Library of Congress Catalogs

Subject Catalog

1979

Volume 4 Chelt–Color F

LIBRARY OF CONGRESS WASHINGTON 1980

CHELTENHAM, ENG.

—ANTIQUITIES

Saville, Alan.
Pre-Regency Cheltenham : an archaeological survey : report / compiled by Alan Saville for Cheltenham Art Gallery and Museum. — Cheltenham : Cheltenham Art Gallery and Museum, 1975.
29 p., 12 leaves of plates : maps (some col.) ; 30 cm. GB76-00377
Bibliography: p. 27-29.
ISBN 0-905157-00-1 : £0.75
1. Cheltenham, Eng.—Antiquities. 2. England—Antiquities. I. Cheltenham Art Gallery and Museum. II. Title.
DA690.C48S28 942.4'16 76-363867
76[r79]rev MARC

CHELYZBINSK, RUSSIA (PROVINCE)

V boevykh ri͡adakh leninskoĭ partii : Cheli͡ab. obl. org. KPSS v t͡sifrakh, 1917-1977 : [Sbornik / Sost. Z. V. Shestakova]. — Cheli͡abinsk : I͡Uzh.-Ural. kn. izd-vo, 1978.
239 p., [6] leaves of plates : col. ill. ; 20 cm. USSR78
Includes bibliographical references.
0.45rub
1. Kommunisticheskai͡a partii͡a Sovetskogo Soi͡uza—Statistics. 2. Chelyzbinsk, Russia (Province) I. Shestakova, Zoi͡a Vasil'evna.
JN6598.K7V1248 79-380402
79 MARC

—BIOGRAPHY—ADDRESSES, ESSAYS, LECTURES

(Tak dobyvalas' Pobeda)
Так добывалась Победа : Воспоминания бойцов и командиров 63-й гвард. Челяб. добровольч. танковой бригады : [Сб. статей / Сост. А. А. Соколов, А. А. Золотов]. — 2-е изд., дораб. — Челябинск : Юж.-Урал. кн. изд-во, 1978.
240 p., [8] leaves of plates : ill. ; 20 cm. USSR 78
Includes bibliographical references.
0.70rub
1. World War, 1939-1945—Tank warfare—Addresses, essays, lectures. 2. World War, 1939-1945—Personal narratives, Russian—Addresses, essays, lectures. 3. World War, 1939-1945—Regimental histories — Russia — 63. gvardeĭskai͡a tankovai͡a Cheli͡abinsko-Petrakovskai͡a dobrovol'cheskai͡a brigada—Addresses, essays, lectures. 4. Russia (1923- U. S. S. R.). Armii͡a. 63. gvardeĭskai͡a tankovai͡a Cheli͡abinsko-Petrakovskai͡a dobrovol'cheskai͡a brigada—Biography—Addresses, essays, lectures. 5. Chelyabinsk, Russia (Province)—Biography—Addresses, essays, lectures. I. Sokolov, Aleksandr Aleksandrovich. II. Zolotov, Aleksandr Afanas'evich.
D793.T34 1978 79-410103

CHEMICAL ABSTRACTS

Campbell, Malcolm B M
The use of Chemical abstracts : a descriptive guide to the organisation, development, and use of Chemical abstracts and its associated indexes / by Malcolm Campbell. — Nathan, Australia : Griffith University Library, 1977.
vi, 37 p. : diagrs. ; 30 cm. Aus***
ISBN 0-86857-023-0 : $3.50
1. Chemical abstracts. I. Chemical abstracts. II. Title.
QD1.C25 540'.72 79-301328
79 MARC

CHEMICAL ADDITIVES IN FOOD
see Food additives

CHEMICAL AFFINITY

see also Chemical reaction, Rate of

—CONGRESSES

Theory and practice in affinity techniques / edited by P. V. Sundaram and F. Eckstein. — London ; New York : Academic Press, 1978.
x, 246 p. : ill. ; 24 cm. GB***
Includes bibliographical references and index.
ISBN 0-12-677150-2
1. Affinity chromatography—Congresses. 2. Chemical affinity—Congresses. 3. Proteins—Affinity labeling—Congresses. I. Sundaram, P. V. II. Eckstein, F.
QP519.9.A35T46 574.1'9285 78-71274
79 MARC

CHEMICAL ANALYSIS see Chemistry, Analytic; and subdivision Analysis under particular subjects

CHEMICAL APPARATUS

see also Centrifuges; Glass blowing and working; Lasers in Chemistry

American Water Works Association.
Guidelines for the selection of laboratory instruments. — [Denver] : American Water Works Association, c1979.
v, 77 p. : diagrs. ; 23 cm. — (AWWA [manual] ; no. M15)
1. Chemical apparatus. I. Title. II. Series: American Water Works Association. AWWA manual ; no. M15.
TD491.A49 no. M15a 628.1'08 s 79-102538
[QD53] 79 MARC

CHEMICAL BONDS

see also Intermolecular forces

Companion, Audrey L
Chemical bonding/ Audrey L. Companion. — 2d ed. — New York : McGraw-Hill, 1979.
x, 179 p. : ill. ; 21 cm.
Includes bibiographical references and index.
ISBN 0-07-012383-7 : $7.95. ISBN 0-07-012379-9 pbk. : $5.95
1. Chemical bonds. I. Title.
QD461.C63 1979 541'.224 78-17081
78 MARC

Carbon-carbon bond formation / edited by Robert L. Augustine. — New York : M. Dekker, c1979-
v. : ill. ; 24 cm. — (Techniques and applications in organic synthesis)
Includes bibliographical references and index.
ISBN 0-8247-6787-X
1. Chemistry, Organic—Synthesis. 2. Chemical bonds. I. Augustine, Robert L., 1932-
QD262.C29 547'.2 78-31483
78 MARC

Gimarc, Benjamin M
Molecular structure and bonding : the qualitative molecular orbital approach / Benjamin M. Gimarc. — New York : Academic Press, 1979.
ix, 224 p. : ill. ; 24 cm.
Includes bibliographical references and index.
ISBN 0-12-284150-6
1. Molecular structure. 2. Chemical bonds. I. Title.
QD461.G56 541'.22 78-3336
78 MARC

I͡At͡simirskiĭ, Konstantin Borisovich.
(Khimicheskai͡a svi͡az')
Химическая связь : [Учебник для хим. специальностей ун-тов] / К. Б. Яцимирский, В. К. Яцимирский. — Киев : Вища школа, 1975.
303 p. : diagrs. ; 22 cm. USSR 75
Bibliography : p. [300]
0.80rub
1. Chemical bonds. I. I͡At͡simirskiĭ, Vitaliĭ Konstantinovich, joint author. II. Title.
QD461.I 18 79-413424

Ladd, Marcus Frederick Charles.
Structure and bonding in solid state chemistry / M. F. C. Ladd. — Chichester, Eng. : Ellis Horwood ; New York : Halsted Press, 1979.
326 p. : ill. ; 24 cm. — (Ellis Horwood series in chemical science) GB***
Bibliography: p. [308]-310.
Includes index.
ISBN 0-470-26597-3 (Halsted Press) : £13.00
1. Solid state chemistry. 2. Chemical structure. 3. Chemical bonds. I. Title. II. Series.
QD478.L32 1979 541'.224 78-41289
79 MARC

Lee, John David.
A new concise inorganic chemistry / by J. D. Lee. — 3d ed. — New York : Van Nostrand Reinhold Co., 1977.
xxi, 505 p. : ill. ; 24 cm.
Editions for 1964 and 1965 published under title: Concise inorganic chemistry.
Bibliography: p. 490.
Includes index.
ISBN 0-442-30169-3
1. Chemistry, Physical and theoretical. 2. Chemical bonds. I. Title. II. Title: Concise inorganic chemistry.
QD453.2.L43 1977 546 76-57657
76 MARC

The Nature of the surface chemical bond / editors, T. N. Rhodin, G. Ertl. — Amsterdam ; New York : North-Holland Pub. Co. ; New York : sole distributed for the U.S.A. and Canada, Elsevier North-Holland, 1979.
xiv, 405 p. : ill. ; 24 cm.
Includes bibliographies and indexes.
ISBN 0-444-85053-8
1. Chemical bonds. 2. Surface chemistry. I. Rhodin, Thor N. II. Ertl, Gerhard.
QD461.N37 541'.224 78-25676
78 MARC

Sanderson, Robert Thomas, 1912-
Chemical bonds in organic compounds / R. T. Sanderson. — Scottsdale, AZ : Sun and Sand Pub. Co., c1976.
73 p. ; 23 cm.
Cover title.
1. Chemical bonds. 2. Chemistry, Physical organic. I. Title.
QD461.S34 547'.1'224 78-325511
79 MARC

Voronkov, Mikhail Grigor'evich.
The siloxane bond : physical properties and chemical transformations / M. G. Voronkov, V. P. Mileshkevich, and Yu. A. Yuzhelevskii ; translated from Russian by John Livak. — New York : Consultants Bureau, c1978.
xii, 493 p. : ill. ; 24 cm. — (Studies in Soviet science : Physical sciences)
Translation of Siloksanovai͡a svi͡az'.
Includes bibliographical references and index.
ISBN 0-306-10940-9
1. Siloxanes. 2. Chemical bonds. I. Mileshkevich, Vladimir Petrovich, joint author. II. I͡Uzhelevskiĭ, I͡Uliĭ Abramovich, joint author. III. Title. IV. Series.
QD412.S6V6713 547'.08 78-16675
78 MARC

—CONGRESSES

Simpozium po khimicheskoĭ svi͡azi v poluprovodnikakh, Minsk? 1967.
Symposium on the Chemical Bond in Semiconductors, May 28-June 3 : theses of the reports. — Minsk : "Science and Engineering", 1967.
102 p. ; 20 cm.
At head of title: Academy of Sciences of the USSR. Department of Physicochemistry and Inorganic Material Technology. Scientific Board on Physics and Chemistry of Semiconductors. Academy of Sciences of the BSSR. Institute of Solids and Semiconductors.
1. Semiconductors—Congresses. 2. Solid state chemistry—Congresses. 3. Chemical bonds—Congresses. I. Akademii͡a nauk SSSR. Otdelenie fiziko-khimii i tekhnologii neorganicheskikh materialov. II. Akademii͡a nauk SSSR. Nauchnyĭ sovet po fizike i khimii poluprovodnikov. III. Akademii͡a navuk BSSR, Minsk. Instytut fiziki tsverdaha tsela i pai͡upravadnikoŭ.
QC610.9.S57 1967 537.6'22 78-321911
79 MARC

CHEMICAL BURNS see Burns, Chemical

CHEMICAL CARRIERS (TANKERS)

Corkhill, Michael, 1947-
Chemical tankers : the ships and the market / by Michael Corkhill ; diagrams drawn by Brian Mehl. — London : Fairplay Publications Ltd, 1976.
[7], 113 p., [40] leaves : ill., plans ; 30 cm. — (A Fairplay report) GB78-05977
Bibliography: p. 113.
ISBN 0-905045-02-5 : £50.00
1. Chemical carriers (Tankers). 2. Chemicals—Transportation. 3. Chemical industries. I. Title.
VM455.3.C67 387.24'5 78-314133
78 MARC

CHEMICAL COMPOSITION OF THE EARTH
see Geochemistry

CHEMICAL EFFECTS OF RADIATION see Radiation chemistry

CHEMICAL ELEMENTS

see also Actinide elements; Allotropy; Nucleosynthesis; Periodic law; Superheavy elements; Technetium; Trace elments

—ANALYSIS

Hamilton, E I
The chemical elements and man : measurements—perspectives—applications / by E. I. Hamilton. — Springfield, Ill. : Thomas, c1979.
xix, 490 p. : ill. ; 24 cm. — (American lecture series ; no. 1021)
Includes bibliographical references and index.
ISBN 0-398-03732-9
1. Chemicals—Physiological effect. 2. Body composition. 3. Health. 4. Chemical elements—Analysis. I. Title.
QP531.H28 612'.0152 78-5409
78 MARC

CHEMICAL ENGINEERING

see also Chemical plants; Chemical processes; Chemistry, Technical; Enzymes—Industrial applications; Fluidization; Geothermal engineering; Mechanical engineering; Metallurgy; Mixing

Aerstin, Frank.
Applied chemical process design / Frank Aerstin and Gary STreet ; with a foreword by K. D. Timmerhaus. — New York : Plenum Press, c1978.
xvii, 294 p. : ill. ; 26 cm.
Includes bibliographical references and index.
ISBN 0-306-31088-0
1. Chemical processes. 2. Chemical engineering. I. Street, Gary, joint author. II. Title.
TP155.7.A35 660.2'81 78-9104
78 MARC

Froment, Gilbert F
Chemical reactor analysis and design / Gilbert F. Froment, Kenneth B. Bischoff. — New York : Wiley, c1979.
xxxix, 765 p. : ill. ; 24 cm.
Includes bibliographical references and indexes.
ISBN 0-471-02447-3
1. Chemical reactors. 2. Chemical reactions. 3. Chemical engineering. I. Bischoff, Kenneth B., joint author. II. Title.
TP157.F76 660'.29'9 78-12465
78 MARC

Hobler, Tadeusz.
Dyfuzyjny ruch masy i absorbery / Tadeusz Hobler. — Wyd. 2 zm. i uzup. — Warszawa : Wydaw-a Naukowo-Techniczne, 1976 (Warsz. : DRP)
612, ⌊2⌋ p. : ill. ; 25 cm. — (Inżynieria chemiczna) P76-4293
Bibliography: p. ⌊604⌋-606.
Includes index.
zł130.00
1. Mass transfer. 2. Diffusion. 3. Chemical engineering. I. Title.
TP156.M3H6 1976 79-413543
79 MARC

Holland, Charles Donald.
Fundamentals of chemical reaction engineering / Charles D. Holland, Rayford G. Anthony. — Englewood Cliffs, N.J. : Prentice-Hall, c1978.
xiv, 541 p. : ill. ; 24 cm. — (Prentice-Hall international series in the physical and chemical engineering sciences)
Includes bibliographical references and index.
ISBN 0-13-335596-9
1. Chemical engineering. 2. Chemical reactions. I. Anthony, Rayford Gaines, 1935- joint author. II. Title.
TP145.H64 660.2'9'9 78-18372
78 MARC

Pohorecki, Ryszard.
Kinetyka i termodynamika procesów inżynierii chemicznej / Ryszard Pohorecki, Stanisław Wroński. — Wyd. 1. — Warszawa : Wydaw-a Naukowo-Techniczne, 1977 (Pozn. : PZGMK)
542, ⌊1⌋ p. : ill. ; 25 cm. P78-1252
Errata slip inserted.
Includes bibliographies and index.
zł70.00
1. Chemical reaction, Rate of. 2. Thermodynamics. 3. Chemical engineering. I. Wroński, Stanisław, Dr., joint author. II. Title.
QD502.P64 78-388752
79 MARC

Rehm, Thomas R
SI for AIChE : a guide for the implementation of the international system of units by the chemical engineering profession / sponsored by the AIChE Metrication Committee ; prepared by Thomas R. Rehm. — New York : American Institute of Chemical Engineers, c1979.
iii, 11 p. ; 29 cm.
Includes bibliographical references.
1. Chemical engineering. 2. Metric system. I. AmericanInstitute of Chemical Engineers. Metrication Committee. II. Title.
TP149.R43 389'.152'02466 79-112032
00 MARC

Sandler, Stanley I 1940-
Chemical and engineering thermodynamics / Stanley I. Sandler. — New York : Wiley, c1977.
xviii, 587 p. : ill. ; 24 cm.
Includes bibliographical references and index.
ISBN 0-471-01774-4
1. Thermodynamics. 2. Chemical engineering. I. Title.
QD504.S25 541'.369 77-1312
77⌊r79⌋rev MARC

Shang-hai hua kung hsüeh yüan.
(Chi ch'u hua hsüeh kung ch'eng)
基础化学工程 / 上海化工学院. — 第1版. — 上海：上海科学技术出版社，1978-
v. : ill. ; 26 cm.
$1.65

1. Chemical engineering. I. Title.
TP155.S444 1978 78-842168

—APPARATUS AND SUPPLIES

see also Heat exchangers; Plate towers

Climax Molybdenum Company.
Stainless steel and the chemical industry / ⌊prepared by Climax Molybdenum Company⌋. — New York : Committee of Stainless Steel Producers, American Iron and Steel Institute, ⌊1966⌋
140 p. : ill. ; 28 cm.
Bibliography: p. 140.
1. Chemical engineering—Apparatus and supplies. 2. Steel, Stainless. I. American Iron and Steel Institute, New York. Committee of Stainless Steel Producers. II. Title.
TP157.C55 1966 660.2'83 78-325524
79 MARC

Pikoń, Jerzy.
Aparatura chemiczna : ⌊podręcznik dla studentów wyższych szkół technicznych⌋ / Perzy Pikoń. — Wyd. 1. — Warszawa : Państ. Wydaw. Naukowe, 1978 (Wroc. : WDN)
1073, ⌊2⌋ p., ⌊4⌋ leaves of plates : ill. ; 25 cm. P78-6875
Bibliography: p. 1072-1073.
zł100.00
1. Chemical engineering—Apparatus and supplies. I. Title.
TP157.P49 79-352099
79 MARC

—APPARATUS AND SUPPLIES—CORROSION

Mihalcu, Mihail.
Coroziunea şi combaterea ei în industria chimică / Mihail Mihalcu, Mihaela Drăgănoiu. — Bucureşti : Editura Tehnică, 1978.
454 p., ⌊2⌋ leaves of plates : ill. (some col.) ; 21 cm. R79-685
Three fold. tables inserted.
Bibliography: p. 350-369.
lei24.00
1. Chemical engineering—Apparatus and supplies—Corrosion. I. Drăgănoiu, Mihaela, joint author. II. Title.
TP157.M52 79-359937
*79 MARC

—CONGRESSES

Chemical engineering in a hostile world : Eurochem conference, 20-24 June 1977, National Exhibition Centre, Birmingham, England : organised in association with the Institution of Chemical Engineers. — ⌊Farnham⌋ : ⌊Conference Communication⌋, ⌊1977⌋
393 p. in various pagings : ill. ; 30 cm. GB78-17569
Cover title.
"Eurochem conference papers"
Includes bibliographical references.
£42.00
1. Chemical engineering—Congresses. 2. Chemical engineering—Environmental aspects—Congresses. I. Institution of Chemical Engineers, London.
TP5.C46 333.8'09'047 79-301629
79 MARC

Conference on Applied Chemistry, Unit Operations, and Processes, 3d, Veszprém, Hungary, 1977.
Proceedings of the 3rd Conference on Applied Chemistry, Unit Operations, and Processes, Veszprém, 29-31 August 1977. — ⌊s.l.⌋ : Magyar Kémikusok Egyesülete, ⌊1977?⌋
716 p. : ill. ; 23 cm. H***
Includes bibliographical references and index.
1. Chemical engineering—Congresses. I. Magyar Kémikusok Egyesülete.
TP5.C56 1977 660.2'84 79-315117
79 MARC

International Symposium on Chemical Reaction Engineering, 4th, Heidelberg, 1976.
Chemical reaction engineering : proceedings of the 4. internat./6. Europ. symposium, Heidelberg, Fed. Rep. Germany, 6-8 April 1976 / Europ. Fed. of Chem. Engineering, the American Chem. Soc. ... ; Publ. by: DECHEMA, Dt. Ges. für Chem. Apparatewesen e. V., Frankfurt am Main. — Frankfurt am Main : DECHEMA, 1976.
664 p. : graphs ; 21 cm. GFR77-B4
Includes bibliographies.
1. Chemical engineering—Congresses. I. European Symposium on Chemical Reaction Engineering, 6th, Heidelberg, 1976. II. European Federation of Chemical Engineering. III. Title.
TP5.I67 1976 660.2'8 78-315634
78 MARC

—COSTS—ADDRESSES, ESSAYS, LECTURES

Modern cost engineering : methods and data / compiled and edited by Chemical engineering. — New York : Chemical Engineering, McGraw-Hill, c1979.
ix, 579 p. : ill. ; 29 cm.
A collection of articles which were originally published in Chemical engineering, 1969-1978.
Includes bibliographical references.
ISBN 0-07-099907-4 (pbk.) : $27.50
1. Chemical engineering—Costs—Addresses, essays, lectures. I. Chemical engineering.
TP155.M6 658.1'55 78-23937
78 MARC

—DICTIONARIES

De Vries, Louis, 1885-
Wörterbuch der Chemie und der chemischen Verfahrenstechnik / L. De Vries, H. Kolb. — 2., überarb. und erw. Aufl. — Weinheim ; New York : Verlag Chemie, 1978-
v. ; 25 cm. GFR78-A
Added t.p. in English with title: Dictionary of chemistry and chemical engineering.
CONTENTS: Bd. 1. Deutsch/Englisch.

ISBN 0-89573-010-0 (v. 1) : DM220 (v. 1)
1. Chemistry—Dictionaries. 2. Chemical engineering—Dictionaries. 3. English language—Dictionaries—German. 4. Chemistry—Dictionaries—German. 5. Chemical engineering—Dictionaries—German. 6. German language—Dictionaries.—English. I. Kolb, Helga, joint author. II. Title. III. Title: Dictionary of chemistry and chemical engineering.
QD5.D47 1978 540'.3 79-346865
79 MARC

Hu, Jun-sheng.
(Tsui hsin shih yung Ying Han hua kung ming tz'u tz'u tien)
最新實用英漢化工名詞辭典 = The newest practical English-Chinese chemical dictionary / 胡潤生，吳美莉合編. — 台北：樂羣出版公司：大一書局總經銷，民國66 ⌊1977⌋
190 p. ; 22 cm.
Colophon title: Tsui hsin shih yung Ying Han hua kung ming ti tz'u tien.
NT$70.00

1. Chemical engineering—Dictionaries. 2. English language—Dictionaries—Chinese. I. Wu, Mei-li, joint author. II. Title. III. Title: The newest practical English-Chinese chemical dictionary. IV. Title: Shih yung Ying Han hua kung ming t'zu t'zu tien.
TP9.H68 79-841149

—DICTIONARIES—GERMAN

De Vries, Louis, 1885-
Wörterbuch der Chemie und der chemischen Verfahrenstechnik / L. De Vries, H. Kolb. — 2., überarb. und erw. Aufl. — Weinheim ; New York : Verlag Chemie, 1978-
v. ; 25 cm. GFR78-A
Added t.p. in English with title: Dictionary of chemistry and chemical engineering.
CONTENTS: Bd. 1. Deutsch/Englisch.

ISBN 0-89573-010-0 (v. 1) : DM220 (v. 1)
1. Chemistry—Dictionaries. 2. Chemical engineering—Dictionaries. 3. English language—Dictionaries—German. 4. Chemistry—Dictionaries—German. 5. Chemical engineering—Dictionaries—German. 6. German language—Dictionaries.—English. I. Kolb, Helga, joint author. II. Title. III. Title: Dictionary of chemistry and chemical engineering.
QD5.D47 1978 540'.3 79-346865
79 MARC

—DICTIONARIES—RUSSIAN

Dresden. Technische Universität. Institut für Angewandte Sprachwissenschaft. Entwicklungsstelle für Fachwörterbücher.
Russisch-deutsches Wörterbuch der Chemie und chemischen Technik : mit etwa 65 000 Fachbegriffen / hrsg. vom Inst. für Angewandte Sprackwiss. d. Techn. Univ. Dresden ; ⌊zsgest. von d. wissenschaftl. Mitarb. d. Entwicklungsstelle für Fachwörterbücher im Bereich Lexikologie d. Fachsprachen d. Inst. für Angewandte Sprachwiss.⌋. — 3., überarb. u. erg. Aufl. — Berlin : Verlag Technik, 1977.
832 p. ; 21 cm. GDR77-A
Added t.p. in Russian: Russko-nemetškiĭ slovar' po khimii i khimicheskoĭ tekhnologii.
Another 3d ed. issued under title: Chemie und chemische Technik, Russisch-Deutsch.

1. Chemistry—Dictionaries—Russian. 2. Chemical engineering—Dictionaries—Russian. 3. Russian language—Dictionaries—German. I. Title. II. Title: Russko-nemetškiĭ slovar' po khimii i khimicheskoĭ tekhnologii.
QD5.D7 1977 540'.3 78-382767
79 MARC

—DOCUMENTATION

Peck, Theodore P
Chemical industries : a guide to information sources / Theodore P. Peck. — Detroit : Gale Research Co., c1979.
xxi, 595 p. ; 23 cm. — (Management information guide ; v. 29)
Includes index.
ISBN 0-8103-0829-0
1. Chemical engineering—Information services. 2. Chemical industries—Information services. 3. Chemical engineering—Documentation. 4. Chemical industries—Documentation. I. Title. II. Series.
TP155P37 660'.07 76-6891
76 MARC

CHEMICAL ENGINEERING (Continued)

—ENVIRONMENTAL ASPECTS—CONGRESSES

Chemical engineering in a hostile world : Eurochem conference, 20-24 June 1977, National Exhibition Centre, Birmingham, England : organised in association with the Institution of Chemical Engineers. — ⌊Farnham⌋ : ⌊Conference Communication⌋, ⌊1977⌋
393 p. in various pagings : ill. ; 30 cm. GB78-17569
Cover title.
"Eurochem conference papers"
Includes bibliographical references.
£42.00
1. Chemical engineering—Congresses. 2. Chemical engineering—Environmental aspects—Congresses. I. Institution of Chemical Engineers, London.
TP5.C46 333.8'09'047 79-301629
79 MARC

—INFORMATION SERVICES

Peck, Theodore P
Chemical industries : a guide to information sources / Theodore P. Peck. — Detroit : Gale Research Co., c1979.
xxi, 595 p. ; 23 cm. — (Management information guide ; v. 29)
Includes index.
ISBN 0-8103-0829-0
1. Chemical engineering—Information services. 2. Chemical industries—Information services. 3. Chemical engineering—Documentation. 4. Chemical industries—Documentation. I. Title. II. Series.
TP155P37 660'.07 76-6891
76 MARC

—LABORATORY MANUALS

(Rukovodstvo k prakticheskim zanfātifām v laboratorii protšessov i apparatov khimicheskoĭ tekhnologii)
Руководство к практическим занятиям в лаборатории процессов и аппаратов химической технологии : допущено в качестве учеб. пособия для студентов химико-технологических специальностей вузов / под ред. П. Г. Романкова. — Изд. 4. перер. и доп. — Ленинград : Химия, 1975.
254 p. : ill. ; 23 cm. USSR***
Second ed. (1964) and 3d ed. (1969) published under title: Rukovodstvo k prakticheskim zanfātifām v laboratorii po protšessam i apparatam khimicheskoĭ tekhnologii.
Includes bibliographies.
0.68rub
1. Chemical engineering—Laboratory manuals. I. Romankov, Petr Gri- gor'evich.
TP161.R83 1975 79-413507

—MATHEMATICAL MODELS

Boiarinov, Anatoliĭ Ivanovich.
(Metody optimizatšii v khimicheskoĭ tekhnologii)
Методы оптимизации в химической технологии : ⌊учеб. пособие для хим.-технол. специальностей вузов⌋ / А. И. Бояринов, В. В. Кафаров. — Изд. 2-е, перераб. и доп. — Москва : Химия, 1975.
574 p. : ill. ; 22 cm. — (Химическая кибернетика) USSR 75
Series romanized: Khimicheskaiā kibernetika.
Includes bibliographies and index.
1.39rub
1. Chemical engineering—Mathematical models. 2. Mathematical optimization. I. Kafarov, Viktor Viācheslavovich, joint author. II. Title.
TP149.B68 1975 79-412160

—NOTATION

Austin, D G
Chemical engineering drawing symbols / D. G. Austin. — London : G. Godwin ; New York : J. Wiley, 1979.
xv, 96 p. : ill. ; 31 cm. GB***
Bibliography: p. ⌊92⌋
Includes index.
ISBN 0-470-26601-5 (Halstead) : £6.80
1. Chemical engineering—Notation. 2. Engineering drawings. I. Title.
TP149.A97 1979 604'.2'6602148 79-317439
79 MARC

—PERIODICALS

Arhiv za rudarstvo i tehnologiju. Technologica acta. g. 6– Tuzla, Institut za rudarska i hemijsko-tehnološka istraživanja, 1968–
v. ill. 29 cm. quarterly (irregular)
Continues: Arhiv za tehnologiju.
"Časopis Instituta za rudarska i hemijsko-tehnološka istraživanja, Tehnološkog fakulteta i Rudarskog fakulteta Sarajevskog univerziteta."
In Serbo-Croatian with summaries in English, French, German, or Russian.

Vols. 6– issued by Institut za rudarska i hemijsko-tehnološka istraživanja, Tehnološki fakultet, and Rudarski fakultet, both faculties of Sarajevski univerzitet.
1. Mining engineering—Periodicals. 2. Chemical engineering—Periodicals. I. Institut za rudarska i hemijsko-tehnološka istraživanja. II. Sarajevo. Univerzitet. Tehnološki fakultet. III. Sarajevo. Univerzitet. Rudarski fakultet. IV. Title: Technologica acta.
TN4.A714 79-642410
MARC-S

Arhiv za tehnologiju. Technologica acta. g. 1-5. Tuzla, Institut za rudarska i hemijsko-tehnološka istraživanja, 1963–1967.
5 v. ill. 28 cm. quarterly (irregular)
"Časopis Instituta za rudarska i hemijsko-tehnološka istraživanja, Tehnološkog fakulteta i Rudarskog fakulteta Sarajevskog univerziteta."
English, French, German, Russian, or Serbo-Croation, with summaries in one of the other languages.

Issued by Institut za rudarska i hemijsko-tehnološka istraživanja, Tehnološki fakultet, and Rudarski fakultet, both faculties of Sarajevski univerzitet.
Continued by: Arhiv za rudarstvo i tehnologiju.
1. Mining engineering—Periodicals. 2. Chemical engineering—Periodicals. I. Institut za rudarska i hemijsko-tehnološka istraživanja. II. Sarajevo. Univerzitet. Tehnološki fakultet. III. Sarajevo. Univerzitet. Rudarski fakultet. IV. Title: Technologica acta.
TN4.A714 79-642412
MARC-S

(Chonghap hwahak)
綜合化學
⌊서울⌋ 한국종합화학공업주식회사
v. ill. 26 cm. semiannual.

1. Chemical engineering—Periodicals. I. Han'guk Chonghap Hwahak Kongŏp Chusik Hoesa.
TP1.C66 79-641827
(MARC-S)

(Hwahak konghak)
화학공학 Hwahak konghak.
⌊서울⌋ 한국화학공학회
v. ill. 26 cm. bimonthly.
Official journal of the Korean Institute of Chemical Engineers.
English or Korean.

1. Chemical engineering—Periodicals. I. Han'guk Hwahak Konghakhoe.
TP1.H88 79-642161
(MARC-S)

(Kagaku kōgaku rombun shū)
化学工学論文集 Kagaku kogaku ronbunshu.
⌊東京⌋ 化学工学協会
v. ill. 26 cm. bimonthly.
In Japanese, with summaries in English.

1. Chemical engineering—Periodicals. I. Kagaku Kōgaku Kyōkai. II. Title: Kagaku kogaku ronbunshu.
TP1.K26 78-647603
(MARC-S)

Sprawy chemików.
Katowice ⌊Zarząd Główny Związku Zawodowego Chemików⌋
v. ill. 44 cm. biweekly.
Vols. for journal of Związek Zawodowy Chemikow.
Microfilm. Washington, Library of Congress, Photoduplication Service, reels. 35 mm.
1. Chemical engineering—Periodicals. I. Związek Zawodowy Chemików.
Microfilm 05025 TP 78-649888
[TP1] MARC-S

—PERIODICALS—INDEXES

Combined 1974 index for the journals Sulphur, Nitrogen, and Phosphorus & potassium. — London : British Sulphur Corporation, 1975.
40 p. ; 30 cm. GB***
Cover title.
1. Chemical engineering—Periodicals—Indexes. I. British Sulphur Corporation, ltd. II. Sulphur. III. Nitrogen. IV. Phosphorus and potassium.
Z7914.C4C64 016.661 78-315218
⌊TP145⌋ 78 MARC

—PROBLEMS, EXERCISES, ETC.

Laurence, Guy.
Problèmes résolus de génie chimique / par Guy Laurence ; préf. de Jean Talbot. — Paris : Eyrolles, 1975.
275 p. : ill. ; 25 cm. F***
55.00F
1. Chemical engineering—Problems, exercises, etc. I. Title.
TP168.L38 660.2'028 78-391244
79 MARC

—SAFETY MEASURES

Chemical Industries Association.
Chemsafe : a manual of the chemical industry scheme for assistance in freight emergencies / ⌊Chemical Industries Association⌋. — 2nd ed. — London : C.I.A., 1976.
⌊3⌋, 33 p. : forms ; 21 cm. GB77-04203
Bibliography: p. 31.
ISBN 0-900623-07-1
1. Chemical engineering—Safety measures. 2. Chemicals—Transportation. I. Title.
TP149.C486 1976 614.8'312 79-303097
79 MARC

—SAFETY MEASURES—BIBLIOGRAPHY

Institutul Central de Documentare Tehnică.
Protecția muncii în industria chimică : bibliografie. — București : Institutul Central de Documentare Tehnică, 1968.
44 p. ; 30 cm.
Includes index.
1. Chemical engineering—Safety measures—Bibliography. I. Title.
Z7914.C4I57 1968 79-379256
⌊TP149⌋ 79 MARC

—SAFETY MEASURES—CONGRESSES

Hazards with chemicals : symposium organised by the Auckland Branch of the New Zealand Institute of Chemistry Inc., 23rd April, 1975. — ⌊Auckland : New Zealand Institute of Chemistry, 1975⌋
⌊98⌋ leaves : ill., plans ; 30 cm. NZ75-7
Proceedings of the symposium held at the University of Auckland School of Engineering.
Bibliography: leaves ⌊96⌋-⌊98⌋
1. Chemical engineering—Safety measures—Congresses. I. New Zealand Institute of Chemistry. Auckland Branch. II. Auckland, N.Z. University. School of Engineering.
TP149.H39 614.8'3 79-306460
79 MARC

—VOCATIONAL GUIDANCE

Hagerty, D Joseph.
Opportunities in chemical engineering / D. Joseph Hagerty, Earl R. Gerhard, Charles A. Plank. — Skokie, Ill. : VGM Career Horizons, c1979.
ix, 146 p. : ill. ; 21 cm.
Bibliography: p. 134-135.
1. Chemical engineering—Vocational guidance. I. Gerhard, Earl Robert, 1922- joint author. II. Plank, Charles A., joint author. III. Title.
TP186.H33 660.2'023 78-60548
79 MARC

—CHINA

U.S. Delegation in Pure and Applied Chemistry.
Chemistry and chemical engineering in the People's Republic of China : a trip report of the U.S. Delegation in Pure and Applied Chemistry / edited by John D. Baldeschwieler. — Washington : American Chemical Society, 1979.
xix, 266 p. : ill. ; 24 cm.
Includes index.
ISBN 0-8412-0501-9
1. Chemistry—China. 2. Chemical engineering—China. I. Baldeschwieler, John D., 1933- . II. Title.
QD18.C5U54 1979 540'.951 79-11217
79 MARC

—JAPAN—HISTORY

(Nihon no kagaku hyakunenshi)
日本の化学百年史：化学と化学工業の歩み / 日本化学会編. — 東京：東京化学同人, 1978.
xiii, 1302 p., ⌊2⌋ leaves of plates : ill. ; 27 cm. Ja 78-10104
"日本化学会創立百周年記念"
Includes bibliographical references.
¥23000

1. Chemistry—Japan—History. 2. Chemical engineering—Japan—History. I. Nippon Kagakukai.
QD18.J3N53 78-812433

CHEMICAL ENGINEERING (Continued)

—NEW JERSEY—HISTORY—
ADDRESSES, ESSAYS, LECTURES

Bernstein, Jeremy, 1924-
Science and the human imagination : Albert Einstein : papers and discussions / by Jeremy Bernstein and Gerald Feinberg, including two lectures on New Jersey's contributions to the nation in chemical industry and chemical education : papers and discussions / by Henry B. Hass and A. K. Bose ; edited by Charles Angoff. — Rutherford [N.J.] : Fairleigh Dickinson University Press, 1978.
94 p. ; 22 cm. — (The Leverton lecture series ; 5)
ISBN 0-8386-2223-2 : $4.50
1. Einstein, Albert, 1879-1955—Addresses, essays, lectures. 2. Physicists—United States—Biography—Addresses, essays, lectures. 3. Physics—History—Addresses, essays, lectures. 4. Chemical engineering—New Jersey—History—Addresses, essays, lectures. 5. Chemistry—Study and teaching—New Jersey—Addresses, essays, lectures. I. Feinberg, Gerald, 1933- joint author. II. Hass, Henry Bohn, 1902- New Jersey's contributions to the nation in chemical industry and chemical education. 1978. III. Title. IV. Series: Leverton lecture series ; 5.
GC16.E5B46 530'.092'4 77-92565
78 MARC

CHEMICAL ENGINEERING LABORATORIES

see also Engineering experiment stations

CHEMICAL ENGINEERING RESEARCH

—JAPAN—PERIODICALS

Tōkyō Kōgyō Shikenjo.
(Tōkyō Kōgyō Shikenjo yōran)
東京工業試験所要覧
[東京 東京工業試験所]
v. ill. 26-30 cm. annual.
Other title : Activities of the National Chemical Laboratory for Industry.
Japanese and English (Japanese only)
1. Tōkyō Kōgyō Shikenjo—Periodicals. 2. Chemical engineering research—Japan—Periodicals. I. Tōkyō Kōgyō Shikenjo. Activities of the National Chemical Laboratory for Industry.
TP194.J3T64a 75-799305
(MARC-S)
rev

CHEMICAL ENGINEERS

see also Chemists

—EMPLOYMENT—INDIA

Gujarat, India (State). Dept. of Technical Education.
Report on engineering and technical manpower, 1974 : chemicals. — Ahmadabad : Dept. of Technical Education, [pref. 1975]
iv, 58 p., [10] leaves of plates : ill. ; 31 cm.
1. Chemical engineers—Employment—India. I. Title.
TP186.G84 1975 331.7'66'020954 77-906294
79 MARC

CHEMICAL EQUILIBRIUM

see also Linear free energy relationship; Phase rule and equilibrium

(Chi pen hua hsüeh p'ing heng yüan li yü ying yung)
基本化學平衡原理與應用 = Applied and principles of chemical equilibrium. — 香港：自然科學出版社, [1978?]
290 p. : ill. : 19 cm.
Includes index.
$12.00

1. Chemical equilibrium. I. Title: Applied and principles of chemical equilibrium.
QD503.C52 78-843408

Gonikberg, Mark Gerfsevich.
Chemical equilibria and reaction rates at high pressures. (Khimicheskoe ravnovesie i skorost' reaktsii pri vysokikh davlieniyakh) [by] M. G. Gonikberg. 2d ed. rev. and augm. Translated from Russian [by Michael Artman] Jerusalem, Published for the National Science Foundation, Washington, by the Israel Program for Scientific Translations; [available from the Office of Technical Services, U.S. Dept. of Commerce, Washington] 1963.
212 p. illus. 24 cm. (NASA TT-F-95)
At head of title: Akademiya nauk SSSR. Institut organicheskoi khimii im. N. D. Zellinskogo. Academy of Sciences of the U.S.S.R. Institute of Organic Chemistry im. N. D. Zelinskii.
Bibliography: p. 194-211.

1. Chemical equilibrium. 2. Chemical reaction, Rate of. I. Title. II. Series: United States. National Aeronautics and Space Administration. NASA technical translation, F-95.
TL507.U745 no. 95 63-61865
[QD503] [r79]rev2 MARC

Polak, Lev Solomonovich.
Neravnovesnaîa khimicheskaîa kinetika i ee primenenie / L. S. Polak. — Moskva : Nauka, 1979.
248 p. : ill. ; 21 cm. USSR79
At head of title: Akademiîa nauk SSSR. Institut neftekhimichiskogo sinteza im. A. V. Topchieva.
Includes bibliographies and index.
2.50rub
1. Chemical reaction, Rate of. 2. Chemical equilibrium. I. Title.
QD502.P66 79-379926
79 MARC

Rossotti, Hazel.
The study of ionic equilibria : an introduction / Hazel Rossotti. — London ; New York : Longman, 1978.
xiv, 194 p. : ill. ; 22 cm.
Includes bibliographical references and index.
ISBN 0-582-44175-7 : £4.25
1. Chemical equilibrium. 2. Ionic solutions. I. Title.
QD503.R67 541.392 77-26048
77 MARC

—CHARTS, DIAGRAMS, ETC.

Kragten, J
Atlas of metal-ligand equilibria in aqueous solution / J. Kragten ; translation editor, Mary Masson. — Chichester, Eng. : E. Horwood ; New York : Halsted Press, 1978.
781 p. : chiefly ill. ; 26 cm. — (Ellis Horwood series in analytical chemistry)
Includes bibliographical references and index.
ISBN 0-470-99309-X (Halsted)
1. Ligands—Charts, diagrams, etc. 2. Coordination compounds—Charts, diagrams, etc. 3. Chemical equilibrium—Charts, diagrams, etc. I. Title.
QD474.K7 1978 541'.2242 77-12168
77 MARC

—PERIODICALS

Fluid phase equilibria. v. 1-
Sept. 1977-
Amsterdam, Elsevier Scientific Pub. Co.
v. ill. 27 cm. quarterly.
"An international journal."
1. Thermodynamics—Periodicals. 2. Fluid dynamics—Periodicals. 3. Chemical equilibrium—Periodicals.
QD504.F55 541'.369 78-646208
MARC-S

—TABLES

International Union of Pure and Applied Chemistry. Commission on Equilibrium Data.
Equilibrium constants of liquid-liquid distribution reactions / compiled by Y. Marcus, A. S. Kertes and E. Yanir. — London : Butterworths, 1974-
v. ; 25 cm. GB74-23311
Pt. 3 has imprint: Pergamon Press, Oxford.
Includes bibliographical references.
CONTENTS: Introduction and pt. 1. Organophosphorus extractants.—pt. 2. Alkylammonium salt extractants.—pt. 3. Compound forming extractants, solvating solvents and inert solvents.

ISBN 0-408-70570-1 (v. 1) : £5.00 (v. 1) £3.00 (v. 2)
1. Chemical equilibrium—Tables. 2. Liquids—Tables, etc. I. Marcus, Yizhak. II. Kertes, A. S. III. Yanir, E.
QD503.I57 1974 546'.3 75-329680
75[r79]rev MARC

CHEMICAL EVOLUTION

Eigen, Manfred.
The hypercycle, a principle of natural self-organization / M. Eigen, P. Schuster. — Berlin ; New York : Springer-Verlag, 1979.
viii, 92 p. : ill. ; 27 cm.
Bibliography: p. 89-90.
Includes index.
ISBN 0-387-09293-5
1. Chemical evolution. 2. Genetic translation. I. Schuster, Peter, 1941- joint author. II. Title.
QH371.E33 575 79-1315
79 MARC

Grassé, Pierre Paul, 1895-
Biologie moléculaire mutagenèse et évolution / par Pierre-P. Grassé. — Paris ; New York : Masson, 1978.
viii, 117 p. : ill. ; 24 cm. — (Collection Les Grands Problèmes de la biologie ; monographie 13) F***
Bibliography: p. [105]-111.
Includes indexes.
ISBN 2-225-49203-4 : 70.00F
1. Chemical evolution. 2. Mutagenesis. I. Title.
QH371.G69 577 79-349986
79 MARC

CHEMICAL FIRE RETARDANTS
see Fireproofing agents

CHEMICAL GENETICS see Biochemical genetics

CHEMICAL GEOLOGY see Geochemistry; Mineralogy, Determinative; Mineralogical chemistry; Rocks—Analysis

CHEMICAL INDICATORS see Indicators and test-papers

CHEMICAL INDUSTRIES

see also Chemical plants; Chemical workers; Paint industry and trade; and names ofspecific industries

Corkhill, Michael, 1947-
Chemical tankers : the ships and the market / by Michael Corkhill ; diagrams drawn by Brian Mehl. — London : Fairplay Publications Ltd, 1976.
[7], 113 p., [40] leaves : ill., plans ; 30 cm. — (A Fairplay report) GB78-05977
Bibliography: p. 113.
ISBN 0-905045-02-5 : £50.00
1. Chemical carriers (Tankers). 2. Chemicals—Transportation. 3. Chemical industries. I. Title.
VM455.3.C67 387.24'5 78-314133
78 MARC

Lanz, Kurt, 1919-
Weltreisender in Chemie / Kurt Lanz. — 1. Aufl. — Düsseldorf ; Wien : Econ-Verlag, 1978.
528 p. : ill. (some col.) ; 22 cm. GFR79-A
Includes indexes.
Bibliography: p. 528.
ISBN 3-430-15895-8 : DM38.00
1. Farbwerke Hoechst AG. 2. Chemical industries. 3. Voyages and travels. 4. Lanz, Kurt, 1919- . 5. Chemists—Germany, West—Biography. I. Title.
HD9654.9.F33L36 79-360712
79 MARC

Shibamura, Yōgo, 1907-
(Nihon kagaku kōgyō to kokusai kyōsōryoku)
日本化学工業と国際競争力 / 柴村羊五著. — 東京：東洋経済新報社, 昭和38 [1963]
viii, 214 p. ; 19 cm.

1. Chemical industries—Japan. 2. Chemical industries. 3. Competition. I. Title.
HD9657.J32S46 78-816165

Wei, James, 1930-
The structure of the chemical processing industries : function and economics / J. Wei, T. W. F. Russell, M. W. Swartzlander. — New York : McGraw-Hill, c1979.
xiii, 397 p. : ill. ; 24 cm. — (McGraw-Hill chemical engineering series)
Includes bibliographies and index.
ISBN 0-07-068985-7
1. Chemical industries. 2. Industrial organization (Economic theory) I. Russell, T. W. F., 1934- joint author. II. Swartzlander, M. W., joint author. III. Title.
HD9650.5.W39 338.6 77-28209
77 MARC

—ABSTRACTS—PERIODICALS

Chemical market abstracts. v. 61-69, no. 8; Jan. 1969-Aug. 1977. Cleveland, Ohio, Predicasts, inc. [etc.]
9 v. 28 cm. monthly.
Continues: CMA. Chemical market abstracts.
Merged with: EMA, equipment market abstracts, ISSN 0098-4779, to form: PROMT. Predicasts overview of markets and technology, ISSN 0161-8032.
Key title: Chemical market abstracts (1969), ISSN 0009-2606
1. Chemical industries — Abstracts — Periodicals. 2. Chemical industries—United States—Periodicals. I. Key title.
HD9651.1.C44 338.4'7'66005 78-646641
MARC-S

CHEMICAL INDUSTRIES
—ABSTRACTS—PERIODICALS
(Continued)

PROMT, Predicasts overview of markets and technology. v. 69, no. 9–
Sept. 1977–
Cleveland, Ohio, Predicasts, inc.
v. 28 cm. monthly.
Formed by the union of: Chemical market abstracts (1969), ISSN 0009–2606, and: EMA, equipment market abstracts, ISSN 0098–4779.
Key title: PROMT. Predicasts overview of markets and technology, ISSN 0161–4556
1. Chemical industries—Abstracts—Periodicals. 2. Electronic industries—Abstracts—Periodicals. 3. Machinery—Trade and manufacture—Abstracts—Periodicals. 4. Energy industries—Abstracts—Periodicals. 5. Manufactures—Abstracts—Periodicals. I. Title: Predicasts overview of markets and technology.
HD9650.1.P16 338.4'05 78–646645
MARC-S

—AUTOMATION

Curievici, Ion.
Automatizări în industria chimică / Ion Curievici. — Bucureşti : Editura Didactică şi Pedagogică, 1966.
392 p. : ill. ; 25 cm.
At head of title: Ministerul Învăţămîntului. Institutul Politehnic Iaşi.
Bibliography: p. ₍386₎-388.
1. Chemical industries—Automation. I. Title.
TP149.C78 79-340831
79 MARC

—BY-PRODUCTS

Kharlampovich, Georgiĭ Dmitrievich.
(Bezotkhodnye tekhnologicheskie protsessy v khimicheskoĭ promyshlennosti)
Безотходные технологические процессы в химической промышленности / Г. Д. Харлампович, Р. И. Кудряшова. — Москва : Химия, 1978.
277 p. : ill. ; 22 cm. USSR 78
Includes bibliographies and index.
1.30rub
1. Chemical industries—By-products. I. Kudriâshova, Rimma Ivanovna, joint author. II. Title.
TP995.K47 79–407577

—CONGRESSES

The Commercial development of chemical engineering projects : a symposium organised by the Nottingham Centre (Midlands Branch) of the Institution of Chemical Engineers. — Rugby : The Institution, 1978.
217 p. in various pagings : ill ; 22 cm. — (Institution of Chemical Engineers symposium series ; no. 55 ISSN 0307-0492) GB79-01980
Includes bibliographical references.
ISBN 0-85295-112-4 : £10.00
1. Chemical industries—Congresses. I. Institution of Chemical Engineers, London. Midlands Branch. Nottingham Group. II. Series: Institution of Chemical Engineers, London. I. Chem. E. symposium series ; no. 55.
HD9650.1.C65 658.4'04 79-317997
79 MARC

—DOCUMENTATION

Peck, Theodore P
Chemical industries : a guide to information sources / Theodore P. Peck. — Detroit : Gale Research Co., c1979.
xxi, 595 p. ; 23 cm. — (Management information guide ; v. 29)
Includes index.
ISBN 0-8103-0829-0
1. Chemical engineering—Information services. 2. Chemical industries—Information services. 3. Chemical engineering—Documentation. 4. Chemical industries—Documentation. I. Title. II. Series.
TP155P37 660'.07 76-6891
76 MARC

—ENVIRONMENTAL ASPECTS

Nakamura, Seiji, 1916–
(Gendai gijutsu ron no kadai)
現代技術論の課題：産業技術と医療技術 / 中村静治著. — 東京：青木書店, 1978.
xiv, 288 p. ; 20 cm. Ja 78–16629
¥1600
1. Technological innovations—Japan. 2. Industry—Social aspects. 3. Medical technology—Social aspects. 4. Chemical industries—Environmental aspects. I. Title.
HC465.T4N27 78–812820

—INFORMATION SERVICES

Peck, Theodore P
Chemical industries : a guide to information sources / Theodore P. Peck. — Detroit : Gale Research Co., c1979.
xxi, 595 p. ; 23 cm. — (Management information guide ; v. 29)
Includes index.
ISBN 0-8103-0829-0
1. Chemical engineering—Information services. 2. Chemical industries—Information services. 3. Chemical engineering—Documentation. 4. Chemical industries—Documentation. I. Title. II. Series.
TP155P37 660'.07 76-6891
76 MARC

—LAW AND LEGISLATION—JAPAN

Japan. Laws, statutes, etc.
₍Kagaku busshitsu no shinsa oyobi seizō tō no kisei ni kansuru hōritsu. English₎
The chemical substances control law in Japan : law concerning the examination and regulation of manufacture, etc., of chemical substances. — ₍Tokyo₎ : Chemical Products Safety Division, Basic Industries Bureau, Ministry of International Trade & Industry, ₍1975₎
79 p. ; 21 cm. Ja77-28
1. Chemicals—Law and legislation—Japan. 2. Chemical industries—Law and legislation—Japan. I. Title.
344'.52'042 78-324401
•79 MARC

—LAW AND LEGISLATION—UNITED STATES

Bureau of National Affairs, Washington, D.C.
Chemical regulation reporter : reference file. — Washington : Bureau of National Affairs, c1977, c1977-
v. ; 30 cm.
Loose-leaf for updating.
1. Chemicals—Law and legislation—United States. 2. Hazardous substances—Law and legislation—United States. 3. Chemical industries—Law and legislation—United States. I. Title.
KF3958.A15B87 Suppl 2 344'.73'047 79-112036
79 MARC

—PERIODICALS

(Hwahak kwa kongŏp ŭi chinbo)
화학과공업의진보 Progress in chemistry & chemical industry.
서울 대한화학회
v. ill. 26 cm.
English or Korean.
1. Chemistry—Periodicals. 2. Chemical industries—Periodicals. I. Daehan Hwahakhwoe. II. Title: Progress in chemistry & chemical industry.
QD1.H85 79-611817
(MARC-S)

—SAFETY MEASURES

National Safety Council.
Personnel safety in chemical and allied industries / ₍prepared with the assistance and cooperation of the Chemical Section of the Industrial Conference of the National Safety Council₎. — Chicago : The Council, c1979.
viii, 190 p. ; 23 cm.
Includes bibliographical references and index.
ISBN 0-87912-125-4
1. Chemical industries—Safety measures. I. National Safety Council. Chemical Section. II. Title.
HD7269.C45N37 1979 614.8'52 78-58303
00 MARC

—SAFETY MEASURES
—HANDBOOKS, MANUALS, ETC.

Huang, Wei-lun.
(Hua hsüeh kung yeh chung tu fang chih shou ts'e)
化學工業中毒防治手册 / 黃偉倫編著. — 香港：香港萬里書店, 1972.
2, 2, 152 p. ; 18 cm.
1. Chemical industries—Safety measures—Handbooks, manuals, etc. I. Title.
TP149.H78 79–837592

—STATISTICS—PERIODICALS

L'Industrie chimique. The Chemical industry. 1973/74– ₍Paris₎ Organisation de coopération et de développement économiques.
v. 27 cm. annual.
Continues: Organisation for Economic Co-operation and Development. Chemical industry.
English and French.
1. Chemical industries—Statistics—Periodicals. 2. Chemicals—Manufacture and industry—Statistics—Periodicals. I. Organization for Economic Cooperation and Development. II. Title: Chemical industry.
HD9650.1.O74 338.4'7'66005 79–646187
MARC-S

—AUSTRIA—WASTE DISPOSAL

Begert, Axel Nikolaus Albert.
Ein Beitrag zur Reinigung des Abwassers eines Chemiefaserwerkes, eines chemischen Betriebes und einer Kokerei / A₍xel Nikolaus Albert₎ Begert. — Wien : Techn. Univ. Wien, Inst. f. Wasserversorgung, Abwasserreinigung u. Gewässerschutz, 1978.
vii, 125, 94, 79 p. : ill., graphs ; 30 cm. (Wiener Mitteilungen, Wasser, Abwasser, Gewässer ; Bd. 24). — (Wiener Mitteilungen, Wasser, Abwasser, Gewässer ; Bd. 24) Au79-8-308
An enlargement of the author's thesis, Technische Universität, Vienna, 1974.
Includes bibliographies.
Not in trade.
1. Chemical industries—Austria—Waste disposal. 2. Sewage—Purification. I. Title. II. Series: Wasser-Abwasser-Gewässer ; Bd. 24.
TD203.W27 Bd. 24 79-369945
₍TD899.C5₎ •79 MARC

—BRAZIL—PIAUÍ

Núcleo de Assistência Gerencial do Piauí.
Diagnóstico das indústrias químicas do Piauí : óleos vegetais e derivados e refinarias de cera de carnaúba / Núcleo de Assistência Gerencial do Piauí—NAG-PI. — Teresina : O Núcleo, 1976.
104 p. : ill. ; 22 cm.
On cover: CEBRAE, SIC, BEP, NAG.
Bibliography: p. 103-104.
1. Chemical industries—Brazil—Piauí. I. Title.
HD9655.B83P526 79-372161
79 MARC

—BULGARIA

Medarov, Vladimir.
Development of the chemical industry in the People's Republic of Bulgaria / Vladimir Medarov ; ₍Engl. transl. Petko Drenkov₎. — ₍Sofia₎ : Sofia-Press, 1978.
51 p., ₍16₎ leaves of plates : ill. (some col.) ; 19 cm. Bu78-2560
1. Chemical industries—Bulgaria. I. Title.
HD9656.B82M4213 338.4'7'660094977 79-307840
•79 MARC

—CHINA

Ch'i, Ju-shan, 1877-1962.
(Chung-kuo ku yu ti hua hsüeh kung i)
中國固有的化學工藝 / 齊如山著. — 台北：中國新聞出版公司, 民國45 ₍1956₎
5, 77 p. ; 19 cm. — (中國百科小叢書)
Series romanized: Chung-kuo pai k'o hsiao ts'ung shu.
1. Chemical industries—China. 2. Cottage industries—China. I. Title.
HD9657.C52C44 79–836825

—CZECHOSLOVAKIA

Czechoslovakia. Ministerstvo chemického průmyslu.
Seznam výrobků pro předkládání požadavků na dodávky centralisovaných, decentralisovaných a ústředně neplánovaných chemických surovin a výrobků na rok 1955. Praha ₍1955?₎
47 p. 21 cm.
At head of title: Ministerstvo chemického průmyslu. Hlavní správa odbytu.
1. Chemical industries—Czechoslovakia. I. Title.
TP65.3.C93 1955 59-38616
₍r79₎rev MARC

Czechoslovakia. Ministerstvo chemického průmyslu.
Thematické úkoly Ministerstva chemického průmyslu pro vynálezce a zlepšovatele, na rok 1954. ₍Praha₎ Techniký odbor Ministerstva chemického průmyslu, 1954.
20 p. 15 cm.
Cover title.
1. Chemical industries—Czechoslovakia. I. Title.
HD9656.C96C9 1954 59-19438
₍r79₎rev MARC

CHEMICAL INDUSTRIES (Continued)

—EUROPE—DIRECTORIES

Chemical Data Services, London.
Chemical company profiles Western Europe / ⌊Chemical Data Services⌋. — 2nd ed. — London : Chemical Data Services, IPC Industrial Press, 1979.
iv, 387 p. : 1 ill. ; 30 cm. GB79-19924
Includes index.
ISBN 0-617-00269-X : £40.00
1. Chemical industries—Europe—Directories. I. Title.
HD9656.A2C48 1979 338.4'7'6600254 79-317509
79 MARC

—EUROPE, EASTERN—COLLECTED WORKS

Sovet ėkonomicheskoĭ vzaimopomoshchi. Postoi͡annai͡a komissii͡a po khimicheskoĭ promyshlennosti.
(Informat͡sionnyĭ bi͡ulleten' po khimicheskoĭ promyshlennosti)
Информационный бюллетень по химической промышленности.
Москва, Совет экон. взаимопомощи, Постоянная комиссия по хим. промышл.,
no. ill. 28 cm.
1. Chemical industries—Europe, Eastern—Collected works. I. Title.
TP55.S685a 79-641821
(MARC-S)

—GERMANY, WEST

West German projects. — ⌊London : Morgan-Grampion⌋, 1977.
43 p. : ill. ; 29 cm. — (A Chemical age survey) GB***
£5.00
1. Chemical industries—Germany, West. I. Series.
TP73.W47 79-316165
79 MARC

—GERMANY, WEST—HISTORY

Die Chemische Industrie in der Bundesrepublik Deutschland : Perspektiven zum 100 jährigen Bestehen des Verbandes der Chemischen Industrie = The chemical industry in the Federal Republic of Germany : perspectives and prospects of the chemical industry : 100 years Verband der Chemischen Industrie / ⌊Chefredaktion, Ernst Koch, Helmut Seidel ; Übersetzungen, Navin Dalal⌋. — ⌊Düsseldorf : Handelsblatt GmbH⌋, 1977.
756 p. in various pagings : ill. (some col.) ; 30 cm. — (Chemische Industrie ; 10/1977, 100. Jahrg. (XXIX) GFR***
Cover title.
Introd. to each section and to each subsection of part 2 in English and German; articles and profiles of companies in German, with summaries in English.
Includes bibliographical references and indexes.
DM48.00
1. Chemical industries—Germany, West—History. I. Verband der Chemischen Industrie. II. Title: The chemical industry in the Federal Republic of Germany. III. Series: Chemische Industrie (Düsseldorf) ; 1977, 10.
TP1.C438 1977, 10 79-358533
⌊HD9656.G38⌋ 79 MARC

—GREAT BRITAIN

Great Britain. Central Office of Information. Reference Division.
Chemicals / ⌊Reference Division, Central Office of Information⌋. — London : H.M. Stationery Off., 1978.
75, ⌊1⌋ p., ⌊4⌋ leaves of plates : ill. ; 24 cm. — (Central Office of Information reference pamphlet ; 151) (British industry today) GB***
Bibliography: p. ⌊76⌋
ISBN 0-11-700933-4 : £1.75
1. Chemical industries—Great Britain. I. Title. II. Series: Great Britain. Central Office of Information. Reference pamphlet ; 151.
HD9652.5.G73 1978 338.4'7'6600941 78-325044
79 MARC

—INDIA—PERIODICALS

India (Republic). Ministry of Chemicals and Fertilisers.
Report - Government of India, Ministry of Chemicals and Fertilisers. 1975/76–1976/77. New Delhi, Govt. of India, Ministry of Chemicals and Fertilisers.
v. 21 cm. annual.
With: India (Republic). Ministry of Petroleum. Report - Government of India, Ministry of Petroleum, supersedes: India (Republic). Ministry of Petroleum and Chemicals. Report - Government of India, Ministry of Petroleum & Chemicals.
1. Fertilizer industry—India—Periodicals. 2. Chemical industries—India—Periodicals.
HD9483.I 4 I 5a 354'.54'008288 78-647958
MARC-S

—INDONESIA

Indonesia. Badan Pimpinan Umum Industri Kimia.
Hasil musjawarah kerdja ke V BPU Industri Kimia dan perusahaan[2] negara dalam lingkungannja di Tretes, pada tg⌊l⌋. 15-7 s/d tgl. 18 Djuli 1965 / Departemen Perindustrian Dasar, Badan Pimpinan Umum Industri Kimia. — ⌊Djakarta⌋ : Badan, ⌊between 1965 and 1967⌋
55 leaves ; 33 cm.
"No. 1190/Dir. Ut/'65."
1. Chemical industries—Indonesia. I. Title.
HD9657.I52I527 1965 79-950368
79 MARC

Indonesia. Badan Pimpinan Umum Industri Kimia.
Pola pembangunan B.P.U. Industri Kimia. — ⌊Djakarta⌋ : Departemen Perindustrian Dasar/Pertambangan, Badan Pimpinan Umum Industri Kimia, Biro Penelitian & Pengembangan, ⌊1962?⌋
152 leaves in various foliations ; 34 cm.
English or Indonesian.
1. Chemical industries—Indonesia. 2. Indonesia. Badan Pimpinan Umum Industri Kimia. I. Title.
HD9657.I52I527 1962 78-950334
79 MARC

Indonesia. Direktorat Djenderal Perindustrian Kimia.
Rentjana kerdja tahun 1968 dan lampiran-lampiran. — ⌊Djakarta⌋ : Republik Indonesia, Departemen Perindustrian Dasar, Ringan dan Tenaga, Direktorat Djendral Perindustrian Kimia, ⌊1968⌋
22, ⌊13⌋ leaves ; 33 cm.
1. Indonesia. Direktorat Djenderal Perindustrian Kimia. 2. Chemical industries—Indonesia. I. Title.
HD9657.I52I53 1968 79-950344
79 MARC

Indonesia. Komando Pelaksana P.N. Industri Kimia.
Hasil rapat kerdja ke I Komando Pelaksana P.N. Industri Kimia beserta P. N.-2 dalam lingkungannja pada tgl. 10 s/d 12 Mei 1966 di Djakarta. — ⌊Djakarta⌋ : Komando, ⌊1966?⌋
41 leaves ; 33 cm.
1. Chemical industries—Indonesia. 2. Government business enterprises—Indonesia. I. Title.
HD9657.I52I54 1966 75-949912
79 MARC

—INDONESIA—PERIODICALS

Indonesia. Badan Pimpinan Umum Industri Kimia.
Laporan tahunan tentang kegiatan[2] B. P. U. Industri Kimia serta P. N.[2] dalam lingkungannja.
⌊Djakarta⌋ Badan Pimpinan Umum Industri Kimia Departemen Perindustrian Dasar.
v. 34 cm.
Continues: Indonesia. Badan Pimpinan Umum Industri Kimia. Laporan tahunan - Badan Pimpinan Umum Industri Kimia.
Cover title, : Laporan tentang kegiatan[2] B. P. U. Industri Kimia serta P. N.[2] dalam lingkungannja.
1. Chemical industries—Indonesia—Periodicals. I. Indonesia. Badan Pimpinan Umum Industri Kimia. Laporan tentang kegiatan[2] B. P. U. Industri Kimia serta P. N.[2] dalam lingkungannja. II. Title.
HD9657.I 5 I 52a 74-940175
MARC-S

—INDONESIA—STATISTICS

Indonesia. Departemen Perindustrian.
Produksi industri kimia. — ⌊Jakarta⌋ : Hubungan Masyarakat, Departemen Perindustrian, ⌊1972⌋
45 p. ; 21 cm.
1. Chemical industries—Indonesia—Statistics. 2. Chemicals—Manufacture and industry—Indonesia—Statistics. I. Title.
HD9657.I52I528 1972 73-940896
79 MARC

—JAPAN

Asahi evening news, Tokyo.
Japan's chemical industry & engineering. ⌊Tokyo⌋ Asahi evening news, international, 1967.
150 p. illus. 34 cm.
Cover title.
"Special supplement."
1. Chemical industries—Japan. 2. Industrial equipment—Japan. I. Title.
HD9656.J32A82 1967 78-313640
78 MARC

Nihon Kagaku Kōgyōhin Yushutsu Kumiai.
Chemical exports & imports of Japan / JCEA, JCIA. — ⌊Tokyo : Japan Chemical Exporters' Association, 1977⌋
26 p. : ill. ; 28 cm. Ja***
Cover title.
1. Chemical industries—Japan. 2. Chemicals—Manufacture and industry—Japan. I. Nihon Kagaku Kōgyōhin Yunyū Kyōkai. II. Title.
HD9657.J32N49 1977 382'.45'6600952 79-315312
79 MARC

Nishiyama, Shigeru, 1927-
The chemical industry of Japan / by Shigeru Nishiyama. — Fairfield, N.J. : C. H. Kline, c1979.
vi, 321 p. : ill. ; 28 cm.
On cover: Nihon no kagaku kōgyō.
1. Chemical industries—Japan. I. Charles H. Kline & Co. II. Title. III. Title: Nihon no kagaku kōgyō.
HD9657.J32N57 338.4'7'6600952 79-108554
79 MARC

Shibamura, Yōgo, 1907-
(Nihon kagaku kōgyō to kokusai kyōsōryoku)
日本化学工業と国際競争力 / 柴村羊五著. — 東京：東洋経済新報社, 昭和38 ⌊1963⌋
viii, 214 p. ; 19 cm.
1. Chemical industries—Japan. 2. Chemical industries. 3. Competition. I. Title.
HD9657.J32S46 78-816165

—KOREA

Chŏng, Ch'ang-yŏng.
The heavy and chemical industry / ⌊author, Chang-Young Jeong⌋. — Seoul : Korean Overseas Information Service, 1977.
58 p. ; 21 cm. — (Korea background series)
1. Metal trade—Korea. 2. Chemical industries—Korea. 3. Industry and state—Korea. I. Korean Overseas Information Service. II. Title. III. Series.
HD9506.K62C5 338.4'09519'5 79-304306
79 MARC

—KOREA (DEMOCRATIC PEOPLE'S REPUBLIC)

Korean chemical industry. — Pyongyang, D.P.R.K. : Foreign Languages Pub. House, 1962.
⌊40⌋ p. : ill. ; 21 cm.
1. Chemical industries—Korea (Democratic People's Republic)
HD9657.K82K67 79-309696
79 MARC

—NETHERLANDS—CATALOGS

Chemfacts, Netherlands. — London : Chemical Data Services, IPC Industrial Press, 1978.
vii, 125, ⌊4⌋ p. : ill., maps ; 30 cm. GB78-12590
Bibliography: p. v.
Includes index.
£40.00
1. Chemical industries—Netherlands—Catalogs. 2. Chemical industries—Netherlands—Directories. I. Chemical Data Services, London.
TP77.C46 382'.456'6009492 79-300876
79 MARC

—NETHERLANDS—DIRECTORIES

Chemfacts, Netherlands. — London : Chemical Data Services, IPC Industrial Press, 1978.
vii, 125, ⌊4⌋ p. : ill., maps ; 30 cm. GB78-12590
Bibliography: p. v.
Includes index.
£40.00
1. Chemical industries—Netherlands—Catalogs. 2. Chemical industries—Netherlands—Directories. I. Chemical Data Services, London.
TP77.C46 382'.456'6009492 79-300876
79 MARC

—NETHERLANDS—EQUIPMENT AND SUPPLIES—STATISTICS—PERIODICALS

Netherlands (Kingdom, 1815-). Centraal Bureau voor de Statistiek. Hoofdafdeling Statistieken van Industrie en Bouwnijverheid.
Fabrieken van machines en apparaten voor voedingsmiddelen-, chemische- en verwante industrieën produktiestatistieken.
's-Gravenhage, Centraal Bureau voor de Statistiek, Hoofdafdeling Statistieken van Industrie en Bouwnijverheid.
v. 29 cm.
Added title page title, : Manufacture of machinery and equipment for food-, chemical- and allied industries production statistics.
Summaries in English.
1. Food industry and trade—Netherlands—Equipment and supplies—Statistics—Periodicals. 2. Chemical industries—Netherlands—Equipment and supplies—Statistics—Periodicals. I. Netherlands (Kingdom, 1815-). Centraal Bureau voor de Statistiek. Hoofdafdeling Statistieken van Industrie en Bouwnijverheid. Manufacture of machinery and equipment for food-, chemical- and allied industries production statistics. II. Title.
HD9015.N4N47b 79-646539
MARC-S

CHEMICAL INDUSTRIES (Continued)

—POLAND

Sosnowska, Alicja.
Kierunki rozwoju chemizacji w polskim przemyśle i jej ekonomiczne uwarunkowania / Alicja Sosnowska. — Warszawa : Szkoła Główna Planowania i Statystyki, 1977.
168 p. ; 24 cm. — (Monografie i opracowania ; 64) P***
Bibliography: p. [149]-167.
zł18.00
1. Chemical industries—Poland. I. Title.
HD9656.P72S67 78-400954
79 MARC

—RUSSIA

Gramoteeva, Lilii͡a Iosifovna.
(Tekhniko-ėkonomicheskie problemy razmeshchenii͡a vazhneĭshikh otrasleĭ khimicheskoĭ promyshlennosti)
Технико-экономические проблемы размещения важнейших отраслей химической промышленности. Москва, "Наука," 1970.
225 p. 20 cm. USSR 70-VKP
At head of title: Академия наук СССР. Госплан СССР. Совет по изучению производительных сил. Л. И. Грамотеева.
0.71rub
1. Chemical industries—Russia. I. Title.
HD9656.R92G7 72-555498

Savinskiĭ, Ėzekiil Simonovich.
Khimizat͡sii͡a narodnogo khozi͡aĭstva i razvitie khimicheskoĭ promyshlennosti / Ė. S. Savinskiĭ. — 2-e izd., pererab. — Moskva : Khimii͡a, 1978.
335 p. ; 22 cm. USSR78
First ed. published in 1972 under title: Khimizat͡sii͡a narodnogo khozi͡aĭstva i proport͡sii razvitii͡a khimicheskoĭ promyshlennosti.
Includes bibliographical references and index.
3.20rub
1. Chemical industries—Russia. I. Title.
HD9656.R92S28 1978 79-368263
79 MARC

United States. Central Intelligence Agency. National Foreign Assessment Center.
Soviet chemical equipment purchases from the West : impact on production and foreign trade. — [Washington] : Central Intelligence Agency, National Foreign Assessment Center, 1978.
v, 34 p. ; 27 cm. — (A Research paper - Central Intelligence Agency, National Foreign Assessment Center)
Cover title.
"Er 78-10554."
1. Chemical industries—Russia. 2. Chemical plants—Russia—Equipment and supplies. 3. Russia—Commerce. I. Title. II. Series: United States. Central Intelligence Agency. National Foreign Assessment Center. A research paper - Central Intelligence Agency, National Foreign Assessment Center.
HD9656.R92U53 1978 338.4'7'681766 79-601092
79 MARC

—RUSSIAN REPUBLIC—BASHKIRIA

Khamzin, Farid Khabibullovich.
(Partkom i rekonstrukt͡sii͡a)
Партком и реконструкция / Ф. Х. Хамзин. — Уфа : Башкирское книжное изд-во, 1976.
78 p. ; 20 cm. USSR***
0.11rub
1. Chemical workers—Russian Republic—Bashkiria. 2. Chemical industries — Russian Republic — Bashkiria. 3. Kommunisticheskai͡a partii͡a Sovetskogo Soi͡uza—Party work. I. Title.
HD8039.C46R937 78-421764

—RUSSIAN REPUBLIC
—KUYBYSHEVSKAYA OBLAST´

(Bol'shai͡a khimii͡a Kuĭbyshevskoĭ oblasti)
Большая химия Куйбышевской области : Сборник документов и материалов (1918–1975 гг.) / Парт. архив Куйбышев. обкома КПСС, Гос. архив Куйбышев. обл. ; [Составитель З. А. Никишина ... et al.]. — Куйбышев : Кн. изд-во, 1977.
420 p. : ill. ; 22 cm. USSR 77
Bibliography: p. 417–[419]
1.50rub
1. Chemical industries—Russian Republic—Kuybyshevskaya oblast'. I. Nikishina, Zoi͡a Andreevna. II. Kommunisticheskai͡a partii͡a Sovetskogo Soi͡uza. Kuĭbyshevskiĭ oblastnoĭ komitet. Partiĭnyĭ arkhiv. III. Gosudarstvennyĭ arkhiv Kuĭbyshevskoĭ oblasti.
HD9656.R93K893 79-407017

—SCANDINAVIA—DIRECTORIES

Chemfacts, Scandinavia : Denmark, Finland, Norway, Sweden. — London : Chemical Data Services, IPC Industrial Press, 1977.
viii, 130, [4] p. : maps ; 30 cm. GB78-38282
Includes index.
£35.00
1. Chemical industries—Scandinavia—Statistics. 2. Chemical industries—Scandinavia—Directories.
HD9656.S32C48 338.4'7'6600948 79-305971
79 MARC

—SCANDINAVIA—STATISTICS

Chemfacts, Scandinavia : Denmark, Finland, Norway, Sweden. — London : Chemical Data Services, IPC Industrial Press, 1977.
viii, 130, [4] p. : maps ; 30 cm. GB78-38282
Includes index.
£35.00
1. Chemical industries—Scandinavia—Statistics. 2. Chemical industries—Scandinavia—Directories.
HD9656.S32C48 338.4'7'6600948 79-305971
79 MARC

—SWEDEN

Khan, Mohammed Sikander.
A study of success and failure in exports : an empirical investigation of the export performance of 165 marked ventures of 83 firms in the chemical and electronics manufacturing industries / by M. Sikander Khan. — Stockholm : Akademilitt., 1978.
260 p. : tables ; 24 cm. S78-8
Thesis—Stockholm.
Bibliography: p. 253-257.
Includes index.
ISBN 9174100459 : kr140.00
1. Chemical industries—Sweden. 2. Electronic industries—Sweden. 3. Export marketing. I. Title.
HD9656.S76K45 382'.6'09485 78-321431
79 MARC

—UKRAINE

Opyt raboty predprii͡atiĭ khimicheskoĭ promyshlennosti USSR, perevedennykh na pi͡atidnevnui͡u rabochui͡u nedeli͡u. — Kiev : [s. n.], 1968.
33 p.
At head of title: Gosudarstvennyĭ planovyĭ komitet Soveta Ministrov USSR. Ukrainskiĭ nauchno-issledovatel'skiĭ institut nauchno-tekhnicheskoĭ informat͡sii i tekhniko-ėkonomicheskikh issledovaniĭ. Ministerstvo khimicheskoĭ promyshlennosti USSR.
L.C. copy on microfilm only.
——— Microfilm. On 1 reel with other items ; 35 mm.
Microfilm Slavic 13016 HD
1. Chemical industries—Ukraine. 2. Hours of labor—Ukraine. I. Ukraine. Derzhavnyĭ planovyĭ komitet. II. Ukraïns'kyĭ naukovo-doslidnyĭ instytut naukovo-tekhnichnoï informatsiï i tekhniko-ekonomichniykh doslidzhen'. III. Ukraine. Ministerstvo khimichnoï promyslovosti.
[HD9656.R93U49] 79–357461

—UNITED STATES

Arskiĭ, Anatoliĭ Konstantinovich.
(Prirodnyĭ gaz v khimicheskoĭ promyshlennosti SShA)
Природный газ в химической промышленности США / А. К. Арский, А. П. Горкин. — Москва : ВНИИЭГазпром, 1968.
50, [4] p. — (Транспорт, хранение и потребление газа)
L. C. copy on microfilm only.
At head of title: Ministerstvo gazovoĭ promyshlennosti. Vsesoi͡uznyĭ nauchno-issledovatel'skiĭ institut ėkonomiki, organizat͡sii proizvodstva, i tekhniko-ėkonomicheskoĭ informat͡sii v gazovoĭ promyshlennosti.
Bibliography: p. [52]
——— ——— Microfilm. On 1 reel with other items ; 35 mm.
Microfilm Slavic 12602 TP
1. Gas, Natural—United States. 2. Chemical industries—United States. I. Gorkin, Aleksandr Pavlovich, joint author. II. Title. III. Series: Transport, khranenie i potreblenie gaza.
[TP350.A68] 78–423439

Frost & Sullivan.
Chemical process equipment market. — New York : Frost & Sullivan, 1979.
vii, 246 p. ; 29 cm.
1. Chemical industries—United States. 2. Chemical processes—Equipment and supplies. 3. Market surveys—United States. I. Title.
HD9651.4.F76 1979a 79-114233
381'.45'6602830973
00 MARC

—UNITED STATES—CONGRESSES

Chemical Marketing Research Association.
The future effect of changing energy values on the CPI : papers presented at the November 1978 meeting of the Chemical Marketing Research Association, held in New Orleans, La., November 28-30, 1978. — Staten Island, N.Y. : The Association, [c1979]
228 p. : ill. ; 22 cm.
$12.50
1. Chemical industries—United States—Congresses. 2. Energy policy—United States—Congresses. I. Title.
HD9651.6.C47 1979 338.4'7'6600973 79-106931
79 MARC

—UNITED STATES—EQUIPMENT AND SUPPLIES

Frost & Sullivan.
The market for pumps and associated controls in the chemical process industry. — New York : Frost & Sullivan, 1978.
v, 309 p. : ill. ; 29 cm.
On spine: CPI pumps & associated controls market.
1. Chemical industries—United States—Equipment and supplies. 2. Pumping machinery. 3. Market surveys—United States. I. Title. II. Title: CPI pumps & associated controls market.
HD9651.5.F76 1978 381'.45'62160973 79-112080
79 MARC

—UNITED STATES—FINANCES

Chemical industry sourcebook : materials / selected by MSS Information Corporation from the journals, indexes, and profiles of Disclosure, inc. — [New York] : The Corporation ; Edison, N.J. : exclusive sales agent, Lasky-Lanouette, c1977.
232 p. ; 23 cm.
Includes extracts of data from current and recent corporate reports filed with the Securities and Exchange Commission.
ISBN 0-8422-4011-X : $50.00
1. Chemical industries—United States—Finances. 2. Chemicals—Manufacture and trade—United States—Finances. I. MSS Information Corporation. II. Disclosure, inc. III. United States. Securities and Exchange Commission.
HD9651.5.C48 338.7'66'02573 77-76597
77 MARC

—UNITED STATES—HISTORY

Yamada, Kiyoshi, 1910–
(Amerika kagaku kōgyō no hatten)
アメリカ化學工業の發展 / 山田清著. — 東京 : 東陽書店, 昭和24 [1949]
9, 219 p. ; 19 cm.
Includes bibliographical references.
1. Chemical industries—United States—History. I. Title.
HD9651.5.Y35 78–815484

—UNITED STATES—PERIODICALS

Chemical market abstracts. v. 61–69, no. 8; Jan. 1969–Aug. 1977. Cleveland, Ohio, Predicasts, inc. [etc.]
9 v. 28 cm. monthly.
Continues: CMA. Chemical market abstracts.
Merged with: EMA, equipment market abstracts, ISSN 0098-4779, to form: PROMT. Predicasts overview of markets and technology, ISSN 0161-8032.
Key title: Chemical market abstracts (1969), ISSN 0009-2606
1. Chemical industries — Abstracts — Periodicals. 2. Chemical industries—United States—Periodicals. I. Key title.
HD9651.1.C44 338.4'7'66005 78–646641
MARC-S

—UZBEKISTAN

Pak, T
(Ėkonomicheskie problemy ėffektivnosti khimicheskoĭ promyshlennosti Uzbekistana)
Экономические проблемы эффективности химической промышленности Узбекистана в период строительства коммунизма / Т. Пак. — Ташкент : Фан, 1976.
73, [3] p. ; 22 cm. USSR***
At head of title: Ministerstvo vysshego i srednego spet͡sial'nogo obrazovanii͡a Uzbekskoĭ SSR. Tashkentskiĭ gosudarstvennyĭ universitet im. V. I. Lenina.
Bibliography: p. 71–[74]
0.51rub
1. Chemical industries—Uzbekistan. I. Title.
HD9656.R93U926 78–421666

CHEMICAL INHIBITORS

see also Antioxidants; Catalysts; Corrosion and anti-corrosives; Enzyme inhibitors

—PATENTS

Robinson, J S 1936-
Corrosion inhibitors : recent developments / J. S. Robinson. — Park Ridge, N.J. : Noyes Data Corp., 1979.
xiii, 306 p. : ill. ; 25 cm. — (Chemical technology review ; no. 132)
Includes indexes.
ISBN 0-8155-0757-7 : $42.00
1. Chemical inhibitors—Patents. 2. Corrosion and anti-corrosives—Patents. I. Title. II. Series.
TP201.R62 620.1'1223 79-14637
79 MARC

CHEMICAL INSTRUMENTS see Chemical apparatus

CHEMICAL KINETICS see Chemical reaction, Rate of

CHEMICAL LABORATORIES

—CONGRESSES

The Modern undergraduate laboratory : innovative techniques / David B. Greenberg and Soon-Jai Khang, editors ; Donald K. Anderson ... [et al.]. — New York : American Institute of Chemical Engineers, 1979.
v, 96 p. : ill. ; 28 cm. — (AIChE symposium series ; no. 183, v. 75)
Includes bibliographical references.
1. Chemical laboratories—Congresses. I. Greenberg, David B., 1928- . II. Khang, Soon-Jai, 1944- . III. Series: American Institute of Chemical Engineers. AIChE symposium series ; no. 183.
QD51.M63 542'.1'07 79-800
79 MARC

—SAFETY MEASURES

Green, Michael E
Safety in working with chemicals / Michael E. Green, Amos Turk. — New York : Macmillan, c1978.
ix, 166 p. : ill. ; 24 cm.
Bibliography: p. 159-161.
Includes index.
ISBN 0-02-346420-8
1. Chemical laboratories—Safety measures. I. Turk, Amos, joint author. II. Title.
QD51.G73 542 78-5122
78 MARC

—EUROPE—DIRECTORIES

Pauwels, J
Inventory of activation analysis facilities available in the European Community to industrial users / J. Pauwels. — [Brussels] : Eurisotop Office, Information and Documentation Service, 1975.
85 p. ; 21 cm. — (Eurisotop Office information booklet ; 99) Be***
At head of title: Commission of the European Communities.
"Series: data-2."
1. Chemical laboratories—Europe—Directories. 2. Radioactivation analysis. I. Title. II. Series: Commission of the European Communities. Eurisotop Office. EURISOTOP Office information booklet ; 99.
TP193.A1P38 543'.088 79-308424
79 MARC

CHEMICAL LASERS

—CONGRESSES

Conference on Organic Lasers, General Telephone & Electronics Laboratories, 1964.
Conference on Organic Lasers at the General Telephone & Electronics Laboratories Inc., May 25, 1964 / sponsored by Office of Naval Research, Washington, D.C., and General Telephone & Electronics Laboratories, Inc., Bayside, New York. — [Washington? : The Office?, 1964]
172 p. : ill. ; 28 cm.
"TR 64-052.6."
1. Chemical lasers—Congresses. I. General Telephone & Electronics Laboratories. II. United States. Office of Naval Research.
TA1690.C66 1964 621.36'6 79-307442
79 MARC

Gasdynamic and chemical lasers : proceedings of the international symposium, 11-15 October 1976, Köln / edited by Martin Fiebig, Helmut Hügel. — Köln-Porz, Germany : DFVLR-Press, [1977?]
xi, 571 p. : ill. ; 23 cm. GFR***
English or German.
Symposium organized by Deutsche Gesellschaft für Luft- und Raumfahrt.
Includes bibliographical references.
DM35.00
1. Gasdynamic lasers—Congresses. 2. Chemical lasers—Congresses. I. Fiebig, Martin, 1932- . II. Hügel, Helmut, 1904- . III. Deutsche Gesellschaft für Luft- und Raumfahrt.
TA1695.G36 621.36'6 78-322415
79 MARC

International Symposium on Gas-Flow and Chemical Lasers, 2d, Von Karman Institute for Fluid Dynamics, 1978.
Gas-flow and chemical lasers : [proceedings] / edited by John F. Wendt. — Washington : Hemisphere Pub. Corp., c1979.
vi, 601 p. : ill. ; 25 cm.
"A Von Karman Institute book."
Includes bibliographical references and index.
ISBN 0-89116-147-3
1. Gasdynamic lasers—Congresses. 2. Chemical lasers—Congresses. I. Wendt, John F., 1936- . II. Title.
TA1695.I54 1978 621.36'6 79-12779
79 MARC

Summer Colloquium on Electronic Transition Lasers 4th, Munich, 1977.
High-power lasers and applications : proceedings of the Fourth Colloquium on Electronic Transition Lasers in Munich, June 20-22, 1977 / editors, K.-L. Kompa and H. Walther. — Berlin ; New York : Springer, 1978.
ix, 228, [2] p. : ill. ; 24 cm. — (Springer series in optical sciences ; v. 9) GFR***
Bibliography: p. [229]-[230]
ISBN 0-387-08641-2
1. Lasers—Congresses. 2. Chemical lasers—Congresses. I. Kompa, Karl L. II. Walther, Herbert, 1935- . III. Title.
TA1673.S9 1977 621.36'6 78-315576
78 MARC

CHEMICAL LITERATURE

Antony, Arthur.
Guide to basic information sources in chemistry / by Arthur Antony. — New York : J. Norton Publishers : distributed by Halsted Press, c1979.
vii, 219 p. ; 22 cm. — (Information resources series)
Includes index.
ISBN 0-400-26587-6
1. Chemical literature. I. Title.
QD8.5.A57 1979 540'.7 79-330
79 MARC

Maizell, Robert Edward, 1924-
How to find chemical information : a guide for practicing chemists, teachers, and students / Robert E. Maizell. — New York : Wiley, c1979.
xxiii, 261 p. : ill. ; 24 cm.
"A Wiley-Interscience publication."
Includes bibliographical references and index.
ISBN 0-471-56531-8
1. Chemical literature. I. Title.
QD8.5.M34 540'.7 78-23222
78 MARC

CHEMICAL MICROSCOPY

Fryer, J R
The chemical applications of transmission electron microscopy / J. R. Fryer. — London ; New York : Academic Press, 1979.
x, 286 p. : ill. ; 24 cm. GB79
Bibliography: p. [259]-278.
Includes index.
ISBN 0-12-269350-7
1. Electron microscope, Transmission. 2. Chemical microscopy. I. Title.
QH212.T7F79 542'.8 78-67894
79 MARC

Stoiber, Richard E
Microscopic identification of crystals / Richard E. Stoiber, Stearns A. Morse. — Huntington, N.Y. : R. E. Krieger Pub. Co., 1979, c1972.
vii, 278 p. : ill. ; 24 cm.
Reprint of the ed. published by Ronald Press Co., New York.
Bibliography: p. 265-271.
Includes index.
ISBN 0-88275-975-2
1. Crystal optics. 2. Chemical microscopy. I. Morse, Stearns Anthony, 1931- joint author. II. Title.
[QD941.S72 1979] 548'.9 79-11715
79 MARC

CHEMICAL MILLING

see also Electrochemical cutting

CHEMICAL MUTAGENESIS

Ames, Bruce N
Environmental chemicals causing cancer and genetic birth defects : developing a strategy to minimize human exposure / Bruce N. Ames. — Berkeley : Published for the California Policy Seminar [by] Institute of Governmental Studies, University of California, 1978.
xi, 28 p. : ill. ; 23 cm. — (MOnograph-California Policy Seminar ; no. 2)
Includes bibliographical references.
ISBN 0-87772-261-7
1. Carcinogenesis. 2. Abnormalities, Human—Etiology. 3. Carcinogens. 4. Carcinogenicity testing. 5. Chemical mutagenesis. 6. Environmentally induced diseases—Prevention. I. Title. II. Series: California Policy Seminar. Monograph - California Policy Seminar ; no. 2.
RC268.5.A44 616.9'94'071 78-31762
79 MARC

Canada. Dept. of National Health and Welfare.
The testing of chemicals for carcinogenicity, mutagenicity and teratogenicity. — [Ottawa] : Health and Welfare Canada, 1975.
iv, 183 p. ; 23 cm. C***
Includes bibliographies.
1. Carcinogenicity testing. 2. Chemicals—Testing. 3. Chemical mutagenesis. 4. Teratogenic agents. I. Title.
RC268.65.C36 1975 615.9 79-317251
79 MARC

Fishbein, Lawrence.
Potential industrial carcinogens and mutagens / Lawrence Fishbein. — Amsterdam ; New York : Elsevier Scientific Pub. Co. ; New York : distributors for the U.S. and Canada, 1979.
x, 534 p. : ill. ; 25 cm. — (Studies in environmental science ; 4)
Includes bibliographical references and index.
ISBN 0-444-41777-X
1. Carcinogens. 2. Chemical mutagenesis. 3. Industrial toxicology. I. Title. II. Series.
[DNLM: 1. Carcinogens, Environmental. 2. Industrial medicine. 3. Mutagens. QZ202.3 F532p]
RC268.6.F57 615.9'02 78-27560
78 MARC

Tsang, Wing-sum.
Metabolic activation of polynuclear aromatic hydrocarbons / by Wing-sum Tsang and Gary W. Griffin. — 1st ed. — Oxford ; New York : Pergamon Press, 1979.
vii, 125 p. : ill. ; 22 cm. GB79
Includes bibliographical references and indexes.
ISBN 0-08-023835-1 : £12.50 ($25.00)
1. Hydrocarbons—Physiological effect. 2. Carcinogenesis. 3. Hydrocarbons—Metabolism. 4. Chemical mutagenesis. I. Griffin, Gary Walter, 1931- joint author. II. Title.
RC268.7.H9T73 1979 616.9'94'071 79-40432
79 MARC

—COLLECTED WORKS

Chemical mutagens. v. 1–
New York, Plenum, 1971–
v. ill. 24 cm.
"Principles and methods for their detection."
Vols. 1– sponsored by the Environmental Mutagen Society.
Key title: Chemical mutagens, ISSN 0093-6855.
1. Chemical mutagenesis—Collected works. I. Environmental Mutagen Society.
QH465.C5C45 575.2'92 79-640909
MARC-S

—CONGRESSES

Assessing chemical mutagens : the risk to humans / edited by Victor K. McElheny, Seymour Abrahamson. — [Cold Spring Harbor, N.Y.] : Cold Spring Harbor Laboratory, 1979.
xiii, 367 p. : ill. ; 23 cm. — (Banbury report ; 1)
Edited proceedings of a meeting sponsored by the Office of Toxic Substances of the U.S. Environmental Protection Agency in May 1978 at Banbury Center, Lloyd Harbor, N.Y.
ISBN 0-87969-200-6 : $30.00
1. Chemical mutagenesis—Congresses. 2. Mutagenicity testing—Congresses. I. McElheny, Victor K. II. Abrahamson, Seymour. III. United States. Environmental Protection Agency. Office of Toxic Substances. IV. Series: Banbury Center. Banbury report ; 1.
QH465.C5A85 615.9 79-998
79 MARC

Chemical mutagenesis in microorganisms and plants : report of symposium held on July 25 to 26, 1972. — Ibaraki-ken, Japan : Institute of Radiation Breeding, 1972.
iii, 96 p. : ill. ; 27 cm. — (Gamma field symposia ; no. 11)
English with summaries in Japanese.
Includes bibliographies.
1. Chemical mutagenesis—Congresses. 2. Microbial mutation—Congresses. 3. Plant mutation—Congresses. I. Hōshasen Ikushujō. II. Title. III. Series.
QH465.A1C43 575.2'92 77-152424
79 MARC

Strategies for short-term testing for mutagens/carcinogens / editor, Byron E. Butterworth. — West Palm Beach, Fla. : CRC Press, c1979.
134 p. : ill. ; 26 cm. — (CRC toxicology series)
Proceedings of the Chemical Industry Institute of Toxicology workshop, Research Triangle Park, N.C., Aug. 11-12, 1977.
Includes bibliographies and index.
ISBN 0-8493-5661-X
1. Chemical mutagenesis—Congresses. 2. Carcinogens—Testing—Congresses. I. Butterworth, Byron E. II. Chemical Industry Institute of Toxicology. III. Series: Chemical Rubber Company, Cleveland. CRC toxicology series.
QH465.C5S77 575.2'92 78-12587
78 MARC

Symposium on Mutagenicity, Carcinogenicity, and Teratogenicity of Chemicals, Maharaja Sayajirao University of Baroda, 1975.
Proceedings of the Symposium on Mutagenicity, Carcinogenicity, and Teratogenicity of Chemicals, M.S. University, Baroda, December 4-6, 1975 / organised by the Basic Sciences Committee II (Biology & Medical), Department of Atomic Energy, Government of India. — [Bombay] : The Symposium, [1975?]
iii, 541, 12 p. : ill. ; 25 cm.
Includes bibliographical references and index.
1. Carcinogenesis—Congresses. 2. Chemical mutagenesis—Congresses. 3. Teratogenic agents—Congresses. I. India (Republic). Dept. of Atomic Energy. Basic Sciences Committee, II (Biology and Medical).
RC268.5.S956 1975 615.9 78-901144
79 MARC

CHEMICAL NERVE TRANSMITTERS see Neurotransmitters

CHEMICAL OCEANOGRAPHY

see also Sea-water—Composition; Water chemistry

Riley, John Price, ed.
Chemical oceanography / edited by J. P. Riley and G. Skirrow. — 2d ed. — London ; New York : Academic Press, 1975-
v. : ill. ; 24 cm. GB***
V. 5- edited by J. P. Riley and R. Chester.
Includes bibliographies and indexes.
ISBN 0-12-588601-2 (v. 1)
1. Chemical oceanography. I. Skirrow, Geoffrey, joint ed. II. Chester, Roy, 1936- joint ed.
GC111.2.R542 1975 551.4'601 74-5679
75[r79]rev2 MARC

—CONGRESSES

Groupement pour l'avancement de la biochimie marine.
Les substances organiques naturelles dissoutes dans l'eau de mer : colloque du Groupement pour l'avancement de la biochimie marine, [Marseille], 18-19 novembre 1976. — Paris : Centre national de la recherche scientifique, 1978.
269 p. : ill. ; 30 cm. — (Actualités de biochimie marine) F***
Summaries in French and English.
Includes bibliographies.
ISBN 2-222-02091-3 : 85.00F
1. Marine biology—Congresses. 2. Biological chemistry—Congresses. 3. Chemical oceanography—Congresses. I. Title. II. Series.
QH91.8.B5G76 1978 574.92 79-352697
79 MARC

—BALTIC SEA

Bloomquist, R Gordon.
Chemical characteristics of interstitial waters from the Southern Baltic Sea / by R. Gordon Bloomquist. — Stockholm : Dept. of Geology, University of Stockholm, 1977.
111 p. : graphs ; 29 cm. S***
"Additions to selected bibliography" published as suppl. (1 p.) and inserted at end.
Errata slip inserted.
Bibliography: p. 77-89.
ISBN 9172221941
1. Chemical oceanography—Baltic Sea. 2. Marine sediments—Baltic Sea. I. Title.
GC571.B57 551.4'613'64 78-318596
78 MARC

—MEXICO, GULF OF—MAPS

Chemistry, primary productivity, and benthic algae of the Gulf of Mexico / S. Z. El-Sayed ... [et al.]. — New York : American Geographical Society, 1972.
1 portfolio (29 p., 6 fold. leaves of plates) : ill., maps ; 41 cm. — (Serial atlas of the marine environment ; fol. 22) (Contribution, Texas A&M University, Department of Oceanography ; no. 502)
Cover title.
Scale of maps 1:9,000,000.
Includes bibliographical references.
1. Chemical oceanography—Mexico, Gulf of—Maps. 2. Algae—Mexico, Gulf of—Maps. 3. Marine biology—Mexico, Gulf of. I. el-Sayed, Sayed Zakaria, 1926- II. Series: American Geographical Society of New York. Serial atlas of the marine environment, fol. 22. III. Series: Texas. A & M University, College Station. Dept. of Oceanography. Contribution ; no. 502.
G1046.C7A4 fol. 22 912'.1'55146008 s 79-306768
[G1107.M4] 79 MARC

CHEMICAL PLANTS

see also Plastics plants

—AUTOMATION

Automatisierte Steuerung kontinuierlicher Produktionsprozesse = Автоматизированное управление непрерывными производственными процессами. — Berlin : Akademie-Verlag, 1975.
xiv, 257 p. : ill. ; 22 cm. — (Kybernetische Methoden in der Chemie und chemischen Technologie) GFR 75-A
Chiefly in German with one contribution in Russian; table of contents also in Russian.
Summaries in English.
Includes bibliographical references.
1. Chemical plants—Automation. I. Title: Avtomatizirovannoe upravlenie nepreryvnymi proizvodstvennymi proßessami. II. Series.
TP155.5.A82 660.2'81 76-479269
(MARC)

Golubiatnikov, Vladimir Alekseevich.
(Avtomatizaßiâ proizvodstvennykh proßessov i ASUP v khimicheskoĭ promyshlennosti)
Автоматизация производственных процессов и АСУП в химической промышленности : [Учебник для хим.-технол. техникумов] / В. А. Голубятников, В. В. Шувалов. — Москва : Химия, 1978.
375 p. : ill. ; 22 cm. — (Автоматизация химических производств) USSR 78
Series romanized: Avtomatizaßiâ khimicheskikh proizvodstv.
Bibliography: p. 366-369.
Includes index.
1.10rub
1. Chemical process control. 2. Chemical plants—Automation. I. Shuvalov, Valeriĭ Vasil'evich, joint author. II. Title.
TP155.75.G649 79-410924

Process flowsheeting / A. W. Westerberg ... [et al.]. — Cambridge ; New York : Cambridge University Press, 1979.
ix, 251 p. : ill. ; 24 cm.
Bibliography: p. [240]-244.
Includes index.
ISBN 0-521-22043-2 : $22.00
1. Chemical process control. 2. Chemical plants—Automation. I. Westerberg, A. W.
TP155.75.P75 660.2'81 78-56182
78 MARC

—DESIGN AND CONSTRUCTION

Hanyang Hwahak Chusik Hoesa.
(Yŏch'ŏn sŏgyu hwahak kongjang kŏnsŏl saŏp kyehoek)
麗川石油化學工場建設事業計劃. — [서울] : 韓洋化學株式會社, 1974.
11 leaves : graph ; 27 cm.
Bound with the author's Yŏch'ŏn sŏgyu hwahak kyenyŏl kongjang kŏnsŏl saŏp kyehoeksŏ. [Sŏul], 1974.
1. Chemical plants—Design and construction. 2. Chemical plants—Korea—Yŏch'ŏn. I. Title. II. Title: Saŏp kyehoek.
TP155.5.H37 1974 79-825509

Hanyang Hwahak Chusik Hoesa.
(Yŏch'ŏn sŏgyu hwahak kyenyŏl kongjang kŏnsŏl saŏp kyehoeksŏ)
麗川石油化學系列工場建設事業計劃書 : 低密度 폴리에틸렌·브이씨엠·이디씨. — [서울] : 韓洋化學株式會社, 1974.
66 leaves : ill. ; 27 cm.
Bound with the author's Yŏch'ŏn sŏgyu hwahak kongjang kŏnsŏl saŏp kyehoek. [Sŏul], 1974.
1. Chemical plants—Design and construction. 2. Chemical plants—Korea—Yŏch'ŏn. I. Title. II. Title: Saŏp kyehoeksŏ.
TP155.5.H37 1974 79-825510

Rose, J C
A guide to project procedure / by J. C. Rose, G. L. Wells, B. H. Yeats. — Rugby, Eng. : Institution of Chemical Engineers, c1978.
50 p., [4] fold p. : diags., tables ; 22 cm. GB***
Bibliography: p. 32.
ISBN 0-85295-108-6 : £3.50
1. Chemical plants—Design and construction. I. Wells, G. L., joint author. II. Yeats, B. H., joint author. III. Title.
TP155.5.R58 79-315952
79 MARC

—DESIGN AND CONSTRUCTION—CONGRESSES

Canadian Chemical Engineering Conference, 27th, Calgary, 1977.
Design of processing plants : presented at 27th Canadian Chemical Engineering Conference, Calgary, Alberta, 23-26 October, 1977. — [Ottawa : Canadian Society for Chemical Engineering, 1977?]
196 p. : ill. ; 28 cm. C***
Includes bibliographical references.
1. Chemical plants—Design and construction—Congresses. I. Canadian Society for Chemical Engineering. II. Title.
TP155.5.C36 1977 660.2'81 78-322158
78 MARC

Symposium on Chemical Process Hazards with Special Reference to Plant Design, 6th, University of Manchester, 1977.
Papers of the Symposium on Chemical Process Hazards with Special Reference to Plant Design, VI : a symposium organised by the North Western Branch of the Institution of Chemical Engineers at the University of Manchester Institute of Science and Technology, 5-7 April 1977, symposium organisers, J. H. Burgoyne ... [et al.]. — Rugby : I. Chem. E. Services for the Institution of Chemical Engineers, 1977.
[8], 152 p. : ill. ; 30 cm. — (Institution of Chemical Engineers symposium series ; no. 49 ISSN 0307-0492) GB77-19507
Includes bibliographical references.
ISBN 0-8429-5100-0 : £12.50
1. Chemicals—Manufacture and industry—Safety measures—Congresses. 2. Chemical plants—Design and construction—Congresses. I. Institution of Chemical Engineers, London. North-Western Branch. II. Series: Institution of Chemical Engineers, London. I. Chem. E. symposium series ; no. 49.
TP149.S97 1977 660.2'804 77-377550
78[r79]rev MARC

—ENVIRONMENTAL ASPECTS

Chizhikov, V A
Meropriâtiâ po sanitarnoĭ okhrane atmosfernogo vozdukha na proizvodstvakh osnovnoĭ khimii / V. A. Chizhikov. — Moskva : TSentr. in-t usovershenstvovaniâ vracheĭ, 1968.
26 p.
At head of title: Ministerstvo zdravookhraneniâ SSSR.
L.C. copy on microfilm only.
——— Microfilm. On 1 reel with other items ; 35 mm.
Microfilm Slavic 13025 TD
1. Chemical plants—Environmental aspects. 2. Air—Pollution. I. Title.
[TD888.C5C48] 79-367068

Kuznetsov, Ivan Efimovich.
(Zashchita vozdushnogo basseĭna ot zagrâzneniâ vrednymi veshchestvami khimicheskikh predpriâtiĭ)
Защита воздушного бассейна от загрязнения вредными веществами химических предприятий / И. Е. Кузнецов, Т. М. Троицкая. — Москва : Химия, 1979.
340 p. : ill. ; 22 cm. USSR 79
Bibliography: p. 330-[334]
Includes index.
1.50rub
1. Chemical plants—Environmental aspects. 2. Air—Pollution. I. Troiţskaiâ, Tat'iâna Mikhaĭlovna, joint author. II. Title.
TD888.C5K89 79-413221

—EQUIPMENT AND SUPPLIES

Evans, Frank L
Equipment design handbook for refineries and chemical plants / Frank L. Evans, Jr. — 2d ed. — Houston, Tex. : Book Division Gulf Pub. Co., c1979-
v. 1 : ill. ; 29 cm.
Includes bibliographies and index.
ISBN 0-87201-254-9 (v. 1) : $24.95
1. Petroleum refineries—Equipment and supplies—Design and construction. 2. Chemical plants—Equipment and supplies. I. Title.
TP690.3.E89 1979 660.2'83 79-50245
79 MARC

Genkin, Avgust Émanuilovich.
Oborudovanie khimicheskikh zavodov : [Uchebnik dlia khim.-tekhnol. i khim.-mekh. tekhnikumov] / A. E. Genkin. — 3. izd., dop. i pererab. — Moskva : Vyssh. shkola, 1978.
272 p. : ill. ; 22 cm. USSR78
Bibliography: p. 269-[270]
0.95rub
1. Chemical plants—Equipment and supplies. I. Title.
TP155.5.G4 1978 79-355307
79 MARC

—EQUIPMENT AND SUPPLIES—CORROSION

Fokin, M N
(Titan i ego splavy v khimicheskoĭ promyshlennosti)
Титан и его сплавы в химической промышленности : справочное пособие / М. Н. Фокин, Ю. С. Рускол, А. В. Мосолов. — Ленинград : Химия, Ленинградское отд-ние, 1978.
198, [2] p. : ill. ; 22 cm. USSR***
Bibliography: p. 185-[199]
0.80rub
1. Titanium—Corrosion. 2. Titanium alloys—Corrosion. 3. Chemical plants—Equipment and supplies—Corrosion. I. Ruskol, IU. S., joint author. II. Mosolov, A. V., joint author. III. Title.
TA480.T54F64 79-409865

IUrchenko, IU F
(Korroziâ svarnykh soedineniĭ v okislitel'nykh sredakh)
Коррозия сварных соединений в окислительных средах / Ю. Ф. Юрченко, Г. И. Агапов. — Москва : Машиностроение, 1976.
150 p. : ill. ; 22 cm. USSR 77
Bibliography: p. 144-[148]
0.56rub
1. Welded joints—Corrosion. 2. Chemical plants—Equipment and supplies—Corrosion. I. Agapov, Gennadiĭ Ignat'evich, joint author. II. Title.
TA492.W4 I 9 78-422199

—EQUIPMENT AND SUPPLIES—DESIGN AND CONSTRUCTION

Vinogradov, Boris Petrovich.
Tekhnologiâ apparatostroeniâ : sredstva mekhanizaßii i avtomatizaßii v apparatostroenii : uchebno-metodicheskoe posobie po speßial'nosti Mashiny i apparaty khimicheskikh proizvodstv / [Vinogradov, Boris Petrovich]. — Moskva : Vysshaia shkola, 1968.
70 p. : ill.
At head of title: Ministerstvo vysshego i srednego speßial'nogo obrazovaniâ RSFSR. Glavnoe upravlenie inzhenerno-tekhnicheskikh vuzov. Moskovskiĭ institut neftekhimicheskoĭ i gazovoĭ promyshlennosti im. I. M. Gubkina.
L.C. copy on microfilm only.
——— Microfilm. On 1 reel with other items ; 35 mm.
Microfilm Slavic 13058 TP
1. Chemical plants—Equipment and supplies—Design and construction. I. Title.
[TP155.5.V54] 79-366974

CHEMICAL PLANTS (Continued)

—EQUIPMENT AND SUPPLIES
—MAINTENANCE AND REPAIR

Grinberg, I͡An Il'ich.
(Slesarnye raboty pri remonte i naladke khimicheskogo oborudovanii͡a)
Слесарные работы при ремонте и наладке химического оборудования : ⌊учеб. пособие для сред. проф. -техн. училищ⌋ / Я. И. Гринберг. — Изд. 2-е, перераб. и доп. — Москва : Высш. школа, 1974.
239 p. : ill. ; 23 cm. — (Профтехобразование : Химическая промышленность) USSR 75
Series romanized: Proftekhobrazovanie : Khimicheskai͡a promyshlennost'.
Bibliography: p. 238.
0.48rub
1. Chemical plants—Equipment and supplies—Maintenance and repair. I. Title.
TP157.G68 1974 78-423516

Shlykov, Aleksandr Ermolaevich.
(Stroitel'no-montazhnye raboty po kapital'nomu remontu osnovnykh fondov)
Строительно-монтажные работы по капитальному ремонту основных фондов : на примере предприятий химической промышленности БССР / А. Е. Шлыков. — Минск : Наука и техника, 1978.
172 p. : ill. ; 21 cm. USSR***
Includes bibliographical references.
1.00rub
1. Chemical plants—Equipment and supplies—Maintenance and repair. I. Title.
TP155.5.S49 79-411390

—EQUIPMENT AND SUPPLIES—
WELDING—CONGRESSES

Fabrication and reliability of welded process plant : an international conference, London, 16-18 November, 1976. — Abington : Welding Institute, c1977-c1978.
2 v. : ill. ; 30 cm. GB***
"Conference technical director: P. T. Houldcroft."
Includes bibliographical references.
CONTENTS: v. 1. Papers.—v. 2. Discussion.
1. Welded joints—Congresses. 2. Welded joints—Testing—Congresses. 3. Chemical plants—Equipment and supplies—Welding—Congresses. 4. Reliability (Engineering)—Congresses. I. Houldcroft, Peter Thomas. II. Title: Welded process plant.
TA492.W4F27 671.5'2'042 78-309854
78⌊r79⌋rev MARC

—MAINTENANCE AND REPAIR

Shklovskiĭ, Ėduard Isaakovich.
Remontnai͡a sluzhba na khimicheskikh predprii͡atii͡akh / Ė. I. Shklovskiĭ, V. S. Vostrikov. — Moskva : Khimii͡a, 1978.
167 p. ; 20 cm. — (Bibliotechka mekhanika khimicheskikh i neftekhimicheskikh proizvodstv) USSR78
Bibliography: p. 167.
0.45rub
1. Chemical plants—Maintenance and repair. I. Vostrikov, Vladimir Semenovich, joint author. II. Title. III. Series.
TP155.5.S48 79-345950
79 MARC

—MANAGEMENT

Kos'min, Nikolaĭ Konstantinovich.
(Rat͡sional'nai͡a organizat͡sii͡a ucheta na predprii͡atii͡akh khimicheskoĭ promyshlennosti)
Рациональная организация учета на предприятиях химической промышленности / Н. К. Косьмин. — Киев : УкрНИИНТИ, 1968.
23 p.
At head of title: Gosudarstvennyĭ planovyĭ komitet Soveta Ministrov USSR. Ukrainskiĭ nauchno-issledovatel'skiĭ institut nauchno-tekhnicheskoĭ informat͡sii i tekhniko-ėkonomicheskikh issledovaniĭ. Ministerstvo khimicheskoĭ promyshlennosti Ukrainskoĭ SSR.
LC copy on microfilm only.
——— Microfilm. On 1 reel with other items ; 35 mm.
Microfilm Slavic 12555 TP
1. Chemical plants—Management. I. Title.
[TP155.5.K67] 78-423647

—PERIODICALS

Institut für Chemieanlagen.
Mitteilungen - Institut für Chemieanlagen.
Berlin. VEB Verlag Technik.
v. 30 cm. quarterly.
1. Chemical plants—Periodicals.
TP155.5.I 57a 660.2'8'005 78-647010
MARC-S

—PIPE LINES

Frost & Sullivan.
The pipe and fittings market in the chemical process industries / Frost & Sullivan. — New York : Frost & Sullivan, c1979.
iii, a-b, 216 p. ; 29 cm.
On spine: Pipes & fittings in chemical process industry (CPI).
1. Pipe industry—United States. 2. Chemical plants—Pipe lines. 3. Pipe-fittings. 4. Market surveys—United States. I. Title.
HD9529.P53U63 381'.45'67183 79-105843
79 MARC

—SAFETY MEASURES

Institution of Chemical Engineers, London. Loss Prevention Panel.
A first guide to loss prevention / ⌊written by members of the Loss Prevention Panel ; edited by W. I. McCrindle⌋. — Rugby ⌊Eng.⌋ : I. of Chemical E. Services for the Institution of Chemical Engineers, c1977, 1978 printing.
ii, 26 p. ; 22 cm. GB***
Cover title.
Includes bibliographical references.
ISBN 0-85295-106-X : £1.80
1. Chemical plants—Safety measures. I. McCrindle, W. I. II. Title.
TP155.5.I58 1977 658.5'6 79-310521
79 MARC

—WASTE DISPOSAL—HYGIENIC
ASPECTS—NEW YORK (STATE)—
NIAGARA FALLS (CITY)

New York (State). Governor's Love Canal Inter-Agency Task Force.
Love Canal, public health time bomb : a special report to the Governor and Legislature / ⌊prepared by the Office of Public Health, Governor's Love Canal Inter-Agency Task Force⌋. — ⌊Albany⌋ : State of New York, Dept. of Health, 1978.
32 p. : ill ; 28 cm.
Cover title.
1. Pollution—New York (State)—Niagara Falls (City) 2. Sanitary landfills—New York (State)—Niagara Falls (City) 3. Chemical plants—Waste disposal—Hygienic aspects—New York (State)—Niagara Falls (City) 4. Pollution—Toxicology—New York (State)—Niagara Falls (City) I. Title.
TD181.N7N46 1978 614.7'6 79-621404
79 MARC

—INDONESIA—SUMATERA SELATAN

Sumatera Selatan. Team Survey Kemungkinan Pendirian Pabrik Asam Semut.
Laporan survey kemungkinan pendirian pabrik asam semut di Propinsi Sumatera Selatan. — ⌊Palembang⌋ : Pemerintah Daerah Propinsi Sumatera Selatan, Team Survey Kemungkinan Pendirian Pabrik Asam Semut di Propinsi Sumatera Selatan, ⌊1972⌋
iv, 19, 6 leaves ; 31 cm.
1. Formic acid. 2. Chemical plants—Indonesia—Sumatera Selatan. I. Title.
TP248.F63S84 1972 72-941525

--KOREA--YŎCH'ŎN

Hanyang Hwahak Chusik Hoesa.
(Yŏch'ŏn sŏgyu hwahak kongjang kŏnsŏl saŏp kyehoek)
麗川石油化學工場建設事業計劃. — ⌊서울⌋ : 韓洋化學株式會社, 1974.
11 leaves : graph ; 27 cm.
Bound with the author's Yŏch'ŏn sŏgyu hwahak kyenyŏl kongjang kŏnsŏl saŏp kyehoeksŏ. ⌊Sŏul⌋, 1974.
1. Chemical plants—Design and construction. 2. Chemical plants—Korea—Yŏch'ŏn. I. Title. II. Title: Saŏp kyehoek.
TP155.5.H37 1974 79-825509

Hanyang Hwahak Chusik Hoesa.
(Yŏch'ŏn sŏgyu hwahak kyenyŏl kongjang kŏnsŏl saŏp kyehoeksŏ)
麗川石油化學系列工場建設事業計劃書 : 低密度폴리에틸렌·브이씨엠·이디씨. — ⌊서울⌋ : 韓洋化學株式會社, 1974.
66 leaves : ill. ; 27 cm.
Bound with the author's Yŏch'ŏn sŏgyu hwahak kongjang kŏnsŏl saŏp kyehoek. ⌊Sŏul⌋, 1974.
1. Chemical plants—Design and construction. 2. Chemical plants—Korea—Yŏch'ŏn. I. Title. II. Title: Saŏp kyehoeksŏ.
TP155.5.H37 1974 79-825510

—ONTARIO—WASTE DISPOSAL

Ball, W L
Emissions of vinyl chloride to the ambient air around manufacturing facilities in Ontario : a summary of three reports produced by Air Resources Branch of the Ontario Ministry of the Environment / prepared by W. L. Ball for the Technology Development Branch, Air Pollution Control Directorate. — Ottawa : Environmental Protection Service, 1977.
iii, 6 p. ; 28 cm. — (Surveillance report EPS ; 5-AP-77-14) C***
Summary also in French.
Bibliography: p. 6.
1. Vinyl chloride—Environmental aspects—Ontario. 2. Air—Pollution—Ontario. 3. Chemical plants—Ontario—Waste disposal. I. Ontario. Air Resources Branch. II. Canada. Air Pollution Control Directorate. Technology Development Branch. III. Title. IV. Series: Canada. Environmental Protection Service. Surveillance report - Environmental Protection Service ; 5-AP-77-14.
TD887.V55B34 79-316110
79 MARC

—RUSSIA—EQUIPMENT AND SUPPLIES

United States. Central Intelligence Agency. National Foreign Assessment Center.
Soviet chemical equipment purchases from the West : impact on production and foreign trade. — ⌊Washington⌋ : Central Intelligence Agency, National Foreign Assessment Center, 1978.
v, 34 p. ; 27 cm. — (A Research paper - Central Intelligence Agency, National Foreign Assessment Center)
Cover title.
"Er 78-10554."
1. Chemical industries—Russia. 2. Chemical plants—Russia—Equipment and supplies. 3. Russia—Commerce. I. Title. II. Series: United States. Central Intelligence Agency. National Foreign Assessment Center. A research paper - Central Intelligence Agency, National Foreign Assessment Center.
HD9656.R92U53 1978 338.4'7'681766 79-601092
79 MARC

—SWEDEN—WASTE DISPOSAL

Sweden. Utredningen rörande omhändertagande och behandling av kemiskt avfall m.m.
Kemiskt avfall-uppgifter och organisation : betänkande / avgivet av Utredningen rörande omhändertagande och behandling av kemiskt avfall m. m. (UKA). — ⌊Stockholm : Jordbruksdepartementet⌋, 19
v. ; 30 cm. S***
"Ds Jo 1973:1."
1. Chemical plants—Sweden—Waste disposal. I. Title.
TD899.C5S93 1973 78-354846
78 MARC

CHEMICAL POLLUTION see Pollution

CHEMICAL PROCESS CONTROL

Badea, L
Modelarea bilanţurilor de materiale ale schemelor complexe / Leonard Badea. — Bucureşti : Editura Tehnică, 1978.
182 p. : ill. ; 25 cm. R79-389
Summary in English.
Bibliography: p. 181-182.
lei19.50
1. Chemical process control. I. Title.
TP155.75.B33 79-363507
•79 MARC

Golubi͡atnikov, Vladimir Alekseevich.
(Avtomatizat͡sii͡a proizvodstvennykh prot͡sessov i ASUP v khimicheskoĭ promyshlennosti)
Автоматизация производственных процессов и АСУП в химической промышленности : ⌊Учебник для хим.-технол. техникумов⌋ / В. А. Голубятников, В. В. Шувалов. — Москва : Химия, 1978.
375 p. : ill. ; 22 cm. — (Автоматизация химических производств) USSR 78
Series romanized: Avtomatizat͡sii͡a khimicheskikh proizvodstv.
Bibliography: p. 366-369.
Includes index.
1.10rub
1. Chemical process control. 2. Chemical plants—Automation. I. Shuvalov, Valeriĭ Vasil'evich, joint author. II. Title.
TP155.75.G649 79-410924

Ostrovskiĭ, Gennadiĭ Markovich.
(Metody optimizat͡sii khimicheskikh reaktorov)
Методы оптимизации химических реакторов. Г. М. Островский, Ю. М. Волин. Москва, Химия, 1967.
247 p. illus. 22 cm.
Bibliography: p. 241-⌊244⌋
1. Chemical reactors. 2. Chemical process control. 3. Mathematical optimization. I. Volin, I͡Uriĭ Markovich. II. Title.
TP157.O88 79-407805

CHEMICAL PROCESS CONTROL (Continued)

Process flowsheeting / A. W. Westerberg ... [et al.]. — Cambridge ; New York : Cambridge University Press, 1979.
ix, 251 p. : ill. ; 24 cm.
Bibliography: p. [240]-244.
Includes index.
ISBN 0-521-22043-2 : $22.00
1. Chemical process control. 2. Chemical plants—Automation. I. Westerberg, A. W.
TP155.75.P75 660.2'81 78-56182
78 MARC

Sklfarskiĭ, Éduard Isaakovich.
(Avtomatizirovannaña sistema fsentralizovannogo kontrolña i upravlenña "Neft'-1")
Автоматизированная система централизованного контроля и управления "Нефть-1" / Э. И. Склярский, Г. И. Грибакин, Л. М. Шиб. — Москва : Химия, 1977.
167, [1] p. : ill. ; 22 cm. USSR***
Bibliography: p. 167-[168]
0.57rub
1. Chemical process control. 2. Remote control. I. Gribakin, Gennadiĭ Ivanovich, joint author. II. Shib, Lfubomir Mikhaĭlovich, joint author. III. Title.
TP155.75.S54 78-421256

Smigelschi, O
Optimizarea proceselor în industria chimică / Octavian Smigelschi, Alexandru Woinaroschy. — Bucureşti : Editura Tehnică, 1978.
423 p., [1] fold. leaf of plates : ill. ; 25 cm. — (Seria Ingineria chimică)
R78-5266
Summary in English.
Includes bibliographies.
lei35.00
1. Chemical process control. I. Woinaroschy, Alexandru, joint author. II. Title.
TP155.75.S57 79-355060
*79 MARC

Trybalski, Zdzisław.
Automatyzacja procesów chemicznych / Zdzisław Trybalski. — Gliwice : PŚ, 1978 (Gliw. : PŚ)
256 p. : ill. ; 24 cm. — (Skrypty centralne wyższych studiów technicznych dla pracujących ; Nr 570/27) P78-6469
At head of title: Politechnika Śląska im. W. Pstrowskiego.
Bibliography: p. [255]-256.
zł29.00
1. Chemical process control. I. Title.
TP155.75.T79 79-350066
79 MARC

—DATA PROCESSING

Lufsenko, Valentin Antonovich.
(Matematicheskoe modelirovanie khimiko-tekhnologicheskikh profsessov na analogovykh vychislitel'nykh mashinakh)
Математическое моделирование химико-технологических процессов на аналоговых вычислительных машинах : лабораторно-практические работы / В. А. Луценко, Л. Н. Финякин. — Москва : Химия, 1975.
355, [1] p. : ill. ; 22 cm. — (Химическая кибернетика) USSR***
"Допущено ... в качестве учебного пособия для студентов химико-технологических специальностей вузов."
Series romanized: Khimicheskaña kibernetika.
Bibliography: p. [336]
0.89rub
1. Chemical process control—Data processing. 2. Chemical process control—Mathematical models. 3. Analog computers. I. Finfakin, Leonid Nikolaevich, joint author. II. Title.
TP155.75.L87 79-413624

(Upravlenie élementarnymi khimicheskimi profsessami i postroenie avtomaticheskikh sistem s primeneniem vychislitel'nykh mashin)
Управление элементарными химическими процессами и построение автоматических систем с применением вычислительных машин / под ред. Бурового, И. А. — Москва : Металлургия, 1967.
324 p. (Сборник научных трудов - Государственный научно-исследовательский институт цветных металлов ; № 25)
At head of title: Ministerstvo fsvetnoĭ metallurgii SSSR.
L.C. copy on microfilm only.
Includes bibliographies.
——— Microfilm. On 1 reel with other items ; 35 mm.
Microfilm Slavic 12747 TN
1. Chemical process control—Mathematical models. 2. Chemical process control—Data processing. I. Burovoĭ, Isaak Abramovich. II. Russia (1923- U. S. S. R.). Ministerstvo fsvetnoĭ metallurgii. III. Series: Moscow. Gosudarstvennyĭ nauchno-issledovatel'skiĭ institut fsvetnykh metallov. Sbornik nauchnykh trudov : no. 25.
[TN758.M62 no. 25] 79-408559
[TP155.75]

—DATA PROCESSING
—CONGRESSES

Chemdata 77, Espoo, Finland, 1977.
Computer applications in the analysis of chemical data and plants. — Princeton, N.J. : Science Press, [1978?]
277 p. : ill. ; 30 cm.
"Plenary and main lectures presented at the Chemdata 77, held in Espoo Korpilampi, Finland, 9-10 June 1977."
Includes bibliographies.
1. Chemical process control—Data processing—Congresses. I. Title.
TP155.75.C44 1977 660.2'8'02854 78-108573
78 MARC

—MATHEMATICAL MODELS

Himmelblau, David Mantner, 1923-
Fault detection and diagnosis in chemical and petrochemical processes / David M. Himmelblau. — Amsterdam ; New York : Elsevier Scientific Pub. Co. ; New York : distributors for the US and Canada, Elsevier North-Holland, 1978.
viii, 414 p. : ill. ; 25 cm. — (Chemical engineering monographs ; v. 8)
Includes bibliographies and index.
ISBN 0-444-41747-8 : $50.00
1. Chemical process control—Mathematical models. I. Title.
TP155.75.H55 660.2'81 78-13953
78 MARC

Lufsenko, Valentin Antonovich.
(Matematicheskoe modelirovanie khimiko-tekhnologicheskikh profsessov na analogovykh vychislitel'nykh mashinakh)
Математическое моделирование химико-технологических процессов на аналоговых вычислительных машинах : лабораторно-практические работы / В. А. Луценко, Л. Н. Финякин. — Москва : Химия, 1975.
355, [1] p. : ill. ; 22 cm. — (Химическая кибернетика) USSR***
"Допущено ... в качестве учебного пособия для студентов химико-технологических специальностей вузов."
Series romanized: Khimicheskaña kibernetika.
Bibliography: p. [336]
0.89rub
1. Chemical process control—Data processing. 2. Chemical process control—Mathematical models. 3. Analog computers. I. Finfakin, Leonid Nikolaevich, joint author. II. Title.
TP155.75.L87 79-413624

Systemverfahrenstechnik / Autoren, G. Gruhn ... [et al.]. — 1. Aufl. — Leipzig : Deutscher Verlag für Grundstoffindustrie, 1976-1978.
2 v. : ill. ; 21 cm. — (Verfahrenstechnik) GDR76-A
Includes bibliographical references and indexes.
CONTENTS: 1. Modellierung und Simulation verfahrenstechnischer Systeme.—2. Bewertung, Optimierung und Synthese verfahrenstechnischer Systeme.
16.00M (v. 1) 13.20M (v. 2)
1. Chemical process control—Mathematical models. I. Gruhn, G.
TP155.7.S96 77-470563
77[r78]rev MARC

(Upravlenie élementarnymi khimicheskimi profsessami i postroenie avtomaticheskikh sistem s primeneniem vychislitel'nykh mashin)
Управление элементарными химическими процессами и построение автоматических систем с применением вычислительных машин / под ред. Бурового, И. А. — Москва : Металлургия, 1967.
324 p. (Сборник научных трудов - Государственный научно-исследовательский институт цветных металлов ; № 25)
At head of title: Ministerstvo fsvetnoĭ metallurgii SSSR.
L.C. copy on microfilm only.
Includes bibliographies.
——— Microfilm. On 1 reel with other items ; 35 mm.
Microfilm Slavic 12747 TN
1. Chemical process control—Mathematical models. 2. Chemical process control—Data processing. I. Burovoĭ, Isaak Abramovich. II. Russia (1923- U. S. S. R.). Ministerstvo fsvetnoĭ metallurgii. III. Series: Moscow. Gosudarstvennyĭ nauchno-issledovatel'skiĭ institut fsvetnykh metallov. Sbornik nauchnykh trudov : no. 25.
[TN758.M62 no. 25] 79-408559
[TP155.75]

CHEMICAL PROCESSES

see also Chemical reactions

Aerstin, Frank.
Applied chemical process design / Frank Aerstin and Gary STreet ; with a foreword by K. D. Timmerhaus. — New York : Plenum Press, c1978.
xvii, 294 p. : ill. ; 26 cm.
Includes bibliographical references and index.
ISBN 0-306-31088-0
1. Chemical processes. 2. Chemical engineering. I. Street, Gary, joint author. II. Title.
TP155.7.A35 660.2'81 78-9104
78 MARC

Churchill, Stuart Winston, 1920-
The interpretation and use of rate data : the rate concept / Stuart W. Churchill. — Rev. print. — Washington : Hemisphere Pub. Corp., c1979.
xxv, 510 p. : ill. ; 24 cm.
Includes bibliographical references and index.
ISBN 0-89116-133-3
1. Chemical processes. 2. Transport theory. 3. Chemical reaction, Rate of. I. Title.
TP155.7.C46 1979 660.2'844 78-23365
78 MARC

Evangelista, John J
Material and energy balances / by John J. Evangelista. — Ed. 1. — Scranton, Pa. : ICS Center for Degree Studies, c1976.
102 p. : ill. ; 27 cm.
"5018."
1. Chemical processes. 2. Thermochemistry. 3. Conservation laws (Physics) I. ICS Center for Degree Studies. II. Title.
TP155.7.E92 660.2 79-300260
79 MARC

Fujita, Shigefumi, 1909-
(Kagaku kōgaku)
化学工学 / [藤田重文], 大山義年著. — 東京 : 岩波書店, 1956-1963.
2 v. : ill. ; 18 cm. — (岩波全書)
Series romanized: Iwanami zensho.
Includes bibliographies and index.
1. Chemical processes. I. Ōyama, Yoshitoshi, 1903- joint author. II. Title.
TP155.7.F84 79-804637

(Kataliz v kipfashchem sloe)
Катализ в кипящем слое / [Мухленов И. П., Анохин В. Н., Проскуряков В. А. ... et al.] ; Под ред. И. П. Мухленова и В. М. Померанцева. — 2-е изд., перераб. — Ленинград : Химия, Ленингр. отд-ние, 1978.
232 p. : ill. ; 22 cm. USSR 78
Bibliography: p. 219-[228]
1.00rub
1. Catalysis. 2. Fluidization. 3. Chemical processes. I. Mukhlenov, Ivan Petrovich. II. Pomeranfsev, V. M.
TP156.C35K39 1978 79-411064

Lebedev, Nikolaĭ Nikolaevich, fl. 1956-
(Khimifa i tekhnologifa osnovnogo organicheskogo i neftekhimicheskogo sinteza)
Химия и технология основного органического и нефтехимического синтеза : [учебник для хим.-технол. специальностей вузов] / Н. Н. Лебедев. — Изд. 2-е, перераб. — Москва : Химия, 1975.
732 p. : ill. ; 22 cm. USSR 75
Bibliography: p. 709-[710]
Includes index.
1.94rub
1. Chemistry, Organic—Synthesis. 2. Petroleum chemicals. 3. Chemical processes. I. Title.
TP247.L42 1975 79-411223

Părăuşanu, Victor.
Procese tehnologice din industria chimică / Victor Părăuşanu. — Bucureşti : Editura Didactică şi Pedagogică, 1978.
ix, 365, [1] p. : ill. ; 25 cm. R79-953
Bibliography: p. 356-[366]
lei19.70
1. Chemical processes. I. Title.
TP155.7.P35 79-368595
*79 MARC

Weiss, S
Thermische Verfahrenstechnik / Autoren, S. Weiss, H.-V. Adolphi, K.-E. Militzer. — 2., bearb. Aufl. — Leipzig : Deutscher Verlag für Grundstoffindustrie, VEB, 1978-
v. : ill. ; 22 cm. — (Verfahrenstechnik) GFR78-A
First ed. entered under Thermische Verfahrenstechnik.
Includes bibliographical references and index.
23.00M (v. 1)
1. Thermodynamics. 2. Chemical processes. I. Adolphi, Hans Velten, joint author. II. Militzer, K. E., joint author. III. Title.
TJ265.T457 1978 621.4'021 79-344198
79 MARC

—ADDRESSES, ESSAYS, LECTURES

(Énergetika promyshlennykh tekhnologicheskikh profsessov)
Энергетика промышленных технологических процессов: Темат. сб. / Ред. А. М. Бакластов, Л. Н. Сидельковский. — Москва : МЭИ, 1977.
114 p. : ill. ; 20 cm. — (Труды Московского энергетического института ; вып. 332) USSR 78
At head of title: Ministerstvo vysshego i srednego spefsial'nogo obrazovanifa SSSR.
Includes bibliographies.
0.41rub
1. Heat engineering—Addresses, essays, lectures. 2. Chemical processes—Addresses, essays, lectures. I. Baklastov, Arseniĭ Mikhaĭlovich. II. Sidel'kovskiĭ, Lazar' Naumovich. III. Series: Moscow. Énergeticheskiĭ institut. Trudy ; vyp. 332.
TJ7.M63 vyp. 332 79-413396
[TJ260]

Solar energy : chemical conversion and storage / edited by Richard R. Hautala, R. Bruce King, and Charles Kutal. — Clifton, N.J. : Humana Press, c1979.
x, 419 p. : ill. ; 24 cm. — (Contemporary issues in science and society)
Based on an American Chemical Society symposium held in Savannah, Georgia, in November, 1978.
Includes bibliographical references and index.
ISBN 0-89603-006-7
1. Solar energy—Addresses, essays, lectures. 2. Chemical processes—Addresses, essays, lectures. 3. Energy storage—Addresses, essays, lectures. I. Hautala, Richard R. II. King, R. Bruce. III. Kutal, Charles. IV. American Chemical Society. V. Series.
TJ810.S4885 621.47 79-87568
79 MARC

—EQUIPMENT AND SUPPLIES

Frost & Sullivan.
Chemical process equipment market. — New York : Frost & Sullivan, 1979.
vii, 246 p. ; 29 cm.
1. Chemical industries—United States. 2. Chemical processes—Equipment and supplies. 3. Market surveys—United States. I. Title.
HD9651.4.F76 1979a 79-114233
381'.45'6602830973
00 MARC

CHEMICAL PROCESSES
—EQUIPMENT AND SUPPLIES
(Continued)

Frost & Sullivan.
Manual control valve market in chemical process industries. — New York : Frost & Sullivan, 1979.
v, 290 p. : ill. ; 29 cm.
Title on spine: Manual valves for the CPI market.
1. Valve industry—United States. 2. Chemical processes—Equipment and supplies. 3. Market surveys—United States. I. Title. II. Title: Manual valves for the CPI market.
HD9529.V33U53 1979 381'.45'62184 79-114183
79 MARC

—MATHEMATICAL MODELS

Krupiczka, Roman.
Optymalizacja procesowa / Roman Krupiczka. — Gliwice : PŚ, 1978 (Gliw : PŚ)
316, ⌊1⌋ p. : ill. ; 25 cm. — (Skrypty uczelniane - Politechnika Śląska im. W. Pstrowskiego ; Nr 748) P78-7747
Bibliography: p. ⌊317⌋
zł27.00
1. Chemical processes—Mathematical models. 2. Mathematical optimization. I. Title.
TP155.7.K78 79-346252
79 MARC

—PERIODICALS

CEC census of buyers in the chemical process industries.
⌊Stamford, Conn., Reinhold Pub. Co.⌋
v. 22 x 28 cm. annual.
Running title, : Chemical engineering catalog census.
Other title, : Census of buyers in the chemical process industries.
Key title: CEC census of buyers in the chemical process industries, ISSN 0190-4760.
1. Chemicals—Manufacture and industry—United States—Directories. 2. Chemical processes—Periodicals. I. Title: Chemical engineering catalog census. II. Title: Census of buyers in the chemical process industries.
HD9651.3.C18 381'.45'66002573 79-640478
MARC-S

CHEMICAL REACTION

see also Phosphorylation

CHEMICAL REACTION, CONDITIONS AND LAWS OF

see also Linear free energy relationship

Bamford, C H
Proton transfer / edited by C. H. Bamford and C. F. H. Tipper. — Amsterdam ; New York : Elsevier Scientific Pub. Co. ; New York : distributors for the U.S. and Canada, Elsevier/North-Holland, 1977.
xi, 261 p. : ill. ; 25 cm. — (Their Comprehensive chemical kinetics ; v. 8)
Includes bibliographical references and index.
ISBN 0-444-41512-2 : $38.00
1. Chemical reaction, Conditions and laws of. 2. Protons.
QD501.B242 vol. 8 541'.39 s 77-23120
77 MARC

Dumas, Geneviève M L
Travaux dirigés de chimie PCEM et premier cycle / Geneviève M.-L. Dumas, Monique Gros, Louise Schriver. — Paris : Hermann, c1978.
2 v. : ill. ; 22 cm. — (Collection Méthodes) F***
Bibliography: v. 1, p. ⌊173⌋
Includes indexes.
CONTENTS: v. 1. Structure, cinétique, thermodynamique.—v. 2. Équilibres.
ISBN (invalid) 2-7056-5882-X (v. 1)
1. Chemical reaction, Conditions and laws of. I. Gros, Monique, joint author. II. Schriver, L., joint author. III. Title.
QD501.D93 541'.39 79-381408
79 MARC

Dynamics of molecular collisions / edited by William H. Miller. — New York : Plenum Press, c1976.
2 v. : ill. ; 26 cm. — (Modern theoretical chemistry ; v. 1-2)
Includes bibliographical references and indexes.
ISBN 0-306-33501-8 : $39.50
1. Molecular dynamics. 2. Chemical reaction, conditions and laws of. I. Miller, William Hughes, 1941- . II. Title. III. Series.
QD461.D95 539'.6 76-12633
76 MARC

Franklin, Joseph Louis, 1906-
Ion-molecule reactions / edited by J. L. Franklin. — Stroudsburg, Pa. : Dowden, Hutchinson & Ross ; New York : distributed worldwide by Academic Press, c1979.
2 v. : ill.; 26 cm. — (Benchmark papers in physical chemistry and chemical physics ; 3)
Includes bibliographical references and indexes.
CONTENTS: pt. 1. Kinetics and dynamics.—pt. 2. Elevated pressures and long reaction times.
ISBN 0-87933-331-6 (pt. 1)
1. Chemical reaction, Conditions and laws of. I. Title.
QD501.F7454 1979 541'.39 78-16358
78 MARC

Gefter, V M
(Raschet sostava i termodinamicheskikh svoĭstv neideal'nykh khimicheski reagirui͡ushchikh gazov)
Расчет состава и термодинамических свойств неидеальных химически реагирующих газов / В. М. Гефтер, В. И. Мика, А. М. Семенов. — Москва : Ин-т высоких температур АН СССР, 1976.
98 p. ; 22 cm. — (Теплофизические свойства веществ ; № 3) USSR***
At head of title: Nauchno-informat͡sionnyĭ t͡sentr po teplofizicheskim svoĭstvam chistykh veshchestv.
Bibliography: p. 49-52.
0.30rub
1. Chemical reaction, Conditions and laws of. 2. Gases—Thermal properties. I. Mika, V. I., joint author. II. Semenov, A. M., joint author. III. Title. IV. Series: Teplofizicheskie svoĭstva veshchestv ; no. 3.
QD501.G342 78-422278

Gilchrist, Thomas Lonsdale.
Organic reactions and orbital symmetry / T. L. Gilchrist and R. C. Storr. — 2d ed. — Cambridge ⌊Eng.⌋ : New York : Cambridge University Press, 1979.
vii, 311 p. : ill. ; 24 cm. — (Cambridge texts in chemistry and biochemistry)
Includes bibliographical references and index.
ISBN 0-521-22014-9
1. Chemistry, Physical organic. 2. Chemical reaction, Conditions and laws of. 3. Molecular orbitals. 4. Symmetry (Physics) I. Storr, R. C., joint author. II. Title. III. Series.
QD476.G54 1979 547'.1'39 78-54578
78 MARC

Karapet'i͡ant͡s, Mikhail Khristoforovich.
(Vvedenie v teorii͡u khimicheskikh prot͡sessov)
Введение в теорию химических процессов : ⌊Учеб. пособие для хим.-технол. специальностей вузов⌋ / М. Х. Карапетьянц. — 2-е изд., перераб. и доп. — Москва : Высш. школа, 1975.
320 p. : ill. ; 21 cm. USSR 75
Bibliography: p. 294-⌊295⌋
0.75rub
1. Chemical reaction, Condition and laws of. I. Title.
QD501.K296 1975 78-421331

Norman, Richard Oswald Chandler.
Mechanisms in organic chemistry : case studies / ⌊by⌋ R. O. C. Norman, M. J. Tomlinson, D. J. Waddington. — London : Mills and Boon, 1978.
110 p. : ill. ; 23 cm. — (Modern chemistry series) GB78-3961
Bibliography: p. ⌊106⌋
Includes index.
ISBN 0-263-06351-8 : £2.50
1. Chemistry, Physical organic. 2. Chemical reaction, Conditions and laws of. I. Tomlinson, Michael John, joint author. II. Waddington, D. J., joint author. III. Title.
QD476.N64 1978 547.1'39 79-305953
79 MARC

Schmalzried, Hermann.
Festkörperthermodynamik : Chemie d. festen Zustandes / von Hermann Schmalzried u. Alexandra Navrotsky. — Berlin : Akademie-Verlag, 1978.
x, 269 p. : ill. ; 25 cm. GDR78-A
Companion volume to H. Schmalzried's Festkörperreaktionen.
Includes bibliographies and indexes.
78.00M
1. Solid state chemistry. 2. Chemical reaction, Conditions and laws of. I. Navrotsky, Alexandra, joint author. II. Title.
QD478.S34 541'.042'1 78-386083
79 MARC

Transition states of biochemical processes / edited by Richard D. Gandour and Richard L. Schowen. — New York : Plenum Press, c1978.
xviii, 616 p. : ill. ; 24 cm.
Includes bibliographical references and indexes.
ISBN 0-306-31092-9
1. Biological chemistry. 2. Chemical reaction, Conditions and laws of. I. Gandour, Richard D. II. Schowen, Richard L.
⌊DNLM: 1. Biochemistry. QU4.3 T772⌋
QP514.2.T7 574.1'92 78-6659
78 MARC

Ulstrup, Jens, 1941-
Charge transfer processes in condensed media / Jens Ulstrup. — Berlin ; New York : Springer-Verlag, 1979.
vii, 419 p. : ill. ; 25 cm. — (Lecture notes in chemistry ; 10)
Includes bibliographical references and indexes.
Includes indexes.
ISBN 0-387-09520-9
1. Chemical reaction, Conditions and laws of. 2. Charge transfer. 3. Quantum chemistry. I. Title.
QD501.U43 541'.39 79-16651
79 MARC

—ADDRESSES, ESSAYS, LECTURES

Gas phase ion chemistry / edited by Michael R. Bowers. — New York : Academic Press, 1979-
v. : ill. ; 24 cm.
Includes bibliographies and index.
ISBN 0-12-120801-X
1. Chemical reaction, Conditions and laws of—Addresses, essays, lectures. 2. Gases, Ionized. I. Bowers, Michael T.
QD501.G324 541'.39 78-19666
79 MARC

—COLLECTED WORKS

Reactive intermediates. v. 1-
New York, John Wiley & Sons, c1978-
v. ill. 24 cm.
"A Wiley-interscience publication."
Key title: Reactive intermediates, ISSN 0190-8375.
1. Chemistry, Physical organic—Collected works. 2. Chemical reaction, Conditions and laws of—Collected works.
QD476.R414 547'.1'39 79-640411
MARC-S

—CONGRESSES

Aspects of mechanism and organometallic chemistry : a volume in honor of Professor Herbert C. Brown / edited by James H. Brewster. — New York : Plenum Press, c1978.
xiii, 347 p. : ill. ; 26 cm.
Proceedings of a symposium held at Purdue Unversity, West Lafayette, Indiana, May 5-6, 1978.
Includes index.
ISBN 0-306-40071-5
1. Chemical reaction, Conditions and laws of—Congresses. 2. Organometallic chemistry—Congresses. 3. Brown, Herbert Charles, 1912- . I. Brewster, James Henry, 1922- . II. Brown, Herbert Charles, 1912- . III. Purdue University, Lafayette, Ind.
QD501.A834 547'.05 78-21684
78 MARC

Nato Advanced Study Institute on Kinetics of Ion-Molecule Reactions, La Baule, 1978.
Kinetics of ion-molecule reactions / edited by Pierre Ausloos. — New York : Plenum Press, published in cooperation with NATO Scientific Affairs Division, c1979.
viii, 508 p. : ill. ; 26 cm. — (Nato advanced study institutes series : Series B, physics ; v. 40)
"Proceedings of the NATO Advanced Study Institute on Kinetics of Ion-Molecule Reactions held at La Baule, France, September 4-15, 1978."
Includes bibliographies and index.
ISBN 0-306-40153-3
1. Chemical reaction, Conditions and laws of—Congresses. 2. Ions—Congresses. 3. Molecules—Congresses. I. Ausloos, Pierre J. II. Title. III. Series.
QD501.N37 1978 541'.39 79-367
79 MARC

CHEMICAL REACTION, RATE OF

see also Linear free energy relationship

Ainsworth, Stanley.
Steady-state enzyme kinetics / Stanley Ainsworth. — Baltimore : University Park Press, c1977.
xii, 255 p. : ill. ; 24 cm.
Includes bibliographies and index.
ISBN 0-8391-1162-2
1. Enzymes. 2. Chemical reaction, Rate of. I. Title.
⌊DNLM: 1. Enzymes. 2. Kinetics. QU135 A297s⌋
QP601.A512 547'.758 77-21435
77⌊r79⌋rev MARC

Caneda, Rodolfo V
Cinética química / por Rodolfo V. Caneda. — Washington : Programa Regional de Desarrollo Científico y Tecnológico, Departamento de Asuntos Científicos, Secretaría General de la Organización de los Estados Americanos, 1978.
vii, 114 p. : ill. ; 23 cm. — (Serie de química ; monografía no. 18)
"78.XX11.B.012-S."
Bibliography: p. 107-109.
$2.00
1. Chemical reaction, Rate of. I. Title. II. Series.
QD502.C36 541'.394 79-106927
79 MARC

Churchill, Stuart Winston, 1920-
The interpretation and use of rate data : the rate concept / Stuart W. Churchill. — Rev. print. — Washington : Hemisphere Pub. Corp., c1979.
xxv, 510 p. : ill. ; 24 cm.
Includes bibliographical references and index.
ISBN 0-89116-133-3
1. Chemical processes. 2. Transport theory. 3. Chemical reaction, Rate of. I. Title.
TP155.7.C46 1979 660.2'844 78-23365
78 MARC

CHEMICAL REACTION, RATE OF (Continued)

Gonikberg, Mark Gerŝevich.
Chemical equilibria and reaction rates at high pressures. (Khimicheskoe ravnovesie i skorost' reaktsii pri vysokikh davlieniyakh) [by] M. G. Gonikberg. 2d ed. rev. and augm. Translated from Russian [by Michael Artman] Jerusalem, Published for the National Science Foundation, Washington, by the Israel Program for Scientific Translations; [available from the Office of Technical Services, U.S. Dept. of Commerce, Washington] 1963.
212 p. illus. 24 cm. (NASA TT-F-95)
At head of title: Akademiya nauk SSSR. Institut organicheskoi khimii im. N. D. Zelinskogo. Academy of Sciences of the U.S.S.R. Institute of Organic Chemistry im. N. D. Zelinskii.
Bibliography: p. 194-211.

1. Chemical equilibrium. 2. Chemical reaction, Rate of. I. Title. II. Series: United States. National Aeronautics and Space Administration. NASA technical translation, F-95.
TL507.U745 no. 95 63-61865
[QD503] [r79]rev2 MARC

Jordan, Peter C
Chemical kinetics and transport / Peter C. Jordan. — New York : Plenum Press, c1979.
xvi, 368 p. : ill. ; 24 cm.
Includes bibliographies and index.
ISBN 0-306-40122-3
1. Chemical reaction, Rate of. 2. Transport theory. I. Title.
QD501.J7573 541'.39 78-20999
78 MARC

Kopanica, Miloslav.
Kinetické metody v chemické analýze : [určeno také stud. stř. i vys. škol chem. / autoři] Miloslav Kopanica, Věra Stará. — 1. vyd. — Praha : SNTL : ČVTS, 1978.
248, [1] p. : ill. ; 21 cm. — (Moderní metody v chemické laboratoři ; sv. 32)
Cz78
Includes bibliographies and index.
Kčs25.00
1. Chemistry, Analytic. 2. Chemical reaction, Rate of. I. Stará, Věra, Ing., joint author. II. Title.
QD75.2.K66 78-382577
*78 MARC

Kuchanov, Semen Il'ich.
(Metody kineticheskikh raschetov v khimii polimerov)
Методы кинетических расчетов в химии полимеров / С. И. Кучанов. — Москва : Химия, 1978.
367 p. : ill. ; 22 cm. USSR 78
Includes bibliographical references and index.
3.80rub
1. Polymers and polymerization. 2. Chemical reaction, Rate of. I. Title.
QD281.P6K78 79-411065

Laboratorium termodynamiki i kinetyki procesów inżynierii chemicznej / Stanisław Wroński [et al.]. — Wyd. 2 popr. — Warszawa : Wydaw-a Politchniki Warszawskiej, 1978 (Warsz : PW)
233, [1] p. : ill. ; 25 cm. P78-1902
Bibliography: p. [234]
zł18.00
1. Chemical reaction, Rate of. 2. Thermodynamics. I. Wroński, Stanisław, Dr.
QD502.L3 1978 78-384149
78 MARC

Laidler, Keith James, 1916-
Theories of chemical reaction rates / Keith J. Laidler. — Huntington, N.Y. : R. E. Krieger Pub. Co., 1979, c1969.
vi, 234 p. : ill. ; 24 cm.
Reprint of the ed. published by McGraw-Hill, New York, in series: McGraw-Hill series in advanced chemistry.
Includes bibliographical references and indexes.
ISBN 0-88275-819-5
1. Chemical reaction, Rate of. I. Title.
[QD502.L34 1979] 541'.394 78-26387
78 MARC

Panchenkov, Georgiĭ Mitrofanovich.
Chemical kinetics and catalysis / G. M. Panchenkov and V. P. Lebedev ; translated from the Russian by G. Leib. — Moscow : Mir Publishers, 1976.
645 p. : diagrs. ; 23 cm. USSR***
Translation of Khimicheskaia kinetika i kataliz.
Bibliography: p. [622]-625.
Includes indexes.
1. Chemical reaction, Rate of. 2. Catalysis. I. Lebedev, Vladimir Petrovich. II. Title.
QD502.P3613 541'.39 79-345776
79 MARC

Pohorecki, Ryszard.
Kinetyka i termodynamika procesów inżynierii chemicznej / Ryszard Pohorecki, Stanisław Wroński. — Wyd. 1. — Warszawa : Wydaw-a Naukowo-Techniczne, 1977 (Pozn. : PZGMK)
542, [1] p. : ill. ; 25 cm. P78-1252
Errata slip inserted.
Includes bibliographies and index.
zł70.00
1. Chemical reaction, Rate of. 2. Thermodynamics. 3. Chemical engineering. I. Wroński, Stanisław, Dr., joint author. II. Title.
QD502.P64 78-388752
79 MARC

Polak, Lev Solomonovich.
Neravnovesnaia khimicheskaia kinetika i ee primenenie / L. S. Polak. — Moskva : Nauka, 1979.
248 p. : ill. ; 21 cm. USSR79
At head of title: Akademiia nauk SSSR. Institut neftekhimichiskogo sinteza im. A. V. Topchieva.
Includes bibliographies and index.
2.50rub
1. Chemical reaction, Rate of. 2. Chemical equilibrium. I. Title.
QD502.P66 79-379926
79 MARC

Poulsen, Knud, 1936-
Kinetics of the Renin system; the basic for determination of the different components of the system. Copenhagen, 1973.
86 p. illus. 25 cm. D***
Thesis—Copenhagen.
Summary in English and Danish.
Bibliography: 71-83.
1. Renin. 2. Angiotensin. 3. Chemical reaction, Rate of. I. Title.
QP609.R38P68 599'.01'9256 78-306210
78 MARC

Ravindran, P V
Comparison of DTA methods for the evaluation of kinetic parameters / by P. V. Ravindran, T. P. Radhakrishnan, and A. K. Sundaram. — Bombay : Bhabha Atomic Research Centre, 1976.
15, [7] p. : graphs ; 28 cm.
At head of title: Govt. of India, Atomic Energy Commission.
"B.A.R.C.-873."
Includes bibliographical references.
1. Thermal analysis. 2. Chemical reaction, Rate of. I. Radhakrishnan, T. P., joint author. II. Sundaram, A. K., joint author. III. India (Republic). Atomic Energy Commission. IV. Bhabha Atomic Research Centre. V. Title.
QD79.T38R38 543'.086 78-984413
79 MARC

Scharpé, S
Bijdrage tot de studie van de enzymatische amylolyse op kwantitatieve en kinetische grondslag / door Simon Scharpé. — Brussel : Paleis der Academiën, 1977.
90 p. : ill. ; 27 cm. — (Verhandelingen van de Koninklijke Academie voor Wetenschappen, Letteren en Schone Kunsten van België, Klasse der Wetenschappen ; jaarg 39, nr. 137) Be77-10
Summary in English.
Bibliography: p. 85-87.
550F
1. Diastase. 2. Amylosysis. 3. Chemical reaction, Rate of. 4. Swine—Physiology. I. Title. II. Series: Vlaamse Academie voor Wetenschappen, Letteren en Schone Kunsten van België. Klasse der Wetenschappen. Verhandelingen ; jaarg. 39, nr. 137.
Q56.V45 jaarg. 39, nr. 137 78-385277
[QP609.A45] *78 MARC

Sieck, L Wayne.
Rate coefficients for ion-molecule reactions : organic ions other than those containing C and H / L. Wayne Sieck. — Washington : U.S. Dept. of Commerce, National Bureau of Standards : for sale by the Supt. of Docs., U.S. Govt. Print. Off., 1979.
iv, 20 p. ; 26 cm. — (Nat. stand. ref. data ser. ; NSRDS-NBS 64)
"National Standard Reference Data System."
Includes bibliographical references.
Supt. of Docs. no.: C 13.48:64
1. Chemical reaction, Rate of. 2. Ions. 3. Molecules. 4. Chemistry, Physical organic. I. Title. II. Series: United States. National Bureau of Standards. National standard reference data series ; NSRDS-NBS 64.
QD502.S52 547'.139 78-10710
78 MARC

Timin, B I
Skorost' khimicheskikh reaktsii, khimicheskoe ravnovesie, kataliz : glavy iz kursa neorganicheskoĭ kimii : [uchebnik] / B. I. Timin. — Sverdlovsk : [s. n.], 1967.
66 p. : ill.
At head of title: Ministerstvo vysshego i srednego spetsial'nogo obrazovaniia RSFSR. Ural'skiĭ politekhnicheskiĭ institut imeni S. M. Kirova.
L.C. copy on microfilm only.
Bibliography: p. 65-66.
——— Microfilm. On 1 reel with other items ; 35 mm.
Microfilm Slavic 13050 QD
1. Chemical reaction, Rate of. I. Title.
[QD502.T48] 79-366968

—CONGRESSES

Fast biochemical reactions in solutions, membranes, and cells : Airlie House, Airlie, Virginia, 2-5 April 1978. — New York : Rockefeller University Press, c1978.
402 p. : ill. ; 27 cm. — (Biophysical discussions) (Biophysical journal ; v. 24, no. 1)
"Proceedings of a meeting sponsored by the Biophysical Society."
Includes bibliographical references and indexes.
ISBN 0-87470-030-2 : $25.00
1. Biological chemistry—Technique—Congresses. 2. Chemical reactions, Rate of—Congresses. I. Biophysical Society. II. Series. III. Series: Biophysical journal ; v. 24, no. 1.
QH505.A1B537 vol. 24 no. 1 78-64444
574.1'91'08 s
[QH345] 79 MARC

National Measurement Laboratory (U.S.)
Summary report on the Workshop on High Temperature Chemical Kinetics, Applications to Combustion Research / D. Garvin ... [et al.], National Measurement Laboratory, National Bureau of Standards ; sponsored by Department of Energy [and] Office of Standard Reference Data, National Bureau of Standards. — [Gaithersburg, Md.] : The Bureau ; Washington : for sale by the Supt. of Docs., U.S. Govt. Print. Off., 1978.
iii, 85 p. ; 27 cm. — (NBS special publication ; 531)
Includes index.
$2.75
1. High temperature chemistry—Congresses. 2. Chemical reaction, Rate of—Congresses. I. Garvin, David, 1923- . II. Workshop on High Temperature Chemical Kinetics: Applications to Combustion Research, National Bureau of Standards, 1977. III. United States. Dept. of Energy. IV. United States. National Bureau of Standards. Office of Standard Reference Data. V. Title. VI. Series: United States. National Bureau of Standards. Special publication ; 531.
QC100.U57 no. 531 602'.1 s 78-600125
[QD515] 79 MARC

Nato Advanced Study Institute on New Applications of Chemical Relaxation Spectrometry and Other Fast Reaction Methods in Solution, University College of Wales, 1978.
Techniques and applications of fast reactions in solution : proceedings of the Nato Advanced Study Institute on New Applications of Chemical Relaxation Spectrometry and Other Fast Reaction Methods in Solution, held at the University College of Wales, Aberystwyth, September 10-20, 1978 / edited by W. J. Gettins and E. Wyn-Jones. — Dordrecht, Holland ; Boston : D. Reidel Pub. Co., c1979.
xi, 608 p. : ill. ; 25 cm. — (NATO advanced study institutes series : Series C, Mathematical and physical sciences ; v. 50)
Includes bibliographical references and index.
ISBN 9027710228

1. Chemical reaction, Rate of—Congresses. 2. Solution (Chemistry)—Congresses. I. Gettins, W. J. II. Wyn-Jones, Evan. III. Title. IV. Series.
QD502.N37 1978 541'.39 79-16650
79 MARC

Structure and dynamics in chemistry : proceedings from symposium held at Uppsala, Sweden, 22-27.9.1977 ed. by P. Ahlberg and L.-O. Sundelöf. — Uppsala : Univ. ; Stockholm : Almqvist & Wiksell international (distr.), 1978.
viii, 339 p. : diagrs., ill. ; 24 cm. — (Symposia Universitatis Upsaliensis annum quingentesimum celebrantis ; 12) S78-26
Includes bibliographical references.
ISBN 9155407935
1. Chemical structure—Congresses. 2. Chemical reaction, Rate of—Congresses. I. Ahlberg, Per. II. Sundelöf, Lars-Olof. III. Series.
QD471.S834 541 79-300559
79 MARC

—MATHEMATICAL MODELS

IAblonskiĭ, Grigoriĭ Semenovich.
(Matematicheskie modeli khimicheskoĭ kinetiki)
Математические модели химической кинетики / Г. С. Яблонский, С. И. Спивак. — Москва : Знание, 1977.
63 p. : ill. ; 20 cm. — (Новое в жизни, науке, технике : Серия Химия ; 11/1977) USSR***
Bibliography : p. 63.
0.11rub
1. Chemical reaction, Rate of—Mathematical models. I. Spivak, Semen Izrailevich, joint author. II. Title. III. Series: Novoe v zhizni, nauke, tekhnike : Seriia Khimiia ; 1977, 11.
QD1.N87 1977, no. 11 78-420763
[QD502]

—TABLES

Buxton, George V
Compilation of rate constants for the reactions of metal ions in unusual valency states / George V. Buxton and Robin M. Sellers. — Washington : U.S. Dept. of Commerce, National Bureau of Standards : for sale by the Supt. of Docs., U.S. Govt. Print. Off., 1978.
v. 68 p. ; 26 cm. — (Nat. stand. ref. data ser., Nat. Bur. of Stand. ; NSRDS-NBS 62)
Includes bibliographical references.
Supt. of Docs. no.: C 13.48:62
1. Chemical reaction, Rate of—Tables, etc. 2. Metal ions—Tables. I. Sellers, Robin M., joint author. II. Title. III. Series: United States. National Bureau of Standards. National standard reference data series ; NSRDS-NBS 62.
QC100.U573 no. 62 602'.1 s 78-816
[QD502] 78 MARC

CHEMICAL REACTIONS

see also Chemical processes; Chemical reactors; Ring formation (Chemistry)

Froment, Gilbert F
Chemical reactor analysis and design / Gilbert F. Froment, Kenneth B. Bischoff. — New York : Wiley, c1979.
xxxix, 765 p. : ill. ; 24 cm.
Includes bibliographical references and indexes.
ISBN 0-471-02447-3
1. Chemical reactors. 2. Chemical reactions. 3. Chemical engineering. I. Bischoff, Kenneth B., joint author. II. Title.
TP157.F76 660'.29'9 78-12465
78 MARC

CHEMICAL REACTIONS (Continued)

Ho, Tse-Lok.
Hard and soft acids and bases principle in organic chemistry / Tse Lok Ho. — New York : Academic Press, 1977.
xii, 209 p. : ill. ; 24 cm.
Includes bibliographical references and indexes.
ISBN 0-12-350050-8
1. Acids. 2. Bases (Chemistry) 3. Chemistry, Organic. 4. Chemical reactions. I. Title.
QD477.H6 547'.1'39 76-13939
77[r79]rev MARC

Holland, Charles Donald.
Fundamentals of chemical reaction engineering / Charles D. Holland, Rayford G. Anthony. — Englewood Cliffs, N.J. : Prentice-Hall, c1978.
xiv, 541 p. : ill. ; 24 cm. — (Prentice-Hall international series in the physical and chemical engineering sciences)
Includes bibliographical references and index.
ISBN 0-13-335596-9
1. Chemical engineering. 2. Chemical reactions. I. Anthony, Rayford Gaines, 1935- joint author. II. Title.
TP145.H64 660.2'9'9 78-18372
78 MARC

Nicholls, David, 1935-
Inorganic chemistry in liquid ammonia / David Nicholls. — Amsterdam ; New York : Elsevier Scientific Pub. Co. ; New York : distributors for the U.S. and Canada, Elsevier/North-Holland, 1979.
x, 238 p. : ill. ; 25 cm. — (Topics in inorganic and general chemistry ; monograph 17)
Includes bibliographical references and index.
ISBN 0-444-41774-5
1. Chemical reactions. 2. Liquid ammonia. I. Title. II. Series.
QD501.N543 541'.39 78-20907
78 MARC

Sykes, Peter.
A guidebook to mechanism in organic chemistry / Peter Sykes. — 4th ed. — New York : Wiley, 1975.
xii, 362 p. : ill. ; 22 cm.
"A Halsted Press book."
Bibliography: p. 345-346.
Includes index.
ISBN 0-470-84101-X
1. Chemistry, Physical organic. 2. Chemical reactions. I. Title.
QD476.S89 1975 547'.1'39 74-12872
75[r79]rev MARC

Sykes, Peter.
Reaktionsmechanismen der organischen Chemie; eine Einführung. Mit einem Geleitwort von A. R. Todd. 5., neubearb. und erw. Aufl. Übers. von Peter Boldt, Hans F. Ebel [und] Peter Hocks. Weinheim/Bergstr., Verlag Chemie, 1972.
viii, 313 p. illus. 21 cm. DM22.00 GDB***
Translation of A guidebook to mechanism in organic chemistry.
Bibliography: p. [303]-305.
ISBN 3-527-25395-5
1. Chemistry, Physical organic. 2. Chemical reactions. I. Title.
QD476.S8915 1972 70-173369
79[r79]rev MARC

Turro, Nicholas J 1938-
Modern molecular photochemistry / Nicholas J. Turro. — Menlo Park, Calif. : Benjamin/Cummings Pub. Co., c1978.
628 p. : ill. ; 25 cm.
Includes bibliographical references and index.
ISBN 0-8053-9353-6
1. Photochemistry. 2. Chemical reactions. 3. Chemistry, Organic. I. Title.
QD601.2.T87 541'.35 78-57151
00 MARC

—COLLECTED WORKS

Organic reactions.
Huntington, N. Y., R. E. Krieger Pub. Co.,
v. 24 cm.
Reprint of the 1942 ed. published by J. Wiley & sons, inc., New York.
1. Chemical reactions—Collected works. 2. Chemistry, Physical organic—Collected works.
QD501.O75 547'.1'39 79-642486
MARC-S

CHEMICAL REACTORS

Froment, Gilbert F
Chemical reactor analysis and design / Gilbert F. Froment, Kenneth B. Bischoff. — New York : Wiley, c1979.
xxxix, 765 p. : ill. ; 24 cm.
Includes bibliographical references and indexes.
ISBN 0-471-02447-3
1. Chemical reactors. 2. Chemical reactions. 3. Chemical engineering. I. Bischoff, Kenneth B., joint author. II. Title.
TP157.F76 660'.29'9 78-12465
78 MARC

Levenspiel, Octave.
The chemical reactor minibook / Octave Levenspiel. — Corvallis, OR : Distributed by OSU Book Stores, c1979.
ca. 250 p. : ill. ; 27 cm.
Includes bibliography.
ISBN 0-88246-068-4
1. Chemical reactors. I. Title.
TP157.L412 541'.39 79-88694
79 MARC

Levenspiel, Octave.
The chemical reactor omnibook / Octave Levenspiel. — Corvallis, OR : Distriubted by OSU Book Stores, c1979.
ca. 600 p. in various pagings : ill. ; 28 cm.
Includes bibliography.
ISBN 0-88246-067-6
1. Chemical reactors. I. Title.
TP157.L413 660.2'9'9 79-88693
79 MARC

Ostrovskiĭ, Gennadiĭ Markovich.
(Metody optimizat͡sii khimicheskikh reaktorov)
Методы оптимизации химических реакторов. Г. М. Островский, Ю. М. Волин. Москва, Химия, 1967.
247 p. illus. 22 cm.
Bibliography : p. 241–[244]
1. Chemical reactors. 2. Chemical process control. 3. Mathematical optimization. I. Volin, I͡Uriĭ Markovich. II. Title.
TP157.O88 79–407805

Respondek, Józef.
Podstawy inżynierii reakcji chemicznych / Józef Respondek. — Wrocław : PW, 1977 (Wroc : PW)
284 p. : ill. ; 24 cm. — (Skrypty - Politechnika Wrocławska) P78-788
Bibliography: p. 281.
zł31.00
1. Chemical reactors. I. Title. II. Series: Politechnika Wrocławska. Skrypty - Politechnika Wrocławska.
TP157.R45 78-390830
79 MARC

—DESIGN AND CONSTRUCTION

Shah, Yatish T
Gas-fluid-solid reactor design / Yatish T. Shah. — London ; New York : McGraw-Hill International Book Co., c1979.
x, 373 p. : ill. ; 25 cm. GB79
Includes bibliographies and index.
ISBN 0-07-056370-5 : $39.50
1. Chemical reactors—Design and construction. I. Title.
TP157.S38 660.2'83 78-40100
78 MARC

CHEMICAL REAGENTS see Chemical tests and reagents

CHEMICAL REDUCTION see Reduction, Chemical; Reduction, Electrolytic

CHEMICAL RESEARCH

—ABSTRACTS—PERIODICALS

Journal of chemical research. Synopses. [Jan.] 1977–
[London] The Chemical Society.
v. ill. 30 cm. monthly.
Contains abstracts of papers appearing in the miniprint edition.
Published on behalf of the Founding Sponsors: Gesellschaft Deutscher Chemiker, Société chimique de France, and The Chemical Society.
Key title: Journal of chemical research. Synopses, ISSN 0308–2342
1. Chemical research—Abstracts—Periodicals. 2. Chemistry—Abstracts—Periodicals. I. Chemical Society, London. II. Gesellschaft Deutscher Chemiker. III. Société chimique de France.
QD40.J64a 540'.5 78–649929
MARC-S

—HISTORY

Zahn, Peter von.
Forschung hat viele Gesichter : Chemie, Pharmazie, Medizin / Peter von Zahn, Ingolf Rheinholz ; [Zeichn., Manfred Popp]. — 1. Aufl. — Düsseldorf ; Wien : Econ-Verlag, 1978.
330 p. : numerous ill. (some col.) ; 31 cm. GFR78-A36
Includes index.
ISBN 3-430-19901-8 : DM68.00
1. Chemical research—History. 2. Pharmaceutical research—History. 3. Medical research—History. I. Rheinholz, Ingolf, joint author. II. Title.
QD40.Z34 79-359756
79 MARC

—PERIODICALS

Journal of chemical research. Miniprint. [Jan.] 1977–
[London] The Chemical Society.
v. ill. 30 cm. monthly.
Contains papers abstracted in the synopses edition.
English, French or German.
Published on behalf of the Founding Sponsors: Gesellschaft Deutscher Chemiker, Société chimique de France, and The Chemical Society.
Key title: Journal of chemical research. Miniprint, ISSN 0308–2350.
1. Chemical research — Periodicals. 2. Chemistry — Periodicals. I. Chemical Society, London. II. Gesellschaft Deutscher Chemiker. III. Société chimique de France.
QD40.J64b 540 78–649930
MARC-S

—PERIODICALS—INDEXES

Journal of chemical research. Indexes. 1977–
[London] The Chemical Society.
v. 30 cm. annual.
Index to: Journal of chemical research. Miniprint, ISSN 0308–2350, and: Journal of chemical research. Synopses, ISSN 0308–2342.
Vols. for 1977– issued by Société chimique de France, the Chemical Society, and Gesellschaft Deutscher Chemiker.
1. Journal of chemical research. Miniprint—Indexes. 2. Journal of chemical research. Synopses—Indexes. 3. Chemical research—Periodicals—Indexes. 4. Chemistry—Periodicals—Indexes. I. Chemical Society, London. II. Gesellschaft Deutscher Chemiker. III. Société chimique de France.
QD40.J64c 540 79–641443
MARC-S

—GERMANY

Gruhn, Werner.
Chemie, technische Chemie, Werkstoffwissenschaften : e. Überblick über Forschung in d. Bundesrepublik Deutschland u. d. DDR / Werner Gruhn, Manfred v. Heimendahl, Hanns Hofmann. — Erlangen : Inst. für Ges. u. Wiss. an d. Univ. Erlangen-Nürnberg, 1978.
135 p. ; 21 cm. — (ABG, Analysen und Berichte aus Gesellschaft und Wissenschaft ; 1978, 6 ISSN 0344-2918) GFR78-B17
Includes bibliographical references.
ISBN 3-88150-034-0 : DM7.00
1. Chemical research—Germany. I. Heimendahl, Manfred von, joint author. II. Hofmann, Hanns, 1923- joint author. III. Title. IV. Series.
H5.A14 1978, no. 6 79-353736
[QD49.G] 79 MARC

—ITALY—CONGRESSES

Dove va la ricerca industriale in Italia : il settore chimico / a cura del Consiglio di fabbrica del Donegani. — Torino : Stampatori, 1979.
314 p. : ill. ; 21 cm. It79-July
"Dal convegno sulla ricerca chimica tenutosi nel 1978 presso l'Istituto Donegani di Novara."
L7000
1. Chemicals—Manufacture and industry—Italy—Congresses. 2. Montecatini Edison—Congresses. 3. Chemical research—Italy—Congresses. I. Istituto di ricerche Donegani di Novara. Consiglio di fabbrica.
HD9656.I82D68 79-381509
•79 MARC

—UNITED STATES—CONGRESSES

Federal regulation and chemical innovation / Christopher T. Hill, editor ; based on a symposium sponsored by the Division of Industrial and Engineering Chemistry at the 176th meeting of the American Chemical Society, Miami Beach, Florida, September 14, 1978. — Washington : American Chemical Society 1979.
xv, 200 p. : ill. ; 24 cm. — (ACS symposium series ; 109 ISSN 0097-6156)
Includes bibliographical references and index.
ISBN 0-8412-0511-6
1. Chemical research—United States—Congresses. 2. Science and state—United States—Congresses. I. Hill, Christopher T. II. American Chemical Society. Division of Industrial and Engineering Chemistry. III. Series: American Chemical Society. ACS symposium series ; 109.
QD47.F4 540'.973 79-15407
79 MARC

CHEMICAL SENSES

see also Chemoreceptors

CHEMICAL STRUCTURE

see also Chemical bonds; Molecular structure; Stereochemistry

Ladd, Marcus Frederick Charles.
Structure and bonding in solid state chemistry / M. F. C. Ladd. — Chichester, Eng. : Ellis Horwood ; New York : Halsted Press, 1979.
326 p. : ill. ; 24 cm. — (Ellis Horwood series in chemical science) GB***
Bibliography: p. ₍308₎-310.
Includes index.
ISBN 0-470-26597-3 (Halsted Press) : £13.00
1. Solid state chemistry. 2. Chemical structure. 3. Chemical bonds. I. Title. II. Series.
QD478.L32 1979 541'.224 78-41289
79 MARC

—ADDRESSES, ESSAYS, LECTURES

(Fiziko-khimicheskie svoĭstva organicheskikh veshchestv i i ikh svi͡az' so stroeniem i mikrostrukturoĭ)
Физико-химические свойства органических веществ и их связь со строением и микроструктурой : сборник трудов / ₍редколлегия, засл. деят. науки РСФСР, проф. В. А. Измаильский ... et al.₎. — Москва : ₍s. n.₎, 1973.
239 p. : ill. ; 22 cm. USSR 74
At head of title: Ministerstvo prosveshcheni͡a RSFSR. Moskovskiĭ gosudarstvennyĭ pedagogicheskiĭ institut imeni V. I. Lenina. Kafedra organicheskoĭ khimii.
Includes bibliographies.
1.00rub
1. Chemistry, Physical organic — Addresses, essays, lectures. 2. Chemical structure—Addresses, essays, lectures. I. Izmail'skiĭ, V. A. II. Moscow. Moskovskiĭ gosudarstvennyĭ pedagogicheskiĭ institut. Kafedra organicheskoĭ khimii.
QD473.F59 79–408062

—CONGRESSES

Structure and dynamics in chemistry : proceedings from symposium held at Uppsala, Sweden, 22-27.9.1977 ed. by P. Ahlberg and L.-O. Sundelöf. — Uppsala : Univ. ; Stockholm : Almqvist & Wiksell international (distr.), 1978.
viii, 339 p. : diagrs., ill. ; 24 cm. — (Symposia Universitatis Upsaliensis annum quingentesimum celebrantis ; 12) S78-26
Includes bibliographical references.
ISBN 9155407935
1. Chemical structure—Congresses. 2. Chemical reaction, Rate of—Congresses. I. Ahlberg, Per. II. Sundelöf, Lars-Olof. III. Series.
QD471.S834 541 79-300559
79 MARC

CHEMICAL STRUCTURE-BIOLOGICAL ACTIVITY RELATIONSHIP (PHARMACOLOGY) see Structure-activity relationship (Pharmacology)

CHEMICAL SYMBOLS see Abbreviations

CHEMICAL SYSTEMS

see also Phase rule and equilibrium

CHEMICAL TANKERS see Chemical carriers (Tankers)

CHEMICAL TECHNOLOGY see Chemistry, Technical

CHEMICAL TESTS AND REAGENTS

see also Indicators andtest-papers; Oxone; Sodium nitroferricyanide

I͡Anson, Ėdgar I͡Ul'evich.
Ditiokarboksilaty v analiticheskoĭ khimii / Ė. I͡U. I͡Anson. — Riga : Zinatne, 1978.
135 p. : ill. ; 22 cm. USSR78
At head of title: Latviĭskiĭ gosudarstvennyĭ universitet im. P. Stuchki.
Bibliography: p. ₍127₎-132.
0.60rub
1. Dithionic acid. 2. Chemistry, Analytic. 3. Chemical tests and reagents. I. Title.
QD77.I33 79-372426
79 MARC

Joint FAO/WHO Expert Committee on Food Additives.
Guide to specifications for general notices, general methods, identification tests, test solutions and other reference materials / Joint FAO/WHO Expert Committee on Food Additives. — Rome : Food and Agriculture Organization of the United Nations, 1978.
v, 91 p. : ill. ; 28 cm. — (FAO food and nutrition paper ; 5) It***
ISBN 9251004439
1. Food additives—Analysis. 2. Chemical tests and reagents. I. Title. II. Series: Food and Agriculture Organization of the United Nations. FAO food and nutrition paper ; 5.
TP455.J63 1978 79-316015
79 MARC

(Magnezon KhS)
Магнезон ХС. — Москва : Всесоюзный научно-исследовательский институт химических реактивов и особо чистых веществ, 1967.
12 p. : ill.
On cover: Assortiment khimicheskikh reaktivov na neorganicheskie iony—Magnezon KhS.
LC copy on microfilm only.
Bibliography: p. 12.
——— ——— Microfilm. On 1 reel with other items ; 35 mm.
Microfilm Slavic 12669 QD
1. Chemical tests and reagents. I. Moscow. Vsesoi͡uznyĭ nauchno-issledovatel'skiĭ institut khimicheskikh reaktivov i osobo chistykh veshchestv. II. Title: Assortiment khimicheskikh reaktivov na neorganicheskie iony— Magnezon KhS.
[QD77.M27] 79–407411

Perrin, Douglas Dalzell.
Organic complexing reagents : structure, behavior, and application to inorganic analysis / D. D. Perrin. — Reprint 1978 w/corrections and new pref. — Huntington, N.Y. : R. E. Krieger Pub. Co., 1978, c1964.
xi, 365 p. : ill. ; 24 cm.
Reprint of the ed. published by Interscience Publishers, New York, which was issued as v. 18 of Chemical analysis.
Includes bibliographical references and index.
ISBN 0-88275-574-9
1. Chemical tests and reagents. 2. Complex compounds. 3. Chemistry, Analytic. I. Title. II. Series: Chemical analysis ; v. 18.
₍QD77.P4 1978₎ 543'.01 77-16083
77 MARC

Petrova, G S
Assortiment reaktivov na kadmiĭ / G. S. Petrova, A. M. Lukin. — Moskva : NIIT-ĖKhim, 1968.
49 p. : ill. — (Khimicheskie reaktivy i preparaty) (Serii͡a Organicheskie reaktivy na neorganicheskie iony)
At head of title: Vsesoi͡uznyĭ nauchno-issledovatel'skiĭ institut khimicheskikh reaktivov i osobo chistykh khimicheskikh veshchestv.
Cover title: Organicheskie reaktivy dli͡a opredeleni͡a neorganicheskikh ionov Cd.
L.C. copy on microfilm only.
Bibliography: p. 43–48.
——— ——— Microfilm. On 1 reel with other items ; 35 mm.
Microfilm Slavic 12937 QD
1. Cadmium—Analysis. 2. Chemical tests and reagents. I. Lukin, Alekseĭ Matveevich, 1900– joint author. II. Title. III. Title: Organicheskie reaktivy dli͡a opredeleni͡a neorganicheskikh ionov Cd. IV. Series. V. Series.
[QD181.C3P47] 79–414423

Welcher, Frank Johnson, 1907-
Organic reagents for copper / by Frank J. Welcher, Erwin Boschmann. — 1st ed. — Huntington, N.Y. : R. E. Krieger, 1979.
xvi, 614 p. ; 24 cm.
Includes bibliographies and indexes.
ISBN 0-88275-440-8
1. Chemical tests and reagents. 2. Chemistry, Organic. 3. Copper—Analysis. I. Boschmann, Erwin, joint author. II. Title.
QD77.W417 546'.652'6 76-18141
76 MARC

—ADDRESSES, ESSAYS, LECTURES

(Reaktivy i osobo chistye veshchestva)
Реактивы и особо чистые вещества : ₍Сборник статей / Ред. совет ... Б. Д. Степин (гл. ред.) ... et al.₎. — Москва : ИРЕА, 1975.
290 p. : ill. ; 22 cm. — (Труды ИРЕА ; вып. 37) USSR 77
At head of title: Vsesoi͡uznyĭ nauchno-issledovatel'skiĭ institut khimicheskikh reaktivov i osobo chistykh khimicheskikh veshchestv.
Includes bibliographies.
1.60rub
1. Chemical tests and reagents—Addresses, essays, lectures. 2. Chemicals—Addresses, essays, lectures. I. Stepin, Boris Dmitrievich. II. Series: Moscow. Vsesoi͡uznyĭ nauchno-issledovatel'skiĭ institut khimicheskikh reaktivov i osobo chistykh veshchestv. Trudy ; vyp. 37.
QD71.V85 vyp. 37 78–421859
[QD77]

—CATALOGS

Chemapol. Fine Chemicals and Pharmaceutical Division.
Reagents and fine chemicals / CHEMAPOL, Fine Chemicals and Pharmaceutical Division. — Praha : The Division, 1964.
833 p., ₍12₎ leaves of plates : ill. ; 21 cm.
1. Chemicals—Catalogs. 2. Chemical tests and reagents—Catalogs. I. Title.
TP202.C49 1964 338.4'7'660025437 79-308388
79 MARC

—CONGRESSES

Vsesoi͡uznai͡a konferent͡sii͡a Organicheskie reagenty v analiticheskoĭ khimii, 4th, Kiev, 1976.
(Organicheskie reagenty v analiticheskoĭ khimii)
Органические реагенты в аналитической химии : Тезисы докл. IV всесоюз. конф., Киев, 21–24 апр. 1976 г. / ₍Ред. коллегия, чл.-кор. АН УССР А. Т. Пилипенко (отв. ред.) ... et al.₎. — Киев : Наук. думка, 1976.
2 v. ; 20 cm. USSR 76
At head of title: Akademii͡a nauk SSSR. Nauchnyĭ sovet po probleme Analiticheskai͡a khimii͡a. Institut geokhimii i analiticheskoĭ khimii im. V. I. Vernadskogo, and other institutions.
Includes bibliographies.
0.50rub (v. 1) 0.50rub (v. 2)

1. Chemical tests and reagents—Congresses. 2. Chemistry, Analytic—Congresses. 3. Chemistry, Organic—Congresses. I. Pilipenko, Anatoliĭ Terent'evich. II. Akademii͡a nauk SSSR. Nauchnyĭ sovet po probleme Analiticheskai͡a khimii͡a. III. Akademii͡a nauk SSSR. Institut geokhimii i analiticheskoĭ khimii.
QD77.V75 1976 79–408786

CHEMICAL WARFARE

see also Gases, Asphyxiating and poisonous—War use

Chemische Kampfstoffe und Schutz vor chemischen Kampfstoffen : Lehrbuch / ₍von e. Autorenkollektiv unter Leitung von Ralf Stöhr ; Autoren, Ralf Stöhr ... ₍et al.₎₎. — 1. Aufl. 1.-6. Tsd. — Berlin : Militärverlag der Dt. Demokrat. Republik, 1977.
445 p. : ill. ; 22 cm. GFR78-A
Bibliography: p. 439-440.
Includes index.
22.00M
1. Chemical warfare. I. Stöhr, Ralf.
UG447.C5225 1977 623.4'59 78-368158
78 MARC

Neutronenbom : wapentuig in opspraak / D. Nauta ... ₍et al.₎. — Deventer : Van Loghum Slaterus, 1978.
50 p. : with ill. ; 20 cm. — (Cahiers bio-wetenschappen en maatschappij ; 4. jaarg., nr. 4) Ne78-19
Cover title.
Bibliography: p. ₍5₎.
ISBN 9060014898 : fl 7.50
1. Neutron bomb. 2. Atomic warfare. 3. Chemical warfare. 4. Biological warfare. I. Nauta, Doede, 1898- . II. Series: Stichting Bio-Wetenschappen en Maatschappij. Cahiers van de Stichting Bio-Wetenschappen en Maatschappij ; 4. jaarg., nr. 4.
UG1282.N48N47 79-344529
•79 MARC

Zimmermann, Dieter.
Grundwissen für die Einsatzkräfte / ₍Autoren, Dieter Zimmermann, Werner Schneider₎. — 1. Aufl. — Berlin : Militärverlag d. Dt. Demokrat. Republic, 1977.
162 p. : ill. ; 19 cm. — (Ausbildungsanleitung der Zivilverteidigung der DDR ; Heft 1) GFR78-A17
Bibliography: p. 162.
1.75M
1. Civil defense. 2. Atomic warfare. 3. Chemical warfare. 4. Biological warfare. I. Schneider, Werner, Major Ingenieur, joint author. II. Title. III. Series.
UA926.Z54 79-371310
79 MARC

—BIBLIOGRAPHY

Robinson, Julian Perry.
Chemical/biological warfare : an introduction and a bibliography / by Julian Perry Robinson. — Los Angeles : California State University, Center for the Study of Armament and Disarmament, c1974.
xiv, 34 p. ; 22 cm. — (Political issues series ; v. 2, no. 2)
1. Chemical warfare—Bibliography. 2. Biological warfare—Bibliography. I. Title. II. Series.
Z6724.C5R62 016.358'34 79-311053
₍UG447₎ 79 MARC

CHEMICAL WORKERS

see also Chemists

—LEGAL STATUS, LAWS, ETC.—SPAIN

Spain. Ministerio de Trabajo.
Industrias químicas. Madrid, Editorial García Enciso, 1966.
345 p. 16 cm. (Colección García Enciso)
"Salario mínimo. Decreto 2.342/1967, del Ministerio de Trabajo, de 21 de septiembre ... sobre salario mínimo interprofesional y bases de cotización para la seguridad social" (4 p.) inserted at end.
1. Chemical workers—Legal status, laws, etc.—Spain. I. Title.
76-233611
77₍r79₎rev MARC

CHEMICAL WORKERS (Continued)

—ITALY—DISEASES AND HYGIENE—CASE STUDIES

Benedetti, Mauro.
La morte colorata : storie di fabbrica / Mauro Benedetti. — 1. ed. — Milano : Feltrinelli, 1978.
160 p. ; 19 cm. — (Franchi narratori ; 28) It78-Aug
L3000
1. Chemical workers—Italy—Diseases and hygiene—Case studies. 2. Industria piemontese coloranti all'anilina. I. Title.
HD7269.C452I833 78-380846
*78 MARC

—NEW ZEALAND—STATISTICS

New Zealand. Dept. of Labour (1954-). Research and Planning Division. Manpower Forecasting Section.
Employment in the manufacture of chemicals and chemical, petroleum, rubber, coal and plastic products. — Wellington : Research and Planning Division, Dept. of Labour, 1975.
25 p. ; 30 cm. — (Ind ; 10) NZ***
1. Chemical workers—New Zealand—Statistics. I. Title. II. Series: New Zealand. Dept. of Labour (1954-). Research and Planning Division. Reports series : Ind ; 10.
HD5718.C45N455 331.1'1 79-317245
79 MARC

—RUSSIAN REPUBLIC—BASHKIRIA

Khamzin, Farid Khabibullovich.
(Partkom i rekonstrukt͡sii͡a)
Партком и реконструкция / Ф. Х. Хамзин. — Уфа : Башкирское книжное изд-во, 1976.
78 p. ; 20 cm. USSR***
0.11rub
1. Chemical workers—Russian Republic—Bashkiria. 2. Chemical industries — Russian Republic — Bashkiria. 3. Kommunisticheskai͡a partii͡a Sovetskogo Soi͡uza—Party work. I. Title.
HD8039.C46R937 78-421764

—RUSSIAN REPUBLIC—VOSKRESENSK (CITY)

Ol'gin, Ol'gert.
(Voskresenskie khimiki)
Воскресенские химики : ⌜О произв. объединении "Минудобрение" им. В. В. Куйбышева⌝ / О. Ольгин. — Москва : Химия, 1977.
83 p. ; 20 cm. — (Передовые коллективы) USSR 77
Series romanized: Peredovye kollektivy.
0.15rub
1. Chemical workers—Russian Republic—Voskresensk (City)
I. Title.
HD8039.C46R963 78-422076

—SWITZERLAND—BASEL-LAND (CANTON)

Hugger, Paul.
Lebensverhältnisse und Lebensweise der Chemiearbeiter im mittleren Fricktal : eine Studie zum sozio-kulturellen Wandel eines ländlichen Gebietes / Paul Hugger. — Basel : ⌜im Selbstverlag⌝, 1976.
iv, 241 p., ⌜7⌝ leaves of plates : ill. ; 21 cm. Sw77-2919
Cover title.
10.00F
1. Chemical workers—Switzerland—Basel-Land (Canton) 2. Basel-Land (Canton)—Social conditions. I. Title. II. Title: Fricktal.
HD8039.C46S934 78-365805
*78 MARC

CHEMICAL WORKS see Chemical plants

CHEMICALS

see also Burns, Chemical; Chemical inhibitors; Fire extinguising agents; Stabilizing agents; and also groups of chemicals and individual chemical substances

Chalmers, Louis.
Chemical specialities : domestic and industrial / Louis Chalmers. — 2nd ed. / revised by Peter Bathe. — London : Godwin, 1978-
v. : ill. ; 25 cm. GB***
First ed. published in 1966 under title: Domestic and industrial chemical specialties.
Includes bibliographies and index.
CONTENTS: v. 1. Soap, detergents, and laundry products. Household pest control.

ISBN 0-7114-3313-5 (v. 1) : £10.00 (v. 1)
1. Chemicals. I. Bathe, Peter. II. Title.
TP200.C36 1978 661 78-317465
78 MARC

Ishii, Raizō, 1899-
(Kagaku shōhin)
化學商品 / 石井賴三著. — 東京 : 新紀元社, 昭和24 ⌜1949⌝
15, 242, 9, ⌜1⌝ p. : ill. ; 22 cm.
Bibliography: p. ⌜10⌝ (3d group)
Includes index.

1. Chemicals. 2. Chemistry, Technical. I. Title.
TP200.I83 78-808557

—ADDRESSES, ESSAYS, LECTURES

Chemicals in ships / edited by L. Kenworthy. — London : published for the Institute of Marine Engineers by Marine Management (Holdings) Ltd. : Forest Grove, Or ; ⌜distributed by⌝ ISBS, c1978.
xiii, 259 p. : ill. ; 22 cm. GB***
Distributor stamped on t.p.
Includes bibliographies and index.
ISBN 0-900976-72-1
1. Ships—Equipment and supplies—Addresses, essays, lectures. 2. Chemicals—Addresses, essays, lectures. I. Kenworthy, Leslie.
VM470.5.C43 660'.02'46238 79-670116
79 MARC

(Reaktivy i osobo chistye veshchestva)
Реактивы и особо чистые вещества : ⌜Сборник статей / Ред. совет ... Б. Д. Степин (гл. ред.) ... et al.⌝. — Москва : ИРЕА, 1975.
290 p. : ill. ; 22 cm. — (Труды ИРЕА ; вып. 37) USSR 77
At head of title: Vsesoi͡uznyĭ nauchno-issledovatel'skiĭ institut khimicheskikh reaktivov i osobo chistykh khimicheskikh veshchestv.
Includes bibliographies.
1.60rub
1. Chemical tests and reagents—Addresses, essays, lectures. 2. Chemicals—Addresses, essays, lectures. I. Stepin, Boris Dmitrievich. II. Series: Moscow. Vsesoi͡uznyĭ nauchno-issledovatel'skiĭ institut khimicheskikh reaktivov i osobo chistykh veshchestv. Trudy ; vyp. 37.
QD71.V85 vyp. 37 78-421859
[QD77]

—CATALOGS

Chemapol. Fine Chemicals and Pharmaceutical Division.
Reagents and fine chemicals / CHEMAPOL, Fine Chemicals and Pharmaceutical Division. — Praha : The Division, 1964.
833 p., ⌜12⌝ leaves of plates : ill. ; 21 cm.
1. Chemicals—Catalogs. 2. Chemical tests and reagents—Catalogs. I. Title.
TP202.C49 1964 338.4'7'660025437 79-308388
79 MARC

European chemical buyer's guide.
London, IPC Business Press.
v. 30 cm.
English, French, German, Italian, and Spanish.
1. Chemicals—Catalogs. 2. Chemicals—Manufacture and industry—Europe—Directories.
TP202.E94 338.4'7'66020254 78-643987
MARC-S

—ENVIRONMENTAL ASPECTS see Pollution

—FIRES AND FIRE PREVENTION

Blum, Paul D
Achtung! Gefährliche Chemikalien : e. Handbuch für d. Feuerwehrmann / Paul D. Blum. — Müchen : König-Industrie-Verlag, 1976.
248 p. ; 21 cm. GFR78-A2
Bibliography: p. 4.
Includes index.
DM49.50
1. Chemicals—Fires and fire prevention. I. Title.
TH9446.C5B55 79-338428
79 MARC

—INFORMATION SERVICES—UNITED STATES

United States. Environmental Protection Agency. Office of Toxic Substances. Chemical Information Division.
Chemical reporting and record-keeping authorities under 15 environmental and consumer acts / prepared by Chemical Information Division, Office of Program Integration and Information, Office of Toxic Substances, Environmental Protection Agency. — Washington : U.S. Environmental Protection Agency, Office of Toxic Substances, 1978 ⌜i.e. 1979⌝
ca. 250 p. in various pagings ; 28 cm.
"EPA-560/3-78-001."
1. Chemicals—Law and legislation—United States. 2. Hazardous substances—Law and legislation—United States. 3. Chemicals—Information services—United States. I. Title.
KF3958.A843 344'.73'0463 79-602550
79 MARC

—LAW AND LEGISLATION

Rehbinder, Eckard.
Das Recht der Umweltchemikalien : Definitionen, Ziele u. Massnahmen : Umweltforschungspl. d. Bundesministers d. Innern, Querschnittsfragen : Forschungsbericht 77-10104004 / von Eckard Rehbinder ; im Auftr. d. Umweltbundesamtes. — Berlin : E. Schmidt, 1978.
227 p. ; 24 cm. — (Berichte - Umwelt Bundes Amt ; 78, 9) GFR79-A
Includes bibliographical references.
ISBN 3-503-01596-5 : DM46.00
1. Hazardous substances—Law and legislation—Germany, West. 2. Chemicals—Law and legislation—Germany, West. 3. Environmental law—Germany, West. 4. Hazardous substances—Law and legislation. 5. Chemicals—Law and legislation. 6. Environmental law. I. Title. II. Series: Germany (Federal Republic, 1949-). Umweltbundesamt. Berichte - Umweltbundesamt ; 78, 9.
79-379843
79 MARC

—LAW AND LEGISLATION—GERMANY, WEST

Rehbinder, Eckard.
Das Recht der Umweltchemikalien : Definitionen, Ziele u. Massnahmen : Umweltforschungspl. d. Bundesministers d. Innern, Querschnittsfragen : Forschungsbericht 77-10104004 / von Eckard Rehbinder ; im Auftr. d. Umweltbundesamtes. — Berlin : E. Schmidt, 1978.
227 p. ; 24 cm. — (Berichte - Umwelt Bundes Amt ; 78, 9) GFR79-A
Includes bibliographical references.
ISBN 3-503-01596-5 : DM46.00
1. Hazardous substances—Law and legislation—Germany, West. 2. Chemicals—Law and legislation—Germany, West. 3. Environmental law—Germany, West. 4. Hazardous substances—Law and legislation. 5. Chemicals—Law and legislation. 6. Environmental law. I. Title. II. Series: Germany (Federal Republic, 1949-). Umweltbundesamt. Berichte - Umweltbundesamt ; 78, 9.
79-379843
79 MARC

—LAW AND LEGISLATION—JAPAN

Japan. Laws, statutes, etc.
⌜Kagaku busshitsu no shinsa oyobi seizō tō no kisei ni kansuru hōritsu. English⌝
The chemical substances control law in Japan : law concerning the examination and regulation of manufacture, etc., of chemical substances. — ⌜Tokyo⌝ : Chemical Products Safety Division, Basic Industries Bureau, Ministry of International Trade & Industry, ⌜1975⌝
79 p. ; 21 cm. Ja77-28
1. Chemicals—Law and legislation—Japan. 2. Chemical industries—Law and legislation—Japan. I. Title.
344'.52'042 78-324401
*79 MARC

—LAW AND LEGISLATION—UNITED STATES

Bureau of National Affairs, Washington, D.C.
Chemical regulation reporter : reference file. — Washington : Bureau of National Affairs, c1977, c1977-
v. ; 30 cm.
Loose-leaf for updating.
1. Chemicals—Law and legislation—United States. 2. Hazardous substances—Law and legislation—United States. 3. Chemical industries—Law and legislation—United States. I. Title.
KF3958.A15B87 Suppl 2 344'.73'047 79-112036
79 MARC

German and American toxic substances legislation and citizen participation in environmental affairs. — Bloomington, Ind. : Institute of German Studies, 1977.
iv, 23 p. ; 28 cm. — (German studies notes) (Institute of German Studies environmental series)
1. Air—Pollution—Law and legislation—Germany, West. 2. Chemicals—Law and legislation—United States. 3. Environmental policy—United States—Citizen participation. 4. Environmental policy—Germany, West—Citizen participation. I. Indiana University, Bloomington. Institute of German Studies. II. Series. III. Series: Indiana University, Bloomington. Institute of German Studies. Environmental series.
KF3593.G4 344'.43'0463 78-622894
78 MARC

Shapo, Marshall S 1936-
A nation of guinea pigs / Marshall S. Shapo. — New York : Free Press, c1979.
xvi, 300 p. ; 25 cm.
Includes bibliographical references and index.
ISBN 0-02-928550-X : $12.95
1. Product safety—United States. 2. Hazardous substances—Law and legislation—United States. 3. Drugs—Law and legislation—United States. 4. Chemicals—Law and legislation—United States. 5. Science and state—United States. I. Title.
KF3945.S53 344'.73'042 78-66977
79 MARC

CHEMICALS
—LAW AND LEGISLATION
—UNITED STATES (Continued)

United States. Congress. Senate. Committee on Commerce, Science, and Transportation. Subcommittee on Science, Technology, and Space.
Toxic substances control act of 1976 : hearing before the Subcommittee on Science, Technology, and Space of the Committee on Commerce, Science, and Transportation, United States Senate, Ninety-sixth Congress, first session ... May 4, 1979. — Washington : U.S. Govt. Print. Off., 1979.
iii, 42 p., 4 fold. leaves of plates ; 24 cm.
"Serial no. 96-25."
1. Chemicals—Law and legislation—United States. 2. Hazardous substances—Law and legislation—United States. 3. Poisons—Law and legislation—United States. I. Title.
KF26.C697 1979c 344'.73'047 79-603101
79 MARC

United States. Congress. Senate. Committee on Environment and Public Works. Subcommittee on Environmental Pollution.
Reauthorization of the Toxic substances control act, Safe drinking water act, and the Marine protection, research, and sanctuaries act : hearing before the Subcommittee on Environmental Pollution of the Committee on Environment and Public Works, United States Senate, Ninety-sixth Congress, first session, March 19, 1979. — Washington : U.S. Govt. Print. Off., 1979.
iii, 147 p. : graph ; 24 cm.
"Serial no. 96-H10."
1. Chemicals—Law and legislation—United States. 2. Hazardous substances—Law and legislation—United States. 3. Drinking water—Law and legislation—United States. 4. Marine resources conservation—Law and legislation—United States. 5. Waste disposal in the ocean—Law and legislation—United States. I. Title.
KF26.E645 1979a 344'.73'046 79-602842
79 MARC

United States. Environmental Protection Agency. Office of Toxic Substances. Chemical Information Division.
Chemical reporting and record-keeping authorities under 15 environmental and consumer acts / prepared by Chemical Information Division, Office of Program Integration and Information, Office of Toxic Substances, Environmental Protection Agency. — Washington : U.S. Environmental Protection Agency, Office of Toxic Substances, 1978 ‹i.e. 1979›
ca. 250 p. in various pagings ; 28 cm.
"EPA-560/3-78-001."
1. Chemicals—Law and legislation—United States. 2. Hazardous substances—Law and legislation—United States. 3. Chemicals—Information services—United States. I. Title.
KF3958.A843 344'.73'0463 79-602550
79 MARC

—LAW AND LEGISLATION—UNITED STATES—CONGRESSES

Toxic Substances Control Conference, 3d, Washington, D.C., 1978.
Toxic substances control volume III : implementing the regulatory program : proceedings of the 3rd Toxic Substances Control Conference, December 4-5, 1978, Washington, D.C. / edited by Marshall Lee Miller. — Washington : Government Institutes, 1979.
viii, 261 p. : ill. ; 23 cm.
Includes bibliographical references.
1. Poisons—Law and legislation—United States—Congresses. 2. Chemicals—Law and legislation—United States—Congresses. 3. Cancer—Law and legislation—United States—Congresses. 4. Carcinogenicity testing—Congresses. 5. Toxicology, Experimental—Congresses. I. Miller, Marshall Lee. II. Title.
KF3958.A75T67 1978 344'.73'047 79-83915
79 MARC

—LAW AND LEGISLATION
—UNITED STATES
—PERIODICALS

United States. Environmental Protection Agency.
Annual report, administration of the Toxic substances control act. 1st- 1977-
‹Washington› U. S. Environmental Protection Agency.
v. 24 cm.
Key title: Annual report, administration of the Toxic substances control act, ISSN 0161-9780
1. Hazardous substances—Law and legislation—United States—Periodicals. 2. Chemicals—Law and legislation—United States—Periodicals. 3. Poisons—Law and legislation—United States—Periodicals. I. Key title.
KF3958.A84 353.007'7 78-643799
MARC-S

—MANUFACTURE AND INDUSTRY

see also Chemical industries; Chemical workers; Chemistry, Technical; Electrochemistry, Industrial; Petroleum chemicals industry

Canada. Dept. of Industry, Trade and Commerce.
Chemicals. — Ottawa : Dept. of Industry, Trade and Commerce, c1977.
35 p. ; 28 cm. — (World market opportunities) C***
ISBN 0-662-01377-8
1. Chemicals—Manufacture and industry. 2. Market surveys. I. Title. II. Series.
HD9650.5.C36 1977 382'.45'6600971 78-319286
79 MARC

Houard, J
Évolution de la competitivité sectorielle des principales économies industrialisées selon l'indicateur synthétique de l'IRES, 1963-1976 / par J. Houard, avec la collaboration de C. Ghymers et de F. G. Prades. — Louvain-la-Neuve : Université catholique de Louvain, Institut de recherches économiques, 1977.
99 leaves : ill. ; 29 cm. — (Bulletin de l'IRES ; no 39-40) Be***
At head of title: Université catholique de Louvain. Institut de recherches économiques. Service de conjoncture.
1. Labor productivity. 2. Labor costs. 3. Competition, International. 4. Steel industry and trade. 5. Chemicals—Manufacture and industry. 6. Textile industry. 7. Metal trade. I. Ghymers, C., joint author. II. Prades, Federico, joint author. III. Université catholique de Louvain. Louvain-la-Neuve, Belgium. Institut de recherches économiques. Service de conjoncture. IV. Title. V. Series: Université catholique de Louvain, Louvain-la-Neuve, Belgium. Institut de recherches économiques. Bulletin de l'IRES ; no 39-40.
HC79.L3H68 333.1'1 78-371378
78 MARC

—MANUFACTURE AND INDUSTRY—SAFETY MEASURES—CONGRESSES

Symposium on Chemical Process Hazards with Special Reference to Plant Design, 6th, University of Manchester, 1977.
Papers of the Symposium on Chemical Process Hazards with Special Reference to Plant Design, VI : a symposium organised by the North Western Branch of the Institution of Chemical Engineers at the University of Manchester Institute of Science and Technology, 5-7 April 1977, symposium organisers, J. H. Burgoyne ... ‹et al.›. — Rugby : I. Chem. E. Services for the Institution of Chemical Engineers, 1977.
‹8›, 152 p. : ill. ; 30 cm. — (Institution of Chemical Engineers symposium series ; no. 49 ISSN 0307-0492) GB77-19507
Includes bibliographical references.
ISBN 0-8429-5100-0 : £12.50
1. Chemicals—Manufacture and industry—Safety measures—Congresses. 2. Chemical plants—Design and construction—Congresses. I. Institution of Chemical Engineers, London. North-Western Branch. II. Series: Institution of Chemical Engineers, London. I. Chem. E. symposium series ; no. 49.
TP149.S97 1977 660.2'804 77-377550
78‹r79›rev MARC

—MANUFACTURE AND INDUSTRY—STATISTICS—PERIODICALS

Ente nazionale idrocarburi.
Chimica.
‹Roma› Ente nazionale idrocarburi.
v. ill. 30 cm.
1. Chemicals—Manufacture and industry—Italy—Statistics—Periodicals. 2. Chemicals—Manufacture and industry—Statistics—Periodicals. I. Title.
HD9656.I 8E55a 338.4'7'6600945 78-646986
MARC-S

L'Industrie chimique. The Chemical industry. 1973/74–
‹Paris› Organisation de coopération et de développement économiques.
v. 27 cm. annual.
Continues: Organisation for Economic Co-operation and Development. Chemical industry.
English and French.
1. Chemical industries—Statistics—Periodicals. 2. Chemicals—Manufacture and industry—Statistics—Periodicals. I. Organization for Economic Cooperation and Development. II. Title: Chemical industry.
HD9650.1.O74 338.4'7'66005 79-646187
MARC-S

United Nations. Economic Commission for Europe.
Annual bulletin of trade in chemical products. Bulletin annuel du commerce des produits chimiques. **Ежегодный бюллетень европейской торговли химическими продуктами.** v. 1– 1974–
New York, United Nations.
v. tables. 28 cm.
English, French, and Russian.
1. Chemicals—Manufacture and industry—Statistics—Periodicals. I. United Nations. Economic Commission for Europe. Bulletin annuel du commerce des produits chimiques. II. United Nations. Economic Commission for Europe. Ezhegodnyĭ bi͡ulleten' evropeĭskoĭ torgovli khimicheskimi produktami. III. Title.
HD9650.1.U54a 382'.45'6600212 77-643654
(MARC-S)

—MANUFACTURE AND INDUSTRY—BELGIUM—DIRECTORIES

Chemfacts, Belgium. — London : Chemical Data Services, IPC Industrial Press, 1977.
vi, 135, ‹4› p. : maps ; 30 cm. GB78-38283
Includes index.
£40.00
1. Chemicals—Manufacture and industry—Belgium—Directories.
TP67.C47 1977 338.4'7'66009493 79-314653
79 MARC

—MANUFACTURE AND INDUSTRY—CANADA—CONGRESSES

Chemical Marketing Research Association.
North American chemical planning, 1977-1985 : papers presented at the November 1977 meeting of the Chemical Marketing Research Associaton, held in Montreal, Canada, November 2-4, 1977. — Staten Island, N.Y. : CMRA, ‹c1978›
175 p. : ill. ; 22 cm.
Includes bibliographical references.
$12.50
1. Chemicals—Manufacture and industry—Canada—Congresses. 2. Chemicals—Manufacture and industry—United States—Congresses. I. Title.
HD9655.C22C48 1978 338.4'7'6600971 78-110756
79 MARC

—MANUFACTURE AND INDUSTRY
—CHINA

Minami Manshū Tetsudō Kabushiki Kaisha. Tenshin Jimusho. Chōsaka.
(Shina ni okeru san, sōda oyobi chisso kōgyō)
支那に於ける酸、曹達及窒素工業 / 南満洲鐵道株式會社天津事務所調査課 ; ‹執筆者 三品頼忠›. — ‹天津› : 同課, 昭和12 ‹1937›
8, 216 p. : ill. ; 22 cm. — (北支經濟資料 ; 第32輯)
Cover title.
"秘"
1. Chemicals—Manufacture and industry—China. 2. Soda industry—China. 3. Nitrogen industries—China. I. Mishina, Yoritada. II. Title. III. Series: Hokushi keizai shiryō ; dai 32-shū.
HD9657.C52M54 1937 79-800102

—MANUFACTURE AND INDUSTRY
—CHINA—MANCHURIA

(Manshū ryūan kōgyō, sōda kōgyō hōsaku)
滿洲硫安工業·曹達工業方策. — ‹新京› : 南滿洲鐵道株式會社經濟調査會, ‹1935›
2, 83, 1, 128 p., ‹1› leaf of plates : ill. ; 23 cm. — (立安調査書類 - 南滿洲鐵道株式會社經濟調査會 ; 第6編第12卷)
Cover title.
"極祕"
CONTENTS: 硫安製造會社設立要綱—曹達製造會社設立要綱
1. Ammonium sulphate. 2. Chemicals—Manufacture and industry—China—Manchuria. 3. Chemicals—Manufacture and industry—Japan. 4. Soda industry—China—Manchuria. I. Minami Manshū Tetsudō Kabushiki Kaisha. Keizai Chōsakai. II. Series: Minami Manshū Tetsudō Kabushiki Kaisha. Keizai Chōsakai. Ritsuan chōsa shorui : dai 6-hen, dai 12-kan.
HD9660.A53C65 79-802061

—MANUFACTURE AND INDUSTRY
—EUROPE

Frost & Sullivan.
Surfactants market in Europe. — New York : Frost & Sullivan, 1979.
xvii, ix, 306 p. ; 29 cm.
On spine: Surfactants market in W. Europe.
Summary in German.
1. Surface active agents. 2. Chemicals—Manufacture and trade—Europe. 3. Market surveys—Europe. I. Title. II. Title: Surfactants market in W. Europe.
HD9660.S853E84 380.1'45'6681094 79-105798
79 MARC

—MANUFACTURE AND INDUSTRY
—EUROPE—DIRECTORIES

Directory of chemical producers, Western Europe. 1978–
‹Menlo Park, Calif.› SRI International.
v. 29 cm. annual.
Issues for 1978– published in 2 vols.
1. Chemicals—Manufacture and industry—Europe—Directories. I. SRI International.
TP12.D485 338.4'7'660025 79-649821
MARC-S

CHEMICALS
—MANUFACTURE AND INDUSTRY
—EUROPE—DIRECTORIES
(Continued)

European chemical buyer's guide.
London, IPC Business Press.
v. 30 cm.
English, French, German, Italian, and Spanish.
1. Chemicals—Catalogs. 2. Chemicals—Manufacture and industry—Europe—Directories.
TP202.E94 338.4'7'66020254 78-643987
MARC-S

—MANUFACTURE AND INDUSTRY—
EUROPE, SOUTHERN

Southern Europe. — ⌜London : Chemical Age, 1977⌝
34 p. : ill. ; 29 cm. — (Chemical age survey) GB***
1. Chemicals—Manufacture and industry—Europe, Southern. I. Chemical age. II. Series.
TP55.S68 79-316054
79 MARC

—MANUFACTURE AND INDUSTRY—
FRANCE

Marthey, Louis.
L'industrie chimique en France / par Louis Marthey. — ⌜Paris⌝ : Documentation française, 1978.
171 p. : ill. ; 24 cm. — (Notes et études documentaires ; no 4454 ISSN 0029-4004) F***
Bibliography: p. 168.
18.00F
1. Chemicals—Manufacture and industry—France. I. Title. II. Series: France. Direction de la documentation. Notes et études documentaires ; no 4454.
D411.F67 no.4454 909.8 s 79-338434
⌜TP71⌝ 79 MARC

—MANUFACTURE AND INDUSTRY—
INDIA—COSTS—CONGRESSES

Seminar on High Cost of Manufacturing Chemicals in India—Reasons & Remedies, Bombay, 1975.
Proceedings of the Seminar on High Cost of Manufacturing Chemicals in India—Reasons & Remedies, Bombay, 1st October 1975. — Bombay : Bombay Regional Office, Indian Chemical Manufacturers Association, ⌜1975 or 1976⌝
114 p. : ill. ; 29 cm.
Organised by the Indian Chemical Manufacturers' Association.
Rs25.00
1. Chemicals—Manufacture and industry—India—Costs—Congresses. I. Indian Chemical Manufacturers' Association. II. Title.
HD9657.I42S42 1975 338.4'3 76-904989
79 MARC

—MANUFACTURE AND INDUSTRY—
INDONESIA—STATISTICS

Indonesia. Departemen Perindustrian.
Produksi industri kimia. — ⌜Jakarta⌝ : Hubungan Masyarakat, Departemen Perindustrian, ⌜1972⌝
45 p. ; 21 cm.
1. Chemical industries—Indonesia—Statistics. 2. Chemicals—Manufacture and industry—Indonesia—Statistics. I. Title.
HD9657.I52I528 1972 73-940896
79 MARC

—MANUFACTURE AND INDUSTRY—
ITALY—CONGRESSES

Dove va la ricerca industriale in Italia : il settore chimico / a cura del Consiglio di fabbrica del Donegani. — Torino : Stampatori, 1979.
314 p. : ill. ; 21 cm. It79-July
"Dal convegno sulla ricerca chimica tenutosi nel 1978 presso l'Istituto Donegani di Novara."
L7000
1. Chemicals—Manufacture and industry—Italy—Congresses. 2. Montecatini Edison—Congresses. 3. Chemical research—Italy—Congresses. I. Istituto di ricerche Donegani di Novara. Consiglio di fabbrica.
HD9656.I82D68 79-381509
•79 MARC

—MANUFACTURE AND INDUSTRY
—ITALY—STATISTICS
—PERIODICALS

Ente nazionale idrocarburi.
Chimica.
⌜Roma⌝ Ente nazionale idrocarburi.
v. ill. 30 cm.
1. Chemicals—Manufacture and industry—Italy—Statistics—Periodicals. 2. Chemicals—Manufacture and industry—Statistics—Periodicals. I. Title.
HD9656.I 8E55a 338.4'7'6600945 78-646986
MARC-S

—MANUFACTURE AND INDUSTRY—
JAPAN

Japan's most advanced industrial fermentation technology and industry / academic review by Dr. K. Yamada ; industrial information by the International Technical Information Institute. — Tokyo : International Technical Information Institute, 1977
1 v. ; 31 cm. Ja***
Spine title: Japan's industrial fermentation.
Looseleaf for updating.
Includes bibliographies and indexes.
$30.00
1. Fermentation. 2. Chemicals—Manufacture and industry—Japan. I. Yamada, Kōichi, 1912- . II. Kaigai Gijutsu Shiryō Kenkyujo. III. Title: Japan's industrial fermentation.
TP534.J3J36 660.2'844'9 78-321160
78⌜r79⌝rev MARC

(Manshū ryūan kōgyō, sōda kōgyō hōsaku)
滿洲硫安工業·曹達工業方策. — ⌜新京⌝ : 南滿洲鐵道株式會社經濟調查會, ⌜1935⌝
2, 83, 1, 128 p., ⌜1⌝ leaf of plates : ill. ; 23 cm. — (立安調查書類 - 南滿洲鐵道株式會社經濟調查會 ; 第6編第12卷)
Cover title.
"極秘"
CONTENTS: 硫安製造會社設立要綱—曹達製造會社設立要綱
1. Ammonium sulphate. 2. Chemicals—Manufacture and industry—China—Manchuria. 3. Chemicals—Manufacture and industry—Japan. 4. Soda industry—China—Manchuria. I. Minami Manshū Tetsudō Kabushiki Kaisha. Keizai Chōsakai. II. Series: Minami Manshū Tetsudō Kabushiki Kaisha. Keizai Chōsakai. Ritsuan chōsa shorui : dai 6-hen, dai 12-kan.
HD9660.A53C65 79-802061

Nihon Kagaku Kōgyōhin Yushutsu Kumiai.
Chemical exports & imports of Japan / JCEA, JCIA. — ⌜Tokyo : Japan Chemical Exporters' Association, 1977⌝
26 p. : ill. ; 28 cm. Ja***
Cover title.
1. Chemical industries—Japan. 2. Chemicals—Manufacture and industry—Japan. I. Nihon Kagaku Kōgyōhin Yunyū Kyōkai. II. Title.
HD9657.J32N49 1977 382'.45'6600952 79-315312
79 MARC

(Tsuioku Morishita Seikichi)
追憶森下清吉 / ⌜編集者 森下清吉追悼文集編集委員会⌝. — 東京 : 森下産業, 昭和52 ⌜1977⌝
vi, 383 p., ⌜4⌝ leaves of plates : ill. ; 22 cm. Ja 78-3480
Not in trade.
1. Morishita, Seikichi, 1899-1975. 2. Executives—Japan—Biography. 3. Chemicals—Manufacture and industry—Japan. I. Morishita, Seikichi, 1899-1975. II. Morishita Seikichi Tsuitō Bunshū Henshū Iinkai.
HD9657.J32M677 79-801697

—MANUFACTURE AND INDUSTRY—
KIRGHIZISTAN—CONGRESSES

Organicheskaia khimiia i puti razvitiia khimicheskikh proizvodstv v Kirgizii : Materialy konf. / ⌜Otv. red. d-r khim. nauk, prof. V. A. Afanas'ev⌝. — Frunze : Ilim, 1976.
132 p. : ill. ; 20 cm. USSR77
At head of title: Akademiia nauk Kirgizskoĭ SSR. Kirgizskoe respublikanskoe pravlenie VKhO im. D. I. Mendeleeva.
1. Chemistry, Organic—Congresses. 2. Chemicals—Manufacture and industry—Kirghizistan—Congresses. I. Afanas'ev, Vitaliĭ Arkad'evich. II. Akademiia nauk Kirgizskoĭ SSR, Frunze. III. Vsesoiuznoe khimicheskoe obshchestvo imeni D. I. Mendeleeva. Kirgizskoe respublikanskoe pravlenie.
QD241.O69 79-385091
79 MARC

—MANUFACTURE AND INDUSTRY
—PACIFIC AREA—CONGRESSES

Market and economic trends for agricultural chemicals in countries bordering the Pacific Ocean ; Business planning ; Chemistry and economics of ocean resources / symposia chairmen, M. Hamamoto ... ⌜et al.⌝. — ⌜Staten Island, N.Y.⌝ : Chemical Marketing and Economics Division of the American Chemical Society, 1979.
494 p. : ill. ; 22 cm. — (Chemical marketing and economics preprints)
"The above symposia, sponsored by the Division of Chemical Marketing and Economics, are to be held at the April 1-6, 1979, Chemical Congress in Honolulu, Hawaii, hosted by the American Chemical Society and the Chemical Society of Japan."
Includes bibliographies.
1. Chemicals—Manufacture and industry—Pacific area—Congresses. 2. Agricultural chemicals—Pacific area—Congresses. 3. Marine resources—Pacific area—Congresses. 4. Market surveys—Pacific area—Congresses. I.
(Continued on next card)
Hamamoto, M. II. American Chemical Society. Division of Chemical Marketing and Economics. III. Series.
HD9657.P172M37 338.4'7'660099 79-84037
00 MARC

—MANUFACTURE AND INDUSTRY—
RUSSIA

ECN Chemical Data Services.
Foreign trade of the USSR, 1971-1973. — London : ⌜ECN Chemical Data Services⌝, ⌜1975⌝
⌜1⌝, 31 p. ; 30 cm. GB76-01012
"Supplement to The Chemical industry of Eastern Europe, third edition."
1. Chemicals—Manufacture and industry—Russia. 2. Russia—Commerce. I. ECN Chemical Data Services. The chemical industry of Eastern Europe. II. Title.
HD9656.R92E15 1975 382'.0947 78-326948
79 MARC

—MANUFACTURE AND INDUSTRY—
SCANDINAVIA—CONGRESSES

European Chemical Marketing Research Association.
European chemical industry : impact of Scandinavia : proceedings of the sixth international conference of the European Chemical Marketing Research Association, Sheraton Hotel, Copenhagen, 9th-11th October 1972 / editor of proceedings, N. A. Perkins. — ⌜London?⌝ : EVAF, ⌜1972?⌝
188 p. : ill. ; 21 cm.
Includes bibliographical references.
1. Chemicals—Manufacture and industry—Scandinavia—Congresses. I. Perkins, Norman Albert. II. Title.
TP88.5.E96 1972 338.4'7'6600948 79-302048
79 MARC

—MANUFACTURE AND INDUSTRY—
SPAIN—DIRECTORIES

Chemfacts, Spain. — London : Chemical Data Services, IPC Industrial Press, 1978.
vi, 138, ⌜4⌝ p. : maps ; 30 cm. GB78-38960
Includes index.
£40.00
1. Chemicals—Manufacture and industry—Spain—Directories. I. Chemical Data Services, London.
TP87.C44 338.4'7'6600946 79-314493
79 MARC

—MANUFACTURE AND INDUSTRY—
UNITED STATES

Frost & Sullivan.
Chemical disinfectants and sanitizers market. — New York : Frost & Sullivan, 1979.
2 v. (xi, 424 p.) : ill. ; 29 cm.
1. Chemicals—Manufacture and industry—United States. 2. Sanitary supply industry—United States. 3. Disinfection and disinfectants. 4. Market surveys—United States. I. Title.
HD9651.4.F76 1979 338.4'7'6681 79-114188
00 MARC

—MANUFACTURE AND INDUSTRY—
UNITED STATES—CONGRESSES

Chemical Marketing Research Association.
Chemicals and plastics growth in the automotive industry in the eighties : papers presented at the February 1978 meeting of the Chemical Marketing Research Association held in Detroit, Michigan, Februrary 7-10, 1978. — New York : The Association, ⌜c1978⌝
226 p. : ill. ; 22 cm.
$12.50
1. Automobile supplies industry—United States—Congresses. 2. Chemicals—Manufacture and industry—United States—Congresses. 3. Plastics in automobiles—Congresses. I. Title.
HD9710.3.U52C49 1978 338.4'7'62920973 78-110753
79 MARC

CHEMICALS
—MANUFACTURE AND INDUSTRY
—UNITED STATES—CONGRESSES
(Continued)

Chemical Marketing Research Association.
North American chemical planning, 1977-1985 : papers presented at the November 1977 meeting of the Chemical Marketing Research Associaton, held in Montreal, Canada, November 2-4, 1977. — Staten Island, N.Y. : CMRA, ͺc1978ͺ
175 p. : ill. ; 22 cm.
Includes bibliographical references.
$12.50
1. Chemicals—Manufacture and industry—Canada—Congresses. 2. Chemicals—Manufacture and industry—United States—Congresses. I. Title.
HD9655.C22C48 1978 338.4'7'6600971 78-110756
79 MARC

Chemical Marketing Research Association.
Trends in performance chemicals : papers presented at the February 1979 meeting of the Chemical Marketing Research Association held in San Francisco, California. — Staten Island, N.Y. : The Association, ͺc1979ͺ
269 p. ; 22 cm.
Includes bibliographical references.
1. Chemicals—Manufacture and industry—United States—Congresses. 2. Market surveys—United States—Congresses. I. Title.
HD9651.4.C54 1979 338.4'7'6600973 79-114005
00 MARC

Chemical Marketing Research Association.
The U.S. chemical industry-government interface : papers presented at the September 1978 meeting of the Chemical Marketing Research Association, Williamsburg, Va., September 24-27, 1978. — Staten Island, N.Y. : The Association, ͺc1978ͺ
95 p. : ill. ; 22 cm.
$12.50
1. Chemicals—Manufacture and industry—United States—Congresses. 2. Industry and state—United States—Congresses. I. Title.
HD9651.6.C47 1978 338.4'7'6600973 79-100658
79 MARC

Petrochemicals—competitive and market relationships ; West Coast petrochemical industry ; External influences on the chemical industry. — ͺStaten Island, N.Y.ͺ : Chemical Marketing and Economics Division of the American Chemical Society, c1978.
224 p. : ill. ; 22 cm. — (Chemical marketing and economics reprints)
Papers presented at 3 symposia held at the 175th ACS meeting, Anaheim, Calif., Mar. 13-15, 1978, and sponsored by the society's Division of Chemical Marketing and Economics.
1. Petroleum chemicals industry—United States—Congresses. 2. Petroleum chemicals industry—Pacific States—Congresses. 3. Chemicals—Manufacture and industry—United States—Congresses. I. American Chemical Society. Division of Chemical Marketing and Economics. II. Series.
HD9579.C33U57 338.4'7'6618040973 78-60436
79 MARC

—MANUFACTURE AND INDUSTRY
—UNITED STATES
—DIRECTORIES

CEC census of buyers in the chemical process industries.
ͺStamford, Conn., Reinhold Pub. Co.ͺ
v. 22 x 28 cm. annual.
Running title, : Chemical engineering catalog census.
Other title, : Census of buyers in the chemical process industries.
Key title: CEC census of buyers in the chemical process industries, ISSN 0190-4760.
1. Chemicals—Manufacture and industry—United States—Directories. 2. Chemical processes—Periodicals. I. Title: Chemical engineering catalog census. II. Title: Census of buyers in the chemical process industries.
HD9651.3.C18 381'.45'66002573 79-640478
MARC-S

—MANUFACTURE AND INDUSTRY—
UNITED STATES—FINANCE

Chemical industry sourcebook : materials / selected by MSS Information Corporation from the journals, indexes, and profiles of Disclosure, inc. — ͺNew Yorkͺ : The Corporation ; Edison, N.J. : exclusive sales agent, Lasky-Lanouette, c1977.
232 p. ; 23 cm.
Includes extracts of data from current and recent corporate reports filed with the Securities and Exchange Commission.
ISBN 0-8422-4011-X : $50.00
1. Chemical industries—United States—Finances. 2. Chemicals—Manufacture and trade—United States—Finances. I. MSS Information Corporation. II. Disclosure, inc. III. United States. Securities and Exchange Commission.
HD9651.5.C48 338.7'66'02573 77-76597
77 MARC

Synthetic fiber and synthetic chemical industry sourcebook, May 1975-April 1976 / materials selected by MSS Information Corporation from the journals, indexes, and profiles of Disclosure, inc. — ͺNew Yorkͺ : The Corporation ; Edison, N.J. : exclusive sales agent, Lasky-Lanouette, c1977.
189 p. ; 23 cm.
"Includes extracts of data from current and recent reports to the SEC."
ISBN 0-8422-4058-6 : $50.00
1. Man-made fibers industry—United States—Finance. 2. Chemicals—Manufacture and industry—United States—Finance. I. MSS Information Corporation. II. Disclosure, inc. III. United States. Securities and Exchange Commission.
HD9855.S95 338.7'61'67740973 77-14741
79 MARC

—MANUFACTURE AND INDUSTRY
—UZBEKISTAN

Turapov, A T
(Ŭzbekistonda khimiia sanoatining rivozhlanishi)
Ўзбекистонда химия саноатининг ривожланиши / А. Туропов. — Тошкент : Ўзбекистон, 1973.
87 p. ; 20 cm. USSR 73-7687
"Ўзбекистон ССР "Билим" жамияти, № 43."
0.06rub
1. Chemicals—Manufacture and industry—Uzbekistan. I. Title.
TP110.U85T87 79-971008

—MANUFACTURE AND INDUSTRY—
ZAMBIA

Zambia. Central Statistical Office.
Chemicals, rubber, and plastic industries. — Lusaka : Central Statistical Office, 1976.
64 p. ; 31 cm. — (Its Industry monographs ; no. 5)
Bibliography: p. 64.
K4.00
1. Chemicals—Manufacture and industry—Zambia. 2. Rubber industry and trade—Zambia. 3. Plastics industry and trade—Zambia. I. Title. II. Series.
HC517.R42A4 1976 no. 5 77-980747
338'.09689'4 s
ͺHD9658.Z35ͺ 78 MARC

—PATENTS

Flick, Ernest W
Household and automotive chemical specialties : recent formulations / Ernest W. Flick. — Park Ridge, N.J. : Noyes Data Corp., 1979.
xxi, 396 p. ; 25 cm.
Includes bibliographical references.
ISBN 0-8155-0751-8 : $32.00
1. Chemicals—Patents. 2. Household supplies—Patents. 3. Automotive supplies—Patents. I. Title.
TP210.F56 660 79-84429
79 MARC

—PHYSIOLOGICAL EFFECT

Hamilton, E I
The chemical elements and man : measurements—perspectives—applications / by E. I. Hamilton. — Springfield, Ill. : Thomas, c1979.
xix, 490 p. : ill. ; 24 cm. — (American lecture series ; no. 1021)
Includes bibliographical references and index.
ISBN 0-398-03732-9
1. Chemicals—Physiological effect. 2. Body composition. 3. Health. 4. Chemical elements—Analysis. I. Title.
QP531.H28 612'.0152 78-5409
78 MARC

—SAFETY MEASURES—CONGRESSES

Toxic chemical and explosives facilities : safety and engineering design : based on a symposium sponsored by the Division of Chemical Health and Safety at the 176th meeting of the American Chemical Society, Miami Beach, Florida, September 11-13, 1978 / Ralph A. Scott, Jr., editor. — Washington : The Society, 1979.
ix, 352 p. : ill. ; 24 cm. — (ACS symposium series ; 96 ISSN 0087-6156)
Includes bibliographical references and index.
ISBN 0-8412-0481-0
1. Explosives—Safety measures—Congresses. 2. Chemicals—Safety measures—Congresses. I. Scott, Ralph A., 1930- . II. American Chemical Society. Division of Chemical Health and Safety. III. Series: American Chemical Society. ACS symposium series ; 96.
TP295.T69 614.8'32 79-9760
79 MARC

—SAFETY MEASURES—HANDBOOKS,
MANUALS, ETC.

Bretherick, L
Handbook of reactive chemical hazards : an indexed guide to published data / L. Bretherick. — 2d ed. — London ; Boston : Butterworth, 1979.
xxviii, 1281 p. ; 24 cm. GB***
Bibliography: p. 1167-1176.
Includes index.
ISBN 0-408-70927-8 : £45.00
1. Hazardous substances—Handbooks, manuals, etc. 2. Chemicals—Safety measures—Handbooks, manuals, etc. I. Title.
T55.3.H3B73 1979 660'.028 78-40367
79 MARC

—STORAGE—CONGRESSES

MariChem 77, Hamburg, 1977.
MariChem 77 : conference on the marine transportation, handling and storage of bulk chemicals, Hamburg, October 11-13, 1977 : ͺproceedingsͺ. — Rickmansworth : Gastech Exhibitions Ltd., 1978.
xiv, 234 p. : ill., ports ; 30 cm. GB78-21024
Includes bibliographies.
ISBN 0-904930-08-4 pbk : £40.00
1. Chemicals—Transportation—Congresses. 2. Chemicals—Storage—Congresses. I. Gastech Exhibitions ltd.
TP201.M34 1977 604.7 78-323392
79 MARC

—TESTING

Canada. Dept. of National Health and Welfare.
The testing of chemicals for carcinogenicity, mutagenicity and teratogenicity. — ͺOttawaͺ : Health and Welfare Canada, 1975.
iv, 183 p. ; 23 cm. C***
Includes bibliographies.
1. Carcinogenicity testing. 2. Chemicals—Testing. 3. Chemical mutagenesis. 4. Teratogenic agents. I. Title.
RC268.65.C36 1975 615.9 79-317251
79 MARC

—TOXICOLOGY see Poisons; Toxicology

—TRANSPORTATION

see also Chemical carriers (Tankers)

Chemical Industries Association.
Chemsafe : a manual of the chemical industry scheme for assistance in freight emergencies / ͺChemical Industries Associationͺ. — 2nd ed. — London : C.I.A., 1976.
ͺ3ͺ, 33 p. : forms ; 21 cm. GB77-04203
Bibliography: p. 31.
ISBN 0-900623-07-1
1. Chemical engineering—Safety measures. 2. Chemicals—Transportation. I. Title.
TP149.C486 1976 614.8'312 79-303097
79 MARC

Corkhill, Michael, 1947-
Chemical tankers : the ships and the market / by Michael Corkhill ; diagrams drawn by Brian Mehl. — London : Fairplay Publications Ltd, 1976.
ͺ7ͺ, 113 p., ͺ40ͺ leaves : ill., plans ; 30 cm. — (A Fairplay report)
GB78-05977
Bibliography: p. 113.
ISBN 0-905045-02-5 : £50.00
1. Chemical carriers (Tankers). 2. Chemicals—Transportation. 3. Chemical industries. I. Title.
VM455.3.C67 387.24'5 78-314133
78 MARC

—TRANSPORTATION—CONGRESSES

MariChem 77, Hamburg, 1977.
MariChem 77 : conference on the marine transportation, handling and storage of bulk chemicals, Hamburg, October 11-13, 1977 : ͺproceedingsͺ. — Rickmansworth : Gastech Exhibitions Ltd., 1978.
xiv, 234 p. : ill., ports ; 30 cm. GB78-21024
Includes bibliographies.
ISBN 0-904930-08-4 pbk : £40.00
1. Chemicals—Transportation—Congresses. 2. Chemicals—Storage—Congresses. I. Gastech Exhibitions ltd.
TP201.M34 1977 604.7 78-323392
79 MARC

CHEMICALS, RUBBER see Rubber chemicals

CHEMICALS, TEXTILE see Textile chemicals

CHEMICULTURE see Hydroponics

CHEMILUMINESCENSE

Bioluminescense and chemiluminescense / edited by Marlene A. DeLuca. — New York : Academic Press, 1978.
xxi, 653 p. : ill. ; 24 cm. — (Methods in enzymology ; v. 57)
Includes bibliographical references and indexes.
ISBN 0-12-181957-4
1. Bioluminescense. 2. Chemiluminescense. I. DeLuca, Marlene A. II. Series.
QP601.M49 vol. 57 574.1'925'08 s 78-13442
ͺQH641ͺ 78 MARC

CHEMIN DES DAMES

Castex, Henri.
L'Affaire du Chemin des Dames / Henri Castex. — Paris : Roblot, 1977.
206 p. ; 24 cm. — (His Les comités secrets) (Collection Cité première) F77-8999
Bibliography: p. 203-204.
35.00F
1. European War, 1914-1918—France. 2. France—Politics and government—1914-1940. 3. Chemin des dames. I. Title. II. Series.
D516.C37 940.3'44 79-361009
79 MARC

CHEMIOTAXIS see Chemotaxis

CHEMISE LACOSTE (FIRM)

—CARICATURES AND CARTOONS

Muntañola, Joaquín.
El cocodrilo Lacoste / Joaquín Muntañola. — [s.l.] : Basi Hnos., [1974]
47 p. : ill. ; 20 cm.
Cartoons.
1. Chemise Lacoste (Firm)—Caricatures and cartoons. 2. Spanish wit and humor, Pictorial. I. Title.
NC1639.M83A4 1974 741.5'946 78-399018
79 MARC

CHEMISORPTION

Tompkins, Frederick Clifford, 1910-
Chemisorption of gases on metals / F. C. Tompkins. — London ; New York : Academic Press, 1978.
xii, 370 p. : ill. ; 24 cm. GB***
Includes bibliographical references and indexes.
ISBN 0-12-694650-7 : £16.80
1. Chemisorption. I. Title.
QD547.T65 1978 541'.3453 77-93207
78 MARC

CHEMISTRY

see also Acids; Alchemy; Assaying; Bases (Chemistry); Biological chemistry; Botanical chemistry; Catalysis; Color; Combustion; Condensation; Cosmochemistry; Crystallization; Crystallography; Decomposition (Chemistry); Dental chemistry; Electrochemistry; Environmental chemistry; Explosives; Fermentation; Fire; Gases—Liquefaction; Geochemistry; Immunochemistry; Pharmacy; Photochemistry; Photographic chemistry; Plasma chemistry; Salts; Sanitary chemistry; Solution (Chemistry); Spectrum analysis; Stereochemistry; Water chemistry

Aminuddin Ressang.
Ilmu kimia sijil persekolahan / oleh Aminuddin Ressang. — Edisi 2. — Kuala Lumpur : Dewan Bahasa dan Pustaka, Kementerian Pelajaran Malaysia, 1976.
xii, 498 p., [2] leaves of plates : ill. ; 22 cm. — (Siri pelajaran menengah DBP ; bil. 87)
Includes index.
$6.00
1. Chemistry. I. Title. II. Series: Dewan Bahasa dan Pustaka. Siri pelajaran menengah ; bil. 87.
QD31.2.A48 1976 77-986772
79 MARC

Bognár, János, Dr.
Elemző kémia : [egyetemi tankönyv] / Bognár János ; sajtó alá rendezte, Szarvas Pál. — Budapest : Tankönyvkiadó, 1977.
551 p. : ill. ; 25 cm. H77
Includes bibliographical references.
ISBN 9631715671 : 57.00Ft
1. Chemistry. I. Title.
QD31.2.B62 79-337950
*79 MARC

Campbell, James Arthur, 1916-
Chemistry, the unending frontier / J. Arthur Campbell ; [illustrators, Barry Age ... et al.]. — Santa Monica, Calif. : Goodyear Pub. Co., c1978.
xii, 532 p. : ill. ; 24 cm.
Includes index.
ISBN 0-87620-133-8
1. Chemistry. I. Title.
QD33.C24 540 77-25214
77 MARC

Ch'u teng hua hsüeh pien hsieh tsu.
(Ch'u teng hua hsüeh)
初等化学 /《初等化学》编写组编. — 第1版. — 上海：上海人民出版社, 1973, 1977 printing.
iii, 246 p., [1] fold. leaf : ill. ; 21 cm.
Includes bibliographical references.
$0.68
1. Chemistry. I. Title.
QD31.2.C49 1973 78-841939

Compton, Charles, 1917-
Inside chemistry / Charles Compton ; [drawings by J & R Services, inc.]. — New York : McGraw-Hill, c1979.
xiv, 569 p. : ill. ; 25 cm.
Includes bibliographies and index.
ISBN 0-07-012350-0 : 16.95
1. Chemistry. I. Title.
QD31.2.C65 540 78-13379
78 MARC

Dickerson, Richard Earl.
Chemical principles / Richard E. Dickerson, Harry B. Gray, Gilbert P. Haight, Jr. — 3d ed. — Menlo Park, Calif. : Benjamin/Cummings Pub. Co., c1979.
1037 p. in various pagings : ill. ; 24 cm.
Includes bibliographies and index.
ISBN 0-8053-2398-8 : $17.95
1. Chemistry. I. Gray, Harry B., joint author. II. Haight, Gilbert Pierce, joint author. III. Title.
QD31.2.D52 1979 540 77-87336
78 MARC

Dickson, Thomas R
Introduction to chemistry / T. R. Dickson. — 3d ed. — New York : Wiley, c1979.
xiii, 465 p. : ill. ; 24 cm.
Includes index.
ISBN 0-471-02223-3
1. Chemistry. I. Title.
QD31.2.D53 1979 540 78-17406
78 MARC

Fessenden, Ralph J 1932-
Chemical principles for the life sciences / Ralph J. Fessenden, Joan S. Fessenden. — 2d ed. — Boston : Allyn and Bacon, c1979.
x, 526 p. : ill. ; 24 cm.
Includes index.
ISBN 0-205-06506-6 : $17.95
1. Chemistry. I. Fessenden, Joan S., joint author. II. Title.
QD31.2.F49 1979 540 78-21279
78 MARC

King, Edward Louis, 1920-
Chemistry / Edward L. King. — 1st ed. — Sausalito, Calif. : Painter Hopkins Publishers, 1979.
xx, 1101 p. : ill. ; 26 cm.
Includes index.
ISBN 0-05-250726-2
1. Chemistry. I. Title.
QD31.2.K54 540 78-25576
78 MARC

Kivinen, Antti.
Kemia / Antti Kivinen, Osmo Mäkitie. — Helsingissä : Otava, [1978]
460 p. ; ill. ; 25 cm. — (Otavan korkeakoulukirjasto) Fi***
Includes index.
ISBN 9511043935
1. Chemistry. I. Mäkitie, Osmo, joint author. II. Title.
QD31.2.K56 78-394994
79 MARC

Kostiner, Edward.
Fundamentals of chemistry / Edward Kostiner, Jesse R. Rea. — New York : Harcourt Brace Jovanovich, c1979.
viii, 472 p. : ill. ; 25 cm.
Includes index.
ISBN 0-15-529430-X
1. Chemistry. I. Rea, Jesse R., joint author. II. Title.
QD31.2.K67 540 78-61298
79 MARC

Latscha, H P 1937-
Chemie : Basiswissen / H. P. Latscha, H. A. Klein. — Berlin ; New York : Springer, 1978-
v. : ill. ; 21 cm. — (Heidelberger Taschenbücher, 193) GFR78-A
Bibliography: p. [411]-413.
CONTENTS: T. 1. Anorganische und allgemeine Chemie.
Includes index.
ISBN 0-387-08630-7
1. Chemistry. I. Klein, Helmut Alfons, 1949- joint author.
QD31.2.L36 540 78-945
78 MARC

March, Jerry, 1929-
General chemistry / Jerry March and Stanley Windwer. — New York : Macmillan, c1979.
xix, 842 p. : ill. ; 29 cm.
Includes index.
ISBN 0-02-375860-0
1. Chemistry. I. Windwer, Stanley, joint author. II. Title.
QD31.2.M365 540 77-28622
78 MARC

Martín, Consuelo.
Químicas / Consuelo Martín Rodríguez. — Pamplona : Ediciones Universidad de Navarra, [1977]
291 p. ; 18 cm. — (Colección Ciencias de la educación) (Guías de los estudios universitarios ; 3)
Bibliography: p. [287]-291.
ISBN 8431302623
1. Chemistry. 2. Chemistry—Study and teaching (Higher) 3. Chemistry—Vocational guidance. I. Title.
QD31.2.M368 79-352367
79 MARC

Meyer, Eugene, 1938-
Introduction to modern chemistry / Eugene Meyer. — Englewood Cliffs, N.J. : Prentice-Hall, c1979.
xvi, 527 p. : ill. ; 24 cm.
Includes index.
ISBN 0-13-488320-9 : $14.95
1. Chemistry. I. Title.
QD31.2.M45 540 78-12664
78 MARC

Miller, Glenn H 1922-
Basic chemistry / Glenn H. Miller, with the assistance of Frederick B. Augustine. — 2d ed. — New York : Harper & Row, c1979.
495 p. in various pagings : ill. ; 23 cm.
Includes index.
ISBN 0-06-385476-7
1. Chemistry. I. Augustine, Frederick B. II. Title.
[QD31.2.M53 1979] 540 78-24062
78 MARC

Mortimer, Charles E
Chemistry : a conceptual approach / Charles E. Mortimer. — 4th ed. — New York : D. Van Nostrand Co., c1979.
xiv, 815 p. : ill. ; 25 cm.
Includes index.
ISBN 0-442-25545-4
1. Chemistry. I. Title.
QD31.2.M65 1979 540 78-65267
79 MARC

Peet, James Henry John.
Fundamentals of chemistry : a chemistry textbook for TEC level 2 / J. H. J. Peet. — London ; New York : Longman, 1978.
238 p. : ill. ; 22 cm. — (Longman technician series : Mathematics and sciences)
Includes bibliographical references and index.
ISBN 0-582-41164-5
1. Chemistry. I. Title.
QD31.2.P43 540 77-26076
77 MARC

Perlman, Philip.
Essentials of modern chemistry / Philip Perlman. — 2d ed. — Woodbury, N.Y. : Barron's Educational Series, c1979.
x, 462 p. : ill. ; 28 cm.
Includes index.
ISBN 0-8120-0646-1 : $5.95
1. Chemistry. I. Title.
QD31.2.P47 1979 540 78-2344
78 MARC

Pilar, Frank L
Chemistry, the universal science / Frank L. Pilar. — Reading, Mass. : Addison-Wesley Pub. Co., c1979.
769 p. : ill. ; 26 cm. — (Addison-Wesley series in chemistry)
Includes bibliographical references and index.
ISBN 0-201-05768-9
1. Chemistry. I. Title.
QD31.2.P55 540 78-55828
78 MARC

Redmore, Fred H
Fundamentals of chemistry / Fred H. Redmore. — Englewood Cliffs, N.J. : Prentice-Hall, c1979.
xix, 711 p. : ill. ; 24 cm.
Includes bibliographies and index.
ISBN 0-13-335158-0 : $17.95
1. Chemistry. I. Title.
QD31.2.R42 540 78-14294
78 MARC

CHEMISTRY (Continued)

Roach, Don, 1936-
Basic college chemistry / Don Roach, Edmund Leddy, Jr. — New York : McGraw-Hill, c1979.
xix, 636 p. : ill. ; 25 cm.
Includes index.
ISBN 0-07-052987-6 : $16.50
1. Chemistry. I. Leddy, Edmund J., joint author. II. Title.
QD31.2.R598 540 78-13382
78 MARC

Rocks, Lawrence.
Developing your chemistry fundamentals / Lawrence Rocks. — Tulsa, Okla. : PPC Books, c1979.
xii, 564 p. : ill. ; 24 cm.
Includes index.
ISBN 0-87814-041-7
1. Chemistry. I. Title.
QD31.2.R63 540 78-52371
79 MARC

Seager, Spencer L
Introductory chemistry : general, organic, biochemistry / Spencer L. Seager, Michael R. Slabaugh. — Glenview, Ill. : Scott, Foresman, c1979.
684 p. : ill. ; 26 cm.
Includes bibliographical references and index.
ISBN 0-673-15026-7 : $16.95
1. Chemistry. I. Slabaugh, Michael R., 1943- joint author. II. Title.
QD31.2.S412 540 78-15936
78 MARC

Sears, Curtis T
Aspects of chemistry for health-related sciences / Curtis T. Sears, Jr., Conrad L. Stanitski. — Englewood Cliffs, N.J. : Prentice-Hall, c1979.
xvi, 409, ix, p. : ill. ; 24 cm.
Includes index.
ISBN 0-13-049262-0 : $16.50
1. Chemistry. I. Stanitski, Conrad L., joint author. II. Title.
[DNLM: 1. Chemistry. QD31.2 S439c]
QD31.2.S42 540'.2'461 78-9015
78 MARC

Sienko, Michell J
Chemistry, principles and applications / Michell J. Sienko, Robert A. Plane. — New York : McGraw-Hill, c1979.
x, 691 p. : ill. ; 24 cm.
Published in 1974 under title: Chemical principles and properties.
Includes bibliographical references and index.
ISBN 0-07-057321-2
1. Chemistry. I. Plane, Robert A., joint author. II. Title.
QD31.2.S558 540 78-11169
78 MARC

Stark, Edward C
Essential chemistry / Edward C. Stark. — Encino, Calif. : Glencoe Pub. Co., c1979.
855 p. in various pagings : ill. ; 25 cm.
Includes index.
ISBN 0-02-478270-X
1. Chemistry. I. Title.
QD31.2.S75 540 78-71681
79 MARC

Timberlake, Karen.
Chemistry / Karen Timberlake. — 2d ed. — New York : Harper & Row, c1979.
582 p. in various pagings : ill. ; 25 cm.
Includes index.
ISBN 0-06-163409-3
1. Chemistry. I. Title.
QD31.2.T55 1979 540 78-27803
78 MARC

Zieliński, Stanisław.
Chemia ogólna i nieorganiczna / Stanisław Zieliński. — Wyd. 1. — Poznań : Wydawn. Naukowe Uniwersytetu im. Adama Mickiewicza, 1978-
v. : ill. ; 25 cm. P***
Bibliography: v. 1, p. [215]
CONTENTS: cz. 1. Chemia ogólna.

zł28.00 (v. 1)
1. Chemistry. I. Title.
QD31.2.Z54 78-399118
79 MARC

—ABSTRACTS—PERIODICALS

Chile. Universidad, Santiago. Facultad de Ciencias Químicas.
Anales de la Facultad de Ciencias Químicas. t. 22/25- 1970/73-
Santiago, Universidad de Chile - Sede Norte.
v. in 26 cm. annual.
"Tesis de químico farmacéuticos, bioquímicos y químicos (Resúmenes)"
Vol. for 1970/73 includes Anales for 1970-71 of the faculty under its earlier name: Facultad de Química y Farmacia.
Continues: Chile. Universidad, Santiago. Facultad de Química y Farmacia. Anales.
English and Spanish.
1. Chemistry, Pharmaceutical—Abstracts—Periodicals. 2. Biological chemistry—Abstracts—Periodicals. 3. Chemistry—Abstracts—Periodicals. I. Chile. Universidad, Santiago. Facultad de Química y Farmacia. Anales.
RS21.C5 78-644597
MARC-S

Journal of chemical research. Synopses. [Jan.] 1977-
[London] The Chemical Society.
v. ill. 30 cm. monthly.
Contains abstracts of papers appearing in the miniprint edition.
Published on behalf of the Founding Sponsors: Gesellschaft Deutscher Chemiker, Société chimique de France, and The Chemical Society.
Key title: Journal of chemical research. Synopses.
ISSN 0308-2342
1. Chemical research—Abstracts—Periodicals. 2. Chemistry—Abstracts—Periodicals. I. Chemical Society, London. II. Gesellschaft Deutscher Chemiker. III. Société chimique de France.
QD40.J64a 540'.5 78-649929
MARC-S

—ADDRESSES, ESSAYS, LECTURES

(Khimiíà)
Химия : [сборник статей / отв. ред. д-р техн. наук проф. В. Н. Крылов]. — Ленинград : ЛИСИ, 1974-
v. : ill. ; 21 cm. — (Сборник трудов ЛИСИ ; № 92,)
USSR 74
At head of title, v. 1- : Ministerstvo vysshego i srednego spetsial'nogo obrazovaniíà RSFSR. Leningradskiĭ inzhenerno-stroitel'nyĭ institut.
Includes bibliographies.
0.70rub (v. 1) varies
1. Chemistry—Addresses, essays, lectures. I. Krylov, Vladimir Nikolaevich, d-r tekhn. nauk. II. Series: Leningrad. Inzhenerno-stroitel'nyĭ institut. Sbornik nauchnykh trudov ; no. 92 [etc.]
TA4.L43 no. 92, etc. 76-531221
[QD39]

Khimiíà i khimicheskaíà tekhonologiíà : [Sbornik stateĭ / Otv. red. kand. tekhn. nauk, dots. V. V. Balandina]. — Novocherkassk : NPI, 1976.
142 p. : ill. ; 20 cm. — (Trudy Novocherkasskogo politekhnicheskogo instituta im. Sergo Ordzhonikidze ; t. 320) USSR76
At head of title: MV i SSO RSFSR.
Includes bibliographies.
1. Chemistry—Addresses, essays, lectures. 2. Chemistry, Technical—Addresses, essays, lectures. I. Balandina, V. V. II. Series: Novocherkassk, Russia. Politekhnicheskiĭ institut. Trudy ; t. 320.
TA7.N65 t. 320 79-382635
[QD39] 79 MARC

Khimiíà i khimicheskaíà tekhnologiíà : [Sb. stateĭ / Redkol. S. I. Smyshlíàev (otv. red.) ... et al.]. — Krasnodar : Kuban. un-t, 1977.
193 p. : ill. ; 20 cm. — (Nauchnye trudy - Kubanskiĭ gosudarstvennyĭ universitet ; vyp. 243) USSR78
At head of title: Ministerstvo vysshego i srednego spetsial'nogo obrazovaniíà RSFSR.
Includes bibliographies.
1. Chemistry—Addresses, essays, lectures. 2. Chemistry, Technical—Addresses, essays, lectures. I. Smyshlíàev, S. I. II. Series: Kubanskiĭ gosudarstvennyĭ universitet. Nauchnye trudy ; vyp. 243.
QD39.K485 79-352096
79 MARC

In memory of H. L. Meerwein. — Berlin ; New York : Springer-Verlag, 1979.
320 p. : ill. ; 25 cm. — (Topics in current chemistry ; v. 80)
Includes bibliographical references and index.
CONTENTS: Brown, H. C. Meerwein and equilibrating carbocations.—Olah, G. A. From boron trifluoride to antimony pentafluoride in search of stable carbocations.—Hogeveen, H. and van Kruchten, E. M. G. A. Wagner-Meerwein rearrangements in long-lived polymethyl substituted bicyclo(3.2.0)heptadienyl cations.—Kirmse, W. Rearrangements of carbocations.
ISBN 0-387-09309-5
1. Chemistry—Addresses, essays, lectures. 2. Meerwein, Hans, 1879- . I. Meerwein, Hans, 1879- . II. Title. III. Series.
QD1.F58 vol. 80 540'.8 s 79-975
[QD39] 79 MARC

Technologia / [zespół redakcyjny Władysław Błaszków, Jerzy Ziobrowski]. — Wyd. 1. — Wrocław : Wydawn. Uczelniane Akademii Ekonomicznej, 1977.
195 p. : ill. ; 24 cm. — (Prace naukowe Akademii Ekonomicznej imienia Oskara Langego we Wrocławiu ; zesz. 118(140)) P***
Summaries in German; table of contents also in German.
Includes bibliographies.
zł30.00
1. Chemistry—Addresses, essays, lectures. I. Błaszków, Władysław. II. Ziobrowski, Jerzy. III. Series: Akademia Ekonomiczna we Wrocławiu. Prace naukowe Akademii Ekonomicznej imienia Oskara Langego we Wrocławiu ; zesz. 118.
HB9.B7522 zesz. 118 78-380560
[QD39] 78 MARC

(Voprosy khimii)
Вопросы химии : [Сборник статей] / Ташк. гос. ун-т им. В. И. Ленина ; [Редколлегия проф. Ш. Т. Талипов (отв. ред.) ... et al.]. — Ташкент : ТашГУ, 1976.
140 p. : ill. ; 19 cm. — (Сборник научных трудов - Ташкентский государственный университет им. В. И. Ленина ; № 513) USSR 77
At head of title: Ministerstvo vysshego i srednego spetsial'nogo obrazovaniíà Uzbekskoĭ SSR.
Series romanized: Sbornik nauchnykh trudov - Tashkentskiĭ gosudarstvennyĭ universitet im. V. I. Lenina.
Includes bibliographies.
0.80rub
1. Chemistry—Addresses, essays, lectures. I. Talipov, Sh. T. II. Tashkend. Universitet.
QD39.V625 79-407227

--APPARATUS see Chemical apparatus

—BIBLIOGRAPHY

see also Chemical literature; Chemical literature

Butman, Donna.
Chemistry for non-science trained librarians and information scientists : a selected annotated bibliography / by Donna Butman. — Silver Spring, Md. : CDB Enterprises, 1979.
ii, 30 leaves ; 29 cm.
Cover title: FLEXIS.
1. Chemistry—Bibliography. I. Title. II. Title: FLEXIS.
Z5521.B95 016.54 79-118702
[QD31.2] 79 MARC

Mexico. Consejo Nacional de Ciencia y Tecnología. Departamento de Sistemas de Información.
Bibliografía básica a nivel profesional, ciencias químicas / elaborado por el Departamento de Sistemas de Información, CONACYT. — 1. ed. — México : Consejo Nacional de Ciencia y Tecnología, 1973, 1976 printing.
vii, 75 leaves ; 28 cm.
Original title: Acervo básico para bibliotecas de escuela de ingeniería química, química industrial y farmacología, lista preliminar.
Spanish, English, French, German, Russian or Italian.
Includes indexes.
1. Chemistry—Bibliography. I. Title.
Z5521.M59 1973 79-371496
[QD31.2] 79 MARC

—BIO-BIBLIOGRAPHY

Novackis, H
Profesors Ernsts Karpovics : personālais literatūras rādītājs / [sastādītāji H. Novackis, D. Paukšēna]. — Rīga : LVU, 1977.
54 p., [1] leaf of plates : port. ; 20 cm. — (Pētera Stučkas Latvijas valsts universitātes zinātnieki) USSR***
At head of title: Latvijas PSR Augstākās un vidējās speciālās izglītības ministrija. Pētera Stučkas Latvijas valsts universitāte. Zinātniskā bibliotēka.
Cover title: E. Karpovics.
Added t.p.: Professor Ėrnst Karpovich.
Latvian and Russian.
Includes index.
0.10rub
1. Karpovics, Ernsts, 1892- . 2. Chemists—Latvia—Biography. 3. Chemistry—Bio-bibliography. I. Paukšēna, D., joint author. II. Riga. Universitāte. Zinātniskā bibliotēka. III. Title. IV. Title: Professor Ėrnst Karpovich.
QD22.K32N68 79-371304
79 MARC

—COLLECTED WORKS

Chemia i inżynieria chemiczna. zesz. 12.-
Poznań [Wydawn. Politechniki Poznańskiej] 1976-
no. ill. 24 cm. (Zeszyty naukowe Politechniki Poznańskiej)
Continues: Chemia techniki zastosowań.
Summaries in English and Russian.
No. 12- issued by Politechnika Poznańska.
1. Chemistry, Technical—Collected works. 2. Chemistry—Collected works. I. Politechnika Poznańska.
TP1.P57a 78-642598
MARC-S

(Khimicheskaíà tekhnologiíà i khimiíà)
Химическая технология и химия.
Рига, Рижский политехнический ин-т [etc.]
no. ill. 20 cm.
"Межвузовский научно-технический сборник."
Began in 1973.
1. Chemistry, Technical — Collected works. 2. Chemistry — Collected works. I. Riga. Politehniskais institūts.
TP1.K4143 79-640870
(MARC-S)

Ostwald, Wilhelm, 1853-1932.
Forschen und Nutzen : Wilhelm Ostwald zur wiss. Arbeit / aus seinen Schriften ausgew., bearb. u. zsgest. anlässlich seines 125. Geburtstages von Günther Lotz, Lothar Dunsch, Uta Kring ; mit e. Geleitw. von Peter Adolf Thiessen u. e. Einf. von Günther Lotz u. Lothar Dunsch. — Berlin : Akademie-Verlag, 1978.
xlviii, 278 p. : 55 ill. (some col.) ; 22 cm. — (Beiträge zur Forschungstechnologie : Sonderband ; 1) GDR78-A4
Includes bibliographical references and indexes.
38.00M
1. Chemistry—Collected works. 2. Ostwald, Wilhelm, 1853-1932. 3. Chemists—Germany—Biography. I. Lotz, Günther. II. Dunsch, Lothar. III. Kring, Uta. IV. Title. V. Series.
QD3.O83 1978 79-364024
79 MARC

CHEMISTRY
—COLLECTED WORKS (Continued)

Plzeň. Pedagogický institut.
Sborník. Biologie, chemie, zeměpis. 3.–
1960–
Praha, Státní pedagogické nakl.
v. illus., maps. 21 cm.
Continues Plzeň. Vyšší pedagogická škola. Sborník. Biologie, chemie.
Summaries in Russian and German.
1. Biology—Collected works. 2. Chemistry—Collected works. 3. Geography—Collected works.
QH301.P57 79–413548

Politechnika Poznańska.
Chemia. zesz. 1–
Poznań, 1962–
no. in v. illus., facsims. 24 cm. (Its Zeszyty naukowe)
Summaries in Russian and English.
Continued by: Chemia techniki zastosowań.
1. Chemistry, Technical — Collected works. 2. Chemistry — Collected works. I. Title.
TP1.P57a 66–54467
MARC–S

Suleĭmanova, Ė T
(S. D. Mekhtiev)
С. Д. Мехтиев : библиография / ₍составители, Э. Т. Сулейманова, М. Р. Мусаев₎. — Баку : Элм, 1976.
168 p. ; 16 cm. — (Деятели науки и культуры Азербайджана)
USSR***
At head of title: Akademii͡a nauk Azerbaĭdzhanskoĭ SSR. Fundamental'nai͡a biblioteka.
Added t. p. in Azerbaijani.
Prefatory matter in Azerbaijani and Russian.
Series romanized: Dei͡ateli nauki i kul'tury Azerbaĭdzhana.
Includes indexes.
1. Mekhtiev, S. D.—Bibliography. 2. Chemists—Azerbaijan—Biography. 3. Chemistry—Bio-bibliography. I. Musaev, M. R., joint author. II. Akademii͡a nauk Azerbaĭdzhanskoĭ SSR, Baku. Fundamental'nai͡a biblioteka.
Z8606.73.S94 79–410682
[QD22.M396]

Uniwersytet Gdański. Wydział Matematyki, Fizyki i Chemii.
Zeszyty naukowe. Chemia. 1–
Gdańsk ₍Wydawn. Uczelniane Uniwersytetu Gdańskiego₎ 1971–
no. illus. 24 cm.
English or Polish.
Summaries in English, French, German, or Polish.
1. Chemistry—Collected works. I. Title: Chemia.
QD1.D2 540'.5 73–640795
MARC–S

Wyższa Szkoła Pedagogiczna w Gdańsku. Wydział Matematyki, Fizyki i Chemii.
Zeszyty naukowe: Matematyka, fizyka, chemia. t. 1–
1961–
₍Gdynia, etc., Wydawn. Morskie₎
v. in ill. 24 cm. annual.
Summaries in English or German.
1. Mathematics—Collected works. 2. Physics—Collected works. 3. Chemistry—Collected works. I. Title. II. Title: Matematyka, fizyka, chemia.
QA1.W96 500.2'05 66–34262
MARC–S

—COMPUTER PROGRAMS

Isenhour, Thomas L
Introduction to computer programming for chemists : FORTRAN / Thomas L. Isenhour, Peter C. Jurs. — 2d ed. — Boston : Allyn & Bacon, c1979.
x, 356 p. : ill. ; 24 cm.
Bibliography: p. 346-347.
Includes index.
ISBN 0-205-05897-3 : $15.95
1. Chemistry—Computer programs. 2. FORTRAN (Computer program language) I. Jurs, Peter C., joint author. II. Title.
QD39.3.C6I83 1979 542'.8 77-14327
77 MARC

—CONGRESSES

Chemistry Symposium, 3d, Aligarh Muslim University, 1972.
Proceedings of the Chemistry Symposium, 1972, Aligarh Muslim University, Aligarh, December 21-23, 1972 / under the auspices of the Department of Atomic Energy, Government of India. — ₍s.l. : s.n., 19
v. : ill. ; 24 cm.
"Under the auspices of the Dept. of Atomic Energy, Govt. of India."
Includes bibliographical references and index.
CONTENTS:
—v. 2. Physical methods in structural inorganic chemistry, actinide chemistry and solid state studies.
1. Chemistry—Congresses. I. India (Republic). Dept. of Atomic Energy.
QD1.C748 1972 540 76-929514
79 MARC

Energy and the chemical sciences : the 1977 Karcher symposium covering the connection between chemistry and the energy problem from several different viewpoints, involving petroleum, natural gas, coal, sunshine, nuclear power and the public policy which guides their use / co-editors and co-chairmen of the symposium, S. D. Christian and J. J. Zuckerman. — Oxford ; New York : Pergamon Press, 1978.
xiv, 144 p. : ill. ; 26 cm. GB78
Includes bibliographical references and indexes.
ISBN 0-08-022094-0 : £8.35 ($15.00)
1. Power resources—Congresses. 2. Chemistry—Congresses. I. Christian, Sherril Duane, 1931- . II. Zuckerman, Jerold J., 1936- .
TJ163.15.E52 1978 621.4 77-30696
78 MARC

ha-Ḥevrah ha-yisr'elit le-khimyah.
Proceedings of the 44th meeting of the Israel Chemical Society and the Israel Institute of Chemical Engineering held at Bar-Ilan University, Ramat-Gan, Israel, June 27-29, 1977. — Ramat Gan : The University, ₍1977?₎
ca. 150 p. in various pagings : ill. ; 24 cm.
Cover title: Abstracts.
Includes bibliographical references and index.
1. ha-Ḥevrah ha-yisr'elit le-khimyah. 2. Chemistry—Congresses. I. ha-Agudah ha-yiśre'elit le-handasah kimit. II. Title.
QD1.H4524 1977 540 79-101771
79 MARC

—DICTIONARIES

Cornubert, Raymond.
Dictionnaire de chimie anglais-français : mots et locutions fréquemment rencontres dans les textes anglais et américains / par R. Cornubert, ... — 6e éd. — Paris : Dunod, 1977.
viii, 212 p. ; 24 cm. — (Collection Dunod technique ISSN 0335-3206)
F78-1867
First-5th editions published under title: Dictionnarie chimique anglais-français.
ISBN 2-04-007334-5 : 65.00F
1. Chemistry—Dictionaries. 2. English language—Dictionaries—French. I. Title.
QD5.C55 1977 540'.3 78-386529
79 MARC

De Vries, Louis, 1885-
Wörterbuch der Chemie und der chemischen Verfahrenstechnik / L. De Vries, H. Kolb. — 2., überarb. und erw. Aufl. — Weinheim ; New York : Verlag Chemie, 1978-
v. ; 25 cm. GFR78-A
Added t.p. in English with title: Dictionary of chemistry and chemical engineering.
CONTENTS: Bd. 1. Deutsch/Englisch.

ISBN 0-89573-010-0 (v. 1) : DM220 (v. 1)
1. Chemistry—Dictionaries. 2. Chemical engineering—Dictionaries. 3. English language—Dictionaries—German. 4. Chemistry—Dictionaries—German. 5. Chemical engineering—Dictionaries—German. 6. German language—Dictionaries.—English. I. Kolb, Helga, joint author. II. Title. III. Title: Dictionary of chemistry and chemical engineering.
QD5.D47 1978 540'.3 79-346865
79 MARC

Duncan, Donald Russell.
English-Esperanto chemical dictionary = Angla-Esperanta kemia vortario / by D. R. Duncan. — 1st ed. — London : Published for British Esperantist Scientific Association by British Esperanto Association, 1956.
56 p. ; 19 cm.
1. Chemistry—Dictionaries. 2. English language—Dictionaries—Esperanto. I. Title. II. Title: Angla-Esperanta kemia vortario.
QD5.D77 540'.3 78-392117
79 MARC

Peristilahan kimia dan farmasi. — Cet. 9. — ₍Bandung₎ : Penerbit Universitas ITB, 1976.
x, 288 p. ; 21 cm.
Compiled by M. M. Purbo-Hadiwidjoyo and S. A. Waworoentoe.
English and Indonesian.
1. Chemistry—Dictionaries. 2. Chemistry, Pharmaceutical—Dictionaries. 3. English language—Dictionaries—Indonesian. I. Purbo Hadiwidjoyo, M. M. II. Waworoentoe, S. A.
QD5.P43 1976 78-940182
79 MARC

(Ying Hua shih yung hua hsüeh ming tz'u shou ts'e)
英華實用化學名詞手册. — 香港：駱駝出版社, 1977.
16, 130 p. ; 21 cm.
$4.20

1. Chemistry—Dictionaries. 2. English language—Dictionaries—Chinese. I. Title: Hua hsüeh ming tz'u shou ts'e.
QD5.Y54 78–842433

—DICTIONARIES—GERMAN

De Vries, Louis, 1885-
Wörterbuch der Chemie und der chemischen Verfahrenstechnik / L. De Vries, H. Kolb. — 2., überarb. und erw. Aufl. — Weinheim ; New York : Verlag Chemie, 1978-
v. ; 25 cm. GFR78-A
Added t.p. in English with title: Dictionary of chemistry and chemical engineering.
CONTENTS: Bd. 1. Deutsch/Englisch.

ISBN 0-89573-010-0 (v. 1) : DM220 (v. 1)
1. Chemistry—Dictionaries. 2. Chemical engineering—Dictionaries. 3. English language—Dictionaries—German. 4. Chemistry—Dictionaries—German. 5. Chemical engineering—Dictionaries—German. 6. German language—Dictionaries.—English. I. Kolb, Helga, joint author. II. Title. III. Title: Dictionary of chemistry and chemical engineering.
QD5.D47 1978 540'.3 79-346865
79 MARC

—DICTIONARIES—MALAY

Universiti Kebangsaan.
Kamus istilah kimia Melayu-Inggeris. — ₍Kuala Lumpur₎ : Universiti Kebangsaan Malaysia, 1970.
xvi, 118 p. ; 33 cm.
Cover title.
English or Malay.
1. Chemistry—Dictionaries—Malay. 2. Malay language—Dictionaries—English. I. Title.
QD5.U54 1970 72-942266
79 MARC

—DICTIONARIES—POLYGLOT

Chung-kuo k'o hsüeh yüan. Tzu jan k'o hsüeh ming tz'u pien ting shih.
(Jih Ying Han hua hsüeh hua kung tz'u hui)
日英汉化学化工詞彙 = Ri Ying Han hua xue huagong cihui / 中国科学院編譯出版委員会名詞室. — 北京：科学出版社, 1959.
iv, 467 p. 18 cm.
Compiled by Chung-kuo k'o hsüeh yüan tzu jan k'o hsüeh ming tz'u pien ting shih under its earlier name.
1. Chemistry—Dictionaries—Polyglot. 2. Japanese language Dictionaries—Polyglot. I. Title. II. Title: Hua hsüeh hua kung tz'u hui. III. Title: Ri Ying Han hua xue huagong cihui.
QD5.C448 74–838870

Tamura, Saburō, 1917-
(Chū-Ei-Nichi kagaku yōgo jiten)
中英日化学用語辞典 = Chinese-English-Japanese glossary of chemical terms = 汉英日化学词江 / 田村三郎, 白鳥冨美子編. — 東京：東方書店, c1977.
xxx, 612 p. ; 22 cm. Ja***
Bibliography: p. x.
Includes indexes.
Y 12000

1. Chemistry—Dictionaries—Polyglot. 2. Dictionaries, Polyglot. I. Shiratori, Fumiko, 1913- joint author. II. Title. III. Title: Chinese-English-Japanese glossary of chemical terms. IV. Title: Han Ying Jin hua hsueh tz'u hui.
QD5.T35 78–811890

—DICTIONARIES—RUSSIAN

Dresden. Technische Universität. Institut für Angewandte Sprachwissenschaft. Entwicklungsstelle für Fachwörterbücher.
Russisch-deutsches Wörterbuch der Chemie und chemischen Technik : mit etwa 65 000 Fachbegriffen / hrsg. vom Inst. für Angewandte Sprackwiss. d. Techn. Univ. Dresden ; ₍zsgest. von d. wissenschaftl. Mitarb. d. Entwicklungsstelle für Fachwörterbücher im Bereich Lexikologie d. Fachsprachen d. Inst. für Angewandte Sprachwiss.₎. — 3., überarb. u. erg. Aufl. — Berlin : Verlag Technik, 1977.
832 p. ; 21 cm. GDR77-A
Added t.p. in Russian: Russko-nemeĭskiĭ slovar' po khimii i khimicheskoĭ tekhnologii.
Another 3d ed. issued under title: Chemie und chemische Technik, Russisch-Deutsch.

1. Chemistry—Dictionaries—Russian. 2. Chemical engineering—Dictionaries—Russian. 3. Russian language—Dictionaries—German. I. Title. II. Title: Russko-nemeĭskiĭ slovar' po khimii i khimicheskoĭ tekhnologii.
QD5.D7 1977 540'.3 78-382767
79 MARC

CHEMISTRY (Continued)

—DICTIONARIES—SPANISH

Carrasco, Cástulo.
Diccionario de medicina, farmacia, veterinaria y química / ₍por₎ Cástulo Carrasco. — Madrid : Garsi, D.L. 1977.
1027 p. ; 19 cm. Sp78-Apr
ISBN 8473910133 : 2500ptas
1. Medicine—Dictionaries—Spanish. 2. Pharmacy—Dictionaries—Spanish. 3. Veterinary medicine—Dictionaries—Spanish. 4. Chemistry—Dictionaries—Spanish. I. Title.
R121.C26 610'.3 78-392328
79 MARC

—EARLY WORKS TO 1800

Glauber, Johann Rudolph, 1604?-1670.
Operis mineralis pars prima ₍-tertia₎ Inventa & publicata in gratiam studiosorum artis chymicae per Johannem Rudolphum Glauberum. Amsterodami, Prostant apud J. Janssonium, 1651-1653.
3 v. 16 cm.
Translation from German of Operis mineralis.
CONTENTS: Pars 1. Ubi docetur separatio auri è filicibus, arena, argillâ, aliisque fossilibus per salis spiritum, quae aliâs eliquari nequeunt. Item Panacea. —Pars 2. De ortu & origine omnium metallorum & mineralium, quo scilicet pacto illa per astra producantur ex aqua & terra corpus sibi suscipiant & multiplici forma formentur.—Pars 3. In qua titulo commentarii in libellum Paracelsi Coelum philosophorum.
1. Chemistry—Early works to 1800. I. Title.
QD27.G6316 79-350811
79 MARC

—EXAMINATIONS, QUESTIONS, ETC.

Hentz, Forrest C 1933-
Grading the advanced placement examination in chemistry, 1977 and 1978 : a report / by Forrest C. Hentz, Jr. — Princeton, N.J. : Advanced Placement Program of the College Board, c1979.
34 p. : ill. ; 28 cm.
$1.00
1. Chemistry—Examinations, questions, etc. 2. Advanced placement programs (Education) 3. Grading and marking (Students) I. College Entrance Examination Board. Advanced Placement Program. II. Title.
QD42.H36 540'.76 79-121343
79 MARC

Hua-nan shih yüan fu chung. Hua hsüeh chiao yen tsu.
(Hua hsüeh fu hsi tzu liao)
化学复习資料 / ₍华南师院附中化学教研组编₎. — ₍香港? : s. n.₎, 1978.
164 p. : ill. ; 19 cm.
Reprint ed.

1. Chemistry—Examinations, questions, etc. I. Title.
QD42.H75 1978 79-840690

Poedjaatmaka, Hadyana.
Bahan-bahan udjian saringan universitas; bagian kimia & ilmu pesawat dengan penjelesaiannja. Kimia dikumpulkan dan disusun oleh Hadyana Poedjaatmaka, ilmu pesawat dikumpulkan dan disusun oleh T. H. M. Pangestu. Bandung, Gelora Masa ₍196-?₎
31, 68 p. 21 cm.
1. Technology—Dictionaries—Indonesian. 2. Mechanics, Applied—Examinations, questions, etc. 3. Chemistry—Examinations, questions, etc. I. Pangestu, T. H. M. II. Title.
TA9.P6 79-950320
79 MARC

—EXPERIMENTS

New Zealand. Education Dept.
Experiments in sixth form chemistry : a teacher's guide. — Wellington : Dept. of Education, 1977.
100 p. : ill. ; 30 cm. NZ***
Includes bibliographical references.
1. Chemistry—Experiments. I. Title.
QD43.N54 1977 542 78-322245
79 MARC

—EXPERIMENTS—JUVENILE LITERATURE

Cherrier, François.
Expériences de chimie amusante / François Cherrier ; photos. de l'auteur ; ouvrage revu par Maurice Finot. — Paris : Hachette, ₍1974₎
90 p. : col. ill. ; 26 cm. F***
ISBN 2-01-300288-1
1. Chemistry—Experiments—Juvenile literature. I. Title. II. Title: Chimie amusante.
QD38.C45 542 78-383739
78 MARC

Cherrier, François.
Fascinating experiments in chemistry / by François Cherrier ; technical consultant, Maurice Finot ; ₍translated by E. W. Egan ; adapted by Dennis Schuchman ; photos. and layout by the author₎. — New York : Sterling Pub. Co., c1978.
96 p. : col. ill. ; 26 cm.
Translation of Expériences de chimie amusante.
Includes index.
SUMMARY: Text and illustrations provide instructions for a variety of chemical experiments including making secret ink, setting off a small volcanic eruption, and growing giant crystals.
ISBN 0-8069-3102-7. ISBN 0-8069-3103-5 lib. bdg.
1. Chemistry—Experiments—Juvenile literature. ₍1. Chemistry—Experiments. 2. Experiments₎ I. Title.
QD38.C4513 542 78-57788
79 MARC

—HANDBOOKS, MANUALS, ETC.

Bernard, Maurice, 1920-
Chimie générale et minérale / Maurice Bernard, ... Florent Busnot, ... — ₍Nouvelle éd. refondue₎. — Paris : Dunod, 1978.
2 v. (xiv, 550 p.) ; 18 cm. — (Collection Aide-mémoire) F78-7872
Bibliography: p. 549-550.
Includes indexes.
CONTENTS: t. 1. Nomenclature, atomistique, données sur les solides et sur les solutions.—t. 2. Constantes physico-chimiques, renseignements pratiques.
ISBN 2-04-002335-6 (v. 1)
1. Chemistry—Handbooks, manuals, etc. I. Busnot, Florent, joint author. II. Title.
QD65.B44 1978 540 78-399128
79 MARC

Lange, Norbert Adolph, 1892-
Lange's Handbook of chemistry / editor, John A. Dean ; formerly compiled and edited by Norbert Adolph Lange. — 12th ed. — New York : McGraw-Hill, c1979.
ca. 1100 p. in various pagings ; 24 cm.
Includes bibliographical references and index.
ISBN 0-07-016191-7 : $28.50
1. Chemistry—Handbooks, manuals, etc. I. Dean, John Aurie, 1921- . II. Title. III. Title: Handbook of chemistry.
QD65.L36 1979 540'.2'02 78-15335
78 MARC

—HISTORY

Debus, Allen G
The chemical philosophy : Paracelsian science and medicine in the sixteenth and seventeenth centuries / Allen G. Debus. — New York : Science History Publications, 1977.
2 v. (xv, 606 p.) : ill. ; 24 cm.
Bibliography: p. ₍555₎-595.
Includes index.
ISBN 0-88202-047-1
1. Chemistry—History. 2. Paracelsus, 1493-1541. 3. Medicine—15th-18th centuries. I. Title.
QD14.D43 540'.9'031 76-1841
76 MARC

McCann, H Gilman, 1942-
Chemistry transformed : the paradigmatic shift from phlogiston to oxygen / H. Gilman McCann. — Norwood, N.J. : Ablex Pub. Corp., 1978.
viii, 179 p. ; 24 cm. — (Modern sociology)
Bibliography: p. 162-169.
Includes indexes.
ISBN 0-89391-004-X
1. Chemistry—History. 2. Science—Social aspects. I. Title. II. Series.
QD11.M23 540'.9'033 78-19173
78 MARC

Raskin, Naum Mikhaĭlovich
Îakov Dmitrievich Zakharov : fizik i khimik konîsa XVIII i nachala XIX v. : 1765-1836 / N. M. Raskin. — Leningrad : Nauka, Leningradskoe otd-nie, 1979.
106 p. ; 20 cm. — (Nauchno-biograficheskaîa seriîa) USSR***
On leaf preceding t.p.: Akademiîa nauk SSSR.
"Osnovnye trudy ÎA. D. Zakharova": p. 104-₍105₎
Includes bibliographical references.
0.20rub
1. Zakharov, Îakov Dmitrievich, 1765-1836. 2. Scientists—Russia—Biography. 3. Physics—History. 4. Chemistry—History.
Q143.Z27R37 79-376202
79 MARC

Sikora, Jan.
Z historii chemii ... / Jan Sikora. — Wyd. 1. — Warszawa : Nasza Księgarnia, 1977 (Gdań. : ZG)
319, ₍1₎ p. : ill., ports. ; 21 cm. P78-1845
Includes bibliographical references.
zł42.00
1. Chemistry—History. I. Title.
QD11.S53 78-378367
78 MARC

Yamaoka, Nozomu, 1892-
(Kagaku shihitsu)
化学史筆 / 山岡望. — 東京 : 内田老鶴圃新社, 1976.
405, ₍31₎ p., ₍6₎ leaves of plates : ill. ; 22 cm. Ja 76-14985
Includes bibliographies and indexes.
¥3800

1. Chemistry—History. I. Title.
QD15.Y35 78-811129

—HISTORY—ADDRESSES, ESSAYS, LECTURES

Yamaoka, Nozomu, 1892-
(Kagakushijin)
化学史塵 : 日本化学会創立百周年慶祝出版 / 山岡望. — 東京 : 内田老鶴圃新社, 1978.
16, 400, 18, 7 p., ₍8₎ leaves of plates : ill. ; 22 cm. Ja 78-18315
Originally published in Kōrutāru, June 1966–July 1974.
Includes index.
¥4800
1. Chemistry—History—Addresses, essays, lectures. 2. Chemists—Biography—Addresses, essays, lectures. 3. Nippon Kagakukai. I. Title.
QD11.Y36 79-800514

—INFORMATION SERVICES

see also National Standard Reference Data System

Chemical Information Review Committee.
Information transfer and use in chemistry : Final report of the Chemical Information Review Committee / J. F. B. Rowland. — London : British Library, 1978.
iv, 33 p. : ill. ; 30 cm. — (British Library research & development reports ; no. 5385 ISSN 0308-2385) GB78-08767
"Appendix": 3 sheets (microfiche) in pocket.
Includes bibliographical references.
ISBN 0-905984-06-4 : £2.50 ($5.00 U.S.)
1. Chemistry—Information services. I. Rowland, J. F. B. II. Title. III. Series: British Library. Research and Development Dept. British Library research & development reports ; no. 5385.
QD8.3.C43 1978 540'.7'2 79-304847
79 MARC

—JUVENILE LITERATURE

Physical science / George A. Williams ... ₍et al.₎. — 2d ed. — New York : Webster Division, McGraw-Hill, c1979.
xi, 468 p. : ill. ; 26 cm. — (Challenges to science)
Includes index.
SUMMARY: A physical science text, stressing an awareness of the environment, with related laboratory activities to lead the student into discovering basic laws and concepts of physics and chemistry.
ISBN 0-07-070415-5
1. Science—Juvenile literature. 2. Physics—Juvenile literature. 3. Chemistry—Juvenile literature. ₍1. Physics. 2. Chemistry₎ I. Williams, George Abiah, 1931- Physical science.
Q163.P577 1979 530 78-12215
78 MARC

—LABORATORIES see Chemical laboratories

—LABORATORY MANUALS

see also Chemistry, Organic—Laboratory manuals; Glass blowing and working

Ch'en, Jui-chu.
(Hua hsüeh shih yen ts'ao tso shou ts'e)
化學實驗操作手冊 / 陳瑞珠編著. — 香港 : 萬里書店, 1978.
2, 100 p. : ill. ; 19 cm.
$5.20

1. Chemistry—Laboratory manuals. I. Title.
QD45.C383 78-841100

Integrated experimental chemistry / David A. Aikens ... ₍et al.₎. — Boston : Allyn and Bacon, c1978.
2 v. (xxii, 752, vi p.) : ill. ; 24 cm. — (The Allyn and Bacon chemistry series)
Includes bibliographical references and index.
CONTENTS: v. 1. Principles and techniques.—v. 2. Laboratory experiments.
ISBN 0-205-05923-6 (v. 1) : $11.95
1. Chemistry—Laboratory manuals. I. Aikens, David A.
QD45.I586 542 77-13690
77₍r79₎rev MARC

Laboratory engineering and manipulations / edited by Edmond S. Perry and Arnold Weissberger. — 3d ed. — New York : Wiley, c1979.
xi, 531 p. : ill. ; 24 cm. — (Techniques of chemistry ; v. 13)
"A Wiley-Interscience publication."
Includes index.
ISBN 0-471-03275-1
1. Chemistry—Manipulation. 2. Chemistry—Laboratory manuals. I. Perry, Edmond S. II. Weissberger, Arnold, 1898- .
QD61.T4 vol. 13 542 s 78-20299
78 MARC

CHEMISTRY
—LABORATORY MANUALS
(Continued)

Mironescu, Mircea.
Tehnica lucrărilor de laborator / Mircea Mironescu, Anton Tutoveanu. — Bucureşti : Editura tehnică, 1967.
332 p. : ill. ; 21 cm.
Bibliography: p. ⌊323⌋
1. Chemistry—Laboratory manuals. I. Tutoveanu, Anton, joint author. II. Title.
QD45.M57 78-390905
79 MARC

Morassi, Roberto.
Tecniche sperimentali moderne del laboratorio chimico / ⌊di⌋ Roberto Morassi, Gian Paolo Speroni ; presentazione del prof. Luigi Sacconi. — ⌊Firenze⌋ : Edizioni scuola universitaria Firenze, 1977.
xx, 546 p. : ill. ; 25 cm. It78-Aug
At head of title: Istituto di chimica generale ed inorganica dell'Università di Firenze.
Bibliography: p. 521-527.
Includes index.
L26000
1. Chemistry—Laboratory manuals. I. Speroni, Gian Paolo, joint author. II. Title.
QD45.M83 542 78-386080
*79 MARC

Pincovschi, Eugen.
Îndrumătorul laborantului chimist / Eugen Pincovschi. — Ediţia a 2-a îmbunătăţită. — Bucureşti : Editura tehnică, 1975.
375 p. : ill. ; 22 cm. R75-4006
Bibliography: p. 343-⌊344⌋
lei17.50
1. Chemistry—Laboratory manuals. I. Title.
QD45.P497 78-393922
*79 MARC

(Rukovodstvo k provedeniiu laboratornykh rabot po obshcheĭ khimii)
Руководство к проведению лабораторных работ по общей химии. — Новосибирск : Новосибирский электротехнический ин-т связи, 1966.
62 p. : ill.
At head of title: Novosibirskiĭ ėlektrotekhnicheskiĭ institut sviazi. Kafedra khimii. Ministerstvo sviazi SSSR.
L.C. copy on microfilm only.
Includes bibliographies.
——— Microfilm. On 1 reel with other items ; 35 mm.
Microfilm Slavic 12620 QD
1. Chemistry—Laboratory manuals. I. Novosibirskiĭ ėlektrotekhnicheskiĭ institut sviazi. Kafedra khimii.
[QD45.R86] 79-410978

—LECTURE EXPERIMENTS see Chemistry—Experiments

—LITERATURE see Chemical literature

—MANIPULATION

see also Distillation, Fractional; Electric furnaces; Glass blowing and working

Laboratory engineering and manipulations / edited by Edmond S. Perry and Arnold Weissberger. — 3d ed. — New York : Wiley, c1979.
xi, 531 p. : ill. ; 24 cm. — (Techniques of chemistry ; v. 13)
"A Wiley-Interscience publication."
Includes index.
ISBN 0-471-03275-1
1. Chemistry—Manipulation. 2. Chemistry—Laboratory manuals. I. Perry, Edmond S. II. Weissberger, Arnold, 1898-
QD61.T4 vol. 13 542 s 78-20299
78 MARC

—MATHEMATICAL MODELS—CONGRESSES

Chemical modeling in aqueous systems : speciation, sorption, solubility, and kinetics : based on a symposium cosponsored by the Society of Environmental Goechemistry and Health and the ACS Division of Enivronmental Chemistry at the 176th meeting of the American Chemical Society, Miami Beach, Florida, September 11-13, 1978 / Everett A. Jenne, editor. — Washington : Society, 1979.
xii, 914 p. : ill. ; 24 cm. — (ACS symposium series ; 93 ISSN 0097-6156)
Includes bibliographies and index.
ISBN 0-8412-0479-9
1. Chemistry—Mathematical models—Congresses. 2. Solution (Chemistry)—Congresses. I. Jenne, Everett A., 1930- . II. Society of Environmental Geochemistry and Health. III. American Chemical Society. Division of Environmental Chemistry. IV. Series: American Chemical Society. ASC symposium series ; 93.
QD39.3.M3C49 541'.3422'0184 79-242
79 MARC

—MATHEMATICS

Goldish, Dorothy M
Basic mathematics for beginning chemistry / Dorothy M. Goldish. — 2d ed. — New York : Macmillan, c1979.
x, 208 p. : ill. ; 24 cm.
Includes index.
ISBN 0-02-344350-2
1. Chemistry—Mathematics. I. Title.
QD39.3.M3G64 1979 510'.2'4541 78-3769
78 MARC

Hashimoto, Yoshirō, 1896–
(Kagaku Sūgaku)
化學數學 / 橋本吉郎, 川端直太郎共著. — 東京 : 三共出版社, 昭和23 ⌊1948⌋
7, 276 p. : ill. ; 19 cm.

1. Chemistry—Mathematics. 2. Calculus. I. Kawabata, Naotarō, joint author. II Title.
QD39.3.M3H37 78–808507

—MATHEMATICS—PROGRAMMED INSTRUCTION

Michels, Leo.
A basic math approach to concepts of chemistry / Leo Michels. — Monterey, Calif. : Brooks/Cole Pub. Co., c1979.
vii, 312 p. : ill. ; 28 cm.
Includes index.
ISBN 0-8185-0273-8 : $7.95
1. Chemistry—Mathematics—Programmed instruction. I. Title.
QD39.3.M3M52 540 78-16259
78 MARC

—NOMENCLATURE

Cahn, Robert Sidney, 1899-
Introduction to chemical nomenclature / R. S. Cahn, O. C. Dermer. — 5th ed. — London ; Boston : Butterworths, 1979.
200 p. : ill. ; 23 cm. GB***
Includes bibliographical references and index.
ISBN 0-408-10608-5
1. Chemistry—Nomenclature. I. Dermer, O. C., 1909- joint author. II. Title.
QD7.C2 1979 540'.1'4 79-40303
79 MARC

Crosland, Maurice P
Historical studies in the language of chemistry / by Maurice P. Crosland. — New York : Dover Publications, 1978.
xxiv, 406 p., ⌊1⌋ leaf of plates : ill. ; 21 cm.
"Unabridged republication of the work originally published by Heinemann Educational Books, Ltd., London in 1962. The author has made corrections and written a new preface to this edition."
Bibliography: p. ⌊355⌋-379.
Includes indexes.
ISBN 0-486-63702-6 : $6.00
1. Chemistry—Nomenclature. I. Title.
QD7.C7 1978 540'.1'4 78-55246
79 MARC

Guyton de Morveau, Louis Bernard, baron, 1737-1816.
Methode der chemischen Nomenklatur für das antiphlogistische System / von Morveau, Lavoisier, Berthollet und de Fourcroy ; mit e. Vorw. von R. Schmitz. — Nachdr. d. Ausg. Wien, 1793. — Hildesheim ; New York : Olms, 1978.
xiv, 365 p. ; 18 cm. GFR78-A
Translation of Méthode de nomenclature chimique.
Includes bibliographical references and index.
ISBN 3-487-06450-2 : DM66.00
1. Chemistry—Nomenclature. I. Lavoisier, Antoine Laurent, 1743-1794, joint author. II. Berthollet, Claude Louis, comte, 1748-1822, joint author. III. Fourcroy, Antoine François de, comte, 1755-1809, joint author. IV. Title.
QD7.G8515 1978 540'.1'4 79-376385
79 MARC

—PERIODICALS

Cluj, Transylvania. Universitatea "Babeş-Bolyai."
Studia Universitatis Babeş-Bolyai. Chemia.
Cluj-Napoca ⌊Universitatea Babeş-Bolyai⌋
v. ill. 24 cm. 2 no. a year.
Continues: Cluj, Transylvania. Universitatea "Babeş-Bolyai." Studia Universitatis Babeş-Bolyai. Series chemia.
English, German, or Romanian.
1. Chemistry—Periodicals. I. Title.
QD1.C789 540'.5 79-646672
MARC-S

Eesti NSV Teaduste Akadeemia.
Eesti NSV Teaduste Akadeemia toimetised. Keemia. Известия Академии наук Эстонской ССР. Химия.
Tallinn, Perioodika,
v. ill. 26 cm. quarterly.
Began in 1978.
In Russian with summaries in English and Estonian.
With: Eesti NSV Teaduste Akadeemia. Eesti NSV Teaduste Akadeemia toimetised. Geoloogia, supersedes: Eesti NSV Teaduste Akadeemia. Toimetised. Keemia, geoloogia.
1. Chemistry—Periodicals. I. Eesti NSV Teaduste Akadeemia. Izvestiia Akademii nauk Ėstonskoĭ SSR. Khimiia.
QD1.E24a 78–648726
(MARC–S)

(Hwahak kwa kongŏp ŭi chinbo)
화학과공업의진보 Progress in chemistry & chemical industry.
서울 대한화학회
v. ill. 26 cm.
English or Korean.

1. Chemistry—Periodicals. 2. Chemical industries—Periodicals. I. Daehan Hwahakhwoe. II. Title: Progress in chemistry & chemical industry.
QD1.H85 79–641817
(MARC–S)

Journal of chemical research. Miniprint. ⌊Jan.⌋ 1977–
⌊London⌋ The Chemical Society.
v. ill. 30 cm. monthly.
Contains papers abstracted in the synopses edition.
English, French or German.
Published on behalf of the Founding Sponsors: Gesellschaft Deutscher Chemiker, Société chimique de France, and The Chemical Society.
Key title: Journal of chemical research. Miniprint, ISSN 0308–2350.
1. Chemical research — Periodicals. 2. Chemistry — Periodicals. I. Chemical Society, London. II. Gesellschaft Deutscher Chemiker. III. Société chimique de France.
QD40.J64b 540 78–649930
MARC–S

Nouveau journal de chimie. v. 1–
Jan. 1977–
⌊Paris⌋ Centre national de la recherche scientifique.
v. ill. 30 cm. bimonthly.
"An interdisciplinary journal."
English or French.
1. Chemistry—Periodicals. I. France. Centre national de la recherche scientifique.
QD1.N848 540'.5 79–649813
MARC–S

Polish journal of chemistry. v. 52–
Warszawa, Państowowe Wydawnictwo Naukowe, 1978–
v. ill. 24 cm. monthly.
At head of title, 1978– : Polish Academy of Sciences, Committee of Chemistry; Polish Chemical Society.
Continues: Roczniki chemii.
1. Chemistry—Periodicals. I. Polska Akademia Nauk. Komitet Nauk Chemicznych. II. Polskie Towarzystwo Chemiczne.
QD1.P6 540'.5 78–646285
MARC–S

Rikagaku Kenkyūjo.
(Rikagaku Kenkyūjo kenkyū nempō)
理化学研究所研究年報 1963-
⌊和光 理化学研究所⌋
v. ill. 26 cm.
"理化学研究所研究報告別冊"
Some issues are numbered by the Japanese calendar.
Other title, 1967- : Annual reports of research activities of the I. P. C. R.
In Japanese.
Some issues are published in 2 vols.

1. Physics—Periodicals. 2. Chemistry—Periodicals. 3. Rikagaku Kenkyūjo—Periodicals. I. Rikagaku Kenkyūjo. Rikagaku Kenkyūjo hōkoku. Bessatsu. II. Rikagaku Kenkyūjo. Annual reports of research activities of the I. P. C. R.
OC1.R543 78–640524
(MARC–S)

SciQuest. v. 52, no. 4–
Apr. 1979–
⌊Easton, PA, American Chemical Society⌋
v. ill. 28 cm. 10 no. a year.
Continues: Chemistry, ISSN 0009–305X.
Key title: SciQuest, ISSN 0190–597X.
1. Chemistry—Periodicals. 2. Science—Periodicals. I. American Chemical Society.
QD1.C74 505 79–643103
MARC–S

CHEMISTRY (Continued)

—PERIODICALS—INDEXES

Journal of chemical research. Indexes. 1977–
[London] The Chemical Society.
v. 30 cm. annual.
Index to: Journal of chemical research. Miniprint, ISSN 0308–2350, and: Journal of chemical research. Synopses, ISSN 0308–2342.
Vols. for 1977– issued by Société chimique de France, the Chemical Society, and Gesellschaft Deutscher Chemiker.
1. Journal of chemical research. Miniprint—Indexes. 2. Journal of chemical research. Synopses—Indexes. 3. Chemical research—Periodicals—Indexes. 4. Chemistry—Periodicals—Indexes. I. Chemical Society, London. II. Gesellschaft Deutscher Chemiker. III. Société chimique de France.
QD40.J64c 540 79–641443
MARC-S

—PHILOSOPHY

Laitko, Hubert.
Kémia és filozófia : a marxista filozófia álláspontja a modern kémia néhány filozófiai kérdésében / Hubert Laitko, Wolf-Dietrich Sprung ; [fordította, Hársing László]. — Budapest : Akadémiai Kiadó, 1975.
130, [1] p. ; 19 cm. — (Korunk tudománya)
Translation of Chemie und Weltanschauung.
Bibliography: p. 129-[131]
ISBN 9630507668 : 16.00Ft
1. Chemistry—Philosophy. 2. Communism and science. I. Sprung, Wolf-Dietrich, joint author. II. Title.
QD6.L3416 78-393863
79 MARC

Makarov, Konstantin Alekseevich.
(Filosofskie voprosy khimii)
Философские вопросы химии : методическое пособие в помощь лектору / К. А. Макаров, Б. М. Суханов. — Ленинград : Об-во Знание РСФСР, 1968.
13, [3] p.
At head of title: Obshestvo "Znanie" RSFSR. Leningradskaía organizaíšiía.
L. C. copy on microfilm only.
Bibliography: p. [15]
——— Microfilm. On 1 reel with other items ; 35 mm.
Microfilm Slavic 12599 QD
1. Chemistry—Philosophy. I. Sukhanov, B. M., joint author. II. Title.
[QD6.M34] 78–423445

—PHYSICAL AND THEORETICAL

P'an, Kuan.
(Wu li hua hsüeh hsin pien)
物理化學新編 / 國立編譯館主編 ; 潘貫編著. — 台北 : 正中書局, 民國67- [1978-
v. : ill. ; 22 cm.
At head of title: Pu pien ta hsüeh yung shu.
Includes index.

1. Chemistry—Physical and theoretical. I. Kuo li pien i kuan, T'ai-pei. II. Title.
QD453.2.P36 78–842615

–POPULAR WORKS

Kotake, Munio, 1894–
(Hishi no mi)
菱の實 / 小竹無二雄著. — 東京 : 葛城書房, 昭和23 [1948]
2, 2, 210 p., [11] leaves of plates : ill. ; 22 cm.

1. Chemistry—Popular works. I. Title.
QD37.K63 79–802876

Sum, N É 1879–1926.
(Uspekhi sovremennoĭ khimii)
Успехи современной химии Н. Э. Сум. — В Ленинграде : П. П. Сойкин, 1926.
79 p. : ill. ; 21 cm.
"Бесплатное приложение к журналу Вестник знания."
Bibliography : p. 78.
1. Chemistry—Popular works. I. Title.
QD37.S95 79–413738

—PROBLEMS, EXERCISES, ETC.

Boni, Elio.
Stechiometria / E. Boni. — [Urbino : AGE, 1976?]
iv, 279 p. : ill. ; 35 cm. It76-Nov
At head of title: Università degli studi di Urbino.
1. Chemistry—Problems, exercises, etc. I. Title.
QD42.B66 540'.76 78-395395
*79 MARC

Butler, Ian S
Relevant problems for chemical principles : With S.I. and conventional units / Ian S. Butler, Arthur E. Grosser. — 3d ed. — Menlo Park, Calif. : Benjamin /Cummings Pub. Co., c1979.
vii, 648 p. : ill. ; 24 cm.
ISBN 0-8053-1587-X
1. Chemistry—Problems, exercises, etc. I. Grosser, Arthur E., 1934- joint author. II. Title.
QD42.B848 1979 540'.76 79-345
79 MARC

Huang, I, fl. 1978–
(Chung hsüeh hua hsüeh hsi t'i chi)
中學化學習題集 / 黃義編. — 香港 : 商務印書館, 1978–1979.
2 v. : ill. ; 27 cm.
$11.00 (v. 1) ; $12.00 (v. 2)

1. Chemistry—Problems, exercises, etc. I. Title.
QD42.H8 79–840261

Urseanu, Florian.
Probleme de chimie şi de tehnologie chimică / Florian Urseanu, Corneliu şanu-Mihăilă, Grigore Bozga. — Bucureşti : Editura Tehnică, 1978.
384 p. : ill. ; 25 cm. R78-4220
Bibliography: p. 372-[373]
lei22.00
1. Chemistry—Problems, exercises, etc. 2. Chemistry, Technical—Problems, exercises, etc. I. Tărăbăşanu-Mihăilă, Corneliu, joint author. II. Bozga, Grigore, joint author. III. Title.
QD42.U77 79-338501
*79 MARC

—PROGRAMMED INSTRUCTION

Programed reviews of chemical principles / Jean D. Lassila ... [et al.]. — 3d ed. — Menlo Park, Calif. : Benjamin/Cummings Pub. Co., c1979.
vi, 746 p. : ill. ; 24 cm.
Includes index.
ISBN 0-8053-6027-1
1. Chemistry—Programmed instruction. I. Lassila, Jean D.
[QD31.2.P76 1979] 540'.7'7 79-381
79 MARC

—RESEARCH see Chemical Research

—STUDY AND TEACHING

Butorina, L B
V Poiskakh nauchnogo postroeniía prepodavaniía khimii po novoĭ programme : tema Khimicheskaía svíàz' / L. B. Butorina. — Moskva : [s. n.], 1968.
7 p.
At head of title: Moskovskiĭ gorodskoĭ otdel narodnogo obrazovaniía. Moskovskiĭ gorodskoĭ institut usovershenstvovaniía uchitel̄eĭ.
L.C. copy on microfilm only.
——— Microfilm. On 1 reel with other items ; 35 mm.
Microfilm Slavic 18057 QD
1. Chemistry—Study and teaching. I. Title.
[QD40.B927] 79–366975

Ćwiczenia z metodyki nauczania chemii dla studentów kierunku nauczycielskiego, stacjonarnego i zaocznego : praca zbiorowa / pod red. Janusza Sławka. — Wyd. 1. — Poznań : Wydawn. Naukowe im. Adama Mickiewicza, 1978.
70 p. : ill. ; 24 cm. P***
At head of title: Uniwersytet im. Adama Mickiewicza w Poznaniu.
Bibliography: p. [67]-68.
zł10.00
1. Chemistry—Study and teaching. I. Sławek, Janusz. II. Uniwersytet im. Adama Mickiewicza w Poznaniu.
QD40.C94 78-400041
79 MARC

Vossen, Herbert.
Kompendium Didaktik Chemie / von Herbert Vossen. — München : Ehrenwirth, 1979.
138 p. ; 20 cm. — (Kompendium Didaktik) GFR79-A
Cover title: Chemie.
Bibliography: p. 131-138.
ISBN 3-431-02054-2 : DM12.00
1. Chemistry—Study and teaching. I. Title. II. Title: Chemie. III. Series.
QD40.V76 79-372722
79 MARC

Zollinger, Heinrich, 1919–
Chemie und Hochschule : Beiträge zum Komplementaritätsdenken in Lehre und Forschung / Heinrich Zollinger. — Basel [etc.] : Birkhäuser, cop. 1978.
163 p. : graphs ; 21 cm. — (Poly ; 11) Sw79-1286
Includes bibliographical references.
ISBN 3-7643-1044-8 : 19.50F
1. Chemistry—Study and teaching. I. Title. II. Series: Poly (Basel) ; 11.
QD40.Z64 79-361641
79 MARC

—STUDY AND TEACHING —PERIODICALS

(Hwahak kyoyuk)
화학교육 Chemical education.
[서울] 대한화학회
v. 26 cm.
In Korean.

1. Chemistry—Study and teaching—Korea—Periodicals. 2. Chemistry—Study and teaching—Periodicals. I. Daehan Hwahakhwoe. II. Title: Chemical education.
QD49.K8H9 78–647726
(MARC-S)

—STUDY AND TEACHING—EGYPT

Labīb, Rushdī.
(Numūw al-mafāhīm al-'ilmīyah)
نمو المفاهيم العلمية / رشدى لبيب. — [القاهرة] : مكتبة الأنجلو المصرية ، 1974.
2, 94 p. ; 24 cm. — (بحوث فى تدريس العلوم ؛ 1)
Bibliography : p. [93]–94.
£E0.45
1. Science—Study and teaching—Egypt. 2. Chemistry—Study and teaching—Egypt. I. Title.
Q183.4.E3L32 74–960829

—STUDY AND TEACHING —KOREA—PERIODICALS

(Hwahak kyoyuk)
화학교육 Chemical education.
[서울] 대한화학회
v. 26 cm.
In Korean.

1. Chemistry—Study and teaching—Korea—Periodicals. 2. Chemistry—Study and teaching—Periodicals. I. Daehan Hwahakhwoe. II. Title: Chemical education.
QD49.K8H9 78–647726
(MARC-S)

—STUDY AND TEACHING—NEW JERSEY—ADDRESSES, ESSAYS, LECTURES

Bernstein, Jeremy, 1924–
Science and the human imagination : Albert Einstein : papers and discussions / by Jeremy Bernstein and Gerald Feinberg, including two lectures on New Jersey's contributions to the nation in chemical industry and chemical education : papers and discussions / by Henry B. Hass and A. K. Bose ; edited by Charles Angoff. — Rutherford [N.J.] : Fairleigh Dickinson University Press, 1978.
94 p. ; 22 cm. — (The Leverton lecture series ; 5)
ISBN 0-8386-2223-2 : $4.50
1. Einstein, Albert, 1879-1955—Addresses, essays, lectures. 2. Physicists—United States—Biography—Addresses, essays, lectures. 3. Physics—History—Addresses, essays, lectures. 4. Chemical engineering—New Jersey—History—Addresses, essays, lectures. 5. Chemistry—Study and teaching—New Jersey—Addresses, essays, lectures. I. Feinberg, Gerald, 1933- joint author. II. Hass, Henry Bohn, 1902- New Jersey's contributions to the nation in chemical industry and chemical education. 1978. III. Title. IV. Series: Leverton lecture series ; 5.
GC16.E5B46 530'.092'4 77-92565
78 MARC

—STUDY AND TEACHING (HIGHER)

Martín, Consuelo.
Químicas / Consuelo Martín Rodríguez. — Pamplona : Ediciones Universidad de Navarra, [1977]
291 p. ; 18 cm. — (Colección Ciencias de la educación) (Guías de los estudios universitarios ; 3)
Bibliography: p. [287]-291.
ISBN 8431302623
1. Chemistry. 2. Chemistry—Study and teaching (Higher) 3. Chemistry—Vocational guidance. I. Title.
QD31.2.M368 79-352367
79 MARC

CHEMISTRY (Continued)

—STUDY AND TEACHING (HIGHER)—CONGRESSES

Unesco International Chemistry Congress, Perth, Australia, 1978.
The role of laboratory teaching in university chemistry courses / edited by J. M. Webb. — Canberra : Australian Government Publishing Service, 1978.
116 p. ; 25 cm. Aus
Organized on behalf of the Australian National Commission for Unesco, and sponsored by the Australian Academy of Science, the Royal Australian Chemical Institute, and the International Union of Pure and Applied Chemistry.
ISBN 0-642-03789-2
1. Chemistry—Study and teaching (Higher)—Congresses. I. Webb, J. M. II. Australian National Commission for the United Nations Educational, Scientific and Cultural Organization. III. Australian Academy of Science. IV. Royal Australian Chemical Institute. V. International Union of Pure and Applied Chemistry. VI. Title.
QD40.U53 1978 540'.7'11 79-314189
00 MARC

—STUDY AND TEACHING (HIGHER)—GERMANY, WEST

Meinel, Christoph.
Die Chemie an der Universität Marburg seit Beginn des 19. Jahrhunderts : e. Beitr. zu ihrer Entwicklung als Hochschulfach / von Christoph Meinel. — Marburg : Elwert [in Komm.], 1978.
xvi, 669 : ill. ; 24 cm. — (Academia Marburgensis ; Bd. 3) GFR79-A
Originally presented as the author's thesis, Marburg, 1977.
Bibliography: p. 638-665.
Includes index.
ISBN 3-7708-0615-8 : DM65.00
1. Chemistry—Study and teaching (Higher)—Germany, West. 2. Marburg. Universität. I. Title. II. Series.
QD49.G33M376 1978 79-359755
79 MARC

—TERMINOLOGY

A Kémiai elnevezés és helyesírás alapjai / Fábián Pál ... [et al.] ; szerk. Erdey-Grúz Tibor, és Fodorné Csányi Piroska. — Budapest : Akadémiai Kiadó, 1977.
332 p. : ill. ; 25 cm. H***
Includes index.
ISBN 9630511266 : 79.00Ft
1. Chemistry—Terminology. I. Erdey-Grúz, Tibor. II. Fodorné Csányi, Piroska.
QD7.K45 78-376808
78 MARC

—VOCATIONAL GUIDANCE

Martin, Consuelo.
Químicas / Consuelo Martin Rodríguez. — Pamplona : Ediciones Universidad de Navarra, [1977]
291 p. ; 18 cm. — (Colección Ciencias de la educación) (Guías de los estudios universitarios ; 3)
Bibliography: p. [287]-291.
ISBN 8431302623
1. Chemistry. 2. Chemistry—Study and teaching (Higher) 3. Chemistry—Vocational guidance. I. Title.
QD31.2.M368 79-352367
79 MARC

—VOCATIONAL GUIDANCE—JUVENILE LITERATURE

Woodburn, John H
Opportunities in chemistry / John H. Woodburn. — Skokie, Ill. : VGM Career Horizons, c1979.
ix, 130 p. : ill. ; 21 cm. — (VGM Career Horizons series)
Bibliography: p. [129]-130.
SUMMARY: Examines the education requirements, necessary personal qualities, and employment opportunities for many jobs related to chemistry.
1. Chemistry—Vocational guidance—Juvenile literature. 2. Chemistry, Technical—Vocational guidance—Juvenile literature. [1. Chemistry—Vocational guidance. 2. Vocational guidance] I. Title.
QD39.5.W64 540'.23 78-58663
79 MARC

—AUSTRALIA—PERIODICALS

Royal Australian Chemical Institute.
Chemistry in Australia. v. 44, no. 7–
July 1977–
[Victoria] Royal Australian Chemical Institute.
v. ill. 30 cm. monthly.
"The proceedings of the Royal Australian Chemical Institute."
Continues: Royal Australian Chemical Institute. PRACI, proceedings of the Royal Australian Chemical Institute.
Key title: Chemistry in Australia, ISSN 0314–4240.
1. Royal Australian Chemical Institute—Periodicals. 2. Chemistry—Australia—Periodicals. I. Title
TP1.R62 660.2'094 79–641475
MARC-S

—AUSTRIA—SOCIETIES, ETC.—DIRECTORIES

Hillisch, Robert.
Leistungsangebot der chemischen Forschungsstätten Österreichs ausserhalb des Unternehmenssektors / [Zsstellung., R. Hillisch]. — Wien : Bundesmin. f. Wissenschaft u. Forschung ; Wien : Bundesmin. f. Handel, Gewerbe u. Industrie ; Wien : Forschungs-Förderungs-Fonds d. Gewerbl. Wirtschaft, 1977.
137 p. ; 21 cm. Au78-6-278
"Suchmatrix": [1] fold. leaf in pocket.
Not in trade.
1. Chemistry—Austria—Societies, etc.—Directories. I. Title.
QD18.A9H54 540'.6'0436 78-365785
*78 MARC

—CHINA

U.S. Delegation in Pure and Applied Chemistry.
Chemistry and chemical engineering in the People's Republic of China : a trip report of the U.S. Delegation in Pure and Applied Chemistry / edited by John D. Baldeschwieler. — Washington : American Chemical Society, 1979.
xix, 266 p. : ill. ; 24 cm.
Includes index.
ISBN 0-8412-0501-9
1. Chemistry—China. 2. Chemical engineering—China. I. Baldeschwieler, John D., 1933- . II. Title.
QD18.C5U54 1979 540'.951 79-11217
79 MARC

—CHINA—HISTORY

Chiang, Lin-ts'ai.
(Chung-kuo ku tai hua hsüeh shih hua)
中国古代化学史话 / 江琳才. — [香港? s. n., 1978]
3, 135 p. : ill. ; 21 cm.
Reprint of the 1978 ed. published by Kuang-tung jen min ch'u pan she.

1. Chemistry—China—History. I. Title.
QD18.C5C54 1978 79-838244

Li, Ch'iao-p'ing, 1897-
The chemical arts of old China / by Li Ch'iao-p'ing ; with a foreword by Tenney L. Davis. 1st AMS ed. New York : AMS Press, 1979, c1948.
viii, 215 p. : ill. ; 23 cm.
Reprint of the ed. published by the Journal of Chemical education, Easton, Pa.
ISBN 0-404-14482-9
1. Chemistry—China—History. I. Title.
QD18.C5L5 1979 540'.951 75-36234
79 MARC

—CZECHOSLOVAKIA—BIBLIOGRAPHY

Šteinerová, Svatava.
Výběrová bibliografie dějin české chemické technologie = Ausgewählte Bibliographie der Geschichte der böhmischen chemischen Technologie / Svatava Šteinerová, Jan Kořan. — Praha : Národní technické muzeum, 1976.
353 p., [19] leaves of plates : ill. ; 23 cm. — (Bibliografie, prameny - Národní technické muzeum v Praze ; 12) Cz***
Text in various languages.
Table of contents also in German.
Bibliography: p. [359]
Includes index.
1. Chemistry—Czechoslovakia—Bibliography. I. Kořan, Jan, joint author. II. Title. III. Title: Ausgewählte Bibliographie der Geschichte der böhmischen chemischen Technologie. IV. Series: Prague. Národní technické muzeum. Bibliografie, prameny ; 12.
Z5525.C9S75 78-380638
[QD18.C9] 78 MARC

—JAPAN—HISTORY

Imoto, Minoru, 1908–
(Nihon no kagaku)
日本の化学：100年のあゆみ / 日本化学会編 ; 井本稔著. — 京都：化学同人, 1978.
v, 207 p. : ill. ; 18 cm. Ja 78–16002
¥350

1. Chemistry—Japan—History. I. Title.
QD18.J3I46 78–816609

(Nihon no kagaku hyakunenshi)
日本の化学百年史：化学と化学工業の歩み / 日本化学会編. — 東京：東京化学同人, 1978.
xiii, 1302 p., [2] leaves of plates : ill. ; 27 cm. Ja 78–10104
"日本化学会創立百周年記念"
Includes bibliographical references.
¥23000

1. Chemistry—Japan—History. 2. Chemical engineering—Japan—History. I. Nippon Kagakukai.
QD18.J3N53 78–812433

—UKRAINE—BIO-BIBLIOGRAPHY

(Vasilii Ivanovich Atroshchenko)
Василий Иванович Атрощенко / Вступ. статья Н. Ф. Киркача ... [et al.] ; Указ. печ. работ [сост.] Н. П. Курган, И. И. Литвиненко. — Киев : Наук. думка, 1978.
78 p., [1] leaf of plates : port. ; 17 cm. — (Biобиблиография ученых УССР) USSR 78
At head of title: Akademiîa nauk USSR.
Series romanized: Biobibliografiîa uchenykh USSR.
Includes indexes.
0.15rub
1. Atroshchenko, Vasiliĭ Ivanovich—Bibliography. 2. Chemistry—Ukraine—Bio-bibliography. 3. Chemists—Ukraine—Biography. I. Kirkach, Nikolaĭ Fedorovich. II. Kurgan, Nelli Pavlovna. III. Litvinenko, Igor' Ivanovich.
Z8046.94.V37 79–410594
[QD22.A74]

CHEMISTRY, ANALYTIC

see also Acetylene reduction assay; Archaeological chemistry; Colorimetric analysis; Electrolysis; Fluorimetry; Instrumental analysis; Metallurgical analysis; Mineralogy, Determinative; Polarograph and polarography; Radioactivation analysis; Separation (Technology); Spectrophotometry; and subdivisions Analysis and Analysis and chemistry under special subjects, e.g. Gases—Analysis; Blood—Analysis and chemistry

Antonovskiĭ, Vilen Lazarevich.
(Analiticheskaîa khimiîa organicheskikh peroksidnykh soedineniĭ)
Аналитическая химия органических пероксидных соединений / В. Л. Антоновский, М. М. Бузланова. — Москва : Химия, 1978.
308 p. : ill. ; 20 cm. USSR 78
Includes bibliographies.
2.00rub
1. Peroxides—Analysis. 2. Chemistry, Analytic. I. Buzlanova, Margarita Mikhaĭlovna, joint author. II. Title.
QD412.O1A56 79–408179

Arima, Junzō.
(Yūki kagōbutsu bunsekihō)
有機化合物分析法 / 有馬純三. — 東京：岩波書店, 昭和9 [1934], 昭和24 [1949] printing.
13, 195, 28 p. : ill. ; 18 cm. — (岩波全書 ; 78)
Colophon inserted.
Series romanized: Iwanami zensho.
Bibliography: p. [2] (1st group)
Includes index.
1. Chemistry, Analytic. 2. Chemistry, Organic. I. Title.
QD271.A73 79–8037[illegible]

Boniface, H J
Analytical notes : a summary of inorganic methods of chemical analysis / H. J. Boniface. — Wolverhampton : Sigma Technical Press, 1978.
[64] p. ; 30 cm. GB78-19162
includes bibliographical references.
ISBN 0-905104-02-1 : £3.50
1. Chemistry, Analytic. 2. Chemistry, Inorganic. I. Title.
QD75.2.B66 545 78-320381
79 MARC

Contemporary topics in analytical and clinical chemistry. v. 1–
New York, Plenum Press [c1977–
v. ill. 24 cm.
Editor: v. 1– D. M. Hercules.
Key title: Contemporary topics in analytical and clinical chemistry, ISSN 0160–3620.
1. Chemistry, Clinical. 2. Chemistry, Analytic. I. Hercules, David M.
RB40.C66 616.07'56'05 79–640979
MARC-S

CHEMISTRY, ANALYTIC (Continued)

Freiser, Henry, 1920-
Ionic equilibria in analytical chemistry / Henry Freiser & Quintus Fernando. — Huntington, N.Y. : R. E. Krieger Pub. Co., 1979, c1963.
xiii, 338 p. : ill. ; 24 cm.
Reprint of the ed. published by Wiley, New York.
Includes bibliographies and index.
ISBN 0-88275-955-8
1. Chemistry, Analytic. 2. Ionic equilibrium. I. Fernando, Quintus, joint author. II. Title.
[QD75.2.F73 1979] 541.392 79-9746
79 MARC

ÎAnson, Édgar ÎUl'evich.
Ditiokarboksilaty v analiticheskoĭ khimii / É. ÎU. ÎAnson. — Riga : Zinatne, 1978.
135 p. : ill. ; 22 cm. USSR78
At head of title: Latviĭskiĭ gosudarstvennyĭ universitet im. P. Stuchki.
Bibliography: p. [127]-132.
0.60rub
1. Dithionic acid. 2. Chemistry, Analytic. 3. Chemical tests and reagents. I. Title.
QD77.I33 79-372426
79 MARC

Kolthoff, Izaak Maurits, 1894-
Treatise on analytical chemistry / edited by I. M. Kolthoff and Philip J. Elving. — 2d ed. — New York : Wiley, c1978-
v. in : ill. ; 25 cm.
"An Interscience publication."
Includes bibliographies and index.
CONTENTS: pt. 1. Theory and practice. v.
ISBN 0-471-03438-X (v. 1)
1. Chemistry, Analytic. I. Elving, Philip Juliber, 1913- joint author. II. Title.
QD75.2.K64 1978 543 78-1707
78 MARC

Kopanica, Miloslav.
Kinetické metody v chemické analýze : [určeno také stud. stř. i vys. škol chem. / autoři] Miloslav Kopanica, Věra Stará. — 1. vyd. — Praha : SNTL : ČVTS, 1978.
248, [1] p. : ill. : 21 cm. — (Moderní metody v chemické laboratoři ; sv. 32) Cz78
Includes bibliographies and index.
Kčs25.00
1. Chemistry, Analytic. 2. Chemical reaction, Rate of. I. Stará, Věra, Ing., joint author. II. Title.
QD75.2.K66 78-382577
*78 MARC

Lîalikov, ÎUriĭ Sergeevich.
(Fiziko-khimicheskie metody analiza)
Физико-химические методы анализа : [учеб. пособие для хим.-технол. специальностей вузов] / Ю. С. Ляликов. — Изд. 5-е, перераб. и доп. — Москва : Химия, 1974.
536 p. : ill. ; 23 cm. USSR 74
Includes bibliographical references and index.
1.34rub
1. Chemistry, Analytic. 2. Chemistry, Physical and theoretical. I. Title.
QD75.L5 1974 78-420883

Mao, Kuang-hsing.
(I ch'i fen hsi)
儀器分析 / 編著者毛光興. — 台北：新學識文教出版中心, 民國67 [1978]
xi, 270 p. : ill. ; 25 cm.
Includes index.
1. Chemical analysis. I. Title.
QD53.M33 78-841236

Nadeinskiĭ, Boris Pavlovich.
(Teoreticheskie obosnovaniîa i raschety v analiticheskoĭ khimii)
Теоретические обоснования и расчеты в аналитической химии. Изд. 3., перер. Допущено в качестве учеб. пособия для химико-технологических вузов и факультетов. Москва, Высшая школа, 1959.
443 p. illus. 23 cm.
First ed. published in 1953 under title: Vvedenie v analiticheskuîu khimiîu.
Includes bibliography.
1. Chemistry, Analytic. 2. Chemistry, Analytic—Problems, exercises, etc. I. Title.
QD75.N27 1959 60-43492 ‡

Nadeinskiĭ, Boris Pavlovich.
(Teoreticheskie obosnovaniîa i raschety v analiticheskoĭ khimii)
Теоретические обоснования и расчеты в аналитической химии. Изд. 2., перер. и доп. Допущено в качестве учеб. пособия для химико-технологических вузов и факультетов. Москва, Советская наука, 1956.
446 p. illus., ports., tables. 23 cm.
First ed. published in 1953 under title: Vvedenie v analiticheskuîu khimiîu.
Errata slip inserted.
——— Microfilm (negative) Made by the Library of Congress.
Negative film in the Library of Congress. Microfilm Slavic 729 AC
1. Chemistry, Analytic. 2. Chemistry, Analytic—Problems, exercises, etc. I. Title.
QD75.N27 1956 57-19721

Perrin, Douglas Dalzell.
Organic complexing reagents : structure, behavior, and application to inorganic analysis / D. D. Perrin. — Reprint 1978 w/corrections and new pref. — Huntington, N.Y. : R. E. Krieger Pub. Co., 1978, c1964.
xi, 365 p. : ill. ; 24 cm.
Reprint of the ed. published by Interscience Publishers, New York, which was issued as v. 18 of Chemical analysis.
Includes bibliographical references and index.
ISBN 0-88275-574-9
1. Chemical tests and reagents. 2. Complex compounds. 3. Chemistry, Analytic. I. Title. II. Series: Chemical analysis ; v. 18.
[QD77.P4 1978] 543'.01 77-16083
77 MARC

Physical methods in modern chemical analysis / edited by Theodore Kuwana. — New York : Academic Press, 1978-
v. : ill. ; 24 cm.
Includes bibliographies and index.
ISBN 0-12-430801-5 (v. 1)
1. Chemistry, Analytic. I. Kuwana, Theodore.
QD75.2.P49 543 77-92242
78 MARC

Pietrzyk, Donald J 1934-
Analytical chemistry / Donald J. Pietrzyk, Clyde W. Frank ; [cover photo by Sigrid Owen]. — 2d ed. — New York : Academic Press, c1979.
xix, 700 p. : ill. ; 25 cm.
ISBN 0-12-555160-6
1. Chemistry, Analytic. I. Frank, Clyde W., joint author. II. Title.
QD75.2.P53 1979 543 77-80792
79 MARC

Popa, Grigore.
Chimie analitică / Gr. Popa, I. A. Paralescu. — București : Editura Didactică și Pedagogică, 1977.
429 p., [1] fold. leaf of plates : ill. ; 25 cm. R78-1279
At head of title: Ministerul Educației și Învățămîntului.
Bibliography: p. 429.
lei20.70
1. Chemistry, Analytic. I. Paralescu, Ion A., joint author. II. Title.
QD75.2.P66 78-375367
*78 MARC

—AUTOMATION—COLLECTED WORKS

Topics in automatic chemical analysis / editors, J. K. Foreman and P. B. Stockwell. — Chichester, Eng. : Ellis Horwood ; New York : Halsted Press, 1979-
v. : ill. ; 24 cm. — (Ellis Horwood series in analytical chemistry) GB***
Includes bibliographical references and index.
ISBN 0-470-26600-7 (v. 1) (Halsted)
1. Chemistry, Analytic—Automation—Collected works. I. Foreman, James K. II. Stockwell, Peter B.
QD75.4.A8T66 543'.0028 79-40239
79 MARC

—CONGRESSES

Zu einigen Problemen der modernen Analytik : [Thesen, Vorträge und Diskussionsbeiträge der Sitzung der Klasse Chemie mit Beteiligung Auswärtiger Mitglieder der Akademie und Gäste am 11. und 12. November 1976 / hrsg. im Auftrage des Präsidenten der Akademie der Wissenschaften der DDR von Heinrich Scheel]. — Berlin : Akademie-Verlag, 1978.
54 p. : ill. ; 22 cm. — (Sitzungsberichte der Akademie der Wissenschaften der DDR ; Jahrg. 1978, Nr. 10/N : Mathematik, Naturwissenschaften, Technik) GDR***
6.00M
1. Chemistry, Analytic—Congresses. I. Scheel, Heinrich. II. Akademie der Wissenschaften der DDR. Klasse Chemie. III. Series: Akademie der Wissenschaften der DDR. Sitzungsberichte der Akademie der Wissenschaften der DDR ; Jahrg. 1978, Nr. 10/N.
QD71.Z8 79-346447
79 MARC

Euroanalysis Conference, 2d, Budapest, 1975.
Reviews on analytical chemistry / presented at the Euroanalysis Conference II held in Budapest, Hungary, 25-30 August, 1975 ; edited by the Working Party on Analytical Chemistry of the Federation of European Chemical Societies ; ed., W. Fresenius. — Budapest : Akadémiai Kiadó, 1977.
253 p. : ill. ; 25 cm. H***
Includes bibliographical references and index.
ISBN 9630510952
1. Chemistry, Analytic—Congresses. I. Fresenius, Wilhelm, 1913- . II. Federation of European Chemical Societies. Working Party on Analytical Chemistry. III. Title.
QD71.E9 1975 543 79-301458
79 MARC

Pittsburgh Conference on Analytical Chemistry and Applied Spectroscopy, 29th, Cleveland, 1978.
The state of the art in analytical chemistry and applied spectroscopy : abstracts, 29th Pittsburgh Conference, February 27-March 3, 1978, Cleveland, Ohio, USA. — [s.l. : s.n., 1978] (Monroeville, PA : ABC Press)
687 p. : ill. ; 22 cm.
Cover title.
Includes bibliographical references and indexes.
1. Chemistry, Analytic—Congresses. 2. Spectrum analysis—Congresses. I. Title.
QD71.P574 1978 543'.085 78-325777
79 MARC

Symposium on Recent Advances in the Analytical Chemistry of Pollutants.
Abstracts - Symposium on Recent Advances in the Analytical Chemistry of Pollutants.
Vienna [Austria, s. n.]
v. 21 cm. annual.
1. Pollution—Congresses. 2. Chemistry, Analytic—Congresses.
TD172.5.S94a 628.5 78-642942
MARC-S

Vsesoiûznaîa konferentsiîa Organicheskie reagenty v analiticheskoĭ khimii, 4th, Kiev, 1976.
(Organicheskie reagenty v analiticheskoĭ khimii)
Органические реагенты в аналитической химии : Тезисы докл. IV всесоюз. конф., Киев, 21–24 апр. 1976 г. / [Ред. коллегия, чл.-кор. АН УССР А. Т. Пилипенко (отв. ред.) ... et al.]. — Киев : Наук. думка, 1976.
2 v. ; 20 cm. USSR 76
At head of title: Akademiîa nauk SSSR. Nauchnyĭ sovet po probleme Analiticheskaîa khimiîa. Institut geokhimii i analiticheskoĭ khimii im. V. I. Vernadskogo, and other institutions.
Includes bibliographies.
0.50rub (v. 1) 0.50rub (v. 2)
1. Chemical tests and reagents—Congresses. 2. Chemistry, Analytic—Congresses. 3. Chemistry, Organic—Congresses. I. Pilipenko, Anatoliĭ Terent'evich. II. Akademiîa nauk SSSR. Nauchnyĭ sovet po probleme Analiticheskaîa khimiîa. III. Akademiîa nauk SSSR. Institut geokhimii i analiticheskoĭ khimii.
QD77.V75 1976 79-408786

—DICTIONARIES—CHINESE

(Fen hsi chi shu tz'u tien)
分析技术词典. — 第1版. — 北京：科学出版社, 1977-
v. : ill. : 19 cm.
CONTENTS: [1] 色谱分析
$0.42 (v. 1)
1. Chemistry, Analytic—Dictionaries—Chinese. 2. Chinese language—Dictionaries—English.
QD131.F46 78-842970

—DICTIONARIES—HUNGARIAN

Analitikai kémiai kislexikon / főszerk., Pungor Ernő, szerk., Buzás Lajosné. — Budapest : Műszaki Könyvkiadó, 1978.
340 p. : ill. ; 24 cm. H78
Includes bibliographical references.
ISBN 9631021815 : 67.00Ft
1. Chemistry, Analytic—Dictionaries—Hungarian. I. Pungor, Ernő. II. Buzás, Lajosné.
QD71.5.A52 79-362736
*79 MARC

—HANDBOOKS, MANUALS, ETC.

Williams, William John.
Handbook of anion determination / W. John Williams. — London ; Boston : Butterworths, 1979.
x, 630 p. : ill. ; 24 cm. GB79
Includes bibliographical references and index.
ISBN 0-408-71306-2 : £45.00
1. Anions—Handbooks, manuals, etc. 2. Chemistry, Analytic—Handbooks, manuals, etc. I. Title.
QD561.W628 543 78-40553
78 MARC

—LABORATORY MANUALS

Instrukcja do ćwiczeń z chemii analitycznej jakościowej / oprac. H[anna] Goszczyńska ... [et al.]. — Wyd. 6 popr. — Gliwice : PŚ, 1977 (Gliw. : PŚ)
123 p. ; 30 cm. — (Skrypty uczelnianc - Politechnika Śląska im. W. Pstrowskiego ; Nr 716) P77-10424
Chiefly tables.
zł6.00
1. Chemistry, Analytic—Laboratory manuals. I. Goszczyńska, Hanna.
QD83.I59 1977 78-369058
78 MARC

CHEMISTRY, ANALYTIC (Continued)

—NOMENCLATURE

Compendium of analytical nomenclature : definitive rules 1977 / prepared for publication by H. M. N. H. Irving, H. Freiser, and T. S. West. — 1st ed. — Oxford ; New York : Pergamon Press, 1978.
viii, 223 p. : ill. ; 28 cm. — (IUPAC publication)
At head of title: International Union of Pure and Applied Chemistry, Analytical Chemistry Division.
Includes bibliographical references and index.
ISBN 0-08-022008-8. ISBN 0-08-022347-8 pbk.
1. Chemistry, Analytic—Nomenclature. 2. Chemistry, Analytic—Terminology. I. Irving, Harry Munroe Napier Hetherington. II. Freiser, Henry, 1920- . III. West, Thomas Summers. IV. International Union of Pure and Applied Chemistry. Analytical Chemistry Division. V. Series: International Union of Pure and Applied Chemistry. IUPAC publication.
QD75.3.C65 1977 543'.001'4 77-8949
77 MARC

—PERIODICALS

(Bunseki)
ぶんせき
〔東京〕 日本分析化学会
no. 26 cm.

1. Chemistry, Analytic—Periodicals. I. Nihon Bunseki-kagakukai.
QD71.B85 79-640655
(MARC-S)

—PROBLEMS, EXERCISES, ETC.

Nadeinskiĭ, Boris Pavlovich.
(Teoreticheskie obosnovaniia i raschety v analiticheskoĭ khimii)
Теоретические обоснования и расчеты в аналитической химии. Изд. 2., перер. и доп. Допущено в качестве учеб. пособия для химико-технологических вузов и факультетов. Москва, Советская наука, 1956.
446 p. illus., ports., tables. 23 cm.
First ed. published in 1953 under title: Vvedenie v analiticheskuiu khimiiu.
Errata slip inserted.

——— Microfilm (negative) Made by the Library of Congress.
Negative film in the Library of Congress.
Microfilm Slavic 729 AC
1. Chemistry, Analytic. 2. Chemistry, Analytic—Problems, exercises, etc. I. Title.
QD75.N27 1956 57-19721

Nadeinskiĭ, Boris Pavlovich.
(Teoreticheskie obosnovaniia i raschety v analiticheskoĭ khimii)
Теоретические обоснования и расчеты в аналитической химии. Изд. 3., перер. Допущено в качестве учеб. пособия для химико-технологических вузов и факультетов. Москва, Высшая школа, 1959.
443 p. illus. 23 cm.
First ed. published in 1953 under title: Vvedenie v analiticheskuiu khimiiu.
Includes bibliography.
1. Chemistry, Analytic. 2. Chemistry, Analytic—Problems, exercises, etc. I. Title.
QD75.N27 1959 60-43492 ‡

—PROGRAMMED INSTRUCTION

Liâlikov, IUriĭ Sergeevich.
(Fiziko-khimicheskie metody analiza)
Физико-химические методы анализа : пособие по программир. опросу : [для преподавателей вузов и сред. спец. учеб. заведений] / Ю. С. Ляликов. — Кишинев : Штиинца, 1974.
226 p. : diagrs. ; 21 cm. USSR 74
At head of title: Ministerstvo narodnogo obrazovaniia Moldavskoĭ SSR. Kishinevskiĭ gosudarstvennyĭ universitet imeni V. I. Lenina.
1.07rub
1. Chemistry, Analytic—Programmed instruction. I. Title.
QD75.2.L52 79-413718

—QUALITATIVE

see also Spectrum analysis

Ankudimova, E V
(Rukovodstvo po kachestvennomu khimicheskomu polumikroanalizu)
Руководство по качественному химическому полумикроанализу / [руководство написано Е. В. Анкудимовой ; под ред. В. И. Петрашень]. — Новочеркасск : Новочеркасский политехнический ин-т, 1967.
230 p. : ill.
At head of title: Ministerstvo vysshego i srednego special'nogo obrazovaniia RSFSR. Novocherkasskiĭ politekhnicheskiĭ institut imeni Sergo Ordzhonikidze. Kafedra analiticheskoĭ khimii.
L.C. copy on microfilm only.

——— Microfilm. On 1 reel with other items ; 35 mm.
Microfilm Slavic 12637 QD
1. Chemistry, Analytic—Qualitative. 2. Microchemistry. I. Petrashen', Vladimir Ivanovich. II. Novocherkassk, Russia. Politekhnicheskiĭ institut. Kafedra analiticheskoĭ khimii. III. Title.
[QD81.A59] 79-408547

Domka, Florian.
Chemiczne metody półmikroanalizy jakościowej / Florian Domka i Aleksandra Binek. — Wyd. 1. — Poznań : Wydawn. Naukowe Uniwersytetu im. Adama Mickiewicza, 1975.
300 p., [1] fold. leaf : ill. ; 25 cm.
Bibliography: p. [293]-294.
zł31.00
1. Chemistry, Analytic—Qualitative. 2. Microchemistry. I. Binek, Aleksandra, joint author. II. Title.
QD81.D6 79-376102
79 MARC

I, Hsien-shih.
(Ting hsing fen hsi)
定性分析 / 執筆者易賢仕 ; 編輯者新學識文教出版中心工專用書編輯委員會. — 台北 : 該中心, 民國67 [1978]
194 p. : ill. ; 25 cm.
"本書係以 Evans 等所著半微量定性分析 (Semimicro qualitive analysis) 一書為藍本"
1. Chemistry, Analytic—Qualitative. I. Hsin hsüeh shih wen chiao ch'u pan chung hsin. Kung chuan yung shu pien chi wei yüan hui. II. Title.
QD81.I2 78-841681

Mello, Ary Ferraz de.
Introdução à análise mineral qualitativa / Ary Ferraz de Mello. — São Paulo : Livraria Pioneira Editora, 1977.
243 p. : ill. ; 22 cm. — (Pioneira-Manuais de estudo)
Bibliography: p. 243.
Cr$90.00
1. Chemistry, Analytic—Qualitative. I. Title.
QD81.M52 544 79-352001
79 MARC

Murashova, Varvara Ivanovna.
(Kachestvennyĭ khimicheskiĭ drobnyĭ analiz)
Качественный химический дробный анализ / В. И. Мурашова, А. Н. Тананаева, Р. Ф. Ховякова. — Москва : Химия, 1976.
279 p. : ill. ; 21 cm. USSR***
Bibliography: p. 276.
0.68rub
1. Chemistry, Analytic—Qualitative. I. Tananaeva, Anna Nikolaevna, joint author. II. Khoviakova, Rimma Fedorovna, joint author. III. Title.
QD81.M87 79-413425

Vogel, Arthur Israel.
Vogel's Practical organic chemistry, including qualitative organic analysis. — 4th ed. / rev. by the following members of the School of Chemistry, Thames (formerly Woolwich) polytechnic, B. S. Furniss ... [et al.]. — London ; New York : Longman, 1978.
xxxv, 1368 p. : ill. ; 24 cm.
Third ed. published in 1956 under title: A text-book of practical organic chemistry, including qualitative organic analysis.
Includes bibliographies and index.
ISBN 0-582-44250-8
1. Chemistry, Organic—Laboratory manuals. 2. Chemistry, Analytic—Qualitative. I. Furniss, Brian Stanley. II. Title. III. Title: Practical organic chemistry, including qualitative organic analysis.
QD261.V63 1978 547 77-23559
77 MARC

Vogel, Arthur Israel.
Vogel's Textbook of Macro and semimicro qualitative inorganic analysis. — 5th ed. / rev. by G. Svehla. — London ; New York : Longman, 1979.
xiii, 605 p. : ill. ; 24 cm.
First-3d ed. published under title: A text-book of qualitative chemical analysis; 4th ed. published under title: A text-book of macro and semimicro qualitative inorganic analysis.
ISBN 0-582-44367-9
1. Chemistry, Analytic—Qualitative. 2. Chemistry, Inorganic. I. Svehla, G. II. Title. III. Title: Textbook of Macro and semimicro qualitative inorganic analysis.
QD81.V6 1979 544 77-8290
77 MARC

—QUALITATIVE—LABORATORY MANUALS

Kemp, William, 1932-
Qualitative organic analysis / William Kemp. — Rev. ed. — London ; New York : McGraw-Hill Book Co., c1979.
xi, 182 p. : ill. ; 23 cm. GB79
Includes bibliographical references and index.
ISBN 0-07-084092-X : £4.85
1. Chemistry, Analytic—Qualitative—Laboratory manuals. 2. Chemistry, Organic—Laboratory manuals. I. Title.
QD271.4.K45 1979 547'.34'0028 78-41217
79 MARC

—QUANTITATIVE

see also Distillation, Fractional; Electrochemical analysis; Volumetric analysis

Fritz, James Sherwood, 1924-
Quantitative analytical chemistry / James S. Fritz, George H Schenk. — 4th ed. — Boston : Allyn and Bacon, c1979.
x, 661 p. : ill. ; 24 cm.
Bibliography: p. [624]-628.
Includes index.
ISBN 0-205-06527-9
1. Chemistry, Analytic—Quantitative. I. Schenk, George H., joint autho[r]. II. Title.
QD101.2.F74 1979 545 78-216[...]
78 MA[RC]

Kenner, C T
Quantitative analysis / Charles T. Kenner, Kenneth W. Busch. — New York : Macmillian, c1979.
xxiii, 623 p. : ill. ; 26 cm.
"A portion of this material has been adapted from Analytical separations and determinations by C. T. Kenner."
Bibliography: p. 583-588.
Includes index.
ISBN 0-02-362490-6
1. Chemistry, Analytic—Quantitative. I. Busch, Kenneth W., joint author. II. Kenner, C. T. Analytical separations and determinations. III. Title.
QD101.2.K46 545 78-4160[...]
78 MARC

Kurs lekfsiĭ po kolichestvennomu analizu / [pod red. V. [...] Petrashen']. — Novocherkassk : Redaktsionno-izdatel'sk[...] otdel NPI, 1968.
109 p. : ill.
At head of title: Ministerstvo vysshego i srednego special'no[go] obrazovaniia SSSR. Novocherkasskiĭ politekhnicheskiĭ institut.
L.C. copy on microfilm only.
——— Microfilm. On 1 reel with other items ; 35 mm.
Microfilm Slavic 18181 QD
1. Chemistry, Analytic—Quantitative. I. Petrashen', Vladim[ir] Ivanovich. II. Novocherkassk, Russia. Politekhnicheskiĭ institu[t]
[QD101.K89] 79-3680[...]

Ma, Tsu Sheng, 1911-
Modern organic elemental analysis / T. S. Ma, Robert C Rittner. — New York : M. Dekker, c1979.
xiv, 518 p. : ill. ; 24 cm.
Includes bibliographical references and indexes.
ISBN 0-8247-6786-1
1. Chemistry, Analytic—Quantitative. 2. Chemistry, Organic. I. Rittner, Robert C., 1924- joint author. II. Title.
QD271.7.M29 547'.34 79-1157[...]
79 MAR[C]

Ma, Tsu Sheng, 1911-
Quantitative analysis of organic mixtures / T. S. Ma, Robert E. Lang. — New York : Wiley, c1979-
v. ; 24 cm.
"A Wiley-Interscience publication."
Includes bibliographical references and indexes.
CONTENTS: pt. 1. General principles.

ISBN 0-471-55800-1
1. Chemistry, Analytic—Quantitative. 2. Chemistry, Organic. I. Lang, Robert E., joint author. II. Title.
QD271.7.M3 547'.35 78-23202
78 MAR[C]

Marczenko, Zygmunt.
Analiza ilościowa : pracownia chemii analitycznej / Zygmun[t] Marczenko ; zadania obliczeniowe przygot. Krzysztof Kasiura. — Wyd. 3. — Warszawa : Wydaw-a Szkolne i Pedagogiczne, 1977 (BielBiała : CDW 5)
243, [1] p. : ill. ; 21 cm. P78-132[...]
Includes index.
zł16.00
1. Chemistry, Analytic—Quantitative. I. Title.
QD101.2.M37 1977 78-38564
78 MAR[C]

CHEMISTRY, ANALYTIC
—QUANTITATIVE (Continued)

Pino Pérez, Francisco.
Gravimetrías y métodos analíticos de separación / ₍por₎ F. Pino Pérez y José M. Cano Pavón. — Sevilla : Publicaciones de la Universidad, 1977.
373 p. : ill., graphs ; 25 cm. Sp78-Feb
Includes bibliographical references and index.
ISBN 8474050677 : 500ptas
1. Chemistry, Analytic—Quantitative. I. Cano Pavón, José Manuel, joint author. II. Title.
QD101.2.P56 545 78-369002
78 MARC

Pozdíunin, E V
(Kratkie zapiski po kolichestvennomu khimicheskomu analizu)
Краткие записки по количественному химическому анализу : пособие для студентов вузов / Е. В. Поздюнин. — Самара : Губиздат., 1924.
88 p. ; 26 cm.
1. Chemistry, Analytic—Quantitative. I. Title.
QD101.P68 79-413887

Skoog, Douglas Arvid, 1918-
Analytical chemistry / Douglas A. Skoog, Donald M. West. — 3d ed. — New York : Holt, Rinehart and Winston, c1979.
xii, 685 p. : ill. ; 25 cm.
Includes bibliographical references and index.
ISBN 0-03-044416-0
1. Chemistry, Analytic—Quantitative. I. West, Donald M., joint author. II. Title.
QD101.2.S55 1979 545 78-14543
78 MARC

Vogel, Arthur Israel.
Vogel's textbook of quantitative inorganic analysis, including elementary instrumental analysis. — 4th ed. / revised by John Bassett ... ₍et al.₎. — London ; New York : Longman, 1978.
xxxi, 925 p. : ill ; 24 cm.
Includes bibliographies and index.
ISBN 0-582-46321-1 : £15.00
1. Chemistry, Analytic—Quantitative. 2. Chemistry, Inorganic. I. Bassett, John, 1924- II. Title.
QD101.2.V63 1978 546 77-5545
77 MARC

Wu-han ta hsüeh, Wuchang, China. Fen hsi hua hsüeh chiao yen shih.
(Hua hsüeh fen hsi)
化学分析 / 武汉大学分析化学教研室编. — 第1版. — 北京：人民教育出版社, 1975, 1976 printing-1977, 1978 printing.
2 v. : ill. ; 21 cm.
$2.65
1. Chemistry, Analytic—Quantitative. I. Title.
QD101.2.W83 1975 77-838288

—QUANTITATIVE —HANDBOOKS, MANUALS, ETC.

Babkov, Aleksandr Vasil'evich.
(Praktikum po obshcheĭ khimii s ėlementami kolichestvennogo analiza)
Практикум по общей химии с элементами количественного анализа : ₍Для мед. ин-тов₎ / А. В. Бабков, Г. Н. Горшкова, А. М. Кононов ; Под ред. А. В. Бабкова. — Москва : Высш. школа, 1978.
168 p. : ill. ; 20 cm. USSR 78
Bibliography: p. ₍159₎
0.25rub
1. Chemistry, Analytic—Quantitative—Handbooks, manuals, etc. I. Gorshkova, Galina Nikolaevna, joint author. II. Kononov, Anatolii Mikhailovich, joint author. III. Title.
QD101.2.B32 79-407181

—STATISTICAL METHODS

Massart, D L
Evaluation and optimization of laboratory methods and analytical procedures : a survey of statistical and mathematical techniques / Desiré L. Massart, Auke Dijkstra, Leonard Kaufman, with contributions by Svante Wold, Bernard Vandeginste, and Yvette Michotte. — Amsterdam ; New York : Elsevier Scientific Pub. Co. ; New York : distributors for the U.S. and Canada, Elsevier North-Holland, 1978.
xvi, 596 p. : ill. ; 25 cm. — (Techniques and instrumentation in analytical chemistry ; v. 1)
Includes bibliographies and index.
ISBN 0-444-41743-5 : $57.75
1. Chemistry, Analytic—Statistical methods. I. Dijkstra, Auke, joint author. II. Kaufman, Leonard, joint author. III. Title. IV. Series.
QD75.4.S8M37 543 78-12667
78 MARC

—STATISTICAL METHODS—CONGRESSES

Statistics / R. F. Hirsch, editor. — Philadelphia : Franklin Institute Press, c1978.
viii, 308 p. : ill. ; 23 cm. — (1977 Eastern Analytical Symposium)
"Based on papers presented at the 1977 Eastern Analytical Symposium in New York City."
Includes bibliographical references and index.
ISBN 0-89168-017-9
1. Chemistry, Analytic—Statistical methods—Congresses. I. Hirsch, Roland F., 1939- II. Series: Eastern Analytical Symposium, 17th, New York, 1977. 1977 Eastern Analytical Symposium series.
QD75.4.S8S74 540'.1'82 78-10508
78 MARC

—TECHNIQUE—HANDBOOKS, MANUALS, ETC.

Knapp, Daniel R
Handbook of analytical derivatization reactions / Daniel R. Knapp. — New York : Wiley, c1979.
xviii, 741 p. : ill. ; 24 cm.
"A Wiley-Interscience publication."
Includes bibliographical references and indexes.
ISBN 0-471-03469-X
1. Chemistry, Analytic—Technique—Handbooks, manuals, etc. 2. Chromatographic analysis—Handbooks, manuals, etc. 3. Mass spectrometry—Handbooks, manuals, etc. I. Title. II. Title: Analytical derivatization reactions.
QD78.K63 543 78-12944
78 MARC

—TERMINOLOGY

Compendium of analytical nomenclature : definitive rules 1977 / prepared for publication by H. M. N. H. Irving, H. Freiser, and T. S. West. — 1st ed. — Oxford ; New York : Pergamon Press, 1978.
viii, 223 p. : ill. ; 28 cm. — (₍IUPAC publication₎)
At head of title: International Union of Pure and Applied Chemistry, Analytical Chemistry Division.
Includes bibliographical references and index.
ISBN 0-08-022008-8. ISBN 0-08-022347-8 pbk.
1. Chemistry, Analytic—Nomenclature. 2. Chemistry, Analytic—Terminology. I. Irving, Harry Munroe Napier Hetherington. II. Freiser, Henry, 1920- III. West, Thomas Summers. IV. International Union of Pure and Applied Chemistry. Analytical Chemistry Division. V. Series: International Union of Pure and Applied Chemistry. IUPAC publication.
QD75.3.C65 1977 543'.001'4 77-8949
77 MARC

CHEMISTRY, ARCHAEOLOGICAL see Archaeological chemistry

CHEMISTRY, ATMOSPHERIC see Atmospheric chemistry

CHEMISTRY, BIOLOGICAL see Biological chemistry

CHEMISTRY, BOTANICAL see Botanical chemistry

CHEMISTRY, CLINICAL

see also Blood—Analysis and chemistry; Enzymes—Analysis; Poisons; Radioactivation analysis

Contemporary topics in analytical and clinical chemistry. v. 1–
New York, Plenum Press ₍c1977–
v. ill. 24 cm.
Editor: v. 1– D. M. Hercules.
Key title: Contemporary topics in analytical and clinical chemistry, ISSN 0160-3620.
1. Chemistry, Clinical. 2. Chemistry, Analytic. I. Hercules, David M.
RB40.C66 616.07'56'05 79-640979
MARC-S

Hodkinson, H M
Biochemical diagnosis of the elderly / H. M. Hodkinson. — London : Chapman and Hall, 1977.
xi, 111 p. : ill. ; 24 cm.
Bibliography: p. 95-100.
Includes index.
ISBN 0-412-13720-8
1. Geriatrics—Diagnosis. 2. Chemistry, Clinical. 3. Function tests (Medicine) I. Title.
₍DNLM: 1. Diagnosis, Laboratory—In old age. WT100 H691b₎
RC953.H58 1977b 618.9'7'0756 77-78950
77₍r79₎rev MARC

Kaplan, Alex, 1910-
Clinical chemistry : interpretation and techniques / Alex Kaplan and LaVerne L. Szabo. — Philadelphia : Lea & Febiger, 1979.
xvii, 440 p. : ill. ; 27 cm.
Includes bibliographical references and index.
ISBN 0-8121-0624-5
1. Chemistry, Clinical. I. Szabo, LaVerne L., joint author. II. Title.
RB40.K38 1979 616.07'56 78-9698
78 MARC

White, Wilma L
Chemistry for the clinical laboratory / Wilma L. White, Marilyn M. Erickson, Sue C. Stevens. — 4th ed. — St. Louis : Mosby, 1976.
ix, 756 p. : ill. ; 27 cm.
First-3d eds. published under title: Chemistry for medical technologists.
First-2d eds. by C. E. Seiverd.
Includes bibliographical references and index.
ISBN 0-8016-5432-7
1. Chemistry, Clinical. I. Erickson, Marilyn M., joint author. II. Stevens, Sue Cassell, 1911- joint author. III. Seiverd, Charles Edward, 1913- Chemistry for medical technologists. IV. Title.
RB40.S46 1976 616.07'56 75-31617
75₍r79₎rev MARC

—COLLECTED WORKS

Bijos, Gerardo Majella.
Contribuições científicos. Rio de Janeiro ₍S.F.Q., i.e. Sociedade de Farmácia e Química de São Paulo₎ 1954.
xii, 275 p. illus. 24 cm.
Includes bibliographies.
1. Pharmacy—Collected works. 2. Pharmacology—Collected works. 3. Chemistry, Clinical—Collected works. 4. Pharmacy, Military—Brazil—Collected works. I. Title.
RS43.B54 78-390913
79 MARC

—CONGRESSES

Arnold O. Beckman Conference in Clinical Chemistry, 1st, Annapolis, 1977.
Clinician and chemist : the relationship of the laboratory to the physician : proceedings of the first Arnold O. Beckman Conference in Clinical Chemistry / editors, Donald S. Young ... ₍et al.₎, executive editor, J. Stanton King. — Washington : American Association of Clinical Chemistry, c1979.
xi, 375 p. : ill. ; 24 cm.
Includes bibliographical references.
ISBN 0-915274-08-6
1. Diagnosis, Laboratory—Philosophy—Congresses. 2. Chemistry, Clinical—Congresses. I. Young, Donald S. II. American Association for Clinical Chemistry. III. Title.
RB37.A1A76 1979 362.1 78-72880
79 MARC

—INSTRUMENTS

Laboratory practice reivews of current techniques and instrumentation / editor Bevan M. Gilpin. — London : United Trade Press, 1977.
82, ₍14₎ p. : ill., plans ; 30 cm. GB78-02289
Cover title.
Includes bibliographies.
ISBN 0-85471-027-2 : £5.00
1. Chemistry, Clinical—Instruments. 2. Medical laboratories—Equipment and supplies. 3. Biological chemistry—Instruments. I. Gilpin, Bevan M. II. Laboratory practice.
RB40.L29 542 78-318157
78 MARC

—OUTLINES, SYLLABI, ETC.

Bold, Alan Maurice.
Clinical chemistry companion / by A. M. Bold & P. Wilding. — Oxford : Blackwell Scientific Publications ; Philadelphia : distributed by Lippincott, 1978.
119 p. ; 19 cm. GB***
Bibliography: p. 11.
Includes index.
ISBN 0-632-00359-6 : £3.75
1. Chemistry, Clinical—Outlines, syllabi, etc. I. Wilding, P., joint author. II. Title.
RB40.B618 616.07'56 79-305855
79 MARC

—PERIODICALS

Journal of clinical chemistry and clinical biochemistry. Zeitschrift für klinische Chemie und klinische Biochemie. v. 14– Jan. 1976–
Berlin, New York, de Gruyter.
v. ill. 30 cm. monthly.
Continues: Zeitschrift für klinische Chemie und klinische Biochemie.
Organ of the Deutsche Gesellschaft für Klinische Chemie, Österreichische Gesellschaft für Klinische Chemie and Schweizerische Gesellschaft für Klinische Chemie.
English or German, with summaries in the other language.
Key title: Journal of clinical chemistry and clinical biochemistry, ISSN 0340-076X
1. Chemistry, Clinical—Periodicals. 2. Biological chemistry—Periodicals. I. Deutsche Gesellschaft für Klinische Chemie. II. Österreichische Gesellschaft für Klinische Chemie. III. Schweizerische Gesellschaft für Klinische Chemie. IV. Title: Zeitschrift für klinische Chemie und klinische Biochemie.
RB40.A1Z4 616.07'56'05 78-645788
MARC-S

CHEMISTRY, DENTAL see Dental chemistry

CHEMISTRY, DIAGNOSTIC see Chemistry, Clinical

CHEMISTRY, ENVIRONMENTAL see Environmental chemistry

CHEMISTRY, FORENSIC

Lowry, William Thomas, 1942-
Forensic toxicology : controlled substances and dangerous drugs / W. T. Lowry and James C. Garriott. — New York : Plenum Press, c1979.
xvii, 445 p. : ill. ; 24 cm.
Bibliography: p. 411-410.
Includes index.
ISBN 0-306-40124-X
1. Chemistry, Forensic. 2. Drugs. 3. Narcotics, Control of—United States. 4. Drugs—Laws and legislation—United States. I. Garriott, James C., joint author. II. Title.
HV8073.L75   614'.19   78-26439
78   MARC

United States. Federal Bureau of Investigation. Laboratory.
Handbook of forensic science, Federal Bureau of Investigation / FBI Laboratory. — Rev. — ⌊Washington⌋ : FBI : for sale by the Supt. of Docs., U.S. Govt. Print. Off., 1978.
i, 133 p. : ill. ; 27 cm.
1. Chemistry, Forensic. 2. Criminal investigation. I. Title.
HV8073.U5 1978   364.12   79-601139
79   MARC

—PERIODICALS

Forensic science international. v. 12– July/Aug. 1978–
Lausanne, Elsevier Sequoia.
v. ill. 22 cm. bimonthly.
"An international journal dedicated to the applications of science to the administration of justice."
Continues: Forensic science (Lausanne), ISSN 0300-9432.
English, French, or German with summaries in English.
1. Medical jurisprudence—Periodicals. 2. Chemistry, Forensic—Periodicals.
RA1001.F65   614'.19   79-641782
MARC-S

CHEMISTRY, INDUSTRIAL see Chemical engineering; Chemistry, Technical

CHEMISTRY, INORGANIC

see also Groups and classes of inorganic compounds, e.g. Acids, Inorganic; Chlorides

Advances in inorganic biochemistry / ed., Gunther L. Eichhorn and Luigi G. Marzilli. — Amsterdam ; New York : Elsevier/-North-Holland ⌊Biomedical Press⌋, 1979-
v. ; 25 cm.   Ne79-20
First ed., edited by G. L. Eichhorn, published in 1973 under title: Inorganic biochemistry.
Includes bibliographical references and index.
ISBN 0-444-00323-1 (v. 1) : fl 50.25 (v. 1)
1. Biological chemistry. 2. Chemistry, Inorganic. I. Eichhorn, Gunther Louis, 1927-   II. Marzilli, Luigi G. III. Eichhorn, Gunther Louis, 1927-   Inorganic biochemistry.
QP531.A33   574.1'921   79-317544
*79   MARC

Araten, David, 1928–
(ha-Kimyah ha-i-organit be-mosdot mada' u-meḥkar be-Yiśrael, ve-ziḳatah le-shimushim tekhnologiyim)
הכימיה האי־אורגאנית במוסדות מדע ומחקר בישראל וזיקתה לשימושים טכנולוגיים / דוד אשבורן (ארטן). — ירושלים : משרד ראש הממשלה, המועצה הלאומית למחקר ופיתוח, 1974.
66 leaves ; 28 cm.
Cover title: Seḳer kimyah i-organit be-Yiśrael.
"NCRD 9–74."
Includes index.
1. Chemistry, Technical—Research—Israel. 2. Chemistry, Inorganic. I. Israel. ha-Mo'atsah ha-le'umit le-meḥkar ule-fituaḥ. II. Title. III. Title: Seker kimyah i-organit be-Yiśrael.
TP177.A72   79–951255

Boniface, H   J
Analytical notes : a summary of inorganic methods of chemical analysis / H. J. Boniface. — Wolverhampton : Sigma Technical Press, 1978.
⌊64⌋ p. ; 30 cm.   GB78-19162
Includes bibliographical references.
ISBN 0-905104-02-1 : £3.50
1. Chemistry, Analytic. 2. Chemistry, Inorganic. I. Title.
QD75.2.B66   545   78-320381
79   MARC

Chemical effects of nuclear transformations in inorganic systems / editors, G. Harbottle and A. G. Maddock. — Amsterdam ; New York : North-Holland Pub. Co. ; New York : sole distributors for the USA and Canada, Elsevier North-Holland, 1979.
viii, 575 p. : ill. ; 23 cm.
Bibliography: p. ⌊505⌋-548.
Includes indexes.
ISBN 0-444-85054-6
1. Radiation chemistry. 2. Chemistry, Inorganic. I. Harbottle, G. II. Maddock, A. G.
QD636.C43   541'.38   78-10479
78   MARC

Marfunin, Arnol'd Sergeevich.
Physics of minerals and inorganic materials : an introduction / A. S. Marfunin ; translated by N. G. Egorova and A. G. Mishchenko. — Berlin ; New York : Springer-Verlag, 1979.
xii, 340 p. : ill. ; 25 cm.
Translation of Vvedenie v fiziku mineralov.
Bibliography: p. ⌊316⌋-333.
Includes index.
ISBN 0-387-08982-9
1. Mineralogical chemistry. 2. Chemistry, Physical and theoretical. 3. Chemistry, Inorganic. I. Title.
QE371.M3713   549   78-23186
78   MARC

New trends in bio-inorganic chemistry / edited by R. J. P. Williams, J. R. R. F. da Silva. — London ; New York : Academic Press, 1978.
489 p. : ill. ; 24 cm. — (Frontiers of knowledge series)   GB***
Includes bibliographical references and index.
ISBN 0-12-755050-X
1. Biological chemistry. 2. Chemistry, Inorganic. I. Williams, Robert Joseph Paton. II. Silva, J. R. R. Fraústo da.
QP531.N47   574.1'921   78-18034
79   MARC

Saitō, Yoshihiko, 1920-
Inorganic molecular dissymmetry / Yoshihiko Saitō. — Berlin ; New York : Springer-Verlag, 1979.
ix, 167 p. : ill. ; 25 cm. — (Inorganic chemistry concepts v. 4)
Bibliography: p. ⌊156⌋-163.
Includes index.
ISBN 0-387-09176-9
1. Molecular theory. 2. Symmetry (Physics) 3. Chemistry, Inorganic. I. Title. II. Series.
QD461.S26   541'.2   78-26270
78   MARC

Tarasov, V   V
(Kinetika ėkstrakt͡sii)
Кинетика экстракции. Москва, ⌊ВИНИТИ⌋, 1974.
119 p. with diagrs. 21 cm. (Итоги науки и техники: Неорганическая химия, т. 4)   USSR 74
At head of title: В. В. Тарасов, Г. А. Ягодин.
On leaf preceding t. p.: Государственный комитет Совета Министров СССР по науке и технике. Академия наук Союза Советских Социалистических Республик. Всесоюзный институст научной и технической информации.
Bibliography: p. 110–⌊118⌋
1. Extraction (Chemistry) 2. Chemistry, Inorganic. I. I͡Agodin, G. A., joint author. II. Title. III. Series: Itogi nauki i tekhniki: Neorganicheskai͡a khimii͡a, t. 4.
QD151.I 85 t. 4   74–343422
[QD63.E88]

Technische anorganische Chemie / ⌊dem Autorenkollektiv gehören an, Gunter Bräutigam ... et al.⌋. — 1. Aufl. — Leipzig : Deutscher Verlag für Grundstoffindustrie, VEB, 1978.
365 p. : ill., graphs ; 24 cm.   GFR79-A4
Table inserted.
Includes bibliographical references and index.
39.80M
1. Chemistry, Technical. 2. Chemistry, Inorganic. I. Bräutigam, Gunter.
TP145.T34   79-372727
79   MARC

Trzebiatowski, Włodzimierz.
Chemia nieorganiczna : podręcznik chemii ogólnej i nieorganicznej dla wydziałów chemicznych politechnik i uniwersytetów / Włodzimierz Trzebiatowski. — Wyd. 8. — Warszawa : Państwowe Wydawn. Naukowe, 1978.
674 p. : ill. ; 23 cm.   P***
Bibliography: p. ⌊637⌋
Includes indexes.
zł110.00
1. Chemistry, Inorganic. I. Title.
QD151.2.T79 1978   79-357340
79   MARC

Vogel, Arthur Israel.
Vogel's textbook of quantitative inorganic analysis, including elementary instrumental analysis. — 4th ed. / revised by John Bassett ... ⌊et al.⌋. — London ; New York : Longman, 1978.
xxxi, 925 p. : ill ; 24 cm.
Includes bibliographies and index.
ISBN 0-582-46321-1 : £15.00
1. Chemistry, Analytic—Quantitative. 2. Chemistry, Inorganic. I. Bassett, John, 1924-   II. Title.
QD101.2.V63 1978   546   77-5545
77   MARC

Vogel, Arthur Israel.
Vogel's Textbook of Macro and semimicro qualitative inorganic analysis. — 5th ed. / rev. by G. Svehla. — London ; New York : Longman, 1979.
xiii, 605 p. : ill. ; 24 cm.
First-3d ed. published under title: A text-book of qualitative chemical analysis; 4th ed. published under title: A text-book of macro and semimicro qualitative inorganic analysis.
ISBN 0-582-44367-9
1. Chemistry, Analytic—Qualitative. 2. Chemistry, Inorganic. I. Svehla, G. II. Title. III. Title: Textbook of Macro and semimicro qualitative inorganic analysis.
QD81.V6 1979   544   77-8290
77   MARC

Williams, Alan Francis, 1950-
A theoretical approach to inorganic chemistry / A. F. Williams. — Berlin ; New York : Springer-Verlag, 1979.
xii, 316 p. : ill. ; 25 cm.
Spine title : Inorganic chemistry.
Includes bibliographies and index.
ISBN 0-387-09073-8
1. Chemistry, Inorganic. 2. Molecular orbitals. 3. Quantum chemistry. I. Title. II. Title: Inorganic chemistry.
QD152.3.W54   546   78-26415
78   MARC

—ADDRESSES, ESSAYS, LECTURES

Inorganic and physical chemistry. — Berlin ; New York : Springer-Verlag, 1978.
239 p. : ill. ; 25 cm. — (Topics in current chemistry ; 77)
Includes bibliographical references and index.
CONTENTS: Bikerman, J. J. Surface energy of solids.—Wiedemann, H. G. and Bayer, G. Trends and applications of thermogravimetry.—Huglin, M. B. Determination of molecular weights by light scattering.
ISBN 0-387-08987-X
1. Chemistry, Physical and theoretical—Addresses, essays, lectures. 2. Chemistry, Inorganic—Addresses, essays, lectures. I. Bikerman, Jacob Joseph, 1898-   Surface energy of solids. 1978. II. Wiedemann, Hans G. Trends and applications of thermogravimetry. 1978. III. Huglin, M. B. Determination of molecular weights by light scattering. 1978. IV. Series.
QD1.F58 vol. 77   540'.8s   78-13471
⌊QD475⌋   78   MARC

(Voprosy fizicheskoĭ khimii v proizvodstve neorganicheskikh veshchestv)
Вопросы физической химии в производстве неорганических веществ / под общей ред. Черненькой Е. И. и Можаровой Т. В. — Харьков : НИОХИМ, 1974.
97, 4 p. : ill. ; 22 cm. — (Труды - Государственный научно-исследовательский и проектный институт основной химии ; т. 34)   USSR 7[illegible]
Includes bibliographies.
1. Chemistry, Physical and theoretical—Addresses, essays, lectures. 2. Chemistry, Inorganic—Addresses, essays, lectures. I. Chernen'kai͡a, E. I. II. Mozharova, T. V. III. Series: Kharkov. Naukovo-doslidnyĭ instytut osnovnoï khimiï. Trudy ; t. 34.
QD1.K48 t. 34   79–40953[illegible]
[QD455]

—COLLECTED WORKS

Politechnika Wrocławska. Instytut Chemii Nieorganicznej i Metalurgii Pierwiastków Rzadkich.
Monografie. nr. 1–
Wrocław, 1971–
no. illus. 24 cm. (Prace naukowe Instytutu Chemii Nieorganicznej i Metalurgii Pierwiastków Rzadkich Politichniki Wrocławskiej) (Prace naukowe Politechniki Wrocławskiej)
Cover title, no. 1–   : Seria Monografie.
1. Chemistry, Inorganic—Collected works. 2. Nonferrous metals—Metallurgy—Collected works. I. Series: Politechnika Wrocławska. Instytut Chemii Nieorganicznej i Metalurgii Pierwiastków Rzadkich. Prace naukowe.
QD1.P566a Subser.   540'.8 s ⌊540⌋   72–61756[illegible]
[QD151.2]   MARC-S

Politechnika Wrocławska. Instytut Chemii Nieorganicznej i Metalurgii Pierwiastków Rzadkich.
Prace naukowe, nr. 1–
Wrocław, 1970–
no. illus. 24 cm. (Prace naukowe Politechniki Wrocławskiej)
1. Chemistry, Inorganic—Collected works. 2. Nonferrous metals—Metallurgy—Collected works.
QD1.P566a subser.   540   78–64996[illegible]
[QD151.2]   MARC-S

Politechnika Wrocławska. Instytut Chemii Nieorganicznej i Metalurgii Pierwiastków Rzadkich.
Studia i materiały. nr. 1–
Wrocław, 1970–
no. illus. 24 cm. (Prace naukowe Instytutu Chemii Nieorganicznej i Metalurgii Pierwiastków Rzadkich Politechniki Wrocławskiej) (Prace naukowe Politechniki Wrocławskiej)
Cover title, no. 1–   : Seria Studia i materiały.
1. Chemistry, Inorganic—Collected works. 2. Nonferrous metals—Metallurgy—Collected works. I. Series: Politechnika Wrocławska. Instytut Chemii Nieorganicznej i Metalurgii Pierwiastków Rzadkich. Prace naukowe.
QD1.P566a subser.   540'.8 s [540]   74–64996[illegible]
[QD151.2]   MARC-S

CHEMISTRY, INORGANIC (Continued)

—CONGRESSES

Inorganic compounds with unusual properties—II : a symposium / sponsored by the Division of Inorganic Chemistry at the inorganic chemistry symposium, Athens, Georgia, February 1-3, 1978 ; R. Bruce King, editor. — Washington : American Chemical Society, 1979.
xii, 418 p. : ill. ; 24 cm. — (Advances in chemistry series ; v. 173 ISSN 0065-2393)
Includes bibliographical references and index.
Includes index.
ISBN 0-8412-0429-2
1. Chemistry, Inorganic—Congresses. I. King, R. Bruce. II. American Chemical Society. Division of Inorganic Chemistry. III. Series.
QD1.A355 no. 173 540'.8 s 78-31900
[QD146] 79 MARC

Latvijas Padomju Socialistiskās Republikas Zinātņu akadēmija. Neorganiskās ķīmijas institūts.
(Pi͡ataia konferenfsii͡a molodykh nauchnykh rabotnikov Instituta neorganicheskoĭ khimii)
Пятая конференция молодых научных работников Института неорганической химии : тезисы докладов / Академия наук Латвийской ССР, Институт неорганической химии ; [редактор Л. Яковлева]. — Рига : Зинатне, 1976.
32 p. ; 20 cm. USSR***
Includes bibliographies and index.
1. Chemistry, Inorganic—Congresses. I. I͡Akovleva, L. II. Title.
QD146.L37 1976 79-410837

—DICTIONARIES

Sittig, Marshall.
Inorganic chemical industry : processes, toxic effluents, and pollution control / Marshall Sittig. — Park Ridge, N.J. : Noyes Data Corp., 1978.
x, 351 p. : ill. ; 25 cm. — (Chemical technology review ; no. 118) (Pollution technology review ; no. 52)
Bibliography: p. 346-347.
Includes indexes.
ISBN 0-8155-0726-7
1. Chemistry, Technical—Dictionaries. 2. Chemistry, Inorganic—Dictionaries. I. Title. II. Series. III. Series: Pollution technology review ; no. 52.
TP9.S57 661'.003 78-61892
79 MARC

—INDEXES—PERIODICALS

Powder diffraction file alphabetical index: inorganic compounds.
Swarthmore, Pa., JCPDS, International Centre for Diffraction Data.
v. 29 cm. annual. (Publication - JCPDS—International Centre for Diffraction Data)
Continues: Powder diffraction file search manual alphabetical listing: inorganic, ISSN 0092-0509.
Key title: Powder diffraction file alphabetical index: inorganic compounds, ISSN 0191-2410.
1. X-rays—Diffraction—Indexes—Periodicals. 2. Powders—Optical properties—Indexes—Periodicals. 3. Chemistry, Inorganic—Indexes—Periodicals. I. JCPDS—International Centre for Diffraction Data. II. Series: JCPDS—International Centre for Diffraction Data. Publication - JCPDS—International Centre for Diffraction Data.
QC482.D5P645 543'.085 79-642626
MARC-S

—MATHEMATICS

Sladkov, Igor' Borisovich.
(Metody priblizhennogo rascheta fiziko-khimicheskikh svoĭstv kovalentnykh neorganicheskikh soedineniĭ)
Методы приближенного расчета физико-химических свойств ковалентных неорганических соединений / И. Б. Сладков ; Под ред. А. Г. Морачевского. — Ленинград : Изд-во ЛГУ, 1978.
104 p. : graphs ; 21 cm. USSR 78
At head of title: Ministerstvo vysshego i srednego spet͡sial'nogo obrazovanii͡a RSFSR.
Bibliography: p. 96-102.
0.90rub
1. Chemistry, Inorganic—Mathematics. I. Morachevskiĭ, Aleksei Georgievich. II. Title.
QD152.3.S58 79-408610

—NOMENCLATURE

Nomenclatura chimiei anorganice : după regulile elaborate de Uniunea Internaţională de Chimie Pură şi Aplicată / sub îngrijirea şi adaptarea dr. D. Purdela ; [adaptarea nomenclaturii I.U.P.A.C. la limba română, sub redacţia prof. dr. doc. ing. Alexandru T. Balaban]. — Bucureşti : Editura Academiei Republicii Socialiste România, 1977.
153 p. ; 21 cm. R78-2
At head of title: Academia Republicii Socialiste România. Secţia de Ştiinţe Chimice. Colectivul de Nomenclatură Chimică.
Summary in English and French.
Includes bibliographical references and index.
lei11.50
1. Chemistry, Inorganic—Nomenclature. I. Purdela, D. II. Academia Republicii Socialiste România. Secţia de Ştiinţe Chimice. III. International Union of Pure and Applied Chemistry.
QD149.N65 78-370667
*78 MARC

—PERIODICALS

Inorganic biochemistry.
Berlin, New York, Springer-Verlag.
v. ill. 25 cm. (Topics in current chemistry)
1. Biological chemistry—Periodicals. 2. Chemistry, Inorganic—Periodicals. I. Series.
QD1.F58 subser. 540'.8 s 78-647910
[QP531] [574.1'921] MARC-S

Journal of inorganic biochemistry. v. 10
Feb. 1979-
New York, Elsevier.
v. ill. 26 cm. 8 no. a year.
Continues: Bioinorganic chemistry, ISSN 0006-3061.
Key title: Journal of inorganic biochemistry, ISSN 0162-0134.
1. Biological chemistry—Periodicals. 2. Chemistry, Inorganic—Periodicals.
QP501.B557 574.1'92'05 79-642923
MARC-S

CHEMISTRY, MEDICAL see Chemistry, Clinical; Chemistry, Pharmaceutical

CHEMISTRY, MEDICAL AND PHARMACEUTICAL

see also Chemistry, Clinical; Disinfection and disinfectants; Drugs; Materia medica; Pharmacy; Poisonous plants; Structure-activity relationship (Pharmacology)

CHEMISTRY, METALLURGIC

Coche, Lucien, 1918-
Angewandte Forschung für die Stahlerzeugung in den Unternehmen, auf nationaler Ebene und in der Europäischen Gemeinschaft / Lucien Coche. Wechselwirkungen zwischen physikalisch-chemischer Grundlagenforschung, theoretischer Metallurgie und grossindustrieller Stahlerzeugung / Ludwig von Bogdandy. — Opladen : Westdeutscher Verlag, c1974.
64 p. : ill. (some col.) ; 24 cm. — (Natur-, Ingenieur- und Wirtschaftswissenschaften : Vorträge - Rheinisch-Westfälische Akademie der Wissenschaften ; N 245) GFR***
Summaries in English and French.
Includes bibliographical references.
1. Steel—Metallurgy. 2. Chemistry, Metallurgic. I. Bogdandy, Ludwig von. Wechselwirkungen zwischen physikalisch-chemischer Grundlagenforschung .... 1974. II. Title. III. Series: Rheinisch-Westfälische Akademie der Wissenschaften. Natur-, Ingenieur- und Wirtschaftswissenschaften : Vorträge ; N 245.
Q49.C95A8 N 245 79-358900
[TN731] 79 MARC

Janiczek, Stanisław.
Materiłoznąwstwo z chemią dla górników / Stanisław Janiczek. — Gliwice : PŚ, 1977 (Gliw. : PŚ)
336 p. : ill. ; 25 cm. — (Skrypty centralne wyższych technicznych studiów dla pracujących ; Nr. 675/23) P77-6103
At head of title: Politechnika Śląska im. W. Pstrowskiego.
zł38.00
1. Materials. 2. Chemistry, Metallurgic. I. Title.
TA403.6.J36 78-366183
78 MARC

Lepinskikh, Boris Mikhaĭlovich.
(Fizicheskai͡a khimii͡a oksidnykh i oksiftoridnykh rasplavov)
Физическая химия оксидных и оксифторидных расплавов / Б. М. Лепинских, А. И. Манаков ; АН СССР, Уральск. науч. центр, Ин-т металлургии. — Москва : Наука, 1977.
190 p. : ill. ; 21 cm. USSR 77
Bibliography : p. 177-188.
1.20rub
1. Slag. 2. Oxides. 3. Chemistry, Metallurgic. I. Manakov, Anatoliĭ Ivanovich, joint author. II. Title.
TN707.L37 79-408888

Parker, Roger Hill.
An introduction to chemical metallurgy / by R. H. Parker. — 2d ed., in SI/metric units. — Oxford [Eng.] ; New York : Pergamon Press, 1978.
xiv, 361 p. : ill. ; 22 cm. — (Pergamon international library of science, technology, engineering, and social studies) (International series on materials science and technology ; v. 26) GB78
Includes bibliographical references and index.
ISBN 0-08-022125-4. ISBN 0-08-022126-2 pbk.
1. Chemistry, Metallurgic. I. Title.
QD132.P37 1978 669'.9 77-30266
77 MARC

Tsybryk, Oleksiĭ Mykolaĭovych.
Fiziko-khimicheskie prot͡sessy v kontaktnoĭ zone metall-forma / A. N. T͡Sibrik ; AN USSR, In-t problem lit'i͡a. — Kiev : Nauk. dumka, 1977.
211 p. : ill. ; 22 cm. USSR77
Bibliography: p. 207-[209]
2.24rub
1. Founding. 2. Liquid metals. 3. Chemistry, Metallurgic. I. Title.
TS233.T846 79-376340
79 MARC

—ADDRESSES, ESSAYS, LECTURES

Fyzikálně-chemické pochody při výrobě oceli : sborník přednášek, květen 1975. — [s.l.] : Česká vědecko-technická společnost, [1975]
333 p. : ill. ; 29 cm. Cz***
Cover title.
At head of title: Česká vědecko-technická společnost and other institutions.
Text in Czech, Slovak, German, or Russian.
Includes bibliographies.
1. Steel—Metallurgy—Addresses, essays, lectures. 2. Chemistry, Metallurgic—Addresses, essays, lectures. I. Česká vědeckotechnická společnost.
TN730.F99 79-339307
79 MARC

—CONGRESSES

Japan-USSR Joint Symposium on Physical Chemistry of Metallurgical Processes, 2d, Tokyo, 1969.
The Second Japan-USSR Joint Symposium on Physical Chemistry of Metallurgical Processes, 1969. — [Tokyo] : Iron and Steel Institute of Japan, [1969]
xvi, 256 p. : ill. ; 27 cm. — (Special report - Iron and Steel Institute of Japan ; no. 10)
Includes bibliographical references.
1. Physical metallurgy—Congresses. 2. Chemistry, Metallurgic—Congresses. I. Nippon Tekkō Kyōkai. II. Series: Nippon Tekkō Kyōkai. Special report - Iron and Steel Institute of Japan ; no. 10.
TN689.2.J36 1969 669'.94 79-300429
79 MARC

Japan-USSR Joint Symposium on Physical Chemistry of Metallurgical Processes, 3d, Moscow, 1971.
The Third Japan-USSR Joint Symposium on Physical Chemistry of Metallurgical Processes, 1971. — [Tokyo] : Iron and Steel Institute of Japan, 1973.
vii, 163 p. : ill. ; 27 cm. — (Special report - Iron and Steel Institute of Japan ; no. 16)
Includes bibliographical references.
1. Physical metallurgy—Congresses. 2. Chemistry, Metallurgic—Congresses. I. Series: Nippon Tekkō Kyōkai. Special report - Iron and Steel Institute of Japan ; no. 16.
TN689.2.J36 1971 669'.94 79-300437
79 MARC

CHEMISTRY, MINERALOGICAL see Mineralogical chemistry

CHEMISTRY, NUCLEAR see Nuclear chemistry

CHEMISTRY, ORGANIC

see also Chemistry, Physical organic; Condensation products (Chemistry); Organic compounds; Organic geochemistry; Organic water pollutants; Radicals (Chemistry); Stereochemistry; Surface active agents;

and also names of classes of organic compounds, e.g. Acids and names of individual organic substances, e.g. Benzene

Abraham, Raymond John, 1933-
Proton and carbon-13 NMR spectroscopy : an integrated approach / [by] R. J. Abraham and P. Loftus. — London ; Philadelphia : Heyden, 1978.
xiii, 230 p. : ill. ; 24 cm. GB78-32139
Includes bibliographies and indexes.
ISBN 0-85501-160-2 : £6.75
1. Nuclear magnetic resonance spectroscopy. 2. Chemistry, Organic. I. Loftus, P., joint author. II. Title.
QD96.N8A27 543'.08 79-302336
79 MARC

CHEMISTRY, ORGANIC (Continued)

Alexandrou, Nikolaos E
(Genikē organikē chēmeia)
Γενικὴ ὀργανικὴ χημεία : δομὴ καὶ μηχανισμοί / Νικολάου Ε. 'Αλεξάνδρου. — Θεσσαλονίκη : [s. n.], 1973.
324 p. : ill. ; 24 cm.
Bibliography : p. [5]–7.
Includes indexes.
1. Chemistry, Organic. I. Title.
QD251.2.A43 78–421691

Arima, Junzō.
(Yūki kagōbutsu bunsekihō)
有機化合物分析法 / 有馬純三. — 東京：岩波書店, 昭和9 [1934], 昭和24 [1949] printing.
13, 195, 28 p. : ill. ; 18 cm. — (岩波全書 ; 78)
Colophon inserted.
Series romanized: Iwanami zensho.
Bibliography: p. [2] (1st group)
Includes index.
1. Chemistry, Analytic. 2. Chemistry, Organic. I. Title.
QD271.A73 79–803740

Avram, Margareta.
Infrared spectroscopy : applications in organic chemistry / Margareta Avram, Gh. D. Mateescu ; translated by Ludmila Bîrlădeanu. — Huntington, N.Y. : R. E. Krieger Pub. Co., 1978, c1972.
527 p. : ill. ; 24 cm.
Translation of Spectroscopia în infraroşu.
Reprint of the ed. published by Wiley-Interscience, New York.
Includes bibliographical references and index.
ISBN 0-88275-711-3
1. Infra-red spectrometry. 2. Chemistry, Organic. I. Mateescu, Gh. D., joint author. II. Title.
[QD272.S6A9513 1978] 547'.34'63 78-16322
78 MARC

Avram, Margareta.
Spectroscopia în infraroşu : aplicaţii în chimia organică / Margareta Avram, Gh. D. Mateescu. — Bucureşti : Editura Tehnică, 1966.
591 p. : ill. ; 25 cm.
Includes bibliographies and index.
1. Infra-red spectrometry. 2. Chemistry, Organic. I. Mateescu, Gh. D., joint author. II. Title.
QD272.S6A95 78-392759
79 MARC

Bioorganic chemistry : a treatise to supplement Bioorganic chemistry, an international journal / edited by E. E. van Tamelen ; [contributors, Michael J. Arslanian ... et al.]. — New York : Academic Press, 1977-1978.
4 v. : ill. ; 24 cm.
Includes bibliographical references and indexes.
CONTENTS: v. 1. Enzyme action.—v. 2. Substrate behavior.—v. 3. Macro- and multimolecular systems.—v. 4. Electron transfer and energy conversion; cofactors; probes.
ISBN 0-12-714301-7 (v. 1)
1. Biological chemistry. 2. Chemistry, Organic. I. Van Tamelen, Eugene E., 1925-
[DNLM: 1. Biochemistry. 2. Chemistry, Organic. QU4 B61597]
QP514.2.B58 574.1'92 76-45994
77[r79]rev MARC

Borowiak, Teresa.
Czynniki określające wzajemne ułożenie elementów strukturalnych w kryształach organicznych / Teresa Borowiak. — Wyd. 1. — Poznań : Uniwersytet im. Adama Mickiewicza, 1975.
88 p., [1] fold. leaf : ill. ; 24 cm. — (Seria Chemia ; Nr. 15)
At head of title: Uniwersytet im. Adama Mickiewicza w Poznaniu. Wydział Matematyki, Fizyki i Chemii.
Summary in English.
Bibliography: p. [81]-88.
zł24.00
1. Crystals. 2. Chemistry, Organic. I. Title. II. Series.
QA1.P682 nr. 15 79-346487
[QD921] 79 MARC

Burgoyne, Edward E
A short course in organic chemistry / Edward E. Burgoyne. — New York : McGraw-Hill, c1979.
xv, 527 p. : ill. ; 24 cm.
Includes index.
ISBN 0-07-009171-4 : $16.50
1. Chemistry, Organic. I. Title.
QD251.2.B83 547 78-17079
78 MARC

Carnduff, John.
An introduction to organic chemistry / John Carnduff. — Chichester [Eng.] ; New York : Wiley, c1978.
xiii, 194 p. : ill. ; 23 cm.
Includes index.
ISBN 0-471-99647-5. ISBN 0-471-99685-8 pbk.
1. Chemistry, Organic. I. Title.
QD251.2.C37 547 78-16664
78 MARC

Fessenden, Ralph J 1932-
Organic chemistry / Ralph J. Fessenden and Joan S. Fessenden. — Boston : Willard Grant Press, c1979
xvi, 1040 p. : ill. ; 27 cm.
Includes index.
ISBN 0-87150-724-2
1. Chemistry, Organic. I. Fessenden, Joan S., joint author. II. Title.
QD251.2.F49 547 78-23768
78 MARC

Fuchs, Philip L 1945-
Carbon-13 NMR based organic spectral problems / Philip L. Fuchs, Charles A. Bunnell. — New York : Wiley, [c1979]
309 p. : ill. ; 28 cm.
Includes bibliographical references.
ISBN 0-471-04907-7
1. Nuclear magnetic resonance spectroscopy. 2. Carbon—Isotopes. 3. Chemistry, Organic. I. Bunnell, Charles A., joint author. II. Title.
QD272.S6F83 543'.08 78-20668
78 MARC

Fuentes Mota, José.
Iniciación al análisis estructural orgánico por espectroscopia de resonancia magnética nuclear / [por] José Fuentes Mota ; revisado por José María Viguera Lobo. — Sevilla : Universidad, 1977 printing.
135 p. : ill. ; 22 cm. — (Publicaciones de la Universidad de Sevilla)
Sp78-Apr
Bibliography: p. 135.
ISBN 8474050634 : 150ptas
1. Nuclear magnetic resonance spectroscopy. 2. Chemistry, Organic. I. Title.
QC762.F83 79-384046
79 MARC

Ho, Tse-Lok.
Hard and soft acids and bases principle in organic chemistry / Tse Lok Ho. — New York : Academic Press, 1977.
xii, 209 p. : ill. ; 24 cm.
Includes bibliographical references and indexes.
ISBN 0-12-350050-8
1. Acids. 2. Bases (Chemistry) 3. Chemistry, Organic. 4. Chemical reactions. I. Title.
QD477.H6 547'.1'39 76-13939
77[r79]rev MARC

Iovu, Mircea.
Chimie organică / Mircea Iovu. — Bucureşti : Editura Didactică şi Pedagogică, 1978.
613 p. : ill. ; 25 cm. R79-919
Bibliography: p. 593-[594]
Includes index.
lei21.70
1. Chemistry, Organic. I. Title.
QD251.2.I58 79-375353
*79 MARC

Jordan, Otto, Dr. phil., Chem.
The technology of solvents. Translated by Alan D. Whitehead for Technical Service Library. London, L. Hill [1938]
xx, 351 p. illus., diagrs., tables. 25 cm.
Pages 348-351, advertising matter.
Translation of Chemische Technologie der Lösungsmittel.
1. Solvents. 2. Chemistry, Technical. 3. Chemistry, Organic. I. Title.
TP247.5.J62 1938 661'.807 38-36462
[r79]rev3 MARC

Jordan, Otto, Dr. phil., Chem.
The technology of solvents. Translated by Alan D. Whitehead for Technical Service Library. [2d ed.] London, L. Hill [1940]
xxix, 361 p. illus., diagrs., tables. 25 cm.
Pages 358-361, advertising matter.
Translation of Chemische Technologie der Lösungsmittel.
1. Solvents. 2. Chemistry, Technical. 3. Chemistry, Organic. I. Title.
TP247.5.J62 1940 661'.807 a 42-1815
New York. Public Libr. [r79]rev2 MARC

(Kachestvennyĭ gazokhromatograficheskiĭ analiz)
Качественный газохроматографический анализ / М. С. Вигдергауз, Л. В. Семенченко, В. А. Езрец, Ю. Н. Богословский. — Москва : Наука, 1978.
243 p. : ill. ; 21 cm. USSR 78
At head of title: Akademiiâ nauk SSSR. Institut organicheskoĭ i fizicheskoĭ khimii im. A. E. Arbuzova.
Includes bibliographies.
1.20rub
1. Gas chromatography. 2. Chemistry, Organic. I. Vigdergauz, Mark Solomonovich. II. Akademiiâ nauk SSSR. Institut organicheskoĭ i fizicheskoĭ khimii.
QD272.C44K32 79–407147

Kakáč, Bohumil.
Handbuch der photometrischen Analyse organischer Verbindungen : Ergänzungsband / Bohumil Kakáč, Zdeněk J. Vejdělek. — Weinheim ; New York : Verlag Chemie, 1977-
v. ; 25 cm. GFR7-7-A49
Includes bibliographical references and indexes.
ISBN 3-527-25707-1 (v. 1) : DM138.00 (V. 1)
1. Photometry. 2. Chemistry, Organic. I. Vejdělek, Zdeněk J., joint author. II. Title.
QD272.P5K3415 Suppl. 78-397590
79 MARC

Levsen, Karsten.
Fundamental aspects of organic mass spectrometry / Karsten Levsen. — 1. Aufl. — Weinheim ; New York : Verlag Chemie, 1978.
xiii, 312 p. : ill. ; 25 cm. — (Progress in mass spectrometry ; v. 4)
GFR78-A
Includes bibliographical references and index.
ISBN 0-89573-009-X (New York) : DM88.00
1. Mass spectrometry. 2. Chemistry, Organic. I. Title. II. Series.
QD272.S6L47 547'.35'33 78-320382
79 MARC

Levy, George C
Nitrogen-15 nuclear magnetic resonance spectroscopy / George C. Levy, Robert L. Lichter. — New York : Wiley, c1979.
ix, 221 p. : ill. ; 24 cm.
"A Wiley-Interscience publication."
Includes bibliographical references and index.
ISBN 0-471-02954-8
1. Nuclear magnetic resonance spectroscopy. 2. Nitrogen—Isotopes. 3. Chemistry, Organic. I. Lichter, Robert L., joint author. II. Title.
QD272.S6L475 547'.308 79-4016
79 MARC

Ma, Tsu Sheng, 1911-
Modern organic elemental analysis / T. S. Ma, Robert C. Rittner. — New York : M. Dekker, c1979.
xiv, 518 p. : ill. ; 24 cm.
Includes bibliographical references and indexes.
ISBN 0-8247-6786-1
1. Chemistry, Analytic—Quantitative. 2. Chemistry, Organic. I. Rittner, Robert C., 1924- joint author. II. Title.
QD271.7.M29 547'.34 79-11573
79 MARC

Ma, Tsu Sheng, 1911-
Quantitative analysis of organic mixtures / T. S. Ma, Robert E. Lang. — New York : Wiley, c1979-
v. ; 24 cm.
"A Wiley-Interscience publication."
Includes bibliographical references and indexes.
CONTENTS: pt. 1. General principles.
ISBN 0-471-55800-1
1. Chemistry, Analytic—Quantitative. 2. Chemistry, Organic. I. Lang, Robert E., joint author. II. Title.
QD271.7.M3 547'.35 78-23202
78 MARC

Miller, John A 1937-
Organic chemistry : concepts and applications / John A. Miller, E. F. Neuzil. — Lexington, Mass. : Heath, c1979.
xv, 573 p. : ill. ; 25 cm.
Includes index.
ISBN 0-669-01885-6
1. Chemistry, Organic. I. Neuzil, Edward Florian, 1930- joint author. II. Title.
QD251.2.M54 547 78-59229
79 MARC

Morrison, James D 1936-
Organic chemistry / James D. Morrison. — Belmont, Calif. : Wadsworth Pub. Co., c1979.
xii, 564 p. : ill. ; 24 cm.
Includes index.
ISBN 0-534-00605-1 : $16.95
1. Chemistry, Organic. I. Title.
QD251.2.M669 547 79-924
79 MARC

Organic spectroscopy : an introduction / S. F. Dyke ... [et al.]. — 2d ed. — London ; New York : Longman, 1978.
290 p. : ill. ; 22 cm. GB78
Includes bibliographies and index.
ISBN 0-582-45076-4 : £4.45
1. Spectrum analysis. 2. Chemistry, Organic. I. Dyke, Stanley Frederick.
QD272.S6O72 1978 547'.308'5 78-40163
78 MARC

Ouhadi, T
Étude des propriétés d'oxydes mixtes solubles en milieu organique / par T. Ouhadi, L. Hocks, J. P. Bioul. — Bruxelles : Palais des académies, 1977.
40 p. : ill. ; 25 cm. — (Mémoires de la classe des sciences - Académie royale de Belgique : Collection in-8° ; 2. sér., t. 42, fasc. 4) Be***
Bibliography: p. [39]-40.
1. Complex compounds. 2. Chemistry, Organic. 3. Catalysis. I. Hocks, L., joint author. II. Bioul, J. P., joint author. III. Title. IV. Series: Académie royale des sciences, des lettres, et des beaux-arts de Belgique. Classe des sciences. Mémoires : Collection in-octavo ; 2. sér., t. 42, fasc. 7.
QD474.O95 547'.1'2242 78-388933
79 MARC

Pavia, Donald L
Introduction to spectroscopy : a guide for students of organic chemistry / Donald L. Pavia, Gary M. Lampman, George S. Kriz, Jr. — Philadelphia : W. B. Saunders Co., 1979.
xii, 367 p. : ill. ; 27 cm. — (Saunders golden sunburst series)
Includes bibliographies and index.
ISBN 0-7216-7119-5
1. Spectrum analysis. 2. Chemistry, Organic. I. Lampman, Gary M., joint author. II. Kriz, George S., joint author. III. Title.
QD272.S6P38 543'.085 77-11348
79 MARC

CHEMISTRY, ORGANIC (Continued)

Polarography of molecules of biological significance / edited by W. Franklin Smyth. — London ; New York : Academic Press, 1979.
xix, 326 p. : ill. ; 24 cm. GB***
Includes bibliographical references and index.
ISBN 0-12-653050-5
1. Voltammetry. 2. Chemistry, Organic. I. Smyth, W. Franklin.
QD272.E4P64 547'.308 78-54541
79 MARC

Remane, Horst.
Massenspektrometrie in der organischen Chemie / Horst Remane, Rainer Herzschuh. — 1. Aufl. — Braunschweig : Vieweg, 1977.
185 p. : ill. ; 18 cm. — (Reihe Wissenschaft) GFR78-A
Includes bibliographical references and index.
ISBN 3-528-06838-8 : DM23.80
1. Mass spectrometry. 2. Chemistry, Organic. I. Herzschuh, Rainer, joint author. II. Title.
QD272.S6R45 78-396098
79 MARC

[Shang-hai shih fan ta hsüeh. Yu chi hua hsüeh chih shih pien hsieh tsu]
(Yu chi hua hsüeh chih shih)
有機化學知識 / 上海師範大學編. — 港1版. — 香港：商務印書館香港分館, 1978.
378 p. : ill. ; 19 cm. — (青年科學叢書)
Reprint of the 1976 ed. published by Shang-hai jen min ch'u pan she. Shanghai under the author : Shang-hai shih fan ta hsüeh yu chi hua hsüeh pien hsieh tsu; issued in series: Ch'ing nien tzu hsüeh ts'ung shu.
1. Chemistry, Organic. I. Shang-hai shih fan ta hsüeh. II. Title. III. Series: Ch'ing nien k'o hsüeh ts'ung shu.
QD251.2.S5 1978 79-837650

Siggia, Sidney.
Quantitative organic analysis via functional groups / Sidney Siggia, J. Gordon Hanna. — 4th ed. — New York : Wiley, c1979.
xii, 883 p. : ill. ; 24 cm.
"A Wiley-Interscience publication."
Includes bibliographical references and index.
ISBN 0-471-03273-5
1. Chemistry, Analytic—Quantitative. 2. Chemistry, Organic. I. Hanna, J. Gordon, joint author. II. Title.
QD271.7.S53 1979 547'.35 78-5940
78 MARC

Szerves vegyipari alapfolyamatok kézikönyve / szerk. Deák Gyula. — Budapest : Műszaki Könyvkiadó, 1978.
1052 p. : ill. ; 25 cm. H78
Errata slip inserted.
Includes bibliographical references and index.
ISBN 9631020746 : 252.00Ft
1. Chemistry, Organic. 2. Chemistry, Technical. I. Deák, Gyula.
TP247.S97 79-354350
*79 MARC

Taylor, Giles Aldred.
Organic chemistry for students of biology and medicine / G. A. Taylor. — 2d ed. — London ; New York : Longman, 1978.
vii, 328 p. : ill. ; 22 cm.
Includes index.
ISBN 0-582-44041-6
1. Chemistry, Organic. 2. Biological chemistry. I. Title.
QD251.2.T39 1978 547 77-2193
77 MARC

Ternay, Andrew L
Contemporary organic chemistry / Andrew L. Ternay, Jr. — 2d ed. — Philadelphia : Saunders, 1979.
xxii, 1277, [68] p. : ill. ; 28 cm. — (Saunders golden sunburst series)
Bibliography: p. 1269-1277.
Includes index.
ISBN 0-7216-8795-4
1. Chemistry, Organic. I. Title.
QD251.2.T47 1979 547 77-27757
79 MARC

Tritium in organic chemistry / edited by E. Buncel and C. C. Lee. — Amsterdam ; New York : Elsevier Scientific Pub. Co. ; New York : distributors for the U.S. and Canada, Elsevier North-Holland, 1978.
xiii, 299 p. : ill. ; 25 cm. — (Isotopes in organic chemistry ; v. 4)
Includes bibliographical references and index.
ISBN 0-444-41741-9
1. Tritium. 2. Chemistry, Organic. I. Buncel, E. II. Lee, Choi Chuck, 1924- III. Series.
QD466.5.H1T74 547'.3 78-13388
78 MARC

Turro, Nicholas J 1938-
Modern molecular photochemistry / Nicholas J. Turro. — Menlo Park, Calif. : Benjamin/Cummings Pub. Co., c1978.
628 p. : ill. ; 25 cm.
Includes bibliographical references and index.
ISBN 0-8053-9353-6
1. Photochemistry. 2. Chemical reactions. 3. Chemistry, Organic. I. Title.
QD601.2.T87 541'.35 78-57151
00 MARC

Urushibara, Yoshiyuki, 1901-
(Yūki kagaku)
有機化學 / 漆原義之. — 東京：岩波書店, 昭和9 [1934], 昭和22- [1947- printing.
v. : ill. ; 18 cm. — (岩波全書 ; 29-)
Colophon inserted.
Series romanized: Iwanami zensho.
1. Chemistry, Organic. I. Title.
QD251.U78 79-803960

Weissermel, Klaus.
Industrial organic chemistry : important raw materials and intermediates / Klaus Weissermel, Hans-Jürgen Arpe ; translated by Alexander Mullen. — 1. Aufl. — Weinheim ; New York : Verlag Chemie, 1978.
xii, 404 p. (20 fold.) ; 25 cm. GFR***
Translation of Industrielle organische Chemie.
Includes bibliographical references and index.
ISBN 0-89573-005-7 (New York) : DM74.00
1. Chemistry, Organic. 2. Chemistry, Technical. I. Arpe, Hans-Jürgen, joint author. II. Title.
TP247.W4413 661'.8 78-390041
78 MARC

Weissermel, Klaus.
Industrielle organische Chemie : bedeutende Vor- u. Zwischenprodukte / Klaus Weissermel, Hans-Jürgen Arpe. — 2., überarb. u. erw. Aufl. — Weinheim ; New York : Verlag Chemie, 1978.
xiv, 430 p. ; 25 cm. GFR78-A
Bibliography: p. 400-407.
ISBN 3-527-25756-X : DM64.00
1. Chemistry, Organic. 2. Chemistry, Technical. I. Arpe, Hans-Jürgen, joint author. II. Title.
TP247.W44 1978 661'.8 78-392826
79 MARC

Welcher, Frank Johnson, 1907-
Organic reagents for copper / by Frank J. Welcher, Erwin Boschmann. — 1st ed. — Huntington, N.Y. : R. E. Krieger, 1979.
xvi, 614 p. ; 24 cm.
Includes bibliographies and indexes.
ISBN 0-88275-440-8
1. Chemical tests and reagents. 2. Chemistry, Organic. 3. Copper—Analysis. I. Boschmann, Erwin, joint author. II. Title.
QD77.W417 546'.652'6 76-18141
76 MARC

Wiberg, Kenneth B
Oxidation in organic chemistry, edited by Kenneth B. Wiberg. New York, Academic Press, 1965-78.
3 v. illus. 24 cm. (Organic chemistry; a series of monographs, v. 5, pts. A-C)
Pts. B-C edited by W. S. Trahanovsky.
Includes bibliographical references.
ISBN 0-12-697250-8 (v. 2)
1. Oxidation. 2. Chemistry, Organic. I. Trahanovsky, Walter S., 1938- II. Title. III. Series.
QD281.O9W5 547'.23 65-26047
[r79]rev2 MARC

Wintner, Claude E
Strands of organic chemistry / Claude E. Wintner. — San Francisco : Holden-Day, c1979.
x, 128 p. : ill. ; 23 cm.
Includes index.
ISBN 0-8162-9661-8
1. Chemistry, Organic. I. Title.
QD251.2.W56 016.547 78-60358
79 MARC

Wu, Chang N
Modern organic chemistry / by C. N. Wu. — New York : Barnes & Noble Books, 1979.
2 v. (590, x p.) : ill. ; 21 cm. — (Barnes & Noble outline series)
ISBN 0-06-460172-2 (v. 1) : $4.95 per vol. ($5.95 Can)
1. Chemistry, Organic. I. Title.
QD251.2.W82 1979 547 79-111865
79 MARC

(Yu chi hua hsüeh)
有機化學 / 執筆者劉熾章 ... [et al.] ; 編輯者新學識文教出版中心工專用書編輯委員會. — 台北：該中心, 民國66- [1977-
v. : ill. ; 25 cm.
1. Chemistry, Organic. I. Liu, Ch'ih-chang. II. Hsin hsüeh shih wen chiao ch'u pan chung hsin. Kung chuan yung shu pien chi wei yüan hui.
QD251.2.Y8 78-841685

—ADDRESSES, ESSAYS, LECTURES

Organic chemistry. — Berlin ; New York : Springer-Verlag, 1979.
173 p. : ill. ; 25 cm. — (Topics in current chemistry ; 79)
Includes bibliographies and index.
ISBN 0-387-09301-X
1. Chemistry, Organic—Addresses, essays, lectures. I. Houk, Kendall N., 1943- Theoretical and experimental insights into cycloaddition reactions. 1979. II. Paquette, Leo A. The development of polyquinane chemistry. 1979. III. Series.
QD1.F58 vol. 79 540'.8 s 79-981
[QD255] 79 MARC

—CONGRESSES

Organicheskai͡a khimii͡a i puti razvitii͡a khimicheskikh proizvodstv v Kirgizii : Materialy konf. / [Otv. red. d-r khim. nauk, prof. V. A. Afanas'ev]. — Frunze : Ilim, 1976.
132 p. : ill. ; 20 cm. USSR77
At head of title: Akademii͡a nauk Kirgizskoĭ SSR. Kirgizskoe respublikanskoe pravlenie VKhO im. D. I. Mendeleeva.
1. Chemistry, Organic—Congresses. 2. Chemicals—Manufacture and industry—Kirghizistan—Congresses. I. Afanas'ev, Vitaliĭ Arkad'evich. II. Akademii͡a nauk Kirgizskoĭ SSR, Frunze. III. Vsesoi͡uznoe khimicheskoe obshchestvo imeni D. I. Mendeleeva. Kirgizskoe respublikanskoe pravlenie.
QD241.O69 79-385091
79 MARC

Vsesoi͡uznai͡a konferent͡sii͡a Organicheskie reagenty v analiticheskoĭ khimii, 4th, Kiev, 1976.
(Organicheskie reagenty v analiticheskoĭ khimii)
Органические реагенты в аналитической химии : Тезисы докл. IV всесоюз. конф., Киев, 21–24 апр. 1976 г. / [Ред. коллегия, чл.-кор. АН УССР А. Т. Пилипенко (отв. ред.) ... et al.]. — Киев : Наук. думка, 1976.
2 v. ; 20 cm. USSR 76
At head of title: Akademii͡a nauk SSSR. Nauchnyĭ sovet po probleme Analiticheskai͡a khimii͡a. Institut geokhimii i analiticheskoĭ khimii im. V. I. Vernadskogo, and other institutions.
Includes bibliographies.
0.50rub (v. 1) 0.50rub (v. 2)
1. Chemical tests and reagents—Congresses. 2. Chemistry, Analytic—Congresses. 3. Chemistry, Organic—Congresses. I. Pilipenko, Anatoliĭ Terent'evich. II. Akademii͡a nauk SSSR. Nauchnyĭ sovet po probleme Analiticheskai͡a khimii͡a. III. Akademii͡a nauk SSSR. Institut geokhimii i analiticheskoĭ khimii.
QD77.V75 1976 79-408786

—EXPERIMENTS

Liao, Ch'ing-chiang, fl. 1957-
(Yu chi hua hsüeh shih yen)
有机化学实验 廖清江编著 南京 江苏人民出版社 1958.
302 p. illus. 21 cm.
1. Chemistry, Organic—Experiments. I. Title.
QD261.L47 C 60-1946 ‡
rev

Mohrig, Jerry R
Laboratory experiments in organic chemistry / Jerry R. Mohrig, Douglas C. Neckers. — 3d ed. — New York : D. Van Nostrand Co., c1979.
xiii, 582 p. : ill. ; 25 cm.
Includes bibliographical references and index.
ISBN 0-442-25471-7
1. Chemistry, Organic—Experiments. I. Neckers, Douglas C., joint author. II. Title.
QD257.5.M63 1979 547'.0028 79-112508
00 MARC

—HANDBOOKS, MANUALS, ETC.

Budde, William L
Organics analysis using gas chromatography/mass spectometry : a techniques & procedures manual / by William L. Budde and James W. Eichelberger. — Ann Arbor, Mich. : Ann Arbor Science Publishers, 1979.
xxvii, 242 p. : ill. ; 24 cm.
Bibliography: p. 229-235.
Includes index.
ISBN 0-250-40318-8
1. Gas chromatography—Handbooks, manuals, etc. 2. Mass spectrometry—Handbooks, manuals, etc. 3. Chemistry, Organic—Handbooks, manuals, etc. I. Eichelberger, James W., joint author. II. Title.
QD272.C44B82 547'.308 79-88484
00 MARC

CHEMISTRY, ORGANIC (Continued)

—HISTORY—ADDRESSES, ESSAYS, LECTURES

Ocherki po istorii organicheskoĭ khimii : ⌈Sbornik stateĭ⌉ / AN SSSR, In-t istorii estestvoznaniîa i tekhniki ; ⌈Otv. red. d-r khim. nauk G. V. Bykov⌉. — Moskva : Nauka, 1977.
198 p. ; 21 cm. USSR77
Includes bibliographies and index.
1.00rub
1. Chemistry, Organic—History—Addresses, essays, lectures. I. Bykov, Georgiĭ Vladimirovich. II. Akademiîa nauk SSSR. Institut istorii estestvoznaniîa i tekhniki.
QD248.O23 79-378488
79 MARC

—LABORATORY MANUALS

Adams, Roger, 1889-1971.
Laboratory experiments in organic chemistry / Roger Adams, John R. Johnson, Charles F. Wilcox. — 7th ed. — New York : Macmillan, c1979.
xviii, 538 p. : ill. ; 24 cm.
First published in 1928 under title: Elementary laboratory experiments in organic chemistry.
Bibliography: p. 505-517.
Includes index.
ISBN 0-02-300590-4
1. Chemistry, Organic—Laboratory manuals. I. Johnson, John Raven, 1900- joint author. II. Wilcox, Charles F., 1930- joint author. III. Title.
QD261.A2 1979 547'.0028 78-4482
78 MARC

Ault, Addison.
Techniques and experiments for organic chemistry / Addison Ault. — 3d ed. — Boston : Allyn and Bacon, c1979.
xiv, 442 p. : ill. ; 28 cm.
Includes bibliographical references and indexes.
ISBN 0-205-06528-7 : $15.95
1. Chemistry, Organic—Laboratory manuals. I. Title.
QD261.A94 1979 547'.0028 78-23920
78 MARC

Fieser, Louis Frederick, 1899-
Organic experiments / Louis F. Fieser, Kenneth L. Williamson. — 4th ed. — Lexington, Mass. : D. C. Heath, c1979.
ix, 418 p. : ill. ; 25 cm.
Includes bibliographical references and indexes.
ISBN 0-669-01688-8
1. Chemistry, Organic—Laboratory manuals. I. Williamson, Kenneth L., joint author. II. Title.
QD261.F5 1979 547'.0028 78-53750
79 MARC

Kemp, William, 1932-
Qualitative organic analysis / William Kemp. — Rev. ed. — London ; New York : McGraw-Hill Book Co., c1979.
xi, 182 p. : ill. ; 23 cm. GB79
Includes bibliographical references and index.
ISBN 0-07-084092-X : £4.85
1. Chemistry, Analytic—Qualitative—Laboratory manuals. 2. Chemistry, Organic—Laboratory manuals. I. Title.
QD271.4.K45 1979 547'.34'0028 78-41217
79 MARC

Loewenthal, H J E
Guide for the perplexed organic experimentalist / ⌈by⌉ H. J. E. Loewenthal. — London ; Philadelphia : Heyden, 1978.
x, 174 p. : ill. ; 20 cm. GB78-39614
Includes bibliographical references and index.
ISBN 0-85501-169-6 : £4.80
1. Chemistry, Organic—Laboratory manuals. I. Title.
QD261.L63 542 79-305525
79 MARC

Mann, Frederick George.
Practical organic chemistry / by Frederick George Mann and Bernard Charles Saunders. — 4th ed. — London ; New York : Longman, 1978, c1960.
xix, 587 p. : ill. ; 22 cm. GB78
Includes index.
ISBN 0-582-44407-1 : £5.95
1. Chemistry, Organic—Laboratory manuals. I. Saunders, Bernard Charles, joint author. II. Title.
QD261.M25 1978 547'.0028 78-40612
78 MARC

Modern experimental organic chemistry / Royston M. Roberts ... ⌈et al.⌉. — 3d ed. — New York : Holt, Rinehart and Winston, c1979.
x, 603 p. : ill. ; 24 cm.
Second ed. published in 1974 under title: An introduction to modern experimental organic chemistry.
Includes bibliographical references and index.
ISBN 0-03-044391-1
1. Chemistry, Organic—Laboratory manuals. I. Roberts, Royston M. II. Title: An introduction to modern experimental organic chemistry.
QD261.I5 1979 547'.0028 78-10752
78 MARC

Nekrasov, Vsevolod Vladimirovich.
(Rukovodstvo k malomu praktikumu po organicheskoĭ khimii)
Руководство к малому практикуму по органической химии / В. В. Некрасов. — 5-е изд., доп. — Москва : Химия, 1975.
328 p. ; ill. ; 22 cm. USSR 76
"Допущено ... в качестве учебного пособия для нехимических специальностей вузов."
Includes index.
1.01rub
1. Chemistry, Organic—Laboratory manuals. I. Title.
QD261.N4 1975 78-421187

Pasto, Daniel J 1936-
Laboratory text for organic chemistry : a source book of chemical and physical techniques / Daniel J. Pasto, Carl R. Johnson. — Englewood Cliffs, N.J. : Prentice-Hall, c1979.
xvi, 542 p. : ill. ; 24 cm.
Includes bibliographies and index.
ISBN 0-13-521302-9
1. Chemistry, Organic—Laboratory manuals. I. Johnson, Carl R., 1937- joint author. II. Title.
QD261.P36 1979 547'.0028 78-16528
78 MARC

Todd, David.
Experimental organic chemistry / David Todd. — Englewood Cliffs, N.J. : Prentice-Hall, c1979.
xxxiii, 347 p. : ill. ; 24 cm.
Includes bibliographical references and index.
ISBN 0-13-294660-2
1. Chemistry, Organic—Laboratory manuals. I. Title.
QD261.T63 547'.0028 78-14730
78 MARC

Vogel, Arthur Israel.
Vogel's Practical organic chemistry, including qualitative organic analysis. — 4th ed. / rev. by the following members of the School of Chemistry, Thames (formerly Woolwich) polytechnic, B. S. Furniss ...⌈et al.⌉. — London ; New York : Longman, 1978.
xxxv, 1368 p. : ill. ; 24 cm.
Third ed. published in 1956 under title: A text-book of practical organic chemistry, including qualitative organic analysis.
Includes bibliographies and index.
ISBN 0-582-44250-8
1. Chemistry, Organic—Laboratory manuals. 2. Chemistry, Analytic—Qualitative. I. Furniss, Brian Stanley. II. Title. III. Title: Practical organic chemistry, including qualitative organic analysis.
QD261.V63 1978 547 77-23559
77 MARC

—NOMENCLATURE

Handbuch zur Anwendung der Nomenklatur organisch-chemischer Verbindungen / hrsg. von Wolfgang Liebscher, unter Mitarb. von P. Becker ... ⌈et al.⌉. — Berlin : Akademie-Verlag, 1979.
xix, 1076 p. : ill. ; 25 cm. GDR79-A
"Errata": ⌈3⌉ p. inserted.
Bibliography: p. 949-963.
Includes indexes.
135.00M
1. Chemistry, Organic—Nomenclature. I. Liebscher, Wolfgang.
QD291.H29 79-379597
79 MARC

—PROBLEMS, EXERCISES, ETC.

Ćwiczenia specjalistyczne z chemii organicznej : skrypt dla studentów IV roku chemii : praca zbiorowa / pod red. W. Z. Antkowiaka i K. Szyftera. — Wyd. 1. — Poznań : Wydawn. Uczelniane Uniwersytetu im. Adama Mickiewicza, 1976.
173 p. : ill. ; 25 cm. P***
At head of title: Uniwersytet im. Adama Mickiewicza w Poznaniu.
Includes bibliographies.
zł17.00
1. Chemistry, Organic—Problems, exercises, etc. I. Antkowiak, Wiesław Z. II. Szyfter, Krzysztof. III. Uniwersytet im. Adama Mickiewicza w Poznaniu.
QD257.C87 79-337925
79 MARC

Gibson, Gerald W
Mastering organic chemistry : a problem-solving approach / Gerald W. Gibson. — Philadelphia : Saunders, c1979.
vii, 563 p. : ill. ; 27 cm. — (Saunders golden sunburst series)
Includes index.
ISBN 0-7216-4113-3
1. Chemistry, Organic—Problems, exercises, etc. I. Title.
QD257.G5 547'.0076 79-105702
79 MARC

(Posobie dlîa samokontrolîa v praktikume po organicheskoĭ khimii)
Пособие для самоконтроля в практикуме по органической химии : ⌈Для фармац. ин-тов и фармац. фак. мед. ин-тов / В. В. Зеленкова, Э. М. Казьмина, Н. И. Калетина, О. Г. Сердюк⌉ ; Под ред. Б. Н. Степаненко. — Москва : Высш. школа, 1978.
272 p. ; 22 cm. USSR 78
0.70rub
1. Chemistry, Organic—Problems, exercises, etc. I. Zelenkova, Vera Vasil'evna. II. Stepanenko, Boris Nikolaevich.
QD257.P67 79-412068

—PROGRAMMED INSTRUCTION

Gordon, John E 1931-
How to succeed in organic chemistry / John E. Gordon. — New York : Wiley, c1979.
xiv, 594 p. : ill. ; 26 cm. — (Wiley self-teaching guides)
Bibliography: p. viii-ix.
Includes index.
ISBN 0-471-03010-4
1. Chemistry, Organic—Programmed instruction. I. Title.
⌈QD251.2.G67⌉ 547'.007'7 78-21496
78 MARC

—SYNTHESIS

see also Biosynthesis; Plastics

Carbon-carbon bond formation / edited by Robert L. Augustine. — New York : M. Dekker, c1979-
v. : ill. ; 24 cm. — (Techniques and applications in organic synthesis)
Includes bibliographical references and index.
ISBN 0-8247-6787-X
1. Chemistry, Organic—Synthesis. 2. Chemical bonds. I. Augustine, Robert L., 1932-
QD262.C29 547'.2 78-31483
78 MARC

Carruthers, W
Some modern methods of organic synthesis / W. Carruthers. — 2d ed. — Cambridge ; New York : Cambridge University Press, 1978.
xii, 532 p. : ill. ; 23 cm. — (Cambridge texts in chemistry and biochemistry)
Bibliography: p. 496-518.
Includes index.
ISBN 0-521-21715-6. ISBN 0-521-29241-7 pbk.
1. Chemistry, Organic—Synthesis. I. Title. II. Series.
QD262.C33 1978 547'.2 77-77735
77 MARC

Freifelder, Morris, 1907-
Catalytic hydrogenation in organic synthesis : procedures and commentary / Morris Freifelder. — New York : Wiley, c1978.
xiv, 191 p. : ill. ; 24 cm.
"A Wiley-Interscience publication."
Includes bibliographical references and index.
ISBN 0-471-02945-9
1. Hydrogenation. 2. Catalysis. 3. Chemistry, Organic—Synthesis. I. Title.
QD281.H8F73 547'.23 78-9458
78 MARC

Gudriniece, Emīlija, 1920-
Organisko savienojumu sintēzes metodes / E. Gudriniece, A. Kārkliņa, I. Strakova. — Rīga : Zvaigzne, 1976.
432 p. ; 22 cm. USSR76-42098
Bibliography: p. 409-⌈410⌉
Includes index.
1.10rub
1. Chemistry, Organic—Synthesis. I. Kārkliņa, Aina, 1941- joint author. II. Strakova, Inta, joint author. III. Title.
QD262.G8 79-378950
*79 MARC

Hajós, Andor.
Complex hydrides and related reducing agents in organic synthesis / by Andor Hajós. — Amsterdam ; New York : Elsevier Scientific Pub. Co. ; New York : distributed by Elsevier/North-Holland, 1979.
398 p. : ill. ; 25 cm. — (Studies in organic chemistry ; 1)
Includes bibliographical references and indexes.
ISBN 0-444-99791-1
1. Chemistry, Organic—Synthesis. 2. Reduction, Chemical. 3. Hydrides. I. Title. II. Series: Studies in organic chemistry (Amsterdam) ; 1.
QD262.H24 547'.2 78-14475
78 MARC

Higashi, Tsuneto.
(Yūki gōsei chūshutsu hō)
有機合成·抽出法 / 東恒人. — 東京：岩波書店，昭和11 ⌈1936⌉，昭和24 ⌈1949⌉ printing.
20, 244, 9 p. : ill. ; 18 cm. — (岩波全書 ; 78)
Colophon inserted.
Series romanized: Iwanami zensho.
Bibliography: p. ⌈4⌉-8 (1st group)
Includes index.
1. Chemistry, Organic—Synthesis. 2. Extraction (Chemistry) I. Title.
QD262.H53 79-803628

Kuo, Chih-liang.
(Ho ch'eng yu chi kung yeh hua hsüeh)
合成有機工業化學 / 郭質良編著. — 臺北：臺灣商務印書館，民國67 ⌈1978⌉
2, 10, 371 p. : ill. ; 22 cm. — (大學叢書)
Series romanized: Ta hsüeh ts'ung shu.
NT$94.00

1. Chemistry, Organic—Synthesis. I. Title.
QD262.K85 79-836467

CHEMISTRY, ORGANIC
—SYNTHESIS (Continued)

Lebedev, Nikolaĭ Nikolaevich, fl. 1956–
(Khimiia i tekhnologiia osnovnogo organicheskogo i neftekhimicheskogo sinteza)
Химия и технология основного органического и нефтехимического синтеза : ₍учебник для хим.-технол. специальностей вузов₎ / Н. Н. Лебедев. — Изд. 2-е, перераб. — Москва : Химия, 1975.
732 p. : ill. ; 22 cm. USSR 75
Bibliography: p. 709–₍710₎
Includes index.
1.94rub
1. Chemistry, Organic—Synthesis. 2. Petroleum chemicals. 3. Chemical processes. I. Title.
TP247.L42 1975 79–411223

Maag, Hans, 1945-
Totalsynthese von Vitamin B12 : Dicyano-Co(III)-Cobyrinsäure-Hexamethylester-f-Amid / vorgelegt von Hans Maag. — Zürich : Juris-Verlag, 1973.
265 p. : ill. ; 23 cm. Sw74-A-735
"Diss. Nr. 5173."
Thesis—Eidgenössische Technische Hochschule, Zürich.
Vita.
Includes bibliographical references.
ISBN 3-260-03588-5 : 48.00F
1. Vitamin B12. 2. Chemistry, Organic—Synthesis. I. Title.
QP772.C9M3 78-393718
*79 MARC

Nefedov, Boris Kostantinovich.
Sintezy organicheskikh soedineniĭ na osnove okisi ugleroda / B. K. Nefedov. — Moskva : Nauka, 1978.
222 p. : ill. ; 21 cm. USSR78
At head of title: Akademiia nauk SSSR. Institut organicheskoĭ khimii im. N. D. Zelinskogo.
Includes bibliographical references.
2.40rub
1. Chemistry, Organic—Synthesis. 2. Carbon monoxide. I. Title.
QD262.N42 79-346343
79 MARC

Norman, Richard Oswald Chandler.
Principles of organic synthesis / R. O. C. Norman. — 2d ed. — London : Chapman and Hall ; New York : Wiley, 1978.
xiii, 800 p. : ill. ; 24 cm.
"A Halsted Press book."
Includes bibliographies and index.
ISBN 0-470-26317-2
1. Chemistry, Organic—Synthesis. I. Title.
QD262.N6 1978 547'.2 78-784
78 MARC

Nucleic acid chemistry : improved and new synthetic procedures, methods, and techniques / edited by Leroy B. Townsend, R. Stuart Tipson. — New York : Wiley, c1978.
2 v. (xv, 1067, 55 p.) : ill. ; 24 cm.
"A Wiley-Interscience publication."
"Successor to volume 1 of Synthetic Procedures in nucleic acid chemistry, edited by W. W. Zorbach and R. S. Tipson."
Includes indexes.
ISBN 0-471-88090-6 (part 1)
1. Chemistry, Organic—Synthesis. 2. Nucleic acids. I. Townsend, Leroy B. II. Tipson, R. Stuart.
QD262.N8 547'.596 77-22816
77 MARC

Organic photochemical syntheses, volume 1 / editor, R. Srinivasan, associate editor T. D. Roberts, with an introductory section on experimental methods in organic photochemical syntheses by R. Srinivasan. — Huntington, N.Y. : R. E. Krieger Pub. Co., 1979, c1971.
x, 108 p. : ill. ; 24 cm.
Reprint of the ed. published by Wiley-Interscience, New York.
Includes bibliographical references and index.
ISBN 0-88275-852-7
1. Chemistry, Organic—Synthesis. 2. Photochemistry. I. Srinivasan, Rangaswamy, 1929- II. Roberts, Thomas D.
₍QD262.O65 1979₎ 547'.1'35 78-27287
78 MARC

(Praktikum po organicheskomu sintezu)
Практикум по органическому синтезу : ₍для пед. ин-тов₎. — Москва : Просвещение, 1974.
239 p. : ill. ; 22 cm. USSR 74
0.54rub
1. Chemistry, Organic—Synthesis.
QD262.P7 79–411224

Rylander, Paul Nels, 1920-
Catalytic hydrogenation in organic syntheses / Paul Rylander. — New York : Academic Press, 1979.
x, 325 p. : ill. ; 24 cm.
Includes bibliographies and index.
ISBN 0-12-605355-3
1. Hydrogenation. 2. Catalysis. 3. Chemistry, Organic—Synthesis. I. Title.
QD281.H8R89 547'.23 79-9711
79 MARC

Sadykov, Abid Sadykovich.
(Sintezy organicheskikh preparatov piridinovogo riada)
Синтезы органических препаратов пиридинового ряда / А. С. Садыков, О. С. Отрощенко, Ю. В. Курбатов. — Ташкент : Фан, 1974.
58 p. ; 22 cm. USSR***
At head of title: Akademiia nauk Uzbekskoĭ SSR. Otdelenie khimiko-tekhnologicheskikh nauk.
Includes bibliographies.
0.35rub
1. Pyridine. 2. Chemistry, Organic—Synthesis. I. Otroshchenko, Ol'ga Sergeevna, joint author. II. Kurbatov, IUriĭ Vasil'evich, joint author. III. Title.
QD401.S134 79–414103

Suvorov, Boris Viktorovich.
Okislitel'nye reaktsii v organicheskom sinteze / B. V. Suvorov, N. R. Bukeĭkhanov. — Moskva : Khimiia, 1978.
197 p. : ill. ; 21 cm. USSR78
Includes bibliographies and index.
0.65rub
1. Oxidation. 2. Chemistry, Organic—Synthesis. I. Bukeĭkhanov, Nurym Raimzhanovich, joint author. II. Title.
QD281.O9S94 79-372542
79 MARC

Szántay, Csaba.
Synthesis of prostaglandins / by Cs. Szántay and L. Novák. — Budapest : Akadémiai Kiadó : distributed by Kultura, 1978.
267 p. : ill. ; 25 cm. — (Recent developments in the chemistry of natural carbon compounds ; v. 8) H***
Includes bibliographical references and indexes.
ISBN 963051303X
1. Prostaglandin synthesis. 2. Chemistry, Organic—Synthesis. I. Novák, Lajos, joint author. II. Title. III. Series.
QP801.P68S9 547'.43'7 78-318873
78 MARC

Transition metal organometallics in organic synthesis / edited by Howard Alper. — New York : Academic Press, 1976-1978.
2 v. : ill. ; 23 cm. — (Organic chemistry, a series of monographs ; v. 33)
Includes bibliographies and indexes.
ISBN 0-12-053101-1 (v. 1)
1. Chemistry, Organic—Synthesis. 2. Organometallic compounds. 3. Transition metal compounds. I. Alper, Howard. II. Series.
QD262.T7 547'.2 75-40604
76₍r79₎rev MARC

—SYNTHESIS—ADDRESSES, ESSAYS, LECTURES

Issledovaniia v oblasti sinteza i kataliza organicheskikh soedineniĭ : ₍sbornik₎ / Pod red. prof. V. G. Kharchenko i prof. IU. N. Usova. — Saratov : Izd-vo Sarat. un-ta, 1975.
68 p. ; 20 cm. USSR75
At head of title: Saratovskiĭ gosudarstvennyĭ universitet im. N. G. Chernyshevskogo. Nauchno-issledovatel'skiĭ institut khimii.
Includes bibliographies.
1. Chemistry, Organic—Synthesis—Addresses, essays, lectures. 2. Catalysis—Addresses, essays, lectures. I. Kharchenko, V. G. II. Usov, IU. N. III. Saratov, Russia (City). Universitet. Nauchno-issledovatel'skiĭ institut khimii.
QD262.I84 79-383930
79 MARC

—SYNTHESIS—COLLECTED WORKS

Macromolecular syntheses. Collective volume. 1– 1977–
New York, Wiley.
v. ill. 24 cm.
"A periodic publication of methods for the preparation of macromolecules."
Key title: Macromolecular syntheses. Collective volume,
ISSN 0163-6189
1. Macromolecules—Collected works. 2. Chemistry, Organic—Synthesis—Collected works.
QD380.M3 547'.7'05 78-648294
MARC-S

—SYNTHESIS—CONGRESSES

Stereoselective synthesis of natural products : proceedings of the seventh workshop conference Hoechst, Schloss Reisensburg, 24-27 September 1978 / editors, W. Bartmann, E. Winterfeldt. — Amsterdam : Excerpta Medica ; New York : sole distributors for the USA and Canada, Elsevier North-Holland, 1979.
216 p. : ill. ; 25 cm. — (Workshop conferences Hoechst ; v. 7)
Includes bibliographical references and indexes.
ISBN 0-444-90067-5 (Elsevier)
1. Chemistry, Organic—Synthesis—Congresses. 2. Natural products—Congresses. I. Bartmann, Wilhelm, 1928- . II. Winterfeldt, Ekkehard. III. Series.
R106.W6 vol. 7 610'.8 s 78-31902
₍QD262₎ 79 MARC

—SYNTHESIS—PROBLEMS, EXERCISES, ETC.

Alcudia, Felipe.
Problemas en síntesis orgánica / ₍por₎ F. Alcudia J. L. García Ruano, P. Prados. — 1. ed. — Madrid : Alhambra, D.L. 1978.
xiii, 204 p. ; 24 cm. — (Vertix 40 : Sección I, Química ; 17) Sp78-Oc
Bibliography: p. 201-204.
ISBN 8420503967 : 580ptas
1. Chemistry, Organic—Synthesis—Problems, exercises, etc. I. García Ruano, José Luis, joint author. II. Prados, Pilar, joint author. III. Title.
QD262.A469 79-366554
79 MARC

—SYNTHESIS—PROGRAMMED INSTRUCTION

Warren, Stuart G
Designing organic syntheses : a programmed introduction to the synthon approach / Stuart Warren. — Chichester ; New York : Wiley, c1978.
285 p. ; 22 cm.
ISBN 0-471-99612-2
1. Chemistry, Organic—Synthesis—Programmed instruction. I. Title.
QD262.W28 547'.2 77-15479
77₍r79₎rev MARC

—TABLES

Pedley, J B
Sussex-N.P.L. computer analysed thermochemical data : organic and organometallic compounds / J. B. Pedley and J. Rylance. — Brighton : University of Sussex, 1977.
9, ca. 200 p. ; 21 x 31 cm. GB***
Cover title.
Chiefly tables.
Includes bibliography.
£9.00
1. Thermochemistry—Tables, etc. 2. Chemistry, Organic—Tables, etc. 3. Organometallic compounds—Tables, etc. I. Rylance, J., joint author. II. Title.
QD511.8.P42 547'.1'36 79-306394
79 MARC

CHEMISTRY, PHARMACEUTICAL

see also Drugs; Materia medica; Pharmacy; Solutions (Pharmacy); Structure-activity relationship (Pharmacology)

Arzneimittelentwicklung : Grundlagen, Strategien, Perspektiven / hrsg. von E. Kutter ; mit Beitr. von V. Austel ... ₍et al.₎. — Stuttgart : Thieme, 1978.
xiii, 194 p. : ill. ; 25 cm. GFR78-A
Includes bibliographical references and index.
ISBN 3-13-563201-6
1. Structure-activity relationship (Pharmacology) 2. Drugs—Physiological effect. 3. Chemistry, Pharmaceutical. I. Kutter, E.
QP906.S75A79 78-397196
79 MARC

Chromatografia na tenkých vrstvách vo farmácii a v klinickej biochémii / ₍aut.₎ Magda Šaršúnová ... ₍et al.₎. — 2., preprac. a dopl. vyd. — Martin : Osveta, 1977.
517 p. : ill. ; 25 cm. — (Farmaceutická edícia ; zv. 13) CzS78
Czech or Slovak.
Includes bibliographies and indexes.
Kčs53.00
1. Drugs—Analysis. 2. Thin layer chromatography. 3. Chemistry, Pharmaceutical. I. Šaršúnová, Magda.
RS189.C57 1977 79-380709
*79 MARC

Lespagnol, Albert.
Chimie des médicaments / par Albert Lespagnol ; avec la collaboration de André Cœur ... ₍et al.₎. — Paris : Entreprise moderne d'édition, ₍1974-1975₎
3 v. : ill. ; 25 cm. F***
Includes bibliographies and index.
ISBN 2-88020-004-0 : 193F (v. 1)
1. Chemistry, Pharmaceutical. I. Title.
RS403.L45 615'.19 75-505409
75₍r79₎rev MARC

Lespagnol, Albert.
Précis de pharmacie chimique usuelle à l'usage des pharmaciens et étudiants en pharmacie : fascicules 1, 2, 3 réunis / Albert Lespagnol, ... ; avec la collaboration des Prof. Charles Lespagnol et Daniel Lesieur. — Paris : Technique et documentation, 1977.
588 p. ; 25 cm. F78-4603
Includes index.
ISBN 2-85206-018-3 : 135F
1. Chemistry, Pharmaceutical. I. Lespagnol, Charles, joint author. II. Lesieur, Daniel, joint author. III. Title.
RS421.L46 615'.19 78-375881
78 MARC

Végh, Antal.
Gyógyszerészi kémia / Végh Antal, Szász György, Takács Mihály. — 2. változatlan kiad. — Budapest : Medicina, 1977.
711 p. ; 24 cm. H***
Includes index.
ISBN 9632404637 : 124.00Ft
1. Chemistry, Pharmaceutical. I. Szász, György, joint author. II. Takács, Mihály, joint author. III. Title.
RS403.V43 1977 78-387905
79 MARC

CHEMISTRY, PHARMACEUTICAL (Continued)

Wagner, Günther, 1925-
Pharmazeutische Chemie / Günther Wagner, Hans Kühmstedt ; ⌐verantwortl. Hrsg. dieses Bd., E. Leibnitz⌐. — 4., bearb. u. erw. Aufl. — Berlin : Akademie-Verlag, 1978.
290 p. : ill. ; 18 cm. — (Wissenschaftliche Taschenbücher ; Bd. 7 : Reihe Chemie) GFR78-A
Includes bibliographical references and index.
12.50M
1. Chemistry, Pharmaceutical. I. Kühmstedt, Hans, joint author. II. Title.
RS403.W28 1978 78-396088
79 MARC

—ABSTRACTS—PERIODICALS

Chile. Universidad, Santiago. Facultad de Ciencias Químicas.
Anales de la Facultad de Ciencias Químicas. t. 22/25- 1970/73-
Santiago, Universidad de Chile - Sede Norte.
v. in 26 cm. annual.
"Tesis de químico farmacéuticos, bioquímicos y químicos (Resúmenes)"
Vol. for 1970/73 includes Anales for 1970-71 of the faculty under its earlier name: Facultad de Química y Farmacia.
Continues: Chile. Universidad, Santiago. Facultad de Química y Farmacia. Anales.
English and Spanish.

1. Chemistry, Pharmaceutical—Abstracts—Periodicals. 2. Biological chemistry—Abstracts—Periodicals. 3. Chemistry—Abstracts—Periodicals. I. Chile. Universidad, Santiago. Facultad de Química y Farmacia. Anales.
RS21.C5 78-644597
MARC-S

—ADDRESSES, ESSAYS, LECTURES

Lespagnol, Albert.
Le Médicament, ange ou démon ... / Albert Lespagnol, ... — Paris : Technique et documentation, 1977.
xi, 248 p. ; 22 cm. F78-720
On cover: La parole est à la défense.
Includes bibliographical references.
ISBN 2-85206-017-5 : 55.00F
1. Pharmacy—Information services—Addresses, essays, lectures. 2. Drugs—Addresses, essays, lectures. 3. Chemistry, Pharmaceutical—Addresses, essays, lectures. I. Title. II. Title: La parole est à la défense.
RS56.L47 615.1 79-338059
79 MARC

—BIBLIOGRAPHY

see also Information storage and retrieval systems—Chemistry, Pharmaceutical

—COLLECTED WORKS

Drug design. Edited by E. J. Ariëns. New York, Academic Press, 1971-
v. illus. 24 cm. (Medicinal chemistry; a series of monographs, v. 11)
Includes bibliographies.
ISBN 0-12-060301-2 (v. 1)
1. Pharmacology—Collected works. 2. Chemistry, Pharmaceutical—Collected works. I. Ariëns, Everhardous Jacobus, ed. II. Series.
RM300.D74 615'.1 72-127678
72⌐r79⌐rev MARC

—CONGRESSES

Cannabinoid analysis in physiological fluids : based on a symposium sponsored by the Division of Analytical Chemistry at the 173rd meeting of the American Chemical Society, New Orleans, Louisiana, March 20-25, 1977 / Joe A. Vinson, editor. — Washington : The Society, 1979.
ix, 2242 p. : ill. ; 24 cm. — (ACS symposium series ; 98 ISSN 0097-6156)
Includes bibliographies and index.
ISBN 0-8412-0488-8
1. Tetrahydrocannabinol—Analysis—Congresses. 2. Tetrahydrocannabinol—Metabolism—Congresses. 3. Chemistry, Pharmaceutical—Congresses. I. Vinson, Joe A. II. American Chemical Society. Division of Analytical Chemistry. III. Series: American Chemical Society. ACS symposium series ; 98.
RS431.T44C36 615'.323'962 79-10934
79 MARC

The Chemistry of radiopharmaceuticals / edited by Ned D. Heindel ... ⌐et al.⌐. — New York : Masson Pub. USA, c1978.
x, 294 p. : ill. ; 24 cm. — (Cancer management)
"Based on a symposium held April 23-24, 1976, at Hahnemann Medical College."
Includes bibliographical references and index.
ISBN 0-89352-019-5
1. Radiopharmaceuticals—Congresses. 2. Chemistry, Pharmaceutical—Congresses. 3. Radioisotope scanning—Congresses. 4. Radiopharmaceuticals—Research—Congresses. I. Heindel, Ned D. II. Hahnemann Medical College and Hospital of Philadelphia. III. Series.
RS431.R34C46 616.07'575 77-94827
78 MARC

Diuretic agents : based on a symposium sponsored by the ACS Division of Medicinal Chemistry at the 174th meeting of the American Chemical Society, Chicago, Illinois, August 29, 1977 / Edward J. Cragoe, Jr., editor. — Washington : The Society, 1978.
ix, 238 p. : ill. ; 24 cm. — (ACS symposium series ; 83 ISSN 0097-6156)
Includes bibliographical references and index.
ISBN 0-8412-0464-0
1. Diuretics and diuresis—Congresses. 2. Chemistry, Pharmaceutical—Congresses. I. Gragoe, Edward J. II. American Chemical Society. Division of Medicinal Chemistry. III. Series: American Chemical Society. ACS symposium series ; 83.
RS431.D58D58 615'.761 78-23405
78 MARC

(Khimii͡a i farmakologii͡a indol'nykh soedinenii)
Химия и фармакология индольных соединений : тезисы докладов IV Всесоюзного коллоквиума, май 1975 г. / ⌐под ред. Г. В. Лазурьевского⌐. — Кишинев : Штиинца, 1975.
107 p. ; 22 cm. USSR***
At head of title: Akademii͡a nauk SSSR. Otdelenie biokhimii, biofiziki i khimii fiziologicheski aktivnykh soedinenii. Akademii͡a nauk MSSR. Institut khimii. Moldavskoe otdelenie VKhO im. D. I. Mendeleeva.
0.79rub

1. Chemistry, Pharmaceutical—Congresses. 2. Indole—Therapeutic use—Congresses. I. Lazur'evskii, Georgii Vasil'evich. II. Akademii͡a nauk SSSR. Otdelenie biokhimii, biofiziki i khimii fiziologicheski aktivnykh soedinenii. III. Akademii͡a de Shtiint͡se a RSSM. Institutul de Kimie. IV. Vsesoi͡uznoe khimicheskoe obshchestvo imeni D. I. Mendeleeva. Moldavskoe otdelenie.
RS421.K45 78-421135

—DATA PROCESSING

Golender, Valerii Efimovich.
Vychislitel'nye metody konstruirovanii͡a lekarstv / V. E. Golender, A. B. Rozenblit ; ⌐Avt. predisl. L. A. Piruzi͡an, V. V. Avidon⌐. — Riga : Zinatne, 1978.
232 p. : ill. ; 21 cm. USSR78
At head of title: Akademii͡a nauk Latviiskoi SSR. Institut organicheskogo sinteza.
Bibliography: p. ⌐211⌐-221.
Includes indexes.
1.20rub
1. Chemistry, Pharmaceutical—Mathematical models. 2. Chemistry, Pharmaceutical—Data processing. 3. Structure-activity relationship (Pharmacology)—Mathematical models. 4. Structure-activity relationship (Pharmacology)—Data processing. I. Rozenbilt, Anatolii Borisovich, joint author. II. Title.
RS403.G64 79-380083
79 MARC

—DATA PROCESSING—CONGRESSES

Symposium on Retrieval of Medicinal Chemical Information, Anaheim, Calif., 1978.
Retrieval of medicinal chemical information / W. Jeffrey Howe, editor, Margaret M. Milne, editor, Ann F. Pennell, editor. — Washington : American Chemical Society, 1978.
viii, 231 p. ; 24 cm. — (ACS symposium series ; 84 ISSN 0097-6156)
"Based on a symposium cosponsored by the Divisions of Computers in Chemistry and Chemical Information at the 175th meeting of the American Chemical Society, California, March 13-17, 1978."
Includes bibliographies and index.
ISBN 0-8412-0465-9
1. Information storage and retrieval systems—Chemistry, Pharmaceutical—Congresses. 2. Chemistry, Pharmaceutical—Data processing—Congresses. I. Howe, William Jeffrey, 1946- II. Milne, Margaret M., 1948- III. Pennell, Ann F., 1946- IV. American Chemical Society. Division of Computers in Chemistry. V. American Chemical Society. Division of Chemical Information. VI. Title. VII. Series: American Chemical Society. ACS symposium series ; 84.
RS421.S93 1978 615'.19'02854 78-21611
78 MARC

—DICTIONARIES

Peristilahan kimia dan farmasi. — Cet. 9. — ⌐Bandung⌐ : Penerbit Universitas ITB, 1976.
x, 288 p. ; 21 cm.
Compiled by M. M. Purbo-Hadiwidjoyo and S. A. Waworoentoe.
English and Indonesian.
1. Chemistry—Dictionaries. 2. Chemistry, Pharmaceutical—Dictionaries. 3. English language—Dictionaries—Indonesian. I. Purbo Hadiwidjoyo, M. M. II. Waworoentoe, S. A.
QD5.P43 1976 78-940182
79 MARC

Sittig, Marshall.
Pharmaceutical manufacturing encyclopedia / Marshall Sittig. — Park Ridge, N.J. : Noyes Data Corp., 1979.
xviii, 702 p. ; 25 cm. — (Chemical technology review ; no. 124)
Includes bibliographies and indexes.
ISBN 0-8155-0739-9 : $72.00
1. Drugs—Dictionaries. 2. Chemistry, Pharmaceutical—Dictionaries. I. Title. II. Series.
RS51.S56 615'.191 79-63053
79 MARC

—HANDBOOKS, MANUALS, ETC.

Riley, Thomas N
A pocket guide to organic medicinal structures / Thomas N. Riley. — Camden, Me. : B. J. Taylor, c1979.
88 p. : ill. ; 16 cm.
1. Chemistry, Pharmaceutical—Handbooks, manuals, etc. 2. Molecular structure—Handbooks, manuals, etc. 3. Drugs—Dictionaries. I. Title.
RS404.5.R54 615'.19 79-63464
00 MARC

—INFORMATION STORAGE AND RETRIEVAL SYSTEMS see Information storage and retrieval systems—Chemistry, Pharmaceutical

—MATHEMATICAL MODELS

Golender, Valerii Efimovich.
Vychislitel'nye metody konstruirovanii͡a lekarstv / V. E. Golender, A. B. Rozenblit ; ⌐Avt. predisl. L. A. Piruzi͡an, V. V. Avidon⌐. — Riga : Zinatne, 1978.
232 p. : ill. ; 21 cm. USSR78
At head of title: Akademii͡a nauk Latviiskoi SSR. Institut organicheskogo sinteza.
Bibliography: p. ⌐211⌐-221.
Includes indexes.
1.20rub
1. Chemistry, Pharmaceutical—Mathematical models. 2. Chemistry, Pharmaceutical—Data processing. 3. Structure-activity relationship (Pharmacology)—Mathematical models. 4. Structure-activity relationship (Pharmacology)—Data processing. I. Rozenbilt, Anatolii Borisovich, joint author. II. Title.
RS403.G64 79-380083
79 MARC

(Matematicheskie metody v khimiobiologii)
Математические методы в химиобиологии / ⌐В. А. Филов ... et al.⌐ ; науч. редактор Г. А. Степанский. — Москва : ВИНИТИ, 1973.
120 p. : ill. ; 22 cm. — (Итоги науки и техники : Серия Фармокология, химиотерапевтические средства, токсикология, проблемы фармакологии ; т. 5) USSR***
On leaf preceding t. p.: Gosudarstvennyi komitet Soveta Ministrov SSSR po nauke i tekhnike. Akademii͡a nauk Soi͡uza Sovetskikh Sot͡sialisticheskikh Respublik. Vsesoi͡uznyi institut nauchnoi i tekhnicheskoi informat͡sii.
Includes bibliographies.
0.65rub

1. Chemistry, Pharmaceutical—Mathematical models. I. Filov, Vladimir Aleksandrovich. II. Stepanskii, Georgii Avraamovich. III. Moscow. Vsesoi͡uznyi institut nauchnoi i tekhnicheskoi informat͡sii. IV. Series: Itogi nauki i tekhniki : Farmokologii͡a, khimioterapevticheskie sredstva, toksikologii͡a, problemy farmakologii ; t. 5.
RM30.I 82 t. 5 79-409911
[RS421]

—PERIODICALS

(Farmat͡sii͡a)
Фармация.
Москва, Изд-во Медицина.
v. ill. 26 cm.
Began in 1952.
Issues for issued by Vsesoi͡uznoe nauchnoe obshchestvo farmat͡sevtov.
1. Pharmacy—Periodicals. 2. Chemistry, Pharmaceutical—Periodicals. I. Vsesoi͡uznoe nauchnoe obshchestvo farmat͡sevtov.
RS1.F264 79-642030
(MARC-S)

CHEMISTRY, PHOTOGRAPHIC see Photographic chemistry

CHEMISTRY, PHYSICAL AND THEORETICAL

see also Adsorption; Allotropy; Atmospheric chemistry; Atomic theory; Atoms; Biochemorphology; Catalysis; Chemical affinity; Chemical bonds; Chemical equilibrium; Chemical evolution; Chemical reaction, Conditions and laws of; Chemical reaction, Rate of; Chemical structure;

Chemistry, Physical organic; Colloids; Crystallization; Crystallography; Electrochemistry; Electrolysis; Gases—Liquefaction; Hydrogen bonding; Instrumental analysis; Mass transfer; Mulecular association; Molecular theory; Nuclear chemistry; Nuclear

physics; Nucleation; Periodic law; Photochemistry; Quantum chemistry; Radiation chemistry; Radiochemistry; Solid state chemistry; Solubility; Solution (Chemistry); Stereochemistry; Surface chemistry; Thermochemistry; Thermodynamics

Adamson, Arthur W
A textbook of physical chemistry / Arthur W. Adamson. — 2d ed. — New York : Academic Press, c1979.
xxv, 953, 23 p. : ill. ; 26 cm.
Includes bibliographical references and index.
ISBN 0-12-044260-4
1. Chemistry, Physical and theoretical. I. Title.
QD453.2.A3 1979 541'.3 79-50783
00 MARC

Alberty, Robert A
Physical chemistry / Robert A. Alberty, Farrington Daniels. — 5th ed. — New York : Wiley, c1979.
xii, 682 p. : ill. ; 24 cm.
Author's names in reverse order in previous editions.
Includes bibliographies and index.
ISBN 0-471-02222-5
1. Chemistry, Physical and theoretical. I. Daniels, Farrington, 1889-1972. Physical chemistry. II. Title.
[.D36 1979.D36 1979] 541'.3 78-14876
78 MARC

Alkali halide vapors : structure, spectra, and reaction dynamics / edited by P. Davidovits and D. L. McFadden ; contributors, J. Berkowitz ... [et al.]. — New York : Academic Press, 1979.
xi, 542 p. ; 24 cm.
Includes bibliographies and indexes.
ISBN 0-12-204250-6
1. Alkali metal halides. 2. Chemistry, Physical and theoretical. I. Davidovits, Paul. II. McFadden, David Lee, 1945-
QD172.A4A44 546'.38 78-11621
78 MARC

Andrews, William A 1930-
Physical science : an introductory study / William A. Andrews, T. J. Elgin Wolfe, John F. Eix. — Scarborough, Ont. : Prentice-Hall of Canada, c1978.
xiii, 444 p. : ill. ; 26 cm. — (Prentice-Hall intermediate science series)
C77-1617-0
Includes index.
ISBN 0-13-671446-3 : $10.75
1. Physics. 2. Chemistry, Physical and theoretical. 3. Physics—Laboratory manuals. 4. Chemistry, Physical and theoretical—Laboratory manuals. I. Wolfe, T. J. E., joint author. II. Eix, John F., 1941- joint author. III. Title. IV. Series.
QC23.A57 500.2 78-319465
79 MARC

Andrews, William A 1930-
Physical science : an introductory study : teacher's guide / William A. Andrews, T. J. Elgin Wolfe, John F. Eix. — Scarborough, Ont. : Prentice-Hall of Canada, c1978.
xix, 529 p. : ill. ; 28 cm. C***
ISBN 0-13-671453-6 : $7.50
1. Physics. 2. Chemistry, Physical and theoretical. 3. Physics—Laboratory manuals. 4. Chemistry, Physical and theoretical—Laboratory manuals. I. Wolfe, T. J. E., joint author. II. Eix, John F., 1941- joint author. III. Title.
QC23.A57 1978b 500.2 79-306943
79 MARC

Barrow, Gordon M
Physical chemistry / Gordon M. Barrow. — 4th ed. — New York : McGraw-Hill, c1979.
xvi, 832 p. : ill. ; 24 cm.
Includes bibliographical references and index.
ISBN 0-07-003825-2 : $23.00.
1. Chemistry, Physical and theoretical. I. Title.
QD453.2.B37 1979 541'.3 78-17080
78 MARC

Bednar, Victoria.
Chimie fizică generală : [manual] / Victoria Bednar, Horst Bednar. — Bucureşti : Editura Didactică şi Pedagogică, 1978.
455 p. : ill. ; 25 cm. R79-2093
At head of title: Ministerul Educaţiei şi Învăţămîntului.
Bibliography: p. 455.
lei21.50
1. Chemistry, Physical and theoretical. I. Bednar, Horst, joint author. II. Title.
QD453.2.B42 79-371406
*79 MARC

Bursa, Stanisław.
Chemia fizyczna / Stanisław Bursa. — Wyd. 1. — Warszawa : Państwowe Wydawn. Naukowe, 1976.
790 p. : ill. ; 24 cm. P***
zł75.00
1. Chemistry, Physical and theoretical. I. Title.
QD453.2.B87 78-421272
79 MARC

Eisenberg, David S
Physical chemistry : with applications to the life sciences / David Eisenberg, Donald Crothers ; artist, Michael Fornalski. — Menlo Park, Calif. : Benjamin/Cummings Pub. Co., c1979.
xxvi, 868 p. : ill. ; 24 cm.
Includes bibliographical references and index.
ISBN 0-8053-2402-X
1. Chemistry, Physical and theoretical. I. Crothers, Donald M., joint author. II. Title.
QD453.2.E37 541'.3 79-12053
79 MARC

Ferri, Antonio, 1912- ed.
Fundamental data obtained from shock-tube experiments. New York, Published for and on behalf of Advisory Group for Aeronautical Research and Development, North Atlantic Treaty Organization, by Pergamon Press, 1961.
xi, 415 p. illus., diagrs. 26 cm. (AGARDograph no. 41)
Includes bibliographies.
1. Shock tubes. 2. Chemistry, Physical and theoretical. I. Title. II. Series.
TL500.N6 no. 41 629.1'08 s 60-14576
[TL567.S4] [r79]rev MARC

Fleury, Maurice B 1934-
Chimie physique : [P.C.E.M., P.C.1, C.P.E.B.H.] / par M.-B. Fleury, ... D. Fleury .... — Paris : Centre de documentation universitaire, 1973.
162 p. : ill. ; 22 cm. F74-12827
29.00F
1. Chemistry, Physical and theoretical. I. Fleury, Danièle, joint author. II. Title.
QD453.2.F54 1973 541'.3 78-393151
79 MARC

Kalous, Vítěz.
Základy fyzikálně chemických metod ; [určeno také stud. odb. škol]. — 2., přeprac. a dopln. vyd. — Praha : SNTL, 1975.
476 p. : ill. ; 21 cm. Cz75
Includes bibliographies and index.
Kčs40.00
1. Chemistry, Physical and theoretical. I. Title.
QD453.2.K36 1975 78-421288
*79 MARC

Kemp, Marwin K 1942-
Physical chemistry : a step-by-step approach / Marwin K. Kemp. — New York : M. Dekker, c1979.
xxii, 1034 p. : ill. ; 26 cm. — (Undergraduate chemistry ; v. 6)
Includes bibliographical references and indexes.
ISBN 0-8247-6640-7
1. Chemistry, Physical and theoretical. I. Title.
QD453.2.K46 541'.1 79-791
79 MARC

Krakow. Akademia Górniczo-Hutnicza. Instytut Metalurgii.
Podstawy chemii fizycznej : dla studentów kierunków hutniczych : teoria, przykłady, zadania / [aut.] Szczepan Chudoba ... [et al.]. — Wyd. 1. — Kraków : AGH, Wydział Metalurgiczny, Instytut Metalurgii, 1976 (Krak. : AGH)
242 p. : ill. ; 24 cm. — (Skrypty uczelniane - Akademia Górniczo-Hutnicza im. S. Staszica w Krakowie ; Nr 563) P77-50
Bibliography: p. [232]
zł17.00
1. Chemistry, Physical and theoretical. I. Chudoba, Szczepan. II. Title.
QD453.2.K7 1976 79-372613
79 MARC

Lee, John David.
A new concise inorganic chemistry / by J. D. Lee. — 3d ed. — New York : Van Nostrand Reinhold Co., 1977.
xxi, 505 p. : ill. ; 24 cm.
Editions for 1964 and 1965 published under title: Concise inorganic chemistry.
Bibliography: p. 490.
Includes index.
ISBN 0-442-30169-3
1. Chemistry, Physical and theoretical. 2. Chemical bonds. I. Title. II. Title: Concise inorganic chemistry.
QD453.2.L43 1977 546 76-57657
76 MARC

Lfâlikov, ÎÛriĭ Sergeevich.
(Fiziko-khimicheskie metody analiza)
Физико-химические методы анализа : [учеб. пособие для хим.-технол. специальностей вузов] / Ю. С. Ляликов. — Изд. 5-е, перераб. и доп. — Москва : Химия, 1974.
536 p. : ill. ; 23 cm. USSR 74
Includes bibliographical references and index.
1.34rub
1. Chemistry, Analytic. 2. Chemistry, Physical and theoretical. I. Title.
QD75.L5 1974 78-420883

Mansfield, David Hutsby.
Principles of physical chemistry / David H. Mansfield. — London : Heinemann Educational Books, 1978.
226 p. : ill. ; 25 cm. GB***
Includes index.
ISBN 0-435-65563-9 : £4.25
1. Chemistry, Physical and theoretical. I. Title.
QD453.2.M35 541.3 78-322694
79 MARC

Marfunin, Arnol'd Sergeevich.
Physics of minerals and inorganic materials : an introduction / A. S. Marfunin ; translated by N. G. Egorova and A. G. Mishchenko. — Berlin ; New York : Springer-Verlag, 1979.
xii, 340 p. : ill. ; 25 cm.
Translation of Vvedenie v fiziku mineralov.
Bibliography: p. [316]-333.
Includes index.
ISBN 0-387-08982-9
1. Mineralogical chemistry. 2. Chemistry, Physical and theoretical. 3. Chemistry, Inorganic. I. Title.
QE371.M3713 549 78-23186
78 MARC

Nicolis, G 1939-
Self-organization in nonequilibrium systems : from dissipative structures to order through fluctuations / G. Nicolis, I. Prigogine. — New York : Wiley, c1977.
xii, 491 p. : ill. ; 24 cm.
"A Wiley-Interscience publication."
Bibliography: p. 475-486.
Includes index.
ISBN 0-471-02401-5
1. Self-organizing systems. 2. Chemistry, Physical and theoretical. 3. Biological chemistry. I. Prigogine, Ilya, joint author. II. Title.
Q325.N5 001.53'3 76-49019
76[r79]rev MARC

Ravich-Shcherbo, Mikhail Iosifovich.
(Fizicheskaîa i kolloidnaîa khimiîa)
Физическая и коллоидная химия : [учебник для мед. ин-тов] / М. И. Равич-Щербо, В. В. Новиков. — Изд. 3-е, испр. и доп. — Москва : Высш. школа, 1975.
254 p. : ill. ; 23 cm. USSR 75
Bibliography : p. [245]
Includes index.
0.62rub
1. Chemistry, Physical and theoretical. 2. Colloids. I. Novikov, Viktorin Vasil'evich, joint author. II. Title.
QD453.R3 1975 78-420885

Sonnenschein, Günter.
Der Einfluss von Hochfrequenzüberlagerungen bei Stossspannungen auf das Durchschlagverhalten von Luft unter Atmosphärendruck bei verschiedenen Elektrodenanordnungen. Köln, Opladen, Westdeutscher Verlag, 1967.
35 p. with illus. 24 cm. (Forschungsberichte des Landes Nordrhein-Westfalen, Nr. 1816) 22.00 GDNB67-A34-510
Issued also as thesis, Technische Hochschule, Aachen.
Bibliography: p. 35.
1. Electrodynamics. 2. Electrodes. 3. Chemistry, Physical and theoretical. I. Title. II. Series.
T7.F65 Nr. 1816 083'.1 s 73-388040
[QC631.3] 79[r79]rev2 MARC

Thiessen, Peter Adolf, 1899-
Erfahrungen, Erkenntnisse, Folgerungen / Peter Adolf Thiessen. — Berlin : Akademie Verlag, 1979.
viii, 263 p. : 2 graphs, port. ; 22 cm. GDR79-A
Includes bibliographical references.
24.00M
1. Chemistry, Physical and theoretical. I. Title.
QD453.2.T55 79-381064
79 MARC

Tuller, Harry L
Electrical properties and defect model of nonstoichiometric and impurity-doped CeO_2 / Harry L. Tuller. — New York : Columbia University, Henry Krumb School of Mines, 1974.
iii, 191 leaves : ill. ; 28 cm.
Cover title.
Originally presented as the author's thesis, Columbia University, 1973.
Includes bibliographical references.
1. Cerium oxide crystals—Electric properties. 2. Point defects. 3. Chemistry, Physical and theoretical. I. Title.
QC611.8.M4T84 1974 546'.412'2 78-324535
79 MARC

Zbiór zadań rachunkowych z chemii fizycznej : praca zbiorowa / pod red. Walerego Miśniakiewicza ; opracowali Zygmunt Dziewięcki ... [et al.]. — Gliwice : Dział Wydawnictw Politechniki Śląskiej, 1975-
v. ; 24 cm. — (Skrypty uczelniane - Politechnika Śląska im. W. Pstrowskiego ; nr. 571)
zł18.00 (v. 1)
1. Chemistry, Physical and theoretical. I. Dziewięcki, Zygmunt. II. Miśniakiewicz, Walery.
QD453.2.Z33 78-399381
79 MARC

CHEMISTRY, PHYSICAL AND THEORETICAL (Continued)

—ADDRESSES, ESSAYS, LECTURES

(Butsurigaku no shimpo)
物理學の進歩 / 監修 仁科芳雄. — 東京：共立出版, 昭和 [19
v. : ill. ; 21 cm.
Includes bibliographies.

1. Solid state physics—Addresses, essays, lectures. 2. Chemistry, Physical and theoretical—Addresses, essays, lectures. 3. Nuclear physics—Addresses, essays, lectures. I. Nishina, Yoshio, 1890-1951.
QC176.2.B88 79-804625

Fiziko-khimicheskie issledovaniia v oblasti stroitel'nykh materialov / pod obshcheĭ red. V. S. Galakhova. — Moskva : Moskovskiĭ inzhenerno-stroit. in-t, 1977.
172 p. : ill. ; 22 cm. — (Sbornik trudov kafedr obshcheĭ khimii, organicheskoĭ, fizicheskoĭ, analiticheskoĭ khimii i fiziki, Moskovskiĭ inzhenerno-stroitel'nyĭ institut im. V. V. Kuĭbysheva ; No. 147) USSR***
At head of title: Ministerstvo vysshego i srednego spetsial'nogo obrazovaniia SSSR.
Includes bibliographies.
1.00rub
1. Building materials—Addresses, essays, lectures. 2. Chemistry, Physical and theoretical—Addresses, essays, lectures. I. Galakhov, V. S. II. Series: Moscow. Inzhenerno-stroitel'nyĭ institut. Sbornik trudov ; no. 147.
TA7.M62A3 no. 147 79-381126
[TA403.8] 79 MARC

For Ilya Prigogine / edited by Stuart A. Rice. — New York : Wiley, c1978.
472 p. : ill. ; 24 cm. — (Advances in chemical physics ; v. 38)
"An Interscience publication."
Includes bibliographical references and index.
CONTENTS: Primas, H. and Müller-Herold, U. Quantum-mechanical system theory.—Spohn, H. and Lebowitz, J. L. Irreversible thermodynamics for quantum systems weakly coupled to thermal reservoirs—Grecos, A. P. Solvable models for unstable states in quantum physics.—Balescu, R. and Résibois, P. Aspects of kinetic theory.—Klimontovich, Y. L. Kinetic theory of plasmas.—Eigen, M. How does information originate?—Nicolis, G., Erneux, T., and Herschkowitz-Kaufman, M. Pattern formation in reacting and diffusing systms.—Hanusse, P., Ross, J., and Ortoleva, P. Instability and far-from-equilibrium states of chemically reacting systems.—Hess, B., Goldbeter, A., and Lefever, R. Temporal, spatial, and functional order in regulated biochemical and cellular systems.—Nazarea, A. D. Electric fields and self-coherent patterns and structures in chemical systems.
ISBN 0-471-03883-0
1. Chemistry, Physical and theoretical—Addresses, essays, lectures. 2. Prigogine, Ilya. 3. Chemists—United States—Biography. I. Prigogine, Ilya. II. Rice, Stuart Alan, 1932- III. Series.
QD453.A27 vol. 38 541'.08 s 79-101799
[QD455] 79 MARC

High pressure chemistry : proceedings of the NATO Advanced Study Institute held in Corfu, Greece, September 24-October 8, 1977 / edited by H. Kelm. — Dordrecht, Holland ; Boston : D. Reidel Pub. Co., c1978.
vii, 600 p. : ill. ; 25 cm. — (NATO advanced study institutes series : Series C, Mathematical and physical sciences ; v. 41)
Includes bibliographical references and indexes.
ISBN 9027709351
1. Chemistry, Physical and theoretical—Addresses, essays, lectures. 2. High pressure (Science)—Addresses, essays, lectures. I. Kelm, H., 1933- II. Series.
QD455.H48 541'.3 78-15304
78 MARC

Inorganic and physical chemistry. — Berlin ; New York : Springer-Verlag, 1978.
239 p. : ill. ; 25 cm. — (Topics in current chemistry ; 77)
Includes bibliographical references and index.
CONTENTS: Bikerman, J. J. Surface energy of solids.—Wiedemann, H. G. and Bayer, G. Trends and applications of thermogravimetry.—Huglin, M. B. Determination of molecular weights by light scattering.
ISBN 0-387-08987-X
1. Chemistry, Physical and theoretical—Addresses, essays, lectures. 2. Chemistry, Inorganic—Addresses, essays, lectures. I. Bikerman, Jacob Joseph, 1898- Surface energy of solids. 1978. II. Wiedemann, Hans G. Trends and applications of thermogravimetry. 1978. III. Huglin, M. B. Determination of molecular weights by light scattering. 1978. IV. Series.
QD1.F58 vol. 77 540'.8s 78-13471
[QD475] 78 MARC

(Voprosy fizicheskoĭ khimii v proizvodstve neorganicheskikh veshchestv)
Вопросы физической химии в производстве неорганических веществ / под общей ред. Черненькой Е. И. и Можаровой Т. В. — Харьков : НИОХИМ, 1974.
97, 4 p. : ill. ; 22 cm. — (Труды - Государственный научно-исследовательский и проектный институт основной химии ; т. 34) USSR 75
Includes bibliographies.
1. Chemistry, Physical and theoretical—Addresses, essays, lectures. 2. Chemistry, Inorganic—Addresses, essays, lectures. I. Chernen'kaia, E. I. II. Mozharova, T. V. III. Series: Kharkov. Naukovo-doslidnyĭ instytut osnovnoï khimiï. Trudy ; t. 34.
QD1.K48 t. 34 79-409531
[QD455]

—BIBLIOGRAPHY

Akademiia nauk URSR, Kiev. Instytut fizychnoï khimiï.
(Ukazatel' opublikovannykh rabot sotrudnikov Instituta fizicheskoĭ khimii im. L. V. Pisarzhevskogo AN USSR)
Указатель опубликованных работ сотрудников Института физической химии им. Л. В. Писаржевского АН УССР : 1927-1976 гг. / [Сост. Б. А. Геллер, Г. П. Корнейчук, М. А. Пионтковская ... et al.]. — Киев : Наук. думка, 1977.
372 p. ; 20 cm. USSR 78
1.20rub
1. Chemistry, Physical and theoretical—Bibliography. I. Geller, Boris Aleksandrovich. II. Title.
Z5524.P6A38 1977 79-407773
[QD453.2]

—COLLECTED WORKS

Prague. Vysoká škola chemicko-technologická.
Sborník Vysoké školy chemicko-technologické v Praze. Fysikální chemie. Сборник Пражского химико-технологического института. Физическая химия. Scientific papers of the Prague Institute of Chemical Technology. Physical chemistry.
N 1-
Praha, Státní pedagogické nakl., 1974-
v. ill. 24 cm.
Title in Russian and English varies slightly.
Czech, English, or German, with summaries in Czech, English, German, and/or Russian.

1. Chemistry, Physical and theoretical—Collected works. I. Prague. Vysoká škola chemicko-technologická. Sbornik Prazhskogo khimiko-tekhnologicheskogo instituta. Fizicheskaia khimiia. II. Prague. Vysoká škola chemicko-technologická. Scientific papers of the Prague Institute of Chemical Technology. Physical chemistry. III. Title: Fysikální chemie.
QD450.P73a 79-646344
(MARC-S)

—CONGRESSES

Société de chemie physique.
Nonlinear behaviour of molecules, atoms, and ions in electric, magnetic, or electromagnetic fields : proceedings of the 31st international meeting of the Société de chemie physique / jointly organized with Associazione italiana di chimica fisica, Deutsche Bunsengesellschaft für physikalische Chemie, Chemical Faraday Division of the Chemical Society, Abbaye de Fontevraud, 25-28 September, 1978 : edited by Louis Néel ; organizing committee, L. Néel, chairman ... [et al.]. — Amsterdam : New York : Elsevier Scientific Pub. Co. ; New York : distribution for the U.S.A. and Canada, Elsevier/North-Holland, 1979.
xx, 515 p. : ill. ; 25 cm.
English or French.

Includes bibliographical references.
ISBN 0-444-41790-7
1. Chemistry, physical and theoretical—Congresses. I. Néel, Louis, 1904- II. Associazione italiana de chimica fisica. III. Deutsche Bunsengesellschaft für Physikalische Chemie. IV. Chemical Society, London. Faraday Division. V. Title.
QD450.S63 1979 541.2 79-10839
79 MARC

—DATA PROCESSING

Zalotai, Lajos.
FORTRAN programok fizikai kémiai feladatok megoldására / Zalotai Lajos, Seres László, Fejes Pál. — Budapest : Műszaki Kiadó, 1978.
440 p. ; 24 cm. H78
Includes bibliographical references and index.
ISBN 9631021610 : 90.00Ft
1. Chemistry, Physical and theoretical—Data processing. 2. FORTRAN (Computer program language) I. Seres, László, joint author. II. Fejes, Pál, joint author. III. Title.
QD455.3.E4Z34 79-371449
*79 MARC

—LABORATORY MANUALS

Andrews, William A 1930-
Physical science : an introductory study / William A. Andrews, T. J. Elgin Wolfe, John F. Eix. — Scarborough, Ont. : Prentice-Hall of Canada, c1978.
xiii, 444 p. : ill. ; 26 cm. — (Prentice-Hall intermediate science series) C77-1617-0
Includes index.
ISBN 0-13-671446-3 : $10.75
1. Physics. 2. Chemistry, Physical and theoretical. 3. Physics—Laboratory manuals. 4. Chemistry, Physical and theoretical—Laboratory manuals. I. Wolfe, T. J. E., joint author. II. Eix, John F., 1941- joint author. III. Title. IV. Series.
QC23.A57 500.2 78-319465
79 MARC

Andrews, William A 1930-
Physical science : an introductory study : teacher's guide / William A. Andrews, T. J. Elgin Wolfe, John F. Eix. — Scarborough, Ont. : Prentice-Hall of Canada, c1978.
xix, 529 p. : ill. ; 28 cm. C***
ISBN 0-13-671453-6 : $7.50
1. Physics. 2. Chemistry, Physical and theoretical. 3. Physics—Laboratory manuals. 4. Chemistry, Physical and theoretical—Laboratory manuals. I. Wolfe, T. J. E., joint author. II. Eix, John F., 1941- joint author. III. Title.
QC23.A57 1978b 500.2 79-306943
79 MARC

Ćwiczenia laboratoryjne z chemii fizycznej : praca zbiorowa / pod red. Józefa Podkówki ; opracowali Jadwiga Dudek ... [et al.]. — Gliwice : Politechnika Śląska, 1977-
v. : ill. ; 24 cm. — (Skrypty uczelniane - Politechnika Śląska im. W. Pstrowskiego ; Nr 705) P***
Includes bibliographies.
zł25.00 (v. 1)
1. Chemistry, Physical and theoretical—Laboratory manuals. I. Podkówka, Józef. II. Dudek, Jadwiga.
QD457.C88 78-377842
78 MARC

Sędzimir, Jerzy.
Ćwiczenia laboratoryjne z chemii fizycznej / Jerzy Sędzimir, Wiesława Ziółkowska. — Wyd. 3. uzup. — Kraków : Akademia Górniczo-Hutnicza, 1977.
iv, 209 p. : ill. ; 21 cm. — (Skrypty uczelniane - Akademia Górniczo-Hutnicza im. S. Staszica w Krakowie ; Nr. 632) P***
At head of title: Akademia Górniczo-Hutnicza im. S. Staszica w Krakowie. Wydział Metali Nieżelaznych. Instytut Metalurgii Metali Nieżelaznych.
Includes bibliographies.
zł17.00
1. Chemistry, Physical and theoretical—Laboratory manuals. I. Ziółkowska, Wiesława, joint author. II. Title.
QD457.S43 1977 78-382191
78 MARC

Vorob'ev, Nikolaĭ Konstantinovich.
(Praktikum po fizicheskoĭ khimii)
Практикум по физической химии : допущено ... в качестве учеб. пособия для студентов химико-технологических специальностей вузов / под общей ред. Н. К. Воробьева. — Изд. 4., перер. и доп. — Москва : Химия, 1975.
366 p. : ill. ; 23 cm. USSR***
Includes bibliographies and index.
0.96rub
1. Chemistry, Physical and theoretical—Laboratory manuals. I. Title.
QD457.V65 1975 78-422836

—PERIODICALS

Tōhoku Daigaku, Sendai, Japan. Genshikaku Rigaku Kenkyū Shisetsu.
(Kakuriken kenkyū hōkoku)
核理研研究報告 Research report of Labotatory of Nuclear Science, Tohoku University.

[仙台] 東北大学理学部[附属]原子核理学研究施設

v. 26 cm.
English or Japanese.
ISSN 0385-2105

1. Nuclear physics—Periodicals. 2. Solid state physics—Periodicals. 3. Chemistry, Physical and theoretical—Periodicals. 4. Nuclear chemistry—Periodicals. I. Tōhoku Daigaku, Sendai, Japan. Genshikaku Rigaku Kenkyū Shisetsu. Research report of Laboratory of Nuclear Science, Tohoku University.
QC770.T58a 539.7'05 78-645859
(MARC-S)

—PROBLEMS, EXERCISES, ETC.

Zbiór zadań z chemii fizycznej : praca zbiorowa / pod red. Marii Kuczyńskiej ; opracowali Zygmunt Dziewięcki ... [et al.]. — Gliwice : Politechnika Śląska im. W. Pstrowskiego, 19
v. : ill. ; 24 cm. P***
Skrypty uczelniane - Politechnika Śląska im. W. Pstrowskiego ; Nr 529)
1. Chemistry, Physical and theoretical—Problems, exercises, etc. I. Kuczyńska, Maria. II. Dziewięcki, Zygmunt.
QD453.2.Z34 79-376306
79 MARC

CHEMISTRY, PHYSICAL ORGANIC

Ballester, Manuel.
Química orgánica física : fundamentos y espectrometrías / Manuel Ballester. — Madrid : Ediciones Pirámide, c1978.
538 p. : ill. ; 24 cm. — (Ciencia y técnica) Sp***
Includes bibliographical references and index.
ISBN 8436800958
1. Chemistry, Physical organic. I. Title.
QD476.B33 79-362501
79 MARC

CHEMISTRY, PHYSICAL ORGANIC (Continued)

Coulson, Charles Alfred.
Hückel theory for organic chemists / C. A. Coulson, Brian O'Leary, and R. B. Mallion. — London ; New York : Academic Press, 1978.
xvi, 182 p. : ill. ; 24 cm. GB***
Bibliography: p. [175]-176.
Includes index.
ISBN 0-12-193250-8
1. Molecular orbitals. 2. Chemistry, Physical organic. I. O'Leary, Brian, joint author. II. Mallion, R. B., joint author. III. Title.
QD461.C663 547'.1'22 77-93202
79 MARC

Gilchrist, Thomas Lonsdale.
Organic reactions and orbital symmetry / T. L. Gilchrist and R. C. Storr. — 2d ed. — Cambridge [Eng.] : New York : Cambridge University Press, 1979.
vii, 311 p. : ill. ; 24 cm. — (Cambridge texts in chemistry and biochemistry)
Includes bibliographical references and index.
ISBN 0-521-22014-9
1. Chemistry, Physical organic. 2. Chemical reaction, Conditions and laws of. 3. Molecular orbitals. 4. Symmetry (Physics) I. Storr, R. C., joint author. II. Title. III. Series.
QD476.G54 1979 547'.1'39 78-54578
78 MARC

Greenberg, Arthur.
Strained organic molecules / Arthur Greenberg, Joel F. Liebman. — New York : Academic Press, 1978.
xi, 406 p. : ill. ; 24 cm. — (Organic chemistry ; v. 38)
Includes bibliographical references and index.
ISBN 0-12-299550-3
1. Chemistry, Physical organic. 2. Molecular structure. I. Liebman, Joel F., joint author. II. Title. III. Series: Organic chemistry, a series of monographs ; v. 38.
QD476.G66 547'.1'3 77-11212
78 MARC

Nasuto, Romuald.
Mechanizm adsorpcji z dwuskładnikowych roztworów niektórych związków organicznych na żelu krzemionkowym : rozprawa habilitacyjna / Romuald Nasuto. — Lublin : Uniwersytet Marii Curie-Skłodowskiej, Wydział Matematyki, Fizyki i Chemii, 1976.
79 p. : ill. ; 21 cm. P77-7176
Bibliography: p. 60-72.
1. Adsorption. 2. Chemistry, Physical organic. I. Title.
QD547.N37 78-384135
78 MARC

Norman, Richard Oswald Chandler.
Mechanisms in organic chemistry : case studies / [by] R. O. C. Norman, M. J. Tomlinson, D. J. Waddington. — London : Mills and Boon, 1978.
110 p. : ill. ; 23 cm. — (Modern chemistry series) GB78-3961
Bibliography: p. [106]
Includes index.
ISBN 0-263-06351-8 : £2.50
1. Chemistry, Physical organic. 2. Chemical reaction, Conditions and laws of. I. Tomlinson, Michael John, joint author. II. Waddington, D. J., joint author. III. Title.
QD476.N64 1978 547.1'39 79-305953
79 MARC

Sanderson, Robert Thomas, 1912-
Chemical bonds in organic compounds / R. T. Sanderson. — Scottsdale, AZ : Sun and Sand Pub. Co., c1976.
73 p. ; 23 cm.
Cover title.
1. Chemical bonds. 2. Chemistry, Physical organic. I. Title.
QD461.S34 547'.1'224 78-325511
79 MARC

Sieck, L Wayne.
Rate coefficients for ion-molecule reactions : organic ions other than those containing C and H / L. Wayne Sieck. — Washington : U.S. Dept. of Commerce, National Bureau of Standards : for sale by the Supt. of Docs., U.S. Govt. Print. Off., 1979.
iv, 20 p. ; 26 cm. — (Nat. stand. ref. data ser. ; NSRDS-NBS 64)
"National Standard Reference Data System."
Includes bibliographical references.
Supt. of Docs. no.: C 13.48:64
1. Chemical reaction, Rate of. 2. Ions. 3. Molecules. 4. Chemistry, Physical organic. I. Title. II. Series: United States. National Bureau of Standards. National standard reference data series ; NSRDS-NBS 64.
QD502.S52 547'.139 78-10710
78 MARC

Sykes, Peter.
A guidebook to mechanism in organic chemistry / Peter Sykes. — 4th ed. — New York : Wiley, 1975.
xii, 362 p. : ill. ; 22 cm.
"A Halsted Press book."
Bibliography: p. 345-346.
Includes index.
ISBN 0-470-84101-X
1. Chemistry, Physical organic. 2. Chemical reactions. I. Title.
QD476.S89 1975 547'.1'39 74-12872
75[r79]rev MARC

Sykes, Peter.
Reaktionsmechanismen der organischen Chemie; eine Einführung. Mit einem Geleitwort von A. R. Todd. 5., neubearb. und erw. Aufl. Übers. von Peter Boldt, Hans F. Ebel [und] Peter Hocks. Weinheim/Bergstr., Verlag Chemie, 1972.
viii, 313 p. illus. 21 cm. DM22.00 GDB***
Translation of A guidebook to mechanism in organic chemistry.
Bibliography: p. [303]-305.
ISBN 3-527-25395-5
1. Chemistry, Physical organic. 2. Chemical reactions. I. Title.
QD476.S8915 1972 70-173369
79[r79]rev MARC

Testa, Bernard.
Principles of organic stereochemistry / Bernard Testa. — New York : M. Dekker, c1979.
xiii, 248 p. : ill. ; 23 cm. — (Studies in organic chemistry ; v. 6)
Includes bibliographical references and indexes.
ISBN 0-8247-6758-6
1. Stereochemistry. 2. Chemistry, Physical organic. I. Title.
QD481.T42 547'.1'223 78-10138
78 MARC

—ADDRESSES, ESSAYS, LECTURES

(Fiziko-khimicheskie svoĭstva organicheskikh veshchestv i i ikh svi͡az' so stroeniem i mikrostrukturoĭ)
Физико-химические свойства органических веществ и их связь со строением и микроструктурой : сборник трудов / [редколлегия, засл. деят. науки РСФСР, проф. В. А. Измаильский ... et al.]. — Москва : [s. n.], 1973.
239 p. : ill. ; 22 cm. USSR 74
At head of title: Ministerstvo prosveshcheni͡a RSFSR. Moskovskiĭ gosudarstvennyĭ pedagogicheskiĭ institut imeni V. I. Lenina. Kafedra organicheskoĭ khimii.
Includes bibliographies.
1.00rub
1. Chemistry, Physical organic — Addresses, essays, lectures. 2. Chemical structure—Addresses, essays, lectures. I. Izmail'skiĭ, V. A. II. Moscow. Moskovskiĭ gosudarstvennyĭ pedagogicheskiĭ institut. Kafedra organicheskoĭ khimii.
QD473.F59 79-408062

(Stroenie i reakt͡sionna͡a sposobnost' organicheskikh soedineniĭ)
Строение и реакционная способность органических соединений : [Сб. статей] : Посвящается 150-летию со дня рождения А. М. Бутлерова / [АН СССР, Казан. фил., Ин-т орган. и физ. химии им. А. Е. Арбузова ; Отв. ред. [и авт. предисл.] Б. А. Арбузов]. — Москва : Наука, 1978.
291 p., [1] leaf of plates : ill. ; 22 cm. USSR 78
Includes bibliographies.
3.00rub
1. Chemistry, Physical organic—Addresses, essays, lectures. 2. Butlerov, Aleksandr Mikhaĭlovich, 1828–1886. I. Butlerov, Aleksandr Mikhaĭlovich, 1828–1886. II. Arbuzov, Boris Aleksandrovich, 1903– III. Akademi͡a nauk SSSR. Institut organicheskoĭ i fizicheskoĭ khimii.
QD476.S75 79–412147

—COLLECTED WORKS

Organic reactions.
Huntington, N. Y., R. E. Krieger Pub. Co.,
v. 24 cm.
Reprint of the 1942 ed. published by J. Wiley & sons, inc., New York.
1. Chemical reactions—Collected works. 2. Chemistry, Physical organic—Collected works.
QD501.O75 547'.1'39 79–642486
MARC-S

Reactive intermediates. v. 1–
New York, John Wiley & Sons, c1978–
v. ill. 24 cm.
"A Wiley-interscience publication."
Key title: Reactive intermediates, ISSN 0190–8375.
1. Chemistry, Physical organic — Collected works. 2. Chemical reaction, Conditions and laws of—Collected works.
QD476.R414 547'.1'39 79–640411
MARC-S

—CONGRESSES

Colston Symposium on the Behaviour of Ions in Macromolecular and Biological Systems, University of Bristol, 1977.
Ions in macromolecular and biological systems : proceedings of the twenty-ninth symposium on the Colston Research Society, held in the University of Bristol, April, 1977 / edited by D. H. Everett and B. Vincent. — Baltimore : University Park Press, 1978.
xiii, 348 p. : ill. ; 25 cm. — (Colston papers ; v. 29)
Includes bibliographical references and index.
ISBN 0-8391-1290-4
1. Ions—Congresses. 2. Macromolecules—Congresses. 3. Membranes (Biology)—Congresses. 4. Biological chemistry—Congresses. 5. Chemistry, Physical organic—Congresses. I. Everett, Douglas Hugh, 1916- II. Vincent, Brian. III. Colston Research Society. IV. Title. V. Series.
QH345.C54 1977 574.1'9283 78-56307
79 MARC

International Symposium on Organic Free Radicals, 2d, Aix, France, 1977.
Radicaux libres organiques : 2nd International Symposium on Organic Free Radicals, Aix-en-Provence, 17-23 juillet 1977. — Paris : Éditions du Centre national de la recherche scientifique, 1978.
549 p. : ill. ; 30 cm. — (Colloques internationaux du Centre national de la recherche scientifique ; no 278) F***
English or French.
Includes bibliographical references.
ISBN 2-222-02281-9 : 175F
1. Radicals (Chemistry)—Congresses. 2. Chemistry, Physical organic—Congresses. I. Title. II. Series: France. Centre national de la recherche scientifique. Colloques internationaux ; no 278.
QD471.I57 1977 547'.1'224 79-348837
79 MARC

Ions in macromolecular and biological systems : proceedings of the twenty-ninth symposium of the Colston Research Society held in the University of Bristol, April 1977 / edited by D. H. Everett and B. Vincent. — Bristol : Scientechnica, 1978.
xiii, 348 p. : ill. ; 24 cm. — (Colston papers ; v. 29) GB***
Includes bibliographical references and index.
ISBN 0-85608-021-7 : £15.00
1. Ions—Congresses. 2. Polymers and polymerization—Congresses. 3. Biological chemistry—Congresses. 4. Chemistry, Physical organic—Congresses. I. Everett, Douglas Hugh, 1916- II. Vincent, B., Ph.D. III. Colston Research Society. IV. Series.
QH345.I66 547'.13'72 79-305540
79 MARC

Organic liquids : structure, dynamics, and chemical properties / edited by A. D. Buckingham, E. Lippert, and S. Bratos. — Chichester, [Eng.] ; New York : Wiley, c1978.
x, 352 p. : ill. ; 24 cm.
"A Wiley-Interscience publication."
Includes bibliographical references and index.
ISBN 0-471-99673-4
1. Chemistry, Physical organic—Congresses. 2. Liquids—Congresses. I. Buckingham, Amyand David. II. Lippert, Ernst. III. Bratos, S.
QD476.O73 547 78-8462
78 MARC

—PROBLEMS, EXERCISES, ETC.

Lambert, Joseph B
Physical organic chemistry, through solved problems / Joseph B. Lambert. — San Francisco : Holden-Day, c1978.
xii, 265 p. : ill. ; 23 cm.
Includes indexes.
ISBN 0-8162-4921-0
1. Chemistry, Physical organic—Problems, exercises, etc. I. Title.
QD476.L35 547'.1'3076 78-60359
79 MARC

CHEMISTRY, QUANTUM see Quantum chemistry

CHEMISTRY, SANITARY see Sanitary chemistry

CHEMISTRY, SOIL see Soil chemistry

CHEMISTRY, SOLID STATE see Solid state cehmistry

CHEMISTRY, SURFACE see Surface chemistry

CHEMISTRY, SYNTHETIC see Chemistry, Organic—Synthesis

CHEMISTRY, TECHNICAL

see also Alloys; Animal products; Biochemical engineering; Bleaching; Ceramics; Chemical engineering; Chemical industries; Chemical reactors; Chemicals—Manufacture and industry; Chemistry, Metallurgic; Cleaning compounds; Concrete —Chemistry; Corrosion and anti-corrosives; Distillation, Destructive; Electrochemistry; Extraction (Chemistry); Food—Analysis; Gums and resins; Gums and resins, Synthetic; High pressure (Technology); Metallurgical analysis; Rubber chemistry; Separation (Technology); Tanning; Textile chemistry; Waste products; Wood—Chemistry; and particular industries and products

Chu, Shu-kung.
(Kung yeh hua hsüeh kai lun)
工業化學概論 / 朱樹恭編. — 臺北 : 臺灣商務印書館, 民國67- [1978-
v. : ill. ; 22 cm. — (大學叢書)
Series romanized: Ta hsüeh ts'ung shu.
Includes bibliographical references.
NT$120.00 (v. 1)

1. Chemistry, Technical. I. Title.
TP145.C48 79-835963

Clausen, Chris A 1940-
Principles of industrial chemistry / Chris A. Clausen III, Guy C. Mattson. — New York : Wiley, c1978.
xiv, 412 p. : ill. ; 24 cm.
"A Wiley-Interscience publication."
Includes bibliographical references and index.
ISBN 0-471-02774-X
1. Chemistry, Technical. I. Mattson, Guy C., joint author. II. Title.
TP145.C67 660.2 78-9450
78 MARC

Ishii, Raizō, 1899-
(Kagaku shōhin)
化學商品 / 石井賴三著. — 東京 : 新紀元社, 昭和24 [1949]
15, 242, 9, [1] p. : ill. ; 22 cm.
Bibliography: p. [10] (3d group)
Includes index.

1. Chemicals. 2. Chemistry, Technical. I. Title.
TP200.I83 78-808557

Jordan, Otto, Dr. phil., Chem.
The technology of solvents. Translated by Alan D. Whitehead for Technical Service Library. London, L. Hill [1938]
xx, 351 p. illus., diagrs., tables. 25 cm.
Pages 348-351, advertising matter.
Translation of Chemische Technologie der Lösungsmittel.
1. Solvents. 2. Chemistry, Technical. 3. Chemistry, Organic. I. Title.
TP247.5.J62 1938 661'.807 38-36462
[r79]rev3 MARC

Jordan, Otto, Dr. phil., Chem.
The technology of solvents. Translated by Alan D. Whitehead for Technical Service Library. [2d ed.] London, L. Hill [1940]
xxix, 361 p. illus., diagrs., tables. 25 cm.
Pages 358-361, advertising matter.
Translation of Chemische Technologie der Lösungsmittel.
1. Solvents. 2. Chemistry, Technical. 3. Chemistry, Organic. I. Title.
TP247.5.J62 1940 661'.807 a 42-1815
New York. Public Libr. [r79]rev2 MARC

Knoblauch, Harald.
Bauchemie / Harald Knoblauch. — Düsseldorf : Werner, 1978.
viii, 292 p. : 69 ill. ; 19 cm. — (Werner-Ingenieur-Texte ; 55 ISSN 0341-0307) GFR78-A
Bibliography: p. 275-278.
Includes index.
ISBN 3-8041-2355-4 : DM34.80
1. Chemistry, Technical. 2. Building materials. I. Title.
TP145.K74 79-339390
79 MARC

Molenda, Jacek.
Chemia techniczna : podręcznik dla studentów szkół wyższych kształcących nauczycieli wychowania technicznego / Jacek Molenda. — Wyd. 2 popr. i uzup. . — Warszawa : Wydawa Naukowo-Techniczne, 1978 (Szczec. : SzZG)
198, [2] p. : ill. ; 25 cm. P78-6661
Bibliography: p. [199]
zł25.00
1. Chemistry, Technical. I. Title.
TP145.M57 1978 79-342627
79 MARC

Szerves vegyipari alapfolyamatok kézikönyve / szerk. Deák Gyula. — Budapest : Műszaki Könyvkiadó, 1978.
1052 p. : ill. ; 25 cm. H78
Errata slip inserted.
Includes bibliographical references and index.
ISBN 9631020746 : 252.00Ft
1. Chemistry, Organic. 2. Chemistry, Technical. I. Deák, Gyula.
TP247.S97 79-354350
•79 MARC

Technische anorganische Chemie / [dem Autorenkollektiv gehören an, Gunter Bräutigam ... et al.]. — 1. Aufl. — Leipzig : Deutscher Verlag für Grundstoffindustrie, VEB, 1978.
365 p. : ill., graphs ; 24 cm. GFR79-A4
Table inserted.
Includes bibliographical references and index.
39.80M
1. Chemistry, Technical. 2. Chemistry, Inorganic. I. Bräutigam, Gunter.
TP145.T34 79-372727
79 MARC

Vajta, László.
Kémiai technológia : mérnökhallgatók számára : [egyetemi tankönyv] / Vajta László, Szebényi Imre. — 3., jav. kiad. — Budapest : Tankönyvkiadó, 1977, c1969.
315 p. : ill. ; 24 cm. H77
Bibliography: p. 308-309.
ISBN 9631725316 : 35.00Ft
1. Chemistry, Technical. I. Szebényi, Imre, joint author. II. Title.
TP145.V27 1977 78-382190
•78 MARC

Weissermel, Klaus.
Industrial organic chemistry : important raw materials and intermediates / Klaus Weissermel, Hans-Jürgen Arpe ; translated by Alexander Mullen. — 1. Aufl. — Weinheim ; New York : Verlag Chemie, 1978.
xii, 404 p. (20 fold.) ; 25 cm. GFR***
Translation of Industrielle organische Chemie.
Includes bibliographical references and index.
ISBN 0-89573-005-7 (New York) : DM74.00
1. Chemistry, Organic. 2. Chemistry, Technical. I. Arpe, Hans-Jürgen, joint author. II. Title.
TP247.W4413 661'.8 78-390041
78 MARC

Weissermel, Klaus.
Industrielle organische Chemie : bedeutende Vor- u. Zwischenprodukte / Klaus Weissermel, Hans-Jürgen Arpe. — 2., überarb. u. erw. Aufl. — Weinheim ; New York : Verlag Chemie, 1978.
xiv, 430 p. ; 25 cm. GFR78-A
Bibliography: p. 400-407.
ISBN 3-527-25756-X : DM64.00
1. Chemistry, Organic. 2. Chemistry, Technical. I. Arpe, Hans-Jürgen, joint author. II. Title.
TP247.W44 1978 661'.8 78-392826
79 MARC

—ADDRESSES, ESSAYS, LECTURES

Industria chimică : [culegere de articole din literatura străină de specialitate]. — București : Consiliul Național pentru Știință și Tehnologie, Institutul Național de Informare și Documentare, 1978.
6 v. : ill. ; 25 cm. — (Materiale, tehnologii și produse noi) R78-426
Includes bibliographies.
lei280
1. Chemistry, Technical—Addresses, essays, lectures. I. Institutul Național de Informare și Documentare Științifică și Tehnică.
TP185.I45 79-338680
•79 MARC

Khimiía i khimicheskaía tekhonologiía : [Sbornik stateĭ / Otv. red. kand. tekhn. nauk, dots. V. V. Balandina]. — Novocherkassk : NPI, 1976.
142 p. : ill. ; 20 cm. — (Trudy Novocherkasskogo politekhnicheskogo instituta im. Sergo Ordzhonikidze ; t. 320) USSR76
At head of title: MV i SSO RSFSR.
Includes bibliographies.
1. Chemistry—Addresses, essays, lectures. 2. Chemistry, Technical—Addresses, essays, lectures. I. Balandina, V. V. II. Series: Novocherkassk, Russia. Politekhnicheskiĭ institut. Trudy ; t. 320.
TA7.N65 t. 320 79-382635
[QD39] 79 MARC

Khimiía i khimicheskaía tekhnologiía : [Sb. stateĭ / Redkol. S. I. Smyshlíaev (otv. red.) ... et al.]. — Krasnodar : Kuban. un-t, 1977.
193 p. : ill. ; 20 cm. — (Nauchnye trudy - Kubanskiĭ gosudarstvennyĭ universitet ; vyp. 243) USSR78
At head of title: Ministerstvo vysshego i srednego spetsial'nogo obrazovaniía RSFSR.
Includes bibliographies.
1. Chemistry—Addresses, essays, lectures. 2. Chemistry, Technical—Addresses, essays, lectures. I. Smyshlíaev, S. I. II. Series: Kubanskiĭ gosudarstvennyĭ universitet. Nauchnye trudy ; vyp. 243.
QD39.K485 79-352096
79 MARC

(Problemy khimii i khimicheskoĭ tekhnologii)
Проблемы химии и химической технологии : Некоторые новые направления хим. технологии и химизации сельск. хоз-ва : Сборник статей : [Посвящается 80-летию акад. С. И. Вольфковича / АН СССР, Секция хим.-технол. и биол. наук, Моск. гос. ун-т им. М. В. Ломоносова ; Ред. коллегия, акад. Н. М. Жаворонков (отв. ред.) ... et al.]. — Москва : Наука, 1977.
309 p. : ill. ; 21 cm. USSR 77
Includes bibliographies.
2.40rub

1. Chemistry, Technical—Addresses, essays, lectures. 2. Agriculture—Addresses, essays, lectures. 3. Vol'fkovich, Semen Isaakovich. I. Vol'fkovich, Semen Isaakovich. II. Zhavoronkov, N. M. III. Akademiía nauk SSSR. Sektsiía khimiko-tekhnologicheskikh i biologicheskikh nauk. IV. Moscow. Universitet.
TP185.P76 78-423159

Technologia / [zespół redakcyjny Władysław Błaszków, Jerzy Ziobrowski]. — Wyd. 1. — Wrocław : Wydawn. Uczelniane Akademii Ekonomicznej, 1976.
239 p. : ill. ; 25 cm. — (Prace naukowe Akademii Ekonomicznej im. Oskara Langego we Wrocławiu ; zesz. 79(101)) P***
Summary in German.
Includes bibliographies.
zł45.00
1. Chemistry, Technical—Addresses, essays, lectures. I. Błaszków, Jerzy. II. Ziobrowski, Jerzy. III. Series: Akademia Ekonomiczna we Wrocławiu. Prace naukowe Akademii Ekonomicznej imienia Oskara Langego we Wrocławiu ; zesz. 79 (101)
HB9.B7522 Zesz. 79 78-372537
[TP185] 78 MARC

Technologia / [zespół redakcyjny Władysław Błaszków, Jerzy Ziobrowski]. — Wyd. 1. — Wrocław : Wydawn. Uczelniane Akademii Ekonomicznej, 1977.
145 p. : ill. ; 25 cm. — (Prace naukowe Akademii Ekonomicznej im. Oskara Langego we Wrocławiu ; zesz. 100 (122)) P***
Summaries in German; table of contents also in German.
200 copies printed.
Includes bibliographies.
zł14.00
1. Chemistry, Technical—Addresses, essays, lectures. I. Błaszków, Władysław. II. Ziobrowski, Jerzy. III. Series: Akademia Ekonomiczna we Wrocławiu. Prace naukowe Akademii Ekonomicznej imienia Oskara Langego we Wrocławiu ; zesz. 100.
HB9.B7522 zesz. 100 78-368118
[TP185] 78 MARC

—COLLECTED WORKS

Chemia i inżynieria chemiczna. zesz. 12.–
Poznań [Wydawn. Politechniki Poznańskiej] 1976–
no. ill. 24 cm. (Zeszyty naukowe Politechniki Poznańskiej)
Continues: Chemia techniki zastosowań.
Summaries in English and Russian.
No. 12– issued by Politechnika Poznańska.
1. Chemistry, Technical—Collected works. 2. Chemistry—Collected works. I. Politechnika Poznańska.
TP1.P57a 78-642598
MARC-S

Chemia techniki zastosowań. zesz. 5.–11. Poznań [Wydawn. Uczelniane Politechniki Poznańskiej] 1967–1972.
7 no. ill. 24 cm. (Zeszyty naukowe Politechniki Poznańskiej)
Continues: Politechnika Poznańska. Chemia.
German or Polish, with summaries in English and Russian.
Issued by Politechnika Poznańska.
Continued by: Chemia i inżynieria chemiczna.
1. Chemistry, Technical—Collected works. I. Politechnika Poznańska.
TP1.P57a 660'.05 72-621484
MARC-S

(Khimicheskaía tekhnologiía i khimiía)
Химическая технология и химия.
Рига, Рижский политехнический ин-т [etc.]
no. ill. 20 cm.
"Межвузовский научно-технический сборник."
Began in 1973.
1. Chemistry, Technical — Collected works. 2. Chemistry — Collected works. I. Riga. Politehniskais institūts.
TP1.K4143 79-640870
(MARC-S)

Politechnika Poznańska.
Chemia. zesz. 1–
Poznań, 1962–
no. in v. illus., facsims. 24 cm. (Its Zeszyty naukowe)
Summaries in Russian and English.
Continued by: Chemia techniki zastosowań.
1. Chemistry, Technical — Collected works. 2. Chemistry — Collected works. I. Title.
TP1.P57a 66-54467
MARC-S

—CONGRESSES

Congrès international de chimie industrielle, 28th, Madrid, 1955.
Memoria : Madrid, 22-31 octubre 1955 / XXVIII Congreso Internacional de Química Industrial. — Madrid : Saez, [1955?]
2 v. (1479 p.) : ill. ; 28 cm.
Includes bibliographies and index.
Spanish, French, Italian, English or German.
CONTENTS: t. 1. Comités. Participantes. Información general. Conferencias plenarias. Comunicaciones de las secciones 1 a 12.—t. 2. Comunicaciones de las secciones 13 a 23. Indices. Documentación industrial y comercial.
1. Chemistry, Technical—Congresses. I. Title.
TP5.C692 1955 78-393808
79 MARC

CHEMISTRY, TECHNICAL
—CONGRESSES (Continued)

Seminar on Chemistry in Industrial Development, Kuala Lumpur, 1975.
Proceedings of the Seminar on Chemistry in Industrial Development, Kuala Lumpur 1975 : a report of the proceedings of the Seminar ... / compiled by M. Mohinder Singh ; organised by the Malaysian Institute of Chemistry and the Royal Institute of Chemistry/Chemical Society (Malaysia Section). — ⌊Kuala Lumpur⌋ : Institut Kimia Malaysia, ⌊1975?⌋
xv, 198 p. : ill. ; 29 cm.
Includes bibliographical references.
1. Chemistry, Technical—Congresses. I. Singh, M. Mohinder. II. Malaysian Institute of Chemistry. III. Royal Institute of Chemistry/Chemical Society (Malaysian Section)
TP5.S45 1975 660.2 77-941568
78 MARC

SPE-AIME International Symposium on Oilfield and Geothermal Chemistry, Houston, Tex., 1979.
1979 International Symposium on Oilfield and Geothermal Chemistry : ⌊papers⌋. — ⌊Dallas : Society of Petroleum Engineers, AIME⌋, c1979.
335 p. : ill. ; 28 cm.
Cover title: 1979 SPE International Symposium on Oilfield and Geothermal Chemistry, Houston, Texas, January 22-24, 1979.
Includes bibliographical references.
1. Oil fields—Congresses. 2. Geothermal engineering—Congresses. 3. Chemistry, Technical—Congresses. I. Society of Petroleum Engineers of AIME.
TN863.S19 1979 622'.33'80154 79-116307
79 MARC

—DICTIONARIES

Sittig, Marshall.
Inorganic chemical industry : processes, toxic effluents, and pollution control / Marshall Sittig. — Park Ridge, N.J. : Noyes Data Corp., 1978.
x, 351 p. : ill. ; 25 cm. — (Chemical technology review ; no. 118) (Pollution technology review ; no. 52)
Bibliography: p. 346-347.
Includes indexes.
ISBN 0-8155-0726-7
1. Chemistry, Technical—Dictionaries. 2. Chemistry, Inorganic—Dictionaries. I. Title. II. Series. III. Series: Pollution technology review ; no. 52.
TP9.S57 661'.003 78-61892
79 MARC

—HANDBOOKS, MANUALS, ETC.

Technisch-chemische Praktikum / von e. Autorenkollektiv unter Federführung von W. Pippel u. K. Iseke ; ⌊dem Autorenkollektiv gehören an, M. Heltzig ... et al.⌋. — 1. Aufl. — Leipzig : Deutscher Verlag für Grundstoffindustrie, VEB, 1977.
431 p. : 127 ill. ; 22 cm. GFR78-A3
Bibliography: p. 406-408.
Includes index.
35.00M
1. Chemistry, Technical—Handbooks, manuals, etc. I. Pippel, Winfried. II. Iseke, K.
TP149.T43 660.2'842 78-378350
78 MARC

—INDEXES

AZ, aide-mémoire : Informations chimie : le premier journal français de la chimie. — ⌊1. éd.⌋. — Paris : EDI, c1977.
271 p. ; 16 cm. F***
"Index analytique des articles et informations publiés ⌊dans Informations chimie⌋ 11.01.1970 au 31.10.1977."
1. Informations chimie—Indexes. 2. Chemistry, Technical—Indexes.
TP1.I6432A19 660.2'05 78-386927
79 MARC

—MATHEMATICAL MODELS—
ADDRESSES, ESSAYS, LECTURES

Modellierung und Optimierung verfahrenstechnischer Systeme = Modelirovanie i optimizatsiia khimiko-tekhnologicheskikh sistem / ⌊verantw. für d. Hrsg. dieses Bd., K. Hartmann⌋. — Berlin : Akademie-Verlag, 1978.
xviii, 335 p. : ill., graphs ; 22 cm. — (Grundlagen der Verfahrenstechnik und chemischen Technologie) GFR78-A27
Chiefly in German with one contribution in English and summaries in English.
Table of contents also in Russian.
Includes bibliographical references.
78.00M
1. Chemistry, Technical—Mathematical models—Addresses, essays, lectures. 2. Mathematical optimization—Addresses, essays, lectures. I. Hartmann, Klaus, Dr. II. Title: Modelirovanie i optimizatsiia khimiko-tekhnologicheskikh sistem. III. Series.
TP149.M7 660.2'81'0184 79-376272
79 MARC

—MATHEMATICS

Floarea, O
Calcule de operații și utilaje din industria chimică / O. Floarea, O. Smigelschi. — București : Editura Tehnică, 1966.
491 p. : ill. ; 25 cm.
Includes bibliographies.
1. Chemistry, Technical—Mathematics. I. Smigelschi, O., joint author. II. Title.
TP149.F55 79-380304
79 MARC

—PERIODICALS

Société financière de transports et d'entreprises industrielles (société anonyme) Brussels.
Bulletin périodique - Société financière de transports et d'entreprises industrielles. 1.–
année; avril 1926–
⌊Bruxelles⌋ Société financière de transports et d'entreprises industrielles.
v. 27 cm. monthly.
1. Electric engineering—Periodicals. 2. Electric railroads—Periodicals. 3. Transportation—Periodicals. 4. Industry—Periodicals. 5. Chemistry, Technical—Periodicals.
TK2.S64 621.3'05 78-646345
MARC-S

—PROBLEMS, EXERCISES, ETC.

Lupu, A
Aplicații și probleme de chimie și tehnologie chimică / Angela Lupu, Gheorghe Țepeș. — București : Editura Didactică și Pedagogică, 1978.
177 p., ⌊1⌋ fold. leaf of plates : ill. ; 20 cm. — (Seria Culegeri de probleme pentru liceele de specialitate) R78-3762
lei10.40
1. Chemistry, Technical—Problems, exercises, etc. I. Țepeș, Gheorghe, joint author. II. Title.
TP168.L86 79-353648
•79 MARC

Urseanu, Florian.
Probleme de chimie și de tehnologie chimică / Florian Urseanu, Corneliu șanu-Mihăilă, Grigore Bozga. — București : Editura Tehnică, 1978.
384 p. : ill. ; 25 cm. R78-4220
Bibliography: p. 372-⌊373⌋
lei22.00
1. Chemistry—Problems, exercises, etc. 2. Chemistry, Technical—Problems, exercises, etc. I. Tărăbășanu-Mihăilă, Corneliu, joint author. II. Bozga, Grigore, joint author. III. Title.
QD42.U77 79-338501
•79 MARC

—RESEARCH

Pogostina, Evgeniia Samuilovna.
(Ėkonomika i organizatsiia nauchnykh issledovaniĭ v khimicheskoĭ promyshlennosti)
Экономика и организация научных исследований в химической промышленности / Е. С. Погостина, П. А. Погудин, Ю. Н. Ширяев. — Москва : Химия, 1978.
175 p. ; 20 cm. USSR***
Includes bibliographical references.
0.45rub
1. Chemistry, Technical—Research. I. Pogudin, Petr Andreevich, joint author. II. Shiriaev, Iuriĭ Nikolaevich, joint author. III. Title.
TP165.P63 78-423401

—RESEARCH—ISRAEL

Araten, David, 1928–
(ha-Kimyah ha-i-organit be-mosdot mada' u-meḥkar be-Yiśrael, ve-ziḳatah le-shimushim tekhnologiyim)
הכימיה האי-אורגאנית במוסדות מדע ומחקר בישראל וזיקתה לשימושים טכנולוגיים / דוד אשבורן (ארטן). — ירושלים : משרד ראש הממשלה, המועצה הלאומית למחקר ופיתוח, 1974.
66 leaves ; 28 cm.
Cover title: Seḳer kimyah i-organit be-Yiśrael.
"NCRD 9–74."
Includes index.
1. Chemistry, Technical—Research—Israel. 2. Chemistry, Inorganic. I. Israel. ha-Mo'atsah ha-le'umit le-meḥkar ule-fituaḥ. II. Title. III. Title: Seker kimyah i-organit be-Yiśrael.
TP177.A72 79–951255

—VOCATIONAL GUIDANCE—JUVENILE LITERATURE

Woodburn, John H
Opportunities in chemistry / John H. Woodburn. — Skokie, Ill. : VGM Career Horizons, c1979.
ix, 130 p. : ill. ; 21 cm. — (VGM Career Horizons series)
Bibliography: p. ⌊129⌋-130.
SUMMARY: Examines the education requirements, necessary personal qualities, and employment opportunities for many jobs related to chemistry.
1. Chemistry—Vocational guidance—Juvenile literature. 2. Chemistry, Technical—Vocational guidance—Juvenile literature. ⌊1. Chemistry—Vocational guidance. 2. Vocational guidance⌋ I. Title.
QD39.5.W64 540'.23 78-58663
79 MARC

CHEMISTRY, TEXTILE see Textile chemistry

CHEMISTRY, THEORETICAL
see Chemistry, Physical and theoretical

CHEMISTRY AS A PROFESSION

see also Chemists

CHEMISTRY IN WARFARE see Chemical warfare

CHEMISTRY OF CONCRETE see Concrete --Chemistry

CHEMISTRY OF FOOD see Food—Analysis; Food—Composition

CHEMISTS

see also Biochemists; Pharmacists

—BIOGRAPHY—ADDRESSES, ESSAYS, LECTURES

Yamaoka, Nozomu, 1892-
(Kagakushijin)
化学史塵：日本化学会創立百周年慶祝出版 / 山岡望. — 東京：内田老鶴圃新社, 1978.
16, 400, 18, 7 p., ⌊8⌋ leaves of plates : ill. ; 22 cm. Ja 78-18315
Originally published in Kōrutāru, June 1966–July 1974.
Includes index.
Y 4800
1. Chemistry—History—Addresses, essays, lectures. 2. Chemists—Biography—Addresses, essays, lectures. 3. Nippon Kagakukai. I. Title.
QD11.Y36 79–800514

—EMPLOYMENT—GERMANY, WEST—STATISTICS
—PERIODICALS

Fonds der Chemischen Industrie zur Förderung von Forschung, Wissenschaft und Lehre.
Statistische Übersichten: Bestand und Bedarf an Chemikern in der chemischen Industrie der Bundesrepublik Deutschland.
Frankfurt am Main, Fonds der Chemischen Industrie.
v. 21–30 cm. (Schriftenreihe des Fonds der Chemischen Industrie)
1. Chemists—Employment—Germany, West—Statistics—Periodicals. I. Title. II. Series: Fonds der Chemischen Industrie zur Förderung von Forschung, Wissenschaft und Lehre. Schriftenreihe des Fonds der Chemischen Industrie.
TP186.F65a 78–643358
MARC-S

—VOCATIONAL GUIDANCE

Royal Institute of Chemistry.
Conditions of employment, guidance for chemists / Royal Institute of Chemistry. — London : The Institute, 1977.
⌊2⌋, 60 p. ; 21 cm. GB78-12551
Cover title.
£2.00
1. Chemists—Vocational guidance. I. Title.
QD39.5.R69 1977 331.2'041'540941 78-318075
78 MARC

—AZERBAIJAN—BIOGRAPHY

Suleĭmanova, Ė T
(S. D. Mekhtiev)
С. Д. Мехтиев : библиография / ⌊составители, Э. Т. Сулейманова, М. Р. Мусаев⌋. — Баку : Элм, 1976.
168 p. ; 16 cm. — (Деятели науки и культуры Азербайджана) USSR***
At head of title: Akademiia nauk Azerbaĭdzhanskoĭ SSR. Fundamental'naia biblioteka.
Added t. p. in Azerbaijani.
Prefatory matter in Azerbaijani and Russian.
Series romanized: Deiateli nauki i kul'tury Azerbaĭdzhana.
Includes indexes.
1. Mekhtiev, S. D.—Bibliography. 2. Chemists—Azerbaijan—Biography. 3. Chemistry—Bio-bibliography. I. Musaev, M. R., joint author. II. Akademiia nauk Azerbaĭdzhanskoĭ SSR, Baku. Fundamental'naia biblioteka.
Z8606.73.S94 79–410682
[QD22.M396]

CHEMISTS (Continued)

—ENGLAND—BIOGRAPHY

Priestley, Joseph, 1733-1804.
Memoirs of Dr. Joseph Priestley to the year 1795, written by himself ; with a continuation to the time of his decease by his son, Joseph Priestley, and observations on his writings by Thomas Cooper and William Christie. — Millwood, N.Y. : Kraus Reprint Co., 1978.
v, 824, x p. ; 21 cm.
Reprint of the 1806 ed. printed by J. Binns, Northumberland, Pa.
"Catalogue of Dr. Priestley's smaller pamphlets and uncollected papers on philosophical subjects": p. 290-293; "A catalogue of books written by Doctor Priestley": p. i-x.
ISBN 0-527-72730-X
1. Priestley, Joseph, 1733-1804. 2. Chemists—England—Biography. I. Priestley, Joseph, 1768-1833.
QD22.P8A33 1978 540'.92'4 78-3422
78 MARC

—FRANCE—BIOGRAPHY

Crosland, Maurice P
Gay-Lussac, scientist and bourgeois / Maurice Crosland. — Cambridge ; New York : Cambridge University Press, 1978.
xvi, 333 p. : port. ; 24 cm.
Bibliography: p. 312-323.
Includes indexes.
ISBN 0-521-21979-5
1. Gay-Lussac, Joseph Louis, 1778-1850. 2. Chemists—France—Biography. I. Title.
QD22.G35C76 540'.92'4 77-91084
78 MARC

Faershteĭn, Moiseĭ Gertŝevich.
(Sharl' Zherar)
Шарль Жерар. 1816–1856. Москва, Наука, 1968.
163 p. with illus., port. 20 cm. (Академия наук СССР. Научно-биографическая серия) USSR 68-11984
At head of title: М. Г. Фаерштейн.
Series romanized: Akademiiâ nauk SSSR. Nauchno-biograficheskaiâ seriiâ.
Bibliography: p. 159–[162]
0.51rub
1. Gerhardt, Charles Frédéric, 1816–1856. 2. Chemists—France—Biography.
QD22.G4F3 78-427586

—GERMANY—BIOGRAPHY

Hoffmann, Klaus.
Otto Hahn : Stationen aus dem Leben eines Atomforschers : Biografie / von Klaus Hoffmann ; mit einem Geleitwort von Manfred von Ardenne. — Berlin : Verlag Neues Leben, c1978.
375 p. : ill., facsims., ports. ; 21 cm. GDR***
Bibliography: p. 375.
Includes index.
13.20M
1. Hahn, Otto, 1879-1968. 2. Chemists—Germany—Biography.
QD22.H2H63 540'.92'4 79-356969
79 MARC

Kritŝman, Viktor Abramovich.
(German Kopp)
Герман Копп / В. А. Крицман, Г. В. Быков. — Москва : Наука, 1978.
159 p. : ill. ; 20 cm. — (Научно-биографическая серия) USSR 78
On leaf preceding t. p.: Akademiiâ nauk SSSR.
Series romanized: Nauchno-biograficheskaiâ seriiâ.
"Библиографический указатель трудов Г. Коппа": p. 149–[156]
Bibliography: p. [157]
0.50rub
1. Kopp, Hermann Franz Moritz, 1817–1892. 2. Chemists—Germany—Biography. I. Bykov, Georgiĭ Vladimirovich, joint author.
QD22.K63K74 78-420635

Kuwada, Satoru, 1898-
(Bikutoru Maiyā no omokage)
ヴィクトル・マイヤーのおもかげ / 桑田智著. — 東京 : 廣川書店, 昭和38 [1963]
viii, 218 p., [4] leaves of plates : ill. ; 19 cm.
Bibliography: p. [207]-209.
Includes index.
1. Meyer, Victor, 1848–1897. 2. Chemists—Germany—Biography. I. Title.
QD22.M48K89 78-809274

Ostwald, Wilhelm, 1853-1932.
Forschen und Nutzen : Wilhelm Ostwald zur wiss. Arbeit / aus seinen Schriften ausgew., bearb. u. zsgest. anlässlich seines 125. Geburtstages von Günther Lotz, Lothar Dunsch, Uta Kring ; mit e. Geleitw. von Peter Adolf Thiessen u. e. Einf. von Günther Lotz u. Lothar Dunsch. — Berlin : Akademie-Verlag, 1978.
xlviii, 278 p. : 55 ill. (some col.) ; 22 cm. — (Beiträge zur Forschungstechnologie : Sonderband ; 1) GDR78-A4
Includes bibliographical references and indexes.
38.00M
1. Chemistry—Collected works. 2. Ostwald, Wilhelm, 1853-1932. 3. Chemists—Germany—Biography. I. Lotz, Günther. II. Dunsch, Lothar. III. Kring, Uta. IV. Title. V. Series.
QD3.O83 1978 79-364024
79 MARC

Otto Hahn : Begr. d. Atomzeitalters : e. Biographie in Bildern u. Dokumenten / mit e. Geleitw. von Reimar Lüst, e. Vorw. von Paul Matussek u. e. Einf. von Walther Gerlach ; hrsg. von Dietrich Hahn. — München : List, 1979.
357 p. : chiefly ill. ; 32 cm. GFR79-A
Bibliography: p. 353.
Includes index.
ISBN 3-471-77841-1 : DM98.00
1. Hahn, Otto, 1879-1968. 2. Chemists—Germany—Biography. I. Hahn, Dietrich, 1946-
QD22.H2O87 79-371691
79 MARC

P'an, Chi-hsing.
(Ko ming hua hsüeh chia Hsiao-lai-ma)
革命化学家肖莱马 / 潘吉星编. — 第1版. — 北京 : 科学出版社, 1978.
ii, 165 p., [1] leaf of plates : port. ; 19 cm.
Includes bibliographical references.
$0.38
1. Schorlemmer, Carl, 1834–1892. 2. Chemists—Germany—Biography. I. Title.
QD22.S43P36 79-835772

—GERMANY, WEST—BIOGRAPHY

Lanz, Kurt, 1919-
Weltreisender in Chemie / Kurt Lanz. — 1. Aufl. — Düsseldorf ; Wien : Econ-Verlag, 1978.
528 p. : ill. (some col.) ; 22 cm. GFR79-A
Includes indexes.
Bibliography: p. 528.
ISBN 3-430-15895-8 : DM38.00
1. Farbwerke Hoechst AG. 2. Chemical industries. 3. Voyages and travels. 4. Lanz, Kurt, 1919- 5. Chemists—Germany, West—Biography. I. Title.
HD9654.9.F33L36 79-360712
79 MARC

—GREAT BRITAIN —BIOGRAPHY

Krivomazov, Aleksandr Nikolaevich.
(Frederik Soddi)
Фредерик Содди : 1877–1956 / А. Н. Кривомазов. — Москва : Наука, 1978.
205 p. : ill. ; 20 cm. — (Научно-биографическая серия) USSR***
Series romanized: Nauchno-biograficheskaiâ seriiâ.
Includes bibliographies and index.
0.75rub
1. Soddy, Frederick, 1877–1956. 2. Chemists—Great Britain—Biography. I. Title.
QD22.S63K74 78-421568

—HUNGARY—BIOGRAPHY

Móra, László.
Korach Mór : a korszerű műszaki kémia úttörője (1888-1975) / Móra László. — Budapest : Budapesti Műszaki Egyetem Könyvtára, 1978.
238 p., [37] leaves of plates : ill. ; 20 cm. — (Műszaki tudománytörténeti kiadványok ; 26 sz) H***
Summary in English, German, Italian and Russian.
Bibliography of works by and about M. Korach: p. 195-220.
Includes index.
ISBN 9634210074
1. Korach, Mór. 2. Chemists—Hungary—Biography. I. Series: Budapest. Műszaki Egyetem. Központi Könyvtár. Műszaki tudománytörténeti kiadványok ; 26 sz.
TP140.K67M67 79-346700
79 MARC

Szabadváry, Ferenc.
Winkler Lajos / Szabadváry Ferenc. — Budapest : Akadémiai Kiadó, 1975.
178 p., [1] leaf of plates : port. ; 15 cm. — (A Múlt magyar tudosai)
Bibliography: p. 155-178.
ISBN 9630505657
1. Winkler, Lajos, 1863-1939. 2. Chemists—Hungary—Biography.
QD22.W493S96 78-399187
79 MARC

Szőkefalvi-Nagy, Zoltán.
Ilosvay Lajos / Szőkefalvi-Nagy Zoltán. — Budapest : Akadémiai Kiadó, 1978.
201 p. : port. ; 14 cm. — (A Múlt magyar tudósai ISSN 0133-1884) H***
Includes bibliographical references.
ISBN 9630514567
1. Ilosvay, Lajos, 1851-1936. 2. Chemists—Hungary—Biography.
QD22.I62S97 78-397266
79 MARC

—INDIA—BIOGRAPHY

New trends in heterocyclic chemistry / edited by R. B. Mitra ... [et al.]. — Amsterdam ; New York : Elsevier Scientific Pub. Co. ; New York : distributors for the United States and Canada, Elsevier/North-Holland, 1979.
xii, 407 p. : ill. ; 25 cm. — (Studies in organic chemistry ; 3)
"This volume has been brought out to commemorate the 60th birthday of Professor Bal Dattatraya Tilak, 26 Sept. 1978."
"Publications of Professor B. D. Tilak": p. 383-395.
Includes bibliographical references and index.
ISBN 0-444-41816-4 : fl 65.00
1. Heterocyclic compounds—Addresses, essays, lectures. 2. Tilak, B. D., 1918- 3. Chemists—India—Biography. I. Mitra, R. B. II. Tilak, B. D., 1918- III. Series: Studies in organic chemistry (Amsterdam) ; 3.
QD400.N48 547'.59 79-13477
79 MARC

—LATVIA—BIOGRAPHY

Alfrēds Ieviņš dzīve un darbā / [atbildīgā redaktore E. Gudriniece]. — Rīga : Zinātne, 1979.
285 p., [9] leaves of plates : ill. ; 21 cm. USSR***
At head of title: Latvijas PSR Zinātņu akadēmija. Neorganiskās ķīmijas institūts. Rīgas politehniskais institūts.
Part of correspondence included in German or Russian.
Includes bibliographies.
1.20rub
1. Ieviņš, Alfrēds. 2. Chemists—Latvia—Biography. I. Gudriniece, Emīlija, 1920- II. Latvijas Padomju Socialistiskās Republikas Zinātņu akadēmija. Neorganiskās ķīmijas institūts. III. Riga. Politehniskais institūts.
QD22.I48A43 79-385421
79 MARC

Novackis, H
Profesors Ernsts Karpovics : personālais literatūras rādītājs / [sastādītāji H. Novackis, D. Paukšēna]. — Rīga : LVU, 1977.
54 p., [1] leaf of plates : port. ; 20 cm. — (Pētera Stučkas Latvijas valsts universitātes zinātnieki) USSR***
At head of title: Latvijas PSR Augstākās un vidējās speciālās izglītības ministrija. Pētera Stučkas Latvijas valsts universitāte. Zinātniskā bibliotēka.
Cover title: E. Karpovics.
Added t.p.: Professor Ërnst Karpovitŝ.
Latvian and Russian.
Includes index.
0.10rub
1. Karpovics, Ernsts, 1892- . 2. Chemists—Latvia—Biography. 3. Chemistry—Bio-bibliography. I. Paukšēna, D., joint author. II. Riga. Universitāte. Zinātniskā bibliotēka. III. Title. IV. Title: Professor Ërnst Karpovitŝ.
QD22.K32N68 79-371304
79 MARC

—POLAND—BIOGRAPHY

Dorabialska, Alicja.
Wojciech Świętosławski, his life and scientific work / by A. Dorabialska ; [English translation by A. Płonka]. — Warszawa : Polish Scientific Publishers, 1974.
98 p., [4] leaves of plates : ill. ; 25 cm. — (Polish men of science)
Includes bibliographical references.
1. Świętosławski, Wojciech, 1881-1968. 2. Chemists—Poland—Biography. I. Series.
QD22.S94D6713 540'.92'4 79-302237
79 MARC

—RUSSIA—BIOGRAPHY

Îur'ev, Îuriĭ Konstantinovich.
Life and work of academician Nikolai Zelinsky / Y. Yuriev, R. Levina ; [translated from the Russian by David Sobolev]. — Moscow : Foreign Languages Pub. House, 1958.
127, [1] p. : ill. ; 20 cm. — (Men of Russian science)
Translation of Zhizn' i deiâtel'nost' akademika Nikolaiâ Dmitrievicha Zelinskogo.
Bibliography: p. 114–[128]
1. Zelinskiĭ, Nikolaĭ Dmitrievich, 1861-1953. 2. Chemists—Russia—Biography. I. Levina, R. ÎA., joint author. II. Title.
QD22.Z4I813 540'.92'4 78-321900
79 MARC

Mendeleeva, Anna Ivanovna, b. 1860.
(Mendeleev v zhizni)
Менделеев в жизни / А. И. Менделеева ; вступ. статья и примечания М. Цявловского. — Москва : Изд. М. и С. Сабашниковыхъ, 1928.
vi, 182 p., [2] leaves of plates : ports. ; 24 cm. — (Записи прошлого : воспоминания и письма)
Series romanized: Zapisi proshlogo : vospominaniiâ i pis'ma.
Includes index.
1. Mendeleev, Dmitriĭ Ivanovich, 1834–1907. 2. Chemists—Russia—Biography. I. Title.
QD22.M43M46 79-413827

Ozarovskaiâ, Ol'ga Ėrastovna.
(D. I. Mendeleev)
Д. И. Менделеев : по воспоминаниям О. Э. Озаровской. — Москва : Федерация, 1929.
163 p., [4] leaves of plates : ill., ports. ; 18 cm.
1. Mendeleev, Dmitriĭ Ivanovich, 1834–1907. 2. Chemists—Russia—Biography. I. Title.
QD22.M43O93 79-413717

CHEMISTS (Continued)

—RUSSIAN REPUBLIC—BIOGRAPHY

Kliūchevich, Aleksandr Solomonovich.
(Flavian Mikhailovich Flavitskii)
Флавиан Михайлович Флавицкий / А. С. Ключевич. — Казань : Изд-во Казан. ун-та, 1978.
127 p. : ill. ; 20 cm. — (Выдающиеся химики Казанского университета) USSR 78
Bibliography : p. 115–[125]
0.52rub
1. Flavitskii, Flavian Mikhailovich, 1848–1917. 2. Chemists—Russian Republic—Biography. I. Series : Vydaiushchiesia khimiki Kazanskogo universiteta.
QD22.F56K55 79–407484

—SWEDEN—BIOGRAPHY

Krook, Hans, 1920-
Jacob Berzelius / Hans Krook. — Stockholm : Rabén & Sjögren, 1979.
143, [1] p., [8] leaves of plates : ill. ; 22 cm. S79
Bibliography: p. 140-141.
Includes index.
ISBN 9129526817
1. Berzelius, Jöns Jakob, friherre, 1779-1848. 2. Chemists—Sweden—Biography.
QD22.B5K76 79-363449
79 MARC

—UKRAINE—BIOGRAPHY

(Vasilii Ivanovich Atroshchenko)
Василий Иванович Атрощенко / Вступ. статья Н. Ф. Киркача ... [et al.] ; Указ. печ. работ [сост.] Н. П. Курган, И. И. Литвиненко. — Киев : Наук. думка, 1978.
78 p., [1] leaf of plates : port. ; 17 cm. — (Биобиблиография ученых УССР) USSR 78
At head of title: Akademiia nauk USSR.
Series romanized: Biobibliografiia uchenykh USSR.
Includes indexes.
0.15rub
1. Atroshchenko, Vasilii Ivanovich—Bibliography. 2. Chemistry—Ukraine—Bio-bibliography. 3. Chemists—Ukraine—Biography. I. Kirkach, Nikolai Fedorovich. II. Kurgan, Nelli Pavlovna. III. Litvinenko, Igor' Ivanovich.
Z8046.94.V37 79–410594
[QD22.A74]

—UNITED STATES—BIOGRAPHY

For Ilya Prigogine / edited by Stuart A. Rice. — New York : Wiley, c1978.
472 p. : ill. ; 24 cm. — (Advances in chemical physics ; v. 38)
"An Interscience publication."
Includes bibliographical references and index.
CONTENTS: Primas, H. and Müller-Herold, U. Quantum-mechanical system theory.—Spohn, H. and Lebowitz, J. L. Irreversible thermodynamics for quantum systems weakly coupled to thermal reservoirs—Grecos, A. P. Solvable models for unstable states in quantum physics.—Balescu, R. and Résibois, P. Aspects of kinetic theory.—Klimontovich, Y. L. Kinetic theory of plasmas.—Eigen, M. How does information originate?—Nicolis, G., Erneux, T., and Herschkowitz-Kaufman, M. Pattern formation in reacting and diffusing systms.—Hanusse, P., Ross, J., and Ortoleva, P. Instability and far-from-equilibrium states of chemically reacting systems.—Hess, B., Goldbeter, A., and Lefever, R. Temporal, spatial, and functional order in regulated biochemical and cellular systems.
—Nazarea, A. D. Electric fields and self-coherent patterns and structures in chemical systems.
ISBN 0-471-03883-0
1. Chemistry, Physical and theoretical—Addresses, essays, lectures. 2. Prigogine, Ilya. 3. Chemists—United States—Biography. I. Prigogine, Ilya. II. Rice, Stuart Alan, 1932- III. Series.
QD453.A27 vol. 38 541'.08 s 79-101799
[QD455] 79 MARC

Neckers, James Winfred, 1902-
The building of a department : chemistry at Southern Illinois University, 1927-1967 / James W. Neckers. — Carbondale : Southern Illinois University Press, c1979.
187 p. ; 23 cm.
Includes index.
ISBN 0-8093-0901-7
1. Neckers, James Winfred, 1902- . 2. Chemists—United States—Biography. 3. Illinois. Southern Illinois University, Carbondale. Dept. of Chemistry and Biochemistry. I. Title.
QD22.N36A34 540'.7'11773994 78-13384
78 MARC

CHEMNYĬ, MYKHAĬLO, 1901-

Chemnyĭ, Mykhaĭlo, 1901–
(Potoptani mriï)
Потоптані мрії / Михайло Чемний. — Детройт : Чемний, 1969.
206 p. : ill. ; 21 cm.
Added t. p. : The trampled dreams.
1. Chemnyĭ, Mykhaĭlo, 1901– 2. Ukrainian Americans—Michigan—Detroit—Biography. 3. Detroit—Biography. I. Title. II. Title: The trampled dreams.
F574.D453C473 79–413421

CHEMODANOV, NIKOLAĬ SERGEEVICH

—ADDRESSES, ESSAYS, LECTURES

(Problemy obshchego i germanskogo iazykoznaniia)
Проблемы общего и германского языкознания : [Сб. статей : Посвящается Н. С. Чемоданову / Редкол. Э. М. Медникова ... et al.]. — Москва : Изд-во МГУ, 1978.
184 p. : port. ; 20 cm. USSR 78
Includes bibliographical references.
1.80rub
1. Linguistics—Addresses, essays, lectures. 2. Germanic languages—Addresses, essays, lectures. 3. Chemodanov, Nikolai Sergeevich—Addresses, essays, lectures. I. Mednikova, Ėsfir' Maksimovna.
P125.P74 79–408131

CHEMORECEPTORS

see also Carotid gland; Smell; Taste

Beets, Muus Gerrit Jan.
Structure-activity relationships in human chemoreception / M. G. J. Beets. — London : Applied Science, 1978.
xii, 408 p. : ill. ; 23 cm. GB***
Bibliography: p. 363-381.
Includes index.
ISBN 0-85334-746-8 : £25.00
1. Chemoreceptors. 2. Structure-activity relationship (Pharmacology) 3. Taste. 4. Smell. I. Title.
QP455.B43 612'.86 78-324346
79 MARC

—CONGRESSES

Morphology and mechanisms of chemoreceptors : proceedings of an international satellite symposium held in Srinagar, Kashmir, India, October 9-12, 1974 / edited by A. S. Paintal. — Delhi : Vallabhbhai Patel Chest Institute, University of Delhi, c1976.
xv, 357 p., [30] leaves of plates : ill. ; 25 cm.
"Sponsored by the National Committee for Physiology of the Indian National Science Academy."
Includes bibliographical references and indexes.
Rs175.00 ($22.00 U.S.)
1. Chemoreceptors—Congresses. 2. Carotid gland—Congresses. 3. Respiration—Regulation—Congresses. I. Paintal, A. S. II. Indian National Science Academy. Committee for Physiology.
QP455.M67 612'.28 78-902398
78 MARC

CHEMOSPHERE

see also Mesosphere

CHEMOSURGERY

Burks, James Willis, 1911-
Dermabrasion and chemical peeling in the treatment of certain cosmetic defects and diseases of the skin / by James W. Burks ; with a foreword by Morris Waisman. — Springfield, Ill. : Thomas, c1979.
xix, 236 p. : ill. ; 24 cm.
Published in 1956 under title: Wire brush surgery in the treatment of certain cosmetic defects and diseases of the skin.
Includes bibliographical references and index.
ISBN 0-398-03778-7
1. Dermabrasion. 2. Chemosurgery. I. Title.
[DNLM: 1. Chemexfoliation. 2. Dermabrasion. WR650 B959w]
RD520.B87 1979 617'.477 78-2704
78 MARC

Mohs, Frederic Edward, 1910-
Chemosurgery : microscopically controlled surgery for skin cancer / by Frederic E. Mohs. — Springfield, Ill. : Thomas, c1978.
xxii, 378 p. : ill. ; 26 cm.
Bibliography: p. 358-364.
Edition of 1956 published under title: Chemosurgery in cancer, gangrene, and infections.
Bibliography: p.
Includes index.
ISBN 0-398-03725-6
1. Skin—Cancer—Surgery. 2. Chemosurgery. 3. Skin—Surgery. 4. Tumors—Surgery. I. Title.
[DNLM: 1. Skin neoplasms—Surgery. 2. Cautery. WR500 M699c]
RD655.M63 1978 616.9'94'77 77-14589
77 MARC

CHEMOTAXIS

Chemotaxis : its biology and biochemistry / editor : E. Sorkin ; contributor, Julius Adler ... [et al.]. — Basel ; New York : S. Karger, 1974.
viii, 467 p. : ill. ; 25 cm. — (Antibiotics and chemotherapy ; v. 19) Sw75-A-647
Includes bibliographies and index.
ISBN 3-8055-1688-6 : 165.00F
1. Chemotaxis. I. Sorkin, Ernst, 1920- . II. Adler, Julius, 1930- . III. Series: Antibiotics and chemotherapy (Basel) ; v. 19.
RM260.A55 vol. 19 615'.58'08 s 75-307066
[QH514] *78 MARC

Taxis and behavior : elementary sensory systems in biology / edited by G. L. Hazelbauer. — London : Chapman and Hall ; New York : Wiley, 1978.
x, 341 p. : ill. ; 24 cm. — (Receptors and recognition : Series B ; v. 5)
"A Halsted Press book."
Includes bibliographies and index.
ISBN 0-470-26483-7
1. Taxes (Biology) 2. Chemotaxis. I. Hazelbauer, G. L. II. Series.
QH514.T39 574.1'8 78-14931
78 MARC

CHEMOTHERAPY

see also Anti-infective agents; Antibacterial agents; Antibiotics; Antitubercular agents; Antiviral agents; Drug utilization; Drugs; Psychopharmacology; and subdivision Chemotherapy under diseases, e.g. Cancer—Chemotherapy; Heart—Chemotherapy

Applied therapeutics for clinical pharmacists / edited by Mary Anne Koda-Kimble, Brian S. Katcher, Lloyd Y. Young. — 2d ed. — San Francisco : Applied Therapeutics, c1978.
944 p. : ill. ; 27 cm.
Includes bibliographies and index.
ISBN 0-915486-02-4 : $31.50
1. Chemotherapy. 2. Pharmacology. I. Koda-Kimble, Mary Anne. II. Katcher, Brian S. III. Young, Lloyd Y.
RM262.A65 1978 615'.58 78-72533
79 MARC

Clinical therapeutics : the forty-seventh Hahnemann symposium / editors, David T. Lowenthal and David A. Major. — New York : Grune & Stratton, c1978.
xiv, 424 p. : ill. ; 27 cm.
Includes bibliographical references and index.
ISBN 0-8089-1080-9
1. Chemotherapy. 2. Pharmacology. I. Lowenthal, David T. II. Major, David A. III. Hahnemann Medical College and Hospital of Philadelphia.
[DNLM: 1. Drug therapy—Congresses. WB330 C643 1977]
RM262.C53 615'.5 78-18899
78 MARC

Hedgecock, Loyd W
Antimicrobial agents [by] Loyd W. Hedgecock. Philadelphia, Lea & Febiger [1967]
xi, 232 p. illus. 26 cm. (Medical technology, 3)
Includes bibliographies.
1. Disinfection and disinfectants. 2. Chemotherapy. I. Title. II. Series: Medical technology (Philadelphia) 1967, 3.
RB37.M42 no. 3 614.4'8 67-31313
[r79]rev MARC

Johns, Marjorie P 1922-
Drug therapy and nursing care / by Marjorie P. Johns ; in consultation with O. James Inashima. — New York : Macmillan, c1979.
x, 694 p. : ill. ; 24 cm.
Bibliography: p. 667-668.
Includes index.
ISBN 0-02-360800-5
1. Chemotherapy. 2. Nursing. I. Title.
[DNLM: 1. Drug therapy—Nursing texts. QV4.3 J65d]
RM262.J64 615'.58'024613 78-17822
78 MARC

Li, Huan-shen.
(Hua hsüeh yao li hsüeh)
化學藥理學 = Chemical pharmacology / 李逸燊, 李匡邦合著. — 台北縣新店鎮：國立中國醫藥研究所, 1975-
v. in : ill. ; 27 cm. — (國立中國醫藥研究所叢書)
In Chinese, with nomenclature also in English.
Series romanized: Kuo li Chung-kuo i yao yen chiu so ts'ung shu.
CONTENTS: 第1編 化學療法 上
1. Pharmacology. 2. Chemotherapy. I. Li, K'uang-pang. II. Title. III. Title: Chemical pharmacology.
RM300.L494 78–842198

CHEMOTHERAPY (Continued)

Manitoba. Drug Standards and Therapeutics Committee.
Manitoba drug standards and therapeutics formulary / [Manitoba Drug Standards and Therapeutics Committee]. — 3d ed. — [Winnipeg : Manitoba Drug Standards and Therapeutics Committee], 1975.
48 p. ; 26 cm. CP75-82190-6
Cover title.
$7.50
1. Drugs—Prescribing—Standards—Manitoba. 2. Drugs—Prices—Manitoba. 3. Chemotherapy. I. Title.
RM138.M35 1975 615'.11'7127 78-312819
78 MARC

Pedersen, Asger, 1919-
Lægemidler. København, Forlaget for Faglitteratur, [1967].
12, 205 p. 25 cm. 43.00 D68-45
Reprinted from Håndbog for sygeplejersker.
1. Drugs. 2. Chemotherapy. I. Title.
RM300.P42 1967 78-391000
79 MARC

Rodman, Morton J
Pharmacology and drug therapy in nursing / Morton J. Rodman, Dorothy W. Smith. — 2d ed. — Philadelphia : Lippincott, c1979.
xiv, 1085 p. : ill. ; 28 cm.
Includes bibliographies and index.
ISBN 0-397-54230-5
1. Pharmacology. 2. Chemotherapy. 3. Drugs. I. Smith, Dorothy W., joint author. II. Title.
[DNLM: 1. Pharmacology—Nursing texts. 2. Drug therapy—Nursing texts. QV4.3 R693p]
RM300.R66 1979 615'.1'02461 79-11342
79 MARC

Turner, Paul.
Clinical pharmacology / Paul Turner, Alan Richens. — 3d ed. — Edinburgh ; New York : Churchill Livingstone ; New York : distributed by Longman, 1978.
254 p. : ill. ; 22 cm. — (Churchill Livingstone medical texts) GB78
Includes indexes.
ISBN 0-443-01759-X : £2.95
1. Pharmacology. 2. Chemotherapy. I. Richens, Alan, joint author. II. Title.
RM300.T86 1978 615'.1 77-30524
77 MARC

Wang, Richard I H 1924-
Practical drug therapy / Richard I. H. Wang, with 28 contributors. — Philadelphia : Lippincott, c1979.
xviii, 693 p. : ill. ; 26 cm.
Includes bibliographies and index.
ISBN 0-397-50420-9
1. Chemotherapy. I. Title.
[DNLM: 1. Drug therapy. WB330 W245p]
RM262.W36 615'.58 79-1306
79 MARC

—COMPLICATIONS AND SEQUELAE
see Drugs—Side effects

—CONGRESSES

Chemotherapy / editor, M. Adolphe. — 1st ed. — Oxford ; New York : Pergamon Press, 1979.
viii, 227 p. : ill. ; 26 cm. — (Advances in pharmacology and therapeutics) (Proceedings of the 7th International Congress of Pharmacology ; v. 10) GB79
ISBN 0-08-023200-0 : £25.00 ($50.00)
1. Chemotherapy—Congresses. I. Adolphe, M. II. Series. III. Series: International Congress of Pharmacology, 7th, Paris, 1978. Proceedings of the 7th International Congress of Pharmacology ; v. 10.
RM21.I58 1978 vol. 10 615'.1'08 s 78-41032
[RM260.2] 79 MARC

International Symposium on Future Trends in Chemotheraphy, Tirrenia, Italy, 1976.
Future trends in chemotherapy-II / edited by A. Bertelli, C. Grassi, M. Neuman ; [sponsored by Italian Society of Clinical Pharmacology, Italian Society of Chemotherapy ; organized by Institute of Pharmacology, University of Pisa]. — [Barcelona] : J. R. Prous ; New York : [distributed by] W. S. Heinman, 1977.
vii, 219 p. : ill. ; 25 cm. — (Drugs under experimental and clinical research ; v. 3, no. 1) Sp***
"Proceedings of an international symposium held in Tirrenia (Pisa) May 15-16, 1976."
Includes bibliographical references.
1. Chemotheraphy—Congresses. I. Bertelli, Aldo. II. Grassi, C. III. Neuman, Maur. IV. Società italiana di farmacologia clinica. V. Società italiana di chemioterapia. VI. Pisa. Università. Istituto di farmacologia. VII. Title. VIII. Series.
RM260.2.I59 1976 615'.58 78-323687
79 MARC

Pharmaceutical medicine / edited by Norman MacLeod ; foreword by Derrick Dunlop. — Edinburgh ; New York : distributed in the United States of America by Longman, 1979.
165 p. : ill. ; 24 cm. GB79
"This symposium follows the proceedings of the Association of Medical Advisers in the Pharmaceutical Industry ... 24 and 25th May 1978 at the Royal College of Physicians."
Includes bibliographies and index.
ISBN 0-443-01950-9 : £12.00
1. Pharmaceutical research—Great Britain—Congresses. 2. Drug trade—Great Britain—Congresses. 3. Chemotherapy—Congresses. I. MacLeod, Norman, M.B. II. Association of Medical Advisers in the Pharmaceutical Industry.
[DNLM: 1. Drug industry—Great Britain. 2. Pharmacy. QV736.3 P533]
RS122.P43 614.3'5 78-41160
79 MARC

Seminář Farmakoterapie zánětů a zánětlivé bolesti, Ostrava, Czechoslovakia, 1976.
Teoretické základy farmakoterapie zánětů : [sborník přednášek semináře Farmakoterapie zánětů a zánětlivé bolesti pořádaného dne 13. října 1976 v Ostravě, Spojenými podniky pro zdravotnickou výrobu SPOFA ... et al.] / kolektiv autorů ; [redakce Jiří Lenfeld]. — Praha : Odbor vědeckých informací, [1978]
195 p. : ill. ; 21 cm. — (Farmakoterapeutické zprávy SPOFA : Supplementum 1/1978) Cz***
Summaries also in English and Russian.
Includes bibliographies.
1. Anti-inflammatory agents—Congresses. 2. Chemotherapy—Congresses. I. Lenfeld, Jiří. II. Spofa. III. Title. IV. Series: Farmakoterapeutické zprávy : Supplementum ; 1978/1.
RM405.S45 1976 79-379206
79 MARC

—DATA PROCESSING—CONGRESSES

IFIP TC-4 Working Conference on Computer Aid to Drug Therapy and to Drug Monitoring, Bern, 1978.
Computer aid to drug therapy and to drug monitoring : proceedings of the IFIP Working Conference on Computer Aid to Drug Therapy and to Drug Monitoring / edited by H. Ducrot ... [et al.] ; sponsored by IFIP Technical Committee 4, Information Processing in Health Care and Biomedical Research, International Federation for Information Processing. — Amsterdam ; New York : North-Holland Pub. Co. ; New York : sole distributors for the U.S.A. and Canada, Elsevier North-Holland, 1978.
xvi, 444 p. : ill. ; 23 cm.
Includes bibliographical references and indexes.
ISBN 0-444-85188-7
1. Chemotherapy—Data processing—Congresses. 2. Drug utilization—Data processing—Congresses. 3. Information storage and retrieval systems—Drugs—Congresses. I. Ducrot, Henry. II. International Federation for Information Processing. Technical Committee 4. III. Title.
RM263.I14 1978 615'.58'02854 78-18294
78 MARC

—EXAMINATIONS, QUESTIONS, ETC.

Clinical pharmacy in ambulatory care / edited by Bruce H. Woolley, Phillip R. Oppenheimer. — [s.l.] : Symposia Specialists, c1978.
ix, 172 p. : ill. ; 24 cm.
"Volume 1, number 10."
Includes bibliographies and index.
ISBN 0-88372-087-6
1. Chemotherapy—Examinations, questions, etc. I. Woolley, Bruce H. II. Oppenheimer, Phillip R.
RM262.C52 615'.58 78-59227
79 MARC

Gerbino, Philip P
Self-assessment of current knowledge in clinical pharmacy : 1,444 multiple choice questions and referenced answers / edited by Philip P. Gerbino. — 2d ed. — Garden City, N.Y. : Medical Examination Pub. Co., 1978.
346 p. : ill. ; 22 cm.
ISBN 0-87488-281-8 : $8.50
1. Pharmacology—Examinations, questions, etc. 2. Chemotherapy—Examinations, questions, etc. I. Title.
RM105.G47 1978 615'.1'076 78-61615
79 MARC

—HANDBOOKS, MANUALS, ETC.

Albanese, Joseph A
Nurses' drug reference / Joseph A. Albanese. — New York : McGraw-Hill, c1979.
ix, 694 p. ; 24 cm.
Bibliography: p. [693]-694.
ISBN 0-07-000765-9. ISBN 0-07-000766-7 pbk.
1. Pharmacology—Handbooks, manuals, etc. 2. Drugs—Handbooks, manuals, etc. 3. Chemotherapy—Handbooks, manuals, etc. 4. Nursing—Handbooks, manuals, etc. I. Title.
[DNLM: 1. Drugs—Nursing texts. QV55.3 N974]
RM300.A487 615'.1'0202 78-23554
78 MARC

Blacque-Belair, Alain.
Dictionnaire médical, clinique, pharmacologique et thérapeutique / Alain Blacque-Belair ; préface du Dr Max Fourestier. — 2e éd. — Paris : Maloine, 1978.
1934 p. ; 18 cm. F78-11589
Includes bibliographical references and index.
ISBN 2-224-00424-9 : 196F
1. Medicine—Handbooks, manuals, etc. 2. Diagnosis—Handbooks, manuals, etc. 3. Chemotherapy—Handbooks, manuals, etc. 4. Therapeutics—Handbooks, manuals, etc. I. Title.
RC55.B56 1978 610'.21'2 78-400720
79 MARC

Handbook of drug therapy / edited by Russell R. Miller, David J. Greenblatt. — New York : Elsevier, c1979.
xxiii, 1126 p. ; 21 cm.
Includes bibliographical references and index.
ISBN 0-444-00329-0 (pbk.)
1. Chemotherapy—Handbooks, manuals, etc. 2. Drugs—Handbooks, manuals, etc. I. Miller, Russell R. II. Greenblatt, David J.
[DNLM: 1. Drug therapy—Handbooks. WB330 H235]
RM262.H26 615'.58'0202 79-10559
79 MARC

—PERIODICALS

Drug therapeutics. 1979–
New York, Elsevier.
v. ill. 24 cm. annual.
"Concepts for physicians."
Key title: Drug therapeutics, ISSN 0163-1705.
1. Chemotherapy—Periodicals. 2. Drugs—Periodicals.
RM260.D77 615'.58'05 79-643483
MARC-S

Facts and comparisons. 1979–
St. Louis, Facts and Comparisons, inc.
v. 26 cm. annual.
"Adapted from Facts and comparisons."
Key title: Facts and comparisons (Annual edition), ISSN 0162-1491.
1. Drugs—Periodicals. 2. Chemotherapy—Periodicals. I. Key title.
RM300.F33 615'.58 79-640262
MARC-S

Pulse of pharmacy. v. [1]–19, no. 3. [Philadelphia, Pa., Wyeth Laboratories, etc.] 1947–1965.
19 v. ill. 29 cm.
Frequency varies.
Issues for Sept. 1947–Nov./Dec. 1951 have no vol. numbering but constitute v. 1–5.
Issued by Wyeth Laboratories (v. 1–6 under the firm's earlier name: Wyeth, inc.)
1. Drugs—Periodicals. 2. Chemotherapy—Periodicals.
I. Wyeth Laboratories.
RM1.P84 615'.1'05 79-641669
MARC-S

—PROGRAMMED INSTRUCTION

Quiring, Julia D
Basic nursing skills : a self-instructional approach / Julia D. Quiring, Robert F. Rubeck. — North Scituate, Mass. : Duxbury Press, c1977.
xiv, 274 p. : ill. ; 24 cm.
Bibliography: p. 274.
ISBN 0-87872-136-3
1. Nursing—Programmed instruction. 2. Asepsis and antisepsis—Programmed instruction. 3. Chemotherapy—Programmed instruction. 4. Vital signs—Programmed instruction. I. Rubeck, Robert F., joint author. II. Title.
[DNLM: 1. Nursing—Programmed texts. WY18 Q8b]
[RT41.Q57] 610.73 77-22972
77 MARC

CHEMUNG RIVER

—FLOODS

see also Corning, N.Y.—Flood, 1972

CH'EN, CHE

—CRITICISM AND INTERPRETATION —ADDRESSES, ESSAYS, LECTURES

Li, Ao, 1935–
(Ma li pu liao ni)
媽離不了你 / 李敖著. — 港3版. — 九龍：文藝書屋, 1974.
5, 232 p. ; 19 cm. — (李敖告別文壇十書 : 3)
Series romanized: Li Ao kao pieh wen t'an shih shu.

1. Ch'en, Che—Criticism and interpretation—Addresses, essays, lectures. I. Title.
PL2877.A6L45 vol. 3 79-835038
[PL2840.C35]

CH'EN, CH'I-MEI, 1877-1916

Shao, Yüan-ch'ung, 1890-1936.
(Ch'en Ying-shih hsien sheng ko ming hsiao shih)
陳英士先生革命小史 / 邵元冲撰. — [台北]: 中國國民黨中央委員會黨史委員會, 民國66 [1977]
21 p., [1] leaf of plates : port. ; 26 cm.
Cover title.
"陳英士先生百年誕辰紀念"
1. Ch'en, Ch'i-mei, 1877-1916. 2. Revolutionists—China—Biography. I. Title.
DS778.C523S5 1977 79-836599

CH'EN, CHING-JUNG

Ch'en, Ching-jung.
(Ch'en Ching-jung su miao hsüan chi)
[陳景容素描選集]. — [台北?: s. n., 197-?]
1 portfolio ([24] leaves of plates : 24 ill.) ; 40 cm.
Cataloged from dealer's slip.
1. Ch'en, Ching-jung. I. Title.
NC350.C48A4 1970z 78-841563

CH'EN, CHÜ-LIANG, 1844-1901

Cheng, Kuan-ying.
(Chen Ch'ien-shan)
陳謙善 鄭冠英著 海外文庫編輯委員會編輯
臺北 海外文庫出版社 民國43 [1954]
25 p. illus. 19 cm. (名人傳之 7)
海外文庫
1. Ch'en, Chü-liang, 1844-1901. 2. Merchants—Philippine Islands—Biography. 3. Merchants—China—Biography.
HF3814.5.C45A33 1954 C 60-1772 ‡
rev

CH'EN, CH'UN, 1483-1544

Ch'en, Ch'un, 1483-1544.
(Ming Ch'en Tao-fu Ku shih shih chiu shou)
明陳道復古詩十九首 / [陳淳]. — 第1版. — 北京: 文物出版社, 1977.
[39] p. (on double leaves) : facsims. ; 42 cm.
In case.
$12.50
1. Ch'en, Ch'un, 1483-1544. 2. Calligraphy, Chinese. 3. Chinese language—Writing, Cursive. I. Ku shih shih chiu shou. 1977. II. Title.
NK3634.C395A4 1977 79-835626

CH'EN, HSIEN-CHANG, 1428-1500

Ch'en, Hsien-chang, 1428-1500.
(Ming Ch'en Hsien-chang tzu shu shih)
明陳獻章自書詩: 上海博物館藏. — 香港: 書譜社, [1977]
[29] p. ; 41 cm. — (書譜叢帖 : 第三輯 ; 15)
Series romanized: Shu p'u ts'ung t'ieh : Ti san chi.
$18.00
1. Ch'en, Hsien-chang, 1428-1500. 2. Calligraphy, Chinese. I. Title.
NK3634.C396A4 1977 78-842164

CH'EN, I, 1901-1972

—ADDRESSES, ESSAYS, LECTURES

(Ch'en I)
陳毅 / 譚震林 ... 等著. — 香港: 廣角鏡出版社: 華風書局發行, 1978.
262 p., [9] leaves of plates : ill. ; 19 cm. — (廣角鏡人物叢書)
Series romanized: Kuang chiao ching jen wu ts'ung shu.
CONTENTS: 譚震林 粟裕 緬懷陳毅同志.—張愛萍 大雪壓青松, 青松挺且直.—楊尚奎 艱難的歲月.—洪伍 陳毅將軍在大江南北的戰鬥生活片斷.—羅建文 憶陳毅對圍棋的關懷.—張茜 陳毅同志詩詞選集編成題後二首.—趙樸初 讀陳毅同志詩詞.—謝冕 無產階級的正氣歌.—陳毅 給羅生特同志的信.—陳毅詩抄.
$8.00
1. Ch'en, I, 1901-1972—Addresses, essays, lectures. 2. Statesmen—China—Biography—Addresses, essays, lectures. I. T'an, Chen-lin.
DS778.C524C47 79-835886

—FICTION

Sung, Sheng-fa.
(Ken sui Ch'en I t'ung chih ta yu chi)
跟随陈毅同志打游击 / 宋生发. — 第1版. — 北京: 解放军文艺社: 新华书店北京发行所发行, 1960, 1978 printing.
93 p., [4] leaves of plates : ill. ; 19 cm.
$0.28
1. Ch'en, I, 1901-1972—Fiction. I. Title.
PL2905.N457K4 79-839482

CH'EN, JU-YÜAN, 16TH CENT. HUNG-LIEN CHAI

Wang, Chih-yung, 1942-
(Tu Liu-ts'ui Ts'ui hsiang meng yü Hung-lien chai san chü ti pi chiao yen chiu)
度柳翠翠鄉夢與紅蓮債三劇的比較研究 / 汪志勇撰. — [台北?: s. n., 1979?]
1, 106 p., [2] fold. leaves ; 21 cm.
Bibliography: p. 101-106.
1. Li, Shou-ch'ing, 13th cent. Tu Liu-ts'ui. 2. Hsü, Wei, 1521-1593. Ts'ui hsiang meng. 3. Ch'en, Ju-yüan, 16th cent. Hung-lien chai. I. Title.
PL2694.L467T838 79-840521

CH'EN, KO

Ch'en, Ko.
(Nan chi hsing)
南極行 / [作者陳戈]. — 臺北: 陳戈, 民國66 [1977]
196 p. : 19 cm.
NT$40.00
1. South Pole. 2. Ch'en, Ko. I. Title.
G850 1976.C48C48 79-835929

CH'EN, KUANG-FU, 1881-1976

Chang, Shou-hsien.
(Ch'en Kuang-fu hsien sheng chuan lüeh)
陳光甫先生傳略 / [上海商業儲蓄銀行編印]. — [台北]: 該行, [1977?]
125 p., [1] leaf of plates : col. port. ; 27 cm.
Cover title.
"張君壽賢...撰述"
1. Ch'en, Kuang-fu, 1881-1976. 2. Bankers—China—Biography. 3. Bankers—Taiwan—Biography. I. Shang-hai shang yeh ch'u hsü yin hang.
HG1552.C45C48 78-841978

CH'EN, KUO-FU, 1892-1951

Hsü, Yung-p'ing, 1912-
(Ch'en Kuo-fu chuan)
陳果夫傳 / 徐詠平著. — 臺北: 正中書局, 民國67 [1978]
2, 970 p., [6] leaves of plates : ill. ; 22 cm.
1. Ch'en, Kuo-fu, 1892-1951. 2. Statesmen—China—Biography. I. Title.
DS778.C526H78 78-841993

CH'EN, MING-JEN

—ADDRESSES, ESSAYS, LECTURES

(Ch'en Ming-jen chih hsi hsiao Chu-ko)
陳明仁智戲小諸葛. — 香港: 集文出版社, 1957.
44 p., [4] leaves of plates : ill. ; 19 cm.
Cover title.
CONTENTS: 陳明仁 愛國一家—陳朗 陳明仁上將訪問記
1. Ch'en, Ming-jen—Addresses, essays, lectures. I. Ch'en, Ming-jen. Ai kuo i chia. 1957. II. Ch'en, Lang, fl. 1957- Ch'en Ming-jen shang chiang fang wen chi. 1957.
DS778.C5264C47 79-840059

CH'EN, PAI-FAN

Ch'en, Pai-fan.
(P'iao yang kuo hai ku hsing yu)
飄洋過海孤星遊 / 陳百範著. — 臺北: 聯合報社: 總經銷聯經出版事業公司, 民國67- [1978-
v. ; 20 cm. — (聯合報叢書) (萬象集刊)
Series 1 romanized: Lien ho pao ts'ung shu. Series 2 romanized: Wan hsiang chi k'an.
NT$40.00 (v. 1)
1. United States—Description and travel—1960- 2. Ch'en, Pai-fan. I. Title.
E169.02.C5816 78-843493

CH'EN, PANG-JUNG, 1913-1976

(Ch'en Pang-jung hsien sheng chi nien chi)
陳邦榮先生紀念集 / [陳邦榮先生治喪委員會編印]. — [台北]: 該會, [1977?]
4, 238 p. : ill. ; 22 cm.
Cover title.
"遺作": p. 203-235.
1. Ch'en, Pang-jung, 1913-1976. 2. Taiwan—Biography. I. Ch'en Pang-jung hsien sheng chih sang wei yüan hui.
CT1828.C52217C45 79-836005

CHEN, PERCY, 1901-

Chen, Percy, 1901-
China called me : my life inside the Chinese Revolution / by Percy Chen. — 1st ed. — Boston : Little, Brown, c1979.
xxi, 423 p., [8] leaves of plates : ill., map (on lining papers ; 25 cm.
Includes index.
ISBN 0-316-13849-5
1. Chen, Percy, 1901- . 2. Statesmen—China—Biography. 3. China—Politics and government—1900- I. Title.
DS778.C5267A33 951.04'092'4 79-13945
79 MARC

CH'EN, SHENG, D. 209 B.C.

—CARTOONS, SATIRE, ETC.

Ssu, Ho.
(Ch'en Sheng, Wu Kuang)
陈胜, 吴广 / 思禾编写 ; 戴敦邦画. — 上海: 上海人民出版社, 1974.
30 p. : col. ill. ; 13 x 15 cm.
Colophon title.
Romanized title on cover: Chen Sheng, Wu Guang.
"连环图画"
$0.14
1. Ch'en, Sheng, d. 209 B.C.—Cartoons, satire, etc. 2. Wu, Kuang, d. 209 B.C.—Cartoons, satire, etc. I. Tai, Tun-pang. II. Title.
DS747.5.C44S68 79-840072

CH'EN, SHENG, D. 209 B.C. (Continued)

—FICTION

Liu, Ya-chou.
(Ch'en Sheng)
陈胜 / 刘亚洲著. — 第1版. — ⌜汉口⌝：湖北人民出版社：湖北省新華書店发行, 1977.
486 p. ; 21 cm.
$1.08

1. Ch'en, Sheng, d. 209 B. C.—Fiction. 2. Peasant uprisings—China—Fiction. 3. China—History—Ch'in dynasty, 221-207 B. C.—Fiction. I. Title.
PL2879.Y3C5 79-840705

CH'EN, TA-CHI, 1886-

—ADDRESSES, ESSAYS, LECTURES

(Ch'en Pai-nien hsien sheng chih chiao wu shih chou nien chi pa chih ta shou chi nien lun wen chi)
陳百年先生執教五十週年暨八秩大壽紀念論文集 / 編輯者國立北京大學校友會, 國立政治大學出版委員會. — 臺北縣木柵鄉：國立政治大學, 民國54 ⌜1965⌝
2, 380 p., ⌜1⌝ leaf of plates : ports. ; 27 cm.
Colophon title.
1. Ch'en, Ta-ch'i, 1886- —Addresses, essays, lectures. I. Ch'en, Ta-ch'i, 1886- II. Kuo li Pei-ching ta hsüeh hsiao yu hui. III. Kuo li cheng chih ta hsüeh. Ch'u pan wei yüan hui.
AC149.C43C45 78-841562

CHEN, TAN-CHENG, 1919

Chen, Tan-cheng, 1919-
(Ch'en Tan-ch'eng hua chi)
陳丹誠畫集. — 臺北：中華民國國立歷史博物館, 民國66 ⌜1977⌝
116 p. : col. ill. ; 29 cm.
Colophon title.
Title in English on p. ⌜4⌝ of cover: Chen Tan-cheng paintings.
English and Chinese.
1. Chen, Tan-cheng, 1919- I. Title.
ND1049.C4775A4 1977 79-836042

CH'EN, TSAI-TAO

—ADDRESSES, ESSAYS, LECTURES

(Ch'en Tsai-tao tzu liao chi)
陳再道資料集：原名「武漢事件資料專輯」. — ⌜香港? : s. n., 1970⌝
52 leaves : ill. ; 23 x 34 cm.
Cover title.
Publication date stamped on p. ⌜4⌝ of cover.
A collection of articles and newspaper clippings reprinted from various sources.
1. Ch'en, Tsai-tao—Addresses, essays, lectures. 2. Generals—China—Biography—Addresses, essays, lectures. 3. China—Politics and government—1949-1976—Addresses, essays, lectures. I. Title: Wu-han shih chien tzu liao chuan chi.
DS778.C53497C46 79-837920

CH'EN, TZU-ANG, 661-702

—CRITICISM AND INTERPRETATION

Ho, Wen-hui.
(Ch'en Tzu-ang kan yü shih chien)
陳子昂感遇詩箋 / ⌜著作者何文匯⌝. — 香港：學津出版社；九龍：發行者學津書店, 1978.
159 p. ; 22 cm.
$30.00

1. Ch'en, Tzu-ang, 661-702—Criticism and interpretation. I. Ch'en, Tzu-ang, 661-702. Poems. 1978. II. Title.
PL2677.C4Z69 79-837724

CHEN, WESO

Chen, Weso.
(Wo ti hu tieh meng)
我的蝴蝶夢 / 陳維壽著. — ⌜台北?⌝：順先出版公司, 民國66 ⌜1977⌝
194 p. : ill. ; 20 cm.
NT$40.00

1. Butterflies—Taiwan. 2. Chen, Weso. 3. Insects—Taiwan. 4. Entomologists—Taiwan—Biography. I. Title.
QL556.T29C494 78-843872

CH'EN, YIN-K'O, 1890-1969

Ch'en, Yin-k'o, 1890-1969.
⌜Ch'en Yin-k'o hsien sheng lun wen chi⌝
陳寅恪先生全集 / 著作人陳寅恪. — 第3次修訂版. — 台北：九思出版有限公司, 民國66 ⌜1977⌝
2 v. : ill. ; 22 cm. — (九思叢書 ; 5)
Colophon title.
Caption title: Ch'en Yin-k'o hsien sheng lun wen chi.
Series romanized: Chiu ssu ts'ung shu.
"陳寅恪先生論文集補編": at end of v. 2.
NT$370.00

1. Ch'en, Yin-k'o, 1890-1969. Ch'en Yin-k'o hsien sheng lun wen chi pu pien. 1977. II. Title.
Title romanized: Ch'en Yin-k'o hsien sheng ch'üan chi.
DS734.9.C49A2 1977 78-843511

Wang, Jung-tsu.
(Shih chia Ch'en Yin-k'o chuan)
史家陳寅恪傳 / 汪榮祖撰. — ⌜臺北縣永和鎮⌝：文海出版社, ⌜1978?⌝
2, 2, 134, 32 p., ⌜1⌝ leaf of plates : ports. ; 22 cm. — (近代中國史料叢刊續編 ⌜i. e.輯⌝ ; 497 : 第五十輯)
Reprint ed.
"附何廣棪 陳寅恪先生著述目錄編年": p. 1-32 (last group)
1. Ch'en, Yin-k'o, 1890-1969. 2. Historians—China—Biography. I. Ho, Kuang-yen. Ch'en Yin-k'o hsien sheng chu shu mu lu pien nien. 1978? II. Title. III. Series: Shen, Yün-lung, comp. Chin tai Chung-kuo shih liao ts'ung k'an hsü chi ; 497.
DS755.S532 vol. 497 79-835748
[DS734.9.C57]

—CRITICISM AND INTERPRETATION

Ho, Kuang-yen.
(Ch'en Yin-k'o hsien sheng i shih shu shih)
陳寅恪先生遺詩述釋 / 述釋人何廣棪. — 台北：九思出版社, 民國66 ⌜1977⌝
34 p. ; 22 cm. — (珠海書院中國文學歷史研究所學會叢刊 ; 7)
Colophon title.
Title in English on p. ⌜4⌝ of cover: Annotation to the poetic works of Ch'en Yin-koh.
Includes bibliographical references.
NT$25.00
1. Ch'en, Yin-k'o, 1890-1969—Criticism and interpretation. I. Title. II. Title: Annotation to the poetic works of Ch'en Yin-koh. III. Series: Chu hai shu yüan Chung-kuo wen hsüeh li shih yen chiu so hsüeh hui. Chu hai shu yüan Chung-kuo wen hsüeh li shih yen chiu so hsüeh hui ts'ung k'an ; 7.
PL2743.Y5Z68 78-842896

CHEN-CHIANG, CHINA

—HISTORY—SOURCES

(Chih-shun Chen-chiang chih)
至順鎮江志：⌜21卷：卷首⌝ / 俞希魯撰；王有立主編. — ⌜台北⌝：台灣華文書局, ⌜1968⌝
2 v. (1358 p.) ; 21 cm. — (中華文史叢史 ; 8 : 第2輯)
Reprint of the Ch'ing T'ung-chih 2 (1862) ed. printed under title: Yüan Chih-shun Chen-chiang chih.
"附校勘記 2卷": p. 1167-1358.

1. Chen-chiang, China—History—Sources. I. Yü, Hsi-lu, 1279-1368. II. Title: Yüan Chih-shun Chen-chiang chih. III. Series: Chung-hua wen shih ts'ung shu ; 8.
DS793.C35C47 1968 77-836467

CHE'N FAMILY

(Ch'en chia ta tsung p'u)
陳家大宗譜 / 編著陳氏大族譜編輯部；主輯莊吳福；主編莊吳圖；助編盧俊華；資料採訪盧有道... ⌜et al.⌝. — ⌜中壢⌝：陳氏大族譜編輯委員會, 民國67 ⌜1978⌝
380, 32 p. : ill. ; 28 cm.
Colophon title.

1. Ch'en family. 2. China—Geneology. I. Chuang, Wu-fu. II. Chuang, Wu-t'u. III. Ch'en shih ta tsu p'u pien chi pu.
CS1169.C475 1978 79-837079

(Ch'en shih tsu p'u)
陳氏族譜 / ⌜編輯莊吳圖⌝. — 桃園縣龍潭鄉：陳氏族譜編輯委員會, 民國66 ⌜1977⌝
332 p. in various pagings : ill. ; 27 cm.
Cover title.
Editions of 1961, 1976 published under title: Ch'en shih ta tsu p'u, and of 1978 published under title: Ch'en chia ta tsung p'u.

1. Ch'en family. 2. China—Geneology. I. Chuang, Wu-t'u. II. Ch'en shih tsu p'u pien chi wei yüan hui.
CS1169.C475 1977a 79-841630

CHEN-HUA, SHIH, 1921-

Chen-hua, Shih, 1921-
(Ts'an hsüeh so t'an)
參學瑣談 / 眞華著. — 再版. — 臺北：天華出版事業股份有限公司, 民國67 ⌜1978⌝
342, ⌜14⌝ p. ; 21 cm. — (天華佛學叢刊 ; 2)
Series romanized: T'ien hua Fo hsüeh ts'ung k'an.
NT$130.00 ($4.00 U.S.)

1. Chen-hua, Shih, 1921- 2. Priests, Buddhist—China—Biography. I. Title.
BQ946.E492A25 1978 79-837125

CHEN-PAO ISLAND

Chang, Hsiao-yen.
(Chen-pao tao shih chien chi fei O kuan hsi)
珍寶島事件及匪俄關係, 一九五六——九六九 / 章孝嚴著. — 臺北：黎明文化事業公司, 民國67 ⌜1978⌝
6, 180, ⌜64⌝ p., ⌜1⌝ leaf of plates : ill. ; 20 cm.
NT$60.00
1. China—Foreign relations—Russia. 2. Russia—Foreign relations—China. 3. China—Boundaries—Russia. 4. Russia—Boundaries—China. 5. Chen-pao Island. I. Title.
DS740.R8C452 78-843090

CHENAVARD, PAUL MARC JOSEPH, 1807-1895

Paul Chenavard et la décoration du Panthéon de Paris en 1848. — Lyon : Musée des beaux-arts, 1977.
98 p. : ill. ; 17 cm. F***
Bibliography: p. 97-98.
1. Chenavard, Paul Marc Joseph, 1807-1895. 2. Paris. Panthéon. I. Lyons. Musée des beaux-arts.
ND553.C48P38 759.4 79-368844
79 MARC

CHENEY FAMILY

Cheney, Leonard Earl, 1908-
The descendants and ancestors of Claudius Franklin Cheney / compiled by Leonard Earl Cheney. — Roswell, N.M. : Roswell Print. Co., 1978.
viii, 232 p. (p. 222-232 blank) : ill. ; 23 cm.
Includes indexes.
1. Cheney family. I. Title.
CS71.C518 1978 929'.2'0973 78-73199
79 MARC

CHENG, CHIH-LUNG, 1604-1661

—FICTION

Chin, Shunshin, 1924-
(Kaze yo kumo yo)
風よ雲よ / 陳舜臣. — 東京 : 講談社, 1978.
2 v. ; 20 cm. Ja 78-9149
¥1200 per vol.

1. Cheng, Chih-lung, 1604-1661—Fiction. I. Title.
PL847.I5K34 78-813538

CHENG, FENG-HSI, 1944- WANG YANG CHUNG TI P'O CH'UAN

Lin, Ch'ing-chieh, 1944-
(Wang yang chung ti i t'iao ch'uan)
汪洋中的一條船 : 電影劇本 / [林清介編劇]. — 新竹 : 楓城出版社, 民國67 [1978]
174 p. : ill. ; 20 cm. — (楓城叢書 ; 27)
Drama.
Series romanized: Feng ch'eng ts'ung shu.
NT$40.00

1. Cheng, Feng-hsi, 1944- Wang yang chung ti p'o ch'uan. II. Title.
PL2878.N1233W3 79-835191

CHENG, FENG-HSI 1944-

—ADDRESSES, ESSAYS, LECTURES

Wei shih san nien)
為師三年 / 著者鄭豐喜[等]. — 第2版. — 臺北 : 鄭豐喜文教基金會出版部 : 總經銷黎明圖書社, 民國67 [1978]
17, 191 p. : ill. ; 20 cm. — (鄭豐喜全集 ; 2)
Colophon title.
NT$30.00
1. Cheng, Feng-hsi, 1944- —Addresses, essays, lectures. 2. Teachers—Taiwan—Biography—Addresses, essays, lectures. I. Cheng, Feng-hsi, 1944- II. Series: Cheng, Feng-hsi, 1944- Cheng Feng-hsi ch'üan chi ; 2.
LA2383.T282C468 1978 79-837801

—CARTOON, SATIRE, ETC. —JUVENILE LITERATURE

Feng, Yu-li.
(Cheng Feng-hsi ti fen tou ku shih)
鄭豐喜的奮鬥故事 : 汪洋中的一條船連環圖 / 主編鄧有立 ; 繪圖江少華. — 臺北 : 聯亞出版社, 民國66 [1977]
128 p. : col. ill. ; 19 cm. — (聯亞特集 ; 2)
Colophon title.
Accompanied with Kuo yü chu yin.
Series romanized: Lien ya t'e chi.
NT$35.00
1. Cheng, Feng-hsi, 1944- —Cartoon, satire, etc.—Juvenile literature. 2. Physically handicapped—Taiwan—Biography—Caricatures and cartoons—Juvenile literature. I. Chiang, Shao-hua. II. Cheng, Feng-hsi, 1944- Wang yang chung ti p'o ch'uan. III. Title.
CT9983.C48T46 78-843496

CHENG, HSIEH, 1693-1765

Cheng, Hsieh, 1693-1765.
(Cheng Pan-ch'iao shu hua hsüan)
鄭板橋書畫選 / 何恭上編選. — 台北 : 藝術圖書公司, 民國67 [1978]
104 p. : chiefly ill. ; 22 cm.
Table of contents and legends also in English.
NT$90.00

1. Cheng, Hsieh, 1693-1765. I. Ho, Kung-shang. II. Title.
ND1049.C488A4 1978 79-836016

Cheng, Hsieh, 1693-1765.
[Hsin hsiu ch'eng huang miao pei chi]
鄭板橋新修城隍廟碑記, 文昌帝君祠記. — 臺北 : 文史哲出版社, 民國64 [1975]
[36] p. : chiefly facsims. ; 39 cm.

1. Cheng, Hsieh, 1693-1765. 2. Calligraphy, Chinese. I. Cheng, Hsieh, 1693-1765. Wen ch'ang ti chün tz'u chi. 1975. II. Title.
Title romanized: Cheng Pan-ch'iao Hsin hsiu ch'eng huang miao pei chi, Wen ch'ang ti chün tz'u chi.
NK3634.C4A4 1975 79-840443

Cheng, Hsieh, 1693-1765.
(Pan-ch'iao jun ko)
板橋潤格 / [鄭燮]. — [蘇州? : s. n., 1977?]
broadside ; 35 x 99 fold. to 26 x 14 cm.
Title from label on pocket.
Rubbing.
"蘇州工藝"
In Case.

1. Cheng, Hsieh, 1693-1765. 2. Painting, Chinese—Prices. I. Title.
ND1049.C488C46 79-840043

CHENG, HSÜAN, 127-200. SHANG SHU CHENG CHU

Ch'en, P'in-ch'ing.
(Shang shu Cheng shih hsüeh)
尚書鄭氏學 / 陳品卿著. — Taipei : 嘉新水泥公司文化基金會, 民國66 [1977]
3, 15, 679 p. ; 22 cm. — (嘉新水泥公司文化基金會研究論文 ; 第305種) (嘉新水泥公司文化基金會叢書)
Series 1 romanized: Chia-hsin shui ni kung ssu wen hua chi chin hui yen chiu lun wen. Series 2 romanized: Chia-hsin shui ni kung ssu wen hua chi chin hui ts'ung shu.
Bibliography: p. 671-679.
1. Shu ching. 2. Cheng, Hsüan, 127-200. Shang shu Cheng chu. I. Shu ching. 1977. II. Cheng, Hsüan, 127-200. Shang shu Cheng chu. 1977. III. Title.
PL2465.Z6C4374 1977 79-837599

CH'ENG, I, 1033-1107. I CHUAN

Lin, I-sheng.
(I-ch'uan i chuan ti ch'u shih che hsüeh)
伊川易傳的處世哲學 / 林益勝撰. — 臺北 : 臺灣商務印書館, 民國67 [1978]
2, 2, 3, 249 p. ; 19 cm. — (人人文庫 ; 特589)
Series romanized: Jen jen wen k'u.
Bibliography: p. 240-249.
NT$30.00
1. Ch'eng, I, 1033-1107. 2. Ch'eng, I, 1033-1107. I chuan. I. Title.
B128.C36I224 78-843261

CH'ENG, I, 1033-1107

Lin, I-sheng.
(I-ch'uan i chuan ti ch'u shih che hsüeh)
伊川易傳的處世哲學 / 林益勝撰. — 臺北 : 臺灣商務印書館, 民國67 [1978]
2, 2, 3, 249 p. ; 19 cm. — (人人文庫 ; 特589)
Series romanized: Jen jen wen k'u.
Bibliography: p. 240-249.
NT$30.00
1. Ch'eng, I, 1033-1107. 2. Ch'eng, I, 1033-1107. I chuan. I. Title.
B128.C36I224 78-843261

CHENG, KUO-YU

(Li Ming-hui san chiao lien ai shih ch'üan chi)
黎明暉三角戀愛史全集 / [編輯者集成出版部]. — [香港? : s. n., 1978?]
15, 72 p. : ill., ports. ; 26 cm.
Cover title.
Colophon title: San chiao lien ai shih ch'üan chi.
Reprint of the 1932 ed. published by Kuang chang shu chü, Hongkong.
1. Li, Ming-hui. 2. Moving-picture actors and actresses—Hong-kong—Biography. 3. Feng, Te-ch'ien, 1906-1932. 4. Cheng, Kuo-yu. I. Chi ch'eng ch'u pan pu. II. Title: San chiao lien ai shih ch'üan chi.
PN2960.H6L5 78-841732

CHENG, SHU-TSU, D. 565?

Cheng, Shu-tsu, d. 565?
(Pei Ch'i Cheng Shu-tsu T'ien-chu shan ming)
北齊鄭述祖天柱山銘. — 台北 : 書法文物館, [1979?]
[80] p. (on double leaves) : all facsims. ; 30 cm.
Title from label on cover.
Colophon inserted.
"中華民[國]書法碑帖"
In case.
1. Cheng, Shu-tsu, d. 565? 2. Calligraphy, Chinese. I. Title. II. Title: T'ien-chu shan ming.
NK3634.C414A4 1979 79-841852

CHENG, TE-K'UN, 1908-

—ART COLLECTIONS—EXHIBITIONS

Céramiques de Tch'ang-Cha : quelques pièces des Han aux T'ang, Ier-Xe siècles, de la collection Mu-fei : [exposition], Musée Cernuschi, septembre-décembre 1976 : [catalogue / V. Elisseeff]. — [Paris] : Musée Cernuschi, [1976]
[34] p. : ill. (some col.) ; 24 cm. F***
1. Pottery, Chinese—To 618—Exhibitions. 2. Pottery, Chinese—T'ang-Five dynasties, 618-960—Exhibitions. 3. Pottery, Chinese—China—Ch'ang-sha—Exhibitions. 4. Ch'ang-sha, China—Antiquities—Exhibitions. 5. Cheng, Te-k'un, 1908- —Art collections—Exhibitions. I. Elisseeff, Vadime. II. Paris. Musée Cernuschi.
NK4166.C45C47 738'.0951'215 79-370387
79 MARC

CH'ENG, T'IEN-KU, 1889-1974

Ch'eng, T'ien-ku, 1889-1974.
(Ch'eng T'ien-ku hui i lu)
程天固回憶錄. — 香港 : 龍門書店, 1978.
2, 3, 3, 539 p. : port. ; 22 cm.
In colophon: Ch'u pan che Ch'eng Ta-min.
$80.00

1. Ch'eng, T'ien-ku, 1889-1974. 2. China—Politics and government—1912-1949. 3. Statesmen—China—Biography. I. Title.
DS778.C5363A38 1978 79-837615

CHENG (MUSICAL INSTRUMENT)

—METHODS

Cheng, Te-yüan.
(Cheng yüeh li lun chi yen tsou)
箏樂理論及演奏 / 鄭德淵著. — 台北 : 中國書局, 1977-
v. : ill. ; 27 cm. — (中國音樂叢書)
Added t. p.: Cheng music of China.
Includes music in number notation.
NT$250.00 (v. 1)

1. Cheng (Musical instrument)—Methods. I. Title. II. Title: Cheng music of China. III. Series: Chung-kuo yin yüeh ts'ung shu.
MT654.C47C54 79-837013

C M

Cheng, Te-yüan.
(Cheng yüeh li lun chi yen tsou)
箏樂理論及演奏 / 鄭德淵著. — 3版修訂. — 香港 : 香港書店, 1977-
v. : ill. ; 27 cm. — (中國音樂叢書)
Added t. p.: Cheng music of China.
Includes music in number notation.
$30.00 ($7.00 U.S.)
1. Cheng (Musical instrument)—Methods. I. Title. II. Series: Chung-kuo yin yüeh ts'ung shu.
MT654.C47C54 1977b 79-837012

CHENG-FENG MEI K'UANG KUNG SSU

Minami Manshū Tetsudō Kabushiki Kaisha. Hokushi Keizai Chōsajo, Peking.
(Seikei, Seihō ryōtankō rōdō gaiyō chōsa hōkoku)
井陘・正豊兩炭礦勞働概要調査報告 / ［執筆者關戸嘉明］. — ［北京］: 滿鐵・北支經濟調査所, 昭和16 ［1941］
4, 85 leaves : 26 cm.
Cover title.
"第一次調査報告書(第六班)"
"極秘" (rubber stamped)

1. Coal miners—China—Hopei (Province) 2. Ching-hsing mei k'uang ku fen yu hsien kung ssu. 3. Cheng-feng mei k'uang kung ssu. I. Sekido, Yoshiaki, 1917- II. Title.
HD8039.M62C639 79-806639

CH'ENG-KUNG LING SHANG. [MOTION PICTURE]

Hsiao-yeh.
(Ching t'ien chiu)
擎天鳩 / 小野著. — 台北 : 文豪出版社 : 總代理學生圖書供應社, 民國67 ［1978］
228 p. : ill. ; 20 cm.
Fiction.
"成功嶺上": p. 101-228.
NT$40.00 ($3.00 U.S.)

1. Ch'eng-kung ling shang. ［Motion picture］ I. Hsiao-yeh. Ch'eng kung ling shang. 1978. II. Title.
PN1997.C4525H8 79-835389

CH'ENG-TE, CHINA (JEHOL PROVINCE)

—ANTIQUITIES—PICTORIAL WORKS

Kishida, Hideto, 1899-
(Nekka iseki)
熱河遺蹟 / 岸田日出刀, 土浦亀城. — 東京 : 相模書房, 昭和15 ［1940］
102 p., ［1］ leaf of plates : all ill. ; 27 cm.

1. Ch'eng-te, China (Jehol Province)—Description and travel—Views. 2. Ch'eng-te, China (Jehol Province)—Antiquities—Pictorial works. 3. China—Antiquities—Pictorial works. I. Tsuchiura, Kameki, 1897- joint author. II. Title.
DS796.C49K57 78-812726

—DESCRIPTION AND TRAVEL —VIEWS

Kishida, Hideto, 1899-
(Nekka iseki)
熱河遺蹟 / 岸田日出刀, 土浦亀城. — 東京 : 相模書房, 昭和15 ［1940］
102 p., ［1］ leaf of plates : all ill. ; 27 cm.

1. Ch'eng-te, China (Jehol Province)—Description and travel—Views. 2. Ch'eng-te, China (Jehol Province)—Antiquities—Pictorial works. 3. China—Antiquities—Pictorial works. I. Tsuchiura, Kameki, 1897- joint author. II. Title.
DS796.C49K57 78-812726

CHENGTU, CHINA

—SOCIAL CONDITIONS

Kessle, Gun.
Vardagskina : Från Stallarnas gata, Gula teglets kvarter i Chengtu stad / Gunn Kessle. — Stockholm : Gidlund, 1978.
149, ［1］ p. : ill. ; 21 cm. S78-21
ISBN 9170212317
1. Chengtu, China—Social conditions. I. Title.
HN740.C45K47 79-340756
79 MARC

CHENNAULT, CLAIRE LEE, 1890-1958

Nalty, Bernard C
Tigers over Asia / Bernard C. Nalty. — 1st ed. — New York : Elsevier-Dutton, c1978.
vii, 182 p. : ill. ; 24 cm. — (Men and battle)
"A Talisman/Parrish book."
Bibliography: p. 175-176.
Includes index.
ISBN 0-525-93007-8
1. World War, 1939-1945—Aerial operations, American. 2. China. K'ung chün. American Volunteer Group. 3. United States. Army Air Forces. 14th Air Force. 4. Chennault, Claire Lee, 1890-1958. I. Title. II. Series.
D790.N33 1978 940.54'49'73 78-56711
78 MARC

CHENOBOSKION MANUSCRIPTS

Fallon, Francis T
The enthronement of Sabaoth : Jewish elements in gnostic creation myths / by Francis T. Fallon. — Leiden : Brill, 1978.
x, 158 p. ; 25 cm. — (Nag Hammadi studies ; v. 10) Ne78-44
Rev. ed. of the author's thesis, Harvard.
Bibliography: p. ［136］-143.
Includes index.
ISBN 9004056831 : fl 56.00
1. Gnosticism—Relations—Judaism. 2. Judaism—Relations—Gnosticism. 3. Creation—Comparative studies. 4. Hypostasis of the Archons. 5. Chenoboskion manuscripts. I. Title. II. Series.
BM536.G54F34 1978 299 79-306229
*79 MARC

Koschorke, Klaus.
Die Polemik der Gnostiker gegen das kirchliche Christentum : unter besonderer Berücksichtigung der Nag-Hammadi-Traktate 'Apokalypse des Petrus' (NHC VII,3) und 'Testimonium Veritatis' (NHC IX,3) / von Klaus Koschorke. — Leiden : Brill, 1978.
xiv, 274 p. ; 25 cm. — (Nag Hammadi studies ; v. 12) Ne78-39
A revision of the author's thesis, Heidelberg, 1976.
Bibliography: p. ［256］-270.
Includes indexes.
ISBN 9004057099 : fl 124.00
1. Gnosticism. 2. Chenoboskion manuscripts. I. Title. II. Series.
BT1390.K67 1978 78-400170
*79 MARC

—CONGRESSES

International Congress of Coptology, 1st, Cairo, 1976.
Nag Hammadi and gnosis : Papers read at the First International Congress of Coptology, (Cairo, December 1976) / ed. by R. McL. Wilson. — Leiden : Brill, 1978.
viii, 178 p. ; 25 cm. — (Nag Hammadi studies ; v. 14) Ne78-49
English, French or German.
Includes bibliographical references and indexes.
ISBN 9004057609 : kr fl 96.00
1. Chenoboskion manuscripts—Congresses. 2. Gnosticism—Congresses. I. Wilson, Robert McLachlen. II. Title. III. Series.
BT1390.I58 1976 299 79-306082
*79 MARC

CHEPELARE, BULGARIA

—HISTORY

Dechov, Vasil.
(Minaloto na Chepelare)
Миналото на Чепеларе : принос за историята на Родопа / Васил Дечов ; ［с предг. от Николай Хайтов］. — 2. изд. — Пловдив : Хр. Г. Данов, 1978-
v. : ill. ; 21 cm. Bu 78-2319
Bibliography : v. 1, p. ［255］
1.52 lv
1. Chepelare, Bulgaria—History. I. Title.
DR98.C47D4 1978 79-407899

CHER, FRANCE (DEPT.)

—CHURCH HISTORY—SOURCES—BIBLIOGRAPHY—CATALOGS

Cher, France (Dept.). Archives.
Série V : cultes 1800-1905 (1940) : répertoire numérique / dressé par Paul Cravayat, archiviste en chef du département ; introd. et bibliographie par Jean-Yves Ribault. — Bourges : Archives départementales du Cher, 1975.
109 p. ; 24 cm. F***
On spine: Répertoire numérique de la série V.
Bibliography: p. 10-45.
Includes index.
ISBN 2-86018-000-1
1. Cher, France (Dept.). Archives—Catalogs. 2. Cher, France (Dept.)—Church history—Sources—Bibliography—Catalogs. I. Cravayat, Paul. II. Ribault, Jean Yves. III. Title: Répertoire numérique de la série V.
CD1215.C4C45 1975 016.944'552 78-385650
78 MARC

—DESCRIPTION AND TRAVEL

Buhot de Kersers, Alphonse, 1835-1897.
Statistique monumentale du département du Cher / A. Buhot de Kersers. — Marseille : Laffitte, 1977.
8 v. : ill. ; 23 cm. F78-8281
Reprint of the 1875-1898 ed. published by A. Morel & Co., Paris.
Three hundred copies printed. No. 298.
Includes bibliographical references and index.
CONTENTS: 1. Canton des Aix-d'Angillon.— 2. Canton de Bourges.—3. Cantons de la Chapelle-d'Angillon, Charenton, Charost, Châteaumeillant et Châteauneuf.—4. Cantons de Le Châtelet, Dun-le-Roi, Graçay, La Gureche, Henrichemont.—5. Cantons de Léré, Levet, Linière, Lury, Mehun.—6. Cantons de Nerondes, Saint-Amand-Montrond, Saint-Martin-d'Auxigny, Sancergues.—7. Cantons de Sancerre, Sancoins, Saulzais-le-Potier, Vailly, Vierzon.—8. Conclusions et table générale.
1400F

1. Monuments—France—Cher (Dept.) 2. Cher, France (Dept.)—Description and travel. I. Title.
DC611.C563B83 1977 944'.552 78-400828
79 MARC

—ECONOMIC CONDITONS—SOURCES—BIBLIOGRAPHY—CATALOGS

Cher, France (Dept.). Archives.
Répertoire numérique de la sous-série 4 Q : fonds de conservation des hypothèques, an VIII-1873 / Archives départementales du Cher ; dressé par Geneviève Bailly, ... ; sous la direction de Jean-Yves Ribault, ... et Jean-Marie Jenn, ... — Bourges : Archives départementales du Cher, 1977.
142 p. : map ; 30 cm. F78-11432
Bibliography: p. 135-136.
ISBN 2-86018-001-X : 15.00F
1. Cher, France (Dept.). Archives—Catalogs. 2. Cher, France (Dept.)—Economic conditons—Sources—Bibliography—Catalogs. I. Bailly, Geneviève. II. Title.
CD1215.C4C45 1977 016.3327 79-347175
79 MARC

—HISTORY

Bers, Günter, 1940-
Die Herrschaft Vierzon. Ein Lehen d. Jülicher Dynasten in Frankreich. (Jülich, Jülicher Geschichtsverein, 1969).
20 p. 21 cm. (Beiträge zur Jülicher Geschichte, Nr. 25) GDR70-B1-231
Cover title.
Includes bibliographical references.
1. Vierzon, France—History. 2. Cher, France (Dept.)—History. 3. Jülich (Duchy)—History. I. Title. II. Series.
BR859.B43 Nr. 25 79-361685
［DC801.V6918］ 79 MARC

CHER, FRANCE (DEPT.). ARCHIVES

—CATALOGS

Cher, France (Dept.). Archives.
Répertoire numérique de la sous-série 4 Q : fonds de conservation des hypothèques, an VIII-1873 / Archives départementales du Cher ; dressé par Geneviève Bailly, ... ; sous la direction de Jean-Yves Ribault, ... et Jean-Marie Jenn, ... — Bourges : Archives départementales du Cher, 1977.
142 p. : map ; 30 cm. F78-11432
Bibliography: p. 135-136.
ISBN 2-86018-001-X : 15.00F
1. Cher, France (Dept.). Archives—Catalogs. 2. Cher, France (Dept.)—Economic conditons—Sources—Bibliography—Catalogs. I. Bailly, Geneviève. II. Title.
CD1215.C4C45 1977 016.3327 79-347175
79 MARC

Cher, France (Dept.). Archives.
Série V : cultes 1800-1905 (1940) : répertoire numérique dressé par Paul Cravayat, archiviste en chef du département introd. et bibliographie par Jean-Yves Ribault. — Bourges : Archives départementales du Cher, 1975.
109 p. ; 24 cm. F*
On spine: Répertoire numérique de la série V.
Bibliography: p. 10-45.
Includes index.
ISBN 2-86018-000-1
1. Cher, France (Dept.). Archives—Catalogs. 2. Cher, France (Dept.) Church history—Sources—Bibliography—Catalogs. I. Cravayat, Paul. II. Ribault, Jean Yves. III. Title: Répertoire numérique de la série V.
CD1215.C4C45 1975 016.944'552 78-3856
78 MA

CHEREMISSES

Kozlova, Klavdiíà Ivanovna.
(Ocherki ėtnicheskoĭ istorii mariĭskogo naroda)
Очерки этнической истории марийского народа / К. И. Козлова. — Москва : Изд-во МГУ, 1978.
344 p. : ill. ; 21 cm. USSR 7
Includes bibliographical references.
3.20rub
1. Cheremisses. 2. Ethnology — Russian Republic — Mariyskay A. S. S. R. I. Title.
DK34.C5K63 79-41285

CHEREMISSES (Continued)

—SOCIAL LIFE AND CUSTOMS

(Baldran-bazar v Kozmodem'ianskom uiezdie)
Балдранъ-базаръ въ Козмодемьянскомъ уѣздѣ. [Переводчикъ чувашскаго языка Спиридонъ Михайловъ. Казань, 1852.]
6 p. 21 cm.
Caption title.
"Изъ Каз. губ. вѣд. 1852 г., № 21."
1. Holidays—Russian Republic—Mariyskaya ASSR. 2. Chuvashes—Social life and customs. 3. Cheremisses—Social life and customs.
GT4856.A3M372 79-412239

CHERENTE INDIANS see Sherente Indians

CHEREVATENKO ALEKSĬ TIKHONOVICH

Cherevatenko, Alekseĭ Tikhonovich.
(Nebo Odessy, 1941-ĭ)
Небо Одессы, 1941-й : [О боевом подвиге 69-го гвард. авиац. полка] / А. Череватенко ; [Лит. запись А. Г. Ключника]. — 3-е изд., перераб. и доп. — Одесса : Маяк, 1978.
176 p., [4] leaves of plates : ill. ; 17 cm. USSR 78
0.40rub
1. Odessa, Battle of, 1941—Personal narratives, Russian. 2. Cherevatenko, Alekseĭ Tikhonovich. 3. Russia (1923- U. S. S. R.). Armiia. 69. gvardeĭskiĭ Odesskiĭ aviatsionnyĭ polk—Biography. 4. World War, 1939-1945—Aerial operations, Russian. 5. Air pilots, Military—Russia—Biography. I. Kliuchnik, Aleksandr Gordeevich. II. Title.
D764.3.O3C52 1978 78-421539

CHERKASOV, NIKOLAĬ KONSTANTINOVICH, 1903-1966

Gerasimov, Iuriĭ Konstantinovich.
(Cherkasov)
Черкасов / Ю. Герасимов, Ж. Скверчинская. — 2-е изд. — Москва : Мол. гвардия, 1977.
348 p., [25] leaves of plates : ill. ; 20 cm. — (Жизнь замечательных людей : Серия биографий ; вып. 8(563)) USSR 77
Series romanized: Zhizn' zamechatel'nykh liudeĭ : Seriia biografiĭ.
Bibliography: p. 346-[347]
1.70rub
1. Cherkasov, Nikolaĭ Konstantinovich, 1903-1966. 2. Actors—Russia—Biography. I. Skverchinskaia, Zhanna Glebovna, joint author.
PN2728.C5G47 1977 78-421157

Hofmann, Heinz, fl. 1959-
Nikolai Tscherkassow. Berlin, Henschelverlag, 1966.
36 p., 16 leaves of illus. 21 cm. DM2.50 GDB***
1. Cherkasov, Nikolaĭ Konstantinovich, 1903- 2. Actors—Russia—Biography.
PN2728.C5H6 67-75945
[r79]rev MARC

CHERKASSKAYA OBLAST, UKRAINE

—ANTIQUITIES

Pidoplichko, Ivan Grigor'evich.
(Mezhirichskie zhilishcha iz kosteĭ mamonta)
Межиричские жилища из костей мамонта / И. Г. Пидопличко ; АН УССР, Ин-т зоологии. — Киев : Наук. думка, 1976.
239 p. : ill. (some col.) ; 20 cm. USSR 76
Summary in English ; legends also in English.
Bibliography : p. 227-[231]
1.98rub
1. Paleolithic period—Ukraine—Cherkasskaya oblast'. 2. Cherkasskaya oblast, Ukraine—Antiquities. I. Title.
GN772.22.R9P5 78-422237

—HISTORY

Honchar, H M
(Zahravy nad Tias'mynom)
Заграви над Тясьмином / Г. М. Гончар. — Київ : Політвидав України, 1978.
214 p., [4] leaves of plates : ports. ; 21 cm. USSR***
0.90rub
1. World War, 1939-1945—Ukraine—Cherkasskaya oblast'. 2. Cherkasskaya oblast', Ukraine—History. I. Title.
D802.R82C473 79-412713

CHERMIANIN, IVAN ALEKSANDROVICH, 1926-

—EXHIBITIONS

Chermianin, Ivan Aleksandrovich, 1926-
(Ivan Chermianin)
Иван Чермянин : персональная юбилейная выставка. — Кемерово : [s. n.], 1976.
46 p. : ill. (1 fold. col.) ; 20 cm. USSR***
At head of title: Kemerovskoe oblastnoe upravlenie kul'tury. Kemerovskoe otdelenie Soiuza khudozhnikov RSFSR. Kemerovskaia oblastnaia kartinnaia galereia.
Fold. col. ill. inserted.
1. Chermianin, Ivan Aleksandrovich, 1926- —Exhibitions. I. Kemerovo, Russia (Province). Oblastnoe upravlenie kul'tury. II. Soiuz khudozhnikov RSFSR. Kemerovskoe otdelenie. III. Kemerovskaia oblastnaia kartinnaia galereia.
N6999.C478A4 1976 79-407450

CHERN, SHIING-SHEN, 1911-

Chern, Shiing-Shen, 1911-
Selected papers / Shiing-Shen Chern. — New York : Springer-Verlag, c1978.
xxxi, 476 p. : ill. ; 25 cm.
Published in conjunction with the International Symposium in Global Analysis and Global Geometry to be held in Berkeley, Calif., June 1979.
Bibliography: p. 467-474.
1. Geometry, Differential—Collected works. 2. Global analysis (Mathematics)—Collected works. 3. Chern, Shiing-Shen, 1911-
QA641.C47 514'.7 78-22093
78 MARC

CHERNETSOV, VLADIMIR SEMENOVICH, 1907-1969

—EXHIBITIONS

Chernetsov, Vladimir Semenovich, 1907-1969.
(Vladimir Chernetsov, 1907-1969)
Владимир Чернецов, 1907-1969 : К семидесятилетию со дня рождения : Каталог выставки / [Составитель каталога И. Б. Ефимович ; Вступит. статья Е. Муриной]. — Москва : Сов. художник, 1977.
[40] p. : ill. ; 21 cm. USSR 77
On leaf preceding t. p.: Moskovskaia organizatsiia Soiuza khudozhnikov RSFSR.
0.36rub
1. Chernetsov, Vladimir Semenovich, 1907-1969—Exhibitions. I. Efimovich, I. B. II. Soiuz khudozhnikov RSFSR. Moskovskaia organizatsiia.
N6999.C48A4 1977 78-423421

CHERNIAEV, MIKHAIL GRIGOR'EVICH, 1828-1898

MacKenzie, David, fl. 1967-
The Lion of Tashkent; the career of General M. G. Cherniaev. [Athens] University of Georgia Press [1974]
xx, 267 p. illus. 25 cm. $11.00
Bibliography: p. 248-258.
ISBN 0-8203-0322-4
1. Cherniaev, Mikhail Grigor'evich, 1828-1898. 2. Russia—History—Alexander II, 1855-1881. 3. Generals—Russia—Biography. I. Title.
DK219.6.C425M3 947.08'092'4 73-81625
74[r79]rev MARC

CHERNIAK, SOFIIA

Cherniak, Sofiia.
(Promin' sered nochi)
Промінь серед ночі : документальна повість / Софія Черняк. — Київ : Політвидав України, 1978.
166 p., [4] leaves of plates : ill. ; 20 cm. USSR***
0.55rub
1. World War, 1939-1945—Personal narratives, Ukrainian. 2. Cherniak, Sofiia. 3. Communists—Ukraine—Biography. I. Title.
D811.5.C443 79-407503

CHERNIGOV

—BIOGRAPHY

TSerkovnyĭ, Mikhail Filippovich.
(Geroev podvigi bessmertny)
Героев подвиги бессмертны : очерки о героях Советского Союза / [М. Ф. Церковный, А. Д. Шиганов, Б. Ф. Юрьев]. — Киев : Политиздат Украины, 1977.
566 p. : ports. ; 21 cm. USSR***
1.10rub
1. Geroi Sovetskogo Soiuza. 2. World War, 1939-1945—Biography. 3. Chernigov—Biography. I. Shiganov, Anatoliĭ Dmitrievich, joint author. II. IUr'ev, Boris Fedorovich, joint author. III. Title.
DK268.A1T77 78-422510

CHERNOMORKA, UKRAINE

—HISTORY

Bond, Anatole.
Deutsche Siedlung am Schwarzen Meer, Lustdorf bei Odessa : [geschichtl. u. sprachl. Studien] / von Anatole Bond. — 1. Aufl. — Marburg/L. : Elwert, 1978.
iv, 186 p. ; 24 cm. — (Deutsche Dialektographie ; Bd. 104) GFR78-A
Bibliography: p. 184-185.
ISBN 3-7708-0576-3
1. Chernomorka, Ukraine—History. 2. Germans in Ukraine. I. Title. II. Series.
DK651.C456B66 947'.717 78-389272
79 MARC

CHERNORIZETS KHRABUR

—FICTION

Rodev, TSoncho.
(Cherniiat konnik)
Черният конник / Цончо Родев. — София : Изд-во отечество, 1978.
236 p. : ill. ; 17 cm. Bu***
0.48 lv
1. Simeon, King of Bulgaria, 868 (ca.)-927—Fiction. 2. Chernorizets Khrabur—Fiction. I. Title.
PG1038.28.O3C5 78-422861

CHERNOV, IURIĬ L'VOVICH, 1935-

—CATALOGS

Chernov, IUriĭ L'vovich, 1935-
(IUriĭ L'vovich Chernov)
Юрий Львович Чернов : заслуженный художник РСФСР, Москва : в бронзе, мраморе, граните / [автор текста Мюда Наумовна Яблонская]. — Москва : Сов. художник, 1978.
[32] p. : col. ill. ; 34 cm. USSR***
Cover title: IU. Chernov.
1.90rub
1. Chernov, IUriĭ L'vovich, 1935- —Catalogs. I. IAblonskaia, Miuda Nakhmanovna.
NB699.C48A4 1978 79-407299

CHERNOVITSKAYA OBLAST', UKRAINE

—HISTORY

Tymoshchuk, Borys Onysymovych.
(Tverdynia na Pruti)
Твердиня на Пруті : з історії виникнення Чернівців : нарис / Б. О. Тимощук. — Ужгород : Карпати, 1978.
86 p. : ill. ; 17 cm. USSR***
Summary in Romanian.
0.30rub
1. Chernovitskaya oblast', Ukraine—History. I. Title.
DK511.C458T95 78-421541

CHERNYSHEVSKIĬ, NIKOLAĬ GAVRILOVICH, 1828-1889. CHTO DELAT'

Naumova, Nelli Naumovna.
(Roman N. G. Chernyshevskogo "Chto delat'?")
Роман Н. Г. Чернышевского "Что делать?" / Н. Наумова. — Изд. 2., доп. — Ленинград : Худож. лит-ра, 1978.
132 p. ; 20 cm. — (Массовая историко-литературная библиотека) USSR***
Series romanized: Massovaia istoriko-literaturnaia biblioteka.
Includes bibliographical references.
0.20rub
1. Chernyshevskiĭ, Nikolaĭ Gavrilovich, 1828-1889. Chto delat'. I. Title.
PG3321.C6C65 1978 79-410271

CHERNYSHEVSKIĬ, NIKOLAĬ GAVRILOVICH, 1828-1889

Melent'ev, IUriĭ Serafimovich.
(Predtechi svobody)
Предтечи свободы / Юрий Мелентьев. — Москва : Правда, 1976.
64 p. ; 17 cm. — (Библиотека Огонек ; № 30) USSR 76
CONTENTS: Предтечи свободы.—Прометей России.
0.06rub
1. Chernyshevskiĭ, Nikolaĭ Gavrilovich, 1828-1889. 2. Intellectuals—Russia—Biography. 3. Revolutionists—Russia—Biography. I. Melent'ev, IUriĭ Serafimovich. Prometeĭ Rossii. 1976. II. Title. III. Series: Biblioteka Ogonek ; no. 30.
DK219.6.C43M44 79-411099

CHERNYSHEVSKIĬ, NIKOLAĬ
GAVRILOVICH, 1829-1889 (Continued)

Pini, O A
(Chernyshevskiĭ v Peterburge)
Чернышевский в Петербурге / О. А. Пини. — Ленинград : Лениздат, 1978.
272 p., [21] leaves of plates : ill. ; 17 cm. USSR 78
Bibliography : p. 269–[271]
1.10rub
1. Chernyshevskiĭ, Nikolaĭ Gavrilovich, 1829–1889. 2. Revolutionists—Russia—Biography. I. Title.
DK219.6.C43P5 79–410361

Shatalov, Stanislav Evgen'evich.
Nikolaĭ Gavrilovich Chernyshevskiĭ : 150 let so dnia rozhdeniia / S. E. Shatalov, A. S. Kurilov. — Moskva : Znanie, 1978.
64 p. ; 20 cm. — (Novoe v zhizni, nauke, tekhnike : Seriia Literatura ; No 11, 1978) USSR78
Includes bibliographical references.
0.11rub
1. Chernyshevskiĭ, Nikolaĭ Gavrilovich, 1828-1889. 2. Authors, Russian—19th century—Biography. 3. Critics—Russia—Biography. 4. Revolutionists—Russia—Biography. I. Kurilov, Aleksandr Sergeevich, joint author. II. Title. III. Series: Novoe v zhizni, nauke, tekhnike : Seriia Literatura ; 1978, 11.
PN35.N6 1978, no. 11 79-377107
[PG3321.C625] 79 MARC

Travushkin, Nikolaĭ Sergeevich.
(Chernyshevskiĭ v gody katorgi i ssylki)
Чернышевский в годы каторги и ссылки / Н. С. Травушкин. — Москва : Худож. лит., 1978.
349 p., [9] leaves of plates : ill. ; 20 cm. USSR 78
Includes bibliographical references and index.
1.20rub
1. Chernyshevskiĭ, Nikolaĭ Gavrilovich, 1828–1889. 2. Revolutionists—Russia—Biography. I. Title.
DK219.6.C43T72 79–410724

—ADDRESSES, ESSAYS, LECTURES

N. G. Chernyshevskiĭ i filosofskaia mysl' na Ukraine : [Sb. stateĭ] / AN USSR, In-t filosofii ; [Redkol. P. T. Manzenko (otv. red.) ... et al.]. — Kiev : Nauk. dumka, 1978.
175 p. : ill. ; 20 cm. USSR79
Includes bibliographical references.
1.20rub
1. Chernyshevskiĭ, Nikolaĭ Gavrilovich, 1828-1889—Addresses, essays, lectures. 2. Philosophy, Ukrainian—Addresses, essays, lectures. I. Manzenko, Petro Tryfonovych. II. Akademiia nauk URSR, Kiev. Instytut filosofii.
B4238.C64N18 79-378715
79 MARC

—AESTHETICS

Solov'ev, Gennadiĭ Arsen'evich.
(Ėsteticheskie vozzreniia Chernyshevskogo)
Эстетические воззрения Чернышевского / Г. А. Соловьев. — 2-е изд., доп. — Москва : Худож. лит., 1978.
421 p., [1] leaf of plates : port. ; 21 cm. USSR 78
First ed. published 1974 under title: Ėsteticheskie vozzreniia Chernyshevskogo i Dobroliubova.
Includes bibliographical references.
1.30rub
1. Chernyshevskiĭ, Nikolaĭ Gavrilovich, 1828–1889—Aesthetics. 2. Criticism—Russia—History. I. Title.
PG2947.C3S6 1978 79–408523

—CORRESPONDENCE

(Poėticheskiĭ kraĭ)
Поэтический край : Н. Г. Чернышевский и груз. общественность : 150 лет со дня рождения Н. Г. Чернышевского : [Сборник / Сост. В. С. Шадури]. — Тбилиси : Мерани, 1978.
204 p. : ill. ; 22 cm. USSR 78
Includes bibliographical references.
1.10rub
1. Chernyshevskiĭ, Nikolaĭ Gavrilovich, 1828–1889—Political and social views. 2. Chernyshevskiĭ, Nikolaĭ Gavrilovich, 1828–1889—Correspondence. 3. Russia — Relations (general) with Georgia (Transcaucasia) 4. Georgia (Transcaucasia)—Relations (general) with Russia. I. Shaduri, Vano Semenovich.
PG3321.C6Z716 79–413074

—FRIENDS AND ASSOCIATES

Bystrova, Nina Mikhaĭlovna Chernyshevskaia, 1896–1975.
(N. G. Chernyshevskiĭ i T. G. Shevchenko)
Н. Г. Чернышевский и Т. Г. Шевченко : воспоминания, заметки, материалы / Н. М. Чернышевская. — Киев : Днiпро, 1978.
133 p., [12] leaves of plates : ill. ; 18 cm. USSR***
Includes bibliographical references.
0.55rub
1. Chernyshevskiĭ, Nikolaĭ Gavrilovich, 1828–1889 — Friends and associates. 2. Shevchenko, Taras Hryhorovych, 1814–1861 — Friends and associates. 3. Russia — Relations (general) with Ukraine. 4. Ukraine — Relations (general) with Russia.
PG3321.C6Z653 1978 79–408259

—HOMES AND HAUNTS

(Nikolaĭ Gavrilovich Chernyshevskiĭ v portretakh, illiustratsiiakh, dokumentakh)
Николай Гаврилович Чернышевский в портретах, иллюстрациях, документах : [Альбом] / Сост. О. А. Пини, А. П. Холина ; Под общ. ред. [и с предисл.] Н. М. Чернышевской. — Ленинград : Просвещение, Ленингр. отд-ние, 1978.
335 p. : ill. ; 27 cm. — (Пособие для учителя) USSR 78
Series romanized : Posobie dlia uchitelia.
Bibliography : p. [333]
3.20rub
1. Chernyshevskiĭ, Nikolaĭ Gavrilovich, 1828–1889 — Portraits. 2. Chernyshevskiĭ, Nikolaĭ Gavrilovich, 1828–1889 — Homes and haunts. 3. Chernyshevskiĭ, Nikolaĭ Gavrilovich, 1828–1889—Illustrations. I. Pini, O. A. II. Kholina, A. P. III. Bystrova, Nina Mikhaĭlovna Chernyshevskaia, 1896–1975.
PG3321.C6Z683 79–410673

—ILLUSTRATIONS

(Nikolaĭ Gavrilovich Chernyshevskiĭ v portretakh, illiustratsiiakh, dokumentakh)
Николай Гаврилович Чернышевский в портретах, иллюстрациях, документах : [Альбом] / Сост. О. А. Пини, А. П. Холина ; Под общ. ред. [и с предисл.] Н. М. Чернышевской. — Ленинград : Просвещение, Ленингр. отд-ние, 1978.
335 p. : ill. ; 27 cm. — (Пособие для учителя) USSR 78
Series romanized : Posobie dlia uchitelia.
Bibliography : p. [333]
3.20rub
1. Chernyshevskiĭ, Nikolaĭ Gavrilovich, 1828–1889 — Portraits. 2. Chernyshevskiĭ, Nikolaĭ Gavrilovich, 1828–1889 — Homes and haunts. 3. Chernyshevskiĭ, Nikolaĭ Gavrilovich, 1828–1889—Illustrations. I. Pini, O. A. II. Kholina, A. P. III. Bystrova, Nina Mikhaĭlovna Chernyshevskaia, 1896–1975.
PG3321.C6Z683 79–410673

—KNOWLEDGE—LITERATURE
—ADDRESSES, ESSAYS, LECTURES

(M. H. Chernyshevs'kyĭ i ukraïns'ka literatura)
М. Г. Чернишевський і українська література : збірник статей / [упорядкування І. Т. Крука]. — Київ : Дніпро, 1978.
219 p. ; 21 cm. USSR***
Includes bibliographical references.
0.95rub
1. Chernyshevskiĭ, Nikolaĭ Gavrilovich, 1828–1889—Knowledge—Literature—Addresses, essays, lectures. 2. Ukrainian literature—History and criticism—Addresses, essays, lectures. 3. Russian literature—19th century—History and criticism—Addresses, essays, lectures. I. Kruk, Ivan Trofimovich.
PG2947.C3M2 79–408360

—POLITICAL AND SOCIAL VIEWS

Nikolaev, Mikhail Pavlovich.
(L. N. Tolstoĭ i N. G. Chernyshevskiĭ)
Л. Н. Толстой и Н. Г. Чернышевский / М. П. Николаев. — Тула : Приок. кн. изд-во, 1978.
144 p. ; 20 cm. USSR 78
Includes bibliographical references.
0.25rub
1. Tolstoĭ, Lev Nikolaevich, graf, 1828–1910—Political and social views. 2. Chernyshevskiĭ, Nikolaĭ Gavrilovich, 1828–1889—Political and social views. I. Title.
PG3415.P6N5 79–410154

(Poėticheskiĭ kraĭ)
Поэтический край : Н. Г. Чернышевский и груз. общественность : 150 лет со дня рождения Н. Г. Чернышевского : [Сборник / Сост. В. С. Шадури]. — Тбилиси : Мерани, 1978.
204 p. : ill. ; 22 cm. USSR 78
Includes bibliographical references.
1.10rub
1. Chernyshevskiĭ, Nikolaĭ Gavrilovich, 1828–1889—Political and social views. 2. Chernyshevskiĭ, Nikolaĭ Gavrilovich, 1828–1889—Correspondence. 3. Russia — Relations (general) with Georgia (Transcaucasia) 4. Georgia (Transcaucasia)—Relations (general) with Russia. I. Shaduri, Vano Semenovich.
PG3321.C6Z716 79–413074

Shabliovs'kyĭ, Ievhen Stepanovych.
(Chernyshevskiĭ i Ukraina)
Чернышевский и Украина / Е. С. Шаблиовский. — Киев : Наук. думка, 1978.
343 p., [8] leaves of plates : ill. ; 20 cm. USSR 78
At head of title: Akademiia nauk Ukrainskoĭ SSR.
Bibliography : p. 323–[334]
Includes index.
2.70rub
1. Chernyshevskiĭ, Nikolaĭ Gavrilovich, 1829–1889—Political and social views. 2. Russia — Relations (general) with Ukraine. 3. Ukraine—Relations (general) with Russia. I. Title.
PG3321.C6Z78 79–408511

—PORTRAITS

(Nikolaĭ Gavrilovich Chernyshevskiĭ v portretakh, illiustratsiiakh, dokumentakh)
Николай Гаврилович Чернышевский в портретах, иллюстрациях, документах : [Альбом] / Сост. О. А. Пини, А. П. Холина ; Под общ. ред. [и с предисл.] Н. М. Чернышевской. — Ленинград : Просвещение, Ленингр. отд-ние, 1978.
335 p. : ill. ; 27 cm. — (Пособие для учителя) USSR 78
Series romanized : Posobie dlia uchitelia.
Bibliography : p. [333]
3.20rub
1. Chernyshevskiĭ, Nikolaĭ Gavrilovich, 1828–1889 — Portraits. 2. Chernyshevskiĭ, Nikolaĭ Gavrilovich, 1828–1889 — Homes and haunts. 3. Chernyshevskiĭ, Nikolaĭ Gavrilovich, 1828–1889—Illustrations. I. Pini, O. A. II. Kholina, A. P. III. Bystrova, Nina Mikhaĭlovna Chernyshevskaia, 1896–1975.
PG3321.C6Z683 79–410673

CHEROKEE AUTOMOBILE

Chilton Book Company.
Chilton's Jeep Wagoneer, Commando & Cherokee, 1966-79 repair & tune-up guide, all models / managing editor, Kerry A. Freeman, senior editor, Richard J. Rivele. — Radnor, Pa. : Chilton Book Co., c1979.
225 p. : ill. ; 26 cm.
First (1974) ed., prepared by the company's Automotive Editorial Dept., published under title: Chilton's repair and tune-up guide: Jeep Wagoneer, Commando, and Cherokee.
On spine: Chilton's repair & tune-up guide for the Jeep Wagoneer, Commando & Cherokee.
Includes index.
ISBN 0-8019-6739-2 : $7.95
1. Cherokee automobile. 2. Commando automobile. 3. Wagoneer automobile. I. Freeman, Kerry A. II. Rivele, Richard J. III. Chilton Book Company. Automotive Editorial Dept. Chilton's repair and tune-up guide: Jeep Wagoneer, Commando, and Cherokee. IV. Title: Jeep Wagoneer, Commando & Cherokee, 1966-79 repair & tune-up guide, all models. V. Title: Chilton's repair & tune-up guide for the Jeep Wagoneer, Commando & Cherokee.
TL215.C47C47 1979 629.28'8'22 79-102577
79 MARC

CHEROKEE INDIAN REMOVAL, 1838 see Cherokee Removal, 1838

CHEROKEE INDIANS

Cherokee nation.
Communication of the delegation of the Cherokee nation the President of the United States, submitting the memorial their national council, with the correspondence between John Ross, principal chief, and certain officers of the rebellious stat Washington, Gibson brothers, printers, 1866.
48 p. 23 cm.
—— —— Microfilm. Microfilm 58191 E
1. Cherokee Indians. 2. Ross, John, Cherokee chief, 1790-1866. 3. United States—History—Civil war 1861-1865—Indian troops.
E99.C5C6 1866 5-22
—— —— Copy 3. Toner Coll. MA
[r79]rev2

Eaton, Rachel Caroline.
John Ross and the Cherokee Indians / by Rachel Caroline Eaton. — 1st AMS ed. — New York : AMS Press, 1978.
153 p. ; 22 cm.
Originally presented as the author's thesis, University of Chicago, 1919.
Reprint of a private ed. distributed in 1921 by the University of Chica Libraries, Chicago.
Bibliography: p. [151]-153.
ISBN 0-404-15526-X
1. Ross, John, Cherokee chief, 1790-1866. 2. Cherokee Inndians. 3. Cherokee Indians—Biography. 4. Indians of North America—Biography. Title.
E99.C5E16 1978 970'.004'97 76-4369
78 MA

Foster, George Everett, 1849-
Se-quo-yah, the American Cadmus and modern Moses / Geo. E. Foster. — New York : AMS Press, 1979.
xviii, 244 p., [6] leaves of plates : ill. ; 19 cm.
Reprint of the 1885 ed. published by the Office of the Indian Rights Association, Philadelphia.
Includes index.
ISBN 0-404-15554-8
1. Sequoyah, 1770?-1843. 2. Cherokee Indians—Biography. 3. Cherokee Indians. 4. Cherokee language—Alphabet. I. Title.
E99.C5S383 1979 970'.004'97 76-437
79 MA

Gilbert, William Harlen, 1904-
The eastern Cherokees / by William Harlen Gilbert, Jr. — AMS ed. — New York : AMS Press, 1978.
p. 169-413 p. : ill. ; 23 cm.
Originally presented as the author's thesis, University of Chicago, 1934.
Reprint of the 1943 ed. published by the Smithsonian Institution, Washington, which was issued as no. 23 of Smithsonian Institution, Bureau of American Ethnology, Anthropological papers, in 133 of its Bulletin.
Bibliography: p. 402-413.
ISBN 0-404-15547-2
1. Cherokee Indians. I. Title. II. Series: United States. Bureau American Ethnology. Anthropological papers ; no. 23. III. Series: United States. Bureau of American Ethnology. Bulletin ; 133.
E99.C5G35 1978 970'.004'97 76-43
78 MA

CHEROKEE INDIANS (Continued)

—ANTIQUITIES

Gleeson, Paul F
Archaeological investigations in the Tellico Reservoir; interim report, 1969. Edited by Paul Gleeson. Contributions by Paul Gleeson ⌐and others. Knoxville, Dept. of Anthropology, University of Tennessee⌐ 1970.
x, 140 p. illus. 28 cm. (University of Tennessee. Dept. of Anthropology. Report of investigations, no. 8)
"Submitted in accordance with Tennessee Valley Authority contract TV-31590A."
Bibliography: p. 135-140.
1. Indians of North America—Little Tennessee Valley, Ga.-Tenn.—Antiquities. 2. Little Tennessee Valley, Ga.-Tenn.—Antiquities. 3. Cherokee Indians—Antiquities. 4. Georgia—Antiquities. 5. Tennessee—Antiquities. I. Tennessee Valley Authority. II. Title. III. Series: University of Tennessee, Knoxville. Dept. of Anthropology. Report of investigations, no. 8.
E78.T3G53 970.3 72-611909
73⌐r79⌐rev MARC

—BIBLIOGRAPHY

Fogelson, Raymond D
The Cherokees, a critical bibliography / Raymond D. Fogelson. — Bloomington : Published for the Newberry Library ⌐by⌐ Indiana University Press, c1978.
x, 98 p. ; 21 cm. — (Bibliographical series)
Includes index.
ISBN 0-253-31346-5
1. Cherokee Indians—Bibliography. I. Title. II. Series.
Z1210.C46F64 016.970'.004'97 78-3254
⌐E99.C5⌐ 78 MARC

—BIOGRAPHY

Adams, Robert G
Nancy Ward, beautiful woman of two worlds / ⌐written by Robert G. Adams⌐. — ⌐Chattanooga⌐ : Hampton House Studios, c1979.
xiv, 109 p. : ill. ; 24 cm.
Bibliography: p. 104-106.
Includes index.
1. Ward, Nancy, d. 1822. 2. Cherokee Indians—Biography. 3. Tennessee—History. I. Title.
E99.C5W26155 970'.004'97 79-87502
00 MARC

Alderman, Pat.
Nancy Ward, Cherokee chieftainess, Dragging Canoe, Cherokee-Chickamauga war chief / by Pat Alderman. — ⌐s.l. : s.n.⌐, c1978 (Johnson City, Tenn. : Overmountain Press)
86 p. : ill. ; 26 cm.
Bibliography: p. 84-86.
SUMMARY: Biographies of two prominent leaders of the Cherokee nation in the 18th century.
1. Ward, Nancy, d. 1822. 2. Dragging Canoe (Cherokee-Chickamauga war chief) 1732-1792. 3. Cherokee Indians—Biography. 4. Cherokee Indians—History. ⌐1. Ward, Nancy, d. 1822. 2. Dragging Canoe (Cherokee-Chickamauga war chief) 1732-1792. 3. Cherokee Indians—Biograpy. 4. Indians of North America—Biography⌐ I. Title.
E99.C5W262 970.'004'97 78-108867
79 MARC

Eaton, Rachel Caroline.
John Ross and the Cherokee Indians / by Rachel Caroline Eaton. — 1st AMS ed. — New York : AMS Press, 1978.
153 p. ; 22 cm.
Originally presented as the author's thesis, University of Chicago, 1919.
Reprint of a private ed. distributed in 1921 by the University of Chicago Libraries, Chicago.
Bibliography: p. ⌐151⌐-153.
ISBN 0-404-15526-X
1. Ross, John, Cherokee chief, 1790-1866. 2. Cherokee Inndians. 3. Cherokee Indians—Biography. 4. Indians of North America—Biography. I. Title.
E99.C5E16 1978 970'.004'97 76-43694
78 MARC

Foster, George Everett, 1849-
Se-quo-yah, the American Cadmus and modern Moses / by Geo. E. Foster. — New York : AMS Press, 1979.
xviii, 244 p., ⌐6⌐ leaves of plates : ill. ; 19 cm.
Reprint of the 1885 ed. published by the Office of the Indian Rights Association, Philadelphia.
Includes index.
ISBN 0-404-15554-8
1. Sequoyah, 1770?-1843. 2. Cherokee Indians—Biography. 3. Cherokee Indians. 4. Cherokee language—Alphabet. I. Title.
E99.C5S383 1979 970'.004'97 76-43709
79 MARC

Moulton, Gary E
John Ross, Cherokee Chief / Gary E. Moulton. — Athens : University of Georgia Press, c1978.
ix, 282 p. : port. ; 24 cm.
Bibliography: p. ⌐257⌐-269.
Includes index.
ISBN 0-8203-0422-0
1. Ross, John, Cherokee Chief, 1790-1866. 2. Cherokee Indians—Biography. 3. Cherokee Indians—History. I. Title.
E99.C5R675 970'.004'97 76-1146
77 MARC

Starr, Emmet.
History of the Cherokee Indians and their legends and folk lore / Emmet Starr. — Tulsa, Okla. : Oklahoma Yesterday, c1979.
680, ⌐104⌐ p. : ill. ; 27 cm.
Reprint of the 1921 ed. published by Warden Co., Oklahoma City.
Includes bibliographical references and indexes.
1. Cherokee Indians—History. 2. Cherokee Indians—Biography. 3. Cherokee Indians—Genealogy. I. Title.
E99.C5S8 1979 970'.004'97 79-112907
00 MARC

—BIOGRAPHY—JUVENILE LITERATURE

Harrell, Sara Gordon.
John Ross / by Sara Gordon Harrell. — Minneapolis : Dillon Press, c1979.
62 p. : ill. ; 24 cm. — (The Story of an American Indian)
SUMMARY: A biography of the Cherokee chief who led his people for more than 40 years, first in an effort to keep their homeland, and later through their greatest trial when they were forced to go west by the United States government.
ISBN 0-87518-173-2 (lib. bdg.) : $5.95
1. Cherokee Indians—History—Juvenile literature. 2. Ross, John, Cherokee chief, 1790-1866—Juvenile literature. 3. Cherokee Indians—Biography—Juvenile literature. ⌐1. Ross, John, Cherokee chief, 1790-1866. 2. Cherokee Indians—Biography. 3. Indians of North America—Biography⌐ I. Title.
E99.C5R674 970'.004'97 78-21042
78 MARC

Marriott, Alice Lee, 1910-
Sequoyah: leader of the Cherokees. Illustrated by Bob Riger. New York, Random House ⌐1956⌐
180 p. illus. 22 cm. (Landmark books)
SUMMARY: A biography of the Cherokee Indian who invented a system of writing for his people.
1. Sequoyah, 1770?-1843—Juvenile literature. 2. Cherokee Indians—Biography—Juvenile literature. 3. Indians of North America—Biography—Juvenile literature. ⌐1. Sequoyah, 1770?-1843. 2. Cherokee Indians—Biography. 3. Indians of North America—Biography⌐ I. Riger, Robert. II. Title.
E99.C5M36 970'.004'97 55-5823
⌐r79⌐rev MARC

—CAPTIVITIES

Marrant, John, b. 1755.
A narrative of the Lord's wonderful dealings with John Marrant, a Black / John Marrant. — New York : Garland Pub., 1978.
v, 38 p. ; 23 cm. — (The Garland library of narratives of North American Indian captivities ; v. 17)
"Taken down from his own relation, arranged, corrected, and published by the Rev. Mr. Aldridge."—1785 ed. t.p.
Reprint of the 1785 ed. printed by Gilbert and Plummer, London.
Issued with the reprints of the 1785 and 1875 editions of Swetland, L. A very remarkable narrative. New York, 1978. The reprint of the 1915 ed. of Merrifield, E. The story of the captivity and rescue from the Indians of Luke Swetland. New York, 1978. The reprints of the 1786 and 1794 editions of M'Donald, P. A surprising account of a captivity and escape. New York, 1978. The reprints of the 1788 and 1794 editions of a Very surprising narrative of a young woman discovered in a rocky-cave, by A. Panther. New York, 1978.
ISBN 0-8240-1641-6 : $29.50
1. Cherokee Indians—Captivities. 2. Marrant, John, b. 1755. 3. Indians of North America—Captivities. 4. Converts—United States—Biography. I. Aldridge, William, 1737-1797. II. Title. III. Series.
E85.G2 vol. 17 973'.04'97 s 77-27409
⌐E99.C5⌐ 77 MARC

M'Donald, Philip.
A surprising account of a captivity and escape / Philip M'Donald and Alexander M'Leon. — New York : Garland Pub., 1978.
11 p. ; 23 cm. — (The Garland library of narratives of North American Indian captivities ; v. 17)
Issued with the reprint of the 1785 ed. of Marrant, J. A narrative of the Lord's wonderful dealings with John Marrant. New York, 1978.
Reprint of the 1794 ed. printed by H. Blake, Keene, N. H., which was issued under title: A surprising account of the captivity and escape of Philip M'Donald & Alexander M'Leod, of Virginia, from the Chickkemogga Indians.
Evans 27248.
ISBN 0-8240-1641-6 : $29.50
1. Cherokee Indians—Captivities. 2. Indians of North America—Captivities. 3. M'Donald, Philip. 4. M'Leod, Alexander. 5. Ohio Valley—Description and travel. 6. Virginia—Biography. I. M'Leod, Alexander, joint author. II. Title. III. Series.
E85.G2 vol. 17 973'.04'97 s 77-27410
⌐E99.C5⌐ 77 MARC

M'Donald, Philip.
A surprising account of a captivity and escape / Philip M'Donald and Alexander M'Leon. — New York : Garland Pub., 1978.
16 p. ; 23 cm. — (The Garland library of narratives of North American Indian captivities ; v. 17)
Issued with the reprint of the 1785 ed. of Marrant, J. A narrative of the Lord's wonderful dealings with John Marrant. New York, 1978.
Reprint of the 1786 ed. printed by Haswell & Russell, Bennington, Vt., which was issued under title: A surprising account, of the captivity and escape of Philip M'Donald, and Alexander M'Leod, of Virginia, from the Chickkemogga Indians, and of their great discoveries in the Western World.
Evans 20472.
ISBN 0-8240-1641-6 : $29.50
1. Cherokee Indians—Captivities. 2. Indians of North America—Captivities. 3. M'Donald, Philip. 4. M'Leod, Alexander. 5. Ohio Valley—Description and travel. 6. Virginia—Biography. I. M'Leod, Alexander, joint author. II. Title. III. Series.
E85.G2 vol. 17 973'.04'97 s 77-27411
⌐E99.C5⌐ 77 MARC

—FICTION

Sat-Okh, 1920-
(Taina starogo Sagamory)
Тайна старого Сагаморы : Повесть / Сат-Окх, А. Расулова ; ⌐Предисл. В. Гуляева ⌐Худож. К. Фадин⌐. — Москва : Мол. гвардия, 1978.
176 p. : ill. ; 16 cm. — (Бригантина) USSR 78
Series romanized: Brigantina.
Includes bibliographical references.
0.70rub
1. Cherokee Indians—Fiction. I. Rasulova, Antonina Leonidovna, joint author. II. Title.
PG3485.8.T2T3 78-420552

—FINANCIAL AFFAIRS

Cherokee Nation. Senate.
⌐Senate bill no. 9 : an act to secure for the Cherokee Nation a loan upon the contract and obligations of the United States for certain portions of the proceeds of the sale of the Cherokee Strip ... — s.l. : s.n., 1849?⌐
3 p. ; 24 cm.
In Cherokee language; printed in Cherokee syllabic characters.
Substituted English title above supplied from English edition issued probably in the same year. Cf. NUC pre-56, NC 0344246.
Cherokee title may be romanized: Galvladi-Dinilawigi unitlvnv 9 goweli : alinegv uwaavehi Tsalagi Ayeli tsutoliquudi ...
Bill vetoed soon after passage by Principal Chief C. J. Harris. Cf. Wardell. A political history.
L.C. copy has a typescript copy of the text of the English-language edition (4 leaves) enclosed.
1. Cherokee Indians—Financial affairs. 2. Cherokee Indians—Land transfers. 3. Cherokee Outlet, Okla. 4. Cherokee language—Texts. I. Title: An act to secure for the Cherokee Nation a loan. II. Title: Galvladi-Dinilawigi unitlvnv sonela goweli.
KF8228.C5A5 1849 79-303047
79 MARC

—GENEALOGY

Starr, Emmet.
History of the Cherokee Indians and their legends and folk lore / Emmet Starr. — Tulsa, Okla. : Oklahoma Yesterday, c1979.
680, ⌐104⌐ p. : ill. ; 27 cm.
Reprint of the 1921 ed. published by Warden Co., Oklahoma City.
Includes bibliographical references and indexes.
1. Cherokee Indians—History. 2. Cherokee Indians—Biography. 3. Cherokee Indians—Genealogy. I. Title.
E99.C5S8 1979 970'.004'97 79-112907
00 MARC

—GOVERNMENT RELATIONS

see also Cherokee Removal, 1838

—HISTORY

see also Cherokee Removal, 1838

Alderman, Pat.
Nancy Ward, Cherokee chieftainess, Dragging Canoe, Cherokee-Chickamauga war chief / by Pat Alderman. — ⌐s.l. : s.n.⌐, c1978 (Johnson City, Tenn. : Overmountain Press)
86 p. : ill. ; 26 cm.
Bibliography: p. 84-86.
SUMMARY: Biographies of two prominent leaders of the Cherokee nation in the 18th century.
1. Ward, Nancy, d. 1822. 2. Dragging Canoe (Cherokee-Chickamauga war chief) 1732-1792. 3. Cherokee Indians—Biography. 4. Cherokee Indians—History. ⌐1. Ward, Nancy, d. 1822. 2. Dragging Canoe (Cherokee-Chickamauga war chief) 1732-1792. 3. Cherokee Indians—Biograpy. 4. Indians of North America—Biography⌐ I. Title.
E99.C5W262 970.'004'97 78-108867
79 MARC

Goodwin, Gary C 1940-
Cherokees in transition : a study of changing culture and environment prior to 1775 / by Gary C. Goodwin. — Chicago : University of Chicago, Dept. of Geography, 1977.
ix, 207 p. : ill. ; 23 cm. — (Research paper - University of Chicago, Department of Geography ; no. 181)
Bibliography: p. 177-207.
ISBN 0-89065-088-8 : $6.00
1. Cherokee Indians—History. 2. Human ecology—Southern States. I. Title. II. Series: Chicago. University. Dept. of Geography. Research paper ; no. 181.
H31.C514 no. 181 970'.004'97 77-2709
⌐E99.C5⌐ 77⌐r79⌐rev MARC

Moulton, Gary E
John Ross, Cherokee Chief / Gary E. Moulton. — Athens : University of Georgia Press, c1978.
ix, 282 p. : port. ; 24 cm.
Bibliography: p. ⌐257⌐-269.
Includes index.
ISBN 0-8203-0422-0
1. Ross, John, Cherokee Chief, 1790-1866. 2. Cherokee Indians—Biography. 3. Cherokee Indians—History. I. Title.
E99.C5R675 970'.004'97 76-1146
77 MARC

CHEROKEE INDIANS
—HISTORY (Continued)

Starr, Emmet.
History of the Cherokee Indians and their legends and folk lore / Emmet Starr. — Tulsa, Okla. : Oklahoma Yesterday, c1979.
680, [104] p. : ill. ; 27 cm.
Reprint of the 1921 ed. published by Warden Co., Oklahoma City.
Includes bibliographical references and indexes.
1. Cherokee Indians—History. 2. Cherokee Indians—Biography. 3. Cherokee Indians—Genealogy. I. Title.
E99.C5S8 1979 970'.004'97 79-112907
00 MARC

—HISTORY—ADDRESSES, ESSAYS, LECTURES

The Cherokee Indian nation : a troubled history / edited by Duane H. King. — Knoxville : University of Tennessee Press, c1979.
xix, 256 p. : ill. ; 24 cm.
Includes bibliographical references and index.
ISBN 0-87049-227-6
1. Cherokee Indians—History—Addresses, essays, lectures. I. King, Duane H.
E99.C5C414 970'.004'97 78-13222
78 MARC

Horton, James H
Our mountain heritage : essays on the natural and cultural history of western North Carolina / by James H. Horton, Theda Perdue, and James M. Gifford ; edited and with an introd. by Clifford R. Lovin. — [Cullowhee? N.C.] : North Carolina Humanities Committee, Western Carolina University, 1979.
131 p. : ill. ; 22 cm.
Includes bibliographies.
1. North Carolina—History, Local—Addresses, essays, lectures. 2. Cherokee Indians—History—Addresses, essays, lectures. 3. Natural history—North Carolina—Addresses, essays, lectures. I. Perdue, Theda, 1949- joint author. II. Gifford, James M., joint author. III. Lovin, Clifford R. IV. North Carolina Humanities Committee. V. Title.
F261.H67 975.6 79-112339
79 MARC

—HISTORY—JUVENILE LITERATURE

Harrell, Sara Gordon.
John Ross / by Sara Gordon Harrell. — Minneapolis : Dillon Press, c1979.
62 p. : ill. ; 24 cm. — (The Story of an American Indian)
SUMMARY: A biography of the Cherokee chief who led his people for more than 40 years, first in an effort to keep their homeland, and later through their greatest trial when they were forced to go west by the United States government.
ISBN 0-87518-173-2 (lib. bdg.) : $5.95
1. Cherokee Indians—History—Juvenile literature. 2. Ross, John, Cherokee chief, 1790-1866—Juvenile literature. 3. Cherokee Indians—Biography—Juvenile literature. [1. Ross, John, Cherokee chief, 1790-1866. 2. Cherokee Indians—Biography. 3. Indians of North America—Biography] I. Title.
E99.C5R674 970'.004'97 78-21042
78 MARC

—JUVENILE LITERATURE

Cunningham, Maggi, 1916-
The Cherokee tale-teller / by Maggi Cunningham ; illustrated by Patrick Des Jarlait. — Minneapolis : Dillon Press, c1978.
158 p. : col. ill. ; 24 cm.
CONTENTS: Atagahi, the wonderful lake.—Selu Corn Woman and the crows.—Deer song.—Princess of the deer.—The monster utlunta.—Desata and the forever boy.—The fire watcher.—The red bird.—Tlanuwa, the great hawk.—The Nunnehi, the gentle people.
ISBN 0-87518-147-3
1. Cherokee Indians—Legends. 2. Cherokee Indians—Juvenile literature. [1. Cherokee Indians—Legends. 2. Indians of North America—Legends] I. Des Jarlait, Patrick, 1921-1972. II. Title.
E99.C5C87 398.2'0973 77-9580
77 MARC

—LAND TRANSFERS

Cherokee Nation. Senate.
[Senate bill no. 9 : an act to secure for the Cherokee Nation a loan upon the contract and obligations of the United States for certain portions of the proceeds of the sale of the Cherokee Strip ... — s.l. : s.n., 1849?]
3 p. ; 24 cm.
In Cherokee language; printed in Cherokee syllabic characters.
Substituted English title above supplied from English edition issued probably in the same year. Cf. NUC pre-56, NC 0344246.
Cherokee title may be romanized: Galvladi-Dinilawigi unitlvnv 9 goweli : alinegv uwaavehi Tsalagi Ayeli tsutoliquudi ...
Bill vetoed soon after passage by Principal Chief C. J. Harris. Cf. Wardell. A political history.
L.C. copy has a typescript copy of the text of the English-language edition (4 leaves) enclosed.

1. Cherokee Indians—Financial affairs. 2. Cherokee Indians—Land transfers. 3. Cherokee Outlet, Okla. 4. Cherokee language—Texts. I. Title: An act to secure for the Cherokee Nation a loan. II. Title: Galvladi-Dinilawigi unitlvnv sonela goweli.
KF8228.C5A5 1849 79-303047
79 MARC

—LAND TENURE

Odell, Marcia Larson, 1945-
Divide and conquer : allotment among the Cherokee / Marcia Larson Odell. — New York : Arno Press, 1979, c1976.
xv, 493 p. : ill. ; 24 cm. — (The Management of public lands in the United States)
Originally presented as the author's thesis, Cornell University, 1975.
Bibliography: p. 474-493.
ISBN 0-405-11347-1
1. Cherokee Indians—Land tenure. 2. Indians of North America—Oklahoma—Land tenure. 3. Land tenure—Oklahoma—Rogers Co. I. Title. II. Series: Management of public lands in the United States.
E99.C5O33 1979 333.1'6 78-56681
78 MARC

—LAND TRANSFERS

see also Cherokee Removal, 1838

—LEGENDS

Cunningham, Maggi, 1916-
The Cherokee tale-teller / by Maggi Cunningham ; illustrated by Patrick Des Jarlait. — Minneapolis : Dillon Press, c1978.
158 p. : col. ill. ; 24 cm.
CONTENTS: Atagahi, the wonderful lake.—Selu Corn Woman and the crows.—Deer song.—Princess of the deer.—The monster utlunta.—Desata and the forever boy.—The fire watcher.—The red bird.—Tlanuwa, the great hawk.—The Nunnehi, the gentle people.
ISBN 0-87518-147-3
1. Cherokee Indians—Legends. 2. Cherokee Indians—Juvenile literature. [1. Cherokee Indians—Legends. 2. Indians of North America—Legends] I. Des Jarlait, Patrick, 1921-1972. II. Title.
E99.C5C87 398.2'0973 77-9580
77 MARC

—SLAVES, OWNERSHIP OF

Littlefield, Daniel F
The Cherokee freedmen : from emancipation to American citizenship / Daniel F. Littlefield, Jr. — Westport, Conn. : Greenwood Press, 1978.
xii, 281 p. : ill. ; 22 cm. — (Contributions in Afro-American and African studies ; no. 40 ISSN 0069-9624)
Bibliography: p. [259]-265.
Includes index.
ISBN 0-313-20413-6 : $18.95
1. Cherokee Indians—Slaves, Ownership of. 2. Afro-Americans—Indian Territory—Relations with Indians. 3. Indians of North America—Indian Territory—Slaves, Ownership of. 4. Freedmen—Indian Territory. I. Title. II. Series.
E99.C5L5 301.45'19'7073 78-53659
78 MARC

Perdue, Theda, 1949-
Slavery and the evolution of Cherokee society, 1540-1866 / Theda Perdue. — 1st ed. — Knoxville : University of Tennessee Press, c1979.
xiv, 207 p. : ill. ; 23 cm.
Bibliography: p. 183-198.
Includes index.
ISBN 0-87049-259-4
1. Cherokee Indians—Slaves, Ownership of. 2. Indians of North America—Slaves, Ownership of. 3. Afro-Americans—Relations with Indians. 4. Slavery in the United States. I. Title.
E99.C5P394 301.45'19'7073 78-16284
78 MARC

CHEROKEE LANGUAGE

—ALPHABET

Foster, George Everett, 1849-
Se-quo-yah, the American Cadmus and modern Moses / by Geo. E. Foster. — New York : AMS Press, 1979.
xviii, 244 p., [6] leaves of plates : ill. ; 19 cm.
Reprint of the 1885 ed. published by the Office of the Indian Rights Association, Philadelphia.
Includes index.
ISBN 0-404-15554-8
1. Sequoyah, 1770?-1843. 2. Cherokee Indians—Biography. 3. Cherokee Indians. 4. Cherokee language—Alphabet. I. Title.
E99.C5S383 1979 970'.004'97 76-43709
79 MARC

—TEXTS

Cherokee Nation. Senate.
[Senate bill no. 9 : an act to secure for the Cherokee Nation a loan upon the contract and obligations of the United States for certain portions of the proceeds of the sale of the Cherokee Strip ... — s.l. : s.n., 1849?]
3 p. ; 24 cm.
In Cherokee language; printed in Cherokee syllabic characters.
Substituted English title above supplied from English edition issued probably in the same year. Cf. NUC pre-56, NC 0344246.
Cherokee title may be romanized: Galvladi-Dinilawigi unitlvnv 9 goweli : alinegv uwaavehi Tsalagi Ayeli tsutoliquudi ...
Bill vetoed soon after passage by Principal Chief C. J. Harris. Cf. Wardell. A political history.
L.C. copy has a typescript copy of the text of the English-language edition (4 leaves) enclosed.

1. Cherokee Indians—Financial affairs. 2. Cherokee Indians—Land transfers. 3. Cherokee Outlet, Okla. 4. Cherokee language—Texts. I. Title: An act to secure for the Cherokee Nation a loan. II. Title: Galvladi-Dinilawigi unitlvnv sonela goweli.
KF8228.C5A5 1849 79-303047
79 MARC

CHEROKEE OUTLET, OKLA.

Cherokee Nation. Senate.
[Senate bill no. 9 : an act to secure for the Cherokee Nation a loan upon the contract and obligations of the United States for certain portions of the proceeds of the sale of the Cherokee Strip ... — s.l. : s.n., 1849?]
3 p. ; 24 cm.
In Cherokee language; printed in Cherokee syllabic characters.
Substituted English title above supplied from English edition issued probably in the same year. Cf. NUC pre-56, NC 0344246.
Cherokee title may be romanized: Galvladi-Dinilawigi unitlvnv 9 goweli : alinegv uwaavehi Tsalagi Ayeli tsutoliquudi ...
Bill vetoed soon after passage by Principal Chief C. J. Harris. Cf. Wardell. A political history.
L.C. copy has a typescript copy of the text of the English-language edition (4 leaves) enclosed.

1. Cherokee Indians—Financial affairs. 2. Cherokee Indians—Land transfers. 3. Cherokee Outlet, Okla. 4. Cherokee language—Texts. I. Title: An act to secure for the Cherokee Nation a loan. II. Title: Galvladi-Dinilawigi unitlvnv sonela goweli.
KF8228.C5A5 1849 79-303047
79 MARC

—BIOGRAPHY—PERIODICALS

Journal of the Cherokee Strip.
Enid, Okla., Association of Sons and Daughters of the Cherokee Strip Pioneers.
v. ill. 28-36 cm.
Semiannual, Sept. 1967-Sept. 1969; annual, Sept. 1970-
Began with Sept. 1959 issue. Cf. New serial titles.
Official publication of the Association of Sons and Daughters of the Cherokee Strip Pioneers.
1. Cherokee Outlet, Okla.—History—Periodicals. 2. Cherokee Outlet, Okla.—Biography—Periodicals. 3. Frontier and pioneer life—Oklahoma—Cherokee Outlet—Periodicals. I. Association of Sons and Daughters of the Cherokee Strip Pioneers.
F702.C42J68a 976.6 78-646260
MARC-S

—HISTORY—PERIODICALS

Journal of the Cherokee Strip.
Enid, Okla., Association of Sons and Daughters of the Cherokee Strip Pioneers.
v. ill. 28-36 cm.
Semiannual, Sept. 1967-Sept. 1969; annual, Sept. 1970-
Began with Sept. 1959 issue. Cf. New serial titles.
Official publication of the Association of Sons and Daughters of the Cherokee Strip Pioneers.
1. Cherokee Outlet, Okla.—History—Periodicals. 2. Cherokee Outlet, Okla.—Biography—Periodicals. 3. Frontier and pioneer life—Oklahoma—Cherokee Outlet—Periodicals. I. Association of Sons and Daughters of the Cherokee Strip Pioneers.
F702.C42J68a 976.6 78-646260
MARC-S

CHEROKEE REMOVAL, 1838

Filler, Louis, 1912- ed.
The removal of the Cherokee Nation: manifest destiny o national dishonor? Edited with an introd. by Louis Filler an Allen Guttmann. Boston, Heath [1962]
113 p. 24 cm. (Problems in Amerian civilization)
Includes bibliography.
1. Cherokee Removal, 1838. I. Guttmann, Allen, joint ed. II. Title. I Series.
E99.C5F5 970'.004'97 62-5179
[r78]rev MAR

CHERRINGTON FAMILY

Cherrington, Dean C 1927-
The Cherrington family history & genealogy / by Dean C. Cherrington. — Scotch Plains, N.J. : Cherrington, 1978.
272, viii p. ; 26 cm.
Includes indexes.
1. Cherrington family. I. Title.
CS71.C5215 1978 929'.2'0973 78-51310
79[r79]rev MARC

CHERRY, JAPANESE see Japanese flowering cherry

CHERRY BROWN ROT see Brown rot of fruit

CHERRY VALLEY REGION, N.Y.

—BIOGRAPHY

Williams, Emily.
Cherry Valley country / text, Emily Williams ; photography, Helen Cardamone ; foreword, David M. Ellis. — 1st ed. — [s.l.] : Williams, c1978.
121 p. : ill. ; 32 cm.
Bibliography: p. 119-121.
$16.95
1. Historic buildings—New York (State)—Cherry Valley region. 2. Historic buildings—New York (State)—Otsego Co. 3. Cherry Valley region, N.Y.—Biography. 4. Otsego Co., N.Y.—Biography. I. Cardamone, Helen. II. Title.
F129.C5W54 974.7'74 78-63381
79 MARC

HERSONESE (CITY)

anilenko, Vitaliĭ Nikolaevich.
(Bashnia Zenona)
Башня Зенона / В. Даниленко, Р. Токарева. — Симферополь : Таврия, 1974.
77, [1] p. : ill. ; 17 cm. — (Археологические памятники Крыма)
USSR 74
Bibliography: p. 76-[78]
0.13rub
1. Chersonese (City) 2. Ukraine—Antiquities. I. Tokareva, Raisa Nikitichna, joint author. II. Title.
DK651.C46D36 79–414618

—HISTORY

hcheglov, Aleksandr Nikolaevich.
(Polis i khora)
Полис и хора / А. Щеглов. — Симферополь : Таврия, 1976.
175 p. : ill. ; 16 cm. — (Археологические памятники Крыма)
USSR 77
Series romanized: Arkheologicheskie pamiatniki Kryma.
Includes bibliographical references.
0.27rub
1. Chersonese (City)—History. I. Title.
DK651.C46S5 78–421114

HERUBIM see Angels

HERUBINI, LUIGI, 1760-1842

Confalonieri, Giulio, 1896-
Cherubini : prigionia di un artista / di Giulio Confalonieri. — [Milano] : Accademia, 1978.
707 p. ; 21 cm. — (I Musicisti) It79-Jan
Earlier ed. (1948) published under title: Prigionia di un artista, il romanzo di Luigi Cherubini.
Bibliography: p. 705-706.
L12000
1. Cherubini, Luigi, 1760-1842. 2. Composers—Italy—Biography.
ML410.C5C6 1978 784'.092'4 79-348639
*79 MARC

ERVEN BRYAG REGION, BULGARIA

—HISTORY

Sanov-Sasho, Simeon.
Edinni v borbata : [spomeni za Chervenobrezhkiia partizanski otriad Diado Vŭlko] / Simeon TSanov-Sasho ; nauch. red. Nediu Nedev. — Sofiia : BZNS, 1978.
262 p., [8] leaves of plates : ill ; 21 cm. Bu78-3125
1.54 lv
1. World War, 1939-1945—Underground movements—Bulgaria—Cherven bryag region—Biography. 2. Bulgaria. Armiia. Chervenobrezhki partizanski otriad Diado Vŭlko. 3. Bulgaria. Armiia. Chervenobrezhki partizanski otriad Georgi Benkovski. 4. World War, 1939-1945—Personal narratives, Bulgarian. 5. TSanov-Sasho, Simeon. 6. Cherven bryag region, Bulgaria—History. 7. Guerrillas—Bulgaria—Cherven bryag region—Biography. I. Nedev, Nediu T. II. Title.
D802.B78C457 79-372401
*79 MARC

ERVES-DE-COGNAC, FRANCE

—RURAL CONDITIONS

a Geste paysanne : la vie au Moyen Age dans les campagnes du Haut-Poitou : documents & enquêtes relatifs au pays de Cherves. Danses, musiques, fêtes et amour dans les campagnes du Haut-Poitou d'après des textes du Moyen-Age et de la Renaissance, témoignages contemporains / choix et commentaires des textes historiques [sic], Francis Rieupeyroux ; enquêtes et réalisations "Les Gens de Cherves". — [Neuville] : Les Gens de Cherves, [1977]
76 p. : ill., music ; 20 x 21 cm. F***
Includes bibliographical references.
40.00F
1. Cherves-de-Cognac, France—Rural conditions. 2. Poitou—Rural conditions. I. Rieupeyroux, Francis. II. Les Gens de Cherves (Society)
HN438.C55G47 309.1'44'6 79-359740
79 MARC

CHERVIN, RONDA

Chervin, Ronda.
Why I am a charismatic, a Catholic explains : reflections on charismatic prayer and the longings of the human heart / Ronda Chervin. — Liguori, MO : Liguori Publications, 1978.
123 p. : ill. ; 18 cm.
ISBN 0-89243-089-3 : $2.95
1. Pentecostalism—Catholic Church. 2. Chervin, Ronda. I. Title.
BX2350.57.C48 248'.2 78-62167
78 MARC

CHERZI, MONGUSH KHOLA-SALOVICH, 1899-1969

Kenin-Lopsan, Mongush Borakhoevich.
Volshebnik—rasskaz o narodnom kamnereze Mongushe Cherzi / Mongush Kenin-Lopsan ; Tuvin. resp. kraeved. muzeĭ im. Aldan-Maadyr. — Kyzyl : Tuvin. kn. izd-vo, 1976.
81 p. : ill. ; 14 cm. USSR76
Bibliography: p. 56-[67]
1. Cherzi, Mongush Khola-Salovich, 1899-1969. 2. Sculptors—Tuvinskaya A.S.S.R.—Biography. I. Title.
NB699.C49K46 79-376865
79 MARC

CHESAPEAKE AND OHIO CANAL

—HISTORY

Unrau, Harlan D
The masonry locks, Chesapeake and Ohio Canal National Historical Park : Maryland, District of Columbia, West Virginia / prepared by Harlan D. Unrau. — Denver : Denver Service Center, Historic Preservation Division, National Park Service, United States Dept. of the Interior, 1978.
xii, 235 p. : ill. ; 27 cm.
Cover title: Chesapeake and Ohio Canal, the masonry locks.
At head of title: Historic structure report, historical data.
Bibliography: p. 231-235.
1. Chesapeake and Ohio Canal—History. 2. Locks (Hydraulic engineering)—Potomac Valley—History. I. United States. National Park Service. Denver Service Center. Historic Preservation Division. II. Title. III. Title: Chesapeake and Ohio Canal, the masonry locks.
TC625.C5U57 627'.1352'09752 78-603463
79 MARC

CHESAPEAKE AND OHIO RAILWAY

—HISTORY—CONGRESSES

The C. and O. in central Virginia : prepared in connection with the 1979 Annual Convention of the Chesapeake and Ohio Historical Society at Charlottesville, Virginia, July 27, 28, & 29, 1979 / edited by Randolph Kean. — Alderson, W. Va. : The Society, c1979.
20 p. : ill. ; 28 cm.
1. Chesapeake and Ohio Railway—History—Congresses. 2. Railroads—Virginia—History—Congresses. I. Kean, Randolph. II. Chesapeake and Ohio Historical Society.
HE2791.C5752 385'.09755 79-114184
00 MARC

CHESAPEAKE BAY

Stevenson, J Court.
Summary of available information on Chesapeake Bay submerged vegetation / by J. Court Stevenson and Nedra M. Confer ; project officer, Burt Brun ; cosponsored by the Maryland Department of Natural Resources and Region 3 Chesapeake Bay Program, U.S. Environmental Protection Agency ; performed for Coastal Ecosystems Team, Office of Biological Services, Fish and Wildlife Service, U.S. Department of the Interior. — [Washington] : The Service, 1978.
xxxii, 335 p. : ill. ; 28 cm.
"FWS/OBS-78/66."
"Contract no. FWS 14-16-0008-2138."
Bibliography: p. 283-335.
1. Aquatic plants—Chesapeake Bay. 2. Aquatic plants—Maryland. 3. Chesapeake Bay. I. Confer, Nedra M., joint author. II. Coastal Ecosystems Team. III. Title.
QK165.S73 581.9'2'147 78-602813
79 MARC

United States. Congress. Senate. Committee on Governmental Affairs. Subcommittee on Governmental Efficiency and the District of Columbia.
Coordination of Federal and State programs affecting Chesapeake Bay : hearing before the Subcommittee on Governmental Efficiency and the District of Columbia of the Committee on Governmental Affairs, United States Senate, Ninety-sixth Congress, first session, January 29, 1979, Annapolis, Md. — Washington : U.S. Govt. Print. Off., 1979.
iv, 322 p. : ill. ; 24 cm.
1. Chesapeake Bay. 2. Chesapeake Bay region. 3. Environmental policy—Chesapeake Bay region. I. Title.
KF26.G6735 1979a 333.9'164'0916347 79-603074
79 MARC

—DESCRIPTION AND TRAVEL

Stone, William T
A cruising guide to the Chesapeake : including the passages from Long Island Sound along the New Jersey coast and inland waterway / by William T. Stone and Fessenden S. Blanchard. — Rev. and updated. — New York : Dodd, Mead, 1978.
xiv, 306 p. : ill. ; 24 cm.
Includes index.
ISBN 0-396-07638-6 : $15.00
1. Boats and boating—Chesapeake Bay. 2. Marinas—Chesapeake Bay—Guide-books. 3. Chesapeake Bay—Description and travel. 4. Chesapeake Bay—Navigation. I. Blanchard, Fessenden Seaver, 1888-1963, joint author. II. Title.
GV776.C47S76 1978 917.55'18'044 78-15229
78 MARC

—NAVIGATION

Stone, William T
A cruising guide to the Chesapeake : including the passages from Long Island Sound along the New Jersey coast and inland waterway / by William T. Stone and Fessenden S. Blanchard. — Rev. and updated. — New York : Dodd, Mead, 1978.
xiv, 306 p. : ill. ; 24 cm.
Includes index.
ISBN 0-396-07638-6 : $15.00
1. Boats and boating—Chesapeake Bay. 2. Marinas—Chesapeake Bay—Guide-books. 3. Chesapeake Bay—Description and travel. 4. Chesapeake Bay—Navigation. I. Blanchard, Fessenden Seaver, 1888-1963, joint author. II. Title.
GV776.C47S76 1978 917.55'18'044 78-15229
78 MARC

—RESEARCH

United States. Congress. House. Committee on Science and Technology. Subcommittee on the Environment and the Atmosphere.
H.R. 12631 : Chesapeake Bay research coordination act of 1978 : hearing before the Subcommittee on the Environment and the Atmosphere of the Committee on Science and Technology, U.S. House of Representatives, Ninety-fifth Congress, second session, August 9, 1978. — Washington : U.S. Govt. Print. Off., 1979.
iii, 108 p. ; 24 cm.
"No. 98."
1. Chesapeake Bay—Research. 2. Research—Law and legislation—United States. I. Title.
KF27.S396 1978e 346'.73'0469164 79-600988
79 MARC

CHESAPEAKE BAY REGION

Hays, Anne M 1933-
Chesapeake kaleidoscope / by Anne M. Hays and Harriet R. Hazleton ; ill. by Mary J. Hazleton. — Cambridge, Md. : Tidewater Publishers, 1975.
viii, 148 p. : ill. ; 23 cm.
Bibliography: p. 137-139.
Includes index.
ISBN 0-87033-214-7
1. Chesapeake Bay region. I. Hazleton, Harriet R., 1913- joint author. II. Title.
F187.C5H38 975.5'18 75-23447
75[r79]rev MARC

United States. Congress. Senate. Committee on Governmental Affairs. Subcommittee on Governmental Efficiency and the District of Columbia.
Coordination of Federal and State programs affecting Chesapeake Bay : hearing before the Subcommittee on Governmental Efficiency and the District of Columbia of the Committee on Governmental Affairs, United States Senate, Ninety-sixth Congress, first session, January 29, 1979, Annapolis, Md. — Washington : U.S. Govt. Print. Off., 1979.
iv, 322 p. : ill. ; 24 cm.
1. Chesapeake Bay. 2. Chesapeake Bay region. 3. Environmental policy—Chesapeake Bay region. I. Title.
KF26.G6735 1979a 333.9'164'0916347 79-603074
79 MARC

—ECONOMIC CONDITIONS

National Planning Association. Center for Economic Projections.
Summary, economic base study, Chesapeake Bay drainage basins / prepared for Department of the Interior, Federal Water Pollution Control Administration, Middle Atlantic Region, and Department of the Army, Baltimore District, Corps of Engineers, in accordance with contract no. PH86-63-92 by National Planning Association, Washington, D.C. — [Washington : The Association], 1967.
iii, 156 p. : ill. ; 27 cm.
Errata slip inserted.
1. Chesapeake Bay region—Economic conditions. 2. Susquehanna River watershed—Economic conditions. I. United States. Federal Water Pollution Control Administration. Middle Atlantic Region. II. United States. Army. Corps of Engineers. Baltimore District. III. Title.
HC107.A12N37 1967 330.9'755'1804 79-311134
79 MARC

CHESHIRE, ENG.

—CHURCH HISTORY—SOURCES—BIBLIOGRAPHY—CATALOGS

Cheshire, Eng. Record Office.
Cheshire Record Office and Chester Diocesan Record Office. — ⌜Chester⌝ : ⌜Cheshire County Council⌝, ⌜1975⌝
38 leaves in various pagings ; 30 cm. GB76-22549
Cover title.
"Compiling a family tree" (sheet (⌜2⌝ sides)) as insert.
ISBN 0-9501590-3-4 : £0.40
1. Cheshire, Eng. Record Office—Catalogs. 2. Cheshire, Eng. (Diocese) —Record Office—Catalogs. 3. Chishire, Eng.—History—Sources—Bibliography—Catalogs. 4. Cheshire, Eng.—Church history—Sources—Bibliography—Catalogs. I. Chester, Eng. (Diocese) Record Office. II. Title.
CD1065.C47C47 1975 016.9427'1 78-317709
78 MARC

—HISTORY

A History of the county of Chester / edited by B. E. Harris. — Oxford ; ⌜New York, etc.⌝ : Oxford University Press for the Institute of Historical Research, 19 -
v. : ill. ; 32 cm. — (The Victoria history of the counties of England) GB79-17997
Includes bibliographical references and index.
ISBN 0-19-722749-X (v. 2) : £45.00 (v. 2)
1. Cheshire, Eng.—History. I. Harris, B. E. II. Series.
DA670.C6H52 942.7'1 79-314972
79 MARC

Hodson, John Howard.
Cheshire, 1660-1780 : Restoration to Industrial Revolution / by Howard Hodson. — Chester : ⌜Eng.⌝ Cheshire Community Council Publications Trust, 1978.
vii, 160 p., ⌜4⌝ leaves of plates : ill. ; 24 cm. — (A History of Cheshire ; v. 9) GB***
Bibliography: p. 155.
Includes index.
ISBN 0-903119-10-2 : £5.95
1. Cheshire, Eng.—History. I. Title. II. Series.
DA670.C6H55 942.7'1'06 79-310395
79 MARC

—HISTORY—SOURCES—BIBLIOGRAPHY—CATALOGS

Cheshire, Eng. Record Office.
Cheshire Record Office and Chester Diocesan Record Office. — ⌜New ed.⌝. — Chester : Cheshire County Council, ⌜1977?⌝
⌜36⌝ leaves in various pagings ; 30 cm. GB78-13232
Caption title.
Cover title: County Record Office and Chester Diocesan Record Office.
£0.40
1. Cheshire, Eng. Record Office—Catalogs. 2. Cheshire, Eng. (Diocese). Record Office—Catalogs. 3. Cheshire, Eng.—History—Sources—Bibliography—Catalogs. I. Chester, Eng. (Diocese). Record Office. II. Title. III. Title: County Record Office and Chester Diocesan Record Office.
CD1065.C47C47 1977 016.9427'1 79-307780
79 MARC

CHESHIRE, ENG. RECORD OFFICE

—CATALOGS

Cheshire, Eng. Record Office.
Cheshire Record Office and Chester Diocesan Record Office. — ⌜Chester⌝ : ⌜Cheshire County Council⌝, ⌜1975⌝
38 leaves in various pagings ; 30 cm. GB76-22549
Cover title.
"Compiling a family tree" (sheet (⌜2⌝ sides)) as insert.
ISBN 0-9501590-3-4 : £0.40
1. Cheshire, Eng. Record Office—Catalogs. 2. Cheshire, Eng. (Diocese) —Record Office—Catalogs. 3. Chishire, Eng.—History—Sources—Bibliography—Catalogs. 4. Cheshire, Eng.—Church history—Sources—Bibliography—Catalogs. I. Chester, Eng. (Diocese) Record Office. II. Title.
CD1065.C47C47 1975 016.9427'1 78-317709
78 MARC

Cheshire, Eng. Record Office.
Cheshire Record Office and Chester Diocesan Record Office. — ⌜New ed.⌝. — Chester : Cheshire County Council, ⌜1977?⌝
⌜36⌝ leaves in various pagings ; 30 cm. GB78-13232
Caption title.
Cover title: County Record Office and Chester Diocesan Record Office.
£0.40
1. Cheshire, Eng. Record Office—Catalogs. 2. Cheshire, Eng. (Diocese). Record Office—Catalogs. 3. Cheshire, Eng.—History—Sources—Bibliography—Catalogs. I. Chester, Eng. (Diocese). Record Office. II. Title. III. Title: County Record Office and Chester Diocesan Record Office.
CD1065.C47C47 1977 016.9427'1 79-307780
79 MARC

CHESHIRE, ENG. (DIOCESE)

—RECORD OFFICE—CATALOGS

Cheshire, Eng. Record Office.
Cheshire Record Office and Chester Diocesan Record Office. — ⌜Chester⌝ : ⌜Cheshire County Council⌝, ⌜1975⌝
38 leaves in various pagings ; 30 cm. GB76-22549
Cover title.
"Compiling a family tree" (sheet (⌜2⌝ sides)) as insert.
ISBN 0-9501590-3-4 : £0.40
1. Cheshire, Eng. Record Office—Catalogs. 2. Cheshire, Eng. (Diocese) —Record Office—Catalogs. 3. Chishire, Eng.—History—Sources—Bibliography—Catalogs. 4. Cheshire, Eng.—Church history—Sources—Bibliography—Catalogs. I. Chester, Eng. (Diocese) Record Office. II. Title.
CD1065.C47C47 1975 016.9427'1 78-317709
78 MARC

CHESHIRE, ENG. (DIOCESE). RECORD OFFICE

—CATALOGS

Cheshire, Eng. Record Office.
Cheshire Record Office and Chester Diocesan Record Office. — ⌜New ed.⌝. — Chester : Cheshire County Council, ⌜1977?⌝
⌜36⌝ leaves in various pagings ; 30 cm. GB78-13232
Caption title.
Cover title: County Record Office and Chester Diocesan Record Office.
£0.40
1. Cheshire, Eng. Record Office—Catalogs. 2. Cheshire, Eng. (Diocese). Record Office—Catalogs. 3. Cheshire, Eng.—History—Sources—Bibliography—Catalogs. I. Chester, Eng. (Diocese). Record Office. II. Title. III. Title: County Record Office and Chester Diocesan Record Office.
CD1065.C47C47 1977 016.9427'1 79-307780
79 MARC

CHESHIRE ACADEMY

—HISTORY

Coleman, Marion Moore, 1900-
The Cheshire Academy : the first twelve decades, 1794-1919 : a memoir / compiled by Marion Moore Coleman. — Cheshire, Conn. : Cherry Hill Books, 1976.
vi, 96 p. : ill. ; 24 cm.
Bibliography: p. 90.
Includes index.
1. Cheshire Academy—History.
LD7501.C35C64 373.746'8 79-311479
79 MARC

CHESHIRE LINES RAILWAY

—HISTORY

Griffiths, R Prys.
The Cheshire Lines Railway / ⌜by⌝ R. Prys Griffiths. — Revised ed. / with additional notes by C. T. Goode. — Blandford Forum : Oakwood Press, 1978.
⌜2⌝, 74 p., ⌜16⌝ p. of plates : ill., coat of arms, facsim., maps, plans ; 22 cm. — (The Oakwood library of railway history ; 5) GB78-32093
£2.40
1. Cheshire Lines Railway—History. I. Goode, Charles Tony. II. Title.
HE3020.C4G7 1978 385'.09427 79-313163
79 MARC

CHESLER, PHYLLIS

Chesler, Phyllis.
With child, a diary of motherhood / Phyllis Chesler. — 1st ed. — New York : Crowell, c1979.
288 p. ; 24 cm.
Bibliography: p. ⌜283⌝-288.
ISBN 0-690-01835-5 : $9.95
1. Pregnancy—Psychological aspects. 2. Childbirth—Psychological aspects. 3. Mother and child. 4. Chesler, Phyllis. 5. Mothers—United States—Biography. 6. Feminists—United States—Biography. I. Title.
RG560.C47 1979 301.41'2'0924 79-7081
79 MARC

CHESNEY, LEE, 1920-

—EXHIBITIONS

Chesney, Lee, 1920-
Lee Chesney—25 years of printmaking / Wayne A. Miyamoto ; with a foreword by Kenneth A. Kerslake. — Orlando : University Presses of Florida, c1978.
vi, 90 p. : ill. ; 23 x 31 cm.
Catalog of a touring retrospective exhibition of prints which opened in Orlando at the Florida Technological University in Nov. 1977.
"A Florida Technological University book."
ISBN 0-8130-0615-5
1. Chesney, Lee, 1920- —Exhibitions. I. Miyamoto, Wayne A., joint author. II. Florida Technological University. III. Title.
NE539.C53A4 1978 769'.92'4 78-13555
78 MARC

CHESS

see also Chess players, Rating of; Chessmen; Pawn (Chess)

Gillam, A J
Starting chess / A. J. Gillam. — London : Batsford, 1978.
127 p. : ill. ; 22 cm. — (Discovering chess series ; book 1) GB***
Previous ed. published under title: Play chess.
ISBN 0-17-341478-2 : £1.45
1. Chess. I. Title. II. Series.
GV1446.G5 1978 794.1'2 78-319652
78 MARC

Keene, Raymond D
Becoming a grandmaster / ⌜by⌝ Raymond Keene. — Londo : Batsford, 1977.
157 p. : ill. ; 22 cm. — (The Club player's library) (Batsford chess book GB78-1855
Includes index.
ISBN 0-7134-0829-4. ISBN 0-7134-0830-8 pbk. : £3.50
1. Chess. 2. Keene, Raymond D. 3. Chess players—Great Britain—Biogra phy. I. Title.
GV1445.K37 794.1 78-32688
79 MAR

Mednis, Edmar, 1937-
How to beat the Russians / by Edmar Mednis. — New York : McKay, c1978.
xiv, 297 p. : ill. ; 21 cm.
ISBN 0-679-13376-3 : $12.95
1. Chess. 2. Chess—Collections of games. I. Title.
GV1445.M35 794.1'2 78-1597
78 MAR

Meinsohn, Francis, 1950-
Je m'améliore / Francis Meinsohn ; dessins de Patrice Do nat ; ⌜publié avec la collaboration de la Fédération française échecs. — Paris : Hatier, 1977.
111 p. : ill. ; 22 cm. — (Jeunes joueurs d'échecs ; 3 ISSN 0335-6841) F78-6
ISBN 2-218-03939-7 : 37.00F
1. Chess. I. Title. II. Series.
GV1445.M37 794.1'2 78-377
78 MA

Négro, Jacques.
Les échecs en 10 ⌜i.e. dix⌝ leçons / Jacques Négro ; préf. de Jean-Pierre Rudin. — ⌜Paris⌝ : Hachette, c1977.
251 p. : ill. ; 20 cm. — (En 10 ⌜i.e. dix⌝ leçons) F***
Includes index.
ISBN 2-01-004182-8
1. Chess. I. Title.
GV1446.N43 794.1 78-39335
79 MAR

Padulli, Giuseppe.
Gli scacchi / Giuseppe Padulli. — 1. ed. — Milano : Mu 1978.
221 p. : ill. ; 18 cm. — (Tascabili Mursia ; 6)
L2000
1. Chess. I. Title.
GV1445.P256 794.1'2 78-382
78 M.

Phillips, Alan, fl. 1978-
The chess teacher / Alan Phillips. — Oxford ⌜Eng.⌝ ; Ne York : Oxford University Press, 1978.
x, 149 p. : ill. ; 21 cm. — (Oxford chess books) GB
Includes index.
ISBN 0-19-217561-0 : £3.95. ISBN 0-19-217560-2 pbk.
1. Chess. I. Title.
GV1445.P42 794.1'07 78-401
78 MA

Pritchard, David Brine.
How to play chess / by D. B. Pritchard. — Toronto : Cole c1975.
236 p. : diagrs. ; 21 cm. C78-1049
Includes index.
$2.95
1. Chess. I. Title.
GV1445.P73 794.1'2 79-31595
79 MA

Pritchard, David Brine.
The right way to play chess / by D. Brine Pritchard ; intr and annotations by Imré Konig. — 8th ed. 1st Barnes & Nob Books ed. — New York : Barnes & Noble Books, 1979, c195
237 p. : ill. ; 21 cm.
Includes index.
ISBN 0-06-463478-7 : $2.95 ($3.75 Can.)
1. Chess. I. König, Imre. II. Title.
GV1445.P74 1979 794.1'2 79-102
79 MA

Soltis, Andrew.
Catalog of chess mistakes / by Andrew Soltis. — New Yo : D. McKay Co., c1979.
213 p. : ill. ; 21 cm.
ISBN 0-679-13250-3
1. Chess. I. Title.
GV1449.5.S64 794.1'2 79-110
79 MA

CHESS (Continued)

Stean, Michael.
Simple chess / Michael Stean. — London ; Boston : Faber, 1978.
116 p. : ill. ; 21 cm. GB***
ISBN 0-571-11215-3 : £3.30. ISBN 0-571-11257-9 pbk.
1. Chess. I. Title.
GV1449.5.S73 794.1'2 78-321183
79 MARC

Webb, Simon.
Chess for tigers / Simon Webb ; with ill. by Edward McLachlan. — Oxford ₍Eng.₎ ; New York : Oxford University Press, 1978.
99 p. : ill. ; 21 cm. GB***
ISBN 0-19-217575-0 : £3.95. ISBN 0-19-217576-9 pbk.
1. Chess. 2. Chess—Psychology. I. Title.
GV1445.W36 794.1'2 78-40204
79 MARC

—ADDRESSES, ESSAYS, LECTURES

Hilsen til Jens Enevoldsen på 70-årsdagen 23. september 1977. — København ₍Skakhuset, Studiestræde 24, 1455 K₎ : Dansk Skakforlag, 1977.
189 p. : ill. ; 21 cm. D78-5
Bibliography of works by Jens Enevoldsen: p. 182-189.
Includes bibliographical references.
CONTENTS: Mester Jens. Et aforistisk selvportræt.—Granberg, N. V. Skak og musik.—Nimzowitsch, A. Hvordan jeg blev stormester.—Larsen, B. Spiller De skak, hr. direktør?—Jansen, B. B. Danske skak-exlibris.—Hage, P. Skakmestre i problemkunstens tjeneste.—Persson, L. Fra timeglas til elektroniske digitalure.—Schmidt, L. C. "Om taktiken paa skakbrættet."—Friedrichsen, P. Hug og parade.—Mortensen, J. Edb og problemskak.—Novrup, S. Mester Jens.
ISBN 8787187043 : kr49.00
1. Enevoldsen, Jens—Addresses, essays, lectures. 2. Chess—Addresses, essays, lectures. 3. Enevoldsen, Jens—Bibliography. I. Enevoldsen, Jens.
GV1445.H49 79-360202
79 MARC

Levy, David N L
Learn chess from the world champions / compiled by David Levy. — 1st ed. — Oxford ₍Eng.₎ ; New York : Pergamon Press, 1979.
vii, 206 p. : ill. ; 27 cm. — (Pergamon chess series) (Pergamon international library of science, technology, engineering and social studies) GB77
ISBN 0-08-021387-1 : £8.00 ($14.50 U.S.)
1. Chess—Addresses, essays, lectures. I. Title.
GV1445.L533 1979 794.1'2 77-30480
77 MARC

—ANECDOTES, FACETIAE, SATIRE, ETC.

Koltanowski, George, 1903-
Chessnicdotes / George Koltanowski ; ₍edited by B. G. Dudley ; line drawings by Ronald Atkinson₎. — Coraopolis, Pa. : Chess Enterprises, 1978-
v. : ill. ; 22 cm.
Includes index.
ISBN 0-931462-01-0 (v. 1)
1. Chess—Anecdotes, facetiae, satire, etc. I. Title.
GV1449.K57 794.1'02'07 78-107872
78 MARC

Marjano, Alek.
Šahovski dezinformator / Alek Marjano. — Beograd : Nova knjiga, 1978.
57 p. : ill. ; 21 cm. — (Biblioteka Savremeni domaći pisci) Yu***
1. Chess—Anecdotes, facetiae, satire, etc. I. Title.
GV1449.M37 79-353983
79 MARC

—BIBLIOGRAPHY

British Library. Lending Division.
Edwin Gardiner chess collection catalogue / by J. E. Vickery and C. A. Webb. — Boston Spa : British Library Lending Division, 1977.
₍2₎, 110 p. ; 30 cm. GB78-38879
ISBN 0-85350-165-3 : £4.00
1. Chess—Bibliography. I. Gardiner, Edwin, 1896- II. Vickery, J. E. III. Webb, C. A. IV. Title.
Z5541.B74 1977 016.7941 79-317700
₍GV1445₎ 79 MARC

Richman, Paul, 1936-
An annotated international bibliography of chess articles in non-chess English language periodicals, 1783-1977 / compiled by Paul Richman. — 1st ed. — Old Bridge, N.J. : Richman, 1978.
v, 70 leaves ; 29 cm.
Cover title.
Includes index.
ISBN 0-9602070-1-5
1. Chess—Bibliography. I. Title.
Z5541.R5 016.7941 78-69856
₍GV1445₎ 79 MARC

—COLLECTIONS OF GAMES

see also Chess—Tournaments

72nd annual US Open Chess Championship, Ventura, California, 8-20 August 1971 / edited by Jack L. Spence. — Nottingham : The Chess Player, ₍1976₎
xxii, 130 p. : ill., ports. ; 19 cm. GB77-01730
Includes index.
£2.70
1. Chess—Tournaments, 1971. 2. Chess—Tournaments—United States. 3. Chess—Collections of games. I. Spence, Jack L., 1926- II. Title.
GV1455.S45 794.1'57 77-361157
77₍r79₎rev MARC

Adams, Jimmy.
Najdorf poisoned pawn / ₍by₎ Jimmy Adams. — Nottingham : "The chess player", 1977.
111 p. : ill. ; 21 cm. — (Sicilian defence) GB78-15136
Includes index.
ISBN 0-906042-07-0 : £2.95
1. Pawn (Chess) 2. Chess—Collections of games. 3. Chess—Openings. I. Title.
GV1451.5.P3A22 794.1'22 78-319649
78 MARC

Adams, Jimmy.
Sicilian Najdorf Polugaevsky variation / ₍by₎ Jimmy Adams. — Nottingham : The Chess Player, 1978.
232 p. : ill. ; 21 cm. GB79-05254
Algebraic notation.
ISBN 0-906042-09-7 : £4.95
1. Chess—Openings. 2. Chess—Collections of games. I. Title.
GV1450.2.A32 794.1'22 79-310915
79 MARC

Alster, Ladislav.
Miniaturní šachová partie / Ladislav Alster. — 1. vyd. — Praha : Olympia, 1978.
141, ₍2₎ p. : ill. ; 20 cm. Cz78
Includes indexes.
Kčs17.00
1. Chess—Collections of games. I. Title.
GV1452.A45 79-356364
*79 MARC

Bath, 1973. — ₍Dallas : Chess Digest, 1973?₎
44 p. : ill. ; 23 cm. — (The Chess player ; no. 164)
Cover title.
1. Chess—Tournaments. 2. Chess—Collections of games. I. Series.
GV1313.C59 no. 164 794.1'08 s 79-314557
₍GV1455₎ 79 MARC

Botvinnik, Mikhail Moiseevich, 1911-
Anatoly Karpov : his road to the world championship / by Mikhail M. Botvinnik ; ₍translated by Kenneth Neat₎. — Oxford ; New York : Pergamon Press, 1978.
xx, 158 p. : ill. ; 21 cm. — (Pergamon chess series) (Pergamon international library of science, technology, engineering, and social studies) GB78
Translation of Tri matcha Anatoliia Karpova.
ISBN 0-08-021139-9. ISBN 0-08-021138-0 pbk. : $5.95
1. Chess—Collections of games. 2. Karpov, Anatoliĭ Evgen'evich, 1951- I. Title.
GV1439.K37.B6713 1978 77-30655
794.1'5
78 MARC

Botvinnik, Mikhail Moiseevich, 1911–
(Polveka v shakhmatakh)
Полвека в шахматах / М. М. Ботвинник. — Москва : Физкультура и спорт, 1978.
272 p. : ill. ; 21 cm. — (Выдающиеся шахматисты мира) USSR 79
Series romanized: Vydaiushchiesia shakhmatisty mira.
1.20rub
1. Botvinnik, Mikhail Moiseevich, 1911– 2. Chess — Collections of games. 3. Chess players—Russia—Biography. I. Title.
GV1439.B6A36 79–411684

Bronshteĭn, David Ionovich, 1924-
The chess struggle in practice : candidates tournament, Zurich 1953 / David Bronstein ; translated by Oscar D. Freedman ; edited by Burt Hochberg. — New York : D. McKay Co., c1978.
xxvii, 499 p. : ill. ; 21 cm.
Translation of Mezhdunarodnyĭ turnir grossmeĭsterov.
Includes indexes.
ISBN 0-679-13064-0 : $17.95
1. Chess—Tournaments, 1953. 2. Chess—Middle games. 3. Chess—Collections of games. I. Hochberg, Burt. II. Title.
GV1455.B713 794.1'57 78-59140
00 MARC

Capablanca, José Raúl, 1888-1942.
Capablanca's Best chess endings / Irving Chernev. — Oxford ₍Eng.₎ ; New York : Oxford University Press, 1978.
x, 288 p. : ill. ; 21 cm. — (Oxford chess books) GB77
Bibliography: p. 286-288.
Includes indexes.
ISBN 0-19-217553-X. ISBN 0-19-217554-8 pbk. : £2.95
1. Chess—End games. 2. Chess—Collections of games. I. Chernev, Irving, 1900- II. Title. III. Title: Best Chess endings.
GV1450.7.C34 794.1'24 77-30392
78 MARC

Capablanca, José Raúl, 1888-1942.
Capablanca's Hundred best games of chess / chosen and annotated by H. Golombek ; with a memoir by J. du Mont. — 1st paperback ed. — New York : McKay, 1978.
xvi, 267 p. : ill. ; 20 cm. — (Tartan books ; 48)
Includes index.
ISBN 0-679-14044-1 : $4.95
1. Capablanca, José Raúl, 1888-1942. 2. Chess—Collections of games. I. Golombek, Harry, 1911- II. Title. III. Title: Hundred best games of chess.
GV1439.C3A265 1978 794.1'5 78-51937
78 MARC

Collins, John W
Maxims of chess / by John W. Collins. — New York : D. McKay, c1978.
xi, 276 p. : ill. ; 21 cm.
Includes indexes.
ISBN 0-679-13066-7 :
1. Chess—Quotations, maxims, etc. 2. Chess—Collections of games. I. Title.
GV1445.C69 794.1 77-18245
77 MARC

Euwe, Machgielis, 1901-
Bobby Fischer—the greatest? / By Max Euwe. — New York : Sterling Pub. Co., c1979.
xiv, 202 p. : ill. ; 23 cm.
Includes index.
ISBN 0-8069-4950-3 : $8.95. ISBN 0-8069-4951-1 lib. bdg. : $8.95
1. Chess players, Rating of. 2. Fischer, Bobby, 1943- 3. Chess—Collections of games. I. Title.
GV1457.5.E97 794.1'092'2 78-66270
79 MARC

Fischer, Bobby, 1943-
Chess games. Edited by Robert G. Wade, Kevin J. O'Connell. With contributions by Leonard Barden ₍and others₎ Garden City, N.Y., Doubleday, 1972.
448 p. illus. 24 cm. $12.50
1. Fischer, Bobby, 1943- 2. Chess—Collections of games. I. Wade, Robert Graham, ed. II. O'Connell, Kevin J., ed. III. Title.
GV1439.F5A29 1972b 794.1'59 70-190393
72₍r79₎rev MARC

Földeák, Walter Árpád.
Chess Olympiads, 1927-1968 / by Árpád Földeák ; translated from the Hungarian by Robert Ejury and Jenö Bochkor ; translation rev. by Peter H. Clarke. — New York : Dover Publications, 1979, c1969.
416 p. : ill. ; 21 cm.
Reprint, with corrections, of the 2d enl. ed. (1969) which was issued under title: Chess Olympiads.
Includes index.
ISBN 0-486-23733-8 : $4.95
1. Chess—Tournaments—History. 2. Chess—Collections of games. I. Title.
GV1455.F583 1979 794.1'57 78-68173
79 MARC

Games of world correspondence chess championships 1-7 / edited by T. D. Harding ; with a foreword by Hans-Werner von Massow. — London : B. T. Batsford, c1979.
152 p. : ill. ; 26 cm. GB***
Foreword in English and German.
Includes indexes.
ISBN 0-7134-2031-6 : £9.50
1. World Correspondence Chess Championship—History. 2. Chess—Collections of games. I. Harding, T. D.
GV1455.G35 794.1'57 79-317705
79 MARC

Hartston, William Roland, 1947-
The battle of Baguio City : Karpov-Korchnoi 1978 / W. R. Hartston. — London : Hutchinson, 1978.
121 p. : ill. ; 21 cm. GB***
ISBN 0-09-134881-1 : £1.95
1. Chess—Tournaments. 2. Chess—Collections of games. 3. Karpov, Anatoliĭ Evgen'evich, 1951- 4. Korchnoĭ, Viktor, 1931- I. Title.
GV1455.H33 794.1'57 79-305230
79 MARC

Heuer, Valter.
Meie Keres : kujunemisaastad / Valter Heuer. — Tallinn : Eesti Raamat, 1977.
381 p. : ill. ; 23 cm. USSR***
2.04rub
1. Keres, Paul, 1916- 2. Chess players—Estonia—Biography. 3. Chess—Collections of games. I. Title.
GV1439.K4H48 78-389256
79 MARC

Karpov, Anatoliĭ Evgen'evich, 1951–
(Deviataia vertikal')
Девятая вертикаль / ₍А. Е. Карпов, А. Б. Рошаль₎. — Москва : Молодая гвардия, 1978.
301 p., ₍16₎ leaves of plates : ill. ; 21 cm. USSR***
1.40rub
1. Karpov, Anatoliĭ Evgen'evich, 1951– 2. Chess—Collections of games. 3. Chess players—Russia—Biography. I. Roshal', Aleksandr, joint author. II. Title.
GV1439.K37A335 79–411688

CHESS

—COLLECTIONS OF GAMES (Continued)

Karpov, Anatoliĭ Evgen'evich, 1951–
(Izbrannye partii, 1969–1977)
Избранные партии, 1969–1977 / Анатолий Карпов. — Москва : Физкультура и спорт, 1978.
256 p. : ill. ; 20 cm. — (Выдающиеся шахматисты мира) USSR 78
Series romanized: Vydaiushchiesia shakhmatisty mira.
1.30rub
1. Karpov, Anatoliĭ Evgen'evich, 1951– 2. Chess—Collections of games. 3. Chess players—Russia—Biography. I. Title.
GV1439.K37A35 79–410789

Keene, Raymond D
Karpov-Korchnoi 1978 : the inside story of the match / ₍by₎ Raymond Keene. — London : Batsford, 1978.
159 p. : ill. ; 20 cm. — (A Batsford chess book) GB78-38616
American ed. published under title: World chess championship.
ISBN 0-7134-1809-5 : £1.50
1. Chess—Tournaments, 1978. 2. Chess—Collections of games. 3. Karpov, Anatoliĭ Evgen'evich, 1951- 4. Korchnoĭ, Viktor, 1931- I. Title.
GV1455.K388 1978b 794.1'57 79-306233
79 MARC

Keene, Raymond D
Korchnoi vs. Spassky : chess crisis / Raymond Keene. — London ; Boston : Allen & Unwin, 1978.
128 p. : ill. ; 20 cm. GB***
Includes index.
ISBN 0-04-794006-9 : £1.95
1. Chess—Tournaments, 1977. 2. Korchnoi, Viktor, 1931- 3. Spasskii, Boris Vasil'evich, 1937- 4. Chess—Collections of games. I. Title.
GV1455.K387 794.1'57 78-313036
78 MARC

Korchnoi, Viktor, 1931-
Korchnoi's 400 best games / Viktor Korchnoi, Robert G. Wade, L. S. Blackstock. — London : Batsford, 1978.
viii, 264 p. : ill. ; 22 cm. GB***
Includes indexes.
ISBN 0-7134-1799-4. ISBN 0-7134-1808-7 pbk. : £\$5.95
1. Korchnoi, Viktor, 1931- 2. Chess—Collections of games. I. Wade, Robert Graham, joint author. II. Blackstock, L. S., joint author. III. Title. IV. Title: 400 best games.
GV1439.K637A35 1978 794.1'092'4 78-323922
78 MARC

Korchnoĭ, Viktor, 1931-
Korchnoi's 400 best games / by Victor Korchnoi, R. G. Wade, L. S. Blackstock. — New York : Arco Pub. Co., c1978.
viii, 264 p. : ill. ; 23 cm.
ISBN 0-668-04698-8 : \$12.95
1. Korchnoĭ, Viktor, 1931- 2. Chess—Collections of games. I. Wade, Robert Graham, joint author. II. Blackstock, L. S., joint author. III. Title. IV. Title: 400 best games.
GV1439.K637A35 1978b 78-16934
794.1'59
78 MARC

Larsen, Bent.
Karpov vs. Korchnoi : World Chess Championship, 1978 / Bent Larsen ; introduced by Michael Stean. — London ; Boston : Unwin Paperbacks, 1978.
143 p. : ill. ; 20 cm. GB***
ISBN 0-04-794008-5 : £1.95
1. Chess—Tournaments, 1978. 2. Chess—Collections of games. 3. Karpov, Anatoliĭ Evgen'evich, 1951- 4. Korchnoĭ, Viktor, 1931- I. Title.
GV1455.L28 794.1'57 79-308377
79 MARC

Magyar sakktörténet. — Budapest : Sport, 197
v. : ill. ; 25 cm. H77
Includes indexes.
CONTENTS:
—Maróczy, G. Világversenyek élén.

ISBN 9632532384 : 55.00Ft (v. 2)
1. Chess—Hungary—History. 2. Chess—Collections of games.
GV1330.H9M33 78-396273
*79 MARC

Mednis, Edmar, 1937-
How to beat the Russians / by Edmar Mednis. — New York : McKay, c1978.
xiv, 297 p. : ill. ; 21 cm.
ISBN 0-679-13376-3 : \$12.95
1. Chess. 2. Chess—Collections of games. I. Title.
GV1445.M35 794.1'2 78-15977
78 MARC

Négro, Jacques.
Karpov-Kortchnoi : championnat du monde 1978 : toutes les parties commentées / Jacques Négro. — ₍Paris₎ : Hachette, 1978.
223 p. : ill. ; 23 cm. F79-3678
ISBN 2-01-005947-6 : 39.00F
1. Chess—Tournaments, 1978. 2. Karpov, Anatoliĭ Evgen'evich, 1951- 3. Korchnoĭ, Viktor, 1931- 4. Chess—Collections of games.
GV1455.N43 794.1'57 79-372971
79 MARC

The New chess player .... — London : Pitman, 1978-
v. : ill. ; 22 cm. GB79-03005
Algebraic notation.
Includes index.
CONTENTS: v. 1. A.

ISBN 0-273-01263-0 (v. 1) : £4.95 (v. 1)
1. Chess—Collections of games.
GV1452.N48 794.1'5 79-316179
79 MARC

Porreca, Giorgio.
La difesa francese / Giorgio Porreca. — Milano : Mursia, 1977.
195 p. : ill. ; 19 cm. — (Biblioteca dello scacchista) (Collana I Giochi) It78-Sept
Includes index.
L4000
1. Chess—Openings. 2. Chess—Collections of games. I. Title.
GV1450.2.P67 794.1'22 79-358409
*79 MARC

Rădulescu, Mihai.
Finaluri complexe în şah : arta marilor campioni : Boris Spasski, Robert Fischer, Anatoli Karpov / Mihai Rădulescu. — ₍Ediţia a 2-a₎. — Bucureşti : "Sport-Turism", 1978.
232 p. : ill. ; 21 cm. R78-5288
Bibliography: p. 227.
lei7.25
1. Chess—End games. 2. Chess—Collections of games. 3. Spasskiĭ, Boris Vasil'evich, 1937- 4. Fischer, Bobby, 1943- 5. Karpov, Anatoliĭ Evgen'evich, 1951- I. Title.
GV1450.7.R33 1978 79-369177
*79 MARC

Shamkovich, Leonid Aleksandrovich.
Chess sacrifices / ₍by₎ Leonid Shamkovich ; ₍translated from the Russian by A. Kalnajs₎. — Nottingham : Chess Player, 1976.
184 p. : ill. ; 21 cm. GB77-33990
Translation of Zhertva v shakhmatakh.
ISBN 0-900928-99-9 : £2.95
1. Chess—Openings. 2. Chess—Collections of games. I. Title.
GV1450.S4413 794.1'2 78-382792
78 MARC

Ştefaniu, Constantin.
Secretele marilor maeştri, Fischer, Gheorghiu, Karpov / Constantin Ştefaniu. — ₍Bucureşti₎ : "Sport-Turism", 1978.
342 p. : ill. ; 22 cm. R78-1823
Bibliography: p. 341.
lei12.00
1. Fischer, Bobby, 1943- 2. Gheorghiu, Florin, 1944- 3. Karpov, Anatoliĭ Evgen'evich, 1951- 4. Chess—Collections of games. I. Title.
GV1439.F5S73 78-386766
*79 MARC

Tālis, Mihails, 1936-
V ogon' ataki / M. N. Tal', ÎA. V. Damskiĭ. — Moskva : Fizkul'tura i sport, 1978.
304 p. : ill. ; 21 cm. — (Vydaiushchiesia shakhmatisty mira) USSR***
1.50rub
1. Tālis, Mihails, 1936- 2. Chess—Collections of games. I. Damskiĭ, ÎAkov Vladimirovich, joint author. II. Title.
GV1439.T3A37 79-372229
79 MARC

Zinser, Sylvain.
La partie italienne / Sylvain Zinser. — Paris : Payot-Diffec, 1978.
166 p. : ill. ; 21 cm. — (Collection d'échecs Payot-Diffec) F***
ISBN 2-228-15310-9
1. Chess—Middle games. 2. Chess—Collections of games. I. Title.
GV1450.3.Z56 794.1'23 78-397454
79 MARC

—COMBINATIONS see Chess —Middle games

—COMPUTER PROGRAMS

see also SARGON (Computer program)

—DATA PROCESSING

Spracklen, Dan.
Sargon, a computer chess program / Dan and Kathe Spracklen. — Rochelle Park, N.J. : Hayden Book Co., c1978.
114 p. : ill. ; 25 cm.
Includes index.
ISBN 0-8104-5155-7 : \$14.95
1. Chess—Data processing. 2. SARGON (Computer program) 3. Zilog Model, Z=80 (Computer)—Programming. 4. Assembler language (Computer program language) I. Spracklen, Kathe, joint author. II. Title.
GV1447.S53 794.1'7 78-64981
79 MARC

—DATA PROCESSING—ADDRESSES, ESSAYS, LECTURES

— **New York : Springer-Verlag, c1978.**
xi, 225 p. : ill. ; 24 cm. — (Texts and monographs in computer science)
Bibliography: p. 215-219.
Includes index.
ISBN 0-387-07957-2
1. Chess—Data processing—Addresses, essays, lectures. I. Frey, Peter, 1942- . II. Title.
GV1318.C45 794.1'7 76-29719
76 MARC

—DICTIONARIES

Brace, Edward R
An illustrated dictionary of chess / ₍by₎ Edward R. Brace ; introduction by Svetozar Gligorić. — London ; New York : Hamlyn, 1977.
320 p. : ill. ; 22 cm. GB78-06276
ISBN 0-600-32920-8 : £3.50
1. Chess—Dictionaries. I. Title.
GV1314.5.B7 794.1'03 78-313043
78 MARC

—EARLY WORKS TO 1800

₍**August II,** Duke of Braunschweig-Lüneburg₎ 1579-1666.
Das schach- oder könig-spiel. Von Gustavo Seleno ₍psued.₎ in vier unterschiedene bücher, mit besonderm fleiss, gründ- und ordentlich abgefasset. Auch mit dienlichen kupferstichen gezieret: desgleichen vorhin nicht aussgangen. Diesem ist zu ende angefüget ein sehr altes spiel genandt rythmomachia. Lipsiæ, Cum privil: Cæsareo, ad Sexennium, CIC ICC XVI.
15 p. l., 495, ₍1₎ p., 1 l. engr. illus. (chess) 3 fold. pl. 29 cm.
Colophon (under woodcut printer's device with monogram H.G.): Leipzig, Gedruckt durch Lorentz Kober. Bey Henning Gross den Jüngern zu finden. Im jahr M.DC.XVI.
Engraved and illustrated t.-p. and special t.-p. for each of the three books and the Rythmomachia.
The plate after p. 216 is signed: J. ab heyden sculpsit.

The order of games follows that of "Il giuoco degli scacchi" by Ruy Lopez, from which Bilguer claims it is largely translated. cf. P. R. von Bilguer, Handbuch des schachspiels, p. 29-30.
Rythmomachia, p. 445-495, has title: Rythmomachia. Ein vortrefflich und uhraltes spiel, dess Pythagoræ: welches Gustavus Selenus ₍pseud.₎ auss des Francisci Barozzii ,.. ins deutsche ubergesetzet ... apud Henningum Gros, Jun. M.DC.XVI.
Author's autograph note of presentation to his mother, dated 25 Jan. 1617, on verso t.-p.
1. Chess—Early works to 1800. I. Barozzi, Francesco, fl. 1550-1590. II nobilissimo et antiquissimo giuoco Pythagoreonominato rythmomachia. II. Heyden, Jacob von der, 1573-1645. III. Title. IV. Title: Rythmomachia.
GV1442.A8 794.1 31-12930
——— Copy Rosenwald Coll. With 13 prelim. leaves, the double-leaf Tabula being bound after p. 12. Bookplate of Baron Landau. MARC
₍r79₎rev

₍**Cessolis, Jacobus de**₎
De ludo scachorum. English. ₍Westminster₎ William Caxton ₍1482?₎
₍84₎ l., the first blank. woodcuts: 24 illus. f°. 27.5 cm.
Leaf ₍2ᵃ₎, line 14: ... an excellent doctour of dyuynyte in the Royame of Fraunce of the Ordre of Thospytal of Saynt Johns of Jherusalem ... hath made a book of the chesse moralysed.
Attributed by William Caxton to Jean de Vignay, whose French version he translated into English.
Hain. Repertorium (with Copinger's Supplement) 4901; Gesamtkat. d. Wiegendr., 6533; Ricci. Census of Caxtons, 18; Goff. Third census, C-414.
L.C. copy imperfect: leaf ₍1₎ wanting; leaf ₍59₎ in facism.
Bookplate of A. Edward Newton.
1. Chess—Early works to 1800. I. Caxton, William, 1422 (ca.)-1491. II. Title. III. Title: A book of the chesse moralysed.
Incun. X.C42 Rosenwald Coll. 48-43293
[GV1442]

₍**Greco, Gioachino, Calabrese**₎ fl. 1640.
The royall game of chesse-play. Sometimes the recreation of the late king, with many of the nobility. Illustrated with almost an hundred gambetts. Being the study of Biochimo the famous Italian. London, H. Herringman, 1656.
9 p.l., 120, ₍2₎ p. front. (port.) 15 cm.
The frontispiece is a portrait of Charles I., engraved by P. Stent.
Tr. by Francis Beale.
Wing G1810.
1. Chess—Early works to 1800. I. Beale, Francis, fl. 1656, tr.
GV1442.G73 5-24312
——— Copy 2. Rosenwald Coll. Bookplate of John Spencer. MARC
₍r79₎rev

Philidor, François André Danican, known as, 1726-1795.
Analyse du jeu des échecs. Nouv. éd., considérablement augmentée, par A. D. Philidor ... Londres, 1777.
302 p. port. 24 cm.
Inscribed to Mr. Rosenwald by Edward Lasker.
1. Chess—Early works to 1800. I. Title.
GV1444.P48 1777 Rosenwald Coll. 5-24315
₍r79₎rev MARC

CHESS (Continued)

—END GAMES

Capablanca, José Raúl, 1888-1942.
Capablanca's Best chess endings / Irving Chernev. — Oxford [Eng.] ; New York : Oxford University Press, 1978.
x, 288 p. : ill. ; 21 cm. — (Oxford chess books) GB77
Bibliography: p. 286-288.
Includes indexes.
ISBN 0-19-217553-X. ISBN 0-19-217554-8 pbk. : £2.95
1. Chess—End games. 2. Chess—Collections of games. I. Chernev, Irving, 1900- II. Title. III. Title: Best Chess endings.
GV1450.7.C34 794.1'24 77-30392
78 MARC

Gillam, A J
Simple checkmates / [by] A. J. Gillam. — London : Batsford, 1978.
128 p. : ill. ; 22 cm. — (Discovering chess series ; book 3) (A Batsford chess book) GB78-20460
ISBN 0-7134-1482-0 : £1.45
1. Chess—End games. I. Title. II. Series.
GV1450.7.G54 794.1'24 79-306529
79 MARC

Rădulescu, Mihai.
Finaluri complexe în şah : arta marilor campioni : Boris Spasski, Robert Fischer, Anatoli Karpov / Mihai Rădulescu. — [Ediţia a 2-a]. — Bucureşti : "Sport-Turism", 1978.
232 p. : ill. ; 21 cm. R78-5288
Bibliography: p. 227.
lei7.25
1. Chess—End games. 2. Chess—Collections of games. 3. Spasskiĭ, Boris Vasil'evich, 1937- 4. Fischer, Bobby, 1943- 5. Karpov, Anatoliĭ Eugen'evich, 1951- I. Title.
GV1450.7.R33 1978 79-369177
*79 MARC

Schuster, Theo.
End games in chess : the essential techniques for winning / by Theo Schuster ; translated [from the German] by Joachim Leibschner. — Guildford : Lutterworth Press, 1977.
64 p. : ill. ; 20 cm. GB78-20462
Translation of Das Endspiel im Schach.
ISBN 0-7188-2326-5 : £1.25
1. Chess—End games. I. Title.
GV1450.7.S3713 794.1'24 79-301871
79 MARC

—HISTORY

Euwe, Machgielis, 1901-
The development of chess style / by M. Euwe ; translated from the Dutch by W. H. Cozens. — 1st paperback ed. — New York : McKay, 1978, c1968.
152 p. : ill. ; 20 cm.
Translation of Veldheerschap op de vierenzestig.
ISBN 0-679-14045-X : $3.95
1. Chess—History. I. Title.
GV1318.E853 1978 794.1'2'09 78-51942
78 MARC

Frank, Joachim A
Lob des königlichen Spiels : oder, Schach-Brevier / Joachim A[rmin] Frank ; [Bildausw. u. Gesamtleitung, Karl Andreas Edlinger]. — Wien ; Berlin : Neff, c1979.
300, [4] p. : ill. ; 22 cm. Au79-13-359
On spine: Schach-Brevier.
Bibliography: p. [302]
ISBN 3-7014-0158-6 : S198.00
1. Chess—History. I. Edlinger, Karl Andreas. II. Title. III. Title: Schach-Brevier.
GV1317.F72 79-379666
*79 MARC

Heydebrand und der Lasa, Tassilo von, 1818-1899.
Zur Geschichte und Literatur des Schachspiels : Forschungen / von T. von der Lasa. — Fotomechan. Neudr. d. Orig.-Ausg. 1897. — Leipzig : Zentralantiquariat d. Dt. Demokrat. Republik, 1978.
vi, 269 p. : ill. ; 21 cm. GFR77-A
Reprint of the ed. published by Von Veit.
Includes bibliographical references.
60.00M
1. Chess—History. I. Title.
GV1317.H45 1978 79-339375
79 MARC

—JUVENILE LITERATURE

Langfield, Paul.
A picture guide to chess / Paul Langfield. — 1st American ed. — Philadelphia : Lippincott, c1976.
48 p. : ill. ; 22 cm.
SUMMARY: Text and illustrations introduce the principles and strategies of chess.
ISBN 0-397-31681-X. ISBN 0-397-31682-8 pbk.
1. Chess—Juvenile literature. [1. Chess] I. Title.
GV1446.L36 1976 794.1'2 75-46637
76 MARC

Love, John F
Chess battle strategies / John Love & John Hodgkins. — New York : Sterling Pub. Co., 1979, c1962.
160 p. : ill. ; 27 cm.
Original ed. published under title: Chess ideas for young players.
SUMMARY: "First lessons in successful chess generalship."
ISBN 0-8069-4952-X
1. Chess—Juvenile literature. [1. Chess] I. Hodgkins, John P., joint author. II. Title.
GV1446.L62 1979 794.1'2 79-65066
79 MARC

Nottingham, Ted.
The children's book of chess / [by] Ted Nottingham, Bob Wade. — London : Batsford, 1978.
111 p. : ill. (some col.), ports. ; 26 cm. GB78-33953
Ill. on lining papers.
ISBN 0-7134-0385-3 : £2.95
1. Chess—Juvenile literature. I. Wade, Robert Graham, joint author. II. Title.
GV1446.N67 1978b 794.1'2 78-326188
79 MARC

Nottingham, Ted.
The children's book of chess / Ted Nottingham, Bob Wade. — New York : Crown Publishers, 1979.
111 p. : ill. ; 26 cm.
SUMMARY: Discusses the pieces, moves, rules, and strategy of chess and youthful champions of the game.
ISBN 0-517-53578-5
1. Chess—Juvenile literature. [1. Chess. 2. Games] I. Wade, Robert Graham, joint author. II. Title.
GV1446.N67 1979 794.1'2 78-19121
78 MARC

Rosenberg, Arthur D
Chess for children and the young at heart / by Arthur D. Rosenberg ; illustrated by Howard Berelson ; photos. by Marianne Groher. — 1st ed. — New York : Atheneum, 1977.
150 p. : ill. ; 25 cm.
SUMMARY: A step-by-step explanation of the basic principles and strategy of chess.
ISBN 0-689-30493-5
1. Chess—Juvenile literature. [1. Chess] I. Berelson, Howard. II. Groher, Marianne. III. Title.
GV1446.R67 794.1'2 75-10963
75 MARC

—MIDDLE GAMES

Bronshteĭn, David Ionovich, 1924-
The chess struggle in practice : candidates tournament, Zurich 1953 / David Bronstein ; translated by Oscar D. Freedman ; edited by Burt Hochberg. — New York : D. McKay Co., c1978.
xxvii, 499 p. : ill. ; 21 cm.
Translation of Mezhdunarodnyĭ turnir grossmeĭsterov.
Includes indexes.
ISBN 0-679-13064-0 : $17.95
1. Chess—Tournaments, 1953. 2. Chess—Middle games. 3. Chess—Collections of games. I. Hochberg, Burt. II. Title.
GV1455.B713 794.1'57 78-59140
00 MARC

Gillam, A J
Simple chess tactics / A. J. Gillam. — London : B. T. Batsford, 1978.
128 p. : ill. ; 22 cm. — (Discovering chess series ; book 2) GB***
ISBN 0-17-341480-4 : £1.45
1. Chess—Middle games. I. Title.
GV1450.3.G54 794.1'23 78-319096
78 MARC

Tāls, Mihails, 1936-
Tal's winning chess combinations the secrets of winning chess combinations described and explained by the Russian Grandmaster Mikhail Tal / by Mikhail Tal and Victor Khenkin ; translated from the Russian by Hanon W. Russell. — New York : Simon and Schuster, c1979.
409 p. : 22 cm.
ISBN 0-671-24262-8
1. Chess—Middle games. I. Khenkin, Viktor, joint author. II. Title.
GV1450.3.T3413 794.1'23 78-31411
78 MARC

Zinser, Sylvain.
La partie italienne / Sylvain Zinser. — Paris : Payot-Diffec, 1978.
166 p. : ill. ; 21 cm. — (Collection d'échecs Payot-Diffec) F***
ISBN 2-228-15310-9
1. Chess—Middle games. 2. Chess—Collections of games. I. Title.
GV1450.3.Z56 794.1'23 78-397454
79 MARC

—MISCELLANEA

Bagnoli, Paolo.
Scacchi : storia, controstoria e altre cose ancora / Paolo Bagnoli. — Milano : Mursia, [1978]
327 p. : ill. ; 19 cm. — (Collana I Giochi) It78-July
Bibliography: p. 315-316.
Includes index.
L5000
1. Chess—Miscellanea. I. Title.
GV1447.B33 78-375688
*78 MARC

—OPENINGS

Adams, Jimmy.
Najdorf poisoned pawn / [by] Jimmy Adams. — Nottingham : "The chess player", 1977.
111 p. : ill. ; 21 cm. — (Sicilian defence) GB78-15136
Includes index.
ISBN 0-906042-07-0 : £2.95
1. Pawn (Chess) 2. Chess—Collections of games. 3. Chess—Openings. I. Title.
GV1451.5.P3A22 794.1'22 78-319649
78 MARC

Adams, Jimmy.
Sicilian Najdorf Polugaevsky variation / [by] Jimmy Adams. — Nottingham : The Chess Player, 1978.
232 p. : ill ; 21 cm. GB79-05254
Algebraic notation.
ISBN 0-906042-09-7 : £4.95
1. Chess—Openings. 2. Chess—Collections of games. I. Title.
GV1450.2.A32 794.1'22 79-310915
79 MARC

Fuller, Maxwell.
Lines with c3 / [by] M. Fuller & L. M. Pickett. — Nottingham : The Chess Player, 1977.
69 p. : ill. ; 21 cm. — (Sicilian defence : 11) GB78-16402
Includes index.
ISBN 0-906042-06-2 : £1.95
1. Chess—Openings. I. Pickett, Leonard M., joint author. II. Title.
GV1450.2.F84 794.1'22 78-320616
78 MARC

Levy, David N L
1 b4 Sokolsky opening / [by] D. N. L. Levy. — Nottingham : Chess Player, 1977.
72 p. : ill. ; 21 cm. GB78-01066
Based on Debiut 1.b2-b4, by A. P. Sokol'skiĭ.
ISBN 0-900928-97-2 : £1.95
1. Chess—Openings. 2. Sokol'skiĭ, Alekseĭ Pavlovich. I. Sokol'skiĭ, Alekseĭ Pavlovich. Debiut 1.b2-b4. II. Title.
GV1450.2.L49 794.1'22 78-319653
78 MARC

Levy, David N L
Benko counter-gambit / [by] David Levy. — London : Batsford, 1978.
127 p. : ill. ; 22 cm. — (Batsford algebraic chess openings) (Batsford chess books) GB78-29859
Includes index.
ISBN 0-7134-1058-2 : £3.95
1. Chess—Openings. I. Title. II. Series.
GV1450.2.L475 794.1'22 79-303720
79 MARC

Levy, David N L
How to play the Sicilian Defence / David Levy and Kevin O'Connell. — London : B. T. Batsford, 1978.
152 p. : ill. ; 23 cm. — (The Club player's library) (Batsford chess books) GB***
Includes index.
ISBN 0-7134-0757-3 : £4.95. ISBN 0-7134-0758-1 pbk.
1. Chess—Openings. I. O'Connell, Kevin J., joint author. II. Title.
GV1450.2.L48 794.1'22 78-319651
78 MARC

MacDonald-Ross, Michael.
Nimzo-Indian defence, Leningrad system ... / [by] Michael MacDonald-Ross. — London : Batsford, 1978.
135 p. : ill. ; 22 cm. — (Batsford algebraic chess openings) (A Batsford chess book) GB78-35492
Includes indexes.
ISBN 0-7134-0947-9 : £5.95
1. Chess—Openings. I. Title. II. Series.
GV1450.2.M33 794.1'22 79-306058
79 MARC

Myers, Hugh E 1930-
Exploring the chess openings / Hugh E. Myers. — Davenport, Iowa : Thinkers' Press, c1978.
102 p. : ill. ; 22 cm.
Includes bibliographical references and indexes.
1. Chess—Openings. I. Title.
GV1450.M93 794.1'22 79-110127
79 MARC

Porreca, Giorgio.
La difesa francese / Giorgio Porreca. — Milano : Mursia, 1977.
195 p. : ill. ; 19 cm. — (Biblioteca dello scacchista) (Collana I Giochi) It78-Sept
Includes index.
L4000
1. Chess—Openings. 2. Chess—Collections of games. I. Title.
GV1450.2.P67 794.1'22 79-358409
*79 MARC

Shamkovich, Leonid Aleksandrovich.
Chess sacrifices / [by] Leonid Shamkovich ; [translated from the Russian by A. Kalnajs]. — Nottingham : Chess Player, 1976.
184 p. : ill. ; 21 cm. GB77-33990
Translation of Zhertva v shakhmatakh.
ISBN 0-900928-99-9 : £2.95
1. Chess—Openings. 2. Chess—Collections of games. I. Title.
GV1450.S4413 794.1'2 78-382792
78 MARC

CHESS

—OPENINGS (Continued)

Sicilian, Lasker-Pelikan ... / ₍by₎ R. G. Wade ... ₍et al.₎. — London : Batsford, 1978.
229 p. : ill. ; 22 cm. — (Batsford algebraic chess openings) (A Batsford chess book) GB78-32363
Includes index.
ISBN 0-7134-1633-5. ISBN 0-7134-1634-3 pbk. : £4.95
1. Chess—Openings. I. Wade, Robert Graham. II. Series.
GV1450.2.S53 794.1'22 79-300111
79 MARC

Williams, Kenneth F
The real American Wilkes-Barre variation two knights defense / by Kenneth F. Williams. — Coraopolis, Pa. : Chess Enterprises, 1979.
58 p. : ill. ; 22 cm.
ISBN 0-931462-03-7
1. Chess—Openings. I. Title.
GV1450.2.W54 794.1'22 79-115382
00 MARC

—PERIODICALS

Chess digest magazine. v. 1–
Jan. 1968–
₍Dallas, s. n.₎
v. ill. 28 cm.
Frequency varies.
Key title: Chess digest magazine, ISSN 0009-3335.
1. Chess—Periodicals. I. Title: Chess digest.
GV1313.C483 794.1'05 79-649818
MARC-S

—PROBLEMS see Chess problems

—PSYCHOLOGICAL ASPECTS

Fuller, Peter.
The champions : the secret motives in games and sport / ₍by₎ Peter Fuller — Revised ed. — London : Allen Lane, 1978.
3-397 p. ; 23 cm. GB78-26256
Bibliography: p. 387-₍397₎
ISBN 0-7139-1077-1 : £6.50
1. Sports—Psychological aspects. 2. Chess—Psychological aspects. 3. Athletes—Biography. I. Title.
GV706.4.F84 1978 796'.01 78-323329
79 MARC

—PSYCHOLOGY

Webb, Simon.
Chess for tigers / Simon Webb ; with ill. by Edward McLachlan. — Oxford ₍Eng.₎ ; New York : Oxford University Press, 1978.
99 p. : ill. ; 21 cm. GB***
ISBN 0-19-217575-0 : £3.95. ISBN 0-19-217576-9 pbk.
1. Chess. 2. Chess—Psychology. I. Title.
GV1445.W36 794.1'2 78-40204
79 MARC

—QUOTATIONS, MAXIMS, ETC.

Collins, John W
Maxims of chess / by John W. Collins. — New York : D. McKay, c1978.
xi, 276 p. : ill. ; 21 cm.
Includes indexes.
ISBN 0-679-13066-7 :
1. Chess—Quotations, maxims, etc. 2. Chess—Collections of games. I. Title.
GV1445.C69 794.1 77-18245
77 MARC

—SACRIFICES see Chess—Middle games

—TOURNAMENTS

Bath, 1973. — ₍Dallas : Chess Digest, 1973?₎
44 p. : ill. ; 23 cm. — (The Chess player ; no. 164)
Cover title.
1. Chess—Tournaments. 2. Chess—Collections of games. I. Series.
GV1313.C59 no. 164 794.1'08 s 79-314557
₍GV1455₎ 79 MARC

Hartston, William Roland, 1947-
The battle of Baguio City : Karpov-Korchnoi 1978 / W. R. Hartston. — London : Hutchinson, 1978.
121 p. : ill. ; 21 cm. GB***
ISBN 0-09-134881-1 : £1.95
1. Chess—Tournaments. 2. Chess—Collections of games. 3. Karpov, Anatoliĭ Evgen'evich, 1951- 4. Korchnoĭ, Viktor, 1931- I. Title.
GV1455.H33 794.1'57 79-305230
79 MARC

Keene, Raymond D
World chess championship : Korchnoi vs. Karpov : the inside story of the match / by Raymond Keene. — New York : Simon and Schuster, 1978.
159 p. : ill. ; 20 cm. — (A Fireside book)
ISBN 0-671-24647-X. ISBN 0-671-24648-8 pbk.
1. Chess—Tournaments. 2. Korchnoĭ, Viktor, 1931- 3. Karpov, Anatoliĭ Evgen'evich, 1951- I. Title.
GV1455.K388 794.1'57 78-16067
78 MARC

—TOURNAMENTS—HISTORY

Földeák, Walter Árpád.
Chess Olympiads, 1927-1968 / by Árpád Földeák ; translated from the Hungarian by Robert Ejury and Jenö Bochkor ; translation rev. by Peter H. Clarke. — New York : Dover Publications, 1979, c1969.
416 p. : ill. ; 21 cm.
Reprint, with corrections, of the 2d enl. ed. (1969) which was issued under title: Chess Olympiads.
Includes index.
ISBN 0-486-23733-8 : $4.95
1. Chess—Tournaments—History. 2. Chess—Collections of games. I. Title.
GV1455.F583 1979 794.1'57 78-68173
79 MARC

—TOURNAMENTS—UNITED STATES

72nd annual US Open Chess Championship, Ventura, California, 8-20 August 1971 / edited by Jack L. Spence. — Nottingham : The Chess Player, ₍1976₎
xxii, 130 p. : ill., ports. ; 19 cm. GB77-01730
Includes index.
£2.70
1. Chess—Tournaments, 1971. 2. Chess—Tournaments—United States. 3. Chess—Collections of games. I. Spence, Jack L., 1926- II. Title.
GV1455.S45 794.1'57 77-361157
77₍r79₎rev MARC

—TOURNAMENTS, 1953

Bronshteĭn, David Ionovich, 1924-
The chess struggle in practice : candidates tournament, Zurich 1953 / David Bronstein ; translated by Oscar D. Freedman ; edited by Burt Hochberg. — New York : D. McKay Co., c1978.
xxvii, 499 p. : ill. ; 21 cm.
Translation of Mezhdunarodnyĭ turnir grossmeĭsterov.
Includes indexes.
ISBN 0-679-13064-0 : $17.95
1. Chess—Tournaments, 1953. 2. Chess—Middle games. 3. Chess—Collections of games. I. Hochberg, Burt. II. Title.
GV1455.B713 794.1'57 78-59140
00 MARC

Bronshteĭn, David Ionovich, 1924-
Zurich international chess tournament, 1953 / by David Bronstein ; translated from the second Russian edition by Jim Marfia. — New York : Dover Publications, c1979.
xviii, 349 p. : ill. ; 24 cm.
Translation of Mezhdunarodnyĭ turnir grossmeĭsterov.
Includes index.
ISBN 0-486-23800-8 : $6.00
1. Chess—Tournaments, 1953. I. Title.
GV1455.B713 1979 794.1'57 78-74881
00 MARC

—TOURNAMENTS, 1971

72nd annual US Open Chess Championship, Ventura, California, 8-20 August 1971 / edited by Jack L. Spence. — Nottingham : The Chess Player, ₍1976₎
xxii, 130 p. : ill., ports. ; 19 cm. GB77-01730
Includes index.
£2.70
1. Chess—Tournaments, 1971. 2. Chess—Tournaments—United States. 3. Chess—Collections of games. I. Spence, Jack L., 1926- II. Title.
GV1455.S45 794.1'57 77-361157
77₍r79₎rev MARC

—TOURNAMENTS, 1977

Keene, Raymond D
Korchnoi vs. Spassky : chess crisis / Raymond Keene. — London ; Boston : Allen & Unwin, 1978.
128 p. : ill. ; 20 cm. GB***
Includes index.
ISBN 0-04-794006-9 : £1.95
1. Chess—Tournaments, 1977. 2. Korchnoi, Viktor, 1931- 3. Spasskiĭ, Boris Vasil'evich, 1937- 4. Chess—Collections of games. I. Title.
GV1455.K387 794.1'57 78-313036
78 MARC

Rakić, Branislav, 1938-
Kako su ratovali : finalni meč kandidata za prvaka sveta / Branislav Rakić. — Beograd : Beogradski izdavačko-grafički zavod, 1978.
200 p. : ill. ; 18 cm. — (Džepna knjiga Beogradskog izdavačko-grafičkog zavoda : Praktična knjiga : Serija Šah) Yu***
On cover: Korčnoj—Spaski, Beograd '77.
30.00Din
1. Chess—Tournaments, 1977. 2. Korchnoĭ, Viktor, 1931- 3. Spasskiĭ, Boris Vasil'evich, 1937- I. Title.
GV1455.R32 78-389425
79 MARC

—TOURNAMENTS, 1978

Keene, Raymond D
Karpov-Korchnoi 1978 : the inside story of the match / ₍by₎ Raymond Keene. — London : Batsford, 1978.
159 p. : ill. ; 20 cm. — (A Batsford chess book) GB78-38616
American ed. published under title: World chess championship.
ISBN 0-7134-1809-5 : £1.50
1. Chess—Tournaments, 1978. 2. Chess—Collections of games. 3. Karpov, Anatoliĭ Evgen'evich, 1951- 4. Korchnoĭ, Viktor, 1931- I. Title.
GV1455.K388 1978b 794.1'57 79-306233
79 MARC

Larsen, Bent.
Karpov vs. Korchnoi : World Chess Championship, 1978 / Bent Larsen ; introduced by Michael Stean. — London ; Boston : Unwin Paperbacks, 1978.
143 p. : ill. ; 20 cm. GB***
ISBN 0-04-794008-5 : £1.95
1. Chess—Tournaments, 1978. 2. Chess—Collections of games. 3. Karpov, Anatoliĭ Evgen'evich, 1951- 4. Korchnoĭ, Viktor, 1931- I. Title.
GV1455.L28 794.1'57 79-308377
79 MARC

Négro, Jacques.
Karpov-Kortchnoi : championnat du monde 1978 : toutes les parties commentées / Jacques Négro. — ₍Paris₎ : Hachette, 1978.
223 p. : ill. ; 23 cm. F79-3678
ISBN 2-01-005947-6 : 39.00F
1. Chess—Tournaments, 1978. 2. Karpov, Anatoliĭ Evgen'evich, 1951- 3. Korchnoĭ, Viktor, 1931- 4. Chess—Collections of games.
GV1455.N43 794.1'57 79-372971
79 MARC

—VARIANTS

see also Chinese chess

—HUNGARY—HISTORY

Magyar sakktörténet. — Budapest : Sport, 197
v. : ill. ; 25 cm. H77
Includes indexes.
CONTENTS:
—Maróczy, G. Világversenyek élén.

ISBN 9632532384 : 55.00Ft (v. 2)
1. Chess—Hungary—History. 2. Chess—Collections of games.
GV1330.H9M33 78-396273
*79 MARC

—UNITED STATES—HISTORY

Korn, Walter.
America's chess heritage : from Benjamin Franklin to Bobby Fischer—and beyond / Walter Korn. — New York : D. McKay Co., c1978.
xv, 302 p. : ill. ; 21 cm.
Includes indexes.
ISBN 0-679-13200-7 : $12.95
1. Chess—United States—History. 2. Chess players—United States—Biography. I. Title.
GV1323.K67 794.1'0973 78-16224
78 MARC

CHESS GAMES see Chess—Collections of games

CHESS MEN see Chessmen

CHESS OPENINGS see Chess—Openings

CHESS PLAYERS

—ESTONIA—BIOGRAPHY

Heuer, Valter.
Meie Keres : kujunemisaastad / Valter Heuer. — Tallinn : Eesti Raamat, 1977.
381 p. : ill. ; 23 cm. USSR***
2.04rub
1. Keres, Paul, 1916- 2. Chess players—Estonia—Biography. 3. Chess—Collections of games. I. Title.
GV1439.K4H48 78-389256
79 MARC

—GREAT BRITAIN—BIOGRAPHY

Keene, Raymond D
Becoming a grandmaster / ₍by₎ Raymond Keene. — London : Batsford, 1977.
157 p. : ill. ; 22 cm. — (The Club player's library) (Batsford chess books) GB78-18552
Includes index.
ISBN 0-7134-0829-4. ISBN 0-7134-0830-8 pbk. : £3.50
1. Chess. 2. Keene, Raymond D. 3. Chess players—Great Britain—Biography. I. Title.
GV1445.K37 794.1 78-326880
79 MARC

—RUSSIA—BIOGRAPHY

Botvinnik, Mikhail Moiseevich, 1911–
(Polveka v shakhmatakh)
Полвека в шахматах / М. М. Ботвинник. — Москва : Физкультура и спорт, 1978.
272 p. : ill. ; 21 cm. — (Выдающиеся шахматисты мира) USSR 79
Series romanized: Vydaûshchiesâ shakhmatisty mira.
1.20rub
1. Botvinnik, Mikhail Moiseevich, 1911– 2. Chess — Collections of games. 3. Chess players—Russia—Biography. I. Title.
GV1439.B6A36 79–411684

Karpov, Anatoliĭ Evgen'evich, 1951–
(Devâtaâ vertikal')
Девятая вертикаль / ₍А. Е. Карпов, А. Б. Рошаль₎. — Москва : Молодая гвардия, 1978.
301 p., ₍16₎ leaves of plates : ill. ; 21 cm. USSR***
1.40rub
1. Karpov, Anatoliĭ Evgen'evich, 1951– 2. Chess—Collections of games. 3. Chess players—Russia—Biography. I. Roshal', Aleksandr, joint author. II. Title.
GV1439.K37A335 79–411688

Karpov, Anatoliĭ Evgen'evich, 1951–
(Izbrannye partii, 1969–1977)
Избранные партии, 1969–1977 / Анатолий Карпов. — Москва : Физкультура и спорт, 1978.
256 p. : ill. ; 20 cm. — (Выдающиеся шахматисты мира) USSR 78
Series romanized: Vydaûshchiesâ shakhmatisty mira.
1.30rub
1. Karpov, Anatoliĭ Evgen'evich, 1951– 2. Chess—Collections of games. 3. Chess players—Russia—Biography. I. Title.
GV1439.K37A35 79–410789

Smyslov, Vasiliĭ Vasil'evich, 1921-
V poiskakh garmonii / V. V. Symslov. — Moskva : Fizkul'tura i sport, 1979.
238 p. : ill. ; 21 cm. — (Vydaûshchiesâ shakhmatisty mira) USSR79
Includes indexes.
1.30rub
1. Smyslov, Vasiliĭ Vasil'evich, 1921- 2. Chess players—Russia—Biography. I. Title.
GV1439.S6A38 79-380035
*79 MARC

—UNITED STATES—BIOGRAPHY

Korn, Walter.
America's chess heritage : from Benjamin Franklin to Bobby Fischer—and beyond / Walter Korn. — New York : D. McKay Co., c1978.
xv, 302 p. : ill. ; 21 cm.
Includes indexes.
ISBN 0-679-13200-7 : $12.95
1. Chess—United States—History. 2. Chess players—United States—Biography. I. Title.
GV1323.K67 794.1'0973 78-16224
78 MARC

CHESS PLAYERS, RATING OF

Elo, Arpad E 1903-
The rating of chessplayers, past and present / by Arpad E. Elo. — New York : Arco Pub., 1978.
206 p. : ill. ; 23 cm.
Bibliography: p. 196-198.
Includes index.
ISBN 0-668-04721-6 : $12.95
1. Chess players, Rating of. 2. Probabilities. I. Title.
GV1457.5.E44 1978 794.1'092'2 78-24077
78 MARC

Euwe, Machgielis, 1901-
Bobby Fischer—the greatest? / By Max Euwe. — New York : Sterling Pub. Co., c1979.
xiv, 202 p. : ill. ; 23 cm.
Includes index.
ISBN 0-8069-4950-3 : $8.95. ISBN 0-8069-4951-1 lib. bdg. : $8.95
1. Chess players, Rating of. 2. Fischer, Bobby, 1943- 3. Chess—Collections of games. I. Title.
GV1457.5.E97 794.1'092'2 78-66270
79 MARC

CHESS PROBLEMS

Magee, James Francis, 1867-
Good companion (Bonus socius); XIIIth century manuscript collection of chess problems; illustrated with coloured frontispiece, and thirty-eight full page reproductions from original manuscript. To which is added a short narrative of Jacobus Cessolis and the two French champions Philidor and La Bourdonnais ... Ed. by James F. Magee, jr., of Philadelphia. Florence, Printed by the Tipografia Giuntina, 1910.
68 p. illuminated front., illus., plates (1 fold.) facsims. (19 double, illuminated) 26 cm.
"Only a limited number of volumes of this edition has been printed. This book is no. 145."
The manuscript has been wrongly attributed to Nicholaus St. Nicholai. cf. p. 6.

Bound in leather with brass bosses and clasp.
1. Chess problems. 2. Cessolis, Jacobus de. 3. Philidor, François André Danican, known as, 1726-1795. 4. La Bourdonnais, Louis Charles Mahé de, 1796?-1840. I. Florence. R. Biblioteca nazionale centrale. Mss. (Banco dei rari, B.A. 6, p. 2, no. 1) II. Title. III. Title: Bonus socius.
GV1451.M17 20-14720
——— Copy 2. Rosenwald coll. MARC
₍r79₎rev

Smullyan, Raymond M
The chess mysteries of Sherlock Holmes / by Raymond Smullyan. — 1st ed. — New York : Knopf, 1979.
xiii, 171 p. : ill. ; 21 cm.
ISBN 0-394-50488-7 : $10.00
1. Chess problems. 2. Holmes, Sherlock (Fiction character) I. Title.
GV1541.S56 1979 794.1'2 79-2101
79 MARC

CHESSMEN

see also Pawn (Chess)

Greygoose, Frank.
Chessmen / Frank Greygoose. — New York : Arco Pub., c1979.
160 p. : ill. ; 25 cm.
Bibliography: p. 155.
Includes index.
ISBN 0-668-04766-6 : $24.95
1. Chessmen. I. Title.
NK4696.G73 688.7'4'1 78-31716
78 MARC

CHEST

see also Heart; Lungs; Respiratory organs

—ATLASES

Lyons, Edward Arthur, 1943-
Color atlas of sectional anatomy : chest, abdomen, and pelvis / E. A. Lyons ; photography, Rob Mathieson ; artist, David Lee. — St. Louis : C. V. Mosby, 1978.
ix, 317 p. : col. ill. ; 25 x 32 cm. & 60 unmounted transparencies (in pocket)
Includes index.
ISBN 0-8016-3052-5
1. Chest—Atlases. 2. Abdomen—Atlases. 3. Pelvis—Atlases. 4. Anatomy, Surgical and topographical—Atlases. I. Mathieson, Rob. II. Lee, David, 1953- III. Title.
QM540.L95 611'.9 78-15746
78 MARC

Wyman, Alvin C
Transverse anatomy of the human thorax, abdomen, and pelvis : an atlas of anatomic, radiologic computed tomographic, and ultrasonic correlation / Alvin C. Wyman, Thomas L. Lawson, Lawrence R. Goodman. — Boston : Little, Brown, 1978.
xv, 135 p. : ill. ; 24 x 32 cm.
Includes index.
ISBN 0-316-96250-3
1. Chest—Atlases. 2. Abdomen—Atlases. 3. Pelvis—Atlases. 4. Tomography—Atlases. I. Lawson, Thomas L., joint author. II. Goodman, Lawrence R., 1943- joint author. III. Title.
QM540.W94 611.9 78-50612
79 MARC

—DISEASES

see also Chest—Radiography; Tuberculosis

Blades, Brian Brewer, 1906-1977, ed.
Blades' Surgical diseases of the chest / edited by Donald B. Effler. — 4th ed. — Saint Louis : Mosby, 1978.
xiv, 839 p. : ill. ; 26 cm.
Includes bibliographies and index.
ISBN 0-8016-0697-7
1. Chest—Surgery. 2. Chest—Diseases. I. Effler, Donald B. II. Title. III. Title: Surgical diseases of the chest.
₍DNLM: 1. Thorax—Surgery. WF980 B632₎
RD536.B53 1978 617'.54 78-7047
00 MARC

—DISEASES—DIAGNOSIS

see also Percussion

—DISEASES—DIAGNOSIS—CASES, CLINICAL REPORTS, STATISTICS

Guter, Marvin.
Chest nuclear medicine case studies : 54 case studies related to chest nuclear medicine / by Marvin Guter and Aldo N. Serafini. — Garden City, N.Y. : Medical Examination Pub. Co., 1979.
170 p. : ill. ; 22 cm.
Includes bibliographies and index.
ISBN 0-87488-083-1
1. Chest—Diseases—Diagnosis—Cases, clinical reports, statistics. 2. Radioisotope scanning—Cases, clinical reports, statistics. 3. Nuclear medicine—Cases, clinical reports, statistics. I. Serafini, Aldo N., joint author. II. Title.
RC941.G87 617'.54'075 78-70442
79 MARC

—DISEASES—RESEARCH—BRAZIL—HISTORY—20TH CENTURY

Silveira, José.
À sombra de uma sigla ou 40 anos de IBIT / José Silveira ; pref. de Edmund Blundi. — Cidade do Salvador, Brasil : ₍s.n.₎, 1977.
199 p., ₍31₎ leaves of plates : ports. ; 23 cm.
1. Instituto Brasileiro para Investigação do Tórax—History—20th century. 2. Tuberculosis—Research—Brazil—History—20th century. 3. Chest—Diseases—Research—Brazil—History—20th century. I. Title.
RC315.B7S54 78-396464
78 MARC

—RADIOGRAPHY

Riggs, Webster.
Pediatric chest roentgenology : recognizing the abnormal / by Webster Riggs, Jr. — St. Louis : W. H. Green, c1979.
xi, 296 p. : ill. ; 24 cm.
Includes bibliographies and index.
ISBN 0-87527-144-8 : $37.50
1. Chest—Radiography. 2. Pediatric radiography. 3. Pediatric respiratory diseases—Diagnosis. I. Title.
RJ433.5.R25R53 618.9'209'7540757 76-19337
79 MARC

—SURGERY

see also Thoracic surgeons

Baumann, Jean.
Notice sur les titres et travaux scientifiques du Dr Jean Baumann. — Paris : Arnette, 1975.
77 p. ; 27 cm. F75-13384
ISBN 2-7184-0152-7 : 100F
1. Chest—Surgery. 2. Surgery. 3. Resuscitation. 4. Anesthesia. 5. Baumann, Jean—Bibliography. I. Title.
RD536.B38 617'.9'08 78-394479
79 MARC

Blades, Brian Brewer, 1906-1977, ed.
Blades' Surgical diseases of the chest / edited by Donald B. Effler. — 4th ed. — Saint Louis : Mosby, 1978.
xiv, 839 p. : ill. ; 26 cm.
Includes bibliographies and index.
ISBN 0-8016-0697-7
1. Chest—Surgery. 2. Chest—Diseases. I. Effler, Donald B. II. Title. III. Title: Surgical diseases of the chest.
₍DNLM: 1. Thorax—Surgery. WF980 B632₎
RD536.B53 1978 617'.54 78-7047
00 MARC

CHEST (Continued)

—SURGERY—ATLASES

Atlas of thoracic surgery / edited by Boris V. Petrovsky ; translated from the Russian by Ludmila Aksenova ; with a foreword to the English ed. by Michale E. DeBakey. — St. Louis : C. V. Mosby Co., 1979.
2 v. (xi, 761 p.) : ill. ; 30 cm.
Translation of Atlas grudnoĭ khirurgii.
ISBN 0-8016-3832-1
1. Chest—Surgery—Atlases. I. Petrovskiĭ, Boris Vasil'evich, 1908-
[DNLM: 1. Thorax—Surgery—Atlases. WF17 A881]
RD536.A7813 617'.54 78-23699
78 MARC

—SURGERY—CASE STUDIES

Thoracic surgery case studies : a compilation of 35 case histories / by Gerard A. Kaiser ... [et al.]. — Garden City, N.Y. : Medical Examination Pub. Co., 1979.
245 p. : ill. ; 22 cm.
Includes bibliographies and index.
ISBN 0-87488-056-4
1. Chest—Surgery—Case studies. I. Kaiser, Gerard A.
RD536.T46 617'.54'0926 78-71345
79 MARC

—SURGERY—COMPLICATIONS AND SEQUELAE

Complications of intrathoracic surgery / edited by A. Robert Cordell and Robert G. Ellison. — 1st ed. — Boston : Little, Brown, c1979.
xv, 415 p. : ill. ; 27 cm.
Includes bibliographies and index.
ISBN 0-316-15732-5
1. Chest—Surgery—Complications and sequelae. I. Cordell, Alfred Robert. II. Ellison, Robert G. III. Title: Intrathoracic surgery.
RD536.C64 617'.54 79-84736
79 MARC

—SURGERY—UNITED STATES

National thoracic surgery manpower study : final report, February 1974 / Lyman A. Brewer, III ... [et al.]. — [s.l. : s.n., 1974] (Los Angeles : Cunningham Press)
xiv, 98 p. : ill. ; 29 cm.
Includes bibliographical references.
1. Chest—Surgery—United States. 2. Thoracic surgeons—United States. 3. Chest—Surgery—United States—Statistics. I. Brewer, Lyman Augustus, 1907-
RD536.N36 362.1'9'7540973 79-310795
79 MARC

—SURGERY—UNITED STATES—STATISTICS

National thoracic surgery manpower study : final report, February 1974 / Lyman A. Brewer, III ... [et al.]. — [s.l. : s.n., 1974] (Los Angeles : Cunningham Press)
xiv, 98 p. : ill. ; 29 cm.
Includes bibliographical references.
1. Chest—Surgery—United States. 2. Thoracic surgeons—United States. 3. Chest—Surgery—United States—Statistics. I. Brewer, Lyman Augustus, 1907-
RD536.N36 362.1'9'7540973 79-310795
79 MARC

CHEST PAIN

see also Angina pectoris

CHEST SURGEONS see Thoracic surgeons

CHESTER, COLBY MITCHELL, 1844-1932

—MANUSCRIPTS—CATALOGS

United States. Library of Congress. Manuscript Division.
Colby Mitchell Chester, William Freeland Fullam, Samuel McGowan, Henry Croskey Mustin: a register of their papers in the Library of Congress. Washington, Library of Congress, 1973.
iii, 4, 6, 7, 5 p. 27 cm.
"Naval Historical Foundation collection."
ISBN 0-8444-0085-8
1. United States—History, Naval—Manuscripts—Catalogs. 2. United States—History, Naval—Sources—Bibliography—Catalogs. 3. Chester, Colby Mitchell, 1844-1932—Manuscripts—Catalogs. 4. Fullam, William Freeland, 1855-1926—Manuscripts—Catalogs. 5. McGowan, Samuel, 1870-1934—Manuscripts—Catalogs. 6. Mustin, Henry Croskey, 1874-1923—Manuscripts—Catalogs. 7. United States. Library of Congress. Manuscript Division—Catalogs. I. Title. II. Title: Naval Historical Foundation collection.
Z1249.N3U5 1973a 016.359 73-2939
[E182] MARC
—— Copy 3. Z663.34.C48
73[r79]rev

CHESTER, NOVA SCOTIA

Women's Institute (Canada) Chester Branch.
History of Chester, 1759-1967. [Lunenburg, N.S., 1967]
165 p. illus., map (on cover), ports. 21 cm. 4.00 C67-3493
Cover title.
Fold. map in pocket.
1. Chester, Nova Scotia. I. Title.
F1039.5.C48W6 1967 971.6'23 77-516387
71[r79]rev MARC

CHESTER BEATTY LIBRARY

—EXHIBITIONS

Chester Beatty Library.
Selected manuscripts from the Chester Beatty Library : exhibited by Princeton University Library, February 28 to April 30, 1967. — [Princeton, N.J. : The Library, 1967]
58 p. : ill. ; 22 cm.
Bibliography: p. 57-58.
1. Manuscripts—Ireland—Dublin—Exhibitions. 2. Chester Beatty Library—Exhibitions. 3. Princeton, N.J.—Exhibitions. I. Princeton University. Library. II. Title.
Z6623.B37C5 1967 016.091'09418'3 79-312971
79 MARC

CHESTER CO., PA.

—BIOGRAPHY

Mac Elree, Wilmer W 1859-
Side lights on the bench and bar of Chester County, by Wilmer W. Mac Elree ... West Chester, Pa., 1918.
2 p.l., 3-435 p. front., illus., plates, ports, maps, facsims. 24 cm.
1. Lawyers—Chester Co., Pa. 2. Judges—Chester Co., Pa. 3. Chester Co., Pa.—Biography.
KF355.C48M3 19-3000
[r79]rev MARC

—GENEALOGY

Emrich, Oran S
Descendants of John George Emmerich, 1730-1800, of Chester County, Pennsylvania / by Oran S. Emrich. — [Washington, D.C.] : Emrich, [1979?]
85 leaves ; 30 cm.
Bibliography: p. 75.
Includes index.
1. Emrich family. 2. Chester Co., Pa.—Genealogy. I. Title.
CS71.E547 1979 929'.2'0973 79-54900
00 MARC

—HISTORY—REVOLUTION, 1775-1783—FICTION

Weitzel, Shirley A
Land o' Goshen : a tale of Chester County, 1776-1778 / by Shirley A. Weitzel ; [design and ill., Susan V. Walker]. — [Philadelphia] : Weitzel ; Valley Forge : may be purchased through Rochambeau Press, c1979.
viii, 279 p. : ill., map ; 24 cm.
One map inserted in pocket.
Includes bibliographical references.
$13.95
1. Chester Co., Pa.—History—Revolution, 1775-1783—Fiction. I. Title.
PZ4.W438 Lan 813'.5'4 79-113837
[PS3573.E432] 79 MARC

CHESTER MENTAL HEALTH CENTER (ILL.)

Illinois. General Assembly. Commission to Visit and Examine State Institutions. Charitable Subcommittee.
Report of the Legislative Commission to Visit and Examine State Institutions / Charitable Subcommittee. — [Springfield] : The Subcommittee, 1978.
22 leaves ; 28 cm.
Cover title.
1. Manteno Mental Health Center (Ill.) 2. Chester Mental Health Center (Ill.) 3. Psychiatric hospitals—Illinois. I. Title.
RC445.I3M364 362.2'1'09773 79-620649
79 MARC

CHESTER PLAYS

Mathews, Godfrey W
The Chester mystery plays / by Godfrey W. Mathews. — Norwood, Pa. : Norwood Editions, 1978.
85 p. ; 23 cm.
Reprint of the 1925 ed. published by E. Howell, Liverpool.
Includes bibliographical references.
ISBN 0-8482-4970-4 lib. bdg. : $12.50
1. Chester plays. 2. Mysteries and miracle-plays, English—History and criticism. I. Title.
[PR644.C4M3 1978] 822'.1 78-18701
78 MARC

CHESTERFIELD, PHILIP DORMER STANHOPE, 4TH EARL OF, 1694-1773

Chesterfield, Philip Dormer Stanhope, 4th earl of, 1694-1773.
Some unpublished letters of Lord Chesterfield / with an introd. by Sidney L. Gulick, Jr. — Norwood, Pa. : Norwood Editions, 1978 [c1937]
84 p. : port. ; 22 cm.
Reprint of the ed. published by University of California Press, Berkeley.
Includes bibliographical references and index.
ISBN 0-8482-0940-0 lib. bdg. : $25.00
1. Young men—Conduct of life. 2. Chesterfield, Philip Dormer Stanhope, 4th Earl of, 1694-1773. I. Gulick, Sidney Lewis, 1902- II. Title.
[BJ1671.C53 1978] 170'.202'232 78-3442
78 MARC

Coxon Roger.
Chesterfield and his critics / by Roger Coxon. — Philadelphia : R. West, 1978.
xii, 328 p. ; 26 cm.
Reprint of the 1925 ed. published by G. Routledge, London.
Includes bibliographical references and index.
ISBN 0-8492-3838-2 : $35.00
1. Chesterfield, Philip Dormer Stanhope, 4th Earl of, 1694-1773. 2. Statesmen—Great Britain—Biography. 3. Authors, English—18th century—Biography. I. Title.
[DA501.C5C6 1977b] 941.07'092'4 78-4570
78 MARC

CHESTERFIELD, ENG.

—GENEALOGY

Lichfield Joint Record Office.
Chesterfield wills and inventories, 1521-1603 / edited by J. M. Bestall and D. V. Fowkes ; with an introduction by David Hey ; a glossary by Rosemary Milward and an index by Barbara Bestall. — Matlock : Derbyshire Record Society, 1977.
xxxiv,348 p. : maps ; 21 cm. — (Publications) - Derbyshire Record Society ; v. 1)
GB78-00606
Catalog of wills and inventories housed in the Lichfield Joint Record Office.
Includes bibliographical references and index.
ISBN 0-9505940-0-8 : £12.00
1. Chesterfield, Eng.—History—Sources. 2. Chesterfield, Eng.—Genealogy. 3. Wills—England—Chesterfield. 4. Inventories of decedents' estates—England—Chesterfield. I. Bestall, J. M. II. Fowkes, Dudley Vincent. III. Title. IV. Series: Derbyshire Record Society. Publications - Derbyshire Record Society ; v. 1.
DA690.C51L52 1977 929'.3425'12 79-303575
79 MARC

—HISTORY—SOURCES

Lichfield Joint Record Office.
Chesterfield wills and inventories, 1521-1603 / edited by J. M. Bestall and D. V. Fowkes ; with an introduction by David Hey ; a glossary by Rosemary Milward and an index by Barbara Bestall. — Matlock : Derbyshire Record Society, 1977.
xxxiv,348 p. : maps ; 21 cm. — (Publications) - Derbyshire Record Society ; v. 1)
GB78-00606
Catalog of wills and inventories housed in the Lichfield Joint Record Office.
Includes bibliographical references and index.
ISBN 0-9505940-0-8 : £12.00
1. Chesterfield, Eng.—History—Sources. 2. Chesterfield, Eng.—Genealogy. 3. Wills—England—Chesterfield. 4. Inventories of decedents' estates—England—Chesterfield. I. Bestall, J. M. II. Fowkes, Dudley Vincent. III. Title. IV. Series: Derbyshire Record Society. Publications - Derbyshire Record Society ; v. 1.
DA690.C51L52 1977 929'.3425'12 79-303575
79 MARC

CHESTERFIELD, MASS.

—BIOGRAPHY

Revolutionary War pension applications of Jesse and Katherine Willcutt. — Livonia, Mich. : Flash Productions, 1975.
6 p. : facsims. ; 28 cm.
Caption title.
Edited by G. A. Calder.
1. Willcutt, Jesse, 1752-1834. 2. United States. Army. Continental Army—Biography. 3. Chesterfield, Mass.—Biography. 4. Willcutt, Katherine. 5. Pensions, Military—United States—Revolution, 1775-1783. I. Willcutt, Jesse, 1752-1834. II. Willcutt, Katherine. III. Calder, George Alexander, 1937-
E263.M4W457 973.3'44'40924 79-312560
79 MARC

CHESTERFIELD CO., VA.

—GENEALOGY

Weisiger, Benjamin B
Chesterfield County, Virginia wills, 1749-1774 / abstracted and compiled by Benjamin B. Weisiger, III. — [Richmond] : Weisiger, c1979.
213 p. ; 28 cm.
Includes index.
1. Wills—Virginia—Chesterfield Co. 2. Chesterfield Co., Va.—Genealogy. I. Title.
F232.C52W44 929'.3755'594 79-102701
79 MARC

CHESTERTON, GILBERT KEITH, 1874-1936

Las Vergnas, Raymond, 1902-
Chesterton, Belloc, Baring / by Raymond Las Vergnas ; translated by C. C. Martindale. — Philadelphia : R. West, 1977.
vi, 153 p. ; 26 cm.
Translation of Portraits anglais.
Reprint of the 1938 ed. published by Sheed & Ward, London.
Includes bibliographical references.
ISBN 0-8492-1516-1 lib. bdg. : $17.50
1. English literature—Catholic authors—History and criticism. 2. Chesterton, Gilbert Keith, 1874-1936. 3. Belloc, Hilaire, 1870-1953. 4. Baring, Maurice, 1874-1945. 5. Authors, English—20th century—Biography. I. Title.
[PR120.C3L32 1977] 820'.9'00912 77-21176
77 MARC

—BIOGRAPHY

Titterton, William Richard.
G. K. Chesterton : a portrait / by W. R. Titterton. — Norwood, Pa. : Norwood Editions, 1978.
235 p. : port. ; 23 cm.
Reprint of the 1947 ed. published by D. Organ, London.
ISBN 0-8482-2646-1 : $20.00
1. Chesterton, Gilbert Keith, 1874-1936—Biography. 2. Authors, English—20th century—Biography. I. Title.
[PR4453.C4Z77 1978] 828'.9'1209 78-3886
78 MARC

—CRITICISM AND INTERPRETATION

Hunter, Lynette.
G. K. Chesterton : explorations in allegory / Lynette Hunter. — New York : St. Martin's Press, 1979.
190 p. ; 23 cm.
Bibliography: p. 179-181.
ISBN 0-312-31492-2 : $18.50
1. Chesterton, Gilbert Keith, 1874-1936—Criticism and interpretation. 2. Allegory.
PR4453.C4Z654 1979 828'.9'1209 79-4643
79 MARC

West, Julius, 1891-1918.
G. K. Chesterton : a critical study / by Julius West. — Norwood, Pa. : Norwood Editions, 1978.
190, [1] p. ; 23 cm.
Reprint of the 1915 ed. published by M. Secker, London.
Bibliography: p. 185-[191]
ISBN 0-8482-2944-4 lib. bdg. : $20.00
1. Chesterton, Gilbert Keith, 1874-1936—Criticism and interpretation.
[PR4453.C4Z85 1978] 828'.9'1209 78-79
78 MARC

—RELIGION AND ETHICS

Denis, Yves.
G. K. Chesterton : paradoxe et catholicisme / Yves Denis. — Paris : Belles Lettres, 1978.
479 p. ; 24 cm. F***
Cover title.
Bibliography: p. 467-478.
Includes index.
80.00F
1. Chesterton, Gilbert Keith, 1874-1936—Religion and ethics. 2. Paradox in literature. 3. Catholic Church—Doctrinal and controversial works.
PR4453.C4Z589 828'.9'1209 79-348582
79 MARC

CHESTS

—ESTONIA

Eesti NSV Riiklik Etnograafiamuuseum.
Eesti rahvapärased kirstud ja vakad / koostanud Tiina Võti. — Tallinn : Valgus, 1974.
71 p., [27] leaves of plates ; 23 cm. USSR***
Summary in German and Russian.
0.62rub
1. Chests—Estonia. 2. Boxes, Ornamental—Estonia. I. Võti, T. II. Title.
NK2725.E3 1974 79-370886
79 MARC

—KOREA

Wickman, Michael.
Korean chests : treasures of the Yi Dynasty / by Michael Wickman. — Seoul : Seoul International Tourist Pub. Co., 1978.
94 p., [1] leaf of plates : ill. (some col.) ; 21 cm.
1. Chests—Korea. 2. Decoration and ornament—Korea—Yi dynasty, 1392-1910. I. Title.
NK2725.W5 749'.3 79-100363
79 MARC

—PENNSYLVANIA

Fabian, Monroe H
The Pennsylvania-German decorated chest / Monroe H. Fabian ; foreword by Frederick S. Weiser. — 1st ed. — New York : Universe Books, c1978.
230 p. : ill. (some col.) ; 26 cm. — ([Publications of the Pennsylvania German Society] ; v. 12)
"A Main Street Press book."
Bibliography: p. 222-225.
Includes index.
ISBN 0-87663-310-6 : $25.00
1. Chests—Pennsylvania. 2. Painted country furniture, Pennsylvania German. I. Title. II. Series: Pennsylvania-German Society. Publications ; 12.
GR110.P4A372 vol. 12 974.8'004'31 s 77-91892
[NK2435.P4] 78 MARC

CHETTY, R. K. SHANMUKHAM, SIR, 1892-

Cettiyār, Pi. Es 1906-
Ariñar Ār. Kē. Es. / Pi. Es. Ceṭṭiyār. — 2. patippu, — Cennai : Kāmanvelt Paplisin Kampeni, 1954.
112 p. ; 19 cm.
In Tamil.
1. Chetty, R. K. Shanmukham, Sir, 1892- 2. Cabinet officers—India—Biography. I. Title.
DS481.C463C47 1954 78-912858
79 MARC

CHEVALLEY GROUPS

Lusztig, George, 1946-
Representations of finite Chevalley groups / by G. Lusztig. — Providence : Published for the Conference Board of the Mathematical Sciences by the American Mathematical Society, c1978.
v, 48 p. ; 26 cm. — (Regional conference series in mathematics ; no. 39)
"Expository lectures from the CBMS regional conference held at Madison, Wisconsin, August 8-12, 1977."
Bibliography: p. [47]-48.
ISBN 0-8218-1689-6
1. Chevalley groups. 2. Finite groups. 3. Representations of groups. I. Conference Board of the Mathematical Science. II. Title. III. Series.
QA1.R33 no. 39 510'.8 s 78-24068
[QA171] 78 MARC

CHEVELLE AUTOMOBILE

General Motors Corporation. Chevrolet Motor Division.
1969 Chevrolet, Chevelle, Camaro, Chevy Nova, and Corvette chassis service manual, and Corvette body service manual. — Detroit : Chevrolet Motor Division, General Motors Corp., c1968.
615 p. in various pagings : ill. ; 28 cm.
1. Chevrolet automobile. 2. Chevelle automobile. 3. Camaro automobile. 4. Nova automobile. 5. Corvette automobile. I. Title: 1969 Chevrolet ...
TL215.C5G25 1968a 629.28'7'22 79-301312
79 MARC

General Motors Corporation. Chevrolet Motor Division.
1969 owner's manual : operation & maintenance instructions for Chevelle, Chevy Nova, Camaro. — 1st ed. — Detroit : Chevrolet Motor Division, General Motors Corporation, 1968.
80 p. : ill.; 14 x 21 cm.
Cover title: 1969 Camaro, Chevelle, Chevy Nova owner's manual.
Includes index.
1. Camaro automobile. 2. Chevelle automobile. 3. Nova automobile. I. Title. II. Title: 1969 Camaro, Chevelle, Chevy Nova owner's manual.
TL215.C33G46 1968 629.22'22 79-303923
79 MARC

General Motors Corportion. Chevrolet Motor Division.
Passenger cars & light duty trucks unit repair manual (overhaul) : 1978 Chevrolet. — [Detroit? : Chevrolet Motors Division, General Motors Corp., c1978]
415 p. in various pagings : ill. ; 28 cm.
Cover title.
Title on spine: 1978 passenger cars & light duty truck unit repair manual.
1. Chevrolet automobile. I. Title. II. Title: 1978 passenger cars & light duty truck unit repair manual.
TL215.C5G25 1978a 629.28'7'22 79-111316
00 MARC

CHEVEREV, ALEKSANDR MIKHAĬLOVICH

(Zheleznye kombrigi)
Железные комбриги : [Докум.-биогр. повести о С. Вострецове, М. Калмыкове, А. Чевереве]. — Уфа : Башк. кн. изд-во, 1978.
332 p. : ill. ; 17 cm. USSR 78
Includes bibliographical references.
0.50rub
1. Russia—History—Revolution, 1917-1921—Biography. 2. Vostretsov, Stepan Sergeevich, 1883-1932. 3. Kalmykov, Mikhail Vasil'evich. 4. Cheverev, Aleksandr Mikhaĭlovich. 5. Bashkiria—Biography.
DK253.Z45 79-413818

CHEVETTE AUTOMOBILE

Chilton Book Company. Automotive Editorial Dept.
Chilton's repair and tune-up guide, Chevette, 1976-77 / prepared by the Automotive Editorial Department, Chilton Book Company. — 1st ed. — Radnor, Pa. : The Company, c1977.
iv, 146 p. : ill. 26 cm.
ISBN 0-8019-6510-1. ISBN 0-8019-6511-X pbk.
1. Chevette automobile. I. Title. II. Title: Repair and tune-up guide, Chevette, 1976-77.
TL215.C49C47 1977 629.28'7'22 76-51724
76[r79]rev MARC

General Motors Corporation. Chevrolet Motor Division.
1976 Chevrolet Chevette shop manual. — Detroit : Chevrolet Motors Division, General Motors Corp., c1976.
500 p. in various pagings : ill. ; 28 cm.
Caption title.
On spine: 1976 Chevette shop manual
"ST-357-76."
1. Chevette automobile. I. Title. II. Title: 1976 Chevette shop manual.
TL215.C49G46 1976 629.28'7'22 78-322581
79 MARC

Jorgensen, Eric.
Chevette, service, repair handbook : all models, 1976-1978 / by Eric Jorgensen ; Jeff Robinson, editor and publisher ; [cover photo by Mike Brown]. — 1st ed. — Los Angeles : Clymer Publications, 1978.
xii, 169 p. : ill. ; 26 cm.
Includes index.
ISBN 0-89287-145-8
1. Chevette automobile. I. Title.
TL215.C48J67 629.28'7'22 78-108298
78 MARC

CHEVIOT HILLS, ENG.

—DESCRIPTION AND TRAVEL

Graham, Frank, 1913-
Wooler, Ford, Chillingham & the Cheviots / by Frank Graham. — Newcastle upon Tyne : Graham, 1976.
48 p. : ill., facsim., maps ; 22 cm. GB76-31407
ISBN 0-85986-066-7 : £0.60
1. Northumberland, Eng.—Description and travel. 2. Cheviot Hills, Eng.—Description and travel. I. Title.
DA670.N8G69 942.8'8 79-300482
79 MARC

CHEVRIER, GENEVIÈVE

—ADDRESSES, ESSAYS, LECTURES

Hommage à Geneviève Chevrier et Alain Geslan : études médiévales / Groupe d'archéologie médiévale, Centre de recherches historiques, École des hautes études en sciences sociales ; [réalisé en collaboration avec Joëlle Burnouf et al.]. — [Strasbourg] : Centre d'archéologie médiévale de Strasbourg, 1975.
146 p. : ill., geneal. table, maps ; 28 cm. — (Cahier - Chantiers d'études médiévales ; no 13) F***
French or Italian.
Includes bibliographies and index.
1. Middle Ages—History—Addresses, essays, lectures. 2. Europe—Antiquities—Addresses, essays, lectures. 3. Chevrier, Geneviève—Addresses, essays, lectures. 4. Geslan, Alain—Addresses, essays, lectures. I. Chevrier, Geneviève. II. Geslan, Alain. III. Burnouf, Joëlle. IV. École des hautes études en sciences sociales. Groupe d'archéologie médiéale. V. Centre d'archéologie médiévale de Strasbourg. VI. Title: Études médiévales. VII. Series: Chantiers d'études médiévales (Association). Cahier - Chantiers d'études médiévales ; no 13.
D119.H64 940.1 78-387767
78 MARC

CHEVROLET AUTOMOBILE

see also Chevrolet trucks;
Corvette automobile

Carroll, William, 1915-
Chevrolet V8 performance guide. — [3d rev. ed.]. — [Oceanside, Calif. : Auto Book Press, 1969]
153 p. : ill. ; 29 cm. — (Bill Carroll's performance engineering handbooks)
Cover title.
"Book 104."
Includes index.
1. Chevrolet automobile. I. Title.
TL215.C5C3 1969 629.2'504 79-306482
79 MARC

Chevrolet tune-up & repair / [edited by Spence Murray and the staff of Specialty Publications Division]. — Los Angeles : Petersen Pub. Co., c1978.
192 p. : ill. ; 28 cm.
Cover title: Petersen's chevrolet tune-up & repair.
ISBN 0-8227-5037-6 : $4.95 ($5.95 Can)
1. Chevrolet automobile. I. Murray, Spencer. II. Petersen Publishing Company. Specialty Publications Division. III. Title: Petersen's chevrolet tune-up & repair.
TL215.C5C48 629.28'7'22 78-62361
78 MARC

General Motors Corporation. Chevrolet Motor Division.
1962 passenger, Chevy II, Corvair, and truck accessories installation reference manual. — [Detroit] : Chevrolet Motor Division, General Motors Corp., c1961.
461 p. : ill. ; 28 cm.
Includes index.
1. Chevrolet automobile. 2. Automobiles—Apparatus and supplies. I. Title.
TL215.C5G25 1961 629.2'7 78-326896
79 MARC

CHEVROLET AUTOMOBILE (Continued)

General Motors Corporation. Chevrolet Motor Division.
1969 Chevrolet, Chevelle, Camaro, Chevy Nova, and Corvette chassis service manual, and Corvette body service manual. — Detroit : Chevrolet Motor Division, General Motors Corp., c1968.
615 p. in various pagings : ill. ; 28 cm.
1. Chevrolet automobile. 2. Chevelle automobile. 3. Camaro automobile. 4. Nova automobile. 5. Corvette automobile. I. Title: 1969 Chevrolet ...
TL215.C5G25 1968a 629.28'7'22 79-301312
79 MARC

General Motors Corporation. Chevrolet Motor Division.
1976 passenger car chassis service and overhaul manual supplement. — Detroit : Chevrolet Motor Division, General Motors Corp., c1976.
423 p. in various pagings : ill. ; 28 cm.
Cover title: 1976 Chevrolet service and overhaul manual supplement.
1. Chevrolet automobile. I. Title. II. Title: 1976 Chevrolet service and overhaul manual supplement.
TL215.C5G25 1976c 629.28'7'22 78-325163
79 MARC

General Motors Corporation. Chevrolet Motor Division.
1977 Chevrolet advance information ship manual : passenger cars and light duty trucks. — ⌈Detroit : Chevrolet Motor Division, General Motors, Corp., c1976⌉
ca. 400 p. in various pagings : ill. ; 28 cm.
Cover title.
On spine: 1977 advance information shop manual.
"ST-356-77."
1. Chevrolet automobile. I. Title.
TL215.C5G25 1976e 629.28'7'22 79-108409
79 MARC

General Motors Corporation. Chevrolet Motor Division.
1977 Chevrolet passenger cars & light duty trucks : unit repair manual (overhaul). — ⌈Detroit : General Motors Corp., c1976⌉
ca. 400 p. in various pagings : ill. ; 28 cm.
Cover title.
"St 333-77."
1. Chevrolet automobile. 2. Chevrolet trucks. I. Title.
TL215.C5G25 1976d 629.28'7'22 78-112822
79 MARC

Vizard, David.
How to rebuild your small-block Chevy : how to find out what needs fixing, how to remove engine, strip & inspect for wear, repair and recondition, assemble, install, and tune up / by David Vizard. — ⌈Tucson, AZ⌉ : H.P. Books, ⌈c1978⌉
168 p. : ill. ; 28 cm.
Cover title.
Includes index.
ISBN 0-912656-66-2 : $5.95
1. Automobiles—Motors—Maintenance and repair. 2. Chevrolet automobile. I. Title.
TL210.V57 629.2'504'028 78-52275
79 MARC

CHEVROLET LUV TRUCK

General Motors Corporation. Chevrolet Motor Division.
Shop manual, Chevrolet-LUV (series 5). — Detroit : Chevrolet Motor Division, General Motors Corp., c1976.
ca. 250 p. in various pagings : ill. ; 28 cm.
On spine: Chevrolet-LUV series 5 shop manual.
Running title: Chevrolet-LUV shop manual.
"ST 351-76."
1. Chevrolet LUV truck. I. Title. II. Title: Chevrolet-LUV series 5 shop manual. III. Title: Chevrolet-LUV shop manual.
TL230.5.C45G36 1976d 629.28'7'3 78-325521
79 MARC

Scott, Ed.
Chevy Luv service-repair handbook, 1972-1978 / by Ed Scott ; Eric Jorgensen, editor ; Jeff Robinson, publisher. — 1st ed. — Arleta, Calif. : Clymer Publications, 1978.
244 p. : ill. ; 26 cm.
Includes index.
ISBN 0-89287-274-8
1. Chevrolet LUV trucks. I. Jorgensen, Eric. II. Clymer Publications. III. Title.
TL230.5.C45S36 629.28'7'22 78-112734
79 MARC

CHEVROLET TRUCKS

Bishop, Mike.
Chevy & GMC 4-wheel drive, 1967-1979 : tune-up, maintenance / by Mike Bishop ; Eric Jorgensen, editor. — 3d ed., rev. / by Mike Bishop to include 1978-1979 models. — Arleta, Calif. : Clymer Publications, c1979.
171 p. : ill. ; 26 cm.
Includes index.
ISBN 0-89287-159-8
1. Chevrolet trucks. 2. GMC trucks. I. Jorgensen, Eric. II. Clymer Publications. III. Title.
TL2305.C45B585 1979 629.28'7'3 79-115750
00 MARC

Bishop, Mike.
Chevy & GMC pickups, 1967-1968, includes suburbans : shop manual / by Mike Bishop ; Eric Jorgensen, editor. — 1st ed. — Arleta, Calif. : Clymer Publications, c1979.
xiv, 265 p., ⌈21⌉ leaves of plates : ill. ; 26 cm.
Includes index.
ISBN 0-89287-301-9
1. Chevrolet trucks. 2. GMC trucks. I. Jorgensen, Eric. II. Clymer Publications. III. Title.
TL230.5.C45B588 629.28'7'3 79-115775
00 MARC

Bishop, Mike.
Chevy & GMC vans : 1967-1978 shop manual / by Mike Bishop ; Eric Jorgensen, editor. — 1st ed. — Arleta, Calif. : Clymer Publications, c1979.
xiv, 265 p., ⌈15⌉ leaves of plates : ill. ; 26 cm.
Includes index.
ISBN 0-89287-300-0
1. Chevrolet trucks. 2. GMC trucks. I. Jorgensen, Eric. II. Clymer Publications. III. Title.
TL230.5.C45B59 629.28'7'3 79-108495
79 MARC

Chevy/GMC pickup repair / ⌈edited by Spence Murray and the staff of the Speciality Publications Division ; photos. by Eric Richman ; cover design by Dick Fischer⌉. — Los Angeles : Petersen Pub. Co., c1978.
ca. 200 p. : ill. ; 28 cm.
Cover title: Petersen's Chevy/GMC pickup repair.
ISBN 0-8227-5038-4 : $4.95 ($5.95 Can)
1. Chevrolet trucks. 2. GMC trucks. I. Murray, Spencer. II. Petersen Publishing Company. Specialty Publications Division. III. Title: Petersen's Chevy/GMC pickup repair.
TL230.5.C45C45 629.28'7'3 78-61768
78 MARC

General Motors Corporation. Chevrolet Motor Division.
1963 Chevrolet truck shop manual. — Detroit, Mich. : Chevrolet Motor Division, General Motors Corp., c1962.
1065 p. in various pagings : ill. ; 28 cm.
"ST 22."
1. Chevrolet trucks. I. Title.
TL230.5.C45G36 1962 629.28'7'4 78-322145
79 MARC

General Motors Corporation. Chevrolet Motor Division.
1965 Chevrolet truck shop manual : supplement. — Detroit : Chevrolet Motor Division, General Motors Corp., c1964.
348 p. in various pagings : ill. ; 28 cm.
1. Chevrolet trucks. I. Title.
TL230.5.C45G36 1964 629.28'7'4 78-325513
79 MARC

General Motors Corporation. Chevrolet Motor Division.
1977 Chevrolet passenger cars & light duty trucks : unit repair manual (overhaul). — ⌈Detroit : General Motors Corp., c1976⌉
ca. 400 p. in various pagings : ill. ; 28 cm.
Cover title.
"St 333-77."
1. Chevrolet automobile. 2. Chevrolet trucks. I. Title.
TL215.C5G25 1976d 629.28'7'22 78-112822
79 MARC

CHEWA (AFRICAN TRIBE)

Ntara, Samuel Yosia, 1904 or 5-
The history of the Chewa (Mbiri ya Achewa) Translated into English by W. S. Kamphandira Jere with comments by Harry W. Langworthy. Edited by Beatrix Heintze. Wiesbaden, F. Steiner, 1973.
xx, 167 p. maps. 24 cm. (Studien zur Kulturkunde, Bd. 31)
Translation with revisions of Mbiri ya Achewa.
Bibliography: p. 157-159.
1. Chewa (African tribe) I. Langworthy, Harry W. II. Heintze, Beatrix, ed. III. Title. IV. Series.
DT864.N8213 968.9'7'004963 73-172940
73⌈r79⌉rev MARC

CHEWING see Mastication

CHEWNING, BILL

Chewning, Bill.
Outward Bound, east Africa : the Loitokitok experience : a pictorial account of an Outward Bound expedition in east Africa / Bill Chewning. — ⌈Leesburg, Va. : LAS Publications, 1975⌉
vii, 80 p. : ill. ; 23 cm.
Cover title.
$5.00
1. Outward Bound Mountain School of Kenya. 2. Africa, East—Description and travel—1951- . 3. Chewning, Bill. I. Title.
GV199.44.K462L633 639'.95'09676 75-18003
79⌈r79⌉rev MARC

CHEYENNE INDIANS

Marquis, Thomas Bailey, 1869-1935.
The Cheyennes of Montana / Thomas B. Marquis ; edited, with an introd. and a biography of the author by Thomas D. Weist. — Algonac, Mich. : Reference Publications, 1978.
297 p. : ill. ; 24 cm.
"The five narratives included herein were published earlier in a slightly different version as Cheyenne and Sioux, the reminiscences of four Indians and a white soldier."
Bibliography: p. 278-283.
Includes index.
ISBN 0-917256-04-2 : $19.50
1. Cheyenne Indians. I. Weist, Thomas D. II. Title.
E99.C53M36 1978 970'.004'97 78-59715
78 MARC

—HISTORY

May, Robin.
Warriors of the West : the story of the fighting Sioux and Cheyennes / ⌈by⌉ Robin May ; illustrated by Harry Bishop ; portraits by Tom Stimpson. — London : Independent Television Books Ltd. ; London : Arrow Books, 1978.
xi, 116 p. : ill., map, ports. ; 18 cm. — (Look-in books) GB78-3347
Includes index.
ISBN 0-09-918030-8 : £0.60
1. Cheyenne Indians—History. 2. Dakota Indians—History. 3. Indians of North America—Wars—1866-1895. I. Title.
E99.C53M39 973.8 79-303727
79 MARC

—MISSIONS

Carter, Joseph LeRoy.
Dian takes to the Indians / by Joseph LeRoy Carter. — ⌈El Reno, OK : Heritage Color Press⌉, 1979.
105 p., ⌈2⌉ leaves of plates : ill. ; 22 cm.
1. Meschberger, Dian Lugenbuehl, 1858-1954. 2. Cheyenne Indians—Missions. 3. Arapaho Indians—Missions. 4. Indians of North America—Oklahoma—El Reno region—Missions. 5. El Reno region, Okla.—Biography. 6. Missionaries—Oklahoma—El Reno region—Biography. I. Title.
E99.C53C37 266'.9'70924 79-84729
79 MARC

—SOCIAL CONDITIONS

Oklahoma. Human Rights Commission.
Report of the April 26, 1977 hearing on Indian civil rights issues in Northwestern Oklahoma. — Oklahoma City : Oklahoma Human Rights Commission, ⌈1977 or 1978⌉
i, 50 p. ; 28 cm.
Includes bibliographies.
1. Indians of North America—Oklahoma—Social conditions. 2. Cheyenne Indians—Social conditions. 3. Arapaho Indians—Social conditions. 4. Indians of North America—Oklahoma—Civil rights. I. Title.
E78.O45O367 1977 301.45'19'70766 78-623622
79 MARC

—WARS—1868-1869

see also Summit Springs, Colo., Battle of, 1869

—WARS, 1876

see also Little Big Horn, Battle of the 1876

Bourke, John Gregory, 1846-1896.
Mackenzie's last fight with the Cheyennes: a winter campaign in Wyoming and Montana. By Captain John G. Bourke ... Governor's Island, N.Y.H., 1890.
cover-title, 44 p. front. (port.) 25 cm.
Reprinted from Journal Military service institution.
LC copy replaced by microfilm.
1. Cheyenne Indians—Wars, 1876. 2. Mackenzie, Ranald Slidell, 1840-1889. 3. United States. Army. Cavalry—Biography. 4. Soldiers—United States—Biography. I. Title.
⌈E83.866.B76⌉ 973.8'2 2-16598
MARC

Rizzi, Leo.
Hoka hey! : l'ultima guerra indiana / Leo Rizzi. — 1. ed. — Milano : Syntesis press, 1978.
121 p. ; 21 cm. It78-Dec
Includes bibliographical references.
L4500
1. Dakota Indians—Wars, 1876. 2. Cheyenne Indians—Wars, 1876. 3. Indians of North America—Wars—1866-1895. I. Title.
E83.876.R57 973.8'2 79-373171
•79 MARC

CHEYENNE LANGUAGE

—GRAMMAR

Leman, Wayne.
Cheyenne grammar notes / by Wayne Leman. — Lame Deer, Mont. : Northern Cheyenne Bilingual Education Program, 1979.
vii, 227 p. ; 28 cm.
Bibliography: p. 227.
1. Cheyenne language—Grammar.
PM795.L4 497'.3 79-115104
00 MARC

CHHATTISGARH, INDIA

—ECONOMIC CONDITIONS

Choubey, H K
Bhilai steel project & its impact on the towns of Chhattisgarh / H. K. Choubey. — 1st ed. — Secunderabad : Indian Institute of Geography, 1975.
xi, 241 p., [4] fold. leaves of plates : ill., maps ; 22 cm. — (Social sciences research publications series ; 1)
Bibliography: p. 238-241.
Rs40.00 ($10.00 U.S.)
1. Steel industry and trade—India—Chhattisgarh. 2. Hindustan Steel, ltd. Bhilai Steel Plant. 3. Chhattisgarh, India—Economic conditions. I. Title. II. Series.
HD9526.I63C473 338.4'7'67209543 78-904363
79 MARC

CHHATTISGARH REGION, INDIA

—BIOGRAPHY

Śarmā, Kamalākānta, 1941-
Chattīsagaṛha ke svātantrya saṅgrāma senāniyoṃ kā paricaya / Kamalākānta Śarmā. — Durga : Jijñāsā Prakāśana, 1973-
v. : ports. ; 22 cm.
In Hindi.
Rs3.00 (v. 1)
1. Chhattisgarh region, India—Biography. 2. Chhattisgarh region, India—Politics and government. I. Title.
DS486.C467S27 79-902547
79 MARC

—POLITICS AND GOVERNMENT

Śarmā, Kamalākānta, 1941-
Chattīsagaṛha ke svātantrya saṅgrāma senāniyoṃ kā paricaya / Kamalākānta Śarmā. — Durga : Jijñāsā Prakāśana, 1973-
v. : ports. ; 22 cm.
In Hindi.
Rs3.00 (v. 1)
1. Chhattisgarh region, India—Biography. 2. Chhattisgarh region, India—Politics and government. I. Title.
DS486.C467S27 79-902547
79 MARC

CHHATTISGARHI POETRY

Gupta, Lakhanalāla.
Sañjhauti ke berā / Lakhana Lāla Gupta. — 1. saṃskaraṇa. — Rāyapura : Lekhaka Sahayogī Prakāśana, 1968.
22 p. ; 19 cm.
In Chhattisgarhi.
Poems.
I. Title.
PK1959.9.G8S2 79-901613
79 MARC

CHHOTU RAM, CHAUDHRI, SIR, RAI BAHADUR

Agnihotri, H L
A profile in courage : a biography of Ch. Chhotu Ram / H. L. Agnihotri, Shiva N. Malik. — 1st ed. — New Delhi : Light & Life Publishers, 1978.
144 p., [5] leaves of plates : ill., geneal. table ; 22 cm.
Includes index.
Rs40.00
1. Chhotu Ram, Chaudhri, Sir, Rai Bahadur. 2. Peasantry—India—Biography. I. Malik, Shiva N., joint author. II. Title.
HD1537.I4A63 78-905954
79 MARC

CH'I, PAI-SHIH, 1863-1957

Ch'i, Pai-shih, 1863–1957.
(Ch'i Huang hua kuai fo)
齊璜畫怪佛 / [齊白石]. — [蘇州? : s. n., 1977?]
Broadside ; 71 x 40 cm. fold. to 27 x 14 cm.
Title from label on envelope.
In case.
Rubbing.
"蘇州工藝"
1. Ch'i, Pai-shih, 1863–1957. I. Title.
ND1049.C5A4 1977 79-840068

Ch'i, Pai-shih, 1863–1957.
(Pai-shih shan weng yin ts'un)
白石山翁印存 / 齊白石著. — 香港 : 宏圖出版社, [1978?]
50 p. : all ill. ; 19 cm.
$4.60
1. Ch'i, Pai-shih, 1863–1957. I. Title.
CD6174.5.C48A35 78-841222
C

Chiang, Hsün.
(Ch'i Pai-shih)
齊白石 : 中國文人畫最後的奇葩 / 蔣勳著. — 臺北 : 雄獅圖書公司, 民國67 [1978]
118 p. : ill. (some col.) ; 22 cm. — (中國畫家叢書 ; 3)
Series romanized: Chung-kuo hua chia ts'ung shu.
"齊白石作品選輯": p. 9–72.
NT$70.00
1. Ch'i, Pai-shih, 1863–1957. 2. Painters—China—Biography. I. Ch'i, Pai-shih, 1863–1957. Ch'i Pai-shih tso p'in hsüan chi. 1978. II. Title. III. Title: Chung-kuo wen jen hua tsui hou ti ch'i p'a.
ND1049.C5C45 79–840504
C

Kim, Yŏng-gi, 1911 (Nov. 30)–
(Chae Paek-sŏk)
齊白石 : 近代中國美術史의巨木白石老人의人間과藝術 / 金晴江 [永基] 編著. — 서울 : 悅話堂, 1978.
170 p., [1] leaf of plates : ill. (some col.) ; 22 cm. — (悅話堂美術選書 ; 13)
Bibliography: p. 31.
₩2200
1. Ch'i, Pai-shih, 1863–1957. I. Title. II. Series: Yŏrhwadang misul sŏnsŏ ; 13.
ND1049.C5K55 79–825151
79 K

CHI, TONG-SIK, 1910-

Chi, Tong-sik, 1910–
(Sinhak ŭi osolkil)
神學의오솔길 / 池東植著. — 서울 : 大韓基督教書會, 1976.
223 p. ; 18 cm. — (現代新書 ; 70)
Series romanized: Hyŏndae sinsŏ.
W500
1. Chi, Tong-sik, 1910– 2. Kidokkyo Taehan Pogŭm Kyohoe—Clergy—Biography. 3. Clergy—Korea—Biography. 4. Theologians—Korea—Biography. I. Title.
BX7990.K538C463 79–826288
K

CHI, YÜN, 1724-1805. SHIH T'UNG HSÜEH FAN

Gagnon, Guy.
(Shih t'ung yü Shih t'ung hsüeh fan t'ung chien)
史通與史通削繁通檢 / [par Guy Gagnon ; avec la collaboration de Emi Gagnon ... et at.]. — Paris : 法蘭西學院漢學研究所, 1977.
2 v. ; 26 cm. — (法蘭西學院漢學研究所漢學通檢提要文獻叢刊 ; 7)
Added t. p.: Concordance combinée du Shitong et du Shitong xiaofan.
Prefatory matter also in French.
Series romanized: Fa-lan-hsi hsüeh yüan Han hsüeh yen chiu so Han hsüeh t'ung chien t'i yao wen hsien ts'ung k'an.
"我們選擇了...四部備要...的通史削繁版本...並以國學基本叢書為藍本"
Includes index (16 p. inserted)
ISBN 2-85757-008-2
1. Liu, Chih-chi, 661–721. Shih t'ung—Concordances. 2. Chi, Yün, 1724–1805. Shih t'ung hsüeh fan—Concordances. 3. China—Historiography—Indexes. I. Gagnon, Emi. II. Liu, Chih-chi, 661–721. Shih t'ung. 1977. III. Chi, Yün, 1724–1805. Shih t'ung hsüeh fan. 1977. IV. Title. V. Title: Concordance combinée du Shitong et du Shitong xiaofan.
DS734.7.L5822G33 79–840253

CHI, YÜN, 1724-1805. YÜEH WEI TS'AO T'ANG PI CHI

Lai, Fang-ling.
(Yüeh wei ts'ao t'ang pi chi chung ti kuan nien shih chieh chi yüan liu ying hsiang)
閱微草堂筆記中的觀念世界及源流影響 / 賴芳伶. — [台北? : s. n.], 民國65 [1976]
76, [1] double leaves ; 27 cm.
"國立台灣大學中文研究所碩士論文"
Reproduced from the typescript.
Bibliography: leaves 75–[77]
1. Chi, Yün, 1724–1805. Yüeh wei ts'ao t'ang pi chi. I. Title.
PL2705.I17Y835 78–843900

CHI, YÜN, 1724-1805

Liu, Wei-tien, 1925–
(Chi Hsiao-lan ch'uan ch'i)
紀曉嵐傳奇 / 東郭先生 [劉維典] 著. — 再版. — 台北 : 石室出版社, 民國67 [1978].
2, 4, 288 p. ; 19 cm.
"石室文史刊書"
NT$50.00
1. Chi, Yün, 1724–1805. 2. Authors, Chinese—Anecdotes, facetiae, satire, etc. I. Title.
PL2705.I17Z75 79–841428
C

CH'I

Tohei, Kōichi, 1920-
Ki in daily life / by Koichi Tohei. — 1st ed. — Tokyo : Ki no Kenkyūkai ; New York : Japan Publications Trading Co., 1978.
136 p. : ill. ; 26 cm. Ja78-31
ISBN 0-87040-436-9 : $8.95
1. Conduct of life. 2. Spiritual exercises. 3. Ch'i. I. Title.
BJ1581.2.T62 796.8'154 79-312137
*79 MARC

CHI-AN HSIEN, CHINA

—HISTORY—SOURCES

(Chi-an fu chih)
吉安府志 : 江西省 / 李興元修 ; 歐陽主生等纂. — 臺1版. — 台北 : 成文出版社, 民國65 [1976]
2 v. (690 p.) : maps ; 27 cm. — (中國方志叢書華中地方 ; 第272號)
"三十六卷"
"據清順治十七年刊本影印"
Each page represents two leaves of the original.
1. Chi-an hsien, China—History—Sources. I. Li, Hsing-yüan. II. Ou-yang, Chu-sheng. III. Series: Chung-kuo fang chih ts'ung shu Hua-chung ti fang ; ti 272 hao.
DS793.C4614C47 1976 78–840892

CHI-LUNG, TAIWAN

—DESCRIPTION—JUVENILE LITERATURE

Huang, Ch'un-ch'ing.
(Chi-lung ho Kao-hsiung)
基隆和高雄 / 著作者黃純青. — 臺北 : 海外文庫出版社, 民國45 [1956]
36 p. : ill. ; 19 cm. — (地方小誌 ; 5) (海外文庫)
Colophon title.
Series 1 romanized: Ti fang hsiao chih ; Series 2 romanized: Hai wai wen k'u.
1. Chi-lung, Taiwan—Description—Juvenile literature. 2. Kao-hsiung, Taiwan (City)—Description—Juvenile literature. I. Title.
DS799.9.C47H8 78–842767

CHI-MO HSIEN, CHINA

—HISTORY—SOURCES

(Chi-mo hsien chih)
即墨縣志 / [總輯林溥 ; 分輯周翕鐄, 黃念昀]. — [臺北?: 即墨縣抗戰戡亂史料編纂委員會], 民國60 [1971]
ca. 600 leaves in various foliations : ill. ; 22 cm.
Reprint of the 1873 ed.

1. Chi-mo hsien, China—History—Sources. I. Lin, P'u, chin shih 1852. II. Chou, Hsi-huang, 19th cent. III. Huang, Nien-yün.
DS793.C4728C47 1971 78-841669

CHI NAN HSÜEH T'ANG

(Chi-nan hsüeh t'ang hsien hsing chang ch'eng)
暨南學堂現行章程. — [香港?: s. n., 1978?]
70 p., [1] fold. leaf of plates : ill. ; 26 cm.
Cover title.
Reprint ed.

1. Chi nan hsüeh t'ang.
LG52.N363C46 1978 79-838566

CHI-NING HSIEN, CHINA

—ECONOMIC CONDITIONS —ADDRESSES, ESSAYS, LECTURES

Chi-ning chi hsing hsieh tso tsu.
(Chi-ning chi hsing)
济宁纪行 / 《济宁纪行》写作组. — [香港?: s. n., 1979?]
157 p. ; 21 cm.
Reprint of the 1978 ed. published by Shan-tung jen min ch'u pan she, Chi-nan.

1. Chi-ning hsien, China—Economic conditions—Addresses, essays, lectures. I. Title.
HC428.C456C46 1979 79-841394

CHI-TS'ANG, SHIH 549-623

Maruyama, Takao.
(Hokke kyōgaku kenkyū josetsu)
法華教学研究序説 : 吉蔵における受容と展開 / 丸山孝雄著. — 京都 : 平楽寺書店, 1978.
22, 523, 29 p. ; 22 cm. Ja 78-26003
Includes bibliographical references and indexes.
"付録 「法華遊意」訳註": p. [405]-523.
¥8000
1. Saddharmapuṇḍarika—Criticism, interpretation, etc. 2. Chi-ts'ang, Shih 549-623. I. Chi-ts'ang, Shih 549-623. Fa hua yu i. Japanese. 1978. II. Title.
BQ2057.M37 79-805791

CHI-TUNG FANG KUNG TZŬ CHIH CHÊNG FU

Takagi, Shōnosuke, 1900–
(Kitō kara Chūka shinseiken e)
冀東から中華新政權へ / 高木翔之助著. — 東京 : 北支那社, 昭和13 [1938]
4, 4, 184 p., [1] leaf of plates : ill. ; 19 cm.
Cover title.

1. China—Politics and government—1912–1949. 2. Hopei, China (Province)—Politics and government. 3. Chi-tung fang kung tzŭ chih chêng fu. I. Title.
DS777.47.T267 79-800467

CHI-YUAN HSIEN, CHINA

—HISTORY—SOURCES

(Chi-yüan hsien chih)
濟源縣志 : 河南省 / 蕭應植纂修. — 臺1版. — 臺北 : 成文出版社, 民國65 [1976]
3 v. (974 p.) : ill. ; 27 cm. — (中國方志叢書 : 華北地方 ; 第492號)
Reprint of the 1761 ed.

1. Chi-yüan hsien, China—History—Sources. I. Hsiao, Ying-chih, 18th cent. II. Series: Chung-kuo fang chih ts'ung shu : Hua-pei ti fang : ti 492 hao.
DS793.C474234C48 1976 79-836349

Library of Congress 79 C

CHIA, I, 200-168 B.C.

Wu, Mei-hui.
(Chia I yen chiu)
賈誼研究 / 吳美惠. — [台北?: s.n., 1978?]
79 p. (on double leaves) ; 27 cm.
Reproduced from typescript.
Thesis (M. A.)—Kuo li T'ai-wan ta hsüeh, 1969.
Bibliography: p. 75–79.

1. Chia, I, 200–168 B. C. 2. Statesmen—China—Biography. I. Title.
DS748.16.C49W8 1978a 78-843487

78 C

CHIA-HO HSIEN, CHINA

—HISTORY—SOURCES

Hsü, Shuo.
(Chih-yüan Chia-ho chih)
至元嘉禾志 / [徐碩撰]. — [臺北 : 臺灣商務印書館, 1978?]
3 v. ; 20 cm. — (四庫全書珍本八集 ; 91–93)
Cover title.

1. Chia-ho hsien, China—History—Sources. I. Title. II. Series: Ssu k'u ch'üan shu chen pen pa chi ; 91–93.
AC149.S7 1978, vol. 91–93 78-842071
[DS793.C5145]

78 C

CHIA-HSIEN, CHINA (SHENSI)

—HISTORY—SOURCES

(Chia-hsien chih)
葭縣志 : 陝西省 / 陳琯修 ; 趙思明纂. — 臺1版. — 台北 : 成文出版社, 民國65 [1976]
282 p., [1] fold. leaf, [1] fold. leaf of plates : map ; 27 cm. — (中國方志叢書 : 華北地方 ; 第535號)
"據民國二十二年石印本影印"

1. Chia-hsien, China (Shensi)—History—Sources. I. Ch'en, Kuan. II. Chao, Ssu-ming. III. Series: Chung-kuo fang chih ts'ung shu : Hua-pei ti fang ; ti 535 hao.
DS793.C47435C47 78-841848

CHIA-I HSIEN, TAIWAN

—COLLECTED WORKS

(Chia-i wen hsien)
嘉義文獻
[嘉義] 嘉義縣政府
no. ill. 21 cm.

1. Chia-i hsien, Taiwan—Collected works. I. Chia-i hsien, Taiwan.
DS799.9.C474C48 79-640697
(MARC-S)
C

—HISTORY—SOURCES

(Chia-i hsien chih)
嘉義縣志 / 監修張炳楠 ; 主修趙璞, 林家駒 ; 纂修賴子清, 賴明初, [張李德和]. — 嘉義 : 嘉義縣政府, 民國64- [1975-
v. : ill., map ; 27 cm.
On v. 7: Chia-i hsien wen hsien wei yüan hui pien yin.
Includes bibliographies.

1. Chia-i hsien, Taiwan—History—Sources. I. Chang, Ping-nan.
DS799.9.C474C47 1975 79-840252

CHIANCIANO, ITALY

—DESCRIPTION—GUIDE-BOOKS

Fumi, Ersilio.
Guida di Montepulciano e dei Bagni di Chianciano / Ersilio Fumi. — Bologna : Atesa, 1978.
146 [i.e. 164] p., [18] leaves of plates (2 fold.) : ill. ; 17 cm. It79-Mar
Reprint of the 1894 ed. published by Fumi e Caleri, Montepulciano.
Includes advertisements (p. 137-146 [i.e. 164]).
Includes index.
1. Montepulciano, Italy—Description—Guide-books. 2. Chianciano, Italy—Description—Guide-books. 3. Bagni di Chianciano, Italy—Description—Guide-books. I. Title.
DG975.M7F85 1978 914.5'58 79-364945
*79 MARC

Stuart, Giorgio.
Guida ai vini ed ai prodotti tipici delle zone limitrofe a Chianciano Terme / Giorgio Stuart. — Montepulciano : Notiziario 468 editore, [1978]
237 p. : ill. ; 20 cm. It***
On spine: 1
Bibliography: p. 237.
L4200
1. Wine and wine making—Italy—Chianciano. 2. Chianciano, Italy—Industries. 3. Chianciano, Italy—Description—Guide-books. I. Title.
HD9385.I82C467 78-392812
79 MARC

—INDUSTRIES

Stuart, Giorgio.
Guida ai vini ed ai prodotti tipici delle zone limitrofe a Chianciano Terme / Giorgio Stuart. — Montepulciano : Notiziario 468 editore, [1978]
237 p. : ill. ; 20 cm. It***
On spine: 1
Bibliography: p. 237.
L4200
1. Wine and wine making—Italy—Chianciano. 2. Chianciano, Italy—Industries. 3. Chianciano, Italy—Description—Guide-books. I. Title.
HD9385.I82C467 78-392812
79 MARC

CHIANG, CH'AO-YÜEH, 1898-

Chiang, Ch'ao-yüeh, 1898-
(Ting ssu sheng ch'i chi)
丁巳聲氣集 / 江山異生〔姜超嶽〕. — 〔台北〕: 三民書局, 民國66 〔1977〕
3, 70 p. : ill. ; 22 cm.

1. Chiang, Ch'ao-yüeh, 1898- 2. China—Officials and employees—Correspondence. I. Title.
DS778.C5373A4 1977 79-841415

C

CHIANG, CH'ING

Li, Hsing-k'o.
The fall of Madam Mao / by Ly Singko. — 1st ed. — New York : Vantage Press, c1979.
136 p. ; 21 cm.
ISBN 0-533-03678-X : $6.50
1. China—Politics and government—1976- . 2. Chiang, Ch'ing. I. Title.
DS779.26.L5173 320.9'51'05 78-54378
79 MARC

—FICTION

Hu-t'ou shan hsia tou Chiang yao pien hsieh tsu.
(Hu-t'ou shan hsia tou Chiang yao)
虎头山下斗江妖 / 《虎头山下斗江妖》编写组编. — 第1版. — 上海 : 上海人民出版社, 1977.
2, 33 p. ; 18 cm.
Colophon title.
$0.10

1. Short stories, Chinese. 2. Chinese fiction—20th century. 3. Chiang, Ch'ing—Fiction. I. Title.
PL2653.H79693 1977c 78-841506

CHIANG, CHING-KUO, 1906-

Chiang, Ching-kuo, 1906-
Calm in the eye of a storm / Chiang Ching-kuo ; 〔translated by Ch'en Shih-Ch'i ; edited by Kan Chu-cheng〕. — Taipei : Li Ming Cultural Enterprise Co., 1978.
iv, 495 p. ; 22 cm.
Translation of Feng yü chung ti ning ching.
Includes index.
NT$200.00 ($5.50 U.S.)
1. Chiang, Ching-kuo, 1906- . 2. Taiwan—Presidents—Biography. I. Kan, Chu-cheng. II. Title.
DS778.C54A29513 951'.249'050924 79-100323
79 MARC

〔**Chiang, Ching-kuo**〕 1906-
(Kung ku tzu chi, ch'iu fa chan, ch'iu sheng li)
鞏固自己, 求發展, 求勝利. — 台北 : 教育部軍訓處, 民國66 〔1977〕
2, 150 p. ; 22 cm. — (軍訓幹部必讀專書 ; 5)
Cover title.
Contains speeches and articles by Chiang Ching-kuo.

I. Title. II. Series: Chün hsün kan pu pi tu chuan shu ; 5.
DS778.C54A563 79-835021

C

Chiang, Ching-kuo, 1906-
〔Selected works〕
蔣主席言論集 / 〔蔣經國〕. — 臺北 : 中央文物供應社, 民國67- 〔1978-
v. ; 22 cm.
CONTENTS: 〔1〕 施政報告.—〔2〕 弘揚憲政光輝.—〔3〕 永生不滅的榮光
NT$120.00
I. Title.
Title romanized: Chiang chu hsi yen lun chi.
DS778.C54A25 1978 78-842586

Chiang, Ching-kuo, 1906-
〔Selected works〕
做一個鋼鐵好漢 / 蔣經國著. — 台北 : 黎明文化事業公司, 民國67 〔1978〕
2, 222 p. ; 19 cm.
NT$40.00

I. Title.
Title romanized: Tso i ko kang t'ieh hao han.
DS778.C54A25 1978c 79-835463

Chiang, Ching-kuo, 1906-
〔Selections〕
蔣院長知勉錄 / 〔恭輯者林大椿〕. — 臺北 : 黎明文化事業公司, 民國66 〔1977〕
7, 4, 374 p. ; 20 cm.
NT$80.00

I. Lin, Ta-ch'un. II. Title.
Title romanized: Chiang yüan chang chih mien lu.
DS778.C54A25 1977c 78-842114

Chiang, Ching-kuo, 1906-
〔Speeches〕
凌霜雪而彌勁 / 蔣經國著. — 台北 : 黎明文化事業公司, 民國66 〔1977〕
4, 244 p. ; 19 cm.
NT$40.00

I. Title.
Title romanized: Ling shuang hsüeh erh mi ching.
DS778.C54A564 1977 78-842111

Chiang, Ching-kuo, 1906-
(Tz'u-hu hsiao ssu)
慈湖孝思 / 蔣經國撰. — 臺北 : 黎明文化事業公司, 民國67 〔1978〕
1, 2, 102 p. ; 19 cm.
CONTENTS: 梅臺思親.—領袖, 慈父, 嚴師.—守父靈一月記
NT$8.00

1. Chiang, Kai-shek, 1887-1975—Funeral and memorial services. 2. Chiang, Ching-kuo, 1906- I. Title.
DS778.C55C44215 1978 78-842948

Chiang, Ching-kuo, 1906-
(Wei ta ti chiao hui)
偉大的教誨 / 著者蔣經國. — 臺北 : 名人出版社, 民國66 〔1977〕
76 p., 〔2〕 leaves of plates : ill. ; 19 cm.
Text accompanied with Kuo yü chu yin.
NT$30.00

1. Chiang, Ching-kuo, 1906- 2. Taiwan—Presidents—Biography. I. Title.
DS778.C54A353 79-838655

Chiang, Ching-kuo, 1906-
(Yu i, shen sheng ti yu i)
友誼, 神聖的友誼 / 蔣經國著. — 臺北 : 黎明文化事業公司, 民國66 〔1977〕
8, 212 p. ; 20 cm.
NT$40.00

1. Chiang, Ching-kuo, 1906- 2. Taiwan—Presidents—Correspondence. I. Title.
DS778.C54A4 1977 78-843004

I yu ch'u pan she.
(Chi wang k'ai lai ti Chiang Ching-kuo hsien sheng)
繼往開來的蔣經國先生 = The yesterday, today, and tomorrow by Chiang Chin-Kuo / 益友出版社編著. — 台北 : 該社, 民國67 〔1978〕
474 p. : ill. ; 22 cm.
Cover title.
In Chinese.
NT$100.00 ($3.00 U.S.)
1. Chiang, Ching-kuo, 1906- 2. Taiwan—Presidents—Biography. 3. Taiwan—Politics and government—1945- I. Title. II. Title: The yesterday, today, and tomorrow.
DS778.C54I2 78-842471

I yu ch'u pan she.
(Chi wang k'ai lai ti Chiang Ching-kuo hsien sheng)
繼往開來的蔣經國先生 / 劉雍熙主編. — 台北 : 益友出版社 : 經銷處泛美圖書公司, 民國67 〔1978〕
478 p. : ill. ; 21 cm.
Cover title.
Another ed. published also in 1978 under the author's name: I yu ch'u pan she.
$3.00 (U.S.)
1. Chiang, Ching-kuo, 1906- 2. Taiwan—President—Biography. I. Liu, Yung-hsi. II. Title.
DS778.C54I2 1978 79-840806
79 C

Li, Yüan-p'ing.
(P'ing fan p'ing tan p'ing shih ti Chiang Ching-kuo hsien sheng)
平凡平淡平實的蔣經國先生 / 李元平著. — 臺北 : 青年戰士報, 民國67 〔1978〕
16, 339 p., 〔54〕 leaves of plates : ill. ; 19 cm. — (青年戰士報叢書)
Cover title.
Series romanized: Ch'ing nien chan shih pao, T'ai-pei. Ch'ing nien chan shih pao ts'ung shu.
NT$60.00
1. Chiang, Ching-kuo, 1906- 2. Taiwan—Presidents—Biography. 3. Taiwan—Politics and government—1945- I. Title.
DS778.C54L52 78-841176
78

Ts'ao, Chü-jen, 1903-1972.
〔Chiang Ching-kuo lun. Japanese〕
蔣経国と台湾 / 曹聚仁著 ; 鈴木博訳. — 東京 : 三一書房, 1978.
236 p. ; 20 cm. Ja 78-11747
¥2000

1. Chiang, Ching-kuo, 1906- 2. Taiwan—Presidents—Biography. 3. Taiwan—Politics and government. I. Title.
Title romanized: Shō Keikoku to Taiwan.
DS778.C54T716 78-817116

J

—ADDRESSES, ESSAYS, LECTURES

Ch'en, Hsiang-chün.
(Chiang Ching-kuo hsien sheng)
蔣經國先生 : 中國民主憲政的導師 / 陳湘君撰. — 臺北 : 白雲文化事業公司, 民國67 〔1978〕
1, 5, 165 p., 〔15〕 leaves of plates : ill. ; 22 cm.
NT$100.00

1. Chiang, Ching-kuo, 1906- —Addresses, essays, lectures. 2. Taiwan—Politics and government—1945- —Addresses, essays, lectures. I. Title.
DS778.C54C43 79-839568

(Chiang Ching-kuo shih tai)
蔣經國時代 / 〔編輯者自立叢書編輯委員會〕. — 臺北 : 自立晚報社, 民國67 〔1978〕
4, 6, 168 p., 〔1〕 leaf of plates : port. ; 20 cm. — (自立晚報叢書)
Series romanized: Tzu li wan pao ts'ung shu.
NT$40.00

1. Taiwan—Politics and government—1945- —Addresses, essays, lectures. 2. Chiang, Ching-kuo, 1906- —Addresses, essays, lectures. I. Tzu li ts'ung shu pien chi wei yüan hui.
DS799.8.C4834 78-842216
78

(Chiang tsung t'ung Ching-kuo hisen sheng ti hua)
蔣總統經國先生的話 / 華僑通訊社恭輯. — 〔台北? : 該社〕, 民國67 〔1978〕
2, 50 p. ; 20 cm.
Cover title.

1. Taiwan—Politics and government—1945- —Addresses, essays, lectures. 2. Chiang, Ching-kuo, 1906- —Addresses, essays, lectures. I. Hua ch'iao t'ung hsün she.
DS799.8.C4838 79-837796

CHIANG, CHING-KUO, 1906- —ADDRESSES, ESSAYS, LECTURES (Continued)

(Ch'üan min yuan wang ti li shih hsing piao hsien)
全民願望的歷史性表現：全民擁戴蔣經國先生競選第六任總統輿論彙編 / 中國國民黨中央文化工作會編印. — [台北]：該會, 民國67 [1978]
2, 10, 420 p. ; 22 cm.
Cover title.

1. Chiang, Ching-kuo, 1906- —Addresses, essays, lectures. I. Chung-kuo kuo min tang. Chung yang wei yüan hui. Wen hua kung tso hui.
DS778.C54C47 79-835646

(Chung-hua min kuo tsung t'ung Chiang Ching-kuo hsien sheng)
中華民國總統蔣經國先生 / [中國出版公司編印]. — [臺北：該公司, 1978]
7 p. : 27 cm.
Cover title.

1. Chiang, Ching-kuo, 1906- —Addresses, essays, lectures. I. Chung-kuo ch'u pan kung ssu.
DS778.C54C48 78-843360
C

T'ao, Hsi-sheng, 1893-
(Li shih ti shih ming, shih tai ti yao ch'iu)
歷史的使命·時代的要求 / 陶希聖演講；中國大陸研究出版社編；[編著者中國大陸問題研究中心]. — 台北：該社, 民國67 [1978]
47 p. ; 20 cm.
Cover title.
NT$20.00
1. Chiang, Ching-kuo, 1906- —Addresses, essays, lectures. 2. Philosophy, Confucian—Addresses, essays, lectures. I. Chung-kuo ta lu yen chiu ch'u pan she. II. Chung-kuo ta lu wen t'i yen chiu chung hsin. III. Title.
DS778.C54T38 79-840665
C

Ts'ai, Chin-hsüan.
(Jen che pu yu)
仁者不憂：蔣經國先生的話 / [編撰者蔡金宣]. — 臺北：照明出版社：總經銷黎明文化事業股份有限公司, 民國67 [1978]
213 p., [1] leaf of plates : ill. ; 22 cm.
NT$200.00

1. Chiang, Ching-kuo, 1906- —Addresses, essays, lectures. I. Title.
DS778.C54T67 78-843362

(Wei kuo hsiao ming, wei min fu wu)
為國效命, 為民服務：蔣經國先生在行政院長任內的施政方針與作法 / [中國出版公司編印]. — [台北]：該公司, [1978?]
2, 24 p. ; 22 cm.

1. Taiwan—Politics and government—1945- —Addresses, essays, lectures. 2. Chiang, Ching-kuo, 1906- —Addresses, essays, lectures. I. Chung-kuo ch'u pan kung ssu.
DS799.8.W44 79-836247

—ANECDOTES

(Che ts'ai shih kung p'u)
這才是公僕 / [編者聯合報編輯部]. — 臺北：聯合報社：總經銷聯經出版事業公司, 民國67 [1978]
2, 5, 249 p. : ill. ; 19 cm. — (聯合報叢書)
Series romanized: Lien ho pao ts'ung shu.
"蔣經國與我們在一起"
NT$40.00

1. Chiang, Ching-kuo, 1906- —Anecdotes. 2. Taiwan—Presidents—Anecdotes, facetiae, satire, etc. I. Lien ho pao.
DS778.C54C42 78-843141

T'ang, Hsing-han.
(Ching-kuo hsien sheng ti ku shih)
經國先生的故事 / 唐興漢編著. — 台北：黎明文化事業公司, 民國67 [1978]
2, 226 p., [20] leaves of plates : ill. ; 19 cm.
NT$40.00

1. Chiang, Ching-kuo, 1906- —Anecdotes. 2. Taiwan—Presidents—Anecdotes, facetiae, satire, etc. I. Title.
DS778.C54T36 78-842037
78

—PORTRAITS, CARICATURES, ETC.

(Chiang chu hsi yü t'ieh lu chien she)
蔣主席與鐵路建設. — [台北縣永和鎮]：中國國民黨台灣區鐵路黨部, 民國67 [1978]
21 p. : ill. ; 27 cm.
Cover title.

1. Railroads—Taiwan—Pictorial works. 2. Chiang, Ching-kuo, 1906- —Portraits, caricatures, etc. I. Chung-kuo kuo ming tang. T'ai-wan ch'ü t'ieh lu tang pu.
HE3290.5.C47 79-837294
C

—QUOTATIONS

Chiang, Ching-kuo, 1906-
[Selections]
蔣院長的懇切話 / [蔣經國]；端木良編選. — 臺北：勵志出版社：總經銷雨辰書報社, 民國67 [1978]
141 p. (p. 140-141 abvertisements) ; 20 cm.
NT$30.00 ($2.00 U.S.)
1. Chiang, Ching-kuo, 1906- —Quotations. I. Tuan-mu, Liang. II. Title.
Title romanized: Chiang yüan chang ti k'en ch'ieh hua.
DS778.C54A25 1978b 78-843284
78

—VIEWS ON PHILOSOPHY

T'ang, Hua.
(Chiang Ching-kuo hsien sheng che hsüeh ssu hsiang)
蔣經國先生哲學思想 / 唐華著. — 臺北：全台文化事業股份有限公司, 民國67 [1978]
14, 22, 712 p., [1] leaf of plates : port. ; 22 cm. — (學術思想名著；第3號)
Bibliography: p. 683-708.
NT$250.00
1. Chiang, Ching-kuo, 1906- —Views on philosophy. 2. Philosophy, Chinese. I. Title. II. Series: Hsüeh shu ssu hsiang ming chu ; ti 3 hao.
DS778.C54T37 79-835440
C

CHIANG, CH'UN-LIN, 1818-1868. SHUI YUN LOU TZ'U

Li, Hsiao-hsüan.
(Chiang Ch'un-lin chi ch'i Shui yün lou shih tz'u)
蔣春霖及其水雲樓詩詞 / 李小萱. — [台北?：s. n.], 民國61 [1972]
177 p. (on double leaves) ; 27 cm.
Reproduced from ms. copy.
"國立台灣大學中文研究所碩士論文"
Bibliography: p. 173-177.

1. Chiang, Ch'un-lin, 1818-1868. Shui yün lou tz'u. 2. Chiang, Ch'un-lin, 1818-1868—Criticism and interpretation. I. Title.
PL2705.I25S534 79-835069

CHIANG, KAI-SHEK, 1887-1975. CHUNG-KUO CHIH MING YÜN

Ch'en, Po-ta, 1905-
(P'ing Chung-kuo chih ming yün)
評「中國之命運」/ 陳伯達著. — [北京?]：新華書店, 1949.
57 p. ; 18 cm.

1. Chiang, Kai-shek, 1887-1975. Chung-kuo chih ming yün. 2. China—Foreign relations—1912-1949. I. Title.
DS775.C531583C45 78-843815

CHIANG, KAI-SHEK, 1887-1975 K'O HSÜEH TI HSÜEH YUNG

Lin, Ta-ch'un.
(K'o hsüeh ti hsüeh yung chih yen chiu)
科學的學庸之研究 / 林大椿編著. — 臺初版. — 臺北：正中書局, 民國67 [1978]
6, 3, 3, 406 p., [1] leaf of plates : facsims. ; 22 cm.
Appendices (p. 341-406): 1. 主要參考書.—2. 次要參考書.—3. 有關學庸之論著.

1. Chiang, Kai-shek, 1887-1975. K'o hsüeh ti hsüeh yung. 2. Ta hsüeh. 3. Chung yung. I. Title.
PL2472.R33 79-835169

CHIANG, KAI-SHEK, 1887-1975

Ch'en, Pin-ho, fl. 1926-
(Lun Chiang Chieh-shih)
論蔣介石 陳彬龢著 上海 太平書局 民國31 [1942]
62 p. 26 cm.
Colophon title.

1. Chiang, Kai-shek, 1887-1975. 2. China—Presidents—Biography. I. Title.
DS778.C55C4355 74-840266

Ch'en, Po-ta, 1905-
(Jen min kung ti Chiang Chieh-shih)
人民公敵蔣介石 / 陳伯達著. — 3版. — [s. n.]：華東新華書店, 1949.
177 p. ; 19 cm.

1. Chiang, Kai-shek, 1887-1975. I. Title.
DS778.C55C436 1949 79-837881

Chiang, Ching-kuo, 1906-
(Feng mu hsiao ssu)
風木孝思 / 蔣經國撰. — 臺北：幼獅文化事業公司, 民國67 [1978]
4, 2, 138 p. ; 22 cm.
NT$15.00

1. Chiang, Kai-shek, 1887-1975. I. Title.
DS778.C55C44197 1978b 79-836596

Chiang, Kai-shek, 1887-1975.
[Selections]
蔣總統遺訓文錄 / 端木良編選. — 臺北縣新店鎮：勵志出版社, 民國66- [1977-
v. ; 19 cm.
CONTENTS: 1. 蔣總統語粹.—2. 蔣總統嘉言.

NT$30.00 ($2.00 U. S.) per vol.
I. Tuan, Mu-liang. II. Title.
Title romanized: Chiang tsung t'ung i hsün wen lu.
DS778.C55A25 1977d 78-840944

CHIANG, KAI-SHEK, 1887-1975 (Continued)

Chiang, Kai-shek, 1887–1975.
[Selections]
蔣總統語粹 / 端木良編選. — 臺北縣新店鎮：勵志出版社, 民國66 [1977]
4, 138 p. ; 19 cm. — (His 蔣總統遺訓文錄 ; 1)
NT$30.00 ($2.00 U. S.)

I. Tuan, Mu-liang. II. Title.
Title romanized: Chiang tsung t'ung yü ts'ui.

DS778.C55A25 1977d, vol. 1 — 78–840951

Chiang, Kai-shek, 1887–1975.
[Selections]
蔣總統嘉言 / 端木良編選. — 臺北縣新店鎮：勵志出版社；臺北：總經銷雨辰書報社, 民國66 [1977]
4, 138 p. ; 19 cm. — (His 蔣總統遺訓文錄 ; 2)
NT$30.00 ($2.00 U. S.)
I. Tuan, Mu-liang. II. Title.
Title romanized: Chiang tsung t'ung chia yen.

DS778.C55A25 1977d, vol. 2 — 78–840950

Chiang, Kai-shek, 1887–1975.
[Selections]
蔣公嘉言選輯. — [高雄?]：中國袖珍出版社；高雄：總經銷心影書店, [1977?]
10, 169 p. ; 19 cm.
NT$15.00

I. Title.
Title romanized: Chiang kung chia yen hsüan chi.

DS778.C55A25 1977e — 78–842548

Chiang, Kai-shek, 1887–1975.
(Tsung ts'ai chung yao hao chao chi yu kuan hsüan ch'uan wen t'i hsün shih chi yao)
總裁重要號召及有關宣傳問題訓示集要 / [蔣介石；中國國民黨中央委員會文化工作會編]. — 增訂本. — 臺北：該會, 民國63 [1974]
42, 386 p. ; 21 cm.

I. Chung-kuo kuo min tang. Chung yang wei yüan hui. Wen hua kung tso hui. II. Title.

DS778.C55A5 1974b — 77–837158

Library of Congress — rev 77 [r78] — C

Chook, Shiu Heng.
Chiang Kai-shek close-up : a personal view / Shiu Heng Chook ; edited by Edward K. Chook. — Oakland, CA. : United California University Press, [1978?]
161 p., [8] leaves of plates : ill. ; 22 cm.
1. Chiang, Kai-shek, 1887-1975. 2. China—Presidents—Biography. I. Chook, Edward K. II. Title.
DS778.C55C496 — 951.04'2'0924 — 78-109073
78 — MARC

Huang, Tsou-sheng.
(Chiang tsung t'ung lun li ssu hsiang chih yen chiu)
蔣總統倫理思想之研究 / 黃奏勝著. — 臺北：中央文物供應社, 民國66 [1977]
2, 4, 250 p. ; 22 cm. — (三民主義學術思想論叢)
Edition of 1975 published under title: Chiang tsung t'ung lun li ssu hsiang yen chiu.
Bibliography: p. 228–249.
NT$120.00
1. Chiang, Kai-shek, 1887–1975. 2. Ethics. I. Title. II. Series: San min chu i hsüeh shu ssu hsiang lun ts'ung.
DS778.C55H85 1977 — 79–836002

Ikeguchi, Ekan, 1936–
(Shō Kaiseki)
蔣介石：その偉大なる生涯 / 池口恵観. — 東京：論稿社, 昭和51 [1976]
215 p. : ill. ; 20 cm. — Ja***
¥1200

1. Chiang, Kai-shek, 1887–1975. 2. China—Presidents—Biography. 3. China—History—Republic, 1912–1949. 4. Japan—Foreign relations—China. 5. China—Foreign relations—Japan. I. Title.
DS778.C55I33 — 78–814965

(She hui hsüeh lun chi)
社會學論集 / [本册主編丘正歐, 蔡漢賢, 龍寶麒]. — 陽明山：中華學術院：發行者華岡出版有限公司, 民國66 [1977]
742, [16] p. : ports. ; 22 cm. — (中華學術與現代文化叢書 ; 第10册)
Includes bibliographical references.
NT$200.00
1. Sociology—Addresses, essays, lectures. 2. Ethnology—Addresses, essays, lectures. 3. Chiang, Kai-shek, 1887–1975. I. Ch'iu, Cheng-ou. II. Ts'ai, Han-hsien. III. Lung, Pao-ch'i. IV. Series: Chung-hua hsüeh shu yü hsien tai wen hua ts'ung shu ; ti 10 ts'e.
HM62.C5S5 — 78–841488

Wu, Man-chün.
(Chiang tsung ts'ai ssu hsiang yen chiu)
蔣總裁思想研究 / 吳曼君著. — 臺北：帕米爾書店, 民國67 [1978]
5, 16, 625 p. ; 22 cm.
Includes bibliographical references.

1. Chiang, Kai-shek, 1887–1975. I. Title.
DS778.C55W85 — 78–842437

—ADDRESSES, ESSAYS, LECTURES

Chang, Ch'ün, 1889?–
(Chih te kuan k'uei lu)
至德管窺錄 / 張羣. — 臺北：中央文物供應社, 民國65 [1976]
2, 128 p. ; 22 cm.
NT$30.00

1. Chiang, Kai-shek, 1887–1975—Addresses, essays, lectures. I. Title.
DS778.C55C423 — 78–842975

(Chung-kuo san chü t'ou)
中國三巨頭：抗戰到勝利的實錄 / 編者曙光編譯社. — 上海：曙光社, 民國34 [1945]
49 p. ; 18 cm.
CONTENTS: 蔣主席領導抗戰.—宋子文院長訪問世界.—何應欽將軍奔馳沙場.

1. Sino-Japanese Conflict, 1937–1945—Addresses, essays, lectures. 2. Chiang, Kai-shek, 1887–1975—Addresses, essays, lectures. 3. Soong, T. V.—Addresses, essays, lectures. 4. Ho, Ying-ch'in, 1889– —Addresses, essays, lectures. I. Shu kuang pien i she.
DS777.53.C58346 — 79–839492

Shen, Ching-hsiung.
(Kao shan yang chih)
高山仰止 / 沈經熊編撰. — 台北：中國佛教會, 民國65 [1976]
117 p. : ill. ; 22 cm.
At head of title: Huai Chiang kung, ssu ming ch'a, hsin i hu kuo ko ming seng.
1. Chiang, Kai-shek, 1887–1975—Addresses, essays, lectures. 2. Temples—China—Addresses, essays, lectures. 3. Priests, Buddhist—China—Political activity—Addresses, essays, lectures. I. Title.
DS778.C55S46 — 79–837544

C

(Tsung t'ung Chiang kung hsing i chuan chi)
總統蔣公行誼專集 / 教育部訓育委員會編. — 台北：該會, 民國66 [1977]
2, 2, 166 p., [1] leaf of plates : port. ; 22 cm.
Cover title.
CONTENTS: 史振鼎 總統蔣公永恆不朽的生命.—李正中 總統蔣公的革命勲業.—李雲漢 總統蔣公早年的讀書生活.—沈若雲 總統蔣公的睿知遠見.—周何 總統蔣公是真正不朽的偉人.—周道濟 總統蔣公逝世的歷史意義.—馬起華 總統蔣公的偉大出於平凡.—張京育 總統蔣公的大智大仁大勇.—鄔昆如 總統蔣公的信仰.—楊崇森 總統蔣公是偉大的愛國者.—趙際良 總統蔣公是現代的完人.—蔣一安 總統蔣公對人類歷史的貢獻.—蔣永敬 總統蔣公怎樣領導抗戰.—儲鴻畢 總統蔣公的立德立功立言

1. Chiang, Kai-shek, 1887–1975—Addresses, essays, lectures. 2. China—Presidents—Biography—Addresses, essays, lectures. I. China (Republic of China). Chiao yü pu. Hsün yü wei yüan hui.
DS778.C55T774 — 79–841413

(Tsung t'ung Chiang kung yü Hua-lien)
總統蔣公與花蓮 / [總編輯楊松德；編輯劉廣元 ... et al.]. — [花蓮縣]：花蓮縣政府, 民國67 [1978]
71 p. : chiefly ill. (some col.) ; 27 cm.
Cover title.

1. Chiang, Kai-shek, 1887–1975—Addresses, essays, lectures. 2. Hua-lien hsien, Taiwan—History—Addresses, essays, lectures. I. Yang, Sung-te. II. Liu, Kuang-yüan.
DS778.C55T794 — 79–838165

—ANNIVERSARIES, ETC., 1978

(Tsung t'ung Chiang kung shih shih san chou nien chi nien chi)
總統蔣公逝世三週年紀念集 / [編輯者總統蔣公逝世三週年紀念集編輯委員會]. — 臺北：近代中國社, 民國67 [1978]
2, 144, 125 p. : ill. ; 23 cm.
Added t. p.: Tribute to President Chiang Kai-shek.
English and Chinese.
"非賣品"
1. Chiang, Kai-shek, 1887–1975—Anniversaries, etc., 1978. I. Tsung t'ung Chiang kung shih shih san chou nien chi nien chi pien chi wei yüan hui. II. Title: Tribute to President Chiang Kai-shek.
DS778.C55T793 — 78–842457

C

—CHRONOLOGY

Ch'en, Pu-lei, 1890–1948.
(Chiang Chieh-shih hsien sheng nien piao)
蔣介石先生年表 / 陳布雷等編；[編輯者傳記文學雜誌社]. — 臺北：傳記文學出版社, 民國67 [1978]
2, 134 p. ; 20 cm. — (傳記文學叢書 ; 68)
Series romanized: Chuan chi wen hsüeh ts'ung shu.
NT$30.00
1. Chiang, Kai-shek, 1887–1975—Chronology. 2. Taiwan—Presidents—Biography. I. Chuan chi wen hsüeh tsa chih she. II. Title.
DS778.C55C438 1978 — 79–835062

—FUNERAL AND MEMORIAL SERVICES

Chiang, Ching-kuo, 1906–
(Feng mu hsiao ssu)
風木孝思 / [蔣經國]. — [台北]：中央文物供應社, [1978]
4, 97 p., [6] leaves of plates : ill. ; 19 cm.
NT$25.00

1. Chiang, Kai-shek, 1887–1975—Funeral and memorial services. I. Title.
DS778.C55C44197 — 79–836000

Chiang, Ching-kuo, 1906–
(Tz'u-hu hsiao ssu)
慈湖孝思 / 蔣經國撰. — 臺北：黎明文化事業公司, 民國67 [1978]
1, 2, 102 p. ; 19 cm.
CONTENTS: 梅臺思親.—領袖, 慈父, 嚴師.—守父靈一月記
NT$8.00
1. Chiang, Kai-shek, 1887–1975—Funeral and memorial services. 2. Chiang, Ching-kuo, 1906– I. Title.
DS778.C55C44215 1978 — 78–842948

(Mei-t'ai ssu ch'in)
梅臺思親. — [台北?：s.n., 1976?]
107 p., [6] leaves of plates : ill. (some col.) ; 29 cm.
Cover title.
CONTENTS: 先君手澤.—母親手諭.—先君紀事手錄.—先君病中隨筆.—梅臺思親.—矢勤矢勇, 毋怠毋忽.—堅此百忍, 奮勵自強.

1. Chiang, Kai-shek, 1887–1975—Funeral and memorial services. I. Chiang, Kai-shek, 1887–1975. II. Chiang, Ching-kuo, 1906–
DS778.C55C44213 1976b — 79–835645

CHIANG, KAI-SHEK, 1887-1975 (Continued)

—MILITARY LEADERSHIP
—ADDRESSES, ESSAYS, LECTURES

Chiang, Wei-kuo, 1916–
(Chiang tsung t'ung yü kuo fang)
蔣總統與國防 / 蔣緯國著 ; 黎明文化公司輯. — 臺北 : 黎明文化事業公司, 民國65 [1976]
2, 152 p., [8] leaves of plates : ill. ; 21 cm.
NT$25.00

1. Chiang, Kai-shek, 1887-1975—Military leadership—Addresses, essays, lectures. 2. Sino-Japanese Conflict, 1937-1945—China—Addresses, essays, lectures. 3. China—Defenses—Addresses, essays, lectures. I. Title.
DS778.C55C4562 78-843151

—PERSONALITY

Kuo, Mang-hsi.
(Tsui kao ling hsiu ti ssu hsiang yü jen ko)
最高領袖的思想與人格 / 郭莽西編著. — [上海?] : 正中出版社 : 總經銷博文書店, [1946?]
2, 2, 115 p. ; 18 cm.

1. Chiang, Kai-shek, 1887-1975—Personality. I. Title.
DS778.C55K87 79-837455

—POLITICAL AND SOCIAL VIEWS

(Chiang tsung t'ung ssu hsiang yen chiu)
蔣總統思想研究 / 國立臺灣師範大學公民訓育學系編印. — [台北] : 該系, 民國65 [1976]
278 p. ; 22 cm. — (公民訓育學系叢刊 ; 1)
Cover title.
Series statement in colophon: Kung min hsün yü ts'ung k'an.
Includes bibliographical references.
1. Chiang, Kai-shek, 1887-1975—Political and social views. I. Kuo li T'ai-wan shih fan ta hsüeh. Kung min hsün yü hsüeh hsi. II. Series: Kuo li T'ai-wan shih fan ta hsüeh. Kung min hsün yü hsüeh hsi. Kung min hsün yü hsüeh hsi ts'ung k'an ; 1.
DS778.C55C46733 78-842422

—POLITICAL AND SOCIAL VIEWS
—ADDRESSES, ESSAYS, LECTURES

(Tsung t'ung Chiang kung cheng chih ssu hsiang kuang po chiang tso chuan chi)
總統蔣公政治思想廣播講座專集. — [台北] : 教育部訓育委員會, 民國67 [1978]
2, 2, 164 p. ; 21 cm.
Cover title.

1. Chiang, Kai-shek, 1887-1975—Political and social views—Addresses, essays, lectures. I. China (Republic of China). Chiao yü pu. Hsün yü wei yüan hui.
DS778.C55T772 79-839948

—QUOTATIONS

Chiang, Kai-shek, 1887-1975.
(Tsung t'ung Chiang kung k'o hsüeh ssu hsiang yen lun chi)
總統蔣公科學思想言論集 / 國家安全會議科學發展指導委員會編印. — [台北] : 該會, 民國67 [1978]
2, 3, 3, 390 p. ; 22 cm.
Cover title.
"總統蔣公科學論著年表": p. 367-390.
includes index.

1. Chiang, Kai-shek, 1887-1975—Quotations. 2. Science—Addresses, essays, lectures. I. China (Republic of China). Kuo chia an ch'üan hui i. K'o hsüeh fa chan chih tao wei yüan hui. II. Title.
DS778.C55A25 1978 79-835746

—RELIGION

[Tang tai Chi-tu chü jen)
當代基督巨人 : 蔣故總統的靈修生活 / 港九基督徒敬悼蔣故總統追思禮拜委員會編印. — [香港] : 該會, 1975.
120 p. : ill. ; 20 cm.
In colophon: Fa hsing che Tung fang hsüeh yüan ch'u pan pu.
"非賣品"

1. Chiang, Kai-shek, 1887-1975—Religion. I. Kang Chiu Chi-tu t'u ching tao Chiang ku tsung t'ung chui ssu li pai wei yüan hui.
DS778.C55T33 79-836536

C

—VIEWS ON ECONOMICS

Wu, Chi-p'ing, fl. 1968–
(Tsung t'ung Chiang kung ching chi ssu hsiang chih yen chiu)
總統蔣公經濟思想之研究 / 吳寄萍. — 台北 : 國立政治大學三民主義研究所, 民國67 [1978]
21 p. ; 27 cm. — (三民主義學術研究專刊 ; 2)
Cover title.
Includes bibliographical references.
1. Chiang, Kai-shek, 1887-1975—Views on economics. I. Title. II. Series: San min chu i hsüeh shu yen chiu chuan k'an ; 2.
DS778.C55W79 79-835714

C

—VIEWS ON EDUCATION

Lo, Hai-hsien.
(Hsüeh chiao tao kuan)
學教道貫 / 羅海賢著. — 臺北 : 文史哲出版社, 民國66 [1977]
[12], 126 p. ; 22 cm.
Reproduced from ms. copy.
Bibliography: p. 123-126.
NT$50.00
1. Confucius—Education. 2. Sun, Yat-sen, 1866-1925—Views on education. 3. Chiang, Kai-shek, 1887-1975—Views on education. 4. Education—Philosophy. I. Title.
B128.C8L62 1977 79-835360

C

—VIEWS ON PHILOSOPHY

Yang, Ju-chou.
(Tsung t'ung Chiang kung chih che hsüeh t'i hsi)
總統蔣公之哲學體系 / 楊汝舟著. — 臺北 : 黎明文化事業公司, 民國67 [1978]
11, 5, 223 p., [1] leaf of plates : ill. ; 22 cm.
Bibliography: p. 217-223.
NT$60.00

1. Chiang, Kai-shek, 1887-1975—Views on philosophy. 2. Philosophy, Chinese. I. Title.
DS778.C55Y36 79-836725

C

—VIEWS ON PUBLIC WELFARE

Shu, Chi-liang.
(Ling hsiu Chiang kung she hui an ch'üan chih tu ti li lun yü shih chien)
領袖蔣公社會安全制度的理論與實踐 / 舒基亮著. — 台北 : 黎明文化事業公司, 民國67 [1978]
8, 14, 500 p. ; 22 cm.
Bibliography: p. 453-476.
NT$100.00
1. Public welfare—Taiwan. 2. Chiang, Kai-shek, 1887-1975—Views on public welfare. I. Title.
HV428.S58 79-835378

CHIANG, KUEI CA. 1155-CA. 1235

—CRITICISM AND INTERPRETATION

Yen, T'ien-yu.
(Nan Sung Chiang Wu p'ai tz'u chih yen chiu)
南宋姜吳派詞之研究 / 顏天佑撰. — [臺北] : 政治大學, 民國63 [1974]
133 double leaves ; 26 cm.
Cover title.
Reproduced from typescript.
"國立政治大學中文研究所研究生論文"
Bibliography: p. 130-133.
1. Tz'u—History and criticism. 2. Chiang, K'uei, ca. 1155-ca. 1235—Criticism and interpretation. 3. Wu, Wen-ying—Criticism and interpretation. I. Title.
PL2343.Y4 78-841148

CHIANG, MEI-LING SUNG, 1897-

Chiang, Mei-ling Sung, 1897–
(Chiang fu jen yen lun chi)
蔣夫人言論集 / [總編纂王亞權, 編纂李萼]. — [台北] : 中華婦女反共聯合會, 民國66 [1977]
2 v. (6, 1556 p., [1] leaf of plates) : port. ; 24 cm.

I. Wang, Ya-ch'üan, 1912– II. Li, O.
DS778.C56A2 1977 78-841814

C

CHIANG, SHAO-TSU, 1876-1895

Chung, Chao-cheng.
(Tan hsin keng keng shu ssu jen)
丹心耿耿屬斯人 : 姜紹祖傳 / [著作者鍾肇政]. — 臺北 : 近代中國出版社 : 總經銷中央文物供應社, 民國66 [1977]
6, 182 p. ; 22 cm. — (近代中國叢書) (先烈先賢傳記叢刊)
$30.00
1. Chiang, Shao-tsu, 1876-1895. 2. Taiwan—History—1895-1945. 3. Revolutionists—Taiwan—Biography. I. Title. II. Series: Chin tai Chung-kuo ts'ung shu. III. Series: Hsien lieh hsien hsien chuan chi ts'ung k'an.
DS799.72.C43C47 79-835733

C

CHIANG, WEI-SHUI, 1891-1931

Huang, Huang-hsiung.
(T'ai-wan ti hsien chih hsien hsüeh che Chiang Wei-shui hsien sheng)
臺灣的先知先覺者蔣渭水先生 / 黃煌雄著. — 臺北 : 總經銷輝煌出版社, 民國65 [1976]
300 p. : ill. ; 20 cm.
Caption title: Chiang Wei-shui hsien sheng.
Bibliography: p. 295-299.
NT$65.00
1. Chiang, Wei-shui, 1891-1931. 2. Statesmen—Taiwan—Biography. 3. Taiwan—History—1895-1945. I. Title. II. Title: Chiang Wei-shui hsien sheng.
DS799.72.C45H8 78-842323

C

Huang, Huang-hsiung.
[T'ai-wan ti hsien chih hsien chüeh che Chiang Wei-shui hsien sheng]
革命家, 蔣渭水 / [黃煌雄著]. — 台北 : 長橋出版社, 民國67 [1978]
300 p. : ill. ; 20 cm.
"本書原名台灣的先知先覺者蔣渭水先生"
NT$70.00

1. Chiang, Wei-shui, 1891-1931. 2. Statesmen—Taiwan—Biography. 3. Taiwan—History—1895-1945. I. Title.
Title romanized: Ko ming chia, Chiang Wei-shui.
DS799.72.C45H8 1978 79-839946

CHIANG, YEE, 1903-

Chiang, Yee, 1903-
(Chan shih Pu-lieh-tien)
戰時不列顛 / 蔣彝著. — [s. l. : s.n., 19--]
23 p. : ill. ; 25 cm.

1. World War, 1939-1945—England. 2. World War, 1939-1945—Personal narratives, Chinese. 3. Chiang, Yee, 1903- 4. Authors, Chinese—20th century—Biography. I. Title.
D759.C42 79-840344

—ADDRESSES, ESSAYS, LECTURES

(Mao chu hsi kuang hui chiao yü ssu hsiang chao liang kung ta ch'ien chin ti tao lu)
毛主席光辉教育思想照亮共大前进的道路 / 江西省革委会政治部教育组编. — [香港? : s.n., 1978?]
2, 99 p. ; 21 cm.

Reprint of the 1972 ed. published by Chiang-hsi jen min ch'u pan she, Nan-ch'ang.

1. Chiang-hsi kung ta—Addresses, essays, lectures. 2. Agricultural colleges—China—Kiangsi (Province)—Addresses, essays, lectures. 3. Education, Cooperative—China—Kiangsi (Province)—Addresses, essays, lectures. 4. Mao, Tse-tung, 1893-1976—Views on education—Addresses, essays, lectures. I. Kiangsi, China (Province). Ko ming wei yüan hui. Cheng chih pu. Chiao yü tsu.
S539.C6C455 1978 79-840287

C

CHIANG-TU, CHINA

—BIOGRAPHY

Chou, Ch'iu-ju.
(Wu-ch'eng chung i lu)
蕪城忠義錄 / 周秋如著. — 屏東 : 中國書局, 民國66 [1977]
82 p. ; 19 cm.
NT$20.00

1. Chiang-tu, China—History. 2. Chiang-tu, China—Biography. I. Title.
DS796.C55C48 78-843214

—DESCRIPTION—VIEWS

(Yang-chou)
扬州 / 扬州地委宣传部编 ; 摄影丁峻 ... [et al.]. — 第1版. — 上海 : 上海人民美術出版社, 1978.
1 portfolio ([11] leaves of plates : all col. ill.) ; 11 x 15 cm.

In case.
$0.59

1. Chiang-tu, China—Description—Views. I. Chung-kuo kung ch'an tang. Yang-chou ti wei. Hsüan ch'uan pu.
DS796.C55Y36 79-837133

—HISTORY

Chou, Ch'iu-ju.
(Wu-ch'eng chung i lu)
蕪城忠義錄 / 周秋如著. — 屏東 : 中國書局, 民國66 [1977]
82 p. ; 19 cm.
NT$20.00

1. Chiang-tu, China—History. 2. Chiang-tu, China—Biography. I. Title.
DS796.C55C48 78-843214

CHIANTI MOUNTAIN REGION

—HISTORY

Flower, Raymond, 1921-
Chianti : the land, the people, and the wine / Raymond Flower. — New York : Universe Books, c1979.
305 p., [8] leaves of plates : ill. ; 25 cm.
Bibliography: p. 295-297.
Includes index.
ISBN 0-87663-330-0
1. Chianti Mountain region—History. 2. Chianti Mountain region—Social life and customs. 3. Wine and wine making—Italy—Chianti Mountain region. 4. Chianti wine. I. Title.
DG975.C47F57 1979 945'.58 78-60736
00 MARC

—RURAL CONDITIONS.

Strumenti di lavoro e oggetti d'uso nel Chianti della mezzadria / [redazione di Renato Stopani, Maurizio Carnasciali]. — Firenze : Salimbeni libreria editrice, [1978]
24, [40] p. : ill. ; 21 cm. — (Quaderno - Centro studi sulla cultura contadina del Chianti, Biblioteca comunale di Radda ; n. 2) It78-Nov
1. Agricultural implements—Italy—Chianti Mountain region—Pictorial works. 2. Farm equipment—Italy—Chianti Mountain region—Pictorial works. 3. Home economics—Italy—Chianti Mountain region—Equipment and supplies—Pictorial works. 4. Agriculture—Italy—Chianti Mountain region. 5. Métayer system—Italy—Chianti Mountain region. 6. Chianti Mountain region—Rural conditions. . I. Stopani, Renato. II. Carnasciali, Maurizio. III. Series: Centro studi sulla cultura contadina del Chianti. Quaderno - Centro studi sulla cultura contadina del Chianti ; n. 2.
S760.I8S77 78-397728
*79 MARC

—SOCIAL LIFE AND CUSTOMS

Flower, Raymond, 1921-
Chianti : the land, the people, and the wine / Raymond Flower. — New York : Universe Books, c1979.
305 p., [8] leaves of plates : ill. ; 25 cm.
Bibliography: p. 295-297.
Includes index.
ISBN 0-87663-330-0
1. Chianti Mountain region—History. 2. Chianti Mountain region—Social life and customs. 3. Wine and wine making—Italy—Chianti Mountain region. 4. Chianti wine. I. Title.
DG975.C47F57 1979 945'.58 78-60736
00 MARC

CHIANTI WINE

Cianferoni, Reginaldo.
Il Chianti classico fra prosperità e crisi / Reginaldo Cianferoni. — 1.ed. — Bologna : Edagricole, 1979.
x, 111 p. : ill. ; 24 cm. It79-July
Includes bibliographical references.
ISBN 8820620138 : L8000
1. Viticulture—Italy—Chianti Mountain region. 2. Chianti wine. I. Title.
SB387.8.I8C52 79-375724
*79 MARC

Flower, Raymond, 1921-
Chianti : the land, the people, and the wine / Raymond Flower. — New York : Universe Books, c1979.
305 p., [8] leaves of plates : ill. ; 25 cm.
Bibliography: p. 295-297.
Includes index.
ISBN 0-87663-330-0
1. Chianti Mountain region—History. 2. Chianti Mountain region—Social life and customs. 3. Wine and wine making—Italy—Chianti Mountain region. 4. Chianti wine. I. Title.
DG975.C47F57 1979 945'.58 78-60736
00 MARC

CHIAO, P'EI-SHU, 1913-1975

—DEATH AND BURIAL

(Chiao P'ei-shu hsien sheng chi nien chi)
焦沛澍先生紀念集. — [台北] : 焦沛澍先生治喪委員會, [1978?]
88 p., [36] leaves of plates : ill. ; 27 cm.
Caption title: Ku Chiao shou hsi chien ch'a kuan P'ei-shu hsien sheng chi nien chi.

1. Chiao, P'ei-shu, 1913-1975—Death and burial. 2. Statesmen—Taiwan—Biography. I. Title: Ku Chiao shou hsi chien ch'a kuan P'ei-shu hsien sheng chi nien chi.
DS799.82.C44C47 78-842395

CHIAO, YÜ-LU, 1922-1964

—JUVENILE FICTION

Ch'eng, I-ju.
(Chiao Yü-lu ho hai tzu men)
焦裕禄和孩子们 / 程逸汝编写 ; 倪绍勇, 张纪平画. — 上海 : 上海人民出版社, 1977.
[28] p. : col. ill. ; 16 cm.
Title also in pinyin on cover: Jiao Yulu he haizimen.
"連環圖畫"
$0 14

1. Chiao, Yü-lu, 1922-1964—Juvenile fiction. I. Ni, Shao-yung. II. Chang, Chi-p'ing. III. Title. IV. Title: Jiao Yulu he haizimen.
PZ10.831.C545C47 79-836645

CHIAO CHI T'IEH LU

Japan. Rikugun. Chintao Shubigun. Minseibu. Tetsudōbu.
(Santō Tetsudō Kaisha ni kansuru chōsa hōkoku)
山東鐵道會社ニ關スル調査報告. — [青島] : 青島守備軍民政部鐵道部, 大正9 [1920]
3, 417 p., [28] fold. leaves of plates : ill. ; 23 cm.
Cover title.

1. Chiao Chi t'ieh lu. 2. Railroads—China—Shantung. I. Title.
HE3290.C4J36 1920 79-802737

CHIAO HUI HSIN PAO

—INDEXES

Bennett, Adrian Arthur.
Research guide to the Chiao-hui hsin-pao (The Church news) 1868-1874 = Chiao hui hsin pao mu lu tao yao / compiled by Adrian A. Bennett. — San Francisco : Chinese Materials Center, 1975.
xvii, 342 p. ; 27 cm. — (Research aids series ; no. 1)
Includes index.
$8.00
1. Chiao hui hsin pao—Indexes. I. Title. II. Series.
PN5370.C55B4 059'.951 76-455108
76[r79]rev MARC

CHIAPAS, MEXICO

—BIOGRAPHY

Gordillo y Ortiz, Octavio.
Diccionario biográfico de Chiapas / Octavio Cordillo y Ortiz. — México : B. Costa-Amic, c1977.
xxiii, 295 p. ; 20 cm.
1. Chiapas, Mexico—Biography. I. Title.
CT557.C47G67 920'.072'7 78-108977
79 MARC

CHIÀPPORI, ALFREDO, 1943-

Alfredo Chiappori / premessa di Giulio Carlo Argan ; saggio introduttivo di Vittorio Fagone ; e inoltre testi di Mario Accolti Gil ... [et al.]. — Milano : Feltrinelli, 1978.
xxiii, 142 p. : ill. ; 23 cm. — (Media ; 1) It78-Nov
Bibliography: p. 141-142.
L8500
1. Chiàppori, Alfredo, 1943- I. Chàppori, Alfredo, 1943- II. Fagone, Vittorio. III. Accolti Gil, Mario.
PN6767.C4Z53 741'.092'4 79-351654
*79 MARC

CHIAPUSO, MANUEL, 1912-

Chiapuso, Manuel, 1912-
Los anarquistas y la guerra en Euskadi : la comuna de San Sebastián / [por] Manuel Chiapuso. — San Sebastián : Txertoa, D.L. 1977.
281 p. : ill. ; 21 cm. — (Colección Ipar haizea) Sp78-Feb
ISBN 8471480360
1. Spain—History—Civil War, 1936-1939. 2. San Sebastián, Spain—History. 3. Anarchism and anarchists—Spain—San Sebastián. 4. Spain—History—Civil War, 1936-1939—Personal narratives. 5. Chiapuso, Manuel, 1912- I. Title.
DP269.27.S2C45 946.081 78-375534
78 MARC

CHIARA, PIERO

—CRITICISM AND INTERPRETATION

Ghidetti, Enrico.
Invito alla lettura di Piero Chiara / Enrico Ghidetti. — Milano : Mursia, [1977]
103 p. ; 17 cm. — (Invito alla lettura : Sezione italiana ; 48) It78-June
Bibliography: p. [9]-96.
Includes indexes.
L2000
1. Chiara, Piero—Criticism and interpretation. I. Title.
PQ4809.H33Z68 853'.9'14 78-388439
*79 MARC

CHIAVENNA, ITALY

—HISTORY

Scaramellini, Guido.
Barbarossa ed Enrico il Leone a Chiavenna / Guido Scaramellini. — Chiavenna : Centro di studi storici valchiavennaschi, 1976.
61 p., [13] leaves of plates : ill. ; 25 cm. — (Quaderni del Centro di studi storici valchiavennaschi ; 5) It77-Sept
Includes bibliographical references.
1. Friedrich I, Barbarossa, Emperor of Germany, 1121-1190. 2. Heinrich der Löwe, Duke of Saxony, 1129-1195. 3. Lombard League—History. 4. Chiavenna, Italy—History. I. Title. II. Series: Centro di studi storici valchiavennaschi. Quaderni del Centro di studi valchiavennaschi ; 5.
DG657.37.S25 943'.024'0924 78-396101
*79 MARC

CHIBA, AYANO, 1889-

(Chiba Ayano, Kurumegasuri)
千葉あやの・久留米絣：正藍染・久留米絣技術保持者会. — 東京：講談社, 1978.
40 p. : ill. ; 36 cm. — (人間国宝シリーズ ; 20) Ja***

1. Chiba, Ayano, 1889- 2. Kurumegasuri Gijutsu Hojishakai. 3. Textile fabrics—Japan. 4. Dyes and dyeing—Textile fibers. 5. Indigo. I. Series: Ningen kokuhō shirizu ; 20.
NK9502.2.C46A4 1978 78-813307

CHIBA, SHŪSAKU, D. 1856

—FICTION

Yamaoka, Sōhachi, 1907-
(Chiba Shūsaku)
千葉周作：[特選時代小説] / 山岡荘八. — 東京：広済堂出版, [昭和52 i.e. 1977]
286 p. ; 19 cm. Ja 78B-89
¥780

1. Chiba, Shūsaku, d. 1856—Fiction.
PL842.A44C5 78-808481

CHIBA, JAPAN (CITY)

—DESCRIPTION—VIEWS

(Shashinshū Meiji Taishō Shōwa Chiba)
写真集明治大正昭和千葉 / 鳥海宗一郎編. — 東京：国書刊行会, 1978.
164 p. : chiefly ill. ; 31 cm. — (ふるさとの想い出 ; 14) Ja 79-4598
¥4800

1. Chiba, Japan (City)—Description—Views. 2. Chiba, Japan (City)—History—Pictorial works. I. Toriumi, Sōichirō, 1932- II. Series: Furusato no omoide ; 14.
DS897.C44S52 79-809101

—HISTORY—PICTORIAL WORKS

(Shashinshū Meiji Taishō Shōwa Chiba)
写真集明治大正昭和千葉 / 鳥海宗一郎編. — 東京：国書刊行会, 1978.
164 p. : chiefly ill. ; 31 cm. — (ふるさとの想い出 ; 14) Ja 79-4598
¥4800

1. Chiba, Japan (City)—Description—Views. 2. Chiba, Japan (City)—History—Pictorial works. I. Toriumi, Sōichirō, 1932- II. Series: Furusato no omoide ; 14.
DS897.C44S52 79-809101

—NIGHTCLUBS, DANCE HALLS, ETC.

Matsuya, Tomihiko, 1936-
(Hasuike nyonin kikigaki shō)
蓮池・女人聞き書抄 / 松谷富彦著. 思い出の花街・吾妻町界隈 / 小池淳達著. — 千葉：吾妻町青蓮会；東京：発売所 スタジオVIC, 1978.
246 p., [2] leaves of plates : ill. ; 22 cm. Ja***
"昭和10年頃の蓮池周辺 吾妻町青蓮会編纂" (1 fold. leaf) inserted.
¥2000
1. Chiba, Japan (City)—Social life and customs. 2. Chiba, Japan (City)—Nightclubs, dance halls, etc. 3. Geishas—Japan—Chiba (City) 4. Matsuya, Tomihiko, 1936- I. Koike, Atsusato, 1929- Omoide no hanamachi Azumachō kaiwai. 1978. II. Title.
DS897.C44M37 79-805136

—SOCIAL LIFE AND CUSTOMS

Matsuya, Tomihiko, 1936-
(Hasuike nyonin kikigaki shō)
蓮池・女人聞き書抄 / 松谷富彦著. 思い出の花街・吾妻町界隈 / 小池淳達著. — 千葉：吾妻町青蓮会；東京：発売所 スタジオVIC, 1978.
246 p., [2] leaves of plates : ill. ; 22 cm. Ja***
"昭和10年頃の蓮池周辺 吾妻町青蓮会編纂" (1 fold. leaf) inserted.
¥2000
1. Chiba, Japan (City)—Social life and customs. 2. Chiba, Japan (City)—Nightclubs, dance halls, etc. 3. Geishas—Japan—Chiba (City) 4. Matsuya, Tomihiko, 1936- I. Koike, Atsusato, 1929- Omoide no hanamachi Azumachō kaiwai. 1978. II. Title.
DS897.C44M37 79-805136

J

CHIBA, JAPAN (PREFECTURE)

—ANTIQUITIES

Chiba-ken Toshi Kōsha.
(Chiba Tōnambu nyū taun)
千葉東南部ニュータウン. — [東京]：日本住宅公団首都圏宅地開発本部, 1975-
v. : ill. ; 26 cm. Ja***
Vol. 3, [pt. 2]: Zuhanhen (205 p.)
"日本住宅公団首都圏宅地開発本部と … 千葉県都市公社との委託契約にもとづく … 埋蔵文化財調査報告"
CONTENTS: 1. 椎名崎古墳群(第1次)—2. 木戸作遺跡(第1次)—3. 有吉遺跡(第1次)
1. Chiba, Japan (Prefecture)—Antiquities. 2. Tombs—Japan—Chiba (Prefecture) 3. Japan—Antiquities. I. Nihon Jūtaku Kōdan. Shutoken Takuchi Kaihatsu Hombu. II. Title.
DS894.49.C4535C468 1975 79-802919

Chiba-ken Toshi Kōsha. Bunkazai Chōsa Jimusho.
(Chiba tōnambu nyū taun)
千葉東南部ニュータウン / [千葉県都市公社文化財調査事務所]. — 東京：日本住宅公団首都圏宅地開発本部, 1975-
v. : ill. ; 26 cm. Ja***
"有吉遺跡構配置図" (1 fold. leaf) inserted in v. 3.
"日本住宅公団首都圏宅地開発本部と … 千葉県都市公社との委託契約"
CONTENTS: 1. 椎名崎古墳群(第1次)—2. 木戸作遺跡(第1次)—3. 有吉遺跡(第1次)
1. Chiba, Japan (City)—Antiquities. 2. Japan—Antiquities. I. Nihon Jūtaku Kōdan. Shutoken Takuchi Kaihatsu Hombu. II. Title.
DS897.C44C48 1975 78-813522

Nakamura, Keiji, 1932-1977.
(Bōsō kofun ronkō)
房総古墳論攷 / 中村恵次著. — [千葉]：故中村恵次氏著作集刊行会, 1978.
237 p. : ill. ; 22 cm. Ja 78-16996
Includes bibliographical references.
List of the author's works: p. [231]-235.
Not in trade.
1. Chiba, Japan (Prefecture)—Antiquities. 2. Tombs—Japan—Chiba (Prefecture) 3. Japan—Antiquities. I. Title.
DS894.49.C4535N34 1978 78-813147

—BIOGRAPHY—ADDRESSES, ESSAYS, LECTURES

(Bōsō jimbutsu shiwa)
房総人物史話：歴史と人物 / 荒川法勝編. — 東京：昭和図書出版, 1978.
185 p. : ill. ; 19 cm. Ja***
¥1300

1. Chiba, Japan (Prefecture)—Biography—Addresses, essays, lectures. I. Arakawa, Hōshō, 1921-
DS894.49.C456B66 79-804131

—CIVILIZATION—PERIODICALS

(Narita: Naritasan Reikōkan Shiryōkan hō)
なりた 成田山霊光館史料館報

成田 成田山霊光館(史料館)
v. ill. 26 cm.

1. Naritasan Reikōkan Shiryōkan—Periodicals. 2. Chiba, Japan (Prefecture)—Civilization—Periodicals. 3. Narita, Japan—Civilization—Periodicals. I. Naritasan Reikōkan Shiryōkan.
DS894.49.C45A2953 78-645921
(MARC-S)
J

—DESCRIPTION AND TRAVEL

Ozawa, Jirōzaemon.
[Kazusa chōsonshi]
上總國町村誌 / [小沢治郎左衛門著]. — 東京：名著出版, 1978.
2 v. ; 22 cm. Ja 78-26430
"上總國輿地全圖" (fold. leaf) inserted.
Photoreprint of the 1889 ed. published under title: Kazusa chosonshi.
¥14000
1. Chiba, Japan (Prefecture)—Description and travel. I. Title.
Title romanized: Kazusa no Kuni chosonshi.
DS894.49.C452O9 1978 79-806840

—INDUSTRIES

Chiba, Japan (Prefecture). Kikakubu. Tōkeika.
(Chiba-ken messhu tōkei hōkoku)
千葉県メッシュ統計報告：昭和50年事業所統計調査編 / [編集責任者 深野一郎]. — [千葉]：千葉県企画部統計課, 昭和53 [1978]
20, 2, 484 p. : ill. ; 30 cm. Ja***
Cover title.
1. Chiba, Japan (Prefecture)—Industries. 2. Labor supply—Japan—Chiba (Prefecture)—Statistics. 3. Industrial statistics—Graphic methods. 4. Grids (Cartography) I. Fukano, Ichirō, 1925- II. Title.
HC463.C45C43 1978 79-809668

—SOCIAL LIFE AND CUSTOMS

Takahashi, Sumihisa.
(Ikite iru minzoku tambō Chiba)
生きている民俗探訪 千葉 / 高橋在久編著. — 東京：第一法規出版, 1978.
174 p., [2] leaves of plates : ill. ; 19 cm. Ja 78-16361
¥1300

1. Chiba, Japan (Prefecture)—Social life and customs. I. Title.
DS894.49.C454T34 78-812112

CHIBA, JAPAN (PREFECTURE). KAIKU GYOGYŌ CHŌSEI IINKAI

Chiba, Japan (Prefecture). Senkyo Kanri Iinkai.
(Shichōson senkyo kekka shirabe, Chiba Kaiku Gyogyō Chōsei Iinkai iin senkyo kekka shirabe)
市町村選挙結果調・千葉海区漁業調整委員会委員選挙結果調：統一地方選挙　昭和50年4月27日執行　市町村選挙　昭和49年4月-昭和52年8月海区漁業調整委員会委員選挙　昭和51年8月4日執行. — [千葉]：千葉県選挙管理委員会, [1977?]
158 p. ; 26 cm. Ja***
Cover title.
Chiefly tables.

1. Elections—Japan—Chiba (Prefecture) 2. Local elections—Japan—Chiba (Prefecture) 3. Chiba, Japan (Prefecture). Kaiku Gyogyō Chōsei Iinkai. I. Title.
JQ1693.C5C47 1977 78-814670
J

CHIBA, JAPAN (PREFECTURE). KEISATSU HOMBU

—HISTORY

Chiba, Japan (Prefecture). Keisatsu Hombu.
(Eikō no rekishi)
栄光の歴史：千葉県警察創立百年記念 / [千葉県警察本部編]. — [千葉]：同本部, [昭和49 i. e. 1974]
80 p. : ill. ; 31 cm. Ja 78-765

1. Police—Japan—Chiba (Prefecture)—History. 2. Chiba, Japan (Prefecture). Keisatsu Hombu—History. I. Title.
HV7827.C48C48 1974 78-810951
J

CHIBA, JAPAN (PREFECTURE) KENGIKAI

—ELECTIONS

Chiba, Japan (Prefecture). Senkyo Kanri Iinkai.
(Senkyo kekka shirabe, Shōwa gojūichinen jūnigatsu itsuka shikkō)
選挙結果調　昭和51年12月5日執行：衆議院議員選挙・最高裁判所裁判官国民審査・千葉県議会議員補欠選挙(鴨川市・天津小湊町選挙区). — [千葉]：千葉県選挙管理委員会, [はしがき　昭和52 i. e. 1977]
409 p., [3] leaves of plates : ill. ; 26 cm. Ja***
Cover title.
1. Elections—Japan—Chiba (Prefecture) 2. Japan. Kokkai. Shūgiin—Elections. 3. Japan. Saikō Saibansho. 4. Judges—Japan. 5. Chiba, Japan (Prefecture). Kengikai—Elections. I. Title.
JQ1693.C5C47 1977a 78-815229
J

CHIBA, JAPAN (PREFECTURE), IN LITERATURE

(Bōsō no bungaku meguri)
房総の文学めぐり / 千葉県高等学校教育研究会国語部会編. — 改訂版. — 東京：富士出版印刷, 昭和52 [1977]
264 p. : ill. ; 17 cm. — (郷土の文学散歩 ; 1) Ja 78-3593
¥400

1. Literary landmarks—Japan—Chiba (Prefecture) 2. Chiba, Japan (Prefecture), in literature. 3. Japanese literature—History and criticism. I. Chiba-ken Kōtō Gakkō Kyōiku Kenkyūkai. Kokugo Bukai. II. Series: Kyōdo no bungaku sampo ; 1.
PL724.2.C5B6 78-811617

CHIBA-KEN SANGYŌ SHIKENJŌ

—HISTORY

Chiba-ken Sangyō Shikenjō.
(Chiba-ken Sangyō Shikenjō rokujūgoshūnen kinen shi)
千葉県蚕業試験場65周年記念誌. — 東金：千葉県蚕業試験場, 昭和49 [1974]
211 p., [3] leaves of plates : ill. ; 22 cm. Ja 75-12004

1. Sericulture—Japan—Chiba (Prefecture)—Abstracts. 2. Chiba-ken Sangyō Shikenjō—History. I. Title.
SF542.C475 1974 79-806772
J

CHIBCHA INDIANS

Friede, Juan.
Los chibchas bajo la dominación española / Juan Friede. — Bogotá : La Carreta, [1974]
286 p., [2] leaves of plates : maps ; 22 cm. — (Serie Libros de bolsillo de La Carreta)
Published in 1960 under title: Descubrimiento del Nuevo Reino de Granada y fundación de Bogotá; in 1966 under title: Invasión del país de los chibchas.
Includes bibliographical references.
1. Chibcha Indians. 2. Indians, Treatment of—Colombia. 3. Colombia—History—To 1800. 4. Bogotá—History. I. Title.
F2270.2.C4F73 980'.004'98 79-383006
79 MARC

—ART

Rojas de Perdomo, Lucía.
Musica, alfarería y decoración / Lucía Rojas de Perdomo. — Bogotá : Ediciones Zazacuabi, c1977.
51 p. : ill. ; 23 cm. — (Colombia : Arte y cultura)
Bibliography: p. 49-50.
1. Chibcha Indians—Pottery. 2. Chibcha Indians—Art. 3. Indians of South America—Colombia—Pottery. 4. Indians of South America—Colombia—Art. I. Title. II. Series.
F2270.2.C4R62 738.3'7 78-396024
79 MARC

—POTTERY

Rojas de Perdomo, Lucía.
Musica, alfarería y decoración / Lucía Rojas de Perdomo. — Bogotá : Ediciones Zazacuabi, c1977.
51 p. : ill. ; 23 cm. — (Colombia : Arte y cultura)
Bibliography: p. 49-50.
1. Chibcha Indians—Pottery. 2. Chibcha Indians—Art. 3. Indians of South America—Colombia—Pottery. 4. Indians of South America—Colombia—Art. I. Title. II. Series.
F2270.2.C4R62 738.3'7 78-396024
79 MARC

—SOCIAL LIFE AND CUSTOMS

Rozo G , José.
Los muiscas : organización social y régimen político / José Rozo Gauta. — Bogotá : Fondo Editorial Suramérica, 1978.
216 p. : diagr. ; 18 cm. — (Colección Historia)
Includes bibliographical references.
1. Chibcha Indians—Social life and customs. 2. Chibcha Indians—Tribal government. 3. Indians of South America—Colombia—Social life and customs. 4. Indians of South America—Colombia—Tribal government. I. Title.
F2270.2.C4R68 301.45'19'80861 79-108159
79 MARC

—TRIBAL GOVERNMENT

Rozo G , José.
Los muiscas : organización social y régimen político / José Rozo Gauta. — Bogotá : Fondo Editorial Suramérica, 1978.
216 p. : diagr. ; 18 cm. — (Colección Historia)
Includes bibliographical references.
1. Chibcha Indians—Social life and customs. 2. Chibcha Indians—Tribal government. 3. Indians of South America—Colombia—Social life and customs. 4. Indians of South America—Colombia—Tribal government. I. Title.
F2270.2.C4R68 301.45'19'80861 79-108159
79 MARC

CHIBCHA LANGUAGE

—GRAMMAR

Lugo, Bernardo de, fl. 1619.
Gramatica en la lengua general del Nuevo Reyno, llamada Mosca / compuesto por fray Bernardo de Lugo. — Madrid : Ediciones Cultura Hispánica del Centro Iberoamericano de Cooperación, 1978.
158 leaves ; 18 cm. Sp***
Issued in case.
Reprint of the 1619 ed. published by B. de Guzmā, Madrid.
"Resurrección de una lengua, introducción a la edición facsimilar de la Gramática chibcha del padre fray Bernardo de Lugo editada en 1619, [por] Manuel Alvar": 47 p. inserted.
Includes bibliographical references.
ISBN 8472321622
1. Chibcha language—Grammar. I. Alvar López, Manuel. Resurrección de una lengua. 1978. II. Title.
PM5811.L8 1978 498 79-111906
79 MARC

CHIBCHAN LANGUAGES

see also Chibcha language

—COLLECTED WORKS

Estudios chibchas. — 1. ed. — Lomalinda, República de Colombia : Editorial Townsend, c1976-
v. ; 23 cm. — (Serie sintáctica ; no. 1-)
On t.p.: Ministerio de Gobierno, Dirección General de Integración y Desarrollo de la Comunidad, División Operativa de Asuntos Indígenas. Instituto Lingüístico de Verano.
Spanish or Chibcha.
Includes bibliographies.
CONTENTS: 1. Tunebo, Ica.
1. Chibchan languages—Collected works. I. Colombia. División Operativa de Asuntos Indígenas. II. Summer Institute of Linguistics. III. Series.
PM5812.A36 79-344521
79 MARC

CHIBEMBA LANGUAGE see Bemba language

CHICAGO

—BARS, SALOONS, ETC.—DIRECTORIES

McCarthy, Dennis B
The great Chicago bar & saloon guide : Chicago's 200 best bars / Dennis B. McCarthy. — 1st ed. — Chicago : Chicago Review Press, c1978.
xii, 158 p. ; 19 cm.
Includes indexes.
ISBN 0-914090-46-1 : $2.95
1. Chicago—Bars, saloons, etc.—Directories. I. Title.
TX950.57.I3M32 647'.95773'11 77-93721
79 MARC

—BIOGRAPHY—EXHIBITIONS

Jachimowicz, Elizabeth.
Eight Chicago women and their fashions, 1860-1929 / Elizabeth Jachimowicz ; color photography by Frederick D. Countiss. — Chicago : Chicago Historical Society, 1978.
65 p. : ill. (some col.) ; 31 cm.
Biographical accompaniment to an exhibition at the Chicago Historical Society.
Bibliography: p. 64.
1. Costume—Illinois—Chicago—Exhibitions. 2. Women—Illinois—Chicago—Biography—Exhibitions. 3. Chicago—Social life and customs—Exhibitions. 4. Chicago—Biography—Exhibitions. I. Chicago Historical Society. II. Title.
GT617.C47J32 391'.2'0977311 78-53885
78 MARC

—CHURCH HISTORY

Ziegenhals, Walter E 1928-
Urban churches in transition : reflections on selected problems and approaches to churches and communities in racial transition based on the Chicago experience / Walter E. Ziegenhals. — New York : Pilgrim Press, c1978.
201 p. ; 21 cm.
Bibliography: p. 195-201.
ISBN 0-8298-0355-6 : $5.95
1. City churches—Illinois—Chicago. 2. Church and race relations—Illinois—Chicago. 3. Chicago—Church history. 4. Chicago—Race relations. I. Title.
BV637.Z53 78-16422
261.8'34'510420977311
78 MARC

CHICAGO (Continued)

—CITY PLANNING—HISTORY

Stampley, James O 1915-
Challenges with changes : a documentary of Englewood / James O. Stampley. — [Chicago] : Stampley, c1979.
ix, 278 p. : ill. ; 22 cm.
Bibliography: p. 277-278.
ISBN 0-9602618-9-3
1. Englewood, Chicago, Ill.—City planning—History. 2. Chicago—City planning—History. 3. Englewood, Chicago, Ill.—Description. 4. Chicago—Description—1951- . 5. Community development, Urban—Illinois. 6. City planning—Illinois. I. Title.
HT168.E53S73 309.2'62'097731 79-117584
00 MARC

—CLUBS

Keiser, R Lincoln, 1937-
The Vice Lords : warriors of the streets / by R. Lincoln Keiser. — Fieldwork ed. — New York : Holt, Rinehart and Winston, c1979.
x, 100 p. : ill. ; 24 cm. — (Case studies in cultural anthropology)
Bibliography: p. 99-100.
ISBN 0-03-045396-8
1. Vice Lords (Organization) 2. Gangs—Illinois—Chicago. 3. Afro-American youth—Illinois—Chicago. 4. Afro-American criminals—Illinois—Chicago. 5. Chicago—Clubs. 6. Chicago—Social conditions. I. Title. II. Series.
F548.9.N4K44 1979 301.18'5 78-16442
78 MARC

—DESCRIPTION—1951-

Stampley, James O 1915-
Challenges with changes : a documentary of Englewood / James O. Stampley. — [Chicago] : Stampley, c1979.
ix, 278 p. : ill. ; 22 cm.
Bibliography: p. 277-278.
ISBN 0-9602618-9-3
1. Englewood, Chicago, Ill.—City planning—History. 2. Chicago—City planning—History. 3. Englewood, Chicago, Ill.—Description. 4. Chicago—Description—1951- . 5. Community development, Urban—Illinois. 6. City planning—Illinois. I. Title.
HT168.E53S73 309.2'62'097731 79-117584
00 MARC

—DWELLINGS

Block, Jean F
Hyde Park houses : an informal history, 1856-1910 / Jean F. Block ; photos. by Samuel W. Block, Jr. — Chicago : University of Chicago Press, 1978.
xii, 156 p., [36] leaves of plates : ill. ; 24 cm.
Bibliography: p. 147-152.
Includes index.
ISBN 0-226-06000-4
1. Chicago. Hyde Park. 2. Chicago—Dwellings. 3. Architecture—Illinois—Chicago. 4. Historic buildings—Illinois—Chicago. I. Block, Samuel W. II. Title.
F548.68.H9B57 977.3'11'04 78-3174
78 MARC

—ECONOMIC CONDITIONS

Chicago. Dept. of Development and Planning.
Industrial Chicago : elements for growth : the report of the Community Economic Development Demonstration Study / prepared by the Department of Development and Planning of the city of Chicago. — Chicago : The Department, 1977.
viii, 90, 39 p. : ill. ; 28 cm.
1. Chicago—Industries. 2. Chicago—Economic conditions. I. Title.
HC108.C4C37 1977 338'.09773'11 78-106877
79 MARC

—FIRE, 1871—PERSONAL NARRATIVES

The Great Chicago fire : in eyewitness accounts and 70 contemporary photographs and illustrations / compiled and edited with an introd. by David Lowe. — New York : Dover Publication, 1979.
87 p. : ill. ; 28 cm.
Text, compiled by M. McIlvaine, was originally published in 1915 under title: Reminiscences of Chicago during the great fire.
ISBN 0-486-23771-0 : $4.00
1. Chicago—Fire, 1871—Personal narratives. I. Lowe, David, 1933- . II. McIlvaine, Mabel, comp. Reminiscences of Chicago during the great fire.
F548.42.G66 1979 977.3'11'04 78-73518
79 MARC

—FIRE, 1958

McBride, Michele, 1945-
The fire that will not die / Michele McBride. — Palm Springs, Calif. : ETC Publications, c1979.
x, 240 p., [3] leaves of plates : ill. ; 24 cm.
ISBN 0-88280-066-3 : $8.95
1. Burns and scalds—Biography. 2. McBride, Michele, 1945- . 3. Chicago—Fire, 1958. 4. Our Lady of the Angels (School)—Fire, 1958. I. Title.
RD96.4.M28 362.1'9'7110926 78-31781
78 MARC

—HAYMARKET SQUARE RIOT, 1886

Zeisler, Sigmund, 1860-1931.
Reminiscences of the anarchist case, by Sigmund Zeisler. [Chicago] Chicago literary club, 1927.
37 p., 1 l. double facsim. 22 cm. (Club papers ... Chicago literary club)
"Edition, five hundred and seventy copies, printed for members of the club in the month of January nineteen hundred and twenty-seven."
——— Microfilm. Microfilm 30307 HX
1. Chicago—Haymarket Square Riot, 1886. I. Title.
HX846.C4Z4 27-4630
[r79]rev MARC

—HISTORY

Holt, Glen E
Chicago, a historical guide to the neighborhoods : the Loop and South Side / Glen E. Holt and Dominic A. Pacyga. — [Chicago] : Chicago Historical Society, 1979.
174 p. : ill. ; 20 x 23 cm.
Bibliography: p. 164-[167]
Includes index.
ISBN 0-913820-07-5
1. Chicago—History. I. Pacyga, Dominic A., joint author. II. Chicago Historical Society. III. Title.
F548.68.A1H64 977.3'11 78-60184
79 MARC

—INDUSTRIES

Chicago. Dept. of Development and Planning.
Industrial Chicago : elements for growth : the report of the Community Economic Development Demonstration Study / prepared by the Department of Development and Planning of the city of Chicago. — Chicago : The Department, 1977.
viii, 90, 39 p. : ill. ; 28 cm.
1. Chicago—Industries. 2. Chicago—Economic conditions. I. Title.
HC108.C4C37 1977 338'.09773'11 78-106877
79 MARC

—MANUFACTURES—DIRECTORIES

Chicago geographic zip-coded edition of the Illinois manufacturers directory.
Chicago, Manufacturers' News, inc.
v. 29 cm.
Key title: Chicago geographic zip-coded edition of the Illinois manufacturers directory, ISSN 0069-3383
1. Chicago—Manufactures—Directories.
T12.C53 338.7'67'025773 78-640713
MARC-S

—MAPS

Lasker, Toy.
Flashmaps, the instant guide to Chicago.
Chappaqua, N. Y., Flashmaps, inc.
v. ill. 21 cm. annual.
Key title: Flashmaps, the instant guide to Chicago, ISSN 0162-4121
1. Chicago—Maps. I. Title.
G1409.C6L3 912'.773'11 77-85321
MARC-S

—OFFICIALS AND EMPLOYEES —SALARIES, ALLOWANCES, ETC.—PERIODICALS

United States. Bureau of Labor Statistics. North Central Regional Office.
Municipal government wage survey, Chicago, Illinois. Chicago, U. S. Dept. of Labor, Bureau of Labor Statistics, North Central Region V.
v. 28 cm. (: North Central regional report)
(: Regional report - Bureau of Labor Statistics, North Central Regional Office)
1. Chicago—Officials and employees—Salaries, allowances, etc.—Periodicals. I. Title. II. Series: North Central regional report. III. Series: United States. Bureau of Labor Statistics. North Central Regional Office. Regional report - Bureau of Labor Statistics. North Central Regional Office.
HD8083.A4N63 331'.0977 s 78-641518
[HD4939.U6] [331.2'81'352077311] MARC-S

—POLICE—HISTORY—19TH CENTURY

Johnson, David Ralph, 1942-
Policing the urban underworld : the impact of crime on the development of the American police, 1800-1887 / David R. Johnson. — Philadelphia : Temple University Press, 1979.
viii, 240 p. ; 22 cm.
Bibliography: p. 233-236.
Includes index.
ISBN 0-87722-148-0 : $15.00
1. Police—United States—History—19th century. 2. Chicago—Police—History—19th century. 3. Philadelphia—Police—History—19th century. 4. Crime and criminals—United States—History—19th century. 5. Crime prevention—United States—History—19th century. I. Title.
HV8138.J58 363.2'09748'11 78-31220
78 MARC

—POLITICS AND GOVERNMENT

League of Women Voters of Chicago.
The key to our local government : Chicago, Cook County, metropolitan area / research, the League of Women Voters of Chicago and Cook County ; [editors, Barbara P. Fiske, Dodie Sugarman, research editor, Patricia Ceperich]. — Newly rev. 4th ed. — Chicago : Citizens Information Service of Illinois, [c1978]
288 p. : ill. ; 23 cm.
Includes index.
$3.95
1. Chicago—Politics and government. 2. Cook Co., Ill.—Politics and government. I. Fiske, Barbara P. II. Sugarman, Dodie. III. Ceprerich, Patricia. IV. League of Women Voters of Cook County. V. Citizens Information Service of Illinois. VI. Title.
JS708.L4 1978 352.0773'1 78-112136
79 MARC

—POOR

McSheehy, William.
Skid Row / by William McSheehy. — Boston : G. K. Hall, c1979.
v, 154 p. : ill. ; 25 cm.
Bibliography: p. 149-151.
Includes index.
ISBN 0-87073-187-4. ISBN 0-87073-181-5 pbk.
1. Tramps—Illinois—Chicago. 2. Chicago—Poor. 3. Chicago—Social conditions. 4. Subculture. I. Title.
HV4506.C5M3 301.2'2 78-11462
78 MARC

—RACE RELATIONS

Ziegenhals, Walter E 1928-
Urban churches in transition : reflections on selected problems and approaches to churches and communities in racial transition based on the Chicago experience / Walter E. Ziegenhals. — New York : Pilgrim Press, c1978.
201 p. ; 21 cm.
Bibliography: p. 195-201.
ISBN 0-8298-0355-6 : $5.95
1. City churches—Illinois—Chicago. 2. Church and race relations—Illinois—Chicago. 3. Chicago—Church history. 4. Chicago—Race relations. I. Title.
BV637.Z53 78-1642[illegible]
261.8'34'510420977311
78 MAR[illegible]

—RESTAURANTS—DIRECTORIES

Restaurants Chicago-style : recipes / collected by James Ward. — Boston : CBI Pub. Co., c1979.
vii, 307 p. : ill. ; 26 cm.
Includes index.
ISBN 0-8436-2116-8 : $14.95
1. Cookery, American—Illinois. 2. Chicago—Social life and customs. 3. Chicago—Restaurants—Directories. I. Ward, James, 1932- .
TX715.R426 641.5 79-996
79 MARC

—RIOT, APRIL 1968

Balbus, Isaac D
The dialectics of legal repression : Black rebels before the American criminal courts / Isaac D. Balbus. — New Brunswick, N.J. : Transaction Books, c1973.
xxiii, 269 p. : ill. ; 23 cm. — (Law and society)
Reprint of the ed. published by Russell Sage Foundation, New York, in series: Publications of the Russell Sage Foundation.
Based on the author's thesis, University of Chicago.
Includes bibliographical references and index.
ISBN 0-87855-609-5 : $4.95
1. Criminal justice, Administration of—United States. 2. Los Angeles—Riots, 1965. 3. Detroit—Riot, 1967. 4. Chicago—Riot, April 1968. I. Title.
[KF9223.B3 1973] 345'.73'05 75-44825
76 MARC

CHICAGO (Continued)

—SEWERAGE

Cain, Louis P
Sanitation strategy for a lakefront metropolis : the case of Chicago / Louis P. Cain. — DeKalb : Northern Illinois University Press, c1978.
xv, 173 p. : ill. ; 29 cm.
Bibliography: p. 162-167.
Includes index.
ISBN 0-87580-064-5
1. Chicago—Water-supply. 2. Chicago—Sewerage. 3. Sewage disposal—Illinois—Chicago. I. Title.
TD225.C5C22 363.6 76-14711
78 MARC

—SOCIAL CONDITIONS

Anderson, Elijah.
A place on the corner / Elijah Anderson. — Chicago : University of Chicago Press, 1978.
x, 237 p. ; 22 cm. — (Studies of urban society)
Bibliography: p. 227-234.
Includes index.
ISBN 0-226-01953-5
1. Small groups—Case studies. 2. Social status—Case studies. 3. Participant observation—Case studies. 4. Chicago—Social conditions. I. Title.
HM133.A55 301.44'0977311 78-1879
78 MARC

Keiser, R Lincoln, 1937-
The Vice Lords : warriors of the streets / by R. Lincoln Keiser. — Fieldwork ed. — New York : Holt, Rinehart and Winston, c1979.
x, 100 p. : ill. ; 24 cm. — (Case studies in cultural anthropology)
Bibliography: p. 99-100.
ISBN 0-03-045396-8
1. Vice Lords (Organization) 2. Gangs—Illinois—Chicago. 3. Afro-American youth—Illinois—Chicago. 4. Afro-American criminals—Illinois—Chicago. 5. Chicago—Clubs. 6. Chicago—Social conditions. I. Title. II. Series.
F548.9.N4K44 1979 301.18'5 78-16442
78 MARC

McSheehy, William.
Skid Row / by William McSheehy. — Boston : G. K. Hall, c1979.
v, 154 p. : ill. ; 25 cm.
Bibliography: p. 149-151.
Includes index.
ISBN 0-87073-187-4. ISBN 0-87073-181-5 pbk.
1. Tramps—Illinois—Chicago. 2. Chicago—Poor. 3. Chicago—Social conditions. 4. Subculture. I. Title.
HV4506.C5M3 301.2'2 78-11462
78 MARC

—SOCIAL LIFE AND CUSTOMS

Restaurants Chicago-style : recipes / collected by James Ward. — Boston : CBI Pub. Co., c1979.
vii, 307 p. : ill. ; 26 cm.
Includes index.
ISBN 0-8436-2116-8 : $14.95
1. Cookery, American—Illinois. 2. Chicago—Soical life and customs. 3. Chicago—Restaurants—Directories. I. Ward, James, 1932-
TX715.R426 641.5 79-996
79 MARC

—SOCIAL LIFE AND CUSTOMS —EXHIBITIONS

Jachimowicz, Elizabeth.
Eight Chicago women and their fashions, 1860-1929 / Elizabeth Jachimowicz ; color photography by Frederick D. Countiss. — Chicago : Chicago Historical Society, 1978.
65 p. : ill. (some col.) ; 31 cm.
Biographical accompaniment to an exhibition at the Chicago Historical Society.
Bibliography: p. 64.
1. Costume—Illinois—Chicago—Exhibitions. 2. Women—Illinois—Chicago—Biography—Exhibitions. 3. Chicago—Social life and customs—Exhibitions. 4. Chicago—Biography—Exhibitions. I. Chicago Historical Society. II. Title.
GT617.C47J32 391'.2'0977311 78-53885
78 MARC

—STATISTICS, MEDICAL

Bourn, Janet MacFarlane.
Incidence of Pfeiffer's bacillus in throats during epidemic and interepidemic periods in Chicago. 1928.
40 l. illus. 29 cm.
Typescript (carbon copy)
Thesis—University of Chicago.
Bibliography: l. 39-40.
1. Hemophilus influenzae—Identification. 2. Throat—Microbiology. 3. Epidemiology. 4. Chicago—Statistics, Medical. I. Title.
QR201.I6B68 78-322597
79 MARC

—TRANSIT SYSTEMS— ENVIRONMENTAL ASPECTS

American Bechtel, inc.
Environmental impact analysis, Chicago Central Area Transit Project / submitted to Chicago Urban Transportation District ⌜by⌝ American Bechtel, Inc. ; subcontractors & consultants, W. V. Rouse Associates, Ltd. ... ⌜et al.⌝. — Chicago : The District, ⌜197-⌝
273 leaves in various foliations : ill. ; 29 cm.
1. Chicago—Transit systems—Environmental aspects. 2. Chicago Central Area Transit Project. I. Chicago Urban Transportation District. II. Title.
TD195.T7A53 1970z 388.4'6'0977311 78-322143
79 MARC

—WATER-SUPPLY

Cain, Louis P
Sanitation strategy for a lakefront metropolis : the case of Chicago / Louis P. Cain. — DeKalb : Northern Illinois University Press, c1978.
xv, 173 p. : ill. ; 29 cm.
Bibliography: p. 162-167.
Includes index.
ISBN 0-87580-064-5
1. Chicago—Water-supply. 2. Chicago—Sewerage. 3. Sewage disposal—Illinois—Chicago. I. Title.
TD225.C5C22 363.6 76-14711
78 MARC

CHICAGO. ART INSTITUTE

(Emaki Jōruri)
繪巻上瑠璃：津山藩松平家伝来(岩佐又兵衛工房) / 著者　辻惟雄，坂田泉，信多純一. — 京都：京都書院，昭和52 ⌜1977⌝
401 p. : ill. (106 col.) ; 44 cm. Ja 77-17912
Issued in a case.
Colophon inserted.
Reproduced from the Atami Bijutsukan copy in 12 rolls, with printed transcript. Also includes photoreproduction of the Art Institute of Chicago copy in 2 rolls.
Includes bibliographical references.
¥120000
1. Emaki Jōruri (Scrolls) 2. Iwasa, Matabē, 1578-1650. 3. Matsudaira family—Art collections. 4. Atami Bijutsukan. 5. Chicago. Art Institute. I. Iwasa, Matabē, 1578-1650. II. Tsuji, Nobuo, 1932- III. Sakata, Izumi. IV. Shinoda, Jun'ichi, 1931- V. Jūnidan sōshi. 1977. VI. Title: Jōruri.
ND1053.5.E4 79-800171

—CATALOGS

Chicago. Art Institute.
The Art Institute of Chicago : 100 masterpieces / ⌜editor and coordinator, Janice J. Feldstein, co-editor, Maureen Smith⌝. — Chicago : The Institute : distributed by Rand McNally, c1978.
159 p. : ill. (some col.) ; 36 cm.
Includes indexes.
$35.00
1. Painting—Illinois—Chicago—Catalogs. 2. Chicago. Art Institute—Catalogs. I. Feldstein, Janice J. II. Smith, Maureen, 1945-
N530.A534 750'.74'017311 78-56322
78 MARC

CHICAGO. BASEBALL CLUB (AMERICAN LEAGUE)

Asinof, Eliot, 1919-
Bleeding between the lines / Eliot Asinof. — 1st ed. — New York : Holt, Rinehart, and Winston, c1979.
203 p. ; 22 cm.
ISBN 0-03-047536-8 : $9.95
1. Talent Associates. 2. Asinof, Eliot, 1919- . 3. Susskind, David, 1920- . 4. Copyright—Broadcasting rights—United States. 5. Television industry—United States. 6. Asinof, Eliot, 1919- Eight men out. 7. Chicago. Baseball club (American League) 8. World Series (Baseball) I. Title.
KF228.T28A85 343'.73'078 78-14181
79 MARC

CHICAGO. BOARD OF TRADE

Lurie, Jonathan, 1939-
The Chicago Board of Trade, 1859-1905 : the dynamics of self-regulation / by Jonathan Lurie. — Urbana : University of Illinois Press, c1979.
xiv, 234 p. : ill. ; 24 cm.
Bibliography: p. ⌜214⌝-232.
Includes index.
ISBN 0-252-00732-8
1. Chicago. Board of Trade. I. Title.
KF1087.C55L87 346'.73'092 78-20881
78 MARC

United States. Congress. House. Committee on Agriculture. Subcommittee on Conservation and Credit.
March wheat futures trading on Chicago Board of Trade : hearing before the Subcommittee on Conservation and Credit of the Committee on Agriculture, House of Representatives, Ninety-sixth Congress, first session, March 28, 1979. — Washington : U.S. Govt. Print. Off., 1979.
iii, 199 p. : graphs ; 23 cm.
"Serial no. 96-I."
1. Wheat trade—United States. 2. Chicago. Board of Trade. I. Title.
KF27.A3226 1979 332.6'44'0973 79-602582
00 MARC

—PERIODICALS

Chicago. Board of Trade.
Statistical annual - Chicago Board of Trade. 1969- ⌜Chicago, Board of Trade⌝
v. ill. 23-26 cm. annual.
Supersedes: Chicago. Board of Trade. Annual report.
Key title: Statistical annual - Chicago Board of Trade, ISSN 0163-5409
1. Chicago. Board of Trade—Periodicals. I. Key title.
HF296.C5453 332.6'44'0977311 78-648247
MARC-S

CHICAGO. CENTER FOR RESEARCH LIBRARIES

—CATALOGS

Chicago. Center for Research Libraries.
Supplementary list of journals accessible through the Center for Research Libraries. — ⌜Chicago⌝ : The Center, ⌜1976⌝
116 p. ; 28 cm.
Cover title.
"This issue supersedes the issue of Feb.-March 1975."
1. Periodicals—Bibliography—Catalogs. 2. Chicago. Center for Research Libraries—Catalogs. I. Title.
Z6945.C525 1976 016.05 78-326120
⌜PN4832⌝ 79 MARC

CHICAGO. FOOT-BALL CLUB (NATIONAL LEAGUE)

Payton, Walter, 1954-
Sweetness / Walter Payton with Jerry B. Jenkins. — Chicago : Contemporary Books, c1978.
xvii, 162 p., ⌜8⌝ leaves of plates : ill. ; 22 cm.
ISBN 0-8092-7544-9
1. Payton, Walter, 1954- . 2. Football players—United States—Biography. 3. Chicago. Foot-ball club (National League) I. Jenkins, Jerry B., joint author. II. Title.
GV939.P39A35 1978 796.33'2'0924 78-7553
78 MARC

CHICAGO. HYDE PARK

Block, Jean F
Hyde Park houses : an informal history, 1856-1910 / Jean F. Block ; photos. by Samuel W. Block, Jr. — Chicago : University of Chicago Press, 1978.
xii, 156 p., ⌜36⌝ leaves of plates : ill. ; 24 cm.
Bibliography: p. 147-152.
Includes index.
ISBN 0-226-06000-4
1. Chicago. Hyde Park. 2. Chicago—Dwellings. 3. Architecture—Illinois—Chicago. 4. Historic buildings—Illinois—Chicago. I. Block, Samuel W. II. Title.
F548.68.H9B57 977.3'11'04 78-3174
78 MARC

CHICAGO. UNIVERSITY. LIBRARY

—HISTORY—19TH CENTURY

Chicago. University. Library.
The Berlin Collection : being a history and exhibition of the books and manuscripts purchased in Berlin in 1891 for the University of Chicago by William Rainey Harper, with the support of nine citizens of Chicago, the Joseph Regenstein Library, April-October, 1979. — ⌜Chicago⌝ : University of Chicago Library, c1979.
iii, 88 p. : ill. ; 23 cm.
1. Chicago. University. Library—History—19th century. 2. Harper, William Rainey, 1856-1906. 3. Calvary, S., & Co., Berlin. 4. Bibliography—Early printed books—Exhibitions. 5. Manuscripts—Illinois—Chicago—Exhibitions. I. Title.
Z733.C535C48 1979 011 79-84896
79 MARC

CHICAGO. UNIVERSITY. PHILIPPINE STUDIES PROGRAM. SELECTED BIBLIOGRAPHY OF THE PHILIPPINES

Houston, Charles Orville, 1918-
Philippine studies; the poor relation in American scholarship [by] Charles O. Houston. Manila, 1957 [i.e. 1960]
20 p. 26 cm.
Cover title.
1. Chicago. University. Philippine Studies Program. Selected bibliography of the Philippines. 2. Philippine Islands—Bibliography. 3. Philippine Islands—Study and teaching—United States. I. Title.
Z3296.H68 78-316541
[DS655] 78 MARC

CHICAGO, BURLINGTON, AND QUINCY RAILROAD

Corbin, Bernard G
Steam locomotives of the Burlington route / by Bernard Corbin, William Kerka. — New York : Bonanza Books, [1978] c1960.
320 p. : ill. ; 29 cm.
ISBN 0-517-26195-2
1. Locomotives—History. 2. Chicago, Burlington and Quincy Railroad. I. Kerka, William, joint author. II. Title.
TJ603.3.C47C67 1978 385'.36'10977 78-10929
78 MARC

Holbrook, Daniel P
Waycars of the Chicago, Burlington & Quincy RR / Daniel P. Holbrook & Steven D. Lorenz. — Danvers, Ma. : Prototype Modeler, [1978]
iii, 107 p. : ill. ; 22 x 28 cm.
1. Cabooses (Railroads) 2. Chicago, Burlington, and Quincy Railroad. I. Lorenz, Steven D., joint author. II. Title.
TF485.H64 625.2'2 78-108792
78 MARC

—EMPLOYEES—BIOGRAPHY

Clayburn, Barbara B 1930-
Prairie stationmaster : the story of one man's railroading career in Nebraska 1917-1963 / by Barbara B. Clayburn. — Detroit : Harlo, c1979.
128 p. : ill. ; 23 cm.
ISBN 0-8187-0034-3
1. Bragg, George Nathan, 1897 or 8-1975. 2. Chicago, Burlington and Quincy Railroad—Employees—Biography. I. Title.
TF140.B6C57 385'.092'4 78-78305
79 MARC

CHICAGO, MILWAUKEE, ST. PAUL, AND PACIFIC RAILROAD

Rehder, Denny.
Grass between the rails : the Waukon, Iowa, branch of the Milwaukee Road / Denny Rehder with Cecil Cook. — Des Moines : Waukon & Mississippi Press, c1972.
208 p. : ill. ; 16 x 23 cm.
1. Chicago, Milwaukee, St. Paul, and Pacific Railroad. 2. Railroads—Iowa—Allamakee Co. I. Cook, Cecil, 1930- joint author. II. Title.
TF25.C5R43 385'.09777'33 72-94745
00 MARC

United States. Congress. House. Committee on Interstate and Foreign Commerce. Subcommittee on Transportation and Commerce.
Milwaukee Railroad bankruptcy : hearing before the Subcommittee on Transportation and Commerce of the Committee on Interstate and Foreign Commerce, House of Representatives, Ninety-fifth Congress, second session ... January 5, 1978. — Washington : U.S. Govt. Print. Off., 1978.
iii, 135 p. ; 24 cm.
"Serial no. 95-90."
1. Chicago, Malwaukee, St. Paul, and Pacific Railroad. 2. Railroad law—United States. 3. Corporate reorganizations—United States. I. Title.
KF27.I5589 1978a 385'.1'0973 78-602163
78 MARC

CHICAGO, NORTH SHORE AND MILWAUKEE RAILRAOD

Central Electric Railfans' Association.
Route of the electroliners. — 2d ed. — Chicago : Central Electric Railfans' Association, c1975.
188 p. : ill. ; 29 cm. — (Bulletin - Central Electric Railfans' Association ; 107)
Continues Interurban to Milwaukee.
ISBN 0-915348-07-1
1. Chicago, North Shore and Milwaukee Railraod. I. Title. II. Series: Central Electric Railfans' Association. CERA bulletin ; 107.
TF725.C43C46 1975 385'.09773'1 79-107263
79 MARC

CHICAGO CENTRAL AREA TRANSIT PROJECT

American Bechtel, inc.
Environmental impact analysis, Chicago Central Area Transit Project / submitted to Chicago Urban Transportation District [by] American Bechtel, Inc. ; subcontractors & consultants, W. V. Rouse Associates, Ltd. ... [et al.]. — Chicago : The District, [197-]
273 leaves in various foliations : ill. ; 29 cm.
1. Chicago—Transit systems—Environmental aspects. 2. Chicago Central Area Transit Project. I. Chicago Urban Transportation District. II. Title.
TD195.T7A53 1970z 388.4'6'0977311 78-322143
79 MARC

CHICAGO HOUSING AUTHORITY

United States. Congress. House. Committee on Government Operations. Manpower and Housing Subcommittee.
The Gautreaux decision and its effect on subsidized housing : hearing before a subcommittee of the Committee on Government Operations, House of Representatives, Ninety-fifth Congress, second session, September 22, 1978. — Washington : U.S. Govt. Print. Off., 1979.
iv, 195 p. : ill. ; 24 cm.
1. Housing—Law and legislation—United States. 2. Discrimination in housing—Law and legislation—United States. 3. Chicago Housing Authority. 4. Gautreaux, Dorothy. I. Title.
KF27.G6678 1978d 363.5 79-602221
79 MARC

—PERIODICALS

Chicago Housing Authority.
Annual report - Chicago Housing Authority.
[Chicago] Chicago Housing Authority.
v. ill. 23 x 31 cm.
Report year ends June 30.
Title varies slightly.
Continued by: Chicago Housing Authority. Chicago Housing Authority annual report.
1. Public housing — Illinois — Chicago — Periodicals. 2. Housing policy—Illinois—Chicago—Periodicals. 3. Chicago Housing Authority—Periodicals.
HD7304.C4C483b 352'.75'0977311 79-640151
MARC-S

Chicago Housing Authority.
Chicago Housing Authority annual report.
[Chicago, Chicago Housing Authority]
v. ill. 28 cm.
Continues: Chicago Housing Authority. Annual report - Chicago Housing Authority.
1. Public housing — Illinois — Chicago — Periodicals. 2. Housing policy—Illinois—Chicago—Periodicals. 3. Chicago Housing Authority—Periodicals. I. Chicago Housing Authority. Annual report.
HD7304.C4C483b 352'.75'0977311 79-640154
MARC-S

CHICAGO-O'HARE INTERNATIONAL AIRPORT

Northeastern Illinois Planning Commission.
O'Hare International Airport, Chicago, Illinois : technical report / prepared for U.S. Department of Housing and Urban Development, U.S. Department of Transportation by Northeastern Illinois Planning Commission. — Chicago : The Commission, 1971.
vi, 118 p. : ill. ; 28 cm.
At head of title: Metropolitan aircraft noise abatement policy study.
"Report no. HUD/DOT IANAP-71-1."
Bibliography: p. 108-109.
1. Airport noise—Illinois—Chicago. 2. Chicago-O'Hare International Airport. I. Title. II. Title: Metropolitan aircraft noise abatement policy study.
TL725.3.N6N67 1971 614.7'8 78-326470
79 MARC

CHICAGO REGION, ILL.

—DESCRIPTION AND TRAVEL—GUIDE-BOOKS

Benton, Chris.
Chicagoland nature trails / by Chris Benton. — Chicago : Contemporary Books, c1978.
xiii, 269 p. : ill. ; 21 cm.
Includes index.
ISBN 0-8092-7662-3
1. Hiking—Illinois—Chicago region—Guide-books. 2. Nature trails—Illinois—Chicago region—Guide-books. 3. Recreation areas—Illinois—Chicago region—Guide-books. 4. Chicago region, Ill.—Description and travel—Guide-books. I. Title.
GV199.42.I32C473 1978 917.73'.2'044 77-91183
78 MARC

CHICAGO SCHOOL OF ECONOMICS

Morgan, Brian, 1947-
Monetarists and Keynesians : their contribution to monetary theory / Brian Morgan. — London : Macmillan, 1978.
183 p. ; 22 cm. GB***
Bibliography: p. 172-177.
Includes indexes.
ISBN 0-333-22508-2 : £6.96. ISBN 0-333-14459-7 pbk.
1. Money. 2. Keynesian economics. 3. Chicago school of economics. I. Title.
HG220.A2M59 1978b 332.4'01 79-301341
79 MARC

Morgan, Brian, 1947-
Monetarists and Keynesians, their contribution to monetary theory / Brian Morgan. — New York : Wiley, 1978.
183 p. : ill. ; 23 cm.
"A Halsted Press book."
Bibliography: p. 172-177.
Includes indexes.
ISBN 0-470-99305-7
1. Money. 2. Keynesian economics. 3. Chicago school of economics. 4. Neoclassical school of economics. I. Title.
HG220.A2M59 1978 332.4'01 77-10905
77 MARC

Purvis, Douglas D
Monetarism : a review / Douglas D. Purvis. — Kingston, Ont. : Institute for Economic Research, Queen's University, 1978.
56 leaves ; 28 cm. — (Discussion paper - Institute for Economic Research, Queen's University ; no. 320 ISSN 0316-5078) C***
Review of Monetarism, published in 1976 by North Holland, Amsterdam.
Bibliography: leaves 50-56.
1. Chicago school of economics. 2. Money. I. Title. II. Series: Queen's University, Kingston, Ont. Institute for Economic Research. Discussion paper ; no. 320.
HB98.2.P87 322.4 79-366500
79 MARC

Robert, Dieter.
Makroökonomische Konzeptionen im Meinungsstreit : zur Auseinandersetzung zwischen Monetaristen u. Fiskalisten / Dieter Robert. — 1. Aufl. — Baden-Baden : Nomos-Verlagsgesellschaft, 1978.
160 p. ; 23 cm. — (Schriften zur monetären Ökonomie ; Bd. 5)
GFR78-A
Bibliography: p. 141-154.
Includes indexes.
ISBN 3-7890-0361-1
1. Monetary policy. 2. Fiscal policy. 3. Chicago school of economics. 4. Keynesian economics. I. Title. II. Series.
HG255.R58 332.4 78-375506
78 MARC

—ADDRESSES, ESSAYS, LECTURES

Monetarismo y capitalismo periférico : el caso de Chile / Jürgen Schuldt, ed. — Lima : Universidad del Pacífico, Centro de Investigación, [1977 or 1978]
141 p. : ill. ; 30 cm. — (Serie Ensayos - Universidad del Pacífico, Centro de Investigación ; no. 13)
Includes bibliographical references.
CONTENTS: Schuldt, J. La "Escuela de Chicago" y los "Chicago Boys" en Chile.—Frank, A. G. Carta abierta a Arnold Harberger.—Friedman, M. Bases para un desarrollo económico de Chile.—Spartacus Youth League. Carta a Milton Friedman.—Friedman, M. Respuesta de Friedman al Spartacus.—Cauas, J. Entrevista al ministro de economía chileno.—Chossudowsky, M. Economía de Chicago a la chilena.—Frank, A. G. Segunda carta abierta a Arnold Harberger y Milton Friedman.—Letelier, O. Chile, dos caras de un mismo modelo.
1. Chile—Economic policy—Addresses, essays, lectures. 2. Chicago school of economics—Addresses, essays, lectures. 3. Monetary policy—Addresses, essays, lectures. 4. Capitalism—Addresses, essays, lectures. I. Schuldt Lange, Juergen. II. Series: Universidad del Pacífico. Centro de Investigación. Serie Ensayos - Universidad del Pacífico, Centro de Investigación ; no. 13.
HC192.M658 79-111638
79 MARC

THE CHICAGO TRIBUNE

Gies, Joseph.
The Colonel of Chicago / by Joseph Gies. — 1st ed. — New York : Dutton, c1979.
261 p., [7] leaves of plates : ill. ; 25 cm.
Bibliography: p. 246-250.
Includes index.
ISBN 0-525-08267-0 : $12.95
1. McCormick, Robert Rutherford, 1880-1955. 2. The Chicago tribune. I. Title.
PN4874.M36G5 070.4'092'4 79-967
79 MARC

CHICANERY (LAW)

—ADDRESSES, ESSAYS, LECTURES

Suekawa, Hiroshi, 1892-1977.
(Fuhō kōi narabini kenri ran'yō no kenkyū)
不法行爲竝に權利濫用の研究 / 末川博著. — 東京 : 岩波書店, 昭和8 [1933]
3, 7, 374 p. ; 23 cm.
Includes bibliographical references.

1. Torts—Addresses, essays, lectures. 2. Chicanery (Law)—Addresses, essays, lectures. 3. Civil rights—Addresses, essays, lectures. I. Title.
79-805839

Suekawa, Hiroshi, 1892-1977.
(Kenri ran'yō no kenkyū)
權利濫用の研究 / 末川博著. — 東京 : 岩波書店, 昭和24 [1949]
4, 6, 309 p. ; 22 cm.
Colophon inserted.
Includes bibliographical references.

1. Chicanery (Law)—Addresses, essays, lectures. 2. Chicanery (Law)—Japan—Addresses, essays, lectures. I. Title.
K579.C48S84 79-805738

—ITALY

Prato, Maurizio.
Il negozio in frode alla legge : diritto e pratica notarile / Maurizio Prato. — Roma : Stamperia nazionale, 1976.
130 p. ; 22 cm. It78-Mar
Includes bibliographical references.
L2650
1. Illegal contracts—Italy. 2. Fraud—Italy. 3. Chicanery (Law)—Italy. I. Title.
346'.45'02 78-387635
*79 MARC

—JAPAN—ADDRESSES, ESSAYS LECTURES

Suekawa, Hiroshi, 1892-1977.
(Kenri ran'yō no kenkyū)
權利濫用の研究 / 末川博著. — 東京 : 岩波書店, 昭和24 [1949]
4, 6, 309 p. ; 22 cm.
Colophon inserted.
Includes bibliographical references.

1. Chicanery (Law)—Addresses, essays, lectures. 2. Chicanery (Law)—Japan—Addresses, essays, lectures. I. Title.
K579.C48S84 79-805738

CHICANERY see Deception

CHICANO COOKERY see Cookery, Mexican American

CHICANO LITERATURE (ENGLISH) see American literature—Mexican-American authors

CHICANO LITERATURE (SPANISH) see Mexican American literature (Spanish)

CHICANOS see Mexican Americans

CHICAS DE CLUB. [MOTION PICTURES]

Grau, Jorge, 1930-
Cantico a unas chicas de club / Jorge Grau. — Barcelona : Ediciones Paulinas, 1975.
302 p., [8] leaves of plates : ill. ; 18 cm. — (Serie Testimonio) (C3S)
Sp76-Feb
ISBN 842850587X : 200ptas
1. Chicas de club. [Motion picture] I. Title.
PN1997.C46423G7 791.43'7 76-474100
*76[r79]rev MARC

CHICHÉ, FRANCE

—HISTORY

—DESCRIPTION—VIEWS

Coudreau, Jean Marie.
Chiché en Poitou / Jean-Marie Coudreau ; ill. de l'auteur. — [Chiché : J. M. Coudreau], c1976.
165 p. : ill. ; 21 cm. F***
Includes bibliographical references.
1. Chiché, France—History. I. Title.
DC801.C5216C68 944'.62 78-390185
79 MARC

CHICHÉN ITZÁ, MEXICO

Folan, William J
Chichén Itzá / William J. Folan ; [fotografía color, Adrián García Valadés, fotografía blanco y negro, Alfredo Gómez Tagle y Adrián García Valadés]. — 1. ed. — México : Ediciones Orto : distribuidor, DICESA, 1977.
62 p., [16] leaves of plates : ill. (some col.) ; 21 cm.
Bibliography: p. 60-62.
1. Chichén Itzá, Mexico. 2. Mayas—Antiquities. I. Title.
F1219.1.Y8F64 917.2'6 78-106603
79 MARC

CHICHESTER, ENG.

—DESCRIPTION—VIEWS

Price, Bernard.
Chichester : the valiant years / by Bernard Price. — London : Phillimore, 1978.
[121] p. : chiefly ill. ; 25 cm. GB***
ISBN 0-85033-322-9 : £4.95
1. Chichester, Eng.—Description—Views. 2. Chichester, Eng.—Social life and customs—Pictorial works. 3. World War, 1939-1945—England—Chichester—Pictorial works. I. Title.
DA690.C53P74 942.2'62 79-313698
79 MARC

—SOCIAL LIFE AND CUSTOMS—PICTORIAL WORKS

Price, Bernard.
Chichester : the valiant years / by Bernard Price. — London : Phillimore, 1978.
[121] p. : chiefly ill. ; 25 cm. GB***
ISBN 0-85033-322-9 : £4.95
1. Chichester, Eng.—Description—Views. 2. Chichester, Eng.—Social life and customs—Pictorial works. 3. World War, 1939-1945—England—Chichester—Pictorial works. I. Title.
DA690.C53P74 942.2'62 79-313698
79 MARC

CHICHIBU, JAPAN

—DESCRIPTION—VIEWS

Minami, Yoshikazu, 1935-
(Chichibu)
秩父 / 写真 南良和 ; 文 井出孫六. — 東京 : 家の光協会, 1978.
181 p. : ill. (some col.) ; 38 cm. Ja 78-7864
¥8500

1. Chichibu, Japan—Description—Views. 2. Chichibu, Japan—History. I. Ide, Magoroku, 1931- II. Title.
DS897.C45M56 78-811399

—HISTORY

Minami, Yoshikazu, 1935-
(Chichibu)
秩父 / 写真 南良和 ; 文 井出孫六. — 東京 : 家の光協会, 1978.
181 p. : ill. (some col.) ; 38 cm. Ja 78-7864
¥8500

1. Chichibu, Japan—Description—Views. 2. Chichibu, Japan—History. I. Ide, Magoroku, 1931- II. Title.
DS897.C45M56 78-811399

CHICHIBU PEASANT UPRISING, 1884

(Chichibu jiken shiryō)
秩父事件史料 浦和 埼玉新聞社出版部 昭和45-53 (1970-78)
5 v. 22 cm. Ja 71-1957 (v. 1)

1. Chichibu Peasant Uprising, 1884. 2. Chichibu Peasant Uprising, 1884—Bibliography. 3. Chichibu Peasant Uprising, 1884—Sources. I. Saitama Shimbun Sha.
DS882.5.C45 71-796714
J

Ide, Magoroku, 1931-
(Chichibu Kommintō)
秩父困民党 / 井出孫六. — 東京 : 講談社, 1979.
245 p. : ill. ; 18 cm. — (講談社現代新書 ; 531) Ja 79-5385
Series romanized: Kōdansha gendai shinsho.
¥390

1. Chichibu Peasant Uprising, 1884. I. Title.
DS882.5.I29 79-809008

(Tama no hyakunen)
多摩の百年 / [著者] 朝日新聞東京本社社会部. — 東京 : 朝日新聞社, 1976.
2 v. : ill. ; 19 cm. Ja 76-12139 (v. 1)
CONTENTS: 1. 悲劇の群像—2. 絹の道
¥920 (per vol.)

1. Tama region, Japan—History. 2. Chichibu Peasant Uprising, 1884. I. Asahi Shimbun Sha. Tōkyō Honsha. Shakaibu.
DS896.7.T34T25 76-804783
rev
6 [r78] J

Toi, Shōzō, 1923-
(Chichibu jiken o aruku)
秩父事件を歩く : 困民党の風土と人 / 戸井昌造. — 東京 : 新人物往来社, 1978.
314 p., [2] leaves of plates : ill. ; 22 cm. Ja 78-25968
¥3000

1. Chichibu Peasant Uprising, 1884. I. Title.
DS882.5.T58 79-803577

—BIBLIOGRAPHY

(Chichibu jiken shiryō)
秩父事件史料 浦和 埼玉新聞社出版部 昭和45-53 (1970-78)
5 v. 22 cm. Ja 71-1957 (v. 1)

1. Chichibu Peasant Uprising, 1884. 2. Chichibu Peasant Uprising, 1884—Bibliography. 3. Chichibu Peasant Uprising, 1884—Sources. I. Saitama Shimbun Sha.
DS882.5.C45 71-796714

Nishino, Tatsukichi, 1916-
(Chichibu Kommintō)
秩父困民党 西野辰吉著 東京 東邦出版社 昭和43 (1968)
222 p. 19 cm. Ja 68-13466
Originally published in Shin Nihon bungaku, Mar. 1954-Feb. 1956.

1. Chichibu Peasant Uprising, 1884—Fiction. I. Title.
PL834.I755C5 1968 77-806933

CHICHIBU PEASANT UPRISING, 1884
(Continued)

—SOURCES

(Chichibu jiken shiryō)
秩父事件史料　浦和　埼玉新聞社出版部　昭和45-53 (1970-78)
5 v. 22 cm. Ja 71-1957 (v. 1)

1. Chichibu Peasant Uprising, 1884. 2. Chichibu Peasant Uprising, 1884—Bibliography. 3. Chichibu Peasant Uprising, 1884—Sources. I. Saitama Shimbun Sha.
DS882.5.C45 71-796714
J

CHICHIBU REGION, JAPAN

—ADDRESSES, ESSAYS, LECTURES

Hara, Zenkyō.
(Oku Chichibu kaiki)
奥秩父回帰 / 原全教. — 東京：河出書房新社, 1978.
iv, 252 p. ; 20 cm. Ja 78-14460
Bibliography: p. 191.
¥1200

1. Chichibu region, Japan—Addresses, essays, lectures. 2. Chichibu region, Japan—Description and travel—Addresses, essays, lectures. 3. Hara, Zenkyō—Addresses, essays, lectures. I. Title.
DS894.49.S249C535 78-812319

—DESCRIPTION AND TRAVEL —ADDRESSES, ESSAYS. LECTURES

Hara, Zenkyō.
(Oku Chichibu kaiki)
奥秩父回帰 / 原全教. — 東京：河出書房新社, 1978.
iv, 252 p. ; 20 cm. Ja 78-14460
Bibliography: p. 191.
¥1200

1. Chichibu region, Japan—Addresses, essays, lectures. 2. Chichibu region, Japan—Description and travel—Addresses, essays, lectures. 3. Hara, Zenkyō—Addresses, essays, lectures. I. Title.
DS894.49.S249C535 78-812319

CHICHIBU-TAMA NATIONAL PARK, JAPAN

(Chichibu Tama)
秩父多摩 / [解説　朝日新聞社；写真　清水武甲]. — 東京：朝日新聞社, 昭和35 [1960]
56 p. : ill. ; 26 cm. — (アサヒ写真ブック ; 101)
Cover title.
In colophon: Henja, Shimizu Bukō.
Series romanized: Asahi shashin bukku.

1. Chichibu-Tama National Park, Japan. I. Shimizu, Bukō, 1913– II. Asahi Shimbun Sha.
DS894.495.C45C45 79-810021

CHICHKOV, ÎURIĬ MIKHAĬLOVICH, 1929-

Chichkov, ÎUriĭ Mikhaĭlovich, 1929-
Pesni nashego serdîsa : razgovor kompozitora s vozhatym o pionerskoĭ pesne / ÎUriĭ Chichkov. — Moskva : Molodaîa gvardiîa, 1976.
111 p. : ill. ; 20 cm. — (Dlîa tekh, kto rabotaet s pionerami)
Songs: p. 43-111.
1. Chichkov, ÎUriĭ Mikhaĭlovich, 1929- 2. Vsesoîuznaîa pionerskaîa organizaîsiîa imeni V. I. Lenina—Songs and music. I. Title.
ML410.C526A3 79-383005
79 MARC

CHICK EMBRYO

Mess, B
Differentiation and function of the hypophyseal-target organ system in chicken embryos, by B. Mess and K. Straznicky. Budapest, Akadémiai Kiadó, 1970.
99 p. illus. 24 cm. (Studia biologica Hungarica 9)
Bibliography: p. 93-98.
1. Hypothalamo-hypophyseal system. 2. Pituitary hormones—Physiological effect. 3. Cell differentiation. 4. Chick embryo. I. Straznicky, K., joint author. II. Title. III. Series: Magyar Tudományos Akadémia, Budapest. Studia biologica, 9.
QP188.H9M47 598.6'1 79-317417
79 MARC

—JUVENILE LITERATURE

The Chicken and the egg / by Oxford Scientific Films ; photos. by George Bernard and Peter Parks. — 1st American ed. — New York : Putnam, 1979.
[30] p. : col. ill. ; 20 x 26 cm.
SUMMARY: Text and photos focus on the behavior, pecking order, and egg production of the most numerous species of bird.
ISBN 0-399-20676-0 : $6.95
1. Chickens—Juvenile literature. 2. Chickens—Behavior—Juvenile literature. 3. Chick embryo—Juvenile literature. [1. Chickens—Habits and behavior] I. Bernard, George, 1949- . II. Parks, Peter, 1942- . III. Oxford Scientific Films.
SF487.5.C44 1979 636.5 78-23663
78 MARC

CHICKASAW CO., IOWA

—MAPS

Rockford Map Publishers, Rockford, Ill.
Land atlas & plat book, Chickasaw County, Iowa : 1977. — Rockford, Ill. : Rockford Map Publishers ; New Hampton, Iowa : distributed by Chickasaw County Auditor, County Courthouse, [c1977]
38 p. : maps ; 28 cm.
Cover title.
Cadastral maps of townships, scale 1:50,688 or 1 1/4 inch equals one mile.
Includes index.
$9.00
1. Chickasaw Co., Iowa—Maps. 2. Real property—Iowa—Chickasaw Co.—Maps. I. Title.
G1433.C55R6 1977 912'.777'315 79-318255
79 MARC

CHICKASAW CO., MISS

—CENSUS, 1860

Felldin, Jeanne Robey.
The 1860 census of Chic[k]asaw County, Mississippi : with an added surname index / [compiled] by Jeanne Robey Felldin and Charlotte Magee Tucker. — Tomball, TX : Genealogical Publications, c1978.
185, vii p. ; 29 cm.
1. Chickasaw Co., Miss.—Census, 1860. 2. Chickasaw Co., Miss.—Genealogy. I. Tucker, Charlotte Magee, joint author. II. United States. Census Office. 8th census, 1860. III. Title.
F347.C4F44 929'.3762'942 79-110270
00 MARC

—GENEALOGY

Felldin, Jeanne Robey.
The 1860 census of Chic[k]asaw County, Mississippi : with an added surname index / [compiled] by Jeanne Robey Felldin and Charlotte Magee Tucker. — Tomball, TX : Genealogical Publications, c1978.
185, vii p. ; 29 cm.
1. Chickasaw Co., Miss.—Census, 1860. 2. Chickasaw Co., Miss.—Genealogy. I. Tucker, Charlotte Magee, joint author. II. United States. Census Office. 8th census, 1860. III. Title.
F347.C4F44 929'.3762'942 79-110270
00 MARC

CHICKASAW TRADE LANGUAGE see Mobilian trade language

CHICKEN BREEDS

see also Old English Game (Chickens); and names of specific breeds, e. g. Leghorns (Poultry; Plymouth Rocks (Poultry)

CHICKENS

see also Chick embryo; Chicks; Cookery (Chicken)

Nesheim, Malden C
Poultry production / Malden C. Nesheim, Richard E. Austic, Leslie E. Card. — 12th ed. — Philadelphia : Lea & Febiger, 1979.
vii, 399 p. : ill. ; 25 cm.
Eleventh ed. (1972) by L. E. Card and M. C. Nesheim.
Includes index.
ISBN 0-8121-0665-2
1. Chickens. 2. Poultry. I. Austic, Richard E., joint author. II. Card, Leslie Ellsworth, 1893- Poultry production. III. Title.
SF487.C28 1979 636.5'08 78-31386
78 MARC

North, Mack O
Commercial chicken production manual / Mack O. North. — 2d ed. — Westport, Conn. : Avi Pub. Co., c1978.
x, 692 p. : ill. ; 26 cm.
Includes index.
ISBN 0-87055-259-7 : $36.00
1. Chickens. I. Title.
SF487.N77 1978 636.5 78-5973
78 MARC

Singer, Andrew.
The backyard poultry book / Andrew Singer ; [illustrated by Colin Spooner]. — New York : Arco Pub. Co., 1979, c1976.
111 p. : ill. ; 21 cm.
Bibliography: p. 107.
Includes index.
ISBN 0-668-04305-9 : $2.95
1. Poultry. 2. Chickens. I. Title.
SF487.S569 1979 636.5 77-3374
77 MARC

—BEHAVIOR—JUVENILE LITERATURE

The Chicken and the egg / by Oxford Scientific Films ; photos. by George Bernard and Peter Parks. — 1st American ed. — New York : Putnam, 1979.
[30] p. : col. ill. ; 20 x 26 cm.
SUMMARY: Text and photos focus on the behavior, pecking order, and egg production of the most numerous species of bird.
ISBN 0-399-20676-0 : $6.95
1. Chickens—Juvenile literature. 2. Chickens—Behavior—Juvenile literature. 3. Chick embryo—Juvenile literature. [1. Chickens—Habits and behavior] I. Bernard, George, 1949- . II. Parks, Peter, 1942- . III. Oxford Scientific Films.
SF487.5.C44 1979 636.5 78-23663
78 MARC

—BREEDING

Uusitalo, Heikki, 1945-
Valintaindeksien rakentaminen kanojen jalostusarvostelua varten / Heikki Uusitalo. — Viikki : Kotieläinten jalostustieteen laitos, Helsingin yliopisto, 1975.
119 p. ; 21 cm. — (Kotieläinjalostuksen tiedote ; 1) Fi75
Errata slip inserted.
Bibliography: p. 115-119.
ISBN 9514505948
1. Chickens—Breeding. I. Title. II. Series.
SF492.U95 78-377876
*78 MARC

—DISEASES

Amaral, Efigénia da Conceição Fernandes Serra do.
Contribuição para o estudo da doença respiratória crónica das aves; valor das associações espiramicina-furaltadona e tilosina-furaltadona. Luanda, Instituto de Investigação Científica de Angola, 1969.
105 p. illus. 25 cm.
Summary in Portuguese, French, and English.
Bibliography: p. 89-101.
1. Chickens—Diseases. 2. Respiratory organs—Diseases. I. Title.
SF995.A53 1969 78-390900
79 MARC

Prastowo, Bambang Probo.
Pemeriksaan pendahuluan terhadap beberapa species tertentu coccidia pada ajam dari Bogor dan Bandung. A preliminary search of certain species of chicken coccidia in Bogor and Bandung, oleh Bambang Probo Prastowo, Gatut Ashadi dan Soeprapto Soekardono. Bogor, Pertjetakan IPB [1967]
8 p. 22 cm. (Institut Pertanian Bogor. Buletin penelitian no. 8, 1967)
Cover title.
In Indonesian; with summary in English.
Bibliography: p. 8.
1. Coccidiosis. 2. Chickens—Disease. 3. Eimeria. I. Ashadi, Gatut, joint author. II. Soekardono, Soeprapto, joint author. III. Title. IV. Series: Institut Pertanian Bogor. Buletin penelitian, 1967, no. 8.
S405.I66 no. 8 79-949590
[SF995.6.C6] 79 MARC

CHICKENS (Continued)

—ECONOMIC ASPECTS—FRANCE

La Gestion technique et économique des élevages de poulettes et pondeuses / Session organisée à Ploufragan, le 21 décembre 1976 ... ITAVI [Institut technique de l'aviculture]. — Paris (28, rue du Rocher, 75008) : ITAVI, 1977, 1978 printing.
37 leaves ; 30 cm. — (Document économique de l'ITAVI) F78-12466
30.00F
1. Chickens—Economic aspects—France. I. Guegan, Y. II. Institut technique de l'aviculture. III. Series: Institut technique de l'aviculture. Document économique de l'ITAVI.
[SF488.F8G45] 338.1'6 79-352653
Shared Cataloging for DNAL 79 MARC

—HATCHERIES

Great Britain. Ministry of Agriculture, Fisheries and Food.
Incubation and hatchery practice / Ministry of Agriculture, Fisheries and Food. — 6th ed. — London : H.M. Stationery Off., 1977.
v, 73 p., [4] leaves of plates : ill. ; 25 cm. — (Bulletin - Ministry of Agriculture, Fisheries and Food ; 148) GB***
Bibliography: p. 72-73.
ISBN 0-11-241448-8 : £1.75
1. Eggs—Incubation. 2. Chickens—Hatcheries. I. Title. II. Series: Great Britain. Ministry of Agriculture, Fisheries and Food. Bulletin ; 148.
S217.A6132 no. 148, 1977 78-321804
630'.8 s
[SF495] 79 MARC

—HOUSING

see also Chickens—Hatcheries

—JUVENILE LITERATURE

The Chicken and the egg / by Oxford Scientific Films ; photos. by George Bernard and Peter Parks. — 1st American ed. — New York : Putnam, 1979.
[30] p. : col. ill. ; 20 x 26 cm.
SUMMARY: Text and photos focus on the behavior, pecking order, and egg production of the most numerous species of bird.
ISBN 0-399-20676-0 : $6.95
1. Chickens—Juvenile literature. 2. Chickens—Behavior—Juvenile literature. 3. Chick embryo—Juvenile literature. [1. Chickens—Habits and behavior] I. Bernard, George, 1949- . II. Parks, Peter, 1942- . III. Oxford Scientific Films.
SF487.5.C44 1979 636.5 78-23663
78 MARC

CHICKS

—EUROPEAN ECONOMIC COMMUNITY COUNTRIES —STATISTICS—PERIODICALS

Statistical Office of the European Communities.
Landbrug: måneds statistik: æg. Landwirtschaft: Monatsstatistik: Eier. Agriculture: monthly statistics: eggs. [Luxembourg, De europæiske fællesskabers statistiske kontor]
v. 30 cm. (Interne meddelelser - De europæiske fællesskabers statistiske kontor)
Danish, Dutch, English, French, German, and Italian.
1. Poultry hatcheries—European Economic Community countries—Statistics—Periodicals. 2. Eggs—European Economic Community countries—Statistics—Periodicals. 3. Egg trade—European Economic Community countries—Statistics—Periodicals. 4. Chicks—European Economic Community countries—Statistics—Periodicals. 5. Poultry industry—European Economic Community countries—Statistics—Periodicals. I. Statistical Office of the European Communities. Landwirtschaft: Monatsstatistik: Eier. II. Statistical Office of the European Communities. Agriculture: monthly statistics: eggs. III. Title. IV. Series: Statistical Office of the European Communities. Interne meddelelser - [De europæiske fællesskabers statistiske kontor]
SF496.E86S7a 78-645848
MARC-S

CHICLAYO, PERU (PROVINCE)

—HISTORY

Ibáñez, Nito.
Salas, Incahuasi, Penachí en la novelesca vida de los brujos : narraciones / de Nito Ibáñez ; [prólogo de José Bracamonte Vera ; portada y dibujos, P. Larzzi Vergara Carrillo]. — [Lima : E. Bracamonte Vera], 1977.
138 p. : ill. ; 25 cm.
S/500.00
1. Witchcraft—Peru—Chiclayo (Province) 2. Chiclayo, Peru (Province)—History. I. Title.
BF1584.P5I2 78-106338
78 MARC

CHICOPEE, MASS.

—HISTORY

McKinstry, Winthrop.
Glimpses of the past / written and compiled by Winthrop McKinstry. — [Chicopee, Mass.] : McKinstry, 1976, c1978.
100 p. (p. 91-95 blank) : ill. ; 29 cm.
Bibliography: p. 99-100.
1. Chicopee, Mass.—History. I. Title.
F74.C58M32 974.4'26 79-109056
79 MARC

CHIDLEY, W. J.

Chidley, W J
Confessions of William James Chidley / edited by S. McInerney. — St. Lucia [Australia] : University of Queensland Press ; Hemel Hempstead, Eng. : distributed by Prentice-Hall International, 1977.
xxxix, 307 p. : ports. ; 22 cm. Aus***
ISBN 0-7022-1491-4
1. Chidley, W. J. 2. Sex—Biography. 3. Hygiene, Sexual. I. McInerney, S., 1946- . II. Title.
HQ21.C49 301.41'58'0924 79-309554
79 MARC

CHIEF JUSTICES see Judges

CHIEFS OF POLICE see Police chiefs

CHIEH, HSI-SSU, 1274-1344

Chieh, Hsi-ssu, 1274–1344.
(Yuan Chieh Hsi-ssu lin Chih-yung chen ts'ao Ch'ien tzu wen)
元揭傒斯臨智永真草千字文. — 香港：書譜社, [1977]
[32] p. : 41 cm. — (書譜叢帖：第3輯：13)
Series romanized: Shu p'u ts'ung t'ieh ti san chi.
Reprint of the 1964 ed. published by Wen wu ch'u pan she, Peking.
1. Chieh, Hsi-ssu, 1274–1344. 2. Calligraphy, Chinese. I. Chih-yung, Shih. II. Chou, Hsing-ssu, d. 521. Ch'ien tzu wen. 1977. III. Title.
NK3634.C417A78 1977 78–842736

CHIEH FANG CHÜN see China (People's Republic of China, 1949-) Chung-kuo jên min chieh fang chün

CHIEMGAU

—DESCRIPTION AND TRAVEL—TOURS

Pause, Walter, 1907-
Münchner Hausberge / Walter Pause. — 14., neubearb. Aufl. — München : BLV-Verlagsgesellschaft, 1978.
181 p. : ill. ; 22 x 25 cm. GFR78-A
Includes index.
ISBN 3-405-12019-5 : DM30.00
1. Bavaria, Upper—Description and travel—Tours. 2. Isartal—Description and travel—Tours. 3. Chiemgau—Description and travel—Tours. 4. Alps, Bavarian—Description and travel—Tours. 5. Mountains—Germany, West—Bavaria, Upper. I. Title.
DD801.B45P328 1978 914.3'36 78-394401
79 MARC

CHIEN-CHEN, 688-763

—ART

Rengyō, fl. 1293–1298.
[Tōsei eden]
東征伝絵巻 / [畫工 蓮行；筆者 忍性；筆師 藤原宣方 ... et al.；編者 小松茂美；執筆者 小松茂美, 菊竹淳一, 小野勝年]. — 東京：中央公論社, 1978.
123 p. : ill. (some col.) ; 36 cm. — (日本絵巻大成；16) Ja 78–18164
Photoreprint of the Tōshōdaiji ms. copy in 5 v., 1298.
Bibliography: p. 123.
¥8800
1. Rengyō, fl. 1293–1298. 2. Chien-chen, 688–763—Art. 3. Ritsu (Sect) in art. 4. Tōshōdaiji, Nara, Japan, in art. 5. Tōshōdaiji, Nara, Japan. I. Komatsu, Shigemi, 1925– II. Kikutake, Jun'ichi, 1939– III. Ono, Katsutoshi. IV. Title. V. Series: Nihon emaki taisei ; 16.
Title romanized: Tōseiden emaki.
ND1059.R46A4 1978 78–816791

CH'IEN, CHEN, 1912-

Ch'ien, Chen, 1912-
(Wo k'an hsi fang)
我看西方 / 錢震著. — 再版. — 臺北：中華日報社, 民國67 [1978]
6, 2, 3, 172 p. ; 19 cm. — (中華日報甲種叢書；59)
Series romanized: Chung-hua jih pao chia chung ts'ung shu.
NT$40.00
1. Europe—Description and travel—1971- 2. United States—Description and travel—1960- 3. Ch'ien, Chen, 1912- I. Title.
D923.C464 79–839596

CH'IEN, CH'IEN-I, 1582-1664. TSU PEN CH'IEN TSENG MU-CHAI SHIH CHU

Chou, Fa-kao.
(Mu-chai shih chu chiao chien)
牧齋詩註校箋 / 周法高著. — [台北]：周法高：經銷處三民書局, 民國67 [1978]
640 p. ; 20 cm.
Chiefly reproduced from ms. copy.
"吾炙集小傳": p. 611–628.
Includes indexes.
NT$350.00
1. Ch'ien, Ch'ien-i, 1582–1664. Tsu pen Ch'ien Tseng Mu-chai shih chu. I. Ch'ien, Ch'ien-i, 1582–1664. Tsu pen Ch'ien Tseng Mu-chai shih chu. II. Teng, Shih, 1878–1898. Wu chih chi hsiao chuan. 1978. III. Title.
PL2701.T83C5 79–841628

CH'IEN, YUNG-MING, 1885-1958

Tu-ch'e, Shih, 1588–1656.
[Nan-lai t'ang shih chi]
王氏輯注南來堂詩集：[4卷：附遺文4首：補編4卷：附錄4卷] / 釋讀徹撰；王培孫輯注；楊家駱主編. — 臺北：鼎文書局, 民國66 [1977]
ca. 300 leaves in various foliations, [4] leaves of plates : ports. ; 22 cm. — (中國學術類編)
"錢新之先生逝世二十周年紀念版"
"蒼雪大師行年考略 陳乃乾撰": 6th prelim. group.
Main text is a reprint of the 1955 ed. published by Nan chung hsiao yu hui, T'ai-pei.
1. Tu-ch'e, Shih, 1588–1656. 2. Authors, Chinese—Biography. 3. Ch'ien, Yung-ming, 1885–1958. I. Wang, P'ei-sun, 1871–1952. II. Ch'en, Nai-ch'ien. Ts'ung-hsüeh ta shih hsing nien k'ao lüeh. 1977. III. Title. IV. Title: Nan-lai t'ang shih chi. V. Series: Chung-kuo hsüeh shu lei pien.
Title romanized: Wang shih chi chu Nan-lai t'ang shih chi.
PL2698.T813A17 1977 78–840975

CHIEN-CH'U, SHIH

Chien-ch'u, Shih.
(Ming t'a T'ang Hsüan-chuang fa shih Chi-kung fa shih t'a ming ho ts'e)
明拓唐玄奘法師基公法師塔銘合册 / [建初書]；主編者洪陸東, 董開章, 李光業；校訂者國立歷史博物館. — [台北]：中國書法學會, 民國63 [1974]
46 p. ; 35 cm.
Colophon title.
Cover title: Ming t'a T'ang Hsüan-chuang fa shih Chi-kung fa shih t'a ming.
Reproduced from rubbings.
1. Chien-ch'u, Shih. 2. Calligraphy, Chinese. I. Hung, Lu-tung. II. Tung, K'ai-chang. III. Li, Kuang-yeh. IV. Title.
NK3634.C423A4 1974 79–837191

CH'IEN-HSIEN, CHINA (SHENSI)

—HISTORY—SOURCES

(Ch'ien-chou hsin chih)
乾州新志：陝西省 / 拜斯呼朗纂修. — 臺1版. — 台北：成文出版社，民國65 ⌈1976⌉
430 p. : ill. ; 27 cm. — (中國方志叢書：華北地方；第549號)
"據清雍正五年刊本影印"

1. Ch'ien-hsien, China (Shensi)—History—Sources. I. Pai-ssu-hu-lang. II. Series: Chung-kuo fang chih ts'ung shu : Hua-pei ti fang ; ti 549 hao.
DS793.C492C48 78-842351

CH'IEN-K'OU TA TUI

Huang, Hsüan-wen.
(Hsiao-ho feng yün)
硝河风云：河南省内黄县千口村史 / 黄烜文. — ⌈香港? : s. n., 1978?⌉
2, 198 p. : ill. ; 21 cm.
Reprint of the 1976 ed. published by Ho-nan jen min ch'u pan she.

1. Ch'ien-k'ou ta tui. I. Title.
HD1492.C52C484 1978 79-837744

CHIERICATO, GIOVANNI, 1633-1717

Beltrame, Guido.
Giovanni Chiericato, oratoriano padovano, ministro e biografo di S. Gregorio Barbarigo : preti e fedeli di S. Gregorio / Guido Beltrame. — Padova : Erredici, 1976.
214 p. : ill. ; 24 cm. It78-Feb
Includes documents in Latin and Italian.
Bibliography: p. 41-45.
Includes indexes.
1. Chiericato, Giovanni, 1633-1717. 2. Catholic Church—Clergy—Biography. 3. Clergy—Italy—Biography. 4. Barbarigo, Gregorio, Saint, Cardinal, 1625-1697. I. Title.
BX4705.C474B44 78-387065
*79 MARC

CHIESA D'ACQUI

—HISTORY—SOURCES

Chiesa d'Acqui.
Le carte medievali della Chiesa d'Acqui / ⌈a cura di⌉ Romeo Pavoni. — Bordighera : Istituto internazionale di studi liguri, 1977.
641 p. ; 25 cm. — (Collana storica dell'oltregiogo ligure ; 2) (Collana storica di fonti e studi ; 22) It79-Jan
In Latin; introductory material in Italian.
Includes bibliographical references and index.
L21200
1. Chiesa d'Acqui—History—Sources. I. Pavoni, Romeo. II. Title.
BX4634.A34C473 79-382063
*79 MARC

CHIESA DEGLI EREMITANI, PADUA

Mantegna, Andrea, 1431-1506.
The frescoes of Mantegna in the Eremitani Church, Padua / text by Giuseppe Fiocco ; with an introd. by Terisio Pignatti. — 2d ed. — Oxford : Phaidon, c1978.
117 p. : chiefly ill. (some col.) ; 38 cm. GB***
Translation of Mantegna, la Capella Ovetari nella Chiesa degli Eremitani.
Bibliography: p. 43.
ISBN 0-7148-1899-2 : £19.95
1. Mantegna, Andrea, 1431-1506. 2. Chiesa degli Eremitani, Padua. I. Fiocco, Giuseppe, 1884- II. Title.
ND623.M3A4 1978a 759.5 78-56684
79 MARC

Mantegna, Andrea, 1431-1506.
Mantegna : la Cappella Ovetari nella Chiesa degli Eremitani / testo di Giuseppe Fiocco ; presentazione di Terisio Pignatti. — 2. ed. — ⌈Milano⌉ : Silvana, ⌈1978⌉
43 p., 25 leaves of plates (3 fold.) : ill. ; 37 cm. It79-Jan
Bibliography: p. 43.
1. Mantegna, Andrea, 1431-1506. 2. Mural painting and decoration—Italy—Padua. 3. Chiesa degli Eremitani, Padua. I. Fiocco, Giuseppe, 1884-
ND623.M3A4 1978 759.5 79-356561
*79 MARC

CHIESA DEL REDENTORE, MASER, ITALY

Basso, Umberto.
La villa e il tempietto dei Barbaro a Maser di Andrea Palladio / Umberto Basso. — Montebelluna : Tip. G. Faggionato, 1976.
49 p. : ill. ; 24 cm. It78-Apr
Includes bibliographical references.
1. Maser, Italy. Villa Barbaro. 2. Chiesa del Redentore, Maser, Italy. I. Title.
DG975.M448B33 1976 78-384464
*78 MARC

CHIESA DELLA DIFESA, SAN VITO DI CADORE, ITALY

Belli, Giuseppe.
La Scuola dei Battuti, la Chiesa della Difesa e la Chiesa di San Floriano in San Vito di Cadore / Giuseppe Belli. — ⌈Belluno⌉ : Istituto bellunese di ricerche sociali e culturali, 1976.
119 p., ⌈8⌉ leaves of plates : ill. ; 24 cm. — (Serie Storia ; 2) It78-June
"Documenti" in Italian or Latin: p. ⌈77⌉-106.
Bibliography: p. 107-110.
1. Scuola dei Battuti, San Vito di Cadore, Italy. 2. Chiesa della difesa, San Vito di Cadore, Italy. 3. Chiesa di San Floriano, San Vito di Cadore, Italy. I. Title. II. Series.
BX814.S383B44 79-353250
*79 MARC

CHIESA DELLA SANTISSIMA TRINITÀ, LIVORNO, ITALY

Museo civico Giovanni Fattori.
I paramenti sacri della Chiesa greco-ortodossa della SS. Trinità di Livorno / a cura di Antonella Capitanio. — Pisa : Giardini, 1978.
36 p. : ill. ; 25 cm. — (Collezione monografica del Museo civico Giovanni Fattori di Livorno ; 2) It78-Oct
1. Ecclesiastical embroidery—History—18th century—Exhibitions. 2. Ecclesiastical embroidery—History—19th century—Exhibitions. 3. Church plate—Exhibitions. 4. Chiesa della Santissima Trinità, Livorno, Italy. I. Capitanio, Antonella. II. Title. III. Series: Museo civico Giovanni Fattori. Collezione monografica del Museo civico Giovanni Fattori di Livorno ; 2.
NK9310.M87 1978 746.9 78-397071
*78 MARC

—HISTORY—SOURCES

Documentazione storico urbanistica sulla comunità greca e sulla Chiesa della SS. Trinità / a cura di Elisabetta De Paz. — Pisa : Giardini, 1978.
43 p. : ill. ; 25 cm. — (Collezione monografica del Museo civico Giovanni Fattori di Livorno ; 3) It78-Oct
Includes bibliographical references.
1. Chiesa della Santissima Trinità, Livorno, Italy—History—Sources. 2. Livorno, Italy—Church history—Sources. 3. Greeks in Livorno, Italy—History—Sources. I. De Paz, Elisabetta. II. Series: Museo civico Giovanni Fattori. Collezione monografica del Museo civico Giovanni Fattori di Livorno ; 3.
BX4711.238.L58C453 78-398314
*79 MARC

CHIESA DI NOSTRA SIGNORA DELLA CONSOLAZIONE, GENOA

Boggero, Franco.
Chiesa di Nostra Signora della Consolazione / a cura di Franco Boggero. — Genova : Sagep, 1977.
32 p. : ill. ; 21 cm. — (Guide di Genova ; 52) It79-Feb
Bibliography: p. 25.
L1000
1. Chiesa di Nostra signora della consolazione, Genoa. I. Series.
DG644.5.C55B63 914.5'182 79-360133
*79 MARC

CHIESA DI SAN BARTOLOMEO IN MODENA

Alfieri, Vittorio.
Guida della Chiesa di S. ⌈i.e. san⌉ Bartolomeo in Modena / a cura di Vittorio Alfieri. — Modena : Tipolitografia editrice Immacolata Concezione, 1976.
78 p. : ill. ; 24 cm. It77-Dec
1. Chiesa di san Bartolomeo in Modena. I. Title.
NA5621.M7993A43 914.5'42 78-376496
*78 MARC

CHIESA DI SAN FLORIANO, SAN VITO DI CADORE, ITALY

Belli, Giuseppe.
La Scuola dei Battuti, la Chiesa della Difesa e la Chiesa di San Floriano in San Vito di Cadore / Giuseppe Belli. — ⌈Belluno⌉ : Istituto bellunese di ricerche sociali e culturali, 1976.
119 p., ⌈8⌉ leaves of plates : ill. ; 24 cm. — (Serie Storia ; 2) It78-June
"Documenti" in Italian or Latin: p. ⌈77⌉-106.
Bibliography: p. 107-110.
1. Scuola dei Battuti, San Vito di Cadore, Italy. 2. Chiesa della difesa, San Vito di Cadore, Italy. 3. Chiesa di San Floriano, San Vito di Cadore, Italy. I. Title. II. Series.
BX814.S383B44 79-353250
*79 MARC

CHIESA DI SAN GIOVANNI DI PIEVE FOSCIANA

—HISTORY

Angelini, Lorenzo.
Una pieve toscana nel Medioevo / Lorenzo Angelini. — Lucca : M. Pacini Fazzi, 1979.
157 p. : ill. ; 24 cm. It79-May
Bibliography: p. 145-147.
Includes index.
L7000
1. Chiesa di San Giovanni di Pieve Fosciana—History. 2. Pieve Fosciana, Italy—Church history. I. Title.
BX4634.P52C483 79-366117
*00 MARC

CHIESA DI SAN NICOLÒ DEL BOSCHETTO, CORNIGLIANO LIGURE, ITALY

Ciliento, Bruno.
Chiesa di San Nicolò del Boschetto / a cura di Bruno Ciliento. — Genova : Sagep, 1977.
16 p. : ill. ; 21 cm. — (Guide di Genova ; n. 45) It79-Feb
Bibliography: p. 16.
L500
1. Chiesa di San Nicolò del Boschetto, Cornigliano Ligure, Italy. I. Series.
DG975.C73853C54 914.5'182 79-360132
*79 MARC

CHIESA DI SANTA MARTA, GENOA

Magnani, Lauro.
Chiesa di Santa Marta / a cura di Lauro Magnani. — Genova : Sagep, 1977.
16 p. : ill. ; 21 cm. — (Guide di Genova ; 51) It79-Feb
Bibliography: p. 13.
L500
1. Chiesa di Santa Marta, Genoa. I. Title. II. Series.
NA5621.G37M33 79-359849
*79 MARC

CHIESA DI SANTO STEFANO, FAENZA

Terzo centenario dell'attuale Chiesa di S. Stefano in Faenza. — ⌈Faenza⌉ : A cura del Comitato parrocchiale per le feste centenarie, 1977.
93 p. : ill. ; 24 cm. It78-June
Cover title: 3. centenario della Chiesa di S. Stefano in Faenza.
Includes bibliographies.
1. Chiesa di Santo Stefano, Faenza.
BX4634.F3S367 78-391058
*78 MARC

CHIESA EVANGELICA VALDESE

Girardet, Franco, 1920-
Il convitto liberante : la ventennale esperienza del Convitto valdese di Torre Pellice / Franco Girardet. — 1. ed. — Rimini ; Firenze : Guaraldi, 1977.
221 p. ; 19 cm. — (Le Frontiere dell'educazione ; 45) It78-Aug
Includes bibliographical references.
L3500
1. Convitto valdese di Torre Pellice—History. 2. Chiesa evangelica valdese. I. Title.
LF3895.T672G57 373.45'12 79-353934
*79 MARC

CHIESA EVANGELICA VALDESE (Continued)

—HISTORY

Peyrot, Giorgio.
Gli evangelici nei loro rapporti con lo Stato dal fascismo ad oggi / Giorgio Peyrot. — Torre Pellice : Società di studi valdesi, [1977]
40 p. : ill. ; 24 cm. — (Monografie edite in occasione del XVII febbraio : Serie italiana) It79-Jan
Includes bibliographical references.
1. Chiesa evangelica valdese—History. 2. Church and state in Italy—History. 3. Italy—Church history—20th century. I. Title.
BX4881.2.P48 79-383118
*79 MARC

CHIETI

—BUILDINGS

Del Bufalo, Alessandro.
I centri antichi di Todi, Monteflavio, Chieti : metodi di analisi storica per il restauro urbano / Alessandro Del Bufalo. — Roma : Bulzoni, 1977.
108 p., [37] leaves of plates : ill. ; 29 cm. It78-Oct
Bibliography: p. 97.
L10000
1. Todi, Italy—Buildings. 2. Monte Flavio, Italy—Buildings. 3. Chieti—Buildings. 4. Historic buildings—Italy—Todi. 5. Historic buildings—Italy—Monte Flavio. 6. Historic buildings—Italy—Chieti. I. Title.
DG975.T63D44 79-347795
*79 MARC

CHIÈZE, JEAN

—EXHIBITIONS

Chièze, Jean.
Rétrospective Jean Chièze en Corse : été 1977, Musée d'ethnographie corse, [Bastia] / [catalogue par Georges Oberti]. — Bastia : Musée d'ethnographie corse, 1977.
113 p. : ill. ; 27 cm. F77-11628
1. Chièze, Jean—Exhibitions. 2. Corsica in art—Exhibitions. I. Oberti, Georges Dominique. II. Musée d'ethnographie corse. III. Title.
N6853.C482A4 1977 760'.092'4 78-378821
78 MARC

CHIFFCHAFF

Schönfeld, Manfred.
Der Weidenlaubsänger, Phylloscopus collybita / von Manfred Schönfeld. — Wittenberg Lutherstadt : Ziemsen, 1978.
136 p. : 58 ill. ; 21 cm. — (Die Neue Brehm-Bücherei ; 511) GFR79-A2
"Gilt als Ringfundmitteilung 1/77 der Vogelwarte Hiddensee."
Erratum slip inserted.
Bibliography: p. 125-132.
Includes index.
15.20M
1. Chiffchaff. 2. Phylloscopus. I. Title.
QL696.P279S3 79-378384
79 MARC

CHIFU, CHINA

—ECONOMIC CONDITIONS

Kōgyō Rengō Kyōkai. Chifū Shibu.
(Chifū jijō)
芝罘事情：昭和十四年版 / 芝罘海軍特別陸戰隊検閲濟；航業聯合協會芝罘支部編. — 青島：同支部, 昭和14 [1939]
6, 247 p., [8] leaves of plates : ill. ; 19 cm.

1. Chifu, China—Economic conditions. 2. Chifu, China—Harbor. I. Title.
HC428.C467K63 1939 79-804018

—HARBOR

Kōgyō Rengō Kyōkai. Chifū Shibu.
(Chifū jijō)
芝罘事情：昭和十四年版 / 芝罘海軍特別陸戰隊検閲濟；航業聯合協會芝罘支部編. — 青島：同支部, 昭和14 [1939]
6, 247 p., [8] leaves of plates : ill. ; 19 cm.

1. Chifu, China—Economic conditions. 2. Chifu, China—Harbor. I. Title.
HC428.C467K63 1939 79-804018

CHIGA, YUKIKO, 1942-

Tanaka, Mariko, 1942-
(Mari to Yuko no ryūgakuki)
マリとユコの留学記：AFS生の記録 / 田中万里子, 値賀由紀子. — 東京：朝日新聞社, 昭和39 [1964]
237 p. : ill. ; 18 cm. — (アサヒ・アドベンチュア・シリーズ ; no. 7)

1. Japanese students in the United States. 2. Students, Interchange of—United States. 3. American Field Service. 4. Tanaka, Mariko, 1942- 5. Chiga, Yukiko, 1942- I. Chiga, Yukiko, 1942- joint author. II. Title.
LB2376.3.J3T36 79-807372

J

CHIGI, FRANCESCO, 1881-1953

Chigi, Francesco, 1881-1953.
Uno sguardo privato : memorie fotografiche / di Francesco Chigi ; a cura di Eva Paola Amendola ; prefazione di Oreste Ferrari. — Torino : G. Einaudi, 1978.
196 p., [6] leaves of plates : ill. ; 22 cm. — (Saggi ; 596) It78-July
1. Photography, Artistic. 2. Italy—Social conditions—Pictorial works. 3. Chigi, Francesco, 1881-1953. I. Amendola, Eva Paola. II. Title.
TR653.C44 1978 770'.92'4 78-392035
*79 MARC

CHIHUAHUA, MEXICO (STATE)

—DESCRIPTION AND TRAVEL

Wampler, Joseph Carson.
La Barranca del Cobre de México : la región y historia de los indios Tarahumaras y de los F.C. Chihuahua al Pacífico : libro con caudal de ilustraciones ... / de Joseph Wampler ; traducción de Lucy Ann Neblett ; fotografías del autor sino en caso notado. — Berkeley, Calif. : [s.n.], c1978.
vii, 100 p., [2] leaves of plates : ill. ; 22 cm.
Translation of New rails to old towns.
Bibliography: p. 99-100.
1. Ferrocarril de Chihuahua al Pacífico. 2. Chihuahua, Mexico (State)—Description and travel. 3. Sinaloa, Mexico (State)—Description and travel. I. Title.
HE2820.F345W3418 78-57149
78 MARC

—DESCRIPTION AND TRAVEL—GUIDE-BOOKS

Hancock, Richard Humphris.
Chihuahua, a guide to the wonderful country / author, Richard H. Hancock : photographer/designer Bill Williams. — [Oklahoma City] : International Training Programs, University of Oklahoma, c1978.
vii, 120 p. : ill. ; 26 cm.
$9.00
1. Chihuahua, Mexico (State)—Description and travel—Guide-books. I. Title.
F1261.H3 917.2'1 79-101349
79 MARC

CHIHUAHUAN DESERT

—GAZETTEERS

Henrickson, James.
A gazetteer of the Chihuahuan Desert region / James Henrickson, Richard M. Straw. — Los Angeles : California State University, 1976.
xxii, 271 p. : maps ; 28 cm.
"A supplement to the Chihuahuan Desert flora [being compiled by M. C. Johnston, and others]"
Bibliography: p. ii-iii.
1. Chihuahuan Desert—Gazetteers. I. Straw, Richard M., joint author. II. Title.
F1262.H46 917.2'1'003 78-622368
79 MARC

CHIKAMATSU, MONZAEMON, 1653-1725. SHINJŪ TEN NO AMIJIMA

(Zōho chūshingura, Shinjū ten no Amijima, Futatsu chōchō kuruwa nikki)
増補忠臣蔵・心中天網島・双蝶々曲輪日記：文楽 / 国立劇場芸能調査室編. — 東京：同調査室, 昭和52 [1977]
62 p. : 21 cm. — (上演資料集 ; 143) Ja***
On cover: Bunraku kōen 1977.9, jōen nempyō, kaisetsu, kata to enshutsu, geidan, kembutsuki.
Includes bibliographies.
1. Bunraku—Addresses, essays, lectures. 2. Zōho chūshingura. 3. Chikamatsu, Monzaemon, 1653-1725. Shinjū ten no Amijima. 4. Takeda, Izumo, 1691-1756. Futatsu chōchō kuruwa nikki. I. Kokuritsu Gekijō. Geinō Chōsashitsu. II. Series: Jōen shiryō shū ; 143.
PN2924.5.K3J64 vol. 143 78-815703
[PN1978.J3]

CHIKAMATSU, MONZAEMON, 1653-1725

—CRITICISM AND INTERPRETATION

Shuzui, Kenji, 1899-
(Chikamatsu)
近松 / 守随憲治著. — 東京：世界評論社, 昭和24 [1949]
191 p., [2] leaves of plates : ill. ; 19 cm. — (世界文學はんどぶっく)
Series romanized: Sekai bungaku handobukku.
Colophon inserted.
Bibliography: p. 167-187.

1. Chikamatsu, Monzaemon, 1653-1725—Criticism and interpretation. I. Title.
PL793.4.Z5S54 79-803196

J

—CRITICISM AND INTERPRETATION—CONGRESSES

Shimpojūmu Chikamatsu to Gendai Ningyōgeki, Osaka, 1977.
(Shimpojūmu Chikamatsu to gendai ningyōgeki)
シンポジウム近松と現代人形劇 / 阪口弘之司会；吉田清治 [et al.] 報告者；[編集 近松と現代人形劇を考える会]. — 川崎：現代人形劇センター, 1978.
128 p., [1] leaf of plates : ill. ; 21 cm. Ja 78-23516
Limited edition: 500 copies printed.
Includes bibliographical references.
¥1200

1. Puppets and puppet plays—Japan—Congresses. 2. Chikamatsu, Monzaemon, 1653-1725—Dramatic production—Congresses. 3. Chikamatsu, Monzaemon, 1653-1725—Criticism and interpretation—Congresses. I. Sakaguchi, Hiroyuki, 1943- II. Yoshida, Seiji, 1931- III. Title. IV. Title: Chikamatsu to gendai ningyōgeki.
PN1978.J3S47 1977 79-803677

—DRAMATIC PRODUCTION—CONGRESSES

Shimpojūmu Chikamatsu to Gendai Ningyōgeki, Osaka, 1977.
(Shimpojūmu Chikamatsu to gendai ningyōgeki)
シンポジウム近松と現代人形劇 / 阪口弘之司会；吉田清治 [et al.] 報告者；[編集 近松と現代人形劇を考える会]. — 川崎：現代人形劇センター, 1978.
128 p., [1] leaf of plates : ill. ; 21 cm. Ja 78-23516
Limited edition: 500 copies printed.
Includes bibliographical references.
¥1200

1. Puppets and puppet plays—Japan—Congresses. 2. Chikamatsu, Monzaemon, 1653-1725—Dramatic production—Congresses. 3. Chikamatsu, Monzaemon, 1653-1725—Criticism and interpretation—Congresses. I. Sakaguchi, Hiroyuki, 1943- II. Yoshida, Seiji, 1931- III. Title. IV. Title: Chikamatsu to gendai ningyōgeki.
PN1978.J3S47 1977 79-803677

CHIKUSA-KU, NAGOYA, JAPAN

—ANTIQUITIES

Nagoya, Japan. Kyōiku Iinkai. Shakai Kyōikubu. Bunkaka.
(Mikage-machi koyōsekigun hakkutsu chōsa hōkoku)
御影町古窯跡群発掘調査報告：千種区御影町 / ［編集 名古屋市教育委員会社会教育部文化課］. — 名古屋：名古屋市教育委員会, 1974.
14 p., 13 leaves of plates : ill. ; 26 cm. — (名古屋市文化財調査報告 ; 3) Ja***
Bibliography: p. 14.
1. Chikusa-ku, Nagoya, Japan—Antiquities. 2. Kilns. 3. Nagoya, Japan—Antiquities. 4. Japan—Antiquities. I. Title. II. Series: Nagoya-shi bunkazai chosa hokoku ; 3.
DS897.N316C456 79-806908

CHIKUZAN HITORITABI. [MOTION PICTURE]

Shindō, Kaneto, 1912-
(Eizō hitoritabi)
映像ひとり旅：映画『竹山ひとり旅』創造の記録 / 新藤兼人. — 東京：未来社, 1978.
281 p. : ill. ; 20 cm. Ja 78-20195
CONTENTS: 映像とは—「竹山ひとり旅」撮影日記—シナリオ「竹山ひとり旅」—高橋竹山との対談—「竹山ひとり旅」の作家意識をめぐって—モスクワ映画祭日記
¥1300
1. Chikuzan hitoritabi. [Motion picture] 2. Takahashi, Chikuzan, 1910- —Drama. I. Shindō, Kaneto, 1912- Chikuzan hitoritabi. 1978. II. Title.
PN1997.C464353E3 79-800280

CHILALO AGRICULTURAL DEVELOPMENT UNIT

Final report on the appraisal of CADU and EPID / by the evaluation team appointed by the Imperial Ethiopian Government and the Swedish International Development Authority. — [s.l. : s.n.], 1974.
iii, 89 p. ; 30 cm.
Cover title.
Errata slip inserted.
1. Chilalo Agricultural Development Unit. 2. Ethiopia. Ya' Eršā ministér. Extension and Project Implementation Dept. I. Ethiopia. II. Styrelsen för internationell utveckling.
HD2124.5.Z9C543 338.1'0963 78-113778
79 MARC

CHILD ABUSE

see also Battered child syndrome; Girls—Crimes against

Fontana, Vincent J
The maltreated child : the maltreatment syndrome in children : a medical, legal, and social guide / by Vincent J. Fontana and Douglas J. Besharov ; with a foreword by Mother Loretto Bernard. — 4th ed. — Springfield, Ill. : Thomas, 1979.
xxii, 170 p. : ill. ; 24 cm.
Bibliography: p. 149-157.
Includes indexes.
ISBN 0-398-03904-6
1. Child abuse. 2. Child abuse—Prevention. I. Besharov, Douglas J., joint author. II. Title.
[DNLM: 1. Child abuse. 2. Child welfare—United States. WA320 F679m]
HV713.F6 1979 362.7'1 78-27727
78 MARC

Inglis, Ruth Langdon, 1927-
Sins of the fathers : a study of the physical and emotional abuse of children / Ruth Inglis. — New York : St. Martin's Press, c1978.
220 p. ; 22 cm.
Bibliography: p. 214-220.
ISBN 0-312-72608-2 : $10.00
1. Child abuse. 2. Parent and child. 3. Child abuse—United States. 4. Child abuse—Great Britain. I. Title.
HV713.I53 362.7'1 78-4004
78 MARC

Kopart, Henning H
Børnemishandling : en kritik af nogle årsagsforklaringer og deres politiske konsekvenser / [af] Henning H. Kopart, Inge M. L. Sørensen. — [København] : [Demos], c1977.
72 p. ; 21 cm. D78-10
Bibliography: p. 71-72.
ISBN 8787337835 pbk.: kr30.00
1. Child abuse. 2. Child abuse—Denmark. I. Sørensen, Inge M. L., joint author. II. Title.
HV713.K66 78-365324
78 MARC

Kroth, Jerome A
Child sexual abuse : analysis of a family therapy approach / by Jerome A. Kroth. — Springfield, Ill. : Thomas, c1979.
x, 200 p. : ill. ; 24 cm.
Bibliography: p. 197-198.
Includes index.
ISBN 0-398-03906-2
1. Child molesting. 2. Child abuse. 3. Family psychotherapy. I. Title.
[DNLM: 1. Child abuse. 2. Sex offenses. 3. Family therapy. WA320 K93c]
RC560.C46K76 362.1'9'68583 79-541
79 MARC

Leulliette, Pierre.
Les enfants martyrs : enquête / Pierre Leulliette. — Paris : Seuil, c1978.
245 p. ; 21 cm. F***
ISBN 2-02-004961-9 : 42.00F
1. Child abuse. I. Title.
HV713.L48 362.7'1 79-350094
79 MARC

Manitoba. Advisory Committee on Battered Children.
Child abuse / Advisory Committee on Battered Children, Margo Buck ... [et al.]. — 5th ed. — [Winnipeg] : Manitoba Dept. of Health and Social Development, 1977.
20 p. ; 23 cm. C***
1. Child abuse. 2. Child abuse—Manitoba. I. Buck, Margo. II. Title.
HV713.M33 1977 362.7'1 78-326222
79 MARC

Willich, Ray.
Shadows of childhood / [by] Ray Willich. — Camberwell, Vic. ; Norwalk, Conn. : Widescope, 1978.
xii, 131 p. ; 23 cm. Aus
"A Visa book."
ISBN 0-86932-059-9. ISBN 0-86932-058-0 pbk.
1. Child abuse. I. Title.
HV873.W5 362.7'1 78-321888
79 MARC

—ADDRESSES, ESSAYS, LECTURES

Il Bambino nella società violenta / [di] autori vari ; a cura di Anna Portoghese. — Assisi : Cittadella editrice, 1978.
193 p. ; 19 cm. — (Obiettivo su ..) It78-June
Includes bibliographical references.
L2800
1. Child welfare—Addresses, essays, lectures. 2. Child abuse—Addresses, essays, lectures. 3. Violence—Addresses, essays, lectures. I. Portoghese, Anna.
HV715.B32 78-384951
78 MARC

Child abuse and violence / edited for the American Orthopsychiatric Association by David G. Gil. — New York : AMS Press, c1979.
xvi, 614 p. : ill. ; 23 cm.
Collection of articles published 1958-75.
Includes bibliographies and index.
ISBN 0-404-16044-1
1. Child abuse—Addresses, essays, lectures. 2. Family—Addresses, essays, lectures. 3. Violence—Addresses, essays, lectures. I. Gil, David G. II. American Orthopsychiatric Association.
RC569.5.C55C45 362.7'8'1971 77-91865
78 MARC

Critical perspectives on child abuse / edited by Richard Bourne, Eli H. Newberger. — Lexington, Mass. : Lexington Books, 1979.
xii, 224 p. ; 24 cm.
Includes bibliographical references and index.
ISBN 0-669-02109-1
1. Child abuse—Addresses, essays, lectures. 2. Child abuse—Prevention—Addresses, essays, lectures. 3. Child abuse—Services—Addresses, essays, lectures. I. Bourne, Richard. II. Newberger, Eli.
HV713.C74 362.7'1 77-18565
78 MARC

El Maltrato a los hijos : el más oculto y menos controlado de todos los crímenes violentos / Giuseppe Amara ... [et al. ; editado por] Jaime Marcovich. — 1. ed. — México : Edicol, 1978.
311 p. : ill. ; 23 cm.
Includes bibliographies.
ISBN 9684080174
1. Child abuse—Addresses, essays, lectures. 2. Battered child syndrome—Addresses, essays, lectures. 3. Child abuse—Mexico—Addresses, essays, lectures. I. Amara, Giuseppe. II. Marcovich, Jaime.
HV713.M27 78-110490
79 MARC

The Maltreatment of children / edited by Selwyn M. Smith. — Baltimore : University Park Press, c1978.
xii, 452 p. ; 24 cm.
Includes bibliographical references and index.
ISBN 0-8391-1220-3
1. Child abuse—Addresses, essays, lectures. I. Smith, Selwyn M.
HV713.M28 362.7'1 78-1698
78 MARC

The Maltreatment of children / edited by Selwyn M. Smith. — Lancaster, Eng. : M.T.P. Press, 1978.
xii, 452 p. : tables ; 24 cm. GB***
Includes bibliographical references and index.
ISBN 0-85200-141-X : £12.95
1. Child abuse—Addresses, essays, lectures. I. Smith, Selwyn M. Smith.
HV713.M28 1978b 362.7'1 79-30434
79 MARC

—DICTIONARIES

Midwest Parent-Child Welfare Resource Center.
Interdisciplinary glossary on child abuse and neglect : legal, medical, social work terms / [prepared by Midwest Parent-Child Welfare Resource Center, Center for Advanced Studies in Human Services, School of Social Welfare, University of Wisconsin-Milwaukee]. — Washington : U.S. Dept. of Health, Education, and Welfare, Office of Human Development Services, Administration for Children, Youth, and Families, Children's Bureau, National Center on Child Abuse and Neglect : for sale by the Supt. of Docs., U.S. Govt. Print. Off., [1978].
xi, 100 p. : ill. ; 27 cm. — (DHEW publication ; no. (OHDS) 78-30137)
Bibliography: p. 91-93.
1. Child abuse—Dictionaries. I. Title. II. Series: United States. Dept. of Health, Education and Welfare. DHEW publication ; no. (OHDS) 78-30137.
HV711.M52 1978 362.7'1 78-601981
79 MARC

—LAW AND LEGISLATION—CALIFORNIA

California. Legislature. Assembly. Committee on Criminal Justice.
Child abuse reporting : hearing, Assembly Committee on Criminal Justice, Los Angeles, November 21, 1978 / Kenneth L. Maddy, chairman ; Terry Goggin, vice chairman. — [Sacramento] : California State Assembly, [1978?]
67 p. ; 28 cm.
Cover title.
"No. 725."
$5.10
1. Child abuse—Law and legislation—California. I. Moody, Kenneth L. II. Goggin, Terry. III. Title.
KFC10.4.C68 1978 344'.794'03271 79-622119
00 MARC

—LAW AND LEGISLATION—GERMANY, WEST—HISTORY

Pfeil, Sigurd, Graf von.
Das Kind als Objekt der Planung : e. kulturhistor. Unters. über Abtreibung, Kindestötung u. Aussetzung / von Sigurd Graf von Pfeil. — Göttingen : Schwartz, 1979.
369 p. ; 24 cm. GFR79-A
Includes bibliographical references.
ISBN 3-509-01096-5
1. Abortion—Law and legislation—Germany, West—History. 2. Infanticide—Germany, West—History. 3. Child abuse—Law and legislation—Germany, West—History. I. Title.
79-367617
79 MARC

—LAW AND LEGISLATION—GREAT BRITAIN

Hall, Jean Graham.
Child abuse, procedure and evidence in juvenile courts / by Jean Graham Hall and Barbara H. Mitchell. — Chichester [Eng.] : B. Rose, c1978.
132 p. ; 19 cm. GB***
Includes index.
Bibliography: p. [8]-[9]
ISBN 0-85992-135-2 : £4.25
1. Child abuse—Law and legislation—Great Britain. 2. Juvenile courts—Great Britain. 3. Evidence (Law)—Great Britain. I. Mitchell, Barbara H., joint author. II. Title.
KD7974.H34 345'.42'025 79-308131
79 MARC

—LAW AND LEGISLATION—PENNSYLVANIA

Pennsylvania. General Assembly. Joint State Government Commission.
Administration of Pennsylvania's child abuse law : a legislative oversight evaluation / conducted on behalf of the Senate Committee on Aging and Youth and the House Committee on Health and Welfare by the staff of the Joint State Government Commission. — Harrisburg : The Commission, 1979.
viii, 96 p. : ill. ; 28 cm.
Includes bibliographical references.
1. Child abuse—Law and legislation—Pennsylvania. I. Pennsylvania. General Assembly. Senate. Committee on Aging and Youth. II. Pennsylvania. General Assembly. House of Representatives. Committee on Health and Welfare. III. Title. IV. Title: Child abuse law.
KFP567.C5A25 1979 353.9'748'008471 79-62253
79 MARC

CHILD ABUSE (Continued)

—LAW AND LEGISLATION—TEXAS

Texas. Legislature. House of Representatives. Select Committee on Child Pornography, Its Related Causes and Control.
Interim report, sixty sixth legislative session / The Select Committee on Child Pornography, Its Related Causes and Control, Texas House of Representatives. — ₍Austin₎ : The Committee, ₍1978?₎
iii leaves, 198 p. ; 28 cm.
Cover title.
1. Child abuse—Law and legislation—Texas. 2. Sex crimes—Texas. 3. Child molesting—Texas. 4. Pornography—Texas.
KFT1211.82.C55 1978 364.1'53'09764 79-623608
00 MARC

—LAW AND LEGISLATION—UNITED STATES

The Abused and neglected child : multi-disciplinary court practice / Douglas J. Besharov ; Michael Baden ... ₍et al.₎. — ₍New York₎ : Practising Law Institute, c1978.
800 p. ; 22 cm. — (Litigation and administrative practice series) (Criminal law and urban problems course handbook series ; no. 104)
"C4-4132."
"Prepared for distribution at the abused and neglected child program, December 1978-February 1979."
Includes bibliographical references.
1. Child abuse—Law and legislation—United States. 2. Juvenile courts—United States. I. Besharov, Douglas J. II. New York (City). Practising Law Institute. III. Series. IV. Series: Criminal law and urban problems course handbook series ; no. 104.
KF9323.A92 344'.73'03271 78-70877
79 MARC

Child abuse and neglect : a compendium of course materials. — ₍Houston, Tex.₎ : National College of District Attorneys, ₍1977₎
387 p. ; 28 cm.
"Originally prepared for presentation at five child abuse seminars under the auspices of the National College of District Attorneys from December 1976 through April 1977."
1. Child abuse—Law and legislation—United States. 2. Child abuse—United States. I. National College of District Attorneys.
KF9323.C5 344'.73'03271 78-317315
78 MARC

United States. Congress. House. Committee on Education and Labor. Subcommittee on Select Education.
Domestic violence : hearings before the Subcommittee on Select Education of the Committee on Education and Labor, House of Representatives, Ninety-fifth Congress, second session, on H.R. 7927 and H.R. 8948 ... March 16 and 17, 1978. — Washington : U.S. Govt. Print. Off., 1978.
vi, 753 p. ; 24 cm.
Bibliography: p. 663-670.
1. Conjugal violence—United States. 2. Wife abuse—United States. 3. Child abuse—Law and legislation—United States. I. Title.
KF27.E373 1978b 344'.73'03282 78-602685
78 MARC

—LAW AND LEGISLATION—UNITED STATES—STATES

Herner and Company, Washington, D.C.
Child abuse and neglect : state reporting laws / ₍developed by Herner and Company₎. — ₍Washington₎ : National Center on Child Abuse and Neglect, U.S. Children's Bureau, Administration for Children, Youth and Families, Office of Human Development Services, U.S. Dept. of Health, Education, and Welfare : ₍for sale by the Supt. of Docs. ; U.S. Govt. Print. Off.₎, 1978.
ii, 34 p. ; 27 cm.
Cover title.
Bibliography: p. 27-30.
1. Child abuse—Law and legislation—United States—States. I. National Center on Child Abuse and Neglect. II. Title.
KF9323.Z95H47 344'.73'03271 78-602512
79 MARC

—LAW AND LEGISLATION—UTAH

Allred, Mark.
Child abuse and neglect : a report to the Forty-second Legislature, containing recommendations of the Joint Social Services Interim Study Committee / by Mark J. Allred. — ₍Salt Lake City₎ : Office of Legislative Research, State of Utah, 1977.
ii, 47 leaves ; 28 cm. — (Research report - Office of Legislative Research, State of Utah ; no. 24)
Includes bibliographical references.
1. Child abuse—Utah. 2. Child abuse—Law and legislation—Utah. I. Utah. Legislature. Joint Social Services Interim Study Committee. II. Title. III. Series: Utah. Office of Legislative Research. Research report - Office of Legislative Research, State of Utah ; no. 24.
HV742.U8A45 362.7'1 78-622525
79 MARC

—PREVENTION

Fontana, Vincent J
The maltreated child : the maltreatment syndrome in children : a medical, legal, and social guide / by Vincent J. Fontana and Douglas J. Besharov ; with a foreword by Mother Loretto Bernard. — 4th ed. — Springfield, Ill. : Thomas, 1979.
xxii, 170 p. : ill. ; 24 cm.
Bibliography: p. 149-157.
Includes indexes.
ISBN 0-398-03904-6
1. Child abuse. 2. Child abuse—Prevention. I. Besharov, Douglas J., joint author. II. Title.
₍DNLM: 1. Child abuse. 2. Child welfare—United States. WA320 F679m₎
HV713.F6 1979 362.7'1 78-27727
78 MARC

—PREVENTION—ADDRESSES, ESSAYS, LECTURES

Critical perspectives on child abuse / edited by Richard Bourne, Eli H. Newberger. — Lexington, Mass. : Lexington Books, 1979.
xii, 224 p. ; 24 cm.
Includes bibliographical references and index.
ISBN 0-669-02109-1
1. Child abuse—Addresses, essays, lectures. 2. Child abuse—Prevention—Addresses, essays, lectures. 3. Child abuse—Services—Addresses, essays, lectures. I. Bourne, Richard. II. Newberger, Eli.
HV713.C74 362.7'1 77-18565
78 MARC

—SERVICES—ADDRESSES, ESSAYS, LECTURES

Critical perspectives on child abuse / edited by Richard Bourne, Eli H. Newberger. — Lexington, Mass. : Lexington Books, 1979.
xii, 224 p. ; 24 cm.
Includes bibliographical references and index.
ISBN 0-669-02109-1
1. Child abuse—Addresses, essays, lectures. 2. Child abuse—Prevention—Addresses, essays, lectures. 3. Child abuse—Services—Addresses, essays, lectures. I. Bourne, Richard. II. Newberger, Eli.
HV713.C74 362.7'1 77-18565
78 MARC

—SERVICES—ALASKA—ANCHORAGE

Mechau, Dorik V
Final evaluation report on shelter for abused & battered women and their children operated by Abused Women's Aid in Crisis (AWAIC), Anchorage, Alsaska / submitted to Dept. of Community and Regional Affairs, Division of Community and Rural Development, by Dorik V. Mechau. — Haines, Or. : Mechau, 1978.
30 p. ; 28 cm.
Cover title.
1. Abused wives—Services for—Alaska—Anchorage. 2. Child abuse—Services—Alaska—Anchorage. I. Alaska. Division of Community and Rural Development. II. Title.
HV6626.M42 362.8'2 79-620896
79 MARC

—SERVICES—MASSACHUSETTS—BOSTON

Massachusetts Society for the Prevention of Cruelty to Children.
Group work as a part of protective services / by the Massachusetts Society for the Prevention of Cruelty to Children ; ₍ill. by Barbara B. Anderson₎. — Englewood, Colo. : Child Protection, American Humane Association, c1976.
iii, 32 p. : ill. ; 23 cm.
1. Massachusetts Society for the Prevention of Cruelty to Children. 2. Child abuse—Services—Massachusetts—Boston. 3. Family social work—Massachusetts—Boston. 4. Social group work. I. Title.
HV743.B692M35 1976 362.7'1 79-108459
79 MARC

—SERVICES—NETHERLANDS—PERIODICALS

Begeleidingskommissie Vertrouwensartsen inzake Kindermishandeling.
Jaarverslag bureaus vertrouwensartsen inzake kindermishandeling.
's-Gravenhage, Begeleidingskommissie Vertrouwensartsen inzake Kindermishandeling.
v. 27 cm.
1. Child abuse—Netherlands—Periodicals. 2. Child abuse—Services—Netherlands—Periodicals. 3. Child abuse—Netherlands—Statistics—Periodicals. I. Title.
HV778.A13a 79-645539
MARC-S

—SERVICES—UNITED STATES

Halperin, Michael, 1941-
Helping maltreated children : school and community involvement / Michael Halperin, with assistance from Pamela Shuman. — St. Louis : Mosby, 1979.
x, 198 p. ; 24 cm.
Bibliography: p. 174-184.
Includes index.
ISBN 0-8016-2020-1
1. Child abuse—Services—United States. 2. Child abuse—United States. 3. Child abuse—United States—Prevention. I. Shuman, Pamela, joint author. II. Title.
HV741.H28 362.7'1 78-31527
78 MARC

Shapiro, Deborah.
Parents and protectors : a study in child abuse and neglect / Deborah Shapiro. — New York : Research Center, Child Welfare League of America, c1979.
xv, 127 p. ; 26 cm.
"G-19."
ISBN 0-87868-139-6
1. Child abuse—United States. 2. Child abuse—Services—United States. I. Title.
HV741.S43 362.7'1 79-88223
79 MARC

Smith, Charles P 1931-
Crisis in the shadows : a preliminary national assessment of child abuse and neglect and the juvenile justice system / Charles P. Smith, David J. Berkman, Warren M. Fraser ; submitted to the U.S. Office of Juvenile Justice and Delinquency Prevention, January 1979 ; prepared by the National Juvenile Justice System Assessment Center of the American Justice Institute. — Sacramento : The Institute, c1979.
xiv, 162 p. ; 29 cm.
Includes bibliographies.
1. Child abuse—Services—United States. 2. Child abuse—United States—Prevention. 3. Juvenile justice, Administration of—United States. I. Berkman, David J., joint author. II. Fraser, Warren M., joint author. III. United States. Law Enforcement Assistance Administration. Office of Juvenile Justice and Delinquency Prevention. IV. National Juvenile Justice System Assessment Center. V. Title.
HV741.S54 362.7'1 79-110961
79 MARC

—SERVICES—UNITED STATES—CONGRESSES

National Conference on Child Abuse and Neglect, 2d, Houston, Tex., 1977.
Child abuse and neglect : issues on innovation and implementation : proceedings of the second annual National Conference on Child Abuse and Neglect, April 17-20, 1977 / edited by Michael L. Lauderdale, Rosalie N. Anderson, Stephen E. Cramer. — ₍Washington : National Center on Child Abuse and Neglect, Children's Bureau, Administration for Children, Youth, and Families, Office of Human Development Services, U.S. Dept. of Health, Education, and Welfare₎, 1978-
v. : ill. ; 27 cm. — (DHEW publication ; no. (OHDS) 78-30147)
Includes bibliographies.
1. Child abuse—United States—Congresses. 2. Child abuse—Services—United States—Congresses. I. Lauderdale, Michael L. II. Anderson, Rosalie N. III. Cramer, Stephen E. IV. National Center on Child Abuse and Neglect. V. Title. VI. Series: United States. Dept. of Health, Education, and Welfare. DHEW publication ; no. (OHDS) 78-30147.
HV881.N35 1977 362.7'1 78-603476
79 MARC

—CALIFORNIA

California. Legislature. Assembly. Committee on Human Resources.
Public hearing on child abuse, its prevention, detection, and treatment programs, San Francisco ... November 29, 1977 ... / Assembly Committee on Human Resources. — ₍Sacramento : Assembly Publications Office, 1977?₎
iii, 211 p. ; 28 cm.
"No. 685."
$5.60
1. Child abuse—California. I. Title: Child abuse, its prevention, detection, and treatment programs.
KFC10.4.H8 1977g 362.7'1 78-623556
79 MARC

—CANADA

Silverman, Peter.
Who speaks for the children? : the plight of the battered child / Peter Silverman. — Don Mills, Ont. : Musson Book Co., 1978.
224 p. ; 22 cm. C***
Includes bibliographical references and index.
ISBN 0-7737-1022-1 : $8.95
1. Child abuse—Canada. I. Title.
HV745.A6S45 362.7'1 79-305621
79 MARC

CHILD ABUSE
—CANADA (Continued)

Van Stolk, Mary.
The battered child in Canada / Mary Van Stolk. — Rev. ed. — Toronto : McClelland and Stewart, c1978.
xii, 178 p. ; 21 cm. C78-1426-X
Includes bibliographical references.
ISBN 0-7710-8712-8 : $7.95
1. Child abuse—Canada. 2. Child welfare—Canada. I. Title.
HV745.A6V3 1978 362.7'1 79-310675
79 MARC

—DENMARK

Kopart, Henning H
Børnemishandling : en kritik af nogle årsagsforklaringer og deres politiske konsekvenser / ₍af₎ Henning H. Kopart, Inge M. L. Sørensen. — ₍København₎ : ₍Demos₎, c1977.
72 p. ; 21 cm. D78-10
Bibliography: p. 71-72.
ISBN 8787337835 pbk.: kr30.00
1. Child abuse. 2. Child abuse—Denmark. I. Sørensen, Inge M. L., joint author. II. Title.
HV713.K66 78-365324
78 MARC

—FRANCE—
SAINT-ANDRÉ-DE-SANGONIS

Guerrin, Charles François.
Morte Pour une messe à l'Espélidou / Charles-François Guerrin. — Nice : A. Lefeuvre, c1978.
222 p., ₍8₎ leaves of plates : ill. ; 24 cm. — (Collection J'accuse) F***
ISBN 2-902639-06-6 : 39.00F
1. Espélidous (Hospital) 2. Mentally handicapped—Institutional care—France—Saint-André-de-Sangonis. 3. Child abuse—France—Saint-André-de-Sangonis. I. Title.
RC450.F8S253 362.7'8'3094484 79-354412
79 MARC

—GERMANY, WEST—ADDRESSES, ESSAYS, LECTURES

Unsere Kinder, unsere Verantwortung : Materialien e. Aktion / hrsg. von Wolfgang Brinkel u. Hermann Polenz. — 1. Aufl. — Freiburg i.Br. : Dreisam-Verlag, 1978.
172 p. ; 19 cm. GFR78-A
Includes the recommendations of the Kommission Anwalt des Kindes Baden-Württemberg.
Includes bibliographical references.
ISBN 3-921472-16-4 : DM7.20
1. Children's rights—Germany, West—Addresses, essays, lectures. 2. Child abuse—Germany, West—Addresses, essays, lectures. I. Brinkel, Wolfgang. II. Polenz, Hermann. III. Kommission Anwalt des Kindes Baden-Württemberg.
HV763.U57 362.7'0943 79-364971
79 MARC

—GREAT BRITAIN

Creighton, Susan J
Child victims of physical abuse : a report on the findings of NSPCC units' registers / by Susan J. Creighton and Peter J. Owtram. — London : National Society for the Prevention of Cruelty to Children, ₍1977₎
₍1₎, 52 p. : ill., forms ; 30 cm. GB78-18150
Cover title.
Bibliography: p. 34-36.
1. Child abuse—Great Britain. 2. Battered child syndrome—Great Britain. 3. Victims of crimes—Great Britain. I. Owtram, Peter J., joint author. II. National Society for the Prevention of Cruelty to Children. III. Title.
HV751.A6C73 362.7'1 79-303682
79 MARC

Inglis, Ruth Langdon, 1927-
Sins of the fathers : a study of the physical and emotional abuse of children / Ruth Inglis. — New York : St. Martin's Press, c1978.
220 p. ; 22 cm.
Bibliography: p. 214-220.
ISBN 0-312-72608-2 : $10.00
1. Child abuse. 2. Parent and child. 3. Child abuse—United States. 4. Child abuse—Great Britain. I. Title.
HV713.I53 362.7'1 78-4004
78 MARC

—ILLINOIS

Rubenstein, Bruce.
Child abuse/neglect systems survey / Bruce Rubenstein. — ₍Springfield₎ : House Democratic Staff, State of Illinois, 1979.
139 p. ; 28 cm.
1. Child abuse—Illinois. 2. Social work with children—Illinois. I. Illinois. General Assembly. House of Representatives. Democratic Staff. II. Title.
HV742.I3R82 362.7'1 79-624227
79 MARC

—ILLINOIS—STATISTICS

Illinois. Dept. of Children and Family Services.
Child abuse reporting, July 1, 1976-June 30, 1977. — ₍Springfield₎ : State of Illinois, Dept. of Children and Family Services, ₍1977₎
7 leaves ; 28 cm.
Tables.
1. Child abuse—Illinois—Statistics. I. Title.
HV742.I3143 1977 362.7'1 78-621288
00 MARC

—INDIANA—CONGRESSES

Indiana Governor's Conference on Child Abuse and Neglect, Indianapolis, 1977.
Proceedings of the Indiana Governor's Conference on Child Abuse and Neglect, Indianapolis, June 12-13, 1977 / Executive Advisory Committee, Otis R. Bowen ... ₍et al.₎ ; sponsored by Midwest Parent-Child Welfare Resource Center ... ₍et al.₎. — ₍Indianapolis : s.n., 1977?₎
40 p. ; 22 cm.
Cover title: Protecting the Hoosier child.
1. Child abuse—Indiana—Congresses. I. Bowen, Otis R. II. Midwest Parent-Child Welfare Resource Center. III. Title: Protecting the Hoosier child.
HV742.I61694 1977 362.7'1 78-623614
79 MARC

—MANITOBA

Manitoba. Advisory Committee on Battered Children.
Child abuse / Advisory Committee on Battered Children, Margo Buck ... ₍et al.₎. — 5th ed. — ₍Winnipeg₎ : Manitoba Dept. of Health and Social Development, 1977.
20 p. ; 23 cm. C***
1. Child abuse. 2. Child abuse—Manitoba. I. Buck, Margo. II. Title.
HV713.M33 1977 362.7'1 78-326222
79 MARC

—MEXICO—ADDRESSES, ESSAYS, LECTURES

El Maltrato a los hijos : el más oculto y menos controlado de todos los crímenes violentos / Giuseppe Amara ... ₍et al. ; editado por₎ Jaime Marcovich. — 1. ed. — México : Edicol, 1978.
311 p. : ill. ; 23 cm.
Includes bibliographies.
ISBN 9684080174
1. Child abuse—Addresses, essays, lectures. 2. Battered child syndrome—Addresses, essays, lectures. 3. Child abuse—Mexico—Addresses, essays, lectures. I. Amara, Giuseppe. II. Marcovich, Jaime.
HV713.M27 78-110490
79 MARC

—MINNESOTA—PERIODICALS

Minnesota. Division of Child Welfare.
Some facts about child abuse in Minnesota, annual report. ₍St. Paul₎ Dept. of Public Welfare, Division of Child Welfare₎
v. 28 cm.
1. Child abuse—Minnesota—Periodicals. I. Title.
HV742.M6M53a 362.7'1 77-646842
MARC-S

—NETHERLANDS—PERIODICALS

Begeleidingskommissie Vertrouwensartsen inzake Kindermishandeling.
Jaarverslag bureaus vertrouwensartsen inzake kindermishandeling.
's-Gravenhage, Begeleidingskommissie Vertrouwensartsen inzake Kindermishandeling.
v. 27 cm.
1. Child abuse—Netherlands—Periodicals. 2. Child abuse—Services—Netherlands—Periodicals. 3. Child abuse—Netherlands—Statistics—Periodicals. I. Title.
HV778.A13a 79-645539
MARC-S

Netherlands (Kingdom, 1815-). Ministerie van Volksgezondheid en Milieuhygiëne.
Jaarverslag vertrouwensartsen inzake kindermishandeling.
₍'s-Gravenhage₎ Ministerie van Volksgezondheid en Milieuhygiëne.
v. 27 cm.
1. Child abuse—Netherlands—Statistics—Periodicals. 2. Child abuse—Netherlands—Periodicals. I. Title.
HV778.A16a 79-646456
MARC-S

—NETHERLANDS—STATISTICS—PERIODICALS

Begeleidingskommissie Vertrouwensartsen inzake Kindermishandeling.
Jaarverslag bureaus vertrouwensartsen inzake kindermishandeling.
's-Gravenhage, Begeleidingskommissie Vertrouwensartsen inzake Kindermishandeling.
v. 27 cm.
1. Child abuse—Netherlands—Periodicals. 2. Child abuse—Services—Netherlands—Periodicals. 3. Child abuse—Netherlands—Statistics—Periodicals. I. Title.
HV778.A13a 79-645539
MARC-S

Netherlands (Kingdom, 1815-). Ministerie van Volksgezondheid en Milieuhygiëne.
Jaarverslag vertrouwensartsen inzake kindermishandeling.
₍'s-Gravenhage₎ Ministerie van Volksgezondheid en Milieuhygiëne.
v. 27 cm.
1. Child abuse—Netherlands—Statistics—Periodicals. 2. Child abuse—Netherlands—Periodicals. I. Title.
HV778.A16a 79-646456
MARC-S

—NORWAY

Skjælaaen, Peer, 1938-
Diagnose, barnemishandling / Peer Skjælaaen ; redigert og bearbeidet av Tone B. Jamholt. — Oslo : Tiden, c1978 (Oslo : Reistad offset)
75 p., ₍2₎ leaves of plates : ill. ; 21 cm. N78-Dec.
Bibliography: p. ₍76₎
ISBN 8210015990 : kr45.00
1. Child abuse—Norway. 2. Battered child syndrome. I. Jamholt, Tone B. II. Title.
HV788.A6S54 79-360688
79 MARC

—ONTARIO

Ontario. Task Force on Child Abuse.
Report of the Task Force on Child Abuse. — ₍Toronto₎ : Ontario Ministry of Community and Social Services, Children's Services Division, 1978.
iv, 24, ₍5₎, 55-65 p. ; 30 cm. C***
Cover title.
Bibliography: p. ₍25₎
1. Child abuse—Ontario. 2. Child abuse—Ontario—Services. I. Ontario. Ministry of Community and Social Services. Children's Services Division.
HV745.O5O5 1978 362.7'1 79-303456
79 MARC

—ONTARIO—SERVICES

Ontario. Task Force on Child Abuse.
Report of the Task Force on Child Abuse. — ₍Toronto₎ : Ontario Ministry of Community and Social Services, Children's Services Division, 1978.
iv, 24, ₍5₎, 55-65 p. ; 30 cm. C***
Cover title.
Bibliography: p. ₍25₎
1. Child abuse—Ontario. 2. Child abuse—Ontario—Services. I. Ontario. Ministry of Community and Social Services. Children's Services Division.
HV745.O5O5 1978 362.7'1 79-303456
79 MARC

—SWEDEN

Rönström, Anitha, 1939-
Barnomsorgen : intentioner och verklighet / Anitha Rönström. — Lund : Studentlitt., 1977.
130 p. ; 23 cm. S77-50/52
Bibliography: p. 124-128.
Includes index.
ISBN 9144142714 : kr31.00
1. Child welfare—Sweden. 2. Children in Sweden. 3. Socialization. 4. Child abuse—Sweden. I. Title.
HV790.A6R63 78-38909
79 MAR

—UNITED STATES

Child abuse and neglect : a compendium of course materials. ₍Houston, Tex.₎ : National College of District Attorneys, ₍197
387 p. ; 28 cm.
"Originally prepared for presentation at five child abuse seminars under auspices of the National College of District Attorneys from December 1 through April 1977."
1. Child abuse—Law and legislation—United States. 2. Child abuse—United States. I. National College of District Attorneys.
KF9323.C5 344'.73'03271 78-3173
78 MA

CHILD ABUSE
—UNITED STATES (Continued)

Geiser, Robert L
Hidden victims : the sexual abuse of children / Robert L. Geiser. — Boston : Beacon Press, c1979.
xi, 191 p. ; 21 cm.
Bibliography: p. 181-182.
Includes index.
SUMMARY: Discusses the possible causes and aftermath of the sexual misuse of children resulting from incidents involving incest, rape, prostitution, and other activities.
ISBN 0-8070-2954-3 : $10.95
1. Child molesting—United States. 2. Child abuse—United States. 3. Incest. [1. Child molesting. 2. Child abuse. 3. Incest] I. Title.
HQ72.U53G43 1979 364.1'53 78-73853
79 MARC

Giovannoni, Jeanne M
Child abuse and neglect : an examination from the perspective of child development knowledge / Jeanne M. Giovannoni, Jonathon Conklin, and Patti Iiyama. — San Francisco : R&E Research Associates, 1978.
iv, 105 p. ; 28 cm.
Includes bibliographies and index.
ISBN 0-88247-514-2 : $10.00
1. Child abuse—United States. I. Conklin, Jonathon, joint author. II. Iiyama, Patti, joint author. III. Title.
HV741.G5 362.7'1 77-90387
79 MARC

Halperin, Michael, 1941-
Helping maltreated children : school and community involvement / Michael Halperin, with assistance from Pamela Shuman. — St. Louis : Mosby, 1979.
x, 198 p. ; 24 cm.
Bibliography: p. 174-184.
Includes index.
ISBN 0-8016-2020-1
1. Child abuse—Services—United States. 2. Child abuse—United States. 3. Child abuse—United States—Prevention. I. Shuman, Pamela, joint author. II. Title.
HV741.H28 362.7'1 78-31527
78 MARC

Herbruck, Christine Comstock.
Breaking the cycle of child abuse / by Christine Comstock Herbruck. — Minneapolis : Winston Press, c1979.
205 p. ; 23 cm.
ISBN 0-03-052691-4
1. Child abuse—United States. 2. Child abuse—United States—Prevention. 3. Parents Anonymous—History. I. Title.
HV741.H47 362.7'1 78-59405
79 MARC

Holder, Wayne M
Understanding child neglect and abuse / by Wayne M. Holder and Patricia Schene ; [ill. by Barbara B. Anderson]. — Englewood, Colo. : Child Protection American Humane Association, c1978.
v, 22 p. : ill. ; 23 cm.
Includes bibliographical references.
1. Child abuse—United States. I. Schene, Patricia, joint author. II. Title.
HV741.H56 362.7'1 79-105526
79 MARC

Inglis, Ruth Langdon, 1927-
Sins of the fathers : a study of the physical and emotional abuse of children / Ruth Inglis. — New York : St. Martin's Press, c1978.
220 p. ; 22 cm.
Bibliography: p. 214-220.
ISBN 0-312-72608-2 : $10.00
1. Child abuse. 2. Parent and child. 3. Child abuse—United States. 4. Child abuse—Great Britain. I. Title.
HV713.I53 362.7'1 78-4004
78 MARC

Shapiro, Deborah.
Parents and protectors : a study in child abuse and neglect / Deborah Shapiro. — New York : Research Center, Child Welfare League of America, c1979.
xv, 127 p. ; 26 cm.
"G-19."
ISBN 0-87868-139-6
1. Child abuse—United States. 2. Child abuse—Services—United States. I. Title.
HV741.S43 362.7'1 79-88223
79 MARC

—UNITED STATES—ADDRESSES, ESSAYS, LECTURES

Child abuse and neglect in residential institutions : selected readings on prevention, investigation, and correction. — [Washington] : National Center on Child Abuse and Neglect, Children's Bureau/Administration for Children, Youth and Families/Office of Human Development Services, U.S. Dept. of Health, Education, and Welfare, 1978.
178 p. in various pagings : ill. ; 27 cm. — (DHEW publication ; no. (OHDS) 78-30160)
1. Child abuse—United States—Addresses, essays, lectures. 2. Children—Institutional care—United States—Addresses, essays, lectures. I. National Center on Child Abuse and Neglect. II. Series: United States. Dept. of Health, Education, and Welfare. DHEW publication ; no. (OHDS) 78-30160.
HV741.C453 362.7'32 78-603525
79 MARC

Focus on services to young children : selected articles from the Journal of home economics. — [Washington] : American Home Economics Association, c1978.
52 p. : ill. ; 28 cm.
Cover title.
ISBN 0-8461-5106-5 : $5.00
1. Day care centers—United States—Addresses, essays, lectures. 2. Child abuse—United States—Addresses, essays, lectures. 3. Nutrition—Addresses, essays, lectures. I. The Journal of home economics. II. American Home Economics Association.
HV854.F6 362.7'08 78-18360
78[r79]rev MARC

—UNITED STATES—CONGRESSES

Interdisciplinary Workshop on Child Abuse, Hershey, Pa., 1977.
Child abuse and neglect : the proceedings of an Interdisciplinary workshop on Child Abuse, April 12-13, 1977, Hershey, PA. — Harrisburg : Commonwealth of Pennsylvania, Dept. of Public Welfare, [1978?].
32 p. ; 28 cm.
Cover title.
"PWPE 30 3-78."
1. Child abuse—United States—Congresses. I. Pennsylvania. Dept. of Public Welfare. II. Title.
HV741.I58 1977 362.7'1 78-623469
79 MARC

National Conference on Child Abuse and Neglect, 2d, Houston, Tex., 1977.
Child abuse and neglect : issues on innovation and implementation : proceedings of the second annual National Conference on Child Abuse and Neglect, April 17-20, 1977 / edited by Michael L. Lauderdale, Rosalie N. Anderson, Stephen E. Cramer. — [Washington : National Center on Child Abuse and Neglect, Children's Bureau, Administration for Children, Youth, and Families, Office of Human Development Services, U.S. Dept. of Health, Education, and Welfare], 1978-
v. : ill. ; 27 cm. — (DHEW publication ; no. (OHDS) 78-30147)
Includes bibliographies.
1. Child abuse—United States—Congresses. 2. Child abuse—Services—United States—Congresses. I. Lauderdale, Michael L. II. Anderson, Rosalie N. III. Cramer, Stephen E. IV. National Center on Child Abuse and Neglect. V. Title. VI. Series: United States. Dept. of Health, Education, and Welfare. DHEW publication ; no. (OHDS) 78-30147.
HV881.N35 1977 362.7'1 78-603476
79 MARC

—UNITED STATES—PREVENTION

Halperin, Michael, 1941-
Helping maltreated children : school and community involvement / Michael Halperin, with assistance from Pamela Shuman. — St. Louis : Mosby, 1979.
x, 198 p. ; 24 cm.
Bibliography: p. 174-184.
Includes index.
ISBN 0-8016-2020-1
1. Child abuse—Services—United States. 2. Child abuse—United States. 3. Child abuse—United States—Prevention. I. Shuman, Pamela, joint author. II. Title.
HV741.H28 362.7'1 78-31527
78 MARC

Helfer, Ray E
Child abuse : a plan for prevention / Ray E. Helfer. — Chicago : National Committee for Prevention of Child Abuse, c1978.
iii, 36 p. ; 22 cm.
Cover title.
Includes bibliographical references.
1. Child abuse—United States—Prevention. I. Title.
HV741.H42 362.7'1 79-108462
79 MARC

Herbruck, Christine Comstock.
Breaking the cycle of child abuse / by Christine Comstock Herbruck. — Minneapolis : Winston Press, c1979.
205 p. ; 23 cm.
ISBN 0-03-052691-4
1. Child abuse—United States. 2. Child abuse—United States—Prevention. 3. Parents Anonymous—History. I. Title.
HV741.H47 362.7'1 78-59405
79 MARC

Smith, Charles P 1931-
Crisis in the shadows : a preliminary national assessment of child abuse and neglect and the juvenile justice system / Charles P. Smith, David J. Berkman, Warren M. Fraser ; submitted to the U.S. Office of Juvenile Justice and Delinquency Prevention, January 1979 ; prepared by the National Juvenile Justice System Assessment Center of the American Justice Institute. — Sacramento : The Institute, c1979.
xiv, 162 p. ; 29 cm.
Includes bibliographies.
1. Child abuse—Services—United States. 2. Child abuse—United States—Prevention. 3. Juvenile justice, Administration of—United States. I. Berkman, David J., joint author. II. Fraser, Warren M., joint author. III. United States. Law Enforcement Assistance Administration. Office of Juvenile Justice and Delinquency Prevention. IV. National Juvenile Justice System Assessment Center. V. Title.
HV741.S54 362.7'1 79-110961
79 MARC

—UTAH

Allred, Mark.
Child abuse and neglect : a report to the Forty-second Legislature, containing recommendations of the Joint Social Services Interim Study Committee / by Mark J. Allred. — [Salt Lake City] : Office of Legislative Research, State of Utah, 1977.
ii, 47 leaves ; 28 cm. — (Research report - Office of Legislative Research, State of Utah ; no. 24)
Includes bibliographical references.
1. Child abuse—Utah. 2. Child abuse—Law and legislation—Utah. I. Utah. Legislature. Joint Social Services Interim Study Committee. II. Title. III. Series: Utah. Office of Legislative Research. Research report - Office of Legislative Research, State of Utah ; no. 24.
HV742.U8A45 362.7'1 78-622525
79 MARC

—WALES—CONGRESSES

Social work practice and non-accidental injury to children : report of the residential workshops held at Abergavenny, January 1977, and Llandudno, March 1977. — [Cardiff] : Welsh Office, Social Work Service, [1977?]
[1], i, 102 p. : ill. ; 30 cm. GB78-14221
Cover title.
Includes bibliographies.
1. Child abuse—Wales—Services—Congresses. 2. Child abuse—Wales—Prevention—Congresses. 3. Child abuse—Wales—Congresses. I. Great Britain. Welsh Office. Social Work Service.
HV751.W34S66 362.7'1 79-303271
79 MARC

—WALES—PREVENTION—CONGRESSES

Social work practice and non-accidental injury to children : report of the residential workshops held at Abergavenny, January 1977, and Llandudno, March 1977. — [Cardiff] : Welsh Office, Social Work Service, [1977?]
[1], i, 102 p. : ill. ; 30 cm. GB78-14221
Cover title.
Includes bibliographies.
1. Child abuse—Wales—Services—Congresses. 2. Child abuse—Wales—Prevention—Congresses. 3. Child abuse—Wales—Congresses. I. Great Britain. Welsh Office. Social Work Service.
HV751.W34S66 362.7'1 79-303271
79 MARC

—WALES—SERVICES—CONGRESSES

Social work practice and non-accidental injury to children : report of the residential workshops held at Abergavenny, January 1977, and Llandudno, March 1977. — [Cardiff] : Welsh Office, Social Work Service, [1977?]
[1], i, 102 p. : ill. ; 30 cm. GB78-14221
Cover title.
Includes bibliographies.
1. Child abuse—Wales—Services—Congresses. 2. Child abuse—Wales—Prevention—Congresses. 3. Child abuse—Wales—Congresses. I. Great Britain. Welsh Office. Social Work Service.
HV751.W34S66 362.7'1 79-303271
79 MARC

CHILD ABUSE PROGRAMS see Child abuse—Services

CHILD ACTORS see Children as actors

CHILD ANALYSIS

Bloch, Dorothy.
"So the witch won't eat me" : fantasy and the child's fear of infanticide / Dorothy Bloch. — Boston : Houghton Mifflin, 1978.
x, 246 p. ; 22 cm.
Includes bibliographical references.
ISBN 0-395-26290-9 : $9.95
1. Child analysis. 2. Psychoanalysis. 3. Fear in children. 4. Fantasy in children. 5. Infanticide. 6. Fear of death. I. Title.
RJ504.2.B57 618.9'28'522 78-13115
78 MARC

Freud, Anna, 1895-
Einführung in die Technik der Kinderanalyse : vier Vorträge am Lehrinstitut der Wiener Psychoanalytischen Vereinigung / von Anna Freud. — Wien : Internationaler Psychoanalytischer Verlag, 1927.
87 p. ; 21 cm.
1. Child analysis. I. Title.
RJ504.2.F73 1927 78-381530
78 MARC

CHILD ANALYSIS (Continued)

Lebovici, Serge.
La connaissance de l'enfant par la psychanalyse / Serge Lebovici, Michel Soulé, avec la collaboration de Simone Decobert et Jeanine Noël. — 3. éd. mise à jour. — ⌐Paris⌐ : Presses universitaires de France, 1977, c1970.
650 p. ; 22 cm. — (Psychanalyse et psychiatrie de l'enfant) (Le Fil rouge ; section 2) F***
Bibliography: p. ⌐601⌐-617.
Includes indexes.
85.00F
1. Child psychology. 2. Child analysis. I. Soulé, Michel, joint author. II. Title.
BF721.L434 1977 155.4 79-344249
79 MARC

Michaud, Ginette.
Écoles buissonnières : pédagogie, psychiatrie, psychanalyse / Ginette Michaud ; préf. de Julia Kirsteva. — Paris : Gauthier-Villars, c1978.
vi, 198 p. ; 24 cm. — (Interférences) F***
Includes bibliographical references.
ISBN 2-04-010356-2
1. School attendance. 2. Mentally ill children. 3. Child analysis. 4. Mentally ill children—Education. I. Title.
LC142.M5 371.9'28 79-357516
79 MARC

Miyatake, Tatsuo.
(Yōji to seihakuji no e ga uttaeru mono)
幼児と精薄児の絵が訴えるもの / 宮武辰夫著. — 名古屋：黎明書房, 1955.
387 p. : ill. (some col.) ; 22 cm.

1. Drawing, Psychology of. 2. Children's art—Japan. 3. Mentally handcapped children—Education—Art. 4. Child analysis. I. Title.
BF723.D7M57 79-801649

Ógida, Hiromoto.
(E ni yoru kosei hakken)
絵による個性発見：描画完成法 / 扇田博元, 畠山忠共著. — 名古屋：黎明書房, 昭和51 ⌐1976⌐
299, 4 p. : ill. ; 20 cm. Ja 76-10597
Bibliography; p. 256-263,
Includes index.
¥2300
1. Drawing, Psychology of. 2. Child analysis. I. Hatakeyama, Tadashi, 1929- joint author. II. Title.
BF723.D7O35 79-805560

Shimoda, Seishi, 1890-
(E ni miru kodomo no shinri)
絵にみる子どもの心理 / 霜田静志著. — ⌐東京⌐：東都書房, 昭和34 ⌐1959⌐
97, ⌐2⌐ p. : ill. (some col.) ; 27 cm.
Bibliography: p. ⌐98⌐

1. Children's art—Japan. 2. Art—Psychology. 3. Child analysis. I. Title.
N352.S52 79-809344

Winnicott, Donald Woods, 1896-
Through paediatrics to psycho-analysis / by D. W. Winnicott ; with an introd. by M. Masud R. Khan. — London : Hogarth Press and the Institute of Psycho-Analysis, 1975.
L, 350 p. : ill. ; 23 cm. — (The International psycho-analytical library ; no. 100) GB***
First published in 1958 under title: Collected papers.
Includes bibliographies and index.
ISBN 0-7012-0399-4 : £5.00
1. Child psychiatry. 2. Child analysis. I. Institute of Psycho-Analysis, London. II. Title. III. Series.
RJ499.W48 1975 618.9'28'9 75-311187
75⌐r79⌐rev MARC

—ADDRESSES, ESSAYS. LECTURES

Il Bambino nella psicoanalisi / a cura di Silvia Vegetti Finzi ; testi di S. Freud ... ⌐et al.⌐. — Bologna : Zanichelli, 1976.
122 p. ; 19 cm. — (Letture di filosofia e scienze umane ; 4) It78-Aug
Bibliography: p. ⌐120⌐-122.
L1800
1. Child analysis—Addresses, essays, lectures. 2. Child analysis—Cases, clinical reports, statistics. I. Vegetti Finzi, Silvia, 1938- . II. Freud, Sigmund, 1856-1939.
RJ504.2.B35 79-340030
*79 MARC

—CASES, CLINICAL REPORTS, STATISTICS

Il Bambino nella psicoanalisi / a cura di Silvia Vegetti Finzi ; testi di S. Freud ... ⌐et al.⌐. — Bologna : Zanichelli, 1976.
122 p. ; 19 cm. — (Letture di filosofia e scienze umane ; 4) It78-Au
Bibliography: p. ⌐120⌐-122.
L1800
1. Child analysis—Addresses, essays, lectures. 2. Child analysis—Cases, clinical reports, statistics. I. Vegetti Finzi, Silvia, 1938- . II. Freud, Sigmund, 1856-1939.
RJ504.2.B35 79-340030
*79 MARC

Sirlin, María Elisabeth.
Una experiencia terapéutica : historia de un grupo de niños de 5 años / María Elisabeth Sirlin. — Buenos Aires : Amorrortu Editores, ⌐1975⌐ c1974.
199 p. ; 21 cm.
Bibliography: p. 197-198.
1. Child analysis—Cases, clinical reports, statistics. I. Title.
RJ504.2.S55 618.9'28'917 77-154641
79 MARC

Winnicott, Donald Woods, 1896-1971.
The Piggle : an account of the psychoanalytic treatment of a little girl / by D. W. Winnicott ; edited by Ishak Ramzy. — London : Hogarth Press, 1978.
xvi, 201 p. : 1 ill. ; 23 cm. — (The International psycho-analytical library ; no. 107) GB***
ISBN 0-7012-0451-6 : £6.50
1. Child analysis—Cases, clinical reports, statistics. I. Ramzy, Ishak. II. Title. III. Series.
RJ504.2.W56 1978 618.9'28'9170926 78-321122
79 MARC

CHILD AND FATHER see Father and child

CHILD AND MOTHER see Mother and child

CHILD AND YOUTH CENTERED INFORMATION SYSTEM see CYCIS (Information retrieval system)

CHILD AUTHORS see Children as authors

CHILD BEHAVIOR see Child psychology; Children—Management

CHILD CARE CENTERS see Child welfare; Day care centers

CHILD CUSTODY see Custody of children

CHILD DEVELOPMENT

see also Child psychology; Children—Growth; Motor ability in children

al-Laqqānī, Fārūq 'Abd al-Ḥamīd.
(Tathqīf al-ṭifl, falsafatuhu wa-ahdāfuh, maṣādiruhu wa-wasā'iluh)
تثقيف الطفل ، فلسفته وأهدافه ، مصادره ووسائله / فاروق عبد الحميد اللقانى . — الاسكندرية : منشأة المعارف ، 1976.
7, 207 p. ; 25 cm.
Bibliography: p. ⌐195⌐-200.
£E1.50
1. Socialization. 2. Child development. I. Title.
HQ783.L36 77-960991

Alasjärvi, Ulla.
Il gioco drammatico ... : proposte pratiche per usare il tempo con i bambini / Ulla Alasjärvi ; grafica di Lucio Diana, Gabriele Vacis, Adriana Zamboni. — Torino : Musolini, 1978.
115 p. : ill. ; 20 x 21 cm. — (Oltre il libro di testo : Manuali) It79-Jan
Bibliography: p. 113.
L5900
1. Creative activities and seat work. 2. Educational games. 3. Drama in education. 4. Child development. I. Title.
LB1537.A37 79-352144
*79 MARC

Ames, Louise Bates.
Your five-year-old : sunny and serene / by Louise Bates Ames and Frances L. Ilg, Gesell Institute of Child Development ; illustrated with photos. by Betty David. — New York : Delacorte Press, c1979.
123 p. : ill. ; 22 cm.
Bibliography: p. 110-116.
Includes index.
ISBN 0-440-09876-9
1. Child development. I. Ilg, Frances Lillian, 1902- joint author. II. Gesell Institute of Child Development, New Haven. III. Title.
HQ772.A474 155.4'23 78-11622
78 MARC

Ames, Louise Bates.
Your six-year-old : defiant but loving / by Louise Bates Ames and Frances L. Ilg, Gesell Institute of Child Development ; illustrated with photos. by Betty David. — New York : Delacorte Press, c1979.
132 p. : ill. ; 22 cm.
Bibliography: p. 121-126.
Includes index.
ISBN 0-440-09877-7
1. Child development. I. Ilg, Frances Lillian, 1902- joint author. II. Gesell Institute of Child Development, New Haven. III. Title.
HQ772.A476 155.4'24 78-11578
78 MARC

Anolli, Luigi.
Lo sviluppo della percezione visiva : contributo alla taratura italiana del DTVP di M. Frostig e analisi del Programma Frostig / ⌐di⌐ Luigi Anolli, Vittorio Cigoli ; introduzione di Giuseppe Girotti. — Firenze : OS, 1978.
326 p. ; 24 cm. It79-Jan
Includes bibliographies.
L7800
1. Visual perception. 2. Visual perception—Testing. 3. Child development. I. Cigoli, Vittorio, joint author. II. Title.
BF241.A58 79-357362
*79 MARC

Baj van a gyerekemmel / ⌐szerk. F. Várkonyi Zsuzsa⌐. — ⌐Budapest⌐ : Gondolat, 1977.
261 p., ⌐8⌐ leaves of plates : ill. (some col.) ; 19 cm. H77
Bibliography: p. 259-⌐260⌐
ISBN 9632805097 : 24.00Ft
1. Handicapped children. 2. Child study. I. Várkonyi, Zsuzsa F.
HQ773.6.B34 78-385271
*78 MARC

Barber, Lucie W
Celebrating the second year of life : a parent's guide for a happy child / Lucie W. Barber and the staff of the Union College Character Research Project. — Birmingham, Ala. : Religious Education Press, c1978.
x, 148 p. ; 20 cm.
Bibliography: p. 132-135.
Includes index.
ISBN 0-89135-015-2
1. Infants. 2. Child development. 3. Children—Management. I. Union College, Schenectady. Character Research Project. II. Title.
HQ774.B33 649'.122 78-21484
78 MARC

Barn och andra människor : en bok för barnskötare / Birgit Andersson-Nyander ... ⌐et al.⌐. — Stockholm : Esselte studium, 1978.
266 p. : ill. ; 24 cm. S78-23
Includes index.
ISBN 9124275557
1. Children—Sweden—Social conditions. 2. Child development. I. Andersson-Nyander, Birgit.
HQ792.S8B37 79-342107
79 MARC

Beaty, Janice J
Skills for preschool teachers / Janice J. Beaty ; ⌐text photos. by Janet Gagnon⌐. — Columbus, Ohio : Merrill, c1979.
xi, 225 p. : ill. ; 25 cm.
Includes bibliographies and index.
ISBN 0-675-08283-8
1. Education, Preschool. 2. Child development. I. Title.
LB1140.2.B413 372.21 78-70405
00 MARC

Bergan, John R 1931-
Child development / John R. Bergan, Ronald W. Henderson. — Columbus, Ohio : Merrill, c1979.
vi, 422 p. : ill. ; 26 cm.
Bibliography: p. 373-401.
Includes index.
ISBN 0-675-08371-0
1. Child development. I. Henderson, Ronald W., joint author. II. Title.
HQ767.9.B45 301.43'1 78-62025
79 MARC

Biere, Bernd Ulrich.
Kommunikation unter Kindern : method. Reflexion u. exemplar. Beschreibung / Bernd Ulrich Biere. — Tübingen : Niemeyer, 1978.
x, 258 p. ; 24 cm. — (Linguistische Arbeiten ; 65 ISSN 0344-6727) GFR78-
A revision of the author's thesis, Tübingen.
Bibliography: p. 196-202.
ISBN 3-484-10313-2
1. Children—Language. 2. Child development. I. Title.
P118.B5 1978 79-3679
79 MAR

CHILD DEVELOPMENT (Continued)

Bigner, Jerry J
Parent-child relations : an introduction to parenting / Jerry J. Bigner. — New York : Macmillan, c1979.
x, 273 p. : ill. ; 26 cm.
Includes bibliographies and indexes.
ISBN 0-02-309820-1
1. Parent and child. 2. Child development. 3. Family. I. Title.
HQ772.B477 301.42'7 78-3572
78 MARC

Bilski, Audrey.
The vital years and your child. ⌊London⌋ Souvenir Press ⌊1973⌋
180 p. illus. 23 cm. £2.20 GB73-10131
Stamped on t.p.: Exclusive distributor, ISBS, inc., Forest Grove, OR.
ISBN 0-285-62088-6
1. Child development. 2. Education, Preschool. I. Title.
⌊DNLM: 1. Child development—Popular works. 2. Child psychology—Popular works. WS105 B599v 1973⌋
HQ772.B478 1973 649'.1 73-595542
Shared Cataloging with DNLM 73⌊r79⌋rev MARC

Biología, psicología y sociología del niño en edad preescolar / Emilia García Manzano ... ⌊et al.⌋ ; coordinación, Jaime Sarramona. — 1. ed. — Barcelona : CEAC, 1978.
187 p. ; 21 cm. — (Colección Educación y enseñanza : Serie Preescolar) Sp***
Bibliography: p. ⌊177⌋-181.
ISBN 8432994014
1. Child development. I. García Manzano, Emilia. II. Sarramona, Jaime. III. Series.
LB1119.B53 78-397733
79 MARC

Bonino, Silvia.
La frustrazione nella dinamica dello sviluppo / Silvia Bonino ; presentazione di Ada Fonzi. — Firenze : Giunti-Barbèra, 1977.
160 p. : ill. ; 21 cm. — (Collezione psicologica) It78-Oct
Includes bibliographies.
L3000
1. Frustration in children. 2. Child development. I. Title.
BF723.F7B66 79-361747
*79 MARC

Bower, T G R
Pathways in development / by T. G. R. Bower. — Paris : Organisation for Economic Co-operation and Development ; ⌊Washington, D.C. : sold by OECD Publications Center⌋, 1977.
28 p. : ill. ; 24 cm. — (Studies in the learning sciences ; 4) F***
At head of title: Centre for Educational Research and Innovation.
Bibliography: p. 27-28.
ISBN 9264117164 : F7.00
1. Child development. 2. Psychological research. I. Centre for Educational Research and Innovation. II. Title. III. Series.
LB1115.B723 155.4'13 79-307602
79 MARC

Breitbach, Irmgard.
Identitätsentwicklung im Kindergartenalter unter den Bedingungen kollektiver Erziehung : d. Analyse dreier Kindergärten in israel. Kibbutzim / Irmgard Breitbach. — München : Minerva-Publikation, 1979.
355 p. ; 21 cm. — (Minerva-Fachserie Pädagogik) GFR79-A
Originally presented as the author's thesis, Marburg, 1978.
Bibliography: p. 345-355.
ISBN 3-597-10071-6 : DM48.00
1. Kindergartens—Israel. 2. Collective settlements—Israel. 3. Child development. 4. Identity (Psychology) I. Title.
LB1438.I75B73 1979 79-371632
79 MARC

Brenig, Ruth.
Pädagogische Probleme im Kindergarten / Ruth Brenig. — 1. Aufl. — Stuttgart : Klett, 1978.
142 p. ; 22 cm. GFR78-A
Bibliography: p. 140-142.
ISBN 3-12-920190-4
1. Kindergarten. 2. Education of children. 3. Child development. I. Title.
LB1167.B68 372.21'8 79-345452
79 MARC

Brooks-Gunn, Jeanne.
He & she : how children develop their sex role identity / Jeanne Brooks-Gunn and Wendy Schempp Matthews. — Englewood Cliffs, N.J. : Prentice-Hall, c1979.
xiv, 338 p. : ill. ; 21 cm. — (A Spectrum book)
Bibliography: p. 300-331.
Includes index.
ISBN 0-13-384388-2
1. Sex role. 2. Child development. I. Matthews, Wendy Schempp, joint author. II. Title.
HQ1075.B76 301.41 79-13187
79 MARC

Bucher, Charles Augustus, 1912-
Physical education for children : movement foundations and experiences / Charles A. Bucher, Nolan A. Thaxton. — New York : Macmillan, c1979.
xiii, 415 p. : ill. ; 24 cm.
Includes bibliographies and index.
ISBN 0-02-316300-3
1. Physical education for children. 2. Child development. 3. Motor learning. 4. Physical education for exceptional children. I. Thaxton, Nolan A., joint author. II. Title.
GV443.B79 372.8'6'044 78-6119
78 MARC

Bühler, Henriette, 1952-
Spielgruppenpädagogik : e. praxisbezogene Einf. / Henriette Bühler, Franziska Fey-Kornau, Karl-Josef Kluge. — München : Kösel, 1978.
156, ⌊1⌋ p. ; 22 cm. — (Kösel Ausbildung) GFR78-A
Bibliography: p. 152-⌊157⌋
ISBN 3-466-30191-2
1. Play. 2. Child development. 3. Play—Psychological aspects. 4. Play—Social aspects. I. Fey-Kornau, Franziska, 1952- joint author. II. Kluge, Karl-Josef, joint author. III. Title.
LB1137.B83 78-394435
79 MARC

Cahoon, Owen W 1937-
Parents and the achieving child / Owen W. Cahoon, Alvin H. Price, A. Lynn Scoresby. — Provo, Utah : Brigham Young University Press, c1979.
vi, 282 p. ; 23 cm.
Includes index.
ISBN 0-8425-0856-2 : $6.95
1. Parent and child. 2. Child development. 3. Home and school. I. Price, Alvin H., 1935- joint author. II. Scoresby, A. Lynn, joint author. III. Title.
HQ772.C32 649'.1 78-32168
79 MARC

Carlson, Carole C
Established in Eden / Carole C. Carlson. — Old Tappan, N.J. : F. H. Revell Co., c1978.
157 p. ; 22 cm.
ISBN 0-8007-0943-8
1. Family. 2. Marriage. 3. Child development. 4. Interpersonal relations. I. Title.
HQ734.C314 301.42 78-14954
78 MARC

Casabianca, Rose Marie de.
L'Éveil social avant trois ans en milieu institutionnel / Rose-Marie de Casabianca .... — Paris : Éditions Fleurus, 1978.
491 p. ; 20 cm. — (Collection Pédagogie psychosociale ; 32) F78-1984
Bibliography: p. 439-466.
Filmography: p. 435-437.
Includes indexes.
ISBN 2-215-00227-1 : 85.00F
1. Socialization. 2. Child development. I. Title.
HQ783.C368 301.15'72 79-363179
79 MARC

Chabauty, Marie Luce.
Les enfants de la maternelle / Marie-Luce Chabauty ; préf. de Louis Leprince-Ringuet. — Paris : Centurion, c1979.
237 p. ; 21 cm. — (Paidoguides ; 19) F***
Bibliography: p. 225-234.
ISBN 2-227-12518-7 : 44.00F
1. Day care centers—France. 2. Child development. I. Title.
HV861.F8C48 362.7'1 79-373641
79 MARC

Ch'en, Chao-ming.
⌊Erh t'ung ti jen ko fa chan yü fu tao⌋
六歲定終身：兒童人格發展與輔導 / 陳照明著. — 第1版. — 新店鎮：巨鷹文化公司, 民國67 ⌊1978⌋
137 p. : 22 cm. — (巨鷹叢書)
Series romanized: Chü ying ts'ung shu.
Includes bibliographical references.
NT$50.00
1. Child development. 2. Parent and child. I. Title. Title romanized: Liu sui ting chung shen.
HQ772.C416 1978 79-835076
Library of Congress 79 C

The Child and its family / edited by Michael Lewis and Leonard A. Rosenblum. — New York : Plenum Press, c1979.
xii, 304 p. : ill. ; 24 cm. — (Genesis of behavior ; v. 2)
Includes bibliographies and indexes.
ISBN 0-306-40112-6
1. Child development. 2. Socialization. 3. Family. I. Lewis, Michael, 1937 (Jan. 10)- II. Rosenblum, Leonard A. III. Series.
⌊DNLM: 1. Infant. 2. Family. 3. Parent-child relations. 4. Sibling relations. W1 GE275 v. 2 / WS105.5.F2 C536⌋
HQ772.C43 155.4'18 78-27020
78 MARC

Chu, Chŏng-il.
(Adong paltarhak)
兒童發達學 / 朱貞一著. — 서울：教文社, 1978.
262 p. : ill. ; 23 cm.
Includes bibliographical references.
W3500
1. Children—Growth. 2. Infants—Growth. 3. Child development. I. Title.
RJ131.C55 79-826417

Cohen, Marilyn A
The developmental resource : behavioral sequences for assessment and program planning / Marilyn A. Cohen and Pamela J. Gross, with contributions by Lynnette Chandler ; ill. by Kathleen E. Anderson. — New York : Grune & Stratton, c1978.
2 v. : ill. ; 26 cm.
Vol. 2 has contributions by T. E. Coggins and R. L. Carpenter.
Includes bibliographies.
ISBN 0-8089-1122-8 (v. 1)
1. Child psychology. 2. Infant psychology. 3. Child development. I. Gross, Pamela J., joint author. II. Title.
BF721.C566 155.4 78-20327
78 MARC

Conger, Flora Stabler.
Child care aide skills / Flora Stabler Conger, Irene B. Rose ; Elizabeth Simpson, consulting editor. — New York : Gregg Division, McGraw-Hill, c1979.
viii, 328 p. : ill. ; 25 cm. — (Careers in home economics)
Includes bibliographical references and index.
ISBN 0-07-012420-5
1. Day care centers. 2. Child development. 3. Children—Care and hygiene. I. Rose, Irene B., joint author. II. Simpson, Elizabeth Jane, 1920. III. Title.
HV851.C67 362.7'1 78-13217
78 MARC

Corella, Laura G
Aprendiendo a ser padres / Laura G. Corella. — Bilbao : Fher, ⌊1975⌋
205 p. ; 20 cm. — (Colección Vida y luz) Sp76-Jan
ISBN 8424304950
1. Child development. 2. Children—Care and hygiene. I. Title.
HQ769.C7414 649'.1 77-467495
77⌊r79⌋rev MARC

Cratty, Bryant J
Perceptual and motor development in infants and children / Bryant J. Cratty. — 2d ed. — Englewood Cliffs, N.J. : Prentice-Hall, c1979.
x, 310 p. : ill. ; 24 cm.
Includes bibliographies and index.
ISBN 0-13-657023-2
1. Child development. 2. Developmental psychobiology. 3. Motor ability in children. 4. Perception in children. I. Title.
⌊DNLM: 1. Child development. 2. Motor activity—In infancy and childhood. 3. Perception—In infancy and childhood. WS105.3 C896p⌋
RJ131.C7 1978 612.6'5 78-7718
78 MARC

Cully, Iris V
Christian child development / Iris V. Cully. — 1st ed. — San Francisco : Harper & Row, c1979.
x, 166 p. ; 21 cm.
Includes bibliographical references and index.
ISBN 0-06-061648-2
1. Christian education of children. 2. Children—Religious life. 3. Child development. I. Title.
BV1475.2.C814 1979 268'.432 78-19507
78 MARC

Denzin, Norman K
Childhood socialization / Norman K. Denzin. — 1st ed. — San Francisco : Jossey-Bass Publishers, 1977.
xiii, 235 p. ; 24 cm. — (The Jossey-Bass social and behavioral science series)
Bibliography: p. 209-225.
Includes index.
ISBN 0-87589-354-6 : $12.95
1. Child development. 2. Socialization. I. Title.
HQ767.9.D46 301.15'72 77-82914
79 MARC

Desarrollo y estimulación del niño : desde su nacimiento hasta los dos años. — Pamplona : Ediciones Universidad de Navarra, c1979.
101 p. : ill. ; 22 cm. — (Colección Ciencias médicas) (Libros de enfermería) Sp***
At head of title: UPEP. Unidad Pediatrica de Estimulación Precoz; Clínica Universitaria, Facultad de Medicina, Universidad de Navarra.
Bibliography: p. ⌊99⌋-101.
ISBN 8431305835
1. Child development. 2. Child development deviations. 3. Motor ability in children. 4. Perception in children. 5. Sensory stimulation.
RJ131.D43 79-108918
00 MARC

Diagram Group.
Child's body : a parent's manual / by the Diagram Group. — New York : Paddington Press : distributed by Grosset & Dunlap, c1977.
ca. 300 p. : ill. ; 28 cm.
Includes bibliography and index.
ISBN 0-448-22187-X : $6.95
1. Children—Care and hygiene. 2. Child development. I. Title.
RJ61.D55 1977 612.6'5 77-12566
77 MARC

Draper, Mary Wanda.
Caring for children / by Mary Wanda Draper and Henry E. Draper. — Rev. — Peoria, Ill. : C. A. Bennett Co., c1979.
560 p. : ill. ; 24 cm. — (A Bennett career book)
"Annotated bibliography of books for children": p. 521-532.
Includes bibliographies and index.
ISBN 0-87002-281-4
1. Child development. 2. Children—Management. 3. Day care aids—United States. I. Draper, Henry E., joint author. II. Title.
HQ769.D644 1979 649'.1 78-52494
79 MARC

CHILD DEVELOPMENT (Continued)

Durojaiye, Susan M
Practical methods for nursery schools / Susan M. Durojaiye. — Ibadan : Oxford University Press Nigeria, 1977.
156 p. : ill. ; 21 cm.
Includes bibliographies.
ISBN 0-19-575313-5
1. Nursery schools. 2. Child development. I. Title.
LB1140.2.D86 372.21'6 79-106637
79 MARC

Dyer, George B
To free a child = La liberación integral del niño / by George B. Dyer ; ⌊Spanish translations by Rosa Maria Icaza⌋. — San Antonio, Tex. : Mexican American Cultural Center, c1979.
20, 20 p. : ill. ; 22 cm.
English and Spanish.
1. Child development. 2. Children's rights. I. Title. II. Title: La liberación del niño.
HQ772.D94 301.43'14 79-84531
79 MARC

L'Enfant. — Paris : R. Laffont ; Lausanne : Grammont, 1976.
144 p. : ill. (some col.) ; 20 cm. — (Bibliothèque Laffont des grands thèmes ; 76 ISSN 0337-064-X) F76-15978
Bibliography: p. ⌊142⌋
ISBN 2-8270-0076-8 : 15.00F
1. Child development.
RJ131.E53 155.4 78-392127
79 MARC

Erny, Pierre.
L'enfant et son milieu en Afrique noire : essais sur l'éducation traditionnelle / Pierre Erny. — Paris : Payot, 1978.
308 p. ; 18 cm. — (Petite bibliothèque Payot ; 342 ISSN 0480-2012) F78-17231
Bibliography: p. ⌊283⌋-301.
Includes indexes.
ISBN 2-228-33420-0 : 20.00F
1. Education—Africa. 2. Blacks—Education—Africa. 3. Child development. 4. Socialization. 5. Africa—Civilization. I. Title.
LA1501.E75 1978 370'.96 79-372973
79 MARC

Feeney, Stephanie.
Who am I in the lives of children? : An introduction to teaching young children / Stephanie Feeney, Doris Christensen ; ⌊photos. by Sally Gale⌋. — Columbus, Ohio : Merrill, c1979.
viii, 279 p. : ill. ; 24 cm.
Includes bibliographies and index.
ISBN 0-675-08391-5
1. Teaching. 2. Child development. 3. Home and school. 4. Teacher-student relationships. I. Christensen, Doris, joint author. II. Title.
LB1715.F38 372.1'1'02 77-92935
79 MARC

Fleischer-Undeutsch, Emma.
Die Wiedervereinigung auf philosophisch-pädagogischer Grundlage; eine Zeitstudie. 2. Ausg. Schorndorf, 1960.
90 p. 21 cm.
——— Nachtrag. Schorndorf, 1964–68.
2 v. (93–166 p.) 21 cm.
Bound with the main work.
1. Education—Philosophy. 2. Child development. I. Title.
LB880.F548W53 1960 77–478465

Fondation pour la recherche sociale.
Grandir de sept à quatorze ans / Fondation pour la recherche sociale, Caisse nationale des allocations familiales. — Paris : Éditions C.N.A.F., 1977.
187 p. ; 23 cm. — (Études C.A.F. ; 20) F78-13030
"Étude ... effectuée sous la direction de Roger Benjamin."
Includes bibliographical references.
20.00F
1. Child development. I. Benjamin, Roger. II. Caisse nationale des allocations familiales. III. Title.
HQ767.9.F65 1977 301.43'14 79-346827
79 MARC

Freinet, Célestin.
Formación de la infancia y de la juventud ⌊por⌋ C. Freinet. La Habana, Ministerio de Educación, Departamento de Publicaciones, 1961.
39 p. 21 cm. (Serie de folletos para maestros, no. 6)
Translation of La formation de l'enfance et de la jeunesse.
1. Child development. I. Title.
LB1117.F72318 78-383791
78 MARC

García, Rosario.
El juego infantil en grupo / trabajo realizado por Rosario García y Rosa María Cañadell. — ⌊Barcelona⌋ : Institut de Ciencies de l'Educació, Universitat de Barcelona, 1978.
64 p. ; 24 cm. — (Col·lecció Documents ; no. A.42) Sp***
Bibliography: p. 64.
1. Play. 2. Child developement. 3. Social groups. I. Cañadell, Rosa María, joint author. II. Title. III. Series.
HQ782.G28 155.4'18 79-350636
79 MARC

Gilmore, John V
A more productive child : guidelines for parents / John V. Gilmore and Eunice C. Gilmore. — Boston : Gilmore Institute, c1978.
xii, 277 p. ; 23 cm.
Bibliography: p. 269-270.
1. Parenting—Psychological aspects. 2. Child development. 3. Scholastic achievement. I. Gilmore, Eunice C., joint author. II. Title.
HQ772.G477 649'.1 78-61689
79 MARC

Goldstein, Bernard, 1925-
Children and work : a study of socialization / Bernard Goldstein, Jack Oldham. — New Brunswick, N.J. : Transaction Books, c1979.
xii, 195 p. ; 24 cm.
Bibliography: p. 183-191.
Includes index.
ISBN 0-87855-285-5 : $12.95
1. Work. 2. Child development. 3. Socialization. 4. Vocational interests. I. Oldham, Jack, 1947- joint author. II. Title.
HQ784.W6G64 301.15'72 78-62894
78 MARC

Guerney, Louise F
Parenting : a skills training manual / Louise F. Guerney. — State College, Penn. : IDEALS, c1978.
vii, 151 p. : ill. ; 23 cm.
ISBN 0-932990-00-2 : $3.75
1. Parenting. 2. Child development. I. Title.
HQ772.G83 649'.1 78-70674
78 MARC

Hersh, Richard.
Promoting moral growth : from Piaget to Kohlberg / Richard H. Hersh, Diana Pritchard Poalitto, Joseph Reimer. — New York : Longman, c1979.
xviii, 270 p. ; 21 cm.
Includes bibliographical references and index.
ISBN 0-582-28057-5 pbk.
1. Moral education. 2. Kohlberg, Lawrence, 1927- . 3. Piaget, Jean, 1896- . 4. Child development. I. Paolitto, Diana Pritchard, joint author. II. Reimer, Joseph, joint author. III. Title.
LC268.H47 370.11'4 78-23963
78 MARC

Herzka, Heinz Stefan.
Das Kind von der Geburt bis zur Schule : Bilderatlas und Texte zur Entwicklung des Kindes / Heinz Stefan Herzka ; ⌊hrsg. unter Mitarb. von Bernardo Ferrari⌋. — 4., erw. Aufl. — Basel : Schwabe, c1978.
278 p. : 260 ill. ; 31 cm. Sw78-2626
Bibliography: p. 263-277.
ISBN 3-7965-0685-2 : 49.50F
1. Child development. I. Ferrari, Bernardo, joint author. II. Title.
RJ131.H44 1978 155.4'22 78-372530
78 MARC

Hofmeister, Alan M
What do I do now? : Practical ways to develop good behavior in your child / Alan M. Hofmeister, Charles Atkinson, Hester Henderson. — Rev. ed. — Niles, Ill. : Argus Communications, c1978.
104 p. : ill. ; 19 cm.
ISBN 0-89505-016-1 : $1.95
1. Child development. 2. Children—Management. I. Atkinson, Charles, 1942- joint author. II. Henderson, Hester, joint author. III. Title.
HQ772.H58 1978 649'.123 78-56863
79 MARC

Ibuka, Masaru, 1908-
Kindergarten is too late! / Masaru Ibuka. — New York : Simon and Schuster, c1977.
183 p. ; 22 cm.
Translation of Yōchien de wa ososugiru.
ISBN 0-671-24036-6
1. Education, Preschool. 2. Child development. I. Title.
LB1140.2.I2913 1977 372.21 78-112
78 MARC

Inoue, Kenji, 1934-
(Kodomo no hattatsu to kankyō)
子どもの発達と環境 / 井上健治. — 東京 : 東京大学出版会, 1979.
v, 241 p. ; 19 cm. — (UP選書 ; 193) Ja 79-5359
Series romanized: UP sensho.
Includes bibliographical references.
¥980
1. Child development. 2. Educational sociology. I. Title.
LB1117.I56 79–808728

Isaacs, Susan.
The inner parent : raising ourselves, raising our children / Susan Isaacs and Marti Keller. — 1st ed. — New York : Harcourt Brace Jovanovich, c1979.
xviii, 234 p. ; 22 cm.
ISBN 0-15-144423-4
1. Parenting—United States—Psychological aspects. 2. Self-actualization (Psychology) 3. Child development. I. Keller, Marti, joint author. II. Title.
HQ755.8.I82 649'.1 78-22256
79 MARC

Kalmar, Jacques M 1913-
La délinquance des maîtres; introduction à l'ontopédagogie ⌊par⌋ Jacques M. Kalmar. Saint-Raphaël, Éditions Les Bardes ⌊1973⌋
256 p. illus. 22 cm. F***
Includes bibliographical references.
1. Kalmar, Jacques M., 1913- . 2. Education—Philosophy. 3. Ontology. 4. Child development. I. Title.
LB880.K23D44 370.1 79-370070
79 MARC

Kampmüller, Otto.
Dein Kind kommt mit dem Zeugnis / Otto Kampmüller. — 10., erw. Aufl. — Wien ; München : Jugend u. Volk, 1975.
72 p. ; 19 cm. Au75-9-188
ISBN 3-7141-5214-8 : S28.00
1. Child development. 2. Parent and child. I. Title.
LB1117.K28 1975 78-398567
*79 MARC

Kaplan, Abraham, 1907-
Freeing your child to learn / by Abraham Kaplan. — 1st ed. — Flushing, N.Y. : Audley Press, c1978.
xi, 116 p. : ill. ; 23 cm.
1. Education, Preschool. 2. Domestic education. 3. Child Development. I. Title.
LB1140.2.K364 372.21 78-73149
79 MARC

Karpova, Sof'îa Nikolaevna.
(Osobennosti orientirovki na slovo u deteĭ)
Особенности ориентировки на слово у детей / С. Н Карпова, И. Н. Колобова. — Москва : Изд-во МГУ, 1978
165 p. : graphs ; 22 cm. USSR 78
Bibliography : p. 160–⌊164⌋
0.90rub
1. Children—Language. 2. Child development. I. Kolobova Irina Nikolaevna, joint author. II. Title.
LB1139.L3K35 79–40918

Kaushall, Philip.
The growing years : a study guide for the televised course Philip Kaushall, Kiki Skagen Munshi. — 2d ed. — New York McGraw-Hill, c1979.
viii, 312 p. : ill. ; 24 cm.
Includes index.
ISBN 0-07-033459-5
1. Child development. I. Skagen, Kiki, joint author. II. The Growin years (Television program)
⌊RJ131.K38 1979⌋ 155.4 78-3119
78 MAR

Kiyānūsh, Maḥmūd.
(Bā farzandān-i khvud bāshīm)
با فرزندان خود باشیــــم / محمود کیانوش. — چاپ 1. —
تهران : سازمان انتشارات اشرفی ، ⌊1977 or 1978⌋ 2536.
121 p. ; 22 cm.
In Persian.
120.00 IR
1. Child development. I. Title.
HQ767.9.K59 79–9710

Kuzma, Kay.
Understanding children / by Kay Kuzma. — Mountain Vie Calif. : Pacific Press Pub. Association, c1978.
157 p. : ill. ; 19 cm. — (Harvest series ; 107)
Bibliography: p. 153-157.
1. Child psychology. 2. Child development. I. Title.
HQ767.9.K89 301.43'1 78-504
79 MA

LaHaye, Beverly.
How to develop your child's temperament / Beverly LaHay — Irvine, Calif. : Harvest House, c1977.
163 p. : ill. ; 21 cm.
ISBN 0-89081-034-6 : $2.95
1. Children—Management. 2. Child development. I. Title.
HQ772.L23 649'.1 76-530
77⌊r79⌋rev MA

Leach, Penelope.
Your baby & child from birth to age five / Penelope Lea photography by Camilla Jessel. — New York : A. A. Kno 1978.
512 p. : ill. ; 24 cm.
First published in London in 1977 under title Baby and child.
Includes index.
ISBN 0-394-40755-5 : $5.95. ISBN 0-394-73509-9 pbk.
1. Infants—Care and hygiene. 2. Children—Care and hygiene. 3. Child velopment. I. Title.
RJ61.L4 1978 649'.122 77-185
78 MA

Leeper, Sarah Hammond, 1912-
Good schools for young children / Sarah Hammond Leepe Dora Sikes Skipper, Ralph L. Witherspoon. — 4th ed. — Ne York : Macmillan, c1979.
vxiii, 588 p. : ill. ; 24 cm.
Bibliography: p. 544-569.
Includes indexes.
ISBN 0-02-369260-X
1. Education, Preschool—United States—Curricula. 2. Child developme 3. Day care centers—United States. I. Skipper, Dora Sikes, joint auth II. Witherspoon, Ralph Leo, 1909- joint author. III. Title.
LB1205.L35 1979 372.21 78-13
78 MA

Lescure, Mireille.
Les carences éducatives : les troubles de la relation à l'au pendant l'enfance et l'adolescence / Mireille Lescure. — To louse : Privat, c1978.
222 p. ; 21 cm. — (Éducateurs) F
Bibliography: p. ⌊215⌋-220.
ISBN 2-7089-7800-4 : 42.00F
1. Social interaction. 2. Child development. 3. Interaction analysis in e cation. I. Title.
HM291.L42 301.11 78-391
79 MA

CHILD DEVELOPMENT (Continued)

Lieberman, Florence.
Social work with children / Florence Lieberman. — New York : Human Sciences Press, c1979.
344 p. ; 21 cm.
Bibliography: p. 311-334.
Includes index.
ISBN 0-87705-255-7
1. Social work with children. 2. Child development. 3. Child welfare. I. Title.
HV713.L54 362.7 78-23287
78 MARC

Ligon, Ernest Mayfield, 1897-
Looking at me : a guide for parents of infant children / Ernest M. Ligon, Lucie W. Barber, Herman J. Williams ; introd. by C. Sybil Waldrop ; illustrated by Karl Cernik. — Nashville : Broadman Press, [1978 or 1979] c1976.
128 p. : ill. ; 29 cm.
First published in 1976 under title: Let me introduce mySelf.
Includes index.
ISBN 0-8054-5633-3
1. Child development. 2. Infants. I. Barber, Lucie W., joint author. II. Williams, Herman J., joint author. III. Title.
HQ774.L53 1978 649'.122 78-72484
00 MARC

Little, Audrey.
Recording and analyzing children's behaviour : a report of the activities of two to three year old children / by Audrey Little. — [Nedlands, W.A.] : [A. Little], [1975?]
38 p. ; 30 cm. Aus
"Under the auspices of the Department of Education of the Australian Government."
Bibliography: leaf 38.
ISBN 0-9596648-0-7
1. Child development. 2. Play groups—Case studies. I. Australia. Dept. of Education. II. Title.
HQ769.L54 301.43'14 79-301877
79 MARC

Liu, Hsiu-chi, 1927-
(P'ei yü yu hsiu ti erh nü)
培育優秀的兒女 / 劉修吉編著. — 香港 : 新力書局, [1978?]
18, 312 p. ; 19 cm.
Cover title.
$7.00

1. Child development. 2. Parent and child. I. Title.
HQ772.L54 79-836832

Losoncy, Lawrence J
When your child needs a hug / Larry Losoncy. — St. Meinrad, Ind. : Abbey Press, c1978.
95 p. : ill. ; 18 cm. — (A When book)
Bibliography: p. 94-95.
ISBN 0-87029-141-6 : $1.95
1. Child development. 2. Children—Management. I. Title.
HQ772.L618 649'.1 78-73016
79 MARC

Lovinger, Sophie L
Learning disabilities and games / Sophie L. Lovinger ; ill. by Shurlee Lee. — Chicago : Nelson-Hall, c1979
143 p. : ill. ; 23 cm.
Bibliography: p. 137-139.
Includes index.
ISBN 0-88229-353-2
1. Play. 2. Child development. 3. Games. 4. Learning disabilities. 5. Perceptual-motor learning. I. Title.
LB1137.L7 155.4'18 78-24619
78 MARC

Meeting children's needs. [Sound recording] New York, Parents' Magazine Educational Enterprises, p1977.
3 cassettes. ½-track. mono. (Conversations about children, set 2)
Interviews, conducted by Bettye Caldwell.
In container (25 cm.)
Bibliography (7 p.) inserted in container.
SUMMARY: Authorities on child health and development and family living present their views on how to meet the physical and psychological needs of children in various settings.
CONTENTS: no. 1. White, B. In homes.—no. 2. Eiduson, B. In non-traditional home settings.—no. 3. Lansburgh, T. In day care.—no. 4. Jackson, P. In schools.—no. 5. Michelson, W. In cities.—no. 6. Palmer, E. Via the media.
1. Children—Care and hygiene. 2. Child development. I. Caldwell, Bettye M. II. Series: Conversations about children. [Sound recording] set 2.
[RJ101] 77-740980
Library of Congress 79 MN

Mercer, Jean.
Small people : how children develop, and what you can do about it / Jean Mercer. — Chicago : Nelson-Hall, c1979.
xiii, 152 p. : ill. ; 23 cm.
Bibliography: p. 141-149.
Includes index.
ISBN 0-88229-318-4
1. Child development. 2. Children—Management. I. Title.
RJ131.M37 649'.1 78-27345
78 MARC

Merleau-Ponty, Maurice, 1908-1961.
Les relations avec autrui chez l'enfant / par Maurice Merleau-Ponty. — Paris : Centre de documentation universitaire, [1975] F***
81 p. ; 22 cm.
ISBN (invalid) 2-7181-1812-3 : 17.50F
1. Socialization. 2. Child development. I. Title.
HQ783.M42 1975 301.15'72 79-365013
79 MARC

Miller, Maureen.
Help your child for life / Maureen Miller. — 1st ed. — Niles, Ill. : Argus Communications, c1978.
228 p. : ill. ; 19 cm.
Bibliography: p. 227-228.
ISBN 0-89505-018-8 : $2.95
1. Child development. 2. Children—Management. I. Title.
HQ772.M446 649'.1 78-63105
79 MARC

Möller, Jannie.
Spelen en leren in het nieuwe basisonderwijs : een beschouwing over de gevolgen van de dreigende verenging van speel- en leermethoden voor kleuters in het nieuwe basisonderwijs / Jannie Möller, Loes Pieters, Wil Schappin. — Culemborg : Tjeenk Willink/Noorduijn, [1979]
118 p. ; 23 cm. Ne79-18
Bibliography: p. 115-118.
ISBN 9011745000 : fl 14.00
1. Education, Preschool. 2. Child development. 3. Play. I. Pieters, Loes, joint author. II. Schappin, Wil, joint author. III. Title.
LB1140.2.M624 79-372931
*79 MARC

Mooney, Bel, 1946-
The year of the child / Bel Mooney. — London : Hutchinson, 1979.
190 p., [4] leaves of plates : ill. ; 23 cm. GB***
ISBN 0-09-138320-X : £5.95
1. Children—Great Britain. 2. Child development. I. Title.
HQ792.G7M66 155.4'092'6 79-317369
79 MARC

Moral development / William Damon, editor. — San Francisco : Jossey-Bass, 1978.
viii, 121 p. ; 24 cm. — (New directions for child development ; no. 2)
Includes bibliographies and index.
1. Child development. 2. Moral education. I. Damon, William, 1944- II. Series.
HQ772.M588 301.15'72 78-61127
79 MARC

Murphy, Liz.
Before school / Liz and Brian Murphy. — London : Cassell, 1978.
192 p. : ill. ; 23 cm. GB***
Bibliography: p. 192.
ISBN 0-304-29791-7 : £4.50
1. Child development. 2. Creative activities and seat work. 3. Perceptual-motor learning. I. Murphy, Brian, 1929- joint author. II. Title.
HQ772.M83 649'.12'3 78-325233
79 MARC

Nisi, Cesare.
Sviluppo infantile e realtà sociale / Cesare Nisi. — Bologna : Pàtron, 1979.
268 p. ; 22 cm. — (Collana di scienze pedagogiche ; 3) It79-July
Includes bibliographical references and index.
L6000
1. Child development. 2. Family. 3. Socialization. I. Title.
HQ767.9.N57 79-378058
*79 MARC

Nord-Larsen, Mogens.
Tabere i skolen : 7 år efter - Losers in school : 7 years later / [af] Mogens Nord-Larsen. — København : Socialforskningsinstituttet : i kommission hos Teknisk Forlag, 1977.
171 p. ; 21 cm. — (Publikation - Socialforskningsinstituttet ; 80) D78-8
Summary in English.
Includes index.
ISBN 8774871307 : kr35.00
1. Problem children—Education—Denmark. 2. Child development. 3. Educational surveys—Denmark. I. Title. II. Series: Socialforskningsinstituttet, Copenhagen. Publikationer ; 80.
LC4801.N68 78-376941
78 MARC

Novoselova, Svetlana Leonidovna.
(Razvitie myshlenii͡a v rannem vozraste)
Развитие мышления в раннем возрасте / С. Л. Новоселова. — Москва : Педагогика, 1978.
159 p., [8] leaves of plates : ill. ; 20 cm. USSR 78
At head of title: Nauchno-issledovatel'skiĭ institut doshkol'nogo vospitanii͡a Akademii pedagogicheskikh nauk SSSR.
Bibliography: p. 155-[158]
0.55rub
1. Thought and thinking. 2. Child development. I. Title.
BF455.N69 79-411221

(Nyūyōji seishin hattatsu kensa)
乳幼兒精神發達検査 / 牛島義友 ... [et al.] 共著. — 東京 : 金子書房, 昭和24 [1949]
4, 331 p., [1] leaf of plates : ill. ; 22 cm. — (兒童研究叢書 ; 2)
Colophon inserted.

1. Psychological tests for children. 2. Child development. 3. Infant psychology—Research. I. Ushijima, Yoshitomo, 1906- II. Series: Jidō kenkyū sōsho ; 2.
BF722.N98 79-809699

Piaget, Jean, 1896-
The child's construction of reality / Jean Piaget ; translated by Margaret Cook. — London : Routledge & Kegan Paul, c1954, 1976 printing.
xiii, 386 p. ; 23 cm.
Translation of La construction du réel chez l'enfant.
ISBN 0-7100-1952-1
1. Child development. 2. Child psychology. I. Title.
LB1131.P48513 1976 155.4 78-314319
78 MARC

Pilling, Doria.
Controversial issues in child development / Doria Pilling and Mia Kellmer Pringle. — London : Elek, 1978.
424 p. ; 23 cm. GB***
Includes bibliographies and index.
ISBN 0-236-40101-7 : £8.95
1. Child development. 2. Education of children. 3. Parent and child. 4. Socially handicapped children. I. Pringle, Mia Lilly Kellmer, joint author. II. Title.
HQ767.9.P54 1978b 301.42'7 78-322387
78 MARC

Pollowy, Anne-Marie.
The urban nest / by Anne-Marie Pollowy. — Stroudsburg, Pa. : Dowden, Hutchinson & Ross, c1977.
xiv, 162 p. : ill. ; 24 cm. — (Community development series ; v. 26)
Includes bibliographies and index.
ISBN 0-87933-235-2
1. City children. 2. Child development. 3. Personal space. I. Title.
HT206.P64 301.36 76-22583
76 MARC

Poulsgaard, Kirsten, 1947-
Socialisering og frigørende pædagogik / [af] Kirsten Poulsgaard. — København : Ejlers, 1978.
233 p. ; 20 cm. — (Christian Ejlers' pædagogiske serie) D78-12/13
A revision of the author's thesis (M.A.).
Bibliography: p. 219-[231]
Includes index.
ISBN 8772414448 : kr74.50
1. Education, Preschool—1965- . 2. Socialization. 3. Child development. I. Title.
LB1140.2.P657 1978 78-387382
79 MARC

Rice, Mary Forman.
Help me learn : a handbook for teaching children from birth to third grade / Mary Forman Rice, Charles H. Flatter ; line drawings by Margaret Cole ; photos. by Judith Goodman. — Englewood Cliffs, N.J. : Prentice-Hall, c1979.
xvii, 398 p. : ill. ; 21 cm. — (A Spectrum book)
Includes bibliographies and index.
ISBN 0-13-386292-5 : $12.95. ISBN 0-13-386284-4 pbk. : $6.95
1. Child development. 2. Parent and child. 3. Learning, Psychology of. 4. Education, Preschool. I. Flatter, Charles H., joint author. II. Title.
LB1115.R445 1979 372.21 78-25781
78 MARC

The Role of the father in child development / edited by Michael E. Lamb. — New York : Wiley, c1976.
xii, 407 p. : ill. ; 24 cm. — (Wiley series on personality processes)
"A Wiley-Interscience publication."
Includes bibliographies and indexes.
ISBN 0-471-51172-2
1. Child development. 2. Father and child. I. Lamb, Michael E., 1953-
[DNLM: 1. Father-Child relations. 2. Personality development. 3. Child development. WS105 R746]
HQ772.R63 301.42'7 76-21778
76[r79]rev MARC

Schuh-Gademann, Lilly.
Erziehung zur Liebesfähigkeit : Materialien u. Konzepte für e. frühkindl. Sexualerziehung / Lilly Gademann. — 2., durchges. Aufl. — Heidelberg : Quelle und Meyer, 1978. GFR78-A
320 p. ; 22 cm.
Bibliography: p. 306-320.
ISBN 3-494-00723-3
1. Sex instruction. 2. Child development. I. Title.
HQ57.2.S38 1978 78-365278
78 MARC

Sidenius, Bente.
At tale sammen med børn : grundrids af en pædagogisk praksis / Bente Sidenius. — København : Hans Reitzel, 1977.
185 p. ; 20 cm. & workbook. D77-35
Workbook by B. Sidenius and K. Pedersen.
Bibliography: p. 184-185.
ISBN 8741231260 : kr62.75
1. Language arts (Preschool)—Denmark. 2. Child development. I. Pedersen, Kirsten. II. Title.
LB1140.5.L3S53 77-570760
78[r79]rev MARC

Skinner, Louise.
Motor development in the preschool years / by Louise Skinner. — Springfield, Ill. : Thomas, c1979.
xvi, 109 p. : ill. ; 23 cm.
Bibliography: p. 93-94.
Discography: p. 94.
Includes index.
ISBN 0-398-03835-X
1. Physical education for children. 2. Perceptual-motor learning. 3. Motor ability in children. 4. Child development. I. Title.
GV443.S55 613.7'042 78-16527
78 MARC

CHILD DEVELOPMENT (Continued)

Sønstabø, Elisabeth Cecilie.
Samlingsstunden i barnehagen : belyst ved observasjon av aktiviteter, emner og barnas oppmerksomhet / Elisabeth Cecilie Sønstabø. — Oslo : Universitetsforl., c1978 (Drammen : Tangen-trykk).
76 p. ; 21 cm. N78-May
Bibliography: p. 53-55.
ISBN 8200051935 : kr34.50
1. Day care centers—Norway. 2. Child development. I. Title.
HV861.N7S63 78-396910
79 MARC

Soziologie der Kindheit / bearb. u. hrsg. von Ernst M. Wallner u. Margret Pohler-Funke. — 1. Aufl. — Heidelberg : Quelle und Meyer, 1978.
192 p. ; 22 cm. — (Soziologie der Gegenwart ; 3) GFR78-A
Bibliography: p. 182-186.
Includes indexes.
ISBN 3-494-00901-5
1. Socialization. 2. Child development. I. Wallner, Ernst Maxim, 1912- . II. Pohler-Funke, Margret. III. Series.
HQ783.S63 301.15'72 78-391481
79 MARC

Stallibrass, Alison.
The self-respecting child : a study of children's play and development / [by] Alison Stallibrass. — Revised ed. — Harmondsworth ; New York : Penguin, 1977.
341 p., [6] p. of plates : ill. ; 18 cm. — (A Pelican book) GB78-10640
Includes index.
Bibliography: p. 329-330.
ISBN 0-14-022023-2 : £1.25
1. Play. 2. Child development. I. Title.
HQ782.S7 1977 155.4'18 78-320851
79 MARC

Stallibrass, Alison.
The self-respecting child : a study of children's play and development / Alison Stallibrass. — New York : Warner Books, c1979.
288 p. : ill. ; 21 cm.
Bibliography: p. 278-279.
Includes index.
ISBN 0-446-87772-7 : $5.95
1. Play. 2. Child development. I. Title.
HQ782.S7 1979 155.4'18 79-106702
79 MARC

Stenbäck, Asser, 1913-
Lapsuus- ja nuoruusiän sukupuolinen kehitys ja kasvatus [kirj.] A. Stenbäck [ja] Lauri Pautola. Helsinki, Otava [1952]
315 p. illus. 21 cm.
1. Sex (Biology) 2. Child development. I. Pautola, Lauri, joint author. II. Title.
QP251.S847 78-393812
79 MARC

Tanner, James Mourilyan.
Education and physical growth; implications of the study of children's growth for educational theory and practice. [1st ed.] London, University of London Press [1961]
144 p. illus. 23 cm.
Includes bibliography.
1. Children in Great Britain—Anthropometry. 2. Children—Growth. 3. Education of children. 4. Child development. I. Title.
LB3425.G7T36 1961 62-51434
[r79]rev MARC

Tanner, James Mourilyan.
Education and physical growth : implications of the study of children's growth for educational theory and practice / J. M. Tanner. — 2d ed. — London : Hodder and Stoughton, 1978.
144 p. : ill. ; 22 cm. — (Unibooks) GB***
Bibliography: p. 127-134.
Includes index.
ISBN 0-340-22484-3 : £2.45
1. Children in Great Britain—Anthropometry. 2. Children—Growth. 3. Education of children. 4. Child development. I. Title.
LB3425.G7T36 1978b 573'.6 78-318319
78 MARC

Tanner, James Mourilyan.
Education and physical growth : implications of the study of children's growth for educational theory and practice / J. M. Tanner. — 2d ed. — New York : International Universities Press, 1978.
144 p. : ill. ; 23 cm.
Bibliography: p. 127-134.
Includes index.
ISBN 0-8236-1531-6
1. Children in Great Britain—Anthropometry. 2. Children—Growth. 3. Education of children. 4. Child development. I. Title.
LB3425.G7T36 1978 573'.6 78-8809
78 MARC

Tanous, Alex, 1926-
Is your child psychic? : A guide for creative parents and teachers / Alex Tanous and Katherine Fair Donnelly. — New York : Macmillan, c1979.
xxiv, 200 p. : ill. ; 22 cm.
Includes bibliographies and index.
ISBN 0-02-616280-6
1. Child development. 2. Psychical research. I. Donnelly, Katherine Fair, joint author. II. Title.
HQ773.T36 1979 155.4 79-226
79 MARC

Taylor, Barbara J
Dear mom and dad : parents and the preschooler / Barbara J. Taylor. — Provo, Utah : Brigham Young University Press, c1978.
147 p. : ill. ; 23 cm.
Includes bibliographies and indexes.
ISBN 0-8425-1231-4 : $5.95
1. Child development. 2. Parent and child. 3. Education, Preschool—1965- . I. Title.
LB1115.T24 372.21 78-5121
78 MARC

Terry, Sheila G
Children, their growth and development / Sheila G. Terry, Jeanette M. Sorrentino, Charles H. Flatter ; Elizabeth Simpson, consulting editor. — New York : Gregg Division, McGraw-Hill, c1979.
vii, 200 p. : ill. ; 23 cm. — (Careers in home economics)
Includes index.
ISBN 0-07-063731-8
1. Child development. I. Sorrentino, Jeanette M., joint author. II. Flatter, Charles H., joint author. III. Title.
RJ131.T47 612.6'5 78-13102
78 MARC

Thompson, Andrew D
When your child learns to choose / Andrew D. Thompson. — St. Meinrad, Ind. : Abbey Press, c1978.
95 p. : ill. ; 18 cm. — (A When book)
Bibliography: p. 91-95.
ISBN 0-87029-145-9 : $1.95
1. Child development. 2. Moral education. I. Title.
HQ772.T52 649'.7 78-73018
79 MARC

Viklund, Monica, 1943-
Barnen behöver oss / Monica Viklund ; [teckningar o. omsl., Helga Henschen]. — Stockholm : Socialstyr. ; Lund : Liber-Läromedel, 1978.
157 p. : ill., tables ; 21 cm. S78-17
Bibliography: p. 153-157.
ISBN 9140045900
1. Children—Sweden. 2. Child development. 3. Child psychology. 4. Child welfare—Sweden. I. Title.
HQ792.S8V54 78-395310
79 MARC

Weiss, Lynn.
Emotionally yours : shaping your child's life / written by Lynn Weiss ; illustrated by Jerry Weiss. — Dallas : I.T. Productions, c1979.
168 p. : ill. ; 22 cm.
$5.95
1. Child development. 2. Children—Management. I. Weiss, Jerry. II. Title.
HQ772.W363 649'.1 79-84406
79 MARC

Westman, Jack C
Child advocacy : new professional roles for helping families / Jack C. Westman. — New York : Free Press, c1979.
xvi, 431 p. ; 25 cm.
Includes bibliographies and indexes.
ISBN 0-02-934540-5 : $16.95
1. Children's rights—United States. 2. Child development.
HV741.W37 1979 362.7'0973 78-24762
79 MARC

(Yōnen no shakaisei)
幼年の社会性：幼年教育のために / 周郷博．井坂行男篇．— 東京：国民図書刊行会．昭和32 [1957]
210 p. : 19 cm.
1. Child development. I. Sugō, Hiroshi, 1907- II. Isaka, Yukio, 1913-
HQ767.9.Y66 79-806930

Yussen, Steven R
Child development : an introduction / Steven R. Yussen, John W. Santrock. — Dubuque, Iowa : W. C. Brown Co., c1978.
xxiii, 647 p. : ill. ; 24 cm.
Bibliography: p. 600-627.
Includes indexes.
ISBN 0-697-06625-8
1. Child psychology. 2. Child development. I. Santrock, John W., joint author. II. Title.
BF721.Y85 155.4 77-86177
79 MARC

Zákostolecká, Jozefa.
Zur Erziehung sozialer Gefühle : aus d. Erfahrungen e. Vorschulpädagogin / Jozefa Zákostolecká ; [aus d. Slowak. übers. von Helmut Barwinek]. — 1. Aufl. — Berlin : Volk und Wissen, 1976.
63 p. : ill. ; 19 cm. GDR76-A
Translation of O výchove spoločenských citov diet'at'a.
Bibliography: p. 62-63.
1.00M
1. Education, Preschool. 2. Child development. 3. Educational sociology. I. Title.
LB1140.2.Z3415 78-376084
78 MARC

Zazzo, Bianka.
Un grand passage : de l'école maternelle à l'école élémentaire / Bianka Zazzo ; préf. de René Zazzo. — 1. éd. — Paris : Presses universitaires de France, c1978.
224 p. ; 22 cm. — (Croissance de l'enfant, genèse de l'homme ; 6) F***
Includes bibliographical references.
ISBN 2-13-035592-7 : 59.00F
1. Child development. 2. Education, Preschool—France. 3. Education, Elementary—France. I. Title. II. Title: De l'école maternelle à l'école élémentaire.
LB1117.Z39 371.21'0944 78-393425
79 MARC

Żebrowska, Maria.
Psychologia rozwojowa dzieci i młodzieży / pod red. Marii Żebrowskiej ; [aut. Zofia Babska ... et al.]. — Wyd. 8. — Warszawa : Państ. Wydaw. Naukowe, 1977, c1975 (Opole : OZGJL)
879, [1] p. : ill. ; 24 cm. P78-2290
Includes bibliographies and indexes.
zł120.00
1. Child psychology. 2. Adolescent psychology. 3. Child development. I. Babska, Zofia. II. Title.
BF721.Z4 1977 78-388491
79 MARC

Ziemska, Maria.
Early child care in Poland / by Maria Ziemska. — London ; New York : Gordon and Breach, 1978.
iii, 148 p. : ill. ; 23 cm. — (Early child development and care ; vol. 5, no. 1/2) (International monographs on early child care ; no. 8) GB***
Cover title.
Bibliography: p. 136-143.
1. Child welfare—Poland. 2. Child development. I. Title. II. Series: International monograph series on early child care ; no. 8.
HV799.P72Z53 362.7'09438 78-324015
79 MARC

—ADDRESSES, ESSAYS, LECTURES

Aspetti cognitivi della socializzazione in età evolutiva / a cura di Augusto Palmonari e Pio E. Ricci Bitti. — Bologna : Il mulino, 1978.
167 p. ; 22 cm. — (Quaderni del Giornale italiano di psicologia ; 3) It78-Aug
Proceedings of a meeting held in Bologna in 1977, organized by the Giornale italiano di psicologia.
Includes bibliographies.
L4000
1. Socialization—Addresses, essays, lectures. 2. Child development—Addresses, essays, lectures. I. Palmonari, Augusto. II. Ricci Bitti, Pio E. III. Giornale italiano di psicologia. IV. Series.
HQ783.A84 79-337963
*79 MARC

Barn / bidrag av Anne-Marit Duve ... [et al.] ; redaksjon : Kari Bruun Wyller, Thomas Chr. Wyller. — Oslo : Gyldendal, 1978 (Oslo : Norbok).
118, [3] p. ; 21 cm. — (Motforestillinger ; 4) N78-July
Bibliography: p. [121]
ISBN 8205108854 : kr89.00
1. Child development—Addresses, essays, lectures. 2. Children—Norway—Addresses, essays, lectures. I. Duve, Anne-Marit, 1919- . II. Wyller, Kari Bruun. III. Wyller, Thomas Christian, 1922- . IV. Series.
HQ772.B379 78-400960
79 MARC

Children communicating : media and development of thought, speech, understanding / Ellen Wartella, editor. — Beverly Hills, Calif. : Sage Publications, c1979.
286 p. : ill. ; 22 cm. — (Sage annual reviews of communications research ; v. 7)
Includes bibliographies.
ISBN 0-8039-1171-8. ISBN 0-8039-1172-6 pbk.
1. Television and children—Addresses, essays, lectures. 2. Child development—Addresses, essays, lectures. 3. Cognition (Child psychology)—Addresses, essays, lectures. 4. Interpersonal communication—Addresses, essays, lectures. I. Wartella, Ellen.
HQ784.T4C5 155.4'18 78-25867
78 MARC

Children in time and space / Kaoru Yamamoto, editor. — New York : Teachers College Press, c1979.
x, 208 p. : ill. ; 23 cm.
Bibliography: p. 187-200.
Includes index.
ISBN 0-8077-2553-6
1. Child development—Addresses, essays, lectures. 2. Yamamoto, Kaoru.
LB1117.C53 301.43'1 79-91
79 MARC

Early adolescents, understanding and nurturing their development / coordinator, John R. Burton ; editor, Monroe D. Cohen ; associate editor for special publications, Bonnie D. Bezon : editorial manager, Lucy Prete Martin ; editorial assistant, Sarah Hadley. — Washington : Association for Childhood Education International, c1978.
36 p. : ill. ; 25 cm.
Includes bibliographical references.
ISBN 0-87173-087-1
1. Adolescence—Addresses, essays, lectures. 2. Child development—Addresses, essays, lectures. I. Burton, John R., 1934- . II. Cohen, Monroe D., 1924- . III. Association for Childhood Education International.
LB1135.E17 301.43'15 78-15408
78 MARC

CHILD DEVELOPMENT
—ADDRESSES, ESSAYS, LECTURES (Continued)

The Education of the young child / edited by David Fontana. — London : Open Books, 1978.
xvi, 335 p. : ill. ; 22 cm. GB***
Includes bibliographies and indexes.
ISBN 0-7291-0132-0 : £4.50
1. Education, Preschool—Addresses, essays, lectures. 2. Child development—Addresses, essays, lectures. I. Fontana, David.
LB1140.2.E325 372 78-319554
79 MARC

(Erh chiao nan chen)
兒教南針 / ⌊編輯者時兆出版社編輯部⌋. — 重訂版. — 香港：時兆出版社, 1977.
2, 2, 191 p. : ill. ; 19 cm.

1. Child development—Addresses, essays, lectures. 2. Parent and child—Addresses, essays, lectures. I. Shih chao ch'u pan she. Pien chi pu.
HQ772.E74 78–841895

Growing up with philosophy / ⌊edited⌋ by Matthew Lipman and Ann Margaret Sharp. — Philadelphia : Temple University Press, 1978.
x, 410 p. ; 24 cm.
Bibliography: p. ⌊405⌋-410.
ISBN 0-87722-118-9 : $20.00
1. Education—Philosophy—Addresses, essays, lectures. 2. Philosophy—Study and teaching (Elementary)—Addresses, essays, lectures. 3. Child development—Addresses, essays, lectures. 4. Motivation in education—Addresses, essays, lectures. I. Lipman, Matthew. II. Sharp, Ann Margaret, 1942-
LB17.G73 107 77-93451
78 MARC

Pei-ching shih fan ta hsüeh. Chiao yü hsi. Hsin li chiao yen shih.
(Hsiao hsüeh sheng nien ling t'e tien yü chiao yü)
小学生年令特点与教育：北京市师范学校试用课本 / ⌊北京师范大学教育系心理教研室主編⌋. — ⌊香港 : s. n., 1979?⌋
120 p. ; 22 cm.
"北京市教育局教材编写组编"
On t. p.: Pei-ching jen min ch'u pan she.
Reprint of the ed. firstly published by Pei-ching jen min ch'u pan she and republished in 1978 by Kuang-tung jen min ch'u pan she.

1. Educational psychology—Addresses, essays, lectures. 2. Child development—Addresses, essays, lectures. I. Peking. Chiao yü chü. Chiao ts'ai pien hsieh tsu. II. Title.
LB1055.P42 1979 79–840647

(Problemy psikhologii lichnosti i sotsial'noĭ psikhologii)
Проблемы психологии личности и социальной психологии : Сб. тр. / Моск. гос. пед. ин-т им. В. И. Ленина ; ⌊Редкол. д-р психол. наук, проф. П. А. Просецкий (отв. ред.) ... et al.⌋. — Москва : МГПИ, 1977.
116 p. ; 20 cm. USSR 78
At head of title: Ministerstvo prosveshcheniia RSFSR.
Includes bibliographies.
0.70rub
1. Educational psychology—Addresses, essays, lectures. 2. Child development—Addresses, essays, lectures. I. Prosetskiĭ, Petr Alekseevich. II. Moscow. Moskovskiĭ gosudarstvennyĭ pedagogicheskiĭ institut.
LB1051.P726 79–408156

Stephens, William N
Our children should be working / by William N. Stephens. — Springfield, Ill. : Thomas, c1979.
ix, 215 p. ; 24 cm.
Includes bibliographical references and index.
ISBN 0-398-03851-1
1. Children—Management—Addresses, essays, lectures. 2. Helping behavior—Case studies. 3. Apprentices—Case studies. 4. Work—Case studies. 5. Child development—Addresses, essays, lectures. I. Title.
HQ772.S69 649'.1 78-11606
78 MARC

Wu, Ching-chi.
(Hsin li yü jen sheng)
心理與人生 / 吳靜吉著. — 臺北：遠流出版社, 民國66 ⌊1977⌋
2, 4, 287 p. ; 20 cm. — (生活叢書 ; 9)
Series romanized: Sheng huo ts'ung shu.
NT$50.00

1. Child development—Addresses, essays, lectures. 2. Personality—Addresses, essays, lectures. I. Title.
HQ772.W8 79–836657

—ADDRESSES, ESSAYS, LECTURES—COLLECTED WORKS

Handbuch der Früh- und Vorschulpädagogik / hrsg. von Rainer Dollase ; ⌊mit Beitr. von E. Barres ... et al.⌋. — 1. Aufl. — Düsseldorf : Pädagogischer Verlag Schwann, 1978.
2 v. ; 23 cm. GFR78-A
Includes bibliographies and indexes.
ISBN 3-590-14356-8 (v. 1)
1. Education, Preschool—Germany, West—Addresses, essays, lectures—Collected works. 2. Child development—Addresses, essays, lectures—Collected works. I. Dollase, Rainer.
LB1342.H33 79-367262
79 MARC

—BIBLIOGRAPHY

Dingle Associates.
An annotated bibliography for child and family development programs / prepared by Dingle Associates, Inc. for Administration for Children, Youth and Families, Office of Human Development Services, U.S. Department of Health, Education, and Welfare. — Washington : U.S. Govt. Print. Off., 1977.
126 p. ; 27 cm. — (DHEW publication ; no. (OHDS) 78-31118)
"Contract no. HEW-105-76-1156."
Includes indexes.
1. Child development—Bibliography. 2. Parenting—Bibliography. 3. Education, Preschool—Bibliography. I. United States. Administration for Children, Youth, and Families. II. Title. III. Series: United States. Dept. of Health, Education, and Welfare. DHEW publication ; no. (OHDS) 78-31118.
Z7164.C5D56 1977 016.30142'7 78-603603
⌊HQ767.85⌋ 79 MARC

International Children's Centre, Paris.
Travaux sur la croissance et le développement de l'enfant normal; études coordonnées par le Centre international de l'enfance, 1968-1972. Paris ⌊1973?⌋
10 p. 27 cm. F***
Cover title.
1. Children—Growth—Bibliography. 2. Child development—Bibliography. I. Title.
Z6671.52.G76157 1973 016.6126'5 78-374369
⌊RJ131⌋ 78 MARC

Lefebvre, Monique.
Techniques d'intervention auprès des enfants d'âge préscolaire : dossier bibliographique ouvert au Centre de documentation du C.A.D.R.E. / par Monique Lefebvre. — Montréal : C.A.D.R.E., 1977.
18 p. ; 28 cm. C78-8384-9
$1.00
1. Education, Preschool—Bibliography. 2. Child development—Bibliography. I. Centre d'animation, de développement et de recherche en éducation. II. Title.
Z5814.P6L44 016.37221 79-359249
⌊LB1140.2⌋ 79 MARC

—CASE STUDIES

Flamant-Paparatti, Danielle.
Emmanuelle : ou, L'enfance au féminin / Danielle et Emmanuelle Flamant-Paparatti ; préf. d'Elena Gianini Belotti. — Paris : Denoël/Gonthier, c1979.
187 p. : ill. ; 22 cm. — (Collection Femme) F***
Text in French.
Includes bibliographical references.
39.00F
1. Mothers and daughters—Case studies. 2. Child development—Case studies. 3. Femininity (Psychology)—Case studies. I. Flamant-Paparatti, Emmanuelle, 1971- joint author. II. Title.
HQ777.F57 301.42'7 79-376704
79 MARC

School en toekomst : beelden van de ontwikkelingsgang van kinderen gedurende de puberteit / door S. Wiegersma ... ⌊et al.⌋ ; met een voorwoord van J. Koekebakker ; in opdracht van het Nederlands Institut voor Praeventieve Geneeskunde. — Haarlem : De Toorts, ⌊1963⌋
250 p. ; 24 cm.
Includes indexes.
1. Children—Netherlands—Case studies. 2. Child development—Case studies. 3. Child psychology—Netherlands—Case studies. I. Wiegersma, S.
HQ792.N4S43 79-344405
79 MARC

—CHARTS, DIAGRAMS, ETC.

International Children's Center, Paris.
Compte rendu de la XIe réunion des équipes chargées des études sur la croissance et le développement de l'enfant normal, Londres 1972. ⌊Paris⌋ Centre international de l'enfance ⌊1973?⌋
249 p. illus. 29 cm. F***
Cover title.
French or English.
Includes bibliographies.
1. Child development—Congresses. 2. Child development—Charts, diagrams, etc.
RJ131.I77 1973 612.6'54 78-373762
78 MARC

—CONGRESSES

Association générale des institutrices des écoles maternelles et classes enfantines.
Actes / du Congrès national de Montpellier ⌊de l'⌋ Association générale des institutrices des écoles maternelles et classes enfantines, 25, 26, 27 et 28 juin 1975 ; ⌊publié par le Centre régional de recherche et de documentation pédagogiques de Montpellier⌋. — Montpellier : Centre régional de recherche et de documentation pédagogiques, 1976.
⌊ii⌋, 187 p. : ill. ; 30 cm. F76-13032
Bibliography: p. 137-138.
25.00F
1. Education, Preschool—France—Congresses. 2. Child development—Congresses.
LB1140.2.A86 1976 372.21'0944 78-382389
78 MARC

Early developmental hazards : predictors and precautions / edited by Frances Degen Horowitz. — Boulder, Colo. : Published by Westview Press for the American Association for the Advancement of Science, 1978.
xvi, 114 p. ; 24 cm. — (AAAS selected symposium ; 19)
Includes bibliographies and indexes.
Based on a symposium presented at the AAAS annual national meetings in Denver, February 1977, and sponsored by the Society for Research in Child Development, in conjunction with Section J (Psychology) of the AAAS.
ISBN 0-89158-084-0 : $12.50
1. Child development deviations—Etiology—Congresses. 2. Child development—Congresses. 3. Infants—Growth—Congresses. I. Horowitz, Frances Degen. II. American Association for the Advancement of Science. III. Society for Research in Child Development. IV. American Association for the Advancement of Science. Sectionon Psychology. V. Series: American Association for the Advancement of Science. AAAS selected symposium ; 19.
RJ135.E25 618.9'2 78-352
78⌊r79⌋rev MARC

Individuation process and biographical aspects of disease / edited by Karl Ernst Schaefer, Uwe Stave, Wolfgang Blankenburg. — Mount Kisco, N.Y. : Futura Pub. Co., 1979.
ix, 233 p. : ill. ; 24 cm. — (A New image of man in medicine ; v. 3)
"Most of the articles in this book are based on presentations made at the meeting on 'Man-centered physiological science and medicine' held at the Gemeinnütziger Krankenhaus in Herdecke, Germany, September 24-28, 1973."
Includes bibliographies.
ISBN 0-87993-117-5
1. Medicine and psychology—Congresses. 2. Individuality—Congresses. 3. Child development—Congresses. 4. Medicine—Philosophy—Congresses. 5. Life cycle, Human—Congresses. I. Schaefer, Karl Ernst, 1912- II. Stave, Uwe, 1923- III. Blankenburg, Wolfgang. IV. Series: New image of man in medicine ; v. 3.
R723.N44 vol. 3 610 s 78-73205
⌊R726.5⌋ 79 MARC

International Children's Center, Paris.
Compte rendu de la XIe réunion des équipes chargées des études sur la croissance et le développement de l'enfant normal, Londres 1972. ⌊Paris⌋ Centre international de l'enfance ⌊1973?⌋
249 p. illus. 29 cm. F***
Cover title.
French or English.
Includes bibliographies.
1. Child development—Congresses. 2. Child development—Charts, diagrams, etc.
RJ131.I77 1973 612.6'54 78-373762
78 MARC

National Conference on the Child in Sport and Physical Activity, Queen's University, Kingston, Ont., 1973.
Child in sport and physical activity : selected papers presented at the national conference workshop, "The child in sport and physical activity," Queen's University, Kingston, Ontario, Canada / edited by J. G. Albinson and G. M. Andrew. — Baltimore : University Park Press, c1976.
viii, 233 p. : ill. ; 24 cm. — (International series on sport sciences ; v. 3)
Includes bibliographies.
ISBN 0-8391-0892-3
1. Child development—Congresses. 2. Sports for children—Congresses. 3. Physical education for children—Congresses. I. Albinson, J. G. II. Andrew, G. M. III. Kingston, Ont. Queen's University. IV. Title. V. Series.
⌊DNLM: 1. Physical education and training—Congresses. 2. Sport medicine—In infancy and childhood—Congresses. QT255 N2754p 1973⌋
RJ131.N32 1973 612'.76 76-1953
76⌊r79⌋rev MARC

National Conference on the Child in Sport and Physical Activity, Queen's University, Kingston, Ont., 1973.
L'enfant dans le sport et dans les activités physiques : découlant de la conférence nationale, Queen's University, Kingston, Ontario du 13 au 18 mai 1973 : comptes rendus, sommaires des ateliers de travail, résolutions / édité par John G. Albinson, George M. Andrew. — Toronto : Conseil canadien de l'enfance et de la jeunesse, ⌊1973⌋
83 p. ; 22 cm. C***
On cover: L'enfant, le sport, l'activité physique.
Published also under title: Child in sport and physical activity.
1. Child development—Congresses. 2. Sports for children—Congresses. 3. Physical education for children—Congresses. I. Albinson, J. G. II. Andrew, G. M. III. Kingston, Ont. Queen's University. IV. Title. V. Title: L'enfant, le sport, l'activité physique.
RJ131.N32 1973a 612'.76 77-470811
77⌊r79⌋rev MARC

CHILD DEVELOPMENT
—CONGRESSES (Continued)

**The Physician and the mental health of the child / ** George Tarjan, monograph series editor. — [Monroe, WI] : American Medical Association, [c1979-
v. ; 23 cm.
Vol. 1: "Based on the proceedings of the AMA Workshops on the Mental Health of Children held November 18-20, 1976 and May 19-21, 1977 in Oakbrook, Illinois."
Vol. 1: "Produced pursuant to contract no. 278-76-0046(SM) with the National Institute of Mental Health of the Alcohol, Drug Abuse, and Mental Health Administration, Department of Health, Education, and Welfare."
Includes bibliographical references.
CONTENTS: 1. Assessing development and treating disorders within a family context.
1. Child mental health—Congresses. 2. Child development—Congresses. 3. Parent and child—Congresses. I. Tarjan, George. II. American Medical Association.
RJ111.P49 618.9'28'9 79-115702
79 MARC

(Seminar gruppy ėkspertov evropeĭskikh sot͡sialisticheskikh stran po razvitii͡u deteĭ rannego vozrasta)
Семинар группы экспертов европейских социалистических стран по развитию детей раннего возраста : Будапешт 21–24 октября 1975 = Seminar der Experten der europäischen sozialistischen Länder für die Entwicklung von Säuglingen und Kleinkindern : Budapest 21–24 oktober 1975 / [под ред. Анна Тардош]. — Будапешт : [s. n., 1975?]
317 p. : ill. ; 29 cm.
German and Russian.
Includes bibliographies.
1. Child development—Congresses. I. Tardos, Anna. II. Title: Seminar der Experten der europäischen sozialistischen Länder für die Entwicklung von Säuglingen und Kleinkindern.
HQ767.82.S45 78–422368

**Social competence in children / ** editors, Martha Whalen Kent, Jon E. Rolf. — Hanover, N.H. : Published for the University of Vermont by the University Press of New England, 1979.
x, 341 p. : ill. ; 24 cm. — (Primary prevention of psychopathology ; v. 3)
Based on papers presented at the 3d Vermont Conference on Primary Prevention of Psychopathology.
Includes bibliographies and indexes.
ISBN 0-87451-153-4
1. Child development—Congresses. 2. Social adjustment—Congresses. 3. Social acceptance—Congresses. 4. Child psychopathology—Prevention—Congresses. I. Kent, Martha Whalen. II. Rolf, Jon E. III. Series: Vermont Conference on the Primary Prevention of Psychopathology. Primary prevention of psychopathology ; v. 3.
[DNLM: 1. Mental disorders—Prevention and control—Congresses. 2. Psychopathology—Congresses. W3 PR945CK]
RC454.V46 1977 vol. 3 616.8'58 s 78-63587
[HQ772] 78 MARC

Tremblay, Marc Adélard.
The social bases of maturity in childhood / Marc-Adelard Tremblay, with the assistance of Vincent Ross. — Toronto, Ontario : Canadian Conference on Children, 1965.
iii, 115, xxix p. ; 29 cm.
Cover title: Les fondements sociaux de la maturation chez l'enfant.
Bibliography: p. [i]-xxix.
1. Child development—Congresses. 2. Socialization—Congresses. 3. Personality and culture—Congresses. I. Ross, Vincent. II. Title. III. Title: Les fondements sociaux de la maturation chez l'enfant.
HQ767.82.T73 155.4'18 78-310167
78 MARC

—CROSS-CULTURAL STUDIES

Werner, Emmy E 1929-
Cross-cultural child development : a view from the planet Earth / Emmy Elisabeth Werner ; [manuscript editor, Elaine Linden ; ill., Lori Gilbo]. — Monterey, Calif. : Brooks/Cole Pub. Co., c1979.
xii, 355 p. : ill. ; 23 cm.
Includes bibliographies and index.
ISBN 0-8185-0332-7
1. Child development—Cross-cultural studies. 2. Underdeveloped areas—Children. I. Title.
RJ131.W37 301.43'1 79-109136
79 MARC

—JUVENILE LITERATURE

Corey, Dorothy.
Tomorrow you can / Dorothy Corey ; pictures by Lois Axeman. — Chicago : A. Whitman, c1977.
[32] p. : col. ill. ; 19 cm. — (A Self-starter book)
SUMMARY: Pictures and brief text reassure small children that soon they will be capable of doing many things: tying shoes, sharing toys, and going off to school.
ISBN 0-8075-8015-5 : $4.25
1. Child development—Juvenile literature. [1. Child development] I. Axeman, Lois. II. Title.
HQ767.9.C67 155.4'23 77-12789
77[r79]rev MARC

—RESEARCH

**International research in early childhood education / ** edited by Maurice Chazan. — Windsor : NFER Pub. Co., 1978.
228 p. ; 22 cm. GB***
On cover: Social Science Research Council, Educational Research Board.
Includes bibliographies and indexes.
ISBN 0-85633-143-0
1. Education, Preschool—Research. 2. Child development—Research. I. Chazan, Maurice. II. Social Science Research Council (Gt. Brit.) Educational Research Board.
LB1140.2.I513 372.21'07'2 78-323291
79 MARC

—STUDY AND TEACHING

Fowler, William, 1921-
Curriculum and assessment guides for infant and child care / William Fowler. — Boston : Allyn and Bacon, c1980.
viii, 188 p. ; 23 cm.
Includes index.
ISBN 0-205-06537-6
1. Child development—Study and teaching. 2. Education, Preschool—Curricula. 3. Day care centers. I. Title.
HQ755.7.F67 372.21 79-15435
79 MARC

—STUDY AND TEACHING—UNITED STATES—DIRECTORIES

Rothenberg, Dianne.
Directory of educational programs for adults who work with children / Dianne Rothenberg, compiler. — Washington : Published by the National Association for the Education of Young Children for the ERIC Clearinghouse on Early Childhood Education, 1979.
97 p. ; 28 cm.
"Prepared by the staff of the ERIC Clearinghouse on Early Childhood Education."
ISBN 0-912674-66-0 : $3.00
1. Child development—Study and teaching—United States—Directories. 2. Education, Preschool—United States—Directories. 3. Adult education—United States—Directories. I. ERIC Clearinghouse on Early Childhood Education. II. Title.
HQ755.7.R67 372.21'025'73 79-64954
00 MARC

—TESTING

see also Psychological tests for children

Chinn, Peggy L 1941-
Child health maintenance : a guide to clinical assessment / Peggy L. Chinn, Cynthia J. Leitch. — 2d ed. — St. Louis : Mosby, 1979.
vii, 157 p. : ill. ; 26 cm.
Bibliography: p. 157.
ISBN 0-8016-0949-6
1. Pediatric nursing. 2. Child development—Testing. 3. Children—Care and hygiene. I. Leitch, Cynthia J., 1933- joint author. II. Title.
[DNLM: 1. Child care. 2. Child development. 3. Child health services. 4. Growth—In adolescence. 5. Pediatric nursing. WY159 C539c]
RJ245.C48 1979 610.73'62 78-11964
78 MARC

Meisels, Samuel J
Developmental screening in early childhood : a guide / by Samuel J. Meisels. — Washington : National Association for the Education of Young Children, [acknowledgments 1978]
viii, 52 p. : ill. ; 20 x 22 cm.
Bibliography: p. 49-51.
ISBN 0-912674-63-6 : $2.50
1. Child development—Testing. I. Title.
LB1131.M444 372.1'2'6 78-70706
79 MARC

Vulpé, Shirley German.
Vulpé assessment battery : developmental assessment, performance analysis, individualized programming : for the atypical child / by Shirley German Vulpé ; with contributions by Ellen I. Rollins and Janet Wilson. — Toronto : National Institute on Mental Retardation, 1977.
vi, 397 p. ; 28 cm. C78-6142-X
"Originally published in 1969 as Home care and management of the mentally retarded child and the Assessment Battery. Revised edition combines both volumes."
Includes bibliographical references.
ISBN 0-919648-07-X
1. Child development—Testing. 2. Mentally handicapped children. 3. Exceptional children. I. Rollins, Ellen I., joint author. II. Wilson, Janet, joint author. III. Title.
HQ773.7.V84 1977 155.4 79-310305
79 MARC

—CANADA—ADDRESSES, ESSAYS, LECTURES

**Childhood and adolescence in Canada / ** edited and with an introd. by K. Ishwaran. — Toronto ; New York : McGraw-Hill Ryerson, c1979.
386 p. ; 23 cm. — (McGraw-Hill Ryerson interdisciplinary studies in family and marriage) C78-1639-4
Includes bibliographies.
ISBN 0-07-082935-7 : $8.95
1. Children—Canada—Addresses, essays, lectures. 2. Child development—Canada—Addresses, essays, lectures. 3. Youth—Canada—Addresses, essays, lectures. I. Ishwaran, Karigoudar. II. Series: Interdisciplinary studies in family and marriage.
HQ792.C3C48 301.43'1'0971 79-318005
79 MARC

—JAPAN

(Atarashii dōtoku no hakken)
新しい道徳の発見：子どもに聞かせる生活の話 / 今井譽次郎編 ; [児童問題研究会]. — 東京：訃信書房, 昭和32 [1957]
8, 314 p. ; 19 cm.

1. Child development—Japan. 2. Education of children. 3. Education—Japan. I. Imai, Takajirō, 1906- II. Jidō Mondai Kenkyūkai.
HQ792.J3A8 79–808764

Isomura, Eiichi, 1903-
(Oya ni mondai ga aru)
親に問題がある：家庭教育に関する六十三條 / 磯村英一, 萱沼素共著. — 東京：刀江書院, 昭和34 [1959]
231 p. ; 19 cm.

1. Parent and child—Japan. 2. Child development—Japan. I. Kayanuma, Hajime, joint author. II. Title.
HQ755.85.I83 79–807322

Itō, Tomonori, 1934-
(Oya)
親：母親、そして父親に問われるもの / 伊藤友宣著. — 大阪：朱鷺書房, 1978.
245 p. : ill. ; 19 cm. Ja 79–3478
¥980

1. Parent and child—Japan. 2. Child development—Japan. I. Title.
HQ755.85.I868 79–807264

J

(Terebikko)
テレビっ子：この新しい世代とつき合うために / 読売新聞社婦人部編. — 東京：文化出版局, 1978.
230 p. : ill. ; 18 cm. — (よつば新書 ; 31) Ja 78–24704
Series romanized: Yotsuba shinsho.
Originally published in Yomiuri shimbun, Jan.–Apr. 1978.
¥600

1. Television and children. 2. Children—Japan. 3. Child development—Japan. I. Yomiuri Shimbun Sha. Fujinbu. II. Yomiuri shimbun.
HQ784.T4T464 79–801814

—NETHERLANDS

**Famine and human development : ** the Dutch hunger winter of 1944-1945 / Zena Stein ... [et al.]. — New York : Oxford University Press, 1975.
xx, 284 p. : graphs ; 24 cm.
Bibliography: p. 259-274.
Includes indexes.
ISBN 0-19-501811-7
1. Fetal malnutrition—Netherlands. 2. Child development—Netherlands. 3. Intellect—Nutritional aspects. 4. Youth—Health and hygiene—Netherlands. 5. Netherlands—History—German occupation, 1940-1945. 6. Netherlands—Famines. I. Stein, Zena.
RG627.6.M34F35 616.3'9 74-83991
75[r79]rev MARC

CHILD DEVELOPMENT (Continued)

—UNITED STATES

United States. Congress. Senate. Committee on Human Resources. Subcommittee on Child and Human Development.
Child care and child development programs, 1977-78 : hearings before the Subcommittee on Child and Human Development of the Committee on Human Resources, United States Senate, Ninety-fifth Congress, first and second sessions ... November 25, 1977, San Francisco, Calif., December 12, 1977, Los Angeles, Calif. — Washington : U.S. Govt. Print. Off., 1978-
v. : ill. ; 24 cm.
1. Children—Care and hygiene—United States. 2. Child development—United States. I. Title.
KF26.H83 1977b 344'.73'0327 79-600556
79 MARC

CHILD DEVELOPMENT DEVIATIONS

see also Developmentally disabled children

Desarrollo y estimulación del niño : desde su nacimiento hasta los dos años. — Pamplona : Ediciones Universidad de Navarra, c1979.
101 p. : ill. ; 22 cm. — (Colección Ciencias médicas) (Libros de enfermería) Sp***
At head of title: UPEP. Unidad Pediatrica de Estimulación Precoz; Clínica Universitaria, Facultad de Medicina, Universidad de Navarra.
Bibliography: p. [99]-101.
ISBN 8431305835
1. Child development. 2. Child development deviations. 3. Motor ability in children. 4. Perception in children. 5. Sensory stimulation.
RJ131.D43 79-108918
00 MARC

Neurodevelopmental problems in early childhood : assessment and management / edited by C. M. Drillien, M. B. Drummond. — Oxford : Blackwell Scientific ; Philadelphia : distributed by J. B. Lippincott, c1977.
ix, 499 p. : ill. ; 25 cm. GB***
Includes bibliographies and index.
ISBN 0-632-00409-6 : $54.50
1. Pediatric neurology. 2. Child development deviations. I. Drillien, Cecil M. II. Drummond, Margaret B.
RJ486.N48 618.9'28 78-320066
78 MARC

—ETIOLOGY—CONGRESSES

Early developmental hazards : predictors and precautions / edited by Frances Degen Horowitz. — Boulder, Colo. : Published by Westview Press for the American Association for the Advancement of Science, 1978.
xvi, 114 p. ; 24 cm. — (AAAS selected symposium ; 19)
Includes bibliographies and indexes.
Based on a symposium presented at the AAAS annual national meetings in Denver, February 1977, and sponsored by the Society for Research in Child Development, in conjunction with Section J (Psychology) of the AAAS.
ISBN 0-89158-084-0 : $12.50
1. Child development deviations—Etiology—Congresses. 2. Child development—Congresses. 3. Infants—Growth—Congresses. I. Horowitz, Frances Degen. II. American Association for the Advancement of Science. III. Society for Research in Child Development. IV. American Association for the Advancement of Science. Sectionon Psychology. V. Series: American Association for the Advancement of Science. AAAS selected symposium ; 19.
RJ135.E25 618.9'2 78-352
78[r79]rev MARC

CHILD DISCIPLINE see Discipline of children

CHILD ENDOWMENT see Family allowances

CHILD HEALTH see Child welfare; Children—Care and hygiene

CHILD HEALTH SERVICES

see also Ambulatory medical care for children; Child mental health services; Immunization of children; Infant health services; Maternal health services; Youth—Medical care

Bisenova, Aklima Bisenovna.
(Ana men balanyñ densaulyghyn saqtau)
Ана мен баланың денсаулығын сақтау / А. Бисенова. — Алматы : Қазақстан, 1968.
27 p.
Kazakh.
L.C. copy on microfilm only.
——— Microfilm. On 1 reel with other items ; 35 mm.
Microfilm Slavic 12710 RG
1. Maternal health services. 2. Child health services. I. Title.
[RG940.B57] 79-970221

—ABSTRACTS

Child & youth services. v. 1-
New York, Haworth Press, 1977-
v. 28 cm. bimonthly.
Key title: Child & youth services, ISSN 0145-935X.
1. Child welfare—Periodicals. 2. Child welfare—Abstracts. 3. Child health services—Periodicals. 4. Child health services—Abstracts. 5. Youth—Services for—Periodicals. 6. Youth—Services for—Abstracts. 7. Adolescence—Periodicals. 8. Adolescence—Abstracts.
HV701.C47 362.7'05 79-649846
MARC-S

—ADMINISTRATION

World Health Organization.
Risk approach for maternal and child health care : a managerial strategy to improve the coverage and quality of maternal and child health/family planning services based on the measurement of individual and community risk. — Geneva : World Health Organization ; [Albany, N.Y. : obtainable from WHO Publications Centre], 1978.
v, 42 p. : ill. ; 30 cm. — (WHO offset publication ; no. 39) Sw***
Includes bibliographical references.
ISBN 9241700394 : 9.00F
1. Maternal health services—Administration. 2. Child health services—Administration. 3. Pregnancy, Complications of. 4. Risk management. I. Title. II. Series: World Health Organization. WHO offset publication ; no. 39.
RG940.W67 1978 362.1'9'82 78-319872
79 MARC

—EVALUATION—CONGRESSES

Chronic childhood illness : assessment of outcome : a conference sponsored by the John E. Fogarty International Center for Advanced Study in the Health Sciences and the National Institute of Child Health and Human Development, National Institutes of Health, Bethesda, Maryland, May 1-2, 1974 / Gilman D. Grave and I. Barry Pless, editors. — Bethesda, Md. : U.S. Dept. of Health, Education, and Welfare, Public Health Service, National Institutes of Health ; Washington : for sale by the Supt. of Docs., U.S. Govt. Print. Off., 1976.
xiv, 238 p. : ill. ; 27 cm. — (Fogarty International Center series on the teaching of preventive medicine ; v. 3) (DHEW publication ; no. (NIH) 76-877)
Includes bibliographies and index.
Supt. of Docs. no.: HE 20.3714:3
$6.10
1. Chronic diseases in children—Congresses. 2. Chronically ill children—Rehabilitation—Congresses. 3. Child health services—Evaluation—Congresses. I. Grave, Gilman D. II. Pless, Ivan B., 1932- III. John E. Fogarty International Center for Advanced Study in the Health Sciences. IV. United States. National Institute of Child Health and Human Development. V. Series: John E. Fogarty International Center for Advanced Study in the Health Sciences. Fogarty International Center series on the teaching of preventive medicine ; v. 3. VI. Series: United States. Dept. of Health, Education, and Welfare. DHEW publication ; no. (NIH) 76-877.
RJ380.C58 618.9'2 76-603560
77[r79]rev MARC

Evaluation of child health services : the interface between research and medical practice / Samuel J. Bosch and Jaime Arias, scientific editors. — Bethesda, Md. : U.S. Dept. of Health, Education, and Welfare, Public Health Service, National Institutes of Health ; Washington : for sale by the Supt. of Docs., U.S. Govt. Print. Off., 1978.
xiii, 291 p. : ill. ; 24 cm. — (DHEW publication ; no. (NIH) 78-1066)
"Proceedings of a conference sponsored by American College of Preventive Medicine ... [et al.] March 7-9, 1977, National Institutes of Health, Bethesda, Maryland."
Includes bibliographies and index.
1. Child health services—Evaluation—Congresses. 2. Child health services—America—Evaluation—Congresses. 3. Children—Care and hygiene—America—Congresses. I. Bosch, Samuel J. II. Arias, Jaime. III. American College of Preventive Medicine. IV. Series: United States. Dept. of Health, Education, and Welfare. DHEW publication ; no. (NIH) 78-1066.
RJ101.E93 362.7'8 79-601520
79 MARC

—LAW AND LEGISLATION—UNITED STATES

United States. Congress. Senate. Committee on Finance. Subcommittee on Health.
Child health assessment act and increased Medicaid funding for Puerto Rico : hearing before the Subcommittee on Health of the Committee on Finance, United States Senate, Ninety-fifth Congress, second session, on H.R. 9434 ... S. 1392 ... August 14, 1978. — Washington : U.S. Govt. Print. Off., 1978 [i.e. 1979]
iv, 267 p. ; 24 cm.
Includes bibliographical references.
1. Medicaid—Law and legislation—Puerto Rico. 2. Child health services—Law and legislation—United States. 3. Medical screening—Law and legislation—United States. I. Title.
KF26.F5538 1978d 344'.73'022 79-601400
79 MARC

United States. Health Services Administration. Bureau of Community Health Services.
Legislative base, maternal and child health programs : (includes 1976 supplement). — Rockville, Md. : U.S. Dept. of Health, Education, and Welfare, Public Health Service, Health Services Administration, Bureau of Community Health Services, 1977.
vii, 82 p. ; 24 cm. — (DHEW publication ; (HSA) 77-5221)
1. Maternal health services—Law and legislation—United States. 2. Child health services—Law and legislation—United States. I. Title. II. Title: Maternal and child health programs. III. Series: United States. Dept. of Health, Education, and Welfare. DHEW publication ; no. (HSA) 77-5221.
KF3728.A84 344'.73'03282 78-601428
79 MARC

—PERIODICALS

Child & youth services. v. 1-
New York, Haworth Press, 1977-
v. 28 cm. bimonthly.
Key title: Child & youth services, ISSN 0145-935X.
1. Child welfare—Periodicals. 2. Child welfare—Abstracts. 3. Child health services—Periodicals. 4. Child health services—Abstracts. 5. Youth—Services for—Periodicals. 6. Youth—Services for—Abstracts. 7. Adolescence—Periodicals. 8. Adolescence—Abstracts.
HV701.C47 362.7'05 79-649846
MARC-S

—UNDERDEVELOPED AREAS see Underdeveloped areas—Child health services

—AMERICA—EVALUATION—CONGRESSES

Evaluation of child health services : the interface between research and medical practice / Samuel J. Bosch and Jaime Arias, scientific editors. — Bethesda, Md. : U.S. Dept. of Health, Education, and Welfare, Public Health Service, National Institutes of Health ; Washington : for sale by the Supt. of Docs., U.S. Govt. Print. Off., 1978.
xiii, 291 p. : ill. ; 24 cm. — (DHEW publication ; no. (NIH) 78-1066)
"Proceedings of a conference sponsored by American College of Preventive Medicine ... [et al.] March 7-9, 1977, National Institutes of Health, Bethesda, Maryland."
Includes bibliographies and index.
1. Child health services—Evaluation—Congresses. 2. Child health services—America—Evaluation—Congresses. 3. Children—Care and hygiene—America—Congresses. I. Bosch, Samuel J. II. Arias, Jaime. III. American College of Preventive Medicine. IV. Series: United States. Dept. of Health, Education, and Welfare. DHEW publication ; no. (NIH) 78-1066.
RJ101.E93 362.7'8 79-601520
79 MARC

—CALIFORNIA

California. Dept. of Finance. Program Evaluation Unit.
Public health services in California provided by local jurisdiction / prepared by State of California, Department of Finance, Program Evaluation Unit ; Robert L. Harris, program budget manager, Bernard P. Donnelly, chief, program evaluation. — [Sacramento] : The Unit, 1977.
xviii, 109 p. ; 28 cm. — (Report - State of California, Department of Finance, Program Evaluation Unit ; no. S77-32)
Includes bibliographical references.
1. Community health services—California. 2. Community health services—California—Finance. 3. Maternal health services—California. 4. Preventive health services—California. 5. Child health services—California. I. Harris, Robert L. II. Donnelly, Bernard P. III. Title. IV. Series: California. Dept. of Finance. Program Evaluation Unit. Report - State of California, Department of Finance, Program Evaluation Unit ; no. S77-32.
RA447.C2C23 1977a 362.1'09794 78-621775
79 MARC

—CZECHOSLOVAKIA

Sziegoleit, Meinhard.
Kinder- und Mütterfürsorge in der CSSR mit besonderer Berücksichtigung der Abtreibung, von Meinhard Sziegoleit. Hrsg. von Hans Harmsen. Hamburg, [Bonn, Büro Bonner Berichte] 1969.
68 p. illus. 29 cm. (Zur Entwicklung und Organisation des Gesundheitswesens in Sowjetrussland, in osteuropäischen Volksdemokratien und in Mitteldeutschland, Bd. 45) GDB70-B3-173
Bibliography: p. 67-68
1. Child health services—Czechoslovakia. 2. Children—Diseases—Czechoslovakia—Statistics. 3. Children—Czechoslovakia—Mortality. 4. Maternal health services—Czechoslovakia. 5. Mothers—Czechoslovakia—Mortality. 6. Abortion—Czechoslovakia—Statistics. I. Title. II. Series.
RJ103.C95S94 79-413683
79 MARC

CHILD HEALTH SERVICES (Continued)

—EUROPE

Silver, George A
Child health, America's future / George A. Silver. — Germantown, Md. : Aspen Systems Corp., 1978.
xviii, 267 p. ; 24 cm.
Includes bibliographical references and index.
ISBN 0-89443-043-2
1. Child health services—United States. 2. Child health services—Europe. I. Title.
[DNLM: 1. Child health services—Standards—United States. WA320 S587c]
RJ102.S54 362.7'8'0973 78-14217
78 MARC

—GERMANY, WEST—CONGRESSES

Unsere Kinder, fit fürs Leben / [hrsg. und überreicht von der Bundesvereinigung für Gesundheitserziehung e.V.]. — Bonn-Bad Godesberg : Bundesvereinigung für Gesundheitserziehung, 1977.
183 p. ; 24 cm. GFR***
On cover: Weltgesundheitstag 1977.
Includes bibliographical references and index.
1. Child health services—Germany, West—Congresses. 2. Children—Care and hygiene—Germany, West—Congresses. I. Bundesvereinigung für Gesundheitserziehung.
RJ103.G3U5 79-356952
79 MARC

—GREAT BRITAIN—CONGRESSES

Reaching the consumer in the ante-natal and child health services : report of conference, 4 April, 1978. — [London : Dept. of Health and Social Security], 1978.
53 p. ; 30 cm. GB***
Cover title.
Arranged jointly by the Child Poverty Action Group and the Dept. of Health and Social Security.
1. Prenatal care—Great Britain—Congresses. 2. Child health services—Great Britain—Congresses. 3. Community health services for children—Great Britain—Congresses. I. Child Poverty Action Group. II. Great Britain. Dept. of Health and Social Security.
RG964.G7R4 79-316023
79 MARC

—HAITI

Bordes, Ary.
For the people, for a change : bringing health to the families of Haiti / Ary Bordes and Andrea Couture. — Boston : Beacon Press, c1978.
xv, 299 p. : ill. ; 21 cm.
Includes index.
ISBN 0-8070-2166-0 : $11.95
1. Maternal health services—Haiti. 2. Birth control—Haiti. 3. Child health services—Haiti. 4. Haiti—Social conditions. 5. Bordes, Ary. 6. Health-officers—Haiti—Biography. I. Couture, Andrea, joint author. II. Title.
RG963.H2B67 1978 362.8'2 77-88372
78 MARC

—INDONESIA—ADDRESSES, ESSAYS, LECTURES

Jakob, Teuku, 1929-
Studi tentang variasi manusia di Indonesia / oleh T. Jacob. Tugas-tugas pokok dalam memelihara kesehatan anak / oleh Ismangoen. — [Yogyakarta] : Universitas Gadjah Mada, 1973.
52 p. ; 24 cm. — (Seri penerbitan pidato pengukuhan - Universitas Gadjah Mada)
Cover title.
Bibliography: p. 19-22.
1. Anthropometry—Indonesia—Addresses, essays, lectures. 2. Fossil man—Indonesia—Addresses, essays, lectures. 3. Children—Diseases—Indonesia—Addresses, essays, lectures. 4. Child health services—Indonesia—Addresses, essays, lectures. I. Ismangoen, R., 1919- Tugas-tugas pokok dalam memelihara kesehatan anak. 1973. II. Title. III. Series: Universitas Gadjah Mada. Seri penerbitan pidato pengukuhan - Universitas Gadjah Mada.
GN58.I46J34 78-942611
79 MARC

—MASSACHUSETTS

University of Massachusetts (System). Institute for Governmental Services.
Report of the Massachusetts health policy study. — [Boston] : Institute for Governmental Services, University of Massachusetts, [1978?]
148 leaves ; 28 cm.
Cover title: Health policy study.
1. Medical policy—Massachusetts. 2. Medical care, Cost of—Massachusetts. 3. Children—Medical examinations—Massachusetts. 4. Child health services—Massachusetts. I. Title. II. Title: Health policy study.
RA395.A4M48 1978 362.1'09744 79-621111
79 MARC

—SWEDEN

Petersson, Owe.
Barns hälsa / av Owe Petersson ; rapport nr. 2 från Barnmiljöutredningen. — Stockholm : [LiberFörlag/Allmänna förlaget], 1975.
120 p. : ill. ; 25 cm. — (Barn ; 2) (Statens offentliga utredningar ; 1975:32) S75-53
ISBN 9138022516 : kr18.00
1. Child health services—Sweden. I. Sweden. Barnmiljöutredningen. II. Title. III. Series. IV. Series: Sweden. Statens offentliga utredningar ; 1975:32.
J406.R15 1975:32 78-390096
[RJ103.S8] *79 MARC

—TUNISIA

Tunisia. Kitābat al-Dawlah lil-Ṣiḥḥah al-'Umūmīyah wa-al-Shu'ūn al-Ijtimā-'īyah.
La protection maternelle et infantile; guide pratique à l'usage des médecins, des sages-femmes, des assistantes sociales et des aides-soignantes. [Tunis, Éditions de "Progrès social," Bulletin du Secrétariat d'État à la santé publique et aux affaires sociales, 1960]
80 p. illus. 25 cm. (Série des monographies techniques)
1. Maternal health services—Tunisia. 2. Child health services—Tunisia. 3. Institut national du P.F. et de la P.M.I. I. Title.
RG966.T8T86 1960 78-970822
79 MARC

—UNITED STATES

Manela, Roger.
Health needs of children / by Roger Manela and Armand Lauffer ; with contributions by Eugene Feingold and Ruben Meyer. — Beverly Hills, Calif. : Sage Publications, published in cooperation with the continuing education program in the human services of the University of Michigan School of Social Work, c1979.
111 p. : ill. ; 22 cm. — (Sage human services guides ; v. 9)
Bibliography: p. 107-109.
ISBN 0-8039-1217-X
1. Children—Care and hygiene. 2. Poor—Health and hygiene. 3. Child health services—United States. 4. Poor—Medical care—United States. I. Lauffer, Armand, joint author. II. Title.
[DNLM: 1. Child health services—United States. 2. Poverty—United States. WA320 M274c]
RJ101.M333 362.7'8 78-26373
78 MARC

Silver, George A
Child health, America's future / George A. Silver. — Germantown, Md. : Aspen Systems Corp., 1978.
xviii, 267 p. ; 24 cm.
Includes bibliographical references and index.
ISBN 0-89443-043-2
1. Child health services—United States. 2. Child health services—Europe. I. Title.
[DNLM: 1. Child health services—Standards—United States. WA320 S587c]
RJ102.S54 362.7'8'0973 78-14217
78 MARC

Task Force on Pediatric Education.
The future of pediatric education : a report / by the Task Force on Pediatric Education. — Evanston, Ill. : The Task Force, [1978 or 1979]
xi, 120 p. ; 23 cm.
Cover title.
Includes bibliographical references.
1. Pediatrics—Study and teaching—United States. 2. Pediatrics—United States. 3. Child health services—United States. 4. Pediatricians—United States—Statistics. I. Title.
RJ84.U6T37 1978 618.9'2'00071 79-114306
79 MARC

—UNITED STATES—ADMINISTRATION

United States. Congress. House. Committee on Interstate and Foreign Commerce. Subcommittee on Oversight and Investigations.
Department of Health, Education, and Welfare's administration of health programs, shortchanging children : report / by the Subcommittee on Oversight and Investigations of the Committee on Interstate and Foreign Commerce, House of Representatives, Ninety-fourth Congress, second session. — Washington : U.S. Govt. Print. Off., 1976.
v, 32 p. ; 24 cm.
At head of title: Subcommittee print.
Includes bibliographical references.
1. Early and Periodic Screening, Diagnosis, and Treatment Program. 2. Child health services—United States—Administration. 3. United States. Dept. of Health, Education, and Welfare. I. Title: Department of Health, Education, and Welfare's administration of health programs ...
RJ102.U54 1976 353.008'47'81 77-600624
77[r79]rev MARC

—UTAH—AUDITING AND INSPECTION

Utah. Office of the Legislative Auditor General.
A performance audit of selected Family Health Services programs / audit performed by audit manager, Roy Dunn, auditor in charge, Susan Elmer, audit staff, Byron Jorgenson. — [Salt Lake City] : Office of Legislative Auditor General, State of Utah, 1978.
iii, 65 p. : graphs ; 29 cm. — (Report to Utah State Legislature ; no. 78-20)
1. Child health services—Utah—Auditing and inspection. 2. Handicapped children—Care and treatment—Utah—Auditing and inspection. 3. Medicine, State—Utah—Auditing and inspection. 4. Children—Dental care—Utah—Emery Co.—Auditing and inspection. 5. Dentistry, State—Utah—Emery Co.—Auditing and inspection. 6. Utah. Family Health Services. I. Dunn, Roy. II. Elmer, Susan. III. Jorgenson, Byron. IV. Title. V. Series.
RJ102.5.U8U83 1978 353.9'792'0084781 79-622586
79 MARC

CHILD LABOR see Children—Employment; Youth—Employment

CHILD MENTAL HEALTH

see also Child mental health services; Child psychiatry; Child psychopathology

Meyers, Joel.
Mental health consultation in the schools / Joel Meyers, Richard D. Parsons & Roy Martin. — 1st ed. — San Francisco : Jossey-Bass Publishers, 1979.
xvi, 246 p. : ill. ; 24 cm. — (The Jossey-Bass social and behavioral science series)
Bibliography: p. 211-237.
Includes index.
ISBN 0-87589-400-3 : $13.95
1. Child mental health. 2. Mental hygiene. I. Parsons, Richard D., joint author. II. Martin, Roy, 1943- joint author. III. Title.
LB3430.M48 371.7'1 79-83567
79 MARC

—ADDRESSES, ESSAYS, LECTURES

Prescriptions for child mental health and education / Arnold P. Goldstein, editor. — New York : Pergamon Press, c1978.
ix, 454 p. ; 24 cm. — (Pergamon general psychology series ; v. 72)
Bibliography: p.
1. Child psychotherapy—Addresses, essays, lectures. 2. Child mental health—Addresses, essays, lectures. I. Goldstein, Arnold P.
RJ504.P7 1978 618.9'28'914 77-14422
77 MARC

—CONGRESSES

The Impact of Piagetian theory : on education, philosophy, psychiatry, and psychology / edited by Frank B. Murray ; contributors, Millie Almy ... [et al.]. — Baltimore : University Park Press, c1979.
xiii, 232 p. ; 24 cm. — (A publication of the Jean Piaget Society)
"Based upon the authors' presentations at the plenary sessions of the fifth annual symposium of the Jean Piaget Society."
Includes bibliographies and indexes.
ISBN 0-8391-1293-9
1. Piaget, Jean, 1896- —Congresses. 2. Cognition in children—Congresses. 3. Knowledge, Theory of—Congresses. 4. Child mental health—Congresses. 5. Educational psychology—Congresses. I. Murray, Frank B. II. Almy, Millie Corinne, 1915- . III. Series: Jean Piaget Society. A publication of the Jean Piaget Society.
[DNLM: 1. Child psychology—Congresses. 2. Philosophy—Congresses. 3. Psychiatry—Congresses. 4. Education—Congresses. WS105.3 I34]
BF723.C5I46 155.4'13'0124 78-13551
78 MARC

The Physician and the mental health of the child / George Tarjan, monograph series editor. — [Monroe, WI] : American Medical Association, [c1979-
v. ; 23 cm.
Vol. 1: "Based on the proceedings of the AMA Workshops on the Mental Health of Children held November 18-20, 1976 and May 19-21, 1977 in Oakbrook, Illinois."
Vol. 1: "Produced pursuant to contract no. 278-76-0046(SM) with the National Institute of Mental Health of the Alcohol, Drug Abuse, and Mental Health Administration, Department of Health, Education, and Welfare."
Includes bibliographical references.
CONTENTS: 1. Assessing development and treating disorders within a family context.
1. Child mental health—Congresses. 2. Child development—Congresses. 3. Parent and child—Congresses. I. Tarjan, George. II. American Medical Association.
RJ111.P49 618.9'28'9 79-115702
79 MARC

CHILD MENTAL HEALTH
—CONGRESSES (Continued)

World Health Organization. Expert Committee on Child Mental Health and Psychosocial Development.
Child mental health and psychosocial development : report of a WHO Expert Committee. — Geneva : WHO ; ⌈Albany : available from WHO Publications Centre USA⌉, 1977.
71 p. ; 20 cm. — (WHO technical report series 613) Sw***
Includes bibliographical references.
ISBN 9241206136 : 7.00F
1. Child mental health—Congresses. 2. Child mental health services—Congresses. I. Title. II. Series: World Health Organization. Technical report series ; 613.
RA8.A25 no. 613 79-316060
⌈RJ11⌉ 79 MARC

—UNITED STATES

Gary, Lawrence E
Support systems in Black communities : implications for mental health services for children and youth / by Lawrence E. Gary. — Washington : Mental Health Research Center, Institute for Urban Affairs and Research, Howard University, 1978.
53 p. ; 28 cm. — (Occasional paper - Institute for Urban Affairs and Research, Howard University ; v. 3, no. 4)
Paper presented at the 1977 annual meeting of the American Association of Psychiatric Services for Children, held in Washington, D.C., Nov. 17, 1977.
Bibliography: p. 43-53.
1. Child mental health—United States. 2. Child mental health services—United States. 3. Afro-Americans—Mental health—Social aspects. I. Title. II. Series: Howard University, Washington, D.C. Institute for Urban Affairs and Research. Occasional paper ; v. 3, no. 4.
RJ111.G37 362.7'8'20973 78-50887
79 MARC

CHILD MENTAL HEALTH SERVICES

see also Child psychiatry; Child psychotherapy

Twiford, Rainer, 1952-
A child with a problem : a guide to the psychological disorders of children / Rainer Twiford. — Englewood Cliffs, N.J. : Prentice-Hall, c1979.
xi, 164 p. ; 21 cm. — (A Spectrum book)
Bibliography: p. 81-84.
Includes index.
ISBN 0-13-130781-9
1. Child psychopathology. 2. Child mental health services. I. Title.
RJ499.T92 1979 618.9'28'9 79-9116
79 MARC

—CONGRESSES

World Health Organization. Expert Committee on Child Mental Health and Psychosocial Development.
Child mental health and psychosocial development : report of a WHO Expert Committee. — Geneva : WHO ; ⌈Albany : available from WHO Publications Centre USA⌉, 1977.
71 p. ; 20 cm. — (WHO technical report series 613) Sw***
Includes bibliographical references.
ISBN 9241206136 : 7.00F
1. Child mental health—Congresses. 2. Child mental health services—Congresses. I. Title. II. Series: World Health Organization. Technical report series ; 613.
RA8.A25 no. 613 79-316060
⌈RJ11⌉ 79 MARC

—DENMARK

Wagner, Marsden, 1930-
Denmark's National family guidance program : a preventive mental health program for children and families / by Marsden Wagner. — Rockville, Md. : U.S. Dept. of Health, Education, and Welfare, Public Health Service, Alcohol, Drug Abuse, and Mental Health Administration, National Institute of Mental Health, 1978.
vi, 65 p. ; 27 cm. — (DHEW publication ; no. (ADM) 77-512)
Translation of Lov om borne-og ungdomsforsorg: p. 40-59.
1. Mental health services—Denmark. 2. Child mental health services—Denmark. I. Denmark. Laws, statutes, etc. Lov om borne- og ungdomsforsorg. English. 1978. II. Title. III. Series: United States. Dept. of Health, Education, and Welfare. DHEW publication ; no. (ADM) 77-512.
RA790.7.D4W34 362.7'8'209489 78-601642
79 MARC

—NETHERLANDS

Kraft, Th B
De jeugdpsychiatrische diensten in Nederland : een sociaal psychiatrische studie / Th. B. Kraft. — Utrecht : Erven J. Bijleveld, c1963.
xvi, 245 p. ; 23 cm.
Bibliography: p. ⌈234⌉-245.
1. Adolescent psychiatry—Netherlands. 2. Child mental health services—Netherlands. I. Title.
RJ503.K72 78-397353
79 MARC

—UNITED STATES

Gary, Lawrence E
Support systems in Black communities : implications for mental health services for children and youth / by Lawrence E. Gary. — Washington : Mental Health Research Center, Institute for Urban Affairs and Research, Howard University, 1978.
53 p. ; 28 cm. — (Occasional paper - Institute for Urban Affairs and Research, Howard University ; v. 3, no. 4)
Paper presented at the 1977 annual meeting of the American Association of Psychiatric Services for Children, held in Washington, D.C., Nov. 17, 1977.
Bibliography: p. 43-53.
1. Child mental health—United States. 2. Child mental health services—United States. 3. Afro-Americans—Mental health—Social aspects. I. Title. II. Series: Howard University, Washington, D.C. Institute for Urban Affairs and Research. Occasional paper ; v. 3, no. 4.
RJ111.G37 362.7'8'20973 78-50887
79 MARC

—UNITED STATES—DIRECTORIES

U. S. facilities and programs for children with severe mental illnesses: directory.
Rockville, Md., U. S. Dept. of Health, Education, and Welfare, Public Health Service, Alcohol, Drug Abuse, and Mental Health Administration, National Institute of Mental Health; ⌈Washington, For sale by the Supt. of Docs., U. S. Govt. Print. Off.⌉
v. 24 cm. (DHEW publication)
Key title: U. S. facilities and programs for children with severe mental illnesses. Directory, ISSN 0160-676X
1. Child mental health services—United States—Directories. I. United States. National Institute of Mental Health. II. Series: United States. Dept. of Health, Education, and Welfare. DHEW publication.
RJ111.U44 362.7'8'20973 78-640895
MARC-S

CHILD MOLESTING

see also Girls—Crimes against

The Forbidden love.
The Normal and abnormal love of children / edited by William Kraemer ; introd. by Philip Nobile. — Kansas City, ⌈Kan.⌉ : Sheed Andrews and McMeel, ⌈1978⌉ c1976.
xiv, 150 p. ; 22 cm.
ISBN 0-8362-6508-4 : $7.95.
1. Pedophilia. 2. Child molesting. I. Kraemer, William Guenther Paul.
RC560.C46F67 1978 616.8'583 78-62205
79 MARC

Kroth, Jerome A
Child sexual abuse : analysis of a family therapy approach / by Jerome A. Kroth. — Springfield, Ill. : Thomas, c1979.
x, 200 p. : ill. ; 24 cm.
Bibliography: p. 197-198.
Includes index.
ISBN 0-398-03906-2
1. Child molesting. 2. Child abuse. 3. Family psychotherapy. I. Title. ⌈DNLM: 1. Child abuse. 2. Sex offenses. 3. Family therapy. WA320 K93c⌉
RC560.C46K76 362.1'9'68583 79-541
79 MARC

Lamontagne, Yves.
L'attentat sexuel contre les enfants : l'apect psychologique, le point de vue légal, la prévention / Yves Lamontagne et Célyne Lacerte-Lamontagne. — ⌈Montréal⌉ : La Presse ; ⌈Saint-Léonard⌉ : distributeur, Nouvelles messageries internationales du livre, ⌈c1977⌉
127 p. ; 21 cm. C***
Bibliography: p. 123-127.
ISBN 0-7777-0189-8
1. Pedophilia. 2. Child molesting. I. Lacerte-Lamontagne, Célyne, joint author. II. Title.
HQ71.L27 301.41'58 78-375954
78 MARC

—NETHERLANDS—CONGRESSES

Conferentie Pedofilie en Samenleving, RAI-Congrescentrum, 1977.
Pedofilie & ⌈i.e. en⌉ samenleving : congresbundel, RAI Amsterdam, 19-3-1977. — ⌈s.l. : NVSH-LWP, 1977⌉
109 p. ; 21 cm. Ne***
Cover title.
1. Pedophilia—Congresses. 2. Pedophilia—Netherlands—Congresses. 3. Child molesting—Netherlands—Congresses. I. Title.
HQ71.C73 1977 78-384032
79 MARC

—TEXAS

Texas. Legislature. House of Representatives. Select Committee on Child Pornography, Its Related Causes and Control.
Interim report, sixty sixth legislative session / The Select Committee on Child Pornography, Its Related Causes and Control, Texas House of Representatives. — ⌈Austin⌉ : The Committee, ⌈1978?⌉
iii leaves, 198 p. ; 28 cm.
Cover title.
1. Child abuse—Law and legislation—Texas. 2. Sex crimes—Texas. 3. Child molesting—Texas. 4. Pornography—Texas.
KFT1211.82.C55 1978 364.1'53'09764 79-623608
00 MARC

—UNITED STATES

Finkelhor, David.
Sexually victimized children / David Finkelhor. — New York : Free Press, c1979.
xii, 228 p. ; 24 cm.
Bibliography: p. 215-221.
Includes index.
ISBN 0-02-910210-3
1. Child molesting—United States. I. Title.
⌈DNLM: 1. Child abuse. 2. Sex offenses. 3. Sex deviation. WM610.3 F499s⌉
HV6626.F56 616.8'583 79-7104
79 MARC

Geiser, Robert L
Hidden victims : the sexual abuse of children / Robert L. Geiser. — Boston : Beacon Press, c1979.
xi, 191 p. ; 21 cm.
Bibliography: p. 181-182.
Includes index.
SUMMARY: Discusses the possible causes and aftermath of the sexual misuse of children resulting from incidents involving incest, rape, prostitution, and other activities.
ISBN 0-8070-2954-3 : $10.95
1. Child molesting—United States. 2. Child abuse—United States. 3. Incest. ⌈1. Child molesting. 2. Child abuse. 3. Incest⌉ I. Title.
HQ72.U53G43 1979 364.1'53 78-73853
79 MARC

CHILD NEGLECT see Child abuse

CHILD PARENT see Parent and child

CHILD PLACING see Adoption; Foster home care

CHILD PSYCHIATRY

see also Adolescent psychiatry; Autism; Child development deviations; Child mental health; Child psychology; Child psychopathology; Child psychotherapy; Mentally handicapped children; Mentally ill children; Psychomotor disorders in children; Psychoses in children; Schizophrenia in children; School phobia

Basic handbook of child psychiatry / Joseph D. Noshpitz, editor-in-chief. — New York : Basic Books, c1979-
v. ; 27 cm.
Includes bibliographies and indexes.
CONTENTS: v. 1. Development.—v. 2. Disturbances in development.

ISBN (invalid) 0-465-00590-X
1. Child psychiatry. I. Noshpitz, Joseph D.
⌈DNLM: 1. Child psychiatry—Handbooks. WS350 B311⌉
RJ499.B33 618.9'28'9 78-7082
78 MARC

Bersihand, Geneviève.
Bettelheim / Geneviève Bersihand. — Champigny-sur-Marne : R. Jauze, c1977.
199 p. ; 22 cm. F***
Bibliography: p. 197-199.
ISBN 2-86214-001-5 : 35.00F
1. Bettelheim, Bruno. 2. Child psychiatry.
RJ499.B475 362.7'8'20924 79-342882
79 MARC

Castets, Bruno.
L'Enfant fou / Bruno Castets. — Nouvelle éd. revue et augmentée. — Paris : Éditions Fleurus, 1978.
251 p., ⌈16⌉ p. of plates : ill. ; 20 cm. — (Collection Pédagogie psychosociale; 10) F78-14390
ISBN 2-215-00223-9 : 50.00F
1. Child psychiatry. 2. Child psychiatry—Cases, clinical reports, statistics. I. Title.
RJ499.C295 1978 618.9'28'9 79-348009
79 MARC

The Child in his family ⌈vol. 1⌉ / edited by E. James Anthony and Cyrille Koupernik. — Huntington, N.Y. : R. E. Krieger Pub. Co., 1979, c1970.
xxxii, 492 p. ; 24 cm.
Reprint of the ed. published by Wiley, New York, as v. 1 of the International yearbook for child psychiatry and allied disciplines.
Includes bibliographies and index.
ISBN 0-88275-863-2
1. Child psychiatry. 2. Parent and child. I. Anthony, Elwyn James. II. Koupernik, Cyrille, 1917- . III. Series: International yearbook for child psychiatry and allied disciplines ; v. 1.
⌈RJ499.C484 1979⌉ 618.9'28'9008 78-31654
78 MARC

CHILD PSYCHIATRY (Continued)

Child psychiatric nursing / Evelyn M. McElroy, guest editor. Implementation of nursing diagnosis / Marjory Gordon, guest editor. — Philadelphia : Saunders, 1979.
vi p., p. 389-573 : ports. ; 24 cm. — (The Nursing clinics of North America ; v. 14, no. 3)
Bibliography: p. 563-569.
Includes index.
1. Nursing. 2. Diagnosis. 3. Child psychiatry. 4. Pscyhiatric nursing. 5. Pediatric nursing. I. Gordon, Marjory. II. McElroy, Evelyn M. III. Title: Implementation of nursing diagnosis. 1979.
[DNLM: 1. Diagnosis—Nursing texts. 2. Nursing care. 3. Psychiatric nursing—In infancy and childhood. W1 NU587G v. 14 no. 3 / WY100.3 I34]
RT1.N77 vol. 14, no. 3 610.73'08 s 79-14981
[RT48] 79 MARC
Shared Cataloging with
DNLM

Children and their parents in a changing world / edited by E. James Anthony and Colette Chiland ; foreword by Erik H. Erikson. — New York : Wiley, c1978.
xxx, 443 p. ; 24 cm. — (Child in his family ; v. 5) (Yearbook of the International Association for Child Psychiatry and Allied Professions ; v. 5)
"A Wiley-Interscience publication."
Includes bibliographies and index.
ISBN 0-471-04432-6
1. Child psychiatry. 2. Child psychology. 3. Family. 4. Social psychiatry. 5. Social change. I. Anthony, Elwyn James. II. Chiland, Colette. III. Series. IV. Series: International Association for Child Psychiatry and Allied Professions. Yearbook ; v. 5.
RJ499.A1C42 vol. 5 618.9'28'9008 s 78-108199
79 MARC

Held, Fritz, writer on psychiatry.
Jugendpsychiatrische Studien. Aus Theorie und Praxis für Ärzte, Psychologen, Pädagogen, Juristen, Sozialarbeiter, Heimerzieher und andere soziale Berufe. (Berlin-Sapndau, Neuwied a. Rh.) Luchterhand (1966)
xi, 131 p. with illus. 21 cm. (Jugend im Blickpunkt) DM14.80
GDB***
1. Child psychiatry. I. Title.
RJ499.H44 618.9'28'9 67-71664
[r79]rev MARC

The Humanization processes : a social, behavioral analysis of children's problems / Robert L. Hamblin ... [et al.]. — Huntington, N.Y. : R. E. Krieger Pub. Co., 1978, c1971.
xxi, 305 p. : graphs ; 24 cm.
Reprint of the ed. published by Wiley, New York.
Bibliography: p. 287-295.
Includes indexes.
ISBN 0-88275-721-0
1. Child psychiatry. 2. Socialization. 3. Socially handicapped children. 4. Aggressiveness in children. 5. Autism. I. Hamblin, Robert Lee, 1927-
[DNLM: 1. Child behavior disorders. 2. Psychoanalysis—In infancy and childhood. WS350 H918 1971a]
[RJ499.H85 1979] 618.92'8'9 78-15049
78 MARC

(Nyūyōjiki no rinshō shinri)
乳幼児期の臨床心理 [辰見敏夫編 東京] 岩崎学術出版社 [1972]
1, 238 p. 22 cm. (児童臨床心理学講座 4) Ja***
Includes bibliographies.

1. Child psychiatry. 2. Infants. I. Tatsumi, Toshio, 1919- ed. II. Series: Jidō-rinshō-shinrigaku kōza, 4.
RJ499.N94 618.9'28'9 76-801861
rev
74 J

Petri, Horst, 1936-
Soziale Schicht und psychische Erkrankung im Kindes- und Jugendalter : e. Erprobungsstudie an e. kinder- u. jugendpsychiatr. Population / Horst Petri, unter Mitarb. von Lutz Rosenberg u. Christine Thate. — Göttingen : Verlag für Med. Psychologie im Verl. Vandenhoeck u. Ruprecht, 1979.
201 p. ; 24 cm. — (Beihefte zur Zeitschrift Gruppenpsychotherapie und Gruppendynamik ; Heft 9) GFR79-A
Bibliography: p. [182]-194.
Includes indexes.
ISBN 3-525-45291-8
1. Child psychiatry. 2. Adolescent psychiatry. I. Rosenberg, Lutz, joint author. II. Thate, Christine, joint author. III. Title. IV. Series: Gruppenpsychotherapie und Gruppendynamik : Beihefte ; Heft 9.
RJ499.P44 79-377755
79 MARC

Stone, Frederick Hope.
Child psychiatry for students / Frederick H. Stone, Cyrille Koupernik ; foreword by L. N. J. Kamp. — 2d ed. — Edinburgh ; New York : Churchill Livingstone ; New York : distributed by Longman, 1978.
111 p. ; 22 cm. GB78
Includes bibliographical references and index.
ISBN 0-443-01797-2 : £2.25
1. Child psychiatry. I. Koupernik, Cyrille, 1917- joint author. II. Title.
[DNLM: 1. Child psychiatry. 2. Mental disorders—In infancy and childhood. WS350 S877c]
RJ499.S797 1978 618.9'28'9 78-40331
78 MARC

Winnicott, Donald Woods, 1896-
Through paediatrics to psycho-analysis / by D. W. Winnicott ; with an introd. by M. Masud R. Khan. — London : Hogarth Press and the Institute of Psycho-Analysis, 1975.
L, 350 p. : ill. ; 23 cm. — (The International psycho-analytical library ; no. 100) GB***
First published in 1958 under title: Collected papers.
Includes bibliographies and index.
ISBN 0-7012-0399-4 : £5.00
1. Child psychiatry. 2. Child analysis. I. Institute of Psycho-Analysis, London. II. Title. III. Series.
RJ499.W48 1975 618.9'28'9 75-311187
75[r79]rev MARC

—ADDRESSES, ESSAYS, LECTURES

Psychiatric problems in medical practice / George U. Balis, editor in chief, Leon Wurmser, Ellen McDaniel, editors, Robert G. Grenell, consulting editor. — Boston : Butterworth Publishers, c1978.
xx, 473 p. ; 24 cm. — (The Psychiatric foundations of medicine ; 6)
Includes bibliographies and index.
ISBN 0-409-95120-X
1. Psychiatry—Addresses, essays, lectures. 2. Child psychiatry—Addresses, essays, lectures. I. Balis, George U. II. Series.
[DNLM: 1. Social sciences. 2. Behavioral sciences. 3. Psychopathology. 4. Psychiatry. WM100.3 P974]
RC458.P775 616.8'9 78-14216
78 MARC

—BIBLIOGRAPHY

Birtchnell, John.
A bibliography in dynamic, familial and social psychiatry / written ... by John Birtchnell and Sheila Hafter Gray. — York : The Society of Clinical Psychiatrists, [1978]
136 p. ; 21 cm. — (S.P.C. reports) GB78-13228
1. Psychiatry—Bibliography. 2. Child psychiatry—Bibliography. 3. Social psychiatry—Bibliography. I. Gray, Sheila Hafter, joint author. II. Series: Society of Clinical Psychiatrists. S.P.C. reports.
Z6664.N5B57 016.6168'9 78-318491
[RC454] 78 MARC

—CASE STUDIES

Castets, Bruno.
L'Enfant fou / Bruno Castets. — Nouvelle éd. revue et augmentée. — Paris : Éditions Fleurus, 1978.
251 p., [16] p. of plates : ill. ; 20 cm. — (Collection Pédagogie psychosociale; 10) F78-14390
ISBN 2-215-00223-9 : 50.00F
1. Child psychiatry. 2. Child psychiatry—Cases, clinical reports, statistics. I. Title.
RJ499.C295 1978 618.9'28'9 79-348009
79 MARC

—CASES, CLINICAL REPORTS, STATISTICS

Yamanaka, Yasuhiro, 1941-
(Shōnenki no kokoro)
少年期の心：精神療法を通してみた影 / 山中康裕著. — 東京：中央公論社, 1978.
viii, 216 p. : ill. ; 18 cm. — (中公新書 ; 515) Ja 78-24236
Series romanized: Chūkō shinsho.
¥380

1. Child psychiatry—Cases, clinical reports, statistics. I. Title.
RJ499.Y35 79-800596

—DICTIONARIES—FRENCH

Lafon, Robert, 1905-
Vocabulaire de psychopédagogie et de psychiatrie de l'enfant / Robert Lafon. — [4. éd. refondue et augm.]. — Paris : Presses universitaires de France, 1979, c1963.
xix, 1060 p. ; 25 cm. F***
On spine: Vocabulaire de psychopédagogie.
182F
1. Child psychiatry—Dictionaries—French. 2. Educational psychology—Dictionaries—French. 3. Child welfare—Dictionaries—French. I. Title. II. Title: Vocabulaire de psychopédagogie.
RJ499.L28 1979 371.9'4'019 79-381763
79 MARC

—VENEZUELA

Bastidas, Arístides, 1924-
Rafael Vegas : reportaje biográfico / [por] Arístides Bastidas. — Barcelona, [etc.] : Ariel, 1978.
268 p. ; 22 cm. — (Horas de Venezuela ; 1) Sp78-Mar
Includes index.
ISBN 8434422005 : 500ptas
1. Vegas, Rafael. 2. Statesmen—Venezuela—Biography. 3. Psychiatrists—Venezuela—Biography. 4. Child psychiatry—Venezuela. I. Series.
F2326.V43B37 79-355574
79 MARC

CHILD PSYCHOANALYSIS see Child analysis

CHILD PSYCHOLOGY

see also Achievement motivation in children; Adolescent psychology; Aggressiveness in children; Anxiety in children; Autonomy in children; Childpsychiatry; Childishness; Children's dreams; Clinical child psychology; Cognition in children; Creative ability in children; Educational psychology;

Emotional problems of children; Emotions in children; Fantasy in children; Fear in children; Frustration in children; Humor in children; Imagination in children; Imitation in children; Infant psychology; Memory in children; Moral development;

Paternal deprivation; Problem solving by children; Psychical research and children; Psychohistory; Psychological tests for children; Reasoning in children; Self-perception in children; Self-reliance in children; Sibling rivalry; Sick children —Psychology

Abe, Saburō, 1898-
(Jidō-shinrigaku kōyo)
兒童心理學綱要 / 安倍三郎著. — 東京 : 協同出版, 昭和23 [1948]
3, 5, 222 p. : 22 cm.
Colophon inserted.
Bibliography: p. 221-222.

1. Child psychology. I. Title.
BF721.A32 79-806891

Ackerman, Paul R 1934-
Signals : what your child is really telling you / Paul R. Ackerman and Murray M. Kappelman. — New York : Dial Press/J. Wade, c1978.
290 p. ; 22 cm.
Includes index.
ISBN 0-8037-8151-2 : $8.95
1. Parent and child. 2. Children—Management. 3. Child psychology. 4. Adolescent psychology. I. Kappelman, Murray, 1931- joint author. II. Title.
HQ755.85.A26 301.42'7 78-15272
78 MARC

Avalle, Vico.
La grammatica del segno infantile : impariamo a leggere i disegni del bambino / Vico Avalle. — [Ivrea] : Priuli & Verlucca, [1975]
237 p. : col. ill. ; 24 cm. It75-Nov
L6000
1. Drawing, Psychology of. 2. Child psychology. 3. Children as artists. I. Title.
BF723.D7A89 79-376775
*79 MARC

Axelrod, Herman C 1904-
Bilingual background and its relation to certain aspects of character and personality of elementary school children / Herman C. Axelrod. — New York : Arno Press, 1978, c1953.
iii, 142 p. ; 24 cm. — (Bilingual-bicultural education in the United States)
Originally presented as the author's thesis, Yeshiva University, 1951.
Bibliography: p. 137-141.
ISBN 0-405-11075-8
1. Education, Bilingual—United States. 2. Biculturalism—United States. 3. Child psychology. 4. Children—Conduct of life. I. Title. II. Series.
LC3731.A93 1978 371.9'7 77-90406
78 MARC

Balloni, Augusto.
Ambiente, ritardo mentale, devianza : deboli di mente si nasce? / [di] Augusto Balloni, Luigi Fadiga. — Bologna : Pàtron, 1978.
136 p. ; 22 cm. — (Collana di studi criminologici ; 2) It78-Apr
Bibliography: p. [129]-134.
Includes index.
L5000
1. Juvenile delinquency. 2. Mental deficiency. 3. Child psychology. I. Fadiga, Luigi, joint author. II. Title.
HV9076.B26 79-373323
*79 MARC

CHILD PSYCHOLOGY (Continued)

Bianco Gablat, Bruna.
Problemi dello sviluppo psicologico / Bruna Bianco Gablat. — 1. ed. — Torino : Paravia, 1977.
190, v p. : ill. ; 23 cm. It78-Sept
Includes bibliographies.
L3500
1. Child psychology. I. Title.
BF721.B4163 79-347515
*79 MARC

Bower, T G R 1941-
Human development / T. G. R. Bower. — San Francisco : W. H. Freeman, c1979.
xiv, 473 p. : ill. ; 25 cm. — (A Series of books in psychology)
Bibliography: p. 434-453.
Includes indexes.
1. Child psychology. I. Title.
[DNLM: 1. Growth. 2. Child development. 3. Cognition. WS105.3 B786h]
BF721.B64 155.4 78-27223
78 MARC

Cunoaşterea elevului : o sinteză a metodelor / Ion Holban coordonator, Ana Gugiuman, Anca Munteanu ... [et al.]. — Bucureşti : Editura Didactica şi Pedagogică, 1978.
338 p. : ill. ; 21 cm. R79-333
At head of title: Institutul de Cercetări Pedagogice şi Psihologice.
Bibliography: p. 333-338.
lei9.45
1. Child study. 2. Educational psychology. I. Holban, Ion. II. Institutul de Cercetări Pedagogice şi Psihologice.
LB1115.C86 79-361606
*79 MARC

Dolto, Françoise Marette.
Lorsque l'enfant paraît / Françoise Dolto. — Paris : Seuil, c1977.
189 p. ; 21 cm. F***
Includes index.
ISBN 2-02-004701-2 : 35.00F
1. Child psychology. 2. Parent and child. 3. Children—Care and hygiene. I. Title.
BF721.D64 649'.1 79-376019
79 MARC

Canestrari, Renzo.
Psicologia fuori programma / Renzo Canestrari. — Bologna : Cappelli, 1978.
316 p. : ill. ; 21 cm. — (Collana di psicologia) It78-May
Bibliography: p. 301-311.
L4500
1. Violence. 2. Psychology, Pathological. 3. Child psychology. 4. Social psychology. I. Title.
RC569.5.V55C36 157'.7'2 78-378635
*78 MARC

Chapman, Arthur Harry, 1924-
Parents talking, kids talking : an original, breakthrough work for all parents who want to discuss adult subjects with children but don't know how / A. H. Chapman. — New York : Putnam, c1979.
245 p. ; 23 cm.
Includes index.
ISBN 0-399-12259-1 : $8.95
1. Parent and child. 2. Children—Management. 3. Adolescent psychology. 4. Child psychology. I. Title.
HQ755.85.C49 1979 301.42'7 78-15271
78 MARC

Chazaud, Jacques.
Tendances nouvelles de la psychanalyse / Jacques Chazaud. — [Paris] : Centurion, c1978.
151 p. ; 21 cm. F***
Includes bibliographical references and index.
ISBN 2-227-02201-9 : 35.00F
1. Psychoanalysis. 2. Psychotherapy. 3. Child psychology. I. Title.
RC504.C49 616.8'917 78-376120
78 MARC

Children and their parents in a changing world / edited by E. James Anthony and Colette Chiland ; foreword by Erik H. Erikson. — New York : Wiley, c1978.
xxx, 443 p. ; 24 cm. — (Child in his family ; v. 5) (Yearbook of the International Association for Child Psychiatry and Allied Professions ; v. 5)
"A Wiley-Interscience publication."
Includes bibliographies and index.
ISBN 0-471-04432-6
1. Child psychiatry. 2. Child psychology. 3. Family. 4. Social psychiatry. 5. Social change. I. Anthony, Elwyn James. II. Chiland, Colette. III. Series. IV. Series: International Association for Child Psychiatry and Allied Professions. Yearbook ; v. 5.
RJ499.A1C42 vol. 5 618.9'28'9008 s 78-108199
79 MARC

Clanché, Pierre.
Le texte libre, écriture des enfants / Pierre Clanché. — Paris : F. Maspero, 1976.
226 p. ; 22 cm. — (Malgré tout) F***
Originally presented as the author's thesis, Bordeaux II.
Includes bibliographical references.
ISBN 2-7071-0820-0 : 35.00F
1. Child psychology. 2. Stream of consciousness fiction. 3. Children as authors. I. Title.
BF721.C545 1976 371.6'23'019 79-349254
79 MARC

Cohen, Marilyn A
The developmental resource : behavioral sequences for assessment and program planning / Marilyn A. Cohen and Pamela J. Gross, with contributions by Lynnette Chandler ; ill. by Kathleen E. Anderson. — New York : Grune & Stratton, c1978.
2 v. : ill. ; 26 cm.
Vol. 2 has contributions by T. E. Coggins and R. L. Carpenter.
Includes bibliographies.
ISBN 0-8089-1122-8 (v. 1)
1. Child psychology. 2. Infant psychology. 3. Child development. I. Gross, Pamela J., joint author. II. Title.
BF721.C566 155.4 78-20327
78 MARC

La Condizione giovanile : introduzione critica alla psicologia degli adolescenti e dei giovani : [scritti / di] G. Lutte ... [et al.]. — Pistoia : Cooperativa Centro documentazione ; Roma : Centro di cultura proletaria della Magliana, 1979.
309 p. ; 21 cm. It79-May
Includes bibliographical references.
L4500
1. Adolescent psychology. 2. Child psychology. I. Lutte, Gérard.
BF724.C58 79-374683
*79 MARC

Craig, Grace J
Child development / Grace J. Craig. — Englewood Cliffs, N.J. : Prentice-Hall, c1979.
xiii, 524 p. : ill. ; 24 cm.
Bibliography: p. 493-506.
Includes indexes.
ISBN 0-13-131250-2 : $14.95
1. Child psychology. I. Title.
[DNLM: 1. Child development. WS105.3 C886c]
BF721.C667 155.4 78-23212
78 MARC

Crow, Lester Donald, 1897-
New approaches to the psychology of childhood and adolescence / Lester D. Crow. — 1st ed. — Hicksville, N.Y. : Exposition Press, c1978.
259 p. ; 24 cm. — (An Exposition-university book)
Bibliography: p. 253-254.
Includes index.
ISBN 0-682-49170-5 : $12.50
1. Child psychology. 2. Adolescent psychology. I. Title.
BF721.C7245 155.4 78-59372
79 MARC

Darroch, Jane.
The formation of the superego. [Sound recording] New York, J. Norton Publishers [1974?] 29157.
1 reel. 3¾ ips. 5 in.
Title from publisher's catalog.
Recorded in 1960.
Originally issued by McGraw-Hill, New York, 75762, in series: Sound seminars.
Duration: 25 min.
SUMMARY: The author considers some of the problems in the development of internal controls in the child.
1. Personality. 2. Child psychology. 3. Superego (Psychology) I. Title. II. Series: Sound seminars. [Sound recording]
[BF723.P4] 78-740878

Davidson, Audrey, 1916-
Phantasy in childhood, by Audrey Davidson and Judith Fay. London, Routledge & Paul [1952]
188 p. 23 cm.
1. Fantasy. 2. Child psychology. I. Fay, Judith, joint author. II. Title.
BF723.F28D38 1952 155.4'13 52-4507
[r79]rev MARC

Davidson, Audrey, 1916-
Phantasy in childhood, by Audrey Davidson and Judith Fay. New York, Philosophical Library [1953, c1952]
viii, 188 p. 22 cm.
Bibliography: p. 187-188.
1. Fantasy. 2. Child psychology. I. Fay, Judith, joint author. II. Title.
BF723.F28D38 1953 155.4'13 53-6559
[r79]rev MARC

Davidson, Audrey, 1916-
Phantasy in childhood, by Audrey Davidson and Judith Fay. Westport, Conn., Greenwood Press [1972, c1952]
viii, 188 p. 22 cm.
Bibliography: p. 187-188.
ISBN 0-8371-6327-7
1. Fantasy. 2. Child psychology. I. Fay, Judith, joint author. II. Title.
[BF723.F28D38 1972] 155.4'13 77-38129
72[r79]rev MARC

Dobson, James C 1936-
The strong-willed child : birth through adolescence / James Dobson. — Wheaton, Ill. : Tyndale House, c1978.
240 p. : ill. ; 22 cm.
Includes bibliographical references.
ISBN 0-8423-0664-1 : $7.95
1. Children—Management. 2. Discipline of children. 3. Child psychology. I. Title.
HQ772.D58 649'.1 77-83559
79 MARC

Donaldson, Margaret C
Children's minds / Margaret Donaldson. — London : Croom Helm, 1978.
156 p. : ill. ; 23 cm.
Bibliography: p. 147-[151]
Includes index.
ISBN 0-85664-759-4 : £5.50
1. Child psychology. 2. Learning, Psychology of. 3. Educational psychology. I. Title.
BF318.D66 155.4'13 79-301348
79 MARC

Elitov, Paula.
Pre-schoolers with special needs : a developmental approach / by Paula Elitov. — Boston : Associated Jewish Community Centers of Greater Boston, [c1978]
vi, 245 p. ; 28 cm.
Includes bibliographies.
$4.00
1. Emotional problems of children. 2. Learning disabilities. 3. Movement disorders in children. 4. Child psychology. 5. Education, Preschool. I. Associated Jewish Community Centers of Greater Boston. II. Title.
RJ499.E414 371.9 79-101689
79 MARC

Elkind, David, 1931-
The child and society : essays in applied child development / David Elkind. — New York : Oxford University Press, 1979.
xii, 304 p. ; 22 cm.
Includes bibliographies and index.
ISBN 0-19-502371-4 : $10.00. ISBN 0-19-502372-2 pbk. : $3.95
1. Child psychology. 2. Socialization. 3. Cognition in children. I. Title.
BF721.E35 155.4'18 78-2758
78 MARC

Ėl'konin, Daniil Borisovich.
(Psikhologiia igry)
Психология игры / Д. Б. Эльконин. — Москва : Педагогика, 1978.
304 p. : port. ; 22 cm. — (Труды действительных членов и членов-корреспондентов Академии педагогических наук СССР) USSR 78
At head of title: Akademiia pedagogicheskikh nauk SSSR.
Series romanized: Trudy deĭstvitel'nykh chlenov i chlenov-korrespondentov Akademii pedagogicheskikh nauk SSSR.
Bibliography: p. 295-[299]
Includes index.
1.70rub
1. Play. 2. Child psychology. I. Title.
BF717.E4 79-412760

Erickson, Karla C
Take time to smell the dandelions / Karla C. Erickson. — Salt Lake City : Bookcraft, 1979.
88 p. ; 24 cm.
ISBN 0-88494-360-7 : $3.95
1. Children—Management. 2. Child psychology. 3. Parent and child. I. Title.
HQ769.E67 301.42'7 78-75283
79 MARC

Fadely, Jack L 1934-
Understanding the Alpha child at home and school : left and right hemispheric function in relation to personality and learning / by Jack L. Fadely and Virginia N. Hosler ; ill. by Debbie Frank. — Springfield, Ill. : Thomas, c1979.
xv, 238 p. : ill. ; 24 cm.
Bibliography: p. 229-230.
Includes index.
ISBN 0-398-03862-7
1. Left and right (Psychology) 2. Cerebral dominance. 3. Personality. 4. Learning disabilities. 5. Child psychology. I. Hosler, Virginia Neemes, 1943- joint author. II. Title.
BF311.F33 155.4'5 78-24687
78 MARC

Faw, Terry.
Schaum's outline of theory and problems of child psychology / by Terry Faw. — New York : McGraw-Hill, c1980.
296 p. : ill. ; 28 cm. — (Schaum's outline series)
On spine: Child psychology.
Includes bibliographies and index.
ISBN 0-07-020110-2
1. Child psychology. I. Title. II. Title: Outline of theory and problems of child psychology. III. Title: Child psychology. IV. Title: Theory and problems of child psychology.
BF721.F3825 155.4 79-10614
79 MARC

Fen, Elisaveta.
Aggression and its interpretation, by Lydia Jackson. With a foreword by Felix Brown. London, Methuen [1954]
237 p. illus. 21 cm.
1. Aggressiveness (Psychology) 2. Child psychology.
BF723.A35F46 152.5'2 54-14392
[r79]rev MARC

Garvey, Chester Roy.
The activity of young children during sleep : an objective study / by Chester Roy Garvey. — Westport, Conn. : Greenwood Press, 1976, c1939.
x, 102 p., [1] leaf of plates : ill. ; 23 cm.
Reprint of the ed. published by the University of Minnesota Press, Minneapolis, which was issued as no. 18 of Monograph series of the University of Minnesota's Institute of Child Welfare.
Bibliography: p. 97-99.
Includes index.
ISBN 0-8371-8073-2
1. Children—Sleep. 2. Sleep movements. 3. Child psychology. I. Title. II. Series: Minnesota. University. Institute of Child Development and Welfare. Monograph series ; no. 18.
[DNLM: WL108 G244a 1939a]
BF723.S45G3 1976 612'.04 76-44474
76 MARC

CHILD PSYCHOLOGY (Continued)

Hatano, Kanji, 1905–
(Gendai shinrigaku to kyōiku)
現代心理學と教育 / 波多野完治著. — 東京：金子書房, 昭和23 [1948]
7, 322 p. ; 19 cm.
Colophon inserted.

1. Child psychology. 2. Educational psychology. I. Title.
LB1115.H32 79–801109

Hatano, Kanji, 1905–
(Jidō shinri to jidō bungaku)
兒童心理と兒童文學 / 波多野完治著. — 東京：金子書房, 昭和25 [1950]
5, 307 p. : ill. ; 22 cm.
Includes index.

1. Child psychology. 2. Children's literature—History and criticism. I. Title.
BF721.H315 78–814922

Hatano, Kanji, 1905–
(Kodomo to wa donna mono ka)
子供とはどんなものか / 波多野完治著. — 東京：眞光社, 昭和23 [1948]
231 p. : ill. ; 18 cm.

1. Child psychology. 2. Education of children. I. Title.
LB1117.H347 79–800964

Hatano, Kanji, 1905–
(Shinkō jidō shinsei ron)
新稿兒童心性論：學習指導の基礎 / 波多野完治著. — 東京：朝倉書店, 昭和27 [1952]
3, 6, 293 p. : ill. ; 22 cm.
Includes bibliographical references.

1. Child psychology. 2. Educational psychology. I. Title. II. Title: Jidō shinsei ron.
LB1115.H33 79–800984

Hepner, Maria.
Schlüssel zur Kinderschrift : Einführung in den Hepner-Schreibtest zur Früherfassung kindlicher Reaktionsformen und seelischer Störungen / Maria Hepner. — Erlenbach-Zürich : E. Rentsch, cop. 1978.
232 p. : 85 ill. (2 col.) ; 23 cm. Sw78-7182
ISBN 3-7249-0502-5 : 36.50F
1. Graphology. 2. Child psychology.. I. Title.
BF905.C5H4 79-356790
79 MARC

Hetherington, Eileen Mavis, 1926-
Child psychology : a contemporary viewpoint / E. Mavis Hetherington, Ross D. Parke. — 2d ed. — New York : McGraw-Hill, c1979.
xxii, 680 p. : ill. ; 24 cm. — (McGraw-Hill series in psychology)
Includes bibliographies and indexes.
ISBN 0-07-028431-8 : $15.95
1. Child psychology. I. Parke, Ross D., joint author. II. Title.
BF721.H418 1979 155.4 78-27624
78 MARC

Holmes, Deborah Lott.
The child, an introduction to developmental psychology / Deborah Lott Holmes, Frederick J. Morrison. — Monterey, Calif. : Brooks/Cole Pub. Co., c1979.
xii, 464 p. : ill. ; 24 cm.
Includes bibliographies and indexes.
ISBN 0-8185-0328-9 : $15.95
1. Child psychology. 2. Adolescent psychology. I. Morrison, Frederick J., joint author. II. Title.
[DNLM: 1. Child psychology. 2. Child development. WS105.3 H749c]
BF721.H593 155.4 78-10711
78 MARC

Isaacs, Susan Sutherland Fairhurst, 1885-1948.
Social development in young children : a study of beginnings / by Susan Isaacs. — 1st AMS ed. — New York : AMS Press, 1979.
xii, 480 p. ; 23 cm.
Reprint of the 1933 ed. published by Harcourt, Brace, New York.
Bibliography: p. 463-467.
Includes indexes.
ISBN 0-404-14557-4
1. Child psychology. 2. Educational psychology. I. Title.
[HQ772.I715 1979] 155.4'18 75-41153
78 MARC

Japan. Mombushō.
(Jidō no rikai to shidō)
兒童の理解と指導 / [著作] 文部省. — 東京：師範学校教科書, 昭和24 [1949]
6, 76 p. ; 21 cm.

1. Child psychology. 2. Education of children. I. Title.
LB1115.J27 1949 78–810960

Jernryd, Elisabeth.
Optimal auktoritets- och propagandaresistens : en sambands- och utvecklingsstudie / Elisabeth Jernryd. — [Lund : Universitet : LiberLäromedel]/Gleerup, 1976.
186 p. ; 24 cm. — (Studia psychologica et paedagogica : Series altera ; 31) S76-23/25
Thesis—Lund.
Extra t.p. with thesis statement inserted.
Summary in English.
Bibliography: p. 178-184.
ISBN 9140042316
1. Child psychology. 2. Adolescent psychology. 3. Autonomy in children. 4. Conformity. I. Title. II. Series: Studia psychologica et paedagogica : Series altera, investigationes ; 31.
BF721.J392 78-377087
*78 MARC

Josselyn, Irene Milliken, 1904-1978.
Psychosocial development of children / Irene M. Josselyn. — 2d ed. — New York : Family Service Association of America, c1978.
192 p. ; 23 cm.
Includes bibliographical references.
ISBN 0-87304-154-2
1. Child psychology. 2. Psychoanalysis. I. Title.
BF721.J6 1978 155.4'18 78-3112
78 MARC

Kleber, Eduard Werner.
Abriss der Entwicklungspsychologie : e. kurze Einf. für Pädagogikstudenten u. Erzieher / Eduard W. Kleber. — 2., überarb. u. erw. Aufl. — Weinheim ; Basel : Beltz, 1978, c1974.
244 p. : ill. ; 21 cm. — (Beltz Studienbuch ; 76) GFR78-A
Bibliography: p. 216-233.
Includes indexes.
ISBN 3-407-51076-4 : DM15.00
1. Child psychology. I. Title.
BF721.K53 1978 155 79-348552
79 MARC

(Kō sureba nobiru yoi ko no me)
こうすれば伸びるよい子の芽 お母さんの教養講座 灯台編集部編 東京 第三文明社 昭和46-51 (1971-76)
2 v. 18 cm. (Todai のホーム・ブックス) (v. 2: 灯台ブックス ; 24) Ja 71-13 (v. 1)
Series 1 romanized: Tōdai no hōmu bukkusu. Series 2 romanized: Tōdai bukkusu.
¥360 (v. 1) ¥560 (v. 2)

1. Child study. I. Tōdai.
BF721.K578 75–802700
rev

Komiyama, Eiichi, 1913–
(Kyōshitsu no shinrigaku)
教室の心理学 / 小見山栄一, 井坂行男, 沢田慶輔共著. — 東京：牧書店, 昭和27 [1952]
2, 4, 191 p. : ill. ; 19 cm.
Originally broadcasted in the program "Kyōshi no jikan" by NHK, Apr. 1951–Mar. 1952.

1. Classroom management. 2. Child psychology. I. Isaka, Yukio, joint author. II. Sawada, Keisuke, 1909– joint author. III. Title.
LB3011.K65 78–817206

Koste, Virginia Glasgow.
Dramatic play in childhood : rehearsal for life / by Virginia Glasgow Koste ; sketches by Susan Russell. — New Orleans : Anchorage Press, 1978.
201 p., [1] leaf of plates : ill. ; 25 cm.
Bibliography: p. 198.
1. Play—Psychological aspects. 2. Drama—Psychological aspects. 3. Child psychology. I. Title.
BF717.K67 155.4'18 78-54331
79 MARC

Kuzma, Kay.
Understanding children / by Kay Kuzma. — Mountain View, Calif. : Pacific Press Pub. Association, c1978.
157 p. : ill. ; 19 cm. — (Harvest series ; 107)
Bibliography: p. 153-157.
1. Child psychology. 2. Child development. I. Title.
HQ767.9.K89 301.43'1 78-50449
79 MARC

Kvols-Riedler, Bill.
Redirecting children's misbehavior / Bill & Kathy Kvols-Riedler. — Boulder, Colo. : R. D. I. C. Publications, c1979.
xvi, 235 p. : ill. ; 23 cm.
Bibliography: p. 227-228.
Includes index.
ISBN 0-933450-00-1
1. Children—Management. 2. Child psychology. I. Kvols-Riedler, Kathy, joint author. II. Title.
HQ772.K86 649'.1 78-68637
79 MARC

Laing, Ronald David.
Conversations with children / [by] R. D. Laing. — London : Allen Lane, 1978.
[1], x, 89 p. ; 23 cm. GB78-29181
ISBN 0-7139-1079-8 : £3.95
1. Child psychology. 2. Children—Language. I. Title.
BF721.L274 1978 155.4'13 79-300488
79 MARC

Landsheere, Viviane de.
On defining educational objectives / by Viviane de Landsheere. — Oxford ; New York : Pergamon Press, 1977.
p. 73-190 : ill. ; 25 cm. — (Evaluation in education ; v. 1, no. 2 ISSN 0145-9228) GB***
Cover title.
Errata slip inserted.
Bibliography: p. 184-190.
1. Learning, Psychology of. 2. Child psychology. 3. Lesson planning. I. Title. II. Series.
LB1051.L17 1977 370.15 79-307721
79 MARC

Lebovici, Serge.
La connaissance de l'enfant par la psychanalyse / Serge Lebovici, Michel Soulé, avec la collaboration de Simone Decobert et Jeanine Noël. — 3. éd. mise à jour. — [Paris] : Presses universitaires de France, 1977, c1970.
650 p. ; 22 cm. — (Psychanalyse et psychiatrie de l'enfant) (Le Fil rouge ; section 2) F***
Bibliography: p. [601]-617.
Includes indexes.
85.00F
1. Child psychology. 2. Child analysis. I. Soulé, Michel, joint author. II. Title.
BF721.L434 1977 155.4 79-344249
79 MARC

Lefèvre, Robert.
De nos jours, doit-on adapter à soi l'enfant ou s'adapter à lui? / Robert Lefèvre. — [Avignon] : Presses universelles, [1976]
209 p. ; 19 cm. F***
25.00F
1. Child psychology. 2. Parent and child. 3. Education of children. I. Title.
BF721.L45 649'.1 79-375735
79 MARC

Lewis, David, 1942-
The secret language of your child : how children talk before they can speak / by David Lewis. — London : Souvenir Press, 1978.
288 p. : ill. ; 23 cm. GB***
ISBN (invalid) 0-285-62358-2 : £4.95
1. Infant psychology. 2. Child psychology. 3. Children—Language. 4. Nonverbal communication. I. Title.
BF723.I6L415 1978 154.4'23 79-303954
79 MARC

Lipsitt, Lewis Paeff, 1929-
Child development / Lewis P. Lipsitt, Hayne W. Reese ; Lyle E. Bourne, Jr, consulting editors. — Glenview, Ill. : Scott, Foresman, [1978] c1979.
152 p. : ill. ; 23 cm.
Bibliography: p. 137-147.
Includes indexes.
ISBN 0-673-05009-2
1. Child psychology. I. Reese, Hayne Waring, 1931- joint author. II. Bourne, Lyle Eugene, 1932- . III. Title.
BF721.L563 155.4 78-16009
78 MARC

Lorrance, Arleen, 1939-
The Love Project / by Arleen Lorrance. — Rev. ed. — San Diego, Calif. : LP Publications, 1978, c1972.
103 p. : ill. ; 22 cm.
ISBN 0-916192-14-8 : $3.50
1. Socially handicapped children—Education—New York (State)—Brooklyn. 2. Child psychology. 3. Social choice. I. Title.
LC4093.B76L67 1978 371.9'67'0974723 78-15162
78 MARC

Meyer, William J
Child psychology, a developmental perspective / William J. Meyer, Jerome B. Dusek. — Lexington, Mass. : Heath, c1979.
xiv, 658 p. : ill. ; 25 cm.
Bibliography: p. 595-635.
Includes indexes.
ISBN 0-669-88971-7
1. Child psychology. I. Dusek, Jerome B., joint author. II. Title.
BF721.M489 155.4 78-69698
79 MARC

Montagner, Hubert.
L'enfant et la communication : comment des gestes, des attitudes, des vocalisations deviennent des messages / Hubert Montagner. — [Paris] : Pernoud/Stock, c1978.
402 p. : ill. ; 22 cm. F***
Bibliography: p. 379-[396]
ISBN 2-234-00289-3 : 49.00F
1. Interpersonal communication. 2. Child psychology. I. Title.
BF723.C57M66 155.4'18 79-341410
79 MARC

CHILD PSYCHOLOGY (Continued)

Mussen, Paul Henry.
Child development and personality / Paul Henry Mussen, John Janeway Conger, Jerome Kagan. — 5th ed. — New York : Harper & Row, c1979.
xi, 579 p. : ill. ; 24 cm.
Includes bibliographies and indexes.
ISBN 0-06-044693-5
1. Child psychology. 2. Personality. I. Conger, John Janeway, joint author. II. Kagan, Jerome, joint author. III. Title.
[DNLM: 1. Child development. WS105.3 M989c]
BF721.M88 1979 155.4'18 78-9840
78 MARC

Mussen, Paul Henry.
The psychological development of the child / Paul Mussen. — 3d ed. — Englewood Cliffs, N.J. : Prentice-Hall, c1979.
xiv, 127 p. : ill. ; 24 cm. — (Prentice-Hall foundations of modern psychology series)
Includes bibliographical references and index.
ISBN 0-13-732420-0 : $10.95
1. Child psychology. I. Title.
[DNLM: 1. Child development. WS105.3 M989p]
BF721.M89 1979 155.4 78-26384
78 MARC

Nakamura, Shigeru, 1912–
(Jidō-shinrigaku nyūmon)
児童心理学入門 / 中村秀. — 東京：有信堂, 昭和29 [1954]
2, 4, 200 p. : ill. ; 18 cm. — (有信堂文庫)
Series romanized: Yūshindō bunko.
Bibliography: p. 189–191.
Includes indexes.

1. Child psychology. I. Title.

BF724.N28 78–816591

J

Naumov, Luk'ı͡an Aleksandrovich.
(Za godom god)
За годом год / Л. Наумов. — Москва : Знание, 1978.
95 p. : ill. ; 16 cm. — (Народный университет : Педагогический факультет ; № 4, 1978) USSR 78
0.14 rub
1. Child psychology. I. Title. II. Series: Narodnyĭ universitet : Pedagogicheskiĭ fakul'tet ; 1978, 4.
AS261.N24 1978 no. 4 78–422231
[LB1051]

Neri, Antonio.
Il bambino trasparente : l'analisi dell'istituzione educativa attraverso il linguaggio del corpo / Antonio Neri. — Milano : Emme, 1977.
168 p. ; 20 cm. — (Scienze dell'educazione) (Il Puntoemme ; 73)
It78-Sept
L2500
1. Child psychology. 2. Children—Language. 3. Nonverbal communication (Psychology) 4. Sex instruction for mentally handicapped children. I. Title.
BF721.N45 79-349129
*79 MARC

Newson, John, 1925-
Toys & playthings in development and remediation / John & Elizabeth Newson. — 1st American ed. — New York : Pantheon Books, c1979.
279 p. : ill. ; 22 cm.
Bibliography: p. [273]-275.
Includes index.
ISBN 0-394-42830-7 : $10.95. ISBN 0-394-73717-2 pbk. : $4.95
1. Play. 2. Child psychology. 3. Handicapped children—Care and treatment. 4. Play therapy. I. Newson, Elizabeth, 1929- joint author. II. Title.
[DNLM: 1. Child development. 2. Play and playthings. WS105.5.P5 N558t]
BF717.N48 1979 155.4'1 78-20419
79 MARC

Nguyễn Khắc Viên.
Ngày thơ : một số suy nghĩ về tâm lý và giáo dục trẻ em / Nguyễn Khắc Viên. — Hà-nội : Phụ Nữ, 1974.
148 p. ; 19 cm.
1. Child psychology. 2. Education. I. Title.
BF721.N518 78-984326
79 MARC

Ōnishi, Kemmei, 1911–
(Jidō-shinrigaku)
児童心理学 / 大西憲明編著. — 東京：ミネルヴァ書房, 昭和27 [1952]
2, 280 p. ; 19 cm. — (教育科学選書 ; 2)
Bibliography: p. 265–280.

1. Child psychology. I. Title. II. Series: Kyōiku kagaku sensho (Tokyo, 1952–) ; 2.

BF724.O53 78–816043

(Oyako kankei no shinri)
親子関係の心理 / 柏木恵子 ... [et al.] 著. — 東京：有斐閣, 1978.
vi, 179, 5 p. : ill. ; 18 cm. — (有斐閣新書) Ja 78–5923
Series romanized: Yūhikaku shinsho.
Includes bibliographies and index.
¥480

1. Mother and child. 2. Child psychology. I. Kashiwagi, Keiko, 1932–

HQ755.85.O93 78–815630

Papalia, Diane E
A child's world : infancy through adolescence / Diane E. Papalia, Sally Wendkos Olds. — 2d ed. — New York : McGraw-Hill, c1979.
xi, 626 p. : ill. ; 25 cm.
Bibliography: p. 580-614.
Includes index.
ISBN 0-07-048458-9
1. Child psychology. 2. Children—Management. I. Olds, Sally Wendkos, joint author. II. Title.
HQ772.P23 1979 301.43'1 78-17198
78 MARC

Paraná, Brazil (State). Universidade Federal
Percepção do espaço urbano / UNESCO, Organização da O.N.U. para Educação, Ciência e Cultura, Universidade Federal do Paraná, Instituto de Pesquisa e Planejamento Urbano de Curitiba. — [Curitiba] : A Universidade, [1978]
150 leaves : ill. ; 30 cm.
"O presente documento foi elaborado pelas equipes da UFPR e IPPUC/PMC, como relatório do projeto' Percepção do espaço urbano/centro,' para o V Simpósio Brasileiro de Psicologia Cognitiva."
Bibliography: leaf 148-150.
1. Geographical perception. 2. Cities and towns—Brazil. 3. Environmental psychology. 4. Child psychology. 5. Curitiba—Description. I. United Nations Educational, Scientific and Cultural Organization. II. Instituto de Pesquisa e Planejamento Urbano de Curitiba. III. Title.
GF125.P37 1978 79-102086
79 MARC

Peck, Johanne.
Young children's behavior : implementing your goals / by Johanne Peck. — Atlanta : Humanics Ltd., c1978.
73 p. : ill. ; 28 cm.
Includes bibliographies.
ISBN 0-89334-015-4
1. Behavior modification. 2. Child psychology. I. Title.
BF637.B4P52 155.4 78-56624
79 MARC

Pereira, Orlindo Manuel Graça Gouveia.
Desenvolvimento psicológico da criança / Orlindo Gouveia Pereira, Jorge, Correia Jesuíno, contribuição de Luís Joyce-Moniz. — Lisboa : Moraes Editores, [1978- , c1976.
v. ; 21 cm. — (Colecção Psicologia e pedagogia)
Includes bibliographies.
CONTENTS: v. 1. História, métodos e factores.
1. Child psychology. 2. Genetic psychology. I. Jesuíno, Jorge Correia, joint author. II. Joyce-Moniz, Luís, joint author. III. Title.
BF721.P387 155.4 79-342761
79 MARC

Piaget, Jean, 1896-
The child's construction of reality / Jean Piaget ; translated by Margaret Cook. — London : Routledge & Kegan Paul, c1954, 1976 printing.
xiii, 386 p. ; 23 cm.
Translation of La construction du réel chez l'enfant.
ISBN 0-7100-1952-1
1. Child development. 2. Child psychology. I. Title.
LB1131.P48513 1976 155.4 78-314319
78 MARC

Plummer, Yolande.
En route pour la vie : étude faite sur le développement de la personnalité humaine / Yolande Plummer. — [Port-au-Prince : Presses nationales d'Haïti, 19-?]
111 p. ; 21 cm.
1. Child psychology. I. Title.
BF721.P584 79-101828
79 MARC

Problémy psychického vývoje a osobnosti / [autor kol. ; předml. Vilém Chmelař]. — 1. vyd. — Praha : Academia, 1978.
343, [1] p., [2] leaves of plates : ill. ; 24 cm. Cz78
"Univ. prof. PhDr. Vilému Chmelařovi, DrSc., členu korespondentu ČSAV, k 85. narozeninám."
Summary in English ; table of contents also in English and Russian.
Bibliography: p. 303-313.
Includes indexes.
Kčs51.00
1. Child psychology. 2. Personality. I. Chmelař, Vilém, 1892-
BF721.P75 79-364573
*79 MARC

Przetacznikowa, Maria.
Podstawy rozwoju psychicznego dzieci i młodzieży / Maria Przetacznikowa. — Wyd. 2. — Warszawa : Wydaw-a Szkolne i Pedagogiczne, 1978 (Bydg. : WSiP)
xvii, 213 p. : ill. ; 20 cm. — (Biblioteka psychologiczna ; 1) P79-3177
Bibliography: p. 207-208.
Includes index.
ISBN 8302001120 : zł35.00
1. Child psychology. I. Title. II. Series.
BF721.P767 1978 79-381775
79 MARC

Przetacznikowa, Maria.
Psychologia wychowawacza, społeczna i kliniczna : materiały pomocnicze dla zakładów kształcenia nauczycieli przedszkoli / Maria Przetacznikowa, Grażyna Makiełło-Jarża. — Wyd. 1. — Warszawa : Wydaw-a Szkolne i Pedagogiczne, 1977 (Łódź : WSiP)
342, [2] p. : ill. ; 22 cm. P78-1529
Bibliography: p. 322-[325]
Includes index.
zł64.00
1. Educational psychology. 2. Child psychology. 3. Education, Preschool. I. Makiełło-Jarża, Grażyna, joint author. II. Title.
LB1051.P7274 79-382473
79 MARC

Pumroy, Donald Keith, 1925-
Modern childrearing : a behavioral approach / Donald K. Pumroy, Shirley S. Pumroy. — Chicago : Nelson-Hall, c1978.
viii, 148 p. : ill. ; 24 cm.
Bibliography: p. 145.
Includes index.
ISBN 0-88229-185-8
1. Children—Management. 2. Child psychology. I. Pumroy, Shirley Ann Spence, 1928- joint author. II. Title.
HQ769.P843 649'.1 77-26964
77 MARC

Robeck, Mildred Coen.
Infants and children : their development and learning / Mildred C. Robeck. — New York : McGraw-Hill, c1978.
x, 534 p. : ill. ; 25 cm.
Includes bibliographical references and index.
ISBN 0-07-053108-0 : $12.95
1. Child psychology. 2. Infant psychology. 3. Learning, Psychology of. I. Title.
BF721.R613 155.4'13 77-25844
77[r79]rev MARC

Sarnoff, Charles A
Latency / Charles Sarnoff. — New York : J. Aronson, c1976.
xii, 402 p. ; 24 cm. — (Classical psychoanalysis and its applications)
Bibliography: p. [387]-394.
Includes index.
ISBN 0-87668-233-6
1. Child psychotherapy. 2. Child psychology. I. Title.
RJ504.S27 618.9'28'914 75-34707
78 MARC

Schlehofer, Jo.
Joy in parenting : parenting skills with preschoolers through adolescents / Jo Schlehofer. — New York : Paulist Press, c1978.
194 p. ; 23 cm.
Bibliography: p. 186-191.
ISBN 0-8091-2125-5 : $3.95
1. Parenting—United States. 2. Child psychology. 3. Parent and child—United States. I. Title.
HQ755.8.S34 649'.1 78-58951
79 MARC

Shaffer, David R
Social and personality development / David R. Shaffer. — Monterey, Calif. : Brooks/Cole Pub. Co., c1979.
xiii, 624 p. : ill. ; 24 cm.
Includes bibliographies and indexes.
ISBN 0-8185-0289-4 : $13.95
1. Child psychology. 2. Personality. 3. Socialization. I. Title.
BF721.S469 155.4'18 78-12032
78 MARC

Shibata, Kaoru, 1934–
(Dōtokusei no hattatsu to kyōiku)
道徳性の発達と教育：人格形成への心理学的提言 / 柴田薫. — 仙台：金港堂出版, 1975.
191 p. : ill. ; 21 cm. Ja 76–3266
Cover title.
Includes bibliographical references and index.
¥1200

1. Moral education. 2. Child psychology. I. Title.

LC273.J3S5 78–811002

Shinagawa, Fujirō, 1916–
(Kyōdai no shinri to michibikikata)
きょうだいの心理と導き方 / 品川不二郎. — 東京：あすなろ書房, 1976.
158 p. ; 19 cm. Ja 76–22755
Bibliography: p. 158.
¥750

1. Child psychology. 2. Brothers and sisters. 3. Education of children. I. Title.

LB1117.S46 78–812009

Singer, Dorothy G
A Piaget primer : how a child thinks / Dorothy G. Singer & Tracey A. Revenson. — New York : New American Library, 1978.
x, 148 p. : ill. ; 21 cm. — (A Plume book)
Bibliography: p. 135-143.
Includes index.
ISBN 0-452-25189-3 : $3.95
1. Cognition in children. 2. Child psychology. 3. Piaget, Jean, 1896- I. Revenson, Tracey A., joint author. II. Title.
BF723.C5S6 155.4'13 78-61227
79 MARC

CHILD PSYCHOLOGY (Continued)

Stone, Lawrence Joseph, 1912-
Childhood & adolescence : a psychology of the growing person / L. Joseph Stone, Joseph Church ; in collaboration with Evelyn Raskin & Alexandria Church ; [cover photo by Elizabeth Crews]. — — 4th ed. — New York : Random House, [1979]
xiv, 590, xx p. : ill. ; 25 cm.
Bibliography: p. 541-564.
Includes index.
ISBN 0-394-32086-7 : $14.95
1. Child psychology. 2. Adolescent psychology. I. Church, Joseph, joint author. II. Title.
BF721.S82 1979 155.4 78-10730
78 MARC

Takahashi, Seiki, 1911-
(Jidō-shinrigaku)
児童心理學：成長と發達を含む / 高橋省己著. — 京都：関書院, 昭和29 [1954]
259 p. : ill. ; 19 cm.
Includes bibliographies.

1. Child psychology. 2. Infant psychology. I. Title.
BF721.T26 78-815443

Takizawa, Takehisa, 1931-
(Kodomo no shikō to ninshiki)
子どもの思考と認識：新しい発達心理学の視点から / [著者 滝沢武久]. — 東京：童心社, 1977
238 p. : 19 cm. — (子どもの文化叢書) Ja 77-10729
Bibliography: p. 237-238.
¥800
1. Child psychology. 2. Thought and thinking. 3. Cognition in children. I. Title. II. Series: Kodomo no benka sosho.
LB1117.T373 79-805247

Thomas, Robert Murray.
Comparing theories of child development / R. Murray Thomas. — Belmont, Calif. : Wadsworth Pub. Co., c1979.
xiv, 507 p. : ill. ; 24 cm.
Bibliography: p. 481-494.
Includes indexes.
ISBN 0-534-00591-8
1. Child psychology. I. Title.
BF721.T456 155.4'01 78-9051
78 MARC

Tsukada, Kōichi.
(Jidō-shinrigaku)
児童心理学 / 塚田紘一. — 東京：めいせい出版, 1977.
384 p. : ill. ; 22 cm. Ja 78-8956
Includes bibliographies and index.
¥2000

1. Child psychology. I. Title.
BF721.T753 78-809452

Ushijima, Yoshitomo, 1906-
(Shōgakusei no shinri)
小學生の心理 / 牛島義友著. — 東京：巖松堂書店, 昭和27 [1952]
2, 5, 425, 7 p. ; 22 cm.
Colophon inserted.
Includes bibliographical references and index.

1. Child psychology. 2. School children. I. Title.
LB1115.U83 78-816237

Várkonyi, Zsuzsa F
Tájékozottság és kompetencia : óvodáskorúak személyiséglélektani vizsgálata / F. Várkonyi Zsuzsa. — Budapest : Akadémiai Kiadó, 1978.
127 p. ; 19 cm. — (Pszichológia a gyakorlatban ; 36. köt) H***
Bibliography: p. 114-118.
ISBN 9630514990 : 15.00Ft
1. Child psychology. 2. Cognition in children. 3. Learning, Psychology of. I. Title. II. Series.
LB1105.P7 Köt. 36 79-376437
[BF721] 79 MARC

Viklund, Monica, 1943-
Barnen behöver oss / Monica Viklund ; [teckningar o. omsl., Helga Henschen]. — Stockholm : Socialstyr. ; Lund : Liber-Läromedel, 1978.
157 p. : ill., tables ; 21 cm. S78-17
Bibliography: p. 153-157.
ISBN 9140045900
1. Children—Sweden. 2. Child development. 3. Child psychology. 4. Child welfare—Sweden. I. Title.
HQ792.S8V54 78-395310
79 MARC

La Voglia di conoscere / a cura di Lucia Lumbelli ; contributi di C. Cecchini ... [et al.]. — Milano : F. Angeli, [1979]
212 p. ; 22 cm. — (Condizionamenti educativi ; 28) It79-Feb
Includes bibliographical references.
L5000
1. Motivation in education. 2. Child Psychology. 3. Lumbelli, Lucia, 1937- I. Lumbelli, Lucia, 1937- . II. Cecchini, Cristina.
LB1065.V63 79-359264
*79 MARC

Watson, Robert Irving, 1909-
Psychology of the child and the adolescent / Robert I. Watson, Henry Clay Lindgren. — 4th ed. — New York : Macmillan, c1979.
xvi, 624 p. : ill. ; 24 cm.
First-3d ed. published under title: Psychology of the child.
Bibliography: p. 585-610.
Includes indexes.
ISBN 0-02-424600-X
1. Child psychology. I. Lindgren, Henry Clay, 1914- joint author. II. Title.
BF721.W33 1979 155.4 78-3423
78 MARC

Weisberger, Eleanor.
Your young child and you : how to manage growing-up problems in the years from one to five / Eleanor Weisberger ; foreword by Benjamin M. Spock. — New, rev. ed. — New York : Dutton, c1979.
xvii, 172 p. ; 21 cm.
"A Sunrise book."
Includes index.
ISBN 0-87690-329-4 : $4.95 ($6.50 Can)
1. Children—Management. 2. Child psychology. I. Title.
HQ769.W43 1979 649'.123 78-22623
79 MARC

Weiss, Rudolf, 1932-
Grundfragen der Entwicklungspsychologie / Rudolf Weiss, mit einem Beitrag v. Hansjörg Walter. — Innsbruck : Univ. Verl. Wagner, c1978.
128 p. ; 24 cm. — (His Pädagogik für Lehrer in der Aus- und Weiterbildung ; T. 5) Au79-4
At head of title: Aus dem Institut für Erziehungswissenschaften der Universität Innsbruck.
"Kurzfassung einer Vorlesung, die im Wintersemester 1978/79 an der Universität Innsbruck gehalten wurde."
Bibliography: p. 123-128.
ISBN 3-7030-0060-0 : S75.00
1. Child psychology. 2. Developmental psychology. I. Walter, Hansjörg, joint author. II. Title. III. Series.
BF721.W36 79-361941
*79 MARC

White, Burton L 1929-
Experience and environment: major influences on the development of the young child [by] Burton L. White and Jean Carew Watts, with Itty Chan Barnett [and others] Englewood Cliffs, N.J., Prentice-Hall [1973-78]
2 v. illus. 24 cm. (Prentice-Hall series in early childhood development)
Vol. 2 by B. L. White, with B. T. Kaban and others.
Vol. 1 lacks series statement.
Bibliography: v. 1, p. 249-250; v. 2, p. 201-204.
ISBN 0-13-294843-5 (v. 1)
1. Child psychology. I. Carew, Jean V., 1936- joint author. II. Title.
BF721.W43 155.4'13 72-10354
72[r79]rev MARC

Wickes, Frances Gillespy, 1875-1967.
The inner world of childhood / Frances G. Wickes ; introd. by C. G. Jung. — Englewood Cliffs, N.J. : Prentice-Hall, [1978] c1966.
xxiv, 280 p. ; 20 cm. — (A Spectrum book ; S-532)
Original ed. published by New American Library, New York.
Includes index.
ISBN 0-13-466045-5 : $10.95. ISBN 0-13-466037-4 pbk. : $5.95
1. Child psychology. 2. Parent and child. I. Title.
BF721.W45 1978 155.4 78-13307
78 MARC

Wolfenstein, Martha, 1911-1976.
Children's humor : a psychological analysis / byMarthaWolfenstein ; forwaord by Alan Dundes. — Bloomington : Indiana University Press, 1978.
224 p. ; 22 cm.
Reprint of the 1954 ed. published by Free Press, Glencoe, Ill.
ISBN 0-253-31351-1. ISBN 0-253-20206-X pbk
1. Wit and humor—Psychology. 2. Child psychology. I. Title.
[PN6149.P5W6 1978] 155.4'1 77-15760
77 MARC

Yokochi, Kiyoshi, 1922-
(Takumashii Nihon no yōji)
たくましい日本の幼児 / 横地清著. — 東京：三一書房, 1977.
217 p. : ill. ; 18 cm. — (三一新書 ; 858) Ja 77-14703
Series romanized: San'ichi shinsho.
¥480

1. Education, Preschool—Japan. 2. Education, Preschool—1965- 3. Child psychology. I. Title.
LB1140.2.Y629 79-806761

Yussen, Steven R
Child development : an introduction / Steven R. Yussen, John W. Santrock. — Dubuque, Iowa : W. C. Brown Co., c1978.
xxiii, 647 p. : ill. ; 24 cm.
Bibliography: p. 600-627.
Includes indexes.
ISBN 0-697-06625-8
1. Child psychology. 2. Child development. I. Santrock, John W., joint author. II. Title.
BF721.Y85 155.4 77-86177
79 MARC

Żebrowska, Maria.
Psychologia rozwojowa dzieci i młodzieży / pod red. Marii Żebrowskiej ; [aut. Zofia Babska ... et al.]. — Wyd. 8. — Warszawa : Państ. Wydaw. Naukowe, 1977, c1975 (Opole : OZGJL)
879, [1] p. : ill. ; 24 cm. P78-2290
Includes bibliographies and indexes.
zł120.00
1. Child psychology. 2. Adolescent psychology. 3. Child development. I. Babska, Zofia. II. Title.
BF721.Z4 1977 78-388491
79 MARC

—ADDRESSES, ESSAYS, LECTURES

Children and the environment / edited by Irwin Altman and Joachim F. Wohlwill. — New York : Plenum Press, c1978.
xv, 300 p. : ill. ; 24 cm. — (Human behavior and environment ; v. 3)
Includes bibliographies and index.
ISBN 0-306-40090-1
1. Child psychology—Addresses, essays, lectures. 2. Environmental psychology—Addresses, essays, lectures. I. Altman, Irwin. II. Wohlwill, Joachim F.
BF353.H85 vol. 3 301.31 s 78-13511
78 MARC

Early cognitive development / edited by John Oates. — New York : Wiley, c1979.
396 p. : ill. ; 22 cm.
"A Halsted Press book."
Includes bibliographies and index.
ISBN 0-470-26431-4
1. Cognition in children—Addresses, essays, lectures. 2. Infant psychology—Addresses, essays, lectures. 3. Child psychology—Addresses, essays, lectures. I. Oates, John, 1946- .
BF723.C5E2 155.4'22 78-17694
78 MARC

Fukaya, Masashi, 1933-
(Gendaikko o sodateru)
現代っ子を育てる 深谷昌志：[編集 富山県教育委員会]. — 富山：同委員会, 昭和52 [1977]
55 p. : 18 cm. — (生涯教育新書 : 13) Ja***

1. Education of children—Japan—Addresses, essays, lectures. 2. Child psychology—Addresses, essays, lectures. I. Title. II. Series: Shōgai kyōiku shinsho : 13.
LB1115.F845 79-806779

(Jidō no shinsei to nōryoku kensa)
兒童の心性と能力檢査 / 波多野完治, 牛島義友編. — 東京：巖松堂, 昭和24 [1949]
6, 2, 244 p. : ill. ; 21 cm. — (教育心理學研究 ; 1)
Colophon inserted.
Includes bibliographical references.

1. Educational psychology—Addresses, essays, lectures. 2. Child psychology—Addresses, essays, lectures. 3. Educational tests and measurements—Addresses, essays, lectures. I. Hatano, Kanji, 1905- II. Ushijima, Yoshitomo, 1906- III. Series: Kyōiku shinrigaku kenkyū : 1.
LB1051.J48 79-802496

Kagan, Jerome.
The growth of the child : reflections on human development / Jerome Kagan. — 1st ed. — New York : Norton, c1978.
283 p. ; 24 cm.
Includes bibliographical references.
ISBN 0-393-01173-0
1. Child psychology—Addresses, essays, lectures. I. Title.
BF721.K15 1978 155.4 78-16008
78 MARC

Lewis, Michael, 1937 (Jan. 10)-
The origins of fear. Edited by Michael Lewis and Leonard A. Rosenblum. New York, Wiley [1974]
xiii, 284 p. illus. 23 cm. (The Origins of behavior, v. 2)
"A Wiley-Interscience publication."
Derived from papers presented and discussed at a conference sponsored by Educational Testing Service, Princeton, N.J.
Includes bibliographies.
ISBN 0-471-53338-6
1. Fear—Congresses. 2. Child psychology—Congresses. I. Rosenblum, Leonard A., joint author. II. Educational Testing Service. III. Title. IV. Series.
[DNLM: 1. Fear—Congresses. W1OR687 v. 2 / WM178 O69 1974]
BF723.F4L48 1974 155.4'18 74-9565
74[r79]rev MARC

CHILD PSYCHOLOGY

—ADDRESSES, ESSAYS, LECTURES (Continued)

Luriia, Alexandr Romanovich, 1902-
The selected writings of A. R. Luria / edited with an introd. by Michael Cole. — White Plains, N.Y. : M. E. Sharpe, c1978.
xxii, 351 p. : ill. ; 24 cm.
Includes bibliographies.
ISBN 0-87332-127-8
1. Child psychology—Addresses, essays, lectures. 2. Neuropsychology—Addresses, essays, lectures. I. Cole, Michael, 1938-
BF721.L82 155.4'08 78-64342
79 MARC

—BIBLIOGRAPHY

Výskumný ústav detskej psychológie a patopsychológie.
Research Institute of Child Psychology and Pathopsychology, 1964-1968 / contributors, Anton Jurovský ... et al. ; editor, Lýdia Jančeková ; translated by Marta Krallová ; cover designed by Emil Bačik. — [Bratislava] : The Institute, [between 1968 and 1971]
74 p. ; 25 cm.
Statement of authorship and imprint stamped on t.p.
"Bibliographical list of publications of the Institute's research workers in the years 1964-1968": p. 41-62.
1. Child psychology—Bibliography. 2. Child psychopatholgy—Bibliography. 3. Výskumný ústav detskey psychológie a patopsychológie—History. I. Jurovský, Anton. II. Jančeková, Lýdia.
Z7204.C5V97 1968 150'.7'204373 79-311309
[BF721] 79[r79]rev MARC

—COLLECTED WORKS

Lecturas de psicología del niño / compilación de Juan Delval. — Madrid : Alianza, c1978.
2 v. : ill. ; 23 cm. — (Alianza universidad : Textos ; 5-6) Sp***
Includes bibliographies.
CONTENTS: v. 1. Las teorías, los métodos y el desarrollo temprano.—v. 2. El desarrollo cognitivo y afectivo del niño y del adolescente.
ISBN 8420689971
1. Child psychology—Collected works. I. Delval, Juan A., 1941-
BF721.L437 155.4'08 78-372963
78[r79]rev MARC

Zazzo, René.
Psychologie de l'enfant et méthode génétique : travaux et conférences / René Zazzo. — [Nouv. éd.]. — Neuchâtel ; Paris : Delachaux & Niestlé, [1977]
316 p. : ill. ; 21 cm. — (His Conduites et conscience ; 1) (Actualités pédagogiques et psychologiques) Sw77-3284
Includes bibliographical references.
ISBN 2-603-00117-5 : 36.00
1. Child psychology—Collected works. I. Title. II. Series. III. Series: Actualités pédagogiques et psychologiques.
BF721.Z35 1977 155.4 78-399805
*79 MARC

—CONGRESSES

International Interdisciplinary Conference on Piagetian Theory and Its Implications for the Helping Professions, 7th, University of Southern California, 1977.
Piagetian theory and its implications for the helping professions : emphasis, social work, and psychological services : proceedings : seventh interdisciplinary conference / co-sponsored by University Affiliated Program, Childrens Hospital of Los Angeles and the University of Southern California Schools of Social Work and Education ; editors, Rama Weizmann ... [et al.]. — Los Angeles : Order from Mail Order Dept., Bookstore, University of Southern California, c1978-
v. : ill. ; 22 cm.
Bibliography: v. 1, p. 403-432.
$12.00
1. Child psychology—Congresses. 2. Mentally ill children—Education—Congresses. 3. Piaget, Jean, 1869- I. Weizmann, Rama. II. Los Angeles. Childrens Hospital. University Affiliated Program. III. Los Angeles. University of Southern California. School of Social Work. IV. Los Angeles. University of Southern California. School of Education. V. Title.
BF721.I52 1977 155.4'13 78-102941
79 MARC

The Normally sick child / cited by Jan van Eys. — Baltimore : University Park Press, c1979.
xiii, 188 p. ; 23 cm.
"Proceedings of a workshop held by the Department of Pediatrics, The University of Texas System Cancer Center, M. D. Anderson Hospital and Tumor Institute, Houston, Texas, on May 6 and 7, 1978."
Includes bibliographies and index.
ISBN 0-8391-1382-X
1. Tumors in children—Psychological aspects—Congresses. 2. Child pscyhology—Congresses. 3. Sick children—Psychology—Congresses. I. Van Eys, Jan. II. Anderson Hospital and Tumor Institute, Houston, Tex. Dept. of Pediatrics.
RC281.C4N67 618.9'29'940019 78-26440
78 MARC

—HANDBOOKS, MANUALS, ETC.

Shinagawa, Takako, 1922-
(Kodomo no shinri shitsuke jiten)
子どもの心理しつけ事典 / 品川孝子著. — 東京：あすなろ書房, 1977.
471 p. ; 21 cm. Ja 77B-5563
Colophon inserted.
¥2500
1. Child psychology—Handbooks, manuals, etc. 2. Children—Management—Handbooks, manuals, etc. I. Title.
LB1115.S58 78-813366

Stein, Sara Bonnett.
A child goes to school / Sara Bonnett Stein ; photos. by Don Connors. — 1st ed. — Garden City, N.Y. : Dolphin Books, 1978.
242 p. : ill. ; 27 cm.
SUMMARY: Separate discussions for youngsters and parents about the problems of starting school and growing up.
ISBN 0-385-12953-X : $12.95. ISBN 0-385-12952-1 pbk. : $5.95
1. Readiness for school—Juvenile literature. 2. First grade (Education)—Juvenile literature. 3. Child psychology—Handbooks, manuals, etc. [1. Kindergarten. 2. Child psychology] I. Connors, Don. II. Title.
LB1132.S74 372.1 77-26528
78 MARC

—HISTORY

Chiari, Silvano.
Introduzione storica alla psicologia dell'età evolutiva nei suoi rapporti col problema educativo / Silvano Chiari. — Roma : Bulzoni, 1976.
118 p. : ill. ; 24 cm. — (Contributi monografici dell'Istituto di psicologia, Facoltà di magistero dell'Università di Siena ; 1) It77-Mar
Bibliography: p. 114-115.
L2500
1. Child psychology—History. 2. Educational psychology—History. I. Title. II. Series: Siena. Università. Istituto di psicologia. Contributi monografici dell'Istituto di psicologia, Università di Siena ; 1.
BF721.C475 79-369350
*79 MARC

—METHODOLOGY

see also Psychological tests for children

Vasta, Ross.
Studying children : an introduction to research methods / Ross Vasta. — San Francisco : W. H. Freeman, c1979.
xi, 212 p. : ill. ; 24 cm. — (A Series of books in psychology)
Includes bibliographies and indexes.
ISBN 0-7167-1067-6. ISBN 0-7167-1068-4 pbk.
1. Child psychology—Methodology. 2. Child psychology—Research. I. Title.
BF722.V37 155.4'072 78-25941
78 MARC

—PERIODICALS

(al-Usrah wa-al-ṭufūlah)
الاسرة والطفولة.
[القاهرة، اللجنـــة العامة لتدريـــب العاملين فى ميادين الاسرة والطفولة]
v. in ill. 20 cm. monthly.
1. Child study — Periodicals. 2. Family — Periodicals. I. al-Lajnah al-ʻĀmmah li-Tadrīb al-ʻĀmilīn fī Mayādīn al-Ṭufūlah wa-al-Usrah.
HQ767.8.U87 78-645729
(MARC-S)

—RESEARCH

Bronfenbrenner, Urie, 1917-
The ecology of human development : experiments by nature and design / Urie Bronfenbrenner. — Cambridge, Mass. : Harvard University Press, 1979.
xv, 330 p. ; 24 cm.
Bibliography: p. 299-319.
Includes index.
ISBN 0-674-22456-6
1. Child psychology—Research. I. Title.
[DNLM: 1. Human development. BF713 B869e]
BF722.B76 155.4 78-27232
78 MARC

Vasta, Ross.
Studying children : an introduction to research methods / Ross Vasta. — San Francisco : W. H. Freeman, c1979.
xi, 212 p. : ill. ; 24 cm. — (A Series of books in psychology)
Includes bibliographies and indexes.
ISBN 0-7167-1067-6. ISBN 0-7167-1068-4 pbk.
1. Child psychology—Methodology. 2. Child psychology—Research. I. Title.
BF722.V37 155.4'072 78-25941
78 MARC

—RESEARCH—DENMARK

Denmark. Socialstyrelsen.
Forskning, forsøgs- og udviklingsarbejde vedrørende børn i førskolealderen / [Socialstyrelsen]. — [København] : Socialstyrelsen, 1977.
62, [12] p. ; 30 cm. D77-44
Cover title.
By G. Berg, I. Bistrup, and F. Hansen.
Includes bibliographical references and index.
1. Children—Research—Denmark. 2. Child psychology—Research—Denmark. 3. Day care centers—Research—Denmark. I. Berg, Gertrud. II. Bistrup, Inge. III. Hansen, Finn, 1946- IV. Title.
HQ767.85.D46 1977 79-343240
79 MARC

—VOCATIONAL GUIDANCE

Wester, William C
Child psychology as a career. [Sound recording] [By] W. C. Wester. New York, J. Norton Publishers [1974?] 15030.
1 reel. 3¾ ips. 5 in.
Title from publisher's catalog.
Recorded in 1968.
Originally issued by McGraw-Hill, New York, 75430, in series: Sound seminars.
Duration: 20 min.
Also issued in cassette.
SUMMARY: Presents material on a career in guidance.
1. Child psychology—Vocational guidance. I. Title. II. Series: Sound seminars. [Sound recording]
[BF722] 78-740986

Williams, Phillip.
Children and psychologists / [by] Phillip Williams. — London : Hodder and Stoughton, 1977.
viii, 88 p. : ill., form, map ; 22 cm. GB77-23274
Expanded version of the author's National Children's Home convocation lecture, 1976.
Bibliography: p. 81-86.
Includes index.
ISBN 0-340-21942-4 : £1.75
1. Child psychology—Vocational guidance. I. Title.
BF721.W48 155.4 79-301349
79 MARC

—JAPAN

Abe, Susumu, 1930-
(Kodomo o shiranai oyatachi)
子どもを知らない親たち：子どもたちに何を教えるか / 阿部進. — 東京：日本経済通信社, 昭和51 [1976]
220 p. : ill. ; 19 cm. — (NKT ブックス) Ja 76-10019
Series romanized: NKT bukkusu.
¥680
1. Child psychology—Japan. 2. Adolescent psychology—Japan. 3. Students—Japan. 4. Parent and child—Japan. 5. Teacher-student relationships. I. Title.
LB1117.A18 79-809194

(Jidō no kyōiku to bunka)
児童の教育と文化 / 小川太郎編. — 京都：ミネルヴァ書房, 1976.
9, 289, 7 p. : ill. ; 22 cm. — (児童問題講座 ; 2) Ja 76-18138
Bibliography: p. 286-289.
Includes index.
¥1800
1. Education—Japan—History. 2. Educational sociology—Japan—History. 3. Child psychology—Japan. I. Ogawa, Tarō, 1907- II. Series: Jidō mondai kōza ; 2.
LA1311.J52 79-809019

—NETHERLANDS—CASE STUDIES

School en toekomst : beelden van de ontwikkelingsgang van kinderen gedurende de puberteit / door S. Wiegersma ... [et al.] ; met een voorwoord van J. Koekebakker ; in opdracht van het Nederlands Institut voor Praeventieve Geneeskunde. — Haarlem : De Toorts, [1963]
250 p. ; 24 cm.
Includes indexes.
1. Children—Netherlands—Case studies. 2. Child development—Case studies. 3. Child psychology—Netherlands—Case studies. I. Wiegersma, S.
HQ792.N4S43 79-344405
79 MARC

CHILD PSYCHOLOGY AS A PROFESSION
see Child psychology—Vocational guidance

CHILD PSYCHOPATHOLOGY

see also Autism; Child psychiatry; Depression in children; Emotional problems of children; Mentally handicapped children; Mentally ill children; Psychoses in children; Schizophrenia in children; Violence in children

Beauchesne, Hervé.
Psychopathologie de l'enfant et de l'adolescent : étude de cas / Hervé Beauchesne. — ⌐Paris⌐ : Presses universitaires de France, c1978.
178 p. ; 21 cm. — (Le Psychologue ; 71) F***
Includes bibliographical references.
45.00F
1. Child psychopathology. 2. Child psychopathology—Cases, clinical reports, statistics. I. Title.
RJ499.B38 618.9'28'9 79-338063
79 MARC

Castets, Bruno.
La loi, l'enfant et la mort : essai de psychopathologie de l'enfant / Bruno Castets ; préf. du docteur Georges Daumezon. — Nouv. éd. rev. et augm. — Paris : Fleurus, c1978.
263 p., ⌐2⌐ leaves of plates : ill. ; 20 cm. — (Collection Pédagogie psychosociale ; 15) F***
ISBN 2-215-00201-8 : 53.00F
1. Child psychopathology. I. Title.
RJ499.C296 1978 79-361048
79 MARC

Knopf, Irwin J
Childhood psychopathology : a developmental approach / Irwin J. Knopf. — Englewood Cliffs, N.J. : Prentice-Hall, c1979.
xi, 497 p. : ill. ; 25 cm.
Includes bibliographies and indexes.
ISBN 0-13-130336-8 : $14.95
1. Child psychopathology. I. Title.
⌐DNLM: 1. Psychopathology—In infancy and childhood. 2. Mental disorders—In infancy and childhood. WS350 K72e⌐
RJ499.K57 618.9'28'907 78-10101
78 MARC

Lang, Jean Louis, of Paris.
Introduction à la psychopathologie infantile : méthodologie, études théoriques et cliniques / Jean-Louis Lang. — Paris : Dunod ; New York : distributed by S.M.P.F. Corp., c1979.
397 p. : ill. ; 24 cm. — (Psychismes) F***
Distributor from label on t.p.
Includes bibliographies and index.
ISBN 2-04-010635-9
1. Child psychopathology. I. Title.
RJ499.L297 618.9'28'582 79-372415
79 MARC

Offer, Daniel.
The psychological world of the juvenile delinquent / by Daniel Offer, Richard C. Marohn, Eric Ostrov. — New York : Basic Books, c1979.
xi, 224 p. ; 24 cm.
Bibliography: p. 203-214.
Includes indexes.
ISBN 0-465-06674-7 : $15.00
1. Juvenile delinquency. 2. Child psychopathology. I. Marohn, Richard C., joint author. II. Ostrov, Eric, 1941- joint author. III. Title.
RJ506.J88O33 618.9'28'58 78-53813
79 MARC

Quay, Herbert Callister, 1927-
Psychopathological disorders of childhood / edited by Herbert C. Quay, John S. Werry. — 2d ed. — New York : Wiley, c1979.
542 p. ; 24 cm.
Includes bibliographies and indexes.
ISBN 0-471-04268-4
1. Child psychopathology. 2. Child psychotherapy. I. Werry, John S., joint author. II. Title.
⌐DNLM: 1. Psychopathology—In infancy and childhood. WS350 P97⌐
RJ499.Q32 1979 618.9'28'9 78-24238
78 MARC

Shanker, Uday.
Problem children : based on case studies of Indian children / by Uday Shanker. — Delhi : Atma Ram, 1978.
303 p., ⌐2⌐ leaves of plates : ill. ; 23 cm.
Bibliography: p. ⌐288⌐-293.
Includes index.
Rs50.00
1. Child psychopathology. I. Title.
RJ499.S489 618.9'28'900954 78-908275
79 MARC

Special problems in child and adolescent behavior / edited by Larry E. Beutler, Richard Greene. — Westport, CT : Technomic Pub. Co., c1978.
404 p. ; 24 cm.
Includes bibliographies and index.
ISBN 0-87762-253-1
1. Behavior therapy. 2. Child psychopathology. 3. Mental deficiency. I. Beutler, Larry E. II. Greene, Richard.
RJ505.B4S63 618.9'28'9 78-56115
78 MARC

Strubbe, W
Anders dan gewoon : beknopte inleiding tot de psychopathologie van het kind / W. Strubbe. — 2e, herz. dr. — Nijkerk : Callenbach, ⌐1978⌐
143 p. ; 22 cm. Ne78-29
Bibliography: p. 133-136.
Includes index.
ISBN 9026601212 : fl 17.50
1. Child psychopathology. I. Title.
RJ499.S82 1978 79-357095
*79 MARC

Twiford, Rainer, 1952-
A child with a problem : a guide to the psychological disorders of children / Rainer Twiford. — Englewood Cliffs, N.J. : Prentice-Hall, c1979.
xi, 164 p. ; 21 cm. — (A Spectrum book)
Bibliography: p. 81-84.
Includes index.
ISBN 0-13-130781-9
1. Child psychopathology. 2. Child mental health services. I. Title.
RJ499.T92 1979 618.9'28'9 79-9116
79 MARC

Vulnerable children / edited by E. James Anthony and Cyrille Koupernik, Colette Chiland ; foreword by Anna Freud ; epilogue by Margaret Mahler. — New York : Wiley, c1978.
xx, 663 p. : graphs ; 25 cm. — (The Child in his family ; v. 4) (Yearbook of the International Association for Child Psychiatry and Allied Professions ; v. 4)
"A Wiley-Interscience publication."
Includes bibliographies and index.
ISBN 0-471-04433-4
1. Child psychopathology. 2. Child psychopathology—Etiology. I. Anthony, Elwyn James. II. Koupernik, Cyrille, 1917- . III. Chiland, Colette. IV. Series. V. Series: International Association for Child Psychiatry and Allied Professions. Yearbook ; v. 4.
RJ499.A1C42 vol. 4 618.9'28'9008 s 78-108200
79 MARC

—ADDRESSES, ESSAYS, LECTURES

Moretti, Giorgio.
Problemi di psicopatologia dell'età evolutiva / Giorgio Moretti. — Ed. ampliata. — Milano : Vita e pensiero, 1979.
129 p. ; 24 cm. — (Corsi universitari) It79-May
Includes bibliographies.
L3500
1. Child psychopathology—Addresses, essays, lectures. I. Title.
RJ499.M656 79-377754
*79 MARC

—BIBLIOGRAPHY

Výskumný ústav detskej psychológie a patopsychológie.
Research Institute of Child Psychology and Pathopsychology, 1964-1968 / contributors, Anton Jurovský ... et al. ; editor, Lýdia Jančeková ; translated by Marta Krallová ; cover designed by Emil Bačik. — ⌐Bratislava⌐ : The Institute, ⌐between 1968 and 1971⌐
74 p. ; 25 cm.
Statement of authorship and imprint stamped on t.p.
"Bibliographical list of publications of the Institute's research workers in the years 1964-1968": p. 41-62.
1. Child psychology—Bibliography. 2. Child psychopatholgy—Bibliography. 3. Výskumný ústav detskej psychológie a patopsychológie—History. I. Jurovský, Anton. II. Jančeková, Lýdia.
Z7204.C5V97 1968 150'.7'204373 79-311309
⌐BF721⌐ 79⌐r79⌐rev MARC

—CASE STUDIES

Beauchesne, Hervé.
Psychopathologie de l'enfant et de l'adolescent : étude de cas / Hervé Beauchesne. — ⌐Paris⌐ : Presses universitaires de France, c1978.
178 p. ; 21 cm. — (Le Psychologue ; 71) F***
Includes bibliographical references.
45.00F
1. Child psychopathology. 2. Child psychopathology—Cases, clinical reports, statistics. I. Title.
RJ499.B38 618.9'28'9 79-338063
79 MARC

Mack, James, 1932-
Psychological examination and report writing / James Mack. — 1st ed. — Hicksville, N.Y. : Exposition Press, c1978.
96 p. ; 21 cm. — (An Exposition-banner book)
Includes bibliographical references.
ISBN 0-682-48964-6 : $7.50
1. Psychodiagnostics. 2. Psychological tests for children. 3. Child psychopathology—Case studies. I. Title.
RJ503.5.M32 618.9'28'9075 78-58553
79 MARC

—ETIOLOGY

Giron, Paule.
La mère empêchée / Paule Giron. — Paris : Seuil, c1978.
118 p. ; 21 cm. — (Libre à elles) F***
ISBN 2-02-004997-X : 28.00F
1. Child psychopathology—Etiology—Cases, clinical reports, statistics. 2. Giron, Hubert, 1956- . 3. Psychiatrists—Malpractice. I. Title.
RJ499.G53 362.7'8'2 79-345918
79 MARC

Vulnerable children / edited by E. James Anthony and Cyrille Koupernik, Colette Chiland ; foreword by Anna Freud ; epilogue by Margaret Mahler. — New York : Wiley, c1978.
xx, 663 p. : graphs ; 25 cm. — (The Child in his family ; v. 4) (Yearbook of the International Association for Child Psychiatry and Allied Professions ; v. 4)
"A Wiley-Interscience publication."
Includes bibliographies and index.
ISBN 0-471-04433-4
1. Child psychopathology. 2. Child psychopathology—Etiology. I. Anthony, Elwyn James. II. Koupernik, Cyrille, 1917- . III. Chiland, Colette. IV. Series. V. Series: International Association for Child Psychiatry and Allied Professions. Yearbook ; v. 4.
RJ499.A1C42 vol. 4 618.9'28'9008 s 78-108200
79 MARC

—LONGITUDINAL STUDIES

Cass, Loretta K
Childhood pathology and later adjustment : the question of prediction / Loretta K. Cass, Carolyn B. Thomas. — New York : Wiley, c1979.
xiv, 268 p. ; 24 cm. — (Wiley series on personality processes)
"A Wiley-Interscience publication."
Bibliography: p. 247-259.
Includes indexes.
ISBN 0-471-04553-5
1. Child psychopathology—Longitudinal studies. 2. Adjustment (Psychology)—Longitudinal studies. 3. Prediction (Psychology)—Longitudinal studies. I. Thomas, Carolyn B., 1927- joint author. II. Title.
⌐DNLM: 1. Affective disturbances—In infancy and childhood—Case studies. 2. Follow-up studies. 3. Personality development—Case studies. WS350.6 C343c⌐
RJ499.C294 618.9'28'909 78-27230
78 MARC

—PREVENTION—CONGRESSES

Social competence in children / editors, Martha Whalen Kent, Jon E. Rolf. — Hanover, N.H. : Published for the University of Vermont by the University Press of New England, 1979.
x, 341 p. : ill. ; 24 cm. — (Primary prevention of psychopathology ; v. 3)
Based on papers presented at the 3d Vermont Conference on Primary Prevention of Psychopathology.
Includes bibliographies and indexes.
ISBN 0-87451-153-4
1. Child development—Congresses. 2. Social adjustment—Congresses. 3. Social acceptance—Congresses. 4. Child psychopathology—Prevention—Congresses. I. Kent, Martha Whalen. II. Rolf, Jon E. III. Series: Vermont Conference on the Primary Prevention of Psychopathology. Primary prevention of psychopathology ; v. 3.
⌐DNLM: 1. Mental disorders—Prevention and control—Congresses. 2. Psychopathology—Congresses. W3 PR945CK⌐
RC454.V46 1977 vol. 3 616.8'58 s 78-63587
⌐HQ772⌐ 78 MARC

CHILD PSYCHOTHERAPY

see also Child analysis

Allen, Frederick Harold, 1890-
Psychotherapy with children / by Frederick H. Allen ; introd. by Harry J. Aponte. — Lincoln : University of Nebraska Press, 1979, c1942.
311 p. ; 20 cm.
Reprint of the ed. published by Norton, New York.
Includes bibliographical references and index.
ISBN 0-8032-1002-7
1. Child psychotherapy. I. Title.
RJ504.A44 1979b 618.9'28'914 79-52647
79 MARC

Drugs and the special child / edited by Michael Jay Cohen. — New York : Gardner Press : distributed by Halsted Press, c1979.
viii, 258 p. : ill. ; 24 cm.
Includes bibliographies and indexes.
ISBN 0-470-99278-6
1. Psychopharmacology. 2. Child psychotherapy. 3. Problem children. 4. Hyperactive children. I. Cohen, Michael Jay.
⌐DNLM: 1. Learning disorders—Drug theray. 2. Child, Exceptional. LC4704 C678d⌐
RJ504.7.D78 618.9'28'5884 77-10151
77 MARC

CHILD PSYCHOTHERAPY (Continued)

Innerhofer, Paul, 1936-
Eltern als Co-Therapeuten : Analyse der Bereitschaft von Müttern zur Mitarbeit bei der Durchführung therapeutischer Programme ihrer Kinder / P. Innerhofer, A. Warnke. — Berlin ; New York : Springer, 1978.
xiv, 182 p. : ill. ; 21 cm. GFR78
ISBN 0-387-08647-1
1. Child psychotherapy. 2. Mother and child. I. Warnke, Andreas, 1945- joint author. II. Title.
RJ504.I56 618.9'28'915 78-1609
78 MARC

Keat, Donald B
Multimodal therapy with children / Donald B. Keat II. — New York : Pergamon Press, c1979.
xi, 171 p. ; 24 cm. — (Pergamon general psychology series ; v. 73)
Bibliography: p. 147-157.
Includes indexes.
ISBN 0-08-022236-6
1. Child psychotherapy. I. Title.
[DNLM: 1. Psychotherapy—In infancy and childhood. WS350.2 K24m]
RJ504.K4 1979 618.9'28'91 78-7044
00 MARC

Krueger, Mark A
Intervention techniques for child care workers / by Mark A. Krueger. — Milwaukee : Franklin Publishers, c1978.
ix, 122 p. : ill. ; 23 cm.
Bibliography: p. 122.
1. Child psychotherapy—Residential treatment. 2. Behavior therapy. 3. Psychiatric aides. I. Title.
RJ504.5.K78 618.9'28'914 79-104744
79 MARC

Mazzetti, Lorenza, 1933-
Il teatro dell'io, l'onirodramma : i bambini drammatizzano a scuola i loro sogni / Lorenza Mazzetti. — Rimini ; Firenze : Guaraldi, c1975.
235 p. : ill. ; 19 cm. — (Le Frontiere dell'educazione ; 25) It75-Oct
L2800
1. Psychodrama. 2. Child psychotherapy. I. Title.
RJ505.P89M39 79-343393
*79 MARC

McAuley, Roger.
Child behaviour problems : an empirical approach to management / Roger McAuley, Patricia McAuley. — 1st American ed. — New York : Free Press, 1978, c1977.
viii, 240 p. : ill. ; 22 cm.
First published in Great Britain under title: Child behaviour problems.
Includes bibliographies and index.
ISBN 0-02-920930-2 : $13.95
1. Behavior therapy. 2. Child psychotherapy. 3. Problem children. I. McAuley, Patricia, joint author. II. Title.
RJ505.B4M3 1978 618.9'28'914 77-94495
78 MARC

Oaklander, Violet.
Windows to our children : a Gestalt therapy approach to children and adolescents / Violet Oaklander. — Moab, Utah : Real People Press, c1978.
335 p. ; 23 cm.
Bibliography: p. 327-335.
ISBN 0-911226-16-8 : $8.00. ISBN 0-911226-17-6 pbk. : $4.50
1. Gestalt therapy. 2. Child psychotherapy. 3. Adolescent psychotherapy. I. Title.
RJ505.G47O24 618.9'28'914 78-10387
78 MARC

Quay, Herbert Callister, 1927-
Psychopathological disorders of childhood / edited by Herbert C. Quay, John S. Werry. — 2d ed. — New York : Wiley, c1979.
542 p. ; 24 cm.
Includes bibliographies and indexes.
ISBN 0-471-04268-4
1. Child psychopathology. 2. Child psychotherapy. I. Werry, John S., joint author. II. Title.
[DNLM: 1. Psychopathology—In infancy and childhood. WS350 P97]
RJ499.Q32 1979 618.9'28'9 78-24238
78 MARC

Resta, Domenico.
Psicoterapia infantile di gruppo / di Domenico Resta. — Milano : Emme, 1977.
178 p. ; 20 cm. — (Il Puntoemme ; 67) It78-Aug
Bibliography: p. 173-175.
L2500
1. Child psychotherapy. 2. Group psychotherapy. I. Title.
RJ504.R47 78-394574
*79 MARC

Santostefano, Sebastian, 1929-
A biodevelopmental approach to clinical child psychology : cognitive controls and cognitive control therapy / Sebastiano Santostefano. — New York : Wiley, c1978.
xvi, 833 p. : ill. ; 26 cm. — (Wiley series on personality processes)
"A Wiley-Interscience publication."
Bibliography: p. 796-813.
Includes index.
ISBN 0-471-75380-7
1. Clinical child psychology. 2. Cognition in children. 3. Psychobiology. 4. Psychological tests for children. 5. Child psychotherapy. I. Title.
[DNLM: 1. Psychiatry—Bibliography. 2. Mental health—Bibliography. ZWM100 G798h]
RJ503.3.S26 618.9'28'9075 78-18485
78 MARC

Sarnoff, Charles A
Latency / Charles Sarnoff. — New York : J. Aronson, c1976.
xii, 402 p. : 24 cm. — (Classical psychoanalysis and its applications)
Bibliography: p. [387]-394.
Includes index.
ISBN 0-87668-233-6
1. Child psychotherapy. 2. Child psychology. I. Title.
RJ504.S27 618.9'28'914 75-34707
78 MARC

Das Symbol im therapeutischen Prozess bei Kindern und Jugendlichen / Ursula Eschenbach (Hrsg.) ; unter Mitarb. von Karl-Franz Daniel ... [et al.]. — 1. Aufl. — Stuttgart : Bonz, 1978.
612 p., [12] leaves of plates : ill. (some col.) ; 22 cm. — (Therapeutische Konzepte der analytischen Psychologie C. G. Jungs ; 1) GFR78-A
"Zitatensammlung zum Begriff 'Symbol' aus dem Gesamtwerk von C. G. Jung, Walter Verlag, Freiburg/Olten": p. 337-608.
Includes bibliographies and index.
ISBN 3-87089-175-0
1. Child psychotherapy. 2. Adolescent psychotherapy. 3. Symbolism (Psychology) 4. Jung, Carl Gustav, 1875-1961. I. Eschenbach, Ursula. II. Series.
RJ504.S88 78-382140
79 MARC

—ADDRESSES, ESSAYS, LECTURES

Prescriptions for child mental health and education / Arnold P. Goldstein, editor. — New York : Pergamon Press, c1978.
ix, 454 p. ; 24 cm. — (Pergamon general psychology series : v. 72)
Bibliography: p.
1. Child psychotherapy—Addresses, essays, lectures. 2. Child mental health—Addresses, essays, lectures. I. Goldstein, Arnold P.
RJ504.P7 1978 618.9'28'914 77-14422
77 MARC

—CASE STUDIES

Kaufman, Barry Neil.
Giant steps / Barry Neil Kaufman. — New York : Coward, McCann & Geoghegan, c1979.
288 p. ; 24 cm.
ISBN 0-698-10956-2 : $9.95
1. Child psychotherapy—Cases, clinical reports, statistics. I. Title.
RJ504.K35 1979 618.9'28'98209 78-10687
78 MARC

—RESIDENTIAL TREATMENT

Edwards, S G T
Treatment in security / by S. G. T. Edwards. — Croydon : Institute for the Study and Treatment of Delinquency, 1976.
14 p. ; 22 cm. GB78-08194
£0.20
1. Juvenile delinquency. 2. Child psychotherapy—Residential treatment. I. Title.
RJ506.J88E38 365'.9426'76 79-302684
79 MARC

Goldstein, Stanley, 1939-
Troubled children/troubled parents : the way out / by Stanley Goldstein. — 1st ed. — New York : Atheneum, 1979.
xiv, 209 p. ; 21 cm.
Includes bibliographical references.
ISBN 0-689-10936-9
1. Emotional problems of children. 2. Parent and child. 3. Mentally ill children. 4. Child psychotherapy—Residential treatment. I. Title.
BF723.E63G64 1979 618.9'28'900240431 78-12622
78 MARC

Horwitz, Robert H
Residential treatment of maladjusted children; a survey and recommendations for Hawaii. Honolulu, Legislative Reference Bureau, University of Hawaii [1956]
80 p. illus. 24 cm. (Legislative Reference Bureau, Territory of Hawaii. Report, 1956, no. 1)
1. Problem children—Hawaiian Islands. 2. Juvenile delinquency—Hawaiian Islands. 3. Child psychotherapy—Residential treatment. I. Title. II. Series: Hawaii. University, Honolulu. Legislative Reference Bureau. Report, 1956, no. 1.
KFH30.H38 1956, no. 1 362.7'4'09599 56-63529
[r79]rev2 MARC

Kramer, Edith.
Art therapy in a children's community : a study of the function of art therapy in the treatment program of Wiltwyck School for Boys / Edith Kramer ; with a foreword by Viola W. Bernard. — New York : Schocken Books, 1977, c1958.
xix, 238 p. : ill. ; 21 cm.
Includes index.
Bibliography: p. 227-231.
ISBN 0-8052-3673-2 : $15.00. ISBN 0-8052-0574-8 pbk. : $8.95
1. Art therapy. 2. Child psychotherapy—Residential treatment. 3. Wiltwyck School for Boys, Esopus, N.Y. I. Title.
[RJ505.A7K72 1977] 618.9'28'915 77-75292
77 MARC

Lambert, Paul, 1911-
The ABCs of child care work in residential care : the Linden Hill manual / Paul Lambert. — New York : Child Welfare League of America, c1977.
76 p. ; 23 cm.
"In use at Linden Hill School, Jewish Board of Guardians, New York."
ISBN 0-87868-165-5
1. Child psychotherapy—Residential treatment. I. Linden Hill School. II. Title.
RJ504.5.L35 616.8'914 77-73022
77[r78]rev MARC

CHILD STUDY see Child development; Child psychology

CHILD SUICIDE see Children--Suicidal behavior

CHILD WELFARE

see also Abandoned children; Child abuse; Child health services; Children—Hospitals; Children—Institutional care; Children—Law; Day care centers; Foster home care; Foundlings; Fresh-air charity;

International Year of the Child, 1979; Juvenile delinquency; Maternal and infant welfare; Mothers' pensions; Runaway children; Youth services

Costin, Lela B
Child welfare : policies and practice / Lela B. Costin. — 2d ed. — New York : McGraw-Hill, c1979.
ix, 469 p. : ill. ; 24 cm.
Includes bibliographies and index.
ISBN 0-07-013206-2
1. Child welfare. 2. Child welfare—United States. I. Title.
HV713.C67 1979 362.7'0973 78-15006
78 MARC

Johansen, Erna Maria, 1911-
Betrogene Kinder : e. Sozialgeschichte d. Kindheit / Erna M. Johansen. — Orig.-Ausg. — Frankfurt am Main : Fischer-Taschenbuch-Verlag, 1978.
267 p. : 43 ill. ; 18 cm. — (Bücher des Wissens) GFR78-A41
Bibliography: p. 259-[263]
Includes indexes.
ISBN 3-596-26622-X : DM7.80
1. Children. 2. Child welfare. 3. Children's rights. I. Title.
HQ781.J62 301.43'14 79-344766
79 MARC

Lieberman, Florence.
Social work with children / Florence Lieberman. — New York : Human Sciences Press, c1979.
344 p. ; 21 cm.
Bibliography: p. 311-334.
Includes index.
ISBN 0-87705-255-7
1. Social work with children. 2. Child development. 3. Child welfare. I. Title.
HV713.L54 362.7 78-23287
78 MARC

Schérer, René.
Une Érotique puérile / René Schérer. — Paris : Éditions Galilée, 1978.
188 p. ; 22 cm. F78-7089
Includes bibliographical references.
ISBN 2-7186-0087-X : 44.00F
1. Child welfare. 2. Children's rights. 3. Youth—Sexual behavior. I. Title.
HV715.S34 301.41'7 79-355835
79 MARC

—ABSTRACTS

Child & youth services. v. 1-
New York, Haworth Press, 1977-
v. 28 cm. bimonthly.
Key title: Child & youth services, ISSN 0145-935X.
1. Child welfare—Periodicals. 2. Child welfare—Abstracts. 3. Child health services—Periodicals. 4. Child health services—Abstracts. 5. Youth—Services for—Periodicals. 6. Youth—Services for—Abstracts. 7. Adolescence—Periodicals. 8. Adolescence—Abstracts.
HV701.C47 362.7'05 79-649846
MARC-S

CHILD WELFARE (Continued)

—ADDRESSES, ESSAYS, LECTURES

Administration, making programs work for children and families / Dorothy W. Hewes, editor ; ⌜cover, Caroline Taylor⌝. — Washington : National Association for the Education of Young Children, c1979.
iv, 243 p. : ill. ; 26 cm.
Articles originally appeared in the journal Young children.
Includes bibliographies and index.
ISBN 0-912674-67-9 : $5.00
1. School management and organization—Addresses, essays, lectures. 2. Preschool education—Addresses, essays, lectures. 3. Child welfare—Addresses, essays, lectures. I. Hewes, Dorothy W. II. Young children.
LB2806.A28 372.21 79-88342
79 MARC

Il Bambino nella società violenta / ⌜di⌝ autori vari ; a cura di Anna Portoghese. — Assisi : Cittadella editrice, 1978.
193 p. ; 19 cm. — (Obiettivo su ..) It78-June
Includes bibliographical references.
L2800
1. Child welfare—Addresses, essays, lectures. 2. Child abuse—Addresses, essays, lectures. 3. Violence—Addresses, essays, lectures. I. Portoghese, Anna.
HV715.B32 78-384951
78 MARC

—BIBLIOGRAPHY

Interamerican Children's Institute. Library.
Bibliografía sobre la protección integral del menor / compilada por la Biblioteca especializada del Instituto como colaboración al XV Congreso Panamericano del Niño, Montevideo, 6-11 junio, 1977. — Montevideo : Instituto Interamericano del Niño, 1977.
ii, v, 199 p. ; 30 cm.
1. Child welfare—Bibliography. I. Title.
Z7164.C5I49 1977 016.3627 78-110230
⌜HV713⌝ 79 MARC

—DICTIONARIES—FRENCH

Lafon, Robert, 1905-
Vocabulaire de psychopédagogie et de psychiatrie de l'enfant / Robert Lafon. — ⌜4. éd. refondue et augm.⌝. — Paris : Presses universitaires de France, 1979, c1963.
xix, 1060 p. ; 25 cm. F***
On spine: Vocabulaire de psychopédagogie.
182F
1. Child psychiatry—Dictionaries—French. 2. Educational psychology—Dictionaries—French. 3. Child welfare—Dictionaries—French. I. Title. II. Title: Vocabulaire de psychopédagogie.
RJ499.L28 1979 371.9'4'019 79-381763
79 MARC

—FEDERAL AID see Federal aid to child welfare

—INFORMATION SERVICES—UNITED STATES

Peat, Marwick, Mitchell, and Company.
CYCIS, a child and youth centered information system : conceptual design / ⌜prepared by Peat, Marwick, Mitchell & Co., Mott-McDonald Associated, inc.⌝ — ⌜New York : Child Welfare League of America, c1978⌝
232 leaes in various foliations : ill. ; 30 cm.
Includes bibliography.
1. CYCIS (Information retrieval system) 2. Child welfare—Information services—United States. 3. Youth services—Information services—United States. I. Mott-McDonald Associates. II. Title.
HV741.P4 1978 029'.9'3627 78-111484
00 MARC

—INFORMATION SERVICES—VIRGINIA

Virginia. Commission for Children and Youth.
An inventory & analysis of State data sources for children & youth, Virginia / Virginia Commission for Children and Youth. — ⌜Richmond⌝ : The Commission, ⌜1977?⌝
156 p. in various pagings : ill. ; 28 cm. — (Its What is the answer, what was the question? ; 1)
Includes bibliographical references.
1. Child welfare—Virginia—Statistics. 2. Child welfare—Information services—Virginia. 3. Youth services—Virginia—Statistics. 4. Youth services—Information services—Virginia. I. Title. II. Series.
HV742.V8V57 1977 362.7'09755 78-622048
79 MARC

—PERIODICALS

Child & youth services. v. 1–
New York, Haworth Press, 1977–
v. 28 cm. bimonthly.
Key title: Child & youth services, ISSN 0145-935X.
1. Child welfare—Periodicals. 2. Child welfare—Abstracts. 3. Child health services—Periodicals. 4. Child health services—Abstracts. 5. Youth—Services for—Periodicals. 6. Youth—Services for—Abstracts. 7. Adolescence—Periodicals. 8. Adolescence—Abstracts.
HV701.C47 362.7'05 79-649846
MARC-S

—RESEARCH—DENMARK—CONGRESSES

Småbørnsforskning i Danmark : rapport fra et tværfagligt forskningsråds seminar 10.-12. marts 1977 / ⌜Lone Heyde har forestået arbejdet med udarbejdelsen⌝. — København : Statens humanistiske Forskningsråd : Statens lægevidenskabelige Forskningsråd : Statens samfundsvidenskabelige Forskningsråd, adresse, Forskningssekretariatet, 1977.
135 p. ; 24 cm. D78-6
Includes bibliographies.
ISBN 8787331365 : not in trade.
1. Child welfare—Research—Denmark—Congresses. 2. Children—Research—Denmark—Congresses. I. Denmark. Statens humanistiske forskningsråd. II. Denmark. Statens lægevidenskabelige forskningsråd. III. Denmark. Statens samfundsvidenskabelige forskningsråd.
HV786.A6S6 78-385609
78 MARC

—RESEARCH —GERMANY, WEST —PERIODICALS

Germany (Federal Republic, 1949–). Bundesministerium für Jugend, Familie und Gesundheit.
Forschung im Geschäftsbereich des Bundesministers für Jugend, Familie und Gesundheit; **Jahresbericht.**
⌜Bonn, Bundesdruckerei⌝
v. illus. 24 cm.
1. Public health—Research—Germany, West—Periodicals. 2. Medical research—Germany, West—Periodicals. 3. Child welfare—Research—Germany, West—Periodicals. 4. Family research—Germany, West—Periodicals. I. Title.
RA440.85.G47a 73-646995
MARC-S

—RESEARCH—UNITED STATES —PERIODICALS

United States. Administration for Children, Youth and Families. Research and Evaluation Division.
Research, demonstration, and evaluation studies.
⌜Washington⌝ Dept. of Health, Education, and Welfare, Office of Human Development Services, Administration for Children, Youth and Families, Research and Evaluation Division.
v. 27 cm. (DHEW publication (OHDS))
Key title: Research, demonstration, and evaluation studies, ISSN 0163-0261
1. Child welfare—Research—United States—Periodicals. I. Title. II. Series: United States. Dept. of Health, Education, and Welfare. DHEW publication.
HV741.U5a 362.7'07'2073 78-645644
MARC-S

—STANDARDS—UNITED STATES

Council on Accreditation of Services for Families and Children.
Provisions for accreditation of agencies serving families and children / Council on Accreditation of Services for Families and Children. — New York : The Council, c1978.
xi, 73 p. ; 26 cm.
Bibliography: p. 66-73.
1. Social service—Standards—United States. 2. Child welfare—Standards—United States. I. Title.
HV95.C67 1978 362.8'2'0973 78-110735
79 MARC

—UNDERDEVELOPED AREAS see Underdeveloped areas—Child welfare

—VOCATIONAL GUIDANCE—GREAT BRITAIN—DIRECTORIES

Hart, Sheila.
Your career with children / by Sheila Hart. — Richmond, Surrey : Careers Consultants Ltd., 1977.
94 p., ⌜8⌝ p. of plates : ill. ; 22 cm. GB78-18148
Bibliography: p. 94.
ISBN 0-85660-026-1. ISBN 0-85660-027-X pbk. : £2.50
1. Vocational guidance—Great Britain—Directories. 2. Child welfare—Vocational guidance—Great Britain—Directories. I. Title.
HF5382.5.G7H37 1977 362.7'023 79-302798
79 MARC

—ALASKA

Alaska. Division of Mental Health & Developmental Disabilities.
Alaska's youth : needs assesment ⌜i.e. assessment⌝ and resource inventory. — ⌜Juneau⌝ : Dept. of Health and Social Services, Division of Mental Health & Developmental Disabilities, 1978.
281 p. : ill. ; 28 cm.
Includes selected papers from Primary prevention: an idea whose time has come, DHEW publication no. (ADM) 77-447.
Includes bibliographies.
1. Social work with youth—Alaska. 2. Child welfare—Alaska. 3. Youth—Alaska—Statistics. 4. Children—Alaska—Statistics. I. Title.
HV1435.A4A45 1978 362.7'09798 78-624036
79 MARC

—AMERICA

Interamerican Children's Institute.
Actividad desarrollada por el Instituto Interamericano del Niño durante los últimos doce años, 1964-1976 / por el director general Rafael Sajón. — Montevideo : El Instituto, 1976.
92 p. : ill. ; 24 cm.
1. Interamerican Children's Institute. 2. Child welfare—America. I. Sajón, Rafael. II. Title.
HV703.I5553 78-394499
79 MARC

Sajón, Rafael.
Menores en situación irregular : aspectos socio legales de su protección / por Rafael Sajón, José Pedro Achard y Ubaldino Calvento. — Montevideo : O.E.A., Instituto Interamericano del Niño, 1973.
40, 16, 3 p. ; 28 cm.
"Trabajo presentado al XIV Congreso Panamericano del Niño, Santiago, Chile del 5 al 11 de agosto.
Bibliography: p. ⌜1⌝-3 (3d group)
1. Child welfare—America. 2. Children—Legal status, laws, etc.—America. 3. Children's rights—America. I. Achard, José P., joint author. II. Calvento, Ubaldino, joint author. III. Interamerican Children's Institute. IV. Title.
HV713.S24 362.7'097 79-373268
79 MARC

—ARABIA

al-Ittiḥād al-'Āmm li-Nisā' al-'Irāq. Sikritārīyat al-Dirāsāt wa-al-Buḥūth.
(al-Takhṭīṭ li-iḥtiyājāt al-ṭufūlah fī al-Khalīj al-'Arabī)
التخطيط لاحتياجات الطفولة في الخليج العربي : دراسة مقدمة الى الدورة الخامسة ، لجنة المراة العربية ، ادارة التنمية الاجتماعية ، جامعة الدول العربية ، المنعقدة في البصرة في ٢٣-٢٦ تشرين الاول ١٩٧٦ / الاتحاد العام لنساء العراق ، سكرتارية الدراسات البحوث . — ⌜s. l.⌝ : السكرتارية ، ⌜1976?⌝
2, 64 leaves ; 26 cm. —
(دراسة ـ الاتحاد العام لنساء العراق ، سكرتارية الدراسات والبحوث ؛ رقم 2)

Includes bibliographical references.
1. Child welfare—Arabia. 2. Child welfare—Iraq. I. Title. II. Series: al-Ittiḥād al-'Āmm li-Nisā' al-'Irāq. Sikritārīyat al-Dirāsāt wa-al-Buḥūth. Dirāsah - al-Ittiḥād al-'Āmm li-Nisā' al-'Irāq, Sikritārīyat al-Dirāsāt wa-al-Buḥūth ; raqm 2.
HV800.A66 I 87 1976 79-970774

—ARGENTINE REPUBLIC

Argentine Republic. Servicio Nacional del Menor.
Informe general de actividades del Servicio Nacional del Menor y la Familia. — ⌜Buenos Aires⌝ : Secretaría de Estado del Menor y la Familia, 1974.
74 p. : ill. ; 31 cm.
1. Argentine Republic. Servicio Nacional del Menor. 2. Child welfare—Argentine Republic. 3. Family social work—Argentine Republic.
HV747.A6A83 1974 362.7'0982 79-364008
79 MARC

—AUSTRALIA

Aspects of school welfare provision. — Canberra : Australian Government Publishing Service, 1977.
xi, 129 p. : charts ; 25 cm. — (Poverty and education series) (Research report - Commission of Inquiry into Poverty) Aus
At head of title: Commission of Inquiry into Poverty.
CONTENTS: 1. Craft, M. School welfare provision in Australia.—2. Tott, P. A. The Red Cross School Volunteer Project.
Includes bibliographical references.
ISBN 0-642-02787-0
1. Child welfare—Australia. 2. Schools—Australia. 3. Volunteers in education. I. Craft, Maurice, 1932- School welfare provision in Australia, 1977. II. Tott, Pamela A. The Red Cross School Volunteer Project. 1977. III. Australia. Commission of Inquiry into Poverty. IV. Series. V. Series: Australia. Commission of Inquiry into Poverty. Research report - Commission of Inquiry into Poverty.
HV802.A9 362.7'0994 79-313928
79 MARC

CHILD WELFARE
—AUSTRALIA (Continued)

Gandevia, Bryan.
Tears often shed : child health and welfare in Australia from 1788 / ₍by₎ Bryan Gandevia with the research assistance of Sheila Simpson. — Rushcutters Bay, N.S.W. : Pergamon Press (Australia), 1978.
151 p. : ill. ; 23 cm.
Aus
Bibliography: p. 148.
Includes index.
ISBN 0-08-023159-4
1. Child welfare—Australia. 2. Children—Care and hygiene. I. Simpson, Sheila. II. Title.
HV802.G36 1978 362.7'0994 78-326343
79 MARC

—AUSTRALIA—CONGRESSES

Australian Child Care Conference, Monash University, 1972.
Australian children 1972, the welfare spectrum : proceedings of the Australian Child Care Conference, Monash University, Melbourne, February 20-25, 1972 / edited by R. J. Lawrence. — Melbourne : Children's Welfare Association of Victoria, 1972.
170 p. : diagrs., tables ; 30 cm. Aus
"Under the auspices of the Children's Welfare Association of Victoria."
Includes bibliographical references.
ISBN 0-9599137-0-X : $3.00
1. Child welfare—Australia—Congresses. I. Lawrence, Robert John, 1931- ed. II. Children's Welfare Association of Victoria. III. Title.
HV802.A6A93 1972 362.7'0994 76-376341
76₍r78₎rev MARC

—AUSTRALIA—VICTORIA

McCaughey, Winsome.
Community child care : resource book for parents and those planning children's services / Winsome McCaughey, Pat Sebastian and the staff of Community Child Care. — Carlton : Greenhouse, 1977.
247 p. : ill. ; 21 x 27 cm. Aus***
Bibliography: p. 246.
ISBN 0-909104-06-9
1. Child welfare—Australia—Victoria. I. Sebastian, Pat, joint author. II. Community Child Care Association. III. Title.
HV802.Z8V525 1977 362.7'09945 78-323784
79 MARC

—BELGIUM—STATISTICS
—PERIODICALS

Statistiek van de kinderen die recht geven op kinderbijslag. Statistique des enfants bénéficiaires d'allocations familiales.
Brussel, Rijksinstituut voor de Sociale Verzekeringen der Zelfstandigen.
v. 30 cm. annual.
Continues: Statistique des enfants bénéficiaires d'allocations familiales.
Dutch or French.
1. Family allowances—Belgium—Statistics—Periodicals. 2. Child welfare—Belgium—Statistics—Periodicals. I. Institut national d'assurances sociales pour travailleurs indépendants.
HD4925.5.B4S73 79-649885
MARC-S

—BRAZIL

Fundação Nacional do Bem-Estar do Menor.
O "menor-problema social" no Brasil e a ação da FUNABEM / Fundação Nacional do Bem-Estar do Menor. — Rio de Janeiro : FUNABEM, 1976.
iv, 142 p. : ill. ; 23 cm.
Bibliography: p. 63-64.
1. Child welfare—Brazil. 2. Fundação Nacional do Bem-Estar do Menor. 3. Marginality, Social—Brazil. I. Title.
HV747.B8F85 1976 362.7'0981 79-104925
79 MARC

—BRAZIL—DUQUE DE CAXIAS

Fundação Nacional do Bem-Estar do Menor. Projeto Integrado da Área Metropolitana do Rio de Janeiro.
Uma abordagem da situação do menor de rua em Duque de Caxias : relatório descritivo, maio-76 / ₍FUNABEM, Diretoria de Programas Integrados, Projeto Integrado da Área Metropolitana do Rio de Janeiro₎. — ₍Rio de Janeiro : O Projeto, 1976?₎
124 leaves : ill. ; 29 cm. — (Documento - FUNABEM, Projeto Integrado da Area Metropolitana do Rio de Janeiro ; no. 4)
1. Abandoned children—Brazil—Duque de Caxias. 2. Child welfare—Brazil—Duque de Caxias. 3. Children—Institutional care—Brazil—Duque de Caxias. I. Title. II. Series: Fundação Nacional do Bem-Estar do Menor. Projeto Integrado da Área Metropolitana do Rio de Janeiro. Documento - FUNABEM, Projeto Integrado da Area Metropolitana do Rio de Janeiro ; no. 4.
HV887.B82D83 362.7'3 79-340486
79 MARC

—BRITISH COLUMBIA

A Survey of adoption and child welfare services to Indians of B.C. — ₍Vancouver? : s.n.₎, 1974.
30, ₍1₎ p. : ill. ; 36 cm. C***
Cover title.
"Report ... requested by the Department of Human Resources from the Union of B.C. Indian Chiefs."
Bibliography: p. ₍40₎
1. Indians of North America—British Columbia—Child welfare. 2. Child welfare—British Columbia. 3. British Columbia—Social policy. I. British Columbia. Dept. of Human Resources. II. Union of British Columbia Indian Chiefs.
E78.B9S94 362.7'09711 78-321878
79 MARC

—CALIFORNIA

California. Dept. of Education.
Final child care pilot study report : a report to the California State Legislature as required by Education Code Section 8280. — Sacramento : California State Dept. of Education, 1978.
728 p. in various pagings ; 30 cm.
Includes bibliographical references.
1. Day care centers—California. 2. Child welfare—California. I. California. Legislature. II. Title.
HV857.C2C35 1978 362.7'1 79-623194
79 MARC

California. Dept. of Finance. Program Evaluation Unit.
Health and welfare eligibility criteria : an inventory / prepared by State of California, Department of Finance, Program Evaluation Unit. — ₍Sacramento₎ : The Department, 1976.
x, 321 p. ; 28 cm. — (Report - State of California, Department of Finance ; no. H631S)
1. Public welfare—California. 2. Child welfare—California. 3. Welfare recipients—California. I. Title. II. Series: California. Dept. of Finance. Program Evaluation Unit. Report - State of California, Department of Finance ; no. H631S.
HV98.C3C35 1976 361.6'2'09794 77-623194
79 MARC

—CANADA

Emberley, Jean.
Children in Canada : a report / prepared by Jean Emberley. — ₍Hamilton, Ont.₎ : McMaster University Press, 1978.
vi, 86 p. : map ; 23 cm. C78-11294-6
Cover title: Today's child ... the world's future.
Prepared for the National Council of Women of Canada.
"The Hospital of Sick Children Foundation, grant no. 78-34."
$3.00
1. Children—Canada. 2. Child welfare—Canada. 3. Children—Care and hygiene—Canada. 4. Education of children. 5. Education—Canada. I. National Council of Women of Canada. II. Title. III. Title: Today's child ... the world's future.
HQ792.C3E43 362.7'0971 79-311504
79 MARC

Task Force on the Child as Citizen.
Admittance restricted : the child as citizen in Canada / ₍Task Force on the Child as Citizen₎. — ₍Ottawa₎ : Canadian Council on Children and Youth, 1978.
172, 195 p. : ill. ; 23 cm. C***
Added t.p.: Interdit aux mineurs.
English and French.
Includes bibliographical references.
$8.95
1. Child welfare—Canada. 2. Children's rights—Canada. I. Canadian Council on Children and Youth. II. Title. III. Title: Interdit aux mineurs.
HV745.A6T37 1978 362.7'0971 79-368442
79 MARC

Van Stolk, Mary.
The battered child in Canada / Mary Van Stolk. — Rev. ed. — Toronto : McClelland and Stewart, c1978.
xii, 178 p. ; 21 cm. C78-1426-X
Includes bibliographical references.
ISBN 0-7710-8712-8 : $7.95
1. Child abuse—Canada. 2. Child welfare—Canada. I. Title.
HV745.A6V3 1978 362.7'1 79-310675
79 MARC

—DENMARK

Denmark. Socialministeriet.
Betænkning vedrørende børneforsorgen og mødrehjælpen, afgivet af det af Socialministeriet under 24, maj 1955 nedsatte udvalg. ₍København₎ 1960.
156 p. 24 cm.
"Betænkning nr. 262."
1. Child welfare—Denmark. 2. Adoption—Denmark. 3. Unwed mothers—Denmark. I. Title.
HV786.A6D46 1960 78-413456
78 MARC

—DENMARK—ADDRESSES, ESSAYS, LECTURES

Forældrenes Børnekommission.
Vore børn er vore børn / Forældrenes Børnekommission. — ₍Varde₎ : ₍Forældrenes Børnekommission₎ ; ₍København₎ : ₍eksp. DBK₎, ₍1978₎.
93 p. : ill. ; 21 cm. D78-10
Bibliography: p. 92.
ISBN 8798065300 : kr15.00
1. Child welfare—Denmark—Addresses, essays, lectures. 2. Family—Denmark—Addresses, essays, lectures. 3. Children—Law—Denmark—Addresses, essays, lectures. I. Title.
HV786.A6F67 1978 78-389093
79 MARC

Når hjemme er et andet sted : debatbog om døgninstitutioner for børn og unge / ₍redaktion, Jens Refsgård ... et al.₎ ; ₍illustrationer, Vibeke Bengtsson₎. — København : ₍Socialarbejdernes Fællesudvalg₎ : salg og distribution, Børn & Unge (Skt. Hans Torv 26, 2200 N), c1978.
160 p. : ill. ; 21 cm. — (SFU-dokumentation ; 3) D78-47
ISBN 8787547066 : kr45.00
1. Children—Institutional care—Denmark—Addresses, essays, lectures. 2. Child welfare—Denmark—Addresses, essays, lectures. 3. Youth services—Denmark—Addresses, essays, lectures. I. Refsgård, Jens. II. Series: Socialarbejdernes fællesudvalg. SFU-dokumentation ; 3.
HV786.A6N37 79-369406
79 MARC

Vores fælles børn? : har vi en børnepolitik? : en debatbog / redigeret af Inge Fischer Møller ; under medvirken af Birgitte Oxdam ... ₍et al.₎. — ₍København₎ : ₍Gyldendal₎, 1978.
87 p. ; 22 cm. — (Gyldendals pædagogiske bibliotek) D78-15
Bibliography: p. 19-₍20₎
ISBN 8701674617 : kr42.50
1. Child welfare—Denmark—Addresses, essays, lectures. 2. Children in Denmark—Addresses, essays, lectures. I. Møller, Inge Fischer. II. Oxdam, Birgitte, 1933-
HV786.A6V67 78-385611
78 MARC

—EUROPE—HISTORY

Dahl, Tove Stang, 1938-
Barnevern og samfunnsvern : om stat, vitenskap og profesjoner under barnevernets oppkomst i Norge / Tove Stang Dahl. — Oslo : Pax, 1978 (Sarpsborg : Sverre Johansen boktr.)
208 p. ; 20 cm. — (Pax ; 555) N78-Apr.
Bibliography: p. 191-200.
Includes index.
ISBN 8253008937 : kr53.00
1. Child welfare—Norway—History. 2. Children—Law—Norway—History. 3. Child welfare—Europe—History. I. Title.
HV788.A6D34 78-385274
78 MARC

—FINLAND

Lastensuojelun Keskusliitto.
Au-lapsi ja hänen vanhempansa : uutta tietoa : kehitys kohti tasa-arvoa, oikeusturva, sosiaalinen tuki, aviottoman äidin vihtoehdot / Lastensuojelun Keskusliitto. — Helsinki : Liitto, 1977.
72 p. ; 21 cm. — (Julkaisu - Lastensuojelun Keskusliitto ; n:o 50 ISSN 0437-2298) Fi***
Includes bibliographical references.
ISBN 9519423036
1. Illegitimacy—Finland. 2. Child welfare—Finland. I. Title. II. Series: Lastensuojelun Keskusliitto. Julkaisu ; n:o 50.
HQ999.F5L37 1977 78-368456
78 MARC

—FINLAND—HELSINKI—HISTORY

Halila, Aimo.
Helsingin kaupungin sosiaalitoimen historia / Aimo Halila. — ₍Helsinki₎ : ₍Helsingin kaupunki₎, 1977.
248 p. : ill. ; 26 cm. — (Helsingin kaupungin julkaisuja ; n:o 30) Fi77
Summary in English and Swedish.
ISBN 9517710542
1. Social service—Finland—Helsinki—History. 2. Child welfare—Finland—Helsinki—History. 3. Helsinki—Charities—History. I. Title.
HV315.5.Z9H444 78-385614
78 MARC

—FRANCE

Meyer, Philippe.
L'enfant et la raison d'État / Philippe Meyer. — Paris : Seuil, c1977.
185 p. ; 18 cm. — (Collection Points) (Série Politique ; Po 88) F***
Includes bibliographical references.
ISBN 2-02-004752-7 : 9.50F
1. Family policy—France. 2. Child welfare—France. I. Title.
HQ624.M46 362.7'1'0944 78-396005
79 MARC

CHILD WELFARE
—FRANCE (Continued)

Protéger l'enfant : problèmes actuels du travail social en milieu judiciaire / ⌐Jean Pinatel, Claude Xuereb, Isabelle Matignon, Henri Daniel Cosnard, etc.⌐ ; sous la direction de Monique Baur-Halbwachs. — Toulouse : Privat, 1977.
172 p. ; 21 cm. — (Agir) F77-11473
ISBN 2-7089-7500-5 : 35.00F
1. Children—Legal status, laws, etc.—France. 2. Child welfare—France. I. Baur-Halbwachs, Monique.
344'.44'0327 79-341231
79 MARC

—GEORGIA

Georgia. General Assembly. Senate. Juvenile Institutions Study Committee.
Report of the Senate Juvenile Institutions Study Committee / Floyd Hudgins, chairman. — Atlanta : General Assembly, 1979.
31 p. ; 36 cm.
1. Problem children—Georgia. 2. Child welfare—Georgia. 3. Children—Institutional care—Georgia. 4. Juvenile courts—Georgia. I. Hudgins, Floyd. II. Title.
HV742.G4G36 1979 362.7'3'09758 79-624487
79 MARC

—GERMANY, WEST

Gernert, Wolfgang.
Jugendhilfe : kurze Einf. in d. sozialpädag. Praxis / Wolfgang Gernert. — 3., neu bearb. Aufl. — München ; Basel : E. Reinhardt, 1978.
222 p. ; 19 cm. — (Uni-Taschenbücher ; 223) GFR78-A
Includes bibliographical references and indexes.
ISBN 3-497-00697-1 : DM16.80
1. Child welfare—Germany, West. 2. Children—Legal status, laws, etc.—Germany, West. I. Title.
HV763.G47 1978 362.7'0943 79-350958
79 MARC

—GERMANY, WEST—BAVARIA
—STATISTICS—PERIODICALS

Bavaria. Staatsministerium für Unterricht und Kultus.
Jugendarbeit in Zahlen.
München, Bayerisches Staatsministerium für Unterricht und Kultus.
v. 30 cm. (Schriften des Bayerischen Staatsministeriums für Unterricht und Kultus. Reihe A : Bildungsstatistik)
"Statistik über Einrichtungen und Personal in der Jugendarbeit in Bayern."
Vols. for issued with: Bayerischer Jugendring.
1. Youth services—Germany, West—Bavaria—Statistics—Periodicals. 2. Child welfare—Germany, West—Bavaria—Statistics—Periodicals. I. Bayerischer Jugendring. II. Title. III. Series: Bavaria. Staatsministerium für Unterricht und Kultus. Schriften. Reihe A : Bildungsstatistik.
HV766.A23a 362.7'0943'3 79-640763
MARC-S

—GERMANY, WEST—HAMBURG
—PERIODICALS

Hamburg. Amt für Jugend.
Landesjugendplan.
Hamburg, Behörde für Schule, Jugend und Berufsbildung, Amt für Jugend.
v. 30 cm. annual.
"Sonderdruck aus dem Amtlichen Anzeiger."
1. Youth—Germany, West—Hamburg—Periodicals. 2. Social work with youth—Germany, West—Hamburg—Periodicals. 3. Child welfare—Germany, West—Hamburg—Periodicals. I. Title.
HQ799.G52H294 301.43'15'0943515 78-643407
MARC-S

—GREAT BRITAIN

Child Benefits Now Campaign.
The great child benefit robbery / ⌐Child Benefits Now Campaign⌐. — ⌐London⌐ : ⌐The Campaign⌐, ⌐1977⌐
30, ⌐8⌐ p. ; 21 cm. GB78-16134
Cover title.
Includes bibliographical references.
£0.30
1. Child welfare—Great Britain. 2. Maternal and infant welfare—Great Britain. I. Title.
HV751.A6C55 1977 362.8'2 78-326358
79 MARC

Child Poverty Action Group.
Children worse off under Labour? / ⌐Child Poverty Action Group⌐ ; ⌐written⌐ by Frank Field. — ⌐London⌐ : ⌐Child Poverty Action Group⌐, 1978.
⌐3⌐, 28 p. ; 21 cm. — (Poverty pamphlet ; 32 ISSN 0306-1868)
GB78-20021
Cover title.
£0.25
1. Child welfare—Great Britain. I. Field, Frank, 1942- . II. Title. III. Series.
HV751.A6C57 1978 362.7'1'0941 79-301403
79 MARC

Field, Frank, 1942-
Priority for children : a Labour success / ⌐by⌐ Frank Field. — London : Child Poverty Action Group, ⌐1978⌐
⌐2⌐, 28 p. : ill. ; 21 cm. — (Poverty pamphlet ; 36 ISSN 0306-1868)
GB78-33664
1. Child welfare—Great Britain. 2. Family policy—Great Britain. I. Title. II. Series.
HV751.A6F52 362.8'2 79-304523
79 MARC

Great Britain. Central Policy Review Staff.
Services for young children with working mothers : report / by the Central Policy Review Staff. — London : H.M. Stationery Off., 1978.
89 p. : ill. ; 30 cm. GB***
ISBN 0-11-630805-2 : £2.75
1. Children of working mothers—Great Britain. 2. Child welfare—Great Britain. 3. Day care centers—Great Britain. I. Title.
HV752.A6G73 1978 362.7'1'0941 79-313843
00 MARC

Wagner, Gillian.
Barnardo / ⌐by⌐ Gillian Wagner. — London : Weidenfeld and Nicolson, 1979.
xv, 344 p., ⌐6⌐ leaves of plates : ill. ; 23 cm. GB***
Bibliography: p. 328-333.
Includes index.
ISBN 0-297-77561-8 : £8.95
1. Barnardo, Thomas John, 1845-1905. 2. Child welfare—Great Britain. 3. Social reformers—Great Britain—Biography.
HV247.B2W34 362.7'3'0924 79-314428
79 MARC

—HUNGARY

Hanák, Katalin.
Társadalom és gyermekvédelem / Hanák Katalin. — Budapest : Akadémiai Kiadó, 1978.
221 p. ; 18 cm. — (Szociológiai tanulmányok ; 19) H78
Includes bibliographical references.
ISBN 9630516810 : 24.00Ft
1. Child welfare—Hungary. I. Title. II. Series.
HV799.H82H36 79-375630
*79 MARC

—ILLINOIS

Illinois. Dept. of Children and Family Services.
1979 plan for Department of Children and Family Services. — ⌐Springfield, Ill. : Bureau of the Budget⌐, 1978-
v. : ill. ; 28 cm. — (Illinois human services plan ; v. 1)
CONTENTS: Phase 1, fiscal years 1977-1979.
1. Child welfare—Illinois. 2. Social work with children—Illinois. 3. Family social work—Illinois. I. Title. II. Series.
HV742.I3I43 1978a 362.7'09773 79-620626
79 MARC

Program plans for children and their families, on file with the Illinois Department of Children and Family Services, fiscal year 1977. — ⌐Springfield⌐ : The Department, ⌐1978⌐
xvi, 792 p. ; 30 cm.
Includes index.
1. Child welfare—Illinois. 2. Family social work—Illinois. 3. Public welfare—Illinois. I. Illinois. Dept. of Children and Family Services.
HV742.I3P75 362.7'1'09773 78-624120
79 MARC

—ILLINOIS—DIRECTORIES

Illinois. Dept. of Children and Family Services.
Program plans for children and their families on file with the Illinois Department of Children and Family Services, fiscal year 1978. — ⌐Springfield⌐ : The Dept., ⌐1978⌐
xix, 1373 p. ; 29 cm.
1. Child welfare—Illinois—Directories. I. Title.
HV742.I3I43 1978 362.7'025'773 78-624119
79 MARC

—ILLINOIS—STATISTICS

Illinois. Dept. of Children and Family Services.
Plan for children and family services, 1978-80 / Illinois Department of Children and Family Services. — ⌐Springfield⌐ : The Dept., ⌐1977⌐
78 p. : ill. ; 28 cm. — (Illinois human services plan ; v. 1)
Cover title.
Title on spine: Children and Family services, 1978-80.
$2.50
1. Child welfare—Illinois—Statistics. 2. Illinois. Dept. of Children and Family Services. I. Title. II. Title: Children and family services, 1978-80. III. Series.
HV742.I3I43 1977a 353.9'773'008471 78-621659
79 MARC

—INDIA—RAJASTHAN

Bedi, M S
Socially handicapped children : a study of their institutional services / M. S. Bedi. — 1st ed. — Jodhpur : Jain Bros., 1978.
xix, 223 p. ; 22 cm.
Includes bibliographical references.
Rs35.00
1. Socially handicapped children—India—Rajasthan. 2. Social work with children—India—Rajasthan. 3. Child welfare—India—Rajasthan. 4. Children—Institutional care—India—Rajasthan. I. Title.
HV800.I53R643 362.7'0954'4 78-905279
79 MARC

—INDONESIA

Indonesia. Treaties, etc. United Nations Children's Fund, June 23, 1975.
Master plan of operations for a programme of services for children in the context of young human resources development in Indonesia, 1975-1979. — ⌐Jakarta : Republic of Indonesia, 1975⌐
xii, 18 leaves ; 29 cm.
1. Child welfare—Indonesia. 2. Youth services—Indonesia. I. United Nations. Children's Fund. Treaties, etc. Indonesia, June 23, 1975. II. Title: Master plan of operations for a programme of services for children ...
HV800.I54I57 1975 362.7'09598 77-940177
79 MARC

—INDONESIA—ADDRESSES, ESSAYS, LECTURES

Konsultasi Periodik Petugas² Taman Penitipan Anak, Temanggung, Indonesia, 1972.
Hasil. Jakarta, Direktorat Kesejahteraan Keluarga dan Masyarakat, 1972.
80 l. 32 cm.
1. Child welfare—Indonesia—Addresses, essays, lectures. 2. Social service—Indonesia—Addresses, essays, lectures.
HV800.I55K57 1972 73-941257
79 MARC

—INDONESIA—HANDBOOKS, MANUALS, ETC.

Indonesia. Direktorat Jenderal Urusan Kesejahteraan Anak, Keluarga dan Masyarakat.
Bimbingan kerja kesejahteraan anak, keluarga, dan masyarakat. — Edisi 2. — Jakarta : Direktorat Jenderal Kesejahteraan Anak, Keluarga dan Masyarakat, Departemen Sosial R.I., ⌐1973⌐
123, ⌐26⌐ p. ; 32 cm.
1. Social service—Indonesia—Handbooks, manuals, etc. 2. Child welfare—Indonesia—Handbooks, manuals, etc. I. Title.
HV401.D53 1973 76-941124
79 MARC

—IRAQ

al-Ittiḥād al-'Āmm li-Nisā' al-'Irāq. Sikritārīyat al-Dirāsāt wa-al-Buḥūth.
(al-Takhṭīṭ li-iḥtiyājāt al-ṭufūlah fī al-Khalīj al-'Arabī)
التخطيط لاحتياجات الطفولة في الخليج العربي : دراسة مقدمة الى الدورة الخامسة ، لجنة المراة العربية ، ادارة التنمية الاجتماعية ، جامعة الدول العربية ، المنعقدة في البصرة في ٢٣ـ٢٦ تشرين الاول ١٩٧٦ / الاتحاد العام لنساء العراق ، سكرتارية الدراسات البحوث . — ⌐s. l.⌐ : السكرتارية ، ⌐1976?⌐
2, 64 leaves ; 26 cm. —
(دراسة ـ الاتحاد العام لنساء العراق ، سكرتارية الدراسات والبحوث ؛ رقم 2)
Includes bibliographical references.
1. Child welfare—Arabia. 2. Child welfare—Iraq. I. Title. II. Series: al-Ittiḥād al-'Āmm li-Nisā' al-'Irāq. Sikritārīyat al-Dirāsāt wa-al-Buḥūth. Dirāsah - al-Ittiḥād al-'Āmm li-Nisā' al-'Irāq, Sikritārīyat al-Dirāsāt wa-al-Buḥūth ; raqm 2.
HV800.A66 I 87 1976 79-970774

—IRELAND

Ireland (Eire). Task Force on Child Care Services.
Interim report to the Tánaiste and Minister for Health / Task Force on Child Care Services. — Dublin : Stationery Off., 1975.
39 p. ; 25 cm. Ir***
£0.25
1. Child welfare—Ireland. I. Ireland (Eire). Tánaiste. II. Ireland (Eire). Dept. of Health.
HV757.A6I77 1975 362.7'09417 78-326363
79 MARC

CHILD WELFARE (Continued)

—JAPAN

Ikegi, Kiyoshi, 1938-
(Gendai no seishōnen)
現代の青少年 / 池木清, 後藤光義, 河上恭雄著. — 東京：総合労働研究所, 1978.
xx, 321 p. : ill. ; 20 cm. — (教育を考えるシリーズ) Ja 78-22771
¥2000

1. Youth—Japan. 2. Child welfare—Japan. 3. Children—Law—Japan. I. Gotō, Mitsuyoshi, 1947- joint author. II. Kawakami, Yasuo, 1945- joint author. III. Title. IV. Series: Kyōiku o kangaeru shirizu.

HQ799.J3I45 78-816649

Kawasaki, Daiji, 1902-
(Senji hoikusho)
戦時保育所：保姆と挺身隊への手引 / 川崎大治著. — 東京：學習社, 昭和20 [1945]
6, 270 p., [2] leaves of plates : ill. ; 19 cm.

1. Day care centers—Japan. 2. Child welfare—Japan. I. Title.

HV861.J3K38 79-804831

Kinjō, Yoshiko, 1902-
(Naha onna ichidaiki)
なはをんな一代記 / 金城芳子. — 那覇：沖縄タイムス社, 1977.
462 p. ; 19 cm. — (タイムス選書 ; 5) Ja***
Series romanized: Taimusu sensho.
Originally published in Okinawa taimusu, Mar. 3-Sept. 23, 1976.
¥1850

1. Kinjō, Yoshiko, 1902- 2. Social workers—Japan—Biography. 3. Child welfare—Japan. I. Title.

HV28.K47A36 79-807643

Morita, Sōichi, 1915-
(Kono chiisaki mono ni)
この小さき者に：少年少女の問題をめぐって / 森田宗一著. — 東京：大日本出版, 昭和25 [1950]
9, 4, 257 p., [1] leaf of plates : ill. ; 19 cm.

1. Child welfare—Japan. 2. Family—Japan. I. Title.

HV800.J3M66 79-808299

(Shikaku shōgai no kyōiku to fukushi)
視覚障害の教育と福祉 / 大山信郎, 佐藤泰正編集. — 東京：日本図書文化協会：発売 図書文化社, 1978.
350 p. ; 22 cm. — (最新心身障害教育・福祉講座 ; 第4巻) Ja 78-15693
Bibliography: p. 340-346.
Includes indexes.
¥3000
1. Blind—Education. 2. Visually handicapped children—Rehabilitation. 3. Child welfare—Japan. I. Ōyama, Nobuo, 1911- II. Satō, Yasumasa, 1929- III. Series: Saishin shinshin shōgai kyōiku fukushi kōza ; dai 4-kan.
HV1631.S53 78-813610

Takeuchi, Yoshimi, 1923-
(Jidō fukushi hō Boshi fukushi hō Boshi hoken hō no kaisetsu)
児童福祉法母子福祉法母子保健法の解説 / 竹内嘉巳. — 新版増補. — 東京：時事通信社, 1978.
17, 671 p. ; 19 cm. Ja 79-3105
Previous editions published under title: Jidō fukushi ho no kaisetsu to un'yō.
¥2200

1. Maternal and infant welfare—Law and legislation—Japan. 2. Maternal health services—Law and legislation—Japan. 3. Child welfare—Japan. I. Japan. Laws, statutes, etc. Jidō fukushi hō. 1978. II. Japan. Laws, statutes, etc. Boshi fukushi hō. 1978. III. Japan. Laws, statutes, etc. Boshi hoken hō. 1978. IV. Title.

79-805236

—JAPAN—ADDRESSES, ESSAYS, LECTURES

(Naku mo no ka)
泣くものか：作文集：子どもの人権10年の証言 / 養護施設協議会編. — 東京：亜紀書房, 1977.
249 p. : ill. ; 19 cm. Ja 77-27417
¥950

1. Children—Japan—Institutional care—Addresses, essays, lectures. 2. Childrens' rights—Japan—Addresses, essays, lectures. 3. Child welfare—Japan—Addresses, essays, lectures. I. Zenkoku Shakai Fukushi Kyōgikai. Yōgo Shisetsu Kyōgikai.

HV1312.N34 79-808623

J

Yamaguchi, Sachio, 1933-
(Gendai no hikō mondai)
現代の非行問題：教育・福祉・司法 / 山口幸男著. — 東京：民衆社, 1978.
228 p. ; 20 cm. Ja 78-14238
List of E. Ferri's works: p. 227-228.
¥1300

1. Juvenile delinquency—Japan—Addresses, essays, lectures. 2. Child welfare—Japan—Addresses, essays, lectures. 3. Children—Law—Japan—Addresses, essays, lectures. I. Title.

HV9207.A5Y294 78-813757

—JAPAN—BIBLIOGRAPHY—CATALOGS

Nihon Shakai Jigyō Daigaku, Tokyo. Toshokan.
(Hoiku kankei bunken mokuroku)
保育関係文献目録 / [編者 日本社会事業大学図書館]. — 東京：同図書館, 1978.
147 p. ; 26 cm. Ja 78-22227
Cover title.
¥800
1. Day care centers—Japan—Bibliography—Catalogs. 2. Education, Preschool—Japan—Bibliography—Catalogs. 3. Child welfare—Japan—Bibliography—Catalogs. 4. Nihon Shakai Jigyō Daigaku, Tokyo. Toshokan—Catalogs. I. Title.

Z7166.N56 1978 79-800236
[HV861]

--JAPAN--COLLECTED WORKS

(Jidō fukushi hō seiritsu shiryō shūsei)
児童福祉法成立資料集成 / 児童福祉法研究会編. — 東京：ドメス出版, 1978-1979.
2 v. ; 22 cm. Ja 79-1119 v. 1
Bibliography: v. 2, p. [863]-873.
Includes indexes.
¥21000 per vol.

1. Children—Law and legislation—Japan—Collected works. 2. Child welfare—Japan—Collected works. I. Jido Fukushiho Kenkyukai.

79-806622

J

--JAPAN--HISTORY--ADDRESSES, ESSAYS, LECTURES

(Sengo hoikusho no rekishi)
戦後保育所の歴史 / 植山つる, 浦辺史, 岡田正章編. — 東京：全国社会福祉協議会, 1978.
7, 353 p. : ill. ; 21 cm. Ja 78-10787
¥1800

1. Day care centers—Japan—History—Addresses, essays, lectures. 2. Child welfare—Japan—History—Addresses, essays, lectures. I. Ueyama, Tsuru. II. Urabe, Hiroshi, 1905- III. Okada, Masatoshi, 1925-

HV800.J3S46 79-803609

—JAPAN—SOCIETIES, ETC.

(Chiiki ni nezashita jidōkan katsudō)
地域にねざした児童館活動 / [編集 朝日生命厚生事業団, 全国社会福祉協議会]. — 東京：全国社会福祉協議会, 昭和50 [1975]
66 p. : ill. ; 21 cm. — (いくせいノート ; 1) Ja 75-22548

1. Community centers—Japan. 2. Playgrounds—Japan. 3. Child welfare—Japan—Societies, etc. I. Asahi Seimei Kōsei Jigyōdan. II. Zenkoku Shakai Fukushi Kyōgikai. III. Series: Ikusei nōto ; 1.

HN46.J3C47 78-814201

—JAPAN—STATISTICS

Japan. Kōseishō. Jidōkyoku.
(Nihon no kodomo to sono katei no jittai)
日本の子どもとその家庭の実態：全国家庭児童調査結果報告 / [厚生省児童局]. — 東京：日本児童福祉協会, 昭和39 [1964]
69 p. : graphs ; 21 cm.
Cover title.

1. Children—Japan—Statistics. 2. Child welfare—Japan—Statistics. 3. Family—Japan—Statistics. I. Title.

HQ792.J3J35 1964 79-803058

--JAPAN--TOKYO

Shakai Fukushi Seisaku Kenkyūkai.
(Tōkyō-to ni okeru minsei iin jidō iin seido no genjō to kongo no arikata)
東京都における民生委員・児童委員制度の現状と今後のあり方：東京都民生局委託研究. — 東京：社会福祉政策研究会, 昭和52 [1977]
128 p. : ill. ; 26 cm. Ja***
Cover title.
1. Public welfare administration—Japan—Tokyo. 2. Public welfare—Japan—Tokyo. 3. Child welfare—Japan—Tokyo. I. Tokyo. Minseikyoku. II. Title.
HV415.T6S45 1977 79-803624

—LATIN AMERICA

Sajón, Rafael.
Protección social y legal del niño de 0 a 5 años / por Rafael Sajón. — Montevideo : O.E.A., Instituto Interamericano del Niño, 1971.
17, [1] p. ; 28 cm.
"15."
Bibliography: p. [18]
1. Child welfare—Latin America. 2. Children—Legal status, laws, etc.—Latin America. I. Interamerican Children's Institute. II. Title.
HV747.L3S24 362.7'098 79-373269
79 MARC

—LATIN AMERICA—CONGRESSES

Seminario Regional Interamericano para Determinar Normas de Protección al Menor en Programas Integrados de Desarrollo, 2d, Montevideo, Uruguay, 1965.
II [i.e. Segundo] Seminario Regional Interamericano para Determinar Normas de Protección al Menor en Programas Integrados de Desarrollo, mayo 2-10 de 1965, Montevideo, Uruguay. — Montevideo : O.E.A., Instituto Interamericano del Niño, 1965.
361 p., [1] leaf of plates : ill. ; 24 cm.
1. Child welfare—Latin America—Congresses. 2. Latin America—Social policy—Congresses. I. Interamerican Children's Institute.
HV709.S44 1965 362.7'098 79-340484
79 MARC

—LATIN AMERICA—HISTORY

Interamerican Children's Institute.
Inter-American Children's Institute, 50th anniversary, 1927-1977. — [Montevideo, Uruguay] : The Institute, [1977]
41 p. : maps ; 23 cm.
1. Interamerican Children's Institute—History. 2. Child welfare—Latin America—History. I. Title.
HV703.I563 362.7'06'28 78-322459
79 MARC

CHILD WELFARE (Continued)

—MARYLAND—STATISTICS —PERIODICALS

Maryland. Juvenile Services Administration.
Annual statistical report. 1974/75–
[Baltimore] Juvenile Services Administration.
v. ill. 28 cm.
Continues: Maryland. Dept. of Juvenile Services. Annual statistical report.
Cover title, 1974/75–1975/76: Annual report - Dept. of Juvenile Services; 1976/77– : Annual report - Juvenile Services Administration.
1. Maryland. Juvenile Services Administration—Periodicals. 2. Child welfare—Maryland—Statistics—Periodicals. 3. Juvenile courts—Maryland—Statistics—Periodicals. 4. Juvenile justice, Administration of—Maryland—Statistics—Periodicals. I. Maryland. Dept. of Juvenile Services. Annual report. II. Maryland. Juvenile Services Administration. Annual report.
HV742.M3M33a 364.36'09752 78-645438
MARC-S

—MASSACHUSETTS

Massachusetts. Committee for Children and Youth.
Report of the Massachusetts Committee on Children and Youth for 1960 White House Conference on Children and Youth. Boston, 1960.
114 p. 29 cm.
1. Child welfare—Massachusetts. I. Golden Anniversary White House Conference on Children and Youth, Washington, D.C., 1960.
HV742.M4M33 1960 362.7'09744 70-222735
69[r79]rev MARC

—MISSOURI

Missouri. Division of Family Services.
Rules for licensing child caring agencies. — Jefferson City, Mo. : Dept. of Social Services, Division of Family Services, 1979.
v, 24 p. ; 28 cm.
Cover title.
1. Children—Legal status, laws, etc.—Missouri. 2. Children—Institutional care—Missouri. 3. Child welfare—Missouri. I. Title. II. Title: Child caring agencies.
KFM8150.C4A32 1979 344'.778'03271 79-625028
79 MARC

—NEBRASKA

Masters, Christopher F
Aid to dependent children : social and economic characteristics of families receiving assistance during March 1977 / prepared by Christopher F. Masters and Division of Research and Statistics. — [Lincoln] : Nebraska Dept. of Public Welfare, [1977?]
63 p. : ill. ; 28 cm.
Cover title: Social and economic characteristics of families receiving assistance during March 1977.
Bibliography: p. 63.
1. Welfare recipients—Nebraska. 2. Child welfare—Nebraska. I. Nebraska. Dept. of Public Welfare. Division of Research and Statistics. II. Title. III. Title: Social and economic characteristics of families receiving assistance during March 1977.
HV98.N2M37 362.7'1 78-622951
79 MARC

—NETHERLANDS

Doek, J E
Wat is een goede raad voor de toekomst? : enige beschouwingen over het ontstaan en de (mogelijke) ontwikkeling van de raden voor de kinderbescherming / J. E. Doek. — Alphen aan den Rijn : Samsom, 1978.
76 p. ; 22 cm. Ne78-34
"Bewerking van de openbare les ... Vrije Universiteit te Amsterdam, 1977.
Bibliography: p. 65-72.
ISBN 9014027796 : fl 14.00
1. Child welfare—Netherlands. 2. Children—Law—Netherlands. I. Title.
HV778.A6D6 79-339042
*79 MARC

Roelfs, Jan.
Kinderbescherming op het gereformeerde erf; het "hoe en waarom" van (her)opvoeding en begeleiding in deze tijd. Samengesteld in opdracht van het Gereformeerd Centrum voor Kinderbescherming van de Stichting Raad voor Gereformeerd Sociale Arbeid. Utrecht, Schotanus & J. Utrecht, 1965.
58 p. illus. 24 cm.
1. Church work with children—Netherlands. 2. Church work with youth—Netherlands. 3. Church work with young adults—Netherlands. 4. Child welfare—Netherlands. 5. Unmarried mothers—Netherlands. 6. Gereformeerde kerken in Nederland—Charities. I. Title.
BV639.C4R63 79-413530
79 MARC

Zandvoort, W J F
Kinderen en de kinderbescherming / W. J. F. Zandvoort. — Amsterdam : Meulenhoff Informatief, [1978]
144 p. ; 20 cm. Ne78-20
Bibliography: p. 140-[144]
ISBN 9029095016 : fl 19.50
1. Child welfare—Netherlands. 2. Children in the Netherlands. I. Title.
HV778.A6Z36 78-375470
*78 MARC

—NETHERLANDS—AMSTERDAM—HISTORY

Fuchs, Johannes Marius.
Opvangen en opvoeden : Lutherse wezenzorg in Amsterdam 1678-1978 / door J. M. Fuchs. — [Amsterdam : Lankamp & Brinkman, 1978]
190 p. : ill. ; 28 cm. Ne78-21
Bibliography: p. 179-181.
Includes index.
fl 45.00
1. Amsterdam—Orphans and orphan-asylums—History. 2. Lutheran Church in Amsterdam—Charities—History. 3. Child welfare—Netherlands—Amsterdam—History. I. Title.
HV1210.A5F8 78-388163
*78 MARC

—NEW YORK (CITY)—CASE STUDIES

Weissman, Harold H
Integrating services for troubled families / Harold H. Weissman. — 1st ed. — San Francisco : Jossey-Bass, 1978.
xviii, 148 p. ; 24 cm. — (The Jossey-Bass social and behavioral science series)
Bibliography: p. 139-142.
Includes index.
ISBN 0-87589-385-6
1. Lower East Side Family Union—History. 2. Child welfare—New York (City)—Case studies. 3. Family social work—New York (City)—Case studies. I. Title.
HV743.N48W44 362.8'2'0973 78-62563
79 MARC

—NEW YORK (CITY)—STATISTICS

Levy, Mel.
Aid to dependent children in New York City : a comparison of large and small households, June 1976 / by Mel Levy. — [New York] : The City of New York, Human Resources Administration, Office of Research and Program Evaluation, 1977.
vii, 43 leaves ; 28 cm.
Cover title.
"PA study no. 2."
1. Child welfare—New York (City)—Statistics. I. Title.
HV743.N48L48 362.7'1 78-326149
79 MARC

—NEW ZEALAND—CONGRESSES

New Zealand Early Childhood Care and Development Convention, 1st, Christchurch, 1975.
Early childhood in New Zealand / editors: Brian O'Rourke and John Clough. — Auckland : Heinemann Educational, 1978.
277 p. ; 26 cm. NZ78-1
"A selection of papers which were presented at the First New Zealand Early Childhood Care and Development Convention, Our Care—Their Future."
Errata slip inserted.
Includes bibliographies.
ISBN 0-86863-259-7 : $11.95
1. Child welfare—New Zealand—Congresses. 2. Parenting—New Zealand—Congresses. 3. Socialization—Congresses. I. O'Rourke, Brian T. II. Clough, John. III. Title.
HV802.5.N48 1975 301.43'14'0993 79-310159
79 MARC

—NORTH DAKOTA—DIRECTORIES

North Dakota. Governor's Committee on Children and Youth.
Directory of services for children and youth : North Dakota, 1977 / compiled by North Dakota Committee on Children and Youth of the Governor's Council on Human Resources. — [Bismarck, N.D.] : The Committee, 1977.
viii, 129 p. ; 22 cm.
Cover title: 1977 directory of services for children and youth.
Includes indexes.
1. Public welfare—North Dakota—Directories. 2. Social service—North Dakota—Directories. 3. Child welfare—North Dakota—Directories. I. Title. II. Title: 1977 directory of services for children and youth.
HV86.N95 1977 362.7'025'784 78-622171
79 MARC

—NORWAY—HISTORY

Dahl, Tove Stang, 1938-
Barnevern og samfunnsvern : om stat, vitenskap og professjoner under barnevernets oppkomst i Norge / Tove Stang Dahl. — Oslo : Pax, 1978 (Sarpsborg : Sverre Johansen boktr.)
208 p. ; 20 cm. — (Pax ; 555) N78-Apr.
Bibliography: p. 191-200.
Includes index.
ISBN 8253008937 : kr53.00
1. Child welfare—Norway—History. 2. Children—Law—Norway—History. 3. Child welfare—Europe—History. I. Title.
HV788.A6D34 78-385274
78 MARC

—OREGON—PERIODICALS

Oregon. Children's Services Division. Staff Development Unit.
Annual report of agency training activities.
[Salem] Children's Services Division, Staff Development Unit.
v. ill. 28 cm.
Continues: Annual report of agency training activities, ISSN 0098-8723, issued by: Oregon. Children's Services Division, Training Unit.
Key title: Annual report of agency training activities, ISSN 0098-8723.
1. Oregon. Children's Services Division. Staff Development Unit. 2. Social work education—Oregon—Periodicals. 3. Child welfare—Oregon—Periodicals. I. Key title.
HV742.O7O65a 353.9'795'00847 78-645414
MARC-S

—QUÉBEC (PROVINCE)

Protection et délinquance : dossiers de la Cour de bien-être social, Québec, 1974 / [données recueillies par] Édith Deleury, Jocelyn Lindsay, Michèle Rivet. — Québec : Presses de l'Université Laval, 1978.
xii, 168 p. ; 23 cm. — (Bibliothèque juridique ; B-2) C78-10958-9
ISBN 0-7746-6872-5 : $12.00
1. Juvenile justice, Administration of—Québec (Province) 2. Child welfare—Québec (Province) I. Deleury, Édith, 1944- . II. Lindsay, Jocelyn, 1944- . III. Rivet, Michèle, 1941- . IV. Québec (Province). Cour de bien-être social. V. Series.
HV9071.P76 364.36'09714 79-373032
79 MARC

—POLAND

Ziemska, Maria.
Early child care in Poland / by Maria Ziemska. — London ; New York : Gordon and Breach, 1978.
iii, 148 p. : ill. ; 23 cm. — (Early child development and care ; vol. 5, no. 1/2) (International monographs on early child care ; no. 8) GB***
Cover title.
Bibliography: p. 136-143.
1. Child welfare—Poland. 2. Child development. I. Title. II. Series: International monograph series on early child care ; no. 8.
HV799.P72Z53 362.7'09438 78-324015
79 MARC

—POLAND—HISTORY

Balcerek, Marian.
Rozwój opieki nad dzieckiem w Polsce w latach 1918-1939 / Marian Balcerek. — Wyd. 1. — Warszawa : Państ. Wydaw. Naukowe, 1978 (Łódź : ŁZG 1)
391, [1] p. ; 20 cm. P78-10509
Errata slip inserted.
Bibliography: p. 361-[382]
Includes index.
zł50.00
1. Child welfare—Poland—History. I. Title.
HV799.P72B34 79-359897
79 MARC

—SCOTLAND

Royal Scottish Society for Prevention of Cruelty to Children.
Child protection (Scotland) : the R.S.S.P.C.C. approach / [Royal Scottish Society for Prevention of Cruelty to Children]. — [Edinburgh] : [The Society], [1978]
[1], 32 p. : 3 forms, map ; 21 cm. GB78-15346
Cover title.
1. Royal Scottish Society for Prevention of Cruelty to Children. 2. Child welfare—Scotland. 3. Children—Law—Scotland. I. Title.
HV703.R6824 179'.2'062411 78-326359
79 MARC

CHILD WELFARE (Continued)

—SOUTH CAROLINA—FINANCE

Patterson, Gary A
Child care : an evaluation of programmatic/financial activities / principal author, Gary A. Patterson, co-authors, Susan P. Agle ... ⌈et al.⌉. — ⌈Columbia, S.C.⌉ : Office of the Governor, Division of Health and Social Development, 1978.
iv, 219, 9 leaves ; 29 cm.
Bibliography: leaves 107-108.
1. Day care centers—South Carolina—Evaluation. 2. Child welfare—South Carolina—Finance. I. South Carolina. Office of the Governor. Division of Health and Social Development. II. Title.
HV857.S6P37 362.7'1 79-620755
79 MARC

—SWEDEN

Bättre barnmiljö / ⌈sammanst. och red. av Margaretha Arnell⌉. — Stockholm : Fören. Sveriges socialchefer, 1977.
160 p. : ill. ; 21 cm. S77-53
1. Child welfare—Sweden. 2. Children—Sweden—Social conditions. I. Arnell, Margaretha, 1929- . II. Föreningen Sveriges socialchefer.
HV790.A6B33 79-382399
79 MARC

Rönström, Anitha, 1939-
Barnomsorgen : intentioner och verklighet / Anitha Rönström. — Lund : Studentlitt., 1977.
130 p. ; 23 cm. S77-50/52
Bibliography: p. 124-128.
Includes index.
ISBN 9144142714 : kr31.00
1. Child welfare—Sweden. 2. Children in Sweden. 3. Socialization. 4. Child abuse—Sweden. I. Title.
HV790.A6R63 78-389094
79 MARC

Sweden. Socialstyrelsen.
Barnomsorgen i siffror : sammanställning av kommunernas barnomsorgsplaner 1977-81, m.m. — Stockholm : Socialstyrelsen, 1977.
62 p. : graphs ; 25 cm. S***
At head of title: Barnomsorgsplanering.
1. Child welfare—Sweden—Statistics. 2. Child welfare—Sweden. I. Title.
HV790.A4 1977 79-339681
79 MARC

Viklund, Monica, 1943-
Barnen behöver oss / Monica Viklund ; ⌈teckningar o. omsl., Helga Henschen⌉. — Stockholm : Socialstyr. ; Lund : Liber-Läromedel, 1978.
157 p. : ill., tables ; 21 cm. S78-17
Bibliography: p. 153-157.
ISBN 9140045900
1. Children—Sweden. 2. Child development. 3. Child psychology. 4. Child welfare—Sweden. I. Title.
HQ792.S8V54 78-395310
79 MARC

Vinterhed, Kerstin, 1941-
Gustav Jonsson på Skå : en epok i svensk barnavård / Kerstin Vinterhed. — Stockholm : Tiden, 1977.
604 p. ; 22 cm. S77-39
Thesis—Stockholm.
Summary in English.
Bibliography: p. 603.
ISBN 9155021409 : kr83.00
1. Jonsson, Gustav, 1907- . 2. Social work with children—Sweden—Biography. 3. Barnbyn Skå. 4. Family social work—Sweden. 5. Child welfare—Sweden. I. Title.
HV887.S8J667 79-356481
79 MARC

—SWEDEN—STATISTICS

Sweden. Socialstyrelsen.
Barnomsorgen i siffror : sammanställning av kommunernas barnomsorgsplaner 1977-81, m.m. — Stockholm : Socialstyrelsen, 1977.
62 p. : graphs ; 25 cm. S***
At head of title: Barnomsorgsplanering.
1. Child welfare—Sweden—Statistics. 2. Child welfare—Sweden. I. Title.
HV790.A4 1977 79-339681
79 MARC

—TAIWAN—CONGRESSES

Ch'üan kuo erh t'ung fu li yeh wu yen t'ao hui, T'ai-pei, Taiwan, 1975.
(Ch'üan kuo erh t'ung fu li yeh wu yen t'ao hui shih lu)
全國兒童福利業務研討會實錄 / 內政部編印. — ⌈臺北⌉ : 該部, 民國64 ⌈1975⌉
85 p. ; 27 cm.
Cover title.
1. Child welfare—Taiwan—Congresses. I. China (Republic of China). Nei cheng pu.
HV800.T3C58 1975 78-842687

—TEXAS

Texas Research League.
A study of the potential capabilities and child rearing practices of parents in active aid to dependent children cases : supplement to Report no. 1 in a study of State welfare programs. — Austin : Texas Research League, ⌈1959?⌉
29 p. ; 23 cm.
Cover title.
1. Child welfare—Texas. 2. Problem family—Texas. I. Texas Research League. Children's services of the Texas Department of Public Welfare. II. Title: A study of the potential capabilities and child rearing practices ...
HV742.T4T45 1959a 362.7'09764 79-312336
79 MARC

—UNITED STATES

Copeland, William C
Roadmap through Title XX / William C. Copeland, Iver A. Iversen. — Washington : CWLA Hecht Institute for State Child Welfare Planning, c1978.
i, 138 p. : diagrs. ; 28 cm. — (His Financing services for children through Title XX and other programs ; manual 5)
ISBN 0-87868-129-9
1. Child welfare—United States. 2. Federal aid to child welfare—United States. 3. Children—Law—United States. I. Iversen, Iver A., joint author. II. Title.
HV741.C653 vol. 5 362.7 s 78-112957
79 MARC

Costin, Lela B
Child welfare : policies and practice / Lela B. Costin. — 2d ed. — New York : McGraw-Hill, c1979.
ix, 469 p. : ill. ; 24 cm.
Includes bibliographies and index.
ISBN 0-07-013206-2
1. Child welfare. 2. Child welfare—United States. I. Title.
HV713.C67 1979 362.7'0973 78-15006
78 MARC

Decker, Celia Anita.
Planning and administering early childhood programs / Celia Anita Decker, John R. Decker. — Columbus, Ohio : C. E. Merrill Pub. Co., c1976.
vi, 266 p. : ill. ; 25 cm.
Includes bibliographies and index.
ISBN 0-675-08623-X
1. Education, Preschool—United States. 2. Day care centers—United States. 3. Child welfare—United States. I. Decker, John R., joint author. II. Title.
LB1140.2.D39 372.21'0973 75-36882
76⌈r79⌉rev MARC

De Lone, Richard H
Small futures : children, inequality, and the limits of liberal reform / by Richard H. de Lone for the Carnegie Council on Children. — 1st ed. — New York : Harcourt Brace Jovanovich, c1979.
xiv, 258 p. ; 24 cm.
Includes bibliographical references and index.
ISBN 0-15-183128-9
1. Children—United States. 2. Children's rights—United States. 3. Child welfare—United States. 4. Equality. 5. Liberalism—United States. I. Carnegie Council on Children. II. Title.
HQ792.U5D44 301.43'1'0973 77-92536
79 MARC

Family Service Association of America.
Proposed provisions for the accreditation of agencies serving families and children : draft #1. — New York : Family Service Association of America, c1977.
81 leaves ; 28 cm.
1. Family social work—United States. 2. Child welfare—United States. 3. Youth services—United States. I. Title. II. Title: Accreditation of agencies serving families and children.
HV699.F32 1977 362.8'2'0973 77-89139
78 MARC

Hosek, James R
Family participation in the AFDC-Unemployed Fathers Program / James R. Hosek. — Santa Monica, CA. : Rand, 1978.
xi, 52 p. : graphs ; 28 cm. — (⌈Report⌉ - Rand Corporation ; R-2316-HEW)
"Prepared under a grant from the U.S. Department of Health, Education, and Welfare."
Bibliography: p. 51-52.
ISBN 0-8330-0062-4
1. Welfare receipients—Employment—United States. 2. Fathers—Employment—United States. 3. Unemployed—United States. 4. Child welfare—United States. I. United States. Dept. of Health, Education, and Welfare. II. Title. III. Series: Rand Corporation. Rand report ; R-2316-HEW.
AS36.R3 R-2316 362.8'2 78-10521
⌈HV85⌉ 78 MARC

Knitzer, Jane.
Children without homes : an examination of public responsibility to children in out-of-home care : final report / written by Jane Knitzer and Mary Lee Allen. — Washington : Children's Defense Fund, c1978.
xv, 282 p. : ill. ; 26 cm.
"This study was conducted by Jane Knitzer, Mary Lee Allen, and Brenda McGowan."
Includes bibliographical references.
1. Child welfare—United States. 2. Children—Institutional care—United States. 3. Foster home care—United States. 4. Adoption—United States. I. Allen, Mary Lee, joint author. II. Washington Research Project. Children's Defense Fund. III. Title.
HV741.K54 362.7'3'0973 78-74230
79 MARC

Lavine, Abe.
Administration of public welfare in the case of Aid to Families with Dependent Children / by Abe Lavine. — ⌈Durham, N.C.⌉ : Institute of Policy Sciences and Public Affairs of Duke University, 1977.
xi, 108 p. ; 28 cm. — (Welfare policy project)
Includes bibliographical references.
1. Public welfare administration—United States. 2. Child welfare—United States. I. Title. II. Series.
HV95.L39 353.008'47'1 79-106626
79 MARC

National Council of Organizations for Children and Youth.
America's children 1976 : a bicentennial assessment : a fact-book about children in poverty, child health problems, changes in American family structure, child care needs, and federal programs serving children. — Washington : National Council of Organizations for Children and Youth, 1976.
87 p. : ill. ; 26 cm.
$4.00
1. Children—United States. 2. Children—United States—Statistics. 3. Child welfare—United States. 4. Social indicators—United States. I. Title.
HQ792.U5N37 1976 362.7'0973 75-42834
79 MARC

Shyne, Ann A 1914-
National study of social services to children and their families : overview : prepared for National Center for Child Advocacy, United States Children's Bureau, Administration on Children, Youth, and Families, Office of Human Development Services, United States Department of Health, Education, and Welfare / Ann W. Shyne, Anita G. Schroeder. — Washington : The Bureau : for sale by the Supt. of Docs. U.S. Govt. Print. Off., 1978.
vi, 26 p. : ill. ; 28 cm. — (DHEW publication ; no. (OHDS) 78-30149)
1. Child welfare—United States. 2. Foster home care—United States. I. Schroeder, Anita G., joint author. II. National Center for Child Advocacy. III. Title. IV. Series: United States. Dept. of Health, Education, and Welfare. DHEW publication ; no. (OHDS) 78-30149.
HV741.S44 362.7'1'0973 79-601088
79 MARC

United States. Congress. House. Committee on Ways and Means. Subcommittee on Public Assistance and Unemployment Compensation.
Definition and measurement of errors in the AFDC program and national AFDC error rate standards : hearings before the Subcommittee on Public Assistance and Unemployment Compensation of the Committee on Ways and Means, House of Representatives, Ninety-fifth Congress, second session, June 14 and 15, 1978. — Washington : U.S. Govt. Print. Off., 1978.
iii, 62 p. ; 24 cm.
"Serial 95-99."
1. Public welfare administration—United States—Quality control. 2. Child welfare—United States. I. Title.
KF27.W3464 1978a 353.008'4 78-603206
79 MARC

United States. Social Security Administration. Office of Research and Statistics.
Aid to families with dependent children : a chartbook. — ⌈Washington⌉ : U.S. Dept. of Health, Education, and Welfare, Social Security Administration, Office of Research and Statistics : for sale by the Supt. of Docs., U.S. Govt. Print. Off., ⌈1978⌉
29 p. : ill. ; 15 x 24 cm. — (HEW publication ; no. (SSA) 78-11721)
1. Father-separated children. 2. Single-parent family—United States. 3. Welfare recipients—United States. 4. Child welfare—United States. I. Title. II. Series: United States. Dept. of Health, Education, and Welfare. DHEW publication ; (SSA) 78-11721.
HV699.U57 1978 362.7'1'0973 78-600048
79 MARC

Young, Leontine R
Wednesday's children : a study of child neglect and abuse / by Leontine Young. — Westport, Conn. : Greenwood Press, 1979, c1964.
195 p. ; 23 cm.
Reprint of the ed. published by McGraw-Hill, New York.
Bibliography: p. 187-191.
Includes index.
ISBN 0-313-20637-6
1. Child welfare—United States. I. Title.
⌈HV741.Y6 1979⌉ 362.7'1 78-12941
78 MARC

—UNITED STATES—ADDRESSES, ESSAYS, LECTURES

Child welfare strategy in the coming years. — ⌈Washington⌉ : U.S. Dept. of Health, Education, and Welfare, Office of Human Development Services, Administration for Children, Youth and Families, Children's Bureau, ⌈1978⌉
465 p. : ill. ; 26 cm. — (DHEW publication ; no. (OHDS) 78-30158)
Includes bibliographies.
1. Child welfare—United States—Addresses, essays, lectures. 2. Youth—Services for—United States—Addresses, essays, lectures. 3. Family social work—United States—Addresses, essays, lectures. I. United States. Children's Bureau. II. Series: United States. Dept. of Health, Education, and Welfare. DHEW publication ; no. (OHDS) 78-30158.
HV741.C5358 362.7'0973 78-603550
79 MARC

CHILD WELFARE
—UNITED STATES—ADDRESSES, ESSAYS, LECTURES (Continued)

Reid, Joseph H
Child welfare perspectives : selected papers of Joseph H. Reid / edited by Ann W. Shyne. — New York : Child Welfare League of America, c1979.
ix, 154 p. ; 24 cm.
"Other publications of Joseph H. Reid": p. 153-154.
Includes bibliographical references.
ISBN 0-87868-133-7
1. Child welfare—United States—Addresses, essays, lectures. 2. Children's rights—United States—Addresses, essays, lectures. I. Shyne, Ann W., 1914- II. Child Welfare League of America. III. Title.
HV741.R44 362.7'08 79-102125
79 MARC

—UNITED STATES—CONGRESSES

Child Welfare Policy Conference, 2d, Washington, D.C., 1978.
Proceedings of the Child Welfare Policy Conference : issues and answers in Region III, March 30-31, 1978 / edited by Barbara J. Stembridge. — Washington : Region III Child Abuse and Neglect Resource Center, Institute for Urban Affairs and Research, Howard University, c1978.
xi, 149 p. : ill. ; 26 cm.
1. Child welfare—United States—Congresses. I. Stembridge, Barbara J. II. Howard University, Washington, D.C. Region III Child Abuse and Neglect Resource Center.
HV741.C5356 1978 362.7'0973 79-102023
79 MARC

—UNITED STATES—DIRECTORIES

A Guide to organizations, agencies, and Federal programs for children / Joyce Lynn, editor. — Washington : ₍Plus Publications₎, 1978.
37 p. ; 28 cm. — (Day care & child development reports ; report 6)
$9.00
1. Child welfare—United States—Directories. I. Lynn, Joyce. II. Series.
HV741.G84 362.7'025'73 79-100379
79 MARC

National Center for Child Advocacy.
Child advocacy programs, 1975. — ₍Washington₎ : U.S. Dept. of Health, Education, and Welfare, Office of Child Development, Office of Human Development, Children's Bureau, National Center for Child Advocacy, ₍1975?₎
118 p. ; 27 cm. — (DHEW publication ; no. (OHD) 76-30082)
1. Child welfare—United States—Directories. 2. Ombudsman—United States—Directories. I. Title. II. Series: United States. Dept. of Health, Education, and Welfare. DHEW publication ; no. (OHD) 76-30082.
HV741.N277 1975 362.7'0973 76-601516
76₍r79₎rev MARC

—UNITED STATES—PERIODICALS

United States. Administration for Children, Youth and Families. Research and Evaluation Division.
The status of children.
₍Washington₎ U. S. Dept. of Health, Education, and Welfare, Office of Human Development Services, Administration for Children, Youth and Families, Research and Evaluation Division.
v. ill. 27 cm. biennial. (DHEW publication (OHDS))
Began with 1975 issue.
Key title: The Status of children, ISSN 0161-4150
1. Child welfare—United States—Periodicals. 2. Child welfare—United States—Statistics—Periodicals. I. Title. II. Series: United States. Dept. of Health, Education, and Welfare. DHEW publication.
HV983.U52a 362.7'0973 78-643260
MARC-S
78

—UNITED STATES—SOCIETIES, ETC.—DIRECTORIES

Child Welfare League of America.
Associate agencies - Child Welfare League of America, inc. 2- ed.; 1978-
New York, Child Welfare League of America.
v. 23 cm. annual.
Continues: Directory of associate agencies - Child Welfare League of America, inc., ISSN 0147-9172.
Key title: Associate agencies - Child Welfare League of America, ISSN 0163-3694
1. Child Welfare League of America. 2. Child welfare—United States—Societies, etc.—Directories. I. Key title.
HV741.C5345a 362.7'025'73 78-646558
MARC-S

—UNITED STATES—STATISTICS—PERIODICALS

United States. Administration for Children, Youth and Families. Research and Evaluation Division.
The status of children.
₍Washington₎ U. S. Dept. of Health, Education, and Welfare, Office of Human Development Services, Administration for Children, Youth and Families, Research and Evaluation Division.
v. ill. 27 cm. biennial. (DHEW publication (OHDS))
Began with 1975 issue.
Key title: The Status of children, ISSN 0161-4150
1. Child welfare—United States—Periodicals. 2. Child welfare—United States—Statistics—Periodicals. I. Title. II. Series: United States. Dept. of Health, Education, and Welfare. DHEW publication.
HV983.U52a 362.7'0973 78-643260
MARC-S
78

—UNITED STATES—STATES

Hutchens, Robert M
Changes in AFDC tax rates, 1967-1971 / Robert M. Hutchens. — ₍Madison₎ : University of Wisconsin—Madison, 1976.
23, ₍2₎ p. ; 28 cm. — (Discussion papers - Institute for Research on Poverty ; 352-76)
Bibliography: p. ₍25₎
1. Welfare recipients—Taxation—United States—States. 2. Child welfare—United States—States. I. Title. II. Series: Wisconsin. University—Madison. Institute for Research on Poverty. Discussion papers ; 352-76.
HJ4652.H955 336.2'42 78-623185
79 MARC

—URUGUAY—HISTORY

Gorlero Bacigalupi, Ruben.
Protección social al menor en el Uruguay, 1800-1935 / Ruben Gorlero Bacigalupi. — Montevideo : Universidad de la República, División Publicaciones y Ediciones, 1978.
220 p. ; 24 cm.
Includes legislation.
Bibliography: p. 217-220.
1. Child welfare—Uruguay—History. I. Title.
HV747.U72G67 362.7'09895 78-112261
79 MARC

—VENEZUELA

Colomine Solarte, Feijoo.
El menor en situación irregular en Venezuela / Feijoo Colomine Solarte. — Caracas : ₍s.n.₎, 1974 ₍i.e. 1976₎
181 p., ₍2₎ leaves of plates : graphs ; 23 cm.
Bibliography: p. 171-175.
1. Child welfare—Venezuela. I. Title.
HV747.V42C64 78-389747
79 MARC

Consejo Venezolano del Niño. Departamento de Metas Generales.
El Consejo Venezolano del Niño / Oficina de Coordinación y Planificación, Departamento de Metas Generales. — Caracas : El Departamento, ₍1976 or 1977₎
46 p., ₍1₎ leaf of plates : diagr. ; 22 cm.
1. Consejo Venezolano del Niño. 2. Child welfare—Venezuela. I. Title.
HV747.V4C6623 362.7'0987 78-100622
79 MARC

—VENEZUELA—STATISTICS

Consejo Venezolano del Niño. Departamento de Estadística.
Series estadísticas, 1959-1972 / Consejo Venezolano del Niño, Oficina de Planificación y Presupuesto, Departamento de Estadística. — Caracas : El Consejo, 1974.
122 p. : graphs ; 28 cm.
Chiefly tables.
1. Child welfare—Venezuela—Statistics. I. Title.
HV747.V42C645 1974 362.7'0987 79-345592
79 MARC

—VIRGINIA

Virginia. Commission for Children and Youth.
Alternative needs assessment techniques for Virginia : a handbook. — ₍Richmond₎ : Virginia Commission for Children and Youth, ₍1977?₎
₍89₎ p. : ill. ; 28 cm.
Bibliography: p. ₍89₎
1. Child welfare—Virginia. I. Title.
HV742.V8V57 1977a 362.7'042'09755 78-622422
79 MARC

Virginia. Commission for Children and Youth.
Innovative programming for children & youth : the state of the art. — ₍Richmond₎ : Virginia Commission for Children and Youth, ₍1977 or 1978₎
90 p. in various pagings ; 28 cm.
"3"
Includes a bibliography.
1. Child welfare—Virginia. 2. Youth services—Virginia. I. Title.
HV742.V8V57 1977b 362.7'09755 78-622049
79 MARC

—VIRGINIA—STATISTICS

Virginia. Commission for Children and Youth.
An inventory & analysis of State data sources for children & youth, Virginia / Virginia Commission for Children and Youth. — ₍Richmond₎ : The Commission, ₍1977?₎
156 p. in various pagings : ill. ; 28 cm. — (Its What is the answer, what was the question? ; 1)
Includes bibliographical references.
1. Child welfare—Virginia—Statistics. 2. Child welfare—Information services—Virginia. 3. Youth services—Virginia—Statistics. 4. Youth services—Information services—Virginia. I. Title. II. Series.
HV742.V8V57 1977 362.7'09755 78-622048
79 MARC

—YUGOSLAVIA—BELGRAD

Samoupravna interesna zajednica društvene brige o deci predškolskog uzrasta Beograda.
Dokumenti / Samoupravna interesna zajednica društvene brige o deci predškolskog uzrasta Beograda. — Beograd : Služba za informacije Stručne službe interesnih zajednica društvenih delatnosti Beograda, 1978.
54 p. ; 24 cm. Yu***
1. Child welfare—Yugoslavia—Belgrad.
HV799.Y83B477 79-370770
79 MARC

CHILD WELFARE LEAGUE OF AMERICA

Child Welfare League of America.
Associate agencies - Child Welfare League of America, inc. 2- ed.; 1978-
New York, Child Welfare League of America.
v. 23 cm. annual.
Continues: Directory of associate agencies - Child Welfare League of America, inc., ISSN 0147-9172.
Key title: Associate agencies - Child Welfare League of America, ISSN 0163-3694
1. Child Welfare League of America. 2. Child welfare—United States—Societies, etc.—Directories. I. Key title.
HV741.C5345a 362.7'025'73 78-646558
MARC-S

CHILDBIRTH

see also Birth (in religion, folk-lore, etc.); Labor (Obstetrics); Natural childbirth

Beels, Christine.
The childbirth book / ₍by₎ Christine Beels ; illustrated by Angela Owen. — London : Turnstone Books, 1978.
256 p. : ill. ; 20 cm. GB78-13608
Includes bibliographical references.
ISBN 0-85500-071-6 : £4.50. ISBN 0-85500-072-4 pbk.
1. Obstetrics—Popular works. 2. Childbirth. I. Title.
RG525.B423 618.4 78-318587
78 MARC

Bittman, Sam, 1943-
Expectant fathers / by Sam Bittman and Sue Rosenberg Zalk ; ill. by F. X. Tobin. — New York : Hawthorn Books, c1978.
xxv, 291 p. : ill. ; 25 cm.
Bibliography: p. 278-283.
Includes index.
ISBN 0-8015-2444-X : $12.95
1. Pregnancy. 2. Childbirth. 3. Fathers. I. Zalk, Sue Rosenberg, joint author. II. Title.
RG525.B63 1978 612.6'3 78-54666
79 MARC

Bordow, Joan Wiener, 1944-
A motherhood book: adventures in pregnancy, birth, and being a mother ₍by₎ Joan Wiener and Joyce Glick. New York, Macmillan ₍1974₎
x, 131 p. illus. 21 cm.
Bibliography: p. 111-114.
ISBN 0-02-625350-X
1. Pregnancy. 2. Childbirth. 3. Infants—Care and hygiene. I. Glick, Joyce, joint author. II. Title.
₍DNLM: 1. Pregnancy—Popular works. WQ150 W423a 1974₎
RG525.B6447 649'.10242 73-2745
73₍r79₎rev2 MARC

CHILDBIRTH (Continued)

Feldman, Silvia.
Choices in childbirth / Silvia Feldman. — New York : Grosset & Dunlap, c1978.
xv, 267 p. : ill. ; 27 cm.
Bibliography: p. [256]-259.
Includes index.
ISBN 0-448-14524-3 : $14.95 ($16.95 Can.). ISBN 0-448-14525-1 pbk.
1. Pregnancy. 2. Childbirth. 3. Maternal health services—United States. I. Title.
RG525.F45 618.2'4 77-95181
78 MARC

Fenlon, Arlene.
Getting ready for childbirth : a guide for expectant parents / Arlene Fenlon, Ellen McPherson, Lovell Dorchak. — Englewood Cliffs, N.J. : Prentice-Hall, 1979.
x, 202 p. : ill. ; 24 cm. — (A Spectrum book)
Includes bibliographies and index.
ISBN 0-13-354803-1. ISBN 0-13-354795-7 pbk.
1. Pregnancy. 2. Childbirth. 3. Obstetrics—Popular works. 4. Infants (Newborn)—Care and hygiene. I. McPherson, Ellen, joint author. II. Dorchak, Lovell, joint author. III. Title.
RG525.F458 1979 618.2'4 79-13442
79 MARC

Fenwick, Peter.
The baby book for fathers : all the expectant husband needs to know / [by] Peter and Elizabeth Fenwick ; illustrated by Leon Olin and Sylvia Gainsford. — London [etc.] : Angus and Robertson, 1978.
vii, 208 p. : ill. ; 21 cm. GB78-22816
Bibliography: p. [194]-196.
Includes index.
ISBN 0-207-95726-6 : £3.95
1. Pregnancy. 2. Childbirth. 3. Fathers. 4. Infants—Care and hygiene. I. Fenwick, Elizabeth, joint author. II. Title.
RG525.F46 618.2'4'0240431 78-325732
79 MARC

Guidez, Guylaine.
Le livre de la future maman / Guylaine Guidez ; préf. d'Annik Beauchamps ; dessins de l'auteur. — [Paris] : Solar, 1975.
333 p. : ill. ; 21 cm. F***
Includes index.
29.95F
1. Pregnancy. 2. Childbirth. 3. Infants—Care and hygiene. I. Title.
RG525.G78 649'.1 78-396135
79 MARC

Isbister, Clair.
Birth of a family : a guide for expectant parents and those who care about children / [by] Clair Isbister. — Melbourne : Thomas Nelson (Australia), 1976.
xii, 260 p. : ill., diagrs. ; 22 cm. Aus
Bibliography: p. 246-249.
Includes index.
ISBN 0-17-005062-9
1. Pregnancy. 2. Childbirth. 3. Infants—Care and hygiene. 4. Parent and child. 5. Family. I. Title.
RG525.I74 1976 618.2'4 77-352652
77[r79]rev MARC

Isbister, Clair.
Birth of a family : a preparation for parenthood / Clair Isbister. — New York : Hawthorn Books, c1978.
x, 252 p. : ill. ; 25 cm.
Bibliography: p. 236-241.
Includes index.
ISBN 0-8015-0653-0 : $8.95
1. Pregnancy. 2. Childbirth. 3. Infants—Care and hygiene. 4. Parent and child. 5. Family.
RG525.I74 1978 618.2'4 78-52878
78 MARC

Journey to motherhood : great writings on pregnancy, from a woman's first love to her baby's first day of life / compiled by Emanuel Martin Greenberg. — New York : St. Martin's Press, c1979.
387 p. ; 24 cm.
ISBN 0-312-44506-7 : $10.00
1. Pregnancy—Literary collections. 2. Childbirth—Literary collections. 3. Pregnancy. 4. Childbirth. I. Greenberg, Emanuel Martin.
PN6071.P62J6 808.8'99287 78-3985
78 MARC

Jurčíková, Vanda.
Než se narodí / Vanda Jurčíková. — 7. vyd. — Praha : Ústav zdravotní výchovy, 1977.
48 p. : col. ill. ; 17 cm. Cz***
1. Pregnancy. 2. Prenatal care. 3. Childbirth. I. Title.
RG525.J87 1977 78-399335
79 MARC

Kieffer, Joyce L
To have, to hold : a parents guide to childbirth & early parenting / by Joyce L. Kieffer ; [design & ill. by Colonial Quill Advertising]. — Harrisburg, Pa. : Training Resource Associates, c1979.
71 p. : ill. ; 22 x 28 cm.
Bibliography: p. 70-71.
ISBN 0-933794-00-2
1. Pregnancy. 2. Childbirth. 3. Infants—Care and hygiene. 4. Parenting. I. Title.
RG525.K49 618.2'4 79-63193
79 MARC

Kramer, Rita.
Giving birth : childbearing in America today / Rita Kramer. — Chicago : Contemporary Books, c1978.
xvii, 236 p. ; 24 cm.
Bibliography: p. 225-227.
Includes index.
ISBN 0-8092-7859-6 : $10.95
1. Childbirth. 2. Obstetrics. 3. Maternal health services. I. Title.
RG652.K73 1978 618.4 78-57440
78 MARC

Nilsson, Lennart, 1922-
Ett barn blir till : en bildskildring av barnets tillblivelse före födelsen och praktiska råd när man väntar barn / foto, Lennart Nilsson ; text, Mirjam Furuhjelm, Axel Ingelman-Sundberg, Claes Wirsén ; teckningar, Bernt Forsblad. — Stockholm : Bonniers, 1976.
112 p. : ill. ; 25 cm. S76-49
Includes index.
ISBN (invalid) 9100413096 : kr112.00
1. Pregnancy. 2. Childbirth. 3. Pregnancy—Pictorial works. 4. Childbirth—Pictorial works. I. Furuholm, Mirjam. II. Ingelman-Sundberg, Axel Gustaf Isidor, 1910- . III. Wirsén, Claes, 1932- . IV. Title.
RG525.N54 79-385456
79 MARC

Wachstein, Alison Ehrlich.
Pregnant moments : the experience of pregnancy and childbirth / photos. and text by Alison Ehrlich Wachstein. — Dobbs Ferry, N.Y. : Morgan & Morgan, c1979.
143 p. : ill. ; 23 x 27 cm.
Bibliography: p. 141-142.
ISBN 0-87100-153-5
1. Pregnancy. 2. Childbirth. 3. Pregnancy—Pictorial works. 4. Childbirth—Pictorial works. I. Title.
RG525.W23 618.2'4 79-88347
00 MARC

—ATLASES

Calera, Ana María.
Atlas de puericultura / Ana M. Calera. — 9. ed. — Barcelona : Ediciones Jover, c1976.
[86] p. : ill. ; 20 cm. — (Colección Atlas : A, Ciencias aplicadas) Sp***
Cover title.
ISBN 8470931377
1. Pregnancy—Atlases. 2. Childbirth—Atlases. 3. Infants—Care and hygiene—Atlases. I. Title.
RG520.C34 1976 78-393949
79 MARC

—JUVENILE LITERATURE

Mørch, Dea Trier, 1941-
Ind i verden / [linoleumssnit og tekst af] Dea Trier Mørch. — [København] : Gyldendal, 1977.
[63] p. : ill. ; 21 x 30 cm. D77-52
ISBN 8701659812 : kr49.75
1. Childbirth—Juvenile literature. I. Title.
RG525.5.M63 79-367461
79 MARC

—LITERARY COLLECTIONS

Journey to motherhood : great writings on pregnancy, from a woman's first love to her baby's first day of life / compiled by Emanuel Martin Greenberg. — New York : St. Martin's Press, c1979.
387 p. ; 24 cm.
ISBN 0-312-44506-7 : $10.00
1. Pregnancy—Literary collections. 2. Childbirth—Literary collections. 3. Pregnancy. 4. Childbirth. I. Greenberg, Emanuel Martin.
PN6071.P62J6 808.8'99287 78-3985
78 MARC

—MISCELLANEA

Curtis, Lindsay R
Pregnant and lovin' it / by Lindsay R. Curtis and Yvonne Coroles ; illustrated by Paul Farber. — Tucson, Ariz. : HPBooks, c1977.
247 p. : ill. ; 23 cm.
Includes index.
ISBN 0-912656-82-4
1. Pregnancy—Miscellanea. 2. Childbirth—Miscellanea. I. Coroles, Yvonne, joint author. II. Title.
RG525.C94 618.2'4 77-8212
78[r79]rev MARC

—PICTORIAL WORKS

Nilsson, Lennart, 1922-
Ett barn blir till : en bildskildring av barnets tillblivelse före födelsen och praktiska råd när man väntar barn / foto, Lennart Nilsson ; text, Mirjam Furuhjelm, Axel Ingelman-Sundberg, Claes Wirsén ; teckningar, Bernt Forsblad. — Stockholm : Bonniers, 1976.
112 p. : ill. ; 25 cm. S76-49
Includes index.
ISBN (invalid) 9100413096 : kr112.00
1. Pregnancy. 2. Childbirth. 3. Pregnancy—Pictorial works. 4. Childbirth—Pictorial works. I. Furuholm, Mirjam. II. Ingelman-Sundberg, Axel Gustaf Isidor, 1910- . III. Wirsén, Claes, 1932- . IV. Title.
RG525.N54 79-385456
79 MARC

Wachstein, Alison Ehrlich.
Pregnant moments : the experience of pregnancy and childbirth / photos. and text by Alison Ehrlich Wachstein. — Dobbs Ferry, N.Y. : Morgan & Morgan, c1979.
143 p. : ill. ; 23 x 27 cm.
Bibliography: p. 141-142.
ISBN 0-87100-153-5
1. Pregnancy. 2. Childbirth. 3. Pregnancy—Pictorial works. 4. Childbirth—Pictorial works. I. Title.
RG525.W23 618.2'4 79-88347
00 MARC

—POETRY

Portugés, Paul Cornel, 1945-
Aztec birth : the turquoise mockingbird of light : English workings / by Paul Portugés ; with Nahuatl as transcribed by Sahagún ; and an afterword by Peter Whigham ; [art by Paul Hobson]. — Santa Barbara, CA : Mudborn Press, 1979.
[23] p. : ill. ; 27 cm. — (Inklings ; 2 ISSN 0190-0234)
English and Nahuatl.
Includes selections from the Historia general de las cosas de Nueva España by Sahagún.
Bibliography: p. [28]
ISBN 0-930012-32-1
1. Aztecs—Poetry. 2. Childbirth—Poetry. 3. Aztecs—Social life and customs. I. Sahagun, Bernardino de, d. 1590. Historia general de las cosas de Nueva España. Selections. 1979. II. Title.
PS3566.O666A97 811'.5'4 79-120182
79 MARC

—PSYCHOLOGICAL ASPECTS

Chesler, Phyllis.
With child, a diary of motherhood / Phyllis Chesler. — 1st ed. — New York : Crowell, c1979.
288 p. ; 24 cm.
Bibliography: p. [283]-288.
ISBN 0-690-01835-5 : $9.95
1. Pregnancy—Psychological aspects. 2. Childbirth—Psychological aspects. 3. Mother and child. 4. Chesler, Phyllis. 5. Mothers—United States—Biography. 6. Feminists—United States—Biography. I. Title.
RG560.C47 1979 301.41'2'0924 79-7081
79 MARC

Oakley, Ann.
Becoming a mother / [by] Ann Oakley. — Oxford : Martin Robertson, 1979.
[7], 328 p. ; 23 cm. GB79-16378
Includes bibliographical references.
ISBN 0-85520-206-8 : £7.95
1. Pregnancy—Psychological aspects. 2. Childbirth—Psychological aspects. 3. Infants—Care and hygiene. 4. Children, First-born. 5. Mothers—Great Britain—Interviews. 6. Housewives—Great Britain—Interviews. I. Title.
RG560.O24 618.2 79-315752
79 MARC

Oxman, Alice, 1946-
Lager maternità : libro-documentario sulle donne e i bambini in venti storie italiane / Alice Oxman. — 2. ed. — Milano : V. Bompiani, 1974.
173 p. ; 22 cm. — (I Mutamenti ; 6) It76-Mar
L3000
1. Pregnancy—Social aspects. 2. Childbirth—Psychological aspects. 3. Mothers—Italy—Biography. I. Title.
RG556.O95 1974 78-399360
*79 MARC

Prince, Joyce.
Minds, mothers, and midwives : the psychology of childbirth / Joyce Prince, Margaret E. Adams ; foreword by Anne Bent. — Edinburgh ; New York : Churchill Livingstone ; New York : distributed by Longman, 1978.
179 p. : ill. ; 22 cm. GB78
Includes bibliographical references and index.
ISBN 0-443-01553-8 : £3.95
1. Pregnancy—Psychological aspects. 2. Childbirth—Psychological aspects. 3. Parent and child. 4. Midwives. I. Adams, Margaret E., 1916- joint author. II. Title.
RG560.P73 612.6'3'019 78-40033
78 MARC

CHILDBIRTH
—PSYCHOLOGICAL ASPECTS
(Continued)

Psychological aspects of gynecology and obstetrics / compiled and edited by Benjamin B. Wolman. — Oradell, N.J. : Medical Economics Co., Books Division, c1978.
xvii, 270 p. ; 27 cm.
Includes bibliographies.
ISBN 0-87489-009-8
1. Gynecology—Psychological aspects. 2. Obstetrics—Psychological aspects. 3. Childbirth—Psychological aspects. 4. Women—Psychology. I. Wolman, Benjamin B.
RG103.5.P77 618'.01'9 79-100594
79 MARC

Whelan, Elizabeth M
The pregnancy experience : the psychology of expectant parenthood / Elizabeth M. Whelan. — 1st ed. — New York : Norton, c1978.
202 p. ; 22 cm.
Bibliography: p. [189]-193.
Includes index.
ISBN 0-393-01179-8
1. Pregnancy—Psychological aspects. 2. Childbirth—Psychological aspects. I. Title.
RG560.W48 1978 618.2'001 78-15397
78 MARC

—STUDY AND TEACHING

Hassid, Patricia.
Textbook for childbirth educators / Patricia Hassid. — Hagerstown, Md. : Medical Dept., Harper & Row, 1978.
xvii, 227 p. : ill. ; 24 cm.
Bibliography: p. 200-204.
Includes index.
ISBN 0-06-141110-8
1. Childbirth—Study and teaching. 2. Natural childbirth—Study and teaching.
RG971.H37 618.4'5 77-13053
77 MARC

Kitzinger, Sheila.
Education and counseling for childbirth / Sheila Kitzinger. — New York : Schocken Books, 1979, c1977.
xvi, 303 p. ; 20 cm.
Bibliography: p. [287]-293.
Includes index.
ISBN 0-8052-0613-2 : $6.95
1. Childbirth—Study and teaching. I. Title.
RG973.K57 1979 618.2'4'0715 78-21114
78 MARC

Sasmor, Jeannette L
Childbirth education : a nursing perspective / Jeannette L. Sasmor. — New York : Wiley, c1979.
xvi, 322 p. : ill. ; 23 cm. — (A Wiley medical publication)
Includes bibliographies and index.
ISBN 0-471-75490-0
1. Childbirth—Study and teaching. 2. Obstetrical nursing. 3. Obstetrical nursing. I. Title.
[DNLM: 1. Obstetrics—Education—Nursing texts. 2. Labor—Nursing texts. 3. Parents—Education—Nursing texts. WY18.3 S252c]
RG973.S37 610.73'678 78-32117
79 MARC

Wiggins, Jayne DeClue, 1923-
Childbearing : physiology, experiences, needs / Jayne DeClue Wiggins. — St. Louis : Mosby, 1979.
x, 178 p. : ill. ; 24 cm.
Bibliography: p. 164-165.
Includes index.
ISBN 0-8016-5543-9
1. Pregnancy. 2. Labor (Obstetrics) 3. Postnatal care. 4. Childbirth—Study and teaching. 5. Obstetrical nursing. I. Title.
[DNLM: 1. Pregnancy. 2. Labor. WQ205 W655c]
RG526.W53 612.6'3 79-10736
79 MARC

—UNITED STATES—HISTORY

Wertz, Richard W
Lying-in : a history of childbirth in America / Richard W. Wertz and Dorothy C. Wertz. — New York : Free Press, c1977.
xii, 260 p. : ill. ; 24 cm.
Bibliography: p. 249-255.
Includes index.
ISBN 0-02-934510-3
1. Obstetrics—United States—History. 2. Childbirth—United States—History. 3. Obstetrics—Social aspects—United States—History. 4. Maternal health services—United States—History. I. Wertz, Dorothy C., joint author. II. Title.
RG518.U5W47 1977 618.4 77-72040
77[r79]rev MARC

Wertz, Richard W
Lying-in : a history of childbirth in America / Richard W. Wertz and Dorothy C. Wertz. — New York : Schocken Books, 1979, c1977.
xii, 260 p. : ill. ; 20 cm. — (Studies in the life of women)
Bibliography: p. 249-255.
Includes index.
ISBN 0-8052-0615-9 : $4.95
1. Obstetrics—United States—History. 2. Childbirth—United States—History. 3. Obstetrics—Social aspects—United States—History. 4. Maternal health services—United States—History. I. Wertz, Dorothy C., joint author. II. Title.
[RG518.U5W47 1979] 618.2'00973 78-26045
78 MARC

—UNITED STATES—STATISTICS

Querec, Linda J
Characteristics of births : United States, 1973-1975 / [Linda Querec and Ernell Spratley]. — Hyattsville, Md. : U.S. Dept. of Health Education and Welfare, Public Health Service, National Center for Health Statistics, 1978.
vi, 49 p. : ill. ; 26 cm. — (Vital and health statistics : Series 21, Data from the National Vital Statistics System ; no. 30) (DHEW publication no. (PHS) 78-1908)
Includes bibliographical references.
ISBN 0-8406-0135-2
1. Childbirth—United States—Statistics. 2. Prenatal care—United States—Statistics. I. Spratley, Ernell, joint author. II. Title. III. Series: United States. National Center for Health Statistics. Vital and health statistics : Series 21, Data from the National Vital Statistics system, Data on natality, marriage, and divorce ; no. 30. IV. Series: United States. Dept. of Health, Education, and Welfare. DHEW publication no. (PHS) 78-1908.
[DNLM: 1. Vital statistical—United States. W2 A N148vu no. 30]
HA211.A3 no. 30 312'.1'73 s 78-15448
[HB915] 78 MARC

CHILDBIRTH, NATURAL see Natural childbirth

CHILDBIRTH AT HOME

Baldwin, Rahima, 1949-
Special delivery : the complete guide to informed birth / Rahima Baldwin. — Millbrae, Calif. : Les Femmes, c1979.
xiii, 169 p. : ill. ; 28 cm.
Bibliography: p. 163-164.
Includes index.
ISBN 0-89087-934-6 : $9.95
1. Natural childbirth. 2. Childbirth at home. I. Title.
RG661.B34 618.4'5 78-61611
78 MARC

Children at birth / by the Hathaways, Jay ... [et al.] ; photos. by Jay Hathaway and many others ; foreword by Robert A. Bradley. — Sherman Oaks, CA : Academy Publications, c1978.
xi, 178 p. : ill. ; 23 cm.
Bibliography: p. 91-92.
ISBN 0-931560-00-4 : $5.95
1. Natural childbirth. 2. Childbirth at home. 3. Family. 4. Brothers and sisters. I. Hathaway, Jay.
RG661.C52 618.4'5 78-53647
79 MARC

—IDAHO—STATISTICS

Barton, Loretta J
Out-of-hospital births in the State of Idaho in 1977 / by Loretta J. Barton and Carol D. H. Harvey. — Boise, Idaho : Center for Research, Grants, and Contracts, Boise State University, 1978.
68 leaves : ill. ; 28 cm.
Bibliography: leaves 43-45.
1. Childbirth at home—Idaho—Statistics. 2. Idaho—Statistics, Vital. I. Harvey, Carol D. H., joint author. II. Boise State University. Center for Research, Grants, and Contracts. III. Title.
RG530.3.U52I333 362.1'9'84 78-623943
79 MARC

CHILDBIRTH IN MIDDLE AGE

The Pregnancy-after-30 workbook : program for safe childbearing, no matter what your age / edited by Gail Sforza Brewer. — Emmaus, Pa. : Rodale Press, c1978.
xiii, 233 p. : ill. ; 28 cm.
Includes bibliographies and index.
ISBN 0-87857-215-5 : $8.95
1. Pregnancy in middle age. 2. Childbirth in middle age. I. Brewer, Gail Sforza.
RG556.6.P73 618.2'4 78-15008
78 MARC

Price, Jane, 1944-
You're not too old to have a baby / Jane Price. — New York : Farrar, Straus, Giroux, c1977.
xii, 147 p. ; 21 cm.
Bibliography: p. 145-147.
ISBN 0-374-29606-5
1. Pregnancy in middle age. 2. Childbirth in middle age. 3. Parent and child. I. Title.
RG556.6.P75 1977 301.42'72'0564 77-23055
77[r79]rev MARC

Price, Jane, 1944-
Your're not too old to have a baby / Jane Price. — New York : Penguin Books, 1978, c1977.
xii, 147 p. ; 18 cm.
Reprint of the 1977 ed. published by Farrar, Straus, Giroux, New York.
Bibliography: p. 145-147.
ISBN 0-14-004910-X pbk.
1. Pregnancy in middle age. 2. Childbirth in middle age. 3. Parent and child. I. Title.
[RG556.6.P75 1978] 612.6'3 78-9876
78 MARC

CHILDBIRTH TEACHERS

—UNITED STATES—DIRECTORIES

Who's who in childbirth education.
[Washington, American Society for Psychoprophylaxis in Obstetrics]
v. 22 cm.
"Directory of the certified teacher and physician members of American Society for Psychoprophylaxis in Obstetrics, Inc."
Key title: Who's who in childbirth education, ISSN 0163-5492
1. Childbirth teachers—United States—Directories. 2. American Society for Psychoprophylaxis in Obstetrics—Directories. I. American Society for Psychoprophylaxis in Obstetrics.
RG973.W47 618.2'4'07 78-648239
MARC-S

CHILDHOOD see Children

CHILDLESSNESS

see also Sterility

CHILDMINDERS see Day care aides

CHILDLESSNESS

Valabrègue, Catherine.
Des enfants, pourquoi? / Catherine Valabrègue. — Paris : Stock, c1978.
167 p. ; 22 cm. — (Stock/femmes) F***
Includes bibliographical references.
ISBN 2-234-00793-3 : 35.00F
1. Childlessness. I. Title.
HQ737.V33 301.43'1 79-338668
79 MARC

CHILDREN, JOHN GEORGE, 1777-1852

Gunther, Albert Edward.
John George Children, F.R.S. (1777-1852) of the British Museum, mineralogist and reluctant Keeper of Zoology / A. E. Gunther. — London : British Museum (Natural History), 1978.
p. 75-108 : ill. ; 25 cm. — (Bulletin of the British Museum (Natural History) : Historical series ; v. 6, no. 4 ISSN 0068-2306) GB***
Cover title.
"Bibliography of the works of J. G. Children": p. 104-105.
Bibliography: p. 105-108.
1. Children, John George, 1777-1852. 2. Zoologists—Great Britain—Biography. 3. Mineralogists—Great Britain—Biography. I. Title. II. Series: British Museum (Natural History). Bulletin : Historical series ; v. 6, no. 4.
QH15.B73 vol. 6, no. 4 574'.08 s 79-313632
[QL31.C533] 79 MARC

CHILDREN

see also Arts and children; Birth order; Boys; Church work with children; City children; Day care centers; Domestic education; Education of children; Eskimos —Children; Exceptional children; Geographical perception in children;

Girls; Infants; Jewish children; Kindergarten; Maoris—Children; Play; Prenatal influences; School children; Sick children; Social work with children; Stepchildren; Television advertising and children; Television and

CHILDREN (Continued)

children; World War, 1939-1945 —Children; Youth; and headings beginning with the word Child

Johansen, Erna Maria, 1911-
Betrogene Kinder : e. Sozialgeschichte d. Kindheit / Erna M. Johansen. — Orig.-Ausg. — Frankfurt am Main : Fischer-Taschenbuch-Verlag, 1978.
267 p. : 43 ill. ; 18 cm. — (Bücher des Wissens) GFR78-A41
Bibliography: p. 259-[263]
Includes indexes.
ISBN 3-596-26622-X : DM7.80
1. Children. 2. Child welfare. 3. Children's rights. I. Title.
HQ781.J62 301.43'14 79-344766
79 MARC

Varga, Domokos.
Kölyökkóstolgató / Varga Domokos. — Budapest : Kozmosz Könyvek, 1974.
221 p. : ill. ; 20 cm.
ISBN 9632110544 : 9.50Ft
1. Children. I. Title.
HQ785.Y37 79-377285
79 MARC

Varga, Domokos.
Kölyökkóstolgató / Varga Domokos. — 2. kiad. — Budpest : [s.n.], 1977.
210 p. : ill. ; 20 cm. — (Kozmosz könyvek)
ISBN 9632111729 : 9.50Ft
1. Children. I. Title.
HQ767.9.V37 1977 78-400043
*79 MARC

—ACCIDENTS see Children's accidents

—ANECDOTES AND SAYINGS

see also Wit and humor, Juvenile

Battaglia, Romano.
Il giardino dei pensieri bambini / Romano Battaglia. — Milano : Le stelle, 1978.
124 p. : col. ill. ; 25 cm. — (Collana Idee nuove) It79-Mar
L5000
1. Children—Anecdotes and sayings. I. Title.
PN6328.C5B3 858'.9'1407 79-359908
*79 MARC

Ephron, Delia.
How to eat like a child, and other lessons in not being a grown-up / Delia Ephron ; with drawings by Edward Koren. — New York : Viking Press, 1978.
87 p. : ill. ; 19 x 23 cm.
ISBN 0-670-38331-7 lib. bdg. : $7.95
1. Children—Anecdotes and sayings. I. Koren, Edward. II. Title.
PN6231.C32E6 818'.5'407 78-19145
78 MARC

Linkletter, Arthur Gordon, 1912-
The new kids say the darndest things! / Art Linkletter ; illustrated by Charles M. Schultz ; introd. by Walt Disney. — Boston : G. K. Hall, 1979, c1978.
xiii, 359 p. : ill. ; 24 cm.
Edition of 1957 published under title: Kids say the darndest things!
"Published in large print."
ISBN 0-8161-6742-7
1. Children—Anecdotes and sayings. 2. Large type books. I. Title.
[PN6328.C5L47 1979] 817'.5'408 79-13791
79 MARC

Richmond, Frank.
Geordie at school / by Frank Richmond. — Newcastle upon Tyne : F. Graham, 1977?
48 p. : ill. ; 22 cm. GB78-00566
ISBN 0-85983-074-8 : £0.50
1. Children—Anecdotes and sayings. 2. Education—Anecdotes, facetiae, satire, etc. I. Title.
PN6328.C5R5 828'.9'1407 79-304769
79 MARC

—ANTHROPOMETRY

see also subdivision Anthropometry under children of specia ethinic groups, e.g. Afro-American children —Anthropometry

Brown, Judith E
Finding the causes of child malnutrition : a community handbook for developing countries / Judith E. Brown and Richard C. Brown. — Atlanta : Task Force on World Hunger, Presbyterian Church in the United States, c1979.
73 p. : ill. ; 23 cm.
Bibliography: p. 56-64.
1. Malnutrition in children. 2. Children—Anthropometry. 3. Nutrition surveys. 4. Underdeveloped areas—Children—Nutrition. I. Brown, Richard Coleman, 1936- joint author. II. Title.
RJ399.M26B75 362.7'8 79-101756
79 MARC

Meredith, Howard Voas, 1903-
Human body growth in the first ten years of life / Howard V. Meredith. — Columbia, S.C. : State Print. Co., c1978.
141 p. : graphs ; 23 cm.
Includes bibliographies and index.
1. Children—Growth. 2. Children—Anthropometry. I. Title.
RJ131.M38 612.6'54 78-63120
78 MARC

Pomerance, Herbert H 1918-
Growth standards in children / Herbert H. Pomerance ; with a contribution by John M. Krall. — Hagerstown, Md. : Medical Dept., Harper & Row, c1979.
xii, 225 p. : graphs ; 27 cm.
Bibliography: p. 215-217.
Includes index.
ISBN 0-06-142210-X
1. Children—Growth. 2. Children—Growth—Longitudinal studies. 3. Children—Anthropometry. I. Title.
[DNLM: 1. Growth. WS103 P779g]
RJ131.P63 612.6'5 78-12460
78 MARC

—ASYLUMS see Children—Institutional care

—BIOGRAPHY

see also Presidents—United States —Children

—BOOKS AND READING see Books and reading for children

—CARE AND HYGIENE

see also Baby sitters; Child health services; Children—Diseases; Children—Nutrition; Children —Sleep; Exercise of children; Health education; Infants—Care and hygiene; Physical education for children; Physical fitness for children; School children—Food; School hygiene; Sports for children

Andriulis, E
Vaiko fiziologija ir mokyklinė higiena / E. Andriulis, V. Kviklys. — Vilnius : Mintis, 1974.
228, [4] p. : ill. ; 22 cm. USSR***
Bibliography: p. [230]
0.67rub
1. Children—Growth. 2. Children—Care and hygiene. 3. School hygiene. I. Kviklys, Vladas, joint author. II. Title.
RJ131.A64 78-366883
78 MARC

Chinn, Peggy L 1941-
Child health maintenance : concepts in family-centered care / Peggy L. Chinn. — 2d ed. — St. Louis : Mosby, 1979.
xiii, 934 p. : ill. ; 26 cm.
Includes bibliographies and index.
ISBN 0-8016-0950-X
1. Pediatric nursing. 2. Children—Care and hygiene.
[DNLM: 1. Child care. 2. Pediatric nursing. 3. Family. 4. Child development. WY159 C5395ca]
RJ245.C49 1979 610.73'62 78-12101
78 MARC

Chinn, Peggy L 1941-
Child health maintenance : a guide to clinical assessment / Peggy L. Chinn, Cynthia J. Leitch. — 2d ed. — St. Louis : Mosby, 1979.
vii, 157 p. : ill. ; 26 cm.
Bibliography: p. 157.
ISBN 0-8016-0949-6
1. Pediatric nursing. 2. Child development—Testing. 3. Children—Care and hygiene. I. Leitch, Cynthia J., 1933- joint author. II. Title.
[DNLM: 1. Child care. 2. Child development. 3. Child health services. 4. Growth—In adolescence. 5. Pediatric nursing. WY159 C5395c]
RJ245.C48 1979 610.73'62 78-11964
78 MARC

Chūbachi, Fujirō, 1893–
(Kodomo no shitsuke to sodatekata)
子供の躾と育て方 / 中鉢不二郎, 小林提樹共著. — 東京 : 佐竹書房, 昭和25 [1950]
244 p. : ill. ; 19 cm.
Colophon inserted.
1. Education of children. 2. Children—Management. 3. Children—Care and hygiene. I. Kobayashi, Teiju, 1908– joint author. II. Title.
LB1115.C56 79–801288

Cohen, Matthew M 1949-
Instructions for parents / Matthew M. Cohen. — New York : Appleton-Century-Crofts, c1979.
vii, 232 p. : ill. ; 28 cm.
ISBN 0-8385-4293-X
1. Children—Diseases. 2. Children—Care and hygiene. I. Title.
RJ61.C67 613'.04'32 79-4282
79 MARC

Conger, Flora Stabler.
Child care aide skills / Flora Stabler Conger, Irene B. Rose ; Elizabeth Simpson, consulting editor. — New York : Gregg Division, McGraw-Hill, c1979.
viii, 328 p. : ill. ; 25 cm. — (Careers in home economics)
Includes bibliographical references and index.
ISBN 0-07-012420-5
1. Day care centers. 2. Child development. 3. Children—Care and hygiene. I. Rose, Irene B., joint author. II. Simpson, Elizabeth Jane, 1920. III. Title.
HV851.C67 362.7'1 78-13217
78 MARC

Corella, Laura G
Aprendiendo a ser padres / Laura G. Corella. — Bilbao : Fher, [1975]
205 p. ; 20 cm. — (Colección Vida y luz) Sp76-Jan
ISBN 8424304950
1. Child development. 2. Children—Care and hygiene. I. Title.
HQ769.C7414 649'.1 77-467495
77[r79]rev MARC

Cunningham, Phyllis Jean.
Nursery nursing / edited by P. J. Cunningham for the British Association for Early Childhood Education ; with 17 line drawings by Audrey Besterman. — 3d ed. — London : Faber, 1977.
214 p. : ill. ; 20 cm. GB***
Bibliography: p. 206-207.
Includes index.
ISBN 0-571-04940-0 : £2.60 ($5.95 U.S.)
1. Children—Care and hygiene. 2. Day care centers. I. British Association for Early Childhood Education. II. Title.
RJ101.C74 1977 649'.1 78-316343
78 MARC

Dete i nezgode : savetovanje Subotica 1975. — Subotica : Izvršno veće Skupštine opštine : Odbor za pripremanje i organizovanje savetovanja na temu "Dete i nezgode," 1975 (Subotica : Minerva)
383 p. ; 20 cm. Yu75-7134
Added t.p.: A gyermek és a balesetek.
Serbo-Croatian and Hungarian.
1. Children—Care and hygiene. I. Title: A gyermek és a balesetek.
RJ101.D53 78-367509
78 MARC

Dolto, Françoise Marette.
Lorsque l'enfant paraît / Françoise Dolto. — Paris : Seuil, c1977.
189 p. ; 21 cm. F***
Includes index.
ISBN 2-02-004701-2 : 35.00F
1. Child psychology. 2. Parent and child. 3. Children—Care and hygiene. I. Title.
BF721.D64 649'.1 79-376019
79 MARC

Diagram Group.
Child's body : a parent's manual / by the Diagram Group. — New York : Paddington Press : distributed by Grosset & Dunlap, c1977.
ca. 300 p. : ill. ; 28 cm.
Includes bibliography and index.
ISBN 0-448-22187-X : $6.95
1. Children—Care and hygiene. 2. Child development. I. Title.
RJ61.D55 1977 612.6'5 77-12566
77 MARC

CHILDREN
—CARE AND HYGIENE (Continued)

Ebrahim, G J
Child care in the tropics / G. J. Ebrahim. — London : Macmillan, 1978.
v, 106 p. : ill. ; 23 cm. — (Macmillan tropical community health manuals) GB78-29401
Includes index.
ISBN 0-333-24038-3 : £5.95
1. Children—Care and hygiene. 2. Pediatric tropical medicine. I. Title.
RJ101.E26 1978 362.7'0913 79-312409
79 MARC

Ebrahim, G J
Practical mother and child health in developing countries / G. J. Ebrahim. — [Revised ed.] — London : Macmillan, 1978.
vi, 130 p. : ill. ; 23 cm. — (Macmillan tropical community health manuals) GB78-29396
Includes index.
ISBN 0-333-24111-8 : £5.95
1. Children—Care and hygiene. 2. Prenatal care. 3. Underdeveloped areas—Child health services. 4. Underdeveloped areas—Maternal health services. I. Title.
RJ101.E27 1978 362.7'8'091724 79-306219
79 MARC

Gandevia, Bryan.
Tears often shed : child health and welfare in Australia from 1788 / [by] Bryan Gandevia with the research assistance of Sheila Simpson. — Rushcutters Bay, N.S.W. : Pergamon Press (Australia), 1978.
151 p. : ill. ; 23 cm.
Aus
Bibliography: p. 148.
Includes index.
ISBN 0-08-023159-4
1. Child welfare—Australia. 2. Children—Care and hygiene. I. Simpson, Sheila. II. Title.
HV802.G36 1978 362.7'0994 78-326343
79 MARC

Gomirato Sandrucci, Maria.
Quando c'è un bambino : manuale di puericultura / [di] Maria Sandrucci, Domenico Castello. — Novara : Istituto geografico De Agostini, 1976.
254 p. : ill. ; 30 cm. It78-Mar
L15500
1. Children—Care and hygiene. 2. Infants—Care and hygiene. I. Castello, Domenico, joint author. II. Title.
RJ61.G56 1976 649'.1 79-342214
*79 MARC

(Hoiku genshō no bunkaronteki tenkai)
保育現象の文化論的展開 / 本田和子，津守真共編；入江礼子，友定啓子，角能清美共著. — 東京：光生館，1977.
177 p. : ill. ; 22 cm. — (人間現象としての保育研究；3) Ja 77-4859
Includes bibliographical references.
¥1400
1. Education, Preschool—1965- 2. Children—Management. 3. Children—Care and hygiene. I. Honda, Kazuko. II. Tsumori, Makoto, 1926- III. Irie, Reiko. IV. Series: Ningen genshō to shite no hoiku kenkyū ; 3.
LB1140.2.H612 78-813577

Howell, Mary C
Healing at home : a guide to health care for children / Mary Howell. — Boston : Beacon Press, c1978.
xvi, 287 p. : ill. ; 20 cm. — (Beacon paperback ; 573)
Bibliography: p. 268-278.
Includes index.
ISBN 0-8070-2368-X : $9.95. ISBN 0-8070-2369-8 pbk. : $5.95
1. Children—Diseases. 2. Children—Care and hygiene. I. Title.
[DNLM: 1. Child care—Popular works. 2. Pediatrics—Popular works. WS113.3 H859h]
RJ61.H85 1978 618.9'2005 77-88329
78 MARC

Hume, Violet Ruth.
Watoto wetu: ulinzi na malezi yao, kimeandikwa kwa Kingereza na V. R. Hume pamoja na J. M. Wigram. Kimefasiriwa kwa Kiswahili na Hyder M. Matano. Nairobi, East African Literature Bureau [1973]
175 p. illus. 22 cm. Sh12.50
1. Children—Care and hygiene. I. Wigram, J. M. II. Title.
RJ61.H896 73-983758
79 MARC

Hutchins, Bobbie.
Child nutrition and health / Bobbie Hutchins. — New York : Gregg Division, McGraw-Hill, c1979.
234 p. : ill. ; 24 cm. — (Careers in home economics)
Includes bibliographical references and index.
ISBN 0-07-031527-2
1. Children—Nutrition. 2. Children—Care and hygiene. I. Title.
RJ206.H87 612'.3 78-21948
78 MARC

Jánošová, Martina.
Dieťa a jeho svet / M. Jánošová a kol. — 1. vyd. — Bratislava : Obzor, 1975.
223, [9] p. : ill. (some col.) ; 30 cm. — (Rodina a spoločnosť ; zv. 1) CzS75
Kčs80.00
1. Children—Care and hygiene. I. Title.
RJ101.J34 78-378101
*78 MARC

Jolly, Hugh.
More commonsense about babies and children / [by] Hugh Jolly. — London : Pelham, 1978.
192 p. ; 21 cm. GB78-21319
Includes index.
ISBN 0-7207-1069-3 : £4.95
1. Children—Care and hygiene. 2. Children—Diseases. 3. Children—Management. I. Title.
RJ61.J68 649.1 79-303949
79 MARC

Leach, Penelope.
Your baby & child from birth to age five / Penelope Leach ; photography by Camilla Jessel. — New York : A. A. Knopf, 1978.
512 p. : ill. ; 24 cm.
First published in London in 1977 under title Baby and child.
Includes index.
ISBN 0-394-40755-5 : $5.95. ISBN 0-394-73509-9 pbk.
1. Infants—Care and hygiene. 2. Children—Care and hygiene. 3. Child development. I. Title.
RJ61.L4 1978 649'.122 77-18558
78 MARC

Leksina, M
(Zhas balany sem'ĭada tărbieleu)
Жас баланы семьяда тәрбиелеу. Алматы, Қазақстан, 1968.
80 p. 14 cm. USSR 68-29960
At head of title: М Лексина.
——— Microfilm. On 1 reel with other items ; 35 mm.
0.04rub
Microfilm Slavic 12711 RJ
1. Children—Care and hygiene. 2. Children—Management. I. Title.
RJ61.L486 72-334411

Linden, Wilhelm zur.
Geburt und Kindheit : Pflege, Ernährung, Erziehung / Wilhelm zur Linden. — 10., überarb. Aufl. — Frankfurt am Main : Klostermann, 1978.
589 p. ; 22 cm. GFR78-A
Bibliography: p. [562]-[571]
Includes index.
ISBN 3-465-01321-2 : DM38.00
1. Infants—Care and hygiene. 2. Children—Care and hygiene. 3. Children—Diseases. I. Title.
RJ61.L597 1978 78-396607
79 MARC

McDonald, Linda.
Everything you need to know about babies / by Linda McDonald ; ill. by Kathryn Wirch. — 1st ed. — Pasadena : Oaklawn Press, 1978.
191 p. : ill. ; 27 cm.
Includes index.
ISBN 0-916198-04-9
1. Infants—Care and hygiene. 2. Children—Care and hygiene. I. Title.
RJ61.M315 649'.1 77-21017
77 MARC

Manela, Roger.
Health needs of children / by Roger Manela and Armand Lauffer ; with contributions by Eugene Feingold and Ruben Meyer. — Beverly Hills, Calif. : Sage Publications, published in cooperation with the continuing education program in the human services of the University of Michigan School of Social Work, c1979.
111 p. : ill. ; 22 cm. — (Sage human services guides ; v. 9)
Bibliography: p. 107-109.
ISBN 0-8039-1217-X
1. Children—Care and hygiene. 2. Poor—Health and hygiene. 3. Child health services—United States. 4. Poor—Medical care—United States. I. Lauffer, Armand, joint author. II. Title.
[DNLM: 1. Child health services—United States. 2. Poverty—United States. WA320 M274c]
RJ101.M333 362.7'8 78-26373
78 MARC

Meeting children's needs. [Sound recording] New York, Parents' Magazine Educational Enterprises, p1977.
3 cassettes. ½-track. mono. (Conversations about children, set 2)
Interviews, conducted by Bettye Caldwell.
In container (25 cm.)
Bibliography (7 p.) inserted in container.
SUMMARY: Authorities on child health and development and family living present their views on how to meet the physical and psychological needs of children in various settings.
CONTENTS: no. 1. White, B. In homes.—no. 2. Eiduson, B. In non-traditional home settings.—no. 3. Lansburgh, T. In day care.—no. 4. Jackson, P. In schools.—no. 5. Michelson, W. In cities.—no. 6. Palmer, E. Via the media.
1. Children—Care and hygiene. 2. Child development. I. Caldwell, Bettye M. II. Series: Conversations about children. [Sound recording] set 2.
[RJ101] 77-740980
MN

Reinisch, Edith H
Health of the preschool child / Edith H. Reinisch, Ralph E. Minear, Jr. — New York : Wiley, c1978.
xii, 256 p. : ill. ; 24 cm.
Bibliography: p. 243-246.
Filmography: p. 247.
Includes index.
ISBN 0-471-60800-9
1. Children—Care and hygiene. 2. Children—Diseases. 3. Education, Preschool—1965- I. Minear, Ralph E., joint author. II. Title.
RJ101.R43 613'.04'32 78-8743
78 MARC

Ruiz, S
Niños en la casa : desde el nacimiento hasta el colegio / [por] S. Ruiz. — Barcelona : Aura, D.L. 1977.
182 p., 3 leaves ; 19 cm. — (Biblioteca Llave) Sp77-July
ISBN 8421401173 : 175ptas
1. Children—Care and hygiene. 2. Infants—Care and hygiene. 3. Pregnancy. I. Title.
HQ769.R768 79-349134
79 MARC

Saitō, Fumio, 1896-
(Haha no ikujisho)
母の育兒書 / 齋藤文雄著. — 東京：主婦之友社，昭和22 [1947]，昭和23 [1948] printing.
2, 14, 404 p. : ill. ; 19 cm.
Includes index.
1. Children—Care and hygiene. 2. Children—Diseases. I. Title.
RJ61.S265 78-817033

T'ai-pei shih chia t'ing chi hua t'ui kuang chung hsin.
(Ch'iao fu shou ts'e)
巧婦手冊. — 台北：台北市家庭計畫推廣中心，[民國65 i. e. 1976序]
68 p. : ill. ; 20 cm.
Cover title.
1. Family life education. 2. Children—Care and hygiene. I. Title.
HQ10.T24 1976 79-837068

Thurmond, Nancy Moore.
Mother's medicine / by Nancy Moore Thurmond. — 1st ed. — New York : Morrow, 1979.
383 p. : ill. ; 24 cm.
Bibliography: p. [375]-376.
Includes index.
ISBN 0-688-03384-9
1. Children—Care and hygiene. 2. Children—Management. I. Title.
RJ61.T53 649.1 78-12552
78 MARC

—CARE AND HYGIENE—CONGRESSES

Gesunde Kinder, unsere Verantwortung / [hrsg. und überreicht von der Bundesvereinigung für Gesundheitserziehung e.V. ; Schriftleitung, Karl Schneider, Heinz Wickert]. — Bonn : Die Bundesvereinigung, 1979.
236 p. ; 24 cm. GFR***
"Weltgesundheitstag 1979."
Includes bibliographies.
1. Children—Care and hygiene—Congresses. 2. Children—Care and hygiene—Germany, West—Congresses. I. Schneider, Karl, Dr. med. Dr. phil. II. Wickert, Heinz. III. Bundesvereinigung für Gesundheitserziehung.
RJ101.G485 79-378350
79 MARC

Der Schulbeginn und die Gesundheit des Kindes : d. Kinder d. ersten beiden Schuljahre im Spannungsfeld von Lernfähigkeit u. Lernanforderungen ; Beitr. zur Gesundheitserziehung aus erziehungswissenschaftl. u. ärztl. Sicht ; Bericht über d. Arbeitstagung vom 3.-5. November 1977 in Bad Kissingen / Veranst. : Arbeitskreis Grundschule e.V., Frankfurt u. Bundesvereinigung für Gesundheitserziehung e.V., Bonn-Bad Godesberg. Unter Beteiligung d. Bundeselternrates für d. Bundesarbeitsgemeinschaft "Eltern, Ärzte u. Lehrer als Anwälte d. Kindes". Verantw. für diesen Bd. : Hans Arno Horn. — Frankfurt am Main : Arbeitskreis Grundschule, 1978.
160 p. : ill. ; 21 cm. — (Beiträge zur Reform der Grundschule : Sonderband ; S 35) GFR78-B18
Includes bibliographies.
1. Readiness for school—Congresses. 2. Education, Primary—Congresses. 3. Children—Care and hygiene—Congresses. I. Horn, Hans Arno. II. Arbeitskreis Grundschule. III. Bundesvereinigung für Gesundheitserziehung. IV. Series.
LB1132.S326 79-353459
79 MARC

—CARE AND HYGIENE—POPULAR WORKS

(Erh t'ung pao chien ch'ang shih)
兒童保健常識 / 丁華華編著. — 第1版. — 臺北：國語日報出版部，民國67 [1978]
2, 6, 240 p. ; 20 cm.
NT$40.00
1. Children—Care and hygiene—Popular works. I. Ting, Hua-hua.
RJ61.E66 78-843994

CHILDREN
—CARE AND HYGIENE
—POPULAR WORKS (Continued)

Huang, Mei-lan.
(Ying erh pao chien shou ts'e)
嬰兒保健手册 / 黃美蘭編著. — 九龍：智學出版社, [1978?]
4, 223 p. : ill. ; 19 cm.
$5.50

1. Children—Care and hygiene—Popular works. I. Title.
RJ61.H875 78-843891

Lin, Chia-jung.
(Hsia i tai)
下一代 / 林佳蓉著. — 三重：世新出版社, 民國65 [1976]
6, 246 p. : ill. ; 20 cm. — (真善美書系 ; 3)
Series romanized: Chen shan mei shu hsi.
NT$35.00

1. Children—Care and hygiene—Popular works. I. Title.
RJ61.L596 78-842922

Wang, Li-ying.
(Tsen yang tai hao hsiao hai)
怎样带好小孩 / [王丽瑛编]. — 第3版. — 北京：人民卫生出版社, 1976.
2, 50 p. : ill. ; 19 cm.
Cover title.
$0.12

1. Children—Care and hygiene—Popular works. 2. Children—Diseases—Popular works. I. Title.
RJ61.W25 1976 78-843053

—CARE AND HYGIENE—AMERICA—CONGRESSES

Evaluation of child health services : the interface between research and medical practice / Samuel J. Bosch and Jaime Arias, scientific editors. — Bethesda, Md. : U.S. Dept. of Health, Education, and Welfare, Public Health Service, National Institutes of Health ; Washington : for sale by the Supt. of Docs., U.S. Govt. Print. Off., 1978.
xiii, 291 p. : ill. ; 24 cm. — (DHEW publication ; no. (NIH) 78-1066)
"Proceedings of a conference sponsored by American College of Preventive Medicine ... [et al.] March 7-9, 1977, National Institutes of Health, Bethesda, Maryland."
Includes bibliographies and index.
1. Child health services—Evaluation—Congresses. 2. Child health services—America—Evaluation—Congresses. 3. Children—Care and hygiene—America—Congresses. I. Bosch, Samuel J. II. Arias, Jaime. III. American College of Preventive Medicine. IV. Series: United States. Dept. of Health, Education, and Welfare. DHEW publication ; no. (NIH) 78-1066.
RJ101.E93 362.7'8 79-601520
79 MARC

—CARE AND HYGIENE—CANADA

Emberley, Jean.
Children in Canada : a report / prepared by Jean Emberley. — [Hamilton, Ont.] : McMaster University Press, 1978.
vi, 86 p. : map ; 23 cm. C78-11294-6
Cover title: Today's child ... the world's future.
Prepared for the National Council of Women of Canada.
"The Hospital of Sick Children Foundation, grant no. 78-34."
$3.00
1. Children—Canada. 2. Child welfare—Canada. 3. Children—Care and hygiene—Canada. 4. Education of children. 5. Education—Canada. I. National Council of Women of Canada. II. Title. III. Title: Today's child ... the world's future.
HQ792.C3E43 362.7'0971 79-311504
79 MARC

—CARE AND HYGIENE—CZECHOSLOVAKIA

Štaifová, Alena.
Zdravotní výchova dětí předškolního věku : metodické podklady pro zdravotníky a pedagogy / Alena Štaifová. — Praha : Ústav zdravot. výchovy, 1976.
47 p. ; 21 cm. — (Metodické podklady pro zdravotní výchovu ; 4) Cz76
On cover: Ministerstvo zdravotnictví.
1. Children—Care and hygiene—Czechoslovakia. 2. Health education—Czechoslovakia. I. Title.
RJ103.C95S7 78-382571
*78 MARC

—CARE AND HYGIENE—ENGLAND—HISTORY—19TH CENTURY—ADDRESSES, ESSAYS, LECTURES

Graham, Sarah.
Little victims : sick children in the Victorian era / by Sarah Graham. — London : Macmillan, 1977.
27 p. : ill., facsims. ; 29 cm. GB78-29402
"A collection of articles ... reprinted from "Nursing times"".
Includes bibliographies.
ISBN 0-333-24168-1 : £0.90
1. Pediatrics—England—History—19th century—Addresses, essays, lectures. 2. Children—Diseases—England—History—19th century—Addresses, essays, lectures. 3. Children—Care and hygiene—England—History—19th century—Addresses, essays, lectures. I. Title.
RJ42.G7G7 362.7'81'0941 79-317810
79 MARC

—CARE AND HYGIENE—GERMANY, WEST—CONGRESSES

Gesunde Kinder, unsere Verantwortung / [hrsg. und überreicht von der Bundesvereinigung für Gesundheitserziehung e.V. ; Schriftleitung, Karl Schneider, Heinz Wickert]. — Bonn : Die Bundesvereinigung, 1979.
236 p. ; 24 cm. GFR***
"Weltgesundheitstag 1979."
Includes bibliographies.
1. Children—Care and hygiene—Congresses. 2. Children—Care and hygiene—Germany, West—Congresses. I. Schneider, Karl, Dr. med. Dr. phil. II. Wickert, Heinz. III. Bundesvereinigung für Gesundheitserziehung.
RJ101.G485 79-378350
79 MARC

Unsere Kinder, fit fürs Leben / [hrsg. und überreicht von der Bundesvereinigung für Gesundheitserziehung e.V.]. — Bonn-Bad Godesberg : Bundesvereinigung für Gesundheitserziehung 1977.
183 p. ; 24 cm. GFR***
On cover: Weltgesundheitstag 1977.
Includes bibliographical references and index.
1. Child health services—Germany, West—Congresses. 2. Children—Care and hygiene—Germany, West—Congresses. I. Bundesvereinigung für Gesundheitserziehung.
RJ103.G3U5 79-356952
79 MARC

—CARE AND HYGIENE—HUNGARY—BUDAPEST—LONGITUDINAL STUDIES

Az 1-60 [i.e. Egytől hatvan] hónapos budapesti gyermekek testi fejlettsége, szociodemográfiai és morbiditási viszonyai (előzetes jelentés / [szerkők, Ács Jolán ... et al. ; szerk., Sárkány Jenő]. — Budapest : Központi Statisztikai Hivatal, Népességtudományi Kutató Intézet, 1977.
237 p. : ill. ; 29 cm. — (A Központi Statisztikai Hivatal Népességtudományi Kutató Intézetének és a Magyar Tudományos Akadémia Demográfiai Bizottságának közleményei ; 45)
Summary in English (37 p.) inserted.
Bibliography p. 101-102.
ISBN 9633401488 : 116.00Ft
1. Children—Care and hygiene—Hungary—Budapest—Longitudinal studies. 2. Children—Hungary—Budapest—Growth—Longitudinal studies. 3. Children—Hungary—Budapest—Anthropometry. 4. Children—Diseases—Hungary—Budapest—Longitudinal studies. I. Ács, Jolán. II. Sárkány, Jenő.

III. Series: Népességtudományos Kutató Intézet. A Központi Statisztikai Hivatal Népességtudományi Kutató Intézetének és a Magyar Tudományos Akadémia Demográfiai Bizottságának közleményei ; 45.
HA1201.A64 no. 45 79-355878
[RJ103.H9] 79 MARC

—CARE AND HYGIENE—RUSSIA

al-Sayyid, Maḥmūd 'Alī.
(al-Nahḍah al-ṣiḥḥīyah al-Rūsīyah)
النهضة الصحية الروسية ، بقلم محمود على السيد. مصر ، مطبعة سمير ، 1943.
117 p. 21 cm.
1. Children—Care and hygiene—Russia. I. Title.
RJ103.R9S27 78-970998

—CARE AND HYGIENE—SWEDEN

Annell, Erik.
Handikappförebyggande hälsokontoll av fyraåringar; förberedande undersökning av särskild utredningsman. Stockholm [Socialdepartementet] 1965.
111 l. 30 cm. ([Sweden] Socialdepartementet. [Publikation] Stencil 1965:2)
Cover title.
1. Children—Care and hygiene—Sweden. 2. Handicapped children—Education—Sweden. I. Title. II. Series.
HV336.A33 1965:2 78-416801
[RJ103.S8] 78 MARC

—CARE AND HYGIENE—UNITED STATES

United States. Congress. Senate. Committee on Human Resources. Subcommittee on Child and Human Development.
Child care and child development programs, 1977-78 : hearings before the Subcommittee on Child and Human Development of the Committee on Human Resources, United States Senate, Ninety-fifth Congress, first and second sessions ... November 25, 1977, San Francisco, Calif., December 12, 1977, Los Angeles, Calif. — Washington : U.S. Govt. Print. Off., 1978-
v. : ill. ; 24 cm.
1. Children—Care and hygiene—United States. 2. Child development—United States. I. Title.
KF26.H83 1977b 344'.73'0327 79-600556
79 MARC

—CHARITIES see Child welfare

—CLOTHING see Children's clothing

—CONDUCT OF LIFE

Axelrod, Herman C 1904-
Bilingual background and its relation to certain aspects of character and personality of elementary school children / Herman C. Axelrod. — New York : Arno Press, 1978, c1953.
iii, 142 p. ; 24 cm. — (Bilingual-bicultural education in the United States)
Originally presented as the author's thesis, Yeshiva University, 1951.
Bibliography: p. 137-141.
ISBN 0-405-11075-8
1. Education, Bilingual—United States. 2. Biculturalism—United States. 3. Child psychology. 4. Children—Conduct of life. I. Title. II. Series.
LC3731.A93 1978 371.9'7 77-90406
78 MARC

Colville, M Josephine.
Growing-up / by M. Josephine Colville. — Boston : St. Paul Editions, c1979.
162 p. : ill. ; 22 cm.
Includes index.
1. Children—Conduct of life. I. Title.
BJ1631.C64 170'.202'22 78-9925
78 MARC

Furuya, Tsunatake, 1908-
(Bokutachi ningen no hokori)
僕たち人間のほこり / 古谷綱武. — 東京：帝國教育會出版部, 昭和17 [1942]
211 p. : ill. ; 19 cm. — (少國民文藝選)

1. Children—Conduct of life. I. Title. II. Series: Shōkokumin bungei sen.
BJ1638.J3F874 78-815354

Fujiwara, Sadamu, 1905-
(Nikushimi yori ai e)
にくしみより愛へ：少年少女のために / 藤原定. — 東京：福村書店, 1953.
5, 117 p. ; 19 cm.

1. Children—Conduct of life—Juvenile literature. I. Title.
BJ1638.J3F84 78-810612

Janney, John R
How to raise happy, godly children / John R. Janney. — 1st ed. — Bridgeton, N.J. : Christian Family Crusades, c1978.
183 p. : ill. ; 22 cm.
Includes bibliographical references.
1. Children—Conduct of life. 2. Children—Religious life. 3. Children—Management. I. Title.
BJ1631.J36 649'.1 78-58575
79 MARC

(Jyā bāyakā̃sa leṅkarē̃ āheta tyā̃sa subodha)
ज्या बायकांस लेंकरें आहेत त्यांस सुबोध = Address to mothers. — 1st ed. — [Bombay : American Mission], 1841.
25 p. ; 19 cm. — (American Mission series ; no. 34)
Caption title.
In Marathi.

1. Children—Conduct of life. 2. Children—Management I. American Marathi Mission. II. Title: Address to mothers.
BJ1638.M37J96 78-984395

CHILDREN
—CONDUCT OF LIFE (Continued)

Rasmussen, Bernard Ingram, 1896-
Character craft, per se : a source book in moral education with a plan / by Bernard Ingram Rasmussen ; drawings by Robert Kutsch. — ‹Angwin, Calif.› : Dupli-Craft Co., c1979.
163 p. : ill. ; 24 cm.
"A book for parents and teachers."
Includes index.
$8.95
1. Moral education. 2. Children—Conduct of life. I. Title.
LC268.R29 370.11'4 79-51459
00 MARC

What do I believe? : answers from the young / compiled and edited by Mary Spain ; analysis by Susan Major ; preface by Jimmy Savile. — London : Shepheard-Walwyn, 1978.
192 p. : ill., port ; 25 cm. GB78-29432
ISBN 0-85683-038-0 : £4.95. ISBN 0-85683-039-9 pbk.
1. Children—Religious life. 2. Children—Conduct of life. 3. Children—England. I. Spain, Mary. II. Major, Susan.
BV4571.2.W47 371.8'1 78-326630
79 MARC

Zaal, Jac N
Sociaal emotioneel gedrag in de klas : de ontwikkeling van een schaal voor het beoordelen van het sociaal emotionele gedrag van leerlingen in kleuter- en basisonderwijs / Jac. N. Zaal. — Groningen : Wolters-Noordhoff, ‹1979›
264 p. ; 24 cm. Ne79-4
Originally presented as the author's thesis, Vrije Universiteit, Amsterdam, under title: Beoordeling van het sociaal emotionele gedrag van leerlingen in het kleuter- en basisonderwijs.
Summary in English.
Bibliography: p. 181-187.
ISBN 9001980716 : fl 43.50
1. Educational tests and measurements—Netherlands—Methodology. 2. Children—Conduct of life. 3. Sociometry. I. Title.
LB3056.N4Z3 1979 370.15'3 79-369511
*79 MARC

—CONDUCT OF LIFE—CASE STUDIES

Scharf, Peter, 1945-
Growing up moral : dilemmas for the intermediate grades / by Peter Scharf, William McCoy, and Diane Ross. — Minneapolis : Winston Press, c1979.
183 p. ; 22 cm.
Bibliography: p. 177-181.
Includes index.
ISBN 0-03-043941-8
1. Moral education—Case studies. 2. Children—Conduct of life—Case studies. I. McCoy, William, 1947- joint author. II. Ross, Diane, joint author. III. Title.
LC268.S37 370.11'4 78-50391
00 MARC

—CONDUCT OF LIFE —JUVENILE LITERATURE

Bickerstaff, George.
My body is a temple / written by George Bickerstaff ; illustrated by Keith Christensen. — Salt Lake City, Utah : Bookcraft, 1978.
‹24› p. : col. ill. ; 30 cm.
SUMMARY: A brother and sister learn the importance of taking care of their body, mind, and spirit.
ISBN 0-88494-354-2
1. Children—Conduct of life—Juvenile literature. ‹1. Christian life. 2. Church of Jesus Christ of Latter-Day Saints. 3. Mormons and Mormonism› I. Christensen, Keith. II. Title.
BJ1631.B45 248'.48'933 78-72565
00 MARC

Hawai Kyōikukai.
(Shūshinsho)
修身書：中等科用 / 布哇教育會編纂. — ‹s. l. : s. n.›, 昭和8 ‹1933›, 昭和15 ‹1940› printing. (‹東京›：内外圖書出版協會)
v. : ill. ; 19 cm.

1. Children—Conduct of life—Juvenile literature. I. Title.
BJ1638.J3H38 1933 79-805657

—COSTUME

see also Children's clothing

—COSTUME—PICTORIAL WORKS

The Way we wore : fashion illustrations of children's wear, 1870-1970 / ‹compiled by› Linda Martin. — New York : Scribner, c1978.
176 p. : chiefly ill. ; 28 cm.
ISBN 0-684-15655-5 : $14.95
1. Children—Costume—Pictorial works. 2. Costume—History—20th century. 3. Costume—History—19th century.
GT1730.M39 391'.07'30904 78-17243
78 MARC

—DEAF—UNITED STATES

Whitehurst, Mary Wood, 1901-
Auditory training for children / Mary Wood Whitehurst. — Washington : Alexander Graham Bell Association for the Deaf, c1978.
v, 103 p. : Ill. ; 26 cm.
ISBN 0-88200-125-6
1. Children—Deaf—United States. 2. Deaf—United States—Means of communication. 3. Hearing aids—United States. I. Title.
HV2483.W47 362.7'8'42 78-71016
00 MARC

—DENTAL CARE—NETHERLANDS

Netherlands (Kingdom, 1815-). Centrale Raad voor de Volksgezondheid.
Gewenste toekomstige tandheelkundige voorzieningen in Nederland : advies van de Centrale raad voor de volksgezondheid. — 's-Gravenhage : Staatsuitgeverij, 1977.
168 p. ; 27 cm. — (Verslagen, adviezen, rapporten - Ministerie van Volksgezondheid en Milieuhygiëne ; 1977, 16) Ne77-32
Cover title.
Bibliography: p. 133-139.
ISBN 9012017491 : fl 17.50
1. Dental care—Netherlands. 2. Preventive dentistry—Netherlands. 3. Children—Dental care—Netherlands. I. Title. II. Series: Netherlands (Kingdom, 1815-). Ministerie van Volksgezondheid en Milieuhygiëne. Verslagen, adviezen, rapporten - Ministerie van Volksgezondheid en Milieuhygiëne ; 1977, 16.
RK52.4.N4N47 1977 78-380545
*78 MARC

—DENTAL CARE—UTAH—EMERY CO. —AUDITING AND INSPECTION

Utah. Office of the Legislative Auditor General.
A performance audit of selected Family Health Services programs / audit performed by audit manager, Roy Dunn, auditor in charge, Susan Elmer, audit staff, Byron Jorgenson. — ‹Salt Lake City› : Office of Legislative Auditor General, State of Utah, 1978.
iii, 65 p. : graphs ; 29 cm. — (Report to Utah State Legislature ; no. 78-20)
1. Child health services—Utah—Auditing and inspection. 2. Handicapped children—Care and treatment—Utah—Auditing and inspection. 3. Medicine, State—Utah—Auditing and inspection. 4. Children—Dental care—Utah—Emery Co.—Auditing and inspection. 5. Dentistry, State—Utah—Emery Co.—Auditing and inspection. 6. Utah. Family Health Services. I. Dunn, Roy. II. Elmer, Susan. III. Jorgenson, Byron. IV. Title. V. Series.
RJ102.5.U8U83 1978 353.9'792'0084781 79-622586
79 MARC

—DICTIONARIES—GERMAN

Kritische Stichwörter zur Kinderkultur / Karl W. Bauer, Heinz Hengst, Hrsg. — München : W. Fink, 1978.
366 p. ; 19 cm. — (Kritische Stichwörter ; 1) GFR***
Includes bibliographies.
ISBN 3-7705-1634-6
1. Children—Dictionaries—German. I. Bauer, Karl W. II. Hengst, Heinz.
HQ767.84.K73 301.43'14'03 79-358466
79 MARC

--DISCIPLINE see Discipline of children

—DISEASES

see also Children—Hospitals; Infants—Diseases; Pediatric hematology; Pediatric otolaryngology; Pediatric respiratory diseases; Pediatric toxicology; Pediatric tropical medicine; and names of individual diseases, e.g. Chicken-pox, Whooping cough, Hypertension ‹Tuberculosis, etc.› in children

Ch'eng-tu Chung i hsüeh yüan.
(Chung i erh k'o hsüeh)
中医儿科学 / 成都中医学院编. — 成都：四川人民出版社, 1976.
3, 2, 244 p. ; 21 cm. — (中医医学丛书 ; 3)
$0.67

1. Children—Diseases. 2. Medicine, Chinese. I. Title. II. Series: Chung i i hsüeh ts'ung shu ; 3.
RJ45.C514 1976 78-843263

Cohen, Matthew M 1949-
Instructions for parents / Matthew M. Cohen. — New York : Appleton-Century-Crofts, c1979.
vii, 232 p. : ill. ; 28 cm.
ISBN 0-8385-4293-X
1. Children—Diseases. 2. Children—Care and hygiene. I. Title.
RJ61.C67 613'.04'32 79-4282
79 MARC

Howell, Mary C
Healing at home : a guide to health care for children / Mary Howell. — Boston : Beacon Press, c1978.
xvi, 287 p. : ill. ; 20 cm. — (Beacon paperback ; 573)
Bibliography: p. 268-278.
Includes index.
ISBN 0-8070-2368-X : $9.95. ISBN 0-8070-2369-8 pbk. : $5.95
1. Children—Diseases. 2. Children—Care and hygiene. I. Title. ‹DNLM: 1. Child care—Popular works. 2. Pediatrics—Popular works. WS113.3 H859h›
RJ61.H85 1978 618.9'2005 77-88329
78 MARC

(Hsiao erh chi fu cheng)
小儿急腹症 / 馬安权主編. — 第1版. — 上海：上海科学技术出版社, 1965.
iv, 265 p. : ill. ; 26 cm. — (医师进修丛书)
Series romanized: I shih chin hsiu ts'ung shu.
Includes bibliographies.

1. Acute abdomen. 2. Children—Diseases. I. Ma, An-ch'üan.
RJ456.A3H73 79-835543

Hsüeh, K'ai, fl. 1488–1505.
(Pao ying ch'üan shu)
保嬰全書 / 薛鎧撰. — 臺北：新文豐出版公司, 民國67 ‹1978›
4 p. (28, 18, 2246 p.) ; 22 cm.
"景自中央圖書館珍藏善本書籍...崇禎沈猶龍閩中刊本."
CONTENTS: 1-2. 薛鎧 卷1-10.—3-4. 薛己 卷11-20.
NT$1000.00

1. Children—Diseases. 2. Medicine, Chinese. I. Hsüeh, Chi, fl. 1506-1566. II. Title.
RJ47.H75 1978 79-837462

Jolly, Hugh.
More commonsense about babies and children / ‹by› Hugh Jolly. — London : Pelham, 1978.
192 p. ; 21 cm. GB78-21319
Includes index.
ISBN 0-7207-1069-3 : £4.95
1. Children—Care and hygiene. 2. Children—Diseases. 3. Children—Management. I. Title.
RJ61.J68 649.1 79-303949
79 MARC

(Kodomo no byōki hyakka)
子供の病気百科：ベビーエイジ / ‹国立小児病院各科専門医執筆›. — 東京：婦人生活社, 昭和51 ‹1976›
528 p. : ill. ; 19 cm. Ja 76B-6076
Colophon inserted.
¥1400

1. Children—Diseases. I. Fujin Seikatsu Sha.
RJ45.K645 79-800019
Shared Cataloging with DNLM
J

Linden, Wilhelm zur.
Geburt und Kindheit : Pflege, Ernährung, Erziehung / Wilhelm zur Linden. — 10., überarb. Aufl. — Frankfurt am Main : Klostermann, 1978.
589 p. ; 22 cm. GFR78-A
Bibliography: p. ‹562›-‹571›
Includes index.
ISBN 3-465-01321-2 : DM38.00
1. Infants—Care and hygiene. 2. Children—Care and hygiene. 3. Children—Diseases. I. Title.
RJ61.L597 1978 78-396607
79 MARC

CHILDREN
—DISEASES (Continued)

Reinisch, Edith H
Health of the preschool child / Edith H. Reinisch, Ralph E. Minear, Jr. — New York : Wiley, c1978.
xii, 256 p. : ill. ; 24 cm.
Bibliography: p. 243-246.
Filmography: p. 247.
Includes index.
ISBN 0-471-60800-9
1. Children—Care and hygiene. 2. Children—Diseases. 3. Education, Preschool—1965- . I. Minear, Ralph E., joint author. II. Title.
RJ101.R43 613'.04'32 78-8743
78 MARC

Saitō, Fumio, 1896–
(Haha no ikujisho)
母の育兒書 / 齋藤文雄著. — 東京：主婦之友社, 昭和22 [1947], 昭和23 [1948] printing.
2, 14, 404 p. : ill. ; 19 cm.
Includes index.

1. Children—Care and hygiene. 2. Children—Diseases. I. Title.
RJ61.S265 78–817033

[Shang-hai Chung i hsüeh yüan]
(Chung i erh k'o hsüeh)
中醫兒科學. — 台北：啓業書局, 民國67 [1978]
3, 289 p. : ill. ; 22 cm.
Previously published by Shang wu yin shu kuan, Hongkong, under the author Shang-hai Chung i hsüeh yüan, in series: Hsin pien Chung i yao chiao ts'ai, with text reset replacing the simplified characters with regular characters.
Includes index.

NT$150.00
1. Children—Diseases. 2. Medicine, Chinese. I. Title.
RJ47.S44 1978 79–836481

Shen, Chin-ao.
(Yu k'o shih mi)
幼科釋謎 / 沈金鰲輯著. — 香港：萬葉出版社；九龍：利通圖書公司發行, [1979?]
4, 215 p. : ill. ; 19 cm.
"本書共分六卷"
Reprint of the 1957 ed. published by Shang-hai wei sheng ch'u pan she, Shanghai.
$10.00
1. Children—Diseases. 2. Medicine, Chinese—Early works to 1800. I. Title.
RJ44.S48 1979 79–840748

Wu, K'o-ch'ien, lexiographer.
(Wu shih erh k'o hsüeh)
吳氏兒科學 / 吳克潛著. — 台北：新文豐出版公司, 民國66 [1977]
2, 16, 393 p. ; 21 cm.
Reprint ed. with original pref. dated 1934.
NT$180.00

1. Children—Diseases. 2. Medicine, Chinese. I. Title.
RJ47.W8 1977 78–842920

—DISEASES—COMPLICATIONS AND SEQUELAE—UNITED STATES

Arnold, Charles B
Could there be a medical basis for the declining SAT scores? : prepared for the Advisory Panel on the Scholastic Aptitude Test Score Decline / Charles B. Arnold ; jointly sponsored by the College Board and Educational Testing Service. — New York : The Board ; Princeton, N.J. : available from College Board Publication Orders, 1977.
8 p. ; 28 cm.
One of 27 papers which supplement On further examination, by the Advisory Panel on the Scholastic Aptitude Test Score Decline of College Entrance Examination Board.
Includes bibliographical references.
$2.00
1. Scholastic aptitude test. 2. Children—Diseases—Complications and sequelae—United States. 3. Birth injuries—Complications and sequelae—United States. I. College Entrance Examination Board. Advisory Panel on the Scholastic Aptitude Test Score Decline. On further examination. II. Title.
LB2353.A76 378.1'6'62 77-154427
78 MARC

—DISEASES—DIAGNOSIS

see also Pediatric radiology

Gädeke, Roland.
Diagnostische und therapeutische Techniken in der Pädiatrie / Roland Gädeke. — 2., neubearb. Aufl. — Berlin ; New York : Springer-Verlag, 1976.
xiii, 191 p. : ill. ; 25 cm.
Bibliography: p. 185.
Includes index.
ISBN 0-387-07595-X
1. Children—Diseases—Diagnosis. 2. Pediatrics. I. Title.
[DNLM: 1. Diagnosis—In infancy and childhood. 2. Therapeutics—In infancy and childhood. WS141 G127d]
RJ50.G33 1976 618.9'2007'5 75-44400
75[r79]rev MARC

Ultrasound in pediatrics / edited by Murray Miskin. — New York : Grune & Stratton, c1979.
viii, 337 p. : ill. ; 24 cm.
Includes bibliographical references and index.
ISBN 0-8089-1117-1
1. Children—Diseases—Diagnosis. 2. Diagnosis, Ultrasonic. I. Miskin, Murray.
[DNLM: 1. Ultrasonics—Diagnostic use. 2. Diagnosis—In infancy and childhood. WS141 U47]
RJ51.U45U47 618.9'2007'54 78-10282
78 MARC

—DISEASES—NURSING see Pediatric nursing

—DISEASES—NUTRITIONAL ASPECTS

Smith, Lendon H 1921-
Feed your kids right : Dr. Smith's program for your child's total health / Lendon Smith. — New York : McGraw Hill, c1979.
ix, 251 p. ; 24 cm.
Bibliography: p. 243-245.
Includes index.
ISBN 0-07-058496-6 : $9.95
1. Children—Nutrition. 2. Children—Diseases—Nutritional aspects. I. Title.
RJ206.S62 613.2 78-23733
78 MARC

—DISEASES—NUTRITIONAL ASPECTS—HANDBOOKS, MANUALS, ETC.

Texas Children's Hospital. Dietary Committee.
Dietary guidelines / written by the Dietary Committee, Texas Children's Hospital, Houston, Texas ; edited by Evelyn J. Lorenzen. — Houston : Gulf Pub. Co., Book Division, c1978.
viii, 182 p. ; 29 cm.
Includes bibliographical references and index.
ISBN 0-87201-176-3
1. Diet therapy for children—Handbooks, manuals, etc. 2. Children—Diseases—Nutritional aspects—Handbooks, manuals, etc. I. Lorenzen, Evelyn J. II. Title.
RJ53.D53T48 1978 615'.854 78-53194
79 MARC

—DISEASES—POPULAR WORKS

Wang, Li-ying.
(Tsen yang tai hao hsiao hai)
怎样带好小孩 / [王丽瑛编]. — 第3版. — 北京：人民卫生出版社, 1976.
2, 50 p. : ill. ; 19 cm.
Cover title.
$0.12

1. Children—Care and hygiene—Popular works. 2. Children—Diseases—Popular works. I. Title.
RJ61.W25 1976 78–843053

—DISEASES—PSYCHOLOGICAL ASPECTS see Sick children —Psychology

—DISEASES—SOCIAL ASPECTS

Psychosocial aspects of pediatric care / edited by Elizabeth Gellert. — New York : Grune & Stratton, c1978.
x, 256 p. : ill. ; 26 cm.
Includes bibliographies and index.
ISBN 0-8089-1091-4
1. Sick children—Psychology. 2. Children—Diseases—Social aspects. I. Gellert, Elizabeth.
[DNLM: 1. Child psychology. 2. Pediatrics. WS105.3 P974]
RJ47.5.P785 618.9'2'001 78-16682
78 MARC

—DISEASES—SOCIAL ASPECTS —CONGRESSES

Perspectieven in de psychosociale zorg voor het zieke kind / onder red. van G. M. H. Veeneklaas, A. Th. Schweizer, G. A. Kohnstamm ; met bijdragen van A. J. M. van Tienen ... [et al.]. — Deventer : Van Loghum Slaterus, [1978]
116 p. ; 22 cm. — (Leven en welzijn) Ne78-12
Includes bibliographies.
ISBN 9060014421 : fl 19.50
1. Sick children—Psychology—Congresses. 2. Children—Diseases—Social aspects—Congresses. 3. Children—Hospital care—Congresses. I. Veeneklaas, G. M. H. II. Schweizer, A. Th. III. Kohnstamm, Geldolph Adriaan, 1908- . IV. Tienen, Aloysius Josef Maria van.
RJ47.5.P38 78-384992
*79 MARC

—DISEASES—CZECHOSLOVAKIA—STATISTICS

Sziegoleit, Meinhard.
Kinder- und Mütterfürsorge in der CSSR mit besonderer Berücksichtigung der Abtreibung, von Meinhard Sziegoleit. Hrsg. von Hans Harmsen. Hamburg, [Bonn, Büro Bonner Berichte] 1969.
68 p. illus. 29 cm. (Zur Entwicklung und Organisation des Gesundheitswesens in Sowjetrussland, in osteuropäischen Volksdemokratien und in Mitteldeutschland, Bd. 45) GDB70-B3-173
Bibliography: p. 67-68
1. Child health services—Czechoslovakia. 2. Children—Diseases—Czechoslovakia—Statistics. 3. Children—Czechoslovakia—Mortality. 4. Maternal health services—Czechoslovakia. 5. Mothers—Czechoslovakia—Mortality. 6. Abortion—Czechoslovakia—Statistics. I. Title. II. Series.
RJ103.C95S94 79-413683
79 MARC

—DISEASES—ENGLAND—HISTORY—19TH CENTURY—ADDRESSES, ESSAYS, LECTURES

Graham, Sarah.
Little victims : sick children in the Victorian era / by Sarah Graham. — London : Macmillan, 1977.
27 p. : ill., facsims. ; 29 cm. GB78-29402
"A collection of articles ... reprinted from "Nursing times"".
Includes bibliographies.
ISBN 0-333-24168-1 : £0.90
1. Pediatrics—England—History—19th century—Addresses, essays, lectures. 2. Children—Diseases—England—History—19th century—Addresses, essays, lectures. 3. Children—Care and hygiene—England—History—19th century—Addresses, essays, lectures. I. Title.
RJ42.G7G7 362.7'81'0941 79-317810
79 MARC

—DISEASES—HUNGARY—BUDAPEST—LONGITUDINAL STUDIES

Az 1-60 [i.e. Egytől hatvan] hónapos budapesti gyermekek testi fejlettsége, szociodemográfiai és morbiditási viszonyai (előzetes jelentés / [szerkők, Ács Jolán ... et al. ; szerk., Sárkány Jenő]. — Budapest : Központi Statisztikai Hivatal, Népességtudományi Kutató Intézet, 1977.
237 p. : ill. ; 29 cm. — (A Központi Statisztikai Hivatal Népességtudományi Kutató Intézetének és a Magyar Tudományos Akadémia Demográfiai Bizottságának közleményei ; 45)
Summary in English (37 p.) inserted.
Bibliography p. 101-102.
ISBN 9633401488 : 116.00Ft
1. Children—Care and hygiene—Hungary—Budapest—Longitudinal studies. 2. Children—Hungary—Budapest—Growth—Longitudinal studies. 3. Children—Hungary—Budapest—Anthropometry. 4. Children—Diseases—Hungary—Budapest—Longitudinal studies. I. Ács, Jolán. II. Sárkány, Jenő.
III. Series: Népességtudományos Kutató Intézet. A Központi Statisztikai Hivatal Népességtudományi Kutató Intézetének és a Magyar Tudományos Akadémia Demográfiai Bizottságának közleményei ; 45.
HA1201.A64 no. 45 79-355878
[RJ103.H9] 79 MARC

—DISEASES—INDONESIA—ADDRESSES, ESSAYS, LECTURES

Jakob, Teuku, 1929-
Studi tentang variasi manusia di Indonesia / oleh T. Jacob. Tugas-tugas pokok dalam memelihara kesehatan anak / oleh Ismangoen. — [Yogyakarta] : Universitas Gadjah Mada, 1973.
52 p. ; 24 cm. — (Seri penerbitan pidato pengukuhan - Universitas Gadjah Mada)
Cover title.
Bibliography: p. 19-22.
1. Anthropometry—Indonesia—Addresses, essays, lectures. 2. Fossil man—Indonesia—Addresses, essays, lectures. 3. Children—Diseases—Indonesia—Addresses, essays, lectures. 4. Child health services—Indonesia—Addresses, essays, lectures. I. Ismangoen, R., 1919- Tugas-tugas pokok dalam memelihara kesehatan anak. 1973. II. Title. III. Series: Universitas Gadjah Mada. Seri penerbitan pidato pengukuhan - Universitas Gadjah Mada.
GN58.I46J34 78-942611
79 MARC

CHILDREN (Continued)

—DISEASES—UNITED STATES —STATISTICS

California. University. School of Public Health.
Health status of children : a review of surveys, 1963-1972 / [prepared by University of California School of Public Health]. — Rockville, Md. : U.S. Dept. of Health, Education, and Welfare, Public Health Service, Health Services Administration, Bureau of Community Health Services ; Washington : for sale by the Supt. of Docs., U.S. Govt. Print. Off., 1978.
vii, 259 p. : ill. ; 28 cm. — (DHEW publication ; no. (HSA) 78-5744)
Report written by Marion Johnson Chabot and others.
Includes bibliographical references.
1. Children—Diseases—United States—Statistics. 2. Health surveys—United States. 3. United States—Statistics, Medical. I. Chabot, Marion Johnson. II. Title. III. Series: United States. Dept. of Health, Education and Welfare. DHEW publication ; no. (HSA) 78-5744.
RJ102.C34 1978 362.7'8'0973 78-603574
79 MARC

Ezzati, Trena M
Ambulatory care utilization patterns of children and young adults : national ambulatory medical care survey, United States, January-December 1975. — Hyattsville, Md. : U.S. Dept. of Health, Education, and Welfare, Public Health Service, Office of the Assistant Secretary for Health, National Center for Health Statistics, 1978.
iv, 43 p. : ill. ; 26 cm. — (Vital and health statistics : Series 13, Data from the National Health Survey ; no. 39) (DHEW publication ; no. (PHS) 78-1790)
Includes bibliographical references.
ISBN 0-8406-0134-4
1. Ambulatory medical care for children—United States—Utilization—Statistics. 2. Physician services utilization—United States—Statistics. 3. Children—Diseases—United States—Statistics. 4. Health surveys—United States.

5. United States—Statistics, Medical. I. United States. National Center for Health Statistics. II. Title. III. Series: United States. National Center for Health Statistics. Vital and health statistics : Series 13, Data from the National Health Survey, Data on health resources utilization ; no. 39. IV. Series: United States. Dept. of Health, Education, and Welfare. DHEW publication ; no. (PHS) 78-1790.
[DNLM: 1. Ambulatory care—In infancy and childhood—Statistics. 2. Ambulatory care—In adolescence—Statistics. 3. Health services—Utilization—United States. 4. Utilization review. W2 A N148vm no. 39 (P)]
RA407.3.A349 no. 39 362.1'1'0973 s 78-606086
[RJ102] 78 MARC

—EDUCATION see Education of children

—EMPLOYMENT

see also Chimmey-sweeps; Hours of labor; Labor supply; Newsboys; Student employment; Sweating system; Youth—Employment

—EMPLOYMENT—CONNECTICUT —NEW BRITAIN

Robinson, Claude Everett, 1900-
Child workers in two Connecticut towns, New Britain and Norwich, by Claude E. Robinson. [New York, N.Y.] National child labor committee, 1929.
44 p., 1 l. tables. 23 cm. ([National child labor committee. Publication no. 350])
"Prepared for the Department of investigation, Charles E. Gibbons, director."
1. Children—Employment—Connecticut—New Britain. 2. Children—Employment—Connecticut—Norwich. I. Title.
[HD6250.U3N2 no. 350] 331.3'1'0973 s e 32-342
[HD6250.U4C7] [r79]rev MARC

—EMPLOYMENT—CONNECTICUT— NORWICH

Robinson, Claude Everett, 1900-
Child workers in two Connecticut towns, New Britain and Norwich, by Claude E. Robinson. [New York, N.Y.] National child labor committee, 1929.
44 p., 1 l. tables. 23 cm. ([National child labor committee. Publication no. 350])
"Prepared for the Department of investigation, Charles E. Gibbons, director."
1. Children—Employment—Connecticut—New Britain. 2. Children—Employment—Connecticut—Norwich. I. Title.
[HD6250.U3N2 no. 350] 331.3'1'0973 s e 32-342
[HD6250.U4C7] [r79]rev MARC

—EMPLOYMENT—DENMARK— HISTORY—JUVENILE LITERATURE

Bolvig, Axel.
Sociale tabere omkring år 1900 : Peter Sabroe og børnene / [af] Axel Bolvig. — [København] : Gyldendal, 1978.
58 p. : ill. ; 22 cm. D78-18
ISBN 8701568515 : kr33.25
1. Children—Employment—Denmark—History—Juvenile literature. 2. Poor—Denmark—History—Juvenile literature. 3. Sabroe, Peter, 1867-1913—Juvenile literature. I. Title.
HD6250.D42B65 78-392736
79 MARC

—EMPLOYMENT—GERMANY— HISTORY—SOURCES

Uhlig, Otto.
Die Schwabenkinder aus Tirol und Vorarlberg / von Otto Uhlig. — Innsbruck : Univ.-Verl. Wagner ; Stuttgart : Aalen : Theiss, 1978.
307 p., [17] leaves of plates : ill. ; 25 cm. — (Tiroler Wirtschaftsstudien, Schriftenreihe der Jubiläumsstiftung der Kammer der Gewerblichen Wirtschaft [für Tirol] ; Folge 34) Au78-15
Bibliography: p. [297]-300.
Includes indexes.
ISBN 3-7030-0052-X : S340.00
1. Children—Austria—Tyrol—History—Sources. 2. Children—Austria—Voralberg—History—Sources. 3. Children—Employment—Germany—History—Sources. 4. Migrant agricultural laborers—Germany—History—Sources. 5. Alien labor, Austrian—Germany—History—Sources. I. Title. II. Series.
HQ799.A92T88 79-337977
*79 MARC

—EMPLOYMENT—GERMANY, WEST

Däubler, Wolfgang.
Schulung und Fortbildung von Betriebsratsmitgliedern und Jugendvertretern nach Paragraph 37 BetrVG : mit e. Anhang zum Personalvertretungsrecht / von Wolfgang Däubler. — 3., überarb. Aufl. — Köln : Bund-Verlag, 1978.
182 p. ; 21 cm. GFR78-A
Bibliography: p. 165-171.
Includes index.
ISBN 3-7663-0230-2 : DM28.00
1. Works councils—Germany, West. 2. Children—Employment—Germany, West. I. Title.
79-343159
79 MARC

Germany (Federal Republic, 1949-). Laws, statutes, etc.
Jugendarbeitsschutzgesetz : Kommentar / begr. von Erich Molitor u. Bernhard Volmer ; fortgef. Claas-Hinrich Germelmann. — 2., völlig neubearb. Aufl. — München : Beck, 1978.
xv, 584 p. ; 23 cm. — (Beck'sche Kommentare zum Arbeitsrecht ; Bd. 15) GFR78-A3
Includes index.
ISBN 3-406-00731-7
1. Children—Employment—Germany, West. I. Molitor, Erich, 1886-1963. II. Volmer, Bernhard. III. Germelmann, Claas-Hinrich. IV. Title.
344'.43'0131 79-350315
79 MARC

Kube, Dietmar, 1941-
Die Koalitionsfreiheit des minderjährigen Arbeitnehmers. [Köln, Spezialdruckerei für Dissertationen Gouder u. Hansen, 1967?]
xxv, 112 p. 21 cm.
Inaug.-Diss.—Cologne.
Vita.
Bibliography: p. vii-xxi.
1. Children—Employment—Germany, West. 2. Freedom of association—Germany, West. I. Title.
77-473987
78 MARC

—EMPLOYMENT—GREAT BRITAIN

Great Britain. Home office.
The law on performances by children: a guide to the Children (Performances) Regulations 1968 and related statutory provisions. London, H.M.S.O., 1968.
iv, 32 p. 25 cm. 4/- B68-25530
ISBN 0-11-340024-1
1. Children—Employment—Great Britain. 2. Children as actors. I. Title.
344'.41'0131 72-180955
72[r78]rev MARC

—EMPLOYMENT—INDIA—DELHI

Indian Council for Child Welfare.
Working children in urban Delhi : a study of their life and work : report submitted to the Department of Social Welfare, Government of India / by the Indian Council for Child Welfare. — New Delhi : The Council, 1977.
173 p., [2] leaves of plates : ill. ; 23 cm.
Includes bibliographical references.
1. Children—Employment—India—Delhi. I. India (Republic). Dept. of Social Welfare. II. Title.
HD6250.I43D444 331.3'1'095456 79-900870
79 MARC

—EMPLOYMENT—ITALY

Legge 285, riflessioni e proposte : consuntivo della prima fase di attuazione della legge in favore dell'occupazione giovanile in Italia. — Roma : Tip. Detti Dario e Ugo, 1978.
275 p. ; 25 cm. — (Quaderni di formazione ISFOL ; n. 48-49) It79-Feb
1. Children—Employment—Italy. 2. Occupational training—Law and legislation—Italy. I. Series: Istituto per lo sviluppo della formazione professionale dei lavoratori. Quaderni di formazione ISFOL ; n. 48-49.
79-354904
*79 MARC

—EMPLOYMENT—ITALY—LECCO— HISTORY

Ballestrero, Maria Vittoria.
Genocidio perfetto : industrializzazione e forza-lavoro nel Lecchese, 1840-1870 / [di] Maria Vittoria Ballestrero, Renato Levrero. — 1. ed. — Milano : Feltrinelli economica, 1979.
177 p. ; 21 cm. — (Le Classi subalterne) (I Nuovi testi ; 179)
It79-May
"Antologia": p. [65]-177.
Includes bibliographical references.
L3000
1. Labor and laboring classes—Italy—Lecco—History. 2. Children—Employment—Italy—Lecco—History. 3. Lecco, Italy—Industries—History. I. Levrero, Renato, 1941- joint author. II. Title. III. Series.
HD8490.L42B34 79-372517
*79 MARC

—EMPLOYMENT—SWEDEN—HISTORY —19TH CENTURY

Ideologi och socialpolitik i 1800-talets Sverige : fyra studier = [Ideology and social policy] / Ingrid Hammerström ... — Uppsala : Univ. ; Stockholm : Almqvist & Wiksell international (distr.), 1978.
xv, 475 p. : diagrs., tables ; 23 cm. — (Studia historica Upsaliensia ; 103 ISSN 0081-6531) S78-27
Summaries in English.
Bibliography: p. 453-463.
Includes index.
ISBN 9155407897 : kr95.00
1. Social service—Sweden—History—19th century. 2. Poor—Sweden—History—19th century. 3. Children—Employment—Sweden—History—19th century. 4. Education, Elementary—Sweden—History—19th century. 5. Medical care—Sweden—History—19th century. 6. Sweden—Social policy. I. Hammarström, Ingrid, 1924- . II. Series.
HV338.I34 79-350393
79 MARC

—EMPLOYMENT—SWEDEN— STOCKHOLM—HISTORY

Utterström, Gustaf.
Fattig och föräldralös i Stockholm på 1600- och 1700-talen = [Arm und elternlos in Stockholm im 17. und 18. Jahrhundert] / av Gustaf Utterström. — Umeå : Univ.-bibl., 1978.
248 p. : diagrs., tables ; 24 cm. — (Umeå studies in the humanities ; 14 ISSN 0345-0155) S78-25
Summary in German.
Errata slip inserted.
Bibliography: p. 242-247.
ISBN 9171740120 : kr75.00
1. Stockholm—Orphans and orphan-asylums. 2. Allmänna barnhuset, Stockholm. 3. Children—Employment—Sweden—Stockholm—History. 4. Stockholm—Poor—History. I. Title. II. Title: Arm und elternlos in Stockholm im 17. und 18. Jahrhundert. III. Series.
HV1240.S8U88 79-346539
79 MARC

—EMPLOYMENT—UNITED STATES

United States. Laws, statutes, etc.
[Fair labor standards act of 1938]
1977 minimum wage law : Fair labor standards act with 1977 amendments. — Washington : Bureau of National Affairs : order from BNA Books, c1977.
15 p., p. 51-55, 101-137 ; 23 cm.
Cover title.
CONTENTS: Explanation of 1977 FLSA amendments.—Joint explanatory statement of Committee of Conference.—Text of Fair labor standards act as amended.
$2.50
1. Labor laws and legislation—United States. 2. Wages—Minimum wage—United States. 3. Overtime—United States. 4. Children—Employment—United States. I. United States. Congress. Conferences Committees, 1977. II. Bureau of National Affairs, Washington, D.C. III. Title.
KF3455.A313A2 1977 344'.73'0121 78-111342
79 MARC

—FICTION

Catalano, Jorge F
Niños / Jorge F. Catalano ; fotoportada, Javier Molina. — [La Paz] : Difusión : distribuidores Gisbert, 1979.
165 p. ; 19 cm.
1. Children—Fiction. I. Title.
PQ7820.C387N5 79-117484
00 MARC

CHILDREN

—FICTION (Continued)

Dandelion clocks : stories of childhood / edited by Alfred Bradley and Kay Jamieson. — London : Joseph, 1978.
[6], 217 p. ; 23 cm. GB78-28934
ISBN 0-7181-1721-2 : £4.95
1. Children—Fiction. 2. Short stories, English. I. Bradley, Alfred, playwright. II. Jamieson, Kay.
PZ1.D17 823'.01 79-310618
[PR1309.C] 79 MARC

Deadly nightshade : strange tales of the dark / edited by Peter Haining. — New York : Taplinger Pub. Co., 1978, c1977.
189 p. ; 21 cm.
ISBN 0-8008-2123-8 : $8.50
1. Horror tales, American. 2. Horror tales, English. 3. Children—Fiction. I. Haining, Peter.
PZ1.D33 823'.0872 77-92767
[PS648.H6] 78 MARC

—FOOD see Children—Nutrition

—GROWTH

Andriulis, E
Vaiko fiziologija ir mokyklinė higiena / E. Andriulis, V. Kviklys. — Vilnius : Mintis, 1974.
228, [4] p. : ill. ; 22 cm. USSR***
Bibliography: p. [230]
0.67rub
1. Children—Growth. 2. Children—Care and hygiene. 3. School hygiene. I. Kviklys, Vladas, joint author. II. Title.
RJ131.A64 78-366883
78 MARC

Ch'en, Mao-liang.
(Chung-hua min kuo ch'ing shao nien chih t'i ko yü t'i neng chuang k'uang fa chan ch'ü shih)
中華民國青少年之體格與體能狀况發展趨勢之研究 / 陳懋良主持；教育部體育司編印. — [台北]：該司, 民國65 [1976]
26 p. : ill. ; 21 cm. — (體育學術研究報告；1)
Summary in Chinese and English.
Bibliography: p. 17.
1. Children—Growth. 2. Physical fitness. 3. Children—Taiwan. I. Title. II. Series: T'i yü hsüeh shu yen chiu pao kao ; 1.
RJ131.C5 312'.6'0951 79-838169

Chu, Chŏng-il.
(Adong paltarhak)
兒童發達學 / 朱貞一著. — 서울：教文社, 1978.
262 p. : ill. ; 23 cm.
Includes bibliographical references.
W3500
1. Children—Growth. 2. Infants—Growth. 3. Child development. I. Title.
RJ131.C55 79-826417

Meredith, Howard Voas, 1903-
Human body growth in the first ten years of life / Howard V. Meredith. — Columbia, S.C. : State Print. Co., c1978.
141 p. : graphs ; 23 cm.
Includes bibliographies and index.
1. Children—Growth. 2. Children—Anthropometry. I. Title.
RJ131.M38 612.6'54 78-63120
78 MARC

Pomerance, Herbert H 1918-
Growth standards in children / Herbert H. Pomerance ; with a contribution by John M. Krall. — Hagerstown, Md. : Medical Dept., Harper & Row, c1979.
xii, 225 p. : graphs ; 27 cm.
Bibliography: p. 215-217.
Includes index.
ISBN 0-06-142210-X
1. Children—Growth. 2. Children—Growth—Longitudinal studies. 3. Children—Anthropometry. I. Title.
[DNLM: 1. Growth. WS103 P779g]
RJ131.P63 612.6'5 78-12460
78 MARC

Smith, David W 1921 (Sept. 24)-
Growth and its disorders : basics and standards, approach and classifications, growth deficiency disorders, growth excess disorders, obesity / by David W. Smith. — Philadelphia : Saunders, 1977.
xiv, : ill. ; 27 cm. — (Major problems in clinical pediatrics ; v. 15)
Includes index.
ISBN 0-7216-8378-9
1. Children—Growth. 2. Growth disorders. I. Title.
RJ131.S57 618.9'2 76-20114
76[r79]rev MARC

Tanner, James Mourilyan.
Education and physical growth; implications of the study of children's growth for educational theory and practice. [1st ed.] London, University of London Press [1961]
144 p. illus. 23 cm.
Includes bibliography.
1. Children in Great Britain—Anthropometry. 2. Children—Growth. 3. Education of children. 4. Child development. I. Title.
LB3425.G7T36 1961 62-51434
[r79]rev MARC

Tanner, James Mourilyan.
Education and physical growth : implications of the study of children's growth for educational theory and practice / J. M. Tanner. — 2d ed. — London : Hodder and Stoughton, 1978.
144 p. : ill. ; 22 cm. — (Unibooks) GB***
Bibliography: p. 127-134.
Includes index.
ISBN 0-340-22484-3 : £2.45
1. Children in Great Britain—Anthropometry. 2. Children—Growth. 3. Education of children. 4. Child development. I. Title.
LB3425.G7T36 1978b 573'.6 78-318319
78 MARC

Tanner, James Mourilyan.
Education and physical growth : implications of the study of children's growth for educational theory and practice / J. M. Tanner. — 2d ed. — New York : International Universities Press, 1978.
144 p. : ill. ; 23 cm.
Bibliography: p. 127-134.
Includes index.
ISBN 0-8236-1531-6
1. Children in Great Britain—Anthropometry. 2. Children—Growth. 3. Education of children. 4. Child development. I. Title.
LB3425.G7T36 1978 573'.6 78-8809
78 MARC

—GROWTH—BIBLIOGRAPHY

International Children's Centre, Paris.
Travaux sur la croissance et le développement de l'enfant normal; études coordonnées par le Centre international de l'enfance, 1968-1972. Paris [1973?]
10 p. 27 cm. F***
Cover title.
1. Children—Growth—Bibliography. 2. Child development—Bibliography. I. Title.
Z6671.52.G76I57 1973 016.6126'5 78-374369
[RJ131] 78 MARC

—GROWTH—CHARTS' DIAGRAMS, ETC.

World Health Organization.
A growth chart for international use in maternal and child health care : guidelines for primary health care personnel. — Geneva : World Health Organization ; [Albany, N.Y. : obtainable from WHO Publications Centre], 1978.
36 p. : ill. ; 30 cm. Sw***
Includes bibliographical references.
ISBN 9241541296
1. Children—Growth—Charts' diagrams, etc. 2. Children—Growth—Testing. 3. Children—Growth—Tables. I. Title.
RJ131.W68 1978 79-316191
79 MARC

—GROWTH—CONGRESSES

Malnutrition, environment, and behavior : new perspectives / edited by David A. Levitsky. — Ithaca, N.Y. : Cornell University Press, 1979.
295 p. : ill. ; 23 cm.
"Originated in the proceedings of the Cornell Conference on Malnutrition and Behavior held November, 1975."
Includes bibliographies and indexes.
ISBN 0-8014-1045-2
1. Malnutrition—Congresses. 2. Children—Growth—Congresses. 3. Developmental psychology—Congresses. I. Levitsky, David A. II. Cornell Conference on Malnutrition and Behavior, Cornell University, 1975.
RC623.M34 616.3'9 78-58016
78 MARC

—GROWTH—LONGITUDINAL STUDIES

Pomerance, Herbert H 1918-
Growth standards in children / Herbert H. Pomerance ; with a contribution by John M. Krall. — Hagerstown, Md. : Medical Dept., Harper & Row, c1979.
xii, 225 p. : graphs ; 27 cm.
Bibliography: p. 215-217.
Includes index.
ISBN 0-06-142210-X
1. Children—Growth. 2. Children—Growth—Longitudinal studies. 3. Children—Anthropometry. I. Title.
[DNLM: 1. Growth. WS103 P779g]
RJ131.P63 612.6'5 78-12460
78 MARC

—GROWTH—TABLES

World Health Organization.
A growth chart for international use in maternal and child health care : guidelines for primary health care personnel. — Geneva : World Health Organization ; [Albany, N.Y. : obtainable from WHO Publications Centre], 1978.
36 p. : ill. ; 30 cm. Sw***
Includes bibliographical references.
ISBN 9241541296
1. Children—Growth—Charts' diagrams, etc. 2. Children—Growth—Testing. 3. Children—Growth—Tables. I. Title.
RJ131.W68 1978 79-316191
79 MARC

—GROWTH—TESTING

World Health Organization.
A growth chart for international use in maternal and child health care : guidelines for primary health care personnel. — Geneva : World Health Organization ; [Albany, N.Y. : obtainable from WHO Publications Centre], 1978.
36 p. : ill. ; 30 cm. Sw***
Includes bibliographical references.
ISBN 9241541296
1. Children—Growth—Charts' diagrams, etc. 2. Children—Growth—Testing. 3. Children—Growth—Tables. I. Title.
RJ131.W68 1978 79-316191
79 MARC

—HEALTH see Children—Care and hygiene

—HISTORY

Voss, Vivian.
Iene miene mutte ... : kinderen van alle tijden / Vivian Voss. — Haarlem : Fibula-Van Dishoeck, c1978.
160 p. : ill. (some col.) ; 30 cm. Ne***
Bibliography: p. 159.
ISBN 9022838552
1. Children—History. 2. Children—Social conditions. I. Title.
HQ781.5.V66 78-375813
78 MARC

—HOSPITAL CARE

American Academy of Pediatrics. Committee on Hospital Care.
Hospital care of children and youth / Committee on Hospital Care. — Evanston, Ill. : American Academy of Pediatrics, c1978.
121 p. : forms ; 22 cm.
Includes index.
1. Children—Hospital care. I. Title.
RJ242.A46 1978 658'.91'3627811 78-53173
79 MARC

Beuf, Ann H 1938-
Biting off the bracelet : a study of children in hospitals / Ann Hill Beuf. — [Philadelphia] : University of Pennsylvania Press, 1979.
ix, 164 p. ; 21 cm.
Bibliography: p. 153-159.
Includes index.
ISBN 0-8122-7766-X
1. Children—Hospital care. 2. Children—Hospitals—Sociological aspects. 3. Hospitals—Sociological aspects. I. Title.
[DNLM: 1. Child, Hospitalized. WS105.5.H7 B566b]
RJ242.B47 362.7'8'11 79-13411
79 MARC

Mutter und Kind im Krankenhaus : e. Situationsbericht aus d. Bundesrepublik Deutschland / Gerd Biermann (Hrsg.). — 1. Aufl. — München ; Basel : E. Reinhardt, 1978.
135 p. (p. 135 advertisement) : graphs ; 22 cm. — (Beiträge zur Psychologie und Soziologie des kranken Menschen ; Bd. 4) GFR78-A
Bibliography: p. 126-130.
Includes indexes.
ISBN 3-497-00817-6
1. Children—Hospital care. 2. Sick children—Family relationships. 3. Mother and child. 4. Sick children—Psychology. 5. Children—Hospital care—Germany, West. I. Biermann, Gerd, 1914- . II. Series.
RJ242.M87 78-381202
79 MARC

CHILDREN (Continued)

—HOSPITAL CARE—ADDRESSES, ESSAYS, LECTURES

Hischer, Erhard, 1925-
Das Kind im Krankenhaus : e. Herausforderung d. Sozialpädagogik / Erhard Hischer. — München : Minerva-Publikation, 1978.
36 p. ; 24 cm. — (Eichstätter Hochschulreden ; 8) GFR78-A
"Als Antrittsvorlesung am 25.1.1977 an der Gesamthochschule Eichstätt gehalten."
Bibliography: p. 33-36.
ISBN 3-597-30008-1 : DM6.80
1. Children—Hospital care—Addresses, essays, lectures. 2. Sick children—Psychology—Addresses, essays, lectures. 3. Educational sociology—Addresses, essays, lectures. 4. Hospital schools—Addresses, essays, lectures. I. Title. II. Series.
RJ242.H58 78-379802
78 MARC

—HOSPITAL CARE—CONGRESSES

Perspectieven in de psychosociale zorg voor het zieke kind / onder red. van G. M. H. Veeneklaas, A. Th. Schweizer, G. A. Kohnstamm ; met bijdragen van A. J. M. van Tienen ... [et al.]. — Deventer : Van Loghum Slaterus, [1978]
116 p. ; 22 cm. — (Leven en welzijn) Ne78-12
Includes bibliographies.
ISBN 9060014421 : fl 19.50
1. Sick children—Psychology—Congresses. 2. Children—Diseases—Social aspects—Congresses. 3. Children—Hospital care—Congresses. I. Veeneklaas, G. M. H. II. Schweizer, A. Th. III. Kohnstamm, Geldolph Adriaan, 1908- . IV. Tienen, Aloysius Josef Maria van.
RJ47.5.P38 78-384992
*79 MARC

—HOSPITAL CARE—HANDBOOKS, MANUALS, ETC.

Hospital for Sick Children. Dept. of Nursing.
Manual of pediatric nursing careplans / Department of Nursing, the Hospital for Sick Children, Toronto, Canada ; edited by U. F. Matthews. — 1st ed. — Boston : Little, Brown, c1979.
xiii, 347 p. ; 28 cm.
Bibliography: p. 9.
Includes index.
ISBN 0-316-37383-4
1. Pediatric nursing—Handbooks, manuals, etc. 2. Children—Hospital care—Handbooks, manuals, etc. I. Matthews, Ursula F. II. Title. III. Title: Pediatric nursing careplans.
RJ245.H67 1979 610.73'62 78-71217
79 MARC

—HOSPITAL CARE—JUVENILE LITERATURE—BIBLIOGRAPHY

Altshuler, Anne.
Books that help children deal with a hospital experience. — Rev. — Rockville, Md. : U.S. Dept. of Health, Education, and Welfare, Public Health Service, Health Services Administration, Bureau of Community Health Services ; Washington : for sale by the Supt. of Docs, U.S. Govt. Print. Off., 1978.
ii, 22 p. ; 15 x 23 cm. — (DHEW publication ; no. (HSA) 78-5224)
Cover title.
Includes indexes.
1. Hospitals—Juvenile literature—Bibliography. 2. Children—Hospital care—Juvenile literature—Bibliography. 3. Children—Preparation for medical care—Juvenile literature—Bibliography. I. United States. Health Services Administration. Bureau of Community Health Services. II. Title. III. Series: United States. Dept. of Health, Education, and Welfare. DHEW publication ; no. (HSA) 78-5224.
Z6675.H75A47 1978 016.3627'8'11 79-602662
[RA963.5] 79 MARC

—HOSPITAL CARE—PSYCHOLOGICAL ASPECTS

Beyond separation : further studies of children in hospital / edited by David Hall and Margaret Stacey. — London ; Boston : Routledge & K. Paul, 1979.
xii, 243 p. ; 23 cm. — (Medicine, illness and society) GB***
Bibliography: p. 219-232.
Includes index.
ISBN 0-7100-0163-0
1. Children—Hospital care—Psychological aspects. 2. Children—Hospital care—Social aspects. 3. Hospitals—Psychological aspects. 4. Hospitals—Sociological aspects. I. Hall, David J. II. Stacey, Margaret.
RJ242.B48 362.7'8'11019 79-40011
79 MARC

—HOSPITAL CARE—SOCIAL ASPECTS

Beyond separation : further studies of children in hospital / edited by David Hall and Margaret Stacey. — London ; Boston : Routledge & K. Paul, 1979.
xii, 243 p. ; 23 cm. — (Medicine, illness and society) GB***
Bibliography: p. 219-232.
Includes index.
ISBN 0-7100-0163-0
1. Children—Hospital care—Psychological aspects. 2. Children—Hospital care—Social aspects. 3. Hospitals—Psychological aspects. 4. Hospitals—Sociological aspects. I. Hall, David J. II. Stacey, Margaret.
RJ242.B48 362.7'8'11019 79-40011
79 MARC

—HOSPITAL CARE—GERMANY, WEST

Mutter und Kind im Krankenhaus : e. Situationsbericht aus d. Bundesrepublik Deutschland / Gerd Biermann (Hrsg.). — 1. Aufl. — München ; Basel : E. Reinhardt, 1978.
135 p. (p. 135 advertisement) : graphs ; 22 cm. — (Beiträge zur Psychologie und Soziologie des kranken Menschen ; Bd. 4) GFR78-A
Bibliography: p. 126-130.
Includes indexes.
ISBN 3-497-00817-6
1. Children—Hospital care. 2. Sick children—Family relationships. 3. Mother and child. 4. Sick children—Psychology. 5. Children—Hospital care—Germany, West. I. Biermann, Gerd, 1914- . II. Series.
RJ242.M87 78-381202
79 MARC

—HOSPITALS

see also Child welfare; Hospital schools

—HOSPITALS—SOCIOLOGICAL ASPECTS

Beuf, Ann H 1938-
Biting off the bracelet : a study of children in hospitals / Ann Hill Beuf. — [Philadelphia] : University of Pennsylvania Press, 1979.
ix, 164 p. ; 21 cm.
Bibliography: p. 153-159.
Includes index.
ISBN 0-8122-7766-X
1. Children—Hospital care. 2. Children—Hospitals—Sociological aspects. 3. Hospitals—Sociological aspects. I. Title.
[DNLM: 1. Child, Hospitalized. WS105.5.H7 B566b]
RJ242.B47 362.7'8'11 79-13411
79 MARC

—HOSPITALS—UNITED STATES—COSTS—STATISTICS

Study to quantify the uniqueness of children's hospitals / conducted by the associated firms of Arthur Andersen Company ... [et al.], submitted to the National Association of Children's Hospitals and Related Institutions, inc. — Wilmington, Del. : The Association, c1978.
xxx, 423 p. ; 29 cm.
1. Children—Hospitals—United States—Statistics. 2. Children—Hospitals—United States—Costs—Statistics. 3. Hospitals—United States—Cost of operation—Statistics. I. Andersen (Arthur) and Company. II. National Association of Children's Hospitals and Related Institutions.
RJ27.S88 362.7'8'110973 78-112721
79 MARC

—HOSPITALS—UNITED STATES—STATISTICS

Study to quantify the uniqueness of children's hospitals / conducted by the associated firms of Arthur Andersen Company ... [et al.], submitted to the National Association of Children's Hospitals and Related Institutions, inc. — Wilmington, Del. : The Association, c1978.
xxx, 423 p. ; 29 cm.
1. Children—Hospitals—United States—Statistics. 2. Children—Hospitals—United States—Costs—Statistics. 3. Hospitals—United States—Cost of operation—Statistics. I. Andersen (Arthur) and Company. II. National Association of Children's Hospitals and Related Institutions.
RJ27.S88 362.7'8'110973 78-112721
79 MARC

—HOURS OF LABOR see Hours of labor

—HYGIENE see Children—Care and hygiene

—INSTITUTIONAL CARE

see also Almshouses; Children—Hospital care; Day care centers; Foster home care; Fresh-air charity; Reformatories

Reinprecht, Hansheinz.
The Hermann Gmeiner book : the story of the SOS-Children's Villages and their founder / by Hansheinz Reinprecht ; [translated by Bertha Meyer]. — 1st ed. — New York : Vantage Press, c1978.
258 p., [8] leaves of plates : ill. ; 22 cm.
Translation of Das Hermann Gmeiner Buch.
ISBN 0-533-03453-1 : $4.95
1. Gmeiner, Hermann. 2. Children—Institutional care. 3. Österreichischer SOS-Kinderdorf-Verein. I. Title.
HV876.G55R4413 362.7'32'0924 79-101151
79 MARC

—INSTITUTIONAL CARE—BIBLIOGRAPHY

Bezuidenhout, Adriana Maria.
Literatuurlys oor pleegsorg = Bibliography on foster care / saamgestel deur A. M. Bezuidenhout. — Pretoria : Departement van Volkswelsyn en Pensioene, 1976.
iii, 36 p. ; 30 cm. — (Reeks Navorsing en inligting - Departement van Volkswelsyn en Pensioene ; 1976:4) SA
Afrikaans and English.
ISBN 0-621-03379-0 : free
1. Adoption—Bibliography. 2. Foster home care—Bibliography. 3. Children—Institutional care—Bibliography. I. Title. II. Title: Bibliography on foster care. III. Series: South Africa. Dept. of Social Welfare and Pensions. Reeks navorsing en inligting ; 1976:4.
Z7164.A23B49 78-392236
[HV875] 79 MARC

—INSTITUTIONAL CARE—EDUCATION—AUSTRIA

Jandl, Karoline.
Chancengerechtigkeit für Heimkinder : eine Untersuchung als Beitrag zur Innovation des Bezugssystems Heim u. Schule / Karoline Jandl. — Wien ; München : Jugend & Volk, c1978.
144 p. ; 21 cm. — (Padagogik der Gegenwart ; 214) Au78-13-2
Bibliography: p. 141-144.
ISBN 3-7141-5368-3 : S148.00
1. Children—Institutional care—Education—Austria. 2. Educational equalization—Austria. 3. Mainstreaming in education. I. Title.
LC4096.A9J36 79-373942
*79 MARC

—INSTITUTIONAL CARE—PERIODICALS

Residential & community child care administration. v. 1- spring 1979-
[New York] Haworth Press.
v. 23 cm. quarterly.
"... the journal of child & youth services management."
Key title: Residential & community care administration, ISSN 0162-1408.
1. Children—Institutional care—Periodicals. 2. Youth—Services for—Periodicals.
HV701.R37 362.7'32'05 79-643708
MARC-S

—INSTITUTIONAL CARE—BRAZIL—DUQUE DE CAXIAS

Fundação Nacional do Bem-Estar do Menor. Projeto Integrado da Área Metropolitana do Rio de Janeiro.
Uma abordagem da situação do menor de rua em Duque de Caxias : relatório descritivo, maio-76 / [FUNABEM, Diretoria de Programas Integrados, Projeto Integrado da Área Metropolitana do Rio de Janeiro]. — [Rio de Janeiro : O Projeto, 1976?]
124 leaves : ill. ; 29 cm. — (Documento - FUNABEM, Projeto Integrado da Area Metropolitana do Rio de Janeiro ; no. 4)
1. Abandoned children—Brazil—Duque de Caxias. 2. Child welfare—Brazil—Duque de Caxias. 3. Children—Institutional care—Brazil—Duque de Caxias. I. Title. II. Series: Fundação Nacional do Bem-Estar do Menor. Projeto Integrado da Área Metropolitana do Rio de Janeiro. Documento - FUNABEM, Projeto Integrado da Area Metropolitana do Rio de Janeiro ; no. 4.
HV887.B82D83 362.7'3 79-340486
79 MARC

CHILDREN (Continued)

—INSTITUTIONAL CARE—DENMARK—ADDRESSES, ESSAYS, LECTURES

Når hjemme er et andet sted : debatbog om døgninstitutioner for børn og unge / ⌐redaktion, Jens Refsgård ... et al.⌐ ; ⌐illustrationer, Vibeke Bengtsson⌐. — København : ⌐Socialarbejdernes Fællesudvalg⌐ : salg og distribution, Børn & Unge (Skt. Hans Torv 26, 2200 N), c1978.
160 p. : ill. ; 21 cm. — (SFU-dokumentation ; 3) D78-47
ISBN 8787547066 : kr45.00
1. Children—Institutional care—Denmark—Addresses, essays, lectures. 2. Child welfare—Denmark—Addresses, essays, lectures. 3. Youth services—Denmark—Addresses, essays, lectures. I. Refsgård, Jens. II. Series: Socialarbejdernes fællesudvalg. SFU-dokumentation ; 3.
HV786.A6N37 79-369406
79 MARC

—INSTITUTIONAL CARE—FRANCE

Verdier, Pierre, 1941-
L'enfant en miettes : l'aide sociale à l'enfance : bilan et perspectives d'avenir / Pierre Verdier. — Toulouse : Privat, c1978.
209 p. ; 21 cm. — (Époque) F***
Bibliography: p. ⌐207⌐-209.
ISBN 2-7089-1235-6 : 46.00F
1. Children—Institutional care—France. 2. Foster home care—France. I. Title.
HV761.A6V46 362.7'32'0944 79-364924
79 MARC

—INSTITUTIONAL CARE—GEORGIA

Georgia. General Assembly. Senate. Juvenile Institutions Study Committee.
Report of the Senate Juvenile Institutions Study Committee / Floyd Hudgins, chairman. — Atlanta : General Assembly, 1979.
31 p. ; 36 cm.
1. Problem children—Georgia. 2. Child welfare—Georgia. 3. Children—Institutional care—Georgia. 4. Juvenile courts—Georgia. I. Hudgins, Floyd. II. Title.
HV742.G4G36 1979 362.7'3'09758 79-624487
79 MARC

—INSTITUTIONAL CARE—GERMANY, WEST

Blandow-Wechsung, Steffi.
Heimerziehung und Devianz : e. Unters. pragmat. Devianztheorie u. Sympathie von Erziehern in ihren Auswirkungen auf Heimkinder / Steffi Blandow-Wechsung. — München : Minerva-Publikation, 1978.
xviii, 419 p. : ill. ; 21 cm. — (Minerva-Fachserie Pädagogik) GFR78-A
Originally presented as the author's thesis, Bremen.
Bibliography: p. 405-419.
ISBN 3-597-10045-7 : DM48.00
1. Children—Institutional care—Germany, West. 2. Children—Institutional care—Germany, West—Hamburg—Case studies. 3. Juvenile delinquency—Germany, West. 4. Deviant behavior. I. Title.
HV887.G4B53 1978 362.7'32'0943 79-350080
79 MARC

Einführung in Theorie und Praxis der Heimerziehung / hrsg. von Heinrich Kupffer ; unter Mitarb. von Eberhard Dyck, Klaus-Rainer Martin, Ulrich Ziethen. — 2., durchges. Aufl. — Heidelberg : Quelle und Meyer, 1978.
163 p. ; 19 cm. — (Uni-Taschenbücher ; 657) GFR78-A
Bibliography: p. 155-160.
Includes index.
ISBN 3-494-02071-X : DM14.80
1. Children—Institutional care—Germany, West. I. Kupffer, Heinrich, 1924-
HV1174.E39 1978 78-389061
79 MARC

Fröhlich, Manfred.
Einführung in die Heimpädagogik / Manfred Fröhlich. — 5., überarb. u. erw. Aufl. — München : Bardtenschlager, 1978.
224 p. ; 21 cm. — (Sozialpädagogische Studienreihe) GFR79-A
Bibliography: p. 215-218.
Includes index.
ISBN 3-7623-0063-1 : DM18.80
1. Handicapped children—Education—Germany, West. 2. Children—Institutional care—Germany, West. I. Title.
LC4036.G4F76 1978 79-369521
79 MARC

Pommerehne-Häss, Barbara, 1945-
Die Reformprogramme für die Heimerziehung : Chancen für e. Demokratisierung d. öffentl. Erziehung? / Barbara Pommerehne-Häss. — Frankfurt am Main ; Bern ; Las Vegas : Lang, 1979.
148 p. : 4 graphs ; 21 cm. — (Beiträge zur Politikwissenschaft ; Bd. 16) GFR79-A
Bibliography: p. 138-148.
ISBN 3-8204-6332-1
1. Socially handicapped children—Education—Germany, West. 2. Children—Institutional care—Germany, West. 3. Open plan schools—Germany, West. I. Title. II. Series.
LC4065.P65 79-384134
79 MARC

Weinschenk, Reinhold.
Geplantes Erziehen im Heim / Reinhold Weinschenk. — Freiburg im Breisgau : Lambertus-Verlag, 1978.
199 p. ; 20 cm. — (Freiburger sozialpädagogische Beiträge ; Bd. 13) GFR79-A
Bibliography: p. 186-193.
Includes indexes.
ISBN 3-7841-0164-X : DM19.80
1. Children—Institutional care—Germany, West. I. Title.
HV887.G4W44 362.7'32'0943 79-359240
79 MARC

—INSTITUTIONAL CARE—GERMANY, WEST—HAMBURG—CASE STUDIES

Blandow-Wechsung, Steffi.
Heimerziehung und Devianz : e. Unters. pragmat. Devianztheorie u. Sympathie von Erziehern in ihren Auswirkungen auf Heimkinder / Steffi Blandow-Wechsung. — München : Minerva-Publikation, 1978.
xviii, 419 p. : ill. ; 21 cm. — (Minerva-Fachserie Pädagogik) GFR78-A
Originally presented as the author's thesis, Bremen.
Bibliography: p. 405-419.
ISBN 3-597-10045-7 : DM48.00
1. Children—Institutional care—Germany, West. 2. Children—Institutional care—Germany, West—Hamburg—Case studies. 3. Juvenile delinquency—Germany, West. 4. Deviant behavior. I. Title.
HV887.G4B53 1978 362.7'32'0943 79-350080
79 MARC

—INSTITUTIONAL CARE—INDIA—RAJASTHAN

Bedi, M S
Socially handicapped children : a study of their institutional services / M. S. Bedi. — 1st ed. — Jodhpur : Jain Bros., 1978.
xix, 223 p. ; 22 cm.
Includes bibliographical references.
Rs35.00
1. Socially handicapped children—India—Rajasthan. 2. Social work with children—India—Rajasthan. 3. Child welfare—India—Rajasthan. 4. Children—Institutional care—India—Rajasthan. I. Title.
HV800.I53R643 362.7'0954'4 78-905279
79 MARC

—INSTITUTIONAL CARE—MISSOURI

Missouri. Division of Family Services.
Rules for licensing child caring agencies. — Jefferson City, Mo. : Dept. of Social Services, Division of Family Services, 1979.
v, 24 p. ; 28 cm.
Cover title.
1. Children—Legal status, laws, etc.—Missouri. 2. Children—Institutional care—Missouri. 3. Child welfare—Missouri. I. Title. II. Title: Child caring agencies.
KFM8150.C4A32 1979 344'.778'03271 79-625028
79 MARC

—INSTITUTIONAL CARE—NEW ZEALAND

McDonald, Dugald Jan.
Perceptions of residential child care / Dugald Jan McDonald. — Christchurch : Dept. of Psychology and Sociology, University of Canterbury, 1977.
viii, 126 p. : ill. ; 29 cm. — (Research project - Dept. of Psychology and Sociology, University of Canterbury ; 27) NZ***
Originally presented as the author's thesis, Massey University.
Bibliography: p. 116-126.
1. Children—Institutional care—New Zealand. I. Title. II. Series: Christchurch, N.Z. University of Canterbury. Dept. of Psychology and Sociology. Research project ; 27.
BF1.C45a no. 27 150'.8 s 79-310215
⌐HV887.N45⌐ 79 MARC

—INSTITUTIONAL CARE—RUSSIAN REPUBLIC—LENINGRAD

Soroka-Rosinskiĭ, Viktor Nikolaevich.
Shkola Dostoevskogo / V. N. Soroka-Rosinskiĭ ; ⌐Vstup. stat'i︠a︡ L. Kabo⌐. — Moskva : Znanie, 1978.
48 p. : ill. ; 20 cm. — (Novoe v zhizni, nauke, tekhnike : Serii︠a︡ Pedagogika i psikhologii︠a︡ ; No 12, 1978 g) USSR78
0.11rub
1. Children—Institutional care—Russian Republic—Leningrad. 2. Problem children—Education—Russian Republic—Leningrad. 3. Orphans and orphan-asylums—Russian Republic—Leningrad. 4. Shkola Dostoevskogo. I. Title. II. Series: Novoe v zhizni, nauke, tekhnike : Serii︠a︡ Pedagogika i psikhologii︠a︡ ; 1978, 12.
HV1215.L52S547 79-371537
79 MARC

—INSTITUTIONAL CARE—SWITZERLAND

Baumgarten, Heinz Hermann.
Ambulante und stationäre Interventions- und Sozialisationshilfen nach schweizerischem Jugendstrafrecht unter besonderer Berücksichtigung der stationären Erziehung männlicher Jugendlicher / Heinz Hermann Baumgarten. — Luzern : Institut für Heilpädagogik, 1978.
201 p. ; 24 cm. — (Formen und Führen ; Heft 30) Sw78-12344
Bibliography: p. 168-180.
Includes indexes.
27.00F
1. Social work with delinquents and criminals—Switzerland. 2. Juvenile delinquency—Switzerland. 3. Children—Institutional care—Switzerland. I. Title. II. Series.
79-350509

—INSTITUTIONAL CARE—UNITED STATES

Knitzer, Jane.
Children without homes : an examination of public responsibility to children in out-of-home care : final report / written by Jane Knitzer and Mary Lee Allen. — Washington : Children's Defense Fund, c1978.
xv, 282 p. : ill. ; 26 cm.
"This study was conducted by Jane Knitzer, Mary Lee Allen, and Brenda McGowan."
Includes bibliographical references.
1. Child welfare—United States. 2. Children—Institutional care—United States. 3. Foster home care—United States. 4. Adoption—United States. I. Allen, Mary Lee, joint author. II. Washington Research Project. Children's Defense Fund. III. Title.
HV741.K54 362.7'3'0973 78-74230
79 MARC

United States. Congress. House. Committee on Education and Labor. Subcommittee on Elementary, Secondary, and Vocational Education.
Part 15, Title I-Neglected and delinquent, and foster children programs : hearing before the Subcommittee on Elementary, Secondary, and Vocational Education of the Committee on Education and Labor, House of Representatives, Ninety-fifth Congress, first session on H.R. 15 to extend for five years certain elementary, secondary, and other education programs, hearings held in Washington, D.C., October 5, 1977. — Washington : U.S. Govt. Print. Off., 1977.
iii, 72 p. : ill. ; 24 cm.
1. Socially handicapped children—Education—Law and legislation—United States. 2. Children—Institutional care—United States. I. Title: Neglected and delinquent, and foster children programs.
KF27.E3364 1977y 344'.73079 79-601067
79 MARC

—INSTITUTIONAL CARE—UNITED STATES—ADDRESSES, ESSAYS, LECTURES

Child abuse and neglect in residential institutions : selected readings on prevention, investigation, and correction. — ⌐Washington⌐ : National Center on Child Abuse and Neglect, Children's Bureau/Administration for Children, Youth and Families/Office of Human Development Services, U.S. Dept. of Health, Education, and Welfare, 1978.
178 p. in various pagings : ill. ; 27 cm. — (DHEW publication ; no. (OHDS) 78-30160)
1. Child abuse—United States—Addresses, essays, lectures. 2. Children—Institutional care—United States—Addresses, essays, lectures. I. National Center on Child Abuse and Neglect. II. Series: United States. Dept. of Health, Education, and Welfare. DHEW publication ; no. (OHDS) 78-30160.
HV741.C453 362.7'32 78-603525
79 MARC

—INSTITUTIONAL CARE—WASHINGTON (STATE)

Washington (State). Division of Community Services.
Child foster care program review / State of Washington, Department of Social and Health Services, Community Services Division. — ⌐Olympia⌐ : The Department, 1976.
129 p. in various pagings ; 28 cm.
1. Children—Institutional care—Washington (State) 2. Foster home care—Washington (State) I. Title.
HV883.W2W37 1976 362.7'33'09797 77-623135
77⌐r79⌐rev MARC

—JUVENILE FICTION

Danois, Jacques.
Les amis du bout du monde / Jacques Danois ; ill. de Jacques Danois. — Bruxelles, r. de Brabant 91 : P. de Méyère, ⌐1977⌐
147 p. : ill. ; 22 cm. Be78-5
242F
1. Children—Juvenile fiction. I. Title.
PZ23.D2173A5 843'.9'14 78-388973
*79 MARC

CHILDREN (Continued)

—JUVENILE LITERATURE

D'Atri, Adriana.
Así es nuestro hermano pequeño / idea Adriana D'Atri ; ilustraciones Ulises Wensell. — Madrid : Altea, D.L. 1978.
32 p. : chiefly col. ill. ; 25 cm. — (Libros para mirar) Sp78-Nov
Cover title.
ISBN 8437213037 : 250ptas
1. Children—Juvenile literature. 2. Family—Juvenile literature. I. Wensell, Ulises. II. Title.
HQ781.D3 79-391167
00 MARC

—LANGUAGE

see also Language acquisition; Language disorders in children

Aitchison, Jean.
The articulate mammal : an introduction to psycholinguistics / Jean Aitchison. — 1st McGraw-Hill paperback ed. — New York : McGraw-Hill, 1978, c1976.
256 p. : ill. ; 21 cm. — (McGraw-Hill paperbacks)
Bibliography: p. [246]-256.
ISBN 0-07-000736-5 : $4.50
1. Psycholinguistics. 2. Children—Language. I. Title.
P37.A37 1978 401'.9 78-6569
78 MARC

Anastasiow, Nicholas J
Oral language : expression of thought / Nicholas Anastasiow. — Updated ed. — Urbana, Ill. : ERIC Clearinghouse on Reading and Communication Skills, National Institute of Education, 1979.
ix, 35 p. ; 23 cm.
Bibliography: p. 33-35.
ISBN 0-87207-840-X
1. Children—Language. 2. Language arts (Primary) I. ERIC Clearinghouse on Reading and Communication Skills. II. Title.
LB1139.L3A483 1979 372.6 79-10263
79 MARC

Bates, Elibabeth.
The emergence of symbols : cognition and communication in infancy / Elizabeth Bates, with the collaboration of Laure Benigni ... [et al.]. — New York : Academic Press, 1979.
xvi, 387 p. ; 24 cm. — (Language, thought, and culture)
Bibliography: p. 371-381.
ISBN 0-12-081540-0
1. Infant psychology. 2. Cognition in children. 3. Children—Language. 4. Symbolism (Psychology) I. Title.
[DNLM: 1. Language developement. 2. Psycholinguistics. WS105.5.C8 E53]
BF723.I6B34 155.4'22 78-20040
79 MARC

Biere, Bernd Ulrich.
Kommunikation unter Kindern : method. Reflexion u. exemplar. Beschreibung / Bernd Ulrich Biere. — Tübingen : Niemeyer, 1978.
x, 258 p. ; 24 cm. — (Linguistische Arbeiten ; 65 ISSN 0344-6727) GFR78-A
A revision of the author's thesis, Tübingen.
Bibliography: p. 196-202.
ISBN 3-484-10313-2
1. Children—Language. 2. Child development. I. Title.
P118.B5 1978 79-367995
79 MARC

Brodskiĭ, Aleksandr Isaakovich.
(Radost' zhizni moeĭ)
Радость жизни моей : Из наблюдений писателя / Александр Бродский ; [Предисл. Н. Забилы]. — Киев : Рад. письменник, 1977.
149 p. : ill. ; 17 cm. USSR 77
0.45rub
1. Children—Language. I. Title.
LB1139.L3B744 78-421910

Brown, Daphne M 1934-
Mother tongue to English : the young child in the multicultural school / Daphne M. Brown. — Cambridge [Eng.] ; New York : Cambridge University Press, 1979.
x, 173 p. ; 24 cm.
Bibliography: p. [168]-169.
Includes index.
ISBN 0-521-21873-X : $8.50. ISBN 0-521-29299-9 pbk. : $4.75
1. Education, Bilingual. 2. Children—Language. 3. Native language and education. I. Title.
LC3715.B76 371.9'7 77-83987
78 MARC

Clezy, Gillian.
Modification of the mother-child interchange in language, speech, and hearing / by Gilliam Clezy, with chapters by Ronald J. Balthazor and Michael J. Cevette ; foreword by Philip S. Dale. — Baltimore : University Park Press, c1979.
xi, 162 p. : ill. ; 23 cm.
Includes bibliographies and index.
ISBN 0-8391-1319-6
1. Speech therapy for children. 2. Children—language. 3. Mother and child. 4. Communicative disorders in children. I. Balthazor, Ronald J., joint author. II. Cevette, Michael J., joint author. III. Title.
RJ496.S7C57 616.8'55 78-23352
78 MARC

Comunicazione verbale e linguaggi visivi : linguistica e audiovisivi nella scuola dell'obbligo. — Torino : Stampatori, 1977.
196 p. : ill. ; 18 cm. — (Didattica ; 5) It78-Nov
Includes bibliographies.
L3000
1. Children—Language. 2. Language arts (Elementary) 3. Audio-visual education—Italy.
LB1139.L3C59 79-356986
*79 MARC

Costal, Cassià.
Quelcom sobre el llenguatge i la intel·ligència del nen petit / Cassià Costal ; pròleg de Maria Pla i Molins. — 1. ed. — Barcelona : Edicions del Mall, 1978.
45 p. ; 21 cm. — (El Tremp ; 2) Sp***
ISBN 8474560101
1. Children—Language. 2. Intelligence levels. I. Title.
LB1139.L3C66 78-377391
78 MARC

Cruttenden, Alan, 1936-
Language in infancy and childhood : a linguistic introduction to language acquisition / Alan Cruttenden. — New York : St. Martin's Press, 1979.
xiv, 193 p. : ill. ; 23 cm.
Bibliography: p. 171-188.
Includes index.
ISBN 0-312-46606-4
1. Children—Language. 2. Reading. I. Title.
P118.C7 1979b 401'.9 78-22106
78 MARC

Dailidienė, Joana.
Ankstyvojo amžiaus vaikų kalbos ypatybės / J. Dailidienė. — Kaunas : Šviesa, 1977.
92, [4] p. ; 22 cm. USSR***
At head of title: Lietuvos TSR Aukštojo ir specialiojo vidurinio mokslo ministerija. Šiaulių K. Preikšo Pedagoginis institutas.
Summary also in Russian.
Bibliography: p. 87-[93]
0.20rub
1. Children—Language. 2. Lithuanian language. I. Title.
LB1139.L3D18 78-381726
78 MARC

Developmental pragmatics / edited by Elinor Ochs, Bambi B. Schieffelin. — New York : Academic Press, c1979.
xv, 437 p. ; 24 cm.
Bibliography: p. 415-429.
Includes index.
ISBN 0-12-524550-5
1. Pragmatics. 2. Children—Language. I. Ochs, Elinor. II. Schieffelin, Bambi B.
[DNLM: 1. Language development. 2. Psycholinquistics. WS105.5.C8 D489]
P99.4.P72D47 401'.9 78-20047
79 MARC

De Villiers, Peter A 1946-
Early language / Peter A. de Villiers, Jill G. de Villiers. — Cambridge : Harvard University Press, 1979.
160 p. : ill. ; 21 cm. — (The developing child series)
Bibliography: p. 157-158.
Includes index.
ISBN 0-674-22140-0. ISBN 0-674-22141-9 pbk.
1. Children—Language. I. De Villiers, Jill G., 1948- joint author. II. Title. III. Series.
P118.D43 401'.9 78-9862
78 MARC

Ehrlich, Stéphane.
Le développement des connaissances lexicales à l'école primaire / par Stéphane Ehrlich, Geneviève Bramaud Du Boucheron, Agnès Florin. — Poitiers : Université de Poitiers, Laboratoire de psychologie, [1978]
387 p. : ill. ; 21 cm. F***
Bibliography: p. 179-182.
60.00F
1. Children—Language. 2. Vocabulary—Study and teaching (Primary) 3. Language arts (Elementary) I. Bramaud Du Boucheron, Geneviève, joint author. II. Florin, Agnès, joint author. III. Title.
LB1139.L3E37 372.6 78-389064
79 MARC

Eneskär, Barbro, 1933-
Children's language at four and six : a longitudinal and multivariable study of language abilities among children / Barbro Eneskär. — Lund : LiberLäromedel/Gleerup, 1978.
69 p. : tables ; 25 cm. — (Studia psychologica et paedagogica : Series altera ; 42 ISSN 0346-5926) S78-15
Thesis—Lund.
Bibliography: p. 67-69.
ISBN 9140046443
1. Children—Language. I. Title. II. Series: Studia psychologica et paedagogica : Series altera, investigationes ; 42.
P118.E53 401'.9 79-302191
79 MARC

Ervin, Jane.
Your child can read and you can help : a book for parents / by Jane Ervin. — 1st ed. — Garden City, N.Y. : Doubleday, 1979.
x, 444 p. ; 22 cm.
Includes bibliographical references and index.
ISBN 0-385-14141-6 : $10.95
1. Reading. 2. Reading readiness. 3. Children—Language. 4. Books and reading for children. I. Title.
LB1050.E75 372.4'1 78-69656
78 MARC

François, Frédéric.
Éléments de linguistique appliqués à l'étude du langage de l'enfant / Frédéric François. — Paris : J.-B. Baillière, c1978.
165 p. ; 28 cm. — (Les Cahiers Baillière ; 45 : Orthophonie ; 6) F***
Bibliography: p. 159.
Includes index.
ISBN 2-7008-0025-7
1. Children—Language. 2. Linguistic analysis (Linguistics) I. Title.
P118.F7 149'.94 78-377052
78 MARC

The Functions of language and cognition / edited by Grover J. Whitehurst, Barry J. Zimmerman. — San Francisco : Academic Press, 1979.
xii, 313 p. : ill. ; 24 cm.
Includes bibliographies and indexes.
ISBN 0-12-747050-6
1. Psycholinguistics. 2. Children—Language. 3. Cognition in children. I. Whitehurst, Grover J. II. Zimmerman, Barry J.
BF455.F86 153 78-20004
79 MARC

García Hoz, Víctor.
Estudios experimentales sobre el vocabulario / Víctor García Hoz. — Madrid : Consejo Superior de Investigaciones Científicas, Instituto de Pedagogía San José de Calasanz, 1977.
122 p. : ill. ; 24 cm. Sp***
Includes bibliographical references.
ISBN 840003712X : 600ptas
1. Spanish language—Study and teaching. 2. Spanish language—Vocabulary. 3. Children—Language. I. Title.
LB1577.S7G35 78-388109
78 MARC

Gimbel, Jørgen.
Kode eller sociolekt? : om socialt bestemte sprogformer i folkeskolens ældste klasser : antaget som licentiatafhandling ved Danmarks Lærerhøjskole / [af] Jørgen Gimbel. — [København] : Danmarks Lærerhøjskole : [eksp. Institut for Dansk Sprog og Litteratur, Danmarks Lærerhøjskole] [Emdrupvej 101, 2400 NV], 1977.
2 v. : ill. ; 30 cm. D78-5
Bibliography: v. 2, p. 16-24.
1. Children—Language. 2. Bernstein, Basil B. 3. Halliday, Michael Alexander Kirkwood. I. Title.
LB1139.L3G493 78-372095
78 MARC

Ginn, Mary E
Speech improvement language development programs : a resource manual for speech and language clinicians / Mary E. Ginn. — [Columbia : Office of Programs for the Handicapped, South Carolina Dept. of Education], 1978.
iii, 60 p. : forms. ; 29 cm.
Bibliography: p. 59-60.
1. Speech—Study and teaching. 2. Children—Language. 3. Language arts. 4. Speech disorders in children. I. South Carolina. Office of Programs for the Handicapped. II. Title.
LB1139.L3G494 407'.1 78-110392
79 MARC

Glebuvienė, Vitolda.
Vaikų ankstyvosios garsinės kalbos raida : metodinė medžiaga / V. Glebuvienė. — Vilnius : PMTI, 1977.
147 p. : ill. ; 21 cm. USSR***
At head of title: Lietuvos TSR Švietimo ministerija. Pedagogikos mokslinio tyrimo institutas.
Summary in English and Russian.
Bibliography: p. 128-130.
0.60rub
1. Children—Language. 2. Lithuanian language—Study and teaching (Preschool)—Lithuania. I. Title.
LB1139.L3G515 78-386478
79 MARC

Greene, George Gordon, 1931-
The development of the language arts from birth through elementary school, by Gordon Greene. Lincoln, Neb., Professional Educators Publications [1973]
79 p. 23 cm. (The Professional education series) $1.75
Bibliography: p. 70-73.
ISBN 0-88224-050-1
1. Children—Language. 2. Language arts (Elementary) I. Title.
LB1139.L3G677 372'.6 73-85548
73[r79]rev MARC

Hillerich, Robert L 1927-
A writing vocabulary of elementary children / by Robert L. Hillerich. — Springfield, Ill. : C. C. Thomas, c1978.
xvi, 295 p. ; 24 cm.
Bibliography: p. xv-xvi.
ISBN 0-398-03814-7
1. English language—Composition and exercises. 2. Vocabulary. 3. Children—Language. I. Title.
LB1576.H44 372.6'1 78-5564
78 MARC

CHILDREN
—LANGUAGE (Continued)

Hoffmann, Ludger.
Zur Sprache von Kindern im Vorschulalter : e. Unters. in 2 Kindergärten aus d. niederdt. Sprachraum / von Ludger Hoffmann. — Köln ; Wien : Böhlau, 1978.
vii, 351 p. : ill. ; 24 cm. — (Niederdeutsche Studien ; Bd. 25) GFR78-A
Originally presented as the author's thesis, Münster, 1977.
Bibliography: p. [340]-351.
ISBN 3-412-03978-0
1. Children—Language. I. Title. II. Series.
LB1139.L3H58 1978 79-343087
79 MARC

Hoffmeister, Walter.
Sprachwechsel in Ost-Lothringen : soziolinguist. Probleme zweisprachiger Schüler / Walter Hoffmeister. — 1. Aufl. — Wiesbaden : Steiner, 1977.
116 p. ; 24 cm. — (Deutsche Sprache in Europa und Übersee ; Bd. 2) GFR77-A
A revision of the author's thesis, Salzburg, 1973.
Bibliography: p. 104-111.
ISBN 3-515-02567-7
1. German language—Dialects—France—Lorraine. 2. Bilingualism—France—Lorraine. 3. Children—Language. 4. Sociolinguistics. I. Title. II. Series.
PF5431.H64 1977 301.2'1 79-360442
79 MARC

Ivić, Ivan.
Čovek kao animal symbolicum : razvoj simboličkih sposobnosti / Ivan Ivić. — Beograd : Nolit, 1978.
331 p. ; 20 cm. — (Biblioteka Sazvežđa ; 60) Yu***
Bibliography: p. [323]-331.
1. Semiotics. 2. Animal communication. 3. Children—Language. 4. Psycholinguistics. I. Title.
P99.I94 78-391643
79 MARC

Karmiloff-Smith, Annette.
A functional approach to child language : a study of determiners and reference / Annette Karmiloff-Smith. — Cambridge [Eng.] ; New York : Cambridge University Press, 1979.
vi, 258 p. ; 24 cm. — (Cambridge studies in linguistics ; no. 24)
Bibliography: p. [242]-251.
Includes index.
ISBN 0-521-22416-0
1. Children—Language. 2. Grammar, Comparative and general—Determiners. 3. Reference (Linguistics) I. Title. II. Series.
P118.K3 401'.9 78-15450
78 MARC

Karpova, Sof'îa Nikolaevna.
(Osobennosti orientirovki na slovo u deteĭ)
Особенности ориентировки на слово у детей / С. Н. Карпова, И. Н. Колобова. — Москва : Изд-во МГУ, 1978.
165 p. : graphs ; 22 cm. USSR 78
Bibliography: p. 160–[164]
0.90rub
1. Children—Language. 2. Child development. I. Kolobova, Irina Nikolaevna, joint author. II. Title.
LB1139.L3K35 79–409186

Klein, Wolfgang.
Linguistik und Didaktik der Kindersprache im Grundschulalter : Unters. zu Konstitution u. Kohäsion von Schülertexten / Wolfgang Klein. — Paderborn : Schöningh, 1978.
287 p. : ill. ; 22 cm. — (Informationen zur Sprach- und Literaturdidaktik ; 18) GFR78-A
A revision of the author's thesis, Bielefeld, 1977.
Bibliography: p. 272-283.
Includes indexes.
ISBN 3-506-74068-7 : DM24.00
1. Children—Language. 2. Text-books—Readability. 3. Readability (Literary style) I. Title.
LB1139.L3K57 1978 79-351313
79 MARC

Laing, Ronald David.
Conversations with children / [by] R. D. Laing. — London : Allen Lane, 1978.
[1], x, 89 p. ; 23 cm. GB78-29181
ISBN 0-7139-1079-8 : £3.95
1. Child psychology. 2. Children—Language. I. Title.
BF721.L274 1978 155.4'13 79-300488
79 MARC

Langevin, Claude, professeur agrégé.
Le langage de votre enfant: comment l'éduquer, le corriger, le développer. Québec, Presses de l'Université Laval [1970] C***
160 p. 21 cm.
Bibliography: p. [159]-160.
1. Children—Language. I. Title.
LB1139.L3L318 372.6 74-19145
79[r79]rev MARC

Lee, Dorris Mary Potter, 1905-
Children and language : reading and writing, talking and listening / Dorris M. Lee, Joseph B. Rubin. — Belmont, Calif. : Wadsworth Pub. Co., c1979.
xxii, 387 p. : ill. ; 25 cm.
Bibliography: p. 363-378.
Includes index.
ISBN 0-534-00686-8
1. Language arts. 2. Children—Language. 3. Communication. I. Rubin, Joseph B., joint author. II. Title.
LB1575.8.L43 372.6 78-27256
78 MARC

Lewis, David, 1942-
The secret language of your child : how children talk before they can speak / by David Lewis. — London : Souvenir Press, 1978.
288 p. : ill. ; 23 cm. GB***
ISBN (invalid) 0-285-62358-2 : £4.95
1. Infant psychology. 2. Child psychology. 3. Children—Language. 4. Nonverbal communication. I. Title.
BF723.I6L415 1978 154.4'23 79-303954
79 MARC

Lewis, David, 1942-
The secret language of your child : how childen talk before they speak / by David Lewis. — New York : St. Martin's Press, c1978.
288 p. : ill. ; 22 cm.
ISBN 0-312-70851-3 : $8.95
1. Infant psychology. 2. Children—Language. 3. Nonverbal communication. I. Title.
BF723.I6L415 155.4'13 78-19411
78 MARC

Lloyd, Gwyneth.
Deprivation and the bilingual child / Gwyneth Lloyd. — Oxford : B. Blackwell, 1977.
x, 159 p. ; 31 cm. — (Studies of infant school children ; 3) GB***
At head of title: Schools Council Research and Development Project in Compensatory Education.
Bibliography: p. 152-155.
Includes index.
ISBN 0-631-16090-6 : £20.00
1. Children—Language. 2. Bilingualism. 3. Compensatory education—England. I. Great Britain. Schools Council. Research and Development Project in Compensatory Education. II. Title. III. Series.
LB1139.L3L53 371.9'67 79-300511
79 MARC

MacWhinney, Brian.
The acquisition of morphophonology / Brian MacWhinney. — Chicago : Published by the University of Chicago Press for the Society for Research in Child Development, 1978.
123 p. : ill. ; 23 cm. — (Monographs of the Society for Research in Child Development ; serial no. 174, v. 43, no. 1-2 ISSN 0037-976X)
Bibliography: p. 117-122.
$7.00
1. Children—Language. 2. Morphophonemics. I. Society for Research in Child Development. II. Title. III. Series: Society for Research in Child Development. Monographs ; v. 43, no. 1-2, serial no. 174.
P118.M33 415 78-108638
78 MARC

Mandel, Michael Peter, 1948-
Sprechen und Verstehen / Michael Peter Mandel. — Frankfurt am Main ; Bern ; Las Vegas : Lang, 1978.
204 p. ; 21 cm. — (Europäische Hochschulschriften : Reihe 11, Pädagogik ; Bd. 59) GFR78-A
Originally presented as the author's thesis, Frankfurt am Main.
Bibliography: p. 195-204.
ISBN 3-261-02542-5 : 39.00F
1. Children—Language. 2. Interaction analysis in education. 3. Oral communication. I. Title. II. Series.
LB1139.L3M348 1978 79-349368
79 MARC

Matthias, Dieter, 1941-
Wahrnehmung und Sprachgebrauch : eine empirische Untersuchung der Filmwirkung auf die Sprache des Kindes / Dieter Matthias. — Köln : Verlag Wissenschaft und Politik, c1978.
iv, 450 p. : ill. ; 21 cm. GFR***
Originally presented as the author's thesis, Pädagogische Hochschule Rheinland, under title Audiovisuelle Faktoren kindlicher Objektbezeichnung.
Bibliography: p. 428-450.
ISBN 3-8046-8545-5
1. Moving-pictures and children. 2. Children—Language. I. Title.
PN1995.9.C45M36 1978 791.43'013 79-367470
79 MARC

Mecham, Merlin J
Language disorders in children : a resource book for speech-language pathologists / by Merlin J. Mecham and Mary Louise Willbrand. — Springfield, Ill. : Thomas, c1979.
vii, 296 p. ; 24 cm.
Includes bibliographies and index.
ISBN 0-398-03865-1
1. Speech disorders in children. 2. Children—Language. I. Willbrand, Mary Louise, joint author. II. Title.
RJ496.S7M42 618.9'28'55 78-10777
78 MARC

Melgar de González, María.
Cómo detectar al niño con problemas del habla / María Melgar de González. — 1. ed. — México : Editorial Trillas, 1976.
44, [59] p. : ill. ; 23 cm.
Bibliography: p. 43-44.
1. Children—Language. 2. Speech, Disorders of. I. Title.
LB1139.L3M42 79-382471
79 MARC

Mendoza Cuba, Aída.
Sustantivos en el habla coloquial de los niños de Cabana / Aída Mendoza Cuba. — Lima : INIDE, 1975.
138 p. ; 29 cm. — (Investigación El Lenguaje del niño peruano hispanohablante : Informe complementario)
On cover: Instituto Nacional de Investigación y Desarrollo de la Educación Augusto Salazar Bondy. Subdirección de Investigaciones Educacionales.
1. Spanish language—Dialects—Peru—Cabana (Pallasca) 2. Spanish language—Noun. 3. Children—Language. I. Title. II. Series.
PC4904.C3M4 467.'9'852 78-108369
79 MARC

Morgan, Jane
Nicknames : their origins and social consequences / Jane Morgan, Christopher O'Neill, Rom Harré. — London ; Boston : Routledge & K. Paul, 1979.
153 p. : ill. ; 23 cm. — (Social worlds of childhood) C***
Bibliography: p. [149]-151.
Includes index.
ISBN 0-7100-0139-8 : £5.95
1. Nicknames. 2. Epithets. 3. Children—Language. 4. Names, Personal. I. O'Neill, Christopher, joint author. II. Harré, Romano, joint author. III. Title. IV. Series.
CS2377.M57 929.4 78-41077
00 MARC

Neri, Antonio.
Il bambino trasparente : l'analisi dell'istituzione educativa attraverso il linguaggio del corpo / Antonio Neri. — Milano : Emme, 1977.
168 p. ; 20 cm. — (Scienze dell'educazione) (Il Puntoemme ; 73) It78-Sept
L2500
1. Child psychology. 2. Children—Language. 3. Nonverbal communication (Psychology) 4. Sex instruction for mentally handicapped children. I. Title.
BF721.N45 79-349129
*79 MARC

Okamoto, Eiichi.
(Yōji ni okeru gengo jōhōryō tesuto no jikkenteki kenkyū)
幼児における言語情報量テストの実験的研究 / [岡本栄一]. — 東京：国立教育研究所, 昭和48 [1973]
63 p. : ill. ; 26 cm. — (国立教育研究所紀要 ; 第82集) Ja***
Cover title.
On p. [4] of cover: An experimental study of language information measurement in children.
Japanese.
Bibliography: p. 63.
1. Children—Language. 2. Japanese language—Ability testing. 3. Kindergartens—Japan—Tokyo. 4. Kindergartens—Japan—Sendai (Miyagi Prefecture) I. Title. II. Title: An experimental study of language information measurement in children. III. Series: Kokuritsu Kyōiku Kenkyūjo, Tokyo. Kokuritsu Kyōiku Kenkyūjo kiyō ; dai 82-shū.
LB1181.O32 79–805334

Ōkubo, Ai, 1921-
(Yōji no kotoba to otona)
幼児のことばとおとな / 大久保愛. — 新版. — 東京：三省堂, 1977.
278 p. : ill. ; 19 cm. — (三省堂選書 ; 17) Ja 77-13623
Series romanized: Sanseidō sensho.
Bibliography: p. 276–278.
¥900
1. Children—Language. I. Title.
LB1139.L3O554 78–812042
*78

Pflaum-Connor, Susanna.
The development of language and reading in the young child. Columbus, Ohio, Merrill [1974]
ix, 226 p. 23 cm. (The Charles E. Merrill comprehensive reading program)
Second ed. published in 1978 under title: The development of language and reading in young children.
Includes bibliographies.
ISBN 0-675-08879-8
1. Children—Language. 2. Reading (Primary) 3. Language arts (Primary) I. Title.
LB1139.L3P43 1974 372.6 73-86011
74[r79]rev MARC

Postman, Neil.
Teaching as a conserving activity / by Neil Postman. — New York : Delacorte Press, c1979.
244 p. ; 22 cm.
Includes bibliographical references and index.
ISBN 0-440-08651-5
1. Education—Philosophy. 2. Postman, Neil. 3. Mass media and children. 4. Children—Language. 5. Audio-visual education. 6. Curriculum planning. I. Title.
LB885.P65T4 371.1'02 79-10728
79 MARC

Prendiamoci le parole : per una psicologia della creatività linguistica : ricerca-didattica di docenti, studenti e operai : Istituto di psicologia sperimentale e sociale, Facoltà di magistero, Università degli studi di Torino / [di] Elena Negro Sancipriano ... [et al.]. — Torino : Books' store, 1977.
101 p. : ill. ; 24 cm. — (Esperienze ; 3) It77-Sept
Bibliography: p. 99-101.
L2000
1. Creativity (Linguistics) 2. Children—Language. I. Negro Sancipriano, Elena. II. Turin. Università. Istituto di psicologia sperimentale e sociale.
P37.5.C74P7 78-381313
*79 MARC

CHILDREN
—LANGUAGE (Continued)

Richards, Jill.
Classroom language : what sort? / Jill Richards. — London ; Boston : Allen & Unwin, 1978.
154 p. ; 22 cm. — (Unwin education books) GB***
Bibliography: p. [145]-151.
Includes index.
ISBN 0-04-372026-9 : $16.25. ISBN 0-04-372027-7 pbk.
1. Children—Language. 2. Language and education. I. Title.
LB1139.L3R47 370.15'2 79-303847
79 MARC

Robinson, William Peter.
Language management in education : the Australian context / W. P. Robinson. — Sydney ; Boston : G. Allen & Unwin, 1978.
188 p. : diagrs. ; 23 cm. Aus***
Bibliography: p. 179-184.
Includes indexes.
ISBN 0-86861-144-1 : $20.25 (U.S.). ISBN 0-86861-152-2 pbk.
1. Bernstein, Basil B. 2. Children—Language. 3. Sociolinguistics. 4. Speech and social status. 5. Students' socio-economic status. 6. Language and education—Australia. I. Title.
P85.B46R6 401'.9 78-53930
78 MARC

Salomone-Levy, Rosemary.
An analysis of the effects of language acquisition context upon the dual language development of non-English dominant students / Rosemary Salomone Levy. — New York : Arno Press, 1978, c1976.
x, 305 p. : ill. ; 24 cm. — (Bilingual-bicultural education in the United States)
Originally presented as the author's thesis, Columbia University, 1976.
Bibliography: p. 227-241.
ISBN 0-405-11095-2 : $25.00
1. Children—Language. 2. Bilingualism. 3. Italian Americans. I. Title. II. Series.
LB1139.L3S2 1978 371.9'7 77-90557
78 MARC

Savage, Hubert William, 1928-
An evaluation of the cooperative English test of effectiveness of expression for use in Ontario / by H. W. Savage. — [Toronto] : Dept. of Educational Research, Ontario College of Education, University of Toronto, 1958.
39 p. ; 23 cm. — (Atkinson study of utilization of student resources, supplementary report ; no. 1)
Bibliography: p. 39.
1. English language—Examinations. 2. English language—Study and teaching (Secondary)—Ontario. 3. Children—Language. I. Title: An evaluation of the cooperative English test of effectiveness of expression ... II. Series.
LB1631.S262 373.1'2'64 78-324209
79 MARC

Schachter, Frances Fuchs, 1930-
Everyday mother talk to toddlers : early intervention / Frances Fuchs Schachter, with Ruth E. Marquis ... [et al.] ; foreword by Herbert Zimiles. — New York : Academic Press, 1979.
xviii, 211 p. : ill. ; 24 cm.
Bibliography: p. 199-205.
Includes index.
ISBN 0-12-621360-7
1. Domestic education—United States. 2. Children—Language. 3. Oral communication. 4. Speech acts (Linguistics) 5. Mother and child. I. Title.
HQ774.5.S3 372.6 78-20049
79 MARC

Schmid-Schönbein, Gisela.
Möglichkeiten und Grenzen des gesteuerten Fremdsprachenerwerbs im Vorschulalter / Gisela Schmid-Schönbein. — 1. Aufl. — München : Minerva-Publikationen, 1978.
xiv, 289 p. ; 21 cm. — (Minerva-Fachserie Pädagogik) GFR78-A
Originally presented as the author's thesis, Pädagogische Hochschule Rheinland, 1977.
Bibliography: p. 275-289.
ISBN 3-597-10034-1 : DM40.00
1. Children—Language. 2. Language and languages—Study and teaching (Preschool) 3. Bilingualism. I. Title.
LB1139.L3S293 1978 79-346159
79 MARC

Schneider, Bruno, 1939-
Sprachliche Lernprozesse : lernpsycholog. u. linguist. Analyse d. Erst- u. Zweitspracherwerbs / Bruno Schneider. — Tübingen : Narr, 1978.
361 p. ; 21 cm. — (Tübinger Beiträge zur Linguistik ; 99 ISSN 0564-7959) GFR78-A
Bibliography: p. 307-361.
ISBN 3-87808-099-9
1. Children—Language. 2. Educational psychology. 3. Language arts—Study and teaching. I. Title.
LB1139.L3S294 79-351259
79 MARC

Schönpflug, Ute, 1940-
Psychologie des Erst- und Zweitspracherwerbs : e. Einführung / Ute Schönpflug. — 1. Aufl. — Stuttgart ; Berlin ; Köln ; Mainz : Kohlhammer, 1977.
176 p. ; 19 cm. — (Urban-Taschenbücher ; Bd. 236) GFR77-A
Bibliography: p. 163-176.
ISBN 3-17-001621-0 : DM10.00
1. Children—Language. 2. Educational psychology. I. Title.
LB1139.L3S299 78-366034
78 MARC

Sign language and language acquisition in man and ape : new dimensions in comparative pedolinguistics / edited by Fred C. C. Peng, editor. — Boulder Colo. : Published by Westview Press for the American Association for the Advancement of Science, [1978]
xxiv, 236 p. ; 24 cm. — (AAAS selected symposium ; 16)
Includes bibliographies.
ISBN 0-89158-445-5 : $16.75
1. Sign language. 2. Children—Language. 3. Animal communication. 4. Apes—Behavior. I. Peng, Fred C. C. II. American Association for the Advancement of Science. III. Series: American Association for the Advancement of Science. AAAS selected symposium ; 16.
HV2474.S54 001.56 78-19807
78 MARC

Spain. Subdirección General de Ordenación Educativa.
El lenguaje en la educación preescolar y ciclo preparatorio : 1.o y 2.o de E.G.B. : Catalán-Castellano / [Subdirección General de Ordenación Educativa]. — Madrid : Ministerio de Educación y Ciencia, Dirección General de Educación Básica, [1978?]
207 p. : ill. ; 22 cm. — (Serie preescolar ; no. 5) (Colección Estudios y experiencias educativas) Sp***
Bibliography: p. 205-207.
ISBN 8436906101 : 200ptas
1. Children—Langauge. 2. Education, Bilingual—Spain. 3. Spanish language—Study and teaching (Elementary) 4. Catalan langauge—Study and teaching (Elementary) I. Title. II. Series. III. Series: Colección Estudios y experiencias educativas.
LB1139.L3S64 1978 79-343171
79 MARC

Streiff, Virginia.
Reading comprehension and language proficiency among Eskimo children / Virginia Streiff. — New York : Arno Press, 1978.
xiv, 268 p. : ill. ; 24 cm. — (Bilingual-bicultural education in the United States)
Bibliography: p. [231]-242.
ISBN 0-405-11097-9 : $22.00
1. Reading comprehension. 2. Children—Language. 3. Eskimos—Children. I. Title. II. Series.
LB1139.L3S715 372.4 77-90559
78 MARC

Structure and development in child language : the preschool years / Marion Potts ... [et al.]. — Ithaca, N.Y. : Cornell University Press, 1979.
237 p. ; 23 cm.
Bibliography: p. 223-229.
Includes index.
ISBN 0-8014-1184-X
1. Children—Language. 2. English language—Syntax. I. Potts, Marion.
P118.S8 401'.9 78-10968
78 MARC

Tough, Joan.
Talking and learning : a guide to fostering communication skills in nursery and infant schools / [written by] Joan Tough [for the] Schools Council Communication Skills in Early Childhood Project. — London : Ward Lock ; Cardiff : Drake Educational Associates, 1977.
viii, 327 p. : ill., plan ; 25 cm. GB78-18183
Bibliography: p. [308]-309.
Includes index.
ISBN 0-7062-3610-6 : £6.00. ISBN 0-7062-3603-3 pbk.
1. Children—Language. 2. Education, Preschool—Great Britain. 3. Oral communication. I. Great Britain. Schools Council. Communication Skills in Early Childhood Project. II. Title.
LB1139.L3T68 372.6 78-323935
79 MARC

Valle de Martínez Pavetti, Elisa E
El vocabulario del escolar / Elisa Estela Valle de Martínez Pavetti. — [Tegucigalpa] : Departamento de Pedagogía y Ciencias de la Educación, UNAH, [1977]
233 p. ; 27 cm.
Includes bibliographical references.
1. Children—Language. 2. Spanish language—Study and teaching (Elementary) I. Honduras. Universidad Nacional Autónoma, Tegucigalpa. Departamento de Pedagogía y Ciencias de la Educación. II. Title.
LB1139.L3V28 79-113676
79 MARC

Verhoeven, Corn
Herinneringen aan mijn moedertaal / Cornelis Verhoeven. — Baarn : Ambo, [1978]
172 p. ; 21 cm. — (Amboboeken) Ne78-13
Includes index.
ISBN 9026304021 : fl 22.50
1. Dutch language—Dialects—Netherlands—Udenhout. 2. Children—Language. 3. Udenhout, Netherlands. I. Title.
PF807.U3V4 79-350248
*79 MARC

Warum weint die Giraffe? / Siegfried Jäger ... [et al.], unter Mitarbeit von Raimund Küchler. — 1. Aufl. — Kronberg/Ts. : Scriptor-Verlag, 1978.
669 p. : graphs ; 21 cm. — (Schichtenspezifischer Sprachgebrauch von Schülern ; Bd. 3) (Monographien Linguistik und Kommunikationswissenschaft ; Bd. 37) GFR78-A
Bibliography: p. 663-669.
ISBN 3-589-20612-8
1. Children—Language. 2. German language—Examinations, question, etc. I. Jäger, Siegfried. II. Series.
P40.5.S63S34 Bd. 3 78-397480
[LB1137.L3] 79 MARC

Weeks, Thelma E
Born to talk / Thelma E. Weeks. — Rowley, Mass. : Newbury House Publishers, c1979.
ix, 147 p. ; 22 cm.
Includes bibliographies and index.
ISBN 0-88377-153-5
1. Children—Language. 2. Nonverbal communication. I. Title.
P118.W37 401'.9 78-23207
78 MARC

—LANGUAGE—ADDRESSES, ESSAYS, LECTURES

Action, gesture, and symbol : the emergence of language / edited by Andrew Lock ; foreword by Jerome Bruner. — London ; New York : Academic Press, 1978.
xiv, 588 p. : ill. ; 23 cm. GB***
Bibliography: p. [557]-581.
Includes index.
ISBN 0-12-454050-3 : $22.00
1. Children—Language—Addresses, essays, lectures. 2. Language and languages—Origin—Addresses, essays, lectures. 3. Animal communication—Addresses, essays, lectures. 4. Nonverbal communication—Addresses, essays, lectures. I. Lock, Andrew.
P118.A17 401'.9 78-52094
79 MARC

Applied linguistics and reading / edited by Robert E. Shafer for the IRA Linguistics and Reading Subcommittee Two. — Newark, Del. : International Reading Association, c1979.
ix, 165 p. ; 22 cm.
Includes bibliographies.
ISBN 0-87207-722-5
1. Reading—Addresses, essays, lectures. 2. Children—Language—Addresses, essays, lectures. 3. Language arts—Addresses, essays, lectures. 4. Applied linguistics—Addresses, essays, lectures. I. Shafer, Robert Eugene, 1925- . II. International Reading Association. Linguistics and Reading Subcommittee.
LB1050.A64 372.6 79-16
79 MARC

Bilingualism and the bilingual child : challenges and problems / edited with an introd. by Francesco Cordasco. — New York : Arno Press, 1978.
ca. 400 p. : ill. ; 24 cm. — (Bilingual-bicultural education in the United States)
Includes bibliographical references.
ISBN 0-405-11108-8 : $35.00
1. Bilingualism—United States—Addresses, essays, lectures. 2. Children—Language—Addresses, essays, lectures. 3. Education, Bilingual—United States—Addresses, essays, lectures. I. Cordasco, Francesco, 1920- II. Series.
P115.5.U5B49 401 77-90569
78 MARC

Kindliche Kommunikation : theoret. Perspektiven, empir. Analysen, methodolog. Grundlagen / hrsg. von Karin Martens. — 1. Aufl. — Frankfurt am Main : Suhrkamp, 1979.
266 p. ; 18 cm. — (Suhrkamp Taschenbücher Wissenschaft ; 272) GFR79-A
Includes bibliographies.
ISBN 3-518-07872-0
1. Children—Language—Addresses, essays, lectures. I. Martens, Karin.
LB1139.L3K556 79-368396
79 MARC

Spracherwerb von 6 bis 16 : linguist., psycholog., soziolog. Grundlagen / Gerhard Augst (Hrsg.). — 1. Aufl. — Düsseldorf : Pädagogischer Verlag Schwann, 1978.
318 p. : ill. ; 21 cm. — (Sprache und Lerner ; Bd. 61) (Schwann Pädagogik) GFR78-A
A revision of selected papers presented at a meeting held at the Gesamthochschule Siegen, spring, 1977.
Includes bibliographies.
ISBN 3-590-14161-1
1. Children—Language—Addresses, essays, lectures. 2. Language arts—Addresses, essays, lectures. I. Augst, Gerhard, 1939-
LB1139.L3S676 78-373818
78[r79]rev MARC

Understanding language through sign language research / edited by Patricia Siple. — New York : Academic Press, 1978.
xiv, 378 p. : ill. ; 24 cm. — (Perspectives in neurolinguistics and psycholinguistics)
Based on the papers presented at the Conference on Sign Language and Neurolinguistics, held in Rochester, N.Y. in Sept. 1976.
Includes bibliographies and index.
ISBN 0-12-646550-9
1. Sign language—Addresses, essays, lectures. 2. Languages—Physiological aspects—Addresses, essays, lectures. 3. Children—Language—Addresses, essays, lectures. 4. Psycholinguistics—Addresses, essays, lectures. I. Siple, Patricia. II. Conference on Sign Language and Neurolinguistics, Rochester, N.Y., 1976.
HV2474.U52 001.56 77-25631
78 MARC

—LANGUAGE—COLLECTED WORKS

Children's language. v. 1–
New York, Gardner Press; distributed by Halstead Press, c1978–
v. ill. 24 cm.
Key title: Children's language, ISSN 0163-2809
1. Children—Language—Collected works.
P118.C47 401'.9 78-646625
MARC-S

CHILDREN (Continued)

—LANGUAGE—CONGRESSES

Pedagogische psycholinguistiek : handelingen van een colloquium georganiseerd to Gent op 25 en 26 november 1976 / M. Spoelders ed. — 9000 (Ghent, (Henri Dunantlaan 1) : Center for Experimental Educational Research, 1977.
284 p. ; 24 cm. Be78-5
Sponsored by the Belgische Vereniging voor Toegepaste Linguïstiek.
Dutch, English, or French.
Includes bibliographies.
285F
1. Children—Language—Congresses. 2. Language arts—Congresses. I. Spoelders, M. II. Belgische Vereniging voor Toegepaste Linguïstiek.
LB1139.L3P36 372.6 78-400900
*79 MARC

Perspectives on Literacy Conference, Minneapolis, 1977.
Perspectives on literacy : proceedings of the 1977 Perspectives on Literacy Conference / edited by Richard Beach, P. David Pearson. — ⌊Minneapolis⌋ : College of Education, University of Minnesota, c1978.
v, 325 p. : ill. ; 21 cm.
Sponsored by the College of Education, University of Minnesota.
Includes bibliographies.
1. Illiteracy—United States—Congresses. 2. Children—Language—Congresses. 3. Language arts—United States—Congresses. I. Beach, Richard. II. Pearson, P. David. III. Minnesota. University. College of Education.
LC151.P47 1979 420'.7 79-101906
79 MARC

Symposium on Language Acquisition, University of Wisconsin—Milwaukee, 1977.
Studies in first and second language acquisition / ⌊edited by⌋ Fred R. Eckman, Ashley J. Hastings. — Rowley, Mass. : Newbury House Publishers, c1979.
xv, 256 p. : ill. ; 23 cm.
Papers presented at the Symposium on Language Acquisition, held Marcy 18-19, 1977 at the University of Wisconsin—Milwaukee.
Bibliography: p. 245-256.
ISBN 0-88377-119-5
1. Children—Language—Congresses. 2. Language and languages—Study and teaching—Congresses. I. Eckman, Fred R. II. Hastings, Ashley J. III. Title.
P118.S93 1977 401'.9 78-27002
78 MARC

Uttering, muttering : collecting, using and reporting talk for social and educational research / compiled by Clem Adelman from contributions by ... ⌊others⌋. — ⌊Reading⌋ : ⌊Bulmershe College of Higher Education⌋, ⌊1978⌋
⌊2⌋, 264 p. ; 30 cm. GB78-24889
Cover title.
Papers originally presented to 2 successive conferences funded by the Social Science Research Council; 1st held in Sept. 1975 at Nottingham University, 2d in Apr. 1976 at Reading University.
Bibliography: p. 257-264.
1. Children—Language—Congresses. 2. Educational research—Congresses. I. Adelman, Clem. II. Social Science Research Council.
LB1139.L3U85 300.72 79-301774
79 MARC

—LANGUAGE—LONGITUDINAL STUDIES

Eneskär, B
Om barns språk vid fyra och sex år : longitudinell undersökning av vissa språkfärdigheter hos barn / Eneskär, B. — ⌊Malmö : Pedagogisk-psykologiska institutionen, Lärarhögskolan⌋, 1977-1978.
5 v. ; 30 cm. — (Pedagogisk-psykologiska problem ; nr. 329, 331-332, 336, 339 ISSN 0346-5004) (Notiser och rapporter från Pedagogisk-psykologiska institutionen, Lärarhögskolan, Malmö) S***
Cover title.
Summaries in English.
Includes bibliographies.
CONTENTS: del 1. Teori, syfte och metoder.—del 2. Studier av uttalet.—del 3. Studier av ordkunskapen.—del 4. Studier av meningsbyggnaden.—del 5. Studier av samband mellan språkfärdigheter.
1. Children—Language—Longitudinal studies. 2. Swedish language—Vocabulary—Longitudinal studies. 3. Educational research—Sweden—Longitudinal studies. I. Title. II. Series: Pedagogisk-psykologiska problem ; nr. 329, ⌊etc.⌋ III. Series: Malmö, Sweden. Lärarhögskolan. Pedagogisk-psykologiska institutionen. Notiser och rapporter från Pedagogisk-psykologiska institutionen, Lärarhögskolan, Malmö.
LB1139.L3E53 78-400706
79 MARC

—LANGUAGE—TESTING

McCabe, Ron.
McCabe's Test handbook : a guide to tests used by speech pathologists, learning disabilities specialists, and special educators / by Ron McCabe. — ⌊Tigard, Or.⌋ : C.C. Publications, c1978.
133 p. ; 24 cm.
Includes index.
1. Children—Language—Testing. 2. Psychological tests for children. I. Title. II. Title: Test handbook.
LB1139.L3M28 371.9'028 79-102482
79 MARC

Sanders, Lois Joan.
Procedure guides for evaluation of speech and language disorders in children / Lois Joan Sanders. — 4th ed. — Danville, IL : Interstate Printers & Publishers, c1979.
206 p. ; 28 cm.
Includes bibliographies.
1. Speech disorders in children—Diagnosis. 2. Language disorders in children—Diagnosis. 3. Children—Language—Testing. I. Title.
RJ496.S7S25 1979 618.9'28'55075 78-71498
00 MARC

Weiner, Frederick F
Phonological process analysis / by Frederick F. Weiner ; ⌊ill. by George V. N. Bullough III and Sheila Humphreys⌋. — Balitmore : University Park Press, c1979.
ix, 181 p. : ill. ; 28 cm.
Bibliography: p. 39.
ISBN 0-8391-1300-5
1. Articulation disorders in children—Diagnosis. 2. Children—Language—Testing. 3. Grammar, Comparative and general—Phonology. I. Title.
RJ496.S7W35 618.9'28'55 78-14199
78 MARC

—LANGUAGE—TESTING—CONGRESSES

International Conference on Frontiers in Language Proficiency and Dominance Testing, 2d, Southern Illinois University, 1978.
Proceedings of the Second International Conference on Frontiers in Language Proficiency and Dominance Testing, held at Southern Illinois University, Carbondale, February 23-25, 1978 / James E. Redden, editor. — Carbondale : Dept. of Linguistics, Southern Illinois University, ⌊1978⌋
vi, 272 p. : ill. ; 28 cm. — (Occasional papers on linguistics ; no. 3)
Bibliography: p. 251-272.
1. Languages, Modern—Examinations—Congresses. 2. Reading—Ability testing—Congresses. 3. Bilingualism—Testing—Congresses. 4. Children—Language—Testing—Congresses. I. Redden, James E. II. Series.
PB36.I59 1978 418'.007'24 78-69842
78 MARC

—LAW

see also Children (International law)

—LAW AND LEGISLATION —JAPAN—COLLECTED WORKS

(Jidō fukushi hō seiritsu shiryō shūsei)
児童福祉法成立資料集成 / 児童福祉法研究会編. — 東京 : ドメス出版, 1978-1979.
2 v. ; 22 cm. Ja 79-1119 v. 1
Bibliography: v. 2. p. ⌊863⌋-873.
Includes indexes.
¥21000 per vol.
1. Children—Law and legislation—Japan—Collected works. 2. Child welfare—Japan—Collected works. I. Jidō Fukushihō Kenkyukai.
79-806622

—LEGAL STATUS, LAWS, ETC.

Belgium. Laws, statutes, etc.
Protection de la jeunesse / sous la direction de Pierre Mahillon. — Bruxelles : F. Larcier, 1978.
lv, 660 p. ; 30 cm. — (Les Novelles) Be***
Bibliography: p. xlix-lv.
Includes index.
1. Children—Legal status, laws, etc.—Belgium. 2. Juvenile courts—Belgium. 3. Children—Legal status, laws, etc. 4. Juvenile courts. I. Mahillon, P. II. Title. III. Series.
346'.493'013 79-361660
79 MARC

Il Codice del minore / ⌊a cura di⌋ Alberto Germano, Francesco Scarcella. — Milano : A. Giuffrè, 1978.
xiv, 2744 p. ; 20 cm. It78-Oct
"Dichiarazioni internazionali" (in French or Italian): p. ⌊3⌋-12.
"Convenzioni internazionali" (in French or Italian): p. ⌊13⌋-80.
Includes jurisprudence.
Includes indexes.
L45000
1. Children—Law—Italy. 2. Children—Law. I. Germanò, Alberto. II. Scarcella, Fancesco.
344'.45'0327 79-337455
*79 MARC

Legal protection of the women and children : analysis of documents of international organisations : contribution of Hungarian Women's Council to the World Congress of Women convened by Women's International Democratic Federation, June 1963 / ⌊written by the Working Group of the Law Commission of the National Council of Hungarian Women ; edited by J. Bokor-Szegő⌋. — ⌊Budapest : Kossuth Kiadó soksz., 1963⌋
60 p. ; 29 cm.
Cover title.
1. Women—Legal status, laws, etc. 2. Children—Law. 3. Women—Legal status, laws, etc.—Hungary. 4. Children—Law—Hungary. I. Bokor, Hanna Szegő. II. World Congress of Women, 5th, Moscow, 1963. III. Magyar Nők Országos Tanácsa. IV. Magyar Nők Országos Tanácsa. Jogi Bizottság. Working Group.
K644.L4 341.48'1 78-318173
78 MARC

Riesgo Menguez, Luis.
Los derechos de los hijos / Luis Riesgo y Carmen Pablo. — Madrid : Editorial Prensa Española, ⌊1979?⌋
247 p. ; 21 cm. Sp***
Includes bibliographical references.
ISBN 8428704864 : 350ptas
1. Children's rights. 2. Children—Legal status, laws, etc. I. Pablo de Riesgo, Carmen, joint author. II. Title.
HQ789.R5 323.4 79-109009
79 MARC

Das Selbstbestimmungsrecht des Jugendlichen im Spannungsfeld von Familie, Gesellschaft und Staat / von Evelyn Kühn ... ⌊et al.⌋. — Bielefeld : Gieseking, c1978.
495 p. ; 22 cm. — (Industriegesellschaft und Recht ; Bd. 10) GFR79-A
One in a series of 3 reports prepared by members of the Forschungsgruppe Recht und Sozialwissenschaften at the Zentrum für Interdisziplinäre Forschung of the Universität Bielefeld and issued as vols. 10-12 of Industriegesellschaft und Recht.
Includes bibliographies and indexes.
ISBN 3-7694-0410-6 : DM125.00
1. Children—Legal status, laws, etc.—Germany, West. 2. Children—Legal status, laws. 3. Parent and child (Law)—Germany, West. 4. Parent and child (Law) 5. Sociological jurisprudence. I. Kühn, Evelyn, 1945- . II. Forschungsgruppe Recht und Sozialwissenschaften. III. Bielefeld. Universität. Zentrum für Interdisziplinäre Forschung. IV. Series.
79-374872
79 MARC

Volpi, Domenico.
I diritti del mio bambino / Domenico Volpi. — Roma : Edizioni paoline, 1979.
105 p. ; 21 cm. It79-July
Bibliography: p. 105.
L3000
1. Children—Legal status, laws, etc. 2. Children—Legal status, laws, etc.—Italy. I. Title.
K639.V64 79-375648
*79 MARC

—LEGAL STATUS, LAWS, ETC.—ADDRESSES, ESSAYS, LECTURES

The Child and the courts / Ian F. G. Baxter, editor of part 2 and Mary A. Eberts, editor of part 1. — Toronto : Carswell Co., 1978.
vi, 429 p. : diagrs. ; 24 cm. — (Carswell's family law series) C***
Includes bibliographical references.
ISBN 0-459-31990-6
1. Parent and child (Law)—Addresses, essays, lectures. 2. Children—Law—Addresses, essays, lectures. 3. Custody of children—Addresses, essays, lectures. 4. Domestic relations courts—Addresses, essays, lectures. I. Baxter, Ian F. G. II. Eberts, Mary A., 1947- . III. Series.
K700.Z9C5 346'.013 78-319449
79 MARC

La Défense des droits de l'enfant : respectons l'enfant et son droit d'être heureux. — ⌊Montréal : Association canadienne pour la santé mentale, Division du Québec, 1978?⌋
160 p. ; 20 cm. C***
Cover title.
1. Children—Legal status, laws, etc.—Addresses, essays, lectures. I. Canadian Mental Health Association. Quebec Division.
HV721.D43 362.7 79-373580
79 MARC

—LEGAL STATUS, LAWS, ETC.—AMERICA

Legislación atinente'a menores en las Américas. Compilación dirigida por José P. Achard. Montevideo, Interamerican Children's Institute, 1959-
v. in 24 cm.
Vol. 2, pt. 3, and v. 3, pt. 1, have title: Legislation pertaining to minors in the Americas.
Text in Spanish; except v. 2, pt. 1, in French or Spanish; v. 2, pt. 3, in English; v. 3, pt. 1, in English and Spanish.
CONTENTS: t.1. América del Sur: fasc. 1. Zona Pacífico, comprende Colombia, Chile, Ecuador y Perú. fasc. 2. Brasil y Venezuela. fasc. 3. Altiplano y Cuenca del Río de la plata, comprende Argentina, Bolivia, Paraguay y Uruguay.—t. 2. América Central y Caribe: fasc. 1. Istmo y Antillas, comprende Haití, Panamá y República Dominicana. fasc. 2. Centroamérica, comprende Costa Rica, El Salvador, Guatemala, Honduras y Nicaragua. fasc. 3. Trinidad-Tobago.—t. 3. América del Norte: fasc. 1. Estados Unidos de América. 2 v. fasc. 2. México.—t. 4. Apéndice 2. Legislación promulgada en los países americanos en el trienio 1958 a 1960.
1. Children—Legal status, laws, etc—America. I. Achard, José P., ed. II. Interamerican Children's Institute. III. Title: Legislation pertaining to minors in the Americas.
346'.72'013 64-35859
⌊r79⌋rev MARC

CHILDREN
—LEGAL STATUS, LAWS, ETC.
—AMERICA (Continued)

Sajón, Rafael.
Menores en situación irregular : aspectos socio legales de su protección / por Rafael Sajón, José Pedro Achard y Ubaldino Calvento. — Montevideo : O.E.A., Instituto Interamericano del Niño, 1973.
40, 16, 3 p. ; 28 cm.
"Trabajo presentado al XIV Congreso Panamericano del Niño, Santiago, Chile del 5 al 11 de agosto.
Bibliography: p. [1]-3 (3d group)
1. Child welfare—America. 2. Children—Legal status, laws, etc.—America. 3. Children's rights—America. I. Achard, José P., joint author. II. Calvento, Ubaldino, joint author. III. Interamerican Children's Institute. IV. Title.
HV713.S24 362.7'097 79-373268
79 MARC

—LEGAL STATUS, LAWS, ETC.—
AUSTRIA

Austria. Laws, statutes, etc.
[Berufsausbildungsgesetz]
Das Berufsausbildungsgesetz in der Fassung der BAG-Novelle 1978 mit Erlässen und Verordnungen ; Kinder- u. Jugendbeschäftigungsgesetz, Land- u. forstwirtschaftl. Berufsausbildungsgesetz ; Bestimmungen des Schulzeit- u. Schulpflichtgesetzes / [Red. v. Dietrich Scherff]. — Wien : Selbstverl. Scherff, [1978]
320 p. ; 17 cm. Au78-22-64
"Als Sondernummer des ARD-Betriebsdienstes."
On cover: Ausgabe betriebswichtiger Gesetze.
Includes index.
S163.00
1. Children—Legal status, laws, etc.—Austria. 2. Vocational education—Law and legislation—Austria. 3. Employees, Training of—Law and legislation—Austria. 4. Agricultural laws and legislation—Austria. I. Scherff, Dietrich. II. ARD-Betriebsdienst. III. Title.
79-350356
•79 MARC

Österreichische Jugendrichtertagung, 11th, Lienz, 1976.
Neue Wege im Jugendrecht. — Wien : Jugendgerichtshof Wien, [1977]
143 p. ; 21 cm. Au78-15
Held Sept. 20-24, 1976.
Includes bibliographical references.
Not in trade.
1. Juvenile delinquency—Austria. 2. Children—Law—Austria. I. Title.
78-383076
•79 MARC

—LEGAL STATUS, LAWS, ETC.
—AUSTRIA—PERIODICALS

Der Österreichische Amtsvormund.
[Wien, Verein der Amtsvormünder Österreichs]
no. 20 cm. bimonthly.
"Fachzeitschrift für Kindschaftsrecht, Vormundschaftswesen und Jugendwohlfahrt."
1. Guardian and ward—Austria—Periodicals. 2. Children—Legal status, laws, etc.—Austria—Periodicals. I. Verein der Amtsvormünder Österreichs.
K15.E76 79-641854
MARC-S

—LEGAL STATUS, LAWS, ETC.—
BELGIUM

Belgium. Laws, statutes, etc.
Protection de la jeunesse / sous la direction de Pierre Mahillon. — Bruxelles : F. Larcier, 1978.
lv, 660 p. ; 30 cm. — (Les Novelles) Be***
Bibliography: p. xlix-lv.
Includes index.
1. Children—Legal status, laws, etc.—Belgium. 2. Juvenile courts—Belgium. 3. Children—Legal status, laws, etc. 4. Juvenile courts. I. Mahillon, P. II. Title. III. Series.
346'.493'013 79-361660
79 MARC

—LEGAL STATUS, LAWS, ETC.—
BOLIVIA

Bolivia. Laws, statutes, etc.
[Código del menor (1975)]
Código del menor. — Ed. oficial. — La Paz : Gaceta Oficial de Bolivia, 1977.
82 p. ; 20 cm. — (Its Codificación Banzer)
1. Children—Legal status, laws, etc.—Bolivia. I. Title. II. Series.
346'.84'013 78-111285
79 MARC

—LEGAL STATUS, LAWS, ETC.
—BRAZIL

Moraes, Oswaldo da Costa.
Legislação especial do menor / Oswaldo da Costa Moraes. — [s.l. : s.n.], 1973 (Porto Alegre [Brasil] : Casa Publicadora Concórdia)
65 p. ; 18 cm.
Includes bibliographical references.
1. Children—Law—Brazil. I. Title.
79-341177
79 MARC

—LEGAL STATUS, LAWS, ETC.
—BRITISH COLUMBIA

Children and the law : a report of a multi-disciplinary seminar / edited by Karen M. Mussell. — Vancouver, B.C. : Legal Publications, The Continuing Legal Education Society of British Columbia, Centre for Continuing Education, The University of British Columbia, 1977.
vii, 89 p. ; 28 cm. C***
Sponsored by the Canadian Bar Association.
"Held January 24 and 25, 1976 at ... Burnaby, B.C."
1. Children—Law—British Columbia. 2. Children—Law—Canada. I. Mussell, Karen M. II. Canadian Bar Association.
KEB190.3.M5C48 344'.711'0327 78-320557
79 MARC

—LEGAL STATUS, LAWS, ETC.—
CALIFORNIA—OUTLINES, SYLLABI, ETC.

D'Embois, E St. Cleening.
The skeptical students workbook of family and juvenile law / E. St. Cleening d'Embois, Krystina Wyzczieki. — Atlanta : Uncommon Lawyers Workshop, 1979.
118 leaves : ill. ; 22 x 28 cm.
Includes bibliographies and index.
1. Domestic relations—United States—Outlines, syllabi, etc. 2. Children—Legal status, laws, etc.—United States—Outlines, syllabi, etc. 3. Domestic relations—California—Outlines, syllabi, etc. 4. Children—Legal status, laws, etc.—California—Outlines, syllabi, etc. I. Wyzczieki, Krystina, joint author. II. Title.
KF505.Z9D45 346'.73'015 79-62889
79 MARC

—LEGAL STATUS, LAWS, ETC.
—CANADA

Canadian Council on Children and Youth.
Legislation related to the needs of children / by Canadian Council on Children and Youth. — [Agincourt, Ont.] : Carswell Co., 1979.
109 p. ; 23 cm. — (Carswell's family law series) C***
"A legislative appendix to Admittance restricted."
"Reprint of an article in volume 6 part 1 of Reports of family law (second series)."
ISBN 0-459-49520-8
1. Children—Legal status, laws, etc.—Canada. 2. Parent and child (Law)—Canada. I. Task Force on the Child as Citizen. Admittance restricted. II. Reports of family law. III. Title. IV. Series.
KE512.C36 346'.71'013 79-315301
79 MARC

Children and the law : a report of a multi-disciplinary seminar / edited by Karen M. Mussell. — Vancouver, B.C. : Legal Publications, The Continuing Legal Education Society of British Columbia, Centre for Continuing Education, The University of British Columbia, 1977.
vii, 89 p. ; 28 cm. C***
Sponsored by the Canadian Bar Association.
"Held January 24 and 25, 1976 at ... Burnaby, B.C."
1. Children—Law—British Columbia. 2. Children—Law—Canada. I. Mussell, Karen M. II. Canadian Bar Association.
KEB190.3.M5C48 344'.711'0327 78-320557
79 MARC

Wilson, Jeffery.
Children and the law / Jeffery Wilson. — Scarborough, Ont. : Butterworths, 1978.
xxxii, 367 p. ; 25 cm. C79-94018-3
Includes bibliographies and indexes.
ISBN 0-409-87760-3 : $30.00
1. Children—Legal status, laws, etc.—Canada. I. Title.
KE512.W54 346'.71'013 79-317430
79 MARC

—LEGAL STATUS, LAWS, ETC.—
DENMARK

Danielsen, Svend.
Myndighedsloven / med kommentarer ved Svend Danielsen. — København : Juristforbundet, 1978.
505 p. ; 21 cm. D78-41
Includes bibliographical references and indexes.
ISBN 8757411905 : kr175.00
1. Children—Legal status, laws, etc.—Denmark. 2. Age (Law)—Denmark. 3. Capacity and disability—Denmark. 4. Domestic relations—Scandinavia—Congresses. I. Denmark. Laws, statutes, etc. Lov om umyndighed og værgemal. 1978. II. Title.
79-342379
79 MARC

—LEGAL STATUS, LAWS, ETC.
—DENMARK—ADDRESSES, ESSAYS LECTURES

Forældrenes Børnekommission.
Vore børn er vore børn / Forældrenes Børnekommission. — [Varde] : [Forældrenes Børnekommission] ; [København] : [eksp. DBK], [1978].
Bibliography: p. 92.
ISBN 8798065300 : kr15.00
1. Child welfare—Denmark—Addresses, essays, lectures. 2. Family—Denmark—Addresses, essays, lectures. 3. Children—Law—Denmark—Addresses, essays, lectures. I. Title.
HV786.A6F67 1978 78-389093
79 MARC

—LEGAL STATUS, LAWS, ETC.—
FRANCE

Néret, Jean Alexis, 1913-
Guide pour l'enfance et l'adolescence en danger / Néret. — Paris : Néret, [1977]
157 p. ; 21 cm. F***
CONTENTS: Actions de prévention.—Aide et assistance à l'enfance.—Délinquance juvénile.—Liberté surveillée.—Nomenclature des services et des associations.
30.00F
1. Children—Law—France. I. Title.
344'.44'0327 78-369046
78 MARC

Protéger l'enfant : problèmes actuels du travail social en milieu judiciaire / [Jean Pinatel, Claude Xuereb, Isabelle Matignon, Henri Daniel Cosnard, etc.] ; sous la direction de Monique Baur-Halbwachs. — Toulouse : Privat, 1977.
172 p. ; 21 cm. — (Agir) F77-11473
ISBN 2-7089-7500-5 : 35.00F
1. Children—Legal status, laws, etc.—France. 2. Child welfare—France. I. Baur-Halbwachs, Monique.
344'.44'0327 79-341231
79 MARC

—LEGAL STATUS, LAWS, ETC.—
GEORGIA

Georgia. Laws, statutes, etc.
Georgia laws relating to minors / Richard J. Anderson, Joanne Jankovic ; research assistants, Nancy Urheim and Peggy Beggs ; Dale F. Ely, consultant in pupil personnel. — 1979 ed. — [New York] : Harcourt Brace Jovanovich, Legal and Professional Publications ; Gardena, CA : distributed by Law Distributors, c1979.
viii, 288 p. ; 28 cm.
Includes index.
1. Children—Legal status, laws, etc.—Georgia. I. Anderson, Richard John, 1929- II. Jankovic, Joanne.
KFG91.M5A3 1979 346'.758'013 79-109688
79 MARC

—LEGAL STATUS, LAWS, ETC.
—GERMANY EAST

Germany (Democratic Republic, 1949-). Laws, statutes, etc.
Jugendhilfe : eine Sammlung von Rechtsvorschriften mit Anmerkungen und Sachregister / hrsg. vom Ministerium für Volksbildung, Abteilung Jugendhilfe und Heimerziehung ; [zusammengestellt und bearb. von einem Bearbeiterkollektiv unter Leitung von Heinz Funke]. — 2., überarb. Aufl. — Berlin : Staatsverlag der Deutschen Demokratischen Republik, 1978.
286 p. ; 24 cm. GDR***
Includes index.
8.00M
1. Children—Law—Germany, East. I. Funke, Heinz. II. Germany (Democratic Republic, 1949-). Ministerium für Volksbildung. Abteilung Jugendhilfe und Heimerziehung. III. Title.
78-391842
79 MARC

—LEGAL STATUS, LAWS, ETC.—
GERMANY, WEST

Abel, Karl.
Grundriss der Jugendhilfe : Leitfaden für d. Aus- u. Fortbildung / bearb. von Karl Abel. — 3., überarb. Aufl. — Köln ; Stuttgart ; Hannover ; Kiel : Deutscher Gemeindeverlag ; Köln ; Stuttgart ; Hannover ; Kiel : Kohlhammer, 1978.
162 p. ; 21 cm. — (Dgt-Skript) GFR78-A
Includes index.
ISBN 3-555-00337-2 : DM14.80
1. Children—Law—Germany, West. I. Title.
344'.43'0327 79-378387
79 MARC

CHILDREN
—LEGAL STATUS, LAWS, ETC.
—GERMANY, WEST (Continued)

Brauser, Hanns.
Jugendarbeitsschutzgesetz : Kommentar für d. Praxis / von Hanns Brauser u. Michael Schoden. — Köln : Bund-Verlag, 1979.
478 p. ; 18 cm. GFR78-A
Includes text of the law.
Bibliography: p. 11.
Includes index.
ISBN 3-7663-0199-3
1. Youth—Employment—Germany, West. 2. Children—Legal status, laws, etc.—Germany, West. I. Schoden, Michael, joint author. II. Germany (Federal Republic, 1949-). Laws, statutes, etc. Jugendarbeitsschutzgesetz. 1979. III. Title.
344'.43'0131 79-342533
79 MARC

Erichsen, Hans Uwe.
Verstaatlichung der Kindeswohlentscheidung? : zur verfassungsrechtl. Bestimmung d. schul. Erziehungsrechts : Vortrag gehalten vor d. Berliner Jur. Ges. am 27. September 1978 / von Hans-Uwe Erichsen. — Berlin ; New York : de Gruyter, 1979.
27 p. ; 23 cm. — (Schriftenreihe der Juristischen Gesellschaft e. V. Berlin ; Heft 57) GFR78-A
Includes bibliographical references.
ISBN 3-11-007951-8 : DM14.00
1. Educational law and legislation—Germany, West. 2. Children—Legal status, laws, etc.—Germany, West. 3. State supervision of teaching—Germany, West. I. Title. II. Series: Juristische Gesellschaft, Berlin. Schriftenreihe ; Heft 57.
79-354005
79 MARC

Frankfurter Kommentar zum Gesetz für Jugendwohlfahrt / von Johannes Münder ... ₍et al.₎. — 1. Aufl. — Weinheim ; Basel : Beltz, 1978.
428 p. ; 19 cm. — (Beltz Bibliothek ; 77) GFR78-A
Includes text of the Jugendwohlfahrtsgesetz.
Bibliography: p. 407-414.
Includes index.
ISBN 3-407-50077-7 : DM21.00
1. Children—Law—Germany, West. I. Münder, Johannes. II. Germany (Federal Republic, 1949-). Laws, statutes, etc. Jugendwohlfahrtsgesetz. 1978.
78-379775
78 MARC

Germany (Federal Republic, 1949-). Laws, statutes, etc.
Jugendwohlfahrtsgesetz : Testausg. mit Nebengesetzen. — 4. Aufl. — Frankfurt/Main : Deutscher Verein für Öffentl. u. Private Fürsorge ; Stuttgart : Grote ₍in Komm.₎, 1977.
178 p. ; 17 cm. — (Kleinere Schriften des Deutschen Vereins für Öffentliche und Private Fürsorge : Heft 38) GFR77-A50
Includes index.
ISBN 3-7745-6378-0
1. Children—Law—Germany, West. I. Title. II. Series: Deutscher Verein für Öffentliche und Private Fürsorge. Kleinere Schriften : Heft 38.
344'.43'0327 78-387218
78 MARC

Gernert, Wolfgang.
Jugendhilfe : kurze Einf. in d. sozialpädag. Praxis / Wolfgang Gernert. — 3., neu bearb. Aufl. — München ; Basel : E. Reinhardt, 1978.
222 p. ; 19 cm. — (Uni-Taschenbücher ; 223) GFR78-A
Includes bibliographical references and indexes.
ISBN 3-497-00697-1 : DM16.80
1. Child welfare—Germany, West. 2. Children—Legal status, laws, etc.—Germany, West. I. Title.
HV763.G47 1978 362.7'0943 79-350958
79 MARC

Hoffmann, Riele, 1926-
Chance im Alltag : Rechtshilfen u. Informationen für d. alleinerziehende Mutter / Riele Hoffmann. — 1. Aufl. — Düsseldorf : Klens, 1977.
72 p. ; 21 cm. GFR78-A44
Includes bibliographical references and index.
ISBN 3-87309-033-3 : DM3.80
1. Maternal and infant welfare—Law and legislation—Germany, West. 2. Children—Legal status, laws, etc.—Germany, West. 3. Single women. I. Title.
79-377904
79 MARC

Hundmeyer, Simon.
Recht für Erzieher / Simon Hundmeyer, Mitarb. Dieter Lücker. — 3., neu bearb. u. erw. Aufl. — München : TR-Verlagsunion, 1977.
185 p. : ill. ; 24 cm. GFR78-A38
"Recht für Erzieher. Aufgabenbogen.": ₍6₎ p. inserted.
Previous ed. published under the title: Telekolleg für Erzieher.
Study guide to a 13-part series broadcast by the Bayerische Rundfunk.
Bibliography: p. 182-183.
Includes index.
ISBN 3-8058-0395-8 : DM18.50
1. Children—Legal status, laws, etc.—Germany, West. 2. Capacity and disability—Germany, West. 3. Guardian and ward—Germany, West. I. Lücker, Dieter, joint author. II. Title.
79-368538
79 MARC

Jugend-Medien-Schutz 1974 ₍i.e. neunzehnhundertvierundsiebzig₎ : Schutz d. Jugend vor Mediengefahren / ₍Bundesprüfstelle für Jugendgefährdende Schriften Bonn-Bad Godesberg₎ : Rudolf Stefen, Hrsg. — Bonn-Bad Godesberg : Bundesprüfstelle für Jugendgefährdende Schriften, 1974.
119 p. ; 21 cm. — (Schriftenreihe der Bundesprüfstelle für Jugendgefährdende Schriften ; Heft 4) GFR76-B19
Includes bibliographical references.
DM2.50
1. Children—Legal status, laws, etc.—Germany, West. 2. Violence in mass media—Law and legislation—Germany, West. 3. Sex in mass media—Law and legislation—Germany, West. I. Stefen, Rudolf. II. Germany (Federal Republic, 1949-). Bundesprüfstelle für Jugendgefährdende Schriften. III. Series: Germany (Federal Republic, 1949-). Bundesprüfstelle für Jugendgefährdende Schriften. Schriftenreihe ; Heft 4.
79-352432
79 MARC

Riedlinger, Raoul.
Minderjährige in der Familienkommanditgesellschaft : ihre Stellung als Gesellschafter u. Mitunternehmer / Raoul Riedlinger. — Frankfurt am Main ; Bern ; Las Vegas : Lang, 1978.
xvii, 262 p. ; 21 cm. — (Bürgerliches Recht, Handels- und Verkehrsrecht ; Bd. 4) GFR78-A
Originally presented as the author's thesis, Frankfurt am Main.
Bibliography: p. 241-254.
Includes indexes.
ISBN 3-261-02653-7
1. Family corporations—Germany, West. 2. Children—Law—Germany, West. I. Title. II. Series.
78-391847
79 MARC

Schmitt-Wenkebach, Rainer.
Das Haftungsrecht in der Jugendarbeit / Rainer Schmitt-Wenkebach. — 1. Aufl. — Neuwied ; Darmstadt : Luchterhand, 1978.
ix, 94 p. ; 21 cm. — (Arbeitsmittel für Studium und Unterricht) GFR78-A
Includes bibliographical references.
ISBN 3-472-55069-4
1. Children—Law—Germany, West. 2. Teachers—Legal status, laws, etc.—Germany, West. 3. Liability (Law)—Germany, West. I. Title.
78-398097
79 MARC

Das Selbstbestimmungsrecht des Jugendlichen im Spannungsfeld von Familie, Gesellschaft und Staat / von Evelyn Kühn ... ₍et al.₎. — Bielefeld : Gieseking, c1978.
495 p. ; 22 cm. — (Industriegesellschaft und Recht ; Bd. 10) GFR79-A
One in a series of 3 reports prepared by members of the Forschungsgruppe Recht und Sozialwissenschaften at the Zentrum für Interdisziplinäre Forschung of the Universität Bielefeld and issued as vols. 10-12 of Industriegesellschaft und Recht.
Includes bibliographies and indexes.
ISBN 3-7694-0410-6 : DM125.00
1. Children—Legal status, laws, etc.—Germany, West. 2. Children—Legal status, laws. 3. Parent and child (Law)—Germany, West. 4. Parent and child (Law) 5. Sociological jurisprudence. I. Kühn, Evelyn, 1945- . II. Forschungsgruppe Recht und Sozialwissenschaften. III. Bielefeld. Universität. Zentrum für Interdisziplinäre Forschung. IV. Series.
79-374872
79 MARC

—LEGAL STATUS, LAWS, ETC.
—GERMANY, WEST
—CONGRESSES

Praxis des neuen Familienrechts : Referate u. Berichte d. grossen Arbeitstagung d. Fachverb. Berliner Stadtvormünder e. V. vom 28. November bis 2. Dezember 1977 in Berlin. — 1. Aufl. — Berlin ; New York : de Gruyter, 1978.
308 p. ; 18 cm. — (Sammlung Göschen ; 2854) GFR78-A
Conference sponsored by the Fachverband Berliner Stadtvormünder.
Includes bibliographical references.
ISBN 3-11-007444-3 : DM19.80
1. Children—Law—Germany, West—Congresses. 2. Domestic relations—Germany, West—Congresses. I. Fachverband Berliner Stadtvormünder.
346'.43'015 78-391885
78 MARC

—LEGAL STATUS, LAWS, ETC.—
GERMANY, WEST—EXAMINATIONS,
QUESTIONS, ETC.

Germany (Federal Republic, 1949-). Bundestag. Rechtsausschuss.
Elterliches Sorgerecht : Sachverständigenanhörung, Gesetzestexte : d. öffentl. Anhörung d. Rechtsausschusses u. d. mitberatenden Ausschusses für Jugend, Familie u. Gesundheit d. Dt. Bundestages vom 12. September 1977 : Entwurf e. Gesetzes zur Neuregelung d. Rechts d. elterl. Sorge : Ausz. aus d. geltenden Recht ₍Hrsg., Dt. Bundestag, Presse- u. Informationszentrum₎. — Bonn : Dt. Bundestag, Presse- u. Informationszentrum, 1978.
227 p. ; 21 cm. — (Zur Sache ; 78, 1 ISSN 0343-8899) GFR78-B21
Includes indexes.
1. Parent and child (Law)—Germany, West—Examinations, questions, etc. 2. Children—Legal status, laws, etc.—Germany, West—Examinations, questions, etc. I. Germany (Federal Republic, 1949-). Bundestag. Ausschuss für Jugend, Familie und Gesundheit. II. Germany (Federal Republic, 1949-). Bundestag. Presse- und Informationszentrum. III. Title. IV. Series: Zur Sache (Bonn) ; 78, 1.
79-375167
79 MARC

—LEGAL STATUS, LAWS, ETC.
—GERMANY, WEST
—NORTH RHINE-WESTPHALIA

Landesjugendring Nordrhein-Westfalen.
Argumente für ein Jugendbildungsgesetz in Nordrhein-Westfalen / ₍Landesjugendring Nordrhein-Westfalen ; Verfasser, Heinz-Jürgen Klinger₎. — 1. Aufl. — Düsseldorf : Landesjugendring NW, 1977.
56 p. ; 21 cm. — (Dokumentation : Analysen, Meinungen, Berichte) GFR***
1. Children—Law—Germany, West—North Rhine-Westphalia. I. Klinger, Heinz-Jürgen. II. Title.
344'.73'0327 78-367557
78 MARC

—LEGAL STATUS, LAWS, ETC.
—GREAT BRITAIN

Bevan, H K
Children act 1975 / by Hugh K. Bevan and Martin L. Parry. — London ; Boston : Butterworths, 1978.
xviii, 307 p. ; 26 cm. GB***
Includes index.
ISBN 0-406-11702-0 : £13.50
1. Parent and child (Law)—Great Britain. 2. Children—Law—Great Britain. I. Parry, Martin L., joint author. II. Title.
KD772.B48 344'.41'0327 78-320159
79 MARC

Feldman, Linda, LL.B.
Care proceedings / Linda Feldman. — London : Oyez Pub., 1978.
xiii, 192 p. ; 22 cm. — (Family and welfare) (Oyez practitioner series) GB***
Bibliography: p. ₍183₎
Includes index.
ISBN 0-85120-344-2 : £6.00
1. Juvenile courts—Great Britain. 2. Parent and child (Law)—Great Britain. 3. Guardian and ward—Great Britain. 4. Children—Law—Great Britain. I. Title. II. Series. III. Series: Oyez practitioner series. IV. Series: Oyez practitioner series.
KD8473.F4 344'.42'03273 78-323119
79 MARC

Great Britain. Home Office.
Part 1 of the Children and young persons act 1969: a guide for courts and practitioners. London, H.M.S.O., 1970.
vii, 63 p. 25 cm. 6/6 B70-27510
ISBN 0-11-340335-6
1. Children—Law—Great Britain. 2. Juvenile courts—Great Britain. I. Title.
346'.41'013 72-178955
72₍r78₎rev MARC

—LEGAL STATUS, LAWS, ETC.
—GREECE, MODERN

Papageōrgiou, Iōannēs K
(Hē koinōnikē prostasia tou paidiou eis tēn hellēnikēn nomothesian)
Ἡ κοινωνικὴ προστασία τοῦ παιδιοῦ εἰς τὴν ἑλληνικὴν νομοθεσίαν / ὑπὸ Ἰωάννου Κ. Παπαγεωργίου. — ₍Ἀθῆναι₎ : Ἔκδ. Ἐθνικοῦ Κέντρου Κοινωνικῶν Ἐρευνῶν, ₍1972₎
39 p. ; 28 cm. — (Κοινωνικὴ ἔρευνα καὶ ἐφαρμογή ; τεῦχος 2)
Bibliography : p. ₍38₎-39.
1. Children—Legal status, laws, etc.—Greece, Modern. I. Ethnikon Kentron Koinōnikōn Ereunōn. II. Title. III. Series: Koinōnikē ereuna kai epharmogē ; teuchos 2.
79-412444

—LEGAL STATUS, LAWS, ETC.
—HUNGARY

Legal protection of the women and children : analysis of documents of international organisations : contribution of Hungarian Women's Council to the World Congress of Women convened by Women's International Democratic Federation, June 1963 / ₍written by the Working Group of the Law Commission of the National Council of Hungarian Women ; edited by J. Bokor-Szegö₎. — ₍Budapest : Kossuth Kiadó soksz., 1963₎
60 p. ; 29 cm.
Cover title.
1. Women—Legal status, laws, etc. 2. Children—Law. 3. Women—Legal status, laws, etc.—Hungary. 4. Children—Law—Hungary. I. Bokor, Hanna Szegö. II. World Congress of Women, 5th, Moscow, 1963. III. Magyar Nők Országos Tanácsa. IV. Magyar Nők Országos Tanácsa. Jogi Bizottság. Working Group.
K644.L4 341.48'1 78-318173
78 MARC

CHILDREN (Continued)

—LEGAL STATUS, LAWS, ETC. —ILLINOIS

Corcoran, James M 1932-
Probate forms for estates of minors, incompetents, and decedents / by James M. Corcoran, Jr. — ₍Evanston, Ill.₎ : Corcoran, c1976.
viii, 423 leaves : forms ; 30 cm.
1. Probate law and practice—Illinois—forms. 2. Children—Legal status, laws, etc.—Illinois. 3. Insanity—Jurisprudence—Illinois. 4. Decedents' estates—Illinois. I. Title.
KFI1344.A65C67 346'.773'050269 79-115591
79 MARC

Illinois. Committee on Youth and the Law.
Report / Committee on Youth and the Law, Illinois Commission on Children. — Springfield : The Committee, 1978.
iv, 76 p. ; 28 cm.
Cover title.
Bibliography: p. 73-76.
1. Status offenders—Legal status, laws, etc.—Illinois. 2. Children—Legal status, laws, etc.—Illinois.
KFI1797.5.A88 345'.773'03 79-622706
79 MARC

—LEGAL STATUS, LAW, ETC—INDIANA

Indiana. Judicial Study Commission. Juvenile Justice Division.
Indiana juvenile code : proposed final draft. — Indianapolis : State of Indiana, Juvenile Justice Division, Judicial Study Commission, 1977.
xii, 293 p. ; 28 cm.
1. Juvenile justice, Administration of—Indiana. 2. Juvenile courts—Indiana. 3. Children—Legal status, law, etc—Indiana. I. Title.
KFI3595.A84 345'.772'08 78-103225
79 MARC

—LEGAL STATUS, LAWS, ETC.— INDONESIA—CONGRESSES

Loka Karya Evaluasi Bimbingan Kemasyarakatan dan Pengentasan Anak, Jakarta, Indonesia, 1976.
Loka Karya Evaluasi Bimbingan Kemasyarakatan dan Pengentasan Anak / diselenggarakan oleh Badan Pembinaan Hukum Nasional bekerja sama dengan Direktorat Jenderal Bina Tuna Warga dan Fakultas Hukum Universitas Padjadjaran, Jakarta, tanggal 9 s/d 11 Agustus 1976. — Cet. 1. — ₍Jakarta₎ : BPHN ; ₍Bandung₎ : diedarkan oleh Binacipta, 1977.
303 p. ; 21 cm.
Includes bibliographical references.
1. Children—Legal status, laws, etc.—Indonesia—Congresses. 2. Juvenile delinquency—Indonesia—Congresses. 3. Juvenile courts—Indonesia—Congresses. I. Badan Pembinaan Hukum Nasional. II. Indonesia. Direktorat Jenderal Bina Tuna Warga. III. Bandung, Indonesia. Universitas Negeri Padjadjaran. Fakultas Hukum.
78-940429
79 MARC

SA

—LEGAL STATUS, LAWS, ETC.—ITALY

Fenizia, Giuseppe.
Appunti di istituzioni di diritto e procedura penale, criminologia e legislazione minorile / Giuseppe Fenizia ; con prefazione del prof. avv. Alfonso Tesauro. — 2. ristampa aggiornata con la Legge 26 luglio 1975, n. 354, Norme sull'ordinamento penitenziario, e il Decreto presidenziale del 29 aprile 1976, n. 431 con il Regolamento. — Padova : CEDAM, 1977.
360 p. ; 25 cm. It78-Aug
Cover title.
L8500
1. Juvenile courts—Italy. 2. Children—Legal status, laws, etc.—Italy. 3. Criminal law—Italy. I. Italy. Laws, statutes, etc. Legge 26 luglio 1975, n. 354. II. Title.
345'.45'08 79-338552
*79 MARC

Il Codice del minore / ₍a cura di₎ Alberto Germano, Francesco Scarcella. — Milano : A. Giuffrè, 1978.
xiv, 2744 p. ; 20 cm. It78-Oct
"Dichiarazioni internazionali" (in French or Italian): p. ₍3₎-12.
"Convenzioni internazionali" (in French or Italian): p. ₍13₎-80.
Includes jurisprudence.
Includes indexes.
L45000
1. Children—Law—Italy. 2. Children—Law. I. Germanò, Alberto. II. Scarcella, Fancesco.
344'.45'0327 79-337455
*79 MARC

Luzzi, Aldo.
La tutela del lavoro dei fanciulli e degli adolescenti : attribuzione degli organi di polizia, con particolare riferimento ai compiti della polizia femminile; prevenzione ed accertamento dei reati; tecnica delle investigazioni / Aldo Luzzi. — Firenze : Laurus, 1976.
87 p. ; 17 cm. It78-Feb
L2400
1. Children—Law—Italy. 2. Youth—Employment—Italy. I. Title.
344'.45'0131 78-379771
*78 MARC

Vitale, Carmine, 1939-
Delinquenza giovanile tra diritto ed educazione : problemi giuridici ed educativi / Carmine Vitale. — Roma : Armando, c1978.
135 p. ; 22 cm. — (Collana medico-pedagogica ; v. 78) It***
Includes bibliographical references.
L3000
1. Juvenile delinquency—Italy. 2. Children—Legal status, laws, etc.—Italy. 3. Crime prevention—Italy. I. Title.
79-355197
79 MARC

Volpi, Domenico.
I diritti del mio bambino / Domenico Volpi. — Roma : Edizioni paoline, 1979.
105 p. ; 21 cm. It79-July
Bibliography: p. 105.
L3000
1. Children—Legal status, laws, etc. 2. Children—Legal status, laws, etc.—Italy. I. Title.
K639.V64 79-375648
*79 MARC

—LEGAL STATUS, LAWS, ETC. —JAPAN

Ikegi, Kiyoshi, 1938-
(Gendai no seishōnen)
現代の青少年 / 池木清, 後藤光義, 河上恭雄著. — 東京 : 総合労働研究所, 1978.
xx, 321 p. : ill. ; 20 cm. — (教育を考えるシリーズ) Ja 78-22771
¥2000
1. Youth—Japan. 2. Child welfare—Japan. 3. Children—Law—Japan. I. Gotō, Mitsuyoshi, 1947- joint author. II. Kawakami, Yasuo, 1945- joint author. III. Title. IV. Series: Kyōiku o kangaeru shirizu.
HQ799.J3I45 78-816649

—LEGAL STATUS, LAWS, ETC. —JAPAN—ADDRESSES, ESSAYS, LECTURES

Yamaguchi, Sachio, 1933-
(Gendai no hikō mondai)
現代の非行問題 : 教育·福祉·司法 / 山口幸男著. — 東京 : 民衆社, 1978.
228 p. ; 20 cm. Ja 78-14238
List of E. Ferri's works: p. 227-228.
¥1300
1. Juvenile delinquency—Japan—Addresses, essays, lectures. 2. Child welfare—Japan—Addresses, essays, lectures. 3. Children—Law—Japan—Addresses, essays, lectures. I. Title.
HV9207.A5Y294 78-813757

—LEGAL STATUS, LAWS, ETC.,—LATIN AMERICA

Sajón, Rafael.
Protección social y legal del niño de 0 a 5 años / por Rafael Sajón. — Montevideo : O.E.A., Instituto Interamericano del Niño, 1971.
17, ₍1₎ p. ; 28 cm.
"15."
Bibliography: p. ₍18₎
1. Child welfare—Latin America. 2. Children—Legal status, laws, etc.—Latin America. I. Interamerican Children's Institute. II. Title.
HV747.L3S24 362.7'098 79-373269
79 MARC

Sajón, Rafael.
Protección social y legal de la niñez / por Rafael Sajón. — Montevideo : O.E.A., Instituto Interamericano del Niño, 1976.
18 p. ; 30 cm.
"25"
1. Children—Legal status, laws, etc.—Latin America. I. Inter-American Children's Institute. II. Title.
346'.8'013 78-105285
79 MARC

—LEGAL STATUS, LAWS, ETC.— MISSOURI

Missouri. Division of Family Services.
Rules for licensing child caring agencies. — Jefferson City, Mo. : Dept. of Social Services, Division of Family Services, 1979.
v, 24 p. ; 28 cm.
Cover title.
1. Children—Legal status, laws, etc.—Missouri. 2. Children—Institutional care—Missouri. 3. Child welfare—Missouri. I. Title. II. Title: Child caring agencies.
KFM8150.C4A32 1979 344'.778'03271 79-625028
79 MARC

—LEGAL STATUS, LAWS, ETC. —NEBRASKA

Nebraska Committee for Children and Youth.
A compendium of Nebraska laws concerning children and youth / sponsored by Nebraska Committee for Children and Youth. — Rev. — ₍Lincoln₎ : The Committee, 1977.
418 p. ; 26 cm.
Previously entered under: Nebraska Committee for Youth.
Includes index.
1. Children—Legal status, laws, etc.—Nebraska. I. Nebraska Committee for Youth. A compendium of Nebraska laws concerning children and youth. II. Title.
KFN91.M5A3976 1977 346'.782'013 79-623137
79 MARC

—LEGAL STATUS, LAWS, ETC.— NETHERLANDS

Bartels, J A C
De persoon van de minderjarige / J. A. C. Bartels. — Deventer : Van Loghum Slaterus, 1979.
xi, 150 p. ; 22 cm. — (Recht en welzijn ; 3) Ne79-4
Bibliography: p. 149-150.
ISBN 9060015401 : fl 19.50
1. Children—Legal status, laws, etc.—Netherlands. I. Title. II. Series.
79-356244
*79 MARC

Doek, J E
Kleine gids bij het kinderrecht : naamgeving, erkenning, adoptie, voogdij, kinderbescherming, pleeggezin / J. E. Doek. — Amersfoort : Bekking, ₍1978?₎
158 p. ; 19 x 9 cm. — (Kleine gidsen) Ne***
Includes index.
ISBN 9061090318
1. Children—Legal status, laws, etc.—Netherlands. I. Title.
79-375649
79 MARC

Doek, J E
Wat is een goede raad voor de toekomst? : enige beschouwingen over het ontstaan en de (mogelijke) ontwikkeling van de raden voor de kinderbescherming / J. E. Doek. — Alphen aan den Rijn : Samsom, 1978.
76 p. ; 22 cm. Ne78-34
"Bewerking van de openbare les ... Vrije Universiteit te Amsterdam, 1977."
Bibliography: p. 65-72.
ISBN 9014027796 : fl 14.00
1. Child welfare—Netherlands. 2. Children—Law—Netherlands. I. Title.
HV778.A6D6 79-339042
*79 MARC

Hammerstein-Schoonderwoerd, W C E 1943-
Jeugd en ouderlijk gezag / W. C. E. Hammerstein-Schoonderwoerd. — Alphen aan den Rijn : H. D. Tjeenk Willink, 1979.
50 p. ; 23 cm. — (Privaatrechtelijke begrippen ; 5) Ne79-13
Includes bibliographical references.
ISBN 9060920449 : fl 13.00
1. Children—Legal status, laws, etc.—Netherlands. 2. Parent and child (Law)—Netherlands. I. Title. II. Series.
79-371075
*79 MARC

Katholieke Hogeschool. Sektie Jeugdrecht.
Omgang met elkaar : onderzoek naar de totstandkoming van omgangsregelingen na scheiding in zuidelijk Nederland / Sektie Jeugdrecht, Katholieke Hogeschool Tilburg ; m. m. v. C. H. L. van Heesch ... ₍et al.₎. — Deventer : Kluwer, c1978.
xiii, 186 p. ; 25 cm. Ne***
Bibliography: p. 152-155.
ISBN 9026810245
1. Children of divorced parents—Netherlands. 2. Parent and child (Law)—Netherlands. 3. Children—Law—Netherlands. 4. Divorce—Netherlands. I. Heesch, C. H. L. van. II. Title.
HQ777.5.K38 1978 301.42'7 78-389835
79 MARC

CHILDREN (Continued)

—LEGAL STATUS, LAWS, ETC.
—NORTHERN IRELAND

Northern Ireland. Children and Young Persons Review Group.
Legislation and services for children and young persons in Northern Ireland : a consultative document for the Children and Young Persons Review Group, June 1977. — Belfast : H.M.S.O., 1977.
98 p. ; 21 cm. Ir77
Includes bibliographical references.
ISBN 0-337-07153-5 : £1.75
1. Children—Law—Northern Ireland. I. Title.
KDE92.C5 362.7'09416 79-307677
*79 MARC

—LEGAL STATUS, LAWS, ETC.
—NORWAY

Norway. Barnelovutvalget.
Lov om barn og foreldre (barneloven) / fra et utvalg (Barnelovutvalget) oppnevnt ved kongelig resolusjon 21. mars 1975 ; utredning avgitt til Justis- og politidepartementet i september 1977. — Oslo : Universitetsforl., 1977.
171 p. ; 28 cm. — (Norges offentlige utredninger ; 1977:35) N77-Nov.
ISBN 8200703819
1. Children—Law—Norway. 2. Parent and child (Law)—Norway. I. Title. II. Series.
78-387392
79 MARC

—LEGAL STATUS, LAWS, ETC.
—NORWAY—HISTORY

Dahl, Tove Stang, 1938-
Barnevern og samfunnsvern : om stat, vitenskap og professjoner under barnevernets oppkomst i Norge / Tove Stang Dahl. — Oslo : Pax, 1978 (Sarpsborg : Sverre Johansen boktr.)
208 p. ; 20 cm. — (Pax ; 555) N78-Apr.
Bibliography: p. 191-200.
Includes index.
ISBN 8253008937 : kr53.00
1. Child welfare—Norway—History. 2. Children—Law—Norway—History. 3. Child welfare—Europe—History. I. Title.
HV788.A6D34 78-385274
78 MARC

—LEGAL STATUS, LAWS, ETC.—
ONTARIO

Ontario. Laws, statutes, etc.
Laws relating to juveniles in Ontario / ⌈compiled by Michael Bennett, James Glover, Dale F. Ely. — 1978 ed. — Gardena, CA : Harcourt Brace Jovanovich Law Group : distributed by Law Distributors, c1978.
v, 255 p. ; 28 cm.
"Endorsed by International Association of Pupil Personnel Workers."
Includes index.
1. Children—Legal status—Ontario. 2. Juvenile justice, Administration of—Ontario. I. Bennett, Michael, 1940- . II. Glover, James, 1947- . III. Ely, Dale F., 1915- . IV. Title.
KEO207.M5A3 1978 346'.713'013 79-109687
00 MARC

Rivard, Glenn, 1953-
Youth and the law / by Glenn Rivard. — 2d ed. — Toronto : Toronto Community Law Program, c1978.
43 p. ; 27 cm. C***
1. Children—Legal status, laws, etc.—Ontario. 2. Children—Legal status, laws, etc.—Ontario—Toronto. I. Title.
KEO207.M5R58 1978 346'.713'013 79-307679
79 MARC

—LEGAL STATUS, LAWS, ETC.—
ONTARIO—TORONTO

Rivard, Glenn, 1953-
Youth and the law / by Glenn Rivard. — 2d ed. — Toronto : Toronto Community Law Program, c1978.
43 p. ; 27 cm. C***
1. Children—Legal status, laws, etc.—Ontario. 2. Children—Legal status, laws, etc.—Ontario—Toronto. I. Title.
KEO207.M5R58 1978 346'.713'013 79-307679
79 MARC

—LEGAL STATUS, LAWS, ETC.—
PENNSYLVANIA

Juvenile Court Judges' Commission.
Juvenile court handbook and directory / Commonwealth of Pennsylvania, Department of Justice, Juvenile Court Judges' Commission. — ⌈Harrisburg, PA⌉ : The Commission, 1977.
ca. 250 p. in various pagings ; 25 cm.
1. Juvenile courts—Pennsylvania. 2. Children—Legal status, laws, etc.—Pennsylvania. 3. Juvenile courts—Pennsylvania—Directories. I. Title.
KFP596.J89 345'.748'08 79-623140
79 MARC

—LEGAL STATUS, LAWS, ETC.—
POLAND

Kulczycki, Michał.
Prawo w ochronie młodocianych / Michał Kulczycki, Jerzy Zduńczyk. — Wyd. 1. — Warszawa : Instytut Wydawniczy CRZZ, 1977.
156 p. ; 17 cm. P***
Includes bibliographical references.
zł23.00
1. Children—Law—Poland. I. Zduńczyk, Jerzy, joint author. II. Title.
78-373917
78 MARC

Lawina, Adam.
Sytuacja prawna młodzieży w PRL / Adam Lawina, Anna Wasilewska. — Wyd. 1. — Warszawa : Młodzieżowa Agencja Wydawnicza, 1978 (Kiel. : PrasZG)
115, ⌈1⌉ p. ; 20 cm. P78-7264
"Wykaz aktów prawnych": p. ⌈113⌉
zł15.00
1. Children—Legal status, laws, etc.—Poland. I. Wasilewska, Anna, joint author. II. Title.
79-339363
79 MARC

Zdańkowski, Baltazar.
Na wczasach bezpiecznie : zbiór przepisów i wytycznych dotyczących ochrony zdrowia i życia dzieci i młodzieży na wczasach / ⌈oprac. Baltazar Zdańkowski ; współprac. Artur Dziak i Bogdan Odyński⌉. — Wyd. 4 popr. i zm. — Warszawa : Młodzieżowa Agencja Wydawnicza, 1978 (Białyst. : BZG)
198, ⌈2⌉ p. ; 17 cm. — (Biblioteczka wczasowa) P78-4115
zł20.00
1. Children—Legal status, laws, etc.—Poland. 2. Summer homes—Law and legislation—Poland. I. Dziak, Artur. II. Odyński, Bogdan. III. Title.
79-367515
79 MARC

—LEGAL STATUS, LAWS, ETC.—
PORTUGAL

Pinho, David Valente Borges de.
Da protecção judiciária dos menores e do Estado : notas, formulários e jurisprudência sobre medidas de prevenção criminal, tutelares cíveis e outras acções com relevo / David Valente Borges de Pinho. — Coimbra : Livraria Almedina, 1978.
173 p. ; 24 cm.
1. Children—Legal status, laws, etc.—Portugal. I. Title.
346'.469'013 79-100004
79 MARC

—LEGAL STATUS, LAWS, ETC.
—RUSSIA

Orlov, Viktor Semenovich, fl. 1958–
(Podrostok i zakon)
Подросток и закон. Пособие для слушателей. Москва, "Знание," 1972.
72 p. 21 cm. USSR 72-VKP
At head of title: Народный университет. Факультет правовых знаний. В. С. Орлов.
Bibliography: p. 70–⌈71⌉
0.13rub
1. Children—Legal status, laws, etc.—Russia. 2. Juvenile delinquency—Russia. I. Title.
72–357448

Orlov, Viktor Semenovich, fl. 1958–
(Pravovaîa okhrana detstva)
Правовая охрана детства : пособие для слушателей / В. С. Орлов, А. М. Белякова, В. С. Пронина. — Москва : Знание, 1975.
190 p. ; 21 cm. — (Народный университет : Факультет правовых знаний) USSR 75
Series romanized: Narodnyĭ universitet : Fakul'tet pravovykh znaniĭ.
Bibliography: p. 187–⌈189⌉
0.42rub
1. Children—Legal status, laws, etc.—Russia. I. Beliâkova, Anna Mikhaĭlovna, joint author. II. Pronina, Vera Sergeevna, joint author. III. Title.
76–512964

—LEGAL STATUS, LAWS ETC.
—SCOTLAND

Royal Scottish Society for Prevention of Cruelty to Children.
Child protection (Scotland) : the R.S.S.P.C.C. approach / ⌈Royal Scottish Society for Prevention of Cruelty to Children⌉. — ⌈Edinburgh⌉ : ⌈The Society⌉, ⌈1978⌉
⌈1⌉, 32 p. : 3 forms, map ; 21 cm. GB78-15346
Cover title.
1. Royal Scottish Society for Prevention of Cruelty to Children. 2. Child welfare—Scotland. 3. Children—Law—Scotland. I. Title.
HV703.R6824 179'.2'062411 78-326359
79 MARC

—LEGAL STATUS, LAWS, ETC.—
SWEDEN

Jacobsson, Ulla.
Ett barns rättigheter / Ulla Jacobsson. — Stockholm : Askild & Kärnekull, 1978.
159 p. ; 22 cm. S78-21
Includes bibliographical references.
ISBN 9170089469
1. Children—Legal status, laws, etc.—Sweden. I. Title.
79-339162
79 MARC

—LEGAL STATUS, LAWS, ETC.
—SWITZERLAND

Schnyder, Bernhard.
Supplement Kindesrecht / von Bernhard Schnyder. — Zürich : Schulthess Polygraphischer Verlag, c1977.
xiii, 101 p. ; 23 cm. Sw***
Supplement to the 9th ed. of Tuor, Peter, 1876-1957. Das Schweizerische Zivilgesetzbuch, published in 1975.
Includes bibliographical references and index.
ISBN 3-7255-1857-2
1. Civil law—Switzerland. 2. Children—Law—Switzerland. I. Tuor, Peter, 1876-1957. Das Schweizerische Zivilgesetzbuch. II. Title. III. Title: Kindesrecht.
79-340248
79 MARC

—LEGAL STATUS, LAWS, ETC.—
SWITZERLAND—CONGRESSES

Berner Tage für die Juristische Praxis, 5th, 1977.
Das neue Kinderecht / Berner Tage für die Juristische Praxis ; ⌈veranstaltet von der⌉ Rechts- und wirtschaftswissenschaftlichen Fakultät der Universität Bern. — Bern : Stämpfli, 1978.
140 p. ; 24 cm. Sw78-6547
Includes bibliographical references.
ISBN 3-7272-9361-6 : 36.00F
1. Children—Legal status, laws, etc.—Switzerland—Congresses. 2. Domestic relations—Switzerland—Congresses. I. Bern. Universität. Rechts- und Wirtschaftswissenschaftliche Fakultät. II. Title.
79-375559
79 MARC

—LEGAL STATUS, LAWS, ETC.
—TAIWAN

Huang, Li-feng.
(Ch'ing shao nien hsing wei chun tse yü fa lü ch'ang shih)
青少年行為準則與法律常識 / ⌈作者黃麗鳳⌉. — 台北：黃麗鳳：總經銷國民圖書服務社, 民國66 ⌈1977⌉
280 p. : 22 cm.
Includes legislation.
Bibliography: p. 142-143.
NT$150.00
1. Youth—Taiwan—Conduct of life. 2. Children—Law—Taiwan. I. Title.
BJ1668.C5H82 78-843375

—LEGAL STATUS, LAWS, ETC.—TEXAS

Barton, Robert R
Texas juvenile officers' manual / by Robert R. Barton. — Kerrville, Tex. : Barton, c1978.
1 v. ; 26 cm.
Loose-leaf for updating.
Includes indexes.
$12.00
1. Juvenile courts—Texas. 2. Juvenile justice, Administration of—Texas. 3. Children—Legal status, laws, etc.—Texas. I. Title.
KFT1796.B37 345'.764'08 79-107818
79 MARC

Texas. Laws, statutes, etc.
Texas laws relating to minors / ⌈compiled by⌉ Roosevelt Miles, Jr., Dale F. Ely. — 1978 ed. — Gardena, CA : Harcourt Brace Jovanovich Law Group : distributed by Law Distributors, c1978.
xx, 336 p. ; 28 cm.
Includes index.
1. Children—Legal status, laws, etc.—Texas. I. Miles, Roosevelt. II. Ely, Dale F. III. Title.
KFT1291.M5A3 1978 346'.764'013 79-114200
79 MARC

CHILDREN (Continued)

—LEGAL STATUS, LAWS, ETC. —UNITED STATES

Copeland, William C
Roadmap through Title XX / William C. Copeland, Iver A. Iversen. — Washington : CWLA Hecht Institute for State Child Welfare Planning, c1978.
i, 138 p. : diagrs. ; 28 cm. — (His Financing services for children through Title XX and other programs ; manual 5)
ISBN 0-87868-129-9
1. Child welfare—United States. 2. Federal aid to child welfare—United States. 3. Children—Law—United States. I. Iversen, Iver A., joint author. II. Title.
HV741.C653 vol. 5 362.7 s 78-112957
79 MARC

United States. Library of Congress. Congressional Research Service. American Law Division.
Constitutional rights of children / prepared for the Subcommittee on the Constitution of the Committee on the Judiciary, United States Senate by the American Law Division, Congressional Research Service of the Library of Congress. — Washington : U.S. Govt. Print. Off., 1978.
ix, 33 p. ; 24 cm.
Prepared by Johnny H. Killian.
At head of title: 95th Congress, 2d session. Committee print.
Includes bibliographical references.
Supt. of Docs. no.: Y4.J89/2:C 43/7
1. Children—Legal status, laws, etc.—United States. I. Killian, Johnny H. II. United States. Congress. Senate. Committee on the Judiciary. Subcommittee on the Constitution. III. Title.
KF479.A25 1978 346'.73'013 79-600857
79 MARC

Weinstein, Noah.
Minors and the law / by Noah Weinstein. — Reno : National Council of Juvenile and Family Court Judges, University of Nevada/Reno, c1978.
iii, 42 p. ; 26 cm. — (Juvenile justice textbook series)
1. Children—Legal status, laws, etc.—United States. I. Title. II. Series.
KF479.Z9W44 346'.73'013 79-100493
79 MARC

Zuckerman, Ruth Jane.
Abortion and the constitutional rights of minors / by Ruth Jane Zuckerman. — New York : American Civil Liberties Union, 1973.
25, [6] leaves : ill. ; 28 cm. — (ACLU reports)
Cover title.
Includes bibliographical references.
1. Children—Law—United States. 2. Abortion—Law and legislation—United States. I. Title. II. Series: American Civil Liberties Union. ACLU reports.
KF479.Z8 346'.73'013 78-320519
79 MARC

—LEGAL STATUS, LAWS, ETC.— UNITED STATES—DIGESTS

Juvenile law digest. v. 10–
Jan. 1978–
[Reno, Nev.] National Council of Juvenile and Family Court Judges.
v. 23 cm. monthly.
Continues: Juvenile court digest, ISSN 0085-2430.
Key title: Juvenile law digest, ISSN 0162-5055.
1. Juvenile delinquency—United States—Digests. 2. Children—Legal status, laws, etc.—United States—Digests. I. National Council of Juvenile and Family Court Judges.
KF9776.3.J88 345'.73'08 78-645492
MARC-S

Weinstein, Noah.
Supreme Court decisions and juvenile justice / by Noah Weinstein. — 2d ed. / by John F. Mendoza. — Reno : National Council of Juvenile and Family Court Judges, University of Nevada/Reno, 1979.
vi, 55 p. ; 26 cm. — (Juvenile justice textbook series ; 102)
1. Juvenile justice, Administration of—United States—Digests. 2. Children—Legal status, laws, etc.—United States—Digests. 3. United States. Supreme Court. I. Mendoza, John F. II. Title. III. Series.
KF9810.A59W43 1979 345'.73'08 79-106182
00 MARC

—LEGAL STATUS, LAWS, ETC. —UNITED STATES —JUVENILE LITERATURE

Englebardt, Leland S
You have a right : a guide for minors / Leland S. Englebardt. — 1st ed. — New York : Lothrop, Lee & Shepard Co., c1979.
128 p. ; 24 cm.
Includes index.
SUMMARY: Discusses the often misunderstood legal rights of young people and points out how these rights vary from state to state.
ISBN 0-688-41893-7
1. Children—Legal status, laws, etc.—United States—Juvenile literature. [1. Law. 2. Children's rights] I. Title.
KF479.Z9E53 346'.73'013 79-4678
79 MARC

Sgroi, Peter P
Blue jeans and black robes : teenagers and the Supreme Court / Peter Sgroi. — New York : J. Messner, c1979.
192 p. ; 22 cm.
Bibliography: p. 187.
Includes index.
SUMMARY: Presents and defends the argument that minors are one of the last groups to receive their rights as evidenced in various Supreme Court decisions.
ISBN 0-671-32884-0
1. Children—Legal status, laws, etc.—United States—Juvenile literature. 2. Civil rights—United States—Juvenile literature. [1. Children's rights. 2. Civil rights. 3. Law]
KF479.Z9S48 346'.73'013 79-115
79 MARC

Swiger, Elinor Porter.
The law in your everyday life / by Elinor Porter Swiger ; illustrated by Mamoru Funai. — Englewood Cliffs, N.J. : Prentice-Hall, 1978.
xxiii, 207 p. : ill. ; 24 cm.
Includes index.
SUMMARY: Discusses how laws are made, enforced, and defined in court, state-by-state differences in law, and several laws pertaining to young people.
ISBN 0-13-526202-X : $8.95
1. Law—United States—Juvenile literature. 2. Children—Law—United States—Juvenile literature. [1. Law. 2. Children's rights] I. Funai, Mamoru. II. Title.
KF387.S96 340'.0973 77-27115
77 MARC

—LEGAL STATUS, LAWS, ETC.— UNITED STATES—OUTLINES, SYLLABI, ETC.

D'Embois, E St. Cleening.
The skeptical students workbook of family and juvenile law / E. St. Cleening d'Embois, Krystina Wyzczieki. — Atlanta : Uncommon Lawyers Workshop, 1979.
118 leaves : ill. ; 22 x 28 cm.
Includes bibliographies and index.
1. Domestic relations—United States—Outlines, syllabi, etc. 2. Children—Legal status, laws, etc.—United States—Outlines, syllabi, etc. 3. Domestic relations—California—Outlines, syllabi, etc. 4. Children—Legal status, laws, etc.—California—Outlines, syllabi, etc. I. Wyzczieki, Krystina, joint author. II. Title.
KF505.Z9D45 346'.73'015 79-62889
79 MARC

—LEGAL STATUS, LAWS, ETC. —UNITED STATES—PERIODICALS

Children's rights report. v. 1–
Sept. 1976–
New York, Juvenile Rights Project, American Civil Liberties Union Foundation.
v. 28 cm. 10 no. a year.
Key title: Children's rights report, ISSN 0190-0242.
1. Children—Legal status, laws, etc.—United States—Periodicals. I. American Civil Liberties Union Foundation. Juvenile Rights Project.
KF479.A15C48 342'.78'087 79-640194
MARC-S

—LEGAL STATUS, LAWS, ETC. —URUGUAY

Uruguay. Laws, statutes, etc.
Código del niño : promulgado por el Poder Ejecutivo el 6 de abril de 1934. — Montevideo : Acali Editorial, 1978.
46 p. ; 24 cm.
1. Children—Law—Uruguay. I. Title.
78-109959
344'.895'032702632
79 MARC

—LEGAL STATUS, LAWS, ETC. —VIRGINIA

Virginia laws relating to minors.
[S. l.] Harcourt Brace Jovanovich Legal and Professional Publications; distributed by Law Distributors.
v. 28 cm. annual.
Key title: Virginia laws relating to minors, ISSN 0198-5496.
1. Children—Legal status, laws, etc.—Virginia. 2. Juvenile courts—Virgina.
KFV2491.M5A298 346'.755'013 79-643623
MARC-S

—LEGAL STATUS, LAWS, ETC. —YUGOSLAVIA

Hirjan, Franjo.
Maloljetnici u krivičnom pravu : materijalnopravni i procesnopravni položaj maloljetnih učinilaca krivičnih djela, krivičnopravna zaštita maloljetnika, kriminalna politika / Franjo Hirjan, Mladen Singer. — Zagreb : Informator, 1978.
510 p. ; 24 cm. — (Pravna biblioteka) Yu***
Bibliography: p. 473-488.
Includes index.
1. Juvenile delinquency—Yugoslavia. 2. Children—Law—Yugoslavia. 3. Criminal law—Yugoslavia. I. Singer, Mladen, joint author. II. Title.
78-387390
79 MARC

—MANAGEMENT

see also Baby sitters; Child abuse; Children's allowances; Children's questions and answers; Discipline of children; Moral education; Parenting; Paternal deprivation; Problem children; Toilet training

Ackerman, Paul R 1934-
Signals : what your child is really telling you / Paul R. Ackerman and Murray M. Kappelman. — New York : Dial Press/J. Wade, c1978.
290 p. ; 22 cm.
Includes index.
ISBN 0-8037-8151-2 : $8.95
1. Parent and child. 2. Children—Management. 3. Child psychology. 4. Adolescent psychology. I. Kappelman, Murray, 1931- joint author. II. Title.
HQ755.85.A26 301.42'7 78-15272
78 MARC

Bágynoi, Attila.
Okos szóval, türelemmel : szülők könyve / Bágynoi Attila, Varjasi Edit ; [ill., Zsoldos Vera]. — Budapest : Medicina, 1978.
230 p. (p. [229]-230 advertisements) : ill. ; 21 cm. H78
ISBN 9632407814 : 30.00Ft
1. Parenting. 2. Children—Management. I. Varjasi, Edit, joint author. II. Title.
HQ755.8.B34 79-366703
*79 MARC

Barber, Lucie W
Celebrating the second year of life : a parent's guide for a happy child / Lucie W. Barber and the staff of the Union College Character Research Project. — Birmingham, Ala. : Religious Education Press, c1978.
x, 148 p. ; 20 cm.
Bibliography: p. 132-135.
Includes index.
ISBN 0-89135-015-2
1. Infants. 2. Child development. 3. Children—Management. I. Union College, Schenectady. Character Research Project. II. Title.
HQ774.B33 649'.122 78-21484
78 MARC

Bartz, Wayne R 1938-
Surviving with kids : a lifeline for overwhelmed parents / Wayne R. Bartz and Richard A. Rasor ; ill. by Dick Pike. — San Luis Obispo, Calif. : Impact Publishers, c1978.
xi, 177 p. : ill. ; 21 cm.
Bibliography: p. 176-177.
ISBN 0-915166-55-0 : $3.95
1. Children—Management. 2. Parent and child. 3. Behavior modification. I. Rasor, Richard A., joint author.
HQ769.B3249 649'.1 78-13328
78 MARC

Baruth, Leroy G
A single parent's survival guide : how to raise the children / Leroy G. Baruth. — Dubuque, Iowa : Kendall/Hunt Pub. Co., c1979.
xiii, 102 p. : ill. ; 23 cm.
Includes bibliographies.
ISBN 0-8403-2053-1
1. Children—Management. 2. Single parents—United States. I. Title.
HQ769.B336 649'.1 79-2388
79 MARC

Birkey, Verna.
A mother's problem solver : imaginative solutions for life's everyday problems / compiled by Verna Birkey and Jeanette Turnquist. — Old Tappan, N.J. : F. H. Revell, c1978.
128 p. ; 21 cm.
ISBN 0-8007-0982-9
1. Children—Management. 2. Family—Religious life. I. Turnquist, Jeanette, joint author. II. Title.
HQ772.B479 301.42'7 78-10142
78 MARC

CHILDREN
—MANAGEMENT (Continued)

Bottel, Helen.
Parents' survival kit : a reassuring guide to living through your child's teen-age years / Helen Bottel. — 1st ed. — New York : Doubleday, 1979.
230 p. ; 22 cm.
1. Adolescence. 2. Children—Management. 3. Parent and child. 4. United States—Social conditions—1960- I. Title.
HQ796.B6846 649'.125 78-1184
78 MARC

Calladine, Carole.
Raising siblings / Carole & Andrew Calladine. — New York : Delacorte Press, c1979.
x, 208 p. ; 22 cm.
ISBN 0-440-07314-6
1. Children—Management. 2. Brothers and sisters. I. Calladine, Andrew, joint author. II. Title.
HQ772.C34 649'.14'3 79-13804
79 MARC

Chapman, Arthur Harry, 1924-
Parents talking, kids talking : an original, breakthrough work for all parents who want to discuss adult subjects with children but don't know how / A. H. Chapman. — New York : Putnam, c1979.
245 p. ; 23 cm.
Includes index.
ISBN 0-399-12259-1 : $8.95
1. Parent and child. 2. Children—Management. 3. Adolescent psychology. 4. Child psychology. I. Title.
HQ755.85.C49 1979 301.42'7 78-15271
78 MARC

Christophersen, Edward R
Little people : guidelines for common sense child rearing / by Edward R. Christophersen. — Lawrence, Kan. : H & H Enterprises, c1977.
ix, 127 p. : ill. ; 24 cm.
Bibliography: p. 127.
1. Children—Management. I. Title.
HQ769.C563 649 79-112461
00 MARC

Chūbachi, Fujirō, 1893-
(Kodomo no shitsuke to sodatekata)
子供の躾と育て方 / 中鉢不二郎, 小林提樹共著. — 東京 : 佐竹書房, 昭和25 [1950]
244 p. : ill. ; 19 cm.
Colophon inserted.
1. Education of children. 2. Children—Management. 3. Children—Care and hygiene. I. Kobayashi, Teiju, 1908- joint author. II. Title.
LB1115.C56 79–801288

Collins, Maurice, 1934-
Kith and kids : self-help for families of the handicapped / by Maurice and Doreen Collins. — London : Souvenir Press (E & A), 1976.
v, 192 p., [4] leaves of plates : ill. ; 23 cm. — (Human horizons series) (A Condor book) GB***
ISBN 0-285-64818-7 : £4.00. ISBN 0-285-64819-5 pbk.
1. Kith and Kids. 2. Mentally handicapped children—Great Britain. 3. Physically handicapped children—Great Britain. 4. Children—Management. I. Collins, Doreen, joint author. II. Title.
HV890.G72L63 1976 362.7'8'3 78-323928
79 MARC

Crary, Elizabeth, 1942-
Without spanking or spoiling : a practical approach to toddler and preschool guidance / by Elizabeth Crary ; [ill. by Ellen B. Ochs]. — 1st ed. — Seattle : Parenting Press, c1979.
102 p. : ill. ; 28 cm.
Includes bibliographies and index.
ISBN 0-9602862-0-9
1. Children—Management. I. Title.
HQ769.C93 649'.1 79-18253
79 MARC

Dobson, James C 1936-
The strong-willed child : birth through adolescence / James Dobson. — Wheaton, Ill. : Tyndale House, c1978.
240 p. ; 22 cm.
Includes bibliographical references.
ISBN 0-8423-0664-1 : $7.95
1. Children—Management. 2. Discipline of children. 3. Child psychology. I. Title.
HQ772.D58 649'.1 77-83559
79 MARC

Doncheva, Vasilka Khristova.
Osobeno dete / Vasilka Doncheva. — 1. izd. — Sofiia : Narodna prosveta, 1978.
117 p. ; 21 cm. — (Biblioteka za roditelia) Bu***
Includes bibliographical references.
0.44lv
1. Children—Management. 2. Juvenile delinquency. I. Title.
HQ796.D59 79-376145
79 MARC

Draper, Mary Wanda.
Caring for children / by Mary Wanda Draper and Henry E. Draper. — Rev. — Peoria, Ill. : C. A. Bennett Co., c1979.
560 p. : ill. ; 24 cm. — (A Bennett career book)
"Annotated bibliography of books for children": p. 521-532.
Includes bibliographies and index.
ISBN 0-87002-281-4
1. Child development. 2. Children—Management. 3. Day care aids—United States. I. Draper, Henry E., joint author. II. Title.
HQ769.D644 1979 649'.1 78-52494
79 MARC

Edmands, Dodie.
The children's astrologer / Dodie & Allan Edmands. — New York : Hawthorn Books, c1978.
143 p. ; 24 cm.
Bibliography: p. 139.
Includes index.
ISBN 0-8015-1227-1 : $7.95
1. Astrology. 2. Children—Management. I. Edmands, Allan, joint author. II. Title.
BF1729.C45E35 133.5'8'6491 78-53406
79 MARC

Erickson, Karla C
Take time to smell the dandelions / Karla C. Erickson. — Salt Lake City : Bookcraft, 1979.
88 p. ; 24 cm.
ISBN 0-88494-360-7 : $3.95
1. Children—Management. 2. Child psychology. 3. Parent and child. I. Title.
HQ769.E67 301.42'7 78-75283
79 MARC

Gould, Shirley, 1917-
The challenge of achievement : helping your child succeed / Shirley Gould. — New York : Hawthorn Books, c1978.
145 p. ; 22 cm.
Bibliography: p. 137-139.
Includes index.
ISBN 0-8015-3385-6 : $7.95
1. Children—Management. 2. Achievement motivation. 3. Success. I. Title.
HQ772.G656 1978 649'.1 78-53400
79 MARC

Grünspun, Haim.
A autoridade dos pais e a educação da liberdade. São Paulo, Edição Sedes Sapientiae, 1968.
116 p. 19 cm. 5.00 LACAP69-3500
Bibliography: p. [115]-116.
1. Children—Management. 2. Parent and child. I. Title.
HQ769.G854 79-438091
79[r79]rev MARC

Heidebrecht, Paul.
Fathering a son / by Paul Heidebrecht and Jerry Rohrbach. — Chicago : Moody Press, c1979.
218 p. : ill. ; 22 cm.
Bibliography: p. 205-218.
ISBN 0-8024-3356-1 : $4.95
1. Boys. 2. Fathers and sons. 3. Children—Management. I. Rohrbach, Jerry, joint author. II. Title.
HQ775.H44 649'.132 78-21871
78 MARC

Hendrick, Gladys West.
My first 300 babies / [Gladys West Hendrick]. — Rev. ed. — Santa Ana, Calif. : Vision House Publishers, 1978.
173 p. : ill. ; 21 cm.
ISBN 0-88449-030-0 : $4.98
1. Infants—Care and hygiene. 2. Discipline of infants. 3. Children—Management. I. Title.
[RJ61.H395 1978] 649'.122 78-50853
79 MARC

Hicks, Mack R
Parent, child, and community : a guide for the middle-class urban family / Mack R. Hicks, Herbert Goldstein, Patricia J. Pearson. — Chicago : Nelson-Hall, 1979.
234 p. : ill. ; 23 cm.
Bibliography: p. 229.
Includes index.
ISBN 0-88229-231-5
1. Children—Management. 2. Middle classes—United States. I. Goldstein, Herbert, 1936- joint author. II. Pearson, Patricia J., joint author. III. Title.
HQ772.H497 649'.1 79-13816
79 MARC

Hofmeister, Alan M
What do I do now? : Practical ways to develop good behavior in your child / Alan M. Hofmeister, Charles Atkinson, Hester Henderson. — Rev. ed. — Niles, Ill. : Argus Communications, c1978.
104 p. : ill. ; 19 cm.
ISBN 0-89505-016-1 : $1.95
1. Child development. 2. Children—Management. I. Atkinson, Charles, 1942- joint author. II. Henderson, Hester, joint author. III. Title.
HQ772.H58 1978 649'.123 78-56863
79 MARC

(Hoiku genshō no bunkaronteki tenkai)
保育現象の文化論的展開 / 本田和子, 津守真共編 ; 入江礼子, 友定啓子, 角能清美共著. — 東京 : 光生館, 1977.
177 p. : ill. ; 22 cm. — (人間現象としての保育研究 ; 3) Ja 77-4859
Includes bibliographical references.
¥1400
1. Education, Preschool—1965- 2. Children—Management. 3. Children—Care and hygiene. I. Honda, Kazuko. II. Tsumori, Makoto, 1926- III. Irie, Reiko. IV. Series: Ningen genshō to shite no hoiku kenkyū ; 3.
LB1140.2.H612 78–813577

Howe, Leland W
How to raise children in a TV world / Leland W. Howe, Bernard Solomon ; illustrated by Beverly Muschlitz. — New York : Hart Pub. Co., c1979.
271 p. : ill. ; 21 cm.
Bibliography: p. [257]-259.
Includes index.
ISBN 0-8055-0395-1 : $6.95
1. Television and children. 2. Children—Management. 3. Television in preschool education—United States. I. Solomon, Bernard, joint author. II. Title.
HQ784.T4H68 649'.5 78-60820
00 MARC

Ichikawa, Eisaku.
(Atarashii sedai ni okuru tadashii kodomo no shitsukekata)
新しい世代におくる正しい子供のしつけ方 : 学校·家庭のために / 市川英作著. — 東京 : 学芸図書出版社, 昭和26 [1951]
126 p., [1] leaf of plates : ill. ; 18 cm.
Colophon inserted.
1. Children—Management. 2. Children—Japan. 3. Education of children. I. Title. II. Title: Tadashii kodomo no shitsukekata.
HQ769.I33 79–809665

Janney, John R
How to raise happy, godly children / John R. Janney. — 1st ed. — Bridgeton, N.J. : Christian Family Crusades, c1978.
183 p. : ill. ; 22 cm.
Includes bibliographical references.
1. Children—Conduct of life. 2. Children—Religious life. 3. Children—Management. I. Title.
BJ1631.J36 649'.1 78-58575
79 MARC

Jolly, Hugh.
More commonsense about babies and children / [by] Hugh Jolly. — London : Pelham, 1978.
192 p. ; 21 cm. GB78-21319
Includes index.
ISBN 0-7207-1069-3 : £4.95
1. Children—Care and hygiene. 2. Children—Diseases. 3. Children—Management. I. Title.
RJ61.J68 649.1 79-303949
79 MARC

(Jyā bāyakā̃sa leṅkarē̃ āheta tyā̃sa subodha)
ज्या बायकांस लेंकरें आहेत त्यांस सुबोध = Address to mothers. — 1st ed. — [Bombay : American Mission], 1841.
25 p. ; 19 cm. — (American Mission series ; no. 34)
Caption title.
In Marathi.
1. Children—Conduct of life. 2. Children—Management. I. American Marathi Mission. II. Title: Address to mothers.
BJ1638.M37J96 78–984395

Kahn, Samuel, 1898-
Child guidance / Samuel Kahn. — New York : Philosophical Library, c1979.
134 p. ; 23 cm.
Bibliography: p. 133-134.
ISBN 0-8022-2230-7 : $10.00
1. Parenting—United States. 2. Children—Management. I. Title.
HQ755.8.K33 649'.1 79-83823
79 MARC

Kohnstamm, Rita.
Meegroeien met uw kind / Rita Kohnstamm. — Amsterdam : Kosmos, 1975.
152 p. ; 20 cm. — (Ouders van nu boek) Ne75-43
ISBN 902150510X : fl 18.50
1. Children—Management. I. Title.
HQ769.K5578 79-343414
*79 MARC

Kvols-Riedler, Bill.
Redirecting children's misbehavior / Bill & Kathy Kvols-Riedler. — Boulder, Colo. : R. D. I. C. Publications, c1979.
xvi, 235 p. : ill. ; 23 cm.
Bibliography: p. 227-228.
Includes index.
ISBN 0-933450-00-1
1. Children—Management. 2. Child psychology. I. Kvols-Riedler, Kathy, joint author. II. Title.
HQ772.K86 649'.1 78-68637
79 MARC

CHILDREN
—MANAGEMENT (Continued)

LaHaye, Beverly.
How to develop your child's temperament / Beverly LaHaye. — Irvine, Calif. : Harvest House, c1977.
163 p. : ill. ; 21 cm.
ISBN 0-89081-034-6 : $2.95
1. Children—Management. 2. Child development. I. Title.
HQ772.L23 649'.1 76-53068
77[r79]rev MARC

Leksina, M
(Zhas balany sem'ı͡ada tărbieleu)
Жас баланы семьяда тәрбиелеу. Алматы, Қазақстан, 1968.
30 p. 14 cm. USSR 68-29960
At head of title: M Лексина.
—— —— Microfilm. On 1 reel with other items ; 35 mm.
0.04rub
Microfilm Slavic 12711 RJ
1. Children—Care and hygiene. 2. Children—Management. I. Title.
RJ61.L486 72-334411

Losoncy, Lawrence J
When your child needs a hug / Larry Losoncy. — St. Meinrad, Ind. : Abbey Press, c1978.
95 p. : ill. ; 18 cm. — (A When book)
Bibliography: p. 94-95.
ISBN 0-87029-141-6 : $1.95
1. Child development. 2. Children—Management. I. Title.
HQ772.L618 649'.1 78-73016
79 MARC

Mamuya, S J
Utu uzima / S. J. Mamuya. — Nairobi : Transafrica, 1975.
vii, 91 p. : ill. ; 22 cm.
Includes index.
1. Marriage. 2. Children—Management. I. Title.
HQ743.M255 77-980013
79 MARC

Matsumoto, Iwao.
(Seikatsu to shitsuke no shidō)
生活と躾の指導 / 松本巖著 ; 佐藤保太郎監修. — 東京 : 西野書店, 昭和25 [1950]
4, 2, 3, 309 p. ; 22 cm. — (教育課程実践叢書 ; 1)
1. Personnel service in elementary education—Japan. 2. Children—Management. 3. Curriculum planning—Japan. 4. Education, Elementary—Japan—Curricula. I. Title. II. Series: Kyōiku katei jissen sōsho ; 1.
LB1027.5.M348 79-806970

Mayle, Peter.
Baby taming / by Peter Mayle ; illustrated by Arthur Robins. — New York : Harmony Books, c1978.
[49] p. : col. ill. ; 27 cm.
ISBN 0-517-53534-3
1. Children—Management. 2. Children—Management—Anecdotes, facetiae, satire, etc. I. Title.
HQ769.M377 1978 649'.1 78-15532
78 MARC

Mercer, Jean.
Small people : how children develop, and what you can do about it / Jean Mercer. — Chicago : Nelson-Hall, c1979.
xiii, 152 p. : ill. ; 23 cm.
Bibliography: p. 141-149.
Includes index.
ISBN 0-88229-318-4
1. Child development. 2. Children—Management. I. Title.
RJ131.M37 649'.1 78-27345
78 MARC

Miller, Maureen.
Help your child for life / Maureen Miller. — 1st ed. — Niles, Ill. : Argus Communications, c1978.
228 p. : ill. ; 19 cm.
Bibliography: p. 227-228.
ISBN 0-89505-018-8 : $2.95
1. Child development. 2. Children—Management. I. Title.
HQ772.M446 649'.1 78-63105
79 MARC

New Zealand. Social Development Council.
Bringing up children in New Zealand : can we do better? — [Wellington] : Social Development Council, 1977.
21 p. ; 30 cm. NZ***
Includes bibliography.
1. Children—Management. 2. Children—New Zealand. I. Title.
HQ772.N492 1977 649'.1'09931 78-323474
79 MARC

Panich, Mikhail Semenovich.
(Khochu poni͡at')
Хочу понять / М. Панич. — Ленинград : Сов. писатель, Ленингр. отд-ние, 1978.
240 p. : port. ; 17 cm. USSR 78
0.70ru
1. Parent and child. 2. Children—Management. I. Title.
HQ755.85.P35 79-407936

Papalia, Diane E
A child's world : infancy through adolescence / Diane E. Papalia, Sally Wendkos Olds. — 2d ed. — New York : McGraw-Hill, c1979.
xi, 626 p. : ill. ; 25 cm.
Bibliography: p. 580-614.
Includes index.
ISBN 0-07-048458-9
1. Child psychology. 2. Children—Management. I. Olds, Sally Wendkos, joint author. II. Title.
HQ772.P23 1979 301.43'1 78-17198
78 MARC

Perrez, Meinrad.
Eltern-Verhaltenstraining : [Für Eltern, Erzieher u. Erwachsenenbildner] / Meinrad Perrez Beate Minsel, Heinz Wimmer. — Salzburg : O. Müller, [1974].
2 v. in 1 ; 24 cm. — (Arbeitsbücher zur psychologischen Schulung)
Au74-18-251
Bibliography: v. 1, p. 65
CONTENTS: 1. Theoretische Einführung.—2. Praktische Übungen
ISBN 3-7013-0498-X : $137.00
1. Children—Management. 2. Behavior modification. 3. Parent and child. I. Minsel, Beate, joint author. II. Wimmer, Heinz, joint author. III. Title.
HQ769.P538 75-569276
*79 MARC

Phillips, Mike, 1946-
Blueprint for raising a child / by Mike Phillips. — Plainfield, N.J. : Logos International, c1978.
xii, 395 p. ; 21 cm.
Bibliography: p. 393-395.
ISBN 0-88270-280-7 : $3.95
1. Children—Management. I. Title.
HQ772.P48 649.1 78-64344
79 MARC

Powell, Barbara, 1929-
How to raise a successful daughter / Barbara Powell. — Chicago : Nelson-Hall, c1979.
xi, 144 p. ; 23 cm.
Includes bibliographical references and index.
ISBN 0-88229-457-1
1. Daughters. 2. Children—Management. I. Title.
HQ777.P68 649'.133 78-16975
78 MARC

Pumroy, Donald Keith, 1925-
Modern childrearing : a behavioral approach / Donald K. Pumroy, Shirley S. Pumroy. — Chicago : Nelson-Hall, c1978.
viii, 148 p. : ill. ; 24 cm.
Bibliography: p. 145.
Includes index.
ISBN 0-88229-185-8
1. Children—Management. 2. Child psychology. I. Pumroy, Shirley Ann Spence, 1928- joint author. II. Title.
HQ769.P843 649'.1 77-26964
77 MARC

Rivers, Caryl.
Beyond sugar and spice : how women grow, learn, and thrive / by Caryl Rivers, Grace Baruch, and Rosalind Barnett. — New York : Putnam, c1979.
333 p. ; 24 cm.
Includes bibliographical references and index.
ISBN 0-399-12164-1 : $10.95
1. Girls. 2. Children—Management. 3. Women—Socialization. I. Barnett, Rosalind C., joint author. II. Baruch, Grace K., joint author. III. Title.
HQ777.R58 649'.133 79-12516
79 MARC

Rutenber, Ralph.
How to bring up 2000 teenagers / Ralph Rutenber. — Chicago : Nelson-Hall, c1979.
228 p. ; 22 cm.
Includes index.
ISBN 0-88229-550-0
1. Adolescence. 2. Children—Management. I. Title.
HQ796.R87 649'.125 78-24060
78 MARC

Rybczyńska, Irena.
Jak być (kochaną) babcią / Irena Rybczyńska. — Wyd. 2 uzup. — Warszawa : Watra, 1977 (Łódź : ŁZG 2)
166, [2] p. ; 20 cm. P78-1779
Errata slip inserted.
zł22.00
1. Grandmothers. 2. Children—Management. I. Title.
HQ759.9.R9 1977 78-400200
79 MARC

Salk, Lee, 1926-
Dear Dr. Salk : answers to your questions about your family / Lee Salk. — 1st ed. — New York : Harper & Row, c1979.
x, 236 p. ; 22 cm.
Includes index.
ISBN 0-06-013768-1 : $8.95
1. Children—Management. 2. Parent and child. 3. Family—United States. 4. Conduct of life. I. Title.
HQ769.S245 301.42'7 78-69629
78 MARC

Shure, Myrna B
Problem-solving techniques in childrearing / Myrna B. Shure, George Spivack. — 1st ed. — San Francisco : Jossey-Bass Publishers, 1978.
xix, 261 p. ; 24 cm. — (The Jossey-Bass social and behavioral science series)
Bibliography: p. 243-254.
Includes index.
ISBN 0-87589-366-X
1. Children—Management. 2. Problem-solving. 3. Parenting. I. Spivack, George, joint author. II. Title.
HQ772.S49 649'.1 77-93677
78 MARC

Sierra Madre Community Nursery School.
Nurturing human growth : an approach for parents and teachers of young children / Sierra Madre Community Nursery School staff and others. — Sierra Madre, Calif. : Sierra Madre Community Nursery School Association, c1978.
vii, 154 p. : ill. ; 22 cm.
Bibliography: p. 143-147.
1. Child development. 2. Children—Management. I. Sierra Madre Community Nursery School Association. II. Title.
HQ772.S497 1978 649'.1 78-111985
79 MARC

Spock, Benjamin McLane, 1903-
Problems of parents / by Benjamin Spock. — Westport, Conn. : Greenwood Press, 1978, c1962.
ix, 308 p. ; 22 cm.
Reprint of the ed. published by Houghton Mifflin, Boston.
Includes index.
ISBN 0-313-21000-4
1. Children—Management. I. Title.
[HQ769.S68 1978] 649'.1 78-13982
78 MARC

Szabó, Pál, Dr.
A játéktól az iskoláig : az óvodáskorról (3-6. év) / Szabó Pál. — [Budapest] : Hazafias Népfront, 1978.
83 p. : ill. ; 20 cm. — (Szülőknek, nevelésről) H78
Bibliography: p. 81-[82]
ISBN 9630911590 : 5.00Ft
1. Children—Management. 2. Games. I. Title. II. Series.
HQ769.S98 79-360393
*79 MARC

Talbot, Nathan Bill, 1909-
Raising children in modern America : what parents and society should be doing for their children / Nathan B. Talbot. — 1st ed. — Boston : Little, Brown, c1976.
x, 180 p. : ill. ; 24 cm.
Bibliography: p. [162]-164.
ISBN 0-316-83136-0
1. Children—Management. 2. Children in the United States. I. Title.
HQ769.T265 649'.1'0973 75-31571
75[r79]rev MARC

Theroux, Rosemary T 1949-
The care of twin children : a common-sense guide for parents / by Rosemary T. Theroux, and Josephine F. Tingley ; edited by Donald M. Keith, Louis G. Keith. — Chicago : Center for Study of Multiple Gestation, 1978.
ix, 119 p. ; 19 cm.
Bibliography: p. 118-119.
Includes index.
ISBN 0-932254-00-4 : 6.75
1. Twins. 2. Children—Management. I. Tingley, Josephine F., 1934- joint author. II. Keith, Louis. III. Keith, Donald M. IV. Title.
HQ777.35.T48 1978 649'.144 78-60823
78 MARC

Thurmond, Nancy Moore.
Mother's medicine / by Nancy Moore Thurmond. — 1st ed. — New York : Morrow, 1979.
383 p. : ill. ; 24 cm.
Bibliography: p. [375]-376.
Includes index.
ISBN 0-688-03384-9
1. Children—Care and hygiene. 2. Children—Management. I. Title.
RJ61.T53 649.1 78-12552
78 MARC

Trimmer, Eric J
The first seven years / [by] Eric Trimmer in association with Elinor Goldschmied. — London : Heinemann, [1978]
3-142 p. : ill. (chiefly col.) ; 27 cm. GB78-39176
Includes index.
ISBN 0-434-79361-2 : £4.95
1. Children—Management. 2. Infants—Care and hygiene. I. Goldschmied, Elinor, joint author. II. Title.
HQ772.T67 649'.1 79-307312
79 MARC

Trimmer, Eric J
The first seven years / Eric Trimmer, in association with Elinor Goldschmied. — New York : St. Martin's Press, 1979, c1978.
142 p. : ill. ; 27 cm.
Includes index.
ISBN 0-312-29305-4 : $8.95
1. Children—Management. 2. Infants—Care and hygiene. I. Goldschmied, Elinor, joint author. II. Title.
HQ772.T67 1979 649'.1 78-64913
79 MARC

CHILDREN
—MANAGEMENT (Continued)

Trofimov, I͡Uriĭ Fedorovich.
(Rabota s sem'i͡ami trudi͡ashchikhsi͡a)
Работа с семьями трудящихся : Воспитание детей и подростков / Ю. Ф. Трофимов. — Москва : Профиздат, 1978.
95 p., [4] leaves of plates : ill. ; 20 cm. USSR 78
0.30rub
1. Vocational guidance—Russia. 2. Children—Management. 3. Communist ethics. 4. Youth—Employment—Russia. I. Title.
HF5382.5.R8T83 78-421007

Weinheimer, Janice Madsen, 1937-
Families are forever ... if I can just get through today! / Janice Madsen Weinheimer. — Salt Lake City : Deseret Book Co., 1979.
141 p. ; 24 cm.
Includes index.
ISBN 0-87747-782-5 : $6.95
1. Weinheimer, Janice Madsen, 1937- . 2. Mothers—United States—Biography. 3. Children—Management. I. Title.
HQ770.W44A34 301.42'7 79-17807
79 MARC

Weisberger, Eleanor.
Your young child and you : how to manage growing-up problems in the years from one to five / Eleanor Weisberger ; foreword by Benjamin M. Spock. — New, rev. ed. — New York : Dutton, c1979.
xvii, 172 p. ; 21 cm.
"A Sunrise book."
Includes index.
ISBN 0-87690-329-4 : $4.95 ($6.50 Can)
1. Children—Management. 2. Child psychology. I. Title.
HQ769.W43 1979 649'.123 78-22623
79 MARC

Weiss, Lynn.
Emotionally yours : shaping your child's life / written by Lynn Weiss ; illustrated by Jerry Weiss. — Dallas : I.T. Productions, c1979.
168 p. : ill. ; 22 cm.
$5.95
1. Child development. 2. Children—Management. I. Weiss, Jerry. II. Title.
HQ772.W363 649'.1 79-84406
79 MARC

Wilt, Joy.
Taming the big bad wolves : how to take the huff and puff out of twelve parenting problems / Joy Wilt. — Waco, Tex. : Word Books, c1979.
209 p. ; 23 cm.
Includes bibliographies.
ISBN 0-8499-0063-8 : $7.95
1. Children—Management. 2. Parent and child. I. Title.
HQ772.W52 649'.122 78-65801
79 MARC

Wood, Samellyn J
Parenting : four patterns of child-rearing / Samellyn J. Wood, Roger S. Bishop, Davene Cohen. — New York : Hart Pub. Co., c1978.
333 p. ; 22 cm.
Bibliography: p. 320-328.
Includes index.
ISBN 0-8055-0362-5 : $10.00
1. Parenting. 2. Children—Management. I. Bishop, Roger S., joint author. II. Cohen, Davene, joint author. III. Title.
HQ755.8.W66 649'.1 77-75707
00 MARC

—MANAGEMENT—ADDRESSES, ESSAYS, LECTURES

Stephens, William N
Our children should be working / by William N. Stephens. — Springfield, Ill. : Thomas, c1979.
ix, 215 p. ; 24 cm.
Includes bibliographical references and index.
ISBN 0-398-03851-1
1. Children—Management—Addresses, essays, lectures. 2. Helping behavior—Case studies. 3. Apprentices—Case studies. 4. Work—Case studies. 5. Child development—Addresses, essays, lectures. I. Title.
HQ772.S69 649'.1 78-11606
78 MARC

—MANAGEMENT—ANECDOTES, FACETIAE, SATIRE, ETC.

Atkinson, Butler M
What Dr. Spock didn't tell us : or, A survival kit for parents : an encyclopedic guide to hitherto uncatalogued afflictions, aberrations, exotic diseases of the American child / by Buddy Atkinson ; illustrated by Whitney Darrow, Jr. — New York : Simon and Schuster, [1979] c1959.
72 p. : ill. ; 24 cm. — (A Fireside book)
Bibliography: p. 69.
Includes index.
ISBN 0-671-24858-8 pbk.
1. Children—Management—Anecdotes, facetiae, satire, etc. I. Title.
PN6231.C3A8 1979 818'.5'407 78-31156
78 MARC

Mayle, Peter.
Baby taming / by Peter Mayle ; illustrated by Arthur Robins. — New York : Harmony Books, c1978.
[49] p. : col. ill. ; 27 cm.
ISBN 0-517-53534-3
1. Children—Management. 2. Children—Management—Anecdotes, facetiae, satire, etc. I. Title.
HQ769.M377 1978 649'.1 78-15532
78 MARC

—MANAGEMENT—CONGRESSES

Raising children in modern America : problems and prospective solutions / Nathan B. Talbot, editor. — 1st ed. — Boston : Little, Brown, c1976.
xiii, 590 p. : ill. ; 24 cm.
Bibliography: p. [497]-499.
Includes index.
ISBN 0-316-83135-2
1. Children—Management—Congresses. 2. Children in the United States—Congresses. I. Talbot, Nathan Bill, 1909- .
HQ769.R17 649'.1'0973 75-19236
75[r79]rev MARC

—MANAGEMENT—HANDBOOKS, MANUALS, ETC.

Kids, day in and day out : a compendium of ideas, recommendations, insights, inspirations, facts, and suggestions, problems and solutions for living with kids every single day / editor, Elisabeth L. Scharlatt ; project director, Christopher Cerf ; research director, Ronnie Shushan ; designer, Beri Greenwald. — New York : Simon and Schuster, c1979.
513 p. : ill. ; 28 cm.
"A Fireside book."
Includes index.
ISBN 0-671-24206-7 : 17.95
1. Children—Management—Handbooks, manuals, etc. I. Scharlatt, Elisabeth Lohman. II. Cerf, Christopher.
HQ772.K46 649'.1 78-11035
78 MARC

Saitama, Japan. Kyōikukyoku. Shakai Kyōikuka.
(Nobiyo Saitamakko)
伸びよ埼玉っ子 / [編集者 埼玉県教育局社会教育課]. — [浦和] : 埼玉県教育委員会, 昭和52 [1977]
102 p. : ill. ; 21 cm. Ja***
1. Domestic education—Japan—Saitama—Handbooks, manuals, etc. 2. Children—Management—Handbooks, manuals, etc. I. Title.
LC37.S25 1977 79-800971

Shinagawa, Takako, 1922-
(Kodomo no shinri shitsuke jiten)
子どもの心理しつけ事典 / 品川孝子著. — 東京 : あすなろ書房, 1977.
471 p. ; 21 cm. Ja 77B-5563
Colophon inserted.
¥2500
1. Child psychology—Handbooks, manuals, etc. 2. Children—Management—Handbooks, manuals, etc. I. Title.
LB1115.S58 78-813366

Stein, Sara Bonnett.
Learn at home the Sesame Street way / by Sara Bonnett Stein in cooperation with children's television workshop. — New York : Simon and Schuster, c1979.
286 p. : ill. ; 26 cm.
Cover title: The Sesame Street way.
Includes index.
ISBN 0-671-22482-4
1. Domestic education—Handbooks, manuals, etc. 2. Children—Management—Handbooks, manuals, etc. 3. Family life education—Handbooks, manuals, etc. 4. Sesame Street. 5. Television in preschool education—Handbooks, manuals, etc. I. Children's Television Workshop. II. Title.
LC37.S73 372.1'3 79-1343
79 MARC

—MANAGEMENT—PROBLEMS, EXERCISES, ETC.

Tuttle, Kenneth C
Better parents, better children : a workbook for a more successful family in 21 days / by Bill C. Marshall and Christina Mae Marshall. — Maplewood, N.J. : Hammond Inc., c1979.
x, 150 p. ; 28 cm.
ISBN 0-8437-3140-0
1. Children—Management—Problems, exercises, etc. I. Marshall, Christina Mae, joint author. II. Title.
HQ769.M298 649'.1 78-17691
78 MARC

—MANAGEMENT—STUDY AND TEACHING

Stewart, Jack C
Counseling parents of exceptional children / Jack C. Stewart. — Columbus, Ohio : Merrill, c1978.
iv, 152 p. ; 23 cm.
Includes bibliographies and indexes.
ISBN 0-675-08386-9
1. Exceptional children. 2. Parent and child. 3. Counseling. 4. Children—Management—Study and teaching. I. Title.
HQ773.5.S73 1978 371.9 77-95122
78 MARC

—MEDICAL CARE

see also Child health serivces

—MEDICAL EXAMINATIONS

see also Child development--Testing

Alexander, Mary Merkel.
Pediatric history taking and physical diagnosis for nurses / Mary M. Alexander, Marie Scott Brown. — 2d ed. — New York : McGraw-Hill, c1979.
viii, 529 p. : ill. ; 23 cm.
First ed. published in 1974 under title: Pediatric physical diagnosis for nurses.
Bibliography: p. 505-508.
Includes index.
ISBN 0-07-001019-6. ISBN 0-07-001018-8 pbk.
1. Children—Medical examinations. 2. Physical diagnosis. 3. Medical history taking. 4. Pediatric nursing. I. Brown, Marie Scott, joint author. II. Title.
[DNLM: 1. Pediatrics—Nursing texts. 2. Physical examination—In infancy and childhood—Nursing texts. 3. Medical records—Nursing texts. WY159 A377p]
RJ50.A4 1979 618.9'2007'54 78-9679
78 MARC

Bernbeck, Rupprecht, 1916-
Vorsorgeuntersuchungen des Bewegungsapparates im Kindesalter : orthopäd. u. neuromotor. Diagnostik / Rupprecht Bernbeck u. Alexander Sinios. — München ; Berlin ; Wien : Urban und Schwarzenberg, 1975.
viii, 120 p. : 86 ill. ; 18 x 26 cm. GFR76-A
Errata slip inserted.
Bibliography: p. 120.
ISBN 3-541-07041-2 : DM28.00
1. Musculoskeletal system—Examination. 2. Children—Medical examinations. 3. Pediatric orthopedia. 4. Neuromuscular diseases in children—Diagnosis. I. Sinios, Alexander, joint author. II. Title.
[DNLM: 1. Bone diseases—In infancy & childhood. 2. Joint diseases—In infancy & childhood. 3. Muscular diseases—In infancy & childhood. 4. Orthopedics—In infancy & childhood. 5. Physical examination—In infancy & childhood. WS270 B524v 1975]
[RJ480.B47] 618.9'27'075 76-675151
Shared Cataloging with DNLM 76[r79]rev MARC

Kawahata, Aiyoshi, 1905-
(Shintai kensa to tairyoku sokutei)
身体検査と体力測定 / 川畑愛義著. — 東京 : 幽書房, 1954.
260, 11 p. : ill. ; 22 cm.
Colophon title: Tairyoku sokutei.
Bibliography: p. 257-260.
1. Physical fitness—Testing. 2. Anthropometry. 3. Children—Medical examinations. I. Title. II. Title: Tairyoku sokutei.
GV436.K35 79-808313

Peteanu, Mihai.
Examinarea medicală în scop de selecţie şi orientare şcolară şi profesională. Ghid. Bucureşti, Editura didactică şi pedagogică, 1971.
132 p. with figs. 20 cm. lei3.15 R71-3410
At head of title: Institutul de Ştiinţe Pedagogice. Filiala Cluj.
Includes bibliographical references.
1. Physical diagnosis. 2. Children—Medical examinations. 3. Vocational guidance—Romania. I. Title.
RC76.P47 79-345301
79 MARC

—MEDICAL EXAMINATIONS
—ILLINOIS—STATISTICS
—PERIODICALS

Illinois. Dept. of Public Health.
Medichek: annual report.
[Springfield] Dept. of Public Health.
v. 28 cm.
Key title: Medichek. Annual report, ISSN 0193-9401.
1. Early and Periodic Screening, Diagnosis, and Treatment Program—Illinois—Statistics—Periodicals. 2. Children—Medical examinations—Illinois—Statistics—Periodicals. 3. Medical screening—Illinois—Statistics—Periodicals. I. Title.
RJ102.5.I 3 I 44a 362.1 79-643708
MARC-S

CHILDREN (Continued)

—MEDICAL EXAMINATIONS—MASSACHUSETTS

University of Massachusetts (System). Institute for Governmental Services.
Report of the Massachusetts health policy study. — [Boston] : Institute for Governmental Services, University of Massachusetts, [1978?]
148 leaves ; 28 cm.
Cover title: Health policy study.
1. Medical policy—Massachusetts. 2. Medical care, Cost of—Massachusetts. 3. Children—Medical examinations—Massachusetts. 4. Child health services—Massachusetts. I. Title. II. Title: Health policy study.
RA395.A4M48 1978 362.1'09744 79-621111
79 MARC

—MEDICAL EXAMINATIONS—UNITED STATES—STATISTICS

Abraham, Sidney.
Total serum cholesterol levels of children, 4-17 years, United States, 1971-1974 / [Sidney Abraham, Clifford L. Johnson, and Margaret D. Carroll.] — Hyattsville, Md. : U.S. Dept. of Health, Education, and Welfare, Public Health Service, Health Resources Administration, National Center for Health Statistics, 1978.
iii, 38 p. : graphs ; 27 cm. — (Vital and health statistics : Series 11, Data from the National Health Survey ; no. 207) (DHEW publication ; (PHS) 78-1655)
Includes bibliographical references.
ISBN 0-8406-0125-5
1. Blood cholesterol—United States—Statistics. 2. Children—Medical examinations—United States—Statistics. 3. Health surveys—United States. I. Johnson, Clifford L., joint author. II. Carroll, Margaret D., joint author. III. Title. IV. Series: United States. National Center for Health Statistics. Vital and health statistics : Series 11, Data from the National Health Survey, Data from the health examination survey ; no. 207. V. Series: United States. Dept. of Health, Education, and Welfare. DHEW publication ; (PHS) 78-1655.
RA407.3.A347 no. 207 312'.0973 s 78-2963
[RJ423] 78 MARC

—MENTAL DISORDERS see Child psychopathology

—MORTALITY

see also Infants—Mortality

Ben-Porath, Yoram.
Fertility and child mortality : issues in the demographic transition of a migrant population : a paper presented at the Universities-National Bureau Conference on Population and Economic Change in Less Developed Countries, Philadelphia, September 1976 / Yoram Ben-Porath. — Rev. — [Santa Monica, Calif. : Rand Corp.], 1976.
72 p. ; 28 cm. — (The Rand paper series ; P-5772)
Bibliography: p. 49-51.
1. Israel—Population. 2. Fertility, Human—Israel. 3. Children—Mortality. I. Title. II. Series: Rand Corporation. Paper ; P-5772.
AS36.R28 no. 5772 081 s 79-304315
[HB3634.A3] 79 MARC

Knodel, John E
Infant and child mortality in Thailand : levels, trends, and differentials as derived through indirect estimation techniques / John Knodel and Apichat Chamratrithirong. — Honolulu : East-West Center, 1978.
vii, 40 p. ; 23 cm. — (Papers of the East-West Population Institute ; no. 57)
Bibliography: p. 39-40.
1. Children—Mortality. 2. Infants—Mortality. 3. Mortality—Thailand. 4. Thailand—Statistics, Vital. I. Apichat Chamratrithirong, 1948- joint author. II. Title. III. Series: East-West Population Institute. Papers of the East-West Population Institute ; no. 57.
HB1323.C5K58 301.32'2 78-26004
78 MARC

McDonald, Peter F
Levels and trends in fertility and childhood mortality in Indonesia / by Peter F. McDonald, Mohammad Yasin, Gavin W. Jones. — Jakarta : Lembaga Demografi, Fakultas Ekonomi, Universitas Indonesia, 1976.
ix, 79 p. : graphs ; 28 cm. — (Indonesian fertility-mortality survey, 1973 ; monograph no. 1)
"LD-FEUI/FM-73/SMP, no. 1/76."
Includes bibliographical references.
1. Fertility, Human—Indonesia. 2. Children—Mortality. 3. Indonesia—Statistics, Vital. I. Yasin, Mohammad, joint author. II. Jones, Gavin W., joint author. III. Series: Survey fertilitas mortalitas Indonesia ; seri monografi no. 1.
HB1057.M32 301.32'1'09598 77-940115
77[r79]rev MARC

Surjit Kaur, 1935-
Wastage of children / Surjit Kaur. — New Delhi : Sterling, c1978.
282 p. : ill. ; 22 cm.
"A Family Planning Foundation book."
Includes bibliographical references and index.
Rs60.00
1. Children—Mortality. 2. Infants—Mortality. 3. Mortality—India. 4. Abortion—India. I. Title.
HB1323.C5S82 301.32'2 78-903795
79 MARC

—NUTRITION

see also Children—Diseases—Nutritional aspects; Infants—Nutrition; Malnutrition in children; Parenteral feeding of children; School children—Food

Children's Foundation.
Working for women and children—the WIC organizing guide / prepared by the WIC advocacy staff, the Children's Foundation. — Washington : The Founation, 1979.
iv, 65 p. : forms ; 28 cm.
$2.50
1. Food relief—United States. 2. Mothers—Nutrition. 3. Children—Nutrition. I. Title.
HV696.F6C49 1979 362.8'2 79-119483
79 MARC

Developmental nutrition / editors, Karen Oliver ... [et al.]. — Columbus, Ohio : Ross Laboratories, c1979.
vi, 105 p. : ill. ; 26 cm.
Includes bibliographical references.
1. Infants—Nutrition. 2. Children—Nutrition. 3. Nutrition. 4. Human growth. I. Oliver, Karen. II. Ross Laboratories, Columbus, Ohio.
RJ216.D48 612'.3 79-65996
79 MARC

Hatfield, Antoinette Kuzmanich.
How to help your child eat right! : A fun cookbook and guide to better nutrition / Antoinette Kuzmanich Hatfield, Peggy Smeeton Stanton. — Rev. and updated. — Washington : Acropolis Books, 1978, c1973.
168 p. : ill. ; 23 cm.
Bibliography: p. 141-145.
Includes indexes.
ISBN 0-87491-252-0 : $8.95. ISBN 0-87491-253-9 pbk. : $4.95
1. Cookery. 2. Children—Nutrition. I. Stanton, Peggy Smeeton, joint author. II. Title.
TX652.H374 641.5'63 78-6300
78 MARC

Hutchins, Bobbie.
Child nutrition and health / Bobbie Hutchins. — New York : Gregg Division, McGraw-Hill, c1979.
234 p. : ill. ; 24 cm. — (Careers in home economics)
Includes bibliographical references and index.
ISBN 0-07-031527-2
1. Children—Nutrition. 2. Children—Care and hygiene. I. Title.
RJ206.H87 612'.3 78-21948
78 MARC

Lantum, Daniel.
Report of the nutrition survey of Pre-school age children of Jakiri and Mbandjock Districts of Cameroon conducted in May 1977 / by Dan. N. Lantum. — [Yaoundé] : Public Health Unit, U.C.H.S., University of Yaoundé, 1978.
vi, 56 leaves : ill. ; 29 cm.
Includes bibliographical references.
1. Nutrition surveys—Cameroon. 2. Children—Nutrition. I. University of Yaoundé. Public Health Unit. II. Title.
TX360.C25L35 362.7'8'6711 78-110217
79 MARC

Lorin, Martin I
The parents' book of physical fitness for children / Martin I. Lorin ; drawings by Nadira Lorin. — 1st ed. — New York : Atheneum, 1978.
viii, 278 p. : ill. ; 22 cm.
Bibliography: p. 269-270.
Includes index.
ISBN 0-689-10922-9 : $9.95
1. Physical fitness for children. 2. Exercise for children. 3. Children—Nutrition. I. Lorin, Nadira. II. Title.
[DNLM: 1. Physical fitness—Popular works. 2. Child care—Popular works. QT255.3 L872p]
RJ61.L86 1978 613.7'04'2 78-3151
78 MARC

(Metodicheskie materialy po voprosam vskarmlivanii͡a i pitanii͡a deteĭ)
Методические материалы по вопросам вскармливания и питания детей. — Москва : Ин-т педиатрии СССР, 1966.
44 p.
At head of title: Ministerstvo zdravookhranenii͡a SSSR. Institut pediatrii AMN SSSR.
L. C. copy on microfilm only.
——— Microfilm. On 1 reel with other items ; 35 mm.
Microfilm Slavic 12594 RJ
1. Children—Nutrition. I. Akademii͡a medit͡sinskikh nauk SSSR, Moscow. Institut pediatrii.
[RJ206.M68] 78-423442

Nutrition assessment and counseling : pregnancy through adolescence / Jill S. Slattery, Gayle Angus Pearson, Carolyn Talley Torre. — New York : Appleton-Century-Crofts, c1979.
xv, 320 p. : ill. ; 24 cm.
Includes bibliographies and index.
ISBN 0-8385-6167-5
1. Children—Nutrition. 2. Youth—Nutrition. 3. Nutrition disorders in children. 4. Pregnancy—Nutritional aspects. I. Slattery, Jill S., 1943- . II. Pearson, Gayle Angus, 1938- . III. Torre, Carolyn Talley, 1946-
RJ206.M32 613.2 79-4457
79 MARC

Nutrition, pre- and postnatal development / edited by Myron Winick. — New York : Plenum Press, c1979.
xx, 496 p. : ill. ; 26 cm. — (Human nutrition ; v. 1)
Includes bibliographies and index.
ISBN 0-306-40132-0
1. Children—Nutrition. 2. Infants—Nutrition. 3. Malnutrition in children—Complications and sequelae. 4. Pregnancy—Nutritional aspects. I. Winick, Myron. II. Series.
[DNLM: 1. Child nutrition. 2. Infant nutrition. 3. Maternal-fetal exchange. 4. Fetus—Growth and development. 5. Growth. QU145.3 H9183 v. 1]
QP141.H78 vol. 1 612'.3'08 s 78-26941
[RJ206] 78 MARC

Nutritional needs of the school age child : elementary eating. — [Winnipeg] : Home Economic Directorate, Manitoba Dept. of Health and Social Development, 1975.
[7] p. : ill. ; 28 cm. C***
Cover title.
1. Children—Nutrition. I. Manitoba. Home Economics Directorate.
RJ206.N82 641.1 78-326239
79 MARC

Powers, Hugh.
Food power : nutrition and your child's behavior / Hugh Powers, James Presley. — New York : St. Martin's Press, c1978.
230 p. : graphs ; 22 cm.
Bibliography: p. 198-208.
ISBN 0-312-29776-9
1. Children—Nutrition. 2. Children—Nutrition—Psychological aspects. 3. Emotional problems of children—Nutritional aspects. I. Presley, James, joint author. II. Title.
RJ206.P68 618.9'28'90654 77-15919
78 MARC

Preschool nutrition education monograph. — Berkeley, CA : National Nutrition Education Clearing House, Society for Nutrition Education, 1978.
48 p. : ill. ; 28 cm. — (NNECH monograph ; no. 3)
Includes bibliographical references.
$4.50
1. Nutrition—Study and teaching (Preschool) 2. Nutrition—Study and teaching (Preschool)—Bibliography. 3. Children—Nutrition. 4. Children—Nutrition—Bibliography. I. Series: National Nutrition Education Clearing House. NNECH monograph ; no. 3.
TX364.P73 372.3'7 79-102729
79 MARC

Smith, Lendon H 1921-
Feed your kids right : Dr. Smith's program for your child's total health / Lendon Smith. — New York : McGraw Hill, c1979.
ix, 251 p. ; 24 cm.
Bibliography: p. 243-245.
Includes index.
ISBN 0-07-058496-6 : $9.95
1. Children—Nutrition. 2. Children—Diseases—Nutritional aspects. I. Title.
RJ206.S62 613.2 78-23733
78 MARC

Thurston, Emory W
The parents' guide to better nutrition for tots to teens (and others!) / Emory W. Thurston. — Rev. and updated. — New Canaan, Conn. : Keats Pub., c1979.
xvi, 162 p. ; 21 cm.
ISBN 0-87983-194-4 : $3.95
1. Children—Nutrition. 2. Nutrition. I. Title.
RJ206.T49 1979 649'.3 76-10041
00 MARC

United States. Federal Trade Commission.
FTC staff report on television advertising to children. — [Washington : Federal Trade Commission] : for sale by the Supt. of Docs., U.S. Govt. Print. Off., 1978
394 p. in various pagings ; 28 cm.
Cover title.
Includes bibliographical references.
1. Television advertising and children—United States. 2. Advertising—Psychological aspects. 3. Youth as consumers—United States. 4. Children—Nutrition. I. Title.
HF6146.T42U55 1978 659.14'3 79-601465
79 MARC

—NUTRITION—ADDRESSES, ESSAYS, LECTURES

Food and nutrition policy in a changing world / edited by Jean Mayer and Johanna T. Dwyer, with Kathryn Dowd and Laura Mayer. — New York : Oxford University Press, 1979.
xii, 300 ; 24 cm.
Includes bibliographical references and index.
ISBN 0-19-502363-3 : $17.95. ISBN 0-19-502364-1 pbk. : $10.95
1. Underdeveloped areas—Nutrition policy—Addresses, essays, lectures. 2. Malnutrition in children—Addresses, essays, lectures. 3. Children—Nutrition—Addresses, essays, lectures. I. Mayer, Jean. II. Dwyer, Johanna T.
[DNLM: 1. Developing countries. 2. Child nutrition. 3. Food. 4. Nutrition. 5. Nutrition disorders. QU145.3 F686]
TX359.F65 362.5 76-42627
78 MARC

CHILDREN (Continued)

—NUTRITION—BIBLIOGRAPHY

Preschool nutrition education monograph. — Berkeley, CA : National Nutrition Education Clearing House, Society for Nutrition Education, 1978.
48 p. : ill. ; 28 cm. — (NNECH monograph ; no. 3)
Includes bibliographical references.
$4.50
1. Nutrition—Study and teaching (Preschool) 2. Nutrition—Study and teaching (Preschool)—Bibliography. 3. Children—Nutrition. 4. Children—Nutrition—Bibliography. I. Series: National Nutrition Education Clearing House. NNECH monograph ; no. 3.
TX364.P73 372.3'7 79-102729
79 MARC

—NUTRITION—LAW AND LEGISLATION—UNITED STATES

United States. Congress. House. Committee on Education and Labor. Subcommittee on Elementary, Secondary, and Vocational Education.
The Child nutrition amendments of 1978 : hearing before the Subcommittee on Elementary, Secondary, and Vocational Education of the Committee on Education and Labor, House of Representatives, Ninety-fifth Congress, second session, on H.R. 12343 and H.R. 11699 ... April 27, 1978. — Washington : U.S. Govt. Off., 1978.
iii, 360 p. : ill. ; 24 cm.
1. Children—Nutrition—Law and legislation—United States. I. Title.
KF27.E3364 1978a 344'.73'0794 78-603164
79 MARC

—NUTRITION—PSYCHOLOGICAL ASPECTS

Powers, Hugh.
Food power : nutrition and your child's behavior / Hugh Powers, James Presley. — New York : St. Martin's Press, c1978.
230 p. : graphs ; 22 cm.
Bibliography: p. 198-208.
ISBN 0-312-29776-9
1. Children—Nutrition. 2. Children—Nutrition—Psychological aspects. 3. Emotional problems of children—Nutritional aspects. I. Presley, James, joint author. II. Title.
RJ206.P68 618.9'28'90654 77-15919
78 MARC

—NUTRITION—UNDERDEVELOPED AREAS see Underdeveloped areas --Children--Nutrition

—PARAPHERNALIA see Children's paraphernalia

—PHOTOGRAPHY see Photography of children

—POETRY

Carlsson, Lennart, 1953-
I närheten : dikter / av Lennart Carlsson. — Stockholm : LT, 1978.
62, [1] p. ; 23 cm. S78-5
ISBN 9136011363 : kr31.00
1. Children—Poetry. I. Title.
PT9876.13.A734I2 78-384543
79 MARC

L'Enfant, précédé de Enfance et poésie / de Maurice Carême, et Ce qui se passait aux Feuillantines vers 1813 / de Victor Hugo ; avec la collaboration de 285 poètes et 28 illustrateurs. — Paris : J. Grassin, 1977.
406 p. : ill. ; 24 cm. — (L'Encyclopédie poétique ; 10) F***
ISBN 2-7055-0941-0 : 120F
1. French poetry—20th century. 2. Children—Poetry. I. Series.
PQ1184.E5 841'.9'108 78-373530
78 MARC

Reich, Hanns.
Die Mutter und ihr Kind / Hanns Reich. — 9. Aufl. — Luzern : Reich, 1975.
64 p. : chiefly ill. ; 24 cm. — (Ein Terra magica Bildband) Sw***
ISBN 3-7243-0002-6
1. Mothers—Portraits. 2. Children—Portraits. 3. Mothers—Poetry. 4. Children—Poetry. I. Title.
TR681.M67R44 1975 779'.24'0924 79-353605
79 MARC

Ünnepi köszöntők anyák napjára és a nemzetközi gyermeknapra / [vál. és az előszót írta Mezei András ; az illusztrációkat készítette Würtz Ádám]. — Budapest : Magyar Vöröskereszt, 1977.
83 p. : col. ill. ; 19 cm. H77
ISBN 9637400494 : 8.00Ft
1. Mothers—Poetry. 2. Hungarian poetry—20th century. 3. Children—Poetry. I. Mezei, András.
PH3164.M6U46 78-395978
•79 MARC

—POETRY—20TH CENTURY

Sal de la tierra : lo esencial es invisible a los ojos, sólo se conoce con el corazón / Patricio Navarro Silva, Benito Baranda Ferrán. — [Santiago, Chile] : Editora Gabriela Mistral, [1977 or 1978]
48 p. ; 20 cm.
CONTENTS: Navarro, P. El infierno es la soledad lugubre de sí mismo.—Baranda, B. Pobre de mí si no predico este mensaje!
1. Children poetry—20th century. I. Navarro Silva, Patricio. El infierno es la soledad lugubre de sí mismo. II. Baranda Ferrán, Benito, 1959- Pobre de mí si no predico este mensaje!
PQ8058.S2 861 78-110135
79 MARC

—PORTRAITS

Baeza, Pepe.
Nens, no ninots / [Pepe Baeza, Joan P. Pastor, Sigfrid Casals]. — [s.l.] : Col·lectiu Document, [1978?]
[72] p. : all ill. ; 21 cm. Sp***
Cover title.
325ptas
1. Photography of children. 2. Children—Portraits. I. Pastor, Joan P., joint author. II. Casals, Sigfrid, joint author. III. Title.
TR681.C5B33 779'.25'0922 79-337032
79 MARC

Children of the past in photographic portraits : an album with 165 prints / edited by Alison Mager. — New York : Dover Publications, 1978.
[94] p. : chiefly ill. ; 24 cm.
ISBN 0-486-23697-8 : $4.00
1. Photography of children. 2. Children—Portraits. I. Mager, Alison.
TR681.C5C48 1978 779'.25 78-55864
79 MARC

Cohen, Morton Norton, 1921-
Lewis Carroll, photographer of children : four nude studies / by Morton N. Cohen. — New York : C. N. Potter : distributed by Crown Publishers, c1979.
31 p., 4 leaves of plates : ill. ; 26 cm.
Includes bibliographical references.
ISBN 0-517-53744-3
1. Children—Portraits. 2. Photography of the nude. 3. Dodgson, Charles Lutwidge, 1832-1898. I. Title.
TR681.C5C63 1979 770'.92'4 78-27872
78 MARC

Plunket, Jean Reasoner.
Faces that won't sit still : celebrated subjects by a prominent portrait artist and how they were captured / by Jean Reasoner Plunket, with Barbara Brandt Ward. — Washington : Acropolis Books, c1978.
201 p. : ill. ; 29 cm.
ISBN 0-87491-261-X : $18.00. ISBN 0-87591-262-8 pbk. : $9.95. ISBN 0-87491-260-1 lim. deluxe ed. : $35.00
1. Plunket, Jean Reasoner. 2. Portrait painters—United States—Biography. 3. Children—Portraits. 4. Portrait drawing—Technique. I. Ward, Barbara Brandt, joint author. II. Title.
NC139.P57A2 1978 741'.092'4 78-14259
78 MARC

Reich, Hanns.
Die Mutter und ihr Kind / Hanns Reich. — 9. Aufl. — Luzern : Reich, 1975.
64 p. : chiefly ill. ; 24 cm. — (Ein Terra magica Bildband) Sw***
ISBN 3-7243-0002-6
1. Mothers—Portraits. 2. Children—Portraits. 3. Mothers—Poetry. 4. Children—Poetry. I. Title.
TR681.M67R44 1975 779'.24'0924 79-353605
79 MARC

Schönberger, Käthe.
Dad Kinderporträt in der Plastik / Käthe Schönberger ; [Fotos von Herbert Schönberger und Beatrix und Hans Böck]. — Erlenbach-Zürich : E. Rentsch, cop. 1978.
[28] leaves of plates : chiefly ill. (all col.) ; 23 cm. Sw78-12677
ISBN 3-7249-0508-4 : 25.00F
1. Schönberger, Käthe. 2. Children—Portraits. 3. Children in art. I. Title.
NB853.S36A4 1978 730'.92'4 79-363992
79 MARC

... to be a child : a collection of photographic essays to commemorate the International Year of the Child on behalf of the International School Psychology Committee / edited by Peter R. Burzynski. — Columbus, Ohio : International School Psychology Committee, c1979.
65 p. : chiefly ill. ; 31 cm.
ISBN 0-917668-03-0
1. Photography of children. 2. Children—Portraits. I. Burzynski, Peter R. II. International School Psychology Committee.
TR681.C5T6 779'.25'0922 79-64136
79 MARC

—PORTRAITS—EXHIBITIONS

Wróblewska, Kamila.
Dawny portret dziecka ze zbiorów polskich : katalog wystawy Olsztyn, maj-październik 1976 rok / [oprac. katalogu Kamila Wróblewska]. — Olsztyn : Muzeum Warmii i Mazur, 1976 (Olszt : OZGSP)
95, [1] p. : ill., ports. ; 25 cm. P76-8695
Includes indexes.
At head of title: Muzeum Warmii i Mazur w Olsztynie.
zł90.00
1. Children—Portraits—Exhibitions. 2. Portrait painting—17th century—Exhibitions. 3. Portrait painting—18th century—Exhibitions. 4. Portrait painting—Poland—Exhibitions. I. Title.
ND1329.3.C45W76 78-383574
78 MARC

—PRAYER-BOOKS AND DEVOTIONS

Tuckett, Gain Ream.
Day's end : bedtime prayers for young children / Guin Ream Tuckett ; illustrated by Joni Fredman. — St. Louis, Mo. : Bethany Press, c1979.
64 p. : col. ill. ; 16 x 18 cm.
SUMMARY: A collection of bedtime prayers for pre-schoolers.
ISBN 0-8272-0611-9 : $4.95
1. Children—Prayer-books and devotions—English. [1. Prayers] I. Fredman, Joni. II. Title.
BV4870.T82 242'.82 78-11565
78 MARC

—PRAYER-BOOKS AND DEVOTIONS—ENGLISH

Coleman, William L
Listen to the animals : devotionals for families with young children / William L. Coleman. — Minneapolis : Bethany Fellowship, inc., c1979.
128 p. : ill. ; 22 cm.
ISBN 0-87123-341-X : $3.50
1. Children—Prayer-books and devotions—English. I. Title.
BV4870.C633 242'.6'2 79-11312
79 MARC

Gospel devotions for children, ages 8-12. — Milwaukee : Northwestern Pub. House, 1978.
195 p. ; 23 cm.
Includes indexes.
1. Children—Prayer-books and devotions—English. I. Northwestern Publishing House, Milwaukee.
BV4870.G7 242'.6'2 78-54769
78 MARC

—PRAYER-BOOKS AND DEVOTIONS—FINNISH

Topi, Päivi.
Taivaan isän heinäsirkat / Päivi Topi ; päällys ja kuvitus, Kaarina Ahonen. — [Helsinki] : SLEY, 1975.
80 p. : ill. ; 20 cm. Fi
ISBN 9516173438
1. Children—Prayer-books and devotions—Finnish. I. Title.
BV4870.T66 78-396693
•79 MARC

—PREPARATION FOR MEDICAL CARE—JUVENILE LITERATURE—BIBLIOGRAPHY

Altshuler, Anne.
Books that help children deal with a hospital experience. — Rev. — Rockville, Md. : U.S. Dept. of Health, Education, and Welfare, Public Health Service, Health Services Administration, Bureau of Community Health Services ; Washington : for sale by the Supt. of Docs, U.S. Govt. Print. Off., 1978.
ii, 22 p. ; 15 x 23 cm. — (DHEW publication ; no. (HSA) 78-5224)
Cover title.
Includes indexes.
1. Hospitals—Juvenile literature—Bibliography. 2. Children—Hospital care—Juvenile literature—Bibliography. 3. Children—Preparation for medical care—Juvenile literature—Bibliography. I. United States. Health Services Administration. Bureau of Community Health Services. II. Title. III. Series: United States. Dept. of Health, Education, and Welfare. DHEW publication ; no. (HSA) 78-5224.
Z6675.H75A47 1978 016.3627'8'11 79-602662
[RA963.5] 79 MARC

—PROTECTION see Child welfare

—PSYCHOLOGICAL TESTING see Psychological tests for children

CHILDREN (Continued)

—PSYCHOLOGY see Child psychology

—RECREATION see Amusements; Creative activities and seat work; Games; Outdoor recreation for children

—RELIGIOUS LIFE

Benson, Dennis C
The ministry of the child / Dennis C. Benson & Stanley J. Stewart. — Nashville : Abingdon, c1979.
142 p. ; 19 cm.
ISBN 0-687-27039-1
1. Children—Religious life. I. Stewart, Stanley J., 1937- joint author. II. Title.
BV4571.2.B46 248'.82 78-12003
78 MARC

Cavalletti, Sofia.
Il potenziale religioso del bambino : descrizione di un'esperienza con bambini da 3 a 6 anni / Sofia Cavalletti. — Roma : Città nuova, 1979.
195 p., [32] leaves of plates : ill. ; 20 cm. It79-Apr
Includes bibliographical references.
L6500
1. Children—Religious life. I. Title.
BV4574.C38 79-376483
*79 MARC

Cully, Iris V
Christian child development / Iris V. Cully. — 1st ed. — San Francisco : Harper & Row, c1979.
x, 166 p. ; 21 cm.
Includes bibliographical references and index.
ISBN 0-06-061648-2
1. Christian education of children. 2. Children—Religious life. 3. Child development. I. Title.
BV1475.2.C814 1979 268'.432 78-19507
78 MARC

[**Elten, Gerhard von**] d. ca. 1484.
Casus utrum liceat pueros ante pubertatis annos ad religionis ingressum inducere. [Cologne, Arnold Ther Hoernen, 1480?]
[12] l. 4°. 21.5 cm.
Caption on leaf [1a]: **Casus per modum questionis proponitur, t est talis.** Text begins: Vtrum liceat pueros.
Attributed to Gerhard von Elten in Neue deut. Biog.
Text signed: Loppo zierrixee humilis seruitor vester.
Hain. Repertorium (with Copinger's Supplement) 7614; Brit. Mus. Cat. (XV cent.) I, p. 208 (IA. 3235); Goff. Third census, G-164.
Rubricated.
1. Vows. 2. Monastic and religious life. 3. Children—Religious life. I. Loppo de Zieriksee, d. 1479. II. Title.
Incun. X.E46 79-414119

Gestwicki, Ronald.
Santa Claus, the Tooth Fairy, and other stories : a child's introduction to religion / by Ronald A. Gestwicki. — 1st ed. — Port Washington, N.Y. : Ashley Books, c1977.
182 p. ; 23 cm.
Bibliography: p. 177-182.
ISBN 0-87949-108-6 : $7.95
1. Children—Religious life. 2. Story-telling (Christian theology) I. Title.
BV4571.2.G47 1977 649'.7 77-80276
78 MARC

Hunt, Gladys M
Honey for a child's heart / Gladys Hunt. — Rev. ed. — Grand Rapids : Zondervan Pub. House, 1978.
182 p. : ill. ; 21 cm.
Bibliography: p. 125-182.
ISBN 0-310-26381-6 : $6.95
1. Books and reading for children. 2. Children—Religious life. I. Title.
Z1037.H945 1978 028.5 78-3488
78 MARC

Janney, John R
How to raise happy, godly children / John R. Janney. — 1st ed. — Bridgeton, N.J. : Christian Family Crusades, c1978.
183 p. : ill. ; 22 cm.
Includes bibliographical references.
1. Children—Conduct of life. 2. Children—Religious life. 3. Children—Management. I. Title.
BJ1631.J36 649'.1 78-58575
79 MARC

Kijm, Jan Marie.
Kinderen vertellen : een vergelijkend onderzoek / J. M. Kijm. — Roermond : Romen, 1961.
202 p. : ill. ; 25 cm.
Includes bibliographical references.
1. Children—Religious life. 2. Moral development. I. Title.
BV4579.D8K54 79-365467
79 MARC

Marxhausen, Joanne.
I am a tree / written by Joanne Marxhausen ; illustrated by Benjamin Marxhausen. — St. Louis : Concordia Pub. House, c1979.
[24] p. : col. ill. ; 26 cm. — (Her I am books)
SUMMARY: Considers the characteristics of trees, compares a tree to the kingdom of God, and provides activities for the reader to get to know a tree.
ISBN 0-570-07952-7 : $1.50
1. Children—Religious life. 2. Trees—Juvenile literature. [1. Trees. 2. Christian life] I. Marxhausen, Benjamin. II. Title. III. Series.
BV4571.2.M342 78-12750
78 MARC

Marxhausen, Joanne.
I am ... people / written by Joanne Marxhausen ; illustrated by Benjamin Marxhausen. — St. Louis : Concordia Pub. House, c1979.
[24] p. : col. ill. ; 26 cm. — (I am books)
SUMMARY: Considers the characteristics of people, explains their role in God's creation, and provides activities for the reader to get to know someone else.
ISBN 0-570-07953-5 : $1.50
1. Children—Religious life. [1. Christian life] I. Marxhausen, Benjamin. II. Title. III. Series.
BV4571.2.M343 78-12755
78 MARC

Marxhausen, Joanne.
I am the sun / written by Joanne Marxhausen ; illustrated by Benjamin Marxhausen. — St. Louis : Concordia Pub. House, c1979.
[24] p. : col. ill. ; 26 cm. — (I am books)
SUMMARY: Considers the characteristics of the sun, observes that the sun is praising God, and provides activities for the reader to get to know the sun.
ISBN 0-507-07950-0 : $1.50
1. Children—Religious life. 2. Sun—Juvenile literature. [1. Sun. 2. Christian life] I. Marxhausen, Benjamin. II. Title. III. Series.
[BV4571.2M344] 78-12759
78 MARC

What do I believe? : answers from the young / compiled and edited by Mary Spain ; analysis by Susan Major ; preface by Jimmy Savile. — London : Shepheard-Walwyn, 1978.
192 p. : ill, port ; 25 cm. GB78-29432
ISBN 0-85683-038-0 : £4.95. ISBN 0-85683-039-9 pbk.
1. Children—Religious life. 2. Children—Conduct of life. 3. Children—England. I. Spain, Mary. II. Major, Susan.
BV4571.2.W47 371.8'1 78-326630
79 MARC

—RELIGIOUS LIFE—JUVENILE LITERATURE

Hanny, Diane.
Closer than my shadow / by Diane Hanny ; illustrated by Richard Henkel. — Plainfield, N.J. : Logos International, c1979.
xvii, 207 p. : ill. ; 21 cm.
ISBN 0-88270-350-1 : $3.95
1. Holy Spirit—Juvenile literature. 2. Children—Religious life—Juvenile literature. I. Title.
BT121.2.H25 248'.82 78-71688
79 MARC

—RESEARCH—DENMARK

Denmark. Socialstyrelsen.
Forskning, forsøgs- og udviklingsarbejde vedrørende børn i førskolealderen / [Socialstyrelsen]. — [København] : Socialstyrelsen, 1977.
62, [12] p. ; 30 cm. D77-44
Cover title.
By G. Berg, I. Bistrup, and F. Hansen.
Includes bibliographical references and index.
1. Children—Research—Denmark. 2. Child psychology—Research—Denmark. 3. Day care centers—Research—Denmark. I. Berg, Gertrud. II. Bistrup, Inge. III. Hansen, Finn, 1946- . IV. Title.
HQ767.85.D46 1977 79-343240
79 MARC

—RESEARCH—DENMARK—CONGRESSES

Småbørnsforskning i Danmark : rapport fra et tværfagligt forskningsråds seminar 10.-12. marts 1977 / [Lone Heyde har forestået arbejdet med udarbejdelsen]. — København : Statens humanistiske Forskningsråd : Statens lægevidenskabelige Forskningsråd : Statens samfundsvidenskabelige Forskningsråd, adresse, Forskningssekretariatet, 1977.
135 p. ; 24 cm. D78-6
Includes bibliographies.
ISBN 8787331365 : not in trade.
1. Child welfare—Research—Denmark—Congresses. 2. Children—Research—Denmark—Congresses. I. Denmark. Statens humanistiske forskningsråd. II. Denmark. Statens lægevidenskabelige forskningsråd. III. Denmark. Statens samfundsvidenskabelige forskningsråd.
HV786.A6S6 78-385609
78 MARC

—RESIDENTIAL CARE see Children—Institutional care

—ROOMS see Children's rooms

—SLEEP

see also Children's dreams

Garvey, Chester Roy.
The activity of young children during sleep : an objective study / by Chester Roy Garvey. — Westport, Conn. : Greenwood Press, 1976, c1939.
x, 102 p., [1] leaf of plates : ill. ; 23 cm.
Reprint of the ed. published by the University of Minnesota Press, Minneapolis, which was issued as no. 18 of Monograph series of the University of Minnesota's Institute of Child Welfare.
Bibliography: p. 97-99.
Includes index.
ISBN 0-8371-8073-2
1. Children—Sleep. 2. Sleep movements. 3. Child psychology. I. Title. II. Series: Minnesota. University. Institute of Child Development and Welfare. Monograph series ; no. 18.
[DNLM: WL108 G244a 1939a]
BF723.S45G3 1976 612'.04 76-44474
76 MARC

—SOCIAL CONDITIONS

Voss, Vivian.
Iene miene mutte ... : kinderen van alle tijden / Vivian Voss. — Haarlem : Fibula-Van Dishoeck, c1978.
160 p. : ill. (some col.) ; 30 cm. Ne***
Bibliography: p. 159.
ISBN 9022838552
1. Children—History. 2. Children—Social conditions. I. Title.
HQ781.5.V66 78-375813
78 MARC

—SOCIETIES AND CLUBS

(Aozora shōnen shōjo gakkō)
青空少年少女学校：地域に子ども会・少年団を / [少年少女組織を育てる全国センター編]. — 東京：同センター, 1977.
149 p. ; 21 cm. Ja 78-9445
¥800

1. Children's clubs—Programs—Addresses, essays, lectures. 2. Children—Japan—Addresses, essays, lectures. I. Shōnen Shōjo Soshiki o Sodateru Zenkoku Sentā.
HQ767.8.A59 79-803688
J

— SPORTS see Sports for children

—STATISTICS

McHale, Magda Cordell.
Children in the world / by Magda Cordell McHale and John McHale, with Guy F. Streatfeild. — Washington : Population Reference Bureau, 1979.
72 p. : ill. ; 21 x 28 cm.
ISBN 0-917136-03-9
1. Children—Statistics. I. McHale, John, joint author. II. Population Reference Bureau, Washington, D.C. III. Title.
HQ767.9.M3 301.43'1 78-24731
79 MARC

—SUICIDAL BEHAVIOR

Inamura, Hiroshi, 1935-
(Kodomo no jisatsu)
子どもの自殺 / 稲村博著. — 東京：東京大学出版会, 1978.
vi, 252 p. ; 19 cm. — (UP選書 ; 188) Ja 78-17462
Series romanized: UP sensho.
Bibliography: p. 251.
¥900

1. Suicide—Japan. 2. Youth—Suicidal behavior. 3. Children—Suicidal behavior. I. Title.
HV6548.J3I53 79-806792

CHILDREN (Continued)

—SURGERY

see also Infants (Newborn)—Surgery; Pediatric anesthesia

Mustardé, John Clark.
Plastic surgery in infancy and childhood / editor, John Clark Mustardé. — 2d ed. — Edinburgh ; New York : Churchill Livingstone ; New York : distributed by Longman, 1979.
xv, 631 p. : ill. ; 26 cm. GB78
Includes bibliographies and index.
ISBN 0-443-01629-1 : £25.00
1. Children—Surgery. 2. Surgery, Plastic. I. Title.
[DNLM: 1. Surgery, Plastic—In infancy and childhood. WO600 P714]
RD137.M87 1978 617'.95 78-40328
78 MARC

Nixon, Harold Homewood.
Surgical conditions in paediatrics / H. Homewood Nixon. — London ; Boston : Butterworths, 1978.
xii, 448 p. : ill. ; 23 cm. — (Postgraduate paediatrics series)
Includes bibliographies and index.
ISBN 0-407-00090-9 : £19.50
1. Children—Surgery. I. Title.
[DNLM: GB78]
RD137.N62 617'.98 77-30676
78 MARC

Pediatric surgery / edited by Mark M. Ravitch ... [et al.]. — 3d ed. — Chicago : Year Book Medical Publishers, c1979.
2 v. (lxxvii, 1622 p., 7 leaves of plates) : ill. ; 29 cm.
Includes bibliographies and index.
ISBN 0-8151-7107-2
1. Children—Surgery. I. Ravitch, Mark M., 1910-
RD137.P42 1979 617'.98 78-25865
78 MARC

Sanders, Bruce, 1943-
Pediatric oral and maxillofacial surgery / Bruce Sanders ; original artwork by Irene Petravicius. — Saint Louis : Mosby, 1979.
xii, 606 p. : ill. ; 26 cm.
Includes bibliographical references and index.
ISBN 0-8016-4308-2
1. Mouth—Surgery. 2. Jaws—Surgery. 3. Children—Surgery. 4. Pedodontics. [1. Surgery—Oral—In infancy and childhood. 2. Maxillofacial injuries—In infancy and childhood. 3. Maxillofacial injuries—Surgery]
[DNLM: WU480 S215p]
RK529.5.C45S26 617'.522 79-11127
79 MARC

Steward, David J
General anesthesia for minor surgery in healthy children / David J. Steward, Robert E. Creighton. — Chicago : Year Book Medical Publishers, 1977.
32 p. ; 23 cm. — (Current problems in anesthesia and critical care medicine ; v. 1, no. 5)
Includes bibliographical references.
1. Pediatric anesthesia. 2. Children—Surgery. 3. Surgery, Minor. I. Creighton, Robert E., joint author. II. Title. III. Series.
RD139.S83 617'.967'98 78-112960
79 MARC

Symposium on pediatric reconstructive plastic surgery / F. X. Paletta, guest editor. — Philadelphia : Saunders, 1977.
vii p., p.p. 477-607 p. : ill. ; 27 cm. — (Clinics in plastic surgery ; v. 4, no. 4)
Includes bibliographies and index.
1. Children—Surgery. 2. Children—Wounds and injuries. 3. Surgery, Plastic. I. Paletta, Francis X. II. Series.
[DNLM: 1. Surgery, Plastic—In infancy and childhood. W1 CL831D v. 4 no. 4 / WO600 S9888]
RD137.S95 617'.95 77-21650
Shared Cataloging with DNLM 77[r79]rev MARC

Wu-han i hsüeh yüan, Hankow, China. Ti erh fu shu i yüan.
(Hsiao erh wai k'o shou shu hsueh)
小儿外科手术学 / 武汉医学院第二附属医院. — 第1版. — [北京] : 人民卫生出版社 : 新华书店北京发行所发行, 1978.
9, 666 p. : ill. ; 27 cm.
$3.75
1. Children—Surgery. I. Title.
RD137.W8 1978 79-836205

—SURGERY—ADDRESSES, ESSAYS, LECTURES

Research in pediatric surgery / editors, P. P. Rickham, W. Ch. Hecker, J. Prévot. — Baltimore : Urban & Schwarzenberg, 1978.
vi, 214 p. : ill. ; 23 cm. — (Progress in pediatric surgery ; v. 12)
Includes bibliographies and index.
ISBN 0-8067-1512-X
1. Surgery, Experimental—Addresses, essays, lectures. 2. Infants (Newborn)—Physiology—Addresses, essays, lectures. 3. Children—Surgery—Addresses, essays, lectures. I. Rickham, Peter Paul. II. Hecker, Waldemar Ch. III. Prévot, J. IV. Series.
RD137.A1P7 vol. 12 617'.98'008 s 78-14280
[RD29] 78 MARC

—SURGERY—CASES, CLINICAL REPORTS, STATISTICS

Pediatric surgery case studies : 68 case histories related to the surgical management of children / by John R. Lilly ... [et. al.]. — Garden City, N.Y. : Medical Examination Pub. Co., 1978.
331 p. : ill. ; 22 cm.
Includes bibliographical references and index.
ISBN 0-87488-069-6
1. Children—Surgery—Cases, clinical reports, statistics. 2. Children—Surgery—Examinations, questions, etc. I. Lilly, John R.
RD137.P44 617'.98'0076 77-94877
78 MARC

—SURGERY—CONGRESSES

Congress of the European Society for Pediatric Neurosurgery, 5th, Stresa, 1976.
Pediatric neurosurgery : Vth Congress of the European Society for Pediatric Neurosurgery, Stresa, September 30-October 2, 1976 / volume editors, R. Villani, M. Giovanelli, and S. M. Gaini. — Basel ; New York : S. Karger, 1977.
viii, 283 p. : ill. ; 23 cm. — (Modern problems in paediatrics ; v. 18)
GFR***
Includes bibliographical references.
ISBN 3-8055-2668-7
1. Nervous system—Surgery—Congresses. 2. Cranial fossa, Posterior—Surgery—Congresses. 3. Hydrocephalus in children—Surgery—Congresses. 4. Children—Surgery—Congresses. 5. Pediatric neurology—Congresses. I. Villani, R. II. Giovanelli, M. III. Gaini, S. M. IV. Title. V. Series: Moderne Probleme der Pädiatrie ; v. 18.
RJ23.M6 vol. 18 617'.48 79-300390
[RD593] 79 MARC

Pediatric neurology and neurosurgery / edited by Richard A. Thompson, John R. Green. — Jamaica, N.Y. : SP Medical & Scientific Books, c1978.
420 p. : ill. ; 24 cm.
"[Sponsored by] Barrow Neurological Institute of St. Joseph's Hospital and Medical Center and the Barrow Neurological Foundation, Phoenix, Arizona."
Includes bibliographical references and index.
ISBN 0-89335-035-4
1. Pediatric neurology—Congresses. 2. Nervous system—Surgery—Congresses. 3. Children—Surgery—Congresses. I. Thompson, Richard Arlen, 1933- . II. Green, John Raymond, 1915- . III. St. Joseph's Hospital, Phoenix, Ariz. Barrow Neurological Institute. IV. Barrow Neurological Foundation.
[DNLM: 1. Nervous System Diseases—In infancy and childhood—Congresses. 2. Neurosurgery—In infancy and childhood—Congresses. WS340 P3694 1977]
RJ486.P34 618.9'28 77-12594
77 MARC

—SURGERY—EXAMINATIONS, QUESTIONS, ETC.

Pediatric surgery case studies : 68 case histories related to the surgical management of children / by John R. Lilly ... [et. al.]. — Garden City, N.Y. : Medical Examination Pub. Co., 1978.
331 p. : ill. ; 22 cm.
Includes bibliographical references and index.
ISBN 0-87488-069-6
1. Children—Surgery—Cases, clinical reports, statistics. 2. Children—Surgery—Examinations, questions, etc. I. Lilly, John R.
RD137.P44 617'.98'0076 77-94877
78 MARC

—SURGERY—GERMANY, WEST—MUNICH—HISTORY—ADDRESSES, ESSAYS, LECTURES

125 [i.e. Hundertfünfundzwanzig] Jahre Münchner Kinderchirurgie / hrsg. von Waldemar Ch. Hecker. — Lübeck : Hansisches Verlagskontor, [1977?]
48 p. : ill. ; 22 cm. — (Documenta pädiatrica ; 3) GFR78-A4
"... Aufsätze und Grussadressen ... vorgetragen auf der Jubiläumsveranstaltung: '125 Jahre Chirurgie im Dr. von Haunerschen Kinderspital - Universitäts-Kinderklinik München - und 90. Geburtstag der Kinderchirurgischen Klinik des Spitals.'"
Bibliography: p. 33.
DM5.50
1. Munich. Universität. Kinderklinik—History—Addresses, essays, lectures. 2. Munich. Universität. Kinderchirurgische Klinik—History—Addresses, essays, lectures. 3. Children—Surgery—Germany, West—Munich—History—Addresses, essays, lectures. 4. Pediatrics—Germany—History—Addresses, essays, lectures. I. Hecker, Waldemar Ch . II. Munich. Universität. Kinderklinik. III. Munich. Universität. Kinderchirurgische Klinik. IV. Series.
RJ28.M955H86 362.7'8'2094336 78-375554
78[r79]rev MARC

—SURGERY—NETHERLANDS

Bos, Johannes Maria.
Hard tegen hart : een sociaal-wetenschappelijk onderzoek onder jongeren die enkele jaren daarvoor werden geopereerd aan een aangeboren hartafwijking / door J. M. Bos. — Amsterdam : Swets & Zeitlinger, 1978.
160 p. ; 24 cm. Ne78-32
Bibliography: p. 143-152.
ISBN 9026502869 : fl 26.00
1. Heart—Surgery—Psychological aspects. 2. Physically handicapped children—Rehabilitation—Netherlands. 3. Sick children—Psychology. 4. Heart—Surgery—Netherlands. 5. Children—Surgery—Netherlands. 6. Pediatric cardiology. I. Title.
RD598.B66 78-396808
*79 MARC

—TRAINING see Children—Management

—TRANSPORTATION see School children—Transportation

—VOCABULARY see Children—Language

—WOUNDS AND INJURIES

see also Battered child syndrome; Burns and scalds in children

Illingworth, Cynthia M
The diagnosis and primary care of accidents and emergencies in children : a manual for the casualty officer and the family doctor / Cynthia M. Illingworth. — Oxford : Blackwell Scientific Publications ; Philadelphia : distributed by J. B. Lippincott, 1978.
viii, 134 p. : ill. ; 22 cm. GB***
Includes bibliographies and index.
ISBN 0-632-00392-8 : £4.25
1. Pediatric emergencies. 2. Children—Wounds and injuries. I. Title.
RJ370.I43 618.9'20025 78-319593
79 MARC

Symposium on pediatric reconstructive plastic surgery / F. X. Paletta, guest editor. — Philadelphia : Saunders, 1977.
vii p., p.p. 477-607 p. : ill. ; 27 cm. — (Clinics in plastic surgery ; v. 4, no. 4)
Includes bibliographies and index.
1. Children—Surgery. 2. Children—Wounds and injuries. 3. Surgery, Plastic. I. Paletta, Francis X. II. Series.
[DNLM: 1. Surgery, Plastic—In infancy and childhood. W1 CL831D v. 4 no. 4 / WO600 S9888]
RD137.S95 617'.95 77-21650
Shared Cataloging with DNLM 77[r79]rev MARC

—WOUNDS AND INJURIES—CONGRESSES

Gesellschaft für Chirurgie der DDR. Sektion Kinderchirurgie.
Das Schädel-Hirn-Trauma im Kindesalter : VI. Symposion mit internationaler Beteiligung der Sektion Kinderchirurgie der Gesellschaft für Chirurgie der DDR vom 30. 9. und 1. 10. 1976 in Jena / [bearb. von Herbert Schickedanz]. — Jena : Friedrich-Schiller-Universität, 1977.
2 v. (391 p.) : ill. ; 21 cm. — (Wissenschaftliche Beiträge der Friedrich-Schiller-Universität Jena) GDR***
5.00M
1. Head—Wounds and injuries—Congresses. 2. Brain—Wounds and injuries—Congresses. 3. Brain-damaged children—Congresses. 4. Children—Wounds and injuries—Congresses. I. Schickedanz, Herbert. II. Title. III. Series: Jena. Universität. Wissenschaftliche Beiträge.
RD521.G47 1977 78-390631
79 MARC

—ALASKA—STATISTICS

Alaska. Division of Mental Health & Developmental Disabilities.
Alaska's youth : needs assesment [i.e. assessment] and resource inventory. — [Juneau] : Dept. of Health and Social Services, Division of Mental Health & Developmental Disabilities, 1978.
281 p. : ill. ; 28 cm.
Includes selected papers from Primary prevention: an idea whose time has come, DHEW publication no. (ADM) 77-447.
Includes bibliographies.
1. Social work with youth—Alaska. 2. Child welfare—Alaska. 3. Youth—Alaska—Statistics. 4. Children—Alaska—Statistics. I. Title.
HV1435.A4A45 1978 362.7'09798 78-624036
79 MARC

—AMERICA—STATISTICS

Interamerican Children's Institute.
Estadísticas básicas sobre menores / O.E.A., Instituto Interamericano del Niño. — Montevideo : El Instituto, 1973.
77 p. : graphs ; 24 cm.
1. Children—America—Statistics. 2. Age distribution (Demography)—America. 3. America—Statistics, Vital. I. Title.
HQ792.A57157 1973 301.43'14'098 79-353204
79 MARC

CHILDREN (Continued)

—ASIA

Bulatao, Rodolfo A 1944-
On the nature of the transition in the value of children / Rudolfo A. Bulatao. — Honolulu : East-West Center, 1979.
xvi, 104 p. : ill. ; 23 cm. — (Current studies on the value of children) (Papers of the East-West Population Institute ; no. 60-A)
Bibliography: p. 99-104.
1. Children—Asia. 2. Family—Asia. 3. Family size—Economic aspects—Asia. 4. Family size—Social aspects—Asia. 5. Fertility, Human—Asia. I. Title. II. Title: Value of children in Asia. III. Series. IV. Series: East-West Population Institute. Papers of the East-West Population Institute ; no. 60-A.
HQ751.B84 301.42'6 79-11835
79 MARC

—AUSTRALIA—PRAHRAN—BIOGRAPHY

Lewis, Brian Bannatyne, 1906-
Sunday at Kooyong Road / [by] Brian Lewis. — Richmond [Aus.] : Hutchinson of Australia, 1976.
182 p. : ill. ; 23 cm.
Aus
ISBN 0-09-130130-0
1. Lewis, Brian Bannatyne, 1906- . 2. Prahran, Australia—Biography. 3. Children—Australia—Prahran—Biography. I. Title.
DU230.P68L484 994.5 78-324509
79 MARC

—AUSTRALIA—TASMANIA—BIOGRAPHY

Sansom, Ruth.
Through the eyes of a child : sketches of a Tasmanian childhood / Ruth Sansom ; decorations by Max Angus. — Hobart, Australia : Cat & Fiddle Press, 1977.
vi, 80 p. : ill. ; 23 cm. Aus***
ISBN (invalid) 0-85853-027-5
1. Sansom, Ruth. 2. Tasmania—Biography. 3. Children—Australia—Tasmania—Biography. I. Title.
CT2808.S27A36 994.604'092'4 79-306174
79 MARC

—AUSTRIA—TYROL—HISTORY—SOURCES

Uhlig, Otto.
Die Schwabenkinder aus Tirol und Vorarlberg / von Otto Uhlig. — Innsbruck : Univ.-Verl. Wagner ; Stuttgart : Aalen : Theiss, 1978.
307 p., [17] leaves of plates : ill. ; 25 cm. — (Tiroler Wirtschaftsstudien, Schriftenreihe der Jubiläumsstiftung der Kammer der Gewerblichen Wirtschaft [für Tirol] ; Folge 34) Au78-15
Bibliography: p. [297]-300.
Includes indexes.
ISBN 3-7030-0052-X : S340.00
1. Children—Austria—Tyrol—History—Sources. 2. Children—Austria—Voralberg—History—Sources. 3. Children—Employment—Germany—History—Sources. 4. Migrant agricultural laborers—Germany—History—Sources. 5. Alien labor, Austrian—Germany—History—Sources. I. Title. II. Series.
HQ799.A92T88 79-337977
*79 MARC

—AUSTRIA—VORALBERG—HISTORY—SOURCES

Uhlig, Otto.
Die Schwabenkinder aus Tirol und Vorarlberg / von Otto Uhlig. — Innsbruck : Univ.-Verl. Wagner ; Stuttgart : Aalen : Theiss, 1978.
307 p., [17] leaves of plates : ill. ; 25 cm. — (Tiroler Wirtschaftsstudien, Schriftenreihe der Jubiläumsstiftung der Kammer der Gewerblichen Wirtschaft [für Tirol] ; Folge 34) Au78-15
Bibliography: p. [297]-300.
Includes indexes.
ISBN 3-7030-0052-X : S340.00
1. Children—Austria—Tyrol—History—Sources. 2. Children—Austria—Voralberg—History—Sources. 3. Children—Employment—Germany—History—Sources. 4. Migrant agricultural laborers—Germany—History—Sources. 5. Alien labor, Austrian—Germany—History—Sources. I. Title. II. Series.
HQ799.A92T88 79-337977
*79 MARC

—BANGLADESH

University of Dacca. Centre for Urban Studies.
The urban poor in Bangladesh : a study on the situation of the urban poor in Bangladesh, with special reference to mothers and children : comprehensive summary report : final draft. — Dacca : Centre for Urban Studies, Dept. of Geography, University of Dacca, 1979.
iv, 108 leaves : map ; 31 cm.
"Centre for Urban Studies research report for UNICEF."
1. Poor—Bangladesh. 2. Family—Bangladesh. 3. Family life surveys—Bangladesh. 4. Mothers—Bangladesh. 5. Children—Bangladesh. I. United Nations. Children's Fund. II. Title.
HC440.8.Z9P627 301.44'1 79-902476
79 MARC

—BANGLADESH—STATISTICS

University of Dacca. Institute of Statistical Research and Training.
Statistical profile of children and mothers in Bangladesh. — [Dacca] : Institute of Statistical Research and Training, University of Dacca, 1977.
iv, 80 p. : map ; 29 cm.
Chiefly tables.
Compiled by the Institute of Statistical Research and Training, University of Dacca, at the request of UNICEF.
1. Bangladesh—Population—Statistics. 2. Bangladesh—Statistics, Vital. 3. Children—Bangladesh—Statistics. 4. Mothers—Bangladesh—Statistics. I. United Nations. Children's Fund. II. Title.
HB3640.6.A3U54 1977 312'.9 77-907181
79 MARC

—BOTSWANA—JUVENILE LITERATURE

Larson, Thomas John.
Dibebeš choice / Thomas J. Larson. — Thompson, Conn. : InterCulture Associates, c1978.
60 p., [1] leaf of plates : ill. ; 21 cm. — (The Africa sketches series)
SUMMARY: The son of a Hambukushu village chief in Botswana must decide whether to earn a living as a miner in South Africa or continue his schooling.
ISBN 0-89253-101-0
1. Hambukushu (African people)—Social life and customs—Juvenile literature. 2. Children—Botswana—Juvenile literature. [1. Hambukushu (Bantu people)—Social life and customs. 2. Botswana—Social life and customs] I. Title. II. Series.
DT797.L37 968'.1 77-94637
00 MARC

—BRAZIL—BIOGRAPHY

Ribeiro, Mozarth.
Luizinho, o menino endiabrado / Mozarth Ribeiro. — Rio de Janeiro : Editora Potengy, 1976.
79 p., [2] leaves of plates : ill. ; 21 cm. — (Coleção Toyo ; no. 4)
1. Luizinho. 2. Brazil—Biography. 3. Children—Brazil—Biography. I. Title.
CT688.L78R5 79-104158
79 MARC

—BRAZIL—RIO DE JANEIRO—NUTRITION

Instituto de Nutrição Annes Dias.
Pesquisa nutricional em escolares do município do Rio de Janeiro / Prefeitura da Cidade do Rio de Janeiro, Secretaria Municipal de Educação e Cultura, Departamento Geral de Educação, Instituto de Nutrição Annes Dias, Secretaria Municipal de Saúde, Departamento Geral de Saúde Pública, Divisão de Medicina Escolar. — Rio de Janeiro : Secretaria Municipal de Educação e Cultura, 1978.
2 v. : ill. ; 30 cm.
"Trabalho apresentado no XI Congresso Internacional de Nutrição, Rio de Janeiro, 1978."
Summary in English.
Includes bibliographies.
1. Diet—Brazil—Rio de Janeiro. 2. Nutrition surveys—Brazil—Rio de Janeiro. 3. Children—Brazil—Rio de Janeiro—Nutrition. I. Rio de Janeiro. Divisão de Medicina Escolar. II. Title.
TX360.B7I57 1978 79-104713
00 MARC

—BULGARIA—BIOGRAPHY

(Pomni tíakhnoto detstvo)
Помни тяхното детство : очерци за деца и юноши, загинали в антифашистката борба / [състав. и ред. Ваня Филипова ; худож. Георги Недялков]. — 8. изд. — София : Нар. младеж, 1978.
294 p. : ports. ; 21 cm. Bu 78-908
1.12 lv
1. World War, 1939-1945—Children. 2. World War, 1939-1945—Bulgaria. 3. World War, 1939-1945—Juvenile literature. 4. World War, 1939-1945—Biography. 5. Bulgaria—Biography. 6. Children—Bulgaria—Biography. I. Filipova, Vanía.
D810.C4P63 1978 78-422948

—CANADA

Emberley, Jean.
Children in Canada : a report / prepared by Jean Emberley. — [Hamilton, Ont.] : McMaster University Press, 1978.
vi, 86 p. : map ; 23 cm. C78-11294-6
Cover title: Today's child ... the world's future.
Prepared for the National Council of Women of Canada.
"The Hospital of Sick Children Foundation, grant no. 78-34."
$3.00
1. Children—Canada. 2. Child welfare—Canada. 3. Children—Care and hygiene—Canada. 4. Education of children. 5. Education—Canada. I. National Council of Women of Canada. II. Title. III. Title: Today's child ... the world's future.
HQ792.C3E43 362.7'0971 79-311504
79 MARC

Roy, Gabrielle Carbotte, 1909-
Ces enfants de ma vie / Gabrielle Roy. — [Montréal] : Stanké, c1977.
212 p. ; 21 cm. C***
ISBN 0-88566-065-X
1. Roy, Gabrielle Carbotte, 1909- —Biography—Teaching career. 2. Novelists, French-Canadian—20th century—Biography. 3. Teachers—Canada—Biography. 4. Children in Canada. I. Title.
PQ3919.R74Z465 843 78-381422
78 MARC

—CANADA—ADDRESSES, ESSAYS, LECTURES

Childhood and adolescence in Canada / edited and with an introd. by K. Ishwaran. — Toronto ; New York : McGraw-Hill Ryerson, c1979.
386 p. ; 23 cm. — (McGraw-Hill Ryerson interdisciplinary studies in family and marriage) C78-1639-4
Includes bibliographies.
ISBN 0-07-082935-7 : $8.95
1. Children—Canada—Addresses, essays, lectures. 2. Child development—Canada—Addresses, essays, lectures. 3. Youth—Canada—Addresses, essays, lectures. I. Ishwaran, Karigoudar. II. Series: Interdisciplinary studies in family and marriage.
HQ792.C3C48 301.43'1'0971 79-318005
79 MARC

—CANADA—BIOGRAPHY—JUVENILE LITERATURE

Grandbois-Paquin, Gabrielle.
La petite fille aux yeux rouges / Gabrielle Grandbois-Paquin ; ill. d'André Bergeron. — Montréal : Fides, c1978.
169 p., [4] leaves of plates : col. ill. ; 21 cm. — (Collection du goéland) C***
ISBN 0-7755-0672-9 : $6.75
1. Grandbois, Bernadette Rousseau—Juvenile literature. 2. Canada—Biography—Juvenile literature. 3. Children in Canada—Biography—Juvenile literature. I. Title.
CT310.G66G7 971.06'092'4 78-394186
78 MARC

—CHINA—KWANGTUN (PROVINCE)

Kokubun, Ichitarō, 1911-
(Senchi no kodomo)
戰地の子供 / 國分一太郎著：清水崑畫. — 東京：中央公論社, 昭和15 [1940]
12, 4, 450 p. : ill. ; 20 cm.

1. Children—China—Kwangtung (Province) 2. Kwangtung, China (Province)—Social conditions. 3. Sino-Japanese Conflict, 1937-1945—Children. I. Shimizu, Kon, 1912- II. Title.
HQ792.C5K64 79-806749

—CZECHOSLOVAKIA—MORTALITY

Sziegoleit, Meinhard.
Kinder- und Mütterfürsorge in der CSSR mit besonderer Berücksichtigung der Abtreibung, von Meinhard Sziegoleit. Hrsg. von Hans Harmsen. Hamburg, [Bonn, Büro Bonner Berichte] 1969.
68 p. illus. 29 cm. (Zur Entwicklung und Organisation des Gesundheitswesens in Sowjetrussland, in osteuropäischen Volksdemokratien und in Mitteldeutschland, Bd. 45) GDB70-B3-173
Bibliography: p. 67-68
1. Child health services—Czechoslovakia. 2. Children—Diseases—Czechoslovakia—Statistics. 3. Children—Czechoslovakia—Mortality. 4. Maternal health services—Czechoslovakia. 5. Mothers—Czechoslovakia—Mortality. 6. Abortion—Czechoslovakia—Statistics. I. Title. II. Series.
RJ103.C95S94 79-413683
79 MARC

—DENMARK

Nellemann, George.
Fremmedarbejderbørn : roepolakkernes efterkommere og deres skæbne i det danske samfund / George Nellemann ; [redaktion, Henning Dehn-Nielsen]. — København : Nationalmuseet : [eksp, Gyldendal], c1977.
76 p. ; 24 cm. — (Working papers - National Museum of Denmark ; 5) D77-32
Includes bibliographical references.
ISBN 8748001414 : kr18.00
1. Denmark—Race relations. 2. Poles in Denmark. 3. Minorities—Denmark. 4. Children—Denmark. I. Title. II. Series: Copenhagen. Nationalmuseet. Working papers - National Museum of Denmark ; 5.
DL142.P6N44 79-346783
79 MARC

CHILDREN (Continued)

—DENMARK—ADDRESSES, ESSAYS, LECTURES

Vores fælles børn? : har vi en børnepolitik? : en debatbog / redigeret af Inge Fischer Møller ; under medvirken af Birgitte Oxdam ... ⌈et al.⌉. — ⌈København⌉ : ⌈Gyldendal⌉, 1978.
87 p. ; 22 cm. — (Gyldendals pædagogiske bibliotek) D78-15
Bibliography: p. 19-⌈20⌉
ISBN 8701674617 : kr42.50
1. Child welfare—Denmark—Addresses, essays, lectures. 2. Children in Denmark—Addresses, essays, lectures. I. Møller, Inge Fischer. II. Oxdam, Birgitte, 1933-
HV786.A6V67 78-385611
78 MARC

—DENMARK—BIOGRAPHY

Christiansen, Valdemar, 1892-1975.
Vi havde en dejlig barndom : optegnelser / af Valdemar Christiansen. — ⌈Hørsholm : Peter Leth Christiansen⌉, 1977.
220 p. : ill. ; 22 cm. D77-44
Not in trade.
1. Christiansen, Valdemar, 1892-1975. 2. Denmark—Biography. 3. Children—Denmark—Biography. I. Title.
CT1278.C49A38 79-337919
79 MARC

—DENMARK—SOCIAL CONDITIONS

Denmark. Børnekommissionen.
Småbørn i Danmark : debat- og studieoplæg / Børnekommissionen ; ⌈fotografier, Julie Rønnow ... et al. ; tegninger, Maja ... et al. ; linoleumssnit, Dea Trier Mørch fra bogen "Vinterbørn" ; forord, tilrættelæggelse, Jakob Vedel-Petersen⌉. — København : Børnekommissionen : ⌈eksp. Socialministeriet⌉ (Slotholmsgade 6, 1216 K), 1977.
72 p. : ill. ; 21 cm. D77-37
Bibliography: p. 72.
Free.
ISBN 8750323016
1. Children—Denmark—Social conditions. 2. Family—Denmark. I. Title.
HQ792.D4D47 1977 79-340226
79 MARC

—DENMARK—VENDSYSSEL—BIOGRAPHY

Barn af Vendsyssel : vendelboere fortæller erindringer fra deres barndom i. Vendsyssel / red. Axel Dalbøge ... ⌈et al.⌉ — ⌈s.l.⌉ : Lions Clubber i Vendsyssel, 1975-1977.
2 v. : ill. ; 26 cm. D76-7
1. Vendsyssel—Biography. 2. Children—Denmark—Vendsyssel—Biography. I. Dalbøge, Axel.
DL271.V4B37 79-345191
79 MARC

—ENGLAND

Freeman, Joan.
Gifted children : their identification and development in a social context / by Joan Freeman. — Lancaster : MTP Press, c1979.
xii, 294 p. ; 24 cm. GB***
Bibliography: p. 281-289.
Includes index.
ISBN 0-85200-250-5 : £6.50
1. Gifted children. 2. Children—England. I. Title.
HQ773.5.F73 301.43'14 79-317507
79 MARC

What do I believe ? : answers from the young / compiled and edited by Mary Spain ; analysis by Susan Major ; preface by Jimmy Savile. — London : Shepheard-Walwyn, 1978.
192 p. : ill, port ; 25 cm. GB78-29432
ISBN 0-85683-038-0 : £4.95. ISBN 0-85683-039-9 pbk.
1. Children—Religious life. 2. Children—Conduct of life. 3. Children—England. I. Spain, Mary. II. Major, Susan.
BV4571.2.W47 371.8'1 78-326630
79 MARC

—ENGLAND—BIOGRAPHY

Elias, Eileen.
On Sundays we wore white / Eileen Elias ; drawings by Debbie Thompson. — London : W. H. Allen, 1978.
263 p. : ill. ; 23 cm. GB***
ISBN 0-491-02143-3 : £4.95
1. Elias, Eileen. 2. England—Biography. 3. Children in England—Biography. I. Title.
CT788.E54A33 942'.083'0924 78-323298
79 MARC

Massey, Victoria.
One child's war / ⌈by⌉ Victoria Massey. — London : British Broadcasting Corporation, 1978.
126 p., plate : 2 ports. ; 24 cm. GB78-20727
ISBN 0-563-17198-7 : £3.30
1. World War, 1939-1945—Personal narratives, English. 2. Massey, Victoria. 3. Children—England—Biography. I. Title.
D811.5.M2917 942.9'084'0924 78-326674
79 MARC

Plant, Ruth.
Nanny and I / Ruth Plant. — London : W. Kimber, 1978.
203 p., ⌈4⌉ leaves of plates : ill. ; 23 cm. GB***
ISBN 0-7183-0495-0 : £4.50
1. Plant, Ruth. 2. England—Biography. 3. Children in England—Biography. I. Title.
CT788.P68A36 942.4'61 78-322803
79 MARC

—ENGLAND—MANCHESTER—BIOGRAPHY—JUVENILE LITERATURE

Bradley, Helen, 1900-
The queen who came to tea / Helen Bradley. — London : Cape, 1978.
32 p. : col. ill. ; 27 x 29 cm. GB***
ISBN 0-224-01545-1 : £3.95
1. Bradley, Helen, 1900- —Juvenile literature. 2. Manchester, Eng.—Biography—Juvenile literature. 3. Children—England—Manchester—Biography—Juvenile literature. I. Title.
DA690.M4B723 759.2 79-301613
79 MARC

—ENGLAND—SALFORD (LANCASHIRE)—BIOGRAPHY

Brereton, Wallace.
Salford boy / by Wallace Brereton. — ⌈Salford⌉ : Salford Local History Society, 1977.
57 p. : ill. ; 21 cm. GB78-13195
1. Brereton, Wallace. 2. Salford, Eng. (Lancashire)—Biography. 3. Children—England—Salford (Lancashire)—Biography. I. Title.
DA690.S157B733 942.7'32 78-322860
79 MARC

—EUROPE—ADDRESSES, ESSAYS, LECTURES

Ina, Takio, 1910-
(Yōroppa no katei to kodomo)
ヨーロッパの家庭と子供 / 衣奈多喜男著. — 東京：新日本教育文化研究所, 昭和24 ⌈1949⌉
106 p. ; 18 cm. — (太陽文庫)
Series romanized: Taiyō bunko.

1. Children—Europe—Addresses, essays, lectures. I. Title.
HQ792.E8I53 79-808651

—FINLAND

Virtanen, Leea.
Children's lore / Leea Virtanen. — ⌈Helsinki⌉ : Suomalaisen Kirjallisuuden Seura, 1978.
100 p., ⌈4⌉ leaves of plates : ill. ; 25 cm. — (Studia Fennica ; 22 ISSN 0085-6835) Fi***
Bibliography: p. ⌈99⌉-100.
ISBN 951717148X
1. Folk-lore and children—Finland. 2. Games—Finland. 3. Children—Finland. I. Title. II. Series.
PH107.S8 no. 22 947.1'008 s 79-316776
⌈GR200⌉ 79 MARC

—FRANCE—BIOGRAPHY

Ruffin, Raymond.
Journal d'un J3 / Raymond Ruffin. — Paris : Presses de la Cité, c1979.
250 p. ; 23 cm. F***
ISBN 2-258-00496-9 : 42.00F
1. World War, 1939-1945—Children. 2. World War, 1939-1945—Personal narratives, French. 3. Ruffin, Raymond. 4. World War, 1939-1945—France. 5. France—History—German occupation, 1940-1945. 6. Children—France—Biography. I. Title.
D810.C4R83 940.53'161'0944 79-376707
79 MARC

—FRANCE—CARE AND HYGIENE

Loux, Françoise.
Le jeune enfant et son corps dans la médecine traditionnelle / Françoise Loux : préf. de Alexandre Minkowski. — Paris : Flammarion, c1978.
276 p., ⌈2⌉ leaves of plates : ill. ; 19 cm. — (La tradition et le quotidien) F***
Bibliography: p. ⌈267⌉-274.
ISBN 2-08-210954-2 : 48.00F
1. Folk-lore of children—France. 2. Folk medicine—France. 3. Children—France—Care and hygiene. I. Title.
GR161.L67 615'.882 79-348510
79 MARC

—FRANCE—HISTORY

Crubellier, Maurice.
L'enfance et la jeunesse dans la société française 1800-1950 / Maurice Crubellier. — Paris : A. Colin, c1979.
389 p. ; 23 cm. — (Collection U : Série Histoire contemporaine) F***
Includes bibliographies.
90.00
1. France—Social conditions. 2. Children—France—History. 3. Educational sociology. I. Title.
HN425.C77 301.43'1'0944 79-374887
00 MARC

—FRANCE—SAINTE-MÈRE-EGLISE—BIOGRAPHY

Duboscq, Geneviève.
Bye bye, Geneviève! / Geneviève Duboscq. — Paris : R. Laffont, c1978.
285 p., ⌈4⌉ leaves of plates : ill. ; 24 cm. — (Collection Vécu) F***
ISBN 2-221-00065-X : 44.00F
1. World War, 1939-1945—France—Sainte-Mère-Eglise. 2. World War, 1939-1945—Personal narratives, French. 3. Duboscq, Geneviève. 4. Sainte-Mère-Eglise, France—History. 5. United States. Army. 82d Division—History. 6. United States. Army. 101st Division—History. 7. Children—France—Sainte-Mère-Eglise—Biography. I. Title.
D762.S25D8 940.53'44'21 78-396390
79 MARC

—GERMANY—BIOGRAPHY

Schmeling, Marianne.
Flee the wolf : the story of a family's miraculous journey to freedom / by Marianne Schmeling. — Norfolk, Va. : Donning, c1978.
290 p., ⌈2⌉ leaves of plates : maps ; 22 cm. — (The Unilaw Library series)
ISBN 0-915442-67-1 : $9.95
1. World War, 1939-1945—Personal narratives, German. 2. Schmeling, E. Marianne. 3. Children—Germany—Biography. I. Title.
D811.5.S32 940.54'82'43 78-24154
78 MARC

—GERMANY, WEST—ADDRESSES, ESSAYS, LECTURES

Kinder in der Bundesrepublik : Materialien, Initiativen, Alternativen / Lottemi Doormann (Hrsg.). — Köln : Pahl-Rugenstein, 1979.
277 p. : ill. ; 19 cm. — (Kleine Bibliothek ; 152) GFR79-A
Includes bibliographical references.
ISBN 3-7609-0431-9
1. Children—Germany, West—Addresses, essays, lectures. 2. Children's rights—Germany, West—Addresses, essays, lectures. 3. Socialization—Addresses, essays, lectures. I. Doormann, Lottemi.
HQ792.G3K56 301.43'14'0943 79-376961
79 MARC

—GERMANY, WEST—ANTHROPOMETRY

Pieper, Ursula.
Anthropometrische Untersuchungen zu Bau und Funktion des kindlichen Körpers / Ursula Pieper, Hans W. Jürgens. — Bremerhaven : Wirtschaftsverlag, 1977.
99 p. ; 21 cm. — (Forschungsbericht - Bundesanstalt für Arbeitsschutz und Unfallforschung, Dortmund ; Nr. 178) GFR***
At head of title: Neue Universität Kiel.
Bibliography: p. 37.
ISBN 3-8331-4007-4
1. Children—Germany, West—Anthropometry. 2. Human growth—Germany, West. I. Jürgens, Hans W., joint author. II. Kiel. Universität. III. Title. IV. Series: Bundesanstalt für Arbeitsschutz und Unfallforschung. Forschungsbericht ; Nr. 178.
GN63.P53 78-389580
79 MARC

CHILDREN (Continued)

—GERMANY, WEST—BIOGRAPHY

Hermansson, Thea, 1931-
Dorte och de samtidas minnen / Thea Hermansson. — Linköping (Lindesbergsg. 10) : Förf., 1978.
222 p. ; 21 cm. S78-11
Autobiographical.
ISBN 9172601620 : kr53.00
1. World War, 1939-1945—Personal narratives, German. 2. Hermansson, Thea, 1931- . 3. Hamburg—History. 4. Children—Germany, West—Biography. 5. World War, 1939-1945—Germany, West—Hamburg. I. Title.
D811.5.H465 78-398425
79 MARC

—GERMANY, WEST—HAMBURG—SOCIAL CONDITIONS

Krause, Hans-Joachim, Dozent.
Kinder in der inneren Stadt : auf Strassen und Plätzen : zum Einfluss der sozialen und städtebauliche[n] Umwelt auf die Lebensverhältnisse und Sozialisationsbedingungen von Kindern in innerstädtischen Gebieten / Auftraggeber, Freie und Hansestadt Hamburg, Baubehörde Landesplanungsamt ; Gutachter, Hans-Joachim Krause, Timm Ohrt, Hille v. Seggern. — Hamburg : Freie und Hansestadt Hamburg, Baubehörde Landesplanungsamt, 1977.
2 v. : ill. (some col.) ; 30 cm. GFR***
Includes bibliographical references.
CONTENTS: T. 1. Interpretation der Daten aus pädagogischer Sicht unter Berücksichtigung städtebaulicher Kategorien.—T. 2. Interpretation der Daten aus städtebaulicher Sicht unter Berücksichtigung pädagogischer Kategorien.
1. Hamburg—City planning. 2. Children—Germany, West—Hamburg—Social conditions. 3. City planning—Germany, West—Case studies. I. Ohrt, Timm, joint author. II. Seggern, Hille von, joint author. III. Title. IV. Title: Auf Strassen und Plätzen.
HT169.G32H285 309.2'62'0943515 79-351961
79 MARC

—GREAT BRITAIN

Mooney, Bel, 1946-
The year of the child / Bel Mooney. — London : Hutchinson, 1979.
190 p., [4] leaves of plates : ill. ; 23 cm. GB***
ISBN 0-09-138320-X : £5.95
1. Children—Great Britain. 2. Child development. I. Title.
HQ792.G7M66 155.4'092'6 79-317369
79 MARC

—GREAT BRITAIN —ANTHROPOMETRY

Tanner, James Mourilyan.
Education and physical growth : implications of the study of children's growth for educational theory and practice / J. M. Tanner. — 2d ed. — London : Hodder and Stoughton, 1978.
144 p. : ill. ; 22 cm. — (Unibooks) GB***
Bibliography: p. 127-134.
Includes index.
ISBN 0-340-22484-3 : £2.45
1. Children in Great Britain—Anthropometry. 2. Children—Growth. 3. Education of children. 4. Child development. I. Title.
LB3425.G7T36 1978b 573'.6 78-318319
78 MARC

—GREAT BRITAIN —HISTORY

Marcus, Leah.
Childhood and cultural despair : a theme and variations in seventeenth-century literature / Leah Sinanoglou Marcus. — Pittsburgh : University of Pittsburgh Press, c1978.
xii, 305 p. : ill. ; 21 cm.
Bibliography: p. 275-295.
Includes index.
ISBN 0-8229-3367-5
1. English poetry—Early modern, 1500-1700—History and criticism. 2. Children in literature. 3. Children (Christian theology) 4. Children in Great Britain—History. I. Title.
PR545.C45M3 821'.009'352 77-15733
78 MARC

—GREAT BRITAIN—HISTORY—JUVENILE LITERATURE

Wilkins, Frances.
Growing up in the Age of Chivalry / Frances Wilkins. — New York : Putnam, [1978] c1977.
72 p. : ill. ; 26 cm.
First published in 1977 under title: Growing up in Tudor times.
Includes index.
SUMMARY: Decribes the family life, working conditions, education, and leisure of children living in Great Britain's cities and countryside during the Tudor period.
ISBN 0-399-20634-5 : $6.95
1. Children—Great Britain—History—Juvenile literature. 2. Great Britain—Social conditions—Juvenile literature. 3. Great Britain—History—Tudors, 1485-1603—Juvenile literature. [1. Great Britain—Social life and customs. 2. Great Britain—History—Tudors, 1485-1603] I. Title.
HQ792.G7W54 1978 941.05 77-84931
78 MARC

—HUNGARY

Csőregh, Eva.
Lakótelepi gyerekek / Csőregh Eva. — Budapest : Közgazd. és Jogi Kiadó, 1978.
259, [1] p. ; 20 cm. H78
Bibliography: p. 257-[260]
ISBN 9632205588 : 24.00Ft
1. Children—Hungary. 2. Housing—Hungary. I. Title.
HQ792.H8C76 79-362439
*79 MARC

—HUNGARY—BUDAPEST—ANTHROPOMETRY

Az 1-60 [i.e. Egytől hatvan] hónapos budapesti gyermekek testi fejlettsége, szociodemográfiai és morbiditási viszonyai (előzetes jelentés / [szerkők, Ács Jolán ... et al. ; szerk., Sárkány Jenő]. — Budapest : Központi Statisztikai Hivatal, Népességtudományi Kutató Intézet, 1977.
237 p. : ill. ; 29 cm. — (A Központi Statisztikai Hivatal Népességtudományi Kutató Intézetének és a Magyar Tudományos Akadémia Demográfiai Bizottságának közleményei ; 45)
Summary in English (37 p.) inserted.
Bibliography p. 101-102.
ISBN 9633401488 : 116.00Ft
1. Children—Care and hygiene—Hungary—Budapest—Longitudinal studies. 2. Children—Hungary—Budapest—Growth—Longitudinal studies. 3. Children—Hungary—Budapest—Anthropometry. 4. Children—Diseases—Hungary—Budapest—Longitudinal studies. I. Ács, Jolán. II. Sárkány, Jenő. III. Series: Népességtudományos Kutató Intézet. A Központi Statisztikai Hivatal Népességtudományi Kutató Intézetének és a Magyar Tudományos Akadémia Demográfiai Bizottságának közleményei ; 45.
HA1201.A64 no. 45 79-355878
[RJ103.H9] 79 MARC

—HUNGARY—BUDAPEST—GROWTH—LONGITUDINAL STUDIES

Az 1-60 [i.e. Egytől hatvan] hónapos budapesti gyermekek testi fejlettsége, szociodemográfiai és morbiditási viszonyai (előzetes jelentés / [szerkők, Ács Jolán ... et al. ; szerk., Sárkány Jenő]. — Budapest : Központi Statisztikai Hivatal, Népességtudományi Kutató Intézet, 1977.
237 p. : ill. ; 29 cm. — (A Központi Statisztikai Hivatal Népességtudományi Kutató Intézetének és a Magyar Tudományos Akadémia Demográfiai Bizottságának közleményei ; 45)
Summary in English (37 p.) inserted.
Bibliography p. 101-102.
ISBN 9633401488 : 116.00Ft
1. Children—Care and hygiene—Hungary—Budapest—Longitudinal studies. 2. Children—Hungary—Budapest—Growth—Longitudinal studies. 3. Children—Hungary—Budapest—Anthropometry. 4. Children—Diseases—Hungary—Budapest—Longitudinal studies. I. Ács, Jolán. II. Sárkány, Jenő. III. Series: Népességtudományos Kutató Intézet. A Központi Statisztikai Hivatal Népességtudományi Kutató Intézetének és a Magyar Tudományos Akadémia Demográfiai Bizottságának közleményei ; 45.
HA1201.A64 no. 45 79-355878
[RJ103.H9] 79 MARC

—INDIA

Pandey, Rama Shankar, 1923-
Child socialization in modernization / Rama S. Pandey. — Bombay : Somaiya Publications, c1977.
x, 191 p. ; 22 cm.
Bibliography: p. [175]-185.
Includes index.
Rs40.00
1. Children in India. 2. Socialization. 3. India—Social conditions—1947- . I. Title.
HQ792.I5P35 301.15'72 77-907006
78 MARC

—INDIA—CALCUTTA METROPOLITAN AREA—ANTHROPOMETRY

India (Republic). Dept. of Anthropology.
A radiological study of skeletal maturity in hand-wrist area of Bengali boys and girls / S. R. Das ... [et al.]. — Calcutta : Anthropological Survey of India, Govt. of India, 1976.
79 p. ; 25 cm. — (Memoir - Anthropological Survey of India ; no. 38)
Bibliography: p. [49]-56.
Rs12.95
1. Children—India—Calcutta metropolitan area—Anthropometry. 2. Hand. 3. Wrist. I. Das, S. R. II. Title. III. Series: India (Republic). Dept. of Anthropology. Memoir ; no. 38.
GN63.I5 1976 573'.6 77-905123
79 MARC

—ITALY—ATTITUDES

Viviani, Cesare.
La pazzia spiegata dai bambini / Cesare Viviani. — Milano : Il formichiere, 1976.
128 p. ; 23 cm. — (Spazio aperto) It77-Mar
L3000
1. Mental illness—Addresses, essays, lectures. 2. Children's writings, Italian. 3. Children—Italy—Attitudes. I. Title.
RC458.V57 79-382215
*79 MARC

—ITALY—SAN BARNABA—BIOGRAPHY

Baroni, Giuseppe, fl. 1978-
Infanzia a Castel San Barnaba / Giuseppe Baroni. — [Cosenza] : Pellegrini, 1978.
163 p. ; 22 cm. It78-Oct
L2000
1. Baroni, Giuseppe, fl. 1978- . 2. San Barnaba, Italy—Biography. 3. Children—Italy—San Barnaba—Biography. I. Title.
DG975.S1837B372 858'.9'1409 79-348524
*79 MARC

—ITALY—SICILY—BIOGRAPHY

Fulco.
Estati felici : un'infanzia in Sicilia / [di] Fulco. — 1. ed. italiana. — Milano : Feltrinelli, 1977.
189 p., [2] leaves of plates : ill. ; 20 cm. It78-Aug
Translation of The happy summer days.
L4500
1. Sicily—Social life and customs. 2. Fulco. 3. Sicily—Nobility—Biography. 4. Children in Sicily—Biography. I. Title.
DG865.6.F8416 945.8'23 78-387799
*78 MARC

—ITALY—SIENA

Pincherle, Mario.
I segni : diluvio unifamiliare / Mario Pincherle. — Ancona : Filelfo, [1978]
223 p. : ill. ; 20 cm. — (L'Avventura dell'uomo ; 7) It78-Nov
L6000
1. Pincherle, Mario. 2. Anthropologists—Italy—Siena—Biography. 3. Children—Italy—Siena. 4. Siena—Biography. I. Title.
CC115.P56A37 78-400672
*79 MARC

—JAPAN

Ichikawa, Eisaku.
(Atarashii sedai ni okuru tadashii kodomo no shitsukekata)
新しい世代におくる正しい子供のしつけ方：学校・家庭のために / 市川英作著. — 東京：学芸図書出版社, 昭和26 [1951]
126 p., [1] leaf of plates : ill. ; 18 cm.
Colophon inserted.
1. Children—Management. 2. Children—Japan. 3. Education of children. I. Title. II. Title: Tadashii kodomo no shitsukekata.
HQ769.I33 79-809665

(Terebikko)
テレビっ子：この新しい世代とつき合うために / 読売新聞社婦人部編. — 東京：文化出版局, 1978.
230 p. : ill. ; 18 cm. — (よつば新書 ; 31) Ja 78-24704
Series romanized: Yotsuba shinsho.
Originally published in Yomiuri shimbun, Jan.-Apr. 1978.
Y600
1. Television and children. 2. Children—Japan. 3. Child development—Japan. I. Yomiuri Shimbun Sha. Fujinbu. II. Yomiuri shimbun.
HQ784.T4T464 79-801814

CHILDREN (Continued)

—JAPAN—ADDRESSES, ESSAYS, LECTURES

(Aozora shōnen shōjo gakkō)
青空少年少女学校：地域に子ども会·少年団を / ［少年少女組織を育てる全国センター編］. — 東京：同センター, 1977.
149 p. ; 21 cm. Ja 78-9445
¥800

1. Children's clubs—Programs—Addresses, essays, lectures. 2. Children—Japan—Addresses, essays, lectures. I. Shōnen Shōjo Soshiki o Sodateru Zenkoku Sentā.
HQ767.8.A59 79-803688

(Naku mo no ka)
泣くものか：作文集：子どもの人権10年の証言 / 養護施設協議会編. — 東京：亜紀書房, 1977.
249 p. : ill. ; 19 cm. Ja 77-27417
¥950

1. Children—Japan—Institutional care—Addresses, essays, lectures. 2. Childrens' rights—Japan—Addresses, essays, lectures. 3. Child welfare—Japan—Addresses, essays, lectures. I. Zenkoku Shakai Fukushi Kyōgikai. Yōgo Shisetsu Kyōgikai.
HV1312.N34 79-808623

—JAPAN—MANAGEMENT

(Chū-kōkōsei no ko o motsu oya e)
中·高校生の子をもつ親へ / 丸岡秀子, 丸木政臣編集. — 東京：ダイヤモンド社, 1979.
vii, 218 p. : ill. ; 20 cm. — (親のための教育学 ; 4) Ja***
¥1200

1. Parent and child—Japan. 2. Children—Japan—Management. 3. High school students—Japan. I. Maruoka, Hideko, 1903?- II. Maruki, Masaomi. III. Series: Oya no tame no kyōikugaku ; 4.
HQ755.85.C5 79-809968

Higuchi, Kiyoyuki, 1909-
(Oyaji taibō ron)
おやじ待望論 / 樋口清之著. — 東京：千曲秀版社, ［昭和52 i. e. 1977］
267 p. ; 20 cm. Ja 77-20161
¥980

1. Fathers—Japan. 2. Children—Japan—Management. I. Title.
HQ756.H5 79-809995

Onosawa, Minoru, 1920-
(Tōsan wa kūki?!)
父さんは空気?!：現代っ子の証言 / 小野澤實著. — 京都：PHP研究所, 1977.
235 p. ; 19 cm. Ja 78-3921
¥880

1. Children—Japan—Management. 2. Parent and child—Japan. 3. Family—Japan. I. Title.
HQ755.85.O44 79-810012

Sekine, Shōichi, 1928-
(Tomobataraki katei kyōiku ron)
共働き家庭教育論 / 関根庄一. — 東京：一光社, 1978.
173 p. ; 18 cm. Ja 79-4549
¥500

1. Parent and child—Japan. 2. Children—Japan—Management. 3. Family—Japan. 4. Mothers—Employment—Japan. I. Title.
HQ769.S445 79-807729

Shinagawa, Fujirō, 1916-
(Hitorikko)
一人っ子：どう育てればいいか / 品川不二郎著. — 東京：日新報道, 昭和51 ［1976］
230 p. : ill. ; 19 cm. Ja 76-16611
¥870

1. Children—Japan—Management. 2. Only child. I. Title.
HQ777.3.S44 79-809680

Shizume, Yasuo, 1925-
(Onna ni ikuji wa makaserarenai)
女に育児はまかせられない / 鎮目恭夫. — 東京：実業之日本社, 1978.
238 p. ; 18 cm. — (実日新書) Ja 79-3171
Series romanized: Jippi shinsho.
¥630

1. Children—Japan—Management. 2. Parent and child—Japan. I. Title.
HQ769.5.S44 79-807710

(Shōgakusei no ko o motsu oya e)
小学生の子をもつ親へ / 丸岡秀子, 丸木政臣編集. — 東京：ダイヤモンド社, 1979.
viii, 213 p. : ill. ; 20 cm. — (親のための教育学 ; 3) Ja***
¥1200

1. Children—Japan—Management. 2. Parent and child—Japan. 3. Education of children. 4. Education—Japan. I. Maruoka, Hideko, 1903?- II. Maruki, Masaomi. III. Series: Oya no tame no kyōikugaku ; 3.
HQ769.S533 79-809488

—JAPAN—PICTORIAL WORKS

Kishimoto, Masayoshi, 1935-
(Bokura no hōkago)
ぼくらの放課後：岸本正義写真集. — 東京：一声社, 1978.
150 p. : chiefly ill. ; 26 cm. Ja 78-25023
¥2000

1. Children—Japan—Pictorial works. 2. School children—Japan—Recreation—Pictorial works. I. Title.
HQ792.J3K55 79-805285

—JAPAN—STATISTICS

Japan. Kōseishō. Jidōkyoku.
(Nihon no kodomo to sono katei no jittai)
日本の子どもとその家庭の実態：全国家庭児童調査結果報告 / ［厚生省児童局］. — 東京：日本児童福祉協会, 昭和39 ［1964］
69 p. : graphs ; 21 cm.
Cover title.

1. Children—Japan—Statistics. 2. Child welfare—Japan—Statistics. 3. Family—Japan—Statistics. I. Title.
HQ792.J3J35 1964 79-803058

—JAPAN—OKINAWA (PREFECTURE)

Kinjō, Kazuhiko.
(Aa Okinawa sen no gakutotai)
嗚呼沖縄戦の學徒隊 / 金城和彦著. — 東京：原書房, 1978.
12, 506 p., ［6］ leaves of plates : ill. ; 20 cm. Ja 78-18086
¥2500

1. World War, 1939-1945—Campaigns—Ryukyu Islands—Okinawa Island. 2. World War, 1939-1945—Children. 3. Children in Okinawa Island. 4. Okinawa Island—History. 5. High school students—Japan—Okinawa (Prefecture) I. Title.
D767.99.O45K56 78-813142

—KENYA—MANAGEMENT

Blount, Ben G 1940-
Aspects of Luo socialization / by Ben G. Blount. — ［Nairobi］ : University of Nairobi, Institute of African Studies, 1971.
15 leaves ; 30 cm. — (Discussion paper - University of Nairobi, Institute of African Studies ; no. 23)
Bibliography: leaf 15.
1. Luo (Nilotic tribe)—Social life and customs. 2. Children—Kenya—Management. I. Title. II. Series: Nairobi. University. Institute of African Studies. Discussion paper - University of Nairobi, Institute of African Studies ; no. 23.
DT433.542.B58 301.15'72'096762 78-108265
79 MARC

—KOREA—CONDUCT OF LIFE

Chŏng, Wŏn-sik.
(Han'guk adong ŭi todŏksŏng paltal)
韓國児童의道德性発達 / 鄭元植. — ［서울］ : 鄭元植, ［1973］
27 leaves ; 26 cm. — (1972년도문교부학술연구조성비에의한연구보고서 : 인문과학계)
Added t. p.: Development of morality in Korean children.
In English with pref. in Korean.
Bibliography: leaf 27.
1. Moral development—Korea. 2. Children—Korea—Conduct of life. 3. Judgement (Ethics) I. Title. II. Title: Development of morality in Korean children. III. Series: Mun'gyobu haksul yŏn'gu chosŏngbi e ŭihan yŏn'gu pogosŏ ; 1972-yŏndo.
BF723.M54C48 79-825953

—MEXICO—ANTHROPOMETRY

Romero, Javier, fl. 1937-
Aspectos psicobiométricos y sociales de una muestra de la juventud mexicana. México, Instituto Nacional de Antropología e Historia, 1956.
63 p. illus. 23 cm. (Dirección de Investigaciones Antropológicas. Publicaciones, 1)
1. Children—Mexico—Anthropometry. 2. Youth—Mexico. I. Title. II. Series: Mexico. Departamento de Investigaciones Antropológicas. Publicaciones, 1.
GN58.M6R65 67-52850
［r79］rev MARC

—NETHERLANDS

Netherlands (Kingdom, 1815-). Centraal Bureau voor de Statistiek.
Het kind in tel : cijfers over kinderen in Nederland in de zeventiger jaren : uitgave ter gelegenheid van het Internationaal Jaar van het kind, 1979 / Centraal Bureau voor de Statistiek. — 's-Gravenhage : Staatsuitgeverij, 1978.
44 p. ; 21 cm. Ne***
Bibliography: p. 40-44.
fl 6.50
1. Children—Netherlands—Statistics. 2. Children—Netherlands. I. Title.
HQ792.N4N46 1978 79-360217
79 MARC

Zandvoort, W J F
Kinderen en de kinderbescherming / W. J. F. Zandvoort. — Amsterdam : Meulenhoff Informatief, ［1978］
144 p. ; 20 cm. Ne78-20
Bibliography: p. 140-［144］
ISBN 9029095016 : fl 19.50
1. Child welfare—Netherlands. 2. Children in the Netherlands. I. Title.
HV778.A6Z36 78-375470
*78 MARC

—NETHERLANDS—CASE STUDIES

School en toekomst : beelden van de ontwikkelingsgang van kinderen gedurende de puberteit / door S. Wiegersma ... ［et al.］ ; met een voorwoord van J. Koekebakker ; in opdracht van het Nederlands Institut voor Praeventieve Geneeskunde. — Haarlem : De Toorts, ［1963］
250 p. ; 24 cm.
Includes indexes.
1. Children—Netherlands—Case studies. 2. Child development—Case studies. 3. Child psychology—Netherlands—Case studies. I. Wiegersma, S.
HQ792.N4S43 79-344405
79 MARC

CHILDREN (Continued)

—NETHERLANDS—JUVENILE LITERATURE

Opzeeland, Bob van.
Waar hoor ik thuis? / geschreven door Bob van Opzeeland, in samenwerking met Robert de Jong en Almar Tjepkema ; met foto's van Thijs Koobs, Maya Pejic en Almar Tjepkema ; met kaartjes van Ineke Lubsen. — Amsterdam : Ploegsma, ⌊1978⌋
79 p. : ill. ; 27 cm. Ne***
ISBN 9021603993
1. Netherlands—Foreign population—Juvenile literature. 2. Minorities—Netherlands—Juvenile literature. 3. Identity (Psychology)—Juvenile literature. 4. Children—Netherlands—Juvenile literature. I. Jong, Robert de, joint author. II. Tjepkema, Almar, joint author. III. Title.
DJ91.O69 79-354615
79 MARC

—NETHERLANDS—STATISTICS

Netherlands (Kingdom, 1815-). Centraal Bureau voor de Statistiek.
Het kind in tel : cijfers over kinderen in Nederland in de zeventiger jaren : uitgave ter gelegenheid van het Internationaal Jaar van het kind, 1979 / Centraal Bureau voor de Statistiek. — 's-Gravenhage : Staatsuitgeverij, 1978.
44 p. ; 21 cm. Ne***
Bibliography: p. 40-44.
fl 6.50
1. Children—Netherlands—Statistics. 2. Children—Netherlands. I. Title.
HQ792.N4N46 1978 79-360217
79 MARC

—NEW YORK (CITY)—PICTORIAL WORKS

New York street kids : 136 photographs / selected by the Children's Aid Society ; with captions by John von Hartz. — New York : Dover Publications, 1978.
106 p., ⌊1⌋ leaf of plates : chiefly ill. ; 28 cm.
ISBN 0-486-23692-7 : $14.50
1. City children—Pictorial works. 2. Children—New York (City)—Pictorial works. I. Von Hartz, John. II. Children's Aid Society, New York.
HT206.N48 1978 301.43'14'097471 78-57823
79 MARC

—NEW ZEALAND

New Zealand. Social Development Council.
Bringing up children in New Zealand : can we do better? — ⌊Wellington⌋ : Social Development Council, 1977.
21 p. ; 30 cm. NZ***
Includes bibliography.
1. Children—Management. 2. Children—New Zealand. I. Title.
HQ772.N492 1977 649'.1'09931 78-323474
79 MARC

—NORWAY—ADDRESSES, ESSAYS, LECTURES

Barn / bidrag av Anne-Marit Duve ... ⌊et al.⌋ ; redaksjon : Kari Bruun Wyller, Thomas Chr. Wyller. — Oslo : Gyldendal, 1978 (Oslo : Norbok).
118, ⌊3⌋ p. ; 21 cm. — (Motforestillinger ; 4) N78-July
Bibliography: p. ⌊121⌋
ISBN 8205108854 : kr89.00
1. Child development—Addresses, essays, lectures. 2. Children—Norway—Addresses, essays, lectures. I. Duve, Anne-Marit, 1919- . II. Wyller, Kari Bruun. III. Wyller, Thomas Christian, 1922- . IV. Series.
HQ772.B379 78-400960
79 MARC

—NORWAY—ATTITUDES

Raundalen, Tora Synøve.
Er du på vår side! : rapport fra barnemiljøet, intervju med 4000 barn, foreldre og lærere / Tora Synøve og Magne Raundalen. — ⌊Oslo⌋ : Universitetsforlaget, c1979.
296 p. : ill. ; 22 cm. N***
Includes bibliographical references.
ISBN 8200271706
1. Children—Norway—Social conditions. 2. Children—Norway—Attitudes. I. Raundalen, Magne, joint author. II. Title.
HQ792.N7R38 79-380615
79 MARC

—NORWAY—BIOGRAPHY

Det Hendte i Bergen : fra overrettssakførerens etterlatte papirer / bearb. av Carl Konow Rieber-Mohn. — ⌊S. l. : s. n.⌋, 1976 (Bergen : Bergmann)
136 p. ; 21 cm. N***
1. Norway—Biography. 2. Children—Norway—Biography. I. Rieber-Mohn, Carl K.
CT1308.Z9H46 79-341298
79 MARC

—NORWAY—SOCIAL CONDITIONS

Raundalen, Tora Synøve.
Er du på vår side! : rapport fra barnemiljøet, intervju med 4000 barn, foreldre og lærere / Tora Synøve og Magne Raundalen. — ⌊Oslo⌋ : Universitetsforlaget, c1979.
296 p. : ill. ; 22 cm. N***
Includes bibliographical references.
ISBN 8200271706
1. Children—Norway—Social conditions. 2. Children—Norway—Attitudes. I. Raundalen, Magne, joint author. II. Title.
HQ792.N7R38 79-380615
79 MARC

—PALESTINE

Keizer, D P R
Het kinderleven in bijbelse tijd : gezondheid en ziekte, geneesmiddelen, spel, ouderliefde en opvoeding, voedsel, kruiden en planten, naamgeving, bijgeloof en amuletten, enz. / D. P. R. Keizer. — Kampen : Kok, ⌊1979⌋
178 p. : with ill. ; 22 cm. Ne79-17
Bibliography: p. 170-173.
Includes index.
ISBN 9024214106 : fl 24.75
1. Children in the Bible. 2. Children—Palestine. I. Title.
BS576.K44 79-379814
*79 MARC

—PERU—LANGUAGE

Mendoza Cuba, Aida.
Sistema fonológico del castellano y variantes regionales / Aida Mendoza, con la colaboración de María del Carmen Cuba. — Lima : Inide, Ministerio de Educación, Instituto Nacional de Investigación y Desarrollo de la Educación Augusto Salazar Bondy, Subdirección de Investigaciones Educacionales, 1976.
152 p. : ill. ; 29 cm. — (Investigación El Lenguaje del niño peruano hispanohablante ; Informe parcial ; no. 3 ⌊pt. 2⌋)
Bibliography: p. 151-152.
1. Spanish language—Dialects—Peru. 2. Spanish language—Dialects—Phonetics. 3. Children—Peru—Language. I. Cuba, María del Carmen. II. Title. III. Series: Investigación El Lenguaje del niño peruano hispanohablante : Informe ; no. 3, ⌊pt. 2⌋.
PC4901.A25 no. 3, pt 2 78-110471
⌊PC4901⌋ 79 MARC

—RHODESIA, SOUTHERN—BIOGRAPHY

Kachengere.
Fear and fun : Rhodesian recollections / by Kachengere (R. H. Lees). — ⌊Axminster⌋ : ⌊The Author⌋, ⌊1978⌋
⌊8⌋, 184 p., plate : port. ; 19 cm. GB79-04752
1. Kachengere. 2. Rhodesia, Southern—Biography. 3. Children—Rhodesia, Southern—Biography. I. Title.
CT1978.K3A33 968.9'1'020924 79-312472
79 MARC

—SWEDEN

Rönström, Anitha, 1939-
Barnomsorgen : intentioner och verklighet / Anitha Rönström. — Lund : Studentlitt., 1977.
130 p. ; 23 cm. S77-50/52
Bibliography: p. 124-128.
Includes index.
ISBN 9144142714 : kr31.00
1. Child welfare—Sweden. 2. Children in Sweden. 3. Socialization. 4. Child abuse—Sweden. I. Title.
HV790.A6R63 78-389094
79 MARC

Sweden. Utbildningsdepartementet. Barnkulturgruppen.
Barnen och kulturen : en rapport / från Barnkulturgruppen. — Stockholm : LiberFörlag, 1978.
300, ⌊2⌋ p. ; 21 cm. — (Publica ISSN 0346-6388) S78-19
Bibliography: p. ⌊301⌋
ISBN 9138041855
1. Children—Sweden. 2. Arts and children—Sweden. I. Title.
HQ792.S8S77 1978 79-343215
79 MARC

Viklund, Monica, 1943-
Barnen behöver oss / Monica Viklund ; ⌊teckningar o. omsl., Helga Henschen⌋. — Stockholm : Socialstyr. ; Lund : Liber-Läromedel, 1978.
157 p. : ill., tables ; 21 cm. S78-17
Bibliography: p. 153-157.
ISBN 9140045900
1. Children—Sweden. 2. Child development. 3. Child psychology. 4. Child welfare—Sweden. I. Title.
HQ792.S8V54 78-395310
79 MARC

—SWEDEN—ATTITUDES

När jag blir stor : 21 barn berättar om framtiden för Lisa Berg och Annika Holm / efterord, Johan Galtung. — Stockholm : Gidlund, 1979.
171 p. : ill. ; 21 cm. S79
ISBN 9170212791
1. Children—Sweden—Attitudes. 2. Children—Sweden—Social conditions. I. Berg, Lisa, 1942- . II. Holm, Annika, 1937- .
HQ792.S8N33 79-382597
79 MARC

—SWEDEN—HISTORY

Ambjörnsson, Ronny, 1936-
Familjeporträtt : essäer om familjen, kvinnan, barnet och kärleken i historien / Ronny Ambjörnsson. — Stockholm : Gidlund, 1978.
188, ⌊2⌋ p. : ill. ; 19 cm. S78
Includes bibliographies.
ISBN 9170212481
1. Family—Sweden—History. 2. Children—Sweden—History. 3. Women—Sweden—Social conditions. I. Title.
HQ647.A6 79-360456
79 MARC

—SWEDEN—SOCIAL CONDITIONS

Barn och andra människor : en bok för barnskötare / Birgit Andersson-Nyander ... ⌊et al.⌋. — Stockholm : Esselte studium, 1978.
266 p. : ill. ; 24 cm. S78-23
Includes index.
ISBN 9124275557
1. Children—Sweden—Social conditions. 2. Child development. I. Andersson-Nyander, Birgit.
HQ792.S8B37 79-342107
79 MARC

Bättre barnmiljö / ⌊sammanst. och red. av Margaretha Arnell⌋. — Stockholm : Fören. Sveriges socialchefer, 1977.
160 p. : ill. ; 21 cm. S77-53
1. Child welfare—Sweden. 2. Children—Sweden—Social conditions. I. Arnell, Margaretha, 1929- . II. Föreningen Sveriges socialchefer.
HV790.A6B33 79-382399
79 MARC

När jag blir stor : 21 barn berättar om framtiden för Lisa Berg och Annika Holm / efterord, Johan Galtung. — Stockholm : Gidlund, 1979.
171 p. : ill. ; 21 cm. S79
ISBN 9170212791
1. Children—Sweden—Attitudes. 2. Children—Sweden—Social conditions. I. Berg, Lisa, 1942- . II. Holm, Annika, 1937- .
HQ792.S8N33 79-382597
79 MARC

Sweden. Familjestödsutredningen.
Daghem för små barn : diskussionspromemoria / från Familjestödsutredningen. — Stockholm : LiberFörlag/Allmänna förl., 1978.
163 p. : diagr., ill. ; 25 cm. S78-26
Bibliography: p. 161-163.
ISBN 9138043076 : kr25.00
1. Day care centers—Sweden. 2. Children—Sweden—Social conditions. I. Title.
HV861.S9S93 1978 79-342082
79 MARC

—SWEDEN—LUND—BIOGRAPHY

Lundberg, Alfred, 1873-1957.
När velocipeden var ung i Lund / Alfred Lundberg. — Lund, ⌊Kulturen, box 1095⌋ : Fören. Gamla Lund, 1976.
130 p. : ill. ; 24 cm. — (Årsskrift - Gamla Lund, Förening för bevarande av stadens minnen ; 59 ISSN 0347-7711) S77-32/33
Autobiographical.
kr25.00
1. Lundberg, Alfred, 1873-1957. 2. Lund—Biography. 3. Children—Sweden—Lund—Biography. I. Title. II. Series: Föreningen Det gamla Lund. Skrifter ; 59.
DL991.L9L854 1976 79-364504
79 MARC

CHILDREN (Continued)

—TAIWAN

Ch'en, Mao-liang.
(Chung-hua min kuo ch'ing shao nien chih t'i ko yü t'i neng chuang k'uang fa chan ch'ü shih)
中華民國青少年之體格與體能狀況發展趨勢之研究 / 陳懋良主持 ; 教育部體育司編印. — ⌊台北⌋ : 該司, 民國65 ⌊1976⌋
26 p. : ill. ; 21 cm. — (體育學術研究報告 ; 1)
Summary in Chinese and English.
Bibliography: p. 17.
1. Children—Growth. 2. Physical fitness. 3. Children—Taiwan. I. Title. II. Series: T'i yü hsüeh shu yen chiu pao kao ; 1.
RJ131.C5 312'.6'0951 79-838169

Ch'en, Yung.
(Tien shih tui erh t'ung sheng huo ying hsiang chih t'iao ch'a yen chiu)
電視對兒童生活影響之調查研究 / 陳墉著. — 臺北 : 臺灣省立臺北師範專科學校, 民國66 ⌊1977⌋
155 p. ; 21 cm. — (教育研究叢書 ; 第42種)
Bibliography: p. 147-150.
1. Television and children. 2. Children—Taiwan. I. Title. II. Series: Chiao yü yen chiu ts'ung shu (T'ai-pei) : ti 42 chung.
HQ784.T4C45 79-835659

—UNITED STATES

see also Italian American children; Mexican American children; Presidents—United States—Children

De Lone, Richard H
Small futures : children, inequality, and the limits of liberal reform / by Richard H. de Lone for the Carnegie Council on Children. — 1st ed. — New York : Harcourt Brace Jovanovich, c1979.
xiv, 258 p. ; 24 cm.
Includes bibliographical references and index.
ISBN 0-15-183128-9
1. Children—United States. 2. Children's rights—United States. 3. Child welfare—United States. 4. Equality. 5. Liberalism—United States. I. Carnegie Council on Children. II. Title.
HQ792.U5D44 301.43'1'0973 77-92536
79 MARC

National Council of Organizations for Children and Youth.
America's children 1976 : a bicentennial assessment : a fact-book about children in poverty, child health problems, changes in American family structure, child care needs, and federal programs serving children. — Washington : National Council of Organizations for Children and Youth, 1976.
87 p. : ill. ; 26 cm.
$4.00
1. Children—United States. 2. Children—United States—Statistics. 3. Child welfare—United States. 4. Social indicators—United States. I. Title.
HQ792.U5N37 1976 362.7'0973 75-42834
79 MARC

—UNITED STATES—ATTITUDES

Listen to us : the Children's express report / edited by Dorriet Kavanaugh ; produced by children's express children and Robert Clampitt. — New York : Workman Pub., ⌊1978⌋
254 p. ; 29 cm.
SUMMARY: Children, ages 7 to 13, candidly discuss their feelings about their families, school, sex, lack of respect for children, and other topics.
ISBN 0-89480-018-3 : $10.95. ISBN 0-89480-017-5 pbk. : $5.95
1. Children—United States—Attitudes. 2. Parent and child. ⌊1. Children—Attitudes. 2. Family problems⌋ I. Kavanaugh, Dorriet. II. Children's express.
HQ772.5.L57 301.43'14 77-94032
78 MARC

—UNITED STATES—PORTRAITS—EXHIBITIONS

Humm, Rosamond Olmsted.
Children in America : a study of images and attitudes : September 30, 1978-May 27, 1979, the High Museum of Art, Atlanta, Georgia / by Rosamond Olmsted Humm. — Atlanta : The Museum, 1978.
72 p. : ill. (some col.) ; 22 x 25 cm.
Bibliography: p. 66-68.
1. Children—United States—Portraits—Exhibitions. 2. Children in art—Exhibitions. I. High Museum of Art. II. Title.
N7644.H84 78-52513
760'.0973'07401558213
79 MARC

—UNITED STATES—STATISTICS

National Council of Organizations for Children and Youth.
America's children 1976 : a bicentennial assessment : a fact-book about children in poverty, child health problems, changes in American family structure, child care needs, and federal programs serving children. — Washington : National Council of Organizations for Children and Youth, 1976.
87 p. : ill. ; 26 cm.
$4.00
1. Children—United States. 2. Children—United States—Statistics. 3. Child welfare—United States. 4. Social indicators—United States. I. Title.
HQ792.U5N37 1976 362.7'0973 75-42834
79 MARC

—VENEZUELA—MONAGAS—SOCIALIZATION

Ruddle, Kenneth.
Education for traditional food procurement in the Orinoco delta / by Kenneth Ruddle and Ray Chesterfield. — Berkeley : University of California Press, 1977.
x, 172 p., ⌊6⌋ leaves of plates : ill. ; 24 cm. — (Ibero-Americana ; 53)
Bibliography: p. 161-165.
Includes index.
ISBN 0-520-09551-0
1. Children—Venezuela—Monagas—Socialization. 2. Children—Education—Venezuela—Monagas. 3. Food habits—Venezuela—Monagas. 4. Human ecology—Venezuela—Monagas. I. Chesterfield, Ray, joint author. II. Title. III. Series.
HQ792.V4R8 301.5'2'09875 75-40954
79 MARC

CHILDREN, ABANDONED see Abandoned children

CHILDREN, ABNORMAL AND BACKWARD see Exceptional children; Handicapped children

CHILDREN, ADOPTED

Popescu, Florica.
Părinţi vitregi, părinţi buni / Florica Popescu. — Bucureşti : Editura Didactică şi Pedagogică, 1976.
55 p. ; 17 cm. — (Pentru părinţi) R77-249
lei1.45
1. Children, Adopted. I. Title.
HV875.P65 78-367330
*78 MARC

—FAMILY RELATIONSHIPS

Weterings, A M 1937-
Het pleeggezin als opvoedingssituatie : een empirisch onderzoek naar de opvoedingsrelatie van voogdijpupillen in een pleeggezin / A. M. Weterings. — Groningen : Brouwer International Publishing, ⌊1977⌋
291 p. ; 24 cm. Ne78-4
Bibliography: p. 274-278.
ISBN 9070198029 : fl 25.00
1. Children, Adopted—Family relationships. 2. Adoption. I. Title.
HV875.W46 78-368449
*78 MARC

—CANADA—BIOGRAPHY

McConnell, Jean, 1947?-
Un homme pleure / Jean McConnell. — ⌊Montréal⌋ : Stanké, c1978.
173 p. : ill. ; 21 cm. C***
ISBN 0-88566-118-4 : $8.25
1. Adoption. 2. McConnell, Jean, 1947- . 3. Children, Adopted—Canada—Biography. I. Title.
HV873.M35 362.7'34'0971 79-347832
79 MARC

—ITALY—FAMILY RELATIONSHIPS—JUVENILE LITERATURE

Antonacci Ciccarelli, Emma, 1904-
Una famiglia per Rocco / Emma Ciccarelli Antonacci. — L'Aquila : Japadre, c1977.
148 p. : ill. ; 19 cm. — (La Spola) It***
Bibliography: p. 147-148.
L1800
1. Abandoned children—Italy—Juvenile literature. 2. Children, Adopted—Italy—Family relationships—Juvenile literature. I. Title.
HV887.I8A57 362.7'34 79-351049
79 MARC

—NETHERLANDS

Nota, Jos A
Kinderen adopteren in Nederland / J. A. Nota. — Alphen aan den Rijn : H. D. Tjeenk Willink, 1979.
viii, 114 p. ; 22 cm. Ne79-13
Bibliography: p. 113.
ISBN 9060920392 : fl 19.50
1. Children, Adopted—Netherlands. 2. Adoption—Netherlands. I. Title.
HV887.N4N67 79-372318
*79 MARC

—NETHERLANDS—ADDRESSES, ESSAYS, LECTURES

Adoptie van buitenlandse kinderen / onder red. van W. H. G. Wolters. — Nijkerk : Callenbach, ⌊1978⌋
165 p. ; 21 cm. — (Introboek) Ne78-50
Includes bibliographies.
ISBN 9026616562 fl 24.50
1. Interracial adoption—Netherlands—Addresses, essays, lectures. 2. Children, Adopted—Netherlands—Addresses, essays, lectures. I. Wolters, W. H. G.
HV887.N4A36 79-355561
*79 MARC

—UNITED STATES

Lifton, Betty Jean.
Lost and found : the adoption experience / Betty Jean Lifton. — New York : Dial Press, c1979.
303 p. ; 24 cm.
Includes bibliographical references and index.
ISBN 0-8037-5036-6 : $9.95
1. Adoptions—United States. 2. Children, Adopted—United States. 3. Children, Adopted—United States—Family relationships. I. Title.
HV875.L458 1979b 929'.1 78-21526
78 MARC

Sheptock, Joanne.
Our growing family / Joanne and Rudy Sheptock, with Wynelle B. Gardner. — Plainfield, N.J. : Logos International, c1979.
xi, 172 p., ⌊4⌋ leaves of plates : ill. ; 22 cm.
ISBN 0-88270-385-4 : $7.95
1. Children, Adopted—United States. I. Sheptock, Rudy, joint author. II. Gardner, Wynelle B., joint author. III. Title.
HV875.S49 362.7'34'0922 79-65214
79 MARC

—UNITED STATES—FAMILY RELATIONSHIPS

Lifton, Betty Jean.
Lost and found : the adoption experience / Betty Jean Lifton. — New York : Dial Press, c1979.
303 p. ; 24 cm.
Includes bibliographical references and index.
ISBN 0-8037-5036-6 : $9.95
1. Adoptions—United States. 2. Children, Adopted—United States. 3. Children, Adopted—United States—Family relationships. I. Title.
HV875.L458 1979b 929'.1 78-21526
78 MARC

CHILDREN, AFRO-AMERICAN see Afro-American children

CHILDREN, ATYPICAL see Exceptional children

CHILDREN, BACKWARD see Mentally handicapped children; Slow learning children

CHILDREN, BLACK

—GREAT BRITAIN

West African families in Britain : a meeting of two cultures / edited by June Ellis ; with contributions by June Ellis, Pat Stapleton, Vivien Biggs. — London ; Boston : Routledge and K. Paul, 1978.
viii, 142 p. ; 23 cm. — (Library of social work) GB***
Includes bibliographies.
ISBN 0-7100-8954-6 : £5.95
1. Blacks—Great Britain. 2. Students, Black—Great Britain. 3. Children, Black—Great Britain. 4. Foster home care—Great Britain. I. Ellis, June. II. Stapleton, Pat. III. Biggs, Vivien.
DA125.N4W47 301.45'19'660941 78-40374
78 MARC

CHILDREN, BLIND

see also Blind—Education; Visually handicapped children

Fraiberg, Selma H
Insights from the blind / ₍by₎ Selma Fraiberg with the collaboration of Louis Fraiberg. — London : Souvenir Press, 1977.
xxii, 297 p. : ill. ; 23 cm. — (Human horizons series) (A Condor book) GB78-05928
Bibliography: p. 289-292.
Includes index.
ISBN 0-285-64848-9 : £5.00. ISBN 0-285-64849-7 pbk.
1. Children, Blind. 2. Blind—Psychology. 3. Infant psychology. I. Title.
HV1598.F69 1977b 362.7'8'41 79-304971
79 MARC

Willoughby, Doris.
A resource guide for parents and educators of blind children / by Doris M. Willoughby. — Baltimore : National Federation of the Blind, c1979.
142 p. ; 23 cm.
"Developed under the sponsorship of the National Federation of the Blind."
Bibliography: p. 139-142.
1. Children, Blind. I. National Federation of the Blind. II. Title.
HV1593.W54 362.7'8'41 79-118061
79 MARC

—CONGRESSES

International Symposium on Problems Concerning the Pre-School Education of Blind Children as Well as Parents' Guidance, Berlin, 1976.
International Symposium on Problems Concerning the Pre-School Education of Blind Children as Well as Parents' Guidance, held in Berlin, May 25-28, 1976 : a documentation / under the auspices of the World Council for the Welfare of the Blind and its European Regional Committee ; organised by the Association of the Blind and Partially Sighted of the GDR. — Berlin : Blinden-und-Sehschwachen-Verband der DDR, ₍1976?₎
213 p. ; 21 cm. — (Series of publications - Association of the Blind and Partially Sighted of the GDR ; no. 25) GDR***
1. Children, Blind—Congresses. 2. Blind—Education—Congresses. I. Blinden-und-Sehschwachen-Verband der DDR. II. Series: Blinden-und-Sehschwachen-Verband der DDR. Schriftenreihe des Blinden-und-Sehschwachen-Verbandes der DDR ; no. 25.
HV1598.I57 1976 371.9'11 79-307089
79 MARC

National Conference for Training Occupational, Physical, and Recreational Therapists in the Area of Deaf-Blind, Dallas, 1975.
Proceedings : training occupational, physical, and recreational therapists in the area of deaf-blind / prepared by Jill M. Gray and Robert Dantona. — Sacramento : California State Dept. of Education, 1977.
vi, 66 p. : ill. ; 28 cm.
"Prepared and published by the Southwestern Region Deaf-Blind Center, California State Department of Education."
"Prepared and published cooperatively by the Texas Education Agency ... and the California State Department of Education."
Includes bibliographies.
1. Blind-deaf—Rehabilitation—Congresses. 2. Blind-deaf—Services for—Congresses. 3. Children, Deaf—Congresses. 4. Children, Blind—Congresses. I. Gray, Jill M. II. Dantona, Robert. III. Southwestern Region Deaf-Blind Center. IV. Texas. Education Agency. V. California. Dept. of Education. VI. Title: Training occupational, physical, and recreational therapists in the area of deaf-blind.
HV1597.N37 1975 362.4'1 78-622076
79 MARC

—ITALY

Conte Oberto, Giuliana.
Handicappati e scuola : il bambino cieco nella scuola di tutti / ₍di₎ Giuliana Conte Oberto, Luciano Paschetta. — Torino : Stampatori, 1978.
225 p. : ill. ; 18 cm. — (Didattica ; 16) It79-Jan
L4000
1. Blind—Education—Italy. 2. Children, Blind—Italy. I. Paschetta, Luciano, joint author. II. Title.
HV1990.C65 371.9'11'045 79-349685
*79 MARC

—JAPAN

Tōkyō-to Shinshin Shōgaisha Fukushi Sentā.
(Mōnyūyōji no yōiku shidō)
盲乳幼児の養育指導：開設10周年記念 / ₍編集 東京都心身障害者福祉センター₎. — 東京：同センター, 1978.
193 p. : ill. ; 26 cm. — (心身障害者の指導技術) Ja 78-18048

1. Children, Blind—Japan. 2. Blind—Education—Japan. I. Title. II. Series: Shinshin shōgaisha no shidō gijutsu.
HV2113.T66 1978 79-800088

—UNITED STATES—CASE STUDIES

Morgan, Barbara Kysor.
Obstacle course / Barbara Kysor Morgan. — San Francisco : Chronicle Books, c1979.
176 p. ; 23 cm.
ISBN 0-87701-141-9
1. Children, Blind—United States—Case Studies. 2. Blind—Education. I. Title.
HV1795.M67 362.7'8'410926 79-16056
79 MARC

CHILDREN, BOOKS AND READING FOR
see Books and reading for children

CHILDREN, CUSTODY OF see Custody of children

CHILDREN, DEAF

see also Deaf—Means of communication

Bloom, Freddy.
Our deaf children : into the 80's / ₍by₎ Freddy Bloom. — 2nd revised ed. — Old Woking : Gresham Books, 1978.
₍7₎, 148 p. : ill. ; 22 cm. GB78-39177
ISBN 0-905418-21-2 : £3.95
1. Children, Deaf. 2. Children, Deaf—Family relationships. I. Title.
HV2380.B57 1978 649'.15'12 79-310651
79 MARC

Denmark. Socialstyrelsen. Revalidering og forsorg.
Det hørehæmmede barn. København, Komiteen for Sundhedsoplysning, 1970.
22 p. illus. 22 cm. (Specialpædagogiske vejledninger) kr2.30
D71-30
1. Children, Deaf. 2. Children, Deaf—Denmark. I. Title.
HV2395.D47 1970 79-375913
79 MARC

Eggermont, J P M
Taalverwerving bij een groep dove kinderen; een experimenteel onderzoek naar de betekenis van een geluidsmethode voor het spraakafzien, door J. P. M. Eggermont. Groningen, J. B. Wolters, 1964.
159 p. 22 cm.
Errata slip inserted.
Summary in English.
Bibliography: p. 149-150.
1. Deaf—Education. 2. Children, Deaf. 3. Deaf—Means of communication. I. Title.
HV2443.E35 78-414919
78 MARC

Gałkowski, Tadeusz S
Psychologia dziecka głuchego : ₍skrypt₎ / Tadeusz Gałkowski Irmina Kunicka-Kaiser, Jadwiga Smoleńska. — Wyd. 2. — Warszawa : Państ. Wydaw. Naukowe, 1978 (Wroc : WDN)
320, ₍2₎ p. : ill. ; 24 cm. P78-3355
Includes bibliographical references.
zł40.00
1. Children, Deaf. 2. Deafness—Psychological aspects. I. Kunicka-Kaiser, Irmina, joint author. II. Smoleńska, Jadwiga, joint author. III. Title.
HV2395.G28 1978 78-400973
79 MARC

Schulte, Klaus.
Fragesätze als sprachdidaktisches Problem bei sprachlosen und spracharmen Kindern : Vorschläge z. Gewinnung e. frühen u. umfassenden interrogativen Kompetenz / Klaus Schulte. — Villingen-Schwenningen : Neckar-Verlag, 1974.
96 p. ; 21 cm. — (Wissenschaftliche Beiträge aus Forschung, Lehre und Praxis zur Rehabilitation behinderter Kinder und Jugendlicher ; WB 9) GFR75-A
Bibliography: p. 87-89.
Includes index.
ISBN 3-7883-0234-8 : DM18.00
1. Deaf—Means of communication. 2. Deaf—Education—Speech. 3. Deaf—Education—German. 4. Children, Deaf. I. Title. II. Series.
HV2483.S37 78-411731
78 MARC

Schulte, Klaus.
Phonembestimmtes Manualsystem (PMS) : Forschungsergebnisse u. Konsequenzen f. d. Artikulation hörgeschädigter Kinder / Klaus Schulte. — Villingen-Schwenningen : Neckar-Verlag, 1974.
212 p. : ill., graphs ; 21 cm. — (Wissenschaftliche Beiträge aus Forschung, Lehre und Praxis zur Rehabilitation behinderter Kinder und Jugendlicher ; WB 12) GFR74-A
Bibliography: p. 206-212.
ISBN 3-7883-0231-8 : DM35.00
1. Deaf—Means of communication. 2. Deaf—Education—Speech. 3. Deaf—Education—German language. 4. Children, Deaf. I. Title. II. Series.
HV2483.S373 78-366144
78 MARC

—ADDRESSES, ESSAYS, LECTURES

Deaf children : developmental perspectives / edited by Lynn S. Liben. — New York : Academic Press, 1978.
xv, 246 p. : ill. ; 24 cm. — (Developmental psychology series)
Includes bibliographies.
ISBN 0-12-447950-2
1. Children, Deaf—Addresses, essays, lectures. I. Liben, Lynn S. II. Series.
₍DNLM: 1. Deafness—In infancy and childhood. 2. Deafness—Rehabilitation. WV271 D489₎
HV2390.D42 155.4'5'12 77-77237
78 MARC

—CONGRESSES

International Conference on Parents and Young Children, Washington University, 1978.
Parent-infant intervention, communication disorders : based on the International Conference on Parents and Young Children held at Washington University, Saint Louis, Missouri, June 20-22, 1978 / edited by Audrey Simmons-Martin, Donald R. Calvert ; sponsored by the Central Institute for the Deaf. — New York : Grune & Stratton, c1979.
xvi, 336 p. : ill. ; 24 cm.
Includes bibliographies and index.
ISBN 0-8089-1185-6
1. Children, Deaf—Congresses. 2. Deaf—Means of communication—Congresses. 3. Parent and child—Congresses. I. Simmons-Martin, Audrey. II. Calvert, Donald R. III. Central Institute for the Deaf. IV. Title.
HV2359.I46 1978 362.7'8'42 79-83870
79 MARC

National Conference for Training Occupational, Physical, and Recreational Therapists in the Area of Deaf-Blind, Dallas, 1975.
Proceedings : training occupational, physical, and recreational therapists in the area of deaf-blind / prepared by Jill M. Gray and Robert Dantona. — Sacramento : California State Dept. of Education, 1977.
vi, 66 p. : ill. ; 28 cm.
"Prepared and published by the Southwestern Region Deaf-Blind Center, California State Department of Education."
"Prepared and published cooperatively by the Texas Education Agency ... and the California State Department of Education."
Includes bibliographies.
1. Blind-deaf—Rehabilitation—Congresses. 2. Blind-deaf—Services for—Congresses. 3. Children, Deaf—Congresses. 4. Children, Blind—Congresses. I. Gray, Jill M. II. Dantona, Robert. III. Southwestern Region Deaf-Blind Center. IV. Texas. Education Agency. V. California. Dept. of Education. VI. Title: Training occupational, physical, and recreational therapists in the area of deaf-blind.
HV1597.N37 1975 362.4'1 78-622076
79 MARC

—EDUCATION—OHIO—PERIODICALS

Vis-à-vis.
Columbus, Ohio Institution for the Education of the Deaf and Dumb ₍etc.₎
v. ill. 45–49 cm. weekly (during the school year)
Continues: Mute's chronicle.
Continued by: The Mute's chronicle.
1. Children, Deaf—Ohio—Periodicals. 2. Children, Deaf—Education—Ohio—Periodicals. I. Ohio. State School for the Deaf, Columbus.
HV2561.O32O388 371.9'12'09771 79–641673
MARC-S

—EDUCATION—OHIO—PERIODICALS
—JUVENILE LITERATURE

The Mute's chronicle.
Columbus, Ohio Institution for the Deaf ₍etc.₎
v. ill. 50 cm.
Weekly, ; weekly (during the school year)
Continues: Vis-à-vis.
1. Children, Deaf — Ohio — Periodicals — Juvenile literature. 2. Children, Deaf—Education—Ohio—Periodicals—Juvenile literature. I. Ohio. State School for the Deaf, Columbus.
HV2561.O32O388 362.7'8'4209771 79–641675
MARC-S

—FAMILY RELATIONSHIPS

Bloom, Freddy.
Our deaf children : into the 80's / ₍by₎ Freddy Bloom. — 2nd revised ed. — Old Woking : Gresham Books, 1978.
₍7₎, 148 p. : ill. ; 22 cm. GB78-39177
ISBN 0-905418-21-2 : £3.95
1. Children, Deaf. 2. Children, Deaf—Family relationships. I. Title.
HV2380.B57 1978 649'.15'12 79-310651
79 MARC

CHILDREN, DEAF
—FAMILY RELATIONSHIPS (Continued)

Luterman, David.
Counseling parents of hearing-impaired children / David Luterman ; foreword by Mark Ross. — 1st ed. — Boston : Little, Brown, c1979.
xxiii, 193 p. ; 23 cm.
Includes bibliographies and index.
ISBN 0-316-53750-0
1. Parenting. 2. Counseling. 3. Children, Deaf—Family relationships. I. Title.
HQ755.8.L87 649'.15'12 79-84419
79 MARC

—JUVENILE LITERATURE

Wolf, Bernard.
Ana y su mundo de silencio / por Bernard Wolf ; traducido del inglés por Sandra Machuca. — Philadelphia : J. B. Lippincott, c1979.
48 p. : ill. ; 23 x 24 cm.
"LLE."
Translation of Anna's silent world.
SUMMARY: Describes the special training and equipment used to help a deaf youngster talk, read, and write.
ISBN 0-397-31847-2
1. Children, Deaf—Juvenile literature. 2. Deaf—Means of communication—Juvenile literature. 3. New York League for the Hard of Hearing. [1. Deaf. 2. Physically handicapped] I. Title.
HV2561.N7W6418 362.4'2 78-26345
78 MARC

—DENMARK

Denmark. Socialstyrelsen. Revalidering og forsorg.
Det hørehæmmede barn. København, Komiteen for Sundhedsoplysning, 1970.
22 p. illus. 22 cm. (Specialpædagogiske vejledninger) kr2.30
D71-30
1. Children, Deaf. 2. Children, Deaf—Denmark. I. Title.
HV2395.D47 1970 79-375913
79 MARC

—GREAT BRITAIN

National Deaf Children's Society.
The deaf school-leaver / [National Deaf Children's Society]. — London : The Society, [1977]
13 p., [4] leaves ; 30 cm. GB77-26980
Cover title.
£0.30
1. Children, Deaf—Great Britain. 2. Deaf—Education—Great Britain. I. Title.
HV2446.N37 1977 371.9'12 79-310160
79 MARC

—OHIO—PERIODICALS

Vis-à-vis.
Columbus, Ohio Institution for the Education of the Deaf and Dumb [etc.]
v. ill. 45–49 cm. weekly (during the school year)
Continues: Mute's chronicle.
Continued by: The Mute's chronicle.
1. Children, Deaf—Ohio—Periodicals. 2. Children, Deaf—Education—Ohio—Periodicals. I. Ohio. State School for the Deaf, Columbus.
HV2561.O32O388 371.9'12'09771 79-641673
MARC-S

—OHIO—PERIODICALS
—JUVENILE LITERATURE

The Mute's chronicle.
Columbus, Ohio Institution for the Deaf [etc.]
v. ill. 50 cm.
Weekly, ; weekly (during the school year)
Continues: Vis-à-vis.
1. Children, Deaf—Ohio—Periodicals—Juvenile literature. 2. Children, Deaf—Education—Ohio—Periodicals—Juvenile literature. I. Ohio. State School for the Deaf, Columbus.
HV2561.O32O388 362.7'8'4209771 79-641675
MARC-S

—UNITED STATES—ADDRESSES, ESSAYS, LECTURES

The Rights of hearing-impaired children / edited by Gary W. Nix. — 1st ed. — Washington : Alexander Graham Bell Association for the Deaf, c1977.
p. 263-351 ; 24 cm. — (The Volta review ; v. 79, no. 5)
Includes bibliographies.
ISBN 0-88200-112-4 : $3.95
1. Children, Deaf—United States—Addresses, essays, lectures. 2. Deaf—Education—United States—Addresses, essays, lectures. 3. Children's rights—United States—Addresses, essays, lectures. I. Nix, Gary W. II. Series.
HV2350.V7 vol. 79, no. 5 77-85240
362.4'2'0973 s
[HV2551] 78 MARC

—UNITED STATES—RECREATION

Boy Scouts of America.
Scouting for the deaf. — North Brunswick, N.J. : Boy Scouts of America, c1978.
48 p. : ill. ; 21 cm.
Cover title.
"Author, Robert Graham Johnson."
Bibliography: p. 35-37.
ISBN 0-8395-3060-0
1. Boy Scouts. 2. Children, Deaf—United States—Recreation. I. Johnson, Robert Graham, 1926- . II. Title.
HS3313.B69 1978a 369.43 78-112699
79 MARC

CHILDREN, EXCEPTIONAL see Exceptional children

CHILDREN, FIRST-BORN

see also Primogeniture

Oakley, Ann.
Becoming a mother / [by] Ann Oakley. — Oxford : Martin Robertson, 1979.
[7], 328 p. ; 23 cm. GB79-16378
Includes bibliographical references.
ISBN 0-85520-206-8 : £7.95
1. Pregnancy—Psychological aspects. 2. Childbirth—Psychological aspects. 3. Infants—Care and hygiene. 4. Children, First-born. 5. Mothers—Great Britain—Interviews. 6. Housewives—Great Britain—Interviews. I. Title.
RG560.O24 618.2 79-315752
79 MARC

CHILDREN, FOLK-LORE OF see Folk-lore of children

CHILDREN, HOSPITALIZED see Children—Hospital care

CHILDREN, ITALIAN AMERICAN see Italian American children

CHILDREN, JEWISH see Jewish children

CHILDREN, MEXICAN AMERICAN see Mexican American children

CHILDREN, PHOTOGRAPHY OF see Photography of children

CHILDREN, RETARDED see Mentally handicapped children; Slow learning children

CHILDREN, RUNAWAY see Runaway children

CHILDREN, SICK see Sick children

CHILDREN, VAGRANT

see also Runaway children

CHILDREN (CHRISTIAN THEOLOGY)

Marcus, Leah.
Childhood and cultural despair : a theme and variations in seventeenth-century literature / Leah Sinanoglou Marcus. — Pittsburgh : University of Pittsburgh Press, c1978.
xii, 305 p. : ill. ; 21 cm.
Bibliography: p. 275-295.
Includes index.
ISBN 0-8229-3367-5
1. English poetry—Early modern, 1500-1700—History and criticism. 2. Children in literature. 3. Children (Christian theology) 4. Children in Great Britain—History. I. Title.
PR545.C45M3 821'.009'352 77-15733
78 MARC

CHILDREN (IN RELIGION, FOLKLORE, ETC.) see Folk-lore of children

CHILDREN (INTERNATIONAL LAW)

—JUVENILE LITERATURE

Agostinelli, Maria Enrica.
On wings of love : the United Nations Declaration of the rights of the child / interpreted and illustrated by Maria Enrica Agostinelli. — New York : Collins, 1979.
[23] p. : col. ill. ; 22 cm.
SUMMARY: An illustrated interpretation of the 10 principles of the United Nations Declaration of the Rights of the Child.
ISBN 0-529-05516-3 : $6.95
1. Children (International law)—Juvenile literature. [1. Children rights] I. United Nations. General Assembly. Declaration of the rights of the child. 1979. II. Title.
K639.A37 341.48'1 78-10318
78 MARC

CHILDREN AND ADULTS

see also Children—Management; Children and strangers; Conflict of generations; Parent and child; Teacher-student relationships

CHILDREN AND ANIMALS

—ADDRESSES, ESSAYS, LECTURES

La Bête et l'enfant. — Paris : Enfances et cultures, 1979.
95 p. : ill. ; 24 cm. — (Enfances et cultures ; no 1) F***
Cover title.
32.00F
1. Children and animals—Addresses, essays, lectures. I. Series.
BF723.A45B47 150 79-374510
79 MARC

CHILDREN AND COMIC BOOKS see Comic books and children

CHILDREN AND DEATH

see also Terminally ill children

Anthony, Elwyn James.
L'enfant devant la maladie et la mort / publié sous la direction de E. James Anthony, Cyrille Koupernik. — Paris : Masson, 1974.
xix, 403 p. ; 24 cm. — (L'Enfant dans la famille ; v. 2) (Livre annuel international de psychiatrie infantile et des professions associées ; v. 2) F***
English ed. has title: The impact of disease and death.
Includes bibliographies.
ISBN 2-225-38808-3
1. Sick children—Psychology. 2. Children and death. 3. Parent and child. I. Koupernik, Cyrille, 1917- joint author. II. Title. III. Series. IV. Series: Livre annuel international de psychiatrie infantile et des professions associées ; v. 2.
RJ499.A1C4214 vol. 2 618.9'28'9008 s 78-376566
[RJ47.5] 78 MARC

CHILDREN AND DEATH (Continued)

Gordon, Audrey K
They need to know : how to teach children about death / Audrey K. Gordon, Dennis Klass. — Englewood Cliffs, N.J. : Prentice-Hall, c1979.
xiii, 274 p. ; 21 cm. — (A Spectrum book)
Includes bibliographies and index.
ISBN 0-13-917104-5. ISBN 0-13-917096-0 pbk.
1. Children and death. 2. Death—Psychological aspects. 3. Death—Study and teaching. I. Klass, Dennis, joint author. II. Title.
BF723.D3G6 155.9'37 79-161
79 MARC

—CONGRESSES

The Child and death / edited by Olle Jane Z. Sahler. — St. Louis : Mosby, 1978.
xviii, 330 p. : ill. ; 23 cm.
Bibliography: p. 260-278.
Includes index.
ISBN 0-8016-4288-4
1. Terminally ill children—Psychology—Congresses. 2. Terminally ill children—Family relationships—Congresses. 3. Children and death—Congresses. I. Sahler, Olle Jane Z., 1944-
[DNLM: 1. Terminal care—In infancy and childhood—Congresses. 2. Death—In infancy and childhood. 3. Attitude to death—Congresses. WS200.3 C536 1977]
RJ47.5.C44 362.7'8'19607 78-18239
78 MARC

CHILDREN AND FOLK-LORE see Folk-lore and children

CHILDREN AND MASS MEDIA see Mass media and children

CHILDREN AND MOVING-PICTURES see Moving-pictures and children

CHILDREN AND POETRY see Poetry and children

CHILDREN AND PSYCHICAL RESEARCH see Psychical research and children

CHILDREN AND STRANGERS

—JUVENILE LITERATURE

Duczman, Linda.
The baby-sitter / words by Linda Duczman ; pictures by Brent Jones. — Milwaukee : Raintree Editions ; Chicago : distributed by Childrens Press, c1977.
30 p. : ill. (some col.) ; 25 cm.
SUMMARY: A child describes his first experience with a baby sitter including his prior fears and anxieties.
ISBN 0-8172-0065-7 lib. bdg. : $4.95
1. Baby sitters—Juvenile literature. 2. Children and strangers—Juvenile literature. [1. Baby sitters] I. Jones, Brent. II. Title.
HQ772.5.D8 649'.1 76-44229
76[r79]rev MARC

CHILDREN AND TELEVISION see Television and children

CHILDREN AS ACTORS

see also Children's plays; Drama in education

Cary, Diana Serra, 1918-
Hollywood's children : an inside account of the child star era / Diana Serra Cary. — Boston : Houghton Mifflin, 1978.
xii, 290 p., [8] leaves of plates : ill. ; 24 cm.
Bibliography: p. [277]-280.
Includes index.
ISBN 0-395-27095-2 : $11.95
1. Cary, Diana Serra, 1918- . 2. Moving-picture actors and actresses—United States—Biography. 3. Children as actors. I. Title.
PN2287.C32A34 791.43'028'0922 78-12585
78 MARC

Great Britain. Home office.
The law on performances by children: a guide to the Children (Performances) Regulations 1968 and related statutory provisions. London, H.M.S.O., 1968.
iv, 32 p. 25 cm. 4/- B68-25530
ISBN 0-11-340024-1
1. Children—Employment—Great Britain. 2. Children as actors. I. Title.
344'.41'0131 72-180955
72[r78]rev MARC

Malkin, Michael R 1943-
Training the young actor / Michael R. Malkin ; photos. by Wayne Nicholls ; ill. by Jim Shubin. — South Brunswick : A. S. Barnes, c1979.
128 p. : ill. ; 25 cm.
Includes index.
ISBN 0-498-01957-8 : $9.95
1. Acting—Study and teaching. 2. Children as actors. 3. Children's plays—Presentation, etc. I. Title.
PN3157.M34 792'.028 76-50201
78 MARC

—JUVENILE LITERATURE

Polsky, Milton E
Today's young stars of stage and screen / by Milton E. Polsky. — New York : F. Watts, 1979.
120 p. : ill. ; 25 cm.
Bibliography: p. 118.
Includes index.
SUMMARY: Spotlights the lives of 12 rising young stars of stage, movies, and television including interviews in which they express their thoughts and attitudes toward their careers.
ISBN 0-531-02885-2
1. Actors—United States—Biography—Juvenile literature. 2. Children as actors—Juvenile literature. [1. Actors and actresses] I. Title.
PN2285.P6 791.43'028'0922 79-13280
79 MARC

CHILDREN AS ARTISTS

see also Children in art; Children's art; Folk art

Avalle, Vico.
La grammatica del segno infantile : impariamo a leggere i disegni del bambino / Vico Avalle. — [Ivrea] : Priuli & Verlucca, [1975]
237 p. : col. ill. ; 24 cm. It75-Nov
L6000
1. Drawing, Psychology of. 2. Child psychology. 3. Children as artists. I. Title.
BF723.D7A89 79-376775
*79 MARC

Brener, Noor Zadé.
Mon atelier magique / Noor-Zadé Brener ; préf. du Pr A. Minkowski. — Paris : R. Laffont, c1978.
253 p. ; 20 cm. — (Collection Réponses) F***
Includes bibliographical references.
44.00F
1. Children as artists. 2. Group work in art. 3. Artists' studios. I. Title.
N351.B73 704'.054 78-382877
78 MARC

Ebert, Wilhelm, 1920-
Zum bildnerischen Verhalten des Kindes in Vor- und Grundschulalter. Ein Beitrag zur Didaktik des Kunstunterrichtes. Ratingen b. Düsseldorf, Henn (1967).
184 p. with illus. 23 cm. 19.60 GDB67-A41-236
Bibliography: p. 170-172.
1. Art—Study and teaching (Elementary) 2. Children as artists. I. Title.
N350.E2 77-581670
79[r79]rev MARC

Green, Gerald.
The artists of Terezin / by Gerald Green. — New York : Hawthorn Books, c1978.
ix, 191 p., [2] leaves of plates : ill. (some col.) ; 26 cm.
Bibliography: p. [193]
ISBN 0-8015-0420-1 : $12.50
1. Prisoners of war as artists. 2. Children as artists. 3. Terezin (Concentration camp) I. Title.
N6833.T47G73 1978b 741.9'437'1 78-108798
78 MARC

Green, Gerald.
The artists of Terezin / Gerald Green. — New York : Schocken Books, 1978.
ix, 191 p., [2] leaves of plates : ill. (some col.) ; 26 cm.
ISBN 0-8052-0609-4 : $7.95
1. Prisoners of war as artists. 2. Children as artists. 3. Terezin (Concentration camp) I. Title.
N6833.T47G73 1978 741.9'437'1 78-59813
78 MARC

A Látás iskolája : tanulmányok a vizuális esztétikai nevelésről / [összeáll., Domonkos Imre]. — [Budapest] : Corvina, c1977.
188 p., [44] leaves of plates : ill. (some col.) ; 22 cm. — (Művészet és elmélet) H77
Includes bibliographical references.
ISBN 9631310000 : 79.00Ft
1. Art—Study and teaching (Elementary) 2. Art—Study and teaching (Secondary) 3. Children as artists. 4. Youth as artists. I. Domonkos, Imre.
N350.L34 78-384122
*78 MARC

Piantoni, Carlo.
72 [i.e. Settantadue] tecniche per le attività espressive / Carlo Piantoni. — Brescia : La scuola, [1978?]
348 p., [32] leaves of plates : ill. ; 26 cm. It78-Nov
Bibliography: p. 345-346.
1. Art—Technique. 2. Children as artists. I. Title.
N7433.P52 702'.8 79-339047
*79 MARC

Séguin-Fontes, Marthe.
Le Second souffle de la créativité / Marthe Séguin-Fontes, ... — Paris : Dessain et Tolra, 1977.
189 p. : ill. ; 22 cm. — (Le Jeu et l'élément créateur) F78-610
ISBN 2-249-20013-0 : 48.00F
1. Children as artists. 2. Youth as artists. 3. Creativity (Literary, artistic, etc.) 4. Perceptual learning. I. Title.
N351.S33 701 78-376824
78 MARC

Strauss, Michaela.
Von der Zeichensprache des kleinen Kindes : Spuren der Menschwerdung / Michaela Strauss ; mit menschenkundlichen Anmerkungen von Wolfgang Schad. — 2. Aufl. — Stuttgart : Verlag Freies Geistesleben, 1977, c1976.
92 p. : ill. (some col.) ; 21 x 23 cm. — (Menschenkunde und Erziehung : Bd. 34) GFR***
Bibliography: p. 91-92.
ISBN 3-7725-0234-0
1. Drawing, Psychology of. 2. Children as artists. I. Schad, Wolfgang. II. Title. III. Series.
BF723.D7S77 1977 78-379208
78 MARC

—CASES, CLINICAL REPORTS, STATISTICS

Selfe, Lorna.
Nadia : a case of extraordinary drawing ability in an autistic child / Lorna Selfe. — London ; New YOrk : Academic Press, 1977.
viii, 137 p. : ill. ; 31 cm. GB***
Bibliography: p. [135]-137.
ISBN 0-12-635750-1
1. Autism—Cases, clinical reports, statistics. 2. Children as artists—Cases, clinical reports, statistics. I. Title.
RJ506.A9S44 1977 618.9'28'982 76-22839
78[r79]rev MARC

—SPAIN

Bisquert Santiago, Adriana.
Las artes plásticas en la escuela / Adriana Bisquert Santiago. — Madrid : Ministerio de Educación y Ciencia, Instituto Nacional de Ciencias de la Educación, 1977.
315 p. : ill. (some col.) ; 18 cm. — (Breviarios de educación ; 4) Sp***
Bibliography: p. 293-296.
Includes index.
ISBN 8436905873 : 300ptas
1. Art—Study and teaching (Elementary)—Spain. 2. Children as artists—Spain. I. Title. II. Series.
N365.S7B57 78-395554
79 MARC

CHILDREN AS AUTHORS

see also Children in literature

Clanché, Pierre.
Le texte libre, écriture des enfants / Pierre Clanché. — Paris : F. Maspero, 1976.
226 p. ; 22 cm. — (Malgré tout) F***
Originally presented as the author's thesis, Bordeaux II.
Includes bibliographical references.
ISBN 2-7071-0820-0 : 35.00F
1. Child psychology. 2. Stream of consciousness fiction. 3. Children as authors. I. Title.
BF721.C545 1976 371.6'23'019 79-349254
79 MARC

Romanciers à treize ans : une expérience pédagogique / [textes réunis et présentés par] Pascal Bouchard, avec la collaboration de Aude Kreisler et Pascale Gruson. — Paris : Denoël-Gonthier, 1978.
261 p. : ill. ; 22 cm. — (Collection Femme) F78-8567
49.00F
1. Children's writings, French—France—Dieppe. 2. French language—Study and teaching—France—Dieppe. 3. Children as authors. I. Bouchard, Pascal. II. Kreisler, Aude. III. Gruson, Pascale.
PQ1109.5.C5R6 840'.8'09282 79-358649
79 MARC

CHILDREN AS CONSUMERS see Youth as consumers

CHILDREN AS MUSICIANS

see also Rhythm bands and orchestras

Lewandowska, Kinga.
Rozwój zdolności muzycznych u dzieci w wieku szkolnym / Kinga Lewandowska. — Wyd. 1. — Warszawa : Wydaw-a Szkolne i Pedagogiczne, 1978 (Bydg : WSiP)
191, [1] p. : ill. ; 21 cm. P78-2999
Summary in English.
Bibliography: p. 178-182.
zł30.00
1. Musical ability—Testing. 2. Children as musicians. I. Title.
ML3838.L49 78-388757
79 MARC

CHILDREN AS POETS see Children as authors

CHILDREN AS SOLDIERS see World War, 1939-1945—Juvenile participants

CHILDREN IN ART

see also Children as artists; Photography of children

Children : a pictorial archive from ninteenth-century sources : 240 copyright-free illustrations for artists and designers / selected by Carol Belanger Grafton. — New York : Dover Publications, 1978.
v, 119 p. : chiefly ill. ; 29 cm. — (Dover pictorial archive series)
Engravings originally published in 1884-99 in The Chatterbox and in 1874 in the British workwoman.
ISBN 0-486-23694-3 : $3.50
1. Children in art. 2. Wood-engraving, English. 3. Wood-engraving—19th century—England. I. Grafton, Carol Belanger.
NE962.C47C47 1978 769'.4'250942 78-59321
79 MARC

Cope, Dawn.
Illustrators of postcards from the nursery / Dawn and Peter Cope. — London : East-West Publications, 1978.
64 p. : ill. (some col.), ports. ; 19 cm. GB***
ISBN 0-904499-05-7
1. Postal cards for children—Great Britain. 2. Nursery rhymes—Illustrations. 3. Children in art. 4. Illustrators—Great Britain—Biography. I. Cope, Peter, joint author. II. Title.
NC1878.C55C66 741.68 79-318396
79 MARC

Dicke, Otto.
De wereld kan alleen maar kleiner worden / Otto Dicke en Hans Bouma. — Baarn : Bosch & Keuning, [1978]
32 p. : ill. ; 32 cm. Ne78-20
ISBN 9024642655 : fl 17.50
1. Dicke, Otto. 2. Children in art. I. Bouma, Hans. II. Title.
NC263.D52A4 1978a 78-376367
*78 MARC

Schönberger, Käthe.
Dad Kinderporträt in der Plastik / Käthe Schönberger ; [Fotos von Herbert Schönberger und Beatrix und Hans Böck]. — Erlenbach-Zürich : E. Rentsch, cop. 1978.
[28] leaves of plates : chiefly ill. (all col.) ; 23 cm. Sw78-12677
ISBN 3-7249-0508-4 : 25.00F
1. Schönberger, Käthe. 2. Children—Portraits. 3. Children in art. I. Title.
NB853.S36A4 1978 730'.92'4 79-363992
79 MARC

—EXHIBITIONS

Humm, Rosamond Olmsted.
Children in America : a study of images and attitudes : September 30, 1978-May 27, 1979, the High Museum of Art, Atlanta, Georgia / by Rosamond Olmsted Humm. — Atlanta : The Museum, 1978.
72 p. : ill. (some col.) ; 22 x 25 cm.
Bibliography: p. 66-68.
1. Children—United States—Portraits—Exhibitions. 2. Children in art—Exhibitions. I. High Museum of Art. II. Title.
N7644.H84 78-52513
760'.0973'07401558213
79 MARC

—POLAND—EXHIBITIONS

Błachowski, Aleksander.
Dziecko w sztuce ludowej : twórczość ludowa dla dziecka : ogólnopolska wystawa pokonkursowa czerwiec-grudzień 1976 : miejsca ekspozycji, Muzeum Etnograficzne w Warszawie, Krakowie i Toruniu / [oprac. katalogu Aleksander Błachowski]. — Toruń : Muzeum Etnograficzne, 1976 (Tor. : ZG)
29, [2] p., [12] leaves of plates : ill. ; 21 cm. P77-9943
zł10.00
1. Children in art—Poland—Exhibitions. 2. Toys—Poland—Exhibitions. 3. Art, Polish—Exhibitions. I. Toruń, Poland. Muzeum Etnograficzne. II. Title.
N7648.P7B58 78-379245
78[r79]rev MARC

CHILDREN IN CITIES see City children

CHILDREN IN ENGLAND

—BIOGRAPHY

Milne, Christopher, 1920-
The enchanted places / Christopher Milne. — Harmondsworth ; New York [etc.] : Penguin, 1976.
183 p., [8] p. of plates : ill., map, ports. ; 18 cm. GB76-32806
ISBN 0-14-003449-8 : £0.70 ($1.95 U.S.)
1. Milne, Alan Alexander, 1882-1956. 2. Milne, Christopher. 3. Authors, English—20th century—Biography. 4. Children in England—Biography. I. Title.
PR6025.I65Z8 1976 828'.9'1209 78-317933
78 MARC

Noel, Diana.
Five to seven : the story of a 1920s childhood / Diana Noel ; with ill. by Eric Stemp. — London : Collins, 1978.
141 p. : ill. ; 22 cm. GB***
ISBN 0-00-211370-8 : £4.95
1. Noel, Diana—Biography—Youth. 2. Authors, English—20th century—Biography. 3. Children in England—Biography. I. Title.
PR6064.O3Z464 828'.9'1409 79-303756
79 MARC

Williamson, Henry, 1897-1977.
The children of Shallowford / [by] Henry Williamson. — [New ed.]. — London : Macdonald and Jane's, 1978.
222 p., [8] p. of plates : ill., ports. ; 23 cm. GB78-33365
ISBN 0-354-04313-7 : £5.50
1. Williamson, Henry, 1897-1977—Biography. 2. Authors, English—20th century—Biography. 3. Children in England—Biography. I. Title.
PR6045.I55Z523 1978 823'.9'12 79-301829
79 MARC

CHILDREN IN FRANCE

—ADDRESSES, ESSAYS, LECTURES

Boulin, Bertrand.
La Charte des enfants / Bertrand Boulin ; construite et révélée au public grâce à Jean-Michel Desjeunes et Philippe Alfonsi ; avec la collaboration de Nicolas Vrain, Alain Chiapello, Michèle Roche, Annie Thomas. — Paris : Stock, 1977.
303 p. ; 20 cm. — (Stock 2 [i.e. deux] : Lutter) F78-625
ISBN 2-234-00766-6 : 30.00F
1. Children in France—Addresses, essays, lectures. 2. Children's rights—France—Addresses, essays, lectures. I. Desjeunes, Jean Michel, joint author. II. Alfonsi, Philippe, joint author. III. Title.
HQ792.F7B68 301.43'14'0944 78-399311
79 MARC

—BIOGRAPHY

Gatard, Marie.
La Guerre, mon père / Marie Gatard. — [Paris] : Mercure de France, 1978.
141, [3] p. ; 21 cm. F78-10976
Bibliography: p. [143]
30.00F
1. World War, 1939-1945—Personal narratives, French. 2. Gatard, Marie. 3. Children in France—Biography. I. Title.
D811.5.G28 940.54'81'44 79-338697
79 MARC

—HISTORY

Entrer dans la vie : naissances et enfances dans la France traditionnelle / présenté par Jacques Gélis, Mireille Laget et Marie-France Morel. — [Paris] : Gallimard, c1978.
245 p., [8] leaves of plates : ill. ; 18 cm. — (Collection Archives ; 72) F***
Includes bibliographical references.
1. Children in France—History. 2. France—Social conditions. I. Gélis, Jacques. II. Laget, Mireille. III. Morel, Marie France.
HQ755.E55 301.43'14'0944 78-399085
79 MARC

CHILDREN IN GREAT BRITAIN

—ANTHROPOMETRY

Tanner, James Mourilyan.
Education and physical growth; implications of the study of children's growth for educational theory and practice. [1st ed.] London, University of London Press [1961]
144 p. illus. 23 cm.
Includes bibliography.
1. Children in Great Britain—Anthropometry. 2. Children—Growth. 3. Education of children. 4. Child development. I. Title.
LB3425.G7T36 1961 62-51434
[r79]rev MARC

Tanner, James Mourilyan.
Education and physical growth : implications of the study of children's growth for educational theory and practice / J. M. Tanner. — 2d ed. — New York : International Universities Press, 1978.
144 p. : ill. ; 23 cm.
Bibliography: p. 127-134.
Includes index.
ISBN 0-8236-1531-6
1. Children in Great Britain—Anthropometry. 2. Children—Growth. 3. Education of children. 4. Child development. I. Title.
LB3425.G7T36 1978 573'.6 78-8809
78 MARC

CHILDREN IN INDONESIA

—CONGRESSES

Badan Koordinasi Nasional untuk Kesejahteraan Keluarga dan Anak.
Mengenali KNAP-PPN 1969 : suatu usaha meneropong anak dan pemuda dari sudut positip dalam rangka pembinaan sumber daja manusia muda. — Djakarta : Badan Koordinasi Nasional untuk Kesedjahteraan Keluarga dan Anak, 1970.
11 p. ; 21 cm.
1. Konperensi Nasional Tentang Anak dan Pemuda Dalam Perentjanaan Pembangunan Nasional, Djakarta, 1969. 2. Children in Indonesia—Congresses. 3. Youth—Indonesia—Congresses. I. Konperensi Nasional Tentang Anak dan Pemuda Dalam Perentjanaan Pembangunan Nasional, Djakarta, 1969. II. Title.
HV800.I55B35 1970 77-986629
79 MARC

Lokakarya Nilai Anak di Indonesia, Universitas Gadjah Mada, 1977.
Lokakarya Nilai Anak di Indonesia : kumpulan kertas kerja / disusun oleh Paul A. Meyer, Simon Andjar Legawan. — Ed. 1. — Yogyakarta : Lembaga Kependudukan, Universitas Gadjah Mada, 1978.
152 p. ; 29 cm. — (Seri kertas kerja lokakarya - Lembaga Kependudukan, Universitas Gadjah Mada ; no. 1)
English or Indonesian.
Includes bibliographical references.
1. Children in Indonesia—Congresses. 2. Family size—Economic aspects—Indonesia—Congresses. I. Meyer, Paul A. II. Legawan, Simon Andjar. III. Series: Universitas Gadjah Mada. Lembaga Kependudukan. Seri kertas kerja lokakarya - Lembaga Kependudukan, Universitas Gadjah Mada ; no. 1.
HQ792.I6L64 1977 301.43'14'09598 79-940365
79 MARC

CHILDREN IN IRAN

—LANGUAGE

Badrah, Farīdūn.
(Vāzhagān-i nivishtārī-i kūdakān-i dabistānī-i Īrān)
واژگان نوشتاری کودکان دبستانی ایران / فریدون بدره‌ای. — [تهران] : فرهنگستان زبان ایران، بنیاد شاهنشاهی فرهنگستانهای ایران، [1352 i. e. 1974]
58, 698 p. ; 25 cm. — (انتشارات فرهنگستان زبان ایران ؛ شمارهٔ 7)
In Persian.
Chiefly tables.
A revision of the author's thesis, Dānishgāh-i Tihrān, 1350.
Bibliography: p. [47]-48 (1st group)
1. Persian language—Word frequency. 2. Children in Iran—Language. I. Title. II. Series: Farhangistān-i Zabān-i Īrān. Intishārāt-i Farhangistān-i Zabān-i Īrān ; shumārah-'i 7.
PK6385.B3 1974 79-970685

—ANTHROPOMETRY

Ente italiano della moda.
Le misure antropometriche della popolazione italiana : l'abbigliamento delle classi giovani dai 6 ai 19 anni / Ente italiano della moda. — Milano : F. Angeli, [1979]
284 p. : ill. ; 22 cm. — (Collana scientifica ; 33) It79-May
Bibliography: p. [279]-284.
ISBN 8820415763 : L6000
1. Children in Italy—Anthropometry. 2. Anthropometry—Italy. I. Title.
GN58.I8E57 1979 573'.6 79-369723
*79 MARC

CHILDREN IN JAPAN

Sarashina, Genzō, 1904-
(Kotan no dōgi)
コタンの童戯 / 更科源蔵. — 東京 : 風書房, 1976.
100 p. : ill. ; 21 cm. Ja***
Limited edition: 300 copies printed. No. 153.
¥2500

1. Play—Japan—Hokkaido. 2. Children in Japan. 3. Ainu—Social life and customs. I. Title.
GT2480.S27 78-812723

CHILDREN IN JAVA

—STATISTICS

Singarimbun, Masri.
Value of children : a study in Java : preliminary tabulations from the Indonesian value of children survey = Nilai anak : hasil penelitian di Jawa : tabulasi pendahuluan dari Survey nilai anak di Indonesia / Masri Singarimbun, Russell K. Darroch, Paul A. Meyer. — 1st ed. — Yogyakarta, : Population Institute, Gadjah Mada University, 1977.
xxxvi, 146 p. ; 33 cm. — (Report series - Population Institute, Gadjah Mada University ; no. 13)
Tables.
English and Indonesian.
Bibliography: p. xxx.
1. Children in Java—Statistics. 2. Family size—Economic aspects—Indonesia—Java—Statistics. 3. Birth control—Indonesia—Java—Statistics. I. Darroch, Russell K., joint author. II. Meyer, Paul A., joint author. III. Title. IV. Title: Nilai anak. V. Series: Universitas Gadjah Mada. Lembaga Kependudukan. Seri laporan - Lembaga Kependudukan, Universitas Gadjah Mada ; no. 13.
HQ792.I6S55 301.42'6'095982 78-941872
79 MARC

CHILDREN IN LA PLATA

—GROWTH—LONGITUDINAL STUDIES

Cusminsky, Marcos.
Estudio longitudinal del crecimiento del niño de 0-1 año en un área del Partido de La Plata / Marcos Cusminsky, Lilia Chaves de Azcona. — La Plata : Provincia de Buenos Aires, Comisión de Investigaciones Científicas, 1974.
29, ⌐3⌐ p. : graphs ; 29 cm. — (Informes - Comisión de Investigaciones Científicas ; 5)
Cover title.
Bibliography: p. ⌐2⌐-⌐6⌐ (2d group)
1. Children in La Plata—Growth—Longitudinal studies. 2. Children in the Argentine Republic—Growth—Longitudinal studies. 3. Infants—Growth. I. Chaves de Azcona, Lilia, joint author. II. Title: Estudio longitudinal del crecimiento del niño de 0-1 año ... III. Series: Comisión de Investigaciones Científicas (Buenos Aires) Informes - Comisión de Investigaciones Científicas ; 5.
GN63.C87 78-358262
78 MARC

CHILDREN IN LAGUNA, PHILIPPINES (PROVINCE)

Boulier, Bryan L
Children and household economic activity in Laguna, Philippines / by Bryan L. Boulier. — ⌐Quezon City?⌐ : Institute of Economic Development and Research, School of Economics, University of the Philippines, 1976.
22, ⌐19⌐ p. ; 29 cm. — (Discussion paper - Institute of Economic Development and Research ; no. 76-19)
Bibliography: p. ⌐29⌐
1. Family size—Economic aspects—Philippine Islands—Laguna (Province) 2. Children in Laguna, Philippines (Province) 3. Time allocation. I. Title. II. Series: Quezon, Philippines. University of the Philippines. Institute of Economic Development and Research. Discussion paper - Institute of Economic Development and Research ; no. 76-19.
HQ766.5.P5B68 301.42'6 79-301306
79 MARC

CHILDREN IN LEBANON

Fāyid, Laylá.
(al-Shahādah wa-al-ḥulm)
الشهادة والحلم : حوارات وثائقية عن تجربة أطفال تل الزعتر / ليلى فايـــد. ـــ ⌐بيروت؟⌐ : الاعلام الموحـــد، م. ت. ف.، منشورات فلسطين الثورة⌐، ؟1977.
86 p. : ill. ; 20 cm.
«في الذكرى ١٢ لانطلاقة الثورة، 1977.»
1. Mukhayyam Tall az Za'tar, Lebanon. 2. Lebanon—History—Civil War, 1975- 3. Children in Lebanon. I. Title.
DS87.F38 79-970935

CHILDREN IN LITERATURE

see also Children as authors

Bethlenfalvay, Marina.
Les Visages de l'enfant dans la littérature française du XIXe siècle : esquisse d'une typologie / Marina Bethlenfalvay. — Genève : Droz, 1979.
148 p. ; 23 cm. — (Histoire des idées et critique littéraire ; v. 176)
Sw79-1987
Bibliography: p. ⌐143⌐-146.
32.00F
1. French literature—19th century—History and criticism. 2. Children in literature. I. Title.
PQ283.B4 840'.9'352 79-371121
79 MARC

Cadogan, Mary.
Women and children first : the fiction of two world wars / by Mary Cadogan and Patricia Craig. — London : Gollancz, 1978.
301 p. : ill., facsims. ; 23 cm. GB78-32549
Bibliography: p. 295.
Includes index.
ISBN 0-575-02418-6 : £7.50
1. English fiction—20th century—History and criticism. 2. Women in literature. 3. Children in literature. 4. European War, 1914-1918—Literature and the war. 5. World War, 1939-1945—Literature and the war. I. Craig, Patricia, joint author. II. Title.
PR888.W6C3 823'.9'12093 78-325411
79 MARC

Cuendet, Simone.
Ramuz et le temps de l'enfance / Simone Cuendet ; éd. sous le patronage de la Fondation C. F. Ramuz. — Chardonne : Éditions Le Cantalou, ⌐S. Cuendet⌐, 1978.
161 p. : ill. ; 21 cm. Sw79-1989
Includes index.
22.00F
1. Ramuz, Charles Ferdinand, 1878-1947—Characters—Children. 2. Children in literature. I. Title.
PQ2635.A35Z5634 79-373593
79 MARC

Marcus, Leah.
Childhood and cultural despair : a theme and variations in seventeenth-century literature / Leah Sinanoglou Marcus. — Pittsburgh : University of Pittsburgh Press, c1978.
xii, 305 p. : ill. ; 21 cm.
Bibliography: p. 275-295.
Includes index.
ISBN 0-8229-3367-5
1. English poetry—Early modern, 1500-1700—History and criticism. 2. Children in literature. 3. Children (Christian theology) 4. Children in Great Britain—History. I. Title.
PR545.C45M3 821'.009'352 77-15733
78 MARC

Song, Ikhwa, 1937-
Die Darstellung des Kindes im frühen Werk Heinrich Bölls : d. Kind in d. Rolle d. moral. Gegenspielers zu e. inhumanen Welt / Ikhwa Song. — Frankfurt/Main ; Bern ; Las Vegas : Lang, 1978.
151 p. ; 21 cm. — (Europäische Hochschulschriften : Reihe 1, Deutsche Literatur und Germanistik ; Bd. 254) GFR78-A
Originally presented as the author's thesis, Frankfurt am Main.
Bibliography: p. 146-151.
ISBN 3-261-02490-9
1. Böll, Heinrich, 1917- —Criticism and interpretation. 2. Children in literature. I. Title. II. Series.
PT2603.O394Z767 1978 78-379502
78 MARC

Stuart, Simon, 1930-
New Phoenix wings : reparation in literature / Simon Stuart. — London ; Boston : Routledge & K. Paul, 1979.
viii, 187 p., ⌐4⌐ leaves of plates : ill. ; 23 cm. GB***
Includes bibliographical references and index.
ISBN 0-7100-0179-7
1. English poetry—19th century—History and criticism. 2. Children in literature. 3. Psychoanalysis and literature. 4. Shakespeare, William, 1564-1616. Much ado about nothing. 5. Creation (Literary, artistic, etc.) I. Reparation in literature. II. Title.
PR589.C45S7 820'.9'3 78-41329
79 MARC

CHILDREN IN MEXICO (CITY)

—GROWTH—LONGITUDINAL STUDIES

Faulhaber, Johanna.
Investigación longitudinal del crecimiento en un grupo de niños caracterizado por su ambiente socioeconómico, su alimentación y su patología / Johanna Faulhaber, en colaboración con María Villanueva. — México : Instituto Nacional de Antropología e Historia, SEP, Departamento de Antropología Física, 1976.
310 p. : ill. ; 27 cm. — (Colección científica : Antropología física ; 26)
Summary in Spanish, English, French and German.
Bibliography: p. 295-298.
1. Children in Mexico (City)—Growth—Longitudinal studies. I. Villanueva, María, joint author. II. Title. III. Series: Mexico. Instituto Nacional de Antropología e Historia. Colección científica - Instituto Nacional de Antropología e Historia ; 26.
GN63.F38 573'.6 78-375443
78 MARC

CHILDREN IN NEPAL

—PERIODICALS

Nepal Bal Sangathan.
(Śrī Pāñca Baḍāmahārānī Sarakārako śubha-janmotsava tathā tṛtīya rāshṭriya bāla divasa)
श्री ५ बडामहारानी सरकारको शुभ-जन्मोत्सव तथा तृतीय राष्ट्रिय बाल दिवस.
⌐काठमाडौं⌐ नेपाल बाल संगठन.
v. ports. 21 cm. annual.
In Nepali.
1. Ratna Rajya Lakshmi Devi Shaha, Badamaharani of Nepal, 1928- —Periodicals. 2. Children in Nepal—Periodicals.
DS495.592.R34N46a S A 68-20786
MARC-S
PL 480: N-56-S

CHILDREN IN POETRY

see also Children—Poetry; Children's poetry

CHILDREN IN THE ARGENTINE REPUBLIC

—GROWTH—LONGITUDINAL STUDIES

Cusminsky, Marcos.
Estudio longitudinal del crecimiento del niño de 0-1 año en un área del Partido de La Plata / Marcos Cusminsky, Lilia Chaves de Azcona. — La Plata : Provincia de Buenos Aires, Comisión de Investigaciones Científicas, 1974.
29, ⌐3⌐ p. : graphs ; 29 cm. — (Informes - Comisión de Investigaciones Científicas ; 5)
Cover title.
Bibliography: p. ⌐2⌐-⌐6⌐ (2d group)
1. Children in La Plata—Growth—Longitudinal studies. 2. Children in the Argentine Republic—Growth—Longitudinal studies. 3. Infants—Growth. I. Chaves de Azcona, Lilia, joint author. II. Title: Estudio longitudinal del crecimiento del niño de 0-1 año ... III. Series: Comisión de Investigaciones Científicas (Buenos Aires) Informes - Comisión de Investigaciones Científicas ; 5.
GN63.C87 78-358262
78 MARC

CHILDREN IN THE BIBLE

Keizer, D P R
Het kinderleven in bijbelse tijd : gezondheid en ziekte, geneesmiddelen, spel, ouderliefde en opvoeding, voedsel, kruiden en planten, naamgeving, bijgeloof en amuletten, enz. / D. P. R. Keizer. — Kampen : Kok, ⌐1979⌐
178 p. : with ill. ; 22 cm. Ne79-17
Bibliography: p. 170-173.
Includes index.
ISBN 9024214106 : fl 24.75
1. Children in the Bible. 2. Children—Palestine. I. Title.
BS576.K44 79-379814
*79 MARC

CHILDREN IN THE IVORY COAST

—JUVENILE LITERATURE

Combet, Sabine.
Fatou au pays des Baoulé / Sabine Combet ; ⌐photos, Gilles Henri⌐. — Dakar : Nouvelles Éditions africaines, c1975.
43 p. : ill. ; 27 cm.
1. Baoulé (African people)—Social life and customs—Juvenile literature. 2. Ivory Coast—Social life and customs—Juvenile literature. 3. Children in the Ivory Coast—Juvenile literature. I. Henri, Gilles, joint author. II. Title.
DT545.42.C65 966.6'8'00496 78-391326
79 MARC

CHILDREN IN THE UKRAINE

—ANTHROPOMETRY

Talko-Hryncewicz, Julian, 1850-1936.
Przyczynek do antropologii dzieci chrześciańskich i żydowskich na Ukrainie / J. Talko-Hryncewicz. — Kraków : Nakł. Akademii Umiejętności, 1908.
60 p. ; 24 cm.
1. Children in the Ukraine—Anthropometry. 2. Jews—Anthropometry. I. Title.
GN63.T34 78-382341
78 MARC

CHILDREN IN THE UNITED STATES

see also Mexican American children; Presidents—United States—Children

Talbot, Nathan Bill, 1909-
Raising children in modern America : what parents and society should be doing for their children / Nathan B. Talbot. — 1st ed. — Boston : Little, Brown, c1976.
x, 180 p. : ill. ; 24 cm.
Bibliography: p. ₍162₎-164.
ISBN 0-316-83136-0
1. Children—Management. 2. Children in the United States. I. Title.
HQ769.T265 649'.1'0973 75-31571
75₍r79₎rev MARC

—ANTHROPOMETRY

Roche, Alex F 1921-
Skeletal maturity of youths 12-17 years : racial, geographic area, and socioeconomic differentials, United States, 1966-1970 / ₍Alex F. Roche, Jean Roberts, and Peter V. V. Hamill₎. — Hyattsville, Md. : U.S. Dept. of Health, Education, and Welfare, Public Health Service, National Center for Health Statistics ; Washington : for sale by the Supt. of Docs., U.S. Govt. Print. Off., 1978.
vi, 98 p. : ill. ; 26 cm. — (Vital and health statistics : Seris 11, Data from the National Health Survey ; no. 167) (DHEW publication ; no. (DHS) 78-1654 no. (PHS) 79-1654)
Includes bibliographical references.
ISBN 0-8406-0121-2
1. Skeletal maturity—United States. 2. Children in the United States—Anthropometry. 3. Health surveys—United States. I. Roberts, Jean, 1918- joint author. II. Hamill, Peter V. V., 1926- joint author. III. Title. IV. Series: United States. National Center for Health Statistics. Vital and health statistics : Series 11, Data from the National Health Survey, Data from the health examination survey ; no. 167. V. Series: United States. Dept. of Health, Education, and Welfare. DHEW publication ; no. (PHS) 79-1654.
₍DNLM: 1. Age determination by skeleton—In adolescence. 2. Bone development—In adolescence. 3. Racial stocks. W2 A N148vk no. 167₎
RA407.3.A347 no. 167 312'.0973 s 78-502
₍GN63₎ 78 MARC

—CONGRESSES

Raising children in modern America : problems and prospective solutions / Nathan B. Talbot, editor. — 1st ed. — Boston : Little, Brown, c1976.
xiii, 590 p. : ill. ; 24 cm.
Bibliography: p. ₍497₎-499.
Includes index.
ISBN 0-316-83135-2
1. Children—Management—Congresses. 2. Children in the United States—Congresses. I. Talbot, Nathan Bill, 1909-
HQ769.R17 649'.1'0973 75-19236
75₍r79₎rev MARC

CHILDREN IN TURKEY

—JUVENILE LITERATURE

Kherdian, David.
The road from home : the story of an Armenian girl / by David Kherdian. — 1st ed. — New York : Greenwillow Books, c1979.
xi, 238 p. : map ; 22 cm.
SUMMARY: A biography of the author's mother concentrating on her childhood in Turkey before the Turkish government deported its Armenian population.
ISBN 0-688-80205-2 : $8.95. ISBN 0-688-84205-4 lib. bdg. : $7.63
1. Kherdian, Veron, 1907- —Juvenile literature. 2. Armenians in Turkey—Biography—Juvenile literature. 3. Children in Turkey—Juvenile literature. 4. Armenian massacres, 1915-1923—Juvenile literature. ₍1. Kherdian, Veron, 1907- . 2. Armenians in Turkey—Biography. 3. Armenian massacres, 1915-1923₎ I. Title.
DR435.A7K47 949.6'1'01 78-72511
78 MARC

CHILDREN IN UNDERDEVELOPED AREAS
see Underdeveloped areas—Children

CHILDREN OF ALCOHOLIC PARENTS

—JUVENILE LITERATURE

Seixas, Judith S
Living with a parent who drinks too much / Judith S. Seixas. — 1st ed. — New York : Greenwillow Books, c1979.
x, 116 p. ; 22 cm.
SUMMARY: Describes alcoholism, alcoholic behavior, and resulting family problems. Advises children of alcoholic parents in dealing with these problems and their own feelings and suggests ways to make life more bearable and productive.
ISBN 0-688-80196-X : $6.95
1. Alcoholism—Juvenile literature. 2. Alcoholics—United States—Family relationships—Juvenile literature. 3. Children of alcoholic parents—Juvenile literature. ₍1. Alcoholism. 2. Alcoholics₎ I. Title.
HV5066.S44 362.8'2 78-11108
78 MARC

—GERMANY, WEST

Stimmer, Franz.
Jugendalkoholismus : e. familiensoziolog. Unters. zur Genese d. Alkoholabhängigkeit männl. Jugendlicher / von Franz Stimmer. — Berlin : Duncker und Humblot, 1978.
192 p. : ill. ; 24 cm. — (Sozialwissenschaftliche Abhandlungen der Görres-Gesellschaft ; Bd. 2) GFR79-A
Bibliography: p. ₍174₎-187.
Includes indexes.
ISBN 3-428-04255-7 : DM58.00
1. Youth and alcohol—Germany, West. 2. Young men—Germany, West. 3. Children of alcoholic parents—Germany, West. 4. Alcoholics—Germany, West—Family relationships. I. Title. II. Series: Görres-Gesellschaft. Sozialwissenschaftliche Abhandlungen der Görres-Gesellschaft ; Bd. 2.
HV5132.S74 362.7'8'2920943 79-353998
79 MARC

—SWEDEN

Börjeson, Bengt, 1932-
Behandlingen : familjeterapi med alkoholskadade föräldrar / Bengt Börjeson. — Stockholm : AWE/Geber, 1977.
184 p. ; 20 cm. S77-50/52
ISBN 9120048823 : kr45.10
1. Alcoholism—Sweden—Treatment. 2. Family social work—Sweden. 3. Barnbyn Skå. 4. Children of alcoholic parents—Sweden. 5. Alcoholics—Sweden—Family relationships. I. Title.
HV5283.S85B63 79-362440
79 MARC

CHILDREN OF ALIEN LABORERS

—EDUCATION—FRANCE

Charlot, Martine.
Mon avenir? Quel avenir? : Témoignages de jeunes immigrés / par Martine Charlot, avec la collaboration d'Annie Lauran et d'Ahmed Ben Dhiab ; dessin d'Ahmed Ben Dhiab. — ₍Paris₎ : Casterman, ₍c1978₎
210 p. : ill. ; 20 cm. — (Collection Orientations E3) F***
Bibliography: p. ₍199₎-207.
ISBN 2-203-20241-6 : 42.00F
1. Children of alien laborers—France. 2. Children of alien laborers—Education—France. I. Lauran, Annie, joint author. II. Ben Dhiab, Ahmed, joint author. III. Title.
HQ792.F7C4 301.43'1'0944 79-359220
79 MARC

Scolarisation des enfants de travailleurs immigrés : vers de nouvelles perspectives : compte rendu du stage C.E.E., Douai, janvier 1976. — ₍Lille₎ : CNDP, CRDP Lille, ₍1978₎
136 p. ; 30 cm. F***
At head of title: Cefisem, Douai, Lille.
Cover title.
1. Children of alien laborers—Education—France. I. Centre de formation et d'information pour la scolarisation des enfants de migrants. II. Title: Vers de nouvelles perspectives.
LC5158.F8S36 371.9'675'0944 79-349999
79 MARC

—EDUCATION—GERMANY, WEST

Akpinar, Ünal.
Pädagogische Arbeit mit ausländischen Kindern und Jugendlichen : Bestandsaufnahme u. Praxishilfen / Ünal Akpinar, Andrés López-Blasco, Jan Vink ; ₍von d. Projektgruppe Modellprogramm für Ausländ. Kinder u. Jugendliche erarb.₎. — München : Juventa-Verlag, 1977.
262 p. ; 23 cm. — (Reihe Deutsches Jugendinstitut : Aktuell) GFR77-A33
Bibliography: p. 249-257.
ISBN 3-7799-0235-4 : DM22.00
1. Children of alien laborers—Education—Germany, West. I. López-Blasco, Andrés, joint author. II. Vink, Jan, joint author. III. Projektgruppe Modellprogramm für Ausländische Kinder und Jugendliche. IV. Title.
LC5158.G3A42 78-383540
78 MARC

Rist, Ray C
Guestworkers in Germany : the prospects for pluralism / Ray C. Rist. : foreword by Irving Louis Horowitz. — New York : Praeger, 1978.
xxii, 263 p. ; 25 cm.
Bibliography: p. 247-258.
Includes index.
ISBN 0-03-040766-4 : $18.95
1. Alien labor—Germany, West. 2. Germany, West—Foreign population. 3. Children of alien laborers—Education—Germany, West. I. Title.
HD8458.A2R58 331.6'2'0943 78-6282
78 MARC

Unsöld, Werner.
Lehrereinstellung und Schülervorurteil : e. empir. Feldstudie über d. Beziehungen zwischen Lehrer- u. Unterrichtsvariablen, Elternvariablen u. Variablen des Kontakts u. Vorurteilen gegenüber Gastarbeitern bei 1118 Schülern d. Grund- u. Hauptschule / Werner Unsöld. — Frankfurt am Main ; Bern ; Las Vegas : Lang, 1978.
367 p. ; 21 cm. — (Europäische Hochschulschriften : Reihe 6, Psychologie ; Bd. 38) GFR78-A
Bibliography: p. 283-309.
ISBN 3-261-02537-9
1. Children of alien laborers—Education—Germany, West. 2. Teachers—Attitudes. 3. Prejudices and antipathies. 4. Interaction analysis in education. I. Title. II. Series.
LC5158.G3U55 79-364638
79 MARC

—EDUCATION—GERMANY, WEST—BREMEN

Bolte, Heinz.
Ausländische Kinder in Bremer Vorbereitungsklassen / Heinz Bolte. — Bremen : Wissenschaftl. Inst. für Schulpraxis, 1977.
336 p. : ill. ; 21 cm. — (Arbeitsberichte - Wissenschafliches Institut für Schulpraxis ; Folge 3) GFR78-B11
A revision of the author's thesis, Bremen.
Bibliography: p. 320-330.
Includes index.
1. Children of alien laborers—Education—Germany, West—Bremen. I. Title. II. Series: Wissenschafliches Institut für Schulpraxis, Bremen. Arbeitsberichte - Wissenschaftliches Institut für Schulpraxis ; Folge 3.
LC5158.G3B64 1977 79-370311
79 MARC

—EDUCATION—GERMANY, WEST—NORTH RHINE-WESTPHALIA

Damanakēs, Michaēl.
Sozialisationsprobleme der griechischen Gastarbeiterkinder in den Grund- und Hauptschulen des Bundeslandes Nordrhein-Westfalen / Michael Damanakis. — Kastellaun : Henn, 1978.
340 p. ; 24 cm. — (Kölner Arbeiten zur Pädagogik) GFR78-A25
Some text in Greek.
Originally presented as the author's thesis, Cologne, 1977.
Bibliography: p. 331-340.
ISBN 3-450-03913-X : DM39.80
1. Children of alien laborers—Education—Germany, West—North Rhine-Westphalia. 2. Socialization. I. Title.
LC5158.G3D35 1978 79-353831
79 MARC

—BERLIN

Berlin (West Berlin).
Bericht zur Lage der Ausländer in Berlin : ressortübergreifende Planung / Berlin ; Hrsg., Der Regierende Bürgermeister von Berlin, Senatskanzlei, Planungsleitstelle. — Berlin : Der Regierende Bürgermeister ; Berlin : Senatskanzlei, Planungsleitstelle, 1978.
ii, 119, 7 p. ; 30 cm. GFR78-B19
1. Alien labor—Berlin. 2. Children of alien laborers—Berlin. I. Title.
HD8457.B47B47 1978 79-378066
79 MARC

CHILDREN OF ALIEN LABORERS (Continued)

—FRANCE

Charlot, Martine.
Mon avenir? Quel avenir? : Témoignages de jeunes immigrés / par Martine Charlot, avec la collaboration d'Annie Lauran et d'Ahmed Ben Dhiab ; dessin d'Ahmed Ben Dhiab. — [Paris] : Casterman, [c1978]
210 p. : ill. ; 20 cm. — (Collection Orientations E3) F***
Bibliography: p. [199]-207.
ISBN 2-203-20241-6 : 42.00F
1. Children of alien laborers—France. 2. Children of alien laborers—Education—France. I. Lauran, Annie, joint author. II. Ben Dhiab, Ahmed, joint author. III. Title.
HQ792.F7C4 301.43'1'0944 79-359220
79 MARC

CHILDREN OF DIVORCED PARENTS

Galper, Miriam, 1940-
Co-parenting : sharing your child equally : a source book for the separated or divored family / Miriam Galper. — Philadelphia : Running Press, c1978.
158 p. ; 23 cm.
Bibliography: p. 151-153.
Includes index.
ISBN 0-89471-038-9 : $8.95. ISBN 0-89471-037-0 lib. bdg. : $15.90. ISBN 0-89471-036-2 pbk. : $4.95
1. Children of divorced parents. 2. Parent and child. I. Title.
HQ777.5.G34 301.42'7 77-29108
78 MARC

—DENMARK—CASE STUDIES

Hold fingrene fra "mor" / Bitten Lolk & Jens Robdrup. — København : Chr. Erichsen, 1977.
121 p. ; 20 cm. D77-27
ISBN 8755505066 : kr32.50
1. Stepparents—Denmark—Attitudes. 2. Parent and child—Case studies. 3. Children of divorced parents—Denmark—Case studies. 4. Stepchildren—Denmark—Case studies. 5. Parent and child (Law)—Denmark. I. Lolk, Bitten. II. Robdrup, Jens.
HQ642.H64 79-343335
79 MARC

—NETHERLANDS

Katholieke Hogeschool. Sektie Jeugdrecht.
Omgang met elkaar : onderzoek naar de totstandkoming van omgangsregelingen na scheiding in zuidelijk Nederland / Sektie Jeugdrecht, Katholieke Hogeschool Tilburg ; m. m. v. C. H. L. van Heesch ... [et al.]. — Deventer : Kluwer, c1978.
xiii, 186 p. ; 25 cm. Ne***
Bibliography: p. 152-155.
ISBN 9026810245
1. Children of divorced parents—Netherlands. 2. Parent and child (Law)—Netherlands. 3. Children—Law—Netherlands. 4. Divorce—Netherlands. I. Heesch, C. H. L. van. II. Title.
HQ777.5.K38 1978 301.42'7 78-389835
79 MARC

—NEW ZEALAND—AUCKLAND

Clay, Marie M
Children of parents who separate / Marie M. Clay and Viviane M. J. Robinson. — Wellington : New Zealand Council for Educational Research, 1978.
xii, 140 p. : ill., form, maps ; 21 cm. — (Bulletin - New Zealand Council for Educational Research ; no. 15) NZ78-5
Bibliography: p. 137-140.
ISBN 0-908567-05-7 : $3.50
1. Children of divorced parents—New Zealand—Auckland. I. Robinson, Viviane M. J., joint author. II. Title. III. Series: New Zealand Council for Educational Research. Bulletin ; no. 15.
LC3501.M3N4 no. 15 370'.78'0931 s 79-303877
[HQ777.5] 79 MARC

—UNITED STATES

Ramos, Suzanne.
The complete book of child custody / Suzanne Ramos. — New York : Putnam, c1979.
330 p. ; 22 cm.
Bibliography: p. 325-330.
ISBN 0-399-12204-4
1. Children of divorced parents—United States. 2. Custody of children—United States. I. Title.
HQ777.5.R35 1979 301.42'8 78-26061
78 MARC

Shepard, Morris A
Divorced dads, their kids, ex-wives, and new lives / Morris A. Shepard and Gerald Goldman. — Radnor, Pa. : Chilton Book Co., c1979.
xi, 154 p. ; 22 cm.
Includes bibliographical references.
ISBN 0-8019-6755-4 : $7.95
1. Divorcees—United States. 2. Fathers. 3. Children of divorced parents—United States. 4. Father and child. I. Goldman, Gerald, joint author. II. Title.
HQ834.S58 1979 301.42'7 78-14637
79 MARC

Vigeveno, H S
Divorce and the children / H. S. Vigeveno and Anne Claire. — Glendale, Calif. : GL Regal Books, c1979.
127 p. ; 21 cm.
Bibliography: p. 125-127.
ISBN 0-8307-0645-3 : $3.25
1. Children of divorced parents—United States. 2. Divorce—United States. 3. Single-parent family—United States. I. Claire, Anne, joint author. II. Title.
HQ777.5.V53 301.42'84 78-67855
79 MARC

Woolley, Persia, 1935-
The custody handbook / Persia Woolley. — New York : Summit Books, c1979.
350 p. ; 22 cm.
Bibliography: p. [339]-340.
Includes index.
ISBN 0-671-40032-0
1. Custody of children—United States. 2. Divorce—United States. 3. Children of divorced parents—United States. 4. Single parents—United States. I. Title.
HQ777.5.W66 301.42'84 79-10862
79 MARC

—UNITED STATES—JUVENILE LITERATURE

White, Ann S
Divorce / by Ann S. White. — New York : F. Watts, 1979.
55 p. : ill. ; 23 cm. — (A First book)
Includes index.
SUMMARY: Discusses the causes and procedures of divorce, the changes it brings to family life, and its effects on children.
ISBN 0-531-02254-4
1. Divorce—United States—Juvenile literature. 2. Children of divorced parents—United States—Juvenile literature. [1. Divorce] I. Title.
HQ834.W57 301.42'84'0973 78-10464
78 MARC

CHILDREN OF GOD

see also Kingdom of God

CHILDREN OF IMMIGRANTS

—EDUCATION—GERMANY, WEST

Mehrländer, Ursula.
Einflussfaktoren auf das Bildungsverhalten ausländischer Jugendlicher : Vergleich italien. u. dt. Auszubildender u. Jungarbeiter / Ursula Mehrländer. — 1. Aufl. — Bonn : Verlag Neue Gesellschaft, 1978.
203 p. ; 24 cm. — (Reihe Arbeit- Forschungsinstitut Friedrich-Ebert-Stiftung) GFR78-A
Errata slip inserted.
Bibliography: p. 155-160.
ISBN 3-87831-280-6 ; DM32.00
1. Children of immigrants—Education—Germany, West. 2. Italians in Germany, West. 3. Educational research—Germany, West. 4. Vocational education—Germany, West. I. Title.
LC3747.G3M43 79-371335
79 MARC

—CANADA—JUVENILE LITERATURE

Come with us : children speak for themselves / [coordinated by Judy McClard and Naomi Wall]. — Toronto : Women's Educational Press, 1978.
119 p. : ill. (some col.) ; 23 x 25 cm. C***
ISBN 0-88961-045-2
1. Children of immigrants—Canada—Juvenile literature. I. McClard, Judy. II. Wall, Naomi.
HQ792.C3C65 301.43'14'0971 79-300569
79 MARC

CHILDREN OF MIGRANT LABORERS

—EDUCATION—LAW AND LEGISLATION—UNITED STATES—STATES

United States. Congress. House. Committtee on Education and Labor. Subcommittee on Elementary, Secondary, and Vocational Education.
Part 17, Title I—State migrant program : hearing before the Subcommittee on Elementary, Secondary, and Vocational Education of the Committee on Education and Labor, House of Representatives, Ninety-fifth Congress, first session on H.R. 15 to extend for five years certain elementary, secondary, and other education programs, hearing held in Washington, D.C., on October 12, 1977. — Washington : U.S. Govt. Print. Off., 1977.
iii, 332 p. ; 24 cm.
1. Children of migrant laborers—Education—Law and legislation—United States—States. 2. Children of migrant laborers—Education—United States—States. I. Title: State migrant program.
KF27.E3364 1977m 344'.73'0791675 79-601069
79 MARC

—EDUCATION—CALIFORNIA

California. Office of Compensatory Education. Special Programs and Support Services Division.
California master plan for migrant education : Elementary and secondary education act of 1965, title I, public law 89-10, as amended by public law 89-750 of 1966 and public law 93-380 of 1974 / prepared by the Office of Compensatory Education, Special Programs and Support Services Division. — Sacramento : California State Dept. of Education, 1976.
viii, 32 p. ; 29 cm.
Bibliography: p. 32.
1. Children of migrant laborers—Education—California. I. Title.
LC5152.C2C32 1976 371.9'675'09794 77-62249
78 MARC

—EDUCATION—COLORADO

Colorado Migrant Child Identification and Recruitment Project.
Colorado Migrant Child Identification and Recruitment Project : field guide / developed by George L. Archuleta, director ; Sherrie L. Archuleta, research assistant. — [La Salle, Colo.] : The Project, 1975.
55 p. : forms ; 29 cm.
1. Children of migrant laborers—Education—Colorado. I. Archuleta, George L. II. Archuleta, Sherrie L.
LC5152.C6C64 371.9'675'09788 78-1027[illegible]
78 MARC

—EDUCATION—EUROPEAN ECONOMIC COMMUNITY COUNTRIES

Commission of the European Communities.
Education of migrant workers' children in the European Community. — [Bruxelles : European Communities-Information ; Washington, D.C. : sold by European Communities-Information], 1975.
13 p. ; 23 cm. — (European documentation ; 1975/1 : School series) Be***
1. Children of migrant laborers—Education—European economic community countries. I. Title. II. Series: European Parliament. European documentation ; 1975/1.
Z699.A1E8 1975, no. 1 309.1'4'055 s 79-309247
[LC5153.E8] 79 MARC

—EDUCATION—FINLAND—CASE STUDIES

The Education of migrant workers and their families : case studies undertaken for the National Commissions of Finland, France, Sweden, Yugoslavia. — Paris : Unesco, c1978.
44 p. : ill. ; 27 cm. — (Educational studies and documents : New series ; no. 27) F***
Includes a bibliography.
ISBN 9231015362
1. Children of migrant laborers—Education—Finland—Case studies. 2. Children of migrant laborers—Education—France—Case studies. 3. Children of migrant laborers—Education—Sweden—Case studies. 4. Children of migrant laborers—Education—Yugoslavia—Case studies. I. Series.
LC5153.F5E34 371.9'675 78-323060
79 MARC

—EDUCATION—FRANCE

La Scolarisation des enfants de migrants : stage d'information et de formation des personnels concernés par la scolarisation des enfants de travailleurs migrants : Roubaix, 25-29 avril 1977. — Lille : CNDP, CRDP, [1978]
iii, 75 p. : ill. ; 30 cm. F***
Cover title.
1. Children of migrant laborers—Education—France. I. Centre régional de documentation pédagogique de Lille.
LC5153.F8S36 78-395352
79 MARC

CHILDREN OF MIGRANT LABORERS (Continued)

—EDUCATION—FRANCE—CASE STUDIES

The Education of migrant workers and their families : case studies undertaken for the National Commissions of Finland, France, Sweden, Yugoslavia. — Paris : Unesco, c1978.
44 p. : ill. ; 27 cm. — (Educational studies and documents : New series ; no. 27) F***
Includes a bibliography.
ISBN 9231015362
1. Children of migrant laborers—Education—Finland—Case studies. 2. Children of migrant laborers—Education—France—Case studies. 3. Children of migrant laborers—Education—Sweden—Case studies. 4. Children of migrant laborers—Education—Yugoslavia—Case studies. I. Series.
LC5153.F5E34 371.9'675 78-323060
79 MARC

—EDUCATION—FRANCE—CONGRESSES

Information et formation des personnels concernés par la scolarisation des enfants de travailleurs migrants dans le second degré (1er cycle et C.E.T.) : stage du 18 au 22 avril 1977. — Lyon : C.R.D.P., ⌊1977⌋
64 p. ; 27 cm. — (Annales du Centre régional de documentation pédagogique de Lyon) F***
On cover: Stage des migrants.
1. Children of migrant laborers—Education—France—Congresses. I. Title: Stage des migrants.
LC5153.F8I53 371.9'675'0944 78-379542
78 MARC

—EDUCATION—GERMANY, WEST

Ottmann, Anton, 1945-
Entwicklung und Evaluation eines Mathematikcurriculums für Ausländerkinder / Anton Ottmann. — Frankfurt am Main ; Bern ; Las Vegas : Lang, 1979.
i, 171 p. : graphs ; 21 cm. — (Studien zur Erziehungswissenschaft ; Bd. 5) GFR79-A
Bibliography: p. 165-171.
ISBN (invalid) 3-261-02701-1
1. Mathematics—Study and teaching (Elementary)—Germany, West. 2. Children of migrant laborers—Education—Germany, West. I. Title.
QA135.5.O86 79-367618
79 MARC

--EDUCATION--NORTH CAROLINA --PERIODICALS

North Carolina. Migrant Education Section.
Migrant education: State evaluation report.
⌊Raleigh⌋ Migrant Education Section, Division of Compensatory Education, Dept. of Public Instruction.
v. ill. 28 cm. annual.
1. Children of migrant laborers—Education—North Carolina—Periodicals. I. Title.
LC5152.N8N67a 371.9'675'09756 79-643678
MARC-S

—EDUCATION—OREGON

Oregon. Dept. of Education.
Oregon state plan for migrant education, 1979. — ⌊Salem : Oregon State Dept. of Education, 1978?⌋
48 leaves ; 28 cm.
1. Children of migrant laborers—Education—Oregon. I. Title.
LC5152.O734 1979 371.9'675'09795 79-620752
79 MARC

—EDUCATION—SOUTH CAROLINA —STATISTICS—PERIODICALS

South Carolina. Dept. of Education. Division of Instruction. Office of Federal Programs. Allocations Section.
ESEA Title I migrant annual evaluation report.
Columbia, South Carolina Dept. of Education, Division of Instruction.
v. 28 cm.
Key title: ESEA Title I migrant annual evaluation report, ISSN 0161-9497
1. Children of migrant laborers—Education—South Carolina—Statistics—Periodicals. I. Title.
LC5152.S6S63b 371.9'675'09757 78-641515
MARC-S

—EDUCATION—UNITED STATES—STATES

United States. Congress. House. Committtee on Education and Labor. Subcommittee on Elementary, Secondary, and Vocational Education.
Part 17, Title I—State migrant program : hearing before the Subcommittee on Elementary, Secondary, and Vocational Education of the Committee on Education and Labor, House of Representatives, Ninety-fifth Congress, first session on H.R. 15 to extend for five years certain elementary, secondary, and other education programs, hearing held in Washington, D.C., on October 12, 1977. — Washington : U.S. Govt. Print. Off., 1977.
iii, 332 p. ; 24 cm.
1. Children of migrant laborers—Education—Law and legislation—United States—States. 2. Children of migrant laborers—Education—United States—States. I. Title: State migrant program.
KF27.E3364 1977m 344'.73'0791675 79-601069
79 MARC

—EDUCATION—UTAH —EVALUATION—PERIODICALS

The Education of migrant workers and their families : case studies undertaken for the National Commissions of Finland, France, Sweden, Yugoslavia. — Paris : Unesco, c1978.
44 p. : ill. ; 27 cm. — (Educational studies and documents : New series ; no. 27) F***
Includes a bibliography.
ISBN 9231015362
1. Children of migrant laborers—Education—Finland—Case studies. 2. Children of migrant laborers—Education—France—Case studies. 3. Children of migrant laborers—Education—Sweden—Case studies. 4. Children of migrant laborers—Education—Yugoslavia—Case studies. I. Series.
LC5153.F5E34 371.9'675 78-323060
79 MARC

Utah. State Board of Education.
Annual evaluation report on migrant education.
Salt Lake City, Office of Utah State Board of Education.
v. 28 cm.
Report year ends June 30.
Cover title, : Utah annual evaluation, summer migrant education.
1. Children of migrant laborers—Education—Utah—Evaluation—Periodicals. I. Utah. State Board of Education. Utah annual evaluation, summer migrant education. II. Title.
LC5152.U8U83a 371.9'675'09792 79-640125
MARC-S

—EDUCATION—YUGOSLAVIA—CASE STUDIES

The Education of migrant workers and their families : case studies undertaken for the National Commissions of Finland, France, Sweden, Yugoslavia. — Paris : Unesco, c1978.
44 p. : ill. ; 27 cm. — (Educational studies and documents : New series ; no. 27) F***
Includes a bibliography.
ISBN 9231015362
1. Children of migrant laborers—Education—Finland—Case studies. 2. Children of migrant laborers—Education—France—Case studies. 3. Children of migrant laborers—Education—Sweden—Case studies. 4. Children of migrant laborers—Education—Yugoslavia—Case studies. I. Series.
LC5153.F5E34 371.9'675 78-323060
79 MARC

—EUROPEAN ECONOMIC COMMUNITY COUNTRIES

Commission of the European Communities.
The children of migrant workers. — Brussels : Commission of the European Communities ; ⌊Washington, D.C. : sold by European Community Information Service⌋, 1977.
53 p. ; 25 cm. — (Collection studies : Education series ; no. 1) Be***
Bibliography: p. 53.
$2.20
1. Migrant labor—European Economic Community countries. 2. Children of migrant laborers—European Economic Community countries. I. Title. II. Series.
HD5856.E8C65 1977 301.43'14 78-317059
79 MARC

CHILDREN OF MILITARY PERSONNEL

—STATISTICS—PERIODICALS

Hawaii. Dept. of Education. Office of the Superintendent. Budget Branch.
Report on federally connected pupils: Hawaii public schools.
⌊Honolulu⌋ Office of the Superintendent, Budget Branch, Dept. of Education, State of Hawaii.
v. 28 cm. annual. (Statistical report - Dept. of Education, State of Hawaii)
Continues: Report on federally connected pupils. Hawaii public schools, ISSN 0090-9440, issued by: Hawaii. Dept. of Education. Office of Planning & Budget.
1. Federal aid to education—Hawaii—Statistics—Periodicals. 2. Children of military personnel—Statistics—Periodicals. I. Title. II. Series: Hawaii. Dept. of Education. Statistical report.
LB2826.H3H33a 379'.122'09969 78-645450
MARC-S

CHILDREN OF WOMEN PRISONERS

—UNITED STATES

McGowan, Brenda G
Why punish the children? : A study of children of women prisoners / Brenda G. McGowan, Karen L. Blumenthal ; foreword by Justine Wise Polier. — Hackensack, N.J. : National Council on Crime and Delinquency, c1978.
xi, 124 p. ; 23 cm.
Bibliography: p. 118-124.
1. Children of women prisoners—United States. I. Blumenthal, Karen L., joint author. II. National Council on Crime and Delinquency. III. Title.
HV8386.U5M3 362.7'3 78-107990
78 MARC

CHILDREN OF WORKING MOTHERS

—SERVICES FOR—GREAT BRITAIN

Trades Union Congress.
The under-fives : report of a TUC working party. — London : T.U.C., ⌊1978⌋
⌊1⌋, 28 p. ; 24 cm. GB78-09441
Cover title.
£0.50
1. Day care centers—Great Britain. 2. Education, Preschool—Great Britain. 3. Children of working mothers—Services for—Great Britain. I. Title.
HV861.G6T72 1978 362.7'1 79-313927
00 MARC

—GREAT BRITAIN

Great Britain. Central Policy Review Staff.
Services for young children with working mothers : report / by the Central Policy Review Staff. — London : H.M. Stationery Off., 1978.
89 p. : ill. ; 30 cm. GB***
ISBN 0-11-630805-2 : £2.75
1. Children of working mothers—Great Britain. 2. Child welfare—Great Britain. 3. Day care centers—Great Britain. I. Title.
HV752.A6G73 1978 362.7'1'0941 79-313843
00 MARC

—GREAT BRITAIN—CONGRESSES

Mothers in employment : papers from a Conference on "Mothers in Employment : Trends and Issues" held at Brunel University, 24-25 May 1976 / edited by Nickie Fonda and Peter Moss. — Uxbridge : Brunel University Management Programme ; ⌊London⌋ : Thomas Coram Research Unit, Institute of Education, University of London, 1976.
⌊3⌋, 141 p. : ill. ; 30 cm. GB78-29317
Includes bibliographies.
ISBN 0-902215-01-9 : £2.50
1. Mothers—Employment—Great Britain—Congresses. 2. Day care centers—Great Britain—Congresses. 3. Children of working mothers—Great Britain—Congresses. I. Fonda, Nickie. II. Moss, Peter G.
HD6055.M67 1976 331.4 79-301836
79 MARC

—INDIA

National Institute of Public Cooperation and Child Development.
Working mother and early childhood education : report of a research sponsored by UNESCO. — New Delhi : National Institute of Public Cooperation and Child Development, 1978.
vi, 120 p. ; 22 cm.
Includes bibliographical references.
Rs25.00
1. Mothers—Employment—India. 2. Children of working mothers—India. 3. Education, Preschool—India. I. Title.
HD6055.N34 1978 331.4'0954 79-901159
79 MARC

CHILDREN OF WORKING PARENTS

see also Children of working mothers

CHILDREN'S ACCIDENTS

see also Children—Wounds andinjuries; Pediatric toxicology

—PREVENTION

Hermand, Marie, 1948-
Secourez votre enfant / Marie Hermand. — Paris : J.-P. Delarge, ⌈c1977⌉
134 p. ; 21 cm. — (Votre santé) F***
Includes index.
ISBN 2-7113-0037-4 : 25.00F
1. Pediatric emergencies. 2. First aid in illness and injury. 3. Children's accidents—Prevention. I. Title.
RJ370.H47 618.9'2002'5 79-337668
79 MARC

CHILDREN'S AID SOCIETY OF WINNIPEG

Duncan, E H
Night-duty social worker / by E. H. Duncan. — 1st ed. — New York : Elsevier/Nelson Books, c1979.
214 p. ; 24 cm.
ISBN 0-525-66630-3 : $8.95
1. Duncan, E. H. 2. Social workers—Manitoba—Winnipeg—Biography. 3. Children's Aid Society of Winnipeg. I. Title.
HV746.W56D86 362.7'092'6 79-4382
79 MARC

CHILDREN'S APPAREL INDUSTRY see Children's clothing industry

CHILDREN'S ALLOWANCES

Davis, Ken, 1932-
Kids & cash : solving a parent's dilemma / by Ken Davis and Tom Taylor. — 1st ed. — La Jolla, Calif. : Oak Tree Publications, c1979.
vii, 280 p. : ill. ; 24 cm.
Includes index.
SUMMARY: Advice to parents on how to teach their children to handle money and understand its value and suggestions to children on ways of earning money.
ISBN 0-916392-26-0
1. Children's allowances. 2. Saving and thrift. 3. Finance, Personal. ⌈1. Finance, personal⌉ I. Taylor, Tom, 1930- joint author. II. Title.
HQ784.S4D38 332'.024 79-13092
79 MARC

Yokosuka-shi Kyōiku Kenkyūjo.
(Kodomo no asobi to kozukai ni tsuite no chōsa)
子どもの遊びとこづかいについての調査 / ⌈編集 横須賀市教育研究所⌉. — 横須賀：同研究所, 1978.
33 p. ; 26 cm. — (紀要 - 横須賀市教育研究所 ; 73集) Ja***
Cover title.
Bibliography: p. 25.
"執筆者 安田龍夫 梅沢光熙"
1. Play—Japan. 2. Games—Japan. 3. Children's allowance. 4. Elementary school students' socio-economic status—Japan. I. Yasuda, Tatsuo. II. Umezawa, Mitsuhiro. III. Title. IV. Series: Yokosuka-shi Kyōiku Kenkyūjo. Kiyō ; 73-shū.
LB1137.Y64 1978 79-808979
J

CHILDREN'S ART

see also Children as artists

An Exceptional view of life : the Easter seal story / written & illustrated by handicapped children ; originated and directed by Edward J. McGrath ; edited by Bob Kraus. — Norfolk Island, Australia : Island Heritage, c1977.
64 p. : col. ill. ; 27 cm. — (An Island Heritage book) Aus***
SUMMARY: Youngsters with a variety of physical disabilities express their feelings about their handicaps through words and art.
1. Handicapped children—Biography. 2. Children's art. 3. Children's writings. ⌈1. Physically handicapped. 2. Children's art. 3. Children's writings⌉ I. McGrath, Edward J. II. Kraus, Bob, 1924-
HV888.E9 1977 362.7'8'40922 77-73684
79 MARC

—CHINA—SHANGHAI

(Shang-hai shao nien erh t'ung hua hsüan)
上海少年儿童画选 / 上海人民出版社⌈编辑⌉. — 上海：該社, 1973.
41 p. : all col. ill. ; 19 cm.
1. Children's art—China—Shanghai. 2. Drawing, Chinese—China—Shanghai. I. Shang-hai jen min ch'u pan she.
N352.S45 78-843165
C

—ENGLAND—CAMBRIDGESHIRE

Moving jewels : an anthology of writing and art / by Cambridgeshire schoolchildren ; ⌈editor Christopher Martin⌉. — ⌈Cambridge⌉ : EARO for the George Edwards Memorial Committee, ⌈1977⌉
80 p. : ill. (some col.), port. ; 21 cm. GB77-33933
ISBN 0-904463-24-9 : £1.50
1. Children's writings, English—England—Cambridgeshire. 2. Children's art—England—Cambridgeshire. I. Martin, Christopher. II. Edwards, George, 1909-1975. III. George Edwards Memorial Committee.
PR1110.C5M6 700'.9426'5 78-322431
79 MARC

—ESTONIA—CATALOGS

Laste loomeröõm : Eesti NSV üldhariduskoolide õpilaste tööd / ⌈koostanud E. Aljasmets⌉. — Tallinn : Valgus, 1978.
159 p. : ill., (some col.) ; 22 cm. USSR***
English, Finnish, German and Russian translations of text (78 p.) inserted.
2.90rub
1. Children's art—Estonia—Catalogs. I. Aljasmets, E.
N352.E75L37 79-355142
79 MARC

—FINLAND

Tapiolan lapset piirtävät ja kirjoittavat. ⌈Julk.⌉ (Lions club Tapiola). ⌈Tapiola, 1968⌉.
63 p. illus. 20 cm. Fi
1. Children's writings, Finnish. 2. Children's art—Finland.
PH342.T36 78-383979
78 MARC

—JAPAN

Miyatake, Tatsuo.
(Yōji to seihakuji no e ga uttaeru mono)
幼児と精薄児の絵が訴えるもの / 宮武辰夫著. — 名古屋：黎明書房, 1955.
387 p. : ill. (some col.) ; 22 cm.
1. Drawing, Psychology of. 2. Children's art—Japan. 3. Mentally handcapped children—Education—Art. 4. Child analysis. I. Title.
BF723.D7M57 79-801649
J

Sasaki, Keizō.
(Kokoro no dezain)
心のデザイン：刺激から創造 / 佐々木敬三著. — 仙台：宝文堂出版販売：宝文堂発売, 昭和52 ⌈1977⌉
266 p. : ill. ; 21 cm. Ja 77-22708
Bibliography: p. 265-266.
Y1500
1. Creative thinking (Education) 2. Children's writings, Japanese. 3. Children's art—Japan. I. Title.
LB1062.S313 79-80€

Shimoda, Seishi, 1890-
(E ni miru kodomo no shinri)
絵にみる子どもの心理 / 霜田静志著. — ⌈東京⌉：東都書房, 昭和34 ⌈1959⌉
97, ⌈2⌉ p. : ill. (some col.) ; 27 cm.
Bibliography: p. ⌈98⌉
1. Children's art—Japan. 2. Art—Psychology. 3. Child analysis. I. Title.
N352.S52 79-80934

Yukawa, Naobumi, 1904-
(E o kaku kodomo)
絵をかく子ども / 湯川尚文. — 東京：誠文堂新光社, 昭和26 ⌈1951⌉
4, 222 p., ⌈3⌉ leaves of plates : ill. ; 19 cm.
1. Art—Study and teaching—Japan. 2. Children's art—Japan. I. Title.
N365.J3Y83 79-80534

—LEBANON

(Aṭfāl Lubnān yarsumūna al-ḥarb)
اطفال لبنان يرسمون الحرب / باشراف سيتا مانوكيان. — بيروت : النجدة الشعبية اللبنانية ، ⌈-197⌉
3 v. : col. ill. ; 20 x 29 cm.
Cover title.
Added cover title: Lebanese children and the war.
English and Arabic.
1. Children's art—Lebanon. 2. War in art. I. Mānūkyān, Sītā. II. Title: Lebanese children and the war.
N352.A87 79-970091

—SPAIN

Algo se rompe : textos libres y dibujos elaborados por niños de 6 a 14 años / ⌈colección animada y coordinada por Antonio Fernández ... et al.⌉. — ⌈Bilbao : Zero ; distribuidor exclusivo, ZYX⌉, c1979.
64 p. : col. ill. ; 28 cm. — (El Trastero ; 1) Sp***
Cover title.
ISBN 8431704950
1. Children's writings, Spanish. 2. Children's art—Spain. I. Fernández, Antonio, fl. 1976-
PQ6173.5.C45A4 79-109090
79 MARC

CHILDREN'S BELONGINGS see Children's paraphernalia

CHILDREN'S BOOKS see Children's literature

CHILDREN'S CHRISTIAN LITERATURE see Christian literature for children

CHILDREN'S CLOTHING

see also Children—Costume; Children's sleepwear; Infants—Clothing

Hotte, Eleanor Boettke.
Self-help clothing : for children who have physical disabiliti / by Eleanor Boettke Hotte. — Rev. ed. — ⌈Chicago⌉ : Nationa Easter Seal Society for Crippled Children and Adults, 1979.
64 p. : ill. ; 23 cm.
Bibliography: p. 62-64.
1. Handicapped—Clothing. 2. Children's clothing. I. Title.
TT648.H67 1979 646.4'7 79-11007
00 MAR

Lawson, Donna.
Kids' clothes for under $5 : how to save money and have yo kids look great by learning to recycle clothes / by Donna La son ; ill. ; by Karen Feldgus ; ⌈interior photos, Barry Wetche — New York : Quick Fox, c1978.
95 p. : ill. ; 28 cm.
ISBN 0-8256-3051-7 : $4.95
1. Children's clothing. 2. Clothing and dress—Remaking. I. Title.
TT635.L38 646.4'06 77-937
78 MA

CHILDREN'S CLOTHING (Continued)

Martensson, Kerstin.
Sew for toddlers / by Kerstin Martensson. — ‹Minneapolis : Kwik-Sew Pattern Co.›, c1979.
134 p., ‹1› fold. leaf of plates : ill. ; 28 cm.
At head of title: Kwik-sew method.
ISBN 0-913212-08-3
1. Children's clothing. 2. Sewing. I. Title. II. Title: Kwik-sew method.
TT635.M38 646.4'06 79-119775
79 MARC

(Osnovy promyshlennoĭ tekhnologii pouzlovoĭ obrabotki verkhnikh sorochek i bel'ia)
Основы промышленной технологии поузловой обработки верхних сорочек и белья / ‹Т. И. Куликова ... et al.›. — Изд. 2., испр. и доп. — Москва : Легкая промышл., 1975.
155, ‹4› p. : ill. ; 22 cm. USSR***
Bibliography : p. ‹157›
0.54rub
1. Shirts, Men's. 2. Blouse's. 3. Underwear. 4. Children's clothing. I. Kulikova, Tamara Ivanovna.
TT612.O8 1975 79-411835

—HISTORY—TWENTIETH CENTURY

Guppy, Alice.
Children's clothes, 1939-1970 : the advent of fashion / ‹by› Alice Guppy. — Poole : Blandford Press for the Pasold Research Fund, 1978.
‹10›, 346 p., ‹8› p. of plates : ill. (some col.) ; 23 cm. GB78-37637
Includes index.
ISBN 0-7137-0896-4 : £9.50
1. Children's clothing—History—Twentieth century. 2. Clothing trade—Great Britain—History—Twentieth century. I. Title.
TT635.G78 391'.07'30941 79-318146
79 MARC

—PATTERN DESIGN

Halpern, Bomie, 1928-
Farm journal's design-and-sew children's clothes / by Bomie Halpern, with Kathryn Larson ; ‹paper dolls, clothes and ill., Rosemary G. Crysler›. — Philadelphia : Farm journal ; Garden City, N.Y. : distributed to the trade by Dolphin Books, c1979.
59 p. : ill. ; 27 cm.
ISBN 0-89795-003-8 : $4.95
1. Children's clothing—Pattern design. I. Larson, Kathryn, joint author. II. Farm journal (Philadelphia, 1956-) III. Title. IV. Title: Design-and-sew children's clothes.
TT640.H34 646.4'3'06 79-1500
79 MARC

Shen, Yang.
(Ts'ai chien hsin fa)
裁剪新法：男裝·女裝·童裝：基本原理 / 沈陽編著. — 香港：大光出版社，1976.
4, 112 p. : ill. ; 19 cm.
$3.50

1. Tailoring—Pattern design. 2. Dressmaking—Pattern design. 3. Children's clothing—Pattern design. I. Title.
TT590.S54 78-841723

CHILDREN'S CLOTHING INDUSTRY

—UNITED STATES—FINANCE

Children's apparel, furniture, and toy industry sourcebook, May 1975-April 1976 / materials selected by MSS Information Corporation from the journals, indexes, and profiles of Disclosure, inc. — ‹New York› : The Corporation ; Edison, N.J. : exclusive sales agent, Lasky-Lanouette, c1977.
157 p. ; 23 cm.
"A guide to corporate reports filed with the Securities and Exchange Commission."
ISBN 0-8422-4028-4 : $50.00
1. Children's clothing industry—United States—Finance. 2. Furniture industry and trade—United States—Finance. 3. Toy industry—United States—Finance. I. MSS Information Corporation. II. Disclosure, inc. III. United States. Securities and Exchange Commission.
HD9940.U4C52 338.7'61'68 77-11746
77 MARC

CHILDREN'S COSTUME see Children—Costume

CHILDREN'S COURTS see Juvenile courts

CHILDREN'S CRUSADE, 1212

—JUVENILE DRAMA

Taylor, Don.
A long march to Jerusalem : a play / Don Taylor. — London ; New York : S. French, c1978.
78 p. ; 22 cm. GB***
ISBN 0-573-05045-7 : £1.10
1. Children's Crusade, 1212—Juvenile drama. I. Title.
PR6070.A914L6 822'.9'14 79-316978
79 MARC

CHILDREN'S DANCES see Dancing—Children's dances

CHILDREN'S DISEASES see Children—Diseases

see also Children—Costume; Children's sleepwear; Infants—Clothing

Cabrer Borràs, Guillem.
La flor romanial / ‹Guillem Cabrer Borràs›. — Barcelona : La Galera, c1978.
24 p. : ill. ; 17 cm. — (Teatre, joc d'equip ; 22) Sp***
Cover title.
ISBN 8424650220 : 60ptas
I. Title.
PZ90.C3C2 79-351033
79 MARC

CHILDREN'S DRAMA, SPANISH

Capmany, María Aurelia, 1918-
El cavaller Tirant / Joanot Martorell, Maria Aurèlia Capmany. — Barcelona : Edicions Don Bosco, c1977.
51 p. : ill. ; 21 cm. — (Col·lecció Teatro edebé ; 1) Sp***
A play adapted from J. Martorell's short story Tirant lo Blanch.
ISBN 842361316X
I. Martorell, Joannot. Tirant lo Blanch. II. Title. III. Series.
PZ90.C3C26 78-391764
79 MARC

CHILDREN'S DREAMS

—CASE STUDIES

Ruspoli, Nancy.
I sogni di Francesco / Nancy Ruspoli. — 1. ed. — Milano : Bompiani, 1979.
123 p. ; 21 cm. It79-Mar
L4500
1. Children's dreams—Case studies. 2. Ruspoli, Francesco. I. Title.
BF1099.C55R87 79-361937
•79 MARC

CHILDREN'S ENCYCLOPEDIAS AND DICTIONARIES

see also subdivision Dictionaries, Juvenile under names of languages or topical headings, e.g. English language—Dictionaries, Juvenile; Science—Dictionaries, Juvenile

Britannica junior encyclopædia for boys and girls / prepared under the supervision of the editors of Encyclopædia Britannica. — Chicago : Encyclopædia Britannica, c1979.
15 v. : ill. (some col.) ; 25 cm.
Includes index.
SUMMARY: A reference set specially prepared for elementary grades, with articles keyed to current school curricula.
ISBN 0-85229-340-2
1. Children's encyclopedias and dictionaries. ‹1. Encyclopedias and dictionaries› I. Encyclopaedia Britannica.
AG5.B8 1979 031 77-94291
79 MARC

Childcraft : the how and why library. — Chicago : World Book-Childcraft International, c1979.
15 v. : ill. (some col.) ; 26 cm.
Includes indexes.
CONTENTS: v. 1. Poems and rhymes.—v. 2. Stories and fables.—v. 3. Children everywhere.—v. 4. World and space.—v. 5. About animals.—5. 6. The green kingdom.—v. 7. How things work.—v. 8. How we get things.— v. 9. Holidays and customs.—v. 10. Places to know.—v. 11. Make and do.—v. 12. Look and learn.—v. 13. Look again.—v. 14. About me.—v. 15. Guide for parents.
ISBN 0-7166-0179-6
1. Children's encyclopedias and dictionaries. ‹1. Encyclopedias and dictionaries›
AG5.C515 1979 081 78-60255
79 MARC

The Illustrated encyclopedia of knowledge. — New York : Marshall Cavendish, c1978.
21 v. (2436 p.) : ill. (some col.) ; 28 cm.
Originally published under title: Tree of knowledge.
Includes index.
1. Children's encyclopedias and dictionaries.
AG6.I38 1978 031 78-12786
78 MARC

Ward Lock Limited.
Ward Lock's Encyclopedia / edited by Harold Boswell Taylor. — London : Ward Lock, 1978.
368 p. : ill. (some col.) ; 27 cm. GB***
ISBN 0-7063-5210-6 : £4.95
1. Children's encyclopedias and dictionaries. I. Boswell-Taylor, Harold. II. Title. III. Title: Encyclopedia.
AG5.W32 1978 032 79-304553
79 MARC

Wiskur, Darrell D
Silver Dollar City's abc words & rhymes / written & illustrated by Darrell Wiskur. — ‹Silver Dollar City, Mo.› : Silver Dollar City, inc., c1977.
‹32› p. : col. ill. ; 22 cm.
1. Children's encyclopedias and dictionaries. I. Title.
AG5.W75 031 78-112644
79 MARC

CHILDREN'S ENCYCLOPEDIAS AND DICTIONARIES, FRENCH

Marchand, Frank.
Premier dictionnaire Nathan : 200 mots illustrés en couleurs / Frank Marchand ... Michèle Barnoud-Maisdon, ... ; illustrations de Monique Gorde, Nicole Lacroix, Henri Mercier. — Paris : F. Nathan, 1977.
95 p. : col. ill. ; 25 cm. — (Éveil aux langages) F78-9585
22.00F
1. Children's encyclopedias and dictionaries, French. I. Barnoud-Maisdon, Michèle, joint author. II. Title.
AG25.M33 034'.1 79-337533
79 MARC

CHILDREN'S ENCYCLOPEDIAS AND DICTIONARIES, GERMAN

Pleticha, Heinrich.
Abenteuer-Lexikon : alles über Motive, Inhalte u. Autoren alter u. neuer Abenteuerbücher / Heinrich Pleticha. — 1. Aufl. — Würzburg : Arena, 1978.
224 p. ; 23 cm. GFR78-A
"Arena Standard Bibliothek."
Includes index.
ISBN 3-401-03846-X : DM19.80
1. Children's encyclopedias and dictionaries, German. I. Title.
AG27.P55 78-397879
79 MARC

CHILDREN'S ENCYCLOPEDIAS AND DICTIONARIES, JAPANESE

(Yōji hyakka)
幼児百科 / 監修 岡田要 ... ‹et al. ; 編集 松村彬夫 ... et al.›. — 東京：学習研究社，昭和36 ‹1961›，昭和39 ‹1964› printing.
7 v. : col. ill. ; 31 cm.
CONTENTS: 1. けもの—2. きとくさばな—3. とり—4. むし—5. さかな—6. りくののりもの—7. そらとうみののりもの

1. Children's encyclopedias and dictionaries, Japanese. I. Okada, Yō, 1891- II. Matsumura, Akio.
AG35.2.Y64 78-811778

CHILDREN'S ENCYCLOPEDIAS AND DICTIONARIES, KOREAN

Kim, Kwang-bae.
(Ŏrini paekkwa sajŏn)
어린이백과사전 / 그림김광배, 박동일 ; 추천 이원수 ; 감수윤극영. — 서울 : 신진출판사, 1978.
255 p. : chiefly col. ill. ; 38 cm.
On cover: The children's picture encyclopedia.
Includes index.
1. Children's encyclopedias and dictionaries, Korean. I. Pak, Tong-il. II. Yun, Kŭg-yŏng. III. Title. IV. Title: The children's picture encyclopedia.
AG35.4.K53 78-826567

CHILDREN'S FURNITURE

Blandford, Percy W
66 children's furniture projects / by Percy W. Blandford. — 1st ed. — Blue Ridge Summit, Pa. : Tab Books, c1979.
444 p. : ill. ; 22 cm.
Includes index.
ISBN 0-8306-9769-1 : $12.95
1. Children's furniture. I. Title.
TT197.5.C5B55 684.1'042 79-17458
79 MARC

Brubaker, Eric.
Make your own toys & children's furniture : a creative handbook for the practical family / by Eric Brubaker. — Toronto : Pagurian Press, c1978.
191 p. : ill. ; 24 cm. C***
ISBN 0-88932-092-6 : $9.95
1. Wooden toy making. 2. Children's furniture. I. Title.
TT174.5.W6B78 745.59'2 79-303451
79 MARC

—CATALOGS

Foa, Linda, 1942-
Kids' stuff / Linda Foa & Geri Brin, photography by Michael Datoli. — 1st ed. — New York : Pantheon Books, 1979.
208 p. : ill. (some col.) ; 28 cm.
ISBN 0-394-41558-2 : $17.95
1. Children's furniture—Catalogs. 2. Children's paraphernalia—Catalogs. I. Brin, Geri, 1947- joint author. II. Datoli, Michael. III. Title.
TT197.5.C5F63 1979 684 78-14343
78 MARC

CHILDREN'S HOMES see **Children—Institutional care**

CHILDREN'S HOSPITALS see **Children—Hospitals**

CHILDREN'S LIBRARIANS

—GREAT BRITAIN—STATISTICS

Library Association. Youth Libraries Group.
Report survey on children's library staffing carried out in April 1975 / edited by Margaret Marshall and Mary Nettlefold for YLG. — [Birmingham, Eng.] : Library Association, Youth Libraries Group, 1976.
2, [14] p. ; 21 cm. — (Pamphlet - Library Association, Youth Libraries Group ; no. 18) GB***
Cover title.
1. Children's librarians—Great Britain—Statistics. I. Marshall, Margaret Richardson. II. Nettlefold, Mary. III. Title. IV. Series: Library Association. Youth Libraries Group. Pamphlet ; no. 18.
Z718.1.L56 no. 18 027.62'5'08 s 79-312714
79 MARC

—NETHERLANDS

Bruyns, R A C
De werksituatie in de jeugdbibliotheek : weergave van de resultaten van een landelijk onderzoek / door R. A. C. Bruyns. — Den Haag : Nederlands Bibliotheek- en Lektuur Centrum, 1978.
59 p. ; 23 cm. — (Bibliotheekonderzoek ; 4) Ne78-40
ISBN 9062527965
1. Libraries, Children's—Netherlands. 2. Children's librarians—Netherlands. 3. Library surveys—Netherlands. I. Title. II. Series.
Z718.1.B826 79-347862
•79 MARC

CHILDREN'S LIBRARIES see **Libraries, Children's; School libraries**

CHILDREN'S LITERATURE

see also **Books and reading for children; Children's encyclopedias and dictionaries; Children's plays; Children's stories; Christian literature for children; Fairy tales; High interest-low vocabulary books; Libraries, Children's; Libraries and schools; Picture-books for children; Slow learning children, Books for; Story-telling;** and subdivision **Juvenile literature** under particular subjects

Aaron, Chester.
Catch Calico! / By Chester Aaron. — 1st ed. — New York : Dutton, c1979.
148 p. ; 22 cm.
SUMMARY: A fourteen-year-old not only suspects his grandfather is dying but is also haunted by the memories of his dead father and a wild cat named Calico.
ISBN 0-525-27551-7
[1. Rabies—Fiction. 2. Grandfathers—Fiction. 3. Fathers and sons—Fiction. 4. Cats—Fiction] I. Title.
PZ7.A12 Cat 1979 [Fic] 78-12319
78 MARC

Abels, Harriette Sheffer.
The creature of Saxony Woods / by Henriette Sheffer Abels ; illustrated by Joann Daley. — Chicago : Childrens Press, c1979.
47 p. : ill. ; 21 cm.
SUMMARY: Spending the summer at a logging operation in the North Woods, Rusty Grant and his friend Len Drover encounter a family of Bigfoot.
ISBN 0-516-03447-2
[1. Sasquatch—Fiction] I. Daley, Joann. II. Title.
PZ7.A1595 Cr [Fic] 78-11305
78 MARC

Abels, Harriette Sheffer.
The haunted cottage / by Harriette Sheffer Abels ; illustrated by Joann Daley. — Chicago : Childrens Press, c1978.
44 p. : ill. ; 21 cm.
SUMMARY: The peace of a family's retreat to an isolated mountain cottage is shattered when a ghost makes its presence known.
ISBN 0-516-03486-3
[1. Ghost stories] I. Daley, Joann. II. Title.
PZ7.A1595 Hau [Fic] 77-16619
77 MARC

Abercrombie, Barbara.
Amanda and Heather, and company / by Barbara Abercrombie ; illustrated by Mimi St. John. — New York : Dandelion Press, 1979.
[31] p. : col. ill. ; 23 cm.
SUMMARY: Two young sisters add sparkle to their parents' dull party.
ISBN 0-89799-086-2
[1. Parties—Fiction] I. St. John, Mimi. II. Title.
PZ7.A1614 Am [E] 78-72117
79 MARC

Adams, Adrienne.
The Christmas party / Adrienne Adams. — New York : Scribner, c1978.
[32] p. : col. ill. ; 26 cm.
SUMMARY: Orson Abbot helps a group of younger rabbits organize a surprise Christmas party for their parents.
ISBN 0-684-15930-9 lib. bdg. : $7.95
[1. Rabbits—Fiction. 2. Christmas stories] I. Title.
PZ7.A194 Ch [E] 78-16230
78 MARC

Addison, Harry.
Write that down for me daddy / by Harry Wayne Addison ; illustrated by Kathy Addison Schendle. — Gretna, La. : Pelican Pub. Co., c1974, 1978 printing.
[60] p. : ill. ; 20 cm.
SUMMARY: A collection of poetry, sayings, and short tales written by a father for his children.
ISBN 0-88289-116-2 : $4.95
[1. Short stories. 2. American poetry] I. Schendle, Kathy Addison. II. Title.
PZ7.A252 Wr 1978 [Fic] 78-9028
78 MARC

Adler, Carole S
The magic of the Glits / C. S. Adler ; illustrated by Ati Forberg. — New York : Macmillan, c1979.
112 p. : ill. ; 22 cm.
SUMMARY: Twelve-year-old Jeremy, spending the summer on Cape Cod, befriends an seven-year-old girl whose mother has just died.
ISBN 0-02-700120-2
[1. Friendship—Fiction] I. Forberg, Ati. II. Title.
PZ7.A26145 Mag [Fic] 78-12149
78 MARC

Adler, Carole S
The silver coach / by C. S. Adler. — New York : Coward, McCann & Geoghegan, c1979.
122 p. ; 22 cm.
SUMMARY: Neither 12-year-old Chris nor her 6-year-old sister look forward to spending the summer with their unknown grandmother in a remote woodland cabin, but the summer holds many surprises for them, not the least of which is a gradual acceptance of their parents' divorce.
ISBN 0-698-20504-9
[1. Grandmothers—Fiction. 2. Divorce—Fiction. 3. Family life—Fiction]
PZ7.A26145 Si 1979 [Fic] 79-10430
79 MARC

Ahlberg, Janet.
Burglar Bill / [by] Janet & Allan Ahlberg. — London : Heinemann, 1977.
[32] p. : col. ill. ; 21 cm. GB77-08118
ISBN 0-434-92500-4 : £2.40
I. Ahlberg, Allan, joint author. II. Title.
PZ7.A2689 Bu 1977b 78-316081
78 MARC

Ahlberg, Janet.
Jeremiah in the dark woods / [by] Janet and Allan Ahlberg. — Harmondsworth : Kestrel Books, 1977.
48 p. : ill. (some col.) ; 23 cm. GB77-30745
ISBN 0-7226-5357-3 : £2.75
I. Ahlberg, Allan, joint author. II. Title.
PZ7.A2689 Je 1977 823'.9'14 78-317597
78 MARC

Ahlberg, Janet.
Jeremiah in the dark woods / Janet and Allan Ahlberg. — 1st American ed. — New York : Viking Press, 1978, c1977.
47 p. : ill (some col.) ; 23 cm.
SUMMARY: A little boy sets out to find the thief of his grandmother's strawberry tarts and meets a number of unusual characters on his search.
ISBN 0-670-40637-6 lib. bdg. : $6.95
[1. Fantasy] I. Ahlberg, Allan, joint author. II. Title.
PZ7.A2689 Je 1978 [Fic] 77-6641
77 MARC

Ahlberg, Janet.
The vanishment of Thomas Tull / Janet and Allan Ahlberg. — New York : Scribner, c1977.
[48] p. : ill. ; 22 cm.
SUMMARY: When Thomas Tull is seven years old, he stops growing and begins shrinking and shrinking and shrinking.
ISBN 0-684-15968-6 : $6.95
[1. Size and shape—Fiction. 2. Humorous stories] I. Ahlberg, Allan. II. Title.
PZ7.A26888 Van [E] 78-15662
78 MARC

Aiken, Joan, 1924-
Mice and Mendelson / Joan Aiken ; illustrated by Babette Cole ; music by John Sebastian Brown. — London : J. Cape, 1978.
124 p. : ill. ; 23 cm. GB***
ISBN 0-224-01615-6 : £3.50
I. Title.
PZ7.A2695 Mh 79-302276
79 MARC

Ainsworth, Ruth.
Look, do, and listen. Illustrated by Bernadette Watts. New York, F. Watts [1969]
159 p. illus. 27 cm. 4.95
SUMMARY: A collection, designed for use with nursery school children, of stories, poems, games, and riddles.
1. Children's literature. [1. Literature—Collections] I. Watts, Bernadette, illus. II. Title.
PZ5.A32 Lo 1969 820'.8 68-31207
[r78]rev MARC

Aitken, Amy.
Ruby! / Story and pictures by Amy Aitken. — Scarsdale, NY : Bradbury Press, c1979.
[32] p. : col. ill. ; 24 cm.
SUMMARY: Anything is possible for Ruby. She decides to be an author or an artist or a movie star or maybe even President.
ISBN 0-87888-144-1 : $7.95
[1. Occupations—Fiction] I. Title.
PZ7.A2836 Ru [E] 78-21283
78 MARC

Alderson, Sue Ann.
Hurry up, Bonnie! / story by Sue Ann Alderson ; illustrated by Fiona Garrick. — Edmonton : Tree Frog Press, c1976.
43 p. : col. ill., ports. ; 18 cm. C77-2025-9
ISBN 0-88967-024-2 : $5.95. ISBN 0-88967-023-4 pbk.
I. Garrick, Fione. II. Title.
PZ7.A3617 Hu 79-300457
79 MARC

Alexander, Lloyd.
The first two lives of Lukas-Kasha / Lloyd Alexander. — 1st ed. — New York : Dutton, c1978.
213 p. ; 24 cm.
SUMMARY: After paying a silver penny to encourage a magician to perform in the town square, a carpenter's helper is conjured to a strange place where the people call him King of Abadan.
ISBN 0-525-29748-0 lib. bdg. : $8.50
[1. Space and time—Fiction] I. Title.
PZ7.A3774 Fi 1978 [Fic] 77-26699
77 MARC

CHILDREN'S LITERATURE (Continued)

Alexander, Sue, 1933-
Marc the Magnificent / by Sue Alexander ; pictures by Tomie de Paola. — New York : Pantheon, c1978.
[32] p. : col. ill. ; 22 cm.
SUMMARY: Marc fantasizes about his future as a world-renowned magician while trying to learn his first magic trick.
ISBN 0-394-83728-2. ISBN 0-394-93728-7 lib. bdg.
[1. Magicians—Fiction] I. De Paola, Thomas Anthony. II. Title.
PZ7.A3784 Mar [E] 78-3285
78 MARC

Alexander, Sue, 1933-
More Witch, Goblin, and Ghost stories / by Sue Alexander ; pictures by Jeanette Winter. — 1st ed. — New York : Pantheon Books, c1978.
60 p. : ill. ; 24 cm. — (An I am reading book)
Sequel to Witch, Goblin, and sometimes Ghost!
SUMMARY: The further adventures of three friends who go on a picnic, learn about honesty, and discover a remedy for insomnia.
ISBN 0-394-83933-1. ISBN 0-394-93933-6 lib. bdg.
[1. Friendship—Fiction] I. Winter, Jeanette. II. Title.
PZ7.A3784 Mo [E] 78-5329
78 MARC

Alexander, Sue, 1933-
Seymour the Prince / by Sue Alexander ; pictures by Lillian Hoban. — New York : Pantheon Books, c1979.
46 p. : ill. ; 24 cm. — (An I am reading book)
SUMMARY: Only one member of the Maple Street Club is unwilling to participate in their play—Seymour simply refuses to kiss the sleeping princess.
ISBN (invalid) 0-394-94141-7
[1. Clubs—Fiction] I. Hoban, Lillian. II. Title.
PZ7.A3784 Se [E] 78-31406
78 MARC

Ali Baba. English.
The story of Ali Baba and the forty theives. New-Haven, Printed for J. Babcock; Sidney's Press, 1820.
31 p. illus. 14 cm.
Caption title: The forty thieves.
I. Title. II. Title: The forty thieves.
PZ6.A398 St 78-315557
78 MARC

Aliki.
The two of them / written and illustrated by Aliki. — New York : Greenwillow Books, c1979.
[30] p. : ill. ; 26 cm.
SUMMARY: Describes the relationship of a grandfather and his granddaughter from her birth to his death.
ISBN 0-688-80225-7 : $7.95
[1. Grandfathers—Fiction. 2. Death—Fiction] I. Title.
PZ7.A397 Tw [E] 79-10161
79 MARC

Allamand, Pascale.
The animals who changed their colors / Pascale Allamand ; English version by Elizabeth Watson Taylor. — 1st U.S. ed. — New York : Lothrop, Lee & Shepard, 1979.
[32] p. : col. ill. ; 20 x 26 cm.
SUMMARY: The polar bear, whale, tortoise, and two crocodiles try to imitate the parrot's beautiful colors, only to discover how impractical they are.
ISBN 0-688-41900-3
[1. Animals—Fiction. 2. Individuality—Fiction] I. Title.
PZ7.A3987 An 1979 [E] 79-196
79 MARC

Allan, Mabel Esther.
The view beyond my father / Mabel Esther Allan. — New York : Dodd, Mead, 1978, c1977.
192 p. ; 22 cm.
SUMMARY: A fifteen-year-old discovers it is her father as well as her blindness that binds her to a narrow world.
ISBN 0-396-07612-2
[1. Fathers and daughters—Fiction. 2. Blind—Fiction. 3. Physically handicapped—Fiction] I. Title.
PZ7.A4 Vi [Fic] 78-7736
78 MARC

Allard, Harry.
Bumps in the night / by Harry Allard ; pictures by James Marshall. — 1st ed. — Garden City, N.Y. : Doubleday, c1979.
32 p. : col. ill. ; 27 cm.
SUMMARY: Dudley Stork enlists his friends to help rid his house of things that go bump in the night.
ISBN 0-385-12942-4 : $4.95
[1. Ghost stories] I. Marshall, James, 1942- . II. Title.
PZ7.A413 Bu [E] 78-22301
78 MARC

Allard, Harry.
I will not go to market today / by Harry Allard ; pictures by James Marshall. — New York : Dial Press, c1979.
[32] p. : col. ill. ; 22 cm.
SUMMARY: Fenimore B. Buttercrunch's repeated attempts to go to market are met with obstacles.
ISBN 0-8037-4019-0 : $6.95
[1. Shopping—Fiction. 2. Animals—Fiction. 3. Humorous stories] I. Marshall, James, 1942- . II. Title.
PZ7.A413 Iac [E] 78-72474
78 MARC

Allen, Judy.
The dream thing / [by] Judy Allen. — London : Hamilton, 1978.
[4], 105 p. ; 23 cm. GB78-13108
ISBN 0-241-89836-6 : £3.25
I. Title.
PZ7.A4273 Dr 1978 78-317397
78 MARC

Allen, Laura Jean.
Ottie and the star / by Laura Jean Allen. — 1st ed. — New York : Harper & Row, c1979.
32 p. : col. ill. ; 22 cm. — (An Early I can read book)
SUMMARY: A little otter encounters a shark, a dolphin, and a starfish while trying to catch a star.
ISBN (invalid) 0-06-020106-X : $6.95
[1. Otters—Fiction. 2. Marine animals—Fiction] I. Title.
PZ7.A4274 Ot 1979 [E] 78-22485
79 MARC

Allen, Linda.
Mr. Simkin's grandma / Linda Allen ; illustrated by Loretta Lustig. — 1st ed. — New York : Morrow, 1979.
[32] p. : ill. ; 21 cm.
SUMMARY: A stranger knocks on Mr. Simkin's door and insists she is his grandmother.
ISBN 0-688-22191-2
[1. Grandmothers—Fiction] I. Lustig, Loretta. II. Title.
PZ7.A4277 Mi [E] 78-20917
78 MARC

Allen, Ted.
Willie the squowse / Ted Allen ; illustrated by Quentin Blake. — New York : Hastings House, 1978, c1973.
57 p. : ill. ; 24 cm.
SUMMARY: Son of a squirrel and a mouse, Willie, resident of an apartment house wall, links the lives of two neighboring families in a very unusual way.
ISBN 0-8038-8086-3 : $5.95
[1. Apartment houses—Fiction] I. Blake, Quentin. II. Title.
PZ7.A4384 Wi 1978 [Fic] 78-1716
78[r79]rev MARC

Allinson, Beverley.
Turkey pops / [Beverley Allinson, Barbara O'Kelly]. — Agincourt, Ont. : Methuen Publications, 1977.
32 p. : ill. (some col.) ; 23 cm. — (Kids like us)
Cover title.
ISBN 0-458-92480-6 : $3.25
I. O'Kelly, Barbara, 1945-, joint author. II. Title.
PZ7.A442 Tu 78-326994
79 MARC

Allison, Rosemary, 1953-
Lazarus laughs / Christiane Duchesne ; adapted by Rosemary Allison. — Toronto : J. Lorimer, 1977.
[12] : col. ill. ; 26 cm. C77-1585-9
Translation of Lazaros olibrius.
ISBN 0-88862-156-6. ISBN 0-88862-157-4
I. Duchesne, Christiane, 1949- Lazaros olibrius. II. Title.
PZ7.A4429 Laz 79-338150
79 MARC

Allison, Rosemary, 1953-
The lonely dragon / Christiane Duchesne ; adapted by Rosemary Allison. — Toronto : J. Lorimer, 1977.
[12] p. : col. ill. ; 27 cm. C***
Translation of Le triste dragon.
ISBN 0-88862-154-X. ISBN 0-88862-155-8 pbk.
I. Duchesne, Christiane, 1949- Le triste dragon. II. Title.
PZ7.A4429 Lo 79-300447
79 MARC

Allison, Rosemary, 1953-
Ms. Beaver travels east / Rosemary Allison and Ann Powell. — Toronto : Women's Press, 1978.
[30] p. : ill. ; 21 cm. C78-1576-2
ISBN 0-88961-057-6
I. Powell, Ann, 1951- joint author. II. Title.
PZ7.A4429 Me 79-310791
79 MARC

Allred, Mary.
The move to a new house / Mary Allred ; illustrated by Paul R. Behrens. — Nashville : Broadman Press, c1978.
32 p. : col. ill. ; 24 cm.
SUMMARY: Todd's worries about moving to a new house are allayed with the help of his parents and of God.
ISBN 0-8054-4252-9
[1. Moving, Household—Fiction. 2. Christian life—Fiction] I. Behrens, Paul R. II. Title.
PZ7.A4434 Mo 78-113092
79 MARC

Ambrus, Victor G
Mishka / Victor Ambrus. — New York : F. Warne, 1978, c1975.
24 p. : col. ill. ; 29 cm.
SUMMARY: Seeking fame with his violin, eight-year-old Mishka joins a circus.
ISBN 0-7232-6150-4 : $6.95
[1. Violinists—Fiction. 2. Circus stories] I. Title.
PZ7.A496 Mi [E] 77-84602
79 MARC

Ames, Mildred.
What are friends for? / Mildred Ames. — New York : Scribner, c1978.
viii, 145 p. ; 22 cm.
SUMMARY: Eleven-year-old Amy finds that her relationship with Michelle tests all of her notions about the true nature and responsibilities of friendship.
ISBN 0-684-15991-0 : $7.95
[1. Friendship—Fiction] I. Title.
PZ7.A5143 Wh [Fic] 78-13747
78 MARC

Amoss, Berthe.
Secret lives / by Berthe Amoss. — 1st ed. — Boston : Little, Brown, c1979.
180 p. ; 22 cm.
"An Atlantic Monthly Press book."
SUMMARY: Twelve-year-old Addie living with her straitlaced aunts in New Orleans unravels the mystery of her dead mother's past.
ISBN 0-316-03740-0
[1. Family life—Fiction. 2. New Orleans—Fiction] I. Title.
PZ7.A5177 Se [Fic] 79-16717
79 MARC

Anastasio, Dina.
A question of time / by Dina Anastasio ; illustrated by Dale Payson. — 1st ed. — New York : Dutton, c1978.
90 p. : ill. ; 23 cm.
SUMMARY: Shortly after moving to a small town in Minnesota, a young girl's curiosity is aroused by four carved dolls resembling her great-grandfather's family.
ISBN 0-525-37990-8 lib. bdg. : $7.50
[1. Mystery and detective stories] I. Payson, Dale. II. Title.
PZ7.A5185 Qe 1978 [Fic] 78-4916
78 MARC

Anderson, Margaret Jean, 1931-
In the circle of time / Margaret J. Anderson. — New York : Knopf : distributed by Random House, c1979.
181 p. ; 22 cm.
SUMMARY: Two children are hurled into the future as a result of their hunt for three 12-foot stones missing from an ancient Scottish stone circle.
ISBN 0-394-84029-1. ISBN 0-394-94029-6 lib. bdg.
[1. Space and time—Fiction. 2. Scotland—Fiction] I. Title.
PZ7.A54397 Ik [Fic] 78-10156
78 MARC

Anderson, Mary, 1929-
F*T*C* and Company / Mary Anderson ; illustrated by Don Sibley. — 1st ed. — New York : Atheneum, 1979.
195 p. : ill. ; 22 cm.
SUMMARY: Freddie the Cat and Emma Pigeon organize an ill-fated company and present "My Fair Lady" starring Rosetta Robin who seeks fame on Broadway.
ISBN 0-689-30673-3
[1. Theater—Fiction. 2. Animals—Fiction] I. Sibley, Don. II. Title.
PZ7.A5444 Fab 1979 [Fic] 78-12097
78 MARC

Anderson, Mary, 1929-
Step on a crack / by Mary Anderson. — 1st ed. — New York : Atheneum, 1978.
180 p. ; 22 cm.
SUMMARY: Sarah suffers from frightening nightmares and compulsive stealing. A friend helps explore the dark areas of her mind, exposing some disturbing facts.
ISBN 0-689-30610-5
[1. Emotional problems—Fiction] I. Title.
PZ7.A5444 St [Fic] 77-22767
77[r79]rev MARC

Angell, Judie.
A word from our sponsor : or, My friend Alfred / by Judie Angell. — Scarsdale, N.Y. : Bradbury Press, c1979.
140 p. ; 22 cm.
SUMMARY: When a 12-year-old boy tries to protect consumers from a dangerous drinking mug, he discovers he must challenge his father's advertising agency.
ISBN 0-87888-142-5 : $6.95
[1. Consumer protection—Fiction. 2. Family life—Fiction. 3. Humorous stories] I. Title.
PZ7.A5824 Wo [Fic] 78-25716
78 MARC

Anglund, Joan Walsh.
The brave cowboy / Joan Walsh Anglund. — 1st Voyager ed. — New York : Harcourt Brace Jovanovich, 1976, c1959.
[40] p. : ill. ; 18 cm. — (A Voyager book ; AVB 104)
SUMMARY: Imaginary adventures with Indians, wild animals, and outlaws keep a little cowboy busy.
ISBN 0-15-614050-0
[1. Cowboys—Fiction] I. Title.
[PZ7.A586 Br 1976] [E] 76-15638
76 MARC

Anglund, Joan Walsh.
Cowboy and his friend / Joan Walsh Anglund. — 1st Voyager ed. — New York : Harcourt Brace Jovanovich, 1976, c1961.
[32] p. : col. ill. ; 18 cm. — (A Voyager book ; AVB 105)
SUMMARY: An imaginary bear can be a friend who goes everywhere you go and does everything you do.
ISBN 0-15-622715-0
[1. Bears—Fiction] I. Title.
PZ7.A586 Co 6 [E] 76-16112
76 MARC

CHILDREN'S LITERATURE (Continued)

Anglund, Joan Walsh.
The Joan Walsh Anglund storybook. — New York : Random House, c1978.
45 p. (p. 45 blank) : col. ill. ; 29 cm.
SUMMARY: A collection of brief stories and poems on a variety of subjects.
ISBN 0-394-83803-3 : $4.95. ISBN 0-394-93803-8 lib. bdg. : $5.99
[1. Short stories. 2. American poetry] I. Title.
PZ7.A586 Jo [E] 78 78-55913 MARC

Anno, Mitsumasa, 1926-
The King's flower / Mitsumasa Anno. — 1st U.S. ed. — New York : Collins, 1979, c1976.
[30] p. : col. ill. ; 29 cm.
SUMMARY: A king discovers that bigger is not always better.
ISBN 0-529-05458-2 : $7.95
[1. Size and shape—Fiction] I. Title.
PZ7.A5875 Ki 1979 [E] 78 78-9596 MARC

Arkle, Phyllis.
Magic in the air / [by] Phyllis Arkle ; illustrated by Mike Cole. — London : Hodder and Stoughton, 1978.
88 p. : ill. ; 20 cm. GB78-28944
ISBN 0-340-22195-X : £1.95
I. Title.
PZ7.A688 Mai 79 79-301703 MARC

Armitage, Ronda.
Don't forget, Matilda! / Ronda and David Armitage. — London : A. Deutsch, 1979.
[32] p. : col. ill. ; 23 cm. GB***
ISBN 0-233-97075-4 : £3.25 ($6.95 U.S.)
I. Armitage, David, joint author. II. Title.
PZ7.A73 Do 79 78-74473 MARC

Armitage, Ronda.
The trouble with Mr Harris / [by] Ronda and David Armitage. — London : Deutsch, 1978.
[32] p. : chiefly col. ill. ; 26 cm. GB78-23761
ISBN 0-233-96963-2 : £3.25
I. Armitage, David, joint author. II. Title.
PZ7.A73 Tr 79 79-305189 MARC

Arnosky, Jim.
Nathaniel / by Jim Arnosky. — Reading, Mass. : Addison-Wesley, c1978.
[48] p. : ill. ; 19 cm.
SUMMARY: The humorous aspects of a man's reclusive life in the country are revealed in three picture stories and two letters.
ISBN 0-201-00171-3
[1. Stories without words. 2. Humorous stories. 3. Country life—Fiction] I. Title.
PZ7.A7384 Nat [Fic] 77 77-13505 MARC

Arrick, Fran.
Steffie can't come out to play / by Fran Arrick. — Scarsdale, N.Y. : Bradbury Press, c1978.
196 p. ; 22 cm.
SUMMARY: Fourteen-year-old Stephanie runs away from her dreary industrial Pennsylvania town to be a model in New York. Instead she meets an expensively-dressed man who puts her to work for him.
ISBN 0-87888-135-2 : $7.95
[1. Runaways—Fiction. 2. Prostitution—Fiction] I. Title.
PZ7.A74335 St [Fic] 78 78-4423 MARC

Arthur, Robert.
Alfred Hitchcock and the three investigators in The mystery of the fiery eye / text by Robert Arthur ; illustrated by Harry Kane. — New York : Random House, 1978, c1967.
viii, 180 p. : ill. ; 20 cm.
SUMMARY: Three junior detectives solve a mystery involving a collection of antique busts, a legacy, a strange letter, and a red ruby.
ISBN 0-394-83770-3 : $1.50
[1. Mystery and detective stories] I. Kane, Harry. II. Title. III. Title: The mystery of the fiery eye.
[PZ7.A744 Ag 1978] [Fic] 78 77-28860 MARC

Arthur, Robert.
Alfred Hitchcock and the three investigators in The secret of terror castle / text by Robert Arthur ; illustrated by Harry Kane. — New York : Random House, 1978, c1964
ix, 179 p. : ill. ; 20 cm.
SUMMARY: The Three Investigators search for an authentic haunted house.
ISBN 0-394-83766-5 : $1.50
[1. Mystery and detective stories] I. Kane, Harry. II. Title. III. Title: The secret of terror castle.
[PZ7.A744 Al 1978] [Fic] 78 77-29129 MARC

Aruego, Jose.
We hide, you seek / by Jose Aruego and Ariane Dewey. — 1st ed. — New York : Greenwillow Books, c1979.
[32] p. : chiefly col. ill. ; 26 cm.
SUMMARY: The reader is invited to find animals hidden in their natural habitat.
ISBN 0-688-80201-X : $7.95. ISBN 0-688-84201-1 lib. bdg. : $7.63
[1. Hide-and-seek—Fiction. 2. Camouflage (Biology)—Fiction. 3. Animals—Habitations—Fiction] I. Dewey, Ariane, joint author. II. Title.
PZ7.A7475 We [E] 78 78-13638 MARC

Asch, Frank.
Moon Bear / by Frank Asch. — New York : Scribner, c1978.
[32] p. : ill. ; 26 cm.
SUMMARY: Worried that the moon is growing smaller each night, Bear decides to do something about it.
ISBN 0-684-15810-8 lib. bdg. : $7.95
[1. Moon—Fiction. 2. Bears—Fiction] I. Title.
PZ7.A778 Mn [E] 78 78-9444 MARC

Asch, Frank.
Sand cake : a Frank Asch bear story. — New York : Parents' Magazine Press, c1978.
[33] p. : col. ill. ; 24 cm.
SUMMARY: Papa Bear uses his culinary skills and a little imagination to concoct a sand cake.
ISBN 0-8193-0985-0. ISBN 0-8193-0986-9 lib. bdg.
[1. Cake—Fiction. 2. Bears—Fiction] I. Title.
PZ7.A778 San [E] 78 78-11183 MARC

Asch, Frank.
Turtle tale / Frank Asch. — New York : Dial Press, c1978.
[32] p. : col. ill. ; 24 cm.
SUMMARY: A young turtle learns how to be a wise old turtle, with a few hard knocks along the way.
ISBN 0-8037-8782-0. ISBN 0-8037-8783-9 lib. bdg. : $6.29
[1. Turtles—Fiction] I. Title.
PZ7.A778 Tu [E] 78 78-51328 MARC

Asheron, Sara.
Little gray mouse goes sailing. Pictures by Claudine Nankivel. Editorial consultant: Lilian Moore. New York, Wonder Books [1965]
60 p. col. illus. 22 cm. (Wonder Books easy reader)
SUMMARY: When he rescues a friend from a swimming pool, a mouse discovers a love for sailing.
[1. Mice—Fiction. 2. Sailing—Fiction] I. Nankivel, Claudine, illus. II. Title.
PZ7.A82 Li [E] [r78]rev 65-14756 MARC

Ashley, Bernard.
A kind of wild justice : a novel / by Bernard Ashley ; ill. by Charle s Keeping. — New York : S. G. Phillips, 1979, c1978.
182 p. : ill. ; 22 cm.
Reprint of the ed. published by Oxford University Press, Oxford.
SUMMARY: Because his father is mixed up with criminals, Ronnie's life becomes one of isolation and fear.
[1. Crime and criminals—Fiction] I. Title.
PZ7.A8262 Ki 1979 [Fic] 78 78-10899 MARC

Ashley, Bernard.
All my men / Bernard Ashley. — New York : S. G. Phillips, 1978, c1977.
159 p. ; 22 cm.
SUMMARY: His parents' move from London to Kent to establish themselves as shopkeepers brings to Paul a difficult period of adjustment in a new school.
ISBN 0-87599-228-5 : $8.95
[1. Moving, Household—Fiction. 2. School stories] I. Title.
PZ7.A8262 Al [Fic] 78 78-12683 MARC

Asimov, Isaac, 1920-
The heavenly host / Isaac Asimov. — Harmondsworth, Eng. ; New York : Puffin Books, 1978, c1975.
55 p. ; 20 cm.
"A shorter and slightly different version ... was published in Boy's life, December 1974."
SUMMARY: Newly arrived on Planet Anderson Two just before Christmas, Jonathan is warned about the dangerous native inhabitants, but an accidental meeting with one of the natives convinces him that they are friendly and peaceful.
ISBN 0-14-031117-3 : $1.95
[1. Science fiction. 2. Christmas stories] I. Colonna, Bernard. II. Title.
[PZ7.A835 He 1978] [Fic] 78 78-4553 MARC

Averill, Esther Holden.
When Jenny lost her scarf. Jenny's adopted brothers. [Sound recording] [By] Esther Averill. Caedmon TC 1608. p1979.
1 disc. 33⅓ rpm. stereo. 12 in.
Children's stories, read by Tammy Grimes.
Duration: 40 min., 13 sec.
Notes by J. Schulman on container.
SUMMARY: Jenny Linsky, a shy little black cat, must figure out a way to get her scarf back from the den of dogs. In the second story, Jenny's owner, Captain Tinker shows that he cares very much for the two homeless cats that Jenny wants him to adopt.
[1. Cats—Fiction] I. Averill, Esther Holden. Jenny's adopted brothers. [Sound recording] 1979. II. Grimes, Tammy. III. Title.
[PZ7] 78-741942

Avi, 1937-
Night journeys / Avi. — New York : Pantheon Books, c1979.
143 p. ; 22 cm.
SUMMARY: Two young indentured servants escape into Pennsylvania in the late 1700's and receive help from an unexpected source.
ISBN 0-394-84116-6
[1. Indentured servants—Fiction. 2. Friends, Society of—Fiction. 3. Pennsylvania—Fiction] I. Title.
PZ7.A953 Ni [Fic] 78 78-10151 MARC

Bach, Alice.
Millicent the Magnificent / by Alice Bach ; pictures by Steven Kellogg. — 1st ed. — New York : Harper and Row, c1978.
43 p. : ill. ; 23 cm.
SUMMARY: When Oliver becomes a pupil of Millicent, a circus bear, his brother Ronald decides to secretly teach himself the same tricks.
ISBN 0-06-020309-9 : $5.95. ISBN 0-06-020312-9 lib. bdg. : $5.79
[1. Bears—Fiction. 2. Twins—Fiction. 3. Acrobats and acrobatics—Fiction] I. Kellogg, Steven. II. Title.
PZ7.B1314 Mi 1978 [E] 78 77-11840 MARC

Bach, Marcus, 1906-
I, Monty / text by Marcus Bach. — 1st ed. — Norfolk Island, Australia : Island Heritage, 1977.
[93] p. : col. ill. ; 27 cm. Aus***
SUMMARY: His metamorphosis from caterpillar into butterfly causes Monty to reflect on the mysteries of life, death, afterlife, rebirth, and love.
$9.95
[1. Butterflies—Fiction. 2. Life—Fiction] I. Title.
PZ7.B1315 Iad [Fic] 79 77-82232 MARC

Bacigalupa, Andrea, 1923-
A good & perfect gift / by Drew Bacigalupa ; illustrated by Jeannie Pear. — Huntington, Ind. : Our Sunday Visitor, c1978.
42 p. : ill. ; 24 cm. — (An OSV read-along book)
SUMMARY: Manuelita's dearest wish is to have a father on Christmas Day and her wish comes true in an unusual way.
ISBN 0-87973-352-7
[1. Christmas stories] I. Pear, Jeannie. II. Title.
PZ7.B13218 Go [E] 79 78-60727 MARC

Bacon, Martha Sherman, 1917-
Moth Manor : a gothic tale / by Martha Bacon ; illustrated by Gail Burroughs. — 1st ed. — Boston : Little, Brown, c1978.
148 p. : ill. ; 21 cm.
"An Atlantic Monthly Press book."
SUMMARY: A young girl takes charge of rescuing a haunted dollhouse and the special dolls that inhabit it from being sold to an antique dealer.
ISBN 0-316-07511-6
[1. Dollhouses—Fiction. 2. Ghost stories] I. Burroughs, Gail. II. Title.
PZ7.B1345 Mo [Fic] 78 78-59680 MARC

Badio, Josephine Greaves.
Naweh and Somah / by Josephine Greaves Badio. — [Monrovia : Division of Printing & Production, Ministry of Education], c1976.
46 p. : ill. ; 22 cm.
I. Title.
PZ7.B1384 Naw 79 79-302214 MARC

Bailey, Paul.
Charlie meets the Goomby : the Land of Urd / created and written by Paul Bailey ; created and illustrated by Dan Wladyka. — [Toronto] : Magook Publishers, 1977.
[32] p. : col. ill. ; 15 x 23 cm. C***
ISBN 0-88869-003-7 : $4.95
I. Wladyka, Dan. II. Title.
PZ7.B1527 Ch 78 78-318020 MARC

Baker, Alan.
Benjamin bounces back / story and pictures by Alan Baker. — 1st American ed. — Philadelphia : Lippincott, c1978.
[32] p. : col. ill. ; 18 x 22 cm.
SUMMARY: Shortsighted Benjamin Hamster goes through a door, whose sign he can't read, and experiences some unusual adventures.
ISBN 0-397-31809-X
[1. Hamsters—Fiction] I. Title.
PZ7.B1688 Bc 1978 [E] 78 78-59689 MARC

Banks, Lynne Reid.
I, Houdini : the autobiography of a self-educated hamster / Lynne Reid Banks ; illustrated by Terry Riley. — London : J. M. Dent, 1978.
119 p. : ill. ; 23 cm. GB***
ISBN 0-460-06873-3 : £3.95
1. Hamsters—Legends and stories. I. Title.
PZ10.3.B2155 Iac 79 79-300504 MARC

Barford, Carol.
Let me hear the music / Carol Barford. — New York : Seabury Press, c1979.
124 p. ; 22 cm.
"A Clarion book."
SUMMARY: A 12-year-old asks herself and her Depression-haunted family some difficult questions about life after experiencing the tragic death of a friend.
ISBN 0-8164-3224-4
[1. Death—Fiction] I. Title.
PZ7.B25035 Le [Fic] 78 78-23966 MARC

Barker, Jane Valentine, 1930-
Adventures in the West / Jane Valentine Barker and Sybil Downing ; [ill. by Robert F. Wilson]. — 1st ed. — Boulder, Colo. : Pruett Pub. Co., c1979.
40 p. : ill. ; 23 cm. — (Colorado heritage series ; book 5)
Bibliography: p. 37.
SUMMARY: Presents four stories from Colorado's history based on incidents involving Zebulon Pike, Lieutenant Fremont and Kit Carson, Fort Misery, and a stampede of buffalo.
ISBN 0-87108-220-9
[1. Colorado—History—To 1876—Fiction. 2. Frontier and pioneer life—Colorado—Fiction. 3. Short stories] I. Downing, Sybil, joint author. II. Wilson, Robert Franklin, 1937- . III. Title. IV. Series.
PZ7.B25052 Ad [Fic] 79 79-52142 MARC

CHILDREN'S LITERATURE (Continued)

Barker, Jane Valentine, 1930-
Mountain treasures / Jane Valentine Barker and Sybil Downing ; ⌜ill., by Dian Diggs and Robert F. Wilson⌝. — 1st ed. — Boulder, Colo. : Pruett Pub. Co., c1978.
44 p. : ill. ; 23 cm. — (Colorado heritage series ; book 1)
Bibliography: p. 44.
SUMMARY: Four stories describe life in Colorado's mountain mining camps of Leadville, Caribou, Ouray, and Cripple Creek in the late 1800's.
ISBN 0-87108-212-8
⌜1. Colorado—History—Fiction. 2. Silver mines and mining—Colorado—Fiction. 3. Mountain life—Colorado—Fiction. 4. Short stories⌝ I. Downing, Sybil, joint author. II. Diggs, Dian. III. Wilson, Robert Franklin, 1937- . IV. Title. V. Series.
PZ7.B25052 Mo ⌜Fic⌝ 79 78-25411 MARC

Barker, Jane Valentine, 1930-
Trappers and traders / Jane Valentine Barker and Sybil Downing ; ⌜ill. by Robert F. Wilson⌝. — 1st ed. — Boulder, Colo. : Pruett Pub. Co., c1979.
36 p., ⌜1⌝ leaf of plates : ill. ; 22 cm. — (Colorado heritage series ; book 3)
Bibliography: p. 35-36.
CONTENTS: Gervais and the bear.—Surprise at the cache.—Juanita's special wish.—Poco's useful trick.
⌜1. Frontier and pioneer life—Fiction. 2. Trapping—Fiction. 3. Short stories⌝ I. Downing, Sybil, joint author. II. Wilson, Robert Franklin, 1937- . III. Title. IV. Series.
PZ7.B25052 Tr ⌜Fic⌝ 79 79-83745 MARC

Barlowe, Wayne Douglas.
Star wars : a pop-up book / paper engineering by Ib Penick ; ill. by Wayne Douglas Barlowe. — New York : Random House, c1978.
⌜16⌝ p. : chiefly col. ill. ; 24 cm.
On spine: The pop-up book of Star wars.
SUMMARY: Pop-up pictures re-create scenes from the movie "Star Wars".
ISBN 0-394-83754-1 : $3.95
⌜1. Star wars. ⌜Motion picture⌝. 2. Science fiction⌝ I. Penick, Ib. II. Star wars. ⌜Motion picture.⌝ III. Title: The pop-up book of Star wars.
PZ7.B2515 St ⌜E⌝ 78 77-90464 MARC

Barrett, John M
The day toys came to Silver Dollar City / written by John Barrett ; illustrated by Rod Ruth. — ⌜Silver Dollar City, Mo : Silver Dollar City, 1978⌝
⌜32⌝ p. : col. ill. ; 22 cm.
SUMMARY: A town with no toys holds a contest to let the children choose a toymaker.
⌜1. Toys—Fiction⌝ I. Ruth, Rod. II. Title.
PZ7.B27519 Day ⌜E⌝ 79 78-112285 MARC

Barrett, John M
Oscar the selfish octopus / by John M. Barrett ; pictures by Joe Servello. — New York : Human Sciences Press, c1978.
⌜32⌝ p. : col. ill. ; 24 cm.
SUMMARY: Oscar regrets his selfish and self-centered ways when only two ocean creatures respond to his plea for help out of a tight situation.
ISBN 0-87705-355-9 : $6.95
⌜1. Conduct of life—Fiction. 2. Selfishness—Fiction⌝ I. Servello, Joe. II. Title.
PZ7.B27518 Os ⌜E⌝ 78 78-18760 MARC

Barrett, John M
Zeke Hatfield and the ghost named Rocky / written by John Barrett ; illustrated by Rod Ruth. — ⌜s.l.⌝ : Silver Dollar City, c1978.
⌜32⌝ p. : col. ill. ; 22 cm.
SUMMARY: Zeke Hatfield achieves remarkable results with his trap designed to catch a candy-thievin' ghost.
⌜1. Humorous stories⌝ I. Ruth, Rod. II. Title.
PZ7.B27518 Ze ⌜E⌝ 79 78-112668 MARC

Bartholic, Edward.
Cricket and sparrow : four stories / by Edward Bartholic ; with pictures by Laura Rader. — New York : Collins, c1979.
58 p. : ill. ; 21 cm.
SUMMARY: In four encounters Cricket and Sparrow meet, pretend, visit, and picnic.
ISBN 0-529-05513-9 (lib. bdg.) : $6.91
⌜1. Crickets—Fiction. 2. Sparrows—Fiction. 3. Friendship—Fiction⌝ I. Rader, Laura. II. Title.
PZ7.B281277 Cr ⌜E⌝ 78 78-13141 MARC

Barton, Byron.
Buzz, buzz, buzz / Byron Barton. — ⌜New York⌝ : Puffin Books, 1979, c1973.
⌜32⌝ p. : col. ill. ; 18 x 23 cm. — (Picture puffin)
SUMMARY: A disastrous chain of events in the farmyard begins when the bee stings the bull.
ISBN 0-14-050307-2 : $1.95
⌜1. Animals—Fiction. 2. Farms—Fiction⌝ I. Title.
⌜PZ7.B2848 Bu 1979⌝ ⌜E⌝ 78 78-18391 MARC

Barton, Byron.
Hester / by Byron Barton. — Harmondsworth, Eng. ; New York : Puffin Books, 1978, c1975.
⌜32⌝ p. : col. ill. ; 18 x 23 cm. — (Picture puffin)
SUMMARY: Dressed too early for her own Halloween party, Hester goes trick-or-treating while she waits for her guests to arrive.
ISBN 0-14-050281-5 : $1.95
⌜1. Halloween—Fiction⌝ I. Title.
PZ7.B2848 He 1978 ⌜E⌝ 78 78-3658 MARC

Bates, Betty, 1921-
My mom, the money nut / Betty Bates. — New York : Holiday House, c1979.
158 p. ; 21 cm.
SUMMARY: A young girl has difficulty understanding and reconciling herself to her mother's materialistic values.
ISBN 0-8234-0347-5
⌜1. Mothers and daughters—Fiction. 2. School stories⌝ I. Title.
PZ7.B29452 My ⌜Fic⌝ 78 78-24213 MARC

Bates, Betty, 1921-
The ups and downs of Jorie Jenkins / Betty Bates. — New York : Holiday House, c1978.
126 p. ; 21 cm.
SUMMARY: A young girl must adjust to her father's serious illness.
ISBN 0-8234-0321-1
⌜1. Family life—Fiction⌝ I. Title.
PZ7.B29452 Up ⌜Fic⌝ 77 77-16698 MARC

Battles, Edith.
What does the rooster say, Yoshio? / Story by Edith Battles : pictures in Toni Hormann. — Chicago : A. Whitman, c1978.
⌜32⌝ p. : col. ill. ; 19 cm.
SUMMARY: A Japanese boy and an American girl comparing the sounds of the animals in their own languages find one animal that says the same thing to both of them.
ISBN 0-8075-3628-8 lib. bdg.
⌜1. Animal sounds—Fiction. 2. Japanese language—Fiction⌝ I. Hormann, Toni. II. Title.
PZ7.B3246 Wh 1978 ⌜E⌝ 78 78-12824 MARC

Baum, Lyman Frank, 1856-1919.
The Land of Oz. ⌜Sound recording⌝ Caedmon TC 1618. p1979.
1 disc. 33⅓ rpm. stereo. 12 in.
Ray Bolger, reader; Don Heckman, music composer and conductor.
Recorded in Hollywood, Calif.
Duration: 1 hr., 1 min., 54 sec.
Biographical notes, with descriptive notes by M. P. Hearn, on container.
SUMMARY: An abridged version of the novel, about a young boy who runs away to the Land of Oz where he and his pumpkinhead creation save the city from its female captors.
⌜1. Fantasy⌝ I. Bolger, Ray. II. Heckman, Don. III. Title.
[PZ7] 78-741953

Baum, Lyman Frank, 1856-1919.
The Wizard of Oz / by L. Frank Baum ; introd. by Ray Bolger ; illustrated by Robert Leydenfrost. — Garden City, N.Y. : Nelson Doubleday, c1978.
x, 153 p., ⌜10⌝ leaves of plates : ill. ; 22 cm.
First published in 1900 under title: The wonderful Wizard of Oz.
SUMMARY: After a cyclone transports her to the land of Oz, Dorothy must seek out the great wizard in order to return home to Kansas. Illustrated with stills from the motion picture.
⌜1. Fantasy⌝ I. Leydenfrost, Robert J., 1925- . II. Title.
PZ7.B327 Wi 1978 ⌜Fic⌝ 79 78-60468 MARC

Baum, Willi.
The golden mountain / Willi Baum ; based on the text by Ernst A. Ekker. — New York : Random House, c1978.
⌜32⌝ p. : col. ill. ; 21 cm. — (The Best book club ever)
Translation of Fred und Tom und der goldene Berg.
SUMMARY: Jack and Pete sell their small farm and head west in search of adventure and gold.
ISBN 0-394-83756-8
⌜1. Gold mines and mining—Fiction. 2. Western stories⌝ I. Ekker, Ernst Alfred. II. Title.
PZ7.B3273 Go 1978 ⌜E⌝ 78 77-85677 MARC

Bawden, Nina, 1925-
Rebel on a rock / by Nina Bawden. — London : Gollancz, 1978.
157 p. ; 21 cm. GB***
ISBN 0-575-02420-8 : £3.20
I. Title.
PZ7.B33 Re 78 78-314294 MARC

Bawden, Nina, 1925-
The robbers / by Nina Bawden. — New York : Lothrop, Lee & Shepard, c1979.
155 p. ; 22 cm.
SUMMARY: Nine-year-old Phillip has always lived with his grandmother, but when his widowed father remarries he finds that he must suddenly adjust to a new way of life with his father and stepmother.
ISBN 0-688-41902-X
⌜1. Grandmothers—Fiction. 2. Fathers and sons—Fiction. 3. Stepmothers—Fiction. 4. Friendship—Fiction. 5. London—Fiction⌝
PZ7.B33 Ro 1979 ⌜Fic⌝ 79 79-4152 MARC

Baxter, Lorna.
The eggchild / Lorna Baxter. — London ; Boston : Faber and Faber, 1978.
157 p. : map ; 21 cm. GB***
ISBN 0-571-11205-6 : £3.50
I. Title.
PZ7.B3376 Eg 823'.9'14 78-314986
⌜PR6052.A848⌝ 78⌜r79⌝rev MARC

Baxter, Lorna.
The eggchild / Lorna Baxter. — 1st ed. — New York : Dutton, 1979, c1978.
157 p. ; 22 cm.
SUMMARY: In a country ruled by four families endowed with special powers, a boy and girl from two of the families find a mysterious Eggchild and feel compelled to protect it and discover its true identity.
ISBN 0-525-29155-5
⌜1. Fantasy⌝ I. Title.
PZ7.B3376 Eg 1979 ⌜Fic⌝ 79 79-16223 MARC

Baylor, Byrd.
Your own best secret place / by Byrd Baylor and Peter Parnall. — New York : Scribner, c1979.
⌜32⌝ p. : col. ill. ; 26 cm.
SUMMARY: Considers the pleasures to be found in one's very own private place, whether it be a hollow in a tree, a sandy gully, or a secret sand dune.
ISBN 0-684-16111-7 (lib. bdg.) : $9.95
I. Parnall, Peter, joint author. II. Title.
PZ7.B3435 Yo ⌜Fic⌝ 78 78-21243 MARC

Beatty, Jerome.
Maria Looney and the Cosmic Circus / by Jerome Beatty, Jr. — New York : Avon Books, 1978.
160 p. : ill. ; 19 cm. — (An Avon Camelot book)
SUMMARY: The arrival of Bill Bailey's circus on the moon means trouble for Maria Looney, especially when it becomes apparent that someone is trying to steal the top-secret designs of the Moon's L-Bomb.
ISBN 0-380-40311-0 : $1.50
⌜1. Science fiction. 2. Circus stories⌝ I. Title.
PZ7.B380542 Maj ⌜Fic⌝ 79 78-56421 MARC

Beatty, Patricia.
Lacy makes a match / Patricia Beatty. — New York : Morrow, 1979.
222 p. ; 22 cm.
SUMMARY: A 13-year-old living in a turn-of-the-century California mining town determines to marry off her adoptive brothers and discover the identity of her real parents.
ISBN 0-688-22183-1. ISBN 0-688-32183-6 lib. bdg.
⌜1. California—Fiction. 2. Western stories⌝ I. Title.
PZ7.B380544 Lac ⌜Fic⌝ 79 79-9813 MARC

Beatty, Patricia.
Wait for me, watch for me, Eula Bee / Patricia Beatty. — 1st ed. — New York : Morrow, c1978.
221 p. ; 22 cm.
SUMMARY: With his father and brother serving in the Confederate Army and the rest of his family murdered in a Comanche raid of their west Texas farm, 13-year-old Lewallen seeks to free himself and his younger sister from their Indian captivity.
ISBN 0-688-22151-3. ISBN 0-688-32151-8 pbk.
⌜1. Indians of North America—Captivities—Fiction. 2. Comanche Indians—Fiction. 3. The West—Fiction⌝ I. Title.
PZ7.B380544 Wai ⌜Fic⌝ 78 78-12782 MARC

Bedini, Silvio A
The spotted stones / by Silvo A. Bedini ; pictures by Richard Erdoes. — New York : Pantheon, c1978.
57 p. : ill. ; 24 cm.
SUMMARY: Two Italian monks who are imprisoned as spies while on a pilgrimage invent a new game to pass the time in a dungeon.
ISBN 0-394-83573-5. ISBN 0-394-93573-X lib. bdg.
⌜1. Dominoes—Fiction. 2. Games—Fiction⌝ I. Erdoes, Richard. II. Title.
PZ7.B38193 Sp ⌜Fic⌝ 00 78-3283 MARC

Beim, Jerrold, 1910-1957.
Sir Halloween. Illustrated by Tracy Sugarman. New York, Morrow, 1959.
48 p. illus. 22 cm. (Morrow junior books)
SUMMARY: Surprised that his costume is not unique, Randy organizes the other knights to stage a make-believe battle at the Halloween party.
⌜1. Halloween—Fiction⌝ I. Sugarman, Tracy, 1921- . II. Title.
PZ7.B38824 Si 1959 ⌜E⌝ ⌜r79⌝rev 59-5000 MARC

Belair, Richard.
Double take : a novel / Richard L. Belair. — New York : Morrow, 1979.
190 p. ; 22 cm.
SUMMARY: Taking a job in a photographer's studio teaches an ungainly high school senior much about photography and himself.
ISBN 0-688-22202-1
⌜1. Photography—Fiction⌝ I. Title.
PZ7.B3887 Do ⌜Fic⌝ 79 79-17056 MARC

Belew, M Wendell.
The dark's a-creepin' / M. Wendell Belew ; illustrated by H. Don Fields. — Nashville : Broadman Press, c1964.
95 p. : col. ill. ; 22 cm.
SUMMARY: After their mother dies and their father is disabled with black lung disease, 11-year-old Chip and his younger sister receive assistance from the minister in their Appalachian community.
ISBN 0-8054-4416-5
⌜1. Christian life—Fiction. 2. Appalachian Mountains—Fiction. 3. Mountain life—Fiction⌝ I. Fields, H. Don. II. Title.
PZ7.B38883 Dar ⌜Fic⌝ 79 77-94678 MARC

CHILDREN'S LITERATURE (Continued)

Benchley, Nathaniel, 1915-
Red fox and his canoe. Pictures by Arnold Lobel. New York, Harper & Row [1964]
62 p. col. illus. 23 cm. (An I can read book)
SUMMARY: Summary; A young Indian boy receives a larger canoe along with some unforseen complications.
[1. Indians of North America—Fiction. 2. Canoes and canoeing—Fiction] I. Lobel, Arnold, illus. II. Title.
PZ7.B4312 Re [E] [r79]rev 64-16650 MARC

Benchley, Nathaniel, 1915-
Sam, the minuteman. Pictures by Arnold Lobel. New York, Harper & Row [1969]
62 p. illus. (part col.) 23 cm. (An I can read history book) 2.50
SUMMARY: An easy-to-read account of Sam and his father fighting as minutemen against the British in the Battle of Lexington.
1. Lexington, Battle of, 1775—Juvenile fiction. 2. Minutemen (Militia)—United States—Juvenile fiction. [1. Lexington, Battle of, 1775—Fiction. 2. Minutemen (Militia)—Fiction. 3. United States—History—Revolution, 1775-1783—Fiction] I. Lobel, Arnold, illus. II. Title.
PZ7.B4312 Sam [E] [r79]rev 68-10211 MARC

Benjamin, Alan.
1000 monsters / by Alan Benjamin ; ill. by Sal Murdocca. — New York : Four Winds Press, c1979.
[10] p. : col. ill. ; 31 x 14 cm.
Cover title.
SUMMARY: Pictures and text dealing with 10 monsters have been cut into three parts so that the reader can flip the various sections to provide many humorous variations.
ISBN 0-590-07636-1 : $3.95
[1. Monsters—Fiction] I. Murdocca, Sal. II. Title.
PZ7.B4346 On [Fic] 79 79-10682 MARC

Berenstain, Stanley, 1923-
The Berenstain bears and the spooky old tree / by Stan and Jan Berenstain. — New York : Beginner Books, c1978.
[40] p. : col. ill. ; 24 cm. — (A Bright and early book ; BE23)
SUMMARY: One by one, three brave little bears have second thoughts about exploring the interior of a spooky old tree.
ISBN 0-394-83910-2 : $3.50. ISBN 0-394-93910-7 lib. bdg.
[1. Bears—Fiction] I. Berenstain, Janice, 1923- joint author. II. Title.
PZ7.B4483 Bek [E] 78 77-93771 MARC

Berenstain, Stanley, 1923-
Papa's pizza : a Berenstain bear sniffy book / Stan and Jan Berenstain. — New York : Random House, c1978.
[21] p. : col. ill. ; 17 cm. — (A Sniffy book)
SUMMARY: Brother and Sister Bear help Papa Bear make a pizza. The reader can smell each ingredient by scratching the appropriate picture.
ISBN 0-394-83922-6 : $2.95
[1. Pizza—Fiction. 2. Cookery—Fiction] I. Berenstain, Janice, 1923- joint author. II. Title.
PZ7.B4483 Pap [E] 00 78-55907 MARC

Beresford, Elisabeth.
Secret magic / [by] Elisabeth Beresford ; illustrated by Caroline Sharpe. — St. Albans : Hart-Davis Educational, 1978.
138 p. : ill. ; 21 cm. GB78-30111
ISBN 0-247-12796-5 : £2.95
I. Title.
PZ7.B4486 Sd 79 79-302277 MARC

Berrisford, Judith Mary, 1921-
Jackie and the pony thieves. / Judith M. Berrisford. — London : Hodder and Stoughton Children's Books, 1978.
112 p. : ill. ; 20 cm. GB***
ISBN 0-340-22196-8 : £2.50
I. Title.
PZ7.B4614 Jab 79 79-300191 MARC

Berson, Harold.
Joseph and the snake / by Harold Berson. — New York : Macmillan, c1979.
[32] p. : col. ill. ; 19 X 24 cm.
SUMMARY: Only a fox's cleverness saves Joseph from being eaten by the hungry and ungrateful snake whose life he saved.
ISBN 0-02-709200-3 lib. bdg. : $6.95
[1. Foxes—Fiction. 2. Snakes—Fiction] I. Title.
PZ7.B4623 Jo [E] 78 78-12317 MARC

Bethancourt, T Ernesto.
Dr. Doom, superstar / T. Ernesto Bethancourt. — New York : Holiday House, c1978.
160 p. ; 22 cm.
SUMMARY: A teenage boy risks his life trying to unwind the threads of an ingenious plot to murder rock superstar, Dr. Doom.
ISBN 0-8234-0333-5
[1. Mystery and detective stories] I. Title.
PZ7.B46627 Dm [Fic] 78 78-6315 MARC

Bethancourt, T Ernesto.
Instruments of darkness / T. Ernesto Bethancourt. — New York : Holiday House, c1979.
159 p. ; 22 cm.
SUMMARY: Strange events involving members of a religious cult headed by a mysterious Rumanian mystic with incredible powers lead authorities to suspect a plot to control the world. Companion volume to "The Mortal Instruments."
ISBN 0-8234-0346-7
[1. Science fiction. 2. Extrasensory perception—Fiction. 3. Cults—Fiction] I. Title.
PZ7.B46627 In [Fic] 78 78-11133 MARC

Bethancourt, T Ernesto.
Nightmare town / by T. Ernesto Bethancourt. — New York : Holiday House, c1979.
158 p. ; 22 cm.
SUMMARY: The residents of Celestial, Arizona and the object of their strange religion are unlike anyone or anything a 16-year-old has ever encountered.
ISBN 0-8234-0366-1 : $7.95
[1. Science fiction] I. Title.
PZ7.B46627 Ni [Fic] 79 79-2091 MARC

Bethell, Jean.
Barney Beagle and the cat. Pictures by Ruth Wood. Editorial consultant: Lilian Moore. New York, Grosset & Dunlap [1965]
60 p. col. illus. 22 cm. (Wonder books easy reader)
SUMMARY: A beagle doesn't want to share his boy with the new addition to the household.
[1. Dogs—Fiction] I. Wood, Ruth, illus. II. Title.
PZ7.B4663 Bar [E] [r78]rev 65-20011 MARC

Bethell, Jean.
Walt Disney Productions' Pete's dragon : based on Walt Disney Productions' full-length cartoon feature film / adaptation by Jean Bethell ; screenplay by Malcolm Marmonstein, based on a story by Seton I. Miller and S. S. Field. — New York : Wonder Books, 1978.
91 p. : ill. ; 21 cm.
SUMMARY: An orphan boy and his dragon have many adventures before they find a suitable family to live with.
ISBN 0-448-16101-X : $0.95
[1. Orphans—Fiction. 2. Dragons—Fiction] I. Disney (Walt) Productions. II. Pete's dragon. [Motion picture] III. Title.
PZ7.B4663 Wal [Fic] 78 77-91860 MARC

Bird, Florence.
Holiday in the woods / Anne Francis [i.e. F. Bird] ; ill., Heather Collins. — Toronto : Clarke, Irwin, 1976.
111 p. : ill. ; 23 cm. C77-1007-5
ISBN 0-7720-1094-3 : $6.95
1. Animals, Legends and stories of. I. Title.
PZ10.3.B525 Ho 1976 78 78-316601 MARC

Bishop, Bonnie.
No one noticed Ralph / by Bonnie Bishop ; illustrated by Jack Kent. — 1st ed. — New York : Doubleday, c1979.
[64] p. : col. ill. ; 24 cm. — (A Reading-on-my-own book)
SUMMARY: An independent parrot flies out into the world to meet people but no one notices him until he spots a fire and saves the day.
ISBN 0-385-12158-X : $4.95. ISBN 0-385-12159-8 lib. bdg.
[1. Parrots—Fiction] I. Kent, Jack, 1920- . II. Title.
PZ7.B5243 No [E] 78 78-18555 MARC

Black, Irma (Simonton) 1906-
The little old man who could not read. Pictures by Seymour Fleishman. Chicago, A. Whitman [1968]
[32] p. illus. (part col.) 24 cm.
SUMMARY: An old toymaker never wanted to learn to read until his wife went away on a visit and he had to do the shopping by himself.
[1. Reading—Fiction] I. Fleishman, Seymour, illus. II. Title.
PZ7.B5288 Li [E] [r79]rev 68-9115 MARC

Blades, Ann.
The cottage at Crescent Beach / written and illustrated by Ann Blades. — [Toronto] : Magook, 1977.
[32] p. : col. ill. ; 23 cm. C***
ISBN 0-88869-000-2 : $4.95
I. Title.
PZ7.B535 Co 78 78-316079 MARC

Blaine, Marge.
Dvora's journey / Marge Blaine ; illustrated by Gabriel Lisomski. — New York : Holt, Rinehart, and Winston, c1976.
126 p. : ill. ; 22 cm.
SUMMARY: After fleeing from Russia in 1904 12-year-old Dvora and her family face unexpected problems.
ISBN 0-03-048306-9
[1. Jews—Fiction. 2. Refugees—Fiction. 3. Family life—Fiction] I. Lisomski, Gabriel. II. Title.
PZ7.B5372 Dv [Fic] 78 78-26349 MARC

Blair, Anne Denton.
Where's Rachel? : Another adventure of Arthur, the White House mouse / by Anne Denton Blair ; with ill. by Carol Watson. — Washington : Acropolis Books, c1978.
45 p. : ill. ; 29 cm.
SUMMARY: When his cousin Rachel runs away to see the White House for herself, Arthur and his cousin Thomas face the ensuing dangers bravely and cleverly.
ISBN 0-87491-264-4 : $7.95
[1. Mice—Fiction. 2. Washington, D.C. White House—Fiction] I. Watson, Carol Stuart. II. Title.
PZ7.B5373 Wh [E] 78 78-12953 MARC

Bloch, Marie Halun.
Displaced person / by Maria Halun Bloch ; decorations by Allen Davis. — 1st ed. — New York : Lothrop, Lee & Shepard, c1978.
191 p. ; 22 cm.
SUMMARY: Fourteen-year-old Stefan, a Ukrainian refugee, struggles to survive as a displaced person in Nazi Germany during the final days of World War II.
ISBN 0-688-41860-0. ISBN 0-688-51860-6 lib. bdg.
1. World War, 1939-1945—Germany—Juvenile fiction. 2. World War, 1939-1945—Ukraine—Juvenile fiction. 3. Ukraine—History—German occupation, 1941-1944—Juvenile fiction. [1. World War, 1939-1945—Germany—Fiction. 2. World War, 1939-1945—Ukraine—Fiction. 3. Ukraine—History—German occupation, 1941-1944—Fiction. 4. World War, 1939-1945—Refugees—Fiction. 5. Refugees—Fiction] I. Davis, Allen. II. Title.
PZ7.B6194 Di [Fic] 78 78-13083 MARC

Blos, Joan W
A gathering of days : a New England girl's journal, 1830-32 : a novel / by Joan Blos. — New York : Scribner, c1979.
144 p. ; 22 cm.
SUMMARY: The journal of a 14-year-old girl, kept the last year she lived on the family farm, records daily events in her small New Hampshire town, her father's remarriage, and the death of her best friend.
ISBN 0-684-16340-3 : $7.95
[1. New Hampshire—Fiction] I. Title.
PZ7.B6237 Gat [Fic] 79 79-16898 MARC

Blue, Rose.
Cold rain on the water / Rose Blue. — New York : McGraw-Hill, c1979.
123 p. ; 21 cm.
SUMMARY: A 16-year-old Russian Jew and his family face challenges and problems in America.
ISBN 0-07-006168-8 : $7.95
[1. Emigration and immigration—Fiction. 2. Russian Americans—Fiction. 3. Family life—Fiction. 4. Jews in the United States—Fiction] I. Title.
PZ7.B6248 Co [Fic] 78 78-23633 MARC

Blue Beard.
A new history of Blue Beard. Written by Gaffer Blackbeard, for the amusement of all good boys and girls. Ornamented with engravings. Albany, Printed by E. & E. Hosford, 1818.
30 p. illus. 11 cm.
The present version derives from Perrault's story. Cf. A. S. W. Rosenbach. Early Amer. children's books, 1933, p. 120.
"The boy and the thief": p. 29-30.
I. Gaffer Blackbeard. II. Perrault, Charles, 1628-1703. La Barbe bleue. III. Title.
PZ6.B625 Ne 1818 78 78-381534 MARC

Bødker, Cecil.
Silas and Ben-Godik / Cecil Bodker ; translated from the Danish by Sheila La Farge. — New York : Delacorte Press /Seymour Lawrence, c1978.
vii, 191 p. ; 21 cm.
Translation of Silas og Ben-Godik.
"A Merloyd Lawrence book."
SUMMARY: Silas and his friend Ben-Godik spend a year traveling by horseback and encountering many strange individuals and harrowing adventures.
ISBN 0-440-07923-3. ISBN 0-440-07924-1 lib. bdg.
I. Title.
PZ7.B635717 Sg [Fic] 78 78-50459 MARC

Bødker, Cecil.
Silas and the runaway coach / Cecil Bodker ; translated from the Danish by Sheila La Farge. — New York : Delacorte Press/S. Lawrence, c1978.
vii, 245 p. ; 21 cm.
"A Merloyd Lawrence book."
Translation of Silas fanger et firspand.
SUMMARY: Follows the further adventures of a firecely independent country boy as he briefly resides with the family of a wealthy urban merchant.
ISBN 0-440-07953-3 : $7.95. ISBN 0-440-07954-3 lib. bdg.
I. Title.
PZ7.B635717 Sk [Fic] 78 78-50465 MARC

Boegehold, Betty Virginia Doyle.
Pippa pops out! / By Betty Boegehold ; illustrated by Cyndy Szekeres. — New York : Knopf : distributed by Random House, c1979.
59 p. : ill. ; 22 cm.
SUMMARY: Four stories relate the further adventures of Pippa Mouse and her animal friends.
ISBN 0-394-84057-7. ISBN 0-394-94057-1 lib. bdg.
[1. Mice—Fiction] I. Szekeres, Cyndy. II. Title.
PZ7.B63572 Pi 1979 [E] 78 78-12491 MARC

CHILDREN'S LITERATURE (Continued)

Bograd, Larry.
Felix in the attic / by Larry Bograd ; illustrated by Dirk Zimmer. — New York : Harvey House, c1978.
[31] p. : ill. ; 25 cm.
SUMMARY: When he is locked up in the attic yet again, Felix is prepared and makes quite a nice place for himself.
ISBN 0-8178-5917-9
[1. Attics—Fiction] I. Zimmer, Dirk. II. Title.
PZ7.B635797 Fe [E] 00 79-100712 MARC

Boles, Paul Darcy, 1916-
A million guitars, and other stories. [1st ed.] Boston, Little, Brown [1968, c1967]
282 p. 21 cm. 4.75
CONTENTS: A million guitars.—The running of the deer.—Swinger.—A few notes about the Arabian oryx.—Treasure.—The charm.—The kindred heart.—The night watch.—A talent for delight.—George the sweet singer.—Wet night in Westmont.—The big beast.
[1. Short stories] I. Title.
PZ7.B63588 Mi [Fic] [r79]rev 68-15386 MARC

Boles, Paul Darcy, 1916-
I thought you were a unicorn, and other stories. [1st ed.] Boston, Little, Brown [1971]
207 p. 21 cm. $5.95
CONTENTS: A verray parfit gentil knight.—The holiday rider.—The somewhere music.—Today is my sister's wedding.—I thought you were a unicorn.—Lucas and Jake.—Miss Rose.—The animal kingdom.—The telephone artist.—The girl at the plaza.—Summer candles.
[1. Short stories] I. Title.
PZ7.B63588 Iat [Fic] 71[r79]rev 70-140477 MARC

Bolliger, Max.
The fireflies : a story / by Max Bolliger ; translated and adapted by Roseanna Hoover ; with pictures by Jiří Trnka. — London : Evans Brothers, 1977, c1970.
[43] p. : col. ill. ; 33 cm. GB***
Based on a Czechoslovakian story: Broučci by J. Karafiát.
ISBN 0-237-44874-2 : £3.50
I. Hoover, Roseanna, tr. II. Trnka, Jiří, 1912- illus. III. Karafiát, Jan, 1846-1929. Broučci. English. 1977. IV. Title.
PZ7.B635915 Fi 1977 78 78-316077 MARC

Bond, Ann Sharpless.
Saturdays in the city / Ann Sharpless Bond ; illustrated by Leonard Shortall. — Boston : Houghton Mifflin, 1979.
147 p. : ill. ; 22 cm.
SUMMARY: Two nine-year-old boys share a series of Saturday adventures in their city.
ISBN 0-395-28376-0
[1. City and town life—Fiction] I. Shortall, Leonard W. II. Title.
PZ7.B6366 Sat [Fic] 79 79-11747 MARC

Bond, Michael.
The great big Paddington book / Michael Bond ; pictures by Ivor Wood. — Glasgow : Collins, 1976.
45 p. : col ill. ; 33 cm. GB***
ISBN 0-00-138075-3
I. Wood, Ivor. II. Title.
PZ7.B6368 Gr 78 78-314312 MARC

Bond, Michael.
Olga takes a bite / Michael Bond. — St. Paul : EMC Corp., 1977.
16 p. : col. ill. ; 20 cm. — (Olga da Polga ; 5)
SUMMARY: The family cat's introduction to Olga the guinea pig proves to be painful.
ISBN 0-88436-460-7
[1. Guinea pigs—Fiction] I. Title.
PZ7.B6368 Oli [E] 77[r79]rev 77-21321 MARC

Bond, Michael.
Paddington at the tower / Michael Bond ; illustrated by Fred Banbery. — London : Collin, 1976.
[29] p. : ill. ; 21 cm. — (Paddington picture book ; 5) GB***
ISBN 0-00-182132-6
I. Banbery, Fred. II. Title.
PZ7.B6368 Pace 1976 79 79-309227 MARC

Bonsall, Crosby Newell.
The goodbye summer / by Crosby Bonsall. — 1st ed. — New York : Greenwillow Books, c1979.
148 p. ; 22 cm.
SUMMARY: Allie hates endings and goodbyes until her friend, Ms. Lenya, helps her learn that some things can't ever be lost.
ISBN 0-688-80202-8 : $6.95. ISBN 0-688-84202-X lib. ed. : $6.67
[1. Friendship—Fiction] I. Title.
PZ7.B64265 Go [Fic] 78 78-23245 MARC

Bonsall, Crosby (Newell)
What spot? [1st ed.] New York, Harper & Row [1963]
64 p. illus. 23 cm. (An I can read book)
SUMMARY: Although his friend tries to discourage him, a walrus is determined to identify a small black spot he spies in the snow.
[1. Walruses—Fiction. 2. Snow—Fiction. 3. Arctic regions—Fiction] I. Title.
PZ7.B64265 Wg [E] [r78]rev 63-8005 MARC

Booth, Eugene, 1940-
At the circus / words by Eugene Booth ; pictures by Derek Collard. — Milwaukee : Raintree Childrens books, c1977.
21 p. : col. ill. ; 24 cm. — (A Raintree spotlight book)
SUMMARY: Uses a circus setting to stimulate such activities as counting, noting visual differences, and making up a story.
ISBN 0-8393-0112-X lib. bdg. : $6.60
[1. Circus] I. Collard, Derek. II. Title.
PZ7.B6467 Ar [E] 77 77-7946 MARC

Booth, Eugene, 1940-
At the fair / words by Eugene Booth ; pictures by Derek Collard. — Milwaukee : Raintree Childrens books, c1977.
21 p. : col. ill. ; 24 cm. — (A Raintree spotlight book)
SUMMARY: Questions encourage the reader to look at, think about, and describe scenes from a fair.
ISBN 0-8393-0114-6 lib. bdg. : $6.60
[1. Fairs] I. Collard, Derek. II. Title.
PZ7.B6467 As [E] 77 77-7961 MARC

Booth, Eugene, 1940-
At the zoo / words by Eugene Booth ; pictures by Derek Collard. — Milwaukee : Raintree childrens books, c1977.
21 p. : col. ill. ; 24 cm. — (A Raintree spotlight book)
SUMMARY: Pictures of animals at the zoo stimulate counting, identifying colors, "finding the mistake," noting visual differences, and other pre-reading activities.
ISBN 0-8393-0107-3 lib. bdg. : $6.60
[1. Zoo animals] I. Collard, Derek. II. Title.
PZ7.B6467 At [E] 77 77-7627 MARC

Booth, Eugene, 1940-
In the air / words by Eugene Booth ; pictures by Derek Collard. — Milwaukee : Raintree Childrens Books, c1977.
21 p. : col. ill. ; 24 cm. — (A Raintree spotlight book)
SUMMARY: Illustrations and simple text encourage thought about things that are flying in the sky.
ISBN 0-8393-0105-7 lib. bdg. : $6.60
[1. Flight. 2. Nature] I. Collard, Derek. II. Title.
PZ7.B6467 Ih [E] 77 77-7984 MARC

Booth, Eugene, 1940-
In the city / words by Eugene Booth ; pictures by Derek Collard. — Milwaukee : Raintree Childrens books, c1977.
21 p. : col. ill. ; 24 cm. — (A Raintree spotlight book)
SUMMARY: Uses a city setting to stimulate such activities as counting, noting visual differences, and making up a story.
ISBN 0-8393-0109-X lib. bdg. : $6.60
[1. City and town life] I. Collard, Derek. II. Title.
PZ7.B6467 Ij [E] 77 77-7949 MARC

Booth, Eugene, 1940-
In the garden / words by Eugene Booth ; pictures by Derek Collard. — Milwaukee : Raintree Childrens books, c1977.
21 p. : col. ill. ; 24 cm. — (A Raintree spotlight book)
SUMMARY: Uses a garden setting to stimulate such pre-reading activities as counting, noting visual differences, and making up a story.
ISBN 0-8393-0115-4 lib. bdg. : $6.60
[1. Gardens] I. Collard, Derek. II. Title.
PZ7.B6467 Il [E] 77 77-7628 MARC

Booth, Eugene, 1940-
In the park / words by Eugene Booth ; pictures by Derek Collard. — Milwaukee : Macdonald-Raintree, c1977.
21 p. : col. ill. ; 24 cm. — (A Raintree spotlight book)
SUMMARY: Questions encourage the reader to look at, think about, and describe scenes from a park.
ISBN 0-8393-0106-5 lib. bdg. : $6.00
[1. Parks] I. Collard, Derek. II. Title.
PZ7.B6467 In [E] 77 77-7622 MARC

Booth, Eugene, 1940-
On the farm / words by Eugene Booth ; pictures by Derek Collard. — Milwaukee : Raintree Childrens books, c1977.
21 p. : col. ill. ; 24 cm. — (A Raintree spotlight book)
SUMMARY: Questions encourage the reader to look at, think about, and describe scenes from a farm.
ISBN 0-8393-0113-8 lib. bdg. : $6.60
[1. Farm] I. Collard, Derek. II. Title.
PZ7.B6467 On [E] 77 77-7965 MARC

Booth, Eugene, 1940-
Under the ground / words by Eugene Booth ; pictures by Derek Collard. — Milwaukee : Raintree Childrens books, c1977.
21 p. : col. ill. ; 24 cm. — (A Raintree spotlight book)
SUMMARY: Pictures and questions stimulate discussion about animals, tunnels, pipes, and even men underground.
ISBN 0-8393-0110-3 lib. bdg. : $6.60
[1. Underground utility lines. 2. Underground construction. 3. Animals. 4. Caves] I. Collard, Derek. II. Title.
PZ7.B6467 Ul [E] 77 77-8037 MARC

Booth, Eugene, 1940-
Under the ocean / words by Eugene Booth ; pictures by Derek Collard. — Milwaukee : Raintree Childrens Books, c1977.
21 p. : col. ill. ; 24 cm. — (A Raintree spotlight book)
SUMMARY: Illustrations and simple text encourage thought about what is happening on the ocean bottom.
ISBN 0-8393-0108-1 lib. bdg. : $6.60
[1. Ocean bottom] I. Collard, Derek. II. Title.
PZ7.B6467 Un [E] 77 77-7983 MARC

Bopp, Joseph B
Herbie Capleenies / by Joseph B. Bopp ; illustrated by Amy Myers. — Reading, Mass. : Addison-Wesley, c1978.
[32] p. : col. ill. ; 24 cm.
SUMMARY: A little boy and his parents reverse their roles.
ISBN 0-201-24721-7 lib. bdg. : $7.95
[1. Parent and child—Fiction] I. Myers, Amy. II. Title.
PZ7.B64746 He [E] 77[r79]rev 77-10715 MARC

Bornstein, Ruth.
The dancing man : story and pictures / by Ruth Bornstein. — New York : Seabury Press, c1978.
[32] p. : ill. ; 24 cm.
"A Clarion book."
SUMMARY: Wearing his pair of special silver shoes, young Joseph sets out to dance with the world.
ISBN 0-8164-3214-7 lib. bdg. : $6.95
[1. Dancing—Fiction] I. Title.
PZ7.B64848 Dan [E] 78 77-29124 MARC

Bornstein, Ruth.
Jim / story and pictures by Ruth Bornstein. — New York : Seabury Press, c1978.
[32] p. : col. ill. ; 21 x 24 cm.
"A Clarion book."
SUMMARY: When Jim the dog searches for his father who has not returned from a day of exploring, he discovers more than he expects.
ISBN 0-8164-3204-X
[1. Dogs—Fiction] I. Title.
PZ7.B64848 Ji [E] 77 77-12712 MARC

Boston, Lucy Maria, 1892-
An enemy at Green Knowe / L. M. Boston ; drawings by Peter Boston. — 1st Voyager/HBJ ed. — New York : Harcourt Brace Jovanovich, 1979, c1964.
196 p. : ill. ; 21 cm. — (A Voyager/HBJ book)
Reprint of the ed. published by Harcourt, Brace & World, New York.
SUMMARY: The inhabitants of Green Knowe become involved with black magic when a modern-day witch attempts to find books of witchcraft supposedly hidden in the old house by a mad alchemist.
ISBN 0-15-628792-7
[1. Witchcraft—Fiction. 2. England—Fiction] I. Boston, Peter. II. Title.
[PZ7.B6497 En 1979] [Fic] 79 78-71151 MARC

Boston, Lucy Maria, 1892-
The stones of Green Knowe / L. M. Boston ; illustrated by Peter Boston. — London : Bodley Head, 1976.
118 p. : ill. ; 21 cm. GB76-12954
ISBN 0-370-11017-X : £2.50
I. Boston, Peter. II. Title.
PZ7.B6497 Sr 78 78-313214 MARC

Boston, Lucy Maria, 1892-
A stranger at Green Knowe / L. M. Boston ; drawings by Peter Boston. — 1st Voyager/HBJ ed. — New York : Harcourt, Brace, Jovanovich, 1979, c1961.
158 p. : ill. ; 20 cm. — (A Voyager/HBJ book)
Reprint of the ed. published by Harcourt, Brace & World, New York.
SUMMARY: A strange friendship develops between an orphaned Chinese refugee who is spending the summer at Green Knowe and a gorilla who has escaped from the London Zoo.
ISBN 0-15-685657-3
[1. Gorillas—Fiction. 2. England—Fiction] I. Boston, Peter. II. Title.
[PZ7.B6497 St 1979] [Fic] 79 78-71150 MARC

Boston, Lucy Maria, 1892-
Treasure of Green Knowe / L. M. Boston ; drawings by Peter Boston. — 1st Voyager/HJB ed. — New York : Harcourt Brace Jovanovich, 1978, c1958.
185 p. : ill. ; 20 cm. — (A Voyager/HBT book)
SUMMARY: A young boy listens to his great-grandmother's tales of Green Knowe as it used to be and, gradually, as past and present blend, he shares the strange adventures of the former inhabitants.
ISBN 0-15-691302-X
[1. Space and time—Fiction. 2. England—Fiction] I. Boston, Peter. II. Title.
[PZ7.B6497 Tr 1978] [Fic] 77 77-16689 MARC

Bottner, Barbara.
Big boss! Little boss! / by Barbara Bottner. — New York : Pantheon Books, c1978.
37 p. : ill. ; 24 cm. — (An I am reading book)
SUMMARY: In these two stories "Big Boss" Penny decides she needs a big sister and "Little Boss" Lizzie hunts for her lost Mickey Mouse ring.
ISBN 0-394-83939-0. ISBN 0-394-93939-5 lib. bdg.
[1. Brothers and sisters—Fiction] I. Title.
PZ7.B6586 Bi [E] 78 78-3281 MARC

CHILDREN'S LITERATURE (Continued)

Bottner, Barbara.
Dumb old Casey is a fat tree / by Barbara Bottner. — 1st ed. — New York : Harper & Row, c1979.
42 p. : ill. ; 23 cm.
SUMMARY: Casey's determination to be a ballet dancer teaches her a valuable secret about hard work.
ISBN 0-06-020616-0 : $6.95. ISBN 0-06-020617-9 lib. bdg. : $6.79
[1. Ballet dancing—Fiction. 2. Weight control—Fiction] I. Title.
PZ7.B6586 Du 1979 [E] 78 78-19474 MARC

Bottner, Barbara.
Messy / written and illustrated by Barbara Bottner. — New York : Delacorte Press, c1979.
[32] p. : col. ill. ; 17 x 22 cm.
SUMMARY: A six-year-old girl has a difficult time being neat.
ISBN 0-440-05492-3
[1. Cleanliness—Fiction] I. Title.
PZ7.B6586 Me [E] 78 78-50420 MARC

Bottner, Barbara.
Myra / by Barbara Bottner. — New York : Macmillan, c1979.
[31] p. : col. ill. ; 14 x 19 cm.
SUMMARY: A dancing class is disrupted when Myra actually changes into the animals the other pupils are imitating.
ISBN 0-02-711740-5
I. Title.
PZ7.B6586 My [E] 78 78-10417 MARC

Boynton, Sandra.
Hester in the wild / by Sandra Boynton. — 1st ed. — New York : Harper & Row, c1979.
[32] p. : col. ill. ; 24 x 26 cm.
SUMMARY: Hester the pig discovers camping has its drawbacks.
ISBN 0-06-020631-4 : $6.95. ISBN 0-06-020654-3 lib. bdg. : $6.79
[1. Pigs—Fiction. 2. Camping—Fiction] I. Title.
PZ7.B6968 He 1979 [E] 78 78-67026 MARC

Bradbury, Bianca.
Mixed-up summer / Bianca Bradbury. — Boston : Houghton Mifflin, 1979.
195 p. ; 22 cm.
SUMMARY: A young girl spends her summer vacillating between marriage to a hardworking, stoic young man and a nursing career.
ISBN 0-395-27816-3 : $7.95
[1. Nursing—Fiction] I. Title.
PZ7.B716 Mi [Fic] 79 78-31792 MARC

Bradbury, Bianca.
Where's Jim now? / Bianca Bradbury. — Boston : Houghton Mifflin, 1978.
174 p. ; 22 cm.
SUMMARY: When Dave's half-brother joins the family after spending time in a rehabilitation institute, he seems unable to stay out of trouble.
ISBN 0-395-27160-6 : $7.95
[1. Brothers and sisters—Fiction] I. Title.
PZ7.B716 Wh [Fic] 78 78-14832 MARC

Bradford, Ann.
The mystery of the missing raccoon / by Ann Bradford and Kal Gezi ; illustrated by Mina Gow McLean. — Elgin, Ill. : Child's World ; Chicago : distributed by Childrens Press, c1978.
32 p. : ill. (some col.) ; 25 cm.
SUMMARY: Five youngsters discover who is repeatedly letting a friendly raccoon out of his cage at the zoo.
ISBN 0-89565-025-8 : $4.95
[1. Raccoons—Fiction. 2. Mystery and detective stories] I. Gezi, Kalil I., joint author. II. McLean, Mina Gow. III. Title.
PZ7.B7226 My [E] 78 78-6455 MARC

Bradley, Michael James, 1944-
The shaping room / Michael Bradley. — New York : Dodd, Mead, c1978.
160 p. ; 21 cm.
SUMMARY: Seeking an escape from his dominating sister, the neighborhood bully, and himself, Stephen Barley learns the shaping of dreams from his sculpture professor.
ISBN 0-396-07540-1
[1. Dreams—Fiction] I. Title.
PZ7.B7254 Sh [Fic] 77[r79]rev 77-16877 MARC

Bradley, Steve.
The candy man / Steve Bradley ; [cover and ill., Warren Lee]. — Chicago : Childrens Press, c1977.
60 p. : ill. ; 21 cm. — (Pacesetters)
SUMMARY: When Peg first meets a fellow traveler to New York, she has no idea he is about to involve her in a deadly game of smuggling.
ISBN 0-516-02152-4
[1. Smuggling—Fiction] I. Lee, Warren. II. Title.
PZ7.B7257 Can [Fic] 78 78-16553 MARC

Bradley, Virginia, 1912-
Bend to the willow / Virginia Bradley. — New York : Dodd, Mead, c1979.
192 p. ; 22 cm.
SUMMARY: In 1935 Ellen Barrow moves to a small town to escape the guilt she feels over her parents' deaths, but instead of being isolated, as she hopes, she becomes involved with the townspeople.
ISBN 0-396-07718-8
[1. Death—Fiction. 2. City and town life—Fiction] I. Title.
PZ7.B7259 Be [Fic] 79 79-52036 MARC

Bram, Elizabeth.
Saturday morning lasts forever / Elizabeth Bram. — New York : Dial Press, c1978.
[32] p. : ill. ; 15 cm.
SUMMARY: While Susan's parents sleep late one Saturday morning she finds various ways to occupy herself.
ISBN 0-8037-7627-6 : $5.95. ISBN 0-8037-7628-4 lib. bdg. : $4.58
I. Title.
PZ7.B7357 Sat [E] 78 78-51318 MARC

Bram, Elizabeth.
There is someone standing on my head / by Elizabeth Bram. — New York : Dial Press, c1979.
[32] p. : col. ill. ; 22 cm.
SUMMARY: A young girl invents an imaginary friend.
ISBN 0-8037-8648-4 : $5.95. ISBN 0-8037-8649-2 lib. bdg. : $5.47
[1. Play—Fiction] I. Title.
PZ7.B7357 Th [E] 78 78-72196 MARC

Brancato, Robin F
Blinded by the light / Robin F. Brancato. — New York : Knopf : distributed by Random House, c1978.
215 p. ; 22 cm.
SUMMARY: A college freshman infiltrates the Light of the World Church in search of her missing brother.
ISBN 0-394-83721-5. ISBN 0-394-93721-X lib. bdg.
1. Cults—Fiction. I. Title.
PZ7.B73587 Bl 813'.5'4 78 78-4583 MARC

Brandenberg, Franz.
Everyone ready? / Franz Brandenberg ; illustrated by Aliki. — 1st ed. — New York : Greenwillow Books, c1979.
55 p. : col. ill. ; 22 cm. — (Greenwillow read-alone)
SUMMARY: The Fieldmouse family take the train to visit Uncle Alfred and Aunt Kate—but not all at once.
ISBN 0-688-80198-6 : $5.95. ISBN 0-688-84198-8 lib. bdg. : $5.71
[1. Mice—Fiction. 2. Humorous stories] I. Aliki. II. Title.
PZ7.B7364 Ev [E] 78 78-13744 MARC

Brandenberg, Franz.
I wish I was sick, too! / By Franz Brandenberg ; illustrated by Aliki. — New York : Puffin Books, 1978, c1976.
[32] p. : col. ill. ; 23 cm.
SUMMARY: Elizabeth envies her brother the pampered treatment he gets when he is sick in bed until she gets sick too.
ISBN 0-14-050292-0 : $1.95
[1. Envy—Fiction. 2. Sick—Fiction] I. Aliki. II. Title.
PZ7.B7364 Iag 1978 [E] 78 78-6113 MARC

Brandt, Lucile, 1900-
Anna Elizabeth, seventeen / by Lucile Long. — Elgin, Ill. : Brethren Press, c1978.
208 p. ; 21 cm.
SUMMARY: Relates the experiences of a seventeen-year-old girl who leaves her home in a Brethren community in eighteenth-century Pennsylvania for the first time to help her aunt care for her nine cousins.
ISBN 0-87178-041-0
[1. Church of the Brethren—Fiction. 2. Pennsylvania Dutch—Fiction] I. Title.
PZ7.B73734 Ar 1978 [Fic] 78 78-9747 MARC

Brandt, Lucile, 1900-
Anna Elizabeth : girl of the plain people / Lucile Long [i.e. L. Brandt] ; ill. by Inez Goughnour. — Elgin, Ill. : Brethren Press, 1975, c1942.
128 p. : ill. ; 18 cm.
SUMMARY: Relates a young girl's struggle for an education in a Brethren community in eighteenth-century Pennsylvania.
ISBN 0-87178-040-2 : $1.50
[1. Church of the Brethren—Fiction. 2. Pennsylvania Dutch—Fiction] I. Goughnour, Inez. II. Title.
[PZ7.B73734 An 10] [Fic] 76[r79]rev 74-29366 MARC

Branscum, Robbie.
For love of Jody / Robbie Branscum ; illustrated by Allen Davis. — 1st ed. — New York : Lothrop, Lee & Shepard Books, c1979.
111 p. : ill. ; 22 cm.
SUMMARY: In the Arkansas hills during a calamity-filled summer of the Depression and Ma's third pregnancy, 12-year-old Frankie overcomes her bitterness toward her mother and jealousy over her younger, mentally retarded sister.
ISBN 0-688-41881-3
[1. Family life—Fiction. 2. Mentally handicapped—Fiction. 3. Depressions—1929—Fiction. 4. Arkansas—Fiction] I. Davis, Allen. II. Title.
PZ7.B73754 Fo [Fic] 79 79-2436 MARC

Branscum, Robbie.
To the tune of a hickory stick / Robbie Branscum. — 1st ed. — Garden City, N.Y. : Doubleday, c1978.
119 p. ; 22 cm.
SUMMARY: Mistreated by their cruel uncle, a backwoods brother and sister find refuge in the schoolhouse where they spend the winter hiding with their teacher and senile grandfather.
ISBN 0-385-13037-6 : $6.95. ISBN 0-385-13038-4 lib. bdg.
[1. Brothers and sisters—Fiction. 2. Arkansas—Fiction] I. Title.
PZ7.B73754 Tm [Fic] 78 77-12844 MARC

Branscum, Robbie.
Toby alone / Robbie Branscum. — 1st ed. — Garden City, N.Y. : Doubleday, c1979.
117 p. ; 19 cm.
SUMMARY: Having driven Johnny Joe off in her grief at Granny's death, Toby, now 14, resolves to win him back. Sequel to "Toby, Granny, and George."
ISBN 0-385-14017-7 : $7.95
[1. Death—Fiction. 2. Mountain life—Fiction. 3. Arkansas—Fiction] I. Title.
PZ7.B73754 Tn [Fic] 79 78-22152 MARC

Branscum, Robbie.
Toby and Johnny Joe / Robbie Branscum. — 1st ed. — Garden City, N.Y. : Doubleday, c1979.
99 p. ; 19 cm.
SUMMARY: Toby and Johnny Joe's blissful marriage is cruelly interrupted by World War II.
ISBN 0-385-13035-X (lib. bdg.) : $7.95
[1. Marriage—Fiction. 2. Mountain life—Fiction. 3. Arkansas—Fiction] I. Title.
PZ7.B73754 Tnd [Fic] 79 78-22153 MARC

Branscum, Robbie.
The ugliest boy / by Robbie Branscum ; illustrated by Michael Eagle. — 1st ed. — New York : Lothrop, Lee & Shepard Co., c1978.
126 p. : ill. ; 22 cm.
SUMMARY: Unexpected romance and a snake cult prompt the teenage son of an Arkansas preacher to reconsider his negative feelings about himself and his family.
ISBN 0-688-41859-7. ISBN 0-688-51859-1 lib. bdg.
[1. Family life—Fiction. 2. Arkansas—Fiction] I. Eagle, Michael. II. Title.
PZ7.B73754 Ug [Fic] 78 78-14388 MARC

Branson, Karen.
The potato eaters / by Karen Branson ; with ill. by Jane Sterrett. — New York : Putnam, c1979.
160 p. : ill. ; 24 cm.
SUMMARY: A family attempts to withstand the hardships brought about by a blight, which strikes Ireland in the 1840's, ruining potato crops.
ISBN 0-399-20678-7 : $8.95
[1. Ireland—Famines—Fiction] I. Sterrett, Jane. II. Title.
PZ7.B73756 Po 1979 [Fic] 78 78-24330 MARC

Breckler, Rosemary, 1920-
Where are the twins? / Rosemary Breckler. — 1st ed. — Philadelphia : Westminster Press, c1979.
96 p. ; 21 cm.
"A Hiway book."
SUMMARY: Two teenagers about to become full-fledged Police Cadets rescue twins lost in a rainstorm and mudslides and perform other helpful deeds too.
ISBN 0-664-32651-X
[1. Adventure stories] I. Title.
PZ7.B744 Wh [Fic] 79 79-10390 MARC

Brenner, Barbara.
Our class presents Ostrich feathers : a play in two acts / story by Barbara Brenner ; pictures by Vera B. Williams, assisted by Evelyn Armstrong. — New York : Parents' Magazine Press, c1978.
[32] p. : ill. ; 19 x 26 cm.
SUMMARY: When the ostrich loses his tail feathers to a hunter, the jungle animals decide to do some hunting of their own.
ISBN 0-8193-0921-4 : $5.95
[1. Jungle animals—Fiction. 2. Hunting—Fiction] I. Williams, Vera B. II. Armstrong, Evelyn. III. Title. IV. Title: Ostrich feathers.
PZ7.B7518 Ou [E] 77 77-24284 MARC

Brězan, Jurij.
Bitwa z wulkimi myšemi a druhe powědki / Jurij Brězan. — 1. nakład. — Budyšin : Domowina, c1977.
76 p. : ill. ; 20 cm. GDR***
5.30M
I. Title.
PZ70.W4B677 78 78-344238 MARC

Bridgers, Sue Ellen.
All together now : a novel / Sue Ellen Bridgers. — New York : Knopf : distributed by Random House, c1979.
238 p. ; 22 cm.
SUMMARY: With her father serving in the Korean War and her mother working, Casey spends her 12th summer visiting her grandparents in their small town where she makes friends with a 33-year-old man with the mind of a 12-year-old.
ISBN 0-394-84098-4. ISBN 0-394-94098-9 lib. bdg.
[1. Grandparents—Fiction] I. Title.
PZ7.B7615 Al 1979 [Fic] 78 78-12244 MARC

Bridgman, Elizabeth P
New dog next door / by Elizabeth Bridgman. — 1st ed. — New York : Harper & Row, c1978.
30 p. : col. ill. ; 15 cm.
SUMMARY: A little boy tells how the neighbors' new dog makes itself at home at his house.
ISBN 0-06-020672-1 : $4.95. ISBN 0-06-020673-X lib. bdg. : $5.79
[1. Dogs—Fiction] I. Title.
PZ7.B7624 Ne [E] 78 77-25667 MARC

CHILDREN'S LITERATURE (Continued)

Briggs, Nancy E
Children's literature through storytelling & drama / Nancy E. Briggs, Joseph A. Wagner. — 2d ed. — Dubuque, Iowa : W. C. Brown Co., c1979.
xiii, 201 p. : ill. ; 23 cm.
First ed. (1970) by J. A. Wagner published under title: Children's literature through storytelling.
On spine: Storytelling & drama.
Includes bibliographical references and indexes.
ISBN 0-697-06212-0
1. Story-telling. 2. Children's literature. I. Wagner, Joseph Anthony, joint author. II. Wagner, Joseph Anthony. Children's literature through storytelling. III. Title. IV. Title: Storytelling & drama.
LB1042.W34 1979 372.6'4 79-50353
79 MARC

Briggs, Raymond.
Fungus the bogeyman / [by] Raymond Briggs. — London : Hamilton, 1977.
41 p. : chiefly col. ill. ; 31 cm. GB77-24560
ISBN 0-241-89553-7 : £2.75
I. Title.
PZ7.B7646 Fu 741.5'942 78-312923
78[r79]rev MARC

Briggs, Raymond.
The snowman / Raymond Briggs. — 1st American ed. — New York : Random House, c1978.
[32] p. : col. ill. ; 31 cm.
SUMMARY: When his snowman comes to life, a little boy invites him home and in return is taken on a flight high above the countryside.
ISBN 0-394-83973-0 : $4.95. ISBN 0-394-93973-5 lib. bdg. : $5.99
[1. Dreams—Fiction. 2. Stories without words. 3. Cartoons and comics]
I. Title.
PZ7.B7646 Sn 1978 [E] 78-55904
78 MARC

Bright, Robert, 1902-
Georgie and the buried treasure / by Robert Bright. — 1st ed. — Garden City, N.Y. : Doubleday, c1979.
[34] p. : ill. ; 21 x 27 cm.
SUMMARY: Georgie the ghost tries to discourage his neighbor from digging for buried treasure.
ISBN 0-385-14626-4 : $6.95
[1. Ghost stories. 2. Buried treasure—Fiction]
PZ7.B765 Gc [E] 78-22305
78 MARC

Bright, Robert, 1902-
My red umbrella. New York, W. Morrrow, 1959.
[32] p. illus. 18 cm.
SUMMARY: A young girl's umbrella grows to accommodate an increasing number of animal friends.
[1. Umbrellas and parasols—Fiction] I. Title.
PZ7.B765 My [E] 59-7928
[r79]rev MARC

Brinsmead, Hesba Fay.
The ballad of Benny Perhaps / [by] Hesba Brinsmead. — Stanmore, N.S.W. : Cassell Australia, 1977.
188 p. ; 23 cm.
Aus
ISBN 0-7269-0449-X
I. Title.
PZ7.B78017 Bal 78-323983
79 MARC

Brisco, Patty.
Raging Rapids / by Patty Briscoe ; [illustrated by Kevin Davidson]. — Mankato, Minn. : Creative Education, c1978.
56 p. : ill. ; 24 cm.
SUMMARY: Temporarily staying with a grandfather he has never met before, Mike decides to run away in aboat that must pass through some "raging rapids."
ISBN 0-87191-683-5
[1. Grandfathers—Fiction. 2. Runaways—Fiction] I. Davidson, Kevin. II. Title.
PZ7.B7803 Rag [Fic] 78-71859
00 MARC

Brisville, Jean Claude, 1922-
Oleg the snow leopard / story by Jean-Claude Brisville ; ill. by Danièle Bour ; translated by Anthea Bell. — London : V. Gollancz, 1978.
[24] p. : col. ill. ; 29 cm. GB***
Translation of Oleg le leopard des neiges.
ISBN 0-575-02557-3 : £2.95
I. Bour, Danièle. II. Title.
PZ7.B78054 Ol 78-326704
79 MARC

Brittain, Bill.
All the money in the world / by Bill Brittain ; ill. by Charles Robinson. — 1st ed. — New York : Harper & Row, c1979.
150 p. : ill. ; 21 cm.
SUMMARY: When Quentin gets his wish for all the money in the world, he gets a pack of troubles too.
ISBN 0-06-020675-6 : $7.95. ISBN 0-06-020676-4 lib. bdg. : $6.79
[1. Money—Fiction] I. Robinson, Charles, 1931- . II. Title.
PZ7.B78067 Al [Fic] 77-25635
78 MARC

Britton, Anna.
Fike's Point / Anna Britton. — 1st American ed. — New York : Coward, McCann & Geoghegan, 1979, c1977.
148 p. ; 22 cm.
SUMMARY: Catherine looks back on her counter-culture life with Mary and Jason, her parents, and Dimsy, her mentally slow brother, born a drug addict, and the events which left her with no family at the age of 14.
0698204743
[1. Family life—Fiction. 2. Hippies—Fiction. 3. Drug abuse—Fiction]
I. Title.
PZ7.B78073 Fi [Fic] 78-11281
78 MARC

Bröger, Achim.
Bruno's journey / [by] Achim Bröger and Gisela Kalow ; translated [from the German] by Caroline Gueritz. — London : Hamilton, 1978.
[28] p. : chiefly col. ill. ; 24 cm. GB78-13916
Translation of Bruno verreist.
American ed. published under title: Bruno takes a trip.
ISBN 0-241-89700-9 : £3.25
I. Kalow, Gisela, joint author. II. Title.
PZ7.B78618 Bv 79-366469
79 MARC

Bröger, Achim.
Little Harry / Achim Bröger ; translation from the German by Elizabeth D. Crawford ; illustrated by Judy Morgan. — 1st U.S. ed. — New York : Morrow, 1979.
189 p. : ill. ; 21 cm.
Translation of Moritzgeschichten.
SUMMARY: A small boy copes with the frustrating limitations of childhood by means of an active imagination which produces fanciful adventures.
ISBN 0-688-22185-8. ISBN 0-688-32185-2 lib. bdg.
[1. Imagination—Fiction. 2. Humorous stories] I. Morgan, Judy. II. Title.
PZ7.B78618 Li 1979 [Fic] 78-26028
78 MARC

Bröger, Achim.
Running in circles / Achim Bröger ; translated from the German by Patricia Crampton. — New York : Morrow, 1977.
224 p. ; 21 cm.
Translation of Kurzschluss.
SUMMARY: After being questioned about assaulting an old man, a fourteen-year-old boy runs away to avoid facing the police, his parents, and his conscience.
ISBN 0-688-22119-X. ISBN 0-688-32119-4 lib. bdg.
[1. Runaways—Fiction] I. Title.
PZ7.B78618 Ru 1977 [Fic] 77-22629
77[r79]rev MARC

Bromley, Dudley.
North to Oak Island / Dudley Bromley ; [cover and ill., David Gore]. — Chicago : Childrens Press, [1979] c1977.
59 p. : ill. ; 21 cm. — (Pacesetters)
SUMMARY: Robbed of all his money and desperate to get to the New World, Will becomes a stowaway on a pirate ship captained by a woman.
ISBN 0-516-02171-0
[1. Pirates—Fiction. 2. Buried treasure—Fiction] I. Gore, David. II. Title.
PZ7.B7873 No 1979 [Fic] 78-15472
78 MARC

Brook, Judy, 1926-
Tim Mouse visits the farm / written and illustrated by Judy Brook. — [Tadworth, Surrey : Windmill Press, 1973, c1968]
[40] p. : ill. (some col.) ; 19 x 23 cm.
"A World's Work children's book."
ISBN 0-437-29202-9 : £1.60
I. Title.
[PZ7.B7896 Tl 1973] 79-109454
00 MARC

Brooks, Jerome.
The big dipper marathon / by Jerome Brooks. — 1st ed. — New York : Dutton, c1979.
134 p. ; 22 cm.
SUMMARY: A fifteen-year-old's visit to relatives in Chicago gives him new insight into his struggle to decide the kind of life he will live as a victim of polio.
ISBN 0-525-26507-4
[1. Physically handicapped—Fiction. 2. Self-acceptance—Fiction. 3. Cousins—Fiction] I. Title.
PZ7.B7953 Bi 1979 [Fic] 78-12485
78 MARC

Brooks, Ron.
Timothy and Gramps / Ron Brooks. — 1st American ed. — Scarsdale, N.Y. : Bradbury Press, c1978.
[32] p. : col. ill. ; 26 cm.
SUMMARY: After his grandfather's visit to his classroom, Timothy enjoys school more.
ISBN 0-87888-139-5 lib. bdg. : $6.95
[1. School stories. 2. Grandfathers—Fiction] I. Title.
PZ7.B7986 Ti 1978 [E] 78-17389
78 MARC

Brown, Beverly Swerdlow.
The myth-adventures of Kraken the sea monster / text by Beverly Swerdlow Brown ; ill. by Carol Louise Koplan. — Ontario, Calif. : Brasch and Brasch, c1978.
[32] p. : col. ill. ; 26 cm.
SUMMARY: A sea monster, finding himself on shore, sets out to learn about his heritage in libraries and museums.
ISBN 0-89554-009-6
[1. Genealogy—Fiction. 2. Sea monsters—Fiction] I. Koplan, Carol Louise. II. Title.
PZ7.B81282 My [Fic] 78-72493
78 MARC

Brown, Bill, 1910-
The hippopotamus that wanted to be a baby. Written by Bill and Rosalie Brown. Illustrated by Haris Petie. New York, Lantern Press [1963]
1 v. (unpaged) illus. 24 cm.
I. Moore, Rosalie, 1910- joint author. II. Title.
PZ7.B81283 Hi 63-17986
[r78]rev MARC

Brown, Fern.
Jockey—or else! / Fern G. Brown ; illustrated by Darrell Wiskur. — Chicago : A. Whitman, c1978.
128 p. : ill. ; 21 cm. — (Pilot books)
SUMMARY: His summer experiences at Lakeside Horse Farm convince Benjy he will never be a jockey, therefore he sets more realistic goals that will still involve him with horses.
ISBN 0-8075-3944-9 lib. bdg. : $4.75
[1. Horses—Fiction] I. Wiskur, Darrell D. II. Title.
PZ7.B81356 Jo [Fic] 78-1715
78 MARC

Brown, Irene Bennett.
Skitterbrain / Irene Bennett Brown. — 1st ed. — Nashville : T. Nelson, c1978.
112 p. ; 21 cm.
SUMMARY: A young pioneer and her spunky companion follow a large cattle herd across the empty prairie in search of a runaway cow.
ISBN 0-8407-6587-8 : $6.95
[1. Frontier and pioneer life—Fiction. 2. Friendship—Fiction] I. Title.
PZ7.B81387 Sk 1978 [Fic] 78-18349
78 MARC

Brown, Jim.
The prime minister's pocket / story by Jim Brown ; drawings by Julie Zangmo. — Vancouver : Blue Mountain Books : [distributed by Intermedia Press], 1978.
56 p. : ill. ; 20 cm. C79-2564-7
ISBN 0-920584-06-3
I. Title.
PZ7.B81423 Pr 79-317211
79 MARC

Brown, Lindsay.
The treasure of Dubarry Castle / by Lindsay Brown ; with illustrations by Val Biro. — London : Hale, 1978.
156 p. : ill. ; 21 cm. GB78-38764
ISBN 0-7091-7164-1 : £2.95
I. Title.
PZ7.B81614 Tr 79-311754
00 MARC

Brown, Marc Tolon.
Arthur's eyes / by Marc Brown. — 1st ed. — Boston : Little, Brown, c1979.
27, [5] p. : col. ill. ; 26 cm.
"An Atlantic-Monthly Press book."
SUMMARY: His friends tease Arthur when he gets glasses, but he soon learns to wear them with pride.
ISBN 0-316-11063-9
[1. Eyeglasses—Fiction. 2. Animals—Fiction] I. Title.
PZ7.B81618 Ap [E] 79-11734
79 MARC

Brown, Marc Tolon.
The cloud over Clarence / Marc Brown. — 1st ed. — New York : Dutton, c1979.
[32] p. : col. ill. ; 21 cm. — (A Unicorn book)
SUMMARY: A special birthday present helps Clarence rid his days of accidents and forgetfulness.
ISBN (invalid) 0-525-30025-2
[1. Friendship—Fiction. 2. Birthdays—Fiction] I. Title.
PZ7.B81618 Cl 1979 [E] 79-4656
79 MARC

Brown, Marc Tolon.
Moose and goose / Marc Brown. — 1st ed. — New York : E. P. Dutton, c1978.
[32] p. : ill. ; 24 cm. — (A Unicorn book)
SUMMARY: Two neighbors resolve their differences about noise.
ISBN 0-525-35175-2
[1. Noise—Fiction. 2. Friendship—Fiction. 3. Geese—Fiction. 4. Moose—Fiction] I. Title.
PZ7.B81618 Mo 1978 [E] 78-6822
78 MARC

Brown, Palmer.
Hickory / by Palmer Brown. — 1st ed. — New York : Harper & Row, c1978.
42 p. : col. ill. ; 23 cm.
SUMMARY: An indoor mouse ventures into the outside world and finds true friendship.
ISBN 0-06-020887-2 : $6.95. ISBN 0-06-020888-0 lib. bdg. : $7.49
[1. Mice—Fiction. 2. Friendship—Fiction] I. Title.
PZ7.B816647 Hi [E] 77-11849
77 MARC

Brown, Pamela Beatrice, 1924-
Every day is market day / [by] Pamela Brown. — London : Hodder and Stoughton, 1977.
120 p. ; 22 cm. GB77-26606
ISBN 0-340-21777-4 : £2.95
I. Title.
PZ7.B81665 Ev 79-307191
79 MARC

CHILDREN'S LITERATURE (Continued)

Brown, Roy.
The cage : a novel / by Roy Brown. — 1st American ed. — New York : Seabury Press, 1977, c1976.
159 p. ; 22 cm.
"A Clarion book."
SUMMARY: A group of teenage boys plan an escape when they realize their imprisonment is part of an experiment to repattern their personalities.
ISBN 0-8164-3198-1 : $6.95
[1. Behavior modification—Fiction] I. Title.
PZ7.B81693 Cag 1977 [Fic] 77-1586
77 MARC

Brown, Roy.
The swing of the gate / by Roy Brown. — New York : Seabury Press, c1978.
108 p. ; 22 cm.
"A Clarion book."
SUMMARY: A young reporter investigating the murder of a call girl uncovers evidence that casts suspicion on his own brother.
ISBN 0-8164-3220-1 : $6.95
[1. Mystery and detective stories] I. Title.
PZ7.B81693 Sw [Fic] 78-6422
78 MARC

Browne, Anthony.
Through the magic mirror / Anthony Browne. — New York : Greenwillow Books, 1977, c1976.
[30] p. : col. ill. ; 22 x 27 cm.
SUMMARY: A young boy steps through his mirror into a world which looks the same but is slightly different.
ISBN 0-688-80064-5. ISBN 0-688-84064-7 lib. bdg.
[1. Fantasy] I. Title.
PZ7.B81984 Th 1977 [E] 76-13024
76 MARC

Browne, Caroline, 1945-
Mrs Christie's farmhouse / Caroline Browne. — London : E. Benn, 1977.
[32] p. : chiefly col. ill. ; 28 cm. GB77-19815
ISBN 0-510-09802-9 : £2.50
I. Title.
PZ7.B81994 Mi 1977 78-306231
78 MARC

Bruna, Dick.
Another story to tell / Dick Bruna. — New York : Methuen, 1978, c1974.
[28] p. : col. ill. ; 16 cm. — (Bruna books)
SUMMARY: A young child plays in newly fallen snow.
ISBN 0-458-92680-9 : $2.50
[1. Snow—Fiction. 2. Stories without words] I. Title.
PZ7.B828416 An 1978 [E] 77-17742
78 MARC

Bruna, Dick.
I can dress myself / Dick Bruna. — New York : Methuen, c1977.
[26] p. : col. ill. ; 17 cm.
SUMMARY: Describes the articles of clothing a brother and sister wear to a party.
ISBN 0-458-93270-1 : $2.50
[1. Clothing and dress—Fiction] I. Title.
PZ7.B828416 Iabm [E] 77-17739
78 MARC

Bruna, Dick.
Miffy in the hospital / Dick Bruna. — New York : Methuen, c1978.
[26] p. : ill. ; 16 cm.
SUMMARY: Relates Miffy's experiences in the hospital.
ISBN 0-458-92700-7 : $2.50
[1. Medical care—Fiction. 2. Hospitals—Fiction. 3. Rabbits—Fiction] I. Title.
PZ7.B828416 Mj [E] 77-17740
78 MARC

Brunhoff, Jean de, 1899–1937.
... Babar and Father Christmas; translated from the French by Merle Haas. New York, Random house, inc. [c1940]
[40] p. col. illus. 37 cm.
Illustrated t.-p. and lining-papers in colors.
Illustrations by the author.
SUMMARY: Babar searches for Father Christmas to ask him to distribute toys to the elephant children just as he does to the children of men.
[1. Elephants—Fiction. 2. Santa Claus—Fiction. 3. Christmas stories] I. Haas, Merle S., tr. II. Title.
PZ7.B82843Baac [E] 40-31350

Brunhoff, Laurent de, 1925-
Babar's mystery / Laurent de Brunhoff. — New York : Random House, c1978.
[29] p. : col. ill. ; 30 cm.
SUMMARY: While vacationing at Celesteville-on-the-Sea, the Babar family matches wits with a bold gang of thieves.
ISBN 0-394-83920-X : $3.95. ISBN 0-394-93920-4 lib. bdg. : $4.99
[1. Elephants—Fiction] I. Title.
PZ7.B82843 Bac [E] 78-55912
78 MARC

Buck, Josephine Darrell.
The turtles tears : a Bermuda fantasy / story by Josephine Darrell Buck ; illustrated by Rhona J. Campbell. — [s.l.: s.n.], c1977. [Hamilton] Bermuda : Island Press)
[16] p. : ill. ; 28 cm. — (Bermuda heritage series ; book 1)
I. Campbell, Rhona J. II. Title. III. Title: A Bermuda fantasy. IV. Series.
PZ7.B8784 Tu 78-107123
78 MARC

Budapest, Zsuzsanna, 1940-
Selene, the most famous bull-leaper on earth / written by Zsuzsanna Budapest ; illustrated by Carol Clement ; edited by Helen Beardwomon. — Baltimore : Diana Press, c1976.
50 p. : ill. ; 27 cm.
SUMMARY: A young acrobat named for the Moon Goddess achieves personal fulfillment by defeating fear and social pressure.
ISBN 0-88447-010-5
[1. Acrobats and acrobatism—Fiction] I. Clement, Carol, 1948- II. Title.
PZ7.B88238 Se [Fic] 76-48224
76 MARC

Budbill, David.
Bones on Black Spruce Mountain / David Budbill. — New York : Dial Press, c1978.
126 p. : maps ; 22 cm.
SUMMARY: Seth and Daniel's camping trip to a lonely mountaintop becomes a journey into a painful past that Daniel must confront.
ISBN 0-8037-0691-X : $6.95. ISBN 0-8037-0699-5 lib. bdg.
[1. Camping—Fiction] I. Title.
PZ7.B8824 Bo [Fic] 77-71520
77 MARC

Bulla, Clyde Robert.
Conquista! / Clyde Robert Bulla & Michael Syson ; illustrated by Ronald Himler. — 1st ed. — New York : Crowell, c1978.
viii, 33 p. : ill. ; 24 cm.
SUMMARY: At the time of Coronado's exploration for a fabled city of gold, a young Indian boy encounters his first horse.
ISBN 0-690-03890-4. ISBN 0-690-03871-2 lib. bdg.
[1. Indians of North America—Fiction. 2. America—Discovery and exploration—Fiction. 3. Horses—Fiction] I. Syson, Michael, joint author. II. Himler, Ronald. III. Title.
PZ7.B912 Co [Fic] 77-26585
78 MARC

Bunting, Anne Eve.
The big find / by Eve Bunting ; [photos by Richard Hutchings]. — Mankato, Mn. : Creative Education, c1978.
56 p. : ill. ; 24 cm.
SUMMARY: Two friends discover how different they really are after they become partners in crime.
ISBN 0-87191-681-9
[1. Friendship—Fiction. 2. Crime and criminals—Fiction] I. Hutchings, Richard. II. Title.
PZ7.B91527 Bk [Fic] 78-12636
78 MARC

Bunting, Anne Eve.
The big red barn / by Eve Bunting ; pictures by Howard Knotts. — 1st ed. — New York : Harcourt Brace Jovanovich, c1979.
31 p. : ill. ; 22 cm. — (A Let me read book)
SUMMARY: A youngster's sense of security is threatened by the arrival of a stepmother and the destruction of the old red barn on her family's farm.
ISBN 0-15-207145-8 : $4.95. ISBN 0-15-611938-2 pbk.
[1. Barns—Fiction. 2. Stepmothers—Fiction. 3. Farm life—Fiction] I. Knotts, Howard. II. Title.
PZ7.B91527 Bm [Fic] 78-12186
78 MARC

Bunting, Anne Eve.
The creature of Cranberry Cove / by Eve Bunting ; illustrated by Scott W. Earle. — St. Paul : EMC Corp., 1976.
39 p. : ill. ; 23 cm.
SUMMARY: While spending the night on an uninhabited island, a Boy Scout troop discovers a skeleton of a large creature whose mate is silently watching the troop's actions.
ISBN 0-88436-300-7. ISBN 0-88436-301-5 pbk.
[1. Monsters—Fiction] I. Earle, Scott W. II. Title.
PZ7.B91527 Cr [Fic] 76-18124
76 MARC

Bunting, Anne Eve.
The demon / by Eve Bunting ; illustrated by Scott W. Earle. — St. Paul : EMC Corp., 1976.
39 p. : ill. ; 23 cm.
SUMMARY: A young boy is determined to break the hold that a demon has on a group of children.
ISBN 0-88436-273-6. ISBN 0-88436-274-4 pbk.
[1. Devil—Fiction] I. Earle, Scott W. II. Title.
PZ7.B91527 Dg [Fic] 76-18125
76 MARC

Bunting, Anne Eve.
The ghost / by Eve Bunting ; illustrated by Scott W. Earle. — St. Paul, Minn. : EMC Corp., 1976.
37 p. : ill. ; 23 cm.
SUMMARY: A young girl is visited by a ghost who warns her of a theft and gives her some insight into her father's remarriage.
ISBN 0-88436-271-X. ISBN 0-88436-272-8 pbk.
[1. Ghost stories. 2. Remarriage—Fiction] I. Scott, Scott W. II. Title.
PZ7.B91527 Gh [Fic] 76-18131
76 MARC

Bunting, Anne Eve.
Maggie the freak / by Eve Bunting ; illustrated by Robert Gadbois. — Mankato, Minn. : Creative Education : Chicago : distributed by Childrens Press, c1978.
30 p. : ill. ; 23 cm.
SUMMARY: Maggie worries that there is no boy in her high school who won't be put off by her athletic ability or her interest in cars.
ISBN 0-87191-633-9 lib. bdg.
[1. Athletes—Fiction. 2. Sex role—Fiction] I. Gadbois, Robert. II. Title.
PZ7.B91527 Mag [Fic] 77-10682
77 MARC

Bunting, Anne Eve.
The mask / written by Eve Bunting ; illustrated by Don Hendricks. — Mankato, Minn. : Creative Education ; Chicago : distributed by Children's Press, c1978.
35 p. : ill. ; 23 cm. — (Creative science fiction)
SUMMARY: When Matt's father purchases an old Chinese mask, strange things begin to happen inside Matthew's head.
ISBN 0-87191-625-8 lib. bdg.
[1. Extrasensory perception—Fiction] I. Hendricks, Donald, 1932- II. Title.
PZ7.B91527 Mas [Fic] 78-5107
78 MARC

Bunting, Anne Eve.
Survival camp! / By Eve Bunting ; illustrated by Robert Gadbois. — Mankato, Minn. : Creative Education ; Chicago : distributed by Childrens Press, [1978]
31 p. : ill. ; 23 cm. — (Creative Education young romance books)
SUMMARY: Survival camp is not Mimi's idea of a vacation but she is determined to meet its challenges.
ISBN 0-87191-631-2
[1. Survival—Fiction. 2. Camping—Fiction] I. Gadbois, Robert. II. Title.
PZ7.B91527 Sv [Fic] 77-10681
77[r79]rev MARC

Bunting, Anne Eve.
The tongue of the ocean / by Eve Bunting ; illustrated by Scott W. Earle. — St. Paul, Minn. : EMC Corp., 1976.
35 p. : ill. ; 23 cm.
SUMMARY: Jo has never believed her grandfather's strange sea stories until the day they become stranded on a fishing boat in the Devil's Triangle.
ISBN 0-88436-302-3. ISBN 0-88436-303-1 pbk.
1. Bermuda Triangle—Fiction. [1. Supernatural—Fiction. 2. Sea stories] I. Earle, Scott W. II. Title.
PZ7.B91527 To [Fic] 76-17624
76 MARC

Burch, Robert, 1925-
Wilkin's ghost / Robert Burch ; illustrated by Lloyd Bloom. — 1st ed. — New York : Viking Press, 1978.
152 p. : ill. ; 22 cm.
SUMMARY: A boy living in rural Georgia in 1935 befriends an accused teenage thief he believes is innocent.
ISBN 0-670-76897-9 : $7.95
[1. Georgia—Fiction. 2. Friendship—Fiction] I. Bloom, Lloyd. II. Title.
PZ7.B91585 Wi 1978 [Fic] 78-6293
78 MARC

Burchard, Peter.
Chinwe / by Peter Burchard ; with drawings by the author. — New York : Putnam, c1979.
127 p. : ill. ; 22 cm.
SUMMARY: Captured by slave traders, a young Ibo girl, her brother and their fellow villagers suffer hardships and cruel treatment on their way to the New World where they are sold as slaves.
ISBN 0-399-20667-1 : $6.95
[1. Slave trade—Fiction. 2. Ibo tribe—Fiction] I. Title.
PZ7.B9159 Cf [Fic] 78-24401
78 MARC

Burman, Ben Lucien, 1895-
High treason at Catfish Bend / Ben Lucien Burman ; illustrated by Alic Caddy. — Harmondsworth, Eng. ; New York : Puffin Books, 1978, c1977.
147 p. : ill. ; 20 cm.
SUMMARY: The further adventures of the residents of Catfish Bend.
ISBN 0-14-031130-0 : $1.95
[1. Animals—Fiction. 2. Fantasy] I. Caddy, Alice. II. Title.
[PZ7.B9268 Hi 1978] [Fic] 78-59685
78 MARC

Burman, Ben Lucien, 1895-
The owl hoots twice at Catfish Bend / Ben Lucien Burman ; ill. by Alice Caddy. — New York : Puffin Books, 1974, c1961.
117 p. : ill. ; 20 cm.
SUMMARY: A city slicker fox joins the animal community at Catfish Bend and becomes dictator.
ISBN 0-14-030397-9 : $1.95
[1. Animals—Fiction] I. Caddy, Alice. II. Title.
[PZ7.B9268 Ow 1974] [Fic] 77-14015
77 MARC

Burningham, John.
Time to get out of the bath, Shirley / [by] John Burningham. — London : Cape, 1978.
[24] p. : chiefly col. ill. ; 20 x 28 cm. GB78-17060
ISBN 0-224-01372-6 : £2.50
I. Title.
PZ7.B936 Ti 79-302763
79 MARC

CHILDREN'S LITERATURE (Continued)

Burningham, John.
Trubloff; the mouse who wanted to play the balalaika. Written and illustrated by John Burningham. New York, Random House [1965, c1964]
[41] p. col. illus. 27 cm.
SUMMARY: A mouse who runs away with gypsies to learn to play the balalaika returns home just in time to save his family.
[1. Mice—Fiction. 2. Balalaika—Fiction] I. Title.
PZ7.B936 Tr 1965 [E] 65-18166
[r78]rev MARC

Burningham, John.
Would you rather / John Burningham. — London : J. Cape, 1978.
[32] p. : col. ill. ; 31 cm. GB***
ISBN 0-224-01635-0 : £3.50
I. Title.
PZ7.B936 Wo 1978 79-302782
79[r79]rev MARC

Burningham, John.
Would you rather ... / John Burningham. — New York : Crowell, c1978.
[32] p. : col. ill. ; 30 cm.
SUMMARY: Presents choices for the reader such as living with a gerbil in a cage or with a rabbit in a hutch and being chased by a crab, by a bull, by a lion, or by wolves.
ISBN 0-690-03917-4 : $8.95. ISBN 0-690-03918-2 lib. bdg. : $8.49
I. Title.
PZ7.B936 Wo [E] 78-7088
78 MARC

Burns, Thomas, fl. 1939-
Terrence O'Hara, by Thomas Burns; illustrations by Reginald Birch. New York, Harcourt, Brace and company [c1939]
3 p.l., 3-155 p. incl. front., illus., plates. 24 cm.
I. Birch, Reginald Bathurst, 1856- illus. II. Title.
PZ7.B9375 Te 39-27849
[r79]rev2 MARC

Burton, Marilee Robin.
The elephant's nest : four wordless stories / by Marilee Robin Burton. — 1st ed. — New York : Harper & Row, c1979.
48 p. : col. ill. ; 16 x 21 cm.
SUMMARY: Four wordless animal fantasies involving flying elephants, mice on the moon, a lion with too many visitors, and a kangaroo with a very full pouch.
ISBN 0-06-020905-4 : $6.95
[1. Animals—Fiction. 2. Stories without words] I. Title.
PZ7.B9534 El 1979 [E] 78-20263
78 MARC

Burton, Virginia Lee, 1909-1968.
The little house / story and pictures by Virginia Lee Burton. — Boston : Houghton Mifflin, [1978] c1942.
40 p. : col. ill. ; 23 x 25 cm. — (Sandpiper books)
SUMMARY: A country house is unhappy when the city, with all its buildings and traffic, grows up around her.
ISBN 0-395-18156-9 (reinforced). ISBN 0-395-25938-X (paperbound) : $2.25
[1. Dwellings—Fiction. 2. Cities and towns—Fiction] I. Title.
PZ7.B954 Li 1978 [E] 79-302678
79 MARC

Butterworth, William E
Under the influence / W. E. Butterworth. — New York : Four Winds Press, c1979.
247 p. ; 22 cm.
SUMMARY: As their friendship deepens, Allan realizes that Keith is an alcoholic whose behavior jeopardizes the well-being of himself and others.
ISBN 0-590-07465-2 : $7.95
[1. Alcoholics—Fiction. 2. Friendship—Fiction] I. Title.
PZ7.B9825 Un [Fic] 78-22127
78 MARC

Byars, Betsy Cromer.
The cartoonist / Betsy Byars. — London : Bodley Head, 1978.
111 p. ; 23 cm. GB***
ISBN 0-370-30104-8 : £2.75
I. Title.
PZ7.B9836 Car 1978b 78-316991
78 MARC

Byars, Betsy Cromer.
Good-bye, Chicken Little / by Betsy Byars. — 1st ed. — New York : Harper & Row, c1979.
101 p. ; 21 cm.
SUMMARY: A boy discovers that he doesn't have to feel personally responsible for his uncle's drowning.
ISBN 0-06-020907-0 : $6.95. ISBN 0-06-020911-9 lib. bdg. : $5.95.
[1. Death—Fiction. 2. Family life—Fiction] I. Title.
PZ7.B9836 Gp 1979 [Fic] 78-19829
78 MARC

Byers, Irene.
Tiptoes and the big race / Irene Byers ; illustrated by Lynette Hemmant. — London : Hodder and Stoughton, 1979.
95 p. : ill. ; 20 cm. GB***
ISBN 0-340-22570-X : £2.25
I. Title.
PZ7.B985 Ti 79-313514
79 MARC

Byfield, Barbara Ninde.
The haunted tower / written and illustrated by Barbara Ninde Byfield. — 1st ed. — Garden City, N.Y. : Doubleday, c1976.
[48] p. : col. ill. ; 16 cm.
SUMMARY: A retired spy and his ghost friend reluctantly become involved with the case of Crown Prince Brulph who disappears on the eve of his coronation.
ISBN 0-385-00450-8 : $5.95. ISBN 0-385-00460-5 lib. bdg.
[1. Mystery and detective stories. 2. Ghost stories] I. Title.
PZ7.B986 Haw [E] 75-11999
75[r78]rev2 MARC

Caldecott, Moyra.
Adventures by leaf light / by Moyra Caldecott ; illustrated by Lyn & Bill Teeple. — La Jolla, Calif. : Green Tiger Press, c1978.
46 p. : ill. (some col.) ; 30 cm.
SUMMARY: A little girl who spends much of her time in the garden soon begins to notice a whole other world peopled with tiny creatures and becomes involved in their activities.
"A Star & elephant book."
[1. Fairies—Fiction. 2. Gardens—Fiction] I. Teeple, Lyn. II. Teeple, Bill. III. Title.
PZ7.C125 Ad 79-105810
79 MARC

Calhoun, Mary.
Cross-country cat / by Mary Calhoun ; illustrated by Erick Ingraham. — 1st ed. — New York : Morrow, 1979.
[40] p. : col. ill. ; 26 cm.
SUMMARY: When he becomes lost in the mountains, a cat with the unusual ability of walking on two legs finds his way home on cross-country skis.
ISBN 0-688-22186-6. ISBN 0-688-32186-0 lib. bdg.
[1. Cats—Fiction. 2. Cross-country skiing—Fiction] I. Ingraham, Erick. II. Title.
PZ7.C1278 Cr [E] 78-31718
78 MARC

Calhoun, Mary.
Houn' dog. Illustrated by Roger Duvoisin. New York, Morrow, 1959.
[32] p. : ill. 26 cm. (Morrow junior books)
SUMMARY: Slyfoot the Fox promises he will ruin Houn' Dog's smeller and beller.
[1. Foxes—Fiction. 2. Dogs—Fiction] I. Duvoisin, Roger Antoine, 1904- . II. Title.
PZ7.C1278 Hs [E] 59-5136
[r79]rev MARC

Calhoun, Mary.
The witch of Hissing Hill. Pictures by Janet McCaffery. New York, W. Morrow, 1964.
[32] p. col. illus. 26 cm.
SUMMARY: After one of her black cats has a yellow kitten, a wicked old witch turns into a loving and good one.
[1. Witches—Fiction. 2. Cats—Fiction] I. McCaffery, Janet. II. Title.
PZ7.C1278 Wi [E] 64-15475
[r79]rev MARC

Calhoun, Mary.
The witch who lost her Shadow / by Mary Calhoun ; ill. by Trinka Hakes Noble. — 1st ed. — New York : Harper & Row, c1979.
30 p. : ill. (some col.) ; 18 x 24 cm.
SUMMARY: A witch is convinced that no cat could take the place of her lost black cat Shadow, until she is pursued by a small striped kitten.
ISBN 0-06-020946-1 : $7.95
[1. Cats—Fiction. 2. Witches—Fiction] I. Noble, Trinka Hakes. II. Title.
PZ7.C1278 Wl 1979 [E] 78-19489
78 MARC

Calkins, Erling, 1917-
Adventure at Beaver Falls / by Erling E. Calkins. — Nashville : Southern Pub. Association, 1979.
96 p. ; 21 cm. — (A Crown book)
SUMMARY: Kevin and Kim are introduced to the teachings of the Seventh-Day Adventists during their stay at their aunt and uncle's Wyoming farm.
ISBN 0-8127-0223-9
[1. Seventh-Day Adventists—Fiction. 2. Christian life—Fiction. 3. Nutrition—Fiction. 4. Farm life—Fiction] I. Title.
PZ7.C1289 Ad [Fic] 79-15601
79 MARC

Callen, Larry.
Sorrow's song / Larry Callen ; illustrated by Marvin Friedman. — 1st ed. — Boston : Little, Brown, c1979.
150 p. : ill. ; 22 cm.
"An Atlantic Monthly Press book."
SUMMARY: Sorrow Nix, a young girl who can't talk, befriends an injured whooping crane and in protecting it comes to better understand freedom and friendship.
ISBN 0-316-12497-4
[1. Mutism—Fiction. 2. Physically handicapped—Fiction. 3. Whooping cranes—Fiction] I. Friedman, Marvin. II. Title.
PZ7.C135 So [Fic] 78-31789
79 MARC

Calloway, Northern J
Northern J. Calloway presents Super-vroomer! / Written by Carol Hall ; illustrated by Sammis McLean ; conceived by Northern J. Calloway. — 1st ed. — Garden City, N.Y. : Doubleday, c1978.
[48] p. : col. ill. ; 29 cm.
SUMMARY: An energetic group of youngsters build what they believe is the winning entry for a go-cart race.
ISBN 0-385-14177-7. ISBN 0-385-14178-5 lib. bdg. : $5.95
[1. Soap box derbies—Fiction] I. Hall, Carol, 1936- II. McLean, Sammis. III. Title. IV. Title: Super vroomer!
PZ7.C135 No [E] 77-26512
78 MARC

Cameron, Eleanor, 1912-
Mr. Bass's planetoid. Illustrated by Louis Darling. [1st ed.] Boston, Little, Brown [1958]
227 p. illus. 20 cm.
I. Title.
PZ7.C143Mi 58-5174 ‡

Cameron, Eleanor, 1912-
A mystery for Mr. Bass. Illustrated by Leonard Shortall. [1st ed.] Boston, Little, Brown [1960]
229 p. illus. 20 cm. (Her A Mushroom Planet book)
I. Title.
PZ7.C143My 60-9344

Campbell, Hope, 1925-
The peak beneath the moon / by Hope Campbell. — New York : Four Winds Press, c1979.
133 p. ; 22 cm.
SUMMARY: An inquisitive 11-year-old tries to find out who is responsible for the strange unfinished tower standing near her house.
ISBN 0-590-07565-9 : $6.95
I. Title.
PZ7.C15414 Paf [Fic] 78-20427
78 MARC

Campbell, John, 1766-1840.
Worlds displayed; for the benefit of young people, by a familiar history of some of their inhabitants. Boston, Lincoln & Edmands, 1811.
70 p. illus. 14 cm.
I. Title.
PZ6.C152 Wo 78-320608
79 MARC

Canfield, Jane White.
Swan cove / by Jane White Canfield ; pictures by Jo Polseno. — 1st ed. — New York : Harper & Row, c1978.
25 p. : col. ill. ; 22 cm. — (An Early I can read book)
SUMMARY: A quiet cove on a small island is the summer home for a pair of swans and their offspring.
ISBN 0-06-020948-8 : $4.95. ISBN 0-06-020949-6 lib. bdg. :
1. Swans—Legends and stories. [1. Swans—Fiction] I. Polseno, Jo. II. Title.
PZ10.3.C164 Sw [E] 77-11832
77 MARC

Carey, M V
Alfred Hitchcock and the three investigators in The mystery of the magic circle / text by M. V. Carey ; based on characters created by Robert Arthur ; illustrated by Jack Hearne. — New York : Random House, c1978.
viii, 150 p. : ill. ; 22 cm. — (Alfred Hitchcock mystery series ; 27)
SUMMARY: Three young sleuths uncover a coven of witches when they search for the missing memoirs of a movie star.
ISBN 0-394-83607-3 : $2.95. ISBN 0-394-93607-8 lib. bdg. : $4.39
[1. Mystery and detective stories. 2. Witches—Fiction] I. Hearne, Jack. II. Title. III. Title: The mystery of the magic circle. IV. Series.
PZ7.C213 Aldfh [Fic] 78-55915
78 MARC

Cargas, Harry J
David's decision : betrayal? Or trust? / by Harry James Cargas. — St. Louis : Concordia Pub. House, c1979.
55 p. : ill. ; 23 cm.
SUMMARY: A nephew of Judas Iscariot determines to regain his family's honor after his uncle's suicide.
ISBN 0-570-07978-0 : $1.95
1. Jesus Christ—Juvenile fiction. [1. Jesus Christ—Fiction. 2. Christian life—Fiction] I. Title.
PZ7.C2144 Dav [Fic] 78-13561
78 MARC

Cargas, Harry J
Simon the crossbearer : a family is affected by their father's chance meeting with the Savior / by Harry James Cargas. — St. Louis : Concordia Pub. House, c1979.
56 p. : ill. ; 23 cm. — (Starlight books)
SUMMARY: Simon the Cyrenean who was forced to carry the cross of Christ begins to see his role as one of honor, not shame.
ISBN 0-570-07977-2 : $1.95
1. Simon of Cyrene—Juvenile fiction. [1. Simon of Cyrene—Fiction] I. Title.
PZ7.C2144 Si [Fic] 78-12731
78 MARC

Carle, Eric.
Watch out! A gaint! / Eric Carle. — New York : Collins-World, c1978.
[28] p. : col. ill. ; 32 cm.
SUMMARY: Two children are caught by a giant who plans to have them for dinner.
ISBN 0-529-05455-8 : $6.95. ISBN 0-529-05456-6 lib. bdg. : $6.91
[1. Giants—Fiction] I. Title.
PZ7.C21476 Wat [E] 78-8244
78 MARC

Carleton, Barbee Oliver.
The secret of Saturday Cove. Illustrated by Charles Geer. [1st ed.] New York, Holt, Rinehart and Winston [1961]
168 p. illus. 22 cm.
SUMMARY: When he faces competition from a newcomer to their Massachusetts town, twelve-year-old David Blake intensifies his own search for treasure long thought to have been hidden by his ancestors during the Revolutionary War.
[1. Buried treasure—Fiction. 2. Massachusetts—Fiction] I. Geer, Charles. II. Title.
PZ7.C2148 Se [Fic] 61-5052
[r78]rev MARC

CHILDREN'S LITERATURE (Continued)

Carlsen, G Robert, 1917- comp.
Insights : themes in literature / G. Robert Carlsen, Anthony Tovatt ; Patricia O. Tovatt. — 3d ed. — New York : Webster Division, McGraw-Hill, c1979.
x, 757 p. : ill. (some col.) ; 25 cm. — (Themes and writers series)
Includes indexes.
SUMMARY: A thematically arranged anthology of poems, short stories, plays, and novellas for the ninth-grade reader.
ISBN 0-07-009859-X
1. Children's literature. [1. Literature—Collections] I. Tovatt, Anthony, joint author. II. Tovatt, Patricia O., 1916- joint author. III. Title.
PZ5.C1828 In 1979 808.8 77 77-17301 MARC

Carlsen, G Robert, 1917- comp.
Perception : themes in literature / [compiled by] G. Robert Carlsen, Ruth Christoffer Carlsen. — 3d ed. — New York : Webster Division, McGraw-Hill, c1979.
x, 630 p. : ill. ; 25 cm. — (Themes and writers series)
Includes indexes.
SUMMARY: A thematically arranged anthology of poems, short stories, plays, and novellas for the eight-grade reader.
ISBN 0-07-009855-7
1. Children's literature. [1. Literature—Collections] I. Carlsen, Ruth Christoffer, joint comp. II. Title.
PZ5.C1828 Pe 1979 808.8 77 77-17316 MARC

Carlson, Dale Bick.
The shining pool / by Dale & Danny Carlson. — 1st ed. — New York : Atheneum, 1979.
138 p. ; 22 cm. — (An Argo book)
SUMMARY: Teenaged Ben is the only person to realize the dangerous meaning of the underwater light which has affected the minds of the other young people in his small town in Maine.
ISBN 0-689-30614-8
[1. Science fiction] I. Carlson, Danny, joint author. II. Title.
PZ7.C21645 Sh [Fic] 78 78-13636 MARC

Carlson, Natalie Savage.
Time for the white egret / by Natalie Savage Carlson. Illustrated by Charles Robinson. — New York : Scribner, c1978.
[32] p. : ill. ; 24 cm.
SUMMARY: A cattle egret, told that time will take care of his problem, goes in search of time.
ISBN 0-684-15990-2 : lib. bdg. : $7.95
[1. Time—Fiction. 2. Cattle egret—Fiction] I. Robinson, Charles, 1931-
PZ7.C2167 Ti [E] 78 78-17226 MARC

Carner, Chas.
Tawny / by Chas Carner ; ill. by Donald Carrick. — New York : Macmillan, c1978.
148 p. : ill. ; 22 cm.
SUMMARY: Twelve-year-old Trey Landry, adjusting to the death of his twin brother, adopts and cares for an injured doe.
ISBN 0-02-716700-3
[1. Death—Fiction. 2. Farm life—Fiction. 3. Deer—Fiction] I. Carrick, Donald. II. Title.
PZ7.C2177 Taw [Fic] 77 77-17411 MARC

Caroselli, Remus F
The mystery cottage in left field / by Remus F. Caroselli. — New York : Putnam, c1979.
140 p. ; 22 cm.
SUMMARY: In the spring of 1929, 12-year-old Jimmy makes friends with the new woman in his Providence, R.I., neighborhood and becomes involved with gangsters who have been searching for her son for a year.
ISBN 0-399-20672-8 : $7.95
[1. Gangs—Fiction] I. Title.
PZ7.C22 My 1979 [Fic] 78 78-24368 MARC

Carrick, Carol.
A rabbit for Easter / by Carol Carrick ; pictures by Donald Carrick. — 1st ed. — New York : Greenwillow Books, c1979.
[32] p. : col. ill. ; 26 cm.
SUMMARY: Paul is entrusted with the class rabbit over Easter Vacation.
ISBN 0-688-80195-1 : $6.95. ISBN 0-688-84195-3 lib. bdg. : $6.67
[1. Rabbits—Fiction. 2. Easter stories] I. Carrick, Donald. II. Title.
PZ7.C2344 Rab [E] 78 78-15647 MARC

Carrick, Carol.
The washout / by Carol Carrick ; pictures by Donald Carrick. — New York : Seabury Press, c1978.
[32] p. : col. ill. ; 23 x 24 cm.
"A Clarion book."
SUMMARY: When a summer storm washes out the road, Christopher decides to row around the lake for —helpand almost gets into serious trouble.
ISBN 0-18-643217-1 lib. bdg. : $6.95
[1. Storms—Fiction] I. Carrick, Donald. II. Title
PZ7.C2344 Was [E] 78 78-8135 MARC

Carrick, Malcolm.
I can squash elephants : a Masai tale about monsters / Malcolm Carrick. — 1st ed. — New York : Viking Press, 1978.
[32] p. : ill. ; 19 x 24 cm.
SUMMARY: A caterpillar in a cave has all the animals outside terrorized by his echoing voice.
ISBN 0-670-38983-8 : $6.95
[1. Animals—Fiction] I. Title.
PZ7.C23452 Iae 1978 [E] 77 77-3455 MARC

Carrick, Malcolm.
"I'll get you!" / By Malcolm Carrick. — 1st ed. — New York : Harper & Row, c1979.
188 p. : ill. ; 21 cm.
SUMMARY: a young boy growing up in post-war London tries to fit in with his neighborhood friends.
ISBN 0-06-021123-7 : $7.95. ISBN 0-06-021124-5 lib. bdg. : $6.79
[1. City and town life—Fiction. 2. England—Fiction] I. Title.
PZ7.C23452 Il 1979 [Fic] 78 78-19490 MARC

Carter, Katharine (Jones) 1905-
Hoppy Long Legs. Illustrated by Elizabeth Rice. Austin, Tex., Steck Co. [1963]
40 p. illus. 24 cm.
SUMMARY: A frog is disappointed with the sound of his song until he discovers that other animals listen for it.
[1. Frogs—Fiction] I. Rice, Elizabeth. II. Title.
PZ7.C246 Ho [E] [r79]rev 63-13047 MARC

Carter, Katharine Jones, 1905-
Willie Waddle. Illustrated by Elizabeth Rice. Austin, Tex., Steck Co. [1959]
40 p. col. illus. 24 cm.
SUMMARY: The barnyard chicks tease a little duck for having big feet, but a day comes when they are grateful that he has them.
[1. Ducks—Fiction] I. Rice, Elizabeth. II. Title.
PZ7.C246 Wi [E] [r79]rev 59-5607 MARC

Carter, Margaret, fl. 1978-
The two-sided dream / Margaret Carter ; illustrated by Cynthia Pickard. — London : J. M. Dent, 1978.
122 p. : ill. ; 23 cm. GB***
ISBN 0-460-06856-3 : £3.50
I. Title.
PZ7.C2464 Tw 78 78-321025 MARC

Casey, Winifred Rosen, 1943-
Cruisin for a bruisin / Winifred Rosen [i.e. W. R. Casey] — New York : Knopf, c1976.
150 p. ; 22 cm.
SUMMARY: The thirteen-year-old daughter of a psychoanalyst describes her growing sexual awareness and her intense relationships with boys and her family.
ISBN 0-394-83291-4
I. Title.
PZ7.C26815 Cr 3 [Fic] 76[r79]rev 76-5488 MARC

Casper, the friendly ghost and Wendy, the good little witch storybook. — [New York] : Grosset & Dunlap, c1978.
[80] p. : col. ill. ; 24 cm.
SUMMARY: Follows the adventures of a small ghost who would rather be friendly than scary and a young witch who refuses to play nasty tricks.
ISBN 0-448-14743-2
[1. Ghost stories. 2. Witches—Fiction]
PZ7.C2684 [E] 78 77-94036 MARC

Castiglia, Julie.
Jill the pill / by Julie Castiglia ; illustrated by Steven Kellogg. — 1st ed. — New York : Atheneum, 1979.
[30] p. : ill. ; 23 cm.
"A Margaret K. McElderry book."
SUMMARY: Patrick finds having a bossy older sister difficult.
ISBN 0-689-50105-6
[1. Brothers and sisters—Fiction] I. Kellogg, Steven. II. Title.
PZ7.C268724 Ji [E] 79 79-12305 MARC

Catling, Patrick Skene.
The Chocolate touch / Patrick Skene Catling ; pictures by Margot Apple. — Newly illustrated ed. — New York : Morrow, 1979, c1952.
126 p. : ill. ; 21 cm.
SUMMARY: A boy acquires a magical gift that turns everything his lips touch into chocolate.
ISBN 0-688-22187-4. ISBN 0-688-32187-9 lib. bdg.
[1. Chocolate—Fiction] I. Apple, Margot. II. Title.
PZ7.C2696 Ch 1979 [Fic] 78 78-31100 MARC

Catton, Bruce, 1899-
Banners at Shenandoah : a story of Sheridan's Fighting Cavalry / Bruce Catton. — Forest Hills, N.Y. : Queens House, 1977, c1955.
254 p. : ill. ; 23 cm.
Reprint of the 1st ed. published by Doubleday, Garden City, N.Y.
SUMMARY: Enlisting in the Union Army, a seventeen-year-old from Michigan ends up in the cavalry under "Fighting Phil" Sheridan headed for Virginia.
ISBN 0-89244-019-8
1. Sheridan, Philip Henry, 1831-1888—Juvenile fiction. [1. Sheridan, Philip Henry, 1831-1888—Fiction. 2. United States—History—Civil War, 1861-1865—Fiction] I. Title.
[PZ7.C272 Ban 5] [Fic] 76 76-40255 MARC

Cauley, Lorinda Bryan.
The animal kids / Lorinda Bryan Cauley. — New York : Putnam, c1979.
[32] p. : chiefly col. ill. ; 23 x 28 cm.
SUMMARY: The party crashers at the animals' party, feeling well-disguised, have a wonderful time.
ISBN 0-399-20677-9 : $8.95
[1. Parties—Fiction. 2. Animals—Fiction] I. Title.
PZ7.C2746 An 1979 [E] 78 78-23632 MARC

Cavanna, Betty, 1909-
Stars in her eyes. New York, Morrow, 1958.
256 p. 21 cm.
SUMMARY: The teenage daughter of a famous entertainer knows she has talent but lacks the professional training and poise necessary for a career of her own.
[1. Performing arts—Fiction. 2. Fathers and daughters—Fiction. 3. Identity—Fiction] I. Title.
PZ7.C286 St [Fic] [r79]rev2 58-9931 MARC

Cazzola, Gus, 1934-
A chisel for Ezekiel / Gus Cazzola. — St. Louis : Concordia Pub. House, c1979.
56 p. : ill. ; 23 cm.
SUMMARY: An apprentice is sent to find Jesus the carpenter to work in his master's shop, but instead, finds Jesus the Savior.
ISBN 0-570-07975-6 : $1.95
1. Jesus Christ—Juvenile literature. [1. Jesus Christ—Fiction] I. Title.
[PZ7.C299 Ch] [Fic] 78 78-12717 MARC

Cebulash, Mel.
The champion's jacket / by Mel Cebulash ; [illustrated by Edward Abrams]. — Mankato, Minn. : Creative Education, [c1978]
54 p. : ill. ; 24 cm.
SUMMARY: A young player examines his schoolyard friendships when he is offered a spot on an organized basketball team.
ISBN 0-87191-687-8
[1. Basketball—Fiction. 2. Friendship—Fiction] I. Title.
PZ7.C2997 Ch [Fic] 79 78-71858 MARC

Chadwick, Roxane.
Don't shoot / Roxane Chadwick ; ill. by Edwin H. Ryan. — Minneapolis : Lerner Publications Co., c1978.
38 p. : ill. ; 23 cm.
SUMMARY: An Eskimo boy determines to devote his life to protecting Arctic polar bears from poachers.
ISBN 0-8225-0706-4 $4.95
[1. Wildlife conservation—Fiction. 2. Eskimos—Fiction. 3. Alaska—Fiction] I. Ryan, Edwin. II. Title.
PZ7.C3472 Do [Fic] 78 78-6101 MARC

Chaikin, Miriam.
I should worry, I should care / by Miriam Chaikin ; drawings by Richard Egielski. — 1st ed. — New York : Harper & Row, c1979.
103 p. : ill. ; 23 cm.
SUMMARY: A young Jewish girl and her family adjust to a new neighborhood and new friends at a time when the radio is telling of Hitler's rise to power in Europe.
ISBN 0-06-021174-1 : $7.95
[1. Jews in the United States—Fiction. 2. Brooklyn—Fiction. 3. Family life—Fiction] I. Egielski, Richard. II. Title.
PZ7.C3487 Ias [Fic] 78 78-19480 MARC

Chambers, Aidan.
Breaktime / Aidan Chambers. — London : Bodley Head, 1978.
138 p. : ill. ; 23 cm. — (A Book for new adults) GB***
ISBN 0-370-30122-6 : £3.25
I. Title.
PZ7.C3557 Br 79 78-325060 MARC

Chambers, Aidan.
Breaktime / Aidan Chambers. — 1st American ed. — New York : Harper & Row, c1978.
180 p. : ill. ; 22 cm.
SUMMARY: A classmate's challenge to prove that literature is related to real life spurs Ditto to leave home during spring vacation and chronicle his life-changing adventures.
ISBN 0-06-021256-X : $7.95. ISBN 0-06-021257-8 lib. bdg.
I. Title.
PZ7.C3557 Br 1978 [Fic] 78 78-19472 MARC

Chaney, Jill.
Canary yellow / [by] Jill Chaney. — London : Dobson, 1977.
144 p. ; 21 cm. GB78-25591
ISBN 0-234-72076-X : £3.50
I. Title.
PZ7.C35966 Can 79 78-325020 MARC

Chapman, Carol.
Ig lives in a cave / by Carol Chapman ; pictures by Bruce Degen. — 1st ed. — New York : Dutton, c1979.
56 p. : ill. ; 24 cm. — (A Fat cat book)
SUMMARY: Though a cave boy, Ig has much in common with little boys of today.
ISBN 0-525-32534-4 : $5.95
[1. Cave dwellers—Fiction. 2. Man, Prehistoric—Fiction. 3. Parent and child—Fiction] I. Degen, Bruce. II. Title. III. Series.
PZ7.C36636 Ig [E] 78 78-12831 MARC

CHILDREN'S LITERATURE (Continued)

Chapman, Elizabeth, 1919-
Marmaduke goes to Morocco / Elizabeth Chapman ; illustrated by Douglas Hall. — London : Hodder and Stoughton, 1979.
107 p. : ill. ; 20 cm. GB***
ISBN 0-340-23173-4 : £2.25
I. Hall, Douglas, 1931- . II. Title.
PZ7.C3666 Map 1979 79-310976
79 MARC

Chapman, Jean.
Supermarket Thursday : six stories / by Jean Chapman ; illustrated by Astra Lācis. — West Melbourne, Vic. : Thomas Nelson (Australia), 1977.
[36] p. : ill. (part col.) ; 19 x 26 cm. Aus
CONTENTS: Supermarket Thursday.—TV special.—The climbing day.—All through the night.—Wet blanket.—Jungle picnic.
ISBN 0-17-005065-3
I. Lācis, Astra. II. Title.
PZ7.C369 Su 78-310761
78 MARC

Charity : or, Nettie's victories : a story for the young / by H. de W. R. — New York : Gen. Prot. Episc. S.S. Union and Church Book Society, 1866.
108 p., [2] leaves of plates : ill. ; 18 cm.
I. H. de W. R. II. R., H. de W.
PZ6.C38 813'.4 78-318176
78 MARC

Charles, Donald.
Fat, fat Calico Cat / written and illustrated by Donald Charles. — Chicago : Childrens Press, c1977.
32 p. : col. ill. ; 25 cm.
SUMMARY: After suffering from eating the wrong kinds of foods, Calico Cat discovers the benefits of a proper diet.
ISBN 0-516-03456-1
[1. Nutrition—Fiction] I. Title.
PZ7.C374 Fat [E] 77-7154
77 MARC

Charlton, Elizabeth, 1937-
Jeremy and the ghost / by Elizabeth Charlton ; illustrated by Celia Reisman. — New York : Dandelion Press, 1979.
[32] p. : col. ill. ; 18 x 23 cm.
SUMMARY: One Halloween Jeremy finds a way to overcome his nickname, Fraidy-Cat.
ISBN 0-89799-118-4
[1. Fear—Fiction. 2. Halloween—Fiction] I. Reisman, Celia. II. Title.
PZ7.C38138 Je [E] 78-72120
79 MARC

Chase, Catherine.
Baby Mouse learns his ABC's / by Catherine Chase ; illustrated by Jill Elgin. — New York : Dandelion Press, c1979.
[32] p. : col. ill. ; 23 cm.
SUMMARY: Baby Mouse learns the alphabet by mastering one letter each day.
ISBN 0-89799-089-7. ISBN 0-89799-001-3 pbk. : $1.50
[1. Alphabet] I. Elgin, Jill. II. Title.
PZ7.C3855 Bab [E] 78-72096
79 MARC

Chase, Catherine.
Duncan McTavish in Switzerland / by Catherine Chase ; illustrated by Robin Moffett. — New York : Dandelion Press, 1978.
[32] p. : col. ill. ; 23 cm.
SUMMARY: While lost in the Swiss Alps, a West Highland white terrier helps a Saint Bernard rescue a group of skiers trapped by an avalanche.
ISBN 0-89799-101-X
[1. Dogs—Fiction. 2. Switzerland—Fiction. 3. Rescue work—Fiction] I. Moffett, Robin. II. Title.
PZ7.C3855 Du 78-64488
79 MARC

Chase, Catherine.
Feet / by Catherine Chase ; illustrated by Susan Heiss. — New York : Dandelion Press, c1979.
[32] p. : col. ill. ; 23 cm. — (A Dandelion first reader)
SUMMARY: Describes the different parts of the foot and all the things one can do with one's feet.
ISBN 0-89799-104-4. ISBN 0-89799-020-X pbk. : $1.50
[1. Foot—Fiction] I. Reiss, Susan. II. Title.
PZ7.C3855 Fe [E] 78-72105
79 MARC

Chase, Catherine.
The mouse in my house / by Catherine Chase ; illustrated by Gail Gibbons. — New York : Dandelion Press, 1979.
[32] p. : col. ill. ; 18 x 23 cm. — (A Dandelion first reader)
SUMMARY: A little girl misses the mouse her mother forces to leave the house, but one day the door is left open.
ISBN 0-89799-126-5. ISBN 0-89799-024-2 pbk. : $1.50
[1. Mice—Fiction] I. Gibbons, Gail. II. Title.
PZ7.C3855 Mo [E] 78-72103
79 MARC

Chase, Catherine.
My balloon / by Catherine Chase ; illustrated by Gail Gibbons. — New York : Dandelion Press, 1979.
[32] p. : col. ill. ; 23 cm. — (A Dandelion first reader)
SUMMARY: A child explores the possibilities in blowing up a balloon.
ISBN 0-89799-127-3. ISBN 0-89799-022-6 pbk. : $1.50
[1. Balloons—Fiction] I. Gibbons, Gail. II. Title.
PZ7.C3855 My [E] 78-72102
79 MARC

Cheatham, Karyn Follis, 1943-
Life on a cool plastic ice floe / by K. Follis Cheatham. — 1st ed. — Philadelphia : Westminster Press, c1978.
180 p. ; 21 cm.
SUMMARY: During a time of growing Indian activism, a young orphan desperately tries to prevent a white family from adopting his sister.
ISBN 0-664-32632-3
[1. Indians of North America—Fiction. 2. Orphans—Fiction] I. Title.
PZ7.C399 Li [Fic] 78-6430
78 MARC

Cheney, Cora.
The mystery of the disappearing cars. Illustrated by Dick Pfahl. New York, Knopf [1964]
146 p. illus. 22 cm.
SUMMARY: Their college scholarships are jeopardized when Windy and Sam are suspected of stealing an antique car which disappeared near the Vermont hotel where they have summer jobs.
[1. Automobiles—Fiction. 2. Vermont—Fiction. 3. Mystery and detective stories] I. Pfahl, Dick. II. Title.
PZ7.C4187 My [Fic] 63-14605
[r79]rev MARC

Chenfeld, Mimi Brodsky.
The house at 12 Rose Street, by Mimi Brodsky. Illustrated by David Hodges. London, New York, Abelard-Schuman [1966]
157 p. illus. 22 cm.
SUMMARY: When a black family moves into a white neighborhood, their twelve-year-old neighbor learns about mob violence, property values, and the importance of scouting honor.
[1. City and town life—Fiction. 2. Race relations—Fiction] I. Hodges, David, illus. II. Title.
PZ7.C4183 Ho [Fic] 67-1918
[r79]rev MARC

Chess, Victoria.
Alfred's alphabet walk / Victoria Chess. — 1st ed. — New York : Greenwillow Books, c1979.
[32] p. : col. ill. ; 26 cm.
SUMMARY: As he encounters animals from A to Z, Alfred learns the letters of the alphabet.
ISBN 0-688-80223-0 : $7.95
[1. Alphabet. 2. Animals—Fiction] I. Title.
PZ7.C42522 Al [E] 79-1185
79 MARC

Chetin, Helen, 1922-
The lady of the strawberries / Helen Chetin ; illustrated by Anita Kunz. — Toronto : PMA Books, 1978.
89 p. : ill. ; 23 cm. C78-1516-9
ISBN 0-88778-183-7 : $7.95
I. Title.
PZ7.C4255 Lad 79-302932
79 MARC

Chew, Ruth.
The magic cave / by Ruth Chew ; illustrated by the author. — New York : Hastings House Publishers, 1978, c1973.
128 p. : ill. ; 22 cm.
Previous ed. published under title: The hidden cave.
SUMMARY: When lightning splits an old oak tree, a brother and sister discover Merlin who has been sleeping inside the tree for many centuries.
ISBN 0-8038-4711-4 : $5.95
[1. Magic—Fiction] I. Title.
PZ7.C429 Mag 1978 [Fic] 78-12972
78 MARC

Chew, Ruth.
The witch's garden / by Ruth Chew ; illustrated by the author. — New York : Hastings House, 1979, c1978.
112 p. : ill. ; 22 cm.
SUMMARY: Susan and Josh discover some surprising things about the garden next door and the little old woman who owns it.
ISBN 0-8038-8093-6 : $5.95
[1. Magic—Fiction. 2. Witches—Fiction] I. Title.
PZ7.C429 Wi 1979 [Fic] 78-26939
78 MARC

Chorao, Kay.
Molly's lies / story and pictures by Kay Chorao. — New York : Seabury Press, c1979.
32 p. : ill. ; 27 cm.
"A Clarion book."
SUMMARY: Challenged by a classmate on the first day of school, Molly stops making up stories and admits her fears.
ISBN 0-8164-3225-2 (lib. bdg.) : $7.50
[1. Honesty—Fiction. 2. Fear—Fiction] I. Title.
PZ7.C4463 Ml [E] 78-12383
78 MARC

Christian, Mary Blount.
The Goosehill Gang and the chocolate cake caper / by Mary Blount Christian ; illustrated by Betty Wind. — St. Louis : Concordia Pub. House, c1976.
32 p. : ill. ; 23 cm.
SUMMARY: Four friends look for a cake thief and a way to help a lonely old lady feel useful.
ISBN 0-570-03606-2 : $0.95
[1. Mystery and detective stories] I. Title.
PZ7.C4528 Gm [E] 76-3638
76 MARC

Christian, Mary Blount.
The Goosehill Gang and the disappearing dues / by Mary Blount Christian ; illustrated by Betty Wind. — St. Louis : Concordia Pub. House, c1976.
32 p. : ill. ; 23 cm.
SUMMARY: The Goosehill Gang suspect the new boy when their club dues disappear, but when they go to his house to confront him they see something sobering.
ISBN 0-570-03607-0 : $0.95
[1. Mystery and detective stories] I. Wind, Betty. II. Title.
PZ7.C4528 Go [E] 76-3634
76 MARC

Christian, Mary Blount.
The Goosehill Gang and the May basket mystery / by Mary Blount Christian ; illustrated by Betty Wind. — St. Louis : Concordia Pub. House, c1978.
32 p. : ill. ; 23 cm.
SUMMARY: Pete decides that the family next door are "weirdos" because they never speak, but learns along with the other members of his gang that the family is not weird but handicapped.
ISBN 0-570-07359-6 : $0.95
[1. Deaf—Fiction. 2. Mutism—Fiction. 3. Physically handicapped—Fiction. 4. Christian life—Fiction] I. Wind, Betty. II. Title.
[PZ7.C4528 Gox] [Fic] 78-2443
78 MARC

Christian, Mary Blount.
The Goosehill Gang and the shadow on the shade / by Mary Blount Christian ; illustrated by Betty Wind. — St. Louis : Concordia, c1978.
32 p. : ill. ; 23 cm.
SUMMARY: The Goosehill Gang helps find a job for a recently paroled convict.
ISBN 0-570-07358-8 : $0.95
[1. Ex-convicts—Fiction. 2. Parole—Fiction] I. Wind, Betty. II. Title.
PZ7.C4528 Gpo [Fic] 78-1291
78 MARC

Christian, Mary Blount.
The Goosehill gang and the stitch-in-time solution / by Mary Blount Christian ; illustrated by Betty Wind. — St. Louis : Concordia Pub. House, c1978.
32 p. : ill. ; 23 cm.
SUMMARY: The "gang" helps Marcus' mother, an invalid, to be useful by involving her talents in a Goosehill "clean-up campaign."
ISBN 0-570-07357-X : $0.95
[1. Physically handicapped—Fiction] I. Wind, Betty. II. Title.
[PZ7.C4528 Gpr] [Fic] 78-702
78 MARC

Christian, Mary Blount.
The Goosehill Gang and the test paper thief / by Mary Blount Christian ; illustrated by Betty Wind. — St. Louis : Concordia Pub. House, c1976.
32 p. : ill. ; 23 cm.
SUMMARY: Tubby is unjustly accused of stealing the answers to a test at school. He knows who really did it and doesn't want to tell.
ISBN 0-570-03608-9 : $0.95
[1. School stories] I. Wind, Betty. II. Title.
PZ7.C4528 Gq [E] 76-3639
76 MARC

Christian, Mary Blount.
The Goosehill Gang and the vanishing sandwich / by Mary Blount Christian ; illustrated by Betty Wind. — St. Louis : Concordia Pub. House, c1976.
32 p. : ill. ; 23 cm.
SUMMARY: When a classmate appears overly tired and carries an empty lunchkit, Beth and the Goosehill Gang take her in hand.
ISBN 0-570-03609-7 : $0.95
[1. School stories] I. Wind, Betty. II. Title.
PZ7.C4528 Gs [E] 76-4809
76 MARC

Christian, Mary Blount.
The lucky man / by Mary Blount Christian ; illustrated by Glen Rounds. — New York : Macmillan, c1979.
63 p. : ill. ; 23 cm. — (Ready-to-read)
SUMMARY: The cumulative catastrophes of hard-luck Felix land him in court where a wise judge's decisions prove him a lucky man.
ISBN 0-02-718270-3
[1. Justice—Fiction] I. Rounds, Glen, 1906- II. Title.
PZ7.C4528 Lu [E] 79-11024
79 MARC

Christian, Mary Blount.
The sand lot / by Mary Blount Christian ; illustrated by Dennis Kendrick. — New York : Harvey House, c1978.
[28] p. : ill. (some col.) ; 20 x 25 cm.
SUMMARY: Baseball suddenly isn't any fun when grownups take over and organize the children's sand lot games.
ISBN 0-8178-5827-X
[1. Baseball—Fiction] I. Kendrick, Dennis. II. Title.
PZ7.C4528 San [E] 77-15894
78 MARC

Christopher, John.
Wild Jack / John Christopher. — New York : Collier Books, 1978, c1974.
147 p. ; 18 cm.
SUMMARY: In twenty-third-century England, young Clive Anderson is imprisoned in an attempt to mold him into a docile member of society but escapes to the Outlands where he is befriended by Wild Jack's outlaw band.
ISBN 0-02-042410-8
[1. Science fiction]
[PZ7.C457 Wi 1978] [Fic] 77-16437
77 MARC

Christopher, Matthew F
Basketball sparkplug. Illustrted by Ken Wagner. [1st ed.] Boston, Little, Brown [1957]
124 p. illus. 20 cm.
SUMMARY: Kim's teammates tease him about singing in the church choir, but they change their tune when the choir helps the Arrows become Small Fry Basketball Champions.
[1. Basketball—Fiction] I. Wagner, Xen. II. Title.
PZ7.C458 Bask [Fic] 57-8038
[r79]rev MARC

CHILDREN'S LITERATURE (Continued)

Christopher, Matthew F
Dirt bike racer / Matt Christopher ; illustrated by Barry Bomzer. — 1st ed. — Boston : Little, Brown, c1979.
149 p. : ill. ; 21 cm.
SUMMARY: Twelve-year-old Ron Baker finds a minibike while scuba diving and, with the help of a former motorcycle rider and racer, restores the bike and enters competition.
ISBN 0-316-13977-7 : $6.95
[1. Minibikes—Fiction. 2. Motorcycle racing—Fiction] I. Bomzer, Barry. II. Title.
PZ7.C458 Dk Fic 79-745
79 MARC

Christopher, Matthew F
The fox steals home / by Matt Christopher ; illustrated by Larry Johnson. — 1st ed. — Boston : Little, Brown, c1978.
178 p. : ill. ; 19 cm.
SUMMARY: Already troubled by his parent's divorce, Bobby Canfield is further distressed when he learns that his father, who has coached him in running bases, intends to move away.
ISBN 0-316-13976-9
[1. Baseball—Fiction. 2. Divorce—Fiction] I. Johnson, Larry, 1949- . II. Title.
PZ7.C458 Fp [Fic] 78-17526
78 MARC

Christopher, Matthew F
Jackrabbit goalie / Matt Christopher ; illustrated by Ed Parker. — 1st ed. — Boston : Little, Brown, c1978.
45 p. : col. ill. ; 22 cm.
SUMMARY: Anxious to make friends in a new town, a young boy lies about his experience as goalie in order to get on the local soccer team.
ISBN 0-316-13975-0
[1. Soccer—Fiction] I. Parker, Edward. II. Title.
PZ7.C458 Jac [Fic] 78-5438
78 MARC

Christopher, Matthew F
Sink it, Rusty. Illus. by Foster Caddell. [1st ed.] Boston, Little, Brown [1963]
138 p. illus. 20 cm.
SUMMARY: Forced to referee rather than play basketball due to the after effects of polio, Rusty believes he will never play again until a new man in town organizes a team and coaches him.
[1. Physically handicapped—Fiction. 2. Basketball—Fiction] I. Caddell, Foster. II. Title.
PZ7.C458 Si [Fic] 63-13457
[r78]rev MARC

Christopher, Matthew F
The team that couldn't lose, by Matt Christopher. Illustrated by Foster Caddell. [1st ed.] Boston, Little, Brown, [1967]
133 p. illus. 20 cm.
SUMMARY: A young, inexperienced football team discovers that its beginner's luck is due to a series of mysterious but successful plays anonymously sent to the coach.
[1. Football—Fiction] I. Caddell, Foster. II. Title.
PZ7.C458 Te [Fic] 67-17550
[r78]rev MARC

Christopher, Matthew F
The twenty-one-mile swim / by Matt Christopher. — 1st ed. — Boston : Little, Brown, c1979.
146 p. ; 21 cm.
SUMMARY: With remarks about his small stature and poor swimming skills ringing in his ears, the son of Hungarian immigrants begins to train for the 21-mile swim across a nearby lake.
ISBN 0-316-13979-3
[1. Swimming—Fiction. 2. Identity—Fiction. 3. Hungarian Americans—Fiction] I. Title.
PZ7.C458 Tv [Fic] 79-15197
79 MARC

Churchill, David.
It, us, and the others / David Churchill. — 1st American ed. — New York : Harper & Row, c1978.
119 p. ; 21 cm.
SUMMARY: Two teenagers, one a handicapped girl and the other a boy in an orphanage, find a special closeness together when they discover a helpless alien being.
ISBN 0-06-021268-3 : $5.95
[1. Science fiction] I. Title.
PZ7.C47226 It 1978 [Fic] 78-19827
78 MARC

Clark, Ann Nolan, 1898-
Secret of the Andes / by Ann Nolan Clark ; with drawings by Jean Charlot. — [New York] : Puffin Books, 1976, c1952.
120 p. : ill. ; 20 cm. — (A Puffin Book)
SUMMARY: An Indian boy who tends llamas in a hidden valley in Peru learns the traditions and secrets of his Inca ancestors.
ISBN 0-14-030926-8 : $1.25
[1. Incas—Fiction. 2. Peru—Fiction] I. Charlot, Jean, 1898- II. Title.
[PZ7.C5296 Se 6] [Fic] 76-39987
76 MARC

Clark, Ann Nolan, 1898-
To stand against the wind / by Ann Nolan Clark. — 1st ed. — New York : Viking Press, 1978.
136 p. ; 22 cm.
Bibliography: p. 134-136.
SUMMARY: A young refugee now living in the United States ponders his memories of life in a Vietnamese village before and during the war.
ISBN 0-670-71773-8 : $7.95
[1. Vietnamese Conflict, 1961-1975—Fiction. 2. Vietnam—Fiction] I. Title.
PZ7.C5296 To 1978 [Fic] 78-5966
78 MARC

Clark, Elizabeth.
Country tales to tell / Elizabeth Clark ; illustrated by Cara Lockhart Smith. — London : Hodder and Stoughton, 1978.
94 p. : ill. ; 22 cm. GB***
CONTENTS: The spindle, the shuttle, and the needle.—The tale of the old iron pot.—The tale of Mrs. Honeycritch and the white rabbit.—Father Sparrow's tug-of-war.— The tale of the harvest bun.— The tale of brave Augustus.—The tale of Mr. and Mrs. Peppercorn and their cuckoo-clock.—Pussy, will you have a sausage?
ISBN 0-340-22512-2 : £2.95
I. Title.
PZ7.C5427 Co 79-303232
79 MARC

Clark, Margaret Goff.
Barney and the UFO / Margaret Goff Clark ; illustrated by Ted Lewin. — New York : Dodd, Mead, c1979.
159 p. : ill. ; 21 cm.
SUMMARY: Barney is afraid to tell his foster parents that he has seen a UFO behind the house even when a hasty promise to a space boy leads him into trouble.
ISBN 0-396-07711-0
[1. Unidentified flying objects—Fiction. 2. Science fiction. 3. Foster home care—Fiction] I. Lewin, Ted. II. Title.
PZ7.C5487 Bar [Fic] 79-52046
79 MARC

Clarke, Pauline, 1921-
The two faces of Silenus. London, Faber and Faber [1972]
162 p. 21 cm. £1.50 B***
ISBN 0-571-09929-7
I. Title.
PZ7.C555 Tw 73-152882
73[r79]rev MARC

Cleary, Beverly.
Ellen Tebbits; illustrated by Louis Darling. New York, Morrow, 1951.
160 p. illus. 21 cm. (Morrow junior books)
SUMMARY: The part of third grade Ellen Tebbits likes most is having Austine Allen for her best friend.
[1. Friendship—Fiction. 2. School stories] I. Darling, Louis. II. Title.
PZ7.C5792 El [Fic] 51-11430
[r79]rev MARC

Cleary, Beverly.
Ramona and her mother / Beverly Cleary ; illustrated by Alan Tiegreen. — New York : Morrow, 1979.
207 p. : ill. ; 21 cm.
SUMMARY: Ramona at 7 1/2 sometimes feels discriminated against by being the youngest in the family.
ISBN 0-688-22195-5
[1. Family life—Fiction. 2. Mothers and daughters—Fiction] I. Tiegreen, Alan. II. Title.
PZ7.C5792 Raif [Fic] 79-10323
79 MARC

Clement, Hal, 1922-
Left of Africa / by Hal Clement ; [ill. & maps by Dany Frolich]. — 1st ed. — New Orleans : Aurian Society Press : distributed by PDA Enterprises, c1976.
160 p., [1] leaf of plates : ill. ; 29 cm.
SUMMARY: A young Phoenician slave becomes involved in the Egyptians' first attempt to circumnavigate Africa.
[1. Africa—Fiction. 2. Explorers—Fiction. 3. Sea stories] I. Frolich, Dany. II. Title.
PZ7.C59018 Le [Fic] 78-107359
78 MARC

Cleveland, David.
The April rabbits / by David Cleveland ; pictures by Nurit Karlin. — New York : Coward, McCann & Geoghegan, c1978.
[32] p. : ill. ; 21 x 23 cm.
SUMMARY: As Robert goes about his daily activities during April, he encounters an ever-increasing number of rabbits.
ISBN 0-698-20463-8
[1. Rabbits—Fiction] I. Karlin, Nurit. II. Title.
PZ7.C5986 Ap [Fic] 78-2044
78 MARC

Clewes, Dorothy.
Missing from home / Dorothy Clewes. — 1st American ed. — New York : Harcourt Brace Jovanovich, 1978, c1975.
150 p. ; 21 cm.
SUMMARY: Two English children living in France devise a scheme to bring their separated parents back together.
ISBN 0-15-254882-3
[1. Family problems—Fiction. 2. Divorce—Fiction. 3. France—Fiction] I. Title.
PZ7.C622 Mi 1978 [Fic] 78-52826
78 MARC

Clewes, Dorothy.
The testing year / [by] Dorothy Clewes, — London : Heinemann, 1977.
[4], 122 p. ; 23 cm. GB78-01212
ISBN 0-434-93288-4 : £3.10
I. Title.
PZ7.C622 Te 78-307726
78 MARC

Clifford, Eth, 1915-
Help! I'm a prisoner in the library / Eth Clifford ; ill. by George Hughes. — Boston : Houghton Mifflin, 1979.
105 p. : ill. ; 22 cm.
SUMMARY: Two girls spend an adventurous night trapped inside the public library during a terrible blizzard.
ISBN 0-395-28478-3
[1. Libraries—Fiction. 2. Blizzards—Fiction] I. Title.
PZ7.C62214 He [Fic] 79-14447
79 MARC

Clifford, Eth, 1915-
The rocking chair rebellion / Eth Clifford. — Boston : Houghton Mifflin, 1978.
147 p. ; 22 cm.
SUMMARY: A teenager chronicles her involvement with the residents of Maple Ridge Home for the Aged and their revolution.
ISBN 0-395-27163-0 : $6.95
[1. Old age—Fiction] I. Title.
PZ7.C62214 Ro [Fic] 78-14834
78 MARC

Clifford, Lucy Lane, d. 1929.
Anyhow stories, moral and otherwise and Wooden Tony, and Anyhow story : reprinted from The last touches / Lucy Lane Clifford ; with a preface for the Garland ed. by Alison Lurie. — New York : Garland Pub., 1977.
x, 146 [23] p., [11] leaves of plates : ill. ; 19 cm. — (Classics of children's literature, 1621-1932)
Reprint of Anyhow stories, moral and otherwise, published in 1882 by Macmillan, London, and of Wooden Tony, published in 1892 by A. & C. Black, London.
Bibliography: p. ix-x.
SUMMARY: A collection of unusual tales with strong psychoanalytic overtones.
ISBN 0-8240-2298-X : $27.00
[1. Conduct of life—Fiction. 2. Short stories] I. Clifford, Lucy Lane, d. 1929. Wooden Tony. 1976. II. Anyhow stories, moral and otherwise. III. Series.
PZ7.C62215 An 4 [Fic] 75-32186
77[r78]rev MARC

Clifford, Sandy.
The smartest person in the world / written and illustrated by Sandy Clifford. — Oakland, Cal. : Parnassus Press, 1979.
[47] p. : ill. ; 17 cm.
SUMMARY: Cleveland knows the answer to every question, even before one's been asked, so his sister Amy devises a plan to keep him from bragging.
ISBN 0-395-28411-2
[1. Questions and answers—Fiction. 2. Brothers and sisters—Fiction] I. Title.
PZ7.C62226 Sm 1979 [E] 79-14138
79 MARC

Clymer, Eleanor Lowenton, 1906-
Horatio goes to the country / Eleanor Clymer ; illustrated by Robert Quackenbush. — 1st ed. — New York : Atheneum, 1978.
63 p. : col. ill. ; 17 cm.
SUMMARY: Everything about the farm visit was awful, until Horatio the cat explored the beauties of a meadow at night.
ISBN 0-689-30649-0 lib. bdg. : $6.95
[1. Cats—Fiction. 2. Country life—Fiction] I. Quackenbush, Robert M. II. Title.
PZ7.C6272 Hk 1978 [E] 78-5137
78 MARC

Coates, Joseph K
The gentle dragon / by Joseph K. Coates. — La Jolla, Calif. : Lane, c1979.
329 p. ; 22 cm.
SUMMARY: Relates the adventures of a very gentle dragon who lived a thousand years ago in the high places of the land now called Japan.
ISBN 0-89882-001-4 : $4.95
[1. Dragons—Fiction] I. Title.
PZ7.C6292 Ge [Fic] 79-84574
79 MARC

Cohen, Barbara.
Bitter herbs and honey / by Barbara Cohen. — New York : Lothrop, Lee & Shepard, c1976.
159 p. ; 22 cm.
SUMMARY: During the early 1900's, a young Jewish girl's loyality to family tradition conflicts with her desire to go to college.
ISBN 0-688-41772-8. ISBN 0-688-51772-2 lib. bdg.
[1. Jews in the United States—Fiction. 2. United States—Social life and customs—1865-1918—Fiction] I. Title.
PZ7.C6595 Bi [Fic] 76-18132
76[r78]rev MARC

Cohen, Dan, 1936-
The case of the battling ball clubs / by Dan Cohen ; pictures by George Overlie. — Minneapolis : Carolrhoda Books, c1979.
[32] p. : col. ill. ; 16 cm. — (Carolrhoda mini-mysteries)
SUMMARY: Before a big game between two rival baseball teams, the manager of one is accused of stealing an autographed picture of Babe Ruth, owned by the other manager.
ISBN 0-87614-101-7
[1. Mystery and detective stories] I. Overlie, George. II. Title.
PZ7.C6598 Cas [Fic] 79-84356
79 MARC

Cohen, Dan, 1936-
The case of the long-losttwin / by Dan Cohen ; pictures by George Overlie. — Minneapolis : Carolrhoda Books, c1979.
[32] p. : col. ill. ; 16 cm. — (The Carolrhoda mini-mysteries)
SUMMARY: When Miss Tromly dies, a woman posing as her twin sister threatens to cheat Miss Tromly's housekeeper out of her inheritance.
ISBN 0-87614-094-0
[1. Mystery and detective stories. 2. Inheritance and succession—Fiction. 3. Twins—Fiction] I. Overlie, George. II. Title.
PZ7.C6598 Cat Fic 79-84357
00 MARC

Cohen, Dan, 1936-
The case of the missing poodle / by Dan Cohen ; pictures by George Overlie. — Minneapolis : Carolrhoda Books, c1979.
[30] p. : ill. (some col.) ; 16 cm. — (Carolrhoda mini-mysteries)
SUMMARY: A poodle disappears from a dog show and sisters Ruthann and Polly help Officer Greenwood find the dognapper.
ISBN 0-87614-095-9
[1. Mystery and detective stories] I. Overlie, George. II. Title.
PZ7.C6599 Cas [Fic] 79-84358
79 MARC

CHILDREN'S LITERATURE (Continued)

Cohen, Dan, 1936-
The case of the runaway rabbit / by Dan Cohen ; pictures by George Overlie. — Minneapolis : Carolrhoda Books, c1979.
[32] p. : col. ill. ; 16 cm. — (Carolrhoda mini-mysteries)
SUMMARY: The search for the biology teacher's missing rabbit reveals a surprising culprit.
ISBN 0-87614-096-7
[1. Mystery and detective stories] I. Overlie, George. II. Title.
PZ7.C6598 Ru [Fic] 79-84359
79 MARC

Cohen, Dan, 1936-
The mystery of the locked door / by Dan Cohen ; pictures by George Overlie. — Minneapolis : Carolrhoda books, c1979.
[32] p. : col. ill. ; 16 cm. — (The Carolrhoda mini-mysteries)
SUMMARY: Two sisters help Officer Greenwood locate a coin stolen from a valuable collection.
ISBN 0-87614-099-1
[1. Mystery and detective stories. 2. Coins—Collectors and collecting—Fiction] I. Overlie, George. II. Title.
PZ7.C6598 Myl [Fic] 79-50786
79 MARC

Cohen, Dan, 1936-
The mystery of the marked money / by Dan Cohen ; pictures by George Overlie. — Minneapolis : Carolrhoda Books, c1979.
[32] p. : col. ill. ; 16 cm. — (Carolrhoda mini-mysteries)
SUMMARY: Two young girls help solve a case of fraud involving marked bills.
ISBN 0-87614-100-9
[1. Mystery and detective stories] I. Overlie, George. II. Title.
PZ7.C6598 Mym Fic 79-50788
00 MARC

Cohen, Dan, 1936-
The mystery of the faded footprint / by Dan Cohen ; pictures by George Overlie. — Minneapolis : Carolrhoda Books, c1979.
[30] p. : ill. (some col.) ; 16 cm. — (Carolrhoda mini-mysteries)
SUMMARY: Two sisters assist their police officer friend in following clues to find who stole a wallet.
ISBN 0-87614-097-5
[1. Mystery and detective stories] I. Overlie, George. II. Title.
PZ7.C6599 My [Fic] 79-84360
79 MARC

Cohen, Miriam.
Lost in the museum / story by Miriam Cohen ; pictures by Lillian Hoban. — 1st ed. — New York : Greenwillow Books, c1979.
[32] p. : col. ill. ; 19 x 21 cm.
SUMMARY: When he and some other first graders get lost in the museum, Jim decides to be brave and go find the teacher.
ISBN 0-688-80187-0 : $7.25. ISBN 0-688-84187-2 lib. bdg. : $6.96
[1. Museums—Fiction] I. Hoban, Lillian. II. Title.
PZ7.C6628 Lo [E] 78-16765
78 MARC

Cohen, Peter Zachary.
Deadly game at Stony Creek / Peter Zachary Cohen ; pictures by Michael J. Deas. — New York : Dial Press, c1978.
107 p. : ill. ; 22 cm.
SUMMARY: A pack of wild killer dogs is hunted down by two teen-age boys.
ISBN 0-8037-1816-0 : $6.95. ISBN 0-8037-1817-9 lib. bdg. : $6.46
[1. Dogs—Fiction. 2. Hunting—Fiction] I. Deas, Michael. II. Title.
PZ7.C663 De [Fic] 78-51772
78 MARC

Coker, Gylbert.
Naptime / Gylbert Coker. — New York : Delacorte Press, c1978.
[32] p. : chiefly ill. ; 16 x 24 cm.
SUMMARY: Seven high-spirited youngsters and their patient teacher prepare for naptime but not without a little horseplay.
ISBN 0-440-06303-5. ISBN 0-440-06304-3 lib. bdg.
[1. Sleep—Fiction. 2. Nursery schools—Fiction] I. Title.
PZ7.C669 Nap [E] 78-50415
78 MARC

Cole, Babette.
Nungu and the Hippopotamus / written and illustrated by Babette Cole. — Johannesburg : Purnell, 1978.
[26] p. : col. ill. ; 24 cm. SA***
ISBN 0-360-00475-X
I. Title.
PZ7.C6734 Nu 1978 79-310789
79 MARC

Cole, Babette.
Nungu and the hippopotamus / written and illustrated by Babette Cole. — New York : McGraw-Hill, 1979.
[28] p. : ill. (some col.) ; 24 cm.
SUMMARY: A young African boy living in a remote village where water is scarce goes on a journey to make a river reappear.
ISBN 0-07-011695-4
[1. Africa—Fiction. 2. Hippopotamus—Fiction] I. Title.
PZ7.C6734 Nu 1979 [E] 78-12382
78 MARC

Cole, Brock.
The king at the door : words and pictures / by Brock Cole. — 1st ed. — Garden City, N.Y. : Doubleday, c1979.
[32] p. : ill. (some col.) ; 22 cm.
SUMMARY: A beggar arrives at the inn declaring he is a king but only manages to convince Little Baggit, the servant.
ISBN 0-385-14718-X : $7.69
[1. Kings, queens, rulers, etc.—Fiction] I. Title.
PZ7.C67342 Ki [E] 78-20064
79 MARC

Collins, David R
If I could, I would / by David R. Collins ; drawings by Kelly Oechsli. — Champaign, Ill. : Garrard Pub. Co., c1979.
46 p. : col. ill. ; 23 cm. — (An Imagination book)
SUMMARY: A child wishes he could arrange to ride behind magical horses, watch monster films, and eat mountains of ice cream with his mother.
ISBN 0-8116-4417-0
[1. Mothers and sons—Fiction] I. Oechsli, Kelly. II. Title.
PZ7.C6955 If [E] 78-27430
78 MARC

Collins, David R
A spirit of giving / written by David R. Collins ; illustrated by Susan Hall. — Nashville : Broadman Press, c1978.
32 p. : col. ill. ; 24 cm.
SUMMARY: After she talks to one of the judges, Maribeth feels differently about her entry in the school Christmas project contest.
ISBN 0-8054-4238-3
[1. Christmas stories] I. Hall, Susan. II. Title.
PZ7.C6955 Sp [E] 78-108429
79 MARC

Collins, Meghan.
Maiden crown / Meghan Collins. — Boston : Houghton Mifflin, 1979.
230 p. : map ; 22 cm.
SUMMARY: Sixteen-year-old Sophie, having left her Russian homeland to marry the Danish king, struggles to preserve a sense of her own value against the court intrigue and double standards of 12th-century Denmark.
ISBN 0-395-28639-5
1. Valdemar I, king of Denmark, 1131-1182—Juvenile fiction. [1. Valdemar I, king of Denmark, 1131-1182—Fiction. 2. Denmark—History—To 1241—Fiction] I. Title.
PZ7.C6967 Mai [Fic] 79-16201
79 MARC

Colman, Hila.
Rachel's legacy / Hila Colman. — 1st ed. — New York : Morrow, 1978.
190 p. ; 21 cm.
SUMMARY: Seven years after her mother dies, 17-year-old Ellie Levine learns her mother left a will, never executed, which could relieve the poverty existence of Ellie and her father during the Depression.
ISBN 0-688-22154-8. ISBN 0-688-32154-2 lib. bdg.
[1. Jews in the United States—Fiction. 2. Inheritance and succession—Fiction. 3. Family life—Fiction] I. Title.
PZ7.C7 Rac [Fic] 78-12783
78 MARC

Colwell, Max.
Full days and pressed pants / [by] Max Colwell ; illustrated by Aubrey Collette. — Adelaide : Rigby, 1978.
160 p. : ill. ; 23 cm. Aus
ISBN 0-7270-0608-8
I. Title.
PZ7.C728 Fu 78-323938
78 MARC

Cone, Molly.
Mishmash and the substitute teacher. Illustrated by Leonard Shortall. Boston, Houghton Mifflin, 1963.
92 p. illus. 22 cm.
SUMMARY: A precocious dog, the school prankster, and a substitute teacher help young Pete learn from his mistakes.
[1. Dogs—Fiction. 2. School stories] I. Shortall, Leonard W., illus. II. Title.
PZ7.C7592 Mk [Fic] 63-18441
[r78]rev MARC

Cone, Molly.
Only Jane. Drawings by Velma Ilsley. New York, T. Nelson [1960]
160 p. illus. 22 cm.
SUMMARY: A 15-year-old overcomes her shyness and attends her first dance during a memorable summer.
I. Ilsley, Velma. II. Title.
PZ7.C7592 On [Fic] 60-7288
[r79]rev MARC

Cone, Molly.
The real dream. Illustrated by Bea Holmes. Boston, Houghton Mifflin, 1964.
139 p. illus. 22 cm.
SUMMARY: A fifteen-year-old girl is certain she has finally met the boy of her dreams.
I. Holmes, Bea, illus. II. Title.
PZ7.C7592 Rc [Fic] 64-12277
[r78]rev MARC

Conford, Ellen.
Anything for a friend / Ellen Conford. — 1st ed. — Boston : Little, Brown, c1979.
180 p. ; 21 cm.
SUMMARY: Having moved frequently in her young life, Wallis is once again the new girl at school and must cope with the difficulties of making new friends.
ISBN 0-316-15308-7
[1. Moving, Household—Fiction. 2. Friendship—Fiction. 3. School stories] I. Title.
PZ7.C7593 Ap [Fic] 78-27843
78 MARC

Conford, Ellen.
Hail, hail, Camp Timberwood / by Ellen Conford ; illustrated by Gail Owens. — 1st ed. — Boston : Little, Brown, c1978.
151 p. : ill. ; 21 cm.
SUMMARY: Relates the experiences of 13-year-old Melanie as she tries to copes with the traumas and pleasures of her first year at summer camp.
ISBN 0-316-15291-9
[1. Camping—Fiction] I. Owens, Gail. II. Title.
PZ7.C7593 Hai [Fic] 78-18715
78 MARC

Conford, Ellen.
We interrupt this semester for an important bulletin / by Ellen Conford. — 1st ed. — Boston : Little, Brown, c1979.
176 p. ; 21 cm.
SUMMARY: Hoping to impress Chip, editor of the high school paper, Carrie takes up investigative reporting and soon finds herself in hot water.
ISBN 0-316-15309-5
[1. Reporters and reporting—Fiction. 2. Journalism—Fiction] I. Title.
PZ7.C7593 We [Fic] 79-9133
79 MARC

Constant, Alberta Wilson.
Does anybody care about Lou Emma Miller? / By Alberta Wilson Constant. — 1st ed. — New York : T.Y. Crowell, c1979.
278 p. ; 21 cm.
SUMMARY: A 15-year-old makes several discoveries about human nature while helping the local suffragettes elect the first woman mayor of Gloriosa, Kansas.
ISBN 0-690-01335-3 : $6.95
[1. Women—Suffrage—Fiction. 2. Kansas—Fiction] I. Title.
PZ7.C766 Do 1979 [Fic] 78-4774
78 MARC

Coombs, Patricia.
Dorrie and the Screebit ghost / by Patricia Coombs. — 1st ed. — New York : Lothrop, Lee & Shepard, c1979.
[48] p. : ill. (some col.) ; 26 cm.
SUMMARY: A little witch conjures up a lively ghost that interferes with Big Witch's séance.
ISBN 0-688-41883-X
[1. Witches—Fiction. 2. Ghost stories] I. Title.
PZ7.C7813 Dn [Fic] 79-4443
79 MARC

Cooney, Caroline B
Safe as the grave / Caroline B. Cooney ; illustrated by Gail Owens. — New York : Coward, McCann & Geoghegan, c1979.
48 p. : ill. ; 22 cm.
SUMMARY: A young girl inadvertently discovers clues leading to a long-lost treasure.
ISBN 0-698-20479-4
[1. Buried treasure—Fiction] I. Owens, Gail. II. Title.
PZ7.C7834 Saf 1979 [Fic] 78-24412
78 MARC

Coontz, Otto.
The quiet house / by Otto Coontz. — 1st ed. — Boston : Little, Brown, c1978.
39 p. : ill. ; 22 cm.
SUMMARY: A lonely dog finds three new friends in a surprising place.
ISBN 0-316-15533-0
[1. Dogs—Fiction. 2. Friendship—Fiction] I. Title.
PZ7.C7845 Qi [E] 78-18347
78[r79]rev MARC

Corbett, Scott.
The discontented ghost / by Scott Corbett. — 1st ed. — New York : Dutton, c1978.
180 p. ; 22 cm. — (A Unicorn book)
SUMMARY: The resident ghost of an English manor house vows to rid the property of its new American owner. A retelling of Oscar Wilde's story "The Canterville Ghost" from the ghost's point of view.
ISBN 0-525-28775-2 : $7.95
[1. Ghost stories] I. Title.
PZ7.C79938 Dk 1978 [Fic] 78-18013
78 MARC

Corbett, Scott.
The donkey planet / by Scott Corbett ; illustrated by Troy Howell. — 1st ed. — New York : Dutton, c1979.
89 p. : ill. ; 22 cm. — (A Unicorn book)
SUMMARY: Two young scientists in disguise carry out a dangerous mission on another planet.
ISBN 0-525-28825-2
[1. Science fiction. 2. Donkeys—Fiction] I. Howell, Troy. II. Title.
PZ7.C79938 Dp 1979 [Fic] 78-11455
78 MARC

Corbett, Scott.
The foolish dinosaur fiasco / by Scott Corbett ; illustrated by Jon McIntosh. — 1st ed. — Boston : Little, Brown, c1978.
46 p. : ill. ; 22 cm.
"An Atlantic Monthly Press book."
SUMMARY: Nick and his dog inadvertently go down Dr. Merlin's magic tunnel and find themselves in Dinosaur Land where Nick is forced to be a dinosaur trainer.
ISBN 0-316-15657-4
[1. Magic—Fiction. 2. Dinosaurs—Fiction] I. McIntosh, Jon. II. Title.
PZ7.C79938 Fo [Fic] 77-16424
77 MARC

Corbett, Scott.
The mysterious Zetabet / Scott Corbett ; illustrated by Jon McIntosh. — 1st ed. — Boston : Little, Brown, c1979.
47 p. : col. ill. ; 22 cm.
"An Atlantic Monthly Press book."
SUMMARY: To find his way out of Zyxland, Zachery Zwicker must meet the challenges of the Zetabet.
ISBN 0-316-15730-9
[1. Alphabet. 2. Humorous stories] I. McIntosh, Jon. II. Title.
PZ7.C79938 Mx [Fic] 78-23243
78 MARC

Corcoran, Barbara.
"Me and you and a dog named Blue" / by Barbara Corcoran. — 1st ed. — New York : Atheneum, 1979.
179 p. ; 22 cm.
SUMMARY: Although she loves her disabled father, a young girl longs to get away from home and is drawn to the idea of going to a baseball training camp for women.
ISBN 0-689-30675-X
I. Title.
PZ7.C814 Md [Fic] 78-13746
78 MARC

CHILDREN'S LITERATURE (Continued)

Cordell, Alexander.
Sea-urchin / ₍by₎ Alexander Cordell. — London : Collins, 1979.
112 p. : map ; 22 cm. GB79-09106
ISBN 0-00-184785-6 : £3.95
I. Title.
PZ7.C81534 Se 79-314412
79 MARC

Coren, Alan, 1938-
Arthur's last stand / Alan Coren ; illustrated by John Astrop. — 1st American ed. — Boston : Little, Brown, c1977.
74 p. : col. ill. ; 22 cm.
SUMMARY: Ten-year-old Arthur employs an unusual technique to save the garrison at Fort Moccasin from an Indian attack.
ISBN 0-316-15742-2
₍1. The West—Fiction. 2. Indians of North America—Fiction₎ I. Astrop, John. II. Title.
PZ7.C81538 As 1977 ₍Fic₎ 79-14052
79 MARC

Coren, Alan, 1938-
Klondike Arthur / Alan Coren ; illustrated by John Astrop. — 1st American ed. — Boston : Little, Brown, ₍1979₎ c1977.
74 p. : ill. ; 22 cm.
SUMMARY: It takes a ten-year-old bard to reform Grizzly Wilkinson, the meanest prospector in the Klondike gold fields.
ISBN 0-316-15733-3 lib. bdg. : $6.95
₍1. Klondike gold fields—Fiction. 2. Gold mines and mining—Fiction. 3. Friendship—Fiction₎ I. Astrop, John. II. Title.
PZ7.C81538 Kl 1979 ₍Fic₎ 78-23176
78 MARC

Corey, Dorothy.
Pepe's private Christmas / written by Dorothy Corey ; pictures by John Wallner. — New York : Parent's Magazine Press, c1978.
₍32₎ p. : col. ill. ; 21 cm.
SUMMARY: Pepe prepares for a visit from Santa Claus—a month in advance.
ISBN 0-8193-0966-4. ISBN 0-8193-0967-2
₍1. Santa Claus—Fiction₎ I. Wallner, John C. II. Title.
PZ7.C815395 Pe ₍E₎ 78-6208
78 MARC

Cormier, Robert.
After the first death / Robert Cormier. — New York : Pantheon Books, c1979.
233 p. ; 22 cm.
SUMMARY: Events of the hijacking of a bus of children by terrorists seeking the return of their homeland are described from the perspectives of a hostage, a terrorist, an Army general involved in the rescue operation, and his son, chosen as the go-between.
ISBN 0-394-84122-0
₍1. Terrorism—Fiction. 2. Fathers and sons—Fiction₎ I. Title.
PZ7.C81634 Af ₍Fic₎ 78-11770
78 MARC

Cornelius, Carol, 1942-
Isabella Wooly Bear Tiger Moth / by Carol Cornelius ; illustrated by Franz Altschuler. — Elgin, Ill. : Child's World ; Chicago : distributed by Childrens Press, c1978.
32 p. : col. ill. ; 25 cm. — (A Concept book)
SUMMARY: A wooly little caterpillar metamorphosizes into a beautiful tiger moth.
ISBN 0-913778-97-4 : $4.95
₍1. Metamorphosis—Fiction₎ I. Altschuler, Franz. II. Title.
PZ10.3.C8157 Is ₍E₎ 77-28423
78 MARC

Cosgrove, Stephen.
Catundra / written by Stephen Cosgrove ; illustrated by Robin James. — Mankato, Minn. : Creative Education, c1978.
₍32₎ p. : col. ill. ; 23 cm.
"A Serendipity book."
SUMMARY: A fat cat loses weight with the help of a friendly mole.
ISBN 0-87191-692-4
₍1. Cats—Fiction. 2. Weight control—Fiction. 3. Friendship—Fiction₎ I. James, Robin. II. Title.
PZ7.C8187 Cat ₍Fic₎ 78-64879
79 MARC

Cosgrove, Stephen.
Hucklebug / written by Stephen Cosgrove ; illustrated by Robin James. — Mankato, Minn. : Creative Education, ₍1978₎ c1975.
₍32₎ p. : col. ill. ; 23 cm. — (A Serendipity book)
SUMMARY: Young Hucklebug runs away from his village rather than help gather food, but soon regrets his decision.
ISBN 0-87191-657-6 (lib. bdg.)
₍1. Community life—Fiction₎ I. James, Robin. II. Title.
PZ7.C8187 Hu 1978 ₍E₎ 78-11453
78 MARC

Cowan, Janice.
The mystery of Castle Hotel : a novel for children / by Janice Cowan. — Ottawa : Borealis Press, 1978.
85 p. ; 19 cm. C***
ISBN 0-919594-56-5 : $5.95
I. Title.
PZ7.C8348 My 79-312613
79 MARC

Cowen, Eve.
Jungle Jenny / Eve Cowen. — Belmont, Calif. : Fearon Pitman Publishers, c1979.
57 p. : ill. ; 18 cm. — (A Pacemaker bestellers book)
SUMMARY: Jenny's arrival in the Brazilian jungle is filled with misunderstandings and terror.
ISBN 0-8224-5365-7
₍1. Brazil—Fiction. 2. Jungle stories₎ I. Title.
₍PZ7.C8357 Ju₎ 78-72331
79 MARC

Craft, Ruth.
The king's collection / story by Ruth Craft ; illustrated by Elisa Trimby. — 1st ed. — Garden City, N.Y. : Doubleday, c1978.
₍40₎ p. : ill. (some col.) ; 18 cm.
SUMMARY: A king who likes to collect many different things finds happiness with a new collection of sounds that he can make by himself.
ISBN 0-385-14664-7 : $6.95
₍1. Collectors and collecting—Fiction. 2. Sounds—Fiction₎ I. Trimby, Elisa. II. Title.
PZ7.C8433 Ki ₍Fic₎ 78-8808
78 MARC

Craft, Ruth.
The king's collection / story by Ruth Craft ; and illustrated by Elisa Trimby. — London ₍etc.₎ : Collins, 1978.
₍40₎ p. : chiefly ill. (some col.) ; 19 cm. GB78-33339
ISBN 0-00-183710-9 : £2.95
I. Trimby, Elisa. II. Title.
PZ7.C8433 Ki 1978b 79-303370
79 MARC

Craig, M Jean.
The dragon in the clock box. Illustrated by Kelly Oechsli. New York, W. W. Norton ₍1962₎
₍48₎ p. illus. 24 cm.
SUMMARY: Joshua's attempts to hatch and raise a baby dragon soon have his whole family interested.
₍1. Dragons—Fiction₎ I. Oechsli, Kelly. II. Title.
PZ7.C84427 Dr ₍E₎ 62-17347
₍r79₎rev MARC

Cresswell, Helen.
Bagthorpes unlimited / Helen Cresswell ; illustrated by Jill Bennett. — London ; Boston : Faber, 1978.
175 p. : ill. ; 21 cm. — (Her The Bagthorpe saga ; pt. 3) GB***
ISBN 0-571-11245-4 : £3.95
I. Title. II. Series.
PZ7.C8645 Bag 1978b 79-314414
79 MARC

Cresswell, Helen.
Bagthorpes unlimited / by Helen Cresswell. — 1st American ed. — New York : Macmillan, 1978.
180 p. ; 22 cm. — (Her The Bagthorpe saga ; pt. 3)
SUMMARY: The competitive Bagthorpes join forces in a rare display of solidarity when Grandma organizes a family reunion.
ISBN 0-02-725430-5
₍1. Family life—Fiction₎ I. Title. II. Series.
PZ7.C8645 Bag 1978 ₍Fic₎ 78-3561
78₍r79₎rev MARC

Cresswell, Helen.
Bagthorpes V. the world : being the fourth part of the Bagthorpe saga / by Helen Cresswell. — 1st American ed. — New York : Macmillan, 1979.
193 p. ; 22 cm. — (Her The Bagthorpe saga ; pt. 4)
SUMMARY: The members of the eccentric Bagthorpe family contend with Mr. Bagthorpe's attempts to be self-sufficient and with Great-aunt Lucy's visit.
ISBN 0-02-725420-8
₍1. Family life—Fiction₎ I. Title. II. Series.
PZ7.C8645 Bai 1979 ₍Fic₎ 79-13260
79 MARC

Cresswell, Helen.
Two hoots in the snow / by Helen Cresswell ; illustrated by Martine Blanc. — New York : Crown Publishers, 1978, c1975.
₍27₎ p. : col. ill. ; 18 cm. — (Her An early reader)
SUMMARY: Two silly owls discover that things are not always what they appear to be.
ISBN 0-517-53495-9
₍1. Owls—Fiction₎ I. Blanc, Martine. II. Title. III. Series.
PZ7.C8645 Tvm 1978 ₍E₎ 78-55546
78 MARC

Cresswell, Helen.
Two Hoots play hide-and seek / by Helen Cresswell ; illustrated by Martine Blanc. — New York : Crown Publishers, 1978, c1977.
₍28₎ p. : col. ill. ; 18 cm. — (Her An early reader)
SUMMARY: Little Hoot tries to interest his owl friends in a game of hide-and-seek.
ISBN 0-517-53282-4
₍1. Owls—Fiction. 2. Hide-and-seek—Fiction₎ I. Blanc, Martine. II. Title. III. Series.
PZ7.C8645 Tw 1978 ₍E₎ 77-15856
77 MARC

Crews, Donald.
Freight train / Donald Crews. — 1st ed. — New York : Greenwillow Books, c1978.
₍22₎ p.: col. ill. ; 22 x 26 cm.
SUMMARY: Brief text and illustrations trace the journey of a colorful train as it goes through tunnels, by cities, and over trestles.
ISBN 0-688-80165-X : $6.95
₍1. Railroads—Trains—Pictorial works. 2. Colors. 3. Picture books₎ I. Title.
PZ7.C8682 Fr ₍E₎ 78-2303
78 MARC

Crook, Beverly.
Fair Annie of Old Mule Hollow / by Beverly Courtney Crook. — New York : McGraw-Hill, c1978.
175 p. : ill. ; 21 cm.
SUMMARY: When Fair Annie, a sheltered mountain girl, inadvertently strays into forbidden McFarr territory her life opens up to tragedy and to love.
ISBN 0-07-014487-7 : $7.95
₍1. Mountain life—Fiction₎ I. Title.
PZ7.C8816 Fai ₍Fic₎ 78-9291
78 MARC

Crosher, G R
The awakening water / G. R. Kesteven ₍i.e. G. R. Crosher₎. — London : Chatto & Windus, 1977.
160 p. ; 21 cm. GB77-09343
ISBN 0-7011-2229-3 : £3.25
I. Title.
PZ7.C8823 Aw 78-314316
78 MARC

Crosher, G R
The awakening water / G. R. Kestewen ₍i.e. G. R. Crosher₎. — New York : Hastings House, 1979, c1977.
160 p. ; 22 cm.
SUMMARY: After the devastation of 1997, 13-year-old Watford Nine John accidentally breaks out of the regimen, imposed by the Party with the surreptitious use of drugs, and decides not to return.
ISBN 0-8038-0471-7 : $6.95
₍1. Science fiction₎ I. Title.
PZ7.C8823 Aw 1979 ₍Fic₎ 78-27186
78 MARC

Crume, Vic.
Napoleon and Samantha : from the Walt Disney Productions' film written by Stewart Raffill / Vic Crume. — New York : Pyramid Books, c1976.
127 p. ; 19 cm.
SUMMARY: Recently orphaned Napoleon, accompanied by his best friend and his pet lion, sets out over 14 miles of rugged mountains seeking a new home.
ISBN 0-515-04140-8 : $1.50
₍1. Orphans—Fiction. 2. Lions—Fiction. 3. Adventure stories₎ I. Napoleon and Samantha. ₍Motion picture₎
PZ7.C8884 Nap 76-27632
79 MARC

Cuneo, Mary Louise Hector.
Inside a sand castle and other secrets / Mary Louise Cuneo ; illustrated by Jan Brett. — Boston : Houghton Mifflin, 1979.
32 p. : col. ill. ; 21 x 22 cm.
SUMMARY: The reader is taken inside a sand castle, a mouse's house, and a pine tree where there is a miniature town.
ISBN 0-395-27805-8
₍1. Fantasy₎ I. Brett, Jan, 1949- II. Title.
PZ7.C91615 In ₍E₎ 78-24000
78 MARC

Cunliffe, John A
Mr Gosling and the great art robbery / story by John Cunliffe ; pictures by William Stobbs. — London : A. Deutsch, 1979.
₍32₎ p. : col. ill. ; 26 cm. GB***
ISBN 0-233-97079-7 : £3.50 ($6.95 U.S.)
I. Stobbs, William. II. Title.
PZ7.C9162 Mh 78-74464
79 MARC

Cunliffe, John A
Mr. Gosling and the runaway chair / story by John Cunliffe ; pictures by William Stobbs. — London : Deutsch, 1978.
₍30₎ p. : ill. ; 26 cm. GB***
ISBN 0-233-96956-X : £2.95
I. Stobbs, William. II. Title.
PZ7.C9162 Mi 78-310204
78 MARC

Cummings, Betty Sue.
Now, Ameriky / Betty Sue Cummings. — 1st ed. — New York : Atheneum, 1979.
175 p. ; 24 cm.
SUMMARY: A young Irish woman, whose family is driven from their land during the potato famine in the 1840's, is sent to America to earn enough money to enable the rest of her family to join her.
ISBN 0-689-30705-5
₍1. United States—Emigration and immigration—Fiction. 2. Irish Americans—Fiction₎ I. Title.
PZ7.C9145 No ₍Fic₎ 79-11750
79 MARC

Cunningham, Bob.
10-5 Alaska skip / by Bob Cunningham ; ill. by Rod and Barbara Furan. — ₍Mankato, Minn.₎ : Crestwood House, c1977.
31 p. : col. ill. ; 24 cm. — (The CB adventures of Neil Hawkins)
SUMMARY: A family climbing on Mt. Olympus in Washington hears an emergency message from Alaska on the skip on their walkie-talkie.
ISBN 0-913940-58-5
₍1. Citizens band radio—Fiction₎ I. Furan, Rodney. II. Furan, Barbara Howell. III. Title. IV. Series: Cunningham, Bob. The CB adventures of Neil Hawkins.
PZ7.C91628 Te ₍Fic₎ 76-51138
79 MARC

Cunningham, Bob.
10-7 for Good Sam / by Bob Cunningham ; ill. by Rod and Barbara Furan. — ₍Mankato, Minn.₎ : Crestwood House, c1977.
31 p. : col. ill. ; 24 cm. — (His The CB adventures of Neil Hawkins)
SUMMARY: When the CB radio rush-hour traffic reports suddenly stop, newspaperman Neil Hawkins decides to investigate the cause.
₍1. Citizens band radio—Fiction₎ I. Furan, Rodney. II. Furan, Barbara Howell. III. Title. IV. Series.
PZ7.C91628 Tf ₍Fic₎ 76-51143
79 MARC

CHILDREN'S LITERATURE (Continued)

Cunningham, Bob.
10-33 emergency / by Bob Cunningham ; ill. by Rod and Barbara Furan. — [Mankato, Minn.] : Crestwood House, c1977.
31 p. : col. ill. ; 24 cm. — (The CB adventures of Neil Hawkins)
SUMMARY: A three-year-old's plea for help mobilizes local CB-radio operators.
ISBN 0-913940-57-7
[1. Citizens band radio—Fiction] I. Furan, Rodney. II. Furan, Barbara Howell. III. Title. IV. Series: Cunningham, Bob. The CB adventures of Neil Hawkins.
PZ7.C91628 Th 79 76-51139 MARC

Cunningham, Bob.
10-200 come on, Smokey! / By Bob Cunningham ; ill. by Rod and Barbara Furan. — [Mankato, Minn.] : Crestwood House, c1979.
31 p. : col. ill. ; 24 cm. — (His The CB adventures of Neil Hawkins)
SUMMARY: CB-radio operators, gathered for a jamboree, join forces to help recover a stolen vehicle.
ISBN 0-913940-55-0
[1. Citizens band radio—Fiction. 2. Mystery and detective stories] I. Furan, Rodney. II. Furan, Barbara Howell. III. Title. IV. Series.
PZ7.C91628 Tj [Fic] 79 76-51142 MARC

Cunningham, Julia.
Tuppenny / by Julia Cunningham. — 1st ed. — New York : Dutton, c1978.
87 p. ; 22 cm. — (A Unicorn book)
SUMMARY: A secretive young girl arrives in a small factory town and flushs out the guilt and hatred festering there.
ISBN 0-525-41572-6
I. Title.
PZ7.C9167 Tu 1978 [Fic] 78 78-7449 MARC

Curley, Daniel.
Hilarion / Daniel Curley ; illustrated by Judith Gwyn Brown. — Boston : Houghton Mifflin, 1979.
88 p. : ill. ; 22 cm.
SUMMARY: At the turn of the century Hilarian joins four other men from Linsk and finds the streets of America paved with gold.
ISBN 0-395-28268-3
[1. Emigration and immigration—Fiction. 2. City and town life—Fiction] I. Brown, Judith Gwyn. II. Title.
PZ7.C929 Hi [Fic] 79 79-11749 MARC

Curry, Jane Louise.
Ghost Lane / Jane Louise Curry. — 1st ed. — New York : Atheneum, 1979.
158 p. : ill. ; 22 cm.
"A Margaret K. McElderry book."
SUMMARY: Three children in an English village face danger as they try to track down the criminals responsible for burglaries in a strange old house.
ISBN 0-689-50129-3
[1. Mystery and detective stories. 2. England—Fiction] I. Title.
PZ7.C936 Gh [Fic] 78 78-73399 MARC

Dahl, Roald.
The enormous crocodile / [by] Roald Dahl ; with pictures by Quentin Blake. — London : Cape, 1978.
[32] p. : col. ill. ; 28 cm. GB78-30118
ISBN 0-224-01579-6 : £2.95
I. Blake, Quentin. II. Title.
PZ7.D1515 En 1978 79 78-323944 MARC

Dahl, Roald.
The enormous crocodile / Roald Dahl ; pictures by Quentin Blake. — New York : Knopf : distributed by Random House, c1978.
[32] p. : col. ill. ; 28 cm.
SUMMARY: The enormous crocodile devises secret plans and a few clever tricks to secure his lunch only to have them foiled by his neighbors.
ISBN 0-394-83594-8. ISBN 0-394-93594-2 lib. bdg.
[1. Crocodiles—Fiction] I. Blake, Quentin. II. Title.
PZ7.D1515 En 1978b [E] 77 77-5081 MARC

Daly, Kathleen N
Cristobalito / Kathleen Daly ; from the Walt Disney Productions' film written by Norman Wright. — New York : Pyramid Books, 1976.
128 p. : ill. ; 19 cm.
Bibliography: p. [7]
SUMMARY: An orphaned stable boy pins his hopes of winning Puerto Rico's All Island Paso Fino Grand Championship on a beautiful palomino named Cristobalito.
ISBN 0-515-04139-4 : $1.50
[1. Horses—Fiction. 2. Puerto Rico—Fiction] I. Cristobalito. [Motion picture] II. Title.
PZ7.D1696 Cr [Fic] 79 76-27624 MARC

Dank, Milton, 1920-
Game's end / Milton Dank. — Philadelphia : Lippincott, c1979.
158 p. ; 21 cm.
SUMMARY: In this sequel to "The Dangerous Game," Charles Marceau, now a second lieutenant in the Free French Army, returns to France and its resistance movement to prepare for the impending Allied invasion.
ISBN 0-397-31821-9
[1. World War, 1939-1945—Underground movements—France—Fiction. 2. France—History—German occupation, 1940-1945—Fiction] I. Title.
PZ7.D228 Gam [Fic] 78 78-12625 MARC

Danziger, Paula, 1944-
Can you sue your parents for malpractice? : a Novel / by Paula Danziger. — New York : Delacorte Press, c1979.
vii, 152 p. ; 21 cm.
SUMMARY: Confused about her life at home and at school, 14-year-old Lauren learns the importance of being her own person.
ISBN 0-440-01050-0
[1. Family life—Fiction. 2. Individuality—Fiction] I. Title.
PZ7.D2394 Can [Fic] 78 78-72856 MARC

Danziger, Paula, 1944-
The pistachio prescription : a novel / by Paula Danziger. — New York : Delacorte Press, c1978.
154 p. ; 21 cm.
SUMMARY: A high school "freshperson" attempts to rise above such inconveniences of life as her older sister, parents, and school.
ISBN 0-440-06936-X
[1. Family life—Fiction. 2. School stories] I. Title.
PZ7.D2394 Pi [Fic] 77 77-86330 MARC

Darling, Kathy.
The Easter bunny's secret / by Kathy Darling ; pictures by Kelly Oechsli. — Champaign, Ill. : Garrard Pub. Co., c1978.
48 p. : col. ill. ; 23 cm. — (A Garrard mystery book)
SUMMARY: Spring Bunny encounters three perplexing problems shortly before the Easter holiday.
ISBN 0-8116-6405-8 lib. bdg. : $4.30
[1. Easter eggs—Fiction. 2. Rabbits—Fiction. 3. Easter stories] I. Oechsli, Kelly. II. Title.
PZ7.D248 Eas [E] 79 78-58521 MARC

Darling, Kathy.
The mystery in Santa's toyshop / by Kathy Darling ; pictures by Lori Pierson. — Champaign, Ill. : Garrard Pub. Co., c1978.
48 p. : col. ill. ; 23 cm. — (A Garrard mystery book)
SUMMARY: As increasing numbers of toys disappear from his workshop, Santa engages a detective to solve the mystery.
ISBN 0-8116-6402-3 lib. bdg. : $4.30
[1. Santa Claus—Fiction. 2. Animals—Fiction. 3. Christmas stories] I. Pierson, Lori. II. Title.
PZ7.D248 My [E] 77 77-19090 MARC

Davidson, Carson.
Fast-talking dolphin / Carson Davidson ; illustrated by Sylvia Stone. — New York : Dodd, Mead, c1978.
128 p. : ill. ; 21 cm.
SUMMARY: When Eric finds a talking dolphin in a trout pond in New England, he uses all his knowledge and skills to save its life.
ISBN 0-396-07696-3
[1. Dolphins—Fiction] I. Stone, Sylvia, S.I. II. Title.
PZ7.D2825 Fas [Fic] 79 79-14271 MARC

Davies, Andrew.
Conrad's war / [by] Andrew Davies. — London [etc.] : Blackie, 1978.
127 p. ; 21 cm. GB78-21625
ISBN 0-216-90533-8 : £3.25
I. Title.
PZ7.D2837 Co 78 78-322013 MARC

Davis, Burke.
Mr. Lincoln's whiskers / by Burke Davis ; illustrated by Douglas Gorsline. — New York : Coward, McCann & Geoghegan, c1978.
33 p. : ill. ; 24 cm.
SUMMARY: At the suggestion of a young girl, Abraham Lincoln grows a beard.
ISBN 0-698-20455-7
1. Lincoln, Abraham, Pres. U.S., 1809-1865—Fiction. [1. Lincoln, Abraham, Pres. U.S., 1809-1865—Fiction] I. Gorsline, Douglas W., 1913- II. Title.
PZ7.D28556 Mi 1978 [Fic] 78 77-29208 MARC

Davis, Douglas F
There's an elephant in the garage / by Douglas F. David ; illustrated by Steven Kellogg. — 1st ed. — New York : Dutton, c1979.
[32] p. : ill. ; 24 x 26 cm.
SUMMARY: The adventures of a little girl and her cat as they hunt rogue elephants in the family garage.
ISBN 0-525-41050-3
[1. Cats—Fiction] I. Kellogg, Steven. II. Title.
PZ7.D28594 Th 1979 [E] 79 79-11378 MARC

Davis, Reda.
Martin's dinosaur. Illustrated by Louis Slobodkin. New York, Crowell [1959]
[40] p. illus. 26 cm.
SUMMARY: While visiting a Welsh castle with his parents, a little boy discovers a dragon and tries to convince him that he is really a dinosaur.
[1. Dragons—Fiction] I. Slobodkin, Louis, 1903- II. Title.
PZ7.D2983 Mar [E] [r78]rev 59-7762 MARC

DeLage, Ida.
ABC pirate adventure / by Ida DeLage ; drawings by Buck Brown. — Champaign, Ill. : Garrard Pub. Co., c1977.
32 p. : ill. ; 23 cm.
SUMMARY: The A to Z adventures of a rollicking crew of pirates who are searching for treasure.
ISBN 0-8116-4352-2 lib. bdg.
[1. Pirates—Fiction. 2. Alphabet] I. Brown, Buck. II. Title.
PZ7.D3698 Aae [E] 77 77-3171 MARC

DeLage, Ida.
ABC Santa Claus / by Ida DeLage ; drawings by Judith Gwyn Brown. — Champaign, Ill. : Garrard Pub. Co., c1978.
32 p. : col. ill. ; 23 cm. — (Her Once upon an ABC)
SUMMARY: Santa Claus prepares for his Christmas Eve journey with a last-minute check of details from A to Z.
ISBN 0-8116-4354-9
[1. Alphabet. 2. Christmas stories] I. Brown, Judith Gwyn. II. Title. III. Series.
PZ7.D3698 Aai [E] 77 77-5629 MARC

DeLage, Ida.
The old witch and her magic basket / by Ida Delage ; illustrated by Ellen Sloan. — Champaign, Ill. : Garrard Pub. Co., c1978.
48 p. : col. ill. ; 23 cm.
SUMMARY: An old witch uses her bag of tricks to cheer a little girl who is sick on Halloween.
[1. Halloween—Fiction. 2. Sick—Fiction] I. Sloan, Ellen. II. Title.
PZ7.D3698 Od [E] 79 78-58520 MARC

DeLage, Ida.
The old witch and the dragon / by Ida DeLage ; illustrated by Unada. — Champaign, Ill. : Garrard Pub. Co., c1979.
48 p. : col. ill. ; 23 cm.
SUMMARY: The Old Witch saves the farmer, the town, and herself from the ravages of the Dragon of Thunder Mountain, who comes down from his mountain every hundred years.
ISBN 0-8116-4064-7 (lib. bdg.) : $4.20
[1. Witches—Fiction. 2. Dragons—Fiction] I. Unada. II. Title.
PZ7.D3698 Of [E] 78 78-11283 MARC

DeLage, Ida.
The old witch and the ghost parade / by Ida DeLage ; illustrated by Jody Taylor. — Champaign, Ill. : Garrard Pub. Co., c1978.
47 p. : col. ill. ; 23 cm.
SUMMARY: Irate at not winning first prize in the Halloween contest, the old witch has her revenge.
ISBN 0-8116-4062-0
[1. Witches—Fiction. 2. Halloween—Fiction] I. Taylor, Jody. II. Title.
PZ7.D3698 Oh [E] 77 77-18185 MARC

DeLage, Ida.
The squirrel's tree party / by Ida Delage ; drawings by Tracy McVay. — Champaign, Ill. : Garrard Pub. Co., c1978.
32 p. : col. ill. ; 23 cm.
SUMMARY: With nothing to do on a rainy day, the squirrel children plan a sunny day party.
ISBN 0-8116-6073-7 : $4.20
[1. Parties—Fiction. 2. Squirrels—Fiction] I. McVay, Tracy. II. Title.
PZ7.D3698 Sq [E] 79 78-58523 MARC

De Paola, Thomas Anthony.
Big Anthony and the magic ring : story and pictures / by Tomie de Paola. — 1st ed. — New York : Harcourt Brace Jovanovich, c1979.
[32] p. : col. ill. ; 29 cm.
SUMMARY: Summary; When Big Anthony borrows Strega Nona's magic ring to turn himself into a handsome man, he gets more trouble than fun.
ISBN 0-15-207124-5
[1. Magic—Fiction] I. Title.
PZ7.D439 Bg [E] 78 78-23631 MARC

De Paola, Thomas Anthony.
Flicks / by Tomie de Paola. — 1st ed. — New York : Harcourt Brace Jovanovich, c1979.
[32] p. : chiefly ill. ; 16 x 24 cm.
SUMMARY: Five short "silent movies" include "Tooth Troubles," "The Birthday Cake," "Rhonda Rolls Along," "The New Baby," and "A Sleep-time Saga."
ISBN 0-15-228487-7
[1. Stories without words. 2. Short stories] I. Title.
PZ7.D439 Fl [E] 79 79-87514 MARC

De Paola, Thomas Anthony.
Four stories for four seasons / stories and pictures by Tomie de Paola. — Englewood Cliffs, N.J. : Prentice-Hall, c1977.
[48] p. : col. ill. ; 24 cm.
SUMMARY: Four friends—a cat, dog, pig, and frog—share the delights of each season.
ISBN 0-13-330175-3 : $7.95
[1. Seasons—Fiction. 2. Friendship—Fiction. 3. Animals—Fiction] I. Title.
PZ7.D439 Fo [E] 76[r79]rev 76-8837 MARC

De Paola, Thomas Anthony.
Oliver Button is a sissy / story and pictures by Tomie de Paola. — 1st ed. — New York : Harcourt Brace Jovanovich, c1979.
[48] p. : col. ill. ; 21 cm.
SUMMARY: His classmates' taunts don't stop Oliver Button from doing what he likes best.
ISBN 0-15-257852-8
[1. Dancing—Fiction. 2. Sex role—Fiction] I. Title.
PZ7.D439 Ol [E] 78 78-12624 MARC

CHILDREN'S LITERATURE (Continued)

De Paola, Thomas Anthony.
Songs of the fog maiden / written and illustrated by Tomie de Paola. — New York : Holiday House, c1979.
[32] p. : col. ill. ; 25 cm.
SUMMARY: The fog maiden sings her songs before and after her nightly travels over the earth.
ISBN 0-8234-0341-6
I. Title.
PZ7.D439 So [E] 78 78-12822 MARC

De Paola, Thomas Anthony.
When everyone was fast asleep / written and illustrated by Tomie de Paola. — [New York] : Puffin Books, 1979, c1976.
[32] p. : col. ill. ; 23 cm. — (Picture puffin)
SUMMARY: When everyone was fast asleep, the Fog Maiden sent a cat to carry two sleeping children off for a fantastical evening highlighted by a palace ball.
ISBN 0-14-050310-2 : $2.25
[1. Fantasy. 2. Night—Fiction] I. Title.
[PZ7.D439 Wh 1979] [E] 79 79-560 MARC

De Regniers, Beatrice Schenk.
Laura's story / written by Beatrice Schenk de Regniers ; illustrated by Jack Kent. — 1st ed. — New York : Atheneum, 1979.
[32] p. : col. ill. ; 24 cm.
SUMMARY: Laura tells her mother a bedtime story about the perilous rescue of a tiny woman.
ISBN 0-689-30677-6
[1. Mothers and daughters—Fiction] I. Kent, Jack, 1920- II. Title.
PZ7.D4417 Lau 1979 [E] 78 78-12623 MARC

De Rico, Ul, 1944-
The rainbow goblins / Ul de Rico. — New York : Thames and Hudson, [1978]
[35] p. : col. ill. ; 35 x 43 cm.
Translation of Die Regenkobolde.
SUMMARY: After seven goblins try to steal it, the Rainbow is careful never again to touch the earth.
ISBN 0-500-95005-9 : $24.95
[1. Rainbow—Fiction] I. Title.
PZ7.D4425 Rai 1978 [E] 79 78-55431 MARC

Deary, Terence.
The Custard Kid / [by] Terry Deary ; illustrations by Charlotte Firmin. — London : A. and C. Black, 1978.
63 p. : ill. ; 22 cm. GB78-15868
ISBN 0-7136-1829-9 : £1.75
I. Firmin, Charlotte. II. Title.
PZ7.D3517 Cu 78 78-318000 MARC

Del Rey, Lester, 1915-
Moon of mutiny. [1st ed.] New York, Holt, Rinehart and Winston [1961]
217 p. 22 cm.
SUMMARY: Charged with insubordination, refusing to follow orders, and endangering the lives of others, a rebellious cadet is dismissed from the Goddard Space Academy and seeks to join the moon colony.
[1. Space flight training—Fiction. 2. Space colonies—Fiction. 3. Science fiction] I. Title.
PZ7.D386 Mo [Fic] [r78]rev 61-14968 MARC

Del Rey, Lester, 1915-
Moon of mutiny / by Lester Del Rey. — Boston : Gregg Press, 1979, c1961.
217 p. ; 22 cm. — (The Gregg Press science fiction series)
Reprint of the ed. published by Holt, Rinehart and Winston, New York.
SUMMARY: Charged with insubordination, refusing to follow orders, and endangering the lives of others, a rebellious cadet is dismissed from the Goddard Space Academy and seeks to join the moon colony.
ISBN 0-8398-2518-8
[1. Space flight training—Fiction. 2. Space colonies—Fiction. 3. Science fiction] I. Title.
[PZ7.D386 Mo 1979] [Fic] 79 79-459 MARC

Del Rey, Lester, 1915-
Outpost of Jupiter. [1st ed.] New York, Holt, Rinehart and Winston [1963]
191 p. 22 cm.
SUMMARY: While he and his father are living on the distant planetoid, Ganymede, a young boy makes contact with aliens from Jupiter.
[1. Science fiction] I. Title.
PZ7.D386 Ou [Fic] [r78]rev 63-10205 MARC

DeLage, Ida.
ABC Easter bunny / by Ida DeLage ; drawings by Ellen Sloan. — Champaign, Ill. : Garrard Pub. Co., c1979.
32 p. : col. ill. ; 23 cm. — (Her Once upon an ABC)
SUMMARY: The Easter bunny makes deliveries to children in the city and in the country.
ISBN 0-8116-4356-5 lib. bdg. : $4.20
[1. Easter stories. 2. Animals—Fiction. 3. Alphabet] I. Sloan, Ellen. II. Title. III. Series.
PZ7.D3698 Aaar [E] 78 78-14829 MARC

DeLage, Ida.
Frannie's flower / by Ida DeLage ; drawings by Ellen Sloan. — Champaign, Ill. : Garrard Pub. Co., c1979.
32 p. : col. ill. ; 23 cm.
SUMMARY: When her grandmother gives her a flower, Frannie takes it to bed with her, has a party with it, and takes it for an airing in her wagon.
ISBN 0-8116-6076-1
[1. Flowers—Fiction. 2. Play—Fiction] I. Sloan, Ellen. II. Title.
PZ7.D3698 Fr [E] 79 79-11724 MARC

DeLage, Ida.
Good morning, lady. Drawings by Tracy McVay. Champaign, Ill., Garrard Pub. Co. [1974]
32 p. col. illus. 23 cm.
SUMMARY: Father Possum peddles his wares to a number of other animals.
ISBN 0-8116-6051-6 (lib. bdg.)
[1. Animals—Fiction] I. McVay, Tracy, illus. II. Title.
PZ7.D3698 Go [E] 73[r79]rev 73-22084 MARC

Delaney, Ned.
Bert and Barney / written and illustrated by Ned Delaney. — Boston : Houghton Mifflin, 1979.
32 p. : col. ill. ; 23 cm.
SUMMARY: Bert, a frog, and his best friend Barney, an alligator, go their separate ways after a disagreement until a near disaster at a dangerous cliff brings them together again.
ISBN 0-395-28377-9
[1. Frogs—Fiction. 2. Alligators—Fiction. 3. Friendship—Fiction] I. Title.
PZ7.D3732 Be [E] 79 79-10633 MARC

Delaney, Ned.
Rufus the Doofus / Ned Delaney. — Boston : Houghton Mifflin, 1978.
32 p. : col. ill. ; 25 cm.
SUMMARY: Rufus is ignored by his classmates until he becomes the holy terror of the class, but a moment of friendship restores him to his own true amiable self.
ISBN 0-395-27153-3 : $6.95
[1. Friendship—Fiction. 2. School stories] I. Title.
PZ7.D3732 Ru [E] 78 78-60493 MARC

Delton, Judy.
The best mom in the world / Judy Delton & Elaine Knox-Wagner ; pictures by John Faulkner. — Chicago : A. Whitman, c1979.
[32] p. : col. ill. ; 24 cm. — (Concept book/level 1)
SUMMARY: A youngster feels angry and betrayed when his doting mother begins asking him to do things for himself.
ISBN 0-8075-0665-6
[1. Mothers and sons—Fiction. 2. Responsibility—Fiction] I. Knox-Wagner, Elaine, joint author. II. Faulkner, John Frink, 1922- III. Title.
PZ7.D388 Be 1979 [E] 78 78-27238 MARC

Delton, Judy.
Brimhall comes to stay / Judy Delton ; ill. by Cyndy Szekeres. — 1st ed. — New York : Lothrop, Lee & Shepard, c1978.
63 p. : col. ill. ; 22 cm. — (A Fun-to-read book)
SUMMARY: Bear welcomes the arrival of his cousin Brimhall to live with him, but Brimhall's eccentric habits soon tax Bear's hospitality.
ISBN 0-688-41863-5. ISBN 0-688-51863-X lib. bdg.
[1. Bears—Fiction. 2. Family life—Fiction] I. Szekeres, Cindy. II. Title.
PZ7.D388 Br [E] 78 78-7524 MARC

Delton, Judy.
Brimhall turns to magic / by Judy Delton ; illustrated by Bruce Degen. — 1st ed. — New York : Lothrop, Lee & Shepard Co., c1979.
62 p. : ill. ; 22 cm. — (A Fun-to-read book)
SUMMARY: Brimhall learns enough magic to make a rabbit appear in a hat, but not enough to make him vanish.
ISBN 0-688-41878-3. ISBN 0-688-51878-8 lib. bdg.
[1. Magic tricks—Fiction. 2. Animals—Fiction. 3. Friendship—Fiction] I. Degen, Bruce. II. Title.
PZ7.D388 Bu [E] 78 78-12141 MARC

Delton, Judy.
Kitty in the middle / Judy Delton ; illustrated by Charles Robinson. — Boston : Houghton Mifflin, 1979.
135 p. : ill. ; 22 cm.
SUMMARY: First love, a madcap mix-up at a stranger's wedding, and exploring a mysterious old house are some of the experiences that make fourth grade exciting for three friends.
ISBN 0-395-28004-4 : $6.95
[1. Friendship—Fiction] I. Robinson, Charles, 1931- II. Title.
PZ7.D388 Ki [Fic] 78-31434 MARC

Delton, Judy.
On a picnic / Judy Delton ; pictures by Mamoru Funai. — 1st ed. — Garden City, N.Y. : Doubleday, c1979.
[64] p. : col. ill. ; 24 cm. — (A Reading-on-my-own book)
SUMMARY: Goose and Gorilla discover they can be friends despite their physical differences.
ISBN 0-385-12944-0 : $4.95.
[1. Individuality—Fiction. 2. Friendship—Fiction. 3. Geese—Fiction. 4. Gorillas—Fiction] I. Funai, Mamoru. II. Title.
PZ7.D388 On [E] 78 78-1240 MARC

Dengler, Marianna.
A pebble in Newcomb's Pond / Marianna Dengler ; illustrated by Kathleen Garry-McCord. — New York : Holt, Rinehart and Winston, c1979.
160 p. : ill. ; 22 cm.
SUMMARY: A 17-year-old girl who suffers from schizophrenia is cured with drug therapy utilizing vitamins and diet.
ISBN 0-03-044641-4
[1. Schizophrenia—Fiction. 2. Mentally ill—Fiction] I. Garry-McCord, Kathleen. II. Title.
PZ7.D4145 Pe [Fic] 78 78-14825 MARC

Devlin, Wende.
Aunt Agatha, there's a lion under the couch! By Wende and Harry Devlin. Princeton, N.J., Van Nostrand [1968]
30 p. col. illus. 29 cm.
SUMMARY: Aunt Agatha tells her nephew to laugh at the lion in the closet and the animal will disappear—but real lions don't vanish.
[1. Lions—Fiction] I. Devlin, Harry, joint author. II. Title.
PZ7.D49875 Au [E] [r79]rev2 68-19181 MARC

Devlin, Wende.
Cranberry Christmas / Wende and Harry Devlin. — New York : Parents' Magazine Press, c1976.
[38] p. : col. ill. ; 26 cm.
SUMMARY: Mr. Whiskers faces a gloomy Christmas until Maggie and her grandmother help him straighten out his house and find the deed to the nearby pond.
ISBN 0-8193-0844-7. ISBN 0-8193-0845-5 lib. bdg.
[1. Christmas stories] I. Devlin, Harry, joint author. II. Title.
PZ7.D49875 Cp [E] 76[r79]rev2 76-2524 MARC

Devlin, Wende.
Cranberry mystery / Wende and Harry Devlin. — New York : Parents' Magazine Press, c1978.
[32] p. : ill. (some col.) ; 26 cm.
SUMMARY: Maggie and Mr. Whiskers go in search of thieves who are stealing many valuable old things in Cranberryport.
ISBN 0-8193-0972-9. ISBN 0-8193-0973-7 lib. bdg.
[1. Mystery and detective stories] I. Devlin, Harry, joint author. II. Title.
PZ7.D49875 Cr [E] 78 78-6219 MARC

Devlin, Wende.
Cranberry Thanksgiving [by] Wende and Harry Devlin. New York, Parents' Magazine Press [1971]
[33] p. col. illus. 27 cm.
SUMMARY: Grandmother's cranberry bread was very special and Maggie took care that no one stole the recipe. Cranberry bread recipe included at the back of the book.
ISBN 0-8193-0499-9 (library binding)
[1. Thanksgiving Day—Fiction] I. Devlin, Harry, joint author. II. Title.
PZ7.D49875 Ct [E] 71[r79]rev2 73-153789 MARC

Devlin, Wende.
How Fletcher was hatched, by Wende and Harry Devlin. New York, Parents' Magazine Press [1969]
[39] p. col. illus. 26 cm. 3.50
SUMMARY: Feeling his mistress has rejected him in preference to her newly hatched chicks, the old hound dog decides he must hatch from an egg and learn to say "peep" to regain favor.
[1. Dogs—Fiction] I. Devlin, Harry, joint author. II. Title.
PZ7.D49875 Ho [E] [r79]rev3 69-12614 MARC

Devlin, Wende.
A kiss for a warthog [by] Wende and Harry Devlin. New York, Van Nostrand Reinhold [1970]
37 p. col. illus. 26 cm. 4.95
SUMMARY: If the warthog destined for the Oldwick zoo had not had a mind of her own, the rival towns of Oldwick and Quimby might never have become friends.
[1. Wart hog—Fiction. 2. Neighborliness—Fiction] I. Devlin, Harry, joint author. II. Title.
PZ7.D49875 Ki [E] 70[r79]rev2 77-90337 MARC

Devlin, Wende.
The Knobby Boys to the rescue, by Wende and Harry Devlin. [New York] Parents' Magazine Press [1965]
1 v. (unpaged) illus. (part col.) 26 cm.
I. Devlin, Harry, joint author. II. Title.
PZ7.D49875 Kn [r79]rev2 65-10465 MARC

Devlin, Wende.
Old Black Witch! By Wende and Harry Devlin. Illustrated by Harry Devlin. Chicago, Encyclopaedia Britannica Press [1963]
32 p. illus. (part col.) 26 cm.
SUMMARY: At first Nicky's and his mother's efforts to turn an old New England house into a tearoom are hampered by an old black witch who haunts the place; then, they find out she can cook.
[1. Witches—Fiction] I. Devlin, Harry, joint author. II. Title.
PZ7.D49875 Ol [E] [r79]rev2 63-18349 MARC

Devlin, Wende.
Old witch and the polka-dot ribbon, by Wende and Harry Devlin. New York, Parents' Magazine Press [1970]
[32] p. col. illus. 27 cm. 3.95
SUMMARY: When the kitchen is finally empty, Old Witch makes her own magic nut cake to enter in the carnival's cake baking contest.
ISBN 0-8193-0418-2
[1. Witches—Fiction] I. Devlin, Harry, joint author. II. Title.
PZ7.D49875 On [E] 70[r79]rev2 74-117565 MARC

CHILDREN'S LITERATURE (Continued)

Devlin, Wende.
Old Witch rescues Halloween! ⌈by Wende and Harry Devlin.⌉ New York, Parents' Magazine Press ⌈1972⌉
⌈41⌉ p. col. illus. 27 cm. $3.95
SUMMARY: Old Witch uses her magic to overcome Mr. Butterbean's edict against Halloween.
ISBN 0-8193-0608-8. ISBN 0-8193-0609-6 (lib. bdg.)
⌈1. Halloween—Fiction. 2. Witches—Fiction⌉ I. Devlin, Harry, joint author. II. Title.
PZ7.D49875 Op ⌈E⌉ 72-676
72⌈r79⌉rev2 MARC

DeWeese, Gene.
Major Corby and the unidentified flapping object / Gene DeWeese. — 1st ed. — Garden City, N.Y. : Doubleday, c1979.
108 p. ; 22 cm. — (A Doubleday signal book)
SUMMARY: Fourteen-year-old Russ encounters a UFO and manages to communicate with its occupant.
ISBN 0-385-13519-X : $5.95
⌈1. Unidentified flying objects—Fiction. 2. Humorous stories⌉ I. Title.
PZ7.D522 Maj ⌈Fic⌉ 78-18557
78 MARC

Deyneka, Anita.
Alexi and the mountain treasure / Anita Deyneka ; ⌈cover ill. by Joe Van Severen, inside ill. by Larry Taylor⌉. — ⌈Elgin, Ill.⌉ : D. C. Cook Pub. Co., c1979.
111 p. : ill. ; 18 cm. — (Chariot books)
SUMMARY: Shortly after moving to a town near the Russian-Chinese border, two young Russian Christians meet a Uighur who translates a mysterious message for them.
ISBN 0-89191-062-X : $1.75
⌈1. Christian life—Fiction. 2. Russia—Fiction. 3. Uigurs—Fiction⌉ I. Title.
⌈PZ7.D54 Ak⌉ ⌈Fic⌉ 78-72835
00 MARC

Dickens, Frank.
Albert Herbert Hawkins, the naughtiest boy in the world, and the space rocket / by Frank Dickens. — 1st ed. in the U.S.A. — Garden City, N.Y. : Doubleday, c1978.
⌈32⌉ p. : col. ill. ; 21 x 27 cm. — (Benn book collection)
SUMMARY: The naughtiest boy in the world skips school to find out if spacemen are afraid of mice.
ISBN 0-385-13327-8 : $5.95
⌈1. Astronauts—Fiction. 2. Fear—Fiction⌉ I. Title.
PZ7.D552 Am 1978 ⌈E⌉ 77-74708
78 MARC

Dickens, Frank.
Albert Herbert Hawkins (the naughtiest boy in the world) and the space rocket / by Frank Dickens. — London : E. Benn, 1978.
⌈32⌉ p. : chiefly col. ill. ; 21 x 27 cm. GB78-17063
ISBN 0-510-22507-1 : £2.50
I. Title.
PZ7.D552 Am 78-318623
78 MARC

Dickinson, Peter, 1927-
Tulku / Peter Dickinson. — 1st ed. — New York : Dutton, 1979.
286 p. ; 22 cm. — (A Unicorn book)
SUMMARY: A thirteen-year-old boy escapes from slaughter by the Boxers in China and joins forces with an English botanist and her escort, traveling with them to Tibet where the power of Buddhist monks transforms the lives of all of them.
ISBN 0-525-41571-8
⌈1. Tibet—Fiction. 2. Buddhism—Fiction. 3. Botany—Fiction. 4. Adventure stories⌉ I. Title.
PZ7.D562 Tu 1979 ⌈Fic⌉ 78-11461
78 MARC

Dicks, Terrance.
The case of the missing masterpiece / ⌈by⌉ Terrance Dicks. — London ⌈etc.⌉ : Blackie, 1978.
128 p. ; 21 cm. GB78-17802
Half t.p. title: The Baker Street Irregulars.
ISBN 0-216-90456-0 : £2.95
I. Title. II. Title: The Baker Street Irregulars.
PZ7.D5627 Cas 1978 78-318756
78 MARC

Dillon, Eilís, 1920-
The shadow of Vesuvius / Eilís Dillon. — London ; Boston : Faber, 1978.
140 p. ; 21 cm. GB78-01277
SUMMARY: A Greek slave of a Roman artisan plans his escape from the city of Pompeii amid ominous signs of the approaching eruption of Mt. Vesuvius.
ISBN 0-571-10913-6 : £3.75
⌈1. Slavery in Rome—Fiction. 2. Rome—Fiction. 3. Pompeii—Fiction⌉ I. Title.
PZ7.D5792 Sh 1978 78-316082
78 MARC

Dionetti, Michelle, 1947-
Thalia Brown and the blue bug / Michelle Dionetti ; illustrated by James Calvin. — Reading, Mass. : Addison-Wesley Pub. Co., c1979.
32 p. : ill. ; 25 cm.
SUMMARY: Feeling neglected by her family, Thalia Brown finds someone else to help her get ready for the Art Fair.
ISBN 0-201-01399-1
⌈1. Family life—Fiction⌉ I. Calvin, James. II. Title.
PZ7.D6214 Th ⌈E⌉ 79-4160
79 MARC

Disney, Walt, 1901-1966.
Walt Disney's Elmer Elephant plus Peculiar penguins. — New York : McKay, ⌈1978⌉
⌈95⌉ p. : col. ill. ; 22 cm.
SUMMARY: Two books originally published separately, one featuring Elmer, a little elephant, the other, a pair of playful penguins named Polly and Peter.
ISBN 0-679-20602-7 : $6.95
⌈1. Elephants—Fiction. 2. Penguins—Fiction⌉ I. Disney, Walt, 1901-1966. Peculiar penguins. 1978. II. Title. III. Title: Walt Disney's Elmer Elephant. IV. Title: Elmer Elephant.
PZ7.D624 Wal 1978 ⌈E⌉ 78-19231
78 MARC

Disney (Walt) Productions.
Walt Disney's Donald Duck and the magic mailbox. — New York : Golden Press, c1978.
⌈24⌉ p. : col. ill. ; 21 cm. — (A Golden look-look book)
SUMMARY: Tired of his dull, old mailbox which seems to bring nothing but bills and advertisements, Donald Duck buys a very unusual mailbox from a traveling antique dealer.
ISBN 0-307-11851-7 : $0.95
⌈1. Mailboxes—Fiction. 2. Magic—Fiction⌉ I. Title. II. Title: Donald Duck and the magic mailbox.
PZ7.D625 Dnm 1978 ⌈E⌉ 78-50650
00 MARC

Disney (Walt) Productions.
Walt Disney's Dumbo, the flying elephant. — 1st American ed. — New York : Random House, c1978.
⌈42⌉ p. : col. ill. ; 25 cm. — (Disney's wonderful world of reading)
"Book club edition."
Based on the story entitled Dumbo, the flying elephant, by H. Aberson and H. Pearl.
SUMMARY: A little elephant uses his enormous ears to become the star of the circus.
ISBN 0-394-84093-3. ISBN 0-394-94093-8 lib. bdg.
⌈1. Elephants—Fiction. 2. Circus stories⌉ I. Aberson, Helen. Dumbo, the flying elephant. II. Title. III. Title: Dumbo, the flying elephant.
PZ7.D625 Du 1978 E 78-112087
79 MARC

Disney (Walt) Productions.
Walt Disney's Li'l Bad Wolf stories. — New York : Golden Press, c1978.
48 p. : col. ill. ; 26 cm. — (A Kid's paperback)
SUMMARY: Presents three adventures involving a young wolf and his wily father.
ISBN 0-307-12078-3 : $1.95
⌈1. Wolves—Fiction. 2. Fathers and sons—Fiction⌉ I. Title. II. Title: Li'l bad wolf stories.
PZ7.D625 Li 1978 ⌈E⌉ 78-50652
79 MARC

Dixon, Franklin W
The firebird rocket / by Franklin W. Dixon. — New York : Grosset & Dunlap, c1978.
180 p. : ill. ; 20 cm. — (The Hardy boys ; 57)
SUMMARY: Two teenage sleuths help their detective father search for a famous rocket scientist whose disappearance endangers the launching of the Firebird Rocket.
ISBN 0-448-08957-2. ISBN 0-448-18957-7 lib. bdg. : $2.50
⌈1. Mystery and detective stories⌉ I. Title.
PZ7.D644 Fh ⌈Fic⌉ 77-76131
78 MARC

Dixon, Franklin W
The sting of the scorpion / by Franklin W. Dixon. — New York : Grosset & Dunlap, c1979.
180 p., ⌈1⌉ leaf of plates : ill. ; 20 cm. — (Hardy boys mystery stories ; 58)
SUMMARY: During their father's investigation of a ruthless gang of terrorists, two young detectives face several adversaries.
ISBN 0-448-08958-0 : $2.95
⌈1. Mystery and detective stories⌉ I. Title.
PZ7.D644 St ⌈Fic⌉ 78-57930
79 MARC

Dockery, Wallene T 1941-
Gabby's Christmas wish / by Wallene T. Dockery ; illustated by Maurine Zook. — Austin, Tex. : Shoal Creek Publishers, c1978.
⌈32⌉ p. : col. ill. ; 28 cm.
SUMMARY: A small bird seeks ways to help his friend who is eager to become a full-grown Christmas tree.
ISBN 0-88319-042-7 : $5.95
⌈1. Christmas trees—Fiction. 2. Birds—Fiction. 3. Friendship—Fiction⌉ I. Zook, Maurine, 1929- II. Title.
PZ7.D6622 Gab ⌈E⌉ 78-11802
78 MARC

Dodgson, Charles Lutwidge, 1832-1898.
Alice's adventures in Wonderland & Through the looking glass / by Lewis Carroll ⌈i.e. C. L. Dodgson⌉ ; with sixty six illustrations by Mervyn Peake. — London : Methuen, 1978.
viii, 221 p. : ill. ; 24 cm. GB78-35569
ISBN 0-416-55280-3 : £5.50
I. Peake, Mervyn Laurence, 1911-1968. II. Dodgson, Charles Lutwidge, 1832-1898. Through the looking glass. III. Title.
PZ7.D684 An 1978c 79-311571
79 MARC

Dodgson, Charles Lutwidge, 1832-1898.
The nursery "Alice," by Lewis Carroll. With a new introd. by Martin Gardner. New York, McGraw-Hill ⌈1966⌉
xi, 56, ⌈7⌉ p. col. illus. 25 cm.
Original t.p. reads: The nursery "Alice," containing twenty coloured enlargements from Tenniel's illustrations to "Alice's adventures in Wonderland." With text adapted to nursery readers by Lewis Carroll. Cover designed and coloured by E. Gertrude Thomson. London, Macmillan and Co., 1890.
Includes the reproductions of the cover designs and the additional material (⌈7⌉ p. at end) from the original ed.
SUMMARY: A little girl falls down a rabbit hole and discovers a strange world inhabited by all sorts of unusual characters.
⌈1. Fantasy⌉ I. Gardner, Martin, 1914- ed. II. Tenniel, John, Sir, 1820-1914, illus. III. Dodgson, Charles Lutwidge, 1832-1898. Alice's adventures in Wonderland. IV. Title.
PZ7.D684 Nu 1966 ⌈Fic⌉ 66-24136
r79⌉rev MARC

Dodson, Susan.
The creep / by Susan Dodson. — New York : Four Winds Press, c1979.
218 p. ; 22 cm.
SUMMARY: After saving a young child from a molester, a 16-year-old girl volunteers to help the police catch the man.
ISBN 0-590-07599-3 : $7.95
⌈1. Child molesting—Fiction. 2. Crime and criminals—Fiction⌉ I. Title.
PZ7.D687 Cr ⌈Fic⌉ 79-1102
79 MARC

The Dog power tower / Beverley Allinson ... ⌈et al. ; ill., Alan Daniel⌉. — New York : Methuen, c1977.
⌈34⌉ p. : col. ill. ; 18 x 26 cm. — (A Mr. Dressup book)
Cover title.
SUMMARY: Searching for a new energy source, Dr. Bernard Boggs unleashes amazing power from the feet of dogs he has dognapped.
ISBN 0-458-93060-1 : $3.95
⌈1. Energy industries—Fiction. 2. Dogs—Fiction⌉ I. Allinson, Beverley. II. Daniel, Alan, 1936-
PZ7.D694 ⌈E⌉ 78-105194
78 MARC

Dolan, Sheila.
The wishing bottle / Sheila Dolan ; illustrated by Leslie Morrill. — Boston : Houghton Mifflin, 1979.
81 p. : ill. ; 22 cm.
SUMMARY: Nora never gives up hope for a pony to fill the barn behind her family's house despite the voices of reason that tell her wishing won't make things so.
ISBN 0-395-28479-1
⌈1. Ponies—Fiction⌉ I. Morrill, Leslie H. II. Title.
PZ7.D698 Wi ⌈Fic⌉ 79-13061
79 MARC

Donohue, Dina.
Mr. Harty comes for Christmas / Dina Donohue ; illustrated by Unada. — Valley Forge, PA. : Judson Press, c1978.
⌈32⌉ p. : ill. ; 24 cm.
SUMMARY: A nine-year-old takes it upon himself to invite a grumpy neighbor to share Christmas with his lively family.
ISBN 0-8170-0771-7 : $1.95
⌈1. Christmas stories. 2. Neighborliness—Fiction. 3. Christian life—Fiction⌉ I. Unada. II. Title.
PZ7.D7223 Mi ⌈E⌉ 78-104871
78 MARC

Douglass, Barbara, 1930-
Skateboard scramble / by Barbara Douglass ; illustrated by Alex Stein. — 1st ed. — Philadelphia : Westminster Press, c1979.
91 p. : ill. ; 21 cm.
SUMMARY: Although she loves skateboarding, Jody is uneasy when her father insists that she participate in a skateboard competition, expecially as she would be competing against her best friend.
ISBN 0-664-32641-2 (lib. bdg.)
⌈1. Skateboarding—Fiction. 2. Friendship—Fiction⌉ I. Stein, Alex. II. Title.
PZ7.D7479 Sk ⌈Fic⌉ 78-12480
78 MARC

Downing, Sybil.
Magic, mystery, and monsters / Sybil Downing and Jane Valentine Barker ; ⌈ill. by Robert F. Wilson⌉. — 1st ed. — Boulder, Colo. : Pruett Pub. Co., c1979.
52 p. : ill. ; 23 cm. — (Colorado heritage series ; book 6)
Bibliography: p. 46-49.
CONTENTS: Wild things.—The mystery box.—Monster bones.—The magic air.
ISBN 0-87108-219-5
⌈1. Colorado—History—Fiction. 2. Short stories⌉ I. Barker, Jane Valentine, 1930- joint author. II. Wilson, Robert Franklin, 1937- III. Title. IV. Series.
PZ7.D7594 Mag ⌈Fic⌉ 79-52143
79 MARC

Downing, Sybil.
Mesas to mountains / Sybil Downing and Jane Valentine Barker ; ⌈ill. by Robert F. Wilson⌉. — 1st ed. — Boulder, Colo. : Pruett Pub. Co., c1979.
47 p. : ill. ; 23 cm. — (Colorado heritage series ; book 4)
Bibliography: p. 42-44.
CONTENTS: Witches and lions.—The lost arrows.—The trick.—Smoking water.
ISBN 0-87108-215-2
⌈1. Indians of North America—Colorado—Fiction. 2. Short stories⌉ I. Barker, Jane Valentine, 1930- joint author. II. Wilson, Robert Franklin, 1937- III. Title. IV. Series.
PZ7.D7594 Me ⌈Fic⌉ 79-107078
79 MARC

Dowsett, Norman C
Ting-a-ling-pa / Norman C. Dowsett. — Pondicherry : Sri Aurobindo Ashram, 1955, 1975 printing.
89 p. : ill. ; 30 cm.
A novel.
Rs9.00
I. Title.
PZ7.D7596 Ti 76-904013
77⌈r79⌉rev MARC

Draper, Cena Christopher.
The worst hound around / Cena C. Draper. — 1st ed. — Philadelphia : Westminster Press, c1979.
115 p. ; 21 cm.
SUMMARY: A young boy, his father, and his dog are living a happy bachelor existence in the Missouri ridgeland until his mother's cousin with three daughters descends on them for a long visit.
ISBN 0-664-32643-9
⌈1. Mountain life—Fiction. 2. Family life—Fiction⌉ I. Title.
PZ7.D7832 Wo ⌈Fic⌉ 78-25687
78 MARC

CHILDREN'S LITERATURE (Continued)

Du Bois, William Pène, 1916-
The Forbidden Forest / ₍by₎ William Pene du Bois. — London : Chatto and Windus, 1978.
56 p. : col. ill. ; 29 cm. GB78-35676
"An Ursula Nordstrom book"
ISBN 0-7011-2396-6 : £2.95
I. Title.
PZ7.D8527 Fo 1978 79-310794
79 MARC

Du Bois, William Pène, 1916-
The forbidden forest / by William Pène du Bois. — 1st ed. — New York : Harper & Row, c1978.
56 p. : col. ill. ; 29 cm.
"An Ursula Nordstrom book."
SUMMARY: Lady Adelaide, a boxing kangaroo, helps to defeat the German army, thus becoming a heroine of the "Great War."
ISBN 0-06-024699-5 : $6.95. ISBN 0-06-024700-2 lib. bdg. : $6.79
₍1. European War, 1914-1918—Fiction. 2. Kangaroos—Fiction₎ I. Title.
PZ7.D8527 Fo ₍Fic₎ 77-25651
78 MARC

Duc, pseud.
Stevie searches for buried treasure. New York, Hart Pub. Co. ₍1968, c1966₎
32 p. col. illus., col. map (on lining papers) 31 cm. 2.95
Published under the author's real name: Bernard Ducourant.
Translation of S.O.S. Stanislas.
SUMMARY: Trying to defend their island and the right to hunt for treasure on it, four young boys wreck their raft and become stranded on a desert island for several weeks.
₍1. Survival—Fiction. 2. Adventure stories₎ I. Title.
₍PZ7.D8549 St 1968₎ ₍Fic₎ 68-29534
₍r79₎rev MARC

Duckling and chick. — Hong Kong : Seagull Pub. Co., ₍1977?₎
12 p. : col. ill. ; 18 x 20 cm.
Cover title.
Translation of Hsiao ya ho hsiao chi.
PZ7.D8577 77-152953
78 MARC

Duncan, Jane.
Janet Reachfar and Chickabird / by Jane Duncan ; pictures by Mairi Hedderwick. — London : Macmillan, 1978.
₍32₎ p. : col. ill. ; 26 cm. GB78-26402
ISBN 0-333-23742-0 : £3.95
I. Hedderwick, Mairi. II. Title.
PZ7.D899 Jam 1978b 78-325019
79 MARC

Duncan, Jane.
Janet Reachfar and Chickabird / by Jane Duncan ; pictures by Mairi Hedderwick. — 1st American ed. — New York : Seabury Press, 1978.
₍32₎ p. : col. ill. ; 26 cm.
"A Clarion book".
SUMMARY: Because of her own carelessness, a young girl must decide the fate of an injured chicken.
ISBN 0-8164-3203-1
₍1. Responsibility—Fiction₎ I. Hedderwick, Mairi. II. Title.
PZ7.D899 Jam 1978 ₍E₎ 77-12709
77 MARC

Duncan, Kathleen M
Crispin's castle / Kathleen M. Duncan. — Grand Rapids : Zondervan Pub. House, c1979.
100 p. ; 21 cm. — (Pathfinder series ; 1)
SUMMARY: An orphan with no hint of his identity, Crispin moves in with the Bernard family, where he feels strongly at home in the shadow of an old English castle and discovers his true identity.
ISBN 0-310-37821-4 : $1.95
₍1. Orphans—Fiction. 2. England—Fiction₎ I. Title. II. Series. III. Series: Pathfinder series (Grand Rapids, 1979-) ; 1.
PZ7.D9116 Cr 1979 ₍Fic₎ 78-21525
78 MARC

Duncan, Kathleen M
Sally and the shepherdess / Kathleen M. Duncan. — Grand Rapids : Zondervan, 1979.
102 p. ; 21 cm. — (Pathfinder series ; 2)
SUMMARY: Sally travels with a family to the mountains in Austria where her artistic talent aids in the recovery of a stolen antique carving.
ISBN 0-310-37801-X
₍1. Family life—Fiction. 2. Christian life—Fiction. 3. Drawing—Fiction₎ I. Title. II. Series: Pathfinder series (Grand Rapids, 1979-) ; 2.
PZ7.D9116 Sal 1979 ₍Fic₎ 78-24324
78 MARC

Duncan, Lois, 1934-
Daughters of Eve / by Lois Duncan. — 1st ed. — Boston : Little, Brown, c1979.
239 p. ; 24 cm.
SUMMARY: A high school teacher uses the guise of feminist philosophy to manipulate the lives of a group of girls with chilling results.
ISBN 0-316-19550-2
₍1. Feminism—Fiction₎ I. Title.
PZ7.D9117 Dau ₍Fic₎ 79-14918
79 MARC

Duncan, Riana.
A tale of ten town mice / ₍by₎ Riana Duncan. — London : Andersen Press ₍etc.₎, 1978.
₍29₎ p. : col. ill. ; 24 cm. GB78-35677
ISBN 0-905478-39-8 : £2.50
I. Title.
PZ7.D913 Tal 1978 79-310972
79 MARC

Dunlop, Eileen.
Fox farm / Eileen Dunlop. — 1st American ed. — New York : Holt, Rinehart and Winston, 1979, c1978.
149 p., ₍1₎ leaf of plates : ill. ; 22 cm.
SUMMARY: Through caring for a stray fox cub, a ten-year-old gradually accepts the fact that, though he too has been abandoned, he does have a place in his foster family.
ISBN 0-03-049051-0
₍1. Foster home care—Fiction. 2. Foxes—Fiction. 3. Scotland—Fiction₎ I. Title.
PZ7.D9214 Fo 1979 ₍Fic₎ 78-14091
78 MARC

Dunne, Mary Collins.
The secret of Captives' Cave / Mary Collins Dunne. — New York : Putnam, c1976.
159 p. ; 21 cm.
SUMMARY: When he moves back to his dead father's birthplace, Dennis tries to uncover the mystery surrounding Captives' Cave which is linked to his ancestors.
ISBN 0-399-20547-0 : $5.95. ISBN 0-399-61036-7 lib. bdg.
₍1. Southern States—Race relations—Fiction. 2. Slavery in the United States—Fiction₎ I. Title.
PZ7.D92174 Se ₍Fic₎ 76-25228
76₍r79₎rev MARC

Durán, Cheli.
Hildilid's night, by Cheli Durán Ryan. Illustrated by Arnold Lobel. New York, Macmillan ₍1971₎
₍30₎ p. illus. (part col.) 18 x 23 cm. $4.50
SUMMARY: An old lady who hates everything about night tries all sorts of methods to chase it away.
₍1. Night—Fiction₎ I. Lobel, Arnold, illus. II. Title.
PZ7.D93 Hi ₍E₎ 75-146627
71₍r79₎rev MARC

Durán, Cheli.
Kindling / by Cheli Durán. — 1st ed. — New York : Greenwillow Books, c1979.
153 p. ; 22 cm.
SUMMARY: Under suspicion of setting fire to some fields, a group of boys from a village in Crete are incarcerated and tortured by officials hoping to implicate the boys' parents.
ISBN 0-688-80199-4 : $6.95
₍1. Crete—Fiction. 2. Prisoners and prisons—Fiction₎ I. Title.
PZ7.D93 Ki ₍Fic₎ 78-23629
78 MARC

Durán, Cheli.
Paz, by Cheli Durán Ryan. Illustrated by Nonny Hogrogian. New York, Macmillan ₍1971₎
37 p. col. illus. 20 cm. $4.95
SUMMARY: A family living on the border of France and Spain find a solution to their divided loyalties when the countries go to war.
₍1. Peace treaties—Fiction₎ I. Hogrogian, Nonny, illus. II. Title.
PZ7.D93 Paz ₍E₎ 73-139444
71₍r79₎rev MARC

Durham, John, fl. 1960-
Me and Arch and the Pest. Illustrated by Ingrid Fetz. New York, Four Winds Press ₍1970₎
96 p. illus. 24 cm. 4.18
SUMMARY: Two boys acquire a German shepherd whose disappearance involves them in a crime.
₍1. Dogs—Fiction₎ I. Fetz, Ingrid, illus. II. Title.
PZ7.D9335 Me ₍Fic₎ 73-105338
70₍r79₎rev MARC

Dutton, June.
Hope is a handful of dreams / by June Dutton ; ill. by Susan Perl. — 1st ed. — San Francisco : Determined Productions, c1978.
ca. 150 p. : col. ill. ; 19 cm.
SUMMARY: Offers various definitions of hope as seen in daily life.
ISBN 0-915696-10-X
₍1. Hope₎ I. Perl, Susan. II. Title.
PZ7.D949 Ho ₍Fic₎ 77-90655
79 MARC

Duvoisin, Roger Antoine, 1904-
Snowy and Woody / Roger Duvoisin. — New York : Knopf : distribution by Random House, c1979.
₍32₎ p. : col. ill. ; 26 cm.
SUMMARY: A polar bear and a brown bear help camouflage each other.
ISBN 0-394-84241-3
₍1. Bears—Fiction. 2. Friendship—Fiction₎ I. Title.
PZ7.D957 Sn 1979 ₍E₎ 79-897
79 MARC

Dwiggins, Don.
The asteroid war / by Don Dwiggins. — Chicago : Childrens Press, c1978.
63 p. : ill. ; 21 cm.
"A Golden Gate junior book."
SUMMARY: The mining operations of a space colony near the asteroid belt are threatened by inhabitants of a distant galaxy.
ISBN 0-516-08807-6
₍1. Science fiction₎ I. Title.
PZ7.D9597 As ₍E₎ 78-5599
78 MARC

Dwiggins, Don.
Jimmy Fox and the Flying Falcon / by Don Dwiggins ; illustrated by Olga Dwiggins. — Chicago : Childrens Press, c1977.
91 p. : ill. ; 21 cm.
"A Golden Gate junior book."
SUMMARY: Although Jimmy has had instruction, his first solo sailplane flight on his 14th birthday is almost more than he can deal with.
ISBN 0-516-08826-2
₍1. Gliding—Fiction₎ I. Dwiggins, Olga. II. Title.
PZ7.D9597 Ji ₍Fic₎ 77-24301
77 MARC

Dygard, Thomas J
Outside shooter / Thomas J. Dygard. — 1st ed. — New York : W. Morrow, 1979.
189 p. ; 21 cm.
SUMMARY: A high school basketball star's personality problems threaten not only his own career but the success of the whole team.
ISBN 0-688-22177-7
₍1. Basketball—Fiction₎ I. Title.
PZ7.D9893 Ou ₍Fic₎ 78-24002
78 MARC

Dyke, John, 1935-
Pigwig / ₍by₎ John Dyke. — London ₍etc.₎ : Methuen, 1978.
₍32₎ p. : chiefly col. ill. ; 30 cm. GB78-34121
ISBN 0-416-87110-0 : £2.95
I. Title.
PZ7.D9895 Pi 79-303231
79 MARC

Dyke, John, 1935-
Pigwig / John Dyke. — New York : Methuen, 1978.
₍32₎ p. : col. ill. ; 30 cm.
SUMMARY: Pigwig decides to win the love of pretty Matilda by improving his appearance and begins to make and wear a variety of unusual hats.
ISBN 0-458-93510-7 : $5.95
₍1. Pigs—Fiction₎ I. Title.
PZ7.D9895 Pi 1978 78-55390
79 MARC

Eager, Edward McMaken.
Magic or not? / Edward Eager ; illustrated by N. M. Bodecker. — 1st Voyager/HBJ ed. — New York : Harcourt, Brace, Jovanovich, 1979, c1959.
190 p. : ill. ; 20 cm. — (A Voyager/HBJ book)
Reprint of the ed. published by Harcourt, Brace & World, New York.
SUMMARY: An old well that might be magic affords an exciting summer for the twins, James and Laura, and helps them to right an ancient wrong.
ISBN 0-15-655121-7
₍1. Magic—Fiction₎ I. Bodecker, N. M. II. Title.
₍PZ7.E115 Maj 1979₎ ₍Fic₎ 78-71152
79 MARC

Earle, Olive Lydia.
Squirrels in the garden, written and illustrated by Olive L. Earle. New York, Morrow, 1963.
63 p. illus. 22 cm.
SUMMARY: An account of two squirrels' activities as they explore their environs, learn survival tactics, prepare for the winter, and mature to adulthood.
1. Squirrels—Legends and stories. ₍1. Squirrels—Fiction₎ I. Title.
PZ10.3.E108 Sq ₍E₎ 62-15762
₍r79₎rev MARC

Edler, Tim.
T-boy in Mossland : story & pictures / by Tim Edler. — Baton Rouge, La. : Little Cajun Books, 1978.
42 p. : ill. ; 28 cm. — (Tim Edler's Tales from the Atchafalaya ; no. 3)
SUMMARY: Through his secret oak tree a Cajun boy enters Mossland where crawfish threaten to boil and eat him.
₍1. Fantasy₎ I. Title.
PZ7.E243 Taj ₍E₎ 78-111707
79 MARC

Edler, Tim.
T-boy the little Cajun / by Tim Edler ; ₍ill. by Van Judice₎. — Baton Rouge, La. : Little Cajun Books, 1978.
29 p. : ill. ; 28 cm. — (Tim Edler's Tales from the Atchafalaya ; no. 1)
SUMMARY: Relates the adventures of a little Cajun boy growing up on the banks of the Bayou Teche in South Louisiana.
ISBN 0-931108-01-2
₍1. Cajuns—Fiction. 2. Louisiana—Fiction₎ I. Judice, Van. II. Title.
PZ7.E243 Tan ₍Fic₎ 78-111728
79 MARC

CHILDREN'S LITERATURE (Continued)

Eichelberger, Rosa Kohler.
Big fire in Baltimore / Rosa Kohler Eichelberger ; illustrated by Rex Schnedier. — 1st ed. — Owings Mills, Md. : Stemmer House, 1979.
204 p. : ill. ; 23 cm.
"A Barbara Holdridge book."
SUMMARY: The desire of a boy to become a Western Union messenger is fulfilled during the conflagration that destroyed much of Baltimore in 1904.
ISBN 0-916144-36-4 : $11.95. ISBN 0-916144-37-2 pbk. : $7.95
1. Baltimore—Fire, 1904—Juvenile fiction. [1. Baltimore—Fire, 1904—Fiction. 2. Fires—Fiction] I. Schneider, Rex. II. Title.
PZ7.E338 Bi 1979 [Fic] 78-31311
78 MARC

Eiseman, Alberta.
Gift from a sheep : the story of how wool is made / by Alberta & Nicole Eiseman ; illustrated by Tracy Sugarman. — 1st ed. — New York : Atheneum, 1979.
52 p. : ill. ; 22 cm.
SUMMARY: Jenny raises a lamb; shears it; cards, spins, washes, and dyes the wool into yarn; knits a poncho; and wins prizes for her spinning and knitting at the county fair. Includes directions for making the poncho.
ISBN 0-689-30707-1
[1. Wool—Fiction] I. Eiseman, Nicole, joint author. II. Sugarman, Tracy, 1921- . III. Title.
PZ7.E345 Gi 1979 [Fic] 79-10629
79 MARC

Eisenstein, Phyllis, 1946-
Born to exile / Phyllis Eisenstein ; ill. by Stephen E. Fabian. — Sauk City, Wis. : Arkham House, 1978.
202 p. : ill. ; 20 cm. — (Arkham House science fiction novel)
SUMMARY: The solitary wanderings of a minstrel with preternatural powers lead him to many adventures before he is ultimately restored to his supernatural antecedents, the Lords of All Power.
ISBN 0-87054-082-3 : $8.95
[1. Fantasy] I. Fabian, Stephen E., 1930- II. Title.
PZ7.E3468 Bo [Fic] 77-78598
79 MARC

Eldridge, Roger.
The shadow of the Gloom-World / Roger Eldridge. — 1st ed. — New York : Dutton, 1978, c1977.
191 p. ; 22 cm.
SUMMARY: Fernfeather questions the barren existence of the folk's underground world and crosses the forbidden barrier to discover the truth about their benefactors and the terrifying Gloom-World.
ISBN 0-525-39235-1 lib. bdg. : $7.95
[1. Fantasy] I. Title.
PZ7.E384 Sh 1978 [Fic] 77-26200
77[r79]rev MARC

Elias, Horace J
The costume party : a new Lori Bigelow story / Story by Horace J. Elias ; illustrated characters created by Lori Bigelow. — [Chicago] : Watertower Books, c1979.
28 p. : col. ill. ; 29 cm.
Running title: A Bigelow bunch story.
SUMMARY: Summer invites her friends to a costume party at which everyone is a character from "The Wizard of Oz."
ISBN (invalid) 0-516-13339-9
[1. Parties—Fiction. 2. Costume—Fiction] I. Bigelow, Lori. II. Title.
PZ7.E395 Co 79-112448
79 MARC

Elias, Horace J
The Flintstones storybook / stories by Horace J. Elias. — Hanna-Barbera authorized ed. — New York : Grosset & Dunlap, c1978.
79 p. : col. ill. ; 24 cm.
SUMMARY: Four brief stories relate the antics of the Flintstone family and their neighbors, the Rubbles.
ISBN 0-448-14744-0
[1. Humorous stories] I. Title.
PZ7.E395 Fl 1978 [E] 77-94035
79 MARC

Ellis, Joyce K
The big split / by Joyce K. Ellis. — Nashville : T. Nelson, c1979.
168 p. : ill. ; 20 cm.
SUMMARY: Rod's newfound faith in God helps heal the wounds of his parent's divorce.
ISBN 0-8407-5672-0 : $3.95
[1. Divorce—Fiction. 2. Christian life—Fiction] I. Title.
PZ7.E4735 Bi [Fic] 78-21262
78 MARC

Epp, Margaret A
Sarah and the lost friendship Epp [sic] / Margaret Epp. — Wheaton, Ill. : Victor Books, c1979.
131 p. : ill. ; 20 cm. — (Prairie adventure series) (A Winner book)
Edition of 1970 published under title: The princess rides a panther.
"Epp" appears in the title by mistake.
SUMMARY: The adventures of a ten-year-old girl and her farm family during a snowy winter.
ISBN 0-88207-483-0 : $1.75
[1. Family life—Fiction. 2. Christian life—Fiction] I. Title. II. Series.
PZ7.E7248 San 1979 78-65203
79 MARC

Epstein, Samuel, 1909-
Hurricane guest, by Sam and Beryl Epstein. Illustrated by Marilyn Miller. New York, Random House [1964]
57 p. col. illus. 24 cm. (The Random House easy to read library, R-33)
SUMMARY: Mike and Tess's plans for their English houseguest are forgotten in the excitement resulting from a ferocious hurricane.
[1. Hurricanes—Fiction] I. Epstein, Beryl (Williams) 1910- joint author. II. Miller, Marilyn, 1925- . III. Title.
PZ7.E7252 Hu [Fic] 64-11174
r79]rev MARC

Erickson, Russell E
Warton and the King of the Skies / by Russell E. Erickson ; pictures by Lawrence Di Fiori. — 1st ed. — New York : Lothrop, Lee & Shepard Co., c1978.
95 p. : ill. ; 22 cm.
SUMMARY: Two toad brothers gain their freedom from a bickering weasel family by teaching them the value of teamwork.
ISBN 0-688-41852-X. ISBN 0-688-51852-4 lib. bdg.
[1. Toads—Fiction. 2. Weasels—Fiction. 3. Brothers and sisters—Fiction] I. Di Fiori, Lawrence. II. Title.
PZ7.E7257 Was [E] 78-4919
78 MARC

Erickson, Russell E
Warton and the traders / by Russell E. Erickson ; pictures by Lawrence Di Fiori. — 1st ed. — New York : Lothrop, Lee & Shepard Co., c1979.
95 p. : ill. ; 22 cm.
SUMMARY: Warton Toad turns the wood rats' trading habits to his own advantage by agreeing to help them trap the dreaded wildcat in exchange for their assistance in rescuing a starving fawn.
ISBN 0-688-41886-4
[1. Toads—Fiction. 2. Animlas—Fiction] I. Di Fiori, Lawrence. II. Title.
PZ7.E7257 Wash [E] 78-25689
78 MARC

Erskine, Jim.
Bert & Susie's messy tale / by Jim Erskine. — New York : Crown Publishers, c1979.
[24] p. : col. ill. ; 14 x 18 cm.
SUMMARY: Bert has problems getting his mother's chrysanthemum to the flower show.
ISBN 0-517-53475-4
[1. Play—Fiction] I. Title.
PZ7.E738 Be 1979 [E] 78-24564
78 MARC

Es ging spazieren vor dem Tor, ein kohlpechrabenschwarzer Mohr
: d. Dritte Welt im Spiegel von Kinder- u. Jugendbüchern : fachübergreifende Unterrichtungsmaterialen für d. Sekundarstufe I / [Bearb.], Jörg Becker ; hrsg. vom Bundesministerium für Wirtschaftl. Zusammenarbeit, Referat Öffentlichkeitsarbeit. — Bonn : Bundesministerium für Wirtschaftl. Zusammenarbeit, Referat Öffentlichkeitsarbeit, 1976.
113 p. : ill. ; 30 cm. — (Schule und Dritte Welt ; Nr. 48) GFR78-B1(
Includes bibliographies.
1. Underdeveloped areas—Education (Elementary) 2. Teaching—Aids and devices. 3. Textbooks. 4. Children's literature. I. Becker, Jörg. II. Germany (Federal Republic, 1949-). Bundesministerium für Wirtschaftliche Zusammenarbeit. Referat Öffentlichkeitsarbeit. III. Series.
LC2608.E8 79-366205
79 MARC

Estes, Eleanor, 1906-
The lost umbrella of Kim Chu / by Eleanor Estes ; illustrated by Jacqueline Ayer. — 1st ed. — New York : Atheneum, 1978.
85 p. : ill. ; 21 cm.
"A Margaret K. McElderry book."
SUMMARY: Nine-year-old Kim Chu who lives in New York City's Chinatown searches for her father's special umbrella that Kim left in an umbrella stand in the library.
ISBN 0-689-50111-0 : $7.95
[1. Umbrellas and parasols—Fiction. 2. Chinese Americans—Fiction. 3. New York (City)—Fiction] I. Ayer, Jacqueline. II. Title.
PZ7.E749 Lo 1978 78-59156
79 MARC

Estes, Eleanor, 1906-
The middle Moffat, by Eleanor Estes, illustrated by Louis Slobodkin. New York, Harcourt, Brace and company [1942]
5 p.l., 3-317 p. incl. plates. front., illus. 21 cm.
SUMMARY: Janey, the middle Moffat, has an imagination that leads her into many difficulties.
[1. Family life—Fiction] I. Slobodkin, Louis, 1903- illus. II. Title.
PZ7.E749 Mi [Fic] 42-36272
r79]rev MARC

Estes, Eleanor, 1906-
Pinky Pye / Eleanor Estes ; illustrated by Edward Ardizzone. — 1st Voyager ed. — New York : Harcourt Brace Jovanovich, 1976, c1958.
192 p. : ill. ; 19 cm. — (A Voyager book ; AVB 100)
SUMMARY: While spending a bird-watching summer on Fire Island, the Pye family acquires a small black kitten that can use a typewriter.
ISBN 0-15-671840-5
[1. Family life—Fiction] I. Ardizzone, Edward, 1900- . II. Title.
[PZ7.E749 Pi 1976] [Fic] 75-31581
75 MARC

Ets, Marie Hall, 1895-
In the forest; story and pictures by Marie Hall Ets. New York, The Viking press, 1944.
[45] p. illus. 19 x 24 xm.
"First published September 1944."
SUMMARY: The animals join a young boy as he walks through the forest blowing his new horn.
[1. Animals—Fiction] I. Title.
PZ7.E855 In [E] 44-7727
——— Copy 2. r79]rev MARC

Evans, Alan, 1930-
Escape at the Devil's Gate / Alan Evans. — London : Hodder and Stoughton, 1978.
111 p. ; 23 cm. GB***
ISBN 0-340-23172-6 : £2.75
I. Title.
PZ7.E8737 Es 79-300185
79 MARC

Evans, Mari, 1923-
Jim flying high / by Mari Evans ; illustrated by Ashley Bryan. — 1st ed. — Garden City, N.Y. : Doubleday, c1979.
[32] p. : col. ill. ; 27 cm.
SUMMARY: A proud flying fish named Jim is rescued from a tree by Olukun, the Chief of the Water.
ISBN 0-385-14129-7 : $7.95
[1. Flying fish—Fiction. 2. Fishes—Fiction] I. Bryan, Ashley. II. Title.
PZ7.E8918 Ji [E] 78-22628
79 MARC

Ewing, Kathryn.
A private matter / Kathryn Ewing ; illustrated by Joan Sandin. — 1st ed. — New York : Harcourt Brace Jovanovich, [1975]
88 p. : ill. ; 21 cm.
SUMMARY: In her longing for a father, nine-year-old Marcy becomes extremely attached to the kind man next door.
ISBN 0-15-263576-9
[1. Single-parent family—Fiction] I. Sandin, Joan, ill. II. Title.
PZ7.E96 Pr [Fic] 74-23673
74[r79]rev MARC

Eyerly, Jeannette.
See Dave run : a novel / by Jeannette Eyerly. — Philadelphia : Lippincott, c1978.
126 p. : port. ; 21 cm.
SUMMARY: Running away from an intolerable home situation, a 15-year-old boy finds that he has nowhere to go and no one to turn to.
ISBN 0-397-31819-7
[1. Runaways—Fiction. 2. Suicide—Fiction] I. Title.
PZ7.E97 Se [Fic] 78-8139
78 MARC

Fanning, Robbie.
100 butterflies / Robbin Fanning. — 1st ed. — Philadelphia : Westminster Pres, c1979.
189 p. ; 21 cm.
SUMMARY: After her father dies, 13-year-old Ristyn spends the summer with her grandmother who affords her more freedom than she has ever known.
ISBN 0-664-32654-4
[1. Death—Fiction. 2. Grandmothers—Fiction] I. Title.
PZ7.F22215 On [Fic] 79-14776
79 MARC

Farrar, Susan Clement.
Samantha on stage / by Susan Clement Farrar ; pictures by Ruth Sanderson. — New York : Dial Press, c1979.
164 p. : ill. ; 22 cm.
SUMMARY: Eleven-year-old Samantha had always been the best in her ballet class, but when she saw the new Russian girl dance she began to wonder who would get the coveted lead in the school's production of the Nutcracker ballet.
ISBN 0-8037-7574-1 : $7.95
[1. Ballet dancing—Fiction. 2. Friendship—Fiction] I. Sanderson, Ruth. II. Title.
PZ7.F2434 Sam [Fic] 78-64958
79 MARC

Farris, Stella.
The magic bubble-pipe / by Stella Farris. — 1st ed. — New York : Harper & Row, 1978.
[28] p. : col. ill. ; 16 cm. — (A Bedtime book)
SUMMARY: Paper pop-ups reveal the strange sights Scott sees when he uses his great-grandfather's pipe in his evening bath.
ISBN 0-06-021878-9 : $3.95
[1. Baths—Fiction. 2. Bubbles—Fiction] I. Title.
PZ7.F2462 Mag [E] 78-108162
78 MARC

Farris, Stella.
The magic castle / by Stella Farris. — 1st ed. — New York : Harper & Row, [1978]
[27] p. : col. ill. ; 17 cm. — (A Bedtime book)
SUMMARY: One night Scott's dreams take him to a castle in a strange land where he meets some extraordinary creatures, as shown in the pop-up illustrations.
ISBN 0-06-021877-0 : $3.95
[1. Dreams—Fiction] I. Title.
PZ7.F2462 Mah [E] 78-108161
79 MARC

CHILDREN'S LITERATURE (Continued)

Feagles, Anita MacRae.
Sophia Scarlotti and Ceecee / Anita MacRae Feagles. — 1st ed. — New York : Atheneum, 1979.
164 p. ; 22 cm.
SUMMARY: Ceecee's 16th summer is one of many changes, all of which affect her relationship with her alter ego, Sophia Scarlotti.
ISBN 0-689-30680-6
[1. Identity—Fiction] I. Title.
PZ7.F295 So [Fic] 78 78-12630 MARC

Feder, Jane.
Beany / by Jane Feder ; pictures by Karen Gundersheimer. — New York : Pantheon Books, c1979.
[32] p. : col. ill. ; 12 x 23 cm.
SUMMARY: A child describes his relationship to his cat.
ISBN 0-394-83734-7. ISBN 0-394-93734-1 lib. bdg.
[1. Cats—Fiction] I. Gundersheimer, Karen. II. Title.
PZ7.F2997 Be 1979 [E] 78 78-10416 MARC

Feder, Paula Kurzband.
Where does the teacher live? / By Paula Kurzband Feder ; pictures by Lillian Hoban. — 1st ed. — New York : Dutton, c1979.
48 p. : col. ill. ; 24 cm. — (A Fat cat book)
SUMMARY: Three children try to discover where their teacher lives.
ISBN 0-525-42586-1
[1. Teachers—Fiction. 2. School stories] I. Hoban, Lillian. II. Title. III. Series.
PZ7.F3113 Wh 1979 [E] 78 78-13157 MARC

Fenner, Carol.
The skates of Uncle Richard / by Carol Fenner ; illustrated by Ati Forberg. — New York : Random House, c1978.
46 p. : ill. ; 24 cm.
SUMMARY: With her Uncle's encouragement a nine-year-old takes the first step toward realizing her dream of becoming a figure skater.
ISBN 0-394-83553-0 : $4.95. ISBN 0-394-93553-5 lib. bdg. : $5.99
[1. Ice-skating—Fiction. 2. Uncles—Fiction] I. Forberg, Ati. II. Title.
PZ7.F342 Sk [Fic] 78 78-55910 MARC

Fernandez, Peter.
A dramatization of The grey king. [Sound recording] Based on the book by Susan Cooper. Newbery Award Records NAR 3099–3100. p1977.
2 discs. 33⅓ rpm. 12 in.
Duration: ca. 1 hr., 34 min., 43 sec.
In container; manual sequence.
Notes on container.
SUMMARY: Visiting in North Wales, Will Stanton is awakened to the fact that he is the youngest of the immortal Old Ones who serve the forces of the Light and that he is destined to find the golden harp, essential to the defeat of the Dark. In his quest he encounters the Grey king, servant of the Dark, and becomes enmeshed in perilous happenings.
[1. Fantasy. 2. Wales—Fiction] I. Cooper, Susan. Grey king. II. Title: Grey king.
[PZ7] 79–740516 MN

Fidler, Kathleen.
The lost cave / [by] Kathleen Fidler. — Glasgow [etc.] : Blackie, 1978.
127 p. : map ; 21 cm. GB78-19618
ISBN 0-216-90465-X : £3.25
I. Title.
PZ7.F452 Lo 1978 78 78-321024 MARC

Filson, Brent.
The puma / by Brent Filson. — 1st ed. — Garden City, N.Y. : Doubleday, c1979.
112 p. ; 22 cm. — (A Doubleday signal book)
SUMMARY: After he is thrown off the team for fighting, 16-year-old Sonny Street seeks to re-establish his position as star wrestler and to close the distance between his father and himself which has been increasing since the death of his mother.
ISBN 0-385-12983-1 : $5.95
[1. Wrestling—Fiction. 2. Fathers and sons—Fiction] I. Title.
PZ7.F49566 Pu [Fic] 78 77-12853 MARC

Fine, Anne.
The summer-house loon / Anne Fine. — 1st U.S. ed. — New York : Crowell, 1979, c1978.
126 p. ; 21 cm.
SUMMARY: A teenage girl and her blind history professor father help a young couple in love solve their biggest problem.
ISBN 0-690-03933-6 : $6.95. ISBN 0-690-03934-4 lib. bdg. : $6.79
[1. Fathers and daughters—Fiction. 2. Blind—Fiction. 3. Physically handicapped—Fiction] I. Title.
PZ7.F495673 Su 1979 [Fic] 78 78-19515 MARC

Firmin, Charlotte.
Hannah's great decision / [by] Charlotte Firmin. — London : Macmillan, 1978.
[25] p. : col. ill. ; 27 cm. GB78-13885
ISBN 0-333-23719-6 : £3.25
I. Title.
PZ7.F498725 Han 1978 78 78-321028 MARC

The First Puffin's pleasure / presented by Kaye Webb & Treld Bicknell. — Harmondsworth, Eng. : Puffin Books ; New York : Penguin Books, 1976.
92 p. : ill. (some col.) ; 29 cm. GB***
SUMMARY: An anthology of short stories, articles, poems, suggested projects, and games.
1. Children's literature. [1. Literature—Collections. 2. Amusements] I. Webb, Kaye. II. Bicknell, Treld.
PZ5.F516 1976 78 78-319515 MARC

Fisher, Paul R 1960-
The ash staff / Paul R. Fisher. — 1st ed. — New York : Atheneum, 1979.
179 p. ; 22 cm.
SUMMARY: When Rhawn, the old scholar, dies, it is restless and temptable Mole who inherits the ash staff, the sword, and the mission.
ISBN 0-689-30708-X
[1. Fantasy] I. Title.
PZ7.F5347 As [Fic] 79 79-11731 MARC

Fisk, Nicholas.
Escape from Splatterbang / [by] Nicholas Fisk. — London : Pelham, 1978.
96 p. ; 21 cm. GB78-18655
ISBN 0-7207-1054-5 : £3.50
I. Title.
PZ7.F548 Es 79 78-319471 MARC

Fisk, Nicholas.
Escape from Splatterbang / Nicholas Fisk. — 1st American ed. — New York : Macmillan, 1979, c1978.
94 p. ; 22 cm.
SUMMARY: Unwittingly abandoned by his parents, Mykl is stranded on a hostile planet, from which escape seems impossible.
ISBN 0-02-735260-9
[1. Science fiction] I. Title.
PZ7.F548 Es 1979 [Fic] 79 79-11494 MARC

Fitzhugh, Louise.
I am five / written and with pictures by Louise Fitzhugh. — New York : Delacorte Press, c1978.
[48 p. : ill. ; 21 cm.
SUMMARY: A little girl describes her activities and her likes and dislikes the day after her fifth birthday.
ISBN 0-440-03952-5 : $4.95. ISBN 0-440-03953-3 lib. bdg. : $4.58
[1. Identity—Fiction] I. Title.
PZ7.F5768 Iam [E] 78 78-50404 MARC

Fitzhugh, Louise.
Sport / Louise Fitzhugh. — New York : Delacorte Press, c1979.
218 p. ; 22 cm.
SUMMARY: Eleven-year-old Sport lives happily with his absent-minded father. His ruthless and wealthy mother suddenly wants his custody.
ISBN 0-440-07886-5
[1. Divorce—Fiction. 2. Remarriage—Fiction. 3. New York (City)—Fiction] I. Title.
PZ7.F5768 Sp [Fic] 79 78-72861 MARC

Flack, Marjorie, 1897-1958.
The story about Ping, by Marjorie Flack and Kurt Wiese. [New York The Viking press, c1933.
[32] p. col. illus. 23 cm.
Illustrated t.-p. in colors.
SUMMARY: A little duck finds adventure on the Yangtze River when he is too late to board his master's houseboat one evening.
[1. Ducks—Fiction. 2. China—Fiction] I. Wiese, Kurt, 1887-1974, illus. II. Title.
PZ7.F593 St [E] [r79]rev 33-29356 MARC

Flanagan, Harley.
Stories & illustrations / by Harley ; introduced by Allen Ginsburg. — [s.l.] : Charlatan Press, [c1976]
[28] p. : col. ill. ; 21 cm.
Cover title.
CONTENTS: The shopkeeper and his donkey.—The sabre tooth tiger.
ISBN 8787594064
[1. Short stories. 2. Children's writings] I. Title.
PZ7.F59828 St [E] 79 79-305058 MARC

Fleischman, Albert Sidney.
The Hey Hey Man / by Sid Fleischman ; illustrated by Nadine Bernard Westcott. — 1st ed. — Boston : Little, Brown, c1979.
31 p. : ill. (some col.) ; 24 cm.
"An Atlantic Monthly Press book."
SUMMARY: A thief steals a farmer's gold but is outwitted by a mischievous wood spirit - the Hey Hey Man.
ISBN 0-316-26001-0
[1. Robbers and outlaws—Fiction] I. Westcott, Nadine Bernard. II. Title.
PZ7.F5992 He [E] 78 78-31702 MARC

Fleischman, Albert Sidney.
Jim Bridger's alarm clock and other tall tales / by Sid Fleischman ; illustrated by Eric Von Schmidt. — 1st ed. — New York : E. P. Dutton, c1978.
56 p. : ill. ; 26 cm. — (A Unicorn book)
SUMMARY: Three tall tales about Jim Bridger and several of his unbelievable discoveries in the wilderness of the West.
ISBN 0-525-32795-9 : $6.95
1. Bridger, James, 1804-1881—Juvenile fiction. [1. Bridger, James, 1804-1881—Fiction. 2. Western stories. 3. Humorous stories] I. Von Schmidt, Eric. II. Title.
PZ7.F5992 Jg 1978 [Fic] 78 78-5854 MARC

Fleischman, Paul.
The birthday tree / by Paul Fleischman ; pictures by Marcia Sewall. — 1st ed. — New York : Harper & Row, c1979.
[32] p. : ill. ; 19 x 20 cm.
SUMMARY: When Jack goes to sea, his parents watch as the tree planted at his birth reflects his fortunes and misfortunes.
ISBN 0-06-021915-7 : $6.95
[1. Trees—Fiction] I. Sewall, Marcia. II. Title.
PZ7.F599233 Bi [E] 78 78-22155 MARC

Fleming, Ian, 1908-1964.
Chitty Chitty Bang Bang; the magical car. Illustrated by John Burningham. New York, Random House [1964]
111 p. illus. 25 cm.
SUMMARY: Two children persuade their father, an inventor, to purchase and restore an old car which turns out to have magical powers.
[1. Automobiles—Fiction. 2. Fantasy] I. Burningham, John. II. Title.
PZ7.F5997 Ch [Fic] [r79]rev 64-21282 MARC

Fletcher, David Leslie.
Mother o'pearl; three tales. Drawings by Susan Obrant. New York, Pantheon Books [1970]
101 p. illus. 24 cm.
CONTENTS: Rose-petal wings.—The young man whose shadow took charge.—Murredin and the wonderful water lily.
ISBN 0-394-90449-4
[1. Short stories] I. Obrant, Susan, illus. II. Title.
PZ7.F628 Mo [Fic] 71[r78]rev2 78-117460 MARC

Floethe, Louise Lee.
Blueberry pie. Pictures by Richard Floethe. New York, Scribner, c1962.
[32] p. col. illus. 26 cm.
SUMMARY: A young boy sets out from his farmhouse to pick blueberries on a hot August day.
[1. Blueberries—Fiction] I. Floethe, Richard. II. Title.
PZ7.F66 Bl [E] [r79]rev 62-9645 MARC

Flora, James.
Grandpa's ghost stories : story and pictures / by James Flora. — 1st ed. — New York : Atheneum, 1978.
[28] p. : col. ill. ; 26 cm.
"A Margaret K. McElderry book."
SUMMARY: Grandpa chooses a stormy night to relate his encounters with several ghoulish characters.
ISBN 0-689-50112-9 : $6.95
[1. Ghost stories] I. Title.
PZ7.F663 Gs [E] 79 78-108947 MARC

Foley, Bernice Williams, 1902-
Spaceships of the ancients / story by Bernice Williams Foley ; drawings by Lee Kolozsi Hoffmann. — Novelty, Ohio : Veritie Press, c1978.
61 p. : ill. ; 24 cm.
SUMMARY: In his dreams a young boy meets a spaceman who takes him back in time to visit ancient civilizations.
ISBN (invalid) 0-915964-04-8
[1. Space and time—Fiction. 2. Science fiction] I. Hoffmann, Lee Kolozsi. II. Title.
PZ7.F7277 Sp 79 78-59116 MARC

CHILDREN'S LITERATURE (Continued)

Foley, Louise Munro.
Tackle 22 / by Louise Munro Foley ; illustrated by John Heinly. — New York : Delacorte Press, c1978.
[48] p. : ill. ; 24 cm.
SUMMARY: A little brother replaces a regular team member who is sick, but the outlook for Saturday's football game is still grim.
ISBN 0-440-08464-4. ISBN 0-440-08465-2 lib. bdg.
[1. Football—Fiction. 2. Brothers and sisters—Fiction] I. Heinly, John, 1932- . II. Title.
PZ7.F732 Tac [E] 78 78-50425 MARC

Fontenot, Mary Alice.
Clovis Crawfish and Etienne Escargot / by Mary Alice Fontenot ; produced and illustrated by Eric Vincent ; musical arrangement by Jeanne and Robert Gilmore. — Lafayette, La. : Acadiana Press, c1979.
[27] p. : col. ill. ; 29 cm.
SUMMARY: The animals of the bayou effect a timely rescue of the snail's eggs, one rainy day.
[1. Animals—Fiction. 2. Louisiana—Fiction] I. Vincent, Eric. II. Title.
PZ7.F73575 Cl [E] 79 79-119985 MARC

Forbes, Bryan, 1926-
International Velvet / [by] Bryan Forbes. — London : Heinemann, 1978.
160 p. ; 23 cm. GB78-25502
ISBN 0-434-26830-5 : £3.90
I. Title.
PZ7.F7485 In 79 79-309642 MARC

Forman, James, 1928-
Freedom's blood / James D. Forman. — New York : F. Watts, 1979.
114, [1] p. ; 25 cm.
Bibliography: p. [115]
SUMMARY: A fictionalized account of the murder of three civil rights workers, Michael Schwerner, James Chaney, and Andrew Goodman, in Mississippi, during the summer of 1964.
ISBN 0-531-02866-6 (lib. bdg.)
[1. Civil rights—Fiction. 2. Mississippi—Fiction. 3. Civil rights workers—Fiction] I. Title.
PZ7.F759 Fr [Fic] 78 78-23265 MARC

Forman, James D
A fine, soft day / James Forman. — New York : Farrar, Straus, Giroux, c1978.
245 p. ; 22 cm.
Includes bibliographical references.
SUMMARY: A sensitive Catholic boy and his family become deeply involved in and emotionally affected by the escalating Protestant-Catholic conflict in Belfast.
ISBN 0-374-32301-1 : $8.95
1. Northern Ireland—History—1969- —Juvenile fiction. [1. Northern Ireland—History—1969- —Fiction] I. Title.
PZ7.F76 Fi 1978 [Fic] 78 78-11127 MARC

Forman, Joan, 1919-
The princess in the tower / Joan Forman ; illustrated by Beryl Sanders. — London : Faber and Faber, 1973.
96 p. : ill. ; 21 cm.
ISBN 0-571-09911-4
I. Sanders, Beryl. II. Title.
PZ7.F764 Pr 79 79-670049 MARC

Foulds, Elfrida Vipont (Brown) 1902-
The lark in the morn, by Elfrida Vipont. [1st ed.] New York, Holt, Rinehart and Winston [1970, c1948]
224 p. 22 cm. 4.95
Sequel: The lark on the wing.
SUMMARY: The "untalented" child of a musical family finally discovers where she fits in.
ISBN 0-03-084274-3
[1. Music—Fiction. 2. Family life—Fiction] I. Title.
PZ7.F825 Lar 5 [Fic] 70[r79]rev 78-98914 MARC

Fradin, Dennis B
Beyond the mountain, beyond the forest / by Dennis Brindell Fradin ; ill. by John Maggard. — Chicago : Childrens Press, c1978.
46 p. : col. ill. ; 24 cm. — (His Early man)
SUMMARY: In the face of the unrelenting cold, a young cave boy decides to go beyond the forest and find a warmer place for his people to live.
ISBN 0-516-03853-2
[1. Cave dwellers—Fiction. 2. Man, Primitive—Fiction. 3. Glacial epoch—Fiction] I. Maggard, John. II. Title. III. Series.
PZ7.F8416 Be [Fic] 78 78-6611 MARC

Fradin, Dennis B
Cave painter / by Dennis Brindell Fradin ; ill., by John Maggard. — Chicago : Childrens Press, c1978.
45 p. : col. ill. ; 25 cm. — (His Early man)
SUMMARY: Eleven-year-old Thor, who does not want to join the other men on the hunt, finds a niche in his community of cave dwellers.
ISBN 0-516-03852-4
[1. Cave dwellers—Fiction. 2. Man, Prehistoric—Fiction] I. Maggard, John. II. Title. III. Series.
PZ7.F8416 Cav [E] 77 77-16623 MARC

Fradin, Dennis B
The new spear / by Dennis Brindell Fradin ; illustrated by Tom Dunnington. — Chicago : Childrens Press, c1979.
45 p. ; col. ill. ; 25 cm. — (His Early man)
SUMMARY: Two young cave girls invent a new kind of spear but can't get the hunters to take their invention seriously.
ISBN 0-516-03854-0
[1. Cave-dwellers—Fiction. 2. Man, Prehistoric—Fiction] I. Dunnington, Tom. II. Title. III. Series.
PZ7.F8416 Ne [E] 78 78-11060 MARC

Fradin, Dennis B
North Star / by Dennis Brindell Fradin ; ill. by William Neebe. — Chicago : Children's Press, c1978.
46 p. : col. ill. ; 24 cm. — (His Early man)
SUMMARY: Nine-year-old Lia, who likes to listen to her grandmother's stories about the stars, finds a way to use her knowledge.
ISBN 0-516-03851-6
[1. Stars—Fiction. 2. Cave dwellers—Fiction. 3. Man, Prehistoric—Fiction] I. Neebe, William. II. Title. III. Series.
PZ7.F8416 No [E] 77 77-26558 MARC

Franco, Marjorie.
So who hasn't got problems? / Marjorie Franco. — Boston : Houghton Mifflin, 1979.
153 p. ; 22 cm.
SUMMARY: A 13-year-old feels she faces nothing but problems the summer a haughty girl moves into the neighborhood, her best friend plans to run away, and she falls in love for the first time.
ISBN 0-395-27814-7 : $6.95
I. Title.
PZ7.F8485 So [Fic] 78 78-31410 MARC

Frank Netherton; or, The talisman. New York, R. Carter, 1859.
201 p. illus. 16 cm. (Carter's fireside library, 2)
PZ6.F8 78 77-363244 MARC

Franzén, Nils Olof.
Agaton Sax and Lispington's grandfather clock / Nils-Olof Franzén ; illustrated by Quentin Blake. — London : A. Deutsch, 1978.
127 p. : ill. ; 21 cm. GB***
ISBN 0-233-96964-0 : £2.95
I. Blake, Quentin. II. Title.
PZ7.F8593 Af 79 79-310790 MARC

Freeman, Barbara Constance.
A pocket of silence / Barbara C. Freeman ; with decorations by the author. — 1st ed. — New York : Dutton, 1978, c1977.
171 p. ; 22 cm.
SUMMARY: For a reason she doesn't understand, Caroline is compelled to return to the town of her childhood on her sixteenth birthday and there meets a stranger with familiar ways and learns the meaning of a recurring dream.
ISBN 0-525-37210-5 : $7.50
[1. Space and time—Fiction] I. Title.
PZ7.F8746 Po 1978 [Fic] 78 78-6392 MARC

Freeman, Bill.
First spring on the Grand Banks / Bill Freeman. — Toronto : J. Lorimer, 1978.
171 p., [8] leaves of plates : ill. ; 21 cm. C78-1396-4
ISBN 0-88862-221-X : $9.95
I. Title.
PZ7.F87464 Fi 79 79-305370 MARC

Freeman, Don.
Bearymore : story and pictures / by Don Freeman. — [New York] : Puffin Books, 1979, c1976.
[32] p. : col. ill. ; 19 x 23 cm.
SUMMARY: A circus bear has trouble hibernating and dreaming up a new act at the same time.
ISBN 0-14-050279-3 : $1.95
[1. Bears—Fiction. 2. Circus stories] I. Title.
PZ7.F8747 Be 1979 [E] 78 78-18281 MARC

Freeman, Don.
Mop top; story and pictures by Don Freeman. New York, Viking Press, 1955.
48 p. illus. 26 cm.
SUMMARY: An almost-six-year-old with a mass of floppy red hair tries to postpone the inevitable trip to Mr. Barberoli's barber shop.
[1. Hair—Care and hygiene—Fiction] I. Title.
PZ7.F8747 Mo [E] [r79]rev 55-14876 MARC

Freeman, Don.
A rainbow of my own. New York, Viking Press [1966]
[32 p.] col. illus. 22 x 24 cm.
SUMMARY: A small boy imagines what it would be like to have his own rainbow to play with.
1. Rainbow—Fiction. I. Title.
PZ7.F8747 Rai [E] [r79]rev 66-13983 MARC

Freeman, Don.
Ski pup, written and illustrated by Don Freeman. New York, Viking Press [1963]
56 p. illus. 29 cm.
SUMMARY: A young Saint Bernard, learning to watch over children in a skiing school in the Swiss Alps, becomes a hero.
[1. Dogs—Fiction. 2. Rescue work—Fiction. 3. Skis and skiing—Fiction. 4. Alps—Fiction] I. Title.
PZ7.F8747 Sk [E] [r79]rev 63-8532 MARC

Fregosi, Claudia.
Are there spooks in the dark? / By Claudia Fregosi. — New York : Four Winds Press, c1977.
[30] p. : col. ill. ; 21 cm.
SUMMARY: Having determined that the "spooks in the dark" are hiding from her, a little girl decides to scare them.
ISBN 0-590-07451-2 : $5.95
[1. Night—Fiction] I. Title.
PZ7.F884 Ar [E] 76[r78]rev 76-9789 MARC

French, Fiona.
Hunt the thimble / [by] Fiona French. — Oxford [etc.] : Oxford University Press, 1978.
[32] p. : chiefly col. ill. ; 29 cm. GB78-20623
SUMMARY: A quiet game of Hunt the Thimble results in a chase through the streets of 17th-century Amsterdam.
ISBN 0-19-279719-0 : £2.50
[1. Play—Fiction. 2. Amsterdam—Fiction] I. Title.
PZ7.F8887 Hu [E] 79 79-313027 MARC

Freschet, Berniece.
Bernard of Scotland Yard / by Berniece Freschet ; pictures by Gina Freschet. — New York : Scribner, c1978.
[48] p. : ill. ; 24 cm.
SUMMARY: An enterprising Bostonian mouse assists his cousin, a Scotland Yard inspector, in apprehending a gang of jewel thieves.
ISBN 0-684-15931-7 lib. bdg. : $7.95
[1. Mice—Fiction. 2. Robbers and outlaws—Fiction] I. Freschet, Gina. II. Title.
PZ7.F88968 Bd [E] 78 78-9584 MARC

Fries, Chloe.
The full of the moon / by Chloe Fries and Richard H. Tomlinson ; [illustrated by E. J. Abrams Associates]. — Mankato, Minn. : Creative Education, c1978.
56 p. : ill. ; 24 cm.
SUMMARY: During their first summer trawling for shrimp together, Juan and his father learn a lot about each other.
ISBN 0-87191-686-X
[1. Fathers and sons—Fiction. 2. Shrimp fisheries—Fiction. 3. Fishing—Fiction] I. Tomlinson, Richard H., joint author. II. Title.
PZ7.F9159 Fu [Fic] 79 78-70805 MARC

Frisinger, Nellie.
Jeff & Jenny and the kidnapping / Nellie Frisinger. — Denver : Accent Books, c1978.
127 p. : ill. ; 20 cm. — (The Jeff & Jenny adventure series)
SUMMARY: The Rollins twins discover valuable lessons in Christian living as they help a young orphan.
ISBN 0-89636-005-9 : $2.25
[1. Christian life—Fiction. 2. Orphans—Fiction] I. Title.
PZ7.F9178 Jb [Fic] 79 78-55337 MARC

Fujikawa, Gyo.
Surprise! Surprise! / Gyo Fujikawa. — [New York] : Grosset & Dunlap, c1978.
[14] p. : col. ill. ; 22 cm.
Cover title.
SUMMARY: A race over the meadow's stone wall ends with a splashy finish for Rabbit and his playmates.
ISBN 0-448-14557-X : $2.50
[1. Animals—Fiction] I. Title.
PZ7.F9513 Su [E] 78 78-103784 MARC

Furminger, Jo, 1932-
Blackbirds and the gift pony / Jo Furminger ; ill. by Susan Hunter. — London : Hodder and Stoughton, 1978.
128 p. : ill. ; 23 cm. GB***
ISBN 0-340-21863-0 : £3.50
I. Title.
PZ7.F96653 Bl 79 79-307192 MARC

CHILDREN'S LITERATURE (Continued)

Gackenbach, Dick.
Crackle Gluck and the sleeping toad / by Dick Gackenbach. — New York : Seabury Press, c1979.
[32] p. : col. ill. ; 22 cm.
"A Clarion book."
SUMMARY: Crackle, the spunky daughter of a Pennsylvania Dutch farmer, becomes suspicious of the pampered sleeping toad that has supposedly brought good luck to the Gluck family for generations.
ISBN 0-8164-3227-9 (lib. bdg.) : $7.95
[1. Toads—Fiction. 2. Humorous stories] I. Title.
PZ7.G117 Cr [E] 78-12635
78 MARC

Gackenbach, Dick.
Pepper and all the legs / by Dick Gackenbach. — New York : Seabury Press, c1978.
[32] p. : col. ill. : 19 x 23 cm.
"A Clarion book."
SUMMARY: Limited by his very short legs, Pepper the dachshund wonders what's —upstairsand finally finds out.
ISBN 0-8164-3221-X lib. bdg. : $6.95
[1. Dogs—Fiction] I. Title.
PZ7.G117 Pe [E] 78-5084
78 MARC

Galbraith, Kathryn Osebold.
Come spring / Kathryn Osebold Galbraith. — 1st ed. — New York : Atheneum, 1979.
198 p. ; 22 cm.
"A Margaret K. McElderry book."
SUMMARY: Reenie hopes that "come spring" her continually moving family will put down roots that will keep them in their new home.
ISBN 0-689-50142-0
[1. Family life—Fiction. 2. Moving, Household—Fiction] I. Title.
PZ7.G1303 Co 1979 [Fic] 79-12311
79 MARC

Galde, Dorothy A
Danger comes to Squirrel Valley / by Dorothy A. Galde. — Chicago : Moody Press, c1979.
125 p. : ill. ; 17 cm.
SUMMARY: The struggles and triumphs of a family of squirrels reveal moral and spiritual truths.
ISBN 0-8024-1175-4 : $1.25
[1. Squirrels—Fiction. 2. Christian life—Fiction] I. Title.
PZ7.G1304 Dan [Fic] 78-24080
78 MARC

Gannett, Ruth Stiles.
My father's dragon; illus. by Ruth Chrisman Gannett. New York, Radom House [1948]
86 p. illus., col. map (on lining-papers) 21 cm.
SUMMARY: A young boy determines to rescue a poor baby dragon who is being used by a group of lazy wild animals to ferry them across the river on Wild Island.
[1. Animals—Fiction. 2. Dragons—Fiction] I. Gannett, Ruth Chrisman.
PZ7.G15 My [E] 48-6527
[r79]rev MARC

Gantos, Jack.
Greedy Greeny / by Jack Gantos & Nicole Rubel. — Garden City, N.Y. : Doubleday, c1979.
[32] p. : col. ill. ; 20 x 24 cm.
SUMMARY: A little monster, having disobeyed his mother by eating the watermelon intended for the family's dessert, suffers for his greed in a subsequent dream.
ISBN 0-385-14685-X : $6.95
[1. Watermelons—Fiction. 2. Dreams—Fiction. 3. Monsters—Fiction] I. Rubel, Nicole, joint author. II. Title.
PZ7.G15334 Gr [E] 78-22630
79 MARC

Gantos, Jack.
The perfect pal / by Jack Gantos and Nicole Rubel. — Boston : Houghton Mifflin, 1979.
30 p. : col. ill. ; 22 x 24 cm.
SUMMARY: Wendell, the pet store salesman, helps Vanessa to find just the right pal for her.
ISBN 0-395-28380-9
[1. Animals—Fiction. 2. Pets—Fiction] I. Rubel, Nicole. II. Title.
PZ7.G15334 Pe [E] 79-12310
79 MARC

Garfield, Leon.
The confidence man / [by] Leon Garfield. — Harmondsworth : Kestrel Books, 1978.
279 p. : map ; 23 cm. GB78-36424
ISBN 0-7226-5407-3 : £3.95
I. Title.
PZ7.G17943 Co 1978 79-306435
79 MARC

Garfield, Leon.
The dumb cake / Leon Garfield ; illustrated by Faith Jaques. — London : Heinemann, 1977.
47 p. : ill. ; 23 cm. — (His Garfield's apprentices ; [9]) GB77-3155:
ISBN 0-434-94040-2 : £2.40
I. Jaques, Faith. II. Title. III. Series: Garfield, Leon. Garfield's apprentices ; 9.
PZ7.G17943 Gar vol. 9 78-326353
79 MARC

Garfield, Leon.
Mirror, mirror / Leon Garfield : illustrated by Antony Maitland. — London : Heinemann, 1976.
48 p. : ill. ; 23 cm. — (His Garfield's apprentices) GB76-16654
£1.90
I. Maitland, Antony, 1935- II. Title.
PZ7.G17943 Mg 78-314314
78 MARC

Garfield, Leon.
Moss and Blister / Leon Garfield ; illustrated by Faith Jaques. — London : Heinemann, 1976.
48 p. : ill. ; 23 cm. — (His apprentices : 3) GB76-34306
ISBN 0-434-94033-X : £1.90
I. Title.
PZ7.G17943 Mo 78-315171
78 MARC

Garfield, Leon.
Tom Titmarsh's devil / Leon Garfield ; illustrated by Faith Jaques. — London : Heinemann, 1977.
47 p. : ill. ; 23 cm. — (His Garfield's apprentices ; [10]) GB77-31554
ISBN 0-434-94039-9 : £2.40
I. Jaques, Faith. II. Title. III. Series: Garfield, Leon. Garfield's apprentices ; 10.
PZ7.G17943 Gar vol. 10 78-326352
79 MARC

Garfield, Leon.
The valentine / Leon Garfield ; illustrated by Faith Jaques. — London : Heinemann, 1977.
47 p. : ill. ; 23 cm. — (His Apprentices : 5) GB77-13384
ISBN 0-434-94036-4 : £2.10
I. Title.
PZ7.G17943 Val 78-315169
78 MARC

Garland, David Betts, 1954-
Oldest Man understands / written and illustrated by David Betts Garland. — New York : Chair Books, c1978.
[24] p. : ill. ; 31 cm.
SUMMARY: Oldest Man overcomes his loneliness with the help of a close friend.
[1. Loneliness—Fiction. 2. Friendship—Fiction] I. Title.
PZ7.G18295 Ol [Fic] 78-62694
79 MARC

Garner, Alan.
The aimer gate / Alan Garner ; etchings by Michael Foreman. — 1st U.S. ed. — New York : Collins, 1979, c1978.
79 p. : ill. ; 21 cm.
SUMMARY: Robert learns that his secret place has another secret the day his soldier uncle comes home to help with the corn harvest.
ISBN 0-529-05506-6 : $6.95
[1. England—Fiction] I. Foreman, Michael, 1938- II. Title.
PZ7.G18417 Ai 1979 [Fic] 78-20964
78 MARC

Garner, Alan.
Granny Reardun / Alan Garner ; etchings by Michael Foreman. — 1st U.S. ed. — New York : Collins + World, 1978, c1977.
61 p. : ill. ; 21 cm.
SUMMARY: Joseph knows that he has to be different from his grandfather, a stone mason whose work is in evidence throughout the village, and the day comes when he must announce his decision.
ISBN 0-529-05505-8 : $6.95
[1. England—Fiction] I. Foreman, Michael, 1938- II. Title.
PZ7.G18417 Gr 1978 [Fic] 78-8141
78 MARC

Garner, Alan.
The moon of Gomrath / Alan Garner. — New York : Collins & World, 1979, c1963.
160 p. ; 22 cm.
Includes bibliographical references.
SUMMARY: Colin, with the help of the wizard Cadellin and the elves, fights to save his sister from the power of the evil witch Morrigan. Sequel to "The Weirdstone of Brisingamen."
ISBN (invalid) 0-00-184503-9 : $6.95
[1. Fantasy] I. Title.
PZ7.G18417 Mo 1978 [Fic] 77-16425
77 MARC

Garner, Alan.
The stone book / Alan Garner ; etchings by Michael Foreman. — New York : Collins, 1978, c1976.
63 p. : ill. ; 21 cm.
SUMMARY: His daughter's request for a book prompts a stonemason to reveal the secret of the stone to her.
ISBN 0-529-05503-1 : $5.95
[1. Fathers and daughters—Fiction] I. Foreman, Michael, 1938- II. Title.
PZ7.G18417 St 1978 [Fic] 78-7965
78 MARC

Garner, Alan.
Tom Fobble's day / Alan Garner ; etchings by Michael Foreman. — 1st U.S. ed. — New York : Collins, 1979, c1977.
72 p. : ill. ; 21 cm.
SUMMARY: When his sled is "Tom Fobbled" by another boy, William's grandfather makes him another from an old oak loom.
ISBN 0-529-05507-4 : $6.95
[1. England—Fiction] I. Foreman, Michael, 1938- II. Title.
PZ7.G18417 To 1979 [Fic] 78-26927
78 MARC

Garner, Alan.
The weirdstone; a tale of Alderley. New York, F. Watts [1961, c1960]
224 p. 21 cm.
"Published in England under the title: The weirdstone of Brisingamen."
I. Title.
PZ7.G18417 Wd 1961 61-10076
[r79]rev MARC

Garner, Alan.
The weirdstone of Brisingamen: a tale of Alderley. London, Collins, 1965.
224 p. map. 20.5 cm. (The Evergreen library) 9/6 B66-17488
I. Title.
PZ7.G18417 We 1965 823'.9'14 74-352456
70[r79]rev MARC

Garner, Alan.
The weirdstone of Brisingamen : a tale of Alderley / Alan Garner. — New York : Collins, 1979, c1960.
224 p. : maps (on lining papers) ; 22 cm.
SUMMARY: A young girl and her brother are catapulted into a battle between good and evil for possession of a magical stone of great power that is contained in her bracelet.
ISBN 0-529-05519-8 : $7.95
[1. Fantasy] I. Title.
PZ7.G18417 We 1979 [Fic] 78-24635
78 MARC

Garrigue, Sheila.
Between friends / by Sheila Garrigue. — Scarsdale, N.Y. : Bradbury Press, c1978.
160 p. ; 22 cm.
SUMMARY: After moving from California to Massachusetts eleven-year-old Jill is eager to make new friends.
ISBN 0-87888-133-6 : $6.95
[1. Moving, Household—Fiction. 2. Friendship—Fiction] I. Title.
PZ7.G1847 Be [Fic] 77-90952
77 MARC

Garrison, Christian.
The dream eater / story by Christian Garrison ; pictires by Diane Goode. — Scarsdale, N.Y.: Bradbury Press, c1978.
[32] p. : col. ill. ; 21 x 27 cm.
SUMMARY: Yukio spares the other villagers from recurring nightmares when he rescues a baku.
ISBN 0-87888-134-4 : $8.95
[1. Dreams—Fiction. 2. Monsters—Fiction] I. Goode, Diane. II. Title.
PZ7.G1848 Dr [E] 78-55213
78 MARC

Gathje, Curtis.
The disco kid / by Curtis Gathje ; photos. by Carole Bertol. — New York : F. Watts, 1979.
88 p. : ill. ; 22 cm. — (A Triumph book)
SUMMARY: A 16-year-old's dream of disco dancing at Planet Earth is realized during an evening he would rather not remember.
ISBN 0-531-02895-X
[1. Disco dancing—Fiction] I. Bertol, Carole. II. Title.
PZ7.G2245 Di [Fic] 79-11495
79 MARC

Gauch, Patricia Lee.
Fridays / Patricia Lee Gauch. — New York : Putnam, c1979.
160 p. ; 22 cm.
SUMMARY: Corey's association with "The Eight" and her participation in their weekly Friday gatherings cause her to examine her values, goals, and friendships.
ISBN 0-399-20703-1 : $7.95
[1. Friendship—Fiction] I. Title.
PZ7.G2315 Fr 1979 [Fic] 79-17047
79 MARC

Geibel, James.
The blond brother / James Geibel. — New York : Putnam, c1979.
201 p. ; 22 cm.
SUMMARY: Racial tension in his new high school undermines Rich Gaskin's efforts to help his team win the state basketball championship.
ISBN 0-399-20653-1 : $7.95
[1. Race relations—Fiction. 2. Basketball—Fiction. 3. School stories] I. Title.
PZ7.G2597 Bl [Fic] 78-16330
78 MARC

Gelman, Rita Golden.
Uncle Hugh : a fishing story / by Rita Golden Gelman and Warner Friedman ; pictures by Eros Kieth. — 1st ed. — New York : Harcourt Brace Jovanovich, c1978.
[48] p. : ill. ; 21 cm.
SUMMARY: A young boy learns the joys of fishing from his Scottish uncle and a legendary fish, both named Uncle Hugh.
ISBN 0-15-292789-1 lib. bdg. : $6.95
[1. Fishing—Fiction] I. Friedman, Warner, joint author. II. Kieth, Eros. III. Title.
PZ7.G2837 Un [E] 78-4918
78 MARC

Gentile, Gennaro L
The mouse in the manger / story by Gennaro L. Gentile ; art by Vernon McKissack. — Notre Dame, Ind. : Ave Maria Press, c1978.
[77] p. : col. ill. ; 18 x 21 cm.
SUMMARY: Oscar the mouse runs away from home after a fight with his father and finds himself in a stable where he befriends a couple and their newborn baby.
ISBN 0-87793-165-8 : $4.95
[1. Friendship—Fiction. 2. Jesus Christ—Nativity—Fiction. 3. Christian life—Fiction. 4. Mice—Fiction] I. McKissack, Vernon. II. Title.
PZ7.G292 Mo [Fic] 78-72944
79 MARC

CHILDREN'S LITERATURE (Continued)

George, Jean Craighead, 1919-
River Rats, inc. / by Jean Craighead George. — 1st ed. — New York : Dutton, c1979.
136 p. ; 22 cm.
SUMMARY: Traveling illegally at night on the Colorado River with a mysterious cargo, two boys are shipwrecked and must depend for survival on their own ingenuity and a wild boy they find living along the river.
ISBN 0-525-38455-3
[1. Survival—Fiction. 2. Feral children—Fiction. 3. Colorado River—Fiction] I. Title.
PZ7.G2933 Ri 1979 [Fic] 78 78-12318 MARC

George, Jean Craighead, 1919-
The wounded wolf / by Jean Craighead George ; pictures by John Schoenherr. — 1st ed. — New York : Harper & Row, c1978.
32 p. : ill. ; 18 x 23 cm.
SUMMARY: As hungry animals close in on an injured wolf, hoping to feed on him after death, help arrives to change the odds.
ISBN 0-06-021949-1 : $6.95. ISBN 0-06-021950-5 lib. bdg. : $5.49
1. Wolves—Legends and stories. [1. Wolves—Fiction] I. Schoenherr, John. II. Title.
PZ10.3.G316 Wo 1978 [E] 78 76-58711 MARC

Geras, Adèle.
The girls in the velvet frame / Adèle Geras. — 1st American ed. — New York : Atheneum, 1979, c1978.
149 p. ; 21 cm.
SUMMARY: In Jerusalem just before World War I five sisters send their photograph to a New York newspaper in hope of contacting their borther from whom they have not heard for one year.
ISBN 0-689-30729-2
[1. Family life—Fiction. 2. Jews—Fiction] I. Title.
PZ7.G29354 Gi 1979 [Fic] 79 79-12352 MARC

German, Tony, 1924-
Tom Penny / Tony German. — Toronto : P. Martin Associates, c1977.
184 p. : ill. ; 21 cm. C77-1516-6
ISBN 0-88778-171-3 : $8.95. ISBN 0-88778-172-1 pbk.
I. Title.
PZ7.G298 To 78 78-316085 MARC

Gerson, Corinne.
Passing through / Corinne Gerson. — New York : Dial Press, c1978.
193 p. ; 22 cm.
SUMMARY: A high school girl trying to cope with her older brother's suicide becomes friends with a handicapped boy who helps her through her grief.
ISBN 0-8037-6842-7 : $7.95
[1. Friendship—Fiction. 2. Physically handicapped—Fiction. 3. Suicide—Fiction] I. Title.
PZ7.G322 Pas [Fic] 78 78-51329 MARC

Gerson, Corinne.
Tread softly / Corinne Gerson. — New York : Dial Press, c1979.
133 p. ; 22 cm.
SUMMARY: A young girl tries to cope with the loss of her parents by inventing an imaginary family.
ISBN 0-8037-9058-9 : $7.95
[1. Family life—Fiction] I. Title.
PZ7.G322 Tr [Fic] 78 78-72199 MARC

Gess, Diane.
Sunshine porcupine / written by Diane Gess ; ill. by Al Lowenheim and the staff of the Lion's Den. — 1st ed. — La Jolla, Calif. : Oak Tree Publications, c1979.
64 p. : col. ill. ; 29 cm.
"Adapted from the animated television program sunshine porcupine."
SUMMARY: The evil and energy-eating Ugli-Unks plan to eat all the energy in egg-producing Eggwood, but Porcupine engages the help of the sun as an alternate power source.
ISBN 0-916392-38-4
[1. Energy conservation—Fiction. 2. Solar energy—Fiction. 3. Eggs—Fiction. 4. Rabbits—Fiction. 5. Porcupines—Fiction] I. Lowenheim, Alfred, 1947- II. Title.
PZ7.G3285 Su [Fic] 79 79-1468 MARC

Gessner, Lynne.
To see a witch / Lynne Gessner. — 1st ed. — Nashville : T. Nelson, c1978.
144 p. ; 21 cm.
SUMMARY: A 12-year-old pre-Columbian Indian boy considered childish by the elders of his tribe finds acceptance as a man when his cousin is accused of being a witch.
ISBN 0-8407-6589-4 : $6.95
[1. Indians of North America—Fiction. 2. Witchcraft—Fiction] I. Title.
PZ7.G329 To [Fic] 78 78-16646 MARC

Gezi, Kalil I
The mystery of the live ghosts / by Kal Gezi and Ann Bradford ; illustrated by Mina McLean. — Elgin, Ill. : Child's World ; Chicago : distributed by Childrens Press, c1978.
32 p. : ill. (some col.) ; 25 cm.
SUMMARY: Five children visit a haunted mansion on Halloween night and are confronted by the mystery of the live ghosts.
ISBN 0-89565-026-6 : $4.95
[1. Halloween—Fiction. 2. Ghost stories] I. Bradford, Ann, joint author. II. McLean, Mina Gow. III. Title.
PZ7.G33902 Mw [Fic] 78 78-8142 MARC

Gezi, Kalil I
The mystery in the secret club house / by Kal Gezi and Ann Bradford ; illustrated by Mina McLean. — Elgin, Ill. : Child's World ; Chicago : distributed by Childrens Press, c1978.
32 p. : ill. (some col.) ; 25 cm.
SUMMARY: Upon entering their secret club house, five youngsters discover they have had a mysterious visitor.
ISBN 0-89565-027-4 : $4.95
[1. Robbers and outlaws—Fiction. 2. Mystery and detective stories] I. Bradford, Ann, joint author. II. McLean, Mina Gow. III. Title.
PZ7.G33902 My [E] 78 78-6418 MARC

Gibbs, Betty L
Miss Independence / Betty L. Gibbs. — Hollywood, Calif. : Ermine Publishers, c1978.
103 p. ; 23 cm.
SUMMARY: A fervently independent young woman begins to wonder if she can integrate her successful career with marriage.
ISBN 0-89343-018-8 : $2.95
I. Title.
PZ7.G33919 Mi [Fic] 79 77-80971 MARC

Gifford, Griselda.
Cass the brave / by Griselda Gifford ; illustrated by Mary Rayner. — London : Gollancz, 1978.
96 p. : ill. ; 21 cm. GB78-32525
ISBN 0-575-02551-4 : £2.95
I. Rayner, Mary. II. Title.
PZ7.C3628 Cas 79 78-326193 MARC

Gilbert, Harriett, 1948-
Running away : a novel / by Harriett Gilbert. — 1st ed. — New York : Harper & Row, c1979.
266 p. ; 21 cm.
SUMMARY: After a series of crises, a 16-year-old girl in an English private school learns to take control of her own life.
ISBN 0-06-021972-6 : $8.95
[1. School stories] I. Title.
PZ7.G3737 Ru 1979 [Fic] 79 79-1937 MARC

Gilman, Dorothy, 1923-
The bells of freedom [by] Dorothy Gilman Butters, illus. by Carol Wilde. Philadelphia, Macrae Smith [1963]
190 p. illus. 22 cm.
1. Massachusetts—History—Revolution, 1775-1783—Juvenile fiction. I. Title.
PZ7.G432 Be 63-19672 [r79]rev MARC

Gilman, Dorothy, 1923-
The calico year [by] Dorothy Gilman Butters. Philadelphia, Macrae Smith [1953]
223 p. 22 cm.
I. Title.
PZ7.G432 Cal 52-14001 [r79]rev MARC

Gilman, Dorothy, 1923-
Carnival gypsy, by Dorothy Gilman Butters. Philadelphia, Macrae-Smith [1950]
217 p. 22 cm.
I. Title.
PZ7.G432 Car 50-7159 [r79]rev MARC

Gilman, Dorothy, 1923-
Enchanted caravan, by Dorothy Gilman Butters; illus. by Janet Smalley. Philadelphia, Macrae-Smith [1949]
207 p. illus. 22 cm.
I. Title.
PZ7.G432 En 49-7952 [r79]rev MARC

Gilman, Dorothy, 1923-
Enchanted caravan. Illustrated by Janet Smalley. New York, Grosset & Dunlap [1954, c1949]
207 p. illus. 20 cm. (Starlight novels for modern girls)
Published under the author's married name: Dorothy Gilman Butters.
I. Title.
[PZ7] 54-8541 [r78]rev MARC

Gilman, Dorothy, 1923-
Four-party line, a junior novel, by Dorothy Gilman Butters. Philadelphia, Macrae Smi..i [1954]
198 p. 22 cm.
"A condensed version ... appeared in Compact magazine under the title: Dial L for love."
I. Title.
PZ7.G432 Fo 54-10361 [r79]rev MARC

Gilman, Dorothy, 1923-
Girl in buckskin, by Dorothy Gilman Butters. Philadelphia, Macrae Smith [1956]
190 p. 21 cm.
I. Title.
PZ7.G432 Gi 56-9969 [r79]rev MARC

Gilman, Dorothy, 1923-
Masquerade, by Dorothy Gilman Butters. Philadelphia, Macrae Smith [1961]
190 p. 22 cm.
I. Title.
PZ7.G432 Mas 61-8303 [r79]rev MARC

Gilman, Dorothy, 1923-
Papa Dolphin's table [by] Dorothy Gilman Butters; illustrated by Kurt Werth. [1st ed.] New York, Knopf, 1955.
88 p. illus. 21 cm.
I. Title.
PZ7.G432 Pap 55-8944 [r79]rev MARC

Gilman, Dorothy, 1923-
Ragamuffin Alley, by Dorothy Gilman Butters. Philadelphia, Macrae Smith [1951]
206 p. 22 cm.
I. Title.
PZ7.G432 Rag 51-13078 [r79]rev MARC

Gilman, Dorothy, 1923-
Ten leagues to Boston Town, by Dorothy Gilman Butters. Philadelphia, Macrae Smith [1962]
187 p. 22 cm.
I. Title.
PZ7.G432 Te 62-19162 [r79]rev MARC

Gilman, Dorothy, 1923-
Witch's silver [by] Dorothy Gilman Butters. Philadelphia, Macrae Smith [1959]
190 p. 22 cm.
I. Title.
PZ7.G432 Wi 59-8239 [r79]rev MARC

Gilson, Jamie.
Dial Leroi Rupert, DJ / Jamie Gilson ; illustrated by John Wallner. — 1st ed. — New York : Lothrop, Lee & Shepard, c1979.
126 p. : ill. ; 22 cm.
SUMMARY: Desperate to pay back 50 dollars they owe, three 12-year-olds form a jazz combo and then receive some surprising assistance from a local disc jockey.
ISBN 0-688-41888-0
[1. Musicians—Fiction. 2. Disc jockeys—Fiction] I. Wallner, John C. II. Title.
PZ7.G4385 Di [Fic] 79 79-4662 MARC

Gilson, Jamie.
Harvey, the beer can king / by Jamie Gilson ; illustrated by John Wallner. — 1st ed. — New York : Lothrop, Lee & Shepard Co., c1978.
127 p. : ill. ; 22 cm.
SUMMARY: A twelve-year-old boy hopes his beer can collection will win him the title of sixth-grade Superkid.
ISBN 0-688-41845-7. ISBN 0-688-51845-1 lib. bdg.
[1. Collectors and collecting—Fiction] I. Wallner, John C. II. Title.
PZ7.G4385 Har [Fic] 78 78-1807 MARC

Gilroy, Tom, 1951-
In Bikole : eight modern stories about life in a West African village / by Tom Gilroy ; with drawings by Monica Vachula. — New York : Knopf : distributed by Random House, c1978.
ix, 83 p. : ill. ; 24 cm.
SUMMARY: Short stories revealing various aspects of life among the Serrers in the West African village of Bikole.
ISBN 0-394-83722-3. ISBN 0-394-93722-8 lib. bdg.
1. Serers—Juvenile fiction. 2. Bikol, Senegal—Juvenile fiction. [1. Serers—Fiction] I. Vachula, Monica. II. Title.
PZ7.G4382 In [Fic] 78 78-3271 MARC

CHILDREN'S LITERATURE (Continued)

Gipson, Frederick Benjamin, 1908-
Curly and the wild boar / by Fred Gipson ; ill. by Ronald Himler. — 1st ed. — New York : Harper & Row, c1979.
88 p. : ill. ; 22 cm.
SUMMARY: A young boy repeatedly puts himself in danger in his determination to kill the wild boar that has destroyed his prize watermelon.
ISBN 0-06-022014-7 : $5.95. ISBN 0-06-022015-5 lib. bdg. : $6.79
[1. Country life—Fiction] I. Himler, Ronald. II. Title.
PZ7.G4393 Cu 1979 [Fic] 77-25644
78 MARC

Gipson, Frederick Benjamin, 1908-
Little Arliss / by Fred Gipson ; ill. by Ronald Himler. — 1st ed. — New York : Harper & Row, c1978.
83 p. : ill. ; 22 cm.
SUMMARY: A small twelve-year old boy's determination to prove he is tough sets him on the trail of a runaway horse.
ISBN 0-06-022008-2 : $5.95. ISBN 0-06-022009-0 lib. bdg. : $5.79
[1. Horses—Fiction. 2. Western stories] I. Himler, Ronald. II. Title.
PZ7.G4393 Li [Fic] 77-17643
78 MARC

Girion, Barbara, 1937-
Joshua, the Czar, and the chicken bone wish / Barbara Girion ; illustrated by Richard Cuffari. — New York : Scribner, c1978.
155 p. : ill. ; 21 cm.
SUMMARY: Fourth grader Josh suffers from being the class "klutz" befriends an elderly gentleman in a home for the aged, becomes a hero, and finally makes a friend and wins his class's acceptance.
ISBN 0-684-15929-5 : $6.95
[1. School stories] I. Cuffari, Richard, 1925- II. Title.
PZ7.G4398 Jo [Fic] 78-15012
78 MARC

Girion, Barbara, 1937-
A tangle of roots / Barbara Girion. — New York : Scribner, c1979.
154 p. ; 22 cm.
SUMMARY: Beth is stunned by her mother's sudden death but soon realizes that she and her father must adjust to their loss and begin a new phase of life.
ISBN 0-684-16074-9 : $7.95
[1. Death—Fiction. 2. Fathers and daughters—Fiction] I. Title.
PZ7.G4398 Tan [Fic] 78-27243
78 MARC

Glazer, Joan I
Introduction to children's literature / Joan I. Glazer, Gurney Williams III. — New York : McGraw-Hill, c1979.
xiii, 737 p. : ill. ; 25 cm.
Includes bibliographies and indexes.
ISBN 0-07-046030-2 : $14.95
1. Books and reading for children. 2. Children's literature. I. Williams, Gurney, 1941- joint author. II. Title.
Z1037.A1G57 028.5 78-12202
78 MARC

Godden, Rumer, 1907-
The rocking horse secret / Rumer Godden ; illustrated by Juliet Stanwell Smith. — 1st American ed. — New York : Viking Press, 1978, c1977.
87 p. : ill. ; 21 cm.
SUMMARY: Terrible changes occur in the house where Tibby's mother works, but although Tibby knows something that may make all the difference, she doesn't feel she can divulge the secret.
ISBN 0-670-60243-4
I. Smith, Juliet Stanwell. II. Title.
PZ7.G54 Ro 1978 [Fic] 77-25489
77 MARC

Goffstein, M B
Family scrapbook / M. B. Goffstein. — 1st ed. — New York : Farrar, Straus, Giroux, 1978.
[48] p. : ill. ; 23 cm.
SUMMARY: Present seven brief episodes in the lives of a curly-haired girl, her little brother, and her parents.
ISBN 0-374-32269-4 : $5.95
[1. Family life—Fiction] I. Title.
PZ7.G5573 Fam [E] 78-51435
78 MARC

Goffstein, M B
My Noah's ark : story and pictures / by M. B. Goffstein. — 1st ed. — New York : Harper & Row, c1978.
[31] p. : ill. ; 19 cm.
SUMMARY: An old woman fondly remembers a lifetime of experiences related to a carved representation of Noah's ark she has had since childhood.
ISBN 0-06-022022-8 : $5.95. ISBN 0-06-022023-6 lib. bdg. : $6.49
[1. Noah's ark—Fiction] I. Title.
PZ7.G5573 Mz [E] 77-25666
78 MARC

Goffstein, M B
Natural history / by M. B. Goffstein. — 1st ed. — New York : Farrar, Straus & Giroux, 1979.
[30] p. : col. ill. ; 19 cm.
SUMMARY: Text and illustrations describe the riches of the earth and how people can promote peace and goodwill by sharing equitably with each other their fellow creatures.
ISBN 0-374-35498-7
[1. Nature—Fiction. 2. Natural resouces—Fiction] I. Title.
PZ7.G5573 Nat 1979 [E] 79-7318
79 MARC

Goffstein, M B
Neighbors / by M. B. Goffstein. — 1st ed. — New York : Harper & Row, c1979.
30 p. : ill. ; 22 cm.
SUMMARY: Relates the year-long attempt of one shy person to befriend her equally shy neighbor.
ISBN (invalid) 0-06-022018-6 : $6.95.
[1. Neighborliness—Fiction. 2. Friendship—Fiction] I. Title.
PZ7.G5573 Ne 1979 [E] 78-19491
78 MARC

Goldman, Susan.
Cousins are special / by Susan Goldman. — Chicago : A. Whitman, 1978.
[32] p. : col. ill. ; 19 cm. — (Self-starter books)
SUMMARY: When Sarah visits Carol Sue, she discovers that they have a special relationship because they are cousins.
ISBN 0-8075-1317-2 : $4.75
[1. Cousins—Fiction. 2. Family—Fiction] I. Title.
PZ7.G5693 Co [E] 78-11924
78 MARC

Goldsmith, Howard.
Invasion, 2200 A.D. / Howard Goldsmith. — 1st ed. — Garden City, N.Y. : Doubleday, c1979.
137 p. ; 22 cm.
SUMMARY: In the 23rd century two rival groups are in constant conflict. Then an evil race from another planet arrives to take over the minds of men in order to enslave them.
ISBN 0-385-14247-1 : $7.95
[1. Science fiction] I. Title.
PZ7.G575 In [Fic] 78-22320
79 MARC

Goldsmith, Howard.
The whispering sea / Howard Goldsmith. — Indianapolis : Bobbs-Merrill, c1976.
131 p. ; 22 cm.
SUMMARY: After discovering the diary of a nineteenth-century suicide victim, two children are drawn into a struggle between the restless spirits inhabiting an old house by the sea.
ISBN 0-672-52199-7
[1. Ghost stories] I. Title.
PZ7.G575 Wh [Fic] 76-11632
76 MARC

Goll, Reinhold Weimar.
The moon twins and the treasure / by Reinhold W. Goll ; illustrated by Russell Hoover. — Elkins Park, Pa. : Elkins Park Press, c1978.
97 p. : ill. ; 23 cm.
SUMMARY: Twins from the Moon Behind the Moon land on Earth in their space capsules and meet two children who are searching for a long lost family treasure.
$6.95
[1. Buried treasure—Fiction. 2. Science fiction] I. Hoover, Russell. II. Title.
PZ7.G5815 Mo [Fic] 78-61141
79 MARC

Goodall, John S
An Edwardian holiday / John S. Goodall. — London [etc.] : Macmillan, 1978.
[64] p. : of col. ill. ; 14 x 19 cm. GB78-43212
ISBN 0-333-24002-2 : £2.50
I. Title.
PZ7.G6018 Ecf 79-303085
79 MARC

Goodall, John S
The surprise picnic / John S. Goodall. — London : Macmillan, 1977.
[57] p. : of col. ill. ; 14 x 18 cm. GB78-03752
Page width varies.
ISBN 0-333-21676-8 : £1.50
I. Title.
PZ7.G6018 Su 1977b 823'.9'14 78-317488
78 MARC

Goodman, Marguerite.
Christmas comes in assorted sizes / by Marguerite Goodman. — 1st ed. — Port Washington, N.Y. : Ashley Books, c1977.
145 p. ; 24 cm.
SUMMARY: After his father's death 15-year-old Tad discovers the newness and power of Christmas while fulfilling his father's last wish.
ISBN 0-87949-111-6 : $6.95
[1. Christmas stories. 2. Death—Fiction] I. Title.
PZ7.G6138 Ch 1977 [Fic] 77-80303
79 MARC

Goodspeed, J M
The ransom note / J. M. Goodspeed ; illustrated by Ray Keane. — Middletown, Conn. : Xerox Education Publications, c1978.
40 p. : ill. ; 22 cm. — (Weekly Reader Children's Book Club edition)
SUMMARY: Searching for the dognapper of their aunt's poodle, Liz and Jake suspect an asthmatic classmate who is not allowed to have a dog.
ISBN 0-88375-216-6
[1. Kidnapping—Fiction. 2. Dogs—Fiction. 3. Asthma—Fiction. 4. Physically handicapped—Fiction] I. Keane, Ray. II. Title.
PZ7.G622 Ran [Fic] 78-51340
79 MARC

Goodwin, Godfrey.
There ain't no angels no more / [by] Godfrey Goodwin. — London : Collins, 1978.
95 p. ; 22 cm. GB78-28958
ISBN 0-00-184837-2 : £2.50
I. Title.
PZ7.G6228 Th 78-323863
79 MARC

Gordon, Clarence, 1835-1920.
Boarding-school days / by Vieux Moustache [i.e. C. Gordon] ; with ill. by F. O. C. Darley and Thomas Nast. — New York : Hurd and Houghton, 1873.
iv, 291 p., [8] leaves of plates : ill. ; 18 cm.
I. Title.
PZ7.G654 Bo 79-302967
79 MARC

Gordon, Sheila.
A monster in the mailbox / by Sheila Gordon ; illustrated by Tony DeLuna. — 1st ed. — New York : Dutton, c1978.
71 p. : ill. ; 23 cm.
SUMMARY: The disappointment Julius experiences when he buys a mail-order monster with his self-earned money is not repeated when he buys a book of monster fairy tales from a used book store.
ISBN 0-525-35150-7 : $6.95
[1. Consumers—Fiction. 2. Reading—Fiction] I. De Luna, Tony. II. Title.
PZ7.G65937 Mo [Fic] 78-4600
78 MARC

Gordon, Shirley.
Grandma zoo / by Shirley Gordon ; pictures by Whitney Darrow, Jr. — 1st ed. — New York : Harper & Row, c1978.
32 p. : ill. ; 21 cm.
SUMMARY: A young boy's grandmother speaks to him with animal voices over the telephone.
ISBN 0-06-022049-X : $6.95. ISBN 0-06-022050-3 lib. bdg. : $7.49
[1. Grandmothers—Fiction] I. Darrow, Whitney, 1909- . II. Title.
PZ7.G6594 Gp 1978 [E] 78-52402
78 MARC

Gould, Lois.
X, a fabulous child's story / Lois Gould ; illustrated by Jacqueline Chwast. — New York : Daughters Pub. Co., [1978]
[52] p. : col. ill. ; 29 cm.
SUMMARY: X is loved by its classmates but despised by adults because n[o] one knows if it is a boy or girl.
ISBN 0-913780-21-9 : $7.50
[1. Sex role—Fiction] I. Chwast, Jacqueline. II. Title.
PZ7.G7352 Xac [E] 77-9498
78 MAR

Goyder, Alice.
Christmas in Catland / Alice Goyder. — New York : Crowell, 1979, c1978.
[24] p. : col. ill. ; 17 cm.
SUMMARY: Tilly and Minnie help the family get ready for Christmas and make secret wishes for the presents they want.
ISBN 0-690-03935-2 : 2.95
[1. Christmas stories. 2. Cats—Fiction. 3. Children's writings] I. Title.
PZ7.G7482 Ch 1979 [E] 78-19517
78 MARC

Goyder, Alice.
Holiday in Catland / Alice Goyder. — New York : Crowell, 1979, c1978.
[24] p. : col. ill. ; 17 cm.
SUMMARY: Four cats take a holiday at the seashore.
ISBN 0-690-03931-X : $2.95. ISBN 0-690-03932-8 lib. bdg. : $6.79
[1. Cats—Fiction. 2. Vacations—Fiction. 3. Children's writings] I. Title.
PZ7.G7482 Ho 1979 [E] 78-19518
78 MARC

Goyder, Alice.
Party in Catland / Alice Goyder. — New York : Crowell, 1979, c1978.
xxii, 538 p. : col. ill. ; 23 cm.
SUMMARY: Two cats, Minnie and Tilly, decide to give a party for all of their friends.
ISBN 0-690-03929-8 : $2.95
[1. Parties—Fiction. 2. Cats—Fiction. 3. Children's writings] I. Title.
PZ7.G7482 Par 1979 [E] 78-19516
78 MARC

Graham, John, 1926-
I love you, mouse / John Graham ; pictures by Tomie d[e] Paola. — 1st Voyager/HBJ ed. — New York : Harcourt Brace Jovanovich, 1978, c1976.
[32] p. : ill. ; 21 cm. — (A Voyager/HBJ book)
SUMMARY: A child imagines the things he would do with various anima[ls] if he were one of them.
ISBN 0-15-644106-3
[1. Animals—Fiction] I. De Paola, Thomas Anthony. II. Title.
[PZ7.G7526 Iad 1978] [E] 78-621
78 MAR

Grant, Gwen.
Private, keep out! / [by] Gwen Grant ; illustrated by Fait[h] Jaques. — London : Heinemann, 1978.
143 p. : ill. ; 21 cm. GB78-2768
ISBN 0-434-94170-0 : £3.20
I. Title.
PZ7.G766754 Pr 78-32495
78 MAR

CHILDREN'S LITERATURE (Continued)

Grant, Louise, 1909-
The fort & the flag / written and illustrated by Louis Grant. — Hanover, N.H. : Regional Center for Educational Training, 1977.
73 p. : ill. (some col.) ; 22 cm. — (Bicentennial historiettes series)
SUMMARY: Relates the separate adventures of a New Hampshire brother and sister as they participate in the Revolution in a different way.
ISBN 0-915892-09-X
1. New Hampshire—History—Revolution, 1775-1783—Juvenile fiction. 2. United States—History—Revolution, 1775-1783—Juvenile fiction. [1. New Hampshire—History—Revolution, 1775-1783—Fiction. 2. United States—History—Revolution, 1775-1783—Fiction] I. Title.
PZ7.G76679 Fo [Fic] 76-40904
76 MARC

Grant, Myrna.
Ivan and the star of David / Myrna Grant ; [illustrated by Jos. E. DeVelasco]. — Wheaton, Ill. : Tyndale House Publishers, c1977.
142 p. : ill. ; 18 cm.
SUMMARY: Ivan and his family use all their ingenuity and courage to help a Jewish girl leave Russia.
ISBN 0-8423-1845-3 : $1.95
[1. Jews in Russia—Fiction. 2. Christian life—Fiction. 3. Russia—Fiction] I. De Velasco, Joe E. II. Title.
PZ7.G7673 Ix [Fic] 77-72439
00 MARC

Gray, Nicholas Stuart.
A wind from nowhere / Nicholas Stuart Gray. — London : Faber, 1978.
155 p. ; 21 cm. GB***
CONTENTS: A wind from nowhere.—Bright silver nothing.—A very near thing.—A touch of oddity.—How mere is moonshine?—A candle for shadows.—Sisters in sorcery.—The gates of Avernus.—Once upon a time there was a chance.
ISBN 0-571-11182-3 : £3.65
I. Title.
PZ7.G78 Wi 1978 823'.9'14 78-321970
78 MARC

Gray, Nigel.
It'll all come out in the wash / by Nigel Gray ; pictures by Edward Frascino. — 1st ed. — New York : Harper & Row, c1979.
32 p. : col. ill. ; 21 cm.
SUMMARY: Dad helps his little girl realize that it's all right not to be perfect.
ISBN 0-06-022067-8 : $6.95
[1. Fathers and daughters—Fiction] I. Frascino, Edward. II. Title.
PZ7.G7813 It 1979 [E] 78-22482
79 MARC

Greaves, Margaret.
A net to catch the wind / by Margaret Greaves ; drawings by Stephen Gammell. — 1st ed. — New York : Harper & Row, c1979.
[47] p. : ill. ; 21 cm.
SUMMARY: A princess shows her father the importance of freedom and love when he tries to capture a silver colt that is dear to her.
ISBN 0-06-022104-6 : $6.95. ISBN 0-06-022105-4 lib. bdg. : $6.79
[1. Unicorns—Fiction. 2. Princesses—Fiction] I. Gammell, Stephen. II. Title.
PZ7.G8 Ne 1979 [E] 78-20265
78 MARC

Gregg, Margaret.
The story of a simple man and other stories / Margaret Gregg ; [illustrated by Innocent Malwa]. — Lusaka : Neczam, 1974, c1971.
76 p. : ill. ; 19 cm. — (Eagle readers : Secondary level)
I. Title.
PZ7.G8538 St 1974 77-980738
79 MARC

Green, Phyllis.
A new mother for Martha / Story by Phyllis Green ; illustrated by Peggy Luks. — New York : Human Sciences Press, c1978.
[32] p. : col. ill. ; 24 cm.
SUMMARY: Faced with her father's remarriage, a first-grader stubbornly insists that her dead mother will return soon.
ISBN 0-87705-330-8
[1. Remarriage—Fiction. 2. Stepmothers—Fiction] I. Luke, Peggy. II. Title.
PZ7.G82615 Ne [E] 78-16731
78 MARC

Green, Phyllis.
Nicky's lopsided, lumpy, but delicious orange / Phyllis Green. — Reading, Mass. : Addison-Wesley, c1978.
111 p. ; 23 cm.
SUMMARY: Thinking that her inadequacies have caused both her father and stepfather to live away from the family motivates Nicky to strive for perfection in all she does.
ISBN 0-201-02590-6 lib. bdg. : $7.95
[1. Family life—Fiction] I. Title.
PZ7.G82615 Ni [Fic] 78-4599
78 MARC

Greene, Constance C
I and Sproggy / Constance C. Greene ; illustrated by Emily McCully. — 1st ed. — New York : Viking Press, 1978.
155 p. : ill. ; 22 cm.
SUMMARY: The meeting between an eleven-year-old boy and Sproggy, his English step-sister who comes to New York City, is less than idyllic, but time, events, and other people help to change that.
ISBN 0-670-38980-3 : $7.95
[1. Brothers and sisters—Fiction. 2. New York (City)—Fiction] I. McCully, Emily Arnold. II. Title.
PZ7.G8287 Iab [Fic] 78-6096
78 MARC

Greene, Shep.
The boy who drank too much / Shep Greene. — 1st ed. — New York : Viking Press, 1979.
149 p. ; 22 cm.
SUMMARY: The efforts of his friends almost fail to stop a lonely and isolated teenager from destroying himself with alcohol.
ISBN 0-670-18381-4
[1. Alcoholism—Fiction. 2. Alcoholics—Fiction] I. Title.
PZ7.G8433 Bo 1979 [Fic] 78-26370
78 MARC

Greene, Vivian.
Chip, oh brother! / By Vivian Greene. — New York : Watts, 1979.
[31] p. : col. ill. ; 23 cm.
SUMMARY: When Chip's new baby brother turns out to be a sister instead, Chip wants to sell her to a friend.
ISBN 0-531-02510-1
[1. Babies—Fiction. 2. Brothers and sisters—Fiction] I. Title.
PZ7.G8436 Ch [E] 79-11969
79 MARC

Greene, Vivian.
Cleveland, the disco king / by Vivian Greene. — New York : F. Watts, 1979.
[32] p. : col. ill. ; 23 cm.
SUMMARY: Cleveland has a crush on Odessa and in order to get her attention and make her like him, he decides to become a fancy dancer.
ISBN 0-531-02513-6
[1. Disco dancing—Fiction] I. Title.
PZ7.G8436 Cl [E] 79-13057
79 MARC

Greene, Vivian.
Oops, Odessa! / by Vivian Greene. — New York : F. Watts, 1979.
[32] p. : col. ill. ; 23 cm.
SUMMARY: Odessa feels very troubled because she can't decide what she wants to be when she grows up. But her friend, Rotunda, helps her with some good advice.
ISBN 0-531-02511-X
[1. Occupations—Fiction] I. Title.
PZ7.G8436 Oo [E] 79-12222
79 MARC

Greene, Vivian.
Rotunda ate a cookie / by Vivian Greene. — New York : Watts, 1979.
[32] p. : col. ill. ; 23 cm.
SUMMARY: Rotunda's brother thinks of a novel way to help his sister diet and improve her memory.
ISBN 0-531-02512-8
[1. Weight control—Fiction. 2. Brothers and sisters—Fiction] I. Title.
PZ7.G8436 Ro [E] 79-11723
79 MARC

Greenwald, Sheila.
All the way to wits' end/ Sheila Greenwald. — 1st ed. — Boston : Little, Brown, c1979.
138 p. : ill. ; 22 cm.
"An Atlantic Monthly Press book."
SUMMARY: Eleven-year-old Drucilla longs to rid herself of her dusty, traditional family heritage but finds the price of conformity more than she wants to pay.
ISBN 0-316-32670-4
[1. Identity—Fiction. 2. Moving, Household—Fiction] I. Title.
PZ7.G852 Al [Fic] 79-15936
79 MARC

Greenwald, Sheila.
The atrocious two / Sheila Greenwald. — Boston : Houghton Mifflin, 1978.
153 p. : ill. ; 22 cm.
SUMMARY: A brother and sister who are "perfect brats" are forced to spend a summer at Aunt Tessie's where they find out some surprising things about themselves and become involved in a jewelry theft.
ISBN 0-395-26451-0 : $6.95
[1. Behavior—Fiction. 2. Mystery and detective stories] I. Title.
PZ7.G852 At [Fic] 78-4538
78 MARC

Griese, Arnold A
The wind is not a river / Arnold A. Griese ; illustrated by Glo Coalson. — 1st ed. — New York : Crowell, c1978.
xi, 108 p. : ill. ; 21 cm.
SUMMARY: As the only ones not captured when the Japanese take over their Aleutian island village during World War II, two children must survive on their own.
ISBN 0-690-03807-0
1. World War, 1939-1945—Campaigns—Aleutian Islands—Juvenile fiction. [1. World War, 1939-1945—Campaigns—Aleutian Islands—Fiction. 2. Aleutian Islands—Fiction] I. Coalson, Glo. II. Title.
PZ7.G8812 Wi 1978 [Fic] 77-5082
77 MARC

Griffiths, Helen.
Grip, a dog story / Helen Griffiths ; illustrated by Douglas Hall. — New York : Holiday House, 1978.
128 p. : ill. ; 21 cm.
SUMMARY: A young boy experiences conflicting loyalties for his father, a breeder of generations of champion fighting dogs, and the playful pup he loves so much.
ISBN 0-8234-0335-1 : $6.95
1. Dogs—Legends and stories. [1. Dogs—Fiction. 2. Fathers and sons—Fiction] I. Hall, Douglas, 1931- . II. Title.
PZ10.3G8808 Gr 1978 [Fic] 78-6819
78 MARC

Griffiths, Helen.
The Kershaw dogs / Helen Griffiths ; illustrated by Douglas Hall. — London : Hutchinson, 1978.
128 p. : ill. ; 21 cm. GB***
ISBN 0-09-132840-3 : £3.50
1. Dogs—Legends and stories. I. Title.
PZ10.3.G8808 Ke 78-319324
79 MARC

Griffiths, Helen.
The last summer : Spain 1936 / Helen Griffiths ; illustrated by Victor Ambrus. — New York : Holiday House, 1979.
151 p. : ill. ; 21 cm.
SUMMARY: A boy and a horse share adventures during the Spanish Civil War.
ISBN 0-8234-0361-0
[1. Spain—History—Civil War, 1936-1939—Fiction. 2. Horses—Fiction] I. Ambrus, Victor G. II. Title.
PZ7.G8837 Las [Fic] 79-10469
79 MARC

Griffiths, Helen.
Witch fear / Helen Griffiths ; illustrated by Victor Ambrus. — London : Hutchinson of London, 1975.
159 p. : ill. ; 21 cm. GB***
ISBN 0-09-124600-8 : £2.95
I. Ambrus, Victor G. II. Title.
PZ7.G8837 Wm 78-312855
78 MARC

Grisé, Jeannette, 1935-
Robert Benjamin and the great blue dog joke / by Jeannette Grisé ; illustrated by Alex Stein. — 1st ed. — Philadelphia : Westminster Press, c1978.
123 p. : ill. ; 21 cm.
Bibliography: p. 119-121.
SUMMARY: Fond of playing practical jokes, a fifth grader finally realizes that it is more fun to make people laugh at you than to embarrass and humiliate them.
ISBN 0-664-32637-4 lib. bdg.
[1. Humorous stories] I. Stein, Alex. II. Title.
PZ7.G8878 Ro [Fic] 78-17006
78 MARC

Grohskopf, Bernice.
Blood & roses / by Bernice Grohskopf. — 1st ed. — New York : Atheneum, 1979.
256 p. ; 22 cm.
SUMMARY: Upset by his mother's plans to remarry, fifteen-year-old Robinson meets the ghost of William Caxton and helps him find in Westminster Abbey a manuscript lost for centuries.
ISBN 0-689-30681-4
1. Caxton, William, 1422 (ca.)-1491—Juvenile fiction. [1. Caxton, William, 1422 (ca.)-1491—Fiction. 2. Ghost stories] I. Title.
PZ7.G8926 Bl [Fic] 78-14583
78 MARC

Grover's Super surprise book : featuring Jim Henson's Muppets / illustrated by Tom Cooke ; paper engineering by Ib Penick. — New York : Random House, c1978.
[14] p. (on double leaves) : chiefly col. ill. ; 24 cm. — (CTW Sesame Street pop-up ; 12)
"Created in cooperation with the Children's Educational Television Workshop."
SUMMARY: Pop-up pictures feature Grover in suprising places and situations.
ISBN 0-394-83841-6 : $3.95
[1. Stories without words] I. Cooke, Tom. II. Penick, Ib. III. Children's Television Workshop. IV. Title: Super surprise book.
PZ7.G9312 [E] 77-93776
79 MARC

Gunther, Louise.
Anna's snow day / by Louise Gunther ; drawings by Paul Frame. — Champaign, Ill. : Garrard Pub. Co., c1979.
46 p. : col. ill. ; 23 cm. — (A For real book)
SUMMARY: When school is called off because of snow, two children in New York City have a wonderful day.
ISBN 0-8116-4310-7 lib. bdg. : $4.30
[1. Snow—Fiction. 2. City and town life—Fiction] I. Frame, Paul, 1913- . II. Title.
PZ7.G9812 An [E] 78-22050
78 MARC

Gunther, Louise.
A tooth for the tooth fairy / by Louise Gunther ; drawings by James Cummins. — Champaign, Ill. : Garrard Pub. Co., c1978.
48 p. : col. ill. ; 23 cm. — (A For real book)
SUMMARY: Rose is convinced the Tooth Fairy won't visit her after she loses the tooth that fell out.
ISBN 0-8116-4308-5 lib. bdg. : $4.50
[1. Teeth—Fiction] I. Cummins, James. II. Title.
PZ7.G9812 To [E] 77-16422
77 MARC

CHILDREN'S LITERATURE (Continued)

Gurney, Nancy.
The king, the mice and the cheese, by Nancy and Eric Gurney. New York, Beginner Books, [1965]
63 p. col. illus. 23 cm.
SUMMARY: A king runs into trouble when he tries to protect his cheese from the palace mice.
[1. Kings, queens, rulers, etc.—Fiction. 2. Mice—Fiction] I. Gurney, Eric, joint author. II. Title.
PZ7.G9816 Ki [E] [r79]rev 65-21212 MARC

Haas, Charlie.
Over the edge / by Charlie Haas and Tim Hunter, based on their screenplay. — 1st ed. — New York : Grove Press, 1979.
164 p., [16] leaves of plates : ill. ; 18 cm. — (A Black cat book ; B-426)
SUMMARY: When schools and recreation facilities are eliminated by cost cuts in what was planned to be an ideal community, teenagers become bored and turn to drugs and vandalism.
ISBN 0-394-17088-1 : $2.25
[1. Juvenile delinquency—Fiction. 2. New towns—Fiction] I. Hunter, Tim, joint author. II. Title.
PZ7.H1123 Ov [Fic] 79 79-51728 MARC

Haas, Dorothy.
The bears upstairs / by Dorothy Haas. — 1st ed. — New York : Greenwillow Books, c1978.
185 p. ; 22 cm.
SUMMARY: A young girl helps two bears rendezvous with the inhabitants of the planet Brun.
ISBN 0-688-80169-2 : $6.95. ISBN 0-688-84169-4 lib. bdg. : $6.67
[1. Fantasy] I. Title.
PZ7.H1124 Be [Fic] 78 78-54683 MARC

Hacks, Peter, 1928-
The bear at the hunters' ball / Peter Hacks ; illustrated by Walter Schmögner. — Reading, Mass. : Addison-Wesley Pub. Co., c1976.
[32] p. : col. ill. ; 18 x 20 cm.
Translation of Der Bär auf dem Försterball.
"An Addisonian Press book."
SUMMARY: Disguised as a hunter, a bear attends the Hunters' Ball where after much drinking, all decide to go bear hunting.
ISBN 0-201-02694-5
[1. Bears—Fiction] I. Schmögner, Walter. II. Title.
PZ7.H1146 Be [E] 76 76-2399 MARC

Hahn, Mary Downing.
The Sara summer / Mary Downing Hahn ; [cover ill. by Ruth Sanderson]. — New York : Houghton Mifflin/Clarion Books, c1979.
152 p. ; 22 cm.
SUMMARY: Forthright, domineering Sara shows Emily how to speak up for herself and what she likes, but as a result Emily finds herself having to take a stand against Sara.
ISBN (invalid) 0-8164-3238-4
[1. Assertiveness (Psychology)—Fiction. 2. Friendship—Fiction] I. Sanderson, Ruth. II. Title.
PZ7.H1256 Sar [Fic] 79 79-12678 MARC

Haigh, Sheila.
Watch for smoke / [by] Sheila Haigh ; illustrated by Trevor Stubley. — London : Methuen, 1978.
v, 128 p. : ill. ; 20 cm. GB78-34898
ISBN 0-416-86380-9 : £2.75
I. Title.
PZ7.H1263 Wat 79 79-303209 MARC

Hale, Irina.
Chocolate Mouse and Sugar Pig, and how they ran away to escape being eaten / written and illustrated by Irina Hale. — 1st American ed. — New York : Atheneum, 1978.
[32] p. : col. ill. ; 24 x 25 cm.
"A Margaret K. McElderry book."
SUMMARY: Upon discovering that they are not to be guests at a party, but part of the menu instead, two confections run away.
ISBN 0-689-50113-7 : $8.95
[1. Confectionery—Fiction] I. Title.
PZ7.H1338 Ch 1978 [E] 78 78-6132 MARC

Hall, Lynn.
The mystery of Pony Hollow / by Lynn Hall ; illustrated by Ruth Sanderson. — Champaign, Ill. : Garrard Pub. Co., c1978.
64 p. : ill. ; 23 cm. — (A Mystery book)
SUMMARY: A young girl living on an Iowa farm finds a mysterious pony's skeleton locked in an abandoned shed and tries to discover where it came from.
ISBN 0-8116-6404-X lib. bdg. : $4.40
[1. Mystery and detective stories] I. Sanderson, Ruth. II. Title.
PZ7.H1458 My [Fic] 78 78-8137 MARC

Hall, Lynn.
The mystery of the lost and found hound / by Lynn Hall ; illustrated by Alan Daniel. — Champaign, Ill. : Garrard Pub. Co., c1979.
64 p. : ill. ; 23 cm. — (A Garrard mystery book)
SUMMARY: A girl and her brother's efforts at tracing the owner of a lost beagle leads them to involvement with dog thieves.
ISBN 0-8116-6408-2
[1. Dogs—Fiction. 2. Mystery and detective stories] I. Daniel, Alan, 1939- . II. Title.
PZ7.H1458 Myl [E] 78 78-27502 MARC

Hall, Lynn.
The mystery of the schoolhouse dog / by Lynn Hall ; illustrated by William Hutchinson. — Champaign, Ill. : Garrard Pub. Co., c1979.
64 p. : ill. ; 23 cm. — (A Mystery book)
SUMMARY: During the renovation of an old one-room schoolhouse, a lonely little boy discovers a mysterious dog that helps him cope with his loneliness.
ISBN 0-8116-6406-6 lib. bdg. : $4.40
[1. Mystery and detective stories] I. Hutchinson, William M. II. Title.
PZ7.H1458 Mz [E] 78 78-12181 MARC

Hallman, Ruth, 1929-
Gimme something, mister! / Ruth Hallman. — 1st ed. — Philadelphia : Westminster Press, c1978.
103 p. ; 21 cm.
"A Hiway book."
SUMMARY: With New Orleans alive with Mardi Gras revelers, a teenager becomes entangled in a plot that includes stolen jewels and voodoo.
ISBN 0-664-32638-2
[1. Robbers and outlaws—Fiction. 2. New Orleans—Fiction. 3. Mardi Gras—Fiction. 4. Voodooism—Fiction] I. Title.
PZ7.H15469 Gi [Fic] 78 78-16721 MARC

Hallman, Ruth, 1929-
Midnight wheels / by Ruth Hallman. — 1st ed. — Philadelphia : Westminister Press, c1979.
91 p. ; 21 cm.
SUMMARY: Always handy with tools, Angie is delighted when she gets a full-time job as a mechanic at Martin's garage but soon begins to suspect that something peculiar is going on.
ISBN 0-664-32650-1
[1. Automobiles—Service stations—Fiction. 2. Automobile mechanics—Fiction. 3. Mystery and detective stories. 4. Sex role—Fiction] I. Tite. II. Title.
PZ7.H15469 Mi [Fic] 79 79-10232 MARC

Hamilton, Dorothy, 1906-
Eric's discovery / Dorothy Hamilton ; illustrated by Betty Wind. — Scottdale, Pa. : Herald Press, 1979.
111 p. : ill. ; 21 cm.
SUMMARY: After witnessing the vandalizing of a church, a young boy doesn't know if he should tell anyone what and whom he saw.
ISBN 0-8361-1902-9 : $4.50
[1. Vandalism—Fiction] I. Wind, Betty. II. Title.
PZ7.H18136 Er [Fic] 79 79-18537 MARC

Hamilton, Dorothy, 1906-
Ken's hideout / Dorothy Hamilton ; illustrated by James Converse. — Scottdale, Pa. : Herald Press, 1979.
84 p. : ill. ; 22 cm.
SUMMARY: Twelve-year-old Ken leaves home when his relationship with his mother deteriorates as a result of her new boyfriend.
ISBN 0-8361-1880-4 : $3.95. ISBN 0-8361-1881-2 pbk. : $2.50
[1. Parent and child—Fiction. 2. Single-parent family—Fiction] I. Converse, James. II. Title.
PZ7.H18136 Kc 1979 [Fic] 78 78-21269 MARC

Hamilton, Dorothy, 1906-
Mari's mountain / Dorothy Hamilton ; illustrated by Esther Rose Graber. — Scottdale, Pa. : Herald Press, 1978.
130 p. : ill. ; 22 cm.
SUMMARY: Abused by her alcoholic father, Mari runs away from home to build a new life for herself in a big city.
ISBN 0-8361-1868-5 : $3.95. ISBN 0-8361-1869-3 pbk. : $2.95
[1. Runaways—Fiction] I. Graber, Esther Rose. II. Title.
PZ7.H18136 Mar [Fic] 78 78-10620 MARC

Hamilton, Elizabeth Verner.
When walls are high / by Elizabeth Verner Hamilton ; illustrated by Laura Peck. — Charleston, S.C. : Tradd Street Press, 1973, c1972.
vii, 151 p. : ill. ; 21 cm.
SUMMARY: Two young girls plan to renovate the garden of a reclusive neighbor with a suspicious past.
[1. Friendship—Fiction] I. Peck, Laura. II. Title.
PZ7.H18145 Wh [Fic] 79 72-93195 MARC

Hamilton, Morse.
My name is Emily / Morse and Emily Hamilton ; pictures by Jenni Oliver. — 1st ed. — New York : Greenwillow Books, c1979.
[32] p. : ill. ; 26 cm.
SUMMARY: Emily and her father play a game to set things right when she returns home after running away.
ISBN 0-688-80181-1 : $6.95. ISBN 0-688-84181-3 lib. bdg. : $6.43
[1. Runaways—Fiction. 2. Brothers and sisters—Fiction. 3. Fathers and daughters—Fiction] I. Hamilton, Emily, joint author. II. Oliver, Jenni. III. Title.
PZ7.H18265 My [E] 78 78-4537 MARC

Hamilton, Virginia.
Justice and her brothers / Virginia Hamilton. — 1st ed. — New York : Greenwillow Books, c1978.
217 p. ; 22 cm.
SUMMARY: An 11-year-old and her older twin brothers struggle to understand their supersensory powers.
ISBN 0-688-80182-X : $7.95. ISBN 0-688-84182-1 lib. bdg. : $7.35
[1. Brothers and sisters—Fiction. 2. Twins—Fiction. 3. Fantasy] I. Title.
PZ7.H1828 Ju [Fic] 78 78-54684 MARC

Hamilton-Paterson, James.
Hostage! / James Hamilton-Paterson. — London : Gollancz, 1978.
160 p. ; 21 cm. GB***
ISBN 0-575-02441-0 : £3.75
I. Title.
PZ7.H18285 Hn 78 78-316084 MARC

Hampton, Ruth E
The girl in the wall / by Ruth E. Hampton. — Chicago : Moody Press, c1978.
224 p. ; 18 cm.
SUMMARY: A college freshman begins to question her faith as she becomes more involved in an unsolved mystery that hovers around the old house where she lives.
ISBN 0-8024-2963-7 : $2.50
[1. Christian life—Fiction. 2. Mystery and detective stories] I. Title.
[PZ7.H1878 Gi] [Fic] 78 78-9805 MARC

Hanlon, Emily.
It's too late for sorry : a novel / by Emily Hanlon. — Scarsdale, N.Y. : Bradbury Press, c1978.
222 p. ; 22 cm.
SUMMARY: Fifteen-year-old Kenny's involvement with the mentally retarded youth on his block brings out the best and the worst in him.
ISBN 0-87888-136-0 : $7.95
[1. Mentally handicapped—Fiction. 2. Friendship—Fiction] I. Title.
PZ7.H1964 It [Fic] 78 78-4422 MARC

Hanlon, Emily.
The swing / By Emily Hanlon. — Scarsdale, N.Y. : Bradbury Press, c1979.
209 p. ; 22 cm.
SUMMARY: An 11-year-old deaf girl and a 13-year-old boy with family problems seek refuge at a swing which has come to have a special meaning for each of them.
ISBN 0-87888-146-8 : $7.95
[1. Friendship—Fiction. 2. Deaf—Fiction. 3. Stepfathers—Fiction] I. Title.
PZ7.H1964 Sw [Fic] 78 78-26400 MARC

Hanson, June Andrea.
Summer of the stallion / June Andrea Hanson ; illustrated by Gloria Singer. — New York : Macmillan, c1979.
108 p. : ill. ; 22 cm.
SUMMARY: Twelve-year-old Janey's experiences as she helps capture and break a wild stallion result in an altered perception of herself and her relationship with her grandfather.
ISBN 0-02-742620-3
[1. Horses—Fiction. 2. Grandfathers—Fiction] I. Singer, Gloria, 1949- . II. Title.
PZ7.H1987 Su [Fic] 78 78-24212 MARC

The Happy sequel; or, The history of Isabella Mordaunt. A tale for young people. New-York Published by W. B. Gilley. Van Winkle & Wiley, printers, 1815.
87 p. front. 15 cm.
L.C. copy imperfect: half title and front. wanting.
PZ6.H214 78 78-315552 MARC

Hardy, Thomas, 1840-1928.
Our exploits at West Poley / Thomas Hardy ; illustrated by John Lawrence ; with an introd. by Richard L. Purdy. — Oxford ; New York : Oxford University Press, 1978.
x, 77 p. : ill. ; 23 cm. — (Oxford illustrated classics) GB78
SUMMARY: While exploring a cave, two boys divert the course of a stream and set off a strange chain of events.
ISBN 0-19-274527-1 : £2.50 $7.95
[1. Caves—Fiction] I. Lawrence, John, 1933- II. Title.
PZ7.H2215 Ou 1978 [Fic] 78 78-40518 MARC

Hare, Norma Q
Wish upon a birthday / by Norma Q. Hare ; drawings by Diane Dawson. — Champaign, Ill. : Garrard Pub. Co., c1979.
46 p. : col. ill. ; 23 cm. — (An Imagination book)
SUMMARY: The first birthday cake ever is baked by a poor cook's helper who cannot afford a present for the princess.
ISBN 0-8116-4418-9 (lib. bdg.) : $4.60
[1. Birthdays—Fiction. 2. Gifts—Fiction. 3. Bakers and bakeries—Fiction] I. Dawson, Diane. II. Title.
PZ7.H2218 Wi [E] 79 79-14596 MARC

CHILDREN'S LITERATURE (Continued)

Harris, Robie H
Don't forget to come back / by Robie H. Harris ; pictures by Tony de Luna. — New York : Knopf : distributed by Random House, c1978.
[40] p. : col. ill. ; 22 cm.
SUMMARY: Annie uses several unsuccessful ploys to keep her parents from going out for the evening, leaving her with a sitter.
ISBN 0-394-83849-1
[1. Behavior—Fiction. 2. Baby sitters—Fiction] I. De Luna, Tony. II. Title.
PZ7.H2436 Do 1978 79 78-3266 MARC

Harris, Rosemary, 1923-
A quest for Orion / Rosemary Harris. — London ; Boston : Faber and Faber, 1978.
233 p. ; 21 cm. GB***
ISBN 0-571-11203-X : £4.50
I. Title.
PZ7.H2437 Qe 1978 78 78-318454 MARC

Hart, Laverne.
Billy Bentson, helper / Laverne Hart ; Bill McPheeters, illustrator. — Nashville : Broadman Press, c1977.
31 p. : col. ill. ; 24 cm.
SUMMARY: A young boy learns how important it is for people to help each other.
ISBN 0-8054-4237-5
[1. Conduct of life—Fiction] I. McPheeters, Bill. II. Title.
PZ7.H2569 Bi [E] 78 78-102280 MARC

Hart, Stephanie.
Is there any way out of sixth grade? / By Stephanie Hart. — New York : Coward, McCann & Geoghegan, c1978.
133 p. ; 23 cm.
SUMMARY: When sixth-grader Sonya and her boarding school classmates are required to recreate a historically accurate version of the 1777 Battle of Princeton, they stage their own version of the battle at night in the snow.
ISBN 0-698-20445-X : $6.95
[1. School stories. 2. Boarding schools] I. Title.
PZ7.H2573 Is [Fic] 78 78-17388 MARC

Haugaard, Erik Christian.
Cromwell's boy / Erik Christian Haugaard. — Boston : Houghton Mifflin, 1978.
214 p. ; 22 cm.
Sequel to A messenger for Parliament.
SUMMARY: A 13-year-old spy's wit and horsemanship make him invaluable to the Cromwell forces as they fight against King Charles.
ISBN 0-395-27203-3 : $7.95
[1. Great Britain—History—Civil War, 1642-1649—Fiction] I. Title.
PZ7.H286 Cr [Fic] 78 78-14392 MARC

Haugaard, Erik Christian.
The rider and his horse. Illustrated by Leo and Diane Dillon. Boston, Houghton Mifflin, 1968.
243 p. illus., map. 23 cm. 3.95
SUMMARY: Involved in the tragedy of the fall of ancient Jerusalem, a young boy, one of the seven survivors of Masada, tells the story of cruelty, disillusionment, and courage.
[1. Mezada (Fortress), Israel—Fiction] I. Dillon, Leo, illus. II. Dillon, Diane, illus. III. Title.
PZ7.H286 Ri [Fic] [r79]rev 68-29900 MARC

Hautzig, Deborah.
The handsomest father / by Deborah Hautzig ; pictures by Muriel Batherman. — New York : Greenwillow Books, c1979.
47 p. : col. ill. ; 22 cm. — (Greenwillow read-alone books)
SUMMARY: Marsha is not eager to have her father come to school on Fathers' Visiting Day.
ISBN 0-688-80214-1 : $5.95
[1. Fathers and daughters—Fiction. 2. School stories] I. Batherman, Muriel. II. Title.
PZ7.H2888 Han [E] 78 78-21277 MARC

Hawkins, Mark.
A lion under her bed / by Mark Hawkins ; illustrated by Jean Vallario. — New York : Holt, Rinehart and Winston, c1978.
[32] p. : col. ill. ; 24 cm.
SUMMARY: At bedtime Molly tells her family about the lion under her bed, but they are quite sure she is just pretending.
ISBN 0-03-040381-2. lib. bdg. : $5.95
[1. Lions—Fiction] I. Vallario, Jean. II. Title.
PZ7.H313518 Li [E] 77 77-18865 MARC

Hayes, Geoffrey.
Patrick comes to Puttyville, and other stories / Geoffrey Hayes. — 1st ed. — New York : Harper & Row, c1978.
118 p. : ill. ; 21 cm.
SUMMARY: Five stories relate the experiences of Mama and Patrick Bear who leave their home in the seafaring town of Catfish Bay and start a new life in the country.
ISBN 0-06-022266-2 : $5.95. ISBN 0-06-022267-0 lib. bdg. : $5.79
[1. Bears—Fiction] I. Title.
PZ7.H31455 Pat 1978 [Fic] 00 77-25668 MARC

Hayes, Leona T
The lost treasures of Peep-Eye Cove : short fiction / by Leona T. Hayes ; illustrated by Lee Budahl. — [Cullowhee, N.C.] : Western Carolina University Mountain Heritage Center, c1978.
68 p. : ill. ; 24 cm. — (Mountain Heritage Day series) (A Western Carolina University heritage publication)
SUMMARY: When she is punished for hunting wildflowers instead of doing her chores, 12-year-old Angy decides to stay up in the mountains with her dog until school is out. Her decision has far-reaching consequences.
[1. Mountain life—Fiction] I. Title. II. Series. III. Series: Western Carolina University. Mountain Heritage Center. A Western Carolina University heritage publication.
PZ7.H31477 Lo [Fic] 00 78-110134 MARC

Hayes, Richard.
The secret army / Richard Hayes. — 1st American ed. — New York : Viking Press, 1979, c1977.
212 p. ; 22 cm.
SUMMARY: An insidious organization infiltrates a school, bent on taking over it and, eventually, the world.
ISBN 0-670-62839-5
[1. Mystery and detective stories. 2. School stories. 3. Children's writings] I. Title.
PZ7.H31487 Se 1979 [Fic] 78 78-11302 MARC

Hayes, Sheila.
The carousel horse / Sheila Hayes. — 1st ed. — Nashville : T. Nelson, c1978.
127 p. ; 21 cm.
SUMMARY: To both wealthy Andrea and Fran, daughter of a cook, the carousel horse symbolizes their budding and unusual friendship.
ISBN 0-8407-6610-6 : $6.95
[1. Friendship—Fiction] I. Title.
PZ7.H314874 Car [Fic] 78 78-15181 MARC

Haywood, Carolyn, 1898-
Eddie the dog holder, written and illustrated by Carolyn Haywood. New York, Morrow, 1966.
187 p. illus. 23 cm.
SUMMARY: The summer Eddie and his friend start a dog portrait business, he holds the subject and she paints the portrait, often with disastrous results.
[1. Dogs—Fiction] I. Title.
PZ7.H31496 Ee [Fic] [r79]rev 66-6357 MARC

Haywood, Carolyn, 1898-
Eddie's menagerie / Carolyn Haywood ; illustrated by Ingrid Fetz. — 1st ed. — New York : Morrow, 1978.
191 p. : ill. ; 23 cm.
SUMMARY: The adventures of Eddie as he gets a job in a pet shop and acquires a hat that soon becomes one of his most prized possessions.
ISBN 0-688-22158-0. ISBN 0-688-32158-5 lib. bdg.
[1. Hats—Fiction. 2. Pet shops—Fiction] I. Fetz, Ingrid. II. Title.
PZ7.H31496 Efm [Fic] 78 78-6519 MARC

Hazen, Barbara Shook.
Gorilla wants to be the baby / by Barbara Shook Hazen ; illustrated by Jacqueline Bardner Smith. — 1st ed. — New York : Atheneum, 1978.
[32] p. : ill. (some col.) ; 26 cm.
SUMMARY: A youngster has a comforting talk with his gorilla who understands the problems of living with a new baby.
ISBN 0-689-30654-7 lib. bdg. : $6.95
[1. Babies—Fiction. 2. Brothers and sisters—Fiction] I. Smith, Jacqueline Bardner. II. Title.
PZ7.H314975 Gq [E] 78 78-5972 MARC

Healey, Larry.
The town is on fire / by Larry Healey. — New York : F. Watts, 1979.
121 p. : map ; 22 cm.
SUMMARY: The discovery that the underground coalfields of a Pennsylvania town are burning is followed by a wave of murders. Three highschool friends get involved in stopping the problems.
ISBN 0-531-02898-4
[1. Pennsylvania—Fiction. 2. Mystery and detective stories] I. Title.
PZ7.H344 To [Fic] 79 79-11354 MARC

Hearne, Betsy Gould.
Home / Betsy Gould Hearne ; illustrated by Trina Schart Hyman. — 1st ed. — New York : Atheneum, 1979.
126 p. : ill. ; 22 cm.
"A Margaret K. McElderry book."
SUMMARY: Megan and Randall's adventuresome and dangerous search for the lost king Brendan takes them to a strange land inhabited by lion-like men.
ISBN 0-689-50114-5
[1. Fantasy] I. Hyman, Trina Schart. II. Title.
PZ7.H3464 Ho [Fic] 78 78-10902 MARC

Hefter, Richard.
No kicks for Dog / written and illustrated by Richard Hefter ; edited by Ruth Lerner Perle. — New York : Holt, Rinehart and Winston, c1978.
[32] p. : col. ill. ; 25 cm. — (Sweet Pickles)
SUMMARY: Dog's self-esteem is so low that he doubts he can do anything well.
ISBN 0-03-042011-3 : $2.95
[1. Dogs—Fiction] I. Perle, Ruth Lerner. II. Title. III. Series: Sweet Pickles series.
PZ7.H3587 Nm [E] 78 78-17009 MARC

Hefter, Richard.
Pig thinks pink / written and illustrated by Richard Hefter ; edited by Ruth Lerner Perle. — New York : Holt, Rinehart, and Winston, 1978.
[32] p. : col. ill. ; 25 cm. — (Sweet Pickles series)
SUMMARY: Pig is positive everything will be all right, even when everything is going all wrong.
ISBN 0-03-042051-2 : $2.95
[1. Pigs—Fiction] I. Title. II. Series.
PZ7.H3587 Pi [E] 78 78-9581 MARC

Hefter, Richard.
Turtle throws a tantrum / written & illustrated by Richard Hefter ; edited by Ruth Lerner Perle. — New York : Holt, Rinehart and Winston, c1978.
[36] p. : col. ill. ; 25 cm. — (Sweet Pickles series)
SUMMARY: Turtle tries every trick in the book to get her own way.
ISBN 0-03-042061-X lib. bdg. : $2.95
[1. Turtles—Fiction] I. Perle, Ruth Lerner. II. Title. III. Series.
PZ7.H3587 Tu [E] 77 77-16319 MARC

Hefter, Richard.
Xerus won't allow it / written and illustrated by Richard Hefter ; edited by Ruth Lerner Perle. — New York : Holt, Rinehart and Winston, c1978.
[32] p. : col. ill. ; 25 cm. — (Sweet Pickles series)
SUMMARY: Xerus is convinced that more rules and regulations will keep order in the town of Sweet Pickles, until she finds that she has to follow the rules too.
ISBN 0-03-042076-8
[1. Squirrels—Fiction] I. Perle, Ruth Lerner. II. Title. III. Series.
PZ7.H3587 Xe [E] 77 77-16321 MARC

Heide, Florence Parry.
Banana twist / Florence Parry Heide. — New York : Holiday House, c1978.
111 p. ; 21 cm.
SUMMARY: Relates the attempts of Jonah D. Krock to elude his oddball neighbor who thinks Jonah has a banana fixation.
ISBN 0-8234-0334-3
[1. Humorous stories] I. Title.
PZ7.H36 Ban [Fic] 78 78-6818 MARC

Heide, Florence Parry.
Face at the Brillstone window / Florence Parry Heide and Roxanne Heide ; cover and front. by Joe Krush. — Chicago : A. Whitman, c1979.
128 p. : ill. ; 21 cm. — (Pilot books)
SUMMARY: Interviews for a research project lead two teenagers to discover the secret of a desperate man.
ISBN 0-8075-2216-3 lib. bdg. : $5.00
[1. Mystery and detective stories] I. Heide, Roxanne, joint author. II. Krush, Joe. III. Title.
PZ7.H36 Fac [Fic] 78 78-15060 MARC

Heide, Florence Parry.
Mystery of the mummy's mask / Florence Parry Heide and Roxanne Heide ; ill. by Seymour Fleishman. — Chicago : A. Whitman, c1979.
127 p. : ill. ; 21 cm. — (A Spotlight Club mystery) (Pilot books)
SUMMARY: The discovery of an ancient mummy mask involves the Spotlight Club detectives with unscrupulous art dealers and danger.
ISBN 0-8075-5384-0 : $5.95
[1. Mystery and detective stories] I. Heide, Roxanne, joint author. II. Fleishman, Seymour. III. Title.
PZ7.H36 Myi 1979 [Fic] 79 78-31728 MARC

Heine, Helme.
The pigs' wedding / written and illustrated by Helme Heine. — 1st American ed. — New York : Atheneum, 1979, [c1978]
[26] p. : col. ill. ; 28 cm.
Translation of Na warte, sagte Schwarte.
"A Margaret K. McElderry book."
SUMMARY: Two pigs marry and celebrate with their friends in splendid fashion.
ISBN 0-689-50127-7 : $7.95
[1. Pigs—Fiction. 2. Weddings—Fiction] I. Title.
PZ7.H3678 Pi 1979 79 78-57691 MARC

Helena, Ann.
The lie / words by Ann Helena ; pictures by Ellen Pizer. — Milwaukee : Raintree Editions, c1977.
30 p. : ill. (some col.) ; 25 cm.
SUMMARY: A young girl tells a lie in order to avoid the consequences of what she has done.
ISBN 0-8172-0958-1 lib. bdg. : $7.35
[1. Honesty—Fiction] I. Pizer, Ellen. II. Title.
PZ7.H3735 Li [E] 77 77-23395 MARC

Heller, Linda.
Lily at the table / by Linda Heller. — New York : Macmillan, c1979.
[32] p. : all ill. ; 22 x 26 cm.
SUMMARY: A reluctant eater transforms the food around her into a fantasy playground.
ISBN 0-02-743530-X
[1. Stories without words. 2. Food—Fiction] I. Title.
PZ7.H37424 Li 1979 [E] 79 79-11415 MARC

CHILDREN'S LITERATURE (Continued)

Henstra, Friso.
Wait and see / written and illustrated by Friso Henstra. — Reading, Mass. : Addison-Wesley Pub. Co., c1978.
[32] p. : col. ill. ; 24 cm.
SUMMARY: Having only wanted to prove himself remarkable, a little man, saddened by the failure of three previous inventions, from sheer need creates something heralded by the world.
ISBN 0-201-03077-2
[1. Inventions—Fiction] I. Title.
PZ7.H3987 Wai [E] 78-1020
78 MARC

Herz, Irene.
Hey! Don't do that! / story by Irene Herz ; illustrated by Lucinda McQueen. — Englewood Cliffs, N.J. : Prentice-Hall, c1978.
[32] p. : col. ill. ; 24 cm.
SUMMARY: A brother and sister go to the playground where they caution a variety of animals about using the swings, slides, and other equipment which is really meant only for children.
ISBN 0-13-387365-X lib. bdg. : $7.95
[1. Playgrounds—Fiction. 2. Animals—Fiction] I. McQueen, Lucinda. II. Title.
PZ7.H43247 He [Fic] 77-27054
77 MARC

Herzig, Alison Cragin.
A word to the wise / Alison Cragin Herzig, Jane Lawrence Mali ; illustrated by Martha Perske. — 1sd ed. — Boston : Little, Brown, c1978.
166 p. : ill. ; 22 cm.
SUMMARY: The thesaurus stolen from their teacher alters the lives of eight fifth-graders in a special reading group.
ISBN 0-316-35898-3
[1. Vocabulary—Fiction. 2. School stories] I. Mali, Jane Lawrence, joint author. II. Perske, Martha. III. Title.
PZ7.H432478 Wo [Fic] 78-17447
78 MARC

Heuck, Sigrid.
The stolen apples / Sigrid Heuck. — New York : Hart Pub. Co., c1978.
[30] p. : col. ill. ; 24 cm.
Translation of Pony, Bär und Apfelbaum.
SUMMARY: A rebus tale relating the adventures of a pony and a bear who go in search of the apples stolen from an apple tree.
ISBN 0-8055-1231-4 : $5.95
[1. Rebuses. 2. Animals—Fiction] I. Title.
PZ7.H438 St [E] 78-54176
00 MARC

Hickman, Martha Whitmore, 1925-
I'm moving [by] Martha Whitmore Hickman. Illustrated by Leigh Grant. Nashville, Abingdon Press [1975, c1974]
[31] p. col. illus. 27 cm.
SUMMARY: William explains what he has brought with him and what he had to leave behind when his family moved to a new town.
ISBN 0-687-18685-4
[1. Moving, Household—Fiction] I. Grant, Leigh, ill. II. Title.
PZ7.H53143 Im [E] 74-3064
74[r79]rev MARC

Hickman, Martha Whitmore, 1925-
My friend William moved away / Martha Whitmore Hickman ; illustrated by Bill Myers. — Nashville : Abingdon, c1979.
[32] p. : col. ill. ; 27 cm.
SUMMARY: Jimmy wonders what he will do now that his best friend has moved away.
ISBN 0-687-27540-7 : $6.95
[1. Friendship—Fiction. 2. Moving, Household—Fiction] I. Myers, Bill, 1940- II. Title.
PZ7.H53143 My [E] 78-24319
78 MARC

Hightower, Florence.
Dreamworld Castle / Florence Hightower. — Boston : Houghton Mifflin, 1978.
214 p. ; 22 cm.
SUMMARY: In need of a friend, Phoebe is flattered by the overtures of a wealthy schoolmate and becomes part of her world of lies and even criminal behavior.
ISBN 0-395-27158-4 : $7.95
I. Title.
PZ7.H544 Dr [Fic] 78-14562
78 MARC

Hildick, Edmund Wallace.
The case of the phantom frog / by E. W. Hildick ; illustrated by Lisl Weil. — New York : Macmillan, c1979.
121 p. : ill. ; 22 cm. — (A McGurk mystery)
SUMMARY: The McGurk detectives go undercover to unmask the demon behind a phantom "werefrog" and expose a "supernatural" hoax.
ISBN 0-02-743840-6
[1. Mystery and detective stories] I. Weil, Lisl. II. Title.
PZ7.H5463 Cauf [Fic] 78-10836
78 MARC

Hildick, Edmund Wallace.
The case of the secret scribbler / E. W. Hildick ; illustrated by Val Biro. — London : Hodder and Stoughton, 1978.
95 p. : ill. ; 21 cm. — (A McGurk mystery) GB***
ISBN 0-340-22516-5 : £2.60
I. Biro, Val, 1921- II. Title.
PZ7.H5463 Cav 1978b 78-325823
79 MARC

Hiller, Catherine.
Argentaybee and the Boonie / by Catherine Hiller ; illustrated by Cyndy Szekeres. — New York : Coward, McCann & Geoghegan, c1979.
[32] p. : col. ill. ; 26 cm.
SUMMARY: Emily's imaginary playmates say goodbye when Emily starts school.
ISBN 0-698-20441-7 : $7.50
[1. Play—Fiction. 2. Imagination—Fiction] I. Szekeres, Cyndy. II. Title.
PZ7.H5578 Ar [E] 77-15643
77 MARC

Hillert, Margaret.
Play ball / Margaret Hillert ; illustrated by Dick Martin. — Chicago : Follett Pub. Co., c1978.
31 p. : col. ill. ; 21 cm. — (A Follett just beginning-to-read book)
SUMMARY: Two boys look for just the right type of ball with which to play.
ISBN 0-695-30879-3. ISBN 0-695-40879-8 lib. bdg.
[1. Balls (Sporting goods)—Fiction] I. Martin, Dick, 1933- II. Title.
PZ7.H558 Pl [E] 78-106127
79 MARC

Himler, Ronald.
Wake up, Jeremiah / Ronald Himler. — 1st ed. — New York : Harper & Row, c1979.
[25] p. : chiefly col. ill. ; 22 x 24 cm.
SUMMARY: A small boy runs to a hilltop to greet the rising sun and home again to wake his parents.
ISBN 0-06-022323-5 : $8.95
[1. Sun—Fiction. 2. Day—Fiction] I. Title.
PZ7.H5684 Wak [E] 77-25679
79 MARC

Hinton, Nigel.
Getting free / Nigel Hinton. — 1st ed. — Nashville : T. Nelson, c1978.
189 p. ; 21 cm.
SUMMARY: Felling pressures from their families, two English teenagers run away together when they discover they are expectant parents.
ISBN 0-8407-6615-7 : $6.95
[1. Unmarried mothers—Fiction. 2. Unmarried fathers—Fiction. 3. Runaways—Fiction. 4. England—Fiction] I. Title.
PZ7.H5974 Ge [Fic] 78-14833
78 MARC

Hirsh, Marilyn.
Potato pancakes all around : a Hanukkah tale / written and illustrated by Marilyn Hirsh. — New York : Bonim Books, c1978.
[34] p. : col. ill. ; 24 cm.
SUMMARY: A wandering peddler teaches the villagers how to make potato pancakes from a crust of bread.
ISBN 0-88482-762-3 lib. bdg. : $6.95
[1. Hanukkah—Fiction] I. Title.
PZ7.H598 Po [E] 78-17927
78 MARC

The History and adventures of Little Eliza, a companion to Little William; illustrated with a series of elegant figures. Philadelphia, Published and sold wholesale by W. Charles, and may be had of all the booksellers, J. Bioren, printer, 1815.
[8] p. 8 plates. 14 cm.
In verse.
Cover dated 1814.
Companion to The history and adventures of Little William.
PZ6.H58 78-320130
79 MARC

The History and adventures of Little Henry, exemplified in a series of figures. London, Printed by D. N. Shury, for S. and J. Fuller, 1810.
20 p. illus. 13 cm.
The illustrations are hand-colored cutouts with interchangeable parts. Laid in the L.C. copy are 7 headless figures, 1 head, and 3 hats (2 from another book?)
Issued in a cardboard case with same printed text as on the printed wrappers.
PZ6.H622 1810 78-313114
78 MARC

The History of Little Fanny, exemplified in a series of figures. The 6th ed. London, S. and J. Fuller, 1810.
15 p. illus. 13 cm.
The illustrations are hand-colored cutouts with interchangeable parts. Laid in the L.C. copy are 7 headless figures, 1 head, and 3 hats.
Issued in a cardboard case with same printed text as on the printed wrappers.
PZ6.H628 1810 79-311528
79 MARC

The History of Little Red Riding-Hood. In verse ... 2nd American ed. Philadelphia, M. Charles, 1821.
12 p. 12 (i.e. 11) col. pl. 13 cm.
Plates colored by hand.
Cover-title: The renowned history of Little Red Riding-Hood.
Based on an earlier prose translation of Perrault's Le Petit Chaperon Rouge.
I. Perrault, Charles, 1628-1703. Le Petit Chaperon rouge. II. Little Red Riding Hood.
PZ6.H62828 1821 21-3047
[r79]rev MARC

The History of Solomon Serious and his dog Pompey. — London : J. Fairburn, [182-?]
35, [1] p., [1] fold. leaf of plates : col. ill. ; 14 cm.
At head of title: Fairburn's edition.
Imprint reads: London, Printed and published by J. Fairburn, 110 Minories; price sixpence.
"The Hermit" (a prose paraphrase of T. Parnell's poem): p. [28]-34.
"The Return" (a story set in America; a fragment?): p. 35-[36]
Attached to L.C. copy, in lieu of front cover, is a leaf with a mounted col. ill. identical with that used for frontispiece in a 1797 edition.
I. The Hermit. 182- II. The Return. 182- III. Title: Fairburn's edition.
PZ6.H6285 1820z 79-301035
79 MARC

Hoadley, H M
The green parakeets : a Trinidad legend / told by H. M. Hoadley ; illustrated by Al Ramsawa[c]k. — [Port-of-Spain : Columbus Publishers, 1972]
16 p. : col. ill. ; 24 cm. — (A Doctor Bird book)
Cover title.
I. Ramsawack, Al. II. Title.
PZ8.1.H67 Gr 79-303329
79 MARC

Hoban, Lillian.
Arthur's prize reader / by Lillian Hoban. — 1st ed. — New York : Harper & Row, c1978.
63 p. : col. ill. ; 23 cm. — (An I can read book)
SUMMARY: Although Arthur loses the Super Chimp Club contest, his pupil, sister Violet, wins the first grade reading competition and a prize for them both.
ISBN 0-06-022379-0 : $4.95. ISBN 0-06-022380-4 lib. bdg. : $5.79
[1. Brothers and sisters—Fiction. 2. Reading—Fiction] I. Title.
PZ7.H635 Ay 1978 [E] 77-25637
78 MARC

Hoban, Russell.
The dancing tigers / Russell Hoban ; pictures by David Gentleman. — London : J. Cape, 1979.
[32] p. : col. ill. ; 27 cm. GB***
ISBN 0-224-01374-2 : £2.95
I. Gentleman, David. II. Title.
PZ7.H637 Dan 79-313316
79 MARC

Hoban, Russell.
The twenty-elephant restaurant / by Russell Hoban ; with pictures by Emily Arnold McCully. — 1st ed. — New York : Atheneum, 1978.
[40] p. : ill. (some col.) ; 23 cm.
SUMMARY: A man builds a new table sturdy enough for an elephant to dance on and works up to a restaurant with twenty dancing elephants.
ISBN 0-689-30593-1
[1. Elephants—Fiction. 2. Restaurants, lunch rooms, etc.—Fiction] I. McCully, Emily Arnold. II. Title.
PZ7.H637 Tw [E] 77-1056
77 MARC

Hodges, Cyril Walter, 1909-
Plain Lane Christmas / C. Walter Hodges. — London : J. M. Dent, 1978.
[32] p. : ill. (some col.) ; 26 cm. GB***
ISBN 0-460-06846-6 : £2.95
I. Title.
PZ7.H663 PK 1978b 78-321009
79 MARC

Hodges, Cyril Walter, 1909-
Plain Lane Christmas / text and pictures by C. Walter Hodges. — 1st American ed. — New York : Coward, McCann & Geoghegan, 1978.
[32] p. : ill. (some col.) ; 26 cm.
SUMMARY: The residents of Plain Lane fight City Hall and urban renewal with their community Christmas festival.
ISBN 0-698-20454-9
[1. Urban renewal—Fiction. 2. Community life—Fiction. 3. Christmas stories] I. Title.
PZ7.H663 Pk 1978 [Fic] 77-17233
77 MARC

Hodgetts, Blake Christopher, 1967-
Dream of the dinosaurs / by Blake Christopher Hodgetts ; pictures by Victoria Hodgetts. — 1st ed. — Garden City, N.Y. : Doubleday, c1978.
[48] p. : ill. ; 32 cm.
SUMMARY: A young boy meets millions of dinosaurs in his vivid dreams.
ISBN 0-385-12138-5 : $6.95. ISBN 0-385-12139-3 lib. bdg.
[1. Dinosaurs—Fiction. 2. Dreams—Fiction. 3. Children's writings] I. Hodgetts, Victoria. II. Title.
PZ7.H6646 Dr [Fic] 77-82949
78 MARC

Hoff, Sydney, 1912-
Henrietta, circus star / by Syd Hoff. — Champaign, Ill. : Garrard Pub. Co., c1978.
48 p. : col. ill. ; 23 cm. — (An imagination book)
SUMMARY: Henrietta the chicken sneaks in to see the circus for the first time and unexpectedly becomes one of the performers.
ISBN 0-8116-4413-8 : $4.30
[1. Circus stories. 2. Chickens—Fiction] I. Title.
PZ7.H672 Haw [E] 78-58524
MARC

CHILDREN'S LITERATURE (Continued)

Hoff, Sydney, 1912-
Henrietta goes to the fair / by Syd Hoff. — Champaign, Ill. : Garrard Pub. Co., c1979.
48 p. : col. ill. ; 23 cm. — (An Imagination Book)
SUMMARY: Farmer Gray chooses Winston the pig to represent his farm at the fair, but it is Henrietta the hen who wins a blue ribbon.
ISBN 0-8116-4416-2 (lib. bdg.) : $4.30
[1. Fairs—Fiction. 2. Domestic animals—Fiction] I. Title.
PZ7.H672 Hb [E] 78 78-22051 MARC

Hoff, Sydney, 1912-
Slugger Sal's slump / story and pictures by Syd Hoff. — New York : Windmill Books, c1979.
48 p. : ill. ; 22 cm.
SUMMARY: As Slugger Sal slips deeper into a slump, he wonders if he is washed up as a baseball player.
ISBN 0-525-61590-3 : $5.95
[1. Baseball—Fiction] I. Title.
PZ7.H672 Sn 1979 [E] 78 78-26338 MARC

Holden, Edith, 1871-1920.
The hedgehog feast / [illustrations] by Edith Holden ; words by Rowena Stott. — London : Joseph [etc.], 1978.
[25] p. : chiefly ill. (some col.) ; 25 cm. GB78-21631
ISBN 0-7181-1693-3 : £2.95
I. Stott, Rowena. II. Title.
[PZ7.H6975 He 1978b] 79 79-303832 MARC

Holding, James.
Mr. Moonlight and Omar. Illustrated by Aliki. [New York] Morrow, 1963.
[32] p. illus. 26 cm.
SUMMARY: A young Moroccan boy devises a clever plan to help his father plow their barley field.
[1. Camels—Fiction. 2. Donkeys—Fiction. 3. Morocco—Fiction] I. Aliki. II. Title.
PZ7.H706 Mi [E] [r79]rev 63-7200 MARC

Holding, James.
Weekly Reader Children's Book Club presents The ugliest dog in the world / by James Holding, Jr. ; pictures by Marilyn Miller. — Middletown, Conn. : Xerox Education Publications, c1979.
31 p. : col. ill. ; 23 cm. — (Weekly Reader Children's Book Club)
SUMMARY: The ugliest dog in the world seeks the secret of success.
ISBN 0-88375-219-0
[1. Dogs—Fiction] I. Miller, Marilyn, 1925- II. Title. III. Title: Ugliest dog in the world.
PZ7.H706 Ug [Fic] 79 78-78365 MARC

Holl, Adelaide.
Small Bear and the secret surprise / by Adelaide Holl ; drawings by Tien. — Champaign, Ill. : Garrard Pub. Co., c1978.
47 p. : ill. ; 23 cm. — (Small Bear adventures)
SUMMARY: Two bear cubs frantically search for the "secret surprise" that has wandered off into the forest.
ISBN 0-8116-4455-3 lib. bdg. : $4.20
[1. Bears—Fiction. 2. Brothers and sisters—Fiction] I. Tien. II. Title.
PZ7.H7078 Sh [E] 77 77-17204 MARC

Holl, Adelaide.
Small Bear solves a mystery / by Adelaide Holl ; drawings by Lorinda Cauley. — Champaign, Ill. : Garrard Pub. Co., c1979.
46 p. : col. ill. ; 23 cm. — (Her Small Bear adventures)
SUMMARY: In search of an apple thief, Small Bear and his friends find a hungry old bear whom they decide to adopt.
ISBN 0-8116-4456-1 lib. bdg. : $4.20
[1. Old age—Fiction. 2. Bears—Fiction] I. Cauley, Lorinda Bryan. II. Title.
PZ7.H7078 Sis [E] 78 78-16727 MARC

Holland, David.
Secret of the old church / written by Dave Holland, in collaboration with Will Day ; illustrated by Gordan Willman. — St. Louis : Concordia Pub. House, c1978.
103 p. : ill. ; 23 cm. — (Bro-kee series)
SUMMARY: While visiting in Georgia, the McCall children use Biblical clues to solve a mystery dating back to the Civil War.
ISBN 0-570-07763-X : $2.95
[1. Christian life—Fiction. 2. Mystery and detective stories] I. Day, Will,. II. Willman, Gordon. III. Title.
PZ7.H70823 Se [Fic] 77 77-13776 MARC

Holland, Isabelle.
Dinah and the green fat kingdom / Isabelle Holland. — Philadelphia : Lippincott, c1978.
189 p. ; 21 cm.
SUMMARY: A 12-year-old girl deals with problems of overweight and family relationships.
ISBN 0-397-31818-9
[1. Weight control—Fiction. 2. Family life—Fiction] I. Title.
PZ7.H7083 Di [Fic] 78 78-8612 MARC

Holling, Holling Clancy.
Tree in the trail, written and illustrated by Holling Clancy Holling. Boston, Houghton Mifflin company, 1942.
[63] p. illus. (part col.; incl. maps) 29 x 24 cm.
SUMMARY: A struggling cottonwood sapling becomes a landmark to travelers, a peace-medicine tree, and after its death in 1834, a yoke which is used on the trail to Santa Fe.
[1. Trees—Fiction. 2. Santa Fe Trail—Fiction] I. Title.
PZ7.H725 Tr [Fic] [r79]rev2 42-25811 MARC

Holman, Felice.
The murderer / by Felice Holman. — New York : Scribner, c1978.
151 p. ; 22 cm.
SUMMARY: A young Jewish boy encounters numerous problems growing up during the Depression in a Pennsylvania mining town.
ISBN 0-684-15904-X : $7.95
[1. Jews in the United States—Fiction. 2. Pennsylvania—Fiction] I. Title.
PZ7.H7325 Mu [Fic] 78 78-14266 MARC

Holmes, Efner Tudor.
Carrie's gift / Efner Tudor Holmes ; illustrated by Tasha Tudor. — New York : Collins+World, 1978.
[26] p. : col. ill. ; 19 x 20 cm.
SUMMARY: Carrie's attempts to befriend the dour old man on the neighboring farm are resisted until they save her dog from a trap together.
ISBN 0-529-05428-0 : $5.95. ISBN 0-529-05429-9 lib. bdg. : $6.49
[1. Friendship—Fiction. 2. Country life—Fiction] I. Tudor, Tasha. II. Title.
PZ7.H735 Car [Fic] 78 78-8452 MARC

Holzenthaler, Jean.
My feet do / by Jean Holzenthaler ; photos. by George Ancona. — 1st ed. — New York : Dutton, c1979.
31 p. : ill. ; 24 cm. — (A Fat cat book)
SUMMARY: Describes the many things one's feet can do.
ISBN 0-525-35485-9
[1. Foot] I. Ancona, George. II. Title. III. Series.
PZ7.H7436 Mw 1979 [E] 78 78-13143 MARC

Holzenthaler, Jean.
My hands can / Jean Holzenthaler ; pictures by Nancy Tafuri. — 1st ed. — New York : Dutton, c1978.
[31] p. : col. ill. ; 21 x 23 cm.
SUMMARY: Describes the varied activities of a pair of hands.
ISBN 0-525-35490-5 lib. bdg.
[1. Hand] I. Tafuri, Nancy. II. Title.
PZ7.H7436 My [E] 78 78-5325 MARC

Homer, Larona C 1909-
Blackbeard the pirate and other stories of the Pine Barrens / by Larona Homer ; illustrated by William Sauts Bock. — Wallingford, Pa. : Middle Atlantic Press, c1979.
96 p. : ill. ; 24 cm.
Bibliography: p. 96.
SUMMARY: Short stories set in the Pine Barrens of New Jersey, featuring both real people and legendary figures who were connected with the area.
ISBN 0-912608-10-2
[1. Pine Barrens—Fiction. 2. Short stories] I. Bock, William Sauts, 1939- II. Title.
PZ7.H745 Bl [Fic] 79 79-13270 MARC

Hooker, Ruth.
The kidnapping of Anna / Ruth Hooker and Carole Smith ; illustrated by George Armstrong. — Chicago : A. Whitman, c1979.
128 p. : ill. ; 21 cm. — (Pilot books)
SUMMARY: A brother and sister uncover a sinister kidnapping plot when their family moves into a strange house over Christmas vacation.
ISBN 0-8075-4176-1
[1. Kidnapping—Fiction. 2. Mystery and detective stories] I. Smith, Carole, joint author. II. Armstrong, George Douglas, 1927- III. Title.
PZ7.H7654 Ki 1979 [Fic] 79 78-32051 MARC

Hooks, William H
Crossing the line / by William H. Hooks. New York : Knopf : distributed by Random House, c1978.
124 p. ; 22 cm.
SUMMARY: Describes the comingof age of an 11-year-old boy on his father's farm in the South in the 1930's and his relationships with his extended family and the black sharecroppers who work the farm.
ISBN 0-394-83938-2. ISBN 0-394-93938-7 lib. bdg.
[1. Family life—Fiction. 2. Race relations—Fiction. 3. Southern states—Fiction] I. Title.
PZ7.H7664 Cr 1978 [Fic] 00 78-3269 MARC

Hoover, H E
The lost star / H. M. Hoover. — 1st ed. — New York : Viking Press, 1979.
150 p. ; 24 cm.
SUMMARY: While on an archaeological expedition to Balthor, a young astrophysicist stumbles upon the Lumpies—gray smiling creatures with a strange secret.
ISBN 0-670-44129-5
[1. Science fiction] I. Title.
PZ7.H77048 Lo [Fic] 78 78-25718 MARC

Hope, Laura Lee.
The Bobbsey twins in a TV mystery show / by Laura Lee Hope. — New York : Grosset & Dunlap, c1978.
179 p. : ill. ; 20 cm. — (Bobbsey twins adventure series ; 71)
SUMMARY: While in California appearing in a television show, the Bobbsey twins stumble upon a mystery involving a key, a missing dog, and some photographs.
ISBN 0-448-08071-0 : $2.50
[1. Twins—Fiction. 2. Actors and actresses—Fiction. 3. Mystery and detective stories] I. Title.
PZ7.H772 Bogn [Fic] 79 77-76127 MARC

Hopkins, Lee Bennett.
Wonder wheels : a novel / by Lee Bennett Hopkins. — New York : Knopf, c1979.
163 p. ; 22 cm.
SUMMARY: The relationship of a star of the local roller rink and a sensitive girl from a neighboring town comes to an abrupt and tragic end.
ISBN 0-394-84125-5. ISBN 0-394-94125-X pbk.
[1. Roller skating—Fiction] I. Title.
PZ7.H7754 Wo 1979 [Fic] 78 78-11774 MARC

Hoppe, Joanne.
April spell / by Joanne Hoppe. — New York : F. Warne, c1979.
127 p. ; 24 cm.
SUMMARY: The warnings of a spiritualist and several bizarre events cause a teenager to believe she is being pursued by a vengeful spirit.
ISBN 0-7232-6161-X
[1. Spiritualism—Fiction. 2. Supernatural—Fiction] I. Title.
PZ7.H7786 Ap [Fic] 78 78-23848 MARC

Hoppe, Joanne.
The lesson is murder / Joanne Hoppe. — 1st ed. — New York : Harcourt Brace Jovanovich, c1977.
viii, 149 p. ; 21 cm.
SUMMARY: The humanities class of elite Scarborough High School is thrown into a panic as its members seem to be the prey of a crazed murderer.
ISBN 0-15-244899-3
[1. Mystery and detective stories] I. Title.
PZ7.H7786 Le [Fic] 79[r79]rev 76-54216 MARC

Hopper, Nancy J
Secrets / by Nancy J. Hopper. — 1st ed. — New York : Elsevier/Nelson Books, c1979.
138 p. ; 21 cm.
SUMMARY: An eighth grade girl, severely traumatized after witnessing her father's death, is not believed when she reveals to her psychiatrist and her English teacher that she and a classmate are targeted for kidnapping.
ISBN 0-525-66621-4 : $7.95
[1. Kidnapping—Fiction] I. Title.
PZ7.H7792 Se 1979 [Fic] 79 79-4667 MARC

Horsfall, Robert Bruce, 1869-1948.
Bluebirds seven : paintings / by R. Bruce Horsfall ; text by Carra E. Horsfall. — 1st ed. — Portland : Audubon Society of Portland, Oregon, 1978.
66 p. : col. ill. ; 27 cm.
SUMMARY: Brief text and 33 watercolors depict the activities of a bluebird family.
ISBN 0-931686-03-2
1. Western bluebird—Legends and stories. [1. Bluebirds—Fiction] I. Horsfall, Carra Elisabeth Hunting. II. Audubon Society of Portland, Oregon. III. Title.
PZ10.3.H788 Bl [E] 78 78-51360 MARC

Hough, Richard Alexander, 1922-
Miaow / Bruce Carter [i.e. R. A. Hough]. — London : J. M. Dent, 1978.
124 p. ; 23 cm. GB***
ISBN 0-460-06852-0 : £3.50
I. Title.
PZ7.H8142 Mi 79 78-325821 MARC

Howe, Deborah.
Bunnicula : a rabbit tale of mystery / by Deborah and James Howe ; illustrated by Alan Daniel. — 1st ed. — New York : Atheneum, 1979.
xii, 98 p. : ill. ; 22 cm.
SUMMARY: Though scoffed at by Harold the dog, Chester the cat tries to warn his human family that their foundling baby bunny must be a vampire.
ISBN 0-689-30700-4
[1. Rabbits—Fiction. 2. Vampires—Fiction. 3. Mystery and detective stories] I. Howe, James, 1935- joint author. II. Daniel, Alan, 1939- III. Title.
PZ7.H836 Bu 1979 [Fic] 78 78-11472 MARC

Hsi, Kao.
Walo Hamlet / text by Hsi Kao ; drawings by Sun Pin, Li Hou-mi, Feng Chih-chung. — 1st ed. — Peking : Foreign Languages Press, 1977.
112 p. : ill. ; 26 cm.
I. Sun, Pin. II. Li, Hou-mi. III. Feng, Chih-chung. IV. Title.
PZ7.H85615 Wal 79 78-112041 MARC

CHILDREN'S LITERATURE (Continued)

Hsiao, Jen.
Capture the old bald eagle : story / by Huang Cheng ; adaptation by Hsiao Jen ; drawings by Kan Wu-yen. — 1st ed. — Peking : Foreign Languages Press, 1977.
59 p. : col. ill. ; 25 cm.
$2.00 (H.K.)
I. Huang, Cheng. Huo cho lao t'u ying. II. Kan, Wu-yen. III. Title.
PZ7.H8562 Cap 78-102591
79 MARC

Hsü, Ying.
A real good holiday / Hsu Ying ; ₍ill. by Hsiao Yu-lei and Ho Pao-chuan₎. — 1st ed. — Peking : Foreign Languages Press, 1977.
255 p. : ill. ; 19 cm.
Translation of Hsiang yang yüan ti ku shih.
Label mounted on p. ₍3₎ of cover: Universal Book Co., Hong Kong.
$3.40 (HK)
I. Title.
PZ7.H85655 Re 78-371461
78 MARC

Hu, Ch'iu, 1918-
The rainbow road. ₍Translated by Tang Sheng₎ Peking, Foreign Languages Press, 1959.
157 p. illus. 21 cm.
Translation of Wu ts'ai lu.
I. Title.
PZ7.H8566 Rai 70-259729
79₍r79₎rev MARC

Hubbard, Ralph.
Queer person / by Ralph Hubbard ; illustrated by Harold von Schmidt. — Medora, N.D. : Theodore Roosevelt Nature and History Association, 1978.
336 p., ₍8₎ leaves of plates : ill. ; 21 cm.
SUMMARY: Relates the experiences of an outcast deaf-mute Indian boy as he grows to adulthood and eventually becomes a great leader.
ISBN 0-9601652-0-7
₍1. Deaf—Fiction. 2. Mutism—Fiction. 3. Physically handicapped—Fiction. 4. Indians of North America—Fiction₎ I. Von Schmidt, Harod., 1893- . II. Title.
PZ7.H862 Qe 1978 78-52836
79 MARC

Hubbard, Ralph.
The wolf song / by Ralph Hubbard ; illustrated by Langdon Kihn. — Medora, N.D. : Theodore Roosevelt Nature and History Association, 1978.
287 p. : ill. ; 21 cm.
Reprint of the 1935 ed. published by Doubleday, Doran, Garden City, N.Y.
SUMMARY: A series of disasters befall a Kiowa Indian family when the father breaks his medicine vow.
ISBN 0-9601652-1-5
₍1. Kiowa Indians—Fiction. 2. Indians of North America—Fiction₎ I. Kihn, W. Langdon. II. Title.
PZ7.H862 Wo 1978 ₍Fic₎ 78-52830
79 MARC

Huber-Rogoz, Viorica, 1927-
Petrolache Făt-Frumos / Viorica Huber ; ilustraţii de Liviu Ghigorţ. — ₍Bucureşti₎ : Editura Tineretului, 1962.
₍24₎ p. : col. ill. ; 26 cm.
I. Title.
PZ90.R6H84 77-479743
77₍r79₎rev MARC

Hubka, Betty.
Octavius. Written and illustrated by Betty Hubka. Austin, Tex., Steck Co. ₍1963₎
48 p. col. illus. 24 cm.
SUMMARY: A shy and lonely octopus who lives at the bottom of the sea learns how to make friends.
₍1. Friendship—Fiction. 2. Octopus—Fiction₎ I. Title.
PZ7.H8632 Oc ₍E₎ 63-13045
₍r79₎rev MARC

Huddy, Delia.
Time piper / Delia Huddy. — 1st American ed. — New York : Greenwillow Books, 1979, c1976.
247 p. ; 22 cm.
SUMMARY: A beautiful and remote village girl follows Luke when he leaves for London to work with the inventor of a revolutionary time machine.
ISBN 0-688-80212-5 : $7.95. ISBN 0-688-84212-7 lib. bdg. : $7.63
₍1. Science fiction₎ I. Title.
PZ7.H864 Ti 1979 ₍Fic₎ 78-24339
78 MARC

Hughes, Monica.
The ghost dance caper / ₍by₎ Monica Hughes. — London : H. Hamilton, 1978.
₍6₎, 122 p. ; 23 cm. GB78-30954
ISBN 0-241-10021-6 : £3.25
I. Title.
PZ7.H87364 Gh 1978 78-323936
79 MARC

Hughes, Monica.
The tomorrow city / ₍by₎ Monica Hughes. — London : Hamilton, 1978.
₍5₎, 137 p. ; 23 cm. GB78-17069
ISBN 0-241-89887-0 : £3.50
I. Title.
PZ7.H87364 To 79-305191
79 MARC

Hughes, Shirley.
Moving Molly / Shirley Hughes. — London : Bodley Head, 1978.
₍32₎ p. : ill. ; 19 x 25 cm. GB***
ISBN 0-370-30125-0 : £2.95
I. Title.
PZ7.H87395 Mo 78-326197
79 MARC

Huggler, Thomas E
Westwind Woods / by Thomas E. Huggler ; art director, Ben Graham, illustrated by James Caulfield, cover ill., Dietmar Krumrey ; foreword, Ben East. — Lansing : Michigan United Conservation Clubs, c1978.
132 p. : ill. ; 23 cm.
SUMMARY: Vignettes of animals who reside in Westwind Woods at various times thoroughout the year are interwoven with the story of a community's efforts to protect its wildlife.
1. Animals, Legends and stories of. ₍1. Animals—Fiction. 2. Wildlife conservation—Fiction. 3. Ecology—Fiction₎ I. Caulfield, James. II. Michigan United Conservation Clubs. III. Title.
PZ10.3.H866 We ₍Fic₎ 78-106385
79 MARC

Hugo, Leon.
High Caste / Leon Hugo. — Johannesburg : Perskor, 1976.
185 p. ; 22 cm. SA77
ISBN 0-628-00905-4 : R5.95
I. Title.
PZ7.H8743 Hi 78-314959
78 MARC

Hunt, Christine.
The blue adder stone / Christine Hunt. — London : H. Hamilton, 1978.
122 p. ; 23 cm. GB***
ISBN 0-241-89757-2 : £3.50
I. Title.
PZ7.H9127 Bl 1978 78-316083
78 MARC

Hunter, Evan, 1926-
Find the Feathered Serpent / by Evan Hunter. — Boston : Gregg Press, 1979, c1952.
xi, 207 p. ; 22 cm. — (The Gregg Press science fiction series)
Reprint of the ed. published by Winston, Philadelphia.
SUMMARY: A 16-year-old and his companion travel by time machine to a Yucatan of more than a thousand years ago to search for the origins of the maya god, Kukulcan.
ISBN 0-8398-2519-6
1. Kukulcan—Juvenile fiction. ₍1. Kukulcan—Fiction. 2. Mayas—Fiction. 3. Science fiction₎ I. Title.
PZ7.H91663 Fi 1979 ₍Fic₎ 79-9914
79 MARC

Hurwitz, Johanna.
Aldo Applesauce / Johanna Hurwitz ; illustrated by John Wallner. — New York : Morrow, 1979.
127 p. : ill. ; 21 cm.
SUMMARY: When he and his family move to the suburbs, Aldo has difficulty finding new friends.
ISBN 0-688-22199-8
₍1. Moving, Household—Fiction. 2. Friendship—Fiction₎ I. Wallner, John C. II. Title.
PZ7.H9574 Al ₍Fic₎ 79-16200
79 MARC

Hurwitz, Johanna.
New neighbors for Nora / Johanna Hurwitz ; illustrated by Susan Jeschke. — 1st ed. — New York : Morrow, 1979.
78 p. : ill. ; 21 cm.
"Morrow eagle library edition."
SUMMARY: The further adventures of an inquisitive seven-year-old who lives in a New York City apartment building.
ISBN 0-688-22173-4. ISBN 0-688-32173-7 lib. bdg.
₍1. Apartment houses—Fiction. 2. City and town life—Fiction. 3. Brothers and sisters—Fiction. 4. Neighborliness—Fiction₎ I. Jeschke, Susan. II. Title.
PZ7.H9574 Ne ₍Fic₎ 78-12631
78 MARC

Hutchins, Pat, 1942-
Happy birthday, Sam / by Pat Hutchins. — 1st ed. — New York : Greenwillow Books, c1978.
₍32₎ p. : col. ill. ; 26 cm.
SUMMARY: Sam's birthday brings a solution to several of his problems.
ISBN 0-688-80160-9. ISBN 0-688-84160-0 lib. bdg.
₍1. Birthdays—Fiction₎ I. Title.
PZ7.H96165 Hap ₍E₎ 78-1295
78 MARC

Hutchins, Pat, 1942-
One-eyed Jake / by Pat Hutchins. — 1st ed. — New York : Greenwillow Books, c1979.
₍32₎ p. : col. ill. ; 21 x 26 cm.
SUMMARY: A greedy pirate plunders one ship too many.
ISBN 0-688-80183-8 : $7.25
₍1. Pirates—Fiction₎ I. Title.
PZ7.H96165 On ₍E₎ 78-18346
78 MARC

Ichikawa, Satomi.
Suzanne and Nicholas at the market / ₍by₎ Satomi Ichikawa / ₍language consultant Henry Pluckrose ; English text translated from the French by Denise Sheldon₎. — London ; New York : F. Watts, 1977.
4-30 p. : chiefly col. ill. ; 29 cm. GB78-09871
Translation of Suzette et Nicolas au marché.
SUMMARY: While doing the daily shopping in their French village, Suzanne and Nicholas learn how various things they have purchased are produced before they arrive at the market.
ISBN 0-85166-669-8 : £2.50
₍1. Markets—Fiction. 2. Shopping—Fiction. 3. France—Fiction₎ I. Title.
PZ7.I16 Sr ₍E₎ 79-300394
79 MARC

I'll take care of the crocodiles / adapted from a story by Stefan Mählqvist ; illustrated by Tord Nygren. — 1st American ed. — New York : Atheneum, 1978.
₍26₎ p. : col. ill. ; 27 cm.
Adapted from S. Mählqvist's Inte farligt pappa, krokodilerna klarar jag.
"A Margaret K. McElderry book."
SUMMARY: Not at all sleepy at bedtime, Eric finds himself, his father, and the bed floating through the picture in his room and confronted with one dangerous adventure after another.
ISBN 0-689-50124-2 : $2.95
₍1. Sleep—Fiction. 2. Fantasy₎ I. Nygren, Tord, 1936- . II. Mählqvist, Stefan. Inte farligt pappa, krokodilerna klarar jag. III. Title.
PZ7.I3 1978 ₍E₎ 78-45642
79 MARC

Immel, Mary Blair.
River of wind / by Mary Blair Immel. — Nashville : Aurora Publishers, c1977.
151 p. : ill. ; 22 cm.
SUMMARY: In mid-19th-century Indiana, a young boy help his father realize his dream of building and sailing a balloon.
ISBN 0-87695-208-2 : $5.95
₍1. Balloons—Fiction₎ I. Title.
PZ7.I34 Ri ₍Fic₎ 77-14643
79 MARC

Ingram, Tom.
The outcasts / ₍by₎ Tom Ingram. — London : Hutchinson, 1977.
158 p. ; 21 cm. GB77-28996
ISBN 0-09-131650-2 : £3.25
I. Title.
PZ7.I55 Ou 79-300189
79 MARC

Inkiow, Dimiter.
Me and Clara and Casimir the cat / by Dimiter Inkiow ; illustrated by Traudl and Walter Reiner ; translated from the German by Paula McGuire. — 1st ed. — New York : Pantheon Books, c1979.
77 p. : ill. ; 20 cmm.
Translation of Ich und Klara und der Kater Kasimir.
SUMMARY: The adventures of a brother and sister with each other, their parents, and their pets.
ISBN 0-394-84124-7
₍1. Brothers and sisters—Fiction₎ I. Reiner, Gertraud. II. Reiner, Walter. III. Title.
PZ7.I56 Me ₍E₎ 78-31316
78 MARC

Inkiow, Dimiter.
Me and my sister Clara / by Dimiter Inkiow ; illustrated by Traudl and Walter Reiner ; translated from the German by Paula McGuire. — 1st ed. — New York : Pantheon Books, c1979.
84 p. : ill. ; 20 cm.
Translation of Ich und meine Schwester Klara.
SUMMARY: The adventures of a young brother and sister as they interact with their parents and other relatives.
ISBN 0-394-84123-9
₍1. Brothers and sisters—Fiction₎ I. Reiner, Gertraud. II. Reiner, Walter. III. Title.
PZ7.I56 Mf ₍E₎ 78-31310
78 MARC

Irwin, Hadley.
The Lilith summer / by Hadley Irwin. — Old Westbury, N.Y. : Feminist Press, c1979.
109 p. ; 24 cm.
SUMMARY: A 12-year-old girl relates her experiences during the summer she spends as a companion to a 77-year-old woman.
ISBN 0-912670-51-7
₍1. Old age—Fiction₎ I. Title.
PZ7.I712 Li ₍Fic₎ 78-24379
78 MARC

Isadora, Rachel.
Ben's trumpet / by Rachel Isadora. — 1st ed. — New York : Greenwillow Books, c1979.
₍32₎ p. : ill. ; 21 x 26 cm.
SUMMARY: Ben wants to be a trumpeter, but plays only an imaginary instrument until one of the musicians in a neighborhood night club discovers his ambition.
ISBN 0-688-80194-3 : $6.95
₍1. Musicians—Fiction₎ I. Title.
PZ7.I763 Be ₍E₎ 78-12885
78 MARC

CHILDREN'S LITERATURE (Continued)

Ishii, Momoko, 1907-
The dolls' day for Yoshiko, by Momoko Ishii. Translated by Yone Mizuta. Illustrated by Mamoru Funai. Chicago, Follett Pub. Co. [1966]
94 p. illus. 22 cm.
Translation of Sangatsu hina no tsuki.
SUMMARY: A ten-year-old Japanese girl's wish to have O-hina dolls for the Peach Blossom Festival comes true in a most unusual way.
[1. Japan—Fiction. 2. Dolls—Fiction] I. Funai, Mamoru, illus. II. Title.
PZ7.I776 Do [Fic] 66-13714
[r78]rev2 MARC

Iverson, Genie.
I want to be big / by Genie Iverson ; pictures by David McPhail. — 1st ed. — New York : Dutton, c1979.
[32] p. : ill. ; 21 x 23 cm. — (A Unicorn book)
SUMMARY: A little girl wishes she were big and able to do all sorts of things but not big enough to have to take care of herself all the time.
ISBN 0-525-32539-5 lib. bdg. : $6.95
[1. Growth—Fiction] I. McPhail, David M. II. Title.
PZ7.I945 Iae 1979 [E] 78-31101
78 MARC

Jack and the beanstalk.
Jack and the bean stalk. — [London] : Darton & Co., [between 1843 and 1862]
[12] p. (on double leaves) : col. ill. ; 19 cm. — (Darton's moveable books)
Cover title.
At head of title: Moveable. One shilling.
Each illustration includes one movable part.
"Mr. Griffin, the original inventor of moveable books for children, arranges and fixes the whole of Darton & Co.'s instructive moveable children's books."
I. Griffin, Mr. II. Title: Moveable Jack and the bean stalk. III. Series.
PZ6.J122 Ja 1843 79-311527
79 MARC

Jacobs, Frank.
Alvin Steadfast on Vernacular Island / Frank Jacobs ; drawings by Edward Gorey. — 1st Taplinger ed. — New York : Taplinger Pub. Co., 1979.
64 p. : ill. ; 23 cm.
SUMMARY: A ten-year-old explorer and a famour discoverer work to overcome the evil forces of the Bitter Grudge, the Purple Rage, the Ugly Rumor and the arch villain, the Doubt.
ISBN 0-8008-0173-3 : $5.95
[1. Humorous stories] I. Gorey, Edward St. John. II. Title.
PZ7.J15225 Al 1979 [Fic] 78-20692
79 MARC

Jacobs, William Jay.
Mother, Aunt Susan and me : the first fight for women's rights / William Jay Jacobs. — New York : Coward, McCann & Geoghegan, c1979.
61 p. : ill. ; 24 cm.
Bibliography: p. 61.
SUMMARY: Sixteen-year-old Harriot Stanton highlights the activities of her mother and her mother's friend Susan B. Anthony in their effort to win equal rights for women.
ISBN 0-698-20480-8
1. Stanton, Elizabeth Cady, 1815-1902—Juvenile fiction. 2. Anthony, Susan Brownell, 1820-1906—Juvenile fiction. [1. Stanton, Elizabeth Cady, 1815-1902—Fiction. 2. Anthony, Susan Brownell, 1920-1906—Fiction. 3. Feminists—Fiction] I. Title.
PZ7.J15256 Mo [Fic] 78-25715
78 MARC

Janice.
Mr. and Mrs. Button's wonderful watchdogs / by Janice ; illustrated by Roger Duvoisin. — 1st ed. — New York : Lothrop, Lee & Shepard, c1978.
[32] p. : col. ill. ; 26 cm.
SUMMARY: Mr. and Mrs. Button have a difficult time finding a fierce watchdog because each one they bring home takes on the happy spirit of their friendly household.
ISBN 0-688-41848-1. ISBN 0-688-51848-6 lib. bdg.
[1. Dogs—Fiction] I. Duvoisin, Roger Antoine, 1904- II. Title.
PZ7.J2426 Mk [E] 78-8451
78[r79]rev MARC

Jaroch, F A
The tornado / characters created by Mike Jaroch, written by F. A. "Randy" Jaroch, J. Chipp Jaroch, and Timothy D. Jaroch ; illustrated by Lois Axeman. — Chicago : Childrens Press, c1977.
31 p. : col. ill. ; 25 cm. — (The Adventures of the Sneeky Sneekers)
SUMMARY: Magical powers from mysteriously acquired sneakers enable four brothers to halt a tornado.
ISBN 0-516-03405-7
[1. Tornadoes—Fiction] I. Jaroch, J. Chipp, joint author. II. Jaroch, Timothy D., joint author. III. Axeman, Lois. IV. Title.
PZ7.J292 To [Fic] 77-4844
77 MARC

Jaroch, F A
Washout at Liberty Valley / characters created by Mike Jaroch ; written by F. A. "Randy" Jaroch, J. Chipp Jaroch, and Timothy D. Jaroch ; illustrated by Lois Axeman. — Chicago : Childrens Press, c1978.
31 p. : col. ill. ; 25 cm. — (The Adventures of the Sneeky Sneekers)
SUMMARY: Four brothers who possess extraordinary powers when wearing their magical sneakers help save three families camped near a breaking dam.
ISBN 0-516-03407-3 (lib. bdg.)
[1. Camping—Fiction] I. Jaroch, Mike. II. Jaroch, J. Chipp, joint author. III. Jaroch, Timothy D., joint author. IV. Axeman, Lois. V. Title.
PZ7.J292 Was [E] 78-5466
78 MARC

Jarrell, Randall, 1914-1965.
The bat-poet / by Randall Jarrell ; pictures by Maurice Sendak. — New York : Collier Books, 1977, c1964.
42 p. : ill. ; 23 cm.
SUMMARY: A bat who can't sleep days makes up poems about the woodland creatures he now perceives for the first time.
ISBN 0-02-043910-5
[1. Bats—Fiction. 2. Animals—Fiction] I. Sendak, Maurice. II. Title.
PZ7.J295 Bat 1977 [Fic] 76-17823
76[r79]rev MARC

Jennings, Michael.
The bears who came to breakfix / by Michael Jennings ; illustrated by Tom Dunnington. — Chicago : Childrens Press, c1977.
32 p. : col. ill. ; 25 cm.
SUMMARY: A young boy has two unexpected guests for breakfast.
ISBN 0-516-03411-1 (lib. bdg.)
[1. Bears—Fiction] I. Dunnington, Tom. II. Title.
PZ7.J42987 Be [E] 77-5078
77 MARC

Jensen, Virginia Allen.
What's that? / Virginia Allen Jensen, Dorcas Woodbury Haller. — New York : Collins, 1978, c1977.
23 p. (on double leaves) : ill. ; 20 cm.
Cover title.
Translation of Hvad er det.
SUMMARY: Four friends attempt to find a hidden playmate. Features raised, textured illustrations.
ISBN 0-529-05500-7 : $9.95
1. Blind, Books for the. [1. Blind, Books for the. 2. Size and shape—Fiction. 3. Textures—Fiction] I. Haller, Dorcas Woodbury, joint author. II. Title.
PZ7.J45 Wh 1978 [E] 78-8613
78 MARC

Jeschke, Susan.
Angela and Bear / by Susan Jeschke. — New York : Holt, Rinehart, and Winston, c1979.
[40] p. : ill. ; 29 cm.
SUMMARY: With the help of her magic crayons, Angela acquires what she wants most—a real bear.
ISBN 0-03-044511-6
[1. Magic—Fiction. 2. Bears—Fiction] I. Title.
PZ7.J553 An [E] 78-15224
78 MARC

Jeschke, Susan.
Sidney. [1st ed.] New York, Holt, Rinehart and Winston [1975]
[30] p. illus. 20 x 24 cm.
SUMMARY: Sidney is surprised how easily he succeeds in fooling the foxes and making them believe he is one of them.
ISBN 0-03-013536-2
[1. Chickens—Fiction. 2. Foxes—Fiction] I. Title.
PZ7.J553 Si [E] 74-13188
74[r79]rev MARC

Jezard, Alison.
Albert goes treasure-hunting / by Alison Jezard ; illustrated by Margaret Gordon. — London : V. Gollancz, 1978.
59 p. : ill. ; 24 cm. GB***
ISBN 0-575-02443-7 : £2.50
I. Gordon, Margaret, 1939- II. Title.
PZ7.J57 Ao 78-316086
78 MARC

Johnson, Crockett, 1906-
Harold's ABC; story and pictures. New York, Harper & Row, 1963.
[64] p. illus. 16 cm. (Purple crayon books)
SUMMARY: Harold takes a trip to the moon and comes home again simply by drawing his way through the letters of the alphabet with a purple crayon.
[1. Alphabet] I. Title.
PZ7.J63162 Har [E] 63-14444
[r78]rev2 MARC

Johnson, Dorothy M
Buffalo woman / Dorothy M. Johnson. — New York : Dodd, Mead, c1977.
247 p. ; 22 cm.
SUMMARY: A fictionalized account of life with the Oglala Sioux from 1820 to 1877 as seen through the eyes of the woman Whirlwind.
ISBN 0-396-07423-5
1. Oglala Indians—Juvenile fiction. [1. Oglala Indians—Fiction. 2. Indians of North America—Fiction] I. Title.
PZ7.J63167 Bu [Fic] 76-53436
76[r79]rev MARC

Johnston, Norma.
The crucible year / by Norma Johnston. — 1st ed. — New York : Atheneum, 1979.
166 p. ; 24 cm.
SUMMARY: In 1963 after her father dies, Elizabeth finds her world changing and expanding as she does volunteer work in Dominica and enters public rather than private high school.
ISBN 0-689-30683-0
I. Title.
PZ7.J6453 Cr [Fic] 78-12173
78 MARC

Johnston, Norma.
A mustard seed of magic / Norma Johnston. — 1st ed. — New York : Atheneum, 1977.
184 p. ; 24 cm.
SUMMARY: Tish wants to write well and she wants to be of help to her friends, but first she has to understand herself.
ISBN 0-689-30587-7
I. Title.
PZ7.J6453 Mu [Fic] 77-1214
77 MARC

Johnston, Tony.
Four scary stories / by Tony Johnston ; pictures by Tomie de Paola. — New York : Putnam, c1978.
[32] p. : col. ill. ; 27 cm.
SUMMARY: An imp, a goblin, a scalawag, and a boy tell each other scary stories.
ISBN 0-399-20614-0 : $7.95
[1. Fairies—Fiction] I. De Paola, Thomas Anthony. II. Title.
PZ7.J6478 Fo [E] 77-13027
77[r79]rev MARC

Johnston, Tony.
Happy birthday, Mole & Troll / by Tony Johnston ; drawings by Cyndy Szekeres. — New York : Putnam, c1979.
63 p. : ill. ; 23 cm. — (A See and read storybook)
SUMMARY: Mole and Troll learn several lessons about friendship in these four stories.
ISBN 0-399-61137-1 : $6.29
[1. Friendship—Fiction. 2. Moles (Animals)—Fiction. 3. Trolls—Fiction] I. Szekeres, Cyndy. II. Title.
PZ7.J6478 Hap [E] 78-25717
78 MARC

Johnston, Tony.
Little mouse nibbling / by Tony Johnston ; drawings by Diane Stanley. — New York : Putnam, c1979.
[32] p. : ill. (some col.) ; 20 cm.
SUMMARY: Though lonely in her little house, a little mouse is too shy to step outside and find a friend until she meets a cricket.
ISBN 0-399-20673-6 : $7.95
[1. Friendship—Fiction. 2. Mice—Fiction. 3. Crickets—Fiction. 4. Christmas stories] I. Stanley, Diane. II. Title.
PZ7.J6478 Li 1979 [E] 78-24265
78 MARC

Jones, Cordelia.
The view from the window / Cordelia Jones. — London : Deutsch, 1978.
143 p. ; 21 cm. GB78-29403
ISBN 0-233-97004-5 : £3.50
I. Title.
PZ7.J683 Vi 78-326198
79 MARC

Jones, Diana Wynne.
Drowned Ammet / Diana Wynne Jones. — 1st American ed. — New York : Atheneum, 1978, c1977.
255 p. ; 22 cm.
SUMMARY: When his protest against the tyrannical government fails, a young boy escapes, with two other children, to the mysterious Holy Islands where they learn the identity and the power of two folk figures celebrated by their countrymen.
ISBN 0-689-30620-2
I. Title.
PZ7.J684 Dr 1978 [Fic] 77-21108
77[r79]rev MARC

Jones, Diana Wynne.
The spellcoats / by Diana Wynne Jones. — 1st American ed. — New York : Atheneum, 1979.
249 p. ; 22 cm. — (An Argo book)
SUMMARY: Tanqui discovers she has the only means to conquer the evil Kankredin who threatens her own people and the Heathens who have invaded prehistoric Dalemark.
ISBN 0-689-30712-8
[1. Fantasy] I. Title.
PZ7.J684 Sp 1979 [Fic] 79-10421
79 MARC

Jones, Harold, 1904-
There & back again / [by] Harold Jones. — Oxford [etc.] : Oxford University Press, 1977.
[32] p. : col. ill. ; 20 cm. GB77-30771
ISBN 0-19-279715-8 : £1.75
I. Title.
PZ8.9.J76 Th 1977b 78-315172
78 MARC

Jones, McClure.
Cast down the stars / McClure Jones. — New York : Holt, Rinehart and Winston, c1978.
186 p. ; 24 cm.
SUMMARY: When the lands of the villagers are threatened by nomadic barbarian tribes, Glory, a starcaster, finds that she needs more than scientific knowledge to make a decision that will affect the people that she loves most.
ISBN 0-03-042501-8
[1. Science fiction] I. Title.
PZ7.J7215 Cas [Fic] 78-4277
78 MARC

CHILDREN'S LITERATURE (Continued)

Judd, Denis, 1938-
Return to Treasure Island / Denis Judd. — New York : St. Martin's Press, 1979, c1978.
209 p. : map ; 23 cm.
SUMMARY: During a reunion with Jim Hawkins, Long John Silver tells of his tireless quest to recover the remaining treasure from Treasure Island.
ISBN 0-312-67912-2 : $8.95
[1. Buried treasure—Fiction. 2. Pirates—Fiction] I. Title.
PZ7.J878 Re [Fic] 78-61014
79 MARC

Kaatz, Evelyn.
Race car driver / Evelyn Kaatz. — 1st ed. — Boston : Little, Brown, c1979.
55 p. : ill. ; 25 cm.
SUMMARY: Greg enters Skip Barber's Racing School to increase his skills in automobile racing.
ISBN 0-316-47751-6
[1. Automobile racing—Fiction] I. Title.
PZ7.K114 Rac [Fic] 79-14766
79 MARC

Kahl, Virginia.
Small white cat / by Virginia Kahl. — New York : Scribner, c1979.
[32] p. : col. ill. ; 20 x 24 cm.
SUMMARY: Although there is only one small white cat in their village, the residents convince inquiring authorities that each household has a cat of its own.
ISBN 0-684-16097-8
[1. Cats—Fiction] I. Title.
PZ7.K1228 Wh [E] 78-31278
78 MARC

Kalan, Robert.
Blue sea / by Robert Kalan ; illustrated by Donald Crews. — 1st ed. — New York : Greenwillow Books, c1979.
[24] p. : col. ill. ; 21 x 26 cm.
SUMMARY: Several fishes of varying size introduce space relationships and size differences.
ISBN 0-688-80184-6 : $7.25
[1. Size and shape—Fiction. 2. Space perception—Fiction. 3. Fishes—Fiction] I. Crews, Donald. II. Title.
PZ7.K123475 Bl [E] 78-18396
78 MARC

Kamm, Josephine.
Runaways / Josephine Kamm. — London : Hodder and Stoughton, 1978.
159 p. ; 22 cm. GB***
ISBN 0-340-20972-0 : £3.50
I. Title.
PZ7.K127 Ru 78-322014
78 MARC

Kane, Betty.
Ow and the crystal clear / by Betty Kane ; with ill. by Tracy Holcomb. — San Rafael, Calif. : Dawne-Leigh Publications, c1978.
28 p. : col. ill. ; 28 cm.
SUMMARY: A naughty gnome who enjoys pinching people changes his ways.
ISBN 0-89742-008-X : $4.95
[1. Fairies—Fiction] I. Holcomb, Tracy. II. Title.
PZ7.K1275 Ow [E] 78-11753
78 MARC

Kaplow, Robert.
Two in the city / Robert Kaplow. — Boston : Houghton Mifflin, 1979.
146 p. ; 22 cm.
SUMMARY: David decides to skip college and live with Stacey in New York City, but soon faces more problems than he's anticipated.
ISBN 0-395-27813-9 : $6.95
[1. New York (City)—Fiction] I. Title.
PZ7.K1295 Tw [Fic] 78-31649
78 MARC

Keene, Carolyn, pseud.
Mystery of Crocodile Island / by Carolyn Keene. — New York : Grosset & Dunlap, c1978.
180 p. : ill. ; 20 cm. — (Her Nancy Drew mystery stories ; 55)
SUMMARY: A young detective's attempts to uncover a group of poachers on Crocodile Island in Florida involve her with kidnapping, reptiles, enemy boats, and a sinister racket.
ISBN 0-448-09555-6 : $2.95
[1. Mystery and detective stories] I. Title.
PZ7.K23 Nan no. 55 [Fic] 77-76128
79 MARC

Keene, Carolyn, pseud.
The Nancy Drew sleuth book : clues to good sleuthing / by Carolyn Keene. — New York : Grosset & Dunlap, c1979.
152 p. : ill. ; 21 cm.
SUMMARY: In solving a series of mysteries, Nancy Drew and her Detective Club study techniques of criminal investigation involving handwriting, fingerprints, codes, and moulages.
ISBN 0-448-15459-5 : $4.95
[1. Mystery and detective stories. 2. Criminal investigation—Fiction] I. Title.
PZ7.K23 Nat [Fic] 78-58209
79 MARC

Keene, Carolyn, pseud.
The witch's omen / by Carolyn Keene. — New York : Grosset & Dunlap, c1979.
179 p. : ill. ; 20 cm. — (The Dana girls mystery stories ; 17)
SUMMARY: After an uninvited witch interrupts their Halloween party, two teenage detectives unravel a mystery involving an international gang and contraband cargo at a waterfront.
ISBN (invalid) 0-04-489097-X : $2.95
[1. Mystery and detective stories] I. Title.
PZ7.K23 Dap no. 17 [Fic] 78-57929
00 MARC

Keeping, Charles.
Miss Emily and the bird of make-believe / [by] Charles Keeping. — London : Hutchinson, 1978.
[29] p. : col. ill. ; 21 cm. GB78-34902
ISBN 0-09-133980-4 : £3.50
I. Title.
PZ7.K2512 Mi 79-303230
79 MARC

Keats, Ezra Jack.
Pet show! New York, Macmillan [1972]
[36] p. col. illus. 21 x 23 cm. $4.95
SUMMARY: When he can't find his cat to enter in the neighborhood pet show, Archie must do some fast thinking to win a prize.
[1. Pets—Stories] I. Title.
PZ7.K2253 Pc [E] 78-317338
78[r78]rev MARC

Keene, Carolyn, pseud.
The riddle of the frozen fountain / by Carolyn Keene. — New York : Grosset & Dunlap, c1972.
173 p. : ill. ; 20 cm. — (The Dana girls mystery stories ; 2)
SUMMARY: When asked to trace the writer of a threatening note, the Dana sisters become enmeshed in a mystery concerning a bronze fountain.
ISBN 0-448-09082-1 : $2.50
[1. Mystery and detective stories] I. Title.
PZ7.K23 Dap no. 2 [Fic] 70-180991
78[r78]rev MARC

Keller, Beverly.
Pimm's place / by Beverly Keller ; pictures by Jacqueline Chwast. — New York : Coward, McCann & Geoghegan, c1978.
63 p. : ill. (some col.) ; 22 cm. — (Break-of-day books)
SUMMARY: A timorous young boy teaches his rowdy cousins the value of being quiet and using one's imagination.
ISBN 0-698-30689-9
[1. Imagination—Fiction] I. Chwast, Jacqueline. II. Title.
PZ7.K2813 Pi 1978 [Fic] 77-27053
77 MARC

Kellerhals-Stewart, Heather.
Rory and the whooping crane / by Heather Kellerhals-Stewart. — Ottawa : Borealis Press, 1977.
vi, 66 p. ; 23 cm. C***
Bibliography: p. 66.
I. Title.
PZ7.K2815 Ro 79-303840
79 MARC

Kelley, Leo P
King of the stars / Leo P. Kelley ; [cover and ill., Rick Guidice]. — Belmont, Calif. : Fearon Pitman Publishers, c1979.
59 p. : ill. ; 18 cm. — (The Galaxy 5 series) (A Pacemaker book)
SUMMARY: As they near Star 84 of Galaxy 5, Don, Steve, and Ellen are moved ahead in space to Planet 1 by the alien Holgo and find themselves locked in fierce battle with King Yorro's silver soldiers.
ISBN 0-8224-3206-4
[1. Science fiction] I. Guidice, Rick. II. Title.
[PZ7.K2817 St] [Fic] 78-68231
00 MARC

Kelley, Leo P
On the red world / Leo P. Kelley. — Belmont, Calif. : Fearon Pitman Publishers, c1979.
60 p. : ill. ; 18 cm. — (The Galaxy 5 series) (A Pacemaker book)
SUMMARY: Steve and Ellen find themselves on a strange planet—half never-ending day, half never-ending night—inhabited by savage robots looking for humans to destroy.
[1. Science fiction] I. Title.
[PZ7.K2817 On] [Fic] 78-68227
00 MARC

Kelley, Leo P
Star gold / Leo P. Kelley. — Belmont : Fearon Pitman Publishers, c1979.
60 p. : ill. ; 18 cm. — (A Pacemaker bestsellers book)
SUMMARY: Brett Kinkaid, serving a life sentence on Earth, the prison planet, is offered freedom if he rescues a kidnapped young woman from the star Alba.
ISBN 0-8224-5369-X
[1. Science fiction] I. Title.
[PZ7.K2817 St] [Fic] 78-72333
79 MARC

Kelley, Leo P
Vacation in space / Leo P. Kelley ; [cover and ill., Rick Guidice]. — Belmont, Calif. : Fearon Pitman Publishers, c1979.
60 p. : ill. ; 18 cm. — (The Galaxy 5 series) (A Pacemaker book)
SUMMARY: While the Voyager's three-person crew rests on the vacation satellite, a beautiful woman schemes to take over their spaceship.
ISBN 0-8224-3203-X
[1. Science fiction] I. Guidice, Rick. II. Title.
PZ7.K28136 Vac [Fic] 78-68232
79 MARC

Kelley, Leo P
Where no sun shines / Leo P. Kelley. — Belmont, Calif. : Fearon Pitman Publishers, c1979.
60 p. : ill. ; 18 cm. — (The Galaxy 5 series) (A Pacemaker book)
SUMMARY: Three crew members of the spaceship Voyager are sold as toys to the giant aliens of the planet Zandar.
ISBN 0-8224-3205-6
[1. Science fiction] I. Title.
[PZ7.K28136 Wh] [Fic] 78-68229
79[r79]rev MARC

Kelley, True.
The Mouses' terrible Christmas / True Kelley & Steven Lindblom. — 1st ed. — New York : Lothrop, Lee & Shepard, c1978.
63 p. : ill. ; 22 cm. — (A Fun-to-read book)
SUMMARY: It's Christmas at the Mouses' where holiday havoc is a family tradition.
ISBN 0-688-41856-2. ISBN 0-688-51856-7 lib. bdg.
[1. Christmas stories. 2. Mice—Fiction. 3. Humorous stories] I. Lindblom, Steven, joint author. II. Title.
PZ7.K2824 Mo [Fic] 78-6995
00 MARC

Kellogg, Steven.
The mystery of the magic green ball / story and pictures by Steven Kellogg. — New York : Dial Press, c1978.
[32] p. : ill. (some col.) ; 17 cm.
SUMMARY: Timmy's long, frustrating search for his favorite green ball ends with a surprising turn of events.
ISBN 0-8037-6214-3 : $4.95. ISBN 0-8037-6215-1 lib. bdg. : $4.58
[1. Balls (Sporting goods)—Fiction] I. Title.
PZ7.K292 Mx [E] 78-51322
78 MARC

Keneally, Thomas.
Ned Kelly and the city of the bees / Thomas Keneally ; illustrated by Stephen Ryan. — London : J. Cape, 1978.
120 p. : ill. ; 23 cm. GB***
ISBN 0-224-01600-8 : £3.50
I. Title.
PZ7.K36 Ne 78-325824
79 MARC

Kennedy, Richard.
The dark princess / written by Richard Kennedy ; illustrated by Donna Diamond. — New York : Holiday House, c1978.
[32] p. : ill. (some col.) ; 24 cm.
SUMMARY: A princess whose radiant beauty blinds her and all who look upon her finds only one man willing to submit to her test for suitors the court fool.
ISBN 0-8234-0329-7
[1. Princesses—Fiction] I. Diamond, Donna. II. Title.
PZ7.K385 Dar [E] 78-1548
78 MARC

Kennedy, Richard.
The leprechaun's story / by Richard Kennedy ; illustrated by Marcia Sewall. — 1st ed. — New York : Dutton, c1979.
[40] p. : ill. ; 21 x 26 cm. — (A Unicorn book)
SUMMARY: A tradesman meets a leprechaun and is determined that he won't be tricked out of a pot of gold.
ISBN 0-525-33472-6
[1. Leprechauns—Fiction] I. Sewall, Marcia. II. Title.
PZ7.K385 Le 1979 [E] 79-11410
79 MARC

Kennedy, Richard.
The lost kingdom of Karnica / Richard Kennedy ; illustrated by Uri Shulevitz. — San Francisco : Sierra Club Books/Scribner's, c1979.
[32] p. : col. ill. ; 26 cm.
SUMMARY: Life in the Kingdom of Karnica is never the same after Farmer Erd unearths a valuable red stone in his field.
ISBN 0-684-16164-8 (lib. bdg.) : $8.95
[1. Conduct of life—Fiction] I. Shulevitz, Uri. II. Title.
PZ7.K385 Lo [E] 78-32052
79 MARC

Kennedy, Richard.
The mouse god / by Richard Kennedy ; illustrated by Stephen Harvard. — 1st ed. — Boston : Little, Brown, c1979.
32 p. : col. ill. ; 24 cm.
"An Atlantic Monthly Press book."
SUMMARY: A vain and lazy barnyard cat makes a coat of mouse skins to protect his fur while chasing mice. The mice consider him their god and trust him to send them to heaven.
ISBN 0-316-48904-2 : $6.95
[1. Cats—Fiction. 2. Mice—Fiction] I. Harvard, Stephen. II. Title.
PZ7.K385 Mo [E] 78-11731
78 MARC

Kennedy, Richard, 1932-
Come again in the spring / by Richard Kennedy ; pictures by Marcia Sewall. — 1st ed. — New York : Harper & Row, c1976.
47 p. : ill. ; 19 x 20 cm.
SUMMARY: An old man tries to outsmart Death with the help of the birds who come to his cabin.
ISBN 0-06-023128-9. ISBN 0-06-023129-7 lib. bdg. : $4.79
[1. Death—Fiction] I. Sewall, Marcia. II. Title.
PZ7.K385 Cm [Fic] 76-3830
76[r79]rev MARC

CHILDREN'S LITERATURE (Continued)

Kennedy, Richard, 1932-
The contests at Cowlick / by Richard Kennedy ; illustrated by Marc Simont. — 1st ed. — Boston : Little, Brown, [1975]
[48] p. : col. ill. ; 22 cm.
"An Atlantic-Monthly Press book."
SUMMARY: Wally outwits a gang of outlaws and saves the town of Cowlick.
[1. Western stories. 2. Humorous stories] I. Simont, Marc, ill. II. Title.
PZ7.K385 Co [Fic] 74-23566
74[r79]rev MARC

Kennedy, Richard, 1932-
Oliver Hyde's dishcloth concert / by Richard Kennedy ; illustrated by Robert Andrew Parker. — 1st ed. — Boston : Little Brown, c1977.
47 p. : ill. ; 20 x 21 cm.
"An Atlantic Monthly Press book."
SUMMARY: A fiddler who went into seclusion after his wife's death agrees to play at a wedding. When he finds that only sacks of grain are around to listen, he goes in search of company.
ISBN 0-316-48179-3
[1. Death—Fiction] I. Parker, Robert Andrew. II. Title.
PZ7.K385 Ol [Fic] 76-15980
76[r79]rev MARC

Kennedy, Richard, 1932-
The rise and fall of Ben Gizzard / by Richard Kennedy ; illustrated by Marcia Sewall. — 1st ed. — Boston : Little, Brown, c1978.
41 p. : ill. ; 22 cm.
"An Atlantic Monthly Press book."
SUMMARY: When he least expects it, a cunning swindler's good fortune runs out.
ISBN 0-316-48903-4 lib. bdg. : $6.95
1. The West—Fiction. [1. Humorous stories] I. Sewall, Marcia. II. Title.
PZ7.K385 Ri [Fic] 78-1816
78[r79]rev MARC

Kent, Alexander.
Midshipman Bolitho and the Avenger / Alexander Kent. — 1st American ed. — New York : Putnam, 1978.
143 p. ; 21 cm.
SUMMARY: Home from the sea on leave in Cornwall, a 17-year-old midshipman becomes involved with smuggling, murder, and "wrecking."
ISBN 0-399-20652-3 : $6.95
[1. Cornwall, Eng.—Fiction. 2. Sea stories] I. Title.
PZ7.K412 Mi 1978 [Fic] 78-9127
78 MARC

Kent, Jack, 1920-
Floyd, the tiniest elephant / by Jack Kent. — 1st ed. — Garden City, N.Y. : Doubleday, c1979.
[32] p. : col. ill. ; 20 x 24 cm.
SUMMARY: A lonely elephant the size of a mouse runs away from the circus because no one ever notices him.
ISBN 0-385-14099-1 : $6.95. ISBN 0-385-14100-9 lib. bdg.
[1. Elephants—Fiction. 2. Mice—Fiction] I. Title.
PZ7.K414 Fj [E] 78-4708
78 MARC

Kent, Jack, 1920-
Jack Kent's Cindy Lou and the witch's dog. — New York : Random House, c1978.
[36] p. : all col. ill. ; 21 cm. — (A Random House pictureback) (The best book club ever)
SUMMARY: Cindy Lou does her best to keep track of a dog full of magic spells.
ISBN 0-394-83745-2
[1. Magic—Fiction. 2. Dogs—Fiction. 3. Hiccups—Fiction. 4. Cartoons and comics] I. Title. II. Title: Cindy Lou and the witch's dog.
PZ7.K414 Jab [E] 77-86144
78 MARC

Kent, Jack, 1920-
Piggy Bank Gonzales / Jack Kent. — New York : Parents' Magazine Press, c1978.
[33] p. : col. ill. ; 24 cm.
SUMMARY: A piggy bank, bored with the sameness of every day, discovers that life outside his quiet Mexican home can be too exciting.
ISBN 0-8193-0997-4. ISBN 0-8193-0998-2 lib. bdg.
[1. Coin banks—Fiction. 2. Mexico—Fiction] I. Title.
PZ7.K414 Pi [E] 78-23515
78 MARC

Kent, Jack, 1920-
Socks for supper / Jack Kent. — New York : Parents' Magazine Press, c1978.
[32] p. : col. ill. ; 24 cm.
SUMMARY: When a poor couple exchange socks for cheese and milk, they receive more than expected.
ISBN 0-8193-0964-8. ISBN 0-8193-0965-6 lib. bdg.
[1. Friendship—Fiction] I. Title.
PZ7.K414 So [E] 78-6224
78 MARC

Kent, Jack, 1920-
Supermarket magic / Jack Kent. — New York : Random House, c1978.
[22] p. : col. ill. ; 16 cm. — (A Sniffy book)
SUMMARY: Wilma Witch turns people into food at a supermarket. The reader can experience several fragrances by scratching and sniffing treated pages.
ISBN 0-394-83921-8 : $2.95
[1. Supermarkets—Fiction. 2. Witches—Fiction] I. Title.
PZ7.K414 Su [E] 78-55908
78 MARC

Kepes, Juliet.
Cock-a-doodle-do / by Juliet Kepes. — New York : Pantheon Books, c1978.
[40] p. : col. ill. ; 23 cm.
SUMMARY: Protective of a young chick she plans to eat later, a tigress changes her plans for dinner when the young rooster becomes her protector.
ISBN 0-394-83494-1. ISBN 0-394-93494-6 lib. bdg.
[1. Animals—Fiction] I. Title.
PZ7.K437 Co [E] 76-44433
76 MARC

Kessler, Leonard P 1920-
Do you have any carrots? / By Leonard Kessler ; drawings by Lori Pierson. — Champaign, Ill. : Garrard Pub. Co., c1979.
32 p. : col. ill. ; 23 cm.
SUMMARY: As they travel around the neighborhood looking for carrots, two little rabbits discover that other animals prefer to eat different things.
ISBN 0-8116-6074-5 lib. bdg. : $4.20
[1. Rabbits—Fiction. 2. Food—Fiction] I. Pierson, Lori. II. Title.
PZ7.K484 Do [E] 78-23708
78 MARC

Kessler, Leonard P 1920-
The pirates' adventure on Spooky Island / by Leonard Kessler. — Champaign, Ill. : Garrard Pub. Co., c1979.
48 p. : col. ill. ; 23 cm. — (An Imagination book)
SUMMARY: Cowardly Captain Ben and his sidekick, a parrot, lead their pirate crew on an adventure to Spooky Island.
ISBN 0-8116-4414-6 lib. bdg. : $4.30.
[1. Pirates—Fiction] I. Title.
PZ7.K484 Pi [E] 78-12379
78 MARC

Kessler, Leonard P 1920-
Tricks for treats on Halloween / by Leonard Kessler ; drawings by Tom Eaton. — Champaign, Ill. : Garrard Pub. Co., c1979.
32 p. : col. ill. ; 23 cm.
SUMMARY: Strange things happen when a witch, a dragon, and a ghost go trick-or-treating.
ISBN 0-8116-6075-3 (lib. bdg.)
[1. Halloween—Fiction] I. Easton, Tom. II. Title.
PZ7.K484 Tr [E] 78-21942
78 MARC

Kibbe, Pat.
The hocus-pocus dilemma / Pat Kibbe ; illustrated by Dan Jones. — New York : Knopf : distributed by Random House, c1979.
124 p. : ill. ; 22 cm.
SUMMARY: Convinced she has psycical powers, a ten-year-old girl makes predictions for the members of her large family which seem to come true.
ISBN 0-394-84058-5. ISBN 0-394-94058-X lib. bdg.
[1. Family life—Fiction] I. Jones, Dan, 1948- II. Title.
PZ7.K529 Ho [Fic] 78-9906
78 MARC

Kidd, Ronald.
That's what friends are for / Ronald Kidd. — 1st ed. — Nashville : T. Nelson, c1978.
127 p. ; 21 cm.
SUMMARY: Two inseparable thirteen-year-old friends confront death when one's illness is diagnosed as leukemia.
ISBN 0-8407-6614-9 : $6.95
[1. Death—Fiction. 2. Leukemia—Fiction. 3. Friendship—Fiction] I. Title.
PZ7.K5315 Th 1978 [Fic] 78-14255
78 MARC

Kilbourne, Frances.
Overnight adventure / by Frances Kilbourne ; Illustrated by Ann Powell. — Toronto : Women's Press, c1977.
[24] p. : all col. ill. ; 21 cm. C***
ISBN 0-88961-054-1. ISBN 0-88961-047-9 pbk
I. Powell, Ann, 1951- II. Title.
PZ7.K5544 Ov 79-300454
79 MARC

Kilpatrick, Christopher H
Kirkmouse / Christopher H. Kilpatrick ; ill. by Jeff MacNelly. — McLean, Va. : EPM Publications, c1979.
[64] p. : ill. ; 14 x 21 cm.
SUMMARY: Kirkmouse tries to think of a way to make money for the destitute New England church in which he lives.
ISBN 0-914440-28-4 : $5.95
[1. Mice-Fiction. 2. Christmas stories] I. MacNelly, Jeff. II. Title.
PZ7.K559 Ki [Fic] 79-13229
79 MARC

Kincl, Kay Owens, 1955-
Mandy's laughing book / written and illustrated by Kay Owens Kincl. — Austin, Tex. : Shoal Creek Publishers, c1978.
[16] p. : col. ill. ; 28 cm.
SUMMARY: Many discovers that everyone, including her animal friends, has a different kind of laugh.
ISBN 0-88319-040-0 : $4.95
[1. Laughter—Fiction. 2. Animals—Fiction] I. Title.
PZ7.K5654 Man [E] 78-12950
78 MARC

The Kind hearted hen. — Hong Kong : Seagull Pub. Co., [1977?]
12 p. : col. ill. ; 18 x 20 cm.
Cover title.
Translation of Hao hsin ch'ang ti mu chi.
PZ7.K5657 77-152961
78 MARC

King, Cynthia.
The year of Mr. Nobody / by Cynthia King ; illustrated by Malcolm Carrick. — 1st ed. — New York : Harper & Row, c1978.
60 p. : ill. ; 23 cm.
SUMMARY: A little boy gradually outgrows his dependence on his imaginary friend.
ISBN 0-06-023132-7 : $5.95. ISBN 0-06-023133-5 lib. bdg. : $5.79
I. Carrick, Malcolm. II. Title.
PZ7.K5758 Ye 1978 [E] 77-11830
77 MARC

Kingman, Lee.
Head over wheels / Lee Kingman. — Boston : Houghton Mifflin, 1978.
186 p. ; 22 cm.
SUMMARY: After an automobile accident, 17-year-old identical twins, Terry and Kerry, must face the fact that Terry will never walk again.
ISBN 0-395-27202-5 : $7.95
[1. Twins—Fiction. 2. Physically handicapped—Fiction. 3. Family problems—Fiction] I. Title.
PZ7.K595 He [Fic] 78-15650
78 MARC

Kipling, Rudyard, 1865-1936.
Just so stories for little children / by Rudyard Kipling. — New York : Weathervane Books : distributed by Crown Publishers, 1978.
210 p. : ill. ; 24 cm.
Reprint of the 1902 ed. published by McMillan, London.
SUMMARY: Twelve stories about animals, insects, and other subjects include "How the Camel Got His Hump," "The Butterfly That Stamped," and "How the Alphabet Was Made."
ISBN 0-517-26655-5
[1. Animals—Fiction. 2. Jungle stories. 3. Short stories] I. Title.
PZ7.K632 Jw 1978 [Fic] 78-23639
78 MARC

Kippax, Janet.
The Gypsy / Janet Kippax. — Zonderavian ed. — Grand Rapids : Zondervan Pub. House, 1979, c1972.
90 p. : 20 cm. — (Pathfinder series ; 3)
SUMMARY: Four brothers, in search of a toy airplane, find a runaway orphan with a broken ankle in the unused railroad tunnel hideout of an escaped convict.
ISBN 0-310-37811-7
[1. Orphans—Fiction. 2. Crime and criminals—Fiction. 3. Christian life—Fiction] I. Title. II. Series: Pathfinder series (Grand Rapids, 1979-) ; 3.
PZ7.K6326 Gy 1979 [Fic] 78-24363
78 MARC

Kirk, Barbara.
Grandpa, me and our house in the tree / by Barbara Kirk. — New York : Macmillan, c1978.
[48] p. : ill. ; 21 x 26 cm.
SUMMARY: A small boy and his grandfather share a special relationship despite the old man's illness.
ISBN 0-02-750750-5
[1. Grandfathers—Fiction] I. Title.
PZ7.K6338 Gr [Fic] 78-9564
78 MARC

Kirkpatrick, Doris.
Honey in the rock / by Doris Kirkpatrick. — 1st ed. — New York : Elsevier/Nelson Books, c1979.
218 p. ; 21 cm.
SUMMARY: During the autumn of 1936, a 16-year-old girl begins to see signs that life in her small village in Vermont is drastically changing.
ISBN 0-525-66643-5 : $8.95
[1. Vermont—Fiction] I. Title.
PZ7.K6353 Ho [Fic] 79-4094
79 MARC

Klein, Norma, 1938-
Mom, the Wolf Man, and me. [New York] Pantheon Books [1972]
x, 128 p. 22 cm. $4.50
SUMMARY: An eleven-year-old girl describes her life and relationship with her mother who has never married.
ISBN 0-394-82470-9. ISBN 0-394-92470-3 (lib. bdg.)
[1. Mother and daughters—Fiction. 2. Unmarried mothers—Fiction. 3. Single-parent family—Fiction] I. Title.
PZ7.K678345 Mo [Fic] 72-260
72[r79]rev MARC

Klein, Norma, 1938-
Tomboy / by Norma Klein. — New York : Four Winds Press, c1978.
115 p. ; 21 cm.
SUMMARY: Somewhat apprehensive about growing up, 10-year-old Antonia tries to cope with changes in herself and her friends.
ISBN 0-590-07521-7 : $5.95
[1. Friendship—Fiction. 2. City and town life—Fiction] I. Title.
PZ7.K678345 To [Fic] 78-4337
78 MARC

CHILDREN'S LITERATURE (Continued)

Klein, Norma, 1938-
Visiting Pamela / by Norma Klein ; pictures by Kay Chorao. — New York : Dial Press, c1979.
[32] p. : ill. ; 25 cm.
SUMMARY: At first Carrie is reluctant to visit Pamela's house, but she soon learns how much fun it can be.
ISBN 0-8037-9307-3 : $5.95. ISBN 0-8037-9308-1 lib. bdg. : $5.47
[1. Fear—Fiction] I. Chorao, Kay. II. Title.
PZ7.K678345 Vi [E] 78 78-72203 MARC

Knight, Hilary.
The circus is coming : a picture parade starring 240 performers in 20 scenes / by Hilary Knight. — New York : Golden Press, c1978.
[44] p. : col. ill. ; 27 cm.
SUMMARY: A circus parade highlights the colorful and exciting acts to come.
ISBN 0-307-13737-6 : $3.95
[1. Circus stories] I. Title.
PZ7.K7375 Ci 79 78-68425 MARC

Koffler, Camilla.
Look who's talking, by Ylla [pseud.] Story by Crosby Newell Bonsall. Planned by Charles Rado. Designed by Luc Bouchage. New York, Harper, c1962.
31 p. illus. 30 cm.
SUMMARY: Only the baby leopard knows why the new ostrich doesn't talk to the other animals.
[1. Animals—Fiction] I. Bonsall, Crosby Newell. II. Title.
PZ7.K8195 Lo [E] [r78]rev 62-15793 MARC

Könner, Alfred.
Jolli, by Alfred Konner. Illustrated by Irmhild and Helman Proft. [Translation by Regina Waldman] Minneapolis, Lerner Publications Co. [1967]
[22] p. col. illus. 24 cm.
SUMMARY: Jolli, the duck, is hungry but she doesn't want the dog's bone, the hen's worm, the cat's mouse, or anyone else's food.
[1. Ducks—Fiction. 2. Animals—Fiction] I. Proft, Irmhild, illus. II. Proft, Helman, illus. III. Title.
PZ7.K8183 Jo [E] [r79]rev2 67-14951 MARC

Kooiker, Leonie.
The magic stone / Leonie Kooiker ; translation from the Dutch by Richard and Clara Winston ; illustrated by Carl Hollander. — 1st ed. — New York : Morrow, 1978.
224 p. : ill. ; 21 cm.
SUMMARY: Chris finds a magic stone and becomes the focus of interest for an association of witches, one of whom wants him to become her successor.
ISBN 0-688-22143-2 : $6.95. ISBN 0-688-32143-7 (lib. bdg.)
[1. Witchcraft—Fiction. 2. Witches—Fiction] I. Hollander, Carl. II. Title.
PZ7.K83557 Mag [Fic] 78 78-1713 MARC

Korman, Gordon.
This can't be happening at Macdonald Hall! / Gordon Korman ; illustrated by Affie Mohammed. — Richmond Hill, Ont. : Scholastic-TAB Publications, c1978.
124 p. : ill. ; 20 cm. C***
"INC 343."
SUMMARY: As they seem to be at the bottom of every prank at Macdonald Hall, the headmaster decides to separate roommates Bruno and Boots, but to little avail.
[1. School stories. 2. Humorous stories. 3. Children's writings] I. Title.
PZ7.K8369 Th [Fic] 79 79-306172 MARC

Kotzwinkle, William.
The ants who took away time / by William Kotzwinkle ; pictures by Joe Servello. — 1st ed. — Garden City, N.Y. : Doubleday, c1978.
[64] p. : ill. (some col.) ; 22 cm.
SUMMARY: A tribe of giant ants steal the Great Timepiece and cause time to stand still throughout the world.
ISBN 0-385-12367-1. ISBN 0-385-12368-X lib. bdg.
[1. Science fiction] I. Servello, Joe. II. Title.
PZ7.K855 An 3 [E] 78 78-17833 MARC

Kotzwinkle, William.
Dream of dark harbor / by William Kotzwinkle ; ill. by Joe Servello. — 1st ed. — Garden City, N.Y. : Doubleday, c1979.
46 p. : ill. ; 25 cm.
SUMMARY: A farm boy who sets out to see the world battles a storm at the edge of the sea and encounters a crew of ghosts rising from the water.
ISBN 0-385-12486-4 : $6.95. ISBN 0-385-12487-2 lib. bdg.
[1. Ghost stories] I. Servello, Joe. II. Title.
PZ7.K855 Dr [Fic] 78 78-1244 MARC

Kotzwinkle, William.
The nap master / by William Kotzwinkle ; with pictures by Joe Servello. — 1st ed. — New York : Harcourt Brace Jovanovich, c1979.
[32] p. : col. ill. ; 21 x 26 cm.
SUMMARY: Herman refuses to take a nap until the Nap Master takes him to the Land of Dreams.
ISBN 0-15-256704-6 : $7.95. ISBN 0-15-665325-7 pbk. : $2.50
[1. Sleep—Fiction. 2. Dreams—Fiction] I. Servello, Joe. II. Title.
PZ7.K855 Nap [E] 78 78-12178 MARC

Knotts, Howard.
The lost Christmas / story and pictures by Howard Knotts. — 1st ed. — New York : Harcourt Brace Jovanovich, c1978.
[32] p. : ill. ; 22 cm. — (A Let me read book)
SUMMARY: A small boy is taken ill on Christmas and in his delirium goes on a fantastic journey.
ISBN 0-15-249361-1 lib. bdg.
[1. Dreams—Fiction. 2. Christmas stories] I. Title.
PZ7.K7615 Lo [E] 78 78-1903 MARC

Knudson, R Rozanne, 1932-
Zanboomer / by R. R. Knudson. — 1st ed. — New York : Harper & Row, c1978.
183 p. : ill. ; 21 cm.
SUMMARY: Zan Hagen separates her shoulder playing baseball and finds cross-country running a very satisfactory substitute.
ISBN 0-06-023217-X : $6.95. ISBN 0-06-023218-8 lib. bdg. : $6.79
[1. Baseball—Fiction. 2. Running—Fiction] I. Title.
PZ7.K785 Zas 1978 [Fic] 78 77-11831 MARC

Krahn, Fernando.
Robot-bot-bot / Fernando Krahn. — 1st ed. — New York : Dutton, c1979.
[30] p. : all ill. ; 18 x 19 cm.
SUMMARY: A family's new robot is all work and no play until the daughter takes matters into her own hands.
ISBN 0-525-38545-2
[1. Robots—Fiction. 2. Stories without words] I. Title.
PZ7.K8585 Ro 1979 [E] 78 78-21959 MARC

Krasilovsky, Phyllis.
The man who tried to save time / by Phyllis Krasilovsky ; pictures by Marcia Sewall. — 1st ed. — Garden City, N.Y. : Doubleday, c1979.
[64] p. : col. ill. ; 24 cm. — (A Reading-on-my-own book)
SUMMARY: A man drastically reorganizes his daily routines to save time, only to come to a startling realization.
ISBN 0-385-12998-X
[1. Time—Fiction] I. Sewall, Marcia. II. Title.
PZ7.K865 Mar [E] 78 77-74304 MARC

Kraus, Robert, 1925-
The good mousekeeper / by Robert Kraus ; drawings by Hilary Knight. — 1st ed. — New York : Windmill Books, c1977.
32 p. : ill. ; 26 cm.
SUMMARY: A mousekeeper relates how she takes care of her mice.
ISBN 0-525-61559-8
[1. Mice—Fiction] I. Knight, Hilary. II. Title.
PZ7.K868 Gn 1977 [E] 77 77-2222 MARC

Krensky, Stephen.
Castles in the air, and other tales / Stephen Krensky ; drawings by Warren Lieberman. — 1st ed. — New York : Atheneum, 1979.
66 p. : ill. ; 22 cm.
SUMMARY: Five clichés woven into stories. Included are "Too Clever for Words," "Barrel of Fun," "Fine Kettle of Fish," and "The Last Straw."
ISBN 0-689-30684-9
[1. Humorous stories] I. Lieberman, Warren. II. Title.
PZ7.K883 Cas [Fic] 78 78-11836 MARC

Kroll, Steven.
The candy witch / by Steven Kroll ; illustrated by Marylin Hafner. — New York : Holiday House, c1979.
[32] p. : col. ill. ; 22 cm.
SUMMARY: Because her good deeds are not noticed, a small witch steals every trick-or-treat bag in town on Halloween.
ISBN 0-8234-0359-9
[1. Confectionery—Fiction. 2. Witches—Fiction. 3. Halloween—Fiction] I. Hafner, Marylin. II. Title.
PZ7.K9225 Can [E] 79 79-10141 MARC

Kroll, Steven.
Fat magic / by Steven Kroll ; illustrated by Tomie de Paola. — New York : Holiday House, c1978.
[32] p. : col. ill. ; 25 cm.
SUMMARY: When the court magician puts a spell on him that makes him grow very fat, Prince Timothy takes some drastic measure to remedy the situation.
ISBN 0-8234-0327-0 lib. bdg. : $6.95
[1. Magic—Fiction. 2. Humorous stories] I. De Paola, Thomas Anthony. II. Title.
PZ7.K9225 Fat [E] 78 77-28736 MARC

Kroll, Steven.
Space cats / by Steven Kroll ; illustrated by Friso Henstra. — New York : Holiday House, c1979.
[48] p. : col. ill. ; 21 cm.
SUMMARY: Perry befriends an unusual cat who turns out to be a visitor from a far-distant planet and invites Perry to return there with him.
ISBN 0-8234-0339-4 (lib. bdg.) : $5.95
[1. Cats—Fiction. 2. Science fiction] I. Henstra, Friso. II. Title.
PZ7.K9225 Sp [Fic] 78 78-15164 MARC

Krumpet, Susie.
Dreamie Jeannie of Krumpetville / written and created by Susie and Stevie Krumpet ; illustrated by SPS Studios. — Boulder, Colo. : Character Imprints, 1979.
[32] p. : col. ill. ; 23 cm.
SUMMARY: Dreamie Jeannie spends so much time daydreaming that she nearly misses out on the one thing she wants most.
ISBN 0-89853-003-2 : $2.50
[1. Imagination—Fiction] I. Krumpet, Stevie. II. SPS Studios. III. Title.
PZ7.K9462 Dr [E] 79 78-74897 MARC

Krumpet, Susie.
Itzi Bitzi Mitzi of Krumpetville / written and created by Susie and Stevie Krumpet ; illustrated by SPS Studios. — Boulder, Colo. : Character Imprints, 1979, c1978.
[32] p. : col. ill. ; 23 cm.
SUMMARY: A tiny girl is unhappy with her size till she is able to experience what it would be like to be bigger.
ISBN 0-89853-001-6 : $2.50
[1. Self-acceptance—Fiction. 2. Size and shape—Fiction] I. Krumpet, Stevie, joint author. II. SPS Studios. III. Title.
PZ7.K9462 It [E] 79 78-72021 MARC

Kuzma, Kay.
The Kim, Kari, and Kevin storybook / by Kay Kuzma ; art by Elfred Lee. — Mountain View, Calif. : Pacific Press Pub. Association, c1979.
126 p. : col. ill. ; 23 cm.
SUMMARY: Stories based on incidents in the lives of the author's children illustrate 13 character traits, such as faithfulness, perseverance, honesty, thoughtfulness, enthusiasm, orderliness, and humility.
[1. Family life—Fiction. 2. Conduct of life—Fiction] I. Lee, Elfred. II. Title.
PZ7.K973 Ki [Fic] 79 78-61115 MARC

Ladd, Elizabeth Grosgrove.
Meg's mysterious island. Illustrated by Mary Stevens. New York, Morow, 1963.
218 p. illus. 21 cm.
SUMMARY: A young man and his spunky little sister hope to escape the gossips of a small town in Maine by agreeing to spend the winter in an isolated cabin on a remote island in West Penobscot Boy.
[1. Maine—Fiction. 2. Islands—Fiction. 3. Brothers, and sisters—Fiction. 4. Mystery and detective stories] I. Stevens, Mary, 1920-1966. II. Title.
PZ7.L137 Mg 1963 [Fic] [r79]rev 63-14205 MARC

Ladd, Elizabeth Crosgrove.
A mystery for Meg. Illustrated by Mary Stevens. New York, Morrow, 1962.
189 p. illus. 21 cm.
SUMMARY: Meg's stay with her older brother on an island off the Maine coast is punctuated by mystery and suspicion.
[1. Mystery and detective stories. 2. Maine—Fiction] I. Stevens, Mary, 1920-1966. II. Title.
PZ7.L137 My [Fic] [r79]rev 62-14012 MARC

Ladouceur, Jeanne Marie.
Entre Vénus et Mars / texte, Jeanne-Marie Ladouceur ; illustrations, Anne Gaud. — Paris (57, Av. de Suffren, 75007) : Éditions du Mascaret, 1978.
[16] p. : col. ill. ; 19 x25 cm. F79-1584
28.00F
I. Gaud, Anne. II. Title.
PZ23.L136E5 843'.9'14 00 79-371294 MARC

Lagerlöf, Selma Ottiliana Lovisa, 1858-1940.
Christ legends and other stories / Selma Lagerlöf [translated by Velma Swanston Howard]. — London : Floris Books, 1977.
204 p. : ill. ; 22 cm. GB***
CONTENTS: The holy night.—The emperor's vision.—The wise men's well.—Bethlehem's children.—The flight into Egypt.—In Nazareth.—In the temple.—Saint Veronica.—Robin Redbreast.—The bird's nest.—The Christmas rose.—The Sacred flame.
ISBN 0-903540-06-1 : £3.95
1. Jesus Christ—Juvenile fiction. [1. Jesus Christ—Fiction. 2. Short stories] I. Title.
PZ7.L1429 Ch 1977 [Fic] 79 78-324165 MARC

Lance, James W
The golden trout / [by] James W. Lance ; illustrated by Chris Riordan. — West Melbourne, Vic. : Thomas Nelson (Australia), 1977.
55 p. : ill. ; 24 cm. Aus
ISBN 0-17-005137-4
I. Title.
PZ7.L2216 Go 79 79-304624 MARC

CHILDREN'S LITERATURE (Continued)

Landis, James David.
The sisters impossible / by James David Landis. — New York : Knopf : distributed by Random House, c1979.
171 p. ; 22 cm.
SUMMARY: Lily's beginning ballet classes draw her unexpectedly closer to her seemingly haughty older sister, already an advanced dancer.
ISBN 0-394-84190-5
[1. Ballet dancing—Fiction. 2. Brothers and sisters—Fiction] I. Title.
PZ7.L23173 Si 1979 [Fic] 78-32148
79 MARC

Lang, Othmar Franz, 1921-
If you are silenced, I will speak for you / Othmar Franz Lang ; translated from the German by Eva Mayer. — 1st English language ed. — New York : Collins Publishers, 1978, c1975.
183 p. ; 21 cm.
Translation of Wenn du verstummt, werde ich sprechen.
SUMMARY: Relates the experiences of several concerned young people who join a West German group of Amnesty International.
ISBN 0-529-05432-9 : $6.95
[1. Amnesty International—Fiction] I. Title.
PZ7.L255 If [Fic] 78-59819
78 MARC

Langner, Nola.
Freddy, my grandfather / by Nola Langner. — New York : Four Winds Press, c1979.
[32] p. : ill. ; 26 cm.
SUMMARY: A child describes her Hungarian-born grandfather.
ISBN 0-590-07577-2 : $6.95
[1. Grandfathers—Fiction] I. Title.
PZ7.L268 Fr [E] 79-9926
79 MARC

Lareuse, Jean.
Devils in the castle / Jean Lareuse. — New York : Scribner, c1979.
[32] p. : col. ill. ; 26 cm.
SUMMARY: The devils who live in the girl's boarding school attic simply will not be chased away, until the boys from a neighboring school arrive to help.
ISBN 0-684-16100-1 (lib. bdg.) : $8.95
[1. Boarding schools—Fiction. 2. School stories] I. Title.
PZ7.L3208 De [E] 79-692
79 MARC

Larson, Glen A
The Hardy boys and Nancy Drew meet Dracula / by Glen A. Larson and Michael Sloan ; novelization by Stratemeyer Syndicate. — New York : Grossett & Dunlap, c1978.
109 p. : ill. ; 23 cm.
Based on the Universal Television series, Hardy boys/Nancy Drew mysteries. Developed for television by Glen A. Larson. Based on the Hardy boys books by Franklin W. Dixon and the Nancy Drew books by Carolyn Keene. Adapted from the episodes The Hardy boys and Nancy Drew meet Dracula (pts. 1 and 2).
SUMMARY: When their father disappears while investigating art thefts, the Hardy boys follow his trail to Dracula's castle in Transylvania where they are helped in their search by Nancy Drew.
ISBN 0-448-16196-6 : $3.95
[1. Mystery and detective stories] I. Sloan, Michael, 1946- joint author. II. Dixon, Franklin W. III. Keene, Carolyn, pseud. IV. Stratemeyer Syndicate. V. Title.
PZ7.L32388 Har [Fic] 78-52863
79 MARC

Lasky, Kathryn.
My island grandma / by Kathryn Lasky ; pictures by Emily McCully. — New York : F. Warne, c1979.
[32] p. : col. ill. ; 21 cm.
SUMMARY: A little girl describes all the things she does with her grandmother during the summers they spend on an island.
ISBN 0-7232-6159-8 (lib. bdg.) : $7.95
[1. Grandmothers—Fiction. 2. Islands—Fiction] I. McCully, Emily Arnold. II. Title.
PZ7.L3274 My [E] 78-12489
78 MARC

Laviolette, Emily A
A little island princess / Emily A. Laviolette. — Fernwood, P.E.I. : E. Harrison & Associates, c1975.
85 p. : ill. ; 23 cm. C76-227-4
I. Title.
PZ7.L4177 Li 78-316069
78 MARC

Lawhead, Terry.
The ferry story / by Terry Lawhead ; illustrated by Paula Richards. — Seattle : Pacific Search Press, c1978.
66 p. : ill. ; 18 x 26 cm.
SUMMARY: Tired of his dull, routine life, a ferry boat joins a group of whales migrating south.
ISBN 0-914718-33-9 : $4.95
[1. Ferries—Fiction. 2. Whales—Fiction] I. Richards, Paula. II. Title.
PZ7.L41845 Fe [E] 78-16743
78 MARC

Lawler, Pat.
My brother's place : a novel / Pat Lawler. — New York : Pantheon Books, c1978.
201 p. ; 22 cm.
SUMMARY: Totally unalike, a fifteen-year old and the older brother he idolizes find the bond between them strained by the difficult task of growing up.
ISBN 0-394-83846-7. ISBN 0-394-93846-1 lib. bdg.
[1. Brothers and sisters—Fiction] I. Title.
PZ7.L4188 My [Fic] 78-3288
78 MARC

Lawrence, Louise, 1943-
Star Lord / by Louise Lawrence. — 1st ed. — New York : Harper & Row, c1978.
170 p. ; 21 cm.
SUMMARY: A Welsh family is caught up in the struggle between the supernatural powers of a mountain and the technological powers of a star lord who crashes into it.
ISBN 0-06-023776-7 : $6.95. ISBN 0-06-023777-5 lib. bdg. : $7.49
[1. Supernatural—Fiction. 2. Wales—Fiction] I. Title.
PZ7.L4367 St [Fic] 77-25674
78 MARC

L'Engle, Madeleine.
Camilla Dickinson, a novel. New York, Simon and Schuster [1951]
245 p. 21 cm.
SUMMARY: Fifteen-year-old Camilla gains new maturity through her relationship with her best friend's brother and the growing realization that her parents are fallible individuals.
[1. Parent and child—Fiction. 2. New York (City)—Fiction] I. Title.
PZ7.L5385 Can [Fic] 51-3595
[r78]rev2 MARC

Le-Tan, Pierre.
Happy birthday, Oliver! / by Pierre Le-Tan. — New York : Random House, c1978.
[32] p. : col. ill. ; 21 cm.
SUMMARY: Excited about his birthday, Oliver is disappointed that no one seems to remember it.
ISBN 0-394-83788-6 : $3.95
[1. Birthdays—Fiction. 2. Dogs—Fiction] I. Title.
PZ7.L566 Hap [E] 78-21762
78 MARC

Le-Tan, Pierre.
Timothy's dream book / story and pictures by Pierre Le-Tan. — 1st ed. — New York : Farrar, Straus and Giroux, 1978.
[30] p. : col. ill. ; 17 cm.
SUMMARY: Just before falling asleep, Timothy thinks about what he wants to be when he grows up.
ISBN 0-374-37598-4 : $5.95
[1. Occupations—Fiction. 2. Dreams—Fiction] I. Title.
PZ7.L566 Ti 78-60457
79 MARC

Leach, Christopher, 1925-
In apple country / Christopher Leach. — London : Dent, 1978.
120 p. ; 23 cm. GB***
ISBN 0-460-06864-4 : £3.75
I. Title.
PZ7.L459 In 78-322065
78 MARC

Leach, Michael, 1940-
Don't call me orphan! / Michael Leach. — 1st ed. — Philadelphia : Westminster Press, c1979.
94 p. ; 21 cm.
SUMMARY: A kindly priest and a lonely little boy help a 13-year-old resident of the Bethlehem Home for Boys recapture his self-confidence.
ISBN 0-664-32645-5
[1. Orphans—Fiction] I. Title.
PZ7.L4613 Do [Fic] 78-26928
78 MARC

Leaf, VaDonna Jean.
Robbie and the stolen minibike / VaDonna Jean Leaf. — Carol Stream, Ill. : Creation House, c1978.
180 p. : ill. ; 22 cm.
SUMMARY: A confused, rebellious teenager, who lives with his crippled grandfather, discovers what it means to be part of God's family.
ISBN 0-88419-127-3 : $2.95
[1. Christian life—Fiction. 2. Grandfathers—Fiction] I. Title.
PZ7.L46314 Ro [Fic] 77-78849
79 MARC

Lee, Benjamin.
It can't be helped / Benjamin Lee. — 1st American ed. — New York : Farrar, Straus, Giroux, 1979, c1976.
155 p. ; 22 cm.
SUMMARY: After his father dies and his mother enters a mental hospital, Max Orloff goes to live with his relatives in London.
ISBN 0-374-33648-2 : $7.95
I. Title.
PZ7.L5116 It 1979 [Fic] 79-1183
79 MARC

Lee, Benjamin.
It can't be helped / Benjamin Lee. — London [etc.] : Bodley Head, 1976.
149 p. ; 23 cm. — (A Book for new adults) GB76-35457
ISBN 0-370-10956-2 : £2.75
I. Title.
PZ7.L5116 It 79-300395
79 MARC

Lee, Beverly Haskell.
The secret of Van Rink's cellar / Beverly Haskell Lee. — Minneapolis : Lerner Publications Co., c1979.
174 p. : map ; 22 cm.
SUMMARY: While searching for the ghost that haunts the house in which their mother is a maid and dressmaker for an English officer's wife, Sarah and Stephen become secretly involved in the Revolutionary War effort.
ISBN 0-8225-0763-3 : $7.95
[1. Mystery and detective stories. 2. Spy stories. 3. United States—History—Revolution, 1775-1783—Fiction] I. Title.
PZ7.L5118 Se 1979 [Fic] 79-52909
79 MARC

Lee, Josephine, 1921-
The fabulous manticora / Josephine Lee. — New York : John Day Co., 1976, c1973.
186 p. ; 21 cm.
SUMMARY: A London boy becomes entangled in the evil of the valuable antique manticora he steals in order to buy his mother a home away from the tenement they live in.
ISBN 0-381-99619-0 : $6.95
[1. Antiques—Fiction. 2. Family life—Fiction] I. Title.
PZ7.L5126 Fab 5 [Fic] 75-25784
75[r78]rev MARC

Lee, Melicent Humason, 1889-1943.
Indians of the oaks / by Melicent Humason Lee (Shah-Ti', "Tall Cedar") ; ill. by Leslie W. Lee. — New ed. — Ramona, Calif. : Acoma Books, 1978.
244 p. : ill. ; 21 cm.
Consists of 2 works, Indians of the oaks, originally published in 1931 under title: The Indians and I, and Secrets of the trail.
SUMMARY: In the first of these two tales a young white boy relates his experiences living with the Kumeyaay, also known as Diegueño, Indians of Southern California and in the second a young girl from the same tribe learns a new kind of medicine when she falls ill and is cared for by a white doctor.
ISBN 0-916552-17-9
[1. Diegueño Indians—Fiction. 2. Indians of North America—Fiction] I. Lee, Melicent Humason, 1889-1943. Secrets of the trail. 1978. II. Title.
PZ7.L5142 In 1978 [Fic] 78-110859
79 MARC

Lee, Tanith.
East of midnight / Tanith Lee. — New York : St. Martin's Press, 1978, c1977.
175 p. ; 22 cm.
SUMMARY: Having exchanged identities, a ruler's consort and a slave discover they seem doomed to die in their new roles.
ISBN 0-312-22494-X
[1. Fantasy] I. Title.
PZ7.L5149 Eas 1978 [Fic] 77-15867
78 MARC

Leech, Jay, 1931-
Bright Fawn and me / by Jay Leech and Zane Spencer ; illustrated by Glo Coalson. — 1st ed. — New York : Crowell, c1979.
[32] p. : col. ill. ; 21 x 23 cm.
SUMMARY: An Indian girl considers her younger sister a pest, but doesn't like other children to call her one.
ISBN 0-690-03937-9 : $6.95
[1. Brothers and sisters—Fiction. 2. Cheyenne Indians—Fictions. 3. Indians of North America—Fiction] I. Spencer, Zane, 1935- joint author. II. Coalson, Glo. III. Title.
PZ7.L5152 Br 1979 [E] 78-19215
78 MARC

Leeson, Robert.
Challenge in the dark / [by] Robert Leeson ; illustrated by Jim Russell. — London : Collins, 1978.
93 p. : ill. ; 21 cm. GB78-16468
ISBN 0-00-184065-7 : £2.50
I. Title.
PZ7.L5158 Ch 78-318423
78 MARC

Leeson, Robert.
Silver's revenge / Robert Leeson. — New York : Collins, 1979, c1978.
196 p. ; 22 cm.
SUMMARY: A sequel to Robert Louis Stevenson's "Treasure Island" relating the events of a second voyage by the "Hispaniola" to Treasure Island in search of adventure, riches, and vengeance.
ISBN 0-529-05530-9 : $7.95
[1. Buried treasure—Fiction. 2. Pirates—Fiction. 3. Adventure stories] I. Title.
PZ7.L5158 Si 1979 [Fic] 78-26952
78 MARC

Leister, Mary.
Wee green witch / Mary Leister ; illustrated by Elaine Arnold. — 1st ed. — Owings Mills, Md. : Stemmer House Publishers, c1978.
30 p. : ill. (some col.) ; 20 cm.
"A Barbara Holdridge book."
SUMMARY: A tiny witch with a temper and a housing problem learns a lesson about the laws of nature.
ISBN 0-916144-30-5 : $9.95. ISBN 0-916144-31-3 pbk. : $5.95
[1. Witches—Fiction] I. Arnold, Elaine. II. Title.
PZ7.L5365 We [E] 78-12380
78 MARC

Leja, Magda.
The boy from the skyscraper / Magda Leja ; translated by Jerzy Brodzi ; illustrated by Tomasz Bogacki. — Warszawa : Krajowa Agencja Wydawnicza, 1977.
14 p. : col. ill. ; 20 cm. P***
Translation of Chłopiec z wieżowca.
I. Bogacki, Tomasz. II. Title.
PZ7.L53725 Bo 79-357753
79 MARC

LeRoy, Gen.
Cold feet / by Gen Le Roy. — 1st ed. — New York : Harper & Row, c1979.
217 p. ; 22 cm.
SUMMARY: Struggling to take charge of her own life, Geneva poses as a boy and ends up getting involved with a gambling ring.
ISBN 0-06-023778-3 : $6.95. ISBN 0-06-023779-1 lib. bdg. : $7.89
[1. Identity—Fiction. 2. Gambling—Fiction] I. Title.
PZ7.L5615 Co [Fic] 77-25642
79 MARC

CHILDREN'S LITERATURE (Continued)

Lesikin, Joan.
Down the road / written and illustrated by Joan Lesikin. — Englewood Cliffs, N.J. : Prentice-Hall, c1978.
32 p. : ill. ; 24 cm.
SUMMARY: A snake and a box turtle set out to find a dwelling place but whatever is suitable for one isn't good for the other.
ISBN 0-13-218909-7 : $5.95
[1. Dwellings—Fiction. 2. Animals—Fiction] I. Title.
PZ7.L56247 Do [E] 78-7864
78 MARC

Lester, Helen.
Cora copycat / by Helen Lester. — 1st ed. — New York : Dutton, c1979.
[32] p. : col. ill. ; 21 x 23 cm.
SUMMARY: An exasperating mimic is left almost speechless after an encounter with an escaped Wild Woolly Wurgal.
ISBN 0-525-28241-6 (lib. bdg.)
[1. Behavior—Fiction] I. Title.
PZ7.L56285 Co 1979 [E] 79-10028
79 MARC

Leverich, Kathleen.
The hungry fox and the foxy duck / written by Kathleen Leverich ; Paul Galdone drew the pictures. — New York : Parents' Magazine Press, c1978.
[41] p. : col. ill. ; 24 cm.
SUMMARY: A fox in pursuit of a wise little duck learns the hard way why she lives exactly where she does.
ISBN 0-8193-0987-7
[1. Ducks—Fiction. 2. Foxes—Fiction] I. Galdone, Paul. II. Title.
PZ7.L5744 Du [E] 78-11215
78 MARC

Levy, Elizabeth.
Frankenstein moved in on the fourth floor / by Elizabeth Levy ; illustrated by Mordicai Gerstein. — 1st ed. — New York : Harper & Row, c1979.
57 p. : ill. ; 24 cm.
SUMMARY: Robert and Sam suspect their weird new neighbor is really Frankenstein.
ISBN 0-06-023810-0 : $6.95
[1. Apartment houses—Fiction] I. Gerstein, Mordicai. II. Title.
PZ7.L5827 Fr 1979 [Fic] 78-19830
79 MARC

Levy, Elizabeth.
The tryouts / by Elizabeth Levy ; illustrated by Jacquie Hann. — New York : Four Winds Press, c1979.
102 p. : ill. ; 22 cm.
SUMMARY: A group of eighth-graders take matters into their own hands when one of their friends fails to make the new co-ed basketball team.
ISBN 0-590-07487-3 : $6.95
[1. Basketball—Fiction] I. Hann, Jacquie. II. Title.
PZ7.L5827 Tr [Fic] 78-22125
78 MARC

Lewis, Milton, fl. 1958-
David White, crime reporter. New York, Dodd, Mead, 1958.
235 p. 21 cm. (Dodd, Mead career books)
I. Title.
PZ7.L5876 Dav 58-9901
[r79]rev MARC

Lexau, Joan M
I hate Red Rover / by Joan M. Lexau ; pictures by Gail Owens. — 1st ed. — New York : Dutton, c1979.
55 p. : ill. ; 24 cm. — (A Fat cat book)
SUMMARY: Jill does poorly with games at school until sharing her problem with Grampa helps her find a solution.
ISBN 0-525-32527-1
[1. Games—Fiction. 2. School stories. 3. Grandfathers—Fiction] I. Owens, Gail. II. Title. III. Series.
PZ7.L5895 Iah 1979 [E] 78-12873
78 MARC

Likhanov, Al'bert Anatol'evich.
The maze : a story about boys / Albert Likhanov ; [translated from the Russian by Fainna Glagoleva ; illustrated by Yuri Ivanov]. — Moscow : Progress Publishers, 1977.
288 p. : ill. ; 21 cm. USSR***
Translation of Labirint.
Second part of author's trilogy: Semeinye obstoiatel'stva.
I. Title.
PZ7.L6268 Maz 79-344318
79 MARC

Lin, Hung-ju.
Tracks in the grass / [written by Lin Hung-ju ; ill. by Tsui Ju-chuo and Kao Pao-sheng] — Peking : Foreign Languages Press, 1977.
[40] p. : col. ill. ; 22 cm.
I. Tsui, Ju-chuo. II. Kao, Pao-sheng. III. Title.
PZ7.L645 Tr 78-316389
78 MARC

Lin, Wan-tsui.
Little pals / ill. by Lin Wan-tsui. — Peking : Foreign Languages Press, 1974.
[37] p. : all col. ill. ; 21 cm.
I. Title.
PZ7.L646 Li 79-300426
79 MARC

Lingard, Joan.
Odd girl out / by Joan Lingard. — 1st ed. — New York : Elsevier/Nelson Books, 1979, c1978.
187 p. ; 21 cm.
First published in 1978 under title: The gooseberry.
SUMMARY: Fourteen-year-old Ellen, already troubled by her height, finds life even more difficult when her mother decides to remarry.
ISBN 0-525-66629-0 : $7.95
[1. Remarriage—Fiction. 2. Edinburgh—Fiction] I. Title.
PZ7.L6626 Od 1979 [Fic] 79-4672
79 MARC

Linklater, Andro.
Amazing Maisie and the cold porridge brigade / [by] Andro Linklater ; illustrated by Joanna Carey. — London : Methuen, 1978.
72 p. : ill., map (on lining papers) ; 25 cm. GB78-23765
ISBN 0-416-58380-6 : £3.50
I. Title.
PZ7.L66284 Am 79-303086
79 MARC

Lionni, Leo, 1910-
Geraldine, music mouse / by Leo Lionni. — New York : Pantheon Books, c1979.
[31] p. : col. ill. ; 28 cm.
SUMMARY: After nibbling an enormous piece of parmesan cheese into the shape of a gaint mouse holding a flute, Geraldine hears music for the first time.
ISBN 0-394-84238-3. ISBN 0-394-94238-8 lib. bdg.
[1. Music—Fiction. 2. Mice—Fiction] I. Title.
PZ7.L6634 Ge [E] 79-932
79 MARC

Lipkind, William, 1904-
The Christmas bunny, by Will [pseud.] and Nicolas [pseud. 1st ed.] New York, Harcourt, Brace [1953]
[49] p. illus. 29 cm.
SUMMARY: A young boy joins in the forest animals' Christmas Eve merrymaking and meets a special brown rabbit.
[1. Animals—Fiction. 2. Christmas stories] I. Mordvinoff, Nicolas, 1911- joint author. II. Title.
PZ7.L6637 Ch [E] 53-7873
[r78]rev MARC

Little, Jack.
Thunder egg : an adventure story / by Jack Little. — Boulder, Colo. : Altair Press, c1978.
115 p. ; 19 cm.
SUMMARY: Certain events during their hike through the Oregon Cascades convince a 14-year-old and his extraordinary dog that their thunder egg has magical properties.
$2.50
[1. Magic—Fiction. 2. Camping—Fiction] I. Title.
PZ7.L7222 Th 77-94288
79 MARC

Little, Mary E
Old Cat and the kitten / Mary E. Little. — 1st ed. — New York : Atheneum, 1979.
119 p. : ill. ; 22 cm.
SUMMARY: When his family moves, Joel knows he can't take his adopted tomcat along nor turn him loose again. The solution is a hard one, but the only one.
ISBN 0-689-30693-8
[1. Cats—Fiction. 2. Pets—Fiction] I. Title.
PZ7.L723 Ol [Fic] 78-11415
78 MARC

Little Red Riding Hood.
Adventures of Little Red Riding Hood. — Albany ed. — Albany : Sprague & Co., [between 1855 and 1863]
8 p. : col. ill. ; 18 cm.
Cover title.
Caption title: History of Little Red Riding Hood.
In verse.
I. Title. II. Title: History of Little Red Riding Hood.
PZ6.L738 Ad 1855 78-315524
78 MARC

Little Red Riding Hood.
History of Little Red Riding Hood. — [s.l. : s.n., 184-?]
8 p. : col. ill. ; 20 cm.
Caption title.
In verse.
L.C. copy has traces of printed yellow wrappers.
I. Title.
PZ6.L738 Hi 1840z 78-315558
78 MARC

Little Red Riding Hood.
Little Red Riding-Hood. — New York : J. B. Jansen, 1824.
8 p. : ill. (some col.) ; 87 mm.
PZ6.L738 Li 1824 78-315523
78 MARC

Little Red Riding Hood.
[Little Red Riding Hood / edited by F. Summerly [i.e. Sir H. Cole]. — London : J. Cundall, 1843]
23 p., [3] leaves of plates : ill. ; 17 cm. — (The Home treasury)
L.C. copy imperfect: t.p. wanting.
Title from Brit. Mus. cat.
I. Cole, Henry, Sir, 1808-1882.
PZ6.L738 Li 1843 78-315529
78 MARC

Little Red Riding Hood.
Little Red Riding Hood / illuminated with ten pictures [by W. H. Thwaites]. — New York : H. W. Hewet, publisher, [1855?]
28 p., [1] leaf of plates : ill. (1 col.) ; 18 cm. — (Hewet's illuminated household stories for little folks ; v. 4)
Imprint from cover. On t.p.: New York, H. W. Hewet, engraver and printer.
"Entered according to act of Congress in the year 1855 ... in the clerk's office of the District Court of the United States for the Southern District of New York."
——— Another issue. [ca. 1857] On t.p.: New York, Loomis & Co., engravers & printers, Hewet imprint retained on cover.
PZ6.L38Li 1857
I. Thwaites, William H.
PZ6.L738 Li 1855 78-315522
78 MARC

Little Red Riding Hood.
Little Red Riding Hood. — London : Dean & Son, [1865?]
[8] leaves : col. ill. ; 26 cm. — (Deans' new scenic books ; no. 1)
Cover title.
The mounted hand-colored illustrations are made to be raised to a position perpendicular to the plane of the text.
I. Series.
PZ6.L738 Li 1865 398.2 77-355254
77[r78]rev MARC

Little Red Riding Hood. Spanish.
La Caperucita Roja. — Nueva York : D. Appleton, [1864?]
[14] p. : col. ill. ; 27 cm.
Cover title.
Three ill. signed by W. Momberger.
"Entered according to act of Congress, in the year 1864 ... in the clerk's office of the District Court of the United States for the Southern District of New York."
Pages [1] and [14] mounted on inside of cover.
I. Momberger, W., b. 1829. II. Title.
PZ6.L738 Ca 78-371816
78 MARC

Liu, Ch'i, fl. 1974-
Two peacocks / written by Chen Wei and Peng Hua ; adapted by Liu Chi and others ; colour woodcuts by Chiang Tieh-feng, Chia Kuo-chung and Wan Chiang-lin. — Peking : Foreign Language Press, 1975.
[36] p. : col. ill. ; 21 cm.
Translation of Liang chih hsiao k'ung ch'üeh.
I. Ch'en, Wei. Liang chih hsiao k'ung ch'üeh. II. Chiang, T'ieh-feng. III. Chia, Kuo-chung. IV. Wan, Ch'iang-lin. V. Title.
PZ7.L7393 Tw 77-551597
77[r79]rev MARC

Lively, Penelope, 1933-
The voyage of QV 66 / Penelope Lively ; illustrated by Harold Jones. — London : Heinemann, 1978.
172 p. : ill. ; 23 cm. GB***
ISBN 0-434-94898-3 : £3.10
I. Jones, Harold, 1904- . II. Title.
PZ7.L7397 Vo 1978 79-301751
79 MARC

Lively, Penelope, 1933-
The voyage of QV 66 / Penelope Lively ; illustrated by Harold Jones. — 1st ed. — New York : Dutton, 1979, c1978.
172 p. : ill. ; 22 cm.
SUMMARY: In an England devoid of People a group of animal friends embarks on a voyage to the London Zoo to learn the identity of one of their members who resembles People but is small and has fur.
ISBN 0-525-42120-3
[1. Animals—Fiction. 2. England—Fiction] I. Jones, Harold, 1904- . II. Title.
PZ7.L7397 Vo 1979 [Fic] 78-12098
78 MARC

Lloyd, Errol.
Nini at carnival / Errol Lloyd. — New York : Crowell, 1979, c1978.
[26] p. : col. ill. ; 21 x 27 cm.
SUMMARY: A young Jamaican girl is without a costume for Carnival until a bigger girl, in the guise of a fairy godmother, makes some deft transformation.
ISBN 0-690-03891-7 : $4.95. ISBN 0-690-03892-5 lib. bdg. : $5.79
[1. Carnival—Fiction. 2. Jamaica—Fiction] I. Title.
PZ7.L776 Ni 1979 [E] 78-4776
78 MARC

Lobel, Arnold.
Grasshopper on the road / by Arnold Lobel. — 1st ed. — New York : Harper & Row, c1978.
62 p. : col. ill. ; 23 cm. — (An I can read book)
SUMMARY: As Grasshopper sets out to follow a road, he meets some unusual characters.
ISBN 0-06-023961-1 : $4.95. ISBN 0-06-023962-X lib. bdg. : $4.79
[1. Animals—Fiction] I. Title.
PZ7.L7795 Gp 1978 [E] 77-25653
78 MARC

CHILDREN'S LITERATURE (Continued)

Lobel, Arnold.
How the rooster saved the day/ Arnold Lobel ; pictures by Anita Lobel. — ⌜New York⌝ : Puffin Books, 1979, c1977.
⌜32⌝ p. : col. ill. ; 18 x 23 cm.
SUMMARY: Because night protects his identity, a thief tries to prevent a clever rooster from crowing up the morning.
ISBN 0-14-050309-9 : $1.95
⌜1. Roosters—Fiction⌝ I. Lobel, Anita. II. Title.
⌜PZ7.L7795 Hq 1979⌝ ⌜E⌝ 79 79-565 MARC

Lobel, Arnold.
A treeful of pigs / Arnold Lobel ; pictures by Anita Lobel. — 1st ed. — New York : Greenwillow Books, c1979.
⌜32⌝ p. : col. ill. ; 21 x 26 cm.
SUMMARY: A farmer's wife uses drastic measures to get her husband to abandon his lazy ways.
ISBN 0-688-80177-3 : $7.95. ISBN 0-688-84177-5 lib. bdg. : $7.35
⌜1. Laziness—Fiction⌝ I. Lobel, Anita. II. Title.
PZ7.L7795 Tr ⌜E⌝ 78 78-1810 MARC

Loeper, John J
The golden dragon : by clipper ship around the Horn / John J. Loeper ; illustrated with old prints and paintings. — 1st ed. — New York : Atheneum, 1978.
61, ⌜4⌝ p. : ill. ; 24 cm.
Bibliography: p. ⌜65⌝
SUMMARY: The account of a young boy's journey by clipper ship from New York, around Cape Horn, to San Francisco in 1850.
ISBN 0-689-30658-X
⌜1. Voyages to the Pacific coast—Fiction. 2. Sailing—Fiction. 3. Sea stories⌝ I. Title.
PZ7.L8266 Go 1978 ⌜Fic⌝ 78 78-5085 MARC

Logie, Ray, 1933-
The houseless mouse / by Ray Logie ; illustrated by Catharine MacKenzie. — Vancouver : Eforbez Enterprises, 1977.
32 p. : ill. ; 22 cm. — (Karpet series) C77-2168-9
ISBN 0-88976-014-4
I. MacKenzie, Catharine. II. Title.
PZ7.L8283 Ho 79 79-300451 MARC

Logue, Christopher, 1926-
The magic circus / Wayne Anderson ; story told by Christopher Logue. — London : J. Cape, 1979.
⌜32⌝ p. : col. ill. ; 27 cm. GB***
ISBN 0-224-01555-9
I. Anderson, Wayne. II. Title.
PZ7.L8287 Mag 79 79-311796 MARC

Lööf, Jan, 1940-
Uncle Louie's fantastic sea voyage / Jan Lööf. — New York : Random House, c1977.
⌜32⌝ p. : col. ill. ; 21 cm.
Translation of Soren Skrot.
SUMMARY: A boy, his uncle, and an old man who dabbles in magic set sail for Africa and become shipwrecked in a surprising place.
ISBN 0-394-83860-2 : $0.95. ISBN 0-394-83704-5 book club. ISBN 0-394-93860-7 lib. bdg.
⌜1. Voyages and travels—Fiction⌝ I. Title.
PZ7.L8264 Un 1977 ⌜E⌝ 77 77-82080 MARC

Lord, Beman.
The trouble with Francis. Picture by Arnold Spilka. New York, H. Z. Walck, 1958.
54 p. illus. 21 cm.
SUMMARY: Unhappy with his name, Francis Harrison Ward III searches for a nickname.
⌜1. Names, Personal—Fiction. 2. Nicknames—Fiction⌝ I. Spilka, Arnold. II. Title.
PZ7.L877 Tr ⌜E⌝ ⌜r79⌝rev 58-10297 MARC

Lovejoy, Jack.
The rebel witch / Jack Lovejoy ; illustrated by Judith Gwyn Brown. — 1st ed. — New York : Lothrop, Lee & Shepard, c1978.
251 p. : ill. ; 22 cm.
SUMMARY: With the powerful Wand of the Necromancers in her possession, an apprentice witch tries to prevent the restoration of the days of the dark lore.
ISBN 0-688-41868-6. ISBN 0-688-51868-0 lib. bdg.
⌜1. Witchcraft—Fiction⌝ I. Brown, Judith Gwyn. II. Title.
PZ7.L95604 Re ⌜Fic⌝ 78 78-18378 MARC

Lovelace, Maud Hart, 1892-
Betsy and Tacy go downtown / Maud Hart Lovelace ; illustrated by Lois Lenski. — 1st Harper trophy ed. — New York : Harper & Row, 1979, c1943.
180 p. : ill. ; 20 cm. — (A Harper trophy book)
Originally published under title: Down town, a Betsy-Tacy story.
SUMMARY: The further adventures of 12-year-old Betsy, Tacy, and Tib as they explore the world beyond their neighborhood and discover the public library, see a real play, and make friends with the owners of the downtown hotel.
ISBN 0-06-440098-0 : $1.95
⌜1. Friendship—Fiction. 2. City and town life—Fiction. 3. Minnesota—Fiction⌝ I. Lensky, Lois, 1893- II. Title.
PZ7.L9561 Bbe 1979 ⌜Fic⌝ 79 79-110884 MARC

Low, Alice.
The witch who was afraid of witches / by Alice Low ; pictures by Karen Gundersheimer. — 1st ed. — New York : Pantheon Books, c1978.
⌜40⌝ p. : col. ill. ; 24 cm.
SUMMARY: A little witch finds a way to out-witch her two bossy sisters.
ISBN 0-394-83718-5. ISBN 0-394-93718-X lib. bdg.
⌜1. Witches—Fiction. 2. Brothers and sisters—Fiction⌝ I. Gundersheimer, Karen. II. Title.
PZ7.L9595 Wi ⌜E⌝ 78 78-5856 MARC

Low, Joseph, 1911-
The devil himself : story and pictures / by Joseph Low. — New York : McGraw-Hill, c1978.
⌜40⌝ p. : ill. ; 22 cm.
SUMMARY: A dog who likes to dress in people's clothing gives his owner quite a scare.
ISBN 0-07-038795-8
⌜1. Dogs—Fiction⌝ I. Title.
PZ7.L9598 De ⌜E⌝ 77 77-17772 MARC

Luger, H C
The elephant tree / H. C. Luger. — 1st ed. — New York : Viking Press, 1978.
115 p. ; 22 cm.
SUMMARY: Lost together in the California desert, two tough kids from hostile gangs are forced to acknowledge their mutual dependence.
ISBN 0-670-29173-0
⌜1. Survival—Fiction. 2. Deserts—Fiction⌝ I. Title.
PZ7.L97813 El ⌜Fic⌝ 77 77-16416 MARC

Lukešová, Milena.
The little girl and the rain / Milena Lukešová ; illustrated by Jan Kudláček ; translated from the Czech by Stuart Amor. — 1st English language ed. — New York : Holt, Rinehart and Winston, c1978.
⌜33⌝ p. : col. ill. ; 25 x 26 cm.
Translation of Holčička a déšť.
SUMMARY: A little girl finds the rain to be a benevolent playmate.
ISBN 0-03-021146-8 : $6.50
⌜1. Rain and rainfall—Fiction⌝ I. Kudláček, Jan, 1928- II. Title.
PZ7.L9783 Li 1978 ⌜E⌝ 77 77-1090 MARC

Lynds, Dennis, 1924-
Alfred Hitchcock and the three investigators in The mystery of the deadly double : based on characters created by Robert Arthur / text by William Arden ⌜i.e. D. Lynds⌝ ; illustrated by Herb Mott. — New York : Random House, c1978.
x, 147 p. : ill. ; 22 cm. — (Alfred Hitchcock mystery series ; 28)
SUMMARY: The Three Investigators foil a plot to kidnap the son of a political leader in an African colony.
ISBN 0-394-83902-1 : $2.95. ISBN 0-394-93902-6 lib. bdg. : $4.39
⌜1. Mystery and detective stories. 2. Kidnapping—Fiction⌝ I. Arthur, Robert. II. Mott, Herb. III. Title. IV. Title: The mystery of the deadly double. V. Series.
PZ7.L984 Ahe ⌜Fic⌝ 78 78-55960 MARC

Mace, Elisabeth.
The Rushton inheritance / Elisabeth Mace. — London : A. Deutsch, 1978.
160 p. ; 21 cm. GB***
ISBN 0-233-96969-1 : £3.50
I. Title.
PZ7.M15845 Ru 78 78-318421 MARC

Mace, Elisabeth.
The Rushton inheritance / Elisabeth Mace. — 1st U.S. ed. — Nashville : T. Nelson, c1978.
173 p. ; 21 cm.
SUMMARY: The young Rushtons find their daily life disrupted by a boy ghost who appears from another time in search of a family secret.
ISBN 0-8407-8611-4 : $6.95
⌜1. Space and time—Fiction. 2. England—Fiction⌝ I. Title.
PZ7.M15845 Ru 1978 ⌜Fic⌝ 78 78-13559 MARC

MacGibbon, Jean.
Three's company / Jean MacGibbon. — London : Chatto & Windus, c1978.
127 p. ; 21 cm. GB***
ISBN 0-7011-2295-1 : £3.25
I. Title.
PZ7.M16753 Th 79 79-300450 MARC

MacGregor, Ellen.
Miss Pickerell and the supertanker / by Ellen MacGregor and Dora Pantell ; illustrated by Charles Geer. — New York : McGraw-Hill, c1978.
157 p. : ill. ; 21 cm.
SUMMARY: Miss Pickerell sets out to have a restful vacation but finds herself involved in the problem of a leaky supertanker and its pollution of the sea.
ISBN 0-07-044588-5 : $6.95
⌜1. Oil pollution of rivers, harbors, etc.—Fiction⌝ I. Pantell, Dora, joint author. II. Geer, Charles. III. Title.
PZ7.M1698 Mhc ⌜Fic⌝ 78 78-8241 MARC

MacKellar, William.
The silent bells / William Mackellar ; illustrated by Ted Lewin. — New York : Dodd, Mead, c1978.
96 p. : ill. ; 21 cm.
SUMMARY: A Young Swiss girl dreams of the day the cathedral bells, which no one has ever heard, will break their long silence when a special gift is presented at the crèche on Christmas Eve.
ISBN 0-396-07618-1
⌜1. Christmas stories. 2. Switzerland—Fiction⌝ I. Lewin, Ted. II. Title.
PZ7.M196 Sl ⌜Fic⌝ 78 78-7744 MARC

MacKellar, William.
The soccer orphans / William MacKellar. — New York : Dodd, Mead, c1979.
157 p. ; 22 cm
SUMMARY: The fact that football overshadows all other sports at Orkney High doesn't discourage a ragtag group from playing soccer.
ISBN 0-396-07667-X
⌜1. Soccer—Fiction. 2. School stories⌝ I. Title.
PZ7.M196 So ⌜Fic⌝ 78 78-22435 MARC

MacKellar, William.
The witch of Glen Gowrie / William MacKellar ; illustrated by Ted Lewin. — New York : Dodd, Mead, c1978.
144 p. : ill. ; 21 cm.
SUMMARY: A young Scottish boy doesn't believe in witches until he meets the old woman who lives in Glen Gowrie with her many animals.
ISBN 0-396-07531-2 : $5.95
⌜1. Witches—Fiction. 2. Scotland—Fiction⌝ I. Lewin, Ted. II. Title.
PZ7.M196 Wi ⌜Fic⌝ 77 77-16864 MARC

MacLachlan, Patricia.
The sick day / by Patricia MacLachlan ; pictures by William Pène du Bois. — ⌜New York⌝ : Pantheon Books, c1979.
⌜32⌝ p. : col. ill. ; 22 cm.
SUMMARY: Emily's father takes care of everything the day she has a "stomach ache in her head."
ISBN 0-394-83876-9
⌜1. Sick—Fiction. 2. Fathers and daughters—Fiction⌝ I. Du Bois, William Pène, 1916- II. Title.
PZ7.M2225 Si ⌜E⌝ 78 78-11686 MARC

Macklem, Michael.
Jacques the woodcutter / text by Michael Macklem ; ill. by Ann Blades. — ⌜Ottawa : Oberon Press, c1977⌝
⌜36⌝ p. : col. ill. ; 26 cm. C***
ISBN 0-88750-239-3
I. Blades, Ann. II. Title.
PZ7.M2175 Jac 79 79-300448 MARC

MacLeod, Charlotte.
King Devil / by Charlotte MacLeod. — 1st ed. — New York : Atheneum, 1978.
212 p. ; 25 cm.
SUMMARY: The last thing Lavinia thinks she will encounter when she goes to live with rich cousin Zilphia in 1908 is mystery and romance.
ISBN 0-689-30659-8
⌜1. Mystery and detective stories⌝ I. Title.
PZ7.M22494 Ki 1978 ⌜Fic⌝ 78 78-5981 MARC

Maddox, Bill, 1938-
Rags and Patches / Bill Maddox and Harold Beeson. — Chicago : Follett Pub. Co., c1978.
192 p. ; 23 cm.
SUMMARY: Relates the adventures of 13-year-old Danny Ragsdale and his dog Patches as they search for Danny's father in south Texas.
ISBN 0-695-80966-0 : $7.95. ISBN 0-695-40966-2 lib. bdg. : $7.98
⌜1. Adventure stories. 2. Dogs—Fiction⌝ I. Beeson, Harold, 1939- joint author. II. Title.
PZ7.M2559 Rag ⌜Fic⌝ 78⌜r79⌝rev 78-3218 MARC

Madison, Winifred.
A portrait of myself : a novel / by Winifred Madison. — New York : Random House, c1979.
239 p. ; 22 cm.
SUMMARY: Though possessing artistic ability, sixteen-year-old Catherine, suffering from a broken home, socialization problems, and a teacher's disfavor, finally attempts suicide.
ISBN 0-394-84021-6 : $7.95.
⌜1. Artists—Fiction. 2. Teachers—Fiction. 3. Suicide—Fiction⌝ I. Title.
PZ7.M2652 Po ⌜Fic⌝ 78 78-13897 MARC

Maestro, Betsy.
Lambs for dinner / by Betsy and Giulio Maestro. — New York : Crown Publishers, c1978.
32 p. : col. ill. ; 24 cm.
SUMMARY: Mr. Wolf keeps insisting he would like to have the lambs for dinner, but Mama and her four little lambs are naturally wary.
ISBN 0-517-53380-4
⌜1. Sheep—Fiction. 2. Wolves—Fiction⌝ I. Maestro, Giulio. II. Title.
PZ7.M267 Lam 1978 ⌜E⌝ 78 78-4407 MARC

CHILDREN'S LITERATURE (Continued)

Maestro, Giulio.
Leopard and the noisy monkeys / by Giulio Maestro. — 1st ed. — New York : Greenwillow Books, c1979.
39 p. : col. ill. ; 22 cm. — (Greenwillow read-alone)
SUMMARY: Twenty monkeys seeking temporary accomodations cause a sleepless night in the jungle.
ISBN 0-688-80221-4 : $5.95
[1. Jungle stories. 2. Sleep—Fiction. 3. Animals—Fiction] I. Title.
PZ7.M268 Ld [E] 78 78-31701 MARC

Maguire, Michael.
Mylor : the kidnap / Michael Maguire ; illustrated by Nick Armstrong. — London : W. H. Allen, 1978.
191 p. : ill. ; 21 cm. — (A Longbow children's book) (A Longbow book) GB***
ISBN 0-491-02345-6 : £3.25
I. Title.
PZ7.M2763 My 79 78-323984 MARC

Mahy, Margaret.
The great piratical rumbustification & The librarian and the robbers / Margaret Mahy ; with pictures by Quentin Blake. — London : J. M. Dent, 1978.
62 p. : ill. ; 24 cm. GB***
ISBN 0-460-06871-7 : £2.95
I. Mahy, Margaret. The librarian and the robberrs. 1978. II. Blake, Quentin. III. Title.
PZ7.M2773 Gr 79 78-325675 MARC

Major, Beverly.
The magic pizza / by Beverly Major ; ill. by Leonard Shortall. — Englewood Cliffs, N.J. : Prentice-Hall, c1978.
[48] p. : ill. ; 22 cm.
SUMMARY: While waiting for his mother at the pizza parlor, a young boy helps capture a dognapper.
ISBN 0-13-545202-3 : $5.95
[1. Pizza—Fiction] I. Shortall, Leonard W. II. Title.
PZ7.M2798 Mag [Fic] 77 77-26993 MARC

Major, Kevin.
Hold fast / Kevin Major. — Toronto : Clarke, Irwin & Co., c1978.
170 p. ; 21 cm. C***
ISBN 0-7720-1175-3
I. Title.
PZ7.M2814 Ho 78 78-318465 MARC

Manes, Stephen, 1949-
The boy who turned into a TV set / by Stephen Manes ; drawings by Michael Bass. — New York : Coward, McCann & Geoghegan, c1979.
32 p. : ill. ; 19 x 22 cm.
SUMMARY: Ogden Pettibone watched television so much that he finally turned into a set with the picture appearing on his stomach.
ISBN 0-698-20491-3
[1. Television—Fiction] I. Bass, Michael, 1947- II. Title.
PZ7.M31264 Bo 1979 [Fic] 78 78-31436 MARC

Manes, Stephen, 1949-
Hooples on the highway / by Stephen Manes ; illustrated by Merle Peek. — New York : Coward, McCann & Geoghegan, c1978.
127 p. : ill. ; 23 cm.
SUMMARY: A seemingly simple automobile trip to Philadelphia is fraught with adventures for the Hoople family.
ISBN 0-698-20459-X
[1. Vacations—Fiction. 2. Family life—Fiction] I. Peek, Merle. II. Title.
PZ7.M31264 Ho [Fic] 78 78-1710 MARC

Mann, Vivienne.
Harry the hedgehog / written by Vivienne Mann ; illustrated by Cheryl Pelavin. — Tadworth : World's Work, 1978.
[32] p. : col. ill. ; 22 cm. GB78-37294
"A World's Work children's book."
ISBN 0-437-57085-1 : £2.70
I. Pelavin, Cheryl. II. Title.
PZ7.M31514 Har 79 79-310793 MARC

Mantinband, Gerda.
Bing bong bang and fiddle dee dee / by Gerda Mantinband ; illustrated by Anne Rockwell. — 1st ed. — Garden City. N.Y. : Doubleday, c1979.
57 p. : col. ill. ; 25 cm. — (A Reading-on-my-own book)
SUMMARY: The old man's fiddling and the old woman's banging drive all their animals away, but some secret music practice makes everything right again.
ISBN 0-385-14211-0 : $4.95. ISBN 0-385-14212-9 lib. bdg.
I. Rockwell, Anne F. II. Title.
PZ7.M3184 Bi [E] 78 78-18562 MARC

Manus, Willard, 1930-
Sea treasure. Illustrated by Lee J. Ames. [1st ed.] Garden City, N.Y., Doubleday [1961]
139 p. illus. 22 cm. (A Signal book)
SUMMARY: Bill Bryan risks his diving career and his life to recover sunken gold.
[1. Skin diving—Fiction. 2. Buried treasure—Fiction] I. Ames, Lee J. II. Title.
PZ7.M3193 Se [Fic] [r78]rev 61-12555 MARC

Marangell, Virginia J
Gianna mia / Virginia J. Marangell. — New York : Dodd, Mead, c1979.
224 p., [1] leaf of leaves : port. ; 22 cm.
SUMMARY: The youngest of the Dellesantos chronicles the lives of the family in their Italian community of New Haven from 1937 to 1953.
[1. Family life—Fiction. 2. Italian Americans—Fiction] I. Title.
PZ7.M326 Gi [Fic] 79 79-52047 MARC

Marbach, Ethel.
The cabbage moth and the shamrock / written by Ethel Marbach ; designed and illustrated by Michael Hague. — La Jolla, Calif. : Star and Elephant Books, c1978.
[32] p. : col. ill. ; 19 cm.
SUMMARY: Separated after a summer spent enjoying each other's company, a cabbage moth and a shamrock are reunited by an old lady who appreciates their beauty.
[1. Moths—Fiction. 2. Plants—Fiction] I. Hague, Michael. II. Title.
PZ7.M327 Cab [E] 00 79-105790 MARC

Marcuse, Katherine.
The devil's workshop / Katherine Marcuse ; drawings by Paul Zepelinski. — Nashville : Abingdon, c1979.
157 p. : ill. ; 23 cm.
SUMMARY: In the mid 1400's in Mainz, Germany, a 12-year-old boy becomes apprenticed to Johann Gutenberg who, because his work is strange and different, is believed by the townspeople to be in league with the devil.
ISBN 0-687-10506-4
1. Gutenberg, Johann, 1397?-1468—Juvenile fiction. [1. Germany—Middle Ages, 843-1517—Fiction. 2. Gutenberg, Joahnn, 1397?-1468—Fiction. 3. Printing—History—Fiction] I. Zepelinski, Paul.
PZ7.M33 De [Fic] 78 78-24121 MARC

Marger, Mary Ann.
Winner at the Dub-Dub Club / by Mary Ann Marger. — 1st ed. — New York : Elsevier/Nelson Books, c1979.
72 p. ; 22 cm.
SUMMARY: Fifteen-year-old Gary, a fine, untaught tennis player, becomes involved with the countryclub tennis group and the city team and then must choose between them.
ISBN 0-525-66634-6 : $7.95
[1. Tennis—Fiction] I. Title.
PZ7.M33545 Wi 1979 [Fic] 79 79-1178 MARC

Marino, Barbara Pavis, 1948-
Eric needs stitches / Barbara Pavis Marino ; photos. by Richard Rudinski. — Reading Mass. : Addison-Wesley, c1979.
28 p. : ill. ; 21 x 26 cm.
SUMMARY: Although he is afraid, Eric goes to the hospital emergency room to get stitches in his knee after a bad fall.
ISBN 0-201-04401-3
[1. Hospitals—Fiction. 2. Medical care—Fiction] I. Rudinski, Richard. II. Title.
PZ7.M33878 Er [E] 78 78-31694 MARC

Marino, Dorothy Bronson, 1912-
Buzzy Bear's busy day, by Dorothy Marino. New York, F. Watts [1965]
[33] p. col. illus. 26 cm.
SUMMARY: A bear who claims there's nothing to do plays busily after breakfast.
[1. Play—Fiction. 2. Boredom—Fiction. 3. Bears—Fiction] I. Title.
PZ7.M3388 Bu [E] [r79]rev 65-19463 MARC

Mark, Jan.
The ennead / Jan Mark. — 1st American ed. — New York : Crowell, c1978.
306 p. ; 21 cm.
SUMMARY: Colonists from a dying Earth find a new solar system similar to the one they left behind and proceed to repeat the same destructive pattern in their new environment until near disaster makes them change their living pattern.
ISBN 0-690-03872-0 : $7.95. ISBN 0-690-03873-9 lib bdg. : $5.79
[1. Science fiction] I. Title.
PZ7.M33924 En 1978 E 00 77-27669 MARC

Mark, Jan.
Thunder and lightnings / Jan Mark ; ill. by Jim Russell. — 1st ed. — New York : Crowell, 1979, c1976.
181 p. : ill. ; 21 cm.
SUMMARY: Shortly after his family moves to the Norfolk countryside, Andrew is befriended by a local boy with a passion for airplanes.
ISBN 0-690-03901-8 : $7.95
[1. Airplanes—Fiction. 2. England—Fiction] I. Russell, Jim, 1933- II. Title.
PZ7.M33924 Th 1979 [Fic] 78 78-4778 MARC

Mark, Jan.
Under the autumn garden / by Jan Mark ; illustrated by Judith Gwyn Brown. — 1st U.S. ed. — New York : Crowell, 1979, c1977.
211 p. : ill. ; 21 cm.
SUMMARY: A young boy's efforts to excavate a historical relic from a neighboring garden for a school project prove futile until, too late for the project deadline, he makes a remarkable find.
ISBN 0-690-03903-4 : $6.95
[1. Excavations (Archaeology)—England—Fiction. 2. England—Fiction] I. Brown, Judith Gwyn. II. Title.
PZ7.M33924 Un 1979 [Fic] 78 78-4779 MARC

Marko, Katherine.
God, when will I ever belong? / written by Katherine D. Marko ; illustrated by Michael Hackett. — St. Louis : Concordia Pub. House, c1979.
78 p. : ill. ; 21 cm.
SUMMARY: After a sequence of foster homes in which she was unhappy, 12-year-old Jeanie finds herself in a new one she dislikes for the same reasons.
ISBN 0-570-03624-0
[1. Foster home care—Fiction. 2. Christian life—Fiction. 3. Adoption—Fiction] I. Hackett, Michael. II. Title.
[PZ7.M3397 Go] [Fic] 79 79-12244 MARC

Marshall, Emily.
Bertha's journal / by Emily Marshall. — Boston : Congregational Sabbath School and Pub. Society, c1868.
387 p., [2] leaves of plates : ill. ; 18 cm.
I. Congregational Publishing Society. II. Title.
PZ6.M358 Be 78 78-102994 MARC

Marshall, James, 1942-
George and Martha, one fine day / James Marshall. — Boston : Houghton Mifflin, 1978.
46 p. : col. ill. ; 22 cm.
SUMMARY: Five new episodes in the friendship of the two hippopotamuses.
ISBN 0-395-27154-1 : $5.95
[1. Friendship—Fiction. 2. Hippopotamus—Fiction] I. Title.
PZ7.M35672 Gd [E] 78 78-60494 MARC

Marshall, James, 1942-
Portly McSwine / James Marshall. — Boston : Houghton Mifflin, 1979.
38 p. : col. ill. ; 21 cm.
SUMMARY: Despite all attempts to reassure him, Portly McSwine frets about the National Snout Day celebration he has planned.
ISBN 0-395-28003-6 : $5.95
[1. Pigs—Fiction. 2. Parties—Fiction] I. Title.
PZ7.M35672 Po [E] 78 78-24814 MARC

Martin, Albert Y
Secret spy / Albert Martin ; [illustrator, Warren Lee]. — Belmont, Calif. : Fearon Pitman Publishers, c1979.
56 p. : ill. ; 18 cm. — (A Pacemaker bestellers book)
SUMMARY: A spy secretly exchanges tape recorders with Jeff Avery as they pass through customs and then desperately tries to recover his machine.
ISBN 0-8224-5368-1
[1. Spy stories] I. Lee, Warren. II. Title.
[PZ7.M356758 Se] [Fic] 79 78-72332 MARC

Martin, Anna.
Star magicians / Anna Martin ; [ill. by Harry Rossit]. — San Rafael, Calif. : Leswing Press, c1978.
79 p. : ill. ; 19 cm. — (Her Star knights)
SUMMARY: A young Magician and his friend attempt to rescue an entire race of underwater people from the clutches of an evil power known as the black brilliance.
[1. Science fiction] I. Rossit, Harry. II. Title. III. Series: Martin, Anna. Star knights.
PZ7.M356759 St [Fic] 78 78-105807 MARC

Martin, Anna.
Star peril / Anna Martin ; [ill. by Harry Rossit]. — San Rafael, Calif. : Leswing Press, c1978.
96 p. : ill. ; 19 cm. — (Star knights)
SUMMARY: A beautiful Tortonian girl must battle with the evil power of the black brilliance in order to bring an isolated tribe into the planet Arrack's ring of tribes.
[1. Science fiction] I. Rossit, Harry. II. Title. III. Series.
PZ7.M356759 Su [Fic] 78 78-105806 MARC

Maschler, Fay.
T. G. and Moonie go shopping / story by Fay Maschler ; pictures by Sylvie Selig. — 1st ed. in the United States of American. — Garden City, N.Y. : Doubleday, c1978.
[32] p. : col. ill. ; 22 cm.
SUMMARY: T. G. and Moonie Adkins go on a shopping spree to buy things for their new house.
ISBN 0-385-14147-5. ISBN 0-385-14148-3 lib. bdg. : $5.95
[1. Shopping—Fiction. 2. Cats—Fiction] I. Selig, Sylvie. II. Title.
PZ7.M369 Taf 1978 [E] 77 77-18451 MARC

CHILDREN'S LITERATURE (Continued)

Maschler, Fay.
T. G. and Moonie go shopping / Sylvie Selig ; story told by Fay Maschler. — London : J. Cape, 1978.
[32] p. : chiefly col. ill. ; 22 cm. GB***
ISBN 0-224-01567-2 : £1.95
I. Selig, Sylvie. II. Title.
PZ7.M369 Taf 1978b 79-308340
79 MARC

Maschler, Fay.
T. G. and Moonie have a baby / story by Fay Maschler ; pictures by Sylvie Selig. — 1st ed. in the U.S.A. — Garden City, N.Y. : Doubleday, 1979.
[32] p. : col. ill. ; 22 cm.
SUMMARY: The arrival of a cat couple's first baby coincides with a visit from helpful cousin Fraser.
ISBN 0-385-15332-5 : $5.95
[1. Babies—Fiction. 2. Cats—Fiction] I. Selig, Sylvie. II. Title.
PZ7.M369 Tafh 1979 [E] 79-3951
79 MARC

Maschler, Fay.
T. G. and Moonie move out of town / story by Fay Maschler ; pictures by Sylvie Selig. — 1st ed. in the U.S.A. — Garden City, N.Y. : Doubleday, [1978] c1977.
[32] p. : col. ill. ; 22 cm.
SUMMARY: Finding city life disagreeable, two cats leave their apartment and seek a home in the country.
ISBN 0-385-14145-9. ISBN 0-385-14146-7 lib. bdg.
[1. Moving, Household—Fiction. 2. Animals—Fiction] I. Selig, Sylvie. II. Title.
PZ7.M369 Tag 1978 [E] 77-15344
77 MARC

Massie, Diane Redfield.
Chameleon was a spy / story and pictures by Diane Redfield Massie. — 1st ed. — New York : Crowell, c1979.
[40] p. : ill. (some col.) ; 24 cm.
SUMMARY: Chameleon becomes a spy in order to recover the formula for the world's best pickles.
ISBN 0-690-03909-3 : $6.95. ISBN 0-690-03910-7 lib. bdg. : $6.79
[1. Anoles—Fiction. 2. Pickles—Fiction. 3. Spy stories] I. Title.
PZ7.M42385 Ch 1979 [Fic] 78-19510
78 MARC

Masterman-Smith, Virginia.
The treasure trap / Virginia Masterman-Smith ; illustrated by Roseanne Litzinger. — New York : Four Winds Press, c1979.
200 p. : ill. ; 22 cm.
SUMMARY: Billy Beak's new neighbor convinces him to help her search a long-missing millionaire's mansion for treasure.
ISBN 0-590-07615-9 : $7.95
[1. Mystery and detective stories] I. Litzinger, Roseanne. II. Title.
PZ7.M42387 Tr [Fic] 79-11722
79 MARC

Mathews, Louise.
Gator pie / by Louise Mathews ; pictures by Jeni Bassett. — New York : Dodd, Mead, c1979.
[32] p. : col. ill. ; 19 x 24 cm.
SUMMARY: Two alligators consider dividing their pie into halves, thirds, fourths, eighths, and hundredths.
ISBN 0-396-07694-7
[1. Fractions—Fiction. 2. Division—Fiction. 3. Alligators—Fiction] I. Bassett, Jeni. II. Title.
PZ7.M4253 Gat [E] 79-1512
79 MARC

Mathieu, Joseph.
The Sesame Street mix or match storybook : over two hundred thousand funny combinations : featuring Jim Henson's Muppets / illustrated by Joe Mathieu. — New York : Random House, c1977.
[9] leaves (incl. cover) : chiefly col. ill. ; 31 x 14 cm.
Cover title.
Published in conjunction with Children's Television Workshop, New York.
SUMMARY: Provides the reader with 48 story elements on split pages that can be combined to create tales featuring Sesame Street characters.
ISBN 0-394-83547-6
[1. Picture books] I. Sesame Street. II. Children's Television Workshop. III. Title.
PZ7.M4275 Se [E] 77-70853
79 MARC

Matthews, Ellen, 1950-
The trouble with Leslie / by Ellen Matthews ; illustrated by Unada Gliewe. — 1st ed. — Philadelphia : Westminister Press, c1979.
109 p. : ill. ; 21 cm.
SUMMARY: Despite being in charge of his three-year-old sister and the house while his mother goes to college, Eric finds his summer isn't wasted.
ISBN 0-664-32653-6
[1. Family life—Fiction. 2. Brothers and sisters—Fiction. 3. Summer—Fiction] I. Unada. II. Title.
PZ7.M43176 Tr [Fic] 79-12348
79 MARC

Mattingley, Christobel.
The jetty / Christobel Mattingley ; illustrated by Gavin Rowe. — London : Hodder and Stoughton, c1978.
95 p. : ill. ; 21 cm. GB***
ISBN 0-340-22521-1 : £2.95
I. Title.
PZ7.M43543 Je 78-318755
78 MARC

Mayer, Mercer, 1943-
Little monster at home / by Mercer Mayer. — New York : Golden Press, c1978.
[24] p. : col. ill. ; 21 cm. — (A Golden look-look book)
SUMMARY: A little monster describes his house and some of his family's activities.
ISBN 0-307-11846-0 : $0.95
[1. Dwellings—Fiction. 2. Family life—Fiction. 3. Monsters—Fiction] I. Title.
PZ7.M462 Lg [E] 77-90843
79 MARC

Mayer, Mercer, 1943-
Little monster at school / by Mercer Mayer. — New York : Golden Press, c1978.
[24] p. : col. ill. ; 21 cm. — (A Golden look-look book)
SUMMARY: Little Monster describes his day in school.
ISBN 0-307-11845-2 : $0.95
[1. School stories. 2. Monsters—Fiction] I. Title.
PZ7.M462 Lgs [E] 77-90845
79 MARC

Mayer, Mercer, 1943-
Little monster at work / by Mercer Mayer. — New York : Golden Press, c1978.
45 p. : col. ill. ; 27 cm.
SUMMARY: Traveling with his grandfather, a little monster discovers the variety of occupations in his community.
ISBN 0-307-13736-8 : $3.95.
[1. Occupations—Pictorial works] I. Title.
PZ7.M462 Lgw 77-95315
79 MARC

Mayer, Mercer, 1943-
Little monster's neighborhood / by Mercer Mayer. — New York : Golden Press, c1978.
[24] p. : col. ill. ; 21 cm. — (A Golden look-look book)
SUMMARY: A little monster describes the people, activities, and building in his town.
ISBN 0-307-11849-5 : $0.95
[1. City and town life—Fiction. 2. Monsters—Fiction] I. Title.
PZ7.M462 Lh [E] 77-90844
79 MARC

Mays, Lucinda.
The other shore / by Lucinda Mays. — 1st ed. — New York : Atheneum, 1979.
223 p. ; 25 cm.
SUMMARY: In 1911 Gabriella lives two lives, one in an American high school and the other with her immigrant family in Little Italy on New York's Lower East Side.
ISBN 0-689-30717-9
[1. Emigration and immigration—Fiction. 2. Italian Americans—Fiction. 3. New York (City)—Fiction] I. Title.
PZ7.M47379 Ot [Fic] 79-10448
79 MARC

Mazer, Harry.
The war on Villa Street ; a novel / by Harry Mazer. — New York : Delacorte Press, c1978.
182 p. ; 22 cm.
SUMMARY: A teen-age boy tries to come to terms with various aspects of his life, his relationship with his often drunk and abusive father, the menace of a local bully and his gang, and his love of running.
ISBN 0-440-09349-X
[1. Fathers and sons—Fiction. 2. Running—Fiction] I. Title.
PZ7.M47397 War [Fic] 78-50454
78 MARC

McAulay, Sara.
In search of the petroglyph / Sara McAulay. — New York : Coward, McCann & Geoghegan, c1978.
170 p. : map ; 23 cm.
SUMMARY: A young girl living in the foothills of the Sierra Madre determines to locate the cave which she believes contains a caving made by Indians.
ISBN 0-698-20470-0
[1. Caves—Fiction. 2. The West—Fiction] I. Title.
PZ7.M1214 In 1978 [Fic] 78-9290
78 MARC

McCaffrey, Anne.
Dragondrums / by Anne McCaffrey. — 1st ed. — New York : Atheneum, 1979.
240 p. : map ; 25 cm. — (An Argo book)
SUMMARY: When his boy soprano voice begins to change, Piemur is drafted by Masterharper Robinton to help with political work and is sent on missions that lead him into unusual and sometimes dangerous adventures.
ISBN 0-689-30685-7
[1. Fantasy] I. Title.
PZ7.M122834 Dm [Fic] 78-11318
78 MARC

McCann, Sean.
Shooting stars / Sean McCann ; illustrated by Barry Raynor. — London : Hodder and Stoughton, 1978.
92 p. : ill. ; 21 cm. GB***
ISBN 0-340-22995-0 : £2.80
I. Title.
PZ7.M1253 Sh 78-322902
79 MARC

McClenathan, Louise.
My mother sends her wisdom / by Louise McClenathan ; illustrated by Rosekrans Hoffman. — 1st ed. — New York : Morrow, 1979.
[32] p. : ill. (some col.) ; 21 x 24 cm.
SUMMARY: A peasant woman uses her wisdom to outwit the local money lender who is expecting to take over her farm.
ISBN 0-688-22193-9. ISBN 0-688-32193-3 lib. bdg.
[1. Russia—Fiction] I. Hoffman, Rosekrans. II. Title.
PZ7.M1327 My [E] 79-164
79 MARC

McCloskey, Robert, 1914-
Homer Price / by Robert McCloskey. — New York : Puffin Books, 1978, c1943.
149 p. : ill. ; 20 cm. — (A Puffin book)
SUMMARY: Six episodes in the life of Homer Price including one in which he and his pet skunk capture four bandits and another about a donut machine on the rampage.
ISBN 0-14-030927-6 : $1.95
[1. Humorous stories] I. Title.
PZ7.M1336 Ho 1978 [Fic] 76-39988
76 MARC

McCord, Jean.
Turkeylegs Thompson / by Jean McCord. — 1st ed. — New York : Atheneum, 1979.
242 p. ; 22 cm.
SUMMARY: Betty Ann's childhood ends abruptly when her father abandons the family and her mother goes to work, leaving her responsible for her younger brother and sister.
ISBN 0-689-30686-5
[1. Brothers and sisters—Fiction. 2. Family problems—Fiction. 3. Death—Fiction] I. Title.
PZ7.M136 Tu [Fic] 78-12174
78 MARC

McDaniel, Suellen R
Serpent treasure / Suellen R. McDaniel. — Winston-Salem, N.C. : J. F. Blair, c1978.
129 p. ; 21 cm.
SUMMARY: The fascinating legend of the lost Aztec civilization and its golden treasure draws 16-year-old Chris and his father to a rocky Texas canyon.
ISBN 0-89587-007-X : $7.95
[1. Aztecs—Fiction. 2. Indians of Mexico—Fiction. 3. Buried treasure—Fiction. 4. Mystery and detective stories] I. Title.
PZ7.M1415 Se [Fic] 78-24288
78 MARC

McGinnis, Lila Sprague.
Secret of the porcelain cats / Lila S. McGinnis ; illustrated by Larry Frederick. — Chicago : A. Whitman, c1978.
127 p. : ill. ; 21 cm. — (Pilot books)
SUMMARY: Two cousins camping with a disabled veteran on an island in Lake Erie discover mysterious activities next to their campsite.
ISBN 0-8075-7288-8 lib. bdg.
[1. Mystery and detective stories] I. Frederick, Larry. II. Title.
PZ7.M16775 Se [Fic] 78-15225
78 MARC

McGivern, Maureen Daly, 1921-
Seventeenth summer; illus. by Jay Robinson. New York, Dodd, Mead, 1948.
255 p. col illus. 21 cm.
SUMMARY: The summer after high school graduation 17-year-old Angie finds herself in love for the first time.
[1. Love—Fiction. 2. Wisconsin—Fiction] I. Robinson, Jay. II. Title.
PZ7.M1678 Se 1948 [Fic] 48-10238
[r79]rev2 MARC

McIlwrath, Maureen Mollie Hunter McVeight, 1922-
The third eye / Mollie Hunter. — 1st ed. — New York : Harper & Row, c1979.
276 p. ; 21 cm.
SUMMARY: Jinty's intuitive sense, her third eye, brings her close to the old Earl as he struggles with the Ballingford doom.
ISBN 0-06-022676-5. ISBN 0-06-022677-3 lib. bdg. $7.89
[1. Scotland—Fiction] I. Title.
PZ7.M18543 Tg [Fic] 78-22159
78 MARC

McInerny, Ralph M
Quick as a dodo / by Ralph McInerny ; with ill. by Pam Butterworth. — New York : Vanguard Press, c1978.
116 p. : ill. ; 22 cm.
SUMMARY: An unusual bird relates how he located his relatives, invented a monetary system, appeared on television, and disproved the notion that dodoes are dumb.
ISBN 0-8149-0794-6 : $6.95
[1. Dodo—Fiction. 2. Birds—Fiction] I. Butterworth, Pam. II. Title.
PZ7.M18545 Qi 1978 [Fic] 77-93301
79[r79]rev MARC

McKillip, Patricia A
Harpist in the wind / by Patricia McKillip. — 1st ed. — New York : Atheneum, 1979.
256 p. ; 24 cm. — (An Argo book)
The 3d vol. in a trilogy, the 1st of which is The riddle-master of Hed; the 2d, Heir of sea & fire.
SUMMARY: In the midst of conflict and unrest the Prince of Hed solves the puzzle of his future when he learns to harp the wind, discovers who the shape changers are, and understands his own relationship to Deth, harpist of the wizard Ohm.
ISBN 0-689-30687-3
[1. Fantasy] I. Title.
PZ7.M19864 Har 1979 [Fic] 78-11410
78 MARC

CHILDREN'S LITERATURE (Continued)

McLean, Andrew, 1946-
The riverboat crew / Andrew and Janet McLean. — Melbourne ; New York : Oxford University Press, 1978.
[32] p. : chiefly col. ill. ; 25 cm. Aus***
SUMMARY: A steamboat captain and the ship's stoker swap jobs to prove whose job is more important.
ISBN 0-19-550571-9 : $5.75
[1. Paddle steamers—Fiction. 2. Steamboats—Fiction. 3. Occupations—Fiction] I. McLean, Janet, 1946- joint author. II. Title.
PZ7.M22417 Ri 79-313028
79 MARC

McMillan, Bruce.
The alphabet symphony : an ABC book / by Bruce McMillan. — New York : Greenwillow Books, c1977.
[32] p. : ill. ; 26 cm.
SUMMARY: Photographs of an orchestra reveal different letters of the alphabet.
ISBN 0-688-80112-9. ISBN 0-688-84112-0 lib. bdg.
[1. Alphabet. 2. Musical instruments] I. Title.
PZ7.M2278 Al [E] 77-5491
77[r79]rev MARC

McNaught, Harry.
Muppets in my neighborhood : featuring Jim Henson's Muppets / Harry McNaught, illustrator. — [New York] : Random House, c1977.
[14] p. : chiefly col. ill. ; 20 cm.
Cover title.
At head of title: CTW Sesame Street.
SUMMARY: Sesame Street characters introduce familiar occupations.
ISBN 0-394-83593-X : $2.50
[1. Occupations] I. Henson, Jim. II. Children's Television Workshop. III. Sesame Street. IV. Title.
PZ7.M23253 Mu 77-74472
79 MARC

McNaughton, Colin.
The great zoo escape / [by] Colin McNaughton. — London : Heinemann, 1978.
[32] p. : chiefly col. ill. ; 21 cm. GB78-34903
ISBN 0-434-94989-2 : £2.90
I. Title.
PZ7.M23256 Gr 1978 79-303217
79 MARC

McNaughton, Colin.
The great zoo escape / Colin McNaughton. — 1st American ed. — New York : Viking Press, 1979, c1978.
[32] p. : col. ill. ; 21 cm.
SUMMARY: When a tropical bird from a desert island is captured and taken away to a zoo, he is helped by his brave friends.
ISBN 0-670-35145-8
[1. Zoological gardens—Fiction. 2. Animals—Fiction. 3. Friendship—Fiction] I. Title.
PZ7.M23256 Gr 1979 [E] 78-5138
78 MARC

McNaughton, Colin.
The rat race : the amazing adventures of Anton B. Stanton / by Colin McNaughton. — London : E. Benn, 1978.
[33] p. : chiefly col. ill. ; 18 x 22 cm. GB78-08651
ISBN 0-510-22513-6 : £2.25
I. Title.
PZ7.M23256 Rat 1978 78-313221
78 MARC

McNaughton, Colin.
Walk rabbit walk / story by Colin McNaughton and Elizabeth Attenborough ; pictures by Colin McNaughton. — 1st American ed. — New York : Viking Press, 1977.
[32] p. : col. ill. ; 21 cm.
SUMMARY: Rabbit enjoys the fact that he can get anywhere quicker if he walks.
ISBN 0-670-74918-4 lib. bdg.
[1. Rabbits—Fiction. 2. Walking—Fiction] I. Attenborough, Elizabeth, joint author. II. Title.
PZ7.M23256 Wal 1977b [E] 76-58504
77[7903] MARC

McPhail, David M
Captain Toad and the motorbike / written and illustrated by David McPhail. — 1st ed. — New York : Atheneum, 1978
[32] p. : col. ill. ; 26 cm. — (A Margaret K. McElderry book)
SUMMARY: A retired toad who wants more than anything to ride a motorcycle surprisingly wins a cross-country race the first time he does.
ISBN 0-689-50118-8 : $9.95
[1. Toads—Fiction. 2. Motorcycling—Fiction] I. Title.
PZ7.M2427 Cap [E] 78-59155
79 MARC

McPhail, David M
Henry Bear's park / by David McPhail. — Harmondsworth, Eng. ; New York : Puffin Books, 1978, c1976.
47 p. : ill. ; 23 cm.
SUMMARY: Henry Bear takes good care of the park that his father bought before taking off on a balloon trip.
ISBN 0-14-050291-2 : $1.95
[1. Bears—Fiction] I. Title.
[PZ7.M2427 He 1978] [E] 78-59683
78 MARC

McPhail, David M
Mistletoe / by David McPhail. — 1st ed. — New York : Dutton, c1978.
[32] p. : col. ill. ; 19 cm. — (A Unicorn book)
SUMMARY: Mickey's disappointment in the horse Santa brings her for Christmas only lasts a little while.
ISBN 0-525-35040-3
[1. Christmas stories] I. Title.
PZ7.M2427 Mi 1978 [E] 78-8372
78 MARC

McPhail, David M
Stanley, Henry Bear's friend / by David McPhail. — 1st ed. — Boston : Little, Brown, c1979.
47 p. : ill. ; 20 cm.
"An Atlantic Monthly Press book."
SUMMARY: A naive young raccoon leaves his unhappy home to make his way in the world.
ISBN 0-316-56318-8 : $6.95
[1. Raccoons—Fiction] I. Title.
PZ7.M2427 St 1979 [E] 78-11467
78 MARC

McPhail, David M
The train / by David McPhail. — [New York] : Puffin Books, 1979, c1977.
[32] p. : col. ill. ; 23 cm.
Reprint of the ed. published by Little, Brown, Boston.
SUMMARY: During the night while everyone is asleep, a young boy takes a ride on his toy train.
ISBN 0-14-050302-1 : $1.95
[1. Railroads—Trains—Fiction] I. Title.
[PZ7.M2427 Tr 1979] [E] 78-17987
78 MARC

McPhail, David M
Where can an elephant hide? / By David McPhail. — 1st ed. — Garden City, N.Y. : Doubleday, c1979.
[32] p. : col. ill. ; 20 x 24 cm.
SUMMARY: The jungle animals help Morris the elephant conceal himself from approaching hunters.
ISBN 0-385-12940-8 : $6.95
[1. Elephants—Fiction. 2. Animals—Fiction] I. Title.
PZ7.M2427 Wh [E] 78-31131
79 MARC

McVicar, Catherine.
The grass beyond the door / [by] Catherine McVicar ; illustrated by Pat Marriott. — London : Methuen, 1978.
111 p. : ill., map (on lining papers) ; 21 cm. GB78-33347
ISBN 0-416-58370-9 : £3.45
I. Title.
PZ7.M2503 Gr 79-303389
79 MARC

Melton, David.
The one and only autobiography of Ralph Miller, the dog who knew he was a boy : an hilarious thriller / written and illustrated by David Melton. — Independence, Mo. : Independence Press, c1979.
90 p. : ill. ; 22 cm.
SUMMARY: Although he looks like a dog, Ralph Miller knows he is a boy. But how can he prove it?
ISBN 0-8309-0233-3
[1. Dogs—Fiction] I. Title.
PZ7.M5164 On [Fic] 79-10795
79 MARC

Menius, Opal.
No escape / by Opal Menius. — 1st ed. — New York : Elsevier/Nelson Books, c1979.
129 p. ; 21 cm.
SUMMARY: Seventeen-year-old Colin Mackifield vows to learn why his father expressed the hope that Colin would avenge him and then committed suicide.
ISBN 0-525-66623-0 : $6.95
[1. Suicide—Fiction. 2. Satanism—Fiction. 3. Witchcraft—Fiction] I. Title.
PZ7.M528 No [Fic] 79-4536
79 MARC

Mercer, Charles E
The Battlestar Galactica storybook / [based on the film by Glen A. Larson and the novel by Glen A. Larson and Robert Thurston ; story adapted by Charles Mercer]. — New York : Putnam, c1979.
[60] p. : col. ill. ; 29 cm.
SUMMARY: Humans, having been driven by the Cylons from the last surviving colonies in the universe and hoping to find Earth, their last chance for help, must battle the destructive Cylons again from their craft in space.
ISBN 0-399-20683-3 lib. bdg. : $4.95
[1. Science fiction] I. Larson, Glen A. Battlestar Galactica. II. Battlestar Galactica. [Motion picture] III. Title.
PZ7.M5327 Bat 1979 [Fic] 79-198
79 MARC

Meshover, Leonard.
The monkey that went to school / photos. by Eve Hoffmann ; story by Leonard Meshover and Sally Feistel. — Chicago : Follett Pub. Co., c1978.
64 p. : chiefly ill. ; 21 x 26 cm.
SUMMARY: A spider monkey spends three days in a classroom of young children.
ISBN 0-695-40878-X. ISBN 0-695-30878-5 pbk.
1. Spider monkeys—Legends and stories. [1. Spider monkeys—Fiction. 2. Monkeys—Fiction] I. Feistel, Sally, joint author. II. Hoffmann, Eve. III. Title.
PZ10.3.M5629 Mo [E] 77-88636
79 MARC

Miklowitz, Gloria D
Did you hear what happened to Andrea? / Gloria D. Miklowitz. — New York : Delacorte Press, c1979.
168 p. ; 22 cm.
SUMMARY: Fifteen-year-old Andrea is raped while hitchhiking home from the beach and finds it difficult to cope with her own emotional reactions and those of her family and friends.
ISBN 0-440-01923-0
[1. Rape—Fiction] I. Title.
PZ7.M593 Di [Fic] 78-72972
79 MARC

Miklowitz, Gloria D
Save that raccoon! / By Gloria D. Miklowitz ; pictures by St. Tamara. — 1st ed. — New York : Harcourt Brace Jovanovich, c1978.
[48] p. : col. ill. ; 22 cm. — (A Let me read book)
SUMMARY: After escaping from a devastating forest fire, Raccoon finds a new home.
ISBN 0-15-270241-5 : $5.95. ISBN 0-15-679476-4 pbk.
1. Raccoons—Legends and stories. [1. Raccoons—Fiction. 2. Forest fires—Fiction] I. St. Tamara. II. Title.
PZ10.3M5778 Sav [E] 77-12031
77 MARC

Miles, Betty.
The trouble with thirteen / Betty Miles. — New York : Knopf ; distributed by Random House, c1979.
108 p. ; 22 cm.
SUMMARY: Twelve-year-old Annie is unwilling to face some major changes in her life.
ISBN 0-394-83930-7. ISBN 0-394-93930-1 lib. bdg.
[1. Friendship—Fiction. 2. Death—Fiction] I. Title.
PZ7.M594 Tr [Fic] 78-31678
78 MARC

Miles, Miska.
Jenny's cat / by Miska Miles ; illustrated by Wendy Watson. — 1st ed. — New York : Dutton, c1979.
39 p. : ill. ; 23 cm. — (A Unicorn book)
SUMMARY: Lonely in their new town, Jenny is delighted when a stray cat comes to their house, but her mother doesn't want the cat to stay.
ISBN 0-525-32746-0 : $6.95
[1. Moving, Household—Fiction. 2. Cats—Fiction] I. Watson, Wendy. II. Title.
PZ7.M5944 Je 1979 [E] 79-11501
79 MARC

Miles, Miska.
Mouse Six and the happy birthday / by Miska Miles ; illustrated by Leslie Morrill. — 1st ed. — New York : Dutton, c1978.
[32] p. : ill. ; 23 cm. — (A Unicorn book)
SUMMARY: Mouse Six inadvertently leads the community of animals to discover the secret of his mother's birthday.
ISBN 0-525-35230-9
[1. Birthdays—Fiction. 2. Mice—Fiction] I. Morrill, Leslie H. II. Title.
PZ7.M5944 Mo 1978 [E] 78-7517
78 MARC

Miles, Patricia.
A disturbing influence / [by] Patricia Miles. — London : H. Hamilton, 1978.
[6], 131 p. ; 23 cm. GB78-34127
ISBN 0-241-10022-4 : £3.75
I. Title.
PZ7.M5948 Di 1978 79-300186
79 MARC

Miles, Patricia.
A disturbing influence / by Patricia Miles. — 1st U.S. ed. — New York : Lothrop, Lee & Shepard Co., c1979.
160 p. ; 22 cm.
SUMMARY: Questions about his future seem all important to Andrew during the summer of his 16th birthday, until he is drawn into a mystery involving a rock festival, forged bills, a runaway neighbor, and a deserted farmhouse.
ISBN 0-688-41911-9
[1. England—Fiction] I. Title.
PZ7.M5948 Di 1979 [Fic] 79-14234
79 MARC

Miller, Edna.
Mousekin's Christmas Eve. Story and pictures by Edna Miller. Englewood Cliffs, N.J., Prentice-Hall [1965]
[32] p. col. illus. 24 cm.
SUMMARY: A whitefoot mouse, searching for a new home on Christmas Eve, finds a peaceful and quiet place in a crèche.
[1. Christmas stories. 2. Mice—Fiction] I. Title.
PZ7.M6128 M 1 [E] 65-25244
[r79]rev MARC

Miller, Edna.
Mousekin's golden house. Story and picures by Edna Miller. Englewood Cliffs, N.J., Prentice-Hall [1964]
[32] p. col. illus. 24 cm.
SUMMARY: A whitefoot mouse makes a home for the winter in a jack-o'-lantern discarded after Halloween.
[1. Mice—Fiction. 2. Pumpkin—Fiction] I. Title.
PZ7.M6128 Mo [E] 64-16429
[r79]rev MARC

CHILDREN'S LITERATURE (Continued)

Miller, Ruth, 1921-
The city rose / by Ruth White Miller. — New York : McGraw-Hill, c1977.
171 p. ; 21 cm.
SUMMARY: After a fire destroys her family, Dee goes to live with an aunt and uncle whose house has a strangely deserted room formerly occupied by a girl no one will talk about.
ISBN 0-07-041950-7 lib. bdg. : $5.72
[1. Orphans—Fiction] I. Title.
PZ7.M6334 Ci [Fic] 76-54949
76[r79]rev MARC

Mills, Therese.
The boy who ran away and other stories / by Therese Mills ; [illustrated by D. E. Williams]. — [s.l. : s.n., c1976] (Port of Spain, Trinidad : College Press)
63 p. : ill. ; 21 cm.
CONTENTS: The boy who ran away.—Joey and Charlie.—A dog called Minty.—The beachgirl.—Christmas customs.
I. Williams, D. E. II. Title.
PZ7.M63986 Bo 78-109646
79 MARC

Milord, Sue.
Maggie and the goodbye gift / written and illustrated by Sue and Jerry Milord. — 1st ed. — New York : Lothrop, Lee & Shepard, c1979.
[33] p. : ill. ; 26 cm.
SUMMARY: Moving to a new town is a lonely experience for Maggie's family until Maggie tries out their goodbye gift.
ISBN 0-688-41912-7
[1. Moving, Household—Fiction. 2. Gifts—Fiction. 3. Family life—Fiction] I. Milord, Jerry, joint author. II. Title.
PZ7.M6444 Mag [E] 79-12424
79 MARC

Milton, Hilary H
Mayday! Mayday! / By Hilary Milton. — New York : F. Watts, 1979.
152 p. ; 22 cm.
SUMMARY: After their airplane crashes, two young people must find their way down a mountain in order to bring help to the survivors.
ISBN 0-531-02890-9
[1. Survival—Fiction] I. Title.
PZ7.M6447 May [Fic] 78-24007
78 MARC

Minarik, Else Holmelund.
Little bear. Pictures by Maurice Sendak. New York, Harper [1957]
63 p. illus. 23 cm. (An I can read book)
SUMMARY: Little Bear's four adventures include taking a trip to the moon and having a birthday party.
[1. Bears—Fiction. 2. Mothers and sons—Fiction] I. Sendak, Maurice. II. Title.
PZ7.M652 Li [E] 57-9263
——— Copy 2. [r79]rev MARC

Mitchison, Naomi Haldane, 1897-
The Brave nurse and other stories / by Naomi Mitchison ; drawings by Polly Loxton. — Cape Town ; New York : Oxford University Press, 1977.
27 p. : ill. ; 22 cm. — (Leopard books ; no. 6D) SA77
"Prepared with the co-operation of the Ministry of Education, Botswana."
CONTENTS: The brave nurse.—A good pot needs a good potter.—For her country.—Letsei and Sekgwari.
ISBN 0-19-570093-7 : R0.60
I. Title.
PZ7.M699 Br 79-306398
79 MARC

Moat, John.
Bartonwood / John Moat ; ill. by the author. — London : Chatto and Windus, 1978. GB***
116 p. : ill. ; 21 cm.
ISBN 0-7011-2312-5 : £3.25
I. Title.
PZ7.M7127 Bar 78-321022
78 MARC

Modell, Frank.
Seen any cats? / Frank Modell. — 1st ed. — New York : Greenwillow Books, c1979.
[32] p. : col. ill. ; 26 cm.
SUMMARY: Milton and Marvin look for cats to use in a circus of their own to earn money for the real circus.
ISBN 0-688-80229-X : $6.95
[1. Cats—Ficiton. 2. Circus stories] I. Title.
PZ7.M714 Se [E] 79-11607
79 MARC

Modell, Frank.
Tooley! Tooley! / Frank Modell. — New York : Greenwillow Books, c1979.
[28] p. : col. ill. ; 26 cm.
SUMMARY: The reward for finding a lost dog sends two young boys to the movies.
ISBN 0-688-80092-0. ISBN 0-688-84092-2 lib. bdg.
[1. Dogs—Fiction] I. Title.
PZ7.M714 To [E] 76-49645
76 MARC

Moeri, Louise.
The girl who lived on the Ferris wheel / by Louise Moeri. — 1st ed. — New York : Dutton, c1979.
117 p. ; 22 cm.
SUMMARY: Til realizes with increasing urgency that her divorced mother's violently abusive behavior is getting more and more out of control.
ISBN 0-525-30659-5
[1. Child abuse—Fiction. 2. Mentally ill—Fiction] I. Title.
PZ7.M7214 Gi 1979 [Fic] 79-11359
79 MARC

Molarsky, Osmond.
A different ball game / by Osmond Molarsky ; illustrated by James Zingarelli. — New York : Coward, McCann & Geoghegan, c1979.
48 p. : ill. ; 23 cm.
SUMMARY: A soccer enthusiast helps a young Chilean boy, living in the United States with his aunt, understand what it means to be arrested and falsely accused of a crime.
ISBN 0-698-20475-1
[1. Juvenile delinquency—Fiction. 2. Soccer—Fiction] I. Zingarelli, James. II. Title.
PZ7.M7317 Di [Fic] 78-23793
78 MARC

Molarsky, Osmond.
Robbery in right field / Osmond Molarsky ; illustrated bby Rob Sauber. — New York : H. Z. Walck, c1978.
52 p. : ill. ; 24 cm.
SUMMARY: Young Cynthia Rose, the Little League's best player, does a little coaching with tremendous results.
ISBN 0-8098-6425-8 : $6.95
[1. Baseball—Fiction. 2. Sex role—Fiction] I. Sauber, Rob. II. Title.
PZ7.M7317 Ro [Fic] 78-5204
78 MARC

Moncure, Jane Belk.
A beach in my bedroom / by Jane Belk Moncure ; illustrated by Helen Endres. — Elgin, Ill. : Child's World ; Chicago : distributed by Childrens Press, c1978.
30 p. : col. ill. ; 23 cm. — (Her Creative dramatics)
SUMMARY: Disappointed because she cannot go to the beach, Lori imagines a beach in her bedroom.
ISBN 0-89565-005-3 : $4.95. ISBN 0-89565-038-X pbk.
[1. Beaches—Fiction. 2. Play—Fiction] I. Endres, Helen. II. Title.
PZ7.M739 Be [E] 77-12960
77 MARC

Moncure, Jane Belk.
Birds, baboons, and barefoot bears / by Jane Belk Moncure ; illustrated by Helen Endres. — Elgin, Ill. : Child's World ; Chicago :
30 p. : col. ill. ; 23 cm. — (Her Creative dramatics)
SUMMARY: Imagines how it feels to fly like a bird, swing from the trees like a baboon, or swim like a whale.
ISBN 0-89565-006-1 : $4.95. ISBN 0-89565-039-8
[1. Play—Fiction] I. Endres, Helen. II. Title. III. Series.
PZ7.M739 Bi [E] 77-13025
77 MARC

Moncure, Jane Belk.
The Four magic boxes / by Jane Belk Moncure ; illustrated by Helen Endres. — Elgin, Ill. : Child's World ; Chicago : distributed by Childrens Press, c1978.
30 p. : ill. ; 23 cm. — (Her Creative dramatics)
SUMMARY: Jim and his new friends turn packing boxes into a house, train, fire engine, airplane, hospital, and ship.
ISBN 0-89565-007-X : $4.95. ISBN 0-89565-040-1 pbk.
[1. Play—Fiction] I. Endres, Helen. II. Title. III. Series.
PZ7.M739 Fo [E] 77-12959
77 MARC

Moncure, Jane Belk.
Magic monsters look for colors / by Jane Belk Moncure ; illustrated by Diana Magnuson. — Elgin, Ill. : Child's World ; Chicago : distributed by Childrens Press, c1979.
31 p. : col. ill. ; 25 cm. — (Magic monster series)
SUMMARY: The magic monsters search for different colors around town.
ISBN 0-89565-056-8 (lib. bdg.) : $5.50
[1. Colors—Fiction] I. Magnuson, Diana. II. Title. III. Title: Look for colors. IV. Series.
PZ7.M739 Mag [E] 78-23792
78 MARC

Moncure, Jane Belk.
Magic monsters look for shapes / by Jane Belk Moncure ; illustrated by Diana Magnuson. — Elgin, Ill. : Child's World ; Chicago : distributed by Childrens Press, c1979.
32 p. : col. ill. ; 25 cm. — (Magic monster series)
SUMMARY: The magic monsters introduce the basic shapes and find representative examples.
ISBN 0-89565-057-6 : $5.50
[1. Size and shape—Fiction. 2. Monsters—Fiction] I. Magnuson, Diana. II. Title. III. Series.
PZ7.M739 Maj [E] 78-21529
78 MARC

Moncure, Jane Belk.
My baby brother needs a friend / by Jane Belk Moncure ; illustrated by Frances Hook. — Elgin, Ill. : Child's World ; Chicago : distributed by Childrens Press, c1979.
30 p. : col. ill. ; 27 cm.
SUMMARY: A young girl describes her relationship with her brother who is not yet one year old.
ISBN 0-89565-019-3 (lib. bdg.) : $5.50
[1. Brothers and sisters—Fiction. 2. Babies—Fiction] I. Hook, Frances. II. Title.
PZ7.M739 Mybc [E] 78-21935
78 MARC

Moncure, Jane Belk.
My c sound box / by Jane Belk Moncure ; illustrated by Linda Sommers. — Elgin, Ill. : Child's World ; Chicago : distributed by Childrens Press, c1979.
30 p. : col. ill. ; 20 x 24 cm. — (Her Sound box books)
SUMMARY: A little girl fills her sound box with many words beginning with the letter "c."
ISBN 0-89565-053-3 : $5.50
[1. Alphabet] I. Sommers, Linda. II. Title. III. Series.
PZ7.M739 Myc [E] 78-23638
78 MARC

Moncure, Jane Belk.
My d sound box / by Jane Belk Moncure ; illustrated by Linda Sommers. — Elgin, Ill. : Child's World ; Chicago : distributed by Childrens Press, c1978.
29 p. : col. ill. ; 20 x 24 cm. — (Her Sound box books)
SUMMARY: A little girl fills her sound box with many words that begin with the letter "d".
ISBN 0-89565-044-4 : $4.95
[1. Alphabet] I. Sommers, Linda. II. Title. III. Series.
PZ7.M739 Myd [E] 78-8450
78 MARC

Moncure, Jane Belk.
My g sound box / by Jane Belk Moncure ; illustrated by Linda Sommers. — Elgin, Ill. : Child's World ; Chicago : distributed by Childrens Press, c1979.
31 p. : col. ill. ; 20 x 24 cm. — (Her Sound box books)
SUMMARY: A little girl fills her sound box with many words that begin with the letter "g".
ISBN 0-89565-053-3 (lib. bdg.) : $5.50
[1. Alphabet] I. Sommers, Linda. II. Title. III. Series.
PZ7.M739 Myg [E] 78-22037
78 MARC

Moncure, Jane Belk.
My j sound box / by Jane Belk Moncure ; illustrated by Linda Sommers. — Elgin, Ill. : Child's World ; Chicago : distributed by Childrens Press, c1979.
29 p. : col. ill. ; 20 x 24 cm. — (Her Sound box books)
SUMMARY: A little boy fills his sound box with many words beginning with the letter "j."
ISBN 0-89565-049-5 (lib. bdg.) : $5.50
[1. Alphabet] I. Sommers, Linda. II. Title. III. Series.
PZ7.M739 Myj [E] 78-23178
78 MARC

Moncure, Jane Belk.
My k sound box / by Jane Belk Moncure ; illustrated by Linda Sommers. — Elgin, Ill. : Child's World ; Chicago : distributed by Childrens Press, c1979.
29 p. : col. ill. ; 20 x 24 cm. — (Her Sound box books)
SUMMARY: A little boy fills his sound box with many words beginning with the letter "k."
ISBN 0-89565-050-9 : $5.50
[1. Alphabet] I. Sommers, Linda. II. Title. III. Series.
PZ7.M739 Myk [E] 78-22034
78 MARC

Moncure, Jane Belk.
My l sound box / by Jane Belk Moncure ; illustrated by Linda Sommers. — Elgin, Ill. : Child's World ; Chicago : distributed by Childrens Press, c1978.
29 p. : col. ill. ; 20 x 24 cm. — (Her Sound box books)
SUMMARY: A little boy fills his sound box with many words beginning with the letter "l".
ISBN 0-89565-045-2 : $4.95
[1. Alphabet] I. Sommers, Linda. II. Title. III. Series.
PZ7.M739 Myl [E] 78-8373
78 MARC

Moncure, Jane Belk.
My m sound box / by Jane Belk Moncure ; illustrated by Linda Sommers. — Elgin, Ill. : Child's World ; Chicago : distributed by Childrens Press, c1979.
29 p. : col. ill. ; 20 x 24 cm. — ([Her Sound box books])
SUMMARY: A little boy fills his sound box with many words that begin with the letter "m".
ISBN 0-89565-051-7 (lib. bdg.) : $5.50
[1. Alphabet] I. Sommers, Linda. II. Title. III. Series.
PZ7.M739 Mym [E] 78-24458
78 MARC

Moncure, Jane Belk.
My n sound box / by Jane Belk Moncure ; illustrated by Linda Sommers. — Elgin, Ill. : Child's World ; Chicago : distributed by Childrens Press, c1979.
29 p. : col. ill. ; 20 x 24 cm. — (Her Sound box books)
SUMMARY: A little girl fills her sound box with many words beginning with the letter "n".
ISBN 0-89565-054-1 (lib. bdg.) : $5.50
[1. Alphabet] I. Sommers, Linda. II. Title. III. Series.
PZ7.M739 Myn [E] 78-22053
78 MARC

Moncure, Jane Belk.
My p sound box / by Jane Belk Moncure ; illustrated by Linda Sommers. — Elgin, Ill. : Child's World ; Chicago : distributed by Childrens Press, c1978.
29 p. : col. ill. ; 20 x 24 cm. — (Her Sound box books)
SUMMARY: A little girl fills her sound box with many words that begin with the letter "p."
ISBN 0-89565-047-9 : $4.95
[1. Alphabet] I. Sommers, Linda. II. Title. III. Series.
PZ7.M739 Myp [E] 78-7841
78 MARC

CHILDREN'S LITERATURE (Continued)

Moncure, Jane Belk.
My r sound box / by Jane Belk Moncure ; illustrated by Linda Sommers. — Elgin, Ill. : Child's World ; Chicago : distributed by Childrens Press, c1978.
29 p. : col. ill. ; 20 x 24 cm. — (Her Sound box books)
SUMMARY: A little boy fills his sound box with many words that begin with the letter "r."
ISBN 0-89565-048-7 : $4.95
[1. Alphabet] I. Sommers, Linda. II. Title. III. Series.
PZ7.M739 Myr [E] 78 78-7842 MARC

Moncure, Jane Belk.
My w sound box / by Jane Belk Moncure ; illustrated by Linda Sommers. — Elgin, Ill. : Child's World ; Chicago : distributed by Childrens Press, c1978.
29 p. : col. ill. ; 20 x 24 cm. — (Her Sound box books)
SUMMARY: A little girl fills her sound box with many words beginning with the letter "w".
ISBN 0-89565-046-0 : $4.95
[1. Alphabet] I. Sommers, Linda. II. Title. III. Series.
PZ7.M739 Myw [E] 78 78-8614 MARC

Moncure, Jane Belk.
Skip aboard a space ship / by Jane Belk Moncure ; illustrated by Helen Endres. — Elgin, Ill. : Child's World ; Chicago : distributed by Childrens Press, c1978.
30 p. : col. ill. ; 23 cm. — (Her Creative dramatics)
SUMMARY: Two boys spending the night in a tree house pretend they are taking a trip to the moon.
ISBN 0-89565-009-6 lib. bdg. : $4.95
[1. Space flight to the moon—Fiction. 2. Play—Fiction] I. Endres, Helen. II. Title. III. Series.
PZ7.M739 Sk [E] 77 77-12958 MARC

Moncure, Jane Belk.
Tick, tock, the popcorn clock / by Jane Belk Moncure ; illustrated by Helen Endres. — Elgin, Ill. : Child's World ; Chicago : distributed by Childrens Press, c1978.
29 p. : col. ill. ; 23 cm. — (Her Creative dramatics)
SUMMARY: Children enact the cycle of popcorn growing from its seeds into tall stalks, under the care of the farmer, and finally being picked and thrown into the popcorn pot for eating.
ISBN 0-89565-010-X lib. bdg. : $4.95
[1. Popcorn—Fiction] I. Endres, Helen. II. Title. III. Series.
PZ7.M739 Ti [E] 77 77-13120 MARC

Montgomery, Elizabeth Rider.
"Seeing" in the dark / by Elizabeth Rider Montgomery ; drawings by Troy Howell. — Champaign, Ill. : Garrard Pub. Co., 1979.
44 p. : ill. (some col.) ; 23 cm. — (A For real book)
SUMMARY: A blind child slowly adjusts to life at her new school and dramatically proves the success of that adjustment during a school fire.
ISBN 0-8116-4312-3 (lib. bdg.) : $4.60
[1. Blind—Fiction. 2. Physically handicapped—Fiction. 3. School stories] I. Howell, Troy. II. Title.
PZ7.M765 Sg [E] 79 79-11736 MARC

Montgomery, Lucy Maud, 1874-1942.
Jane of Lantern Hill / L. M. Montgomery. — [Toronto] : McClelland and Stewart, 1977, c1937.
297 p. ; 20 cm. — (Canadian favourites) C78-4271-9
ISBN 0-7710-6301-6 : $3.95
I. Title. II. Series.
PZ7.M768 Jan 1977 79 79-305238 MARC

Montresor, Beni.
Bedtime! / Beni Montresor. — 1st ed. — New York : Harper & Row, c1978.
[32] p. : chiefly col. ill. ; 24 x 28 cm.
SUMMARY: A little boy who is not ready for bedtime runs to hide in a magic castle and finds some unusual friends there.
ISBN 0-06-024353-8
[1. Sleep—Fiction. 2. Fantasy] I. Title.
PZ7.M769 Be E 00 78-108581 MARC

Moore, Gerald Ernest.
The cuckoo who flew backwards / [by] Gerald Moore. — London : Bachman & Turner, 1977.
71 p. : ill. (some col.) ; 23 cm. GB78-02769
ISBN 0-85974-064-1 : £2.95
I. Title.
PZ7.M7837 Cu 78 78-308052 MARC

Moore, Marvin, 1937-
Witnesses through trial / by Marvin Moore. — Nashville : Southern Pub. Association, c1979.
128 p. ; 17 cm. — (Orion)
SUMMARY: Eight fictionalized accounts of Christian men, women, and children who adhered to their faith through times of adversity.
ISBN 0-8127-0216-6
[1. Chritian life—Fiction. 2. Seventh-Day Adventists—Fiction] I. Title.
[PZ7.M78716 Wi] [Fic] 78 78-24294 MARC

Moore, Ruth Nulton.
Mystery of the lost treasure / Ruth Nulton Moore ; illustrated by James L. Converse. — Scottdale, Pa. : Herald Press, 1978.
177 p. : ill. ; 22 cm.
SUMMARY: Two youngsters follow clues from an old diary hoping to find a treasure buried years ago on their aunt's Pennsylvania farm.
ISBN 0-8361-1870-7. ISBN 0-8361-1871-5 pbk.
[1. Buried treasure—Fiction. 2. Pennsylvania—Fiction. 3. Mystery and detective stories] I. Converse, James. II. Title.
PZ7.M7878 My [Fic] 78 78-11748 MARC

Moore, Ruth Nulton.
Tomás and the talking birds / Ruth Nulton Moore ; illustrated by Esther Rose Graber. — Scottdale, Pa. : Herald Press, 1979.
115 p. : ill. ; 22 cm.
SUMMARY: A Puerto Rican boy has trouble adjusting to his new life in a Pennsylvania steel town. When he asked to teach some English words to a parrot, Tomás learns the difficult language himself.
ISBN 0-8361-1873-1 : $3.95. ISBN 0-8361-1874-X pbk. : $2.95
[1. Puerto Ricans in the United States—Fiction] I. Graber, Esther Rose. II. Title.
PZ7.M7878 To [Fic] 78 78-23509 MARC

Morey, Walter.
Sandy and the rock star / by Walt Morey. — 1st ed. — New York : Dutton, c1979.
171 p. ; 22 cm.
SUMMARY: Running away from his overly protective parents and restricted life, a teenage singing star finds himself on a remote island where he befriends a tame but frightened cougar that he determines to protect from a vicious hunter.
ISBN 0-525-38785-4
[1. Pumas—Fiction. 2. Hunting—Fiction. 3. Islands—Fiction] I. Title.
PZ7.M816 San 1979 [Fic] 78 78-12375 MARC

Morgenroth, Barbara.
Impossible Charlie / by Barbara Morgenroth ; pictures by Velma Ilsley. — 1st ed. — New York : Atheneum, 1979.
153 p. : ill. ; 22 cm.
SUMMARY: A horse that seemed ideal when Jackie got him turns out to have too much of a mind of his own.
ISBN 0-689-30718-7
[1. Horses—Fiction] I. Ilsley, Velma. II. Title.
PZ7.M82669 Im [Fic] 79 79-10447 MARC

Morgenroth, Barbara.
Last junior year / Barbara Morgenroth. — 1st ed. — New York : Atheneum, 1978.
179 p. ; 22 cm.
SUMMARY: Seventeen-year-old Kim has her heart set on joining the United States Equestrian Team, but finds she must make alternative plans for her future.
ISBN 0-689-30663-6
[1. Horses—Fiction] I. Title.
PZ7.M82669 Las 1978 [Fic] 78 78-2750 MARC

Morgenroth, Barbara.
Tramps like us / Barbara Morgenroth. — 1st ed. — New York : Atheneum, 1979.
145 p. ; 22 cm.
SUMMARY: Oppressed by the restrictions and provincial standards of her conventional school and hometown, a gifted girl finds the incentive to seek her individuality through her relationship with a musically dedicated boy from a neighboring town.
ISBN 0-689-30690-3
[1. Individuality—Fiction] I. Title.
PZ7.M82669 Tr 1979 [Fic] 78 78-13739 MARC

Morrison, Bill, 1935-
Louis James hates school / written and illustrated by Bill Morrison. — Boston : Houghton Mifflin, 1978.
32 p. : col. ill. ; 21 x 22 cm.
SUMMARY: A young boy learns the value of formal education when he quits school to find a job.
ISBN 0-395-27156-8 : $6.95
[1. Dropouts—Fiction] I. Title.
PZ7.M82925 Lo [E] 78 78-60496 MARC

Morrow, Charlotte.
The rain woman / [by] Charlotte Morrow. — Glasgow [etc.] : Blackie, 1978.
117 p. ; 23 cm. GB78-19626
ISBN 0-216-90380-7 : £3.25
I. Title.
PZ7.M8454 Rai 78 78-321026 MARC

Morton, Jane, 1931-
Running scared / by Jane Morton. — 1st ed. — New York : Elsevier/Nelson Books, c1979.
118 p. ; 21 cm.
SUMMARY: Stealing a car, the latest of 15-year-old Dave's misdeeds, gets him sent to Juvenile Hall where he meets Pat, a counselor determined to help Dave.
ISBN 0-525-66631-1 : $6.95
I. Title.
PZ7.M8467 Ru [Fic] 79 79-1495 MARC

Moskin, Marietta D
Day of the blizzard / Marietta Moskin ; drawings by Stephen Gammell. — New York : Coward, McCann & Geoghegan, c1978.
79 p. : ill. ; 18 cm.
SUMMARY: Eager to accomplish an errand for her sick mother, Katie braves one of New York City's severest snowstorms in history, the blizzard of 1888.
ISBN 0-698-20468-9
[1. Blizzards—Fiction. 2. New York (City)—Fiction] I. Gammell, Stephen. II. Title.
PZ7.M849 Day [Fic] 78 78-6461 MARC

Moss, Charles Norman.
The wreck of the Pied Piper / C. N. Moss. — Grand Rapids : Zondervan Pub. House, 1979, c1973.
136 p. ; 21 cm. — (Pathfinder series ; 4)
SUMMARY: While on summer vacation, three youths unearth the truth about a stolen jewel on an old wreck when the thief comes back to claim his treasure.
ISBN 0-310-37831-1
[1. Mystery and detective stories] I. Title. II. Series.
PZ7.M8519 Wr 1979 [Fic] 78 78-23264 MARC

Moss, Elaine.
Polar / story by Elaine Moss ; pictures by Jeannie Baker. — London : Deutsch, 1975.
[32] p. : col. ill. ; 26 cm. GB75-31339
SUMMARY: A teddy bear's exploits in his toboggan bring him injury, but also the attention of kind friends.
ISBN 0-233-96695-1 : $5.95 (U.S.)
[1. Dolls—Fiction. 2. Tobogganing—Fiction] I. Baker, Jeannie. II. Title.
PZ8.9.M75 Po [E] 79 78-74459 MARC

Muir, Frank.
What-a-mess / Frank Muir ; illustrated by Joseph Wright. — London : E. Benn ; Toronto : distributed by General Pub. Co., c1977.
30 p. : ill. ; 28 cm. GB***
ISBN 0-510-22509-8 : £1.95
I. Wright, Joseph. II. Title.
PZ7.M883 Wh 1977 78 78-314313 MARC

Muir, Frank.
What-a-mess, the good / Frank Muir ; illustrated by Joseph Wright. — 1st ed. in the U.S.A. — Garden City, N.Y. : Doubleday, c1978.
[32] p. : col. ill. ; 28 cm.
"The Benn book collection."
SUMMARY: A well-meaning puppy decides to show everyone how good he can be by ridding his owners' home of ants.
ISBN 0-385-14699-X : $5.95. ISBN 0-385-14700-7 lib. bdg.
[1. Dogs—Fiction] I. Wright, Joseph. II. Title.
PZ7.M883 Wk 1978 [Fic] 78 78-8807 MARC

Muir, Frank.
What-a-mess, the Good / Frank Muir ; illustrated by Joseph Wright. — London : E. Benn, 1978.
[32] p. : col. ill. ; 28 cm. GB***
ISBN 0-510-12400-3 : £2.50
I. Wright, Joseph. II. Title.
PZ7.M883 Wk 1978b 79 79-307174 MARC

Munves, James.
The treasure of Diogenes Sampuez / James Munves. — New York : Four Winds Press, c1979.
186 p. ; 22 cm.
SUMMARY: When, after his father's death, his uncle tries to take the house and horse away from the family, 14-year-old Diogenes, his sister, and a friend travel across the Colombian mountains to seek help from an older brother.
ISBN 0-590-07384-2 : $7.95
[1. Colombia—Fiction] I. Title.
PZ7.M93 Tr [Fic] 78 78-21768 MARC

Murphy, Shirley Rousseau.
The flight of the fox / Shirley Rousseau Murphy ; drawings by Don Sibley based on original designs by Richard Cuffari. — 1st ed. — New York : Atheneum, 1978.
164 p. : ill. ; 22 cm.
SUMMARY: With the help of his friends, Rory, a wandering kangaroo rat, restores a model of a Fairey Fox airplane to fly on his travels and foils a flock of starlings that has overrun the town.
ISBN 0-689-30662-8
I. Sibley, Don. II. Cuffari, Richard, 1925- . III. Title.
PZ7.M956 Fl 1978 [Fic] 78 78-5436 MARC

Murphy, Shirley Rousseau.
The wolf bell / by Shirley Rousseau Murphy. — 1st ed. — New York : Atheneum, 1979.
182 p. : map ; 24 cm.
SUMMARY: With the help of the wolves and the slave Jerthon, Ramad must search out the Runestone for the future good of the entire planet of Ere.
ISBN 0-689-30692-X
[1. Fantasy] I. Title.
PZ7.M956 Wo [Fic] 78 78-10415 MARC

CHILDREN'S LITERATURE (Continued)

Murray, Frances.
White Hope / Frances Murray. — London : Hodder and Stoughton, 1978.
124 p. ; 23 cm. GB***
ISBN 0-340-22994-2 : £3.25
I. Title.
PZ7.M9575 Wh 78-325957
79 MARC

Myers, Amy.
I know a monster / written and illustrated by Amy Myers. — Reading, Mass. : Addison-Wesley, c1979.
[31] p. : ill. ; 16 x 19 cm.
SUMMARY: Each child sitting in a circle adds to the story of a monster.
ISBN 0-201-04990-2
[1. Monsters—Fiction] I. Title.
PZ7.M9813 Iad [E] 78-11692
78 MARC

Myers, Bernice.
Little John Bear in the big city / by Bernice Myers. — New York : Four Winds Press, c1978.
32 p. : col. ill. ; 22 cm.
SUMMARY: Little John Bear experiences some typical big city experiences as he travels there to borrow a fur coat from his friend.
ISBN 0-590-07601-9 : $5.95
[1. City and town life—Fiction. 2. Bears—Fiction] I. Title.
PZ7.M9817 Li [E] 78-15595
78 MARC

Myers, Walter Dean, 1937-
It ain't all for nothin' / Walter Dean Myers. — 1st ed. — New York : Viking Press, 1978.
217 p. ; 22 cm.
SUMMARY: A young black boy must decide whether to go along with his father, who is a thief, or reject his father's way of life and risk losing him.
ISBN 0-670-40301-6 : $8.95
[1. Fathers and sons—Fiction] I. Title.
PZ7.M992 It 1978 [Fic] 78-57516
78 MARC

Myller, Rolf.
How big is a foot? [1st ed.] New York, Atheneum, 1962.
[36] p. illus. 23 cm.
SUMMARY: Thrown in jail because the bed he made for the Queen is too small, an apprentice comes up with a more accurate way of measuring size.
[1. Measuring—Fiction] I. Title.
PZ7.M995 Ho [E] 62-10252
[r78]rev MARC

Myra, Harold Lawrence, 1939-
Easter bunny, are you for real? / Written by Harold Myra ; illustrated by Dwight Walles. — Nashville : T. Nelson, c1979.
32 p. : col. ill. ; 28 cm.
SUMMARY: Two children learn about the origins of the Easter bunny and other symbols and customs associated with Easter.
ISBN 0-8407-5148-6 : $4.95
[1. Easter stories] I. Walles, Dwight. II. Title.
PZ7.M9954 Eas [Fic] 78-21268
78 MARC

The Mystery of the rebellious robot / illustrated by Mark Corcoran. — New York : Random House, c1979.
[32] p. : col. ill. ; 21 cm. — (Star wars)
SUMMARY: Luke Skywalker and his companions are puzzled by the sudden bizarre behavior of their robots and the malfunction of their machines.
ISBN 0-394-84086-0 : $1.25. ISBN 0-394-94086-5 lib. bdg. : $3.99
[1. Science fiction] I. Corcoran, Mark. II. Series.
PZ7.M9973 [Fic] 78-19703
79 MARC

Nagel, Shirley, 1922-
Escape from the tower / by Shirley Nagel ; [illustrated by Rudy Obero]. — Mankato, Minn. : Creative Education, c1978.
55 p. : ill. ; 24 cm.
SUMMARY: A mistreated servant girl for the head jailer at the Tower of London becomes involved in the daring escape plot of a prominent prisoner.
ISBN 0-87191-685-1
[1. England—Fiction. 2. London. Tower—Fiction] I. Obero, Rudy. II. Title.
PZ7.N135 Es [Fic] 78-71860
00 MARC

Narell, Irena.
Joshua, fighter for Bar Kochba / by Irena Narell. — Pebble Beach, CA : Akiba Press, c1978.
ix, 172 p. ; 22 cm.
SUMMARY: In 132 A.D. a young man follows the leadership of Bar Kochba in the second Hebrew insurrection against the Romans.
$5.95
1. Bar Kokba, fl. 130—Juvenile fiction. [1. Bar Kokba, fl. 130—Fiction. 2. Palestine—History—70-638—Fiction] I. Title.
PZ7.N16 Jo [Fic] 78-55959
79 MARC

Naylor, Phyllis Reynolds.
The witch herself / Phyllis Reynolds Naylor ; illustrated by Gail Owens. — 1st ed. — New York : Atheneum, 1978.
164 p. : ill. ; 22 cm.
The third vol. of a trilogy; the first vol. is Witch's sister and the second is Witch water.
SUMMARY: As more people fall under Mrs. Tuggle's spells, Lynn and Mouse grow more desperate in their efforts to find a way to stop the old woman's evil.
ISBN 0-689-30664-4 : $7.95
[1. Witchcraft—Fiction] I. Owens, Gail. II. Title.
PZ7.N24 Wk 1978 [Fic] 78-5437
78 MARC

Neal, Berniece Roer.
Chicken / by Berniece Roer Neal ; illustrated by Jeffrey K. Lindberg. — New York : Dandelion Press, 1979.
[32] p. : col. ill. ; 23 cm.
SUMMARY: A young boy is convinced he is a coward because when faced with danger, he is afraid.
ISBN 0-89799-092-7
[1. Courage—Fiction] I. Lindberg, Jeffrey K. II. Title.
PZ7.N248 Ch 78-72119
79 MARC

Needle, Jan.
My mate Shofiq / Jan Needle. — London : Deutsch, 1978.
174 p. ; 21 cm. GB***
ISBN 0-233-96987-X : £3.50
I. Title.
PZ7.N33 My 78-322117
79 MARC

Needle, Jan.
The size spies / by Jan Needle ; illustrated by Roy Bentley. — London : Deutsch, 1978.
156 p. : ill. ; 21 cm. GB***
ISBN 0-233-97003-7 : £3.50
I. Title.
PZ7.N33 Si 79-305218
79 MARC

Nelson, Darrel.
Little Millard Mustardseed / Darrel Nelson. — Elgin, IL : D. C. Cook Pub. Co., c1979.
[48] p. : col. ill. ; 18 x 23 cm. — (Chariot books)
SUMMARY: Millard attends a seed gathering hoping to find an important task suitable for a tiny seed like himself.
ISBN 0-89191-063-8 : $2.95
[1. Mustard seed—Fiction. 2. Seeds—Fiction. 3. Size and shape—Fiction] I. Title.
PZ7.N43375 Li [E] 78-72834
00 MARC

Nelson, Gladys Tirrell.
War drums at Eden Prairie : story and illustrations / by Gladys Tirrell Nelson. — Saint Cloud, Minn. : North Star Press, c1976.
115 p. : ill. ; 23 cm.
SUMMARY: A family living in Minnesota during the mid-1800's is threatened by an Indian uprising.
ISBN 0-87839-023-5
[1. Indians of North America—Fiction. 2. Minnesota—Fiction. 3. Frontier and pioneer life—Fiction] I. Title.
PZ7.N4338 War [Fic] 78-104895
78 MARC

Ness, Evaline.
Marcella's guardian angel / written and illustrated by Evaline Ness. — New York : Holiday House, c1979.
[40] p. : col. ill. ; 21 cm.
SUMMARY: Marcella's guardian angel persuades her to play an interesting games.
ISBN 0-8234-0343-2 (lib. bdg.) : $6.95
[1. Behavior—Fiction] I. Title.
PZ7.N4384 Mar [E] 78-10414
78 MARC

Newcomb, Ellsworth.
Three came riding. Illustrated by Cristina F. Christensen. New York, Norton [1964]
82 p. illus. 24 cm.
SUMMARY: Following times of hardship and disappointment, good fortune comes to José, his sister, and mother on Three Kings Day.
[1. Christmas stories. 2. Family life—Puerto Rico—Fiction] I. Christensen, Cristina F., illus. II. Title.
PZ7.N4729 Th [Fic] 64-17535
[r78]rev MARC

Newman, Marjorie.
Wilkins, the armchair cat / Marjorie Newman ; illustrated by Clare Beaton. — London : A. & C. Black, 1978.
[25] p. : col. ill. ; 27 cm. GB***
ISBN 0-7136-1860-4 : £2.50
I. Beaton, Clare. II. Title.
PZ7.N484 Wi 78-323937
78 MARC

Newman, Shirlee Petkin.
Tell me, Grandma, tell me, Grandpa / story by Shirlee P. Newman ; pictures by Joan Drescher. — Boston : Houghton Mifflin, 1979.
30 p. : col. ill. ; 20 x 24 cm.
SUMMARY: As a youngster hears about her parents when they were young, she turns her imagination loose.
ISBN 0-395-27815-5 : $6.95
[1. Fathers—Fiction. 2. Mothers—Fiction. 3. Humorous stories] I. Drescher, Joan E. II. Title.
PZ7.N486 Te [E] 78-23711
78 MARC

Newton, Suzanne.
Reubella and the Old Focus Home / Suzanne Newton. — 1st ed. — Philadelphia : Westminister Press, c1978.
197 p.; 21 cm.
SUMMARY: Three vivacious elderly ladies turn life in a sleepy southern town upside down.
ISBN 0-664-32635-8
[1. Old age—Fiction] I. Title.
PZ7.N4875 Re [Fic] 78-18336
78 MARC

Nichlas, Ruth, 1948-
The left-handed spirit / Ruth Nichols. — 1st ed. — New York : Atheneum, 1978.
220 p. ; 22 cm.
"A Margaret K. McElderry book."
SUMMARY: Endowed with the power to cure sickness, a young woman in 2nd-century Rome is kidnapped and taken to China by the Chinese ambassador who desperately needs a healer.
ISBN 0-689-50120-X
[1. Kidnapping—Fiction. 2. Rome—History—Marcus Aurelius, 161-180—Fiction. 3. China—History—Han dynasty, 202 B.C.-220 A.D.—Fiction] I. Title.
PZ7.N5418 Le [Fic] 78-5573
78 MARC

Nichols, Ruth, 1948-
The left-handed spirit / Ruth Nichols. — Toronto : Macmillan of Canada, c1978.
220 p. ; 22 cm. C***
ISBN 0-7705-1701-3 : $9.95
I. Title.
PZ7.N5418 Le 1978b 79-305515
79 MARC

Nicklaus, Carol.
Harry the hider / by Carol Nicklaus. — New York : F. Watts, 1979.
[32] p. : col. ill. ; 22 cm. — (An Easy-read story book)
SUMMARY: Miranda can't think of a way to involve her cat, Harry, in the neighborhood circus, but Harry comes up with an idea himself.
ISBN 0-531-02381-8. ISBN 0-531-02298-6 lib. bdg.
[1. Cats—Fiction] I. Title.
PZ7.N5583 Har [E] 78-10250
78 MARC

Niland, Deborah.
ABC of monsters / Deborah Niland. — New York : McGraw-Hill, [1978] c1976.
[30] p. : col. ill. ; 15 x 22 cm.
SUMMARY: A monster party features annoying apes, drawing dragons, and ogling ogres.
ISBN 0-07-046560-6 lib. bdg. : $3.95
[1. Monsters—Fiction. 2. Alphabet] I. Title.
PZ7.N586 Aab 1978 [E] 78-7939
78 MARC

Nivola, Ruth.
The messy rabbit / by Ruth Nivola ; pictures by Claire Nivola. — New York : Pantheon Books, c1978.
34 p. : col. ill. ; 22 cm.
SUMMARY: A messy rabbit finally learns the advantages of being neat.
ISBN 0-394-83764-9. ISBN 0-394-93764-3 lib. bdg.
[1. Cleanliness—Fiction] I. Nivola, Claire. II. Title.
PZ7.N645 Me [E] 78-6018
78 MARC

Nixon, Joan Lowery.
The butterfly tree / by Joan Lowery Nixon ; illustrated by James McIlrath. — Huntington, Ind. : Our Sunday Visitor, [c1979]
[30] p. : col. ill. ; 29 cm. — (An OSV read-along book)
SUMMARY: As her great-grandmother nears the end of her life, young Jennifer gives her a special gift.
ISBN 0-87973-355-1
[1. Grandmothers—Fiction. 2. Death—Fiction. 3. Monarch butterfly—Fiction. 4. Butterflies—Fiction] I. McIlrath, James. II. Title.
PZ7.N65 Bu [E] 78-71264
79[r79]rev MARC

Nixon, Joan Lowery.
The halloween mystery / Joan Lowery Nixon ; pictures by Jim Cummins. — Chicago : A. Whitman, 1979.
32 p. : ill. ; 21 cm. — (First read-alone mysteries)
SUMMARY: When they are left with a mysterious, witchlike babysitter on Halloween, Susan and Mike try to figure out whether she really has magical powers.
ISBN 0-8075-3136-7 : $4.95
[1. Mystery and detective stories. 2. Halloween—Fiction] I. Cummins, James. II. Title.
PZ7.N65 Hal 1979 [E] 79-166
79 MARC

Nixon, Joan Lowery.
The kidnapping of Christina Lattimore / Joan Lowery Nixon. — 1st ed. — New York : Harcourt Brace Jovanovich, c1979.
179 p. ; 21 cm.
SUMMARY: A teenage girl is kidnapped, but when freed, is accused of masterminding the scheme to extort money from her wealthy grandmother.
ISBN 0-15-242657-4
[1. Kidnapping—Fiction. 2. Mystery and detective stories] I. Title.
PZ7.N65 Ki [Fic] 78-20570
78 MARC

CHILDREN'S LITERATURE (Continued)

Nixon, Joan Lowery.
Muffie Mouse and the busy birthday / by Joan Lowery Nixon ; illustrated by Geoffrey Hayes. — New York : Seabury Press, c1978.
46 p. : ill. ; 21 cm.
"A Clarion book."
SUMMARY: Muffie displays her imagination in four episodes that take place on her mother's birthday.
ISBN 0-8164-3213-9
[1. Birthdays—Fiction. 2. Mice—Fiction] I. Hayes, Geoffrey. II. Title.
PZ7.N65 Mn [E] 78 77-28866 MARC

Nixon, Joan Lowery.
The New Year's mystery / Joan Lowery Nixon ; pictures by Jim Cummins. — Chicago : A. Whitman, 1979.
31 p. : ill. ; 21 cm. — (First read-alone mysteries)
SUMMARY: Susan and Mike use their detective skills to figure out what happened to the missing ingredients for their New Year's party refreshments.
ISBN 0-8075-5592-4 : $4.95
[1. Mystery and detective stories. 2. New Year—Fiction] I. Cummins, James. II. Title.
PZ7.N65 Ne [E] 79 79-172 MARC

Noble, June.
Two homes for Lynn / June Noble ; illustrated by Yuri Salzman. — New York : Holt, Rinehart and Winston, c1979.
[32] p. : ill. (some col.) ; 24 cm.
SUMMARY: Lynn's make-believe friend helps her adjust to her two homes after her parents' divorce.
ISBN 0-03-046186-3 (lib. bdg.) : $6.95
[1. Divorce—Fiction] I. Salzman, Yuri. II. Title.
PZ7.N669 Tw [E] 78 78-14090 MARC

The Noel Streatfeild weekend story book : an anthology. — London : J. M. Dent, 1977.
153 p. : ill. ; 14 cm. GB***
CONTENTS: Morpurgo, M. Sam's duck.—Johnstone, H. The first winter of a filly.—Aiken, J. Who would not sing for Lycidas?—Cousins, P. Battle of the allies.—Farmer, P. A tall story.—Farjeon, A. Song of the captured princess.—Williams, U. M. Family feeling.—Macleod, I. The battle.—Streatfeild, N. Green silk.—Samjar, R. The desert.—Treadgold, M. The cornfield.—Hughes, T. A moon-lily.—Williams, J. The question box.—Patience, I. Mr. Plintyloolooloock-ton.—Smith, S. S. The burning mountain.—Willcox, L. Thoughts.—Mahy, M. The man who lived in a tree.—Marjoram, J. The wounded tree.—Cooper, L. The difficult one.—Thompson, S. Cold.—Simpson, J. M. Dragon marmalade.—MacGill, P. The farmer's boy.—Roberts, C. G. D. Gentleman in feathers.—Barton, S. Heron.
ISBN 0-460-06766-4 : £3.50
1. Children's literature. 2. Children's writings, English. I. Streatfeild, Noel.
PZ5.N54 820'.8'09282 79 79-309618 MARC

Nolan, Madeena Spray.
My daddy don't go to work / Madeena Spray Nolan ; ill. by Jim LaMarche. — Minneapolis : Carolrhoda Books, c1978.
[32] p. : ill. ; 20 cm.
SUMMARY: A family is troubled by Daddy's unemployment.
ISBN 0-87614-093-2
[1. Family life—Fiction. 2. Unemployed—Fiction] I. LaMarche, Jim. II. Title.
PZ7.N69 My 1978 [E] 78 78-53640 MARC

Noodles.
The phantom athlete : a diary of great events in my life / by your friend Noodles ; illustrated by Ted Rand. — New York : Holt, Rinehart, and Winston, c1979.
[32] p. : col. ill. ; 17 cm.
SUMMARY: A versatile athlete enumerates his outstanding feats.
ISBN 0-03-047346-2
[1. Sports stories] I. Rand, Ted. II. Title.
PZ7.N738 Ph 1979 E 00 79-117430 MARC

Norton, Andre.
The defiant agents / Andre Norton. — Boston : Gregg Press, 1978 i.e. 1979, c1962.
190 p. ; 21 cm. — (Her The time traders series)
Reprint of the ed. published by Ace Books, New York.
SUMMARY: A group of space agents run into opposition while attempting to carry out its mission of colonizing a planet.
ISBN 0-8398-2423-6
[1. Science fiction. 2. Space and time] I. Title. II. Series.
[PZ7.N82 De 1979] [Fic] 78 78-20923 MARC

Norton, Andre.
Galactic derelict / Andre Norton. — Boston : Gregg Press, 1978 i.e. 1979, c1959.
192 p. ; 21 cm. — (Her The time traders series)
Reprint of the ed. published by Ace Books, New York.
SUMMARY: Three men undertake a perilous journey to investigate an alien spaceship wrecked during prehistoric times.
ISBN 0-8398-2422-X
[1. Science fiction] I. Title. II. Series.
[PZ7.N82 Gal 1979] [Fic] 78 78-26944 MARC

Norton, Andre.
Key out of time / Andre Norton. — Boston : Gregg Press, 1978 i.e. 1979, c1963.
188 p. ; 21 cm. — (Her The time traders series)
Reprint of the ed. published by Ace Books, New York.
SUMMARY: Ross Murdock, Gordon Ashe, and other members of the Terran settlement team travel back in time to investigate a civilization that once flourished on Hawaika, a world of shallow seas and archipelagoes.
ISBN 0-8398-2424-6
[1. Space and time—Fiction. 2. Science fiction] I. Title. II. Series.
[PZ7.N82 Ke 1978] [Fic] 78 78-27244 MARC

Norton, Andre.
Seven spells to Sunday / Andre Norton and Phyllis Miller. — 1st ed. — New York : Atheneum, 1979.
136 p. ; 22 cm.
"A Margaret K. McElderry book."
SUMMARY: Two children receive messages in an old mailbox in a vacant lot and are plunged into the strangest week of their lives.
ISBN 0-689-50134-X
[1. Fantasy] I. Miller, Phyllis, joint author. II. Title.
PZ7.N82 Se [Fic] 78 78-24362 MARC

Norton, Andre.
Star Ka'ats and the plant people / Andre Norton and Dorothy Madlee ; illustrated by Jean Jenkins. — New York : Walker, 1979.
122 p. : ill. ; 22 cm.
SUMMARY: Follows the adventures of two children and a super race of cats as they rescue a group of plant people.
ISBN 0-8027-6342-1 : $6.95. ISBN 0-8027-6343-X lib. bdg. : $6.95
[1. Science fiction] I. Madlee, Dorothy H., joint author. II. Jenkins, Jean. III. Title.
PZ7.N82 Ssy 1979 [Fic] 78 78-67177 MARC

Norton, Andre.
The time traders / Andre Norton. — Boston : Gregg Press, 1978 ie. 1979, c1958.
xiv, 220 p. ; 21 cm. — (Her The time traders series)
Reprint of the ed. published by Ace Books, New York.
SUMMARY: In an effort to discover the source of the alien technology the Russians are using, Ross Murdock travels back in time to Britain in the year 2000 B.C. to uncover the Russian base of operations.
ISBN 0-8398-2421-1
[1. Space and time—Fiction. 2. Science fiction] I. Title. II. Series.
[PZ7.N82 Ti 1979] [Fic] 78 78-26925 MARC

Nöstlinger, Christine.
Marrying off mother / [by] Christine Nostlinger. ; translated by Anthea Bell. — London : Andersen Press, 1978.
140 p. ; 21 cm. GB***
Translation of Ein Mann für Mama.
ISBN 0-905478-38-X : £2.95
I. Title.
PZ7.N672 Mar 1978 79 79-300704 MARC

Noyes, Beppie.
Mosby, the Kennedy Center cat / by Beppie Noyes. — Washington : Acropolis Books, c1978.
x, 127 p. : ill. ; 24 cm.
SUMMARY: The chief ratcatcher at a performing arts center enjoys a very private and culturally rich life.
ISBN 0-87491-263-6 : $7.95
[1. Cats—Fiction. 2. United States. John F. Kennnedy Center for the Performing Arts—Fiction] I. United States. John F. Kennedy Center for the Performing Arts. II. Title.
PZ7.N9615 Mo [Fic] 78 78-12640 MARC

Oakley, Graham.
The church mice at bay / Graham Oakley. — London : Macmillan, 1978.
36 p. : col. ill. ; 21 x 27 cm. GB***
ISBN 0-333-23235-6 : £3.95
I. Title.
PZ7.O1048 Cfd 1978 79 79-313029 MARC

Oakley, Graham.
The church mice at bay / Graham Oakley. — 1st American ed. — New York : Atheneum, 1979, c1978.
[36] p. : col. ill. ; 21 x 26 cm.
SUMMARY: Samson the church cat and all the church mice devise a campaign to rid the vicarage of the substitute vicar.
ISBN 0-689-30629-6
[1. Mice—Fiction] I. Title.
PZ7.O1048 Cec 1979 [E] 79 78-62260 MARC

Oana, Kay D
The little dog who wouldn't be / illustrated by Robert Russo ; written by Kay D. Oana. — Fayetteville, Ga. : Oddo Pub., c1978.
32 p. : col. ill. ; 25 cm.
SUMMARY: A plain brown dog is convinced that he would be happier if only he could be something else.
ISBN 0-87783-150-5
[1. Dogs—Fiction. 2. Identity—Fiction] I. Russo, Robert. II. Title.
PZ7.O1055 Li 79 77-18351 MARC

Oana, Kay D
Robbie and the raggedy scarecrow / written by Kay D. Oana ; illustrated by Jackie Stephens. — Fayetteville, Ga. : Oddo Publ., c1978.
32 p. : col. ill. ; 25 cm.
SUMMARY: A young boy thinks of a way to prevent his grandmother's cherry tree from being chopped down.
ISBN 0-87783-154-8
[1. Scarecrows—Fiction] I. Stephens, Jackie. II. Title.
PZ7.O1055 Ro 79 77-18349 MARC

Oana, Kay D
Shasta and the shebang machine / written by Kay D. Oana ; illustrated by Jackie Stephens. — Fayetteville, Ga. : Oddo Pub., c1978.
32 p. : col. ill. ; 25 cm.
SUMMARY: Shasta the kitten wants very much to make her mistress's new typewriter go clickety-click shebang.
ISBN 0-87783-152-1
[1. Typewriters—Fiction. 2. Cats—Fiction] I. Stephens, Jackie. II. Title.
PZ7.O1055 Sh [E] 79 77-18350 MARC

O'Brine, Manning, 1913-1977.
Dagger before me. London, Hammond, Hammond [1957]
254 p. 19 cm. (A Cloak and dagger mystery)
I. Title.
PZ5.O137 Dag 57-35807
[PR6065.B75] [r79]rev MARC

O'Connor, Jane.
Yours til Niagara Falls, Abby / by Jane O'Connor ; illustrated by Margaret Apple. — New York : Hastings House, [1979]
xiv, 141 p. ; 24 cm.
Includes bibliographical references.
SUMMARY: Ten-year-old Abby faces two months of summer camp without her best friend.
ISBN 0-394-50551-4
[1. Camping—Fiction. 2. Friendship—Fiction] I. Apple, Margaret. II. Title.
PZ7.O222 Yo [Fic] 79 79-19782 MARC

Okimoto, Jean Davies.
My mother is not married to my father / Jean Davies Okimoto. — New York : Putnam, 1979.
109 p. ; 21 cm.
SUMMARY: Eleven-year-old Cynthia and her six-year-old sister try to adjust to their parents' separation and divorce.
ISBN 0-399-20664-7 : $7.95
[1. Divorce—Fiction] I. Title.
PZ7.O415 My 1979 [Fic] 78 78-26924 MARC

Olds, Helen Diehl, 1895-
Kate can skate. Illustrated by Carol Beech. New York, Knopf [1960]
[33] p. illus. 23 cm.
SUMMARY: Seven-year-old Kate who lives in the country where there are no sidewalks learns to skate at her grandmother's house in the city.
[1. Roller skating—Fiction] I. Beech, Carol. II. Title.
PZ7.O46 Kat [E] [r79]rev 60-5513 MARC

Oliver, Symmes Chadwick, 1928-
Mists of dawn / by Chad Oliver. — Boston : Gregg Press, 1979, c1952.
xv, 208 p. : ill. ; 21 cm. — (The Gregg Press science fiction series) (Reprint of the ed. published by Holt, Rinehart and Winston, Philadelphia, in series: A Science fiction novel)
SUMMARY: A 17-year-old makes an unplanned trip through space and time to Europe 50,000 years ago where Neanderthal and Cro-Magnon man engage in conflict for survival.
ISBN 0-8398-2520-X
[1. Space and time—Fiction. 2. Science fiction] I. Title.
[PZ7.O494 Mi 1979] [Fic] 79 79-1151 MARC

Oneal, Zibby.
Turtle and snail / by Zibby Oneal ; pictures by Margot Tomes. — Philadelphia : Lippincott, c1979.
47 p. : col. ill. ; 24 cm. — (A Lippincott I-like-to read book)
SUMMARY: Five episodes in which a turtle and a snail demonstrate their fondness for and loyalty to each other.
ISBN 0-397-31829-4 (lib. bdg.) : $6.95
[1. Snails—Fiction. 2. Turtles—Fiction. 3. Friendship—Fiction] I. Tomes, Margot. II. Title.
PZ7.O552 Tu [E] 78 78-14826 MARC

O'Reilly, Edward.
Brown pelican at the pond / by Edward O'Reilly ; illustrated by Florence Strange. — San Rafael, Calif. : Manzanita Press, c1979.
[32] p. : ill. ; 15 x 21 cm.
SUMMARY: A young boy adopts an ailing pelican knowing that someday the bird must be returned to its flock.
ISBN 0-931644-01-1 : $5.95
[1. Pelicans—Fiction. 2. Children's writings] I. Strange, Florence. II. Title.
PZ7.O628 Br [E] 79 78-58689 MARC

CHILDREN'S LITERATURE (Continued)

Oreshnik, A F
The Demeter Star / A. F. Oreshnik. — Belmont, Calif. : Fearon-Pitman Publishers, c1977.
60 p. : ill. ; 18 cm. — (A Pacemaker bestseller book)
SUMMARY: On his first ocean dive, Mike discovers a sunken cargo ship that he later learns may be carrying teasure in silver.
ISBN 0-8224-5270-4
[1. Buried treasure—Fiction. 2. Skin diving—Fiction] I. Title.
[PZ7.O6316 De 1977] [Fic] 77-77157
79 MARC

Orgel, Doris.
A certain magic / Doris Orgel. — New York : Dial Press, c1976.
176 p. : ill. ; 22 cm.
SUMMARY: The discovery of an old copy book belonging to her aunt leads eleven-year-old Jenny on an unusual search into the past.
ISBN 0-8037-5405-1 : $6.95
I. Title.
PZ7.O632 Ce [Fic] 75-9204
75[r79]rev MARC

Orgel, Doris.
The devil in Vienna / by Doris Orgel. — New York : Dial Press, c1978.
246 p. ; 22 cm.
SUMMARY: A Jewish girl and the daughter of a Nazi have been best friends since they started school, but in 1938 the 13-year-olds find their close relationship difficult to maintain.
ISBN 0-8037-1920-5 : $7.95
[1. Jews in Austria—Fiction. 2. Friendship—Fiction. 3. Austria—History—1938-1945—Fiction] I. Title.
PZ7.O632 De [Fic] 78-51319
78 MARC

Ostrovsky, Vivian.
Mumps! / story by Vivian Ostrovsky ; illustrated by Rose Ostrovosky. — New York : Holt, Rinehart, and Winston, 1978, c1976.
[28] p. : col. ill. ; 23 x 27 cm.
Translation of Caxumba!
SUMMARY: A young girl shares her mumps with her family.
ISBN 0-03-042126-8 lib. bdg. : $6.95
[1. Mumps—Fiction. 2. Sick—Fiction] I. Ostrovsky, Rose. II. Title.
PZ7.O854 Mu 1978 [E] 78-4276
78 MARC

Owen, Dilys.
Leo possessed / Dilys Owen. — London : F. Muller, 1975.
128 p. ; 23 cm. GB***
ISBN 0-584-31162-5 : £2.35
I. Title.
PZ7.O9697 Le 823'.9'14 75-322089
75[r79]rev MARC

Owen, Dilys.
Leo possessed / Dilys Owen ; ill. by Stephen Gammell. — 1st ed. — New York : Harcourt Brace Jovanovich, [1979] c1975.
150 p. : ill. ; 21 cm.
SUMMARY: When her family moves into an old Georgian townhouse, a 13-year-old English girl is distrubed by haunting presences.
ISBN 0-15-244897-7
[1. Ghost stories] I. Gammell, Stephen. II. Title.
PZ7.O9697 Le 1979 [Fic] 78-20572
78 MARC

Oxenbury, Helen.
The queen and Rosie Randall / Helen Oxenbury from an idea by Jill Buttfield-Campbell. — 1st ed. — New York : Morrow, 1979, c1978.
[33] p. : ill. (some col.) ; 25 cm.
SUMMARY: A little girl serves as the Queen's social adviser and acts as joint hostess at a royal tea party.
ISBN 0-688-22171-8. ISBN 0-688-32171-2 lib. bdg.
[1. Parties—Fiction. 2. Humorous stories] I. Buttfield-Campbell, Jill. II. Title.
PZ7.O975 Qe 1979 [E] 78-10375
78 MARC

Packard, Edward, 1931-
The third planet from Altair / Edward Packard ; illustrated by Barbara Carter. — Philadelphia : Lippincott, c1979.
96 p. : ill. ; 21 cm.
SUMMARY: The reader, en route to the third planet from Altair to seek the source and meaning of extraterrestrial messages, is given choices to make determining the course of the spaceship and the survival of its crew.
ISBN 0-397-31827-8 : $6.95
[1. Science fiction] I. Carter, Barbara, 1952- II. Title.
PZ7.P1245 Th [Fic] 78-23998
78 MARC

Palacios, Argentina.
The knight and the squire : a retelling of the adventures of Don Quixote and Sancho Panza, based on Cervantes, Don Quixote de la Mancha / by Argentina Palacios ; illustrated by Ray Cruz. — 1st ed. — Garden City, N.Y. : Doubleday, c1979.
96 p. : ill. ; 27 cm.
SUMMARY: An abridged version of the adventures of a Spanish country gentlemen, considered mad, and his companion who set out as knights of old to right wrongs and punish evil.
ISBN 0-385-12433-3 : $7.95
[1. Knights and knighthood—Fiction. 2. Spain—Fiction] I. Cruz, Ray. II. Cervantes Saavedra, Miguel de, 1547-1616. Don Quixote. III. Title.
PZ7.P1753 Kn [Fic] 78-20091
78 MARC

Palmer, C Everard.
A dog called Houdini / C. Everard Palmer ; illustrated by Maurice Wilson. — London : Deutsch, 1978.
100 p. : ill. ; 21 cm. GB***
ISBN 0-233-96985-3 : £3.25
1. Dogs—Legends and stories. I. Title.
PZ10.3.P1814 Do 823'.9'14 78-321833
79 MARC

Palmer, Michele.
Zoup soup / by Michele Palmer ; ill. by Janine Gugler. — Storrs, Conn. : Rocking Horse Press, c1978.
[24] p. : ill. ; 11 x 17 cm.
SUMMARY: After years of potato soup one night and tomato soup the next, Mr. and Mrs. Zoup vary their menu.
ISBN 0-932306-00-4
[1. Soups—Fiction] I. Gugler, Janine. II. Title.
PZ7.P18558.Zo [E] 78-66342
79 MARC

Panek, Dennis.
Catastrophe cat : story and pictures / by Dennis Panek. — Scarsdale, N.Y. : Bradbury Press, c1978.
[34] p. : chiefly col. ill. ; 29 cm.
SUMMARY: A cat named Catastrophe demonstrates how he got his name and why it is appropriate.
ISBN 0-87888-130-1 : $7.95
[1. Cats—Fiction] I. Title.
PZ7.P1887 Cau [E] 77-90951
79 MARC

Panek, Dennis.
Catastrophe Cat at the zoo : story and pictures / by Dennis Panek. — Scarsdale, N.Y. : Bradbury Press, c1979.
[32] p. : chiefly col. : ill. ; 29 cm.
SUMMARY: Catastrophe Cat gets off a city bus hoping to snooze under a tree, but starts a romp through the zoo instead.
ISBN 0-87888-147-6 : $7.95
[1. Cats—Fiction. 2. Zoological gardens—Fiction] I. Title.
PZ7.P1887 Cat [E] 78-26369
78 MARC

Pape, Donna Lugg.
Doghouse for sale / by Donna Lugg Pape ; drawings by Tom Eaton. — Champaign, Ill. : Garrard Pub. Co., [1979]
40 p. : col. ill. ; 23 cm. — (An Imagination books)
SUMMARY: Freckles the dog fixes up his house in order to attract prospective buyers, but then has second thoughts.
ISBN 0-8116-4415-4 lib. bdg. : $4.20
[1. Dogs—Fiction] I. Eaton, Tom. II. Title.
PZ7.P1978 Do [E] 78-11685
78 MARC

Pape, Donna Lugg.
Where is my little Joey? / By Donna Lugg Pape ; drawings by Tom Eaton. — Champaign, Ill. : Garrard Pub. Co., c1978.
44 p. : col. ill. ; 23 cm. — (An Imagination book)
SUMMARY: A kangaroo's empty pouch has several different uses until her lost baby returns.
ISBN 0-8116-4411-1 lib. bdg. : $4.30
[1. Kangaroos—Fiction] I. Eaton, Tom. II. Title.
PZ7.P1978 Wh [E] 78-1022
78 MARC

Parenteau, Shirley.
Blue hands, blue cloth : story / by Shirley Parenteau ; pictures by Ann Toulmin-Rothe. — Chicago : Childrens Press, c1978.
31 p. : col. ill. ; 24 cm.
SUMMARY: A young African girl is impatient to learn to be an indigo dyer in the tradition of her mother and grandmother.
ISBN 0-516-03414-6
[1. Dyes and dyeing—Fiction. 2. Africa—Fiction] I. Toulmin-Rothe, Ann. II. Title.
PZ7.P2167 Bl [E] 77-16538
77 MARC

Parenteau, Shirley.
Secrets of scarlet / story by Shirley Parenteau ; pictures by Ann Toulmin-Rothe. — Chicago : Children'S Press, c1979.
31 p. : col. ill. ; 24 cm.
SUMMARY: A French girl looks for a way to save her village whose "scarlet bug" industry is in danger.
ISBN 0-516-03614-9
[1. Dyes and dyeing—Fiction. 2. France—Fiction] I. Toulmin-Rothe, Ann. II. Title.
PZ7.P2167 Se [E] 78-23797
78 MARC

Parish, Peggy.
Amelia Bedelia helps out / Peggy Parish ; pictures by Lynn Sweat. — New York : Greenwillow Books, c1979.
63 p. : col. ill. ; 22 cm. — (Greenwillow read-alone)
SUMMARY: Amelia Bedelia shows her niece Effie Lou how to follow instructions to the letter as they dust the potato bugs and sew seeds.
ISBN 0-688-80231-1 : $5.95
[1. Humorous stories] I. Sweat, Lynn. II. Title.
PZ7.P219 Ap [E] 79-11729
79 MARC

Parish, Peggy.
Be ready at eight / by Peggy Parish ; pictures by Leonard Kessler. — New York : Macmillan, c1979.
63 p. : ill. (some col.) ; 23 cm. — (Ready-to-read)
SUMMARY: Absent-minded Miss Molly tries to remember why she tied a string around her finger—and why all her friends plan to see her at eight.
ISBN 0-02-769830-0
[1. Memory—Fiction. 2. Birthdays—Fiction] I. Kessler, Leonard P., 1920- II. Title.
PZ7.P219 Bc [E] 78-11847
78 MARC

Parish, Peggy.
Zed and the monsters / by Peggy Parish ; Paul Galdone drew the pictures. — 1st ed. — Garden City, N.Y. : Doubleday, c1979.
[64] p. : col. ill. ; 25 cm. — (A Reading-on-my-own book)
SUMMARY: Zed, a lazy backwoods boy, meets four monsters and cleverly outwits them.
ISBN 0-385-12948-3 : $4.95. ISBN 0-385-12949-1 lib. bdg.
[1. Monsters—Fiction] I. Galdone, Paul. II. Title.
PZ7.P219 Ze [E] 77-15876
77 MARC

Park, Ruth.
Come danger, come darkness / [by] Ruth Park. — Sydney : Hodder and Stoughton, 1978.
162 p. : ill. ; 22 cm. Aus
ISBN 0-340-22746-X
I. Title.
PZ7.P2214 Co 79-300453
79 MARC

Parker, Lois M
Brothers of the Longhouse / Lois M. Parker ; [editor, Thomas A. Davis ; ill., Terry Crews]. — Washington : Review and Herald Pub. Association, [1979]
124 p. : ill. ; 21 cm.
SUMMARY: Two Quaker boys are forcibly taken from Penn's young settlement of Philadelphia and made guests of the Onondaga Indians for a reason no one will reveal to them.
[1. Onondaga Indians—Fiction. 2. Indians of North America—Pennsylvania—Fiction. 3. Pennsylvania—History—Colonial period, ca. 1600-1775—Fiction. 4. Frontier and pioneer life—Pennsylvania—Fiction] I. Davis, Thomas A. II. Crews, Terry, 1950- III. Title.
PZ7.P227 Br [Fic] 79-13190
79 MARC

Parker, Nancy Winslow.
The ordeal of Byron B. Blackbear / by Nancy Winslow Parker ; with drawings by the author. — 1st ed. — New York : Dodd, Mead, c1979.
[70] p. : ill. (some col.) ; 19 x 24 cm.
Includes bibliography.
SUMMARY: A world-famous scientist's study of a hibernating bear produces surprising results.
ISBN 0-396-07642-4 lib. bdg. : $6.95
[1. Bears—Fiction. 2. Scientists—Fiction] I. Title.
PZ7.P2274 Or [E] 78-12140
78 MARC

Parkinson, Ethelyn M
Rupert Piper & the boy who could knit / Ethelyn M. Parkinson ; illustrated by Jim Padgett. — Nashville : Abingdon, c1979.
159 p. : ill. ; 23 cm.
SUMMARY: Everyone in Wakefield has problems. Maybe Shirley, the boy who loves to knit, has some answers.
ISBN 0-687-36654-2
[1. Humorous stories] I. Padgett, Jim. II. Title.
PZ7.P2393 Rum [Fic] 78-24001
78 MARC

Parr, Jhon.
Jim Tweed / John Parr. — Winnipeg, Man. : Queenston House Pub. Co., c1978.
303 p. ; 22 cm. C***
ISBN 0-919866-12-3
I. Title.
PZ7.P2444 Ji 78-318453
78 MARC

Pascal, Francine.
My first love & other disasters / Francine Pascal. — 1st ed. — New York : Viking Press, 1979.
186 p. ; 22 cm.
SUMMARY: Fourteen-year-old Victoria attracts the attention of the boy she likes, but discovers her life is still full of problems.
ISBN 0-670-49952-8
I. Title.
PZ7.P2678 My 1979 [Fic] 78-25720
78 MARC

Paterson, Diane, 1946-
The bathtub ocean / Diane Paterson. — New York : Dial Press, c1979.
[32] p. : col. ill. ; 14 x 19 cm.
SUMMARY: Henry discovers deep sea adventures in his bathtub.
ISBN 0-8037-0460-7
[1. Baths—Fiction. 2. Imagination—Fiction] I. Title.
PZ7.P2727 Bat [E] 78-72517
78 MARC

Paton Walsh, Jill, 1937-
A chance child / Jill Paton Walsh. — London : Macmillan, 1978.
158 p. ; 21 cm. GB***
ISBN 0-333-23833-8 : £3.95
I. Title.
PZ7.P2735 Cg 1978b 79-311212
79 MARC

CHILDREN'S LITERATURE (Continued)

Paton Walsh, Jill, 1937-
A chance child / Jill Paton Walsh. — 1st American ed. — New York : Farrar, Straus and Giroux, 1978.
185 p. ; 22 cm.
SUMMARY: Compelled to search for his half brother Creep who some people insist is nonexistent, Christopher locates Parliamentary Papers containing Nathaniel Creep's personal narrative of working conditions during the Industrial Revolution 100 years earlier.
ISBN 0-374-31236-2 : $7.95
[1. Children—Employment—Fiction. 2. Great Britain—History—19th century—Fiction. 3. Space and time—Fiction] I. Title.
PZ7.P2735 Cg 1978 [Fic] 78 78-21521 MARC

Paton Walsh, Jill, 1937-
Children of the fox / Jill Paton Walsh ; pictures by Robin Eaton. — New York : Farrar, Straus and Giroux, 1978.
115 p. : ill. ; 22 cm.
CONTENTS: Crossing to Salamis.—The walls of Athens.—Persian gold.
ISBN 0-374-31242-7 : $7.95
1. Themistokles, ca. 524-ca. 459 B.C.—Juvenile fiction. [1. Themistokles, ca. 524-ca. 459 B.C.—Fiction. 2. Greece—History—Persian Wars, 500-449 B.C.—Fiction. 3. Short stories] I. Title.
PZ7.P2735 Ch 1978 [Fic] 78 78-8138 MARC

Paton Walsh, Jill, 1937-
Crossing to Salamis / Jill Paton Walsh ; illustrated by David Smee. — London : Heinemann, 1977.
43 p. ; 21 cm. — (Long ago children books) GB77-19054
ISBN 0-434-94925-6 : £2.10
1. Themistokles, ca. 524-ca. 459 B.C.—Juvenile fiction. I. Title.
PZ7.P2735 Cr 79 79-313026 MARC

Paton Walsh, Jill, 1937-
Crossing to Salamis / Jill Paton Walsh ; illustrated by David Smee. — London : Heinemann, 1977.
43 p. : ill. ; 21 cm. — (Long ago children books) GB77-19054
ISBN 0-434-94925-6 : £2.10
1. Themistokles, ca. 524-ca. 459 B.C.—Juvenile fiction. I. Title.
PZ7.P2735 Cr 79 79-313026 MARC

Paton Walsh, Jill, 1937-
The emperor's winding sheet. [1st American ed.] New York, Farrar, Straus and Giroux [1974]
273 p. 21 cm.
SUMMARY: An English boy, shipwrecked, hungry, and lost, finds his way into the court of Constantine where he is interpreted as a symbol of good luck and, as such, ordered to be kept always near the king.
ISBN 0-374-32160-4
1. Byzantine Empire—History—Constantine XI Dragases, 1448-1453—Juvenile fiction. [1. Byzantine Empire—History—Constantine XI Dragases, 1448-1453—Fiction. 2. Istanbul—Siege, 1453—Fiction. 3. Turkey—History—Ottoman Empire, 1288-1918—Fiction] I. Title.
PZ7.P2735 Em 1974 [Fic] 74[r79]rev 73-90970 MARC

Paton Walsh, Jill, 1937-
Fireweed. Harmondsworth, Penguin, 1972.
127 p. 18 cm. (Puffin books) £0.20 B72-29514
SUMMARY: Two teen-age runaways who refuse to be evacuated from London struggle to survive the blitz of 1940.
ISBN 0-14-030560-2
1. London—Bombardment, 1940—Juvenile fiction. [1. London—Bombardment, 1940—Fiction. 2. World War, 1939-1945—Great Britain—Fiction] I. Title.
PZ7.P2735 Fi 1972 [Fic] 73[r79]rev2 73-330950 MARC

Paton Walsh, Jill, 1937-
The walls of Athens / Jill Paton Walsh ; illustrated by David Smee. — London : Heinemann, 1977.
41 p. : ill. ; 21 cm. — (Long ago children books) GB***
ISBN 0-434-94931-0 : £2.50
1. Themistokles, ca. 524-ca. 459 B.C.—Juvenile fiction. I. Smee, David. II. Title.
PZ7.P2735 Wal 79 79-313025 MARC

Patterson, Lillie.
The grouchy Santa / by Lillie Patterson ; drawings by Lou Cunette. — Champaign, Ill. : Garrard Pub. Co., c1979.
39 p. : col. ill. ; 23 cm. — (First holiday books)
SUMMARY: Mr. Grouch doesn't like Christmas and doesn't even believe in Santa Claus until Santa takes him around the world on Christmas Eve.
ISBN 0-8116-7254-9 lib. bdg. : $4.20
[1. Christmas stories] I. Cunette, Lou. II. Title. III. Series.
PZ7.P2768 Gr [E] 78 78-21936 MARC

Patterson, Lillie.
The jack-o'lantern trick / by Lillie Patterson ; drawings by William Hutchinson. — Champaign, Ill. : Garrard Pub. Co., 1979.
38 p. : col. ill. ; 23 cm. — (First holiday books)
SUMMARY: When unfriendly Indians threaten them and their home, two pioneer girls use their Halloween jack-o-lanterns to good advantage.
ISBN 0-8116-7250-6 (lib. bdg.) : $4.48
[1. Halloween—Fiction. 2. Frontier and pioneer life—Fiction] I. Hutchinson, William M. II. Title. III. Series.
PZ7.P2768 Jac [E] 78 78-11307 MARC

Patterson, Lillie.
Jenny, the Halloween spy / by Lillie Patterson ; drawings by Diane Dawson. — Champaign, Ill. : Garrard Pub. Co., c1979.
37 p. : col. ill. ; 23 cm.
SUMMARY: Jenny's curiosity leads her into a world of fairies and elves and on a wild Halloween ride with the Devil and his headless hounds.
ISBN 0-8116-7251-4 lib. bdg. : $4.20
[1. Halloween—Fiction] I. Dawson, Diane. II. Title.
PZ7.P2768 Je [E] 78 78-11538 MARC

Paulsen, Gary.
The night the white deer died / Gary Paulsen. — 1st ed. — Nashville : T. Nelson, c1978.
96 p. ; 21 cm.
SUMMARY: A teenage girl and an old Indian are brought together by the same haunting dream.
ISBN 0-8407-6616-5 : $6.50
[1. Indians of North America—Fiction. 2. Dreams—Fiction] I. Title.
PZ7.P2843 Ni [Fic] 78 78-12641 MARC

Pearce, Ann Philippa.
The battle of Bubble and Squeak / Philippa Pearce ; illustrated by Alan Baker. — London : Deutsch, 1978.
88 p. : ill. ;21 cm. GB***
ISBN 0-233-96986-1 : £2.50
I. Title.
PZ7.P3145 Bat 1978 78 78-318736 MARC

Pearce, Ann Philippa.
The shadow-cage, and other tales of the supernatural / Philippa Pearce ; illustrated by Janet Archer. — Harmondsworth : Kestrel Books, 1977.
142 p. : ill. ; 23 cm. GB77-15555
CONTENTS: The shadow-cage.—Miss Mountain.—Guess.—At the river-gates.—Her father's attic.—The running-companion.—Beckoned.—The dear little man with his hands in his pockets.—The dog got them.—The strange illness of Mr. Arthur Cook.
1. Supernatural—Juvenile fiction. I. Archer, Janet. II. Title.
PZ7.P3145 Sh 1977 79 79-314268 MARC

Pearson, Jeanne.
Pony of the Sioux, by M. J. Pearson. Illustrated by Carl Pfeufer. [1st ed.] Garden City, N.Y., Doubleday, 1961.
143 p. illus. 22 cm. (A Signal book)
I. Title.
PZ7.P32313 Pr [r78]rev 61-9756 MARC

Pearson, Jeanne.
Ride the red-eyed wind / by M. J. Pearson ; illustrated by Charles E. Pearson. — Minneapolis : Dillon Press, c1978.
175 p. : ill. ; 22 cm.
SUMMARY: In the summer of 1862 a young Sioux warrior returns to the land of his people in southern Minnesota to follow a vision which comes to pass in an unexpected way.
ISBN 0-87518-157-0 : $8.95
1. Dakota Indians—Wars—1862-1865—Juvenile fiction. [1. Dakota Indians—Wars—1862-1865—Fiction. 2. Indians of North America—Wars—1862-1865—Fiction] I. Pearson, Charles. II. Title.
PZ7.P32313 Ri [Fic] 78 78-2206 MARC

Pearson, Susan.
Everybody knows that! / By Susan Pearson ; pictures by Diane Paterson. — New York : Dial Press, c1978.
[32] p. : col. ill. ; 24 cm.
SUMMARY: A young girl decides to teach her playmate a lesson after he lets her know she can't do certain things because she's a girl.
ISBN 0-8037-2417-9 : $6.95. ISBN 0-8037-2418-7 lib. bdg. : $7.45
[1. Sex role—Fiction] I. Paterson, Diane, 1946- . II. Title.
PZ7.P323316 Ev [E] 78 78-51311 MARC

Pearson, Susan.
Monday I was an alligator / by Susan Pearson ; pictures by Sal Murdocca. — Philadelphia : Lippincott, c1979.
40 p. : col. ill. ; 24 cm. — (A Lippincott I-like-to-read book)
SUMMARY: Every day Emily's imaginative play brings despair to a different member of her family.
ISBN 0-397-31830-8 (lib. bdg.) : $6.95
[1. Behavior—Fiction. 2. Play—Fiction] I. Murdocca, Sal. II. Title.
PZ7.P323316 Ml [E] 78 78-23618 MARC

Peck, Richard.
Father figure : a novel / by Richard Peck. — 1st ed. — New York : Viking Press, 1978.
192 p. ; 22 cm.
SUMMARY: After being a father figure for years, Jim and his younger brother are reunited with their divorced father and Jim is forced to find a new role for himself.
ISBN 0-670-30930-3
[1. Brothers and sisters—Fiction. 2. Fathers and sons—Fiction] I. Title.
PZ7.P338 Fat 1978 [Fic] 78 78-7909 MARC

Peck, Robert Newton.
Basket case / Robert Newton Peck. — Garden City, N.Y. : Doubleday, c1979.
94 p. ; 22 cm.
SUMMARY: Higbee Hartburn, Graffiti Prep's foremost prankster, goes to great lengths to provide the school with a victorious basketball team.
ISBN 0-385-14361-3 : $5.95. ISBN 0-385-14362-1 lib. bdg.
[1. Basketball—Fiction. 2. School stories. 3. Humorous stories] I. Title.
PZ7.P339 Bas [Fic] 78 78-60298 MARC

Peck, Robert Newton.
Clunie / Robert Newton Peck. — New York : Knopf : distributed by Random House, c1979.
124 p. ; 22 cm.
SUMMARY: A teenage boy risks his own popularity to give friendship and support to a retarded girl who is harassed by her classmates.
ISBN 0-394-84166-2
[1. Friendship—Fiction. 2. Mentally handicapped—Fiction] I. Title.
PZ7.P339 Cl 1979 [Fic] 78 78-24335 MARC

Peck, Robert Newton.
Hub / Robert Newton Peck ; illustrated by Ted Lewin. — New York : Knopf : distributed by Random House, c1979.
104 p. : ill. ; 22 cm.
SUMMARY: When their teacher enters the annual Chump's Landing Overland Obstacle Bicycle Race, Horrace Hubert and his best friend Spooner Jitwell do everything they can to make sure she wins the silver trophy.
ISBN 0-394-83968-4. ISBN 0-394-93968-9 lib. bdg.
[1. Bicycle racing—Fiction] I. Lewin, Ted. II. Title.
PZ7.P339 Hu 1979 [Fic] 78 78-11763 MARC

Peck, Robert Newton.
Mr. Little / Robert Newton Peck ; illustrated by Ben Stahl. — 1st ed. — Garden City, N.Y. : Doubleday, c1979.
87 p. : ill. ; 22 cm.
SUMMARY: Greatly disappointed that the celebrated Miss K will not be their teacher, Finley Streeter and Stanley Dragavich pull pranks on the new teacher in their small town.
ISBN 0-385-13657-9 : $6.95
[1. Humorous stories] I. Stahl, Ben, 1910- . II. Title.
PZ7.P339 Mk [Fic] 78 78-22347 MARC

Peck, Robert Newton.
Trig sees red / by Robert Newton Peck ; illustrated by Pamela Johnson. — 1st ed. — Boston : Little, Brown, c1978.
59 p. : ill. ; 21 cm.
SUMMARY: An official Junior G-man takes matters into her own hands when Clodburg's only uniformed policemen policeman is replaced by a traffic light.
ISBN 0-316-69656-0 lib. bdg. : $6.95
[1. Humorous stories] I. Johnson, Pamela. II. Title.
PZ7.P339 Tt [Fic] 78 78-18348 MARC

Peet, Bill.
Cowardly Clyde / Bill Peet. — Boston : Houghton Mifflin, 1979.
38 p. : col. ill. ; 26 cm.
SUMMARY: For a war horse, Clyde is an abysmal coward, but he finally decides that even if he isn't brave he can at least act brave.
ISBN 0-395-27802-3 : $8.95
[1. Courage—Fiction. 2. Horses—Fiction] I. Title.
PZ7.P353 Co [E] 78 78-24343 MARC

Pegram, Lorna.
Blood and fire / by Lorna Pegram. — London : Gollancz, 1978.
192 p. ; 21 cm. GB78-30980
ISBN 0-575-02410-0 : £4.95
I. Title.
PZ7.P3583 Bl 79 79-318347 MARC

Peña de Bordas, Virginia de.
La princesa de los cabellos platinados ; La eracra de oro / Virginia de Peña de Bordas ; [presentación, Sócrates Nolasco, Fabio Mota]. — [Santo Domingo] : Editora Taller, [1978]
44 p., [1] leaf of plates : port. ; 30 cm.
I. Peña de Bordas, Virginia de. La eracra de oro. 1978. II. Title.
PZ73.P463 79 78-111080 MARC

Perl, Lila.
Don't ask Miranda / Lila Perl. — New York : Seabury Press, c1979.
164 p. ; 21 cm.
"A Clarion book."
SUMMARY: When Miranda realizes she is expected to cheat to help her candidate win the school election, she must make some painful choices.
ISBN 0-8164-3229-5
[1. Elections—Fiction. 2. School stories] I. Title.
PZ7.P432 Do [Fic] 78 78-23835 MARC

Perl, Susan, 1922-
Susan Perl's Color wheel / by Peter Robinson. — New York : Platt & Munk, c1979.
[20] p. : col. ill. ; 32 cm. — (A Cricket book)
SUMMARY: Describes the six colors of the rainbow and lists objects that display these colors.
ISBN 0-448-46529-9 : $1.95
[1. Color] I. Robinson, Peter, 1951- . II. Title. III. Title: Color wheel.
PZ7.P4323 Co 1979 [E] 79 78-64439 MARC

Peterson, Esther Allen.
Frederick's alligator / Esther Allen Peterson ; illustrated by Susanna Natti. — New York : Crown Publishers, c1979.
[32] p. : col. ill. ; 19 x 21 cm.
SUMMARY: Frederick has always claimed to have wild animals in his house. Then one day he really does become the owner of a baby alligator.
ISBN 0-517-53597-1
[1. Alligators—Fiction. 2. Imagination—Fiction] I. Natti, Susanna. II. Title.
PZ7.P4434 Fr [E] 78 78-15597 MARC

CHILDREN'S LITERATURE (Continued)

Peterson, Jeanne Whitehouse.
That is that / by Jeanne Whitehouse Peterson ; pictures by Deborah Ray. — 1st ed. — New York : Harper & Row, c1979.
[32] p. : col. ill. ; 21 x 26 cm.
SUMMARY: When their father leaves home, Emma Rose and Meko devise their own ways of remembering and understanding.
ISBN 0-06-024708-8 : $7.95. ISBN 0-06-024709-6 (lib. bdg.) : $7.89
[1. Single-parent family—Fiction] I. Ray, Deborah. II. Title.
PZ7.P4444 Th 1979 [Fic] 77-25676
78 MARC

Pevsner, Stella.
And you give me a pain, Elaine / Stella Pevsner. — New York : Seabury Press, c1978.
182 p. ; 22 cm.
"A Clarion book."
SUMMARY: Andrea feels her parents unfairly focus most of their attention on her trouble-making older sister.
ISBN (invalid) 0-18-643219-8
[1. Brothers and sisters—Fiction. 2. Family life—Fiction] I. Title.
PZ7.P44815 An [Fic] 78-5857
78 MARC

Peyton, K M pseud.
Prove yourself a hero / K. M. Peyton. — 1st American ed. — New York : Collins+World, 1978, c1977.
182 p. ; 22 cm.
SUMMARY: Cycling home after his guitar lesson, Jonathan becomes a kidnapping victim.
ISBN 0-529-05452-3 : $6.95
[1. Kidnapping—Fiction] I. Title.
PZ7.P4483 Pr 1978 [Fic] 78-18802
78 MARC

Pfeffer, Susan Beth, 1948-
Awful Evelina / by susan Beth Pfeffer ; pictures by Diane Dawson. — Chicago : A Whitman, 1979.
[32] p. : ill. (some col.) ; 24 cm.
SUMMARY: Meredith, dreading an encounter with her cousin who always hits her and steps on her toes, fantasizes some imaginative disasters that might prevent their visit.
ISBN 0-8075-0494-7 : $5.75
[1. Cousins—Fiction] I. Dawson, Diane. II. Title.
PZ7.P44855 Aw 1979 [E] 79-108
79 MARC

Pfeffer, Susan Beth, 1948-
Starring Peter and Leigh : a novel / Susan Beth Pfeffer. — New York : Delacorte Press, c1979.
200 p. ; 22 cm.
SUMMARY: When her mother remarries, 16-year-old Leigh abandons her acting career and tries to lead the life of a normal teenager. She is coached by her 17-year-old stepbrother, homebound with hemophilia.
ISBN 0-440-08226-9
[1. Remarriage—Fiction. 2. Actors and actresses—Fiction. 3. Hemophilia—Fiction] I. Title.
PZ7.P44855 St [Fic] 78-72855
79 MARC

Phelan, Terry Wolfe.
The S. S. Valentine / by Terry Wolfe Phelan ; illustrated by Judy Glasser. — New York : Four Winds Press, c1979.
40 p. : ill. ; 22 cm.
SUMMARY: A new classmate adds a unique touch to the class St. Valentine's Day play.
ISBN 0-590-07578-0 : $5.95
[1. Physically handicapped—Fiction. 2. St. Valentine's Day—Fiction. 3. School stories] I. Glasser, Judy. II. Title.
PZ7.P4489 Sac [Fic] 79-10037
79 MARC

Phelan, Terry Wolfe.
The week mom unplugged the TVs / by Terry Wolfe Phelan ; illustrated by Joel Schick. — New York : Four Winds Press, c1979.
36 p. : ill. ; 22 cm.
SUMMARY: A parental ban on all television viewing for one week stimulates the curiosity of three youngsters.
ISBN 0-590-07561-6 : $5.95
[1. Play—Fiction. 2. Television—Fiction. 3. Family life—Fiction] I. Schick, Joel. II. Title.
PZ7.P4489 We [Fic] 78-12180
78 MARC

Phillips, Charles Fox.
The Hoity-Toity Mouse, and other bayou tales / Charles Fox Phillips. — 1st ed. — Garden City, N.Y. : Doubleday, c1979.
48 p. : ill. (some col.) ; 27 cm.
CONTENTS: The Hoity-Toity Mouse.—Fleas and cockleburrs.—Looking is free.—Rooting out bad feelings.—The cows who came to tea.
ISBN 0-385-14546-2 : $6.95
[1. Animals—Fiction] I. Title.
PZ7.P5294 Ho [E] 78-22643
79 MARC

Phipson, Joan.
The cats / Joan Phipson. — London ; New York : Macmillan, 1976.
[4], 168, [1] p. ; 21 cm. GB76-33548
ISBN 0-333-19230-3 : £2.95
I. Title.
PZ7.P55204 Cat 78-322281
79 MARC

Phipson, Joan.
Fly free / Joan Phipson. — 1st ed. — New York : Atheneum, 1979.
134 p. ; 22 cm.
"A Margaret K. McElderry book."
SUMMARY: Anxious to go on a school trip to a remote island, two Australian boys with divergent philosophies decide to earn the neccessary money by trapping wild animals—a scheme that badly misfires when one of them becomes involved with unscrupulous poachers.
ISBN 0-689-50149-8
[1. Hunting—Fiction. 2. Conservation of wild animals—Fiction. 3. Australia—Fiction] I. Title.
PZ7.P55204 Fh [Fic] 79-14661
79 MARC

Phipson, Joan.
Keep calm / Joan Phipson. — London : Macmillan, 1978.
166 p. ; 21 cm. GB***
ISBN 0-333-23840-0 : £3.50
I. Title.
PZ7.P55204 Ke 79-303318
79 MARC

Phipson, Joan.
When the city stopped / Joan Phipson. — 1st ed. — New York : Atheneum, 1978.
181 p. ; 22 cm.
"A Margaret K. McElderry book."
SUMMARY: When a series of strikes and labor disputes immobilize the Australian city where they live, a young boy and his sister join their cleaning lady and her crippled husband and try to make their way out of an increasingly hostile environment.
ISBN 0-689-50121-8
[1. Survival—Fiction. 2. Australia—Fiction] I. Title.
PZ7.P55204 Wh [Fic] 78-6930
78 MARC

Pierce, Mary Cunningham Fitzgerald, 1908-
Weekly Reader Children's Book Club presents The rescue / by Mary Cunningham [i.e. M. C. F. Pierce] ; illustrated by Marilyn Miller. — Middletown, Conn. : Xerox Education Publication, c1978.
59 p. : ill. ; 21 cm. — (Weekly Reader Children's Book Club edition)
Cover title: The rescue.
SUMMARY: A family's vacation at an isolated mountain cabin turns into a desperate attempt to escape a raging flood.
ISBN 0-88375-215-8
[1. Floods—Fiction. 2. Survival—Fiction] I. Miller, Marilyn, 1925- . II. Title. III. Title: Rescue.
PZ7.P6145 We [Fic] 78-52132
79 MARC•

Pierce, Robert, 1915-
The day of the wind, rain, and snow : stories and pictures / by Robert Pierce. — New York : Golden Press, c1978.
72 p. : chiefly col. ill. ; 26 cm. — (A Kid's paperback)
SUMMARY: Extremes in the weather cause suspense and excitement for several children in this collection of three stories.
ISBN 0-307-13434-2 : $2.95
[1. Rain and rainfall—Fiction. 2. Snow—Fiction. 3. Winds—Fiction. 4. Short stories] I. Title.
PZ7.P6146 Day [E] 77-80629
78 MARC

Pierson, Jan.
The mystery of Five Finger Island / by Jan Pierson ; [ill. by Ralph Furmanski]. — Wheaton, Ill. : Tyndale House Publishers, c1979.
123 p. : ill. ; 18 cm.
SUMMARY: Blake and Jennifer explore Five Finger Island hoping to find clues to their father's strange disappearance.
ISBN 0-8423-4663-5 (pbk.) : $1.95
[1. Christian life—Fiction. 2. Mystery and detective stories] I. Furmanski, Ralph. II. Title.
PZ7.P618 My [Fic] 78-63616
79 MARC

Pinkwater, Daniel Manus, 1941-
Alan Mendelsohn, the boy from Mars / by Daniel M. Pinkwater. — 1st ed. — New York : Dutton, c1979.
248 p. ; 22 cm.
SUMMARY: Leonard's life at his new junior high is just barely tolerable until he becomes friends with the unusual Alan and with him shares an extraordinary adventure.
ISBN 0-525-25360-2
[1. Science fiction. 2. Humorous stories] I. Title.
PZ7.P6335 Al 1979 [Fic] 78-12052
78 MARC

Pinkwater, Daniel Manus, 1941-
The last guru / written and illustrated by Daniel M. Pinkwater. — New York : Dodd, Mead, c1978.
127 p. : ill. ; 22 cm.
SUMMARY: One harmless bet on a horse race starts a chain of events that turns a 12-year-old into one of the world's richest people and a spiritual guru.
ISBN 0-396-07636-X
[1. Humorous stories] I. Title.
PZ7.P6335 Las [Fic] 78-14260
78 MARC

Pinkwater, Daniel Manus, 1941-
Pickle creature / by Daniel M. Pinkwater. — New York : Four Winds Press, c1979.
32 p. : col. ill. ; 20 x 24 cm.
SUMMARY: Instead of the pickle his grandmother asked him to get at the supermarket, Conrad brings home a pickle creature instead.
ISBN 0-590-07579-9 : $7.95
[1. Pickles—Fiction] I. Title.
PZ7.P6335 Pi [E] 78-11157
78 MARC

Pinkwater, Daniel Manus, 1941-
Return of the moose / written and illustrated by Daniel M. Pinkwater. — New York : Dodd, Mead, [1979]
64 p. : ill. ; 24 cm.
Sequel to Blue moose.
SUMMARY: The blue moose who helps Mr. Breton run his restaurant writes what he is sure is the greatest book ever written by man or moose.
ISBN 0-396-07674-2
[1. Moose—Fiction] I. Title.
PZ7.P6335 Re [Fic] 78-22434
78 MARC

Place, Marian Templeton.
The boy who saw Bigfoot / Marian T. Place. — New York : Dodd, Mead, c1979.
96 p. ; 21 cm.
SUMMARY: A ten-year-old boy, placed once again with new foster parents, becomes involved in a search for Bigfoot.
ISBN 0-396-07644-0
[1. Sasquatch—Fiction. 2. Foster home care—Fiction. 3. Washington (State)—Fiction. 4. Forests and forestry—Fiction] I. Title.
PZ7.P69 Bp [Fic] 78-23199
78 MARC

Platt, Kin.
Chloris and the weirdos / Kim Platt. — Scarsdale, N.Y. : Bradbury Press, c1978.
231 p. ; 22 cm.
SUMMARY: A 13-year-old chronicles life with a mixed-up sister, a twice-divorced mother, and a boyfriend who is an ace skateboarder.
ISBN 0-87888-137-9 : $7.95
[1. Family problems—Fiction. 2. Divorce—Fiction] I. Title.
PZ7.P7125 Cm [Fic] 78-55214
78 MARC

Platt, Kin.
Dracula, go home! / By Kin Platt ; ill. by Frank Mayo. — New York : F. Watts, 1979.
87 p. : ill. ; 22 cm. — (A Triumph book)
SUMMARY: Several eerie events at his aunt's hotel prompt Larry to investigate a guest who strongly resembles Dracula.
ISBN 0-531-01464-9
[1. Mystery and detective stories. 2. Dracula—Fiction] I. Mayo, Frank. II. Title.
PZ7.P7125 Dr [Fic] 78-11335
78 MARC

Pleasence, Donald.
Scouse the mouse / by Donald Pleasence ; pictures by Gerald Potterton. — London : New English Library, 1977.
48 p. : col. ill. ; 30 cm. GB77-32269
ISBN 0-450-03216-7 pbk. : £2.50
I. Potterton, Gerald. II. Title.
PZ7.P717 Sc 78-303273
78 MARC

Polese, Carolyn.
Something about a mermaid / by Carolyn Polese ; illustrated by Gail Owens. — 1st ed. — New York : Dutton, c1978.
27 p. : ill. ; 22 cm.
SUMMARY: A mermaid discovered by Janie and her family at the seashore goes home with them to their city apartment.
ISBN 0-525-39590-3 lib. bdg. : $6.95
[1. Mermaids—Fiction] I. Owens, Gail. II. Title.
PZ7.P7526 So 1978 [Fic] 77-15585
77 MARC

Pollock, Bruce.
Me, Minsky & Max / Bruce Pollock. — Boston : Houghton Mifflin, 1978.
242 p. ; 22 cm.
SUMMARY: An eighteen-year-old film buff hitchhikes to San Francisco during the summer of 1965 with his friend Minsky and true love Max.
ISBN 0-395-26455-3 : $7.95
I. Title.
PZ7.P759 Me [Fic] 78-15599
78 MARC

Pollock, Penny.
Ants don't get Sunday off / by Penny Pollock ; illustrated by Lorinda Bryan Cauley. — New York : Putnam, c1978.
45 p. : ill. (some col.) ; 23 cm. — (A See and read storybook)
SUMMARY: Anya, a hard-working ant, longs for adventure but gets more than she bargained for when a heavy rainstorm disturbs the ant nest.
ISBN 0-399-61129-0 lib. bdg. : $5.69
[1. Ants—Fiction] I. Cauley, Lorinda Bryan. II. Title.
PZ7.P765 An 1978 [Fic] 78-8283
78 MARC

Polushkin, Maria.
Mother, Mother, I want another / by Maria Polushkin ; illustrated by Diane Dawson. — New York : Crown Publishers, c1978.
[32] p. : col. ill. ; 21 cm.
SUMMARY: Anxious to get her son to sleep, Mrs. Mouse goes off to find what she thinks he wants.
ISBN 0-517-53401-0
[1. Mice—Fiction. 2. Sleep—Fiction] I. Dawson, Diane. II. Title.
PZ7.P7695 Mo 1978 [E] 78-5443
78 MARC

CHILDREN'S LITERATURE (Continued)

Pomerantz, Charlotte.
The downtown fairy godmother / Charlotte Pomerantz ; illustrated Susanna Natti. — Reading, Mass. : Addison-Wesley, c1978.
42 p. : ill. ; 23 cm.
SUMMARY: When Olivia wishes for a fairy godmother, she encounters a Fairy Godmother Grade C, still in training and limited in powers, who takes her on her rounds in New York City.
ISBN 0-201-05858-8
[1. Magic—Fiction] I. Natti, Susanna. II. Title.
PZ7.P77 Do [Fic] 78 78-1014 MARC

Pontiflet, Ted.
Poochie / by Ted Pontiflet ; pictures by Mahiri Fufuka. — New York : Dial Press, c1978.
40 p. : ill. ; 25 cm.
SUMMARY: The day one of their cops-and-robbers games leads to a fire in the basement is a day Poochie, Wahoo, and Jessie Jr. would rather forget.
ISBN 0-8037-7029-4 : $7.50. ISBN 0-8037-7030-8 lib. bdg. : $5.45
[1. Friendship—Fiction] I. Fufuka, Mahiri. II. Title.
PZ7.P775 Po [E] 78 78-51321 MARC

Postma, Lidia.
The witch's garden / Lidia Postma. — New York : McGraw-Hill, 1979, c1978.
[25] p. : col. ill. ; 29 cm.
Translation of De heksen-tuin.
SUMMARY: Although they fear the woman who lives in the little red brick house, a group of children accept her invitation to share the secrets of her garden.
ISBN 0-07-050535-7 lib. bdg.
[1. Magic—Fiction. 2. Witches—Fiction] I. Title.
PZ7.P83945 Wi [E] 78 78-11414 MARC

Potter, Beatrix, 1866-1943.
A treasury of Peter Rabbit and other stories : a collection of classics / by Beatrix Potter. — New York : Watts, 1978.
ca. 300 p. : col. ill. ; 21 cm.
CONTENTS: The tale of Peter Rabbit.—The tale of Benjamin Bunny.—The tale of Squirrel Nutkin.—The tale of two bad mice.—The tale of Jeremy Fisher.
[1. Animals—Fiction] I. Title.
PZ7.P85 Tr 1978 [E] 78 78-105544 MARC

Potter, Marian.
The shared room / by Marian Potter. — New York : Morrow, 1979.
192 p. ; 21 cm.
SUMMARY: Despite her grandmother's opposition, a 10-year-old reestablishes contact with her long-institutionalized mother.
ISBN 0-688-22209-9
[1. Mentally ill—Fiction. 2. Mothers and daughters—Fiction. 3. Grandmothers—Fiction] I. Title.
PZ7.P853 Sh [Fic] 79 79-18012 MARC

Powell, Ann, 1951-
Strange Street / by Ann Powell. — Toronto : Kids Can Press, c1975.
[30] p. : ill. ; 14 x 22 cm. C***
I. Title.
PZ7.P875 St 76[r79]rev 76-372031 MARC

Powers, Bill.
A test of love / by Bill Powers ; photos. by Bill Aron. — New York : F. Watts, 1979.
90 p. : ill. ; 22 cm. — (A Triumph book)
SUMMARY: A pregnant teenager is ashamed to tell her parents and although her college boyfriend loves her he is not ready for marriage.
ISBN 0-531-02888-7
[1. Unmarried mothers—Fiction] I. Aron, Bill. II. Title.
PZ7.P8834 Te 1979 [Fic] 78 78-12913 MARC

Poynter, Margaret.
Crazy Minnie / by Margaret Poynter ; [ill. by David Sherrod]. — Mankato, Minn. : Creative Education, c1978.
56 p. : ill. ; 24 cm.
SUMMARY: After her parents are killed in an accident, Karen is sent to Los Angeles to live with her grandmother who is called "Crazy Minnie" by the kids in the area.
ISBN 0-87191-680-0 (lib. bdg.)
[1. Orphans—Fiction. 2. Grandmothers—Fiction] I. Sherrod, David. II. Title.
PZ7.P885 Cr [Fic] 78 78-12887 MARC

Prishvin, Mikhail Mikhailovich, 1873-1954.
The sun's storehouse / M. Prishvin ; drawings by V. Kirillov ; [translated from the Russian by Ivy Litvinov]. — Moscow : Progress Publishers, 1975.
121 p. : col. ill. ; 29 cm. USSR***
Translation of Kaldovaîa solnîsa.
1.03rub
I. Kirillov, V. II. Title.
PZ7.P938 Su 1975 78 78-371139 MARC

Provensen, Alice.
The year at Maple Hill Farm / Alice and Martin Provensen. — 1st ed. — New York : Atheneum, 1978.
[32] p. : col. ill. ; 33 cm.
"A Jonathan Cape book."
SUMMARY: Describes the seasonal changes on a farm and surrounding countryside throughout the year.
ISBN 0-689-30642-3. ISBN 0-689-20494-9 lib. bdg. : $5.69
[1. Farm life—Fiction. 2. Seasons—Fiction] I. Provensen, Martin, joint author. II. Title.
PZ7.P9457 Ye 1978 [E] 77 77-18518 MARC

Quackenbush, Robert M
Detective Mole and the seashore mystery / story and pictures by Robert Quackenbush. — 1st ed. — New York : Lothrop, Lee & Shepard, c1979.
64 p. : col. ill. ; 22 cm. — (A Fun-to-read book)
SUMMARY: With cool-headed logic, Detective Mole tracks down the robber of Captain Bill's valuable pearl.
ISBN 0-688-41917-8. ISBN 0-688-51917-2 lib. bdg.
[1. Mystery and detective stories. 2. Animals—Fiction] I. Title.
PZ7.Q16 Df [E] 79 79-14431 MARC

Quackenbush, Robert M
Detective Mole and the Tip-Top mystery : story and pictures / by Robert Quackenbush. — 1st ed. — New York : Lothrop, Lee & Shepard Co., c1978.
62 p. : col. ill. ; 22 cm.
SUMMARY: When mysterious mishaps drive guests away from the Tip-Top Inn, Detective Mole investigates.
ISBN 0-688-41858-9. ISBN 0-688-51858-3 lib. bdg.
[1. Mystery and detective stories. 2. Animals—Fiction] I. Title.
PZ7.Q16 Di [E] 78[r79]rev 78-6082 MARC

Quackenbush, Robert M
Moose's store / story and pictures by Robert Quackenbush. — 1st ed. — New York : Lothrop, Lee & Shepard, c1979.
63 p. : col. ill. ; 22 cm. — (A Fun-to-read books)
SUMMARY: Country storekeeper Moose shows that he can be pushed just so far when his friends take over and try to modernize the store.
ISBN 0-688-41884-8. ISBN 0-688-51884-2 lib. bdg.
[1. Moose—Fiction. 2. Friendship—Fiction] I. Title.
PZ7.Q16 Ml [E] 78 78-21534 MARC

Quackenbush, Robert M
The most welcome visitor / by Robert Quackenbush. — 1st ed. — New York : Windmill Books, c1978.
32 p. : ill. ; 26 cm.
SUMMARY: Fred Horny Toad endures visiting relatives all year long, but on Christmas Day, after the visit of Santa Claus, the relatives are all anxious to leave.
ISBN 0-525-61587-3 : $7.95
[1. Reptiles—Fiction. 2. Holidays—Fiction. 3. Santa Claus—Fiction. 4. Christmas stories] I. Title.
PZ7.Q16 Mo 1978 [E] 78 78-16627 MARC

Quackenbush, Robert M
Pete Pack Rat and the Gila Monster Gang : story and pictures / by Robert Quackenbush. — 1st ed. — New York : Lothrop, Lee & Shepard Co., c1978.
64 p. : col. ill. ; 22 cm. — (Fun-to-read books)
SUMMARY: The Gila Monster Gang robs the Pebble Junction bank and kidnaps Sheriff Sally Gopher, but Pete Pack Rat's ingenuity saves the day.
ISBN 0-688-41838-4. ISBN 0-688-51838-9 lib. bdg.
[1. Western stories. 2. Animals—Fiction] I. Title.
PZ7.Q16 Pf [E] 77[r79]rev 77-90535 MARC

Rabe, Berniece.
The orphans / Berniece Rabe. — 1st ed. — New York : Dutton, c1978.
184 p. ; 22 cm.
SUMMARY: An orphaned twin brother and sister living in rural Missouri during the Depression finally find a permanent home.
ISBN 0-525-36450-1 lib. bdg. : $7.95
[1. Orphans—Fiction. 2. Twins—Fiction. 3. Country life—Fiction] I. Title.
PZ7.R105 Or 1978 [Fic] 78 78-9418 MARC

Rabinowitz, Sholom, 1859-1916.
Holiday tales of Sholom Aleichem / selected and translated from the Yiddish by Aliza Shevrin ; pictures by Thomas di Grazia. — New York : Scribner, c1979.
xii, 145 p. : ill. ; 21 cm.
CONTENTS: Really a Sukkah!—Benny's luck.—A ruined Passover.—The esrog.—The goldspinners.—The Passover exiles.—The first commune.
ISBN 0-684-16118-4
1. Fasts and feasts—Judaism—Juvenile fiction. [1. Fasts and feast—Judaism—Fiction. 2. Jews in the Ukraine—Fiction. 3. Short stories] I. Shevrin, Aliza. II. Di Grazia, Thomas. III. Title.
PZ7.R110 Ho 1979 839'.09'33 79 79-753 MARC

Radford, William Lewis, 1931-
Flying doctor. Illus. by Adrienne Moore. [1st ed. Nairobi] East African Pub. House [1969]
100 p. illus. 19 cm. (Safari adventure books, 5) 4/95
I. Title.
PZ7.R1215 Fl 73[r79]rev 73-156839 MARC

Radley, Gail.
The night Stella hid the stars / Gail Radley ; illustrated by John Wallner. — New York : Crown, c1978.
[32] p. : col. ill. ; 29 cm.
SUMMARY: After millions of years of dusting the stars all day to have them ready for the night sky, Stella decides she needs time to enjoy the day.
ISBN 0-517-53256-5
[1. Stars—Fiction. 2. Night—Fiction] I. Wallner, John C. II. Title.
PZ7.R1223 Ni 1978 [E] 78 78-55545 MARC

Radu, Kenneth.
The mouse who came to dinner / story by Kenneth Radu ; illustrations by Diane Radu. — Ottawa : Borealis Press, 1978.
39 p. : ill. ; 22 x 26 cm. C***
ISBN 0-919594-66-2
I. Radu, Diane L. II. Title.
PZ7.R1225 Mo 79 79-300445 MARC

Rae, Gwynedd.
The Mary Plain omnibus / Gwynedd Rae ; illustrated by Irene Williamson. — London [etc.] : Routledge and Kegan Paul, 1976.
[8], 458 p. : ill. ; 20 cm. GB76-26142
CONTENTS: Mostly Mary.—All Mary.—Mary Plain in town.—Mary Plain on holiday.
ISBN 0-7100-8437-4 : £3.95
1. Children's stories, English. [1. Bears—Fiction] I. Williamson, Irene. II. Title.
PZ7.R1227 Mar [Fic] 77[r79]rev 77-364253 MARC

Rae, John, 1931-
Return to the Winter Palace / John Rae ; illustrated by Susan Edwards. — London : Hodder and Stoughton, 1979.
128 p. : ill. ; 22 cm.
ISBN 0-340-21778-2 : £3.50
I. Title.
PZ7.R123 Re 79 79-310974 MARC

Ramos, Francie.
The land of health / [story by Francie Ramos ; edited by W. Karl Parker ; illustrated by Bill Hall]. — Fort Worth, Tex. : Parker Chiropractic Research Foundation, c1977.
[39] p. : col. ill. ; 29 cm.
SUMMARY: A youngster learns the five main secrets of good health while on his way to the chiropractor office.
[1. Health—Fiction. 2. Chiropractic—Fiction] I. Parker, W. Karl. II. Hall, Bill. III. Parker Chiropractic Research Foundation. IV. Title.
PZ7.R1425 Lan [Fic] 79 79-114692 MARC

Raphael, Elaine.
Donkey and Carlo / by Elaine Raphael and Don Bolognese. — 1st ed. — New York : Harper & Row, c1978.
31 p. : ill. ; 22 cm.
SUMMARY: Carlo becomes so involved at the market that he forgets his promise to Donkey.
ISBN 0-06-020552-0 $6.95. ISBN 0-06-020553-9 lib. bdg. : $6.79
[1. Donkeys—Fiction. 2. Friendship—Fiction] I. Bolognese, Don, joint author. II. Title.
PZ7.R1812 Do 1978 [E] 78 77-25660 MARC

Ray, Mary, 1932-
The voice of Apollo. Illustrated by John Cooper. London, J. Cape [1964]
190 p. illus. 21 cm.
I. Title.
PZ7.R21015 Vo [r78]rev 65-1905 MARC

Rayner, Mary.
Mr and Mrs Pig's evening out / [written and illustrated by] Mary Rayner. — London [etc.] : Macmillan, 1976.
[32] p. : chiefly col. ill. ; 27 cm. GB76-12968
ISBN 0-333-19371-7 : £2.95
I. Title.
PZ7.R2315 Mi 78 78-313129 MARC

Razzi, Jim.
Don't open this box! Words and pictures by James Razzi. New York, Parents' Magazine Press [1973]
[39] p. col. illus. 22 cm.
SUMMARY: Harvey's curiosity finally tempts him to ignore the "Don't open this box" sign on the box he finds on his doorstep.
ISBN 0-8193-0669-X. ISBN 0-8193-0670-3 (lib. bdg.)
[1. Boxes—Fiction] I. Title.
PZ7.R233 Do [E] 72[r79]rev 72-10219 MARC

Read, Beryl J
The runaway girl / Beryl J. Read. — Grand Rapids : Zondervan Pub. House, 1979, c1976.
89 p. ; 21 cm. — (Pathfinder series)
SUMMARY: Frustrated by life in England with her stern uncle, an 11-year-old stows away on an Australia-bound ship in order to search for her father.
ISBN 0-310-37851-6 : $1.95
[1. Stowaways—Fiction. 2. Voyages and travels—Fiction] I. Title. II. Series: Pathfinder series (Grand Rapids)
PZ7.R2335 Ru 1979 [Fic] 78 78-31367 MARC

CHILDREN'S LITERATURE (Continued)

Reed, Tom.
Melissa on parade : written and illustrated / by Tom Reed. — Scarsdale, N.Y. : Bradbury Press, c1979.
[40] p. : col. ill. ; 29 cm.
SUMMARY: While everyone else is watching the parade, Melissa spots a fleeing bank robber.
ISBN 0-87888-143-3 : $7.95
[1. Parades—Fiction. 2. Robbers and outlaws—Fiction] I. Title.
PZ7.R2534 Me [E] 78 78-26339 MARC

Rees, David, 1936-
The Exeter blitz / David Rees. — London : Hamilton, 1978.
[vi], 116 p. ; 23 cm. GB***
ISBN 0-241-89759-9 : £3.25
1. Exeter, Eng.—Bombardment, 1942—Juvenile fiction. I. Title.
PZ7.R25468 Ex 78 78-314296 MARC

Rees, David, 1936-
Risks / David Rees. — 1st U.S. ed. — Nashville : T. Nelson, c1977.
91 p. ; 21 cm.
SUMMARY: Two 15-year-old boys split up while hitchhiking on a weekend trip to London in order to facilitate catching rides. One of them never reaches their destination.
ISBN 0-8407-6605-X : $5.95
[1. Hitchhiking—Fiction. 2. Death—Fiction. 3. Friendship—Fiction] I. Title.
PZ7.R25468 Ri 1977 [Fic] 78 78-1006 MARC

Reeves, James.
The James Reeves story book / illustrated by Edward Ardizzone. — London : Heinemann, 1978.
208 p. : ill. ; 24 cm. GB78-33350
ISBN 0-434-95897-2 : £3.95
I. Ardizzone, Edward, 1900- . II. Title.
PZ7.R257 Jam 79 79-300187 MARC

Reid, James, 1907-
Peter, Peter, pumpkin eater. Story and illus. by James Reid. Philadelphia, Fortress Press [1970]
[32] p. col. illus. 33 cm.
SUMMARY: Three children visit a pumpkin farm to select pumpkins for Halloween.
[1. Halloween—Fiction] I. Title.
PZ7.R269 Pe [E] 70[r79]rev 76-16532 MARC

Reig, June.
Diary of the boy king, Tut Anhk Amen / by June Reig ; illustrated by the author. — New York : Scribner, c1978.
viii, 147 p. : ill. ; 22 cm.
SUMMARY: A fictionalized diary kept by Tutankhamen during his ninth year, the year he became King of Egypt, about 1334 B.C.
ISBN 0-684-15992-6 : $7.95
1. Tutankhamen, King of Egypt—Juvenile fiction. [1. Tutankhamen, King of Egypt—Fiction. 2. Kings, queens, rulers, etc.—Fiction. 3. Egypt—Civilization—To 332 B.C.—Fiction] I. Title.
PZ7.R27383 Di [Fic] 79 78-71467 MARC

Reilly, Robert T
Rebels in the shadows / by Robert T. Reilly. — Pittsburgh : University of Pittsburgh Press, 1979, c1962.
x, 179 p. : map ; 21 cm.
Reprint of the 1962 ed. published by Bruce Pub. Co., Milwaukee.
SUMMARY: The violent activities of the Molly Maguires in the Pennsylvania coal fields during the 1870's leave their mark on a family of Irish immigrants.
ISBN 0-8229-5304-5
[1. Coal mines and mining—Fiction. 2. Molly Maguires—Fiction. 3. Pennsylvania—History—1865- —Fiction] I. Title.
[PZ7.R2746 Rc 1979] [Fic] 78 78-66069 MARC

Reinach, Jacquelyn.
Happy birthday Unicorn / written by Jacquelyn Reinach ; illustrated by Richard Hefter ; edited by Ruth Lerner Perle. — New York : Holt, Rinehart, and Winston, c1978.
[36] p. : col. ill. ; 25 cm. — (Sweet pickles series)
SUMMARY: Unicorn changes everyone's conception of a proper birthday party for a senior citizen in a unique way.
ISBN 0-03-042066-0 (lib. bdg.) : $2.95
[1. Unicorns—Fiction. 2. Birthdays—Fiction] I. Hefter, Richard. II. Perle, Ruth Lerner. III. Title. IV. Series.
PZ7.R2747 Hap [E] 78 78-9585 MARC

Reinach, Jacquelyn.
Jackal wants everything / written by Jacquelyn Reinach ; illustrated by Richard Hefter ; edited by Ruth Lerner Perle. — New York : Holt, Rinehart and Winston, c1978.
[36] p. : col. ill. ; 25 cm. — (Sweet Pickles series)
SUMMARY: Jackal wants everything he sees—and then some more.
ISBN 0-03-042036-9
[1. Jackals—Fiction] I. Hefter, Richard. II. Perle, Ruth Lerner. III. Title. IV. Series.
PZ7.R2747 Jac [E] 77 77-16324 MARC

Reinach, Jacquelyn.
Ms. Klondike / written by Jacquelyn Reinach ; illustrated by Richard Hefter; edited by Ruth Lerner Perle. — 1st ed. — New York : Viking Press, c1977.
[34] p. : col. ill. ; 20 x 24 cm.
SUMMARY: A taxicab driver faces stiff opposition in her new job because people think she is just another dumb woman driver.
ISBN 0-670-49510-7 : $6.50
[1. Taxicab drivers—Fiction. 2. Sex role—Fiction] I. Hefter, Richard. II. Title.
PZ7.R71969 Ms [E] 77[r79]rev 77-4649 MARC

Reinach, Jacquelyn.
Octopus protests / written by Jacquelyn Reinach ; illustrated by Richard Hefter ; edited by Ruth Lerner Perle. — New York : Holt, Rinehart, and Winston, c1978.
[36] p. : col. ill. ; 25 cm. — (Sweet Pickles series)
SUMMARY: Octopus is not afraid to speak out to make the town a better place to live.
ISBN 0-03-042046-6 lib. bdg. : $2.95
[1. Octopus—Fiction. 2. City and town life—Fiction] I. Hefter, Richard. II. Perle, Ruth Lerner. III. Title. IV. Series.
PZ7.R2747 Oc [E] 78 78-54981 MARC

Reit, Seymour.
Bugs Bunny goes to the dentist / by Seymour Reit ; pictures by Lou Cunette. — New York : Golden Press, c1978.
[24] p. : col. ill. ; 21 cm. — (A Golden look-look book)
SUMMARY: Bugs Bunny visits the dentist hoping to be inspired to write the last line of his Happy Tooth Week poem.
ISBN 0-307-11843-6 : $0.95
[1. Teeth—Care and hygiene—Fiction. 2. Rabbits—Fiction] I. Cunette, Lou. II. Title.
PZ7.R2785 Bs [E] 78 77-88808 MARC

Renick, Marion.
The big basketball prize. Illustrated by Paul Galdone. New York, Scribner, 1963.
[32] p. col. illus. 27 cm.
SUMMARY: Fearing he won't outgrow his awkwardness in time to win the basketball foul-shooting contest, a young boy takes matters into his own hands.
[1. Basketball—Fiction] I. Galdone, Paul. II. Title.
PZ7.R288 Bi [Fic] [r79]rev 63-19908 MARC

Ressner, Philip.
Dudley Pippin's summer / Philip Ressner ; pictures by Ben Shecter. — 1st ed. — New York : Harper & Row, c1979.
40 p. : ill. ; 24 cm.
SUMMARY: Eleven short stories recount Dudley Pippin's discoveries and adventures during a summer holiday.
ISBN 0-06-024887-4 : $6.89
[1. Vacations—Fiction] I. Shecter, Ben. II. Title.
PZ7.R323 Dx 1979 [Fic] 78 78-19831 MARC

Rey, Hans Augusto, 1898-
Cecily G. and the 9 monkeys, by H. A. Rey. Boston, Houghton Mifflin company, 1942.
31, [1] p. col. illus. 30 cm.
SUMMARY: A lonely giraffe teams up with nine playful monkeys.
[1. Giraffes—Fiction. 2. Monkeys—Fiction] I. Title.
PZ7.R33 Ce [E] [r78]rev 42-20276 MARC

Rhiner, Gladys.
Jimmy goes camping / Gladys Rhiner ; illustrated by John Peterson. — Nashville : Broadman Press, c1978.
32 p. : col. ill. ; 24 cm.
SUMMARY: Jimmy and his family reaffirm their faith in God while on a camping trip.
ISBN 0-8054-4255-3
[1. Christian life—Fiction. 2. Camping—Fiction] I. Peterson, John Lawrence, 1924- . II. Title.
PZ7.R34 Jk [E] 79 78-113038 MARC

Rice, Eve.
New blue shoes : story and pictures / by Eve Rice. — [New York] : Puffin Books, 1979, 1975.
[32] p. : col. ill. ; 23 cm. — (Picture puffin)
SUMMARY: A little girl goes shopping with her mother for new shoes which she insists must be blue.
ISBN 0-14-050316-1 : $1.95
[1. Shoes and boots-Fiction] I. Title.
[PZ7.R3622 Ne 1979] [E] 79 79-582 MARC

Rice, Eve.
Oh, Lewis! Story and pictures / by Eve Rice. — [New York] : Puffin Books, 1979, c1974.
32 p. : col. ill. ; 18 x 23 cm. — (Picture puffin)
SUMMARY: Lewis has a lot of trouble with his winter clothing.
ISBN 0-14-050257-2 : $1.95
[1. Winter—Fiction. 2. Clothing and dress—Fiction] I. Title.
[PZ7.R3622 Oh 1979] [E] 78 78-17990 MARC

Rice, Hugo.
The remarkable feat of King Caboodle / Hugo Rice ; illustrations by Charlotte Firmin. — London : A. & C. Black, c1978.
62 p. : ill. ; 23 cm. GB***
ISBN 0-7136-1862-0 : £2.50
I. Firmin, Charlotte. II. Title.
PZ7.R3625 Re 79 78-326196 MARC

Richardson, Arleta.
More stories from grandma's attic / Arleta Richardson. — Elgin, IL : D. C. Cook Pub. Co., c1979.
141 p. : ill. ; 18 cm. — (Chariot books)
Sequel to In grandma's attic.
SUMMARY: Relates the further adventures of a young girl growing up during the late 19th century on a farm in Michigan.
ISBN 0-89191-131-6 : $1.95
[1. United States—Social life and customs—1865-1918—Fiction] I. Title.
PZ7.R3942 Mo [Fic] 79 78-73125 MARC

Richardson, Henry Handel, pseud.
The getting of wisdom / Henry Handel Richardson. — London : Heinemann, 1977.
[6], 233 p. ; 20 cm. GB***
ISBN 0-434-95950-2 : £2.90
I. Title.
PZ7.R3945 Ge 1977 79 78-319322 MARC

Richmond, Legh, 1772-1827.
The dairyman's daughter; an authentic and interesting narrative, in five parts. Newark, Published and sold by W. Tuttle, J. Tuttle & Co. printers, 1815.
72 p. 15 cm.
I. Title.
PZ6.R415 Da 1815 79 78-324777 MARC

Richter, Alice Numeroff.
You can't put braces on spaces / Alice Richter & Laura Joffe Numeroff ; illustrated by Laura Joffe Numeroff. — 1st ed. — New York : Greenwillow Books, c1979.
54 p. : col. ill. ; 22 cm. — (Greenwillow read-alone)
SUMMARY: His younger brother enviously watches as Neil has braces fitted on his teeth.
ISBN 0-688-80190-0 : $5.95. ISBN 0-688-84190-2 lib. bdg. $5.71
[1. Orthodontics—Fiction. 2. Teeth—Care and Hygiene—Fiction] I. Numeroff, Laura Joffe, joint author. II. Title.
PZ7.R41527 Yo [E] 78 78-16367 MARC

Ricks, Charlotte Hall.
Look at me / Charlotte Hall Ricks ; illustrated by Annie Gusman. — Boston : Houghton Mifflin, 1979.
32 p. : col. ill. ; 21 cm.
SUMMARY: Mother is too busy to pay attention to Catherine's antics until Catherine, in exasperation, carries out her threat to disappear.
ISBN 0-395-28480-5
[1. Mother and child—Fiction] I. Gusman, Annie.
PZ7.R41615 Lo [E] 79 79-14463 MARC

Ritchie, Jo-An.
Jonie goes to academy / by Jo-An Ritchie. — Nashville : Southern Pub. Association, c1979.
124 p. ; 17 cm.
SUMMARY: Jonie's first year at the academy brings friendship, happy times, and a continuing strong faith in God.
ISBN 0-8127-0193-3
[1. Boarding schools—Fiction. 2. School stories. 3. Christian life—Fiction] I. Title.
[PZ7.R51165 Jl] [Fic] 78 78-21271 MARC

Ritchie, Jo-An.
Jonie graduates / by Jo-An Ritchie. — Nashville : Southern Pub. Association, c1979.
128 p. ; 17 cm. — (Orion)
SUMMARY: With God's guidance 17-year-old Jonie works her way through senior year at an Adventist boarding school.
ISBN 0-8127-0201-8
[1. Boarding schools—Fiction. 2. School stories. 3. Seventh-Day Adventists—Fiction. 4. Christian life—Fiction] I. Title.
[PZ7.R51165 Jm] [Fic] 78 78-27431 MARC

Roach, Marilynne K
Presto : or, The adventures of a turnspit dog / written and illustrated by Marilynne K. Roach. — Boston : Houghton Mifflin, 1979.
vi, 148 p. : ill. ; 22 cm.
SUMMARY: A young dog escapes the drudgery of an inn's turnspit to become a part of the colorful street life of 18th-century London.
ISBN 0-395-28269-1
[1. Dogs—Fiction. 2. London—Fiction] I. Title.
PZ7.R528 Pr [Fic] 79 79-11746 MARC

Robb, Brian, 1913-
The last of the centaurs / Brian Robb. — [London] : A. Deutsch, [1979]
[32] p. : col. ill. ; 26 cm. GB***
ISBN 0-233-97000-2 : £3.75 ($6.95 U.S.)
I. Title.
PZ7.R5317 Las 79 79-315475 MARC

CHILDREN'S LITERATURE (Continued)

Robinson, Jan M
The story of Warple / by Jan Robinson ; illustrated by Jack Jewell. — Nashville : Aurora Publishers, c1978.
[32] p. : col. ill. ; 21 x 26 cm.
SUMMARY: Warple a slightly warped guitar, awaits his chance to prove he can play beautiful music.
ISBN 0-87695-209-0
[1. Guitar—Fiction. 2. Musical instruments—Fiction] I. Jewell, Jack. II. Title.
PZ7.R5662 St [E] 77-15907
78 MARC

Robinson, Joan G
The dark house of the sea witch / by Joan G. Robinson. — 1st American ed. — New York : Coward, McCann & Geoghegan, 1979.
128 p. ; 22 cm.
SUMMARY: On their own while their parents and housekeeper are away, Meg and her brother make the acquaintance of a neighbor they suspect is a witch.
ISBN 0-698-20494-8
[1. Brothers and sisters—Fiction] I. Title.
PZ7.R5664 Dar [Fic] 79-10845
79 MARC

Robinson, Marileta.
Mr. Goat's bad good idea / by Marileta Robinson ; illustrated by Arthur Getz. — 1st ed. — New York : Crowell, c1979.
39 p. : col. ill. ; 21 cm.
SUMMARY: Three episodes featuring Mr. Goat, Grandfather Sheep, Jerry Prairie Dog, and other Navajo friends.
ISBN 0-690-03862-3. ISBN 0-690-03864-X lib. bdg.
[1. Animals—Fiction. 2. Navaho Indians—Fiction. 3. Indians of North America—Fiction] I. Getz, Arthur. II. Title.
PZ7.R56747 Mi [E] 77-26601
78 MARC

Robinson, Rony.
A time of bears / Rony Robinson. — London : Macmillan, 1978.
161 p. ; 21 cm. GB***
ISBN 0-333-24773-6 : £3.95
I. Title.
PZ7.R56757 Ti 79-314992
79 MARC

Robison, Nancy.
The lizard hunt / Nancy Robison ; illustrated by Lynn Munsinger. — 1st ed. — New York : Lothrop, Lee & Shepard, c1979.
62 p. : ill. ; 22 cm. — (A Fun-to-read book)
SUMMARY: Talley leads her brother into a canyon while hunting lizards. When they become lost, Greg leads them to safety.
ISBN (invalid) 0-688-41875-9
[1. Lizards—Fiction. 2. Hunting—Fiction. 3. Lost children—Fiction] I. Munsinger, Lynn. II. Title.
PZ7.R5697 Li [E] 78-24374
78 MARC

Robison, Nancy.
The mystery at Hilltop Camp / by Nancy L. Robison ; pictures by Ethel Gold. — Champaign, Ill. : Garrard Pub. Co., c1979.
47 p. : col. ill. ; 23 cm. — (A Garrard mystery book)
SUMMARY: Patty and her friends determine to apprehend the culprit who is stealing milk from the refrigerator of Hilltop Camp.
ISBN 0-8116-6407-4 (lib. bdg.)
[1. Mystery and detective stories] I. Gold, Ethel. II. Title.
PZ7.R5697 My [E] 78-21523
78 MARC

Robison, Nancy.
On the balance beam / Nancy Robison ; illustrated by Rondi Anderson. — Chicago : A. Whitman, c1978.
64 p. : ill. ; 21 cm.
SUMMARY: Andrea's dream of becoming an accomplished gymnast is threatened by her own impulsiveness and a jealous member of her gym team.
ISBN 0-8075-6067-7 lib. bdg. : $4.50
[1. Gymnastics—Fiction] I. Anderson, Rondi. II. Title.
PZ7.R5697 On 1978 [Fic] 78-15228
78 MARC

Robison, Nancy.
The other place / by Nancy L. Robison ; illustrated by Joan Drescher. — New York : Walker, 1978.
72 p. : ill. ; 22 cm.
SUMMARY: Following her mother's death, Elena and her father relocate in a remote valley that brings tranquility and happiness to her father and loneliness and a sense of foreboding to Elena.
ISBN 0-8027-6340-5 : $6.95. ISBN 0-8027-6341-3 lib. bdg. : $6.85
[1. Science fiction] I. Drescher, Joan E. II. Title.
PZ7.R56754 Ot [Fic] 78-56287
78 MARC

Robison, Nancy.
Space hijack! / Nancy Robison ; illustrated pictures by Edward Frascino. — 1st ed. — New York : Lothrop, Lee & Shepard Co., c1979.
63 p. : col. ill. ; 22 cm. — (A Fun-to-read book)
SUMMARY: When a mysterious passenger hijacks their Moon shuttle, Mark and Ted use the difference in gravity to escape from the see-through space people that the hijacker controls.
ISBN 0-688-41897-X. ISBN 0-688-51897-4 lib. bdg.
[1. Science fiction] I. Frascino, Edward. II. Title.
PZ7.R5697 Sp [E] 78-24218
78 MARC

Robles, Al.
Looking for Ifugao Mountain = Paghahanap sa Bundok ng Ifugao / written by Al Robles ; translated into Pilipino by Alfredo Carigma ; illustrated by Jim Dong. — San Francisco : Children's Book Press, c1977.
20, [4] p. : col. ill. ; 25 cm. — (Fifth world tales = Ika-limang pandaigdig na mga istoriya)
SUMMARY: Caught up in an old man's vision of the past, a young Filipino American begins a search for his own heritage.
ISBN 0-89239-012-3
[1. Filipino Americans—Fiction. 2. Tagalog language—Readers] I. Dong, Jim. II. Title. III. Title: Paghahanap sa Bundok ng Ifugao.
PZ90.T27R6 [Fic] 77-78104
79 MARC

Rockwell, Anne F
Blackout / by Anne & Harlow Rockwell. — New York : Macmillan, c1979.
63 p. : col. ill. ; 23 cm. — (Ready-to-read)
SUMMARY: Dan and his family face hardship and danger when an ice storm causes a 3-day failure in their town.
ISBN 0-02-777610-7 lib. bdg. : $6.95
[1. Natural disasters—Fiction. 2. Survival—Fiction] I. Rockwell, Harlow, joint author. II. Title.
PZ7.R5943 Bl [E] 78-12185
78 MARC

Rockwell, Anne F
The bump in the night / by Anne Rockwell. — 1st ed. — New York : Greenwillow Books, c1979.
55 p. : col. ill. ; 22 cm. — (Greenwillow read-alone)
SUMMARY: Fearless Toby the tinker repairs a ghost and reaps the reward.
ISBN 0-688-80180-3 : $5.95. ISBN 0-688-84180-5 lib. bdg. : $5.71
[1. Tinkers—Fiction. 2. Ghost stories] I. Title.
PZ7.R5943 Bs [E] 78-16272
78 MARC

Rockwood, Roy, pseud.
Bomba, the jungle boy / by Roy Rockwood. — New York : Grosset & Dunlap, c1978.
180 p. ; 20 cm. — (Bomba series ; no. 1)
SUMMARY: Separated from his parents since childhood, a young boy, who has been brought up in the jungle by an old naturalist, begins the long search for his true identity.
ISBN 0-448-14701-7 : $2.50
[1. Jungle stories. 2. Adventure stories] I. Title.
PZ7.R598 Bod 77-84109
79 MARC

Rockwood, Roy, pseud.
Bomba the jungle boy : the Moving Mountain / by Roy Rockwood. — New York : Grosset & Dunlap, c1978.
vi, 180 p. ; 20 cm. — (Bomba series ; no. 2)
SUMMARY: Searching for his long-lost parents, 14-year-old Bomba embarks on a perilous journey to a distant volcano.
ISBN 0-448-14702-5 : $2.50
[1. Jungle stories] I. Title.
PZ7.R598 Bov 1978 [Fic] 77-84110
79 MARC

Romanek, Enid Warner.
Teddy / written and illustrated with etchings by Enid Warner Romanek. — New York : Scribner, c1978.
[32] p. : ill. ; 26 cm.
SUMMARY: Teddy visits the park, takes a nap, paints a picture, and at the end of the day, retires for the night with some favorite toys.
ISBN 0-684-15811-6 lib. bdg. : $6.95
[1. Toys—Fiction] I. Title.
PZ8.9.R68 Te [E] 78-23177
78 MARC

Rose, Gerald.
"Ahhh!" said Stork / Gerald Rose. — London : Faber, 1977.
[14] p. : chiefly col. ill. ; 25 cm. GB***
I. Title.
PZ7.R7156 Ah 78-314317
78 MARC

Rose, Gerald.
Watch out! / [by] Gerald Rose. — Harmondsworth : Kestrel Books, 1978.
[28] p. : chiefly col. ill. ; 21 cm. — (A Minnow book) GB78-3804?
ISBN 0-7226-5474-X : £1.95
I. Title.
PZ7.R7156 Wat 79-310792
79 MARC

Roth, Arthur J 1925-
The yucky monster / by Arthur Roth ; illustrated by Tom O'Sullivan. — New York : Dandelion Press, c1979.
[29] p. : col. ill. ; 23 cm.
SUMMARY: Convinced that there is a monster in his room, a little boy asks his father to help him get rid of it.
ISBN (invalid) 0-89799-146-6
[1. Monsters—Fiction. 2. Fathers and sons—Fiction] I. O'Sullivan, Tom. II. Title.
PZ7.R727 Yu [E] 78-72112
00 MARC

Roy, Ron, 1940-
Awful Thursday / by Ron Roy ; pictures by Lillian Hoban. — [New York] : Pantheon Books, c1979.
38 p. : ill. ; 24 cm. — (An I am reading book)
SUMMARY: Jack feels responsible for ruining a school tape recorder and dreads telling Miss Hope what happened.
ISBN 0-394-84003-8
[1. Honesty—Fiction] I. Hoban, Lillian. II. Title.
PZ7.R8139 Aw [E] 78-14049
78 MARC

Roy, Ron, 1940-
Three ducks went wandering / by Ron Roy ; pictures by Paul Galdone. — New York : Seabury Press, c1979.
[32] p. : col. ill. ; 23 cm.
"A Clarion book."
SUMMARY: Blind luck protects three little ducks when they venture out of the barnyard.
ISBN 0-8164-3231-7
[1. Ducks—Fiction] I. Galdone, Paul. II. Title.
PZ7.R8139 Tj [E] 78-12629
78 MARC

Roy, Thomas Albert.
The curse of the turtle / Thomas Albert Roy ; illustrated by Rex Backhaus-Smith. — 1st American ed. — New York : Collins + World, 1978, c1977.
184 p. : ill. ; 23 cm.
SUMMARY: After a series of tragedies befall Jimmy's family, whose home is built on sacred Aborigine land, he and his friend, Tajurra, realize the survival of Canadera depends on reconciliation between black and white.
ISBN 0-529-05502-3 : $6.95
[1. Australian aborigines—Fiction. 2. Australia—Fiction] I. Backhaus-Smith, Rex. II. Title.
PZ7.R8144 Cu 1978 [Fic] 78-6413
78 MARC

Rubinstein, Robert E
Who wants to be a hero! / Robert E. Rubinstein. — New York : Dodd, Mead, c1979.
158 p. ; 21 cm.
SUMMARY: After a "heroic" rescue of an old man from thugs, Jason himself becomes a frequent victim of the attackers, which causes him to wonder about the wisdom of his actions.
ISBN 0-396-07737-4
[1. Crime and criminals—Fiction] I. Title.
PZ7.R83135 Wh [Fic] 79-16103
79 MARC

Ruck-Pauquèt, Gina.
Oh, that koala! / By Gina Ruck-Pauquet ; illustrated by Anna Mossakowska. — New York : McGraw-Hill, 1979.
[28] p. : col. ill. ; 30 cm.
Translation of Oh la la koala, originally published under title: Der Koala-Bär.
SUMMARY: A merry little koala almost forfeits his new home in a sedate bear community because of a rowdy sleepwalk.
ISBN 0-07-054192-2
[1. Koalas—Fiction] I. Title.
PZ7.R84 Oh [E] 78-10838
78 MARC

Ruckman, Ivy.
Encounter / by Ivy Ruckman. — 1st ed. — Garden City, N.Y. : Doubleday, c1978.
215 p. ; 22 cm.
SUMMARY: A young boy's encounter with an unidentified flying object drastically alters his life.
ISBN 0-385-14150-5 : $7.95. ISBN 0-385-14151-3 lib. bdg.
[1. Unidentified flying objects—Fiction] I. Title.
PZ7.R844 En [Fic] 78-16736
78 MARC

Ruckman, Ivy.
Melba the Brain / Ivy Ruckman ; illustrated by Ruth Van Sciver. — 1st ed. — Philadelphia : Westminster Press, c1979.
132 p. : ill. ; 21 cm.
SUMMARY: A fifth-grade genius' experiments with spliced energies acquaint her with an unusual cat and hurtle her light years beyond Earth to a planet ruled by animals.
ISBN 0-664-32655-2
[1. Science fiction. 2. Animals—Fiction] I. Van Sciver, Ruth. II. Title.
PZ7.R844 Me [Fic] 79-19326
79 MARC

Ruffins, Reynold.
My brother never feeds the cat / Reynold Ruffins. — New York : Scribner, c1979.
[32] p. : col. ill. ; 20 x 24 cm.
SUMMARY: A little girl tells everyone she meets about all the work she does and all that her brother doesn't do.
ISBN 0-684-16211-3 : $7.95
[1. Brothers and sisters—Fiction] I. Title.
PZ7.R88546 My [E] 78-31548
79 MARC

Russ, Joanna, 1937-
Kittatinny : a tale of magic / Joanna Russ ; illustrated by Loretta Li. — 1st ed. — New York : Daughters Pub. Co., c1978.
92 p. : ill. ; 28 cm.
SUMMARY: A woman warrior, a sleeping beauty, a baby satyr, and a mermaid encourage a young girl into accepting free womanhood.
ISBN 0-913780-24-3 : $5.00
[1. Fantasy] I. Li, Loretta. II. Title.
PZ7.R9 Ki [Fic] 77-94981
78 MARC

Ryan, John, 1921-
Doodle's homework / or, The fuddi-duddi-dodo's great mathematical experiment / story and pictures by John Ryan. — London : A Deutsch, 1978.
[32] p. : col. ill. ; 26 cm. GB***
ISBN 0-233-96998-5 : £3.50
I. Title.
PZ7.R954 Do 79-303084
79 MARC

CHILDREN'S LITERATURE (Continued)

Sachar, Louis. 1954-
Sideways stories from Wayside School / by Louis Sachar ; illustrated by Dennis Hockerman. — Chicago : Follett Pub. Co., c1978.
141 p. : ill. ; 23 cm.
SUMMARY: Humorous episodes from the classroom on the thirtieth floor of Wayside School, which was accidentally built sideways with one classroom on each story.
ISBN 0-695-80964-4 : $7.95. ISBN 0-695-40964-6 lib. bdg. : $7.98
[1. School stories. 2. Humorous stories] I. Hockerman, Dennis. II. Title.
PZ7.S1185 Si [Fic] 78-3213
78 MARC

Sachs, Marilyn.
A secret friend / Marilyn Sachs. — Garden City, N.Y. : Doubleday, c1978.
111 p. ; 22 cm.
SUMMARY: Ten-year-old Jessica loses her best friend, and in struggling to win her back finds out some important things about friendship.
ISBN 0-385-13569-6 : $5.95. ISBN 0-385-13570-X lib. bdg. : $6.95
[1. Friendship—Fiction] I. Title.
PZ7.S1187 Se [Fic] 77-25606
78 MARC

Sachs, Marilyn.
A summer's lease / Marilyn Sachs. — 1st ed. — New York : Dutton, c1979.
124 p. ; 22 cm.
SUMMARY: A teenage girl's driving ambition to be a writer prevents her from forming normal friendships. A favorite teacher tries to help her.
ISBN 0-525-40480-5
[1. Authorship—Fiction. 2. Teachers—Fiction] I. Title.
PZ7.S1187 Su 1979 [Fic] 78-12486
78 MARC

Sage, Alison.
The ogre's banquet / [Alison Sage ; ill., Gian Calvi] — 1st ed. in the U.S.A. — Garden City, N.Y. : Doubleday, c1978.
[31] p. : col. ill. ; 30 cm.
"The Benn book collection."
SUMMARY: When he is called upon to serve the local ogre, the seventh son of a seventh son charms him with his flute.
ISBN 0-385-14360-5 : $5.95. ISBN 0-385-14462-8 lib. bdg.
[1. Ghouls and ogres—Fiction] I. Calvi, Gian. II. Title.
PZ7.S1303 Og 1978 [E] 77-18648
78 MARC

Salkey, Andrew.
Hurricane / Andrew Salkey. — Oxford ; New York : Oxford University Press, 1979.
89 p. ; 21 cm. GB***
SUMMARY: Thirteen-year-old Joe and his nine-year-old sister face their first hurricane in Kingston, Jamaica with a mixture of excitement and curiosity that gradually turns to fear.
[1. Hurricanes—Fiction. 2. Jamaica—Fiction] I. Title.
PZ7.S1526 Hu 1979 [Fic] 79-317572
79 MARC

Sallis, Susan.
An open mind : a novel / by Susan Sallis. — 1st ed. — New York : Harper & Row, c1978.
139 p. ; 21 cm.
SUMMARY: Torn by conflicting loyalties and emotions, a teen-age boy finally comes to terms with his father's pending remarriage with the help of a spastic boy.
ISBN 0-06-025162-X : $7.95. ISBN 0-06-025163-8 lib. bdg. : $7.49
[1. Fathers and sons—Fiction. 2. Physically handicapped—Fiction. 3. Divorce—Fiction] I. Title.
PZ7.S1533 Op 1978 [Fic] 77-25678
78 MARC

Salus, Naomi Panush.
My daddy's mustache / yby Naomi Panush Salus ; pictures by Tomie de Paola. — 1st ed. — Garden City, N.Y. : Doubleday, c1979.
[32] p. : ill. ; 22 cm.
SUMMARY: A little boy learns why his father won't shave off his mustache.
ISBN 0-385-13188-7 : $5.95. ISBN 0-385-13189-5 lib. bdg.
[1. Mustache—Fiction] I. De Paola, Thomas Anthony. II. Title.
PZ7.S1553 My [E] 78-68363
78 MARC

Salzman, Yuri.
Hope you're feeling better / Yuri Salzman. — 1st ed. — New York : Harper & Row, c1979.
[32] p. : col. ill. ; 20 cm.
SUMMARY: Morry the bear finds being a camp counselor is more exciting than he expected.
ISBN 0-06-025164-6 : $7.95. ISBN 0-06-025165-4 lib. bdg. : $6.79
[1. Camping—Fiction. 2. Bears—Fiction] I. Title.
PZ7.S1556 Ho 1979 [E] 77-25655
78 MARC

Sampson, Fay.
Half a welcome / [by] Fay Sampson. — London : Dobson, 1977.
107 p. ; 21 cm. GB78-00559
ISBN 0-234-72031-X : £2.50
I. Title.
PZ7.S162 Hal 78-317999
78 MARC

Sandburg, Carl, 1878-1967.
The wedding procession of the rag doll and the broom handle and who was in it / by Carl Sandburg ; pictures by Harriet Pincus. — 1st Voyager/HBJ ed. — New York : Harcourt Brace Jovanovich, 1978.
[32] p. : col. ill. ; 20 x 25 cm. — (A Voyager/HBJ book)
SUMMARY: The rag doll and the broom handle marry and have a grand wedding procession that includes the Easy Ticklers, the Chubby Chubbs, and the Sleepyheads.
ISBN 0-15-695487-7
[1. Weddings—Fiction] I. Pincus, Harriet. II. Title.
[PZ7.S1965 We 1978] [E] 78-7912
78 MARC

Sanger, Richard Harlakenden, 1905-
Holly's hurricane / by Richard Sanger ; illustrated by Joy Seidler. — Martha's Vineyard, Mass. : Tashmoo Press, c1976.
[42] p. : ill. (some col.) : 21 x 28 cm.
SUMMARY: Two little girls, one outgoing, the other shy and retiring, visit their grandparents on Martha's Vineyard and have many adventures including coping with a full-fledged hurricane.
[1. Hurricanes—Fiction. 2. Martha's Vineyard, Mass.—Fiction. 3. Islands—Fiction] I. Seidler, Joy. II. Title.
PZ7.S228 Ho [Fic] 76-40749
79 MARC

Sargent, Sarah.
Edward Troy and the witch cat / Sarah Sargent ; illustrated by Emily McCully. — Chicago : Follett Pub. Co., c1978.
144 p. : ill. ; 23 cm.
SUMMARY: Relates the experiences of Edward Troy the spring and summer he is 9 years old.
ISBN 0-695-40876-3 (lib. bdg.)
[1. Cats—Fiction. 2. Friendship—Fiction. 3. City and town life—Fiction. 4. Afro-Americans—Fiction] I. McCully, Emily Arnold. II. Title.
PZ7.S2479 Ed [Fic] 77-89686
79 MARC

Savitt, Sam.
The Dingle Ridge fox, and other stories / written and illustrated by Sam Savitt. — New York : Dodd, Mead, c1978.
122 p. : ill. ; 22 cm.
SUMMARY: A collection of short stories about a variety of animals and their relationship with each other and with humans.
ISBN 0-396-07614-9
1. Animals, Legends and stories of. [1. Animals—Fiction. 2. Short stories] I. Title.
PZ10.3.S259 Di [Fic] 78-7739
78 MARC

Savitz, Harriet May.
Run, don't walk : a novel / by Harriet May Savitz. — New York : Watts, 1979.
122 p. ; 22 cm.
SUMMARY: When a teenager returns to school in a wheelchair after a debilitating accident, all she wants is to be left alone. However, a handicapped activist insists she join his crusade.
ISBN 0-531-02897-6
[1. Physically handicapped—Fiction] I. Title.
PZ7.S2664 Ru [Fic] 79-14612
79 MARC

Say, Allen.
The ink-keeper's apprentice / by Allen Say. — New York : Harper & Row, c1979.
185 p. ; 21 cm.
SUMMARY: A 14-year-old boy lives on his own in Tokyo and becomes apprenticed to a famous Japanese cartoonist.
ISBN 0-06-025208-1 : $6.95. ISBN 0-06-025209-X lib. bdg. : $6.79
[1. Cartoonists—Fiction. 2. Artists—Fiction. 3. Japan—Fiction] I. Title.
PZ7.S2744 In 1979 [Fic] 78-20264
78 MARC

Scarry, Richard.
Richard Scarry's Little bedtime book. — 1st American ed. — New York : Random House, c1978.
[48] p. : col. ill. ; 15 cm. — (Richard Scarry's Best little books ever)
Stories adapted from Richard Scarry's Funniest storybook ever.
SUMMARY: Four short stories include "The Three Fishermen," The Accident," "The Three Sitters," and "The Unlucky Day."
ISBN 0-394-83967-6 : $0.95
[1. Animals—Fiction. 2. Short stories] I. Scarry, Richard. Funniest storybook ever. II. Title. III. Title: Little bedtime book.
PZ7.S327 Rks 1978 [E] 78-55298
79 MARC

Scarry, Richard.
Richard Scarry's Little counting book. — 1st American ed. — New York : Random House, c1978.
[48] p. : chiefly col. ill. ; 15 cm.
SUMMARY: Willy Bunny counts items from 1 to 20.
ISBN 0-394-83966-8 : $0.95
[1. Counting. 2. Animals—Fiction] I. Title. II. Title: Little counting book.
PZ7.S327 Skt 1979 [E] 78-55299
79 MARC

Scarry, Richard.
Richard Scarry's Lowly Worm sniffy book. — New York : Random House, c1978.
[22] p. : col. ill. ; 17 cm. — (A Sniffy book)
SUMMARY: Lowly Worm and Huckle the cat explore some smells of the four seasons of the year. The reader can experience the same fragrances by scratching and sniffing treated pages.
ISBN 0-394-83778-9 : $2.95
[1. Seasons—Fiction] I. Title. II. Title: Lowly worm sniffy book.
PZ7.S327 Rky [E] 77-90054
78 MARC

Scarry, Richard.
Richard Scarry's Mr. Fixit and other stories. — 1st American ed. — New York : Random House, c1978.
[48] p. : col. ill. ; 15 cm. — (Richard Scarry's best little books ever)
"Adapted from Richard Scarry's ... Funniest storybook ever."
SUMMARY: Four short stories include "Mr. Fixit," "Ma Pig's New Car," "Absent-minded Mr. Rabbit," and "Tanglefoot."
ISBN 0-394-83965-X : $0.95
[1. Animals—Fiction. 2. Short stories] I. Title. II. Title: Mr. Fixit and other stories.
PZ7.S327 Rli 1978 [E] 78-55300
00 MARC

Schaffer, Ulrich, 1942-
Zilja's secret plan / by Ulrich Schaffer ; illustrated by Takashi Shoji. — 1st U.S. ed. — Grand Rapids : Eerdmans, 1978.
[32] p. : col. ill. ; 25 cm.
Translation of Siljas geheimer Plan.
SUMMARY: A young girl convinces her fellow townspeople to conserve the natural resources of their beautiful valley.
ISBN 0-8028-3514-7 lib. bdg. : $3.95
[1. Conservation of natural resources—Fiction] I. Shoji, Takashi. II. Title.
PZ7.S3336 Zi [E] 78-6404
78 MARC

Schell, Mildred.
The tiniest Christmas star / by Masahiro Kasuya ; adapted by Mildred Schell. — Valley Forge, PA : Judson Press, [1979]
[24] p. : col. ill. ; 22 x 30 cm.
SUMMARY: A tiny star carries a message of love to Jesus on the night he is born.
ISBN 0-8170-0822-5 : $5.95
[1. Stars—Fiction. 2. Christmas stories] I. Kasuya, Masahiro. II. Title.
PZ7.S3432 Ti 1979 [E] 79-116818
79 MARC

Schellie, Don, 1932-
Kidnapping Mr. Tubbs / Don Schellie. — New York : Four Winds Press, c1978.
182 p. ; 22 cm.
SUMMARY: Two teenagers sneak a very elderly ex-cowboy out his rest home for a visit to the ranch where he spent his life. Their trip is full of unusual adventures and Mr. Tubbs proves a surprising traveling compainon.
ISBN 0-590-07542-X : $6.95
[1. Old age—Fiction. 2. Arizona—Fiction] I. Title.
PZ7.S3433 Ki [Fic] 78-6153
78 MARC

Schermer, Judith.
Mouse in house / Judith Schermer. — Boston : Houghton Mifflin, 1979.
32 p. : col. ill. ; 21 x 26 cm.
SUMMARY: A family nearly tears the house apart trying to oust a mouse but their daughter saves the day with an inventive scheme.
ISBN 0-395-27801-5 : $6.95
[1. Family life—Fiction. 2. Mice—Fiction] I. Title.
PZ7.S3439 Mo [E] 78-24406
78 MARC

Schick, Alice.
Santaberry and the Snard / Alice and Joel Schick. — Philadelphia : Lippincott, c1979.
47 p. : col. ill. ; 24 cm. — (A Lippincott I-like-to-read book)
SUMMARY: A nearsighted Snard causes problems for Santa Claus on Christmas Eve.
ISBN 0-397-31824-3 (lib. bdg.) : $6.95
[1. Santa Claus—Fiction. 2. Animals—Fiction] I. Schick, Joel, joint author. II. Title.
PZ7.S3443 San [E] 78-23796
78 MARC

Schick, Eleanor, 1942-
One summer night / Eleanor Schick. — New York : Greenwillow Books, c1977.
[32] p. : col. ill. ; 26 cm.
SUMMARY: Laura unknowingly triggers a chain of neighborhood events the night she dances instead of going to bed.
ISBN 0-688-80072-6. ISBN 0-688-84072-8 lib. bdg. : $6.95
[1. Summer—Fiction. 2. City and town life—Fiction] I. Title.
PZ7.S3445 On [E] 76-25199
76 MARC

Schick, Eleanor, 1942-
Summer at the sea / Eleanor Schick. — 1st ed. — New York : Greenwillow Books, c1979.
56 p. : col. ill. ; 22 cm. — (Greenwillow read-alone)
SUMMARY: A young girl returns to the city at summer's end and misses her activities at the beach.
ISBN 0-688-80116-1. ISBN 0-688-84116-3 lib. bdg.
[1. Seashore—Fiction] I. Title.
PZ7.S3445 Su [E] 77-3026
77 MARC

Schleier, Curt, 1944-
You'd better not tell! / By Curt Schleier. — 1st ed. — Philadelphia : Westminster Press, c1979.
93 p. ; 21 cm.
SUMMARY: Living in a neighborhood where the police are the enemy, a young boy must make a difficult decision when he finds evidence that his best friend's brother is a mugger.
ISBN 0-664-32646-3
[1. Crime and criminals—Fiction. 2. City and town life—Fiction] I. Title.
PZ7.S3469 Yo [Fic] 78-20921
78 MARC

CHILDREN'S LITERATURE (Continued)

Schulman, Janet.
Camp KeeWee's secret weapon / by Janet Schulman pictures by Marylin Hafner. — 1st ed. — New York : Greenwillow Books, c1979.
63 p. : col. ill. ; 22 cm. — (Greenwillow read-alone)
SUMMARY: Jill's softball talent helps her find her niche at camp.
ISBN 0-688-80185-4 : $5.95. ISBN 0-688-84185-6 lib. bdg. : $5.71
[1. Camping—Fiction. 2. Softball—Fiction] I. Hafner, Marylin. II. Title.
PZ7.S3866 Cam [E] 78-16742
78 MARC

Schulman, Janet.
The great big dummy / Janet Schulman ; pictures by Lillian Hoban. — 1st ed. — New York : Greenwillow books, c1979.
30 p. : col. ill. ; 22 cm. — (Greenwillow read-alone)
SUMMARY: When Anna's friends can't play with her she creates a playmate of her own.
ISBN 0-688-80208-7 : $5.95. ISBN 0-688-84208-9 lib. bdg. : $5.71
[1. Friendship—Fiction] I. Hoban, Lillian. II. Title.
PZ7.S3866 Gr [E] 78-27728
78 MARC

Schulz, Charles M
A Charlie Brown Thanksgiving, [by] Charles M. Schulz. New York, Random House c1974.
[44] p. : col. ill. ; 25 cm.
"A Charlie Brown special."
SUMMARY: Charlie Brown is in a dilemma when Peppermint Pattie invites herself and several friends to his house for Thanksgiving dinner.
ISBN 0-394-83047-4
[1. Thanksgiving Day—Fiction] I. Title.
PZ7.S38877 Ch 741'.5'973 74-17004
74 MARC

Schulz, Charles M
It's your first kiss, Charlie Brown / Charles M. Schulz. — New York : Random House, c1978.
[44] p. : col. ill. ; 25 cm.
"A Charlie Brown special."
SUMMARY: Charlie Brown panics when selected to escort the homecoming queen to the big post-football game dance.
ISBN 0-394-83955-2 : $3.50. ISBN 0-394-93955-7 lib. bdg. : $4.39
[1. School stories] I. Title.
PZ7.S38877 Iu [E] 78-7460
78 MARC

Schulz, Charles M
Race for your life, Charlie Brown / Charles M. Schulz. — 1st ed. — New York : Holt, Rinehart, and Winston, c1978.
[126] p. : col. ill. ; 29 cm.
Based on the feature film with the same title, written and created by Charles M. Schulz.
SUMMARY: Charlie Brown and his friends compete in a raft race against a group of bullies who use some unsportsmanlike conduct.
ISBN 0-03-042646-4 : $10.95.
[1. Boats and boating—Fiction] I. Title.
PZ7.S38877 Rac [E] 77-15208
78 MARC

Schultz, Gwen M
The blue valentine / Gwen Schultz ; pictures by Elizabeth Coberly. — Newly illustrated ed. — New York : Morrow, 1979, c1965.
53 p. : ill. ; 20 cm.
SUMMARY: New to her school, Cindy strives to make a contact with her teacher by sending her a valentine in her favorite color.
ISBN 0-688-22176-9. ISBN 0-688-32176-3 lib. bdg.
[1. St. Valentine's Day—Fiction] I. Coberly, Elizabeth. II. Title.
PZ7.S3875 Bl 1979 [E] 78-12184
78 MARC

Schurch, Maylan.
Mark, Legion, and the little red schoolhouse / Maylan Schurch. — Washington : Review and Herald Pub. Association, c1978.
126 p. : ill. ; 21 cm.
SUMMARY: A family searches for a church that closely adheres to the Bible.
[1. Seventh-Day Adventists—Fiction. 2. Christian life—Fiction] I. Title.
PZ7.S3966 Mar [Fic] 77-88347
78 MARC

Schwartz, Sheila.
Growing up guilty : a novel / by Sheila Schwartz. — New York : Pantheon Books, c1978.
177 p. ; 22 cm.
SUMMARY: An intelligent, naive, and lonely girl in Brooklyn grows up in the late 1930's, coping with first love, communist ideology, and changing ideas of her self-perception.
ISBN 0-394-83606-5. ISBN 0-394-93606-X lib. bdg.
I. Title.
PZ7.S4113 Gr [Fic] 78-7720
78 MARC

Schweninger, Ann.
A dance for three / Ann Schweninger. — New York : Dial Press, c1979.
[32] p. : col. ill. ; 20 cm.
SUMMARY: In three wordless stories, three friends try to make the moon appear, a bewildered gentleman has difficulties wooing a young lady, and a musician proves he has more than musical talent.
ISBN 0-8037-1629-X : $6.95
[1. Stories without words. 2. Short stories] I. Title.
PZ7.S41263 Dan [E] 78-72197
78 MARC

Scoppettone, Sandra.
Happy endings are all alike : a novel / by Sandra Scoppettone. — 1st ed. — New York : Harper & Row, c1978.
202 p. ; 21 cm.
SUMMARY: Small town prejudices emerge when a love affair between two teenage girls is revealed.
ISBN 0-06-025239-1 : $6.95. ISBN 0-06-025240-5 lib. bdg. : $6.79
[1. Lesbians—Fiction. 2. Rape—Fiction. 3. Prejudices—Fiction]
PZ7.S4136 Hap 1978 [Fic] 78-2976
78 MARC

Seed, David.
Stream runner / David Seed. — New York : Four Winds Press, c1979.
185 p. ; 22 cm.
SUMMARY: Leif Collins is in no hurry to grow up as he enjoys trout fishing and swimming with his friends in his northern California town.
ISBN 0-590-07568-3 : $7.95
[1. Fishing—Fiction. 2. Friendship—Fiction] I. Title.
PZ7.S4512 St [Fic] 78-21769
78 MARC

Segal, Joyce.
It's time to go to bed / by Joyce Segal ; pictures by Robin Eaton. — 1st ed. — Garden City, N.Y. : Doubleday, c1979.
[32] p. : col. ill. ; 20 x 24 cm.
SUMMARY: Presents the numerous familiar and usually unsuccessful excuses used to delay bedtime.
ISBN 0-385-14034-7 : $6.95
[1. Sleep—Fiction. 2. Animals—Fiction] I. Eaton, Robin. II. Title.
PZ7.S4526 It [E] 78-20284
78 MARC

Ségur, Sophie Rostopchine, comtesse de, 1799-1874.
The angel inn = (L'auberge de l'ange-gardien) / the Comtesse de Ségur ; translated by Joan Aiken ; illustrated by Pat Marriott. — 1st ed. — Owings Mills, Md. : Stemmer House Publishers, 1978, c1976.
231 p., [8] leaves of plates : ill. (some col.) ; 20 cm.
"A Barbara Holdridge book."
SUMMARY: A soldier who finds two abandoned boys entrusts them to the care of the innkeepers of the Guardian Angel, thus starting the boys and their benefactors on a series of adventures.
ISBN 0-916144-28-3 : $9.95. ISBN 0-916144-29-1 pbk. : $6.95
[1. Hotels, motels, etc.—Fiction] I. Marriott, Pat, 1920- . II. Title.
[PZ7.S454 An 1978] [Fic] 78-12784
78 MARC

Shafer, Ted.
The great silver air fleet / Ted Shafer ; [ill. by John Hunt ; photos. by Ed Lawless ; Douglas DC-3 twin-engine transport by Ray Hernandez ; other flying model airplanes by Ted Shafer]. — [Mill Valley, Calif.] : Merry Makers Press, [c1978]
131 p. : ill. ; 22 cm.
SUMMARY: Follows the escapades of the Great Silver Air Fleet as they thrill their radio audience with their many adventures.
$5.95
[1. Air pilots—Fiction. 2. Radio serials—Fiction. 3. Adventure stories] I. Hunt, John, 1947- . II. Title.
PZ7.S5276 Gr [Fic] 78-112714
00 MARC

Shapiro, Irwin, 1911-
The hungry ghost mystery / by Irwin Shapiro ; Illustrated by William Hutchinson. — Champaign, Ill. : Garrard Pub. Co., c1978.
63 p. : ill. ; 23 cm. — (A Garrard mystery book)
SUMMARY: The evidence of a ghost in their new house seems conclusive until two boys discover something else in an old graveyard.
ISBN 0-8116-6403-1
[1. Mystery and detective stories] I. Hutchinson, William. II. Title.
PZ7.S5295 Hu [E] 78-5841
78 MARC

Sharmat, Marjorie Weinman.
The 329th friend / by Marjorie Weinman Sharmat ; ill. by Cyndy Szekeres. — New York : Four Winds Press, c1979.
[48] p. : col. ill.; 19 x 26 cm.
SUMMARY: Bored with his own company, Emery Raccoon invites 328 guests to lunch but finds that none of them have time to listen to him.
ISBN 0-590-07558-6 : $8.95
[1. Friendship—Fiction. 2. Animals—Fiction] I. Szekeres, Cyndy. II. Title.
PZ7.S5299 Tk [E] 78-21770
78 MARC

Sharmat, Marjorie Weinman.
Griselda's new year / by Marjorie Weinman Sharmat ; pictures by Normand Chartier. — New York : Macmillan, c1979.
63 p. : ill. (some col.) ; 23 cm. — (Ready-to-read)
SUMMARY: Griselda Goose attempts to carry out her New Year's resolutions, but her good deeds backfire.
ISBN 0-02-782420-9
[1. Geese—Fiction. 2. New Year—Fiction] I. Chartier, Normand. II. Title.
PZ7.S5299 Gr [E] 79-11375
79 MARC

Sharmat, Marjorie Weinman.
Mitchell is moving / by Marjorie Weinman Sharmat ; pictures by Jose Aruego & Ariane Dewey. — New York : Macmillian, c1978.
47 p. : col. ill. ; 23 cm. — (Ready-to-read)
SUMMARY: A dinosaur's exuberance about moving cools considerably when he realizes how much he misses his next-door friend.
ISBN 0-02-782410-1
[1. Moving, Household—Fiction. 2. Friendship—Fiction. 3. Dinosaurs—Fiction] I. Aruego, Jose. II. Dewey, Ariane. III. Title.
PZ7.S5299 Mi 1978 [E] 78-6816
78 MARC

Sharmat, Marjorie Weinman.
Mr. Jameson and Mr. Phillips / by Marjorie Weinman Sharmat ; pictures by Bruce Degen. — 1st ed. — New York : Harper & Row, c1979.
48 p. : col. ill. ; 26 cm.
SUMMARY: Two friends go searching for a special place away from the pushing and shoving of the crowded cities, but find it harder to get away than they thought.
ISBN 0-06-025528-5. ISBN 0-06-025529-3 lib. bdg. : $7.89
[1. Population—Fiction. 2. Privacy, Right of—Fiction] I. Degen, Bruce. II. Title.
PZ7.S5299 Mh 1979 [Fic] 77-25665
78 MARC

Sharmat, Marjorie Weinman.
Mooch the messy meets Prudence the neat / vy Marjorie Weinman Sharmat ; pictures by Ben Shecter. — New York : Coward, McCann & Geoghegan, c1978.
63 p. : ill. ; 22 cm. — (A Break-of-day book)
SUMMARY: A very messy rat worries about his friendship with his extremely neat new neighbor.
ISBN 0-698-30703-8
[1. Cleanliness—Fiction. 2. Rats—Fiction. 3. Friendship—Fiction] I. Shecter, Ben. II. Title.
PZ7.S5299 Mnm 1978 [E] 77-29049
78 MARC

Sharmat, Marjorie Weinman.
Octavia told me a secret / by Marjorie Weinman Sharmat ; illustrated by Roseanne Litzinger. — New York : Four Winds Press, c1979.
[32] p. : col. ill. ; 20 x 24 cm.
SUMMARY: A child delights in sharing a secret with her friend but can't resist toying with the idea of swapping or telling it.
ISBN 0-590-07605-1 (lib. bdg.) : $6.95
[1. Secrets—Fiction. 2. Friendship—Fiction] I. Litzinger, Roseanne. II. Title.
PZ7.S5299 Oc * [E] 79-10425
79 MARC

Sharmat, Marjorie Weinman.
Scarlet Monster lives here / by Marjorie Weinman Sharmat ; illustrated by Denis Kendrick. — 1st ed. — New York : Harper & Row, c1979.
64 p. : col. ill. ; 23 cm. — (An I can read book)
SUMMARY: New in the neighborhood, Scarlet Monster keeps herself so busy preparing for guests she doesn't have time to realize they are waiting to visit.
ISBN 0-06-025526-9 : $6.95. ISBN 0-06-025527-7 lib. bdg. : $5.79
[1. Neighborliness—Fiction. 2. Moving, Household—Fiction. 3. Monsters—Fiction] I. Kendrick, Denis. II. Title.
PZ7.S5299 Sc 1979 [E] 78-19484
78 MARC

Sharmat, Marjorie Weinman.
Uncle Boris and Maude / by Marjorie Weinman Sharmat ; pictures by Sammis McLean. — 1st ed. — Garden City, N.Y. : Doubleday, c1979.
62 p. : color ill. ; 25 cm. — (A Reading-on-my-own book)
SUMMARY: Maude and Uncle Boris discover a cure for their boredom.
ISBN 0-385-12946-7 : $4.95.
[1. Boredom—Fiction. 2. Moles (Animals)—Fiction] I. McLean, Sammis. II. Title.
PZ7.S5299 Un [E] 78-18565
78 MARC

Sharmat, Marjorie Weinman.
Thornton, the worrier / by Marjorie Weinman Sharmat ; illustrated by Kay Chorao. — New York : Holiday House, c1978.
[32] p. : ill. ; 21 cm.
SUMMARY: Thornton Rabbit worries about everything until he meets an old man who worries about nothing.
ISBN 0-8234-0328-9 lib. bdg. : $5.95
[1. Worry—Fiction] I. Chorao, Kay. II. Title.
PZ7.S5299 Th [E] 78-1286
78 MARC

Sharmat, Mitchell.
Reddy Rattler and pictures Easy Eagle / by Mitchell Avell Sharmat ; pictures by Marc Simont. — 1st ed. — Garden City, N.Y. : Doubleday, c1979.
63 p. : col. ill. ; 24 cm. — (A Reading-on-my-own book)
SUMMARY: An ambitious rattlesnake is discouraged about his future until his eagle friend helps him make use of his natural talents.
ISBN 0-385-14215-3 : $4.95. ISBN 0-385-14216-1 lib. bdg.
[1. Rattlesnakes—Fiction. 2. Eagles—Fiction] I. Simont, Marc. II. Title.
PZ7.S52992 Re [E] 78-20924
78 MARC

CHILDREN'S LITERATURE (Continued)

Sharp, Margery, 1905-
Bernard into battle : a Miss Bianca story / Margery Sharp ; ill. by Leslie Morrill. — 1st American ed. — Boston : Little, Brown, c1978.
87 p. : ill. ; 21 cm.
SUMMARY: Miss Bianca's faithful lieutenant, Bernard, directs a hazardous operation that repulses the army of rats infesting the regions under the Ambassador's cellar.
ISBN 0-316-78326-9 : $7.95
[1. Mice—Fiction] I. Morrill, Leslie H. II. Title.
PZ7.S5315 Bc 1978 [Fic] 78-11332
78 MARC

Shea, George.
Big bad Ernie / by George Shea ; [photos. by Ted Williams]. — Mankato, MN : Creative Education, c1978.
56 p. : ill. ; 24 cm.
SUMMARY: No one denies that Ernie is a skilled athlete but can he withstand the pressure when his baseball team reaches the playoffs?
ISBN 0-87191-679-7
[1. Baseball—Fiction] I. Williams, Ted C., 1930- II. Title.
PZ7.S5375 Bi 78-12632
78 MARC

Shea, George.
I died here / George Shea ; [illustrator, Bill Shields]. — Belmont, Calif. : Fearon Pitman Publishers, c1979.
60 p. ill. ; 18 cm. — (A Pacemaker bestellers book)
SUMMARY: Haunting dreams lead a young man to a small Greek village where he uncovers a murder.
ISBN 0-8224-5364-9
[1. Mystery and detective stories] I. Shields, Bill. II. Title.
[PZ7.S5375 Iac] [Fic] 78-72329
79 MARC

Shea, George.
Nightmare Nina / by George Shea ; [illustrated by Dan Hanrahan]. — Mankato, Minn. : Creative Education, c1978.
56 p. : ill. ; 24 cm.
SUMMARY: A girl whose nightmares often come true dreams of a kidnapped boy and soon all the details of that dream actually come true.
ISBN 0-87191-682-7
[1. Dreams—Fiction. 2. Kidnapping—Fiction. 3. Mystery and detective stories] I. Hanrahon, Dan. II. Title.
PZ7.S5375 Ni [Fic] 78-71857
00 MARC

Shearer, John, 1947-
Billy Jo Jive and the case of the missing pigeons / by John Shearer ; pictures by Ted Shearer. — New York : Delacorte Press, c1978.
42 p. : ill. ; 24 cm.
SUMMARY: Young detectives search the streets for a pigeon thief.
ISBN 0-440-00567-1. ISBN 0-440-00568-X lib. bdg.
[1. Mystery and detective stories. 2. City and town life—Fiction] I. Shearer, Ted. II. Title.
PZ7.S5379 Bf [E] 78-50409
78 MARC

Shecter, Ben.
The river witches / Ben Shecter. — 1st ed. — New York : Harper & Row, c1979.
180 p. : ill. ; 22 cm.
SUMMARY: A young boy becomes caught in the struggle between good witches and bad witches.
ISBN 0-06-026843-3
[1. Witchcraft—Fiction] I. Title.
PZ7.S5382 Ri [Fic] 75-25397
78 MARC

Sheedy, Alexandra Elizabeth.
She was nice to mice : the other side of Elizabeth I's character never before revealed by previous historians / by Alexandra Elizabeth Sheedy ; illustrated by Jessica Ann Levy. — New York : McGraw-Hill, [1975]
95 p. : ill. ; 19 cm.
SUMMARY: The memoirs of a literary mouse living at the court of Elizabeth I reveal the public and private life of the Queen and her courtiers.
ISBN 0-07-056515-5 : $5.95. ISBN 0-07-056516-3 lib. bdg. : $5.83
1. Elizabeth, Queen of England, 1533-1603—Juvenile fiction. [1. Elizabeth, Queen of England, 1533-1603—Fiction. 2. Great Britain—History—Elizabeth, 1558-1603—Fiction. 3. Mice—Fiction. 4. Children's writings] I. Levy, Jessica Ann. II. Title.
PZ7.S53825 Sh [Fic] 75-9960
75[r79]rev MARC

Sherwood, Mary Martha (Butt) 1775-1851.
The history of Little Henry and his bearer. 2d American, from the 2d London ed. Hartford, Hudson and Co. printers, 1817.
32 p. 17 cm.
Caption title: Little Henry and his bearer.
I. Title. II. Title: Little Henry and his bearer.
PZ6.S554 Hi 1817 78-316918
78 MARC

Shreve, Susan Richards.
Family secrets : five very important series / by Susan Shreve ; illustrated by Richard Cuffari. — New York : Knopf : distributed by Random House, c1979.
56 p. : ill. ; 22 cm.
SUMMARY: Eight-year-old Sammy tries to come to terms with several difficult situations including the death of his dog, the divorce of his aunt and uncle, the suicide of his best friend's brother, coping with his terminally ill grandmother, and cheating on a school test.
ISBN 0-394-83896-3. ISBN 0-394-93896-8 lib. bdg.
[1. Family Problems—Fiction] I. Cuffari, Richard, 1925- II. Title.
PZ7.S55915 Fam 1979 [Fic] 78-12471
78 MARC

Shub, Elizabeth.
Seeing is believing / by Elizabeth Shub ; pictures by Rachel Isadora. — 1st ed. — New York : Greenwillow Books, c1979.
63 p. : ill. ; 22 cm. — (Greenwillow read-alone)
SUMMARY: Two stories about Tom and the piskies, "The Leprechaun's Trick" and "Pisky Mischief."
ISBN 0-688-80211-7 : $5.95
[1. Leprechauns—Fiction] I. Isadora, Rachel. II. Title.
PZ7.S5592 Se [E] 78-12378
78 MARC

Shura, Mary Francis.
The Barkley Street six-pack / Mary Francis Shura ; illustrated by Gene Sparkman. — New York : Dodd, Mead, c1979.
159 p. : ill. ; 21 cm.
SUMMARY: Shy, plain Jane is delighted with her new best friend, but this association costs her the friendship of all the other kids on Barkley Street.
ISBN 0-396-07714-5
[1. Friendship—Fiction] I. Sparkman, Gene. II. Title.
PZ7.S55983 Bar [Fic] 79-52043
79 MARC

Shura, Mary Francis.
The gray ghosts of Taylor Ridge / Mary Francis Shura ; illustrated by Michael Hampshire. — New York : Dodd, Mead, c1978.
128 p. : ill. ; 21 cm.
SUMMARY: Nat's pesty little sister, Nan, blackmails him into going out to Taylor Ridge where, an old man tells them, gray ghosts come "chinking" after dark.
ISBN 0-396-07526-6 : $5.95
[1. Ghost stories] I. Hampshire, Michael A. II. Title.
PZ7.S55983 Gr [Fic] 77-16861
77 MARC

Shura, Mary Francis.
Mister Wolf and me / Mary Francis Shura ; illustrated by Konrad Hack. — New York : Dodd, Mead, c1979.
126 p. : ill. ; 21 cm.
SUMMARY: Thirteen-year-old Miles desperately tries to prove his German shepherd has not been killing a farmer's sheep.
ISBN 0-396-07666-1
[1. Dogs—Fiction] I. Hack, Konrad. II. Title.
PZ7.S55983 Mi [Fic] 78-22432
79 MARC

Silverberg, Robert.
Time of the great freeze. [1st ed.] New York, Holt, Rinehart and Winston [1964]
192 p. 22 cm.
SUMMARY: During the receding Fifth Ice Age seven men expelled from underground New York in 2650 and one deserter of that isolated colony attempt to travel to London, where contact has been made with other people.
[1. Survival—Fiction] I. Title.
PZ7.S5858 Ti [Fic] 64-14345
[r79]rev MARC

Simmons, Diane, 1939-
Joanna, the crowing hen of Bethel : a little peasant girl meets the boy Jesus and follows his advice / by Diane Simmons. — St. Louis : Concordia Pub. House, c1979.
56 p. : ill. ; 23 cm.
SUMMARY: A young girl who is often hostile and rebellious changes her attitudes after she meets 12-year-old Jesus on the way to Jerusalem.
ISBN 0-570-07976-4 : $1.95
1. Jesus Christ—Juvenile fiction. [1. Jesus Christ—Fiction] I. Title.
PZ7.S59165 Jo [Fic] 78-12726
78 MARC

Simon, Marcia L
A special gift / Marcia L. Simon. — 1st ed. — New York : Harcourt Brace Jovanovich, c1978.
132 p. ; 21 cm.
SUMMARY: A young boy tries to cope with the difficulties of developing his skills as a dancer and at the same time maintain his identity as an athlete.
ISBN 0-15-277865-9
[1. Ballet dancing—Fiction. 2. Sex role—Fiction] I. Title.
PZ7.S60519 Sp [Fic] 78-4329
78 MARC

Simon, Norma.
We remember Philip / by Norma Simon ; pictures by Ruth Sanderson. — Chicago : A. Whitman, c1979.
[32] p. : ill. ; 24 cm. — (Concept books)
SUMMARY: Sam and his classmates seek a way to express their feelings about the accidental death of their teacher's son.
ISBN 0-8075-8709-5 lib. bdg. : $4.50
[1. Death—Fiction. 2. Grief—Fiction. 3. School stories] I. Sanderson, Ruth. II. Title.
PZ7.S6053 We [Fic] 78-11691
78 MARC

Singer, Marilyn.
It can't hurt forever / by Marilyn Singer ; ill. by Leigh Grant. — 1st ed. — New York : Harper & Row, c1978.
186 p. : ill. ; 21 cm.
SUMMARY: Eleven-year-old Ellie describes her experiences during twelve days of hospitalization for heart surgery.
ISBN 0-06-025681-8 : $7.95. ISBN 0-06-025682-6 lib. bdg. : $7.49
[1. Hospitals—Fiction. 2. Heart—Surgery—Fiction] I. Grant, Leigh. II. Title.
PZ7.S6172 It [Fic] 77-25657
78 MARC

Singer, Marilyn.
Pickle plan / by Marilyn Singer ; pictures by Steven Kellogg. — 1st ed. — New York : Dutton, c1978.
[40] p. : ill. ; 22 cm.
SUMMARY: A young girl who feels no one is interested in her devises three plans for attracting attention.
ISBN 0-525-37021-8 : $5.95
[1. Individuality—Fiction] I. Kellogg, Steven. II. Title.
PZ7.S6172 Pi 1978 [E] 78-4603
78 MARC

Sivers, Brenda.
The snailman / Brenda Sivers ; illustrated by Shirley Hughes. — 1st ed. — Boston : Little, Brown, c1978.
118 p. : ill. ; 21 cm.
SUMMARY: Recently moved to a country village, Timothy is bored and lonely until he meets the eccentric snailman who is feared and avoided by the rest of the townspeople.
ISBN 0-316-79118-0
[1. Friendship—Fiction. 2. Country life—Fiction] I. Hughes, Shirley. II. Title.
PZ7.S6237 Sn [Fic] 78-9297
78 MARC

Skármeta, Antonio.
Chileno! / Antonio Skármeta ; translation from the Spanish by Hortense Carpentier. — New York : Morrow, 1979.
92 p. ; 21 cm.
SUMMARY: A 14-year-old boy grows up in Berlin after his family is exiled from their native Chile.
ISBN 0-688-22213-7
[1. Exiles—Fiction] I. Title.
PZ7.S6257 Ch [Fic] 79-17540
79 MARC

Skolsky, Mindy Warshaw.
Carnival and kopeck and more about Hannah / by Mindy Warshaw Skolsky ; illustrated by Karen Ann Weinhaus. — 1st ed. — New York : Harper & Row, c1979.
74 p. : ill. ; 23 cm.
SUMMARY: When Hannah and her grandmother have their first quarrel, Hannah learns that living near someone you love somtimes makes problems.
ISBN 0-06-025686-9 : $7.95
[1. Grandmothers—Fiction] I. Weinhaus, Karen Ann. II. Title.
PZ7.S62836 Car 1979 [Fic] 77-25643
78 MARC

Skurzynski, Gloria.
Martin by himself / Gloria Skurzynski ; illustrated by Lynn Munsinger. — Boston : Houghton Mifflin, 1979.
36 p. : col. ill. ; 21 cm.
SUMMARY: Martin doesn't like coming home to an empty house when his mother goes back to work.
ISBN 0-395-28271-3
[1. Mothers—Employment—Fiction] I. Munsinger, Lynn. II. Title.
PZ7.S6287 Mar [E] 79-11748
79 MARC

Slaatten, Evelyn.
In the captain's shoes / Evelyn Slaatten. — New York : Watts, 1978.
122 p. ; 22 cm.
SUMMARY: An orphaned boy goes to live with his unusual grandfather, subsequently sharing many adventures and learning the meaning of friendship.
ISBN 0-531-02215-3
[1. Grandfathers—Fiction. 2. Friendship—Fiction] I. Title.
PZ7.S6288 In [Fic] 78-1986
78 MARC

Slater, Jim, 1929-
Goldenrod / Jim Slater ; illustrated by Christopher Chamberlain. — London : J. Cape, 1978.
118 p. : ill. ; 23 cm.
ISBN 0-224-01629-6 : £3.95 GB***
I. Title.
PZ7.S6293 Go 78-326195
79 MARC

Sleator, William.
Into the dream / by William Sleator ; illustrated by Ruth Sanderson. — 1st ed. — New York : Dutton, c1979.
137 p. : ill. ; 22 cm.
SUMMARY: When two youngsters realize they are having the same frightening dream, they begin searching for an explanation for this mysterious coincidence.
ISBN 0-525-32583-2
[1. Extrasensory perception—Fiction. 2. Psychokinesis—Fiction. 3. Unidentified flying objects—Fiction] I. Sanderson, Ruth. II. Title.
PZ7.S6313 In 1979 [Fic] 78-11825
78 MARC

CHILDREN'S LITERATURE (Continued)

Sleator, William.
Once, said Darlene / by William Sleator ; pictures by Steven Kellogg. — 1st ed. — New York : Dutton, c1979.
55 p. : ill. ; 24 cm. — (A Fat cat book)
SUMMARY: Darlene's stories sound unbelievable but she insists they are all true.
ISBN 0-525-36410-2
[1. Fantasy] I. Kellogg, Steven. II. Title. III. Series.
PZ7.S6313 On 1979 [E] 78 78-12643 MARC

Sloan, Glenna Davis, 1930-
Spotlight on Liz / Glenna Davis Sloan. — Toronto : Macmillan of Canada, 1977.
124 p. ; 18 cm. — (A Canadian topliner) C77-1089-X
ISBN 0-7705-1444-8 : $1.65
I. Title.
PZ7.S63313 Sp 79 78-321037 MARC

Sloan, Michael, 1946-
The haunted house / by Michael Sloan & Flight to nowhere / by James Henerson ; novelizations by Stratemeyer Syndicate. — New York : Grosset & Dunlap, c1978.
110 p. : ill. ; 23 cm.
SUMMARY: In the first of these two stories, the Hardy brothers help their father track down an amnesia victim who has access to important NATO secrets and in the second, the brothers foil a plot to do away with a famous rock star.
Stories based on the Universal Television series, Hardy boys/Nancy Drew mysteries. Developed for television by Glen A. Larson. Based on the Hardy boys books by Franklin W. Dixon. The haunted house adapted from the episode The house on possessed hill. Flight to nowhere adapted from the episode The flickering torch mystery.
ISBN 0-448-16197-4 : $3.95
[1. Mystery and detective stories] I. Larson, Glen A. II. Dixon, Franklin W. III. Henerson James. Flight to nowhere. 1978. IV. Stratemeyer Syndicate. V. Title.
PZ7.S63314 Hau [Fic] 00 78-52858 MARC

Slote, Alfred.
My trip to Alpha I / by Alfred Slote ; illustrated by Harold Berson. — Philadelphia : Lippincott, c1978.
94 p. : ill. ; 21 cm.
SUMMARY: Voya-code is the most sophisticated form of interplanetary travel but a young boy finds out it is not without dangers.
ISBN 0-397-31810-3
[1. Science fiction] I. Berson, Harold. II. Title.
PZ7.S635 Mzt [Fic] 78 78-6463 MARC

Smaridge, Norah.
Peter's tent. Illustrated by Brinton Turkle. New York, Viking Press [1965]
[32] p. col. illus. 21 cm.
SUMMARY: Peter makes a tent for his very own private place, but enjoys sharing it with friends.
[1. Tents—Fiction] I. Turkle, Brinton, illus. II. Title.
PZ7.S6392 Pe [E] [r79]rev 65-18146 MARC

Smith, Doris Buchanan.
Dreams & drummers / Doris Buchanan Smith. — 1st ed. — New York : Crowell, c1978.
180 p. ; 21 cm.
SUMMARY: Growing up is not quite so easy as she'd expected for a teenage girl in a small Georgia town.
ISBN 0-690-01381-7 : $6.95. ISBN 0-690-03843-7 lib. bdg. : $6.79
[1. Georgia—Fiction] I. Title.
PZ7.S64474 Dr 1978 [Fic] 78 77-26590 MARC

Smith, Dorothy Gladys, 1896-
The midnight kittens / Dodie Smith ; illustrated by Anne and Janet Grahame Johnstone. — London : Allen, 1978.
119 p. : ill. ; 25 cm. — (A Longbow children's book) GB***
ISBN 0-491-02403-7 : £3.25
I. Title.
PZ7.S644747 Mi 79 79-300397 MARC

Smith, Jim, 1920-
The Frog Band and Durrington Dormouse / Jim Smith. — 1st American ed. — Boston : Little, Brown, [1978] c1977.
[32] p. : col. ill. ; 30 cm.
SUMMARY: The Frog Band comes to the aid of Durrington Dormouse who wakes up in the 20th century following a Rip Van Winkle sleep to find his family being evicted from his property.
ISBN 0-316-80155-0 lib. bdg. : $6.95
[1. Animals—Fiction. 2. Space and time—Fiction] I. Title.
PZ7.S6515 Fr 1978 [Fic] 79 77-10549 MARC

Smith, Jim, 1920-
The Frog Band and the mystery of Lion Castle / Jim Smith. — 1st American ed. — Boston : Little, Brown, c1978.
[32] p. : col. ill. ; 30 cm.
SUMMARY: The Frog Band and the eminent detective, Alphonse le Flic, combine their efforts to solve the mystery of Lion Castle.
ISBN 0-316-80161-5 : $7.95
[1. Animals—Fiction. 2. Mystery and detective stories] I. Title.
PZ7.S6515 Ft 1978 [E] 79 78-61426 MARC

Smith, Pauline Coggeshall, 1908-
Brush fire! / a journalist's guide to the most needed information sources and contacts / — 1st ed. — Philadelphia : Westminster Press, c1979.
96 p. ; 21 cm.
"A Hiway book."
SUMMARY: While house sitting for his shop teacher in the California hills, a teenage boy learns a great deal about responsibility, resourcefulness, and courage when a motorcycle gang attacks the house and starts a brush fire.
ISBN 0-664-32639-0
[1. Gangs—Fiction. 2. Fires—Fiction] I. Title.
PZ7.S65752 Br [Fic] 78 78-14768 MARC

Smith, Theresa (Kalab)
Dilly Dally, written and illustrated by Theresa Kalab Smith. Austin, Tex.,Steck Co. [1961]
48 p. illus. 22 cm.
SUMMARY: A little duck who does not like to swim must call upon that skill to rescue a friend from drowning.
[1. Ducks—Fiction] I. Title.
PZ7.S65947 Di [E] [r79]rev 61-8168 MARC

Smith, Wendy.
Haddock McCraddock : a picture story / by Wendy Smith. — 1st ed. in the United States of America. — Garden City, N.Y. : Doubleday, c1978.
[38] p. : col. ill. ; 28 cm.
"The Benn book collection."
SUMMARY: What can a penniless spoon player who yodels, carves wooden animals, and has six toes on his left foot do for a living?
ISBN 0-385-14656-6 : $5.95. ISBN 0-385-14657-4 lib. bdg.
[1. Musicians—Fiction] I. Title.
PZ7.S6615 Had 1978 [E] 78 78-7136 MARC

Snyder, Carol.
Ike and Mama and the block wedding / by Carol Snyder ; drawings by Charles Robinson. — New York : Coward, McCann & Geoghegan, c1979.
80 p. : ill. ; 21 cm.
SUMMARY: Rosie Weinstein is getting married on Sunday but not without a little help from the residents of East 136th Street.
ISBN 0-698-20461-1
[1. Weddings—Fiction. 2. Neighborliness—Fiction. 3. Jews in the United States—Fiction. 4. New York (City)—Fiction] I. Robinson, Charles, 1931- . II. Title.
PZ7.S68517 Ih 1979 [Fic] 78 78-11702 MARC

Solomon, Joan, fl. 1978-
Berron's tooth / [by] Joan Solomon, photographs by Richard Harvey. — London : H. Hamilton, 1978.
[32] p. : col. ill. ; 22 cm. GB78-24659
ISBN 0-241-89783-1 : £2.25
I. Harvey, Richard. II. Title.
PZ7.S696 Be 79 79-305217 MARC

[Somerville, Elizabeth]
The history of little Phœbe, and The reclaimed child. Adorned with cuts. Hartford : Printed by Sheldon & Goodwin. <Stereotyped by J. F. & C. Starr.> [ca. 1815]
30 p. incl. front., illus. 10 cm.
Illustrated paper cover.
I. Title.
PZ6.S74 Hi [r78]rev 21-17593 MARC

Spangenberg, Judith Dunn.
The little goat / story by Judy Dunn ; photos. by Phoebe Dunn. — New York : Random House, c1978.
[32] p. : col. ill. ; 21 cm. — (A Random House picture book)
SUMMARY: Andy receives a little goat as a birthday present. He names him Sam and they soon become best friends.
ISBN 0-394-83872-6. ISBN 0-394-93872-0 lib. bdg. ISBN 0-394-83871-8 best book club
[1. Goats—Fiction. 2. Pets—Fiction] I. Dunn, Phoebe. II. Title.
PZ7.S7365 Li 1978 [E] 78 77-91658 MARC

Speed, Eric.
The midnight rally / by Eric Speed. — New York : Grosset & Dunlap, c1978.
180 p. ; 20 cm. — (Wynn and Lonny racing series ; 6)
SUMMARY: When they become stunt men for a movie company, two automobile racers solve a mystery of long standing.
ISBN 0-448-14558-8
[1. Stunt driving—Fiction. 2. Automobile racing—Fiction. 3. Mystery and detective stories] I. Title.
PZ7.S7417 Mi [Fic] 79 77-89962 MARC

Spence, Eleanor.
A candle for Saint Anthony / Eleanor Spence. — Oxford [Eng.] ; New York : Oxford University Press, 1977.
139 p. ; 23 cm. GB77
SUMMARY: As fifteen-year-old Justin's feelings toward Rudi shift from intense hostility to strong friendship, his friends, astonished at the reversal, begin to question the nature of the relationship.
ISBN 0-19-271415-5 £2.95
[1. Friendship—Fiction] I. Title.
PZ7.S7475 Can [Fic] 77 77-9976 MARC

Spence, Eleanor.
The October child / Eleanor Spence ; illustrated by Malcolm Green. — London : Oxford University Press, 1976.
[7], 151 p. : ill. ; 23 cm. GB76-24692
ISBN 0-19-271384-1 : £3.25
I. Title.
PZ7.S7475 Oc 79 79-315200 MARC

Sperry, Armstrong, 1897-
Black Falcon, a story of piracy and Old New Orleans; written and illustrated by Armstrong Sperry. [1st ed.] Philadelphia, J. C. Winston Co. [1949]
v, 218 p. illus. 22 cm.
SUMMARY: During the War of 1812 a young boy joins the forces of the buccaneer Jean Laffite.
1. Laffite, Jean, 1782-1854—Juvenile fiction. [1. Laffite, Jean, 1782-1854—Fiction. 2. United States—History—War of 1812—Fiction] I. Title.
PZ7.S749 Bl [Fic] [r79]rev2 49-11410 MARC

Spier, Peter.
Bored—nothing to do! / By Peter Spier. — 1st ed. — Garden City, N.Y. : Doubleday, c1978.
[44] p. : col. ill. ; 22 x 28 cm.
SUMMARY: On a lazy afternoon, two bored brothers keep themselves busy by building and flying an airplane.
ISBN 0-385-13177-1 : $6.95. ISBN 0-385-13178-X lib. bdg.
[1. Boredom—Fiction. 2. Airplanes—Fiction] I. Title.
PZ7.S7544 Bo [E] 78 77-20726 MARC

Spillane, Frank Morrison, 1918-
The day the sea rolled back / Mickey Spillane ; illustrated by Maroto. — 1st ed. — New York : Windmill Books, c1979.
138 p. : ill. ; 22 cm.
SUMMARY: A man and his 12-year-old son lose all hope of finding a sunken treasure in the Caribbean. Then one day a strange thing happens to the sea.
ISBN 0-525-61589-X : $7.95
[1. Buried treasure—Fiction. 2. Caribbean area—Fiction] I. Title.
PZ7.S7548 Day [Fic] 78 78-7855 MARC

St. George, Judith, 1931-
The Halloween pumpkin smasher / by Judith St. George ; illustrated by Margot Tomes. — New York : Putnam, c1978.
[46] p. : ill. (some col.) ; 24 cm.
SUMMARY: Mary Grace and her imaginary friend Nellie plan to find out who has been smashing all the jack-o'-lanterns the week before Halloween.
ISBN 0-399-20617-5
[1. Halloween—Fiction. 2. Mystery and detective stories] I. Tomes, Margot. II. Title.
PZ7.S142 Hal 1978 [E] 77 77-26294 MARC

St. George, Judith, 1931-
The halo wind / Judith St. George. — New York : Putnam, c1978.
156 p. : map ; 22 cm.
SUMMARY: A young girl begins to wonder if a Chinook companion is causing or alleviating hardships a group of pioneers experiences on its way to Oregon.
ISBN 0-399-20651-5 : $7.95
[1. Overland journeys to the Pacific—Fiction. 2. Chinook Indians—Fiction. 3. Indians of North America—Fiction] I. Title.
PZ7.S142 Han 1978 [Fic] 78 78-16640 MARC

St. John, Patricia Mary, 1920-
The secret at Pheasant Cottage / by Patricia St. John. — Chicago : Moody Press, c1978.
160 p. : ill. ; 18 cm.
SUMMARY: Raised by her grandparents, 12-year-old Lucy starts a search for her father that takes her from a prison in England to the Mediterranean coast of Spain.
ISBN 0-8024-7683-X : $1.95
[1. Fathers and daughters—Fiction. 2. Family life—Fiction] I. Title.
[PZ7.S143 Se] [Fic] 78 78-24384 MARC

Stack, Jerome.
The adventures of Little and Big Whiskers (two rats in case you didn't know) / by Jerome Stack ; ill. by Jack Billups. — San Diego, Calif. : Reward Books, c1976.
153 p. : ill. ; 22 cm.
Spine title: Little and Big Whiskers.
SUMMARY: Two pet rats who have lived quietly for five years in a cage suddenly begin to converse with the family that owns them.
$4.95
[1. Rats—Fiction. 2. Pets—Fiction] I. Billups, Jack. II. Title. III. Title: Little and Big Whiskers.
PZ7.S77517 Ad [E] 78 76-55083 MARC

CHILDREN'S LITERATURE (Continued)

Stahl, HildaAnn.
Elizabeth Gail and the mystery at the Johnson farm / Hilda Stahl. — Wheaton, Ill. : Tyndale House, c1978.
128 p. ; 18 cm. — (Elizabeth Gail series ; no. 1)
SUMMARY: Eleven-year-old Libby is happy in her new foster home where she has found Jesus as her Savior, but it is clear that someone wants her to leave.
ISBN 0-8423-0720-6 (pbk.) : $1.95
[1. Foster home care—Fiction. 2. Family life—Fiction. 3. Christian life—Fiction] I. Title.
PZ7.S78244 El [Fic] 78-66373
79 MARC

Stahl, HildaAnn.
Elizabeth Gail and the secret box / Hilda Stahl. — Wheaton, Ill. : Tyndale House, c1979.
124 p. ; 18 cm. — (Elizabeth Gail series ; no. 2)
SUMMARY: Happy in her new foster home where she has found Jesus as her Savior, Libby is troubled when her father, whom she doesn't remember, sends her gifts for her 12th birthday.
ISBN 0-8423-0721-4 : $1.95
[1. Foster home care—Fiction. 2. Family life—Fiction. 3. Christian life—Fiction] I. Title.
PZ7.S78244 En [Fic] 78-63097
79 MARC

Stanek, Muriel Novella, 1915-
Growl when you say R / Muriel Stanek ; pictures by Phil Smith. — Chicago : A. Whitman & Co., c1979.
[32] p. : ill. ; 24 cm. — (Concept books/level 2)
SUMMARY: Robbie overcomes his speech problem with the help of a skillful teacher, hard work, and the discovery that he is not alone.
ISBN 0-8075-3074-3 : $5.25
[1. Speech—Disorders—Fiction] I. Smith, Phil. II. Title.
PZ7.S78637 Gr 1979 [E] 79-171
79 MARC

Stauffer, Anne.
The Redcoats return : sequel to Friendly rebels / by Anne Stauffer. — [s.l. : s.n.], c1978 (Rockland, Me. : Courier-Gazette)
261 p. : map (on lining paper) ; 22 cm.
SUMMARY: After 40 years of peace, war returns to southern Maine when the British capture a tiny peninsula during the War of 1812.
$8.95
1. Maine—History—War of 1812—Juvenile fiction. [1. Maine—History—War of 1812—Fiction. 2. United States—History—War of 1812—Fiction] I. Title.
PZ7.S806 Re 78-107106
79 MARC

Staynes, Jill.
Out of that world / Jill Staynes. — London ; Boston : Faber & Faber, 1979.
138 p. ; 21 cm. GB79-06101
SUMMARY: The inhabitants of the Knowe at Longbarrow seek 15-year-old Bertie's help as they face the destruction of their ancient knoll.
ISBN 0-571-11302-8 : £3.95
[1. England—Fiction. 2. Space and time—Fiction] I. Title.
PZ7.S808 Ou [Fic] 79-670152
79 MARC

Stearns, Pamela.
The fool and the dancing bear / by Pamela Stearns ; illustrated by Ann Strugnell. — 1st ed. — Boston : Little, Brown, c1979.
167 p. : ill. ; 21 cm.
"An Atlantic Monthly Press book."
SUMMARY: A lovesick king, his court jester, and a bear embark on a quest to lift the curse of a spiteful queen.
ISBN 0-316-81171-8
[1. Fantasy] I. Strugnell, Ann. II. Title.
PZ7.S8124 Fo [Fic] 78-26965
78 MARC

Steele, Mary Q
Mrs. Gaddy and the ghost / by Wilson Gage [i.e. M. Q. Steele] ; pictures by Marylin Hafner. — 1st ed. — New York : Greenwillow Books, c1979.
55 p. : col. ill. ; 22 cm. — (Greenwillow read-alone)
SUMMARY: Mrs. Gaddy finds a ghost in the kitchen can be good company.
ISBN 0-688-80179-X : $5.95. ISBN 0-688-84179-1 lib. bdg. : $5.71
[1. Ghost stories] I. Hafner, Marylin. II. Title.
PZ7.S8146 Mk [E] 78-16366
78 MARC

Steele, Mary Q
The owl's kiss : three stories / by Mary Q. Steele. — 1st ed. — New York : Greenwillow Books, c1978.
113 p. : ill. ; 22 cm.
CONTENTS: The owl's kiss.—The last great snake.—Dita's story.
ISBN 0-688-80174-9 : $6.95. ISBN 0-688-84174-0 lib. bdg. : $6.43
[1. Fantasy] I. Title.
PZ7.S8146 Ow [Fic] 78-1983
78 MARC

Steele, William O 1917-
The lone hunt / William O. Steele ; illustrated by Paul Caldone. — 1st Voyager ed. — New York : Harcourt Brace Jovanovich, 1976, c1956.
175 p. : ill. ; 19 cm. — (A Voyager book ; AVB 101)
SUMMARY: A young Tennessee boy in the early 1800's goes on a buffalo hunt to prove his manhood.
ISBN 0-15-652983-1
[1. Hunting—Fiction. 2. Tennessee—Fiction] I. Galdone, Paul. II. Title.
[PZ7.S8148 Lo 7] [Fic] 75-29489
75 MARC

Steele, William O 1917-
The magic amulet / William O. Steele. — 1st ed. — New York : Harcourt Brace Jovanovich, c1979.
114 p. ; 21 cm.
SUMMARY: Left to die by his prehistoric family band, a wounded young hunter must find and join a new group if he is to survive.
ISBN 0-15-250427-3
[1. Man, Prehistoric—Fiction] I. Title.
PZ7.S8148 Mag [Fic] 78-20573
78 MARC

Steele, William O 1917-
The war party / by William O. Steele ; ill. by Lorinda Bryan Cauley. — 1st ed. — New York : Harcourt Brace Jovanovich, c1978.
63 p. : ill. ; 22 cm. — (A Let me read book)
SUMMARY: Despite his preparations a young Indian warrior's first battle is not what he expected.
ISBN 0-15-294789-2. ISBN 0-15-694697-1 pbk.
[1. Indians of North America—Fiction. 2. War—Fiction] I. Cauley, Lorinda Bryan. II. Title.
PZ7.S8148 War [E] 78-52815
78 MARC

Steig, William, 1907-
Farmer Palmer's wagon ride : story and pictures / by William Steig. — New York : Puffin Books, 1978, c1974.
[30] p. : col. ill. ; 23 cm. — (Picture Puffin)
Originally published by Farrar, Straus, Giroux, New York.
SUMMARY: The wagon ride from town is so hazardous that Farmer Palmer, a pig, and Ebenezer, an ass, barely make it home again.
ISBN 0-14-050267-X : $2.25
[1. Pigs—Fiction. 2. Donkeys—Fiction. 3. Humorous stories] I. Title.
PZ7.S8177 Far 1978 [E] 77-29063
78 MARC

Steig, William, 1907-
Tiffky Doofky / William Steig. — 1st ed. — New York : Farrar, Straus, and Giroux, 1978.
[32] p. : col. ill. ; 27 cm.
SUMMARY: Tiffky Doofky, a kindly garbage collector, spends his day waiting for a fortune teller's prophesy to come true.
ISBN 0-374-37542-9 lib. bdg. : $7.95
[1. Fortune telling—Fiction] I. Title.
PZ7.S8177 Ti 1978 [E] 78-19657
78 MARC

Steinberg, Fannie.
Birthday in Kishinev / Fannie Steinberg ; illustrated by Luba Hanuschak. — 1st ed. — Philadelphia : Jewish Publication Society of America, 1978.
79 p. : ill. ; 22 cm.
SUMMARY: Sarah believes her 12th birthday to be the happiest day of her life in Kishinev, Russia, but it becomes the most frightening as well as the Jewish population is brutally attacked in a pogrom.
ISBN 0-8276-0111-5 : $5.95
1. Kishinev Massacre, 1903—Juvenile fiction. [1. Kishinev Massacre, 1903—Fiction. 2. Jews in Russia—History—Fiction. 3. Russia—History—Nicholas II, 1894-1917—Fiction] I. Hanuschak, Luba. II. Title.
PZ7.S8218 Bi Fic 78-56546
00 MARC

Steiner, Jörg.
Rabbit Island / text, Jörg Steiner ; pictures, Jörg Müller ; translation by Ann Conrad Lammers. — 1st American ed. — New York : Harcourt Brace Jovanovich, 1978.
[32] p. : col. ill. ; 33 cm.
Translation of Die Kanincheninsel.
SUMMARY: The adventures of two rabbits who escape from the rabbit factory.
[1. Rabbits—Fiction] I. Müller, Jörg. II. Title.
PZ7.S8262 Rab 1978 [E] 78-1512
78 MARC

Steinmetz, Leon.
Clocks in the woods / story and pictures by Leon Steinmetz. — 1st ed. — New York : Harper & Row, c1979.
32 p. : col. ill. ; 26 cm.
SUMMARY: The forest animals realize that the sun is a more reliable clock than the clocks sold them by the greedy porcupine.
ISBN 0-06-025649-4 : $7.95. ISBN 0-06-025650-8 lib. bdg. : $7.89
[1. Forest animals—Fiction. 2. Clocks and watches—Fiction] I. Title.
PZ7.S8277 Cl 1979 [E] 77-11859
77 MARC

Steinmetz, Leon.
The story of Ricky, the royal dwarf / written and illustrated by Leon Shtainmets. — 1st ed. — New York : Harper & Row, c1976.
48 p. : ill. ; 24 cm.
SUMMARY: Ricky, the court dwarf, suffers silently when people laugh at him until his music shows them that he isn't a toy.
ISBN 0-06-025647-8 : $4.95. ISBN 0-06-025648-6 lib. bdg.
[1. Dwarfs—Fiction] I. Title.
PZ7.S8277 St [Fic] 75-21671
76[r79]rev MARC

Stern, Mark.
It's a dog's life / Mark Stern. — London : Cape, 1978.
[24] p. : col. ill. ; 27 cm. GB78-01788
ISBN 0-224-01568-0 : £2.50
I. Title.
PZ7.S83884 It 1978b 78-321008
78 MARC

Stern, Mark.
It's a dog's life / Mark Stern. — 1st American ed. — New York : Atheneum, 1978.
[24] p. : col. ill. ; 26 cm.
"A Jonathan Cape book."
SUMMARY: A dog's efforts to make a fortune in dog food turn him into a virtual prisoner.
ISBN 0-689-30643-1
[1. Dogs—Fiction] I. Title.
PZ7.S83884 It 1978 [E] 77-25494
77 MARC

Stevens, Clifford J
Wild dogs of Chongdo / Clifford Stevens ; illustrated by James McIlrath. — Huntington, Ind. : Our Sunday Visitor, c1979.
79 p. : ill. ; 21 cm.
SUMMARY: A 14-year-old North Korean refugee plots his escape from a Communist prison camp set in a crater-like depression on a mountain and surrounded by wild dogs.
ISBN 0-87973-301-2 : $1.95
[1. Prisoners of war—Fiction. 2. Dogs—Fiction. 3. Escapes—Fiction. 4. Korean War, 1950-1953—Fiction] I. McIlrath, James. II. Title.
PZ7.S8438 Wi [Fic] 79-84745
79 MARC

Stevenson, James, 1929-
Could be worse! / James Stevenson. — [New York] : Puffin Books, 1979, c1977.
[32] p. : col. ill. ; 18 x 23 cm. — (Picture puffin)
Reprint of the ed. published by Greenwillow Books, New York.
SUMMARY: Everything is always the same at Granpa's house, even the things he says—until one unusual morning.
ISBN 0-14-050285-8 : $2.25
[1. Dreams—Fiction] I. Title.
[PZ7.S84748 Co 1979] [E] 78-18280
78 MARC

Stevenson, James, 1929-
Fast friends : two stories / by James Stevenson. — 1st ed. — New York : Greenwillow Books, c1979.
64 p. : col. ill. ; 23 cm. — (Greenwillow read-alone)
CONTENTS: Murray and Fred.—Thomas and Clem.
SUMMARY: In the first of two easy-to-read stories a turtle and a snail make some friends with the help of a skateboard. In the second a mouse and a snail learn, through trial and error, how to be friends.
ISBN 0-688-80197-8 : $6.95. ISBN 0-688-84197-X lib. bdg. : $6.67
[1. Friendship—Fiction. 2. Animals—Fiction. 3. Short stories] I. Title.
PZ7.S84748 Fas [E] 78-14828
78 MARC

Stevenson, James, 1929-
Monty / by James Stevenson. — 1st ed. — New York : Greenwillow Books, c1979.
[32] p. : col. ill. ; 26 cm.
SUMMARY: When Monty the alligator goes on vacation, a rabbit, duck, and frog look for another way to cross the river to school.
ISBN 0-688-80209-5 : $7.25. ISBN 0-688-84209-7 lib. bdg. : $6.96
[1. Alligators—Fiction. 2. Animals—Fiction. 3. Cartoons and comics] I. Title.
PZ7.S84748 Mo [E] 78-11409
78 MARC

Stevenson, James, 1929-
Winston, Newton, Elton, and Ed / James Stevenson. — 1st ed. — New York : Greenwillow Books, c1978.
56 p. : col. ill. ; 22 cm. — (Greenwillow read-alone)
SUMMARY: Two stories—the first featuring three pugnacious walruses, the second, a stranded penguin.
ISBN 0-688-80152-8. ISBN 0-688-84152-X lib. bdg.
[1. Walruses—Fiction. 2. Penguins—Fiction. 3. Short stories] I. Title.
PZ7.S84748 Wk [E] 77-17139
77 MARC

Stevenson, Robert Louis, 1850-1894.
Treasure Island / Robert Louis Stevenson ; drawings by Mervyn Peake. — New York : Schocken Books, 1979, c1976.
222 p. : ill. ; 20 cm.
Reprint of the 1976 ed. published my Methuen, London.
SUMMARY: While going through the possessions of a deceased guest who owed them money, the mistress of the inn and her son find a treasure map that leads to a pirate fortune as well as great danger.
ISBN 0-8052-0620-5
[1. Buried treasure—Fiction. 2. Pirates—Fiction. 3. Adventure stories] I. Peake, Mervyn Laurence, 1911-1968. II. Title.
[PZ7.S8482 Tr 1979] [Fic] 78-74610
79 MARC

Stewart, Robert S
The daddy book. Pictures by Don Madden. New York, American Heritage Press [1972]
[48] p. illus. 27 cm. $3.95
SUMMARY: Describes the various activities of fathers at home and away.
ISBN 0-07-061347-8. ISBN 0-07-061348-6 (lib. bdg)
[1. Fathers—Fiction] I. Madden, Don, 1927- illus. II. Title.
PZ7.S8498 Dad [E] 72-20
72[r79]rev MARC

Stiff, Peter.
Tommy goes home / by Peter Stiff. — Salisbury, Rhodesia : Jacaranda Press, 1977.
232 p. : ill. ; 26 cm.
ISBN 0-7974-0279-9 : Rh$7,98
1. Lions—Legends and stories. I. Title.
PZ4.S8546 To 823'.9'14 78-315602
[PR9390.9.S75] 78 MARC

CHILDREN'S LITERATURE (Continued)

Stiles, Martha Bennett.
The star in the forest : a mystery of the Dark Ages / by Martha Bennett Stiles. — New York : Four Winds Press, c1979.
xiv, 206 p., [1] leaf of plates : maps ; 22 cm.
SUMMARY: Near the end of the sixth century, the daughter of a Gallic lord witnesses what she believes to be the murder of her brother and the beginning of a plot to usurp her father's title.
ISBN 0-590-07537-3 : $8.95
[1. Gaul—Fiction. 2. France—History—To 987—Fiction] I. Title.
PZ7.S8557 Ss [Fic] 78 78-22284 MARC

Stiles, Martha Bennett.
Tana and the useless monkey / by Martha Bennett Stiles. — 1st ed. — New York : Elsevier/Nelson Books, c1979.
85 p. ; 21 cm.
SUMMARY: Tana tries to demonstrate the usefulness of her mischievous pet monkey.
ISBN 0-525-66625-4 : $6.95
[1. Monkeys—Fiction. 2. Pets—Fiction] I. Title.
PZ7.S8557 Tan 1979 [Fic] 79 79-4655 MARC

Stoddard, Richard Henry, 1825-1903.
The story of Little Red Riding Hood / told in verse by Richard Henry Stoddard ; illustrated by Alfred Fredericks ; engraved in colours by Bobbett and Hooper. — New York : J. C. Gregory, c1864.
[8] p., [4] leaves of plates : col. ill. ; 27 cm.
I. Fredericks, Alfred. II. Little Red Riding Hood. III. Title.
PZ6.S83 St 78 78-315527 MARC

Stoddard, Sandol.
Curl up small [by] Sandol Stoddard Warburg. Illustrated by Trina Schart Hyman. [Boston, Houghton Mifflin] 1964.
32 p. col. illus. 24 cm.
SUMMARY: Describes how the world appears to a baby fish, a bird, a bear, and a child who go exploring with their mothers.
[1. Size and shape—Fiction] I. Hyman, Trina Schart, illus. II. Title.
PZ7.S868 Cu [E] [r79]rev 64-10725 MARC

Stoddard, Sandol.
I like you. Illustrated by Jacqueline Chwast. Boston, Houghton Mifflin, 1965.
48 p. illus. 14 cm.
Published under the author's earlier name: Sandol Stoddard Warburg.
I. Chwast, Jacqueline, illus. II. Title.
PZ7.S868 Iac [r79]rev 65-11020 MARC

Stolz, Mary Slattery, 1920-
Cider days / by Mary Stolz. — 1st ed. — New York : Harper & Row, c1978.
130 p. ; 21 cm.
"An Ursula Nordstrom book."
SUMMARY: A young girl's persistent overtures to a new neighbor result in friendship.
ISBN 0-06-025837-3 : $6.95. ISBN 0-06-025838-1 lib. bdg. : $6.79
[1. Friendship—Fiction. 2. Family life—Fiction. 3. Vermont—Fiction] I. Title.
PZ7.S875854 Ci 1978 [Fic] 78 77-25652 MARC

Stolz, Mary Slattery, 1920-
Go and catch a flying fish / Mary Stolz. — 1st ed. — New York : Harper & Row, c1979.
213 p. ; 22 cm.
"An Ursula Nordstrom book."
SUMMARY: Taylor, Jem, and B. J. each react differently to the breakup of their parents' marriage.
ISBN 0-06-025867-5 : $7.95. ISBN 0-06-025868-3 lib. bdg. : $7.89
[1. Family problems—Fiction] I. Title.
PZ7.S875854 Gn 1979 [Fic] 79 78-21785 MARC

Stone, Josephine Rector.
Praise all the moons of morning / by Josephine Rector Stone. — 1st ed. — New York : Atheneum, 1979.
172 p. ; 22 cm. — (An Argo book)
SUMMARY: While recording the enslavement of her space mission crew by inhabitants of the plants Ix-thlan, an exo-biologist is projected into the future where she meets descendants of the crew who have known nothing but slavery.
ISBN 0-689-30697-0
[1. Science fiction]
PZ7.S87787 Pr 1979 [Fic] 78 78-12633 MARC

Stone, Josephine Rector.
Those who fall from the Sun / Josephine Rector Stone. — 1st ed. — New York : Atheneum, 1978.
153 p. ; 22 cm.
SUMMARY: Separated from her family group during a hurried flight from a chaotic Earth, a teen-age girl finds that her heretofore unappreciated prophetic vision helps to guide the vessel she is traveling in to a small, hospitable planet.
ISBN 0-689-30615-6
[1. Science fiction] I. Title.
PZ7.S87787 Th 1978 [Fic] 78 78-5139 MARC

Storr, Catherine.
Kate and the island / Catherine Storr ; illustrated by Gareth Floyd. — London : Faber and Faber, c1972, 1974 printing.
87 p. : ill. ; 21 cm.
ISBN 0-571-10119-4
I. Title.
PZ7.S8857 Kat 79 79-670055 MARC

Storr, Catherine.
Winter's end / Catherine Storr. — London : Macmillan, 1978.
183 p. ; 21 cm. GB***
ISBN 0-333-24778-7 : £3.95
I. Title.
PZ7.S8857 Wi 79 79-300190 MARC

Storr, Catherine.
Winter's end / by Catherine Storr. — 1st American ed. — New York : Harper & Row, c1979.
206 p. ; 21 cm.
SUMMARY: An old family house has a mysterious effect on a group of teenagers staying there.
ISBN 0-06-026069-6 : $6.95. ISBN 0-06-026072-6 lib. bdg. : $6.79
[1. Supernatural—Fiction. 2. England—Fiction] I. Title.
PZ7.S8857 Wi 1979 [Fic] 78 78-19488 MARC

Strang, Celia.
Foster Mary / by Celia Strang. — New York : McGraw-Hill, c1979.
162 p. ; 21 cm.
SUMMARY: Mary's dreams for her foster children include a warm house, regular meals and no more traveling.
ISBN 0-07-061996-4 : $7.95
[1. Foster home care—Fiction. 2. Migrant labor—Fiction] I. Title.
PZ7.S89729 Fo [Fic] 78 78-23630 MARC

Strasser, Todd.
Angel dust blues : a novel / by Todd Strasser. — New York : Coward, McCann & Geoghegan, c1979.
203 p. ; 22 cm.
SUMMARY: Feeling increasingly alienated from his successful and affluent parents, a 17-year-old boy becomes involved in drug dealing.
ISBN 0-698-20485-9
[1. Drug abuse—Fiction] I. Title.
PZ7.S899 An 1979 [Fic] 78 78-31735 MARC

Streatfield, Noel.
Meet the Maitlands / [by] Noel Streatfield ; illustrated by Antony Maitland. — London : W. H. Allen, 1978.
[5], 118 p. : ill. ; 21 cm. GB78-40561
ISBN 0-491-02036-8 : £3.25
I. Title.
PZ7.S914 Me 79 79-303176 MARC

Stren, Patti.
Sloan and Philamina : or, How to make friends with your luch / by Patti Stren. — 1st ed. — New York : Dutton, c1979.
[42] p. : ill. (some col) ; 21 x 23 cm. — (A Unicorn book)
SUMMARY: The anteaters pressure Sloan to give up his friend Philamina, an ant. He has no intention of doing so and must resort to drastic measures.
ISBN 0-525-39485-0
[1. Anteaters—Fiction. 2. Ants—Fiction. 3. Friendship—Fiction] I. Title. II. Title: How to make friends with your lunch.
PZ7.S9163 Sl 1979 [E] 78 78-10377 MARC

Strete, Craig.
When grandfather journeys into winter / Craig Kee Strete; illustrated by Hal Frenck. — 1st ed. — New York : Greenwillow Books, c1979.
86 p. : ill. ; 22 cm.
SUMMARY: A young Indian boy struggles to accept his grandfather's rapidly approaching death.
ISBN 0-688-80193-5 : $6.95
[1. Death—Fiction. 2. Grandfathers—Fiction. 3. Indians of North America—Fiction] I. Frenck, Hal. II. Title.
PZ7.S9164 Wh [Fic] 78 78-14830 MARC

Stubbs, Joanna.
Happy bear's day / Joanna Stubbs. — London : A. Deutsch, 1978.
[30] p. : chiefly ill. (some col.) ; 26 cm. GB***
ISBN 0-233-96999-3 : £3.25 ($6.50 U.S.)
I. Title.
PZ7.S9375 Hap 79 78-74461 MARC

Sudbery, Rodie.
Somewhere else / Rodie Sudbery. — London : A. Deutsch, 1978.
143 p. ; 21 cm. GB***
ISBN 0-233-96974-8 : £3.25
I. Title.
PZ7.S9434 Sm 1978 78 78-317396 MARC

Sudbery, Rodie.
A tunnel with problems / Rodie Sudbery. — London : Deutsch, 1979.
112 p. ; 21 cm. GB***
ISBN 0-233-97073-8 : £2.95
I. Title.
PZ7.S9434 Tu 823'.9'14 79 79-317364 MARC

Sugita, Yutaka, 1930-
The flower family / by Yutaka Sugita. — New York : McGraw-Hill, [1975] c1972.
[28] p. : col. ill. ; 21 x 30 cm.
SUMMARY: Flowers bloom all around the world bringing cheer, beauty, and happiness wherever they are found.
ISBN 0-07-061768-6 : $5.95. ISBN 0-07-061769-4 (lib. bdg.) $5.72
[1. Flowers—Fiction] I. Title.
PZ7.S944 Fl 5 [E] 75[r79]rev2 75-20082 MARC

Sumiko.
Kittymouse / Sumiko. — [1st American ed.]. — New York : Harcourt Brace Jovanovich, [1979, c1978]
[30] p. : col. ill. ; 30 cm.
Translation of Okashino neko no hanashi.
Idea and original story from M. Norton's The Kitten who thought he was a mouse.
SUMMARY: Adopted as a kitten by a family of mice, a cat believes herself to be a mouse until she comes in contact with two loving children.
ISBN 0-15-243028-8 : $7.95
[1. Cats—Fiction. 2. Mice—Fiction] I. Norton, Miriam. The kitten who thought he was a mouse. II. Title.
PZ7.S9539 Ki 1979 [E] 78 78-16716 MARC

Supraner, Robyn.
Sam Sunday and the strange disappearance of Chester Cats / story by Robyn Supraner ; pictures by Robert Tallon. — New York : Parents' Magazine Press, c1978.
[37] p. : col. ill. ; 24 cmn.
SUMMARY: The best detective on the police force searches for Mrs. Cats' missing son.
ISBN 0-8193-0993-1
[1. Mystery and detective stories. 2. Cats—Fiction] I. Tallon, Robert, 1939- . II. Title.
PZ7.S9652 Sam [E] 78 78-11062 MARC

Sutcliff, Rosemary.
Song for a dark queen / [by] Rosemary Sutcliff. — London Pelham, 1978.
176 p. : map ; 21 cm. GB78-2164
ISBN 0-7207-1060-X : £3.75
1. Boadicea, Queen, d. 62—Juvenile fiction. 2. Great Britain—History—Roman period, 55 B.C.-449 A.D.—Juvenile fiction. I. Title.
PZ7.S966 So 1978 78 78-32129 MAR

Sutton, Eve.
Moa hunter / Eve Sutton ; illustrated by Bernard Brett. — London : Hamilton ; Auckland : distributed by Hodder and Stoughton, 1978.
112 p. : ill. ; 24 cm. GB***
ISBN 0-241-89697-5 : £3.75
I. Title.
PZ7.S96817 Mo 1978 823 78 78-317491 MARC

Sutton, Jane.
What should a Hippo wear? / by Jane Sutton ; illustrated by Lynn Munsinger. — Boston : Houghton Mifflin, 1979.
32 p. : col. ill. ; 21 x 22 cm.
SUMMARY: For the jungle dance, Bertha, a hippopotamus, invents an eye catching outfit only to discover that her date, Fred, invited her for her plain old everyday self.
ISBN 0-395-27800-7 : $6.95
[1. Hippopotamus—Fiction. 2. Jungle stories. 3. Self-acceptance—Fiction] I. Munsinger, Lynn. II. Title.
PZ7.S96824 Wh [E] 78 78-24005 MARC

Swetnam, Evelyn.
Yes, my darling daughter / by Evelyn Swetnam ; ill. by Laurie Harden. — New York : Harvey House, c1978.
166 p. : ill. ; 23 cm.
SUMMARY: An 11-year-old is slow to accept the love of her foster family.
ISBN 0-8178-5782-6
[1. Foster home care—Fiction] I. Harden, Laurie. II. Title.
PZ7.S97489 Ye [Fic] 78 77-85134 MARC

Swindells, Robert.
The very special baby / by Robert Swindells ; illustrated by Victor Ambrus. — 1st American ed. — Englewood Cliffs, N.J. : Prentice-Hall, 1978, c1977.
44 p. : ill. ; 22 cm.
SUMMARY: A camel born with three humps helps three wise men carry gifts to a newborn king.
ISBN 0-13-941633-1 : $5.95
[1. Camels—Fiction. 2. Jesus Christ—Nativity—Fiction] I. Ambrus, Victor G. II. Title.
PZ7.S9818 Ve 1978 [E] 78 78-639 MAR

CHILDREN'S LITERATURE (Continued)

Symons, Geraldine.
Crocuses were over, Hitler was dead / by Geraldine Symons. — 1st American ed. — Philadelphia : Lippincott, c1977.
158 p. ; 21 cm.
Originally published in Great Britian 1977 under title: Now and then.
SUMMARY: While visiting an old English manor, a young girl is befriended by one of its former inhabitants, a soldier killed during World War II.
ISBN 0-397-31814-6
[1. Space and time—Fiction. 2. England—Fiction] I. Title.
PZ7.S9846 Cr 1977 [Fic] 78 78-8751 MARC

Talbot, Charlene Joy.
An orphan for Nebraska / Charlene Joy Talbot. — 1st ed. — New York : Atheneum, 1979.
208 p. ; 22 cm.
SUMMARY: Orphaned on the journey to America in 1872, a young Irish boy finally makes his way to Nebraska where he goes to work for a newspaper editor and learns to do the work of a printer's devil.
ISBN 0-689-30698-9
[1. Orphans—Fiction. 2. Frontier and pioneer life—Fiction. 3. Nebraska—Fiction] I. Title.
PZ7.T1418 Or 1979 [Fic] 78 78-12179 MARC

Talbot, Charlene Joy.
The Great Rat Island adventure / Charlene Joy Talbot ; drawings by Ruth Sanderson. — 1st ed. — New York : Atheneum, 1977.
164 p. : ill. ; 22 cm.
SUMMARY: A young boy accompanies his preoccupied ornithologist father on a summer expedition to study terns on an island sanctuary and finds himself involved in a dangerous adventure when a hurricane strikes.
ISBN 0-689-30596-6
[1. Hurricanes—Fiction. 2. Islands—Fiction. 3. Fathers and sons—Fiction] I. Sanderson, Ruth. II. Title.
PZ7.T1418 Gr [Fic] 77[r79]rev 77-1055 MARC

Tallon, Robert, 1939-
Little Cloud / by Robert Tallon. — New York : Parents' Magazine Press, c1978.
[32] p. : col. ill. ; 24 cm.
SUMMARY: A small cloud defies a mountain and brings rain to a parched valley.
ISBN 0-8193-0981-8. ISBN 0-8193-0982-6 lib. bdg.
[1. Clouds—Fiction] I. Title.
PZ7.T157 Li [E] 78 78-11061 MARC

Taylor, Ronald Ditz, 1948-
The book where Michael meets the Royal Street elves and learns about whales and whale oil, the electric light, the ostrich, and the two-headed sea serpent / by R. D. Taylor ; ill. by Ellen Turner. — Austin, Tex. : Curbstone Press, c1978.
69 p. : ill. ; 26 cm.
Spine title: The Royal Street elves.
SUMMARY: A youngster becomes concerned about the fate of the Royal Street elves.
ISBN 0-931604-00-1 : $9.95. ISBN 0-931604-01-1 pbk. : $6.95
[1. Fairies—Fiction. 2. New Orleans—Fiction] I. Turner, Ellen. II. Title. III. Title: The Royal Street elves.
PZ7.T2174 Bo Fic 79 78-55985 MARC

Taylor, Elisabeth Russell.
The loadstone / [by] Elisabeth Russell Taylor ; illustrations Jan Mitchener. — London : Abelard, 1978.
3-50 p. : ill. ; 23 cm. GB78-31037
ISBN 0-200-72562-9 : £1.95
I. Title.
PZ7.T21263 Lo 79 78-325958 MARC

Taylor, Mark.
The case of the missing kittens / Mark Taylor ; illustrated by Graham Booth. — 1st ed. — New York : Atheneum, 1978.
[48] p. : ill. ; 17 x 19 cm.
At head of title: The secret life of Angus.
SUMMARY: Angus the dog does some detective work to locate Mrs. Friday's missing kittens.
ISBN 0-689-30627-X
[1. Dogs—Fiction. 2. Cats—Fiction] I. Booth, Graham. II. Title.
PZ7.T2172 Cas 1978 [E] 78 78-4908 MARC

Teague, Kathleen.
What happened to Hector? Drawings by Ric Del Rossi. Champaign, Ill., Garrard Pub. Co. [1974]
32 p. col. illus. 23 cm.
SUMMARY: Hector the worm is put into a flower pot by mistake.
ISBN 0-8116-6050-8 (lib. bdg.)
[1. Worms—Fiction] I. Del Rossi, Ric, illus. II. Title.
PZ7.T21938 Wh [E] 73[r79]rev 73-22078 MARC

Teibl, Margaret.
Davey come home / by Margaret Teibl ; pictures by Jacqueline Bardner Smith. — 1st ed. — New York : Harper & Row, c1979.
60 p. : ill. ; 21 cm.
SUMMARY: A young boy yearns for a mother figure to make their home a welcoming place for him and his father.
ISBN 0-06-026135-8 : $7.95
[1. Single-parent family—Fiction. 2. Fathers and sons—Fiction] I. Smith, Jacqueline Bardner. II. Title.
PZ7.T22966 Dav 1979 [Fic] 79 78-22477 MARC

Telemaque, Eleanor Wong.
It's crazy to stay Chinese in Minnesota / Eleanor Wong Telemaque. — 1st ed. — Nashville : T. Nelson, c1978.
118 p. ; 21 cm.
SUMMARY: A seventeen-year-old Chinese American in Minnesota and her family tread a balance between the Far East and Middle West.
ISBN 0-8407-6613-0 : $6.95
[1. Chinese Americans—Fiction. 2. Minnesota—Fiction] I. Title.
PZ7.T236 It [Fic] 78 78-14054 MARC

Tempos / [edited by] Jo M. Stanchfield, Harvey R. Granite ; consultant, Millard H. Black. — Boston : Houghton Mifflin, c1979.
248 p. : ill. ; 23 cm. — (Vistas)
Includes bibliographical references.
SUMMARY: A collection of short stories, plays, poems, and topical articles on a variety of subjects.
ISBN 0-395-25227-X
1. Children's literature. [1. Literature—Collections] I. Stanchfield, Jo M. II. Granite, Harvey R. III. Series.
PZ5.T2968 810'.8 78 77-83336 MARC

Tennant, Veronica.
On stage, please : a story / by Veronica Tennant ; ill. by Rita Briansky. — New York : Holt, Rinehart, and Winston, 1979, c1977.
176 p. : ill. ; 22 cm.
SUMMARY: A young girl begins serious training for a career as a professional dancer.
ISBN 0-03-049306-4
[1. Ballet dancing—Fiction] I. Briansky, Rita. II. Title.
PZ7.T2639 On 1979 [Fic] 79 79-4819 MARC

Terris, Susan.
Tucker and the horse thief / Susan Terris. — New York : Four Winds Press, c1979.
188 p. ; 22 cm.
SUMMARY: A 12-year-old girl, posing as a boy and abused and neglected by a father who has brought her to California to help him get rich in the aftermath of the gold rush, finds both warmth and uneasiness in a friendship with a Jewish boy who does not suspect she is a girl.
ISBN 0-590-07626-4 : $7.95
[1. Friendship—Fiction. 2. California—Fiction] I. Title.
PZ7.T276 Tu [Fic] 79 79-12810 MARC

Tester, Sylvia Root, 1939-
Sometimes I'm afraid / by Sylvia Root Tester ; illustrated by Frances Hook. — Elgin, Ill. : Child's World ; Chicago : distributed by Children's Press, c1979.
30 p. : col. ill. ; 27 cm.
SUMMARY: A young child describes some of her fears and how she deals with them.
ISBN 0-89565-021-5 (lib. bdg.) : $5.50
[1. Fear—Fiction] I. Hook, Frances. II. Title.
PZ7.T288 So [E] 78 78-23262 MARC

Tester, Sylvia Root, 1939-
We laughed a lot, my first day of school / by Sylvia Root Tester ; illustrated by Frances Hook. — Elgin, Ill. : Child's World ; Chicago : distributed by Childrens Press, c1979.
32 p. : col. ill. ; 27 cm.
SUMMARY: Scared to go to his first day of kindergarten, Juan discovers it's not so bad—he even laughs a lot.
ISBN 0-89565-020-7 (lib. bdg.) : $5.50
[1. Kindergarten—Fiction] I. Hook, Frances. II. Title.
PZ7.T288 We [E] 78 78-10900 MARC

Thackray, Patricia.
Big Bird gets lost / by Patricia Thackray ; pictures by Carol Nicklaus. — [Racine, Wis.] : Western Pub. Co., c1978.
[28] p. : col. ill. ; 25 cm. — (A Golden scratch & sniff book)
"A Sesame Street/Golden Press book."
SUMMARY: Big Bird loses his way while shopping for bird seed for his nest-warming party. The reader can experience some of the fragrances Big Bird encounters on his trip by scratching and sniffing treated pages.
ISBN 0-307-13524-1
[1. Shopping—Fiction] I. Nicklaus, Carol. II. Title.
PZ7.T298 Bi [E] 78 77-93657 MARC

Thiele, Colin.
The sknuks / text by Colin Thiele ; drawings by Mary Milton. — Adelaide : Rigby, 1977.
[32] p. : col. ill. ; 26 cm. — (Rigby opal books) Aus
ISBN 0-7270-0386-0
I. Milton, Mary. II. Title.
PZ7.T354 Sk 78 78-316078 MARC

Thiele, Colin.
Storm Boy / by Colin Thiele ; pictures by John Schoenherr. — 1st American ed. — New York : Harper & Row, [1978] c1963.
61 p. : ill. ; 21 cm.
SUMMARY: A boy living on a desolate beach in Australia rescues a baby pelican that becomes his constant companion.
ISBN 0-06-026133-1 : $6.95. ISBN 0-06-026134-X lib. bdg. : $7.89
[1. Pelicans—Fiction. 2. Friendship—Fiction. 3. Australia—Fiction] I. Schoenherr, John. II. Title.
PZ7.T354 St 1978 [Fic] 78 77-25675 MARC

Thomas, Ianthe, 1951-
Hi, Mrs. Mallory! / By Ianthe Thomas ; pictures by Ann Toulmin-Rothe. — 1st ed. — New York : Harper & Row, c1979.
48 p. : ill. ; 22 cm.
SUMMARY: A young girl describes her special friendship with her neighbor, Mrs. Mallory.
ISBN 0-06-026128-5 : $7.95. ISBN 0-06-026129-3 lib. bdg. : $6.79
[1. Friendship—Fiction] I. Toulmin-Rothe, Ann. II. Title.
PZ7.T36693 Hi 1979 [E] 78 78-3013 MARC

Thomas, Piri, 1928-
Stories from El Barrio / by Piri Thomas. — New York : Knopf, c1978.
viii, 141 p. ; 22 cm.
CONTENTS: The three mosquiteers.—La peseta.—The konk.—Mighty Miguel.—The blue wings and the Puerto Rican knights.—Coney Island.—Putting it on for Juanita.—Amigo brothers.
ISBN (invalid) 0-394-83568-7
1. Spanish Americans in New York (City)—Juvenile fiction. [1. Spanish Americans in New York (City)—Fiction. 2. Short stories] I. Title.
PZ7.T3693 St [Fic] 78 78-3287 MARC

Thompson, Jean, 1933-
Don't forget Michael / Jean Thompson ; illustrated by Margot Apple. — New York : Morrow, 1979.
64 p. : ill. ; 21 cm.
SUMMARY: Four episodes in which seven-year-old Michael copes with being the smallest and quietest in a large, clamorous family.
ISBN 0-688-22196-3
[1. Family life—Fiction] I. Apple, Margot. II. Title.
PZ7.T371595 Do [E] 79 79-16637 MARC

Thornton, J Kaye.
Ghosts are here to stay / by J. Kaye Thornton ; with ill. by Pamela J. Brown. — Fayette, AL : Southern Style Creations, c1978.
92 p., [2] leaves of plates : ill. ; 22 cm.
SUMMARY: Uncle Shane delights the MacIvy children with stories of weird happenings and restless ghosts.
[1. Ghost stories] I. Brown, Pamela J. II. Title.
PZ7.T3948 Gh [Fic] 79 78-56770 MARC

Three kittens. — Hong Kong : Seagull Pub. Co., [1977?]
12 p. : col. ill. ; 18 x 20 cm.
Cover title.
Translation of San chih hsiao mao.
PZ7.T414 78 77-152949 MARC

Thwaite, Ann.
The chatterbox / story by Ann Thwaite ; pictures by Glenys Ambrus. — London : Deutsch, 1978.
[16] p. : ill. (some col.) ; 26 cm. GB***
ISBN 0-233-96967-5 : £3.25
I. Ambrus, Glenys. II. Title.
PZ7.T43 Ch 78 78-318424 MARC

Tierney, Frank M
Silly Sally and the little pumpkin / by Frank M. Tierney ; illus. by Wendy Irvine. — Ottawa : Borealis Press, 1978.
18 p. : ill. ; 21 x 25 cm. C***
I. Irvine, Wendy. II. Title.
PZ7.T443 Sf 79 79-300446 MARC

Tierney, Hanne.
Where's your baby brother, Becky Bunting? / By Hanne Tierney ; illustrated by Paula Winter. — 1st ed. — Garden City, N.Y. : Doubleday, c1979.
[32] p. : col. ill. ; 27 cm.
SUMMARY: Having a difficult time adjusting to her new baby brother, a little girl trades him away for a bike.
ISBN 0-385-08653-9 : $5.95. ISBN 0-385-08654-7 lib. bdg.
[1. Brothers and sisters—Fiction. 2. Babies—Fiction] I. Winter, Paula. II. Title.
PZ7.T444 Wh [E] 76 76-54474 MARC

Timmermans, Gommaar, 1930-
The little white hen and the emperor of France. [Reading, Mass.] Addison-Wesley [1976] c1973.
[27] p. col. illus. 28 cm.
SUMMARY: The adventures of Blanche, the little white hen who, drafted into the French army by the Emperor, participates in a war against the Russians.
Translation of Henne Blanche Soldat des Kaisers.
"An Addisonian Press book."
[1. Chickens—Fiction. 2. War—Fiction] I. Title.
PZ7.T486 Li [Fic] 74[r78]rev 74-9759 MARC
ISBN 0-201-09332-4 (lib. bdg.)

CHILDREN'S LITERATURE (Continued)

Tison, Annette.
Coloured pig town : a take another look book with a colour wheel / Annette Tison and Talus Taylor. — ⌐London⌐ : F. Warne, 1975.
⌐13⌐ p. : col. ill. ; 28 cm. GB***
ISBN 0-7232-1807-2
I. Taylor, Talus, joint author.
PZ7.T518 Co 78-315748
78 MARC

Titus, Eve.
Anatole and the Pied Piper / by Eve Titus ; pictues by Paul Galdone. — New York : McGraw-Hill, c1979.
32 p. : ill. (some col.) ; 26 cm.
SUMMARY: Anatole the mouse is called upon to be a hero when an ambitious musician concocts a scheme involving 24 schoolmice at a picnic.
ISBN 0-07-064897-2 : $7.95
⌐1. Mice—Fiction. 2. France—Fiction⌐ I. Galdone, Paul. II. Title.
PZ7.T543 Ao ⌐E⌐ 78-23513
78 MARC

Tobias, Tobi.
Liquid or solid? / By Tobi Tobias ; illustrated by Gene Sharp. — Chicago : Children's Press, c1977.
32 p. : col. ill. ; 25 cm. — (His That's a good question!)
SUMMARY: Simple yet thought provoking questions which stem from one basic question—is it liquid or solid?
ISBN 0-516-03537-1
⌐1. Liquids. 2. Solids⌐ I. Sharp, Gene, 1923- . II. Title.
PZ7.T56 Li ⌐E⌐ 77-4845
77 MARC

Tobias, Tobi.
Where does it come from? / By Tobi Tobias ; illustrated by Sharon Elzaurdia. — Chicago : Childrens Press, c1977.
31 p. : col. ill. ; 25 cm. — (That's good a question!)
SUMMARY: Simple yet thought provoking questions which stem from one basic question—where does it come from?
ISBN 0-516-03667-X
I. Elzaurdia, Sharon. II. Title.
PZ7.T56 Wh ⌐E⌐ 77-4846
77 MARC

Todd, Herbert Eatton.
The roundabout horse / story by H. E. Todd ; pictures by Val -Biro. — London : Hodder and Stoughton, 1978.
⌐24⌐ p. : col. ill. ; 25 cm. GB***
ISBN 0-340-22510-6 : £2.80
I. Biro, Val, 1921- . II. Title.
PZ7.T5664 Ro 79-300396
79 MARC

Tolle, Jean Bashor.
The Great Pete Penney / Jean Bashor Tolle. — 1st ed. — New York : Atheneum, 1979.
90 p. ; 22 cm.
"A Margaret K. McElderry book."
SUMMARY: An 11-year-old pitcher's unusual friendship with a tiny invisible man is responsible for her team's very memorable season.
ISBN 0-689-50145-5
⌐1. Baseball—Fiction⌐ I. Title.
PZ7.T57447 Gr ⌐Fic⌐ 79-14603
79 MARC

Tomalin, Ruth.
Gone away / Ruth Tomalin. — London ; Boston : Faber & Faber, 1979.
158 p. ; 21 cm. GB***
SUMMARY: Francie's first term at a school in Alchester is filled with many new experiences including a ghost.
ISBN 0-571-11342-7 : £4.25 ($8.95)
⌐1. Ghost stories. 2. School stories. 3. England—Fiction⌐ I. Title.
PZ7.T583 Go ⌐Fic⌐ 79-670253
79 MARC

Tompert, Ann.
The clever princess / by Ann Tompert ; illustrated by Patricia Riley. — Chapel Hill, N.C. : Lollipop Power, c1977.
39 p. : ill. ; 26 cm.
SUMMARY: Princess Lorna passes more tests than were required of her father before he became King and ascends to the throne of the kingdom.
ISBN 0-914996-15-0. ISBN 0-914996-13-4 pbk.
⌐1. Sex role—Fiction. 2. Princesses—Fiction⌐ I. Riley, Patricia. II. Title.
PZ7.T598 Cl ⌐Fic⌐ 77-72559
77⌐r79⌐rev MARC

Tompert, Ann.
Three foolish tales / by Ann Tompert ; illustrated by Diane Dawson. — New York : Crown Publishers, c1979.
48 p. : col. ill. ; 22 cm.
SUMMARY: Raccoon and Skunk attempt to trick Fox out of his umbrella by telling tales of foolish animals, but Fox sets them straight about who is really foolish.
ISBN 0-517-53595-5
⌐1. Animals—Fiction⌐ I. Dawson, Diane. II. Title.
PZ7.T598 Th 1979 ⌐E⌐ 78-12637
78 MARC

Todd, Joan Marlow.
A child's Swiss family Robinson / in a new version by Joan Marlow Todd ; from the classic original by Johann Wyss ; illustrated by Violette Diserens. — New York : Hart Pub. Co., c1978.
95 p. : ill. ; 22 cm.
SUMMARY: An abridged version of the fortunes of a shipwrecked Swiss family as they adapt to life on a deserted island with abundant animal and plant life.
ISBN 0-8055-1263-2 : $5.95
⌐1. Family life—Fiction. 2. Survival—Fiction⌐ I. Diserens, Violette. II. Wyss, Johann David, 1743-1818. The Swiss family Robinson. III. Title.
PZ7.T5666 Ch ⌐Fic⌐ 78-105151
79 MARC

Topelius, Zakarias, 1818-1898.
The tomten in Åbo castle / Z. Topelius ; ⌐transl. from the Swedish original text by Frank R. Southerington. — ⌐Åbo⌐ : Sydvästkusten, 1977.
58 p. : ill. ; 17 cm. Fi77
Translation of Tomtegubben i Åbo slott.
ISBN 9519020039
I. Title.
PZ7.T6442 To 79-351035
79 MARC

Torgersen, Don Arthur, 1934-
The girl who tricked the troll / by Don Arthur Torgersen ; ill. by Tom Dunnington. — Chicago : Childrens Press, c1978.
31 p. : col. ill. ; 25 cm.
SUMMARY: Clever Karin thinks of a way to drive an enraged troll back into the forest.
ISBN 0-516-03465-0
⌐1. Trolls—Fiction⌐ I. Dunnington, Tom. II. Title.
PZ7.T6444 Gi ⌐E⌐ 77-17198
77 MARC

Torgersen, Don Arthur, 1934-
The scariest night in Troll Forest / by Don Arthur Torgersen ; ill. by Tom Dunnington. — Chicago : Childrens Press, c1979.
31 p. : col. ill. ; 25 cm.
SUMMARY: Several children decide not to follow the safe path on their hike through Troll Forest and consequently experience a very scary night.
ISBN 0-516-03613-0
⌐1. Trolls—Fiction⌐ I. Dunnington, Tom. II. Title.
PZ7.T6444 Sc ⌐E⌐ 78-13563
78 MARC

Torgersen, Don Arthur, 1934-
The troll who lived in the lake / by Don Arthur Torgersen ; ill. by Tom Dunnington. — Chicago : Childrens Press, c1978.
31 p. : col. ill. ; 25 cm.
SUMMARY: The troll living at the bottom of the lake finally decides to do something about the children who are taking all the fish out of his lake and leaving a lot of debris behind.
ISBN 0-516-03631-9
⌐1. Trolls—Fiction. 2. Pollution—Fiction⌐ I. Dunnington, Tom. II. Title.
PZ7.T6444 Tr ⌐Fic⌐ 78-3592
78 MARC

Travers, Pamela L 1906-
Mary Poppins comes back, by P. L. Travers. Illustrated by Mary Shepard. 1st Voyager ed. New York, Harcourt Brace Jovanovich ⌐1975, c1963⌐
xiii, 268 p. illus. 20 cm. (A Voyager book, AVB 93)
SUMMARY: Mary Poppins comes back on the end of a kite string, stays with the Banks family for a while, and then disappears on a merry-go-round horse.
⌐1. Fantasy⌐ I. Shepard, Mary, 1909- illus. II. Title.
⌐PZ7.T689 Mas 8⌐ ⌐Fic⌐ 74-17258
74 MARC

Travers, Pamela L 1906-
Mary Poppins in the park / by P. L. Travers ; illustrated by Mary Shepard. — 1st Voyager ed. — New York : Harcourt Brace Jovanovich, 1976, c1952.
xii, 235 p. : ill. ; 19 cm. — (A Voyager book ; AVB 102)
SUMMARY: More adventures of Mary Poppins, Michael, Jane, the twins and baby Annabel.
ISBN 0-15-657690-2
⌐1. Fantasy⌐ I. Shepard, Mary, 1909- . II. Title.
⌐PZ7.T689 Masp 15⌐ ⌐Fic⌐ 75-30526
75 MARC

Travers, Pamela L 1906-
Mary Poppins opens the door / by P. L. Travers ; illustrated by Mary Shepard and Agnes Sims. — 1st Voyager ed. — New York : Harcourt Brace Jovanovich, 1976, c1943.
xiii, 239 p. : ill. ; 19 cm. — (A Voyager book ; AVB 103)
SUMMARY: Mary Poppins returns to the Banks family in a rocket and involves the Banks children in more magical adventures including those with Peppermint Horses, the Marble Boy, and the Cat that Looked at the King.
ISBN 0-15-657692-9
⌐1. Fantasy⌐ I. Shepard, Mary, 1909- . II. Sims, Agnes. III. Title.
⌐PZ7.T689 Mat 10⌐ ⌐Fic⌐ 75-30697
75 MARC

Trease, Geoffrey, 1909-
Red comet; a tale of travel in the U.S.S.R., by Geoffrey Trease. Moscow, Co-operative publishing society of foreign workers in the U.S.S.R., 1936.
211, ⌐1⌐ p. illus. 19 cm.
Maps on lining-papers.
I. Title.
PZ7.T6895 Rc 37-10700
⌐r79⌐rev2 MARC

Trease, Geoffrey, 1909-
The seas of morning / ⌐by⌐ Geoffrey Trease ; illustrated by David Smee. — Harmondsworth : Puffin Books, 1976.
176 p. : ill. ; 19 cm. GB76-33557
SUMMARY: In the late fifteenth century, a young boy travels to Rhodes to find out more about the Order of St. John and becomes involved in a dangerous adventure.
ISBN 0-14-030891-1 : £0.65
1. Rhodes (City)—Siege, 1480—Juvenile fiction. ⌐1. Rhodes (City)—Siege, 1480—Fiction. 2. Knights of Malta—Fiction⌐ I. Title.
PZ7.T6895 Sd ⌐Fic⌐ 78-321289
78 MARC

Trevor, Elleston.
Badger's Beech / Elleston Trevor ; with drawings by Leslie Atkinson. — Charter House Publishers ; ⌐New York : distributed by Two Continents⌐, Charter House Publishers, c1978.
151 p. : ill. ; 21 cm. — (The woodlander series)
"First published 1948."
SUMMARY: A rhyme in an ancient chest hints at buried gold for those who follow its directions.
ISBN 0-8202-5015-5. ISBN 0-8202-5005-8 pbk.
⌐1. Animals—Fiction⌐ I. Atkinson, Leslie. II. Title.
PZ7.T7328 Bad 1978 ⌐Fic⌐ 78-18487
78 MARC

Trevor, Elleston.
Badger's moon / Elleston Trevor ; illustrated by Leslie Atkinson. — Nashville : Charter House ; New York : distributed by Two Continents, c1978.
156 p., ⌐4⌐ leaves of plates : ill. ; 21 cm. — (The Woodlander series)
SUMMARY: The Woodlanders make an unexpected trip to the moon in the Brown Wizard's rocket ship.
ISBN 0-8202-5016-3
⌐1. Space flight to the moon—Fiction. 2. Animals—Fiction⌐ I. Atkinson, Leslie. II. Title.
PZ7.T7328 Bag 1978 ⌐Fic⌐ 78-18484
00 MARC

Trevor, Elleston.
Sweethallow Valley / Elleston Trevor ; with drawings by Leslie Atkinson. — Nashville : Charter House Publishers ; New York : distributed by Two Continents, c1978.
222 p., ⌐4⌐ leaves of plates : ill. ; 21 cm. — (The Woodlander series)
SUMMARY: Twenty Gnomes come to Deep Wood to make mischief, but the Woodlander animals convince them to do good instead.
ISBN 0-8202-5018-X. ISBN 0-8202-5008-2 pbk. : $2.95
⌐1. Animals—Fiction. 2. Fairies—Fiction⌐ I. Atkinson, Leslie. II. Title.
PZ7.T7328 Sw 1978 ⌐Fic⌐ 78-18490
79 MARC

Tricolici, Chiril.
Campionii : ⌐povestiri⌐ / Chiril Tricolici. — ⌐Ediția a 2-a / coperta de Horia Constantinescu⌐. — ⌐București⌐ : Editura Ion Creangă, ⌐1978⌐
463 p. ; 20 cm. R78-4347
Cover title.
CONTENTS: Marele premiu.—Ultima variantă.
lei10.50
1. Automobile racing—Juvenile fiction. I. Title.
PZ90.R6T68 1978 79-361152
*79 MARC

Trivelpiece, Laurel.
During water peaches / Laurel Trivelpiece. — Philadelphia : Lippincott, c1979.
160 p. ; 21 cm.
SUMMARY: A seventten-year-old's job with a World War II project employing Mexican nationals reveals facets of life that influence her future plans.
ISBN 0-397-31831-6
⌐1. Family problems—Fiction. 2. World War, 1939-1945—Fiction⌐ I. Title.
PZ7.T738 Du ⌐Fic⌐ 78-14393
78 MARC

Tudor, Bethany.
Samuel's tree house / Bethany Tudor. — New York : Collins, 1979.
⌐25⌐ p. : col. ill. ; 16 cm.
SUMMARY: A duckling who wants to build a tree house all by himself accepts some help from his friends.
ISBN 0-529-05435-3 : 5.95
⌐1. Toys—Fiction⌐ I. Title.
PZ7.T8227 San ⌐E⌐ 78-12087
78 MARC

CHILDREN'S LITERATURE (Continued)

Tully, John, 1923-
Johnny and the Yank / ⌜by⌝ John Tully. — London : Methuen, 1978.
190 p. ; 21 cm. GB78-34138
ISBN 0-416-86170-9 : £3.75
I. Title.
PZ7.T8232 Jo 1978 79-308886
79 MARC

Turco, Lewis.
Murgatroyd and Mabel / by Wesli Court ⌜i.e. L. Turco⌝ ; ill. by Robert Michaels. — Oswego, N.Y. : Mathom Pub. Co., c1978.
⌜40⌝ p. : col. ill. ; 16 x 23 cm.
SUMMARY: Two caterpillars, anxiously awaiting metamorphosis, are suprised at the results when it actually occurs.
ISBN 0-930000-06-4
⌜1. Caterpillars—Fiction. 2. Butterflies—Fiction⌝ I. Michaels, Robert, 1926- . II. Title.
PZ7.T837 Mu ⌜E⌝ 79-101735
79 MARC

Turkle, Brinton.
Obadiah the Bold; story and pictures by Brinton Turkle. New York, Viking Press ⌜1965⌝
⌜36⌝ p. col. illus. 18 x 23 cm.
SUMMARY: A young boy's desire to be a pirate is quelled by his brothers and sisters during a game.
⌜1. Friends, Society of—Fiction. 2. Nantucket, Mass.—Fiction⌝ I. Title.
PZ7.T847 Ob ⌜E⌝ 65-13350
MARC

Turkle, Brinton.
Rachel and Obadiah / by Brinton Turkle. — 1st ed. — New York : Dutton, c1978.
⌜31⌝ p. : col. ill. ; 18 x 23 cm.
SUMMARY: Both Rachel and Obadiah want to earn some money by carrying the news of the next ship's arrival to Nantucket.
ISBN 0-525-38020-5 lib. bdg. : $7.95
⌜1. Nantucket, Mass.—Fiction. 2. Friends, Society of—Fiction⌝ I. Title.
PZ7.T847 Rac 1978 ⌜E⌝ 77-15661
77 MARC

Turner, Gladys T 1935-
Autobiography of Tammy : a life full of love and fun / ⌜written by Gladys T. Turner ; illustrated by Frederick M. Finney⌝. — Dayton, Ohio : Challenge Press, c1978.
56 p. : ill. ; 22 cm.
SUMMARY: A cat describes her birth, siblings, parents, growth and development during her early years, and her adulthood.
ISBN 0-89421-007-6 : $8.95. ISBN 0-89421-009-2 lib. bdg. : $9.95
1. Cats—Legends and stories. ⌜1. Cats—Fiction⌝ I. Finney, Frederick Marshall. II. Title.
PZ10.3.T8927 Au ⌜Fic⌝ 78-1317
78 MARC

Tzu, Ch'iang.
Gathering mushrooms : story and drawings / by Tzu Chiang, Chiu Yi, and Ku Ping. — 1st ed. — Peking : Foreign Languages Press, 1977.
⌜72⌝ p. : col. ill. ; 19 cm.
Translation of Ts'ai mo ku.
I. Ch'iu, I., joint author. II. Ku, P'ing, joint author. III. Title.
PZ7.T996 Gat 79-354766
79 MARC

Uchida, Yoshiko.
Journey home / Yoshiko Uchida ; illustrated by Charles Robinson. — 1st ed. — New York : Atheneum, 1978.
131 p. : ill. ; 22 cm.
"A Margaret K. McElderry book."
SUMMARY: After their release from an American concentration camp, a Japanese-American girl and her family try to reconstruct their lives amidst strong anti-Japanese feelings which breed fear, distrust, and violence.
ISBN 0-689-50126-9 : $7.95
⌜1. Japanese-Americans—Fiction. 2. Prejudices—Fiction⌝ I. Robinson, Charles, 1931- . II. Title.
PZ7.U25 Jn 1978 78-8792
79 MARC

Uchida, Yoshiko.
Sumi's prize. Pictures by Kazue Mizumura. New York, Scribner ⌜1964⌝
⌜48⌝ p. col. illus. 24 cm.
SUMMARY: Seven-year-old Sumi longs to do something well enough to win a prize. Opportunity arises when she enters the kite flying contest in her village.
⌜1. Kites—Fiction. 2. Japan—Fiction⌝ I. Mizumura, Kazue, illus. II. Title.
PZ7.U25 Su ⌜E⌝ 64-19650
⌜r79⌝rev MARC

Union is strength. — Hong Kong : Seagull Pub. Co., ⌜197-?⌝
12 p. : col. ill. ; 18 x 20 cm.
$1.80
PZ7.U4412 77-152950
79 MARC

Uttley, Alison, 1884-
From spring to spring : stories of the four seasons / Alison Uttley ; chosen by Kathleen Lines ; illustrated by Shirley Hughes. — London : Faber and Faber, 1978.
131 p. : ill. ; 23 cm. GB***
CONTENTS: Mrs Mimble and Mr Bumble Bee.—The three flowers of fortune.—Green shoes.—The rainbow.—The cornfield.—John Barleycorn.—Country mice meet church mouse.—Sam Pig's trousers.—The wind in a frolic.—The snow goose.—Star-shine.—The seven sleepers.—The Queen Bee.
ISBN 0-571-11144-0
I. Lines, Kathleen. II. Title.
PZ7.U72 Fr 79-670056
79 MARC

Vallejo, Doris.
The boy who saved the stars / by Doris Vallejo ; illustrated by Boris Vallejo. — New York : O'Quinn Studios, c1978.
32 p. : col. ill. ; 22 x 26 cm.
SUMMARY: A young boy on a faraway planet must find a way to stop the stars from disappearing from the sky.
ISBN 0-931064-05-8 : $5.95
⌜1. Science fiction⌝ I. Vallejo, Boris. II. Title.
PZ7.V254 Bo ⌜E⌝ 79-101629
79 MARC

Van Horn, William.
Twitchtoe, the beastfinder / by William Van Horn. — 1st ed. — New York : Atheneum, 1978.
36 p. : col. ill. ; 25 x 29 cm.
SUMMARY: Victor Twitchtoe is an expert at finding the lazy beasts hiding from work in the village, but he almost meets defeat from Snort, who causes giant winds and knocks down chimneys to gain attention.
ISBN 0-689-30670-9
I. Title.
PZ7.V3253 Tw ⌜E⌝ 78-3679
78 MARC

Van Leeuwen, Jean.
Seems like this road goes on forever / Jean Van Leeuwen. — New York : Dial Press, c1979.
214 p. ; 22 cm.
SUMMARY: A 17-year-old minister's daughter who feels her world is crashing down suffers an emotional breakdown which eventually forces her to make some difficult decisions.
ISBN 0-8037-7687-X : $7.95
⌜1. Emotional problems—Fiction. 2. Family life—Fiction. 3. Christian life—Fiction⌝ I. Title.
PZ7.V3273 Se ⌜Fic⌝ 78-72201
79 MARC

Van Leeuwen, Jean.
Tales of Oliver Pig / Jean Van Leeuwen ; pictures by Arnold Lobel. — New York : Dial Press, c1979.
64 p. : col. ill. ; 23 cm. — (Dial easy-to-read)
SUMMARY: Five adventures of Oliver Pig with his family.
ISBN 0-8037-8736-7 (lib. bdg.) : $5.89
⌜1. Pigs—Fiction. 2. Family life—Fiction⌝ I. Lobel, Arnold. II. Title.
PZ7.V3273 Tal ⌜E⌝ 79-4276
79 MARC

Van Steenwyk, Elizabeth.
Mystery at Beach Bay / by Elizabeth Van Steenwyk ; photos. by Lee Hooper. — Mankato, MN : Creative Education, c1978.
56 p. : ill. ; 24 cm.
SUMMARY: Two young sleuths set out to unravel the mystery surrounding the Toy Soldier Store and its proprietor.
ISBN 0-87191-676-2
⌜1. Mystery and detective stories⌝ I. Hooper, H. Lee. II. Title.
PZ7.V358 My ⌜Fic⌝ 78-12639
78 MARC

Van Steenwyk, Elizabeth.
Ride to win / by Elizabeth Van Steenwyk ; ⌜photos. by George Long Photography⌝. — Mankato, MN : Creative Education, c1978.
56 p. : ill. ; 24 cm.
SUMMARY: Nancy's decision to take up endurance riding brings her hard work and much satisfaction.
ISBN 0-87191-677-0
⌜1. Horsemanship—Fiction⌝ I. George Long Photography. II. Title.
PZ7.V358 Ri ⌜Fic⌝ 78-12377
78 MARC

Van Steenwyk, Elizabeth.
Rivals on ice / Elizabeth Van Steenwyk ; illustrated by Rondi Anderson. — Chicago : A. Whitman, c1979.
63 p. : ill. ; 21 cm. — (Springboard books)
SUMMARY: As two young figure-skaters grow to like each other, they are forced into a dilemma: should friends compete against each other?
ISBN 0-8075-7071-0 : $4.95
⌜1. Ice skating—Fiction. 2. Friendship—Fiction⌝ I. Anderson, Rondi. II. Title.
PZ7.V358 Rl ⌜Fic⌝ 78-31102
78 MARC

Van Woerkom, Dorothy.
Donkey Ysabel / by Dorothy O. Van Woerkom ; pictures by Normand Chartier. — New York : Macmillan, c1978.
47 p. : col. ill. ; 23 cm. — (Ready-to-read)
SUMMARY: Donkey Ysable is upset when she is replaced by a car but proves to her farmer owner that she is still superior to an automobile.
ISBN 0-02-791280-9
⌜1. Donkeys—Fiction⌝ I. Chartier, Normand, 1945- . II. Title.
PZ7.V39 Do ⌜Fic⌝ 78-5140
78 MARC

Vasiliu, Mircea.
A day at the beach / by Mircea Vasiliu. — New York : Random House, c1977.
⌜32⌝ p. : col. ill. ; 21 cm. — (A Random House pictureback)
SUMMARY: A family spends the day at the beach where they swim, build sand castles, explore the tide pools, and watch birds.
ISBN 0-394-83475-5 : 0.95. ISBN 0-394-93475-X lib. bdg. : $3.99
⌜1. Seashore—Fiction⌝ I. Title.
PZ7.V446 Day ⌜E⌝ 76-24169
78⌜r79⌝rev MARC

Važa-P'šavela, 1861-1915.
The lofty mountains ⌜by⌝ Vazha-Pshavela. Tbilisi, Pub. House Ganatleba, 1972.
61 p. illus. 27 cm. (Georgian literature) 0.78rub USSR***
Translated from the Georgian.
I. Title.
PZ7.V483 Lo 1972 75-304641
75⌜r79⌝rev MARC

Veglahn, Nancy.
Dance of the planets : the universe of Nicolaus Copernicus / by Nancy Veglahn ; illustrated by George Ulrich. — New York : Coward, McCann & Geoghegan, c1979.
63 p. : ill. ; 22 cm.
Bibliography: p. 63.
SUMMARY: Nicolaus Copernicus' inquiring mind probes the secrets of the universe during a time when such behavior is considered heretical.
ISBN 0-698-30693-7 lib. bdg.
1. Copernicus, Nicolaus, 1473-1543—Juvenile fiction. ⌜1. Copernicus, Nicolaus, 1473-1543—Fiction. 2. Astronomers—Fiction⌝ I. Ulrich, George. II. Title.
PZ7.V487 Dan 1979 ⌜Fic⌝ 78-8615
78 MARC

Vergara, René, 1916-
Más allá del crimen / René Vergara ; selección y prólogo de Alfonso Calderón. — Santiago : Editorial Nascimento, 1978.
302 p. ; 19 cm.
I. Title.
PQ8098.32.E75M3 863 78-109799
79 MARC

Very, Lydia Louisa Anna, 1823-1901.
Red Riding Hood / ⌜written & designed by Lydia L. Very⌝. — Boston, Mass. : L. Prang, ⌜1863?⌝
⌜16⌝ p. : col. ill. ; 18 cm.
Title from illustrated paper cover.
Volume cut in the shape of a standing Little Red Riding Hood.
"Entered according to act of Congress, in the year 1863 ... in clerk's office of the District Court of Massachusetts."
In verse.
I. Little Red Riding Hood.
PZ6.V4 Re 78-315528
78 MARC

Vigna, Judith.
Anyhow, I'm glad I tried / story and pictures by Judith Vigna. — Chicago : A. Whitman, c1978.
⌜32⌝ p. : ill. (some col.) ; 24 cm.
SUMMARY: A child treats a disagreeable classmate with kindness and is glad she makes an attempt at friendliness even though she feels her effort is in vain.
ISBN 0-8075-0378-9
⌜1. Kindness—Fiction⌝ I. Title.
PZ7.V67 An ⌜E⌝ 78-12883
78 MARC

Viorst, Judith.
Alexander, who used to be rich last Sunday / Judith Viorst ; illustrated by Ray Cruz. — 1st ed. — New York : Atheneum, 1978.
⌜32⌝ p. : ill. ; 20 x 26 cm.
SUMMARY: Alexander and his money are quickly parted yet he realizes there are many things a dollar can do.
ISBN 0-689-30602-4 lib. bdg. : $6.95
⌜1. Finance, Personal—Fiction⌝ I. Cruz, Ray. II. Title.
PZ7.V816 Am 1978 ⌜E⌝ 77-1579
77 MARC

Vogel, Ilse Margret.
Farewell, Aunt Isabell / by Ilse-Margret Vogel ; pictures by the author. — 1st ed. — New York : Harper & Row, c1979.
54 p. : ill. ; 22 cm.
SUMMARY: Two young sisters try to make their mentally ill aunt happy so that she will get well.
ISBN 0-06-026317-2 : $6.95
⌜1. Mentally ill—Fiction⌝ I. Title.
PZ7.V867 Far 1979 ⌜Fic⌝ 79-1785
79 MARC

Vogt, Esther Loewen.
Harvest gold / Esther Loewen Vogt ; illustrated by Seymour Fleishman. — Elgin, Ill. : D.C. Cook Pub. Co., c1978.
128 p. : ill. ; 18 cm.
Sequel to Turkey Red.
SUMMARY: The further experiences of Martha and her family as they face the hardships and the rewards of life on the Kansas prairie in the 1870's.
ISBN 0-89191-105-7 : $1.95
⌜1. Mennonites—Fiction. 2. Frontier and pioneer life—Fiction. 3. Kansas—Fiction⌝ I. Fleishman, Seymour. II. Title.
⌜PZ7.V8678 Har⌝ ⌜Fic⌝ 77-87257
78 MARC

CHILDREN'S LITERATURE (Continued)

Von Canon, Claudia.
The moonclock / Claudia Von Canon. — Boston : Houghton Mifflin, 1979.
ix, 159 p. : ill. ; 22 cm.
Bibliography: p. 159.
SUMMARY: The correspondence to and from a Viennese couple during 1683 chronicles domestic and political affairs including the Turks' siege of their city.
ISBN 0-395-27810-4 : $6.95
1. Vienna—Siege, 1683—Juvenile fiction. [1. Vienna—Siege, 1683—Fiction. 2. Austria—History—Leopold I, 1657-1705—Fiction] I. Title.
PZ7.V889 Mo [Fic] 79 79-1076 MARC

Waber, Bernard.
The snake : a very long story / Bernard Waber. — Boston : Houghton Mifflin, c1978.
[48] p. : col. ill. ; 15 x 24 cm.
SUMMARY: A snake goes on a very long trip only to face a big surprise at the end.
ISBN 0-395-27157-6 : $7.95
[1. Snakes—Fiction] I. Title.
PZ7.W113 Sn [E] 78 78-17991 MARC

Wagner, Jenny.
Aranea : a story about a spider / story by Jenny Wagner ; ill. by Ron Brooks. — 1st American ed. — Scarsdale, N.Y. : Bradbury Press, 1978, c1975.
[34] p. : ill. ; 26 cm.
SUMMARY: An industrious spider spends her days and nights spinning perfect webs.
ISBN 0-87888-138-7 : $6.95
[1. Spiders—Fiction] I. Brooks, Ron. II. Title.
PZ7.W12425 Ar 1978 [E] 78 78-55212 MARC

Wahl, Jan.
Carrot Nose / by Jan Wahl ; pictures by James Marshall. — 1st ed. — New York : Farrar, Straus, Giroux, 1978.
[39] p. : col. ill. ; 21 cm.
SUMMARY: A bunny is given a carrot nose when he displeases the carrot elf by refusing to eat any more carrots.
ISBN 0-374-31122-6 : $6.95
[1. Rabbits—Fiction. 2. Carrots—Fiction] I. Marshall, James, 1942- . II. Title.
PZ7.W1266 Car 1978 [E] 79 78-60514 MARC

Wahl, Jan.
The Norman Rockwell storybook / told by Jan Wahl. — New York : Simon and Schuster, [1979], c1969.
[46] p. : col. ill. ; 31 cm.
SUMMARY: Twenty, one-page, humorous stories illustrated with paintings by Norman Rockwell.
ISBN 0-671-25102-3
[1. Humorous stories. 2. Short stories] I. Rockwell, Norman, 1894-1978. II. Title.
PZ7.W1266 No 1979 [Fic] 79 79-14462 MARC

Wahl, Jan.
The teeny, tiny witches / by Jan Wahl ; illustrated by Margot Tomes. — New York : Putnam, c1979.
[48] p. : ill. (some col.) ; 23 cm.
SUMMARY: Relates the adventures of a family of teeny tiny witches as they search for a home.
ISBN 0-399-20682-5 : $6.95
[1. Witches—Fiction] I. Tomes, Margot. II. Title.
PZ7.W1266 Te 1979 [E] 78 78-15657 MARC

Wahl, Jan.
Who will believe Tim Kitten? / By Jan Wahl ; illustrated by Cyndy Szekeres. — New York ; Pantheon Books, c1978.
54 p. : ill. ; 22 cm. — (A Read aloud/read alone book)
SUMMARY: Tim Kitten has his great-grandmother's talent for telling tales, but no one believes his stories.
ISBN 0-394-83666-9. ISBN 0-394-93666-3 lib. bdg.
[1. Cats—Fiction] I. Szekeres, Cyndy. II. Title.
PZ7.W1266 Wh [E] 77 77-12956 MARC

Wahl, Jan.
Youth's magic horn : seven stories / by Jan Wahl. — 1st ed. — Nashville : T. Nelson, c1978.
127 p. ; 21 cm.
CONTENTS: Jackal, wolf, and fox.—A nun's afternoon off.—The walk past Anny's place.—The valedictorian.—The scalp.—Heaven bound.—At the crossing.
ISBN 0-8407-6582-7
[1. Short stories] I. Title.
PZ7.W1266 Yo [Fic] 78 77-29127 MARC

Waide, Jan, 1952-
Jennifer / written and illustrated by Jan Waide. — Austin, Tex. : Shoal Creek Publishers, c1978.
[32] p. : col. ill. ; 28 cm.
SUMMARY: A maple tree makes friends with a little boy and learns from him that some humans want nothing from the woods but to enjoy their peace and beauty.
ISBN 0-88319-039-7 : $5.95
[1. Trees—Fiction] I. Title.
PZ7.W128 Je [E] 78 78-12948 MARC

Waldman, Guido.
I went to school one morning / Guido Waldman ; illustrated by Michael Charlton. — London : Bodley Head, 1978.
[24] p. : col. ill. ; 22 cm. GB***
ISBN 0-370-30076-9 : £2.50
I. Charlton, Michael, fl. 1968- . II. Title.
PZ7.W1453 Iad 78 78-318422 MARC

Waldman, Jeri.
Pitcher in left field / by Jeri and Neil Waldman. — Englewood Cliffs, N.J. : Prentice-Hall, c1979.
63 p. : ill. ; 22 cm.
SUMMARY: A frustrated pitcher's resentment at playing left field causes him to make so many errors he brings down the whole team.
ISBN 0-13-676411-8 : $5.95
[1. Baseball—Fiction] I. Waldman, Neil, joint author. II. Title.
PZ7.W1459 Pi 1979 [Fic] 79 74-2327 MARC

Walsh, Ellen Stoll.
Brunus and the new bear / Ellen Stoll Walsh. — 1st ed. — Garden City, N.Y. : Doubleday, c1979.
[32] p. : col. ill. ; 22 cm.
SUMMARY: Benjamin's stuffed bear is jealous when a new bear arrives in the mail.
ISBN 0-385-14660-4 : $7.95. ISBN 0-385-14661-2 lib. bdg.
[1. Jealousy—Fiction. 2. Toys—Fiction] I. Title.
PZ8.9.W153 Br [E] 78 78-22361 MARC

Walt Disney Productions presents Button soup. — Book club ed. — New York : Random House, c1975.
[44] p. : col. ill. ; 24 cm. — (Disney's wonderful world of reading)
SUMMARY: Daisy tricks her stingy Uncle Scrooge into making enough soup for the whole town—using just one button.
ISBN 0-394-82562-4 : $2.50. ISBN 0-394-92562-9 lib. bdg.
I. Disney (Walt) Productions. II. Title: Button soup.
PZ7.W16897 1975 [E] 74 74-28376 MARC

Walt Disney Productions presents Donald Duck, mountain climber. — 1st American ed. — New York : Random House, c1978.
[42] p. : col. ill. ; 25 cm. — (Disney's wonderful world of reading)
Translation of Anders And som bjergbestiger.
"Book Club edition."
SUMMARY: Donald Duck, after much encouragement from his nephews, agrees to go mountain climbing with a guide named Goofy.
ISBN 0-394-84078-X
[1. Mountaineering—Fiction. 2. Ducks—Fiction] I. Disney (Walt) Productions. II. Title: Donald Duck, mountain climber.
PZ7.W168972 1978 79 78-112190 MARC

Walt Disney Productions presents Hiawatha, the brave hunter. — Book club ed., 1st American ed. — New York : Random House, c1979.
[42] p. : col. ill. ; 25 cm. — (Disney's wonderful world of reading)
SUMMARY: Little Hiawatha has a busy and exciting day in the forest with his animal friends.
ISBN 0-394-94234-5 (lib. bdg.)
1. Hiawatha, Iroquois Indian—Juvenile fiction. [1. Hiawatha, Iroquois Indian—Fiction. 2. Indians of North America—Fiction] I. Title: Hiawatha, the brave hunter.
PZ7.W1689723 1979 [E] 00 79-115115 MARC

Walt Disney Productions presents Lucky puppy. — 1st American ed., Book club ed. — New York : Random House, c1978.
[42] p. : col. ill. ; 25 cm. — (Disney's wonderful world of reading)
Translation of Dalmatineren plet og dronningens.
"Suggested by the Walt Disney motion picture One hundred and one dalmatians, based on The hundred and one dalmatians, by Dodie Smith."
SUMMARY: Lucky Puppy and his dad aae instrumental in recovering the queen's stolen jewels.
ISBN 0-394-84117-4
[1. Dogs—Fiction. 2. Mystery and detective stories] I. One hundred and one dalmatians. [Motion picture] II. Disney (Walt) Productions. III. Title: Lucky puppy.
PZ7.W1689724 [E] 79 79-106706 MARC

Walt Disney Productions presents The love bug. — New York : Random House, c1979.
[42] p. : col. ill. ; 25 cm. — (Disney's wonderful world of reading ; 45)
SUMMARY: Jim and his little white car, Herby, continue to enter races despite the sneaky tricks another driver plays on them.
ISBN 0-394-84139-5 : $3.50
[1. Automobile racing—Fiction] I. Disney (Walt) Productions. II. Title. III. Title: Love bug.
PZ7.W168974 [E] 78 78-13896 MARC

Walt Disney's the jungle book. New York, Random House [c1974]
[42] p. col. ill. 25 cm. (Disney's wonderful world of reading)
SUMMARY: The animal friends of the boy raised by wolves escort him out of the jungle to keep him safe from the tiger who seeks to kill him.
ISBN 0-394-82560-8. ISBN 0-394-92560-2 (lib. bdg.)
[1. Jungle stories] I. Kipling, Rudyard, 1865-1935. The jungle book. II. The jungle book.
PZ10.3.W18 [E] 74 74-3407 MARC

Wangerin, Walter.
The book of the dun cow / by Walter Wangerin, Jr. — 1st ed. — New York : Harper & Row, 1978.
ix, 241 p. ; 21 cm.
SUMMARY: Good struggles with Evil as Chauntecleer the rooster fights against the mysterious Wyrm.
ISBN 0-06-026346-6 : $6.95. ISBN 0-06-026347-4 lib. bdg. : $6.79
[1. Fantasy] I. Title.
PZ7.W1814 Bo 1978 [Fic] 78 77-25641 MARC

Warach, Marie Norkin.
I like red / written and illustrated by Marie Norkin Warach. — New York : Dandelion Press, 1979.
[31] p. : col. ill. ; 18 x 23 cm.
SUMMARY: Two youngsters describe their favorite colors.
ISBN 0-89799-116-8. ISBN 0-89799-002-1 pbk. : $1.50
[1. Color—Fiction] I. Title.
PZ7.W184 Iac [E] 79 78-72123 MARC

Ward, Leila.
I am eyes : ni macho / words by Leila Ward ; pictures by Nonny Hogrogian. — 1st ed. — New York : Greenwillow Books c1978.
[32] p. : ill. (some col.) ; 21 x 26 cm.
SUMMARY: An African child wakes to the morning marvels of her native land.
ISBN 0-668-80161-7 : $7.95. ISBN 0-688-84161-9 lib. bdg. : $7.35
[1. Africa—Fiction] I. Hogrogian, Nonny. II. Title.
PZ7.W2143 Iad [E] 78 78-1314 MARC

Ward, Mary Frances.
Fast as the wind : a story of Curtin Village in the 1860's / by Mary Frances Ward ; ill. by Nancy H. Strailey. — Curtin, Pa. : Roland Curtin Foundation for the Preservation of Eagle Furnace, 1978.
vi, 42 p. : ill. ; 23 cm.
SUMMARY: Portrays the life of a fictional family in Curtin Village in the Allegheny Mountains of central Pennsylvania in 1868.
[1. Pennsylvania—Social life and customs—Fiction. 2. United States—Social life and customs—1865-1918—Fiction] I. Strailey, Nancy H. II. Title.
PZ7.W2152 Fas [Fic] 78 78-105162 MARC

Wartski, Maureen Crane, 1940-
My brother is special / by Maureen Crane Wartski. — 1st ed. — Philadelphia : Westminster Press, c1979.
153 p. ; 21 cm.
SUMMARY: Eighth grader Noni Harlow hopes to solve several problems by entering her younger, mentally retarded brother in the Special Olympics, but on the morning of the competition Kip must be so heavily sedated that it appears he won't be able to run the relay.
ISBN 0-664-32644-7
[1. Mentally handicapped—Fiction. 2. Family life—Fiction. 3. Special Olympics—Fiction] I. Title.
PZ7.W2582 My [Fic] 78 78-23999 MARC

Waters, John Frederick, 1930-
Summer of the seals / by John F. Waters ; illustrated by Mike Eagle. — New York : F. Warne, c1978.
94 p. : ill. ; 23 cm.
SUMMARY: A teenage boy and girl befriend an old hermit and attempt to discover who is shooting the seals along their stretch of Maine coast.
ISBN 0-7232-6155-5 : $6.95
[1. Harbor seal—Fiction. 2. Seals (Animals)—Fiction. 3. Maine—Fiction] I. Eagle, Michael. II. Title.
PZ7.W2643 Su [Fic] 78 78-6712 MARC

Watson, Jane Werner, 1915-
The case of the semi-human beans / by Jane Werner Watson. — New York : Coward, McCann & Geoghegan, c1979.
144 p. ; 22 cm.
SUMMARY: Two boys' observation of some beans which seem almost human lead them into some near-fatal adventures, experiences with a spacecraft, and unsettling interviews with Air Force investigators of UFO's.
ISBN 0-698-20476-X
[1. Science fiction. 2. The West—Fiction] I. Title.
PZ7.W3274 Cas [Fic] 78 78-24320 MARC

Watson, Nancy Dingman.
Sugar on snow. Illus. by Aldren A. Watson. New York, Viking Press [1964]
43 p. col. illus. 21 x 25 cm.
SUMMARY: Peter, Cammie, and Cait make maple syrup in the spring, just in time for Cammie's birthday party.
[1. Maple syrup—Fiction] I. Watson, Aldren Auld, 1917- . II. Title.
PZ7.W328 Su [E] [r79]rev 64-21474 MARC

Watson, Nancy Dingman.
What does A begin with? Pictures by Alden A. Watson. New York, Knopf [1956]
[33] p. illus 21 x 26 cm.
SUMMARY: Peter and Linda identify objects around their far, that begin with each letter of the alphabet.
[1. Alphabet. 2. Farms—Fiction] I. Watson, Aldren Auld, 1917- . II. Title.
PZ7.W328 Wf [E] [r79]rev 56-8901 MARC

CHILDREN'S LITERATURE (Continued)

Watson, Pauline.
Wriggles, the little wishing pig / by Pauline Watson ; pictures by Paul Galdone. — New York : Seabury Press, c1978.
[32] p. : col. ill. ; 23 cm.
"A Clarion book."
SUMMARY: What's a little pig to do when his wish for a mouth like an alligator, wings like a pelican, and legs like a crane comes true?
ISBN 0-8164-3216-3 lib. bdg. : $7.95
[1. Pigs—Fiction. 2. Wishes—Fiction] I. Galdone, Paul. II. Title.
PZ7.W3285 Wr [E] 78 78-5855 MARC

Watson, Simon.
The new red bike, and other stories for the very young / [by] Simon Watson ; illustrated by Charlotte Voake. — London : Heinemann, 1978.
75 p. : ill. ; 23 cm. GB78-13115
ISBN 0-434-97165-0 : £2.90
I. Voake, Charlotte. II. Title.
PZ7.W3294 Ne 78 78-321967 MARC

Watson, Wendy.
Has winter come? / By Wendy Watson. — London : Collins ; New York : Collins & World, c1978.
[28] p. : col. ill. ; 19 x 21 cm.
SUMMARY: Although the children don't recognize the faint smell of winter in the air, a woodchuck family begins preparing for long snowy nights.
ISBN 0-529-05439-6 : $6.95. ISBN 0-529-05441-8 (lib. bdg.) : $6.91
[1. Woodchuck—Fiction. 2. Winter—Fiction] I. Title.
PZ7.W332 Has [E] 77 77-18863 MARC

Watson, Wendy.
Lollipop / by Wendy Watson. — New York : Puffin Books, 1978, c1976.
[31] p. : col. ill. ; 23 cm. — (Picture puffin)
SUMMARY: Bunny goes through a lot before he finally gets his lollipop.
ISBN 0-14-050264-5 pbk. : $1.95
[1. Rabbits—Fiction] I. Title.
[PZ7.W332 Lo 1978] [E] 77 77-28340 MARC

Watts, Marjorie-Ann.
Crocodile medicine / Marjorie-Ann Watts. — New York : F. Warne, 1978, c1977.
[32] p. : col. ill. ; 25 cm.
SUMMARY: Crocodile isn't the easiest patient in the hospital nor are his friends the best-behaved visitors, but he does make life in the children's ward much more fun.
ISBN 0-7232-6154-7 (lib. bdg.) : $6.95
[1. Hospitals—Fiction. 2. Crocodiles—Fiction] I. Title.
PZ7.W342 Cr 1978 79 78-52992 MARC

Watts, Marjorie-Ann.
Crocodile plaster / Marjorie-Ann Watts. — London : A. Deutsch, 1978.
[32] p. : col. ill. ; 25 cm. GB***
ISBN 0-233-96962-4 : £2.95
I. Title.
PZ7.W342 Ct 1978 79 78-323864 MARC

Way, Irene.
Armada quest / Irene Way. — Grand Rapids : Zondervan Pub. House, 1979, c1976.
102 p. ; 21 cm. — (Pathfinder series)
SUMMARY: Newly moved to a Scottish island, a pair of twins runs into danger as they race with a thief to solve the mystery of the lost treasure ship, the San Salvador.
ISBN 0-310-37841-9
[1. Mystery and detective stories. 2. Scotland—Fiction] I. Title. II. Series: Pathfinder series (Grand Rapids)
PZ7.W3512 Ar 1979 [Fic] 78 78-26401 MARC

Weber, Judith.
Lights, camera, cats! / Judith Eichler Weber ; illustrated by Pat Grant Porter. — New York : Lothrop, Lee & Shepard Co., c1978.
127 p. : ill. ; 22 cm.
SUMMARY: A 12-year-old girl auditions five cats for a television commercial advertising a brand of cat food that is the top rival of her father's company.
ISBN 0-688-41867-8. ISBN 0-688-51867-2 lib. bdg.
[1. Cats—Fiction. 2. Family life—Fiction] I. Porter, Patricia Grant. II. Title.
PZ7.W388 Li [Fic] 78 78-17083 MARC

Weintraub, Linda, 1940-
Runaway ! / Linda Weintraub. — 1st ed. — New York : Ballantine Books, 1978.
294 p. ; 22 cm.
SUMMARY: Three runaways in search of love finally find it in Hollywood with the help of a bachelor who yearns for a family.
ISBN 0-345-27720-1 : $7.95
[1. Runaways—Fiction] I. Title.
PZ4.W42259 Ru [Fic] 78 78-5259 MARC
[PS3573.E3972]

Weiss, Ellen.
Millicent Maybe / by Ellen Weiss. — New York : F. Watts, 1979.
[32] p. : col. ill. ; 22 cm. — (An Easy-read story book)
SUMMARY: Indecisive Millicent, who has trouble making up her mind, discovers what she thinks is a permanent solution—only to experience real chaos.
ISBN 0-531-02382-6. ISBN 0-531-02299-4 lib. bdg.
[1. Decision making—Fiction] I. Title.
PZ7.W4472 Mi [E] 78 78-13144 MARC

Weldrick, Valerie.
Time sweep / Valerie Weldrick ; illustrated by Ron Brooks. — 1st ed. — New York : Lothrop, Lee & Shepard Co., [1978] c1976.
157 p. : ill. ; 22 cm.
SUMMARY: A boy is mysteriously transported from present-day Sydney, Australia, to the London of 1862.
ISBN 0-688-41869-4. ISBN 0-688-51869-9 lib. bdg.
[1. Space and time—Fiction. 2. England—Fiction] I. Brooks, Ron. II. Title.
PZ7.W4493 Ti 1978 [Fic] 78 78-11837 MARC

Wells, Rosemary.
Stanley & Rhoda / by Rosemary Wells. — New York : Dial Press, c1978.
40 p. : col. ill. ; 22 cm.
SUMMARY: In three episodes a brother and sister deal with Rhoda's untidy room, a bee sting, and a baby sitter.
ISBN 0-8037-8248-9 : $5.95. ISBN 0-8037-8249-7 lib. bdg. : $6.46
[1. Brothers and sisters—Fiction] I. Title.
PZ7.W46843 St [E] 78 78-51874 MARC

Wensell, Ulises.
Come to our house / Ulises Wensell ; text adapted by Charlotte Herman and D. Francis ; ill., Ulises Wensell. — Chicago : Rand McNally, c1978.
[60] p. : chiefly col. ill. ; 29 cm.
Text adapted from the author's Así somos nosotros and Así son papá y mamá.
SUMMARY: A young brother and sister describe their family and their daily activities.
ISBN 0-528-82442-2 : $5.95
[1. Family life—Fiction] I. Herman, Charlotte. II. Haas, Dorothy. III. Wensell, Ulises. Así somos nosotros. IV. Wensell, Ulises. Así son papá y mamá. V. Title.
PZ7.W46926 Co [E] 79 78-57836 MARC

Wentworth, Ruth Starbuck.
The first Nantucket tea party / newly edited by Meredith Marshall Brenizer ; with new ill. by Mary Thomas Curtis. — Taunton, Mass., : W. S. Sullwold Pub., c1978.
[63] p. : ill. ; 23 x 18 cm.
SUMMARY: A young girl living in mid-18th-century Nantucket describes how her family first brewed the tea brought to them from the Orient by her cousin.
ISBN 0-88492-022-4
[1. Nantucket, Mass.—Fiction. 2. Tea—Fiction] I. Brenizer, Meredith Marshall. II. Title.
PZ7.W4697 Fi [Fic] 79 78-62476 MARC

Wesker, Arnold, 1932-
Fatlips : a story for children / by Arnold Wesker ; illustrations by Oscar Zarate. — London : Writers and Readers Publishing Cooperative, 1978.
3-93 p. : ill. ; 21 cm. GB78-32545
ISBN 0-904613-72-0 : £2.95
I. Zarate, Oscar, 1942- . II. Title.
PZ7.W49 Fat 1978b 79 79-303371 MARC

Wesker, Arnold, 1932-
Fatlips : a story for children / by Arnold Wesker ; ill. by Oscar Zarate. — New York : Harper & Row, c1978.
93 p. : ill. ; 21 cm.
SUMMARY: A mysterious old man with magical powers reveals his secrets to a family that has befriended him on a ship.
ISBN 0-06-026388-1 : $6.95. ISBN 0-06-026389-X lib. bdg.
[1. Magic—Fiction] I. Zarate, Oscar. II. Title.
PZ7.W49 Fat [Fic] 79 78-109064 MARC

West, Carl.
Dark wing / by Katherine MacLean and Carl West. — 1st ed. — New York : Atheneum, 1979.
242 p. ; 25 cm. — (An Argo book)
SUMMARY: Travis's illegal medical knowledge, his experiences, and his friendships enable him to help Iron Phedon, a persecuted boy from another planet, and also to help himself to what he really wants.
ISBN 0-689-30668-1
[1. Science fiction] I. MacLean, Katherine, joint author. II. Title.
PZ7.W5173 Dar [Fic] 78 78-10837 MARC

Westall, Robert.
The devil on the road / Robert Westall. — London : Macmillan, 1978.
247 p. : [1] ill. ; 21 cm. GB***
ISBN 0-333-25352-3 : £3.95
I. Title.
PZ7.W51953 De 79 79-303216 MARC

Westall, Robert.
The Devil on the road / Robert Westall. — 1st American ed. — New York : Greenwillow Books, [1979] c1978.
248 p. ; 22 cm.
SUMMARY: While seeking shelter from a sudden rainstorm in an old barn, a young motorcyclist finds himself catapulted into a mid-17th century England troubled by witch hunts.
ISBN 0-688-80227-3 : $7.95
[1. Space and time—Fiction. 2. Witchcraft—Fiction. 3. England—Fiction] I. Title.
PZ7.W51953 De 1979 Fic 79 79-10427 MARC

Westall, Robert.
The machine gunners / by Robert Westall. — New York : Greenwillow Books, 1976.
186 p. ; 22 cm.
SUMMARY: After an air raid, a group of English children find a German machine gun and hide it from adults who are looking for it.
ISBN 0-688-80055-6. ISBN 0-688-84055-8 lib. bdg.
[1. World War, 1939-1945—Great Britain—Fiction. 2. Great Britain—History—20th century—Fiction] I. Title.
PZ7.W51953 Mac 3 [Fic] 76 76-13630 MARC

Westbrook, Dante.
Daxius / story by Dante Westbrook ; ill. by Dewey Crumpler. — San Francisco : Children's Book Press, c1977.
23 p. : col. ill. ; 25 cm. — (Fifth world tales)
SUMMARY: Relates the adventures of Daxius, son of a black panther and a part-panther human woman, and how he eventually becomes the ruler of a faraway planet known as Black Panther Land.
ISBN 0-89239-014-X
[1. Panthers—Fiction. 2. Fantasy. 3. Children's writings] I. Crumpler, Dewey. II. Title.
PZ7.W51957 Dax [Fic] 79 77-83857 MARC

Wheeler, Penny Estes.
With long life / Penny Estes Wheeler. — Nashville : Southern Pub. Association, c1978.
127 p. ; 21 cm.
SUMMARY: Fourteen stories illustrating Seventh-Day Adventist principles.
ISBN 0-8127-0192-5
[1. Seventh-Day Adventists—Fiction. 2. Christian life—Fiction] I. Title.
PZ7.W5664 Wi [Fic] 78 78-13748 MARC

Whelan, Gloria.
A time to keep silent / by Gloria Whelan. — New York : Putnam, c1979.
127 p. ; 22 cm.
SUMMARY: After her mother's death, 13-year-old Clair refuses to speak and her minister father moves them from their comfortable existence to the wilds of northern Michigan to start a small mission and to begin a new life.
ISBN 0-399-20693-0 : $7.95
[1. Country life—Fiction] I. Title.
PZ7.W5718 Ti 1979 [Fic] 79 79-11608 MARC

White, John, 1924 (Mar. 5)-
The Tower of Geburah : a children's fantasy / John White. — Downers Grove, Ill. : Inter-Varsity Press, c1978.
402 p. : ill. ; 21 cm.
SUMMARY: Left at home by themselves one snowy evening, three youngsters enter the magical kingdom of Anthropos.
ISBN 0-87784-560-3 : $5.95
[1. Fantasy] I. Title.
PZ7.W5837 To [Fic] 79 78-6363 MARC

White, Mary.
Dominic / [by] Mary White. — Sydney : Methuen of Australia, 1977.
182 p. ; 21 cm. Aus
ISBN 0-454-00028-6
I. Title.
PZ7.W5839 Do 79 79-300188 MARC

White, Robb, 1909-
Fire storm : a novel / by Robb White. — 1st ed. — Garden City, N.Y. : Doubleday, c1979.
111 p. ; 22 cm.
SUMMARY: A raging forest fire in the National Parks area of the Sierras traps a forest ranger and a young boy he suspects is an arsonist.
ISBN 0-385-14630-2 : $6.95. ISBN 0-385-14631-0 lib. bdg.
[1. Forest fires—Fiction] I. Title.
PZ7.W5844 Fi [Fic] 79 78-72186 MARC

White, Robin, 1928-
The troll of crazy mule camp / by Robin White ; illustrated by Susan Reisse. — New York : Dandelion Press, 1979.
[32] p. : col. ill. ; 23 cm.
SUMMARY: Shaman gives up his life as a troll and joins a human family.
ISBN 0-89799-142-7
[1. Trolls—Fiction] I. Reiss, Susan. II. Title.
PZ7.W58442 Tr [E] 79 78-72140 MARC

CHILDREN'S LITERATURE (Continued)

White, Wallace, 1930-
One dark night / by Wallace White ; photos. by Bill Aron. — New York : F. Watts, 1979.
90 p. : ill. ; 22 cm. — (A Triumph book)
SUMMARY: Greg visits his widowed uncle, sheriff of a small southern town, and is confused and frightened by the man's changed personality.
ISBN 0-531-02896-8
[1. Uncles—Fiction. 2. Alcoholism—Fiction] I. Aron, Bill. II. Title.
PZ7.W58487 On [Fic] 79-10842
79 MARC

Widney, Gaylord G
Bunny Snowhite visits Mr. McDougal's garden / by Gaylord G. Widney. — San Diego : Grossmont Press, c1977.
37 p. : col. ill. ; 23 cm.
SUMMARY: A young bunny frequents Mr. McDougal's vegetable patch until he is ensnared in the gardener's trap.
ISBN 0-913182-89-3 : $4.95
[1. Rabbits—Fiction] I. Title.
PZ7.W6348 Bu [E] 77-73162
78[r79]rev MARC

Wild, Robin.
The mouse who stole a zoo [by] Robin and Jocelyn Wild. [1st American ed.] New York, Coward-McCann & Geoghegan [1972]
[32] p. col. illus. 27 cm. $4.95
SUMMARY: Hating to see the zoo animals living in cages, a clever mouse devises a scheme to set them free.
[1. Zoo animals—Fiction] I. Wild, Jocelyn, joint author. II. Title.
PZ7.W64576 Mo 1972 [E] 72-76704
73[r79]rev MARC

Wild, Robin.
Spot's dogs and the Alley Cats / Robin and Jocelyn Wild. — 1st American ed. — Philadelphia : Lippincott, c1979.
[32] p. : col. ill. ; 25 cm.
SUMMARY: Two rival gangs play mean tricks and do good deeds in "A Trip to the Sea," "A Day at the Fair," and "The Christmas Party."
ISBN 0-397-31841-3
[1. Dogs—Fiction. 2. Cats—Fiction] I. Wild, Jocelyn, joint author. II. Title.
PZ7.W64576 Sp 1979 [Fic] 78-23836
78 MARC

Wildsmith, Alan.
The Northern phantom / Alan Wildsmith. — [London] : Deutsch, [1978]
176 p. ; 21 cm. GB***
ISBN 0-233-97002-9 : £3.95
I. Title.
PZ7.W6468 No 1978 78-325692
79 MARC

Wildsmith, Brian.
Hunter and his dog / Brian Wildsmith. — Oxford ; New York : Oxford University Press, 1979.
[30] p. : chiefly col. ill. ; 29 cm. GB***
"Though trained to retrieve birds for his master, a dog finds that he is too kind-hearted for such work."
ISBN 0-19-279725-5 : £2.95 ($7.95)
[1. Dogs—Fiction. 2. Ducks—Fiction. 3. Kindness—Fiction. 4. Hunting—Fiction. 5. Wildlife conservation—Fiction] I. Title.
PZ7.W647 Hu [E] 78-41181
00 MARC

Wiles, Julian.
The Tradd Street Follies / by Julian Wiles ; with ill. by the author. — Charleston, S.C. : Tradd Street Press, c1978.
101 p. : ill. ; 15 x 23 cm.
SUMMARY: A group of children help P. G. T. Pinckney, an elderly gentleman who does magic and other interesting things, save his house from the ladies of the Historical Society who are determined to buy it for their new headquarters.
[1. City and town life—Fiction. 2. Charleston, S.C.—Fiction] I. Title.
PZ7.W6475 Tr [Fic] 76-45729
79 MARC

Wilkerson, Don.
Hell-bound / Don Wilkerson, with David Manuel. — Orleans, Mass. : Rock Harbor Press, c1978.
199 p. ; 18 cm.
SUMMARY: The street ministry of Teen Challenge offers a worldly bank robber, a glue-sniffing arsonist, and a bored, alcoholic teenager a new view of life.
$2.25
[1. Christian life—Fiction] I. Manuel, David, joint author. II. Title.
PZ7.W64836 He [Fic] 78-60735
00 MARC

Willard, Barbara.
The country maid / Barbara Willard. — London : Hamilton, 1978.
177 p. ; 23 cm. GB***
ISBN 0-241-89936-2 : £3.75
I. Title.
PZ7.W6547 Cr 1978 78-318451
78 MARC

Willard, Barbara.
The gardener's grandchildren / Barbara Willard ; ill. by Gordon King. — New York : McGraw-Hill, 1979, c1978.
143 p. : ill. ; 21 cm.
SUMMARY: Two children become curious about the departure of their grandfather's employer, who left their "Garden Island" off Scotland years ago after instructing the gardener to faithfully keep the island beautiful until his return.
ISBN 0-07-070291-8 : $7.95
[1. Mystery and detective stories. 2. Islands—Fiction] I. King, Gordon, 1939- II. Title.
PZ7.W6547 Gar 1979 [Fic] 78-23637
78 MARC

Willard, Nancy.
The island of the grass king : the further adventures of Anatole / by Nancy Willard ; ill. by David McPhail, decorative letters by John O'Connor. — 1st ed. — New York : Harcourt Brace Jovanovich, c1979.
120 p. : ill. ; 21 cm.
SUMMARY: Anatole embarks on a fantastical journey to the island where the wild fennel grows.
ISBN 0-15-239082-0
[1. Fantasy] I. Title.
PZ7.W6553 Is [Fic] 78-20574
78 MARC

Willard, Nancy.
Simple pictures are best / story by Nancy Willard ; pictures by Tomie de Paola. — 1st Voyager/HBJ ed. — New York : Harcourt Brace Jovanovich, 1978.
[32] p. : col. ill. ; 21 x 25 cm. — (A Voyager/HBJ book)
SUMMARY: A shoemaker and his wife being photographed for their wedding anniversary keep adding items to the picture despite the photographer's admonition that "Simple pictures are best."
ISBN 0-15-682625-9
[1. Photography—Fiction] I. De Paola, Thomas Anthony. II. Title.
[PZ7.W6553 Si 1978] [E] 78-6424
78 MARC

William, Thomas.
Prince Donkey : a Christmas story / by Thomas William ; illustrated by Marty Gooder. — Anchorage : Dow Publications, c1977.
30 p., [1] leaf of plates : ill. ; 23 cm.
SUMMARY: Driven away from the royal stables where he served the king's horses, a lonely donkey finds happiness serving a poor, young couple.
[1. Donkeys—Fiction. 2. Christmas stories] I. Gooder, Marty. II. Title.
PZ7.W6555 Pr [Fic] 79-101686
79 MARC

Williams, Barbara.
Jeremy isn't hungry / by Barbara Williams ; pictures by Martha Alexander. — 1st ed. — New York : Dutton, c1978.
[32] p. : col. ill. ; 21 x 22 cm.
SUMMARY: A youngster's attempts to amuse and feed his baby brother almost end in disaster.
[1. Brothers and sisters—Fiction] I. Alexander, Martha G. II. Title.
PZ7.W65587 Je 1978 [E] 78-4924
78 MARC

Williams, Barbara.
Whatever happened to Beverly Bigler's birthday? / By Barbara Williams ; illustrated by Emily McCully. — 1st ed. — New York : Harcourt Brace Jovanovich, c1979.
[64] p. : ill. (some col.) ; 22 cm. — (A Let me read book)
SUMMARY: Seven-year-old Beverly fears her birthday has been totally eclipsed by her sister's wedding.
ISBN 0-15-295286-1. ISBN 0-15-696083-4 pbk.
[1. Birthdays—Fiction] I. McCully, Emily Arnold. II. Title.
PZ7.W65587 Wh [Fic] 78-20575
78 MARC

Williams, Barbara.
Where are you, Angela von Hauptmann, now that I need you? / By Barbara Williams. — New York : Holt, Rinehart, and Winston, c1979.
192 p. ; 22 cm.
SUMMARY: When a talkative sixth-grader arrives at Walt Whitman Elementary in the spring of 1939, she has an unusual effect on her classmates.
ISBN 0-03-050606-9
[1. School stories] I. Title.
PZ7.W65587 Wk [Fic] 79-1069
79 MARC

Williams, Jay, 1914-
The city witch & the country witch / by Jay Williams ; pictures by Ed Renfro. — New York : Macmillan, c1979.
[43] p. : col. ill. ; 26 cm.
SUMMARY: A country and a city witch trade places for two weeks.
ISBN 0-02-793050-5
[1. Witches—Fiction] I. Renfro, Ed. II. Title.
PZ7.W666 Ci [E] 78-11333
78 MARC

Williams, Jay, 1914-
The magic grandfather / by Jay Williams ; illustrated by Gail Owens. — New York : Four Winds Press, c1979.
149 p. : ill. ; 22 cm.
SUMMARY: An 11-year-old discovers that not only is his seemingly ne'er-do-well grandfather a bona fide sorcerer but he too may have an untapped talent for magic.
ISBN 0-590-07588-8 : $7.95
[1. Magic—Fiction] I. Owens, Gail. II. Title.
PZ7.W666 Mai [Fic] 78-22285
78 MARC

Williams, Ursula Moray, 1911-
Bogwoppit / Ursula Moray Williams ; illustrated by Shirle[y] Hughes. — London : H. Hamilton, 1978.
158 p. : ill. ; 23 cm. GB**
ISBN 0-241-89760-2 : £3.75
I. Title.
PZ7.W6683 Bm 78-31910
78 MAR

Willoughby, Elaine Macmann.
Mystery of the lobster thieves / by Elaine Macmann Willoughby ; illustrated by Raymond Keane. — Middletown, Conn : Xerox Education Publications, c1978.
56 p. : ill. ; 21 cm. — (Weekly Reader Children's Book Club edition)
SUMMARY: When Annie and Wallace move to an island just off the coas[t] of New Hampshire, they solve a mystery that has been puzzling the local resi[-]dents for a long time.
ISBN 0-88375-217-4
[1. Mystery and detective stories] I. Keane, Raymond. II. Title.
PZ7.W6837 My [Fic] 78-52140
79 MARC

Wilson, David Henry, 1937-
Elephants don't sit on cars / by David Henry Wilson ; illustrated by Patricia Drew. — London : Chatto and Windus, 1977.
95 p. : ill. ; 21 cm. GB77-2154
ISBN 0-7011-2273-0 : £2.50
I. Title.
PZ7.W6905 El 79-31179
79 MAR

Wilson, Eric H
Vancouver nightmare ; illustrated by Tom McNeely. / Eric Wilson. — Toronto : Clarke, Irwin, 1978.
C78-1254-2
ISBN 0-7720-1234-2 : $6.95
I. Title.
PZ7.W6935 Van 79-306037
79 MAR

Wilson, Merna.
Python cave / by Merna Wilson ; illustrated by Irene Duggan. — [Gwelo] : Modern Publications, 1977.
76 p. : ill. ; 21 cm.
I. Title.
PZ7.W6974 Py 79-111627
79 MARC

Wingham, Peter.
The toy box / [by] Peter Wingham. — London : Heinemar[n], 1978.
[32] p. : col. ill. ; 20 cm. GB78-373
ISBN 0-434-97276-2 : £2.95
I. Title.
PZ8.9.W69 To 79-3109
79 MA

Winn, Alison.
Charlie's iron horse / Alison Winn ; illustrated by Val Biro. London : Hodder and Stoughton, 1979.
158 p. : ill. ; 22 cm. GB*
ISBN 0-340-22194-1 : £2.25
I. Title.
PZ7.W7297 Ch 79-3109
79 MA

Winthrop, Elizabeth.
Are you sad, Mama? / By Elizabeth Winthrop ; illustrated b[y] Donna Diamond. — 1st ed. — New York : Harper & Row, c1979.
24 p. : col. ill. ; 18 cm.
SUMMARY: A little girl tries to cheer up her sad mother.
ISBN 0-06-026539-6 : $5.95
[1. Sadness—Fiction. 2. Emotions—Fiction. 3. Mothers and daughters—Fiction] I. Diamond, Donna. II. Title.
PZ7.W768 Ar 1979 [E] 77-2566
78 MAR

Winthrop, Elizabeth.
Journey to the bright kingdom / by Elizabeth Winthrop illustrated by Charles Mikolaycak. — New York : Holiday House, c1979.
[47] p. : ill. ; 25 cm.
SUMMARY: The mice of Kakure-sato grant a blind woman temporary sig[ht] during the journey she and her kind daughter make to their magical undergrou[nd] kingdom.
ISBN 0-8234-0357-2 (lib. bdg.) : $7.95
[1. Blind—Fiction. 2. Physically handicapped—Fiction. 3. Mice—Fiction 4. Japan—Fiction] I. Mikolaycak, Charles. II. Title.
PZ7.W768 Jo [E] 78-2326
78 MAR

Winthrop, Elizabeth.
Knock, knock, who's there? / Elizabeth Winthrop. — Ne[w] York : Holiday House, c1978.
192 p. ; 22 cm.
SUMMARY: Two brothers and their mother each try in their own way come to terms with the father's death and in the process learn to cope with disease that threatens to cripple all of them.
SUMMARY: Two brothers and their mother each try in their own way come to terms with the father's death and in the process learn to cope w[ith] a disease that threatens to cripple all of them.
ISBN 0-8234-0337-8
[1. Family problems—Fiction. 2. Death—Fiction. 3. Alcoholism—F[ic]tion] I. Title.
PZ7.W768 Kn [Fic] 78-69
00 MA

CHILDREN'S LITERATURE (Continued)

Winthrop, Elizabeth.
Marathon Miranda / Elizabeth Winthrop. — New York : Holiday House, c1979.
155 p. ; 21 cm.
SUMMARY: Miranda feels left out of everything until she meets a jogger in training for a marathon. Despite her apprehension about a sudden attack of asthma, Miranda starts to run.
ISBN 0-8234-0349-1
[1. Running—Fiction. 2. Asthma—Fiction. 3. Physically handicapped—Fiction. 4. Adoption—Fiction] I. Title.
PZ7.W768 Mar [Fic] 78-20615
78 MARC

Wise, Robert A
Mystery of Tanglefoot Island, by Robert Wise. Illus. by Paul Snyder. St. Paul, EMC Corp. 1974.
42 p., illus. ; 23 cm. (His Sea Wolf mysteries)
SUMMARY: When their boat is damaged during a storm, a group of youngsters and their Indian guide land on an island that seems to be haunted.
ISBN 0-88436-144-6 (lib. bdg.). ISBN 0-88436-145-4 (pbk.)
[1. Mystery and detective stories] I. Snyder, Paul, 1923- illus. II. Title.
PZ7.W777 Mz [Fic] 74-16077
74 MARC

Wise, Robert A
Mystery of Totem Pole Inlet, by Robert Wise. Illus. by Paul Snyder. St. Paul, EMC Corp., 1974.
43 p., ill., 23 cm. (His Sea Wolf mysteries)
SUMMARY: When Eddie is late for an appointment, two friends set off by boat to look for him at an Indian burial village.
ISBN 0-88436-138-1 (lib. bdg.). ISBN 0-88436-139-X (pbk.)
[1. Mystery and detective stories] I. Snyder, Paul, 1923- illus. II. Title.
PZ7.W777 Mzh [Fic] 74-16078
74 MARC

Wiseman, Bernard.
The lucky runner / by Bernard Wiseman. — Champaign, Ill. : Garrard Pub. Co., c1979.
48 p. : ill. ; 23 cm. — (A Forreal book)
SUMMARY: When he wins the track championship, fleet-footed Buddy Barnes believes his lucky socks were responsible until he discovers that he wasn't wearing them during the race.
ISBN 0-8116-4313-1 (lib. bdg.) : $4.60
[1. Track and field—Fiction] I. Title.
PZ7.W7802 Lu [E] 79-14215
79 MARC

Wiseman, Bernard.
Morris tells Boris Mother Moose stories and rhymes / written and illustrated by Bernard Wiseman. — New York : Dodd, Mead, c1979.
[48] p. : col. ill. ; 24 cm.
SUMMARY: In order to help Boris the Bear fall asleep, Morris the Moose tells his friend some very familiar stories.
ISBN 0-396-07693-9 (lib. bdg.) : $5.95
[1. Sleep—Fiction. 2. Moose—Fiction. 3. Bears—Fiction] I. Title.
PZ7.W7802 Mt [E] 79-694
79 MARC

Wiseman, Bernard.
Quick quackers / by Bernard Wiseman. — Champaign, Ill. : Garrard Pub. Co., c1979.
32 p. : col. ill. ; 23 cm.
SUMMARY: A communication gap exists between three ducks and a baby parrot who has trouble pronouncing the letter "r".
ISBN 0-8116-6077-0
[1. English language—Pronunciation—Fiction. 2. Parrots—Fiction. 3. Ducks—Fiction] I. Title.
PZ7.W7802 Qi [E] 79-14216
79 MARC

Witter, Evelyn.
The Mystery of animal haven / by Evelyn Witter ; illustrated by Lois Axeman. — Chicago : Childrens Press, c1979.
31 p. : col. ill. ; 25 cm.
SUMMARY: Roger's knowledge of animals becomes particularly useful when he is kidnapped.
ISBN 0-516-03554-1
[1. Kidnapping—Fiction. 2. Mystery and detective stories] I. Axeman, Lois. II. Title.
PZ7.W78434 Mk [E] 78-13564
78 MARC

Witter, Evelyn.
The mystery of the red brick house / by Evelyn Witter ; illustrated by Lois Axeman. — Chicago : Childrens Press, c1978.
30 p. : col. ill. ; 25 cm.
SUMMARY: Jenny and a friend discover the reason why her family's newly purchased house seems haunted.
ISBN 0-516-03548-7 (lib. bdg.)
[1. Mystery and detective stories] I. Axeman, Lois. II. Title.
PZ7.W78434 Mw [Fic] 77-20061
77 MARC

Witter, Evelyn.
The mystery of the missing bicycles / by Evelyn Witter ; illustrated by Lois Axeman. — Chicago : Childrens Press, c1978.
30 p. : col. ill. ; 25 cm.
SUMMARY: Quick thinking helps three youngsters when they stumble upon a mystery in an old carriage house.
ISBN 0-516-03540-1
[1. Mystery and detective stories] I. Axeman, Lois. II. Title.
PZ7.W78434 Mu [Fic] 78-5106
78 MARC

Witter, Evelyn.
The mystery of the music in the night / by Evelyn Witter ; illustrated by Lois Axeman. — Chicago : Childrens Press, c1979.
31 p. : col. ill. ; 25 cm.
SUMMARY: Three youngsters determine to locate the source of mysterious music coming from the Gray Stone Church Cemetery.
ISBN 0-516-03555-X (lib. bdg.)
[1. Kites—Fiction. 2. Mystery and detective stories] I. Axeman, Lois. II. Title.
PZ7.W78434 Mu [E] 78-13637
78 MARC

Witter, Evelyn.
The mystery of the red-eyed camel pin / by Evelyn Witter ; illustrated by Lois Axeman. — Chicago : Childrens Press, c1979.
29 p. : col. ill. ; 25 cm.
SUMMARY: Jenny's loss and recovery of her family's precious pin leads to a new friendship.
ISBN 0-516-03556-8
[1. Mystery and detective stories. 2. School stories] I. Axeman, Lois. II. Title.
PZ7.W78434 Mx [Fic] 78-12376
78 MARC

Wittman, Sally.
A special trade / by Sally Wittman ; pictures by Karen Gundersheimer. — 1st ed. — New York : Harper & Row, c1978.
[32] p. : col. ill. ; 15 cm.
SUMMARY: As the years go by a little girl is able to help an old man as he helped her when she was very young.
ISBN 0-06-026553-1 : $5.95. ISBN 0-06-026554-X lib. bdg. : $7.89
[1. Friendship—Fiction. 2. Old age—Fiction] I. Gundersheimer, Karen. II. Title.
PZ7.W78444 Sp 1978 [E] 77-25673
78 MARC

Wojciechowska, Maia, 1927-
Shadow of a bull / [by] Maia Wojciechowska. — South Melbourne, Vic. : Globe Books, 1975.
117 p. ; 19 cm. Aus
ISBN 0-333-17563-8
I. Title.
PZ7.W8182 Sh 1975 78-319211
79 MARC

Wolde, Gunilla, 1939-
Betsy and the doctor / Gunilla Wolde. — 1st American ed. — New York : Random House, 1978, c1977.
[23] p. : col. ill. ; 17 cm. — (Betsy books)
Translation of Emma hos doktorn.
SUMMARY: Betsy's fall from a tree results in a visit to the hospital emergency room to see a doctor.
ISBN 0-394-83782-7 : $2.50. ISBN 0-394-93782-1 lib. bdg.
[1. Hospitals—Fiction. 2. Medical care—Fiction] I. Title.
PZ7.W81855 Bce 1978 78-50057
79 MARC

Wolde, Gunilla, 1939-
Betsy's fixing day / Gunilla Wolde. — 1st American ed. — New York : Random House, 1978, c1977.
[24] p. : col. ill. ; 17 cm. — (Betsy books)
SUMMARY: A little girl fixes three of her toys and bandages her younger brother's hurt finger.
ISBN 0-394-83781-9 : $2.50. ISBN 0-394-93781-3 lib. bdg. : $3.69
[1. Repairing—Fiction] I. Title.
PZ7.W81855 Bi 1978 [E] 78-50056
79 MARC

Woldin, Beth Weiner, 1955-
Benjamin's perfect solution : story and pictures / by Beth Weiner Woldin. — New York : F. Warne, c1978.
[31] p. : ill. ; 17 x 21 cm.
SUMMARY: Benjamin's attempts to acquire procupine quills are unsuccessful because he overlooks one simple fact.
ISBN 0-7232-6160-1
[1. Opossums—Fiction. 2. Identity—Fiction] I. Title.
PZ7.W81856 Be [E] 78-12245
78 MARC

Wolkstein, Diane.
White wave : a chinese tale / Diane Wolkstein ; [illustrated by] Ed Young. — 1st ed. — New York : Crowell, c1979.
[32] p. : ill. ; 27 cm.
SUMMARY: A young Chinese farmer discovers a generous goddess living in a snail shell.
ISBN 0-690-03893-3 : $6.95. ISBN 0-690-03894-1 lib. bdg. : $6.79
[1. Shells—Fiction. 2. China—Fiction] I. Young, Ed. II. Title.
PZ7.W8344 Wh 1979 [E] 78-4781
78 MARC

Wolman, Judith.
David's new baby / by Judith Wolman ; illustrated by Isobel Read. — New York : Dandelion Press, 1979.
[32] p. : col. ill. ; 23 cm.
SUMMARY: David looks forward to the new baby, but when little Tommy comes home from the hospital David is not so sure he likes having the baby brother.
ISBN 0-89799-099-4. ISBN 0-89799-016-1 pbk. : $1.50
[1. Babies—Fiction. 2. Brothers and sisters—Fiction] I. Read, Isobel. II. Title.
PZ7.W836 Dav [E] 78-72116
79 MARC

Wolman, Judith.
Duncan McTavish in Paris / by Judith Wolman ; illustrated by Robin Moffett. — New York : Dandelion Press, c1979.
[32] p. : col. ill. ; 23 cm.
SUMMARY: A brave terrier from Scotland, on vacation in Paris, helps a French poodle in distress.
ISBN 0-89799-100-1. ISBN 0-89799-011-0 pbk. : $1.50
[1. Dogs—Fiction. 2. Paris—Fiction] I. Moffett, Robin. II. Title.
PZ7.W836 Du [E] 78-72111
79 MARC

Wolman, Judith.
Jenny's birthday party / by Judith Wolman ; illustrated by Isobel Read. — New York : Dandelion Press, 1979.
[30] p. : col. ill. ; 18 x 23 cm.
SUMMARY: Jenny plans to have a big party for her 7th birthday but then finds out that her friend Joey is having a birthday party on the same day.
ISBN 0-89799-117-6
[1. Birthdays—Fiction. 2. Parties—Fiction] I. Read, Isobel. II. Title.
PZ7.W836 Je 78-72114
79 MARC

Wolman, Judith.
Lizzie and the Tooth Fairy / by Judith Wolman ; illustrated by Karen Sukoneck. — New York : Dandelion Press, 1979.
[32] p. : col. ill. ; 23 cm.
SUMMARY: Lizzie gets more than a visit from the Tooth Fairy who takes her to a castle in the sky where there are lots and lots of teeth.
ISBN 0-89799-123-0
[1. Tooth Fairy—Fiction. 2. Teeth—Fiction] I. Sukoneck, Karen. II. Title.
PZ7.W836 Li [E] 78-72115
00 MARC

Wood, James Playsted, 1905-
Chase scene / by James Playsted Wood. — 1st ed. — New York : Elsevier/Nelson Books, c1979.
170 p. ; 21 cm.
SUMMARY: Two teenagers help an escapee from a hospital for the criminally insane elude the police and prove his innocence.
ISBN 0-525-66640-0 : $7.95
[1. Friendship—Fiction] I. Title.
PZ7.W8496 Ch [Fic] 79-11399
79 MARC

Wood, Joyce, 1928-
Grandmother Lucy goes on a picnic / story by Joyce Wood ; pictures by Frank Francis. — Cleveland : Collins + World, 1976, c1970.
[32] p. : col. ill. ; 26 cm.
SUMMARY: Grandmother Lucy takes a picnic basket down to the river with a little friend and together they have a wonderful time.
ISBN 0-529-05288-1 lib. bdg. : $5.95
[1. Picnicking—Fiction] I. Francis, Frank. II. Title.
PZ7.W84964 Gs 1976 [E] 76-80
76[r79]rev MARC

Wood, Kay.
Big Bird's shape book / [by Kay Wood]. — 1st American ed. — New York : Random House, c1977.
[26] p. : col. ill. ; 10 x 12 cm.
SUMMARY: Two Sesame Street characters collect different shapes they see everyday.
ISBN 0-394-83263-9
[1. Size and shape—Fiction] I. Title.
PZ7.W84966 Bi 1977 [E] 75-39343
79 MARC

Wood, Phyllis Anderson.
Get a little lost, Tia / by Phyllis Anderson Wood. — 1st ed. — Philadelphia : Westminster Press, c1978.
173 p. ; 21 cm.
SUMMARY: With his father dead and mother working, a teenage boy is given the responsibility of looking after his thirteen-year-old sister, a task that proves exasperating.
ISBN 0-664-32636-6
[1. Brothers and sistesr—Fiction] I. Title.
PZ7.W854 Ge [Fic] 78-17762
78 MARC

The Wookiee storybook / illustrated by Patricia Wynne. — New York : Random House, c1979.
[40] p. : col. ill. ; 24 cm. — (Star wars)
SUMMARY: Lumpy's descent to the 'nother world on his jungle planet to hunt for special berries for his father's homecoming becomes a scary adventure.
ISBN 0-394-84256-1 : $2.95
[1. Science fiction] I. Wynne, Patricia. II. Series.
PZ7.W8813 [E] 79-1747
79 MARC

CHILDREN'S LITERATURE (Continued)

Woolley, Catherine.
Where is Squirrel? / By Jane Thayer [i.e. C. Woolley] ; illustrated by Bari Weissman. — 1st ed. — New York : Morrow, 1979.
[32] p. : col. ill. ; 21 cm.
SUMMARY: Skunk leads the animals in a search for Squirrel who has disappeared from the garden and woods.
ISBN 0-688-22192-0
[1. Animals—Fiction. 2. Mystery and detective stories] I. Weissman, Bari. II. Title.
PZ7.W882 Wg [E] 78 78-31611 MARC

Wosmek, Frances.
Mystery of the eagle's claw / by Frances Wosmek. — 1st ed. — Philadelphia : Westminster Press, c1979.
131 p. ; 21 cm.
SUMMARY: When her adoptive parents die, a young girl is sent to live with an old aunt who owns an inn in Maine and finds herself in a dangerous situation when she discovers a conspiracy to defraud her aunt.
ISBN 0-664-32640-4
[1. Mystery and detective stories] I. Title.
PZ7.W8884 My [Fic] 78 78-12221 MARC

Wright, Dare.
Edith and Midnight / Dare Wright. — 1st ed. — Garden City, N.Y. : Doubleday, c1978.
[48] p. : ill. ; 32 cm.
SUMMARY: A doll and a teddy bear attempt to capture and tame their dream horse.
ISBN 0-385-14155-6 : $5.95. ISBN 0-385-14156-4 lib. bdg.
[1. Toys—Fiction] I. Title.
PZ8.9.W875 Eg [E] 78 77-27767 MARC

Wright, Samuel.
The music machine : a fantasy story from Agapeland (from the pages of the ancient manuscripts) / Samuel Wright ; [illustrated by Mark D. Pendergrass]. — Minneapolis : Bethany Fellowship, c1979.
142 p. : ill. ; 20 cm.
SUMMARY: Steve and Nancy inadvertently enter Agapeland where they meet The Conductor and discover the amazing music machine.
ISBN 0-87123-707-5 : $2.95
[1. Fantasy] I. Pendergrass, Mark D. II. Title.
PZ7.W9585 Mu [Fic] 79 79-9200 MARC

Wrightson, Patricia.
The dark bright water / Patricia Wrightson. — 1st American ed. — New York : Atheneum, 1979, c1978.
223 p. : map ; 22 cm.
"A Margaret K. McElderry book."
SUMMARY: Tribal elders urge a young Australian Aborigine to investigate the strange events occurring in the interior of their vast continent.
ISBN 0-689-50122-6
[1. Australian aborigines—Fiction. 2. Australia—Fiction] I. Title.
PZ7.W9593 Dar 1979 [Fic] 78 78-8793 MARC

Yaffe, Alan.
Sid & Sol / story by Arthur Yorinks [i.e. Alan Yaffe] ; pictures by Richard Egielski. — 1st ed. — New York : Farrar, Straus and Giroux, 1977.
[32] p. : ill. ; 25 cm.
SUMMARY: Short Sid sets off to save the world from the giant Sol.
ISBN 0-374-36904-6
[1. Giants—Fiction] I. Egielski, Richard. II. Title.
PZ7.Y128 Si [E] 77[r79]rev 77-24126 MARC

Yaffe, Arthur.
The magic meatballs / by Alan Yaffe ; pictures by Karen Born Andersen. — New York : Dial Press, c1979.
[40] p. : ill. ; 22 x 28 cm.
SUMMARY: Marvin meets a man who gives him special, magical chopped meat. Now he must find out if the magical meat will help solve his problems.
ISBN 0-8037-5139-7 : $5.95
[1. Magic—Fiction. 2. Humorous stories] I. Andersen, Karen Born. II. Title.
PZ7.Y128 Mag [E] 79 78-72472 MARC

Yang, Chao-lin.
At the ferry-boat landing : story / by Chin Hsin ; adapted by Yang Chao-lin ; drawings by Ho Yu-men. — 1st ed. — Peking : Foreign Languages Press, 1977.
70 p. : col. ill. ; 19 x 27 cm.
$3.00 (H.K.)
I. Chin Hsin. At the ferry-boat landing. II. Ho, Yu-men. III. Title.
PZ7.Y193 At 79 78-108308 MARC

Yeoman, John.
The young performing horse / Story by John Yeoman ; pictures by Quentin Blake. — New York : Parents' Magazine Press, 1978, c1977.
[32] p. : col. ill. ; 26 cm.
SUMMARY: A brother and sister and their unusually talented horse journey to London to seek their fortunes as actors.
ISBN 0-8193-0970-2. ISBN 0-8193-0971-0 lib. bdg.
[1. Horses—Fiction] I. Blake, Quentin. II. Title.
PZ7.Y418 Yo 1978 [E] 78 78-6304 MARC

Yolen, Jane H
The giants go camping / by Jane Yolen ; pictures by Tomie de Paola. — New York : Seabury Press, c1979.
46 p. : col. ill. ; 21 cm.
"A Clarion book."
SUMMARY: Five giants share the fun and surprises of a camping trip to the mountains.
ISBN 0-8164-3223-6 (lib. bdg.) : $6.95
[1. Giants—Fiction. 2. Camping—Fiction] I. De Paola, Thomas Anthony. II. Title.
PZ7.Y78 Gk [E] 78 78-17928 MARC

Yolen, Jane H
No bath tonight / by Jane Yolen ; pictures by Nancy Winslow Parker. — 1st ed. — New York : Crowell, c1978.
[32] p. : col. ill. ; 21 x 22 cm.
SUMMARY: A small boy refuses to take a bath until his grandmother shows him how to make kid tea.
ISBN 0-690-03881-X. ISBN 0-690-03882-8 lib. bdg.
[1. Baths—Fiction. 2. Grandmothers—Fiction] I. Parker, Nancy Winslow. II. Title.
PZ7.Y78 No [E] 78 77-26605 MARC

Yolen, Jane H
Spider Jane / Jane Yolen ; pictures by Stefen Bernath. — New York : Coward, McCann & Geoghegan, c1978.
61 p. : ill (some col.) ; 23 cm. — (A Break-of-day book)
SUMMARY: Four episodes about a spider and a fly who become friends and decide to live together.
ISBN 0-698-30696-1 lib. bdg.
[1. Spiders—Fiction. 2. Flies—Fiction. 3. Friendship—Fiction] I. Bernath, Stefen. II. Title.
PZ7.Y78 Sp [E] 77 77-17193 MARC

Yolen, Jane H
The simple prince / Jane Yolen ; pictures by Jack Kent. — New York : Parents' Magazine Press, c1978.
[33] p. : col. ill. ; 24 cm.
SUMMARY: A prince who has tired of his life at court aspires to lead the simple life, but finds it much harder than he imagined.
ISBN 0-8193-0960-5. ISBN 0-8193-0961-3 lib. bdg.
[1. Princes—Fiction] . I. Kent, Jack, 1920- . II. Title.
PZ7.Y78 Si [E] 78 78-6118 MARC

[Yonge, Charlotte Mary], 1823-1901.
The history of Sir Thomas Thumb / by the author of "The heir of Redcliffe," "Heartsease," "The little duke," &c. &c. [i.e. C. M. Yonge] ; illustrated by J. B. — Edinburgh : T. Constable, 1855.
142 p. : ill. ; 22 cm.
I. Blackburn, Jane. II. Heir of Redcliffe, Author of. III. B., J. IV. J. B. V. Tom Thumb. VI. Title.
PZ6.Y6 Hi 78 78-315554 MARC

York, Carol Beach.
Revenge of the dolls / by Carol Beach York. — 1st ed. — New York : Elsevier/Nelson Books, c1979.
103 p. ; 21 cm.
SUMMARY: Her uneasy feeling about her aunt's collection of grotesque homemake dolls causes Alice to be particularly suspicious of a doll given to a mischevious cousin.
ISBN 0-525-66632-X : $6.95
[1. Dolls—Fiction. 2. Mystery and detective stories] I. Title.
PZ7.Y82 Re [Fic] 79 79-10761 MARC

Youldon, Gillian.
Colors / by Gillian Youldon. — London ; New York : F. Watts, 1979.
[25] p. : col. ill. ; 20 cm. GB***
SUMMARY: With pages split horizontally, the reader can mix the heads, bodies, and legs of 11 animals, each of a different color.
ISBN 0-531-02387-7
[1. Color. 2. Animals] I. Title.
PZ7.Y827 Co [E] 79 79-63326 MARC

Yü, T'ien-tz'u.
Wily wolf / edited by Yu Tien-tzu ; text by Chiang Chin-cheng ; ill. by Chen Yung-chen. — Peking : Foreign Languages Press, 1976.
[28] p. : col. ill. ; 19 cm.
Colophon title.
Title also in Chinese: Chiao hua ti ta hui lang.
Translation of Hui yao wei pa ti lang.
I. Chiang, Chin-ch'eng. II. Ch'en, Yung-chen. III. Title. IV. Title: Chiao hua ti ta hui lang.
PZ7.Y8964 Wi 79 79-365468 MARC

Zabel, Jennifer, 1947-
Miss Priscilla's secret / written by Jennifer Zabel ; illustrated by Christopher Masters. — London : F. Warne, 1978.
[25] p. : col. ill. ; 26 cm. GB78-31041
ISBN 0-7232-2117-0 : £2.95
I. Masters, Christopher, 1944- . II. Title.
PZ7.Z15 Mi 79 79-300184 MARC

Zalben, Jane Breskin.
Norton's nightime / story and pictures by Jane Breskin Zalben. — New York : Collins, 1979.
[24] p. : col. ill. ; 16 cm.
SUMMARY: A raccoon tries to overcome his fear of the dark by imagining what some of his braver friends would do.
ISBN (invalid) 0-592-05430-2 : $5.95
[1. Fear—Fiction. 2. Night—Fiction. 3. Raccoons—Fiction. 4. Animals—Fiction] I. Title.
PZ7.Z254 No [E] 78 78-12144 MARC

Zalben, Jane Breskin.
Penny and the captain : tales and pictures / by Jane Breskin Zalben. — New York : Collins + World, c1978.
60 p. : ill. ; 21 cm.
CONTENTS: The birthday gift.—The meeting.—All alone.—The card game.—Penny and the captain.
ISBN 0-529-05424-8. ISBN 0-529-05425-6 lib. bdg.
[1. Penguins—Fiction] I. Title.
PZ7.Z254 Pe 1978 [E] 76 76-54826 MARC

Zalben, Jane Breskin.
Will you count the stars without me? / story and watercolor by Jane Breskin Zalben. — 1st ed. — New York : Farrar, Straus and Giroux, 1979.
[28] p. : col. ill. ; 24 cm.
SUMMARY: Saddened by the separation from his mate who has gone in search of food, Saba learns to survive with the help of friends who keep him busy.
ISBN 0-374-38433-9 : $7.95
[1. Individuality—Fiction. 2. Animals—Fiction. 3. Islands—Fiction] I. Title.
PZ7.Z254 Wi [E] 78 78-16069 MARC

Zander, Hans, 1937-
The harmonica man / written and illustrated by Hans Zander. — Richmond Hill, Ont. : Scholastic-Tab Publications ; [New York : distributed by Scholastic Book Services], 1976.
[32] p. : col. ill. ; 18 x 22 cm. C***
SUMMARY: The harmonica man brings happiness to a little boy and teaches him to play the harmonica.
[1. Harmonica—Fiction] I. Title.
PZ7.Z2546 Har [E] 77[r79]rev 77-370157 MARC

Zeï, Alkē.
The sound of the dragon's feet / by Alki Zei ; translated from the Greek by Edward Fenton. — 1st ed. — New York : Dutton, c1979.
113 p. ; 22 cm.
Translation of Konta stis ragies.
SUMMARY: The time spent with her "revolutionary" tutor opens 10-year-old Sasha's eyes to more of life in turn-of-the-century Russia than her somewhat sheltered existence had previously allowed her to experience.
ISBN 0-525-39712-4
[1. Russia—Fiction. 2. Fathers and daughters—Fiction] I. Title.
PZ7.Z32 So 1979 [Fic] 79 79-14917 MARC

Zemach, Kaethe.
The beautiful rat / written and illustrated by Kaethe Zemach. — New York : Four Winds Press, c1979.
[31] p. : col. ill. ; 24 cm.
SUMMARY: Mr. and Mrs. Rat search for someone very great to be their daughter's husband.
ISBN 0-590-07584-5 : $7.95
[1. Marriage—Fiction. 2. Rats—Fiction]
PZ7.Z413 Be [E] 79 79-11752 MARC

Zhitkov, Boris Stepanovich, 1882-1938.
How I hunted the little fellows / Boris Zhitkov ; translated from the Russian by Djemma Bider ; illustrated by Paul O. Zelinsky. — New York : Dodd, Mead, c1979.
[64] p. : ill. ; 22 cm.
SUMMARY: Fascinated by the life-like miniature of a steamer on his grandmother's shelf, a young boy becomes convinced that there are little people inside and, ignoring the admonition not to touch, decides to find out for himself.
ISBN 0-396-07692-0
[1. Russia—Fiction] I. Zelinsky, Paul O. II. Title.
PZ7.Z48 Ho [Fic] 79 79-11738 MARC

Zimmerman, Andrea Griffing.
Yetta, the trickster / by Andrea Griffing Zimmerman ; pictures by Harold Berson. — New York : Seabury Press, c1978.
45 p. : col. ill. ; 21 cm.
"A Clarion book."
SUMMARY: Yetta loves to play tricks on people until one day the joke is on her.
ISBN 0-8164-3218-X
[1. Practical jokes—Fiction] I. Berson, Harold. II. Title.
PZ7.Z618 Ye [E] 78 78-6164 MARC

Zimnik, Reiner.
Billy's balloon ride. English version by Richard Whittingham. Northbrook, Ill., Hubbard Press [1974]
[32] p. illus. 32 cm.
SUMMARY: Determined to fly, a little boy ties balloons to his bed and travels over many lands.
ISBN 0-8331-0033-5
[1. Flight—Fiction] I. Whittingham, Richard. II. Title.
PZ7.Z64 Bi [E] 74 74-6041 MARC

CHILDREN'S LITERATURE (Continued)

Zindel, Paul.
The undertaker's gone bananas / Paul Zindel. — 1st ed. — New York : Harper & Row, c1978.
239 p. ; 21 cm.
SUMMARY: Two teenagers believe a neighbor, an undertaker, has murdered his wife but can't convince anyone else.
ISBN 0-06-026845-X : $6.95. ISBN 0-06-026846-8 lib. bdg. : $6.79
[1. Crime and criminals—Fiction. 2. Mystery and detective stories] I. Title.
PZ7.Z647 Un 1978 [Fic] 78-54606
78 MARC

Zion, Gene.
Harry and the lady next door. Pictures by Margaret Bloy Graham. New York, Harper [1960]
62 p. illus. 23 cm. (An I can read book)
SUMMARY: Harry the dog goes to fantastic lengths to make his neighbor stop singing.
[1. Singing—Fiction. 2. Dogs—Fiction] I. Graham, Margaret Bloy. II. Title.
PZ7.Z66 Han [E] 60-9452
[r78]rev MARC

Zion, Gene.
The sugar mouse cake. Pictures by Margaret Bloy Graham. New York, Scribner [1964]
[48] p. col. illus. 27 cm.
SUMMARY: With the help of a mouse a cook's assistant bakes a cake judged the best in the kingdom and thus becomes the King's Chief Pastry Cook.
[1. Baking—Fiction. 2. Mice—Fiction] I. Graham, Margaret Bloy. II. Title.
PZ7.Z66 Sr [E] 64-22756
[r79]rev MARC

Zuliani, Vilma.
Randy robin finds a friend / by Vilma Zuliani ; illustrated by Robert de Claire. — Lawrence, Mass. : Hermitage Press, c1977.
[31] p. : col. ill. ; 23 x 29 cm.
On dust jacket: Distributed by Follett Publishing Co. [Chicago]
SUMMARY: Randy's the robin's bold and fearless spirit is almost his undoing.
ISBN 0-89441-002-4 : $4.95
[1. Robins—Fiction] I. De Claire, Robert. II. Title.
PZ7.Z8 Ran [E] 77-74629
78[r79]rev MARC

—ADDRESSES, ESSAYS, LECTURES

Almendros, Herminio, 1898-
A propósito de "La edad de oro"; notas sobre literatura infantil. La Habana, Instituto Cubano del Libro, 1972.
312 p. 19 cm. (Gente nueva)
1. Martí, José, 1853-1895. La edad de oro. 2. Children's literature—Addresses, essays, lectures. I. Title.
PQ7389.M2E332 1972 73-217721
79[r79]rev MARC

—AUTHORSHIP

From writers to students : the pleasures and pains of writing / edited by M. Jerry Weiss for the IRA Literature for Adolescents Committee. — Newark, Del. : International Reading Association, c1979.
vi, 113 p. ; 22 cm.
ISBN 0-87207-724-1
1. Children's literature—Authorship. 2. Children's literature, American—History and criticism. 3. Authors, American—20th century—Interviews. I. Weiss, Morton Jerome. II. International Reading Association. Literature for Adolescents Committee.
PN147.5.F7 808'.068 79-11695
79 MARC

—BIBLIOGRAPHY

see also Bibliography—Bibliography—Children's literature; Bibliography—Children's books issued in series; School libraries; School libraries—Book lists

Buckinghamshire County Library. School Library Service.
Outstanding fiction for young people, 1970-1976 / Buckinghamshire County Library, School Library Service. — [Aylesbury] : The Library, [1978]
[2], 29 p. : ill. ; 21 cm. GB78-34264
Text on back cover.
ISBN 0-86059-021-6 : £1.00
1. Children's literature—Bibliography. I. Title.
Z1037.9.B8 1978 028.52 79-303628
[PN1009.A1] 79 MARC

Building a children's literature collection. — Rev. ed. — Middletown, Conn. : CHOICE, 1978.
v, 42 p. ; 28 cm. — (CHOICE bibliographical essay series ; no. 5)
Includes indexes.
CONTENTS: Quimby, H. B. A suggested basic reference collection for academic libraries.—Weber, R. A suggested basic collection of children's books.
ISBN 0-914492-04-7 : $4.75
1. Reference books—Children's literature—Bibliography. 2. Children's literature—Bibliography. I. Quimby, Harriet B. A suggested basic reference collection for academic libraries. 1978. II. Weber, Rosemary. A suggested basic collection of children's books. 1978.
Z1037.1.B84 1978 028.52 78-15886
[PN1009.A1] 78 MARC

Fundação Nacional do Livro Infantil e Juvenil.
Bibliografia analítica da literatura infantil e juvenil publicada no Brasil, 1965-1974 / Fundação Nacional do Livro Infantil e Juvenil (FNLIJ). — São Paulo : Edições Melhoramentos, c1977.
384 p. ; 20 cm.
"Trabajo realizado sob o patrocínio do Instituto Nacional de Estudos e Pesquisas Educacionais (INEP) ... em convênio com o Instituto Nacional do Livro, Ministério da Educação e Cultura."
"Bibliografia sobre literatura infantil e juvenil": p. 359-384.
Includes indexes.
1. Children's literature—Bibliography. 2. Brazil—Imprints. I. Title.
Z1684.5.F86 1977 028.52'0981 79-101308
[PN1009.A1] 79 MARC

Gillespie, John Thomas, 1928-
Best books for children, preschool through the middle grades / edited by John T. Gillespie and Christine B. Gilbert. — New York : Bowker, 1978.
x, 541 p. ; 26 cm.
Includes indexes.
ISBN 0-8352-1104-5
1. Children's literature—Bibliography. I. Gilbert, Christine Bell, 1909- joint author. II. Title.
Z1037.G48 028.52 78-10649
[PN1009.A1] 78 MARC

Home and school study and reading guides, The new book of knowledge. — [New York : Grolier, c1977]
86, 115 p. ; 25 cm.
Cover title.
CONTENTS: Home and school study guide.—Home and school reading guide.
1. The New book of knowledge—Indexes. 2. Children's literature—Bibliography. I. The New book of knowledge.
AG5.B64 1977 Suppl. 031 78-108097
78 MARC

Hunt, Abby Campbell.
The world of books for children : a parent's guide / by Abby Campbell Hunt. — New York : Sovereign Books, c1979.
xii, 242 p. : ill. ; 21 cm. — (A Sovereign book)
Includes bibliographies and indexes.
ISBN 0-671-18383-4. ISBN 0-671-18350-8 pbk.
1. Children's literature—Bibliography. I. Title. II. Series.
Z1037.A1H86 028.52 78-9723
[PN1009.A1] 78 MARC

James, Philip Brutton, 1901-
Children's books of yesterday / by Philip James ; edited by C. Geoffrey Holme. — Detroit : Gale Research Co., 1976.
128 p. : ill. ; 29 cm.
At head of title: The Studio special autumn number.
"The illustrations ... have been in great part selected from the exhibition of illustrated books for children held at the Victoria and Albert Museum in 1932."
Includes bibliographical references.
ISBN 0-8103-4135-2
1. Children's literature—Bibliography. 2. Illustrated books, Children's—Bibliography. I. Studio international. II. Title.
Z1037.A1J3 1976 028.52 79-174059
75 MARC

(Jidō yomimono to dokusho shidō)
児童読物と読書指導 / 責任編集 亀井勝一郎 ... [et al.]. — 東京 : 牧書店, 昭和30 [1955]
2, 3, 273, 7, 2 p. : ill. ; 22 cm. — (読書指導講座 ; 9)
Includes bibliographies and index.
1. Books and reading for children. 2. Children's literature—History and criticism. 3. Children's literature—Bibliography. I. Kamei, Katsuichirō, 1907-1966. II. Series: Dokusho shidō kōza ; 9.
Z1037.J6 79-804690

Lassam, Christine.
Stories and settings, Americas, Caribbean, Arctic, and Antarctic : a handbook for teachers and librarians / selected and annotated by Christine Lassam. — London : National Book League, 1978.
32 p. ; 21 cm. GB78-33487
ISBN 0-85353-273-7 : £0.50
1. Children's literature—Bibliography. 2. America—Juvenile fiction—Bibliography. I. Title.
Z1037.9.L33 028.52 79-302121
[PN1009.A1] 79 MARC

Lassam, Christine.
Stories and settings : Europe and the U.S.S.R. : a handbook for teachers and librarians / selected and annotated by Christine Lassam. — London : National Book League, [1978?]
53 p. ; 21 cm. GB***
£0.50
1. Children's literature—Bibliography. 2. Europe—Fiction—Bibliography. 3. Europe—Children's literature—Bibliography. I. National Book League, London. II. Title.
Z1037.9.L34 028.52 79-302187
[PN1009.A1] 79 MARC

Lassam, Christine.
Stories and settings, Scotland, Wales and Ireland : a handbook for teachers and librarians / selected and annotated by Christine Lassam. — London : National Book League, 1978.
28 p. ; 21 cm. GB78-34266
ISBN 0-85353-272-9 : £0.50
1. Children's literature—Bibliography. 2. Scotland—Juvenile fiction—Bibliography. 3. Wales—Juvenile fiction—Bibliography. 4. Ireland—Juvenile fiction—Bibliography. I. Title.
Z1037.9.L345 028.52 79-308495
[PN1009.A1] 79 MARC

Library Association of Australia. Children's Libraries Section. New South Wales Division.
The new classics : a selection of award-winning children's books / [prepared by a sub-committee of members of the New South Wales Divisions of the School and Children's Libraries Sections of the Library Association of Australia]. — Sydney : Library Association of Australia, Children's Libraries Section & School Libraries Section, New South Wales Divisions, 1975.
64 p. : ill. ; 21 cm. Aus
Bibliography: p. 52-53.
Includes index.
ISBN 0-909915-34-2
1. Children's literature—Bibliography. 2. Literary prizes. I. Library Association of Australia. School Libraries Section. New South Wales Division. II. Title.
Z1037.A2L52 1975 028.52 79-303874
[PN1009.A1] 79 MARC

Marshall, Margaret Richardson.
Children's paperbacks, 5-11 / selected and annotated by Margaret Marshall. — London : National Book League [for] the School Library Association, 1978.
24 p. ; 21 cm. GB78-32664
ISBN 0-85353-304-0 : £0.65
1. Children's literature—Bibliography. 2. Bibliography—Paperback editions. 3. Books—Prices—Great Britain. I. Title.
Z1037.9.M375 028.52 78-325652
[PN1009.A1] 79 MARC

Meacham, Mary, 1946-
Information sources in children's literature : a practical reference guide for children's librarians, elementary school teachers, and students of children's literature / Mary Meacham. — Westport, Conn. : Greenwood Press, 1978.
xvii, 256 p. ; 24 cm. — (Contributions in librarianship and information science ; no. 24 ISSN 0084-9243)
Bibliography: p. [237]-249.
Includes index.
ISBN 0-313-20045-9
1. Bibliography—Bibliography—Children's literature. 2. Children's literature—Bibliography. 3. Children's literature—Book reviews—Periodicals—Bibliography. 4. Books and reading for children—Bibliography. 5. Libraries, Children's—Bibliography. I. Title. II. Series.
Z1037.A1M4 028.52 77-91107
78 MARC

Miller, Bertha E Mahony, comp.
Contemporary illustrators of children's books / compiled by Bertha E. Mahony and Elinor Whitney. — Detroit : Gale Research Co., 1978, c1930.
viii, 135 p. : ill. ; 24 cm.
Reprint of the ed. published by The Bookshop for Boys and Girls, Women's Industrial Union, Boston.
"A Firenze book."
ISBN 0-8103-4308-8
1. Illustrators—Biography. 2. Illustration of books—20th century. 3. Illustrated books, Children's. 4. Children's literature—Bibliography. I. Whitney, Elinor, joint author. II. Title.
NC965.M58 1978 741'.092'2 78-17018
78 MARC

Modern realistic stories for children and young people : international bibliography / [editor, Rudolf Majonica]. — [Munich] : Arbeitskreis für Jugendliteratur, c1978.
119 p. ; 21 cm. GFR***
1. Children's literature—Bibliography. 2. Bibliography—Bibliography—Children's literature. 3. Children's literature—History and criticism—Bibliography. I. Majonica, Rudolf.
Z1037.M8 028.52 79-305273
[PN1009.A1] 79 MARC

Moss, Elaine.
Children's books of the year, 1977 / selected and annotated by Elaine Moss. — London : H. Hamilton [for] the National Book League and the British Council, 1978.
141 p. : ill. ; 22 cm. GB78-25784
Includes indexes.
ISBN 0-241-89876-5 : £2.25
1. Children's literature—Bibliography. I. National Book League, London. II. Great Britain. British Council. III. Title.
Z1035.A2M66 1978 028.52 79-312160
[PN1009.A1] 79 MARC

CHILDREN'S LITERATURE —BIBLIOGRAPHY (Continued)

Reading is Fundamental, inc.
RIF's guide to book selection / Reading is Fundamental. — 1st ed. — Washington : RIF, 1970-
v. ; 28 cm.
Vol. 4 has title: RIF's guide to developing a program.
1. Children's literature—Bibliography. 2. Minorities—United States—Juvenile literature—Bibliography. I. Title.
Z1037.A1R39 1970 028.52 73-603512
79 MARC

Scotland / [compiled by four Scottish children's librarians for the committee of the Scottish Branch of the Youth Libraries Group of the Library Association ; editor Mary Nettlefold]. — Birmingham : Library Association, Youth Libraries Group, 1974.
[2], 21 p. : ill. ; 22 cm. — (Storylines ; no. 5) GB75-05370
Cover title.
Annotated bibliography.
ISBN 0-85356-147-7 : £0.30
1. Children's literature—Bibliography. 2. Scotland—Fiction—Bibliography. 3. Scotland—Juvenile literature—Bibliography. I. Nettlefold, Mary. II. Library Association. Youth Libraries Group. Scottish Branch. III. Series.
Z1037.9.S38 028.52 79-302914
[PN1009.A1] 79 MARC

Spirt, Diana L
Introducing more books : a guide for the middle grades / by Diana L. Spirt. — New York : Bowker, 1978.
xv, 240 p. ; 24 cm.
Includes indexes.
ISBN 0-8352-0988-1 : $13.95
1. Children's literature—Bibliography. 2. Books and reading for children. I. Title.
Z1037.S758 028.52 78-13490
[PN1009.A1] 78 MARC

Toronto Public Libraries. Boys and Girls Services.
Books for boys and girls, being a list of two thousand books which the librarians of the Boys and Girls Division of the Toronto Public Library deem to be of definite and permanent interest, with annotations and descriptions. [Toronto] Boys and Girls House, Public Library of Toronto, Canada [1927]
3 p.l., 290 p., 1 l. 20 cm.
Edited by Miss Lillian Smith, head of the Boys and Girls House and of that division of the library.
1. Children's literature—Bibliography. I. Title.
Z1037.T72 1927 28-16205
[PN1009.A1] [r79]rev4 MARC

Toronto Public Libraries. Boys and Girls Services.
Books for boys and girls, edited by Lillian H. Smith; prepared at Boys and Girls House, Toronto Public Libraries. [Toronto] Ryerson Press [1940]
x, 367 p. illus. 20 cm.
"The compiling ... has been a co-operative effort in which all members of the staff of the Boys and Girls Division ... have shared."—Pref.
"First edition, 1927 ... second edition, 1940."
1. Children's literature—Bibliography. I. Smith, Lillian Helena, 1887- ed. II. Title.
Z1037.T72 1940 028.52 40-27465
[PN1009.A1] [r79]rev4 MARC

Toronto Public Libraries. Boys and Girls Services.
Books for boys and girls. [3rd ed.] Edited by Jean Thomson, head of the Boys and Girls Division, Toronto Public Library. With an introd. by Charles R. Sanderson. Toronto, Ryerson Press [1954]
297 p. illus. 21 cm.
——— A supplement 1953/1958. With an introd. by H. C. Campbell. Toronto, Ryerson Press [1960, c1959]
116 p. illus. 22 cm.
Z1037.T72 1954 Suppl.
1. Children's literature—Bibliography. I. Thomson, Jean, ed. II. Title.
Z1037.T72 1954 028.52 54-13337 rev 3 ‡
[PN1009.A1] (MARC)

Toronto Public Libraries. Boys and Girls Services.
Books for boys and girls. Edited by Marguerite Bagshaw, head of Boys and Girls Services, Toronto Public Library, assisted by Doris Scott. With an introd. by Henry C. Campbell. [4th ed.] Toronto, Ryerson Press [1966]
301 p. illus. 21 cm.
1. Children's literature—Bibliography. I. Bagshaw, Marguerite, ed. II. Scott, Doris, ed. III. Title.
Z1037.T72 1966 028.52 67-5200
[PN1009.A1] [r79]rev MARC

Toronto Public Libraries. Boys and Girls Services.
Books for boys and girls. Toronto, Boys and Girls Division, Toronto Public Libraries, 1966.
[6] p. 22 cm. CB67-192
1. Children's literature—Bibliography. I. Title.
Z1037.T72 1966b 028.52 68-142541
[PN1009.A1] [r79]rev MARC

Toronto Public Libraries. Boys and Girls Services.
A chronicle of Boys and Girls House and a selected list of recent additions to the Osborne collection of early children's books, 1542-1910, and the Lillian H. Smith collection, 1911-1963. [Toronto, 1964]
29 p. illus. 26 cm.
"Prepared to commemorate the opening of the new Boys and Girls House, Toronto Public Library, Toronto, Canada, May, 1964."
1. Children's literature—Bibliography. I. Title.
Z1037.T73 66-96408
[PN1009.A1] [r79]rev2 MARC

Warren, Dorothea.
Fiction, verse and legend : an annotated guide to a selection of imaginative literature for middle and secondary schools / [by] Dorothea Warren. — 2nd ed., extensively revised. — Oxford : School Library Association, 1978.
[8], 58 p. : ill. ; 21 cm. — (School Library Association book lists)
GB78-25705
Includes index.
ISBN 0-900641-34-7 : £2.50
1. Children's literature—Bibliography. 2. School libraries—Book lists. I. Title. II. Series: School Library Association. Book lists.
Z1037.W29 1978 028.52 79-304850
[PN1009.A1] 79 MARC

White, Virginia L
Books about children's books : an annotated bibliography / compiled by Virginia L. White and Emerita S. Schulte. — Newark, Del. : International Reading Association, 1979.
48 p. ; 22 cm. — (IRA annotated bibliography series)
Includes index.
ISBN 0-87207-333-5
1. Bibliography—Bibliography—Children's literature. 2. Children's literature—Bibliography. 3. Children's literature—History and criticism—Bibliography. 4. Books and reading for children—Bibliography. I. Schulte, Emerita S., joint author. II. Title. III. Series: International Reading Association. IRA annotated bibliography series.
Z1037.A1W59 028.52 79-17553
[PN1009.A1] 79 MARC

—BIBLIOGRAPHY—CATALOGS

Emporia State University. William Allen White Memorial Library.
The May Massee collection : creative publishing for children, 1923-1963, a checklist : William Allen White Library, Emporia State University, Emporia, Kansas / edited by George V. Hodowanec ; annotations prepared by Jeanne Frederickson. — Limited 1st ed. — [Emporia] : The University, 1979.
xviii, 316 p. : ill. ; 24 cm.
Includes indexes.
ISBN 0-934068-00-3 : $25.00
1. Children's literature—Bibliography—Catalogs. 2. Emporia State University. William Allen White Memorial Library. 3. Massee, May. 4. Children's literature—Publishing—United States—History—20th century. I. Hodowanec, George V., 1935- . II. Frederickson, Jeanne. III. Title.
Z1038.E46E46 1979 028.52 79-4868
[PN1009.A1] 79 MARC

—BIBLIOGRAPHY—PERIODICALS

Children's Book Council, New York.
Children's books: awards & prizes.
New York, Children's Book Council, inc.
v. 19–28 cm. biennial.
Continues: Children's books. Awards and prizes, ISSN 0069-3472, issued by: Westchester Library System.
Key title: Children's books. Awards & prizes, ISSN 0069-3472
1. Children's literature—Bibliography—Periodicals. 2. Literary prizes—Periodicals. I. Title.
Z1037.A2W47a 028.52 76-357347
MARC-S

Westchester Library System.
Children's books: awards and prizes. 1960/61–
[New York, Children's Book Council, inc.]
v. 19–24 cm. annual.
At head of title, 1960/61– : Children's Book Council, inc.
Vols. for 1961/62– called also supplements.
Continued by: Children's books. Awards & prizes, ISSN 0069-3472, issued by: Children's Book Council, New York.
Key title: Children's books. Awards and prizes, ISSN 0069-3472
1. Children's literature—Bibliography—Periodicals. 2. Literary prizes—Periodicals. I. Children's Book Council, New York. II. Title.
Z1037.A2W47a 028.52 68-4275
MARC-S

—BIO-BIBLIOGRAPHY

De översätts för barn : porträtt av 15 utländska barnboksförfattare / red., Carl-Agnar Lövgren. — Lund : Bibliotekstjänst, 1977.
159 p. : ill. ; 19 cm. — (Btj-serien ; 86) S***
Includes bibliographies.
ISBN 9170181497
1. Children's literature—History and criticism. 2. Children's literature—Bio-bibliography. 3. Authors—Biography. I. Lövgren, Carl-Agnar, 1929-
PN1009.A1D39 78-395884
79 MARC

De skrev för barn : porträtt av 15 klassiska barnboksförfattare / red., Carl-Agnar Lövgren. — Lund : Bibliotekstjänst, 1977.
172 p. : ill. ; 19 cm. — (Btj-serien ; 89) S***
Includes bibliographies and index.
ISBN 9170181543
1. Children's literature—Bio-bibliography. 2. Authors—Biography. I. Lövgren, Carl-Agnar, 1929-
PN497.D4 78-397610
79 MARC

Twentieth-century children's writers / with a preface by Naomi Lewis ; editor D. L. Kirkpatrick. — London [etc.] : Macmillan, 1978.
xvi, 1507 p. ; 24 cm. GB78-33981
ISBN 0-333-23414-6 : £17.50
1. Children's literature—History and criticism. 2. Children's literature—Bio-bibliography. I. Kirkpatrick, D. L.
PN1009.A1T9 1978 820'.9'9282 79-315893
79 MARC

Twentieth-century children's writers / with a pref. by Naomi Lewis ; editor, D. L. Kirkpatrick. — New York : St. Martin's Press, 1978.
xvi, 1507 p. ; 24 cm.
Includes bibliographies.
ISBN 0-312-82413-0 : $40.00
1. Children's literature—History and criticism. 2. Children's literature—Bio-bibliography. I. Kirkpatrick, D. L.
PN1009.A1T9 1979 809'.89282 79-110314
79 MARC

Ward, Martha Eads.
Authors of books for young people : supplement to the second edition / by Martha E. Ward and Dorothy A. Marquardt. — Metuchen, N.J. : Scarecrow Press, 1979.
v, 302 p. ; 22 cm.
ISBN 0-8108-1159-6
1. Children's literature—Bio-bibliography. I. Marquardt, Dorothy A., joint author. II. Title.
PN497.W3 1979 809'.89282 78-16011
78 MARC

—BOOK REVIEWS

Nemoto, Masayoshi, 1942-
(Shōwa jidō bungaku annai)
昭和児童文学案内 / 根本正義著. — 東京 : 高文堂出版社, 昭和51 [1976]
97 p. : 18 cm. — (高文堂新書) Ja 76-19585
Series romanized: Kōbundō shinsho.
"児童文学研究文献目録": p. 72-96.
Y 350
1. Children's literature. Japanese—Bio-Bibliography. 2. Children's literature—Book reviews. I. Title.
Z1037.8.J3N45 78-811951
[PN1009.J3]

—BOOK REVIEWS—PERIODICALS—BIBLIOGRAPHY

Meacham, Mary, 1946-
Information sources in children's literature : a practical reference guide for children's librarians, elementary school teachers, and students of children's literature / Mary Meacham. — Westport, Conn. : Greenwood Press, 1978.
xvii, 256 p. ; 24 cm. — (Contributions in librarianship and information science ; no. 24 ISSN 0084-9243)
Bibliography: p. [237]-249.
Includes index.
ISBN 0-313-20045-9
1. Bibliography—Bibliography—Children's literature. 2. Children's literature—Bibliography. 3. Children's literature—Book reviews—Periodicals—Bibliography. 4. Books and reading for children—Bibliography. 5. Libraries, Children's—Bibliography. I. Title. II. Series.
Z1037.A1M4 028.52 77-91107
78 MARC

—CONGRESSES

Children's Literature Association.
Proceedings of the fifth annual conference of the Children's Literature Association, Harvard University, March, 1978 / edited by Margaret P. Esmonde and Priscilla A. Ord. — Villanova, Pa. : Villanova University, 1979.
122 p. ; 27 cm.
Includes bibliographical references.
1. Children's literature—Congresses. I. Esmonde, Margaret P. II. Ord, Priscilla A.
PN1009.A1C51384 1978 809'.89282 79-103606
79 MARC

Loughborough International Seminar on Children's Literature, 10th, Woudschoten, 1977.
Contributions / edited by the Dept. Book and Youth of the Dutch Centre for Libraries and Literature. — The Hague : Nederlands Bibliotheek en Lektuur Centrum, 1978.
117 p. ; 23 cm. Ne***
Includes bibliographies.
ISBN 9062527167
1. Children's literature—Congresses. I. Nederlands Bibliotheek- en Lektuurcentrum. Bureau Boek en Jeugd.
PN1009.A1L64 1977 809'.89282 79-300831
79 MARC

CHILDREN'S LITERATURE —CONGRESSES (Continued)

National Seminar on Children's Literature, Frankston State College, 1975.
Readings in children's literature : proceedings of the National Seminar on Children's Literature at Frankston State College / edited by Moira Robinson. — Frankston, Vic. : Frankston State College, 1977.
v, 291 p. ; 21 cm. Aus
"A Frankston State College publication."
Includes bibliographies.
ISBN 0-9597053-0-9
1. Books and reading for children—Congresses. 2. Books and reading for youth—Congresses. 3. Children's literature—Congresses. I. Robinson, Moira. II. Title.
Z1037.A1N34 1975 028.52 78-322420
79 MARC

Pacific Rim Conference on Children's Literature, 1st, University of British Columbia, 1976.
One ocean touching : papers from the First Pacific Rim Conference on Children's Literature / edited by Sheila A. Egoff. — Metuchen, N.J. : Scarecrow Press, 1979.
viii, 252 p. ; 23 cm.
Includes bibliographies and index.
ISBN 0-8108-1199-5
1. Children's literature—Congresses. 2. Children's literture, Canadian—Congresses. I. Egoff, Sheila A. II. Title.
PN1009.A1P28 1976 028.5 78-31308
78 MARC

Variety is king : aspects of fiction for children : a symposium / compiled by Norman Culpan and Clifford Waite. — Oxford : School Library Association, 1977.
vii, 173 p. : ill. ; 25 cm. GB78-23785
Includes bibliographies.
ISBN 0-900641-32-0 : £5.00
1. Children's literature—Congresses. 2. Books and reading for children—Congresses. I. Culpan, Norman. II. Waite, Clifford Anthony.
PN1009.A1V34 028.5 79-307803
79 MARC

Viewpoints : an international sharing of ideas about children, books, and reading : papers presented at a conference at Kutztown State College, Kutztown, Pennsylvania, July 5-8, 1976 / co-sponsored by Division of Library Science, Kutztown State College and the International Institute for Children's Literature and Reading Research, Vienna, Austria. — Kutztown : The College, ₍1976₎
108 p. ; 21 cm.
Includes bibliographical references.
1. Children's literature—Congresses. 2. Books and reading for children—Congresses. I. Kutztown State College. Division of Library Science. II. International Institute for Children's Juvenile and Popular Literature.
PN1009.A1V48 809'.89282 79-300957
79 MARC

—EXHIBITIONS

Börsenverein des Deutschen Buchhandels. Arbeitsgemeinschaft Antiquariat.
Schöne alte Kinderbücher : e. Ausstellung d. Arbeitsgemeinschaft Antiquariat im Börsenverein d. Dt. Buchhandels : Frankfurter Buchmesse, 18.-23. Oktober 1978 / ₍Ausstellung u. Katalogred., Raimund Kitzinger, Christine Pressler, Ingeborg Ramseger. — Frankfurt/Main : Buchhändler-Vereinigung, 1978.
96 p. : ill. (some col.) ; 22 cm. GFR78-A
Bibliography: p. 93-96.
Includes index.
1. Bibliography—Rare books—Exhibitions. 2. Children's literature—Exhibitions. 3. Frankfurt am Main—Exhibitions. I. Kitzinger, Raimund. II. Pressler, Christine. III. Ramseger, Ingeborg. IV. Frankfurt am Main. Messe. V. Title.
Z1029.B63 79-355778
79 MARC

Monschein, Johanna.
Europäische Kinderbücher vom 15. bis zum 19. Jahrhundert : Ausstellung im Prunksaal d. Österr. Nationalbibliothek, 17. Mai bis 14. Sept. 1979 / ₍Verf. des Kataloges, Johanna Monschein ; Photos, Alfred Janderka₎. ₍Wien : Österr. Nationalbibliothek, 1979₎
199 p. : ill. (some col.) ; 26 cm. Au79-13-11
Bibliography: p. 191-192.
Includes indexes.
S350.00
1. Children's literature—Exhibitions. 2. Bibliography—Rare books—Exhibitions. I. Vienna. Nationalbibliothek. II. Title.
Z1037.A1M76 79-382943
₍PN1009.A1₎ 79 MARC

Les Plus beaux livres d'images du monde ... : Bibliothèque municipale de Mulhouse, du 16 octobre au 6 novembre 1976. — Mulhouse : Bibliothèque municipale, 1976.
20 p. : ill. ; 22 cm. F77-12683
On cover: Une sélection de 50 livres d'images.
"Partiellement extrait du 'Bulletin d'analyses de livres pour enfants.'" — Bibl. de la Fr.
1. Children's literature—Exhibitions. 2. Mülhausen, Alsace—Exhibitions. I. Mülhausen, Alsace. Bibliothèque municipale. II. Title: Une sélection de cinquante livres d'images. III. Title: Cinquante livres d'images.
Z1037.9.P58 78-384624
₍PN1009.A1₎ 78 MARC

—HISTORY AND CRITICISM

Børnebogen i klassekampen : artikler om marxistisk litteratur for børn / Finn Barlby, red. — København : Tiderne Skifter, ₍1975₎
265 p. : ill. ; 20 cm. D76-6
Includes bibliographical references.
kr42.00
1. Children's literature—History and criticism. 2. Communism and literature. I. Barlby, Finn.
PN1009.Z6B6 78-345084
•78 MARC

Bressan, Tullio.
Non ammazzate Pinocchio / Tullio Bressan. — Napoli : Guida, 1979.
280 p. ; 18 cm. — (I Tascabili ; 58) It79-June
Bibliography: p. ₍18₎-24.
Includes index.
L4500
1. Children's literature—History and criticism. I. Title.
PN1009.A1B76 79-383984
•79 MARC

Campen, S G van.
De onderste plank : het jeugdboek nader bekeken / S. G. van Campen en C. van der Burg. — 2e herz. en uitgebreide dr. — Zeist : Uitgeverij NIB, ₍1977₎
183 p. : ill. ; 21 cm. Ne77-45
Bibliography: p. 160-168.
Includes index.
ISBN 9027500762 : fl 19.75
1. Children's literature—History and criticism. I. Burg, C. van der, joint author. II. Title.
PN1009.A1C314 1977 78-391669
•79 MARC

Correia, João David Pinto.
Literatura juvenil, paraliteratura : problemas de delimitação de conceitos e aspectos didáctico-pedagógicos : propostas para um melhor aproveitamento pedagógico das Leituras dos jovens / João David Pinto Correia. — Lisboa : Livraria Novidades Pedagógicas, 1978.
61 p. : diagrs. ; 21 cm. — (Colecção Biblioteca prática do professor ; 6)
Includes bibliographical references.
1. Children's literature—History and criticism. 2. Children's literature, Portuguese—History and criticism. I. Title.
PN1009.A1C665 79-112194
79 MARC

Furuta, Taruhi, 1927-
(Gendai jidō bungaku ron)
現代児童文学論：近代童話批判 / 古田足日. — 東京：くろしお出版, 1959.
253 p. ; 19 cm.

1. Children's literature, Japanese—History and criticism. 2. Children's literature—History and criticism. I. Title.
PN1009.J3F78 79–803361

Hatano, Kanji, 1905–
(Jidō shinri to jidō bungaku)
兒童心理と兒童文學 / 波多野完治著. — 東京：金子書房, 昭和25 ₍1950₎
5, 307 p. : ill. ; 22 cm.
Includes index.

1. Child psychology. 2. Children's literature—History and criticism. I. Title.
BF721.H315 78–814922

(Jidō yomimono to dokusho shidō)
児童読物と読書指導 / 責任編集 亀井勝一郎 ... ₍et al.₎. — 東京：牧書店, 昭和30 ₍1955₎
2, 3, 273, 7, 2 p. : ill. ; 22 cm. — (読書指導講座 ; 9)
Includes bibliographies and index.

1. Books and reading for children. 2. Children's literature—History and criticism. 3. Children's literature—Bibliography. I. Kamei, Katsuichirō, 1907–1966. II. Series: Dokusho shidō kōza ; 9.
Z1037.J6 79–804690

De översätts för barn : porträtt av 15 utländska barnboksförfattare / red., Carl-Agnar Lövgren. — Lund : Bibliotekstjänst, 1977.
159 p. : ill. ; 19 cm. — (Btj-serien ; 86) S***
Includes bibliographies.
ISBN 9170181497
1. Children's literature—History and criticism. 2. Children's literature—Bio-bibliography. 3. Authors—Biography. I. Lövgren, Carl-Agnar, 1929-
PN1009.A1D39 78-395884
79 MARC

Petroni, Mario.
Effetto infanzia : viaggio nell'immaginario / Mario Petroni. — Roma : A. Armando, 1978.
190 p. : ill. ; 20 cm. — (Controcampo ; 26) It78-Aug
Bibliography: p. ₍6₎
L4000
1. Children's literature—History and criticism. I. Title.
PN1009.A1P47 78-390684
•79 MARC

Rosenberg, Fulvia.
La Famille dans les livres pour enfants / Fulvia Rosenberg. — Paris : Magnard : Éditions l'École, 1976.
158 p. ; 17 x 18 cm. — (Collection Lecture en liberté) F78-15545
Originally presented as the author's thesis, École pratique des hautes études, Paris, 1969, under title: La famille et les relations familiales dans les livres pour enfants.
Bibliography: p. ₍127₎-136.
ISBN 2-210-46110-4 (Magnard). ISBN 2-211-85537-7 (l'École)
1. Children's literature—History and criticism. 2. Family in literature. I. Title.
PN1009.Z6R67 1976 809'.89282 79-380201
79 MARC

Scherf, Walter, 1920-
Strukturanalyse der Kinder- und Jugendliteratur : Bauelemente u. ihre psycholog. Funktion / Walter Scherf. — Bad Heilbrunn/Obb. : Klinkhardt, 1978.
200 p. ; 21 cm. — (Schriften des Arbeitskreises für Jugendliteratur) GFR78-A
"Die Erstfassung dieses Buches ... stützt sich auf dänische Beispiele und erscheint 1978 unter dem Titel 'Handling og spaending. Om estetisk og psykologisk kritiddaf børnelitteratur.'"
Includes bibliographical references and indexes.
ISBN 3-7815-0357-7 : DM19.80
1. Children's literature—History and criticism. I. Title. II. Series: Arbeitskreis für Jugendliteratur. Schriften des Arbeitskreises für Jugendliteratur.
PN1009.A1S348 78-396164
79 MARC

Schultz Cazenueve de Mantovani, Fryda, 1912-
La torre en guardia : lógica del mito en la infancia y en los pueblos / Fryda Schultz de Mantovani. — ₍Buenos Aires₎ : Editorial Plus Ultra, ₍c1978₎
94 p. ; 23 cm. — (Comunicación)
1. Children's literature—History and criticism. 2. Books and reading for children. I. Title.
PN1009.Z6S36 79-100870
79 MARC

Sexism in children's books : facts, figures & guidelines / edited by the Children's Rights Workshop ; ₍ill. by Lee Robinson₎. — London : Writers and Readers Pub. Cooperative, 1976.
56 p. : ill. ; 21 cm. — (Papers on children's literature ; no. 2) GB***
Includes bibliographies.
ISBN 0-904613-22-4 : £0.60
1. Children's literature—History and criticism. 2. Sex role in literature. I. Children's Rights Workshop.
PN1009.Z6S49 1976 820'.9'35 78-314103
78 MARC

Simons, Ludo.
Geschiedenis van het jeugdboek : twee bijdragen / door L. Simons, W. Goeman-Van Randen. — 2e dr. — Den Haag : Nederlands Bibliotheek en Lektuur Centrum, 1976.
18 p. : ill. ; 16 x 22 cm. — (Buiten het boekje ; 1)
Ne76
Bibliography: p. 18.
ISBN 9062525016 : fl 4.95
1. Children's literature—History and criticism. I. Goeman-Van Randen, W. II. Title. III. Series.
PN1009.A1S393 79-346057
79 MARC

Twentieth-century children's writers / with a preface by Naomi Lewis ; editor D. L. Kirkpatrick. — London ₍etc.₎ : Macmillan, 1978.
xvi, 1507 p. ; 24 cm. GB78-33981
ISBN 0-333-23414-6 : £17.50
1. Children's literature—History and criticism. 2. Children's literature—Bio-bibliography. I. Kirkpatrick, D. L.
PN1009.A1T9 1978 820'.9'9282 79-315893
79 MARC

Twentieth-century children's writers / with a pref. by Naomi Lewis ; editor, D. L. Kirkpatrick. — New York : St. Martin's Press, 1978.
xvi, 1507 p. ; 24 cm.
Includes bibliographies.
ISBN 0-312-82413-0 : $40.00
1. Children's literature—History and criticism. 2. Children's literature—Bio-bibliography. I. Kirkpatrick, D. L.
PN1009.A1T9 1979 809'.89282 79-110314
79 MARC

—HISTORY AND CRITICISM —ADDRESSES, ESSAYS, LECTURES

The Cool web : the pattern of children's reading / ₍edited by₎ Margaret Meek, Aidan Warlow, Griselda Barton. — 1st American ed. — New York : Atheneum, 1978, c1977.
427 p. ; 23 cm.
Bibliography: p. 399-418.
Includes index.
ISBN 0-689-10834-6
1. Children's literature—History and criticism—Addresses, essays, lectures. 2. Books and reading for children—Addresses, essays, lectures. I. Meek, Margaret. II. Warlow, Aidan. III. Barton, Griselda.
PN1009.A1C654 1978 809 77-3852
77 MARC

CHILDREN'S LITERATURE
—HISTORY AND CRITICISM
—ADDRESSES, ESSAYS, LECTURES
(Continued)

(Nikātī dar bārah-'i adabīyāt-i kūdakān)
نکاتی در بارهٔ ادبیات کودکان. — ‹تهران›: شورای کتاب کودک،
.2536 [1978]
51 p. ; 21 cm.
In Persian.
1. Children's literature—History and criticism—Addresses, essays, lectures. I. Shūrā-yi Kitāb-i Kūdak.
PN1009.A1N5 79-970414

Ozeki, Iwaji, 1896-
(Jidō bungaku no riron to jissai)
兒童文学の理論と実際 / 尾関岩二著. — 東京：関書院, 昭和24 [1949]
5, 216 p. ; 19 cm.
Colophon inserted.
1. Children's literature, Japanese—History and criticism—Addresses, essays, lectures. 2. Children's literature—History and criticism—Addresses, essays, lectures. I. Title.
PN1009.J3O9 78-811067

—HISTORY AND CRITICISM—
BIBLIOGRAPHY

Modern realistic stories for children and young people : international bibliography / [editor, Rudolf Majonica]. — [Munich] : Arbeitskreis für Jugendliteratur, c1978.
119 p. ; 21 cm. GFR***
1. Children's literature—Bibliography. 2. Bibliography—Bibliography—Children's literature. 3. Children's literature—History and criticism—Bibliography. I. Majonica, Rudolf.
Z1037.M8 028.52 79-305273
[PN1009.A1] 79 MARC

Vakliteratuur over jeugdliteratuur, jeugdbibliotheken en schoolbibliotheken / onder red. van Alice Gerardts ; met medew. van Jannie Daane ... [et al.]. — 2e dr. — Den Haag : Nederlands Bibliotheek en Lektuur Centrum, 1976.
38 p. ; 16 x 21 cm. — (Buiten het boekje ; 3)
Ne77
Includes index.
ISBN 9062525032 : fl 4.95
1. Libraries, Children's—Bibliography. 2. School libraries—Bibliography. 3. Books and reading for children—Bibliography. 4. Children's literature—History and criticism—Bibliography. I. Gerardts, Alice. II. Daane, Jannie. III. Series.
Z718.1.V33 79-349114
79 MARC

White, Virginia L
Books about children's books : an annotated bibliography / compiled by Virginia L. White and Emerita S. Schulte. — Newark, Del. : International Reading Association, 1979.
48 p. ; 22 cm. — (IRA annotated bibliography series)
Includes index.
ISBN 0-87207-333-5
1. Bibliography—Bibliography—Children's literature. 2. Children's literature—Bibliography. 3. Children's literature—History and criticism—Bibliography. 4. Books and reading for children—Bibliography. I. Schulte, Emerita S., joint author. II. Title. III. Series: International Reading Association. IRA annotated bibliography series.
Z1037.A1W59 028.52 79-17553
[PN1009.A1] 79 MARC

—HISTORY AND CRITICISM
—PERIODICALS

La Revue des livres pour enfants. no 56-
oct. 1977-
[Paris] Joie par les livres.
no. ill. 23 cm. 6 no. a year.
Continues: Bulletin d'analyses de livres pour enfants.
1. Children's literature—History and criticism—Periodicals.
I. Joie par les livres (Association)
PN1009.A1B84 028.52 79-649812
MARC-S

—INDEXES

Sarkissian, Adele.
Children's authors and illustrators : an index to biographical dictionaries / Adele Sarkissian, editor. — 2d ed. — Detroit : Gale Research Co., c1978.
xxiii, 251 p. ; 24 cm. — (Gale biographical index series ; no. 2)
Earlier ed. by D. La Beau.
ISBN 0-8103-1079-1
1. Authors—Biography—Indexes. 2. Illustrators—Biography—Indexes. 3. Children's literature—Indexes. I. La Beau, Dennis. Children's authors and illustrators. II. Title.
Z1037.A1L2 1978 016.809 78-16868
78 MARC

—LIBRARY RESOURCES—NEW YORK
(CITY)

Safford, Barbara Ripp.
Children's literature : a guide to research collections in the Libraries of Columbia University and Union Theological Seminary / Barbara Ripp Safford and Sharyl G. Smith. — New York : School of Library Service, The University, 1979.
36 p. ; 22 cm.
Includes index.
1. Children's literature—Library resources—New York (City) 2. Columbia University. Libraries. 3. New York (City). Union Theological Seminary. Library. I. Smith, Sharyl G., joint author. II. Title.
Z1038.N7S23 028.52 79-112035
[PN1009.A1] 79 MARC

—MARKETING

Duke, Judith S
Children's books and magazines : a market study / by Judith S. Duke. — White Plains, N.Y. : Knowledge Industry Publications, c1979.
iv, 236 p. ; 24 cm. — (Communications library)
Bibliography: p. 233.
Includes index.
ISBN 0-914236-17-2 : $24.95
1. Children's literature—Publishing—United States. 2. Children's literature—Marketing. I. Title. II. Series.
Z479.D83 78-24705
658.8'09'0705730973
79 MARC

—PUBLISHING

Hürlimann, Bettina, 1909-
Sieben Häuser : Aufzeichnungen einer Bücherfrau / Bettina Hürlimann. — Zürich ; München : Artemis-Verlag, 1976.
247 p., [10] leaves of plates : ill. ; 22 cm. Sw76-4581
34.50F
1. Hurlimann, Bettina, 1909- 2. Publishers and publishing—Switzerland—Biography. 3. Children's literature—Publishing. I. Title.
Z430.H85A34 1976 78-390168
*79 MARC

Turow, Joseph.
Getting books to children : An exploration of publisher-market relations / Joseph Turow. — Chicago : American Library Association, 1978.
vi, 137 p. ; 23 cm. — (ALA studies in librarianship ; no. 7)
Based on the author's thesis, University of Pennsylvania.
Bibliography: p. 135-137.
ISBN 0-8389-0276-6 : $12.50
1. Children's literature—Publishing. 2. Libraries and publishing. 3. Booksellers and bookselling. I. Title. II. Series: American Library Associatioin. ALA studies in librarianship ; no. 7.
Z286.C48T87 658.8'09'070573 78-24103
78 MARC

—PUBLISHING—EUROPE

Martin, Helen, 1889-1943.
Methods of distributing children's books employed in typical cities of certain selected European countries, 1931-1932 : progress report / presented by Helen Martin. — 1933.
leaves a-r, 175, xxix leaves ; 29 cm.
Typescript (carbon copy)
1. Libraries, Children—Europe. 2. Books and reading for children—Europe. 3. Children's literature—Publishing—Europe. I. Title: Methods of distributing children's books ...
Z718.1.M28 1933 79-306973
79 MARC

—PUBLISHING—GERMANY—HISTORY
—19TH CENTURY—EXHIBITIONS

Deutsche Kinderbücher des 18. Jahrhunderts : ein Beitrag zur Vorbereitung einer Bibliographie alter deutscher Kinderbücher : Ausstellung der Herzog August Bibliothek Wolfenbüttel 3. November 1978 bis 10. Januar 1979 / [Katalog, Georg Ruppelt, Ingrid Nutz und Michael Schlosser, unter Mitarbeit von Dorothea Balzer und Verona Borchardt]. — Wolfenbüttel : Die Bibliothek, 1978.
82 p. : ill. ; 26 cm. — (Ausstellungskataloge der Herzog-August-Bibliothek ; Nr. 25) GFR***
Includes index.
ISBN 3-88373-003-3
1. Children's literature—Publishing—Germany—History—19th century—Exhibitions. 2. Children's literature, German—Exhibitions. 3. Wolfenbüttel—Exhibitions. I. Ruppelt, Georg. II. Nutz, Ingrid. III. Schlosser, Michael. IV. Wolfenbüttel. Herzog-August-Bibliothek. V. Series: Wolfenbüttel. Herzog-August-Bibliothek. Ausstellungskataloge ; Nr. 25.
Z1035.3.D52 028.52'0943 79-368979
[PN1009.G3] 79 MARC

—PUBLISHING—GHANA—CONGRESSES

Seminar on Writing and Production of Literature for Children, University of Ghana, 1976.
Seminar on Writing and Production of Literature for Children : [selected papers]. — Legon : University of Ghana, Dept. of Library and Archival Studies, 1976-
v. ; 28 cm.
Includes bibliographical references.
1. Children's literature—Publishing—Ghana—Congresses. 2. Books and reading for children—Ghana—Congresses. 3. Children's literature, Ghanaian—Authorship—Congresses. I. Ghana. University, Legon. Dept. of Library and Archival Studies.
Z470.G45S45 1976 070.5'73'096 78-313652
78 MARC

—PUBLISHING—UNITED STATES

Duke, Judith S
Children's books and magazines : a market study / by Judith S. Duke. — White Plains, N.Y. : Knowledge Industry Publications, c1979.
iv, 236 p. ; 24 cm. — (Communications library)
Bibliography: p. 233.
Includes index.
ISBN 0-914236-17-2 : $24.95
1. Children's literature—Publishing—United States. 2. Children's literature—Marketing. I. Title. II. Series.
Z479.D83 78-24705
658.8'09'0705730973
79 MARC

—PUBLISHING—UNITED STATES—
HISTORY—20TH CENTURY

Emporia State University. William Allen White Memorial Library.
The May Massee collection : creative publishing for children, 1923-1963, a checklist : William Allen White Library, Emporia State University, Emporia, Kansas / edited by George V. Hodowanec ; annotations prepared by Jeanne Frederickson. — Limited 1st ed. — [Emporia] : The University, 1979.
xviii, 316 p. : ill. ; 24 cm.
Includes indexes.
ISBN 0-934068-00-3 : $25.00
1. Children's literature—Bibliography—Catalogs. 2. Emporia State University. William Allen White Memorial Library. 3. Massee, May. 4. Children's literature—Publishing—United States—History—20th century. I. Hodowanec, George V., 1935- . II. Frederickson, Jeanne. III. Title.
Z1038.E46E46 1979 028.52 79-4868
[PN1009.A1] 79 MARC

—PUBLISHING—UNITED STATES—
HISTORY—20TH CENTURY—
BIBLIOGRAPHY

Gottlieb, Robin.
Publishing children's books in America, 1919-1976 : an annotated bibliography / Robin Gottlieb. — 1st ed. — New York : Children's Book Council, c1978.
viii, 195 p. : ill. ; 23 cm.
Includes index.
1. Children's literature—Publishing—United States—History—20th century—Bibliography. I. Children's Book Council, New York. II. Title.
Z472.G67 78-113505
016.6588'09'0705730973
79 MARC

—STORIES, PLOTS, ETC.—
ILLUSTRATIONS—EXHIBITIONS

Paris. Bibliothèque Forney.
Quelques héros de l'enfance au pays de leurs images : exposition, Bibliothèque Forney, du 30 mars-3 juin 1979. — Paris : La Bibliothèque, [1979]
p. A-K, 41 p. : ill. ; 18 x 20 cm. F***
ISBN 2-7012-0562-X : 15.00F
1. Illustrated books, Children's—Exhibitions. 2. Children's literature—Stories, plots, etc.—Illustrations—Exhibitions. I. Title.
NC965.P3 1979 741.64'2'07404361 79-115670
79 MARC

—STUDY AND TEACHING

Huck, Charlotte S
Children's literature in the elementary school / Charlotte S. Huck. — 3d ed., updated. — New York : Holt, Rinehart, and Winston, c1979.
xiii, 788 p. : ill. ; 24 cm.
Includes bibliographies and indexes.
ISBN 0-03-046086-7
1. Children's literature—Study and teaching. I. Title.
LB1575.H8 1979 372.6'4 78-21081
78 MARC

CHILDREN'S LITERATURE —STUDY AND TEACHING (Continued)

Somers, Albert B 1939-
Response guides for teaching children's books / Albert B. Somers, Janet Evans Worthington. — Urbana, Ill. : National Council of Teachers of English, c1979.
vi, 119 p. ; 23 cm.
Includes bibliographical references.
ISBN 0-8141-4086-6 : $5.00
1. Children's literature—Study and teaching. 2. Books and reading for children. I. Worthington, Janet Evans, 1942- joint auhor. II. Title.
LB1575.S56 372.6'4 79-18
79 MARC

—STUDY AND TEACHING—JAPAN

(Ehon no ataekata)
絵本の与え方：幼児の文学教育 / 西郷竹彦編. — 東京：明治図書出版, 1968.
218 p. : ill. ; 19 cm. — (シリーズ·現代幼児教育新書 ; 2)
Series romanized: Shirizu gendai yōji kyōiku shinsho.

1. Children's literature—Study and teaching—Japan. 2. Picture -books for children. I. Saigō, Takehiko, 1920–
LB1179.E36 78–817189

—STUDY AND TEACHING—JAPAN —ADDRESSES, ESSAYS, LECTURES

Takamori, Kuniaki, 1930–
(Jidō bungaku to kokugo kyōiku)
児童文学と国語教育 / 高森邦明著. — 東京：鳩の森書房, 昭和51 [1976]
223 p. ; 22 cm. Ja 77–4869
Includes bibliographical references.
¥1200

1. Language arts (Elementary)—Japan—Addresses, essays, lectures. 2. Children's literature—Study and teaching—Japan—Addresses, essays, lectures. I. Title.
LB1577.J3T27 78–812587

—STUDY AND TEACHING (HIGHER)—SWEDEN

Kursplaner i barn- och ungdoms- litteratur vid Litteraturvetenskapliga institutionerna i Stockholm och Uppsala samt förteckningar över Barnlitteraturprojektets utgivning. — Uppsala : Avd. för litteratursociologi, Litteraturturvetenskapliga inst., [1978]
44 leaves ; 30 cm. — (Meddelande från Avd. för litteratursociologi vid Litteraturvetenskapliga institutionen i Uppsala ; nr. 1977:2 ISSN 0345-6889)
S78-4
Caption title.
"Projektet Barn- och ungdomslitteratur i Sverige."
Danish, English, Norwegian or, Swedish.
Bibliography: leaves 37-38.
kr7.00
1. Children's literature—Study and teaching (Higher)—Sweden. 2. Universities and colleges—Sweden—Curricula. I. Stockholm. Universitetet. Litteraturvetenskaplige institutionen. II. Uppsala. Universitet. Litteraturvetenskapliga institutionen. III. Series: Uppsala. Universitet. Litteraturvetenskapliga institutionen. Avdelningen för litteratursociologi. Meddelanden ; nr. 1977:2.
PN1009.A1K85 78-400593
79 MARC

—STUDY AND TEACHING (SECONDARY)

Schwartz, Sheila.
Teaching adolescent literature : a humanistic approach / Sheila Schwartz. — Rochelle Park, N.J. : Hayden Book Co., c1979.
216 p. ; 23 cm. — (Hayden English education series)
Includes bibliographies and index.
ISBN 0-8104-6036-X
1. Children's literature—Study and teaching (Secondary). 2. Books and reading for youth. I. Title.
PN1009.A1S353 809'.89283 79-2042
79 MARC

CHILDREN'S LITERATURE, AFRICAN

Atanda, F A
Ewì : fún àwọn ọmọdé / lati ọwó F. A. Atanda. — Ibadan : Onibonoje Press & Book Industries, 1974.
iv, 72 p. : ill. ; 18 cm.
I. Title.
PZ90.Y6A83 78-385752
79 MARC

Katalambulla, F H H 1942-
Lawalawa na hadithi nyingine / F. H. H. Katalambulla. — Dar es Salaam : East African Literature Bureau, [1976]
93 p. ; 19 cm.
CONTENTS: Lawalawa.—Ugeni wa wasiwasi.—Tafadhali niadhibu.
I. Title.
PZ90.S94K35 77-980184
78 MARC

Katalambulla, F H H 1942-
Mirathi / F. H. H. Katalambulla. — Nairobi : East African Literature Bureau, 1976.
48 p. ; 19 cm.
I. Title.
PZ90.S94K36 77-980014
78 MARC

Katalambulla, F H H 1942-
Pendo pevu / F. H. H. Katalambulla. — Kampala : East African Literature Bureau, 1976.
54 p. ; 19 cm.
I. Title.
PZ90.S94K37 77-980012
78 MARC

Malya, S
Babu Simulia / Simoni Malya. Dar es Salaam : Tanzania Pub. House, 1975.
50 p. : ill. ; 19 cm.
I. Title.
PZ90.S94M25 76-980334
78 MARC

Mbelwa, C M
Donda ndugu / H. C. M. Mbelwa. — Dar es Salaam : Tanzania Pub. House, 1973.
81 p. ; 19 cm.
I. Title.
PZ90.S94M29 77-980179
78 MARC

Mwenegoha, Hamza A K
Shujaa wa vijana / kimeandikwa na H. A. K. Mwenegoha. — Nairobi : Foundation Books for Tanzania Library Services, c1976.
44 p. : ill. ; 22 cm.
I. Title.
PZ90.S94M9 77-980031
79 MARC

Ngahyoma, Ngalimecha.
Kijiji chetu / Ngalimecha Ngahyoma. — Dar es Salaam : Tanzania Pub. House, 1975.
45 p. ; 19 cm. — (Michezo ya kuigiza ; 7)
I. Title.
PZ90.S94N45 77-980200
78 MARC

Ngomoi, John.
Ndoto ya Ndaria / J. Ngomoi. — Dar es Salaam : Tanzania Pub. House, 1976.
102 p. ; 19 cm.
I. Title.
PZ90.S94N47 77-980177
78 MARC

Njai, Mary Wanjiru, 1948-
Mapenzi gani haya? / Mary Wanjiru Njai ; illustrated by Isaac M. Thuku. — Nairobi : Njogu Gitene Publications, [1976?]
63 p. : ill. ; 19 cm.
Sh6.00
I. Title.
PZ90.S94N5 77-980187
78 MARC

Njoroge, Lizzie N
Mbũri yetagwo Njirũ / Lizzie N. Njoroge ; mbica icoretwo nĩ V. Murray Ngoima. — Nairobi : East African Pub. House, 1976.
21 p. : ill. ; 19 cm.
I. Title.
PZ90.K495N5 77-980759
79 MARC

Omari, Cuthbert K
Maswali waulizayo watoto wadogo / C. K. Omari na I. M. Omari. — Kampala : East African Literature Bureau, 1976.
32 p. ; 19 cm.
I. Omari, I. M., joint author. II. Title.
PZ90.S94O42 77-980767
79 MARC

Warwick, Francis.
Mlariba = Usurer / [Francis Warwick]. — 1st ed. — Karatina, Kenya : Gakaara Book Service, 1971.
48 p. : ill. ; 22 cm.
Based on Shakespeare's Merchant of Venice.
I. Shakespeare, William, 1564-1616. The merchant of Venice. II. Title. III. Title: Usurer.
PZ90.S94W28 77-983562
78 MARC

CHILDREN'S LITERATURE, ALBANIAN

Gjokutaj, Tasim.
Keci gë s'bëri detyrat : tregim në vargje / Tasim Gjokutaj ; [piktore Dolli Gjinali]. — Tiranë : Mihal Duri, 1975.
[20] p. : col. ill. ; 16 x 24 cm. — (Biblioteka e fatosit)
L1.50
I. Gjinali, Dolli. II. Title.
PZ90.A4G54 77-515967
78[7901] MARC

Karaulli, Jonuz.
Iriqi, lepuri dhe gjarperi : tregime në vargje / Jonuz Karaulli ; [piktor I. Shehu]. — Tiranë : Naim Frashëri, 1974.
[18] p. : col. ill. ; 16 x 24 cm. — (Biblioteka e yllkave)
L1.00
I. Shehu, Inajet. II. Title.
PZ90.A4K345 77-515968
78[7810] MARC

Malltezi, Xhevat.
Skalitësit e lirisë : roman / Xhevat Malltezi. — Tiranë : Naim Frashëri, 1976.
144 p. ; 17 cm. — (Biblioteka e pionierit)
L2.00
I. Title.
PZ90.A4M265 77-556532
78[r79]rev2 MARC

—HISTORY AND CRITICISM

Dedja, Bedri.
Burimet e letërsisë shqipe për fëmijë / Bedri Dedja. — Prishtinë : Rilindja, 1978.
122 p. ; 19 cm. — (Biblioteka Hejza) Yu***
15.00Din
1. Children's literature, Albanian—History and criticism. I. Title.
PN1009.A65D4 79-366664
79 MARC

—HISTORY AND CRITICISM—ADDRESSES, ESSAYS, LECTURES

Hoxha, Hysni.
Kritika / Hysni Hoxha. — Prishtinë : Rilindja, 1977.
221 p. ; 17 cm. — (Biblioteka Spektri) Yu***
15.00Din
1. Albanian poetry—20th century—History and criticism—Addresses, essays, lectures. 2. Children's literature, Albanian—History and criticism—Addresses, essays, lectures. 3. Literature—Addresses, essays, lectures. I. Title.
PG9607.H68 79-354555
79 MARC

CHILDREN'S LITERATURE, AMERICAN

Busy signal / [selected by] William D. Sheldon, Nina Woessner, José Gallegos. — Boston : Allyn and Bacon, c1978.
vi, 183 p. : ill. ; 21 cm. — (Breakthrough!)
SUMMARY: A collection of essays, biographies, short stories, and poems.
1. Children's literature, American. [1. Literature—Collections. 2. Short stories] I. Sheldon, William B. II. Woessner, Nina C. III. Gallegos, José, 1921-
PZ5.B98 [Fic] 77-90577
79 MARC

CHILDREN'S LITERATURE, AMERICAN (Continued)

Monsters and wild creatures / prepared by the Bank Street College of Education ; senior editor, Irma Simonton Black ... ⌊et al.⌋. — Boston : Houghton Mifflin, c1979.
47 p. : col. ill. ; 24 cm. — (Houghton Mifflin reading series)
SUMMARY: A collection of short stories, poems, and folktales, which includes a glossary and pronunciation guide.
1. Children's literature, American. ⌊1. Literature—Collections⌋ I. Black, Irma Simonton, 1906- . II. Bank Street College of Education, New York.
PZ5.M815 78-60563
79 MARC

The Peanut butter hamster, and other animal tails / edited by Grace Fox Anderson ; illustrated by Richard Johnson. — Wheaton, Ill. : Victor Books, c1979.
132 p. : ill. ; 20 cm. — (A Winner book)
SUMMARY: Presents short stories and poems featuring animals in a religious setting.
ISBN 0-88207-484-9 : $1.75
1. Animals—Literary collections. 2. Christian life—Literary collections. 3. Children's literature, American. ⌊1. Animals—Literary collections. 2. Christian life—Literary collections. 3. American literature—Collections⌋ I. Fox, Grace. II. Johnson, Richard, 1953-
PZ5.P37 ⌊Fic⌋ 78-65204
79 MARC

Sheldon, William D
The big ones-2 / William D. Sheldon, Nina Woessner. — 2d ed. — Boston : Allyn and Bacon, c1978.
v, 195 p. : ill. ; 21 cm. — (Breakthrough)
First ed. (1969) by N. Woessner and W. D. Sheldon.
SUMMARY: A collection of essays, biographies, short stories, and poems.
1. Children's literature, American. ⌊1. Literature—Collections. 2. Short stories⌋ I. Woessner, Nina C. The big ones. II. Title.
PZ5.S518 Bi 1978 ⌊Fic⌋ 78-50387
79 MARC

The Sesame Street library : with Jim Henson's Muppets / ⌊written by Michael Frith, et al. ; illustrated by Mel Crawford, et al. ; photos. by Charles P. Rowan⌋. — ⌊New York⌋ : Children's Television Workshop : distributed by Funk & Wagnalls, c1978.
12 v. : col. ill. ; 29 cm.
"Created in cooperation with the Children's Television Workshop, producers of Sesame Street."
SUMMARY: A multivolume collection of short stories, games, craft projects, and other activities. Each volume introduces several letters of the alphabet and a number from 1 to 12.
ISBN 0-8343-0009-5 (v. 1)
1. Children's literature, American. ⌊1. Literature—Collections. 2. Alphabet. 3. Counting⌋ I. Frith, Michael K. II. Crawford, Mel. III. Rowan, Charles, 1933- . IV. Children's Television Workshop. V. Sesame Street.
PZ5.S475 ⌊E⌋ 79-101209
79 MARC

Views from nature. New York, American tract society ⌊1863?⌋
80 p. incl. front., illus. 21 cm.
1. Children's literature, American. I. American tract society.
PZ6.V6 1863 810'.8'003 42-35743
⌊r79⌋rev MARC

Wheelock, Warren, comp.
Beyond the block 2 / ⌊compiled by⌋ Warren H. Wheelock, William D. Sheldon. — 2d ed. — Boston : Allyn and Bacon, c1978.
v, 127 p. : ill. ; 21 cm. — (Breakthrough!)
SUMMARY: A collection of essays, biographies, short stories, and poems.
1. Children's literature, American. ⌊1. Literature—Collections. 2. Short stories⌋ I. Sheldon, William D., joint comp. II. Title.
PZ5.W5 Bf 1978 ⌊Fic⌋ 77-90628
79 MARC

—BIBLIOGRAPHY

McCulloch, Lou W
Children's books of the 19th century / Lou W. McCulloch ; photo ill. by Thomas R. McCulloch. — Des Moines, Iowa : Wallace-Homestead Book Co., c1979.
152 p. : ill. ; 24 cm.
Includes bibliographies and index.
ISBN 0-87069-239-9 : $9.95
1. Children's literature, English—History and criticism. 2. Children's literature, American—History and criticism. 3. Children's literature, English—Bibliography. 4. Children's literature, American—Bibliography. I. McCulloch, Thomas R. II. Title.
PN1009.A1M233 820'.9'9282 77-87948
79 MARC

Minnesota. University. Children's Literature Research Collections.
Girls series books : a checklist of hardback books published 1900-1975. — Minneapolis : Children's Literature Research Collections, University of Minnesota Libraries, 1978.
ix, 121 p. ; 29 cm.
Bibliography: p. 120-121.
1. Bibliography—Children's books issued in series. 2. Children's literature, American—Bibliography. 3. Books and reading for girls. I. Title.
Z1232.M56 1978 028.52 78-623834
⌊PN1009.A1⌋ 79 MARC

—CONGRESSES

Society & children's literature : papers presented on research, social history, and children's literature at a symposium sponsored by the School of Library Science, Simmons College, and the Committee on National Planning for Special Collections of the Children's Services Division of the American Library Association, May 14-15, 1976 / edited by James H. Fraser. — Boston : D. R. Godine, 1978.
ix, 209 p. ; 24 cm.
Includes bibliographical references and index.
ISBN 0-87923-236-6 : $15.00. ISBN 0-8389-3213-4 pbk.
1. Children's literature, American—Congresses. I. Fraser, James Howard, 1934- . II. Simmons College, Boston. School of Library Science. III. American Library Association. Committee on National Planning for Special Collections of Children's Books.
PN1009.A1S63 028.5 77-94110
78 MARC

—HISTORY AND CRITICISM

From writers to students : the pleasures and pains of writing / edited by M. Jerry Weiss for the IRA Literature for Adolescents Committee. — Newark, Del. : International Reading Association, c1979.
vi, 113 p. ; 22 cm.
ISBN 0-87207-724-1
1. Children's literature—Authorship. 2. Children's literature, American—History and criticism. 3. Authors, American—20th century—Interviews. I. Weiss, Morton Jerome. II. International Reading Association. Literature for Adolescents Committee.
PN147.5.F7 808'.068 79-11695
79 MARC

McCulloch, Lou W
Children's books of the 19th century / Lou W. McCulloch ; photo ill. by Thomas R. McCulloch. — Des Moines, Iowa : Wallace-Homestead Book Co., c1979.
152 p. : ill. ; 24 cm.
Includes bibliographies and index.
ISBN 0-87069-239-9 : $9.95
1. Children's literature, English—History and criticism. 2. Children's literature, American—History and criticism. 3. Children's literature, English—Bibliography. 4. Children's literature, American—Bibliography. I. McCulloch, Thomas R. II. Title.
PN1009.A1M233 820'.9'9282 77-87948
79 MARC

—MISCELLANEA

Woolman, Bertha.
The Caldecott Award : the winners and the honor books / by Bertha Woolman. — Minneapolis : T. S. Denison, c1978.
54 p. ; 28 cm.
Bibliography: p. 49-52.
Includes index.
SUMMARY: Explains the history and significance of the Caldecott Award and includes a list of the winning and honor books, brief biographies of their illustrators, and questions about the books.
ISBN 0-513-01648-1
1. Caldecott medal books. 2. Children's literature, American—Miscellanea. 3. Illustrators—Biography. ⌊1. Caldecott medal books. 2. Literature—History and criticism. 3. Illustrators⌋ I. Title.
Z1037.A2W65 028.5'07'9 78-67194
⌊PN1009.A1⌋ 79 MARC

Woolman, Bertha.
The Newbery award : these are winners / by Bertha Woolman. — Minneapolis : T. S. Denison, c1978.
78 p. ; 29 cm.
Bibliography: p. 54-78.
SUMMARY: Explains the history and the significance of the Newbery Award and includes a list of the winning and honor books, brief biographies of the authors, and questions about the books.
ISBN 0-513-01650-3
1. Children's literature, American—Miscellanea. 2. Newbery medal books. ⌊1. Newbery Award. 2. Literature—History and criticism. 3. Authors, American⌋ I. Title.
PN1009.A1W6 028.5'07'9 78-67193
79 MARC

—KANSAS—BIO-BIBLIOGRAPHY

Shadle, Leona M
Kansas authors of books for children and young people, 1850-1975 / Leona M. Shadle. — ⌊s.l. : s.n.⌋, c1978 (Manhattan, KS : Ag Press)
208 p. (p. 205-208 blank) ; 22 cm.
Bibliography: p. 203-204.
1. Children's literature, American—Kansas—Bio-bibliography. 2. Authors, American—Kansas—Bio-bibliography. I. Title.
Z1232.S5 016.8109'9282 78-75122
⌊PS283.K2⌋ 79 MARC

CHILDREN'S LITERATURE, ARABIC

—SUDAN—BIBLIOGRAPHY

Nadīm, Hikmat.
(Bibliyūjrāfiyā adab al-aṭfāl fī al-Sūdān)
ببليوجرافيا ادب الاطفال فى السودان / جمع حكمت نديم ؛ اعداد وتصنيف عثمان عوض عثمان. — ⌊الخرطوم⌋ : ادارة النشر الثقافى ، مصلحة الثقافة ، ⌊1976?⌋
133 p. ; 24 cm.
Includes index.
1. Children's literature, Arabic—Sudan—Bibliography. I. 'Uthmān, 'Uthmān 'Awaḍ. II. Title.
Z1037.8.A7N3 77-982074
[PN1009.A68]

CHILDREN'S LITERATURE, ARGENTINE

Domínguez, María Alicia, 1908-
Cocotón : poema escénico en dos actos y dos cuadros / María Alicia Domínguez ; ⌊ilus., Kitty Loréfice de Passalia⌋. — Buenos Aires : Editorial Plus Ultra, ⌊c1977⌋
63 p. : ill. ; 21 cm. — (Colección El Escenario)
I. Title. II. Series.
PZ77.D63 79-371762
79 MARC

Gori, Gastón.
Y además, era pecoso / Gastón Gori. — S⌊anta⌋ Fe, ⌊Argentina⌋ : Librería Nuevo Best-Seller, ⌊1977⌋
122 p. ; 21 cm.
I. Title.
PZ73.G64 863 78-110722
79 MARC

Murillo, José.
Brunita / José Murillo. — Buenos Aires : Ediciones Pespir, ⌊1977⌋
101 p. : ill. ; 20 cm.
I. Title.
PZ73.M86 863 78-111255
79 MARC

Ocampo, Silvina.
El tobogán : segunda parte de El cofre volante / Silvina Ocampo ; ilustradora, Beatriz Bolster. — ⌊1. ed.⌋. — Buenos Aires : A. Estrada, ⌊1975⌋
⌊19⌋ p. : col. ill. ; 26 cm. — (Colección Cuentos para seguir contando)
Sequel to El cofre volante.
I. Bolster, Beatriz. II. Title. III. Series.
PZ73.O2 863 79-353639
79 MARC

Vettier, Adela.
Cinco diálogos fantásticos / Adela Vettier ; ilus. de Juan R. Carvajal. — Buenos Aires : Conjunta Editores, c1977.
92 p. : ill. ; 23 cm.
On cover: Para niños de 12 a 80 años.
CONTENTS: La puerta de la señora Roen.—El paraguas.—La armadura.—La mano del manto celeste.—La mujer que lava las tazas de porcelana.
I. Title.
PZ77.V4 862 79-348908
79 MARC

CHILDREN'S LITERATURE, AUSTRALIAN

—EXHIBITIONS

Australian children's book exhibition, 30 September-15 October 1978 : ⌊held at the⌋ Library & Resources Centre, Commonwealth Institute. — ⌊London⌋ : ⌊The Centre⌋, ⌊1978⌋.
⌊1⌋, 21 p. ; 30 cm. GB79-00891
Cover title.
£0.20
1. Children's literature, Australian—Exhibitions. 2. London—Exhibitions. I. London. Commonwealth Institute. Library and Resources Centre.
Z1037.A97 028.52 79-313050
⌊PN1009.A1⌋ 79 MARC

Gurbanova, M
(Alagīalyn sărchălăr)
Ajaгјалын сәрчәләр / М. Гурбанова ; ⌊рәссаму Р. Јанбулатовдур⌋. — Бакы : Көнчлик, 1968.
⌊15⌋ p.
Azerbaijani.
L.C. copy on microfilm only.
"Мәктәб јашы чатмамыш ушаглар үчүн."
——— Microfilm. On 1 reel with other items ; 35 mm.
Microfilm Slavic 12708 PZ
I. Īanbulatov, R. II. Title.
[PZ90.A9G87] 79-970217

CHILDREN'S LITERATURE, BENGALI

Bose, Dakshina Ranjan.
(Yuddhe gelena Giyam)
যুদ্ধে গেলেন গিয়ং / দক্ষিণারঞ্জন বসু. — কলিকাতা : ফরোয়ার্ড পাবলিশিং কনসার্ন, 1384 [1977]
86 p. : ill. ; 23 cm.
In Bengali.
"লেখকের অন্যান্য বই": p. [4]
Rs5.00
I. Title.
PZ90.B4B615 78-902088

Guha, Buddhadev, 1936-
(Ṛjudāra saṅge jaṅgale)
ঋজুদার সঙ্গে জঙ্গলে / বুদ্ধদেব গুহ. — 1. সংস্করণ. — কলিকাতা : আনন্দ পাবলিশার্স, 1973, 1974 printing.
88 p. : ill. ; 22 cm.
In Bengali.
Rs5.00
1. Title.
PZ90.B4G82 78-911771

Mukhopadhyay, Sirshendu, 1936-
Manojadera adbhuta bāṛi / Śīrshendu Mukhopādhyāya. — 1. saṃskaraṇa. — Kalikātā : Ānanda, 1978.
137 p. : ill. ; 20 cm.
In Bengali.
A novel.
Rs6.00
I. Title.
PZ90.B4M87 78-913197
79 MARC

CHILDREN'S LITERATURE, BOLIVIAN

Aguiar, Adonias, 1915-
Fora da pista / Adonias Filho. — Rio de Janeiro : Civilização Brasileira, 1978.
100 p. ; 21 cm. — (Coleção Vera Cruz ; v. 263)
Cr$60.00
I. Title.
PZ83.A42 78-379632
78 MARC

Almeida, Márcio.
Gagarino, o menino bimundi / Márcio Almeida ; [ilustrações, Sebastião Laguardia]. — 1. ed. — Belo Horizonte : Interlivros, 1975.
65 p. : ill. ; 23 cm. — (Coleção Saci ; 3)
I. Title. II. Series.
PZ83.A545 78-383036
78 MARC

JaimesFreyre, Raúl, 1884-1970.
Evangelio infantil, versospara los niños ... La Paz, Bolivia [Editorial "boliviana", 1933?]
[16] p. 19 cm. (Biblioteca de "apuntes")
I. Title. II. Series.
PZ74.3.J3 861 43-30181
[r79]rev MARC

Leuzinger, Maria Alice do Nascimento e Silva.
O diário de Marcos Vinicius / Maria Alice do Nascimento e Silva Leuzinger ; [ilustrações, J. L. Mileno]. — Rio de Janeiro : Livraria J. Olympio Editora, 1977.
105 p. : ill. ; 21 cm.
"Em convênio com o Instituto Nacional do Livro."
Cr$22.00
I. Title.
PZ83.L475 79-103552
79 MARC

Musatti, Jeanete.
Sonhos do Pondji / Jeanete Musatti. — São Paulo : M. Ohno, 1979.
[52] p. : col. ill. ; 27 cm.
Cr$350.00
I. Title.
PZ83.M9 79-114545
79 MARC

Rector, Mônica.
Bimba pensa na fazenda / Mônica Rector ; [capa e ilustrações, Virgílio Veloso]. — [Belo Horizonte] : Editora Comunicação, 1977.
38 p. : ill. ; 20 cm.
"Em convênio com o Instituto Nacional do Livro - Ministério da Educação e Cultura."
Cr$10.00
I. Title.
PZ83.R36 869'.3 79-365531
79 MARC

Rezende Filho, José, 1929-1977.
Tonico / José Rezende Filho ; [ilus., Iranildo Alves]. — 2. ed. — São Paulo : Editora Ática, [1978]
95 p. : ill. ; 21 cm. — (Série Vaga-lume)
Cr$17.00
I. Title.
PZ83.R44 1978 78-385255
79 MARC

Rocha, Antonio.
Fura-Nuvens / Antonio Rocha ; [capa de Branca de Castro]. — Belo Horizonte : Editora Itatiaia, 1978.
114 p. : ill. ; 21 cm. — (Coleção Sapo Sapeca ; v. 1)
Cr$70.00
I. Title. II. Series: Rocha, Antonio. Coleção Sapo Sapeca ; v. 1.
PZ83.R64 79-117520
79 MARC

Vasconcelos, José Mauro de.
O menino invisível / José Mauro de Vasconcelos ; capa e ilustrações do autor. — São Paulo : Edições Melhoramentos, c1978.
25 p. : ill. ; 27 cm.
Cr$40.00
I. Title.
PZ83.V3 869'.3 79-353643
79 MARC

Werneck, Leny.
O velho que foi embora / Leny Werneck ; ilustrações de Regina Yolanda. — 2. ed. — São Paulo : Edições Melhoramentos, 1977.
48 p. : col. ill. ; 20 cm.
Cr$40.00
I. Title.
PZ83.W45 1977 79-108236
79 MARC

CHILDREN'S LITERATURE, BULGARIAN

Denev, Mladen Zhekov, 1931-
(Kŭsni izpiti)
Късни изпити : разкази за деца / Младен Денев ; [худож. Петър Петров]. — София : Бълг. писател, 1978.
101 p. : ill. ; 21 cm. Bu 78-2061
0.58 lv
I. Title.
PZ70.B8D44 79-409073

Dichev, Stefan, 1920-
(Sŭdbonosnata misiia)
Съдбоносната мисия : новите приключения на Рали : роман : [за юноши] / Стефан Дичев ; [худож. Лиляна Дичева]. — 1. изд. — София : Отечество, 1978 (Враца : печ. В. Александров)
[144] p. : ill. ; 21 cm. Bu 78-2758
0.62 lv
I. Title.
PZ70.B8D53 79-408161

Krŭsteva, Olga Stefanova.
(Edna godina bez znachenie)
Една година без значение : повест / Олга Кръстева. — 1. изд. — София : Изд-во отечество, 1978.
189 p. : ill. ; 24 cm. Bu***
0.68 lv
I. Title.
PZ70.B8K76 79-409549

Radostni ochi : sbornik prikazki, razkazi i stikhotvoreniia : [za nach. uchilishtna vŭzrast / redaktori Dariia Tabakova, Ivan Krŭstev]. — Sofiia : Bŭlg. khudozhnik, 1978.
160 p. : chiefly col. ill. ; 32 cm. Bu78-3096
CONTENTS: Rusafov, G. V koziata kŭshturka.—Ovadiia, D. Nosht v planinata.—Veselinov, G. Gorski tramvaĭ.—Danchev, S. Radostni ochi.—TSenov, B. Otkradnatata diniia.—Bozhilov, K. Vinagi shte ima prikazki.—Konstantinova, K. Snezhka.—Chakŭrov, S. Pŭrviiat urok.—Borov, S. Istinski priiateli.
2.58 lv
1. Children's literature, Bulgarian. I. Tabakova, Dariia. II. Krŭstev, Ivan, 1921-
PZ70.B7R3 79-378502
*79 MARC

Staĭnova, Anna Madzharova, 1894-
(Taka kaza Benkovski)
Така каза Бенковски : повест за юноши / Анна Каменова [i. e. А. Стайнова]. — София : Български писател, 1978.
92 p. : ill. ; 21 cm. Bu***
0.56 lv
1. Benkovski, Georgi, 1843-1876—Juvenile fiction. I. Title.
PZ70.B8S778 79-408160

Stŭpov, Petŭr Ivanov, 1910-
(Vsichki na orŭzhie)
Всички на оръжие : повест за деца и юноши / Петър Стъпов ; [худож. Мая Чолакова]. — София : Бълг. писател, 1978.
165 p. : ill. ; 21 cm. Bu 78-2763
0.77 lv
I. Title.
PZ70.B8S9 79-408918

Vaptsarova, Raĭna.
(Bak'o)
Бакьо : случки из детските и юношески години на Никола Йонков Вапцаров / Райна Вапцарова ; [худож. Георги Недялков]. — 3. изд. — София : Нар. младеж, 1978.
244 p. : ill. ; 21 cm. — (Библиотека Любими книги и герои ; № 86) Bu 78-791
Series romanized: Biblioteka Liubimi knigi i geroi.
0.93 lv
1. Vaptsarov, Nikola Ionkov, 1909-1942—Juvenile fiction.
I. Title.
PZ70.B8V27 1978 79-408678

Zidar, Pavle.
Glavne osebe na potepu / [avtor] Pavle Zidar ; [pogovor Berta Golob ; ilustriral Rudi Skočir ; opremil Dušan Klun]. — Ljubljana : Mladinska knjiga, 1975 (Ljubljana : "Jože Moškrič")
206 p. : ill. ; 19 cm. — (Knjižnica Sinjega galeba ; 186)
I. Title.
PZ70.S6Z5 78-385254
78 MARC

—BIBLIOGRAPHY—PERIODICALS

(Detska literatura i izkustva)
Детска литература и изкуства. 1976-
София, Дом на литературата и изкуствата за деца и юноши.
v. 21 cm. annual.
"Библиографски преглед."
1. Children's literature, Bulgarian—Bibliography—Periodicals. 2. Arts, Bulgarian—Bibliography—Periodicals. 3. Education—Bulgaria—Bibliography—Periodicals. I. Dom na literaturata i izkustvata za detsa i iunoshi.
Z1037.6.D427 79-641766
(MARC-S)

CHILDREN'S LITERATURE, BURMESE

E Maung, U, 1905-
(Tịrithsan pon pyin)
တိရစ္ဆာန်ပုံပြင်. ဦးအေးမောင် ရေးသားပြုစုသည်. Rangoon, ဘာသာပြန်စာပေအသင်း [1955]
72 p. col. illus. 18 cm.
In Burmese.
1. Animals, Legends and stories of. I. Title.
PZ90.B8E2 78-929194

Hla, U, Ludu.
(Ạ pyi pyi hmạ pon pyin myā)
အပြည်ပြည်မှပုံပြင်များ. [ရေးသူ] လူထုဦးလှ. [မန္တလေး, ကိုပါ, 1962]
239 p. col. illus. 18 cm.
In Burmese.
1. Tales. I. Title.
PZ90.B8H5 S A 65-2256

(Thạma nạdi nat mīn wathtụ)
သမာနဒီနတ်မင်း ဝတ္ထု = The king of the Golden River. — Rangoon : Burma Christian Literature Society, 1954.
40 p. : ill. ; 19 cm.
In Burmese.
A free adaptation of John Ruskin's The king of the Golden River.
"Reprint."
I. Ruskin, John, 1819-1900. The king of the Golden River. II. Title: The king of the Golden River.
PZ90.B8T47 1954 76-984756

—CONGRESSES

Kạlēi Lu Nge Sa Pei Hnī Hnāw Hpạle Pwē, Rangoon, 1973.
(Kạlēi lu nge sa pei)
ကလေးလူငယ်စာပေ, စာတန်း လေးစောင် / အုန်းကျော် ... [et al.]. — ရန်ကုန် : စာပေဗိမာန်, 1974.
2 v. (320 p.) ; 22 cm.
In Burmese.
Includes bibliographical references.
K3.00 per vol.
1. Children's literature, Burmese—Congresses. I. Ōn Kyaw, U. II. Title.
PN1009.B85K3 1973 77-985210

CHILDREN'S LITERATURE, CANADIAN

—BIBLIOGRAPHY

Frayne, June, 1952-
Print for young Canadians : a bibliographical catalogue of Canadian fiction for children, from 1825-1920 / prepared by June Frayne, Jennifer Laidley, and Henry Hadeed. — Toronto : [s.n.], 1975.
x, 80 p. : ill. ; 18 cm. C***
Bibliography: p. 79-80.
1. Children's literature, Canadian—Bibliography. 2. Canadian fiction—Bibliography. I. Laidley, Jennifer, joint author. II. Hadeed, Henry, joint author. III. Title.
Z1037.F844 028.52 76-372173
[PN1009.A1] 76[r79]rev2 MARC

Writers' Development Trust. Ontario Work Group.
Action/adventure / prepared by the Ontario Work Group, Barry Dickson ... [et al.]. — [s.l.] : Writers' Development Trust ; [Toronto : sold by Teachers and Writers Education Project, 197-]
200 p. ; 21 cm. — (A Resource guide for the teaching of Canadian literature) C***
Includes bibliographies.
$2.00
1. Canadian literature—Outlines, syllabi, etc. 2. Adventure stories, Canadian—Bibliography. 3. Canadian literature—Bibliography. 4. Children's literature, Canadian—Bibliography. I. Dickson, Barry, 1939- . II. Title. III. Series.
PR9185.5.A37W7 1970z 810 78-325338
79 MARC

—CONGRESSES

Pacific Rim Conference on Children's Literature, 1st, University of British Columbia, 1976.
One ocean touching : papers from the First Pacific Rim Conference on Children's Literature / edited by Sheila A. Egoff. — Metuchen, N.J. : Scarecrow Press, 1979.
viii, 252 p. ; 23 cm.
Includes bibliographies and index.
ISBN 0-8108-1199-5
1. Children's literature—Congresses. 2. Children's literture, Canadian—Congresses. I. Egoff, Sheila A. II. Title.
PN1009.A1P28 1976 028.5 78-31308
78 MARC

CHILDREN'S LITERATURE, CATALAN

Grau, Florència.
La bruixeta sense escombra / [guió i escenificació, Florència Grau]. — Barcelona : La Galera, c1978.
24 p. : il. ; 17 cm. — (Teatre joc d'equip ; 23) Sp***
Cover title.
ISBN 8424650239
I. Title.
PZ90.C3G7 79-351052
79 MARC

—BIBLIOGRAPHY

Omnium Cultural. Delegació d'Ensenyament Català. Equip Pedagogic.
Guia de material per a l'ensenyament i de lectura infantil i juvenil 1977-1978 / elaborada per l'Equip Pedagògic de la Delegació d'Ensenyament Català d,Omnium Cultural. — [Barcelona] : Omnium Cultural, [1977?]
19 p. : il. ; 21 cm. Sp***
1. Children's literature, Catalan—Bibliography. 2. Catalan imprints. I. Title.
Z1037.7.O47 1977 78-400347
[PN1009.S8] 79 MARC

CHILDREN'S LITERATURE, CHINESE

Chang, Cheng.
(Hsiao-lien-tzu)
小莲子 / 张正著 ; [插图文义]. — [香港? : s. n., 1978?]
51 p. : ill. ; 21 cm.
Fiction.
Reprint of the 1st (1978) ed. published by Shao nien erh t'ung ch'u pan she, Shanghai.
I. Wen, I. II. Title.
PZ10.831.C4613H7 79-837984

Chang, Yen-hsün, 1925-
(Liang ken ts'ao)
兩根草 / 張彥勳著. — 臺南 : 閩道出版社, 民國62 [1973]
2, 100 p. : ill. ; 19 cm. — (兒童小叢書 ; 5)
English title on added t. p.: History of two kids.
Juvenile literature.
Text accompanied with Chu yin fu hao.
Series romanized: Erh t'ung hsiao ts'ung shu.
I. Title. II. Title: History of two kids.
PZ10.831.C465L5 79-841620

Ch'en, Chao-hsiang, fl. 1978-
(Ma chüeh ho hu li)
麻雀和狐狸 / 陈兆祥, 李恒辰绘. — 第1版. — 北京 : 人民美術出版社 : 发行者新华书店北京发行所, 1978.
22 p. : chiefly col. ill. ; 13 cm.
"根据保加利亚民間故事改编."
"連環圖畫."
$0.11
I. Li, Heng-ch'en, joint author. II. Title.
PZ10.831.C48M3 79-838179

Ch'en, Hsiang-yin.
(Shui shih hao p'eng yu)
誰是好朋友 / 陳相因文 ; 李林圖. — [南投縣] : 臺灣省政府教育廳 ; 臺北 : 發行者臺灣書店, 民國56 [1967]
36 p. : col. ill. ; 21 cm. — (中華兒童叢書)
Text accompani. d with Kuo yü chu yin.
Series romanized: Chung-hua erh t'ung ts'ung shu.
I. Li, Lin, fl. 1967- II. Title.
PZ10.831.C514S5 78-840613

Ch'eng, I-ju.
(Chiao Yü-lu ho hai tzu men)
焦裕禄和孩子们 / 程逸汝编写 ; 倪绍勇, 张纪平画. — 上海 : 上海人民出版社, 1977.
[28] p. : col. ill. ; 16 cm.
Title also in pinyin on cover: Jiao Yulu he haizimen.
"連環圖畫"
$0.14
1. Chiao, Yü-lu, 1922-1964—Juvenile fiction. I. Ni, Shao-yung. II. Chang, Chi-p'ing. III. Title. IV. Title: Jiao Yulu he haizimen.
PZ10.831.C545C47 79-836645

(Ch'i i ti shih chieh)
奇異的世界. — 香港 : 新雅兒童教育出版社, 1975-1976 [v. 1, 1976]
5 v. in 1. : col. ill. ; 18 x 20 cm.
Cover title.
CONTENTS: [1] 王珀 動物的語言—[2] 薛思新 自然建築師—[3]-[5] 青川 山林趣獵 好幫手 綠色的世界冠軍
$2.20 per vol.
I. Wang, P'o, fl. 1976-
PZ10.831.C55 78-842302

(Ch'ing niao)
青鳥. — 香港 : 大衆圖書公司, [1978?]
10 p. : col. ill. ; 18 x 19 cm. — (兒童故事彩色畫 ; 第2輯, [4])
Bound with Kuo wang ti hsin i. Hsiang-kang, 1978?
$1.80
I. Series: Erh t'ung ku shih ts'ai se hua ; ti 2 chi, 4.
PZ10.831 79-840727

Ch'iu, Hung.
(Feng ku niang)
風姑娘 / 編寫秋紅 ; 插圖劉兆源. — 香港 : 新雅兒童教育出版社, 1978.
16 p. : col. ill. ; 27 cm. — (自然的故事 ; 第一輯)
Bound with the author's Hsiao shui ti. Hsiang-kang, 1978.
$3.50
I. Liu, Chao-yüan. II. Title. III. Series: Tzu jan ti ku shih.
PZ10.831.C6196H74 79-837531
[PZ10.831.C6196]

Ch'iu, Hung.
(Hsiao shui ti)
小水滴 / 編寫秋紅 ; 插圖劉兆源. — 香港 : 新雅兒童教育出版社, 1978.
16 p. : col. ill. ; 27 cm. — (自然的故事 ; 第一輯)
Bound with the author's Ying t'ao shu. Hsiang-kang, 1978. Feng ku niang. Hsiang-kang, 1978. Huo miao ti ti. Hsiang-kang, 1978.
$3.50
I. Liu, Chao-yüan. II. Title. III. Series: Tzu jan ti ku shih.
PZ10.831.C6196H74 79-83753

Ch'iu, Hung.
(Huo miao ti ti)
火苗弟弟 / 編寫秋紅 ; 插圖劉兆源. — 香港 : 新雅兒童教育出版社, 1978.
16 p. : col. ill. ; 27 cm. — (自然的故事 ; 第1輯)
Bound with the author's Hsiao shui ti. Hsiang-kang, 1978.
$3.50
I. Liu, Chao-yüan. II. Title. III. Series: Tzu jan ti ku shih.
PZ10.831.C6196H74 79-837530
[PZ10.831.C6196]

Ch'iu, Hung.
(Ying t'ao shu)
櫻桃樹 / 編寫秋紅 ; 插圖劉兆源. — 香港 : 新雅兒童教育出版社, 1978.
16 p. : col. ill. ; 27 cm. — (自然的故事 ; 第一輯)
Bound with the author's Hsiao shui ti. Hsiang-kang, 1978.
$3.50
I. Liu, Chao-yüan, II. Title. III. Series: Tzu jan ti ku shih.
PZ10.831.C6196H74 79-837532
[PZ10.831.C6196]

Chu, Ch'uan-yü.
(Tzu yu hao lieh ch'e)
自由號列車 / 朱傳譽文 ; 王寧生圖. — [南投縣] : 臺灣省政府教育廳 ; 臺北 : 發行者臺灣書店, 民國55 [1966]
36 p. : col. ill. ; 21 cm. — (中華兒童叢書)
Text accompanied with Kuo yü chu yin.
Series romanized: Chung-hua erh t'ung ts'ung shu.
I. Wang, Ning-sheng. II. Title.
PZ10.831.C6343T9 78-840615

Chu, Jen-min.
(I t'iao hai chi yü)
一条海鲫鱼 = Yi tiao hai ji yu / [朱人民编画]. — [第1版]. — [上海 : 上海人民出版社, 1976]
[20] p. : col. ill. ; 13 x 19 cm.
Cover title.
"連環圖畫"
$0.11
I. Title. II. Title: Yi tiao hai ji yu.
PZ10.831.C6344I2 79-836663

Fan-ch'ang hsien, China. Wen chiao chü.
(Miao erh ch'ing ch'ing)
苗儿青青 / 繁昌县文教局. — [香港? : s. n., 1978?]
68 p. : ill. ; 21 cm.
Reprint of the 1977 ed. published by An-hui jen min ch'u pan she, Ho-fei, in series: Shao nien erh t'ung wen i t'u wu.
CONTENTS: 鲁安 蜜蜂的歌.—唐宗力 心愿.—操利群 纪念.—刘人云 "1"比"0"以后.—曹晓乔 苗儿青青.—徐承执 两根笛子.
I. Title.
PZ10.831.F29M5 1978 79-835178

Fang, Ming.
(Yu erh ch'ang shih hua ts'ung)
幼兒常識畫叢 / [繪圖者方明]. — [香港 : 中外出版社 ; 九龍 : 發行者時代圖書有限公司, 1977?]
5 v. in 1 : all col. ill. ; 22 cm.
Spine title.
CONTENTS: 我們住的地球.—世界真美麗.—世界奇景.—各國人民.—動物
I. Title.
PZ10.831.F32Y9 79-836678

CHILDREN'S LITERATURE, CHINESE (Continued)

(Ha-mai)
哈迈. — 香港：海鷗出版公司, [1977?]
[96] p. : ill. ; 19 cm.
$3.00

PZ10.841.H3 78–842303

Hai, Hsiao.
(Hung hung ti Yü-hua shih)
红红的雨花石 / 海笑著；[华三川插图]. — 第1版. — 上海：少年儿童出版社：新華書店上海发行所发行, 1978.
271 p. : ill. ; 19 cm.
Juvenile literature.
$0.48

I. Hua, San-chuan. II. Title.
PZ10.831.H276H8 79–840593

Ho, Chih, fl. 1973–
(Hsiao shan ying)
小山鷹 何芷 [著 广州] 广东人民出版社 [1973]
92 p. illus. 19 cm. $0.21
Juvenile literature.

I. Title.
PZ10.831.H6H74 73–896209

Ho, Chih, fl. 1973–
(Hsiao shan ying)
小山鹰 / 何芷. — 第2版. — [广州]：广东人民出版社, 1977.
137 p. : ill. ; 19 cm. — (少年文艺丛书)
Children's fiction.
Title in Pinyin on cover: Xiao Shanying.
Series romanized: Shao nien wen i ts'ung shu.
$0.29
I. Title. II. Title: Xiao Shanying.
PZ10.831.H6H74 1977 79–835630

Ho, Chih, fl. 1973–
(T'ieh chiang ti erh tzu)
铁匠的儿子 / 何芷；[插图王维宝]. — 第2版, 修订本. — [广州]：广东人民出版社, 1977.
191 p. : ill. ; 19 cm. — (少年文艺丛书)
Title in Pinyin on cover: Tiejiang de erzi.
Fiction.
Series romanized: Shao nien wen i ts'ung shu.
$0.38
I. Wang, Wei-pao. II. Title. III. Title: Tiejiang de erzi.
PZ10.831.H6T5 1977 79–835325

Ho, I.
(P'ao hsiao ti Shih-yu ho)
咆哮的石油河 / 贺宜. — 第1版. — 北京：中國少年兒童出版社, 1978.
274 p. : ill. ; 19 cm.
$0.53

1. Wang, Chin-hsi, d. 1970—Juvenile fiction. I. Title.
PZ10.831.H616P3 79–836052

Ho, Lin-k'en.
(Hsiao ku shou)
小鼓手 / 何林墾文；高山嵐圖. — [南投縣]：臺灣省政府教育廳；臺北：發行者臺灣書店, 民國57 [1968]
44 p. : ill. (some col.) ; 21 cm. — (中華兒童叢書)
Text accompanied with Kuo yü chu yin.
Series romanized: Chung-hua erh t'ung ts'ung shu.

I. Kao, Shan-lan. II. Title.
PZ10.831.H62H75 78–841376

Ho, Tzu.
(Erh t'ung hsiao shuo chi)
兒童小説集 / 何紫著. — 香港：山邊公司出版部, 1977–
v. ; 19 cm.
Cover title.
Juvenile literature.
CONTENTS: 1. 培培和小鴿子.—2. 我想飛. —4. 老師不要走.
$3.00 per vol.
I. Title.
PZ10.831.H64E74 78–843447

Ho, Tzu.
(Lao shih pu yao tsou)
老師不要走 / 何紫著. — 香港：山邊公司出版部, 1977.
3, 99 p. ; 19 cm. — (His 兒童小説集 ; 4)
Juvenile literature.
Series romanized: Erh t'ung hsiao shuo chi.
$3.00
I. Title.
PZ10.831.H64E74 vol. 4 78–843448

Ho, Tzu.
(P'ei-p'ei ho hsiao ko tzu)
培培和小鴿子 / 何紫著. — 香港：山邊公司出版部, 1977.
3, 84 p. ; 19 cm. — (His 兒童小説集 ; 1)
Juvenile literature.
Series romanized: Erh t'ung hsiao shuo chi.
$3.00
I. Title.
PZ10.831.H64E74 vol. 1 78–843446

Ho, Tzu.
(T'un se ti yu i)
褪色的友誼 / 何紫著. — 香港：山邊公司, 1977.
3, 91 p. ; 19 cm. — (His 兒童小説集 ; 3)
Juvenile literature.
Series romanized: Erh t'ung hsiao shuo chi.
$3.00
I. Title.
PZ10.831.H64E74 vol. 3 79–835458

Ho, Tzu.
(Wo hsiang fei)
我想飛 / 何紫著. — 香港：山邊公司出版部, 1977.
3, 87 p. ; 19 cm. — (His 兒童小説集 ; 2)
Juvenile literature.
Series romanized: Erh t'ung hsiao shuo chi.
$3.00
I. Title.
PZ10.831.H64E74 vol. 2 78–843449

(Hsiao ch'i ch'e)
小汽車. — 香港：大衆圖書公司, [1978?]
10 p. : col. ill. ; 18 x 19 cm. — (兒童故事彩色畫 ; 第2輯, [5])
Bound with Kuo wang ti hsin i. Hsiang-kang, 1978?
$1.80

I. Series: Erh t'ung ku shih ts'ai se hua ; ti 2 chi, 5.
PZ10.831.K88 79–840698

(Hsiao fei hsia)
小飛俠. — 香港：大衆圖書公司, [1978?]
10 p. : col. ill. ; 18 x 19 cm. — (兒童故事彩色畫 ; 第2輯, [6])
Bound with Kuo wang ti hsin i. Hsiang-kang, 1978?
$1.80

I. Series: Erh t'ung ku shih ts'ai se hua ; ti 2 chi, 6.
PZ10.831.K88 79–840699

(Hsiao jen kuo ho ta jen kuo)
小人國和大人國. — 香港：藝美圖書公司, 1978.
61 p. : ill. ; 19 cm. — (世界名著兒童故事精選)
Bound with Pai-hsüeh kung chu. Hongkong. 1978.
$1.80

I. Series: Shih chieh ming chu erh t'ung ku shih ching hsüan.
PZ10.831.P32 79–837267

(Hsiao lü Pan-pi)
小鹿班比. — 香港：大衆圖書公司, [1978?]
10 p. : col. ill. ; 18 x 19 cm. — (兒童故事彩色畫 ; 第2輯, [8])
Bound with Kuo wang ti hsin i. Hsiang-kang, 1978?
$1.80

I. Series: Erh t'ung ku shih ts'ai se hua ; ti 2 chi, 8.
PZ10.831.K88 79–840697

(Hsiao ma i li hsien chi)
小螞蟻歷險記. — 香港：海鷗出版公司, [1978?]
6 v. in 1 : col. ill. ; 18 x 19 cm.

PZ10.831.H6639 78–842685

Hsieh, Chu-sheng.
(Chu Te t'ung chih ti ku shih ho ch'uan shuo)
朱德同志的故事和传说 / 谢储生；[顾炳鑫插图]. — 第1版. — 上海：上海人民出版社, 1977.
2, 116 p., [2] leaves of plates : ill. ; 19 cm. — (少年儿童文艺读物)
Series romanized: Shao nien erh t'ung wen i tu wu.
$0.24

1. Chu, Te, 1886–1976—Juvenile fiction. I. Title.
PZ10.831.H6715C47 78–842531

Hsieh, Ping-ying, 1906–
(Lin Lin)
林琳 / 謝冰瑩文；楊震夷圖. — [南投縣]：臺灣省政府教育廳；臺北：發行者臺灣書店, 民國55 [1966]
88 p. : ill. (some col.) ; 21 cm. — (中華兒童叢書)
Juvenile literature.
Text accompanied with Kuo yü chu yin.
Series romanized: Chung-hua erh t'ung ts'ung shu.
I. Yang, Chen-i. II. Title.
PZ10.831.H6718L5 78–841380

Hsieh, Wan-ying, 1902–
(Hsiao chü teng)
小桔灯 / 冰心[謝婉瑩]. — 第1版. — 北京：人民文学出版社：新華书店北京发行所, 1978.
3, 259 p. ; 19 cm.
Literature.
$0.49

I. Title.
PZ10.831.H6725H76 79–838041

Hu, Ch'i, 1918–
(Lü se ti yüan fang)
绿色的远方 / 胡奇著. — [再次修订]. — 北京：中國少年兒童出版社：新华書店北京发行所发行, 1978.
2, 204 p., [4] leaves of plates : ill. (some col.) ; 19 cm.
Juvenile literature.
$0.51

I. Title.
PZ10.831.H75 1978 79–836298

CHILDREN'S LITERATURE, CHINESE (Continued)

Hu, Ch'i, 1918–
(Shen huo)
神火　胡奇著　温勇雄插圖　北京　中國少年兒童出版社 1957.
64 p. illus. 19 cm.

I. Title.
PZ10.831.H752　　C 58-5296

Hu, Ch'i, 1918–
(Wu ts'ai lu)
五彩路　胡奇著　林凡插圖　北京　中國少年兒童出版社　1957.
127 p. illus. 19 cm.

I. Title.
PZ10.831.H754　　C 58-5297

(Hu k'ou chen ch'a chi)
虎口偵察記. — 第1版. — 上海：上海人民出版社, 1977.
167 p. : ill. ; 19 cm. — (少年儿童文艺读物)
Series romanized: Shao nien erh t'ung wen i tu wu.
CONTENTS: 贾晓晨　虎口侦察记.—黎汝清　竹斗笠的秘密.—文然·尤华　找红军.—林阿绵, 高鑫　石柱子的绝招.—李学中　鱼鹰姑娘.—曲山山妹.—王希平　老槐树的秘密.—徐汝成　湖边少年.—娇健　铁虎.—苏淮军, 石宏　火红的战旗.—张方文　观察哨.—今新　小猎人的礼物.—潘新宁　桂桂.—牛云祥　铁道旁的枪声
$0.30
PZ10.831.H773　　79-840527

Hua, Hsia-ling.
(I mao ch'ien)
一毛錢 / 華霞菱文；張悦珍圖. — ［南投縣］：臺灣省政府教育廳；臺北：發行者臺灣書店, 民國56 ［1967］
36 p. : col. ill. ; 21 cm. — (中華兒童叢書)
Text accompanied by Kuo yü chu yin.
Series romanized: Chung-hua erh t'ung ts'ung shu.

I. Chang, Yüeh-chen. II. Title.
PZ10.831.H777I2　　78-841373

Hua, Hsia-ling.
(Lao kung kung te hua yüan)
老公公的花園 / 華霞菱文；陳壽美圖. — ［南投縣］：臺灣省政府教育廳；臺北：發行者臺灣書店, 民國54 ［1965］
35 p. : col. ill. ; 21 cm. — (中華兒童叢書)
Text accompanied with Kuo yü chu yin.
Series romanized: Chung-hua erh t'ung ts'ung shu.

I. Ch'en, Shou-mei. II. Title.
PZ10.831.H777L3　　78-840612

Hua, Hsia-ling.
(Wu ts'an jih chi)
午餐日記 / 華霞菱文；高山嵐圖. — ［南投縣］：臺灣省政府教育廳；臺北：發行者臺灣書店, 民國55 ［1966］
36 p. : ill. (some col.) ; 21 cm. — (中華兒童叢書)
Text accompanied with Kuo yü chu yin.
Series romanized: Chung-hua erh t'ung ts'ung shu.

I. Kao, Shan-lan. II. Title.
PZ10.831.H777W8　　78-841387

I, T'ao.
(Suan shu ho hsiang jih k'uei)
算術和向日葵：独幕兒童剧 / 以滔編剧. — ［香港?：s. n., 1978?］
12 p. ; 18 cm.
Cover title.
Children's one act play.
Reprint of the 1956 ed. published by Ch'ang-chiang wen i ch'u pan she, Hankow.

I. Title.
PZ10.87.I16S9 1978　　78-840854

(Ko hsüeh ho t'i ch'in)
格雪和提琴. — 香港：大衆圖書公司, ［1978?］
10 p. : col. ill. ; 18 x 19 cm. — (兒童故事彩色畫；第2輯, ［3］)
Bound with Kuo wang ti hsin i. Hsiang-kang, 1978?
$1.80

I. Series: Erh t'ung ku shih ts'ai se hua ; ti 2 chi, 3.
PZ10.831.K88　　79-840695

Kuan, Hua.
(Lao hu ho hei hsiung)
老虎和黑熊 / 管樺著；徐琛, 黃景插圖. — 第1版. — 北京：中國少年兒童出版社, 1956.
47 p. : ill. ; 19 cm.

I. Title.
PZ10.831.K796L3　　78-841752

(Kuo wang ti hsin i)
國王的新衣. — 香港：大衆圖書公司, ［1978?］
10 p. : col. ill. ; 18 x 19 cm. — (兒童故事彩色畫；第2輯, ［1］)
Bound with Meng yu hsien ching. Hsiang-kang, 1978? Ko hsüeh ho t'i ch'in. Hsiang-kang, 1978? Ch'ing liao. Hsiang-kang, 1978? Hsiao ch'i ch'e. Hsiang-kang, 1978? Hsiao fei hsia. Hsiang-kang, 1978? Hsiao fei hsiang. Hsiang-kang, 1978? Hsiao lü Pan-pi. Hsiang-kang, 1978?
$1.80

I. Series: Erh t'ung ku shih ts'ai se hua ; ti 2 chi, 1.
PZ10.831.K88　　79-840701

Li, Li-wen.
(A-hui ti ch'i yü)
阿灰的奇遇 / 李麗雯文；趙國宗圖. — ［南投縣］：臺灣省政府教育廳；臺北：發行者臺灣書店, 民國56 ［1967］
35 p. : col. ill. ; 21 cm. — (中華兒童叢書)
Text accompanied with Kuo yü chu yin.
Series romanized: Chung-hua erh t'ung ts'ung shu.

I. Chao, Kuo-tsung. II. Title.
PZ10.831.L476A2　　78-841383

Li, Li-ying.
(Liu lang ti hsiao hei mao)
流浪的小黑貓 / 黎麗瑛文；林蒼筤圖. — ［南投縣］：臺灣省政府教育廳；臺北：發行者臺灣書店, 民國56 ［1967］
36 p. : col. ill. ; 21 cm. — (中華兒童叢書)
Text accompanied with Kuo yü chu yin.
Series romanized: Chung-hua erh t'ung ts'ung shu.

I. Lin, Ts'ang-lang. II. Title.
PZ10.831.L477L5　　78-841377

Liang, Hsüeh-cheng.
(T'ai-pei ti huang hun)
台北的黄昏 / 梁学政著；［范一辛插图］. — 第1版. — 上海：少年儿童出版社：新華書店上海发行所发行, 1978.
91 p. : ill. ; 19 cm.
Juvenile literature,
CONTENTS: 在海岸上盼望.—台北的黄昏.—台湾街头.—拾破烂的孩子.—战斗.—阿爸一定会回来的.
$0.19
I. Fan, I-hsin. II. Title.
PZ10.831.L517T3　　79-837945

Lin, Chung-lung.
(Mei-yü ho hsiao kou)
美玉和小狗 / 林鐘隆文；張國雄圖. — ［南役縣］：臺灣省政府教育廳；臺北：發行者臺灣書店, 民國54 ［1965］
36 p. : ill. (some col.) ; 21 cm. — (中華兒童叢書)
Juvenile literature.
Text accompanied with Kuo yü chu yin.
Series romanized: Chung-hua erh t'ung ts'ung shu.
I. Chang, Kuo-hsiung. II. Title.
PZ10.831.L55M4　　78-840611

Lin, Hai-yin.
(Ts'ai chia lao wu)
蔡家老屋 / 林海音文；曾璧圖. — ［南投縣］：臺灣省政府教育廳；臺北：發行者臺灣書店, 民國55 ［1966］
36 p. : col. ill. ; 21 cm. — (中華兒童叢書)
Text accompanied with Kuo yü chu yin.
Series romanized: Chung-hua erh t'ung ts'ung shu.

I. Tseng, Pi. II. Title.
PZ10.831.L556T7　　78-840616

Lin, Liang.
(Hui shuo hua ti niao)
會説話的鳥 / 林良文；林友竹圖. — ［南投縣］：臺灣省政府教育廳；臺北：發行者臺灣書店, 民國57 ［1968］
32 p. : col. ill. ; 21 cm. — (中華兒童叢書)
Text accompanied with Kuo yü chu yin.
Series romanized: Chung-hua erh t'ung ts'ung shu.

I. Lin, Yu-chu. II. Title.
PZ10.831.L572H8　　78-840609

Lin, Ming-chung.
(I ko kuo chung sheng ti hsüeh lei ku shih)
一個國中生的血淚故事 / 著者林銘忠. — 臺北：東方出版社, 民國66 ［1977］
213 p. ; 20 cm.
Colophon title.
Text accompanied with Kuo yü chu yin.
NT$25.00

I. Title.
PZ10.831.L5723I2　　78-843247

Liu, Ch'i, fl. 1974–
(Liang chih hsiao k'ung ch'üeh)
两只小孔雀 / 陈炜, 彭华供稿；刘绮等改编；蒋铁峰, 贾国中, 万强麟绘制. — 昆明：云南人民出版社, 1974.
［38］ p. : chiefly col. ill. ; 22 cm.
Juvenile literature.
$0.36

I. Ch'en, Wei. II. P'eng, Hua. III. Chiang, T'ieh-feng. IV. Title.
PZ10.831.L583L5　　75-839717

Lo, Chün-ch'ing.
(Hsiao hsiung ti t'ou k'uei)
小熊的頭盔 / 編寫羅俊青；插圖陳棪文. — 香港：新雅兒童教育出版社, 1978.
23 p. : col. ill. ; 19 x 22 cm. — (科學發明的故事；第1輯, ［1］)
Bound with the author's Huo yao fa ming liao. Hsiang-kang, 1978. Tso ch'e lun. Hsiang-kang, 1978. P'iao liang ti po li. Hsiang-kang, 1978.
$3.50

I. Ch'en, Yen-wen. II. Title. III. Series: K'o hsüeh fa ming ti ku shih ; ti 1 chi, 1.
PZ10.831.L64H86　　79-837781

Lo, Chün-ch'ing.
(Huo yao fa ming liao)
火藥發明了 / 編寫羅俊青；插圖陳棪文. — 香港：新雅兒童教育出版社, 1978.
23 p. : col. ill. ; 19 x 22 cm. — (科學發明的故事；第1輯, ［2］)
Bound with the author's Hsiao hsiung ti t'ou k'uei. Hsiang-kang, 1978. Tso ch'e lun. Hsiang-kang, 1978. P'iao liang ti po li. Hsiang-kang, 1978.
$3.50
I. Ch'en, Yen-wen. II. Title. III. Series: K'o hsüeh fa ming ti ku shih ; ti 1 chi, 2.
PZ10.831.L64H86　　79-837778

Lo, Chün-ch'ing.
(P'iao liang ti po li)
漂亮的玻璃 / 編寫羅俊青；插圖陳棪文. — 香港：新雅兒童教育出版社, 1978.
23 p. : col. ill. ; 19 x 22 cm. — (科學發明的故事；第1輯, ［4］)
Bound with the author's Huo yao fa ming liao. Hsiang-kang, 1978. Hsiao hsiung ti t'ou k'uei. Hsiang-kang, 1978. Tso ch'e lun. Hsiang-kang, 1978.
$3.50
I. Ch'en, Yen-wen. II. Title. III. Series: K'o hsüeh fa ming ti ku shih ; ti 1 chi, 4.
PZ10.831.L64H86　　79-837780

CHILDREN'S LITERATURE, CHINESE (Continued)

Lo, Chün-ch'ing.
(Tso ch'e lun)
做車輪 / 編寫羅俊青；插圖陳棪文. — 香港：新雅兒童教育出版社, 1978.
23 p. : col. ill. ; 19 x 22 cm. — (科學發明的故事；第1輯, ₍3₎)
Bound with the author's Huo yao fa ming liao. Hsiang-kang, 1978. Hsiao hsiung ti t'ou k'uei. Hsiang-kang, 1978. P'iao liang ti po li. Hsiang-kang, 1978.
$3.50
I. Ch'en, Yen-wen. II. Title. III. Series: K'o hsüeh fa ming ti ku shih ; ti 1 chi, 3.
PZ10.831.L64H86 79-837779

Lo, Te-chen.
(Li Yung ta pao)
黎勇打豹 / 罗德祯. — 修订本. — ₍广州₎：广东人民出版社, 1977.
186 p. : ill. ; 19 cm.
Juvenile literature.
Title in Pinyin on cover: Liyong da bao.
$0.37
I. Title. II. Title: Liyong da bao.
PZ10.831.L68L5 1977 79-835644

Lou, Tzu-k'uang.
(Hsiang t'u shen hua)
鄉土神話 / 婁子匡文；曹俊彥圖. — ₍南投縣₎：臺灣省政府教育廳；臺北：發行者臺灣書店, 民國57 ₍1968₎
64 p. : ill. (some col.) ; 21 cm. — (中華兒童叢書)
Text accompanied with Kuo yü chu yin.
Series romanized: Chung-hua erh t'ung ts'ung shu.
I. Ts'ao, Chün-yen. II. Title.
PZ10.831.L7H75 78-841381

Lu, Chih-lo.
(Sung ya tzu li hsien chi)
松伢子历险记 / 鲁之洛. — ₍香港?：s. n., 1978?₎
194 p. : ill. ; 21 cm.
Reprint of the 1977 ed. published by Kuang-tung jen min ch'u pan she, in series: Shao nien wen i ts'ung shu.
I. Title.
PZ10.831.L73S86 1978 78-842530

Lu, Feng.
(Hsiao ch'ung pien ch'eng chen chu)
小蟲變成珍珠 / 魯風編著. — 香港：上海書局, 1976.
27 p. : ill. ; 19 cm.
Juvenile literature.
$1.30
I. Title.
PZ10.831.L74H7 78-840594

Lu, Feng.
(Lung wang ch'ing chiu)
龍王請酒 / 魯風編著. — 香港：上海書局, 1976.
25 p. : ill. ; 20 cm.
$1.20
I. Title.
PZ10.831.L74L8 78-841672

Lu, Feng.
(Lung wang ch'ing chiu)
龍王請酒 / 魯風編著. — 香港：上海書局, 1976.
25 p. : ill. ; 20 cm.
$1.20
I. Title.
PZ10.831.L74L8 78-841672

Lu, Feng.
(Ts'ung ming ti mu t'ung)
聰明的牧童 / 魯風編. — 香港：上海書局, 1975.
24 p. : ill. ; 19 cm.
$0.80
I. Title.
PZ10.831.L74T7 78-840600

Lu, Feng.
(Tzu yu ti mei kuei)
自由的玫瑰 / 魯風編. — 香港：上海書局, 1975.
25 p. : ill. ; 19 cm.
Juvenile literature.
$0.80
I. Title.
PZ10.831.L74T98 78-841713

Lu, K'o.
(Shang yeh pan ti tung wu)
上夜班的动物 = Shang ye ban de dong wu / ₍鲁克写；陈永镇画₎. — ₍第1版₎. — ₍上海：上海人民出版社, 1977₎
₍24₎ p. : col. ill. ; 15 cm.
Cover title.
$0.10
I. Ch'en, Yung-chen. II. Title. III. Title: Shang ye ban de dong wu.
PZ10.831.L747S5 78-840811

Lu, Ping.
(Ts'ung hsiao tuan lien shen t'i hao)
从小锻炼身体好 / 学前街幼儿园原作；鲁兵改写；吴僦声画. — 第1版. — 上海：上海人民出版社, 1976.
₍26₎ p. : col. ill. ; 19 cm.
Colophon title.
Cover title: Cong xiao duan lian shen ti hao.
"連環圖畫"
$0.12
I. Wu, Ching-lu. II. Hsüeh ch'ien chieh yu erh yüan. III. Title.
PZ10.831.L77T7 78-840607

Lu, Ssu, fl. 1975–
(Hsiao ho shui li kung)
小河水立功 / 魯思編. — 香港：上海書局, 1976.
25 p. : ill. ; 20 cm.
CONTENTS: 坐着滑梯旅行.—巨大的墙.—小河水立功
$1.20
I. Title.
PZ10.831.L825H7 78-841673

Lu, Yang-lieh.
(Wu tu pao t'ung)
雾都报童 / 陆扬烈, 冰夫. — 新1版. — 上海：少年儿童出版社, 1978.
216 p., ₍8₎ leaves of plates : ill. ; 19 cm.
Juvenile literature.
Title in Pinyin on cover: Wudu baotong.
$0.41
I. Ping-fu, pseud., joint author. II. Title. III. Title: Wudu baotong.
PZ10.831.L85W8 79-835512

(Lu-pin-hsün p'iao liu chi)
魯濱遜飄流記. — 香港：藝美圖書公司, 1978.
55 p. : ill. ; 19 cm. — (世界名著兒童故事精選)
Bound with Pai-hsüeh kung chu. Hongkong. 1978.
$1.80
I. Series: Shih chieh ming chu erh t'ung ku shih ching hsüan.
PZ10.831.P32 79-837263

(Lü yeh hsien tsung)
綠野仙蹤. — 香港：藝美圖書公司, 1978.
65 p. : ill. ; 19 cm. — (世界名著兒童故事精選)
Bound with Pai-hsüeh kung chu. Hongkong. 1978.
$2.00
I. Series: Shih chieh ming chu erh t'ung ku shih ching hsüan.
PZ10.831.P32 79-837265

(Mao chu hsi pao kuo ti wa)
毛主席抱过的娃：少年儿童朗诵诗选. — ₍香港?：s. n., 1978?₎
2, 123 p. ; 21 cm.
Reprint of the 1977 ed. published by Pei-ching jen min ch'u pan she, Peking.
1. Children's poetry, Chinese. 2. Mao, Tse-tung, 1893–1976, in fiction, drama, poetry, etc.—Juvenile poetry.
PZ10.843.M36 1978 79-837944

(Meng yu hsien ching)
夢遊仙境：兒童故事彩色畫. — 香港：大衆圖書公司, ₍1976?₎
10 p. : col. ill. ; 19 cm.
$1.80
I. Dodgson, Charles Lutwidge, 1832–1898. Alice in Wonderland.
PZ10.84.M4 78-842755

(Meng yu hsien ching)
夢遊仙境. — 香港：大衆圖書公司, ₍1978?₎
10 p. : col. ill. ; 18 x 19 cm. — (兒童故事彩色畫；第2輯, ₍2₎)
Bound with Kuo wang ti hsin i. Hsiang-kang, 1978?
$1.80
I. Series: Erh t'ung ku shih ts'ai se hua ; ti 2 chi, 2.
PZ10.831.K88 79-840702

(Min-hao-sheng ch'i yu chi)
敏豪生奇遊記. — 香港：藝美圖書公司, 1978.
66 p. : ill, ; 19 cm. — (世界名著兒童故事精選)
Bound with Pai-hsüeh kung chu. Hongkong. 1978.
$2.30
I. Series: Shih chieh ming chu erh t'ung ku shih ching hsüan.
PZ10.831.P32 79-837262

(Mu ou ch'i yü chi)
木偶奇遇記. — 香港：藝美圖書公司, 1978.
75 p. : ill. ; 19 cm. — (世界名著兒童故事精選)
Bound with Pai-hsüeh kung chu. Hongkong. 1978.
$2.50
I. Series: Shih chieh ming chu erh t'ung ku shih ching hsüan.
PZ10.831.P32 79-837261

Nakayama, Mayumi, 1948–
(Koronda kabasan)
ころんだかばさん / 仲山真弓·文；大貫明美·絵. — 東京：講談社, 1978.
31 p. : col. ill. ; 24 cm. Ja 78-27
¥690
I. Ōnuki, Akemi, 1958– II. Title.
PZ49.63.N35 78-814172

CHILDREN'S LITERATURE, CHINESE (Continued)

Ono, Chūkō, 1909–
(Namida no hakuchō)
涙の白鳥：少女小説 / [小野忠孝著]. — 東京：講談社, 昭和24 [1949]
189 p. ; 19 cm.

I. Title.
PZ49.31.O492 78-809898

Ōno, Michio, 1919–
(Machiya no ko)
町屋の子 / 大野道雄. — 東京：中央出版社, 昭和23 [1948]
2, 179 p. : ill. ; 18 cm.

I. Title.
PZ49.31.O499 78-809894

(Pai-hsüeh kung chu)
白雪公主. — 香港：藝美圖書公司, 1978.
48 p. : ill. ; 19 cm. — (世界名著兒童故事精選)
Bound with Mu ou ch'i yü chi. Hongkong. 1978. Min-hao-sheng ch'i yu chi. Hongkong. 1978. Lu-pin-hsün p'iao liu chi. Hongkong. 1978. Shen teng. Hongkong. 1978. Hsiao jen kuo ho ta jen kuo. Hongkong. 1978. Lü yeh hsien tsung. Hongkong. 1978.
$1.80

I. Series: Shih chieh ming chu erh t'ung ku shih ching hsüan.
PZ10.831.P32 79-837228

P'an, Hsi-chen.
(Mai niu chi)
賣牛記 / 潘琦君[希真]文；林顯宗圖. — [南投縣]：臺灣省政府教育廳；臺北：發行者臺灣書店, 民國55 [1966]
75 p. : ill. (some col.) ; 21 cm. — (中華兒童叢書)
Juvenile literature.
Text accompanied with Kuo yü chu yin.
Series romanized: Chung-hua erh t'ung ts'ung shu.
I. Lin, Hsien-tsung. II. Title.
PZ10.831.P33M3 78-841385

P'an, I-ta.
(Ch'en Yü-ch'eng ti ku shih)
陈玉成的故事 / 潘益大, 张广智. — [香港?：s. n., 1978?]
2, 2, 111 p. : ill. ; 22 cm.
Title in Pinyin on cover: Chenyucheng de gushi.
Reprint of the 1977 ed. published by Shang-hai jen min ch'u pan she, Shanghai, in series: Shao nien erh t'ung li shih tu wu.
1. Ch'en, Yü-ch'eng, 1837–1862—Juvenile fiction. I. Chang, Kuang-chih, joint author. II. Title.
PZ10.831.P34C47 1978 78-842529

P'an, Sui.
(Ai p'iao liang ti hu tieh)
愛漂亮的蝴蝶 / 潘遂文；陳壽美圖. — [南投縣]：臺灣省政府教育廳；臺北：發行者臺灣書店, 民國57 [1968].
32 p. : col. ill. ; 21 cm. — (中華兒童叢書)
Text accompanied with Kuo yü chu yin.
Series romanized: Chung-hua erh t'ung ts'ung shu.

I. Ch'en, Shou-mei. II. Title.
PZ10.831.P35A36 78-841707

(Pao t'ung)
報童：六场儿童剧 / 邵冲飞 ... [et al.]. — 第1版. — [鄭州]：河南人民出版社；[鄭州?]：河南省新華書店发行, 1978.
92 p. ; 19 cm.
$0.20

I. Shao, Ch'ung-fei.
PZ10.87.P36 79-837857

P'eng, Hsin-ken.
(I k'o hsiao shui ti)
一颗小水滴 = Yike xiao shui di：科学童话 / 彭鑫根著. — 第1版. — 上海：上海人民出版社, 1977.
63 p. : ill. ; 19 cm.
Juvenile literature.
Text in characters.
$0.14
I. Title. II. Title: Yike xiao shui di.
PZ10.831.P38I2 78-840893

Pi, P'u, 1923–
(Nan wang ti chia ch'i)
難忘的假期 / 畢璞文；夏祖明圖. — [南投縣]：臺灣省政府教育廳；臺北：發行者臺灣書店, 民國56 [1967]
76 p. : ill. (some col.) ; 21 cm. — (中華兒童叢書)
Text accompanied with Kuo yü chu yin.
Series romanized: Chung-hua erh t'ung ts'ung shu.

I. Hsia, Tsu-ming. II. Title.
PZ10.831.P5N3 78-840610

P'u, Fei.
(Lieh-ts'un ti hai tzu)
猎村的孩子 / 普飞写；詹同画. — 第1版. — 上海：上海人民出版社, 1975.
[28] p. : col. ill. ; 16 cm.
Title in romanization on cover: Liedun de haizi.
$0.12

I. Chan, T'ung. II. Title.
PZ10.831.P8L5 78-842304

Shang-hai shih I-miao ch'ü chung hsin hsiao hsüeh.
(Wo men ti hsiao kung ch'ang)
我們的小工場 / 上海市邑庙区中心小学編寫；[張之凡繪圖]. — 第1版. — 上海：少年兒童出版社, 1956.
20 p. : ill. ; 19 cm.

I. Chang, Chih-fan. II. Title.
PZ10.831.S458W6 1956 78-842306

Shang-hai ti liu tien piao ch'ang. Kung jen li lun tsu.
(Wang Hsiao-po ch'i i ti ku shih)
王小波起义的故事 / 上海第六电表厂工人理论组编；[颜梅华, 颜志强插图]. — [香港?：s. n., 1978]
2, 155 p., [1] leaf of plates : ill. ; 21 cm.
Reprint of the 1978 ed. published by Shao nien erh t'ung ch'u pan she, Shanghai, in series: Shao nien erh t'ung li shih tu wu.
1. Wang, Hsiao-po, d. 994—Juvenile fiction. I. Yen, Mei-hua. II. Yen, Chih-ch'iang. III. Title.
PZ10.831.S4615W3 1978 78-842873

Shao, Hsien, 1934–
(Tsai yang kuang hsia)
在陽光下 / 邵僩文；張英超圖. — [南投縣]：臺灣省政府教育廳；臺北：發行者臺灣書店, 民國56 [1967]
72 p. : ill. (some col.) ; 21 cm. — (中華兒童叢書)
Text accompanied with Kuo yü chu yin.
Series romanized: Chung-hua erh t'ung ts'ung shu.

I. Chang, Ying-ch'ao. II. Title.
PZ10.831.S463T74 78-841382

Shao nien erh t'ung ch'u pan she, Shanghai.
(Na-cha)
哪吒 / 少年兒童出版社節选；[董天野繪圖]. — 第1版. — 上海：該社, 1957.
30 p. : ill. ; 19 cm.
"哪吒是从...封神演义中節选出来的...只有个别深奥的字句作了修改，为了使故事能够很好地衔接，个别情節上作了些删節"

I. Feng shen yen i. II. Title.
PZ10.831.S4639N3 1957 78-841698

(Shen teng)
神燈；阿里巴巴與四十大盜. — 香港：藝美圖書公司, 1978.
49 p. : ill. ; 19 cm. — (世界名著兒童故事精選)
Bound with Pai-hsüeh kung chu. Hongkong. 1978.
$1.80

I. Title: A-li-pa-pa yü ssu shih ta tao. II. Series: Shih chieh ming chu erh t'ung ku shih ching hsüan.
PZ10.831.P32 79-837264

Shen, Ting-i.
(Chi hsin)
寄信 / 申丁伊編；劉兆源繪. — 香港：新雅兒童教育出版社, 1979.
16 p. : col. ill. ; 27 cm. — (社會的工作；第1輯, [2])
Colophon title.
Bound with the author's Chiu huo. Hsiang-kang, 1979.
$3.50
I. Liu, Chao-yüan. II. Title. III. Series: She hui ti kung tso ; ti 1 chi, 2.
PZ10.831.S473C5 79-841131

Shen, Ting-i.
(Chiu huo)
救火 / 申丁伊編；劉兆源繪. — 香港：新雅兒童教育出版社, 1979.
16 p. : col. ill. ; 27 cm. — (社會的工作；第1輯, [1])
Colophon title.
Bound with the author's Chi hsin. Hsiang-kang, 1979. Tso fei chi. Hsiang-kang, 1979. I yüan li. Hsiang-kang, 1979.
$3.50
I. Liu, Chao-yüan. II. Title. III. Series: She hui ti kung tso ; ti 1 chi, 1.
PZ10.831.S473C5 79-841132

Shen, Ting-i.
(Chung tzu ti ku shih)
種子的故事 / 申丁伊編；劉兆源繪. — 香港：新雅兒童教育出版社, 1979.
16 p. : col. ill. ; 27 cm. — (社會的工作；第2輯, [2])
Colophon title.
Bound with the author's Tso mien pao. Hsiang-kang, 1979. Mien hua chih ch'eng i fu. Hsiang-kang, 1979. Mu ts'ai ti li yung. Hsiang-kang, 1979.
$3.50
I. Liu, Chao-yüan. II. Title. III. Series: She hui ti kung tso ; ti 2 chi, 2.
PZ10.831.S473T7 79-841125

Shen, Ting-i.
(I yüan li)
醫院裏 / 申丁伊編；劉兆源繪. — 香港：新雅兒童教育出版社, 1979.
16 p. : col. ill. ; 27 cm. — (社會的工作；第1輯, [4])
Colophon title.
Bound with the author's Chiu huo. Hsiang-kang, 1979.
$3.50
I. Liu, Chao-yüan. II. Title. III. Series: She hui ti kung tso ; ti 1 chi, 4.
PZ10.831.S473C5 79-841128

Shen, Ting-i.
(Mien hua chih ch'eng i fu)
棉花製成衣服 / 申丁伊編；劉兆源繪. — 香港：新雅兒童教育出版社, 1979.
16 p. : col. ill. ; 27 cm. — (社會的工作；第2輯, [3])
Colophon title.
Bound with the author's Tso mien pao. Hsiang-kang, 1979. Chung tzu ti ku shih. Hsiang-kang, 1979. Mu ts'ai ti li yung. Hsiang-kang, 1979.
$3.50
I. Liu, Chao-yüan. II. Title. III. Series: She hui ti kung tso, ti 2 chi, 3.
PZ10.831.S473T7 79-841124

Shen, Ting-i.
(Mu ts'ai ti li yung)
木材的利用 / 申丁伊編；劉兆源繪. — 香港：新雅兒童教育出版社, 1979.
16 p. : col. ill. ; 27 cm. — (社會的工作；第2輯, [4])
Colophon title.
Bound with the author's Tso mien pao. Hsiang-kang, 1979. Chung tzu ti ku shih. Hsiang-kang, 1979. Mien hua chih ch'eng i fu. Hsiang-kang, 1979.
$3.50
I. Liu, Chao-yüan. II. Title. III. Series: She hui ti kung tso ; ti 2 chi, 4.
PZ10.831.S473T7 79-841127

CHILDREN'S LITERATURE, CHINESE (Continued)

Shen, Ting-i.
(Tso fei chi)
坐飛機 / 申丁伊；劉兆源繪. — 香港：新雅兒童教育出版社，1979.
16 p. : col. ill. ; 27 cm. — (社會的工作；第1輯，[3])
Colophon title.
Bound with the author's Chiu huo. Hsiang-kang, 1979.
$3.50
I. Liu, Chao-yüan. II. Title. III. Series: She hui ti kung tso, ti 1 chi, 3.
PZ10.831.S473C5 79–841130

Shen, Ting-i.
(Tso mien pao)
做麵包 / 申丁伊編；劉兆源繪. — 香港：新雅兒童教育出版社，1979.
16 p. : col. ill. ; 27 cm. — (社會的工作；第2輯，[1])
Colophon title.
Bound with the author's Chung tzu ti ku shih. Hsiang-kang, 1979. Mien hua chih ch'eng i fu. Hsiang-kang, 1979. Mu ts'ai ti li yung. Hsiang-kang, 1979.
$3.50
I. Liu, Chao-yüan. II. Title. III. Series: She hui ti kung tso ; ti 2 chi, 1.
PZ10.831.S473T7 79–841126

(Shuo ta hua ti hsiao hui t'u)
説大話的小灰兔. — 香港：海鷗出版公司，1975.
25 p. : col. ill. ; 18 cm.
$1.80
PZ10.842.S5 78–842754

Su, Shang-yao.
(T'i teng ti nü hai)
提燈的女孫 / 蘇尚耀文；廖未林圖. — [南投縣]：臺灣省政府教育廳；臺北：發行者臺灣書店，民國55 [1966]
36 p. : col. ill. ; 21 cm. — (中華兒童叢書)
Text accompanied with Kuo yü chu yin.
Series romanized: Chung-hua erh t'ung ts'ung shu.
I. Liao, Wei-lin. II. Title.
PZ10.831.S745T5 78–842305

Su, Yün-ch'ing.
(Yu ch'ing)
友情 / 蘇雲青文；馮鍾睿圖. — [南投縣]：臺灣省政府教育廳；臺北：發行者臺灣書店，民國56 [1967]
72 p. : ill. (some col.) ; 21 cm. — (中華兒童叢書)
Text accompanied with Kuo yü chu yin.
Series romanized: Chung-hua erh t'ung ts'ung shu.
I. Feng, Chung-jui. II. Title.
PZ10.831.S75Y8 78–841374

Sun, Ken-hsi.
(Ho t'ao ti ku shih)
核桃的故事 / 孙根喜著；[插图朱延龄]. — 第1版. — 上海：上海人民出版社，1977.
81 p. : ill. ; 19 cm. — (红小兵故事丛书) (少年儿童文艺读物)
Series 1 romanized: Hung hsiao ping ku shih ts'ung shu. Series 2 romanized: Shao nien erh t'ung wen i tu wu.
$0.18
I. Chu, Yen-ling. II. Title.
PZ10.831.S8H6 78–843254

Takakura, Teru, 1891–
(Emma Daiō)
エンマ大王 / [著者 タカクラ・テル]. — 東京：文化評論社，1948.
9, 154 p. : ill. ; 19 cm. — (タカクラ・テルえんげき集)
Juvenile literature.
I. Title. II. Series: Takakura, Teru, 1891– Takakura Teru engeki shū.
PL839.A525E47 78–816035

Ts'ung-ts'ung.
(Hsiao Tse-tang ta ch'ai lang)
小泽当打豺狼 / 聪聪写；施大畏画. — 上海：上海人民出版社，1975.
[23] p. : col. ill. ; 13 x 19 cm.
Colophon title.
"連環圖畫"
$0.13
I. Shih, Ta-wei. II. Title.
PZ10.843.T75H75 78–842752

Tsutsui, Keisuke, 1918–
(Onēsan to issho)
おねえさんといっしょ / 筒井敬介. — 東京：角川書店，1978.
297 p. ; 15 cm. — (読んで聞かせるお話；2) (角川文庫；4072)
Ja 78–20296
Series 2 romanized: Kadokawa bunko.
¥300
I. Title. II. Series: Yonde kikaseru ohanashi ; 2.
PZ49.31.T797 78–814887

Wei, Chin-chih.
(Chung-kuo ku tai yü yen)
中國古代寓言 / 魏金枝编写；程十发绘图. — 修订第3版. — 上海：少年儿童出版社：新華書店上海发行所发行，1978.
9, 168 p. : ill. ; 18 cm.
$0.35
1. Fables, Chinese. I. Ch'eng, Shih-fa. II. Title.
PZ10.842.W4C5 1978 79–841625

(Wo ti t'u hua ku shih shu)
我的圖畫故事書. — 香港：上海書局，1976.
2 v. : col. ill. ; 22 cm.
Title from spine.
10 separate works bound in 2.
Juvenile literature.
CONTENTS: 1. 小貓釣魚 小雞找圍 小羊不哭了 小賴兒失蹤了 布穀鳥叫錯了—2. 蝴蝶和小鏡子 謝謝小花貓 貓頭鷹打「壞人」 黃汽球 「慈善」的白螞蟻
$30.00
PZ10.831.W65 78–842294

Wu, Lü.
(She tao ti mi mi)
蛇岛的秘密 / 伍律著. — 第3版. — 北京：中國少年兒童出版社：新华书店北京发行所发行，1978.
99 p. : ill. ; 19 cm.
Title in Pinyin on cover: She dao de mi mi.
Juvenile literature.
$0.21
I. Title. II. Title: She dao de mi mi.
PZ10.831.W79S5 1978 79–840236

Wu, Yeh, fl. 1967–
(Chia chia chiu)
家家酒 / 吳葉文；蘇新田圖. — [南投縣]：臺灣省政府教育廳；臺北：發行者臺灣書店，民國56 [1967]
36 p. : col. ill. ; 21 cm. — (中華兒童叢書)
Juvenile literature.
Text accompanied with Kuo yü chu yin.
Series romanized: Chung-hua erh t'ung ts'ung shu.
I. Su, Hsin-t'ien. II. Title.
PZ10.831.W86C5 78–841384

(Wu wan nien i ch'ien ti k'o jen)
五万年以前的客人 / [插图周宪彻]. — 第1版. — 北京：人民文学出版社：新华书店北京发行所发行，1978.
176 p. : ill. ; 19 cm.
Title in Pinyin on p. [4] of cover: Wu wan nian yi qian de ke ren.
CONTENTS: 郑文光 黑宝石.—干止 到人造月亮去.—郑文光 飞上天去的小猴子.—干止 失踪的哥哥.—郭以实 孙悟空大闹原子世界.—童恩正 五万年以前的客人.—萧建亨 布克的奇遇
$0.36
I. Chou, Hsien-ch'e. II. T'ung, En-cheng. Wu wan nien i ch'ien ti k'o jen. 1978.
PZ10.831.W88 79–840101

Yang, Shang-k'uei.
(Ch'en I tsai Kan nan)
陈毅在赣南 / 杨尚奎. — 北京：中國少年兒童出版社：新华书店北京发行所发行，1978.
2, 94 p., [3] leaves of plates : ill. ; 19 cm.
$0.21
1. Ch'en, I, 1901–1972—Juvenile fiction. I. Title.
PZ10.831.Y325C4 79–840003

Yang, Tsung-chen, 1919–
(Wang en fu i ti lang)
忘恩負義的狼 / 孟瑤[楊宗珍]文；林雨樓圖. — [南投縣]：臺灣省政府教育廳；臺北：發行者臺灣書店，民國57 [1968]
36 p. : col. ill. ; 21 cm. — (中華兒童叢書)
Text accompanied with Kuo yü chu yin.
Series romanized: Chung-hua erh t'ung ts'ung shu.
I. Lin, Yü-lou. II. Title.
PZ10.831.Y33W3 78–841388

Yeh, Hsin, fl. 1977–
(Shen yeh ma t'i sheng)
深夜马蹄声 / 叶辛. — 第1版. — 上海：上海人民出版社，1977.
2, 252 p. : ill. ; 19 cm. — (少年儿童文艺读物)
Juvenile literature.
Series romanized: Shao nien erh t'ung wen i tu wu.
$0.47
I. Title.
PZ10.831.Y366S5 78–842709

Yen, I-yen.
(Hsiao ma kuan ho Ta-p'i-hsüeh shu shu)
小马倌和"大皮靴"叔叔 / 颜一烟著. — 第1版. — 北京：中国少年兒童出版社：新华书店北京发行所发行，1959, 1978 printing.
155 p. : ill. ; 19 cm.
Title in Pinyin on cover: Xiaomaguan he "Dapixue" shushu.
Juvenile literature.
$0.30
I. Title. II. Title: Xiaomaguan he "Dapixue" shushu.
PZ10.831.Y385H7 79–837618

Yen, Yu-mei.
(Po li hai)
玻璃海 / 嚴友梅文；張悦珍圖. — [南投縣]：臺灣省政府教育廳；臺北：發行者臺灣書店，民國57 [1968]
76 p. : ill. (some col.) ; 21 cm. — (中華兒童叢書)
Text accompanied with Kuo yü chu yin.
Series romanized: Chung-hua erh t'ung ts'ung shu.
I. Chang, Yüeh-chen. II. Title.
PZ10.831.Y44P6 78–841375

Yen, Yu-mei.
(Wan chü ch'uan)
玩具船 / 嚴友梅文；黃昌惠圖. — [南投縣]：臺灣省政府教育廳；臺北：發行者臺灣書店，民國57 [1968]
36 p. : col. ill. ; 21 cm. — (中華兒童叢書)
Text accompanied with Kuo yü chu yin.
Series romanized: Chung-hua erh t'ung ts'ung shu.
I. Huang, Ch'ang-hui. II. Title.
PZ10.831.Y44W3 78–841390

Yen, Yu-mei.
(Wan chü ti chia ch'i)
玩具的假期 / 嚴友梅文；曹俊彥圖. — [南投縣]：臺灣省政府教育廳；臺北：發行者臺灣書店，民國57 [1968]
36 p. : ill. (some col.) ; 21 cm. — (中華兒童叢書)
Juvenile literature.
Text accompanied with Kuo yü chu yin.
Series romanized: Chung-hua erh t'ung ts'ung shu.
I. Ts'ao, Chün-yen. II. Title.
PZ10.831.Y44W33 78–841379

CHILDREN'S LITERATURE, CHINESE (Continued)

Yoshiya, Nobuko, 1896–1973.
(Ano michi kono michi)
あの道この道：少女小説 / 吉屋信子著. — 東京：東和社, 昭和23 ⌊1948⌋
2 v. ; 19 cm. — (吉屋信子少女小説選集)
Fiction.
Series romanized: Yoshiya Nobuko shōjo shōsetsu senshū.

I. Title.
PZ49.31.Y64 78–809868

Yü, Li-ch'ing, 1945–
(Erh ko pai shou)
兒歌百首 / 喻麗清編. — 台北：爾雅出版社, 民國67 ⌊1978⌋
151 p. : ill. ; 20 cm. — (爾雅叢書 : 44)
Series romanized: Erh ya ts'ung shu.

I. Title.
PZ10.843.Y75E7 79–836413

Yü, Shen-ssu.
(K'uai le chung ch'iu)
快樂中秋 / 于愼思文 ; 林友竹圖. — ⌊南投縣⌋：臺灣省政府教育廳 ; 臺北：發行者臺灣書店, 民國57 ⌊1968⌋
36 p. : col. ill. ; 21 cm. — (中華兒童叢書)
Text accompanied with Kuo yü chu yin.
Series romanized: Chung-hua erh t'ung ts'ung shu.

I. Lin, Yu-chu. II. Title.
PZ10.831.Y79K8 78–841712

Yü, T'ien-tz'u.
(Hui yao wei pa ti lang)
会摇尾巴的狼 / 虞天慈改编 ; 姜金城配诗 ; 陈永镇绘. — 上海：上海人民出版社, 1975.
⌊27⌋ p. : col. ill. ; 16 cm.
$0.12

I. Chiang, Chin-ch'eng. II. Ch'en, Yung-chen. III. Title.
PZ10.831.Y83H8 76–837463

Yüan, Ching.
(Hsiao hei ma ti ku shih)
小黑马的故事 / 袁静著. — 第1版. — ⌊北京⌋ 中国少年兒童出版社：新华书店北京发行所发行, 1958, 1978 printing.
2, 127 p., ⌊4⌋ leaves of plates : ill. ; 19 cm.
Fiction.

I. Title.
PZ10.831.Y835H7 1958 79–83739

—HISTORY AND CRITICISM

Hsü, I-tsung.
(Wo kuo erh t'ung wen hsüeh ti yen chin yü chan wang)
我國兒童文學的演進與展望 / 許義宗著. — 臺北：許義宗, 民國65 ⌊1976⌋
2, 38 p. : ill. ; 21 cm. — (兒童文學研究叢書)
Series romanized: Erh t'ung wen hsüeh yen chiu ts'ung shu.
Bibliography: p. 36–37.
NT$10.00

1. Children's literature, Chinese—History and criticism. I. Title.
PN1009.C5H78 78–841266

Price, R F
What Chinese children read about : serial picture books & foreigners / ⌊by⌋ R. F. Price. — Bundoora, Vic. : Centre for Comparative and International Studies in Education, School of Education, La Trobe University, 1976.
52 p. : ill. ; 30 cm. — (Occasional paper - Centre for Comparative and International Studies in Education) Aus
English and Chinese.
Includes bibliographical references.
ISBN 0-85816-089-7
1. Children's literature, Chinese—History and criticism. 2. Comic books, strips, etc.—China—History and criticism. 3. Aliens in literature. I. Title. II. Series: Centre for Comparative and International Studies in Education. Occasional paper - Centre for Comparative and International Studies in Education.
PN1009.C5P7 895.1'3'509352 79-304146
79 MARC

CHILDREN'S LITERATURE, COLOMBIAN

Fornaguera de Roda, María.
Fortacha y otros cuentos / María Fornaguera ; ⌊ilus. de Juan Antonio Roda⌋. — ⌊Bogotá⌋ : Instituto Colombiano de Cultura, c1978.
63 p., ⌊13⌋ leaves of plates : ill. ; 28 cm. — (Colección Literatura infantil ; 1)
CONTENTS: Fortacha.—El canto de miría.—Cuento de diciembre.—La veloz.—El mal viaje de Candela.—Próspero y Eugenia.
I. Roda, Juan Antonio. II. Title. III. Series.
PZ73.F587 863 79-104213
79 MARC

CHILDREN'S LITERATURE, CROATIAN

—HISTORY AND CRITICISM

Turjačanin, Zorica.
Ključevi zlatnih vrata / Zorica Turjačanin. — Banjaluka : Glas, 1978.
99 p. ; 22 cm. Yu***
Includes bibliographical references.
1. Children's literature, Serbian—History and criticism. 2. Children's literature, Croatian—History and criticism. I. Title.
PN1009.S4T8 79-357685
79 MARC

CHILDREN'S LITERATURE, CUBAN

Devetach, Laura, 1936-
Monigote en la arena / ⌊Laura Devetach⌋ ; ilus., Muñoz Bachs. — ⌊La Habana : Casa de las Américas, 1975⌋
79 p. : col. ill. ; 29 cm. — (Obras para niños y jóvenes)
CONTENTS: Monigote en la arena.—Un elefante ocupa mucho espacio.—Historia de Ratita.—Los picaflores de cola roja.—El garbanzo peligroso.—Cuento que cuento ...
I. Title. II. Series.
PZ73.D46 79-350229
79 MARC

Reyes, Carlos José.
Globito manual / Carlos José Reyes ; ilus., Armando Millares. — La Habana : Casa de las Américas, ⌊1977⌋
⌊30⌋ p. : col. ill. ; 29 cm. — (Obras para niños y jóvenes)
Issued with the author's El hombre que escondió el sol y la luna. La Habana, 1977.
I. Millares, Armando, 1938- II. Title. III. Series.
PZ77.R38 79-104873
79 MARC

Reyes, Carlos José.
El hombre que escondió el sol y la luna / Carlos José Reyes ; ilus., Justo Luis. — La Habana : Casa de las Américas, ⌊1977⌋
⌊34⌋ p. : col. ill. ; 29 cm. — (Obras para niños y jóvenes)
Issued with the author's Globito manual. La Habana, 1977.
I. Justo Luis, 1945- II. Title. III. Series.
PZ77.R38 79-104875
79 MARC

CHILDREN'S LITERATURE, CZECH

Bezděková, Zdeňka.
Děvčátko Zdena a moudrý pes : ⌊pro čtenáře od 9 let⌋ / Zdeňka Bezděková ; ⌊il. Zdeněk Filip⌋. — 1. Vyd. — Praha : Albatros, 1977.
163, ⌊3⌋.p., ⌊8⌋ leaves of plates : ill. (some col.) ; 25 cm. Cz77
Kčs26.00
I. Title.
PZ70.C9B428 79-343840
*79 MARC

Borská, Ilona.
Ve čtvrtek budeme dospělí / Ilona Borská ; ⌊il. Josef Žemlička⌋. — 1. vyd. — Praha : Albatros, 1977.
125, ⌊2⌋ p. : ill. ; 22 cm. — (Edice Veronika) Cz78
Kčs17.00
I. Title.
PZ70.C9B628 78-375390
*78 MARC

Čtvrtek, Václav.
Franta a já : ⌊pro čtenáře od 7 let⌋ / Václav Čtvrtek ; ⌊il. Jitka Kolínská⌋. — 1. vyd. — Praha : Albatros, 1978.
85, ⌊3⌋ p., ⌊4⌋ leaves of plates : ill. (some col.) ; 19 x 21 cm. — (Edice Jiskřičky) Cz78
Kčs20.00
I. Title.
PZ70.C9C812 78-378400
*78 MARC

Dědková, Jarmila.
Místo pro Danu : dívčí román / Jarmila Dědková ; ⌊kresby Olga Čechová⌋. — 1. vyd. — Brno : Blok, 1978.
134 p. : ill. ; 21 cm. Cz78
Kčs16.00
I. Title.
PZ70.C9D39 79-343665
*79 MARC

Frýd, Norbert.
Květovaný kůň : básně, hry a rýmovačky : ⌊pro malé čtenáře⌋ / Norbert Fryd. — 1. vyd. — Praha : Albatros, 1975.
54, ⌊2⌋ p., ⌊5⌋ leaves of plates : col. ill. ; 21 cm. Cz75
Kčs15.00
I. Title.
PZ70.C9F78 79-359819
*79 MARC

Hurban, Jozef Miloslav, 1817-1888.
Olejkár / Jozef Miloslav Hurban ; ⌊dosl. nap. Július Noge ; pozn. a vysvetl. sprac. Viera Bosáková ; prebal a väzbu navrhol a graf. upr. Vladislav Rostoka⌋. — 3. vyd. — Bratislava : Tatran, 1977.
224 p., ⌊4⌋ leaves of plates : ill. ; 18 cm. — (Čítanie študujúcej mládeže ; zv. 96) CzS77
CONTENTS: Olejkár.—Od Silvestra do Troch kráľov.
Kčs14.00
I. Hurban, Jozef Miloslav, 1817-1888. Od Silvestra do Troch kráľov. 1977. II. Title.
PZ70.S46H8 1977 78-362748
*78 MARC

Jirásek, Alois, 1851-1930.
Filozofská historie : ⌊pro čtenáře od 12 let⌋ / Alois Jirásek ; ⌊text k vyd. připr. a vysvětl. opatřila Františka Havlová ; doslov Zdeněk Nejedlý ; il. Karl Müller⌋. — 8. vyd. Albatrosu. — Praha : Albatros, 1977.
157, ⌊2⌋ p., ⌊8⌋ leaves of plates : col. ill. ; 21 cm. — (Zlatoroh) Cz78
"Psáno roku 1877 (Světozar)"
Previous editions published under title: Filosofská historie.
I. Havlová, Františka. II. Title.
PZ70.C9J544 1977 79-343750
*79 MARC

Lhotová, Dagmar.
Kocour Vavřinec na vodě i ve vzduchu / ⌊autoři⌋ Dagmar Lhotová, Z. K. Slabý ; ⌊barev. il.⌋ Věra Faltová. — 1. vyd. — Praha : Olympia, 1977.
55, ⌊1⌋ p. : col. ill. ; 22 x 24 cm. Cz77
Kčs31.00
I. Slabý, Z. K., joint author. II. Faltová, Věra, joint author. III. Title.
PZ70.C9L48 78-375384
*78 MARC

Pecháček, Jaroslav.
Ostrov tisíce vůní / Jaroslav Pecháček. ; ⌊il. Jindřich Kovařík⌋. — 1. vyd. — Praha : Albatros, 1978.
301, ⌊3⌋ p. : ill. ; 21 cm. — (Knihy odvahy a dobrodružství ; sv. 146) Cz78
Kčs26.00
1. Cuba—History—Revolution, 1895-1898—Juvenile fiction. I. Title.
PZ70.C9P37 78-378396
*78 MARC

Peychlová, Bohumíra.
Dobrodružství kočky Vločky : aneb, Všude dobře, doma nejlíp : skoromuzikál pro mládež o osmi obrazech / Bohumíra Peychlová ; ⌊hudba Zdeněk Mašek ; texty písní napsala Bohumíra Peychlová⌋. — Praha : Dilia, 1975.
64, ⌊1⌋ leaves ; 29 cm. Cz76
I. Title.
PZ70.C9P412 79-378692
*79 MARC

Šajner, Donát.
Takový nádherný kus cesty : ⌊pro děti od 12 let⌋ / Donát Šajner ; ⌊il. Jaroslav Lukavský⌋. — 1. vyd. — Praha : Albatros, 1975.
72, ⌊2⌋ p. : ill. ; 21 cm. — (Klub mladých čtenářů) Cz75
Kčs11.00
I. Title.
PZ70.C9S22 79-347015
*79 MARC

Šmahelová, Helena.
Bratr a sestra : ⌊pro čtenáře od 12 let⌋ / Helena Šmahelová ; ⌊il. Helena Chvojková⌋. — 2. vyd. V nakl. Albatros 1. vyd. — Praha : Albatros, 1977.
212 p. : ill. (some col.) ; 20 cm. — (Střelka ; sv. 128) Cz78
Kčs20.00
I. Title.
PZ70.C9S547 1977 78-369184
*78 MARC

Štíplová, Ljuba.
Co si povídají zvířátka v noci : ⌊pro děti od 5 let⌋ / Ljuba Štíplová ; ⌊barev. il.⌋ Mirko Hanák. — 1. vyd. — Praha : Albatros, 1976.
28, ⌊3⌋ p. : col. ill. ; 26 cm. — (Edice Mezinárodní série) Cz77
Kčs18.00
I. Hanák, Mirko. II. Title.
PZ70.C9S787 78-399526
*79 MARC

Toonder, Marten, 1912-
Tom Poes en de toverfluit. Grammofoonplaat: Harry Geelen ⌊Amsterdam, Ring, c1969⌋
⌊28⌋ p. illus. (part col.) and phonodisc (2 s. 18 cm. 33 1/3 rpm. stereo) in pocket. 20 x 21 cm. Ne***
I. Geelen, Harry. II. Title.
PZ16.3.T637 74-340505
79⌊r79⌋rev MARC

CHILDREN'S LITERATURE, CZECH (Continued)

Vrána, Jan.
Čaroděj Mordydy / na motivy rus. lid. pohádek vypráví Jan Vrána ; ⌊il. z díla Josefa Lady vybral a k vyd. připravil J. Vrána⌋. — 1. vyd. — Praha : Albatros, 1978.
86, ⌊4⌋ p. : ill. (some col.) ; 20 cm. — (Edice Z pohádky do pohádky) Cz78
Kčs18.00
I. Title.
PZ70.C9V68 79-338148
*79 MARC

—HISTORY AND CRITICISM

Chaloupka, Otakar.
Children's books in Czechoslovakia / Otakar Chaloupka. — Prague : Orbis Press Agency, 1978.
85 p., ⌊8⌋ leaves of plates : ill. ; 19 cm. Cz***
Kčs8.00
1. Children's literature, Czech—History and criticism. 2. Children's literature, Slovak—History and criticism. I. Title.
PN1009.C9C44 891.8'6'099282 79-303345
79 MARC

—HISTORY AND CRITICISM—CONGRESSES

Literatúra pre predškolský vek a jej využitie v materských školách : zborník referátov z pracovného seminára v Modre 30.-9.-1.10.1976 ... / ⌊konferenciu viedla a vyd. zborníka referátov pripravila Lýdia Kyselová⌋. — 1. vyd. — Bratislava : Mladé letá, 1978.
147 p. ; 21 cm. — (Otázky detskej literatúry) CzS***
Czech or Slovak.
Bibliography: p. 140-141.
Kčs13.50
1. Children's literature, Czech—History and criticism—Congresses. 2. Children's literature, Slovak—History and criticism—Congresses. 3. Education, Pre-school—Czechoslovakia—Congresses. I. Kyselová, Lýdia.
PN1009.C9L5 79-363666
79 MARC

CHILDREN'S LITERATURE, DANISH

Arre, Svend.
Menneskebørn / tekst og tegninger af Svend Arre. — ⌊Skævinge⌋ ⌊Hulvejen, Lille Lyngby, 3320⌋ : ⌊Svend Arre⌋, c1977.
⌊32⌋ p. : ill. ; 30 cm. D78-23
Cover title.
"I 500 nummererede eksemplarer ... dette er nr. 404."
kr38.00
I. Title.
PZ54.3.A7 78-401102
79 MARC

Bak Rasmussen, Hedvig.
Hulens hemmelighed / ⌊af⌋ Hedvig Bak Rasmussen ; ⌊illustrationer af forfatteren⌋. — Fredericia : Frimodt, 1977.
166 p. : ill. ; 20 cm. D78-11
ISBN 8774460609 : kr39.50
I. Title.
PZ53.B326H8 79-359044
79 MARC

Belert, Keld.
Søstrene Musehjerte / ⌊af⌋ Keld Belert ; illustrationer, Ole Fick. — ⌊København⌋ : Borgen, c1978.
213 p. : ill. ; 22 cm. D78-34
ISBN 8741847199 : kr75.00
I. Title.
PZ53.B39S6 79-343671
79 MARC

Bødker, Cecil.
Jerutte besøger Hundejens / Cecil Bødker. — Viborg : Arena, 1977.
78 p. ; 21 cm. D77-51
ISBN 8774055747 : kr28.50
I. Title.
PZ53.B56J4 79-356027
79 MARC

Boesdal, Nis.
Gry / ⌊af⌋ Nis Boesdal ; illustreret af Bodil Nygård. — Århus : Husets Forlag, cop. 1978.
⌊62⌋ p. : ill. ; 21 cm. D78-23
ISBN 8787582368 : kr30.00
I. Nygård, Bodil. II. Title.
PZ53.B565G7 78-387999
79 MARC

Esmann, Karin.
Kvalitet? for børn / ⌊af⌋ Karin Esmann ; ⌊bogens illustrationer er lavet af Carsten Poulsen⌋. — 1. udgave. — Tranehuse : GMT, 1978.
110 p. : ill. ; 22 cm. — (Opvækst og opdragelse ; 3) D79-1/2
Bibliography: p. 107-110.
ISBN 8773300810 : kr35.00
1. Books and reading for children—Denmark. 2. Children's literature, Danish. I. Title. II. Series.
Z1037.5.E84 79-359486
79 MARC

Halby, Lis.
Det er osse lige meget / ⌊af⌋ Lis Halby ; ⌊fotografier, Mogens Steen Jacobsen⌋. — ⌊København⌋ : Fremad, 1977.
116 p. : ill. ; 20 cm. D78-6
ISBN 8755708862 : kr34.75
I. Title.
PZ53.H26E7 79-363054
79 MARC

Haller, Bent.
Indianeren / ⌊af⌋ Bent Haller ; tegninger af Palle Bregnhøi. — ⌊København⌋ : Borgen, c1978.
99 p. : ill. ; 22 cm. D78-25
ISBN 8741847083 : kr44.00
I. Bregnhøi, Palle. II. Title.
PZ53.H27I5 79-358480
79 MARC

Hare, Kirsten.
Sus og Søren : ind i stuen og ud ad døren / ⌊text af⌋ Kirsten Hare ; illustreret af Thora Lund. — ⌊Herning⌋ : Specialpædagogisk forlag, 1978.
36 p. : ill. ; 18 cm. — (Her Sus og Søren ; 1. bog) D78-30
ISBN 8787420740 : kr32.00
I. Lund, Thora. II. Title.
PZ53.H35S97 vol. 1 78-393311
79 MARC

Hare, Kirsten.
Sus og Søren—og pakken : over, under og bagved broen / ⌊text af⌋ Kirsten Hare ; illustreret af Thora Lund. — ⌊Herning⌋ : Specialpædagogisk forlag, 1978.
36 p. : ill. ; 18 cm. — (Her Sus og Søren ; 2. bog) D78-30
ISBN 8787420767 : kr32.00
I. Lund, Thora. II. Title.
PZ53.H35S97 vol. 2 79-350818
79 MARC

Hare, Kirsten.
Sus og Søren / ⌊text af⌋ Kirsten Hare ; illustreret af Thora Lund. — ⌊Herning⌋ : Specialpædagogisk forlag, 1978-
v. : ill. ; 18 cm. D78-30
CONTENTS: 1. bog. Sus og Søren.—2. bog Sus og Søren—og pakken.

ISBN 8787420742 (v. 1) : kr32.00 (v. 1)
I. Lund, Thora. II. Title.
PZ53.H35S97 78-393310
79 MARC

Holst, Kirsten.
Baggårds-rødderne / ⌊af⌋ Kirsten Holst ; tegninger af Dorte Salicath. — ⌊København⌋ : Gyldendal, 1978.
173 p. : ill. ; 21 cm. — (LL-serien) D78-15
ISBN 8701589210 : kr56.50
I. Salicath, Dorte. II. Title.
PZ53.H62B34 78-391741
79 MARC

Ispen, Henning.
Mistanken / ⌊af⌋ Henning Ispen. — ⌊København⌋ : Sesam, 1979.
80 p. ; 21 cm. — (Danske forfattere fortæller) D79-30
ISBN 8773245054 : kr44.75
I. Title.
PZ53.I6M5 79-383341
79 MARC

Jensen, Virginia Allen.
Ida og døren / tekst af Virginia Allen Jensen ; tegninger af Joan Sandin. — ⌊København⌋ : Gyldendal, 1978.
⌊28⌋ p. : ill. ; 17 cm. D78-45
ISBN 8701671219 : kr17.75
I. Sandin, Joan. II. Title.
PZ53.J375I3 79-358507
79 MARC

Johansson, Magnus, 1921-
Den røde strømer / ⌊af⌋ Magnus Johansson. — ⌊København⌋ : Borgen, c1978.
149 p. ; 20 cm. D78-46
ISBN 8741848101 : kr48.40
I. Title.
PZ53.J628R6 79-345628
79 MARC

Juni, Djorn.
Fridag i solskin : digte / af Djorn Juni ; ⌊vignetter af Kirsten Clemann⌋. — Nexø : Gornitzka's Bogtrykkeri : sælges gennem Keramikståuan ⌊Strandgade, 3730⌋, 1977.
⌊35⌋ p. : ill. ; 30 cm. D78-20
ISBN 8787709007 : kr28.00
I. Title.
PZ54.3.J8 78-401119
79 MARC

Larsen, Inger, 1917-
Runes lange rejse : en fortælling om en drengs oplevelser på en vandring fra Stockholm til København / ⌊af⌋ Inger Larsen. — København : Branner og Korch, 1978.
139 p. ; 20 cm. D78-35
ISBN 8741147650 : kr42.75
I. Title.
PZ53.L26Ru 78-390446
79 MARC

Lööf, Jan, 1940-
Søren Skrot / af Jan Lööf ; overs. af Jens Peder Agger. — Copenhagen : Carlsen, 1976.
⌊32⌋ p. : ill. ; 26 cm. D***
Translation of Skrot-Nisse.
I. Title.
PZ53.L7S65 78-400767
79 MARC

Nielsen, Torben, 1918-
Landsbydreng—røgterdreng / ⌊af⌋ Torben Nielsen ; illustreret af Leif Bøgild. — ⌊Malling⌋ : Tommeliden, 1978.
88 p. : ill. ; 21 cm. D79-3
Continues the author's En landsbydrengs dagbog.
ISBN 8785184969 : kr36.45
I. Title.
PZ53.N53L3 79-369561
79 MARC

Nørlund, Mogens Christian.
Slavepigen Rosita : en sandfærdig beretning fra Columbia / ⌊af⌋ Mogens Chr. Nørlund. — ⌊København⌋ : Branner & Korch, 1978.
156 p. ; 20 cm. D78-48
ISBN 8741146727
I. Title.
PZ53.N66S5 79-345626
79 MARC

Olsen, Ib Spang.
Thor og hammeren / ⌊tekst og tegninger⌋, Ib Spang Olsen. — ⌊København⌋ : Gyldendal, 1978.
⌊32⌋ p. : ill. ; 26 cm. D79-11
ISBN 8701771116 : kr42.50
1. Thor—Juvenile fiction. I. Title.
PZ53.O55T47 79-384325
79 MARC

Olsen, Jørgen M 1912-
Krypilin / Jørgen M. Olsen ; teknað hevur Birgit Søndergård. — Tórshavn : Føroya Skúlabókagrunnur, 1976.
53 p. : ill. ; 21 cm. D77
Kr22.00
I. Title.
PZ90.F3O4 79-357839
79 MARC

Olsen, Jørgen M 1912-
Pisuvarðin / Jørgen M. Olsen ; teknað hevur Birgit Søndergård. — Tórshavn : Føroya Skúlabókagrunnur, 1976.
44 p. : ill. ; 21 cm. D77
Kr22.00
I. Title.
PZ90.F3O43 79-371532
79 MARC

Petersen, Palle, 1943-
Bo på Vesterbro / af Palle Petersen ; illustreret af Ullrich Rössing. — København : Gad, 1978.
56 p. : ill. ; 19 cm. — (Rundt om os) D78-31
ISBN 8712777269 : kr19.50
I. Rössing, Ullrich. II. Title.
PZ53.P45B6 78-401123
79 MARC

Rasmussen, Bent, 1934-
Den dobbelte sol / ⌊af⌋ Bent Rasmussen. — ⌊København⌋ : Borgen, 1978.
106 p. ; 20 cm. D78-15
ISBN 8741846885 : kr49.00
I. Title.
PZ53.R27D6 78-397371
79 MARC

Rasmussen, Bent.
Rapserne / af Bent Rasmussen ; illustreret af Kirsten Hoffmann. — København : Gad, 1978.
47 p. : ill. ; 19 cm. — (Rundt om os) D78-31
ISBN 8712777277 : kr19.50
I. Hoffmann, Kirsten. II. Title.
PZ53.R2698R3 78-396551
79 MARC

CHILDREN'S LITERATURE, DANISH (Continued)

Reif, Ole.
Fra onsdag til mandag / ₍af₎ Ole Reif. — København : Branner og Korch, 1978.
168 p. ; 20 cm. D78-18
ISBN 8741146735 : kr43.75
I. Title.
PZ53.R37F7 79-338160
79 MARC

Schmidt, Birgit, 1947-
Må jeg låne din bog? / tekst og tegninger af Birgit Schmidt. — ₍København₎ : Fremad, 1978.
26 p. : ill. ; 21 x 25 cm. D78-32
ISBN 8755709311 : kr29.75
I. Title.
PZ53.S287M3 79-348028
79 MARC

Wedelius, Marie.
Hilde : Ørvads søster : børne- og ungdomsbog fra sagatiden / ₍af₎ Marie Wedelius ; ₍tegning, Werner Møller₎. — ₍Haderslev₎ ₍Rosenvænget 21, 6100₎ : Afondo, 1978.
127 p. ; 21 cm. D78-46
ISBN 8799716178 : kr72.15
I. Title.
PZ53.W38H5 79-345627
79 MARC

Weinreich, Torben.
Det sker igen / ₍af₎ Torben Weinreich ; illustreret af Ulf Rössing. — 1. oplag. — ₍Aarhus₎ : ₍Modtryk₎, 1978.
77 p. : ill. ; 19 cm. — (Modpol ; 4) D78-33
ISBN 8787620596 : kr30.00
I. Rössing, Ulf. II. Title.
PZ54.3.W4 78-384681
79 MARC

—20TH CENTURY —HISTORY AND CRITICISM

Ørn, B
Conni og bolighajen : produktion, distribution, konsumtion / B. Ørn. — Grenå : GMT, 1976.
112 p. : ill. ; 22 cm. — (Pædagogiske arbejdsmapper ; 15) D77-7
Includes bibliographies.
ISBN 8773300330 : kr25.00
1. Knudsen, Per Holm. Conni og bolighajen. 2. Children's literature, Danish—20th century—History and criticism. 3. Literature and society. I. Title.
PT8176.21.N813C636 78-367548
78 MARC

—HISTORY AND CRITICISM

Belert, Keld.
Vold og modvold : analyser af dansk ungdomslitteratur i slutningen af 1970'erne / ₍af₎ Keld Belert. — Gentofte : Tellerup ; ₍København₎ : ₍eksp. DBK₎, c1978.
126 p. ; 20 cm. D78-29
ISBN 8785191183 : kr59.00
1. Children's literature, Danish—History and criticism. 2. Violence in literature. I. Title.
PN1009.D3B4 78-396163
79 MARC

CHILDREN'S LITERATURE, DUTCH

Bouhuys, Mies.
Het dorp van Maarten / Mies Bouhuys ; met tekeningen van Pieter Kok. — Amsterdam : Kosmos, ₍1978₎
55 p. : ill. ; 20 x 27 cm. Ne78-26
ISBN 9021506904 : fl 19.90
I. Kok, Pieter. II. Title.
PZ13.B6184 79-359830
•79 MARC

Christophers, Ewald, 1922-
Der 6. ₍i.e. sechste₎ Erdteil : auf d. Spuren d. Ostfriesen / Ewald Christophers ; mit Zeichn. von Peter Wirsing. — Bremerhaven : Wirtschaftsverlag NW, Verl. für Neue Werbung, 1977.
176 p. : ill. ; 24 cm. GFR77-A
ISBN 3-920320-90-5
1. East Friesland—Juvenile fiction. I. Title.
PZ13.C47 833'.9'14 79-364808
79 MARC

De Klerk, Willem Abraham, 1917-
Die Gésel van Namaland / W. A. de Klerk. — 2. uitg. — Johannesburg : Perskor, 1976.
197 p. ; 22 cm. SA***
ISBN 0-628-01093-1
I. Title.
PZ13.D46 1976 79-348240
79 MARC

Diekmann, Miep.
Wiele wiele stap / Miep Diekmann ; met tekeningen van The Tjong Khing. — Amsterdam : Querido, 1977.
32 p. : with ill. ; 21 cm. Ne77-46
ISBN 9021413337 : fl 5.90
I. The, Tjong Khing. II. Title.
PZ13.D56 78-393640
•79 MARC

Du Plessis, Izak David, 1900-
Die Maboersa en ander Kaapse stories / opgeteken en oorvertel deur I. D. du Plessis ; met illustrasies deur Chris Lovell. — Kaapstad : Human & Rousseau, 1978.
72 p. : ill. ; 22 cm. SA***
"Oorspronklike uitgawe as deel van Uit die Slamse buurt ..."
ISBN 0-7981-0793-6
I. Lovell, Chris. II. Title.
PZ13.D86 1978 79-371534
79 MARC

Evenhuis, Gertie.
De tram is geel, het gras is groen / Gertie Evenhuis ; met tekeningen van Piet Klaasse. — ₍Amsterdam, Langestraat 61₎ : Commissie voor de Collectieve Propaganda van het Nederlandse Boek, 1978.
68 p. : with ill. ; 25 cm. Ne78-43
"Uitgegeven ... ter gelegenheid van de Kinderboekenweek."
ISBN 9010022404 : fl 3.00
I. Klaasse, Piet. II. Title.
PZ13.E865 79-359624
•79 MARC

Eykman, Karel.
De allerallersterkste / Karel Eykman, Joep Bertrams. — Odijk : Sjaloom, c1977.
₍31₎ p. : ill. ; 16 x 22 cm. Ne***
ISBN 906249031X
I. Bertrams, Joep, joint author. II. Title.
PZ13.E927 79-338173
79 MARC

Hildebrand, A D
P. P. Koekelberg en zijn woonwagen / A. D. Hildebrand. — Hoorn : Westfriesland, ₍1977₎
112 p. : ill. ; 22 cm. Ne77-50
ISBN 9020512978 : fl 14.90
I. Title.
PZ13.H4995 79-338169
•79 MARC

Hofman, Wim.
Het tweede boek over Wim / Wim Hofman. — Bussum : Van Holkema & Warendorf, c1978.
137 p. : ill. ; 21 cm. Ne***
ISBN 9026915187
I. Title.
PZ13.H554 79-359102
79 MARC

Pelgrom, Els.
De kinderen van het Achtste Woud / Els Pelgrom ; met tekeningen van Peter van Straaten. — Amsterdam : Kosmos, c1977.
204 p. : ill. ; 23 cm. Ne***
ISBN 9021506548
I. Straaten, Peter van. II. Title.
PZ13.P36 78-396188
79 MARC

Pieters, Roger, 1922-
De muis / Roger Pieters. — Leuven : De Clauwaert, ₍1977₎
152 p. ; 21 cm. Be***
ISBN 9063060696
I. Title.
PZ13.P49 79-338170
79 MARC

Polet, Sybren.
Een heel klein mannetje en andere sprookjes / Sybren Polet ; met illustraties van Pieterjan Grasmayer. — Utrecht/Antwerpen : A. W. Bruna, 1978.
255 p. : with ill. ; 25 cm. — (Een Bruna jeugdboek) Ne78-42
CONTENTS: 1. Een heel klein mannetje.—2. De man die een hoofd groter was.—3. Klein Kareltje wordt Keizer.—De steen.
ISBN 9022945324 : fl 24.90
I. Grasmayer, Pieterjan. II. Title.
PZ13.P557 78-399469
79 MARC

Sauwer, Monika.
Wilbert's lunchpakket / Monika Sauwer. — Amsterdam : Bert Bakker, 1978.
55 p. : with ill. ; 20 cm. Ne78-46
ISBN 9060195841 : fl 10.00
I. Title.
PZ16.3.S2 79-343833
•79 MARC

Stander, Eric.
My eie seestorieboek / saamgestel deur Eric Stander ; illustraties deur Alida Bothma. — Pretoria : D. Retief, ₍1979?₎
40 p. : ill. ; 23 cm. SA***
ISBN 0-909230-89-7
I. Bothma, Alida. II. Title.
PZ13.S687 79-363149
79 MARC

Terpstra, Pieter, 1919-
Heimwee naar het eiland. ₍Door₎ Peter Terpstra. Den Haag, Voorhoeve, ₍1971₎.
147 p. with illus. 20 cm. fl 4.90 Ne71-2
I. Title.
PZ13.T43 72-863245
79₍r79₎rev2 MARC

Theissen-Pouderoyen, Maries.
Door het rozenpoortje / Maries Theissen-Pouderoyen. — Zeist : Vrij Geestesleven, ₍1977₎
159 p. : ill. ; 15 x 22 cm. Ne77-31
Bibliography: p. 159.
ISBN 9060380746 : fl 17.50
1. Children's literature, Dutch. I. Title.
PZ11.T47 79-367015
•79 MARC

Van Schalkwyk, Nickey.
Die geheim van die bergvolk / deur Nickey van Schalkwyk ; illustrasies deur Alida Bothma. — Pretoria : Retief, 1978.
124 p. : ill. ; 22 cm. SA***
ISBN 0-909230-86-2
I. Bothma, Alida. II. Title.
PZ13.V3319 79-345610
79 MARC

Van Schalwyk, Nickey.
Ngonyama keer terug : die avonture van Kristof en die slaweseun, Kobo / deur Nickey van Schalkwyk. — Pretoria : D. Retief, 1978.
116 p. : ill. ; 23 cm. SA***
ISBN 0-909230-81-1
I. Title.
PZ13.V33194 79-363665
79 MARC

CHILDREN'S LITERATURE, ENGLISH

The Noel Streatfeild birthday story book : an anthology. — London : Dent, 1976.
158 p. : ill. ; 25 cm. GB76-27886
ISBN 0-460-06667-6 : £2.95
1. Children's literature, English. I. Streatfeild, Noel.
PZ5.N527 823'.9'14 79-317764
79 MARC

—BIBLIOGRAHY

Gawith, Gwen.
Children's paperbacks 11 to 16 / selected, annotated, and illustrated by Gwen Gawith. — London : National Book League ₍for₎ the School Library Association, 1977.
44 p. : ill. ; 21 cm. GB78-32663
ISBN 0-85353-303-2 : £0.85
1. Children's literature, English—Bibliography. 2. Bibliography—Paperback editions. I. Title.
Z1037.9.G38 028.52 79-301143
₍PN1009.A1₎ 79 MARC

Lassam, Christine.
Stories and settings : England : a handbook for teachers and librarians / selected and annotated by Christine Lassam. — London : National Book League, c1977.
48 p. ; 21 cm. GB***
ISBN 0-85353-262-1 : £0.50
1. England—Juvenile literature—Bibliography. 2. Children's literature, English—Bibliography. I. Title.
Z2014.5.L37 016.942 78-324689
₍DA27.5₎ 79 MARC

McCulloch, Lou W
Children's books of the 19th century / Lou W. McCulloch ; photo ill. by Thomas R. McCulloch. — Des Moines, Iowa : Wallace-Homestead Book Co., c1979.
152 p. : ill. ; 24 cm.
Includes bibliographies and index.
ISBN 0-87069-239-9 : $9.95
1. Children's literature, English—History and criticism. 2. Children's literature, American—History and criticism. 3. Children's literature, English—Bibliograhy. 4. Children's literature, American—Bibliography. I. McCulloch, Thomas R. II. Title.
PN1009.A1M233 820'.9'9282 77-87948
79 MARC

CHILDREN'S LITERATURE, ENGLISH (Continued)

—EXHIBITIONS

Kloet, Christine A
After "Alice" : a hundred years of children's reading in Britain / material selected and catalogued by Christine A. Kloet on behalf of the Youth Libraries Group of the Library Association. — London : Library Association, c1977.
63 p., ⌊3⌋ leaves of plates : ill. ; 25 cm. GB***
"An exhibition to celebrate the Centenary of the Library Association, jointly organised by the Association and the Victoria and Albert Museum at the Museum of Childhood ... 29 September 1977 to 15 January 1978."
Bibliography: p. 57.
Includes index.
ISBN 0-85365-740-8 : £2.00
1. Children's literature, English—Exhibitions. 2. London—Exhibitions. I. Library Association. Youth Libraries Group. II. Library Association. III. Victoria and Albert Museum, South Kensington. IV. Museum of Childhood. V. Title.
Z1037.K59 028'.5 78-317600
⌊PN1009.A1⌋ 78 MARC

—HISTORY AND CRITICISM

Laski, Marghanita, 1915-
Mrs. Ewing, Mrs. Molesworth, and Mrs. Hodgson Burnett / by Marghanita Laski. — Norwood, Pa. : Norwood Editions, 1977.
121 p. ; 23 cm.
Reprint of the 1950 ed. published by A. Barker, London, in series: the English novelists.
"List of Mrs. Ewing's works": p. 97-107. "List of Mrs. Molesworth's works": p. 108-112. "List of Mrs. Hodgson Burnett's works": p. 113-117.
Includes index.
ISBN 0-8482-1574-5 lib. bdg. : $15.00
1. English fiction—Women authors—History and criticism. 2. Ewing, Juliana Horatio Gatty, 1841-1885—Criticism and interpretation. 3. Molesworth, Mary Louisa Stewart, 1842-1921—Criticism and interpretation. 4. Burnett, Frances Hodgson, 1849-1924—Criticism and interpretation. 5. Children's literature, English—History and criticism. I. Title. II. Series: The English novelists (London).
PR115.L34 1977 823'.8'09 77-16574
77 MARC

McCulloch, Lou W
Children's books of the 19th century / Lou W. McCulloch ; photo ill. by Thomas R. McCulloch. — Des Moines, Iowa : Wallace-Homestead Book Co., c1979.
152 p. : ill. ; 24 cm.
Includes bibliographies and index.
ISBN 0-87069-239-9 : $9.95
1. Children's literature, English—History and criticism. 2. Children's literature, American—History and criticism. 3. Children's literature, English—Bibliography. 4. Children's literature, American—Bibliography. I. McCulloch, Thomas R. II. Title.
PN1009.A1M233 820'.9'9282 77-87948
79 MARC

CHILDREN'S LITERATURE, FINNISH

Backlén, Marianne.
Tyttö tanssii / Marianne Backlén. — Porvoo ; Helsinki : Juva : WSOY, 1978 (Porvoo : WSOY:n graafiset laitokset)
187, ⌊1⌋ p. ; 19 cm. Fi78-4
ISBN 9510085073 : Fmk29.00
I. Title.
PZ90.F5B3 79-346514
79 MARC

Erhomaa, Ester.
Minä, Aina Ilona ... : romaani / Ester Erhomaa. — Helsingissä : Otava, ⌊1978⌋
190 p. ; 21 cm. Fi78
ISBN 9511048716
I. Title.
PZ90.F5E68 79-368232
79 MARC

Hyvärinen, Aarne.
Kiljuva seurapiiri / runot, Aarne Hyvärinen ; kuv., Anuliina Mäki. — ⌊Oulu⌋ : Pohjoinen, 1977.
56 p. : ill. ; 21 x 24 cm. Fi78
ISBN 9519099697 : Fmk20.00
I. Mäki, Anuliina. II. Title.
PZ90.F5H9 78-400762
79 MARC

Isotalo, Kaarlo.
Ellimaarit / Kaarlo Isotalo. — Hämeenlinna : Karisto, 1978.
109 p. ; 20 cm. Fi***
ISBN 9512312573
I. Title.
PZ90.F5I7 79-346516
79 MARC

Jalo, Merja.
Ponisaaren arvoitus : tyttöromaani / Merja Jalo. — ⌊Espoo⌋ : Weilin + Göös, 1978.
144 p. ; 20 cm. Fi78-4
ISBN 9513516628 : Fmk22.00
I. Title.
PZ90.F5J235 79-338168
79 MARC

Karjalainen, Kastehelmi.
Ollaanpas vaan. Kuvat piirsi Sirkku. Hämeenlinna, A. A. Karisto ⌊1966⌋
169 p. illus. 21 cm.
I. Title.
PZ90.F5K295 75-212253
77⌊r79⌋rev MARC

Krohn, Leena.
Ihmisen vaatteissa : kertomus kaupungilta / Leena Krohn ; ⌊kansikuva ja kuvitus tekijän⌋. — Helsinki : Tammi, 1976.
157 p. : ill. ; 21 cm. Fi
ISBN 9513039021 : Fmk29.00
I. Title.
PZ90.F5K7339 77-554329
*77⌊r79⌋rev MARC

Kujala, Yrjö.
Kesä Luotsisaaressa / Yrjö Kujala. — ⌊Helsinki⌋ : SLEY ⌊Suomen luterilainen evankeliumiyhdistys⌋, 1977 (Joensuu : Pohjois-Karjalan kp.)
80 p. : ill. ; 21 cm. Fi77
I. Title.
PZ90.F5K7398 78-371911
78 MARC

Kuoppala, Helvi.
Sininen soitto. Satuja. Kuv. Paula Kuoppala. Lapua, Herättäjä, 1971.
55, (1) p. illus. 18 cm. Fi
I. Title.
PZ90.F5K822 78-348223
78 MARC

Kuurinmaa, Aira.
Leikisti / Aira Kuurinmaa : Kuvittanut, Ulla Vaajakallio. — Porvoo : Söderström, 1978.
110 p. : ill. ; 21 cm. — (WSOY-nuortenkirjat) Fi***
ISBN 9510087327
I. Title.
PZ90.F5K88 78-401127
79 MARC

Lehikoinen, Urho.
Oskarin kujeet. Toim. Urho Lehikoinen. Kuvitus ja päällys: Olavi Hurmerinta. Helsinki, Kirjayhtymä, 1972.
75 p. illus. 25 cm. Fmk16.00 Fi
I. Hurmerinta, Olavi, illus. II. Title.
PZ90.F5L347 78-408542
ISBN 9512601109 78 MARC

Lipsonen, Leo, 1952-
Aaveita Salojärvellä / Leo Lipsonen. — ⌊Hämeenlinna⌋ : Karisto, 1977.
140 p. ; 20 cm. — (Kariston nuortenkirjasto) Fi78-4
ISBN 9512311755
I. Title.
PZ90.F5L47 79-346518
79 MARC

Mickwitz, Camilla.
Jason : tarina tavallisesta pienestä pojasta / Camilla Mickwitz. — Helsinki : Wilin & Göös, 1975.
30 p. : ill. ; 22 x 26 cm. Fi76
ISBN 9513513912 : Fml18.00
I. Title.
PZ90.F5M456 79-357127
*79 MARC

Mickwitz, Camilla.
Jasonin kesä / kert. ja kuv., Camilla Mickwitz. — Helsinki : Weilin + Göös, 1976.
⌊34⌋ p. : ill. ; 21 x 26 cm. Fi77-Mar.
ISBN 9513514943 : Fmk19.00
I. Title.
PZ90.F5M457 79-357128
79 MARC

Mikkanen, Raili.
Ikkunaton talo / Raili Mikkanen. — Helsinki : Otava, 1977.
123 p. ; 20 cm. — (Otavan nuortenkirjoja) Fi77Dec.
ISBN 9511045083 : Fmk22.00
I. Title.
PZ90.F5M47 79-356307
79 MARC

Mikkola, Marja-Leena.
Anni Manninen / Marja-Leena Mikkola ; kuv., Kaarina Kaila. — Helsinki : Otava, 1977 (Keuruu : Otavan painolaitokset)
190 p. : ill. ; 21 cm. Fi78-Jan.
ISBN 9511046063 : Fmk28.00
I. Title.
PZ90.F5M52 79-351625
79 MARC

Niskanen, Vuokko, 1933-
Hiekkalan lapset matkustavat : romaani juuri lukemaan oppineille / Vuokko Niskanen ; kansi ja kuvat, Sanna Hietala. — Jyväskylä : Gummerus, 1978.
66 p. : ill. ; 17 cm. Fi78-10
ISBN 9512015994 : Fmk22.00
I. Title.
PZ90.F5N48 79-356312
79 MARC

Numminen, Mauri Antero.
Jänukset maailmankartalle : seikkailukertomus lapsille / M. A. Numminen. — Helsinki : Kirjayhtymä, 1977 (Helsinki : Yhteiskp.)
73 p. : ill. ; 21 cm. Fi78
ISBN 951261412X : Fmk24.00
I. Title.
PZ90.F5N76 79-345638
79 MARC

Pakkanen, Kaija.
Huilu hilpeä : loruja ja saturunoja / Kaija Pakkanen ; ⌊kuv., Pirkko Kilpelä⌋. — Porvoo ; Helsinki : Söderström, c1976.
47 p. : ill. ; 19 cm. — (WSOY nuortenkirjat) Fi76Nov.
ISBN 9510077402 : Fmk22.00
I. Title.
PZ90.F5P284 79-356319
*79 MARC

Palvo, Aila.
Mökkerö / Aila Palvo ; kansi, kuv. ja taitto, Tuula Savolainen. — Helsinki : Kirjayhtymä, 1978 (Vaasa : Vaasa oy).
30 p. : ill. ; 25 cm. Fi78-5
ISBN 9512614375 : Fmk7.80
I. Title.
PZ90.F5P33 79-369562
79 MARC

Parkkinen, Jukka.
Korppi ja kumppanit / Jukka Parkkinen ; kuvittanut, Jyrki Vuori. — Porvoo : Söderström, 1978.
86 p. : ill. ; 25 cm. Fi***
ISBN 9510087807
I. Title.
PZ90.F5P34 79-348909
79 MARC

Polva, Anni.
Tiinalle otetaan pikkusisko / Anni Polva. — Hämeenlinna : Karisto, 1978.
155 p. ; 18 cm. Fi***
ISBN 9512312395
I. Title.
PZ90.F5P744 78-401126
79 MARC

Puputti, Marja-Liisa.
Tuli palaa apua! / Marja-Liisa Puputti ; kuvittanut, Ulla Vaajakallio. — Porvoo : Söderström, c1978.
56 p. : ill. ; 20 cm. Fi***
ISBN 9510087785
I. Title.
PZ90.F5P854 79-359611
79 MARC

Pyyhtiä, Pekka Antero.
Mitäs me mieliharmit : seikkailukertomus 10-100 vuotiaille pojille / Pekka Antero Pyyhtiä. — Hämeenlinna : Karisto, 1978.
128 p. ; 20 cm. — (Kariston nuortenkirjasto) Fi78-4
ISBN 951231200X : Fmk16.00
I. Title.
PZ90.F5P87 79-346515
79 MARC

Saari, P M comp.
Nastajuttu. Poikien kaskukirja. Valikoinut P. M. Saari. Helsinki, Otava, 1970.
63 p. illus. 18 cm. Fmk3.75 Fi
I. Title.
PZ90.F5S117 78-348225
78 MARC

Saarto, Tuula, 1942-
Junnu / Tuula Saarto. — Helsingissä : Otava, ⌊1977⌋
112 p. ; 20 cm. — (Otavan nuortenkirjoja) Fi77Dec.
ISBN 9511043595 : Fmk22.00
I. Title.
PZ90.F5S1197 79-356025
79 MARC

Samulinen, Mikko.
Tunari / kirjoittanut ja kuvittanut Mikko Samulinen. — Helsinki : Sanoma, c1978.
46 p. : ill. ; 27 cm. Fi***
ISBN 9519133933
I. Title.
PZ90.F5S259 79-355563
79 MARC

CHILDREN'S LITERATURE, FINNISH (Continued)

Seppälä, Arto.
Kättä päälle, Vasco da Gama / Arto Seppälä. — Porvoo; Helsinki ; Juva : WSOY, 1978 (Porvoo : (WSOY:n graafiset laitokset)
139, [1] p. ; 19 cm. — (Nuorten toivekirjasto ; 248) Fi78-10
ISBN 9510087718 : Fmk25.00
I. Title.
PZ90.F5S37 79-358484
79 MARC

Soini, Tuulikki.
Nukkumatin satuja / Tuulikki Soini ; kansi ja kuv., Valpuri Pihlajamäki. — Jyväskylä : Gummerus, 1977.
96 p. : ill. ; 28 cm. Fi78
ISBN 9512014262 : Fmk23.00
I. Pihlajamäki, Valpuri. II. Title.
PZ90.F5S585 78-400755
79 MARC

Tanninen, Oili.
Milli / Oili Tanninen. — Helsinki : Otava, 1976.
[32] p. : ill. ; 17 x 20 cm. Fi76Nov.
ISBN 9511040448 : Fmk16.00
I. Title.
PZ90.F5T312 79-358794
*79 MARC

Wendelin, Martta.
Metsänväen suojatti / Martta Wendelin, Elsa Virkkunen. — [Helsinki] : Kirjapaja, 1977 (Jyväskylä : Gummerus)
54 p., [1] p. : ill. ; 27 cm. Fi78-2
ISBN 9516211798
I. Virkkunen, Elsa, joint author. II. Title.
PZ90.F5W4 78-400763
79 MARC

CHILDREN'S LITERATURE, FRENCH

15 [i.e. Quinze] chasses au trésor : [histoires / Bertrand Solet et al.] ; ill. de Georges Pichard. — [Paris] : Gautier-Languereau, [1974]
217 p. : ill. ; 22 cm. — (Série 15)
1. Treasure-trove—Juvenile fiction.
PZ21.Q52217 79-381084
79 MARC

15 [i.e. Quinze] histories de rugby / sélection des textes par Bertrand Solet ; ill. de Alain d'Orange. — [Paris] : Gautier-Languereau, c1978.
211 p. : ill. ; 22 cm. — (Série 15 [i.e. Quinze])
CONTENTS: Hughes, T. La première partie.—Llewellyn, R. Une journée de fête.—Jauréguy, A. L'époque héroïque.—Solet, B. Yves ... et les copains.—Barran, R. Rugby gazogène.—Prouteau, G. La passion d'André Jalabert.—Barran, R. Seul contre trente.—Barran, R. Celui qui venait des îles Fidji.—Solet, B. Les élections à Pipérade.—Barran, R. Au cœur du village.—Barran, R. Rugby, capitale Béziers.—Solet, B. Télé-rugby.—Barran, R. Astre, comme son nom l'indique.—Barran, R. A nous le grand chelem.—Patal, D. Un tallon très précieux.
ISBN 2-217-04072-3
1. Rugby football—Juvenile fiction. I. Solet, Bertrand.
PZ21.Q555 79-381083
79 MARC

15 [i.e. Quinze] récits de la préhistoire / [Clément Borgal, J.-H. Rosny, aîné] ; ill. de Alain d'Orange. — [Paris] : Gautier-Languereau, c1978.
214 p. : ill. ; 22 cm. — (Série 15 [i.e.Quinze]) F***
1. Man, Prehistoric—Juvenile fiction. I. Borgal, Clément. II. Rosny, J. H., aîné, pseud. of J. H. H. Boëx, 1856-1940.
PZ21.Q577 79-380889
79 MARC

Alain, Jean Claude.
Les Enfants de Dublin : chronique irlandaise / Jean-Claude Alain. — Paris : Éditions G.P., 1977.
219 p. ; 21 cm. — (Grand angle ; 26) F77-8884
ISBN 2-261-00203-3 : 15.00F
1. Ireland—History—Civil War, 1922-1923—Juvenile fiction. I. Title.
PZ23.A47E5 843'.9'14 79-358710
79 MARC

Aubry, Cécile.
Au secours, Poly! / Cécila Aubry ; illustrations d'Annie-Claude Martin. — Paris : Hachette, 1978.
155 p. : ill. (part col.) ; 25 cm. — (Collection Vermeille) F78-20684
ISBN 2-01-005037-1 : 28.00F
I. Title.
PZ23.A834A8 79-381082
79 MARC

Aubry, Cécile.
Les vacances de Poly / Cécile Aubry ; ill. d'Annie-Claude Martin. — [Paris] : Hachette, c1978.
158 p. : ill. (some col.) ; 26 cm. — (Collection Vermeille) F***
ISBN 2-01-004592-0
I. Title.
PZ23.A834V3 843'.9'14 79-348364
79 MARC

Bardot, Brigitte.
Noonoah, le petit phoque blanc / Brigitte Bardot ; [illustré par] Daniel Dollfuss. — [Paris] : Grasset et Fasquelle, 1978.
[24] p. : col. ill. ; 23 cm. — (Grasset jeunesse) F79-1218
ISBN 2-246-00574-4 : 25.00F
I. Dollfuss, Daniel. II. Title.
PZ23.B334N6 843'.9'14 79-372792
79 MARC

Barokas, Bernard, 1950-
Le plus bel âge de la vie / Bernard Barokas. — Paris : Duculot, c1978.
155 p. ; 19 cm. — (Travelling ; 39) Be***
ISBN 2-8011-0204-0
I. Title.
PZ23.B339P55 843'.9'14 79-367192
79 MARC

Bazin, Maryvonne.
La Rose bleue / texte de Maryvonne Bazin ; illustré par Boiry. — Paris : Grasset, 1977.
[24] p. : col. ill. ; 23 cm. — (Grasset jeunesse) F78-12861
ISBN 2-246-00506-X : 25.00F
I. Title.
PZ23.M3513R67 79-361311
79 MARC

Bosco, Henri, 1888-1976.
Bras de fer / Henri Bosco ; imagé par Romain Simon. — Tournai : Casterman, [c1977]
57 p. : col. ill. ; 24 cm. — (Plaisir des contes) Be***
ISBN 2-203-13324-4
I. Simon, Romain, 1915- II. Title.
PZ23.B632B7 1977 843'.9'12 78-385795
78 MARC

Brunhoff, Laurent de, 1925-
Babar et l'arbre de Noël / Laurent de Brunhoff. — Paris : Hachette, 1976.
23 p. : col. ill. ; 22 cm. — (Gentil coquelicot : ISSN 0338-1048) F76-16278
ISBN 2-01-003016-8 : 5.00F
I. Title.
PZ23.B78B326 77-479754
77[r79]rev MARC

Butor, Michel.
Chacun son cadeau / texte de Michel Butor ; ill. de T. Kitada. — Paris : Hachette, c1977.
[28] p. : chiefly col. ill. ; 25 cm. — (Le Vert paradis) F***
ISBN 2-01-004200-X
I. Kitada, Takushi. II. Title.
PZ23.B87C5 78-363390
78 MARC

Cailloux, André.
François et l'oiseau du Brésil, suivi de Tombé des étoiles / André Cailloux. — [Montréal] : Leméac, [c1977]
151 p. : ill. ; 18 cm. — (Théâtre pour enfants) C***
ISBN 0-7761-9906-4
I. Cailloux, André. Tombé des étoiles. 1977. II. Title: François et l'oiseau du Brésil.
PZ27.C28 78-366455
78 MARC

Čapek, Josef, 1887-1945.
Histoire de la [lettre] que le [chat] et le [chien] écrivirent à leurs amies, les petites [filles] Images de M. Chevallier. Thème de J. Čapek. Traduit du tchèque par L. Hirsch. Paris, Flammarion, 1970.
24 p. col. illus. 18 x 21 cm. (Albums du père Castor) 2.90F F***
I. Chevallier, Micheline, illus. II. Title.
PZ23.C25H5 74-571355
79[r79]rev MARC

Čapek, Josef, 1887-1945.
Histoire de la [lettre] que le [chat] et le [chien] écrivirent à leurs amies, les petites [filles] Images de M. Chevallier. Thème de J. Čapek. Traduit du tchèque par L. Hirsch. Paris, Flammarion, 1970.
24 p. col. illus. 18 x 21 cm. (Albums du père Castor) 2.90F F***
I. Chevallier, Micheline, illus. II. Title.
PZ23.C25H5 74-571355
79[r79]rev MARC

Chica.
Célestine va au marché / Chica. — Paris : Hachette, c1978.
[20] p. : chiefly col. ill. ; 25 cm. — (Les Poussins Hachette) F***
ISBN 2-01-005043-6
I. Title.
PZ23.C53C4 843'.9'14 78-400839
79 MARC

Ciravégna, Nicole.
César, poisson marseillais / texte de Nicole Ciravégna ; imagé par Alain Leray. — Paris : Magnard, c1977.
31 p. : col. ill. ; 29 cm. — (Grand carré : Formule 1) F***
Cover title.
ISBN (invalid) 2-210-99813-3
I. Leray, Alain. II. Title.
PZ23.C56C4 843'.9'14 79-372803
79 MARC

Clancier, Georges Emmanuel, 1914-
L'enfant de neige / Georges-Emmanuel Clancier ; imagé par Jean-Claude Luton. — [Tournai] : Casterman, [c1978]
61 p. : col. ill. ; 24 cm. — (Plaisir des contes) Be***
ISBN 2-203-13325-2
I. Title.
PZ23.C597E5 79-371755
79 MARC

Cohen, Mathew, 1942-
The leaves of Louise / Matt Cohen ; illustrated by Rikki. — Toronto : McClelland and Stewart, c1978.
30 p. : ill. ; 26 cm. C***
ISBN 0-7710-2190-9 : $7.95
I. Rikki. II. Title.
PZ27.C6627Le 78-319264
78 MARC

Conte, Tonino.
Brun l'ours et Le Bois / histoire de Tonino Conte ; illustré par Alain Letort. — Paris : J.-P. Delarge, 1978.
[27] p. : col. ill. ; 29 cm. F79-2889
ISBN 2-7113-0112-5 : 38.00F
I. Letort, Alain. II. Title.
PZ23.C727B7 843'.9'14 79-373113
79 MARC

Corriveau, Monique Chouinard.
La petite fille du printemps / Monique Corriveau ; ill. de Louise Pomminville. — Montréal : Fides, c1978.
124 p., [4] leaves of plates : col. ill. ; 21 cm. — (Collection du goéland) C79-2019-X
ISBN 0-7755-0705-9 : $7.75
I. Title.
PZ23.C756P4 1978 79-371750
79 MARC

Corriveau, Monique Chouinard.
Le Wapiti : roman / Monique Corriveau ; ill. de Melinda Wilson. — Montréal : Fides, c1978.
177 p., [4] leaves of plates : col. ill. ; 21 cm. — (Collection du goéland) C***
ISBN 0-7755-0685-0
1. Canada—History—To 1763 (New France)—Juvenile fiction. I. Title.
PZ23.C756W3 1978 843 79-352600
79 MARC

Dailly, Claudine.
Le voyage du train : théâtre musical pour une fugue en forme de conte / écrit, mis en paroles et en musique par Claudine Dailly ; ill., Yvonne Wastchenko-Cattier. — Paris : J. C. Simoën, c1978.
61 p. : ill., music ; 21 cm. — (Collection mise en pièces) F***
29.00F
I. Title.
PZ27.D27 78-398594
78 MARC

Danois, Jacques.
Les amis du bout du monde / Jacques Danois ; ill. de Jacques Danois. — Bruxelles, r. de Brabant 91 : P. de Méyère, [1977]
147 p. : ill. ; 22 cm. Be78-5
242F
1. Children—Juvenile fiction. I. Title.
PZ23.D2173A5 843'.9'14 78-388973
*79 MARC

Daudet, Alphonse, 1840-1897.
La chèvre de M. Seguin ; Le secret de maître Cornille / Alphonse Daudet ; illustrés par May Angeli. — Paris : Hachette, c1977.
[26] p. : col. ill. ; 32 cm. — (Le Jardin des rêves) F***
ISBN 2-01-004252-2
I. Angeli, May. II. Daudet, Alphonse, 1840-1897. Le secret de maître Cornille. 1977. III. Title: La chèvre de M. Seguin.
PZ23.D27C52 1977 843'.8 79-372836
79 MARC

Daudet, Alphonse, 1840-1897.
L'élixir du révérend père Gaucher, suivi de Les sauterelles / Alphonse Daudet ; images d'Élisabeth Ivanovsky. — Tournai : Casterman, c1977.
31 p. : col. ill. ; 31 cm. — (L'Age d'or) Be***
ISBN 2-203-13124-1
I. Ivanowsky, Élisabeth, 1910- II. Daudet, Alphonse, 1840-1897. Les sauterelles. 1977. III. Title: L'élixir du révérend père Gaucher.
PZ23.D27E4 1977 843'.8 79-372802
79 MARC

Daveluy, Paule.
La maison des vacances / Paule Daveluy ; ill. de Lise Thérien. — [Montréal] : Fides, c1977.
137 p., [4] leaves of plates : col. ill. ; 21 cm. — (Her Une année du tonnerre ; 1) (Collection du Goéland) C***
ISBN 0-7755-0662-1 : $6.45
I. Title. II. Series.
PZ23.D276M3 843 78-370083
78 MARC

CHILDREN'S LITERATURE, FRENCH
(Continued)

Delahaye, Gilbert, 1923-
Martine prend le train / Gilbert Delahaye, Marcel Marlier. — ₍Tournai₎ : Casterman, c1978.
20 p. : col. ill. ; 26 cm. — (Collection Farandole) Be***
ISBN 2-203-10128-8
I. Marlier, Marcel, joint author. II. Title.
PZ23.D313M34 79-380347
79 MARC

Denis, Jean C
Oncle Ernest & ₍i.e. et₎ les ravis / dessins, Jean C. Denis ; texte, Martin Veyron. — ₍Tournai₎ : Casterman, c1978.
27 p. : col. ill. ; 25 cm. — (Funambule) Be***
ISBN 2-203-10911-4
I. Veyron, Martin. II. Title.
PZ23.D3594O5 843'.9'14 79-361186
79 MARC

du Pavillon, Berthe.
Les jeux de Lisbeth et d'Estelle / Berthe du Pavillon. — ₍s.l. : s.n., 1975₎ (Vacoas, Ile Maurice : Impr. Petite Ourse)
85 p. : ill. ; 18 cm.
I. Title.
PZ23.D7865J4 76-980488
78 MARC

Dubouquet, Amélie.
La Maison des collines / Amélie Dubouquet. — Caen (142, rue Basse, B.P. 3050, 14018) : L. O. Four, 1976.
278 p. : ill. ; 22 cm. — (Large vision) F78-2524
49.00F
I. Title.
PZ23.D734M3 1976 843'.9'14 78-372379
78 MARC

Dumas, Philippe.
Laura sur la route / Philippe Dumas. — Paris : L'École des loisirs, c1978.
₍60₎ p. : col. ill. ; 22 cm. — (Joie de lire) F***
ISBN 2-211-05347-5 : 19.50F
I. Title.
PZ23.D7823L3 79-374082
79 MARC

Durousseau, Serge.
Franck et le héron cendré / texte de Serge Durousseau ; imagé par Rachel Alexandre Cohn. — Paris : Magnard, 1976.
28 p. : col. ill. ; 29 cm. — (Les Albums Grand carré) F78-5638
ISBN 2-210-99810-7 : 13.65F
I. Alexandre Cohn, Rachel. II. Title.
PZ23.D83F7 843'.9'14 79-372804
79 MARC

Escarpit, Robert, 1918-
Les reportages de Rouletabosse / Robert Escarpit ; ill. de l'auteur. — Paris : Magnard, c1978.
183 p., ₍8₎ leaves of plates : ill. (some col.) ; 21 cm. — (Collection Fantasia ; 18) F***
I. Title.
PZ23.E67R4 79-371752
79 MARC

Essen, Anne van der.
La souris s'en va-t-en guerre / par Anne van der Essen et Étienne Delessert. — ₍Paris₎ : Gallimard, c1978.
₍48₎ p. : ill. ; 22 cm. — (Aux couleurs de la vie) F***
I. Delessert, Étienne, joint author. II. Title.
PZ23.E736S68 843'.9'14 79-342686
79 MARC

Farge, Monique.
Par ici les souris! / Monique Farge ; illustré par Luis Camps. — Tournai : Casterman, c1978.
21 p. : col. ill. ; 26 cm. — (Les Aventures des pouscaloufs) Be***
ISBN 2-203-12201-3
I. Title.
PZ23.F24P3 741.5'9493 79-372789
79 MARC

Farge, Monique.
Toutes voiles dehors! / écrit par Monique Farge ; ill. par Luis Camps. — ₍Paris₎ : Casterman, c1978.
21 p. : col. ill. ; 26 cm. — (Les Aventures des pouscaloufs) F***
ISBN 2-203-12204-8
I. Camps, Luis. II. Title.
PZ23.F24T6 741.5'9493 79-372669
79 MARC

Féraud, Marie.
Anne ici, Sélima là-bas / Marie Féraud. — Paris : Duculot, c1978.
159 p. ; 19 cm. — (Travelling ; 40) Be***
ISBN 2-8011-0202-4
I. Title.
PZ23.F338A5 79-371753
79 MARC

Fournel, Paul, 1947-
Les Aventures très douces de Timothée le rêveur / texte Paul Fournel ; illustrations Claude Lapointe, ₍Gabriel Garcia₎. — Paris : Hachette, 1978.
113 p. : ill. ; 26 cm. — (La Bouteille à l'encre)
ISBN 2-01-005698-1 : 34.00F
I. Title. II. Title: Timothée le rêveur.
PZ23.F634A95 843'.9'14 79-365218
79 MARC

Gabalda, Jacques.
Histoires et légendes de tous les temps / Jacques Gabalda et René Beaulieu ; ill. de Jacques Fromont. — ₍Paris₎ : Hachette, c1978.
112 p. : col. ill. ; 27 cm. F***
ISBN 2-01-005087-8
I. Beaulieu, René, joint author. II. Title.
PZ23.G23H5 843'.9'14 79-348358
79 MARC

Gabalda, Jacques.
Histoires, récits et légendes / Jacques Gabalda et René Beaulieu ; ill. de Jacques Fromont. — ₍Paris₎ : Hachette, c1978.
114 p. : col. ill. ; 27 cm. F***
ISBN 2-01-005086-X
I. Beaulieu, René, joint author. II. Title.
PZ23.G23H52 843'.9'14 79-348360
79 MARC

Giono, Jean, 1895-1970.
Le petit garçon qui avait envie d'espace / Jean Giono ; dessiné par Gilbert Raffin. — ₍Paris₎ : Gallimard, c1978.
₍28₎ p. : col. ill. ; 20 cm. — (Enfantimages) F***
I. Raffin, Gilbert. II. Title.
PZ23.G47P47 1978 843'.9'12 79-368203
79 MARC

Gorde, Monique.
Les jumeaux jouent sur la plage / Monique Gorde. — ₍Paris₎ : Hachette, c1978.
₍21₎ p. : col. ill. ; 25 cm. — (Collection Les Poussins : Série verte) F***
I. Title.
PZ23.G68J8 79-342684
79 MARC

Gorde, Monique.
Les jumeaux rangent leur chambre / Monique Gorde. — Paris : Hachette, c1978.
₍19₎ p. : col. ill. ; 25 cm. — (Les Poussins Hachette) F***
ISBN 2-01-004782-6
I. Title.
PZ23.G68J827 843'.9'14 79-372390
79 MARC

Gorde, Monique.
Les Jumeaux se déguisent / Monique Gorde. — Paris : Hachette, c1978.
₍19₎ p. : col. ill. ; 25 cm. — (Les Poussins Hachette) F***
ISBN 2-01-005601-9
I. Title.
PZ23.G68J83 843'.9'14 79-372391
79 MARC

Grenier, Christian.
Les cascadeurs du temps / Christian Grenier ; ill. de Jean Revelin. — Paris : Magnard, c1977.
189 p., ₍7₎ leaves of plates : ill. (some col.) ; 21 cm. — (Fantasia ; 30) F***
I. Title.
PZ23.G745 Cas 843'.9'14 79-347292
79 MARC

Grenier, Christian.
Il y a deux soleils chez les Tortupatons / texte de Christian Grenier ; imagé par Frédéric Clément. — Paris : Magnard, c1977.
30 p. : col. ill. ; 29 cm. — (Grand carré : 1 ₍i.e. un₎) F***
Cover title.
ISBN 2-210-99813-1
I. Clément, Frédéric, fl. 1975- II. Title.
PZ23.G745I4 843'.9'14 79-372790
79 MARC

Gripari, Pierre.
Nanasse et Gigantet : conte en forme d'échelle / de Pierre Gripari ; illustré par Jean-Luc Allart. — Paris : Grasset, 1977.
₍23₎ p. : col. ill. ; 26 cm. — (Grasset jeunesse) F78-12877
ISBN 2-246-00501-9 : 28.00F
I. Allart, Jean Luc. II. Title.
PZ23.G747N3 843'.9'14 79-372389
79 MARC

Held, Jacqueline.
Le crocodile aspirateur/ texte de Jacqueline et Claude Held ; imagé par Françoise Purrey. — ₍Paris₎ : Magnard, ₍c1977₎
27 p. : col. ill. ; 29 cm. — (Grand carré : Formule 0) F***
Cover title.
ISBN 2-210-99706-2
I. Held, Claude Ludwig. II. Purrey, Françoise. III. Title.
PZ23.H335C7 79-360329
79 MARC

Held, Jacqueline.
Mais où est donc Arsinoë? / texte de Jacqueline Held ; imagé par André Vial. — Paris : Magnard, 1976.
31 p. : col. ill. ; 29 cm. — (Les Albums grand carré) F78-5649
ISBN 2-210-99803-4 : 13.65F
I. Vial, André. II. Title.
PZ23.H335M3 843'.9'14 79-372791
79 MARC

Helft, Claude.
Le Voyage / texte de Claude Helft ; illustrations de Renée Koch. — Paris (180, rue Lafayette, 75010) : d'Au, 1977.
₍26₎ p. : col. ill. ; 28 cm. F78-3464
ISBN 2-85936-105-7 : 31.00F
I. Koch, Renée. II. Title.
PZ23.H337V68 843'.9'14 79-345664
79 MARC

Hélias, Pierre Jakez.
Comment un Breton devint roi d'Angleterre / conte de Per Jakez Hélias ; dessins de Claude Kerfriden. — Paris : Éditions G.P., ₍c1976₎
₍34₎ p. : col. ill. ; 27 cm. F***
ISBN 2-261-00192-4 : 35.00F
I. Kerfriden, Claude. II. Title.
PZ23.H339C6 891.6'8'33 79-373112
79 MARC

Joubert, Jean.
Le voyage à Poudrenville : conte-poème / de Jean Joubert ; images de Danièle Bour. — 1. éd. — ₍Paris₎ : J.-P. Delarge, 1977.
₍24₎ p. : col. ill. ; 18 cm. F***
ISBN 2-7113-0070-6
I. Bour, Danièle. II. Title.
PZ23.J67V6 841'.9'14 79-372993
79 MARC

Kahn, Michèle.
Boucles d'Or va au bal / ₍texte de₎ Michèle Kahn, ₍illustrations de₎ Henriette Munière. — Paris : Hachette, 1978.
₍21₎ p. : col. ill. ; 25 cm. — (Collection Les Poussins : Série bleue) F79-599
ISBN 2-01-004996-9 : 10.00F
I. Munière, Henriette. II. Title.
PZ23.K26B6 843'.9'14 79-372386
79 MARC

Kahn, Michèle.
Moi et les autres / Michèle Kahn ; illustré par Arnaud Laval ... ₍et al.₎. — Paris : Bordas, c1978.
95 p. : col. ill. ; 26 cm. F***
ISBN 2-04-010187-X
I. Laval, Arnaud. II. Title.
PZ23.K26M6 843'.9'14 79-365981
79 MARC

Laboulaye, Édouard de, 1811-1883.
Nouveaux contes bleus / Édouard Laboulaye ; ill. de Janine Godin. — Paris : Flammarion-Jeunesse, c1961.
152 p. : col. ill. ; 22 cm.
"10"
CONTENTS: Zerbin le farouche.—Le pacha berger.—Perlino.—La sagesse des nations.—Le château de la vie.
I. Title.
PZ23.L12N6 1961 79-338100
79 MARC

Laroque, Robert, 1889-
Les habits rouges : roman / Robert de Roquebrune ₍i.e. R. Laroque₎ ; ill. de Josée Guberek. — Montréal : Fides, c1978.
137 p., ₍4₎ leaves of plates : col. ill. ; 21 cm. — (Collection du goéland) C***
ISBN 0-7755-0684-2 : $7.75
I. Canada—History—Rebellion, 1837-1838—Juvenile fiction. I. Title.
PZ23.L194H3 1978 843 79-358441
79 MARC

Laude, André.
Animalphabet / André Laude ; ill. de J. Ghin. — Paris : Éditions Saint-Germain-des-Prés, c1977.
38 p. : ill. ; 19 cm. — (Collection L'Enfant, la poésie) F***
Poems.
ISBN 2-243-00189-6 : 20.00F
1. Animals—Juvenile poetry. I. Title.
PZ24.3.L35 841'.9'14 78-363391
78 MARC

L'Hermite, Simone.
Ursule Gazon / ₍Simone l'Hermite₎. — Paris : Magnard, c1977.
31 p. : chiefly col. ill. ; 29 cm. — (Grand carré : Formule O) F***
Paging begins and ends on covers.
ISBN 2-210-99705-4
I. Title.
PZ23.L577U7 78-398273
79 MARC

CHILDREN'S LITERATURE, FRENCH (Continued)

Lorenzo, Charles.
Contes et récits / Charles Lorenzo. — La Prairie, Qué[bec] : Entreprises culturelles, 1978-
v. ; 22 cm. C79-2691-0
CONTENTS: recueil 1. L'ombre du cimetière. Le cadavre de l'étang. La mystérieuse adolescente. La cachette impénétrable. La rivale.

$5.50 (v. 1)
PZ23.L74C6 79-371756
79 MARC

Marais, Jean, 1913-
Contes / de Jean Marais ; illustration ... de l'auteur. — Paris : A. Michel, 1978.
145 p. : ill. ; 24 cm. F78-20842
ISBN 2-226-00733-4 : 29.00F
I. Title.
PZ23.M255C6 79-371751
79 MARC

Martin-Chauffier, Simone.
L'Autre chez les corsaires / Simone Martin-Chauffier ; illustrations de Arias Crespo. — Paris : Éditions G.P., 1977.
186 p. : ill. (some col.) ; 21 cm. — (Souveraine ; 354) F78-8791
ISBN 2-261-00368-4 : 12.00F
I. Title.
PZ23.M427A9 843'.9'14 79-348742
79 MARC

Obaldia, René de.
Chez moi : innocentine / de René de Obaldia ; illustrée par Letizia Galli. — Paris : Grasset, 1977.
[24] p. : col. ill. ; 23 cm. — (Grasset jeunesse) F78-12888
ISBN 2-246-00503-5 : 28.00F
I. Title.
PZ24.3.O2 79-361312
79 MARC

Pont, Gabriel.
Les crêpes de l'avent : nouvelle / Gabriel Pont ; ill. Marie-Paule Gailland. — Sierre : Château Ravire ; Saint-Maurice : diffusion Œuvres Saint-Augustin, achevé d'impr. 1978.
[38] p., [8] leaves of col. plates ; 22 cm. Sw78-11967
15.00F
I. Title.
PZ23.P623C7 843'.9'14 79-365985
79 MARC

Pulicani, Suzanne.
Les vacances marseillaises de Pistou / Suzanne Pulicani ; ill. de Patrice Harispe. — Paris : Magnard, c1977.
183 p., [8] leaves of plates : ill. (some col.) ; 21 cm. — (Fantasia ; 25)
F***
I. Title.
PZ23.P8V3 843'.9'14 79-358439
79 MARC

Rabier, Benjamin Armand, 1869-1939
Gédéon, chef de brigands / Benjamin Rabier. — Paris : Garnier frères, c1978.
47 p. : col. ill. ; 32 cm. F***
Reprint of the 1931 ed. published by Garnier frères, Paris.
ISBN 2-7050-0139-5
I. Title.
PZ23.R18G4 1978 843'.8 79-373111
79 MARC

Rabier, Benjamin Armand, 1869-1939.
Gédéon en Afrique / par Benjamin Rabier. — Paris : Garnier, c1977.
[48] p. : col. ill. ; 32 cm. F***
ISBN 2-7050-0117-4
I. Title.
PZ26.3.R15 1977 843'.9'12 78-371033
78 MARC

Remise, Jac.
Marie-Stéphanie et les quatre saisons / une histoire photographiée et racontée par Jac Remise. — [Paris] : Flammarion, 1975.
[46] p. : col. ill. ; 29 cm. F76-1189
48.00F
I. Title.
PZ24.9.R38 78-401112
79 MARC

Salembier, Maïlé Ch.
Un petit éléphant de rien du tout / Maïlé-Ch. Salembier. — Tournai : Casterman, c1978.
20 p. : col. ill. ; 26 cm. — (Collection Farandole) Be***
ISBN 2-203-10440-6
I. Title.
PZ23.S252P4 843'.9'14 79-372788
79 MARC

Selve, Francine de.
Le pommier doux / Francine de Selve ; ill. de Patrice Harispe. — Paris : Magnard, c1976.
184 p., [8] leaves of plates : ill. (some col.) ; 21 cm. — (Fantasia ; 21)
F***
ISBN 2-210-99021-1
I. Title.
PZ23.S48P65 843'.9'14 79-354398
79 MARC

Tersac, Hélène.
Petit Jean et le marchand de sable / texte d'Hélène Tersac ; illustrations de Renate Magnier. — [Paris] : Grasset, 1977.
[24] p. : col. ill. ; 20 cm. — (Album trois pommes pour les enfants hauts comme trois pommes) (Grasset jeunesse) F78-2043
ISBN 2-246-00502-7 : 22.00F
I. Magnier, Renate. II. Title.
PZ23.T45P4 78-400788
79 MARC

Testa, Fulvio.
La courte échelle / texte et images de Fulvio Testa. — 1. éd. — [Paris] : J.-P. Delarge, c1977.
[25] p. : ill. ; 18 cm. F***
ISBN 2-7113-0069-2
I. Title.
PZ23.T47C6 1977 843'.9'14 79-364354
79 MARC

Thériault, Yves.
Le ru d'Ikoué : roman / Yves Thériault ; ill. de Michelle Poirier. — Montréal : Éditions Fides, c1963, 1977 printing.
120 p., [4] leaves of plates : ill. ; 21 cm. — (Collection du Goéland)
C77-10178-X
ISBN 0-7755-0631-1
I. Title.
PZ23.T48R8 843 78-376130
78 MARC

Vérité, Marcelle.
Mirliton, le poulain / Marcelle Vérité, Romain Simon. — Paris : Gautier-Languereau, [c1977]
44 p. : col. ill. ; 20 cm. — (Premiers livres ISSN 0338-8670) F***
ISBN 2-217-23030-1
I. Simon, Romain, 1915- joint author. II. Title.
PZ26.3.V445 78-363387
78 MARC

—BIBLIOGRAPHY

Centre régional de documentation pédagogique de Lille.
Littérature et presse de jeunesse : bibliographie / Centre national de documentation pédagogique, Centre régional de documentation pédagogique de Lille. — Lille : CRDP, 1977.
35 p. ; 30 cm. F***
Cover title.
1. Books and reading for children—France—Bibliography. 2. Bibliography—Bibliography—Children's literature, French. 3. Children's literature, French—Bibliography. 4. Press, Juvenile—France—Bibliography. 5. Libraries, Children's—France. I. Title.
Z1037.2.C46 1977 028.52 78-372412
78 MARC

—HISTORY AND CRITICISM

Kaeppelin-Billaudot, Marie L'or.
Il sera une fois : essai sur le contenu des livres pour enfants / Marie-L'or Kaeppelin-Billaudot. — Claix : La Pensée sauvage, c1978.
149 p. : ill. ; 22 cm. — (Espaces féminins) F***
Bibliography: p. [147]-149.
ISBN 2-85919-011-2 : 32.00F
1. Children's literature, French—History and criticism. I. Title.
PN1009.F8K3 840.9'9282 78-390549
79 MARC

CHILDREN'S LITERATURE, GERMAN

Askenazy, Ludvik.
Wo die goldene Schildkröte tanzt / Ludvik Askenazy ; [Ill. von Dieter Wiesmüller]. — Aarau [etc.] : Sauerländer, c1977.
98 p. : ill. ; 22 cm. Sw78-1653
ISBN 3-7941-1603-8 : 16.80F
I. Title.
PZ33.A835 78-371908
78 MARC

Baum, Willi.
Fred und Tom und die Indianer / erdacht und gemalt von Willi Baum ; erzählt von Ernst A. Ekker. — 1. Aufl. — Ravensburg : O. Maier, 1978.
[32] p. : col. ill. ; 26 cm. GFR***
ISBN 3-473-33846-X
I. Ekker, Ernst Alfred. II. Title.
PZ33.B31123 79-371760
79 MARC

Der Boss und andere Geschichten / ausgew. von Weert Flemmig, Hans May und Hans-Heinrich Strube ; [Ill., Frantz Wittkamp]. — [Orig.-Ausg.]. — Gütersloh : Gütersloher Verlagshaus Mohn, 1977.
124 p. ; 19 cm. — (Gütersloher Taschenbücher ; 223 : Siebenstern)
GFR77-A
Includes bibliographical references.
ISBN 3-579-03983-0 : DM4.80
1. Children's literature, German. 2. Conduct of life—Literary collections. I. Flemmig, Weert. II. May, Hans, Pastor. III. Strube, Hans-Heinrich.
PZ31.B64 833'.9'1080355 79-362305
79 MARC

Bouhuys, Mies.
Een boot in het koren / Mies Bouhuys ; ill. Francien van Westering. — Bentveld-Aerdenhout : Landshoff, c1977.
96 p. : ill. ; 25 cm. Ne78
ISBN 9062100430 : fl 24.50
I. Westering, Francien van. II. Title.
PZ13.B6178 78-385027
79 MARC

Brandes, Sophie, 1943-
Hauptsache, jemand hat dich lieb : Kinderroman / Sophie Brandes ; mit Bildern von Sophie Brandes. — Weinheim : Beltz & Gelberg, c1976.
158 p. : ill. ; 20 cm. GFR***
ISBN 3-407-80513-6
I. Title.
PZ33.B6984 78-361981
78 MARC

Brender, Irmela, 1935-
Die Kinderfamilie / Irmela Brender ; [Illustrationen, Heinz Edelmann]. — München : C. Bertelsmann, c1976.
127 p. : ill. ; 22 cm. GFR***
ISBN 3-570-07688-1
I. Title.
PZ33.B7435 78-375386
78 MARC

Bruyn, Günter de.
Tristan und Isolde / nach Gottfried von Strassburg neu erzählt von Günter de Bruyn ; [farb. Ill. von Peter Nagengast]. — München : Kindler, 1975.
134 p. : ill. (col.) ; 24 cm. GFR76-A
ISBN 3-463-00628-6 : DM20.00
1. Tristan—Romances—Juvenile fiction. I. Gottfried von Strassburg, 13th cent. Tristan und Isolde. II. Title.
PZ34.1.B78 78-400789
79 MARC

Danella, Utta.
Die Tränen vom vergangenen Jahr : Roman / Utta Danella. — München : Bertelsmann, 1978.
191 p. ; 22 cm. GFR78-A
ISBN 3-570-03171-3 : DM16.80
I. Title.
PZ33.D24 78-400786
79 MARC

David, Kurt, 1924-
Der Bär mit dem Vogel auf dem Kopf : Geschichten u. Bilder aus d. Mongolei / erzählt von Kurt David ; gemalt von Gerhard Gossmann. — 1. Aufl. — Berlin : Kinderbuchverlag, 1977.
134 p. : numerous ill. (some col.) ; 25 cm. GDR77-A
12.80M
1. Mongolia—Social life and customs—Juvenile fiction. I. Gossmann, Gerhard. II. Title.
PZ33.D27 78-384434
78 MARC

Der dicke Karpfen Kilobald / von Luis Murschetz. — Zürich : Diogenes Verlag, cop. 1978.
[26] p. : chiefly col. ill. ; 28 cm. — (Ein Diogenes Kinderbuch)
Sw78-10867
ISBN 3-257-00600-4 : 16.80F
I. Title.
PZ36.3.M823 78-400799
79 MARC

Der Fliegende Robert : Erzählungen, Bilder, Gedichte, Märchen, Zukunftsgeschichten, Rätsel, Fotos, Kinderzeitung / [Autoren und Illustratoren, Max Barthel ... et al.] ; hrsg. von Hans-Joachim Gelberg. — Weinheim : Beltz & Gelberg, c1977.
333 p. : ill. (some col.) ; 25 cm. — (Jahrbuch der Kinderliteratur ; 4)
GFR***
ISBN 3-407-80530-6
1. Children's literature, German. I. Barthel, Max, 1893-1975. II. Gelberg, Hans-Joachim.
PZ31.F57 78-368202
78 MARC

Fix, Philippe.
Serafin gegen Serafin / Text von Alain Grée ; aus dem Französischen von Sabine Ibach. — Zürich : Diogenes, c1971.
28 p. : ill. ; 31 cm. — (Ein Diogenes Kinderbuch)
Translation of Sérafin contre Sérafin.
I. Grée, Alain. II. Title.
PZ33.F4866 78-378314
78 MARC

CHILDREN'S LITERATURE, GERMAN
(Continued)

Fuchshuber, Annegert.
Die Geschichte vom Bombardon und vom Konrad, der es fand und wieder verlor / von Annegert Fuchshuber. — Stuttgart : Thienemann, c1977.
[28] p. : numerous ill. (all col.) ; 24 cm. GFR***
"Ein Thienemann Bilderbuch."
ISBN 3-522-41410-1
I. Title.
PZ33.F7723 78-369178
78 MARC

Gabel, Wolfgang, 1942-
Der Aufstieg der Familie Kohlbrenner / Wolfgang Gabel ; Fotomontagen von Hans Mende. — Neunkirchen im Odenwald : Anrich, c1976.
148 p. : ill. ; 21 cm. GFR***
ISBN 3-920110-28-5
I. Title.
PZ33.G26 78-361980
78 MARC

Gast, Lise, 1908-
Unser kleiner Esel Jan. [Von] Lise Gast. Bilder von Rolf Rettich. Schrift von Julius Kim. Bayreuth, Loewe, 1969.
59 p. illus. 24 cm. 5.90 GDNB69-A23-337
I. Rettich, Rolf, illus. II. Title.
PZ33.G333 75-482842
77[r78]rev MARC

Gast, Lise, 1908-
Was wird aus Regine? Reutlingen, Ensslin & Laiblin [1953]
79 p. illus. 19 cm. (Kleine Ensslin-Bücher, 33)
I. Title.
PZ33.G334 53-39701
[r78]rev MARC

Gernhardt, Almut, 1940-
Mit dir sind wir vier / Bilder von Almut Gernhardt ; mit Geschichten von Robert Gernhardt. — Frankfurt am Main : Insel Verlag, c1976.
[29] p. : numerous col. ill. ; 23 cm. — (Ein Insel-Bilderbuch) GFR***
I. Gernhardt, Robert, 1937- II. Title.
PZ36.3.G46 78-367944
78 MARC

Grasso, Mario, fl. 1975-
Die Geschichte vom Jäger, der alle Tiere auf der ganzen Welt töten wollte : 10 Geschichten / erzählt und ill. von Mario Grasso. — Basel [Bachlettenstrasse 39] : J. Fink, cop. 1975.
31 p. : numerous ill. ; 30 cm. Sw78-9839
14.80F
I. Title.
PZ36.3.G66 78-400785
79 MARC

Grasso, Mario, fl. 1975-
Metamorfosia / Mario Grasso. — Basel : Sphinx-Verlag, 1976.
[36] p. : col. ill. ; 37 cm. Sw77-307
ISBN 3-85914-120-1 : 38.00F
I. Title.
PZ33.G737 77-559970
*77[r79]rev MARC

Grünler, Jörg.
Knacks, sagte das Ei, die ganze Luft ist voller Bilder / Jörg Grünler, Hoschy Tiessler ; hrsg. von Harald Hohenacker und illustriert von Dschinghi Weisse. — Stuttgart : Deutsche Verlags-Anstalt, c1976.
96 p. : ill. ; 25 cm. GFR***
ISBN 3-421-02492-8
I. Tiessler, Hoschy, joint author. II. Hohenacker, Harald. III. Weisse, Dschinghi. IV. Title.
PZ33.G8474 78-361979
78 MARC

Hacks, Peter, 1928-
Meta Morfoss und Ein Märchen für Claudias Puppe / Peter Hacks ; [Ill. von Gisela Neumann]. — 1. Aufl. — Berlin : Kinderbuchverlag, 1975.
59 p. : col. ill. ; 15 cm. — (Die kleinen Trompeterbücher ; 109)
GDR74-A
1.75M
I. Neumann, Gisela. II. Title.
PZ33.H1067 833'.9'14 79-375826
79 MARC

Hauptmann, Tatjana.
Ein Tag im Leben der Dorothea Wutz / von Tatjana Hauptmann. — Zürich : Diogenes-Verlag, cop. 1978.
[13] leaves : all col. ill. ; 30 x 33 cm. — (Ein Diogenes Kinderbuch)
Sw78-10859
ISBN 3-257-00601-2 : 19.80F
I. Title.
PZ36.3.H33 78-400797
79 MARC

Heuck, Sigrid.
Petah Eulengesicht / Sigrid Heuck ; illustriert von Monika Cronshagen. — München : C. Bertelsmann, c1977.
105 p. : ill. ; 22 cm. GFR***
ISBN 3-570-07691-1
I. Cronshagen, Monika. II. Title.
PZ33.H326 833'.9'12 79-364021
79 MARC

Hoerning, Martina.
Eine Geschichte ohne Ende : u. andere Szenen mit Liedern f. d. Unterstufe, nach Text oder aus d. Stegreif, im Kreis oder auf d. Bühne zu spielen / Martina u. Konrad Hoerning. — Leipzig : Zentralhaus für Kulturarbeit, 1974.
53 p. : music ; 21 cm. — (Laientheater) (Kindertheater) GDR74-A
"Die Spiele wurden angeregt durch das Heft 'Little plays' von S. B. Saks, Moskau, 1962."
2.20M
I. Hoerning, Konrad, joint author. II. Title.
PZ37.H62 78-371910
78 MARC

Hoffman, Felix.
Der Bärenhäuter : ein Märchenbilderbuch im Faksimiledruck nach den Brüdern Grimm / von Felix Hoffmann. — 1. Aufl. — Aarau : Sauerländer, 1978, c1977.
[32] p. : col. ill. ; 31 cm. Sw***
ISBN 3-7941-1660-7
I. Grimm, Jakob Ludwig Karl, 1785-1863. Der Bärenhäuter. 1978. II. Title.
PZ33.H6185 833'.9'14 79-374779
79 MARC

Hoffmann, Hans.
Beim ersten Mal gelingt nicht alles : Geschichten aus Icksdorf / Hans Hoffmann. — 1. Aufl. — Mainz : Matthias-Grünewald-Verlag, 1977.
96 p. : ill. ; 20 cm. GFR77-A
ISBN 3-7867-0614-X
I. Title.
PZ33.H6275 78-371909
78 MARC

Horváth, Maria, fl. 1968-
Wenn ich gross bin; Kinderwünsche aufgeschrieben von Maria Horváth, mit Bildern von Annegert Fuchshuber. München, A. Betz [1968]
[27] p. col. illus. 22 cm. (Bilderbücher der Sechs)
I. Fuchshuber, Annegert, illus. II. Title.
PZ33.H658 76-365632
79[r79]rev MARC

Janosch.
Das grosse Janosch Buch : Geschichten und Bilder. — 21.-30. Tausend. — Weinheim : Beltz & Gelberg, 1977, c1976.
293 p. : ill. (some col.) ; 25 cm. GFR***
ISBN 3-407-80514-4
I. Title.
PZ33.J3348 1977 78-361976
78 MARC

Kalendergeschichten unserer Zeit : ein Hausbuch zum Jahreslauf : Geschichten, Lieder und Gedichte, Merkwürdiges und Merkenswertes zum Lesen und Vorlesen, Aufschreiben und Spielen / hrsg. von Barbara Bartos-Höppner ; grafische Gesamtgestaltung von Aiga Rasch. — 1. Aufl. — Würzburg : Arena, 1977.
260 p. : ill. (chiefly col.) ; 21 cm. GFR***
ISBN 3-401-03807-9
1. Children's literature, German. I. Bartos-Höppner, Barbara, 1923- II. Rasch, Aiga.
PZ31.K34 78-361975
78 MARC

Kaulbach, Hermann, 1846-1909.
Bilderbuch / Text von Adelheid Stier. — Bearb. Ausg. — München : Deutscher Taschenbuch Verlag, 1977.
53 p. : ill. ; 19 cm. — (Dtv Junior ; 7290) GFR***
ISBN 3-423-07290-3 : DM6.80
I. Stier, Adelheid, 1852-1942.
PZ33.K35 1977 78-366803
78 MARC

Kleberger, Ilse.
Damals mit Kulicke : vom Luftschutzkeller bis zum schwarzen Markt / Ilse Kleberger ; mit Zeichnungen von Kossatz. — 1. Aufl. — Berlin : Arani, 1978.
103 p. ; 19 cm. GFR78-A
ISBN 3-7605-8532-9
I. Kossatz, Hans, 1901- II. Title.
PZ33.K4867 79-359028
79 MARC

Korschunow, Irina.
Steffis roter Luftballon : Geschichten für die Kleinsten / Irina Korschunow ; mit Bildern von Edith Schindler. — Zürich [etc.] : Benziger, cop. 1978.
74 p. : numerous ill. (some col.) ; 22 cm. Sw78-10314
ISBN 3-545-30024-2 : 15.80F
I. Schindler, Edith. II. Title.
PZ33.K59657 79-352033
79 MARC

Kreye, Walter.
Die Leseratte und die Wasserratte / Walter Kreye ; mit Bildern von Mette Newth. — Mönchaltorf [etc.] : Nord-Süd-Verlag, cop. 1978.
[28] p. : chiefly col. ill. ; 30 cm. — (Ein Nord-Süd Bilderbuch)
Sw78-11417
ISBN 3-85825-108-9 : 16.80F
I. Newth, Mette. II. Title.
PZ36.3.K64 79-351893
79 MARC

Kreye, Walter Arthur.
Dascha sagenhaft : wie Gräfin Emma d. klein Kerl mang Roland seine Füsse inne Gange brachte u. so d. Bremer zu ihre Bürgerweide kamen / Walter A. Kreye ; und Volker Ernsting. — Bremen : Schmalfeldt, 1977.
[47] p. : col. ill. ; 28 cm. GFR78-A
Text by W. A. Kreye; illustrations by V. Ernsting.
ISBN 3-921749-02-6 : DM19.80
I. Ernsting, Volker. II. Title.
PZ33.K6457 79-360330
79 MARC

Kunert, Günter, 1929-
Jeder Wunsch ein Treffer / Günter Kunert ; illustriert von Heinz Edelmann. — Velber/Hannover : G. Middelhauve, c1976.
76 p. : ill. ; 19 cm. GFR***
ISBN 3-7876-8160-4
I. Edelmann, Heinz, fl. 1967- II. Title.
PZ33.K798 79-383037
79 MARC

Ludwig, Helmut.
Kampf um den Südpol : 2 Tatsachenberichte / von Helmut Ludwig. — Wuppertal : Brockhaus, 1976.
125 p. ; 15 cm. — (Kleine R. Brockhaus Bücherei ; Bd. 118)
GFR77-A25
ISBN 3-417-00565-5 : DM2.50
1. Scott, Robert Falcon, 1868-1912—Juvenile fiction. 2. South Pole—Juvenile fiction. I. Title.
PZ33.L812 79-383236
79 MARC

Morgenstern, Christian, 1871-1914.
Klein Irmchen : Kindergedichte ; Klaus Burrmann der Tierweltphotograph / Christian Morgenstern ; [Ill. von Katharina Eisleben]. — Basel : Zbinden, 1978.
93 p. : ill. (some col.) ; 21 cm. — (His Sämtliche Dichtungen ; Bd. 16) (Aus mitteleuropäischem Geistesleben) Sw78-8531
Includes index.
ISBN 3-85989-165-0 : 27.00F
I. Morgenstern, Christian, 1871-1914. Klaus Burrmann der Tierweltphotograph. II. Title.
PZ34.3.M63 1978 78-389258
79 MARC

Müller, Jörg.
Die Kanincheninsel / Text, Jörg Steiner ; Bilder, Jörg Müller. — 2. Aufl. — Aarau : Sauerländer, cop. 1977.
[32] p. : chiefly ill. (all col.) ; 33 cm. Sw78-1074
ISBN 3-7941-1636-4 : 18.80F
I. Steiner, Jörg. II. Title.
PZ36.3.M794 1977 78-368240
78 MARC

Munchausen. German. Selections.
Die wunderbaren Reisen und Abenteuer des Freiherrn von Münchhausen, wie er dieselben bei einer Flasche im Zirkel seiner Freunde selbst zu erzählen pflegte : die Landabenteuer / [Ill. von] Binette Schroeder ; Textbearb. von Peter Nickl. — Mönchaltorf [etc.] : Nord-Süd-Verlag, cop. 1977.
52 p. : numerous col. ill. ; 33 cm. — (Ein Nord-Süd Bilderbuch)
Sw78-8579
"Der Textbearbeitung dienten die 3. Raspe'sche Auflage (1786) und die 2. Bürger'sche Auflage (1788) als Vorlage."
ISBN 3-85825-074-0 : 28.00F
I. Schroeder, Binette. II. Nickl, Peter. III. Title.
PZ33.M852 1977 79-349474
79 MARC

Ruck-Pauquèt, Gina.
Das grosse Buch von Gina Ruck-Pauquèt : Geschichten, Spässe u. Gedichte / ill. von Edith Schindler. — 1. Aufl. — Ravensburg : Maier, 1978.
217 p. : ill. ; 24 cm. GFR78-A
Bibliography of the author's works: p. 214-216.
ISBN 3-473-37320-6 : DM24.00
I. Schindler, Edith. II. Title.
PZ33.R699 79-352032
79 MARC

Säuberli, Jack.
Köbi & [i.e. und] Lisi : eine Zwillingsgeschichte / ausgedacht und gezeichnet von Jack Säuberli. — Horgen : Bosbo Verlag, c1977.
[28] p. : chiefly ill. (all col.) ; 17 x 20 cm. — (Ein Bosbo Kinderbuch)
Sw78-4217
12.00F
I. Title.
PZ33.S186 833'.9'14 78-381233
79 MARC

Schmid, Eleonore.
Mein Kätzchen Sebastian / Geschichte und Bilder, Eleonore Schmid ; [Text, Margrit Rüedi-Mattes nach einer Idee von Eleonore Schmid]. — Mönchaltorf [etc.] : Nord-Süd-Verlag, cop. 1978.
[28] p. : chiefly ill. (col.) ; 30 cm. — (Ein Nord-Süd Bilderbuch)
Sw78-11424
ISBN 3-85825-109-7 : 16.80F
I. Rüedi-Mattes, Margrit. II. Title.
PZ36.3.S3918 79-345525
79 MARC

CHILDREN'S LITERATURE, GERMAN (Continued)

Scholey, Arthur.
Der Wunschvogel : eine Geschichte / von Arthur Scholey ; mit Bildern von Bernadette. — Mönchaltorf ⌈etc.⌉ : Nord-Süd-Verlag, cop. 1978.
⌈32⌉ p. : col. ill. ; 19 x 25 cm. Sw78-11425
ISBN 3-85825-107-0 : 16.80F
I. Watts, Bernadette. II. Title.
PZ36.3.S4156 79-346817
79 MARC

Schlote, Wilhelm.
Briefe an Sarah / Wilhelm Schlote ; Text von Elisabeth Borchers. — 1. Aufl. — Frankfurt am Main : Insel Verlag, 1977.
⌈32⌉ p. : col. ill. ; 31 cm. — (Ein Insel-Bilderbuch) GFR***
ISBN 3-458-05067-1
I. Borchers, Elisabeth. II. Title.
PZ33.S4635 78-361984
78 MARC

Schlote, Wilhelm.
Die Zeichenstunde / Wilhelm Schlote ; Text von Elisabeth Borchers. — Frankfurt am Main : Insel Verlag, c1977.
⌈31⌉ p. : chiefly col. ill. ; 31 cm. — (Ein Insel-Bilderbuch) GFR***
ISBN 3-458-05039-6
I. Borchers, Elisabeth. II. Title.
PZ33.S46376 78-366454
78 MARC

Schlote, Wilhelm.
Uaaaahhh / Wilhelm Schlote. — Hamburg : Oetinger, c1977.
⌈24⌉ p. : chiefly ill. (all col.) ; 18 x 25 cm. GFR***
Cover title.
ISBN 3-7891-6151-9
I. Title.
PZ36.3.S3915 78-368530
78 MARC

Siegentaler, Peter.
Da staunte Noachs Tochter : Geschichten von Menschen u. Tieren für kleine u. grosse Leute / Peter Siegentaler, Fulvio Testa. — 1. Aufl. — Düsseldorf : Patmos-Verlag, 1978.
112 p. : ill. ; 21 cm. GFR78-A
ISBN 3-491-79028-X : DM14.80
I. Testa, Fulvio, joint author. II. Title.
PZ36.3.S513 78-400790
79 MARC

Stahel, Nelly.
Der kleine Rothut / ein Bilderbuch von Nelly Stahel ; mit Ill. von Bruno Raimondi. — Zürich-Erlenbach ⌈etc.⌉ : E. Rentsch, c1978.
⌈24⌉ p. : numerous ill. (all col.) ; 22 x 29 cm. Sw78-7371
ISBN 3-7249-0504-1 : 17.50F
I. Raimondi, Bruno. II. Title.
PZ33.S7192 79-383078
79 MARC

Tidick, Markus Joachim.
Pit und Petty an Bord : Segelferien auf der Ostsee / Markus Joachim Tidick ; illustriert von Kurt Schmischke. — Oldenburg : G. Stalling, c1977.
87 p. : ill. ; 26 cm. GFR***
ISBN 3-7979-2527-1
I. Schmischke, Kurt. II. Title.
PZ33.T498 79-371757
79 MARC

Tolkien, John Ronald Reuel, 1892-1973.
Farmer Giles of Ham = Die Geschichte vom Bauern Giles und dem Drachen / J. R. R. Tolkien ; ⌈Übers., Angela Uthe-Spencker ; ill., Pauline Diana Baynes⌉. — 9.-12. Tsd. — München : Deutscher Taschenbuch-Verlag, 1976.
143 p. : ill. ; 18 cm. — (Dtv zweisprachig ; 9073) (Edition Langewiesche-Brandt) GFR76-A
English and German in parallel texts.
ISBN 3-423-09073-1 : DM4.80
I. Title. II. Title: Die Geschichte vom Bauern Giles und dem Drachen.
PZ34.T65 1976 823'.9'12 79-365907
79 MARC

Walbert, Helmut, 1937-
Der Riesenvogel / Helmut Walbert ; illustriert von Leo Leonhard. — München : C. Bertelsmann, c1976.
109 p. : ill. ; 22 cm. GFR***
ISBN 3-570-05146-3
I. Leonhard, Leo. II. Title.
PZ36.3.W29 78-361977
78 MARC

Weiss, Walter W
Die Rache der Kachinas : ein Indianerbuch / Walter Weiss ; ⌈sämtl. Aufn. stammen vom Autor⌉. — Wien ; München : Jugend & Volk, c1977.
191 p. : col. ill. ; 25 cm. Au77-21-429
ISBN 3-7141-1510-2 : S178.00
1. Navaho Indians—Juvenile fiction. I. Title.
PZ33.W34787 78-380127
*78 MARC

Wordel, Ria.
Allerhands vun allerhands Deere : ov künnten dat och Minsche sin?! / Ria Wordel. — Köln : Greven-Verlag, 1978.
78 p. ; 21 cm. GFR78-A
ISBN 3-7743-0163-8
I. Title.
PZ36.3.W67 833'.9'14 79-366347
79 MARC

—EXHIBITIONS

Deutsche Kinderbücher des 18. Jahrhunderts : ein Beitrag zur Vorbereitung einer Bibliographie alter deutscher Kinderbücher : Ausstellung der Herzog August Bibliothek Wolfenbüttel 3. November 1978 bis 10. Januar 1979 / ⌈Katalog, Georg Ruppelt, Ingrid Nutz und Michael Schlosser, unter Mitarbeit von Dorothea Balzer und Verona Borchardt⌉. — Wolfenbüttel : Die Bibliothek, 1978.
82 p. : ill. ; 26 cm. — (Ausstellungskataloge der Herzog-August-Bibliothek ; Nr. 25) GFR***
Includes index.
ISBN 3-88373-003-3
1. Children's literature—Publishing—Germany—History—19th century—Exhibitions. 2. Children's literature, German—Exhibitions. 3. Wolfenbüttel—Exhibitions. I. Ruppelt, Georg. II. Nutz, Ingrid. III. Schlosser, Michael. IV. Wolfenbüttel. Herzog-August-Bibliothek. V. Series: Wolfenbüttel. Herzog-August-Bibliothek. Ausstellungskataloge ; Nr. 25.
Z1035.3.D52 028.52'0943 79-368979
⌈PN1009.G3⌉ 79 MARC

—HISTORY AND CRITICISM

Hanxleden, Rüdiger von.
Bilderbücher für das Vorschulalter : e. Analyse d. Erfahrungswelt d. Hauptpersonen / Rüdiger von Hanxleden. — 1. Aufl. — Lollar : Achenbach, 1978.
381 p. ; 20 cm. — (Edition 2000 ⌈i.e. Zweitausend⌉ ; Bd. 63 : Dissertation) GFR78-A
Bibliography: p. 375-381.
ISBN 3-87958-163-0 : DM38.00
1. Children's literature, German—History and criticism. 2. Picture-books for children. I. Title.
PN1009.G3H24 830'.9'9282 79-364239
79 MARC

Schmidt, Joachim, fl. 1968-
Volksdichtung und Kinderlektüre in der ersten Hälfte des 19. Jahrhunderts / Joachim Schmidt. — 1. Aufl. — Berlin : Kinderbuchverlag, 1977.
166 p. ; 24 cm. — (Studien zur Geschichte der deutschen Kinder- und Jugendliteratur ; 3) GDR78-A
Includes bibliographical references and index.
4.80M
1. Children's literature, German—History and criticism. 2. Literature and folk-lore. 3. Folk-lore and children. I. Title. II. Series.
PN1009.Z6S3 78-379249
78 MARC

Wegehaupt, Heinz.
Vorstufen und Vorläufer der deutschen Kinder- und Jugendliteratur bis in die Mitte des 18. Jahrhunderts / Heinz Wegehaupt. — 1. Aufl. — Berlin : Kinderbuchverlag, 1977.
127 p. ; 24 cm. — (Studien zur Geschichte der deutschen Kinder- und Jugendliteratur ; 1) GDR78-A
Bibliography: p. 109-119.
Includes index.
4.80M
1. Children's literature, German—History and criticism. I. Title. II. Series.
PN1009.G3W43 78-396896
79 MARC

CHILDREN'S LITERATURE, GHANAIAN

—AUTHORSHIP—CONGRESSES

Seminar on Writing and Production of Literature for Children, University of Ghana, 1976.
Seminar on Writing and Production of Literature for Children : ⌈selected papers⌉. — Legon : University of Ghana, Dept. of Library and Archival Studies, 1976-
v. ; 28 cm.
Includes bibliographical references.
1. Children's literature—Publishing—Ghana—Congresses. 2. Books and reading for children—Ghana—Congresses. 3. Children's literature, Ghanaian—Authorship—Congresses. I. Ghana. University, Legon. Dept. of Library and Archival Studies.
Z470.G45S45 1976 070.5'73'096 78-313652
78 MARC

Tarsoulē, Geōrgia.
(Hoi philoi mas ta zōa)
Οἱ φίλοι μας τὰ ζῶα· διηγήματα γι' ἀγόρια καὶ κορίτσια. 'Αθῆναι, Ν. 'Αλικιώτης ⌈1961⌉
66 p. illus. 25 cm. (Τὸ Βιβλίο τοῦ παιδιοῦ, 86)
I. Title.
PZ90.G7T3 64-43703

CHILDREN'S LITERATURE, HEBREW

Ashman, Ya'akov.
(Lulu)
לולו / יעקב אשמן ; ציורים, אלישבע נדל־לנדאו. — תל־אביב : הדר, 1978.
96 p. : ill. ; 21 x 23 cm.
Children's stories.
I. Nadel-Landa'u, Elisheva'. II. Title.
PZ90.H3A862 79-951894

Atlas, Yehuda, 1937-
(Yeha-yeled ha-zeh hu ani)
והילד הזה הוא אני / יהודה אטלס ; ציורים, דני קרמן. — ירושלים : כתר ירושלים, c1977.
⌈48⌉ p. : ill. ; 24 cm. — (כתר־לי)
Vocalized.
Series romanized: Keter-li.
I. Kerman, Danni. II. Title.
PZ90.H3A875 78-952039

Gal, Abraham.
(Yalde 708)
ילדי תש"ח : סיפורים לילדים / אברהם גל ; ⌈איורים, נעה⌉. — תל־אביב : אל"ף, ⌈1978⌉
38 p. : ill. ; 22 cm.
I. Title.
PZ90.H3G29 79-951046

Goldszmit, Henryk, 1878-1942.
⌈Król Maciuś Pierwszy. Hebrew⌉
המלך מתיא הראשון / יאנוש קורצ'ק ⌈i. e., הנריק גולדשמיט⌉ ; עברית, אורי אורלב. — ירושלים : בית הוצאה כתר, c197-.
v. : ill. ; 24 cm.
On verso of t. p.: Król Maciuś Pierwszy.
I. Title.
Title romanized: Melekh Matya ha-rishon.
PZ90.H3G5795 1979 79-951775

József, Attila, 1905-037.
Gyöngy a csillag : ⌈versek⌉ / József Attila ; ⌈válogatta és az utószót írta Fodor András ; Würtz Ádám rajzaival⌉. — Budapest : Móra Kiadó, 1977.
75 p. : ill. ; 23 cm. H77
ISBN 9631109194 : 12.00Ft
I. Fodor, András, 1929- II. Würtz, Ádám. III. Title.
PZ90.H8J68 1977 78-395376
*79 MARC

Kohen, Malkah, 1908-
(ha-Dubon Shinzoli ve-tsipor ha-zahav)
הדבון שינזלי וצפור הזהב : ספור לילדים / מלכה כהן. — תל־אביב : אל"ף, ⌈1978⌉
45 p. : ill. ; 22 cm.
Vocalized.
I. Title.
PZ90.H3K62 78-952251

Levanon, No'omi.
(Shenayim anashim)
שנים אנשים : שירים לילדים / נעמי לבנון ; ציורים, מרטין הולט. — תל־אביב : אל"ף, ⌈1977⌉
48 p. : ill. ; 16 x 23 cm.
Vocalized.
I. Holt, Martin. II. Title.
PZ90.H3L425 78-952092

Ron-Feder, Galilah.
(Kamah medinat Yiśrael)
קמה מדינת ישראל / גלילה רון־פדר ; ⌈ציורים, שלום (סול) בסקין⌉. — תל־אביב : הוצאת מלוא, c1977.
96 p. : ill. ; 22 cm. — (סדרת ילדים אלמונים)
Series romanized: Sidrat yeladim almonim.
"מהדורת ידיעות אחרונות."
1. Israel-Arab War, 1948-1949—Juvenile fiction. I. Title.
PZ90.H3R6145 78-952091

Singer, Isaac Bashevis, 1904-
(Shelumi'el ish-Ḥelem ve-'od sipurim)
שלומיאל איש־חלם ועוד סיפורים / יצחק בשבים־זינגר ; ⌈תירגם מאנגלית, צבי ארד ; ציורים, מאת מרגוט צמח⌉. — תל אביב : עם עובד, c1977.
80 p. : ill. ; 23 cm.
On verso of t. p.: When Shlemiel went to Warsaw and other stories.
English translation has title: When Shlemiel went to Warsaw & other stories.
Short stories.
CONTENTS: מודית הפיקה וליזר הקמצן.—צרצור ופזיזה.—רבי ליב והמכשפה קוניגונדה.—חכמי־חלם והמפתח של גנגדל.—שלומיאל הסוחר.—מר עצל ובתו דלות.—חלומו של מנשה.—שלומיאל איש־חלם יוצא לורשה.
I. Title.
PZ90.H3S4915 79-951421

CHILDREN'S LITERATURE, HEBREW (Continued)

—HISTORY AND CRITICISM

Deshe, Michael.
(Ḥinukh li-ḳeri'ah ṭovah)
חינוך לקריאה טובה : על ספרות ילדים ועל בעיות הקריאה החפשית : מיועד למחנכים ולהורים / מיכאל דשא. — תל־אביב : הוצאת יזרעאל, [c1978.
279, [1] p. ; 22 cm.
Bibliography: p. [280]
1. Books and reading for children. 2. Children's literature, Hebrew—History and criticism. I. Title.
Z1037.A1D4 1978 79-951097

CHILDREN'S LITERATURE, HINDI

—BIBLIOGRAPHY

Agrawal, Suren, 1928-
(Hindī meṃ bāla sāhitya, svatantratā prāpti ke paścāt)
हिन्दी में बाल साहित्य, स्वतन्त्रता प्राप्ति के पश्चात्. Children literature in Hindi. [लेखक] सुरेन अग्रवाल, जगदीश चन्द अग्रवाल [तथा] जगदीश लाल. [1. संस्करण] दिल्ली, रचना राकेश, 1972-
v. 19 cm. Rs18.00 (v. 1)
In Hindi.
CONTENTS: 1. 1947-1960.
1. Children's literature, Hindi—Bibliography. I. Aggarwal, J. C., joint author. II. Jagadīśa Lāla, 1932- joint author. III. Title. IV. Title: Children literature in Hindi.
Z1037.8.H5A36 73-907741
[PN1009.H5]

CHILDREN'S LITERATURE, HUNGARIAN

Április 4 : műsorfüzet felszabadulásunk 25. évfordulójára 5-8. osztályosoknak / [Gaál Albert összeállítása. — Budapest : Móra, 1970.
102 p. ; 18 cm. — (Iskolai színpad)
3.50Ft
1. Hungary—History—1945- —Literary collections. 2. Children's literature, Hungarian. I. Gaál, Albert.
PZ90.H8A65 79-349050
79 MARC

Arany, László, 1844-1898.
A kis ködmön : [képes mesekönyv] / Arany László ; Heinzelmann Emma rajzaival. — Budapest : Móra, 1977.
[24] p. : col. ill. ; 22 x 24 cm. H77
"Öt éven felülieknek."
ISBN 9631108732 : 24.00Ft
I. Heinzelmann, Emma. II. Title.
PZ90.H8A73 1977 79-343670
•79 MARC

Balázs, Béla, 1884-1949.
Halálugrás a szakadékon át : [történelmi regény az ifjúság számára] / Balázs Béla ; [Kondor Lajos rajzaival]. — Budapest : Móra Ferenc Könyvkiadó, 1978.
230 p. : ill. ; 19 cm. H78
ISBN 9631113019 : 12.00Ft
1. Spain—History—Civil War, 1936-1939—Juvenile fiction. I. Title.
PZ90.H8B198 1978 79-352039
•79 MARC

Devecseri, Gábor.
Állatkerti útmutató : [gyermekversek] / Devecseri Gábor ; Borsos Miklós rajzaival. — Budapest : Móra, 1977.
37 p. : numerous col. ill. ; 25 cm. H77
ISBN 9631108457 : 22.00Ft
I. Borsos, Miklós, 1906- II. Title.
PZ90.H8D48 78-375385
•78 MARC

Fehér, Klára.
Lesz nekem egy szigetem : [gyermekregény] / Fehér Klára ; [Demjén Zsuzsa rajzaival]. — 3. kiad. — Budapest : Móra, 1977, c1972.
74 p. : ill. ; 24 cm. H77
Sequel to Én sose kapok levelet.
ISBN 963110768X : 11.00Ft
I. Title.
PZ90.H8F393 1977 78-400473
•79 MARC

Hárs, László.
Buci királyfi összes meséje / Hárs László ; [Heinzelmann Emma rajzaival]. — Budapest : Móra, 1977.
226 p. : col. ill. ; 25 cm. H77
ISBN 9631107221 : 32.00Ft
I. Heinzelmann, Emma. II. Title.
PZ90.H8H27 78-366452
•78 MARC

Hevesi, Lajos, 1843-1910.
Jelky András kalandjai / Hevesi Lajos ; [átdolgozta Czibor János ; Kondor Lajos rajzaival]. — Budapest : Móra F. Könyvkiadó, 1977, c1955.
304 p. : ill. ; 20 cm. H***
"Tizenkét éven felülieknek."
ISBN 963110785X : 20.00Ft
1. Jelky, András, 1730-1783—Juvenile fiction. I. Czibor, János. II. Title.
PZ90.H8H48 1977 79-343837
79 MARC

Horváth, Péter, 1951-
A sárga táltos / Horváth Péter ; [Bokros László rajzaival]. — Budapest : Móra, 1977.
58 p. : col. ill. ; 23 cm. H77
"Hat éven felülieknek."
ISBN 9631108252 : 6.00Ft
I. Title.
PZ90.H8H64 78-384433
•78 MARC

Karczag, György.
Idegen páncélban : történelmi regény / Karczag György ; [Nagy András rajzaival]. — Budapest : Móra, 1977.
517 p. : ill. ; 20 cm. H77
ISBN 9631108937 : 39.00Ft
I. Title.
PZ90.H8K32 78-375388
•78 MARC

Kormos, István.
Kerek eszetendő / Kormos István versei ; Reich Károly rajzai. — Budapest : Móra, 1977.
[14] leaves : numerous col. ill. ; 21 x 28 cm. H77
ISBN 9631107337 : 29.00Ft
I. Reich, Károly. II. Title.
PZ90.H8K59 78-370527
•78 MARC

Kormos, István.
A pöttyös zöld lovacska bátyja / Kormos István meséje ; Reich Károly rajzaival. — Budapest : Móra, 1977.
20 p. : numerous col. ill. ; 28 cm. H77
ISBN 9631107051 : 29.00Ft
I. Reich, Károly. II. Title.
PZ90.H8K593 78-370531
•78 MARC

Lipták, Gábor.
Ezüsthíd : [mesék, mondák, történetek] / Lipták Gábor ; [Gaál Éva rajzaival]. — Budapest : Móra, 1977.
385 p. : ill. ; 20 cm. H77
CONTENTS: Ezüsthíd Füreden.—A tihanyi varázsfüvek.—A három tallér.—Éjszaka a kolostorban.—A szepezdi csutora.—Az elrabolt menyasszony.—A vérkúti borbély.
ISBN 9631109216 : 33.00Ft
1. Balaton Lake, Hungary—Juvenile fiction. I. Title.
PZ90.H8L5 79-345445
•79 MARC

Martić, Andjelka.
Radowanowy dar / Andjelka Martić ; ilustrował Stevo Binički ; [přełožił Helmut Kaltšmit]. — W Budyšinje : Domowina, 1966.
[6] p. : col. ill. ; 26 cm.
Cover title.
Translated from Serbo-Croatian.
I. Binički, Stevo. II. Title.
PZ70.W4M37 79-367802
79 MARC

Marton Pálné Homok, Erzsébet.
A faluban szól a nóta : [regény] / Martonné Homok Erzsébet ; [Balogh Péter rajzaival]. — Budapest : Móra, 1978.
283 p. : ill. ; 19 cm. H78
ISBN 963111144X : 15.00Ft
I. Title.
PZ90.H8M29 79-345277
•79 MARC

Nagy, László, 1925-
Csodafiu-szarvas : [versek] / Nagy László ; [az utószót Pomogáts Béla írta ; Kass János rajzaival]. — Budapest : Móra, 1977.
123 p. : ill. ; 22 cm. H77
ISBN 9631110486 : 14.00Ft
I. Title.
PZ90.H8N35 78-378404
•78 MARC

Pándy, Lajos.
A csodálatos nagykendő : perbetei mesék / Pándy Lajos ; [Balogh Péter rajzaival]. — Budapest : Móra, 1977.
146 p. : ill. ; 19 cm. H77
ISBN 9631109062 : 13.00Ft
I. Title.
PZ90.H8P3 78-400777
•79 MARC

Petrolay, Margit.
A szél nótája; mesék. Budapest, Móra Könyvkiadó, 1968.
106 p. illus (part col.) 20 cm. 12.00
I. Title.
PZ90.H8P43 71-228068
79[r79]rev2 MARC

Pintér, Pálné.
A csodakút tündére : balatoni mesék / Pintér Pálné ; [eredeti szöveget gyerekek számára feldolgozta Kiss Anna ; Tamás Noémi rajzaival]. — Budapest : Móra, 1977.
97 p. : ill. ; 20 cm. H77
ISBN 9631108821 : 12.00Ft
I. Kiss, Anna, 1939- II. Title.
PZ90.H8P57 79-351036
•79 MARC

Piros betűk : ünnepi műsorok általános iskolásoknak. — Budapest : Népművelési Propaganda Iroda, 1973.
159 p. ; 23 cm. — (Színjátszók kiskönyvtára ; 158. sz)
21.00Ft
1. Holidays—Hungary—Juvenile literature. 2. Children's literature, Hungarian.
PZ90.H8P59 79-365043
79 MARC

Szentiványi, Jenő.
A postának mennie kell / Szentiványi Jenő ; [Hegedüs István rajzaival]. — Budapest : Móra, 1977.
107 p. : col. ill. ; 25 cm. H77
ISBN 9631103129 : 42.00Ft
I. Hegedüs, István, 1932- II. Title.
PZ90.H8S8654 79-338164
•79 MARC

Szép, Ernő, 1884-1953.
Egyszer volt Budán kutyavásár : [mesék] / Szép Ernő ; Gyulai Liviusz rajzaival. — 2. kiad. — Budapest : Móra, 1977.
[8] leaves : numerous col. ill. ; 29 cm. H77
CONTENTS: Egyszer volt Budán kutyavásár.—Róka fogta csuka.
ISBN 9631105121 : 24.00Ft
I. Szép, Ernő, 1884-1953. Róka fogta csuka. 1977. II. Guylai, Liviusz. III. Title.
PZ90.H8S866 1977 78-378403
•78 MARC

Sző, fon, nem takács, mi az? : találós kérdések gyerekeknek / [válogatta és összeállította Varga Ferencné ; Bálint Endre rajzaival]. — Budapest : Móra, 1975.
130, [1] p. : ill. ; 24 cm.
"Forrásmunkák": p. 129-[131]
ISBN 9631102998 : 16.00Ft
I. Varga, Ferencné. II. Bálint, Endre, 1914-
PZ90.H8S877 79-364338
79 MARC

Tamkó Sirató, Károly, 1905-
Szélkiáltó : [gyermekversek] / Tamkó Sirató Károly ; Kass János rajzaival. — Budapest : Móra, 1977.
33 p. : col. ill. ; 19 cm. H77
ISBN 9631109577 : 24.00Ft
I. Kass, János. II. Title.
PZ90.H8T324 79-365044
•79 MARC

Thury, Zsuzsa.
Mostohatestvérek : regény / Thury Zsuzsa ; [Reich Károly rajzaival]. — 5. kiad. — Budapest : Móra, 1977, c1960.
299 p. : ill. ; 19 cm. H77
"Tizenkét éven felülieknek."
ISBN 9631108430 : 17.50Ft
I. Title.
PZ90.H8T5 1977 78-401121
•79 MARC

Tűz, tűz, lobogó : nemzeti ünnepeink, március 15., március 21., április 4., május 1., augusztus 20., november 7 / [válogatta, szerk., és az előszót írta Vasy Géza ; Balogh Péter rajzaival]. — Budapest : Móra, 1977.
224 p. : ill. ; 19 cm. H77
ISBN 9631108910 : 18.00Ft
1. Children's literature, Hungarian. 2. Holidays—Hungary—Literary collections. I. Vasy, Géza.
PZ90.H8T8 78-401117
•79 MARC

—BIBLIOGRAPHY

Magyar gyermek- és ifjúsági irodalom 1965-1975 : válogatás / [főszerkesztő, Bereczky László, a kötetet összeállította és szerk. Benkő Attiláné]. — [Budapest] : Magyar Könyvkiadók és Könyvterjesztők Egyesülése, 1978.
187 p. ; 24 cm. H***
On cover: Nemzetközi Gyermekév 1979.
ISBN 9632010582
1. Chidren's literature, Hungarian—Bibliography. I. Bereczky, László. II. Benkő, Attiláné.
Z1037.6.M33 79-382644
[PZ90.H8] 79 MARC

CHILDREN'S LITERATURE, HUNGARIAN (Continued)

—HISTORY AND CRITICISM

Petrolay, Margit.
Gondolatok a gyermekirodalomról / Petrolay Margit. — Budapest : Tankönyvkiadó, 1978.
175 p. ; 20 cm. H78
ISBN 9631736636 : 8.00Ft
1. Children's literature, Hungarian—History and criticism. I. Title.
PN1009.H8P47 79-356397
*79 MARC

CHILDREN'S LITERATURE, ICELANDIC

Gunnar M Magnúss.
Bærinn á ströndinni : saga úr lífi Jóns Guggusonar / Gunnar M. Magnúss. — Reykjavík : Setberg, 1974.
135 p. : ill. ; 22 cm.
I. Title.
PZ56.G75 75-546316
79[r79]rev MARC

CHILDREN'S LITERATURE, INDONESIAN

Harsajanta.
Kota didasar laut : kumpulan dongeng untuk remadja / Harsajanta. — Djakarta : Widjaya, [19-
v. : ill. ; 21 cm.
I. Title.
PZ90.I5H334 79-949552
79 MARC

Kechendai, Edward, 1932-
Sekeda cherita pandak / ditulis Edward Kechendai ; gambar di gaga Krisno Jitab. — Kuching : Borneo Literature Bureau, 1970.
iv, 84 p. : ill. ; 19 cm.
I. Title.
PZ90.I2K4 70-943752
79 MARC

Mufty, A Rahim.
Hang Tuah / ditjeritakan kembali oleh A. Rahim Mufty ; gambar² oleh Dahlan Djazh. — [Djakarta] : Balai Pustaka, 1962.
16 p. : ill. ; 15 x 24 cm.
I. Djazh, Dahlan. II. Hikayat Hang Tuah. III. Title.
PZ90.I5M74 78-949789
79 MARC

CHILDREN'S LITERATURE, IRISH

Dukes, Bríd, 1947-
Filleann an feall / Bríd Dukes. — Baile Átha Cliath : Foilseacháin Náisiúnta, 1977.
98 p. : ill. ; 19 cm. Ir***
£1.50
I. Title.
PZ90.I7D8 79-343578
79 MARC

Heymans, Margriet.
Pipín, Grágálaí agus an naíonán mór / Margriet Heymans ; Eoghan Ó Colla a chuir an leagan Gaeilge in oiriúint do leanaí 6-9 mbliana d'aois. — Baile Átha Cliath : [Oifig an tSoláthair], c1977.
[24] p. : col. ill. ; 22 x 20 cm. Ir***
Translated from the Dutch.
£0.60
I. Title.
PZ90.I7H43 78-372092
78 MARC

Ní Shúilleabháin, Siobhán.
Mé féin agus Síle / Siobhán Ní Shúilleabháin. — Baile Átha Cliath : Oifig an tSoláthair : le ceannach díreach ón Oifig Dhíolta Foilseachán Rialtais, c1978.
77 p. ; 24 cm. Ir***
I. Title.
PZ90.I7N48 79-380762
79 MARC

Wolde, Gunilla, 1939-
Tomás agus a chuid éadaigh / Gunilla Wolde ; Treasa Ní Ailpín [a d'aistrigh]. — [Baile Átha Cliath] : Oifig an tSoláthair, [c1978]
[24] p. : col. ill. ; 16 cm. Ir***
I. Title.
PZ90.I7W6 79-365060
79 MARC

CHILDREN'S LITERATURE, ITALIAN

Clama, Nike.
Fu così ... : leggende / Nike Clama. — 1. ed. — Trieste : Società artistico letteraria, 1975.
93 p. ; 21 cm. — (Il Timavo ; ser. I, 22) It77-Oct
I. Title.
PZ43.C69 78-394523
*79 MARC

D'Arzo, Silvio.
Il pinguino senza frac : racconto per ragazzi / [di] Silvio D'Arco ; con illustrazioni di A. Manfredi. — Reggio Emilia : Prandi, 1978.
28 p. : ill. ; 26 cm. It79-Oct
150 Copies printed. No. 13.
I. Manfredi, Alberto, 1930- II. Title.
PZ46.3.D34 79-371761
*79 MARC

Giusti, Mario.
Lo scacciapensieri : scioglilingua, indovinelli, passerotti, ninne-nanne, filastrocche, canzoncine, cantilene, giochi e novelline popolari / Mario Giusti. — Roma : Edizioni paoline, 1978.
237 p. : col. ill. ; 30 cm. It78-Nov
L8000
I. Title.
PZ48.G5 1978 858'.9'1409 79-345270
*79 MARC

Manzi, Alberto.
Orzowei / Alberto Manzi. — Milano : Bompiani, 1977.
229 p., [8] leaves of plates : ill. ; 24 cm. It78-July
L5000
I. Title.
PZ43.M274 78-378304
*78 MARC

Moravia, Alberto, 1907-
Un miliardo di anni fa ... / Alberto Moravia ; illustrazioni di Gloria Leonetti. — Torino : Stampatori, 1979.
68 p. : ill. (some col.) ; 21 cm. — (Collana Il Cerchio) It79-July
L3800
I. Title.
PZ43.M643 79-379718
*79 MARC

Moravia, Alberto, 1907-
Quando Ba Lena era tanto piccola / Alberto Moravia ; illustrazioni di Diana Saputi. — Teramo : Lisciani & Zampetti, [1978]
[23] p. : ill. ; 24 cm. — (C'era non c'era ; 13) It79-Jan
L2300
I. Saputi, Diana, joint author. II. Title.
PZ43.M644 853'.9'12 79-354827
*79 MARC

Quintavalle, Ruggero R Y
La città difficile : una storia di contestazione e violenza / Ruggero Quintavalle. — 1. ed. — Milano : Massimo, 1978.
191 p. ; 20 cm. — (Collana Verdi anni ; n. 59) It***
L2400
I. Title.
PZ43.Q5 78-378395
78 MARC

Salvadeo, Guido.
Cuori lanciati nell'azzurro / Guido Salvadeo. — Brescia : La Scuola, 1946.
228 p., [8] leaves of plates : ill. ; 22 cm. — (Incontro al sole)
I. Title.
PZ46.S3 78-378875
78 MARC

Sissa, Pietro, 1916-
Pelandrone e la fattoria / Pietro Sissa ; illustrazioni di Jacovitti. — 1. ed. — Milano : Rizzoli, 1977.
170 p. : ill. ; 20 cm. — (La BUR dei bambini ; 6) (Biblioteca universale Rizzoli) It78-Jan
L3800
I. Jacovitti, Benito. II. Title.
PZ43.S527 78-394515
*79 MARC

Tábori, Piroska, 1892-1947.
Le avventure di Tino Tappo / Piroska Tábori ; [traduzione di Emilia Villoresi e Filippo Faber]. — Milano : Genio, 1934.
207 p. : ill. (some col.) ; 25 cm.
I. Title.
PZ43.T3 78-378876
78 MARC

Vergani, Viscardo.
La storia di Cirino : ovvero, Il Cuore dei nostri tempi e dei nostri ragazzi raccontato dal maestro di Mirteto / Viscardo Vergani. — Bologna : Cappelli, 1978.
125 p. : ill. ; 26 cm. It79-Mar
L3500
I. Title. II. Title: Il cuore dei nostri tempi e dei nostri ragazzi ...
PZ43.V43 853'.9'14 79-365536
*79 MARC

—BIBLIOGRAPHY

Il Segnalibro : manuale del bibliotecario per ragazzi / [a cura del] Centro didattico nazionale di studi e documentazione. — Firenze : Edizioni del Centro didattico nazionale di studi e documentazione, 1976.
312 p. : ill. ; 24 cm. It78-Apr
Includes indexes.
L6000
1. Children's literature, Italian—Bibliography. 2. Books and reading for children. I. Florence. Centro didattico nazionale di studi e documentazione.
Z1037.4.S43 1976 78-376085
*78 MARC

—EXHIBITIONS

Mostra del libro italiano per l'infanzia e la gioventù = Italienische Kinder-und Jugendbuch-Ausstellung : organizzata dalla Biblioteca internazionale per la gioventù di Monaco di Baviera e dal Ministero degli affari esteri della Repubblica italiana : Monaco di Baviera, 20 ottobre-10 novembre 1977, Internationale Jugendbibliothek-Kaulbachstrasse, 11a. : catalogo. — [s.l. : s.n., 1977] (Firenze : Tip. fiorentina)
lxiv, 116 p. ; 24 cm. It***
1. Children's literature, Italian—Exhibitions. 2. Munich—Exhibitions. I. Munich. Internationale Jugendbibliothek. II. Italy. Ministero degli affari esteri. III. Title: Italienische Kinder-und Jugendbuch-Ausstellung.
Z1037.4.M68 850'.9'9282 79-339257
[PN1009.I8] 79 MARC

—HISTORY AND CRITICISM

Andrianopoli Cardullo, Mariacarla.
Letture e modelli comportamentali : motivi moralistici ed esemplaristici nei libri italiani di lettura per l'infanzia, la fanciullezza e le famiglie (1830-1914) / Mariacarla Andrianopoli Cardullo. — Genova : Tilgher, 1977.
166 p. ; 24 cm. It79-Jan
Bibliography: p. [147]-159.
Includes index.
L4700
1. Children's literature, Italian—History and criticism. I. Title.
PN1009.I8A5 850'.9'9282 79-366220
*79 MARC

Pedullà, Walter.
La letteratura emarginata : i narratori giovani degli anni '70 / Walter Pedullà, Silvana Castelli, Stefano Giovanardi. — Roma : Lerici, 1978.
231 p. ; 21 cm. — (La Rivista ; 1) It***
Cover title.
Includes bibliographical references.
L5000
1. Children's literature, Italian—History and criticism. I. Castelli, Silvana, joint author. II. Giovanardi, Stefano, joint author. III. Title.
PN1009.I8P4 850'.9'9282 79-354181
79 MARC

CHILDREN'S LITERATURE, JAPANESE

Chiba, Shōzō, 1892-
(Umi no hayabusa)
海の隼：少年少女小説 / 千葉省三作. — 東京：中文館, 昭和23 [1948]
3, 258 p. : 19 cm.
Colophon inserted.

I. Title.
PZ49.31.C46 79–801314

Chin, Chin.
(Genki na kogoitachi)
げんきな小ごいたち / 文・金近 ; え・揚善子, 丁榕臨. — 北京：外文出版社, 1961.
[33] p. : col. ill. ; 21 cm.
Juvenile literature.

I. Yang, Shan-tsu. II. Ting, Jung-lin. III. Title.
PZ49.63.C47 79–805840

CHILDREN'S LITERATURE, JAPANESE (Continued)

Gotō, Ryūji, 1943-
(Shiokaze no gakkō)
潮風の学校 / 後藤竜二著；高橋透絵. — 東京：新日本出版社, 1978.
174 p. : ill. ; 22 cm. — (新日本少年少女の文学) Ja 78-14
Series romanized: Shin Nihon shōnen shōjo no bungaku.
¥980

I. Takahashi, Tōru, 1939- II. Title.
PZ49.31.G683 78-811432

(Hama wa bokura no kyōshitsu da)
浜はぼくらの教室だ / 三重・子どもと教師の文学の会編；こさかしげる絵. — 東京：ポプラ社, 1978.
165 p., [1] leaf of plates : ill. ; 22 cm. — (先生のとっておきの話；21) Ja 79-4
Series romanized: Sensei no totteoki no hanashi.
¥680
1. Children's stories, Japanese—Japan—Mie (Prefecture) I. Kosaka, Shigeru, 1925- II. Mie Kodomo to Kyōshi no Bungaku no Kai.
PZ49.2.H315 79-807409

Haruyama, Yukio.
(Haha no higan)
母の悲願 / 春山雪夫. — 東京：偕成社, 昭和25 [1950]
250 p. : ill. ; 19 cm.
Fiction.

I. Title.
PZ49.31.H378 78-814689

Hirata, Ryōichi, 1931-
(Kurabete mitara)
くらべてみたら / 平田良一作；岡田昌子絵. — 東京：金の星社, 1978.
[25] p. : col. ill. ; 26 x 27 cm. — (しぜんのくにえほん) Ja 78-39

I. Okada, Masako. II. Title. III. Series: Shizen no kuni ehon.
PZ49.31.H538 78-816663

Iba, Nantetsu, 1902-
(Arashio no wakōdo)
荒潮の若人 / 伊波南哲著. — 東京：講談社, 昭和18 [1943]
184 p. : ill. ; 19 cm.

1. Fishermen—Japan—Itoman—Juvenile fiction. I. Title.
PZ49.31.I15 79-801543

Imanishi, Sukeyuki, 1923-
(Shimabara no eshi)
島原の絵師 / 今西祐行. — 東京：小峰書店, 1978.
190 p. : ill. ; 19 cm. — (文学のひろば；6) Ja 79-9
Colophon inserted.
¥800
1. Yamada, Emosaku, fl. 1614-1638—Juvenile fiction. 2. Shimabara, Japan—History—Juvenile fiction. I. Title. II. Series: Bungaku no hiroba ; 6.
PZ49.31.I447 79-807492

Iwasaki, Chihiro, 1918-1974.
(Watashi no ehon)
わたしのえほん / いわさきちひろ. — 東京：新日本出版社, 1978.
23 p. : ill. ; 16 cm. Ja 78-7977
¥500

I. Title.
PZ49.31.I814 1978 78-811446

Kako, Satoshi, 1926-
(Yamura no Yasuke)
矢村のヤ助 / かこさとし. — 東京：偕成社, 1978.
[48] p. : col. ill. ; 26 cm. — (かこさとし語り絵本；1) Ja 78-40
¥880

I. Title. II. Series: Kako, Satoshi, 1926- Kako Satoshi katari ehon ; 1.
PZ49.31.K2792 78-816411

Kambe, Junkichi, 1920-
(Genroku no shiroi shio)
元禄の白い塩 / 神戸淳吉・[作；小島直・画]. — 東京：金の星社, 1975.
190 p., [1] leaf of plates : ill. ; 22 cm. — (現代・創作児童文学；6) Ja 75-49b
Series romanized: Gendai sōsaku jidō bungaku.
¥850

1. Japan—History—Akō Vendetta, 1703—Juvenile fiction. I. Title.
PZ49.31.K235 79-803520

Katō, Teruji, 1934-
(Yowamushi Torata)
よわむし虎太 / 加藤輝治作；中山正美絵. — 京都：PHP研究所, 1978.
157 p., [2] leaves of plates : ill. ; 22 cm. Ja 78-28
¥950

I. Nakayama, Masami, 1914- II. Title.
PZ49.31.K283 78-814198

Kawamura, Takashi, 1931-
(Kōya no tabibitotachi)
広野の旅人たち：新十津川物語 / 川村たかし. — 東京：偕成社, 1978.
252 p. : ill. ; 21 cm. — (偕成社の創作文学) Ja 78-44
Juvenile literature.
Series romanized: Kaiseisha no sōsaku bungaku.
¥950

I. Title.
PZ49.31.K2945 79-800747

Kawatake, Shigetoshi, 1889-1967.
(Matsuō, Umeō, Sakuramaru)
松王・梅王・櫻丸：菅原傳授手習鑑 / 竹田, 並木, 三好原作；河竹繁俊譯. — 東京：童話春秋社, 昭和24 [1949]
2, 188 p. ; 19 cm. — (少年少女世界名作文庫)
Colophon inserted.

I. Sugawara denju tenarai kagami. II. Title. III. Series: Shōnen shōjo sekai meisaku bunko.
PZ49.31.K296 78-816443

Kishida, Kyōko, 1930-
(Hitomishiri na irie)
ひとみしりな入江 / [岸田今日子]. — 東京：大和書房, 1978.
100 p. : ill. ; 22 cm. Ja 78-26593
"岸田今日子ファンタジー"
CONTENTS: ミッシェル—沼で—悲恋—七匹目の仔山羊—いつもの夏—ひとみしりな入江—縞が似合う女—風が…—絵日記—いばら姫またはねむり姫
¥980
I. Title.
PZ49.31.K5468 79-805405

Kogure, Masao, 1939-
(Machi ni midori no kaze ga fuku)
町にみどりの風がふく / 木暮正夫・作；下河辺史子・画. — 東京：童心社, 1978.
125 p. : ill. ; 22 cm. — (現代童話館；16) Ja 78-25
Series romanized: Gendai dōwakan.
¥780

I. Shimokawabe, Fumiko, 1955- II. Title.
PZ49.31.K5777 78-811474

Konomi, Hikaru.
(Makkakka na musume ga mado kara nozoite iru)
まっかっかなむすめがまどからのぞいている：[なぞなぞのえほん] / 木乃美光編；正田壌画. — 東京：福音館書店, 1973.
31 p. : col. ill. ; 26 cm. — (こどものとも；213号) Ja***
Cover title.
On p. [4] of cover: The book of riddles.
Series romanized: Kodomo no tomo.
I. Shōda, Jō. II. Title. III. Title: Nazonazo no ehon. IV. Title: The book of riddles.
PZ49.31.K5935 79-802177

Madokoro, Hisako, 1938-
(Nyugu nyugu puchu puchu)
にゅぐにゅぐぷちゅぷちゅ / 間所ひさこ文；河本祥子絵. — 東京：講談社, 1978.
31 p. : col. ill. ; 24 cm. Ja 78-27
¥690

I. Kōmoto, Sachiko, 1940- II. Title.
PZ49.31.M213 78-814121

Matsutani, Miyoko, 1926-
(Asobimasho)
あそびましょ / 松谷みよ子；丸木俊 え. — 東京：偕成社, 1977.
[20] p. : col. ill. ; 22 cm. — (あかちゃんのわらべうた；1) Ja***
¥580

I. Maruki, Toshi. II. Title. III. Series: Akachan no warabeuta ; 1.
PZ49.43.M365 78-814310

Matsutani, Miyoko, 1926-
(Fūchan to Chacha)
ふうちゃんとチャチャ / 松谷みよ子・さく；遠藤てるよ・え. — 東京：金の星社, 1978.
31 p. : ill. (some col.) ; 25 cm. — (新しいえほん) Ja 78-40
Series romanized: Atarashii ehon.
¥850

I. Endō, Teruyo. II. Title.
PZ49.31.M2874 78-816645

Matsutani, Miyoko, 1926-
(Ōkami no mayuge)
おおかみの眉毛 / 松谷みよ子. — 東京：講談社, 1978.
211 p. : ill. ; 15 cm. — (講談社文庫) Ja 78-25568
Series romanized: Kōdansha bunko.
Juvenile literature.
¥260

I. Title.
PZ49.31.M312 79-802189

Matsutani, Miyoko, 1926-
(Yamamba no nishiki)
やまんばのにしき / ぶん まつたにみよこ；え せがわやすお. — 東京：ポプラ社, 昭和47 [1972]
[40] p. : col. ill. ; 27 cm. — (むかしむかし絵本；2) Ja***

I. Segawa, Yasuo. II. Title. III. Series: Mukashi mukashi ehon ; 2.
PZ49.41.M35 1972 79-802420

(Monogatari Niitsu no mukashibanashi)
ものがたり にいつの昔話 / [編集者 水野興二郎, 宮崎進]. — 新津：新津青年会議所20周年記念出版委員会, 1977.
176 p., [6] leaves of plates : ill. (some col.) ; 22 cm. Ja 78-14
¥1800

1. Tales, Japanese—Japan—Niitsu. I. Mizuno, Kōjirō, 1941- II. Miyazaki, Susumu, 1947- III. Title: Niitsu no mukashibanashi.
PZ49.41.M59 78-816339

CHILDREN'S LITERATURE, JAPANESE (Continued)

Nishimaki, Kayako.
(Fun fun nandaka ii nioi)
ふんふん なんだかいいにおい / えとぶん にしまきかやこ. — 東京：こぐま社, 1977.
[44] p. : col. ill. ; 20 x 22 cm. Ja***
¥800

I. Title.
PZ49.31.N48 78-810119

Ogawa, Mimei, 1882-1961.
(Mimei dōwa shū)
未明童話集 / 小川未明著 ; 初山滋挿画 ; 武井武雄装幀. — 東京：丸善, 昭和2- [1927- 昭和6- [1931- printing.
v. : ill. ; 23 cm.

I. Hatsuyama, Shigeru, 1897- II. Title.
PZ49.2.O43 78-814520

Ogawa, Mimei, 1882-1961.
(Teihon Ogawa Mimei dōwa zenshū)
定本小川未明童話全集. — 東京：講談社, 1976-1978.
16 v. : ill. ; 22 cm. Ja 77-10 (v. 3)
¥2700 per vol.

I. Title.
PZ49.31.O332 1977 77-810546

Ōhashi, Ayumi.
(Sayonara sankaku mata kite shikaku)
さよならさんかくまたきてしかく / [著者 大橋歩]. — 東京：大和書房, 1976.
205 p. : ill. ; 18 cm. Ja 76B-6605
¥790

I. Title.
PZ49.31.O335 78-812959

Ōno, Michio, 1919-
(Akai rampu)
赤いランプ / 大野道雄著. — 東京：中央出版社, 昭和22 [1947]
2, 166 p. : ill. ; 19 cm.

I. Title.
PZ49.31.O49897 78-811395

Ōno, Tetsurō.
(Nousagi to kāsan)
のうさぎとかあさん / 大野哲郎作：中島保彦絵. — 東京：ポプラ社, 1974.
61 p. : ill. (some col.) ; 23 cm. — (カラー版・創作えばなし ; 19) Ja 74-40
Series romanized: Karaban sosaku ebanashi.
¥800

I. Nakajima, Yasuhiko, 1921- II. Title.
PZ49.31.O535 76-806858

Ōtomo, Noriko, 1948-
(Mō sugu onichan)
もうすぐおにいちゃん / 大友のり子 さく ; 大友康夫 え. — 東京：童心社, 1978.
31 p. : col. ill. ; 21 x 23 cm. — (えほん・ぼくらはきょうだい ; 2) Ja 78-39
¥780

I. Ōtomo, Yasuo. II. Title. III. Series: Ehon bokura wa kyōdai ; 2.
PZ49.31.O84 78-816336

Ōtomo, Yasuo.
(Momoko to Gorō no okurimono)
ももことごろうのおくりもの / 大友康夫 さく. — 東京：童心社, 1978.
39 p. : col. ill. ; 21 x 23 cm. — (えほん・ぼくらはきょうだい ; 1) Ja 78-39
¥780

I. Title. II. Series: Ehon bokura wa kyōdai ; 1.
PZ49.31.O845 78-816337

Saijō, Yaso, 1892-1970.
(Shōjo junjō shishū)
少女純情詩集 / 西條八十著. — 東京：講談社, 昭和26 [1951]
266 p., [10] leaves of plates : ill. ; 16 cm.
Poems.

I. Title.
PZ49.43.S24 79-803612

Saitō, Ryūsuke, 1917-
(Ten no akauma)
天の赤馬 / 斎藤隆介 作 ; 滝平二郎 画. — 東京：岩崎書店, 1977, 1978 printing.
269 p., [1] leaf of plates : ill. (some col.) ; 22 cm. — (創作児童文学 ; 4) Ja***
Series romanized: Sōsaku jidō bungaku.
¥1200

I. Takidaira, Jirō, 1921- II. Title.
PZ49.31.S315 78-815851

Sakurai, Nobuo.
(Oshaberi shiroiruka)
おしゃべり白いるか / 桜井信夫・文 ; 芝美千世・画. — 東京：小学館, 1978.
43 p. : ill. ; 21 cm. — (小学館のノンフィクション童話 ; 5) Ja 78-33
Series romanized: Shōgakukan no nonfikushon dōwa.
¥430

I. Shiba, Michiyo. II. Title.
PZ49.63.S32 78-814371

Samukawa, Kōtarō, pseud.
(Shin'ya no bōken)
深夜の冒険 / 寒川光太郎. — 東京：偕成社, 昭和25 [1950]
281 p. ; 19 cm.

I. Title.
PZ49.31.S376 79-800985

Satō, Hachirō, 1903-1973.
(Ashinami soroete)
足なみ揃えて / サトウ・ハチロー. — 東京：全音楽譜出版社, 昭和23 [1948]
181 p. ; 18 cm.
CONTENTS: 足なみ揃へて—昭和の一休—足軽の先祖

I. Title.
PZ49.31.S395 78-811397

Satō, Kōroku, 1874-1949.
(Kōgan bidan)
紅顔美談 / 佐藤紅緑著. — 東京：まひる書房, 昭和23 [1948]
235 p. ; 19 cm.
Fiction.

I. Title.
PZ49.31.S412 79-803650

Segawa, Yasuo.
(Ōsamu kosamu)
おおさむこさむ：わらべうた / 瀬川康男え. — 東京：福音館書店, 1972.
27 p. : col. ill. ; 26 cm. — (こどものとも ; 190号)
Cover title.
On p. [4] of cover: The winter wind, old nursery rhymes from Japan.
Series romanized: Kodomo no tomo.

I. Title. II. Title: The winter wind, old nursery rhymes from Japan.
PZ49.43.S4 79-802190

Seta, Teiji, 1916-
(Ma no ii ryōshi)
まのいいりょうし / 瀬田貞二再話：赤羽末吉画. — 東京：福音館書店, 1973.
31 p. : col. ill. ; 26 cm. Ja***
On p. [4] of cover: A lucky hunter, old Japanese tale.
"こどものとも増刊号"

I. Akaba, Suekichi. II. Kodomo no tomo. III. Title. IV. Title: A lucky hunter, old Japanese tale.
PZ49.41.S443 79-802176

Shōgenji, Haruko, 1914-
(Dare kashira dare kashira)
だれかしら だれかしら / 生源寺美子・文；安井淡・絵. — 東京：講談社, 1978.
32 p. : col. ill. ; 24 cm. Ja 78-27
¥690

I. Yasui, Tan. II. Title.
PZ49.31.S559 78-814151

Tachihara, Erika, 1937-
(Yōseitachi)
妖精たち / 立原えりか. — 東京：サンリオ出版, 昭和49 [1974]
129 p. : ill. ; 18 cm. Ja***
¥680

I. Title.
PZ49.31.T278 79-804400

Tachihara, Erika, 1937-
(Yōseitachi no kōrigashi)
妖精たちの氷菓子 / 立原えりか. — 東京：角川書店, 1978.
214 p., [1] leaf of plates : ill. ; 20 cm. Ja 78-7505
CONTENTS: 小鳥が降る日—雪をよぶ鳥—恋唄—青い睡蓮—思いちがえて—羽の首飾り—海のまんなかにある館—星を見る窓—しあはせのたね—妖精たちの氷菓子
¥960

I. Title.
PZ49.31.T279 79-803627

Takezaki, Yūhi, 1923-
(Otōsan no daijōbu)
おとうさんのだいじょうぶ / 作 竹崎有斐；絵 岩村和朗. — 京都：PHP研究所, 1978.
61 p. : ill. ; 23 cm. — (PHPおはなしひろばシリーズ) Ja 78-28
¥880

I. Iwamura, Kazuo, 1937- II. Title. III. Series: PHP ohanashi hiroba shirizu.
PZ49.31.T3344 78-812850

Takikawa, Takashi, 1948-
(Ōmu wa ōmu)
おうむはおうむ：[こどものうた] / 滝川たかし・詩；[松本伸次・イラストレーション]. — 高松：讃文社, 1973.
[24] p. : ill. ; 26 cm. Ja 74-13
¥500

I. Title.
PZ49.43.T344 79-806856

CHILDREN'S LITERATURE, JAPANESE (Continued)

Tanikawa, Shuntarō, 1931–
(Sekai wa hiroshi)
せかいはひろし / 谷川俊太郎作 ; 和田誠絵. — 東京 : あかね書房, 1978.
[48] p. : col. ill. ; 22 cm. — (新絵本 ; 3)   Ja 78-45
Juvenile literature.
Series romanized: Shin'ehon.
¥780

I. Wada, Makoto, 1936–   II. Title.
PZ49.31.T3443   79-802465

Tanikawa, Shuntarō, 1931–
(Yuri no uta)
由利の歌 / 谷川俊太郎 ; え・長新太, 山口はるみ, 大橋歩. — 東京 : すばる書房, 1977.
[39] p. : chiefly ill. ; 24 cm.   Ja***
Cover title.
Juvenile literature.
Colophon inserted.
¥2800
I. Chō, Shinta, 1927–   II. Yamaguchi, Harumi.   III. Ōhashi, Ayumi.   IV. Title.
PL862.A54Y8   78-815483

Taniuchi, Rokurō, 1921–
(Nemunoki)
ねむの木 = My deepest love to a silk tree : 谷内六郎詩画集 / 序 宮城まり子. — 長野 : あすか書房 ; 東京 : 発売 教育書籍, 1977.
157 p. : col. ill. ; 26 cm.   Ja 78-11296
In Japanese.
"「ねむの木学園」児童画8点収録"
¥6000
I. Title.   II. Title: My deepest love to a silk tree.
PZ49.43.T36   78-810139

(Tembō Nihon no jidō bungaku)
展望日本の児童文学 / [編者 大藤幹夫]. — 東京 : 双文社出版, 1978.
297 p. ; 21 cm.   Ja 78-15875
Includes bibliographies.
¥1300

1. Children's literature, Japanese.   2. Children's literature, Japanese—Bio-bibliography.   I. Ōfuji, Mikio.   II. Nihon no jidō bungaku.
PZ49.2.T4   79-803570

Tōge, Hyōta, 1931–
(Asuka no ishi no monogatari)
明日香の石の物語 / 峠兵太. — 川崎 : 緑陰書房, 昭和52 [1977]
133 p. ; 19 cm.   Ja 77-27189
¥800

I. Title.
PZ49.31.T582   79-804347

Tokita, Shirō, 1943–
(Urashima Tarō)
うらしまたろう / 時田史郎再話 ; 秋野不矩画. — 東京 : 福音館書店, 1974.
31 p. ; 20 x 27 cm. — (日本傑作絵本シリーズ)   Ja***
Series romanized: Nihon kessaku ehon shirizu.

I. Akino, Fuku, 1908–   II. Title.
PZ49.41.T59   79-802585

Tsubota, Jōji, 1890–
(Tsubota Jōji sakuhin shū)
坪田譲治作品集 / 大石真編. — 東京 : 角川書店, 昭和51– [1976–   ]
v. : ill. ; 15 cm. — (角川文庫 ; 3618, 3649   )
Ja 76B-1765 (v. 1)
Juvenile literature.
Series romanized: Kadokawa bunko.
CONTENTS: 1. 風の中の子供—2. 子供の四季
¥300 (v. 1)

PL839.S85T78   79-802084

Tsukiji, Fujiko.
(Haruka nari tomo no yobu koe)
遥かなり友の呼ぶ声 / 築地藤子. — 東京 : 古今書院, 昭和25 [1950]
245 p. ; 19 cm.
CONTENTS: 芍薬譜—紫のネッカチーフ—香港の薔薇—ボルネオの歌—美しきカプアスの流れ—カカトワの島をたづねて—白楊の蔭に

I. Title.
PZ49.22.T75   78-813864

Tsutsui, Keisuke, 1918–
(Kachikachiyama no sugu soba de)
かちかち山のすぐそばで / 文 筒井敬介 ; 絵 瀬川康男. — 東京 : フレーベル館, 昭和47 [1972]
126 p. ; 23 cm. — (フレーベルどうわ文庫 ; 1)   Ja 73-3
Series romanized: Furēberu dōwa bunko.

I. Segawa, Yasuo, ill.   II. Title.
PZ49.63.T8   74-809667

rev

Uchiyama, Kensho, 1899–
(Hoiku no tame no shinwa kyōzai shūsei)
保育のための神話教材集成 / 内山憲尚, 伊東挙位著. — [東京] : 白眉学芸社, 昭和43 [1968]
252 p. : ill. ; 19 cm.

1. Tales, Japanese.   2. Mythology, Japanese—Juvenile literature.   3. Education, Preschool—Japan.   I. Itō, Kyoi, 1901–   joint author.   II. Title.   III. Title: Shinwa kyōzai shūsei.
PZ49.41.U25   79-803704

Yokoyama, Yoshiko, 1930–
(Boku no mandara)
ぼくのまんだら / 横山佳子著. — 東京 : たいまつ社, 1978.
143 p. : ill. ; 20 cm.   Ja 78-25340
¥1200

I. Title.
PZ49.31.Y555   79-803691

—BIO-BIBLIOGRAPHY

Nemoto, Masayoshi, 1942–
(Shōwa jidō bungaku annai)
昭和児童文学案内 / 根本正義著. — 東京 : 高文堂出版社, 昭和51 [1976]
97 p. ; 18 cm. — (高文堂新書)   Ja 76-19585
Series romanized: Kōbundō shinsho.
"児童文学研究文献目録": p. 72-96.
¥550

1. Children's literature, Japanese—Bio-Bibliography.   2. Children's literature—Book reviews.   I. Title.
Z1037.8.J3N45   78-811951
[PN1009.J3]

J

—HISTORY AND CRITICISM

Furuta, Taruhi, 1927–
(Gendai jidō bungaku ron)
現代児童文学論 : 近代童話批判 / 古田足日. — 東京 : くろしお出版, 1959.
253 p. ; 19 cm.

1. Children's literature, Japanese—History and criticism.   2. Children's literature—History and criticism.   I. Title.
PN1009.J3F78   79-803361

Kan, Tadamichi, 1909–
(Jidenteki jidō bunka shi) .
自伝的児童文化史 / 管忠道. — 東京 : ほるぷ総連合ほるぷ教育開発研究所, 1978–
v. : ill. ; 19 cm. — (ほるぷ叢書 ; 2–   )   Ja 78-19219 (v. 1)
Series romanized: Horupu sōsho.
"「子どもの本棚」の一九七二年三月号から三年数か月にわたって連載されたものを土台にして補筆、改稿したもの"
CONTENTS: [1] 戦前・戦中期編
¥1600 (v. 1)

1. Children's literature, Japanese—History and criticism.   2. Mass media and children—Japan.   3, Kan, Tadamichi, 1909–   4. Critics—Japan—Biography.   I. Title.   II. Series: Horupu sōsho ; 2 [etc.]
PN1009.J3K25   79-802655

*79   J

Namekawa, Michio, 1906–
(Shōkokumin bungaku shiron)
少國民文學試論 / 滑川道夫著. — 東京 : 帝國教育會出版部, 昭和17 [1942]
2, 2, 228 p. ; 19 cm

1. Children's literature, Japanese—History and criticism—Addresses, essays, lectures.   I. Title.
PN1009.J3N326   79-803183

Yagi, Noboru.
(Taishū bungei zushi)
大衆文芸図誌 : 装釘・挿絵にみる昭和ロマンの世界 / 八木昇. — 東京 : 新人物往来社, 昭和52 [1977]
130 p. : ill. ; 31 cm.   Ja 77-5243
Includes indexes.
¥4800
1. Book covers—Japan.   2. Illustration of books—Japan.   3. Japanese fiction—20th century—History and criticism.   4. Children's literature, Japanese—History and criticism.   I. Title.
NC973.5.J3Y33   78-816075

—HISTORY AND CRITICISM —ADDRESSES, ESSAYS, LECTURES

Namekawa, Michio, 1906–
(Shōkokumin bungaku shiron)
少國民文學試論 / 滑川道夫著. — 東京 : 帝國教育會出版部, 昭和17 [1942]
2, 2, 228 p. ; 19 cm

1. Children's literature, Japanese—History and criticism—Addresses, essays, lectures.   I. Title.
PN1009.J3N326   79-803183

Ozeki, Iwaji, 1896–
(Jidō bungaku no riron to jissai)
児童文学の理論と実際 / 尾関岩二著. — 東京 : 関書院, 昭和24 [1949]
5, 216 p. ; 19 cm.
Colophon inserted.

1. Children's literature, Japanese—History and criticism—Addresses, essays, lectures.   2. Children's literature—History and criticism—Addresses, essays, lectures.   I. Title.
PN1009.J3O9   78-811067

Torigoe, Shin, 1929–
(Sengo jidō bungaku e no shōgen)
戦後児童文学への証言 / 鳥越信著. — 東京 : 理論社, 1978.
2 v. ; 19 cm. — (児童文学セミナー)   Ja 78-12416
Series romanized: Jidō bungaku seminā.
¥1200 per vol.

1. Children's literature, Japanese—History and criticism—Addresses, essays, lectures.
PN1009.J3T648   78-812613

J

CHILDREN'S LITERATURE, KIRGHIZ

Gor'kiĭ, Maksim, 1868–1936.
[Pesnâ o burevestnike. Kirghiz]
Бороон кабарчысы жөнүндө ыр ; Шумкар жөнүндө ыр / М. Горький. — Фрунзе : Мектеп, 1967.
17 p.
On cover : Èki yr.
Fiction.
Kirghiz.
"Тестиер балдар үчүн."
L.C. copy on microfilm only.
—— —— Microfilm. On 1 reel with other items ; 35 mm.
Microfilm Slavic 12709 PZ
I. Gor'kiĭ, Maksim. Pesnâ o sokole. Kirghiz. 1967. II. Title.
Title romanized: Boroon kabarchysy zhônûndô yr.
[PZ90.K5G67 1967] 79–970218

CHILDREN'S LITERATURE, KOREAN

Cho, Tae-hyŏn, 1939-
(Pŏmbawigol ŭi mae)
범바위골의매 : 조대현창작동화집. — 서울 : 경지사, 1976.
242 p. : ill. (some col.) ; 22 cm.
Children's literature.
CONTENTS: 돌아온토끼.—흥부.—우렁이.—아기노루의모험.—종달새와소년.—양회다리와승이할아버지.—동물농장의산토끼.—황새.—범바위골의매.
₩800
I. Title.
PZ50.52.C47 79–825687

Ch'oe, Hyo-sŏp.
(Ich'ŏn-ch'ilsibinyŏn)
2072년 / 최효섭지음. — 서울 : 대한기독교서회, 1972.
230 p. : ill. ; 19 cm. — (세계명작공상소설선집 ; 2) (새벗문고)
Fiction.
In colophon: The year 2072.
Series 2 romanized: Saebŏt mun'go.
I. Title. II. Title: The year 2072. III. Series: Segye myŏngjak kongsang sosŏl sŏnjip ; 2.
PZ50.531.C545 79–825759

Chŏng, Yong-wŏn, 1944-
(Kohyang kŭ yetkang)
고향그옛강 : 정용원시집 / [지은이정용원]. — 서울 : 시문학사, 1978.
132 p. : ill. ; 22 cm.
Poems.
₩1500
I. Title.
PZ50.543.C49 79–826026

Kim, Yong-ik.
(Salmŭi kil)
삶의길 : 중편소설 / 김룡익 ; 그림 최일. — [평양] : 금성청년출판사, 1977.
198 p. : ill. ; 19 cm.
Fiction.
₩0.30
I. Title.
PZ50.531.K49 78–826552

Kim, Yŏng-il, 1914-
(Ŭn pangul kkot)
은방울꽃 / 김영일지음. — 서울 : 대한기독교서회, 1969.
173 p. : ill. ; 19 cm. — (한국신작동화선집 ; 10) (새벗문고)
On verso of t. p.: Lily of the valley.
Series 2 romanized: Saebŏt mun'go.
I. Title. II. Title: Lily of the valley. III. Series: Han'guk sinjak tonghwa sŏnjip ; 10.
PZ50.52.K54 79–825479

Ku, Hye-yŏng, 1931-
(Ŏndŏk e punŭn param)
언덕에부는바람 / 구혜영지음. — 서울 : 성바오로출판사, 1977.
226 p. : ill. ; 19 cm.
₩650
I. Title.
PZ50.52.K8 79–825076

Lee, Hyun Joo.
(Usŭm ŭi ch'ong)
웃음의총 / 이현주지음. — 서울 : 대한기독교서회, 1970.
165 p. : ill. ; 19 cm. — (한국신작동화선집 ; 5) (새벗문고)
English title in colophon: A pistol roaring with laughter.
Series 2 romanized: Saebŏt mun'go.
I. Title. II. Title: A pistol roaring with laughter. III. Series: Han'guk sinjak tonghwa sŏnjip ; 5.
PZ50.52.L4 79–825681

Ŏ, Hyo-sŏn.
(Inhyŏng ŭi nunmul)
인형의눈물 : 어효선동화집 / [지은이어효선] ; 그림·김정. — 서울 : 일지사, 1978.
191 p. : ill. ; 22 cm.
Children's literature.
₩1000
I. Title.
PZ50.52.O23 79–825399

Pak, Hwa-mok, 1923-
(Chŏnyŏk nol ch'ŏrŏm)
저녁놀처럼 / 박화목지음. — 서울 : 대한기독교서회, 1970.
211 p. : ill. ; 19 cm. — (한국신작동화선집 ; 4) (새벗문고)
English title in colophon: Like the evening glow.
Series 2 romanized: Saebŏt mun'go.
I. Title. II. Title: Like the evening glow. III. Series: Han'guk sinjak tonghwa sŏnjip ; 4.
PZ50.531.P33 79–825911

Pak, Sŭng-il.
(Moksa ŭi adŭl)
목사의아들 : 박승일동화집 / [저자박승일]. — 서울 : 목양사 : 총판교문사, 1977.
202 p. : ill. ; 21 cm.
₩1000
I. Title.
PZ50.531.P336 79–825937

Pak, Sŭng-il.
(P'ari wangguk ŭi mudohoe)
파리왕국의무도회 : 동화집 / 박승일지음. — 서울 : 세종문화사, 1976.
290 p. : ill. ; 21 cm.
Children's stories.
₩1500
I. Title.
PZ50.531.P337 78–826717

Pang, Ki-hwan, 1921-
(Kkot param punŭn chip)
꽃바람부는집 / 방기환지음. — 서울 : 大韓基督教書會, 1969.
167 p. : ill. ; 19 cm. — (한국아동문학신작선집 ; 7) (새벗문고)
Fiction.
On verso of t. p.: The house in flower dust.
Series 2 romanized: Saebŏt mun'go.
I. Title. II. Title: The house in flower dust. III. Series: Han'guk adong munhak sinjak sŏnjip ; 7.
PZ50.531.P36 79–825628

Park, Kyung Chong.
(Turisŏman anŭn pimil)
둘이서만아는비밀 / 박경종지음. — 서울 : 大韓基督教書會, 1969.
168 p. : ill. ; 19 cm. — (한국신작동화선집 ; 2) (새벗문고)
Fiction.
On verso of t. p.: The secret only two know.
Series 2 romanized: Saebŏt mun'go.
I. Title. II. Title: The secret only two know. III. Series: Han'guk sinjak tonghwa sŏnjip ; 2.
PZ50.531.P37 79–825629

Ryu, Young-Hi.
(Nat to pam to ŏmnun kot)
낮도밤도없는곳 / 유영희지음. — 서울 : 대한기독교서회, 1971.
296 p. : ill. ; 19 cm. — (세계아동종교문학선집 ; 4) (새벗문고)
Fiction.
In colophon: The hell and heaven.
Series 2 romanized: Saebŏt mun'go.
I. Title. II. Title: The hell and heaven. III. Series: Segye adong chonggyo munhak sŏnjip ; 4.
PZ50.531.R95 79–825758

Yi, Cha-ŭng.
(Kojŏn sosŏl p'yŏn)
고전소설편 / [편저리자웅 ; 편집원안정환]. — 평양 : 아동도서출판사, 1964-
v. : ill. ; 22 cm. — (조선아동문학문고 ; 5)
I. An, Chŏng-hwan. II. Title. III. Series: Chosŏn adong munhak mun'go ; 5.
PZ50.52.Y47 79–82562[illegible]

Yi, Chong-hwan, 1920-
(Ŏndŏk wi ŭi p'urŭn chip)
언덕위의푸른집 / 이종환지음. — 서울 : 大韓基督教書會, 1969, 1971 printing.
191 p. : ill. ; 19 cm. — (한국아동문학신작선집 ; 6) (새벗문고)
Fiction.
On verso of t. p.: The blue house on the hill.
Series 2 romanized: Saebŏt mun'go.
I. Title. II. Title: The blue house on the hill. III. Series: Han'guk adong munhak sinjak sŏnjip ; 6.
PZ50.531.Y48 79–825630

Yi, Mun-hŭi, 1933-
(Ttŏdori pyŏl)
떠돌이별 / 이문희지음. — 서울 : 대한기독교서회, 1973.
145 p. : ill. ; 19 cm. — (새벗문고 ; 41)
Fiction.
Series romanized: Saebŏt mun'go.
I. Title.
PZ50.531.Y53 79–82594[illegible]

Yi, Wŏn-su, 1911-
(Hosu sok ŭi odumak chip)
호수속의오두막집 : 李元壽童話小說集. — 서울 : 세종문화사, 1975.
310 p. : ill. ; 22 cm.
Fiction.
Colophon inserted.
I. Title.
PZ50.531.Y54 78–826908

Yi, Yŏng-hŭi, 1931-
(Pyŏlnim ŭl sarang han iyagi)
별님을사랑한이야기 : 젊은이를위한사랑의동화 / 李寧熙著. — 서울 : 甲寅出版社, 1978.
127 p. : ill. ; 20 cm.
Short stories.
CONTENTS: 투명나비의집.—분홍조개껍질의사랑.—불개가있는마당.—민들레가보낸편지.—금가락지.—가슴에꽃을가꾸는짐승.—기다림이라는이름의꽃씨.—꽃과벌레의토론.—비둘기의여행.—불로초.—아름다운너무나아름다운.—별님네전화번호.—별이열리는나무.—잉어동.—별님을사랑한이야기.
₩800
I. Title.
PZ50.52.Y5 79–825168

CHILDREN'S LITERATURE, KOREAN (Continued)

Yu, Kyŏng-hwan, 1936-
(Onui kage)
오누이가개 / 유경환지음. — 서울 : 대한기독교서회, 1971, 1974 printing.
216 p. : ill. ; 19 cm. — (한국신작동화선집 ; 8) (새벗문고)
On verso of t. p.: Brother and sister shop.
Series 2 romanized: Saebŏt mun'go.

I. Title. II. Title: Brother and sister shop. III. Series: Han'guk sinjak tonghwa sŏnjip ; 8.
PZ50.52.Y78 79-825680

—HISTORY AND CRITICISM

Yi, Chae-ch'ŏl, 1931-
(Han'guk hyŏndae adong munhaksa)
韓國現代兒童文學史 / 李在徹著. — 서울 : 一志社, 1978.
598 p., [2] leaves of plates : ill. ; 23 cm.
Includes bibliographical references and index.
₩7000

1. Children's literature, Korean—History and criticism. I. Title.
PN1009.K6Y518 79-826266

Yi, Chae-ch'ŏl, 1931-
(Han'guk hyŏndae adong munhaksa ŭi yŏn'gu)
韓國現代兒童文學史의研究 / 李在徹. — [大邱?] : 李在徹, 1968.
327 p. ; 25 cm.
"1968年度文教部學術研究助成費에依한研究報告書(人文科學系)"
Bibliography: p. 324-327.

1. Children's literature, Korean—History and criticism. I. Title.
PN1009.K6Y52 79-826287

—HISTORY AND CRITICISM —PERIODICALS

(Adong munhak p'yŏngnon)
兒童文學評論 第1- 卷; 1976年여름-
[서울 兒童文學評論社]
v. ill. 21 cm. quarterly.

1. Children's literature, Korean—History and criticism—Periodicals. 2. Children's literature, Korean—Periodicals.
PN1009.K6A36 78-647587 (MARC-S)

—PERIODICALS

(Adong munhak p'yŏngnon)
兒童文學評論 第1- 卷; 1976年여름-
[서울 兒童文學評論社]
v. ill. 21 cm. quarterly.

1. Children's literature, Korean—History and criticism—Periodicals. 2. Children's literature, Korean—Periodicals.
PN1009.K6A36 78-647587 (MARC-S)

CHILDREN'S LITERATURE, LATVIAN

Vāczemnieks, Laimonis.
Noslēpumu grāmata : stāsti bērniem / Laimonis Vāczemnieks ; [mākslinieks Edgars Ozoliņš]. — Rīga : Liesma, 1978.
361 p. : ill. ; 23 cm. USSR***
CONTENTS: Līvsalas zēni.—Atdodiet mammu!— Veco vraku noslēpums.
0.90rub
I. Title.
PZ90.L4V26 79-380775
79 MARC

—HISTORY AND CRITICISM

Stikāne, Anastāsija.
Latviešu bērnu literatūra : [mācību līdzeklis] / A. Stikāne. — Rīga : Zvaigzne, 1977.
213 p. : ill. ; 21 cm. USSR77-31891
Bibliography: p. 205-[211]
0.66rub
1. Children's literature, Latvian—History and criticism. I. Title.
PN1009.L3S76 78-382861
78 MARC

CHILDREN'S LITERATURE, LITHUANIAN

Avyžius, Jonas, 1922-
Aštuonetas iš Trepsės namų / Jonas Avyžius ; iliustravo Lelija Bičiūnaitė. — Vilnius : Vaga, 1976.
143 p. : col. ill. ; 27 cm. USSR77-5773
"Jaunesniam mokykliniam amžiui."
1.48rub
I. Bičiūnaitė, Lelija. II. Title.
PZ90.L5A9 78-368871
78 MARC

Isokas, Gediminas.
Pavasario žibintai / Gediminas Isokas. — Vilnius : Vaga, 1978.
187 p. : ill. ; 17 cm. USSR***
"Jaunesniam mokykliniam amžiui."
0.35rub
I. Title.
PZ90.L5I8 79-345609
79 MARC

Krėvė-Mickevičius, Vincas, 1882-1954.
Bobulės vargai / Vincas Krėvė. — 2. leidimas. — Vilnius : Vaga, 1977.
193 p. : ill. ; 18 cm. USSR***
"Vidutiniam mokykliniam amžiui."
CONTENTS: Bobulės vargai.—Antanuko rytas.—Galvažudys.—Silkės.—Skerdžius.—Gilšė.—Milžinkapis.
0.43rub
I. Title.
PZ90.L5K72 1977 78-369176
78 MARC

Lietuvių vaikų proza / [sudaryt. V. Auryla]. — Vilnius : Vaga, 1977.
644 p. : ill. ; 23 cm. USSR***
3.30rub
1. Children's literature, Lithuanian. I. Auryla, Vincas, 1923-
PZ90.L5L47 79-343841
79 MARC

Marukas, K
Sunku būti geram / K. Marukas ; iliustravo Virginija Martinkėnaitė-Stankevičienė. — Vilnius : Vaga, 1978.
131 p. : col. ill. ; 23 cm. USSR***
"Ikimokykliniam ir jaunesniam mokykliniam amžiui."
0.80rub (cellophane binding)
I. Title.
PZ90.L5M274 79-352583
79 MARC

Palčinskaitė, Violeta, 1943-
Namai namučiai / Violeta Palčinskaitė ; iliustravo Marija Ladigaitė. — Vilnius : Vaga, 1977.
142 p. : col. ill. ; 22 cm. USSR78-22346
"Ikimokykliniam ir jaunesniam mokykliniam amžiui."
0.95rub
I. Ladigaitė, Marija. II. Title.
PZ90.L5P28 79-352552
79 MARC

Pikturna, Ignas, 1924-
Ko jūra ūžia / Ignas Pikturna. — Vilnius : Vaga, 1977.
85 p. : ill. ; 17 cm. USSR***
Short stories.
"Jaunesniam ir vidutiniam mokykliniam amžiui."
0.15rub
I. Title.
PZ90.L5P47 79-343664
79 MARC

Pocius, Algirdas, 1930-
Laiškas ant šalpusnio lapo / Algirdas Pocius. — Vilnius : Vaga, 1977.
102 p. : col. ill. ; 22 cm. USSR***
"Jaunesniam ir vidutiniam mokykliniam amžiui."
0.74rub
I. Title.
PZ90.L5P56 78-400757
79 MARC

Sokas, Juozas.
Gegutė ieško auklės / Juozas Sokas. — Vilnius : Vaga, 1978.
234 p. : ill. ; 17 cm. USSR***
"Vidutiniam mokykliniam amžiui."
0.40rub
I. Title.
PZ90.L5S56 79-345276
79 MARC

Tarasenka, Petras.
Užburti lobiai / Petras Tarasenka. — 3. leidimas. — Vilnius : Vaga, 1978.
170 p. : ill. ; 18 cm. USSR***
"Vidutiniam mokykliniam amžiui."
0.35rub (cellophane binding)
I. Title.
PZ90.L5T3 1978 78-389424
79 MARC

CHILDREN'S LITERATURE, MALAYAN

Baharuddin Zainal.
Hikayat Bakhtiar / di-usahakan oleh Baharudin bin Zainal. — Chet. 1. — Kuala Lumpur : Dewan Bahasa dan Pustaka, Kementerian Pelajaran Persekutuan Tanah Melayu, 1963.
xi, 142 p. ; 20 cm. — (Siri perpustakaan sekolah DBP ; bil. 12) (Bunga rampai sastera lama ; 3)
I. Hikayat Bakhtiar. 1963. II. Series: Dewan Bahasa dan Pustaka. Siri perpustakaan sekolah ; bil. 12. III. Series: Bunga rampai sastra lama ; 3.
PZ90.M25B28 78-950466
79 MARC

Matlob, 1937-
Anggun Cik Tunggal, putera malang / oleh Pawang Ana dan Raja Haji Yahya ; diceriterakan kembali oleh Matlob. — Kuala Lumpur : Fajar Bakti, 1975.
61 p. : ill. ; 22 cm. — (Siri bacaan Fajar Bakti : peringkat 2)
Based on Hikayat Anggun Cik Tunggal.
ISBN 0-19-580869-X
I. Ana, Pawang. II. Yahya, Raja Haji. III. Hikayat Anggun Cik Tunggal. IV. Title.
PZ90.M25M239 77-941259
78 MARC

Mohamed Naim Daipi, 1950-
Tangan-tangan yang menghancurkan / oleh Mohamed Naim Daipi, Mohd. Latiff Mohd. — Cet. 1. — Singapura : Solo Enterprises, 1977.
84 p. : ill. ; 19 cm. — (Siri novel kanak-kanak ; bk. 102)
I. Latiff Mohd., Mohd., 1950- joint author. II. Title.
PZ90.M25M6 78-943308
79 MARC

CHILDREN'S LITERATURE, MALAYSIAN

Hood bin Musa.
Rampaian Pandawa Lima / di-usahakan oleh Hood bin Musa. — Chet. 1. — Kuala Lumpur : Dewan Bahasa dan Pustaka, Kementerian Pelajaran, 1964.
xiii, 117 p., [2] leaves of plates : ill. ; 19 cm. — (Bunga rampai sastera lama ; 7) (Siri perpustakaan sekolah DBP ; bil. 35)
I. Hikayat Pandawa Lima. 1964. II. Title. III. Series. IV. Series: Dewan Bahasa dan Pustaka. Siri perpustakaan sekolah ; bil. 35.
PZ90.M25H6 78-950421
78 MARC

Ravaoarivelo, Elisabeth.
Voahirana / [Ravaoarivelo Elisabeth no namboatra ity tantara ity amin'ny teny Malagasy ; Razafintsalama Noël no nandoko ny sary]. — Fianarantsoa : Librairie Ambozontany, [1973]
38 p. : col. ill. ; 27 cm.
I. Razafintsalama, Noël. II. Title.
PZ90.M23R34 75-980244
79 MARC

Rubaidin Siwar.
Ekspedisi / Rubaidin Siwar. — Cet. 1. — Kuala Lumpur : Dewan Bahasa dan Pustaka, Kementerian Pelajaran Malaysia, 1974.
vii, 104 p. : ill. ; 19 cm. — (Siri perpustakaan sekolah DBP ; bil. 141)
CONTENTS: Lubang maut.—Hati waja.—Ribut salji.—Raksaksa Pulau Komodo.—Si profesor gila.—Menyeberang angkasa.—Bahaya di bawah laut.—Titian salji.—Mengembara ke dunia gelap.—Bola api.—Gergasi putih yang sepi.—Misi.
$2.20
I. Title. II. Series: Dewan Bahasa dan Pustaka. Siri perpustakaan sekolah ; bil. 141.
PZ90.M25R784 1974b 78-949721
79 MARC

CHILDREN'S LITERATURE, MARATHI

Purandare, Balavanta Moreśvara, 1922-
(Śelāra khiṇḍa)
शेलार खिंड / प्रस्तावना, सुमित्राराजे भोसले ; चित्रकार भाऊसाहेब आचरेकर ; लेखक बाबासाहेब पुरंदरे. — पुणें : महाराजा शिवछत्रपति प्रतिष्ठान प्रकाशन ; नागपूर : वितरक साहित्य प्रसार केंद्र, [1976?]
200 p. : ill. (some col.) ; 29 cm.
In Marathi
A novel.
Rs105.00
1. Shivaji, Raja, 1627-1680—Juvenile fiction. I. Acharekar, Murlidhar R., 1907- II. Title
PZ90.M35P8 78-901469

CHILDREN'S LITERATURE, MOLDAVIAN

Deleanu, Liviu, 1911–1967.
(De la mik la mare)
Де ла мик ла маре / Ливиу Делеану ; алкэтуире де Б. Делеану. — Кишинэу : Лумина, 1968.
255 p. ; 17 cm. — (Библиотека школарулуй ; 8)
Series romanized: Biblioteka shkolarului.
0.45rub
I. Title.
PZ90.M7D4 1968 78–415042

Drutsė, Ion, 1928–
(Frunze de dor)
Фрунзе де дор : повестире / Ион Друцэ. — Кишинэу : Лумина, 1966.
212 p. : ill. ; 17 cm.
"Пентру школаа медие."
I. Title.
PZ90.M7D7 78–415036

CHILDREN'S LITERATURE, NORWEGIAN

Armand, Yngvil.
—og ingen skal være aleine : noveller / Yngvil Armand ; tegninger av Kjersti Scheen. — Oslo : Gyldendal, 1978 (Oslo : Norbok)
133 p. : ill. ; 20 cm. — (Les mer bokene) N78-Nov.
CONTENTS: Lene og Liv.—Jenni.—To fortellinger om Eline.—De voksnes fest.—Katten.—Tore, seks år.—Sykle ute aleine.—Demonstrasjon.—Britas dagbok.—Joggen.—Tomas.—Ingen skal være aleine.
I. Scheen, Kjersti. II. Title.
PZ54.5.A75 78-401120
79 MARC

Asmervik, Sverre, 1942-
Og så kom vi til Norge ... / Sverre Asmervik. — Oslo : Gyldendal, c1977.
166 p. ; 19 cm. — (Treff-bokene) N***
ISBN 8205106460
I. Title.
PZ54.5.A76 79-359104
79 MARC

Brodtkorb, Reidar.
Skogen av siv og strå / Reidar Brodtkorp ; illustrert av Tor Morisse. — ₍Oslo₎ : Damm, 1978 (Stavanger : Dreyer aksjeselskap).
P. ₍3₎-187 : ill. ; 23 cm. — (Damms illustrerte barneboker) N78-Nov.
ISBN 8251795427 : kr39.50
I. Morisse, Tor. II. Title.
PZ54.5.B687 79-342005
79 MARC

Dybvad, Øyvind, 1908-
Håkon Hestekar gjennom eld og vatn / tekst, Øyvind Dybvad ; illustrert av Egil Torin Næsheim. — ₍Oslo₎ : Noregs boklag, c1977.
70 p. : ill. ; 26 cm. N77-Nov.
ISBN 8252203604 : kr40.00
I. Næsheim, Egil Torin. II. Title.
PZ54.5.D92 78-400765
79 MARC

Ekman, Fam, 1946-
Hva skal vi gjøre med lille Jill? / ₍av₎ Fam Ekman. — Oslo : Cappelen, 1976 (Oslo : Reistad offset)
₍24₎ leaves : ill. ; 19 cm. N76-Mar.
ISBN 8202034582 : kr25.00
I. Title.
PZ54.5.E483 78-375448
78 MARC

Ekman, Fam, 1946-
Kyss katta / Fam Ekman. — Oslo : Cappelen, 1977 (Oslo : Grøndahl & Søn tr.)
₍24₎ leaves : ill. ; 20 cm. N77-Apr.
ISBN 8202036496 : kr27.00
I. Title.
PZ54.5.E485 78-375449
78 MARC

Gaarder, Inger Margrethe, 1926-
Du er stor gutt nå, Nikolai! / Inger Margrethe Gaarder. — ₍Oslo₎ : Cappelen, 1978 (Oslo : Joh.Nordahls tr.)
109 p. ; 19 cm. — (Delfin-bokene) N78-Oct.
Sequel to Nikolai begynner på skolen.
ISBN 8202040256 : kr29.00
I. Title.
PZ54.5.G237 79-338172
79 MARC

Hanssen, Arvid, 1932-
Den vonde vinteren : forteljing frå Nord-Noreg i 1930-åra / Arvid Hanssen. — ₍Larvik₎ : Norsk barneblad, 1975.
213 p. ; 20 cm. N75-Nov.
kr28.00
I. Title.
PZ54.5.H2577 76-519534
*79₍r79₎rev MARC

Heggland, Johannes.
Det pantsette landet / Johannes Heggland. — Larvik : Norsk barneblads forl., 1977 (Larvik : Ant. Anderssens tr.)
172 p. ; 20 cm. N78-May
kr36.00
I. Title.
PZ54.5.H396 78-400769
79 MARC

Heie, Sigbjørn, 1945-
Framfor det ukjende / Sigbjørn Heie ; illustrert av Solveig Muren Sanden. — Larvik : Norsk barneblads forl., 1977 (Larvik : Elo trykk).
96 p. : ill. ; 20 cm. N78-May
kr28.00
I. Muren Sanden, Solveig. II. Title.
PZ54.5.H43 79-338163
79 MARC

Hopp, Zinken, 1905-
Papegøyen / fortalt av Zinken Hopp ; tegnet av Reidar Johan Berle. — Bergen : Eide, 1978 (Bergen : Bergens Tidende-J. W. Eide)
₍15₎ leaves : ill. ; 26 cm. N
ISBN 8251401291 : kr33.50
I. Berle, Reidar Johan. II. Title.
PZ54.5.H67 79-343669
79 MARC

Høydal, Aslaug, 1916-
Gaupus / Aslaug Høydal ; illustrert av Agnete Erichsen. — Oslo : Fonna, 1978 (Sandnes : Dales offset).
29 p., ₍1₎ leaf : ill. ; 25 cm. N78-Nov.
ISBN 8251300428 : kr40.00
I. Erichsen, Agnete. II. Title.
PZ54.5.H524 79-343839
79 MARC

Jåstad, Birger, 1906-
Kongeørnungen / Birger Jåstad. — Oslo : Noregs boklag, 1978 (Drammen : Orion trykkservice)
47 p. ; 20 cm. N78-Sept.
ISBN 8252203922 : kr40.00
I. Title.
PZ54.5.J28 78-389603
79 MARC

Jonsmoen, Ola, 1932-
Farleg ferd for fire / Ola Jonsmoen. — Oslo : Noregs boklag, 1978 (Oslo : EKB boktr.).
112 p. ; 21 cm. N78-Sept.
ISBN 8252203884 : kr40.00
I. Title.
PZ54.5.J577 78-390515
79 MARC

Kleiva, Ivar, 1903-
Vågar / Ivar Kleiva. — Oslo : Damm, 1978 (Oslo : Norbok)
119 p. ; 19 cm. N78-Sept.
ISBN 8251795435 : kr36.00
I. Title.
PZ54.5.K514 78-390846
79 MARC

Leer-Salvesen, Paul, 1951-
Sprekker i muren : skisser fra et landsfengsel / Paul Leer-Salvesen. — Oslo : Gyldendal, 1978 (Oslo : Norbok)
134 p., ₍1₎ leaf ; 20 cm. — (Treffbokene) N78-Nov.
ISBN 8205112002 : kr35.00
I. Title.
PZ54.5.L38 79-337790
79 MARC

Lund, Reidar, 1938-
Trond Travel / Reidar Lund ; tegninger av Kjersti Scheen. — ₍Oslo₎ : Damm, 1978 (Fjellhamar : O.K.-trykk)
p. ₍5₎-174, ₍1₎ leaf : ill. ; 20 cm. — (Damms barnebibliotek) N78-Nov.
ISBN 8251795494 : kr38.50
I. Scheen, Kjersti. II. Title.
PZ54.5.L8 78-400772
79 MARC

Newth, Philip.
Benjamins borg / ₍av₎ Philip Newth ; ill. av Mette Newth. — ₍Oslo₎ : Tiden, c1976.
₍10₎ leaves : ill. ; 22 cm. N76-Sept.
ISBN 8210012126 : kr23.00
I. Newth, Mette. II. Title.
PZ54.5.N464 79-358512
79 MARC

Norås, Thor G
Per 4 ₍i.e. fire₎ år og vennene hans / Thor G. Norås ; tegninger av Kjersti Scheen. — Oslo : Gyldendal, 1977 (Oslo : Norbok).
75 p. : ill. ; 19 cm. — (Les mer-bøkene) N77-Oct.
ISBN 8205106517 : kr36.00
I. Scheen, Kjersti. II. Title.
PZ54.9.N6 78-375451
78 MARC

Ranum, Edith, 1922-
"... men du kan ikke flykte ..." / Edith Ranum. — Oslo : Aschehoug, 1978 (₍Trondhjem₎ : Aktietr. i Trondhjem).
109 p. ; 20 cm. — (Ung i dag) N78-Oct.
ISBN 8203095003 : kr25.00
I. Title.
PZ54.5.R35 78-400770
79 MARC

Rønningen, Bjørn, 1937-
Tinius / Bjørn Rønningen ; ₍illustrasjoner, Egil Torin Næsheim₎. — Oslo : Luther, 1978 (₍Oslo₎ : Furuset boktr.)
136 p. : ill. ; 23 cm. N78-Oct.
ISBN 8253151454 : kr47.00
I. Næsheim, Egil Torin. II. Title.
PZ54.5.R588 79-338171
79 MARC

Sandbeck, Vidar, 1918-
Ferie i Hamnehagan / Vidar Sandbeck ; illustrert av Rune J. Andersson. — ₍Oslo₎ : Cappelen, 1978 (₍Oslo₎ : Joh. Nordahls tr.)
72 p., ₍1₎ leaf : ill. ; 19 cm. — (Delfinbokene) N78-Oct.
ISBN 8202040272 : kr29.00
I. Andersson, Rune J., 1936- II. Title.
PZ54.5.S24 79-342231
79 MARC

Skrede, Torill Thorson.
Skyld ikke på oss! / Torill Thorson Skrede ; illustrert av Freda Magnussen. — ₍Oslo₎ : Damm, 1978 (Halden : E. Sem)
p. ₍5₎-100, ₍1₎ leaf : ill. ; 21 cm. N78-Nov.
ISBN 8251794641 : kr36.50
1. Gipsies—Norway—Juvenile fiction. I. Magnussen, Freda. II. Title.
PZ54.5.S514 78-404128
79 MARC

Skulstad, Terje, 1946-
Første tørn / Terje Skulstad. — Oslo : Cappelen, 1977.
102 p. ; 20 cm. — (Delfin-bokene) N77-Oct.
ISBN 8202037883 : kr23.00
I. Title.
PZ54.5.S517 79-358482
79 MARC

Slapgard, Bjarne, 1901-
Vollstad-borna : forteljing frå okkupasjonstida (1940-1945) / Bjarne Slapgard. — Trondheim : Rune, 1976.
106 p. ; 21 cm. N77-Mar.
ISBN 8252301118 : kr30.00
1. Norway—History—German occupation, 1940-1945—Juvenile fiction. I. Title.
PZ54.5.S544 79-339811
79 MARC

Sørlle, Steinar.
Jørgen og uteliggeren / Steinar Sørlle. — Oslo : Gyldendal, 1978 (Oslo : Norbok)
127 p. ; 20 cm. — (Alle vi-bokene) N78-Nov.
ISBN 8205112541 : kr30.00
I. Title.
PZ54.5.S555 79-358483
79 MARC

Tenfjord, Johanne Marie Giæver, 1918-
Tre trylleord : en julefortelling fra Troms—om å være redd / Jo Tenfjord ; ill. av Reidar Johan Berle. — Bergen : Eide, c1976.
18 leaves : ill. ; 26 cm. N76-Nov.
ISBN 8251401224 : kr28.00
I. Berle, Reidar Johan. II. Title.
PZ54.5.T48 79-367458
79 MARC

Toonder, Marten, 1912-
Tom Poes en de toverfluit. Grammofoonplaat: Harry Geelen ₍Amsterdam, Ring, c1969₎
₍28₎ p. illus. (part col.) and phonodisc (2 s. 18 cm. 33 1/3 rpm. stereo) in pocket. 20 x 21 cm. Ne***
I. Geelen, Harry. II. Title.
PZ16.3.T637 74-340505
79₍r79₎rev MARC

Tvedt, Inger Vonheim.
Tone / Inger Vonheim Tvedt ; illustrert av Borghild Rud. — ₍Bodø₎ : Nordnorsk forfatterlag ; ₍Oslo₎ : Tiden, c1978 (Oslo : Reistad offset)
66 p., ₍1₎ leaf : ill. ; 21 cm. N78-Sept.
ISBN 8210016016 : kr32.00
I. Rud, Borghild. II. Title.
PZ54.5.T86 78-389604
79 MARC

Wikström, Olga, 1905-
Vart är du på väg, Durra? / Olga Wikström. — Stockholm : AWE/Geber, 1978.
114 p. ; 19 cm. S78-26
ISBN 9120057261
I. Title.
PZ59.W53V3 78-400771
79 MARC

CHILDREN'S LITERATURE, NORWEGIAN (Continued)

—BIBLIOGRAPHY

Moen, Kjersti.
Barne- og ungdomslitteraturen 1970-76 : et utvalg gruppert etter emner med handlingsreferat og vurdering / Kjersti Moen, Solvår Østebø, Anne Mari Aamelfot. — Oslo : Cappelen, 1977 ₍s.l.₎ : Reistad offset)
₍4₎ leaves, 131 p. ; 23 cm. N78-Jan
Includes indexes.
ISBN 8202014719 : kr29.00
1. Children's literature, Norwegian—Bibliography. I. Østebø, Solvår, joint author. II. Aamelfot, Anne Mari, joint author. III. Title.
Z1037.8.N7M63 78-366379
₍PN1009.N6₎ 78 MARC

CHILDREN'S LITERATURE, PERSIAN

Kashkūlī, Mah'dukht.
(Afsānah-i bārān dar Īrān)
افسانه باران در ایران / نوشتهٔ مه‌دخت کشکولی ؛ نقاشی از حسین محجوبی. — چاپ 1. — ₍تهران₎ : رادیو تلویزیون آموزشی ، سازمان رادیو تلویزیون ملی ایران ، ₍1977₎ 2535.
29 p. : col. ill., facsims. ; 28 cm.
In Persian.
«ویژه نوجوانان.»
I. Maḥjūbī. Ḥusayn. II. Title.
PZ90.P4K25 79–970979

—HISTORY AND CRITICISM
—ADDRESSES, ESSAYS, LECTURES

Maḥjūb, Muḥammad Ja'far.
(Intikhāb va inṭibāq-i manābi'-i adab-i Fārsī barā-yi tadvīn-i kitāb'hā-yi kūdakān va nawjavānān)
انتخاب و انطباق منابع ادب فارسی برای تدوین کتاب‌های کودکان و نوجوانان / محمد جعفر محجوب. — ₍تهران₎ : شورای کتاب کودک ، ₍1978₎ 2536.
47 p. ; 22 cm.
In Persian.
1. Books and reading for children—Iran—Addresses, essays, lectures. 2. Children's literature, Persian—History and criticism—Addresses, essays, lectures. I. Title.
Z1037.8.P47M34 79–970426

CHILDREN'S LITERATURE, PERUVIAN

Díaz Herrera, Jorge.
Parque de leyendas / Jorge Díaz Herrera. — Lima : Instituto Nacional de Cultura, 1977.
77 p. : ill. ; 18 cm.
"29"
I. Title.
PZ73.D52 863 79-348722
79 MARC

Vega Herrera, César.
Pasakón / César Vega Herrera ; ₍ill., Maruja García Naranjo₎. — ₍Lima?₎ : Vega Herrera, ₍between 1974 and 1977₎
24 p. : ill. ; 21 cm.
I. Title.
PZ73.V37 863 79-366305
79 MARC

CHILDREN'S LITERATURE, POLISH

Bahdaj, Adam.
Czarne sombrero / Adam Bahdaj ; ₍il. Jerzy Moskal₎. — Wyd. 3. — Katowice : Śląsk, 1977 (Warsz. : DSP)
262, ₍2₎ p. : ill. ; 21 cm. P78-2467
zł21.00
I. Title.
PZ69.B28C9 1977 79-374052
79 MARC

Bahdaj, Adam.
Pilot i ja / Adam Bahdaj ; il. ₍kolor.₎ Danuta Konwicka. — Wyd. 4. — Warszawa : Nasza Księgarnia, 1977 (Łódź : ŁZG 2)
₍24₎ p. ; 26 cm. — (Moje książeczki) P78-1828
zł13.00
I. Konwicka, Danuta. II. Title.
PZ69.B28P54 1977 78-388666
79 MARC

Broniewska, Janina.
Historia toczonego dziadka i malowanej babki / Janina Broniewska ; ilustrowała Maria Sołtyk. — Wyd. 5. — Warszawa : Nasza księgarnia, 1975.
144 p. : ill. ; 17 cm. P***
zł13.00
I. Sołtyk, Maria. II. Title.
PZ69.B65H54 1975 78-417809
78 MARC

Broszkiewicz, Jerzy.
Wielka, większa i największa / Jerzy Broszkiewicz ; ₍il. Gabriel Rechowicz₎. — Wyd. 11. — Warszawa : Nasza Księgarnia, 1978 (Łódź : ŁZG 1)
265, ₍3₎ p. : ill. ; 20 cm. P78-6765
Errata slip inserted.
zł20.00
I. Title.
PZ69.B72W5 79-347237
79 MARC

Brzoza, Jan, 1900-
Dzieci / Jan Brzoza. — Wyd. 4. — Katowice : Śląsk, 1978 (Ciesz. : CDW)
261, ₍3₎ p. ; 20 cm. P78-6767
zł30.00
I. Title.
PZ69.B78D9 1978 79-358561
79 MARC

Dąbrowska, Krystyna.
Ucieczka z domu pod Ołymką / Krystyna Dąbrowska ; il. Emilia Freudenreich. — Wyd. 1. — Warszawa : Krajowa Agencja Wydawnicza, 1978 (Lub. : LZGPKWN)
100, ₍4₎ p. : ill. ; 19 cm. P78-5957
zł20.00
I. Title.
PZ69.D24U3 79-341546
79 MARC

Domańska, Antonina.
Paziowie króla Zygmunta : opowiadania obyczajowe na tle dawnych wieków / Antonina Domańska ; il. Antoni Gawiński. — Wyd. 7. — Warszawa : Czytelnik, 1978 (Łódź : ŁZG 1)
215, ₍4₎ p. : ill. ; 21 cm. P78-6983
zł17.00
I. Gawiński, Antoni. II. Title.
PZ69.D6P3 1978 79-345530
79 MARC

Falski, Marian.
Elementarz / Marian Falski : ilustrował Janusz Grabiański. — Wyd. 2. — Warszawa : Wydawnictwa Szkolne i Pedagogiczne, 1976.
159 p. : col. ill. ; 21 x 23 cm. P***
zł40.00
I. Grabiański, Janusz. II. Title.
PZ69.F3E4 1976 78-370510
78 MARC

Goldszmit, Henryk, 1878-1942.
Fragmenty utworów / Janusz Korczak ₍i.e. Henryk Goldszmit₎ ; wybrała Danuta Stępniewska ; il. ₍kolor.₎ Małgorzata Różańska ; ₍posłowie Kazimierz Dębnicki₎. — Wyd. 1. — Warszawa : Nasza Księgarnia, 1978 (Krak. : DNar)
244, ₍4₎ p. : col. ill. ; 25 cm. P78-6569
zł75.00
I. Stępniewska, Danuta. II. Title.
PZ69.G6F7 1978 78-398298
79 MARC

Goldszmit, Henryk, 1878-1942.
Król Maciuás na wyspie bezludnej / Janusz Korczak ; il. ₍kolor.₎ Jerzy Srokowski. — Wyd. 6. — Warszawa : Nasza Księgarnia, 1977 (Krak : DNar 2)
146, ₍2₎ p., ₍6₎ leaves of plates : col. ill. ; 29 cm. P78-1832
I. Srokowski, Jerzy. II. Title.
PZ69.G6K7 1977 78-380131
78 MARC

Goldszmit, Henryk, 1878-1942.
Król Maciuś Pierwszy / Janusz Korczak ; il. ₍kolor.₎ Jerzy Srokowski ; ₍wstępem opatrzył Igor Newerly₎. — Wyd. 6. — Warszawa : Nasza Księgarnia, 1977 (Krak : DNar 2)
210, ₍2₎ p., ₍11₎ leaves of plates : col. ill., port. ; 29 cm. P78-1833
zł90.00
I. Srokowski, Jerzy. II. Title.
PZ69.G6K72 1977 78-380132
78 MARC

Goldszmit, Henryk, 1878-1942.
Król Maciuś Pierwszy / Janusz Korczak ₍i.e. Henryk Goldszmit₎ ; il. ₍kolor.₎ Waldemar Andrzejewski ; wstęp Igor Newerly. — Wyd. 7. — Warszawa : Nasza Księgarnia, 1978 (Krak. : DNar)
292, ₍4₎ p. : ill. (some col.) ; 25 cm. P78-8543
zł70.00
I. Andrzejewski, Waldemar. II. Title.
PZ69.G6K72 1978 79-354558
79 MARC

Grabowski, Jan.
Czarna owieczka / Jan Grabowski ; ilustrowała Maria Orłowska-Gabryś. — Wyd. 3. — Warszawa : Nasza Księgarnia, 1978, c1966.
34 p. : col. ill. ; 26 cm. — (Lektura szkolna ; 1) P***
"Opowiadanie ze zbioru Kochany zwierzyniec."
zł20.00
I. Orłowska-Gabryś, Maria. II. Title.
PZ69.G68C9 1978 79-350935
79 MARC

Groński, Ryszard Marek.
Katar żyrafy / Ryszard Marek Groński ; il. ₍kolor.₎ Maria Uszacka. — Wyd. 2. — Warszawa : Nasza Księgarnia, 1977 (Łódź : ŁZG 2)
₍12₎ p. : col. ill. ; 16 cm. — (Poczytaj mi mamo) P78-1831
Children's story.
zł6.00
I. Uszacka, Maria. II. Title.
PZ69.G714K3 1977 79-373690
79 MARC

Józefacka, Maria.
Kto sieje wiatr : powieść dla młodzieży / Maria Józefacka. — Wyd. 2. — Lublin : Wydaw. Lubelskie, 1978 (Rzesz. : RzZG)
378, ₍2₎ p., ₍6₎ leaves of plates : ill. ; 20 cm. P78-3082
Part 2 of the author's trilogy, the 1st of which is: Karkołomna gra, and the 3d: To co najpiękniejsze.
zł37.00
I. Title.
PZ69.J68K7 1978 78-396578
79 MARC

Józefacka, Maria.
To co najpiękniejsze : powieść dla młodzieży / Maria Józefacka ; ₍il. Zofia Szulc₎. — Wyd. 1. — Lublin : Wydaw. Lubelskie, 1977 (Lub : LZGPKWN)
263, ₍1₎ p. : ill. ; 19 cm. P77-10750
zł28.00
I. Szulc, Zofia.
PZ69.J68T6 78-373305
78 MARC

Kern, Ludwik Jerzy.
Mądra poduszka : wiersze dla dzieci / Ludwik Jerzy Kern ; il. ₍kolor.₎ Zbigniew Rychlicki. — Wyd. 2. — Warszawa : Nasza Księgarnia, 1977 (Pozn. : PZGMK)
107, ₍1₎ p. : col. ill. ; 34 cm. P78-2470
zł74.00
I. Rychlicki, Zbigniew. II. Title.
PZ69.K4M3 1977 78-400697
79 MARC

Konopnicka, Maria, 1842-1910.
Choinka / Marji Konopnickiej ; zdobiła graficznie Marja Dziewulska. — Warszawa : W Drukarni i litografji Jan Cotty, 1928.
₍10₎ p. : col. ill. ; 20 cm.
Cover title.
1. Christmas trees—Juvenile poetry. I. Dziewulska, Maria. II. Title.
PZ69.K6C5 1928 79-372385
79 MARC

Korczakowska, Jadwiga.
Zabawa w chowanego / Jadwiga Korczakowska. — Wyd. 1. — Warszawa : Ludowa Spółdzielnia Wydawnicza, 1978 (Inowr. : ZW CZSR)
93, ₍3₎ p. ; 18 cm. P78-7426
Short stories.
CONTENTS: Zabawa w chowanego.—Dziki Królik.—Zakład.—Mały.
zł13.00
I. Title.
PZ69.K62Z3 79-350938
79 MARC

Kowalewska, Maria.
Zwariowane podwórko / Maria Kowalewska ; il. Krystyna Michałowska. — Wyd. 1. — Warszawa : Ludowa Spółdzielnia Wydawnicza, 1976 (Olszt : OZGSP)
185 p. : ill. ; 18 cm. P77-1338
zł20.00
I. Title.
PZ69.K66 78-366283
78 MARC

Kowalewski, Stanisław, 1918-
Maja / Stanisław Kowalewski. — Wyd. 1. — Warszawa : Młodzieżowa Agencja Wydawnicza, 1977 (Białyst : BZG)
139, ₍1₎ p. ; 17 cm. — (Portrety ; 13) P78-3083
zł13.00
I. Title.
PZ69.K635M3 78-398832
79 MARC

Kownacka, Maria.
Wiatrak profesora Biedronki / Maria Kownacka, Jan E. Kucharski ; il. Konstanty M₍aria₎ Sopoćko. — Wyd. 4. — Warszawa : Nasza Księgarnia, 1978 (Pozn. : PZGMK)
174, ₍2₎ p. ; 22 cm. P78-1597
Errata slip inserted.
zł25.00
I. Kucharski, Jan Edward, joint author. II. Title.
PZ69.K64W5 1978 78-401096
79 MARC

Kubiak, Tadeusz.
Kocham ciebie, mamo / Tadeusz Kubiak ; il. ₍kolor.₎ Krzysztof Zeydler-Zborowski. — Wyd. 1. — Warszawa : Krajowa Agencja Wydawnicza, 1977 (Gdań. : ZG)
₍28₎ p. : ill. ; 28 cm. P77-9505
Poems.
zł16.00 (pbk)
I. Title.
PZ69.K76K6 78-380333
79 MARC

CHILDREN'S LITERATURE, POLISH
(Continued)

Kuczyński, Maciej.
Kask / Maciej Kuczyński. — Wyd. 1. — Warszawa : Czytelnik, 1978 (Olszt. : OZGSP)
130, [2] p. : ill. ; 20 cm. P78-8346
zł16.00
I. Title.
PZ69.K8K3 79-348441
79 MARC

Lengren, Zbigniew.
Gałgankowy skarb / Zbigniew Lengren ; il. [kolor.] aut. — Wyd. 1. — Lublin : Wydaw. Lubelskie, 1978 (Lub. : LZGPKWN)
[12] p. : col. ill. ; 15 x 21 cm. P78-3731
Cover title.
zł6.50
I. Title.
PZ69.L38G3 79-349739
79 MARC

Lesman, Jan, 1898-
Brzechwa dzieciom / Jan Brzechwa [i.e. Jan Lesman] ; il. [kolor] Jan Marcin Szancer. — Wyd. 11. — Warszawa : Nasza Księgarnia, 1978 (Krak. : DNar)
124, [4] p. : col. ill. ; 22 cm. — (Klasyka dziecięca) P78-6587
zł50.00
I. Szancer, Jan Marcin. II. Title. III. Series.
PZ69.L4B7 1978 78-399525
79 MARC

Lesman, Jan, 1898-
ZOO / Jan Brzechwa [i.e. J. Lesman] ; il. [kolor.] Teresa Wilbik. — Wyd. 2. — Warszawa : Nasza Księgarnia, 1977 (Łódź : ŁZG 2)
[8] p. : col. ill. ; 19 x 21 cm. P78-1829
Cover title.
zł20.00
I. Wilbik, Teresa. II. Title.
PZ69.L4Z6 79-339607
79 MARC

Leśmian, Bolesław, pseud.
Przygody Sindbada Żeglarza / Bolesław Leśmian ; oprac. graf. Janusz Stanny. — Wyd. 6. — Warszawa : Czytelnik, 1977 (Krak : DNar)
205, [3] p., [8] leaves of plates : ill. (some col.) ; 25 cm. P78-1368
I. Stanny, Janusz. II. Title.
PZ69.L45P7 1977 78-381102
79 MARC

Liskowacki, Ryszard.
Wracamy do domu : powieść dla młodzieży. / Ryszard Liskowacki ; il. Szymon Kobyliński. — Wyd. 1. — Poznań : Wydaw. Poznańskie, 1978 (Pozn : PZGMK)
249, [3] p., [3] leaves of plates : ill., (some col.) ; 21 cm.
zł40.00
1. World War, 1939-1945—Juvenile fiction. I. Kobyliński, Szymon. II. Title.
PZ69.L58W7 78-399844
79 MARC

Lothamer, Henryk.
Król z Babskich Łąk / Henryk Lothamer ; il. Maria Mackiewicz. — Wyd. 1. — Warszawa : Nasza Księgarnia, 1977 (Gdań : ZG)
199, [1] p. : ill. ; 20 cm. P78-1595
zł25.00
I. Mackiewicz, Maria. II. Title.
PZ69.L64K7 78-398833
79 MARC

Makuszyński, Kornel, 1884?-1953.
Bajka o wawelskim smoku / Kornel Makuszyński i Marian Walentynowicz. — Wyd. nicejskie. — [s.l.] : Nakł. Związku Polaków we Francji, 1941.
[32] p. : chiefly col. ill. ; 27 cm.
I. Walentynowicz, Marian, joint author. II. Title.
PZ69.M3B3 1941 78-400764
79 MARC

Michałowska, Mira.
Ania i Ewa, i Marian, i Rusak, i inni ... / Mira Michałowska ; il. [kolor.] Janusz Stanny. — Wyd. 2. — Warszawa : Krajowa Agencja Wydawnicza, 1977 (Gdań. : ZG)
55, [5] p. : col. ill. ; 28 cm. P78-2865
zł25.00
I. Stanny, Janusz. II. Title.
PZ69.M43A5 1977 78-400756
79 MARC

Michałowska, Mira.
Trzy po trzy z sąsiadami : nowe spotkanie z Marcinem / przygot. Mira Michałowska ; il. [kolor.] Jan Bokiewicz ... [et al.] — Wyd. 1. — Warszawa : Krajowa Agencja Wydawnicza, 1978 (Gdańsk : ZG)
60 p. : col. ill. ; 19 x 22 cm. P78-7850
zł20.00
I. Title.
PZ69.M43T7 79-347238
79 MARC

Mrówczyński, Bolesław.
Cień Montezumy : opowieść o Indianinie-tułaczu i pułkowniku Luisie de Tulancingo, polskim powstańcu / Bolesław Mrówczyński ; il. Bohdan Zieleniec. — Wyd. 5. — Warszawa : Ludowa Spółdzielnia Wydawnicza, 1978 (Inowr. : ZW CZSR)
434, [6] p. : ill. ; 20 cm. — (Biblioteka młodych) P78-3732
zł55.00
1. Mexico—History—European intervention, 1861-1867—Juvenile fiction. I. Title.
PZ69.M7C5 1978 78-391398
79 MARC

Nawrocka-Dońska, Barbara.
Historia życia czarnego kota / Barbara Nawrocka-Dońska ; il. Teresa Wilbik. — Wyd. 2. — Warszawa : Nasza Księgarnia, 1978 (Gdań. : ZG)
178, [6] p. : ill. ; 22 cm. P78-2866
zł20.00
I. Wilbik, Teresa. II. Title.
PZ69.N3H5 1978 78-396579
79 MARC

Nienacki, Zbigniew.
Pan Samochodzik i Fantomas / Zbigniew Nienacki ; [il. Magdalena Kapuścińska]. — Wyd. 3. — Łódź : Wydaw. Łódzkie, 1978 (Łódź : ŁZG 1)
338, [1] p. : ill. ; 20 cm. P78-3956
zł39.00
I. Title.
PZ69.N474P3 1978 78-400696
79 MARC

Niziurski, Edmund.
Naprzód, wspaniali! / Edmund Niziurski. — Wyd. 3. — Warszawa : Młodzieżowa Agencja Wydawnicza, 1977 (Białyst. : BZG)
298, [1] p. ; 21 cm. P78-2666
zł40.00
I. Title.
PZ69.N5N3 1977 78-378869
78 MARC

Nowacka, Ewa.
Może nie, może tak / Ewa Nowacka ; il. Krystyna Michałowska. — Wyd. 2. — Warszawa : Nasza Księgarnia, 1978 dr. ukończ. 1977 (Gdań : ZG)
222, [2] p. ; ill. ; 20 cm. — (Klub siedmiu przygód) P78-2667
zł25.00
I. Michałowska, Krystyna. II. Title.
PZ69.N6M6 1977 79-338457
79 MARC

Okaryna i inne opowiadania / [Hanna Januszewska ... et al.] ; il. [kolor.] Maria Orłowska-Gabryś. — Wyd. 1. — Warszawa : Krajowa Agencja Wydawnicza, 1977 (Łódź : PrasZG)
84, [4] p. : col. ill. ; 21 cm. P78-6163
CONTENTS: Januszewska, H. Okaryna.—Pollakówna, J. Złoty deszcz.—Michałowska, M. Wyobraźcie sobie.—Korczakowska, J. Ulica dobrych zdarzeń.—Korda, P. Raki.
zł36.00
1. Children's literature, Polish. I. Januszewska, Hanna. II. Orłowska-Gabryś, Maria.
PZ69.A2O36 79-349738
79 MARC

Owsianko, Bronisław.
Ucieczka z Pawiaka : opowieść o działaczach KPP [Komunistycznej Partii Polski] / Bronisław Owsianko. — Wyd. 2. — Warszawa : Wydaw-a Szkolne i Pedagogiczne, 1977 (Wroc. : WZG)
69, [3] p. : ill. ; 17 cm. — (Biblioteczka historyczna : Cykl Dzieje ruchu robotniczego) P78-1844
zł12.00
1. Zdziarski, Mirosław, 1892- —Juvenile fiction. I. Title.
PZ69.O88U3 1977 78-381105
79 MARC

Petecki, Bohdan.
Prosto w gwiazdy / Bohdan Petecki ; [il. Henryk Krzysztof Smoleński]. — Wyd. 1. — Katowice : Śląsk, 1978 (Byt. : ZGK 5)
251, [1] p. : ill. ; 20 cm. P78-2243
zł21.00
I. Title.
PZ69.P4P7 79-371500
79 MARC

Porazińska, Janina.
Nad wiślaną wodą / Janina Porazińska ; il. [kolor.] Zbigniew Rychlicki. — Wyd. 2. — Warszawa : Nasza Księgarnia, 1976 (Łódź : ŁZG 2)
[2], 175, [5] p. : col. ill. ; 27 cm. P77-1667
zł60.00
I. Rychlicki, Zbigniew. II. Title.
PZ69.P6N3 1976 78-378402
78 MARC

Szczepańska, Nora, 1914-
Odwaga / Nora Szczepańska ; il. Andrzej Strumiłło. — Wyd. 2. — Poznań : Wydaw. Poznańskie, 1977 (Warsz. : DRP)
169, [3] p. ; 21 cm. — (Biblioteka młodych) P77-3882
zł18.00
I. Strumiłło, Andrzej. II. Title.
PZ69.S917O3 1977 78-367035
78 MARC

Szczepańska, Nora, 1914-
Zemsta Karibu / Nora Szczepańska ; il. Zygmunt Zaradkiewicz. — Wyd. 2. — Warszawa : Krajowa Agencja Wydawnicza, 1978 (Szcz. : SzZG)
173, [3] p. : ill. ; 21 cm. P78-6164
zł25.00
I. Title.
PZ69.S917Z4 1978 78-399524
79 MARC

Szczygieł, Jerzy.
Powódź / Jerzy Szczygieł. — Wyd. 3. — Lublin : Wydaw. Lubelskie, 1978 (Lub. : LZGPKWN)
227, [1] p. : ill. ; 20 cm. P78-3737
zł24.00
I. Title.
PZ69.S92P68 1978 79-351855
79 MARC

Szymański, Edward.
Niespodzianka / Edward Szymański ; il. Jan S[amuel] Miklaszewski. — Wyd. 3. — Warszawa : "Czytelnik", 1978 (Krak : DW)
[24] p. : col. ill. ; 20 cm. P78-2033
zł20.00
I. Miklaszewski, Jan S. II. Title.
PZ69.S98N5 1978 78-381379
78 MARC

Zakrzewska, Wanda.
Wakacje z braciszkiem / Wanda Zakrzewska. — Wyd. 1. — Warszawa : Instytut Wydawniczy Pax, 1975.
113 p. ; 20 cm. p***
zł25.00
I. Title.
PZ69.Z25W3 79-376787
79 MARC

Zamojski, Ferdynand.
Chłopiec znad Bzury / Ferdynand Zamojski. — Wyd. 1. — Warszawa : Wydaw. Min. Obrony Narodowej, 1978 (Warsz : WojsZG)
274, [2] p. ; 19 cm. P78-2244
zł20.00
1. Kutno, Battle of, 1939—Juvenile fiction. I. Title.
PZ69.Z27C5 79-338099
79 MARC

Zobaczyć inaczej : wybór opowiadań / [aut.] Lech Borski ... [et al. ; il. Witold Rębkowski]. — Wyd. 1. — Warszawa : Młodzieżowa Agencja Wydawnicza, 1978 (Białyst : BZG)
217, [3] p. : ill. ; 20 cm. P78-7189
CONTENTS: Borski, L. Co ma but do samochodu.—Wieczerska, J. Bywa i zła pogoda. —Boglar, K. Semafory.—Nowacka, E. Coś, co o nich wiem.—Gawęcka-Bisko, K. Sorymina.—Szczygieł, J. Ścieżka.—Minkowski, A. Zaczynasz być mężczyzną.—Siemieński, M. Pięćminut, ja i matka.
zł32.00
I. Borski, Lech. II. Rębkowski, Witold.
PZ69.A2Z6 79-346270
79 MARC

—HISTORY AND CRITICISM

Frycie, Stanisław.
Literatura dla dzieci i młodzieży w latach 1945-1970 / Stanisław Frycie. — Wyd. 1. — Warszawa : Wydawn. Szkolne i Pedagogiczne, 1978-
v. : ill. ; 21 cm. p***
Includes bibliographical references and index.
CONTENTS: t. 1. Proza.

zł80.00 (v. 1)
1. Children's literature, Polish—History and criticism. I. Title.
PN1009.P6F7 78-389712
78 MARC

Z dziejów literatury dla dzieci i młodzieży. [Redaktor naukowy: Izabela Lewańska]. Opole, 1970.
99 p. 25 cm. (Zeszyty naukowe Wyższej Szkoły Pedagogicznej w Opolu) (Historia literatury, 7)
Includes bibliographical references.
1. Children's literature, Polish—History and criticism. I. Kaniowska-Lewańska, Izabela, ed. II. Series: Oppeln. Wyższa Szkoła Pedagogiczna. Historia literatury, 7.
PG7003.H57 no. 7 74-231571
[PN1009.P6] 79[r79]rev MARC

CHILDREN'S LITERATURE, PORTUGUESE

--BIBLIOGRAPHY--PERIODICALS

Bibliografia de literatura infantil em língua portuguêsa. 1953–
São Paulo, Secretaria de Cultura, Departamento de Bibliotecas Infanto-Juvenís [etc.]
v. ill. 23–26 cm.
Vols. for 1953–55 called also 1.–2. ed.; called supplemento.
Vols. for 1953– issued by the Divisão de Bibliotecas Infanto-Juvenís, Departamento de Cultura; by the division under a variant name: Divisão de Bibliotecas Infanto-Juvenís e Cinema Educativo; Departamento de Bibliotecas Infanto-Juvenís, São Paulo.
1. Children's literature, Portuguese — Bibliography — Periodicals. I. São Paulo, Brazil (City). Departamento de Cultura. Divisão de Bibliotecas Infanto-Juvenís. II. São Paulo, Brazil (City). Departamento de Bibliotecas Infanto-Juvenís.
Z1037.7.B53 79–641321
[PN1009.P65] MARC–S

CHILDREN'S LITERATURE, PORTUGUESE (Continued)

Sérgio, António, 1883-1969.
Os dez anõezinhos da Tia Verde-Água e outros contos / António Sérgio ; ilustrações de Luís Filipe de Abreu. — 1. ed. — Lisboa : Sá da Costa Editora, 1978.
55 p. : ill. ; 23 cm. — (Obras infantis de António Sérgio ; v. 1)
CONTENTS: Os dez anõezinhos da Tia Verde-Água.—O ratão pelado.—A cabana dos lobos.
I. Title. II. Series: Sérgio, António, 1883-1969. Obras infantis de António Sérgio ; v. 1.
PZ83.S43 1978 — 79-113224
79 — MARC

—HISTORY AND CRITICISM

Correia, João David Pinto.
Literatura juvenil, paraliteratura : problemas de delimitação de conceitos e aspectos didáctico-pedagógicos : propostas para um melhor aproveitamento pedagógico das Leituras dos jovens / João David Pinto Correia. — Lisboa : Livraria Novidades Pedagógicas, 1978.
61 p. : diagrs. ; 21 cm. — (Colecção Biblioteca prática do professor ; 6)
Includes bibliographical references.
1. Children's literature—History and criticism. 2. Children's literature, Portuguese—History and criticism. I. Title.
PN1009.A1C665 — 79-112194
79 — MARC

Grips,
Una festa pro Antonio! : gö in tschinch parts / Grips ; versiun rumantscha da Clo Duri Bezzola. — [Ittigen : Tista Murk], 1977.
67 p. ; 21 cm. — (La Scena ; n. 55) — Sw***
10.00F
I. Title.
PZ90.R3G7 — 79-347469
79 — MARC

CHILDREN'S LITERATURE, ROMANIAN

Anton, Costache, 1930-
În soarele de april : [schițe] / Costache Anton ; [desenul copertei de Vintilă Mihăescu]. — București : Editura Ion Creangă, 1977.
199 p. ; 17 cm. — (Biblioteca contemporană) — R77-2765
lei2.75
I. Title.
PZ90.R6A57 — 78-368527
*78 — MARC

Băjenaru, Grigore, 1907-
Cișmigiu et Comp. : [roman] / Grigore Băjenaru. — Ediția a 2-a / [portretul autorului de Silvan ; copertă de Gheorghe Marinescu]. — București : Editura Ion Creangă, 1978.
237 p. ; 21 cm. — R79-538
lei4.50
I. Title.
PZ90.R6B258 1978 — 79-372189
*79 — MARC

Băran, Tudor.
Casa destinelor : [roman] / Tudor Băran ; [desenul copertei, Gh. Marinescu]. — București : Editura Ion Creangă, 1978.
286 p. ; 17 cm. — R78-4589
lei4.50
I. Title.
PZ90.R6B27 — 79-372191
*79 — MARC

Berindei, Ruxandra.
Povestiri de lună plină / Ruxandra Berindei ; [coperta, Done Stan]. — [București] : Editura Eminescu, 1978.
218 p. ; 20 cm. — R78-2548
lei5.00
I. Title.
PZ90.R6B444 — 79-338177
*79 — MARC

Căldăraru, Emilia.
Lîngă păsări, lîngă vînt : [povestiri] / Emilia Căldăraru ; [desenul copertei de Vintilă Mihăescu]. — București : Editura Ion Creangă, 1978.
213 p. ; 17 cm. — (Biblioteca contemporană) — R78-2528
lei2.75
I. Title.
PZ90.R6C32 — 79-343832
*79 — MARC

Cristescu, Maria Luiza.
Roman din dragoste : (povești de adormit copiii) / Maria-Luiza Cristescu ; [coperta, Done Stan]. — [București] : "Albatros", 1977.
157 p. : ill. ; 17 cm. — R78-189
lei4.75
I. Title.
PZ90.R6C76 — 78-368210
*78 — MARC

Drumeș, Mihail.
Povestiri despre cutezători / Mihail Drumeș ; [copertă și portrete de Florin Ionescu]. — București : Editura Ion Creangă, 1977.
235 p. : ill. ; 24 cm. — R78-613
lei9.25
1. Explorers—Juvenile fiction. 2. Air pilots—Juvenile fiction. I. Title.
PZ90.R6D73 — 78-366412
*78 — MARC

Iuteș, Gica, 1925-
Măr roșu ca focul : [roman] / Gica Iuteș ; [desenul copertei, Kalab Francisc]. — București : Editura Ion Creangă, 1978.
306 p. ; 17 cm. — R79-767
lei5.00
I. Title.
PZ90.R6I9 — 79-371058
*79 — MARC

Manoliu, Emanoil.
Bulgăre de foc : [roman științifico-fantastic] / Emanoil Manoliu ; [coperta de Pusztai Georgeta]. — București : Editura Ion Creangă, 1978.
314 p. ; 20 cm. — R78-1488
lei6.75
I. Title.
PZ90.R6M24 — 79-343835
*79 — MARC

Munteanu, Francisc.
Roșcovanul : [roman] / Francisc Muntean ; [desenul copertei de Kalab Francisc]. — București : Editura Ion Creangă, 1979.
214 p. ; 17 cm. — R79-1447
lei3.25
I. Title.
PZ90.R6M85 — 79-380749
*79 — MARC

Sadoveanu, Ion Marin.
Taurul mării : [roman] / Ion Marin Sadoveanu ; cuvînt înainte de Ovidiu Papadima ; [coperta colecției de Done Stan ; desenul copertei și ilustrațiile de Contantin Baciu]. — București : Editura Ion Creangă, 1977.
207 p. : ill. ; 21 cm. — (Biblioteca pentru toți copiii ; 53) — R77-5052
lei8.50
1. Istria, Romania—History—Juvenile fiction. I. Title.
PZ90.R6S23 1977 — 78-369185
*78 — MARC

Sadoveanu, Mihail, 1880-1961.
Dumbrava minunată : [povestiri] / Mihail Sadoveanu. — București : Editura Ion Creangă, 1978.
124 p. ; 17 cm. — (Biblioteca școlarului ; 96) — R78-4599
CONTENTS: Dumbrava minunată.—Domnu' Trandafir.—Povestea cu iepurii.—Ianoș năzdrăvan.—Ursul.—Un om năcăjit.—La Mestecănei.—Hultanul.—În pădurea Petrișorului.—Mergînd spre Hîrlău.—Făt-Frumos măzărean.
lei3.00
I. Title.
PZ90.R6S252 1978 — 79-372190
*79 — MARC

Sîntimbreanu, Mircea.
Recreația mare : [povestiri] / Mircea Sîntimbreanu. — Ediția a 3-a / [ilustrații de Iurie Darie]. — București : Editura Ion Creangă, 1978.
362 p. : ill. ; 24 cm. — R78-3856
lei13.00
I. Title.
PZ90.R6S455 1978 — 79-338567
*79 — MARC

Vestea cea mare : provestiri istorice / cuvînt înainte de Dumitru Almaș ; [coperta de Stan Done ; desenul copertei și planșa de Romeo Voinescu]. — București : Editura Ion Creangă, 1978.
310 p. ; 21 cm. — (Biblioteca pentru toți copiii ; 59) — R78-3844
lei11.50
1. Romania—History—Juvenile fiction.
PZ90.R6V45 — 78-396147
*79 — MARC

Vrabie, Gheorghe.
Gruia lui Novac : repovestire în proză de Gheorghe Vrabie ; [gravuri în lemn de Mircea Dumitrescu]. — [București] : Editura Ion Creangă, 1977.
69 p. : col. ill. ; 34 cm. — R78-1087
lei10.50
I. Title.
PZ90.R6V67 — 78-400761
*79 — MARC

—BIBLIOGRAPHY

Gafița, Viniciu, 1926-
Bibliografie de literatură română pentru copii / Viniciu Gafița ; [ilustrația copertei, Gh. Marinescu ; coperta seriei, Val Munteanu]. — București : Editura Ion Creangă, 1978.
270 p. ; 17 cm. — R78-1684
lei3.50
1. Children's literature, Romanian—Bibliography. I. Title.
Z1037.8.R6G33 — 78-378863
[PN1009.R6] — *78 — MARC

CHILDREN'S LITERATURE, RUSSIAN

Aksakov, Sergeĭ Timofeevich, 1791–1859.
(Detskie gody Bagrova-vnuka)
Детские годы Багрова-внука : [Для детей] / С. Т. Аксаков ; [Вступ. статья В. Солоухина]. — Чебоксары : Чуваш. кн. изд-во, 1978.
320 p. : ill. ; 20 cm. — (Золотые родники) — USSR 78
Series romanized: Zolotye rodniki.
1.60rub
I. Title.
PZ62.A38 1978 — 79–413075

Amlinskiĭ, Vladimir Il'ich, 1935–
(Spor s sud'boĭ)
Спор с судьбой : [Очерки : Для детей] / Владимир Амлинский. — Москва : Молодая гвардия, 1977.
159 p. : ill. ; 20 cm. — USSR 77
0.25rub
I. Title.
PZ63.A495 — 78–420583

Arzhentinski, Ivan.
(Malkiiat toreador)
Малкият тореадор : роман : [за средна и горна училищна възраст] / Иван Аржентински ; [худож. Петър Кръстев]. — 1. изд. — София : Нар. младеж, 1977.
152 p. : ill. ; 20 cm. — Bu 77–2785
0.42 lv
I. Title.
PZ70.B8A79 — 79–408509

Ăsănov, Fănil.
(Uramansy malaĭy)
Урмансы малайы : повесть / Фәнил Әсәнов. — Өфө : Башкортостан китап нәшриәте, 1966.
94 p. : ill. ; 20 cm.
I. Title.
PZ90.B36A8 — 74–237294

Balter, Boris.
(Do svidaniia, mal'chiki!)
До свидания, мальчики! : Повесть / Борис Балтер ; [Худож. В. И. Левинсон]. — Москва : Сов. писатель, 1978.
264 p. : ill. ; 16 cm. — USSR 78
0.85rub
I. Title.
PZ63.B329 1978 — 79–409605

Batrov, Aleksandr Mikhaĭlovich.
(Vremia sinego tsveta)
Время синего цвета : Повесть и новеллы / Александр Батров ; [Худож. Г. В. Гармидер]. — Одесса : Маяк, 1978.
262 p. : ill. ; 17 cm. — USSR 78
0.85rub
I. Title.
PZ63.B34396 — 79–407124

Belozerov, Timofeĭ Maksimovich.
(Volshebnyĭ pososhok)
Волшебный посошок : стихи и рассказы / Тимофей Белозеров ; рисунки Г. Акулова. — Москва : Детская лит-ра, 1976.
142 p. : ill. ; 21 cm. — USSR***
0.42rub
I. Akulov, Georgiĭ Valentinovich, 1931– II. Title.
PZ64.3.B433 — 78–415037

Dobkin, Vladimir Iosifovich.
(Konets fanernogo chemodana)
Конец фанерного чемодана : Повесть, рассказы : [Для сред. и ст. школ. возраста] / В. Добкин. — Москва : Мол. гвардия, 1978.
159 p. : ill. ; 16 cm. — USSR 78
CONTENTS: Прогулка за воспоминаниями.—Конец фанерного чемодана.—Неистовый Каблуков.—Кошка, виляющая хвостом.
0.30rub
I. Title.
PZ63.D583 — 78–422405

Dodgson, Charles Lutwidge, 1832–1898.
[Alice's adventures in Wonderland. Russian]
Приключения Алисы в Стране Чудес ; Зазеркалье и про то, что увидела там Алиса : [Сказки] / Льюис Кэрролл [i. e. C. L. Dodgson] ; Перевод с англ. А. Щербакова ; [Предисл. Ю. Кагарлицкого ; Ил. М. Митурича]. — Москва : Худож. лит., 1977.
303 p. : ill. (some col.) ; 20 cm. — USSR 77
4.80rub
I. Dodgson, Charles Lutwidge, 1832–1898. Through the looking-glass and what Alice found there. Russian. 1977. II. Title.
Title romanized: Prikliucheniia Alisy v Strane Chudes.
PZ64.D584 1977 — 78–423598

Ershov, Iakov Alekseevich, 1915–
(Malaia Tamanskaia)
Малая Таманская. Докум. повесть. [Для детей]. — Москва, Изд-во ДОСААФ, 1972.
142 p. with illus. 20 cm. — USSR 72–VKP
At head of title: Яков Ершов.
0.22rub
1. Brianskaia Malaia Tamanskaia mal'chisheskaia diviziia—Juvenile fiction. I. Title.
PZ63.E744 — 79–413579

CHILDREN'S LITERATURE, RUSSIAN (Continued)

Frolov, Leonid Anatol'evich.
Gruzdi-ryzhiki : Rasskazy [Dlia sred. vozrasta] / Leonid Frolov ; [Ris. A. Tambovkina]. — Moskva : Det. lit., 1978.
159 p. : ill. ; 20 cm. USSR78
CONTENTS: Gruzdi-ryzhiki.—Fonarik.—Stakan kiselia.—Nechaiannyĭ klad.—Serezhkina premiia.—Tishkiny kotiata.—Kliuchi.—Ziablevaia vspashka.
0.40rub
I. Title.
PZ63.F794 79-365958
79 MARC

Gaustov, Ivan Konstantinovich.
(Lesnaia skazka)
Лесная сказка : [рассказы натуралиста] / Иван Гаустов. — Ставрополь : Кн. изд-во, 1973.
75 p. : ill. ; 16 cm. USSR 73
0.18rub
1. Animals, Legends and stories of. I. Title.
PZ66.3.G28 79–411240

Hu, Ch'i, 1918–
[Wu ts'ai lu. Russian]
Дорога счастья. Пекин, Изд-во лит-ры на иностранных языках, 1960.
155 p. illus. 21 cm.
At head of title: Ху Ци.
I. Title.
Title romanized: Doroga schast'ia.
PZ63.H8 76–201238

Ivanov, Sergeĭ Anatol'evich.
(Ol'ga Iakovleva)
Ольга Яковлева : Повесть : [Для мл. школьного возраста] / Сергей Иванов ; Рис. Е. Медведева. — Москва : Дет. лит., 1976.
158 p. : ill. ; 20 cm. USSR 76
0.49rub
I. Title.
PZ63.I 877 78–414840

Kal'ma, N pseud.
(Vozmozhno, ikh zovut inache)
Возможно, их зовут иначе; [повесть для среднего и старшего школьного возраста. Москва] Молодая гвардия, 1963.
77 p. illus. 20 cm.
At head of title: Н. Кальма.
I. Title.
PZ63.K21523 68–39762

Kobets-Filimonava, Alena Ryhoraŭna, 1932–
(Stoĭ, mashinist!)
Стой, машинист! : повесть / Елена Кобец-Филимонова ; рисунки В. Штаркина ; [перевод с белорусского автора]. — Москва : Детская лит-ра, 1976.
174 p. : ill. ; 21 cm. USSR***
0.43rub
I. Title.
PZ63.K5877 78–421940

Milne, Alan Alexander, 1882–1956.
(Stikhi dlia deteĭ)
Стихи для детей / А. А. Мильн и Т. С. Элиот ; в переводах-переложениях Якова Бергера. — Лондон : [s. n.], 1978.
16 p. : ill. ; 21 cm. GB***
Cover title.
I. Eliot, Thomas Stearns, 1888–1965. II. Berger, Iakov. III. Title.
PZ64.3.M57 1978 79–413301

Montvila, Vytautas.
Į plačią žemę / Vytautas Montvila. — Vilnius : Vaga, 1977.
110 p. ; 14 cm. USSR***
"Vyresniam mokykliniam amžiui."
0.24rub
I. Title.
PZ90.L5M66 78-394514
79 MARC

Multhaupt, Hermann, 1937-
Die silbernen Ballschuhe : e. Erzählung / von Hermann Multhaupt. — Regensburg : Habbel, 1978.
28 p. : ill. ; 19 cm. GFR78-A
ISBN 3-7748-0358-7
I. Title.
PZ33.M8515 79-358787
79 MARC

Nechaev, I 1906–1941.
(Belyĭ karlik)
Белый карлик; рассказ. Рисунки Д. Павлинова. Москва, Гос. изд-во детской лит-ры, 1943.
19 p. illus. 20 cm. (Библиотека научной фантастики и приключений)
"Для среднего и старшего возраста."
Series romanized: Biblioteka nauchnoĭ fantastiki i prikliuchenii.
I. Title.
PZ63.N364 1943 52–48784

(Parus–77)
Парус–77 : Сб. лит.-худож. и публицист. произведений для подростков / [Редкол. Алексин А. Г., лауреат премии Ленинского комсомола ... et al.]. — Москва : Мол. гвардия, 1977.
367 p. : ill. ; 23 cm. USSR 78
1.30rub
1. Children's literature, Russian. I. Aleksin, Anatoliĭ Georgievich.
PZ61.P34 78–420862

Pushyk, Stepan Hryhorovych.
(Pero zolotoho ptakha)
Перо золотого птаха : повість / Степан Пушик. — Ужгород : Карпати, 1978.
271 p. : ill. ; 21 cm. USSR***
1.10rub
I. Title.
PZ70.U72P8 79–411374

Sakharnov, Sviatoslav Vladimirovich.
Belye kity : Puteshestviia i prikliucheniia : Povesti [Dlia ml. shkol. vozrasta] / S. Sakharnov ; [Ris. M. Belomlinskogo]. — Leningrad : Lenizdat, 1978.
415 p., [1] leaf of plates : ill. (some col.) ; 21 cm. USSR79
CONTENTS: Puteshestvie na Trigle.—Trepangolovy.—Dom pod vodoĭ.—Fotorasskaz: Na Tikhom okeane.—Fotorasskaz: Na ostrove Kuba.—Belye kity.—Doroga na Bagamoĭo.—Os'minogi za steklom.
1.00rub
1. Sea stories—Juvenile fiction. I. Title.
PZ63.S332 79-379290
79 MARC

Semenov-Spasskiĭ, Leonid Grigor'evich.
Berillovoe ozherel'e : Rasskazy ob Antarktide : [dlia sred. i st. vozrasta] / L. Semenov-Spasskiĭ. — Leningrad : Det. lit., Leningr. otd-nie, 1978.
127 p. : ill. ; 22 cm. USSR78
0.35rub
1. Antarctic regions—Juvenile fiction. I. Title.
PZ63.S3827 79-371531
79 MARC

Semushkin, Tikhon Zakharovich.
(Prikliucheniia Aĭvama)
Приключения Айвама / Тихон Семушкин ; художник Юрий Копейко. — Москва : Малыш, 1978.
92 p. : col. ill. ; 22 cm. USSR***
"Для младшего школьного возраста."
0.90rub
I. Kopeĭko, Iuriĭ. II. Title.
PZ63.S3855 79–411462

Smirnova-Maksheeva, Tat'iana.
(Skazka O Russkom Soldate, Krylatom Bese i Tsarevne Elene)
Сказка О Русском Солдате, Крылатом Бесе и Царевне Елене / Т. А. Смирнова-Макшеева. — [s. l. : s. n.], 1973 (Paris : P. I. U. F.)
122 p. : ill. ; 22 cm. F***
I. Title.
PZ64.S62 79–414060

Sysoev, Vsevolod Petrovich.
(Amurskie zverolovy)
Амурские звероловы : Повесть и рассказы / Всеволод Сысоев ; [Худож. Г. А. Палкин]. — Хабаровск : Кн. изд-во, 1977.
144 p. : ill. ; 20 cm. USSR 78
CONTENTS: Амурские звероловы.—Лебединая песня.—Черный лис. — Голубой песец. — Клыкастый зверь. — Рысь. — Снежный баран.—Красный волк.—Амурский горал.—Дальневосточный лесной кот.—Крылатая белка.—Ход до смерти.
0.50rub
1. Animals, Legends and stories of. I. Title.
PZ66.3.S9 79–412106

Važa-P'šavela, 1861–1915.
(Dumy malen'kogo pastushka)
Думы маленького пастушка. [Перевод Е. Гогоберидзе] Художник Мераб Бердзенишвили. Тбилиси, Накадули, 1966.
[12] p. illus. 21 cm.
At head of title: Важа Пшавела.
"Для детей младшего школьного возраста."
I. Berdzenishvili, Merab. II. Title.
PZ63.V3258 1966 67–51378

Vnukov, Nikolaĭ Andreevich.
"Sverre" zovet na pomoshch' : Povestvovanie v trekh ėpizodakh : [Dlia sred. i st. vozrasta] / Nikolaĭ Vnukov ; [Ris. E. Anosova]. — Leningrad : Det. lit., Leningr. otd-nie, 1978.
127 p. : ill. ; 20 cm. USSR78
0.45rub
1. World War, 1939-1945—Juvenile fiction. I. Title.
PZ63.V5433 79-371518
79 MARC

Zheliĭkhovskaia, Vera Petrovna, 1835–1896.
Мое отрочество. 4. изд. Съ рисунками С. С. Соломко. С. Петербургъ, Издание А. Ф. Девріена [1903]
2 p. l., 322 p. illus., port. 23 cm.
At head of title: В. П. Желиховская.
I. Title.
PZ63.Z52 1903 Russian Imperial Coll. 33–38051

--BIBLIOGRAPHY--HISTORY

(Istoriia bibliografii detskoĭ literatury, 1900—okt. 1917 gg.)
История библиографии детской литературы, 1900–окт. 1917 гг. : Учеб. пособие для студентов фак. универс. б-к / [К. А. Алексахина, Н. Н. Гутман, Н. А. Кужелева ... et al. ; Сост. Н. А. Кужелева] ; Ленингр. гос. ин-т культуры им. Н. К. Крупской. — Ленинград : [s. n.], 1976.
149 p. ; 21 cm. USSR 77
At head of title: Ministerstvo kul'tury RSFSR.
Includes bibliographical references.
1. Children's literature, Russian—Bibliography—History. I. Aleksakhina, K. A. II. Kuzheleva, N. A. III. Leningradskiĭ gosudarstvennyĭ institut kultury.
Z2491.A2 I 87 79–414354
[PN1009.R8]

—HISTORY AND CRITICISM

O'Dell, Felicity Ann.
Socialisation through children's literature : the Soviet example / Felicity Ann O'Dell. — Cambridge [Eng.] ; New York : Cambridge University Press, 1978.
x, 278 p. ; 23 cm. — (Soviet and East European studies)
Bibliography: p. [269]-274.
Includes index.
ISBN 0-521-21968-X
1. Children's literature, Russian—History and criticism. 2. Socialization. I. Title. II. Series.
PN1009.R8O3 301.15'7'0947 77-91093
78 MARC

CHILDREN'S LITERATURE, SERBIAN

Krklec, Gustav, 1899-
Pod Gupčevom lipom / tekst Gustav Krklec ; likovni prilozi Ivan Rabuzin. — Zagreb : Mladost, 1977.
[20] p. : col. ill. ; 32 cm. — (Velike slikovnice) Yu***
I. Rabuzin, Ivan, 1921- II. Title.
PZ70.S42K73 78-378398
78 MARC

—HISTORY AND CRITICISM

Turjačanin, Zorica.
Ključevi zlatnih vrata / Zorica Turjačanin. — Banjaluka : Glas, 1978.
99 p. ; 22 cm. Yu***
Includes bibliographical references.
1. Children's literature, Serbian—History and criticism. 2. Children's literature, Croatian—History and criticism. I. Title.
PN1009.S4T8 79-357685
79 MARC

CHILDREN'S LITERATURE, SINHALESE

Hērat, Piyal B
(Raja daruvō)
රජ දරුවෝ / පියල් බී. හේරත්. — කොළඹ : පුබුදු මුද්‍රණ ශිල්පියෝ සහ ප්‍රකාශකයෝ, [1977]
203 p. ; 20 cm.
In Sinhalese.
A novel.
Rs8.00
I. Title.
PZ90.S5H4 78–911123

CHILDREN'S LITERATURE, SLOVAK

Bodenek, Ján.
Leto na Rovniach : [pre čitateľov od 9 rokov] / Ján Bodenek ; [ilustr. Viera Bombová]. — 1. vyd. — Bratislava : Mladé letá, 1975.
131, [5] p. : ill. ; 21 cm. CzS75
Kčs17.00
I. Title.
PZ70.S46B64 78-387819
•79 MARC

Dônčová, Zlata.
Agnesa a Tizian / Zlata Dônčová ; [ilustr. Mária Podhradská]. — 2. vyd. — Bratislava : Mladé letá, 1978.
200 p. : ill. ; 18 cm. — (Čajka) CzS78
Kčs16.00
I. Title.
PZ70.S46D66 1978 79-339503
•79 MARC

CHILDREN'S LITERATURE, SLOVAK (Continued)

Vansová, Terézia Medvecká, 1857-1942.
Julinkin prvý bál / Terézia Vansová ; ⌈výber zost., text upr., dosl., pozn. a slovník nap. Margita Valehrachová-Matulayová ; ilustr. Jozef Baláž⌉. — 5. novoupr. vyd. — Bratislava : Mladé letá, 1977.
146 p. : ill. ; 21 cm. CzS77
Kčs20.00
I. Title.
PZ70.S46V33 1977 78-396548
*79 MARC

Vansová, Terézia Medvecká, 1857-1942.
Sirota Podhradských / Terézia Vansová ; ⌈ilustr. Osvald Klapper⌉. — 5. vyd. — Bratislava : Mladé letá, 1977.
291 p. : ill. ; 18 cm. — (Čajka) CzS77
Kčs20.00
I. Title.
PZ70.S46V34 1977 78-396547
*79 MARC

—BIBLIOGRAPHY

Sliacky, Ondrej.
Bibliografia literatúry pre deti a mládež, 1778-1917 / Ondrej Sliacky. — 1. vyd. — Bratislava : Mladé letá, 1977.
307 p. ; 21 cm. — (Otázky detskej literatúry) CzS78
Bibliography: p. 305-306.
Includes indexes.
Kčs19.00
1. Children's literature, Slovak—Bibliography. I. Title.
Z1037.6.S538 79-359886
⌈PN1009.C9⌉ *79 MARC

—HISTORY AND CRITICISM

Chaloupka, Otakar.
Children's books in Czechoslovakia / Otakar Chaloupka. — Prague : Orbis Press Agency, 1978.
85 p., ⌈8⌉ leaves of plates : ill. ; 19 cm. Cz***
Kčs8.00
1. Children's literature, Czech—History and criticism. 2. Children's literature, Slovak—History and criticism. I. Title.
PN1009.C9C44 891.8'6'099282 79-303345
79 MARC

—HISTORY AND CRITICISM —CONGRESSES

Literatúra pre predškolský vek a jej využitie v materských školách : zborník referátov z pracovného seminára v Modre 30.-9.-1.10.1976 ... / ⌈konferenciu viedla a vyd. zborníka referátov pripravila Lýdia Kyseľová⌉. — 1. vyd. — Bratislava : Mladé letá, 1978.
147 p. ; 21 cm. — (Otázky detskej literatúry) CzS***
Czech or Slovak.
Bibliography: p. 140-141.
Kčs13.50
1. Children's literature, Czech—History and criticism—Congresses. 2. Children's literature, Slovak—History and criticism—Congresses. 3. Education, Pre-school—Czechoslovakia—Congresses. I. Kyseľová, Lýdia.
PN1009.C9L5 79-363666
79 MARC

CHILDREN'S LITERATURE, SPANISH

Alborch, Francesc.
Miles gloriosus (el militar fanfarró) : adaptació lliure per a teatre de l'obra de Plaute. — Barcelona : Edicions Don Bosco, c1977.
48 p. : ill. ; 17 cm. — (Col.lecció Teatre edebé ; no. 6) Sp***
A play.
ISBN 8423613550
I. Plautus, Titus Maccius. Miles gloriosus. II. Title. III. Series.
PZ90.C3A4 78-389221
78 MARC

Amo y Gili, Montserrat del.
La torre / Montserrat del Amo ; dibujos de Miguel Angel Pacheco. — Valladolid : Miñón, ⌈1975⌉
32 p. : col. ill. ; 25 cm. Sp75-June
ISBN 843550414X
I. Pacheco, Miguel Angel. II. Title.
PZ73.A4955 76-452150
*76⌈r79⌉rev2 MARC

Anglada y Sarriera, Dolores.
El més petit de tots / text i dibuixos de Lola Anglada i Sarriera. — ⌈Barcelona⌉ : Alta Fulla, ⌈c1978⌉
80 p. : col. ill. ; 19 cm. Sp***
Reprint of the 1937 ed. published by Comissariat de Propaganda de la Generalitat de Catalunya.
ISBN 8485403037
I. Title.
PZ90.C3A613 1978 78-381551
78 MARC

Armijo, Consuelo.
Más bataultos / Consuelo Armijo ; ilustrado por Jordi Ciuró. — 1. ed. — Barcelona : Editorial Juventud, 1978.
70 p. : col. ill. ; 24 cm. — (Colección La Hora del cuento ; 2) Sp***
ISBN 8426115136
I. Ciuró, Jordi, joint author. II. Title.
PZ73.A68 79-363049
79 MARC

Barceló, Joan, 1955-
Científicament s'ha demostrat ... / Joan Barceló i Cullerés. — Barcelona : Edicions Don Bosco, c1977.
47 p. : ill. ; 17 cm. — (Col·lecció Teatre edebé ; no. 7) Sp***
A play.
ISBN 8423613569
I. Title. II. Series.
PZ90.C3B27 79-338176
79 MARC

Barceló, Xesc.
El duc Meu-Meu / Xesc Barceló ; música de Joan Albert Amargós ; pròleg de Josep A. Codina. — 1. ed. — Mataró : S. Serres, 1978.
89 p. ; 17 cm. (Edicions Robrenyo e teatre de tots els temps : Sèrie infantil ; 3) Sp***
ISBN 8474660033
I. Title.
PZ90.C3B29 79-354212
79 MARC

Batiste, Jaume.
Els tres mosqueters : versió escènica lliure de la novel.la d'Alexandre Dumas. — Barcelona : Edicions Don Bosco, c1977.
79 p. : ill. ; 17 cm. — (Col.lecció Teatre edebé ; no. 8) Sp***
A play.
ISBN 8423613593
I. Dumas, Alexandre, 1802-1870. Les trois mousquetaires. II. Title. III. Series.
PZ90.C3B3 78-389222
78 MARC

Benet i Jornet, Josep Maria, 1940-
El somni de Bagdad / Josep M. Benet i Jornet. — Barcelona : Edicions Don Bosco, c1977.
63 p. ; ill. ; 17 cm. — (Col·lecció teatre edebé ; no. 2) Sp***
A play.
Bibliography: p. 11-12.
ISBN 8423613119
I. Title. II. Series.
PZ90.C3B4 79-338162
79 MARC

Blasco Casanovas, Joan.
Els dos vailets del castell de Montsó / Joan Blasco Casanovas ; amb il·lustracions de Llucià Navarro. — Barcelona : Edicions La Galera, c1977.
117 p. : ill. ; 22 cm. — (Els Grumets de La Galera) Sp***
ISBN 8424640306
1. Jaime I, King of Aragon, 1208-1276—Juvenile fiction. I. Navarro, Llucià. II. Title.
PZ90.C3B55 78-365378
78 MARC

Bright, Robert, 1902-
Mi paraguas rojo. Traducido del inglés por Marion H. Redfield. New York, W. Morrow ⌈1968⌉
⌈32⌉ p. col. illus. 18 cm.
Translation of My red umbrella.
SUMMARY: A young girl's umbrella grows to accomodate an increasing number of animal friends.
⌈1. Umbrellas and parasols—Fiction. 2. Spanish language—Readers⌉ I. Title.
PZ73.B68 ⌈E⌉ 68-20836
⌈r79⌉rev MARC

Coquard, Lluis, 1922-
Rabilinda, la ratita presumida : fantasía infantil en 3 actos divididos en 10 cuadros / Luis Coquard. — ⌈s.l. : s.n., 1978⌉ (Barcelona : Ingro)
51 p. : music ; 17 cm. — (Teatro infantil ; 5) Sp***
ISBN 8440051107
PZ77.C574 862'.6'4 79-368104
79 MARC

Donato, Mário.
Tiro de guerra : novela juvenil / Mário Donato. — Rio de Janeiro : Livraria . Olympio Editora, 1978.
viii, 87 p. : port. ; 22 cm.
"Em convênio com o Instituto Nacional do Livro."
Cr$35.00
I. Title.
PZ83.D66 79-105288
79 MARC

Font, Manuel.
La capseta dels reis d'Orient : i altres narracions / Manuel Font ; il·lustracions de Francesc Salvà. — 1. ed. — Montserrat : Publicacions de l'Abadia, 1978.
109 p. : ill. ; 20 cm. — (La Xarxa ; 15) Sp***
ISBN 8472023028
I. Title.
PZ90.C3F64 79-363040
79 MARC

Gandolfi Herrero, Arístides, 1890-
Nuestros muchachos / Alvaro Yunque ⌈i.e. A. Gandolfi Herrero⌉. — Buenos Aires : Plus Ultra, c1976.
78 p. ; 20 cm. — (Colección juvenil)
CONTENTS: Nuestros muchachos.—El vivo y el zonzo.—El vigilante y los hippies.—Yuyo es secuestrado.—Por qué quiero a mi ciudad?—La Noche Buena de Cirilo.—Uberto Sonio, precoz suicida.—Dos terribles pistoleros.—El gordo se venga.—Alcohol y agua.
I. Title.
PZ73.G236 863 79-338134
79 MARC

García Sánchez, José Luis.
La niña invisible / ⌈idea y texto, J. L. García Sánchez y M. A. Pacheco ; ilus., Ulises Wensell⌉. — ⌈Madrid⌉ : Ediciones Altea, ⌈c1978⌉
32 p. : col. ill. ; 31 cm. — (His Los Derechos del niño ; 10) Sp***
ISBN 8437213398
I. Pacheco, Miguel Angel, joint author. II. Wensell, Ulises. III. Title. IV. Series.
PZ73.G362 863'.6'4 79-363728
79 MARC

García Sánchez, José Luis.
La niña sin nombre / ⌈idea y texto, J. L. García Sánchez y M. A. Pacheco ; ilus., Asun Balzola⌉. — ⌈Madrid⌉ : Ediciones Altea, ⌈c1978⌉
32 p. : col. ill. ; 31 cm. — (His Los Derechos del niño ; 3) Sp***
ISBN 8437213444
I. Pacheco, Miguel Angel, joint author. II. Balzola, Asun. III. Title. IV. Series.
PZ73.G3622 863'.6'4 79-363729
79 MARC

García Sánchez, José Luis.
El niño gigante / ⌈idea y texto, J. L. García Sánchez y M. A. Pacheco ; ilus., Carme Solé⌉. — ⌈Madrid⌉ : Ediciones Altea, ⌈c1978⌉
32 p. : col. ill. ; 31 cm. — (His Los Derechos del niño ; 9) Sp***
ISBN 8437213460
I. Pacheco, Miguel Angel, joint author. II. Solé, Carme. III. Title. IV. Series.
PZ73.G363 863'.6'4 79-363734
79 MARC

García Sánchez, José Luis.
El niño llorón / ⌈idea y texto, J. L. García Sánchez y M. A. Pacheco ; ilus., Carme Solé⌉. — ⌈Madrid⌉ : Ediciones Altea, ⌈c1978⌉
32 p. : col. ill. ; 31 cm. — (His Los Derechos del niño ; 4) Sp***
ISBN 8437213436
I. Pacheco, Miguel Angel, joint author. II. Solé, Carme. III. Title. IV. Series.
PZ73.G3632 863'.6'4 79-363736
79 MARC

García Sánchez, José Luis.
El niño que tenía dos ojos / ⌈idea y texto, J. L. García Sánchez y M. A. Pacheco ; ilus., Ulises Wensell⌉. — ⌈Madrid⌉ : Ediciones Altea, ⌈c1978⌉
32 p. : col. ill. ; 31 cm. — (His Los Derechos del niño ; 5) Sp***
ISBN 8437213452
I. Pacheco, Miguel Angel, joint author. II. Wensell, Ulises. III. Title. IV. Series.
PZ73.G3633 863'.6'4 79-363730
79 MARC

García Sánchez, José Luis.
El niño y el robot / ⌈idea y texto, J. L. García Sánchez y M. A. Pacheco ; ilus., Asun Balzola⌉. — ⌈Madrid⌉ : Ediciones Altea, ⌈c1978⌉
32 p. : col. ill. ; 31 cm. — (His Los Derechos del niño ; 6) Sp***
ISBN 8437213401
I. Pacheco, Miguel Angel, joint author. II. Balzola, Asun. III. Title. IV. Series.
PZ73.G3634 863'.6'4 79-363735
79 MARC

García Sánchez, José Luis.
Los niños de los cuentos / ⌈idea y texto, J. L. García Sánchez y M. A. Pacheco ; ilus., Nella Bosnia⌉. — ⌈Madrid⌉ : Ediciones Altea, ⌈c1978⌉
32 p. : col. ill. ; 31 cm. — (His Los Derechos del niño ; 1) Sp***
ISBN 8437213487
I. Pacheco, Miguel Angel, joint author. II. Bosnia, Nella. III. Title. IV. Series.
PZ73.G364 863'.6'4 79-363596
79 MARC

García Sánchez, José Luis.
Los niños que no eran como niños / ⌈idea y texto, J. L. García Sánchez y M. A. Pacheco ; ilus., Karin Schubert⌉. — ⌈Madrid⌉ : Ediciones Altea, ⌈c1978⌉
32 p. : col. ill. ; 31 cm. — (His Los Derechos del niño ; 2) Sp***
ISBN 8437213428
I. Pacheco, Miguel Angel, joint author. II. Schubert, Karin. III. Title. IV. Series.
PZ73.G3642 863'.6'4 79-363732
79 MARC

CHILDREN'S LITERATURE, SPANISH (Continued)

García Sánchez, José Luis.
Los niños que no tenían escuelas / ₍idea y texto, J. L. García Sánchez y M. A. Pacheco ; ilus., Nella Bosnia₎. — ₍Madrid₎ : Ediciones Altea, ₍c1978₎
32 p. : col. ill. ; 31 cm. — (His Los Derechos del niño ; 7) Sp***
ISBN 843721341X
I. Pacheco, Miguel Angel, joint author. II. Bosnia, Nella. III. Title. IV. Series.
PZ73.G3643 863'.6'4 79-363733
79 MARC

García Sánchez, José Luis.
El pueblo que se quedó sin niños / ₍idea y texto, J. L. García Sánchez y M. A. Pacheco ; ilus., Karin Schubert₎. — ₍Madrid₎ : Ediciones Altea, ₍c1978₎
32 p. : col. ill. ; 31 cm. — (His Los Derechos del niño ; 8) Sp***
ISBN 8437213479
I. Pacheco, Miguel Angel, joint author. II. Schubert, Karin. III. Title. IV. Series.
PZ73.G365 863'.6'4 79-363595
79 MARC

Giménez Frontín, J L 1943-
Historia del pequeño chamán / J. L. Giménez-Frontín ; dibujos de Gonzalo Tena. — Barcelona : Lumen, c1978.
103 p. : ill. ; 18 cm. — (Colección Grandes autores ; 36) Sp***
ISBN 8426430368
I. Tena, Gonzalo, ill. II. Title.
PZ73.G49 79-371749
79 MARC

Gómez-Fernández, Eduardo, 1914-
Trece cuentos para niños / Eduardo Gómez-Fernández ; ₍ilustraciones de Mercedes Graña de Iriarte₎. — ₍s.l. : s.n.₎, 1978 (s.l.₎ : Master Typesetting of Puerto Rico)
107 p. : ill. ; 22 cm.
CONTENTS: María Cabra y sus dos cabritos.—El duendecillo travieso.—El girasol vanidoso.—El bien y el mal.—Lindo, un gato astuto.—La araña y la cucaracha.—El cantar del coquí.—El árbol de pino y la bruja.—Por que tienen escamas los peces.—Rex y Sultán.—El conejito Boni.—El cangrejo y las olas.—El cocuyo y el hada.
₍1. Animals—Fiction. 2. Nature—Fiction. 3. Short stories. 4. Spanish langauge—Readers₎ I. Title.
PZ73.G58 ₍Fic₎ 78-107513
79 MARC

Lázaro, Angel, 1900-
Santiago de Cuba / Angel Lázaro ; ilus. de José Francisco Aguirre. — ₍Madrid₎ : Aguilar, c1966.
91 p. : col. ill. ; 23 cm. — (Colección El Globo de colores : Relatos históricos)
1. United States—History—War of 1898—Juvenile fiction. I. Aguirre, José Francisco, painter.
PZ73.L35 863 79-370930
79 MARC

León, María Teresa.
Cervantes, el soldado que nos enseñó a hablar / María Teresa León. — Madrid : Altalena, c1978.
193 p. : ports. ; 22 cm. — (Mundo ancho y propio) Sp***
ISBN 8474750032
1. Cervantes Saavedra, Miguel de, 1547-1616, in fiction, drama, poetry, etc.—Juvenile literature. I. Title.
PZ73.L396 78-385206
79 MARC

Martí Pol, Miquel.
Hivern / Miquel Martí Pol ; ₍imatge₎ Escola Cooperativa El Puig. — 1. ed. — Sant Boi de Llobregat : Edicions del Mall, 1978.
₍32₎ p. : chiefly col. ill. ; 24 cm. — (El Petit mall) Sp***
"Extret del poema Les quatre estacions del llibre El poble."
ISBN 8474560195
1. Winter—Juvenile poetry. I. Escola cooperativa El Puig. II. Title.
PZ90.C3M3 79-371288
79 MARC

Martí Pol, Miquel.
Primavera / Miquel Martí i Pol ; ₍imatge₎ Escola Cooperativa El Puig. — 1. ed. — Sant Boi de Llobregat : Edicions del Mall, 1978.
₍44₎ p. : chiefly col. ill. ; 24 cm. — (El Petit mall) Sp***
"Extret del poema Les quatre estacions del llibre El poble."
ISBN 8474560209
1. Spring—Juvenile poetry. I. Escola cooperativa El Puig. II. Title.
PZ90.C3M33 79-371287
79 MARC

Medina, Roberto Nicolás.
Érase un viejo pirata : un pirata y un loro en dos travesías y un hundimiento / Roberto Nicolas Medina. — Buenos Aires : Editoria Plus Ultra, c1977.
95 p. : ill. ; 20 cm. — (Colección El Escenario ; 7)
I. Title. II. Series.
PZ77.M38 862 78-109761
79 MARC

Pannunzio, Martha Azevedo, 1938-
Veludinho / Martha Azevedo Pannunzio. — s.l. : Livraria J. Olympio Editora, 1978.
vii, 52 p. : ill. ; 21 cm.
"Em convênio com o Instituto Nacional do Livro."
Cr$20.00
I. Title.
PZ83.P26 869'.8 79-104642
79 MARC

Sáenz E , Carlos Luis.
El Abuelo Cuentacuentos / Carlos Luis Sáenz. — San José : Editorial Costa Rica, 1974.
92 p. : ill. ; 28 cm.
I. Title.
PZ73.S25 863 79-371022
79 MARC

Soto, Apuleyo.
El buey de los cuernos de oro : (farsa musical infantil) / ₍por₎ Apuleyo Soto. — Madrid : San Pío X, 1978.
85 p. : ill. ; 17 cm.
ISBN 8472210928 : 90ptas
I. Title.
PZ77.S63 79-110569
79 MARC

Soto, Apuleyo.
Doña Noche : (comedia musical infantil) / ₍por₎ Apuleyo Soto. — Madrid : San Pío X, 1978.
79 p. : ill. ; 17 cm. Sp78-Jul
ISBN 8472210901 : 120ptas
I. Title.
PZ77.S64 79-110564
79 MARC

Suqué i d'Espona, Carme.
Un conte del mar / Carme Suqué i d'Espona ; il·lustracions de Joan A. Vallvé. — 1. ed. — Barcelona : Publicacions de l'Abadia de Montserrat, 1978.
97 p. : ill. ; 20 cm. — (Col·lecció La Xarxa ; 14) Sp***
ISBN 8472023001
I. Title.
PZ90.C3S9 79-358270
79 MARC

Tallon, Robert, 1939-
ABCDEFGHIJKLMNOPQRSTUVWXYZ in English and Spanish. ₍New York₎ Lion Press ₍1969₎
₍57₎ p. (chiefly col. illus.) 21 x 26 cm.
SUMMARY: Illustrations for each letter of the alphabet with words in both English and Spanish from airplane/avión to zigzag/ziszás.
On spine: ABC's in English and Spanish.
₍1. Alphabet books. 2. Alphabet books, Spanish₎ I. Title. II. Title: ABC's in English and Spanish.
PZ73.T3 ₍468₎ 76-86987
77₍r78₎rev MARC

Tapia, Luis de, 1871-1937.
Zapaquilda de viaje / texto de Luis de Tapia ; dibujos de Marco. — Madrid : Editorial "Saturnino Calleja", ₍19—?₎
₍32₎ p. : ill. (some col.); 28 cm. — (Cuentos de Calleja en colores : Séptima serie, Historias cómicas de animales, plantas y cosas ; ₍2₎)
I. Marco, Fernando. II. Title. III. Series: Cuentos de Calleja en colores ; ₍2₎
PZ71.C8 7. ser. no. 2 79-358055
₍PZ73₎ 79 MARC

Tubau, Núria.
Any de neu, any de Déu / ₍guió i escenificació, Núria Tubau ; coberta i l·lustracions, M. Eulàlia Sariola₎. — Barcelona : La Galera, c1977.
22 p. : ill. ; 17 cm. — (Teatre, joc d'equip ; 20) Sp***
Cover title.
ISBN 8424650204
I. Sariola, M. Eulàlia. II. Title.
PZ90.C3T8 79-367808
79 MARC

Vallverdú, José.
En Mir l'Esquirol / Josep Vallverdú ; amb il·lustracions de Joan Andreu Vallvé. — Barcelona : Ed. La Galera, c1978.
112 p. : ill. ; 22 cm. — (Els Grumets de la Galera) Sp***
ISBN 8424640322
1. Spain—History—Arab period, 711-1492—Juvenile fiction. I. Vallvé, Joan Andreu. II. Title.
PZ90.C3V35 79-338165
79 MARC

Vergés, Oriol.
La ciutat sense muralles / Oriol Vergés ; amb il·lustracions d'Isidre Monés. — Barcelona : Edicions La Galera, 1978.
101 p. : ill. ; 22 cm. — (Els Grumets de la Galera) Sp***
ISBN 8424640314
1. Barcelona—History—Juvenile fiction. I. Title.
PZ90.C3V47 78-389185
79 MARC

—BIBLIOGRAPHY

Demerson, Paula de.
Esbozo de biblioteca de la juventud ilustrada (1740-1808) / Paula de Demerson. — Oviedo : Universidad, Cátedra Feijoo, Facultad de Filosofía y Letras, 1976.
140 p. ; 25 cm. — (Textos y estudios del siglo XVIII ; no. 5) Sp***
Includes bibliographical references.
ISBN 8460068978
1. Children's literature, Spanish—Bibliography. I. Title. II. Series.
Z1037.7.D45 028.52 77-554102
77₍r79₎rev MARC

New York (City). Public Library.
Libros en español : an annotated list of children's books in Spanish / sponsored by South Bronx Project. — New York : New York Public Library, 1977, cover 1978.
72 p. ; 23 cm.
Prepared by M. Arnoldson, P. B. White, C. Behrmann, and M. K. Conwell. Revision of the 1971 ed. by M. K. Conwell.
English and Spanish.
Includes index.
1. Children's literature, Spanish—Bibliography. I. Arnoldson, Marta. II. Conwell, Mary K. Libros en español. III. Title.
Z1037.7.N48 1978 79-106543
₍PN1009.S8₎ 79 MARC

Quins llibres han de llegir els nens? / Seminari de Bibliografia Infantil de Rosa Sensat. — 1. ed. — Barcelona : Publicaciones de Rosa Sensat, 1977.
181 p. ; 24 cm. Sp***
Includes indexes.
ISBN (invalid) 84850008014 : 300ptas
1. Children's literature, Spanish—Bibliography. 2. Books and reading for children—Spain—Congresses. I. Rosa Sensat. Seminari de Bibliografia Infantil.
Z1037.7.Q84 79-383139
₍PN1009.S8₎ 79 MARC

Schon, Isabel.
Books in Spanish for children and young adults : an annotated guide = Libros infantiles y juveniles en español : una guía anotada / by Isabel Schon. — Metuchen, N.J. : Scarecrow Press, 1978.
xi, 153 p. ; 23 cm.
English and Spanish.
Includes indexes.
ISBN 0-8108-1176-6
1. Children's literature, Spanish—Bibliography. 2. Children's literature, Spanish American—Bibliography. I. Title. II. Title: Libros infantiles y juveniles en español.
Z2694.5.S35 028.52 78-10429
₍PN1009.S8₎ 78 MARC

—BIBLIOGRAPHY—PERIODICALS

Proyecto Leer.
Proyecto LEER bulletin. no. 1-
spring 1968-
Washington, Proyecto LEER.
no. 29 cm.
Frequency varies.
1. Children's literature, Spanish—Bibliography—Periodicals. 2. Spanish imprints—Bibliography. 3. Spanish Americans in the United States—Periodicals. I. Proyecto Leer. Bulletin. II. Title.
Z1037.7.P76a 028.52'05 78-644391
[PN1009.S8] MARC-S

CHILDREN'S LITERATURE, SPANISH AMERICAN

—BIBLIOGRAPHY

Schon, Isabel.
Books in Spanish for children and young adults : an annotated guide = Libros infantiles y juveniles en español : una guía anotada / by Isabel Schon. — Metuchen, N.J. : Scarecrow Press, 1978.
xi, 153 p. ; 23 cm.
English and Spanish.
Includes indexes.
ISBN 0-8108-1176-6
1. Children's literature, Spanish—Bibliography. 2. Children's literature, Spanish American—Bibliography. I. Title. II. Title: Libros infantiles y juveniles en español.
Z2694.5.S35 028.52 78-10429
₍PN1009.S8₎ 78 MARC

CHILDREN'S LITERATURE, SWEDISH

Åberg, Lars, 1940-
Lektajm med Trazan och Banarne / ord o. bilder, Lasse Åberg alias "Trazan". — Stockholm : Askild & Kärnekull, 1977.
48 p. : ill. ; 29 cm. S78-13
ISBN 9170088705 : kr37.00
I. Title.
PZ59.A14L4 78-400768
79 MARC

CHILDREN'S LITERATURE, SWEDISH (Continued)

Anderson, Kjell Ivan, 1935-
Den gamla cykeln / Kjell Ivan Anderson. — Göteborg : Zinderman, 1978.
[20] p. : ill. ; 16 cm. S78-8
ISBN 9152802353 : kr16.00
I. Title.
PZ59.A54G3 78-390512
79 MARC

Andersson, Christina, 1936-
Jakob Dunderskägg berättar mera. Omslag och illus. av författaren. Stockholm, Saga, 1970.
109, (1) p. illus. 20 cm. (Saga, 575) kr11.50 S70-46
I. Title.
PZ59.A56J27 77-553036
79[r79]rev MARC

Andersson, Christina, 1936-
Jakob Dunderskäggs sällsamma släktträff / Christina Andersson. — Stockholm : AWE/Geber, 1978.
110, [1] p. : ill. ; 21 cm. S78-18
ISBN 9120056621
I. Title.
PZ59.A56J29 78-401129
79 MARC

Beckman, Gunnel.
Oskuld / Gunnel Beckman. — Stockholm : Bonnier, 1978.
179, [1] p. ; 20 cm. S78-23
ISBN 9100423726 : kr35.00
I. Title.
PZ59.B34O8 79-356311
79 MARC

Beckman, Gunnel.
Varför just Eva? / Gunnel Beckman ; ill. av Inger Eriksson. — 1. uppl. — Lund : Cavefors, 1978.
98, [1] p. : ill. ; 21 cm. — (Lättläst för vuxna) S78-11
ISBN 9150406620 : kr28.00
I. Title.
PZ59.B34V35 78-382218
78 MARC

Berger, Yngve, 1934-
Ligga i lä / Yngve Berger. — Stockholm : Raben & Sjögren, 1978.
146 p., [1] p. ; 22 cm. — (Ung nu) S78-24
ISBN 9129514894 : kr30.00
I. Title.
PZ59.B384L5 79-337404
79 MARC

Bexell, Eva, 1945-
Kalabalik hos morfar prosten / Eva Bexell ; [omsl. o. ill. av Karin Stjernholm-Raeder]. — Stockholm : Bonnier, 1978.
85, [2] p. : ill. ; 21 cm. — (Bonniers illustrerade barnböcker) S78-2(
Sequel to Prostens barnbarn.
ISBN 9100421235
I. Stjernholm-Raeder, Karin. II. Title.
PZ59.B45K3 79-338155
79 MARC

Blom, K , Arne, 1946-
Stora tågkalabaliken : [1:a boken om Anders] / K. Arne Blom ; ill. av Ursula Wilby. — [Malmö] : [Hemmets journal (bokförl.)], 1978.
91 p. : ill. ; 25 cm. — (Hemmets barnbibliotek ; nr. 1) S79
ISBN 9173006726
I. Title.
PZ59.B49S8 79-365957
79 MARC

Brattström, Inger, 1920-
Jungfrurna : bilder ur 30-talet / Inger Brattström. — Stockholm : Rabén & Sjögren, 1978.
121, [1] p. ; 21 cm. — (Allåldersbok) S78-26
ISBN 9129521793
I. Title.
PZ59.B64J8 79-351051
79 MARC

Engqvist, Hans Erik, 1934-
Rymmarna / Hans Erik Engqvist. — Stockholm : Bonnier, 1978.
169, [1] p. ; 19 cm. — (Bonniers ungdomsböcker) S78-20
ISBN 9100423491
I. Title.
PZ59.E49R9 78-393626
78 MARC

Falk, Ann Mari.
Fredriks dröm / Ann Mari Falk. — Stockholm : Rabén & Sjögren, 1977.
122 p. ; 21 cm. S77-41
ISBN 9129498910 : kr27.00
I. Title.
PZ59.F3F7 78-385028
78 MARC

Falk, Ann Mari.
Räkna med bråk / Ann Mari Falk. — Stockholm : Rabén & Sjögren, c1978.
124 p. ; 23 cm. — (Ung nu) S***
ISBN 9129517893
I. Title.
PZ59.F3R3 79-338156
79 MARC

Franzén, Nils Olof.
Agaton Sax och den mörklagda ljusmaskinen / Nils-Olof Franzén. — Stockholm : Bonnier, 1978.
150, [2] p. ; 19 cm. — (Bonniers ungdomsböcker) S78-31
ISBN 9100423262
I. Title.
PZ59.F7A26 79-369560
79 MARC

Geijerstam, Brita af, 1902-
Tycker du som jag / Brita af Geijerstam ; [omsl. o. ill. av Tord Nygren]. — Stockholm : Bonnier, 1978.
61 p. : ill. ; 19 cm. S78-20
ISBN 9100423246
I. Nygren, Tord, 1936- II. Title.
PZ60.3.G44 79-338161
79 MARC

Gripe, Camilla, 1947-
Spöken Hjälp! / Camilla Gripe ; [omsl. o. ill. av Måd Olsson]. — Stockholm : Bonnier, 1978.
103, [1] p. : ill. ; 20 cm. — (Bonniers illustrerade barnböcker) S78-20
ISBN 9100423661
I. Olsson, Måd. II. Title.
PZ59.G68S6 79-343651
79 MARC

Gripe, Maria, 1923-
Att vara Elvis / Maria Gripe ; teckningar av Harald Gripe. — Stockholm : Bonnier, 1977.
156, [1] p. : ill. ; 21 cm. — (Bonniers illustrerade barnböcker) S77-46
ISBN 9100421618 : kr35.00
I. Gripe, Harald, 1921- II. Title.
PZ59.G7A8 78-375450
78 MARC

Gustavsson, Anita, 1945-
Jens / Anita Gustavsson. — Stockholm : B. Wahlström, 1978.
222 p. ; 21 cm. S78-31
ISBN 913241773X
I. Title.
PZ59.G86J4 79-350902
79 MARC

Hambe, Alf, 1931-
Molom—en värld i Belums tecken / Alf Hambe ; ill. av Knut Tiberg. — Stockholm : Atlantis, 1978.
110 p. ; 24 cm. S79
ISBN 9174860224
I. Title.
PZ59.H315M6 79-366626
00 MARC

Hasso, Signe, 1915-
Hoppa hoppa hage och andra rim / Signe Hasso ; ill. av Jerry Roupe. — Stockholm : Askild & Kärnekull, 1978.
[53] p. : ill. ; 19 x 20 cm. S79
ISBN 9170089728
I. Title.
PZ60.3.H35 79-369563
79 MARC

Hellberg, Hans-Eric, 1927-
Björn med trollhatten / Hans-Erik Hellberg ; teckningar av Tord Nygren. — Stockholm : Bonnier, 1978.
138, [2] p. : ill. ; 21 cm. — (Bonniers illustrerade barnböcker) S78-3
ISBN 9100424137
I. Nygren, Tord, 1936- II. Title.
PZ59.H35B5 1978 79-350815
79 MARC

Hermansson, Ritva, 1936-
Mirja flyttar till Sverige / Ritva Hermansson. — Stockholm : Prisma, 1978.
96, [1] p. ; 21 cm. S78-21
ISBN 9151812177
I. Title.
PZ59.H39M5 79-343834
79 MARC

Jacobson, Gun, 1930-
Salamander / Gun Jacobson. — Stockholm : Bonnier, 1978.
161, [2] p. ; 20 cm. S78-33
ISBN 9100423785
I. Title.
PZ59.J22S3 79-343667
79 MARC

Jakobsson, Greta, 1932-
Barn på Bränd Jons gård / Greta Jakobsson ; ill. av Kaisa Bäcklund. — Malung : Dalaförl., 1978.
171, [1] p. : ill. ; 19 cm. S78-25
ISBN 9185014117
I. Bäcklund, Kaisa. II. Title.
PZ59.J25B3 79-350177
79 MARC

Lange, Dagmar.
Jan och Katja jagar en kista / av Maria Lang [i.e. D. Lange]. — Stockholm : Norstedt, 1977.
148 p. ; 21 cm. S77-42
ISBN 9117722829 : kr25.00
I. Title.
PZ59.L27J3 79-367473
79 MARC

Lindborg-Sylvest, Rigmor, 1912-
Pojken Per / Rigmor Lindborg-Sylvest ; ill., Vera Mulder. — 1. uppl. — Stockholm : Författarförl., 1978.
42 p. : ill. ; 24 cm. S78-8
ISBN 9170542287 : kr26.00
I. Mulder, Vera. II. Title.
PZ59.L44P6 78-390847
79 MARC

Lindell, Lasse, 1951-
Tid för uppbrott / Lasse Lindell. — Stockholm : AWE/Geber, 1979.
172 p. ; 19 cm. S79
ISBN 9120057385
I. Title.
PZ59.L4515T5 79-370675
00 MARC

Löfgren, Ulf.
Harlekin / text o. bild av Ulf Löfgren. — Stockholm : Awe/-Geber, 1978.
[24] p. : ill. ; 24 x 28 cm. S78-21
ISBN 9120056664
I. Title.
PZ59.L55H37 79-356757
79 MARC

Lundgren, Max, 1937-
Boken om Birthe : en kärleksroman / Max Lundgren. — Stockholm : Bonnier, 1978.
117, [1] p. ; 20 cm. S78-34
ISBN 9100424153
I. Title.
PZ59.L8B6 79-343666
79 MARC

Mählqvist, Stefan.
Inte farligt pappa, krokodilerna klarar jag / text, Stefan Mählqvist ; bild, Tord Nygren. — [Stockholm] : Raben & Sjögren, c1977.
[25] p. : ill. ; 26 cm. S***
ISBN 9129486378
I. Nygren, Tord, 1936- II. Title.
PZ59.M224I5 79-339136
79 MARC

Mählqvist, Stefan.
Kom in i min natt kom in i min dröm / text, Stefan Mählqvist ; bild, Tord Nygren. — Stockholm : Rabén & Sjögren, 1978.
[26] p. : ill. ; 27 cm. S79
ISBN 9129522552
I. Nygren, Tord, 1926- II. Title.
PZ59.M224K6 79-365067
79 MARC

Martin, Bengt, 1933-
Bengt och kärleken / Bengt Martin. — Stockholm : Rabén + Sjögren, 1978.
127, [1] p. ; 23 cm. — (Ung nu) S78-26
ISBN 9129514932
I. Title.
PZ59.M2513B4 79-367457
79 MARC

Mickwitz, Camilla.
Jason och Arga Agnes / berättad o. ill. av Camilla Mickwitz. — Bromma : Opal, 1978.
[31] p. : ill. ; 21 x 26 cm. S78-22
ISBN 9172701374
I. Title.
PZ59.M53J3 78-400766
79 MARC

Nilsdotter, Katarina, 1944-
Helen, Anna och Caroline : en saga / berättad av Katarina Nilsdotter ; till akvarellmålningar av Carl Larsson. — [Stockholm] : Zeus, [1976], c1975.
[30] p. : ill. ; 21 x 30 cm. S77-14/16
ISBN 9174040073 : kr19.50
I. Larsson, Carl Olof, 1853-1919. II. Title.
PZ59.N478H4 1976 78-361081
78 MARC

Nordin, Carl-Gustaf, 1915-
Cecilia och morfar : bildberättelser / av Carl-Gustaf Nordin och Cecilia 4 år. — Bjästa : CeWe-förl./CeWe-pool, 1978.
58, [2] p. : ill. ; 22 x 30 cm. S79-3
I. Cecilia. II. Title.
PZ59.N53C4 79-371773
00 MARC

CHILDREN'S LITERATURE, SWEDISH (Continued)

Norrlid, Carl-Anders, 1938-
Judit Helena och andra berättelser / Carl-Anders Norrlid. — Stockholm : Rabén & Sjögren, 1977.
131 p. ; 23 cm. — (Ung nu) S77-42
CONTENTS: Ge dig, gosse.—Ann-Mari står i den stora hallen.—Johanna.—Sekunda vara.—Judit Helena.
ISBN 9129501997 : kr30.00
I. Title.
PZ59.N66J8 79-358506
79 MARC

Peterson, Hans.
De fyra hemlösa / Hans Peterson ; [omsl. o. teckningar, Stig Södersten]. — Stockholm : AWE/Geber, 1977.
176 p. : ill. ; 22 cm. S77-42
ISBN 9120046987 : kr31.00
I. Södersten, Stig. II. Title.
PZ59.P45F9 78-384679
79 MARC

Peterson, Hans.
Jag, Alberta / Hans Peterson. — Stockholm : AWE/Geber, 1978.
132 p. ; 19 cm. S78-21
ISBN 9120056796
I. Title.
PZ59.P45J28 78-390848
79 MARC

Peterson, Hans.
Kära Alberta / Hans Peterson. — Stockholm : AWE/Geber, 1977.
130 p. ; 19 cm. S77-39
Sequel to Alberta.
ISBN 9120056540 : kr26.50
I. Title.
PZ59.P45K3 79-367474
79 MARC

Peterson, Lars, 1944-
"Snöret" / Lars Peterson. — Stockholm : Gidlund, 1978.
162 p. ; 19 cm. S79
ISBN 9170212430
I. Title.
PZ59.P46S6 79-369559
79 MARC

Pettersson, Allan Rune, 1936-
Frankensteins faster : ett fasansfullt påhitt / Allan Rune Pettersson. — Stockholm : Bonnier, 1978.
151 p. ; 19 cm. S78-20
ISBN 9100423165
I. Title.
PZ59.P5F7 78-390514
79 MARC

Rimås, Lena, 1952-
Lill-semester på Blåholmen / Lena o. Sven-Otto Rimås ; ill. av Tanja Perskaia-Wisén. — Stockholm : Norman, 1978.
63, [1] p. : ill. ; 18 cm. — (Den Kristna barnbokringen) S78-20
ISBN 9153660048
I. Rimås, Sven-Otto, 1949- joint author. II. Perskaia-Wisén, Tanja. III. Title.
PZ59.R537L5 78-395287
79 MARC

Sandberg, Inger.
Var är Labolinas Millimina? / Inger o. Lasse Sandberg. — Stockholm : Rabén & Sjögren, 1978.
[32] p. : ill. ; 29 cm. S78-26
ISBN 9129508851
I. Sandberg, Lasse, joint author. II. Title.
PZ59.S25V3 79-371519
79 MARC

Sandman Lilius, Irmelin, 1936-
Skeppet Flygande Gedda / Irmelin Sandman Lilius ; omslag och illustr. av förf. — Stockholm : Bonnier, 1976.
127 p. : ill. ; 21 cm. S76-47
ISBN 9100413461 : kr26.50
I. Title.
PZ59.S285S54 78-381661
79 MARC

Schlossman, Sonia, 1942-
Sofia och Sebastian / Sonia Schlossman ; ill. av Rie Hägerdal. — Stockholm : Prisma, 1978.
76, [1] p. : ill. ; 21 cm. S78-22
ISBN 9151812274
I. Hägerdal, Rie. II. Title.
PZ59.S42S66 79-338159
79 MARC

Söderhjelm, Kai.
Över gränsen / Kai Söderhjelm. — Stockholm : Rabén & Sjögren, 1978.
114, [1] p. ; 22 cm. S78-26
ISBN 9129514975
1. World War, 1939-1945—Norway—Juvenile fiction. 2. World War, 1939-1945—Sweden—Juvenile fiction. I. Title.
PZ59.S59O3 79-338167
79 MARC

Sundh, Kerstin, 1912-
Önske-klövern / Kerstin Sundh ; ill., Ylva Källström. — Stockholm : AWE/Geber, 1978.
141, [3] p. : ill. ; 22 cm. S78-31
ISBN 9120056834
I. Källström, Ylva. II. Title.
PZ59.S88O4 79-343668
79 MARC

Wildh, Gunborg, 1919-
Det trodde Jenny och jag / Gunborg Wildh ; [omsl. o. ill. av Tor Morisse]. — Stockholm : Bonnier, 1978.
103 p. : ill. ; 21 cm. — (Bonniers illustrerade barnböcker) S78-20
ISBN 9100423394
I. Morisse, Tor. II. Title.
PZ59.W536T7 79-342239
79 MARC

Wolde, Gunilla, 1939-
Emma Värtemot / Gunilla Wolde. — [Stockholm] : Almqvist & Wiksell Förlag, 1974.
[24] p. : ill. ; 17 cm. S***
ISBN 9120042213
I. Title.
PZ59.W57E54 77-567636
78[r79]rev MARC

—BIBLIOGRAPHY

Andersson, Christina.
Boktips : 300 bra böker för barn och ungdom / valda av Christina Andersson och Kerstin Rydsjö. — Lund : Bibliotekstjänst, 1977.
63 p. ; 19 cm. — (Btj-serien ; 90) S***
Includes index.
ISBN 9170181551
1. Children's literature, Swedish—Bibliography. I. Rydsjö, Kerstin, joint author. II. Title.
Z1037.5.A53 78-375859
[PN1009.S9] 78 MARC

Koldenius, Malin, 1945-
För de yngsta : ett urval bilderböcker, ramsor, sagor och berättelser / av Malin Koldenius. — Lund : Biblitekstjänst, 1978.
76 p. ; 19 cm. — (Btj-serien ; 93 ISSN 0523-4646) S78-18
Rev. ed. of Småbarnsböcker by K. Auraldsson.
Includes indexes.
ISBN 9170181632
1. Children's literature, Swedish—Bibliography. I. Auraldsson, Kerstin, 1931- Småbarnsböcker. II. Title.
Z1037.5.A95 1978 79-359952
[PN1009.S9] 79 MARC

Loman, Anna, 1923-
Vad kommer sen? : Barnboksserier aktuella 1977. — 5. omarb. uppl. — Lund : Bibliotekstjänst, 1977.
80 p. ; 21 cm. — (Btj-serien ; 87) S***
Includes index.
ISBN 9170181527
1. Children's literature, Swedish—Bibliography. I. Bibliotekstjänst. II. Title.
Z1037.5.L65 1977 78-394470
[PN1009.S9] 79 MARC

Nilson, Margot.
Låt oss läsa högt : ett intressegrupperat urval böcker för åk 3 9 / av Margot Nilson. — 2. omarb. uppl. — Lund : Bibliotekstjänst, 1977.
65 p. ; 19 cm. — (Btj-serien ; 85) S***
Includes indexes.
ISBN 9170181462
1. Children's literature, Swedish—Bibliography. I. Title.
Z1037.5.N48 1977 78-375855
[PN1009.S9] 78 MARC

—BIO-BIBLIOGRAPHY

De skriver för barn : porträtt av fjorton svenska barnboksförfattare / red., Carl-Agnar Lövgren. — 2. omarb. uppl. — Lund : Bibliotekstjänst, 1978.
176 p. ; 19 cm. — (Barn- och ungdomsförfattare ; 1 ISSN 0348-2839) S78-18
Includes bibliographies.
ISBN 9170181640
1. Children's literature, Swedish—History and criticism. 2. Children's literature, Swedish—Bio-bibliography. 3. Authors, Swedish—Biography. I. Lövgren, Carl-Agnar, 1929- II. Series.
PN1009.S9S55 1978 79-349075
79 MARC

—HISTORY AND CRITICISM

De skriver för barn : porträtt av tolv svenska barnboksförfattare / red., Carl-Agnar Lövgren. — Lund : Bibliotekstjänst, 1974.
120 p. : ports. ; 19 cm. — (Btj-serien ; 64) S74-53
Includes bibliographies.
ISBN 9170181071 : kr38.30
1. Authors, Swedish—Biography. 2. Children's literature, Swedish—History and criticism. I. Lövgren, Carl-Agnar, 1929-
PT9305.D4 75-530716
78[r78]rev MARC

De skriver för barn : porträtt av fjorton svenska barnboksförfattare / red., Carl-Agnar Lövgren. — 2. omarb. uppl. — Lund : Bibliotekstjänst, 1978.
176 p. ; 19 cm. — (Barn- och ungdomsförfattare ; 1 ISSN 0348-2839) S78-18
Includes bibliographies.
ISBN 9170181640
1. Children's literature, Swedish—History and criticism. 2. Children's literature, Swedish—Bio-bibliography. 3. Authors, Swedish—Biography. I. Lövgren, Carl-Agnar, 1929- II. Series.
PN1009.S9S55 1978 79-349075
79 MARC

De skriver för ungdom : porträtt av svenska ungfomsförfattare / red., Carl Agnar Lövgren. — 2. omarb. uppl. — Lund : Bibliotekstjänst, c1978-
v. ; 19 cm. — (Barn- och ungdoms- författare ; 2) S***
Includes bibliographies.
ISBN 9170181683
1. Children's literature, Swedish—History and criticism. 2. Children's literature, Swedish—Bio-bibliography. I. Lövgren, Carl-Agnar. II. Series.
PN1009.S9S56 1978 79-363055
79 MARC

CHILDREN'S LITERATURE, SWISS

Tidlund.
Jag minns en dag / Ture Gerdes [i.e. Tidlund] ; [omsl. o. ill., Lennart Karlsson]. — Göteborg : Korpen, 1978.
79, [1] p. : ill. ; 24 cm. S78-11
ISBN 9173741566 : kr30.00
I. Title.
PZ59.T57J3 79-338157
79 MARC

CHILDREN'S LITERATURE, UKRAINIAN

Dodgson, Charles Lutwidge, 1832–1898.
[Alice's adventures in Wonderland. Ukrainian]
Аліса в Країні Чудес : казка : для молодшого шкільного віку / Льюіс Керролл [i. e. Charles Lutwidge Dodgson] **; переклад з англійської Галини Бушиної ; малюнки англійського художника Джона Тенніла**]. — **Київ : Веселка**, 1976.
142 p. : ill. ; 17 cm. — (Бібліотечна серія) USSR***
Series romanized: Bibliotechna serifa.
0.28rub
I. Tenniel, Sir John, 1820–1914. II. Title.
Title romanized: Alisa v Kraïni Chudes.
PZ70.U72D6 1976 78–408425

CHILDREN'S LITERATURE, URDU

Aṣarī, 'Azīz.
('Ālī par kyā guzrī)
عالی پر کیا گزری : ایک ناول بچوں کے لیے / عزیز اثری. —
لاہور : فیروزسنز, 1977.
199 p. : ill. ; 18 cm.
In Urdu.
A novel.
Rs4.50
I. Title.
PZ90.U7A8 78–931029

CHILDREN'S LITERATURE, VIETNAMESE

Phạm Hổ.
Những quả chùm rượu / Phạm Hổ. — Hà-nội : Kim Đồng, 1970.
21 p. : ill. ; 19 cm.
Short stories.
1. Vietnamese Conflict, 1961-1975—Juvenile fiction. I. Title.
PZ90.V5P46 78-984329
79 MARC

CHILDREN'S LITERATURE, WELSH

Lloyd, John Selwyn.
Trysor Bryniau Caspar / [gan] J. Selwyn Lloyd ; arluniwyd gan Catherine Thomas. — Llandysul : Gwasg Gomer, 1976.
99 p. : ill. ; 21 cm. GB78-28998
ISBN 0-85088-481-0 pbk : £1.10
I. Title.
PZ90.W4L57 78-398299
79 MARC

CHILDREN'S LITERATURE, YUGOSLAV

Dragojević, Danijel.
Bajka o vratima / Danijel Dragojević ; [ilustracije Boris Dogan]. — Zagreb : Naprijed, 1975.
[20] p. : col. ill. ; 27 cm. — (Biblioteka Zelena žirafa) Yu***
I. Dogan, Boris, 1923- II. Title.
PZ70.S42D72 78-343538
78 MARC

Jovanović, Jovan, known as Zmaj, 1833-1904.
Da čudne radosti / J. Jovanović Zmaj ; [antologiju pesama za decu J. Jovanovića Zmaja priredio Husein Tahmiščić]. — Gornji Milanovac : Dečje novine, 1977.
114 p. : ill. (some col.) ; 22 cm. Yu***
I. Tahmiščić, Husein. II. Title.
PZ70.S42J59 1977 78-366282
78 MARC

—HISTORY AND CRITICISM—ADDRESSES, ESSAYS, LECTURES

Šijan, Dane, 1922-
Savremeni dječji pisci Jugoslavije : intervjui / Dane Šijan ; [likovno opremio Šimo Klaić]. — Zagreb : Stylos, 1975 (Zagreb : Zadružna štampa)
553 p., [29] leaves of plates : ports. ; 23 cm. — (Biblioteka Savremenici) Yu76-7138
Includes bio-bibliographical notes of the authors.
1. Children's literature, Yugoslav—History and criticism—Addresses, essays, lectures. 2. Authors, Yugoslav—20th century—Interviews. I. Title.
PN1009.Y8S5 79-350266
79 MARC

CHILDREN'S LITURGICS

—CATHOLIC CHURCH

Hilliard, Dick.
The Lord blesses me : center celebrations of God's word for children and young families / by Dick Hilliard ; illustrated by George Collopy. — Saratoga, CA. : Resource Publications, c1978.
152 p. : ill. ; 28 cm.
Includes indexes.
ISBN 0-89390-005-2
1. Children's liturgics—Catholic Church. I. Title.
[BX2045.C55H54] 264'.02 78-61308
00 MARC

CHILDREN'S LITURGIES

see also Children—Prayer-books and devotions

—CATHOLIC CHURCH

Children's liturgies : seventy-four Eucharistic liturgies, prayer services, and penance services, designed for primary, middle, and junior high children / edited by Bernadette Kenny. — New York : Paulist Press, c1977.
xi, 161 p. ; 28 cm.
"Selected from the first two years of Service resources for pastoral ministry."
SUMMARY: A collection of seventy-four fully planned liturgies for religious holidays and special occasions throughout the year.
ISBN 0-8091-2030-5 : $5.95
1. Children's liturgies—Catholic Church. [1. Catholic Church. Liturgy and ritual] I. Kenny, Bernadette. II. Catholic Church. Liturgy and ritual. III. Service.
BX2045.C55C47 264'.02 77-74582
79 MARC

CHILDREN'S MASS

Matthews, Edward.
Celebrating mass with children : a commentary on the Directory for masses with children / Edward Matthews. — New York : Paulist Press, 1978, c1975.
192 p. ; 22 cm.
"Directory for masses with children": p. 173-192.
Includes bibliographical references.
ISBN 0-8091-2160-3 : $6.95
1. Children's mass. 2. Mass—Celebration. I. Catholic Church. Congregatio pro Cultu Divino. Directorium de missis cum pueris. English. 1978. II. Title.
BX2230.2.M33 1978 264'.0203 78-58564
79 MARC

CHILDREN'S MUSEUMS

see also Museums and schools

CHILDREN'S NIGHTWEAR see Children's sleepwear

CHILDREN'S OPERAS see Operas, Juvenile

CHILDREN'S PARAPHERNALIA

see also Children's clothing; Children's furniture; Children's literature; Games; Infants' supplies; Toys

—CATALOGS

Foa, Linda, 1942-
Kids' stuff / Linda Foa & Geri Brin, photography by Michael Datoli. — 1st ed. — New York : Pantheon Books, 1979.
208 p. : ill. (some col.) ; 28 cm.
ISBN 0-394-41558-2 : $17.95
1. Children's furniture—Catalogs. 2. Children's paraphernalia—Catalogs. I. Brin, Geri, 1947- joint author. II. Datoli, Michael. III. Title.
TT197.5.C5F63 1979 684 78-14343
78 MARC

CHILDREN'S PARTIES

Wolfsohn, Reeta Bochner.
Successful children's parties / Reeta Bochner Wolfsohn (Bunky the Clown). — New York : Arco Pub. Co., c1979.
104 p. : ill. ; 24 cm.
Includes index.
ISBN 0-668-04607-4 : $7.95
1. Children's parties. I. Title.
GV1205.W64 793.2'1 78-18421
78 MARC

—JUVENILE LITERATURE

Fiarotta, Phyllis.
Confetti : the kid's make-it yourself, do-it yourself party book / by Phyllis Fiarotta and Noel Fiarotta. — New York : Workman Pub., c1978.
223 p. : ill. ; 22 cm.
SUMMARY: Instructions for twenty-two parties, including birthday, holiday, Mother's Day, and new baby, with details for making invitations, decorations, favors, and food.
ISBN 0-89480-049-3. ISBN 0-89480-050-7 pbk.
1. Handicraft—Juvenile literature. 2. Cookery—Juvenile literature. 3. Children's parties—Juvenile literature. [1. Parties] I. Fiarotta, Noel, joint author. II. Title.
TT160.F472 790.19'22 78-7121
78 MARC

The Mickey Mouse birthday book. — New York : Random House, c1978.
[42] p. : col. ill. ; 24 cm. — (Disney's wonderful world of reading ; 44)
SUMMARY: Minnie gives Mickey a surprise party and the reader learns how she handled invitations, place cards, food, and games.
ISBN 0-394-83963-3 : $3.50
1. Children's parties—Juvenile literature. 2. Birthdays—Juvenile literature. [1. Parties. 2. Birthdays] I. Disney (Walt) Productions.
GV1205.M52 793.2'1 78-55911
78 MARC

CHILDREN'S PERIODICALS, DANISH

Cederborg, Else.
Ungdomsblade / Else Cederborg. — [København] : Munksgaard, 1977.
43 p. : ill. ; 30 cm. — (Trivi) D77-12
Includes index.
ISBN 8716022319 : kr29.00
1. Children's periodicals, Danish. I. Title.
PN5287.J8C4 78-386463
79 MARC

CHILDREN'S PERIODICALS, ENGLISH

The Best of Chums / edited by Philip Warner. — London : Cassell, 1978.
191 p. : ill. ; 32 cm. GB***
ISBN 0-304-30059-4 : £7.50
I. Warner, Philip. II. Chums.
AP201.C65 052 79-300886
79 MARC

CHILDREN'S PERIODICALS, HUNGARIAN

Kisdobosok évkönyve.
[Budapest] Ifjúsági Lapkiadó Vállalat.
v. ill. 24 cm.
AP215.H8K57 75-643299
MARC-S

CHILDREN'S PERIODICALS, JAPANESE

Shū, Hajime, 1930-
(Karasu no takaramono)
からすのたからもの / 周はじめ　さく ; 小松崎邦雄　え. — 東京 : 福音館書店, c1972.
27 p. : col. ill. ; 19 x 27 cm. — (こどものとも ; 192号)
Cover title.
On p. [4] of cover: The crows and their treasures.
I. Komatsuzaki, Kunio. II. Title. III. Title: The crows and their treasures. IV. Series: Kodomo no tomo ; 192-gō.
AP215.J3K6 no. 192 79-804259
[PZ49.63]

Uchida, Risako, 1928-
(Babayagā no shiroi tori)
ババヤガーのしろいとり : ロシア民話 / 内田莉莎子再話 ; 佐藤忠良画. — 東京 : 福音館書店, 1973.
31 p. : col. ill. ; 27 cm. — (こどものとも ; 212号) Ja***
On p. [4] of cover: Babayaga and her white birds.
I. Satō, Chūryō, 1912- II. Title. III. Title: Babayaga and her white birds. IV. Series: Kodomo no tomo ; 212-gō.
AP215.J3K6 no. 212 79-804403
[PZ49.41]

Uchida, Risako, 1928-
(Jūni no tsuki no okurimono)
12のつきのおくりもの : スロバキア民話 / 内田莉莎子再話 ; 丸木俊画. — 東京 : 福音館書店, c1971.
27 p. : col. ill. ; 27 cm. — (こどものとも ; 189号)
Cover title.
On p. [4] of cover: Twelve months.
I. Maruki, Toshi. II. Title. III. Title: Twelve months. IV. Series: Kodomo no tomo ; 189-gō.
AP215.J3K6 no. 189 79-804416
[PZ49.41]

CHILDREN'S PERIODICALS, KOREAN

(Sonyŏn)
少年　第1-　年; 隆熙2年11月-　[Nov. 1908-
[서울　文學資料原文社 1977-
v. ill. 22 cm. monthly.
. Reprint of a periodical published by Sinmunkwan.

AP215.K6S642 79-640732
(MARC-S)

CHILDREN'S PERIODICALS, MALAGASY (Continued)

Vankona. –50/51.
[Tananarive] Ministère de l'éducation nationale [etc., –1974?]
no. ill. 22 cm.
French or Malagasy.
No. issued by the Direction de l'enseignement primaire; –50/51 by the Ministère de l'éducation nationale.
No more published?
I. Malagasy Republic. Direction de l'enseignement primaire. II. Malagasy Republic. Ministère de l'éducation nationale.
AP215.M3V36 79–642219
MARC-S

CHILDREN'S PLAYS

see also Bible plays; Children as actors; College and school drama

Bailey, Carolyn Sherwin, 1875-1961.
Plays for the children's hour : an American childhood presentation / by Carolyn Sherwin Bailey. — Great Neck, N.Y. : Core Collection Books, 1978.
257 p. ; 20 cm.
SUMMARY: Discusses staging and producing plays and includes a variety of plays, with some for holidays and special occasions.
Reprint of the 1931 ed. published by M. Bradley Co., Springfield, Mass.
ISBN 0-8486-2032-1
1. Children's plays. [1. Plays] I. Title.
PN6120.A5B215 1978 808.82'41 77-94332
79 MARC

Brown, Abbie Farwell, d. 1927.
The lantern, and other plays for children / by Abbie Farwell Brown. — Great Neck, N.Y. : Cora Collection Books, 1978.
152 p. ; 20 cm. — (One act plays in reprint)
Reprint of the 1928 ed. published by Houghton, Mifflin, Boston.
CONTENTS: The lantern.—Rhœcus.—The wishing moon.—The little shadows.
ISBN 0-8486-2033-X
1. Children's plays. [1. Plays] I. Title. II. Series.
PN6120.A5B68 1978 812'.5'2 77-94333
79 MARC

Murray, John, 1923-
Fifteen plays for teen-agers : a collection of one-act royalty-free comedies and mysteries / by John Murray. — Boston : Plays, inc., c1979.
v, 353 p. ; 22 cm.
SUMMARY: A collection of royalty-free, one-act mysteries and comedies for teenage actors with production notes.
ISBN 0-8238-0227-2 : $8.95
1. Amateur theatricals. 2. Children's plays. [1. Plays] I. Title.
PN6120.A5M817 812'.041 78-16588
78 MARC

Watts, Irene N
A blizzard leaves no footprints ; Listen to the drum ; Patches ; The rainstone : four children's plays / by Irene N. Watts. — 1st ed. — Toronto : Playwrights Co-op, 1978.
39 p. ; 28 cm. C79-3899-4
ISBN 0-88754-096-1 : $3.50
1. Children's plays. I. Title.
PR9199.3.W384B5 812'.5'4 79-314564
00 MARC

—PRESENTATION, ETC.

Luzzati, Emanuele, 1921-
Facciamo insieme teatro / [di] Emanuele Luzzati, Tonino Conte. — Torino : Einaudi, 1977.
166 p. : ill. ; 20 cm. — (Ragazzi) (Gli Struzzi ; 133) It77-Aug
L3000
1. Children's plays—Presentation, etc. 2. Theater and society. I. Conte, Tonino, joint author. II. Title. III. Series.
PN6041.S85 no. 133 79-383741
[PN315] *79 MARC

Malkin, Michael R 1943-
Training the young actor / Michael R. Malkin ; photos. by Wayne Nicholls ; ill. by Jim Shubin. — South Brunswick : A. S. Barnes, c1979.
128 p. : ill. ; 25 cm.
Includes index.
ISBN 0-498-01957-8 : $9.95
1. Acting—Study and teaching. 2. Children as actors. 3. Children's plays—Presentation, etc. I. Title.
PN3157.M34 792'.028 76-50201
78 MARC

Pierini, Mary Paul Francis.
Children's theater in elementary school / Sister Paul Francis Pierini ; art, Sister Roberta Carson ; photography, George F. Perry. — New York : Seabury Press, 1977.
84 p. : ill. ; 28 cm.
"A crossroad book".
ISBN 0-8164-0342-2 : $3.95
1. Children's plays—Presentation, etc. 2. Drama in education. I. Title.
PN3157.P5 372.6'6 76-54641
76 MARC

—PRESENTATION, ETC. —JUVENILE LITERATURE

Murayama, Tomoyoshi, 1901-
(Watakushitachi no geki)
私たちの劇 / 村山知義著. — 東京：國民図書刊行会, 昭和24 [1949]
121 p., [1] leaf of plates : ill. ; 19 cm. — (教室文庫)
Series romanized: Kyōshitsu bunko.
Includes index.
1. College and school drama—Juvenile literature. 2. Children's plays—Presentation, etc.—Juvenile literature. I. Title.
PN3175.M8 78–811105

—PRESENTATION, ETC.—DENMARK

Teater for børn / redigeret af Lis Vibeke Kristensen ; i samarbejde med Eva Marie Møller. — [København] : Kulturministeriets arbejdsgruppe om børn og kultur : [eksp. DBK], 1978.
64 p. : ill. ; 21 cm. — (Børn, kultur, samfund ; 2) D78-16
Includes bibliographical references.
ISBN 8750324632 : kr6.20
1. Children's plays—Presentation, etc.—Denmark. 2. Amateur theatricals—Denmark. I. Kristensen, Lis Vibeke, 1953- II. Møller, Eva Marie. III. Denmark. Arbejdsgruppen om børn og kultur. IV. Series.
PN3169.D4T4 79-381465
79 MARC

—PRESENTATION, ETC.—DENMARK—ADDRESSES, ESSAYS, LECTURES

Børneoffentlighed : socialistisk børneteater : muligheder for en frigørende opdragelse : en antologi / red. af Projektgruppe Børneteater. — København : Tiderne Skifter, c1977.
271 p. : ill. ; 20 cm. D77-17
Most of the articles translated from German.
Includes bibliographical references.
CONTENTS: Langsted, J. Perspektiver i et socialistisk teaterarbejde med børn og unge.—Benjamin, W. Program for et proletarisk børneteater.—Lacis, A. Om "Program for et proletarisk børneteater".—Hoernle, E. Teaterspil som udtryksmulighed og som hvervested.—Schedler, M. Opfordring om at genopdage det tidlige kommunistiske børneteater.—Paul, A. Eventyrteater og emancipation.—Paris-Ebert, W., Paris, V. og Riemeyer, E. Tre faser i arbejdet i Kindertheater im Märkischen Viertel.—Fest.—Maier, H., Praml, W., Ring, R. Om æstetisk opdragelse sammen med lærlinge.—Richard, J. Om Lederstyret rollespil med arbejderbørn i skolealderen. Schedler, M. Perspektiver i et nyt børneteater.—Frommlet, W. et al. Marlermodellen—Bram, A., Dolberg, H. og Møller, C. H. Muligheder i Danmark.
1. Children's plays—Presentation, etc.—Denmark—Addresses, essays, lectures. 2. Children's plays—Presentation, etc.—Germany—Addresses, essays, lectures. 3. Theater and socialism—Addresses, essays, lectures. I. Projektgruppe Børneteater.
PN3157.B58 79-381462
79 MARC

—PRESENTATION, ETC. —GERMANY

Studienseminar Kinder- und Jugendtheater, Berlin, 1978.
Kinder- und Jugendtheater zwischen Tradition und Emanzipation : Grusswort, Programm, Thesen, Analysen, Dokumentation zum Studienseminar Kinder- und Jugendtheater 1978 / [verantwortlich, Gotthard Welzel]. — Bonn : Theater Rundschau, 1978.
63 p. : ill. ; 21 cm. — (Schriftenreihe des Bundes der Theatergemeinden ; Heft 11) GFR***
Sponsored by the Bund der Theatergemeinden and held May 25-28, 1978.
Includes index.
1. Children's plays—Presentation, etc.—Germany. I. Welzel, Gotthard. II. Bund der Theatergemeinden. III. Title. IV. Series: Bund der Theatergemeinden. Schriftenreihe des Bundes der Theatergemeinden ; Heft 11.
PN3157.S8 1978 78-393326
79 MARC

—PRESENTATION, ETC.—GERMANY—ADDRESSES, ESSAYS, LECTURES

Børneoffentlighed : socialistisk børneteater : muligheder for en frigørende opdragelse : en antologi / red. af Projektgruppe Børneteater. — København : Tiderne Skifter, c1977.
271 p. : ill. ; 20 cm. D77-17
Most of the articles translated from German.
Includes bibliographical references.
CONTENTS: Langsted, J. Perspektiver i et socialistisk teaterarbejde med børn og unge.—Benjamin, W. Program for et proletarisk børneteater.—Lacis, A. Om "Program for et proletarisk børneteater".—Hoernle, E. Teaterspil som udtryksmulighed og som hvervested.—Schedler, M. Opfordring om at genopdage det tidlige kommunistiske børneteater.—Paul, A. Eventyrteater og emancipation.—Paris-Ebert, W., Paris, V. og Riemeyer, E. Tre faser i arbejdet i Kindertheater im Märkischen Viertel.—Fest.—Maier, H., Praml, W., Ring, R. Om æstetisk opdragelse sammen med lærlinge.—Richard, J. Om Lederstyret rollespil med arbejderbørn i skolealderen. Schedler, M. Perspektiver i et nyt børneteater.—Frommlet, W. et al. Marlermodellen—Bram, A., Dolberg, H. og Møller, C. H. Muligheder i Danmark.
1. Children's plays—Presentation, etc.—Denmark—Addresses, essays, lectures. 2. Children's plays—Presentation, etc.—Germany—Addresses, essays, lectures. 3. Theater and socialism—Addresses, essays, lectures. I. Projektgruppe Børneteater.
PN3157.B58 79-381462
79 MARC

--PRESENTATION, ETC. -- HONGKONG

(Erh t'ung hsi chü lun wen chi)
兒童戲劇論文集：紀念兒童劇場開幕. 第1輯. — [香港? : s. n., 1978?]
32 p., [1] leaf of plates : ill. ; 21 cm.
Cover title.
No more published?
Reprint of the 1940 ed. published by Hsiang-kang Chung-hua Chi-tu chiao ch'ing nien hui.
CONTENTS: 王永葳　寫在前面.—曾昭森　兒童戲劇運動在香港的意義.—胡春冰　兒童戲劇運動講話.—郭步陶　兒童戲劇運動的大前題.—陳丹林　一份豐厚的兒童恩物.—狄睎　兒童劇場與兒童教育.—李殊倫　兒童劇場與兒童演劇運動.—黃慶雲　關于兒童劇本的一個討論.—吳其敏　兒童劇的製作問題.—盧秉良為兒童劇場成立致小朋友書.—Oglebey, H. M.　什麼叫做教育戲劇　王永葳譯
1. Children's plays—Presentation, etc.—Hongkong.
PN3169.H6E7 1978 79–836823

—PRESENTATION, ETC.—HUNGARY

Műsorok gyermekszinjátszó együttesek részére. — [Nyiregyháza] : Szabolcs-Szatmár Megyei Művelődési Központ, 1977.
60 p. ; 29 cm. — (Szinjátszó módszertani füzetek ; 3. sz) H77
1. Children's plays—Presentation, etc.—Hungary. 2. Children's plays, Hungarian—Bibliography.
PN3157.M83 79-346594
*79 MARC

—PRESENTATION, ETC.—ITALY

Giampaola, Giuseppe.
Scuola e teatro / Giuseppe Giampaola. — [Firenze] : Nuovedizioni E. Vallecchi, [1975?]
253 p. ; 19 cm. It76-Oct
Includes bibliographical references.
L2000
1. Drama in education. 2. Children's plays—Presentation, etc.—Italy. 3. College theater—Italy. 4. Theater and society—Italy. I. Title.
PN3191.I8G5 792'.0945 79-360237
*79 MARC

Mamprin, Lodovico.
Tre dialoghi sull'animazione / [di] L. Mamprin, G. R. Morteo, L. Perissinotto. — Roma : Bulzoni, [1977]
282 p. ; 21 cm. — (Biblioteca di cultura ; 114) It78-Jan
Bibliography: p. 277-282.
L9000
1. Drama in education. 2. Children's plays—Presentation, etc.—Italy. 3. Expression. 4. Group relations training. I. Morteo, Gian Renzo, joint author. II. Perissinotto, Loredana, joint author. III. Title.
PN3171.M375 371.33'2 79-353960
*79 MARC

—PRESENTATION, ETC.—NORWAY

Norsk kulturråd. Utvalg til utredning av barneteaterarbeid i Norge.
Utredning om barneteaterarbeid i Norge : innstilling fra Norsk kulturråds utvalg til utredning av barneteaterarbeid i Norge. — Oslo : Norsk kulturråd, c1976.
107 p. ; 23 cm. N77-Nov.
Bibliography: p. [98]-100.
ISBN 8270810142
1. Children's plays—Presentation, etc.—Norway. I. Title.
PN3157.N6 1976 79-359561
79 MARC

CHILDREN'S PLAYS, AMERICAN

Henderson, Nancy Wallace.
Celebrate America : a baker's dozen of plays / by Nancy Henderson ; illustrated by Paul Frame. — New York : J. Messner, c1978.
128 p. : ill. ; 22 cm.
CONTENTS: Soul force.—Hail the lucky year.—Legend for our time.—"John Muir, Earth-planet, universe."—Honor the brave.—Casey at the bat.—Moonlife, 2069.—Come to the fair.—Little Turtle.—The land we love.—M.D. in petticoats.—Keeping Christmas merry.—Popcorn whoppers.
ISBN 0-671-32907-3
1. Children's plays, American. [1. Plays] I. Frame, Paul, 1913- II. Title.
PS3558.E489C4 812'.5'4 78-15169
78 MARC

CHILDREN'S PLAYS, BENGALI (Continued)

Modabber, Mohammad.
Eka thali nāṭaka / Mohāmmada Modābbera. — Ḍhākā : Śirina Presa, 1978.
76 p. : ill. ; 23 cm.
On verso of t.p.: Ek tholi natok.
In Bengali.
Plays.
CONTENTS: Dukhu Miẏā.—Duṭi greneḍa.—Cā̃dera deśe piẏā.—Minaṭura kīrtti.—Katbela.—Kula.
Tk6.00
I. Title.
PZ90.B4M564 79-900325
79 MARC

CHILDREN'S PLAYS, BOLIVIAN

Díaz Villamil, Antonio.
Retamita ; Muñecas de bazar ; Crisol ; El pequeño estafeta / Antonio Díaz Villamil. — La Paz : Librería Editorial Juventud, 1977.
168 p. ; 19 cm. — (Colección Ayer y hoy)
1. Children's plays, Bolivian. 2. War of the Pacific, 1879-1884—Juvenile drama. I. Title.
PZ77.D5 862 78-102475
78 MARC

CHILDREN'S PLAYS, BRAZILIAN

Teatro infantil : três peças. — Rio de Janeiro : Serviço Nacional de Teatro, Ministério da Educação e Cultura, 1975.
72 p. ; 21 cm. — (Coleção Prêmios ; 2)
CONTENTS: Santos, M. L. C. M. T. Estória da moça preguiçosa.—Filgueiras, R. M. Papo de anjo.—Peduto, L. M. José de Maria (Auto de Natal)
1. Children's plays, Brazilian. I. Santos, Maria de Lourdes Cavalcanti Martini Teixeira dos, 1928- Estória da moça preguiçosa. 1975. II. Filgueiras, Ricardo Mack. Papo de anjo. 1975. III. Peduto, Luiz Menezes. José de Maria. 1975. IV. Brazil. Serviço Nacional de Teatro. V. Series.
PZ87.T37 869'.2 79-358843
79 MARC

CHILDREN'S PLAYS, CHINESE

Shu, Ch'ing-ch'un, 1898–1966.
(Ch'ing wa ch'i shou)
青蛙騎手：儿童歌剧 / 老舍[舒慶春]. — [香港? : s. n., 1978?]
24 p. : ill. : 26 cm.
Drama.
Reprint ed.

1. Children's plays, Chinese. I. Title.
PZ10.87.S55C47 1978 79–835008

CHILDREN'S PLAYS, FRENCH

Morency, Pierre.
Tournebire et le Malin Frigo ; Les écoles de bon Bazou / Pierre Morency. — [Montréal] : Leméac, [c1978]
153 p. : ill. ; 18 cm. C***
ISBN 0-7761-9907-2 : $5.50
1. Children's plays, French. I. Morency, Pierre. Les écoles de bon Bazou. 1978. II. Title.
PZ27.M65 842 78-376354
78 MARC

Pinguilly, Yves.
Les Vagabonds du miroir ... : [Le Havre, Maison de la culture, octobre 1977] / Yves Pinguilly ; illustration Gérard Esquerre. — Paris (2, rue des Petits-Pères, 75002) : J.-C. Simoën, 1977.
63 p. : ill. ; 21 cm. — (Collection Mise en pièces) F78-10739
24.00F
1. Children's plays, French. I. Title.
PZ27.P5 842'.9'14 78-392661
79 MARC

CHILDREN'S PLAYS, HUNGARIAN

—BIBLIOGRAPHY

Műsorok gyermekszinjátszó együttesek részére. — [Nyíregyháza] : Szabolcs-Szatmár Megyei Művelődési Központ, 1977.
60 p. ; 29 cm. — (Szinjátszó módszertani füzetek ; 3. sz) H77
1. Children's plays—Presentation, etc.—Hungary. 2. Children's plays, Hungarian—Bibliography.
PN3157.M83 79-346594
•79 MARC

CHILDREN'S PLAYS, ITALIAN

Vastapane, Sergio.
Garibaldi e la spedizione dei mille : drammatizzazione per fanciulli / Sergio Vastapane, Claudio Corio. — Leumann (Torino) : Elle Di Ci, 1978.
37 p. : ill. ; 19 cm. — (Collana Espressione fanciulli) It***
"Supplemento a Armonia di voci 2/1978."
ISBN 8801127642
1. Garibaldi, Giuseppe, 1807-1882—Juvenile drama. I. Corio, Claudio, joint author. II. Armonia di voci. III. Title.
PZ47.V34 78-383311
79 MARC

CHILDREN'S PLAYS, JAPANESE

(Gakkōgeki kyakuhon shū)
学校劇脚本集：低学年用 / [落合聰三郎 ... et al. 共著]. — 東京：学習研究社, 昭和25 [1950]
58 p. : ill. ; 18 cm. — (学研教育シリーズ ; 第5集)

1. Children's plays, Japanese. 2. College and school drama. I. Ochiai, Sōzaburō, 1910- II. Series: Gakken kyōiku shirizu ; dai 5-shū.
PL768.A35G3 79–809833

(Maŭm sok ŭi mongnan kkot)
마음속의목란꽃 : 학생작품집. — [평양] : 금성청년출판사, 1977.
199 p. : ill. ; 19 cm.
₩0.20

1. Children's writings, Korean—Korea (Democratic People's Republic) I. Kŭmsŏng Ch'ŏngnyŏn Ch'ulp'ansa.
PL973.2.C45M3 79–826450

CHILDREN'S PLAYS, LITHUANIAN

Žilinskaitė, Vytautė, 1933-
Stebuklingas apsiaustas : pjesės lėlių teatrui / Vytautė Žilinskaitė. — Vilnius : Vaga, 1977.
92 p. ; 17 cm. USSR***
"Jaunesniam ir vidutiniam mokykliniam amžiui."
CONTENTS: Paršiukas ir parašiutas.—Paskutinė lapė.—Stebuklingas apsiaustas.
0.13rub
I. Žilinskaitė, Vytautė, 1933- Paršiukas ir parašiutas. 1977. II. Žilinskaitė, Vytautė, 1933- Paskutinė lapė. 1977. III. Title.
PZ90.L5Z53 79-343662
79 MARC

CHILDREN'S PLAYS, RAETO-ROMANCE

Fasser, Carl.
Ot gōs per scolars / Carl Fasser. — [Chur] : Lia rumantscha, 1977.
44 p. ; 21 cm. — (La Scena ; Nr. 52) Sw78-2777
6.00F
1. Children's plays, Raeto-Romance. I. Title.
PZ90.R3F26 79-349028
79 MARC

Fasser, Carl.
Parablas dramatisadas per nossas scoulas / Carl Fasser. — [Chur] : Lia rumantscha, 1977.
48 p. ; 21 cm. — (La Scena ; Nr. 51) Sw78-2778
6.30F
1. Children's plays, Raeto-Romance. I. Title.
PZ90.R3F28 78-386219
79 MARC

Fasser, Carl.
Teatrins per las scoulas bassas / Carl Fasser. — [Chur] : Lia rumantscha, 1977.
24 p. : music ; 21 cm. — (La Scena ; nr. 53) Sw78-2779
4.50F
1. Children's plays, Raeto-Romance. I. Title.
PZ90.R3F3 78-380079
78 MARC

CHILDREN'S PLAYS, ROMANIAN

Toți copiii pe scenă! : [Piese de teatru pentru copii / de Alecu Popovici, Virgil Stoenescu, Viniciu Gafița, et al. ; coperta colecției și desenul, E. Munteanu]. — București : Editura Ion Creangă, 1977.
317 p. ; 17 cm. — (Micul actor) R78-813
lei6.25
1. Children's plays, Romanian.
PZ90.R6T64 78-375389
•78 MARC

CHILDREN'S PLAYS, RUSSIAN

—HISTORY AND CRITICISM

Khalfin, IUliĭ Anatol'evich.
(Vospitanie garmonieĭ)
Воспитание гармонией / Ю. А. Халфин. — Москва : Знание, 1978.
95 p. : ill. ; 16 cm. — (Народный университет : Педагогический факультет ; № 1, 1978) USSR 78
0.14rub
1. Amateur theatricals—Russia—History and criticism. 2. Children's plays, Russian—History and criticism. 3. Moral education (Elementary) I. Title. II. Series: Narodnyĭ universitet : Pedagogicheskiĭ fakul'tet ; 1978, no. 1.
AS261.N24 1978, no. 1 79–410427
[PN3167.R8]

—PRESENTATION, ETC.
—BIBLIOGRAPHY

(Sovetskiĭ teatr dli͡a deteĭ, 1918–1972)
Советский театр для детей, 1918–1972 : Указ. лит. /Сост. Э. И. Зерницкая, В. С. Мещерякова, Н. Д. Самойлова, Н. В. Шашкова]. — Москва : ГБЛ, 1978.
3 v. ; 21 cm. USSR 78
At head of title: Ministerstvo kul'tury SSSR. Gosudarstvenni͡a biblioteka SSSR imeni V. I. Lenina. Gosudarstvenni͡a t͡sentral'nai͡a teatral'nai͡a biblioteka.
Includes indexes.
1.98rub
1. Theater—Russia—Bibliography. 2. Children's plays, Russian—Presentation, etc.—Bibliography. I. Zernit͡skai͡a, Ė. I. II. Moscow. Publichnai͡a biblioteka. III. Gosudarstvennai͡a t͡sentral'nai͡a teatral'nai͡a biblioteka.
Z2504.D7S68 79–409090

CHILDREN'S PLAYS, SPANISH

Efak, Guillem F H d'.
El dimoni Cucarell / Guillem d'Efak. — Barcelona : Edicions Don Bosco, c1977.
55 p. : ill. ; 20 cm. — (Col·lecció Teatre edebé per a nois i noies ; 5) Sp***
ISBN 8423613631
I. Title. II. Series.
PZ90.C3E37 78-401125
79 MARC

CHILDREN'S POETRY

see also Children's songs; Counting out rhymes; Lullabies; Nursery rhymes; Poetry and children

Always begin where you are : themes in poetry and song / edited by Walter Lamb. — New York : McGraw-Hill, c1979.
xi, 276 p. : ill. ; 24 cm.
Includes index.
ISBN 0-07-035921-0
1. Children's poetry. I. Lamb, Walter, 1950-
PN6109.97.A45 808.81 78-32108
79 MARC

Boynton, Sandra.
Hippos go berserk / written and illustrated by Sandra Boynton. — Boston : Little, Brown, [1979] c1977.
[31] p. : col. ill. ; 15 x 19 cm.
SUMMARY: Larger and larger groups of hippos join a lone hippopotamus for a night-time party.
ISBN 0-316-10488-4
[1. Hippopotamus—Fiction. 2. Counting. 3. Stories in rhyme] I. Title.
PZ8.3.B7 Hi 1979 [E] 79-16134
79 MARC

Cole, William, 1919- ed.
The birds and the beasts were there : animal poems / selected by William Cole ; woodcuts by Helen Siegl. — Cleveland : Collins + World, [1977], c1963.
320 p. : ill. ; 24 cm.
Reprint of the ed. published by World Pub. Co., Cleveland.
Includes indexes.
SUMMARY: Approximately 300 poems, old and new, about real and impossible animals of all kinds.
ISBN 0-529-03742-4 (lib. bdg.) : $6.91
1. Animals—Juvenile poetry. 2. Children's poetry. [1. Animals—Poetry. 2. Poetry—Collections] I. Siegl, Helen. II. Title.
[PN6110.A7C65 1977] 808.81'9'36 77-3586
77 MARC

CHILDREN'S POETRY (Continued)

Cole, William, 1919- ed.
Poems of magic and spells / edited by William Cole ; illustrated by Peggy Bacon. — Cleveland : Collins + World, ⌈1977⌉ c1960.
224 p. : ill. ; 25 cm.
Includes index.
ISBN 0-529-03587-1 (lib. bdg.) : $6.91
1. Children's poetry. I. Bacon, Peggy, 1895- . II. Title.
⌈PN6110.C4C63 1977⌉ 821'.008'0352 77-2938
77 MARC

Fyleman, Rose, 1887-1957, ed.
Here we come a'piping / edited by Rose Fyleman. — Great Neck, N.Y. : Granger Book Co., 1978.
83 p. : ill. ; 20 cm. — (Granger poetry library)
Reprint of the 1937 ed. published by F. A. Stokes, New York.
SUMMARY: A collection of poetry by various authors grouped under such categories as "Morning and Night," "Country Things," and "Wind and Weather."
ISBN 0-89609-085-X
1. Children's poetry. ⌈1. Poetry—Collections⌉ I. Title.
PN6109.97.F9 1978 808.81 77-94813
00 MARC

Friskey, Margaret, 1901-
Indian Two Feet and the ABC moose hunt / by Margaret Friskey ; illustrated by John Hawkinson. — Chicago : Childrens Press, c1977.
32 p. : col. ill. ; 25 cm.
SUMMARY: A young Indian goes hunting for his first moose.
ISBN 0-516-03500-2
⌈1. Indians of North America—Fiction. 2. Stories in rhyme. 3. Alphabet⌉ I. Hawkinson, John, 1912- . II. Title.
PZ8.3.F9179 In ⌈E⌉ 77-4467
77 MARC

Gaige, Grace, 1881- comp.
Recitations : old and new, for boys and girls / by Grace Gaige ; introd. by Christopher Morley. — Great Neck, NY : Granger Book Co., 1979.
xxiii, 466 p. ; 20 cm. — (Granger poetry library)
Reprint of the 1924 ed. published by Appleton, New York.
Includes indexes.
SUMMARY: A collection of poems on such topics as animals, birds, patriotism, and selected holidays that are suitable for recitation.
ISBN 0-89609-112-0
1. Recitations. 2. Children's poetry. ⌈1. Recitations. 2. Poetry—Collections⌉ I. Title. II. Series.
PN4271.G16 1979 821'.008 78-73486
79 MARC

A Galaxy of verse / selected and with a commentary by Louis Untermeyer ; with editorial assistance by Bryna Ivens Untermeyer. — New York : M. Evans ; Philadelphia, c1978.
224 p. : ill. ; 23 cm.
Includes indexes.
SUMMARY: Story poems, animal poems, love poems, nonsense poems, — limericksfrom such great poets as Keats, Browning, Dickinson, Yeats, Frost, Eliot, and the Bible.
ISBN 0-87131-258-1 : $6.95
1. Children's poetry. ⌈1. Poetry—Collections⌉ I. Untermeyer, Louis, 1885-1977.
PN6110.C4G25 808.81 78-1255
78 MARC

Gernhardt, Almut, 1940-
One more makes four / pictures by Almut Gernhardt ; with verses by Robert Gernhardt ; translated from the German by Elizabeth Watson Taylor. — London : Cape, 1978.
⌈32⌉ p. : col. ill. ; 23 cm. GB78-18677
Translation of Mit der sind wir vier.
Poems.
ISBN 0-224-01577-X : £2.95
1. Humorous poetry, German. I. Gernhardt, Robert, 1937- . II. Title.
PZ8.3.G322 On 1978 831'.9'14 79-301024
79 MARC

Ghost poems / edited by Daisy Wallace ; illustrated by Tomie de Paola. — New York : Holiday House, c1979.
30 p. : ill. ; 24 cm.
SUMMARY: Presents 17 poems about ghosts.
ISBN 0-8234-0344-0
1. Ghosts—Juvenile poetry. 2. Children's poetry. ⌈1. Ghosts—Poetry. 2. American poetry—Collections. 3. English poetry—Collections⌉ I. Wallace, Daisy. II. De Paola, Thomas Anthony.
PN6109.97.G5 811'.008'0375 78-11028
78 MARC

Go with the poem : a new collection / chosen by Lilian Moore. — New York : McGraw-Hill, ⌈1979⌉
125 p. ; 24 cm.
Includes index.
SUMMARY: A collection of some 90 poems written by such outstanding 20th-century poets as William Carlos Williams, Robert Frost, and Lillian Morrison.
ISBN 0-07-042880-8 : $7.95
1. Children's poetry. ⌈1. Poetry—Collections⌉ I. Moore, Lilian.
PN6109.97.G6 811'.5'208 78-8393
78 MARC

Graham, Carolyn.
Jazz chants for children : rhythms of American English through chants, songs and poems / Carolyn Graham : ill. by Anna Veltfort. — New York : Oxford University Press, 1979.
xi, 84 p. : ill. ; 28 cm.
SUMMARY: A collection of chants, songs, and poems designed to teach the rhythms of spoken American English primarily to students of English as a second language.
ISBN 0-19-502496-6
1. English language—Text-books for foreigners. 2. English language—Spoken English—Juvenile literature. 3. Children's songs—Texts. 4. Children's poetry. ⌈1. English language—Textbooks for foreigners⌉ I. Veltfort, Anna. II. Title.
PE1128.G65 784'.3 78-11858
78 MARC

Hoban, Tana.
One little kitten / by Tana Hoban. — 1st ed. — New York : Greenwillow Books, c1979.
⌈24⌉ p. : chiefly ill. ; 26 cm.
SUMMARY: Brief rhyming text and photographs follow a kitten as it explores its surroundings.
ISBN 0-688-80222-2 : $6.95
⌈1. Cats—Fiction. 2. Stories in rhyme⌉ I. Title.
PZ8.3.H6517 On ⌈E⌉ 78-31862
79 MARC

Illustrated poems for children : a special collection / illustrated by Keystyna Stasiak ; with an introd. by Miriam Peterson. — New York : Rand McNally, ⌈1977⌉ c1973.
160 p. : ill. (some col.) ; 31 cm.
Includes indexes.
SUMMARY: 159 poems include well-known selections by Tennyson, Stevenson, Carroll, Housman, Wordsworth, Dickinson, Frost, Lear, and many other famous poets.
ISBN 0-528-88584-7 : $7.95
1. Children's poetry. ⌈1. Poetry—Collections⌉ I. Stasiak, Krystyna.
PN6110.C4I4 1977 821'.008 77-22105
77 MARC

Livermore, Elaine.
Three little kittens lost their mittens / written and illustrated by Elaine Livermore. — Boston : Houghton Mifflin, 1979.
30 p. : ill. ; 23 cm.
SUMMARY: Three little kittens lose their mittens while playing and the reader must help find them.
ISBN 0-395-28379-5
⌈1. Stories in rhyme. 2. Cats—Fiction⌉ I. Title.
PZ8.3.L747 Th ⌈E⌉ 79-12709
79 MARC

Lovejoy, Mary Isabella, 1894- comp.
Nature in verse : a poetry reader for children / compiled by Mary I. Lovejoy. — Great Neck, NY : Granger Book Co., 1979.
xiv, 305 p. : ill. ; 20 cm. — (Granger poetry library)
Reprint of the 1895 ed. published by Silver, Burdett, Boston.
SUMMARY: A collection of poetry celebrating the beauty of nature throughout the year.
ISBN 0-89609-116-3
1. Nature—Juvenile poetry. 2. Children's poetry. ⌈1. Nature—Poetry. 2. Poetry—Collections⌉ I. Title. II. Series.
PN6110.N2L6 1979 821'.008'036 78-73490
00 MARC

The Man in the Moon as he sails the sky and other moon verse / collected and illustrated by Ann Schweninger. — New York : Dodd, Mead, c1979.
48 p. : col. ill. ; 20 cm.
SUMMARY: Poems about the moon from such sources as Mother Goose, Vachel Lindsay, Edward Lear, William Wordsworth, and R. L. Stevenson.
ISBN 0-396-07741-2
1. Moon—Juvenile poetry. 2. Children poetry. ⌈1. Moon—Poetry. 2. Poetry—Collections⌉ I. Schweninger, Ann.
PN6110.M6M3 821'.008'032 79-52051
79 MARC

McClintock, Marshall, 1906-
Stop that ball! By Mike McClintock. Illustrated by Fritz Siebel. New York, Beginner Books; distributed by Random House, c1959.
62 p. illus. 24 cm. (⌈Beginner books⌉ B-10)
SUMMARY: A young boy chases his ball all over town.
⌈1. Stories in rhyme. 2. Balls (Sporting goods)—Fiction⌉ I. Siebel, Fritz. II. Title.
PZ8.3.M1325 ST ⌈E⌉ 59-9741
⌈r78⌉rev2 MARC

Pavey, Peter.
One dragon's dream / by Peter Pavey. — 1st American ed. — Scarsdale, N.Y. : Bradbury Press, 1979, c1978.
⌈32⌉ p. : col. ill. ; 21 x 29 cm.
SUMMARY: The reader can count items from 1 to 10 included in dragon's dream.
ISBN 0-87888-148-4 : $7.95
⌈1. Counting. 2. Dreams—Fiction. 3. Dragons—Fiction. 4. Stories in rhyme⌉ I. Title.
PZ8.3.P2737 On 1979 ⌈E⌉ 79-12008
79 MARC

Poèmes de partout et de toujours / recueillis par Paulette Lequeux. — Paris : A Colin, c1978.
159, ⌈1⌉ p. ; 20 cm. F***
Bibliography: p. 157-⌈160⌉
29.50F
1. Children's poetry. I. Lequeux-Gromaire, Paulette.
PN6109.97.P63 808.81 79-366607
79 MARC

Scarpa, Roque Esteban, 1914-
El árbol deshojado de sonrisas : poemas / por Roque Esteban Scarpa. — Santiago, Chile : Editorial Nascimento, 1977.
242 p. : ill. ; 27 cm.
I. Title.
PQ8097.S4A9 861 78-109823
79 MARC

De Vette hertog : en andere volksverhalen van de Veluwe / (op) getekend door D. J. W. Teding van Berkhout. — Zutphen : Walburg Pers, 1978.
88 p. : with ill. ; 22 cm. Ne
Bibliography: p. 88.
ISBN 9060113047 : fl 15.00
1. Folk-lore—Netherlands—Veluwe—Juvenile poetry. I. Teding van Berkhout, D. J. W.
PZ14.3.V4 79-337718
*79 MARC

—BIBLIOGRAPHY

Haviland, Virginia, 1911-
Children & poetry : a selective, annotated bibliography / compiled by Virginia Haviland, Head, Children's Literature Center, and William Jay Smith, Consultant in Poetry, 1968-70. — 2d ed., rev. — Washington : Library of Congress : for sale by the Supt. of Docs., U.S. Govt. Print. Off., 1979.
xii, 84 p. : ill. ; 19 x 22 cm.
Includes index.
ISBN 0-8444-0267-2
1. Childrens' poetry—Bibliography. I. Smith, William Jay, 1918- joint author. II. United States. Library of Congress. Children's Literature Center. III. Title.
Z1037.H36 1979 028.52 78-57071
⌈PN1009.A1⌉ MARC
——— Copy 3. Z663.292.H34 1978
78

CHILDREN'S POETRY, AFRIKAANS

Klein verseboek / saamgestel deur D. J. Opperman. — 7. druk. — Kaapstad : Tafelberg, 1976.
184 p. : ill. ; 22 cm. SA***
Sequel to Kleuter-verseboek.
Bibliography: p. 181-184.
ISBN 0-624-00828-2
1. Children's poetry, Afrikaans. I. Opperman, Diederik Johannes.
PZ14.3.K5 1976 79-382595
79 MARC

CHILDREN'S POETRY, AMERICAN

Barnes, Walter, 1880-
The children's poets : analyses and appraisals of the greatest English and American poets for children : for use in normal schools, library schools, and homes / by Walter Barnes. — Folcroft, Pa. : Folcroft Library Editions, 1978 ⌈c1924⌉
vi, 264 p. ; 23 cm.
Reprint of the 1925 ed. published by World Book Co., Yonkers-on-Hudson, N.Y.
Bibliography: p. 241-258.
Includes index.
ISBN 0-8414-1736-9 lib. bdg. : $30.00
1. Children's poetry, English—History and criticism. 2. Children's poetry, American—History and criticism. 3. Children's poetry, English. 4. Children's poetry, American. I. Title.
PR508.C6B3 1978 821'.009 78-13147
78 MARC

A Child's treasury of verse / compiled by Eleanor Doan ; illustrated by Nancy Munger. — Grand Rapids : Zondervan Pub. House, c1977.
207 p. : ill. ; 26 cm.
Includes indexes.
SUMMARY: Selection topics include beauty, bits of wisdom, friends, growing up, fun and nonsense, and talking to God.
ISBN 0-310-23800-5
1. Children's poetry, American. ⌈1. American poetry—Collections⌉ I. Doan, Eleanor Lloyd, 1914- . II. Munger, Nancy.
PN6110.C4C538 811'.008'0352 77-2142
77⌈r79⌉rev MARC

Cock Robin.
Cock Robin's death and funeral. Boston ⌈T. and J. Fleet, ca. 1780⌉
32 p. illus. 96 mm.
Bristol B5059.
"The catalogue and characters of the letters of the alphabet ...": p. 11-24.
1. Children's poetry, English. 2. Children's poetry, American. 3. Bibliography—Microscopic and miniature editions—Specimens. I. Title.
PZ6.C644 Fl 78-321612
79 MARC

CHILDREN'S POETRY, AMERICAN (Continued)

Dinosaurs and beasts of yore : poems / verses selected by William Cole ; illustrated by Susanna Natti. — Cleveland : Collins, 1979.
62 p. : ill. ; 22 cm.
SUMMARY: A collection of humorous poetry about extinct animals.
ISBN 0-529-05511-2 : $8.95
1. Extinct animals—Juvenile poetry. 2. Dinosauria—Juvenile poetry. 3. Children's poetry, American. 4. Children's poetry, English. 5. Children's poetry, Canadian. ⌈1. Dinosaurs—Poetry. 2. Extinct animals—Poetry. 3. Humorous poetry. 4. Poetry—Collections⌉ I. Cole, William, 1919- . II. Natti, Susanna.
PS595.E93D5 821'.9'1408 78-31619
79 MARC

Frank, Josette, 1893- comp.
Poems to read to the very young / selected by Josette Frank ; illustrated by Dagmar Wilson. — New York : Random House, c1977.
⌈32⌉ p. : col. ill. ; 21 cm. — (The Best book club ever) (A Randon House pictureback)
ISBN 0-394-83705-3
1. Children's poetry, American. 2. Children's poetry, English. I. Wilson, Dagmar. II. Title.
PS593.C45F7 1977 821'.008 77-82083
79 MARC

Go to bed! : A book of bedtime poems / selected by Lee Bennett Hopkins ; illustrated by Rosekrans Hoffman. — New York : Knopf ; distributed by Random House, c1979.
27 p. : ill. ; 17 x 21 cm.
SUMMARY: An anthology of poems depicting the pleasant and sometimes unpleasant aspects of bedtime.
ISBN 0-394-83869-6. ISBN 0-394-93869-0 lib. bdg.
1. Night—Juvenile poetry. 2. Sleep—Juvenile poetry. 3. Children's poetry, American. ⌈1. Night—Poetry. 2. Sleep—Poetry⌉ I. Hopkins, Lee Bennett. II. Hoffman, Rosekrans.
PS595.N54G6 811'.008 78-3575
78 MARC

I sing the song of myself : an anthology of autobiographical poems / edited by David Kherdian. — 1st ed. — New York : Greenwillow Books, c1978.
xiii, 146 p. ; 24 cm.
Includes indexes.
SUMMARY: A collection of autobiographical poems by modern American poets.
ISBN 0-688-80172-2 : $7.95. ISBN 0-688-84172-4 lib. bdg. : $6.43
1. American poetry—20th century. 2. Children's poetry, American. ⌈1. American poetry—Collections⌉ I. Kherdian, David.
PS613.I24 811'.5'408 78-5807
78 MARC

Juvenile poems; or, The alphabet in verse. Designed for the entertainment of all good little boys and girls. Ornamented with engravings. Albany, Printed by E. and E. Hosford, 1810.
31 p. illus. 11 cm.
Pagination includes front cover and recto of back cover, with both covers illustrated.
"The story of a Moor and a Spaniard" (in prose): p. 29-31.
1. Alphabet rhymes. 2. Children's poetry, American. I. The Story of a Moor and a Spaniard. 1810.
PZ6.J993 78-325058
79 MARC

Juvenile poems. Wendell, Mass., Printed and sold by J. Metcalf, 1828.
16 p. illus. 10 cm.
Illustrated covers.
L.C. copy imperfect: p. 13-14 mutilated.
CONTENTS: Pork and beef.—What is veal.—What is mutton.—The value of order.—Bricks.—To a little girl that liked to look in the glass.—The Bee.—The Beggar.—The Nest.—Days, hours, and minutes.—Mamma and the baby.
1. Children's poetry, American.
PZ6.J99 21-3861
⌈r79⌉rev MARC

Juvenile poems : to be read and learned. — Providence ⌈R.I.⌉ : G. P. Daniels, 1839.
16 p. : ill. ; 11 cm.
In L.C. copy, some of the illustrations have been hand-colored.
"Illustrations also on inside wrappers."—NUC pre-56. L.C. copy lacks the wrappers.
CONTENTS: Coasting.—A Wicked thing.—My little sister.—Anna and her chicken.—My kite.—Stop! Stop!—The Little beggar girl.—Lucy and her lamb.—Cruel kitty.—The Alphabet.—The Boy and the bird.—The Little bird's answer.
1. Children's poetry, American.
PZ6.J994 811'.3'08 78-325057
79 MARC

Littledale, Freya, 1929-
I was thinking : poems / by Freya Littledale ; pictures by Leonard Kessler. — 1st ed. — New York : Greenwillow Books, c1979.
46 p. : col. ill. ; 22 cm. — (Greenwillow read-alone)
SUMMARY: Brief poems describing the world from a child's point of view.
ISBN 0-688-80217-6 : $5.95
1. Children's poetry, American. ⌈1. American poetry⌉ I. Kessler, Leonard P., 1920- . II. Title.
PS3562.784 811'.5'4 79-4062
79 MARC

The Moon's the North Wind's cooky : night poems / selected and illustrated by Susan Russo. — 1st ed. — New York : Lothrop, Lee & Shepard, c1979.
36 p. : ill. ; 18 cm.
SUMMARY: A collection of 14 images of night by such poets as Nikki Giovanni, Louis Untermeyer, and Felice Holman.
ISBN 0-688-41879-1. ISBN 0-688-51879-6 lib. bdg.
1. Night—Juvenile poetry. 2. Children's poetry, American. 3. Children's poetry, English. ⌈1. Night—Poetry. 2. American poetry—Collections⌉ I. Russo, Susan.
PS595.N54M66 811'.008 78-32178
79 MARC

Oh, such foolishness! : Poems / selected by William Cole ; pictures by Tomie de Paola. — Philadelphia : Lippincott, c1978.
96 p. : ill. ; 24 cm.
Includes indexes.
SUMMARY: A collection of humorous and nonsense poems.
ISBN 0-397-31807-3
1. Humorous poetry, American. 2. Humorous poetry, English. 3. Children's poetry, American. 4. Children's poetry, English. 5. Nonsense-verses. ⌈1. Humorous poetry. 2. Poetry—Collections. 3. Nonsense verses⌉ I. Cole, William, 1919- . II. De Paola, Thomas Anthony.
PS595.H8O36 811'.08 78-1622
78 MARC

Rosenbloom, Joseph.
Silly verse (and even worse) / Joseph Rosenbloom ; ill. by Joyce Behr. — New York : Sterling Pub. Co., c1979.
128 p. : ill. ; 22 cm.
Includes index.
SUMMARY: Contains more than 200 short, humorous poems from anonymous sources.
ISBN 0-8069-4600-8 : $4.95
1. Children's poetry, American. 2. Humorous poetry, American. ⌈1. Humorous poetry. 2. American poetry—Collections⌉ I. Behr, Joyce. II. Title.
PS591.C45R6 811'.07 78-66322
79 MARC

Sergeant, Howard, 1914- comp.
The Two Continents book of children's verse / edited by Howard Sergeant. — New York : Two Continents Pub. Group, ⌈1979?⌉
286 p. : ill. ; 23 cm.
Published in 1972 under title: Evans book of children's verse.
Includes indexes.
SUMMARY: A collection containing humorous poems by English and American poets.
ISBN 0-8467-0238-X : $7.95
1. Children's poetry, English. 2. Children's poetry, American. ⌈1. English poetry—Collections. 2. American poetry—Collections⌉ I. Title. II. Title: Book of children's verse.
PN6110.C4S47 1979 821'.008 76-47173
76 MARC

Starbird, Kaye.
The covered bridge house and other poems / by Kaye Starbird ; illustrated by Jim Arnosky. — New York : Four Winds Press, c1979.
53 p. : ill. ; 24 cm.
SUMMARY: A collection of 30 poems about a variety of childhood experiences.
ISBN 0-590-07544-6 : $6.95
1. Children's poetry, American. ⌈1. American poetry⌉ I. Arnosky, Jim. II. Title.
PS3537.T2396C6 811'.5'4 79-11418
79 MARC

CHILDREN'S POETRY, ARGENTINE

Galimberti de Guacci, Haydée.
Canto al amanecer : cantarina / Haydée Galimberti de Guacci. — ⌈City Bell, República Argentina⌉ : Ediciones Baires, 1975.
89 p. : ill. ; 23 cm.
1. Children's poetry, Argentine. I. Title.
PZ74.3.G28 861 78-109864
79 MARC

CHILDREN'S POETRY, AUSTRALIAN

Turner, Ian.
Cinderella dressed in yella / compiled and edited by Ian Turner, June Factor, Wendy Lowenstein ; with an interpretative essay by Ian Turner. — 2d ed. — Richmond ⌈Australia⌉ : Heinemann Educational Australia, 1978.
ix, 173 p. : ill. ; 26 cm. Aus***
Bibliography: p. 166.
Includes index.
ISBN 0-85859-155-3
1. Counting-out rhymes—Australia. 2. Children's poetry, Australian. 3. Folk-lore and children—Australia. I. Factor, June. II. Lowenstein, Wendy J. III. Title.
GR485.T87 1978 398.8 79-307869
79 MARC

CHILDREN'S POETRY, BENGALI

Māhamuda Haka, 1948–
⌈Chu mantara chu⌉
মাহমুদ হকের ছু মন্তর ছু. — সিলেট : ছড়া পরিষদ, 1976.
⌈24⌉ p. : ill. ; 13 x 19 cm.
In Bengali.
Rhymes.
On verso of t.p.: Prakāśaka Śāmasula Karīma Kayesa.
Tk5.00
I. Title: Chu mantara chu.
Title romanized: Māhamuda Hakera Chu mantara chu.
PZ90.B4M27 1976 77–903334

—BANGLADESH

(Bāṃlādeśera charā)
বাংলাদেশের ছড়া / মহফিল হক ও নূরুল ইসলাম সম্পাদিত. — ⌈Rangpur : রংপুর কবিসভা'র ; ঢাকা : পরিবেশক নওরোজ কিতাবিস্তান, 1977⌉
⌈64⌉ p. ; 22 cm.
In Bengali.
Tk8.00
1. Children's poetry, Bengali—Bangladesh. I. Haque, Mahfil. II. Isalāma, Nurula.
PZ90.B4B27 78–911578

CHILDREN'S POETRY, BURMESE

Nụ Yin, 1916–
(Hnīn ngon wei)
နှင်းငုံဝေ, ကလေးကဗျာများ / နုယဉ်. — ရန်ကုန် : ရွှေပန်းချီ စာပေ, 1974.
156 p. : ill. ; 17 cm.
In Burmese.
Poems.
K3.00
I. Title.
PZ90.B8N8 77–986326

Tin Mō.
(Ka lē kabya pān uyyin)
ကလေးကဗျာပန်းဥယျာဉ် / တင်မိုး ; ပန်းချီ, ပေါ်ဦးသက်. — 1. အကြိမ်. — ရန်ကုန် : မုတ်သုန်စာပေ : ထုတ်ဝေသူ သန်းဌေးစာပေ, ⌈pref. 1976⌉
152 p. : ill. ; 26 cm.
In Burmese.
Poems.
K12.00
I. Paw Oo Thet. II. Title.
PZ90.B8T53 78–920001

CHILDREN'S POETRY, CANADIAN

Dinosaurs and beasts of yore : poems / verses selected by William Cole ; illustrated by Susanna Natti. — Cleveland : Collins, 1979.
62 p. : ill. ; 22 cm.
SUMMARY: A collection of humorous poetry about extinct animals.
ISBN 0-529-05511-2 : $8.95
1. Extinct animals—Juvenile poetry. 2. Dinosauria—Juvenile poetry. 3. Children's poetry, American. 4. Children's poetry, English. 5. Children's poetry, Canadian. ⌈1. Dinosaurs—Poetry. 2. Extinct animals—Poetry. 3. Humorous poetry. 4. Poetry—Collections⌉ I. Cole, William, 1919- . II. Natti, Susanna.
PS595.E93D5 821'.9'1408 78-31619
79 MARC

Downie, Mary Alice, comp.
The wind has wings : poems from Canada / compiled by Mary Alice Downie & Barbara Robertson ; illustrated by Elizabeth Cleaver. — Toronto : Oxford University Press, 1978, c1968.
95 p. : col. ill. ; 25 cm. C78-6884-X
Includes index.
ISBN 0-19-540287-1 : $5.95
1. Canadian poetry. 2. Children's poetry, Canadian. I. Robertson, Barbara. II. Cleaver, Elizabeth, 1939- . III. Title.
PR9195.25.D6 1978 811'.008 79-308081
79 MARC

CHILDREN'S POETRY, CHINESE

Kitamura, Yoriharu.
(Okera no negai)
おけらのねがい：[北村順治少年詩集：きたむら・よりはる第二子ども詩集] / きたむら・よりはる・さく. — 横浜：北村, 昭和49 [1974]
27 p. ; 13 x 19 cm. Ja***
Poems.

I. Title.
PZ49.43.K55 78–814123

Lu, Ssu, fl. 1975-
(San chih chiao ao ti hsiao mao)
三隻驕傲的小貓 / 魯思編. — 香港：上海書局, 1975.
20 p. : ill. ; 19 cm.
Juvenile literature.
$0.80

I. Title.
PZ10.842.L85S2 78–841873

(Ta chia lai tuan lien)
大家来锻炼 = Da jia lai duan lian / [本社编；龚韵文, 吴克柔画]. — 上海：上海人民出版社, 1974.
[24] p. : col. ill. ; 15 cm.
Cover title.
Juvenile literature.
$0.10

1. Children's poetry, Chinese. I. K'ung, Yün-wen. II. Wu, K'o-jou. III. Shang-hai jen min ch'u pan she.
PZ10.843.T29 78–842753

(Ta hung hua)
大红花：革命儿歌. — [香港?：s. n., 1978?]
3, 70 p. : ill. ; 20 cm.
Reprint of the 1975 ed. published by Ho-pei jen min ch'u pan she, Shih-chia-chuang.

1. Children's poetry, Chinese.
PZ10.843.T35 1978 78–842455

Wang, Jung-tzu.
(T'ung hua ch'eng)
童話城 / 王蓉子文；趙國宗圖. — [南投縣]：臺灣省政府教育廳；臺北：發行者臺灣書店, 民國56 [1967]
60 p. : ill. (some col.) ; 21 cm. — (中華兒童叢書)
Text accompanied with Kuo yü chu yin.
Series romanized: Chung-hua erh t'ung ts'ung shu.

1. Children's poetry, Chinese. I. Chao, Kuo-tsung. II. Title.
PZ10.843.W33T8 78–841709

Wang, Shu-t'ien.
(Ts'ao yüan hsin ko)
草原新歌 / 王树田. — [香港?：s. n., 1978?]
3, 117 p., [4] leaves of plates : ill. ; 20 cm.
Children's poetry.
Reprint of the 1977 ed. published by Jen min wen hsüeh ch'u pan she, Peking.

I. Title.
PZ10.843.W37T75 1978 78–842638

Yeh, Hsin.
(Kao kao ti Miao ling)
高高的苗岭 / 叶辛. — 第1版. — 上海：上海人民出版社, 1977.
131 p. : ill. ; 19 cm. — (少年儿童文艺读物)
Fiction.
Series romanized: Shao nien erh t'ung wen i tu wu.
$0.32

I. Title.
PZ10.831.Y365K3 78–842338

Yü, Shen-ssu.
(Hsia yü t'ien)
下雨天 / [于]慎思文；周春江圖. — [南投縣]：臺灣省政府教育廳；臺北：發行者臺灣書店, 民國56 [1967]
36 p. : col. ill. ; 21 cm. — (中華兒童叢書)
Text accompanied with Kuo yü chu yin.
Series romanized: Chung-hua erh t'ung ts'ung shu.

1. Children's poetry, Chinese. I. Chou, Ch'un-chiang. II. Title.
PZ10.843.Y8H7 78–841708

—CHINA—NANKING

Wang, Su-ts'un.
(Nan-ching hua)
南京話 / 王素存編纂. — [台北?：s. n., 1977?]
2 v. ; 27 cm.
Caption title.
Reproduced from ms. copy on double leaves.

1. Proverbs, Chinese—China—Nanking. 2. Children's poetry, Chinese—China—Nanking. 3. Chinese language—Dialects—China—Nanking. I. Title.
PN6519.C5W36 78–840812

CHILDREN'S POETRY, CZECH

—HISTORY AND CRITICISM—CONGRESSES

Poezie pro děti, tradice a současnost : referáty a diskusní příspěvky z pracovního semináře Společnosti přátel knihy pro mládež, Praha. 17. května 1976. — [s.l. : s.n., 1976?]
39 p. ; 21 cm. Cz***
Summary in German and Russian.
1. Children's poetry, Czech—History and criticism—Congresses. I. Společnost přátel knihy pro mládež.
PN1009.S5P57 79-370832
79 MARC

CHILDREN'S POETRY, DANISH

Ditlevsen, Tove Irma Margit, 1918-1976.
Til en lille pige : efterladte digte / [af] Tove Ditlevsen. — [København] : Gyldendal, 1978.
75 p. ; 21 cm. D78-45
ISBN 8701738224 : kr69.00. ISBN 8701738216 pbk.
I. Title.
PZ54.3.D5 1978 79-347046
79 MARC

Sigsgaard, Jens.
Da oldefar Mads var ulydig / [af] Jens Sigsgaard ; illustreret af Oskar Pletsch. — København : Gad, 1978.
31 p. : ill. ; 21 cm. D78-25
ISBN 8712806501 : kr14.20
I. Pletsch, Oskar. II. Title.
PZ53.S4D32 78-394517
79 MARC

Verroen, Dolf.
Maak dat je weg komt / Dolf Verroen ; met illustraties van The Tjong Khing. — Den Haag : Leopold, c1978.
61 p. : ill. ; 23 cm. Ne***
ISBN 9025846823
I. The, Tjong Khing. II. Title.
PZ14.3.V38 79-370176
79 MARC

CHILDREN'S POETRY, ENGLISH

Ahlberg, Janet.
Cops and robbers / [by] Janet & Allan Ahlberg. — London : Heinemann, 1978.
[32] p. : col. ill. ; 21 cm. GB78-30108
ISBN 0-434-92501-2 : £2.90
I. Ahlberg, Allan, joint author. II. Title.
PZ8.3.A278 Co 1978 78-325674
79 MARC

Ahlberg, Janet.
Cops and robbers / Janet & Allan Ahlberg. — 1st American ed. — New York : Greenwillow Books, c1978.
[32] p. : col. ill. ; 21 cm.
SUMMARY: A band of thieves plot to steal the children's toys before they are received on Christmas.
ISBN 0-688-80178-1 : $7.25. ISBN 0-688-84178-3 lib. bdg.
[1. Stories in rhyme. 2. Robbers and outlaws—Fiction. 3. Christmas stories] I. Ahlberg, Allan, joint author. II. Title.
PZ8.3.A278 Co [E] 78-5354
78 MARC

Ahlberg, Janet.
Each peach pear plum : an "I spy" story / Janet and Allan Ahlberg. — [1st American ed.] — New York : Viking Press, [1979] c1978.
[32] p. : col. ill. ; 20 x25 cm. — (An I-spy-book)
SUMMARY: Rhymed text and illustrations invite the reader to play "I Spy" with a variety of Mother Goose and other folklore characters.
ISBN 0-670-28705-9 : $8.95
[1. Stories in rhyme] I. Ahlberg, Allan, joint author. II. Title.
PZ8.3.A278 Eac 1979 [E] 78-16726
78 MARC

Allingham, William, 1824-1889.
Robin Redbreast, and other verses / by William Allingham ; pictures by Helen Allingham ... [et al.]. — Great Neck, N.Y. : Core Collection Books, 1979.
xi, 113 p. : ill. ; 20 cm.
Reprint of the 1930 ed. published by Macmillan, New York, in series: The Little library.
SUMMARY: A collection of poems grouped under such headings as "Playing Alone," "Playmates and Pets," and "Fairies and Elves."
ISBN 0-8486-0007-X
1. Children's poetry, English. [1. English poetry] I. Allingham, Helen Paterson, 1848-1926. II. Title.
PR4004.A5A77 1979 821'.8 78-74507
79 MARC

Asch, Frank.
Country pie / Frank Asch. — 1st ed. — New York : Greenwillow Books, c1979.
32 p. : ill. ; 24 cm.
SUMMARY: Fourteen poems describe various aspects of country life.
ISBN 0-688-80188-9 : $6.95. ISBN 0-688-84188-0 lib. bdg. : $6.67
1. Country life—Juvenile poetry. [1. Country life—Poetry. 2. American poetry] I. Title.
PZ8.3.A7 Co 811'.5'4 78-14837
78 MARC

Ashford, Ann.
If I found a wistful unicorn / by Ann Ashford ; illustrated by Bill Drath. — 1st ed. — Atlanta : Peachtree Publishers, c1978.
[39] p. : col. ill. ; 28 cm.
SUMMARY: If his friend answers yes to any of his questions, then a boy will know his friend loves him.
ISBN 0-931948-00-2
[1. Friendship—Fiction. 2. Stories in rhyme] I. Drath, Bill. II. Title.
PZ8.3.A76 If [E] 78-59094
79 MARC

Atwood, Margaret Eleanor, 1939-
Up in the tree / Margaret Atwood. — Toronto : McClelland and Stewart, [1978]
[30] p. : chiefly col. ill. ; 26 cm. C***
ISBN 0-7710-0824-4 : $6.95
I. Title.
PZ8.3.A923 Up 78-319301
78 MARC

Barnes, Walter, 1880-
The children's poets : analyses and appraisals of the greatest English and American poets for children : for use in normal schools, library schools, and homes / by Walter Barnes. — Folcroft, Pa. : Folcroft Library Editions, 1978 [c1924]
vi, 264 p. ; 23 cm.
Reprint of the 1925 ed. published by World Book Co., Yonkers-on-Hudson, N.Y.
Bibliography: p. 241-258.
Includes index.
ISBN 0-8414-1736-9 lib. bdg. : $30.00
1. Children's poetry, English—History and criticism. 2. Children's poetry, American—History and criticism. 3. Children's poetry, English. 4. Children's poetry, American. I. Title.
PR508.C6B3 1978 821'.009 78-13147
78 MARC

The Batsford book of light verse for children / edited by Gavin Ewart ; Nicolas Bentley drew the pictures. — London : Batsford, 1978.
160 p. : ill. ; 23 cm. GB78-33985
ISBN 0-7134-0916-9 : £3.50
1. English poetry. 2. Humorous poetry, English. 3. Children's poetry, English. I. Ewart, Gavin.
PR1175.B273 821'.08 79-307917
79 MARC

Bemelmans, Ludwig, 1898-1962.
Madeline's rescue; story and pictures by Ludwig Bemelmans. New York, Viking Press, 1953.
56 p. illus. 31 cm.
SUMMARY: A hound rescues a schoolgirl from the Seine, becomes a beloved school pet, is chased away by the trustees, and returns with a surprise.
[1. Dogs—Fiction. 2. Paris—Fiction. 3. Stories in rhyme] I. Title.
PZ8.3.B425 Mah 3 [E] 53-8709
[r78]rev2 MARC

Chase, Catherine.
See the fly fly / by Catherine Chase ; illustrated by Bari Weissman. — New York : Dandelion Press, 1979.
[32] p. : col. ill. ; 23 cm. — (Dandelion first reader)
SUMMARY: A pesky fly will not disappear.
ISBN 0-89799-133-8
[1. Flies—Fiction. 2. Stories in rhyme] I. Weissman, Bari. II. Title.
PZ8.3.C387 Se [E] 78-72104
79 MARC

CHILDREN'S POETRY, ENGLISH (Continued)

Chase, Edith Newlin.
Small window panes : verses for children / by Edith Newlin Chase ; pictures by Janet Hendry Brown. — Alstead, N.H. : Chase, c1978.
29 p. : ill. ; 25 cm.
SUMMARY: A collection of 20 poems with a variety of themes.
$4.00
[1. American poetry] I. Brown, Janet Hendry. II. Title.
PZ8.3.C388 Sm 811'.5'4 78-111719
79 MARC

Children in the wood.
The affecting history of the children in the wood. Hartford, Printed by J. Babcock, 1796.
30, [1] p. illus. 11 cm.
Evans 29955.
In prose.
"Story of three robbers": p. [23]-24; selections in verse: p. [25]-30.
Title page of L.C. copy follows p. 30; front. and p. [31] are mounted on inside of covers.
1. Children's stories, English. 2. Children's poetry, English. I. Title.
PZ6.C436 1796 78-319667
79 MARC

Clifton, Lucille, 1936-
Everett Anderson's nine month long / by Lucille Clifton ; ill. by Ann Grifalconi. — New York : Holt, Rinehart, and Winston, c1978.
[31] p. : ill. ; 21 x 22 cm.
SUMMARY: A small boy and his family anticipate the birth of their newest member.
ISBN 0-03-043536-6
[1. Babies—Fiction. 2. Brothers and sisters—Fiction. 3. Stories in rhyme]
I. Grifalconi, Ann. II. Title.
PZ8.3.C573 Evk [E] 78-4202
78 MARC

Cock Robin.
Cock Robin's death and funeral. Boston [T. and J. Fleet, ca. 1780]
32 p. illus. 96 mm.
Bristol B5059.
"The catalogue and characters of the letters of the alphabet ...": p. 11-24.
1. Children's poetry, English. 2. Children's poetry, American. 3. Bibliography—Microscopic and miniature editions—Specimens. I. Title.
PZ6.C644 Fl 78-321612
79 MARC

Corey, Dallas.
The Christmas legend of Monkey Joe / by Dallas Corey. — Hermitage, Tenn. : Monkey Joe Enterprises, c1978.
37 p. : col. ill. ; 21 x 28 cm.
SUMMARY: When a witch kidnaps Santa Claus, Monkey Joe takes over the Christmas Eve deliveries.
$2.95
[1. Stories in rhyme. 2. Santa Claus—Fiction. 3. Monkeys—Fiction]
I. Title.
PZ8.3.C818 Ch [E] 78-65591
79 MARC

Craft, Ruth.
Carrie Hepple's garden / Ruth Craft ; illustrated by Irene Haas. — 1st American ed. — New York : Atheneum, 1979.
[32] p. : col. ill. ; 24 cm.
"A Margaret K. McElderry book."
SUMMARY: Three children venture into eccentric Carrie Hepple's garden in search of their ball and learn some strange and unexpected things.
ISBN 0-689-50099-8 (lib. bdg.) $9.95
[1. Stories in rhyme] I. Haas, Irene. II. Title.
PZ8.3.C842 Car 1979 [E] 78-397
78 MARC

Crichton Smith, Iain.
River, river : poems for children / by Iain Crichton Smith. — Loanhead : Macdonald Publishers, 1978.
40 p. : ill. ; 22 cm. GB78-34798
ISBN 0-904265-19-6
I. Title.
PZ8.3.C87 Ri 821'.9'14 79-308124
79 MARC

Dinosaurs and beasts of yore : poems / verses selected by William Cole ; illustrated by Susanna Natti. — Cleveland : Collins, 1979.
62 p. : ill. ; 22 cm.
SUMMARY: A collection of humorous poetry about extinct animals.
ISBN 0-529-05511-2 : $8.95
1. Extinct animals—Juvenile poetry. 2. Dinosauria—Juvenile poetry. 3. Children's poetry, American. 4. Children's poetry, English. 5. Children's poetry, Canadian. [1. Dinosaurs—Poetry. 2. Extinct animals—Poetry. 3. Humorous poetry. 4. Poetry—Collections] I. Cole, William, 1919- . II. Natti, Susanna.
PS595.E93D5 821'.9'1408 78-31619
79 MARC

Dodd, Lynley.
The nickle nackle tree / Lynley Dodd. — London : Hamilton, 1976.
[35] p. : chiefly col. ill. ; 21 cm. GB76-31674
ISBN 0-241-89330-5 : £1.80
I. Title.
PZ8.3.D637 Ni 1976 78-319099
78 MARC

Dodson, Fitzhugh, 1923-
I wish I had a computer that makes waffles— : teaching your child with modern nursery rhymes / written by Fitzhugh Dodson ; illustrated by Al Lowenheim. — 1st ed. — La Jolla, Calif. : Oak Tree Publications, c1978.
80 p. : col. ill. ; 22 x 28 cm.
Includes bibliographical references and index.
SUMMARY: A collection of contemporary nursery rhymes with tips for writing original rhymes and a note to parents.
ISBN 0-916392-27-9
1. Nursery rhymes, American. [1. Nursery rhymes] I. Lowenheim, Alfred, 1947- . II. Title.
PZ8.3.D669 Iah 811'.5'4 78-13178
78 MARC

Driz, Ovseĭ Ovseevich, 1908-1971.
The boy and the tree : poems / by Ovsei Driz ; illustrated by Victor Pivovarov ; translated by Joachim Neugroschel. — 1st American ed. — Englewood Cliffs, N.J. : Prentice-Hall, 1978.
31 p. : col. ill. ; 24 cm.
Translation of Mal'chik i derevo, a collection of children's poems originally written in Yiddish.
SUMMARY: A collection of children's poems written by a Yiddish poet in the Soviet Union.
ISBN 0-13-080929-2
[1. Yiddish poetry. 2. Russian poetry] I. Neugroschel, Joachim. II. Pivovarov, Viktor. III. Title.
PZ8.3.D835 Bo 1978 839'.09'13 78-4119
78 MARC

Dunstan, Peggy.
Sunflowers and sandcastles : a selection of poems for children / written by Peggy Dunstan ; illustrated by Mark Dunstan. — Wellington : Millwood Press, 1977.
37 p. : chiefly ill. (part col.) ; 24 cm. NZ77-10
$6.50
I. Dunstan, Mark. II. Title.
PZ8.3.D927 Su 821 78-670092
78 MARC

Farber, Norma.
Small wonders : poems / by Norma Farber ; woodcuts by Kazue Mizumura. — New York : Coward, McCann & Geoghegan, c1979.
31 p. : ill. ; 23 cm.
SUMMARY: A collection of more than 20 poems about noise, the sun, flowers, and other topics.
ISBN 0-698-20484-0
[1. American poetry] I. Mizumura, Kazue. II. Title.
PZ8.3.F224 Sm 811'.5'4 78-31282
78 MARC

Farber, Norma.
There goes feathertop! / By Norma Farber ; illustrated by Marc Brown. — 1st ed. — New York : Dutton, c1979.
[32] p. : ill. ; 22 cm. — (A Unicorn book)
"A changeling version of Nathaniel Hawthorne's 'Feathertop.'"
SUMMARY: When the witch's scarecrow comes to life, Miss Polly Gookin takes a shine to him.
ISBN 0-525-29667-0
[1. Scarecrows—Fiction. 2. Stories in rhyme] I. Brown, Marc Tolon. II. Hawthorne, Nathaniel, 1804-1864. Feathertop. III. Title.
PZ8.3.F224 Tg [E] 78-12096
78 MARC

Farber, Norma.
There once was a woman who married a man / Norma Farber ; drawings by Lydia Dabcovich. — Reading, Mass. : Addison-Wesley, c1978.
[32] p. : ill. ; 24 cm.
SUMMARY: No matter how hard she tries, a silent man's wife can't get him to speak.
ISBN 0-201-01947-7
[1. Stories in rhyme] I. Dabcovich, Lydia. II. Title.
PZ8.3.F224 Th [Fic] 78-1036
78 MARC

Feelings, Tom.
Something on my mind / Tom Feelings ; words by Nikki Grimes. — New York : Dial Press, 1978.
[32] p. : ill. ; 27 cm.
SUMMARY: Poems expressing the hopes, fears, joys, and sorrows of growing up.
ISBN 0-8037-8229-2 : $7.50. ISBN 0-8037-8225-X lib. bdg. : $7.28.
[1. American poetry] I. Grimes, Nikki. II. Title.
PZ8.3.F32 So 811'.5'4 77-86266
77 MARC

Frank, Josette, 1893- comp.
Poems to read to the very young / selected by Josette Frank ; illustrated by Dagmar Wilson. — New York : Random House, c1977.
[32] p. : col. ill. ; 21 cm. — (The Best book club ever) (A Randon House pictureback)
ISBN 0-394-83705-3
1. Children's poetry, American. 2. Children's poetry, English. I. Wilson, Dagmar. II. Title.
PS593.C45F7 1977 821'.008 77-82083
79 MARC

From morn to midnight / [illustrated by] Satomi Ichikawa ; children's verses chosen by Elaine Moss. — London : Heinemann, 1977.
[32] p. : chiefly col. ill. ; 23 x 25 cm. GB77-25837
ISBN 0-434-94363-0 : £2.90
1. English poetry. 2. Children's poetry, English. I. Ichikawa, Satomi. II. Moss, Elaine.
PR1175.F76 1977 821'.008 79-311784
00 MARC

Frost, Robert, 1874-1963.
Stopping by woods on a snowy evening / Robert Frost ; illustrated by Susan Jeffers. — 1st ed. — New York : Dutton, [1978]
[32] p. : ill. ; 27 cm.
SUMMARY: Illustrations of wintry scenes accompany each line of the well-known poem.
ISBN 0-525-40115-6
[1. Winter—Poetry. 2. American poetry] I. Jeffers, Susan. II. Title.
PZ8.3.F937 St 1978 811'.5'2 78-8134
78 MARC

Fyleman, Rose, 1877-1957, ed.
Pipe & drum / edited by Rose Fyleman. — Great Neck, N.Y. : Core Collection Books, 1979.
88 p. : ill. ; 23 cm.
Reprint of the 1940 ed.
SUMMARY: Presents a collection of poems by a variety of English and American authors grouped under such topics as "Over Hill, Over Dale," "The World of Waters," and "Bells on Earth."
ISBN 0-8486-0010-X
1. Children's poetry, English. [1. English poetry—Collections. 2. American poetry—Collections] I. Title.
[PR1175.F9 1979] 821'.008 78-74510
79 MARC

Gág, Wanda, 1893-1946.
The ABC bunny / by Wanda Gág ; hand lettered by Howard Gág. — New York : Coward, McCann & Geoghegan, 1978, c1933.
[40] p. : ill. ; 28 cm.
Includes "ABC song" (words and music)
SUMMARY: Follows a little bunny as it scampers through the alphabet.
ISBN 0-698-20465-4 : $1.95
1. Alphabet rhymes. [1. Stories in rhyme. 2. Alphabet. 3. Rabbits—Fiction] I. Gág, Howard. II. Title.
PZ8.3.G1213 Aac 1978 [E] 78-110970
79 MARC

Gefen, Jonathan.
Songs Anat loves best / by Jonathan Gefen ; illustrated by Eric Smith ; translated from the Hebrew by Margaret Soifer and Rebekah Soifer Ben-Yitzhak. — Tel Aviv : Dvir, c1974.
46 p. : col. ill. ; 25 cm.
Translation of Shirim she-'Anat ohevet bi-meyuḥad.
I. Smith, Eric. II. Title.
PZ8.3.G27 So 892.4'1'6 75-326047
78 MARC

Geisel, Theodor Seuss, 1904-
Happy birthday to you! By Dr. Seuss [pseud.] New York, Random House [1959]
[57] p. illus. 28 cm.
SUMMARY: Describes a birthday celebration in Katroo presided over by the Birthday Bird.
[1. Birthdays—Fiction. 2. Stories in rhyme] I. Title.
PZ8.3.G276 Hap [E] 59-8475
[r79]rev MARC

[Geisel, Theodor Seuss] 1904-
McElligot's pool, written and illustrated by Dr. Seuss [pseud.] New York, Random House [1947]
[56] p. illus. (part col.) 29 cm.
SUMMARY: A boy imagines the rare and wonderful fish he might catch in McElligot's pool.
[1. Fishing—Fiction. 2. Stories in rhyme] I. Title.
PZ8.3.G276 Mac [E] 47-4895
MARC

Geisel, Theodor Seuss, 1904-
Ten apples up on top! By Theo. LeSieg [pseud.] Illustrated by Roy McKie. [New York] Beginner Books [1961]
59 p. illus. 24 cm.
SUMMARY: A lion, a dog, and a tiger balance apples on their heads.
[1. Counting. 2. Animals—Fiction. 3. Stories in rhyme] I. McKié, Roy. II. Title.
PZ8.3.G276 Te [E] 61-7068
[r79]rev2 MARC

Goldthwaite, John.
Roll call : the story of Noah's Ark & the world's first losers / words by John Goldthwaite ; pictures by Henri Galeron. — [New York] : Harlin Quist, c1978.
[24] p. : ill. ; 23 cm.
SUMMARY: An alphabetical list of the animals Noah forgot to invite aboard the ark.
ISBN 0-8252-7482-6 : $5.95
[1. Noah's ark—Fiction. 2. Animals—Fiction. 3. Stories in rhyme] I. Galeron, Henri. II. Title.
PZ8.3.G58 Ro [E] 78-70570
00 MARC

Hoberman, Mary Ann.
A house is a house for me / Mary Ann Hoberman ; illustrated by Betty Fraser. — 1st ed. — New York : Viking Press, 1978.
[48] p. : col. ill. ; 29 cm.
SUMMARY: Lists in rhyme the dwellings of various animals, and things.
ISBN 0-670-38016-4
[1. Dwellings—Fiction. 2. Stories in rhyme] I. Fraser, Betty. II. Title.
PZ8.3.H66 Ho 1978 [E] 77-15518
77 MARC

CHILDREN'S POETRY, ENGLISH (Continued)

Hobzek, Mildred.
We came a-marching ... 1, 2, 3 / by Mildred Hobzek ; illustrated by William Pene du Bois. — New York : Parents' Magazine Press, c1978.
[38] p. : col. ill. ; 24 cm.
SUMMARY: Three marching soldiers determine to see the birds in a nest on top of a tall tree. The recurring refrain gives the marching count in twelve different languages.
ISBN 0-8193-0974-5. ISBN 0-8193-0975-3 lib. bdg.
[1. American poetry] I. Du Bois, William Pène, 1916- . II. Title.
PZ8.3.H665 We 398.8 78-7793
78 MARC

Holland, Marion, 1908-
A big ball of string, written and illustrated by Marion Holland. [New York] Beginner Books; distributed by Random House [1958]
64 p. illus. 24 cm. (Beginner books, 5)
SUMMARY: After winding a large ball of string, a young boy has fun finding ways of using it.
[1. Stories in rhyme] I. Title.
PZ8.3.H699 Bi [E] 58-11964
[r79]rev2 MARC

Holland, Marion, 1908-
A big ball of string, written and illustrated by Marion Holland. [New York] Beginner Books; distributed by Random House [1958]
64 p. illus. 24 cm. (Beginner books, 5)
SUMMARY: After winding a large ball of string, a young boy has fun finding ways of using it.
[1. Stories in rhyme] I. Title.
PZ8.3.H699 Bi [E] 58-11964
[r79]rev MARC

Howard, Jean G
Of mice and mice / by Jean G. Howard ; with drawings by the author. — Cranberry Isles, Me. : Tidal Press, c1978.
[48] p. : ill. ; 15 x 30 cm.
SUMMARY: A mouse begins a new life and family when he becomes a young girl's pet.
ISBN 0-930954-03-3. ISBN 0-930954-04-1 de luxe
[1. Stories in rhyme. 2. Mice—Fiction. 3. Pets—Fiction] I. Title.
PZ8.3.H818 Of [E] 78-50486
79 MARC

Krauss, Ruth.
Somebody spilled the sky / Ruth Krauss ; illustrated by Eleanor Hazard. — 1st ed. — New York : Greenwillow Books, c1979.
32 p. : ill. (some col.) ; 26 cm.
SUMMARY: Sixteen poems of feelings, thoughts, and behavior of childhood.
ISBN 0-688-80186-2: $6.95. ISBN 0-688-84186-4 lib. bdg. : $6.67
[1. American poetry] I. Hazard, Eleanor Lanahan. II. Title.
PZ8.3.K865 So 811'.5'4 78-14306
78 MARC

Larcom, Lucy, 1824-1893.
Childhood songs : [poems] / by Lucy Larcom. — Great Neck, N.Y. : Granger Book Co., 1978.
202 p. : ill. ; 20 cm.
Reprint of the 1874 ed. published by Houghton Mifflin, Boston.
SUMMARY: Fifty-four poems with widely varying themes.
ISBN 0-89609-069-8
[1. American poetry] I. Title.
PZ8.3.L323 Ch 1978 811'.4 78-102394
79 MARC

Lopshire, Robert.
Put me in the zoo. [New York] Beginner Books; distributed by Random House [1960]
58 p. illus. 24 cm. (Beginner books, B-17)
SUMMARY: A large, spotted animal discovers he really belongs in a circus, not a zoo.
[1. Animals—Fiction. 2. Stories in rhyme] I. Title.
PZ8.3.L862 Pu [E] 60-13494
[r79]rev MARC

Lubran, Alfred.
I spy with my little eye : a little book for little folk. — London ([4 Stradella Rd., SE24 9HA]) : Narbulla Agency [for the] Society of Private Presses [and the] Private Libraries Association, 1975.
[16] p. : ill. ; 11 cm. GB78-31631
Text by Alfred Lubran.
I. Title.
PZ8.3.L96 Iac 821'.9'14 79-306434
79 MARC

Macaulay, Thomas Babington Macaulay, Baron, 1800-1859.
Horatius / by T. B. Macaulay ; [ill. by Nicholas Garland]. — London : G. Duckworth, 1977.
79 p. : ill. ; 28 cm. GB***
ISBN 0-7156-1272-7 : £3.95
1. Horatius Cocles—Juvenile poetry. I. Garland, Nicholas. II. Title.
PZ8.3.M118 Ho 1977 821'.8 78-319116
78 MARC

Marilue.
Bobby Bear's Christmas / story and ill. by Marilue. — Fayetteville, Ga. : Oddo Pub., c1978.
32 p. : col. ill. ; 24 cm.
SUMMARY: Bobby Bear and his family get ready for Christmas Day.
ISBN 0-87783-143-2
[1. Bears—Fiction. 2. Christmas stories. 3. Stories in rhyme] I. Title.
PZ8.3.M3914 Bm 77-83628
79 MARC

Marilue.
Bobby Bear's Thanksgiving / story and ill. by Marilue. — Fayetteville, Ga. : Oddo Pub., c1978.
32 p. : col. ill. ; 24 cm.
SUMMARY: Bobby Bear and his friends search for a turkey for their Thanksgiving celebration.
ISBN 0-87783-142-4
[1. Thanksgiving Day—Fiction. 2. Animals—Fiction. 3. Stories in rhyme] I. Title.
PZ8.3.M3914 Bt [E] 77-83623
79 MARC

Mayer, Mercer, 1943-
Little monster's bedtime book / by Mercer Mayer. — New York : Golden Press, c1978.
[24] p. : col. ill. ; 21 cm. — (A Golden look-look book)
SUMMARY: A collection of humorous poems introduces 15 unusual monsters.
ISBN 0-307-11848-7 : $0.95
[1. Monsters—Poetry. 2. American poetry] I. Title.
PZ8.3.M4527 Li 811'.5'4 77-90981
79 MARC

McClintock, Marshall, 1906-
A fly went by, by Mike McClintock. Illustrated by Fritz Siebel. [New York] Beginner Books; distributed by Random House [1958]
62 p. illus. 24 cm. (Beginner books, 3)
SUMMARY: A sheep with its foot caught in a tin can sets off a chase with a fly in the lead.
[1. Stories in rhyme. 2. Animals—Fiction] I. Siebel, Fritz. II. Title.
PZ8.3.M1325 Fl [E] 58-9018
[r79]rev MARC

McKié, Roy.
Snow [by] Roy McKié and P. D. Eastman. [New York] Beginner Books [1962]
61 p. illus. 24 cm. (Beginner books, B-27)
SUMMARY: Two children enjoy sledding, skiing, making tracks, building a snowman, and other activities in the snow.
[1. Snow—Fiction. 2. Stories in rhyme] I. Eastman, Philip D., joint author. II. Title.
PZ8.3.M223 Sn [E] 62-15114
[r79]rev MARC

Miss Browne : the story of a superior mouse / illustrated by Madeline Hall. — New York : Hart Pub. Co., c1978.
24 p. : col. ill. ; 25 cm.
SUMMARY: Relates in rhymed text the events of a visit by an elegant city mouse to her simple country cousins.
ISBN 0-8055-1262-4 : $3.95
[1. Mice—Fiction. 2. Stories in rhyme] I. Hall, Madeline.
PZ8.3.M66 [E] 78-106900
78 MARC

Mitchell, Cynthia.
Playtime / [by] Cynthia Mitchell ; with pictures by Satomi Ichikawa. — London : Heinemann, 1978.
[32] p. : col. ill. ; 16 x 21 cm. GB78-38777
ISBN 0-434-94364-9 : £2.50
1. Play—Juvenile poetry. I. Ichikawa, Satomi. II. Title.
PZ8.3.M676 Pl 1978 79-310971
79 MARC

Moncure, Jane Belk.
If a dinosaur came to dinner / by Jane Belk Moncure ; illustrated by Helen Endres. — Elgin, Ill. : Child's World ; Chicago : distributed by Childrens Press, c1978.
30 p. : col. ill. ; 23 cm. — (Her Creative dramatics)
SUMMARY: A child imagines having a pet dinosaur.
ISBN 0-89565-008-8 lib. bdg. : $4.95
[1. Play—Fiction. 2. Dinosaurs—Fiction. 3. Stories in rhyme] I. Endres, Helen. II. Title. III. Series.
PZ8.3.M72 If [E] 77-12957
77 MARC

The Moon's the North Wind's cooky : night poems / selected and illustrated by Susan Russo. — 1st ed. — New York : Lothrop, Lee & Shepard, c1979.
36 p. : ill. ; 18 cm.
SUMMARY: A collection of 14 images of night by such poets as Nikki Giovanni, Louis Untermeyer, and Felice Holman.
ISBN 0-688-41879-1. ISBN 0-688-51879-6 lib. bdg.
1. Night—Juvenile poetry. 2. Children's poetry, American. 3. Children's poetry, English. [1. Night—Poetry. 2. American poetry—Collections] I. Russo, Susan.
PS595.N54M66 811'.008 78-32178
79 MARC

Moss, Howard, 1922-
Tigers & other lilies / Howard Moss ; drawings by Frederick Henry Belli. — 1st ed. — New York : Atheneum, 1977.
40 p. : ill. ; 24 cm.
SUMMARY: Brief humorous poems inspired by plants with animals in their names.
ISBN 0-689-30592-3
1. Plants—Juvenile poetry. 2. Humorous poetry, American. [1. Plants—Poetry. 2. Humorous poetry. 3. American poetry] I. Belli, Frederick Henry. II. Title.
PZ8.3.M845 Ti 811'.5'4 77-1998
77[r79]rev MARC

Mother Goose.
Kate Greenaway's Mother Goose : or, The old nursery rhymes. — New York : Gramercy Pub. Co., [1978]
55 p. : col. ill. ; 27 cm.
First published in 1881 by G. Routledge, London, New York, under title: Mother Goose.
SUMMARY: A collection of 44 traditional nursery rhymes with drawings of life in Victorian times.
ISBN 0-517-26289-4 lib. bdg.
1. Nursery rhymes. [1. Nursery rhymes] I. Greenaway, Kate, 1846-1901. II. Title.
PZ8.3.M85 Gr 1978 398.8 78-61430
78 MARC

Noodles.
The false-face detective / thought up and put down by Noodles ; illustrated by Sal Murdocca. — New York : Holt, Rinehart, and Winston, c1979.
[32] p. : col. ill. ; 17 cm.
SUMMARY: Seldoon the detective, using a variety of disguises, searches for a cap snatcher.
ISBN 0-03-047356-X
[1. Mystery and detective stories. 2. Stories in rhyme] I. Murdocca, Sal. II. Title.
PZ8.3.N738 Fal [E] 79-117432
79 MARC

Noodles.
How to catch a ghost / thought up and written down by Noodles ; illustrated by Michael Foreman. — New York : Holt, Rinehart, and Winston, c1979.
[32] p. : col. ill. ; 16 cm.
SUMMARY: Brief instructions in rhyme for catching a ghost.
ISBN 0-03-047366-7
[1. Ghost stories. 2. Stories in rhyme] I. Foreman, Michael, 1938- . II. Title.
PZ8.3.N738 Ho E 79-117444
00 MARC

Noodles.
The night of the ooley bugs : a story / by Noodles ; with goblins by James Endicott. — New York : Holt, Rinehart, and Winston, c1979.
[32] p. : col. ill. ; 16 cm.
SUMMARY: Describes in rhyme the scary prowl of the ooley bugs on Halloween night.
ISBN 0-03-047371-3
[1. Halloween—Fiction. 2. Stories in rhyme] I. Endicott, James R. II. Title.
PZ8.3.N738 Ni [E] 79-117448
79 MARC

Noodles.
This little thing / by another little thing—Noodles ; illustrated by Peter Lippman. — New York : Holt, Rhinehart, and Winston, c1979.
[32] p. : col. ill. ; 16 cm.
SUMMARY: A small creature that jumps and eats a lot grows enormously after eating a pot of skunk cabbage stew.
ISBN 0-03-047361-6
[1. Stories in rhyme] I. Lippman, Peter J. II. Title.
PZ8.3.N738 Th [E] 79-117445
79 MARC

Oh, such foolishness! : Poems / selected by William Cole ; pictures by Tomie de Paola. — Philadelphia : Lippincott, c1978.
96 p. : ill. ; 24 cm.
Includes indexes.
SUMMARY: A collection of humorous and nonsense poems.
ISBN 0-397-31807-3
1. Humorous poetry, American. 2. Humorous poetry, English. 3. Children's poetry, American. 4. Children's poetry, English. 5. Nonsense-verses. [1. Humorous poetry. 2. Poetry—Collections. 3. Nonsense verses] I. Cole, William, 1919- . II. De Paola, Thomas Anthony.
PS595.H8O36 811'.08 78-1622
78 MARC

Pahz, Jim.
Robin sees a song / Jim & Cheryl Pahz ; [illustrated by Cheryl Pahz]. — Silver Spring, Md. : National Association of the Deaf, c1977.
[51] p. : ill. (some col.) ; 29 cm. — (People potential series ; 2)
SUMMARY: A deaf girl discovers that songs can be sung in many different ways.
[1. Deaf—Fiction. 2. Physically handicapped—Fiction. 3. Stories in rhyme] I. Pahz, Cheryl, joint author. II. Title.
PZ8.3.P148 Ro [E] 76-17710
78[r79]rev MARC

Park, Ruth.
Roger Bandy / [by] Ruth Park ; illustrated by Deborah and Kilmeny Niland. — Adelaide : Rigby, 1977.
32 p. : col. ill. ; 27 cm. Aus
ISBN 0-7270-0466-2
I. Niland, Deborah. II. Niland, Kilmeny. III. Title.
PZ8.3.P164 Ro 78-314960
78[r79]rev MARC

Peet, Bill.
Ella. Written and illustrated by Bill Peet. Boston, Houghton Mifflin, 1964.
48 p. col. illus. 26 cm.
SUMMARY: A spiteful, conceited elephant runs away from the circus and immediately regrets her course of action.
[1. Elephants—Fiction. 2. Stories in rhyme] I. Title.
PZ8.3.P2764 El [E] 63-20703
[r79]rev2 MARC

CHILDREN'S POETRY, ENGLISH (Continued)

Perle, Ruth Lerner.
The fisherman and his wife, with Benjy and Bubbles / adapted by Ruth Lerner Perle and Susan Horowitz ; illustrated by Giulio Maestro. — New York : Holt, Rinehart, and Winston, c1979.
30 p. : col. ill. ; 25 cm. — (Read with me)
Adapted from the Grimm brothers' Vom Fischer und seiner Frau.
SUMMARY: A rhymed retelling of the classic tale of the fisherman and his greedy wife, with the addition of Benjy the bunny and Bubbles the cat.
ISBN 0-03-044971-5 : $2.95
[1. Fairy tales. 2. Folklore—Germany. 3. Stories in rhyme] I. Horowitz, Susan, joint author. II. Maestro, Giulio. III. Grimm, Jakob Ludwig Karl, 1785-1863. Vom Fischer und seiner Frau. IV. Title. V. Series: Read with me series.
PZ8.3.P423 Fi 398.2 78-55629
78 MARC

Preston, Edna Mitchell.
The temper tantrum book / Edna Mitchell Preston ; illustrated by Rainey Bennett. — New York : Puffin Books, 1978, c1969.
[36] p. : col. ill. ; 23 cm. — (Picture puffin)
SUMMARY: Relates in rhyme several issues between parent and child that provoke temper tantrums.
ISBN 0-14-050181-9 : $2.25
[1. Behavior—Fiction. 2. Stories in rhyme] I. Bennett, Rainey. II. Title.
[PZ8.3.P913 Te 1978] [E] 76-29003
76 MARC

Rhodes, Irma G
Flashbacks : poems for children / by Irma G. Rhodes and Enid R. Peschel ; [ill. by Malcolm F. McKesson]. — Limited 1st ed. — Georgetown, Calif. : Dragon's Teeth Press, c1978.
62 p. (p. 61 blank) : ill. ; 22 cm. — (Living poets series ; no. 19)
SUMMARY: A collection of more than 50 poems celebrating the joys of childhood.
$3.50
[1. American poetry] I. Peschel, Enid Rhodes, joint author. II. McKesson, Malcolm F. III. Title.
PZ8.3.R347 Fl 811'.5'4 78-106002
78 MARC

Rice, James, 1934-
Gaston goes to Texas / written and illustrated by James Rice. — Gretna, La. : Pelican Pub. Co., 1978.
[32] p. : ill. (some col.) ; 29 cm.
SUMMARY: Gaston the alligator gets hurled by a Louisiana hurricane into west Texas where he becomes the first alligator cowboy.
ISBN 0-88289-204-5 : $6.95
[1. Alligators—Fiction. 2. Cowboys—Fiction. 3. Stories in rhyme] I. Title.
PZ8.3.R36 Gar [Fic] 78-12490
78 MARC

Roffey, Maureen.
Rhyming Nell : a picture book / by Maureen Roffey ; verses by Bernard Lodge. — London : Bodley Head, 1979. GB***
[24] p. : col. ill. ; 25 cm.
ISBN 0-370-30124-2 : £2.95
I. Lodge, Bernard. II. Title.
[PZ8.3.R63 Rh 1979b] 398.8 79-309916
79 MARC

Roffey, Maureen.
Rhyming Nell : a picture book / by Maureen Roffey ; verses by Bernard Lodge. — 1st U.S. ed. — New York : Lothrop, Lee & Shepard Co., c1979.
[28] p. : col. ill. ; 25 cm.
SUMMARY: A mischievous witch named Nell is transformed by her rhyming spell.
ISBN 0-688-41898-8
[1. Witches—Fiction. 2. Stories in rhyme] I. Lodge, Bernard. II. Title.
PZ8.3.R63 Rh 1979 [E] 78-21188
78 MARC

Sergeant, Howard, 1914- comp.
The Two Continents book of children's verse / edited by Howard Sergeant. — New York : Two Continents Pub. Group, [1979?]
286 p. : ill. ; 23 cm.
Published in 1972 under title: Evans book of children's verse.
Includes indexes.
SUMMARY: A collection containing humorous poems by English and American poets.
ISBN 0-8467-0238-X : $7.95
1. Children's poetry, English. 2. Children's poetry, American. [1. English poetry—Collections. 2. American poetry—Collections] I. Title. II. Title: Book of children's verse.
PN6110.C4S47 1979 821'.008 76-47173
76 MARC

Seuss, Dr.
I can read with my eyes shut / by Dr. Seuss. — New York : Beginner Books, c1978.
[40] p. : col. ill. ; 24 cm.
SUMMARY: The Cat in the Hat takes Young Cat in tow to show him the fun he can get out of reading.
ISBN 0-394-83912-9 : $3.50. ISBN 0-394-93912-3 lib. bdg. : $4.39
[1. Stories in rhyme. 2. Reading—Fiction] I. Title.
PZ8.3.G276 Iae [E] 78-7193
78 MARC

The Speaking toybook : reproducing the voices of the cock, the goat, the cat, the bird, the lamb, and the cuckoo. — [s.l. : s.n., 189-?]
[16] p. : 6 col. ill. ; 19 x 23 cm.
Poems.
Pages [1] and [16] are mounted inside the free front cover and on the top of a book-shaped box. The appropriate "voice" is sounded by pulling one of 6 cords attached to the box.
"Printed in Germany."
1. Children's poetry, English. 2. Animals—Juvenile poetry. 3. Toy and movable books—Specimens.
PR3991.A1S646 821'.8'08036 79-316862
79 MARC

Stevenson, Robert Louis, 1850-1894.
Selections from A child's garden of verses / by Robert Louis Stevenson ; illustrated by Elizabeth Ehrlich. — New York : Dandelion Press, 1979.
[32] p. : col. ill. ; 23 cm.
SUMMARY: Presents selections from a collection of poems evoking the world and feeling of childhood.
ISBN 0-89799-134-6. ISBN 0-89799-055-2 pbk. : $1.50
[1. English poetry] I. Ehrlich, Elizabeth. II. Title.
PZ8.3.S848 Se 821'.8 78-72100
79 MARC

Sweet Betsy from Pike : a song from the gold rush days / illustrated by Robert Andrew Parker. — 1st ed. — New York : Viking Press, 1978.
[32] p. : ill. ; 19 x 24 cm.
SUMMARY: Betsy crosses the plains with Ike to look for gold in California.
ISBN 0-670-68632-8 lib. bdg. : 7.95
[1. Folk songs, American] I. Parker, Robert Andrew.
PZ8.3.S9953 1978 [E] 77-13924
77 MARC

This little pig-a-wig, and other rhymes about pigs / chosen by Lenore Blegvad ; illustrated by Erik Blegvad. — 1st ed. — New York : Atheneum, 1978.
[32] p. : ill. (some col.) ; 21 cm.
"A Margaret K. McElderry book."
SUMMARY: A collection of both well- and lesser-known English and American rhymes about pigs.
ISBN 0-689-50110-2
1. Nursery rhymes. 2. Pigs—Juvenile poetry. [1. Nursery rhymes. 2. Pigs—Poetry] I. Blegvad, Lenore. II. Blegvad, Erik.
PZ8.3.T294 1978 398.8 78-7015
00 MARC

Thomas, Patricia.
"There are rocks in my socks," said the ox to the fox / by Patricia Thomas ; illustrated by Mordicai Gerstein. — 1st ed. — New York : Lothrop, Lee & Shepard Co., c1979.
[37] p. : ill. (some col.) ; 26 cm.
SUMMARY: An ox gets himself into ridiculous situations made worse when he listens to the advice of his friend the fox.
ISBN 0-688-41851-1
[1. Stories in rhyme] I. Gerstein, Mordicai. II. Title.
PZ8.3.T316 Th [E] 78-16632
78 MARC

Willington, Louisa Penn.
Aunt Louisa's Rip Van Winkle / Louisa Penn Willington. — New York : Hart Pub. Co., c1977.
24 p. : ill. ; 29 cm.
Poem based on the story by W. Irving.
SUMMARY: A poem based on the Washington Irving tale of the man who falls asleep in the Catskill Mountains and doesn't wake up for twenty years.
ISBN 0-8055-0357-9 : $3.95
[1. Folklore—United States. 2. Stories in rhyme] I. Irving, Washington, 1783-1859. Rip Van Winkle. II. Title. III. Title: Rip Van Winkle.
PZ8.3.W6796 Au 811'.5'4 s 77-79198
78[r79]rev MARC

Wilson, Dorothy, 1920-
Morning [noon, night] Illustrated by W. K. Plummer. New York, Grosset & Dunlap, 1963.
3 v. col. illus. 11 cm.
I. Title. II. Title: Noon. III. Title: Night.
PZ8.3.W696 Mo 63-18966
[r78]rev MARC

Worth, Valerie.
Still more small poems / by Valerie Worth ; Pictures by Natalie Babbitt. — 1st ed. — New York : Farrar, Straus, and Giroux, 1978.
41 p. : ill. ; 20 cm.
SUMMARY: A collection of brief poems about a variety of everyday things.
ISBN 0-374-37258-6 : $5.95
[1. American poetry] I. Babbitt, Natalie. II. Title.
PZ8.3.W913 St 1978 811'.5'4 78-11739
78 MARC

Wright, Kit.
Arthur's sister / [by] Kit Wright ; illustrated by Eileen Browne. — London : Methuen, 1978.
[16] p. : chiefly col. ill. ; 15 x 21 cm. GB78-34004
ISBN 0-416-86050-8 : £1.30. ISBN 0-416-86060-5 pbk.
I. Browne, Eileen. II. Title.
PZ8.3.W936 Ar 79-303238
79 MARC

Younger, Damon, 1950-
The tiniest candle / written by Damon Younger ; illustrated by Joyce Kitchell. — 1st ed. — La Jolla, Calif. : Oak Tree Publications, c1978.
[64] p. : col. ill. ; 25 cm.
Poem.
SUMMARY: A verse retelling of the events surrounding the birth of Jesus.
ISBN 0-916392-25-2
1. Jesus Christ—Nativity—Juvenile poetry. [1. Jesus Christ—Nativity—Poetry. 2. American poetry] I. Kitchell, Joyce. II. Title.
PZ8.3.Y86 Ti 811'.5'4 78-13709
78 MARC

Yun, Sŏk-chung, 1911-
Half past four : poems for children / by Suk-Joong Yoon ; translated from the Korean by Francis Taewon Yoon & David D. Lapham ; illustrated by Jung Kim. — Los Angeles : F. T. Yoon Co., c1978.
94 p. : ill. ; 18 x 20 cm.
English and Korean.
SUMMARY: A collection of 80 poems including lullabies, play songs, and scenes of family life and nature by a major Korean children's poet.
ISBN 0-931168-01-5 : $7.95
[1. Korean poetry] I. Yoon, Francis Taewon. II. Lapham, David D. III. Kim, Jung. IV. Title.
PZ8.3.Y96 Haf 891.7'1'3 78-51351
79 MARC

Zolotow, Charlotte Shapiro, 1915-
River winding / poems by Charlotte Zolotow ; pictures by Kazue Mizumura. — 1st ed. — New York : Crowell, [1978]
[40] p. : ill. ; 23 cm.
SUMMARY: An illustrated collection of short poems about bedtime, rivers, seasons, and other topics.
ISBN 0-690-03866-6 : $5.95. ISBN 0-690-03867-4 lib. bdg. : $5.79
[1. American poetry] I. Mizumura, Kazue. II. Title.
PZ8.3.Z6 Ri 1978 811'.5'4 77-27670
78 MARC

Zuliani, Vilma.
I believe in make believe / by Vilma Zuliani ; illustrated by Robert de Claire. — [Lawrence, Mass. : Hermitage Press, c1977]
[47] p. : ill. (some col.) ; 22 x 29 cm.
On dust jacket: Distributed by Follett Publishing Co. [Chicago]
SUMMARY: A collection of 16 illustrated poems including the saga of "Terrible Teddy the Toad" and "Mike the Mighty Monster."
ISBN 0-89441-001-6 : $4.95
[1. American poetry] I. De Claire, Robert. II. Title.
PZ8.3.Z8 Iac 811'.5'4 77-74630
78 MARC

—HISTORY AND CRITICISM

Barnes, Walter, 1880-
The children's poets : analyses and appraisals of the greatest English and American poets for children : for use in normal schools, library schools, and homes / by Walter Barnes. — Folcroft, Pa. : Folcroft Library Editions, 1978 [c1924]
vi, 264 p. ; 23 cm.
Reprint of the 1925 ed. published by World Book Co., Yonkers-on-Hudson, N.Y.
Bibliography: p. 241-258.
Includes index.
ISBN 0-8414-1736-9 lib. bdg. : $30.00
1. Children's poetry, English—History and criticism. 2. Children's poetry, American—History and criticism. 3. Children's poetry, English. 4. Children's poetry, American. I. Title.
PR508.C6B3 1978 821'.009 78-13147
78 MARC

CHILDREN'S POETRY, FINNISH

Lennä, lennä leppäkerttu : tuttuja lastenrunoja ja -loruja / koonnut Helyi Karahka ; kuy. Eila Kaustia ja Annikki Puurunen. — 4. p. — Helsinki : Otaya, 1976.
117 p. : ill. ; 21 cm. Fi77Jan
ISBN 9511022679
1. Children's poetry, Finnish. I. Karahka, Helvi. II. Kaustia, Eila. III. Puurunen, Annikki.
PZ90.F5L434 1976 79-349032
79 MARC

CHILDREN'S POETRY, FRENCH

Chausteur, Marie Andrée.
Brindilles : poèmes pour les enfants / Marie-Andrée Chausteur ; illustrés par Thibault Jacquet, Laurent Brihaye et par J.-J. Bauduin. — Paris : Éditions Arcam, c1979.
55 p. : ill. ; 21 cm. — (Le Miroir poétique) F***
ISBN 2-902312-72-5 : 25.00F
1. Children's poetry, French. I. Title.
PQ2663.H363B7 841'.9'14 79-386666
79 MARC

Menanteau, Pierre.
Chansons venues par la fenêtre / Pierre Menanteau. — Paris : Éditions Saint-Germain-des-Prés, c1978.
62 p. ; 19 cm. — (Collection L'Enfant, la poésie) F***
ISBN 2-243-00668-5 : 19.00F
1. Children's poetry, French. I. Title.
PZ24.3.M4 841'.9'14 79-340520
79 MARC

CHILDREN'S POETRY, FRENCH (Continued)

La Récitation pour tous : [textes / choisis] par J. B. Abessolo et F. Meyo-Bibang. — Libreville : Ministère de l'éducation nationale et de la recherche scientifique, Institut pédagogique national, [between 1973 and 1976]
172 leaves : ill. ; 29 cm.
Cover title.
1. French language—Text-books for foreigners--African. 2. Children's poetry, French. 3. African poetry (French). I. Abessolo, Jean Baptiste. II. Meyo-Bibang, Frédéric.
PC2129.A5R4 448'.2'496 78-101894
79[r79]rev MARC

Roy, Claude, 1915-
Nouvelles enfantasques : des poèmes et des images pour les lecteurs de plusiers âges entre 4 et 104 ans / imaginés et imagés par Claude Roy. — [Paris] : Gallimard, 1978.
121 p. : ill. ; 22 cm. — (Collection 1000 [i.e. Mille] soleils) F78-14598
18.70F
I. Title.
PZ24.3.R65 78-400796
79 MARC

CHILDREN'S POETRY, GERMAN

Hoffmann von Fallersleben, August Heinrich, 1798-1874.
Kinderlieder / August Heinrich Hoffmann von Fallersleben ; hrsg. von Lionel von Donop ; mit e. Nachw. von Jürgen Dittmar. — Nachdr. d. Ausg. Berlin, 1877. — Hildesheim ; New York : Olms, 1976.
vii, 318 p. ; 16 cm. — (Volkskundliche Quellen : [Reihe] 8, Volkslied) GFR76-A
Reprint of the ed. published by Grote'sche Verlagsbuchhandlung.
Includes index.
ISBN 3-487-05681-X : DM35.80
1. Children's poetry, German. I. Title. II. Series.
PZ34.3.H647 1976 78-385489
79 MARC

Mein erstes Gedichtbuch : 100 Gedichte für Schule und Haus : eingeleitet, ausgewählt und für den Unterricht geordnet / von Nikolaus Britz. — Wien : W. Braumüller, c1978.
vii, 124 p. ; 21 cm. Au***
Bibliography: p. 123-124.
ISBN 3-7003-0172-3
1. Children's poetry, German. I. Britz, Nikolaus.
PZ34.3.M357 821'.008 79-352030
79 MARC

Sieh mal einer guck! : fünfundfünfzig Bildgeschichten, zwei Daumenkinos und eine Katzenschwanzparade / hrsg. von Viktor Christen. — Oldenburg : G. Stalling, c1977.
88 p. : ill. (some col.) ; 25 cm. GFR***
Includes index.
1. Children's poetry, German. I. Christen, Viktor.
PZ34.3.S47 78-358110
78 MARC

CHILDREN'S POETRY, ITALIAN

Pin pidìn : poeti d'oggi per i bambini / a cura di Antonio Porta e Giovanni Raboni. — 1. ed. — Milano : Feltrinelli, 1978.
127 p. ; 24 cm. It78-Dec
Includes bio-bibliographical sketches of the authors.
L3000
1. Children's poetry, Italian. 2. Italian poetry—20th century. I. Porta, Antonio, 1935- . II. Raboni, Giovanni, 1932-
PZ44.3.P5 851'.9'108 79-365970
*79 MARC

CHILDREN'S POETRY, JAPANESE

Abe, Tomoji, 1903-
(Jinkō teien)
人工庭園 / 阿部知二. — 東京 : 講談社, 昭和29 [1954]
245 p. ; 19 cm.
Short stories.
CONTENTS: 人工庭園—隠者と人々—二つの死—少年

I. Title.
PL822.B4J56 78-812299

(Okāsan)
おかあさん : 詩集 / [監修 高田敏子, 吉野弘]. — 東京 : ハナマルキ味噌出版部 : 発売 大和書房, 1978.
217 p. : ill. ; 18 cm. Ja 78-5008
¥580

1. Mothers—Juvenile poetry. 2. Children's poetry, Japanese. 3. Children's writings, Japanese. 4. Japanese poetry—20th century. I. Takada, Toshiko, 1916- II. Yoshino, Hiroshi, 1926-
PL757.4.C45O37 79-805754

Saotome, Katsumoto, 1932-
火の瞳 早乙女勝元・作 鈴木義治・絵 東京 講談社 昭和44 (1969)
218 p. illus. 21 cm. ¥480 Ja 69-27

I. Suzuki, Yoshiharu, illus. II. Title.
Title romanized: Hi no hitomi.
PZ49.31.S3774 77-811548

(Wasure-enu jidōshi)
忘れえぬ児童詩 / 日本作文の会編. — 東京 : 民衆社, 1977.
2 v. ; 20 cm. Ja 78-5490
Bibliography: v. 2, p. 261-262.
¥1200 per vol.

1. Children's poetry, Japanese. 2. Children's writings, Japanese. 3. Children's poetry, Japanese—History and criticism. I. Nihon Sakubun no Kai.
PL757.4.C45W37 78-811721

Watanabe, Megumi, 1958-
(Yowamushi obake)
よわむしおばけ / わたなべめぐみ作 ; 原ゆたか絵. — 東京 : 理論社, 1978.
91 p. : ill. (some col.) ; 23 cm. — (理論社のどうわ) Ja 78-27
Cover title.
Series romanized: Riron Sha no dōwa.
¥880

I. Hara, Yutaka, 1953- II. Title.
PZ49.31.W2997 78-812393

—HISTORY AND CRITICISM

(Wasure-enu jidōshi)
忘れえぬ児童詩 / 日本作文の会編. — 東京 : 民衆社, 1977.
2 v. ; 20 cm. Ja 78-5490
Bibliography: v. 2, p. 261-262.
¥1200 per vol.

1. Children's poetry, Japanese. 2. Children's writings, Japanese. 3. Children's poetry, Japanese—History and criticism. I. Nihon Sakubun no Kai.
PL757.4.C45W37 78-811721

CHILDREN'S POETRY, KOREAN

Yi, Sang-hyŏn.
(Saenggak hanŭn sonyŏn)
생각하는소년 : 이상현동시집 / [지은이이상현]. — 서울 : 일지사, 1978.
153 p. : ill. ; 21 cm.
Poems.
₩800

I. Title.
PZ50.543.Y46 79-825593

CHILDREN'S POETRY, LITHUANIAN

Skripka, Vytautas.
Balandžio rytas / Vytautas Skripka ; [iliustravo Rima Stasiūnaitė]. — Vilnius : Vaga, 1977.
41 p. : col. ill. ; 16 cm. USSR***
Poems.
"Vidutiniam mokykliniam amžiui."
0.10rub
I. Title.
PZ90.L5S53 79-343663
79 MARC

CHILDREN'S POETRY, MOLDAVIAN

(Epurashul albastru)
Епурашул албастру : дин литература универсалэ / алкэтуире ши традучере де Спиридон Вангели ; презентаре графикэ Емил Килдеску. — Кишинэу : Лумина, 1968.
72 p., [4] leaves of plates : ill. (some col.) ; 21 cm.
"Пентру школа примарэ."
1. Children's poetry, Moldavian. I. Vangeli, Spiridon.
PZ90.M7E6 79-407427

CHILDREN'S POETRY, NORWEGIAN

Francke, Ellen, 1943-
Ingen har bruk for oss : dikt / Ellen Francke ; [med tegninger av Runa Førde]. — Oslo : Gyldendal, 1978 (Oslo : Hestholms boktr.)
95 p. : ill. ; 21 cm. N78-Dec
ISBN 8205113025 : kr58.00
I. Title.
PZ54.9.F7 79-354102
79 MARC

CHILDREN'S POETRY, POLISH

Badalska, Wiera.
Pocztówka z podróży / Wiera Badalska ; ilustrowała Bożena Truchanowska. — 1. wyd. — Warszawa : Nasza Księgarnia, 1978.
[8] p. : col. ill. ; 20 x 21 cm. P***
Cover title.
zł20.00
1. Egypt—Description and travel—Juvenile poetry. I. Truchanowska, Bożena. II. Title.
PZ69.B24P6 79-341200
79 MARC

Chotomska, Wanda.
Muzyka pana Chopina / Wanda Chotomska ; ilustrował Wiesław Majchrzak. — Wyd. 1. — Warszawa : Nasza Księgarnia, 1978.
36 p. : col. ill. ; 25 cm. P***
zł27.00
1. Chopin, Fryderyk Franciszek, 1810-1849—Juvenile poetry. I. Majchrzak, Wiesław. II. Title.
PZ69.C5M8 79-350934
79 MARC

CHILDREN'S POETRY, ROMANIAN

Vulpe, Marius.
Vitralii pionierești : [versuri] / Marius Vulpe ; [copertă și ilustrații, Th. Bogoi]. — București : Editura Ion Creangă, 1978.
34 p. : ill. ; 23 cm. R78-5349
lei2.00
1. Children's poetry, Romanian. I. Title.
PC840.32.U47V5 79-377068
*79 MARC

CHILDREN'S POETRY, SERBIAN

Jato čudesnih ptica : zbirka pesama sedam pesnika / Nedeljko Terzić ... [et al.]. — Sremska Mitrovica : Dom omladine "1. novembar", 1977.
73 p., [7] leaves of plates : ill. ; 24 cm. — (Mala biblioteka Nova knjiga ; br. 1) Yu***
Includes biobibliographical notes of the authors.
1. Children's poetry, Serbian. I. Terzić, Nedeljko, 1949-
PZ70.S4J33 78-395479
79 MARC

CHILDREN'S POETRY, SPANISH

Antología de poetas españoles / [selección] Francisco Alvarez Allende, Alfonso García Rodríguez. — León : Nebrija, D.L. 1978.
218 p. : ill. ; 25 cm. — (Colección Biblioteca) Sp78-Oct
Basque, Catalan, Gallegan, or Spanish.
ISBN 8439122004 : 300ptas
1. Spanish poetry. 2. Children's poetry, Spanish. I. Alvarez Allende, Francisco. II. García Rodríguez, Alfonso.
PQ6176.A534 861'.008 79-374843
79 MARC

CHILDREN'S POETRY, SWEDISH

Peterson, Hans.
Jag, Alberta / Hans Peterson. — Stockholm : AWE/Geber, 1978.
132 p. ; 19 cm. S78-21
ISBN 9120056796
I. Title.
PZ59.P45J28 78-390848
79 MARC

CHILDREN'S POETRY, URUGUAYAN

Figueredo, Alvaro.
ABC del gallito verde / Alvaro Figueredo ; prólogo de Arturo Sergio Visca ; carátula e ilus., Elsa Carafí de Marchand. — Pan de Azúcar, R.O. del Uruguay : [Comisión Pro Edición de Obras de Alvaro Figueredo], 1977.
90 p. : ill. ; 19 cm.
1. Children's poetry, Uruguayan. I. Title.
PZ74.3.F53 78-107645
79 MARC

CHILDREN'S POETRY, VENEZUELAN

Mendoza Sagarzazu, Beatriz.
Tarea de vacaciones / Beatriz Mendoza Sagarzazu ; portada e ill., Lola Altamira. — ⌐s.l. : s.n., 1977⌐ (Caracas : Artegrafía)
⌐38⌐ p. : ill. ; 21 x 22 cm.
1. Children's poetry, Venezuelan. I. Altamira, Lola. II. Title.
PZ74.3.M37 861 79-366301
79 MARC

CHILDREN'S POETRY, YIDDISH

Young voices from the ghetto : a collection of children's and young people's poetry written in the ghettos of World War II / translation and introd. by Judith Berkowitz and Eve Edelman. — Waltham, Mass. : Dept. of Near Eastern and Judaic Studies, Brandeis University, 1979.
60 p. ; 29 cm.
"These poems were selected from two collections ... Lider fun di ghettos un lagern, collected by Sh. Kaczerginski, and Dos lid fun ghetto-zamlung."
Cover title: Young voices from the ghetto = Yunge koyleš fun geto.
Poems in English and Yiddish.
1. Yiddish poetry. 2. Children's poetry, Yiddish. 3. Holocaust, Jewish (1939-1945)—Poetry. 4. Yiddish poetry—Translations into English. 5. English poetry—Translations from Yiddish. I. Berkowitz, Judith. II. Edelman, Eve. III. Title: Yunge koyleš fun geto.
PJ5126.Y6 839'.09'1308 79-115532
79 MARC

CHILDREN'S POETRY, YUGOSLAV

Jokanović-Toumin, Dean.
Reponje / ⌐ilustracije⌐ Dean Jokanović-Toumin ; ⌐pjesme Dubravko Horvatić⌐. — Zagreb : Naprijed, 1975.
⌐20⌐ p. : col. ill. ; 27 cm. — (Biblioteka Zelena žirafa) Yu***
I. Horvatić, Dubravko, 1939- II. Title.
PZ70.S42J53 78-343540
78 MARC

CHILDREN'S PRESS see Press, Juvenile

CHILDREN'S QUESTIONS AND ANSWERS

Spencer, Douglas Barry, 1946-
Questions kids ask for those who care to listen / by Douglas Barry Spencer ; foreword by Bill Cosby. — Toronto : Simon & Pierre Pub. Co., c1977.
94 p. : ill. ; 21 cm. C***
Bibliography: p. 94.
ISBN 0-88924-061-2. ISBN 0-88924-060-4 pbk.
1. Children's questions and answers. I. Title.
HQ784.Q4S64 301.43'1 78-309926
78 MARC

CHILDREN'S REFERENCE BOOKS see Children's encyclopedias and dictionaries

CHILDREN'S RIGHTS

see also Children—Law; International Year of the Child, 1979

Dyer, George B
To free a child = La liberación integral del niño / by George B. Dyer ; ⌐Spanish translations by Rosa Maria Icaza⌐. — San Antonio, Tex. : Mexican American Cultural Center, c1979.
20, 20 p. : ill. ; 22 cm.
English and Spanish.
1. Child development. 2. Children's rights. I. Title. II. Title: La liberación del niño.
HQ772.D94 301.43'14 79-84531
79 MARC

Johansen, Erna Maria, 1911-
Betrogene Kinder : e. Sozialgeschichte d. Kindheit / Erna M. Johansen. — Orig.-Ausg. — Frankfurt am Main : Fischer-Taschenbuch-Verlag, 1978.
267 p. : 43 ill. ; 18 cm. — (Bücher des Wissens) GFR78-A41
Bibliography: p. 259-⌐263⌐
Includes indexes.
ISBN 3-596-26622-X : DM7.80
1. Children. 2. Child welfare. 3. Children's rights. I. Title.
HQ781.J62 301.43'14 79-344766
79 MARC

Riesgo Menguez, Luis.
Los derechos de los hijos / Luis Riesgo y Carmen Pablo. — Madrid : Editorial Prensa Española, ⌐1979?⌐
247 p. ; 21 cm. Sp***
Includes bibliographical references.
ISBN 8428704864 : 350ptas
1. Children's rights. 2. Children—Legal status, laws, etc. I. Pablo de Riesgo, Carmen, joint author. II. Title.
HQ789.R5 323.4 79-109009
79 MARC

Schérer, René.
Une Érotique puérile / René Schérer. — Paris : Éditions Galilée, 1978.
188 p. ; 22 cm. F78-7089
Includes bibliographical references.
ISBN 2-7186-0087-X : 44.00F
1. Child welfare. 2. Children's rights. 3. Youth—Sexual behavior. I. Title.
HV715.S34 301.41'7 79-355835
79 MARC

—ADDRESSES, ESSAYS, LECTURES

Having children : philosophical and legal reflections on parenthood : essays / edited for the Society for Philosophy and Public Affairs by Onora O'Neill and William Ruddick. — New York : Oxford University Press, 1979.
vi, 362 p. ; 24 cm.
Includes bibliographical references.
ISBN 0-19-502412-5 : $5.50
1. Parenthood—Moral and religious aspects—Addresses, essays, lectures. 2. Children's rights—Addresses, essays, lectures. I. O'Neill, Onora, 1941- . II. Ruddick, William, 1931- . III. Society for Philosophy and Public Affairs.
HQ755.8.H38 301.42'7 78-2951
78 MARC

—JUVENILE LITERATURE

Archer, Jules.
Who's running your life? : A look at young people's rights / Jules Archer. — 1st ed. — New York : Harcourt Brace Jovanovich, c1979.
x, 168 p., ⌐6⌐ leaves of plates : ill. ; 22 cm.
Bibliography: p. 157-162.
Includes index.
SUMMARY: Discusses the rights of children and young adults, including those concerning family life, school, working, and personal freedom.
ISBN 0-15-296058-9
1. Children's rights—Juvenile literature. 2. Children's rights—United States—Juvenile literature. ⌐1. Children's rights. 2. Law⌐ I. Title.
HV713.A7 362.7 78-20566
79 MARC

—AMERICA

Sajón, Rafael.
Menores en situación irregular : aspectos socio legales de su protección / por Rafael Sajón, José Pedro Achard y Ubaldino Calvento. — Montevideo : O.E.A., Instituto Interamericano del Niño, 1973.
40, 16, 3 p. ; 28 cm.
"Trabajo presentado al XIV Congreso Panamericano del Niño, Santiago, Chile del 5 al 11 de agosto."
Bibliography: p. ⌐1⌐-3 (3d group)
1. Child welfare—America. 2. Children—Legal status, laws, etc.—America. 3. Children's rights—America. I. Achard, José P., joint author. II. Calvento, Ubaldino, joint author. III. Interamerican Children's Institute. IV. Title.
HV713.S24 362.7'097 79-373268
79 MARC

—CANADA

Task Force on the Child as Citizen.
Admittance restricted : the child as citizen in Canada / ⌐Task Force on the Child as Citizen⌐. — ⌐Ottawa⌐ : Canadian Council on Children and Youth, 1978.
172, 195 p. : ill. ; 23 cm. C***
Added t.p.: Interdit aux mineurs.
English and French.
Includes bibliographical references.
$8.95
1. Child welfare—Canada. 2. Children's rights—Canada. I. Canadian Council on Children and Youth. II. Title. III. Title: Interdit aux mineurs.
HV745.A6T37 1978 362.7'0971 79-368442
79 MARC

—FRANCE—ADDRESSES, ESSAYS, LECTURES

Boulin, Bertrand.
La Charte des enfants / Bertrand Boulin ; construite et révélée au public grâce à Jean-Michel Desjeunes et Philippe Alfonsi ; avec la collaboration de Nicolas Vrain, Alain Chiapello, Michèle Roche, Annie Thomas. — Paris : Stock, 1977.
303 p. ; 20 cm. — (Stock 2 ⌐i.e. deux⌐ : Lutter) F78-625
ISBN 2-234-00766-6 : 30.00F
1. Children in France—Addresses, essays, lectures. 2. Children's rights—France—Addresses, essays, lectures. I. Desjeunes, Jean Michel, joint author. II. Alfonsi, Philippe, joint author. III. Title.
HQ792.F7B68 301.43'14'0944 78-399311
79 MARC

—GERMANY, WEST—ADDRESSES, ESSAYS, LECTURES

Kinder in der Bundesrepublik : Materialien, Initiativen, Alternativen / Lottemi Doormann (Hrsg.). — Köln : Pahl-Rugenstein, 1979.
277 p. : ill. ; 19 cm. — (Kleine Bibliothek ; 152) GFR79-A
Includes bibliographical references.
ISBN 3-7609-0431-9
1. Children—Germany, West—Addresses, essays, lectures. 2. Children's rights—Germany, West—Addresses, essays, lectures. 3. Socialization—Addresses, essays, lectures. I. Doormann, Lottemi.
HQ792.G3K56 301.43'14'0943 79-376961
79 MARC

Unsere Kinder, unsere Verantwortung : Materialien e. Aktion / hrsg. von Wolfgang Brinkel u. Hermann Polenz. — 1. Aufl. — Freiburg i.Br. : Dreisam-Verlag, 1978.
172 p. ; 19 cm. GFR78-A
Includes the recommendations of the Kommission Anwalt des Kindes Baden-Württemberg.
Includes bibliographical references.
ISBN 3-921472-16-4 : DM7.20
1. Children's rights—Germany, West—Addresses, essays, lectures. 2. Child abuse—Germany, West—Addresses, essays, lectures. I. Brinkel, Wolfgang. II. Polenz, Hermann. III. Kommission Anwalt des Kindes Baden-Württemberg.
HV763.U57 362.7'0943 79-364971
79 MARC

--JAPAN--ADDRESSES, ESSAYS, LECTURES

(Naku mo no ka)
泣くものか：作文集：子どもの人権10年の証言 / 養護施設協議会編. — 東京：亜紀書房, 1977.
249 p. : ill. ; 19 cm. Ja 77-27417
¥950

1. Children—Japan—Institutional care—Addresses, essays, lectures. 2. Childrens' rights—Japan—Addresses, essays, lectures. 3. Child welfare—Japan—Addresses, essays, lectures. I. Zenkoku Shakai Fukushi Kyōgikai. Yōgo Shisetsu Kyōgikai.
HV1312.N34 79-808623

—MADAGASCAR

Fifanampiana Malagasy. Komisiona Maharitra ho amin' ny Fiarovana ny Zon' ny Vehivavy sy ny Ankizy.
Commission permanente pour la défense des droits de la femme et de l'enfance = Komisiona Maharitra ho amin' ny Fiarovana ny Zon' hy Vehivavy sy ny Ankizy : 1969-1974. — Tananarive : Fifanampiana Malagasy, ⌐1974?⌐
10 p., ⌐6⌐ leaves of plates : ill. ; 28 cm.
Cover title.
French and Malagasy.
1. Women's rights—Madagascar. 2. Children's rights—Madagascar. I. Title. II. Title: Komisiona Maharitra ho amin' ny Fiarovana ny Zon' ny Vehivavy sy ny Ankizy.
HQ1799.5.F54 1974 323.4 79-342421
79 MARC

—NORTH CAROLINA

Thomas, Mason P
Kids and cops : law enforcement services for children in North Carolina / Mason P. Thomas, Jr. and L. Lynn Hogue. — ⌐Chapel Hill⌐ : Institute of Government, University of North Carolina at Chapel Hill, 1974, c1978.
122 p. : ill. ; 28 cm.
"Statutory material updated 1977."
Bibliography: p. 122.
1. Police services for juveniles—North Carolina. 2. Juvenile justice, Administration of—North Carolina. 3. Children's rights—North Carolina. I. Hogue, L. Lynn, joint author. II. University of North Carolina at Chapel Hill. Institute of Government. III. Title.
HV8079.25.T48 1978 363.23 79-624394
79 MARC

—UNITED STATES

Summerhill, Louise.
The story of Birthright : the alternative to abortion / by Louise Summerhill. — Kenosha, Wis. : Prow Books, c1973.
134 p. ; 18 cm.
ISBN 0-913382-06-X : $1.95
1. Birthright (Organization) 2. Abortion—United States—Moral and religious aspects. 3. Children's rights—United States. I. Title.
HQ766.5.U5S8 301 72-96117
79 MARC

—UNITED STATES—ADDRESSES, ESSAYS, LECTURES

Reid, Joseph H
Child welfare perspectives : selected papers of Joseph H. Reid / edited by Ann W. Shyne. — New York : Child Welfare League of America, c1979.
ix, 154 p. ; 24 cm.
"Other publications of Joseph H. Reid": p. 153-154.
Includes bibliographical references.
ISBN 0-87868-133-7
1. Child welfare—United States—Addresses, essays, lectures. 2. Children's rights—United States—Addresses, essays, lectures. I. Shyne, Ann W., 1914- . II. Child Welfare League of America. III. Title.
HV741.R44 362.7'08 79-102125
79 MARC

—UNITED STATES—CONGRESSES

Children's rights : contemporary perspectives / Patricia A. Vardin and Ilene N. Brody, editors. — New York : Teachers College Press, c1979.
xvii, 182 p. ; 23 cm.
"Outgrowth of a Conference on Children's Rights and Child Advocacy held at Teachers College, Columbia University, in June 1977."
Includes bibliographical references.
ISBN 0-8077-2550-1
1. Children's rights—United States—Congresses. I. Vardin, Patricia A. II. Brody, Ilene N., 1954- . III. Institute on Children's Rights and Child Advocacy, 2d, Teachers College, Columbia University, 1977.
HQ789.C46 301.43'1'0973 78-12584
78 MARC

CHILDREN'S RIGHTS (Continued)

—UNITED STATES—JUVENILE LITERATURE

Archer, Jules.
Who's running your life? : A look at young people's rights / Jules Archer. — 1st ed. — New York : Harcourt Brace Jovanovich, c1979.
x, 168 p., [6] leaves of plates : ill. ; 22 cm.
Bibliography: p. 157-162.
Includes index.
SUMMARY: Discusses the rights of children and young adults, including those concerning family life, school, working, and personal freedom.
ISBN 0-15-296058-9
1. Children's rights—Juvenile literature. 2. Children's rights—United States—Juvenile literature. [1. Children's rights. 2. Law] I. Title.
HV713.A7 362.7 78-20566
79 MARC

De Lone, Richard H
Small futures : children, inequality, and the limits of liberal reform / by Richard H. de Lone for the Carnegie Council on Children. — 1st ed. — New York : Harcourt Brace Jovanovich, c1979.
xiv, 258 p. ; 24 cm.
Includes bibliographical references and index.
ISBN 0-15-183128-9
1. Children—United States. 2. Children's rights—United States. 3. Child welfare—United States. 4. Equality. 5. Liberalism—United States. I. Carnegie Council on Children. II. Title.
HQ792.U5D44 301.43'1'0973 77-92536
79 MARC

Westman, Jack C
Child advocacy : new professional roles for helping families / Jack C. Westman. — New York : Free Press, c1979.
xvi, 431 p. ; 25 cm.
Includes bibliographies and indexes.
ISBN 0-02-934540-5 : $16.95
1. Children's rights—United States. 2. Child development.
HV741.W37 1979 362.7'0973 78-24762
79 MARC

CHILDREN'S ROOMS

see also Children's furniture; Nurseries

Children's spaces : 50 architects & designers create environments for the young / [edited by] Molly & Norman McGrath ; foreword by Ivan Chermayeff. — 1st ed. — New York : Morrow, c1978.
222 p. : ill. (some col.) ; 29 cm.
ISBN 0-688-03329-6
1. Children's rooms. 2. Interior decoration. I. McGrath, Molly. II. McGrath, Norman.
NK2117.C4C47 747'.77 78-15324
78 MARC

The Rights of hearing-impaired children / edited by Gary W. Nix. — 1st ed. — Washington : Alexander Graham Bell Association for the Deaf, c1977.
p. 263-351 ; 24 cm. — (The Volta review ; v. 79, no. 5)
Includes bibliographies.
ISBN 0-88200-112-4 : $3.95
1. Children, Deaf—United States—Addresses, essays, lectures. 2. Deaf—Education—United States—Addresses, essays, lectures. 3. Children's rights—United States—Addresses, essays, lectures. I. Nix, Gary W. II. Series.
HV2350.V7 vol. 79, no. 5 77-85240
362.4'2'0973 s
[HV2551] 78 MARC

CHILDREN'S SERMONS

Francis, Dorothy Brenner.
The boy with the blue ears and 49 other object lessons for children / Dorothy Brenner Francis. — Nashville : Abingdon, c1979.
112 p. ; 20 cm.
ISBN 0-687-03908-8
1. Object-teaching. 2. Children's sermons. I. Title.
BV4315.F638 268'.432 79-9899
79 MARC

Johnson, Samuel Lawrence.
The mouse's tale and other children's sermons / S. Lawrence Johnson. — Nashville : Abingdon, c1978.
125 p. ; 19 cm.
SUMMARY: Forty-eight brief story/sermons provide lessons for living.
ISBN 0-687-27255-6
1. Children's sermons. [1. Sermons. 2. Christian life] I. Title.
BV4315.J637 251'.53 78-858
78 MARC

Moore, Harvey Daniel, 1942-
The droopy flower mystery, and other object lessons for children / Harvey D. & Patsie S. Moore. — Nashville : Abingdon, c1979.
127 p. ; 20 cm.
ISBN 0-687-11190-0
1. Children's sermons. I. Moore, Patsie S., joint author. II. Title.
BV4315.M623 352'.5'3 78-13172
78 MARC

Sherer, Michael L
Stories for special days : messages for children on the lesser festivals / Michael L. Sherer. — Minneapolis : Augsburg Pub. House, c1978.
71 p. ; 20 cm.
SUMMARY: Presents 30 brief sermons that introduce the Lesser Festivals including the conversion of St. Paul, the Visitation, and Reformation Day.
ISBN 0-8066-1667-9 : $2.95
1. Children's sermons. [1. Sermons. 2. Christian life. 3. Fasts and feasts] I. Title.
BV4315.S524 252'.53 78-51342
79 MARC

CHILDREN'S SLEEPWEAR

—FIRES AND FIRE PREVENTION

United States. Congress. House Committee on the Judiciary. Subcommittee on Administrative Law and Governmental Relations.
Reimbursement for losses incurred by Government banning of Tris : hearings before the Subcommittee on Administrative Law and Governmental Relations of the Committee on the Judiciary, House of Representatives, Ninety-fifth Congress, second session ... June 14 and 15, 1978. — Washington : U.S. Govt. Print. Off., 1978.
iv, 382 p. : ill. ; 24 cm.
"Serial no. 55."
Includes bibliographical references.
1. Damages—United States. 2. Children's sleepwear—Fires and fire prevention. 3. Dibromopropanol phosphate—Toxicology. I. Title.
KF27.J832 1978b 342'.73'088 79-600671
79 MARC

CHILDREN'S SONGS

see also Games with music; Kindergarten—Music; Lullabies; Nursery rhymes; Operas, Juvenile

Raposo, Joseph G
The Sesame Street song book / words and music by Joe Raposo and Jeffrey Moss ; arrangements by Sy Oliver ; illustrated by Loretta Trezzo. — New York : Simon and Schuster, [1978] c1971.
126 p. : ill. ; 28 cm. (A Fireside book)
"Published in conjunction with Children's Television Workshop."
SUMMARY: Words and music for thirty-six songs from the television production "Sesame Street."
ISBN 0-671-21036-X ; 0-671-24208-3 pbk. : $4.95
1. Children's songs. [1. Songs] I. Moss, Jeffrey, joint composer. II. Trezzo, Loretta. III. Children's Television Workshop. IV. Title.
[M1997.R212S5 1978] 784.6'24'06 78-6845

—INDEXES

Peterson, Carolyn Sue, 1938-
Index to children's songs : a title, first line, and subject index / compiled by Carolyn Sue Peterson and Ann D. Fenton. — New York : H. W. Wilson Co., 1979.
xviii, 318 p. ; 26 cm.
ISBN 0-8242-0638-X
1. Children's songs—Indexes. 2. Songs—Indexes. I. Fenton, Ann D., joint author. II. Title.
ML128.S3P48 016.7846'24 79-14265
79 MARC

—TEXTS

Graham, Carolyn.
Jazz chants for children : rhythms of American English through chants, songs and poems / Carolyn Graham : ill. by Anna Veltfort. — New York : Oxford University Press, 1979.
xi, 84 p. : ill. ; 28 cm.
SUMMARY: A collection of chants, songs, and poems designed to teach the rhythms of spoken American English primarily to students of English as a second language.
ISBN 0-19-502496-6
1. English language—Text-books for foreigners. 2. English language—Spoken English—Juvenile literature. 3. Children's songs—Texts. 4. Children's poetry. [1. English language—Textbooks for foreigners] I. Veltfort, Anna. II. Title.
PE1128.G65 784'.3 78-11858
78 MARC

CHILDREN'S SONGS, CHINESE

—TEXTS

Chu, Chieh-fan.
(Chung-kuo erh ko)
中國兒歌 / 朱介凡編著. — 臺北 : 純文學出版社, 民國66 [1977]
4, 7, 426 p. ; 22 cm. — (純文學叢書)
Series romanized: Ch'un wen hsüeh ts'ung shu.
Bibliography: p. 418-420.
"本書編者其他著述": p. 425-426.

1. Children's songs, Chinese—Texts. I. Title.
PZ10.843.C587C58 78-842485

CHILDREN'S SONGS, GERMAN

—TEXTS

Eia popeia et cetera : e. Sammlung alter Wiegenleider aus d. Volk / Anita Albus ; [Bilder, Anita Albus ; musikal. Bearb., Friedrich Kur]. — Frankfurt am Main : Insel-Verlag, 1978.
71 p. : col. ill., music ; 27 cm. GFR79-A
Includes index.
DM120.00
1. Children's songs, German—Texts. 2. Folk-songs, German—Texts. I. Albus, Anita, 1942- . II. Kur, Friedrich.
ML54.6.E4 784.6'24'06 79-360580
79 MARC

CHILDREN'S SONGS, HUNGARIAN

Borsai, Ilona.
Magyar népi gyermekjátékok / [irták, Borsai Ilona, Hajdu Gyula, Igaz Mária ; szerk., Keszler Mária]. — 2. kiad. — Budapest : Tankönyvkiadó, 1977.
128 p. : ill. ; 24 cm. — (Ének-zene szakköri füzetek ; 2) H77
Includes music.
Includes bibliographical references and index.
ISBN 9631722732 : 10.00Ft
1. Games—Hungary. 2. Children's songs, Hungarian. I. Hajdu, Gyula, 1914- , joint author. II. Igaz, Mária, joint author. III. Keszler, Mária. IV. Title.
GV1204.68.H9B67 1977 79-346431
*79 MARC

Kivirágzott a diófa : népi gyermekjátékok / [a kötet anyagát válogatta és a játékokat feldolgozta Haider Edit ; szerk. Kovács Ágnes ; Matkó Katalin rajzaival]. — Budapest : Móra, 1977.
149, [1] p. : col. ill. ; 24 cm. H77
Includes music.
Bibliography: p. [150]
ISBN 9631108406 : 24.00Ft
1. Games—Hungary. 2. Children's songs, Hungarian. I. Haider, Edit. II. Kovács, Ágnes.
GV1204.68.H9K58 79-346432
*79 MARC

CHILDREN'S SONGS, JAPANESE

(Bokura no uta)
ぼくらの歌 : 児童作詞作曲集 / 玉川学園編. — 町田 : 玉川大学出版部, 昭和49 [1974]
79 p. ; 21 cm. Ja 75-1
Colophon inserted.
For 1-2 voices; in part with piano acc.
¥800

1. Children's songs, Japanese. I. Tamagawa Gakuen. Machida, Japan.
M1997.B7534 78-815193
[M1812]

Takeda, Tadashi, 1930-
(Warabeuta meruhen)
わらべ唄めるへん : こどもの遊戯に民俗文化を訪ねて / 武田正著. — 東京 : 明文社, 1976.
362 p. : ill. ; 19 cm. Ja 76-8864
Bibliography: p. 351-354.
Includes index.
¥1300

1. Play—Japan. 2. Children's songs, Japanese. 3. Singing games. I. Title.
GV1204.77.T34 78-812391

—JAPAN—KOBE

Mifune, Kiyoshi, 1913-
(Kōbe no asobi to asobiuta)
神戸の遊びと遊び歌 : 大正・昭和の兵庫かいわい / 三船清著. — 神戸 : のじぎく文庫 : 発売元神戸新聞出版センター, 昭和52 [1977]
239 p. : ill. ; 19 cm. Ja 78-2994
Originally published in Kōbe shimbun, Apr. 1976-Sept. 1977, under title: Hyōgo no asobi to asobiuta.
¥1500
1. Play—Japan—Kobe. 2. Children's songs, Japanese—Japan—Kobe. I. Title.
GT2480.M53 78-809514

--JAPAN--TO-YAMA (PREFECTURE) --TEXTS

Kawabata, Tsuka.
(Moshiogusa)
もしおぐさ : 富山県の方言・唄 / 川端つか. — 黒部 : 川端, 1978-
v. ; 26 cm. Ja 79-4338 (v. 1)
Cover title.
Errata slip inserted in v. 1.
1. Japanese language—Dialects—Japan—Toyama (Prefecture)—Glossaries, vocabularies, etc. 2. Children's songs, Japanese—Japan—Toyama (Prefecture)—Texts. 3. Folk-songs, Japanese—Japan—Toyama (Prefecture)—Texts. I. Title.
PL693.T63K3 79-809085

CHILDREN'S STORIES

see also Fairy tales; Missionary stories; Story-telling

And everywhere, children! : An international story festival / selected by the Literature Committee, Association for Childhood Education International. — 1st ed. — New York : Greenwillow Books, c1979.
xvi, 266 p. ; 24 cm.
SUMMARY: Thirteen stories and excerpts from longer books which reflect life in the country of their setting.
ISBN 0-688-80215-X : $8.95
1. Children's stories. [1. Short stories] I. Association for Childhood Education International. Literature Committee.
PZ5.A52 [Fic] 78-25932
78 MARC

—BIBLIOGRAPHY

Griffiths, Vivien.
Ghostly encounters / [compiled by Vivien Griffiths, Carole Barlow]. — London : National Book League, [1978?]
24 p. ; 22 cm. — (Pamphlet - Library Association, Youth Libraries Group ; no. 19) GB***
ISBN 0-85365-930-3 : £0.55 ($1.80 U.S.)
1. Ghost stories—Bibliography. 2. Children's stories—Bibliography. I. Barlow, Carole, joint author. II. Title. III. Series: Library Association. Youth Libraries Group. Pamphlet ; no. 19.
Z718.1.L56 no. 19 027.62'5'08 s 79-311417
[ZS917.G45] 79 MARC
[PN3435]

—DICTIONARIES

Fisher, Margery Turner, 1913-
Who's who in children's books : a treasury of the familiar characters of childhood / Margery Fisher. — 1st ed. — New York : Holt, Rinehart and Winston, c1975.
399 p. : ill. ; 26 cm.
Includes indexes.
ISBN 0-03-015091-4
1. Children's stories—Dictionaries. 2. Characters and characteristics in literature—Dictionaries. I. Title.
PN1009.A1F575 809'.89282 75-5463
75[r79]rev2 MARC

—HISTORY AND CRITICISM

Seki, Hideo, 1912-
(Jidō bungaku ron)
児童文学論 / 関英雄著. — 東京 : 新評論社, 昭和30 [1955]
274 p. ; 18 cm. — (教育新書 ; 8)
Series romanized: Kyōiku shinsho.

1. Children's stories, Japanese—History and criticism. 2. Children's stories—History and criticism. I. Title.
PN1009.J3S43 79-802849

—HISTORY AND CIRITICISM—CONGRESSES

Baśń i dziecko / wstęp, wybór i oprac. wypowiedzi Halina Skrobiszewska. — Wyd. 1. — Warszawa : Ludowa Spółdzielnia Wydawnicza, 1978 (Olszt. : OZGSP)
357, [3] p. ; 20 cm. P79-1662
"Praca oparta na materiałach z sesji naukowo-literackiej pod hasłem 'Funkcja baśni w wychowaniu i kształtowaniu osobowości dziecka' zorganizowanej przez Związek Literatów Polskich w Warszawie w grudniu 1973 r."
Includes bibliographies.
Includes index.
zł68.00
1. Children's stories—History and ciriticism—Congresses. 2. Tales—History and criticism—Congresses. I. Skrobiszewska, Halina. II. Związek Literatów Polskich.
PN1009.Z6B25 79-380859
79 MARC

—INDEXES

Frend, P M
Junior fiction index / [compiled by] Patricia M. Frend. — 3rd ed. — London : Association of Assistant Librarians, 1977.
44 p. ; 24 cm. GB78-35038
£4.70
1. Children's stories—Indexes. I. Title.
Z1037.F85 1977 028.52 79-312843
[PN1009.A1] 79 MARC

Yonkers, N. Y. Public Library. Children's Services.
A guide to subjects & concepts in picture book format / Yonkers Public Library Children's Services. — 2d ed. — Dobbs Ferry, N.Y. : Oceana Publications, 1979.
163 p. : ill. ; 24 cm.
Bibliography: p. [159]-163.
ISBN 0-379-20276-X : $10.00
1. Picture books for children—Indexes. 2. Children's stories—Indexes. 3. Subject headings, Children's literature. I. Title.
Z1037.Y65 1979 028.52 78-31811
79 MARC

CHILDREN'S STORIES, AMERICAN

Nature stories for tiny folk. Springfield, Mass., McLoughlin brothers, inc. [c1934]
[252] p. illus. 14 cm. (The little big books)
1. Children's stories, American.
PZ5.N215 1934 34-10381
[r78]rev MARC

The Sesame street bedtime storybook : featuring Jim Henson's Muppets / written by Tony Geiss ... [et al.] ; illustrated by Tom Cooke ... [et a.]. — New York : Random House, c1978.
64 p. : col. ill. ; 29 cm.
Published in conjunction with Children's Television Workshop.
SUMMARY: A collection of 12 read-aloud stories featuring the Sesame Street Muppets.
ISBN 0-394-83843-2 : $3.95. ISBN 0-394-93843-7 lib. bdg. : $5.99
1. Children's stories, American. [1. Short stories] I. Geiss, Tony. II. Cooke, Tom. III. Children's Television Workshop. IV. Bedtime storybook.
PZ5.S47 [E] 77-93774
78 MARC

Walt Disney's treasury of children's classics / edited by Darlene Geis. — New York : H. N. Abrams, 1978.
306 p. : ill. (some col.) ; 29 cm.
SUMMARY: Retells seventeen well-known fairy tales, folk tales, and short stories illustrated with scenes from Walt Disney films. Includes a behind-the-scenes look at the production of animated films.
ISBN 0-8109-0812-3
1. Children's stories, American. 2. Children's stories, English. 3. Fairy tales. 4. Tales. 5. Animation (Cinematography)—Juvenile literature. [1. Short stories. 2. Fairy tales. 3. Folklore. 4. Animation (Cinematography)] I. Geis, Darlene. II. Disney (Walt) Productions. III. Title: Treasury of children's classics.
PZ5.W167 [Fic] 78-3529
78 MARC

CHILDREN'S STORIES, CHINESE

Ch'ang, Wen.
(I mei yin pi)
一枚銀幣 : 兒童故事 / 昌雯編. — 香港 : 大光出版社, 1977.
92 p. : ill. ; 19 cm.
Cover title.
Juvenile literature.
CONTENTS: 一枚銀幣.—兩張馬戲票.—「聖誕老人」的禮物.—一瓶牛奶.—「小霸王」受到的教訓.—母與子.—父與子.—捉金雀.—山羊與奶油餅.—天鵝湖畔三兄妹.—小黃尼勇救小羊.
$3.00

1. Children's stories, Chinese. I. Title.
PZ10.831.C4643I17 78-842757

Chiang, Chu-chün.
(Lao-ts'an yu chi li ti ku shih)
老殘遊記裏的故事 / 蔣竹君改寫 ; 陳雄繪圖. — 第1版. — 臺北 : 國語日報出版部, 民國64 [1975]
2, 152 p. : ill. ; 20 cm. — (中國古典文學名著)
At head of title: Chung-kuo ku tien ming chu.
Text accompanied with Chu yin fu hao.
Series romanized: Chung-kuo ku tien wen hsüeh ming chu.
NT$30.00
1. Children's stories, Chinese. I. Liu, E, 1857-1909. Lao-ts'an yu chi. II. Title.
PZ10.831.C562.L3 79-835717

(Ch'ing pao kua)
情报瓜 / 广西军区政治部编. — [南宁?] : 广西人民出版社, 1977.
240 p. : ill. ; 19 cm.
$0.45

1. Children's stories, Chinese. I. China (People's Republic of China, 1949-). Chung-kuo jen min chieh fang chün. Kuang-hsi chün ch'ü. Cheng chih pu.
PZ10.831.C6193 1977 79-838726

(Ch'ing pao kua)
情报瓜 / 广西军区政治部编. — [香港? : s. n., 1978?]
240 p. : ill. ; 22 cm.
Reprint of the 1977 ed. published by Kuang-hsi jen min ch'u pan she.

1. Children's stories, Chinese. I. China (People's Republic of China, 1949-). Chung-kuo jen min chieh fang chün. Kuang-hsi chün ch'ü. Cheng chih pu.
PZ10.831.C6193 1978 79-838725

(Erh ko)
儿歌 / 如东县革委会教育局编. — [香港? : s. n., 1978?]
2, 13, 334 p. : ill. ; 22 cm.
At head of title: yu erh chiao hsüeh ts'an k'ao ts'ai liao.
Reprint of the 1977 ed. published by Chiang-su jen min ch'u pan she.

1. Children's poetry, Chinese. I. Ju-tung hsien, China, Ko wei hui. Chiao yü chü.
PZ10.843.E7 1978 79-836822

(Hai tao shao nien)
海島少年. — 香港 : 港華出版社, 1978.
49 p. : col. ill. ; 19 cm. — (圖畫故事叢書)
Cover title.
Series romanized: T'u hua ku shih ts'ung shu.
CONTENTS: 雜技團的小狗.—海島少年.—奇異的花.—笨公鷄.—永康的生日.
$2.00

1. Children's stories, Chinese.
PZ10.831.H32 79-836650

Ho-pei jen min ch'u pan she.
(T'ui p'iao)
退票 ; 果园小主人 / 本社改编 ; 朱妍, 李耀华画. — 第1版. — [石家荘?] : 河北人民出版社, 1977.
[28] p. : ill. ; 86 x 188 mm.
"連環圖書"
$0.08
1. Children's stories, Chinese. I. Chu, Yen, fl. 1977- II. Li, Yao-hua. III. Ho-pei jen min ch'u pan she. Kuo yüan hsiao chu jen. 1977. IV. Title. V. Title: Kuo yüan hsiao chu jen.
PZ10.831.H655T8 79-835249

(Kuang jung chün chien shang ti ying hsiung chan shih)
光荣军舰上的英雄战士 / [上海市出版革命组编辑]. — [香港? : s. n., 1978?]
2, 2, 136 p. : ill. ; 21 cm.
At head of title: Hung hsiao ping hsüeh hsi Mao Tse-tung ssu hsiang fu chu tu wu.
Reprint of the 1970 ed. published by Shang-hai shih ch'u pan ko ming tsu, Shanghai.

1. Children's stories, Chinese. I. Shang-hai shih ch'u pan ko ming tsu.
PZ10.831.K798 1978 79-837556
s 79 C

(Pao ku hu hsü hung)
苞谷胡须红 = Baogu huxu hong / [蒋咸美, 潘昌仁等著]. — 第1版. — [南宁] : 广西人民出版社, 1975.
91 p. : ill. ; 19 cm.
CONTENTS: 潘昌仁 小喜子智擒熊二拐.—蒋咸美 苞谷胡须红.—潘昌仁 阿葵.—蒋咸美 小铁生三教黄老头.—邓幸涛 雏燕展翅.

1. Children's stories, Chinese. I. Chiang, Hsien-mei. II. P'an, Ch'ang-jen. III. Title: Baogu huxu hong.
PZ10.831.P36 78-842953
78 C

(P'u-t'ao-kou ti shao nien)
葡萄沟的少年 / 通化市《葡萄沟的少年》创编组. — [香港? : s. n., 1978?]
141 p. : ill. ; 21 cm.
"这是一本富有长白山区特色的少年儿童故事集"
Reprint of the 1977 ed. published by Chi-lin jen min ch'u pan she.

1. Children's stories, Chinese. I. T'ung-hua shih P'u-t'ao-kou ti shao nien ch'uang pien tsu.
PZ10.831.P84 1978 78-842656

CHILDREN'S STORIES, CHINESE (Continued)

(Ta yen nan fei ti shih hou)
大雁南飞的时候 / 江阴县儿童文学创作组. — [香港?: s. n., 1978?]
142 p. : ill. ; 21 cm.
Reprint of the 1977 ed. published by Shang-hai jen min ch'u pan she. Shanghai, issued in series: Shao nien erh t'ung wen i tu wu.
CONTENTS: 程玮 大雁南飞的时候.—陈光潜, 赵雲芬 急浪飞舟.—丁阿虎 会走路的肥料.—春晖 大雨哗哗.—孙方亮 十万个为什么和小弹弓.—金鑫, 文陶 小斌管水.—春晖 两只白洛克.—顾红 虹.—唐麒 蛙鼓声声.—程玮 桃树林.
1. Children's stories, Chinese. I. Chiang-yin hsien erh t'ung wen hsüeh ch'uang tso tsu.
PZ10.831.T2 78–842405

(T'e pieh jen wu)
特别任务 / 郑胤飞等著. — [香港?: s. n., 1978?]
68 p. : ill. ; 22 cm.
Reprint of the 1977 ed. published by Shang-hai jen min ch'u pan she, Shanghai.
1. Children's stories, Chinese. I. Cheng, Yin-fei.
PZ10.831.T4 1978 78–841336

(Tiao yü)
钓鱼. — [香港?: s. n., 1977]
97 p. : ill. ; 22 cm.
Reprint of the 1977 ed. published by Che-chiang jen min ch'u pan she.
CONTENTS: 王昌盛, 蒋陈华原作；王柏强, 张林泉改写 小莲学农.—云涛 大地作业.—陈志超 钓鱼.—瞿俊伟 三号订前.—张林泉, 俞汉创作；嘉兴县文化馆革命故事创作组修改 看瓜.—松岭 弹子的故事.—周策 浪里鲭大爷
1. Children's stories, Chinese.
PZ10.831.T45 1977 79–836387

CHILDREN'S STORIES, CUBAN

González, Omar, 1950-
Nosotros los felices / Omar González ; ilus., Roberto Fabelo. — La Habana : Casa de las Américas, [1978 or 1979]
68 p. : ill. ; 22 cm. — (Obras para niños y jóvenes)
CONTENTS: El güinal de los muertos.—De la gracia de mi padre, los juegos y los recién llegados de la guerra.—La muerte de un general.—Los sucesos de Varadero y las ocurrencias de tía Celina.—El presente.—Historia de dos amigos.—Tirando con morteros.—Cruce de cartas.
1. Children's stories, Cuban. I. Title. II. Series.
PZ73.G59 79-116702
79 MARC

CHILDREN'S STORIES, ENGLISH

Children in the wood.
The affecting history of the children in the wood. Hartford, Printed by J. Babcock, 1796.
30, [1] p. illus. 11 cm.
Evans 29955.
In prose.
"Story of three robbers": p. [23]-24; selections in verse: p. [25]-30.
Title page of L.C. copy follows p. 30; front. and p. [31] are mounted on inside of covers.
1. Children's stories, English. 2. Children's poetry, English. I. Title.
PZ6.C436 1796 78-319667
79 MARC

The Entertaining, moral, and religious repository; containing, upwards of three score separate performances, all of which are written in a simple yet pleasing stile ... for the amusement and instruction of the youth of both sexes. Published by a society in Great-Britain, instituted for the beneficent purpose of aiding the intention of His Majesty, as expressed in his royal proclamation for the suppression of vice and immorality. Elizabeth-Town, Printed by S. Kollock, for C. Davis, 1799.
324 p. 17 cm.
Evans 35298. Probably a companion vol. to Evans 35297 (same title; designated v. 1)
Previously published in 1798, with "In two volumes. Vol. II" on t. p. (Evans 35296, v. 2) In the present ed. the same sheets are used for p. [99]-324.
CONTENTS: Contents originally published in England as Cheap repository tracts.
1. Christian life—Juvenile fiction. 2. Children's stories, English. I. Cheap repository.
PZ6.E63 1799 78-317619
ᐩ78 MARC

Manes, Stephen, 1949-
Mule in the mail / by Stephen Manes ; pictures by Mary Chalmers. — New York : Coward, McCann & Geoghegan, c1978.
46 p. : ill. ; 24 cm.
SUMMARY: A toy mule longs for a home but undergoes some fearful adventures before that wish becomes reality.
ISBN 0-698-20453-0 lib. bdg. : $6.95
[1. Toys—Fiction. 2. Christmas stories] I. Chalmers, Mary, 1927- II. Title.
PZ8.9.M28 Mu [Fic] 77-26193
77 MARC

Rae, Gwynedd.
The Mary Plain omnibus / Gwynedd Rae ; illustrated by Irene Williamson. — London [etc.] : Routledge and Kegan Paul, 1976.
[8], 458 p. : ill. ; 20 cm. GB76-26142
CONTENTS: Mostly Mary.—All Mary.—Mary Plain in town.—Mary Plain on holiday.
ISBN 0-7100-8437-4 : £3.95
1. Children's stories, English. [1. Bears—Fiction] I. Williamson, Irene. II. Title.
PZ7.R1227 Mar [Fic] 77-364253
77[r79]rev MARC

Stoker, Bram, 1847-1912.
Under the Sunset / by Bram Stoker ; with ill. by W. Fitzgerald and W. V. Cockburn. — 1st American ed. — North Hollywood, Calif. : Newcastle Pub. Co., 1978.
ix, 190 p., [5] leaves of plates : ill. ; 22 cm. — (The Newcastle Forgotten Fantasy library ; v. 17)
CONTENTS: Under the Sunset.—The Rose Prince.—The invisible giant.—How 7 went mad.—Lies and lilies.—The castle of the king.—The wondrous child.
ISBN 0-87877-116-6 : $3.95
1. Children's stories, English. [1. Short stories] I. Fitzgerald, William. II. Cockburn, W. V. III. Title.
PZ7.S8745 Un 1978 [Fic] 78-113034
79 MARC

Stories from Asia today : a collection for young readers. — 1st ed. — New York : Weatherhill/Heibonsha, 1979-
v. : ill. ; 24 cm. — (Asian copublication programme series four ; book 1-)
"Sponsored by Asian Cultural Centre for Unesco."
SUMMARY: A collection of contemporary short stories from 14 Asian countries.
ISBN 0-8348-1038-7
1. Children's stories, Oriental—Translation into English. 2. Children's stories, English—Translation from Oriental languages. [1. Asia—Fiction. 2. Short stories] I. Asian Cultural Centre for Unesco.".
PZ5.S8819 1979 [Fic] 79-19059
79 MARC

Walt Disney's treasury of children's classics / edited by Darlene Geis. — New York : H. N. Abrams, 1978.
306 p. : ill. (some col.) ; 29 cm.
SUMMARY: Retells seventeen well-known fairy tales, folk tales, and short stories illustrated with scenes from Walt Disney films. Includes a behind-the-scenes look at the production of animated films.
ISBN 0-8109-0812-3
1. Children's stories, American. 2. Children's stories, English. 3. Fairy tales. 4. Tales. 5. Animation (Cinematography)—Juvenile literature. [1. Short stories. 2. Fairy tales. 3. Folklore. 4. Animation (Cinematography)] I. Geis, Darlene. II. Disney (Walt) Productions. III. Title: Treasury of children's classics.
PZ5.W167 [Fic] 78-3529
78 MARC

—TRANSLATION FROM ORIENTAL LANGUAGES

Stories from Asia today : a collection for young readers. — 1st ed. — New York : Weatherhill/Heibonsha, 1979-
v. : ill. ; 24 cm. — (Asian copublication programme series four ; book 1-)
"Sponsored by Asian Cultural Centre for Unesco."
SUMMARY: A collection of contemporary short stories from 14 Asian countries.
ISBN 0-8348-1038-7
1. Children's stories, Oriental—Translation into English. 2. Children's stories, English—Translation from Oriental languages. [1. Asia—Fiction. 2. Short stories] I. Asian Cultural Centre for Unesco.".
PZ5.S8819 1979 [Fic] 79-19059
79 MARC

CHILDREN'S STORIES, FAROESE

Kjeld, Jákup M 1901-1965.
Lítla Gunvør og kálvurin / J. M. Kjeld. — Tórshavn : Føroya Lærarfelag, 1976.
162 p. : ill. ; 20 cm. D77
Kr30.00
1. Children's stories, Faroese. I. Title.
PZ90.F3K58 1976 79-371533
79 MARC

CHILDREN'S STORIES, FRENCH

15 [i.e. Quinze] histoires de Provence / ill. de Alain d'Orange. — [Paris] : Gautier-Languereau, c1977.
220 p. : ill. ; 22 cm. — (Série 15 [i.e. Quinze]) F***
CONTENTS: Daudet, A. Mon moulin.—Vérité, M. Le prisonnier des barbaresques.—Vérité, M. Grenouilles-charbon.—Daudet, A. L'élixir du révérend père Gaucher.—Vérité, M. La mort de Clanclan.—Vérité, M. L'étoile du berger.—Vérité, M. Les trois bonheurs de Mariette.—Vérité, M. Le barbaroto de Cucuron.—Gasquet, M. Le blé de Noël.—Gasquet, M. Le "décoconnage" au mas.—Vérité, M. Bélugue et le mistral.—Vérité, M. Pauvre Cantaloup.—Vérité, M. La bravade des Espagnols.—Vérité, M. Pichelune.—Mauron, M. La Provence et son histoire.
ISBN 2-217-04070-7
1. Children's stories, French. 2. Provence—Juvenile fiction.
PZ21.Q553 843'.008 78-376266
78 MARC

CHILDREN'S STORIES, GERMAN

Die Kopfkissen-Gans und andere Geschichten von grossen Dichtern für kleine Leute / Vorw., Peter Härtling ; Bilder, Mario Grasso. — Frauenfeld [etc.] : Huber, cop. 1978.
190 p. : ill. (some col.) ; 28 cm. Sw78-11414
ISBN 3-7193-0620-8 : 24.80F
1. Children's stories, German. I. Härtling, Peter, 1933- . II. Grasso, Mario, fl. 1975-
PZ33.K592 79-345526
79 MARC

Kremer, Eva Maria, 1928-
Jesus hat viele Gesichter : wie Kinder in fernen Ländern Jesus erleben / Eva-Maria Kremer, Werner Zurfluh ; [Ill., Mona Helle-Ineichen]. — Luzern ; München : Rex-Verlag, c1975.
100 p. : ill. ; 22 cm. Sw75-A-7776
ISBN 3-7252-0279-6 : 18.00F
1. Children's stories, German. 2. Christian life—Juvenile literature. I. Zurfluh, Werner, joint author. II. Title.
PZ33.K6446 79-383696
*79 MARC

Mein Pferd kann fliegen : vierundzwanzig Geschichten für kleine und grosse Leser / hrsg. von Erika Klopp. — 1. Aufl. — Berlin : E. Klopp, c1975.
201 p. : ill. ; 20 cm. GFR***
CONTENTS: Bailey, B. Jimmy hat Glück.—Betke, L. Die Fahrt nach Jevensand.—Blyton, E. Hupf und Schlurr.—Dietz, L. Das Ende der grossen Katze.—Gast, L. Hände im Stall.—Heyst, I. van. Onkel Lügenpeter.—Höcker, K. Meine Schwester ist ein Held.—Inkiow, D. Das Ei, das Sonnenstrahlen isst.—Italiaander, R. Prahlen zahlt sich nicht aus.—Kleberger, I. Die Kunst, ein Mann zu werden.—Laurey, H. Serena, die Halbhexe.—Maran, T. Eine Frage der Höflichkeit.—Metka, D. Gedichte.—Neumeister, A. Der Mann mit dem Goldknopf.—Orgel-Köhne, L. u. A. Ein Hund nach Mass.—Patchett, M. Wenn mich der Mond weckte.—Reitmann, E. Kille und Kalle.—Richard, A. Das Geschenk des Meeres.—Saxegaard, A. Ein Zimmer bei der Schwiegermutter.—Schilling, M. Nanook, der Eisär.—Seelhorst, G. Der Löwe Fridolin.—Süssmann, C. Das schwarze Schäfchen.—Wahlstedt, V. Rettungseifer.—Klopp, E. Mein Pferd kann fliegen.
ISBN 3-7817-1030-0
1. Children's stories, German. I. Klopp, Erika.
PZ31.M39 79-358786
79 MARC

Mit mir nicht! : Geschichten / ausgew. von Weert Flemmig, Hans May und Hans-Heinrich Strube. — 1. Aufl. — Gütersloh : Gütersloher Verlagshaus Mohn, 1978.
128 p. : ill. ; 19 cm. — (Gütersloher Taschenbücher ; 283 : Siebenstern) GFR78-A
ISBN 3-579-03683-1 : DM5.80
1. Children's stories, German. I. Flemmig, Weert. II. May, Hans, Pastor. III. Strube, Hans-Heinrich.
PZ31.M58 78-400793
79 MARC

Die Räuber gehen baden : Gute-Nacht-Geschichten von A bis Z / illustriert von 52 Grafikern ; hrsg. von Beate Hanspach und Fred Rodrian. — 1. Aufl. — Berlin : Der Kinderbuchverlag, c1977.
220 p. : col. ill. ; 25 cm. GDR***
12.80M
1. Children's stories, German. I. Hanspach, Beate. II. Rodrian, Fred.
PZ31.R3 78-395480
79 MARC

—TRANSLATIONS FROM UKRAINIAN

Wilde Steppe, Abenteuer : Kosakengeschichten / hrsg. von Anna-Halja Horbatsch ; [Aus dem Urkainischen von Anna-Halja Horbatsch ; illustriert von Kurt Schmischke]. — Göttingen : W. Fischer, c1974.
160 p. : ill. ; 21 cm. GFR***
CONTENTS: Berdnyk, O. Das Scharmützel.—Schewtschuk, W. Die Rivalen.—Hinaus aufs blaue Meer.—Sturmnacht auf dem Schwarzen Meer.—Malyk, W. Die Lehrstunde.—Kulisch, P. Am Schandpfahl Kosakentreue.—Wowtschok, M. Ein geheimnisvoller Bote.
ISBN 3-439-00854-1
1. Cossacks—Juvenile fiction. 2. Children's stories, German—Translations from Ukrainian. 3. Children's stories, Ukrainian—Translations into German. I. Horbatsch, Anna-Halja (Lutziak), 1924-
PZ31.W53 78-370529
78 MARC

CHILDREN'S STORIES, GREEK

Tarsoulē, Geōrgia, ed.
(Ta prōta mou paramythia)
Τὰ πρῶτα μου παραμύθια. Συλλογὴ καὶ διασκευὴ Γεωργίας Ταρσούλη. Εἰκονογράφηση Ἑλένης Περάκη-Θεοχάρη. Ἀθῆναι, Ν. Ἀλικιώτης [c1954]
[32] p. illus. (part col.) 21 cm. (Τὰ Βιβλία τῆς Θείας Λένας, 2. [i. e. 1.] σειρά, ἀριθ. 3)
1. Children's stories, Greek. I. Perakē-Theocharē, Helenē, illus. II. Title.
PZ90.G7T32 64–51098

CHILDREN'S STORIES, JAPANESE

(Harapeko sensō)
はらぺこ戦争 / 東京・子どもと教師の文学の会編 ; 高橋透絵. — 東京 : ポプラ社, 昭和51 [1976]
165 p., [1] leaf of plates : ill. ; 22 cm. — (先生のとっておきの話 ; 5) Ja 76-20
Series romanized: Sensei no totteoki no hanashi.
¥680

1. Children's stories, Japanese. I. Takahashi, Tōru, 1939- II. Tōkyō Kodomo to Kyōshi no Bungaku no Kai.
PZ49.2.H32 78-808498

*78 J

(Kanashii gochisō)
かなしいごちそう / [編者 代田昇, 増村王子]. — 東京 : 岩崎書店, 1974.
130 p., [1] leaf of plates : ill. ; 23 cm. — (母が子に語るとっておきの話 ; 1) Ja***

1. Children's stories, Japanese. I. Shirota, Noboru, 1924- II. Masumura, Kimiko, 1913- III. Series: Haha ga ko ni kataru totteoki no hanashi ; 1.
PZ49.2.K35 79-805902

79 J

—HISTORY AND CRITICISM

Seki, Hideo, 1912-
(Jidō bungaku ron)
児童文学論 / 関英雄著. — 東京 : 新評論社, 昭和30 [1955]
274 p. ; 18 cm. — (教育新書 ; 8)
Series romanized: Kyōiku shinsho.

1. Children's stories, Japanese—History and criticism. 2. Children's stories—History and criticism. I. Title.
PN1009.J3S43 79-802849

—JAPAN—GIFU (PREFECTURE)

(Hazukashikatta hadakambo)
はずかしかったはだかんぼ / 岐阜・子どもと教師の文学の会編 ; 若山憲絵. — 東京 : ポプラ社, 1978.
149 p. : ill. ; 22 cm. — (先生のとっておきの話 ; 19) Ja 78-32
Series romanized: Sensei no totteoki no hanashi.
¥680

1. Children's stories, Japanese—Japan—Gifu (Prefecture) I. Gifu Kodomo to Kyōshi no Bungaku no Kai.
PZ49.2.H33 78-814386

--JAPAN--IWATE

(Ōkina wasuremono)
大きなわすれもの / 岩手・子どもと教師の文学の会編 ; 石倉欣二絵. — 東京 : ポプラ社, 1978.
165 p. : ill. ; 22 cm. — (先生のとっておきの話 ; 18) Ja 78-28
Series romanized: Sensei no totteoki no hanashi.
CONTENTS: 吉田繁紀 大きなわすれもの—大槻忠士 ほやほやパンツ—加藤義男 なぐられ損のびんた—柴田貴美子 ばくはつしたたまご—今松桂子 草とりの思い出—宍戸春雄 うさぎを売った日—菅野礼子 カエルなんかしらないよ—堀合英行 勇太のランドセル—福島敬次郎 オシッコとかくれんぼ
¥680

1. Children's stories, Japanese—Japan—Iwate. I. Ishikura, Kinji, 1937- II. Iwate Kodomo to Kyōshi no Bungaku no Kai.
PZ49.2.O47 78-813438

—JAPAN—KAGOSHIMA (PREFECTURE)

(Hitoribotchi no Midori)
ひとりぼっちのみどり / 鹿児島・子どもと教師の文学の会編 ; 大古尅己絵. — 東京 : ポプラ社, 1978.
165 p., [1] leaf of plates : ill. ; 22 cm. — (先生のとっておきの話 ; 20) Ja 78-40
Series romanized: Sensei no totteoki no hanashi.
CONTENTS: ひとりぼっちのみどり—圭太郎のラッパ—赤い自転車の思い出—ヘビ—劇にでなくても—赤いくじゃくの羽の着物—たき火
1. Children's stories, Japanese—Japan—Kagoshima (Prefecture) I. Ōfuru, Katsumi, 1936- II. Kagoshima Kodomo to Kyōshi no Bungaku no Kai.
PZ49.2.H5 78-816372

—JAPAN—MIE (PREFECTURE)

(Hama wa bokura no kyōshitsu da)
浜はぼくらの教室だ / 三重・子どもと教師の文学の会編 ; こさかしげる絵. — 東京 : ポプラ社, 1978.
165 p., [1] leaf of plates : ill. ; 22 cm. — (先生のとっておきの話 ; 21) Ja 79-4
Series romanized: Sensei no totteoki no hanashi.
¥680
1. Children's stories, Japanese—Japan—Mie (Prefecture) I. Kosaka, Shigeru, 1925- II. Mie Kodomo to Kyōshi no Bungaku no Kai.
PZ49.2.H315 79-807409

—KOREA (DEMOCRATIC PEOPLE'S REPUBLIC)

(Ch'ŏt palchaguk)
첫발자국 : 단편소설집. — [평양] : 금성청년출판사, 1976.
287 p. : ill. ; 21 cm.
CONTENTS: 정인학 해빛받는날.—김정 아버지의얼굴.—박춘삼 날개를활짝펼치라.—김정 일요일.—박현 유격대원의딸.—리태렬 바람부는날.—문희준 다리.—리준길 우리가맡은일.—리성지 첫분공.—김성웅 영웅을찾는소년.—리준길 속이큰아이.—리보흠 선녀산의무지개.—리관숙 은백색19호.—윤경수 우리는친형제.—리림수 해님의나라.
₩0.40

1. Children's stories, Korean—Korea (Democratic People's Republic) I. Kŭmsŏng Ch'ŏngnyŏn Ch'ulp'ansa.
PZ50.52.C48 79-825077

(Kkach'i ne saejiptŭri nal)
까치네새집들이날 : 동화집. — [평양] : 사로청출판사, 1974.
96 p. : col. ill. ; 21 cm.
CONTENTS: 차용구 까치네새집들이날.—김신복 점백이와얼룩이.—김우경 물방울.—최낙서 하늘로날아오른꽃발.—허원길 빈도토리깍지를받은베쾌지.—고수정 문어의각을뜬산호마을동무들.

1. Children's stories, Korean—Korea (Democratic People's Republic) I. Saroch'ŏng Ch'ulp'ansa.
PZ50.52.K56 79-826360

(K'ŭna k'ŭn sarang)
크나큰사랑. — 東京 : 학우서방, 1975.
123 p., [12] leaves of plates : ill. ; 19 cm. — (학생문고)
Series romanized: Haksaeng mun'go.

1. Children's stories, Korean—Korea (Democratic People's Republic) I. Hagu Sŏbang, Tokyo.
PZ50.52.K85 79-826382

(Ppalgan kŭmbungŏ)
빨간금붕어 : 동화집. — [평양] : 금성청년출판사, 1977.
95 p. : col. ill. ; 26 cm.
CONTENTS: 최낙서 [illegible].—[illegible] 빨간금붕어.—김우경 [illegible].—최낙서 [illegible].—전종심 꽃시계.—허원길 [illegible].—배풍 [illegible].—김우경 [illegible].
1. Children's stories, Korean—Korea (Democratic People's Republic) I. Kŭmsŏng Ch'ŏngnyŏn Ch'ulp'ansa.
PZ50.52.P7 79-825963

(Ttŭgŏun ŭnjŏng)
뜨거운은정 : 단편소설집. — [평양] : 금성청년출판사, 1977.
182 p. ; 21 cm.
CONTENTS: 박춘삼 샘줄기.—리림수 특별반.—김정 영원한불빛.—김익철 소년단인사.—박현 붉은넥타이.—최병환 꼬마척후병.—박춘삼 아버지.—최병환 낚시터의불빛.
₩0.45
1. Children's stories, Korean—Korea (Democratic People's Republic) 2. Kim, Il-sŏng, 1912- —Juvenile fiction. I. Kŭmsŏng Ch'ŏngnyŏn Ch'ulp'ansa.
PZ50.52.T77 79-825419

CHILDREN'S STORIES, ORIENTAL

Stories from Asia today : a collection for young readers. — 1st ed. — New York : Weatherhill/Heibonsha, 1979-
v. : ill. ; 24 cm. — (Asian copublication programme series four ; book 1-)
"Sponsored by Asian Cultural Centre for Unesco."
SUMMARY: A collection of contemporary short stories from 14 Asian countries.
ISBN 0-8348-1038-7
1. Children's stories, Oriental—Translation into English. 2. Children's stories, English—Translation from Oriental languages. [1. Asia—Fiction. 2. Short stories] I. Asian Cultural Centre for Unesco.".
PZ5.S8819 1979 [Fic] 79-19059
79 MARC

—TRANSLATION INTO ENGLISH

Stories from Asia today : a collection for young readers. — 1st ed. — New York : Weatherhill/Heibonsha, 1979-
v. : ill. ; 24 cm. — (Asian copublication programme series four ; book 1-)
"Sponsored by Asian Cultural Centre for Unesco."
SUMMARY: A collection of contemporary short stories from 14 Asian countries.
ISBN 0-8348-1038-7
1. Children's stories, Oriental—Translation into English. 2. Children's stories, English—Translation from Oriental languages. [1. Asia—Fiction. 2. Short stories] I. Asian Cultural Centre for Unesco.".
PZ5.S8819 1979 [Fic] 79-19059
79 MARC

CHILDREN'S STORIES, ROMANIAN

Cinste şi omenie : [culegere de povestiri, proverbe, maxime şi cugetări / coperta şi ilustraţiile de Gh. Cernoeanu]. — Bucureşti : Editura Ion Creangă, 1976.
131 p. : ill. ; 17 cm. — (Prima mea bibliotecă ; 49) R77-586
lei2.25
1. Children's stories, Romanian.
PZ90.R6C53 78-368537
78 MARC

CHILDREN'S STORIES, SLOVAK

Pozdrav z Bratislavy : [sborník současné slov. prózy pro mládež : pro čtenáře od 13 let / uspoř., předml. a medailónky autorů naps. Lýdie Kyselová ; ze slov. originálů přel. kol. ; il. kol.]. — 1. vyd. — Praha : Albatros, 1977.
192, [3] p. : ill. (some col.) ; 25 cm. — (Klub mladých čtenářů) Cz78
Kčs25.00
1. Children's stories, Slovak. 2. Slovak fiction—20th century. I. Kyselová, Lydia.
PZ70.S45P6 79-338149
*79 MARC

CHILDREN'S STORIES, UKRAINIAN

—TRANSLATIONS INTO GERMAN

Wilde Steppe, Abenteuer : Kosakengeschichten / hrsg. von Anna-Halja Horbatsch ; [Aus dem Urkainischen von Anna-Halja Horbatsch ; illustriert von Kurt Schmischke]. — Göttingen : W. Fischer, c1974.
160 p. : ill. ; 21 cm. GFR***
CONTENTS: Berdnyk, O. Das Scharmützel.—Schewtschuk, W. Die Rivalen.—Hinaus aufs blaue Meer.—Sturmnacht auf dem Schwarzen Meer.—Malyk, W. Die Lehrstunde.—Kulisch, P. Am Schandpfahl Kosakentreue.—Wowtschok, M. Ein geheimnisvoller Bote.
ISBN 3-439-00854-1
1. Cossacks—Juvenile fiction. 2. Children's stories, German—Translations from Ukrainian. 3. Children's stories, Ukrainian—Translations into German. I. Horbatsch, Anna-Halja (Lutziak), 1924-
PZ31.W53 78-370529
78 MARC

CHILDREN'S STORIES, VIETNAMESE

Hai bàn tay chiến sĩ : tập truyện và ký chọn lọc về đề tài kháng chiến chống Pháp, 1945-1954. — Hà-nội : Kim Đồng, 1970.
250 p. : ill. ; 19 cm.
1. Children's stories, Vietnamese. 2. Indochinese War, 1946-1954—Juvenile fiction.
PZ90.V5H3 78-984262
79 MARC

Hai làng Tà pình và Đông hía : tập truyện kháng chiến chống thực dân Pháp / Hoài An ... ₍et al.₎ ; bìa và minh họa của Hà Quang Phương. — Hà-nội : Kim Đồng, 1977.
173 p. : ill. ; 19 cm.
Short stories.
0đ45
1. Children's stories, Vietnamese. 2. Vietnam—History—1945-1975—Juvenile fiction. I. Hoài An. II. Hà Quang Phương.
PZ90.V5H34 79-984217
79 MARC

CHILDREN'S THEATER see Children's plays—Presentation, etc.

CHILDREN'S VILLAGES see Children —Institutional care

CHILDREN'S WITH AND HUMOR see Children—Anecdotes and sayins; Wit and humor, Juvenile

CHILDREN'S WRITINGS

see also School prose

An Exceptional view of life : the Easter seal story / written & illustrated by handicapped children ; originated and directed by Edward J. McGrath ; edited by Bob Kraus. — Norfolk Island, Australia : Island Heritage, c1977.
64 p. : col. ill. ; 27 cm. — (An Island Heritage book) Aus***
SUMMARY: Youngsters with a variety of physical disabilities express their feelings about their handicaps through words and art.
1. Handicapped children—Biography. 2. Children's art. 3. Children's writings. ₍1. Physically handicapped. 2. Children's art. 3. Children's writings₎ I. McGrath, Edward J. II. Kraus, Bob, 1924-
HV888.E9 1977 362.7'8'40922 77-73684
79 MARC

Morgan, Alice P
If you ask me ...! / By Alice P. Morgan ; ill. by Robert Duffy. — ₍Troy, Ala. : Troy State University Press, c1978.
₍32₎ leaves : ill. ; 28 cm.
"The questions and answers in this book were taken from a thesis written by the author ...."
ISBN 0-916624-23-4
1. Cognition in children. 2. Children's writings.
BF723.C5M67 500.9 78-62901
79 MARC

CHILDREN'S WRITINGS, AMERICAN

see also School verse, American

Let the children speak / edited by Nancy Welbourn ... ₍et al.₎. — ₍s.l.₎ : Teacher Education Projects, ₍1979?₎
vii, 125 p. : ill. ; 28 cm.
1. Children's writings, American. 2. American literature—Minority authors. I. Welbourn, Nancy.
PS508.C5L47 810'.8'09282 79-109426
00 MARC

—OHIO

They came to school dressed like flowers : an anthology, 1977 / edited by Robert Fox. — ₍Columbus?₎ : Published by the Ohio Foundation on the Arts for the Ohio Arts Council, c1978.
170 p. : ill. ; 22 cm.
SUMMARY: Poems by elementary and high school students in Ohio that resulted from the Ohio Poetry-in-the-Schools Program.
1. Children's writings, American—Ohio. 2. School verse, American—Ohio. 3. American poetry—Ohio—Bio-bibliography. 4. American poetry—20th century—Bio-bibliography. ₍1. American poetry—Collections. 2. Children's writings₎ I. Fox, Robert, 1943- . II. Ohio Arts Council.
PS571.O3T4 811'.5'408 78-112784
79 MARC

—PACIFIC STATES

Miracle finger : a book of works / by children ages two to fifteen ; with notes by parent and teacher poets ₍Dick Bakken ... et al.₎. — Portland, Or. : Salted feathers, 1975.
80 p. : ill. ; 28 cm. — (Salted feathers ; no. 11)
1. Children's writings, American—Pacific States. I. Bakken, Dick. II. Series.
PS591.S3M5 810'.8'09282 79-303814
79 MARC

—TENNESSEE—MEMPHIS

Artworks : an anthology of workshop poetry / David Spicer, editor ; ₍shell drawings by Marvina Wyatt₎. — Memphis : Brooks Memorial Art Gallery, c1979.
72 p. ; 21 cm.
ISBN 0-918518-10-5 : $3.95
1. American poetry—Tennessee—Memphis. 2. American poetry—20th century. 3. Children's writings, American—Tennessee—Memphis. I. Spicer, David. II. Brooks Memorial Art Gallery, Memphis, Tenn.
PS559.M4A77 811'.5'408 78-13298
78 MARC

CHILDREN'S WRITINGS, CATALAN

—SPAIN—CALDAS DE MONTBUY

Recull dels treballs guanyadors del 5 ₍i.e. quint₎ concurs literari local celebrat a Caldes de Montbui el dia 12 de Març del 1972. Caldes de Montbui ₍Associació Cultural₎ (1972)
48 p. 18 cm. Sp73-Apr
Cover title.
Prose and Poetry.
"Conferència pronunciada per Joan Alavedra en l'homenatge dedicat per la vila de Caldes de Montbui al mestre Enric Morera.": p. ₍33₎-47.
1. Children's writings, Catalan—Spain—Caldas de Montbuy. 2. Children's writings, Spanish—Spain—Caldas de Montbuy. 3. Morera, Enric, 1865-1942—Addresses, essays, lectures. 4. Musicians—Spain—Catalonia—Biography—Addresses, essays, lectures. I. Associació Cultural de Caldes de Montbui.
PC3927.R4 79-345294
79 MARC

CHILDREN'S WRITINGS, DUTCH

Kinderen schrijven de Koningin. — ₍'s-Gravenhage : Staatsuitgeverij, 1973₎
₍96₎ p. : ill. ; 14 cm. Ne***
ISBN 9012002184
1. Children's writings, Dutch. 2. Dutch letters.
PT5537.K5 79-412096
79 MARC

CHILDREN'S WRITINGS, ENGLISH

Night ride and sunrise : an anthology of new poems / edited and introduced by Edward Lowbury. — Aberystwyth, Wales : Celton Poetry, c1978.
72 p. ; 22 cm. GB***
Prepared on behalf of the British Migraine Association.
ISBN 0-86053-018-3 : £3.99. ISBN 0-86053-019-1 pbk.
1. English poetry—20th century. 2. Children's writings, English. I. Lowbury, Edward Joseph Lister. II. British Migraine Association.
PR1110.C5N54 821'.9'1408 78-322257
79 MARC

The Noel Streatfeild weekend story book : an anthology. — London : J. M. Dent, 1977.
153 p. : ill. ; 14 cm. GB***
CONTENTS: Morpurgo, M. Sam's duck.—Johnstone, H. The first winter of a filly.—Aiken, J. Who would not sing for Lycidas?—Cousins, P. Battle of the allies.—Farmer, P. A tall story.—Farjeon, A. Song of the captured princess.—Williams, U. M. Family feeling.—Macleod, I. The battle.—Streatfeild, N. Green silk.—Samjar, R. The desert.—Treadgold, M. The cornfield.—Hughes, T. A moon-lily.—Williams, J. The question box.—Patience, I. Mr. Plintylooloolockton.—Smith, S. S. The burning mountain.—Willcox, L. Thoughts.—Mahy, M. The man who lived in a tree.—Marjoram, J. The wounded tree.—Cooper, L. The difficult one.—Thompson, S. Cold.—Simpson, J. M. Dragon marmalade.—MacGill, P. The farmer's boy.—Roberts, C. G. D. Gentleman in feathers.—Barton, S. Heron.
ISBN 0-460-06766-4 : £3.50
1. Children's literature. 2. Children's writings, English. I. Streatfeild, Noel.
PZ5.N54 820'.8'09282 79-309618
79 MARC

—ENGLAND—CAMBRIDGESHIRE

Moving jewels : an anthology of writing and art / by Cambridgeshire schoolchildren ; ₍editor Christopher Martin₎. — ₍Cambridge₎ : EARO for the George Edwards Memorial Committee, ₍1977₎
80 p. : ill. (some col.), port. ; 21 cm. GB77-33933
ISBN 0-904463-24-9 : £1.50
1. Children's writings, English—England—Cambridgeshire. 2. Children's art—England—Cambridgeshire. I. Martin, Christopher. II. Edwards, George, 1909-1975. III. George Edwards Memorial Committee.
PR1110.C5M6 700'.9426'5 78-322431
79 MARC

CHILDREN'S WRITINGS, FINNISH

Tapiolan lapset piirtävät ja kirjoittavat. ₍Julk.₎ (Lions club Tapiola). ₍Tapiola, 1968₎.
63 p. illus. 20 cm. Fi
1. Children's writings, Finnish. 2. Children's art—Finland.
PH342.T36 78-383979
78 MARC

CHILDREN'S WRITINGS, FRENCH

see also School verse, French

Bing, Élisabeth.
Et je nageai jusqu'à la page : vers un atelier d'écriture / Élisabeth Bing. — ₍Paris₎ : Éditions des Femmes, 1976.
326 p. : ill. (some col.) ; 18 cm. F76-9677
ISBN 2-7210-0041-1 : 35.00F
1. Creative writing (Elementary). 2. Children's writings, French. I. Title.
LB1576.B493 372.6'34 78-384694
78 MARC

—FRANCE—DIEPPE

Romanciers à treize ans : une expérience pédagogique / ₍textes réunis et présentés par₎ Pascal Bouchard, avec la collaboration de Aude Kreisler et Pascale Gruson. — Paris : Denoël-Gonthier, 1978.
261 p. : ill. ; 22 cm. — (Collection Femme) F78-8567
49.00F
1. Children's writings, French—France—Dieppe. 2. French language—Study and teaching—France—Dieppe. 3. Children as authors. I. Bouchard, Pascal. II. Kreisler, Aude. III. Gruson, Pascale.
PQ1109.5.C5R6 840'.8'09282 79-358649
79 MARC

—FRANCE—SEINE (DEPT.)

Poètes dans les écoles : ₍textes / présentés par₎ Daniel Habrekorn. — Vanves : Thot, c1978.
159 p. : ill. ; 22 cm. F***
Bibliography: p. 159.
33.00F
1. Children's writings, French—France—Seine (Dept.) 2. School verse, French—France—Seine (Dept.) I. Habrekorn, Daniel.
PQ1109.5.C5P67 841'.9'1408 79-360771
79 MARC

CHILDREN'S WRITINGS, GERMAN

Ich übe die Verteidigung : 37 Texte von Jugendlichen / hrsg. von Irene Johanson. — Stuttgart : Urachhaus, 1976.
104 p. ; 21 cm. GFR77-A
ISBN 3-87838-208-1 : DM10.00
1. German literature—20th century. 2. Children's writings, German. I. Johanson, Irene.
PT1141.I25 838'.9'1407 79-376321
79 MARC

CHILDREN'S WRITINGS, ITALIAN

Viviani, Cesare.
La pazzia spiegata dai bambini / Cesare Viviani. — Milano : Il formichiere, 1976.
128 p. ; 23 cm. — (Spazio aperto) It77-Mar
L3000
1. Mental illness—Addresses, essays, lectures. 2. Children's writings, Italian. 3. Children—Italy—Attitudes. I. Title.
RC458.V57 79-382215
*79 MARC

CHILDREN'S WRITINGS, JAPANESE

(Mitsubachi no ko)
みつばちの子：青森県黒石小学校四年生の生活記録 / 鈴木喜代春編. — 東京：東洋書館, 1952.
232 p. ; 19 cm.

1. Children's writings, Japanese. I. Suzuki, Kiyoharu, 1925–
PL757.4.C45M57 78–811477

CHILDREN'S WRITINGS, JAPANESE (Continued)

(Okāsan)
おかあさん：詩集 / [監修 高田敏子，吉野弘]. — 東京：ハナマルキ味噌出版部：発売 大和書房，1978.
217 p. : ill. ; 18 cm. Ja 78-5008
¥580

1. Mothers—Juvenile poetry. 2. Children's poetry, Japanese. 3. Children's writings, Japanese. 4. Japanese poetry—20th century. I. Takada, Toshiko, 1916- II. Yoshino, Hiroshi, 1926-
PL757.4.C45O37 79-805754

Sasaki, Keizō.
(Kokoro no dezain)
心のデザイン：刺激から創造 / 佐々木敬三著. — 仙台：宝文堂出版販売：宝文堂発売，昭和52 [1977]
266 p. : ill. ; 21 cm. Ja 77-22708
Bibliography: p. 265-266.
¥1500

1. Creative thinking (Education) 2. Children's writings, Japanese. 3. Children's art—Japan. I. Title.
LB1062.S313 79-80(

(Shōgakkō dōtoku jisaku shiryō senshū)
小学校道徳自作資料選集 / 井上治郎，伴三男編. — 東京：明治図書出版，1975-
v. ; 19 cm. Ja 75-8740 (v. 1)
¥1800 (v. 1)

1. Moral education (Elementary)—Japan. 2. Children's writings, Japanese. I. Inoue, Jirō, 1923- II. Ban, Mitsuo.
LC315.J3S53 78-813077

(Wasure-enu jidōshi)
忘れえぬ児童詩 / 日本作文の会編. — 東京：民衆社，1977.
2 v. ; 20 cm. Ja 78-5490
Bibliography: v. 2, p. 261-262.
¥1200 per vol.

1. Children's poetry, Japanese. 2. Children's writings, Japanese. 3. Children's poetry, Japanese—History and criticism. I. Nihon Sakubun no Kai.
PL757.4.C45W37 78-811721
*78 J

—CALIFORNIA

(Dainisei Nihongo sakubun shū)
第二世日本語作文集 / [編纂 南加日本語學園協會]. — Los Angeles：同協會，昭和14 [1939]
154 p., [1] leaf of plates : ill. ; 19 cm.
Cover title.

1. Children's writings, Japanese—California. 2. School prose. I. Nanka Nihongo Gakuen Kyōkai.
PL756.C5D34 78-813039
78 J

CHILDREN'S WRITINGS, KOREAN

(Il hanŭn aidŭl)
일하는아이들 / 李五德. — 서울：靑年社，1978.
291 p. : ill. ; 19 cm.
₩1000

1. Children's writings, Korean. I. Yi, O-dŏk, 1925-
PL975.8.C5I4 79-825111

(Parŭn kŭl koun kkum)
바른글고운꿈：교사와아동이엮은 / [엮은이충북글짓기지도회]. — 서울：세종문화사，1975-
v. : ill. ; 22 cm.

1. Korean language—Study and teaching (Elementary)—Korea—Ch'ungch'ŏng-pukto. 2. Korean language—Composition and exercises. 3. Children's writings, Korean. I. Ch'ungbuk Kŭl Çhikki Chidohoe.
LB1577.K6P37 79-826515

CHILDREN'S WRITINGS, RAETO-ROMANCE

—ITALY—FRIULI-VENEZIA GIULIA

Concors di un compit par furlàn in tes scuelis elementârs e mèdiis : Gurizze, Pordenon, Udin : 1975-1976 / [a cura di Giuseppe Bergamini]. — Udine : Societât filologjche furlane, 1977.
45 p. ; 24 cm. It***
1. Children's writings, Raeto-Romance—Italy—Friuli-Venezia Giulia. I. Bergamini, Giuseppe.
PC947.6.C6 859'.9 79-371841
79 MARC

CHILDREN'S WRITINGS, ROMANIAN

Zburătorii : culegere de versuri şi proză / îngrijită de prof. Gh. Izbăşescu ; Cenaclul literar "Zburătorul" ; Casa Pionierilor Gh. Gheorghiu-Dej. — [Gheorghe Gheorghiu Dej] : Consiliul Judeţean al Organizaţiei Pionierilor, 1976.
107 p. : ill. ; 22 cm. R77-4415
lei9.80
1. Children's writings, Romanian. I. Izbăşescu, Gh. II. Organizaţia Pionierilor din Republica Socialistă România. Consiliul Judeţean Bacău.
PC830.5.C5Z2 78-370525
*78 MARC

—ROMANIA—ALBA (JUDEŢ)

Recunoştinţă vie patriei şi partidului iubit : [culegere de versuri ale pionierilor din judeţul Alba ; cuvînt înainte de Ion Sângereanu ; coperta, Nicolae Haţegan]. — Alba Iulia : [s.n.], 1974.
80 p. : ill. ; 20 cm. R75-518
At head of title: Comitetul Judeţean Alba al U.T.C. Consiliul Judeţean Alba al Organizaţiei Pionierilor.
lei4.15
1. Children's writings, Romanian—Romania—Alba (Judeţ) 2. Patriotic poetry, Romanian—Romania—Alba (Judeţ) I. Uniunea Tineretului Comunist. Comitetul Judeţean Alba. II. Organizaţia Pionierilor din Republica Socialistă România. Consiliul Judeţean Alba.
PC830.5.C5R4 78-393919
*79 MARC

—ROMANIA—BISTRIŢA-NĂSĂUD

Ascendenţă : culegere de versuri şi proză. — (Ediţia a 3-a) / [au colaborat, Silviu Ignat, Nicolae Rebreanu, Leonida Dănilă ... ; cuvînt înainte, Valer Cîmpean ; coperta, prof. Domiţian Barteş]. — Bistriţa : Comitetul judeţean al U.T.C., 1974.
119 p. ; 20 cm. R75-901
At head of title: Uniunea Tineretului Comunist. Comitetul Judeţean Bistriţa-Năsăud.
lei12.00
1. Children's writings, Romanian—Romania—Bistriţa-Năsăud. 2. Romanian literature—Romania—Bistriţa-Nasaud. I. Uniunea Tineretului Comunist. Comitetul Judeţean Bistriţa-Năsăud.
PC830.5.C5A7 78-393914
*79 MARC

—ROMANIA—MEHEDINTI

Din inimi tinereşti : [din creaţia uteciştilor din judeţul Mehedinţi / cuvînt înainte de Pavel Ciobanu]. — [Drobeta Turnu-Severin] : Uniunea Tineretului Comunist, Comitetul Judeţean Mehedinţi, 1977.
58 p. : ill. ; 21 cm. — (Cîntarea României) R78-619
lei5.80
1. Children's writings, Romanian—Romania—Mehedinţi. I. Uniunea Tineretului Comunist. Comitetul Judeţean Mehedinţi.
PC830.5.C5D5 78-371586
*78 MARC

—ROMANIA—SIBIU (JUDEŢ)

Anii împlinirilor : plachetă de versuri ale elevilor din judeţul Sibiu, dedicată celei de a 30-a aniversări a eliberării şi Congresului al XI-lea a P.C.R. — Sibiu : [s.n.], 1974.
60 p. : ill. ; 21 cm. R75-1869
At head of title: Uniunea Tineretului Comunist. Comitetul Judeţean Sibiu.
lei3.00
1. Children's writings, Romanian—Romania—Sibiu (Judeţ) 2. Patriotic poetry, Romanian—Romania—Sibiu (Judeţ) I. Uniunea Tineretului Comunist. Comitetul Judeţean Sibiu.
PC830.5.C5A4 78-393915
*79 MARC

Raze de soare : creaţii literare închinate anului jubiliar. — Sibiu : [s.n.], 1974.
69 p. ; 17 cm. R75-3587
At head of title: Consiliul Jud· ;ean Sibiu al Organizaţiei Pionierilor.
Includes bibliographical references.
lei7.20
1. Children's writings, Romanian—Romania—Sibiu (Judeţ) 2. Romanian literature—Romania—Sibiu (Judeţ) I. Organizaţia Pionierilor din Republica Socialistă România. Consiliul Judeţean Sibiu.
PC830.5.C5R33 78-393943
*79 MARC

—ROMANIA—TIMIŞ

Muguri în lumină : [culegere literară din creatiile pionierilor din judeţul Timiş / sub îngrijirea, prof. Mirela Frigioiu, prof. Iulia Milin ; coperta şi ilustraţiile, prof. Tiberiu Simon, prof. Margareta Mureşan]. — [Timişoara] : Consiliul Judeţean Timiş al Organizaţiei Pionierilor, 1977.
58 p. : ill. ; 20 cm. R78-3843
lei6.00
1. Children's writings, Romanian—Romania—Timiş. 2. Romanian literature—Romania—Timiş. I. Organizaţia Pionierilor din Republica Socialistă România. Consiliul Judeţean Timiş.
PC830.5.C5M8 79-351375
*79 MARC

—ROMANIA—VASLUI (JUDEŢ)

Imn sentimental : culegere de versuri ale elevilor dedicată celei de a XXX-a aniversări a eliberării patriei de sub dominaţia fascistă şi celui de al XI-lea Congres al P.C.R. — Vaslui : Uniunea Tineretului Comunist, Comitetul Judeţean Vaslui, 1974.
51 p. ; 20 cm. R75-2623
lei5.20
1. Children's writings, Romanian—Romania—Vaslui (Judeţ) 2. Patriotic poetry, Romanian—Romania—Vaslui (Judeţ) I. Uniunea Tineretului Comunist. Comitetul Judeţean Vaslui.
PC830.5.C5I4 78-393916
*79 MARC

"La ce visează mugurii" : versuri scrise de copii / lucrare realizată cu sprijinul profesorilor, Elena Condrea, Mihai Gheorghe, Teodor Pracsiu ; [coperta şi vignetele au fost executate de pionierii, Daniela Pintilie, Liviu Neamţu, prof. Eugen Iftenie]. — Vaslui : [s.n.], 1973.
94 p. : ill. ; 20 cm. R74-2778
At head of title: Organizaţia Pionierilor din Republica Socialistă România. Consiliul Judeţean Vaslui.
1. Children's writings, Romanian—Romania—Vaslui (Judeţ) 2. Romanian poetry—Romania—Vaslui (Judeţ) I. Organizaţia Pionierilor din Republica Socialistă România. Consiliul Judeţean Vaslui.
PC830.5.C5C4 78-393920
*79 MARC

CHILDREN'S WRITINGS, RUSSIAN

—HISTORY AND CRITICISM

Levin, Vadim Aleksandrovich.
(Vospitanie tvorchestva)
Воспитание творчества / В. А. Левин. — Москва : Знание, 1977.
63 p. ; 16 cm. — (Новое в жизни, науке, технике : Серия Педагогика и технология ; № 7, 1977)
Includes bibliographical references.
0.11rub
1. Poetry and children. 2. Children's writings, Russian—History and criticism. I. Title. II. Series: Novoe v zhizni, nauke, tekhnike : Serii͡a Pedagogika i tekhnologii͡a ; 1977, no. 7.
PN1085.L48 79-407719

CHILDREN'S WRITINGS, SPANISH

Algo se rompe : textos libres y dibujos elaborados por niños de (a 14 años / [colección animada y coordinada por Antonio Fernández ... et al.]. — [Bilbao : Zero ; distribuidor exclusivo, ZYX], c1979.
64 p. : col. ill. ; 28 cm. — (El Trastero ; 1) Sp***
Cover title.
ISBN 8431704950
1. Children's writings, Spanish. 2. Children's art—Spain. I. Fernández, Antonio, fl. 1976-
PQ6173.5.C45A4 79-109090
79 MARC

—SPAIN—CALDAS DE MONTBUY

Recull dels treballs guanyadors del 5 [i.e. quint] concurs literari local celebrat a Caldes de Montbui el dia 12 de Març del 1972. Caldes de Montbui [Associació Cultural] (1972)
48 p. 18 cm. Sp73-Apr
Cover title.
Prose and Poetry.
"Conferència pronunciada per Joan Alavedra en l'homenatge dedicat per la vila de Caldes de Montbui al mestre Enric Morera.": p. [33]-47.
1. Children's writings, Catalan—Spain—Caldas de Montbuy. 2. Children's writings, Spanish—Spain—Caldas de Montbuy. 3. Morera, Enric, 1865-1942—Addresses, essays, lectures. 4. Musicians—Spain—Catalonia—Biography—Addresses, essays, lectures. I. Associació Cultural de Caldes de Montbui.
PC3927.R4 79-345294
79 MARC

CHILDREN'S WRITINGS, SWISS

Ein Bernerschädel ist nicht fon Blastigg : [Berner Kinder schreiben über Bern]. — Bern : Benteli, c1975.
70 p. ; 22 cm. Sw75-A-6461
ISBN 3-7165-0019-4 : 9.80F
1. Children's writings, Swiss. 2. Bern—Poetry. 3. Swiss poetry (German)—20th century. 4. Swiss poetry (German)—Switzerland—Bern.
PT1109.C5B4 79-384285
*79 MARC

CHILE

—ANTIQUITIES

Brown, John Macmillan, 1846-1935.
The riddle of the Pacific / by J. Macmillan Brown. — 1st AMS ed. — New York : AMS Press, 1979.
xii, 312 p., [62] leaves of plates : ill. ; 24 cm.
Reprint of the 1924 ed. published by T. F. Unwin, London.
Includes index.
ISBN 0-404-14205-2
1. Easter Island—Antiquities. 2. Sepulchral monuments—Easter Island. 3. Polynesians in Easter Island. 4. Chile—Antiquities. I. Title.
F3169.B87 1979 996'.18 75-35177
78 MARC

Echaiz, René León.
Prehistoria de Chile Central / René León Echaiz. — Buenos Aires : Editorial Francisco de Aguirre, c1976.
x, 122 p., [11] leaves of plates : ill. ; 19 cm. — (Biblioteca Antártica (pocket) ; 24)
Bibliography: p. 113-114.
Includes index.
1. Indians of South America—Chile—Antiquities. 2. Chile—Antiquities. I. Title.
F3069.E3 1976 78-110911
79 MARC

Núñez A , Lautaro.
Sobre el comienzo de la agricultura prehistórica en el Norte de Chile : estudio de arqueología andina / Lautaro Núñez A. — La Paz : Centro de Investigaciones Arqueológicas en Tiwanaku, 1972.
24 p. : ill. ; 22 cm. — (Publicación - Centro de Investigaciones Arqueológicas en Tiwanaku ; no. 4)
Cover title.
Bibliography: p. 19-20.
1. Indians of South America—Chile—Agriculture. 2. Indians of South America—Chile—Antiquities. 3. Chile—Antiquities. I. Title. II. Series: Centro de Investigaciones Arqueológicas en Tiwanaku. Publicación ; no. 4.
F3319.1.T55C43 no. 4 79-366980
[F3069.3.A3] 79 MARC

—ANTIQUITIES—PERIODICALS

Chungará. no. 1/2- año 1972/73-
Arica [Chile] Universidad del Norte, Depto. de Antropología.
no. ill. 27 cm.
1. Indians of South America—Chile—Arica (Dept.)—Antiquities—Periodicals. 2. Arica, Chile (Dept.)—Antiquities—Periodicals. 3. Excavations (Archaeology)—Chile—Arica (Dept.)—Periodicals. 4. Chile—Antiquities—Periodicals. I. Universidad del Norte. Departamento de Antropología.
F3069.1.A74C47 983'.1 76-647090
MARC-S

—ARMED FORCES

Rivas, Fernando.
Las fuerzas armadas de Chile : un caso de penetración imperialista / Fernando Rivas Sánchez, Elisabeth Reimann Weigert. — 1. ed. — México : Ediciones 75, 1976.
xiii, 295 p. ; 18 cm.
Bibliography: p. 283-291.
1. Chile—Armed Forces. 2. Chile—History—Coup d'etat, 1973. 3. Military assistance, American—Chile. 4. Military education—Chile. I. Reiman, Elisabeth, joint author. II. Title.
UA622.R58 78-103997
79 MARC

—BIBLIOGRAPHY

Oppenheimer, Robert, 1942-
Chile / compiled by Robert Oppenheimer. — Los Angeles : Latin American Studies Center, California State University, [1977?]
vi, 91 p., [1] leaf of plates : ill. ; 27 cm. — (Latin America bibliography series ; #6)
Cover title.
1. Chile—Bibliography. I. Title. II. Series.
Z1701.O66 016.983 79-620691
[3058] 79 MARC

—BIBLIOGRAPHY—CATALOGS

Biblioteca José A. Echeverría.
Bibliografía de Chile : Biblioteca José A. Echeverría, Casa de las Américas. — La Habana : Editorial Orbe, 1977.
195 p. ; 23 cm.
Errata slip inserted.
1. Chile—Bibliography—Catalogs. 2. Biblioteca José A. Echeverría—Catalogs. I. Title.
Z1701.B53 1977 016.983 78-106029
[F3058] 79 MARC

—BIOGRAPHY

Presa Casanueva, Rafael de la.
Venida y aporte de los españoles a Chile independiente / Rafael de la Presa Casanueva. — Santiago de Chile : [s.n.], 1978.
686 p. : port. ; 27 cm.
1. Spaniards in Chile—History. 2. Chile—Biography. 3. Chile—Relations (general) with Spain. 4. Spain—Relations (general) with Chile. 5. Chile—History. I. Title.
F3285.S7P73 983'.004'61 78-109784
79 MARC

Subercaseaux Browne, Julio, b. 1870.
Reminiscencias / Julio Subercaseaux Browne ; prólogo del autor. — Santiago, Chile : Editorial Nascimento, 1976.
322 p. ; 18 cm. — (Biblioteca popular Nascimento)
1. Subercaseaux Browne, Julio, b. 1870. 2. Chile—Biography. I. Title.
CT698.S9A37 983'.06'0924 77-150291
78 MARC

—BOUNDARIES

Chile. Embajada. U.S. Cultural Dept.
Chilean Antarctica, Easter Island, and Juan Fernandez Archipelago. — [Washington : Cultural Dept., Embassy of Chile, 1978]
27 p. : ill. ; 28 cm. — (Chilean cultural panorama ; 3)
Cover title.
1. Easter Island. 2. Juan Fernández (Islands) 3. Antarctic regions. 4. Chile—Boundaries. I. Title. II. Series.
F3169.C55 1978 996'.18 79-104906
79 MARC

—BOUNDARIES—ARGENTINE REPUBLIC

Argentina, Chile : present situation on the border controversy in the southern region = Estado actual de las diferencias sobre limites en la zona Austral. — Washington : Embassy of the Argentine Republic, 1978.
v, 24 p. : map ; 23 cm.
Cover title.
English and Spanish.
1. Argentine Republic—Boundaries—Chile. 2. Chile—Boundaries—Argentine Republic. I. United States. Embassy. Argentine Republic. II. Title.
F2986.A67 341.42 79-100419
79 MARC

Argentina en el Atlántico, Chile en el Pacífico / Isaac F. Rojas, Arturo L. Medrano. — [Buenos Aires : Nemont Ediciones, c1978]
47 [i.e. 57] p. : maps ; 20 cm.
Cover title.
CONTENTS: Rojas, I. F. La cuestión del Beagle y de las islas argentinas de la zona austral usurpadas por Chile.—Medrano, A. Origenes y desarrollo del problema del Beagle.
1. Beagle Channel, Chile. 2. Argentine Republic—Boundaries—Chile. 3. Chile—Boundaries—Argentine Republic. I. Rojas, Isaac Francisco, 1906- Cuestión del Beagle y de las islas argentinas de la zona austral usurpadas por Chile. 1978. II. Medrano, Arturo L. Origenes y desarrollo del problema del Beagle. 1978.
F2986.A68 341.42 79-109669
79 MARC

Controversia en la región del Canal Beagle : laudo arbitral. — Ed. bilingüe con notas y documentos adicionales. — [Santiago de Chile] : República de Chile, 1977.
xv, 498 p. : maps (4 fold., col. in pocket) ; 26 cm.
Added t.p.: Controversy concerning the Beagle Channel region.
Cover title: Canal Beagle laudo-award, 1977.
On spine: Arbitraje Canal Beagle. Beagle Channel Arbitration.
Spanish and English.
Includes text of Award of Her Britannic Majesty's Government pursuant to the agreement for arbitration (compromiso) of a controversy between the Argentine Republic and the Republic of Chile concerning the region of the Beagle Channel.
Includes bibliographical references.
1. Argentine Republic—Boundaries—Chile. 2. Chile—Boundaries—Argentine Republic. 3. Beagle Channel, Chile. I. Chile. II. Great Britain. Award of Her Britannic Majesty's Government pursuant to the agreement for arbitration (compromiso) of a controversy between the Argentine Republic and the Republic of Chile concerning the region of the Beagle Channel. 1977. III. Title: Controversy concerning the Beagle Channel region. IV. Title: Canal Beagle laudo-award, 1977. V. Title: Arbitraje Canal Beagle. VI. Title: Beagle Channel arbitration.
JX1531.Z7C43 341.42 78-390596
79 MARC

Fitte, Ernesto J
Los límites con Chile / Ernesto J. Fitte. — Buenos Aires : Editorial Plus Ultra, 1977.
84 p. ; 23 cm.
Includes bibliographical references.
1. Argentine Republic—Boundaries—Chile. 2. Chile—Boundaries—Argentine Republic. 3. Beagle Channel, Chile. I. Title.
F2851.F536 911'.82 78-109693
79 MARC

Great Britain.
Award of Her Britannic Majesty's Government pursuant to the agreement for arbitration (compromiso) of a controversy between the Argentine Republic and the Republic of Chile concerning the region of the Beagle Channel. — London : H.M.S.O., 1977.
[6], a-f, xvii, 152 p. & atlas (4 fold. leaves of plates : col. maps) ; 25 cm.
GB77-31058
"Reproduces the text of the original signed Report and Decision of the Court."
Includes bibliographical references.
ISBN 0-11-591649-0 : £15.00
1. Argentine Republic—Boundaries—Chile. 2. Chile—Boundaries—Argentine Republic. 3. Beagle Channel, Chile. I. Title.
JX1531.Z7C438 1977 341.42 78-322534
79 MARC

Rodríguez Zía, Jorge L
De mar a mar : el fallo del Beagle / Jorge L. Rodríguez Zía. — [Buenos Aires] : Editorial Moharra : distribución en Gran Bs. As., Distribuidora Rubbo, [1978]
[32] p. : maps ; 29 cm. — (Temas argentinos trascendentales ; 1)
Bibliography: p. [25]
$800.00
1. Argentine Republic—Boundaries—Chile. 2. Chile—Boundaries—Argentine Republic. 3. Beagle Channel, Chile. I. Title. II. Series.
F2851.R64 341.42 78-109871
79 MARC

Salguero, Ramón.
Todo sobre el Beagle / Ramón Salguero. — Buenos Aires : El Cid Editor, [1978]
82 p. : ill. ; 23 cm. — (Colección Geopolítica ; 7)
1. Beagle Channel, Chile. 2. Argentine Republic—Boundaries—Chile. 3. Chile—Boundaries—Argentine Republic. I. Title. II. Series.
F2986.S24 983'.6 79-109660
79 MARC

—BOUNDARIES—ARGENTINE REPUBLIC—ADDRESSES, ESSAYS, LECTURES

La Argentina en el Beagle y Atlántico Sur / [J. S. Campobassi ... et al. ; coordinación, Isaac Francisco Rojas]. — [Buenos Aires] : Nemont Ediciones : distribución exclusiva, Editorial Diagraf, [1978-
v. : maps ; 20 cm.
1. Argentine Republic—Boundaries—Chile—Addresses, essays, lectures. 2. Chile—Boundaries—Argentine Republic—Addresses, essays, lectures. 3. Beagle Channel, Chile—Addresses, essays, lectures. I. Campobassi, José Salvador.
F2851.A66 78-110223
79 MARC

El Conflicto del Beagle / Juan E. Guglialmelli y otros. — Buenos Aires : El Cid Editor, [1978]
237 p. : maps ; 23 cm. — (Colección Geopolítica ; 8)
Articles previously published in Estrategia.
Includes documents.
Includes bibliographical references.
CONTENTS: Bianchi y von Kirch, A. El Canal de Beagle: fábula y especulación.—Rizzo Romano, A. H. Aspectos legales del arbitraje solicitado por Chile en la cuestión del Beagle.—Bianchi y von Kirch, A. El Beagle en los Pactos de julio.—Pillet, G. La cuestión del Canal de Beagle.—Capitán Piedrabuena. Sin información sobre el Beagle?—Rizzo Romano, A. H. La Constitución nacional y el Acuerdo arbitral argentino-chileno-británico de 1971.—Orstein, R. M. Consideraciones técnicas sobre el límite austral del mar argentino.—Ruggeri, A. Canal de Beagle, algunas reflexiones sobre el laudo arbitral.—Carrió, G. R. Una crítica al laudo arbitral dictado en el caso del Beagle.—Gobbi, H. J. Problemas australes argentino-chilenos.—Montañez Santiago, J. M. La cuestión al oriente del Cabo de Hornos.—Díaz Loza, F. Geopolítica de Chile.—Guglialmelli, J. E. El área meridional del Atlántico suroccidental, la geopolítica de Chile y el laudo del Beagle.—Guglialmelli, J. E. Cuestión del Beagle, negociación directa o diálogo de armas.
1. Beagle Channel, Chile—Addresses, essays, lectures. 2. Argentine Republic—Boundaries—Chile—Addresses, essays, lectures. 3. Chile—Boundaries—Argentine Republic—Addresses, essays, lectures. I. Guglialmelli, Juan Enrique. II. Bianchi y von Kirch, A. III. Series.
F2986.C65 983'.6 79-108232
79 MARC

—BOUNDARIES—ARGENTINE REPUBLIC—BIBLIOGRAPHY

Matijevic, Nicolás.
Bibliografía sobre el Canal de Beagle : primera contribución / Nicolás Matijevic ; palabras liminares por Camilo H. Rodríguez Berrutti. — Bahía Blanca, Argentina : Universidad Nacional del Sur, Departamento de Ciencias Sociales, Centro de Documentación Patagónica, [1978 or 1979]
iii, 26 leaves ; 25 cm.
Includes index.
1. Beagle Channel, Chile—Bibliography. 2. Chile—Boundaries—Argentine Republic—Bibliography. 3. Argentine Republic—Boundaries—Chile—Bibliography. I. Title.
Z1634.B42M37 016.983'6 79-109145
[F2986] 79 MARC

CHILE (Continued)

—BOUNDARIES—BOLIVIA

Botelho Gosálvez, Raúl, 1917-
Breve historia del litoral boliviano ; La mision Jaimes Freyre en Santiago / Raúl Botelho Gosálvez. — La Paz : Ediciones Populares Ultima Hora, 1978.
151 p. : maps ; 18 cm. — (Colección Litoral boliviano) (Biblioteca popular boliviana de Ultima Hora)
"La Negociación charaña" (p. [127]-151) includes documents.
Bibliography: p. 67-[75]
1. Bolivia—Boundaries—Chile. 2. Chile—Boundaries—Bolivia. 3. Bolivia—Foreign relations—Chile. 4. Chile—Foreign relations—Bolivia. 5. Jaimes Freyre, Ricardo, 1868-1933. I. Botelho Gosálvez, Raúl, 1917- La misión Jaimes Freyre en Chile. 1978. II. Title. III. Series.
F3139.B78 327.84'083 79-105491
79 MARC

Canelas O , Amado.
Bolivia : mito y realidad de su enclaustramiento / Amado Canelas Orellana. — 1 ed. — Lima : Canelas Orellana, 1977.
266 p. ; 21 cm.
S/600.00
1. Bolivia—Boundaries. 2. Bolivia—Boundaries—Chile. 3. Chile—Boundaries—Bolivia. I. Title.
F3341.B7C36 911'.84 78-110291
79 MARC

Oblitas Fernández, Edgar.
Bolivia y su derecho al Pacífico / Edgar Oblitas Fernández. — La Paz : Ediciones Populares Ultima Hora, 1978.
93, 35 p., [2] leaves of plates : maps ; 18 cm. — (Colecció Litoral boliviano) (Biblioteca popular boliviana de Ultima Hora)
"Jeografía náutica de Bolivia por Ramón Vidal Gormaz": p. [3]-35 (2d group)
Includes bibliographical references.
1. Bolivia—Boundaries—Chile. 2. Chile—Boundaries—Bolivia. 3. War of the Pacific, 1879-1884—Sources. 4. Pilot guides—Bolivia. I. Vidal Gormaz, Ramón, b. 1836. Jeografía náutica de Bolivia. 1978. II. Title. III. Series.
F3139.O24 983'.061 79-104710
79 MARC

—CHURCH HISTORY

Condamines, Carlos.
L'Église catholique au Chili : complicité ou résistance? / Charles Condamines ; [publié par le Centre international d'information et de documentation sur l'Église conciliaire]. — Paris : L'Harmattan, 1977.
274 p. ; 22 cm. F77-9080
Bibliography: p. 258-268.
ISBN 2-85802-018-3 : 45.00F
1. Catholic Church in Chile—History. 2. Chile—Church history. I. Title.
BX1468.2.C66 261.7'0983 79-343063
79 MARC

Donoso Loero, Teresa.
Los cristianos por el socialismo en Chile / Teresa Donoso Loero. — 3. ed. aumentada y complementada con abundante material gráfico y documentación. — Santiago de Chile : Editorial Vaitea, 1976, c1975.
314 p. : ill. ; 24 cm.
Previous eds. published in 1975 or 1976 under title: Historia de los cristianos por el socialismo en Chile.
"Esta obra ha sido desglosada de la Colección Ciencia política, no. 5, de Editorial Vaitea."
Bibliography: p. 307-308.
Includes index.
1. Catholic Church in Chile—History. 2. Socialism and Catholic Church. 3. Chile—Church history. I. Title.
BX1468.2.D66 1976 79-360994
79 MARC

—CHURCH HISTORY—SOURCES

Curatos en Bolivia, 1800-1923 : documentos / Rosario Jiménez [compiladora]. — Lima : [s.n.], 1978.
153 p. ; 33 cm. — (Fuentes de historia social americana ; 4) (Biblioteca andina)
1. Parishes—Bolivia—History—Sources. 2. Bolivia—Church history—Sources. 3. Parishes—Chile—History—Sources. 4. Chile—Church history—Sources. 5. Parishes—Peru—History—Sources. 6. Peru—Church history—Sources. I. Jiménez, Rosario. II. Series. III. Series: Biblioteca andina.
BX1464.A3C87 79-100132
79 MARC

—CIVILIZATION

Barros Lezaeta, Luis.
El modo de ser aristocrático : el caso de la oligarquía chilena hacia 1900 / Luis Barros Lezaeta, Ximena Vergara Johnson ; prólogo de Tomás Moulian. — 1. ed. — Santiago : Ediciones Aconcagua, [1978]
190, [7] p. ; 19 cm. — (Colección Lautaro)
Bibliography: p. 187-[191]
1. Upper classes—Chile—History. 2. Aristocracy—History. 3. Chile—Civilization. I. Vergara Johnson, Ximena, joint author. II. Title.
HT653.C44B37 78-113643
79 MARC

—COMMERCE—DIRECTORIES

Chile. Oficina de Planificación Agrícola.
Directorio de productos exportables chilenos : agrícolas, pecuarios, forestales, del mar / República de Chile, Ministerio de Agricultura, Oficina de Planificación Agrícola, ODEPA. — Santiago de Chile : ODEPA, 1976-77 [i.e. 1977?]-
v. : graphs ; 27 cm.
Vol. 1: chiefly tables.
1. Chile—Commerce—Directories. 2. Produce trade—Chile. 3. Forest products—Chile. 4. Marine resources—Chile. I. Title.
HF3413.C48 1977 382'.6'02583 78-109647
79 MARC

—COMMERCE—HISTORY

Muñoz G , Oscar.
Estado e industrialización en el ciclo de expansión del salitre / Oscar Muñoz. — [Santiago, Chile] : Corporación de Investigaciones Económicas para Latinoamérica, 1977.
50 p. ; 22 cm. — (Estudios Cieplan ; 6)
Includes bibliographical references.
1. Chile—Industries—History. 2. Industry and state—Chile—History. 3. Chile—Commerce—History. I. Title. II. Series: Corporación de Investigaciones Económicas para Latinoamérica. Estudios CIEPLAN ; 6.
HC192.M84 338'.0983 79-358743
79 MARC

--COMMERCE--PERIODICALS

Asociación Latinoamericana de Libre Comercio.
Chile: estadísticas de comercio exterior, exportaciones. 1961/69–
[Montevideo] Asociación Latinoamericana de Libre Comercio.
v. 28 cm. (Serie A - Asociación Latinoamericana de Libre Comercio)
1. Chile — Commerce — Periodicals. I. Asociación Latinoamericana de Libre Comercio. Estadísticas de comercio exterior, exportacions. II. Title. III. Series: Asociación Latinoamericana de Libre Comercio. Serie A - Asociación Latinoamericana de Libre Comercio (Chile)
HF3411.A74a 79-645726
MARC-S

—CONSTITUTIONAL HISTORY

Actas constitucionales : antecedentes y textos actualizados al 20 marzo de 1977 / preparado por Eduardo Soto Kloss y Gustavo Fiamma Olivares. — Santiago : Editorial Jurídica de Chile, [1976]
83 p. ; 24 cm.
At head of title: República de Chile.
"Separata del Anuario de derecho administrativo, v. 1, 1975/1976."
CONTENTS: Metas u objetivos fundamentales para la nueva Constitución política de la República, 26-11-1973.—Declaración de principios del Gobierno de Chile, 11-3-1974.—Informe del Sr. Presidente de la Comisión de Reforma Constitucional al cumplir ésta un año de trabajo, 8-11-1974.—Objetivo nacional de Chile, 23-12-1975.—Actas constitucionales: DL 1319 (9-1-1976) DL1551 (13-911976) DL 1552 (13-9-1976) DL 1553 (13-9-1976)
1. Chile—Constitutional law. 2. Chile—Constitutional history. I. Soto Kloss, Eduardo. II. Fiamma O., Gustavo. III. Chile. Laws, statutes, etc.
342'.83 78-110023
79 MARC

Bravo Lira, Bernardino.
Chile 1925-1932 [i.e. mil novecientos veinticinco-mil novecientos treinta y dos] : de la nueva Constitución al nuevo régimen de gobierno / Bernardino Bravo Lira. — Santiago : Contraloría General de la República, 1977.
p. 20-53 ; 27 cm.
Includes bibliographical references.
1. Chile—Politics and government—1920- . 2. Chile—Constitutional history. I. Title.
JL2626 1925.B722 320.9'81'06 79-104973
79 MARC

Novoa Monreal, Eduardo.
Vía legal hacia el socialismo? : El caso de Chile, 1970-1973 / Eduardo Novoa Monreal. — Caracas : Editorial Jurídica Venezolana, 1978.
136 p. ; 20 cm. — (Colección Monografías políticas ; no. 1)
Includes bibliographical references.
Bs30.00
1. Law—Chile—History and criticism. 2. Chile—Constitutional history. 3. Law and socialism. 4. Chile—Politics and government—1970- . I. Title. II. Series.
340'.0983 78-113584
79 MARC

Rogers Sotomayor, Jorge.
La incógnita constitucional / Jorge Rogers Sotomayor. — Santiago de Chile : Distribuidora Universitaria Chilena, 1978.
195 p. ; 19 cm.
Cover title: Chile, la incógnita constitucional.
1. Chile—Constitutional law. 2. Chile—Constitutional history. I. Title. II. Title: Chile, la incógnita constitucional.
342'.83 78-112388
79 MARC

—CONSTITUTIONAL LAW

Actas constitucionales : antecedentes y textos actualizados al 20 marzo de 1977 / preparado por Eduardo Soto Kloss y Gustavo Fiamma Olivares. — Santiago : Editorial Jurídica de Chile, [1976]
83 p. ; 24 cm.
At head of title: República de Chile.
"Separata del Anuario de derecho administrativo, v. 1, 1975/1976."
CONTENTS: Metas u objetivos fundamentales para la nueva Constitución política de la República, 26-11-1973.—Declaración de principios del Gobierno de Chile, 11-3-1974.—Informe del Sr. Presidente de la Comisión de Reforma Constitucional al cumplir ésta un año de trabajo, 8-11-1974.—Objetivo nacional de Chile, 23-12-1975.—Actas constitucionales: DL 1319 (9-1-1976) DL1551 (13-911976) DL 1552 (13-9-1976) DL 1553 (13-9-1976)
1. Chile—Constitutional law. 2. Chile—Constitutional history. I. Soto Kloss, Eduardo. II. Fiamma O., Gustavo. III. Chile. Laws, statutes, etc.
342'.83 78-110023
79 MARC

Rogers Sotomayor, Jorge.
La incógnita constitucional / Jorge Rogers Sotomayor. — Santiago de Chile : Distribuidora Universitaria Chilena, 1978.
195 p. ; 19 cm.
Cover title: Chile, la incógnita constitucional.
1. Chile—Constitutional law. 2. Chile—Constitutional history. I. Title. II. Title: Chile, la incógnita constitucional.
342'.83 78-112388
79 MARC

Silva Bascuñán, Alejandro.
La experiencia chilena sobre referedum y su valoración / Alejandro Silva Bascuñán. — [Santiago : Instituto de Ciencia Política] Universidad Católica de Chile, 1977.
29 leaves ; 27 cm. — (Cuadernos del Instituto de Ciencia Política ; no. 19)
1. Referendum—Chile. 2. Chile—Constitutional law. I. Title. II. Series: Santiago de Chile. Universidad Católica. Instituto de Ciencia Política. Cuadernos del Instituto de Ciencia Política ; no. 19.
JF493.C54S54 78-107558
79 MARC

—CONSTITUTIONAL LAW—CASES

Chile. Tribunal Constitucional.
Tribunal Constitucional de la República de Chile. — [Santiago] : Editorial Jurídica de Chile, 1972.
212 p. ; 19 cm.
"II."
CONTENTS: Constitución política de la República de Chile, capítulo VI, Tribunal Constitucional y Tribunal Calificador de Elecciones.—Auto acordado, Estatuto jurídico sobre organización y funcionamiento del Tribunal Constitucional y régimen de su personal.—Auto acordado, Estatuto jurídico sobre procedimiento aplicable ante el Tribunal Constitucional.—Segundo grupo de fallos (VI al XVII) dictados por el Tribunal Constitucional de la República de Chile.
1. Chile. Tribunal Constituticional. 2. Chile—Constitutional law—cases. I. Title.
342'.83'00269 79-354560
79 MARC

—DESCRIPTION AND TRAVEL—1951-

Popović, Ljuba.
Kondori lete nad Andima / Ljuba Popović. — Gornji Milanovac : Dečje novine, 1976 (Kragujevac : "Nikola Nikolić")
77 p. : ill. ; 24 cm. Yu76-10049
1. Chile—Description and travel—1951- . 2. Indians of South America—Chile. 3. Popović, Ljuba. I. Title.
F3064.P66 79-357409
79 MARC

—ECONOMIC CONDITIONS

Stallings, Barbara.
Class conflict and economic development in Chile, 1958-1973 / Barbara Stallings. — Stanford, Calif. : Stanford University Press, 1978.
xviii, 295 p. : graphs ; 24 cm.
Includes bibliographical references and index.
ISBN 0-8047-0978-5 : $18.50
1. Chile—Economic policy. 2. Chile—Economic conditions. 3. Chile—Politics and government—1920- . 4. Chile—Social conditions. 5. Social conflict. I. Title.
HC192.S7 330.9'83'064 77-89181
79 MARC

—ECONOMIC CONDITIONS—1918-1970

Cunill Grau, Pedro.
Geografía de Chile : nuevo texto para la educación media / Pedro Cunill Grau. — 7. ed. corr. — [Santiago] : Editorial Universitaria, [1978, c1970]
557, [1] p. : ill. ; 28 cm.
Bibliography: p. 551-[558]
1. Chile—Economic conditions—1918-1970. 2. Chile—Social conditions. 3. Physical geography—Chile. I. Title.
HC192.C89 1978 78-109841
79 MARC

CHILE (Continued)

—ECONOMIC CONDITIONS—1970-

Castelijn, J J J W
Chili / J. J. J. W. Castelijn. — 's-Gravenhage : Staatsuitgeverij, 1978.
50 p. : ill. ; 24 cm. — (Landendocumentatie ; 1977, nr. 4) Ne
Bibliography: p. 50.
ISBN 9012019842 : fl 10.00
1. Chile—Economic conditions—1970- . 2. Chile—Social conditions. I. Title.
HC192.C236 79-339646
*79 MARC

Chile. Oficina de Planificación Nacional.
Posibles áreas de interés para inversiones extranjeras / ⌈República de Chile, Presidencia de la República, Oficina de Planificación Nacional, ODEPLAN⌉. — ⌈Santiago de Chile⌉ : La Oficina, 1978.
2 v. : ill. ; 26 cm.
Bibliography: v. 1, p. 60.
CONTENTS: ⌈1⌉ Descripción geo-económica de Chile.—⌈2⌉ Proyectos de inversión.
1. Chile—Economic conditions—1970- . 2. Infrastructure (Economics)—Chile. 3. Investments, Foreign—Chile. I. Title.
HC191.A4 1978a 79-105662
79 MARC

Le Chili d'Allende : témoignages de la vie quotidienne. — Montréal : Éditions coopératives A. St-Martin, c1978.
291 p. ; 21 cm. C78-11890-1
$3.00
1. Chile—Economic conditions—1970- . 2. Chile—Social conditions —1970- . 3. Chile—Politics and government—1970-
HC192.C523 983'.064 79-357738
79 MARC

Reiman, Elisabeth.
Chile, antecedentes para un análisis / Elisabeth Reimann Weigert, Fernando Rivas Sánchez. — La Habana : Editorial de Ciencias Sociales, 1977.
453 p. : ill. ; 18 cm. — (Ediciones políticas)
1. Chile—Politics and government—1970- . 2. Chile—Economic conditions—1970- . 3. Chile—Social conditions—1970- . I. Rivas, Fernando, joint author. II. Title.
F3100.R43 309.1'83'064 78-110436
79 MARC

Rieben, Pierre.
Chili : un an après le coup d'état / Pierre Rieben. — Lausanne : Cedips, ⌈1974⌉
110 p. : ill. ; 21 cm. Sw74-5985
Includes bibliographical references.
1. Chile—Politics and government—1970- . 2. Chile—Economic conditions—1970- . I. Title. II. Title: Un an après le coup d'état.
F3100.R53 320.9'83'064 78-358591
*78 MARC

Seers, Dudley.
Chile : the rule of the Chicago boys / ⌈by⌉ Dudley Seers. — London : World University Service (UK), 1978.
8 p. ; 21 cm. GB78-34354
Includes bibliographical references.
ISBN 0-9505721-4-4 : £0.20
1. Chile—Economic conditions—1970- . 2. Chile—Economic policy. I. Title.
HC192.S38 330.9'83'064 79-301611
79 MARC

La Vida diaria en Chile bajo la Junta / Colectivo. — Salamanca : Ediciones Sígueme, 1978.
138 p. ; 18 cm. — (Pedal : 77) Sp***
ISBN 8430107088 : 190ptas
1. Chile—Economic conditions—1970- . 2. Chile—Social conditions —1970- . 3. Chile—Politics and government—1970-
HC192.V52 983'.064 78-375628
78 MARC

—ECONOMIC CONDITIONS—1970- —ADDRESSES, ESSAYS, LECTURES

Chile, ein schwieriger Weg / ⌈Hrsg., Hanns-Seidel-Stiftung e.V.⌉. — München : Olzog, 1978.
119 p. ; 23 cm. — (Politische Studien : Sonderheft ; 1978, 1 ISSN 0032-3462)
GFR78-A50
Bibliography: p. ⌈118⌉-119.
DM5.80
1. Chile—Politics and government—1970- —Addresses, essays, lectures. 2. Chile—Economic conditions—1970- —Addresses, essays, lectures. I. Hanns-Seidel-Stiftung.
JL2681.C5 320.9'83'064 79-350141
79 MARC

Chile 1970-73 : economic development and its international setting : self-criticism of the Unidad Popular government's policies / edited by S. Sideri, in collaboration with B. Evers. — The Hague ; Boston : Martinus Nijhoff, 1979.
xxvi, 400 p. ; 25 cm. — (Series on the development of societies ; v. 4)
Ne***
Papers originally presented at an Apr. 1976 seminar held at the Institute of Social Studies in The Hague, the Netherlands; then rev. for presentation at a 2d seminar held in July 1976.
Includes bibliographical references and index.
ISBN 9024721989
1. Chile—Economic conditions—1970- —Addresses, essays, lectures. 2. Chile—Economic policy—Addresses, essays, lectures. 3. Chile—Foreign economic relations—Addresses, essays, lectures. I. Sideri, S., 1934- . II. Evers, Ben. III. Series.
HC192.C49 330.9'83'064 79-313430
79 MARC

—ECONOMIC POLICY

Chile. Laws, statutes, etc.
Normas legales y reglamentarias relativas a la planificación nacional y regional. — ⌈Santiago de Chile⌉ : Oficina de Planificación Nacional, ⌈1977⌉
148 p. ; 28 cm.
CONTENTS: Ley no. 16.635.—Decreto Supremo no. 239, de 1972.—Decreto Supremo no. 1.391, de 1967.—Normas constitucionales y legales aplicables a las Secretarías Regionales de Planificación y Coordinación.
1. Chile. Oficina de Planificación Nacional. 2. Chile—Economic policy. I. Chile. Oficina de Planificación Nacional. II. Title.
78-395139
79 MARC

Chile. Oficina de Planificación Nacional.
Plan nacional indicativo de desarrollo, 1978-1983 / ⌈República de Chile, Presidencia de la República, Oficina de Planificación Nacional, ODEPLAN⌉. — ⌈Santiago⌉ : ODEPLAN, ⌈1978⌉
137 p. ; 26 cm.
1. Chile—Economic policy. 2. Chile—Social policy. I. Title.
HC191.A4 1978 309.1'83'064 78-111486
79 MARC

Muñoz G , Oscar.
La ciencia económica en la perspectiva de los problemas de la sociedad chilena / Oscar Muñoz Gomá. — Santiago, Chile : Corporación de Investigaciones Económicas para Latinoamérica, 1977.
20 leaves ; 27 cm. — (Apuntes CIEPLAN ; no. 9)
Includes bibliographical references.
1. Economics—Chile—History. 2. Economics—History. 3. Chile—Economic policy. I. Title. II. Series: Corporación de Investigaciones Económicas para Latinoamérica. Apuntes CIEPLAN ; no. 9.
HB123.C5M84 79-105547
79 MARC

Palacios, Jorge, 1926-
Chile : un ensayo de compromiso histórico / Jorge Palacios. — ⌈Barcelona : Editorial 7 1/2, 1978?⌉
272 p. ; 24 cm. Sp***
Includes bibliographical references.
ISBN 8485411048
1. Unidad Popular. 2. Chile—Politics and government—1970- . 3. Chile—Economic policy. 4. Communism—Chile. I. Title.
JL2698.U5P34 320.9'83'064 79-346877
79 MARC

Programa Regional del Empleo para América Latina y el Caribe.
Políticas de estabilización y empleo : el caso de Chile, 1973-1975 / Organización Internacional del Trabajo, OIT, Programa Mundial del Empleo, Programa Regional del Empleo para América Latina y el Caribe (P.R.E.A.L.C.). — ⌈Santiago de Chile?⌉ : P.R.E.A.L.C., 1977.
ix, 238 p. ; 28 cm. — (Monografía - PREALC ; 5)
"Esta monografía fue escrita por el señor Alberto Bastías."
"Original: español"
Includes bibliographical references.
1. Chile—Economic policy. 2. Economic stabilization. 3. Labor supply—Chile. 4. Consumption (Economics)—Chile. 5. National income—Chile. I. Bastías, Alberto. II. Title. III. Series: Programa Regional del Empleo para América Latina y el Caribe. Monografía - PREALC ; 5.
HC192.P77 1977 78-113864
79 MARC

Seers, Dudley.
Chile : the rule of the Chicago boys / ⌈by⌉ Dudley Seers. — London : World University Service (UK), 1978.
8 p. ; 21 cm. GB78-34354
Includes bibliographical references.
ISBN 0-9505721-4-4 : £0.20
1. Chile—Economic conditions—1970- . 2. Chile—Economic policy. I. Title.
HC192.S38 330.9'83'064 79-301611
79 MARC

Stallings, Barbara.
Class conflict and economic development in Chile, 1958-1973 / Barbara Stallings. — Stanford, Calif. : Stanford University Press, 1978.
xviii, 295 p. : graphs ; 24 cm.
Includes bibliographical references and index.
ISBN 0-8047-0978-5 : $18.50
1. Chile—Economic policy. 2. Chile—Economic conditions. 3. Chile—Politics and government—1920- . 4. Chile—Social conditions. 5. Social conflict. I. Title.
HC192.S7 330.9'83'064 77-89181
79 MARC

Vega M , Julio.
Introducción a la juseconomía / Julio Vega M. — Santiago de Chile : Editorial Andrés Bello, c1972.
215 p. ; 23 cm.
Bibliography: p. 209-215.
1. Economic policy. 2. Chile—Economic policy. I. Title.
K3820.V43 78-383720
78 MARC

—ECONOMIC POLICY—ADDRESSES, ESSAYS, LECTURES

Chile 1970-73 : economic development and its international setting : self-criticism of the Unidad Popular government's policies / edited by S. Sideri, in collaboration with B. Evers. — The Hague ; Boston : Martinus Nijhoff, 1979.
xxvi, 400 p. ; 25 cm. — (Series on the development of societies ; v. 4)
Ne***
Papers originally presented at an Apr. 1976 seminar held at the Institute of Social Studies in The Hague, the Netherlands; then rev. for presentation at a 2d seminar held in July 1976.
Includes bibliographical references and index.
ISBN 9024721989
1. Chile—Economic conditions—1970- —Addresses, essays, lectures. 2. Chile—Economic policy—Addresses, essays, lectures. 3. Chile—Foreign economic relations—Addresses, essays, lectures. I. Sideri, S., 1934- . II. Evers, Ben. III. Series.
HC192.C49 330.9'83'064 79-313430
79 MARC

Monetarismo y capitalismo periférico : el caso de Chile / Jürgen Schuldt, ed. — Lima : Universidad del Pacífico, Centro de Investigación, ⌈1977 or 1978⌉
141 p. : ill. ; 30 cm. — (Serie Ensayos - Universidad del Pacífico, Centro de Investigación ; no. 13)
Includes bibliographical references.
CONTENTS: Schuldt, J. La "Escuela de Chicago" y los "Chicago Boys" en Chile.—Frank, A. G. Carta abierta a Arnold Harberger.—Friedman, M. Bases para un desarrollo económico de Chile.—Spartacus Youth League. Carta a Milton Friedman.—Friedman, M. Respuesta de Friedman al Spartacus.—Cauas, J. Entrevista al ministro de economía chileno.—Chossudowsky, M. Economía de Chicago a la chilena.—Frank, A. G. Segunda carta abierta a Arnold Harberger y Milton Friedman.—Letelier, O. Chile, dos caras de un mismo modelo.
1. Chile—Economic policy—Addresses, essays, lectures. 2. Chicago school of economics—Addresses, essays, lectures. 3. Monetary policy—Addresses, essays, lectures. 4. Capitalism—Addresses, essays, lectures. I. Schuldt Lange, Juergen. II. Series: Universidad del Pacífico. Centro de Investigación. Serie Ensayos - Universidad del Pacífico, Centro de Investigación ; no. 13.
HC192.M658 79-111638
79 MARC

—ECONOMIC POLICY—HISTORY

Falabella, Gonzalo.
Desarrollo de la planificación en Chile y Perú : carácter y condicionamientos / Gonzalo Falabella. — ⌈Lima⌉ : Pontificia Universidad Católica del Perú, Departamento de Ciencias Sociales, Area de Sociología, 1975.
53, 10 p. ; 29 cm. — (Serie Publicaciones previas - Centro de Investigaciones Sociales, Económicas, Políticas y Antropológicas ; no. 14) (Publicaciones CISEPA)
Includes bibliographical references.
1. Chile—Economic policy—History. 2. Peru—Economic policy—History. I. Title. II. Series: Centro de Investigaciones Sociales, Económicas, Políticas y Antropológicas. Publicaciones previas - Centro de Investigaciones Sociales, Económicas, Políticas y Antropológicas ; no. 14.
HC192.F3 78-374734
78 MARC

—EXILES—BIOGRAPHY

Martín, José Carlos.
Manuel Pardo en Chile / José Carlos Martín. — Lima : ⌈s.n.⌉, 1978.
111 p., ⌈5⌉ leaves of plates : ports. ; 24 cm.
"Apéndice" (p. ⌈79⌉-109) includes documents.
Bibliography: p. ⌈110⌉-111.
1. Pardo y Lavalle, Manuel, Pres. Peru, 1834-1878. 2. Peru—Presidents—Biography. 3. Chile—Exiles—Biography. I. Title.
F3447.P25M3 78-106462
78 MARC

—FOREIGN ECONOMIC RELATIONS—ADDRESSES, ESSAYS, LECTURES

Chile 1970-73 : economic development and its international setting : self-criticism of the Unidad Popular government's policies / edited by S. Sideri, in collaboration with B. Evers. — The Hague ; Boston : Martinus Nijhoff, 1979.
xxvi, 400 p. ; 25 cm. — (Series on the development of societies ; v. 4)
Ne***
Papers originally presented at an Apr. 1976 seminar held at the Institute of Social Studies in The Hague, the Netherlands; then rev. for presentation at a 2d seminar held in July 1976.
Includes bibliographical references and index.
ISBN 9024721989
1. Chile—Economic conditions—1970- —Addresses, essays, lectures. 2. Chile—Economic policy—Addresses, essays, lectures. 3. Chile—Foreign economic relations—Addresses, essays, lectures. I. Sideri, S., 1934- . II. Evers, Ben. III. Series.
HC192.C49 330.9'83'064 79-313430
79 MARC

—FOREIGN OPINION, AMERICAN

Moraes, Maria Teresa.
Images of Chile in the U.S. press / by Maria Teresa Moraes and George A. Lawton. — Toronto : Brazilian Studies, ⌈1974⌉
57 p. ; 22 cm. — (Studies - Brazilian Studies ; 4) C***
"Ref. NLR 740318."
Includes bibliographical references.
1. Foreign news—United States. 2. Chile—Foreign opinion, American. I. Lawton, George A., joint author. II. Title. III. Series: Brazilian Studies (Firm). Studies - Brazilian Studies ; 4.
PN4888.F69M6 070.4'49'983064 79-314849
79 MARC

CHILE (Continued)

—FOREIGN OPINION, WEST GERMAN

Jungsozialisten. Landesorganisation Hamburg. Arbeitskreis Internationales.
Die Welt, unabhängige Tageszeitung für Deutschland über Chile / Jungsozialisten in der SPD Landesorganisation Hamburg, Arbeitskreis Internationales. — Hamburg : Arbeitskreis Internationales, 1973.
15, xiv p. ; 21 cm. GFR74-B3
1. Chile—Politics and government—1970- . 2. Chile—Foreign opinion, West German. 3. Public opinion—Germany, West. 4. Die Welt, Hamburg. I. Title.
F3100.J86 1973 320.9'83 76-475189
76[r79]rev MARC

—FOREIGN RELATIONS

Naudón de la Sotta, Carlos.
El nuevo escenario latinoamericano : un enfoque global / Carlos Naudón de la Sota. — Santiago : Instituto de Ciencia Política, 1978.
49 leaves ; 26 cm. — (Cuadernos del Instituto de Ciencia Política ; no. 26)
Includes bibliographical references.
1. Latin America—Economic conditions—1945- . 2. Latin America—Social conditions—1945- . 3. Latin America—Foreign relations. 4. Chile—Foreign relations. I. Title. II. Series: Santiago de Chile. Universidad Católica. Instituto de Ciencia Política. Cuadernos del Instituto de Ciencia Política ; no. 26.
HC125.N38 309.1'8'003 78-112063
79 MARC

—FOREIGN RELATIONS—TREATIES

Tratados, convenciones y arreglos internacionales de Chile, 1810-1977. — Santiago de Chile : Ministerio de Relaciones Exteriores, Dirección de Documentación, Departamento de Tratados, 19
v. : maps ; 27 cm.
CONTENTS:
—[. 3. Tratados bilaterales. v. 2. Chile-Argentina.
1. Chile—Foreign relations—Treaties. I. Chile. Departamento de Tratados. II. Chile. Treaties, etc.
JX546 1977.T7 341'.0266'83 79-102320
79 MARC

—FOREIGN RELATIONS—BOLIVIA

Botelho Gosálvez, Raúl, 1917-
Breve historia del litoral boliviano ; La mision Jaimes Freyre en Santiago / Raúl Botelho Gosálvez. — La Paz : Ediciones Populares Ultima Hora, 1978.
151 p. : maps ; 18 cm. — (Colección Litoral boliviano) (Biblioteca popular boliviana de Ultima Hora)
"La Negociación charaña" (p. [127]-151) includes documents.
Bibliography: p. 67-[75]
1. Bolivia—Boundaries—Chile. 2. Chile—Boundaries—Bolivia. 3. Bolivia—Foreign relations—Chile. 4. Chile—Foreign relations—Bolivia. 5. Jaimes Freyre, Ricardo, 1868-1933. I. Botelho Gosálvez, Raúl, 1917- La misión Jaimes Freyre en Chile. 1978. II. Title. III. Series.
F3139.B78 327.84'083 79-105491
79 MARC

—FOREIGN RELATIONS —UNITED STATES

Artamonov, Aleksandr Aleksandrovich.
(Taĭnye tropy "ITT")
Тайные тропы "ИТТ" / А. Артамонов. — Москва : Политиздат, 1978.
60 p. : facsims. ; 20 cm. — (Владыки капиталистического мира) USSR***
Series romanized: Vladyki kapitalisticheskogo mira.
0.10rub
1. International Telephone and Telegraph Corporation. 2. Chile—Politics and government—1970- 3. United States—Foreign relations—Chile. 4. Chile—Foreign relations—United States. I. Title.
HE8846.I 64A77 79-411910

—HISTORY

Loveman, Brian.
Chile : the legacy of Hispanic capitalism / Brian Loveman. — New York : Oxford University Press, 1979.
xi, 429 p. : maps ; 22 cm. — (Latin American histories)
Bibliography: p. 371-419.
Includes index.
ISBN 0-19-502518-0
1. Chile—History. I. Title.
F3081.L68 983'.064 78-13965
78 MARC

Presa Casanueva, Rafael de la.
Venida y aporte de los españoles a Chile independiente / Rafael de la Presa Casanueva. — Santiago de Chile : [s.n.], 1978.
686 p. : port. ; 27 cm.
1. Spaniards in Chile—History. 2. Chile—Biography. 3. Chile—Relations (general) with Spain. 4. Spain—Relations (general) with Chile. 5. Chile—History. I. Title.
F3285.S7P73 983'.004'61 78-109784
79 MARC

—HISTORY—TO 1810

Zudaire Huarte, Eulogio.
Don Agustín de Jáuregui y Aldecoa / [por] Eulogio Zudaire. — Pamplona : Diputación Foral de Navarra, Institución Principe de Viana, Consejo Superior de Investigaciones Científicas, 1978-
v. : ill. ; 24 cm. — (Colección Historia) Sp78-June
Bibliography: p. 235-238.
Includes indexes.
CONTENTS: 1. Presidente, gobernador y capitán general del reino de Chile.
ISBN 8423501337 : 650ptas (v. 1)
1. Jáuregui y Aldecoa, Agustín de, Viceroy of Peru, 1708-1784. 2. Viceroys—Peru—Biography. 3. Chile—History—To 1810. 4. Peru—History—To 1820.
F3444.J38Z83 79-340921
79 MARC

—HISTORY—1565-1810

Anadón, José.
Pineda y Bascuñán, defensor del araucano : vida y escritos de un criollo chileno del siglo XVII / José Anadón. — [Santiago de Chile] : Editorial Universitaria, c1977.
287 p., 9 leaves of plates : ill. ; 23 cm.
"Versión ampliada y revisada de la tesis doctoral que se presentó a la Universidad de Michigan, Ann Arbor, en 1974."
Bibliography: p. 259-269.
Includes index.
1. Núñez de Pineda y Bascuñán, Francisco, 1607?-1676. 2. Soldiers—Chile—Biography. 3. Spain. Ejército—Biography. 4. Chile—History—1565-1810. 5. Araucanian Indians—Captivities. I. Title.
F3091.A6 983'.03'0924 78-110816
79 MARC

—HISTORY—1810-

Alvear Godoy, Aníbal, 1898-
Por los caminos de Chile : ayer y hoy, hechos históricos y anécdotas, 1810-1976 / Aníbal Alvear Godoy. — 2. ed., corr. y ampliada. — [s.l. : s.n., 1977] (Santiago de Chile : Gráficos Corporación)
257 p. : ill. ; 25 cm.
1. Alvear Godoy, Aníbal, 1898- . 2. Generals—Chile—Biography. 3. Chile. Ejército—Biography. 4. Chile—History—1810- . I. Title.
F3099.A5 1977 983'.064'0924 78-110204
79 MARC

—HISTORY—WAR OF INDEPENDENCE, 1810-1824—NAVAL OPERATIONS

Cox Balmaceda, Ricardo.
La gesta de Cochrane / Ricardo Cox Balmaceda. — 1. ed. — Buenos Aires : Editorial Francisco de Aguirre, 1976, c1975.
x, 378 p., [17] leaves of plates : ill. ; 20 cm. — (Biblioteca Francisco de Aguirre ; 58) (Colección Nación latinoamericana ; 3)
Bibliography: p. 359-361.
Includes index.
1. Chile—History—War of Independence, 1810-1824—Naval operations. 2. Dundonald, Thomas Cochrane, 10th earl of, 1775-1860. I. Title. II. Series: Colección Nación latinoamericana ; 3.
F3094.C75 1976 983'.04 78-109344
79 MARC

—HISTORY—1824-1920

see also War of the Pacific, 1879-1884

—HISTORY—INSURRECTION OF 1859

Bahamonde Silva, Mario, 1910-
El caudillo de Copiapó : Copiapó, 1859 / Mario Bahamonde. — Santiago, Chile : Editorial Nascimento, 1977.
70 p. ; 19 cm.
1. Chile—History—Insurrection of 1859. 2. Gallo Goyenechea, Pedro León, 1830-1877. 3. Revolutionists—Chile—Biography. 4. Copiapó, Chile—History. I. Title.
F3095.B16 983'.05 79-341829
79 MARC

—HISTORY—WAR WITH PERU, 1879-1884

see War of the Pacific, 1879-1884

—HISTORY—REVOLUTION, 1891—FICTION

Vicuña Subercaseaux, Benjamín, 1876-
Zozobras / Benjamin Vicuña Subercaseaux, (Tatin) ; prólogos de Jaime Mendoza y C. T. Robinet. — Santiago, Chile : Editorial Nascimento, 1978.
116 p. ; 18 cm. — (Biblioteca popular Nascimento)
1. Chile—History—Revolution, 1891—Fiction. I. Title.
PQ8097.V5617Z4 863 79-100055
79 MARC

—HISTORY—1920- —FICTION

Bálek, František.
Katedrála hrůzy / František Bálek. — 1. vyd. — Praha : NV, 1977.
203 p. : ill. ; 20 cm. — (Stopa ; sv. 15) Cz77
Kčs14.00
1. Chile—History—1920- —Fiction. I. Title.
PG5039.12.A36K37 79-348561
*79 MARC

—HISTORY—COUP D'ÉTAT, 1973

Garcés, Joan E
Allende et l'expérience chilienne / Joan Garcés. — Paris : Presses de la Fondation nationale des sciences politiques, 1976.
285 p. : graphs ; 24 cm. — (Cahiers de la Fondation nationale des sciences politiques ; no 207) F77-4973
Translation of Allende y la experiencia chilena.
Includes bibliographical references and index.
ISBN 2-7246-0371-0 : 76.00F
1. Chile—Politics and government—1970- . 2. Chile—History—Coup d'état, 1973. I. Title. II. Series: Fondation nationale des sciences politiques. Cahiers ; no. 207.
H31.F6 no. 207 320'.08 s 78-390056
[F3100.G35] 79 MARC

Mader, Julius, 1928-
Instruction 37/57 [i. e. siebenunddreissig-siebenundfünfzig] : Tatsachen u. Hintergründe d. Putsches in Chile 1973 / Julius Mader ; [Zeichn., Wilhelm Kaufmann]. — 1. Aufl. 1.30. Tsd. — Berlin : Militärverlag der Dt. Demokrat. Republik, VEB, 1974.
142 p. : ill.; 21 cm. GFR74-A
Includes bibliographical references.
2.70M
1. Chile—History—Coup d'état, 1973. I. Title.
F3100.M315 75-552764
79 MARC

Rivas, Fernando.
Las fuerzas armadas de Chile : un caso de penetración imperialista / Fernando Rivas Sánchez, Elisabeth Reimann Weigert. — 1. ed. — México : Ediciones 75, 1976.
xiii, 295 p. ; 18 cm.
Bibliography: p. 283-291.
1. Chile—Armed Forces. 2. Chile—History—Coup d'etat, 1973. 3. Military assistance, American—Chile. 4. Military education—Chile. I. Reiman, Elisabeth, joint author. II. Title.
UA622.R58 78-103997
79 MARC

—HISTORY—COUP D'ÉTAT, 1973—CAUSES

Sakhnin, Arkadiĭ Ĭakovlevich.
Ognennaia zemlia / Arkadiĭ Sakhnin. — Moskva : Sov. Rossiia, 1978.
223 p. ; 21 cm. USSR79
Second ed. published in 1976 under title: Tovarishch Chili.
0.60rub
1. Chile—History—Coup d'état, 1973—Causes. 2. United States. Central Intelligence Agency. I. Sakhnin, Arkadiĭ Ĭakovlevich. Tovarishch Chili. II. Title.
F3100.S24 1978 79-381835
79 MARC

—HISTORY—COUP D'ÉTAT, 1973 —CONGRESSES

International Commission of Enquiry into the Crimes of the Military Junta in Chile.
First session of the International Commission of Enquiry into the Crimes of the Military Junta in Chile : Dipoli, Finland, March 21-24, 1974 : documents / edited by the International Commission of Enquiry into the Crimes of the Military Junta in Chile. — Helsinki : Secretariat, [1974?]
47 p. ; 20 cm. Fi***
Statutes of the International Commission: p. 8-[11]
Errata slip inserted.
1. Chile—History—Coup d'état, 1973—Congresses. 2. Civil rights—Chile—Congresses. 3. Political prisoners—Chile—Congresses. 4. Atrocities—Chile—Congresses.
F3100.I56 1974 323.4'0983 79-306646
*79 MARC

CHILE (Continued)

—HISTORY—COUP D'ÉTAT, 1973—DRAMA

Guzmán, Patricio.
La batalla de Chile : la lucha de un pueblo sin armas / Patricio Guzmán ; prólogos Marta Harnecker, Julio García Espinosa. — Pamplona : Peralta, D.L. 1977.
232 p. : ill. ; 21 cm. — (Libros Hiperión ; 24) Sp78-Feb
ISBN 8485272102
1. Chile—History—Coup d'état, 1973—Drama. I. La Batalla de Chile. ⌜Motion picture⌝
PN1997.B343 862 78-385978
*79 MARC

—HISTORY—COUP D'ÉTAT, 1973—MISCELLANEA

Chile si, Junta no! : Agit-Prop-Theater u. Bildergeschichte / ⌜hrsg. von d. Initiative z. Gründung e. Vereinigung Sozialist. Kulturschaffender u.d. Arbeitskollektiv Kämpfende Kunst im Oberbaumverlag⌝. — 1.-3. Tsd. — Berlin : Oberbaumverlag, Verlag f. Politik u. Ökonomie ; Berlin : Universum-Buch-Dienst ⌜Auslieferung f.d. BRD u. Westberlin⌝, 1974.
23, 16 p. : ill., map, ; 24 cm. — (Kämpfende Kunst) GFR75-A
DM2.50
1. Chile—Politics and government—1970- —Miscellanea. 2. Chile—History—Coup d'état, 1973—Miscellanea. I. Initiative zur Gründung einer Vereinigung Sozialistischer Kulturschaffender. II. Arbeitskollektiv Kämpfende Kunst.
F3100.C4815 320.9'83'064 79-352793
79 MARC

Karrasko Moĭîa, Rolando, 1929-
("Voennoplennye" v Chili)
"Военнопленные" в Чили / Роландо Карраско. — Москва : Изд-во агентства печати Новости, 1977.
191 p. : port. ; 21 cm. — (Библиотечка АПН) USSR 77
Series romanized: Bibliotechka APN.
0.50rub
1. Chile—History—Coup d'état, 1973—Personal narratives. 2. Concentration camps—Chile. 3. Karrasko Moĭîa, Rolando, 1929- 4. Journalists—Chile—Biography. I. Title.
F3100.K39 78-421471

—HISTORY—COUP D'ÉTAT, 1973—SOURCES

The GDR's fervent solidarity with the courageous Chilean people.
Berlin, Panorama DDR ⌜1973?⌝
14 p. illus. 20 cm. GDR***
1. Chile—History—Coup d'état, 1973—Sources. 2. Allende Gossens, Salvador, 1908-1973. 3. Germany, East—Relations (general) with Chile—Sources. 4. Chile—Relations (general) with East Germany—Sources.
F3100.G17 301.29'43'1083 79-306238
79 MARC

—HISTORY—ADDRESSES, ESSAYS, LECTURES

Estampas de Chile : (de la población hispánica a la experiencia allendista). — ⌜Madrid⌝ : Speiro, 1974.
142 p. ; 22 cm. Sp75-Mar
Includes bibliographical references.
CONTENTS: Vallet de Goytisolo, J. La fundación de Chile hispánico, ciudades y cabildos.—Elías de Tejada, F. El reino de Chile en el padre Alonso de Ovalle (1601-1655).—Ibáñez Santamaría, G. La política en Chile desde 1810 a la caída de Allende.—Cáceres C., C. F. La economía en Chile durante el periodo de la Unidad Popular.—Widow, J. A. Significado de un despertar.—Crenzet, M. Reflexiones desde Europa tras los acontecimientos de Chile.—G., A. R. Apéndice documental y comentarios.
1. Chile—History—Addresses, essays, lectures.
F3081.E8 983'.064 79-356874
*79 MARC

—HISTORY, NAVAL

López Urrutia, Carlos.
Breve historia naval de Chile / Carlos López Urrutia ; prólogo de Sergio Aguirre Mac-Kay. — 1. ed. — Buenos Aires : Editorial Francisco de Aguirre, 1976.
xv, 166 p. : ill. ; 19 cm. — (Biblioteca Antartica (pocket) ; 30)
Bibliography: p. 149-152.
Includes index.
1. Chile—History, Naval. I. Title.
F3082.L65 983 78-109932
79 MARC

—INDUSTRIES

Meller, Patricio.
Pequeña y gran indústria : generación de empleo y sectores claves / Patricio Meller, Manuel Marfán. — Santiago de Chile : Corporación de Investigaciones Económicas para Latinoamérica, 1977.
51 p. ; 22 cm. — (Estudios CIEPLAN ; 20)
"Presentado a la Tercera Jornada de Economistas realizada en Viña del Mar."
Bibliography: p. 49-51.
1. Chile—Industries. 2. Industries, Size of. 3. Labor supply—Chile. I. Marfán, Manuel, joint author. II. Series: Corporación de Investigaciones Económicas para Latinoamérica. Estudios CIEPLAN ; 20.
HC192.M45 338'.0983 78-107288
79 MARC

Programa Regional del Empleo para América Latina y el Caribe.
Pequeña y gran industria : generación de empleo y sectores claves : un comentario / Organización Internacional del Trabajo, OIT, Programa Mundial del Empleo, Programa Regional del Empleo para América Latina y el Caribe (P.R.E.A.L.C.). — ⌜Santiago, Chile⌝ : P.R.E.A.L.C., 1978.
30 p. ; 28 cm. — (Documento de trabajo - PREALC ; 126)
Prepared by Norberto E. García.
Includes bibliographical references.
1. Meller, Patricio. Pequeña y gran industria. 2. Industries, Size of. 3. Labor supply—Chile. 4. Chile—Industries. I. García, Norberto E. II. Title. III. Series: Programa Regional del Empleo para América Latina y el Caribe. Documento de trabajo - PREALC ; 126.
HC192.P77 1978 338'.0983 79-100823
79 MARC

—INDUSTRIES—HISTORY

Muñoz G , Oscar.
Estado e industrialización en el ciclo de expansión del salitre / Oscar Muñoz. — ⌜Santiago, Chile⌝ : Corporación de Investigaciones Económicas para Latinoamérica, 1977.
50 p. ; 22 cm. — (Estudios Cieplan ; 6)
Includes bibliographical references.
1. Chile—Industries—History. 2. Industry and state—Chile—History. 3. Chile—Commerce—History. I. Title. II. Series: Corporación de Investigaciones Económicas para Latinoamérica. Estudios CIEPLAN ; 6.
HC192.M84 338'.0983 79-358743
79 MARC

—JUVENILE LITERATURE

Sabella, Andrés.
Chile, fértil provincia ... / Andrés Sabella. — 4. ed., corr. y aumentada. — Santiago : Editorial Nascimento, 1976.
119 p. ; 19 cm.
CONTENTS: El caliche.—Guía del Norte.—Estampas de niños.—Grabados en un árbol.—Choapinos.—La sombra del camino.—Retratos de araucanos y Elogio de la trutruca.
1. Chile—Juvenile literature. I. Title.
F3058.5.S23 1976 78-101192
78 MARC

—LAW see Law—Chile

—MAPS

Chile. Instituto Geográfico Militar.
Atlas escolar de Chile : con la microregionalización del país — 2. ed., 1976. — ⌜Santiago⌝ : Instituto Geográfico Militar, 1975.
48 p. : col. ill., col. maps ; 34 cm.
1. Chile—Maps. I. Title.
G1750.C485 1975 912'.83 78-395956
79 MARC

Chile. Instituto Geográfico Militar.
Atlas escolar de Chile : con la microrregionalización del país. — 3. ed., 1978. — ⌜Santiago⌝ : Instituto Geográfico Militar, 1978.
48 p. : col. ill., col. maps ; 34 cm.
1. Chile—Maps. I. Title.
G1750.C485 1978 912'.83 78-395957
79 MARC

—MAPS, EARLY—TO 1800

Sociedad de Promoción de Valores Naciónales.
Cartografía colonial chilena : mapas antiguos de Chile / creación de Sociedad de Promoción de Valores Nacionales. — ⌜s.l.⌝ : La Sociedad, 1973.
1 portfolio (⌜8⌝ leaves : chiefly maps (some col.) ; 55 cm. — (Legajo - Sociedad de Promoción de Valores Nacionales ; 1)
Bibliography: leaf ⌜3⌝
1. Chile—Maps, Early—To 1800. 2. Maps, Early—Facsimile. I. Title. II. Series: Sociedad de Promoción de Valores Nacionales. Legajo - Sociedad de Promoción de Valores Nacionales ; 1.
G1750.S6 1973 78-379684
78 MARC

—POLITICS AND GOVERNMENT

Heise González, Julio.
Evolución del pensamiento democrático en Chile / Julio Heise González ; discurso de recepción, Juan Gómez Millas. — Santiago, Chile : Editorial Universitaria, ⌜1977⌝
30 p. ; 27 cm.
At head of title: Instituto de Chile, Academia de Ciencias Sociales, Políticas y Morales.
1. Political science—Chile—History. 2. Democracy—History. 3. Chile—Politics and government. I. Gómez Millas, Juan. II. Title.
JA84.C5H43 320.9'83 79-105520
79 MARC

Rodríguez Grez, Pablo.
Democracia liberal o democracia orgánica? / Pablo Rodríguez Grez. — ⌜Santiago de Chile⌝ : Universidad Católica de Chile, 1977.
27 leaves ; 26 cm. — (Cuadernos del Instituto de Ciencia Política ; no. 12)
E°42.00
1. Democracy. 2. Political parties. 3. Communist parties. 4. Chile—Politics and government. I. Title. II. Series: Santiago de Chile. Universidad Católica. Instituto de Ciencia Política. Cuadernos del Instituto de Ciencia Política ; no. 12.
JC423.R6154 320.9'83'064 79-342807
79⌜7906⌝ MARC

—POLITICS AND GOVERNMENT—1810-1824

Heise González, Julio.
Años de formación y aprendizaje políticos, 1810-1833 / Julio Heise González. — ⌜Santiago de Chile⌝ : Editorial Universitaria, ⌜c1978⌝
286 p. ; 24 cm.
Errata slip inserted.
Bibliography: p. 279-286.
1. Chile—Politics and government—1810-1824. 2. Chile—Politics and government—1824-1920. I. Title.
F3094.H47 78-113788
79 MARC

—POLITICS AND GOVERNMENT—1824-1920

Heise González, Julio.
Años de formación y aprendizaje políticos, 1810-1833 / Julio Heise González. — ⌜Santiago de Chile⌝ : Editorial Universitaria ⌜c1978⌝
286 p. ; 24 cm.
Errata slip inserted.
Bibliography: p. 279-286.
1. Chile—Politics and government—1810-1824. 2. Chile—Politics and government—1824-1920. I. Title.
F3094.H47 78-11378
79 MAR

—POLITICS AND GOVERNMENT—1920-

Bravo Lira, Bernardino.
Chile 1925-1932 ⌜i.e. mil novecientos veinticinco-mil novecientos treinta y dos⌝ : de la nueva Constitución al nuevo régimen de gobierno / Bernardino Bravo Lira. — Santiago : Contraloría General de la República, 1977.
p. 20-53 ; 27 cm.
Includes bibliographical references.
1. Chile—Politics and government—1920- . 2. Chile—Constitutional history. I. Title.
JL2626 1925.B722 320.9'81'06 79-104973
79 MARC

Bravo Lira, Bernardino.
Régimen de gobierno y partidos políticos en Chile, 1924-1973 / Bernardino Bravo Lira. — Santiago de Chile : Editorial Jurídica de Chile, ⌜1978⌝
320 p. : graphs ; 24 cm.
Includes bibliographical references.
1. Political parties—Chile—History. 2. Chile—Politics and government—1920- I. Title.
JL2698.A1B72 329.9'83 79-108565
79 MARC

Caviedes L , César.
The politics of Chile : a sociogeographical assessment / Cesa Caviedes. — Boulder, Colo. : Westview Press, 1979.
xii, 357 p. : ill. ; 24 cm. — (Westview special studies in Latin America)
Bibliography: p. 325-346.
Includes index.
ISBN 0-89158-365-3
1. Chile—Politics and government—1920- . 2. Elections—Chile. 3. Social Classes—Chile. I. Title.
JL2631.C37 301.5'92'0983 78-2384
78 MAR

Stallings, Barbara.
Class conflict and economic development in Chile, 1958-1973 / Barbara Stallings. — Stanford, Calif. : Stanford Universi Press, 1978.
xviii, 295 p. : graphs ; 24 cm.
Includes bibliographical references and index.
ISBN 0-8047-0978-5 : $18.50
1. Chile—Economic policy. 2. Chile—Economic conditions. 3. Chile—Politics and government—1920- . 4. Chile—Social conditions. 5. Social conflict. I. Title.
HC192.S7 330.9'83'064 77-8918
79 MAR

Valenzuela, Arturo, 1944-
The breakdown of democratic regimes, Chile / Arturo Valenzuela. — Baltimore : Johns Hopkins University Press, c1978.
xv, 140 p. : ill. ; 23 cm.
Includes bibliographical references and index.
ISBN 0-8018-2010-3 pbk.
1. Chile—Politics and government—1920- . I. Title.
JL2611.V3 320.9'83'064 78-58
78 MAR

CHILE (Continued)

—POLITICS AND GOVERNMENT— 1920-1970

Alexander, Robert Jackson, 1918-
The tragedy of Chile / Robert J. Alexander. — Westport, Conn. : Greenwood Press, 1978.
xiii, 509 p. ; 24 cm. — (Contributions in political science ; no. 8 ISSN 0147-1066)
Bibliography: p. ₍483₎-493.
Includes index.
ISBN 0-313-20034-3 : $29.95
1. Chile—Politics and government—1970- . 2. Chile—Politics and government—1920-1970. I. Title. II. Series.
F3100.A34 320.9'83'064 77-91101
78 MARC

—POLITICS AND GOVERNMENT—1970-

Alexander, Robert Jackson, 1918-
The tragedy of Chile / Robert J. Alexander. — Westport, Conn. : Greenwood Press, 1978.
xiii, 509 p. ; 24 cm. — (Contributions in political science ; no. 8 ISSN 0147-1066)
Bibliography: p. ₍483₎-493.
Includes index.
ISBN 0-313-20034-3 : $29.95
1. Chile—Politics and government—1970- . 2. Chile—Politics and government—1920-1970. I. Title. II. Series.
F3100.A34 320.9'83'064 77-91101
78 MARC

Artamonov, Aleksandr Aleksandrovich.
(Taĭnye tropy "ITT")
Тайные тропы "ИТТ" / А. Артамонов. — Москва : Политиздат, 1978.
60 p. : facsims. ; 20 cm. — (Владыки капиталистического мира)
USSR***
Series romanized: Vladyki kapitalisticheskogo mira.
0.10rub
1. International Telephone and Telegraph Corporation. 2. Chile—Politics and government—1970- 3. United States—Foreign relations—Chile. 4. Chile—Foreign relations—United States. I. Title.
HE8846.I 64A77 79-411910

Bernatowicz-Bierut, Grażyna.
Chile: Rewolucja parlamentarna? ₍Wyd. 1.₎ Warszawa, Państwowe Wydawn. Naukowe, 1973.
200 p. illus. 20 cm. zł24.00
At head of title: Grażyna Bernatowicz, Zbigniew Dobosiewicz.
Bibliography: p. 196-197.
1. Chile—Politics and government—1970- . I. Dobosiewicz, Zbigniew.
F3100.B47 74-200032
79₍r79₎rev MARC

Chilekommittén, Stockholm.
De "försvunna" politiska fångarna : repressionens utveckling i Chile 1973-76 / ₍Chilekommittén₎. — ₍Stockholm₎ : ₍Komm.₎, 1977.
29, ₍3₎ p. : ill. ; 21 cm. — (Chilekommitténs småskriftserie ; nr. 5)
S78-29
1. Political prisoners—Chile. 2. Chile—Politics and government—1970- . I. Title. II. Series: Chilekommittén, Stockholm. Chilekommitténs småskriftserie ; nr. 5.
HV9598.C44 1977 79-366022
79 MARC

Le Chili d'Allende : témoignages de la vie quotidienne. — Montréal : Éditions coopératives A. St-Martin, c1978.
291 p. ; 21 cm. C78-11890-1
$3.00
1. Chile—Economic conditions—1970- . 2. Chile—Social conditions—1970- . 3. Chile—Politics and government—1970- .
HC192.C523 983'.064 79-357738
79 MARC

Cusack, David F
Revolution and reaction : the internal dynamics of conflict and confrontation in Chile / by David F. Cusack. — ₍Denver₎ : University of Denver, Graduate School of International Studies, ₍c1977₎
xi, 146 p., ₍1₎ leaf of plates : ill. ; 23 cm. — (Studies in conflict ; book 3) (Monograph series in world affairs ; v. 14)
Includes bibliographical references.
1. Chile—Politics and government—1970- . 2. Chile—Relations (general) with foreign countries. I. Title. II. Series. III. Series: Denver. University. Social Science Foundation. Monograph series in world affairs ; v. 14.
F3100.C87 320.9'83'064 78-305174
78₍r79₎rev MARC

Dorfman, Ariel.
Culture et résistance au Chili / Ariel Dorfman. — Grand-Saconnex : Institut d'action culturelle, ₍1978₎
43 p. : ill. ; 24 cm. — (Document IDAC ; 15) Sw78-2301
1. Chile—Politics and government—1970- . 2. Politics and culture—Chile. I. Title. II. Series: Institut d'action culturelle. IDAC document ; 15.
F3100.D668 320.9'83'064 79-354587
79 MARC

Enriquez, Carlos, 1944-
Le Parti communiste italien et l'expérience chilienne / Carlos Enriquez. — ₍Montréal₎ : Université du Québec à Montréal, Dép. de science politique, 1977.
36 leaves ; 28 cm. — (Notes de recherche-Université du Québec à Montréal, Département de science politique ; no. 5 ISSN 0707-1833) C79-974-9
Includes bibliographical references.
Free
1. Partito comunista italiano. 2. Chile—Politics and government—1970- I. Title. II. Series: Université du Québec à Montréal. Département de science politique. Notes de recherche - Université du Québec à Montréal, Département de science politique ; no 5.
JN5657.C63E57 329.9'45 79-356716
79 MARC

Falabella G , Gonzalo.
Clase, partido y Estado : la CUT en el gobierno de la Unidad Popular / Gonzalo Falabella. — ₍Lima₎ : Pontificia Universidad Católica del Perú, Departamento de Ciencias Sociales, Area de Sociología, 1975.
61 p. ; 29 cm. — (Serie Publicaciones previas - Centro de Investigaciones Sociales, Económicas, Políticas y Antropológicas ; no. 10) (Publicaciones CISEPA)
Includes bibliographical references.
1. Central Unica de Trabajadores de Chile. 2. Unidad Popular. 3. Chile—Politics and government—1970- . 4. Trade-unions—Chile. I. Title. II. Series: Centro de Investigaciones Sociales, Económicas, Políticas y Antropológicas. Publicaciones previas - Centro de Investigaciones Sociales, Económicas, Políticas y Antropológicas ; no. 10.
HD8292.C43F34 331.88'0983 79-344158
79 MARC

Garcés, Joan E
Allende et l'expérience chilienne / Joan Garcés. — Paris : Presses de la Fondation nationale des sciences politiques, 1976.
285 p. : graphs ; 24 cm. — (Cahiers de la Fondation nationale des sciences politiques ; no 207) F77-4973
Translation of Allende y la experiencia chilena.
Includes bibliographical references and index.
ISBN 2-7246-0371-0 : 76.00F
1. Chile—Politics and government—1970- . 2. Chile—History—Coup d'état, 1973. I. Title. II. Series: Fondation nationale des sciences politiques. Cahiers ; no. 207.
H31.F6 no. 207 320'.08 s 78-390056
₍F3100.G35₎ 79 MARC

Jungsozialisten. Landesorganisation Hamburg. Arbeitskreis Internationales.
Die Welt, unabhängige Tageszeitung für Deutschland über Chile / Jungsozialisten in der SPD Landesorganisation Hamburg, Arbeitskreis Internationales. — Hamburg : Arbeitskreis Internationales, 1973.
15, xiv p. ; 21 cm. GFR74-B3
1. Chile—Politics and government—1970- . 2. Chile—Foreign opinion, West German. 3. Public opinion—Germany, West. 4. Die Welt, Hamburg. I. Title.
F3100.J86 1973 320.9'83 76-475189
76₍r79₎rev MARC

Novoa Monreal, Eduardo.
Vía legal hacia el socialismo? : El caso de Chile, 1970-1973 / Eduardo Novoa Monreal. — Caracas : Editorial Jurídica Venezolana, 1978.
136 p. ; 20 cm. — (Colección Monografías políticas ; no. 1)
Includes bibliographical references.
Bs30.00
1. Law—Chile—History and criticism. 2. Chile—Constitutional history. 3. Law and socialism. 4. Chile—Politics and government—1970- . I. Title. II. Series.
340'.0983 78-113584
79 MARC

Palacios, Jorge, 1926-
Chile : un ensayo de compromiso histórico / Jorge Palacios. — ₍Barcelona : Editorial 7 1/2, 1978?₎
272 p. ; 24 cm. Sp***
Includes bibliographical references.
ISBN 8485411048
1. Unidad Popular. 2. Chile—Politics and government—1970- . 3. Chile—Economic policy. 4. Communism—Chile. I. Title.
JL2698.U5P34 320.9'83'064 79-346877
79 MARC

Puente, Mario.
Chile, enseñanzas y balance de un proceso revolucionario / Mario Puente. — ₍Lima₎ : Taller de Estudios Políticos, Programa Académico de CCSS, Universidad Catolica del Perú, ₍1975?₎
51 p. ; 28 cm. — (Serie Procesos políticos latinoamericanos ; no. 3)
1. Chile—Politics and government—1970- . 2. Unidad Popular. I. Title. II. Series.
F3100.P84 320.9'83'064 78-106584
79 MARC

Reiman, Elisabeth.
Chile, antecedentes para un análisis / Elisabeth Reimann Weigert, Fernando Rivas Sánchez. — La Habana : Editorial de Ciencias Sociales, 1977.
453 p. : ill. ; 18 cm. — (Ediciones políticas)
1. Chile—Politics and government—1970- . 2. Chile—Economic conditions—1970- . 3. Chile—Social conditions—1970- . I. Rivas, Fernando, joint author. II. Title.
F3100.R43 309.1'83'064 78-110436
79 MARC

Rieben, Pierre.
Chili : un an après le coup d'état / Pierre Rieben. — Lausanne : Cedips, ₍1974₎
110 p. : ill. ; 21 cm. Sw74-5985
Includes bibliographical references.
1. Chile—Politics and government—1970- . 2. Chile—Economic conditions—1970- . I. Title. II. Title: Un an après le coup d'état.
F3100.R53 320.9'83'064 78-358591
*78 MARC

Spence, Jack.
Search for justice : neighborhood courts in Allende's Chile / Jack Spence. — Boulder, Colo. : Westview Press, 1979.
xi, 206 p. ; 23 cm. — (A Westview replica edition)
Bibliography: p. 199-206.
ISBN 0-89158-279-7
1. Courts—Chile. 2. Chile—Politics and government—1970- . 3. Allende Gossens, Salvador, 1908-1973. I. Title.
347'.83'01 78-21446
78 MARC

Valenzuela, Arturo, 1944-
Il crollo della democrazia in Cile / Arturo Valenzuela. — 1. ed. — ₍Torino₎ : Centro di ricerca e documentazione Luigi Einaudi, 1977.
xv, 89 p. ; 21 cm. — (Quaderni di Biblioteca della libertà ; n. 4)
It78-June
Translation of From established democracy to polarization and dictatorship: Chile.
L2300
1. Chile—Politics and government—1970- . I. Title. II. Series.
F3100.V3216 79-340936
*79 MARC

La Vida diaria en Chile bajo la Junta / Colectivo. — Salamanca : Ediciones Sígueme, 1978.
138 p. ; 18 cm. — (Pedal : 77) Sp***
ISBN 8430107088 : 190ptas
1. Chile—Economic conditions—1970- . 2. Chile—Social conditions—1970- . 3. Chile—Politics and government—1970- .
HC192.V52 983'.064 78-375628
78 MARC

World University Service. United Kingdom Council.
The present situation in Chile and Argentina / ₍World University Service (UK)₎. — London : W.U.S. (U.K.), 1976.
₍2₎, 21 p. ; 22 cm. GB77-16745
Bibliography: p. 21.
1. Education, Higher—Chile. 2. Education, Higher—Argentine. Republic. 3. Chile—Politics and government—1970- . 4. Argentine Republic—Politics and government—1955- . I. Title.
LA548.W67 1976 309.1'82'06 79-317734
79 MARC

—POLITICS AND GOVERNMENT—1970- —ADDRESSES, ESSAYS, LECTURES

Chile, ein schwieriger Weg / ₍Hrsg., Hanns-Seidel-Stiftung e.V.₎. — München : Olzog, 1978.
119 p. ; 23 cm. — (Politische Studien : Sonderheft ; 1978, 1 ISSN 0032-3462)
GFR78-A50
Bibliography: p. ₍118₎-119.
DM5.80
1. Chile—Politics and government—1970- —Addresses, essays, lectures. 2. Chile—Economic conditions—1970- —Addresses, essays, lectures. I. Hanns-Seidel-Stiftung.
JL2681.C5 320.9'83'064 79-350141
79 MARC

Cox Balmaceda, Ricardo.
La crisis democrática / Ricardo Cox Balmaceda. — ₍Santiago, Chile₎ : Editorial Andrés Bello, ₍1978₎
389 p. ; 19 cm.
CONTENTS: Desarrollo y mentalidad subdesarrollada.—Los prejuicios norteamericanos.—Estados Unidos y Rusia ante el mañana.—Fisonomía de la postguerra.—Democracia, auge o decadencia?—El colapso democrático en Chile.—La defensa chilena frente al marxismo.—Marxismo, nacionalismo y régimen militar.—El movimiento de animación espiritual de la democracia universal.
1. Democracy—Addresses, essays, lectures. 2. Chile—Politics and government—1970- —Addresses, essays, lectures. 3. Communism—Chile—Addresses, essays, lectures. I. Title.
JC423.C735 320.9'83'064 79-107439
79 MARC

One thousand days of revolution : CP Chile leaders on lessons of the events in Chile. — Prague : Peace and Socialism International Publishers, 1978.
148 p. ; 19 cm. — (Marxism-Leninism and our time) Cz***
Includes bibliographical references.
Kčs6.00
1. Chile—Politics and government—1970- —Addresses, essays, lectures. 2. Communism—Chile—Addresses, essays, lectures.
F3100.O53 983'.064 79-314284
79 MARC

Posadas, J
Chile : conclusiones del golpe militar y la resistencia del proletariado : recopilación de artículos del camarada Posadas que componen este folleto cuyo título es del Comité de Redacción / J. Posadas. — ₍Montevideo₎ : Ediciones Revista Marxista Latinoamericana, 1973.
94 p. ; 20 cm.
CONTENTS: Significado histórico de la táctica, la política y el programa para aplastar la contrarrevolución en Chile.—Hay que defender incondicionalmente la revolución chilena.—Conclusiones del golpe transitorio y del heroísmo del proletario y las masas chilenas.—Conclusiones de la derrota transitoria de la revolución socialista chilena.—Conclusiones sobre las experiencias en Chile en el golpe transitorio de la contrarrevolución.—Llamado del Secretariado Internacional de la IV Internacional.
1. Chile—Politics and government—1970- —Addresses, essays, lectures. 2. Communism—Chile—Addresses, essays, lectures. I. Title.
F3100.P68 320.9'83'064 79-367752
79 MARC

CHILE (Continued)

—POLITICS AND GOVERNMENT—1970-
—MISCELLANEA

Chile si, Junta no! : Agit-Prop-Theater u. Bildergeschichte / ₍hrsg. von d. Initiative z. Gründung e. Vereinigung Sozialist. Kulturschaffender u.d. Arbeitskollektiv Kämpfende Kunst im Oberbaumverlag₎. — 1.-3. Tsd. — Berlin : Oberbaumverlag, Verlag f. Politik u. Ökonomie ; Berlin : Universum-Buch-Dienst ₍Auslieferung f.d. BRD u. Westberlin₎, 1974.
23, 16 p. : ill., map, ; 24 cm. — (Kämpfende Kunst) GFR75-A
DM2.50
1. Chile—Politics and government—1970- —Miscellanea. 2. Chile—History—Coup d'état, 1973—Miscellanea. I. Initiative zur Gründung einer Vereinigung Sozialistischer Kulturschaffender. II. Arbeitskollektiv Kämpfende Kunst.
F3100.C4815 320.9'83'064 79-352793
79 MARC

—POLITICS AND GOVERNMENT
—1970- —PICTORIAL WORKS

Jacques, André.
Chili, un peuple brode sa vie et ses luttes / André Jacques ; ₍publié par le Comité intermouvements auprès des évacués₎. — Paris (176, rue de Grenelle, 75007) : CIMADE, 1977.
₍86₎ p. : ill. ; 21 cm. F78-10649
Cover title.
On spine: Arpilleras du Chili.
45.00F
1. Embroidery—Chile—History—20th century. 2. Chile in art. 3. Chile—Politics and government—1970- —Pictorial works. I. Title. II. Title: Arpilleras du Chili.
NK9234.A1J32 746.3'983 79-345847
79 MARC

—POLITICS AND GOVERNEMENT
—1970— —SOURCES

Corvalán L , Luis.
(Nas zhdut novye bitvy)
Нас ждут новые битвы : Избр. статьи и речи : ₍Пер. с исп.₎ / Луис Корвалан ; ₍Сост. И. Е. Рыбалкин₎. — Москва : Политиздат, 1978.
551 p., ₍1₎ leaf of plates : port. ; 22 cm. USSR 78
Includes bibliographical references.
1.20rub
1. Chile—Politics and government—1970- —Sources. 2. Partido Comunista de Chile—History—Sources. I. Title.
F3100.C73 79-409422

—RELATIONS (GENERAL) WITH EAST
GERMANY—SOURCES

The GDR's fervent solidarity with the courageous Chilean people. Berlin, Panorama DDR ₍1973?₎
14 p. illus. 20 cm. GDR***
1. Chile—History—Coup d'état, 1973—Sources. 2. Allende Gossens, Salvador, 1908-1973. 3. Germany, East—Relations (general) with Chile—Sources. 4. Chile—Relations (general) with East Germany—Sources.
F3100.G17 301.29'43'1083 79-306238
79 MARC

—RELATIONS (GENERAL) WITH
FOREIGN COUNTRIES

Cusack, David F
Revolution and reaction : the internal dynamics of conflict and confrontation in Chile / by David F. Cusack. — ₍Denver₎ : University of Denver, Graduate School of International Studies, ₍c1977₎
xi, 146 p., ₍1₎ leaf of plates : ill. ; 23 cm. — (Studies in conflict ; book 3) (Monograph series in world affairs ; v. 14)
Includes bibliographical references.
1. Chile—Politics and government—1970- . 2. Chile—Relations (general) with foreign countries. I. Title. II. Series. III. Series: Denver. University. Social Science Foundation. Monograph series in world affairs ; v. 14.
F3100.C87 320.9'83'064 78-305174
78₍r79₎rev MARC

—RELATIONS (GENERAL) WITH SPAIN

Presa Casanueva, Rafael de la.
Venida y aporte de los españoles a Chile independiente / Rafael de la Presa Casanueva. — Santiago de Chile : ₍s.n.₎, 1978.
686 p. : port. ; 27 cm.
1. Spaniards in Chile—History. 2. Chile—Biography. 3. Chile—Relations (general) with Spain. 4. Spain—Relations (general) with Chile. 5. Chile—History. I. Title.
F3285.S7P73 983'.004'61 78-109784
79 MARC

—RURAL CONDITIONS

Collins, Joseph, 1945-
Agrarian reform and counter-reform in Chile / Joseph Collins. — San Francisco : Institute for Food and Development Policy, c1979.
17 p. : ill. ; 22 cm.
$1.00
1. Land reform—Chile. 2. Chile—Rural conditions. I. Institute for Food and Development Policy. II. Title.
HD508.C64 333.3 79-116348
79 MARC

—SOCIAL CONDITIONS

Castelijn, J J J W
Chili / J. J. J. W. Castelijn. — 's-Gravenhage : Staatsuitgeverij, 1978.
50 p. : ill. ; 24 cm. — (Landendocumentatie ; 1977, nr. 4) Ne
Bibliography: p. 50.
ISBN 9012019842 : fl 10.00
1. Chile—Economic conditions—1970- . 2. Chile—Social conditions. I. Title.
HC192.C236 79-339646
*79 MARC

Cunill Grau, Pedro.
Geografía de Chile : nuevo texto para la educación media / Pedro Cunill Grau. — 7. ed. corr. — ₍Santiago₎ : Editorial Universitaria, ₍1978, c1970₎
557, ₍1₎ p. : ill. ; 28 cm.
Bibliography: p. 551-₍558₎
1. Chile—Economic conditions—1918-1970. 2. Chile—Social conditions. 3. Physical geography—Chile. I. Title.
HC192.C89 1978 78-109841
79 MARC

Mattelart, Armand.
Atlas social de las comunas de Chile / Armand Mattelart. — Santiago de Chile : Editorial del Pacífico, ₍c1965₎
126 p. : ill. ; 27 cm. — (Colección Estudios regionales) (Publicación - Centro de Investigaciones Sociológicas, Universidad Católica de Chile)
1. Cost and standard of living—Chile—Statistics. 2. Cost and standard of living—Chile—Maps. 3. Chile—Social conditions. I. Title. II. Series. III. Series: Santiago de Chile. Universidad Católica. Centro de Investigaciones Sociológicas. Publicación.
HD7014.M37 78-106515
79 MARC

Stallings, Barbara.
Class conflict and economic development in Chile, 1958-1973 / Barbara Stallings. — Stanford, Calif. : Stanford University Press, 1978.
xviii, 295 p. : graphs ; 24 cm.
Includes bibliographical references and index.
ISBN 0-8047-0978-5 : $18.50
1. Chile—Economic policy. 2. Chile—Economic conditions. 3. Chile—Politics and government—1920- . 4. Chile—Social conditions. 5. Social conflict. I. Title.
HC192.S7 330.9'83'064 77-89181
79 MARC

—SOCIAL CONDITIONS—1970-

Chile. Oficina de Planificación Nacional.
Social report : summary, 1976 / Oficina de Planificación Nacional. — ₍Santiago₎ : La Oficina, ₍1976₎
22 leaves ; 27 cm.
1. Chile—Social conditions—1970- . 2. Chile—Social policy. I. Title.
HN293.5.C47 1976 309.1'83'064 79-303063
79 MARC

Le Chili d'Allende : témoignages de la vie quotidienne. — Montréal : Éditions coopératives A. St-Martin, c1978.
291 p. ; 21 cm. C78-11890-1
$3.00
1. Chile—Economic conditions—1970- . 2. Chile—Social conditions—1970- . 3. Chile—Politics and government—1970- .
HC192.C523 983'.064 79-357738
79 MARC

Reiman, Elisabeth.
Chile, antecedentes para un análisis / Elisabeth Reimann Weigert, Fernando Rivas Sánchez. — La Habana : Editorial de Ciencias Sociales, 1977.
453 p. : ill. ; 18 cm. — (Ediciones políticas)
1. Chile—Politics and government—1970- 2. Chile—Economic conditions—1970- 3. Chile—Social conditions—1970- . I. Rivas, Fernando, joint author. II. Title.
F3100.R43 309.1'83'064 78-110436
79 MARC

La Vida diaria en Chile bajo la Junta / Colectivo. — Salamanca : Ediciones Sígueme, 1978.
138 p. ; 18 cm. — (Pedal : 77) Sp***
ISBN 8430107088 : 190ptas
1. Chile—Economic conditions—1970- . 2. Chile—Social conditions—1970- . 3. Chile—Politics and government—1970- .
HC192.V52 983'.064 78-375628
78 MARC

—SOCIAL CONDITIONS—1970-
—PERIODICALS

Chile. Oficina de Planificación Nacional.
Informe social.
₍Santiago₎ Oficina de Planificación Nacional.
v. 27 cm. semiannual.
1. Chile—Social policy—Periodicals. 2. Chile—Social conditions—1970- —Periodicals. I. Title.
HN292.5.O35a 309.1'83'064 78-648684
MARC-S

—SOCIAL POLICY

Chile. Oficina de Planificación Nacional.
Plan nacional indicativo de desarrollo, 1978-1983 / ₍República de Chile, Presidencia de la República, Oficina de Planificación Nacional, ODEPLAN₎. — ₍Santiago₎ : ODEPLAN, ₍1978₎
137 p. ; 26 cm.
1. Chile—Economic policy. 2. Chile—Social policy. I. Title.
HC191.A4 1978 309.1'83'064 78-111486
79 MARC

Chile. Oficina de Planificación Nacional.
Social report : summary, 1976 / Oficina de Planificación Nacional. — ₍Santiago₎ : La Oficina, ₍1976₎
22 leaves ; 27 cm.
1. Chile—Social conditions—1970- . 2. Chile—Social policy. I. Title.
HN293.5.C47 1976 309.1'83'064 79-303063
79 MARC

—SOCIAL POLICY—PERIODICALS

Chile. Oficina de Planificación Nacional.
Informe social.
₍Santiago₎ Oficina de Planificación Nacional.
v. 27 cm. semiannual.
1. Chile—Social policy—Periodicals. 2. Chile—Social conditions—1970- —Periodicals. I. Title.
HN292.5.O35a 309.1'83'064 78-64868
MARC-

—STATISTICS—PERIODICALS

Germany (Federal Republic, 1949–). Statistisches Bundesamt.
Länderkurzberichte: Chile.
Stuttgart, W. Kohlhammer.
v. 30 cm. (19– –77: Allgemeine Statistik des Auslandes) (1977– : Statistik des Auslandes)
Title varies slightly.
1. Chile—Statistics—Periodicals. I. Title. II. Series: Statistik des Auslandes.
HA1004.G45a 318.3 76-61671
MARC-

CHILE. EJÉRCITO

—BIOGRAPHY

Alvear Godoy, Aníbal, 1898-
Por los caminos de Chile : ayer y hoy, hechos históricos y anécdotas, 1810-1976 / Aníbal Alvear Godoy. — 2. ed., corr. y ampliada. — ₍s.l. : s.n., 1977₎ (Santiago de Chile : Gráficos Corporación)
257 p. : ill. ; 25 cm.
1. Alvear Godoy, Aníbal, 1898- . 2. Generals—Chile—Biography. 3. Chile. Ejército—Biography. 4. Chile—History—1810- . I. Title
F3099.A5 1977 983'.064'0924 78-11020
79 MAR

CHILE. OFICINA DE PLANIFICACIÓN
NACIONAL

Chile. Laws, statutes, etc.
Normas legales y reglamentarias relativas a la planificación nacional y regional. — ₍Santiago de Chile₎ : Oficina de Planificación Nacional, ₍1977₎
148 p. ; 28 cm.
CONTENTS: Ley no. 16.635.—Decreto Supremo no. 239, de 1972.—Decreto Supremo no. 1.391, de 1967.—Normas constitucionales y legales aplicables a las Secretarías Regionales de Planificación y Coordinación.
1. Chile. Oficina de Planificación Nacional. 2. Chile—Economic policy. I. Chile. Oficina de Planificación Nacional. II. Title.
78-39513
79 MAR

CHILE. TRIBUNAL CONSTITUTICIONAL

Chile. Tribunal Constitucional.
Tribunal Constitucional de la República de Chile. — ₍Santiago₎ : Editorial Jurídica de Chile, 1972.
212 p. ; 19 cm.
"II."
CONTENTS: Constitución política de la República de Chile, capítulo VI, Tribunal Constitucional y Tribunal Calificador de Elecciones.—Auto acordado, Estatuto jurídico sobre organización y funcionamiento del Tribunal Constitucional y régimen de su personal.—Auto acordado, Estatuto jurídico sobre procedimiento aplicable ante el Tribunal Constitucional.—Segundo grupo de fallos (VI al XVII) dictados por el Tribunal Constitucional de la República de Chile.
1. Chile. Tribunal Constitutional. 2. Chile—Constitutional law—cases. I. Title.
342'.83'00269 79-354560
79 MARC

CHILE. UNIVERSIDAD, SANTIAGO

—HISTORY—ADDRESSES, ESSAYS, LECTURES

Imágenes de la Universidad de Chile. — Santiago : Universidad de Chile, 1977.
253 p. : ill. ; 37 cm.
"Con motivo de conmemorarse el 19 de noviembre el centésimo trigésimo quinto aniversario de la solemne instalación de la Casa de Bello."
1. Chile. Universidad, Santiago—History—Addresses, essays, lectures. I. Chile. Universidad, Santiago.
LE36.C52I45 378.83'3 79-342391
79 MARC

CHILE. UNIVERSIDAD, SANTIAGO. ESCUELA DE MEDICINA

—HISTORY—19TH CENTURY

Orrego Luco, Augusto, 1848-1933.
Recuerdos de la Escuela / Augusto Orrego Luco ; prólogo de Leonidas Aguirre Mac-Kay ; semblanza del autor por Amelia Orrego Cifuentes. — Buenos Aires : Editorial Francisco de Aguirre, c1976.
xxiv, 176 p., ₍17₎ leaves of plates : ill. ; 21 cm. — (Colección Historia de la medicina ; 2) (Biblioteca Francisco de Aguirre ; 59)
Includes index.
1. Chile. Universidad, Santiago. Escuela de Medicina—History—19th century. I. Title. II. Series.
R762.C47O77 1976 78-109726
79 MARC

CHILE IN ART

Jacques, André.
Chili, un peuple brode sa vie et ses luttes / André Jacques ; ₍publié par le Comité intermouvements auprès des évacués₎. — Paris (176, rue de Grenelle, 75007) : CIMADE, 1977.
₍86₎ p. : ill. ; 21 cm. F78-10649
Cover title.
On spine: Arpilleras du Chili.
45.00F
1. Embroidery—Chile—History—20th century. 2. Chile in art. 3. Chile—Politics and government—1970- —Pictorial works. I. Title. II. Title: Arpilleras du Chili.
NK9234.A1J32 746.3'983 79-345847
79 MARC

CHILEAN CLOVER see Alfalfa

CHILEAN COOKERY see Cookery, Chilean

CHILEAN DRAMA

Under this heading, without subdivision, are entered individual plays and collections of plays by one author irrespective of whether they are comedies or tragedies

Díaz, Jorge.
Mata a tu prójimo como a ti mismo / Jorge Díaz. — Madrid : Ediciones Cultura Hispánica, 1977.
45 p. ; 22 cm. — (Colección de teatro) Sp***
ISBN 8472321371
I. Title.
PQ8098.14.I12M3 862 78-376128
78 MARC

Gómez-Correa, Enrique, 1915-
Mandrágora, rey de gitanos / Enrique Gómez-Correa. — Santiago de Chile : Ediciones Mandrágora, 1954.
83 p. ; 20 cm.
"Quinientos ejemplares." No. 214.
I. Title.
PQ8097.G615M29 862 79-378939
79 MARC

Morgan, Patricia, fl. 1938-1948.
Teatro: Búscame entre las estrellas, La tarde llega callada ₍comedias. Santiago de Chile₎ Editorial Tegualda ₍1948₎
169 p. illus. 21 cm.
Cover title.
I. Title: Búscame entre las estrellas. II. Title: La tarde llega callada.
a 50-484
New York. Public Libr. ₍r78₎rev MARC

Vadell, Jaime.
Bienaventurados los pobres / de Jaime Vadell y José Manuel Salcedo y la coparticipación de David Benavente. — Santiago ₍Chile₎ : Ediciones Aconcagua, ₍1978₎
73 p. ; 19 cm. — (Colección Mistral)
On t.p.: CENECA de la Fundación Civitas.
1. Hurtado Cruchaga, Alberto, 1901-1952—Drama. I. Salcedo, José Manuel, joint author. II. Benavente, David, joint author. III. Title.
PQ8098.32.A3B5 862 78-111064
79 MARC

CHILEAN FICTION

Under this heading, without subdivision, are entered individual works and collections by one author

Agurto Chamorro, Jorge.
Un extraño en la clínica y otros cuentos / Jorge Agurto. — Santiago, Chile : Editorial Nascimento, 1978.
151 p. ; 19 cm.
CONTENTS: Pasajero de la eternidad.—La muerte ante el espejo.—Veta ignorada.—Embrujo de amor.—Retazos de una vida.—Un extraño en la clínica.—Calle 103, no. 12.—El extraño personaje.—Los humildes.—El caballero del monóculo.—El último viaje.—De cómo se originó la brisa.—El último pollo.—La ciudad perdida.—La sombra del otro.
I. Title.
PQ8098.1.G87E9 863 79-107176
79 MARC

Alcalde Ibáñez, Luciano.
Vida y amor / ₍por₎ Luciano Alcalde. — Madrid : Afrodisio Aguado, D.L. 1977.
286 p. ; 17 cm. Sp77-Sept
200ptas
I. Title.
PQ8097.A732V5 78-361613
78 MARC

Arancibia Bonilla, Guillermo.
Pijes, rotos, y pirquineros : cuentos / Guillermo Arancibia Bonilla. — ₍Santiago de Chile₎ : Arancibia Hnos., 1978.
107 p. ; 20 cm.
I. Title.
PQ8098.1.R27P5 863 79-107070
79 MARC

Arenas, Braulio.
Los sucesos del Budi : novela de Braulio Arenas. — Santiago de Chile : Ediciones Aconcagua, 1978.
207 p. ; 19 cm. — (Colección Mistral)
I. Title.
PQ8097.A76S8 79-115727
79 MARC

Bahamonde Silva, Mario, 1910-
Derroteros y cangalla : cuentos / Mario Bahamonde. — Santiago : Editorial Nascimento, 1978.
312 p. ; 19 cm.
I. Title.
PQ8097.B2D4 863 78-109931
79 MARC

Barrenechea, Julio, 1910-
El compadre Mucho Gasto / Julio Barrenechea. — Santiago, Chile : Editorial Nascimento, 1978.
122 p. ; 19 cm.
I. Title.
PQ8097.B33C65 863 79-348283
79 MARC

Barros, Alvaro, 1931-
Lennox, Nueva, Picton / Alvaro Barros. — Santiago : Editorial Nascimento, 1978.
147 p. ; 19 cm.
CONTENTS: Gualaia.—La hora del yámana.—Las minas del Confín.—Juana Carlito.—Río Toro.—No la vide más.—Así éramos nosotros.—Fresia Argentina.—La cueca y la muerte.—Puedo seguir esperando.—El mensaje.—Sólo un besito.—San Juan pa San Juan.—Si disparas, te mato.
I. Title.
PQ8098.12.A78L4 863 79-107896
79 MARC

Blanco, Guillermo.
Dulces chilenos / ₍por₎ Guillermo Blanco. — Barcelona ₍etc.₎ : Pomaire, 1977 printing.
226 p., 1 leaf ; 22 cm. Sp78-Feb
ISBN 8428602069 : 250ptas
I. Title.
PQ8097.B39D8 863 79-368925
*79 MARC

Blest Gana, Alberto, 1830-1920.
Martín Rivas / ₍por₎ Alberto Blest Gana. — 1. ed. — Barcelona : Vosgos, 1977.
428 p., 1 leaf ; 18 cm. — (Temas girasol) (Colección Grandes maestros) Sp77-Sept
70ptas
I. Title. II. Series.
PQ8097.B5M3 1977 863 79-348734
79 MARC

Bustos Arratia, Myriam.
Tribilín prohibido y otras vedas : cuentos / Myriam Bustos Arratia. — Santiago : Editorial Nascimento, 1978.
142 p. ; 19 cm.
I. Title.
PQ8098.12.U76T74 863 79-108575
79 MARC

Carreño Latorre, Héctor.
A veces los jueces : cuentos / Héctor Carreño Latorre. — Santiago, Chile : Editorial Nascimento, 1977.
143 p. ; 19 cm.
CONTENTS: Cada uno.—Una pasajera.—El disparo.—El informe.—La vista vendada.—La audiencia.—No.
I. Title.
PQ8097.C243A63 863 78-105971
78 MARC

Carreño Latorre, Héctor.
A veces no es delito : novela / Héctor Carreño Latorre. — Santiago, Chile : Editorial Nascimento, 1978.
399 p. ; 19 cm.
I. Title.
PQ8097.C243A64 863 79-107309
79 MARC

Castro Palma, Baltazar.
Ha almorzado la gente? / Baltazar Castro. — ₍1. ed.₎ — ₍Santiago de Chile₎ : Editorial del Pacífico : ₍distribuído por Distribuidora Alonso Ovalle, 1978₎
316 p. ; 19 cm.
I. Title.
PQ8097.C29H3 863 79-107301
79 MARC

Collados Baines, Alberto.
Ostras con mostaza / Alberto Collados Baines. — Santiago, Chile : Editorial Nascimento, 1977.
49 p. ; 18 cm.
I. Title.
PQ8098.13.O4O8 863 78-105965
78 MARC

Délano, Poli, 1936-
En este lugar sagrado / Poli Délano. — 1. ed. — Barcelona : Editorial Grijalbo, c1977.
255 p. ; 20 cm. — (Best sellers Grijalbo) Sp***
I. Title.
PQ8098.14.E4E5 863 79-108792
79 MARC

Délano, Poli, 1936-
Lo primero es un morral : notas de un viaje al Africa / Poli Délano. — ₍1. ed.₎. — ₍Santiago, Chile₎ : Pineda Libros, 1972.
100 p. ; 19 cm.
I. Title.
PQ8098.14.E4P7 863 79-339413
79 MARC

Donoso, José, 1924-
Casa de campo / José Donoso. — 1. ed. — Barcelona : Seix Barral, 1978.
498 p. ; 20 cm. — (Biblioteca breve ; 439 : Novela) Sp***
ISBN 8432203440
I. Title.
PQ8097.D617C3 79-383970
79 MARC

Echeverría Y , Alfonso.
Conservemos el asombro / Alfonso Echeverría. — Santiago, Chile : Editorial Nascimento, 1978.
190 p. ; 19 cm.
I. Title.
PQ8097.E175C58 863 79-108090
79 MARC

Edwards, Jorge.
Los convidados de piedra / Jorge Edwards. — 1. ed. — Barcelona : Seix Barral, 1978.
364 p. ; 20 cm. — (Biblioteca breve ; 436 : Novela) Sp***
ISBN 8432202278
I. Title.
PQ8098.15.D9C6 863 78-377494
78 MARC

Edwards Bello, Joaquín, 1888-1968.
El bisabuelo de piedra / Joaquín Edwards Bello. — Santiago : Editorial Nascimento, 1978.
154 p. ; 19 cm.
1. Bello, Andrés, 1781-1865, in fiction, drama, poetry, etc. I. Title.
PQ8097.E2B5 1978 863 79-107306
79 MARC

CHILEAN FICTION (Continued)

Edwards Cruchaga, Isabel.
El cajón de las cosas perdidas : cuentos / Isabel Edwards Cruchaga. — Santiago, Chile : Editorial Nascimento, 1978.
200 p. ; 19 cm.
I. Title.
PQ8098.15.D94C3 863 79-103964
79 MARC

García, Pablo, 1919-
La tarde en que ardió la bahía / Pablo García. — Santiago de Chile : Editorial Nascimento, 1979.
175 p. ; 19 cm.
I. Title.
PQ8097.G372T3 79-115881
79 MARC

Lafourcade, Enrique, 1927-
Frecuencia modulada / Enrique Lafourcade. — Caracas : Monte Avila Editores, c1977.
408 p. ; 18 cm. — (Colección Continentes)
I. Title.
PQ8097.L235F7 1977 863 79-101242
79 MARC

Latorre Uribe, Marina.
Cuál es el Dios que pasa? : Novela / Marina Latorre. — Santiago : Editorial Nascimento, 1978.
111 p. ; 19 cm.
Includes bibliographical references.
I. Title.
PQ8098.22.A8C8 863 79-107899
79 MARC

Lavín Cerda, Hernán, 1939-
Los tormentos del hijo / Hermán Lavín Cerda. — 1. ed. — México : J. Mortiz, 1977.
120 p. ; 20 cm. — (Nueva narrativa hispánica)
I. Title.
PQ8098.22.A9T6 79-116282
79 MARC

Leighton Illanes, Marcela.
Vendimia en diciembre / Marcela Leighton Illanes. — Santiago : Editorial Nascimento, 1978.
208 p. ; 19 cm.
I. Title.
PQ8098.22.E37V4 863 78-109912
79 MARC

León Alvarado, Carlos.
Algunos días ... : crónicas / Carlos León ; fotografía, Juan Hernández. — ₍Valparaíso, Chile₎ : Ediciones Universitarias de Valparaíso, ₍c1977₎
131 p., ₍11₎ leaves of plates : ill. ; 19 cm.
I. Title.
PQ8097.L38A8 868 78-106098
78 MARC

Lezaeta, Gabriela.
Incendiaron la escuela / Gabriela Lezaeta. — Santiago, Chile : Editorial Nascimento, 1978.
133 p. ; 20 cm.
I. Title.
PQ8098.22.E96I5 863 78-110694
79 MARC

Merani, María.
La parábola del loto ciego : cuentos de la otra realidad : segunda etapa / María Merani. — Santiago de Chile : ₍Ediciones Barcelona₎, 1978.
127 p., ₍1₎ leaf of plates : port. ; 20 cm.
I. Title.
PQ8098.23.E68P3 1978 863 79-107721
79 MARC

Mihovilovic R , Nicolás.
En el último mar del mundo / Nicolás Mihovilovic. — 1. ed. — Santiago de Chile : Zig-Zag, 1978.
248 p. ; 19 cm. — (Libros de bolsillo Zig-Zag)
I. Title.
PQ8098.23.I45E46 863 79-104063
79 MARC

Montero Abt, Antonio.
Nos vemos en Santiago / Antonio Montero Abt. — Santiago, Chile : Arancibia Hnos., 1978.
190 p. ; 19 cm.
I. Title.
PQ8098.23.O55N65 863 78-109809
79 MARC

Neruda, Pablo, 1904-1973.
Huellas de dolor y esperanza / Neruda, Guillén y León Felipe ; selección de Domingo Arteaga P. — 1. ed. — México : Editores Mexicanos Unidos, 1976.
277 p. : ill. ; 17 cm.
I. Guillén, Nicolás, 1902- joint author. II. Camino Galicia, León Felipe, 1884-1968, joint author. III. Arteaga P., Domingo. IV. Title.
PQ8097.N4H8 1976 861 78-106506
78 MARC

Palma, Clemente, 1872-1944.
Cuentos malévolos / Clemente Palma. — París : Librería P. Ollendorff, ₍1912₎
xv, 276 p. ; 20 cm.
I. Title.
PQ8497.P23C8 1912 79-341498
79 MARC

Rodríguez A , Héctor.
Una noche con el diablo : cuentos / Héctor Rodríguez A. — Santiago : Editorial Nascimento, 1977.
199 p. ; 19 cm.
CONTENTS: Una noche con el diablo.—La píldora.—Boutique fantastique.—Diez años, cuatro segundos.—Eva, mi primer amor.—Aló Tierra! Aló Marte!—Medios de comunicación.—Vestido de rojo.
I. Title.
PQ8098.28.O297N6 863 78-110814
79 MARC

Rojas, Manuel, 1896-1973.
El colocolo y otros cuentos / Manuel Rojas. — La Habana : Ediciones Huracán, Editorial Arte y Literatura, 1977.
269 p. ; 18 cm.
I. Title.
PQ8097.R782C6 1977 863 78-108992
79 MARC

Rosales, Justo Abel, 1855-1896.
La negra Rosalía : o, El club de los picarones ; Los amores del diablo en Alhué / Justo Abel Rosales ; prólogo de Jaime Mendoza. — Santiago de Chile : Editorial Nascimento, 1978.
418 p. ; 18 cm. — (Biblioteca popular Nascimento)
I. Rosales, Justo Abel, 1855-1896. Los amores del diablo en Alhué. 1978. II. Title.
PQ8097.R862N44 1978 863 78-113530
79 MARC

Rosenrauch, Erich.
La burra / Erich Rosenrauch V. — Santiago, Chile : Editorial del Pacífico, ₍1978₎
125 p. ; 20 cm.
"500 ejemplares."
I. Title.
PQ8097.R88B8 863 79-104797
79 MARC

Ruiz-Tagle Gandarillas, Carlos.
Cuentos de Santiago / Carlos Ruiz-Tagle. — Santiago, Chile : Editorial Nascimento, 1978.
179 p. ; 19 cm.
I. Title.
PQ8097.R94C8 863 78-109808
79 MARC

Szmulewicz, Efraim.
Forja de hombre / Efraín Szmulewicz. — Santiago, Chile : Ediciones Rumbo, c1978.
207 p. ; 20 cm.
I. Title.
PQ8097.S95F6 863 79-107307
79 MARC

Vicuña Subercaseaux, Benjamín, 1876-
Zozobras / Benjamín Vicuña Subercaseaux, (Tatin) ; prólogos de Jaime Mendoza y C. T. Robinet. — Santiago, Chile : Editorial Nascimento, 1978.
116 p. ; 18 cm. — (Biblioteca popular Nascimento)
1. Chile—History—Revolution, 1891—Fiction. I. Title.
PQ8097.V5617Z4 863 79-100055
79 MARC

Wally.
Gabriela : novela / Wally. — Santiago de Chile : Ediciones Aire Libre, 1973.
120 p. ; 18 cm.
I. Title.
PQ8098.33.A3G3 863 79-378940
79 MARC

Yankas, Lautaro, 1902-
El vado de la noche : novela de la raza / Lautaro Yankas. — 1. ed. — Santiago de Chile : Quimantú, 1972.
187 p. ; 18 cm. — (Colección Cordillera : Narrativa)
I. Title.
PQ8097.Y3V3 863 79-113138
79 MARC

CHILEAN LITERATURE

Under this heading, without subdivision, are entered individual works of creative writing not clearly indicated to be dramatic works, fiction or poetry, and collections of works of mixed genre by individual authors

Aldunate, María Elena.
Del cosmos las quieren vírgenes / Elena Aldunate. — Santiago de Chile : Zig-Zag, 1977.
91 p. ; 19 cm.
I. Title.
PQ8097.A733D4 863 78-105964
78 MARC

Carrión Canales, José Francisco, 1938-
Tatuajes íntimos / José Fco. Carrión Canales. — Santiago, Chile : Ediciones Barcelona, ₍1978₎
92 p. : port. ; 19 cm.
I. Title.
PQ8098.13.A773T3 79-115665
00 MARC

Gómez-Correa, Enrique, 1915-
La noche al desnudo / Enrique Gómez-Correa. — Santiago de Chile : Ediciones Mandrágora, 1945.
106 p. ; 20 cm.
"Quinientos cinco ejemplares. Ejemplar no. 353."
I. Title.
PQ8097.G615N6 79-382462
79 MARC

Jobet, Jorge.
El principio del fin / Jorge Jobet. — Santiago, Chile : Editorial Nascimento, 1978.
259 p. ; 19 cm.
I. Title.
PQ8098.2.O2P7 861 79-339278
79 MARC

Lihn, Enrique.
Lihn y Pompier / ₍Enrique Lihn₎. — 1. ed. — Santiago : Dpto. de Estudios Humanísticos, Facultad de Ciencias Físicas y Matemáticas, Universidad de Chile, 1978.
₍50₎ p. : ill. ; 22 x 31 cm.
Issued in case.
I. Title.
PQ8097.L47L5 868 78-112366
79 MARC

Lihn, Enrique.
La orquesta de cristal / Enrique Lihn. — Buenos Aires : Editorial Sudamericana, c1976.
153 p. ; 18 cm. — (Colección "El Espejo")
I. Title.
PQ8097.L47O7 78-377928
78 MARC

Moreno, Inés.
Al umbral de la luz / Inés Moreno. — ₍s.l.₎ : Ediciones Guilla de Navío, ₍between 1976 and 1979₎
51 p. ; 23 cm.
I. Title.
PQ8098.23.O765A79 79-116501
79 MARC

Weinstein, Luis.
Fábulas abiertas / Luis Weinstein. — Santiago : Editorial Nascimento, 1978.
110 p. ; 19 cm.
I. Title.
PQ8098.33.E38F3 868 79-104191
79 MARC

Yrarrázaval, Renato.
Por la cerradura del espejo / Renato Yrarrázaval ; ₍portada, Jaime Villaseca₎. — Santiago, Chile : Editorial Nascimento, 1979.
107 p. ; 19 cm.
I. Title.
PQ8098.35.R3P6 79-115878
79 MARC

—20TH CENTURY

Antología atacameña : poesía y cuento / compilación y notas de Alfredo Aranda. — Santiago : Editorial Nascimento, 1978.
260 p. ; 19 cm.
1. Chilean literature—Chile—Atacama Desert. 2. Chilean literature—20th century. I. Aranda, Alfredo.
PQ8091.A7A5 860 79-104427
79 MARC

—BIO-BIBLIOGRAPHY

Szmulewicz, Efraim.
Diccionario de la literatura chilena / Efraín Szmulewicz ; prólogo de Roque Esteban Scarpa. — Santiago : Selecciones Lautaro : distribuidores, Librería Libertad, 1977.
563 p. ; 24 cm.
Bibliography: p. 535-563.
1. Chilean literature—Bio-bibliography. I. Title.
Z1710.S95 860 78-104231
₍PQ7911₎ 78 MARC

—PERIODICALS

Literatura chilena en exilio. v. 1- (no. 1-); enero 1977-
Los Angeles, Calif., Ediciones de la Frontera.
v. 28 cm. quarterly.
1. Chilean literature—Periodicals.
PQ7900.L58 860 78-646096
MARC-S

CHILEAN LITERATURE (Continued)

—TRANSLATIONS INTO RUSSIAN

Neruda, Pablo, 1904–1973.
[Selected works. Russian]
Местожительство—земля : стихи разных лет / Пабло Неруда ; перевод с испанского. — Москва : Прогресс, 1977.
422 p. : ill. ; 23 cm. USSR***
3.40rub
I. Title.
Title romanized: Mestozhitel'stvo—zemlia.
PQ8097.N4A57 1977 79-407231

—CHILE—ATACAMA DESERT

Antología atacameña : poesía y cuento / compilación y notas de Alfredo Aranda. — Santiago : Editorial Nascimento, 1978.
260 p. ; 19 cm.
1. Chilean literature—Chile—Atacama Desert. 2. Chilean literature—20th century. I. Aranda, Alfredo.
PQ8091.A7A5 860 79-104427
79 MARC

—CHILE—ATOCAMA—HISTORY AND CRITICISM

Avalos Ansieta, Benigno.
Panorama espiritual de una provincia, tercera región / por Benigno Avalos Ansieta. — [La Serena, Chile : Editorial de Norte, 1977?]
164 p. ; 19 cm.
1. Atacama, Chile—Intellectual life. 2. Chilean literature—Chile—Atocama—History and criticism. I. Title.
F3131.A82 78-381153
78 MARC

CHILEAN NEWSPAPERS

Blanco, Guillermo.
Los incidentes de Riobamba y Pudahuel en tres diarios chilenos / Guillermo Blanco. — 1. ed. — [Santiago] : Instituto Chileno de Estudios Humanísticos, [1977]
174 p. ; 22 cm.
1. Chilean newspapers. 2. Riobamba, Ecuador—Church history. 3. Catholic Church in Chile. I. Title.
PN5047.R45B57 079.83 79-367799
79 MARC

CHILEAN PERIODICALS (GENERAL)

Hoy. v. 1- 1/7 jun. 1977-
[Santiago de Chile, Araucaria Ltda.]
v. ill. 29 cm. weekly.
AP63.H774 056'.1 78-646252
MARC-S

CHILEAN POETRY

Under this heading, without subdivision, are entered individual works and collections by one author

Aciares, Vicente.
Hermana jungla / Vicente Aciares. — [s.l. : s.n.], 1977 (Santiago, Chile : Editorial Universitaria)
91 p. ; 19 cm.
I. Title.
PQ8098.1.C5H4 861 78-109898
79 MARC

Adriazola, Ximena.
Tiempo detenido : poemas / Ximena Adriazola. — Santiago de Chile : Ediciones Orto, 1977.
47 leaves ; 19 cm.
"200 ejemplares .... No. 162."
I. Title.
PQ8097.A57T5 861 78-110751
79 MARC

Aguirre, Berta.
Afrodita / [Berta Aguirre Vigouroux]. — Santiago de Chile : Grupo Fuego de la Poesiá, 1977.
57 p. ; 22 cm.
I. Title.
PQ8098.1.G84A7 861 78-110007
79 MARC

Arenas, Braulio.
Una mansión absolutamente espejo deambula insomne por una mansión absolutamente imagen / Braulio Arenas. — [Valparaíso] : Ediciones Universitarias de Valparaíso, [1978?]
86 p. ; 19 cm. — (Colección Cruz del sur)
I. Title.
PQ8097.A76M3 861 78-110215
79 MARC

Bansart, André.
Prólogo para una primavera : poesías / Andrés Bansart. — [Santiago, Chile] : Ediciones Camilo Henríquez, [1971 or 1972]
71 p. ; 18 cm.
I. Title.
PQ8098.12.A5P7 861 78-401145
79 MARC

Castellano Girón, Hernán, 1937-
El automóvil celestial = L'automobile celestiale / Hernán Castellano Girón ; Mario Lunetta presenta e traduce. — Bari : Gea, 1977.
94 p. ; 20 cm. — (Proiezioni ; 1) It***
Poems.
L3000
I. Title. II. Title: L'automobile celestiale.
PQ8098.13.A8A9416 861 79-351379
79 MARC

Correa, Raúl, 1922-
Ancestral : poemas terrestres / Raúl Correa. — Vicuña [Chile] : Correa, 1978.
39 p. ; 19 cm.
Alternate pages blank.
I. Title.
PQ8098.13.O724A84 861 79-107478
79 MARC

Corvalán, Stella.
Amphion / Stella Corvalán. — Montevideo : [s.n.], 1949.
78 p. : ill. ; 20 cm.
I. Title.
PQ8097.C68A8 861 78-111718
79 MARC

Cravero Lizama, Herbart.
Exploración al subconsciente : poemas / Herbart Cravero Lizama. — Santiago, Chile : Impr. Fco. Carrión, 1978.
142 p. ; 28 cm.
I. Title.
PQ8098.13.R3E95 861 78-109679
79 MARC

Cruchaga de Walker, Rosa.
Bajo la piel del aire / Rosa Cruchaga de Walker. — Santiago, Chile : Editorial Nascimento, 1978.
70 p. ; 19 cm.
I. Title.
PQ8098.13.R8B3 861 78-109765
79 MARC

Domarchi V , Edilberto, 1924-
El viejo aarmonio / Edilberto Domarchi V. — [Santiago, Chile] : Arancibia Hnos., 1977.
76 p. ; 19 cm. — (Ediciones del "Grupo Fuego")
"Referencias críticas y bibliográficas acerca de la obra de E. Domarchi": p. 68-72.
"Comentarios literarios de Edilberto Domarchi, años 1967 a 1976": p. 73-74.
I. Title.
PQ8098.14.O4V5 861 78-110022
79 MARC

Durán, Harold.
El túnel / Harold Durán. — Santiago de Chile : Ediciones Bolt, [1978 or 1979]
33 p. ; 19 cm. — (Colección Grupo literario ñuble)
I. Title.
PQ8098.14.U7T8 861 79-107416
79 MARC

Ferrero, Mario, 1920-
Picasso a cuatro manos / Mario Ferrero. — 1. ed. — Madrid : Ediciones Rialp, 1978.
60 p. ; 18 cm. — (Adonais ; 353) Sp***
Poems.
ISBN 8432119520
I. Title.
PQ8097.F46P5 861 78-382027
78 MARC

Flores Alvarez, Roberto.
Añañucas : poemas del Norte Verde / por Roberto Flores Alvarez. — La Serena [Chile : Editorial de Norte], 1977.
78 p. ; 21 cm.
I. Title.
PQ8098.16.L58A83 861 79-104005
79 MARC

Flores V , Julio.
Mis días lejanos : poemas / Julio Flores. — Valparaíso : [s.n.], 1978.
65 p. ; 19 cm.
I. Title.
PQ8098.16.L6M5 861 79-107074
79 MARC

Frugone, Anita.
A la tierra vuelvo, y otros poemas / Anita Frugone. — Santiago de Chile : [s.n.], 1978.
54 p. ; 20 cm.
I. Title.
PQ8098.16.R8A63 861 79-107303
79 MARC

García-Díaz, Eugenio, 1930-
Los escritos del otoño : poesía / Eugenio García-Díaz. — Santiago, Chile : Ediciones del Grupo Fuego de la Poesía, 1978.
76 p. ; 20 cm.
I. Title.
PQ8098.17.A73E8 861 79-107636
79 MARC

García Lorca, Federico, 1898-1936.
Antología poética ; Doña Rosita la soltera / Federico García Lorca ; introd., selección y análisis de Angel Rodríguez González. — 1. ed. — Santiago, Chile : Santillana, 1976.
xxiii, 196 p. ; 21 cm. — (Biblioteca escolar literaria ; no. 23)
ISBN 8439948085
I. García Lorca, Federico, 1898-1936. Doña Rosita la soltera. 1976. II. Rodríguez González, A. III. Title. IV. Series.
PQ6613.A763A6 1976 868'.6'209 78-103661
79 MARC

Godoy Alcayaga, Lucila, 1889-1957.
Gabriela Mistral / presentada por Lelia Garreaud y Mario Céspedes. — San José : Ministerio de Cultura, Juventud y Deportes, Departamento de Publicaciones, 1977.
230 p. ; 19 cm. — (Serie Pensamiento de América ; no. 9)
I. Garreaud, Lelia. II. Céspedes, Mario. III. Title. IV. Series.
PQ8097.G6A6 1977 861 78-103653
78 MARC

Gómez-Correa, Enrique, 1915-
Mother-darkness / Enrique Gómez-Correa ; [English version by Susana Wald ; collages by Ludwig Zeller]. — Toronto : Oasis, 1975.
[30] p. : ill. ; 23 cm. C***
Poems.
Limited ed. of 270 numb. copies, no. 62.
I. Wald, Susana. II. Title.
PQ8097.G615A25 861 78-375348
78 MARC

González Figueroa, Raúl.
El espejo de los días : poesía / Raúl González Figueroa. — Santiago, Chile : Editorial Nascimento, 1978.
74 p. ; 19 cm.
I. Title.
PQ8097.G674E8 861 78-113815
79 MARC

González Urízar, Fernando.
Al sur del ayer / Fernando González-Urízar. — Santiago, Chile : Editorial Nascimento, 1978.
334 p. ; 19 cm.
I. Title.
PQ8097.G678A79 861 78-113813
79 MARC

González Urízar, Fernando.
Domingo de pájaros / Fernando González-Urízar. — [Santiago, Chile] : Editorial Pedro de Valdivia, 1977.
146 p. ; 23 cm.
I. Title.
PQ8097.G678D6 861 78-105999
78 MARC

Gonzáles Urízar, Fernando.
Tañedor de lluvias : poesías / de Fernando González-Urízar. — Santiago de Chile : Edición Aconcagua, 1978.
179 p. ; 19 cm. — (Colección Mistral)
PQ8097.G678T3 79-116032
79 MARC

Guzmán Cruchaga, Juan, 1896-
Sed / Juan Guzmán Cruchaga. — Valparaíso : Ediciones Universitarias de Valparaíso, c1978.
79 p. ; 19 cm. — (Colección Cruz del sur)
I. Title.
PQ8097.G86S4 861 79-107310
79 MARC

Hernández Jara, Fernando.
El mensaje de los peces / Fernando Hernández Jara. — [Santiago de Chile] : Ediciones "Rissú Valeria", 1976.
[37] p. ; 19 cm.
Errata slip inserted.
I. Title.
PQ8098.18.E69M46 861 79-356015
79 MARC

Ibáñez Langlois, Pilar.
Cómplice / Pilar Ibáñez Langlois. — Santiago, Chile : Editorial Nascimento, 1978.
174 p. ; 19 cm.
I. Title.
PQ8098.19.B36C65 861 79-104423
79 MARC

Jobet, Jorge.
Contacto en Norteamérica : [poesías] / Jorge Jobet. — Santiago, Chile : Editorial Nascimento, 1978.
2 v. ; 19 cm.
1. United States—Poetry. I. Title.
PQ8098.2.O2C6 861 79-107168
79 MARC

CHILEAN POETRY
(Continued)

Larraín Pinedo, Ricardo.
14 ₍i.e. Catorce₎ poemas / Ricardo Larraín Pinedo ; prólogo de Andrés Bansart. — ₍Santiago de Chile : s.l., 1978?₎
₍17₎ leaves ; 22 cm.
I. Title.
PQ8098.22.A69C3 861 79-107382
79 MARC

Lavín Cerda, Hernán, 1939-
Ciegamente los ojos, 1962-1976 / Hernán Lavín Cerda. — 1. ed. — México : Dirección General de Publicaciones, Universidad Nacional Autónoma de México, 1977.
302 p. ; 19 cm. — (Colección Poemas y ensayos)
I. Title.
PQ8098.22.A9C5 861 79-100101
79 MARC

Martínez, Cesáreo.
Cinco razones puras para comprometerse, con la huelga / Cesáreo Martínez. — ₍Lima?₎ : Ediciones Quipu, ₍1978₎
₍15₎ p. ; 23 cm. — (Poema coyuntural ; no. 2)
1. Peru—History—1968- —Poetry. I. Title. II. Series.
PQ8498.23.A742C5 861 78-112198
79 MARC

Massone, Juan Antonio.
Alguien hablara por mi silencio / Juan Antonio Massone. — 1. ed. — Santiago de Chile : Ediciones Nueva Linea, 1978.
64 p. ; 20 cm. — (Colección Poesía ; no. 3)
I. Title.
PQ8098.23.A77A78 861 79-345358
79 MARC

Matte, Ester.
Las leyes del viento / Ester Matte. — Santiago, Chile : Editorial Nascimento, 1977.
103 p. ; 19 cm.
I. Title.
PQ8098.23.A78L4 861 78-105720
78 MARC

Mesa Seco, Manuel Francisco.
Carro de fuego; poema y coro fantástico en 21 cantos. ₍Linares? Chile, 1961₎
x, 65 p. illus. 27 cm.
I. Title.
PQ8098.23.E75C3 67-41670
₍r79₎rev MARC

Mesa Seco, Manuel Francisco.
Ruinas y transparencias / Manuel Francisco Mesa Seco. — Santiago, Chile : Editorial Nascimento, 1978.
90 p. ; 19 cm.
I. Title.
PQ8098.23.E75R8 861 79-342737
79 MARC

Neruda, Pablo, 1904-1973.
Crepusculario : poemas (1920-1923) / ₍por₎ Pablo Neruda. — 1. ed. — Barcelona ₍etc.₎ : Seix Barral, 1977.
100 p., 1 leaf ; 20 cm. — (Biblioteca breve ; 400 : Poesía) Sp77-Oct
ISBN 8432203203
I. Title.
PQ8097.N4C7 1977 861 79-369907
79 MARC

Neruda, Pablo, 1904-1973.
Estravagario / ₍por₎ Pablo Neruda. — 1. ed. — Barcelona, ₍etc.₎ : Seix Barral, 1977.
181 p. ; 20 cm. — (Biblioteca breve ; 408 : Poesía) Sp78-July
250ptas
I. Title.
PQ8097.N4E83 1977 861 79-353956
79 MARC

Neruda, Pablo, 1904-1973.
Poemas inmortales / Pablo Neruda ; ₍prólogo y selección de Jaime Concha₎. — 1. ed. — Santiago de Chile : Quimantú, 1971.
147 p. ; 18 cm. — (Colección Quimantú para todos ; 5)
I. Concha, Jaime. II. Title.
PQ8097.N4A17 1971 861 79-345588
79 MARC

Neruda, Pablo, 1904-1973.
Tercer libro de las odas / Pablo Neruda. — Buenos Aires : Editorial Losada, 1957.
205 p. ; 24 cm. — (Poetas de España y América)
I. Title.
PQ8097.N4A17 1957 861 79-358046
79 MARC

Neruda, Pablo, 1904-1973.
Tercera residencia, 1935-1945 / Pablo Neruda. — 1. ed. — Barcelona : Seix Barral, 1977.
103 p. ; 20 cm. — (Biblioteca breve ; 403 : Poesía) Sp***
ISBN 843220322X
I. Title.
PQ8097.N4T44 1977 861 78-392290
79 MARC

Neruda, Pablo, 1904-1973.
Twenty love poems and a song of despair / Pablo Neruda ; translated by W. S. Merwin. — New York : Penguin Books, 1978.
68 p. ; 18 cm. — (The Penguin poets)
Translation of Veinte poemas de amor y una canción desesperada.
Parallel Spanish texts and English translation.
Bibliography: p. 66-67.
ISBN 0-14-042205-6
I. Title.
₍PQ8097.N4V413 1978₎ 861 76-29006
76 MARC

Neruda, Pablo, 1904-1973.
Los versos más populares de Pablo Neruda : edición del cincuentenario. — ₍Santiago de Chile₎ : Editora Austral, ₍1954?₎
48 p. : port. ; 21 cm.
I. Title.
PQ8097.N4A6 1954 861 79-354143
79 MARC

Neruda, Pablo, 1904-1973.
Viajes / Pablo Neruda. — Santiago de Chile : Nascimento, 1955.
212 p., ₍1₎ leaf of plates : ill. ; 28 cm.
CONTENTS: Viaje al corazón de Quevedo.—Viaje por las costas del mundo.—Viaje al Norte.—Viaje de vuelta.—El esplendor de la tierra.
I. Title.
PQ8097.N4V48 861 79-348671
79 MARC

Los Poetas del mar : antología poética / selección y prólogo de Hugo Montes Brunet. — ₍Santiago₎ : Editorial Andrés Bello, ₍1978₎
156 p. ; 19 cm. — (Biblioteca Andrés Bello ; no. 14)
Includes bibliographical references.
1. Chilean poetry. 2. Sea poetry. I. Montes Brunet, Hugo.
PQ8061.S4P6 861 79-348278
79 MARC

Rodríguez-Cano, Adolfo.
Humanidad : poemas / Adolfo Rodríguez-Cano. — Santiago de Chile : ₍s.n.₎, 1977.
105 p. ; 19 cm.
I. Title.
PQ8097.R7H8 861 78-105719
79 MARC

Silva Acevedo, Manuel.
Mester de Bastardía / Manuel Silva Acevedo. — ₍Santiago de Chile?₎ : Ediciones El Viento en la Llama, 1977.
₍50₎ leaves ; 22 cm. — (Colección Poemas y relatos del país más largo del mundo)
I. Title. II. Series.
PQ8098.29.I495M47 861 79-100265
79 MARC

Teillier, Jorge.
Para un pueblo fantasma / Jorge Teillier. — Valparaíso : Ediciones Universitárias de Valparaíso, c1978.
138 p. ; 19 cm. — (Colección Cruz del Sur)
I. Title.
PQ8098.3.E4P29 861 78-113216
79 MARC

Valenzuela M , Fernando.
Desclaveterías / Fernando Valenzuela M. — Santiago de Chile : ₍s.n.₎, 1978.
109 p. ; 20 cm.
I. Title.
PQ8098.32.A45D39 861 79-104072
79 MARC

Valle, Rosamel del, 1900-1965.
Antología / Rosamel del Valle ; prólogo de Humberto Díaz Casanueva ; selección de Juan Sánchez Peláez. — Caracas : Monte Avila, ₍1976₎
191 p. ; 21 cm. — (Colección Altazor)
CONTENTS: Mirador.—País blanco y negro.—Poesía.—Orfeo.—El joven olvido.—Fuegos y ceremonias.—La visión comunicable.—El corazón escrito.—El sol es un pájaro cautivo en el reloj.—Adiós enigma tornasol.
I. Sánchez Peláez, Juan.
PQ8097.V266A83 1976 861 78-113694
79 MARC

Vial, Sara.
Al oído del viento / Sara Vial. — ₍Valparaíso, Chile₎ : Ediciones Universitarias de Valparaíso, ₍1977₎
70 p. : port. ; 19 cm. — (Colección Cruz del Sur)
I. Title.
PQ8098.32.I33A78 861 78-105854
79 MARC

Wally.
Sólo poemas / Wally. — ₍Santiago? Chile₎ : Ediciones Renovación, ₍1973₎
44 p. ; 19 cm. — (Ediciones del Grupo Fuego de la Poesía)
I. Title.
PQ8098.33.A3S64 79-381179
79 MARC

Zambelli, Hugo, 1927-
De la mano del tiempo / Hugo Zambelli. — ₍Valparaíso, Chile₎ : Ediciones Universitarias de Valparaíso, ₍c1978₎
60 p. ; 19 cm. — (Colección Cruz del Sur)
I. Title.
PQ8098.36.A4D4 861 78-110996
79 MARC

—20TH CENTURY

Poesía para el camino : antología / Unión de Escritores Jóvenes. — Santiago de Chile : Ediciones Nueva Universidad, Pontificia Universidad Católica de Chile, Vicerrectoría de Comunicaciones, 1977.
107 p. ; 19 cm.
1. Chilean poetry—20th century. I. Unión de Escritores Jóvenes, Santiago de Chile.
PQ8058.P62 78-111978
79₍r79₎rev MARC

—TRANSLATIONS INTO FRENCH

Neruda, Pablo, 1904-1973.
Chant général / Pablo Neruda ; traduit de l'espagnol par Claude Couffon. — ₍Paris₎ : Gallimard, c1977.
578 p. ; 21 cm. — (Du monde entier) F***
Translation of Canto general.
I. Title.
PQ8097.N4C1714 1977 861 79-374840
79 MARC

Viñuela, Francisco, 1947-
Exil transitoire : poèmes = ₍Exilio transitorio : poemas₎ / Francisco Viñuela ; traduits par Johanne Garneau-Lassonde. — Montréal : Éditions Nouvelles Frontiéres, c1977.
57 p., ₍7₎ leaves of plates : ill. ; 22 cm. C78-4305-7
Spanish texts with French translations on opposite pages.
$3.00
I. Garneau-Lassonde, Johanne. II. Title. III. Title: Exilio transitorio.
PQ8098.32.I58E914 861 78-397079
78 MARC

CHILEAN PORTRAIT PAINTING see
Portrait painting, Chilean

CHILI

Chili-lover's cook book : chili recipes and recipes with chiles / compiled by Al Fischer and Mildred Fischer. — Phoenix, Ariz. : Golden West Publishers, c1978.
131 p. : ill. ; 22 cm.
Includes index.
ISBN 0-914846-06-X
1. Chili con carne. 2. Chili. I. Fischer, Al, 1923- . II. Fischer, Mildred, 1924-
TX749.C48 641.8'2 78-16382
78 MARC

CHILI CON CARNE

Chili-lover's cook book : chili recipes and recipes with chiles / compiled by Al Fischer and Mildred Fischer. — Phoenix, Ariz. : Golden West Publishers, c1978.
131 p. : ill. ; 22 cm.
Includes index.
ISBN 0-914846-06-X
1. Chili con carne. 2. Chili. I. Fischer, Al, 1923- . II. Fischer, Mildred, 1924-
TX749.C48 641.8'2 78-16382
78 MARC

CHILIASM see Millennium

CHILLI see Chili

CHILLIDA, EDUARDO, 1924-

Fullaondo, Juan Daniel.
Oteiza y Chillida en la moderna historiografía del arte / Juan Daniel Fullaondo. — Bilbao : Editorial La Gran Enciclopedia Vasca, c1976.
109 p., ₍22₎ leaves of plates : ill. ; 20 cm. Sp***
ISBN 8424802594
1. Chillida, Eduardo, 1924- . 2. Oteiza, Jorge de, 1908- . 3. Sculpture, Modern—20th century—Historiography. I. Title.
NB813.C45F83 78-387305
79 MARC

CHILLING FISH AT SEA see Fishing boats —Refrigeration

CHILLS AND FEVER see Malaria

CHILOÉ, CHILE (PROVINCE)

—SOCIAL LIFE AND CUSTOMS

Cárdenas Tabies, Antonio.
Usos y costumbres de Chiloé / Antonio Cárdenas Tabies. — Santiago : Editorial Nascimento, 1978.
222 p. ; 19 cm.
Bibliography: p. 213-214.
1. Chiloé, Chile (Province)—Social life and customs. I. Title.
F3146.C35 983'.5 78-109911
79 MARC

CHILOSTOMATA, FOSSIL see
Cheilostomata, Fossil

CHILTERN HILLS, ENG.

—DESCRIPTION AND TRAVEL

Pigram, R J
Around the historic Chilterns / [by] Ron Pigram. — Tunbridge Wells : Midas Books, 1978.
128 p. : ill., map ; 23 cm. GB78-38083
ISBN 0-85936-110-1 : £3.25
1. Chiltern Hills, Eng.—Description and travel. I. Title.
DA670.C62P53 914.25 79-303465
79 MARC

—DESCRIPTION AND TRAVEL—GUIDE-BOOKS

Bull, Leonard.
A king's highway : (some trackways of the Chilterns) / by Leonard Bull. — London ; New York : Regency Press, 1978.
87 p. : maps ; 23 cm. GB78-19694
£3.00
1. Chiltern Hills, Eng.—Description and travel—Guide-books. 2. Bull, Leonard. I. Title.
DA670.C62B84 914.25 79-300039
79 MARC

Hammond, Reginald J W
Complete Thames and Chilterns / [by] Reginald J. W. Hammond. — London : Ward Lock, 1977.
192 p. : ill., maps ; 20 cm. — (Red guide) GB78-01253
Includes index.
ISBN 0-7063-5502-4 : £3.95
1. Thames Valley—Description and travel—Guide-books. 2. Chiltern Hills—Description and travel—Guide-books. 3. Hiking—England—Chiltern Hills—Guide-books. I. Title.
DA670.T2H33 914.22'04'857 79-307330
79 MARC

CHILUBA LANGUAGE see Luba language

CHILWA, LAKE, MALAWI AND MOZAMBIQUE

Lake Chilwa : studies of change in a tropical ecosystem / edited by Margaret Kalk, A. J. McLachlan & C. Howard-Williams. — The Hague ; Boston : W. Junk, 1979.
xviii, 462 p. ; 25 cm. — (Monographiae biologicae ; v. 35)
Papers written by participants in the Lake Chilwa Co-ordinated Research Project, University of Malawi, 1966-1976.
Includes bibliographies and indexes.
ISBN 9061930871
1. Lake ecology—Chilwa, Lake, Malawi and Mozambique. 2. Chilwa, Lake, Malawi and Mozambique. I. Kalk, Margaret. II. McLachlan, A. J. III. Howard-Williams, C. IV. University of Malawi. Lake Chilwa Co-ordinated Research Project. V. Series.
QP1.P37 vol. 35 574'.08 s 79-16288
[QH195.M47] 79 MARC

CHIMAERA (GENETICS) see Mosaicism

CHIMING see Change ringing

CHIMING CLOCKS

—NETHERLANDS—GELDERLAND

Jong, Rinus de.
De klokkenspelkunst in Gelderland 1945-1975 / door Rinus de Jong. — Zutphen : Walburg Pers, 1975.
24 p. : ill. ; 22 cm. — (Walburg reeks ; 7) Ne75-9
ISBN 9060113721 : fl 5.00
1. Chiming clocks—Netherlands—Gelderland. I. Title.
TS543.N4J66 78-415657
*78 MARC

CHIMNEY-SWEEPS

—DENMARK

Meister, Knud.
Op gennem en skorsten / [af] Knud Meister. — [København] : Skorstensfejerlauget, [1978]
48 p. : ill. ; 23 cm. D78-28
"Udgivet af Skorstensfejerlauget af 11. februar 1778 ved dets 200 års jubilæum."
Not in trade.
1. Chimney-sweeps—Denmark. 2. Skorstensfejerlauget af 11. februar 1778. I. Skorstensfejerlauget af 11. februar 1778. II. Title.
HD8039.C482D425 78-389077
79 MARC

CHIMNEYPIECES see Fireplaces

CHIMNEYS

see also Spark arresters

Disertori, Andrea.
Il libro dei camini e dei caminetti / Andrea Disertori, Anna Necchi Disertori. — Milano : G. De Vecchi, c1976.
189 p. : ill. ; 28 cm. It***
Bibliography: p. 188.
1. Fireplaces. 2. Chimneys. I. Necchi, Anna, joint author. II. Title.
NA3050.D56 729'.39 78-368790
78 MARC

—DESIGN AND CONSTRUCTION

Bán, Lajos.
Kéményépítés / Bán Lajos, Bauer György. — Budapest : Műszaki Kiadó, 1977.
206 p. : ill. ; 20 cm. — (Ipari szakkönyvtár ISSN 0324-217K) H77
ISBN 9631021858 : 21.00Ft
1. Chimneys—Design and construction. I. Bauer, György, joint author. II. Title.
TH2281.B36 78-376070
*78 MARC

Jackson, Kenneth R
A guide to chimney design / Kenneth R. Jackson. — Guildford : IPC Science and Technology Press, 1978.
vii, 143 p., [3] fold. leaves of plates : ill. ; 22 cm. GB78-07671
Includes index.
ISBN 0-902852-62-0 : £11.00
1. Chimneys—Design and construction. I. Title.
TH2281.J32 697'.8 79-303013
79 MARC

—DESIGN AND CONSTRUCTION—CONGRESSES

Chimney Design Symposium and Exhibition, 2d, Edinburgh, 1976.
2nd Chimney Design Symposium and Exhibition : organised by [the] University of Edinburgh, Centre for Industrial Consultancy and Liaison, March 1976. — Edinburgh : The Centre, 1976.
3 v. (v, 462 p.) : ill. ; 30 cm. GB76-17608
Includes bibliographical references.
ISBN 0-903042-04-5 : £10.00
1. Chimneys—Design and construction—Congresses. I. Edinburgh. University. Centre for Industrial Consultancy and Liaison.
TH2281.C55 1976 697'.8 76-373372
76[r79]rev MARC

CHIMPANZEES

Badham, Molly.
Chimps with everything : the story of Twycross Zoo / Molly Badham, with Maureen Lawless ; foreword by Gerald Durrell. — London : W. H. Allen, 1979.
143 p., [16] leaves of plates : ill. ; 23 cm. GB***
ISBN 0-491-02158-5 : £4.95
1. Twycross Zoo. 2. Chimpanzees. 3. Zoo aniamls. I. Lawless, Maureen, joint author. II. Title.
QL76.5.G72T953 636.08'899 79-319247
79 MARC

Brewer, Stella Margaret.
The chimps of Mt. Asserik / Stella Brewer. — 1st American ed. — New York : Knopf : distributed by Random House, 1978.
xiii, 302 p. : ill. ; 24 cm.
Includes index.
ISBN 0-394-49449-0
1. Chimpanzees. 2. Mammals—Senegal—Assirik, Mount. I. Title.
QL737.P96B73 1978 599'.884 77-20359
78 MARC

—BEHAVIOR

Desmond, Adrian J 1947-
The ape's reflexion / Adrian J. Desmond. — New York : Dial Press/J. Wade, c1979.
288 p., [12] leaves of plates : ill. ; 24 cm.
Bibliography: p. [245]-247.
Includes index.
ISBN 0-8037-0674-X
1. Chimpanzees—Psychology. 2. Chimpanzees—Behavior. 3. Gorillas—Psychology. 4. Gorillas—Behavior. 5. Mammals—Psychology. 6. Mammals—Behavior. I. Title.
QL737.P96D47 156'.3 79-13480
79 MARC

Sabater Pí, Jorge.
El chimpancé y los orígenes de la cultura / J. Sabater Pí. — Barcelona : Promoción Cultural, 1978.
124 p. : ill. ; 21 cm. — (Temas antropologicos ; 2) Sp***
Bibliography: p. 109-114.
Includes index.
ISBN 8473530365
1. Chimpanzees—Behavior. 2. Social behavior in animals. 3. Social evolution. 4. Mammals—Behavior. I. Title. II. Series.
QL737.P96S2 79-359704
79 MARC

—JUVENILE LITERATURE

Whitlock, Ralph.
Chimpanzees / Ralph Whitlock. — Milwaukee : Raintree Children's Books, 1977.
57 p. : col. ill. ; 24 cm. — (Animals of the world)
Bibliography: p. 56.
Includes index.
SUMMARY: Introduces the monkey family, particularly the wild chimpanzees of the central African forests.
ISBN 0-8172-1077-6 lib. bdg. ISBN 0-8467-0400-5 pbk.
1. Chimpanzees—Juvenile literature. [1. Chimpanzees] I. Title. II. Series.
QL737.P96W45 1977 599'.884 77-13965
77 MARC

—PSYCHOLOGY

Desmond, Adrian J 1947-
The ape's reflexion / Adrian J. Desmond. — New York : Dial Press/J. Wade, c1979.
288 p., [12] leaves of plates : ill. ; 24 cm.
Bibliography: p. [245]-247.
Includes index.
ISBN 0-8037-0674-X
1. Chimpanzees—Psychology. 2. Chimpanzees—Behavior. 3. Gorillas—Psychology. 4. Gorillas—Behavior. 5. Mammals—Psychology. 6. Mammals—Behavior. I. Title.
QL737.P96D47 156'.3 79-13480
79 MARC

Okano, Tsuneya, 1926-
(Chimpanji no chinō)
チンパンジーの知能 / 岡野恒也著. — 東京 : ブレーン出版, 昭和53 [1978]
12, 4, 214, iv p. : ill. ; 19 cm. Ja***
Bibliography: p. 210-214.
Includes indexes.
¥1500
1. Chimpanzees—Psychology. 2. Animal intelligence. I. Title.
QL737.P96O38 79-803513

CHIN, CHIEN, D. 1795

Pyŏn, In-sŏk.
(Sago chŏnsŏ Chosŏn saryo ŭi yŏn'gu)
四庫全書朝鮮史料의研究 / 卞麟錫著. — 慶山 : 嶺南大學校出版部 ; 서울 : 普及處 太學社, 1977.
877 p. : facsims ; 23 cm.
"四庫全書의朝鮮史資料 附·金簡傳" (in facsims.): p. 83-877.
₩10000
1. Korea—History—Sources. 2. Korea—Description and travel—To 1900. 3. Chin, Chien, d. 1795. 4. Ssu k'u ch'uan shu. I. Title.
DS901.62.P9 78-826323

CH'IN, KUAN, 1049-1100

Ho, Chin-lan.
(Su Tung-p'o yü Ch'in Shao-yu)
蘇東坡與秦少游 / 何金蘭. — [臺北 : s. n., 1970?]
133, [1] double leaves ; 27 cm.
"國立臺灣大學中國文學研究所五十九學年度畢業論文"
Reproduced from ms. copy.
Bibliography: leaves 131-[134]
1. Su, Shih, 1036-1101. 2. Ch'in, Kuan, 1049-1100. 3. Authors, Chinese—Biography. I. Title.
PL2685.Z5H6 1970 78-842809

CHIN, LÜ-HSIANG, 1232-1303
—CRITICISM AND INTERPRETATION

Ho, Shu-chen.
(Chin Lü-hsiang ti sheng p'ing chi ching hsüeh)
金履祥的生平及經學 / 何淑貞. — [台北? : s. n.], 民國64 [1975]
2 v. (436 p. on double leaves) ; 27 cm.
"國立台灣大學中文研究所博士論文"
Cover title.
Reproduced from typescript copy.
Bibliography: p. 421-436.
1. Chin, Lü-hsiang, 1232-1303—Criticism and interpretation. I. Title.
PL2687.C555Z68 79-835426

CHIN. LÜ-HSIANG, 1232-1303

—CRITICISM AND INTERPRETATION

CHIN, SHUNSHIN, 1924-

Chin, Shunshin, 1924-
(Pekin no tabi)
北京の旅 / 陳舜臣. — 東京 : 平凡社, 1978.
3, 247 p., [17] leaves of plates : ill. ; 20 cm. Ja 79-97
¥1700

1. Peking—Description. 2. Chin, Shunshin, 1924- I. Title.
DS795.C528 79-805911

—JOURNEYS—CHINA—SINKIANG

Chin, Shunshin, 1924-
(Shiruku Rōdo no tabi)
シルクロードの旅 / 陳舜臣. — 東京 : 平凡社, 昭和52 [1977]
333 p., [18] leaves of plates : ill. ; 20 cm. Ja 78-562
¥1600

1. Chin, Shunshin, 1924- —Journeys—China—Sinkiang. 2. Sinkiang—Description and travel. 3. Authors, Japanese—20th century—Biography. I. Title.
PL847.I5Z518 78-809367

CH'IN, YÜEH-JEN, 5TH CENTURY B. C.
NAN CHING

Huang, Wei-san.
(Nan ching chih yao)
難經知要 / 黃維三著. — 臺北縣新店鎮 : 國立中國醫藥研究所, 民國58 [1969], 民國66 [1977] printing.
11, 2, 10, 128 p., [1] fold. leaf of plates : ill. ; 21 cm.
NT$40.00
1. Ch'in, Yüeh-jen, 5th cent. B. C. Nan ching. 2. Medicine, Chinese—Early works to 1800. I. Ch'in, Yüeh-jen, 5th cent. B. C. Nan ching. 1977. II. Title.
R127.1.C483H83 79-840804

CHIN, YUNG

—ADDRESSES, ESSAYS, LECTURES

(Chin Yung yü Ming pao)
金庸與明報. — [香港? : s. n., 1977]
40 leaves : ill. ; 22 x 31 cm.
Cover title.
Date of publication stamped on p. [4] of cover.
A collection of articles and newspaper clippings reprinted from various sources.

1. Chin, Yung—Addresses, essays, lectures. 2. Authors—China—Biography—Addresses, essays, lectures. 3. Ming pao—Addresses, essays, lectures.
PL2848.Y8Z6 78-842390

CH'IN-AN HSIEN, CHINA

–HISTORY–SOURCES

(Ch'in-an chih)
秦安志 : 甘肅省 / 胡讃[i. e.纘]宗纂修. — 臺1版. — 台北 : 成文出版社, 民國65 [1976]
364 p. ; 27 cm. — (中國方志叢書 : 華北地方 ; 第559號)
Reprint of the 1535 ed.

1. Ch'in-an hsien, China—History—Sources. I. Hu, Tsuan-tsung, 1480-1560. II. Series: Chung-kuo fang chih ts'ung shu : Hua-pei ti fang ; ti 559 hao.
DS793.C5148C45 1976 79-838048

CH'IN DYNASTY, 221-207 B. C. see China —History—Ch'in dynasty 221-207 B. C.

CHIN DYNASTY, 265-419 see China--History --Chin dynasty, 265-419

CHIN FAMILY

Chin, T'ae-jun, 1925-
(Hyŏlt'ong kwa kamun)
血統과家門 / 秦泰俊著. — [서울] : 弘人文化社, 1976.
229 p., [5] leaves of plates : ill. ; 19 cm.
Bibliography: p. 225-227.
₩2000

1. Chin family. 2. Korea—Genealogy. I. Title.
CS3012.Z8C452 79-826324

CHIN HSI CHUNG HSÜEH

(Chin hsi hsüeh ch'ao ch'üan pu chen hsiang)
金禧學潮全部真相 : 香港教育界醜聞大揭秘. — [香港? : s. n., 1978?]
32 p. : ill. ; 27 cm.
Cover title.
$1.00

1. Chin hsi chung hsüeh.
LG52.H663C47 78-842510

CHIN LANGUAGES

see also Tibeto-Burman languages

CH'IN SHIH-HUANG-TI, EMPEROR OF CHINA, 259-210 B. C.

Li, T'ang.
(Ch'in Shih-huang)
秦始皇 / 李唐. — 臺景印初版. — 臺北 : 河洛圖書出版社, 民國67 [1978]
3, 3, 102 p. ; 22 cm. — (中國歷史小叢書 ; 005)
Series romanized: Chung-kuo li shih hsiao ts'ung shu.
Reprint of the 1962 ed. published by Hsiang-kang hung yeh shu chü, Hong Kong, in series: Li shih hsiao ts'ung shu.
Pbk. : NT$35.00
1. China—History—Ch'in dynasty, 221-207 B. C. 2. Ch'in Shih-huang-ti, Emperor of China, 259-210 B. C.
DS747.5.L52 1978 79-837023

CHIN SHIH TI I CHIEN YÜ

The First Peking Prison. — [Peking? : s.n.], 1916.
[50] p. : chiefly ill. ; 16 x 24 cm.
Cover title.
Title also in Chinese: Ching shih ti i chien yü t'u.
1. Chin shih ti i chien yü. I. Title: Ching shih ti i chien yü t'u.
HV9820.P442C443 365'.951'156 79-309823
79 MARC

CHIN STATE, BURMA

Carey, Bertram Sausmarez, 1864-
The Chin Hills : a history of the people, our dealings with them, their customs and manners, and a gazetteer of their country / by Bertram S. Carey and H. N. Tuck. — Calcutta : Firma KLM on behalf of Tribal Research Institute, Aizawl, 1976.
2 v. : ill., maps ; 25 cm.
Reprint of the 1896 ed. published by the Govt. of Burma, Rangoon.
On social life and customs of the Chins, a tribal people of the Indo-Burmese border.
1. Chin State, Burma. 2. Chin tribes. I. Tuck, Henry Newman, joint author. II. Tribal Research Institute. III. Title.
DS530.8.C46C37 1976 959.1 77-911140
79 MARC

CHIN TRIBES

Carey, Bertram Sausmarez, 1864-
The Chin Hills : a history of the people, our dealings with them, their customs and manners, and a gazetteer of their country / by Bertram S. Carey and H. N. Tuck. — Calcutta : Firma KLM on behalf of Tribal Research Institute, Aizawl, 1976.
2 v. : ill., maps ; 25 cm.
Reprint of the 1896 ed. published by the Govt. of Burma, Rangoon.
On social life and customs of the Chins, a tribal people of the Indo-Burmese border.
1. Chin State, Burma. 2. Chin tribes. I. Tuck, Henry Newman, joint author. II. Tribal Research Institute. III. Title.
DS530.8.C46C37 1976 959.1 77-911140
79 MARC

—POLITICS AND GOVERNMENT

Dạw Sūn, Ma.
(Chīn ạmyō thā myā ị ne che hsạnkyin yŏi thạmāing)
ချင်းအမျိုးသားများ၏နေထိုင်မှုဆန့်ကျင်ရေးသမိုင်း. — ဟားခါးမြို့, ချင်းပြည်နယ်, ဒေါ်ဒေါ်စွန်း, 1976.
16, 288 p. : ill. ; 18 cm.
In Burmese.
Bibliography: p. 283-[285]
K12.00

1. Chin tribes—Politics and government. 2. Burma—Politics and government. I. Title.
DS528.2.C45D38 77-986485

CHINA

(Chūkyō mondai no rikai)
中共問題の理解 / 田中直吉 ... [et al.] 共著. — 東京 : 文教書院, 昭和37 [1962]
177 p. : ill. ; 19 cm. — (現代文化シリーズ ; 2)

1. China. I. Tanaka, Naokichi, 1907- II. Series: Gendai bunka shirizu ; 2.
DS706.C5484 78-813808

Goodnow, Frank Johnson, 1859-1939.
China, an analysis / Frank J. Goodnow. — New York : Arno Press, 1979, c1926.
viii, 279 p. ; 21 cm. — (Johns Hopkins University Press reprints)
Reprint of the ed. published by Johns Hopkins Press, Baltimore.
ISBN 0-405-10603-3 : $22.00
1. China. 2. China—Civilization. I. Title.
DS736.G6 1979 951 78-19330
78 MARC

Hinton, Harold C ed.
Major topics on China and Japan : a handbook for teachers / edited by Harold C. Hinton and Marius B. Jansen. — Westport, Conn. : Greenwood Press, 1979.
xi, 236 p. : map ; 29 cm.
Reprint of the 1957 ed. published by Institute of Pacific Relations, New York.
Bibliography: p. [229]-323.
ISBN 0-313-20657-0
1. China. 2. Japan. I. Jansen, Marius B. II. Title.
[DS706.H53 1979] 951 78-10673
78 MARC

Moore, Joanna.
China in pictures / by Joanna Moore. — New York : Sterling Pub. Co., c1974.
64 p. : ill. ; 26 cm. — (Visual geography series)
Includes index.
SUMMARY: Introduces the geography, history, government, people, and economy of this Asian country whose size is exceeded only by the Soviet Union and Canada.
ISBN 0-8069-1170-0. ISBN 0-8069-1171-9 lib bdg.
1. China. [1. China] I. Title.
DS711.M63 1974 951 77-356449
77[r79]rev2 MARC

CHINA (Continued)

The People's Republic of China : a handbook / edited by Harold C. Hinton. — Boulder, Colo. : Westview Press, 1979.
xvii, 443 p. : ill. ; 24 cm.
Includes bibliographies and index.
ISBN 0-89158-419-6 : $30.00
1. China. I. Hinton, Harold C.
DS706.P38 1979 951.05 78-10306
78 MARC

(Saishin Shina taikan)
最新支那大観 / [編輯者 佐藤定勝]. — 東京：誠文堂新光社, 昭和12 [1937], 昭和13 [1938] printing.
2, 8, 713 p., [11] leaves of plates : ill. ; 27 cm.

1. China. I. Satō, Sadakatsu, 1904- II. Title: Shina taikan.
DS706.S217 79-800979

(Seinen tairiku tokuhon)
青年大陸讀本 / 松村磊序文；藤谷重雄, 福田義郎共編. — 東京：元宇館：發賣所 上田屋書店, 昭和15 [1940]
3, 334 p., [2] leaves of plates : ill. ; 21 cm.

1. China. I. Fujitani, Shigeo. II. Fukuda, Yoshirō.
DS706.S386 78-815228

(Shina)
支那：消夏漫筆筆 / 中山成太郎譯[編]. — 東京：有斐閣書房, 大正5 [1916]
2, 3, 6, 619 p. ; 23 cm.
"第一章...「ニコラウス, ポスト」氏の報告に依り, 第二章...「グルンヘルド」氏支那海港殖民地論に依り, 第三章...第四章...第五章...「フランケー」氏東亞新氣運に依り...抄譯"

1. China. 2. China—Foreign population. 3. Exterritoriality. 4. Tibet. 5. Mongolia (Inner Mongolia) I. Nakayama, Seitarō.
DS706.S48 78-810619

(Shina oyobi Manshūkoku gensei chiri)
支那及滿洲國現勢地理 / 支那地理歷史大系刊行會編. — 東京：白揚社, 昭和16 [1941]
2 v. : ill. ; 19 cm. — (支那地理歷史大系；第1-2卷)

1. China. 2. Manchuria. 3. Mongolia (Inner Mongolia) I. Shina Chiri Rekishi Taikei Kankōkai. II. Series: Shina chiri rekishi taikei : dai 1-2-kan.
DS735.S4854 78-813612

Uchida, Ryōhei, 1874-1937.
(Shina kan)
支那觀 / 內田良平著. — 東京：黑龍會, 大正2 [1913]
3, 3, 127, 7 p. ; 23 cm.

1. Japan—Foreign relations—China. 2. China—Foreign relations—Japan. 3. China. I. Title.
DS849.C6U26 78-817226

Wang, Su-hsin.
(Chin hsiu Chung-hua)
錦繡中華 / 編者王素心. — 台南：青溪出版社：總經銷東風圖書供應社, 民國67 [1978]
281, [13] leaves of plates : ill. ; 21 cm. — (東風新刊；1)
English and Chinese.
Series romanized: Tung feng hsin k'an.
NT$100.00

1. China. I. Title.
DS706.W329 915.1 04'5 79-840728

Yung, Wing, 1828-1912.
My life in China and America / Yung Wing. — New York : Arno Press, 1978 [c1909]
vi, 286 p. : port. ; 24 cm. — (The Asian experience in North America : Chinese and Japanese)
Reprint of the ed. published by H. Holt, New York.
Includes index.
ISBN 0-405-11301-3
1. Yung, Wing, 1828-1912. 2. China. 3. Education—China. 4. Chinese in the United States—Education. I. Title. II. Series: Asian experience in North America : Chinese and Japanese.
DS763.Y8A3 1978 951'.03'0924 78-54850
78 MARC

—ADDRESSES, ESSAYS, LECTURES

(Chūgoku konjaku)
中国今昔 / 山内恭彦 ... [et al.]. — 東京：二玄社, 昭和51 [1976]
207 p. : ill. ; 19 cm. — (どう考えるか；8) Ja 76-11674
Series romanized: Dō kangaeru ka.
¥750

1. China—Addresses, essays, lectures. I. Yamanouchi, Takahiko, 1902-
DS706.C532 78-812455

Perspectives on a changing China : essays in honor of Professor C. Martin Wilbur on the occasion of his retirement / edited by Joshua A. Fogel and William T. Rowe. — Boulder, Colo. : Westview Press, 1979.
xiv, 325 p. ; 25 cm. — (Westview special studies on China and East Asia)
"Bibliography of the publications of C. Martin Wilbur": p. 307-314.
Includes bibliographical references and index.
CONTENTS: Harrison, J. P. An appreciation.—Fogel, J. A. and Rowe, W. T. Introduction.—Prospect: Lang, O. Two visions of the future.—Chinese institutions and institutional change: Lee, R. H. G. Frontier politics in the southwestern Sino-Tibetan borderlands during the Ch'ing dynasty. Wou, O. Y. K. The political kin unit and the family origin of Ch'ing local officials. Rowe, W. T. Urban control in late imperial China. Kennedy, T. L. Mausers and the opium trade. DeFrancis, J. Mao Tse-tung and writing reform.—Retrospects: Fogel, J. A. On the rediscovery of the Chinese past. Li, Y. The vicissitudes of Chinese Communist historiography.
ISBN 0-89158-091-3 : $18.75
1. China—Addresses, essays, lectures. 2. Wilbur, Clarence Martin, 1908- . I. Wilbur, Clarence Martin, 1908- . II. Fogel, Joshua A. III. Rowe, William T.
DS706.P45 1979 951 79-626
79 MARC

—ADMINISTRATIVE AND POLITICAL DIVISIONS

Chao, Ch'üan-ch'eng, fl. 1955-
(Ch'ing tai ti li yen ko piao)
清代地理沿革表 / 趙泉澄著. — [台北縣永和鎮]：文海出版社, [1979?]
[14], 204 p., [16] fold. leaves ; 22 cm. — (近代中國史料叢刊續編 i. e. 輯；628：第六十三輯)
Reprint of the ed. with pref. dated 1940.
Includes bibliographical references and index.
1. China—Historical geography. 2. China—Administrative and political divisions. I. Title. II. Series: Shen, Yün-lung, comp. Chin tai Chung-kuo shih liao ts'ung k'an hsü chi ; 628.
DS755.S532 vol. 628 79-840821
[DS706.5]

China (People's Republic of China, 1949-). Kung an pu.
(Chung-hua jen min kung ho kuo hsing cheng ch'ü hua chien ts'e)
中华人民共和国行政区划简册：截至一九七七年底的区划 / 中华人民共和国公安部编. — 第1版. — 北京：地图出版社, 新华书店北京发行所发行, 1978.
134 p. : 19 cm.
Includes indexes.
$0.35.

1. China—Administrative and political divisions. I. Title.
79-836770
79 C

—ANECDOTES, FACETIAE, SATIRE, ETC.

(Goen)
語園 / [桃華老人撰]；吉田幸一編. — 東京：古典文庫, 昭和53 [1978]
164, 60 p. : ill. ; 17 cm. — (古典文庫；377) Ja*
Series romanized: Koten bunko.
"古活字版複製"
Not in trade.

1. Anecdotes—China. 2. China—Anecdotes, facetiae, etc. I. Ichijō, Kanera, 1402-1481. II. Yoshida, Kōichi, 1909-
PN6267.C5G6 79-804191

(Minu yo no tomo)
見ぬ世の友：[5冊 / 編者 吉田幸一]. — 東京：古典文庫, 昭和53 [1978]
ca. 300 p. : ill. ; 17 cm. — (古典文庫；378) Ja***
Series romanized: Koten bunko.
Photoreprint of the 1658 ed.
Not in trade.

1. Anecdotes—China. 2. China—Anecdotes, facetiae, satire, etc. I. Yoshida, Kōichi, 1909-
PN6267.J3M5 79-804702

—ANTIQUITIES

Akiyama, Shingo.
(Kodai Chūgoku no doki)
古代中国の土器 / 秋山進午. — 東京：平凡社, 1978.
141 p. : ill. (some col.) ; 27 cm. — (陶磁大系；33) Ja 78-27096
Bibliography: p. 134-137.
¥1700

1. Pottery, Chinese—To 618. 2. China—Antiquities. I. Title. II. Series: Tōjī taikei ; 33.
NK4167.T63 vol. 33 79-803711
[NK4165.2]

An-yang, China. Wen hua chü.
(Yin hsü)
殷墟：奴隶社会的一个缩影 / 河南省安阳市文化局编. — 第1版. — 北京：文物出版社, 1976.
96 p. : ill. ; 19 cm.
$0.30

1. An-yang, China—Antiquities. 2. China—Antiquities. I. Title.
DS796.A55A5 1976 78-841163

Australian Art Exhibitions Corp.
The Chinese exhibition : a selection of recent archaeological finds of the People's Republic of China. — Melbourne : National Gallery of Victoria, 1977.
128 p. : ill. (part col.) ; 26 cm. Aus
Presented by the Australian Art Exhibitions Corporation and exhibited at the National Gallery of Victoria, the Art Gallery of New South Wales, and the Art Gallery of South Australia.
ISBN 0-7241-0029-6
1. Art objects, Chinese—Exhibitions. 2. China—Antiquities. I. Victoria, Australia. National Gallery, Melbourne. II. Sydney. Art Gallery of New South Wales. III. South Australia. Art Gallery. IV. Title.
NK1068.A95 1977 709'.51'0740994 79-300220
79 MARC

Chi-lin ta hsüeh, Ch'ang-ch'un, China. Li shih hsi. K'ao ku chuan yeh.
(Kung nung k'ao ku chi ch'u chih shih)
工农考古基础知识 / 吉林大学历史系考古专业, 河北省文物管理处. — 第1版. — 北京：文物出版社, 1978.
7, 233 p. : ill. ; 27 cm.
$1.20
1. Archaeology—Methodology. 2. China—Antiquities. I. Hopei, China (Province). Wen wu kuan li ch'u. II. Title.
CC75.C44 1978 79-836123

(Chung-hua wen wu ching hua)
中華文物精華 = The Chinese art treasures / 編輯者中華世界資料供應出版社編輯部. — Hong Kong：中華新聞與世界報導國際通訊社：總代理香港華僑文化中心, c1976.
681 p. : chiefly ill. (some col.) ; 39 cm.
English and Chinese.
$300.00 ($65.00 U.S.)
1. Art, Chinese. 2. China—Antiquities. I. Chung-hua shih chieh tzu liao kung ying ch'u pan she. Pien chi pu. II. Title: The Chinese art treasures.
N7340.C5385 709'.51 79-837170

(Chung-kuo ch'ing t'ung ch'i)
中國青銅器 = Ancient Chinese bronzes / [文物出版社编辑出版]. — [第1版]. — [北京：該社, 1976-1977]
5 portfolios (50 leaves of plates : 50 col. ill) ; 16 cm.
Title from portfolio.
Legends in English and Chinese.
In 5 cases.
$0.65 per portfolio.
1. Bronzes, Chinese. 2. China—Antiquities. I. Wen wu ch'u pan she, Peking. II. Title: Ancient Chinese bronzes.
NK7983.A1C476 78-842142

CHINA

—ANTIQUITIES (Continued)

Creel, Herrlee Clessner, 1905-
Studies in early Chinese culture, first series / by Herrlee Glessner Creel. — Philadelphia : Porcupine Press, 1978.
xxii, 266 p. ; 22 cm. — (Perspectives in Asian History ; no. 3)
Reprint of the 1937 ed. published by Waverly Press, Baltimore, which was issued as no. 3 of the American Council of Learned Societies' Studies in Chinese and related civilizations.
Bibliography: p. 255-262.
Includes index.
ISBN 0-87991-601-X : $15.00
1. China—Civilization. 2. China—Antiquities. I. Title. II. Series. III. Series: American Council of Learned Societies Devoted to Humanistic Studies. Studies in Chinese and related civilizations ; no. 3.
DS723.C73 1978 951 78-14504
78 MARC

Hagiwara, Shōzō, 1914-
(Tonkō sembutsudō)
敦煌千仏洞 / 萩原正三編著. — 東京 : シルクロード, 1978.
158 p. : ill. ; 26 cm. Ja 78-27092
"蘇瑩輝著「敦煌学概要」...を基本として論述したもの"

1. Tun-huang, Kansu—Antiquities. 2. China—Antiquities. I. Su, Ying-hui. Tung-huang hsüeh kai yao. II. Title.
DS793.T8H33 79-804288

Hibino, Takeo, 1914-
[Me de miru Chūgoku no rekishi. Chinese]
用眼睛看的中國歷史 / [日比野丈夫著述] ; 李方中編著. — 九龍 : 田園書店, [1978?]
2, 3, 230 p. : ill. (some col.) ; 20 cm.
Translation of Me de miru Chūgoku no rekishi, originally published in 1976 by Mu t'ung ch'u pan she, Taipei; issued as 14 in series Mu yüan ts'ung k'an.
$15.00

1. China—History. 2. China—Antiquities. I. Title.
Title romanized: Yung yen ching k'an ti Chung-kuo li shih.
DS735.H513 1978 79-835593

Ho-pei sheng po wu kuan.
(Ho-pei ch'u t'u t'ao tz'u)
河北出土陶瓷 / 河北省博物馆编. — 第1版. — 北京 : 文物出版社, 1977.
1 portfolio (10 leaves of plates : 10 col. ill.) ; 16 cm.
Colophon title.
In case.
$0.55
1. Pottery, Chinese—China—Hopei (Province) 2. Porcelain, Chinese—China—Hopei (Province) 3. Hopei, China (Province)—Antiquities. 4. China—Antiquities. I. Title.
NK4166.H66H6 1977 79-835648

Ho-pei sheng po wu kuan.
(Kao-ch'eng T'ai-hsi Shang tai i chih)
藁城台西商代遺址 / 河北省博物馆, 文管处台西考古队, 河北省藁城县台西大队理论小组编. — 第1版. — 北京 : 文物出版社, 1977.
91 p. : ill. ; 19 cm.
$0.30
1. China—History—Shang dynasty, 1766-1122 B.C. 2. T'ai-hsi ts'un, China—Antiquities. 3. China—Antiquities. I. Wen kuan ch'u T'ai-hsi k'ao ku tui. II. Kao-ch'eng hsien, China (Hopei). T'ai-hsi ta tui. Li lun hsiao tsu. III. Title.
DS744.H66 1977 78-843255

Hsin-chiang Wei-wu-erh tzu chih ch'u po wu kuan.
(Hsin-chiang li shih wen wu)
新疆历史文物 / 新疆维吾尔自治区博物馆编. — 第1版. — 北京 : 文物出版社 : 新華書店发行, 1977, t. p. 1978.
131 p. : ill. (some col.) ; 21 cm.
$0.60

1. Sinkiang—Antiquities. 2. China—Antiquities. I. Title.
DS793.S62H774 1977 79-840267

Kuang-hsi Chuang tsu tzu chih ch'ü. Wen wu kuan li wei yüan hui.
(Kuang-hsi ch'u t'u wen wu)
广西出土文物 / 广西壮族自治区文物管理委员会编. — 第1版. — 北京 : 文物出版社 : 新华书店发行, 1978.
6, 23 p., [68] leaves of plates : ill. (some col.) ; 36 cm.
Bibliography: p. 23 (2d group)
$35.00

1. Art objects, Chinese—China—Kwangsi (Province) 2. Kwangsi, China (Province)—Antiquities. 3. China—Antiquities. I. Title.
NK1069.K85K82 1978 79-837243

Lovell, Hin-cheung.
Report on the investigation of kiln-sites of Ju-type ware and Chün ware in Lin-ju hsien, Honan / by Fêng Hsien-ming ; [translator, Hin-cheung Lovell]. — [London] : Victoria & Albert Museum, c1968.
18 p., [3] leaves of plates : ill. ; 29 cm. — (Chinese translations ; no. 3)
An abstract translated from a report originally published in Wen wu, no. 8, 1964 under title: Ho-nan sheng Lin-ju hsien Sung tai Ju yao i chih tiao ch'a.
1. Lin-ju hsien, China—Antiquities. 2. China—Antiquities. 3. Celadon ware—China—Lin-ju hsien. 4. Chün ware—China—Lin-ju hsien. I. Feng, Hsien-ming. Ho-nan sheng Lin-ju hsien Sung tai Ju yao i chih tiao ch'a. II. Title. III. Series.
DS793.L464L68 931 78-326476
79 MARC

Ma-wang-tui Han mu po shu cheng li hsiao tsu.
(Chan kuo tsung heng chia shu)
戰國縱横家書 / 马王堆汉墓帛书整理小组编. — 第1版. — 北京 : 文物出版社, 1976.
201 p. : ill. ; 21 cm. — (马王堆汉墓帛书)
Includes bibliographical references.
$0.75

1. China—History—Warring States, 403-221 B.C.—Sources. 2. Manuscripts, Chinese. 3. China—Antiquities. I. Title. II. Series: Ma-wang-tui Han mu po shu.
DS747.2.M33 1976 79-836137

Maspero, Henri, 1883-1945.
China in antiquity / by Henri Maspero ; translated by Frank A. Kierman, Jr. — Amherst : University of Massachusetts Press, 1978.
xxxii, 527 p. : maps ; 24 cm.
Translation of La Chine antique.
Bibliography: p. [493]-497.
"Bibliography of the writings of Henri Maspero": p. [498]-511.
Includes index.
ISBN 0-87023-193-6
1. China—History—To 1643. 2. China—Civilization. 3. China—Antiquities. I. Title.
DS745.M313 951'.01 77-26756
77 MARC

Nan-ching po wu yüan.
(Nan-ching po wu yüan ts'ang tz'u ch'i)
南京博物院藏瓷器 = Ceramics from the collection of the Nanking Museum. — 北京 : 文物出版社, 1977.
1 portfolio ([10] leaves of plates : all col. ill.) ; 16 cm.
Descriptive legends in English and Chinese.
In case.
$2.50

1. Pottery, Chinese. 2. China—Antiquities. I. Title. II. Title. Ceramics from the collection of the Nanking Museum.
NK4165.N3 1977 78-843654

(Nei Meng-ku wen wu tzu liao hsüan chi)
内蒙古文物資料选輯 / 内蒙古文物工作队編. — 香港 : 神州圖書公司, [1978?]
18, 215 p., [32] leaves of plates : ill. ; 26 cm.
Reprint of the 1964 ed. published by Nei Meng-ku jen min ch'u pan she, Kuei-sui.

1. Mongolia (Inner Mongolia)—Antiquities. 2. China—Antiquities. I. Mongolia (Inner Mongolia). Wen wu kung tso tui.
DS793.M7N44 1978 79-836186

Pai k'o ch'u pan she.
(Chung-kuo ch'u t'u wen wu ching hua)
中國出土文物精華 / 百科出版社編印. — 香港 : 該社 ; 九龍 : 發行利源書報社, 1978.
50 p. : chiefly ill. (some col.) ; 22 cm. — (百科叢書)
Series romanized: Pai k'o hua ts'ung.
$3.00

1. China—Antiquities. 2. Art objects, Chinese. I. Title.
DS715.P34 1978 79-836811

79 C

(Seiiki shutsudo mokkan)
西域出土木簡 : 展大 / 編集 飯島稲太郎 [春敬]. — 東京 : 書藝文化新社, 昭和45 [1970], 昭和49 [1974] printing.
[34] p. (on double leaves) ; 34 cm. & explanatory text. — (和漢墨寶選集 ; 第6巻)
Ja***
Cataloged from colophon (inserted).
Folded book, issued in case.
¥2700

1. China—Antiquities. 2. Inscriptions, Chinese. 3. Inscriptions—China. 4. Chinese language—Writing—History. I. Iijima, Shunkei, 1906- II. Series: Wa-Kan bokuhō senshū ; dai 6-kan.
ND1457.J33W3 vol. 6 78-814353
[DS715]

Shang-hai po wu kuan. Chung-kuo yüan shih she hui ts'an k'ao t'u chi pien chi hsiao tsu.
(Chung-kuo yüan shih she hui ts'an k'ao t'u chi)
中国原始社会参考图集 / 上海博物馆中国原始社会参考图集编辑小组. — 第1版. — 上海 : 上海人民出版社, 1977.
74 p. : ill. ; 19 x 27 cm.
$1.10

1. China—Antiquities. 2. Man, Prehistoric—China. I. Title.
DS715.S45 1977 78-841497

(Ti chen yü ti chen k'ao ku)
地震与地震考古 / 孟繁兴 ... [et al.]. — 第1版. — 北京 : 文物出版社, 1977.
185 p. : ill. ; 19 cm.
Includes bibliographical references.
$0.55

1. Earthquakes—China. 2. China—Antiquities. I. Meng, Fan-hsing.
QE537.2.C6T53 79-835466

Wei, Chü-hsien, 1898-
(Ku ch'i wu hsüeh)
古器物學 / [衛聚賢]. — 新竹 : 說文社書社, [1978?-
v. : ill. ; 22 cm.
Colophon title.
CONTENTS: 第1冊. 石器時代.
NT$80.00 (v. 1)

1. China—Antiquities. 2. Archaeology—Methodology. 3. Implements, utensils, etc.—China. I. Title.
DS715.W43 79-835662

Weidenreich, Franz, 1873-1948.
Giant early man from Java and south China / Franz Weidenreich. — 1st AMS ed. — New York : AMS Press, 1978.
134 p., [6] leaves of plates : ill. ; 26 cm.
Reprint of the 1945 ed., which was issued as v. 40, pt. 1, of Anthropological papers of the American Museum of Natural History.
Bibliography: p. 125-129.
ISBN 0-404-15978-8
1. Fossil man—Indonesia—Java. 2. Fossil man—China. 3. Java—Antiquities. 4. China—Antiquities. 5. Indonesia—Antiquities. I. Title. II. Series: American Museum of Natural History, New York. Anthropological papers ; v. 40, pt. 1.
GN282.5.W44 1978 573'.3 76-44798
78 MARC

—ANTIQUITIES—ADDRESSES, ESSAYS, LECTURES

New archaeological finds in China : discoveries during the cultural revolution. — 2d ed., rev. — Peking : Foreign Languages Press, 1973-78.
2 v. : ill. (some col.) ; 19 cm.
Vol. 2, 1st ed.
1. China—Antiquities—Addresses, essays, lectures.
DS715.N48 1973 931 76-364761
76[r79]rev MARC

(Shui wen, sha mo, huo shan k'ao ku)
水文, 沙漠, 火山考古 / 重庆市博物馆等. — 第1版. — 北京 : 文物出版社, 1977.
186 p., [9] leaves of plates (1 fold.) : ill. ; 19 cm.
"这里的文章曾在《文物》《考古》杂志上发表过"
Includes bibliographical references.
CONTENTS: 略谈长江上游水文考古.—长江上游宜渝段历史枯水调查.—从石刻题记看长江上游的历史洪水.—从红桝河上的古城废墟看毛乌素沙漠的变迁.—乌兰布和沙漠的考古发现和地理环境的变迁.—风沙威胁不可怕, "榆林三迁"是谣传.—隋唐胜州榆林城的发现.—貧下中农发现古代文物证实大同火山群是死火山.—大同火山的活动时代.—关于大同火山问题的座谈纪要.
$0.70

1. China—Antiquities—Addresses, essays, lectures. I. Ch'ung-ch'ing shih po wu kuan. II. Wen wu.
DS715.S58 79-836095

(Wen wu tzu liao ts'ung k'an)
文物资料丛刊 / 文物编辑委员会编. — 第1版. — 北京 : 文物出版社, 1977-
v. : ill. ; 27 cm.
$1.30 (v. 1)

1. China—Antiquities—Addresses, essays, lectures. I. Wen wu pien chi wei yüan hui.
DS715.W46 78-843659

CHINA (Continued)

—ANTIQUITIES—CATALOGS

Ch'en, Meng-chia, 1911-
[Mei ti kuo chu i chieh lüeh ti wo kuo Yin Chou t'ung ch'i t'u lu. Japanese & Chinese]
殷周青銅器分類図録 = A corpus of Chinese bronzes in American collections / 陳夢家編；松丸道雄改編. — 東京：汲古書院, 1977.
2 v. (42, 1188 p.) : ill. ; 27 cm. Ja 77-6053
Reprint of the 1962 ed. published by K'o hsüeh ch'u pan she; with new introduction by M. Matsumaru.
"殷周青銅器の形態 陳夢家": p. [15]-42.
¥38000
1. Bronzes, Chinese—To 221 B. C.—Catalogs. 2. Bronzes—United States—Catalogs. 3. China—Antiquities—Catalogs. I. Matsumaru, Michio. II. Title. III. Title: A corpus of Chinese bronzes in American collections.
Title romanized: In Shū seidōki bunrui zuroku.
NK7983.22.C4516 1977 79-800345

Finsterbusch, Käte.
Zur Archäologie der Pei-Ch'i- (550-577) und Sui-Zeit (581-618) : mit e. Fundkatalog / Käte Finsterbusch. — Wiesbaden : Steiner, 1976.
ix, 147, 83 [i.e. 84] p. : numerous ill., map ; 29 cm. — (Münchener ostasiatische Studien : Sonderreihe ; Bd. 1) GFR76-A
Bibliography: p. [127]-134.
ISBN 3-515-02017-9 : DM98.00
1. China—Antiquities—Catalogs. 2. China—History—Northern Ch'i dynasty, 550-577. 3. China—History—Sui dynasty, 581-618. I. Title. II. Series.
DS715.F56 78-382300
79 MARC

Paris. Musée Cernuschi.
Formes chinoises : centenaire de Victor Segalen, 1878-1919 : [exposition], Musée Cernuschi, 17 novembre 1978-11 février 1979. — [Paris : Le Musée, 1978?]
[91] p. : ill. (some col.) ; 24 cm. F***
At head of title: Ville de Paris.
1. China—Antiquities—Catalogs. 2. Art, Chinese—Catalogs. 3. Segalen, Victor, 1878-1919. I. Title.
DS715.P37 1978 931 79-369121
79 MARC

—ANTIQUITIES—EXHIBITIONS

(Chūka Jimmin Kyōwakoku shutsudo bumbutsu ten)
中華人民共和国出土文物展：東京展 1978年1月2日-2月26日 西武美術館 / 主催 日本中国文化交流協会 ... [et al.]. — [東京]：西武美術館, c1978.
169 p. : chiefly ill. (some col.) ; 26 cm. Ja***
Catalog of an exhibition.
List of articles also in English with caption title: Exhibition of archaeological finds of the People's Republic of China.
Bibliography: p. 159-161.
1. China—Antiquities—Exhibitions. I. Seibu Bijutsukan. II. Nitchū Bunka Kōryū Kyōkai.
DS715.C533 78-815300

Chung-kuo ch'u t'u wen wu Hsiang-kang chan lan wei yüan hui.
(Chung-hua jen min kung ho kuo ch'u t'u wen wu chan lan)
中華人民共和国出土文物展覽：一九七八年四月·香港 = The exhibition of archaeological finds of the People's Republic of China : April 1978, Hong Kong / 中国出土文物香港展覽委員會編印. — Hong Kong : 該會, [1978]
[103] p. : chiefly ill. (some col.) ; 26 cm.
Explanations in Chinese and English.
1. China—Antiquities—Exhibitions. 2. Art, Chinese—Exhibitions. I. Title. II. Title: The exhibition of archaeological finds of the People's Republic of China. III. Title: Ch'u t'u wen wu chan lan.
DS715.C536 1978 78-841950

Chung-kuo ch'u t'u wen wu Hsiang-kang chan lan wei yüan hui.
(Chung-kuo ch'u t'u wen wu Hsiang kang chan lan)
中國出土文物香港展覽 = Exhibition of archaeological finds of China : Hong Kong, 1978 / 中國出土文物香港展覽委員會. — Hong Kong : 該會, [1978]
[66] p. : ill. ; 26 cm.
Cover title.
Chinese and English.
1. China—Antiquities—Exhibitions. 2. Art, Chinese—Exhibitions. I. Title. II. Title: Exhibition of archaelogical finds of China, Hong Kong, 1978.
DS715.C536 1978a 78-841949

--ANTIQUITIES--PERIODICALS

Chinese studies in archeology. v. 1- summer 1979-
[White Plains, N. Y., M. E. Sharpe]
v. ill. 23 cm. quarterly.
"A journal of translations."
Key title: Chinese studies in archeology, ISSN 0192-9628.
1. China—Antiquities—Periodicals.
DS715.C46 931'.005 79-643207
MARC-S

—ANTIQUITIES—PICTORIAL WORKS

Kishida, Hideto, 1899-
(Nekka iseki)
熱河遺蹟 / 岸田日出刀, 土浦龜城. — 東京：相模書房, 昭和15 [1940]
102 p., [1] leaf of plates : all ill. ; 27 cm.
1. Ch'eng-te, China (Jehol Province)—Description and travel—Views. 2. Ch'eng-te, China (Jehol Province)—Antiquities—Pictorial works. 3. China—Antiquities—Pictorial works. I. Tsuchiura, Kameki, 1897- joint author. II. Title.
DS796.C49K57 78-812726

—ARMED FORCES

Daini Kokuminkai. Henshūbu.
(Ugoku Hokushi)
動く北支 / [編者 第二國民會編輯部；校閲 高橋坦]. — 東京：第二國民會出版部：發賣所 栗田書店, 昭和12 [1937]
2, 4, 277 p., [9] leaves of plates : ill. ; 19 cm.
1. China—Politics and government—1937-1949. 2. China—Foreign relations—Japan. 3. Japan—Foreign relations—China. 4. China—Armed Forces. 5. Marco Polo Bridge Incident, 1937. I. Title.
DS777.53.D27 1939 78-815478

Fan, Chi-Yuan, 1934-
The paralysis and reconstruction of the central leadership of Peiping's armed forces / [Fan Chih-yuan]. — Taipei : Chung Hwa Information Service, [1978?]
16 p. ; 20 cm.
1. China—Armed Forces. I. Title.
UA837.F36 355.3'31'0951 78-103824
78 MARC

(Pei-ching, Chin-men chün ch'i ch'i kuan)
北京, 金門軍器奇觀. — 香港：新時代出版社, [1976?]
34 p. : ill. ; 26 cm.
Cover title.
1. China—Armed Forces. 2. Taiwan—Armed Forces.
UA835.P35 78-843812

Semenov, Georgiĭ Gavrilovich, 1908-
(Tri goda v Pekine)
Три года в Пекине : записки военного советника / Г. Г. Семенов. — Москва : Наука, 1978.
294 p. : ill. ; 23 cm. USSR***
At head of title: Akademiíà nauk SSSR. Institut Dal'nego Vostoka.
1.90rub
1. Military assistance, Russian—China. 2. China—Armed Forces. 3. Semenov, Georgiĭ Gavrilovich, 1908- 4. Russians in China. 5. Generals—Russia—Biography. 6. Russia (1923- U. S. S. R.). Armiíà—Biography. I. Title.
DS777.55.S436 79-411690

—ARMED FORCES—BIOGRAPHY

Liu, Chung-k'ang, 1916-
(Li yin hung chi)
笠蔭鴻跡 / 劉仲康著. — 臺北：禹甸文化事業有限公司, 民國66 [1977]
6, 11, 464 p. ; 22 cm.
NT$120.00
1. Tai, Li, 1896-1946. 2. Generals—China—Biography. 3. China—Armed Forces—Biography. I. Title.
DS778.T26L58 79-836136

--ARMED FORCES--ORGANIZATION --HISTORY

Hsü, Shih-chung.
(Li tai kuan chih ping chi k'o chü chih ch'ang shih)
歷代官制兵制科舉制常识 / 徐師中編. — 第1版. — 澳門：爾雅出版社, 1977, 1978 printing.
ii, 278 p. ; 19 cm.
$10.00 (H.K.)
1. Civil service positions—China—History. 2. China—Armed Forces—Organization—History. I. Title.
JQ1512.Z1H77 79-837502

—BIBLIOGRAPHY

Chung yang t'u shu kuan, T'ai-pei. Kuan shu ku.
Unesco documents; an annotated bibliography with an index to documents related to China. Based on the collection of the National Central Library. Edited by Government Documents Division, National Central Library. Chief: Daniel T. C. Chang. Assistants: Jocelin Tsai [and] Richard Chin. Taipei, National Central Library, 1970.
130, xxvii, iii p. 27 cm.
On cover: Kuo li chung yang t'u shu kuan ts'ang Lien-ho-kuo chiao yü k'o hsüeh wen hua tsu chih tzu liao mu lu.
1. United Nations Educational, Scientific and Cultural Organization—Bibliography. 2. China—Bibliography. I. Chang, Daniel T. C. II. Tsai, Jocelin. III. Chin, Richard. IV. Title. V. Title: Kuo li chung yang t'u shu kuan ts'ang Lien-ho-kuo chiao yü k'o hsüeh wen hua tsu chih tzu liao mu lu.
Z6483.U5C58 016.9151 74-163609
78[r78]rev MARC

Kobylinski, Hanna, 1907-
Litteratur om Kina, Taiwan og Hong Kong, Korea og Mongoliet : nyt udvalg : annoteret bibliografi / ved Hanna Kobylinski, Huy Dong Shin og Vagn Plenge. — København : Mellemfolkeligt Samvirke : [eksp. DBK], 1978.
100 p. ; 21 cm. — (MS bibliografier) D78-27
Includes indexes.
ISBN 8770282420 : kr23.00
1. China—Bibliography. 2. Taiwan—Bibliography. 3. Hongkong—Bibliography. 4. Korea—Bibliography. 5. Mongolia (Mongolian People's Republic)—Bibliography. 6. East Asia—Bibliography. I. Shin, Huy Dong, joint author. II. Plenge, Vagn, joint author. III. Title. IV. Series: Mellemfolkeligt samvirke. MS bibliografier.
Z3106.K58 78-386757
[DS706] 79 MARC

Kumar, Brij Kishore, 1938-
China through Indian eyes : a select bibliography, 1911-1977 / B. K. Kumar. — Delhi : Concept Pub. Co., 1978.
140 p. ; 22 cm.
Includes indexes.
Rs60.00 ($12.00 U.S.)
1. China—Bibliography. 2. China—Study and teaching—India—Bibliography. I. Title.
Z3106.K84 016.951 78-904380
[DS706] 78 MARC

Lowe, Chuan-hua, 1902-
Notable books on Chinese studies : a selected, annotated, and subject-divided bibliographic guide / compiled and edited, with introductory essays, by C. H. Lowe ; foreword by Jesse H. Shera. — Taipei : China Print. ; Menlo Park, Calif. : obtainable from C. H. Lowe, c1978.
ix, 228 p., [1] leaf of plates : ill. ; 24 cm.
English and Chinese.
1. China—Bibliography. I. Title.
Z3106.L68 016.951 78-107687
[DS706] 79 MARC

Nihon Kokusai Mondai Kenkyūjo, Tokyo. Gendai Chūgoku Kenkyū Bukai.
(Chūgoku kankei Eibun rombun mokuroku)
中国関係英文論文目録：『国際年報』1969年版抜刷. — [東京]：日本国際問題研究所現代中国研究部会, 1973.
p. 686-744 ; 21 cm. Ja 75-4812
Cover title.
1. China—Bibliography. I. Kokusai nempō. II. Title.
Z3106.N48 1973 78-813036
[DS706]

CHINA

—BIBLIOGRAPHY (Continued)

Oslo. Norske Nobelinstitutt. Biblioteket.
Folkerepublikken China etter 1957 : litteratur i Nobelinstituttets bibliotek, 2. utg. : suppl. / [à]jourført av Tone Lise Rogstad]. — Oslo : Nobelinstituttets bibliotek, 1978.
14 leaves ; 30 cm. N***
Cover title.
1. China—Bibliography. I. Rogstad, Tone Lise. II. Title.
Z3106.O8 1971 Suppl. 016.951 79-343972
[DS706] 79 MARC

Tanis, Norman E
China in books : a basic bibliography in Western language / compiled by Norman E. Tanis, David L. Perkins, Justine Pinto. — Greenwich, Conn. : Jai Press, 1979.
xvi, 328 p. ; 24 cm. — (Foundation in library and information science ; v. 4)
Includes index.
ISBN 0-89232-071-0 : $35.00
1. China—Bibliography. I. Perkins, David, 1939- joint author. II. Pinto, Justine, joint author. III. Title. IV. Series.
Z3106.T35 016.951 77-24396
[DS706] 79 MARC

—BIBLIOGRAPHY—CATALOGS

Cornell University. Libraries.
The catalog of the Wason Collection on China and the Chinese, Cornell University Libraries / compiled and edited by Paul P. W. Cheng. — Washington : Center for Chinese Research Materials, Association of Research Libraries, [1978-
v. : port. ; 39 cm. — (Bibliographical series - Center for Chinese Research Materials, Association of Research Libraries ; no. 17a)
Includes index.
CONTENTS: pt. 1. Serials catalog.
1. China—Bibliography—Catalogs. 2. Chinese imprints—Catalogs. 3. Wason, Charles William, 1854-1918—Library—Catalogs. I. Cheng, Paul P. W. II. Title. III. Series: Association of Research Libraries. Center for Chinese Research Materials. Bibliographical series ; no. 17a-.
Z3109.C67 1978 015.51 79-113498
[DS706] 00 MARC

San Jose State University. Library.
East Asian language works in the San Jose State University Library / compiled by Jo Bell Whitlatch and Simon P. J. Chen, with the assistance of Clayton D. Trost. — San Jose : San Jose State University Library, 1975.
xi, 95, 20, 3 p. ; 28 cm. — (Publication - San Jose State University Library ; B-12)
Includes indexes.
1. Japan—Bibliography—Catalogs. 2. China—Bibliography—Catalogs. 3. Korea—Bibliography—Catalogs. 4. San Jose State University. Library—Catalogs. I. Whitlatch, Jo Bell. II. Chen, Simon P. J. III. Trost, Clayton D. IV. Title. V. Series: San Jose State University. Library. Publication - San Jose State University Library ; B-12.
Z3306.S26 1975 015.52 76-621152
[DS806] 79 MARC

—BIBLIOGRAPHY—PERIODICALS

United States. Dept. of State. Office of External Research.
Foreign affairs research special papers available, People's Republic of China. 1971/75-
[Washington] Office of External Research, Bureau of Intelligence and Research, U. S. Dept. of State.
v. 27 cm. annual. (Department of State publication)
Vols. for 1971/75 cumulative from June 1971–Dec. 1975.
Key title: Foreign affairs research special papers available, People's Republic of China, ISSN 0163-7681
1. China—Bibliography—Periodicals. I. Title. II. Series: United States. Dept. of State. Publication.
Z3106.U44a 016.951'005 78-648286
[DS706] MARC-S

—BIOGRAPHY

Angus, Fay.
The white pagoda / Fay Angus. — Wheaton, Ill. : Tyndale House, c1978.
192 p. : ill. ; 22 cm.
ISBN 0-8423-8213-5 : $3.95
1. Angus, Fay. 2. China—Biography. 3. China—Social life and customs. I. Title.
CT1828.A53A38 951.04'2'0924 77-83555
00 MARC

Biographical dictionary of Republican China. Howard L. Boorman, editor; Richard C. Howard, associate editor. New York, Columbia University Press, 1967-79.
5 v. map. 28 cm. $25.00 (v. 3)
Bibliography: v. 4, p. [97]-418, by J. K. H. Cheng.
Vol. 5: "A personal name index, by Janet Krompart."
ISBN 0-231-08957-0 (v. 3)
1. China—History—Republic, 1912-1949—Biography. 2. China—Biography. I. Boorman, Howard L., ed. II. Howard, Richard C., ed. III. Cheng, Joseph K. H.
DS778.A1B5 920'.051 67-12006
[r79]rev2 MARC

Chang, Chung-fu, 1902-
(Mi wang chi)
迷惘集 / 張忠紱著. — 臺北縣永和鎮：文海出版社, 民國67 [1978]
2, 4, 258 p. ; 22 cm. — (近代中國史料叢刊續編 [i. e.輯] ; 530 : 第五十三輯)
Reprint of the 1968 ed.
1. Chang, Chung-fu, 1902- 2. China—Biography. I. Title. II. Series: Shen, Yün-lung, comp. Chin tai Chung-kuo shih liao ts'ung k'an hsü chi ; 530.
DS755.S532 vol. 530 79-835353
[CT1828.C47]

Chao, Hung-te.
(Wan li shuang hung chi)
萬里雙鴻記 趙鴻德著 台北 民國48 [1959]
246 p. illus. 19 cm.
1. Chao, Hung-te. 2. China—Biography. I. Title.
CT1828.C497A38 1959 C 60-680 ‡

Chao, Hung-te.
(Wan li shuang hung chi)
萬里雙鴻記 / 趙鴻德著. — 3版. — 臺北：趙鴻德：總經銷傳眞書店, 民國65 [1976]
272 p. : ill. ; 20 cm. — (傳眞叢書 ; 3)
Series romanized: Ch'uan chen ts'ung shu.
NT$45.00
1. Chao, Hung-te. 2. China—Biography. I. Title.
CT1828.C497A38 1976 78-843724

Ch'en, Chiung-hsü.
(Wan ku liu fang)
萬古流芳 / 陳烱旭編. — 三重：世新出版社, 民國65 [1976]
255 p. : ill. ; 22 cm.
NT$40.00
1. China—Biography. I. Title.
DS734.C513 79-835142

Cheng, Ch'ien.
(Sung jen sheng tsu k'ao shih li)
宋人生卒考示例 / 鄭騫撰. — 景美[臺北縣]：華世出版社, 民國66 [1977]
11, 162 p. ; 21 cm.
NT$60.00
1. China—History—Sung dynasty, 960-1279—Biography. 2. China—Biography. I. Title.
DS751.C455 79-837165
79 C

Ch'ien, Mu, 1895-
(Huang-ti)
黃帝 / 錢穆著. — 臺初版. — 臺北：東大圖書公司：總經銷三民書局, 1978.
3, 5, 116 p. ; 22 cm. — (創海叢刊 : 歷史)
Series romanized: Ts'ang hai ts'ung k'an : Li shih.
"堯舜禹湯文武周公附"
NT$35.00
1. China—Biography. I. Title.
DS734.C566 1978 78-843320
78 C

Ch'ien, Pao-t'ang.
(Li tai ming jen sheng tsu lu)
歷代名人生卒錄 / 錢保塘撰. — 台北：廣文書局, 民國67 [1978]
2 v. (795 p.) ; 22 cm. — (史料七編)
Reprint ed.
NT$750.00
1. China—Biography. I. Title. II. Series: Shih liao ch'i pien.
DS734.C568 1978 79-838671

Ch'in, Hsiang-jui.
(Li tai wei jen chuan)
歷代偉人傳 / 秦祥瑞編著. — 台南：大夏出版社, 民國67 [1978]
2, 262 p. ; 19 cm.
NT$35.00
1. China—Biography. I. Title.
DS734.C5747 79-836561

Ch'in, Hsiang-jui.
(Chung-kuo wei jen chuan chi)
中國偉人傳記 / 秦祥瑞編著. — 台南：大夏出版社, 民國66 [1977]
6, 454 p. : ill. ; 20 cm.
CONTENTS: 民族英雄傳.—歷代偉人傳記.
NT$70.00
1. China—Biography. I. Title.
DS734.C574 78-843217
C

(Ch'in ting Sheng ch'ao hsün chieh chu ch'en lu)
欽定勝朝殉節諸臣錄 / [舒赫德, 于敏中]. — [台北：台灣]商務[印書館, 1976?]
3 v. ; 20 cm. — (四庫全書珍本六集 ; 105-107)
Cover title.
Reproduced from Wen yüan ko copy.
1. China—History—Ming dynasty, 1368-1644—Biography. 2. China—Biography. I. Shu-ho-te, 1711-1777. II. Yü, Min-chung, 1714-1779. III. Title: Sheng ch'ao hsün chieh chu ch'en lu. IV. Series: Ssu k'u ch'üan shu. Ssu k'u ch'üan shu chen pen liu chi ; 105-107.
AC149.S7 1976, vol. 105-107 78-843855
[DS753.5.C455]

(Ch'ing nien ying hsiung ti ku shih)
青年英雄的故事 / 本社編. — [香港? : s. n., 1978]
1, 201 p. : ill. ; 21 cm.
Reprint of the 1978 ed published by Chung-kuo ch'ing nien ch'u pan she, Peking.
1. China—Biography. 2. Communists—China—Biography. I. Chung-kuo ch'ing nien ch'u pan she, Peking.
DS778.A1C535 1978 79-838011

Chow, Ching Lie, 1936-
Journey in tears : memory of a girlhood in China / by Chow Ching-li ; translated from the French by Abby Israel. — New York : McGraw-Hill, c1978.
262 p., [4] leaves of pplates : ill. ; 24 cm.
Translation of Le palanquin des larmes.
ISBN 0-07-010818-8 : $11.95
1. Chow, Ching Lie, 1936- . 2. China—Social life and customs. 3. China—Biography. I. Title.
CT1828.C544A3413 951.05'092'4 78-18171
78 MARC

Chu, K'ung-chang, 1842-1919.
(Hsien-feng i lai kung ch'en pieh chuan)
咸豐以來功臣別傳 / 朱孔彰撰. — 台北：廣文書局, 民國67 [1978]
2 v. (552 p.) ; 22 cm. — (史料七編)
Reprint of the 1898 ed. published by Chien hsüeh lu, in series: Chien hsüeh lu ts'ung shu.
NT$400.00
1. China—History—Hsien-feng, 1850-1861—Biography. 2. China—History—T'ung-chih, 1861-1875—Biography. 3. China—Biography. I. Title. II. Sereis: Shih liao ch'i pien.
DS760.8.C486 1978 78-841108
C

(Chung-kuo li tai ming jen chuan)
中國歷代名人傳 / [編者中外文史研究社]. — 香港：文昌書店, [1978?]
847 p. in various pagings ; 19 cm.
Cover title.
$20.00
1. China—Biography. I. Chung wai wen shih yen chiu she.
DS734.C6255 79-840088

CHINA

—BIOGRAPHY (Continued)

Fan, Yin-nan.
(Tang tai Chung-kuo ssu ch'ien ming jen lu)
當代中國四千名人錄 / 編纂者樊蔭南. — Hong Kong：波文書局, 1978.
9, 460 p. ; 20 cm.
Title in English on verso of t. p.: Who's who in contemporary China (Four thousand persons)
"據1936年版增訂版重印"
Edition of 1931 published under title: Tang tai Chung-kuo ming jen lu.
1. China—History—1912-1937—Biography. 2. China—Biography. I. Title. II. Title: Who's who in contemporary China.
DS778.A1F36 1978 79-835014

Hsia, Ch'eng-ming.
(Sheng hsüeh chi)
升學記 / 夏成明著. — [台北]：中國大陸災胞救濟總會, [1977?]
2, 2, 462 p. : 22 cm.
Cover title.

1. Hsia, Ch'eng-ming. 2. China—Biography. I. Title.
CT1828.H69A37 79-836047

Hsiao, Ping-nan.
(Wo yao k'ung su)
我要控訴 / 蕭炳南口述；楊昆南筆記. — [台北]：中國大陸災胞救濟總會, [1978?]
1, 2, 36 p. ; 20 cm.
Cover title.

1. Hsiao, Ping-nan. 2. China—Biography. I. Yang, K'un-nan. II. Title.
CT1828.H74A34 79-839573

(Hsien tai Chung-kuo ssu hsiang chia)
現代中國思想家 / 龔自珍 ... [et al.]；[編撰委員王曉波 ... et al.]. — 台北：巨人出版社, 民國67 [1978]
8 v. ; 22 cm.
NT$180.00 per vol.

1. China—Biography. I. Kung, Tzu-chen, 1792-1841. II. Wang, Hsiao-po.
CT1826.H75 79-837489

(Huan yao nu li tso chan)
还要努力作战：无产阶级专政下继续革命的先进战士. — [香港?：s. n., 1979?]
2, 136 p. ; 21 cm.
Reprint of the 1975 ed. published by Pei-ching jen min ch'u pan she.

1. Communists—China—Biography. 2. China—Biography. I. Title: Wu ch'an chieh chi chuan cheng hsia chi hsü ko ming ti hsien chin chan shih.
DS778.A1H82 1979 79-840095

(K'ai kuo ying hsiung chuan)
開國英雄傳 / [編著者孟銓, 潘朝陽]. — 台北：莊嚴出版社, 民國67 [1978]
6 v. ; 19 cm. — (古典新刊 ; 31-36)
CONTENTS: [1] 孟銓, 潘朝陽 秦前漢編.—[2] 漢水, 孟銓 後漢三國編.—[3] 劉吳 唐朝編.—[4] 端木鐸 宋朝編.—[5] 潘朝陽 元朝編.—[6] 李侑森, 詹順裕 明清編.
NT$315.00

1. China—Biography. I. Meng, Ch'üan. II. P'an, Chao-yang. III. Series: Ku tien hsin k'an ; 31-36.
DS734.K35 79-837833

Ku, Wei-wei.
(Shih liu sui, Pei-p'ing)
16歲, 北平 / 古威威著. — 臺北：名人出版社, 民國67 [1978]
201 p. ; 20 cm.
NT$38.00

1. Ku, Wei-wei. 2. China—Biography. I. Title.
CT1828.K84A347 78-843329

Kuo fu chi nien kuan.
(Chung-hua wei jen hua hsiang hsü chi)
中華偉人畫像續輯 / 國父紀念館珍藏；[主編國父紀念館文教活動科]. — 台北：該局, 民國67 [1978]
50, [19] leaves : 50 col. ports. ; 27 cm.
Added t. p.: Portraits of the greats of China, a sequel.
English and Chinese.
NT$320.00 ($8.00 U.S.)
1. China—Biography. 2. China—Biography—Portraits. I. Kuo fu chi nien kuan. Kuan li ch'u. Wen chiao huo tung tsu. II. Title. III. Title: Portraits of the greats of China, a sequel.
DS734.K825 1978 920'.051 79-837169

(Kuo kung feng yün ming jen lu)
國共風雲名人錄 / 李約等著. — 香港：廣角鏡出版社, 19
v. : ill. ; 19 cm. — (廣角鏡人物叢書)
Series romanized: Kuang chiao ching jen wu ts'ung shu.
$5.50 (v. 3)

1. China—History—Republic, 1912-1949—Biography. 2. China—History—1949-1976—Biography. 3. China—Biography. I. Li, Yüeh.
DS778.A1K87 78-842847

Li, Han-ch'ing.
(Li Han-ch'ing hsien sheng hui i lu)
李漢青先生回憶錄. — [s. l. : s. n.], 民國66 [1977] (台北：達文文具印刷有限公司)
161 p. : ill. ; 22 cm.

1. Li, Han-ch'ing. 2. China—Biography. I. Title.
CT1828.L348A34 79-835226

Mamo, David, 1948-
Dizionario storico-biografico della Cina moderna / [di] David Mamo, Laurie Upson. — Firenze : Vallecchi, 1977.
xxvii, 458 p. ; 20 cm. — (Il Pellicano : Storia) It78-Sept
Bibliography: p. [449]-455.
Includes index.
L8000
1. China—History—Republic, 1912-1949—Biography. 2. China—History—1949-1976—Biography. 3. China—Biography. I. Title.
DS778.A1M35 920'.051 78-396318
*79 MARC

(Mo kuei hsing hsiang)
魔鬼形像. — 台北：大陸資料出版社：總經銷新亜出版社, 民國56 [1967]
9, 257 p. : ill. ; 19 cm.
Colophon title.
CONTENTS: 鄧小平—陶鑄—羅瑞卿—陳雲—王光美—朱德—薄一波—陳毅

1. China—History—1949-1976—Biography. 2. China—Biography. 3. Communists—China—Biography. I. Ta lu tzu liao ch'u pan she.
DS778.A1M58 78-843817

Moriya, Hiroshi, 1932-
(Chūgokujin no kōshōjutsu)
中国人の交渉術：世渡りの才覚とは / 守屋洋著. — 東京：産業能率短期大学出版部, 1978.
202 p. : map ; 19 cm. Ja 79-3227
¥950

1. China—History—Warring States, 403-221 B. C.—Biography. 2. China—Biography. 3. Diplomacy. I. Title.
DS747.2.M67 79-805946
*79 J

Moriya, Hiroshi, 1932-
(Hasha no senryaku to ketsudan)
覇者の戦略と決断 / 守屋洋著. — 東京：産業能率短期大学出版部, 1978.
218 p. : ill. ; 19 cm. Ja 78-23422
¥980

1. China—History—Spring and autumn period, 722-481 B. C.—Biography. 2. China—Kings and rulers—Biography. 3. China—Biography. I. Title.
DS747.15.M67 79-809183

(Ningen no kyōshi, Tōyō hen)
人間の教師 東洋編 / 古川哲史編. — 東京：大阪教育図書, 昭和34 [1959]
264 p. : ill. ; 18 cm.

1. China—Biography. 2. India—Biography. I. Furukawa, Tesshi, 1912-
DS734.N56 78-815271

Okuzaki, Hiroshi, 1935-
(Chūgoku kyōshin jinushi no kenkyū)
中國郷紳地主の研究 / 奥崎裕司著. — 東京：汲古書院, 1978.
15, 599, 9, 57 p., [1] fold. leaf of plates : ill. ; 22 cm. Ja***
Includes bibliographical references and index.
¥8000

1. Upper classes—China. 2. China—Gentry. 3. Land tenure—China. 4. China—Intellectual life. 5. Yüan, Huang, chin shih 1586. 6. China—Biography. I. Title.
HT653.C6O48 79-800227

(Shao-hsing shih pa nien t'ung nien hsiao lu)
紹興十八年同年小錄. 中州人物考 / [孫奇逢撰]. — [台北：臺灣]商務[印書館, 1978?]
2 v. ; 20 cm. — (四庫全書珍本八集 ; 82-83)
Cover title.
Photoreprint ed. of two works, both of the Wen Yüan Ko copy.
1. China—Registers. 2. Civil service—China. 3. China—History—Sung dynasty, 960-1279—Biography. 4. China—Biography. 5. China—History—Ming dynasty, 1368-1644—Biography. 6. Honan, China (Province)—Biography. I. Sun, Ch'i-feng, 1585-1675. Chung chou jen wu k'ao. 1978?. II. Title: Chung chou jen wu k'ao. III. Series: Ssu k'u ch'uan shu. Ssu k'u ch'uan shu chen pen pa chi ; 82-83.
AC149.S7 1978, vol. 82-83 78-841087
[JQ1507]

Shen, Tsung-han, 1895-
(Shen Tsung-han wan nien wen lu)
沈宗瀚晚年文錄. — 臺北：傳記文學雜誌社, 民國68 [1979]
4, 2, 350 p., [1] leaf of plates : port. ; 21 cm. — (傳記文學叢刊 ; 50)
Series romanized: Chuan chi wen hsüeh ts'ung k'an.
NT$140.00

1. Shen, Tsung-han, 1895- 2. China—Biography. I. Title.
CT1828.S47A374 79-841041

Shih tsu ta ch'üan.
排韻增廣事類氏族大全. — [台北：臺灣]商務[印書館, 1978?]
6 v. ; 20 cm. — (四庫全書珍本八集 ; 138-143)
Cover title.
Caption and running title: Shih tsu ta ch'üan.
Photoreprint ed. of the Wen Yüan Ko copy.
1. China—History—Sung dynasty, 960-1279—Biography. 2. China—Biography. I. Title. II. Series: Ssu k'u ch'üan shu. Ssu k'u ch'üan shu chen pen pa chi ; 138-143.
Title romanized: P'ai yün tseng kuang shih lei shih tsu ta ch'üan.
AC149.S7 1978, vol. 138-143 78-841132
[DS751.5]

Sin, Yong-gyun.
(Chungguk taeryuk ŭi maek)
中國大陸의脈：그時代와人物 / 申容均, 鄭甲龍共著. — 서울：鄉書閣, 1977.
362 p. : ill. ; 20 cm.
Errata slip inserted.
₩1500

1. China—Biography. I. Chŏng, Kam-nyŏng, joint author. II. Title.
DS734.S56 79-825124

Ssu-ma, Lu, 1919-
[Tou cheng shih pa nien. Japanese]
聖地延安：失われし祖国よ・青春よ / 司馬璐著；矢内茂訳. — 東京：生活社, 1955.
262 p. : ill. ; 19 cm.
On verso of t. p.: My strife in eighteen years.

1. China—Politics and government—1937-1949. 2. Ssu-ma, Lu, 1919- 3. China—Biography. I. Title.
Title romanized: Seichi En'an.
DS777.47.S7516 79-800646

CHINA

—BIOGRAPHY (Continued)

Ting, Wang.
(Hua Kuo-feng Chi Teng-k'uei ho hsin ch'i ti i tai)
華國鋒紀登奎和新起的一代 / 丁望著. — 香港：明報月刊社, 1977.
10, iv, 191 p. : ill. ; 21 cm. — (現代中國研究叢刊：傳記類；1) (明報月刊叢書；10)
Title in English in colophon and on cover Hua Kuo-feng, Chi Teng-kuei and a new generation.
Series 2 romanized: Ming pao yüeh k'an ts'ung shu.
Includes index.
Bibliography: p. 181–184.
$18.00 ($4.50 U. S.)

1. China—History—1976- —Biography. 2. China—Biography. 3. Communists—China—Biography. 4. Statesmen—China—Biography. I. Title. II. Title: Hua Kuo-feng, Chi Teng-kuei and a new generation. III. Series: Hsien tai Chung-kuo yen chiu ts'ung k'an : Chuan chi lei ; 1.
DS779.28.T56 78–842374

Vogau, Boris Andreevich, 1894-1937.
Une femme russe en Chine : [Journal de Xénia Mikhaïlovna] / [par] B. Pilniak [i.e. B. A. Vogau] ; [collab.] A. Rogozina ; trad., préf. et notes de Jocelyne de Maack. — Lausanne : L'Age d'homme, 1976.
127 p. ; 21 cm. — (Contemporains) Sw77-4338
19.00F
1. China—Politics and government—1912-1949. 2. Biespalykh, Xénia Mikhaïlovna. 3. China—Biography. 4. Russia—Biography. I. Rogozina, A., joint author. II. Title.
DS775.V6414 320.9'51'04 78-400604
*79 MARC

Wang, Feng-i, b. 1864.
(Wang kung Feng-i yen hsing lu)
王公鳳儀言行錄 / 鄭子東講述；[記錄者鄭宜時]. — [台北]：佛教印經處, 民國65 [1976]
463 p. in various pagings : port. ; 22 cm.
Cover title.

1. Wang, Feng-i, b. 1864. 2. China—Biography. I. Wang, Lü-jen. II. Cheng, I-shih. III. Title.
CT1828.W316A38 78–843681

Wang, Shih-chen.
(Min kuo jen hao hua chuan)
民國人豪畫傳 / 王世禎主編. — 台南：青溪出版社：總經銷東風圖書供應社, 民國66 [1977]
328, 12 p. : ill. ; 22 cm. — (東風新刊；2)
Cover title varies slightly.
Reprint ed.?
Series romanized: Tung feng hsin k'an.
English and/or Chinese.
NT$80.00
1. China—History—Republic, 1912-1949—Biography. 2. China—Biography. I. Title.
DS778.A1W329 1977 79–835035
79 C

Wang, Te-i.
(Chung-kuo li tai ming jen nien p'u tsung mu)
中國歷代名人年譜總目 / 王德毅編. — 台北：華世出版社, 民國68 [1979]
27, 328, 63, [8] p. ; 22 cm. — (檢索叢刊：甲種；2)
Includes indexes.
NT$300.00

1. China—Biography. 2. China—Biography—Bibliography. I. Title. II. Series: Chien so ts'ung k'an : chia chung ; 2.
DS734.W32 79–838195

(Wei ch'ung kao li hsiang wang wo fen tou ti jen)
为崇高理想忘我奋斗的人. — [香港? : s. n., 1978?]
80 p. ; 21 cm.
Reprint of the 1975 ed. published by Fu-chien jen min ch'u pan she.
CONTENTS: 万里征途不歇脚.—珠穆朗玛一青松.—"爱民模范"洛桑单增.—"毛主席教育出来的好军医".—"干社会主义, 就需要他这种人!".—帕米尔高原上的共产主义战士杜洪亮.—火红的青春.—扎根农村干革命, 身背药箱献青春.

1. China—Biography.
CT1823.W43 1978 79–838645

(Wei pei san chung)
魏碑三種. — [上海]：上海古籍书店, [1978]
[52] p. ; 27 cm.
CONTENTS: 魏驃騎將軍冀州刺史元診墓誌銘.—魏寧遠將軍元君墓誌銘.—魏江陽王次妃石夫人墓誌銘.
$5.50

1. Calligraphy, Chinese. 2. Epitaphs—China. 3. China—History—Northern Wei dynasty, 386-534—Biography. 4. China—Biography.
NK3634.A2W45 79–835870

Who's who in China. 6th ed. Shanghai, China Weekly Review [1950]
263 p. ports. 25 cm.
Title also in Chinese.
1. China—History—Republic, 1912-1949—Biography. 2. China—Biography. I. China monthly review.
DS778.A1W443 1950 920'.051 51-1089
[r79]rev MARC

Who's who in China, containing the pictures and biographies of some of China's political, financial, business and professional leaders. 2d ed. Shanghai, Millard's review [1920]
2 p.l., 314, iv p. illus. (ports.) 26 cm.
"The material incorporated in this book has been reproduced practically without change from the pages of Millard's Review during the past three years."—Pref., signed M. C. [i.e. John B.] Powell, H. K. Tong, editors.
1. China—History—Republic, 1912-1949—Biography. 2. China—Biography. I. Millard's review of the Far East. II. Powell, John Benjamin, 1888-, ed. III. Tong, Hollington Kong, 1887-1971, joint ed.
DS778.A1W443 1920 21-10280
[r79]rev MARC

Who's who in China ... containing the pictures and biographies of China's best known political, financial, business and pofessional men ... 3d ed. Shanghai, The China weekly review [1926]
2 p.l., 972 [1], 1-25, 11 p. illus. (ports.) 24 cm.
"Practically all of these pictures and biographies have appeared originally in the pages of the China weekly review during the past eight years."—Pref., signed M. C. [i.e. John B.] Powell, editor.
"All of these biographical sketches have appeared in the pages of the China weekly review since January 1927."-Pref., signed J. B. P. [i.e. John B. Powell]
1. China—History—Republic, 1912-1949—Biography. 2. China—Biography. I. China monthly review. II. Powell, John Benjamin, 1888- ed.
DS778.A1W443 1926 26-24595
[r79]rev2 MARC

Wu, Hsi.
(Chung-kuo li tai wei jen shih lüeh)
中國歷代偉人史略 / [著作者吳熹]. — [s. l.]：雋思出版社, [1978?]
2, 154 p. ; 19 cm. — (僑務委員會. 中華函授學校補充教材；47) (雋思文化叢書)
Series 2 romanized: Chün ssu wen hua ts'ung shu.

1. China—Biography. I. Title. II. Series: China (Republic of China). Ch'iao wu wei yüan hui. Ch'iao wu wei yüan hui. Chung hua han shou hsüeh hsiao pu ch'ung chiao ts'ai ; 47.
DS734.W77 79–836369

Wu, Yung-chiu, 1919–
(Li tai sheng hsien yen hsing lu)
歷代聖賢言行錄 / 宋瑞[吳詠九]著. — 臺北：中央文物供應社, 民國67 [1978]
6, 166 p. ; 22 cm.
NT$30.00

1. China—Biography. 2. Conduct of life—Quotations, maxims, etc. I. Title.
DS734.W79 79–835133

—BIOGRAPHY—ANECDOTES, FACETIAE, SATIRE, ETC.

Ch'ai, Fu.
(Chung-kuo hsien tai ming jen lien ai shih)
中國現代名人戀愛史 / 柴甫著. — 台北縣新莊鎮：綠園出版社, 民國67 [1978]
192 p. : ill. ; 20 cm.
NT$50.00

1. China—Biography—Anecdotes, facetiae, satire, etc. I. Title.
DS778.A1C42 78–843087

Ch'en, Hao-yang, 1918–
(Chūgoku bijo den)
中国美女伝 / 陳浩洋著. — [s. l. : s. n.], 民国62 [1973] (台北：文和印刷有限公司印刷)
3, 5, 290 p. : ill. ; 19 cm.

1. China—Biography—Anecdotes, facetiae, satire, etc. 2. Women—China—Biography—Anecdotes, facetiae, satire, etc. I. Title.
DS734.C516 78–816084

Ch'en, K'o-wei.
[Chung kung t'e ch'üan chieh chi sheng huo cheng hsiang]
中共高級幹部的生活 / 陳克煒編. — 九龍：阿爾泰出版社, 1977.
2, 212 p., [2] leaves of plates : ill. ; 19 cm.
Reprint of the ed. published by Li ming wen hua shih yeh kung ssu, Taipei under title: Chung kung t'e ch'üan chieh chi sheng huo cheng hsiang; with intro. deleted.
$6.50 ($1.40 U.S.)
1. China—History—1949-1976—Biography—Anecdotes, facetiae, satire, etc. 2. China—Biography—Anecdotes, facetiae, satire, etc. I. Title.
Title romanized: Chung kung kao chi kan pu ti sheng huo.
DS778.A1C464 1977 78–843450

(Chiang hsiang feng yün)
將相風雲 / [編譯者河洛圖書出版社]. — 臺初版. — 臺北：該社, 民國67 [1978]
2, 7, 237 p. ; 22 cm.
At head of title: Chung-kuo chiang hsiang ku shih.
"本書所記載的十八位將相...從春秋時代迄三國時代"

1. China—Biography—Anecdotes, facetiae, satire, etc. I. Ho lo t'u shu ch'u pan she.
DS734.C534 78–842677

(Chung-kuo jen ti chih hui)
中國人的智慧. — [s. l.]：蘇宿出版社；高雄：心影書店, [1977?]
6, 184 p. ; 19 cm.
NT$30.00

1. China—History—Anecdotes, facetiae, satire, etc. 2. China—Biography—Anecdotes, facetiae, satire, etc.
DS736.C539 78–843641

(Hao chieh lieh chuan)
豪傑列傳. — 台北：偉文圖書出版社, 民國67 [1978]
101 p. ; 19 cm. — (中國偉人傳記
NT$25.00

1. China—Biography—Anecdotes, facetiae, satire, etc. I. Series: Chung-kuo wei jen chuan chi.
DS734.H29 79–837508

Hui-lu.
(Chung-kuo li shih jen wu man t'an)
中國歷史人物漫談 / 晦廬編著. — 香港：上海書局, 1978.
2, 67 p. ; 19 cm
$2.20

1. China—Biography—Anecdotes, facetiae, satire, etc. I. Title.
DS734.H88 79–836799

(Luan shih ying hsiung p'u)
亂世英雄譜. — 臺北：河洛圖書出版社, 民國66 [1977]
190 p., [57] leaves of plates : ill. ; 22 cm.
At head of title: Chung-kuo li shih jen wu ku shih.

1. China—Biography—Anecdotes, facetiae, satire, etc.
DS734.L79 78–843433

CHINA

—BIOGRAPHY—ANECDOTES, FACETIAE, SATIRE, ETC. (Continued)

Ma, Pin.
(Chung-kuo li tai ming jen i shih)
中國歷代名人軼事 / 南宮博[馬彬]著 ; 白羽圖. — 台南 : 大夏出版社, 民國67 [1978]
2 v. : ill. ; 20 cm.
NT$70.00

1. China—Biography—Anecdotes, facetiae, satire, etc. I. Title.
DS734.M199 79-835362

Yü, Ling.
(Chung-kuo min tsu ying hsiung ku shih)
中國民族英雄故事 / 編者俞凌. — 香港 : 王氏出版社 : 發行者南洋圖書公司, 1977.
1, 139 p. : ill. ; 19 cm.
Short stories.
CONTENTS: 魯曹沫收復失地.—藺相如完璧歸趙.—澠池會不辱國體.—為祖國廉藺交歡.—申包胥秦庭乞師.—報國仇臥薪嘗膽.—火牛陣田單復齊.—博浪沙張良行刺.—漢蘇武牧羊北海.—趙充國屯田保邊.—岳武穆盡忠報國.—文天祥殺身成仁.—戚繼光勦平倭寇.—史可法為國犧牲.—鄭成功死守台灣.

1. China—Biography—Anecdotes, facetiae, satire, etc. I. Title.
DS734.Y75 78-842738

--BIOGRAPHY--BIBLIOGRAPHY

Wang, Te-i.
(Chung-kuo li tai ming jen nien p'u tsung mu)
中國歷代名人年譜總目 / 王德毅編. — 台北 : 華世出版社, 民國68 [1979]
27, 328, 63, [8] p. ; 22 cm. — (檢索叢刊 : 甲種 ; 2)
Includes indexes.
NT$300.00

1. China—Biography. 2. China—Biography—Bibliography. I. Title. II. Series: Chien so ts'ung k'an : chia chung ; 2.
DS734.W32 79-838195

—BIOGRAPHY—INDEXES

Chung, Hua.
(Shih chi jen ming so yin)
史記人名索引 / 鍾華編. — 第1版. — 北京 : 中華書局, 1977.
4, 267 p. ; 21 cm.
"中華書局在一九五九年出版了《史記》 點校本 本索引就是根據點校本編製的"
Includes index.
$0.72
1. Ssu-ma, Chien, ca. 145–ca. 86 B. C. Shih chi—Indexes. 2. China—History—Early to 1766 B. C.—Indexes. 3. China—History—1766 B. C.–220 A. D.—Indexes. 4. China—Biography—Indexes. I. Ssu-ma, Chien, ca. 145–ca. 86 B. C. Shih chi. II. Title.
DS735.A2S612 78-841309

Umehara, Kaoru, 1934–
(Zoku Shiji tsugan chōhen jimmei sakuin)
續資治通鑑長編人名索引 / 梅原郁編. — 京都 : 同朋舍, 1978.
7, 697 p. ; 22 cm. — (東洋史研究資料叢刊) Ja 78-20683
"底本としては光緒7年(1881)浙江書局刊本を使用"
¥14000
1. Li, T'ao, 1115–1184. Hsü Tzu chih t'ung chien ch'ang pien—Indexes. 2. China—History—Sung dynasty, 960–1279—Indexes. 3. China—History—Sung dynasty, 960–1279—Biography—Indexes. 4. China—Biography—Indexes. I. Li, T'ao, 1115–1184. Hsü Tzu chih t'ung ch'ang pien. II. Title. III. Series: Tōyōshi kenkyū shiryō sōkan.
DS751.U43 78-816027

--BIOGRAPHY--PORTRAITS

Kuo fu chi nien kuan.
(Chung-hua wei jen hua hsiang hsü chi)
中華偉人畫像續輯 / 國父紀念館珍藏 ; [主編國父紀念館文教活動科]. — 台北 : 該局, 民國67 [1978]
50, [19] leaves : 50 col. ports. ; 27 cm.
Added t. p.: Portraits of the greats of China, a sequel.
English and Chinese.
NT$320.00 ($8.00 U.S.)
1. China—Biography. 2. China—Biography—Portraits. I. Kuo fu chi nien kuan. Kuan li ch'u. Wen chiao huo tung tsu. II. Title. III. Title: Portraits of the greats of China, a sequel.
DS734.K825 1978 920'.051 79-837169

—BOUNDARIES

China (Republic of China). Kuang fu ta lu she chi yen chiu wei yüan hui.
(Kung fei pien chiang hsing cheng ching chi wen chiao ts'o shih chih yen chiu)
共匪邊疆行政·經濟·文教措施之研究 / 光復大陸設計研究委員會研擬. — [台北] : 該會, 民國65 [1976]
2, 204 p. ; 22 cm. — (邊疆研究叢書)
Cover title.
1. China—Boundaries. 2. China—Race relations. I. Title. II. Series: Pien chiang yen chiu ts'ung shu.
DS737.C4884 1976 78-841058

China (Republic of China). Kuang fu ta lu she chi yen chiu wei yüan hui.
(Ta lu kuang fu hou pien fang chien she chih yen chiu)
大陸光復後邊防建設之研究 / 光復大陸設計研究委員會研擬. — [台北] : 該會, 民國65 [1976]
3, 100 p. ; 22 cm. — (邊疆研究叢書)
Cover title.

1. China—Defences. 2. China—Boundaries. 3. Fortifications—China. I. Title. II. Series: Pien chiang yen chiu ts'un, shu.
UA835.C446 1976 79-837134

--BOUNDARIES--ADDRESSES, ESSAYS, LECTURES

(Hsi-nan pien chiang wen t'i yen chiu pao kao)
西南邊疆問題研究報告 / 華中大學中國文學系研究室編. 第1期. — [大理] : 華中大學庶務處, 民國31 [1942]
98 double leaves in various foliations ; 27 cm.
No more published.
Lithoprinted.
CONTENTS: 羅常培 貢山俅語初探.—游國恩 說洱海.—游國恩 火把節考.—包鷺賓 民家非白國後裔考.—包鷺賓 蒙氏滅南詔說.—萬先法 史記西南夷傳考疏.
1. China—Boundaries—Addresses, essays, lectures. I. Hua-chung ta hsüeh, Wuchang, China. Chung-kuo wen hsüeh hsi. Yen chiu shih.
DS737.H68 79-839917

—BOUNDARIES—RUSSIA

Chang, Hsiao-yen.
(Chen-pao tao shih chien chi fei O kuan hsi)
珍寶島事件及匪俄關係, 一九五六—一九六九 / 章孝嚴著. — 臺北 : 黎明文化事業公司, 民國67 [1978]
6, 180, [64] p., [1] leaf of plates : ill. ; 20 cm.
NT$60.00
1. China—Foreign relations—Russia. 2. Russia—Foreign relations—China. 3. China—Boundaries—Russia. 4. Russia—Boundaries—China. 5. Chen-pao Island. I. Title.
DS740.R8C452 78-843090

China (Republic of China). Kuang fu ta lu she chi yen chiu wei yüan hui.
(Fei O pien chieh cheng chih wen t'i chih yen chiu)
匪俄邊界爭執問題之研究 / 光復大陸設計研究委員會研擬. — [台北] : 該會, 民國65 [1976]
2, 52 p. ; 22 cm. — (邊疆研究叢書)
Cover title.
1. China—Foreign relations—Russia. 2. Russia—Foreign relations—China. 3. China—Boundaries—Russia. 4. Russia—Boundaries—China. I. Title. II. Series: Pien chiang yen chiu ts'ung shu.
DS740.5.R8C5512 1976 78-843066

Chung-kuo she hui k'o hsüeh yüan. Chin tai shih yen chiu so.
(Sha O ch'in Hua shih)
沙俄侵华史 / 中国社会科学院近代史研究所. — 第1版. — [北京] : 人民出版社, 1978–
v. : ill. ; 21 cm.
Includes bibliographies.
$0.90 (v. 1)
1. China—Foreign relations—Russia. 2. Russia—Foreign relations—China. 3. China—Boundaries—Russia. 4. Russia—Boundaries—China. I. Title.
DS740.5.R8C558 1978 78-843452

Ginsburgs, George.
The Sino-Soviet territorial dispute, 1949-64 / George Ginsburgs, Carl F. Pinkele. — New York : Praeger, 1978.
viii, 1435 p. ; 25 cm.
Bibliography: p. 139-145.
ISBN 0-03-027096-0 : $15.95
1. China—Boundaries—Russia. 2. Russia—Boundaries—China. I. Pinkele, Carl F., joint author. II. Title.
DS740.5.R8G48 1978 911'.51 78-19458
78 MARC

Olgin, V
On the Soviet-Chinese border : questions and answers / Vladimir Olgin. — Moscow : Novosti Press Agency Pub. House, 1978.
63 p. ; 17 cm. — (The Soviet viewpoint) USSR***
Includes bibliographical references.
0.25rub
1. Russia—Foreign relations—China. 2. China—Foreign relations—Russia. 3. China—Boundaries—Russia. 4. Russia—Boundaries—China. I. Title. II. Series.
DK68.7.C5O4313 327.47'051 79-314784
79 MARC

—BOUNDARIES—RUSSIA—ADDRESSES, ESSAYS, LECTURES

(Po huang yen chih tsao che)
驳谎言制造者 / [人民出版社编辑出版]. — 第1版. — [北京] : 該社, 1977.
89 p. : ill. ; 21 cm.
Includes bibliographical references.
$0.22

1. China—Boundaries—Russia—Addresses, essays, lectures. 2. Russia—Boundaries—China—Addresses, essays, lectures. I. Jen min ch'u pan she, Peking.
DS740.5.R8P6 78-843028

—CHURCH HISTORY

Bergeron, Marie Ina.
Le Christianisme en Chine : approches et stratégies / Marie-Ina Bergeron. — Lyon : Chalet, 1977.
155 p. : maps ; 21 cm. F78-17147
Includes bibliographical references.
ISBN 2-7023-0292-0 : 33.00F
1. Catholic Church in China—History. 2. China—Church history. 3. Missions—China. I. Title.
BX1665.B39 266'.00951 79-373931
79 MARC

—CIVILIZATION

Anghel, Paul, 1921-
O clipă în China : eseu despre eternitatea culturii chineze / Paul Anghel. — Bucureşti : "Sport-Turism", 1978.
235 p. ; 20 cm. R78-5258
lei7.50
1. China—Civilization. I. Title.
DS721.A62 79-368585
*79 MARC

Cheng, Te-k'un, 1908–
(Chung-hua min tsu wen hua shih lun)
中華民族文化史論 / 鄭德坤著. — 香港 : 天地圖書有限公司, 1978.
3, 195 p., [11] leaves of plates : ill. ; 22 cm.
Added t. p.: Man and culture in China.
$15.00

1. China—Civilization. I. Title. II. Title: Man and culture in China.
DS721.C4773 79-838591

Chin, Yao-chi.
(Ts'ung ch'uan t'ung tao hsien tai)
從傳統到現代 / 金耀基著. — 增訂1版. — 台北 : 時報文化出版事業有限公司, 民國67 [1978]
327 p. ; 20 cm. — (時報書系 ; 118)
Series romanized: Shih pao shu hsi.
"附錄 胡適與中國現代化運動": p. 267–285.
Includes bibliographical references.
NT$55.00
1. China—Civilization. I. Title.
DS721.C4878 1978 79-836986

CHINA

—CIVILIZATION (Continued)

Chou, Ching-wen, 1909–

(Chung-hua min tsu tsai i ch'ieh min tsu ko ming tou cheng chung ti ling tao ti wei)

中華民族在一切民族革命鬥爭中的領導地位 / 周鯨文著. — 香港：時代批評社, 民國28 [1939]

2, 2, 108 p. ; 22 cm. — (時代叢書 : 政治組 ; 第1種)

Series romanized: Shih tai ts'ung shu : Cheng chih tsu.
Includes bibliographical references.

1. China—Civilization. 2. Nationalism. I. Title.

DS721.C54 1939 77–838507

Chuang yen ch'u pan she. Pien chi pu.

(Kuo hsüeh ch'ien t'an)

國學淺談 / [本社編輯部]. — 臺北：莊嚴出版社, 民國67 [1978]

2, 9, 189 p. ; 19 cm. — (古典新刊 ; 20)

NT$40.00

1. China—Civilization. I. Title. II. Series: Ku tien hsin k'an ; 20.

DS721.C57215 1978 79–839105

(Chūgoku bummei no genzō)

中国文明の原像 / 宮川寅雄, 関野雄, 長広敏雄編. — 東京：日本放送出版協会, 1978.

2 v. : ill. ; 19 cm. — (放送ライブラリー ; 21–22) Ja 78–21773

Series romanized: Hōsō raiburari.
Bibliography: v. 2, p. 259.
¥900 per vol.

1. China—Civilization. I. Miyagawa, Torao, 1908– II. Sekino, Takeshi, 1915– III. Nagahiro, Toshio, 1905–

DS721.C57218 78–815599

(Chung-hua wen hua ch'ang shih)

中華文化常識. — 九龍：百靈出版社, [1978?]

2, 252 p. ; 19 cm.

Reprint ed.
$7.50

1. China—Civilization.

DS721.C57224 1978 78–843089

Chung-hua wen hua fu hsing yün tung t'ui hsing wei yüan hui.

(Chung-hua wen hua fu hsing yün tung ti shih chien yü chan wang)

中華文化復興運動的實踐與展望 / 中華文化復興運動推行委員會編. — 第2次增訂. — [台北]：該會, 民國66 [1977]

10, 4, 238 p., [1] leaf of plates : col. ill. ; 22 cm.

1. Taiwan—Social life and customs. 2. China—Civilization. I. Title.

DS799.4.C457 1977 78–843716

Cotterell, Yong Yap.

Chinese civilization : from the Ming revival to Chairman Mao / Yong Yap & Arthur Cotterell. — New York : St. Martin's Press, 1977.

256 p. : ill. (some col.) ; 26 cm.
Bibliography: p. 247-249.
Includes index.
ISBN 0-312-13381-2
1. China—Civilization. I. Cotterell, Arthur, joint author. II. Title.

DS721.C838 1977b 951 77-24816
77[r79]rev MARC

Creel, Herrlee Clessner, 1905-

Studies in early Chinese culture, first series / by Herrlee Glessner Creel. — Philadelphia : Porcupine Press, 1978.

xxii, 266 p. ; 22 cm. — (Perspectives in Asian History ; no. 3)
Reprint of the 1937 ed. published by Waverly Press, Baltimore, which was issued as no. 3 of the American Council of Learned Societies' Studies in Chinese and related civilizations.
Bibliography: p. 255-262.
Includes index.
ISBN 0-87991-601-X : $15.00
1. China—Civilization. 2. China—Antiquities. I. Title. II. Series. III. Series: American Council of Learned Societies Devoted to Humanistic Studies. Studies in Chinese and related civilizations ; no. 3.

DS723.C73 1978 951 78-14504
78 MARC

Dawson, Raymond Stanley.

The Chinese experience / [by] Raymond Dawson. — London : Weidenfeld and Nicolson, 1978.

xxv, 318 p., [24] p. of plates : ill., facsim., map, ports. ; 25 cm. — (History of civilisation) GB78-36519
Includes index.
ISBN 0-297-77403-4 : £12.50.
1. China—Civilization. I. Title.

DS721.D358 951 79-314486
79 MARC

Enoki, Kazuo, 1913–

(Tōzai bummei no kōryū)

東西文明の交流 / 榎一雄. — 東京：講談社, 昭和52 [1977]

182 p. : ill. (some col.) ; 27 cm. — (図説中国の歴史 ; 11) Ja***

Bibliography: p. 174–176.
Includes index.
¥2400

1. China—Civilization. 2. East and West. I. Title. II. Series: Zusetsu Chūgoku no rekishi ; 11.

DS735.Z86 vol. 11 78–812202
[DS721]

J

Feng, Hu-hsiang, 1948–

(Wen hua che hsüeh mien mien kuan)

文化哲學面面觀 / 馮滬祥著. — 台北縣新店鎮：先知出版社 ; 台北：總經銷先知書卡禮物中心, 1976.

337 p. ; 20 cm. — (先知叢書 ; 36)

Title on verso of t. p.: Essays on the cultural philosophy.
Series romanized: Hsien chih ts'ung shu.
Includes bibliographical references.
NT$85.00
1. Civilization—Philosophy. 2. China—Civilization. I. Title.

CB19.F38 78–840890

FitzGerald, Charles Patrick, 1902-

Ancient China / by Patrick FitzGerald. — Oxford : Elsevier-Phaidon, 1978.

151 p. : ill. (chiefly col.), col. maps, col. plans ; 29 cm. — (The Making of the past) GB78-39380
Bibliography: p. [137]
Includes index.
ISBN 0-7290-0065-6 : £4.95
1. China—Civilization. I. Title. II. Series.

DS721.F54 931 79-314492
79 MARC

Goodnow, Frank Johnson, 1859-1939.

China, an analysis / Frank J. Goodnow. — New York : Arno Press, 1979, c1926.

viii, 279 p. ; 21 cm. — (Johns Hopkins University Press reprints)
Reprint of the ed. published by Johns Hopkins Press, Baltimore.
ISBN 0-405-10603-3 : $22.00
1. China. 2. China—Civilization. I. Title.

DS736.G6 1979 951 78-19330
78 MARC

Howard, Leslie G

Can there be a Chinese church / Leslie G. Howard ; Chinese translation by Fung Yue Ling. — Hong Kong : International Book Services, [1978]

54, 65 p., [2] leaves of plates : ill. ; 21 cm.
Added t.p.: Neng fou yu i ko chen cheng ti Chung-kuo chiao hui.
English and Chinese.
Errata slip inserted.
1. Missions—China. 2. China—Religion. 3. China—Civilization. I. Title. II. Title: Neng fou yu i ko chen cheng ti Chung-kuo chiao hui.

BV3415.2.H68 266'.00951 79-374213
79 MARC

Hsiang, Ta, 1900–

(Chung wai chiao t'ung hsiao shih)

中外交通小史 / 向達著. — [香港? : s. n., 1978?]

2, 2, 112 p. ; 19 cm.

Cover title.
Reprint of the ed. published by Shang wu yin shu kuan, Shanghai, in series: Wan yu wen k'u : Ti i chi, i ch'ien chung.

1. China—Relations (general) with foreign countries. 2. China—Civilization. I. Title.

DS740.4.H716 1978 79–835981

Ku, Chieh-kang, 1893–

[Han tai hsüeh shu shih lüeh. Japanese]

中国古代の学術と政治 / 顧頡剛著 ; 小倉芳彦 ... [et al.] 共訳. — 東京：大修館書店, 1978.

vi, 260 p. ; 20 cm. — (中国叢書) Ja 79–2923

Includes bibliographical references.
¥1600

1. Philosophy, Chinese. 2. China—Civilization. I. Title. II. Series: Chūgoku sōsho (Tokyo, 1978–)

Title romanized: Chūgoku kodai no gakujutsu to seiji.

B126.K7816 79–806934

Ku, Hung-ming.

[The spirit of the Chinese people. Japanese]

支那人の精神 / 辜鴻銘著 ; 魚返善雄譯. — 東京：目黒書店, 昭和15 [1940], 昭和19 [1944] printing.

213, [17] p., [1] leaf of plates : ill. ; 19 cm.

Added t. p.: The spirit of the Chinese people.
"附錄 暴民崇拜教-大戦とその出路": p. 183–206.
1. China—Civilization. 2. National characteristics, Chinese. 3. European War, 1914–1918—Causes. I. Title.

Title romanized: Shinajin no seishin.

DS721.K816 78–814782

Li, Chiu-jui.

(Chung-hua wen hua hsing t'ai shih kai shu)

中華文化形態史概述 / 李九瑞著. — 臺北：順先出版公司, 民國66 [1977]

4, 14, 340 p. ; 22 cm.

1. China—Civilization. I. Title.

DS721.L54323 79–835384

Liang, Shu-ming, 1893–

(Chung-kuo wen hua yao i)

中國文化要義 / [著者梁漱溟]. — 臺北：問學出版社, 民國66 [1977]

5, 2, 332 p., [1] leaf of plates : port. ; 22 cm.

Reprint ed.
Includes bibliographical references.
NT$70.00 (pbk.)

1. China—Civilization. I. Title.

DS721.L549 1977 78–841952

Maspero, Henri, 1883-1945.

China in antiquity / by Henri Maspero ; translated by Frank A. Kierman, Jr. — Amherst : University of Massachusetts Press, 1978.

xxxii, 527 p. : maps ; 24 cm.
Translation of La Chine antique.
Bibliography: p. [493]-497.
"Bibliography of the writings of Henri Maspero": p. [498]-511.
Includes index.
ISBN 0-87023-193-6
1. China—History—To 1643. 2. China—Civilization. 3. China—Antiquities. I. Title.

DS745.M313 951'.01 77-26756
77 MARC

Nakayama, Kyūshirō, 1874-1961.

(Saikin Shina no seiji oyobi bunka)

最近支那の政治及文化 / 中村[山]久四郎著. — 東京：三省堂, 大正11 [1922]

2, 5, 285 p. ; 18 cm. — (日本百科叢書)

Series romanized: Nihon hyakka sōsho.

1. China—Politics and government—1912–1937. 2. China—Civilization. I. Title.

DS777.44.N37 79–803670

(San min chu i yü Chung-hua wen hua)

三民主義與中華文化 / 吳經熊等著 ; [編輯者中央文物供應社]. — 臺北：該社, 民國67 [1978]

4, 2, 228 p. ; 22 cm.

NT$40.00

1. Sun, Yat-sen, 1866–1925. San min chu i. 2. China—Civilization. I. Wu, Ching-hsiung, 1899– II. Chung yang wen wu kung ying she.

DS777.A567S287 79–836477

79 C

(Shina taikan to saikan)

支那大観と細観 / 久米邦武, 永井柳太郎合著. — 東京：新日本社, 大正6 [1917]

8, 10, 114, 280 p. ; 20 cm.

CONTENTS: 永井柳太郎 支那大観—久米邦武 「支那大観」の細観

1. China—Civilization. 2. Nagai, Ryūtarō, 1881–1944. Shina taikan. 3. China—Description and travel—1901–1948. 4. Nagai, Ryūtarō, 1881–1944. I. Nagai, Ryūtarō, 1881–1944. Shina taikan. 1917. II. Kume, Kunitake, 1839–1931. Shina taikan no saikan. 1917.

DS721.S477 79–801365

CHINA

—CIVILIZATION (Continued)

T'an, Cheng-pi.
(Kuo hsüeh kai lun hsin pien)
國學概論新編 / 譚正璧編. — 臺景印初版. — 臺北：河洛圖書出版社, 民國67 [1978]
2, 5, 4, 240 p. ; 22 cm. — (河洛文庫 ; 56)
Series romanized: Ho lo wen k'u.
Reprint of the 1936 ed. published by Pei hsin shu chü, Shanghai.

1. China—Civilization. I. Title.
DS721.T26 1978 79-839109

Wang, Ching-chih, 1916-
(Kuo hsüeh tao tu)
國學導讀 / 王靜芝, 蔡與濟合著. — 4版. — 台北縣新莊鎮：輔仁大學文學院, 民國66 [1977]
1, 4, 302 p. ; 22 cm. — (輔仁大學文學院叢書)
NT$90.00

1. China—Civilization. I. Ts'ai, Hsing-chi, joint author. II. Title. III. Series: Fu jen ta hsüeh, Hsin-chuang, Taiwan. Wen hsüeh yüan. Fu jen ta hsüeh wen hsüeh yüan ts'ung shu.
DS721.W33373 1977 79-839562

Yin, Hai-kuang.
(Chung-kuo wen hua ti chan wang)
中國文化的展望 / 殷海光著. — 九龍：文藝書屋, 1969.
2 v. ; 21 cm.
Title in English in colophon: Reappraisal of cultural change in modern China.
Reprint of the 1966 ed. published by Wen hsing shu tien, T'ai-pei.
Includes bibliographical references.
1. China—Civilization. I. Title. II. Title: Reappraisal of cultural change in modern China.
DS721.Y45 1969 79-840818

—CIVILIZATION—1949-

(Chūgoku o shiru)
中国を知る / 朝日新聞外報部. — 東京：朝日新聞社, 1978.
256 p. : ill. ; 19 cm. Ja 78-19614
"一九七七年の初めから同年末まで朝日新聞の国際面に...連載した記事に、かなりの部分を書き加え...たもの"
¥800

1. China—Civilization—1949- I. Asahi Shimbun Sha. Gaihōbu. II. Asahi shimbun, Tokyo.
DS724.C468 78-814852

Egashira, Kazuma, 1925-
(Chūgoku no bunka to kyōiku)
中国の文化と教育 / 江頭数馬. — [東村山]：教育社：東京：販売 教育社出版サービス, [1978]
165 p. : ill. ; 18 cm. — (時事問題解説 ; 76. 入門新書) Ja 78-21006
Cover title.
Series 2 romanized: Nyūmon shinsho.
¥400

1. China—Civilization—1949- 2. Education—China. I. Title. II. Series: Jiji mondai kaisetsu ; 76.
DS724.E35 79-801545

Ishikawa, Shō, 1928-
(Shin Chūgoku jijō)
新·中国事情：近代化をめざす政治·経済から大衆の暮しぶりまで / 石川昌著. — 東京：日本実業出版社, 1979.
238 p. : ill. ; 19 cm. Ja 79-4618
¥980

1. China—Civilization—1949- I. Title.
DS779.2.I8 79-806364

Mooneyham, Walter Stanley, 1926-
China, a new day / by W. Stanley Mooneyham. — Plainfield, N.J. : Logos International, c1979.
241 p. ; 18 cm.
ISBN 0-88270-364-1 : $2.50
1. Christianity—China. 2. China—Civilization—1949- 3. China—Politics and government—1976- I. Title.
BR1285.M55 275.1 79-83789
79 MARC

—CIVILIZATION—1949- —ADDRESSES, ESSAYS, LECTURES

The China difference / edited by Ross Terrill. — 1st ed. — New York : Harper & Row, c1979.
xiv, 335 p. : ill. ; 24 cm.
Includes index.
ISBN 0-06-014249-9 : $12.95
1. China—Civilization—1949- —Addresses, essays, lectures. I. Terrill, Ross.
DS724.C453 1979 951.05 78-2179
78 MARC

Shu, Ch'ing-yü.
(Chung-kuo wen hua wang na li ch'ü)
中國文化往哪裡去 / 孫慶餘著. — 三重：慧龍文化公司；大甲鎮 [台中縣]：總經銷文星書店, 民國67 [1978]
226 p. ; 19 cm.
NT$48.00

1. China—Civilization—1949- —Addresses, essays, lectures. I. Title.
DS724.S9 78-843772

Wang, Chang-ling.
(Kung fei wen hua cheng ts'e)
共匪文化政策 / 王章陵講；教育部社會教育司編. — [台北]：該司, 民國65 [1976]
36 p. : port. ; 19 cm. — (文化講座專集 ; 58)
Includes bibliographical references.

1. China—Civilization—1949- —Addresses, essays, lectures. I. Title. II. Series: Wen hua chiang tso chuan chi ; 58.
DS724.W35 78-843190

—CIVILIZATION—ADDRESSES, ESSAYS, LECTURES

Ching-k'ung, Shih.
(Hsing chien Chung-hua min tsu pai hsing tsung tz'u ch'u i)
興建「中華民族百姓宗祠」芻議 / [淨空]. — 台北：華藏法施會, 民國66 [1977]
6, 16 p. ; 26 cm.
Colophon title.
"文化復興與宗教": p. 1-16 (last group)
"非賣品"
1. China—Genealogy—Addresses, essays, lectures. 2. China—Civilization—Addresses, essays, lectures. I. Ching-k'ung, Shih. Wen hua fu hsing yü tsung chiao. 1977. II. Title.
CS1162.C48 78-843619

Hu, Lan-ch'eng, 1906-
(Shan ho sui yüeh)
山河歲月 / 胡蘭成著. — [香港? : s. n., 1978?]
10, 295 p. ; 19 cm.
Cover title.
Literature.
Reprint ed.

1. China—Civilization—Addresses, essays, lectures. I. Title.
DS721.H6976 1978 79-836101

Kao, Ming, 1909-
(Kao Ming lun hsüeh tsa chu)
高明論學雜著 / 高明著. — 臺北：黎明文化事業公司, 民國67 [1978]
4, 2, 317 p. ; 22 cm. — (國學叢書)
Series romanized: Kuo hsüeh ts'ung shu.
NT$70.00

1. China—Civilization—Addresses, essays, lectures. I. Title.
DS721.K1757 79-839931

Kōsaka, Jun'ichi, 1915-
(Pekin Daigaku ninen)
北京大学二年 / 香坂順一. — 東京：竜渓書舎, 1976.
348 p. ; 20 cm. — (日中関係選書 ; 2) Ja 76-8746
Series romanized: Nitchū kankei sensho.
¥1900

1. Education—China—1949- —Addresses, essays, lectures. 2. Pei-ching ta hsüeh—Addresses, essays, lectures. 3. China—Civilization—Addresses, essays, lectures. I. Title.
LA1131.82.K67 78-812438

Kuwabara, Takeo, 1904-
(Chūgoku to tsukiau hō)
中国とつきあう法 / 桑原武夫, 加藤周一. — 東京：潮出版社, 1978.
194 p. ; 20 cm. Ja 78-10317
¥880

1. China—Civilization—Addresses, essays, lectures. 2. Civilization—Addresses, essays, lectures. I. Katō, Shūichi, 1919- joint author. II. Title.
DS721.K93 78-812158

Li, Ko.
(Chung-kuo wen hua shih hua)
中國文化史話 / 黎戈等編著. — 香港：商務印書館, 1978-
v. : ill. ; 19 cm. — (中學生文庫)
Series romanized: Chung hsüeh sheng wen k'u.
$4.00 (v. 1)

1. China—Civilization—Addresses, essays, lectures. I. Title.
DS721.L5453 78-842622

Uchiyama, Kanzō, 1885-1959.
(Ryanpentō)
両辺倒 / 内山完造著. — 東京：乾元社, 昭和28 [1953]
303 p. ; 19 cm.

1. China—Civilization—Addresses, essays, lectures. I. Title.
DS721.U269 79-804664

Yü, Ta-ch'eng.
(Chung-kuo wen hua ching shen)
中國文化精神 / 于大成講；教育部社會教育司編印. — [台北]：該司, 民國66 [1977]
14 p. : port. ; 19 cm. — (文化講座專集 ; 114)

1. China—Civilization—Addresses, essays, lectures. I. Title. II. Series: Wen hua chiang tso chuan chi ; 114.
DS721.Y82 78-842573

Yü, Ta-ch'eng.
(Wen hua chien she yü kuo chia hsien tai hua)
文化建設與國家現代化 / 于大成講；教育部社會教育司編印. — [台北]：該司, 民國65 [1976]
18 p. : port. ; 20 cm. — (文化講座專集 ; 68)

1. China—Civilization—Addresses, essays, lectures. I. Title. II. Series: Wen hua chiang tso chuan chi ; 68.
DS724.Y83 78-841443

—CIVILIZATION—JUVENILE LITERATURE

Knox, Robert.
Ancient China / [Robert Knox ; editor, Frances Clapham ; illustrators, Brian and Constance Dear, Richard Hook]. — New York : Warwick Press, 1979.
44 p. : ill.(some col.) ; 28 cm. — (Modern knowledge library)
Includes index.
SUMMARY: A brief, illustrated history of China from 1500 B.C. to 907 A.D.
ISBN 0-531-09111-2
1. China—Civilization—Juvenile literature. [1. China—Civilization] I. Clapham, Francis M. II. Dear, Brian. III. Dear, Constance. IV. Hook, Richard. V. Title.
DS721.K595 1979 931 78-63101
79 MARC

—CIVILIZATION—MISCELLANEA

Brelet-Rueff, Claudine.
Les étonnantes techniques chinoises / Claudine Brelet-Rueff. — [Paris : Retz, c1977]
191 p. : ill. ; 19 cm. — (Les Encyclopédies du mieux être) F***
Bibliography: p. [190]-191.
ISBN 2-7256-0266-1 : 44.00F
1. Health. 2. Medicine, Chinese. 3. China—Civilization—Miscellanea. I. Title.
RA776.5.B73 615'.8'0951 78-376072
78 MARC

CHINA (Continued)

—CIVILIZATION—OCCIDENTAL INFLUENCES

Chu, Chieh-ch'in, comp.
(Chung hsi wen hua chiao t'ung shih i ts'ui)
中西文化文[i. e. 交]通史譯粹 / 朱傑勤譯. — 臺北：華世出版社, 民國63 [1974]
2, 2, 302 p. ; 21 cm.
Reprint of the 1939 ed. published by Chung-hua shu chü, in series: Li shih ts'ung shu.
Includes bibliographical references.

1. China—Civilization—Occidental influences. I. Title.
DS721.C568 1974 79-835311

(Gendai Shina shisō no shomondai)
現代支那思想の諸問題 / 神谷正男譯編. — 東京：生活社, 昭和15 [1940]
5, 12, 271, 16, 21 p. : 22 cm.
Bibliography: p. 257-271.
Includes indexes.
"郭湛波氏の「近五十年中國思想史」...蔡尚志氏の論文「近三十年來の中國思想界」を加えて本文となし、外に二·三の参考論文を附録として編篡飜譯した"
1. China—Intellectual life. 2. China—Civilization—Occidental influences. 3. Learning and scholarship—China. I. Kamiya, Masao, 1910-1972. II. Kuo, Chan-po. Chin wu shih nien Chung-kuo ssu hsiang shih. Japanese. Selections. 1940. III. Ts'ai, Shang-ssu, 1905- Kin sanjūnenrai no Chūgoku shisōkai. 1940.
DS721.G36 78-815255

—CIVILIZATION—PERIODICALS

(K'ung-tzu hsüeh shuo yü Chung-hua wen hua)
孔子學説與中華文化 第[1]- 集 63- 年 [1974- 台北] 教育部社會教育司
v. 19 cm. annual.
"國學研究會講演專集"

1. Confucius—Societies, periodicals, etc. 2. China—Civilization—Periodicals. I. Kuo hsüeh yen chiu hui. II. China (Republic of China). Chiao yü pu. She hui chiao yü ssu.
B128.C8K844 77-836450 (MARC-S)

(Shūkan Tōyōgaku)
集刊東洋學
[仙台] 東北大学中国文史哲研究会
no. 21 cm.
Organ of Chūgoku Bunshitetsu Kenkyūkai.
Other title : Chinese and Oriental studies.
In Japanese.

1. China—Civilization—Periodicals. 2. East Asia—Civilization—Periodicals. I. Chūgoku Bunshitetsu Kenkyūkai. II. Title: Chinese and Oriental studies. III. Title: Tōyōgaku.
DS701.S46 79-640536 (MARC-S)

--CIVILIZATION--PHILOSOPHY

Ogura, Yoshihiko, 1927-
(Gyakuryū to junryū)
逆流と順流：わたしの中国文化論 / 小倉芳彦著. — 東京：研文出版, 1978.
325 p. ; 20 cm. — (研文選書 ; 3) Ja 79-3226
Series romanized: Kenbun sensho.
¥1800

1. China—Civilization—Philosophy. 2. China—Historiography. I. Title.
DS721.O436 79-805676

—COMMERCE

Babu, V Vittal.
China trade manual / V. Vithal Babu. — [New Delhi] : Vithal Babu, [pref. 1978]
ix, 61 p. : map ; 23 cm.
Rs20.00
1. China—Commerce. 2. China—Commercial policy. I. Title.
HF3836.5.B3 382'.0951 78-911934
79 MARC

Hsiao, Gene T
The foreign trade of China : policy, law, and practice / [by] Gene T. Hsiao. — Canberra : Australian National University Press, 1977.
xxi, 291 p. ; 25 cm. Aus
Bibliography: p. 263-277.
Includes index.
ISBN 0-7081-1342-7
1. China—Commerical policy. 2. China—Commerce. 3. Commercial law—China. I. Title.
HF1604.H77 1977 382'.0951 78-314253
78 MARC

Japan. Keizai Kikakuchō. Chōsakyoku. Kaigai Chōsaka.
(Chūgoku no taigai bōeki)
中国(本土)の対外貿易：東西貿易と日中貿易の推移と特徴. — [東京]：経済企画庁調査局海外調査課, 昭和39 [1964]
153 p. : graphs ; 25 cm.
Cover title.
"研究資料64年1号" on label mounted on cover.
Includes bibliographical references.
Errata slip inserted.

1. China—Commerce. 2. China—Commerce—Japan. 3. Japan—Commerce—China. I. Title.
HF3836.5.J36 1964 79-802864

Japan. Keizai Kikakuchō. Chōsakyoku. Kaigai Chōsaka.
(Chūgoku no taigai bōeki kōzō no jittai to mondaiten)
中国の対外貿易構造の実態と問題点. — [東京]：経済企画庁調査局海外調査課, [1961]
284 p. ; 25 cm.
Cover title.
Chiefly tables.
"経企調外昭36第3号[4月26日]"
Errata slip inserted.
1. China—Commercial policy. 2. China—Commerce. I. Title.
HF1604.J363 1961 79-803087

Kitto, N
Jebsen & Co. — [Hong Kong : Grant Advertising International, 197-]
88 p., [2] leaves of plates : ill. ; 26 cm.
Title also in Chinese: Chieh-ch'eng yang hang.
1. Jebsen & Co. 2. China—Commerce. 3. Hongkong—Commerce. I. Grant Advertising International.
HF3834.K57 382'.06'55125 78-324950
78 MARC

Minami Manshū Tetsudō Kabushiki Kaisha. Tenshin Jimusho. Chōsaka.
(Kita Shina kōwan jijō)
北支那港灣事情. — [天津]：南滿洲鐵道株式會社天津事務所調査課, 昭和11 [1936]
3, 128 p. ; 23 cm. — (北支經濟資料 ; 第4輯)
Cover title.
1. Harbors—China—Hopei (Province) 2. Harbors—China—Shantung. 3. China—Commerce. I. Title. II. Series: Hokushi keizai shiryō ; dai 4-shū.
HE559.C5M55 1936 78-815367

Schultz, Jeffrey.
China's foreign trade corporations ... organization and personnel / by Jeffrey Schultz ; executive editor, Nicholas H. Ludlow. — Washington : National Council for US-China Trade, 1979.
339 p. : ill. ; 29 cm.
Bibliography: p. 31-32.
Includes index.
1. China—Commerce. 2. Foreign trade promotion—China. 3. China—Registers. I. Ludlow, Nicholas H. II. National Council for United States-China Trade. III. Title.
HF3836.5.S38 354'.51'00827 78-71596
00 MARC

Ting, Kuo-chen, fl. 1950- ed.
(Hsin kung shang ts'an k'ao tzu liao)
新工商參考資料 丁國楨編 北京 五十年代出版社 1950.
2, 5, 238 p. tables. 18 cm.

1. China—Economic policy—1949-1976. 2. China—Industries—1949-1976. 3. China—Commerce. I. Title.
HC427.9.T57 C 64-1039

Ting, Kuo-chen, fl. 1950- ed.
(Hsin kung shang ts'an k'ao tzŭ liao)
新工商参考資料 丁國楨編 增訂本 北京 五十年代出版社 1952.
2, 6, 313 p. tables. 19 cm.

1. China—Economic policy—1949-1976. 2. China—Industries—1949-1976. 3. China—Commerce. I. Title.
HC427.9.T57 1952 C 60-5309

Uchiyama, Kiyoshi, 1887-
(Bōekijō yori mitaru Shina fūzoku no kenkyū)
貿易上ヨリ見タル支那風俗之研究 / 内山清著. — 上海：日々新聞社：賣捌所 日本堂, 大正4 [1915]
8, 17, 538 p., [16] leaves of plates : ill. ; 23 cm.
Originally published in Nichinichi shimbun, Jan.-July 1915.

1. China—Social life and customs. 2. China—Commerce. I. Title. II. Title: Shina fūzoku no kenkyū.
DS721.U29 78-808535

United States. Bureau of Foreign and Domestic Commerce.
Commercial handbook of China / Julean Arnold. — New York : Garland Pub., 1979.
2 v. : ill. ; 24 cm. — (The Modern Chinese economy)
Reprint of the 1919-1920 ed. printed by the Govt. Print. Off., Washington, which was issued as no. 84 of the Bureau of Foreign and Domestic Commerce's Misclaneous series.
Bibliography: v. 2, p. 460-466.
ISBN 0-8240-4264-6 : $100.00
1. China—Commerce. 2. China—Commerce—United States. 3. United States—Commerce—China. 4. China—Economic conditions—1912-1949. I. Arnold, Julean Herbert, 1875-1946. II. Title. III. Series. IV. Series: United States. Bureau of Foreign and Domestic Commerce. Miscellaneous series ; no. 84.
HF3836.U52 1979 382'.0951 78-24800
79 MARC

United States. Central Intelligence Agency. National Foreign Assessment Center.
China demand for foreign grain / National Foreign Assessment Center. — Washington : The Center, 1979.
iii, 8 p. : ill. ; 27 cm. — (A research paper - National Foreign Assessment Center)
"ER 79-10073."
1. Grain trade—China. 2. China—Commerce. I. Title. II. Series: United States. Central Intelligence Agency. National Foreign Assessment Center. Research paper - Central Intelligence Agency, National Foreign Center.
HD9046.C62U53 1979 382'.41'310951 79-601777
79 MARC

United States. Central Intelligence Agency. National Foreign Assessment Center.
China : real trends in trade with non-Communist countries since 1970 / National Foreign Assessment Center. — [Washington] : The Center, 1977.
iv, 43 p. : graphs ; 27 cm. — (A research paper - National Foreign Assessment Center)
Cover title.
"ER77-10477."
1. China—Commerce. 2. Prices—China. I. Title. II. Series: United States. Central Intelligence Agency. National Foreign Assessment Center. Research paper - Central Intelligence Agency, National Foreign Assessment Center.
HF3836.5.U54 1977 382'.0951 77-604456
78[r79]rev MARC

Yang, Tao-lin, 1856-1932.
(Kuang-hsü t'ung shang lieh piao)
光緒通商列表 / 揚楷[i. e. 楊道霖]製. — [台北縣永和鎮]：文海出版社, [1977?]
134 p. ; 22 cm. — (近代中國史料叢刊續編[i. e. 輯] ; 479 : 第48輯)
Reprint ed. with original pref. dated 1886.

1. China—Commerce. I. Title. II. Series: Shen, Yün-lung, comp. Chin tai Chung-kuo shih liao ts'ung k'an hsü chi ; 479.
DS755.S532 vol. 479 78-841626
[HF3831]

—COMMERCE—ADDRESSES, ESSAYS, LECTURES

Wei, O.
(Chung kung tui wai mao i chih ching chi fen hsi)
中共對外貿易之經濟分析 / 魏蕚講 ; 教育部社會教育司編印. — [台北]：該司, 民國66 [1977]
42 p. : port. ; 20 cm. — (文化講座專集 ; 100)
Includes bibliographical references.

1. China—Commerce—Addresses, essays, lectures. 2. China—Commercial policy—Addresses, essays, lectures. I. Title. II. Series: Wen hua chiang tso chuan chi ; 100.
HF3836.5.W44 78-840818

CHINA (Continued)

—COMMERCE—HANDBOOKS, MANUALS, ETC.

Terry, Edith.
China's foreign trade organizations, domestic corporations, and ministries / compiled by Edith Terry. — [Washington] : National Council for US-China Trade, 1979.
68 p. : ill. ; 23 x 11 cm.
Cover title: Pocket guide to China's foreign trade organizations, domestic corporations, and ministries.
1. China—Commerce—Handbooks, manuals, etc. 2. China—Industries—Handbooks, manuals, etc. 3. Foreign trade promotion—China—Handbooks, manuals, etc. I. National Council for United States-China Trade. II. Title. III. Title: Pocket guide to China's foreign trade organizations, domestic corporations, and ministries.
HF3837.T47 382'.025'51 79-88564
79 MARC

—COMMERCE—HISTORY

Ch'en, Wen-shih.
(Ming Hung-wu Chia-ching chien ti hai chin cheng ts'e)
明洪武嘉靖間的海禁政策 / 陳文石. — 台北：國立臺灣大學文學院, 1966.
3, 7, 176, 2 p. ; 22 cm. — (國立臺灣大學文史叢刊 ; 20)
Title in English in Colophon: The sea-faring prohibition policy in the early Ming Dynasty.
Text in Chinese, with summary in English.
Includes bibliographical references.
1. China—Foreign relations—To 1912. 2. China—History—Ming dynasty, 1368–1644. 3. China—Commerce—History. 4. Pirates—China. I. Title. II. Title: The sea-faring prohibition policy. III. Series: T'ai-wan ta hsüeh, T'ai-pei. Kuo li T'ai-wan ta hsüeh wen shih ts'ung k'an ; 20.
DS753.C5115 78–843024

Ecsedy, Ildikó.
Nomádok és kereskedők Kína határain / Ecsedy Ildikó. — Budapest : Akadémiai Kiadó, 1979.
237 p. ; 18 cm. — (Kőrösi Csoma kiskönyvtár ; 16) H***
Includes bibliographical references.
ISBN 9630517698 : 23.00Ft
1. China—Commerce—History. 2. China—Foreign relations. 3. Nomads—China—History. 4. Sinkiang—History. I. Title.
HF3835.E27 79-381930
79 MARC

—COMMERCE—PERIODICALS

United States. Industry and Trade Administration. Office of East-West Country Affairs.
China's economy and foreign trade. 1977/78–
[Washington] U. S. Dept. of Commerce, Industry and Trade Administration, Office of East-West Country Affairs; for sale by the Supt. of Docs., U. S. Govt. Print. Off.
v. 27 cm. annual.
Continues: Chinese economy and foreign trade perspective, ISSN 0147-6351.
Key title: China's economy and foreign trade, ISSN 0192-5555.
1. China — Commerce — Periodicals. 2. China — Economic conditions—1976— —Periodicals. 3. China—Commerce—United States—Periodicals. 4. United States — Commerce — China — Periodicals. I. Title.
HF3836.5.U53a 382'.0951 79-643027
MARC-S

—COMMERCE—ASIA, SOUTHEASTERN

Japan. Gaimushō. Keizaikyoku. Keizai Chōsashitsu.
(Chūkyō shōhin no Tōnan Ajia shinshutsu)
中共商品の東南アジア進出：その英連邦商品への影響. — [東京]：外務省経済局経済調査室, 昭和34 [1959]
67 p. ; 21 cm. — (在外公館調査報告 ; 第26号)
Cover title.
1. China—Commerce—Asia, Southeastern. 2. Asia, Southeastern—Commerce—China. I. Title. II. Series: Zaigai kōkan chōsa hōkoku ; dai 26-gō.
HF3838.A7842J36 1959 79–802976

—COMMERCE—BRAZIL —PERIODICALS

Brazil. Serviço de Estatística Econômica e Financeira.
Comércio exterior do Brasil com a China.
Rio de Janeiro, Impr. Nacional.
v. 27 cm.
1. Brazil—Commerce—China—Periodicals. 2. China—Commerce—Brazil—Periodicals. I. Title.
HF3408.C5B7a 79-645294
MARC-S

—COMMERCE—GREAT BRITAIN

Correspondence respecting the revision of the Treaty of Tien-Tsin
: presented to the House of Lords by command of Her Majesty, in pursuance of their address dated March 24, 1871. — San Francisco : Chinese Materials Center, 1975.
viii, 467 p., [5] fold. leaves of plates : maps ; 27 cm. — ([Reprint series] - Chinese Materials Center, inc. ; [no. 26])
Spine title: Revision of the Treaty of Tien-Tsin.
Reprint of the 1871 ed. printed by Harrison, London, which was issued as China no. 5 and as no. 389 of Papers by command, c.
1. Great Britain—Foreign relations—China. 2. China—Foreign relations—Great Britain. 3. Great Britain—Commerce—China. 4. China—Commerce—Great Britain. 5. Unequal treaties. I. Title: Revision of the Treaty of Tien-Tsin. II. Series: Chinese Materials Center. Reprint series - Chinese Materials Center ; no. 26. III. Series: Great Britain. Foreign Office. China ; no. 5. IV. Series: Great Britain. Parliament. Papers by command, c ; 389.
JX637 1975.C67 1975 327.41'051 78-324368
79 MARC

Greenberg, Michael.
British trade and the opening of China, 1800-42 / by Michael Greenberg. — New York : Monthly Review Press, [1979]
xii, 238 p. ; 20 cm.
Reprint of the 1951 ed. published by University Press, Cambridge, Eng., issued in series: Cambridge studies in economic history.
Bibliography: p. [226]-232.
Includes index.
ISBN 0-85345-497-3
1. Great Britain—Commerce—China. 2. China—Commerce—Great Britain. 3. Opium trade—China. 4. China—History—War of 1840-1942. I. Title.
[HF3508.C5G73 1979] 382'.41'051 79-12206
79 MARC

—COMMERCE—JAPAN

Japan. Keizai Kikakuchō. Chōsakyoku. Kaigai Chōsaka.
(Chūgoku no taigai bōeki)
中国(本土)の対外貿易：東西貿易と日中貿易の推移と特徴. — [東京]：経済企画庁調査局海外調査課, 昭和39 [1964]
153 p. : graphs ; 25 cm.
Cover title.
"研究資料64年1号" on label mounted on cover.
Includes bibliographical references.
Errata slip inserted.

1. China—Commerce. 2. China—Commerce—Japan. 3. Japan—Commerce—China. I. Title.
HF3836.5.J36 1964 79-802864

(Nitchū bōeki antei kakudai e no michisuji)
日中貿易, 安定拡大への道すじ：長期取決めをめぐる諸条件. — 東京：日中経済協会, 1978.
408, 5 p. : ill. ; 25 cm. — (中国経済関係調査報告書) (日中経報 ; no. 107) Ja 78–19906
Cover title.
"52-調-14"
Includes bibliographical references.
¥1290
1. China—Commerce—Japan. 2. Japan—Commerce—China. 3. China—Commercial policy. I. Nitchū Keizai Kyōkai. II. Series: Chūgoku keizai kankei chōsa hōkokusho. III. Series: Nitchū Keizai Kyōkai. Nitchūkei hō ; no. 107.
HF3838.J3N56 78–814347

Remer, Charles Frederick, 1889-
A study of Chinese boycotts with special reference to their economic effectiveness / C. F. Remer ; with the assistance of William B. Palmer. — New York : Arno Press, 1979, c1933.
xii, 306 p., [1] leaf of plates : graph ; 21 cm. — (Johns Hopkins University Press reprints)
Reprint of the ed. published by Johns Hopkins Press, Baltimore.
Bibliography: p. 293-295.
Includes index.
ISBN 0-405-10620-3 : $22.00
1. Boycott—China. 2. China—Commercial policy. 3. China—Commerce—Japan. 4. Japan—Commerce—China. I. Title.
HF1604.R45 1979 382'.0951 78-19287
78 MARC

—COMMERCE—MALAYSIA

Wong, John, 1939-
The political economy of Malaysia's trade relations with China / by John Wong. — Singapore : Institute of Southeast Asian Studies, 1974.
31 leaves ; 28 cm. — (Occasional paper - Institute of Southeast Asian Studies ; no. 20)
Includes bibliographical references.
$3.00
1. Malaysia—Commerce—China. 2. China—Commerce—Malaysia. I. Title. II. Series: Institute of Southeast Asian Studies. Occasional paper ; no. 20.
HF3800.6.Z7C488 382'.0951'0595 78-301603
78[r79]rev MARC

—COMMERCE—RUSSIA

Sladkovskiĭ, Mikhail Iosifovich.
(Istorii͡a torgovo-ėkonomicheskikh otnosheniĭ narodov Rossii s Kitaem (do 1917 g.))
История торгово-экономических отношений народов России с Китаем (до 1917 г.) Москва, "Наука," 1974.
439 p. 22 cm. USSR 74
At head of title: Академия наук СССР. Институт Дальнего Востока. М. И. Сладковский.
Continued by the author's Istorii͡a torgovo-ėkonomicheskikh otnosheniĭ SSSR s Kitaem (1917–1974)
Bibliography: p. 357–[368]
2.52rub
1. Russia—Commerce—China. 2. China—Commerce—Russia. I. Title.
HF3628.C6S59 382'.0947'051 74–328748

—COMMERCE—UNITED STATES

Kamm, John
Report of the National Council for US-China Trade agricultural machinery delegation to China, May 19-June 2, 1978 / John Kamm. — Washington : The Council, c1978.
80 leaves ; 28 cm.
Cover title.
1. Agricultural machinery—Trade and manufacture—United States. 2. Agricultural machinery—Trade and manufacture—China. 3. United States—Commerce—China. 4. China—Commerce—United States. I. National Council for United States-China Trade. II. Title.
HD9486.U4K35 631.3'0951 79-113017
00 MARC

Shao, Hsi, fl. 1952-
(Shih chiu shih chi Mei-kuo tui Hua ya p'ien ch'in lüeh)
十九世紀美國對華鴉片侵略 / 紹溪著. — [香港? : s. n., 1978?]
1, 110 p. ; 21 cm.
Reprint of the 1952 ed. published by Sheng huo, Tu shu, Hsin chih san lien shu tien, Peking.
Includes bibliographical references.
1. Opium trade—China. 2. United States—Commerce—China. 3. China—Commerce—United States. I. Title.
HV5816.S48 1978 79–836295

United States. Bureau of Foreign and Domestic Commerce.
Commercial handbook of China / Julean Arnold. — New York : Garland Pub., 1979.
2 v. : ill. ; 24 cm. — (The Modern Chinese economy)
Reprint of the 1919-1920 ed. printed by the Govt. Print. Off., Washington, which was issued as no. 84 of the Bureau of Foreign and Domestic Commerce's Misclaneous series.
Bibliography: v. 2, p. 460-466.
ISBN 0-8240-4264-6 : $100.00
1. China—Commerce. 2. China—Commerce—United States. 3. United States—Commerce—China. 4. China—Economic conditions—1912-1949. I. Arnold, Julean Herbert, 1875-1946. II. Title. III. Series. IV. Series: United States. Bureau of Foreign and Domestic Commerce. Miscellaneous series ; no. 84.
HF3836.U52 1979 382'.0951 78-24800
79 MARC

United States. Congress. Senate. Committee on Agriculture, Nutrition, and Forestry. Subcommittee on Foreign Agricultural Policy.
Agricultural trade with the People's Republic of China and Taiwan : hearing before the Subcommittee on Foreign Agricultural Policy of the Committee on Agriculture, Nutrition, and Forestry, United States Senate, Ninety-sixth Congress, first session, Tuesday, March 13, 1979. — Washington : U.S. Govt. Print. Off., 1979.
iii, 28 p. ; 24 cm.
1. Farm produce—United States. 2. United States—Commerce—China. 3. China—Commerce—United States. I. Title.
KF26.A3549 1979 382'.41'0973 79-602850
79 MARC

United States. Industry and Trade Administration.
Doing business with China. — Rev. Mar. 1979. — Washington : U.S. Dept. of Commerce, Industry and Trade Administration : for sale by the Supt. of Docs., U.S. Govt. Print. Off., 1979.
iii, 44 p. : ill. ; 26 cm.
Bibliography: p. 23-25.
1. United States—Commerce—China. 2. China—Commerce—United States. 3. China—Description and travel—1949- I. Title.
HF3128.U53 1979 346'.51'07 79-601897
00 MARC

—COMMERCE—UNITED STATES —PERIODICALS

United States. Industry and Trade Administration. Office of East-West Country Affairs.
China's economy and foreign trade. 1977/78–
[Washington] U. S. Dept. of Commerce, Industry and Trade Administration, Office of East-West Country Affairs; for sale by the Supt. of Docs., U. S. Govt. Print. Off.
v. 27 cm. annual.
Continues: Chinese economy and foreign trade perspective, ISSN 0147-6351.
Key title: China's economy and foreign trade, ISSN 0192-5555.
1. China — Commerce — Periodicals. 2. China — Economic conditions—1976— —Periodicals. 3. China—Commerce—United States—Periodicals. 4. United States — Commerce — China — Periodicals. I. Title.
HF3836.5.U53a 382'.0951 79-643027
MARC-S

CHINA (Continued)

—COMMERCE—UNITED STATES—STATISTICS

Sino-US trade statistics 1977 / prepared for members of the National Council for US-China Trade. — Washington : The Council, 1978.
44 p. ; 28 cm. — (Special report- National Council for US-China Trade ; no. 22)
Cover title.
1. United States—Commerce—China—Statistics. 2. China—Commerce—United States—Statistics. I. National Council for United States-China Trade. II. Series: National Council for United States-China Trade. Special report - National Council for United States-China Trade ; no. 22.
HF3128.S57 382'.0951'073 79-116267
00 MARC

—COMMERCIAL POLICY

Akeno, Yoshio.
(Chūgoku no taigai keizai kōryū no tenkai)
中国の対外経済交流の展開 / 明野義夫著. — 東京 : 大東文化大学東洋研究所, 昭和53 [1978]
264 p. : graphs ; 22 cm. Ja***
Includes bibliographical references.
¥3500

1. China—Commercial policy. 2. China—Economic policy—1976- I. Title.
HF1604.A43 78-816803

Babu, V Vittal.
China trade manual / V. Vithal Babu. — [New Delhi] : Vithal Babu, [pref. 1978]
ix, 61 p. : map ; 23 cm.
Rs20.00
1. China—Commerce. 2. China—Commercial policy. I. Title.
HF3836.5.B3 382'.0951 78-911934
79 MARC

Hsiao, Gene T
The foreign trade of China : policy, law, and practice / [by] Gene T. Hsiao. — Canberra : Australian National University Press, 1977.
xxi, 291 p. ; 25 cm. Aus
Bibliography: p. 263-277.
Includes index.
ISBN 0-7081-1342-7
1. China—Commerical policy. 2. China—Commerce. 3. Commercial law—China. I. Title.
HF1604.H77 1977 382'.0951 78-314253
78 MARC

Japan. Keizai Kikakuchō. Chōsakyoku. Kaigai Chōsaka.
(Chūgoku no taigai bōeki kōzō no jittai to mondaiten)
中国の対外貿易構造の実態と問題点. — [東京] : 経済企画庁調査局海外調査課, [1961]
284 p. ; 25 cm.
Cover title.
Chiefly tables.
"経企調外昭36第3号[4月26日]"
Errata slip inserted.
1. China—Commercial policy. 2. China—Commerce. I. Title.
HF1604.J363 1961 79-803087

McMillan, M Sean.
Attitudes of the Peoples Republic of China to international commercial trade (1979) / by M. Sean McMillan. — [Chicago] : American Bar Association, c1979.
28, 11, 44, [5] p. ; 23 cm. — (Monograph - Section of International Law ; 3)
Includes bibliographical references.
1. Foreign trade regulation—China. 2. Commercial law—China. 3. China—Commercial policy. I. Title. II. Series: American Bar Association. Section of International Law. Monograph - American Bar Association, Section of International Law ; 3.
346'.51'07 79-117905
79 MARC

(Nitchū bōeki antei kakudai e no michisuji)
日中貿易, 安定拡大への道すじ : 長期取決めをめぐる諸条件. — 東京 : 日中経済協会, 1978.
408, 5 p. : ill. ; 25 cm. — (中国経済関係調査報告書) (日中経報 ; no. 107)
Ja 78-19906
Cover title.
"52-調-14"
Includes bibliographical references.
¥1290
1. China—Commerce—Japan. 2. Japan—Commerce—China. 3. China—Commercial policy. I. Nitchū Keizai Kyōkai. II. Series: Chūgoku keizai kankei chōsa hōkokusho. III. Series: Nitchū Keizai Kyōkai. Nitchūkei hō ; no. 107.
HF3838.J3N56 78-814347

Remer, Charles Frederick, 1889-
A study of Chinese boycotts with special reference to their economic effectiveness / C. F. Remer ; with the assistance of William B. Palmer. — New York : Arno Press, 1979, c1933.
xii, 306 p., [1] leaf of plates : graph ; 21 cm. — (Johns Hopkins University Press reprints)
Reprint of the ed. published by Johns Hopkins Press, Baltimore.
Bibliography: p. 293-295.
Includes index.
ISBN 0-405-10620-3 : $22.00
1. Boycott—China. 2. China—Commercial policy. 3. China—Commerce—Japan. 4. Japan—Commerce—China. I. Title.
HF1604.R45 1979 382'.0951 78-19287
78 MARC

Wei, O
The foreign trade policy of the Chinese Communist regime / by Wei O. — [Taipei] : World Anti-Communist League, China Chapter, 1978.
33 p. ; 19 cm. — (English publications of the APACL/ROC ; pamphlet xxiv, 209)
Cover title.
Includes bibliographical references.
$0.50 (U.S.)
1. China—Commercial policy. I. Title.
HF1604.W44 382'.3'0951 78-323919
79 MARC

—COMMERCIAL POLICY—ADDRESSES, ESSAYS, LECTURES

China's finance and trade : a policy reader / Gordon Bennett, editor. — White Plains, N.Y. : M. E. Sharpe, c1978.
xiii, 250 p. ; 23 cm.
"Selections from Chinese economic studies."
Bibliography: p. 247-249.
ISBN 0-87332-115-4
1. China—Economic policy—1949-1976—Addresses, essays, lectures. 2. Finance—China—Addresses, essays, lectures. 3. China—Commercial policy—Addresses, essays, lectures. I. Bennett, Gordon A. II. Chinese economic studies.
HC427.9.C56543 330.9'51'05 77-99080
78 MARC

Wei, O.
(Chung kung tui wai mao i chih ching chi fen hsi)
中共對外貿易之經濟分析 / 魏蕚講 ; 教育部社會教育司編印. — [台北] : 該司, 民國66 [1977]
42 p. : port. ; 20 cm. — (文化講座專集 ; 100)
Includes bibliographical references.

1. China—Commerce—Addresses, essays, lectures. 2. China—Commercial policy—Addresses, essays, lectures. I. Title. II. Series: Wen hua chiang tso chuan chi ; 100.
HF3836.5.W44 78-840818

C

—COMMERICAL TREATIES

China. Treaties, etc., 1875-1908 (Ch'ing Te-tsung)
[Chung O ting li Chi-lin mu chih ho t'ung. Japanese]
伐木ニ關スル露(東支鐵道會社)支(吉林省)間ノ協定. — [新京 : 滿鐵經濟調査會, 1934]
8 leaves ; 27 cm. — (北鐵林區接收及經營ニ關スル對策案參考資料 ; 第2)
Cover title.
At head of title: Gokuhi (rubber stamped)
1. Lumber trade—China—Heilungkiang (Province) 2. Chinese Eastern Railway. 3. China—Commercial treaties. 4. Russia—Commercial treaties. I. Russia. Treaties, etc., 1894-1917 (Nicholas II) II. Chinese Eastern Railway. III. Heilungkiang, China (Province) IV. Title. V. Series: Hokutetsu rinku sesshū oyobi keiei ni kansuru taisakuan sankō shiryō ; dai 3.
Title romanized: Batsuboku ni kansuru Ro-Shi, Kokuryūkō-shō, kan no kyōtei.
K4611.L86C57416 79-803803

China. Treaties, etc., 1875-1908 (Ch'ing Te-tsung)
[Hei-lung chiang t'ieh lu kung ssu fa mu ho t'ung. Japanese]
伐木ニ關スル露(東支鐵道會社)支(黑龍江省)間ノ協定. — [新京 : 滿鐵經濟調査會, 1934?]
8 leaves : 27 cm. — (北鐵林區接收及經營ニ關スル對策案參考資料 ; 第3)
Cover title.
At head of title: Gokuhi (rubber stamped)
1. Lumber trade—China—Kirin (Province) 2. Chinese Eastern Railway. 3. China—Commercial treaties. 4. Russia—Commercial treaties. I. Russia. Treaties, etc., 1894-1917 (Nicholas II) II. Chinese Eastern Railway. III. Kirin, China (Province) IV. Title. V. Series: Hokutetsu rinku sesshū oyobi keiei ni kansuru taisakuan sankō shiryō : dai 2.
Title romanized: Batsuboku ni kansuru Ro-Shi, Kitsurin-sho, kan no kyotei.
K4611.L86C57316 79-803802

—CONSTITUTIONAL LAW

Adjitorop, Jusuf.
Integrasi kekuasaan politik dan sistim hukum dgn revolusi di Tiongkok Rakjat / Jusuf Adjitorop. — [Djakarta] : Jajasan Pendidikan dan Kebudajaan Baperki Djakarta, 1964.
52 p. : ill. ; 22 cm.
Includes bibliographical references.
1. China—Politics and government—1949-1976. 2. China—Constitutional law. I. Title.
JQ1508.A35 76-949952
79 MARC

Ch'iao, Pao-t'ai.
(Chung-hua min kuo hsien fa yü wu wu hsien ts'ao chih pi chiao yen chiu)
中華民國憲法與五五憲草之比較研究 / 喬寶泰著. — 臺北 : 中央文物供應社, 民國67 [1978]
2 v. (2, 6, 18, 900 p.) ; 22 cm. — (三民主義學術思想論叢)
Bibliography: p. 884-900.
Appendices: (p. 847-883): 1. 中華民國憲法.—2. 中華民國憲法草案(五五憲草)
1. China—Constitutional law. I. China. Constitution. 1978. II. Title. III. Series: San min chu i hsüeh shu ssu hsiang lun ts'ung.
79-837616

China. Constitution.
The Constitution of the Republic of China / English translation by Charles C. H. Wan. — Nanking : Chinese Ministry of Information, 1947.
20 p. ; 22 cm.
"An unofficial translation of the newly promulgated Constitution."
1. China—Constitutional law. I. Title.
342'.51'023 79-111359
79 MARC

China (People's Republic of China, 1949-). Constitution, 1978.
(Chung-hua jen min kung ho kuo hsien fa)
中华人民共和国宪法. 关于修改宪法的报告 / 叶剑英. — 第1版. — [北京?] : 人民出版社, 1978.
66 p. : 23 cm.
"一九七八年三月五日中华人民共和国第五届全国人民代表大会第一次会议通过"
$1.50
(Continued on next card)
1. China—Constitutional law. I. Yeh, Chien-ying. Kuan yü hsiu kai hsien fa ti pao kao. 1978. II. Title.
79-835901

China (People' Republic of China, 1949-) Constitution, 1978.
The Constitution of the People's Republic of China : adopted on March, 5, 1978 by the Fifth National People's Congress of the People's Republic of China at its first session. — 1st ed. — Peking : Foreign Languages Press, 1978.
41 p. ; 22 cm.
1. China—Constitutional law. I. Title.
342'.51'023 78-110754
79 MARC

Chou, Ching-wen, 1909-
(Chung-kuo hsien cheng wen t'i yen chiu)
中國憲政問題研究 周鯨文著 香港 時代批評社 民國29 [1940]
1, 3, 106 p. 23 cm. (時代叢書 政治組第2種)
Series romanized: Shih tai ts'ung shu : Cheng chih tsu.

1. China—Constitutional law. I. Title.
C 66-725

(Chūgoku Zenjindai, Seikyō Zenkoku Kaigi tokushūgō)
中国全人代·政協全国会議特集号 / [編集者 ラヂオプレス]. — 東京 : 同プレス, 1978.
183, 10 p. ; 25 cm. Ja***
Cover title.
"中華人民共和国憲法": p. 76-92.
"政協会議規約": p. 120-127.
Includes index.
1. China—Politics and government—1976- 2. China—Constitutional law. 3. China—Registers. I. China (People's Republic of China, 1949-). Ch'üan kuo jen min tai piao ta hui. II. Chung-kuo jen min cheng chih hsieh shang hui i. III. Rajio Puresu.
JQ1503 1978.C47 79-805288

Kuroda, Ryōichi, 1911-
(Kempō kōwa)
憲法講話 : 入門篇 / 黒田了一. — 京都 : 三和書房, 昭和30 [1955]
2, 5, 270 p. ; 22 cm.
Includes bibliographical references.

1. Japan—Constitutional law. 2. China—Constitutional law. 3. Constitutional law—Study and teaching. I. Title.
79-800957

CHINA

—CONSTITUTIONAL LAW (Continued)

Lei, Chen, 1897-
(Chien ch'a yüan chih chiang lai)
監察院之將來 雷震著 台北 自由中國出版社 民國42 [1953]
5, 6, 173 p. 19 cm. (自由中國社叢書之 23)
Includes bibliographical references.

1. China (Republic of China). Chien ch'a yüan. 2. China—Constitutional law. I. Title.
JQ1530.5.L44 C 66-1055

Liu, Ch'ing-po.
(Chung kung hsien fa chih yen chiu)
中共憲法之研究 / 劉清波著. — Taipei : 中國大陸問題研究所, 1971.
2, 4, 220 p., [5] fold. leaves ; 21 cm.
Added t.p.: The Constitution of the People's Republic of China.
Table of contents also in English.
Includes bibliographical references.
Appendices (p. 187-220): 1. 中共政權一九五四年憲法全文.—2. 中共九屆二中全會憲法修改草案全文.
1. China—Constitutional law. I. China (People's Republic of China, 1949-). Constitution. 1971. II. Chung-kuo kung ch'an tang. Chung yang wei yüan hui. Chung-hua jen min kung ho kuo hsien fa hsiu kai ts'ao an. 1971. III. Title. IV. Title: The Constitution of the People's Republic of China.
78-842223

Tyau, Min-ch'ien T Z 1888-
China's new constitution and international problems / by Min-ch'ien T. Z. Tyau. — Reprint ed. — Arlington, Va. : University Publications of America, 1976.
xv, 286 p. ; 24 cm. — (China studies) (Studies in Chinese government and law)
Reprint of the 1918 ed. published by Commercial Press, Shanghai.
ISBN 0-89093-064-3
1. China—Constitutional law. 2. China—Politics and government—1900-1949. 3. China—Foreign relations. 4. European War, 1914-1918—China. I. Title.
342'.51'029 76-55090
78 MARC

--CONSTITUTIONAL LAW--ADDRESSES, ESSAYS, LECTURES

(Hsin shih ch'i chih kuo ti tsung chang ch'eng)
新时期治国的总章程 / 余先予, 周尚文, 陆世友等编写. — [香港?] : s. n., 1978?]
132 p. ; 21 cm.
Reprint of the 1st (1978) ed. published by Shang-hai jen min ch'u pan she, Shanghai.

1. China—Constitutional law—Addresses, essays, lectures. I Yü, Hsien-yü. II. Chou, Shang-wen. III. Lu, Shih-yu.
79-837935
C

—COURT AND COURTIERS

Der Ling, Princess.
Two years in the forbidden city / by the Princess Der Ling. — San Francisco : Chinese Materials Center, 1977 [c1911]
ix, 383 p., [17] leaves of plates : ill. ; 22 cm. — (Reprint series - Chinese Materials Center ; no. 68)
Reprint of the 1912 ed. published by Moffat, Yard, New York.
1. Tz'u-hsi, Empress dowager of China, 1835-1908. 2. China—Kings and rulers—Biography. 3. China—Court and courtiers. I. Title. II. Series: Chinese Materials Center. Reprint series - Chinese Materials Center ; no. 68.
DS763.T8D43 1977 951'.03 78-306763
78 MARC

Lo, Lung-chih.
(T'ang tai ti hou fei yü wai ch'i)
唐代的后妃與外戚 / 羅龍治著. — 臺北 : 桂冠圖書有限公司, 民國67 [1978]
2, 8, 178 p. ; 20 cm. — (桂冠系列)
Series romanized: Kuei kuan hsi lieh.
Bibliography: p. 172-178.
NT$50.00

1. China—History—T'ang dynasty, 618-907. 2. China—Court and courtiers. I. Title.
DS749.3.L62 79-839962

—COURT AND COURTIERS
—ANECDOTES, FACETIAE, SATIRE, ETC.

Wen, Chi.
(Feng chien shih tai ti yin wei)
封建時代的淫穢 / 文輯. — [香港?] : 華學研究社, [1977?]
4, 123 p. ; 19 cm.
"古稀筆記"

1. China—Court and courtiers—Anecdotes, facetiae, satire, etc. I. Title.
78-841438
DS727.W44

—COURT AND COURTIERS
—FICTION

Hao-shang-k'o.
(Li tai ti wang yü nü jen)
歷代帝王與女人 / 濠上客編著. — 高雄 : 勝夫書局, 民國66 [1977]
3 v. ; 20 cm.
NT$120.00

1. China—Court and courtiers—Fiction. I. Title.
PL2861.A78L5 79-838199

—DEFENSES

China (Republic of China). Kuang fu ta lu she chi yen chiu wei yüan hui.
(Ta lu kuang fu hou pien fang chien she chih yen chiu)
大陸光復後邊防建設之研究 / 光復大陸設計研究委員會研擬. — [台北] : 該會, 民國65 [1976]
3, 100 p. ; 22 cm. — (邊疆研究叢書)
Cover title.

1. China—Defences. 2. China—Boundaries. 3. Fortifications—China. I. Title. II. Series: Pien chiang yen chiu ts'un, shu.
UA835.C446 1976 79-837134

(Chung Su chan huo)
中蘇戰火. — 香港 : 海星出版印刷中心, [1979?]
[32] p. : ill. ; 26 cm.
Cover title.
$2.00

1. China—Defenses. 2. Russia—Defenses.
UA835.C55 79-839915

Middleton, Drew, 1913-
The Duel of the giants : China and Russia in Asia / by Drew Middleton. — New York : Scribner, c1978.
viii, 241 p. ; 22 cm.
Includes index.
ISBN 0-684-15785-3 : $8.95
1. China—Foreign relations—Russia. 2. Russia—Foreign relations—China. 3. China—Defenses. I. Title.
DS740.5.R8M48 327.51'047 78-9305
78 MARC

—DEFENSES—ADDRESSES, ESSAYS, LECTURES

Chiang, Wei-kuo, 1916-
(Chiang tsung t'ung yü kuo fang)
蔣總統與國防 / 蔣緯國著 ; 黎明文化公司輯. — 臺北 : 黎明文化事業公司, 民國65 [1976]
2, 152 p., [8] leaves of plates : ill. ; 21 cm.
NT$25.00

1. Chiang, Kai-shek, 1887-1975—Military leadership—Addresses, essays, lectures. 2. Sino-Japanese Conflict, 1937-1945—China—Addresses, essays, lectures. 3. China—Defenses—Addresses, essays, lectures. I. Title.
DS778.C55C4562 78-843151

—DESCRIPTION AND TRAVEL

Anderson, Aeneas, fl. 1802.
[A narrative of the British embassy to China, in the years 1792, 1793, and 1794. Japanese]
マカートニー奉使記 / イーニアス・アンダースン著 ; 加藤憲市譯. — 東京 : 筑摩書房, 昭和22 [1947]
8, 317 p., [1] leaf of plates : port. ; 21 cm.
1. China—Description and travel. 2. Great Britain—Foreign relations—China. 3. China—Foreign relations—Great Britain. 4. Macartney, George, Macartney, earl, 1737-1806. I. Title.
Title romanized: Makātoni hōshiki.
DS708.A5416 79-800799

Ku, Tsu-yü, 1624-1680.
[Ku shih tu shih fang yü chi yao ching lüeh. Selections]
顧氏讀史方輿紀要精略 / 顧祖禹撰. — 台北 : 老古出版社, 民國67 [1978]
310, 318, [13] p. ; 22 cm. — (正統謀略學叢編 : 初輯 ; 第18冊)
"精略之京省每章序文, 取雖達通甫先生註釋為準, 京省內容則取辛鍾靈先生輯本"
"附: 顧炎武 天下郡國利病書論略" p. 1-318 (2d group)
1. China—Historical geography. 2. China—History, Military. 3. China—Description and travel. I. Ku, Yen-wu, 1613-1682. T'ien hsia chün kuo li ping shu. Selections. 1978. II. Title. III. Series: Cheng t'ung mou lüeh hsüeh hui pien : Ch'u chi ; ti 18 ts'e.
Title romanized: Ku shih tu shih fang yü chi yao ching lüeh.
DS706.5.K822 1978 79-840130

The Times atlas of China / [editors and chief contributors, P. J. M. Geelan, D. C. Twitchett ; cartography by Kyobunkaku Limited, John Bartholomew & Son Limited, and Hunting Surveys Limited]. — London : Times Books, 1974.
xl, 145, 27 p. : chiefly col. maps ; 39 cm. GB75-00895
Includes index.
ISBN 0-7230-0118-9 : £25.00
1. China—Maps. 2. China—Description and travel. 3. Names, Geographical—China. I. Geelan, P. J. M. II. Twitchett, Denis Crispin. III. Kyobunkaku. IV. Bartholomew (John) and Son, ltd. V. Hunting Surveys, ltd. VI. Title: Atlas of China.
G2305.T47 1974 912'.51 75-313716
75[r79]rev2 MARC

Wang, Kuo-wei, 1877-1927.
[Ch'ang-ch'un-chen-jen hsi yu chi chu] 校注者王國維 蒙古律例 [上·下卷] [12卷] 臺北 廣文書局 民國61 [1972]
長春眞人西遊記校注
4, 146, 129 p. 22 cm. (史料四編) NT$200.00
Colophon title.
Caption title: Ch'ang-ch'un-chen-jen hsi yu chi chu.
Reprint ed.
1. Li, Chih-ch'ang, 1193-1278. Ch'ang-chun-chen-jen hsi yu chi. 2. China—Description and travel. 3. Mongolia—Description and travel. 4. Asia, Central—Description and travel. 5. Ch'iu, Ch'ang-ch'un, 1148-1227. 6. Law—Mongolia—Sources. I. Li, Chih-ch'ang, 1193-1278. Ch'ang-ch'un-chen-jen hsi yu chi. 1972. II. Title. III. Title: Meng-ku lü li. IV. Series: Shih liao ssu pien.
Title romanized: Ch'ang-ch'un-chen-jen hsi yu chi chiao chu.
DS707.L53W36 1972 73-897702

—DESCRIPTION AND TRAVEL—TO 1900

Fortune, Robert, 1813-1880.
Three years' wanderings in the northern provinces of China / Robert Fortune. — New York : Garland Pub., 1979.
xiv, 406 p. : ill. ; 23 cm. — (The Modern Chinese economy)
Reprint of the 1847 ed. published by J. Murray, London.
ISBN 0-8240-4253-0 : $40.00
1. China—Description and travel—To 1900. 2. Fortune, Robert, 1813-1880. I. Title. II. Series.
DS709.F765 1979 951'.03 78-74307
79 MARC

Gherardini, Giovanni Battista, 1652?-1723.
Relation du voyage fait à la Chine sur le vaisseau l'Amphitrite, en l'année 1698, par Gio Ghirardini. Paris, N. Pepie, 1700.
94 p. 18 cm.
1. China—Description and travel—To 1900. 2. Gherardini, Giovanni Battista, 1652?-1723. 3. Amphitrite (Ship) I. Title.
DS708.G48 79-348832
79 MARC

Huc, Évariste Régis, 1813-1860.
Souvenirs d'un voyage dans la Tartarie, le Thibet, et la Chine pendant les années 1844, 1845, et 1846 / par M. Huc. — Paris : Librairie d'Adrien le Clerc, 1850.
2 v. : fold. map ; 22 cm.
Includes bibliographical references.
1. China—Description and travel—To 1900. 2. Huc, Évariste Régis, 1813-1860. I. Title.
DS709.H84 1850 915.1'04'3 78-390195
79 MARC

CHINA (Continued)

--DESCRIPTION AND TRAVEL--TO 1900--ADDRESSES, ESSAYS, LECTURES

(Shan hai ching hsin lun)
山海經新論 / 淩純聲等. — Taipei：東方文化書局複刊, [1978?]
203 p. ; 21 cm. — (國立北京大學·中國民俗學會民俗叢書 ; 142 : 第八輯) (東方文叢)
Added t. p.: Recent studies of the Shan hai ching.
Series 2 romanized: Tung fang wen ts'ung.
A collection of articles reprinted from various sources.
1. Shan hai ching—Addresses, essays, lectures. 2. China—Description and travel—To 1900—Addresses, essays, lectures. I. Ling, Ch'un-sheng. II. Title: Recent studies of the Shan hai ching. III. Series: Kuo li Pei-ching ta hsüeh, Chung-kuo min su hsüeh hui min su ts'ung shu ; 142.
GR335.K87 vol. 142 79-837557
[DS709]

--DESCRIPTION AND TRAVEL --TO 1900--COLLECTED WORKS

Chêng, Kuang-tsu, b. 1775.
(Chou chü so chih)
舟車所至 鄭光祖編 [臺北] 正中書局 民國51 [1962]
884 p. illus., maps. 22 cm.
Facsimile reproduction of 道光23 (1843) 刊本
CONTENTS: 寧古塔紀畧—朝鮮志—隨鑾紀恩—出塞紀畧—塞北紀聞—西域舊聞—烏魯木齊雜詩—伊犂日記—金川舊事—維西見聞—西藏紀聞—容美紀游—海國聞見—中山傳信—採硫日紀—臺灣使槎錄—海島逸志—海錄
1. China—Description and travel—To 1900—Collected works. 2. Asia—Description and travel—Collected works. I. Title.
DS708.C5 1843a C 63-1638

Cheng, Kuang-tsu, b. 1775.
(Chou chü so chih)
舟車所至 / 青玉山房居士[鄭光祖]輯. — [台北縣永和鎮]：文海出版社, [1978?]
2 v. (882 p.) : ill. ; 22 cm. — (近代中國史料叢刊續編 [i. e. 輯] ; 511 : 第五十二輯)
Reprint of the 1843 ed.
1. China—Description and travel—To 1900—Collected works. 2. Asia—Description and travel—Collected works. I. Title. II. Series: Shen, Yün-lung, comp. Chin tai Chung-kuo shih liao ts'ung k'an hsü chi ; 511.
DS755.S532 vol. 511 79-837939
[DS708]

—DESCRIPTION AND TRAVEL—TO 1900 —VIEWS

Beers, Burton F
China in old photographs, 1860-1910 / Burton F. Beers. — New York : Scribner, c1978.
160 p. : ill. ; 28 cm.
Selected from the collection of the Museum of the American China Trade in Milton, Mass.
ISBN 0-684-15800-2 : $17.50
1. China—Description and travel—To 1900—Views. 2. China—Description and travel—1901-1948—Views. I. Museum of the American China Trade. II. Title.
DS709.B43 951'.0022'2 78-13220
78 MARC

—DESCRIPTION AND TRAVEL— 1901-1948

Easton, Robert Olney.
Guns, gold, and caravans : the extraordinary life and times of Fred Meyer Schroder, frontiersman and soldier of fortune, in California, Mexico, Alaska, and China, includes his discovery of the mysterious pyramids of Shensi and rescue of the Boy Emperor / Robert Easton. — 1st ed. — Santa Barbara, Calif. : Capra Press, 1978.
256 p. : ill. ; 24 cm.
Bibliography: p. 252-256.
Includes index.
ISBN 0-88496-122-2 : $11.95
1. Schroder, Fred. 2. Pioneers—United States—Biography. 3. Soldiers of fortune—Biography. 4. China—Description and travel—1901-1948. I. Title.
CT275.S3455E2 909.81'092'4 77-25022
77 MARC

Gotō, Asatarō, 1881-1945.
(Shina ryokō tsū)
支那旅行通 / 後藤朝太郎. — 東京：四六書院：発売所 三省堂, 昭和5 [1930]
2, 5, 4, 211 p., [12] leaves of plates : ill. ; 20 cm. — (通叢書)
Series romanized: Tsū sōsho.
Bibliography: p. 210.
1. China—Description and travel—1901-1948. 2. Gotō, Asatarō, 1881-1945. I. Title.
DS710.G664 78-816235

Hiramatsu, Takashi.
(Shina no jissai o mite chōki kensetsu o kataru)
支那の實際を見て長期建設を語る / 平松卓著. — 東京：國際同盟通信社, 昭和13 [1938]
140 p., [1] leaf of plates : ill. ; 19 cm.
1. China—Description and travel—1901-1948. 2. China—Politics and government—1937-1949. 3. Hiramatsu, Takashi. I. Title. II. Title: Chōki kensetsu o kataru.
DS710.H69 79-804115

Kao, Ch'ing-hsin.
(Shan ho lien)
山河戀 / 著者高清心. — 台北：金陵圖書有限公司, 民國67 [1978]
6, 2, 134, 108 p. ; 20 cm. — (中國遊記大觀 : 1) (金陵文庫)
Series 2 romanized: Chin-ling wen k'u.
"第二[i. e.一]部 北平回憶錄 第一[i. e.二]部 大陸風光集."
NT$48.00
1. Kao, Ch'ing-hsin—Homes and haunts—China. 2. China—Description and travel—1901-1948. 3. Authors, Chinese—20th century—Biography. 4. Kao, Ch'ing-hsin—Journeys—China. I. Title. II. Series: Chung-kuo yu chi ta kuan ; 1.
PL2772.O12Z474 79-837848

Kobayashi, Shūichi.
(Nanshi o yuku)
南支を征く / 小林秀一著. — 東京：潮文閣, 昭和15 [1940]
190 p., [1] leaf of plates : ill. ; 19 cm.
1. Sino-Japanese Conflict, 1937-1945—Personal narratives, Japanese. 2. China—Description and travel—1901-1948. 3. Kobayashi, Shūichi. I. Title.
DS777.53.K565 79-808137

Nakamura, Kōya, 1885-1970.
(Shina o yuku)
支那を行く / 中村孝也著. — [東京]：講談社, 昭和17 [1942]
5, 7, 467 p. : ill. ; 19 cm.
1. China—Description and travel—1901-1948. 2. Nakamura, Kōya, 1885-1970. I. Title.
DS710.N33 78-811048

Nakayama, Shōzen, 1905-1967.
(Shanhai kara Hokuhei e)
上海から北平へ / [著作者 中山正善]. — 改訂版. — 丹波：天理教道友社, 昭和14 [1939]
2, 3, 427 p., [32] leaves of plates : ill. ; 24 cm.
Colophon inserted.
Bibliography: p. 2-3 (2d group)
1. China—Description and travel—1901-1948. 2. Nakayama, Shōzen, 1905-1967. I. Title.
DS710.N335 78-816509

Noo, Tetsumi, 1913-
(Mō Taku-tō no kuni ni ikite)
毛澤東の国に生きて——歸国者の生活記錄—乃生哲巳著 東京 真珠社 昭和28 [1953]
6, 2, 254 p. illus., map. 19 cm.
Colophon inserted.
1. China—Description and travel—1901-1948. 2. China—Description and travel—1949-1975. 3. Noo, Tetsumi, 1913- I. Title.
DS710.N59 J 59-3005

Nozaki, Keisuke.
(Shina jihen senseki angya)
支那事變戰跡行脚 / 野崎圭介著. — 東京：聚文館, 昭和14 [1939]
10, 706 p. ; 17 cm.
1. Sino-Japanese Conflict, 1937-1945. 2. China—Description and travel—1901-1948. 3. Nozaki, Keisuke. I. Title.
DS777.53.N69 78-816515

Sasaki, Tōichi, 1886-1955.
(Nampō kakumei seiryoku no jissō to sono hihan)
南方革命勢力の實相と其の批判 / 佐々木到一著. — [s. l. : s. n., 自序 昭和2 i. e. 1927]
2, 4, 8, 418+ p., [2] leaves of plates : ill. ; 19 cm.
LC copy imperfect: all after p. 418 wanting.
1. China—History—1912-1937. 2. China—Description and travel—1901-1948. 3. Sasaki, Tōichi, 1886-1955. I. Title.
DS777.44.S27 79-804155

Scott, Munroe, 1927-
McClure : the China years of Dr. Bob McClure : a biography / by Munroe Scott. — Toronto : Canec Pub. and Supply House, c1977.
x, 409 p. : map (on lining paper) ; 24 cm. C***
ISBN 0-919000-12-6
1. McClure, Robert Baird, 1900- 2. Missionaries, Medical—China—Biography. 3. China—Description and travel—1901-1948.
R722.32.M28S34 266'.0092'4 78-315222
78 MARC

(Shina taikan to saikan)
支那大觀と細觀 / 久米邦武, 永井柳太郎合著. — 東京：新日本社, 大正6 [1917]
8, 10, 114, 280 p. ; 20 cm.
CONTENTS: 永井柳太郎 支那大觀—久米邦武 「支那大觀」の細觀
1. China—Civilization. 2. Nagai, Ryūtarō, 1881-1944. Shina taikan. 3. China—Description and travel—1901-1948. 4. Nagai, Ryūtarō, 1881-1944. I. Nagai, Ryūtarō, 1881-1944. Shina taikan. 1917. II. Kume, Kunitake, 1839-1931. Shina taikan no saikan. 1917.
DS721.S477 79-801365

Watanabe, Minojirō, 1869-1924.
(Rōtaikoku no sanga)
老大國の山河：余と朝鮮及支那 / 渡邊巳之次郎著. — 東京：金尾文淵堂, 大正10 [1921]
10, 433 p., [10] leaves of plates : ill. ; 19 cm.
1. China—Description and travel—1901-1948. 2. Korea—Description and travel—1900-1953. 3. Watanabe, Minojirō, 1869-1924. I. Title.
DS710.W355 79-808194

79 J

Wu, Chia-ch'ing, 1901-
(Shan shui ch'ing hui chi)
山水清暉集 伍稼青著 [臺北] 臺灣商務印書館 [民國58 i. e. 1969]
2, 3, 2, 158 p. 18 cm. (人人文庫 1025-1026) 12.00
Series romanized: Jen jen wen k'u.
1. China—Description and travel—1901-1948. I. Title.
DS710.W88 78-838776

Wu, Chia-ch'ing, 1901-
(Yu tsung ch'u ch'u)
遊踪處處 伍稼青著 [臺北] 臺灣商務印書館 [民國61 i. e. 1972]
2, 4, 4, 205 p. 19 cm. (人人文庫 特225) NT$20.00
Essays.
Series romanized: Jen jen wen k'u.
1. China—Description and travel—1901-1948. 2. Taiwan—Description and travel. I. Title.
DS710.W89 73-897008

CHINA (Continued)

—DESCRIPTION AND TRAVEL —1901-1948—ADDRESSES, ESSAYS, LECTURES

Funatsu, Yūsuke, 1878-1940.
(Enkyō kashin)
燕京佳信：船津輪助の北京通信 明治三十五年～三十六年 / 船津喜助編；小川博注. — 鳩ケ谷：船津, 昭和53 ⌊1978⌋
v, 276 p., ⌊4⌋ leaves of plates : ill. ; 22 cm. Ja***
1. Peking—Description and travel—Addresses, essays, lectures. 2. China—Description and travel—1901-1948—Addresses, essays, lectures. 3. Funatsu, Yūsuke, 1878-1940—Addresses, essays, lectures. I. Funatsu, Kisuke, 1904- II. Ogawa, Hiroshi, 1929- III. Title.
DS795.F86 1978 79-801628

Ishii, Baku, 1886?-1962.
(Kōgun imon Hokushi kara Chūshi e)
皇軍慰問北支から中支へ / 石井漠著. — 東京：日本教育資料刊行會, 昭和14 ⌊1939⌋
12, 152 p., ⌊8⌋ leaves of plates : ill. ; 19 cm.
1. Sino-Japanese Conflict—1937-1945—Personal narratives, Japanese. 2. China—Description and travel—1901-1948—Addresses, essays, lectures. 3. Soldiers—China—Recreation—Addresses, essays, lectures. 4. Ishii, Baku, 1886?-1962—Addresses, essays, lectures. I. Title.
DS777.53.I7824 79-804810

Shōgakkō Kyōin Manshū oyobi Shina Shisatsudan, 1st, 1939.
(Daiikkai shōgakkō kyōin Manshū oyobi Shina shisatsu hōkoku)
第一回小學校教員滿洲及支那視察報告 / ⌊第一回小學校教員滿洲及支那視察團著⌋. — ⌊東京⌋：外務省文化事業部, 昭和15 ⌊1940⌋
2, 275 p., ⌊2⌋ leaves of plates : ill. ; 22 cm.
Cover title.
1. Manchuria—Description and travel—Addresses, essays, lectures. 2. China—Description and travel—1901-1948—Addresses, essays, lectures. I. Japan. Gaimushō. Bunka Jigyōbu. II. Title.
DS782.3.S53 1939 78-815433

Tōa Dōbun Shoin Daigaku, Shanghai. Dai 36-kisei. Ryokōshi Hensan Iinkai.
(Dairyokōki)
大旅行紀 / 東亞同文書院第三十六期生 ⌊旅行誌編纂委員會⌋著. — 東京：同亞同文會業務部, 昭和15 ⌊1940⌋
1, 2, 432 p., ⌊2⌋ leaves of plates : ill. ; 19 cm.
1. China—Description and travel—1901-1948—Addresses, essays, lectures. 2. Asia, Southeastern—Description and travel—Addresses, essays, lectures. I. Title.
DS710.T673 1940 79-805177

—DESCRIPTION AND TRAVEL —1901-1948—VIEWS

Beers, Burton F
China in old photographs, 1860-1910 / Burton F. Beers. — New York : Scribner, c1978.
160 p. : ill. ; 28 cm.
Selected from the collection of the Museum of the American China Trade in Milton, Mass.
ISBN 0-684-15800-2 : $17.50
1. China—Description and travel—To 1900—Views. 2. China—Description and travel—1901-1948—Views. I. Museum of the American China Trade. II. Title.
DS709.B43 951'.0022'2 78-13220
78 MARC

—DESCRIPTION AND TRAVEL—1949-

Maitraye Devi, 1914-
(Acenā Cīna)
অচেনা চীন / মৈত্রেয়ী দেবী. — 1. সংস্করণ. — কলিকাতা : আনন্দ পাবলিশার্স, 1978.
150 p., ⌊4⌋ leaves of plates : ill. ; 22 cm.
In Bengali.
Rs15.00
1. Maitraye Devi, 1914- —Journeys—China. 2. China—Description and travel—1949- 3. Authors, Bengali—20th century—Biography. I. Title.
PK1718.M24244Z47 78-905893

Tito u Kini / ⌊uređivački odbor Nikola Ban ... et al. ; autori fotografija Miloš Rašeta, Foto Tanjug, Joco Znidaršić ; urednik Boža Aleksić⌋. — Beograd : Jugoslovenska revija, 1978.
86 p. : col. ill. ; 26 cm. Yu***
Issued also in English and Slovenian.
CONTENTS: Tito u Kini.—Smederevac, M. Jugoslavija i Kina.—Šaranović, M. Narodna republika Kina.
300.00Din
1. Tito, Josip Broz, Pres. Yugoslavia, 1892- . 2. Visits of state—China. 3. China—Description and travel—1949- . I. Aleksić, Božo. II. Smederevac, Milivoje. Jugoslavija i Kina. 1978. III. Šaranović, Mihailo, 1935- Narodna Republika Kina. 1978.
DR327.C5T57 78-394413
79 MARC

United States. Industry and Trade Administration.
Doing business with China. — Rev. Mar. 1979. — Washington : U.S. Dept. of Commerce, Industry and Trade Administration : for sale by the Supt. of Docs., U.S. Govt. Print. Off., 1979.
iii, 44 p. : ill. ; 26 cm.
Bibliography: p. 23-25.
1. United States—Commerce—China. 2. China—Commerce—United States. 3. China—Description and travel—1949- I. Title.
HF3128.U53 1979 346'.51'07 79-601897
00 MARC

—DESCRIPTION AND TRAVEL —1949- —ADDRESSES, ESSAYS, LECTURES

Niigata-ken Seinen Kaigai Hakendan, 7th, 1976.
(Chūgoku kenshū no tabi)
中国研修の旅：第7回新潟県青年海外派遣団報告書. — 新潟：新潟県, 昭和51 ⌊1976⌋
127 p. : ill. ; 26 cm. Ja 76-13
Cover title.
1. China—Description and travel—1949- —Addresses, essays, lectures. 2. Agriculture—China—Addresses, essays, lectures. I. Title.
DS711.N52 1976 76-806980

(P'arŏgin kwa ŭi taehwa)
8億人과의對話：現地에서본中國大陸 / 李泳禧編譯. — 서울：創作과批評社, 1977.
370 p. ; 21 cm. — (創批新書 ; 18)
Series romanized: Ch'angbi sinsŏ.
₩1700
1. China—Description and travel—1949- —Addresses, essays, lectures. I. Yi, Yŏng-hŭi, 1929-
DS711.P318 78-826477

—DESCRIPTION AND TRAVEL—1949-1975

Bredsdorff, Jan.
Revolution tur/retur / ⌊af⌋ Jan Bredsdorff ; ⌊fotos, Stine Heger og Jan Bredsdorff⌋. — ⌊København⌋ : Information, 1978.
274 p. : ill. ; 20 cm. D78-47
ISBN 8787498782 : kr95.00
1. China—Description and travel—1949-1975. 2. China—Description and travel—1976- . 3. Bredsdorff, Jan. I. Title.
DS711.B735 79-342446
79 MARC

Hevi, Emmanuel John.
⌊An African student in China. Chinese⌋
一個留學北平的外國學生 / 約翰·哈維著；古原譯. — 九龍：友聯出版社, 1965.
4, 108 p. ; 19 cm.
1. African students in China. 2. China—Description and travel—1949-1976. 3. Hevi, Emmanuel John. I. Title.
Title romanized: I ko liu hsüeh Pei-p'ing ti wai kuo hsüeh sheng.
LA1133.7.H412 79-837053

Irving, Esther Bertha.
Journey into China / Esther Irving. — 1st ed. — Foxton : Peacestar Publications, 1976.
127 p. : ill., map ; 22 cm. NZ77-2
Includes index.
$5.95
1. China—Description and travel—1949-1975. I. Title.
DS711.I73 915.1'04'50924 79-301482
79 MARC

Koning, Hans, 1921-
Love and hate in China. ⌊1st ed.⌋ New York, McGraw-Hill ⌊1966⌋
150 p. illus. (on lining papers) 22 cm.
1. China—Description and travel—1949-1975. I. Title.
DS711.K66 915.1045 66-19286
⌊r79⌋rev MARC

Koning, Hans, 1921-
Love and hate in China. London, Cape, 1967.
⌊4⌋, 150 p. 21 cm. 21/- B67-7774
1. China—Description and travel—1949-1975. I. Title.
DS711.K66 1967 915.1'04'5 67-91672
⌊r79⌋rev MARC

Ling, Chiang.
(Chung-kuo ti li chih shih)
中國地理知識 / 凌江編著. — 香港：商務印書館, 1977.
328 p. : ill. ; 21 cm.
$10.00
1. China—Description and travel—1949-1975. I. Title.
DS711.L56 78-841535
C

Mizuki, Akira.
(Horobi to tanjō)
滅びと誕生：中国·西欧·日本見聞録 / 水木明著. — 島田：山田富久, 1975.
210 p. : ill. ; 21 cm. Ja 76-12670
Errata slips inserted.
Bibliography: p. 205-208.
1. China—Description and travel—1949-1975. 2. Europe—Description and travel—1945-1970. 3. Japan—Description and travel—1945- 4. Mizuki, Akira. I. Title.
DS711.M546 79-801721

Noo, Tetsumi, 1913-
(Mō Taku-tō no kuni ni ikite)
毛澤東の国に生きて——歸国者の生活記錄—乃生哲巳著 東京 真珠社 昭和28 ⌊1953⌋
6, 2, 254 p. illus., map. 19 cm.
Colophon inserted.
1. China—Description and travel—1901-1948. 2. China—Description and travel—1949-1975. 3. Noo, Tetsumi, 1913- I. Title.
DS710.N59 J 59-3005
rev ⌊r78⌋

Robinson, Joan, 1903-
Joan Robinson-reports from China, 1953-76. — London : Anglo-Chinese Educational Institute, 1977.
131 p. ; 22 cm. GB78-27081
Errata slip inserted.
£1.00
1. China—Description and travel—1949-1975. 2. Robinson, Joan, 1903- I. Title.
DS711.R55 951.05 79-301413
79 MARC

Schell, Orville.
China : in the People's Republic / Orville Schell. — London : Gollancz, 1978, c1977.
ix, 271 p., ⌊12⌋ leaves of plates : ill. ; 21 cm. GB***
American ed. published under title: In the People's Republic.
ISBN 0-575-02434-8 : £5.95
1. China—Description and travel—1949-1975. 2. Schell, Orville. I. Title.
DS711.S288 1978b 951.05 78-313776
78 MARC

—DESCRIPTION AND TRAVEL —1949-1975—OUTLINES, SYLLABI, ETC.

Wang, Chiung-sheng.
(Ti li chih shih t'i kang)
地理知識提綱 / 王烱聲編著；陳碩礎校訂. — ⌊香港?：s.n., 196-?⌋
ca. 150 leaves : ill. ; 22 x 30 cm.
Reprint of the 1957 ed. published by Chung-kuo ch'u pan she, Hongkong, issued in series: Chung hsüeh ko k'o fu hsi ts'ung shu.
Each leaf represents 2 pages of the original.
1. Physical geography—Outlines, syllabi, etc. 2. China—Description and travel—1949-1975—Outlines, syllabi, etc. I. Title.
GB55.5.W36 1960z 78-843085

CHINA (Continued)

--DESCRIPTION AND TRAVEL--1949-1975--VIEWS

(Chūgoku no fūkei)
中国の風景 /「中国撮影」編集部編. — 北京：外文出版社, 1962.
27 p. : all ill. (some col.) ; 24 x 26 cm.

1. China—Description and travel—1949-1975—Views. I. Chung-kuo she ying pien chi pu.
DS711.C51864 79–805123

(Midori no daichi ni yūgi no haru)
緑の大地に友誼の春 = 迎来春色増友誼 = a blossoming springtime of friendship：池田会長日中友好の記録 / ⌊編集撮影 原山正征⌋. — 東京：聖教新聞社, 昭和53 ⌊1978⌋
135 p. : all ill. (some col.) : 31 cm. Ja***
¥2800
1. Ikeda, Daisaku—Journeys—China—Pictorial works. 2. China—Description and travel—1949–1975—Views. I. Harayama, Masayuki. II. Title: Ying lai ch'un se tseng yu i. III. Title: A blossoming springtime of friendship.
BQ8449.I387M5 79–804482

—DESCRIPTION AND TRAVEL—1976-

Bredsdorff, Jan.
Revolution tur/retur / ⌊af⌋ Jan Bredsdorff ; ⌊fotos, Stine Heger og Jan Bredsdorff⌋. — ⌊København⌋ : Information, 1978.
274 p. : ill. ; 20 cm. D78-47
ISBN 8787498782 : kr95.00
1. China—Description and travel—1949-1975. 2. China—Description and travel—1976- . 3. Bredsdorff, Jan. I. Title.
DS711.B735 79-342446
79 MARC

Buffetti, Maria.
Cina quotidiana : un'estate cinese / di Maria Buffetti. — Roma : Trevi editore, 1978.
204 p. : ill. ; 22 cm. — (I Doni ; 8) It79-Jan
L12000
1. China—Description and travel—1976- . 2. Buffetti, Maria. I. Title.
DS712.B83 951.05'022'2 79-361723
•79 MARC

China travel : Zhengzhou, Luoyang, Anyang, Linxian / edited by China International Travel Service. — Beijing : China Travel and Tourism Press, 1977.
113 p., ⌊3⌋ leaves of plates : col. ill. ; 21 cm.
1. China—Description and travel—1976- I. Chung-kuo kuo chi lü hsing she.
DS712.C482 951.05 78-112922
79 MARC

China view : first NASW study tour / ChaunceyA. Alexander and Sally J. Alexander, editors. — Washington : National Association of Social Workers, c1979.
xxiv, 357 p. : ill. ; 19 cm.
Bibliography: p. 312-317.
ISBN (invalid) 0-571-01081-X
1. Social service—China. 2. China—Description and travel—1976- . 3. National Association of Social Workers. I. Alexander, Chauncey A. II. Alexander, Sally J.
HV418.C44 361'.951 79-64536
00 MARC

La Chine telle que l'ont vue dix syndicalistes. — Paris : Éditions du Centenaire, 1977.
137 p., ⌊4⌋ p. of plates : ill. ; 22 cm. F78-3676
15.00F
1. China—Description and travel—1976-
DS712.C483 951.05 78-395636
79 MARC

(Chung-kuo hsin lü yu)
中國新旅遊. — 香港：風格出版社；九龍：發行利源書報社, ⌊1978?⌋
46 p. : ill. ; 27 cm.
Cover title.
$2.00

1. China—Description and travel—1976-
DS712.C495 79–836602

Gille, Hans Werner, 1928-
Das Antlitz Chinas / Hans-Werner Gille. — Wels : Welsermühl, c1976.
256 p. : ill. (some col.) ; 30 cm. Au78-22
Bibliography: p. 252-255.
ISBN 3-85339-128-1 : S460.00
1. China—Description and travel—1976- . 2. Gille, Hans Werner, 1928- . I. Title.
DS712.G54 951.05 79-338477
•79 MARC

Gómez Pompa, Arturo.
Viaje a China / Arturo Gómez-Pompa. — Xalapa, México : Instituto de Investigaciones sobre Recursos Bióticos, 1978.
iv, 159, ⌊4⌋ p., ⌊5⌋ leaves of plates : ill. ; 28 cm.
Bibliography: p. ⌊161⌋
$125.00
1. China—Description and travel—1976- . 2. Gómez Pompa, Arturo. I. Title.
DS712.G65 915.1'04'5 78-113821
79 MARC

Gustafsson, Lars, 1936-
Kinesisk höst / Lars Gustafsson ; reseteckningar av Sven Ljungberg. — Stockholm : Norstedt, 1978.
227, ⌊1⌋ p. : ill. ; 21 cm. S78-26
Bibliography: p. 227-⌊228⌋
ISBN 9117811627
1. Gustafsson, Lars, 1936- —Journeys—China. 2. China—Description and travel—1976- . 3. Authors, Swedish—20th century—Biography. I. Ljungberg, Sven, 1913- II. Title.
PT9876.17.U8Z468 78-399654
79 MARC

Hägglöf, Gunnar, 1904-
Kina som jag såg det / Gunnar Hägglöf. — Stockholm : Norstedt, 1978.
191 p., ⌊2⌋ leaves of plates : ill. ; 22 cm. S***
Bibliography: p. 185-190.
Includes index.
ISBN 9117832829
1. China—Description and travel—1976- . 2. Hägglöf, Gunnar, 1904- . I. Title.
DS712.H33 78-399647
79 MARC

Honig, Albert M
China today : sin or virtue? / Albert M. Honig. — 1st ed. — Hicksville, N.Y. : Exposition Press, c1978.
vi, 98 p., ⌊1⌋ leaf of plates : ill. ; 22 cm.
Bibliography: p. 98.
ISBN 0-682-49086-5 : $7.00
1. China—Description and travel—1976- . 2. Honig, Albert M. I. Title.
DS712.H66 915.1'04'5 78-53758
78 MARC

Huang, Chiu-shun.
(Chung-kuo ti li kai lun)
中國地理概論 / 黃就順編著. — 香港：上海書局, 1978.
6, 501 p. : ill. ; 21 cm.
$15.00

1. Physical geography—China. 2. China—Description and travel—1976- I. Title.
GB316.H827 79–835433

Jamīluzzamān.
(Cistān-i Cīn)
پیستان چین : ایک متجسس سفرنامہ اپریل تا جون ۱۹۷۶ء / از جمیل الزماں. — اشاعت 1. — لاہور : نیشنل بک فاؤنڈیشن، 1977.
240 p. ; 22 cm.
In Urdu.
Rs20.00

1. China—Description and travel—1976- 2. Jamīluzzamān. I. Title.
DS712.J35 78–930541

Kaurinkoski, Kaarina.
Kiinalaista arkipäivää / Kaarina ja Jaakko Kaurinkoski. — Helsinki : Kirjayhtymä, ⌊1978⌋
207 p. ⌊8⌋ leaves of plates : ill. ; 21 cm. Fi***
ISBN 951261538X
1. China—Description and travel—1976- . 2. Kaurinkoski, Kaarina. I. Kaurinkoski, Jaakko, 1934- joint author. II. Title.
DS712.K38 79-349700
79 MARC

Kiyooka, Takayuki, 1922–
(Geijutsuteki na akushu)
藝術的な握手：中國旅行の同想 / 清岡卓行. — 東京：文芸春秋, 1978.
358 p. ; 22 cm. Ja 78–17174
¥2500

1. China—Description and travel—1976– 2. Kiyooka, Takayuki, 1922– I. Title.
DS712.K58 78–814157

Ljungberg, Sven, 1913-
Kina : flitens land / bilder och text av Sven Ljungberg. — Stockholm : Alba, 1979.
138, ⌊5⌋ p. : ill. ; 23 cm. S79
ISBN 9174581880
1. China—Description and travel—1976- . 2. Ljungberg, Sven, 1913- . I. Title.
DS712.L56 79-360222
79 MARC

Luttwak, Edward.
⌊Seeing China plain. Chinese⌋
平實的看中國大陸 / ⌊拉瓦克作；李玉成譯⌋. — ⌊臺北⌋：黃不撓印贈, ⌊1977⌋
20 p. ; 19 cm.
Cover title.
At head of title: Ching ch'ing nin.
"原載「評論雜誌」"
1. China—Description and travel—1976– 2. Luttwak, Edward. I. Title.
Title romanized: P'ing shih ti k'an Chung-kuo ta lu.
DS712.L8712 79–840064

Masi, Edoarda, 1927-
Per la Cina : confuciani e proletari / Edoarda Masi. — 1. ed. — Milano : Mondadori, 1978.
492 p. ; 20 cm. — (Saggi ; 109) It78-Nov
Includes index.
L5500
1. China—Description and travel—1976- . 2. Masi, Edoarda, 1927- . I. Title.
DS712.M37 79-340400
•79 MARC

Meiner, Louis Michel.
La Chine au présent / L.-M. Meiner. — Paris : Klincksieck, c1978.
167 p., ⌊4⌋ leaves of plates : ill. ; 21 cm. F***
ISBN 2-252-02097-0 : 40.00F
1. China—Description and travel—1976- 2. Meiner, Louis Michel. I. Title.
DS712.M44 915.1'04'5 79-364514
79 MARC

Nimmergut, Jörg, 1939-
Westwind weht durch China : als Tourist im China von heute / Jörg u. Ulla Nimmergut. — München : Heyne, 1978.
254 p., ⌊16⌋ leaves of plates : ill., map ; 18 cm. — (Heyne Bücher ; Nr. 7072) (Heyne Sachbuch) GFR79-A
Includes index.
ISBN 3-453-00897-9 : DM6.80
1. China—Description and travel—1976- . 2. Nimmergut, Jörg, 1939- . I. Nimmergut, Ulla, joint author. II. Title.
DS712.N55 951.05 79-354748
79 MARC

Nitchū Yūkō Tochigi Kemmin no Tsubasa Hōchūdan.
(Asu e no Chūgoku)
あすへの中国：日中友好栃木県民の翼訪中団報告書：自1978年5月15日～至1978年5月26日 / ⌊編集者 「日中友好栃木県民の翼」実行委員会報告書編集委員会⌋. — ⌊宇都宮⌋：同実行委員会, 1978.
103 p. : ill. ; 26 cm. Ja***
Cover title.
Accompanied by "一衣帯水の隣国·中国を訪ねて：音で綴る県民の翼訪中団の足跡" (3 discs, 33 1/3 r.p.m. mono. 20 in.) in pocket.

1. China—Description and travel—1976– 2. Nitchū Yūkō Tochigi Kemmin no Tsubasa Hōchūdan. I. Nitchū Yūkō Tochigi Kemmin no Tsubasa Jikkō Iinkai. Hōkokusho Henshū Iinkai. II. Title.
DS712.N59 1978 79–807340
79 J

Paris. École normale supérieure. Centre de sciences humaines.
Choses vues en République populaire de Chine : rapport sur le voyage effectué ... en septembre 1974 / École normale supérieure, Centre de sciences humaines. — 2e éd. — Paris : Presses de l'École normale supérieure, 1976.
85 p., ⌊4⌋ leaves of plates : ill. ; 27 cm. F77-17327
ISBN 2-7288-0029-4 : 18.00F
1. China—Description and travel—1976- . I. Title.
DS712.P37 1976 309.1'51'05 78-390356
79 MARC

Petersen, Gwenn Boardman.
Across the bridge to China / by Gwenn Boardman Petersen. — 1st ed. — New York : Elsevier/Nelson Books, c1979.
243 p. : ill. ; 24 cm.
Includes index.
ISBN 0-525-66622-2 : $8.95
1. China—Description and travel—1976- . 2. Petersen, Gwenn Boardman. I. Title.
DS712.P47 1979 915.1'04'5 79-4171
79 MARC

CHINA

—DESCRIPTION AND TRAVEL
—1976- (Continued)

Salisbury, Charlotte Y
China diary, after Mao / Charlotte Y. Salisbury. — New York : Walker, 1979.
ix, 214 p. : map ; 24 cm.
Includes index.
ISBN 0-8027-0621-5 : $9.95
1. China—Description and travel—1976- 2. Salisbury, Charlotte Y. I. Title.
DS712.S24 1979 915.1'04'5 78-70932
00 MARC

Samuel, Yvon.
Chine 79 [i.e. soixante dix-neuf] : une brèche dans la grande muraille / Yvon Samuel. — Paris : J.-C. Simoën, c1979.
194 p. ; 21 cm. F***
ISBN 2-7313-0105-8 : 44.00F
1. China—Description and travel—1976- 2. Samuel, Yvon. I. Title.
DS712.S25 951.05 79-380295
79 MARC

Tschudi, Aadel Brun, 1909-
Gjensyn med Kina / Aadel Brun Tschudi. — Oslo : Grøndahl, 1978 (Lillestrøm : Lillestrøm industritrykk)
189 p. : ill. ; 26 cm. N78-Nov.
ISBN 8250403320 : kr88.00
1. China—Social conditions—1976- . 2. China—Description and travel—1976- . 3. Missions—Norway—History. 4. Missions—China—History. I. Title.
HN733.5.T7 79-341862
79 MARC

United States. Central Intelligence Agency.
China city brief[s]. — Washington : [Available from] Document Expediting Project, Exchange and Gift Division, Library of Congress, [197-]-
v. in : maps ; 27 cm.
Cover title.
1. Cities and towns—China. 2. China—Description and travel—1976- I. Title.
DS737.U54 1970z 951 79-600827
79 MARC

La Vie en jaune : sept jeunes giscardiens en Chine populaire / Dominique Bussereau, Jean-Pierre Raffarin, Henri Giscard d'Estaing, Benoît Roger-Vasselin ... [etc.]. — Paris : Stock, 1977.
222 p. ; 22 cm. F78-1054
ISBN 2-234-00731-3 : 38.00F
1. China—Description and travel—1976- I. Bussereau, Dominique.
DS712.V53 915.1'1'045 78-383710
78 MARC

Wieviorka, Annette.
L'écureuil de Chine / Annette Wieviorka. — Paris : Presses d'aujourd'hui, c1979.
211 p. ; 21 cm. F***
42.00F
1. China—Description and travel—1976- 2. Wieviorka, Annette. I. Title.
DS712.W54 951.05 79-373270
79 MARC

Yamaki, Shigeru, 1908–
(Shakaishugi Chūgoku mita mama)
社会主義中国みたまま：ソ連邦と対比して / 山木茂. — [広島]：山新, 1978.
27 p. ; 22 cm. Ja 78-14483
Cover title.
¥300
1. China—Description and travel—1976– 2. Yamaki, Shigeru, 1908– I. Title.
DS712.Y35 78-814669

Zenkoku Nōkyō Yakushokuin Remmei Dainiji Yūkō Hōchūdan.
(Jimmin Chūgoku o tazunete)
人民中国を訪ねて：全国農協役職員連盟第二次友好訪中団報告書. — 東京：全国農業協同組合役職員連盟, 1978.
117 p. : ill. ; 21 cm. Ja 78-12884
¥400
1. China—Description and travel—1976– I. Title.
DS712.Z46 1978 78-812992

—DESCRIPTION AND TRAVEL
—1976- —ADDRESSES, ESSAYS LECTURES

Hataraku Fujin no Yūkō Hōchūdan.
(Chūgoku no onnatachi o tazunete)
中国の女たちを訪ねて：一九七七年·夏. — 東京：働く婦人の友好訪中団, 1978.
74 p. : ill. ; 26 cm. Ja 78-21105
Cover title.
¥380
1. China—Description and travel—1976– —Addresses, essays, lectures. 2. Women—China—Social conditions—Addresses, essays, lectures. I. Title.
DS712.H37 1978 78-816638

Hui-lu.
(Chung-kuo shan ch'uan chi sheng)
中國山川紀勝 / 晦廬著. — 香港：上海書局, 1978.
6, 232 p. ; 19 cm.
$5.90
1. China—Description and travel—1976– —Addresses, essays, lectures. 2. Taiwan—Description and travel—Addresses, essays, lectures. I. Title.
DS712.H84 78-842207

Nitchū Yūkō Dainiji Hōchūdan.
(Yottsu no gendaika ni susumu Chūgoku)
"四つの現代化"に進む中国：四人組を粉砕して：1977年9月5日～21日 / 日中友好第二次訪中団. — 東京：中国旅行読者の会, 1977.
206 p. : ill. ; 21 cm. Ja 78-14482
Spine title: Chūgoku ryokō.
¥800
1. China—Description and travel—1976– —Addresses, essays, lectures. I. Title. II. Title: Chūgoku ryokō.
DS712.N567 1977 78-816959

—DESCRIPTION AND TRAVEL—1976-
—GUIDE-BOOKS

De Keijzer, Arne J
The China guidebook : a traveler's guide to the People's Republic of China / Arne J. de Keijzer, Fredric M. Kaplan ; with an introd. by Esther Gollobin. — 1979/80 ed., 1st ed. — Fair Lawn, N.J. : Eurasia Press, c1979.
304 p., [1] fold. leaf of plates : ill. ; 21 cm.
Bibliography: p. 280-283.
Includes index.
ISBN 0-397-01358-2. ISBN 0-397-01359-0 pbk.
1. China—Description and travel—1976- —Guide-books. I. Kaplan, Fredric M., joint author. II. Title.
DS712.D4 1979 915.1'04'5 79-37
79 MARC

De Keijzer, Arne J
JAL guide to the People's Republic of China / Arne J. de Keijzer, with Fredric M. Kaplan. — 1st ed. — Fair Lawn, N.J. : Eurasia Press, c1978.
x, 214 p. : ill. ; 18 cm.
Bibliography: p. 205-207.
Includes index.
ISBN 0-932030-01-7
1. China—Description and travel—1976- —Guide-books. I. Kaplan, Fredric M., joint author. II. Japan Air Lines. III. Title. IV. Title: Guide to the People's Republic of China.
DS705.D44 915.1'04'5 78-58817
79 MARC

Fodor's People's Republic of China. 1979–
New York, D. McKay.
v. ill. 21 cm. (Fodor's modern guides)
Editors: 1979– R. C. Fisher, L. Brown.
Key title: Fodor's People's Republic of China, ISSN 0192-2378.
1. China—Description and travel—1976– —Guide-books. I. Fisher, Robert C. II. Brown, Leslie Ruth. III. Title: People's Republic of China. IV. Title: China.
DS712.F62 915.1'04'5 79-642969
MARC-S

Lai, Tien-chang.
Visiting China : a cultural guide / T. C. Lai. — Hong Kong : Swindon Book Co., c1979.
248 p., [2] leaves of plates : ill. ; 22 cm.
1. China—Description and travel—1976- —Guide-books. I. Title.
DS705.L34 915.1'04'5 79-109151
79 MARC

Saunders, Hilliard.
The complete travel guide to China / by Hilliard Saunders. — Seal Beach, CA : China Pub. Co., c1979.
vii, 179 p., [1] leaf of plates : col. ill. ; 22 cm.
Bibliography: p. 174-176.
Includes index.
ISBN 0-89260-142-6 : $6.95
1. China—Description and travel—1976- —Guide-books. I. Title.
DS705.S28 915.1'04'5 79-63530
79[r79]rev MARC

—DESCRIPTION AND TRAVEL—1976-
—VIEWS

Cross, Mary.
Behind the Great Wall : a photographic essay on China / black and white photos. by Mary Cross, color photos. by Theodore Cross ; text by Mary and Theodore Cross. — New York : Atheneum, 1979.
ca. 150 p. : ill. ; 32 cm.
ISBN 0-689-11018-9 : $25.00
1. China—Description and travel—1976- —Views. 2. Cross, Mary. 3. Cross, Theodore L., 1924- I. Cross, Theodore L., 1924- joint author. II. Title.
DS712.C76 779'.9'95105 79-1742
79 MARC

Selected photographs from China. — 1st ed. — Peking : Foreign Languages Press, 1977.
100 p. : chiefly ill. ; 26 cm.
1. China—Description and travel—1976- —Views.
DS712.S44 779'.9'95105 79-100615
79 MARC

Stanger, D L
Communist China, democratic India / introd., Richard C. Gripp ; photography and comment, D. L. Stanger. — [San Diego : Overdive High Communications Corp., c1978]
180 p. : chiefly col. ill. ; 25 x 32 cm.
On spine: China, India.
1. China—Description and travel—1976- —Views. 2. India—Description and travel—1947- —Views. I. Title. II. Title: China, India.
DS712.S7 779'.9'951 78-112134
79 MARC

--DESCRIPTION AND TRAVEL
--COLLECTED WORKS

T'ao, T'ien-ts'e.
(Chung-kuo yu chi ta kuan)
中國遊記大觀 / 編著陶天策. — 台北：金陵圖書有限公司, 民國 [19
v. : ill. ; 20 cm. — (中國各地區旅遊「文學之旅」總集 ; -2) (金陵文庫)
Series 2 romanized: Chin-ling wen k'u.
NT$55.00 ($3.00 U. S.) v. 2.
1. China—Description and travel—Collected works. I. Title. II. Series: Chung-kuo ko ti ch'ü lü yu wen hsüeh chih lü tsung chi ; 2.
DS706.T35 79-838646

—DICTIONARIES AND ENCYCLOPEDIAS

Chi, Chia-ling.
(Feng su t'ung i chiao chu)
風俗通義校注 / 李嘉玲撰. — [臺北? : s. n.], 民國65 [1976]
121 p. ; 27 cm.
Cover title.
Reproduced from ms. copy.
"國立台灣師範大學國文研究所碩士論文"
1. Ying, Shao, fl. 189-194. Feng su t'ung i. 2. China—Dictionaries and encyclopedias. I. Ying, Shao, fl. 189-194. Feng su t'ung i. Selections. 1976. II. Title.
DS705.C447 1976 76-840979

Encyclopedia of China today / [edited by] Fredric M. Kaplan, Julian M. Sobin, Stephen Andors ; introd. by John S. Service. — 1st ed. — Fair Lawn, N.J. : Eurasia Press ; New York : distributed in the U.S.A. by Harper & Row, c1979.
xvi, 336 p. : ill. ; 29 cm.
Bibliography: p. 305-320.
Includes index.
ISBN 0-06-012256-0 : $20.00
1. China—Dictionaries and encyclopedias. I. Kaplan, Frederic M. II. Sobin, Julian M. III. Andors, Stephen, 1938-
DS705.E54 1979 951.05'03 77-6886
77 MARC

Hsiao, I, 508-554.
(Chin-lou-tzu)
金樓子 6卷 梁元帝[蕭繹]撰 謝章鋌校 [台北] 世界書局 [1959]
1 v. (various leaves) 19 cm. (中國思想名著之一)
世界文庫 四部刊要
At head of title: 景印國立中央圖書館珍藏鈔永樂大典本
1. China—Dictionaries and encyclopedias. I. Hsieh, Chang-t'ing, 1820-1903, ed. II. Title.
DS705.H77 1959 77-837114

CHINA

—DICTIONARIES AND ENCYCLOPEDIAS (Continued)

Hsü, Tê-p'ing, 1940–
(Chin-lou-tzu chiao chu)
金樓子校注　許德平著　[臺北　嘉新水泥公司文化基金會　民國58 i. e. 1969]
5, 2, 270 p.　21 cm.　(嘉新水泥公司文化基金會研究論文第103種)
Series statement in colophon: 嘉新水泥公司文化基金會叢書
Bibliography: p. 266–270.
1. Hsiao, I, 508–554. Chin-lou-tzŭ. 2. China—Dictionaries and encyclopedias. I. Hsiao, I, 508–554. Chin-lou-tzŭ. 1969. II. Title.
D705.H78　75–843575

Kondō, Haruo, 1914–
(Chūgoku gakugei daijiten)
中国学芸大事典 / 近藤春雄. — 東京：大修館書店, 1978.
5, 1000 p. ; 27 cm.　Ja 78–26440
Includes bibliographies.
¥17000
1. Chinese philology—Dictionaries—Japanese. 2. China—Dictionaries and encyclopedias. I. Title.
PL1071.K6　79–801491

Liu, Chou, 516?–567?
(Liu tzu)
劉子 / [劉晝撰；袁孝政注]. 意林 / [馬總撰]. — [台北?]：中國子學名著集成編印基金會, [1977?]
3, 3, 3, 555 p. : 22 cm. — (中國子學名著集成 ; 091 : 珍本初編　雜家子部)
"劉子十卷...明萬曆壬辰蔣以化孝昌刊本, 朱藍批校, 清沈韵珩手書題記"
"意林五卷...清乾隆四十七年武英殿聚珍本...並輯有意林逸文"
1. China—Dictionaries and encyclopedias. I. Yüan, Hsiao-cheng. II. Ma, Tsung, d. 823. I lin. 1977. III. Title. IV. Title: I lin. V. Series: Chung-kuo tzu hsüeh ming chu chi ch'eng ; 091.
B128.L58A1　1977　78–842623
78　C

Schicht, Peter, 1935-
China / [Photos] Peter Schicht ; [Text] Ulla und Johannes Küchler. — Luzern [etc.] : C. J. Bucher, cop. 1978.
220 p. : numerous ill. (some col.) ; 31 cm.　Sw78-12744
Bibliography: p. 220.
Includes index.
ISBN 3-7658-0269-7 : 88.00F
1. China—Description and travel—1976-　—Views. I. Küchler, Ulla, 1942-　II. Küchler, Johannes. III. Title.
DS712.S34　951.05　79-352877
79　MARC

Ying, Shao, fl. 189–194.
(Feng su t'ung i)
風俗通義　[10卷]　附佚文　[6卷　應劭著　中法漢學研究所編輯　北京　中法漢學研究所　來薰閣書店代售]　民國32 [1943]
10, 4, 150 p.　27 cm.　(中法漢學研究所通檢叢刊之3)
Added t. p.: Le Fong sou t'ong yi, avec un appendice contenant les fragments du texte perdu cités en d'autres ouvrages.
——通檢　[中法漢學研究所編輯　北京　中法漢學研究所　來薰閣書店代售]　民國32 [1943]
xx, 97 p.　27 cm.　(中法漢學研究所通檢叢刊之3)　Bound with the main work.
Added t. p.: Index du Fong sou t'ong yi.
1. China—Dictionaries and encyclopedias. I. Chung Fa Han hsüeh yen chiu so. II. Title. III. Title: Le Fong sou t'ong yi. IV. Title: Index du Fong sou t'ong yi. (Series: Chung Fa Han hsüeh yen chiu so. Chung Fa Han hsüeh yen chiu so t'ung chien ts'ung k'an chih 3)
DS705.Y55　1943　C 68–2019
rev [r79]　C

Ying, Shao, fl. 189–194.
(Feng su t'ung i)
風俗通義 / [應劭撰]. 金樓子 / [梁元帝撰]. — [臺北?]：中國子學名著集成編印基金會, [1978?]
2, 3, 3, 570 p. ; 22 cm. — (中國子學名著集成 ; 90 : 珍本初編雜家子部)
Reprint ed.
1. China—Dictionaries and encyclopedias. I. Hsiao, I, 508–554. Chin-lou-tzu. 1978? II. Series: Chung-kuo tzu hsüeh ming chu chi ch'eng ; 90.
DS705.Y56　1978　78–842882

—DIRECTORIES

China. Yu cheng tsung chü.
(Chung-kuo t'ung yu ti fang wu chan chih)
中國通郵地方物產誌 / 交通部郵政總局編輯. — 上海：商務印書館, 民國26 [1937]
ca. 1100 p. in various pagings : ill. ; 27 cm.
Chiefly tables.
1. China—Directories. 2. Commercial products—China. I. Title.
DS705.C46　1937　79–837428

—DIRECTORIES—TELEPHONE

China Phone Book Company.
The China phone book & address directory. — [Hongkong] : The China Phone Book Co., c1978.
130 p. ; 26 cm.
1. China—Directories—Telephone. I. Title.
DS705.C52　1978　915.1'0025　79-101073
MARC

—ECONOMIC CONDITIONS

Crow, Carl, 1883–1945.
[Four hundred million customers. Japanese]
支那で成功する道 / カール・クロウ著；柴田賢一譯. — 東京：高山書院, 昭和15 [1940]
7, 8, 289 p. ; 19 cm.
1. China—Social life and customs. 2. China—Social conditions. 3. China—Economic conditions. 4. Crow, Carl, 1883–1945. I. Title.
Title romanized: Shina de seikōsuru michi.
DS721.C9316　79–800152

Menzel, Ulrich.
Wirtschaft und Politik im modernen China : e. Sozial- u. Wirtschaftsgeschichte von 1842 bis nach Maos Tod / Ulrich Menzel. — Opladen : Westdeutscher Verlag, 1978.
230 p. ; 19 cm.　GFR79-A
Bibliography: p. 222-230.
ISBN 3-531-11460-3 : DM19.80
1. China—Economic conditions. 2. China—Polities and government. I. Title.
HC427.7.M46　79-355874
79　MARC

—ECONOMIC CONDITIONS—TO 1644

Chung, K'uai-ming.
(T'ang tai Tung-nan ti ch'ü ching chi k'ai fa ti yen chiu)
唐代東南地區經濟開發的研究 / 鍾快鳴撰. — [台中? : s. n.], 民國66 [1977]
114 p. (on double leaves) : maps ; 27 cm.
Cover title.
Reproduced from ms. copy.
"東海大學歷史研究所碩士論文"
Bibliography: p. 107–114.
1. China—Economic conditions—To 1644. I. Title.
HC427.6.C557　1977　78–843904

—ECONOMIC CONDITIONS—1644-1912

Katsudō no Nihon Dōnin.
(Nanshin no fugen)
南清之富源 / [活動の日本同人著]. — 東京：隆文館, 明治38 [1905]
4, 6, 184 p., [7] leaves of plates : ill. ; 19 cm. — (海外富源叢書)
1. China—Economic conditions—1644–1912. I. Title. II. Series: Kaigai fugen sōsho.
HC427.7.K37　1905　79–802944

Wu, Chieh, fl. 1958-　ed.
[Chung-kuo chin tai kuo min ching chi shih. Japanese]
中国近代経済史 / 呉杰[傑]編；大塚恒雄, 陳繼昌訳. — 東京：角川書店, 1978.
457 p. ; 22 cm.　Ja 78–13681
Includes bibliographical references and indexes.
¥3500
1. China—Economic conditions—1644–1912. I. Title.
Title romanized: Chūgoku kindai keizai shi.
HC427.7.W816　78–811619

--ECONOMIC CONDITIONS--1644-1912--SOURCES

(Chung-kuo ching chi shih liao, Ch'ing tai pien)
中國經濟史料·清代編 / 楊家駱主編. — 臺北：鼎文書局, 民國66 [1977]
9, 740 p. ; 26 cm. — (中國史料系編)
CONTENTS: 清史稿食貨志.—清實錄經濟史料.
1. China—Economic conditions—1644–1912—Sources. I. Yang, Chia-lo. II. Chao, Erh-hsün, 1844–1927, ed. Ch'ing shih kao. Selections. 1977. III. Ta Ch'ing li ch'ao shih lu. Selections. 1977. IV. Series: Chung-kuo shih liao hsi pien.
HC427.7.C54　79–837479

—ECONOMIC CONDITIONS—1912-1949

Akaike, Nō, b. 1879.
(Shina jihen to Yudayajin)
支那事変と猶太人 / 赤池濃著. — 東京：政経書房, 昭和14 [1939]
4, 104 p. ; 19 cm. — (国際秘密力研究叢書 ; 第4冊)
Series romanized: Kokusai himitsu ryoku kenkyū sōsho.
1. China—Politics and government—1937–1945. 2. China—Economic conditions—1912–1949. 3. Sino-Japanese Conflict, 1937–1945. 4. Jews, in China. I. Title.
DS777.53.A7366　78–815789

(Chung-kuo chan hou ching chi wen t'i yen chiu)
中國戰後經濟問題研究 / 方顯廷著[i. e. 編]. — 上海：商務印書館, 民國35 [1946]
257, [5] p. ; 21 cm.
1. China—Economic conditions—1912–1949. I. Fang, Hsien-t'ing, 1902–
HC427.8.C54　79–835587

Hawes, Grace M
The Marshall Plan for China : Economic Cooperation Administration 1948-1949 / Grace M. Hawes. — Cambridge, Mass. : Schenkman Pub. Co., c1977.
138 p. : ill. ; 24 cm.
Bibliography p. [134]-138.
ISBN 0-87073-373-7
1. Economic assistance, American—China. 2. China—Economic conditions—1912-1949. 3. United States. Economic Cooperation Administration. I. Title.
HC427.8.H39　338.91'51'073　76-40300
76[r79]rev　MARC

Hennicke, Peter.
Die entwicklungstheoretischen Konzeptionen Mao Tse-tungs : histor. Grundlagen u. sozialökonom. Bedingungen d. Entwicklungspolitik d. Volksrepublik China (1927-1957) / Peter Hennicke. — München : Minerva-Publikationen, 1978.
xlix, 608 ; 21 cm. — (Minerva-Fachserie Wirtschafts- und Sozialwissenschaften)　GFR78-A
Originally presented as the author's thesis, Bremen, under title: Historische Grundlagen und sozialökonomische Bedingungen der Entwicklungspolitik der Volksrepublik China ....
Bibliography: p. 594-608.
ISBN 3-597-10017-1 : DM78.00
1. China—Economic conditions—1912-1949. 2. China—Economic conditions—1949-1976. I. Title.
HC427.8.H46　1978　78-391441
79　MARC

Manshū Nichinichi Shimbun Sha.
(Hokushi keizai taikan)
北支經濟大観 / [編輯人　磯部和夫]. — 大連：滿洲日日新聞社, 昭和13 [1938]
6, 234 p. ; 19 cm.
1. China—Economic conditions—1912–1949. I. Title.
HC427.8.M23　1938　79–802436

Nearing, Scott, 1883-
Whither China? : An economic interpretation of recent events in the Far East / by Scott Nearing. — Westport, Conn. : Hyperion Press, 1977, c1927.
225 p. ; 23 cm. — (China studies)
Reprint of the ed. published by International Publishers, New York.
Bibliography: p. 211-214.
Includes index.
ISBN 0-88355-385-6 : $16.50
1. China—Economic conditions—1912-1949. 2. China—Industries. 3. China—Social conditions. 4. China—History—1900-　. 5. China—Foreign relations. I. Title. II. Series: China studies (Westport, Conn.)
HC427.N38　1977　330.9'51'05　75-39029
77　MARC

CHINA

—ECONOMIC CONDITIONS—1912-1949 (Continued)

United States. Bureau of Foreign and Domestic Commerce.
Commercial handbook of China / Julean Arnold. — New York : Garland Pub., 1979.
2 v. : ill. ; 24 cm. — (The Modern Chinese economy)
Reprint of the 1919-1920 ed. printed by the Govt. Print. Off., Washington, which was issued as no. 84 of the Bureau of Foreign and Domestic Commerce's Miscellaneous series.
Bibliography: v. 2, p. 460-466.
ISBN 0-8240-4264-6 : $100.00
1. China—Commerce. 2. China—Commerce—United States. 3. United States—Commerce—China. 4. China—Economic conditions—1912-1949. I. Arnold, Julean Herbert, 1875-1946. II. Title. III. Series. IV. Series: United States. Bureau of Foreign and Domestic Commerce. Miscellaneous series ; no. 84.
HF3836.U52 1979 382'.0951 78-24800
79 MARC

—ECONOMIC CONDITIONS —1912-1949—ADDRESSES, ESSAYS, LECTURES

Chūgoku Sōryokusen Kenkyūjo.
(Chūgoku tōsei keizai no sōgō kenkyū)
中國統制經濟の綜合研究 / 中國總力戰研究所編著. — 上海：中支那經濟年報刊行會, 昭和19 ₍1944₎
9, 228 p. ; 22 cm. — (中國戰爭經濟 ; 第1輯)
1. Sino-Japanese Conflict, 1937-1945—Economic aspects—China—Addresses, essays, lectures. 2. China—Economic conditions—1912-1949—Addresses, essays, lectures. I. Title. II. Series: Chūgoku sensō keizai ; dai 1-shū.
HC427.8.C52 1944 79-808766

(Tairiku keiei no senkusha o kataru)
大陸經營の先驅者を語る / 森田英亮編. — 東京：天佑書房出版部, 昭和15 ₍1940₎
2, 2, 233 p. ; 19 cm. — (評判叢書 ; 1)
Series romanized: Hyōban sōsho.
1. Japan—Foreign economic relations—China—Addresses, essays, lectures. 2. China—Foreign economic relations—Japan—Addresses, essays, lectures. 3. China—Economic conditions—1912-1949—Addresses, essays, lectures. I. Morita, Hidesuke.
HF1602.15.C6T34 79-801694

Umwälzung einer Gesellschaft : zur Sozialgeschichte der chinesischen Revolution (1911-1949) / hrsg. von Richard Lorenz. — 1. Aufl., Erstausg. — Frankfurt am Main : Suhrkamp Verlag, 1977.
511 p. ; 18 cm. — (Edition Suhrkamp ; 870) GFR***
Bibliography: p. 495-501.
1. China—Economic conditions—1912-1949—Addresses, essays, lectures. 2. China—Politics and government—1912-1949—Addresses, essays, lectures. 3. Labor and laboring classes—China—Addresses, essays, lectures. 4. Agriculture—China—Addresses, essays, lectures. I. Lorenz, Richard.
HC427.9.U44 330.9'51'04 78-393231
79 MARC

—ECONOMIC CONDITIONS —1912-1949—YEARBOOKS

(Chūgoku nenkan)
中國年鑑 宜統3- 年 ₍1911-
₍台北 天一出版社 民國62- i. e. 1973-
v. 27 cm.
Reprint of a yearbook with original title: Shina nenkan, published 1911-14 in Tokyo by Tōa Dōbunkai Chōsa Hensanbu.
Title in Chinese: Chung-kuo nien chien.
1. China—Economic conditions—1912-1949—Yearbooks. 2. China—Politics and government—1912-1949—Yearbooks. I. Tōa Dōbunkai. II. Title: Shina nenkan. III. Title: Chung-kuo nien chien.
HC426.C5192 78-643208
(MARC-S)

—ECONOMIC CONDITIONS—1949-1976

Breth, R M
Mao's China : a study of socialist economic development / ₍by₎ R. M. Breth. — Melbourne : Longman Cheshire, 1977.
viii, 178 p. : ill., portrs. ; 22 cm. Aus
Bibliography: p. 171-176.
Includes index.
ISBN 0-582-68443-9
1. China—Economic conditions—1949-1976. I. Title.
HC427.9.B68 330.9'51'05 78-320054
79 MARC

Chen, Ting Kai.
Die Volksrepublik China : Nord u. Süd in d. Entwicklung / Ting Kai Chen, mit d. Beitr., Ausgewählte Kommunen, regionale Gegensätze und Differenzierungen (aus, K. Buchanan, The transformation of the Chinese earth) u. China nach Mao von Harry Hamm. — 1. Aufl. — Stuttgart : Klett, 1977.
219 p. : ill., maps ; 23 cm. — (Länderprofile, geographische Strukturen, Daten, Entwicklungen) GFR78-A2
Edition for 1974 published under title: Nord- und Südchina.
Bibliography: p. 215-219.
ISBN 3-12-928050-2 : DM28.00
1. China—Economic conditions—1949-1976. 2. China—Social conditions—1949-1976. 3. Physical geography—China. I. Buchanan, Keith M., joint author. II. Hamm, Harry, joint author. III. Title. IV. Series.
HC427.9.C52172 1977 309.1'51'05 79-355023
79 MARC

China (People's Republic of China, 1949-). Ch'üan kuo jen min tai piao ta hui.
(Wu nien lai ti ts'ai cheng ching chi kung tso)
五年來的財政經濟工作：在第一屆全國人民代表大會第一次會議上代表們關於政府工作報告的發言選輯 / ₍著者陳雲等₎. — ₍香港?：s. n., 1978?₎
93 p. : 20 cm.
Reprint of the 1955 ed. published by Ts'ai cheng ching chi ch'u pan she, Peking.
1. China—Economic conditions—1949-1976. I. Ch'en, Yün, 1904- II. Title.
HC427.9.A49 1978 79-835582

Ching, Ch'ih.
(Tzu pen chu i kung shang yeh ti she hui chu i kai tsao)
资本主义工商业的社会主义改造 / 景池. — 第1版. — 上海：上海人民出版社, 1976.
80 p. ; 19 cm.
Includes bibliographical references.
$0.16
1. China—Economic conditions—1949- 2. China—Industries—1949- I. Title.
HC427.9.C56548 79-835450

Dragičević, Ivo.
Kina između siromaštva i industrijskog razvitka / Ivo Dragičević. — Zagreb : Stvarnost, 1978.
407 p., ₍28₎ leaves of plates : ill. (some col.) ; 23 cm. Yu***
Includes bibliographical references and indexes.
1. China—Economic conditions—1949-1976. I. Title.
HC427.9.D65 79-352350
79 MARC

Hennicke, Peter.
Die entwicklungstheoretischen Konzeptionen Mao Tse-tungs : histor. Grundlagen u. sozialökonom. Bedingungen d. Entwicklungspolitik d. Volksrepublik China (1927-1957) / Peter Hennicke. — München : Minerva-Publikationen, 1978.
xlix, 608 ; 21 cm. — (Minerva-Fachserie Wirtschafts- und Sozialwissenschaften) GFR78-A
Originally presented as the author's thesis, Bremen, under title: Historische Grundlagen und sozialökonomische Bedingungen der Entwicklungspolitik der Volksrepublik China ....
Bibliography: p. 594-608.
ISBN 3-597-10017-1 : DM78.00
1. China—Economic conditions—1912-1949. 2. China—Economic conditions—1949-1976. I. Title.
HC427.8.H46 1978 78-391441
79 MARC

Howe, Christopher.
China's economy : a basic guide / ₍by₎ Christopher Howe. — London : Elek, 1978.
xxxvii, 248 p. : ill., maps ; 25 cm. GB78-23231
Includes bibliographical references and index.
ISBN 0-236-40115-7 : £7.50
1. China—Economic conditions—1949-1976. I. Title.
HC427.9.H66 1978b 330.9'51'05 79-301285
79 MARC

Hung, Yu-chiao.
An appraisal of the Maoists' "fourth five-year plan" / ₍by Hung Yu-chiao₎. — Taipei : Chung Hwa Information Service, ₍1978?₎
22 p. ; 19 cm.
1. China—Economic policy—1949-1976. 2. China—Economic conditions—1949-1976. I. Title.
HC427.9.H898 338.951 78-103851
78 MARC

Lardy, Nicholas R
Economic growth and distribution in China / Nicholas R. Lardy. — Cambridge ₍Eng.₎ ; New York : Cambridge University Press, 1978.
x, 244 p. ; 24 cm.
Bibliography: p. 213-237.
Includes index.
ISBN 0-521-21904-3 : $23.00
1. China—Economic conditions—1949- 2. China—Economic policy—1949- I. Title.
HC427.9.L36 330.9'51'05 77-27508
77 MARC

Lefebvre, Alain.
Le Système économique chinois / Alain Lefebvre. — Toulouse : Privat, 1978.
216 p. : map ; 23 cm. F78-20004
Bibliography: p. ₍189₎-199.
ISBN 2-7089-2367-6 : 47.00F
1. China—Economic conditions—1949-1976. 2. China—Economic conditions—1976- 3. China—Economic policy—1949-1976. 4. China—Economic policy—1976- I. Title.
HC427.9.L42 330.9'51'05 79-347381
79 MARC

Li, Mien.
(Hsin Chung-kuo ching chi wen t'i chiang hua)
新中國經濟問題講話 / 李勉著. — ₍香港?：s. n., 1978?₎
2, 114 p. : 21 cm.
Reprint of the 1949 ed. published by Chih yüan shu chü, Hongkong.
1. China—Economic conditions—1949-1976. I. Title.
HC427.9.L526 1978 79-835586

Liang, Jen-ts'ai.
(Hua nan ti ch'ü ching chi ti li)
华南地区经济地理：广东, 广西, 福建 / 梁仁彩, 黄勉, 申維丞編写 ; 孫敬之主編. — ₍s. l. : s. n., 1977?₎
vi, 147 leaves, ₍12₎ leaves of plates : ill. ; 32 cm.
Reprint of the 1959 ed. published by K'o hsüeh ch'u pan she, Peking, and issued in series: Chung-kuo k'o hsüeh yüan Chung-hua ti li chih ching chi ti li ts'ung shu ; 4.
Includes bibliographical references.
1. China—Economic conditions—1949- I. Huang, Mien, joint author. II. Shen, Wei-ch'eng, joint author. III. Sun, Ching-chih. IV. Title. V. Series: Chung-kuo k'o hsüeh yüan. Chung-hua ti li chih pien chi pu. Chung-kuo k'o hsüeh yüan Chung-hua ti li chih ching chi ti li ts'ung shu ; 4.
HC427.9.L553 1977 79-839621

Lin, Terng-Yaw, 1946-
Wirtschaftsverwaltung der VR China / vorgelegt von Terng-Yaw Lin. — ₍s. l. : s. n., 1978₎ (Köln : R. J. Hundt)
204 p. ; 21 cm. GFR***
Thesis—Cologne.
Vita.
Bibliography: p. 184-204.
1. China—Economic policy—1949-1976. 2. China—Economic policy—1976- 3. China—Economic conditions—1949-1976. 4. China—Economic conditions—1976- I. Title.
HC427.9.L568 330.9'51'05 79-357608
79 MARC

Mader, Jean.
Mille questions, mille réponses sur la République populaire de Chine / Jean Mader ; ₍publié par l'Association des élèves de l'Institut supérieur de gestion₎. — ₍Paris₎ (B, rue de Lota, 75116) : ₍Association des élèves de l'Institut supérieur de gestion₎, 1977.
126 p., ₍20₎ p. of col. plates ; 21 cm. F78-1824
30.00F
1. China—Economic conditions—1949-1976. 2. China—Social conditions. I. Title.
HC427.9.M15 330.9'51'05 79-343311
79 MARC

Maeda, Hisao, 1919-
(Chūgoku no keizai)
中国の経済：四つの近代化 / 前田寿夫. — ₍東村山₎：教育社 ; 東京：販売 教育社出版サービス, ₍1978₎
168 p. : ill. ; 18 cm. — (時事問題解説 ; no. 59) (入門新書)
Ja 78-23806
Cover title.
Series 2 romanized: Nyūmon shinsho.
¥400
1. China—Economic policy—1949-1976. 2. China—Economic conditions—1949-1976. I. Title. II. Series: Jiji mondai kaisetsu ; 59.
HC427.9.M16 79-800064

*79 J

Private enterprise in China : a catalogue of expropriation. — ₍s.l. : s.n.₎, 1956.
19 p. ; 24 cm.
Cover title.
Includes bibliographical references.
1. Business enterprises—China. 2. China—Economic conditions—1949- 3. Government ownership—China.
HC427.9.P68 79-311609
79 MARC

CHINA

—ECONOMIC CONDITIONS—1949-1976 (Continued)

Sun, Ching-chih, ed.

(Hua-tung ti ch'ü ching chi ti li)

华东地区經济地理：上海·江苏·安徽·浙江 / 孙敬之主編；胡序威 ... [et al.] 編写；[編著者中国科学院中华地理志編輯部]. — 第1版. — 北京：科学出版社：新华书店总經售, 1959.

vi, 153 p., [4] leaves. — (中国科学院中华地理志經济地理丛書；5)

Photocopy. vi, 153 p. on 158 leaves. 32 x 23 cm.

1. China—Economic conditions—1949-1976. I. Hu, Hsü-wei, writer on economic geography. II. Chung-kuo k'o hsüeh yüan. Chung-hua ti li chih pien chi pu. III. Title. IV. Series: Chung-kuo k'o hsüeh yüan. Chung-hua ti li chih pien chi pu. Chung-kuo k'o hsüeh yüan Chung-hua ti li chih ching chi ti li ts'ung shu ; 5.

HC427.9.S86 1959a 79-839994

Tomiyama, Eikichi, 1926-

(Chūgoku keizai zusetsu)

中国経済図説 / 富山栄吉，須藤正親編著. — 東京：東研，1978.

vii, 326 p. : ill. ; 22 cm. Ja 79-601

Bibliography: p. 322-326.

¥3200

1. China—Economic conditions—1949-1976. I. Sudō, Masachika, 1941- joint author. II. Title.

HC427.9.T594 79-803693

United States. Central Intelligence Agency. National Foreign Assessment Center.

China, economic indicators / National Foreign Assessment Center. — Washington : The Center : may purchase from Photoduplication Service, Library of Congress, 1978.

iii, 50 p., [1] fold. leaf of plates : ill. ; 27 cm. — (A Reference aid - National Foreign Assessment Center)

"ER 78-10750."

Chiefly tables.

1. Economic indicators—China. 2. China—Economic conditions—1949-1976. 3. China—Economic conditions—1976- 4. China—Statistics. I. Title.

HC427.9.U462 1978 330.9'51'05 79-602325
79 MARC

United States. Central Intelligence Agency. National Foreign Assessment Center.

China : post-Mao search for civilian industrial technology / National Foreign Assessment Center. — [Washington] : The Center, 1979.

viii, 17 p. : graphs ; 27 cm. — (A research paper - Central Intelligence Agency, National Foreign Assessment Center)

"ER 79-10020U."

1. Technology—China. 2. China—Economic conditions—1949- I. Title. II. Series: United States. Central Intelligence Agency. National Foreign Assessment Center. A research paper - Central Intelligence Agency, National Foreign Assessment Center.

T27.C5U54 1979 338.4'7'60951 79-601761
79 MARC

—ECONOMIC CONDITIONS —1949-1976—ADDRESSES, ESSAYS, LECTURES

Ch'ai, Sung-lin, 1934-

(Ts'ung ching chi kuan tien k'an kung ch'an chu i li lun p'o ch'an)

從經濟觀點看共產主義理論破產 / 柴松林講；教育部社會教育司編印. — [台北]：該司，民國65 [1976]

18 p. : port. ; 20 cm. — (文化講座專集；85)

1. Communism—Addresses, essays, lectures. 2. Taiwan—Economic conditions—Addresses, essays, lectures. 3. China—Economic conditions—1949-1976—Addresses, essays, lectures. I. Title. II. Series: Wen hua chiang tso chuan chi ; 85.

HX44.C39 78-841437
78 C

Chin, Hsiung-pai, 1903-

(Chūkyō no jūdai mondai)

中共の十大問題 / 金雄白；本郷賀一訳. — 東京：時事通信社：発売所 同社出版局，昭和38 [1963]

443 p. ; 19 cm.

1. China—Politics and government—1949-1976. 2. China—Economic conditions—1949-1976. 3. Communism—China. I. Title.

DS777.55.C446638 78-814380

Communication and development in China / Godwin Chu ... [et al.]. — Honolulu : East-West Center, 1976.

xiv, 105 p. ; 23 cm. — (Communication monographs ; 1)

"Four papers from thefirst advanced summer seminar (1974) on 'Communication and Development'."

Bibliography: p. 105.

1. China—Economic conditions—1949- —Addresses, essays, lectures. 2. Communication—China—Addresses, essays, lectures. I. Chu, Godwin C., 1927- . II. Series: Communication monographs (Honolulu) ; 1.

HC427.9.C62 330.9'51'05 76-45831
76 MARC

Die Wirtschaft Osteuropas und der VR China : 1970-1980 : Bilanz u. Perspektiven / Hans-Hermann Höhmann (Hrsg.). — 1. Aufl. — Stuttgart ; Berlin ; Köln ; Mainz : Kohlhammer, 1978.

380 p. ; 24 cm. GFR78-A

Includes bibliographical references and index.

ISBN 3-17-004517-2

1. Europe, Eastern—Economic conditions—Addresses, essays, lectures. 2. Europe, Eastern—Economic policy—Addresses, essays, lectures. 3. China—Economic conditions—1949-1976—Addresses, essays, lectures. 4. China—Economic policy—1949-1976—Addresses, essays, lectures. I. Höhmann, Hans-Hermann.

HC244.W576 78-393273
79 MARC

—ECONOMIC CONDITIONS —1949-1976—HANDBOOKS, MANUALS, ETC.

Bydžovský, Zdeněk.

ČLR-Čínska lidová republika / [autor] Zdeněk Bydžovský a kol. — 1. vyd. — Praha : Pressfoto, 1978.

292, [3] p. : ill. ; 21 cm. — (Obchodně ekonomické sborníky : Řada SZ ; sv. 40) Cz78

Bibliography: p. 291-292.

Kčs22.00

1. China—Handbooks, manuals, etc. 2. China—Economic conditions—1949- —Handbooks, manuals, etc. 3. China—Foreign economic relations—Czechoslovakia—Handbooks, manuals, etc. 4. Czechoslovakia—Foreign economic relations—China—Handbooks, manuals, etc. I. Title. II. Title: Čínská lidová republika.

DS777.55.B92 79-358341
*79 MARC

—ECONOMIC CONDITIONS—1976-

Bettelheim, Charles.

Questions sur la Chine après la mort de Mao Tsé-toung / Charles Bettelheim. — Paris : F. Maspero, 1978.

153 p. ; 20 cm. — (Économie et socialisme ; 35) F***

Includes bibliographical references.

ISBN 2-7071-1026-4 : 25.00F

1. China—Economic conditions—1976- 2. China—Politics and government—1976- 3. Socialism in China. I. Title.

HC427.92.B47 335.43'0951 78-387336
79 MARC

Lefebvre, Alain.

Le Système économique chinois / Alain Lefebvre. — Toulouse : Privat, 1978.

216 p. : map ; 23 cm. F78-20004

Bibliography: p. [189]-199.

ISBN 2-7089-2367-6 : 47.00F

1. China—Economic conditions—1949-1976. 2. China—Economic conditions—1976- . 3. China—Economic policy—1949-1976. 4. China—Economic policy—1976- I. Title.

HC427.9.L42 330.9'51'05 79-347381
79 MARC

Lin, Terng-Yaw, 1946-

Wirtschaftsverwaltung der VR China / vorgelegt von Terng-Yaw Lin. — [s. l. : s. n., 1978] (Köln : R. J. Hundt)

204 p. ; 21 cm. GFR***

Thesis—Cologne.

Vita.

Bibliography: p. 184-204.

1. China—Economic policy—1949-1976. 2. China—Economic policy—1976- . 3. China—Economic conditions—1949-1976. 4. China—Economic conditions—1976- I. Title.

HC427.9.L568 330.9'51'05 79-357608
79 MARC

Nitchū Keizai Kyōkai.

(Shindankai o mukaeta Chūgoku keizai tembō)

新段階を迎えた中国経済展望. — 東京：日中経済協会，1978.

194, 5 p. ; 25 cm. — (中国経済関係調査報告書) (日本経報；no. 106) Ja 78-19895

Cover title.

"52-調-13"

Includes bibliographical references.

¥690

1. China—Economic conditions—1976- 2. Economic forecasting—China. 3. China—Economic policy—1976- I. Title. II. Series: Chūgoku keizai kankei chōsa hōkokusho. III. Series: Nitchū Keizai Kyōkai. Nitchūkei hō ; no 106.

HC427.92.N57 1978 78-814373

United States. Central Intelligence Agency. National Foreign Assessment Center.

China, economic indicators / National Foreign Assessment Center. — Washington : The Center : may purchase from Photoduplication Service, Library of Congress, 1978.

iii, 50 p., [1] fold. leaf of plates : ill. ; 27 cm. — (A Reference aid - National Foreign Assessment Center)

"ER 78-10750."

Chiefly tables.

1. Economic indicators—China. 2. China—Economic conditions—1949-1976. 3. China—Economic conditions—1976- 4. China—Statistics. I. Title.

HC427.9.U462 1978 330.9'51'05 79-602325
79 MARC

United States. Congress. House. Committee on International Relations. Subcommittee on Asian and Pacific Affairs.

A new realism : factfinding mission to the People's Republic of China, July 3-13, 1978 : report / by the Subcommittee on Asian and Pacific Affairs to the Committee on International Relations, U.S. House of Representatives. — Washington : U.S. Govt. Print. Off., 1978.

vii, 132 p. ; 24 cm.

At head of title: 95th Congress, 2d session. Committee print.

Includes bibliographical references.

1. United States—Relations (general) with China. 2. China—Relations (general) with the United States. 3. China—Economic conditions—1976- I. United States. Congress. House. Committee on International Relations. II. Title.

E183.8.C5U48 1978 301.29'73'051 79-600630
79 MARC

—ECONOMIC CONDITIONS—1976- —ADDRESSES, ESSAYS, LECTURES

China im Spiegel der Weltpresse : Dokumentation ausgew. Beitr. u. Daten zu polit. u. wirtschaftl. Entwicklungen 1977 u. 1978 / Wilhelm G. Franken, unter Mitarb. von Detlev Passarge, Marlies Prigge, Hubert-Günter Striefler. — Hamburg : Verlag Weltarchiv, 1978.

iv, 820 p. ; 29 cm. — (Veröffentlichungen des HWWA-Institut für Wirtschaftsforschung-Hamburg : Aktuelle Fachinformation) GFR79-A

"Abgeschlossen: 31.10.1978."

English, French, German, or Italian.

Research carried out by HWWA-Institut für Wirtschaftsforschung-Hamburg, Hauptabteilung Dokumentation, Bibliothek und Archive.

Bibliography: p. 803-817.

ISBN 3-87895-179-5 : DM68.00

1. China—Politics and government—1976- —Addresses, essays, lectures. 2. China—Economic conditions—1976- —Addresses, essays, lectures. 3. China—Foreign opinion—Addresses, essays, lectures. I. Franken, Wilhelm G. II. HWWA-Institut für Wirtschaftsforschung-Hamburg. Hauptabteilung Dokumentation, Bibliothek und Archive.

DS779.26.C4735 320.9'51'05 79-369082
79 MARC

Chinese economy post-Mao : a compendium of papers / submitted to the Joint Economic Committee, Congress of the United States. — Washington : U.S. Govt. Print. Off., 1978-

v. : ill. ; 24 cm.

At head of title: 95th Congress, 2d session. Joint committee print.

Includes bibliographical references.

CONTENTS: v. 1. Policy and performance.

1. China—Economic policy—1976- —Addresses, essays, lectures. 2. China—Economic conditions—1976- —Addresses, essays, lectures. I. United States. Congress. Joint Economic Committee.

HC427.9.C56546 330.9'51'05 79-600609
79 MARC

(Hua chu hsi tai ling wo men pan hao she hui chu i ts'ai mao)

华主席带领我们办好社会主义财贸 / 湖南省革命委员会财贸办公室编. — 第1版. — [北京]：中国财政经济出版社，1978.

132 p. ; 21 cm.

$0.29

1. China—Economic conditions—1976- —Addresses, essays, lectures. I. Hunan, China (Province). Ko ming wei yüan hui. Ts'ai mao pan kung shih.

HC427.92.H8 79-836868

(Ta lan "ssu jen pang" sheng ch'an li ta chieh fang)

打烂"四人帮"生产力大解放 / 上海第三十三棉纺织厂等编写. — [香港? : s. n., 1978?]

56 p. ; 21 cm.

Reprint of the 1977 ed. published by Shang-hai jen min ch'u pan she, Shanghai.

1. China—Economic conditions—1976- —Addresses, essays, lectures. I. Shang-hai ti san shih san mien fang chih ch'ang.

HC427.92.T3 1978 79-837819

—ECONOMIC CONDITIONS—1976- —CASE STUDIES

Erasmus Universiteit. China Studiegroep.

Hoe ver is China? : protocollen en artikelen naar aanleiding van de studiereis naar de Volksrepubliek China in Oktober 1977 van de China Studiegroep van de Erasmus Universiteit te Rotterdam / organisatie van de reis De China Studiegroep, in samenwerking met het Bureau Buitenlandse Betrekkingen van de Erasmus Universiteit Rotterdam ; comité van aanbeveling voor de reis, H. C. Bos ... [et al.]. — [3000 DR] Rotterdam, (Postbus 1738) : [Bureau Buitenlandse Betrekkingen], 1978.

x, 619 p. : with ill. ; 21 cm. Ne78-34

Includes bibliographies.

fl 17.50

1. China—Social conditions—Case studies. 2. China—Economic conditions—1976- —Case studies. 3. Medical care—China—Case studies. I. Bos, Hendricus Cornelis, 1926- II. Title.

HN737.E72 1978 79-342561
*79 MARC

CHINA (Continued)

—ECONOMIC CONDITIONS—1976-
—PERIODICALS

United States. Industry and Trade Administration. Office of East-West Country Affairs.
China's economy and foreign trade. 1977/78-
[Washington] U. S. Dept. of Commerce, Industry and Trade Administration, Office of East-West Country Affairs; for sale by the Supt. of Docs., U. S. Govt. Print. Off.
v. 27 cm. annual.
Continues: Chinese economy and foreign trade perspective, ISSN 0147-6351.
Key title: China's economy and foreign trade, ISSN 0192-5555.
1. China — Commerce — Periodicals. 2. China — Economic conditions—1976- —Periodicals. 3. China—Commerce—United States—Periodicals. 4. United States — Commerce — China — Periodicals. I. Title.
HF3836.5.U53a 382'.0951 79-643027
MARC-S

—ECONOMIC CONDITIONS—
CONGRESSES

Conference on Modern Chinese Economic History, T'ai-pei, 1977.
Conference on Modern Chinese Economic History, August 26-29, 1977 : [proceedings]. — Taipei : Institute of Economics, Academia Sinica, [1977?]
vi, 735 p. : ill. ; 26 cm.
Title also in Chinese: Chung-kuo chin tai ching chi shih hui i.
Includes bibliographical references.
1. China—Economic conditions—Congresses. I. Chung yang yen chiu yüan. Ching chi yen chiu so.
HC426.C65 1977 330.9'51'05 79-301773
79 MARC

Conference on Modern Chinese Economic History, T'ai-pei, 1977.
Modern Chinese economic history : proceedings of the Conference on Modern Chinese Economic History, Academia Sinica, Taipei, Taiwan, Republic of China, August 26-29, 1977 / edited by Chi-ming Hou, Tzong-shian Yu. — Taipei : Institute of Economics, Academia Sinica ; Seattle : distributed by University of Washington Press, c1979.
xiv, 9-678 p. : graphs ; 26 cm.
Includes bibliographical references.
1. China—Economic conditions—Congresses. I. Hou, Chi-ming, 1924- . II. Yü, Tsung-hsien. III. Title.
HC426.C65 1977a 330.9'51'05 79-114995
79 MARC

Studies in Chinese society / edited by Arthur P. Wolf ; contributors, Emily M. Ahern ... [et al.]. — Stanford, Calif. : Stanford University Press, 1978.
xii, 372 p. : ill. ; 23 cm.
Papers, from conferences held between 1966 and 1973, which were first published in the seven volume series: Studies in Chinese society.
Includes bibliographical references and index.
ISBN 0-8047-1006-6
1. China—Social conditions—Congresses. 2. China—Economic conditions—Congresses. I. Wolf, Arthur P. II. Ahern, Emily M.
HN733.S8 309.1'51 78-62272
00 MARC

—ECONOMIC CONDITIONS
—PERIODICALS

Minami Manshū Tetsudō Kabushiki Kaisha. Sōmubu. Chōsaka.
(Chōsa shiryō)
調査資料　第10-　輯
[大連]　南滿洲鐵道株式會社總務部調査課　[大正8-　i. e. 1919-
v. 22 cm.
Earlier vols. classified separately in L. C.
Superseded by: Minami Manshū Tetsudō Kabushiki Kaisha. Shomubu. Chōsaka. Chōsa shiryō.

1. China—Economic conditions—Periodicals. 2. China Politics and government—Periodicals. 3. East Asia—Periodicals.
HC426.M562a 75-791578
(MARC-S)

—ECONOMIC POLICY

Kojima, Seiichi, 1895-
(Nihon senji keizai ron)
日本戰時經濟論 / 小島精一著. — 東京：春秋社：發賣所　松柏館, 昭和14 [1939]
6, 277 p. ; 23 cm.

1. Japan—Economic policy. 2. Industrial mobilization—Japan. 3. China—Economic policy. 4. Sino-Japanese Conflict, 1937-1945—Economic aspects—East Asia. I. Title.
HC462.8.K62157 79-802787

Matsumoto, Shigekazu, 1934-
China's economic policy towards Asia, 1949-72 / Shigekazu Matsumoto. — Tokyo : Institute of Developing Economies : [sold by] Maruzen Co., c1977.
34 p. ; 24 cm. — (IDE special paper ; no. 8) Ja***
Cover title.
Includes bibliographical references.
$2.00 (U.S.)
1. China—Foreign economic relations—Asia, Southeastern. 2. Asia, Southeastern—Foreign economic relations—China. 3. China—Economic policy. I. Title. II. Series: Ajia Keizai Kenkyūjo, Tokyo. IDE special paper ; no. 8.
HF1604.Z5A73 338.91'51'05 78-315379
78 MARC

--ECONOMIC POLICY--1949-
--ADDRESSES, ESSAYS, LECTURES

(Chung-kuo hsien tai hua ti pu fa)
中國現代化的步伐：學聯第五屆中國周論文集 = China towards modernization : collected essays for the 5th China Week. — 九龍：香港專上學生聯會, 1977.
507 p. : ill. ; 21 cm.
Chinese or English.
$12.00
1. China—Economic policy—1949- —Addresses, essays, lectures. 2. Education—China—Addresses, essays, lectures. 3. Science—China—Addresses, essays, lectures. I. Hsiang-kang chuan shang hsüeh sheng lien hui. II. Title: Hsüeh lien ti wu chieh Chung-kuo chou lun wen chi. III. Title: China towards modernization.
HC427.9.C5824 338.951 79-837123

—ECONOMIC POLICY—1949-1976

Ch'ai, Mao.
(Tsen yang jen shih wo kuo hsien chieh tuan ti shang p'in chih tu)
怎样认识我国现阶段的商品制度 / 柴茂. — [香港? : s. n., 1978?]
12 p. ; 20 cm.
Reprint of the 1975 ed. published by Pei-ching jen min ch'u pan she, in series Cheng chih ching chi hsüeh chiang hua.
Includes bibliographical references.
1. Commercial products—China. 2. China—Economic policy—1949-1976. I. Title. II. Series: Cheng chih ching chi hsüeh chiang hua.
HF1040.9.C6C44 1978 79-837279

Christiansen, Eigil, 1945-
Kina—industri, plan, menneske / [af] Eigil Christiansen og Kjeld A. Larsen ; [Kort og diagrammer er udført af Mogens Skjoldager]. — [København] : Gyldendal, 1978.
103 p. : ill. ; 30 cm. — (Regionalgeografiske analyser) D78-21
Bibliography: p. 92-93.
Includes index.
ISBN 8701588311 : kr68.50
1. China—Industries—1949-1976. 2. China—Economic policy—1949- I. Larsen, Kjeld Allan, 1942- joint author. II. Title. III. Series.
HC427.9.C58116 78-380758
79 MARC

Hung, Yu-chiao.
An appraisal of the Maoists' "fourth five-year plan" / [by Hung Yu-chiao]. — Taipei : Chung Hwa Information Service, [1978?]
22 p. ; 19 cm.
1. China—Economic policy—1949-1976. 2. China—Economic conditions—1949-1976. I. Title.
HC427.9.H898 338.951 78-103851
78 MARC

Kondō, Yasuo, 1899-
(Gendai Chūgoku keizai ron)
現代中国経済論 / 近藤康男著. — 東京：農山漁村文化協会, 1978.
6, 256 p. ; 22 cm. Ja 78-21001
Bibliography: p. 245-247.
¥2200

1. China—Economic policy—1949-1976. 2. China—Economic policy—1976- I. Title.
HC427.9.K59 78-815929

*78 J

Lardy, Nicholas R
Economic growth and distribution in China / Nicholas R. Lardy. — Cambridge [Eng.] ; New York : Cambridge University Press, 1978.
x, 244 p. ; 24 cm.
Bibliography: p. 213-237.
Includes index.
ISBN 0-521-21904-3 : $23.00
1. China—Economic conditions—1949- . 2. China—Economic policy—1949- I. Title.
HC427.9.L36 330.9'51'05 77-27508
77 MARC

Lefebvre, Alain.
Le Système économique chinois / Alain Lefebvre. — Toulouse : Privat, 1978.
216 p. : map ; 23 cm. F78-20004
Bibliography: p. [189]-199.
ISBN 2-7089-2367-6 : 47.00F
1. China—Economic conditions—1949-1976. 2. China—Economic conditions—1976- . 3. China—Economic policy—1949-1976. 4. China—Economic policy—1976- I. Title.
HC427.9.L42 330.9'51'05 79-347381
79 MARC

Lin, Terng-Yaw, 1946-
Wirtschaftsverwaltung der VR China / vorgelegt von Terng-Yaw Lin. — [s. l. : s. n., 1978] (Köln : R. J. Hundt) GFR***
204 p. ; 21 cm.
Thesis—Cologne.
Vita.
Bibliography: p. 184-204.
1. China—Economic policy—1949-1976. 2. China—Economic policy—1976- 3. China—Economic conditions—1949-1976. 4. China—Economic conditions—1976- I. Title.
HC427.9.L568 330.9'51'05 79-357608
79 MARC

Maeda, Hisao, 1919-
(Chūgoku no keizai)
中国の経済：四つの近代化 / 前田寿夫. — [東村山]：教育社；東京：販売　教育社出版サービス, [1978]
168 p. : ill. ; 18 cm. — (時事問題解説 ; no. 59) (入門新書)
Ja 78-23806
Cover title.
Series 2 romanized: Nyūmon shinsho.
¥400
1. China—Economic policy—1949-1976. 2. China—Economic conditions—1949-1976. I. Title. II. Series: Jiji mondai kaisetsu ; 59.
HC427.9.M16 79-800064

*79 J

Ting, Kuo-chen, fl. 1950- ed.
(Hsin kung shang ts'an k'ao tzu liao)
新工商參考資料　丁國楨編　北京　五十年代出版社　1950.
2, 5, 238 p. tables. 18 cm.

1. China—Economic policy—1949-1976. 2. China—Industries—1949-1976. 3. China—Commerce. I. Title.
HC427.9.T57 C 64-1039

Ting, Kuo-chen, fl. 1950- ed.
(Hsin kung shang ts'an k'ao tzŭ liao)
新工商參考資料　丁國楨編　增訂本　北京　五十年代出版社　1952.
2, 6, 313 p. tables. 19 cm.

1. China—Economic policy—1949-1976. 2. China—Industries—1949-1976. 3. China—Commerce. I. Title.
HC427.9.T57 1952 C 60-5309
rev
[r78]

—ECONOMIC POLICY—1949-1976—
ADDRESSES, ESSAYS, LECTURES

China's finance and trade : a policy reader / Gordon Bennett, editor. — White Plains, N.Y. : M. E. Sharpe, c1978.
xiii, 250 p. ; 23 cm.
"Selections from Chinese economic studies."
Bibliography: p. 247-249.
ISBN 0-87332-115-4
1. China—Economic policy—1949-1976—Addresses, essays, lectures. 2. Finance—China—Addresses, essays, lectures. 3. China—Commercial policy—Addresses, essays, lectures. I. Bennett, Gordon A. II. Chinese economic studies.
HC427.9.C56543 330.9'51'05 77-99080
78 MARC

Chinese economic planning : translations from Chihua ching-chi / edited with an introd. by Nicholas R. Lardy ; translated by K. K. Fung. — White Plains, N.Y. : M. E. Sharpe, 1978.
xii, 268 p. ; 24 cm.
Includes bibliographical references.
ISBN 0-87332-117-8
1. China—Economic policy—1949- —Addresses, essays, lectures. I. Lardy, Nicholas R. II. Chi hua ching chi.
HC427.9.C56545 1978 330.9'51'05 78-52292
78 MARC

CHINA

—ECONOMIC POLICY—1949-1976
—ADDRESSES, ESSAYS, LECTURES
(Continued)

(Kuan yü kung shang yeh ti cheng ts'e)
關於工商業的政策 / 陳伯達等著. — ₍香港? : s. n., 1978?₎
94 p. ; 22 cm.
Reprint of the 1949 ed. published by Chung-kuo ch'u pan she, Hongkong, in series: Chieh fang ts'ung shu : ti 1 chi, ti 3 chung.

1. China—Economic policy—1949-1976—Addresses, essays, lectures. I. Ch'en, Po-ta, 1905-
HC427.9.K815 1978 79-836555

Mao, Tse-tung, 1893-1976.
(Chung-kuo chiang yao ch'u hsien i ko ta yao chin)
中国将要出现一个大跃进 ; 关于农业机械化问题的一封信 / 毛泽东. — 香港 : 生活, 讀書, 新知三联书店香港分店, 1977.
5 p. ; 21 cm.

1. China—Economic policy—1949-1976—Addresses, essays, lectures. 2. Farm mechanization—China—Addresses, essays, lectures. I. Mao, Tse-tung, 1893-1976. Kuan yü nung yeh chi hsieh hua wen t'i ti i feng hsin. 1977. II. Title.
HC427.9.M224 1977 78-843862

Die Wirtschaft Osteuropas und der VR China : 1970-1980 : Bilanz u. Perspektiven / Hans-Hermann Höhmann (Hrsg.). — 1. Aufl. — Stuttgart ; Berlin ; Köln ; Mainz : Kohlhammer, 1978.
380 p. ; 24 cm. GFR78-A
Includes bibliographical references and index.
ISBN 3-17-004517-2
1. Europe, Eastern—Economic conditions—Addresses, essays, lectures. 2. Europe, Eastern—Economic policy—Addresses, essays, lectures. 3. China—Economic conditions—1949-1976—Addresses, essays, lectures. 4. China—Economic policy—1949-1976—Addresses, essays, lectures. I. Höhmann, Hans-Hermann.
HC244.W576 78-393273
79 MARC

—ECONOMIC POLICY—1949-1976
—MISCELLANEA

(Jen min kung she ching chi cheng ts'e wen ta)
人民公社經济政策問答 / 广东人民出版社編輯. — 第1版. — 广州 : 該社, 1958.
28 p. ; 19 cm. — (为人民公社宣传参考材料)
Colophon title.
Series romanized: Pan jen min kung she hsüan ch'uan ts'an k'ao ts'ai liao.

1. China—Economic policy-1949-1976—Miscellanea. 2. Communes (China)—Miscellanea. I. Kuang-tung jen min chu pan she, Canton, China.
HC427.9.J364 78-840579

78 C

—ECONOMIC POLICY—1976-

Akeno, Yoshio.
(Chūgoku no taigai keizai kōryū no tenkai)
中国の対外経済交流の展開 / 明野義夫著. — 東京 : 大東文化大学東洋研究所, 昭和53 ₍1978₎
264 p. : graphs ; 22 cm. Ja***
Includes bibliographical references.
¥3500

1. China—Commercial policy. 2. China—Economic policy—1976- I. Title.
HF1604.A43 78-816803

Ch'ing-shan.
(She hui chu i chien she tsung lu hsien chiang hua)
社会主义建設总路线讲话 / 青山. — ₍香港? s. n., 1979?₎
149 p. ; 21 cm.
Reprint of the 1978 ed. published by Kuang-tung jen min ch'u pan she.

1. China—Economic policy—1976- I. Title.
HC427.92.C47 1979 79-838664

Kondō, Yasuo, 1899-
(Gendai Chūgoku keizai ron)
現代中国経済論 / 近藤康男著. — 東京 : 農山漁村文化協会, 1978.
6, 256 p. ; 22 cm. Ja 78-21001
Bibliography: p. 245-247.
¥2200

1. China—Economic policy—1949-1976. 2. China—Economic policy—1976- I. Title.
HC427.9.K59 78-815929

Ku, Huan-chang.
(T'an t'an fa chan kuo min ching chi tsung fang chen)
谈谈发展国民经济总方针 / 顾焕章, 章宗礼. — ₍香港? : s. n., 1978?₎
2, 71 p. ; 21 cm.
Reprint of the 1978 ed. published by Chiang-su jen min ch'u pan she.

1. China—Economic policy—1976- I. Chang, Tsung-li, joint author. II. Title.
HC427.92.K79 1978 79-837657

Lefebvre, Alain.
Le Système économique chinois / Alain Lefebvre. — Toulouse : Privat, 1978.
216 p. : map ; 23 cm. F78-20004
Bibliography: p. ₍189₎-199.
ISBN 2-7089-2367-6 : 47.00F
1. China—Economic conditions—1949-1976. 2. China—Economic conditions—1976- . 3. China—Economic policy—1949-1976. 4. China—Economic policy—1976- I. Title.
HC427.9.L42 330.9'51'05 79-347381
79 MARC

Lin, Terng-Yaw, 1946-
Wirtschaftsverwaltung der VR China / vorgelegt von Terng-Yaw Lin. — ₍s. l. : s. n., 1978₎ (Köln : R. J. Hundt)
204 p. ; 21 cm. GFR***
Thesis—Cologne.
Vita.
Bibliography: p. 184-204.
1. China—Economic policy—1949-1976. 2. China—Economic policy—1976- . 3. China—Economic conditions—1949-1976. 4. China—Economic conditions—1976- I. Title.
HC427.9.L568 330.9'51'05 79-357608
79 MARC

Maeda, Isao, 1942-
(Shinnittetsu Chūgoku kensetsutai)
新日鉄中国建設隊 : 日中巨大プロジェクトはいかに組み上げられたか / 前田勲著. — 東京 : こう書房, 1978.
218 p. : ill. ; 19 cm. — (Kou business) Ja 79-1271
Bibliography: p. 218.
¥980
1. China—Economic policy—1976- 2. Technical assistance, Japanese—China. 3. Shin Nihon Seitetsu Kabushiki Kaisha. I. Title.
HC427.92.M33 79-804224

1 J

Nitchū Keizai Kyōkai.
(Shindankai o mukaeta Chūgoku keizai tembō)
新段階を迎えた中国経済展望. — 東京 : 日中経済協会, 1978.
194, 5 p. ; 25 cm. — (中国経済関係調査報告書) (日本経報 ; no. 106) Ja 78-19895
Cover title.
"52-調-13"
Includes bibliographical references.
¥690

1. China—Economic conditions—1976- 2. Economic forecasting—China. 3. China—Economic policy—1976- I. Title. II. Series: Chūgoku keizai kankei chōsa hōkokusho. III. Series: Nitchū Keizai Kyōkai. Nitchūkei hō ; no 106.
HC427.92.N57 1978 78-814373

*78 J

Nitchū Keizai Kyōkai Kaihō Henshūsha Yūkō Hōchūdan.
(Chūgoku no seiji to keizai, sen-kyūhyaku-shichijūshichinen hachigatsu)
中国の政治と経済—1977年8月 : 日中経済協会会報編集者友好訪中団報告書. — 東京 : 同協会, 1978.
119 p. ; 25 cm. — (日中経報 ; no. 104) Ja 78-19892
Cover title.
1. China—Economic policy—1976- 2. China—Politics and government—1976- I. Title. II. Series: Nitchū Keizai Kyōkai. Nitchūkei hō ; no. 104.
HC427.92.N58 1978 78-814261

(Sen-kyūhyaku-shichijūshichinen Chūgoku keizai kankei shuyō ronsetsu shū)
1977年中国経済関係主要論説集 : 付 経済関係論説・記事索引. — 東京 : 日中経済協会, 1978.
307, 5 p. : ill. ; 25 cm. — (日中経報 ; no. 105) Ja 78-19896
Cover title.
"52-調-12"

1. China—Economic policy—1976- 2. China—Economic policy—1976- —Periodicals—Indexes. I. Nitchū Keizai Kyōkai. II. Series: Nitchū Keizai Kyōkai. Nitchūkei hō ; no. 105.
HC427.92.S48 78-814354

Teng, Hsiao-p'ing, 1900-
Programm für Chinas Zukunft : Dèng Xiǎopíngs Dokumente zur Lage der Nation auf dem Höhepunkt des Machtkampfes 1975 : Übersetzung und Kommentar / Martin Krott. — Hamburg : Institut für Asienkunde, 1978.
126 p. : ill. ; 21 cm. — (Mitteilungen des Instituts für Asienkunde Hamburg ; Nr. 92) GFR***
Summary in English.
Includes bibliographical references.
ISBN 3-921469-41-4 : DM12.00
1. China—Politics and government—1976- 2. China—Economic policy—1976- . 3. Science and state—China. I. Krott, Martin, 1953- . II. Title. III. Series: Institut für Asienkunde. Mitteilungen ; Nr. 92.
DS1.I55 Nr. 92 78-376858
₍DS779.26₎ 78 MARC

Teng, Hsiao-p'ing, 1904-
Programm für Chinas Zukunft : Dèng Xiǎopíngs Dokumente zur Lage der Nation auf dem Höhepunkt des Machtkampfes 1975 : Übersetzung und Kommentar / Martin Krott. — 2. erw. Aufl. — Hamburg : Institut für Asienkunde, 1978.
143 p. ; 21 cm. — (Mitteilungen des Instituts für Asienkunde Hamburg ; Nr. 95) GFR***
Summary in English.
Includes bibliographical references.
ISBN 3-921469-45-7
1. China—Politics and government—1976- 2. China—Economic policy—1976- . 3. Science and state—China. I. Krott, Martin, 1953- . II. Title. III. Series: Institut für Asienkunde. Mitteilungen ; Nr. 95.
DS1.I55 Nr. 95 950 s 79-372086
₍DS799.26₎ 79 MARC

United States. Central Intelligence Agency. National Foreign Assessment Center.
China, in pursuit of economic modernization / Central Intelligence Agency, National Foreign Assessment Center. — ₍Washington₎ : The Center, 1978.
ii, 17 p. ; 27 cm. — (A research paper - Central Intelligence Agency, National Foreign Assessment Center)
Cover title.
"ER 78-10680."
Includes bibliographical references.
1. China—Economic policy—1976- I. Title. II. Series: United States. Central Intelligence Agency. National Foreign Assessment Center. A research paper - Central Intelligence Agency, National Foreign Assessment Center.
HC427.92.U54 1978 330.9'51'05 79-601094
79 MARC

—ECONOMIC POLICY—1976- —
ADDRESSES, ESSAYS, LECTURES

China (People's Republic of China, 1949-). Kuo wu yüan. Ts'ai mao hsiao tsu. Li lun tsu.
(Po ch'ih ssu jen pang ti hui she hui chu i shang p'in sheng ch'an ti fan tung miu lun)
驳斥"四人帮"诋毁社会主义商品生产的反动谬论 / 国务院财贸小组理论组. — 第1版. — 北京 : 人民出版社, 1978.
27 p. ; 20 cm.
Includes bibliographical references.
$0.10

1. China—Economic policy—1976- —Addresses, essays, lectures. I. Title.
HC427.92.C46 1978 79-836162

79 C

Chinese economy post-Mao : a compendium of papers / submitted to the Joint Economic Committee, Congress of the United States. — Washington : U.S. Govt. Print. Off., 1978-
v. : ill. ; 24 cm.
At head of title: 95th Congress, 2d session. Joint committee print.
Includes bibliographical references.
CONTENTS: v. 1. Policy and performance.
1. China—Economic policy—1976- —Addresses, essays, lectures. 2. China—Economic conditions—1976- —Addresses, essays, lectures. I. United States. Congress. Joint Economic Committee.
HC427.9.C56546 330.9'51'05 79-600609
79 MARC

CHINA

—ECONOMIC POLICY—1976-
—ADDRESSES, ESSAYS, LECTURES
(Continued)

(Hsien ch'i shang yeh hsüeh Ta-ch'ing, hsüeh Ta-chai ch'ün chung yün tung hsin kao ch'ao)
掀起商业学大庆，学大寨群众运动新高潮 / 商业部学大庆学大寨办公室编. — 第1版. — 北京：中国财政经济出版社，1977.
54 p. : 19 cm.
$0.13
1. China—Economic policy—1976- —Addresses, essays, lectures. I. China (People's Republic of China, 1949-). Shang yeh pu. Hsüeh Ta-ch'ing hsüeh Ta-chai pan kung shih.
HC427.92.H77 78-843388

(Kuan ch'e chih hsing an lao fen p'ei ti she hui chu i yüan tse)
貫彻执行按劳分配的社会主义原则 / 《人民日报》特约评论员. — 第1版. — 北京：人民出版社，1978.
35 p. ; 21 cm.
Includes bibliographical references.
$0.12
1. China—Economic policy—1976- —Addresses, essays, lectures. 2. Socialism—Addresses, essays, lectures. I. Jen min jih pao, Peking.
HC427.92.K8 79-836365

—ECONOMIC POLICY
—1976- —PERIODICALS
—INDEXES

(Sen-kyūhyaku-shichijūshichinen Chūgoku keizai kankei shuyō ronsetsu shū)
1977年中国経済関係主要論説集：付 経済関係論説・記事索引. — 東京：日中経済協会，1978.
307, 5 p. ; ill. ; 25 cm. — (日中経報 ; no. 105) Ja 78-19896
Cover title.
"52-調-12"
1. China—Economic policy—1976- 2. China—Economic policy—1976- —Periodicals—Indexes. I. Nitchū Keizai Kyōkai. II. Series: Nitchū Keizai Kyōkai. Nitchūkei hō ; no. 105.
HC427.92.S48 78-814354

--ECONOMIC POLICY--ADDRESSES, ESSAYS, LECTURES

(Tairiku shinsakusen to Nisshi keizai sen)
大陸新作戰と日支經濟戰 / 中國總力戰研究所編著. — 上海：中支那經濟年報刊行會，昭和19 [1944]
4, 169, 2 p. ; 22 cm. — (中國戰爭經濟 ; 第2輯)
1. China—Economic policy—Addresses, essays, lectures. 2. Sino-Japanese Conflict, 1937-1945—Economic aspects—China—Addresses, essays, lectures. 3. Sino-Japanese Conflict, 1937-1945—Economic aspects—Japan—Addresses, essays, lectures. I. Chūgoku Sōryokusen Kenkyūjo. II. Series: Chūgoku sensō keizai ; dai 2-shū.
HC427.8.T353 79-804939

—EMIGRATION AND IMMIGRATION

Ch'ên, Ta, 1896-
Emigrant communities in south China : a study of overseas migration and its influence on standards of living and social change / by Ta Chen ; English version edited by Bruno Lasker. — New York : AMS Press, 1978.
xvi, 287 p. ; 23 cm.
Reprint of the 1940 ed. published by the Institute of Pacific Relations, New York and issued under the auspices of the China Institute of Pacific Relations International research series.
Includes bibliographical references and index.
ISBN 0-404-59515-4
1. China—Social life and customs. 2. China—Emigration and immigration. I. Lasker, Bruno, 1880-1965. II. Title. III. Series: Institute of Pacific Relations. International research series.
DS721.C476 1978 301.32'9'51 75-30052
78 MARC

Li, Tien-Lu.
Congressional policy of Chinese immigration / Tien-Lu Li. — New York : Arno Press, 1978 [c1916]
132 p. ; 24 cm. — (The Asian experience in North America : Chinese and Japanese)
Reprint of the ed. printed for the author by the pub. house of the Methodist Episcopal Church, South Nashville, Tenn.
Originally presented as the author's thesis, Vanderbilt University, 1916.
Bibliography: p. [126]-132.
ISBN 0-405-11280-7
1. United States—Emigration and immigration. 2. China—Emigration and immigration. 3. Chinese in the United States. I. Title. II. Series: Asian experience in North America, : Chinese and Japanese.
JV6874.L5 1978 325'.251'0973 78-54824
78 MARC

Taiwan Ginko. Chōsaka.
(Nan'yō ni okeru Kakyō)
南洋ニ於ケル華僑(支那移住民) 附・爲替關係 臺灣銀行總務部調査課 [臺北] 臺灣銀行 大正3 [1914]
172 p. illus. 22 cm.
1. Chinese in Southeastern Asia. 2. China—Emigration and immigration. I. Title.
DS732.T25 1914 J 58-4065 ‡

—EMIGRATION AND IMMIGRATION—HISTORY

Andracki, Stanislaw.
Immigration of orientals into Canada, with special reference to Chinese / Stanislaw Andracki. — New York : Arno Press, 1978.
ii, 227, [10] p. ; 24 cm. — (The Asian experience in the North America : Chinese and Japanese)
Reprint of the author's thesis, McGill University, 1958.
Bibliography: p. [231]-[234].
ISBN 0-405-11262-9
1. Canada—Emigration and immigration—History. 2. China—Emigration and immigration—History. I. Title. II. Series: Asian experience in North america : Chinese and Japanese.
JV7285.C5A6 1978 325'.251'0971 78-54806
78 MARC

Zo, Kil Young.
Chinese emigration into the United States, 1850-1880 / Kil Young Zo. — New York : Arno Press, 1978, c1971.
226 p. : map ; 24 cm. — (The Asian experience in North America : Chinese and Japanese)
Reprint of the author's thesis, Columbia.
Bibliography: p. 207-224.
ISBN 0-405-11302-1
1. United States—Emigration and immigration—History. 2. China—Emigration and immigration—History. 3. Chinese in the United States—History. I. Title. II. Series: Asian experience in North America : Chinese and Japanese.
JV6874.Z6 1978 301.32'8'51073 78-54851
78 MARC

—EMPERORS see China—Kings and rulers

--EMPRESSES--BIOGRAPHY

Kamada, Shigeo, 1909-1969.
[Ryo Taikō. Chinese]
呂太后：農民皇帝和他的妻 / 鎌田重雄著；李君奭譯. — 彰化：專心企業有限公司，民國66 [1977]
9, 194 p. ; 20 cm. — (專心文庫 : 第一輯)
Series romanized: Chuan hsin wen k'u : Ti 1 chi.
NT$35.00
1. Han Lü-hou, Empress of China, 241-180 B. C. 2. Han Kao-tsu, Emperor of China, 247-195 B. C. 3. China—Empresses—Biography. 4. Emperors—China—Biography. I. Title.
Title romanized: Lü t'ai hou.
DS748.16.H356K3512 79-838749

—FAMINES

Lu, Tseng-yü, 17th cent.
(Ch'in ting K'ang chi lu)
欽定康濟錄 / [陸曾禹作；倪國璉錄]. — [台北：台灣]商務[印書館，1976?]
2 v. : ill. ; 20 cm. — (四庫全書珍本六集 ; 130-131)
Cover title.
Reproduced from Wen yüan ko copy.
"欽定武英殿聚珍版程式 金簡撰": last groups of v. 2.
1. China—Famines. 2. Disaster relief—China. 3. Printing, Practical. 4. Printing—History—China. I. Ni, Kuo-lien, chin shih 1730. II. Chin, Chien, d. 1795. Wu ying tien chü chen pan ch'eng shih. 1976? III. Title. IV. Title: K'ang chi lu. V. Series: Ssu k'u ch'üan shu. Ssu k'u ch'üan shu chen pen liu chi ; 130-131.
AC149.S7 1976, vol. 130-131 78-843856
[HC430.F3L78]

—FOREIGN ECONOMIC RELATIONS

(Saikin ni okeru Ō-Beijin no Shina-kan)
最近ニ於ケル歐米人ノ支那觀. — [東京]：外務省情報部，大正2-3 [1913-1914]
6 v. in 1 ; 23 cm.
1. China—Politics and government—1912-1937. 2. China—Foreign economic relations. 3. China—Foreign opinion. I. Japan. Gaimushō. Jōhōbu.
DS774.S3 79-802817

Vashchilko, Alekseĭ ÎAkovlevich.
China and the developing countries : political gambling and trade / Alexei Vashchilko. — Moscow : Novosti Press Agency Pub. House, 1978.
44 p. ; 17 cm. USSR***
Includes bibliographical references.
0.15rub
1. China—Foreign economic relations. 2. Underdeveloped areas—Foreign economic relations. I. Title.
HF1604.V3713 382'.0951'01724 79-303901
79 MARC

—FOREIGN ECONOMIC RELATIONS
—ADDRESSES, ESSAYS, LECTURES

Nakahira, Ryō.
(Shina wa dō naru)
支那はどうなる / 中平亮著. — 東京：東洋研究會，昭和7 [1942]
ii, 149 p. ; 19 cm. — (His 東洋研究叢書 ; 3)
Series romanized: Tōyō kenkyū sōsho.
1. China—History—19th century—Addresses, essays, lectures. 2. China—History—1912-1949—Addresses, essays, lectures. 3. China—Foreign economic relations—Addresses, essays, lectures. I. Title.
DS755.N54 79-808145

—FOREIGN ECONOMIC RELATIONS
—ASIA, SOUTHEASTERN

Matsumoto, Shigekazu, 1934-
China's economic policy towards Asia, 1949-72 / Shigekazu Matsumoto. — Tokyo : Institute of Developing Economies : [sold by] Maruzen Co., c1977.
34 p. ; 24 cm. — (IDE special paper ; no. 8) Ja***
Cover title.
Includes bibliographical references.
$2.00 (U.S.)
1. China—Foreign economic relations—Asia, Southeastern. 2. Asia, Southeastern—Foreign economic relations—China. 3. China—Economic policy. I. Title. II. Series: Ajia Keizai Kenkyūjo, Tokyo. IDE special paper ; no. 8.
HF1604.Z5A73 338.91'51'05 78-315379
78 MARC

—FOREIGN ECONOMIC RELATIONS
—CZECHOSLOVAKIA
—HANDBOOKS, MANUALS, ETC.

Bydžoyský, Zdeněk.
ČLR-Čínska lidová republika / [autor] Zdeněk Bydžovský a kol. — 1. vyd. — Praha : Pressfoto, 1978.
292, [3] p. : ill. ; 21 cm. — (Obchodně ekonomické sborníky : Řada SZ ; sv. 40) Cz78
Bibliography: p. 291-292.
Kčs22.00
1. China—Handbooks, manuals, etc. 2. China—Economic conditions—1949- —Handbooks, manuals, etc. 3. China—Foreign economic relations—Czechoslovakia—Handbooks, manuals, etc. 4. Czechoslovakia—Foreign economic relations—China—Handbooks, manuals, etc. I. Title. II. Title: Čínská lidová republika.
DS777.55.B92 79-358341
*79 MARC

—FOREIGN ECONOMIC RELATIONS
—JAPAN

Haeoe Kyŏngje Yŏn'guso.
(Il-Chunggong muyŏk i uri nara e mich'inŭn yŏnghyang kwa kŭ taech'aek)
日・中共貿易이우리나라에미치는影響과그對策. — [서울]：海外經濟研究所，1975.
176 p. ; 27 cm. — (海外經濟報告 ; 75-5, 通卷第61號)
Chiefly tables.
1. China—Foreign economic relations—Japan. 2. Japan—Foreign economic relations—China. 3. Korea—Commercial policy. I. Title. II. Series: Haeoe kyŏngje pogo ; tonggwan che 61-ho.
HC427.9.H26 1975 79-826352

Nitchū Keizai Kyōkai.
(Hendōsuru kokusai keizai to Nitchū keizai kankei)
変動する国際経済と日中経済関係. — 東京：日中経済協会，1978.
175, 5 p. : graphs ; 25 cm. — (中国経済関係調査報告書) (日中経報 ; no. 108) Ja 78-19903
Cover title.
"52-調-15"
Includes bibliographies.
¥630
1. Japan—Foreign economic relations—China. 2. China—Foreign economic relations—Japan. 3. International economic relations. I. Title. II. Series: Chūgoku keizai kankei chōsa hōkokusho. III. Series: Nitchū Keizai Kyōkai. Nitchūkei hō ; no. 108.
HF1602.15.C6N57 1978 78-814319

CHINA (Continued)

—FOREIGN ECONOMIC RELATIONS —JAPAN—ADDRESSES, ESSAYS, LECTURES

(Jihenka no tairiku keizai seisaku o kataru)
事変下の大陸經済政策を語る / ［朝日新聞經済部編］. — 大阪：朝日新聞社, 昭和13 ［1938］
2, 6, 287 p., ［4］ leaves of plates : ports. ; 18 cm.
"本社は…北支、中支、滿洲、朝鮮の各現地に於て…座談會を開き、本紙經濟面に連載"
1. Japan—Foreign economic relations—China—Addresses, essays, lectures. 2. China—Foreign economic relations—Japan—Addresses, essays, lectures. 3. Japan—Foreign economic relations—Korea—Addresses, essays, lectures. 4. Korea—Foreign economic relations—Japan—Addresses, essays, lectures. I. Asahi Shimbun Sha. Osaka Honsha. Keizaibu. II. Asahi shimbun, Osaka. III. Title: Tairiku keizai seisaku o kataru.
HF1602.5.C6J54 79–806126

(Tairiku keiei no senkusha o kataru)
大陸經營の先驅者を語る / 森田英亮編. — 東京：天佑書房出版部, 昭和15 ［1940］
2, 2, 233 p. ; 19 cm. — (評判叢書 ; 1)
Series romanized: Hyōban sōsho.
1. Japan—Foreign economic relations—China—Addresses, essays, lectures. 2. China—Foreign economic relations—Japan—Addresses, essays, lectures. 3. China—Economic conditions—1912–1949—Addresses, essays, lectures. I. Morita, Hidesuke.
HF1602.15.C6T34 79–801694

—FOREIGN ECONOMIC RELATIONS —JAPAN—PERIODICALS

Nitchū Keizai Kyōkai.
(Nitchū Keizai Kyōkai kaihō)
日中経済協会会報
［東京］ 日中経済協会
no. ill. 26 cm. monthly.
1. China—Foreign economic relations—Japan—Periodicals. 2. Japan—Foreign economic relations—China—Periodicals. 3. Nitchū Keizai Kyōkai—Periodicals.
HF1604.Z4J366 79–649649 (MARC-S)

—FOREIGN ECONOMIC RELATIONS —RUSSIA

Sladkovskiĭ, Mikhail Iosifovich.
(Istorii͡a torgovo-ėkonomicheskikh otnosheniĭ SSSR s Kitaem, 1917–1974)
История торгово-экономических отношений СССР с Китаем, 1917–1974 / М. И. Сладковский ; АН СССР, Ин-т Дальнего Востока. — Москва : Наука, 1977.
368 p. ; 22 cm. USSR 77
Continues the author's Istorii͡a torgovo-ėkonomicheskikh otnosheniĭ narodov Rossii s Kitaem (do 1917 g.)
Bibliography: p. 320–325.
2.00rub
1. Russia—Foreign economic relations—China. 2. China—Foreign economic relations—Russia. I. Title.
HF1558.15.C5S52 78–422251

—FOREIGN ECONOMIC RELATIONS—SUDAN

Ali, Ali Abdalla.
The Sudan and the People's Republic of China : an appraisal of the trade and aid relations / by Ali Abdalla Ali. — Khartoum : Economic and Social Research Council, National Council for Research, 1975, cover 1976.
35 leaves ; 28 cm. — (Bulletin - Economic and Social Research Council ; no. 29)
1. Sudan—Foreign economic relations—China. 2. China—Foreign economic relations—Sudan. 3. Economic assistance, Chinese—Sudan. I. Title. II. Series: Majlis al Abḥāth al-Iqtiṣādīyah wa-al-Ijtimāʻīyah. Bulletin - Economic and Social Research Council ; no. 29.
HF1611.8.Z4C492 382'.0951'0624 78-980272
79 MARC

—FOREIGN ECONOMIC RELATIONS—UNITED STATES

Swannack-Nunn, Susan.
US business and East-West industrial cooperation prospects for the PRC / Susan Swannack-Nunn. — Washington : National Council for US-China Trade, c1978.
iv, 92 leaves ; 28 cm. — (Reference series ; 3-RS 3)
Cover title.
Bibliography: leaves 90-92.
1. East-West trade (1945-) 2. United States—Foreign economic relations—China. 3. China—Foreign economic relations—United States. 4. Foreign trade promotion—United States. I. National Council for UnitedStates-China Trade. II. Title. III. Series: Reference series (Washington) ; 3-RS 3.
HF1411.S89 382'.0951'073 79-112995
00 MARC

—FOREIGN OPINION

(Saikin ni okeru Ō-Beijin no Shina-kan)
最近ニ於ケル歐米人ノ支那觀. — ［東京］：外務省情報部, 大正2-3 ［1913-1914］
6 v. in 1 ; 23 cm.
1. China—Politics and government—1912–1937. 2. China—Foreign economic relations. 3. China—Foreign opinion. I. Japan. Gaimushō. Jōhōbu.
DS774.S3 79–802817

—FOREIGN OPINION—ADDRESSES, ESSAYS, LECTURES

China im Spiegel der Weltpresse : Dokumentation ausgew. Beitr. u. Daten zu polit. u. wirtschaftl. Entwicklungen 1977 u. 1978 / Wilhelm G. Franken, unter Mitarb. von Detlev Passarge, Marlies Prigge, Hubert-Günter Striefler. — Hamburg : Verlag Weltarchiv, 1978.
iv, 820 p. ; 29 cm. — (Veröffentlichungen des HWWA-Institut für Wirtschaftsforschung-Hamburg : Aktuelle Fachinformation) GFR79-A
"Abgeschlossen: 31.10.1978."
English, French, German, or Italian.
Research carried out by HWWA-Institut für Wirtschaftsforschung-Hamburg, Hauptabteilung Dokumentation, Bibliothek und Archive.
Bibliography: p. 803-817.
ISBN 3-87895-179-5 : DM68.00
1. China—Politics and government—1976- —Addresses, essays, lectures. 2. China—Economic conditions—1976- —Addresses, essays, lectures. 3. China—Foreign opinion—Addresses, essays, lectures. I. Franken, Wilhelm G. II. HWWA-Institut für Wirtschaftsforschung-Hamburg. Hauptabteilung Dokumentation, Bibliothek und Archive.
DS779.26.C4735 320.9'51'05 79-369082
79 MARC

—FOREIGN POPULATION

(Shina)
支那：消夏漫筆筆 / 中山成太郎譯［編］. — 東京：有斐閣書房, 大正5 ［1916］
2, 3, 6, 619 p. ; 23 cm.
"第一章…「ニコラウス, ポスト」氏の報告に依り, 第二章…「グルンヘルド」氏支那海港殖民地論に依り, 第三章…第四章…第五章…「フランケー」氏東亞新氣運に依り…抄譯"
1. China. 2. China—Foreign population. 3. Exterritoriality. 4. Tibet. 5. Mongolia (Inner Mongolia) I. Nakayama, Seitarō.
DS706.S48 78–810619

—FOREIGN RELATIONS

Ch'en, Pin-ho, fl. 1926–
(Ti kuo chu i ch'in lüeh Chung-kuo shih)
帝國主義侵略中國史 陳彬龢著 湯彬華編輯
上海 世界書局 民國16 ［1927］
2, 132 p. 20 cm.
1. China—Foreign relations. 2. East Asia—Politics and government. I. Title.
DS740.4.C35337 C 67–948

Ch'en, Shih-ch'ang.
(Chung-kuo yü men hu k'an fang)
中國與門戶開放：清末民初外患史 / 陳世昌著. — 台北：海國書局, 民國66 ［1977］
11, 420 p. ; 22 cm.
At head of title: Ta chuan yung shu.
Title on spine: Ch'ing mo min ch'u wai huan shih.
Bibliography: p. 361–367.
NT$140.00
1. China—Foreign relations. 2. China—History—Ch'ing dynasty, 1644–1912. 3. China—History—1912–1937. I. Title. II. Title: Ch'ing mo min ch'u wai huan shih.
DS740.4.C35338 78–841899

Ch'ien, I-shih, 1893–1937.
(Chung-kuo wai chiao shih)
中國外交史 / 錢亦石著. — ［香港?：s. n., 1977?］
2, 5, 242 ［i. e. 247］ p. ; 21 cm.
Reprint of the 1947 ed. published by Sheng huo shu tien, Shanghai.
1. China—Foreign relations. I. Title.
DS740.4.C35426 1977 78–841287

China (Republic of China). Kuang fu ta lu she chi yen chiu wei yüan hui.
(Lieh ch'iang ch'in lüeh wo kuo chih yen chiu)
列強侵略我國之研究 / 光復大陸設計研究委員會研擬. — ［台北］：該會, 民國65 ［1976］
3, 32, 4, 48 p. ; 22 cm. — (邊疆研究叢書)
Cover title.
CONTENTS: 近百年來日本與西歐列強侵略我領土損害我主權之研究.—俄國侵略我領土損害我主權之研究.
1. China—Foreign relations. 2. China—History—19th century. 3. China—History—1900– I. Title. II. Series: Pien chiang yen chiu ts'ung shu.
DS740.4.C354518 1976 78–841329

Clements, Paul Henry, 1884-
An outline of the politics and diplomacy of China and the Powers, 1894-1902, by Paul H. Clements ... New York, 1914.
49 p. 24 cm.
Vita.
Submitted in 1914 as author's thesis for Ph.D. Replaced by thesis with title "The Boxer rebellion, a political and diplomatic review, 1915," and degree granted in 1916.
1. China—Foreign relations. I. Title.
DS740.4.C4 14-12896
Columbia Univ. Libr. ［r79］rev MARC

Ecsedy, Ildikó.
Nomádok és kereskedők Kína határain / Ecsedy Ildikó. — Budapest : Akadémiai Kiadó, 1979.
237 p. ; 18 cm. — (Kőrösi Csoma kiskönyvtár ; 16) H***
Includes bibliographical references.
ISBN 9630517698 : 23.00Ft
1. China—Commerce—History. 2. China—Foreign relations. 3. Nomads—China—History. 4. Sinkiang—History. I. Title.
HF3835.E27 79-381930
79 MARC

Inoguchi, Takashi.
(Gaikō taiyō no hikaku kenkyū)
外交態様の比較研究：中国・英国・日本 / 猪口孝著. — 東京：巌南堂書店, 1978.
5, 351 p. ; 22 cm. Ja 79–4426
Bibliography: p. 314–349.
¥4500
1. East Asia—Foreign relations. 2. China—Foreign relations. 3. World politics—19th century. 4. World politics—20th century. I. Title.
DS740.6.I56 79–807406

Keeton, George Williams, 1902-
The development of extraterritoriality in China / by G. W. Keeton. — ［s.l. : s.n., 1978?］
2 v. in 4 ; 27 cm.
Reprint of the 1928 ed. published by Longmans, Green, London.
Bibliography: v. 2 (i.e. 4), p. ［399］-402.
Includes index.
1. Exterritoriality. 2. Justice, Administration of—China. 3. Courts—China. 4. China—Foreign relations. I. Title.
JX1570.K4 1978 341.48'8 78-105475
78 MARC

Li, Chien-nung.
The political history of China, 1840-1928. Translated and edited by Ssu-yu Teng and Jeremy Ingalls. Stanford, Calif., Stanford University Press ［1967, c1956］
xii, 545 p. maps. 22 cm.
"SP60."
Translation of Chung-kuo chin pai nien chêng chih shih.
Bibliography: p. 515-528.
1. China—Politics and government—19th century. 2. China—Politics and government—1900-1949. 3. China—Foreign relations. I. Têng, Ssŭ-yü, 1906- ed. II. Ingalls, Jeremy, 1911- ed. III. Title.
DS755.L4313 1967 951'.03 67-26648
［r78］rev2 MARC

CHINA

—FOREIGN RELATIONS (Continued)

Nearing, Scott, 1883-
Whither China? : An economic interpretation of recent events in the Far East / by Scott Nearing. — Westport, Conn. : Hyperion Press, 1977, c1927.
225 p. ; 23 cm. — (China studies)
Reprint of the ed. published by International Publishers, New York.
Bibliography: p. 211-214.
Includes index.
ISBN 0-88355-385-6 : $16.50
1. China—Economic conditions—1912-1949. 2. China—Industries. 3. China—Social conditions. 4. China—History—1900- . 5. China—Foreign relations. I. Title. II. Series: China studies (Westport, Conn.)
HC427.N38 1977 330.9'51'05 75-39029
77 MARC

Tyau, Min-ch'ien T Z 1888-
China's new constitution and international problems / by Min-ch'ien T. Z. Tyau. — Reprint ed. — Arlington, Va. : University Publications of America, 1976.
xv, 286 p. ; 24 cm. — (China studies) (Studies in Chinese government and law)
Reprint of the 1918 ed. published by Commercial Press, Shanghai.
ISBN 0-89093-064-3
1. China—Constitutional law. 2. China—Politics and government—1900-1949. 3. China—Foreign relations. 4. European War, 1914-1918—China. I. Title.
342'.51'029 76-55090
78 MARC

Yu, Se-hŭi.
(Chuhan Migun kamch'uk ch'ŏlsu e ttarŭn Chung So ŭi tae Hanbando chŏngch'aek chŏnmang)
駐韓美軍減縮撤収에따른中·蘇의對韓半島政策展望 / 研究執筆責任 柳世熙 ; 刊行責任 李壽一. — [서울] : 國土統一院調査研究室, 1977.
36 p. ; 26 cm. — (周邊環境與件)
"국통정77-10-1187"
Includes bibliographical references.
1. Military assistance, American—Korea. 2. China—Foreign relations. 3. Russia—Foreign relations. 4. Korea (Democratic People's Republic)—Foreign relations. I. Yi, Su-il. II. Title. III. Series: Chubyŏn hwan'gyŏng yŏkŏn.
UA853.K6Y78 79-825304

K

Yu, Se-hŭi.
(Hanbando p'yŏnghwa chŏngch'ak e taehan Chung-So ŭi chagyong pyŏnsu)
韓半島平和定着에대한中蘇의作用変数 / 研究執筆責任 柳世熙 ; 刊行責任 朴鏞德. — [서울] : 國土統一院調査研究室, 1976.
65 p. ; 26 cm. — (周辺勢力調査研究)
"국통조76-12-98."
Includes bibliographical references.
1. China—Foreign relations. 2. Russia—Foreign relations. 3. Korea—Foreign relations. 4. Korea (Democratic People's Republic)—Foreign relations. I. Pak, Yong-dŏk. II. Title. III. Series: Chubyŏn seryŏk chosa yŏn'gu.
DS777.55.Y7636 79-825320

—FOREIGN RELATIONS—TO 1912

Chan Lau, Kit-ching.
Anglo-Chinese diplomacy in the careers of Sir John Jordan and Yüan Shih-kai / Chan Lau Kit-ching. — [Hong Kong] : Hong Kong University Press, 1978.
viii, 168 p. ; 23 cm.
Bibliography: p. [144]-156.
Includes index.
ISBN (invalid) 0-9622090-1-09
1. Great Britain—Foreign relations—China. 2. China—Foreign relations—Great Britain. 3. Great Britain—Foreign relations—1901-1910. 4. Great Britain—Foreign relations—1910-1936. 5. China—Politics and government—1900-1949. 6. China—Foreign relations—To 1912. 7. China—Foreign relations—1912-1949. 8. Yüan, Shih-k'ai, 1859-1916. 9. Jordan, John Newell, Sir, 1852- . I. Title.
DA47.9.C6C45 327.41'051 79-106848
79 MARC

Ch'en, Wen-shih.
(Ming Hung-wu Chia-ching chien ti hai chin cheng ts'e)
明洪武嘉靖間的海禁政策 / 陳文石. — 台北 : 國立臺灣大學文學院, 1966.
3, 7, 176, 2 p. ; 22 cm. — (國立臺灣大學文史叢刊 ; 20)
Title in English in Colophon: The sea-faring prohibition policy in the early Ming Dynasty.
Text in Chinese, with summary in English.
Includes bibliographical references.
1. China—Foreign relations—To 1912. 2. China—History—Ming dynasty, 1368-1644. 3. China—Commerce—History. 4. Pirates—China. I. Title. II. Title: The sea-faring prohibition policy. III. Series: T'ai-wan ta hsüeh, T'ai-pei. Kuo li T'ai-wan ta hsüeh wen shih ts'ung k'an ; 20.
DS753.C5115 78-843024

—FOREIGN RELATIONS—1912-1949

Chan Lau, Kit-ching.
Anglo-Chinese diplomacy in the careers of Sir John Jordan and Yüan Shih-kai / Chan Lau Kit-ching. — [Hong Kong] : Hong Kong University Press, 1978.
viii, 168 p. ; 23 cm.
Bibliography: p. [144]-156.
Includes index.
ISBN (invalid) 0-9622090-1-09
1. Great Britain—Foreign relations—China. 2. China—Foreign relations—Great Britain. 3. Great Britain—Foreign relations—1901-1910. 4. Great Britain—Foreign relations—1910-1936. 5. China—Politics and government—1900-1949. 6. China—Foreign relations—To 1912. 7. China—Foreign relations—1912-1949. 8. Yüan, Shih-k'ai, 1859-1916. 9. Jordan, John Newell, Sir, 1852- . I. Title.
DA47.9.C6C45 327.41'051 79-106848
79 MARC

Ch'en, Po-ta, 1905-
(P'ing Chung-kuo chih ming yün)
評「中國之命運」/ 陳伯達著. — [北京?] : 新華書店, 1949.
57 p. ; 18 cm.
1. Chiang, Kai-shek, 1887-1975. Chung-kuo chih ming yün. 2. China—Foreign relations—1912-1949. I. Title.
DS775.C531583C45 78-843815

Sakamaki, Teiichirō.
(Shina bunkatsu ron)
支那分割論 : 附袁世凱 / [著者 酒巻貞一郎]. — 東京 : 啓成社, 大正2 [1913]
584, 204, [35] p., [5] leaves of plates : maps ; 23 cm.
1. China—History—1912-1937. 2. China—Foreign relations—1912-1949. 3. China—History—Revolution, 1911-1912. 4. Revolutions—China. 4. Yüan, Shih-k'ai, 1859-1916. 6. China—Presidents—Biography. I. Title. II. Sakamaki, Teiichirō. En Seigai. 1913.
DS774.S332 79-809704

Washington, D. C. Conference on the Limitation of Armament, 1921-1922.
(Hua fu hui i yüeh chang hui lu)
華府會議約章彙錄. — [台北? : s. n., 1978?]
2 v. ; 27 cm.
Spine title.
English and Chinese.
Reprint ed.
1. China—Foreign relations—1912-1949. 2. China—Foreign relations—Treaties. I. Title.
DS775.W37 1978 79-840111

Yao, Sung-ling, 1900-
(Jui En-shih shih Hua chi yao)
芮恩施使華記要 / 姚崧齡著. — 臺北 : 傳記文學出版社, 民國60 [1971]
2, 103 p., [1] leaf of plates : ill. ; 19 cm. — (傳記文學叢書 ; 46)
Series romanized: Chuan chi wen hsüeh ts'ung shu.
"本書係姚先生根據芮所撰使華記一書編譯而成"
1. China—Foreign relations—1912-1949. 2. Reinsch, Paul Samuel, 1869-1923. I. Reinsch, Paul Samuel, 1869-1923. An American diplomat in China. II. Title.
DS775.Y34 79-836572

—FOREIGN RELATIONS—1949-

Holbraad, Carsten.
Superpowers and international conflict / Carsten Holbraad. — New York : St. Martin's Press, 1979.
viii, 178 p. ; 23 cm.
Includes bibliographical references and index.
ISBN 0-312-77674-8
1. World politics—1945- 2. United States—Foreign relations—1945- . 3. Russia—Foreign relations—1945- 4. China—Foreign relations—1949- I. Title.
D843.H64 1979 327'.09'045 79-9942
79 MARC

Kokusai Jōsei Kenkyūkai.
(Chūkyō naigai seisaku no kihon hōshin)
中共内外政策の基本方針. — 東京 : 国際情勢研究会, 昭和41 [1966]
37 p. ; 25 cm.
Cover title.
1. China—Foreign relations—1949- 2. China—Politics and government—1949-1976. I. Title.
DS777.55.K615 1966a 79-803468

—FOREIGN RELATIONS—1949-1976

Ch'en, Yü-ch'ing, 1918-
(Fen sui kung fei t'ung chan yin mou k'ai ch'uang fan kung fu kuo hsin chü)
粉碎共匪統戰陰謀開創反共復國新局 / 陳裕清著 ; 海外出版社編印. — 1st ed. — 臺北 : 該社, 1973.
18, 28 p. ; 18 cm.
English and Chinese.
Added t. p.: Chinese communist united front scheme vs. ROC's mainland recovery mission.
1. China—Foreign relations—1949-1976. 2. China—Politics and government—1949-1976. 3. Taiwan—Foreign relations. 4. Taiwan—Politics and government—1945- I. Title. II. Title: Chinese communist united front scheme vs. ROC's mainland recovery mission.
DS777.55.C44517 78-841699

Clark, Gregory, 1936-
In fear of China [by] Gregory Clark. [Melbourne] Lansdowne [1967]
ix, 219 p. maps. 23 cm. $4.95 Aust. Aus67-3047
Includes bibliographical references.
1. China—Foreign relations—1949-1976. I. Title.
DS777.55.C49 1967 327.51 68-79303
[r79]rev MARC

Clark, Gregory, 1936-
In fear of China, [by] Gregory Clark. London, Cresset P., 1968.
xi, 219 p. 4 maps. 23 cm. 30/- B69-03008
Map on lining papers.
Includes bibliographies.
ISBN 0-214-66767-7
1. China—Foreign relations—1949-1976. I. Title.
DS777.55.C49 1968 327.51 73-402997
69[r79]rev MARC

Clark, Gregory, 1936-
(Kokusai seiji to Chūgoku)
国際政治と中国 オーストラリア外交から見る 松本繁一訳 [東京 アジア経済研究所 アジア経済出版会発売 1970]
xiv, 366 p. maps. 18 cm. (アジアを見る眼 36)(アジア経済研究所出版物通巻第938号) ¥500 Ja***
Translation of In fear of China.
Includes bibliographical references.
1. China—Foreign relations—1949-1976. I. Title. II. Series: Ajia o miru me. III. Series: Ajia Keizai Kenkyūjo, Tokyo. Ajia Keizai Kenkyūjo shuppanbutsu, tsūkan dai 938-gō.
DS740.4.C39 1970 79-796610

Kang, Chong-Sook.
China in Waffen : d. Rüstungs- u. Abrüstungspolitik d. Volksrepublik China von 1969 bis zum Tod Mao-Tse-Tungs / Chong-Sook Kang. — Frankfurt/Main : Haag und Herchen, 1979.
143 p. ; 21 cm. — (Internationale Beziehungen ; Bd. 2 ISSN 0171-4414) GFR79-A
Summary in English.
Bibliography: p. 125-143.
ISBN 3-88129-208-X : DM19.80
1. China—Military policy. 2. China—Foreign relations—1949-1976. 3. China—Foreign relations—1976- 4. Munitions—China. 5. Arms control. I. Title. II. Series.
UA835.K36 79-376036
79 MARC

Kao, Hsiang-kao, 1905-
(Chung-kung tui wai kuan hsi chih fa chan)
中共對外關係之發展 / 著者高向杲. — 臺初版. — 臺北 : 正中書局, 民國67 [1978]
4, 2, 10, 482 p., [1] fold. leaf of plates : maps ; 22 cm. — (中國大陸研究叢書)
Colophon title.
Series romanized: Chung-kuo ta lu yen chiu ts'ung shu.
Includes bibliographical references.
1. China—Foreign relations—1949-1976. I. Title.
DS777.55.K236 79-838072

C

Lieberthal, Kenneth.
The foreign policy debate in Peking as seen through allegorical articles, 1973-1976 / Kenneth Lieberthal. — Santa Monica, Calif. : Rand Corp., 1977.
v, 42 p. ; 28 cm. — (The Rand paper series ; P-5768)
Cover title.
Includes bibliographical references.
1. China—Foreign relations—1949-1976. I. Title. II. Series.
AS36.R28 no. 5768 081 s 77-155636
[DS777.55] 78 MARC

CHINA

—FOREIGN RELATIONS—1949-1976 (Continued)

Pak, Hong-min.
(Chunggong oegyo chŏngch'aek kyŏlchŏng kwajŏng kwa Mo T'aek-tong ihu tae Hanbando chŏngch'aek)
中共外交政策決定過程과毛澤東以後対韓半島政策 / 研究執筆責任 朴洪眠 ; 刊行責任 羅昌柱. — [서울] : 國土統一院調査研究室, 1977.
44 p. ; 26 cm. — (周邊環境與件 : 中共)
"국통조 77-1-1104."
1. China—Foreign relations—1949- 2. China—Foreign relations—Korea. 3. Korea—Foreign relations—China. I. Na, Ch'ang-ju. II. Title. III. Series: Chubyŏn hwan'gyŏng yŏkŏn.
DS777.55.P268 78-826281

Sergeeva, Ėleonora Arkad'evna.
(Vneshnfafa politika i vneshnepoliticheskafa propaganda rukovodstva KPK)
Внешняя политика и внешнеполитическая пропаганда руководства КПК / Э. А. Сергеева. — Москва : Наука, 1978.
190 p. ; 20 cm. USSR 78
Bibliography: p. 181-[189]
0.65rub
1. China—Foreign relations—1949-1976. I. Title.
DS777.55.S438 79-411037

Sutter, Robert G
Chinese foreign policy after the cultural revolution, 1966-1977 / Robert G. Sutter. — Boulder, Colo. : Westview Press, 1978.
x, 176 p. ; 24 cm. — (Westview special studies on China and East Asia)
Bibliography: p. [167]-169.
Includes index.
ISBN 0-89158-342-4
1. China—Foreign relations—1949-1976. I. Title.
DS777.55.S88 327.51 77-7018
77[r79]rev MARC

United States. Library of Congress. Foreign Affairs and National Defense Division.
Recognizing the People's Republic of China : the experience of Japan, Australia, France, and West Germany : report / prepared for the Subcommittee on Asian and Pacific Affairs of the Committee on Foreign Affairs, U. S. House of Representatives by the Foreign Affairs and National Defense Division, Congressional Research Service, Library of Congress. — Washington : U. S. Govt. Print. Off., 1979.
xiii, 33 p. ; 24 cm.
At head of title: 96th Congress, 1st session. Committee print. Prepared by L. S. Christopher and R. G. Sutter.
1. China—Foreign relations—1949-1976. 2. China—Foreign relations—1976- I. Christopher, Luella Sue. II. Sutter, Robert G. III. United States. Congress. House. Committee on Foreign Affairs. Subcommittee on Asian and Pacific Affairs. IV. Title.
DS777.55.U55 1979 327.51 79-602528
00 MARC

—FOREIGN RELATIONS —1949-1976—ADDRESSES, ESSAYS, LECTURES

(Chung kung meng yu yü ti jen)
中共盟友與敵人. — 香港 : 香港中國文史哲出版社, [1978?]
32 p. : ill. ; 29 cm.
Cover title.
$2.00
1. China—Foreign relations—1949-1976—Addresses, essays, lectures.
DS777.55.C468633 78-843966

Yeh, Po-t'ang.
(Mao kung wai chiao cheng ts'e lun)
毛共外交政策論 / 葉伯棠著. — 臺北 : 幼獅文化事業公司, 民國66 [1977]
2, 382 p. ; 22 cm.
Includes bibliographical references.
NT$100.00
1. China—Foreign relations—1949-1976—Addresses, essays, lectures. I. Title.
DS777.55.Y366 78-842602

—FOREIGN RELATIONS —1949-1976 —COLLECTED WORKS

Fitzgerald, Stephen.
China and the world / [by] Stepnen FitzGerald. — Canberra : Australian National University Press, 1977.
x, 126 p. ; 18 cm. — (Contemporary China papers ; no. 11) Aus
ISBN 0-7081-0224-7
1. China—Foreign relations—1949-1976—Collected works. 2. China—Foreign relations—1976- —Collected works. I. Title.
DS777.55.F544 327.51 78-312808
78 MARC

—FOREIGN RELATIONS—1949-1976—SOURCES

China and the three worlds : a foreign policy reader / edited by King C. Chen. — White Plains, N.Y. : M. E. Sharpe, c1979.
xii, 383 p. ; 24 cm.
Bibliography: p. 379-381.
ISBN 0-87332-118-9. ISBN 0-87332-134-0 pbk.
1. China—Foreign relations—1949-1976—Sources. I. Chen, King C., 1926-
DS777.55.C446736 327.51 78-51973
79 MARC

—FOREIGN RELATIONS—1976-

Johansen, Jahn Otto.
Kinas utenrikspolitikk / Jahn Otto Johansen, Erik Solheim, Sigurd Allern. — Oslo : Oktober, 1979 (Oslo : Duplotrykk)
95 p. ; 19 cm. N79-May
"Diskusjonsmøte arrangert av Opplysningsutvalget i Oslo AKP(m-l) 8. februar 1979."
Includes bibliographical references.
ISBN 8270949795 : kr35.00
1. China—Foreign relations—1976- I. Solheim, Erik, joint author. II. Allern, Sigurd, joint author. III. Arbeidernes kommunistparti (marxist-leninistene) Opplysningsutvalget i Oslo. IV. Title.
DS779.27.J63 79-375017
79 MARC

Kang, Chong-Sook.
China in Waffen : d. Rüstungs- u. Abrüstungspolitik d. Volksrepublik China von 1969 bis zum Tod Mao-Tse-Tungs / Chong-Sook Kang. — Frankfurt/Main : Haag und Herchen, 1979.
143 p. ; 21 cm. — (Internationale Beziehungen ; Bd. 2 ISSN 0171-4414) GFR79-A
Summary in English.
Bibliography: p. 125-143.
ISBN 3-88129-208-X : DM19.80
1. China—Military policy. 2. China—Foreign relations—1949-1976. 3. China—Foreign relations—1976- 4. Munitions—China. 5. Arms control. I. Title. II. Series.
UA835.K36 79-376036
79 MARC

Kim, Samuel S 1935-
China, the United Nations, and world order / Samuel S. Kim. — Princeton, N.J. : Princeton University Press, c1979.
xviii, 581 p. ; 25 cm.
Bibliography: p. 531-566.
Includes index.
ISBN 0-691-07599-9. ISBN 0-691-10076-4 pbk.
1. China—Foreign relations—1976- 2. United Nations—China. 3. World politics—1975-1985. I. Title.
DS779.27.K55 327.51 78-51174
78 MARC

Moya, R
China del rojo al amarillo : la teoría de los tres mundos / R. Moya. — [Barcelona] : Editorial 7 1/2, [1978?]
246 p. ; 21 cm. — (Controversias de nuestro tiempo) Sp***
Includes bibliographical references.
1. China—Foreign relations—1976- 2. World politics—1975-1985. 3. Communism. I. Title.
DS779.27.M69 78-381822
78 MARC

Sarandova, Zhivka.
Peking's foreign policy after Mao Tse-Tung / Zhivka Sarandova ; [translated by S. Kostov]. — [Sofia] : Sofia Press, 1978.
49 p. ; 19 cm. Bu78-794
Includes bibliographical references.
1. China—Foreign relations—1976- I. Title.
DS779.27.S2713 327.51 79-313462
79 MARC

United States. Library of Congress. Foreign Affairs and National Defense Division.
Recognizing the People's Republic of China : the experience of Japan, Australia, France, and West Germany : report / prepared for the Subcommittee on Asian and Pacific Affairs of the Committee on Foreign Affairs, U. S. House of Representatives by the Foreign Affairs and National Defense Division, Congressional Research Service, Library of Congress. — Washington : U. S. Govt. Print. Off., 1979.
xiii, 33 p. ; 24 cm.
At head of title: 96th Congress, 1st session. Committee print. Prepared by L. S. Christopher and R. G. Sutter.
1. China—Foreign relations—1949-1976. 2. China—Foreign relations—1976- I. Christopher, Luella Sue. II. Sutter, Robert G. III. United States. Congress. House. Committee on Foreign Affairs. Subcommittee on Asian and Pacific Affairs. IV. Title.
DS777.55.U55 1979 327.51 79-602528
00 MARC

Yahuda, Michael B
China's role in world affairs / Michael B. Yahuda. — London : Croom Helm, c1978.
298 p. ; 22 cm. GB***
Bibliography: p. 286-288.
Includes index.
ISBN 0-85664-269-X : £10.95
1. China—Foreign relations—1976- I. Title.
DS779.27.Y33 327.51 78-323720
79 MARC

Yahuda, Michael B
China's role in world affairs / Michael B. Yahuda. — New York : St. Martin's Press, c1978.
298 p. ; 23 cm.
Bibliography: p. 286-288.
Includes index.
ISBN 0-312-13358-8 : $18.50
1. China—Foreign relations—1976- I. Title.
DS779.27.Y33 1978b 327.51 78-19218
79 MARC

Yin, Ch'ing-yao.
Chinese Communist foreign policy and trends / by Yin Ching-yao. — [Taipei?] : World Anti-Communist League, China Chapter, 1977.
64, x p. ; 19 cm. — (English publications of the APACL/ROC : Pamphlets ; 207)
Cover title.
Includes bibliographical references.
$0.70
1. China—Foreign relations—1976- I. World Anti-Communist League. China Chapter. II. Title.
DS779.27.Y56 327.51 78-109090
78 MARC

Yin, Ch'ing-yao.
(Chung kung ti tui wai cheng ts'e yü tung hsiang)
中共的對外政策與動向 / [尹慶耀撰]. — [台北] : 世界反共聯盟中華民國分會, 民國66 [1977]
46 p. ; 19 cm.
Cover title.
Includes bibliographical references.
1. China—Foreign relations—1976- I. Title.
DS779.27.Y57 78-843394

—FOREIGN RELATIONS —1976- —COLLECTED WORKS

Fitzgerald, Stephen.
China and the world / [by] Stepnen FitzGerald. — Canberra : Australian National University Press, 1977.
x, 126 p. ; 18 cm. — (Contemporary China papers ; no. 11) Au
ISBN 0-7081-0224-7
1. China—Foreign relations—1949-1976—Collected works. 2. China—Foreign relations—1976- —Collected works. I. Title.
DS777.55.F544 327.51 78-31280
78 MARC

—FOREIGN RELATIONS—ADDRESSES, ESSAYS, LECTURES

Hou, Li-ch'ao.
(Mei Jih O san ti tsai Chung-kuo)
美日俄三帝在中國 / 侯立朝著. — 臺北 : 博學出版社 : 總經銷傳眞書店, 民國66 [1977]
8, 223 p., [4] leaves of plates : ill. ; 20 cm. — (中國人抗外史 ; 5)
CONTENTS: 中美關係黃皮書.—中日敵友關係百年論定.—俄共與中共的歷史關係.—中共一面倒馬列化徹底失敗大事記(一九一七—一九七三)
1. China—Foreign relations—Addresses, essays, lectures. 2. Taiwan—Foreign relations—Addresses, essays, lectures. I. Title. II Series: Chung-kuo jen k'ang wai shih ; 5.
DS740.4.H63 78-84378

—FOREIGN RELATIONS—TREATIES

(Chung-kuo pu p'ing teng t'iao yüeh chih fei ch'u)
中國不平等條約之廢除 / [編著者王世杰, 胡慶育]. — [台北] : 蔣總統對中國及世界之貢獻叢編編纂委員會 : 總經銷中央文物供應社, 民國56 [1967]
2, 4, 460 p. ; 27 cm.
"蔣總統對中國及世界之貢獻"
Includes bibliographical references.
1. China—Foreign relations—Treaties. 2. Unequal treaties. I Wang, Shih-chieh, 1891- II. Hu, Ch'ing-yü.
JX926 1967.C48 79-83617

Heine, Irwin M
China's maritime agreements : prepared for members of the National Council for US-China Trade / Irwin Millard Heine. — Washington : The Council, 1977.
68 p. : ill. ; 28 cm. — (Special report - National Council for US-China Trade ; no. 21)
1. Maritime law—China. 2. China—Foreign relations—Treaties. I. National Council for United States-China Trade. II. Title. III. Series: National Council for United States-China Trade. Special report - National Council for US-China trade ; no. 21.
341.7'566'026451 77-84946
79 MARC

CHINA

—FOREIGN RELATIONS—TREATIES (Continued)

Washington, D. C. Conference on the Limitation of Armament, 1921-1922.
(Hua fu hui i yüeh chang hui lu)
華府會議約章彙錄. — ₍台北?：s. n., 1978?₎
2 v. ; 27 cm.
Spine title.
English and Chinese.
Reprint ed.
1. China—Foreign relations—1912-1949. 2. China—Foreign relations—Treaties. I. Title.
DS775.W37 1978 79-840111

—FOREIGN RELATIONS—AFRICA

Chung, Kuan.
A case study of Chinese Communist subversive activity in Africa : relations between Red China and Ghana, 1964-1966 / by Kuan Chung. — ₍Taipei₎ : World Anti-Communist League, China Chapter, 1977.
37, 8 p. ; 20 cm. — (Publications of the APACL/ROC ; pamphlet 23, no. 195)
Cover title.
$0.50
1. China—Foreign relations—Ghana. 2. Ghana—Foreign relations—China. 3. China—Foreign relations—Africa. 4. Africa—Foreign relations—China. I. Title.
DS740.5.G45C46 327.51'0667 77-152741
79 MARC

Greig, Ian.
The Communist challenge to Africa : an analysis of contemporary Soviet, Chinese, and Cuban policies / by Ian Greig. — Sandton : Southern African Freedom Foundation, 1977.
384 p. ; 22 cm. SA78
Bibliography: p. 337-339.
Includes index.
ISBN 0-908395-19-1 : R9.90
1. Africa—Foreign relations—Russia. 2. Russia—Foreign relations—Africa. 3. Africa—Foreign relations—Cuba. 4. Cuba—Foreign relations—Africa. 5. Africa—Foreign relations—China. 6. China—Foreign relations—Africa. 7. Communism—Africa. I. Title.
DT38.9.R8G73 1977b 327'.096 79-310858
79 MARC

—FOREIGN RELATIONS—ALBANIA

Everlasting friendship : from the visit of the delegation of the Party of Labor and Government of the People's Republic of Albania to the People's Republic of China, from April 26 to May 11, 1966. — Tirana : "Naim Frashëri" Pub. House, 1966.
85 p. ; 17 cm.
1. Albania—Foreign relations—China. 2. China—Foreign relations—Albania. 3. Visits of state—China.
DR953.C5E93 327.4965'057 79-305821
79 MARC

—FOREIGN RELATIONS—ASIA

(Chūkyō no jittai to Nitchū shinkankei no arikata)
中共の実態と日中新関係の在り方：中国総合研究会・第二十一次研究報告. — 東京：中国総合研究会, 昭和43 ₍1968₎
376 p. ; 18 x 25 cm.
Cover title.
Not in trade.
1. China—Politics and government—1949-1976. 2. China—Foreign relations—Asia. 3. Asia—Foreign relations—China. I. Chūgoku Sōgō Kenkyūkai.
DS777.55.C465762 78-813750

United States. Library of Congress. Foreign Affairs and National Defense Division.
China and Asia—an analysis of China's recent policy toward neighboring states : report / by the Foreign Affairs and National Defense Division, Congressional Research Service, Library of Congress preceded by a State Department report on normalization negotiations with China ; prepared for the Subcommittee on Asian and Pacific Affairs of the Committee on Foreign Affairs, U.S. House of Representatives. — Washington : U.S. Govt. Print. Off., 1979.
vii, 54 p. ; 24 cm.
At head of title: 96th Congress, 1st session. Committee print.
Supt. of Docs. no.: Y 4.F76/1:C44/14
1. Asia—Foreign relations—China. 2. China—Foreign relations—Asia. 3. United States—Foreign relations—China. 4. China—Foreign relations—United States. I. United States. Congress. House. Committee on Foreign Affairs. Subcommittee on Asian and Pacific Affairs. II. Title.
DS33.4.C5U54 1979 327.51'05 79-601691
79 MARC

—FOREIGN RELATIONS—ASIA, SOUTHEASTERN

Shee, Poon Kim.
ASEAN and the People's Republic of China / by Shee Poon Kim. — ₍Singapore₎ : Institute of Humanities and Social Sciences, College of Graduate Studies, Nanyang University, 1976.
16, 2 leaves ; 30 cm. — (Occasional paper series - Institute of Humanities and Social Sciences, Nanyang University ; no. 53)
Includes bibliographical references.
1. Asia, Southeastern—Foreign relations—China. 2. China—Foreign relations—Asia, Southeastern. 3. Association of South East Asian Nations. I. Title. II. Series: Nanyang University. Institute of Humanities and Social Sciences. Occasional paper series - Institute of Humanities and Social Sciences, Nanyang University ; no. 53.
DS518.15.S47 327.59'051 78-940462
79 MARC

—FOREIGN RELATIONS—CAMBODIA—ADDRESSES, ESSAYS, LECTURES

(Je lieh huan ying Chien-pu-chai chan yu)
热烈欢迎柬埔寨战友波尔布特书记率领的民主柬埔寨党政代表团访问中国. — 第1版. — ₍北京₎：人民出版社, 1977.
2, 48 p., ₍8₎ leaves of plates (1 fold.) : ill. ; 21 cm.
$0.30
1. China—Foreign relations—Cambodia—Addresses, essays, lectures. 2. Cambodia—Foreign relations—China—Addresses, essays, lectures.
DS740.5.C15J4 79-836784

—FOREIGN RELATIONS—EUROPE

Ivory Coast. Service pédagogique. Section histoire.
Expansion européenne et l'extrême orient : La Chine (1838-1911) : documents à l'usage des professeurs des lycées, collèges, c. complémentaires de la République de Cote d'Ivoire / République de Côte d'Ivoire, Ministère de l'éducation nationale, Section générale des études et des programmes, Service pédagogique, Section histoire. — ₍Abidjan₎ : La Section, ₍1964 or 1965₎
2 v. : ill. ; 27 cm. — (Document - République de Côte d'Ivoire, Service pédagogique, Section histoire ; no 172, 183)
Cover title: Documents d'histoire : l'extrême orient.
Bound with the author's Le Japon de 1716 à 1867. Abidjan, 1964 or 1965.
1. Europe—Foreign relations—China. 2. China—Foreign relations—Europe. I. Title. II. Title: Documents d'histoire. III. Series: Ivory Coast. Service pédagogique. Section histoire. Document - République de Côte d'Ivoire, Service pédagogique, Section histoire ; no 172, 183.
D34.C6I93 1964 908 s 79-379073
79 MARC

—FOREIGN RELATIONS—FRANCE

Barlow, Jeffrey G
Sun Yat-sen and the French, 1900-1908 / Jeffrey G. Barlow. — Berkeley : Center for Chinese Studies, Institute of East Asian Studies, University of California, ₍c1979₎
93 p. : map ; 23 cm. — (China research monograph ; no. 14)
Includes bibliographical references.
ISBN 0-912966-19-X : $4.00
1. China—Foreign relations—France. 2. France—Foreign relations—China. 3. Sun, Yat-sen, 1866-1925. I. Title. II. Series: China research monographs ; no. 14.
DS740.5.F8B35 327.51'044 79-620017
79 MARC

—FOREIGN RELATIONS—GHANA

Chung, Kuan.
A case study of Chinese Communist subversive activity in Africa : relations between Red China and Ghana, 1964-1966 / by Kuan Chung. — ₍Taipei₎ : World Anti-Communist League, China Chapter, 1977.
37, 8 p. ; 20 cm. — (Publications of the APACL/ROC ; pamphlet 23, no. 195)
Cover title.
$0.50
1. China—Foreign relations—Ghana. 2. Ghana—Foreign relations—China. 3. China—Foreign relations—Africa. 4. Africa—Foreign relations—China. I. Title.
DS740.5.G45C46 327.51'0667 77-152741
79 MARC

—FOREIGN RELATIONS—GREAT BRITAIN

Anderson, Aeneas, fl. 1802.
₍A narrative of the British embassy to China, in the years 1792, 1793, and 1794. Japanese₎
マカートニー奉使記 / イーニアス・アンダースン著；加藤憲市譯. — 東京：筑摩書房, 昭和22 ₍1947₎
8, 317 p., ₍1₎ leaf of plates : port. ; 21 cm.
1. China—Description and travel. 2. Great Britain—Foreign relations—China. 3. China—Foreign relations—Great Britain. 4. Macartney, George, Macartney, earl, 1737-1806. I. Title.
Title romanized: Makātoni hōshiki.
DS708.A5416 79-800799

Chan Lau, Kit-ching.
Anglo-Chinese diplomacy in the careers of Sir John Jordan and Yuän Shih-kåi / Chan Lau Kit-ching. — ₍Hong Kong₎ : Hong Kong University Press, 1978.
viii, 168 p. ; 23 cm.
Bibliography: p. ₍144₎-156.
Includes index.
ISBN (invalid) 0-9622090-1-09
1. Great Britain—Foreign relations—China. 2. China—Foreign relations—Great Britain. 3. Great Britain—Foreign relations—1901-1910. 4. Great Britain—Foreign relations—1910-1936. 5. China—Politics and government—1900-1949. 6. China—Foreign relations—To 1912. 7. China—Foreign relations—1912-1949. 8. Yüan, Shih-k'ai, 1859-1916. 9. Jordan, John Newell, Sir, 1852- I. Title.
DA47.9.C6C45 327.41'051 79-106848
79 MARC

Correspondence respecting the revision of the Treaty of Tien-Tsin : presented to the House of Lords by command of Her Majesty, in pursuance of their address dated March 24, 1871. — San Francisco : Chinese Materials Center, 1975.
viii, 467 p., ₍5₎ fold. leaves of plates : maps ; 27 cm. — (₍Reprint series₎ - Chinese Materials Center, inc. ; ₍no. 26₎)
Spine title: Revision of the Treaty of Tien-Tsin.
Reprint of the 1871 ed. printed by Harrison, London, which was issued as China no. 5 and as no. 389 of Papers by command, c.
1. Great Britain—Foreign relations—China. 2. China—Foreign relations—Great Britain. 3. Great Britain—Commerce—China. 4. China—Commerce—Great Britain. 5. Unequal treaties. I. Title: Revision of the Treaty of Tien-Tsin. II. Series: Chinese Materials Center. Reprint series - Chinese Materials Center ; no. 26. III. Series: Great Britain. Foreign Office. China ; no. 5. IV. Series: Great Britain. Parliament. Papers by command, c ; 389.
JX637 1975.C67 1975 327.41'051 78-324368
79 MARC

Waung, W S K 1938-
The controversy : opium and Sino-British relations, 1858-1887 / W. S. K. Waung. — Hong Kong : Lung Men Press, c1977.
xii, 282 p. : maps (on lining papers) ; 27 cm.
Bibliography: p. 263-274.
Includes index.
1. Opium trade—China—History. 2. Narcotics, Control of—International cooperation—History. 3. China—Foreign relations—Great Britain. 4. Great Britain—Foreign relations—China. I. Title.
HV5816.W36 363.4'5'0951 78-318904
78 MARC

—FOREIGN RELATIONS—HUNZA

Müller-Stellrecht, Irmtraud.
Hunza und China (1761-1891) : 130 Jahre e. Beziehung u. ihre Bedeutung für d. wirtschaftl. u. polit. Entwicklung Hunzas im 18. u. 19. Jh. / von Irmtraud Müller-Stellrecht. — Wiesbaden : Steiner, 1978.
139 p. ; 24 cm. — (Beiträge zur Südasienforschung ; Bd. 44) GFR78-A
Includes some text and summary in English.
Bibliography: p. 124-133.
ISBN 3-515-02799-8
1. China—Foreign relations—Hunza. 2. Hunza—Foreign relations—China. I. Title. II. Series.
DS740.5.H86M83 78-381684
79 MARC

—FOREIGN RELATIONS—INDOCHINA—ADDRSSES, ESSAYS, LECTURES

(Betonamu Chūgoku Kambojia no kankei to shakaishugi to o kangaeru)
ベトナム・中国・カンボジアの関係と社会主義とを考える / 本多勝一編. — 東京：朝日新聞社, 1979.
286 p. : ill. ; 20 cm. Ja 79-4428
¥1100
1. Indochina—Politics and government—Addresses, essays, lectures. 2. Cambodian-Vietnamese Conflict, 1977- —Addresses, essays, lectures. 3. China—Foreign relations—Indochina—Addresses, essays, lectures. 4. Indochina—Foreign relations—China—Addresses, essays, lectures. 5. Socialism in Indochina—Addresses, essays, lectures. I. Honda, Katsuichi, 1933-
DS550.B47 79-808819

—FOREIGN RELATIONS—JAPAN

Chu, Shao-hsien.
Peiping's policy and tactics toward Japan / ₍Chu Shao-hsien₎. — Taipei : Chung Hwa Information Service, ₍1978?₎
25 p. ; 20 cm.
Includes bibliographical references.
1. China—Foreign relations—Japan. 2. Japan—Foreign relations—China. I. Title.
DS740.5.J3C519313 327.51'052 78-103812
79 MARC

CHINA

—FOREIGN RELATIONS—JAPAN (Continued)

Chung-kuo kuo min tang. Chung yang wei yüan hui. Wen hua kung tso hui.
(Jih-pen cheng fu ying ming pien shih fei)
日本政府應明辨是非 / 中國國民黨中央委員會文化工作會編印. — [臺北] : 該會, 民國61 [1972]
2, 130 p. ; 22 cm. — (我們的叢書 ; 8)
Cover title.
Series romanized: Wo men ti ts'ung shu.
1. China—Foreign relations—Japan. 2. Japan—Foreign relations—China. I. Title.
DS740.5.J3C542 1972 79-838691

Daini Kokuminkai. Henshūbu.
(Ugoku Hokushi)
動く北支 / [編者 第二國民會編輯部 ; 校閲 高橋坦]. — 東京 : 第二國民會出版部 : 發賣所 栗田書店, 昭和12 [1937]
2, 4, 277 p., [9] leaves of plates : ill. ; 19 cm.
1. China—Politics and government—1937-1949. 2. China—Foreign relations—Japan. 3. Japan—Foreign relations—China. 4. China—Armed Forces. 5. Marco Polo Bridge Incident, 1937. I. Title.
DS777.53.D27 1939 78-815478

Furui, Yoshimi, 1903-
(Nitchū jūhachinen)
日中十八年 : 一政治家の軌跡と展望 / 古井喜實. — 東京 : 牧野出版, 1978.
243 p., [2] leaves of plates : port. ; 19 cm. Ja 78-16736
¥1600
1. Japan—Foreign relations—China. 2. China—Foreign relations—Japan. 3. China—History—1949-1976. I. Title.
DS849.C6F78 78-814478

Gurūpu Nijūichi.
(Nitchū kankei)
日中関係 : 平和条約と国際環境 / グループ21. — [東村山] : 教育社 ; 東京 : 販売 教育社出版サービス, [1978]
144 p. : ill. ; 18 cm. — (時事問題解説 ; 113) (入門新書) Ja 78-25757
Cover title.
Series 2 romanized: Nyūmon shinsho.
¥400
1. Japan—Foreign relations—China. 2. China—Foreign relations—Japan. I. Title. II. Series: Jiji mondai kaisetsu ; 113.
DS849.C6G87 1978 79-803880

Hasegawa, Satoru, 1896-1969.
(Meirō shin Shina no kensetsu)
明朗新支那の建設 / 長谷川了著. — 東京 : 第二国民会出版部, 昭和13 [1938]
2, 4, 299 p. ; 19 cm.
1. China—Politics and government—1937-1949. 2. China—Foreign relations—Japan. 3. Japan—Foreign relations—China. 4. Russia—Foreign relations. 5. Communist strategy. I. Title.
DS777.53.H337 78-816935

Hayashi, Kyūjirō, 1882-1964.
(Manshū jihen to Hōten sōryōji)
満州事変と奉天総領事 : 林久治郎遺稿 / 林久治郎著. — 東京 : 原書房, 1978.
3, 204, 7 p. ; 20 cm. Ja 78-12863
¥1500
1. Manchuria—History—1931-1945. 2. Japan—Foreign relations—China. 3. China—Foreign relations—Japan. I. Title.
DS784.2.H35 1978 78-813325

J

Honda, Kumatarō, 1874-1948.
(Ōshū jōsei to Shina jihen)
歐洲情勢と支那事變 / 本多熊太郎著. — 東京 : 千倉書房, 昭和14 [1939]
3, 8, 284 p. ; 19 cm.
1. World politics, 1933-1945. 2. Japan—Foreign relations—China. 3. China—Foreign relations—Japan. I. Title.
D727.H59 78-817157

Hsü, Tung-fan, 1886-
(K'ang cheng Lu an chih wen chang)
抗爭魯案之文章 / 著作者徐東藩 ; 校閲者郁悼童, 顧旭侯. — 上海 : 山東外交協會江蘇分會 : 代售處商務印書總分館, 民國9 [1920]
66, [14] p., [1] leaf of plates : ports. ; 24 cm.
Colophon title.
1. China—Foreign relations—Japan. 2. Japan—Foreign relations—China. 3. European War, 1914-1918—Territorial questions—China—Shantung. I. Title.
DS740.5.J3H79 78-843821

Ikeguchi, Ekan, 1936-
(Shō Kaiseki)
蔣介石 : その偉大なる生涯 / 池口恵観. — 東京 : 論稿社, 昭和51 [1976]
215 p. : ill. ; 20 cm. Ja***
¥1200
1. Chiang, Kai-shek, 1887-1975. 2. China—Presidents—Biography. 3. China—History—Republic, 1912-1949. 4. Japan—Foreign relations—China. 5. China—Foreign relations—Japan. I. Title.
DS778.C55I33 78-814965

Inoue, Masaji, 1876-
(Kyojin Arao Sei)
巨人荒尾精 : 附・十二烈士傳 / [著作者 井上雅二]. — 東京 : 東亜同文会, 昭和11 [1936]
220, [19]., [5] leaves of plates : ill. ; 22 cm.
1. Arao, Sei, 1859-1896. 2. Sinologists—Japan—Biography. 3. Japan—Foreign relations—China. 4. China—Foreign relations—Japan. 5. Japan—Politics and government—1868-1912. I. Title.
DS734.9.A7I55 78-816106

Itō, Hirobumi, 1841-1909.
(T'ai-wan ko jang Chung Jih t'an p'an mi hua lu)
台灣割讓中日談判秘話錄 : 伊藤博文・李鴻章一問一答. — 再版. — 臺北 : 西南書局, 民國64 [1975]
162 p., [2] leaves of plates : ill. ; 22 cm. — (學術叢書 ; 43)
"李鴻章手記"
Chinese and Japanese.
Series romanized: Hsüeh shu ts'ung shu.
1. Chinese-Japanese War, 1894-1895—Diplomatic history. 2. Japan—Foreign relations—China. 3. China—Foreign relations—Japan. 4. Taiwan—History—To 1895. I. Li, Hung-chang, 1823-1901. II. Title. III. Title: Chung Jih t'an p'an mi hua.
DS765.I93 1975 79-836789

79 C

Jain, Rajendra Kumar.
China and Japan, 1949-1976 / R. K. Jain. — Atlantic Highlands, N.J. : Humanities Press, 1977.
xix, 336 p. ; 22 cm.
Bibliography: p. [323]-327.
Includes index.
ISBN 0-391-00749-1
1. China—Foreign relations—Japan. 2. Japan—Foreign relations—China. I. Title.
DS740.5.J3J34 1977 327.51'052 77-70008
77[r79]rev MARC

Jōno, Hiroshi.
(Chūgokugaku genri)
中国学原理 / 城野宏. — 東京 : 原書房, 1978.
214 p. ; 20 cm. Ja 78-18115
¥1300
1. China—Study and teaching. 2. China—Politics and government—1949-1976. 3. Japan—Foreign relations—China. 4. China—Foreign relations—Japan. I. Title.
DS734.95.J66 78-813203

*78 J

Katogawa, Kōtarō, 1909-
(Chūgoku to Nihon Rikugun)
中国と日本陸軍 / 加登川幸太郎. — 東京 : 圭文社, 1978.
2 v. : ill. ; 19 cm. Ja 78-23983
Bibliography: v. 2, p. 272-273.
¥1500 per vol.
1. Japan—Foreign relations—China. 2. China—Foreign relations—Japan. 3. Japan. Rikugun—History. I. Title.
DS849.C6K273 79-800999

Kobayashi, Jun'ichirō.
(Hijōji kyūkoku gaikō taisaku)
非常時救國外交對策 : 附 日支問題に就て帝國外交の驚くべき失敗の眞相 / 小林順一郎著. — 東京 : 自衛社 : 大賣捌元 東京堂, 昭和8 [1933]
119, 56 p., [1] leaf of plates : port. ; 23 cm.
1. Japan—Foreign relations—1912-1945. 2. League of Nations. 3. Japan—Foreign relations—China. 4. China—Foreign relations—Japan. I. Title. II. Title: Gaikō taisaku.
DS888.5.K58 79-801111

Liao, T'ieh-chün.
(T'ang-ku hsieh ting hou ti Hua pei chü shih)
塘沽協定後(一九三三-一九三六)的華北局勢 / 廖鐵軍編著. — 臺北 : 驚聲文物供應公司, 民國66 [1977]
4, 106 p. ; 22 cm.
Bibliography: p. 100-106.
NT$40.00
1. China—Foreign relations—Japan. 2. Japan—Foreign relations—China. I. Title.
DS740.5.J3L53 79-837260

Lin, Wen-lung.
(Sansen to kensetsu Kahoku)
参戰と建設華北 / 林文龍著. — 東京 : 聯合出版社, 昭和18 [1943]
2, 2, 148 p., [1] leaf of plates : port. ; 19 cm.
Cover title.
1. World War, 1939-1945—China. 2. China—Politics and government—1937-1945. 3. China—Foreign relations—Japan. 4. Japan—Foreign relations—China. I. Title.
D767.3.L55 79-804166

Mishima, Yasuo.
(Shina jihen no honshitsu to mitōshi)
支那事變の本質と見透し : 國際新情勢と日本の立場 / 三島康夫著. — 東京 : 實業之日本社, 昭和14 [1939]
5, 230 p. ; 19 cm.
1. Sino-Japanese Conflict—1937-1945. 2. Japan—Foreign relations—China. 3. China—Foreign relations—Japan. 4. Japan—Foreign relations—1912-1945. I. Title.
DS777.53.M553 78-815469

Nakamura, Nobuyasu.
(Taiheiyō to rekkyō kaigun)
太平洋と列强海軍 / 中村伸康著. — 東京 : 高山書院, 昭和13 [1938]
6, 11, 246 p., [2] leaves of plates : ill. ; 19 cm.
1. Japan—Foreign relations—China. 2. China—Foreign relations—Japan. 3. Japan—Foreign relations—1912-1945. 4. Pan-Pacific relations. 5. Sea-power. I. Title.
DS849.C6N333 79-804176

(Nihon fashizumu no keisei)
日本ファシズムの形成 / 江口圭一編. — 東京 : 日本評論社, 1978.
6, 3, 302 p. ; 19 cm. — (体系・日本現代史 ; 1) Ja 79-1402
Includes bibliographical references.
1. Fascism—Japan—History. 2. Japan—History—1926-1945. 3. Japan—Foreign relations—China. 4. China—Foreign relations—Japan. I. Eguchi, Keiichi, 1932- II. Series: Taikei Nihon gendai shi ; 1.
DS888.25.N53 79-807182

*79 J

(Nissō bōeki to Genkō)
日宋貿易と元寇 / [編集 第二アートセンター]. — 東京 : 毎日新聞社, 昭和54 [1979]
187 p. : ill. (some col.) ; 27 cm. — (図説人物海の日本史 ; 2) Ja***
Includes bibliographies.
¥1900
1. Japan—History—Heian period, 794-1185. 2. Pirates—Japan—History. 3. Japan—History, Naval—To 1868. 4. Japan—Foreign relations—China. 5. China—Foreign relations—Japan. 6. Japan—History—Attempted Mongol invasions, 1274-1281. I. Daini Āto Sentā. II. Mainichi Shimbun Sha. III. Series: Zusetsu jimbutsu umi no Nihon shi ; 2.
DS856.3.N57 79-805978

CHINA

—FOREIGN RELATIONS—JAPAN (Continued)

Nohara, Shirō, 1903–
(Chūgoku kakumei to Dai Nippon Teikoku)
中国革命と大日本帝国 / 野原四郎著. — 東京: 研文出版, 1978.
iii, 282 p. ; 20 cm. — (研文選書 ; 2) Ja 78-24003
Series romanized: Kenbun sensho.
Includes bibliographical references.
¥1800
1. China—History—1900– 2. China—Foreign relations—Japan. 3. Japan—Foreign relations—China. I. Title.
DS774.N63 79-800112

Takagi, Rikurō, 1880–
(Nikka kōyūroku)
日華交友錄 / 高木陸郎著. — 東京: 救護會出版部, 昭和18 [1943]
6, 6, 236 p. ; 18 cm.
1. China—Politics and government—1912-1949. 2. China—Foreign relations—Japan. 3. Japan—Foreign relations—China. I. Title.
DS774.T26 79-801739

Uchida, Ryōhei, 1874-1937.
(Shina kan)
支那觀 / 內田良平著. — 東京: 黑龍會, 大正2 [1913]
3, 3, 127, 7 p. ; 23 cm.
1. Japan—Foreign relations—China. 2. China—Foreign relations—Japan. 3. China. I. Title.
DS849.C6U26 78-817226

Ueji, Tatsunori, 1932–
(Senkaku Rettō to Takeshima)
尖閣列島と竹島: 中国·韓国との領土問題 / 上地龍典. — [東村山]: 教育社; 東京: 販売 教育社出版サービス, [1978]
158 p. : ill. ; 18 cm. — (時事問題解説 ; 95)(入門新書) Ja 78-23753
Cover title.
Series 2 romanized: Nyūmon shinsho.
Bibliography: p. 158.
¥400
1. Japan—Foreign relations—China. 2. China—Foreign relations—Japan. 3. Senkaku Islands—International status. 4. Japan—Foreign relations—Korea. 5. Korea—Foreign relations—Japan. 6. Take-shima. 7. Japan—Boundaries. I. Title. II. Series: Jiji mondai kaisetsu ; 95.
DS849.C6U33 79-801087

Willoughby, Westel Woodbury, 1867-1945.
Japan's case examined / Westel W. Willoughby. — New York : Arno Press, 1979, c1940.
x, 237 p. ; 22 cm. — (Johns Hopkins University Press reprints)
Reprint of the ed. published by the Johns Hopkins Press, Baltimore.
Includes index.
ISBN 0-405-10635-1 : $18.00
1. Japan—Foreign relations—China. 2. China—Foreign relations—Japan. 3. United States—Foreign relations—Japan. 4. Japan—Foreign relations—United States. 5. United States—Foreign relations—China. 6. China—Foreign relations—United States. 7. Eastern question (Far East) I. Title.
DS849.C6W52 1979 327.52'051 78-19305
78 MARC

Young, Carl Walter, 1902-1939.
Japan's special position in Manchuria. New York, AMS Press [1971]
xxxiv, 412 p. 23 cm. (His Japan's jurisdiction and international legal position in Manchuria (v. 1))
Bibliography: p. 393-401.
ISBN 0-404-07072-8
1. Japan—Foreign relations—China. 2. China—Foreign relations—Japan. 3. Manchuria—International status. 4. Railroads—China—Manchuria. I. Title. II. Series.
DS849.C6Y63 1971 301.29'52'0518 71-175937
72[r79]rev MARC

Young, Carl Walter, 1902-1939.
Japan's special position in Manchuria / C. Walter Young. — New York : Arno Press, 1979, c1931.
xxxiv, 412 p. : map ; 21 cm. — (Johns Hopkins University Press reprints)
Reprint of the ed. published by Johns Hopkins Press, Baltimore, in the author's series: Japan's jurisdiction and international legal position in Manchuria, v. 1.
Bibliography: p. 393-401.
Includes index.
ISBN 0-405-10636-X : $28.00
1. Japan—Foreign relations—China. 2. China—Foreign relations—Japan. 3. Manchuria—International status. 4. Railroads—China—Manchuria. I. Title. II. Series: Young, Carl Walter, 1902-1939. Japan's jurisdiction and international legal position in Manchuria ; v. 1.
[DS849.C6Y63 1979] 341.2'9 78-19322
78 MARC

—FOREIGN RELATIONS—JAPAN —ADDRESSES, ESSAYS, LECTURES

Hirano, Reiji, 1897–
(Mōko no uta)
蒙古の唄 / 平野零兒著. — 東京: 平原社, 昭和11 [1936]
244 p. ; 19 cm.
1. Manchuria—History—1931-1945—Addresses, essays, lectures. 2. Mongolia (Outer Mongolia)—History—Addresses, essays, lectures. 3. China—Foreign relations—Japan—Addresses, essays, lectures. 4. Japan—Foreign relations—China—Addresses, essays, lectures. I. Title.
DS783.7.H45 79-801414

Ho, Ying-ch'in, 1889–
(Jih fei kuan hsi yü Chung kung yin mou)
日匪關係與中共陰謀 / 著述者何應欽. — 臺北: 海外出版社, 民國61 [1972]
2, 52 p., [1] leaf of plates : ill. ; 19 cm.
Colophon title.
1. China—Foreign relations—Japan—Addresses, essays, lectures. 2. Japan—Foreign relations—China—Addresses, essays, lectures. I. Title.
DS740.5.J3H59 78-843822

(Kōa kenkoku undō no riron to shuchō)
興亞建國運動の理論と主張 / 上海日本總領事館特別調査班譯. — [s. l. : s. n.], 昭和17 [1942]
8, 6, 564 p. ; 21 cm.
"民國二十八年(昭和十四年)秋より約一ヶ年の間に興亞建國運動の幹部同志によって發表された論文を邦譯編輯したもの"
1. China—Politics and government—1937-1945—Addresses, essays, lectures. 2. Sino-Japanese Conflict, 1937-1945—Collaborationists—Addresses, essays, lectures. 3. China—Foreign relations—Japan—Addresses, essays, lectures. 4. Japan—Foreign relations—China—Addresses, essays, lectures. I. Japan. Sōryōjikan, Shanghai.
DS777.53.K54 78-815769

Shimizu, Yoshitarō, 1899–
(Dōran no Shina to shōrai)
動乱の支那と將來 / 清水芳太郎著. — 福岡: 清水芳太郎先生後援會本部, 昭和12 [1937]
6, 189 p., [5] leaves of plates : ill. ; 19 cm.
Cover title.
1. Japan—Politics and government—1912-1945—Addresses, essays, lectures. 2. Japan—Foreign relations—China—Addresses, essays, lectures. 3. China—Foreign relations—Japan—Addresses, essays, lectures. 4. World politics—1933-1945—Addresses, essays, lectures. I. Title.
DS888.2.S 79-803970
J

Suzuki, Teiichi, 1888–
(Suzuki Teiichi Shi danwa sokkiroku)
鈴木貞一氏談話速記録 / 日本近代史料研究会, 木戸日記研究会編. — 東京: 日本近代史料研究会, 昭和 [19
v. ; 25 cm. — (日本近代史料叢書 ; B-4) Ja 75-14 (v. 2)
1. Suzuki, Teiichi, 1888– —Addresses, essays, lectures. 2. Statesmen—Japan—Biography—Addresses, essays, lectures. 3. Japan—Politics and government—1912-1945—Addresses, essays, lectures. 4. Japan—Foreign relations—China—Addresses, essays, lectures. 5. China—Foreign relations—Japan—Addresses, essays, lectures. I. Kido Nikki Kenkyūkai. II. Nihon Kindai Shiryō Kenkyūkai. III. Title. IV. Series: Nihon kindai shiryō sōsho ; B-4.
DS885.5.S86A36 79-803566
*79 J

Wang, Kuang-t'i, 1918–
(Chung Jih kuan hsi erh shih wu nien)
中日關係二十五年 / 司馬桑敦[王光逖]著. — 臺北: 聯合報社: 總經銷聯經出版事業公司, 民國67 [1978]
5, 14, 422 p. ; 21 cm. — (聯合報叢書)
Series romanized: Lien ho pao ts'ung shu.
NT$160.00
1. China—Foreign relations—Japan—Addresses, essays, lectures. 2. Japan—Foreign relations—China—Addresses, essays, lectures. 3. Taiwan—Foreign relations—Japan—Addresses, essays, lectures. 4. Japan—Foreign relations—Taiwan—Addresses, essays, lectures. I. Title.
DS740.5.J3W313 78-841077

—FOREIGN RELATIONS—JAPAN —FOREIGN OPINION CHINESE

Lo, Kang.
[Wo men tui Jih Wang hsieh ting ying yu chih jen shih. Japanese]
日汪協定に對する吾人の認識 / [羅剛]. — [上海]: 大陸調査會上海調査室, 昭和15 [1940]
26 p. (on double leaves) ; 26 cm. — (大調政治資料 ; 第3輯)
Cover title.
Translation of article published in Shang-hai chou pao, v. 2, no. 26, Dec. 7, 1940.
1. China—Foreign relations—Japan—Foreign opinion, Chinese. 2. Japan—Foreign relations—China—Foreign opinion, Chinese. 3. Public opinion—China. 4. China—Politics and government—1912-1949. I. Title. II. Series: Tairiku Chōsakai. Shanhai Chōsashitsu. Taichō seiji shiryō ; dai 3-shū.
Title romanized: Nichi-Ō kyōtei ni taisuru gojin no ninshiki.
DS740.5.J3L616 78-812948

—FOREIGN RELATIONS—JAPAN —PICTORIAL WORKS

(Manshū jihen zengo)
満州事変前後: 孤立への道: 昭和元年-10年. — 東京: 毎日新聞社, 1975.
258 p. : chiefly ill. ; 28 cm. — (一億人の昭和史 ; 1) Ja***
Cover title.
"特別企画"王道楽土"建国から滅亡まで"
¥1000
1. Mukden Incident, 1931—Pictorial works. 2. Japan—Foreign relations—China—Pictorial works. 3. China—Foreign relations—Japan—Pictorial works. 4. Japan—History—1926-1945—Pictorial works. I. Mainichi Shimbun Sha. II. Series: Ichiokunin no Shōwa shi ; 1.
DS888.15.I25 vol. 1 79-800024
[DS783.8]

—FOREIGN RELATIONS—JAPAN —SOURCES

(Nihon mondai bunken shū)
日本問題文献集 / 中国世界知識社編, 国民文庫編集部訳. — 東京: 大月書店, 1955.
319 p. ; 19 cm.
"北京の世界知識社編「日本問題文件彙編」の構成にそのまゝしたがい, 可能なかぎり原文によって訳出"
1. Japan—Foreign relations—1945– —Sources. 2. Japan—Politics and government—1945– —Sources. 3. China—Foreign relations—Japan—Sources. 4. Japan—Foreign relations—China—Sources. I. Shih chieh chih shih she, Peking.
DS889.5.N53 79-801953

—FOREIGN RELATIONS—KOREA

Pak, Hong-min.
(Chunggong oegyo chŏngch'aek kyŏlchŏng kwajŏng kwa Mo T'aek-tong ihu tae Hanbando chŏngch'aek)
中共外交政策決定過程과毛澤東以後対韓半島政策 / 研究執筆責任 朴洪眠; 刊行責任 羅昌柱. — [서울]: 國土統一院調査研究室, 1977.
44 p. ; 26 cm. — (周邊環境興件 : 中共)
"국통조 77-1-1104."
1. China—Foreign relations—1949– 2. China—Foreign relations—Korea. 3. Korea—Foreign relations—China. I. Na, Ch'ang-ju. II. Title. III. Series: Chubyŏn hwan'gyŏng yŏkŏn.
DS777.55.P268 78-826281

—FOREIGN RELATIONS—KOREA —SOURCES

(Ch'ao t'ien lu)
朝天錄: 明代中韓關係史料選輯 / 權近等著; 本社編輯部選輯. — 台北: 珪庭出版社, 民國67 [1978]
4 v. (1737 p.) ; 22 cm. — (中韓關係史料輯要 ; 2)
Reprint ed.
NT$2200.00
1. Korea—Foreign relations—China—Sources. 2. China—Foreign relations—Korea—Sources. I. Kwŏn, Kŭn, 1352-1409. II. Kuei t'ing ch'u pan she. Pien chi pu. III. Series: Chung Han kuan hsi shih liao chi yao ; 2.
DS910.2.C5C48 79-837729

CHINA

—FOREIGN RELATIONS
—KOREA—SOURCES (Continued)

Tongmum hwigo. Chinese.

同文彙考 / 鄭昌順等編纂. — 台北：珪庭出版社, 民國67- [1978-
v. ; 22 cm. — (中韓關係史料輯要 ; 4)
"收錄文件自清順治元年至乾隆五十二年, 朝鮮仁祖至正祖十一年, 西元一六四四——七八七年.
Reprint ed.
1. Korea—Foreign relations—China—Sources. 2. China—Foreign relations—Korea—Sources. I. Chŏng, Ch'ang-sun, b. 1727. II. Title. III. Series: Chung Han kuan hsi shih liao chi yao ; 4.
Title romanized: T'ung wen hui k'ao.
DS910.2.C5T6512 1978 79-837766

—FOREIGN RELATIONS—NEAR EAST

Shichor, Yitzhak.
The Middle East in China's foreign policy, 1949-1977 / Yitzhak Shichor. — Cambridge ; New York : Cambridge University Press, 1979.
xiii, 268 p. ; 23 cm. — (International studies)
Bibliography: p. [250]-259.
Includes index.
ISBN 0-521-22214-1 : $21.00
1. Near East—Foreign relations—China. 2. China—Foreign relations—Near East. I. Title.
DS63.2.C5S52 327.51'056 78-58801
78 MARC

—FOREIGN RELATIONS—PAKISTAN

Rais, Rasul Bux.
China and Pakistan : a political analysis of mutual relations / Rasul Bux Rais. — Lahore : Progressive Publishers, 1977.
iv, 163 p. ; 21 cm.
Bibliography: p. 156-163.
Rs20.00
1. Pakistan—Foreign relations—China. 2. China—Foreign relations—Pakistan. I. Title.
DS383.5.C5R34 327.549'1'051 78-930572
79 MARC

—FOREIGN RELATIONS—RUSSIA

Borisov, Oleg Borisovich.
The Soviet Union and the Manchurian revolutionary base : (1945-1949) / O. Borisov ; [translated from the Russian by David Fidlon]. — Moscow : Progress Publishers, c1977.
275 p. ; 18 cm. USSR***
Translation of Sovetskiĭ Soiūz i Man'chzhurskaia revoliūtsionnaia baza.
Includes bibliographical references.
1. Manchuria—History—1945- . 2. China—History—Civil War, 1945-1949. 3. Russia—Foreign relations—China. 4. China—Foreign relations—Russia. I. Title.
DS784.2.B6713 951.04'2 78-380143
78 MARC

Chang, Hsiao-yen.
(Chen-pao tao shih chien chi fei O kuan hsi)
珍寶島事件及匪俄關係, 一九五六——九六九 / 章孝嚴著. — 臺北：黎明文化事業公司, 民國67 [1978]
6, 180, [64] p., [1] leaf of plates : ill. ; 20 cm.
NT$60.00
1. China—Foreign relations—Russia. 2. Russia—Foreign relations—China. 3. China—Boundaries—Russia. 4. Russia—Boundaries—China. 5. Chen-pao Island. I. Title.
DS740.R8C452 78-843090

Ch'en, Pin-ho, fl. 1926-
(Chung Su fu chiao wen t'i)
中蘇復交問題 / 陳彬龢. — [香港? : s. n., 1978?]
60 p. ; 18 cm.
Reprint of the 1933 ed. published by Liang yu t'u shu yin shua kung ssu, Shanghai, issued in series: I chiao ts'ung shu ; 51 chung.
Includes bibliographical references.
1. China—Foreign relations—Russia. 2. Russia—Foreign relations—China. I. Title.
DS740.5.R8C468 1978 79-837976

China (Republic of China). Kuang fu ta lu she chi yen chiu wei yüan hui.
(Fei O pien chieh cheng chih wen t'i chih yen chiu)
匪俄邊界爭執問題之研究 / 光復大陸設計研究委員會研擬. — [台北]：該會, 民國65 [1976]
2, 52 p. ; 22 cm. — (邊疆研究叢書)
Cover title.
1. China—Foreign relations—Russia. 2. Russia—Foreign relations—China. 3. China—Boundaries—Russia. 4. Russia—Boundaries—China. I. Title. II. Series: Pien chiang yen chiu ts'ung shu.
DS740.5.R8C5512 1976 78-843066

Chu, Wen-lin.
The current situation and prospect of the Russo-Chinese Communist relationship / [by Chu Wen-lin]. — Taipei : Chung Hwa Information Service, [1978?]
18 p. ; 20 cm.
1. China—Foreign relations—Russia. 2. Russia—Foreign relations—China. I. Title.
DS740.5.R8C552713 327.51'047 78-103819
78 MARC

Chung-kuo she hui k'o hsüeh yüan. Chin tai shih yen chiu so.
(Sha O ch'in Hua shih)
沙俄侵华史 / 中国社会科学院近代史研究所. — 第1版. — [北京]：人民出版社, 1978-
v. : ill. ; 21 cm.
Includes bibliographies.
$0.90 (v. 1)
1. China—Foreign relations—Russia. 2. Russia—Foreign relations—China. 3. China—Boundaries—Russia. 4. Russia—Boundaries—China. I. Title.
DS740.5.R8C558 1978 78-843452

Fejtö, François, 1909-
Chine—U.R.S.S. : de l'alliance au conflit : 1950-1977 / François Fejtö. — Nouvelle éd., complétée et mise à jour. — Paris : Éditions du Seuil, 1978.
479 p. ; 18 cm. — (Points) (Politique ; 60) F78-11842
Bibliography: p. [472]-473.
Includes index.
ISBN 2-02-000364-3 : 16.00F
1. China—Foreign relations—Russia. 2. Russia—Foreign relations—China. I. Title.
DS740.5.R8F4 1978 327.51'047 79-339324
79 MARC

Heinzig, Dieter.
Sowjetische Militärberater bei der Kuomintang : 1923-1927 / Dieter Heinzig. — 1. Aufl. — Baden-Baden : Nomos-Verlagsgesellschaft, 1978.
339 p. ; 23 cm. — (Osteuropa und der internationale Kommunismus ; Bd. 1) GFR78-A
Includes glossary in Chinese with transliteration.
Bibliography: p. 295-309.
Includes index.
ISBN 3-7890-0371-9
1. China—Foreign relations—Russia. 2. Russia—Foreign relations—China. 3. Chung-kuo kuo min tang. I. Title. II. Series.
DS740.5.R8H43 327.47'051 79-341654
79 MARC

Heinzig, Dieter.
Sowjetisch-chinesische Beziehungen nach Mao Tse-tungs Tod / Dieter Heinzig. — Köln : Bundesinst. für Ostwissenschaftl. u. Internat. Studien, 1978.
36 p. ; 30 cm. — (Berichte des Bundesinstituts für Ostwissenschaftliche und Internationale Studien ; 1978, 35) GFR79-B1
Cover title.
Summary also in English.
Includes bibliographical references.
1. China—Foreign relations—Russia. 2. Russia—Foreign relations—China. I. Title. II. Series: Germany (Federal Republic, 1949-). Bundesinstitut für Ostwissenschaftliche und Internationale Studien. Berichte ; 1978, 35.
HX15.G468 1978, nr. 35 309.1'171'7 s 79-358652
[DS740.5.R8] 79 MARC

Hsü, Hsiang-t'ao.
(Chan hou Su-lien tui Hua cheng ts'e chih yen chiu)
戰後蘇聯對華政策之研究 / 許湘濤撰. — 民國65 [1976]
[14], 248 p. (on double leaves) ; 27 cm.
Reproduced from typescript.
Thesis (M. A.)—Kuo li cheng chih ta hsüeh Tung-ya yen chiu so.
Bibliography: leaves 232-248.
1. Russia—Foreign relations—China. 2. China—Foreign relations—Russia. I. Title.
DK68.7.C5H75 1976b 79-841844

Hu, Wei-chen.
(Ts'ung Mo-ssu-k'o kuang po k'an Su O tui Hua cheng ts'e)
從莫斯科廣播看蘇俄對華政策 / 著者胡為眞. — 臺北：國立政治大學國際關係研究中心, 民國67 [1978]
3, 153 p. ; 22 cm.
Colophon title.
Bibliography: p. 145-153
NT$40.00 ($1.50 U.S.)
1. Russia—Foreign relations—China. 2. China—Foreign relations—Russia. 3. Propaganda, Russian. I. Title.
DK68.7.C5H8 79-841420

Lieberthal, Kenneth.
Sino-Soviet conflict in the 1970s : its evolution and implications for the strategic triangle / Kenneth G. Lieberthal ; a report prepared for Director of Net Assessment, Office of the Secretary of Defense. — Santa Monica, CA : Rand, 1978.
xv, 197 p. ; 28 cm. — ([Report] - Rand Corporation ; R-2342-NA)
Bibliography: p. 193-197.
ISBN 0-8330-0049-7
1. Russia—Foreign relations—China. 2. China—Foreign relations—Russia. I. United States. Dept. of Defense. Director of Net Assessment. II. Title. III. Series: Rand Corporation. Rand report ; R-2342-NA.
AS36.R3 R-2342 081 s 78-13529
[DK68.7.C5] 78 MARC

Louis, Victor E
The coming decline of the Chinese Empire / Victor Louis : with a dissenting introd. by Harrison E. Salisbury. — New York : Times Books, c1979.
xxii, 198 p., [1] leaf of plates : map ; 22 cm.
Includes index.
ISBN 0-8129-0819-8
1. Russia—Foreign relations—China. 2. China—Foreign relations—Russia. I. Title.
DK68.7.C5L67 1979 327.47'051 78-20686
79 MARC

Marcy, Sam, 1911-
The class character of the USSR : an answer to the theory of "Soviet social-Imperialism" / by Sam Marcy. — New York : World View Publishers, 1977.
71 p. : ill. ; 21 cm.
Includes an appraisal of the History of USSR-China relations, by V. Copeland.
Includes bibliographical references.
ISBN 0-89567-009-7 : $1.00
1. Social classes—Russia. 2. Russia—Foreign relations—China. 3. China—Foreign relations—Russia. 4. World politics—20th century. I. Copeland, Vincent. History of USSR-China relations. 1977. II. Title.
HN530.Z9S637 327.47'051 78-15436
78 MARC

Middleton, Drew, 1913-
The Duel of the giants : China and Russia in Asia / by Drew Middleton. — New York : Scribner, c1978.
viii, 241 p. ; 22 cm.
Includes index.
ISBN 0-684-15785-3 : $8.95
1. China—Foreign relations—Russia. 2. Russia—Foreign relations—China. 3. China—Defenses. I. Title.
DS740.5.R8M48 327.51'047 78-9305
78 MARC

Minami Manshū Tetsudō Kabushiki Kaisha. Chōsaka.
(Rokoku senryō zengo ni okeru Dairen oyobi Ryojun)
露國占領前後ニ於ケル大連及旅順. — [大連]：南滿洲鐵道株式會社調查課, [緒言 明治44 i. e. 1911]
4, 61 p. ; 23 cm.
Cover title.
1. Dairen—Description. 2. Port Arthur—Description. 3. Russia—Foreign relations—China. 4. China—Foreign relations—Russia. I. Title.
DS796.D3M52 1911 78-817217

Nakajima, Mineo, 1936-
(Chū-So tairitsu to gendai)
中ソ対立と現代：戦後アジアの再考察 / 中嶋嶺雄. — 東京：中央公論社, 1978.
298 p. ; 21 cm. — (叢書国際環境) Ja 79-3092
Includes bibliographical references and indexes.
¥1700
1. China—Foreign relations—Russia. 2. Russia—Foreign relations—China. 3. East Asia—Foreign relations. I. Title. II. Series: Sōsho kokusai kankyō.
DS740.5.R8N246 79-805615

Olgin, V
On the Soviet-Chinese border : questions and answers / Vladimir Olgin. — Moscow : Novosti Press Agency Pub. House, 1978.
63 p. ; 17 cm. — (The Soviet viewpoint) USSR***
Includes bibliographical references.
0.25rub
1. Russia—Foreign relations—China. 2. China—Foreign relations—Russia. 3. China—Boundaries—Russia. 4. Russia—Boundaries—China. I. Title. II. Series.
DK68.7.C5O4313 327.47'051 79-314784
79 MARC

Ramachandran, K N 1938-
Power and ideology : Sino-Soviet dispute : an over view / K. N. Ramachandran and S. K. Ghosh. — New Delhi : Young Asia Publications, 1977.
ii, 243 p. ; 22 cm.
Includes bibliographical references and index.
Rs48.00 ($15.00 U.S.)
1. China—Foreign relations—Russia. 2. Russia—Foreign relations—China. I. Ghosh, S. K., joint author. II. Title.
DS740.5.R8R3 327.51'047 78-900708
79 MARC

CHINA

—FOREIGN RELATIONS—RUSSIA
(Continued)

Shih, Ta.
(Sha O ch'in Hua chien shih)
沙俄侵華簡史 / 史達著. — 香港 : 香港朝陽出版社, 1977.
88 p. ; 19 cm.
Includes bibliographical references.
$2.00

1. China—Foreign relations—Russia. 2. Russia—Foreign relations—China. I. Title.
DS740.5.R8S369 1977 79-838741

Wang, Sheng.
(O ti ch'in Hua ts'e lüeh chih yen chiu)
俄帝侵華策略之研究 / 王昇, 曹敏合著. — 臺北 : 黎明文化事業公司, 民國66 [1977]
2, 6, 264 p., [2] fold. leaves of plates ; 22 cm. — (共黨問題研究叢書)
Series romanized: Kung tang wen t'i yen chiu ts'ung shu.
NT$70.00

1. China—Foreign relations—Russia. 2. Russia—Foreign relations—China. I. Ts'ao, Min, joint author. II. Title.
DS740.5.R8W34 78-841173

Wun, Kin Wah.
A documentary analysis of the Sino-Soviet relations : the border negotiations / by Wun Kin-wah. — [Singapore] : Institute of Humanities and Social Sciences, College of Graduate Studies, Nanyang University, 1976.
20 leaves ; 28 cm. — (Occasional paper series - Institute of Humanities and Social Sciences, Nanyang University ; no. 18)
"A part of the author's Ph.D. thesis entitled The politics of conflict and defence: a case study into the Sino-Soviet relations in 1969-1972."
Includes bibliographical references.
1. China—Foreign relations—Russia. 2. Russia—Foreign relations—China. I. Title. II. Series: Nanyang University. Institute of Humanities and Social Sciences. Occasional paper series - Institute of Humanities and Social Sciences, Nanyang University ; no. 18.
DS740.5.R8W86 327.51'047 78-940356
79 MARC

—FOREIGN RELATIONS
—RUSSIA—ADDRESSES, ESSAYS, LECTURES

(Fen sui Su hsiu she hui ti kuo chu i fa tung ch'in lüeh chan cheng ti ch'iang tao "li lun")
粉碎苏修社会帝国主义发动侵略战争的强盗"理论". — [香港? : s. n., 1977?]
54 p. ; 17 cm.
Cover title.
Reprint of the 1969 ed. published by Kuang-tung jen min ch'u pan she. in series: Chan pei chiao yü ts'an k'ao tzu liao ; 4.
CONTENTS: "国际专政论"是社会帝国主义的强盗"理论".—粉碎新沙皇的"有限主权论".—撕掉苏修所谓"侵略定义"的画皮.—新沙皇向亚洲扩张的新步骤.—苏修"经济一体化新阶段"的侵略实质.

1. China—Foreign relations—Russia—Addresses, essays, lectures. 2. Russia—Foreign relations—China—Addresses, essays, lectures. I. Series: Chan pei chiao yü ts'an k'ao tzu liao ; 4.
DS740.5.R8F45 1977 79-838217

(Nien liu nien ti Su-lien yü Chung-kuo)
廿六年的蘇聯與中國 / 新華社編. — [福州?] : 新華書店, 1949.
28 p. ; 18 cm.
Cover title: Erh shih liu nien ti Su-lien yü Chung-kuo.

1. China—Foreign relations—Russia—Addresses, essays, lectures. 2. Russia—Foreign relations—China—Addresses, essays, lectures. I. Hsin hua t'ung hsün she. II. Title: Erh shih liu nien ti Su-lien yü Chung-kuo.
DS740.5.R8N43 78-843073

—FOREIGN RELATIONS—RUSSIA—SOURCES

(Chung O hui i ts'an k'ao wen chien)
中俄會議參考文件 中俄交涉公署編 [臺北縣永和鎮] 文海出版社 [民國60 i. e. 1971]
2 v. (816 p.) 20 cm. (紀念中華民國建國六十週年史料彙刊 87)
Reprint ed.
CONTENTS: 第1類 中俄舊約節要—第2類 中俄問題往來文件—第3類 中俄臨時協定—第4類 中俄協商草案—第5類 蘇俄與各國所訂協約—蘇維埃社會共和國聯邦憲法

1. China—Foreign relations—Russia—Sources. 2. Russia—Foreign relations—China—Sources. I. China. Chung O chiao she kung shu. II. China. Treaties, etc. III. Russia (1917- R. S. F. S. R.). Treaties, etc. IV. Series: Chi nien Chung-hua min kuo chien kuo liu shih chou nien shih liao hui k'an 87.
DS774.C397 vol. 87 73-897895
[DS740.5.R8]

Russko-kitaĭskie otnoshenii͡a v XVIII [i.e. vosemnadt͡satom] veke : materialy i dokumenty / [sostavlenie i obrabotka teksta N. F. Demidova i V. S. Mi͡asnikov ; otv. redaktor L. S. Tikhvinskiĭ]. — Moskva : Nauka, 1978-
USSR***
v. ; 27 cm.
On leaf preceding t.p., v. 1- : Akademii͡a nauk SSSR. Institut Dal'nego Vostoka, and other organizations.
Includes bibliographical references and indexes.
CONTENTS: t. 1. 1700-1725.

7.40rub (v. 1)
1. Russia—Foreign relations—China—Sources. 2. China—Foreign relations—Russia—Sources. I. Demidova, Natal'i͡a Fedorovna. II. Mi͡asnikov, Vladimir Stepanovich. III. Tikhvinskiĭ, Sergeĭ Leonidovich. IV. Akademii͡a nauk SSSR. Institut Dal'nego Vostoka.
DK68.7.C5R83 79-375545
79 MARC

—FOREIGN RELATIONS—SWITZERLAND

Dubois, Howard, 1947-
Die Schweiz und China / Howard Dubois. — Bern ; Frankfurt am Main ; Las Vegas : P. Lang, c1978.
153 p. ; 23 cm. — (Schweizer asiatische Studien : Studienhefte ; Bd. 1)
Sw***
Bibliography: p. 145-153.
ISBN 3-261-03105-0
1. Switzerland—Foreign relations—China. 2. China—Foreign relations—Switzerland. I. Title. II. Series.
DQ76.S9D8 78-382301
79 MARC

—FOREIGN RELATIONS—UNITED STATES

Chang, King-yuh.
Washington-Peiping relations : three levels of analysis / by King-yuh Chang. — [Taipei] : World Anti-communist League, China Chapter, Asian Peoples' Anti-communist League, Republic of China, 1978.
36 p. ; 19 cm. — (Pamphlet - Asian Peoples' Anti-communist League ; 217)
Cover title.
Includes bibliographical references.
$0.40
1. United States—Foreign relations—China. 2. China—Foreign relations—United States. I. Title.
E183.8.C5C415 327.73051 79-109437
00 MARC

Downen, Robert L
The Taiwan pawn in the China game : Congress to the rescue / by Robert L. Downen ; with a foreword by Ray S. Cline. — Washington : Center for Strategic and International Studies, Georgetown University, c1979.
v, 80 p. ; 22 cm.
Includes bibliographical references.
ISBN 0-89206-007-7
1. United States—Foreign relations—China. 2. China—Foreign relations—United States. 3. United States—Foreign relations—Taiwan. 4. Taiwan—Foreign relations—United States. I. Georgetown University, Washington, D.C. Center for Strategic and International Studies. II. Title.
E183.8.C5D67 327.73'051 79-88334
00 MARC

Fairbank, John King, 1907-
China perceived : images and policies in Chinese-American relations / John K. Fairbank. — New York : Vintage Books, 1976, c1974.
xx, 245, xi p. : ill. ; 21 cm.
Includes bibliographical references and index.
ISBN 0-394-71863-1
1. United States—Foreign relations—China. 2. China—Foreign relations—United States. I. Title.
[E183.8.C5F27 1976] 327.51'073 76-9046
76[r79]rev MARC

Fairbank, John King, 1907-
The United States and China / John King Fairbank. — 4th ed. — Cambridge, Mass. : Harvard University Press, 1979.
xxiv, 606 p. : ill. ; 22 cm. — (The American foreign policy library)
Bibliography: p. [481]-580.
Includes index.
ISBN 0-674-92435-5
1. China—History. 2. United States—Foreign relations—China. 3. China—Foreign relations—United States. I. Title. II. Series.
DS735.F3 1979 327.51'073 78-13667
78 MARC

Hou, Li-ch'ao.
(Ti kuo chu i mai pan chu i)
帝國主義買辦主義 / 侯立朝著. — 臺北 : 博學出版社 : 總經銷傳眞書店, 民國66 [1977]
6, 249 p., [4] leaves of plates : ill. ; 20 cm. — (中國人抗外史 ; 6)
Includes bibliographical references.
NT$55.00

1. United States—Foreign relations—China. 2. China—Foreign relations—United States. I. Title. II. Series: Chung-kuo jen k'ang wai shih ; 6.
E183.8.C5H723 79-837151

Hunt, Michael H
Frontier defense and the open door : Manchuria in Chinese-American relations, 1895-1911 / Michael H. Hunt. — New Haven : Yale University Press, 1973.
xiv, 281 p., [2] leaves of plates : ill. ; 24 cm. — (Yale historical publications/-miscellany ; 95)
Bibliography: p. 267-274.
Includes index.
ISBN 0-300-01616-6
1. United States—Foreign relations—China. 2. China—Foreign relations—United States. 3. Manchuria—History. 4. Manchuria—Commerce—United States. 5. United States—Commerce—Manchuria. I. Title. II. Series: Yale historical publications : Miscellany ; 95.
E183.8.C5H85 327.51'073 73-77152
79 MARC

Kintner, William Roscoe, 1915-
A matter of two Chinas : the China-Taiwan issue in U.S. foreign policy / William R. Kintner & John F. Copper. — Philadelphia : Foreign Policy Research Institute, c1979.
vii, 127 p. ; 23 cm.
Includes bibliographical references.
1. United States—Foreign relations—China. 2. China—Foreign relations—United States. 3. United States—Foreign relations—Taiwan. 4. Taiwan—Foreign relations—United States. I. Copper, John Franklin, joint author. II. Title.
E183.8.C5K565 327.73'051 79-4254
79 MARC

Kuan, Chung, 1940-
(Mei fei chien chiao yü kuo chia ch'ien t'u)
美匪建交與國家前途 / 關中著. — 臺北 : 自由青年社, 民國68 [1979]
3, 2, 212 p. ; 22 cm.
NT$70.00

1. United States—Foreign relations—China. 2. China—Foreign relations—United States. 3. Taiwan—Foreign relations. I. Title.
E183.8.C5K75 79-840506

Kuo, Wu-hsiung.
(Mei-kuo ch'uan t'ung wai chiao cheng ts'e yüan tse yü tui Hua cheng ts'e)
美國傳統外交政策原則與對華政策 / 郭武雄著. — 臺北 : 臺灣商務印書館, 民國67 [1978]
2, 1, 2, 96 p. ; 19 cm. — (人人文庫 ; 2373-2374)
Series romanized: Jen jen wen k'u.
Appended with two historical policy statements in English by U. S. Dept. of State.
Bibliography: p. 89-96.
NT$18.00

1. United States—Foreign relations—China. 2. China—Foreign relations—United States. 3. United States—Foreign relations. I. Title.
E183.8.C5K86 79-836585

Kuz'min, Vi͡acheslav Vasil'evich.
(Kitaĭ v strategii amerikanskogo imperializma)
Китай в стратегии американского империализма / В. В. Кузьмин. — Москва : Междунар. отношения, 1978.
USSR 78
271 p. ; 21 cm.
Includes bibliographical references.
1.20rub
1. United States—Foreign relations—China. 2. China—Foreign relations—United States. 3. World politics—1965-1975. 4. World politics—1975-1985. I. Title.
E183.8.C5K89 79-412287

Li, Ling-Chün.
Washington-Peiping relations and future developments / [by Li Ling-Chun]. — Taipei : Chung Hwa Information Service, [1978?]
9 p. ; 19 cm.
1. United States—Foreign relations—China. 2. China—Foreign relations—United States. I. Title.
E183.8.C5L4413 327.73'051 78-103828
79 MARC

Liang, Chin-tung, 1890-
Patrick Hurley, the China mediator / by Chin-tung Liang. — Jamaica, N.Y. : Center of Asian Studies, St. John's University, [1977?]
37 leaves ; 28 cm. — (St. John's papers in Asian studies ; no. 23)
Includes bibliographical references.
1. China—Politics and government—1937-1945. 2. United States—Foreign relations—China. 3. China—Foreign relations—United States. 4. Hurley, Patrick Jay, 1883-1963. I. Title. II. Series.
DS777.53.L465 320.9'51'042 79-311171
79 MARC

Liu, Ta-jen, fl. 1936-
A history of Sino-American diplomatic relations, 1840-1974 / by Ta Jen Liu. — Taipei : China Academy, 1978.
iii, 645 p. ; 22 cm. — (Chinese culture series ; 2-2)
Title also in Chinese: Chung Mei wai chiao kuan hsi shih.
Bibliography: p. [599]-618.
Includes index.
NT$420.00 ($11.00 U.S.)
1. United States—Foreign relations—China. 2. China—Foreign relations—United States. I. Title. II. Title: Chung Mei wai chiao kuan hsi shih. III. Series.
E183.8.C5L498 327.51'073 78-112427
79 MARC

CHINA
—FOREIGN RELATIONS
—UNITED STATES (Continued)

May, Gary, 1944-
China scapegoat, the diplomatic ordeal of John Carter Vincent / Gary May ; introd. by John K. Fairbank. — Washington : New Republic Books, 1979.
370 p., [12] leaves of plates : ill. ; 24 cm.
Bibliography: p. 295-305.
Includes index.
ISBN 0-915220-49-0 : $15.95
1. Vincent, John Carter, 1900- . 2. China—History—Civil War, 1945-1949. 3. United States—Foreign relations—China. 4. China—Foreign relations—United States. 5. Diplomats—United States—Biography. I. Title.
E748.V564M39 327'.2'0973 79-4129
79 MARC

McKee, Delber L 1923-
Chinese exclusion versus the open door policy, 1900-1906 : clashes over China policy in the Roosevelt era / Delber L. McKee. — Detroit : Wayne State University Press, 1977.
905/1(d
292 p. : ill. ; 24 cm.
Bibliography: p. 277-283.
Includes index.
ISBN 0-8143-1565-8
1. United States—Foreign relations—China. 2. China—Foreign relations—United States. 3. Chinese in the United States. 4. United States—Emigration and immigration. 5. United States—Foreign relations—1901-1909. 6. Roosevelt, Theodore, Pres. U.S., 1858-1919. I. Title.
E183.8.C5M29 327.73'051 76-47024
76[r79]rev MARC

(Pai kung meng yen)
白宮夢魘 / 張陵華, 丁義鳴編著. — [臺北]：張陵華：總經銷自由青年社, 民國68 [1979]
406 p., [14] leaves of plates : ill. ; 22 cm.
NT$100.00 ($6.00 U.S.)

1. United States—Foreign relations—Taiwan. 2. Taiwan—Foreign relations—United States. 3. United States—Foreign relations—China. 4. China—Foreign relations—United States. I. Chang, Ling-hua. II. Ting, I-ming.

E183.8.T3P34 79-841089

Schaller, Michael, 1947-
The U.S. crusade in China, 1938-1945 / Michael Schaller. — New York : Columbia University Press, 1979.
xiii, 364 p. ; 24 cm.
Bibliography: p. 345-355.
Includes index.
ISBN 0-231-04454-2
1. United States—Foreign relations—China. 2. China—Foreign relations—United States. 3. China—Politics and government—1937-1945. 4. World War, 1939-1945—China. I. Title.
E183.8.C5S325 940.53'22'73 78-15032
78 MARC

Solomon, Richard H 1937-
China policy and America's public debate : ten arguments in search of normalized U.S.-PRC relations / Richard H. Solomon. — Santa Monica, Calif. : Rand Corp., 1977.
43 p. ; 28 cm. — (The Rand paper series ; P-6048)
Cover title.
Includes bibliographical references.
1. United States—Foreign relations—China. 2. China—Foreign relations—United States. I. Title. II. Series.
AS36.R28 no. 6048 081 s 79-300792
[E183.8.C5] 79 MARC

Tu, Heng-chih.
(Chung Mei kuan hsi yü kuo chi fa)
中美關係與國際法 / 杜蘅之著. — 臺北：臺灣商務印書館, 民國68 [1979]
2, 1, 5, 324 p. ; 19 cm. — (人人文庫；特629)
Series romanized: Jen jen wen k'u.
NT$36.00

1. United States—Foreign relations—China. 2. China—Foreign relations—United States. 3. International law. I. Title.
E183.8.C5T83 79-841442

(T'uan chieh, tzu ch'iang, chiu kuo)
團結·自強·救國 / [主編者幼獅公司編譯部]. — 臺北：該公司, 民國68 [1979]
2, 134 p., [1] leaf of plates : ill. ; 19 cm.
NT$15.00

1. United States—Foreign relations—Taiwan. 2. Taiwan—Foreign relations—United States. 3. United States—Foreign relations—China. 4. China—Foreign relations—United States. I. Yu shih wen hua shih yeh kung ssu. Pien i pu.
E183.8.T3T82 79-841091

United States. Library of Congress. Foreign Affairs and National Defense Division.
China and Asia—an analysis of China's recent policy toward neighboring states : report / by the Foreign Affairs and National Defense Division, Congressional Research Service, Library of Congress preceded by a State Department report on normalization negotiations with China ; prepared for the Subcommittee on Asian and Pacific Affairs of the Committee on Foreign Affairs, U.S. House of Representatives. — Washington : U.S. Govt. Print. Off., 1979.
vii, 54 p. ; 24 cm.
At head of title: 96th Congress, 1st session. Committee print.
Supt. of Docs. no.: Y 4.F76/1:C44/14
1. Asia—Foreign relations—China. 2. China—Foreign relations—Asia. 3. United States—Foreign relations—China. 4. China—Foreign relations—United States. I. United States. Congress. House. Committee on Foreign Affairs. Subcommittee on Asian and Pacific Affairs. II. Title.
DS33.4.C5U54 1979 327.51'05 79-601691
79 MARC

Willoughby, Westel Woodbury, 1867-1945.
Japan's case examined / Westel W. Willoughby. — New York : Arno Press, 1979, c1940.
x, 237 p. ; 22 cm. — (Johns Hopkins University Press reprints)
Reprint of the ed. published by the Johns Hopkins Press, Baltimore.
Includes index.
ISBN 0-405-10635-1 : $18.00
1. Japan—Foreign relations—China. 2. China—Foreign relations—Japan. 3. United States—Foreign relations—Japan. 4. Japan—Foreign relations—United States. 5. United States—Foreign relations—China. 6. China—Foreign relations—United States. 7. Eastern question (Far East) I. Title.
DS849.C6W52 1979 327.52'051 78-19305
78 MARC

Yao, Meng-hsüan.
Peiping's two-face policy toward the United States / [by Yao Meng-hsuan]. — Taipei : Chung Hwa Information Service, [1978?]
20 p. ; 20 cm.
Includes bibliographical references.
1. United States—Foreign relations—China. 2. China—Foreign relations—United States. I. Title.
E183.8.C5Y35 327.51'073 78-103835
79 MARC

Yin, Ch'ing-yao.
A review of the Washington-Peiping "normalization" question / by Yin Ching-yao. — [Taipei] : World Anti-Communist League, China Chapter, Asian Peoples' Anti-Communist League, Republic of China, 1978.
49, xii p. ; 19 cm. — (English publications of the APACL/ROC : Pamphlet ; 214)
Cover title.
$0.60
1. United States—Foreign relations—China. 2. China—Foreign relations—United States. I. World Anti-Communist League. China Chapter. II. Asian Peoples' Anti-Communist League. China. III. Title.
E183.8.C5Y58 327.73'051 78-113255
79 MARC

Yu, Se-hŭi.
(Mi-Chunggong kwan'gye ŭi chŏnmang kwa Hanbando munje e mich'inun yŏnghyang)
美·中共關係의 展望과 韓半島問題에 미치는 影響 / 研究執筆責任柳世熙. — [서울]：國土統一院政策企劃室, 1977.
43 p. ; 26 cm. — (周邊環境研究)
"국통정 77-12-1365"
Includes bibliographical references.
1. China—Foreign relations—United States. 2. United States—Foreign relations—China. 3. Korea—Foreign relations. I. Title. II. Series: Chubyŏn hwan'gyŏng yŏn'gu.
E183.8.C5Y78 78-826434

—FOREIGN RELATIONS—UNITED STATES—ADDRESSES, ESSAYS, LECTURES

About face : the China decision and its consequences / edited by John Tierney, Jr. — New Rochelle, N.Y. : Arlington House, c1979.
444 p. ; 24 cm.
Includes bibliographical references.
ISBN 0-87000-438-7
1. United States—Foreign relations—Taiwan—Addresses, essays, lectures. 2. Taiwan—Foreign relations—United States—Addresses, essays, lectures. 3. United States—Foreign relations—China—Addresses, essays, lectures. 4. China—Foreign relations—United States—Addresses, essays, lectures. I. Tierney, John, 1940-
E183.8.T3A26 327.73'051'249 79-4545
79 MARC

(Fan Pai p'i shu hsüeh hsi ts'ai liao)
反「白皮書」学習材料 / 中共天津市委總學委會編. — [天津]：該會, 1949.
3, 179 p. ; 18 cm.
Cover title.

1. United States. Dept. of State. United States relations with China—Addresses, essays, lectures. 2. United States—Foreign relations—China—Addresses, essays, lectures. 3. China—Foreign relations—United States—Addresses, essays, lectures. I. Chung-kuo kung ch'an tang. T'ien-chin shih wei. Tsung hsüeh wei hui.
E183.8.C5U5333 79-840067

(Futatsu no Chūgoku o tsukuru Amerika no shin'imbō ni hantaisuru)
二つの中国をつくるアメリカの新陰謀に反対する. — 北京：外文出版社, 1962.
124 p. ; 18 cm.
Contains translations of 12 articles originally published in 1961 in various Chinese journals.

1. United States—Foreign relations—China—Addresses, essays, lectures. 2. China—Foreign relations—United States—Addresses, essays, lectures. I. Wai wen ch'u pan she, Peking.
E183.8.C5F82 78-813412

Kuan, Chung, 1940-
(Mei fei kuan hsi ti p'ing ku)
美匪關係的評估 / 關中著. — [台北?]：中央青年工作會, [1978?]
94 p. ; 21 cm.
CONTENTS: 美國人在中國問題上最大可能的作為.—最近美匪關係評述.—鮑大可著「美國對華政策：舊問題與新挑戰」評介.—美國對華政策與我自處之道.—從范錫訪匪談美匪關係的可能發展.—范錫訪匪的因素分析.—從卡特政府對中共政策之演變看范錫北平之行.—范錫訪匪的評估.
1. United States—Foreign relations—China—Addresses, essays, lectures. 2. China—Foreign relations—United States—Addresses, essays, lectures. I. Title.
E183.8.C5K76 79-838609

Li, Chung-kuei.
(Chung Mei kuan hsi yü kuo chia ch'ien t'u)
中美關係與國家前途 / 李鍾桂講；教育部社會教育司編印. — [台北]：該司, 民國66 [1977]
34 p. ; 19 cm. — (文化講座專集；97)
"中華民國六十六年二月八日講於高雄"

1. United States—Foreign relations—Taiwan—Addresses, essays, lectures. 2. Taiwan—Foreign relations—United States—Addresses, essays, lectures. 3. United States—Foreign relations—China—Addresses, essays, lectures. 4. China—Foreign relations—United States—Addresses, essays, lectures. I. China (Republic of China). Chiao yü pu. She hui chiao yü ssu. II. Title. III. Series: Wen hua chiang tso chuan chi ; 97.
E183.8.T3L5 78-841226

Li, Chung-kuei.
(Mei-kuo ta hsüan yü Chung Mei kuan hsi)
美國大選與中美關係 / 李鍾桂講；教育部社會教育司編印. — [台北]：該司, 民國65 [1976]
30 p. (p. 30 blank) ; 20 cm. — (文化講座專集；67)

1. United States—Foreign relations—China—Addresses, essays, lectures. 2. China—Foreign relations—United States—Addresses, essays, lectures. 3. United States—Politics and government—1974-1977—Addresses, essays, lectures. I. Title. II. Series: Wen hua chiang tso chuan chi ; 67.
E183.8.C5L438 79-839133

Li, Chung-kuei.
(Mei-kuo wai chiao yü Chung Mei ch'ien t'u)
美國外交與中美前途 / 李鍾桂講；教育部社會教育司編印. — [台北]：該司, 民國66 [1977]
30 p. ; 20 cm. — (文化講座專集；112)
"中華民國六十六年九月五—六日講於苗栗新竹桃園臺北縣."

1. United States—Foreign relations—1977- —Addresses, essays, lectures. 2. United States—Foreign relations—China—Addresses, essays, lectures. 3. China—Foreign relations—United States—Addresses, essays, lectures. 4. United States—Foreign relations—Taiwan—Addresses, essays, lectures. 5. Taiwan—Foreign relations—United States—Addresses, essays, lectures. I. Title. II. Series: Wen hua chian tso chuan chi ; 112.
E872.L5 78-842144

T'ao, Pai-ch'uan, 1903-
(Tung Ya hao tu)
東亞豪賭 / 陶百川著. — 臺北：聯經出版事業公司, 民國67 [1978]
1, 4, 385 p. ; 22 cm.
NT$140.00
1. United States—Foreign relations—China—Addresses, essays, lectures. 2. China—Foreign relations—United States—Addresses, essays, lectures. 3. United States—Foreign relations—Taiwan—Addresses, essays, lectures. 4. Taiwan—Foreign relations—United States—Addresses, essays, lectures. 5. United States—Foreign relations—1974-1977—Addresses, essays, lectures. I. Title.
E183.8.C5T33 78-843378

Thomson, James Claude, 1931-
An anatomy of Chinese-American relations / James C. Thomson. — [Johannesburg] : South African Institute of International Affairs, 1976.
11 p. ; 30 cm. — (Occasional paper - South African Institute of International Affairs) SA***
Cover title.
Text of an address to the Witwatersrand Branch of the Institute, 1975.
ISBN 0-909239-25-8
1. United States—Foreign relations—China—Addresses, essays, lectures. 2. China—Foreign relations—United States—Addresses, essays, lectures. 3. United States—Foreign relations—1969-1974—Addresses, essays, lectures. I. Title. II. Series: South African Institute of International Affairs. Occasional paper - South African Institute of International Affairs.
E183.8.C5T44 327.73'051 77-357515
77[r79]rev MARC

CHINA

—FOREIGN RELATIONS —UNITED STATES—ADDRESSES, ESSAYS, LECTURES (Continued)

Tu, Heng-chih.
(T'ai hai hsing shih ti chan wang)
台海形勢的展望 / 杜蘅之講；教育部社會教育司編印. — ₍台北₎：該司, 民國66 ₍1977₎
12 p. : port. ; 20 cm. — (文化講座專集；111)
1. United States—Foreign relations—Taiwan—Addresses, essays, lectures. 2. Taiwan—Foreign relations—United States—Addresses, essays, lectures. 3. United States—Foreign relations—China—Addresses, essays, lectures. 4. China—Foreign relations—United States—Addresses, essays, lectures. I. China (Republic of China). Chiao yü pu. She hui chiao yü ssu. II. Title. III. Series: Wen hua chiang tso chuan chi ; 111.
E183.8.T3T8 78-842543

Two Chinese states : U.S. foreign policy and interests / edited by Ramon H. Myers ; introd. by Robert A. Scalapino. — Stanford, Calif. : Hoover Institution Press, c1978.
xii, 84 p. ; 23 cm. — (Hoover international studies) (Hoover Institution publication ; 200)
Includes bibliographical references and index.
ISBN 0-8179-7002-9
1. United States—Foreign relations—China—Addresses, essays, lectures. 2. China—Foreign relations—United States—Addresses, essays, lectures. 3. United States—Foreign relations—Taiwan—Addresses, essays, lectures. 4. Taiwan—Foreign relations—United States—Addresses, essays, lectures. I. Myers, Ramon Hawley, 1929- . II. Series. III. Series: Stanford University. Hoover Institution on War, Revolution and Peace. Publications ; 200.
E183.8.C5T86 327.73'051 78-70389
79 MARC

(Wo men tsen mo pan?)
我們怎麼辦? / 謝禮群, 莊秀鈴合編. — 高雄：心影出版社, 民國68 ₍1979₎
200 p. ; 19 cm.
Cover title.
NT$50.00
1. United States—Foreign relations—Taiwan—Addresses, essays, lectures. 2. Taiwan—Foreign relations—United States—Addresses, essays, lectures. 3. China—Foreign relations—United States—Addresses, essays, lectures. 4. United States—Foreign relations—China—Addresses, essays, lectures. I. Hsieh, Li-ch'ün. II. Chuang, Hsiu-ling.
E183.8.T3W6 79-840650

(Wo men tui Mei fei kuan hsi cheng ch'ang hua ti k'an fa)
我們對「美匪關係正常化」的看法 / 世界反共聯盟中華民國分會, 亞洲人民反共聯盟中華民國總會印行. — ₍台北₎：該會, 民國67 ₍1978₎
43 p. ; 20 cm.
Cover title.
CONTENTS: 蔣總統向美報人指出匪共陰謀.—臺海形勢的展望.—避免戰爭與延遲戰爭.—戰爭險影與世局危機.—論美國與中共的三種關係.
1. United States—Foreign relations—China—Addresses, essays, lectures. 2. China—Foreign relations—United States—Addresses, essays, lectures. I. World Anti-Communist League. China Chapter. II. Asian People's Anti-Communist League. China.
E183.8.C5W6 79-840326

—FOREIGN RELATIONS—UNITED STATES—CONGRESSES

The Future of Chinese-American relations : ₍forum₎ held on March 28, 1979 and sponsored by the American Enterprise Institute for Public Policy Research, Washington, D.C. / John Charles Daly, moderator ; Jonathan B. Bingham ... ₍et al.₎. — Washington : American Enterprise Institute for Public Policy Research, c1979.
35 p. ; 21 cm. — (AEI forum ; 29)
ISBN 0-8447-2156-5 : $3.00
1. United States—Foreign relations—China—Congresses. 2. China—Foreign relations—United States—Congresses. I. Daly, John Charles, 1914- . II. Bingham, Jonathan B. III. American Enterprise Institute for Public Policy Research. IV. Series: American Enterprise Institute for Public Policy Research. AEI public policy forum ; 29.
E183.8.C5F84 327.73'051 79-65540
00 MARC

—FOREIGN RELATIONS—UNITED STATES—SOURCES

American diplomatic and public papers, the United States and China : Series 2, the United States, China, and imperial rivalries, 1861-1893 / Jules Davids, editor. — Wilmington, Del. : Scholary Resources, 1979.
18 v. ; 29 cm.
Bibliography: v. 1, p. xlviii-lvii.
ISBN 0-8420-2136-1 (set) : $995.00
1. United States—Foreign relations—China—Sources. 2. China—Foreign relations—United States—Sources. I. Davids, Jules. II. Title: The United States, China, and imperial rivalries, 1861-1893.
E183.8.C5A72 327.73'051 79-12914
79 MARC

—FOREIGN RELATIONS—VIETNAM

China and Indochina : the realities behind he headlines / ₍editors, Hugh Deane, Esther Gollobin, Richard Pastor₎. — New York : US-China Peoples Friendship Association of New York, ₍1979?₎
23 p. ; 24 cm.
"A China and us publication."
Includes bibliographical references.
$1.00
1. Indochina—History—1945- . 2. Chine—Foreign relations—Vietnam. 3. Vietnam—Foreign relations—China. I. Deane, Hugh. II. Gollobin, Esther. III. Pastor, Richard. IV. U.S.-China Peoples Friendship Association. V. China and us.
DS550.C534 327.51'0597 79-51418
00 MARC

Kazawa, Gō, 1936-
(Chūgoku Betonamu kankei)
中国·ベトナム関係：中越紛争の歴史と国際環境 / 鹿沢剛. — ₍東村山₎：教育社：東京：販売教育社出版サービス, ₍1978₎
172 p. : ill. ; 18 cm. — (時事問題解説：148) (入門新書) Ja 79-3413
Cover title.
Series 2 romanized: Nyūmon shinsho.
¥400
1. Vietnam—Foreign relations—China. 2. China—Foreign relations—Vietnam. 3. Cambodian-Vietnamese Conflict, 1977- 4. Asia, Southeastern—Foreign relations. I. Title. II. Series: Jiji mondai kaisetsu ; 148.
DS556.58.C5K39 79-806774

*79 J

World Council of Peace.
China's hostility to Vietnam : a historical review of Sino-Vietnamese relations. — Helsinki : World Peace Council, ₍1979₎
FI***
26 p. ; 19 cm.
1. Vietnam—Foreign relations—China. 2. China—Foreign relations—Vietnam. I. Title.
DS556.58.C5W67 1979 327.597'051 79-313567
79 MARC

(Yüeh-nan ch'ü kan Hua ch'iao hui kuo chen hsiang)
越南驱赶华侨回国真相. — 香港：生活, 讀書, 新知三联书店香港分店, 1978.
35 p. ; 21 cm.
1. Chinese in Vietnam. 2. China—Foreign relations—Vietnam. 3. Vietnam—Foreign relations—China.
DS556.45.C55Y83 79-835445

79 C

—FOREIGN RELATIONS —VIETNAM—ADDRESSES, ESSAYS, LECTURES

(Chin i pu fa chan Chung Yüeh liang kuo jen min ti wei ta yu i ho chan tou t'uan chieh)
进一步发展中越两国人民的伟大友谊和战斗团结. — 第1版. — 北京：人民出版社, 1973.
2, 59 p., ₍3₎ fold. leaves of plates ; 21 cm.
"越南劳动党和越南民主共和国政府代表团访问中国文件集"
1. China—Foreign relations—Vietnam—Addresses, essays, lectures. 2. Vietnam—Foreign relations—China—Addresses, essays, lectures.
DS740.5.V5C486 79-838137

—FOREIGN RELATIONS—VIETNAM (DEMOCRATIC REPUBLIC, 1946-) —COLLECTED WORKS

Schneider, Eberhard, 1936-
Die sowjetische Vietnampolitik.
₍Köln, Bundesinstitut für Ostwissenschaftliche und Internationale Studien₎
v. 30 cm. (Berichte des Bundesinstitut für ostwissenschaftliche und internationale Studien)
Summaries in English.
1. Vietnam (Democratic Republic, 1946-)—Foreign relations—Russia—Collected works. 2. Russia—Foreign relations—Vietnam (Democratic Republic, 1946-)—Collected works. 3. Vietnam (Democratic Republic, 1946-)—Foreign relations—China—Collected works. 4. China—Foreign relations—Vietnam (Democratic Republic, 1946-)—Collected works. I. Title. II. Series: Germany (Federal Republic, 1949-). Bundesinstitut für Ostwissenschaftliche und Internationale Studien. Berichte.
HX15.G468 subser. 309.1'171'7 s 78-643406
[DS560.69.R9] [327.597'047] MARC-S

—FOREIGN RELATIONS —YUGOSLAVIA

(Chung Nan liang kuo jen min yu i ti hsin fa chan)
中南两国人民友谊的新发展：热烈欢迎铁托总统访问中国. — 第1版. — 北京：人民出版社, 1977.
2, 45 p., ₍8₎ leaves of plates (1 fold) : ill. ; 21 cm.
$0.29
1. Tito, Jospi Broz, Pres. Yugoslavia, 1892- —Journeys—China. 2. Visits of state—China. 3. China—Foreign relations—Yugoslavia. 4. Yugoslavia—Foreign relations—China.
DR359.T5C48 78-842857

—GAZETEERS

(Chūgoku oyobi Manshūkoku jimmei chimei benran)
中國及滿洲國人名地名便覽：英文對照 / ₍編纂者タイムス出版社編輯部₎. — 東京：タイムス出版社, 昭和8 ₍1933₎
277 p. in various pagings ; 19 cm.
1. China—Registers. 2. Manchuria—Registers. 3. China—Gazetteers. 4. Manchuria—Gazetteers. I. Taimusu Shuppansha. Henshubu.
DS705.C5253 79-804239

Heibonsha.
(Chūgoku chizuchō)
中国地図帳 平凡社版 東京 平凡社 1973.
80 p. 29 cm. Ja 73-13801
付：中国地名表記便覧
1. China—Maps. 2. China—Gazetteers. 3. Names, Geographical—China. I. Title.
G2305.H35 1973 73-808398

—GENEALOGY

(Ch'en chia ta tsung p'u)
陳家大宗譜 / 編著陳氏大族譜編輯部；主輯莊吳福；主編莊吳圖；助編盧俊華；資料採訪盧有道... ₍et al.₎. — ₍中壢₎：陳氏大族譜編輯委員會, 民國67 ₍1978₎
380, 32 p. : ill. ; 28 cm.
Colophon title.
1. Ch'en family. 2. China—Geneology. I. Chuang, Wu-fu. II. Chuang, Wu-t'u. III. Ch'en shih ta tsu p'u pien chi pu.
CS1169.C475 1978 79-837079

(Ch'en shih tsu p'u)
陳氏族譜 / ₍編輯莊吳圖₎. — 桃園縣龍潭鄉：陳氏族譜編輯委員會, 民國66 ₍1977₎
332 p. in various pagings : ill. ; 27 cm.
Cover title.
Editions of 1961, 1976 published under title: Ch'en shih ta tsu p'u, and of 1978 published under title: Ch'en chia ta tsung p'u.
1. Ch'en family. 2. China—Geneology. I. Chuang, Wu-t'u. II. Ch'en shih tsu p'u pien chi wei yüan hui.
CS1169.C475 1977a 79-841630

(Yü shih ta tsu p'u)
余氏大族譜 / ₍主編賴振興, 賴信作₎. — 嘉義：余氏族譜編輯委員會, 民國66 ₍1977₎
34, 120, 326, 328 p. ; ill. ; 27 cm.
On spine: Hsin-chu hsien Yü hsing tsung ch'in hui kung yin.
1. Yü family. 2. China—Geneology. I. Lai, Chen-hsing. II. Lai, Hsin-tso.
CS1169.Y8 1977 79-837959

CHINA (Continued)

—GENEALOGY—ADDRESSES, ESSAYS, LECTURES

Ching-k'ung, Shih.
(Hsing chien Chung-hua min tsu pai hsing tsung tz'u ch'u i)
興建「中華民族百姓宗祠」芻議 / [淨空]. — 台北：華藏法施會, 民國66 [1977]
6, 16 p. ; 26 cm.
Colophon title.
"文化復興與宗教": p. 1–16 (last group)
"非賣品"
1. China—Genealogy—Addresses, essays, lectures. 2. China—Civilization—Addresses, essays, lectures. I. Ching-k'ung, Shih. Wen hua fu hsing yü tsung chiao. 1977. II. Title.
CS1162.C48 78–843619

Ho, Ch'i-min.
(Chung ku men ti lun chi)
中古門第論集 / 何啓民著. — 臺北：臺灣學生書局, 民國67 [1978]
iii, 311 p. ; 22 cm.
Includes bibliographical references.
NT$120.00
1. China—Genealogy—Addresses, essays, lectures. 2. Names, Personal—China—Addresses, essays, lectures. I. Title.
CS1166.H6 78–841494

—GENTRY

Okuzaki, Hiroshi, 1935–
(Chūgoku kyōshin jinushi no kenkyū)
中國郷紳地主の研究 / 奥崎裕司著. — 東京：汲古書院, 1978.
15, 599, 9, 57 p., [1] fold. leaf of plates : ill. ; 22 cm. Ja***
Includes bibliographical references and index.
¥8000
1. Upper classes—China. 2. China—Gentry. 3. Land tenure—China. 4. China—Intellectual life. 5. Yüan, Hvang, chin shih 1586. 6. China—Biography. I. Title.
HT653.C6O48 79–800227

—HANDBOOKS, MANUALS, ETC.

Bydžovský, Zdeněk.
ČLR-Čínska lidová republika / [autor] Zdeněk Bydžovský a kol. — 1. vyd. — Praha : Pressfoto, 1978.
292, [3] p. : ill. ; 21 cm. — (Obchodně ekonomické sborníky : Řada SZ ; sv. 40) Cz78
Bibliography: p. 291-292.
Kčs22.00
1. China—Handbooks, manuals, etc. 2. China—Economic conditions—1949- —Handbooks, manuals, etc. 3. China—Foreign economic relations—Czechoslovakia—Handbooks, manuals, etc. 4. Czechoslovakia—Foreign economic relations—China—Handbooks, manuals, etc. I. Title. II. Title: Čínská lidová republika.
DS777.55.B92 79–358341
*79 MARC

—HISTORICAL GEOGRAPHY

Chao, Ch'üan-ch'eng, fl. 1955–
(Ch'ing tai ti li yen ko piao)
清代地理沿革表 / 趙泉澄著. — [台北縣永和鎮]：文海出版社, [1979?]
[14], 204 p., [16] fold. leaves ; 22 cm. — (近代中國史料叢刊續編 i. e. 輯] ; 628 : 第六十三輯)
Reprint of the ed. with pref. dated 1940.
Includes bibliographical references and index.
1. China—Historical geography. 2. China—Administrative and political divisions. I. Title. II. Series: Shen, Yün-lung, comp. Chin tai Chung-kuo shih liao ts'ung k'an hsü chi ; 628.
DS755.S532 vol. 628 79–840821
[DS706.5]

Hibino, Takeo, 1914–
(Shina rekishi chiri)
支那歷史地理 / [日比野丈夫, 小野勝年] ; 支那地理歴史大系刊行會編. — 東京：白揚社, 昭和17 [1942]
2, 4, 2, 404 p. : ill. ; 19 cm. — (支那地理歷史大系 ; 3)
1. China—Historical geography. I. Ono, Katsutoshi, joint author. II. Title. III. Series: Shina chiri rekishi taikei ; 3.
DS706.5.H49 78–812802

Hsü, Chi-yü, 1795–1873.
(Liang Han chih yen pien shih chün k'ao lüeh)
兩漢志沿邊十郡考略 ; 兩漢幽幷涼三州今地考略 / 徐繼畬撰. — 台北：廣文書局, 民國67 [1978]
4, 32, 72 p. ; 22 cm. — (史料七編)
Reprint the ed. with postscript dated 1913.
NT$150.00
1. China—Historical geography. 2. China—History—Han dynasty, 202 B. C.–220 A. D. I. Hsü, Chi-yü, 1795–1873. Liang Han Yu Ping Liang san chou chin ti k'ao lüeh. 1978. II. Title. III. Series: Shih liao ch'i pien.
DS706.5.H72 1978 79–841396

Ku, Tsu-yü, 1624–1680.
[Ku shih tu shih fang yü chi yao ching lüeh. Selections]
顧氏讀史方輿紀要精略 / 顧祖禹撰. — 台北：老古出版社, 民國67 [1978]
310, 318, [13] p. ; 22 cm. — (正統謀略學彙編 : 初輯 ; 第18冊)
"精略之京省每章序文, 取藏達通甫先生註釋為準, 京省內容則取辛鐘靈先生輯本"
"附: 顧炎武 天下郡國利病書論略" p. 1–318 (2d group)
1. China—Historical geography. 2. China—History, Military. 3. China—Description and travel. I. Ku, Yen-wu, 1613–1682. T'ien hsia chün kuo li ping shu. Selections. 1978. II. Title. III. Series: Cheng t'ung mou lüeh hsüeh hui pien : Ch'u chi ; ti 18 ts'e.
Title romanized: Ku shih tu shih fang yü chi yao ching lüeh.
DS706.5.K822 1978 79–840130

Ku, Tsu-yü, 1624–1680.
[Tu shih fang yü chi yao. Selections]
讀史方輿紀要歷代州域形勢 / 顧祖禹撰. — 台灣初版. — 台北：老古出版社, 民國67 [1978]
2 v. (938, [38] p.) ; 22 cm. — (正統謀略學彙編 : 初輯 ; 第19–20冊)
Reprint ed.
1. China—Historical geography. I. Title. II. Series: Cheng t'ung mou lüeh hsüeh hui pien : Ch'u chi : ti 19–20 ts'e.
Title romanized: Tu shih fang yü chi yao li tai chou yü hsing shih.
DS706.5.K832 1978 79–839134

Liu, Wen-ch'i, 1789–1854.
(Ch'u Han chu hou chiang yü chih)
楚漢諸侯疆域志 / 劉文淇撰. — 台北：廣文書局, 民國67 [1978]
74 p. ; 22 cm. — (史料七編)
Reprint of the 1876 ed.
NT$150.00
1. China—Historical geography. 2. China—History—Han dynasty, 202 B. C.–220 A. D. I. Title. II. Series: Shih liao ch'i pien.
DS706.5.L57 1978 79–839919

Mihara, Kazuo.
(Kōtei densetsu no rekishi chiri)
黃帝傳説乃歴史地理 / [著者 三原一雄]. — 東京：陝甘文化研究所, 昭和50 [1975]
36, 32, 612, [11] p. (chiefly on double leaves) : ill. ; 27 cm. Ja 76–6893
Cover title.
On p. [4] of cover: The study of legend on Huang Ti in vieu [i. e. view] of historical geography.
Caption title: Kōtei densetsu no rekishi-chirigakuteki kenkyū.
Photocopy of the author's ms.
Summary in English.
¥35000
1. Huang-ti (Legendary ruler) 2. China—Historical geography. I. Title. II. Title: The study of legend on Huang Ti in view of historical geography.
DS741.M53 1975a 78–816048
*78 J

Ogawa, Takuji, 1870–1941.
(Shina rekishi chiri kenkyū)
支那歷史地理研究 / 小川琢治著. — 京都：弘文堂書房 ; 東京：發賣元 弘文堂東京店, 昭和3–4 [1928–1929]
2 v. : ill. ; 23 cm.
Includes bibliographies and index in v. 2.
1. China—Historical geography. I. Title.
DS706.5.O35 78–812082

—HISTORIOGRAPHY

Chao, I, 1727–1814.
(Nien erh shih cha chi)
廿二史劄記 [36卷] 趙翼著 上海 世界書局 民國36 [1947]
3, 17, 554 p. 18 cm.
Appendix (p. 538–554): 補遺
1. China—Historiography. I. Title.
DS734.7.C456 1947 C 58–5688

Chao, I, 1727–1814.
(Nien erh shih cha chi)
廿二史劄記 [36卷] 附補遺 趙翼撰 北京 商務印書館 1958.
1, 1, 12, 782, 11 p. 19 cm.
1. China—Historiography. I. Title.
DS734.7.C456 1958 C 61–1221

Chao, I, 1727–1814.
(Nien erh shih cha chi)
廿二史劄記 [36卷 附補遺] 趙翼撰 新1版 北京 中華書局 1963.
2 v. 20 cm.
1. China—Historiography. I. Title.
DS734.7.C456 1963 C 65–1163
rev

Chao, I, 1727–1814.
[Nien erh shih cha chi. Japanese & Chinese]
廿二史劄記 / [趙翼著 ; 解說 長澤規矩也]. — [東京]：古典研究会：汲古書院(発売), 昭和48 [1973]
457 p. ; 25 cm. — (和刻本正史 : 別卷 ; 8) Ja 74–2120
Japanese reading marks added to Chinese text.
Photoreprint of the 1862 ed. published by Kenkensha.
¥4000
1. China—Historiography. I. Title. II. Series: Wakokubon seishi : Bekkan ; 8.
Title romanized: Nijūnishi sakki.
DS734.7.C45617 1973 76–811139

Chu, Hsi-tsu, 1879–1945.
(Chung-kuo shih hsüeh t'ung lun)
中國史學通論 / [撰著者朱希祖]. — 臺北：莊嚴出版社, 民國66 [1977]
2, 206 p. ; 21 cm. — (古典大學叢書 ; 1)
"朱希祖先生著作目錄": p. 185–206.
1. China—Historiography. I. Title. II. Series: Ku tien ta hsüeh ts'ung shu ; 1.
DS734.7.C57 1977 78–841209

Dzo, Ching-Chuan.
Sseu-ma Ts'ien et l'historiographie chinoise / Dzo Ching-chuan ; préface, R. Etiemble .... — Paris : Publications orientalistes de France, cop. 1978.
v, 356 p. ; 21 cm. — (POF-études ISSN 0336-3279) F78-17769
Bibliography: p. 316-353.
ISBN 2-7169-0089-2 : 65.00F
1. Ssu-ma, Ch'ien, ca. 145-ca. 86 B.C. 2. China—Historiography. I. Title.
DS734.9.S8D96 931'.007'2024 79-357352
79 MARC

Ku, Chieh-kang, 1893–
(Shih lin tsa shih)
史林雜識 : 初編 / 顧頡剛著. — 第1版. —北京：中華書局, 1963, 1977 printing.
4, 4, 296 p., [6] leaves of plates : ill., maps ; 21 cm.
"附圖(共十幅)"(4 fold. leaves) : inserted.
$1.65
1. China—Historiography. I. Title.
DS734.7.K8 79–835266

CHINA

—HISTORIOGRAPHY (Continued)

Ogura, Yoshihiko, 1927-
(Gyakuryū to junryū)
逆流と順流：わたしの中国文化論 / 小倉芳彦著. — 東京：研文出版, 1978.
325 p. ; 20 cm. — (研文選書 ; 3) Ja 79-3226
Series romanized: Kenbun sensho.
¥1800

1. China—Civilization—Philosophy. 2. China—Historiography. I. Title.
DS721.O436 79-805676

Wang, Erh-min.
(Shih hsüeh fang fa)
史學方法 / 王爾敏. — 臺北：東華書局, 民國66 [1977]
3, 3, 338 p. ; 22 cm. — (東華歷史叢書)
Series romanized: Tung hua li shih ts'ung shu.
Bibliography: p. 325-338.
NT$100.00

1. Historiography. 2. China—Historiography. I. Title.
D13.W338 78-841894

—HISTORIOGRAPHY—INDEXES

Gagnon, Guy.
(Shih t'ung yü Shih t'ung hsüeh fan t'ung chien)
史通與史通削繁通檢 / [par Guy Gagnon ; avec la collaboration de Emi Gagnon ... et at.]. — Paris：法蘭西學院漢學研究所, 1977.
2 v. ; 26 cm. — (法蘭西學院漢學研究所漢學通檢提要文獻叢刊 ; 7)
Added t. p.: Concordance combinée du Shitong et du Shitong xiaofan.
Prefatory matter also in French.
Series romanized: Fa-lan-hsi hsüeh yüan Han hsüeh yen chiu so Han hsüeh t'ung chien t'i yao wen hsien ts'ung k'an.
"我們選擇了...四部備要...的通史削繁版本...並以國學基本叢書為藍本"
Includes index (16 p. inserted)
ISBN 2-85757-008-2

1. Liu, Chih-chi, 661-721. Shih t'ung—Concordances. 2. Chi, Yün, 1724-1805. Shih t'ung hsüeh fan—Concordances. 3. China—Historiography—Indexes. I. Gagnon, Emi. II. Liu, Chih-chi, 661-721. Shih t'ung. 1977. III. Chi, Yün, 1724-1805. Shih t'ung hsüeh fan. 1977. IV. Title. V. Title: Concordance combinée du Shitong et du Shitong xiaofan.
DS734.7.L5822G33 79-840253

—HISTORY

(Chūgoku minshū hanran shi)
中国民衆叛乱史 / 谷川道雄, 森正夫編. — 東京：平凡社, 1978-
v. ; 18 cm. — (東洋文庫 ; 336,) Ja 78-21774 (v. 1)
Series romanized: Tōyō bunko.
Bibliography: v. 1, p. 416.
CONTENTS: 1. 秦～唐
¥1400 (v. 1)

1. China—History. 2. Peasant uprisings—China. I. Tanigawa, Michio, 1925- II. Mori, Masao, 1935-
DS735.C52665 78-814851
J

Chung-kuo wen hua hsüeh yüan. Chung-kuo t'ung shih pien chi wei yüan hui.
(Chung-kuo t'ung shih)
中國通史 / [中國文化學院中國通史編輯委員會編著]；中國文化學院編. — 修訂版. — 臺北：華岡出版有限公司, 民國66 [1977]
2, 2, 14, 506 p. ; 22 cm.
Includes bibliographical references.
NT$150.00
1. China—History. I. Chung-kuo wen hua hsüeh yüan. II. Title.
DS735.C545 1977 78-842574

Fairbank, John King, 1907-
The United States and China / John King Fairbank. — 4th ed. — Cambridge, Mass. : Harvard University Press, 1979.
xxiv, 606 p. : ill. ; 22 cm. — (The American foreign policy library)
Bibliography: p. [481]-580.
Includes index.
ISBN 0-674-92435-5
1. China—History. 2. United States—Foreign relations—China. 3. China—Foreign relations—United States. I. Title. II. Series.
DS735.F3 1979 327.51'073 78-13667
78 MARC

Fan, Wen-lan, 1891-1969.
[Chung-kuo t'ung shih chien pien. Japanese]
中國通史 / 范文瀾著；貝塚茂樹, 陳顯明譯. — 東京：岩波書店, 昭和33- [1958-
v. : ill. ; 22 cm.
Includes bibliographical references.

1. China—History. I. Title.
Title romanized: Chūgoku tsūshi.
DS735.F3317 78-816143

Frankfurth, Horst.
Die Geheimnisse der gelben Götter : Macht u. Liebe in 3000 Jahren chines. Geschichte / Horst Frankfurth. — Stuttgart : Seewald, 1978.
380 p. : ill., maps ; 25 cm. GFR78-A
Bibliography: p. 362-364.
Includes indexes.
ISBN 3-512-00511-X : DM42.00
1. China—History. I. Title.
DS736.F8 951 79-348624
79 MARC

Grousset, René, 1885-1952.
Histoire de la Chine / René Grousset, ... — Paris : Fayard, 1976.
446 p. : maps ; 18 cm. — (Les Grandes études historiques) F77-15291
Includes bibliographical references.
ISBN 2-213-00417-X : 43.00F
1. China—History. I. Title.
DS735.G85 1976 951 78-375347
78 MARC

Hibino, Takeo, 1914-
[Me de miru Chūgoku no rekishi. Chinese]
用眼睛看的中國歷史 / [日比野丈夫著述]；李方中編著. — 九龍：田園書店, [1978?]
2, 3, 230 p. : ill. (some col.) ; 20 cm.
Translation of Me de miru Chūgoku no rekishi, originally published in 1976 by Mu t'ung ch'u pan she, Taipei; issued as 14 in series Mu yüan ts'ung k'an.
$15.00
1. China—History. 2. China—Antiquities. I. Title.
Title romanized: Yung yen ching k'an ti Chung-kuo li shih.
DS735.H513 1978 79-835593

Koutsoukis, Albert J 1933-
Asian response : China and Japan / [by] A. J. Koutsoukis. — East Perth, W.A. : Bookland, 1976.
190 p. : ill., col. maps ; 25 cm. Aus
Bibliography: p. 187-188.
Includes index.
ISBN 0-9597361-1-5
1. China—History. 2. Japan—History. I. Title.
DS736.K65 951 79-310396
79 MARC

Kri͡ukov, Mikhail Vasil'evich.
(Drevnie kitaĭt͡sy—problemy ėtnogeneza)
Древние китайцы—проблемы этногенеза / М. В. Крюков, М. В. Софронов, Н. Н. Чебоксаров. — Москва : Наука, 1978.
342 p., [4] leaves of plates : ill. ; 22 cm. USSR 78
At head of title: Akademii͡a nauk SSSR. Institut Dal'nego Vostoka. Institut ėtnografii im. N. N. Miklukho-Maklai͡a.
Summary in English.
Bibliography: p. 296-[314]
Includes indexes.
2.40rub
1. Ethnology—China. 2. Chinese—History. I. Sofronov, M. V., joint author. II. Cheboksarov, Nikolaĭ Nikolaevich, joint author. III. Title.
DS730.K74 79-409594

Li, Dun Jen, 1920-
The ageless Chinese : a history / by Dun J. Li ; [maps by Jeanyee Wong]. — 2d [i.e. 3d] ed. — New York : Scribner, c1978.
xiii, 629 p. : ill. ; 24 cm.
Bibliography: p. 591-593.
Includes index.
ISBN 0-684-15276-2 : $17.50
1. China—History. I. Title.
DS735.L43 1978 951 78-7787
78 MARC

Milston, Gwendda.
A short history of China / Gwendda Milston. — Stanmore [Australia] : Cassell Australia, 1978.
ix, 407 p. : ill. ; 23 cm. — (Cassell Asian histories) Aus***
Bibliography: p. 375-389.
Includes index.
ISBN 0-7269-5449-7 : $8.95
1. China—History. I. Title.
DS735.M554 951 79-309870
79 MARC

Mitchell, Peter MacVicar, 1938-
China : tradition and revolution / [by] Peter M. Mitchell ; study material prepared by Douglas M. Parker. — London : Edward Arnold, 1977.
vi, 234 p. : ill., maps, ports. ; 23 cm. GB77-30890
Includes bibliographies and index.
ISBN 0-7131-0110-5 : £4.95
1. China—History. I. Parker, Douglas M., joint author. II. Title.
DS736.M54 1977 951 78-315532
78 MARC

Miyazaki, Ichisada, 1901-
(Chūgoku shi)
中国史 / 宮崎市定著. — 東京：岩波書店, 1977-78.
2 v. : maps ; 19 cm. — (岩波全書 ; 295, 303) Ja 77-19759 (v. 1)
Series romanized: Iwanami zensho.
Includes index in v. 2.
¥1100 (v. 1) ¥1300 (v. 2)

1. China—History. I. Title.
DS735.M583 78-800329

Servas, Ploutēs.
(Palĕa kai nea Kina)
Παληὰ καὶ νέα Κίνα / Πλουτὴ Σέρβα. — Ἀθῆνα : Διάλογος, 1977.
257 p. ; 21 cm.
1. China—History. I. Title.
DS735.S46 79-413745

Wiethoff, Bodo.
China / Bodo Wiethoff. — 5. überarb. Neuaufl. — Hannover : Fackelträger, 1978.
178 p. : ill., maps ; 21 cm. — (Edition Zeitgeschehen) GFR78-A49
Bibliography: p. 170-174.
Includes index.
ISBN 3-7716-2083-X : DM16.00
1. China—History. I. Title.
DS735.W64 1978 951 79-367646
79 MARC

Żbikowski, Tadeusz.
Legendy i pradzieje Kraju Środka / Tadeusz Żbikowski. — Wyd. 1. — Warszawa : Iskry, 1978 (Łódź : ŁZG 1)
457, [3] p., [8] leaves of plates : ill. ; 21 cm. P78-10050
Bibliography: p. 455-[458]
zł70.00
1. Legends—China. 2. Mythology, Chinese. 3. China—History. I. Title.
GR335.Z38 79-354605
79 MARC

Zierer, Otto, 1909-
China / [by] Otto Zierer ; [translated from the German by Godfrey Ivins]. — London ; New York [etc.] : Spring Books, 1978.
3-125 p. : ill. (some col.), ports. (some col.) ; 26 cm. — (Concise history of great nations) GB78-35789
On fly leaf: History of China.
ISBN 0-600-32019-7 : £2.95
1. China—History. I. Title. II. Title: History of China. III. Series.
DS736.Z5313 951 79-307291
79 MARC

—HISTORY—TO 1766 B.C.

Ch'en, Fei-lung.
(Shih chi tao yen)
史記導言 / 陳飛龍著. — 臺北：莊嚴出版社, 民國67 [1978]
2, 68 p. ; 20 cm.
Bibliography: p. 67-68.
NT$15.00

1. Ssu-ma, Ch'ien, ca. 145-ca. 86 B. C. Shih chi. 2. China—History—To 1766 B. C. 3. China—History—1766 B. C.-220 A. D. I. Title.
DS735.A2S6688 79-840579

Chou, Hu-lin.
(Ssu-ma Ch'ien yü ch'i shih hsüeh)
司馬遷與其史學 / 周虎林著. — 臺北：文史哲出版社, 民國67 [1978]
2, 4, 338 p. ; 22 cm. — (文史哲學集成)
Series romanized: Wen shih che hsüeh chi ch'eng.
Bibliography: p. 331-338.
NT$100.00
1. Ssu-ma, Ch'ien, ca. 145-ca. 86 B. C. Shih chi. 2. China—History—To 1766 B. C. 3. China—History—1766 B. C.-220 A. D. I. Title.
DS735.A2C6714 79-838629

Ni, Ssu, 1147-1220.
(Pan Ma i t'ung)
班馬異同 / 倪思編. — [台北：臺灣]商務[印書館, 1978?]
3 v. ; 20 cm. — (四庫全書珍本八集 ; 77-79)
Cover title.
Selections of and comparison between passages from Ssu-ma Ch'ien's Shih chi and Pan Ku's Han shu.
Photoreprint ed. of the Wen Yüan Ko copy.
1. Pan, Ku, 32-92. Han shu. 2. China—History—Han dynasty, 202 B. C.-220 A. D. 3. Ssu-ma, Ch'ien, ca. 145-ca. 86. B. C. Shih chi. 4. China—History—Early to 1766 B. C. 5. China—History—1766 B. C.-220 A. D. I. Ssu-ma, Ch'ien, ca. 145-ca. 86 B. C. Shih chi. Selections. 1978. II. Pan, Ku, 32-92. Han Shu. Selections. 1978. III. Title. IV. Series: Ssu k'u ch'uan shu. Ssu k'u ch'uan shu chen pen pa chi ; 77-79.
AC149.S7 1978, vol. 77-79 78-841074
[DS748.P35]

CHINA

—HISTORY—TO 1766 B.C. (Continued)

Shih, Chih-mien.
(Shih chi chung mu chi)
史記冢墓記 / [著作者施之勉]. — 台北：大陸雜誌社, 民國66 [1977]
9, 263 p. ; 18 cm.
NT$30.00

1. Tombs—China. 2. Ssu-ma, Ch'ien, ca. 145–ca. 86 B.C. Shih chi. 3. China—History—To 1766 B.C. 4. China—History—1766 B.C.–220 A.D. I. Ssu-ma, Ch'ien, ca. 145–ca. 86 B.C. Shih chi. 1977. II. Title.
DS719.S53 78–843333

Ssu-ma, Ch'ien, ca. 145–ca. 86 B.C.
[Shih chi]
白話史記 / [司馬遷]. — 臺北：河洛圖書出版社, 民國68 [1979]
3 v. (6, 14, 10, 1769 p.) ; 22 cm.
"本書為集體翻譯, 凡…六十位教授"
NT$500.00
1. China—History—To 1766 B.C. 2. China—History—1766 B.C.–220 A.D. I. Title.
Title romanized: Pai hua Shih chi.
DS735.A2S6 1979 79–840139

Ssu-ma, Ch'ien, ca. 145–ca. 86 B.C.
[Shih chi. Selections]
史記菁華錄 / [司馬遷著]；姚祖恩編著. — 臺北：聯經出版事業公司, 民國66 [1977]
4, 40, 2, 266 p. ; 22 cm.
NT$100.00
1. Ssu-ma, Ch'ien, ca. 145–ca. 86 B.C. Shih chi. 2. China—History—To 1766 B.C. 3. China—History—1766 B.C.–220 A.D. I. Yao, Tsu-en, chin shih 1784. II. Title.
Title romanized: Shih chi ching hua lu.
DS735.A2S63 1977 79–838632

Ssu-ma, Ch'ien, ca. 145–ca. 86 B.C.
[Shih chi. Selections]
史記精華 / 秦同培注釋；宋晶如增訂. — 台南：大夏出版社, 民國67 [1978]
2, 4, 398 p. ; 21 cm.
"作者司馬遷"
"注音·詳註·白話"
"「史記菁華錄」指導大概 朱自清著": p. 381–398.
NT$100.00
1. Ssu-ma, Ch'ien, ca. 145–ca. 86 B.C. Shih chi. 2. China—History—To 1766 B.C. 3. China—History—1766 B.C.–220 A.D. I. Ch'in, T'ung-pei. II. Sung, Ching-ju. III. Chu, Tzu-ch'ing, 1898–1948. Shih chi ching hua lu chih tao ta kai. 1978. IV. Title.
Title romanized: Shih chi ching hua.
DS735.A2S63 1978 79–838631
79 C

Ts'ui, Shih.
(Shih chi t'an yüan)
史記探源 / [崔適]. — 台北縣板橋鎮：廣城出版社, 民國66 [1977]
[141] leaves ; 22 cm.
Reprint of the 1967 ed. published by Kuang wen shu chü, T'ai-pei.
NT$75.00

1. Ssu-ma, Ch'ien, ca. 145–ca. 86 B.C. Shih chi. 2. China—History—To 1766 B.C. 3. China—History—1766 B.C.–220 A.D. I. Title.
DS735.A2S6797 1977 78–843318

—HISTORY—TO 1766 B.C. —INDEXES

Chung, Hua.
(Shih chi jen ming so yin)
史記人名索引 / 鍾華編. — 第1版. — 北京：中華書局, 1977.
4, 267 p. ; 21 cm.
"中華書局在一九五九年出版了《史記》 點校本 本索引就是根據點校本編製的"
Includes index.
$0.72
1. Ssu-ma, Chien, ca. 145–ca. 86 B.C. Shih chi—Indexes. 2. China—History—Early to 1766 B.C.—Indexes. 3. China—History—1766 B.C.–220 A.D.—Indexes. 4. China—Biography—Indexes. I. Ssu-ma, Chien, ca. 145–ca. 86 B.C. Shih chi. II. Title.
DS735.A2S612 78–841309

—HISTORY—TO 1766 B.C. —JUVENILE FICTION

Yang, Tsung-chen.
(Chih shui ho chih kuo)
治水和治國 / 揚宗珍文；洪義男圖. — [南投縣]：臺灣省政府教育廳；臺北：發行者 臺灣書店, 民國60 [1971]
68 p. : ill. ; 21 cm. — (中華兒童叢書)
Series romanized: Chung-hua erh t'ung ts'ung shu.
"編號 61091…年級 六年級"
1. China—History—To 1766 B.C.—Juvenile fiction. I. Hung, I-nan. II. Title.
PZ10.831.Y3298C5 78–842756

—HISTORY—TO 1643

Chang, Tai, 1597–ca. 1676.
(Shih ch'üeh)
史闕：附明紀史闕 / 張岱撰. — 台1版. — 台北：華世出版社, 民國66 [1977]
2 v. (4, 4, 797 p.) ; 22 cm.
Reprint ed. ; with new pref.
NT$300.00 per vol.

1. China—History—To 1643. I. Chang, Tai, 1597–ca. 1676. Ming chi shih ch'üeh. 1977. II. Title.
DS735.A2C38 1977 78–840996
C

Ch'en, Yüan, 1880–1971.
(T'ung chien Hu chu piao wei)
通鑑胡注表微 / 陳垣著. — 香港：廣角鏡出版社, 1978.
1, 2, 411 p., [1] leaf of plates : facsim. ; 21 cm.
Reprint of the 1958 ed. published by K'o hsüeh ch'u pan she, Peking.
Bibliography: p. 405–408.
$20.00

1. Ssu-ma, Kuang, 1019–1086. Tzu chih t'ung chien. 2. China—History—To 1643. I. Title.
DS745.S783C47 1978 79–838174

Maspero, Henri, 1883-1945.
China in antiquity / by Henri Maspero ; translated by Frank A. Kierman, Jr. — Amherst : University of Massachusetts Press, 1978.
xxxii, 527 p. : maps ; 24 cm.
Translation of La Chine antique.
Bibliography: p. [493]-497.
"Bibliography of the writings of Henri Maspero": p. [498]-511.
Includes index.
ISBN 0-87023-193-6
1. China—History—To 1643. 2. China—Civilization. 3. China—Antiquities. I. Title.
DS745.M313 951'.01 77-26756
77 MARC

(Tzu chih t'ung chien chiao k'an piao)
資治通鑑校勘表. — [香港?]：中華書局, [1978?]
68 p. ; 21 cm.

1. Ssu-ma, Kuang, 1019–1086. Tzu chih t'ung chien. 2. China—History—To 1643.
DS745.S783T98 79–835477

—HISTORY—SHANG DYNASTY, 1766–1122 B.C.

Chung, Pai-sheng.
(Pu tz'u chung so chien Yin wang t'ien yu ti ming k'ao)
卜辭中所見殷王田游地名考：兼論田游地名研究方法 / 鐘柏生. — [台北? : s.n.], 民國61 [1972]
191 p. (on double leaves) : maps ; 27 cm.
"國立台灣大學中文研究所碩士論文"
Reproduced from typescript.
Bibliography: p. 189–191.

1. China—History—Shang dynasty, 1766–1122 B.C. 2. Inscriptions, Chinese. 3. Names, Geographical—China. I. Title.
DS744.C45 1972b 79–836185

Ho-pei sheng po wu kuan.
(Kao-ch'eng T'ai-hsi Shang tai i chih)
藁城台西商代遺址 / 河北省博物馆, 文管处台西考古队, 河北省藁城县台西大队理论小组编. — 第1版. — 北京：文物出版社, 1977.
91 p. : ill. ; 19 cm.
$0.30
1. China—History—Shang dynasty, 1766–1122 B.C. 2. T'ai-hsi ts'un, China—Antiquities. 3. China—Antiquities. I. Wen kuan ch'u T'ai-hsi k'ao ku tui. II. Kao-ch'eng hsien, China (Hopei). T'ai-hsi ta tui. Li lun hsiao tsu. III. Title.
DS744.H66 1977 78–843255
C

—HISTORY—SHANG DYNASTY, 1766-1122 B.C.—SOURCES

Keightley, David N
Sources of Shang history : the oracle-bone inscriptions of bronze age China / David N. Keightley. — Berkeley : University of California Press, c1978.
xvii, 281 p., [15] leaves of plates : ill. ; 32 cm.
Bibliography: p. 229-256.
Includes index.
ISBN 0-520-02969-0
1. China—History—Shang dynasty, 1766-1122 B.C.—Sources. 2. Inscriptions, Chinese. 3. Oracle bones. I. Title.
DS744.K44 931 74-29806
79 MARC

—HISTORY—1766 B.C. - 220 A.D.

Ch'en, Fei-lung.
(Shih chi tao yen)
史記導言 / 陳飛龍著. — 臺北：莊嚴出版社, 民國67 [1978]
2, 68 p. ; 20 cm.
Bibliography: p. 67–68.
NT$15.00

1. Ssu-ma, Ch'ien, ca. 145–ca. 86 B.C. Shih chi. 2. China—History—To 1766 B.C. 3. China—History—1766 B.C.–220 A.D. I. Title.
DS735.A2S6688 79–840579

Chou, Hu-lin.
(Ssu-ma Ch'ien yü ch'i shih hsüeh)
司馬遷與其史學 / 周虎林著. — 臺北：文史哲出版社, 民國67 [1978]
2, 4, 338 p. ; 22 cm. — (文史哲學集成)
Series romanized: Wen shih che hsüeh chi ch'eng.
Bibliography: p. 331–338.
NT$100.00
1. Ssu-ma, Ch'ien, ca. 145–ca. 86 B.C. Shih chi. 2. China—History—To 1766 B.C. 3. China—History—1766 B.C.–220 A.D. I. Title.
DS735.A2C6714 79–838629

Ni, Ssu, 1147–1220.
(Pan Ma i t'ung)
班馬異同 / 倪思編. — [台北：臺灣]商務[印書館, 1978?]
3 v. ; 20 cm. — (四庫全書珍本八集 ; 77–79)
Cover title.
Selections of and comparison between passages from Ssu-ma Ch'ien's Shih chi and Pan Ku's Han shu.
Photoreprint ed. of the Wen Yüan Ko copy.
1. Pan, Ku, 32–92. Han shu. 2. China—History—Han dynasty, 202 B.C.–220 A.D. 3. Ssu-ma, Ch'ien, ca. 145–ca. 86. B.C. Shih chi. 4. China—History—Early to 1766 B.C. 5. China—History—1766 B.C.–220 A.D. I. Ssu-ma, Ch'ien, ca. 145–ca. 86 B.C. Shih chi. Selections. 1978. II. Pan, Ku, 32–92. Han Shu. Selections. 1978. III. Title. IV. Series: Ssu k'u ch'uan shu. Ssu k'u ch'uan shu chen pen pa chi ; 77–79.
AC149.S7 1978, vol. 77–79 78–841074
[DS748.P35]
C

Shih, Chih-mien.
(Shih chi chung mu chi)
史記冢墓記 / [著作者施之勉]. — 台北：大陸雜誌社, 民國66 [1977]
9, 263 p. ; 18 cm.
NT$30.00

1. Tombs—China. 2. Ssu-ma, Ch'ien, ca. 145–ca. 86 B.C. Shih chi. 3. China—History—To 1766 B.C. 4. China—History—1766 B.C.–220 A.D. I. Ssu-ma, Ch'ien, ca. 145–ca. 86 B.C. Shih chi. 1977. II. Title.
DS719.S53 78–843333

CHINA

—HISTORY—1766 B.C.-220 A.D. (Continued)

Ssu-ma, Ch'ien, ca. 145-ca. 86 B.C.
[Shih chi]
白話史記 / [司馬遷]. — 臺北：河洛圖書出版社, 民國68 [1979]
3 v. (5, 14, 10, 1769 p.) ; 22 cm.
"本書為集體翻譯, 凡…六十位教授"
NT$500.00
1. China—History—To 1766 B.C. 2. China—History—1766 B.C.-220 A.D. I. Title.
Title romanized: Pai hua Shih chi.
DS735.A2S6 1979 79-840139

Ssu-ma, Ch'ien, ca. 145-ca. 86 B.C.
[Shih chi. Selections]
史記菁華錄 / [司馬遷著] ; 姚祖恩編著. — 臺北：聯經出版事業公司, 民國66 [1977]
4, 40, 2, 266 p. ; 22 cm.
NT$100.00
1. Ssu-ma, Ch'ien, ca. 145-ca. 86 B.C. Shih chi. 2. China—History—To 1766 B.C. 3. China—History—1766 B.C.-220 A.D. I. Yao, Tsu-en, chin shih 1784. II. Title.
Title romanized: Shih chi ching hua lu.
DS735.A2S63 1977 79-838632

Ssu-ma, Ch'ien, ca. 145-ca. 86 B.C.
[Shih chi. Selections]
史記精華 / 秦同培注釋 ; 宋晶如增訂. — 台南：大夏出版社, 民國67 [1978]
2, 4, 398 p. ; 21 cm.
"作者司馬遷"
"注音·詳註·白話"
"「史記菁華錄」指導大概 朱自清著": p. 381-398.
NT$100.00
1. Ssu-ma, Ch'ien, ca. 145-ca. 86 B.C. Shih chi. 2. China—History—To 1766 B.C. 3. China—History—1766 B.C.-220 A.D. I. Ch'in, T'ung-pei. II. Sung, Ching-ju. III. Chu, Tzu-ch'ing, 1898-1948. Shih chi ching hua lu chih tao ta kai. 1978. IV. Title.
Title romanized: Shih chi ching hua.
DS735.A2S63 1978 79-838631

Ts'ui, Shih.
(Shih chi t'an yüan)
史記探源 / [崔適]. — 台北縣板橋鎮：廣城出版社, 民國66 [1977]
[141] leaves ; 22 cm.
Reprint of the 1967 ed. published by Kuang wen shu chü, T'ai-pei.
NT$75.00
1. Ssu-ma, Ch'ien, ca. 145-ca. 86 B.C. Shih chi. 2. China—History—To 1766 B.C. 3. China—History—1766 B.C.-220 A.D. I. Title.
DS735.A2S6797 1977 78-843318

—HISTORY—1766 B.C.-220 A.D.—ADDRESSES, ESSAYS, LECTURES

Kaizuka, Shigeki, 1904-
(Chūgoku kodai no denshō)
中国古代の伝承 / [著者 貝塚茂樹]. — 東京：中央公論社, 1976.
405 p. ; 21 cm. — (貝塚茂樹著作集 ; 第5巻) Ja 76-21325
¥2800
1. China—History—1766 B.C.-220 A.D.—Addresses, essays, lectures. I. Title. II. Series: Kaizuka, Shigeki, 1904- Kaizuka Shigeki chosaku shū ; dai 5-kan.
DS747.2.K34 78-815490

—HISTORY—1766 B.C.-220 A.D.—INDEXES

Chung, Hua.
(Shih chi jen ming so yin)
史記人名索引 / 鍾華編. — 第1版. — 北京：中華書局, 1977.
4, 267 p. ; 21 cm.
"中華書局在一九五九年出版了《史記》點校本 本索引就是根據點校本編製的"
Includes index.
$0.72
1. Ssu-ma, Chien, ca. 145-ca. 86 B.C. Shih chi—Indexes. 2. China—History—Early to 1766 B.C.—Indexes. 3. China—History—1766 B.C.-220 A.D.—Indexes. 4. China—Biography—Indexes. I. Ssu-ma, Chien, ca. 145-ca. 86 B.C. Shih chi. II. Title.
DS735.A2S612 78-841309

—HISTORY—CHOU DYNASTY, 1122-221 B.C.

Chang, I-jen.
(Kuo yü yen chiu)
國語研究 / 張以仁. — [台北?：張以仁], 民國48 [1959]
iii, 138 double leaves ; 27 cm.
Cover title.
"國立台灣大學中文研究所碩士論文"
Reproduced from ms. copy.
1. Tso-ch'iu, Ming. Kuo yü. 2. China—History—Chou dynasty, 1122-221 B.C. 3. Tso-ch'iu, Ming. Tso chuan. I. Title.
DS747.T753C476 1959 78-842846

Ch'iu, Hsin-i.
(Wu teng chüeh shuo yen chiu)
五等爵說研究 / 邱信義. — [台北?：s.n.], 民國59 [1970]
111 double leaves ; 27 cm.
"國立臺灣大學中文研究所碩士論文"
Reproduced from ms. copy.
Bibliography: leaves 106-111.
1. China—History—Chou dynasty, 1122-221 B.C. 2. Feudalism—China. I. Title.
DS747.C46 1970 79-835882

Li, T'ang.
(Chou Wu-wang)
周武王 / 李唐. — 臺景印初版. — 臺北：河洛圖書出版社, 民國67 [1978]
3, 2, 83 p. ; 21 cm. — (中國歷史小叢書 ; 002)
Series romanized: Chung-kuo li shih hsiao ts'ung shu.
Reprint of the 1961 ed. published by Hsiang-kang hung yeh shu chü, Hong Kong, in series: Li shih hsiao ts'ung shu.
Pbk.: NT$35.00
1. China—History—Chou dynasty, 1122-221 B.C. 2. Chou Wu-wang, Emperor of China, 1169-1116 B.C.
DS747.L48 1978 79-836048

Tso-chiu, Ming.
(Kuo yü)
國語 / [上海师范大学古籍整理组校点]. — 上海：上海古籍出版社, 1978.
2 v. (1, 18, 728 p.) ; 21 cm.
"左丘明所作, 全書共二十一卷"
CONTENTS: 上册 卷1至卷9.—下卷 卷10至卷21, 索引.
$2.75
1. China—History—Chou dynasty, 1122-221 B.C. I. Shang-hai shih fan ta hsüeh. Ku chi cheng li tsu. II. Title.
DS747.T75 1978 78-843640

—HISTORY—CHOU DYNASTY, 1122-221 B.C.—CHRONOLOGY

Ch'en, P'an.
(Ch'un ch'iu ta shih piao lieh kuo chüeh hsing chi ts'un mieh piao chuan i)
春秋大事表列國爵姓及存滅表譔異 / [譔述者陳槃]. — 增訂本. — 臺北：中央研究院歷史語言研究所, 民國58 [1969]
7 v. (9, 743 double leaves) in 2 cases : ill. ; 28 cm. — (中央研究院歷史語言研究所專刊 ; 52)
Bibliography: v. 7, leaves 712-733.
Includes index.
1. Ku, Tung-kao, 1679-1759. Ch'un ch'iu ta shih piao. 2. China—History—Chou dynasty, 1122-221 B.C.—Chronology. I. Title. II. Series: Chung yang yen chiu yüan. Li shih yü yen yen chiu so. Chung yang yen chiu yüan li shih yü yen yen chiu so chuan k'an ; 52.
DS747.K763C47 78-842487

Feng, Meng-lung, 1574?-1645? — [Hsin lieh kuo chih] ... [1978]
NT$120.00
1. China—History—Chou dynasty, 1122-221 B.C.—Fiction. I. Ts'ai, Yüan-fang, fl. 1736-1770. II. Ho lo t'u shu ch'u pan she. III. Title.
Title romanized: Tung Chou lieh kuo chih.
PL2698.F4H8 1978 79-837316

—HISTORY—SPRING AND AUTUMN PERIOD, 722-481 B.C.

Ch'en, Hou-yao, 1648-1722.
(Ch'un ch'iu chan kuo i tz'u)
春秋戰國異辭：五十四卷 / 陳厚耀撰. — 臺北：鼎文書局, 民國66 [1977]
ca. 1500 p. in various pagings ; 22 cm. — (中國學術類編)
Reprint ed.
CONTENTS: 春秋戰國通表 上下卷.—春秋戰國異辭 54卷.—摭遺.
1. China—History—Spring and Autumn period, 722-481 B.C. 2. China—History—Warring States, 403-221 B.C. I. Title. II. Series: Chung-kuo hsüeh shu lei pien.
DS747.15.C47 1977 79-838266

Meng, Chung-jen.
(Ch'un ch'iu hsiao ku shih)
春秋小故事 / 孟仲仁編著. — 臺北：出版家文化事業有限公司出版部, 民國67 [1978]
2, 195 p. ; 20 cm. — (出版家叢書 ; 55)
Series romanized: Ch'u pan chia ts'ung shu.
NT$40.00
1. China—History—Spring and Autumn period, 722-481 B.C. I. Title.
DS747.15.M46 79-837403

Sun, Yao.
(Ch'un ch'iu shih tai chih shih tsu)
春秋時代之世族 / 孫曜撰. — 台1版. — [台北]：華世出版社, 民國65 [1976]
2, 192 p. ; 21 cm.
Reprint of the 1931 ed.
NT$50.00
1. China—Nobility—Genealogy. 2. China—History—Spring and Autumn period, 722-481 B.C. I. Title.
CS1164.S9 1976 79-838276

—HISTORY—SPRING AND AUTUMN PERIOD, 722-481 B.C.—BIOGRAPHY

Moriya, Hiroshi, 1932-
(Hasha no senryaku to ketsudan)
覇者の戦略と決断 / 守屋洋著. — 東京：産業能率短期大学出版部, 1978.
218 p. : ill. ; 19 cm. Ja 78-23422
¥980
1. China—History—Spring and autumn period, 722-481 B.C.—Biography. 2. China—Kings and rulers—Biography. 3. China—Biography. I. Title.
DS747.15.M67 79-809183

—HISTORY—WARRING STATES, 403-221 B.C.

Chang, Yang-hao, 1269-1329.
(San shih chung kao)
三事忠告 / [張養浩撰]. 七國考 / [董說撰]. — [臺北：臺灣]商務[印書館, 1979]
4 v. ; 22 cm. — (四庫全書珍本九集 ; 165-168)
Cover title.
"三事忠告四卷…七國攷十四卷"
Photoreprint ed. of the Wen Yüan Ko copy.
1. Local government—China. 2. Civil service ethics. 3. China—Politics and government—To 1643. 4. China—History—Warring States, 403-221 B.C. I. Tung, Shuo, 1620-1686. Ch'i kuo k'ao. 1979. II. Title. III. Series: Ssu k'u ch'üan shu. Ssu k'u ch'üan shu chen pen chiu chi ; 165-168.
AC149.S7 1979, vol. 165-168 79-840282
[JS7353.A3]

Ch'en, Hou-yao, 1648-1722.
(Ch'un ch'iu chan kuo i tz'u)
春秋戰國異辭：五十四卷 / 陳厚耀撰. — 臺北：鼎文書局, 民國66 [1977]
ca. 1500 p. in various pagings ; 22 cm. — (中國學術類編)
Reprint ed.
CONTENTS: 春秋戰國通表 上下卷.—春秋戰國異辭 54卷.—摭遺.
1. China—History—Spring and Autumn period, 722-481 B.C. 2. China—History—Warring States, 403-221 B.C. I. Title. II. Series: Chung-kuo hsüeh shu lei pien.
DS747.15.C47 1977 79-838266

CHINA

—HISTORY—WARRING STATES, 403-221 B.C. (Continued)

Li, T'ang.
(Chan kuo ch'i hsiung)
戰國七雄 / 李唐. — 臺景印初版. — 臺北：河洛圖書出版社, 民國67 [1978]
3, 3, 96 p. ; 21 cm. — (中國歷史小叢書 ; 004)
Series romanized: Chung-kuo li shih hsiao ts'ung shu.
Reprint of the 1962 ed. published by Hsiang-kang hung yeh shu chü, Hong Kong, in series: Li shih hsiao ts'ung shu.
Pbk.: NT$35.00

1. China—History—Warring States, 403-221 B.C. I. Title.
DS747.2.L48 1978 79-836018

Wang, Hsien-ch'ien, 1842-1918.
(Hsien-yü Chung-shan kuo shih piao, chiang yü t'u shuo)
鮮虞中山國事表, 疆域圖説 / 王先謙輯. — 台北：廣文書局, 民國67 [1978]
50 p. (p. 50 blank) : map ; 22 cm. — (史料七編)
Reprint of the 1883 ed. published by Ch'ang-sha Wang shih.
NT$150.00

1. China—History—Warring States, 403-221 B.C. I. Title. II. Series: Shih liao ch'i pien.
DS747.2.W34 1978 79-838580

—HISTORY—WARRING STATES, 403-221 B.C.—ANECDOTES, FACETIAE, SATIRE, ETC.

Chan kuo ts'e.
戰國策 / 劉向集錄. — 第1版. — 上海：上海古籍出版社：新華書店上海發行所發行, 1978.
3 v. (1, 2, 17, 1366 p., [1] leaf of plates) : map ; 21 cm.
$5.10

1. China—History—Warring States, 403-221 B.C.—Sources. 2. China—History—Warring States, 403-221 B.C.—Anecdotes, facetiae, satire, etc. I. Liu, Hsiang, 77?-6? B.C. II. Title.
Title romanized: Chan kuo ts'e.
DS747.2.C375 1978 79-836509

C

Chan kuo ts'e.
戰國策 / 劉向撰；高誘注. — 台灣初版. — 台北：老古出版社, 民國67 [1978]
696, [27] p. ; 22 cm. — (正統謀略學彙編 ; 初輯 ; 第3冊)
Reprint ed.

1. China—History—Warring States, 403-221 B.C.—Sources. 2. China—History—Warring States, 403-221 B.C.—Anecdotes, facetiae, satire, etc. I. Liu, Hsiang, 77?-6? B.C. II. Kao, Yu, fl. 205-212. III. Series: Cheng t'ung mou lüeh hsüeh hui pien : Ch'u chi : ti 3 ts'e.
Title romanized: Chan kuo ts'e.
DS747.2.C375 1978a 79-839139

Meng, Chung-jen.
(Chan kuo hsiao ku shih)
戰國小故事 / 孟仲仁編著. — 臺北：出版家文化事業有限公司出版部, 民國67 [1978]
6, 246 p., [1] leaf of plates : maps ; 19 cm. — (出版家叢書 ; 64)
Series romanized: Ch'u pan chia ts'ung shu.
NT$40.00

1. China—History—Warring States, 403-221 B.C.—Anecdotes, facetiae, satire, etc. I. Title.
DS747.2.M46 79-837331

—HISTORY—WARRING STATES, 403-221 B.C.—BIOGRAPHY

Moriya, Hiroshi, 1932-
(Chūgokujin no kōshōjutsu)
中国人の交渉術：世渡りの才覚とは / 守屋洋著. — 東京：産業能率短期大学出版部, 1978.
202 p. : map ; 19 cm. Ja 79-3227
¥950

1. China—History—Warring States, 403-221 B.C.—Biography. 2. China—Biography. 3. Diplomacy. I. Title.
DS747.2.M67 79-805946

—HISTORY—WARRING STATES, 403-221 B.C.—SOURCES

Chan kuo ts'e.
戰國策 / 劉向集錄. — 第1版. — 上海：上海古籍出版社：新華書店上海發行所發行, 1978.
3 v. (1, 2, 17, 1366 p., [1] leaf of plates) : map ; 21 cm.
$5.10

1. China—History—Warring States, 403-221 B.C.—Sources. 2. China—History—Warring States, 403-221 B.C.—Anecdotes, facetiae, satire, etc. I. Liu, Hsiang, 77?-6? B.C. II. Title.
Title romanized: Chan kuo ts'e.
DS747.2.C375 1978 79-836509

Chan kuo ts'e.
戰國策 / 劉向撰；高誘注. — 台灣初版. — 台北：老古出版社, 民國67 [1978]
696, [27] p. ; 22 cm. — (正統謀略學彙編 ; 初輯 ; 第3冊)
Reprint ed.

1. China—History—Warring States, 403-221 B.C.—Sources. 2. China—History—Warring States, 403-221 B.C.—Anecdotes, facetiae, satire, etc. I. Liu, Hsiang, 77?-6? B.C. II. Kao, Yu, fl. 205-212. III. Series: Cheng t'ung mou lüeh hsüeh hui pien : Ch'u chi : ti 3 ts'e.
Title romanized: Chan kuo ts'e.
DS747.2.C375 1978a 79-839139

Ma-wang-tui Han mu po shu cheng li hsiao tsu.
(Chan kuo tsung heng chia shu)
戰國縱横家書 / 马王堆汉墓帛书整理小组编. — 第1版. — 北京：文物出版社, 1976.
201 p. : ill. ; 21 cm. — (马王堆汉墓帛书)
Includes bibliographical references.
$0.75

1. China—History—Warring States, 403-221 B.C.—Sources. 2. Manuscripts, Chinese. 3. China—Antiquities. I. Title. II. Series: Ma-wang-tui Han mu po shu.
DS747.2.M33 1976 79-836137

Shui-hu-ti Ch'in mu chu chien cheng li hsiao tsu.
(Shui-hu-ti Ch'in mu chu chien)
睡虎地秦墓竹简 / [編者《睡虎地秦墓竹简》整理小組]. — [第1版]. — [北京]：文物出版社, [1977]
7 v. (on double leaves) in case : ill. ; 41 cm.
$280.00
1. China—History—Ch'in dynasty, 221-207 B.C.—Sources. 2. China—History—Warring States, 403-221 B.C.—Sources. 3. Law—China—Sources. 4. Inscriptions, Chinese—China—Shui-hu-ti site. 5. Shui-hu-ti site, China.
DS747.5.S56 78-840921

78 C

—HISTORY—CH'IN DYNASTY, 221-207 B.C.

Li, T'ang.
(Ch'in Shih-huang)
秦始皇 / 李唐. — 臺景印初版. — 臺北：河洛圖書出版社, 民國67 [1978]
3, 3, 102 p. ; 22 cm. — (中國歷史小叢書 ; 005)
Series romanized: Chung-kuo li shih hsiao ts'ung shu.
Reprint of the 1962 ed. published by Hsiang-kang hung yeh shu chü, Hong Kong, in series: Li shih hsiao ts'ung shu.
Pbk.: NT$35.00
1. China—History—Ch'in dynasty, 221-207 B.C. 2. Ch'in Shih-huang-ti, Emperor of China, 259-210 B.C.
DS747.5.L52 1978 79-837023

C

—HISTORY—CH'IN DYNASTY, 221-207 B.C.—BIBLIOGRAPHY

Sakade, Yoshinobu, 1934-
(Shin Kan shisō kenkyū bunken mokuroku)
秦漢思想研究文献目録 / 坂出祥伸編. — 吹田：関西大学出版広報部, 1978.
8, 137 p. ; 26 cm. Ja 78-17058
¥4800

1. China—Intellectual life—Bibliography. 2. China—History—Ch'in dynasty, 221-207 B.C.—Bibliography. 3. China—History—Han dynasty, 202 B.C.-220 A.D.—Bibliography. I. Title.
Z3108.C55S24 78-813240
[DS721]

—HISTORY—CH'IN DYNASTY, 221-207 B.C.—FICTION

Liu, Ya-chou.
(Ch'en Sheng)
陈胜 / 刘亚洲著. — 第1版. — [汉口]：湖北人民出版社：湖北省新華書店发行, 1977.
486 p. ; 21 cm.
$1.08

1. Ch'en, Sheng, d. 209 B.C.—Fiction. 2. Peasant uprisings—China—Fiction. 3. China—History—Ch'in dynasty, 221-207 B.C.—Fiction. I. Title.
PL2879.Y3C5 79-840705

—HISTORY—CH'IN DYNASTY, 221-207 B.C.—SOURCES

Shui-hu-ti Ch'in mu chu chien cheng li hsiao tsu.
(Shui-hu-ti Ch'in mu chu chien)
睡虎地秦墓竹简 / [編者《睡虎地秦墓竹简》整理小組]. — [第1版]. — [北京]：文物出版社, [1977]
7 v. (on double leaves) in case : ill. ; 41 cm.
$280.00
1. China—History—Ch'in dynasty, 221-207 B.C.—Sources. 2. China—History—Warring States, 403-221 B.C.—Sources. 3. Law—China—Sources. 4. Inscriptions, Chinese—China—Shui-hu-ti site. 5. Shui-hu-ti site, China.
DS747.5.S56 78-840921

—HISTORY—THREE KINGDOMS, 220-265—FICTION

[**Lo, Kuan-chung**] ca. 1330-ca. 1400.
[San-kuo chih yen i]
繡像第一才子書 / 毛聲山評. — [s.l.]：經國堂發兑, [順治甲申i.e. 1644序]
20 v. (on double leaves) in 2 cases : ill. ; 25 cm. — (四大奇書 ; 第1種)
Running title: Ti i ts'ai tzu shu.
In 51 chüan, 120 hui.
Later editions published under the author: Lo Kuan-chung and the title: San-kuo chih yen i.
"聖歎外書"
Block print.

1. China—History—Three kingdoms, 220-265—Fiction. I. Mao, Sheng-shan. II. Title. III. Title: Ti i ts'ai tzu shu. IV. Series: Ssu ta ch'i shu : ti 1 chung.
Title romanized: Hsiu hsiang ti i ts'ai tzu shu.
PL2690.S3 1644 79-836164

79 C

—HISTORY—HAN DYNASTY, 202 B.C.-220 A.D.

An, Tso-chang.
(Liu Pang yü Lü Chih)
刘邦与吕雉 / 安作璋. — [香港?：s.n., 1977?]
2, 2, 78 p., [2] fold. leaves of plates : maps ; 22 cm.
Reprint of the 1974 ed. published by Shan-tung jen min ch'u pan she.

1. China—History—Han dynasty, 202 B.C.-220 A.D. 2. Han Kao-tsu, Emperor of China, 247-195 B.C. 3. Han Lü-hou, Empress of China, 241-180 B.C. I. Title.
DS748.A63 1977 79-836328

C

(Chou Ch'en erh shih Han shu pu cheng ho k'an)
周陳二氏漢書補證合刊. — 臺北：鼎文書局, 民國66 [1977]
[17], 987, [12], 208 p. ; 22 cm. — (中國學術類編)
Reprint of two works, with new pref.
CONTENTS: 周壽昌 漢書注校補五十六卷.—陳直 漢書新證二卷.

1. Pan, Ku, 32-92. 2. China—History—Han dynasty, 202 B.C.-220 A.D. I. Chou, Shou-ch'ang, 1814-1884. Han shu chu chiao pu. 1977. II. Ch'en, Chih. Han shu hsin cheng. 1977. III. Series: Chung-kuo hsüeh shu lei pien.
DS748.16.P36C45 79-838256

CHINA

—HISTORY—HAN DYNASTY, 202 B.C.-220 A.D. (Continued)

De Crespigny, Rafe.
Portents of protest in the later Han dynasty : the memorials of Hsiang Kai to Emporor Huan in 166 A.D. / by Rafe de Crespigny. — Canberra : Australian National University Press in association with the Faculty of Asian Studies, Australian National University, 1976.
vii, 124 p. ; 25 cm. — (Oriental monograph series ; no. 19)
Aus
Includes bibliographical references and index.
ISBN 0-7081-0178-X
1. China—History—Han dynasty, 202 B.C.-220 A.D. 2. Omens. 3. Hsiang, K'ai, 2d cent. 4. Scholars—China—Biography. I. Title. II. Series.
DS748.D42 1976 931 78-322713
79 MARC

Hsü, Chi-yü, 1795-1873.
(Liang Han chih yen pien shih chün k'ao lüeh)
兩漢志沿邊十郡考略；兩漢幽幷涼三州今地考略 / 徐繼畬撰. — 台北：廣文書局, 民國67 [1978]
4, 32, 72 p. ; 22 cm. — (史料七編)
Reprint the ed. with postscript dated 1913.
NT$150.00
1. China—Historical geography. 2. China—History—Han dynasty, 202 B.C.-220 A.D. I. Hsü, Chi-yü, 1795-1873. Liang Han Yu Ping Liang san chou chin ti k'ao lüeh. 1978. II. Title. III. Series: Shih liao ch'i pien.
DS706.5.H72 1978 79-841396

Hsü, T'ien-lin, chin shih 1205.
(Hsi Han hui yao)
西漢會要 / 徐天麟撰. — 第1版. — 上海：上海人民出版社, 1977.
2 v. (2, 22, 835 p.) ; 19 cm.
$1.85
1. China—History—Han dynasty, 202 B.C.-220 A.D. I. Title.
DS748.H7 1977 78-840862

Hsü, T'ien-lin, chin shih 1205.
(Tung Han hui yao)
東漢會要 / 徐天麟撰. — 第1版. — 上海. — 上海古籍出版社：新華書店上海發行所發行, 1978.
600, [22] p. ; 19 cm.
"以清江蘇書局翻刻的武英殿為底本"
$1.60
1. China—History—Han dynasty, 202 B.C.-220 A.D. I. Title.
DS748.H72 1978 79-836807

Huang, Lin.
(Yeh hsin chia Lü-hou)
野心家吕后 / 黄霖编写. — 第1版. — 上海：上海人民出版社：新華書店上海所发行, 1977.
69 p. ; 19 cm.
$0.16
1. China—History—Han dynasty, 202 B.C.-220 A.D. 2. Han Lü-hou, Empress of China, 241-180 B.C. I. Title.
DS748.H8 79-836947

Li, T'ang.
(Han Kao-tsu)
漢高祖 / 李唐著. — 臺景印初版. — 臺北：河洛圖書出版社, 民國67 [1978]
3, 2, 106 p. ; 22 cm. — (中國歷史小叢書 ; 006)
Series romanized: Chung-kuo li shih hsiao ts'ung shu.
Reprint of the 1962 ed. published by Hsiang-kang hung yeh shu chü, Hong Kong, in series: Li shih hsiao ts'ung shu.
Pbk. : NT$35.00
1. China—History—Han dynasty, 202 B.C.-220 A.D. 2. Han Kao-tsu, Emperor of China, 247-195 B.C.
DS748.L514 1978 79-838142

Li, T'ang.
(Han Kuang-wu)
漢光武 / 李唐. — 臺景印初版. — 臺北：河洛圖書出版社, 民國67 [1978]
3, 3, 111 p. ; 21 cm. — (中國歷史小叢書 ; 008)
Series romanized: Chung-kuo li shih hsiao ts'ung shu.
Reprint of the 1962 ed. published by Hsiang-kang hung yeh shu chü, Hong Kong, in series: Li shih hsiao ts'ung shu.
Pbk. : NT$35.00
1. China—History—Han dynasty, 202 B.C.-220 A.D. 2. Han Kuang-wu, Emperor of China, 4 B.C.-57 A.D.
DS748.L515 1978 79-836288

Li, T'ang.
(Han Wu-ti)
漢武帝 / 著作者李唐. — 臺景印初版. — 臺北：河洛圖書出版社, 民國67 [1978]
3, 3, 108 p. ; 22 cm. — (中國歷史小叢書 ; 7)
Series romanized: Chung-kuo li shih hsiao ts'ung shu.
Reprint of the 1962 ed. published by Hsiang-kang hung yeh shu chü, Hongkong, in series: Li shih hsiao ts'ung shu.
NT$35.00
1. Han Wu-ti, Emperor of China, 156-87 B.C. 2. Wang, Meng, d. 23. 3. China—History—Han dynasty, 202 B.C.-220 A.D. I. Title.
DS748.16.H36L5 1978 78-842325

Liu, Chi-hua, 1943-
(Wang Fu yü Ch'ien fu lun)
王符與潛夫論 / 劉紀華著. — 台北：世紀書局, 民國66 [1977]
2, 160 p. ; 22 cm.
Bibliography: p. 157-160.
NT$60.00
1. Wang, Fu, ca. 76-ca. 157. Ch'ien fu lun. 2. China—History—Han dynasty, 202 B.C.-220 A.D. I. Title.
DS748.W323L58 79-840266

Liu, Wen-ch'i, 1789-1854.
(Ch'u Han chu hou chiang yü chih)
楚漢諸侯疆域志 / 劉文淇撰. — 台北：廣文書局, 民國67 [1978]
74 p. ; 22 cm. — (史料七編)
Reprint of the 1876 ed.
NT$150.00
1. China—Historical geography. 2. China—History—Han dynasty, 202 B.C.-220 A.D. I. Title. II. Series: Shih liao ch'i pien.
DS706.5.L57 1978 79-839919

Ni, Ssu, 1147-1220.
(Pan Ma i t'ung)
班馬異同 / 倪思編. — [台北：臺灣]商務[印書館, 1978?]
3 v. ; 20 cm. — (四庫全書珍本八集 ; 77-79)
Cover title.
Selections of and comparison between passages from Ssu-ma Ch'ien's Shih chi and Pan Ku's Han shu.
Photoreprint ed. of the Wen Yüan Ko copy.
1. Pan, Ku, 32-92. Han shu. 2. China—History—Han dynasty, 202 B.C.-220 A.D. 3. Ssu-ma, Ch'ien, ca. 145-ca. 86. B.C. Shih chi. 4. China—History—Early to 1766 B.C. 5. China—History—1766 B.C.-220 A.D. I. Ssu-ma, Ch'ien, ca. 145-ca. 86 B.C. Shih chi. Selections. 1978. II. Pan, Ku, 32-92. Han Shu. Selections. 1978. III. Title. IV. Series: Ssu k'u ch'uan shu. Ssu k'u ch'uan shu chen pen pa chi ; 77-79.
AC149.S7 1978, vol. 77-79 78-841074
[DS748.P35]
78 C

Pan, Ku, 32-92.
(Han shu)
漢書 / [班固撰 ; 顏師古注]. — [成都] : 成都書局, 同治10 [1871]
32 v. (on double leaves) in 4 cases ; 36 cm.
Caption and running title: Ch'ien Han shu.
Issued in case with collective title on spine: Ch'ien ssu shih and with individual title: Ch'ien Han shu.
"摹殿本"
Block print.
1. China—History—Han dynasty, 202 B.C.-220 A.D. I. Yen, Shih-ku, 581-645. II. Title. III. Title: Ch'ien Han shu. IV. Title: Ch'ien ssu shih.
DS748.P24 1871 79-841873
79 C

Pan, Ku, 32-92.
(Han shu)
漢書 / 班固撰 ; 顏師古注. — 第1版. — 北京：中華書局：新華書店北京發行所發行, 1962, 1975 printing.
12 v. (4273, [35] p.) ; 21 cm.
"包括十二紀·八表·十志·七十列傳, 凡一百篇…班固死時,《漢書》的八表和《天文志》還沒有完成…他的妹妹班昭參考東觀藏書替他補作"
1. China—History—Han dynasty, 202 B.C.-220 A.D. I. Yen, Shih-ku, 581-645. II. Pan, Chao, 1st cent. III. Title
DS748.P24 1962b 79-841111

Shen, Chan-ju.
(Hsin Mang ch'üan shih)
新莽全史：又名新莽政治大騙局：Hsin Mang, a political trickster / 沈展如著. — 臺北：正中書局, 民國66 [1977]
6, 18, 588 p., [1] leaf of plates : ill. ; 22 cm.
In Chinese with appendices in English.
Appendices (p. 572-588): 1. Explanation.—2. Table of contents.—3. Chronological table of the important events in the life of Wang Mang.
1. Wang, Mang, d. 23. 2. Statesmen—China—Biography. 3. China—History—Han dynasty, 202 B.C.-220 A.D. I. Title. II. Title: Hsin Mang cheng chih ta p'ien chü. III. Title: Hsin Mang, a political trickster.
DS748.16.W36S53 78-842040

Wang, Ying-lin, 1223-1296.
(Han chih k'ao)
漢制考：[4卷] / 撰者王應麟. 遊志續編 / [陶宗儀撰 i.e. 輯]. — 臺北：臺灣商務印書館, 民國62 [1973]
ca. 450 p. ; 20 cm. — (景印岫廬現藏罕傳善本叢刊)
Colophon title.
Reprint of two works; each has separate t. p.
NT$160.00
1. China—History—Han dynasty, 202 B.C.-220 A.D. 2. China—Description and travel. I. T'ao, Tsung-i, fl. 1360-1368. II. Yu chih hsü pien. 1973. III. Title. IV. Series: Ying yin Hsiu-lu hsien ts'ang han ch'uan shan pen ts'ung k'an.
DS748.W35 1973 79-840030

—HISTORY—HAN DYNASTY, 202 B.C.-220 A.D.—ADDRESSES, ESSAYS, LECTURES

Pei-ching ch'i ch'e chih tsao ch'ang. Kung jen li lun yen chiu so.
(Lü-Hou ch'i jen)
吕后其人：吕后及诸吕叛国篡权资料选译 / 北京汽车制造厂工人理论研究所. — 北京：中华书局, 1977.
7, 199 p. ; 19 cm.
$0.42
1. China—History—Han dynasty, 202 B.C.-220 A.D.—Addresses, essays, lectures. 2. Han Lü-hou, Empress of China, 241-180 B.C.—Addresses, essays, lectures. I. Ssu-ma, Ch'ien, ca. 145-ca. 86 B.C. Shih chi. Selections. 1977. II. Pan, Ku, 32-92. Han shu. Selections. 1977. III. Title.
DS748.P4 1977 78-843655

—HISTORY—HAN DYNASTY, 202 B.C.-200 A.D.—ANECDOTES, FACETIAE, SATIRE, ETC.

Lin, Han-ta.
(Ch'ien hou Han ku shih hsin pien)
前后汉故事新编 / 林汉达编写. — 第1版. — 北京：中华书局：新华书店北京发行所发行, 1978.
2 v. (1, 806 p.) : ill. ; 19 cm.
$4.10
1. China—History—Han dynasty, 202 B.C.-220 A.D.—Anecdotes, facetiae, satire, etc. I. Title.
DS748.L54 79-841445

—HISTORY—HAN DYNASTY, 202 B.C.-220 A.D.—BIBLIOGRAPHY

Ma, Hsien-hsing.
(Han shih ts'ai liao yü Han shih lun chu tsung ho mu lu)
漢史材料與漢史論著綜合目錄 = A classified catalogue of articles and books on the history of the Han dynasty / 馬先醒編. — [台北]：中華學術院史學研究所, 民國59 [1970]
vii, 336 p. ; 27 cm. — (史學研究叢書 ; 第2種)
Cover title.
In Chinese with prefatory notes also in English.
Includes index.
(Continued on next card)
1. China—History—Han dynasty, 202 B.C.-220 A.D.—Bibliography. I. Title. II. Title: A classified catalogue of articles and books on the history of the Han dynasty. III. Series: Shih hsüeh yen chiu mu lu ; ti 2 chung.
Z3108.A3M29 016.931 79-841042
[DS748]

CHINA

—HISTORY—HAN DYNASTY, 202 B. C. -220 A. D. —BIBLIOGRAPHY (Continued)

Sakade, Yoshinobu, 1934–
(Shin Kan shisō kenkyū bunken mokuroku)
秦漢思想研究文献目録 / 坂出祥伸編. — 吹田：関西大学出版広報部, 1978.
8, 137 p. ; 26 cm. Ja 78-17058
¥4800

1. China—Intellectual life—Bibliography. 2. China—History—Ch'in dynasty, 221-207 B. C.—Bibliography. 3. China—History—Han dynasty, 202 B. C.-220 A. D.—Bibliography. I. Title.

Z3108.C55S24
[DS721] 78-813240

Library of Congress

—HISTORY—HAN DYNASTY, 202 B. C. -220 A. D. —CARICATURES AND CARTOONS

(Chang Ch'ien)
張騫. — 香港：海鷗出版公司, ［1976?］
81 p. : chiefly ill. ; 18 x 19 cm.
"連環畫故事"
$3.00

1. China—History—Han dynasty, 202 B. C.-220 A. D.—Caricatures and cartoons. 2. Chang, Ch'ien, 2d. cent. B. C.—Cartoons, satire, etc.

DS748.C434 79-839669

—HISTORY—HAN DYNASTY, 202 B. C. -220 A. D. —CHRONOLOGY

(Han chi Hsi Han nien chi ho k'an)
漢紀西漢年紀合刊. — 臺北：鼎文書局, 民國66 ［1977］
2, 218, 1, 495 p. ; 22 cm. — (中國學術類編)
CONTENTS: 荀悅 漢紀.—王益之 西漢年紀.

1. China—History—Han dynasty, 202 B. C.-220 A. D.—Chronology. I. Hsün, Yüeh, 148-209. Han chi. 1977. II. Wang, I-chih, fl. 1196. Hsi Han nien chi. 1977. III. Series: Chung-kuo hsüeh shu lei pien.

DS748.H27 79-838639

—HISTORY—THREE KINGDOMS, 220-265

Ch'en, Shou, 233-297.
(San kuo chih)
三國志 / ［陳壽撰；裴松之注］. — ［成都］：成都書局, 同治10 ［1871］
14 v. (on double leaves) in 2 cases ; 37 cm.
Issued in case with collective title on spine: Ch'ien ssu shih.
Block print.

1. China—History—Three kingdoms, 220-265. I. P'ei, Sung-chih, 372-451. II. Title. III. Title: Ch'ien ssu shih.

DS748.2.C42 1871 79-841848

Li, T'ang.
(San kuo)
三國 / 李唐. — 臺景印初版. — 臺北：河洛圖書出版社, 民國67 ［1978］
3, 3, 125 p. ; 21 cm. — (中國歷史小叢書；009)
Series romanized: Chung-kuo li shih hsiao ts'ung shu.
Reprint of the 1962 ed. published by Hsiang-kang hung yeh shu chü, Hong Kong, in series: Li shih hsiao ts'ung shu.
Pbk. : NT$35.00

1. China—History—Three Kingdoms, 220-265. I. Title.

DS748.2.L39 1978 79-836278

—HISTORY—THREE KINGDOMS, 220-265—DRAMA

(Ch'ün ying hui)
群英会：京剧. — 新1版. — 上海：上海文化出版社, 1957.
111 p. ; 14 cm. — (戏曲小丛书)
Cover title.
"这个本子是由蕭長華… 整理的"
Series romanized: Hsi ch'ü hsiao ts'ung shu.
"本書是根据新文藝出版社出版的"京剧叢刊"本重排出版"

1. China—History—Three kingdoms, 220-265—Drama. I. Hsiao, Ch'ang-hua.

PL2832.5.C55 78-840606

(Kuan Yün-ch'ang hsin chuan)
關雲長新傳 / 曲鳳還 ... ［et al.］編著. — 台南：長河出版社, 民國67 ［1978］
2, 4, 308 p. ; 20 cm. — (長河叢書；16)
Series romanized: Ch'ang ho ts'ung shu.
NT$55.00

1. Kuan, Yü, 160-220—Fiction. 2. China—History—Three kingdoms, 220-265—Fiction. I. Ch'ü, Feng-huan.

PL2832.5.K84 79-835363

Umemoto, Sutezō, 1903–
(Shin Sangokushi)
新・三国志 / ［著者 楳本捨三］. — 東京：秀英書房, 昭和53 ［1978］
2 v. : ill. ; 20 cm. — (楳本捨三著作集；第5-6巻) Ja 78-26620 (v. 1)
Novelization of Ch'en Shou's San-kuo chih.
¥1400 per vol.
1. China—History—Three kingdoms, 220-265—Fiction. I. Ch'en Shou, 233-297. San-kuo chih. Japanese. II. Title. III. Title: Sangokushi. IV. Series: Umemoto, Sutezō, 1903– Umemoto Sutezō chosaku shū ; dai 5-6-kan.

PL840.M38S5 1978 79-804932

Yao, Hsin-nung, 1905–
(Mei jen chi)
美人計 / ［著者姚克 i. e. 姚莘農］. — 臺北：聯經出版事業公司, 民國67 ［1978］
6, 176 p. ; 20 cm. — (姚克戲劇選集)
Drama.
Series romanized: Yao K'o hsi chü hsüan chi.
NT$30.00

1. China—History—Three kingdoms, 220-265—Fiction. I. Title.

PL2923.H73M4 79-836216

—HISTORY—220-618

Lin, Jui-han.
(Wei Chin Nan-pei ch'ao shih)
魏晉南北朝史 / 林瑞翰著. — 台北：至大圖書文具教育用品股份有限公司, 民國66 ［1977］
471 p. ; 22 cm.
Cover title.
NT$360.00

1. China—History—220-618. I. Title.

DS748.2.L45 78-841547

Moriya, Mitsuo, 1915-1966.
(Rikuchō mombatsu no ichikenkyū)
六朝門閥の一研究：太原王氏系譜考 / 守屋美都雄著. — 東京：日本出版協同, 昭和26 ［1951］
2, 150, 2 p. ; 22 cm. — (東洋大學學術叢書)
Colophon inserted.
Includes bibliographical references.
Summary in English with title: A study of the clan-institute in the age of Liu-Chao or six dynasty.
1. China—History—220-618. I. Title. II. Series: Tōyō Daigaku, Tokyo. Tōyō Daigaku gakujutsu sōsho.

DS748.4.M67 78-814419

—HISTORY—220-618—ADDRESSES, ESSAYS, LECTURES

Lin, Jui-han.
(Tu shih ou te)
讀史偶得 / 林瑞翰著. — 臺北：幼獅文化事業公司, 民國66 ［1977］
4, 6, 442 p. ; 22 cm.
NT$100.00

1. China—History—220-618—Addresses, essays, lectures. 2. China—History—Five dynasty, 907-960—Addresses, essays, lectures. 3. China—History—Sung dynasty, 960-1279—Addresses, essays, lectures. I. Title.

DS748.2.L448 79-838669

—HISTORY—CHIN DYNASTY, 265-419

Wang, I-t'ung, 1914–
(Wu ch'ao men ti)
五朝門第 / 王伊同著. — 重刊第1版. — 香港：中文大學出版社, 1978.
2 v. (viii, 331 p., ［43］ leaves of plates (39 fold.)) : 75 Geneal. tables ; 24 cm.
Bibliography: p. ［290］-296.
"附高門權門世系婚姻表": v. 2.
ISBN 962-201-147-0
ISBN 962-201-150-0 pbk.
1. China—Nobility—Genealogy. 2. China—History—Chin dynasty, 265-419. 3. China—History—Southern and northern dynasties, 317-589. I. Title.

CS1164.W36 79-840333

—HISTORY—SOUTHERN AND NORTHERN DYNASTIES, 317-589

Wang, I-t'ung, 1914–
(Wu ch'ao men ti)
五朝門第 / 王伊同著. — 重刊第1版. — 香港：中文大學出版社, 1978.
2 v. (viii, 331 p., ［43］ leaves of plates (39 fold.)) : 75 Geneal. tables ; 24 cm.
Bibliography: p. ［290］-296.
"附高門權門世系婚姻表": v. 2.
ISBN 962-201-147-0
ISBN 962-201-150-0 pbk.
1. China—Nobility—Genealogy. 2. China—History—Chin dynasty, 265-419. 3. China—History—Southern and northern dynasties, 317-589. I. Title.

CS1164.W36 79-840333

—HISTORY—NORTHERN WEI DYNASTY, 386-534

Lu, Yao-tung, 1933–
(Ts'ung P'ing-ch'eng toa Lo-yang)
從平城到洛陽：拓跋魏文化轉變的歷程 / 逯耀東著. — 臺北：聯經出版事業公司, 民國68 ［1979］
1, 285 p. ; 22 cm.
NT$130.00

1. China—History—Northern Wei dynasty, 386-534. I. Title.

DS748.5.L83 79-841633

—HISTORY—NORTHERN WEI DYNASTY, 386-534—BIOGRAPHY

(Wei pei san chung)
魏碑三種. — ［上海］：上海古籍书店, ［1978］
［52］ p. ; 27 cm.
CONTENTS: 魏驃騎將軍冀州刺史元診墓誌銘.—魏寧遠將軍元君墓誌銘.—魏江陽王次妃石夫人墓誌銘.
$5.50

1. Calligraphy, Chinese. 2. Epitaphs—China. 3. China—History—Northern Wei dynasty, 386-534—Biography. 4. China—Biography.

NK3634.A2W45 79-835870

C

—HISTORY—NORTHERN CH'I DYNASTY, 550-577

Finsterbusch, Käte.
Zur Archäologie der Pei-Ch'i- (550-577) und Sui-Zeit (581-618) : mit e. Fundkatalog / Käte Finsterbusch. — Wiesbaden : Steiner, 1976.
ix, 147, 83 ［i.e. 84］ p. : numerous ill., map ; 29 cm. — (Münchener ostasiatische Studien : Sonderreihe ; Bd. 1) GFR76-A
Bibliography: p. ［127］-134.
ISBN 3-515-02017-9 : DM98.00
1. China—Antiquities—Catalogs. 2. China—History—Northern Ch'i dynasty, 550-577. 3. China—History—Sui dynasty, 581-618. I. Title. II. Series.
DS715.F56 78-382300
79 MARC

—HISTORY—SUI DYNASTY, 581-618

Finsterbusch, Käte.
Zur Archäologie der Pei-Ch'i- (550-577) und Sui-Zeit (581-618) : mit e. Fundkatalog / Käte Finsterbusch. — Wiesbaden : Steiner, 1976.
ix, 147, 83 ［i.e. 84］ p. : numerous ill., map ; 29 cm. — (Münchener ostasiatische Studien : Sonderreihe ; Bd. 1) GFR76-A
Bibliography: p. ［127］-134.
ISBN 3-515-02017-9 : DM98.00
1. China—Antiquities—Catalogs. 2. China—History—Northern Ch'i dynasty, 550-577. 3. China—History—Sui dynasty, 581-618. I. Title. II. Series.
DS715.F56 78-382300
79 MARC

CHINA

—HISTORY—SUI DYNASTY, 581-618 (Continued)

Lin, T'ien-wei.
(Sui T'ang shih hsin lun)
隋唐史新論 / 林天蔚. — 臺北：東華書局, 民國67 [1978]
9, 440 p., [1] fold. leaf of plates : ill. ; 22 cm. — (東華歷史叢書)
Series romanized: Tung hua li shih ts'ung shu.
Includes bibliographical references.
NT$120.00
1. China—History—Sui dynasty, 581-618. 2. China—History—T'ang dynasty, 618-907. I. Title.
DS749.2.L53 79-838648

Wright, Arthur F 1913-1976.
The Sui dynasty / Arthur F. Wright. — 1st ed. — New York : Knopf, 1978.
237 p. : maps ; 24 cm.
Bibliography: p. [223]-228.
Includes index.
ISBN 0-394-49276-5 : $15.00
1. China—History—Sui dynasty, 581-618. I. Title.
DS749.2.W74 1978 951'.01 78-54898
78 MARC

—HISTORY—T'ANG DYNASTY, 618-907

Forte, Antonio.
Political propaganda and ideology in China at the end of the seventh century : inquiry into the nature, authors and function of the Tunhuang document S.6502, followed by an annotated translation / Antonino Forte. — Napoli : Istituto universitario orientale, Seminario di studi asiatici, 1976.
xii, 312 p., [18] leaves of plates : facsims. ; 25 cm. It78-May
Bibliography: p. [285]-296.
Includes index.
1. China—Politics and government—To 1643. 2. Buddhism and state—China. 3. Tun-huang manuscripts. 4. China—History—T'ang dynasty, 618-907. I. Title.
DS749.3.F63 951'.01 79-309494
*79 MARC

Herbert, Penelope Ann, 1945-
Under the brilliant emperor : imperial authority in T'ang China as seen in the writings of Chang Chiu-ling / P. A. Herbert. — Canberra : Faculty of Asian Studies in association with Australian National University Press, 1978.
vi, 156 p. ; 25 cm. — (Oriental monograph series ; no. 21) Aus***
Bibliography: p. 138-145.
Includes index.
ISBN 0-7081-1348-6
1. China—History—T'ang dynasty, 618-907. 2. T'ang Hsün-tsung, Emperor of China, 685-762. 3. Chang, Chiu-ling, 678-740. 4. Statesmen—China—Biography. I. Title. II. Series.
DS749.3.H42 951'.01 78-325232
79 MARC

Hsü, Liang-chih.
(I tai hsiung ts'ai Li Shih-min)
一代雄才李世民 / [著作者徐亮之]. — 臺北：河洛圖書出版社, 民國67 [1978]
2, 2, 239 p., [2] leaves of plates : ill. ; 22 cm. — (一代叢書 ; 4)
Reprint ed.
NT$50.00
1. T'ang T'ai-tsung, Emperor of China, 597-649. 2. China—History—T'ang dynasty, 618-907. 3. China—Kings and rulers—Biography. I. Title. II. Series: I tai ts'ung shu ; 4.
DS749.42.T35H76 1978 79-841853

Jaeger, Georgette.
Les Lettrés chinois : poètes T'ang et leur milieu / Georgette Jaeger. — Neuchâtel : Éditions de la Baconnière, cop. 1977.
347 p. ; 21 cm. Sw78-978
Bibliography: p. 339-342.
Includes index.
ISBN 2-8252-0950-3 : 39.00F
1. Chinese poetry—T'ang dynasty, 618-907—History and criticism. 2. China—Intellectual life. 3. China—History—T'ang dynasty, 618-907. I. Title.
PL2321.J3 895.1'3'09 78-368780
78 MARC

Li, Fu-teng.
(T'ang tai chien ch'a chih tu)
唐代監察制度 / 李福登著. — 再版. — 臺南縣永康鄉：私立台南家政專科學校, 民國66 [1977]
3, 5, 206 p. ; 22 cm. — (私立台南家政專科學校叢書 ; 2)
Bibliography: p. 193-206.
NT$40.00
1. Separation of powers—China. 2. Impeachments—China. 3. China—History—T'ang dynasty, 618-907. I. Title. II. Series: T'ai-nan chia cheng chuan k'o hsüeh hsiao. Ssu li T'ai-nan chia cheng chuan k'o hsüeh hsiao ts'ung shu ; 2.
79-840079

Li, T'ang.
(T'ang Ming-huang)
唐明皇 / 李唐. — 臺景印初版. — 臺北：河洛圖書出版社, 民國67 [1978]
3, 4, 170 p., [11] leaves of plates : ill. ; 22 cm. — (中國歷史小叢書 ; 016)
Series romanized: Chung-kuo li shih hsiao ts'ung shu.
Reprint of the 1963 ed. published by Hsiang-kang hung yeh shu chü, Hong Kong, in series: Li shih hsiao ts'ung shu.
Pbk. : NT$35.00
1. China—History—T'ang dynasty, 618-907. 2. T'ang Hsüan-tsung, Emperor of China, 685-762.
DS749.3.L4937 1978 79-837024

Lin, T'ien-wei.
(Sui T'ang shih hsin lun)
隋唐史新論 / 林天蔚. — 臺北：東華書局, 民國67 [1978]
9, 440 p., [1] fold. leaf of plates : ill. ; 22 cm. — (東華歷史叢書)
Series romanized: Tung hua li shih ts'ung shu.
Includes bibliographical references.
NT$120.00
1. China—History—Sui dynasty, 581-618. 2. China—History—T'ang dynasty, 618-907. I. Title.
DS749.2.L53 79-838648

Lo, Lung-chih.
(T'ang tai ti hou fei yü wai ch'i)
唐代的后妃與外戚 / 羅龍治著. — 臺北：桂冠圖書有限公司, 民國67 [1978]
2, 8, 178 p. ; 20 cm. — (桂冠系列)
Series romanized: Kuei kuan hsi lieh.
Bibliography: p. 172-178.
NT$50.00
1. China—History—T'ang dynasty, 618-907. 2. China—Court and courtiers. I. Title.
DS749.3.L62 79-839962

Ma, Ch'i-hua.
(Chen-kuan cheng lun)
貞觀政論 / 馬起華著. — 台北板橋：漢苑出版社, 民國66 [1977]
143, [9] p. ; 22 cm. — (政大政治叢書 ; 1)
At head of title: Ta hsüeh yung shu.
Includes bibliographical references.
1. China—Politics and government. 2. China—History—T'ang dynasty, 618-907. 3. T'ang T'ai-tsung, Emperor of China, 597-649. I. Title. II. Series: Kuo li cheng chih ta hsüeh. Cheng ta cheng chih ts'ung shu ; 1.
JQ1508.M29 79-841071
79 C

Pei-ching ta hsüeh. Fa hsüeh yüan. Chung-kuo ching chi yen chiu shih.
(T'u ti wen t'i)
土地問題 / 國立北京大學法學院中國經濟研究室編 ; 主編人陶希聖. — 臺北：食貨出版社：總代理中央文物供應社, 民國63 [1974]
10, 6, 250 p., [5] leaves of plates : facsims ; 22 cm. — (食貨史學叢書) (唐代經濟史料叢編 ; 第2種)
Cover title: T'ang tai t'u ti wen t'i.
Series 1 romanized: Shih huo shih hsüeh ts'ung shu.
Reprint of the ed. printed in 1937 by Kuo li Pei-ching ta hsüeh ch'u pan tsu ; in series: Chung-kuo ching chi shih liao ts'ung pien : T'ang tai p'ien ; 2.
1. Land use—China. 2. China—History—T'ang dynasty, 618-907. I. T'ao, Hsi-sheng, 1893- II. Title. III. Title: T'ang tai t'u ti wen t'i. IV. Series: T'ang tai ching chi shih liao ts'ung pien ; ti 2 chung.
HD924.P43 1974 79-838043
79 C

Tseng, I-min.
(T'ang tai k'ao k'o chih tu yen chiu)
唐代考課制度研究 / 曾一民著. — 臺北：臺灣商務印書館, 民國67 [1978]
8, 2, 3, 172 p. ; 20 cm. — (岫廬文庫 ; 52)
Series romanized: Hsiu-lu wen k'u.
Bibliography: p. 169-172.
NT$24.00
1. Civil service—China. 2. Employees, Rating of—China. 3. China—History—T'ang dynasty, 618-907. I. Title.
JQ1512.Z13R38 79-840571

Wang, Chung.
(Hsin T'ang shu T'u-fan chuan chien cheng)
新唐書吐蕃傳箋證 / 王忠著. — 香港：神州圖書公司, [1978?]
166 p. ; 19 cm.
Reprint of the 1958 ed. published by K'o hsüeh ch'u pan she, Peking.
1. Ou-yang, Hsiu, 1007-1072. Hsin T'ang shu. 2. China—History—T'ang dynasty, 618-907. 3. Tibet—History. I. Ou-yang, Hsiu, 1007-1072. Hsin T'ang shu. Selections. 1978. II. Title. III. Title: T'u-fan chuan chien cheng.
DS749.3.O8223W36 1978 79-836617

—HISTORY—T'ANG DYNASTY, 618-907—ANECDOTES, FACETIAE, SATIRE, ETC.

Li, Chao, fl. 806-820.
[T'ang kuo shih pu. Japanese & Chinese]
唐國史補 / [李肇撰] ; 長澤規矩也解題. — 東京：出版 古典研究会：発行所 汲古書院, 昭和48 [1973], 昭和53 [1978] printing.
297 p. ; 25 cm. — (和刻本漢籍随筆集 ; 6) Ja***
In Chinese, with Japanese reading marks.
Photoreprint ed.
CONTENTS: 李肇 唐國史補—段成式 西陽雜俎—王仁裕 開元天寶遺事
¥4500
1. China—History—T'ang dynasty, 618-907—Anecdotes, facetiae, satire, etc. 2. Chinese essays—T'ang dynasty, 618-907. I. Nagasawa, Kikuya, 1902- II. Title. III. Series: Wakokubon Kanseki zuihitsu shū ; 6.
Title romanized: Tō kokushi ho.
PL2606.W3 vol. 6 79-809165
[DS749.3]

—HISTORY—T'ANG DYNASTY, 618-907—SOURCES

Sun, Kuo-tung.
(T'ang tai chung yang chung yao wen kuan ch'ien chuan t'u ching yen chiu)
唐代中央重要文官遷轉途徑研究 / 孫國棟. — 香港：龍門書店, 1978.
2, 6, 565 p. : ill. ; 25 cm. — (新亞研究所叢書)
$70.00
1. China—Officials and employees. 2. China—History—T'ang dynasty, 618-907—Sources. I. Title. II. Series: Hongkong. Chinese University. Institute of Advanced Chinese Studies and Research. Hsin ya yen chiu so ts'ung shu.
JQ1512.S88 79-837757

T'ang liu tien.
唐六典：[30卷 / 唐元宗御撰 ; 李林甫等奉敕注]. — [日本東都?]：官板, [天保7 i. e. 1836?]
8 v. (on double leaves) in case ; 26 cm.
Caption title.
Block print.
1. China—Officials and employees. 2. China—History—T'ang dynasty, 618-907—Sources. I. T'ang Hsüan-tsung, Emperor of China, 685-762. II. Li, Lin-fu, d. 752.
Title romanized: T'ang liu tien.
JQ1512.T3 1836 77-836510

T'ang liu tien.
大唐六典 [30卷] 唐玄宗御撰 李林甫奉敕註 台北 文海出版社 民國51 [1962]
522 p. 20 cm.
Facsimile ed.
Each page represents 2 leaves of the original.
1. China—Officials and employees. 2. China—History—T'ang dynasty, 618-907—Sources. I. T'ang Hsüan-tsung, Emperor of China, 685-762. II. Li, Lin-fu, d. 752. III. Title.
Title romanized: Ta T'ang liu tien.
JQ1512.T3 1962 C 63-2101

(T'ang liu tien)
唐六典 / [張九齡等撰]. — [台北：台灣]商務[印書館, 1976?]
3 v. ; 20 cm. — (四庫全書珍本六集 ; 117-119)
Cover title.
Reproduced from Wen yüan ko copy.
1. China—Officials and employees. 2. China—History—T'ang dynasty, 618-907—Sources. I. Chang, Chiu-ling, 678-740. II. Series: Ssu k'u ch'üan shu. Ssu k'u ch'üan shu chen pen liu chi ; 117-119.
AC149.S7 1976, vol. 117-119 78-843860
[JQ1512]

CHINA (Continued)

—HISTORY—FIVE DYNASTY, 907-960—ADDRESSES, ESSAYS, LECTURES

Lin, Jui-han.
(Tu shih ou te)
讀史偶得 / 林瑞翰著. — 臺北：幼獅文化事業公司, 民國66 [1977]
4, 6, 442 p. ; 22 cm.
NT$100.00

1. China—History—220-618—Addresses, essays, lectures. 2. China—History—Five dynasty, 907-960—Addresses, essays, lectures. 3. China—History—Sung dynasty, 960-1279—Addresses, essays, lectures. I. Title.

DS748.2.L448 79-838669

—HISTORY—SUNG DYNASTY, 960-1279

Ahrens-Thiele, Dagmar.
Der Abschluss eines Vertrages: Diplomatic zwischen Sung- und Chin-Dynastie, 1117-1123. Wiesbaden, F. Steiner, 1971.
289 p. fold. map. 24 cm. (Münchener ostasiatische Studien, Bd. 6)
GDB***
Originally presented as the author's thesis, Munich.
Summary in English.
Bibliography: p. 259-264.
1. China—History—Sung dynasty, 960-1279. 2. Ju-chen (Tribe)—History. I. Title. II. Series.
DS751.A58 1971 951'.02 71-595958
79[r79]rev MARC

Chang, T'ien-yu.
(Sung Ming shih yen chiu lun chi)
宋明史研究論集：宋明衰亡時期 / 張天佑撰. — [台北縣永和]：華世出版社, 民國66 [1977]
4, 212 p. ; 21 cm.
Includes bibliographical references.
NT$80.00

1. China—History—Sung dynasty, 960-1279. 2. China—History—Ming dynasty, 1368-1644. I. Title.

DS751.C434 78-841215

Ch'en, I-yen, 1942-
(Pei Sung t'ung chih chieh ts'eng she hui liu tung chih yen chiu)
北宋統治階層社會流動之研究 / 陳義彥著. — Taipei：嘉新水泥公司文化基金會, 民國66 [1977]
2, 2, 222 p. ; 22 cm. — (嘉新水泥公司文化基金會研究論文：第226種) (嘉新水泥公司文化基金會叢書)
Series 1 romanized: Chia-hsin shui ni kung ssu wen hua chi chin hui yen chiu lun wen. Series 2 romanized: Chia-hsin shui ni kung ssu wen hua chi chin hui ts'ung shu.
Bibliography: p. 217-222.

1. Social mobility—China. 2. China—History—Sung dynasty, 960-1279. I. Title.

HN733.C42 79-837494

Huang, K'uan-chung.
(Wan Sung ch'ao ch'en tui kuo shih ti cheng i)
晚宋朝臣對國是的爭議：理宗時代的和戰, 邊防與流民 / [著作者黃寬重]. — 臺北：國立臺灣大學文學院, 1978.
4, 4, 188, 3 p. ; 21 cm. — (國立臺灣大學文史叢刊 ; 50)
Summary in English with title: The controversies of late Sung dynasty officials concerning national affairs.
Bibliography: p. 173-188.

1. China—History—Sung dynasty, 960-1279. I. Title. II. Title: The controversies of late Sung dynasty officials concerning national affairs. III. Series: T'ai-wan ta hsüeh, T'ai-pei. Kuo li T'ai-wan ta hsüeh wen shih ts'ung k'an ; 50.

DS751.H855 79-835737

Li, T'ang.
(Sung Hui-tsung)
宋徽宗 / [著作者李唐]. — 臺北：河洛圖書出版社, 民國67 [1978]
3, 3, 161 p. ; 22 cm. — (中國歷史小叢書 ; 19)
Reprint ed.
Series romanized: Chung-kuo li shih hsiao ts'ung shu.
NT$35.00

1. China—History—Sung dynasty, 960-1279. 2. Sung Hui-tsung, Emperor of China, 1082-1135. I. Title.

DS751.L5418 1978 79-838181

Li, T'ang.
(Sung Kao-tsung)
宋高宗 / [著作者李唐]. — 臺景印初版. — 臺北：河洛圖書出版社, 民國67 [1978]
3, 3, 169 p. ; 22 cm. — (中國歷史小叢書 ; 20) (歷史小叢書)
Series 1 romanized: Chung-kuo li shih hsiao ts'ung shu. Series 2 romanized: Li shih hsiao ts'ung shu.
Reprint of the 1964 ed. published by Hsiang-kang hung yeh shu chü, Hongkong.
NT$35.00

1. China—History—Sung dynasty, 960-1279. 2. Sung Kao—tsung, Emperor of China 1107-1187. I. Title.

DS751.S76L5 1978 79-835704

Li, T'ien-ming.
(Sung Yüan Hsiang Fan chan i chih yen chiu)
宋元襄樊戰役之研究 / 李天鳴撰. — [台北?：李天鳴], 民國62 [1973]
202 double leaves : ill., maps ; 26 cm.
Cover title.
Reproduced from typescript.
"私立中國文化學院史學研究所碩士論文"
Bibliography: leaves 191-202.

1. China—History, Military. 2. Battles—China. 3. China—History—Sung dynasty, 960-1279. 4. China—History—Yüan dynasty, 1260-1368. I. Title.

DS738.L45 1973b 79-835350

Lin, T'ien-wei.
(Sung shih shih hsi)
宋史試析 / 林天蔚著. — 臺北：臺灣商務印書館, 民國67 [1978]
9, 2, 1, 330 p. ; 19 cm. — (岫廬文庫 ; 040)
Includes bibliographical references.
Series romanized: Hsiu lu wen k'u.
NT$36.00

1. China—History—Sung dynasty, 960-1279. I. Title.

DS751.L545 78-843920

Liu, An-shih, 1048-1125.
(Chin yen chi)
盡言集 / [劉安世撰]. — [臺北：臺灣]商務[印書館, 1976]
ca. 500 p. in various pagings ; 20 cm. — (四庫全書珍本六集 ; 097)
Cover title.
Photoreprint ed. of the Wen Yüan Ko copy.

1. China—History—Sung dynasty, 960-1279. I. Title. II. Series: Ssu k'u ch'üan shu. Ssu k'u ch'üan shu chen pen liu chi ; 097.

AC149.S7 1976, vol. 97 79-835381
[DS751]

Liu, Tao-yüan.
(Liang Sung t'ien fu chih tu)
兩宋田賦制度 / 劉道元著. — 臺灣再版. — 臺北：食貨出版社, 民國67 [1978]
3, 5, 190 p. ; 20 cm. — (食貨史學叢書)
Series romanized: Shih huo shih hsüeh ts'ung shu.
Includes bibliographical references.
NT$120.00

1. Land value taxation—China. 2. China—History—Sung dynasty, 960-1279. I. Title.

HJ4398.L6 79-841101

P'an, Sheng-te.
(Pei Sung yü lun yü tang cheng chih yen chiu)
北宋輿論與黨爭之研究 / 潘昇德. — [台北?：s. n., 1978?]
79 double leaves ; 27 cm.
Reproduced from ms. copy.
Originally presented as the author's thesis (M.A.), Kuo li cheng chih ta hsüeh, 1961.
Bibliography: leaves 77-79.
1. China—History—Sung dynasty, 960-1279. 2. Public opinion—China. 3. China—Politics and government—To 1643. I. Title.
DS751.P29 1978 79-835700

Shih, Wen-chi.
(Nan Sung chung hsing ssu chen)
南宋中興四鎮 / 石文濟撰. — [台北?：s. n.], 民國63 [1974]
4 v. (4, 2, 419, [1] double leaves, [8] leaves of plates) : maps ; 27 cm.
"私立中國文化學院史學研究所博士論文"
Reproduced from typescript.
Bibliography: v. 4, leaves 414-[420]

1. China—History—Sung dynasty, 960-1279. I. Title.

DS751.S54 1974 79-835057

(Sung shih)
宋史：華岡版新刊本 / 編纂者中華學術院新刊本宋史編纂委員會；主編蔣復璁宋晞. — 臺北：華岡出版有限公司, 民國
v. ; 26 cm. — (中華大典)
Title in English also in colophon: The history of Sung dynasty.
"新刊本宋史以商務印書館百衲本二十四史之宋史為底本, 由...等十三位學者編纂...包括...主要參考書目, 人名索引...全書十冊"
Series romanized: Chung-hua ta tien.
NT$2,800.00 (v. 1-10)
1. China—History—Sung dynasty, 960-1279. I. Chiang, Fu-ts'ung, 1898- II. Sung, Hsi, 1920- III. Chung-hua hsüeh shu yüan. Hsin k'an pen Sung shih pien tsuan wei yüan hui. IV. Title: The history of Sung dynasty.
DS751.S798 951'.02 79-839119

T'o-t'o, 1313-1355.
(Sung shih)
宋史 / 脱脱等撰. — 第1版. — 北京：中華書局, 1977.
40 v. (8, 144, 14263 p.) ; 21 cm.
"全書...共四百九十六卷"
Includes bibliographical references.
$42.75

1. China—History—Sung dynasty, 960-1279. I. Title.

DS751.T6285 1977 78-842373

(Wang Hsiao-po Li Shun ch'i i tzu liao hui pien)
王小波李順起义资料汇编 / 四川大学历史系王小波李顺起义调查组编. — 第1版. — 成都：四川人民出版社：四川省新華書店发行, 1978.
2, 3, 173 p., [4] leaves of plates : ill. ; 19 cm.
Bibliography: p. 165-169.
Includes index.
$0.39

1. Wang, Hsiao-po, d. 994. 2. Li, Shun, d. 994. 3. Peasant uprisings—China—Szechwan. 4. China—History—Sung dynasty, 960-1279. 5. Revolutionists—China—Biography. I. Ssu-ch'uan ta hsüeh, Chengtu, China. Wang Hsiao-po Li Shun ch'i i tiao ch'a tsu.

DS751.6.W3W36 79-839131

—HISTORY—SUNG DYNASTY, 960-1279—ADDRESSES ESSAYS, LECTURES

Lin, Jui-han.
(Tu shih ou te)
讀史偶得 / 林瑞翰著. — 臺北：幼獅文化事業公司, 民國66 [1977]
4, 6, 442 p. ; 22 cm.
NT$100.00

1. China—History—220-618—Addresses, essays, lectures. 2. China—History—Five dynasty, 907-960—Addresses, essays, lectures. 3. China—History—Sung dynasty, 960-1279—Addresses, essays, lectures. I. Title.

DS748.2.L448 79-838669

—HISTORY—SUNG DYNASTY, 960-1279—BIBLIOGRAPHY

Sung, Hsi, 1920-
(Sung shih yen chiu lun wen)
宋史研究論文與書籍目錄 史學研究目錄第一種 宋晞編 A classified list of Chinese articles and books on the Sung history, by Shee Sung. [陽明山] 中華學術院 中國文化學院史學研究所 民國55 [1966]
ill, 82, ii, 13 p. 27 cm.
Cover title.
In Chinese.
1. China—History—Sung dynasty, 960-1279—Bibliography. I. Title.

Z3108.A3S8 C 67-1027
[DS751]
rev

Sung, Hsi, 1920-
(Sung shih yen chiu lun wen yü shu chi mu lu)
宋史研究論文與書籍目錄 = A classified list of Chinese articles and books on the Sung history：史學研究目錄第一種 / 宋晞編. — 再版. — [台北]：中華學術院 中國文化學院史學研究所, 1976.
2 v. ; 27 cm.
Vol. 2: Reprinted from Chinese culture, vol. 12, no. 4, December, 1971.
1. China—History—Sung dynasty, 960-1279—Bibliography. I. Title. II. Title: A classified list of Chinese articles and books on the Sung history.

Z3108.A3S8 1976 79-835146
[DS751]

CHINA

—HISTORY—SUNG DYNASTY, 960-1279—BIBLIOGRAPHY (Continued)

A **Sung bibliography** = Bibliographie des Sung / initiated by Etienne Balazs ; edited by Yves Hervouet. — Hong Kong : Chinese University Press, c1978.
xxxiv, 598 p. ; 28 cm.
Introduction and instructions to the reader in English and French. Notices in English or French.
Errata slip inserted.
ISBN 9622011586 : $60.00
1. China—Imprints. 2. China—History—Sung dynasty, 960-1279—Bibliography. I. Balazs, Étienne, 1905-1963. II. Hervouet, Yves. III. Title: Bibliographie des Sung.
Z3102.S77 016'.951'02 79-106852
79 MARC

—HISTORY—SUNG DYNASTY, 960-1279—BIOGRAPHY

Cheng, Ch'ien.
(Sung jen sheng tsu k'ao shih li)
宋人生卒考示例 / 鄭騫撰. — 景美[臺北縣]: 華世出版社, 民國66 [1977]
11, 162 p. ; 21 cm.
NT$60.00

1. China—History—Sung dynasty, 960-1279—Biography. 2. China—Biography. I. Title.
DS751.C455 79-837165

(Shao-hsing shih pa nien t'ung nien hsiao lu)
紹興十八年同年小錄. 中州人物考 / [孫奇逢撰]. — [台北：臺灣]商務[印書館, 1978?]
2 v. ; 20 cm. — (四庫全書珍本八集 ; 82-83)
Cover title.
Photoreprint ed. of two works, both of the Wen Yüan Ko copy.
1. China—Registers. 2. Civil service—China. 3. China—History—Sung dynasty, 960-1279—Biography. 4. China—Biography. 5. China—History—Ming dynasty, 1368-1644—Biography. 6. Honan, China (Province)—Biography. I. Sun, Ch'i-feng, 1585-1675. Chung chou jen wu k'ao. 1978?. II. Title: Chung chou jen wu k'ao. III. Series: Ssu k'u ch'uan shu. Ssu k'u ch'uan shu chen pen pa chi ; 82-83.
AC149.S7 1978, vol. 82-83 78-841087
[JQ1507]

Shih tsu ta ch'üan.
排韻增廣事類氏族大全. — [台北：臺灣]商務[印書館, 1978?]
6 v. ; 20 cm. — (四庫全書珍本八集 ; 138-143)
Cover title.
Caption and running title: Shih tsu ta ch'üan.
Photoreprint ed. of the Wen Yüan Ko copy.
1. China—History—Sung dynasty, 960-1279—Biography. 2. China—Biography. I. Title. II. Series: Ssu k'u ch'üan shu. Ssu k'u ch'üan shu chen pen pa chi ; 138-143.
Title romanized: P'ai yün tseng kuang shih lei shih tsu ta ch'üan.
AC149.S7 1978, vol. 138-143 78-841132
[DS751.5]

—HISTORY—SUNG DYNASTY, 960-1279—BIOGRAPHY—INDEXES

Umehara, Kaoru, 1934-
(Zoku Shiji tsugan chōhen jimmei sakuin)
續資治通鑑長編人名索引 / 梅原郁編. — 京都：同朋舎, 1978.
7, 697 p. ; 22 cm. — (東洋史研究資料叢刊) Ja 78-20683
"底本としては光緒7年(1881)浙江書局刊本を使用"
¥14000
1. Li, T'ao, 1115-1184. Hsü Tzu chih t'ung chien ch'ang pien—Indexes. 2. China—History—Sung dynasty, 960-1279—Indexes. 3. China—History—Sung dynasty, 960-1279—Biography—Indexes. 4. China—Biography—Indexes. I. Li, T'ao, 1115-1184. Hsü Tzu chih t'ung ch'ang pien. II. Title. III. Series: Tōyōshi kenkyū shiryō sōkan.
DS751.U43 78-816027

—HISTORY—SUNG DYNASTY, 960-1279—INDEXES

Umehara, Kaoru, 1934-
(Zoku Shiji tsugan chōhen jimmei sakuin)
續資治通鑑長編人名索引 / 梅原郁編. — 京都：同朋舎, 1978.
7, 697 p. ; 22 cm. — (東洋史研究資料叢刊) Ja 78-20683
"底本としては光緒7年(1881)浙江書局刊本を使用"
¥14000
1. Li, T'ao, 1115-1184. Hsü Tzu chih t'ung chien ch'ang pien—Indexes. 2. China—History—Sung dynasty, 960-1279—Indexes. 3. China—History—Sung dynasty, 960-1279—Biography—Indexes. 4. China—Biography—Indexes. I. Li, T'ao, 1115-1184. Hsü Tzu chih t'ung ch'ang pien. II. Title. III. Series: Tōyōshi kenkyū shiryō sōkan.
DS751.U43 78-816027

—HISTORY—SUNG DYNASTY, 960-1279—POETRY

Sommardal, Göran, 1947-
Sång efter ett uppror : [dikter] / Göran Sommardal. — Stockholm : Rabén & Sjögren, 1978.
177, [1] p. ; 20 cm. S78-34
ISBN 9129522870
1. Peasant uprisings—China—Poetry. 2. China—History—Sung dynasty, 960-1279—Poetry. I. Title.
PT9876.29.O49S2 79-360093
79 MARC

—HISTORY—SUNG DYNASTY, 960-1279—SOURCES

Kracke, Edward A
Translations of Sung civil service titles, classification terms, and governmental organ names / by E. A. Kracke, Jr. — 2d rev. ed. — San Francisco : Chinese Materials Center, 1978.
xiv, 35, [21] p. ; 21 cm. — (Asian library series ; no. 11)
Designations also in Chinese.
ISBN 0-89644-526-7
1. Civil service positions—China. 2. China—History—Sung dynasty, 960-1279—Sources. I. Title. II. Series.
JQ1512.Z1K7 1978 354'.51'00103 79-100815
79 MARC

—HISTORY—YÜAN DYNASTY, 1260-1368

Franke, Herbert, 1914-
From tribal chieftain to universal emperor and god : the legitimation of the Yüan dynasty / Herbert Franke. — München : Verlag der Baerischen Akademie der Wissenschaften, 1978.
84 p. ; 23 cm. — (Sitzungsberichte · Bayerische Akademie der Wissenschaften, Philosophisch-Historische Klasse ; Jahrg. 1978, Heft 2 ISSN 0342-5991) GFR***
Includes bibliographical references.
ISBN 3-7696-1489-5
1. Mongols—History. 2. China—History—Yüan dynasty, 1260-1368. 3. Buddhism and state—China. 4. Sovereignty. I. Title. II. Series: Akademie der Wissenschaften, Munich. Philosophisch-Historische Klasse. Sitzungsberichte ; Jahrg. 1978, Heft 2.
DS19.F72 951'.025 79-310210
79 MARC

Li, T'ang.
(Yüan Shih-tsu)
元世祖 / 李唐著. — 香港：香港宏業書局, 1978.
3, 4, 188 p. ; 19 cm. — (歷史小叢書)
Series romanized: Li shih hsiao ts'ung shu.
$5.90

1. China—History—Yüan dynasty, 1260-1368. 2. Kublai Khan, 1216-1294. I. Title.
DS752.L519 79-840636

Li, T'ien-ming.
(Sung Yüan Hsiang Fan chan i chih yen chiu)
宋元襄樊戰役之研究 / 李天鳴撰. — [台北?：李天鳴], 民國62 [1973]
202 double leaves : ill., maps ; 26 cm.
Cover title.
Reproduced from typescript.
"私立中國文化學院史學研究所碩士論文"
Bibliography: leaves 191-202.
1. China—History, Military. 2. Battles—China. 3. China—History—Sung dynasty, 960-1279. 4. China—History—Yüan dynasty, 1260-1368. I. Title.
DS738.L45 1973b 79-835350
79 C

Makino, Shūji, 1931-
(Gendai kōtōkan no taikeiteki kenkyū)
元代勾當官の體系的研究 / 牧野修二著. — 東京：大明堂, 1979.
v, 204 p. ; 22 cm. Ja 79-8100
Originally presented as the author's thesis, Kyūshū Daigaku, 1972.
Includes bibliographical references.
Errata slip inserted.
¥3500
1. China—Officials and employees—History. 2. China—History—Yüan dynasty, 1260-1368. I. Title.
JQ1512.Z1M34 1979 79-810184

Yang, Shu-fan.
(Yüan tai chung yang cheng chih chih tu)
元代中央政治制度 / 楊樹藩著. — 臺北：臺灣商務印書館, 民國67 [1978]
[16], 212 p., [1] fold. leaf ; 19 cm. — (岫廬文庫 ; 035)
Series romanized: Hsiu-lu wen k'u.
NT$36.00

1. China—Politics and government—To 1643. 2. China—History—Yüan dynasty, 1260-1368. I. Title.
JQ1509 1260.Y36 78-842346

Yüan, Chi, 1923-
(Yüan shih lun ts'ung)
元史論叢 / 袁冀著. — 臺北：聯經出版事業公司, 民國67 [1978]
309 p. : map ; 22 cm.
NT$100.00

1. China—History—Yüan dynasty, 1260-1368—Addresses, essays, lectures. I. Title.
DS752.Y79 79-838592

—HISTORY—YÜAN DYNASTY, 1260-1368—BIBLIOGRAPHY

Harayama, Akira.
(Genchō hishi kankei bunken mokuroku)
元朝秘史関係文献目録 / 原山煌編. — 東京：日本モンゴル学会, 1978.
36 p. ; 26 cm. Ja 78-9923
Cover title.
Title also in Mongolian: Mongolyn nuuts tovchoony nomyn zhagsaalt.

1. China—History—Yüan dynasty, 1260-1368—Bibliography. I. Title. II. Title: Mongolyn nuuts tovchoony nomyn zhagsaalt.
Z3108.A3H37 78-814860
[DS752]

—HISTORY—YÜAN DYNASTY, 1260-1368—CHRONOLOGY

Hu, Ts'ui-chung.
(Yüan shih hsü pien)
元史續編 / [胡粹中撰]. — [台北：台灣]商務[印書館, 1976?]
2 v. ; 20 cm. — (四庫全書珍本六集 ; 62-63)
Cover title.
Reproduced from Wen yüan ko copy.
1. China—History—Yüan dynasty, 1260-1368—Chronology. I. Title. II. Series: Ssu k'u ch'üan shu. Ssu k'u ch'üan shu chen pen liu chi ; 62-63.
AC149.S7 1976, vol. 62-63 78-843854
[DS752]

—HISTORY—YÜAN DYNASTY, 1260-1368—SOURCES

Schuh, Dieter.
Erlasse und Sendschreiben mongolischer Herrscher für tibetische Geistliche : e. Beitr. zur Kenntnis d. Urkunden d. tibet. Mittelalters u. ihrer Diplomatik / von Dieter Schuh. — St. Augustin : VGH-Wissenschaftsverlag, 1977.
xxvii, 212 p. ; 30 cm. — (Monumenta tibetica historica : Abteilung 3, Diplomata et epistolae ; Bd. 1) GFR78-A6
Some text in Tibetan.
Habilitationsschrift—Bonn, 1975.
Bibliography: p. 183-193.
ISBN 3-88280-002-X
1. Tibet—History—Sources. 2. China—History—Yüan dynasty, 1260-1368—Sources. 3. Tibet—Relations (general) with China. 4. China—Relations (general) with Tibet. I. Title. II. Series.
DS785.S33 78-380180
79 MARC

—HISTORY—MING DYNASTY, 1368-1644

Chang, T'ien-yu.
(Sung Ming shih yen chiu lun chi)
宋明史研究論集：宋明衰亡時期 / 張天佑撰. — [台北縣永和]：華世出版社, 民國66 [1977]
4, 212 p. ; 21 cm.
Includes bibliographical references.
NT$80.00
1. China—History—Sung dynasty, 960-1279. 2. China—History—Ming dynasty, 1368-1644. I. Title.
DS751.C434 78-841215

CHINA

—HISTORY—MING DYNASTY, 1368-1644 (Continued)

Ch'en, Wen-shih.
(Ming Hung-wu Chia-ching chien ti hai chin cheng ts'e)
明洪武嘉靖間的海禁政策 / 陳文石. — 台北：國立臺灣大學文學院, 1966.
3, 7, 176, 2 p. ; 22 cm. — (國立臺灣大學文史叢刊 ; 20)
Title in English in Colophon: The sea-faring prohibition policy in the early Ming Dynasty.
Text in Chinese, with summary in English.
Includes bibliographical references.
1. China—Foreign relations—To 1912. 2. China—History—Ming dynasty, 1368-1644. 3. China—Commerce—History. 4. Pirates—China. I. Title. II. Title: The sea-faring prohibition policy. III. Series: T'ai-wan ta hsüeh, T'ai-pei. Kuo li T'ai-wan ta hsüeh wen shih ts'ung k'an ; 20.
DS753.C5115 78-843024

Greiner, Peter, fl. 1975-
Die Brokatuniform-Brigade (chin-i wei) der Ming-Zeit von den Anfängen bis zum Ende der T'ien-shun-Periode (1368-1464) / von Peter Greiner. — Wiesbaden : O. Harrassowitz, 1975.
viii, 272 p. ; 24 cm. — (Veröffentlichungen des Ostasien-Instituts der Ruhr-Universität Bochum ; Bd. 12) GFR75-A31
Originally presented as part of the author's thesis, Ruhr-Universität Bochum, 1970.
Bibliography: p. 253-260.
Includes indexes.
ISBN 3-447-01651-5
1. Secret service—China. 2. China—History—Ming dynasty, 1368-1644. I. Title. II. Series: Ruhr-Universität Bochum. Ostasien-Institut. Veröffentlichungen ; Bd. 12.
HV8260.A2G73 75-512239
75[r79]rev2 MARC

Hsüeh, San-ts'ai, chin shih 1586.
(Hsüeh Kung-min kung tsou shu)
薛恭敏公奏疏 / 薛三才撰. — 臺北：偉文圖書出版社, 民國66 [1977]
3 v. ; 22 cm. — (明季史料集珍 ; 第二輯)
Series statement in colophon: Ming tai shih liao chi chen.
"本書係中央圖書館所藏"
Reproduced from ms. copy.
CONTENTS: [1] 卷1-7. 上.—[2] 卷8-14. 下.—[3] 不分卷.
NT$900.00 ($23.81 U.S.)
1. China—History—Ming dynasty, 1368-1644. I. Title. II. Series: Ming chi shih liao chi chen : Ti erh chi.
DS753.H78 79-836697

Hu, Ta-hsün.
(Ming tai t'u ti wen t'i)
明代土地問題 / 胡大恂著. 江都耕地分配 / 吳致華著. — 臺北：成文出版社, 民國66 [1977]
4 p., p. 34347-34946, [1] fold. leaf of plates : ill. ; 22 cm. — (民國二十年代中國大陸土地問題資料 ; 66) (中國地政研究所叢刊)
Cover title.
Each work has a separate t. p.
"地政學院之論文…均為手寫恭楷…乃訂約影印"
1. Land use—China—History. 2. China—History—Ming dynasty, 1368-1644. 3. Land tenure—China—Chiang-tu. I. Wu, Chih-hua. Chiang-tu keng ti fen p'ei. 1977. II. Title. III. Title: Chiang-tu keng ti fen p'ei. IV. Series: Min kuo erh shih nien tai Chung-kuo ta lu t'u ti wen t'i tzu liao ; 66. V. Series: Chung-kuo ti cheng yen chiu so, T'ai-pei. Chung-kuo ti cheng yen chiu so ts'ung k'an.
HD925.M56 vol. 66 78-841771
[HD923]
78 C

Li, T'ang.
(Ming Wu-tsung)
明武宗 / 李唐著. — 香港：香港宏業書局, 1978.
3, 3, 175 p. ; 19 cm. — (歷史小叢書)
Series romanized: Li shih hsiao ts'ung shu.
$5.60
1. China—History—Ming dynasty, 1368-1644. 2. Ming Wu-tsung, Emperor of China, 1491-1521. I. Title.
DS753.L5419 79-838110

Li, T'ang.
(Ta Ming)
大明 / 著作者李唐. — 臺北：河洛圖書出版社, 民國67 [1978]
93 p. : ill., maps ; 22 cm. — (中國歷史小叢書 ; 25)
Series romanized: Chung-kuo li shih hsiao ts'ung shu.
Reprint of the ed. issued in series: Li shih hsiao ts'ung shu.
NT$35.00
1. China—History—Ming dynasty, 1368-1644. I. Title.
DS753.L54194 1978 79-840800

Meng, Sen, 1867-1937.
(Ming tai shih)
明代史 / 孟森撰. — 修訂本. — 台北：華世出版社, 民國66 [1977]
2, 4, 392 p. ; 22 cm.
Reprint ed.
NT$80.00
1. China—History—Ming dynasty, 1368-1644. I. Title.
DS753.M4 1977 78-842711

Pokotilov, Dmitriĭ Dmitrievich.
History of the eastern Mongols during the Ming dynasty from 1368 to 1634 / Dmitriĭ Pokotilov ; translation of the Russian text by Rudolph Loewenthal ; Addenda and corrigenda to Pokotilov's History of the eastern Mongols during the Ming dynasty by Wolfgang Franke. — Philadelphia : Porcupine Press, 1976.
148, 95 p. ; 23 cm. — (Perspectives in Asian history ; no. 1)
Translation of Istoriĭa vostochnykh mongolov v period dinastii Min.
Reprint of the 1st ed. published in 1947-1949 by West China Union University, Chengtu, which was issued as no. 1 and 5, ser. A of Studia Serica monographs.
Includes bibliographical references and indexes.
ISBN 0-87991-602-8
1. Mongolia—History. 2. China—History—Ming dynasty, 1368-1644. I. Franke, Wolfgang, 1912- . II. Title. III. Series. IV. Series: Studia Serica monographs ; ser. A, no. 1, 5.
DS793.M7P5713 1976 951'.7 76-50087
76[r79]rev MARC

Teng, Ming-yen.
(Ming tai chih chien kuo chi ch'i hsing wang)
明代之建國及其興亡 / [著作者鄧明炎]. — 台北：天山出版社, 民國65 [1976]
4, 1, 167 p. : map ; 21 cm.
Includes bibliographical references.
NT$50.00
1. China—History—Ming dynasty, 1368-1644. I. Title.
DS753.T43 78-841218

Yang, Hsüeh-feng.
(Ming tai ti shen p'an chih tu)
明代的審判制度 / 楊雪峯著. — 臺北：黎明文化事業公司, 民國67 [1978]
10, 428 p. : ill. ; 22 cm. — (學術叢書)
Series romanized: Hsüeh shu ts'ung shu.
Bibliography: p. 413-427.
NT$150.00
1. Trial practice—China—History. 2. China—History—Ming dynasty, 1368-1644. I. Title.
79-836316

Yang, Shu-fan.
(Ming tai chung yang cheng chih chih tu)
明代中央政治制度 / 楊樹藩著. — 臺北：臺灣商務印書館, 民國67 [1978]
3, 246, [13] p. ; 20 cm. — (岫廬文庫 ; 36)
Series romanized: Hsiu lu wen k'u.
Includes bibliographical references.
NT$36.00
1. China—Politics and government—To 1643. 2. China—History—Ming dynasty, 1368-1644. I. Title.
JQ1509 1368.Y36 78-841510

Yeh, Hsiang-kao, 1559-1627.
(Lun fei tsou ts'ao)
綸扉奏草 / 葉向高撰. — 台北：偉文圖書出版社, 民國66 [1977]
7 v. (3706 p.) ; 22 cm. — (明季史料集珍 ; 第二輯)
Reprint ed.
CONTENTS: 1-5. 綸扉奏草 30卷.—6-7. 續綸扉奏草 14卷.
NT$2500.00 ($65.88 U.S.)
1. China—History—Ming dynasty, 1368-1644. I. Title. II. Series: Ming chi shih liao chi chen : Ti erh chi.
DS753.Y38 1977 78-843264
78 C

(Yü ting tzu chih t'ung chien kang mu san pien)
御定資治通鑑綱目三編 / [張廷玉等編]. — [台北：台灣]商務[印書館, 1976?]
7 v. ; 20 cm. — (四庫全書珍本六集 ; 64-70)
Cover title.
Reproduced from Wen yüan ko copy.
1. China—History—Ming dynasty, 1368-1644. I. Ch'ing, Kao-tsung, Emperor of China, 1711-1799. II. Chang, T'ing-yü, 1672-1755. III. Title: Tzu chih t'ung chien kang mu san pien. IV. Series: Ssu k'u ch'üan shu. Ssu k'u ch'üan shu chen pen liu chi ; 64-70.
AC149.S7 1976, vol. 64-70 78-843859
[DS753]

Yüeh, Ho-sheng, 17th cent.
(Ts'an wei tzu chi)
餐微子集 / 岳和聲撰. — 臺北：偉文圖書出版社, 民國66 [1977]
2 v. (1064 p.) ; 22 cm. — (明季史料集珍 ; 第2輯)
Series statement in colophon: Ming tai shih liao chi chen.
"本書係中央圖書館所藏"
Reprint ed.
NT$700.00 ($18.44 U.S.)
1. China—History—Ming dynasty, 1368-1644. I. Title. II. Series: Ming chi shih liao chi chen : Ti erh chi.
DS753.Y84 79-836696

—HISTORY—MING DYNASTY, 1368-1644 —ADDRESSES, ESSAYS, LECTURES

From Ming to Ch'ing : conquest, region, and continuity in seventeenth-century China / edited by Jonathan D. Spence and John E. Wills, Jr. — New Haven : Yale University Press, 1979.
xxiv, 413 p. : maps ; 25 cm.
Bibliography: p. 367-396.
Includes index.
ISBN 0-300-02218-2
1. China—History—Ming dynasty, 1368-1644—Addresses, essays, lectures. 2. China—History—Ch'ing dynasty, 1644-1912—Addresses, essays, lectures. I. Spence, Jonathan D. II. Wills, John Elliot, 1936- .
DS753.F74 951 78-15560
78 MARC

(Min Shin shi ronsō)
明清史論叢：中山八郎教授頌寿記念 / [編集 佐久間重男, 山根幸夫]. — 東京：燎原書店, 1977.
16, 381 p., [1] leaf of plates : port. ; 22 cm. Ja 78-10852
"中山八郎教授著作目録": p. 7-10 (1st group)
CONTENTS: 大島立子 元末の農民反乱の背景—細野浩二 耆宿制から里老人制へ—川越泰博 所謂「鄭和行香碑」について—谷口規矩雄 陣友諒の末裔について—間野潜龍 「大学衍義補」の「正朝廷」について—王賢徳 海瑞に関する一考察—岡野昌子 秀吉の朝鮮侵略と中国—佐藤文俊 光山県・麻城県奴変考—森正夫 一六四五年太倉州沙溪鎮における烏龍会の反乱について—谷光隆 明代漕運の一齣—佐久間重男 明代の門攤税と都市商業との関係—山根幸夫 明・清初の華北の市集と紳士・豪民—田尻利 十九世紀中葉江蘇の蚕桑書について—中山八郎 私の学習記録
¥6000
1. China—History—Ming dynasty, 1368-1644—Addresses, essays, lectures. 2. China—History—Ch'ing dynasty, 1644-1912—Addresses, essays, lectures. 3. Nakayama, Hachirō, 1907- I. Sakuma, Shigeo, 1914- II. Yamane, Yukio, 1921- III. Nakayama, Hachirō, 1907-
DS753.M533 78-813401

(Ming mo nung min ch'i i yen chiu lun ts'ung)
明末農民起義研究論叢 / 存萃學社編集. — 第1版. — 香港：大東圖書公司, 1977.
299, 40 p. : ill. ; 27 cm.
Includes bibliographies.
$18.00 (U. S.)
1. China—History—Ming dynasty, 1368-1644—Addresses, essays, lectures. 2. Peasant uprisings—China—Addresses, essays, lectures. I. Ts'un ts'ui hsüeh she.
DS753.M557 78-842625

—HISTORY—MING DYNASTY, 1368-1644—ANECDOTES, FACETIAE, SATIRE, ETC.

(Chiu ch'ao t'an tsuan)
九朝談纂 / 明·不著撰人. — 臺北：偉文圖書出版社, 民國66 [1977]
3 v. (1502 p.) ; 22 cm. — (清代禁燬書叢刊 ; 第一輯)
Photoreprint ed.
NT$1000.00 ($26.40 U. S.)
1. China—Kings and rulers—Anecdotes, facetiae, satire, etc. 2. China—History—Ming dynasty, 1368-1644—Anecdotes, facetiae, satire, etc. I. Series: Ch'ing tai chin hui shu ts'ung k'an ; ti i chi.
DS753.5.C46 1977 78-841445
78 C

Chuang, Lien.
(Ming Ch'ing jen wu yü chang ku)
明清人物與掌故 / 莊練著. — 台北：求精出版社：總代理聯亞出版社, 民國67 [1978]
2, 238 p. ; 20 cm. — (精益叢書 ; 14)
Series romanized: Ching i ts'ung shu.
NT$45.00
1. China—History—Ming dynasty, 1368-1644—Anecdotes, facetiae, satire, etc. 2. China—History—Ch'ing dynasty, 1644-1912—Anecdotes, facetiae, satire, etc. I. Title.
DS753.C565 79-837332

CHINA

—HISTORY—MING DYNASTY, 1368-1644—ANECDOTES, FACETIAE, SATIRE, ETC. (Continued)

Li, Le, chin shih 1568.
(Chien wen tsa chi)
見聞雜記 / 李樂撰. — 臺北：偉文圖書出版社，民國66 [1977]
2 v. (1046 p.) ; 22 cm. — (明季史料集珍 ; 第二輯)
Series title varies slightly.
Reprint of the ed. in the possession Chung-yang t'u shu kuan, Taipei.
NT$650.00 ($17.13 U.S.)
1. China—History—Ming dynasty, 1368-1644—Anecdotes, facetiae, satire, etc. I. Title. II. Series: Ming chi shih liao chi chen : Ti erh chi.
DS753.L5417 1977 79-835065

Su, T'ung-ping.
(Ming Ch'ing shih shih ts'ung t'an)
明清史事叢談 莊練 [蘇同炳] 著 [臺北] 學生書局 [民國61 i. e. 1972]
1, 272 p. 20 cm. (學生書苑5) NT$35.00
1. China—History—Ming dynasty, 1368-1644—Anecdotes, facetiae, satire, etc. 2. China—History—Ch'ing dynasty, 1644-1912—Anecdotes, facetiae, satire, etc. I. Title.
DS753.S93 73-897013

Wu, Yüan-ts'ui, chin shih 1580.
(Lin chü man lu)
林居漫錄 / 伍袁萃撰. — 臺北：偉文圖書出版社，民國66 [1977]
2 v. (701 p.) ; 22 cm. — (清代禁燬書叢刊 ; 第一輯)
Photoreprint ed.
NT$600.00 ($15.81 U.S.)
1. China—History—Ming dynasty, 1368-1644—Anecdotes, facetiae, satire, etc. I. Title. II. Series: Ch'ing tai chin hui shu ts'ung k'an : Ti i chi.
DS753.W85 1977 78-842469

78 C

Yang, Shih-ts'ung, 1597-1648.
(Yü t'ang hui chi)
玉堂薈記 / 楊士聰撰. — 臺北：偉文圖書出版社，民國66 [1977]
221 p. : 22 cm. — (清代禁燬書叢刊 ; 第1輯)
Reprint ed.
NT$300.00 ($8.00 U.S.)
1. China—History—Ming dynasty, 1368-1644—Anecdotes, facetiae, satire, etc. I. Title. II. Series: Ch'ing tai chin hui shu ts'ung k'an : ti 1 chi.
DS753.Y28 1977 79-838596

C

—HISTORY—MING DYNASTY, 1368-1644—BIOGRAPHY

(Ch'in ting Sheng ch'ao hsün chieh chu ch'en lu)
欽定勝朝殉節諸臣錄 / [舒赫德，于敏中]. — [台北：台灣] 商務 [印書館，1976?]
3 v. ; 20 cm. — (四庫全書珍本六集 ; 105-107)
Cover title.
Reproduced from Wen yüan ko copy.
1. China—History—Ming dynasty, 1368-1644—Biography. 2. China—Biography. I. Shu-ho-te, 1711-1777. II. Yü, Min-chung, 1714-1779. III. Title: Sheng ch'ao hsün chieh chu ch'en lu. IV. Series: Ssu k'u ch'üan shu. Ssu k'u ch'üan shu chen pen liu chi ; 105-107.
AC149.S7 1976, vol. 105-107 78-843855
[DS753.5.C455]

79 C

(Shao-hsing shih pa nien t'ung nien hsiao lu)
紹興十八年同年小錄. 中州人物考 / [孫奇逢撰]. — [台北：臺灣] 商務 [印書館，1978?]
2 v. ; 20 cm. — (四庫全書珍本八集 ; 82-83)
Cover title.
Photoreprint ed. of two works, both of the Wen Yüan Ko copy.
1. China—Registers. 2. Civil service—China. 3. China—History—Sung dynasty, 960-1279—Biography. 4. China—Biography. 5. China—History—Ming dynasty, 1368-1644—Biography. 6. Honan, China (Province)—Biography. I. Sun, Ch'i-feng, 1585-1675. Chung chou jen wu k'ao. 1978?. II. Title: Chung chou jen wu k'ao. III. Series: Ssu k'u ch'üan shu. Ssu k'u ch'üan shu chen pen pa chi ; 82-83.
AC149.S7 1978, vol. 82-83 78-841087
[JQ1507]

—HISTORY—MING DYNASTY, 1368-1644—FICTION

Fang, I-chih, 1611-1671.
Bitter gourd : Fang I-chih and the the impetus for intellectural change / Willard J. Peterson. — New Haven : Yale University Press, 1979.
xiv, 228 p. ; 24 cm.
Translation of Ch'i chieh which is a part of Chi ku t'ang wen chi.
Bibliography: p. 203-219.
Includes index.
ISBN 0-300-02208-5
1. Fang, I-chih, 1611-1671, in fiction, drama, poetry, etc. 2. China—History—Ming dynasty, 1368-1644—Fiction. I. Peterson, Willard J. II. Title.
PL2698.F33C5213 1979 895.1'3'4 78-18491
78 MARC

—HISTORY—MING DYNASTY, 1368-1644—JUVENILE LITERATURE

Yang, Tsung-chen, 1919-
(Ta Ming ti kuo)
大明帝國 / 孟瑤 [楊宗珍] 文 ; 沈以正圖. — [南投縣]：臺灣省政府教育廳；臺北：發行者臺灣書店，民國64 [1975]
56 p. : ill. ; 21 cm. — (中華兒童叢書)
Series romanized: Chung-hua erh t'ung ts'ung shu.
1. China—History—Ming dynasty, 1368-1644—Juvenile literature. I. Shen, I-cheng. II. Title.
DS753.Y29 79-840023

—HISTORY—MING DYNASTY, 1368-1644—SOURCES

China. Li pu. K'ao kung ssu.
(Li pu k'ao kung ssu t'i kao)
吏部考功司題稿 / 吏部考功司撰. — 臺北：偉文圖書出版社，民國66 [1977]
3 v. (1643 p.) ; 22 cm. — (明季史料集珍 ; 第二輯)
"起嘉靖二十年十一月訖嘉靖廿二年五月"
Series statement in colophon: Ming tai shih liao chi chen.
"本書係中央圖書館所藏"
Reprint ed.
NT$1000.00 ($26.40 U.S.)
1. Civil service—China—History—Sources. 2. China—History—Ming dynasty, 1368-1644—Sources. 3. Employees, Rating of—China—History—Sources. I. Title. II. Series: Ming chi shih liao chi chen : Ti erh chi.
JQ1512.Z1C49 1977 79-837285

Chou, I, 1506-1569.
(Na-hsi tsou shu)
訥溪奏疏 / [周怡撰]. 兩垣奏議 / [逯中立撰]. 御選明臣奏議. — [台北]：商務 [印書館，1979]
12 v. ; 20 cm. — (四庫全書珍本九集 ; 117-128)
Cover title.
Photoreprint ed. of the Wen Yüan Ko copy.
CONTENTS: 1. 訥溪奏疏 兩垣奏議 御選明臣奏議 卷1.—2-12. 御選明臣奏議 卷2-40.
1. China—History—Ming dynasty, 1368-1644—Sources. I. Lu, Chung-li, chin shih 1589. Liang yüan tsou i. 1979. II. Yü hsüan Ming ch'en tsou i. 1979. III. Title. IV. Series: Ssu k'u ch'üan shu. Ssu k'u ch'üan shu chen pen chiu chi ; 117-128.
AC149.S7 1979, vol. 117-128 79-840245
[DS753]

Hu, Shih-ning, 1469-1530.
(Hu Tuan-min kung tsou i)
胡端敏公奏議 10卷 胡世寧撰 [n. p.] 浙江書局刻 [光緒癸巳(1893)跋]
4 v. (double leaves) in case. 24 cm.
Block print.
1. China—History—Ming dynasty, 1368-1644—Sources.
DS753.H793 1893 C 60-5245

(Ming chi li)
明集禮 / [徐一夔 ... et al.]. — [台北：台灣商務印書館，1978?]
12 v. : ill. ; 20 cm. — (四庫全書珍本八集 ; 113-124)
Cover title.
Reprint ed.; originally issued in 53 chüan.
1. China—Social life and customs. 2. Rites and ceremonies—China. 3. China—History—Ming dynasty, 1368-1644—Sources. I. Hsü, I-k'uei, 14th cent. II. Series: Ssu k'u ch'üan shu. Ssu k'u ch'üan shu chen pen pa chi ; 113-124.
AC149.S7 1978, vol. 113-124 78-841089
[DS721]

—HISTORY—K'ANG-HSI, 1622-1722—SOURCES

(Ch'ing tai Chun-ka-erh shih liao)
清代準噶爾史料：初編 / 莊吉發譯注. — 臺北：文史哲出版社，民國66 [1977]
x, 223 p. ; 22 cm.
Chinese and Manchu language.
NT$120.00
1. China—History—K'ang-hsi, 1622-1722—Sources. 2. Sungaria—History—Sources. I. Chuang, Chi-fa. II. Ch'ing sheng-tsu, Emperor of China, 1654-1722.
DS754.2.C56 79-835023

—HISTORY—TATAR CONQUEST, 1643-1644

Palafox y Mendoza, Juan de, bp., 1600-1659.
History of the conquest of China / Palafox ; introd. by H. S. Bhatia. — New Delhi : Deep & Deep Publications, c1978.
272 p. : ill. ; 23 cm.
Translation of Historia de la conquista de la China por el Tartaro.
Bibliography: p. [269]
Includes index.
Rs65.00
1. China—History—Tatar Conquest, 1643-1644. I. Title.
DS753.P3313 951'.026 78-913186
79 MARC

—HISTORY—CH'ING DYNASTY, 1644-1912

Chang, Hsing-t'ang.
(Meng-ku meng pu ch'i chih ti i i ho yen ko)
蒙古盟部旗制的意義和沿革 / [張興唐編著]. — 節本. — [台北]：藏蒙委員會，民國67 [1978]
59 p. ; 19 cm.
Appendices (p. 34-59): 1. 蒙古盟部旗組織法—2. 蒙古自治辦法原則八項 附表蒙古盟部旗行政區劃表
1. Mongols—History. 2. China—History—Ch'ing dynasty, 1644-1912. I. China. Laws, statutes, etc. Meng-ku meng pu ch'i tsu chih fa. 1978. II. Title.
DS19.C4517 79-838049

Ch'en, Shih-ch'ang.
(Chung-kuo yü men hu k'an fang)
中國輿門戶開放：清末民初外患史 / 陳世昌著. — 台北：海國書局，民國66 [1977]
11, 420 p. ; 22 cm.
At head of title: Ta chuan yung shu.
Title on spine: Ch'ing mo min ch'u wai huan shih.
Bibliography: p. 361-367.
NT$140.00
1. China—Foreign relations. 2. China—History—Ch'ing dynasty, 1644-1912. 3. China—History—1912-1937. I. Title. II. Title: Ch'ing mo min ch'u wai huan shih.
DS740.4.C35338 78-841899

78 C

Linck-Kesting, Gudula.
Ein Kapitel chinesischer Grenzgeschichte, Han und Nicht-Han im Taiwan der Qing-Zeit : 1683-1895 / von Gudula Linck-Kesting. — Wiesbaden : Steiner, 1979.
360 p. : ill. ; 24 cm. — (Münchener ostasiatische Studien ; Bd. 22)
GFR79-A
Summary in English.
Bibliography: p. 336-345.
Includes indexes.
ISBN 3-515-02983-4 : DM28.00
1. Taiwan—Race relations. 2. China—History—Ch'ing dynasty, 1644-1912. I. Title. II. Series.
DS799.42.L56 951'.249'03 79-362019
79 MARC

Lu, Pao-ch'ien.
(Ch'ing tai ssu hsiang shih)
清代思想史 / 陸寶千著. — 台北：廣文書局，民國67 [1978]
1, 1, 481, 6 p. ; 22 cm.
Includes bibliographical references and index.
NT$180.00
1. Learning and scholarship—China. 2. China—History—Ch'ing dynasty, 1644-1912. I. Title.
AZ791.L8 79-841414

CHINA

—HISTORY—CH'ING DYNASTY, 1644-1912 (Continued)

Rawski, Evelyn Sakakida.
Education and popular literacy in Ch'ing China / Evelyn Sakakida Rawski. — Ann Arbor : University of Michigan Press, c1979.
xii, 294 p. : ill. ; 24 cm. — (Michigan studies on China)
Bibliography: p. 257-289.
Includes index.
ISBN 0-472-08753-3 : $16.50
1. Education—China—History. 2. Illiteracy—China—History. 3. China—History—Ch'ing dynasty, 1644-1912. I. Title. II. Series.
LA1131.R35 1979 370'.951 78-15526
78 MARC

—HISTORY—CH'ING DYNASTY, 1644-1912—ADDRESSES, ESSAYS, LECTURES

Ch'en, Chieh-hsien.
(Ch'ing shih tsa pi)
清史雜筆 / 陳捷先著. — 台北：學海出版社, 民國66 [1977]
2 v. ; 22 cm.
Includes bibliographical references.
NT$50.00 per vol.

1. China—History—Ch'ing dynasty, 1644-1912—Addresses, essays, lectures. I. Title.
DS754.C493 79-838751

From Ming to Ch'ing : conquest, region, and continuity in seventeenth-century China / edited by Jonathan D. Spence and John E. Wills, Jr. — New Haven : Yale University Press, 1979.
xxiv, 413 p. : maps ; 25 cm.
Bibliography: p. 367-396.
Includes index.
ISBN 0-300-02218-2
1. China—History—Ming dynasty, 1368-1644—Addresses, essays, lectures. 2. China—History—Ch'ing dynasty, 1644-1912—Addresses, essays, lectures. I. Spence, Jonathan D. II. Wills, John Elliot, 1936- .
DS753.F74 951 78-15560
78 MARC

(Min Shin shi ronsō)
明清史論叢：中山八郎教授頌寿記念 / [編集 佐久間重男, 山根幸夫]. — 東京：燎原書店, 1977.
16, 381 p., [1] leaf of plates : port. ; 22 cm. Ja 78-10852
"中山八郎教授著作目録": p. 7-10 (1st group)
CONTENTS: 大島立子 元末の農民反乱の背景—細野浩二 耆宿制から里老人制へ—川越泰博 所謂「鄭和行香碑」について—谷口規矩雄 陣友諒の末裔について—間野潜龍 「大学衍義補」の「正朝廷」について—王賢徳 海瑞に関する一考察—岡野昌子 秀吉の朝鮮侵略と中国—佐藤文俊 光山県·麻城県奴変考—森正夫 一六四五年太倉州沙渓鎮における烏龍会の反乱について—谷光隆 明代漕運の一齣—佐久間重男 明代の門攤税と都市商業との関係—山根幸夫 明·清初の華北の市集と紳士·豪民—田尻利 十九世紀中葉江蘇の蚕桑書について—中山八郎 私の学習記録
¥6000

1. China—History—Ming dynasty, 1368-1644—Addresses, essays, lectures. 2. China—History—Ch'ing dynasty, 1644-1912—Addresses, essays, lectures. 3. Nakayama, Hachirō, 1907- I. Sakuma, Shigeo, 1914- II. Yamane, Yukio, 1921- III. Nakayama, Hachirō, 1907-
DS753.M533 78-813401

—HISTORY—CH'ING DYNASTY, 1644-1912—ANECDOTES, FACETIAE, SATIRE, ETC.

Chin, Liang, 1878-
[Ssu ch'ao i wen]
道咸同光四朝佚聞 / 金梁撰. — 台北：廣文書局, 民國67 [1978]
118 p. ; 22 cm. — (史料七編)
Reprint of the 1936 ed. printed by Fu-tung yin shua chü, with caption and running title: Ssu ch'ao i wen.
NT$150.00
1. China—History—Ch'ing dynasty, 1644-1912—Anecdotes, facetiae, satire, etc. I. Title. II. Series: Shih liao ch'i pein.
Title romanized: Tao Hsien T'ung Kuang ssu ch'ao i wen.
DS754.C53134 1978 78-841107

Chuang, Lien.
(Ming Ch'ing jen wu yü chang ku)
明清人物與掌故 / 莊練著. — 台北：求精出版社：總代理聯亞出版社, 民國67 [1978]
2, 238 p. ; 20 cm. — (精益叢書 ; 14)
Series romanized: Ching i ts'ung shu.
NT$45.00
1. China—History—Ming dynasty, 1368-1644—Anecdotes, facetiae, satire, etc. 2. China—History—Ch'ing dynasty, 1644-1912—Anecdotes, facetiae, satire, etc. I. Title.
DS753.C565 79-837332

Hsü, Kuo-ying.
(Shih yeh yeh wen)
十葉野聞 許指嚴著 [臺北縣永和鎮] 文海出版社 [民國61 i. e. 1972]
4, 156, 138 p. 20 cm. (近代中國史料叢刊830)
Reprint ed.

1. China—History—Ch'ing dynasty, 1644-1912—Anecdotes, facetiae, satire, etc. I. Title. II. Series: Shen, Yün-lung, comp. Chin tai Chung-kuo shih liao ts'ung k'an 830.
DS755.S53 no. 830 73-897190
[DS754]
rev

I, Tsung-k'uei, 1875-
(Hsin shih shuo)
新世說：[8卷] / 易宗夔撰述. — 台北：廣文書局, 民國66 [1977]
2 v. (634, [14] p.) ; 22 cm. — (國學珍籍彙編)
Reprint of the 1922 printing.
"本書承國立中央研究院惠借影印"
NT$410.00
1. China—History—Ch'ing dynasty, 1644-1912—Anecdotes, facetiae, satire, etc. 2. China—History—Republic, 1912-1949—Anecdotes, facetiae, satire, etc. I. Title. II. Series: Kuo hsüeh chen chi hui pien.
DS754.I443 1977 79-836358

Su, T'ung-ping.
(Ming Ch'ing shih shih ts'ung t'an)
明清史事叢談 莊練[蘇同炳]著 [臺北] 學生書局 [民國61 i. e. 1972]
1, 272 p. 20 cm. (學生書苑5) NT$35.00

1. China—History—Ming dynasty, 1368-1644—Anecdotes, facetiae, satire, etc. 2. China—History—Ch'ing dynasty, 1644-1912—Anecdotes, facetiae, satire, etc. I. Title.
DS753.S93 73-897013
rev

—HISTORY—CH'ING DYNASTY, 1644-1912—HISTORIOGRAPHY

Lo, Ping-mien.
(Ch'ing tai hsüeh shu lun chi)
清代學術論集 / 羅炳綿著. — 臺北：食貨出版社：總代理中央文物供應社, 民國67 [1978]
2, 2, 724 p. ; 22 cm. — (食貨史學叢書)
Series romanized: Shih huo shih hsüeh ts'ung shu.
NT$260.00

1. China—History—Ch'ing dynasty, 1644-1912—Historiography. I. Title.
DS754.L63 79-839499

—HISTORY—CH'ING DYNASTY, 1644-1912—JUVENILE LITERATURE

Yang, Tsung-chen, 1919-
(Ta Ch'ing ti kuo)
大清帝國 / 孟瑶[楊宗珍]文；洪義男圖. — [南投縣]：臺灣省政府教育廳；臺北：發行者臺灣書店, 民國64 [1975]
56 p. : ill. ; 21 cm. — (中華兒童叢書)
Series romanized: Chung-hua erh t'ung ts'ung shu.

1. China—History—Ch'ing dynasty, 1644-1912—Juvenile literature. I. Hung, I-nan. II. Title.
DS754.Y28 78-843803

—HISTORY—CH'ING DYNASTY, 1644-1912—SOURCES

(Ch'in ting ta Ch'ing t'ung li)
欽定大清通禮 / [來保等編纂]. — [台北：台灣]商務[印書館, 1978?]
6 v. ; 20 cm. — (四庫全書珍本八集 ; 125-130)
Cover title.
Photoreprint ed. of the Wen Yüan Ko copy.
1. China—Social life and customs. 2. Rites and ceremonies—China. 3. China—History—Ch'ing dynasty, 1644-1912—Sources. I. Lai, Pao. II. Title: Ta Ch'ing t'ung li. III. Series: Ssu k'u ch'üan shu. Ssu k'u ch'üan shu chen pen pa chi ; 125-130.
AC149.S7 1978, vol. 125-130 78-841106
[DS721]

(Ch'ing shih tzu liao hui pien pu pien)
清史資料彙編補編. — [香港? : s. n., 1979?]
3 v. in 4 ; 22 cm.
Cover title.
Includes bibliographical references.

1. China—History—Ch'ing dynasty, 1644-1912—Sources.
DS754.C53317 79-837730

(Kuo ch'ao shih liao ling shih)
國朝史料零拾 / 羅福頤輯. — [台北縣永和鎮]：文海出版社, [1977?]
202 p. ; 22 cm. — (近代中國史料叢刊續編[i. e.輯] ; 441 : 第45輯)
Reprint ed.

1. China—History—Ch'ing dynasty, 1644-1912—Sources. I. Lo, Fu-i, 1905- II. Series: Shen, Yün-lung, comp. Chin tai Chung-kuo shih liao ts'ung k'an hsü chi ; 441.
DS755.S532 vol. 441 79-839912
[DS754]

O, Hai.
(Fu Miao lu)
撫苗錄 / 鄂海撰. — 台北：廣文書局, 民國67 [1978]
372 p. ; 22 cm. — (史料七編)
Reprint ed.
NT$300.00

1. Miao people. 2. China—History—Ch'ing dynasty, 1644-1912—Sources. 3. China—Race relations. I. Title. II. Series: Shih liao ch'i pien.
DS731.M5O12 1978 78-842307
78 C

—HISTORY—K'ANG-HSI, 1662-1722

Tai, Hsüan-chih.
(Chu San T'ai-tzu an)
朱三太子案 / 戴玄之. — Singapore：南洋大學研究院人文與社會科學研究所, 1977.
15 leaves ; 30 cm. — (研究論文 - 南洋大學研究院人文與社會科學研究所 ; no. 59)
Reproduced from typewritten copy.
Includes bibliographical references.
1. China—History—K'ang-hsi, 1662-1722. 2. Chu San T'ai-tzu, 1633-1708. I. Title. II. Series: Nanyang University. Institute of Humanities and Social Sciences. Occasional paper series - Institute of Humanities and Social Sciences, Nanyang University ; no. 59.
DS754.2.T3 1977 78-940556
78 C

Wu, Silas H L 1929-
Passage to power : K'ang-hsi and his heir apparent, 1661-1722 / Silas H. L. Wu. — Cambridge : Harvard University Press, 1979.
xv, 252 p. : maps ; 24 cm. — (Harvard East Asian series ; 91)
Bibliography: p. 227-240.
Includes index.
ISBN 0-674-65625-3
1. Ch'ing Sheng-tsu, Emperor of China, 1654-1722. 2. China—History—K'ang-hsi, 1662-1722. I. Title. II. Series.
DS754.4.C53W8 951'.03'0924 79-4191
79 MARC

—HISOTRY—YUNG-CHENG, 1723-1735—SOURCES

(Kung chung tang Yung-cheng ch'ao tsou che)
宮中檔雍正朝奏摺 / [編輯者國立故宮博物院]. — 臺北：該院, 民國66- , 1977-
v. ; 27 cm.
Added t. p.: Secret palace memorials of the Yung-cheng period.
"本院現藏雍正朝漢滿文奏摺逾二萬三千件...除請安摺及重錄複本外, 猶得二萬一千餘件...為留存御批真跡, 特以硃墨套色景印."
NT$1,400.00 ($38.00 U.S.) (v. 1-2)
1. China—History—Yung-cheng, 1723-1735—Sources. I. Kuo li ku kung po wu yüan. II. Title: Secret palace memorials of the Yung-cheng period.
DS754.2.K87 79-841049

CHINA (Continued)

—HISTORY—19TH CENTURY

Chang, Yü-fa, 1936–
(Chung-kuo chin tai hsien tai shih)
中國近代現代史 / 張玉法. — 臺北：東華書局，民國67 [1978]
14, 469 p. ; 22 cm.
Bibliography: p. 465–469.
NT$140.00

1. China—History—19th century. 2. China—History—1900– I. Title.
DS755.C354 79–838600

China (Republic of China). Kuang fu ta lu she chi yen chiu wei yüan hui.
(Lieh ch'iang ch'in lüeh wo kuo chih yen chiu)
列強侵略我國之研究 / 光復大陸設計研究委員會研擬. — [台北]：該會，民國65 [1976]
3, 32, 4, 48 p. ; 22 cm. — (邊疆研究叢書)
Cover title.
CONTENTS: 近百年來日本與西歐列強侵略我領土損害我主權之研究.—俄國侵略我領土損害我主權之研究.
1. China—Foreign relations. 2. China—History—19th century. 3. China—History—1900– I. Title. II. Series: Pien chiang yen chiu ts'ung shu.
DS740.4.C354518 1976 78–841329

(Chūgoku kakumei no kiten)
中国革命の起点 / [執筆者 田中正俊 ... et al.]. — 東京：東京大学出版会，1978.
xii, 296 p. ; 21 cm. — (講座中国近現代史 ; 1) Ja 78–11730
Includes bibliographical references.
¥2400

1. China—History—19th century. I. Tanaka, Masatoshi, 1922– II. Series: Kōza Chūgoku kingendai shi ; 1.
DS755.K69 vol. 1 78–813396

Fu tan ta hsüeh, Shanghai. Li shih hsi. Chung-kuo chin tai shih chiao yen tsu.
(Chung-kuo chin tai chien shih)
中国近代简史 / 复旦大学历史系中国近代史教研组编著. — 第2版. — 上海：上海人民出版社，1978.
4, 366 p. : ill. ; 19 cm. — (青年自学丛书)
Series romanized: Ch'ing nien tzu hsüeh ts'ung shu.
Includes bibliographical references.
"附录大事年表": p. 351–366.
$0.62

1. China—History—19th century. 2. China—History—1900– I. Title.
DS755.F8 1978 79–836705
79

Graham, Gerald Sandford, 1903-
The China station : war and diplomacy 1830-1860 / by Gerald S. Graham. — Oxford [Eng.] : Clarendon Press ; New York : Oxford University Press, 1978.
xvi, 444 p. : ill. ; 23 cm. GB***
Bibliography: p. [424]-432.
Includes index.
ISBN 0-19-822472-9 : £15.00
1. China—History—19th century. 2. Great Britain—History, Naval—19th century. 3. Great Britain. Navy—History. I. Title.
DS757.G7 1978 951'.03 78-40070
79 MARC

Histoire de la Chine. Sous la direction de Jean Chesneaux ... Paris, Hatier, 1969-77.
4 v. maps. 24 cm. (Collection d'histoire contemporaine) (Hatier Université) 38.05F (v. 3)
On spine, v. 1; cover title, v. 2-4: La Chine.
Includes bibliographies and indexes.
CONTENTS: 1. Des guerres de l'opium à la guerre franco-chinoise, 1840-1855, par J. Chesneaux et M. Bastid.—2. De la guerre franco-chinoise à la fondation du Parti communiste chinois, 1885-1921, par M. Bastid, M.-C. Bergère, J. Chesneaux.—3. La marche de la Révolution, 1921-1949, par J. Chesneaux et F. Le Barbier.—4. Un nouveau communisme, 1949-1976, par J. Bellassen et al.
ISBN 2-218-03028-4 (v. 3)
1. China—History—19th century. 2. China—History—1900- I. Chesneaux, Jean, ed. II. Title: La Chine.
DS757.5.H57 76-512614
79[r79]rev2 MARC

Hu-nan shih fan hsüeh yüan. Chin tai Chung-kuo shih hua pien hsieh tsu.
(Chin tai Chung-kuo shih hua)
近代中国史话 / 湖南师范学院《近代中国史话》编写组. — 第1版. — 北京：人民出版社，1977.
3, 380 p. : ill. ; 19 cm.
$0.75

1. China—History—19th century. 2. China—History—1900– I. Title.
DS755.H87 1977 78–841957

Kawakami, Tadao, 1915–
(Kindai Chūgoku hyakunenshi)
近代中国百年史 / 川上忠雄著. — 東京：高文堂出版社，1979.
212 p. : maps ; 19 cm. Ja 79–4616
Bibliography: p. 210–212.
Errata slip inserted.
¥1800

1. China—History—19th century. 2. China—History—1900– I. Title.
DS755.K38 79–807173

Matsushima, Shūe, 1871–1935.
(Shinchō matsuro hishi)
清朝末路秘史 / 松島宗衛. — 長崎村(東京市外)：大星社，大正14 [1925]
6, 271 p., [6] leaves of plates : ports. ; 22 cm.
L.C. copy imperfect: all before pref. wanting; title from caption.

1. Yuan, Shih-k'ai, 1859–1916. 2. China—Presidents—Biography. 3. China—History—19th century. I. Title.
DS778.Y84M37 78–814187

—HISTORY—19TH CENTURY—ADDRESSES, ESSAYS, LECTURES

China, reform and revolution : background papers on Chinese history, 1850 to 1949 / edited by H. D. M.-h. Chan. — [Palmerston North, N.Z.] : Massey University, Dept. of History, 1978.
iv, 103 p. ; 21 cm. NZ***
Bibliography: p. 92-101.
1. China—History—19th century—Addresses, essays, lectures. 2. China—History—1900- —Addresses, essays, lectures. I. Chan, H. D. M-h.
DS758.C47 951'.03 79-310525
79 MARC

Liu, Ta-nien.
(Chung-kuo chin tai shih wen t'i)
中国近代史问题 / 刘大年著. — 第2版. — 北京：人民出版社：新華書店发行，1978.
262 p. ; 21 cm.
First ed. published in 1965 under title: Chung-kuo chin tai shih chu wen t'i.
Includes bibliographical references
$0.64
1. China—History—19th century—Addresses, essays, lectures. 2. China—History—1900– —Addresses, essays, lectures. I. Title.
DS754.L588 1978 79–837278

Nakahira, Ryō.
(Shina wa dō naru)
支那はどうなる / 中平亮著. — 東京：東洋研究會，昭和7 [1942]
ii, 149 p. ; 19 cm. — (Hia 東洋研究叢書 ; 3)
Series romanized: Tōyō kenkyū sōsho.

1. China—History—19th century—Addresses, essays, lectures. 2. China—History—1912–1949—Addresses, essays, lectures. 3. China—Foreign economic relations—Addresses, essays, lectures. I. Title.
DS755.N54 79–808145

Wu, Hsiang-hsiang, 1913–
(Li shih yü jen wu)
歷史與人物 / 吳相湘著. — 臺北：東大圖書有限公司：總經銷三民書局，1978.
2, 3, 551 p. ; 22 cm. — (滄海叢刊 : 歷史)
Series romanized: Ts'ang hai ts'ung k'an : Li shih.

1. China—History—19th century—Addresses, essays, lectures. 2. China—History—1900– —Addresses, essays, lectures. I. Title.
DS755.W82 79–839501

—HISTORY—19TH CENTURY —COLLECTED WORKS

(Kōza Chūgoku kingendai shi)
講座中国近現代史 / 野沢豊，田中正俊編集. — 東京：東京大学出版会，1978–
v. ; 21 cm. Ja 78–11730 (v. 1)
¥2400 per vol.

1. China—History—19th century—Collected works. 2. China—History—1900– —Collected works. I. Nozawa, Yutaka. II. Tanaka, Masatoshi, 1922–
DS755.K69 78–813399
J

—HISTORY—19TH CENTURY —SOURCES

(Ch'ing tai tang an shih liao ts'ung pien)
清代档案史料丛编 / 故宫博物院明清档案部编. — 第1版. — 北京：中华书局：新华书店北京发行所，1978–
v. ; 21 cm.
$0.66 (v. 1)

1. Taiping Rebellion, 1850–1964—Sources. 2. China—History—19th century—Sources. I. Ku kung po wu yüan, Peking. Ming Ch'ing tang an pu.
DS759.C425 79–839128

—HISTORY—19TH CENTURY —SOURCES—BIBLIOGRAPHY

Washington (State). University. Modern Chinese History Project.
Guide to the memorials of seven leading officials of nineteenth-century China. Prepared by the staff of the Modern Chinese History Project. Edited by Chung-li Chang and Stanley Spector. Translators: Mort Bobrow [and others] Foreword by Franz Michael. Seattle, University of Washington Press, 1955.
xv, 457 p. 28 cm. (University of Washington publications on Asia)
1. China—History—19th century—Sources—Bibliography. I. Chang, Chung-li, 1919- ed. II. Spector, Stanley, 1924- ed. III. Title. IV. Series: Washington (State). University. University of Washington publications on Asia.
Z3108.A4W3 1955 a 55-9932
[DS755] [r79]rev MARC
Washington. Univ., Seattle. Library

—HISTORY—TAO-KUANG, 1820-1850

Shang-hai shih fan ta hsüeh. Chung-kuo chin tai shih tsu.
(Lin Tse-hsü shih wen hsüan chu)
林则徐诗文选注 / 上海师范大学历史系中国近代史组. — 第1版. — 上海：上海古籍出版社，1978.
7, 3, 331 p., [1] leaf of plates ; 21 cm.
$1.10
1. Lin, Tse-hsü, 1785–1850. 2. China—History—War of 1840–1842. 3. China—History—Tao-kuang, 1820–1850. I. Lin, Tse-hsü, 1785–1850. Selected works. 1978. II. Title.
DS760.9.L5S5 1978 78–841955
C

—HISTORY—WAR OF 1840-1842

Greenberg, Michael.
British trade and the opening of China, 1800-42 / by Michael Greenberg. — New York : Monthly Review Press, [1979]
xii, 238 p. ; 20 cm.
Reprint of the 1951 ed. published by University Press, Cambridge, Eng., issued in series: Cambridge studies in economic history.
Bibliography: p. [226]-232.
Includes index.
ISBN 0-85345-497-3
1. Great Britain—Commerce—China. 2. China—Commerce—Great Britain. 3. Opium trade—China. 4. China—History—War of 1840-1942. I. Title.
[HF3508.C5G73 1979] 382'.41'051 79-12206
79 MARC

CHINA

—HISTORY—WAR OF 1840-1842
(Continued)

Hsia-men ta hsüeh. Ya p'ien chan cheng shih ch'i ti Lin Tse-hsü pien hsieh tsu.
(Ya p'ien chen cheng shih ch'i ti Lin Tse-hsü)
鴉片战争时期的林则徐 / 厦门大学历史系《鸦片战争时期的林则徐》编写组. — [香港? : s. n., 1979?]
2, 80 p., [2] leaves of plates : ill. ; 21 cm.
Reprint of the 1978 ed. published by Fu-chien jen min ch'u pan she, Foochow.
Includes bibliographical references.

1. Lin, Tse-hsü, 1785-1850. 2. Statesmen—China—Biography. 3. China—History—War of 1840-1842. I. Title.
DS760.9.L5H76 1979 79-841169

Masui, Tsuneo, 1907-
(Chūgoku no futatsu no higeki)
中国の二つの悲劇 : アヘン戦争と太平天国 / 増井経夫著. — 東京 : 研文出版, 1978.
247 p. ; 20 cm. — (研文選書 ; 1) Ja 78-24002
Series romanized: Kenbun sensho.
"日本における太平天国関係著作目錄": p. 237-244.
¥1500

1. China—History—War of 1840-1842. 2. Taiping Rebellion, 1850-1864. I. Title.
DS757.5.M37 79-800113

Matsuzaki, Keiji, 1906-
(Ahen sensō)
阿片戦争 : その史實と物語 / 松崎啓次, 小國英雄共著. — 東京 : 高山書院, 昭和17 [1942]
263 p., [10] leaves of plates : ill. ; 19 cm.

1. China—History—War of 1840-1842. I. Oguni, Hideo, 1904- joint author. II. Title.
DS757.5.M38 79-804074

Shang-hai shih fan ta hsüeh. Chung-kuo chin tai shih tsu.
(Lin Tse-hsü shih wen hsüan chu)
林则徐诗文选注 / 上海师范大学历史系中国近代史组. — 第1版. — 上海 : 上海古籍出版社, 1978.
7, 3, 331 p., [1] leaf of plates ; 21 cm.
$1.10
1. Lin, Tse-hsü, 1785-1850. 2. China—History—War of 1840-1842. 3. China—History—Tao-kuang, 1820-1850. I. Lin, Tse-hsü, 1785-1850. Selected works. 1978. II. Title.
DS760.9.L5S5 1978 78-841955

Tan, Chung, 1929-
China and the brave new world : a study of the origins of the Opium War (1840-42) / Tan Chung. — Durham, N.C. : Carolina Academic Press, c1978.
viii, 271 p. ; 22 cm.
Bibliography: p. [246]-256.
Includes index.
ISBN 9890890862 : $12.95
1. China—History—War of 1840-1842. 2. Opium trade—China. I. Title.
DS757.5.T28 951'.03 78-50590
79 MARC

—HISTORY—WAR OF 1840-1842—FICTION

Lloyd, Alan, 1927-
Trade imperial / Alan Lloyd. — 1st American ed. — New York : Coward, McCann & Geoghegan, 1979.
272 p. ; 24 cm.
ISBN 0-698-10970-8 : $10.95
1. China—History—War of 1840-1842—Fiction. I. Title.
PZ4.L7925 Tr 1979 823'.9'14 78-26239
[PR6062.L57] 78 MARC

—HISTORY—HSIEN-FENG, 1850-1861

Tuan, Ch'ang-kuo.
(Kung ch'in wang I-hsin yü Hsien T'ung chih chi ti wai chiao yü cheng chih chiu fen)
恭親王奕訢與咸同之際的外交與政治糾紛 : 1858-1865 / 段昌國. — [台北? : s. n., 1972?]
70 double leaves ; 27 cm.
"國立臺灣大學歷史學研究所(近代史組)六十一學年度碩士畢業論文"
Reproduced from the typescript.
Bibliography: leaves 67-70.
1. China—History—Hsien-feng, 1850-1861. 2. China—History—T'ung-chih, 1861-1875. 3. China—Politics and government—1644-1911. 4. I-hsin, 1833-1898. I. Title.
DS758.T8 79-841141

—HISTORY—HSIEN-FENG, 1850-1861—BIOGRAPHY

Chu, K'ung-chang, 1842-1919.
(Hsien-feng i lai kung ch'en pieh chuan)
咸豐以來功臣別傳 / 朱孔彰撰. — 台北 : 廣文書局, 民國67 [1978]
2 v. (552 p.) ; 22 cm. — (史料七編)
Reprint of the 1898 ed. published by Chien hsüeh lu, in series: Chien hsüeh lu ts'ung shu.
NT$400.00
1. China—History—Hsien-feng, 1850-1861—Biography. 2. China—History—T'ung-chih, 1861-1875—Biography. 3. China—Biography. I. Title. II. Sereis: Shih liao ch'i pien.
DS760.8.C486 1978 78-841108

—HISTORY—T'UNG-CHIH, 1861-1875

Tuan, Ch'ang-kuo.
(Kung ch'in wang I-hsin yü Hsien T'ung chih chi ti wai chiao yü cheng chih chiu fen)
恭親王奕訢與咸同之際的外交與政治糾紛 : 1858-1865 / 段昌國. — [台北? : s. n., 1972?]
70 double leaves ; 27 cm.
"國立臺灣大學歷史學研究所(近代史組)六十一學年度碩士畢業論文"
Reproduced from the typescript.
Bibliography: leaves 67-70.
1. China—History—Hsien-feng, 1850-1861. 2. China—History—T'ung-chih, 1861-1875. 3. China—Politics and government—1644-1911. 4. I-hsin, 1833-1898. I. Title.
DS758.T8 79-841141

79 C

—HISTORY—T'UNG-CHIH, 1861-1875—BIOGRAPHY

Chu, K'ung-chang, 1842-1919.
(Hsien-feng i lai kung ch'en pieh chuan)
咸豐以來功臣別傳 / 朱孔彰撰. — 台北 : 廣文書局, 民國67 [1978]
2 v. (552 p.) ; 22 cm. — (史料七編)
Reprint of the 1898 ed. published by Chien hsüeh lu, in series: Chien hsüeh lu ts'ung shu.
NT$400.00
1. China—History—Hsien-feng, 1850-1861—Biography. 2. China—History—T'ung-chih, 1861-1875—Biography. 3. China—Biography. I. Title. II. Sereis: Shih liao ch'i pien.
DS760.8.C486 1978 78-841108

—HISTORY—1862-1899

Fields, Lanny B
Tso Tsung-t'ang and the Muslims : statecraft in northwest China, 1868-1880 / by Lanny B. Fields. — Kingston, Ont. : Limestone Press, 1978.
viii, 206 p., [4] leaves of plates : ill. ; 24 cm. — (Central Asia series ; no. 1) C***
Originally presented as the author's thesis, Indiana State University.
Bibliography: p. 177-200.
Includes index.
ISBN 0-919642-85-3
1. Tso, Tsung-t'ang, 1812-1885. 2. Statesmen—China—Biography. 3. China—History—1862-1899. 4. Muslims in China. I. Title. II. Series.
DS760.9.T7F53 1978 951'.03'0924 79-309212
79 MARC

—HISTORY—KUANG-HSÜ, 1875-1908

Chao, Ping-lin, b. 1873.
(Kuang-hsü ta shih hui chien)
光緒大事彙鑑 / 趙柏巖 [炳麟] 撰. — 台北 : 廣文書局, 民國67 [1978]
2 v. (530 p.) ; 22 cm. — (史料七編)
NT$380.00

1. China—History—Kuang-hsü, 1875-1908. I. Title. II. Series: Shih liao ch'i pien.
DS764.C485 1978 78-841203

(Giwadan undō)
義和団運動 / [執筆者 高橋孝助 ... et al.]. — 東京 : 東京大学出版会, 1978.
vi, 266 p. : 21 cm. — (講座中国近現代史 ; 2) Ja 78-14481
Includes bibliographical references.
¥2400

1. Boxers. 2. China—History—1900. 3. China—History—Kuang-hsü, 1875-1908. I. Takahashi, Kōsuke. II. Series: Kōza Chūgoku kingendai shi ; 2.
DS755.K69 vol. 2 78-813156
[DS771]

—HISTORY—WAR WITH JAPAN, 1894-1895 see Chinese-Japanese War, 1894-1895

—HISTORY—REFORM MOVEMENT, 1898—ADDRESSES, ESSAYS, LECTURES

Chang, Ho-ling, 1867-1908.
(Pien fa ching wei kung li lun)
變法經緯公例論 / 張鶴齡著 ; 沈雲龍主編. — [臺北縣永和鎮] : 文海出版社, [1977?]
212 p. ; 22 cm. — (近代中國史料叢刊續編 [i. e. 輯] ; 472 : 第48輯)
Reprint ed. with pref. dated 1900.

1. China—History—Reform Movement, 1898—Addresses, essays, lectures. I. Title. II. Series: Shen, Yün-lung, comp. Chin tai Chung-kuo shih liao ts'ung k'an hsü chi ; 472.
DS755.S532 vol. 472 78-841810
[DS764]

—HISTORY—20TH CENTURY

Chung-kuo wen hua hsüeh yüan. Chung-kuo hsien tai shih pien chi wei yüan hui.
(Chung-kuo hsien tai shih)
中國現代史 / [編著者中國文化學院中國現代史編輯委員會] ; 中國文化學院編 ; [總編輯魏汝霖]. — 臺北 : 華岡出版有限公司, 民國66 [1977]
2, 1, 20, 254 p. : 22 cm.
Bibliography: p. 249-253.
NT$100.00
1. China—History—20th century. I. Wei, Ju-lin. II. Chung-kuo wen hua hsüeh yüan. III. Title.
DS774.C485 1977 78-843298

—HISTORY—1900

Chang, Yü-fa, 1936-
(Chung-kuo chin tai hsien tai shih)
中國近代現代史 / 張玉法. — 臺北 : 東華書局, 民國67 [1978]
14, 469 p. ; 22 cm.
Bibliography: p. 465-469.
NT$140.00

1. China—History—19th century. 2. China—History—1900- I. Title.
DS755.C354 79-838600

China (Republic of China). Kuang fu ta lu she chi yen chiu wei yüan hui.
(Lieh ch'iang ch'in lüeh wo kuo chih yen chiu)
列強侵略我國之研究 / 光復大陸設計研究委員會研擬. — [台北] : 該會, 民國65 [1976]
3, 32, 4, 48 p. ; 22 cm. — (邊疆研究叢書)
Cover title.
CONTENTS: 近百年來日本與西歐列強侵略我領土損害我主權之研究.—俄國侵略我領土損害我主權之研究.
1. China—Foreign relations. 2. China—History—19th century. 3. China—History—1900- I. Title. II. Series: Pien chiang yen chiu ts'ung shu.
DS740.4.C354518 1976 78-841329

Duiker, William J 1932-
Cultures in collision : the Boxer Rebellion / William J. Duiker. — San Rafael, CA : Presidio Press, c1978.
xvii, 226 p. : ill. ; 22 cm.
Bibliography: p. 219-221.
Includes index.
ISBN 0-89141-028-7 : $12.95
1. Boxers. 2. Nationalism—China. 3. China—History—1900. I. Title.
DS771.D84 951'.03 77-73550
78 MARC

CHINA
—HISTORY—1900- (Continued)

Fu tan ta hsüeh, Shanghai. Li shih hsi. Chung-kuo chin tai shih chiao yen tsu.
(Chung-kuo chin tai chien shih)
中国近代简史 / 复旦大学历史系中国近代史教研组编著. — 第2版. — 上海：上海人民出版社, 1978.
4, 366 p. : ill. ; 19 cm. — (青年自学丛书)
Series romanized: Ch'ing nien tzu hsüeh ts'ung shu.
Includes bibliographical references.
"附录大事年表": p. 351-366.
$0.62
1. China—History—19th century. 2. China—History—1900- I. Title.
DS755.F8 1978 79-836705

(Giwadan undō)
義和団運動 / [執筆者 高橋孝助 ... et al.]. — 東京：東京大学出版会, 1978.
vi, 266 p. ; 21 cm. — (講座中国近現代史 ; 2) Ja 78-14481
Includes bibliographical references.
¥2400
1. Boxers. 2. China—History—1900. 3. China—History—Kuang-hsü, 1875-1908. I. Takahashi, Kōsuke. II. Series: Kōza Chūgoku kingendai shi ; 2.
DS755.K69 vol. 2 78-813156
[DS771]

Histoire de la Chine. Sous la direction de Jean Chesneaux ... Paris, Hatier, 1969-77.
4 v. maps. 24 cm. (Collection d'histoire contemporaine) (Hatier Université) 38.05F (v. 3)
On spine, v. 1; cover title, v. 2-4: La Chine.
Includes bibliographies and indexes.
CONTENTS: 1. Des guerres de l'opium à la guerre franco-chinoise, 1840-1855, par J. Chesneaux et M. Bastid.—2. De la guerre franco-chinoise à la fondation du Parti communiste chinois, 1885-1921, par M. Bastid, M.-C. Bergère, J. Chesneaux.—3. La marche de la Révolution, 1921-1949, par J. Chesneaux et F. Le Barbier.—4. Un nouveau communisme, 1949-1976, par J. Bellassen et al.
ISBN 2-218-03028-4 (v. 3)
1. China—History—19th century. 2. China—History—1900- I. Chesneaux, Jean, ed. II. Title: La Chine.
DS757.5.H57 76-512614
79r79rev2 MARC

Hu-nan shih fan hsüeh yüan. Chin tai Chung-kuo shih hua pien hsieh tsu.
(Chin tai Chung-kuo shih hua)
近代中国史话 / 湖南师范学院《近代中国史话》编写组. — 第1版. — 北京：人民出版社, 1977.
3, 380 p. : ill. ; 19 cm.
$0.75
1. China—History—19th century. 2. China—History—1900- I. Title.
DS755.H87 1977 78-841957

Kalíuzhnaía, Nina Mikhaĭlovna.
Vosstanie ikhètuaneĭ, 1898-1901 / N. M. Kalíuzhnaía. — Moskva : Nauka, 1978.
363 p. ; 21 cm. USSR79
At head of title: Akademiía nauk SSSR. Institut vostokovedeniía.
Bibliography: p. 335-[350]
Includes indexes.
3.20rub
1. China—History—1900. 2. Boxers. I. Title.
DS771.K3 79-379552
*79 MARC

Kawakami, Tadao, 1915-
(Kindai Chūgoku hyakunenshi)
近代中国百年史 / 川上忠雄著. — 東京：高文堂出版社, 1979.
212 p. : maps ; 19 cm. Ja 79-4616
Bibliography: p. 210-212.
Errata slip inserted.
¥1800
1. China—History—19th century. 2. China—History—1900- I. Title.
DS755.K38 79-807173

Nearing, Scott, 1883-
Whither China? : An economic interpretation of recent events in the Far East / by Scott Nearing. — Westport, Conn. : Hyperion Press, 1977, c1927.
225 p. ; 23 cm. — (China studies)
Reprint of the ed. published by International Publishers, New York.
Bibliography: p. 211-214.
Includes index.
ISBN 0-88355-385-6 : $16.50
1. China—Economic conditions—1912-1949. 2. China—Industries. 3. China—Social conditions. 4. China—History—1900- 5. China—Foreign relations. I. Title. II. Series: China studies (Westport, Conn.)
HC427.N38 1977 330.9'51'05 75-39029
77 MARC

Nohara, Shirō, 1903-
(Chūgoku kakumei to Dai Nippon Teikoku)
中国革命と大日本帝国 / 野原四郎著. — 東京：研文出版, 1978.
iii, 282 p. ; 20 cm. — (研文選書 ; 2) Ja 78-24003
Series romanized: Kenbun sensho.
Includes bibliographical references.
¥1800
1. China—History—1900- 2. China—Foreign relations—Japan. 3. Japan—Foreign relations—China. I. Title.
DS774.N63 79-800112

P'êng, Tse-chou, 1926-
(Chūgoku gendai shi)
中国現代史：五四運動から四人組追放まで / 彭澤周著. — 東京：泰流社, 1978.
vii, 352, xv p., [1] fold. leaf of plates : map ; 19 cm. — (泰流選書) Ja 78-18116
Series romanized: Tairyū sensho.
Bibliography: p. 305-320.
Includes indexes.
¥1300
1. China—History—1900-
DS774.P4316 78-814552

Sladkovskiĭ, Mikhail Iosifovich.
Kitaĭ—osnovnye problemy istorii, èkonomiki, ideologii / M. I. Sladkovskiĭ. — Moskva : Mysl', 1978.
300 p. ; 21 cm. USSR***
Includes bibliographical references.
1.20rub
1. China—History—1900- I. Title.
DS774.S57 79-383756
79 MARC

T'ao, Chü-yin, 1898-
(Pei-yang chün fa t'ung chih shih ch'i shih hua)
北洋軍閥統治時期史話 / 陶菊隐著. — 第1版. — 北京：生活·读书·新知三联书店：新华书店发行, 1957-1959, 1978 printing.
8 v. ; 19 cm.
CONTENTS: 第1册. 辛亥革命前后和第一次南北战争时期(一八九五年至一九一三年).—第2册. 洪宪帝制和护国战争时期(一九一三年至一九一六年).—第3册. 督军团叛变和复辟政变时期(一九一六年六月至一九一七年七月).—第4册. 第二次南北战争时期(一九一七年七月至一九一八年十一月).—第5册. 五四运动和直皖战争时期(一九一九年一月至一九二〇年十二月).—第6册. 联省自治运动和第一次奉直战争时期(一九二一年一月至一九二三年六月).—第7册. 第二次奉直战争和五卅运动时期(一九二三年七月至一九二六年四月).—第8册. 北伐战争时期(一九二六年五月至一九二八年十二月)
$3.72
1. China—History—1900- I. Title.
DS774.T332 1957 79-837212

—HISTORY—1900- —ADDRESSES, ESSAYS, LECTURES

China, reform and revolution : background papers on Chinese history, 1850 to 1949 / edited by H. D. M.-h. Chan. — [Palmerston North, N.Z.] : Massey University, Dept. of History, 1978.
iv, 103 p. ; 21 cm. NZ***
Bibliography: p. 92-101.
1. China—History—19th century—Addresses, essays, lectures. 2. China—History—1900- —Addresses, essays, lectures. I. Chan, H. D. M.-h.
DS758.C47 951'.03 79-310525
79 MARC

Liu, Ta-nien.
(Chung-kuo chin tai shih wen t'i)
中国近代史问题 / 刘大年著. — 第2版. — 北京：人民出版社：新華書店发行, 1978.
262 p. ; 21 cm.
First ed. published in 1965 under title: Chung-kuo chin tai shih chu wen t'i.
Includes bibliographical references
$0.64
1. China—History—19th century—Addresses, essays, lectures. 2. China—History—1900- —Addresses, essays, lectures. I. Title.
DS754.L588 1978 79-837278

Wu, Hsiang-hsiang, 1913-
(Li shih yü jen wu)
歷史與人物 / 吳相湘著. — 臺北：東大圖書有限公司：總經銷三民書局, 1978.
2, 3, 551 p. ; 22 cm. — (滄海叢刊 : 歷史)
Series romanized: Ts'ang hai ts'ung k'an : Li shih.
1. China—History—19th century—Addresses, essays, lectures. 2. China—History—1900- —Addresses, essays, lectures. I. Title.
DS755.W82 79-839501

—HISTORY—1900- —COLLECTED WORKS

(Kōza Chūgoku kingendai shi)
講座中国近現代史 / 野沢豊, 田中正俊編集. — 東京：東京大学出版会, 1978-
v. ; 21 cm. Ja 78-11730 (v. 1)
¥2400 per vol.
1. China—History—19th century—Collected works. 2. China—History—1900- —Collected works. I. Nozawa, Yutaka. II. Tanaka, Masatoshi, 1922-
DS755.K69 78-813399

Li, Liu-ju.
(Liu shih nien ti pien ch'ien)
六十年的變遷 / 李六如著. — 香港：廣角鏡出版社, 1977.
2 v. : ill. ; 21 cm.
Vol. 3 was planned but never published.
$22.00
1. Li, Liu-ju, in fiction, drama, poetry, etc. 2. China—History—1900- —Fiction. I. Title.
PL2780.L5L5 1977 78-839938

—HISTORY—1900- —SOURCES

Morrison, George Ernest, 1862-1920.
The correspondence of G. E. Morrison / edited by Lo Hui-Min. — Cambridge ; New York : Cambridge University Press, 1976-1978.
2 v. : ill. ; 24 cm.
Includes indexes.
CONTENTS: 1. 1895-1912.—2. 1912-1920.
ISBN 0-521-20486-0 (v. 1)
1. China—History—1900- —Sources. 2. Morrison, George Ernest, 1862-1920. 3. Journalists—Correspondence. I. Lo, Hui-min.
DS764.M67 951'.03 78-11354
78 MARC

—HISTORY—HSÜAN-T'UNG, 1908-1912—CHRONOLOGY

Chao, Ping-lin, b. 1873.
(Hsüan-t'ung ta shih chien)
宣統大事鑑 / 趙柄麟 [炳麟] 撰. — 台北：廣文書局, 民國67 [1978]
1, 19, 1 leaves ; 22 cm. — (漢學彙編)
"本書承國立中央研究院惠借影印"
NT$140.00
1. China—History—Hsüan-t'ung, 1908-1912—Chronology. I. Title. II. Series: Han hsüeh hui pien.
DS773.C488 1978 79-837382

—HISTORY—REVOLUTION, 1911-1912

China (Republic of China). Kuo fang pu. Shih cheng chü.
(K'ai kuo chan shih)
開國戰史 / 國防部史政局編纂. — 臺2版. — 臺北：正中書局, 民國65 [1976]
2 v. (16, 914, [64] p., [83] leaves of plates (43 fold.)) : ill. ; 22 cm.
1. China—History—Revolution, 1911-1912. 2. China—History—Republic, 1912-1949. I. Title.
DS773.C5144 1976 78-841813

Chŏng, Se-hyŏn.
(Kŭndae Chungguk minjok undongsa yŏn'gu)
近代中國民族運動史研究：孫文의反清革命과會黨 / 鄭世鉉著. — 서울：一志社, 1977.
454 p. ; 24 cm.
Chinese or Korean.
Includes bibliographical references and index.
₩4500
1. China—History—Revolution, 1911-1912. 2. Political parties—China. 3. Secret societies—China. 4. Sun, Yat-sen, 1866-1925. I. Title.
DS773.C519 78-826561

CHINA

—HISTORY—REVOLUTION, 1911-1912 (Continued)

Hsi, Hsien-te.

(Sun Chung-shan hsien sheng yü ko ming ssu hsiang ti ch'uan po)

孫中山先生與革命思想的傳播 / 習賢德著. — ［台北：s. n.］, 民國67 ［1978］

248 p. (on double leaves) ; 27 cm.

"國立台灣大學三民主義研究所碩士論文"
Bibliography: p. 241-248.

1. Sun, Yat-sen, 1866-1925—Views on revolutions. 2. China—History—Revolution, 1911-1912. I. Title.

DS777.H73 78-842563

Huang, Ta-shou.

(Chung-hua min kuo chih ti tsao)

中華民國之締造 / 著者黃大受. — 臺初版. — 臺北：正中書局；九龍：海外總經銷集成圖書公司, 民國66 ［1977］

1, 3, 129 p. ; 19 cm. — (青年理論叢書：第七輯)

Colophon title.
Series romanized: Ch'ing nien li lun ts'ung shu : Ti ch'i chi.

1. China—History—Revolution, 1911-1912. I. Title.

DS773.H84 79-835444

Nakamura, Tadashi, 1929-

(Shingai kakumei shi kenkyū)

辛亥革命史研究 / 中村義著. — 東京：未来社, 1979.

365, 16 p. ; 22 cm. Ja 79-4617

Caption title: A study of the 1911 revolution in China.
Includes bibliographical references and index.
¥4500

1. China—History—Revolution, 1911-1912. I. Title. II. Title: A study of the 1911 revolution in China.

DS773.N32 79-807196

Sakamaki, Teiichirō.

(Shina bunkatsu ron)

支那分割論：附袁世凱 / ［著者 酒巻貞一郎］. — 東京：啓成社, 大正2 ［1913］

584, 204, ［35］ p., ［5］ leaves of plates : maps ; 23 cm.

1. China—History—1912-1937. 2. China—Foreign relations—1912-1949. 3. China—History—Revolution, 1911-1912. 4. Revolutions—China. 4. Yüan, Shih-k'ai, 1859-1916. 6. China—Presidents—Biography. I. Title. II. Sakamaki, Teiichirō. En Seigai. 1913.

DS774.S332 79-809704

(Shingai kakumei)

辛亥革命 / ［執筆者 久保田文次 ... et al.］. — 東京：東京大学出版会, 1978.

vi, 288 p. ; 21 cm. — (講座中国近現代史；3) Ja 78-17033

Includes bibliographical references.
¥2400

1. China—History—Revolution, 1911-1912. I. Kubota, Bunji, 1936- II. Series: Kōza Chūgoku kingendai shi ; 3.

DS755.K69 vol. 3 78-813180

Yang, Yü-ju, 1877?-

(Hsin hai ko ming hsien chu chi)

辛亥革命先著記 / 楊玉如編. — 重版. — 香港：文化資料供應社, 1978.

2, 285 p. ; 21 cm.
$8.50

1. China—History—Revolution, 1911-1912. I. Title.

DS773.Y3 1978 79-835352

—HISTORY—REVOLUTION, 1911-1912—ADDRESSES, ESSAYS, LECTURES

(Kung-ch'i T'ao-t'ien lun Sun Chung-shan yü Huang Hsing)

宮崎滔天論孫中山與黃興 / 譯著者陳鵬仁. — 臺初版. — 臺北：正中書局, 民國66 ［1977］

270, ［6］ p., ［2］ leaves of plates : ill. ; 19 cm. — (國民文庫；13)

Colophon title.
Series romanized: Kuo min wen k'u.
Includes bibliographical references.

1. Sun, Yat-sen, 1866-1925—Addresses, essays, lectures. 2. Huang, Hsing, 1874-1916—Addresses, essays, lectures. 3. Miyazaki, Tōten, 1870-1922—Addresses, essays, lectures. 4. China—History—Revolution, 1911-1912—Addresses, essays, lectures. I. Miyazaki, Tōten, 1870-1922. II. Ch'en, P'en-jen.

DS777.A595K84 78-843783

(Kung-ch'i T'ao-t'ien yü Chung-kuo ko ming)

宮崎滔天與中國革命 / 陳鵬仁譯著. — 高雄：三信出版社, 民國66 ［1977］

1, 266, ［2］ p., ［4］ leaves of plates : ill. ; 19 cm.

"陳鵬仁其他著書及譯書": p. ［267］-［268］
CONTENTS: 彭澤周 宮崎滔天與中國革命.—宮崎槌子 我對於中國革命的回憶.—陳鵬仁 關於「宮崎滔天全集」第一卷.—陳鵬仁 關於「宮崎滔天全集」第二卷.—陳鵬仁 關於「宮崎滔天全集」第三卷.—陳鵬仁 關於「宮崎滔天全集」第四卷.—陳鵬仁 關於「宮崎滔天全集」第五卷.—武田範之 滔天無賜二士斷髮記.—近藤秀樹 宮崎滔天年譜.—孫中山 國父給宮崎滔天的序函.

1. Miyazaki, Tōten, 1870-1922—Addresses, essays, lectures. 2. Politicians—Japan—Biography—Addresses, essays, lectures. 3. China—History—Revolution, 1911-1912—Addresses, essays, lectures. I. Ch'en, P'eng-jen. II. P'eng, Tse-chou, 1926-

DS884.M5K86 78-842737

—HISTORY—1912-1937

Ajia Seikei Gakkai.

(Kōgun sōkenki no Mō Takutō to Shū Onrai)

紅軍創建期の毛沢東と周恩来：立三路線論考 / ［編集責任者 蜂屋亮子］. — 東京：アジア政経学会：発売所 巌南堂書店, 1978.

6, 190 p. ; 21 cm. — (現代中国研究叢書；15) Ja 78-7624

Includes bibliographical references.
¥1000

1. China—History—1912-1937. 2. Chung-kuo kung ch'an tang—History. 3. Chung-kuo kung nung hung chün—History. 4. Mao, Tse-tung, 1893-1976. 5. Chou, En-lai, 1898-1976. I. Hachiya, Akiko. II. Title. III. Series: Gendai Chūgoku kenkyū sōsho ; 15.

DS777.55.G43 vol. 15 78-816766
[DS777.47]

Ch'en, Shih-ch'ang.

(Chung-kuo yü men hu k'an fang)

中國與門戶開放：清末民初外患史 / 陳世昌著. — 台北：海國書局, 民國66 ［1977］

11, 420 p. ; 22 cm.

At head of title: Ta chuan yung shu.
Title on spine: Ch'ing mo min ch'u wai huan shih.
Bibliography: p. 361-367.
NT$140.00

1. China—Foreign relations. 2. China—History—Ch'ing dynasty, 1644-1912. 3. China—History—1912-1937. I. Title. II. Title: Ch'ing mo min ch'u wai huan shih.

DS740.4.C35338 78-841899

Ch'en, Wei-lun, 1897-

(Huang K'o-ch'iang hsien sheng chuan chi)

黃克強先生傳記 / 陳維綸撰；［編輯者中國國民黨中央委員會黨史委員會］. — 臺北：經銷處中央文物供應社, 民國62 ［1973］

2, 4, 318 p., ［4］ leaves of plates : ill. ; 22 cm.

Includes bibliographical references.

1. Huang, Hsing, 1874-1916. 2. Statesmen—China—Biography. 3. China—History—Revolution, 1911-1912. 4. China—Politics and government—1912-1937. I. Chung-kuo kuo min tang. Chung yang wei yüan hui. Tang shih wei yüan hui. II. Title.

DS778.H85C47 78-842983

Sakamaki, Teiichirō.

(Shina bunkatsu ron)

支那分割論：附袁世凱 / ［著者 酒巻貞一郎］. — 東京：啓成社, 大正2 ［1913］

584, 204, ［35］ p., ［5］ leaves of plates : maps ; 23 cm.

1. China—History—1912-1937. 2. China—Foreign relations—1912-1949. 3. China—History—Revolution, 1911-1912. 4. Revolutions—China. 4. Yüan, Shih-k'ai, 1859-1916. 6. China—Presidents—Biography. I. Title. II. Sakamaki, Teiichirō. En Seigai. 1913.

DS774.S332 79-809704

Sasaki, Tōichi, 1886-1955.

(Nampō kakumei seiryoku no jissō to sono hihan)

南方革命勢力の實相と其の批判 / 佐々木到一著. — ［s. l. : s. n., 自序 昭和2 i. e. 1927］

2, 4, 8, 418+ p., ［2］ leaves of plates : ill. ; 19 cm.

LC copy imperfect: all after p. 418 wanting.

1. China—History—1912-1937. 2. China—Description and travel—1901-1948. 3. Sasaki, Tōichi, 1886-1955. I. Title.

DS777.44.S27 79-804155

—HISTORY—1912-1937—ADDRESSES, ESSAYS, LECTURES

(Lien ho chan hsien lun)

聯合戰綫論 / 謝漢夫, 吳敏. — ［上海：讀書生活出版社］, 1936.

2, 70 p. ; 18 cm.

CONTENTS: 漢夫 聯合戰線論.—吳培敏 怎樣在中國建立救亡聯合戰線

1. China—History—1912-1937—Addresses, essays, lectures. I. Han-fu. II. Wu, Min. Tsen yang tsai Chung-kuo chien li chiu wang lien ho chan hsien. 1936.

DS777.47.L48 78-843801

—HISTORY—1912-1937—ANECDOTES, FACETIAE, SATIRE, ETC.

T'ao, Chü-yin, 1898-

(Tu chün t'uan chuan)

督軍團傳：菊隱叢譚 / 陶菊隱著. — ［香港?：s. n., 1976?］

4, 270 p. ; 19 cm.

Reprint of the 1948 ed. published by Chung-hua shu chü, Shanghai.

1. China—History—1912-1937—Anecdotes, facetiae, satire, etc. I. Title.

DS774.T333 1976 78-843146

—HISTORY—1912-1937—BIOGRAPHY

Fan, Yin-nan.

(Tang tai Chung-kuo ssu ch'ien ming jen lu)

當代中國四千名人錄 / 編纂者樊蔭南. — Hong Kong：波文書局, 1978.

9, 460 p. ; 20 cm.

Title in English on verso of t. p.: Who's who in contemporary China (Four thousand persons)
"據1936年版增訂版重印"
Edition of 1931 published under title: Tang tai Chung-kuo ming jen lu.
1. China—History—1912-1937—Biography. 2. China—Biography. I. Title. II. Title: Who's who in contemporary China.

DS778.A1F36 1978 79-835014

—HISTORY—1912-1937—SOURCES

(I chiu i ch'i nien ting ssu Ch'ing ti fu p'i shih liao hui chi)

1917年丁巳清帝復辟史料彙輯 / 存萃學社編集. — 香港：大東圖書公司, 1977.

224 p., ［3］ leaves of plates : ill., ports. ; 21 cm. — (中國近代史資料叢編；3)

Photoreprint ed.
$18.00 (U. S.)

1. China—History—1912-1937—Sources. I. Ts'un ts'ui hsüeh she. II. Series: Chung-kuo chin tai shih tzu liao ts'ung pien ; 3.

DS777.4.I18 1977 79-837308

—HISTORY—REPUBLIC, 1912-1949

China (Republic of China). Kuo fang pu. Shih cheng chü.

(K'ai kuo chan shih)

開國戰史 / 國防部史政局編纂. — 臺2版. — 臺北：正中書局, 民國65 ［1976］

2 v. (16, 914, ［64］ p., ［83］ leaves of plates (43 fold.)) : ill. ; 22 cm.

1. China—History—Revolution, 1911-1912. 2. China—History—Republic, 1912-1949. I. Title.

DS773.C5144 1976 78-841813

China from the 1911 revolution to liberation / by Jean Chesneaux, Françoise Le Barbier, and Marie-Claire Bergère ; translated from the French by Paul Auster and Lydia Davis, chapter 1 to 3 translated by Anne Destenay. — 1st American ed. — New York : Pantheon Books, c1977.

vi, 372 p. : maps ; 25 cm. — (The Pantheon Asia library)
Translation of chapters 4-6 of De la guerre franco-chinoise à la fondation du parti communiste chinois, 1885-1921, by J. Chesneaux, M. Bastid, and M. C. Bergère, v. 2 of Histoire de la Chine; and Le Chine, la marche de la Révolution, 1921-1949, by Chesneaux and F. Le Barbier, v. 3 of the Histoire.
Includes bibliographies and index.
ISBN 0-394-41193-5 : $17.95. ISBN 0-394-73332-0 pbk. : $6.95
1. China—History—Republic, 1912-1949. I. Chesneaux, Jean. II. Le Barbier, Françoise. III. Bergère, Marie Claire.

DS774.H47213 1977 951.04 77-76494
77 MARC

CHINA

—HISTORY—REPUBLIC, 1912-1949 (Continued)

(Chūgoku kakumei no tenkai)
中国革命の展開 / ［執筆者 姫田光義 ... et al.］. — 東京 : 東京大学出版会, 1978.
vi, 319 p. ; 21 cm. — (講座中国近現代史 ; 5) Ja***
Includes bibliographical references.
¥2400

1. China—History—Republic, 1912-1949. I. Himeta, Mitsuyoshi, 1937- II. Series: Kōza Chūgoku kingendai shi ; 5.
DS755.K69 vol. 5 78-815037
[DS774]

Fang, Fei.
(Hsin Chung-kuo shih tsen yang tan sheng ti)
新中國是怎樣誕生的 / 方非編著. — 香港 : 新知識出版社, 1977-
v. : ill. ; 19 cm.
$4.00 (v. 1)

1. China—History—Republic, 1912-1949. 2. Communism—China. I. Title.
DS774.F35 78-843566

Fuse, Katsuji.
(Shina kokumin kakumei to Hyō Gyokushō)
支那國民革命と馮玉祥 / 布施勝治著. — 東京 : 大阪屋號書店, 昭和4 ［1929］
5, 500 p., ［21］ leaves of plates : ill. ; 20 cm.

1. Feng, Yü-hsiang, 1882-1948. 2. Generals—China—Biography. 3. China—History—Republic, 1912-1949. I. Title.
DS778.F45F87 78-814186

Great Britain. Schools Council. History 13-16 Project.
The rise of Communist China / ［Project team David Sylvester et al.］. — ［Student's book］. — Edinburgh : Holmes McDougall, 1977.
88 p. : ill., col. maps, ports. ; 25 cm. — (Modern world studies) GB77-21638
ISBN 0-7157-1558-5 : £2.05
1. China—History—1949-1976. 2. China—History—Republic, 1912-1949. I. Title. II. Series.
DS777.55.G685 1977 951.04 78-314102
78 MARC

Guillermaz, Jacques.
La Chine populaire / Jacques Guillermaz, ... — 7e éd. mise à jour. — ［Paris］ : PUF, 1977.
126, ［2］ p. : maps ; 18 cm. — (Que sais-je? ; 840) F78-1300
Bibliography: p. ［127］
9.00F
1. China—History—1949-1976. 2. China—History—Republic, 1912-1949. I. Title.
DS777.55.G833 1977 320.9'51'04 79-337042
79 MARC

Ikeguchi, Ekan, 1936-
(Shō Kaiseki)
蔣介石 : その偉大なる生涯 / 池口恵観. — 東京 : 論稿社, 昭和51 ［1976］
215 p. : ill. ; 20 cm. Ja***
¥1200

1. Chiang, Kai-shek, 1887-1975. 2. China—Presidents—Biography. 3. China—History—Republic, 1912-1949. 4. Japan—Foreign relations—China. 5. China—Foreign relations—Japan. I. Title.
DS778.C55I33 78-814965

Keenan, Barry, 1940-
The Dewey experiment in China : educational reform and political power in the early Republic / by Barry Keenan. — Cambridge, Mass. : Council on East Asian Studies, Harvard University : distributed by Harvard University Press, 1977.
xii, 335 p. ; 24 cm. — (Harvard East Asian monographs ; 81)
Bibliography: p. 289-325.
Includes index.
ISBN 0-674-20277-5
1. Dewey, John, 1859-1952. 2. Education—China—History. 3. China—History—Republic, 1912-1949. I. Title. II. Series.
LB875.D5K4 370'.951 77-6682
77［r79］rev MARC

Li, Tsung-jen, 1890-1969.
The memoirs of Li Tsung-jen / Te-kong Tong and Li Tsung-jen. — Boulder, Colo. : Westview Press, 1979.
xxviii, 642 p. ; 24 cm. — (Studies of the East Asian Institute of Columbia University)
Includes index.
ISBN 0-89158-343-2
1. Li, Tsung-jen, 1890-1969. 2. Heads of state—China—Biography. 3. China—History—Republic, 1912-1949. I. Tong, Te-Kong, 1920- . II. Title. III. Series: Columbia University. East Asian Institute. Studies.
DS778.L416A35 1979 951.04'2'0924 78-60155
78 MARC

Wang, Anna, 1907-
［Ich kämpfte für Mao. Chinese］
嫁給革命的中國 / 王安娜著 ; 龍翔譯. — 香港 : 廣角鏡出版社, 1978.
4, 14, 454 p., ［4］ leaves of plates : ill. ; 23 cm.
$14.00

1. Sino-Japanese Conflict, 1937-1945—Personal narratives. 2. Wang, Anna, 1907- 3. China—History—Republic, 1912-1949. I. Title.
DS777.53.W317312 78-843527

—HISTORY—REPUBLIC, 1912-1949—ADDRESSES, ESSAYS, LECTURES

Li, Hsin.
(Chung-kuo hsin min chu chu i ko ming shih chiang hua)
中国新民主主义革命史讲话 / 李新. — ［香港? : s.n., 1978?］
132 p. : maps ; 21 cm.
Reprint of the 1977 ed. published by Kuang tung jen min ch'u pan she.

1. China—History—Republic, 1912-1949—Addresses, essays, lectures. 2. China—History—Civil War, 1945-1949—Addresses, essays, lectures. 3. Communism—China—Addresses, essays, lectures. I. Title.
DS774.L436 1978 78-843259
C

Nakahira, Ryō.
(Shina wa dō naru)
支那はどうなる / 中平亮著. — 東京 : 東洋研究會, 昭和7 ［1942］
ii, 149 p. ; 19 cm. — (His 東洋研究叢書 ; 3)
Series romanized: Tōyō kenkyū sōsho.

1. China—History—19th century—Addresses, essays, lectures. 2. China—History—1912-1949—Addresses, essays, lectures. 3. China—Foreign economic relations—Addresses, essays, lectures. I. Title.
DS755.N54 79-808145

(Revolûtsiíà tysíàcha devíàt'sot dvadtsat' píàtogo—tysíàcha devíàt'sot dvadtsat' sed'mogo gg. v Kitae)
Революция 1925-1927 гг. в Китае / ［отв. редактор Л. П. Делюсин］. — Москва : Наука, 1978.
236 p. ; 22 cm. USSR***
At head of title: Akademiíà nauk SSSR. Institut vostokovedeniíà.
Includes bibliographical references.
1.70rub
1. China—History—Republic, 1912-1949—Addresses, essays, lectures. I. Delíusin, Lev Petrovich. II. Akademiíà nauk SSSR. Institut vostokovedeniíà.
DS777.45.R49 79-411683

Ssu-ma, Ch'ang-feng.
(Chung-kuo chin tai shih chi yao)
中國近代史輯要 / 司馬長風著. — 香港 : 創作書社, 1977.
4, 275 p. ; 21 cm.
Includes bibliographical references.
$12.00

1. China—History—Republic, 1912-1949—Addresses, essays, lectures. 2. China—History—Republic, 1912-1949—Historiography—Addresses, essays, lectures. I. Title.
DS774.S6 78-840860

—HISTORY—REPUBLIC, 1912-1949—ANECDOTES, FACETIAE, SATIRE, ETC.

(Chin tai Chung-kuo cheng t'an ku shih)
近代中國政壇故事. — ［台北 : 新萬象月刊社］, 民國67- ［1978-
v. ; 20 cm. — (新萬象月刊叢書 ; 2)
Cover title: Chung-kuo cheng t'an ku shih.
"名家聯合執筆"
Imprint stamped at the end of text.
Series romanized: Hsin wan hsiang yüeh k'an ts'ung shu.
NT$50.00 (v. 1)
1. China—History—Republic, 1912-1949—Anecdotes, facetiae, satire, etc. I. Title: Chung-kuo cheng t'an ku shih.
DS774.C443 79-835127

I, Tsung-k'uei, 1875-
(Hsin shih shuo)
新世說 : ［8卷］ / 易宗夔撰述. — 台北 : 廣文書局, 民國66 ［1977］
2 v. (634, ［14］ p.) ; 22 cm. — (國學珍籍彙編)
Reprint of the 1922 printing.
"本書承國立中央研究院惠借影印"
NT$410.00
1. China—History—Ch'ing dynasty, 1644-1912—Anecdotes, facetiae, satire, etc. 2. China—History—Republic, 1912-1949—Anecdotes, facetiae, satire, etc. I. Title. II. Series: Kuo hsüeh chen chi hui pien.
DS754.I443 1977 79-836358

Kuo, T'ung.
(Ch'ien shih hsin p'ing)
前事新評 : 張作霖, 西安事變及其他 / 郭桐著. — 香港 : 七十年代雜誌社, 1977.
191 p., ［4］ leaves of plates : ill. ; 19 cm.
$5.00

1. China—History—Republic, 1912-1949—Anecdotes, facetiae, satire, etc. I. Title.
DS777.47.K86 78-840832

T'ang, Lu-sun.
(Ku yüan ch'ing)
故園情 / 唐魯孫著. — 台北 : 時報文化出版事業有限公司, 民國67 ［1978］
4, 310 p., ［3］ leaves of plates : ill. ; 20 cm. — (時報書系 ; 144)
Series romanized: Shih pao shu hsi.
NT$70.00

1. China—History—Republic, 1912-1949—Anecdotes, facetiae, satire, etc. 2. Restaurants, lunch rooms, etc.—China. 3. Cookery, Chinese. I. Title.
DS775.T34 79-840231

Tso, Shun-sheng, 1883-1969.
(Chung-kuo hsien tai ming jen i shih)
中國現代名人軼事 / 左舜生著. — 再版. — 九龍 : 自由出版社 ; 香港 : 總經售處福利書店, 民國41 ［1952］
2, 4, 110 p. ; 18 cm.

1. China—History—Republic, 1912-1949—Anecdotes, facetiae, satire, etc. I. Title.
DS775.T782 1952 79-837406

—HISTORY—REPUBLIC, 1912-1949—BIOGRAPHY

Biographical dictionary of Republican China. Howard L. Boorman, editor; Richard C. Howard, associate editor. New York, Columbia University Press, 1967-79.
5 v. map. 28 cm. $25.00 (v. 3)
Bibliography: v. 4, p. ［97］-418, by J. K. H. Cheng.
Vol. 5: "A personal name index, by Janet Krompart."
ISBN 0-231-08957-0 (v. 3)
1. China—History—Republic, 1912-1949—Biography. 2. China—Biography. I. Boorman, Howard L., ed. II. Howard, Richard C., ed. III. Cheng, Joseph K. H.
DS778.A1B5 920'.051 67-12006
［r79］rev2 MARC

(Kuo kung feng yün ming jen lu)
國共風雲名人錄 / 李約等著. — 香港 : 廣角鏡出版社, 19
v. : ill. ; 19 cm. — (廣角鏡人物叢書)
Series romanized: Kuang chiao ching jen wu ts'ung shu.
$5.50 (v. 3)

1. China—History—Republic, 1912-1949—Biography. 2. China—History—1949-1976—Biography. 3. China—Biography. I. Li, Yüeh.
DS778.A1K87 78-842847
78 C

Mamo, David, 1948-
Dizionario storico-biografico della Cina moderna / ［di］ David Mamo, Laurie Upson. — Firenze : Vallecchi, 1977.
xxvii, 458 p. ; 20 cm. — (Il Pellicano : Storia) It78-Sept
Bibliography: p. ［449］-455.
Includes index.
L8000
1. China—History—Republic, 1912-1949—Biography. 2. China—History—1949-1976—Biography. 3. China—Biography. I. Title.
DS778.A1M35 920'.051 78-396318
*79 MARC

CHINA

—HISTORY—REPUBLIC, 1912-1949
—BIOGRAPHY (Continued)

Wang, Shih-chen.
(Min kuo jen hao hua chuan)
民國人豪畫傳 / 王世禎主編. — 台南：青溪出版社：總經銷東風圖書供應社, 民國66 [1977]
328, 12 p. : ill. ; 22 cm. — (東風新刊 ; 2)
Cover title varies slightly.
Reprint ed.?
Series romanized: Tung feng hsin k'an.
English and/or Chinese.
NT$80.00
1. China—History—Republic, 1912-1949—Biography. 2. China—Biography. I. Title.
DS778.A1W329 1977 79-835035

Who's who in China. 6th ed. Shanghai, China Weekly Review [1950]
263 p. ports. 25 cm.
Title also in Chinese.
1. China—History—Republic, 1912-1949—Biography. 2. China—Biography. I. China monthly review.
DS778.A1W443 1950 920'.051 51-1089
[r79]rev MARC

Who's who in China ... containing the pictures and biographies of China's best known political, financial, business and pofessional men ... 3d ed. Shanghai, The China weekly review [1926]
2 p.l., 972 [1], 1-25, 11 p. illus. (ports.) 24 cm.
"Practically all of these pictures and biographies have appeared originally in the pages of the China weekly review during the past eight years."—Pref., signed M. C. [i.e. John B.] Powell, editor.
"All of these biographical sketches have appeared in the pages of the China weekly review since January 1927."-Pref., signed J. B. P. [i.e. John B. Powell]
1. China—History—Republic, 1912-1949—Biography. 2. China—Biography. I. China monthly review. II. Powell, John Benjamin, 1888- ed.
DS778.A1W443 1926 26-24595
[r79]rev2 MARC

Who's who in China, containing the pictures and biographies of some of China's political, financial, business and professional leaders. 2d ed. Shanghai, Millard's review [1920]
2 p.l., 314, iv p. illus. (ports.) 26 cm.
"The material incorporated in this book has been reproduced practically without change from the pages of Millard's Review during the past three years."—Pref. signed M. C. [i.e. John B.] Powell, H. K. Tong, editors.
1. China—History—Republic, 1912-1949—Biography. 2. China—Biography. I. Millard's review of the Far East. II. Powell, John Benjamin, 1888-, ed. III. Tong, Hollington Kong, 1887-1971, joint ed.
DS778.A1W443 1920 21-10280
[r79]rev MARC

—HISTORY—REPUBLIC, 1912-1949—HISTORIOGRAPHY —ADDRESSES, ESSAYS, LECTURES

Ssu-ma, Ch'ang-feng.
(Chung-kuo chin tai shih chi yao)
中國近代史輯要 / 司馬長風著. — 香港：創作書社, 1977.
4, 275 p. ; 21 cm.
Includes bibliographical references.
$12.00
1. China—History—Republic, 1912-1949—Addresses, essays, lectures. 2. China—History—Republic, 1912-1949—Historiography—Addresses, essays, lectures. I. Title.
DS774.S6 78-840860

—HISTORY—REPUBLIC, 1912-1949—PERIODICALS

Chinese Republican studies newsletter. v. 1- Oct. 1975-
Granville, Ohio [etc., s. n.]
v. 28 cm. semiannual.
Chinese, English, or Japanese.
1. China—History—Republic, 1912-1949—Periodicals.
DS774.C442 951'.005 79-646167
MARC-S

—HISTORY—TSINAN INCIDENT, 1928

(Chi-nan wu san ts'an an)
濟南五三慘案 / 蔣永敬編. — 臺北：正中書局, 民國67 [1978]
4, 8, 8, 368 p. : ill. ; 21 cm. — (中國現代史史料選輯)
1. China—History—Tsinan Incident, 1928. I. Chiang, Yung-ching. II. Series: Chung-kuo hsien tai shih shih liao hsüan chi.
DS777.49.C48 79-841446

—HISTORY—SIAN INCIDENT, 1936—ADDRESSES, ESSAYS, LECTURES

Kuo, Tseng-k'ai.
(Hsi-an shih pien nei mu)
西安事変内幕 / 郭增愷著. — [香港? : s. n., 1972]
[35] leaves ; 22 x 31 cm.
Cover title.
Publication date stamped on p. [4] of cover.
Each page represents 2 leaves of the original.
A collection of articles reprinted from various sources.
"夏遇安. 西安事變秘聞拾遺": leaves [31]-[33]
1. China—History—Sian Incident, 1936—Addresses, essays, lectures. I. Hsia, Yü-an. Hsi-an shih pien mi wen shih i. 1972 II. Title.
DS777.47.K85 78-841901

Lo, Jui-ch'ing, 1906-
(Hsi-an shih pien ho Chou En-lai t'ung chih)
西安事変和周恩来同志 / 罗瑞卿, 吕正操, 王炳南. — 第1版. — 北京：人民出版社：新華書店发行, 1978.
2, 101 p., [3] leaves of plates : ill. ; 21 cm.
Errata slip inserted.
$0.31
1. China—History—Sian Incident, 1936—Addresses, essays, lectures. 2. Chou, En-lai, 1898-1976.—Addresses, essays, lectures. I. Lü, Cheng-ts'ao. II. Wang, Ping-nan. III. Title.
DS777.47.L6 79-839135

Lu, Pi.
(Lun Chang Hsüeh-liang)
論張學良 / 魯泌著. — Hong Kong : 神州圖書公司, [1978?]
7, 132 p. ; 19 cm.
Reprint of the 1948 ed. published by Shih tai p'i p'an she, Hong Kong, in series: Shih tai ts'ung shu : Jen wen tsu.
1. Chang, Hsüeh-liang, 1898- —Addresses, essays, lectures. 2. China—History—Sian Incident, 1936—Addresses, essays, lectures. 3. China—Politics and government—1937-1949—Addresses, essays, lectures. I. Title.
DS778.C495L8 1978 78-842246

—HISTORY—1937-1945

see also Sino-Japanese Conflict, 1937-1945

Kashima, Sōjirō, 1904-
(Chūgoku no saininshiki)
中國の再認識 / 吉田東祐[鹿島宗二郎]著. — [上海?]：東方文化編譯館, 民國34 [1945]
2, 153 p. ; 19 cm.
1. China—History—1937-1945. 2. Nationalism—China. I. Title.
DS777.53.K2938 78-815432

—HISTORY—1937-1945—ANECDOTES, FACETIAE, SATIRE, ETC.

Jung-ma-shu-sheng.
(Ai chiang nan)
哀江南 退職記者[戎馬書生]著 九龍 振華出版社 1960-64.
8 v. 19 cm. (中國近代逸史)
Colophon title.
Series romanized: Chung-kuo chin tai i shih.
1. China—History—1937-1945—Anecdotes, facetiae, satire, etc. I. Title.
DS777.53.J86 1960 C 61-1570 rev

—HISTORY—1945-

Arai, Takeo, 1919-
(Ka Kokuhō den)
華国鋒伝 / 新井宝雄著. — 東京：青年出版社, 1978.
341 p., [2] leaves of plates : ill. ; 20 cm. Ja 78-15779
¥2800
1. Hua, Kuo-feng. 2. Heads of state—China—Biography. 3. China—History—1937-1945. 4. China—History—1945- I. Title.
DS779.29.H8A8 78-812874

—HISTORY—CIVIL WAR, 1945-1949

Borisov, Oleg Borisovich.
The Soviet Union and the Manchurian revolutionary base : (1945-1949) / O. Borisov ; [translated from the Russian by David Fidlon]. — Moscow : Progress Publishers, c1977.
275 p. ; 18 cm. USSR***
Translation of Sovetskiĭ Soiūz i Man'chzhurskaiā revoliūtsionnaiā baza.
Includes bibliographical references.
1. Manchuria—History—1945- . 2. China—History—Civil War, 1945-1949. 3. Russia—Foreign relations—China. 4. China—Foreign relations—Russia. I. Title.
DS784.2.B6713 951.04'2 78-380143
78 MARC

(Chou-shan ch'e t'ui nien ch'i chou nien chi nien t'e k'an)
舟山撤退廿七週年紀念特刊 / 編輯人張行周. — 臺北：舟山同鄉會, 民國66 [1977]
3, 170 p. : ill. ; 22 cm.
Colophon title.
1. Chou-shan Islands, China—History. 2. China—History—Civil War, 1945-1949. I. Chang, Hsing-chou.
DS793.C555C47 1977 79-836121

(Chūgoku kakumei no shōri)
中国革命の勝利 / [執筆者 吉沢南 ... et al.]. — 東京：東京大学出版会, 1978.
v, 305, 6 p. ; 21 cm. — (講座中国近現代史 ; 7) Ja 78-27051
Includes bibliographical references.
¥2400
1. China—History—Civil War, 1945-1949. 2. China—History—1949-1976. I. Yoshizawa, Minami. II. Series: Kōza Chūgoku kingendai shi ; 7.
DS755.K69 vol. 7 79-801655
[DS777.54]

Hung yen ko ming chi nien kuan. Pien hsieh tsu.
(Mao chu hsi fu Ch'ung-ch'ing t'an p'an)
毛主席赴重庆谈判 / 红岩革命纪念馆编写组. — [香港? : s. n., 1978?]
101 p., [4] leaves of plates : ill. ; 22 cm.
Title in Pinyin on cover: Maozhuxi fu Chongqingtanpan.
Reprint of the 1978 ed. published by Ssu-ch'uan jen min ch'u pan she, Chengtu.
1. China—History—Civil War, 1945-1949. 2. Mao, Tse-tung, 1893-1976. I. Title. II. Title: Maozhuxi fu Chongqingtanpan.
DS777.54.H847 1978 79-835702

May, Gary, 1944-
China scapegoat, the diplomatic ordeal of John Carter Vincent / Gary May ; introd. by John K. Fairbank. — Washington : New Republic Books, 1979.
370 p., [12] leaves of plates : ill. ; 24 cm.
Bibliography: p. 295-305.
Includes index.
ISBN 0-915220-49-0 : $15.95
1. Vincent, John Carter, 1900- . 2. China—History—Civil War, 1945-1949. 3. United States—Foreign relations—China. 4. China—Foreign relations—United States. 5. Diplomats—United States—Biography. I. Title.
E748.V564M39 327'.2'0973 79-4129
79 MARC

—HISTORY—CIVIL WAR, 1945-1949—ADDRESSES, ESSAYS, LECTURES

(Ch'uan K'ang lun hsien ching kuo)
川康淪陷經過 / 周開慶編著. — 臺北縣新店鎮：四川文獻研究社, 民國61 [1972]
2, 2, 168 p. ; 18 cm. — (四川文獻研究社叢書)
Series romanized: Ssu-ch'uan wen hsien yen chiu she ts'ung shu.
CONTENTS: 周開慶 川康淪陷經過.—蔣經國 危急存亡之秋.—陶希聖 危難中之奮鬥.—袁暌九 大陸棄守前總統留渝記.—楊森 在渝反共苦鬥記.—孫震 回顧三十八年.—川康最後苦鬥記：節錄胡宗南上將年譜.—盛文 成都保衛戰前後.—鍾樹楠 成都淪陷目擊記.—賀國光 西康淪陷前後.—周君亮 濵變身歷記.
1. Szechwan, China—History—Addresses, essays, lectures. 2. Sikang, China—History—Addresses, essays, lectures. 3. China—History—Civil War, 1945-1949—Addresses, essays, lectures. I. Chou, K'ai-ch'ing.
DS793.S8C66 78-842589

(K'an luan yü chien kuo)
戡亂与建國. — [南京? : s. n., 194-?]
4, 164 p. ; 18 cm.
Cover title.
1. China—History—Civil War, 1945-1949—Addresses, essays, lectures.
DS777.54.K27 78-843145

CHINA

—HISTORY—CIVIL WAR, 1945-1949
—ADDRESSES, ESSAYS, LECTURES
(Continued)

Li, Hsin.
(Chung-kuo hsin min chu chu i ko ming shih chiang hua)
中国新民主主义革命史讲话 / 李新. — [香港?: s. n., 1978?]
132 p. : maps ; 21 cm.
Reprint of the 1977 ed. published by Kuang tung jen min ch'u pan she.

1. China—History—Republic, 1912-1949—Addresses, essays, lectures. 2. China—History—Civil War, 1945-1949—Addresses, essays, lectures. 3. Communism—China—Addresses, essays, lectures. I. Title.

DS774.L436 1978 78-843259

Li, Wei-ch'en.
(Chung-kuo chü shih ti pi jan fa chan)
中國局勢的必然發展 / 李微塵著. — 九龍: 自由出版社, 民國39 [1950]
84 p. ; 19 cm.

1. China—Civil War, 1945-1949—Addresses, essays, lectures. 2. Korean War, 1950-1953—Addresses, essays, lectures. I. Title.

DS777.54.L522 78-840588

Lung, Fei-hu.
(Hsi-pei kao yüan shuai ch'i p'iao)
西北高原帅旗飘 / 龙飞虎. — [香港?: s. n., 1979?]
156 p., [6] leaves of plates : ill. ; 21 cm.
Reprint of the 1978 ed. published by Chieh fang chün wen i she, Peking.

1. China—History—Civil War, 1945-1949—Addresses, essays, lectures. I. Title.

DS777.54.L87 1979 79-841117

—HISTORY—CIVIL WAR, 1945-1949
—ATROCITIES

(Ju tz'u Chung Mei t'e chung chi shu ho tso so)
如此中美特種技術合作所 = The so-called Sino-American Cooperative Institute for Special Technology / 重慶市各界追悼楊虎城將軍暨被難烈士籌備委員會編. Washington : Center for Chinese Research Materials, Association of Research Libraries, 1979.
1, 84 p. on [43] leaves, [1] leaf of plates : ill. ; 29 cm.
Title from caption in preface.
"蔣美特務重慶大屠殺之血錄"
Text in Chinese.
Reprint of the 1950 ed. with added preface in English by the CRM, ARL, Washington, D. C.
$9.40
1. China—History—Civil War, 1945-1949—Atrocities. 2. Communists—China—Biography. 3. Yang, Hu-ch'eng, 1893-1949. 4. Secret service—China. 5. Intelligence service—United States. I. Ch'ung-ch'ing shih ko chieh chui tao Yang Hu-ch'eng chiang chün chi pei nan lieh shih ch'ou pei wei yüan hui. II. Title: So-called Sino-American Cooperative Institute for Special Technology.

DS777.54.J8 1979 79-841152

79 C

—HISTORY—CIVIL WAR, 1945-1949
—CHRONOLOGY

Ti san tz'u kuo nei ko ming chan cheng ta shih yüeh piao.
第三次國共内戰大事月表：一九四五年七月至一九四九年十月. — 香港：文化資料供應社, 1978.
100 p. ; 19 cm.
"本書原名為「第三次國内革命戰爭大事月表」, 一九六一年由北京人民出版社出版"
$4.00
1. China—History—Civil War, 1945-1949—Chronology. I. Title.
Title romanized: Ti san tz'u kuo kung nei chan ta shih yüeh piao.

DS777.54.T48 1978 78-842053

—HISTORY—CIVIL WAR, 1945-1949
—FICTION

Chang, Tung-p'ing.
(Hsiao hsiang tao)
小向导 / 原著粟谦；改编张东平；绘画冯越, 冯远. — 第1版. — 上海：上海人民出版社, 1976.
70 p. : ill. ; 10 x 13 cm.
Title in romanization on cover: Xiao xiang dao.
"連環圖畫"
$0.07

1. China—History—Civil War, 1945-1949—Fiction. I. Su, Ch'ien. II. Feng, Yüeh. III. Title. IV. Title: Hiao siang dao.

PL2837.T8H7 78-843189

(Ch'ang ying sung)
长缨颂：广西民兵革命斗争故事集 / 广西军区政治部编. — 第1版. — [南宁]：广西人民出版社, 1977.
2, 498 p., [8] leaves of plates : col. ill. ; 19 cm.
$1.20
1. Short stories, Chinese. 2. Sino-Japanese Conflict, 1939-1945—Fiction. 3. China—History—Civil War, 1945-1949—Fiction. 4. Kwangsi, China (Province)—History—Fiction. I. China (People's Republic of China, 1949-). Chung-kuo jen min chieh fang chün. Kuang-hsi chün ch'ü. Cheng chih pu.

PL2653.C435 78-843876

(Ch'e lun kun kun)
车轮滚滚：根据同名电影 / 长春电影制片厂供稿；本社改编. — 第1版. — 北京：人民美術出版社, 1976.
188 p. : ill. ; 11 x 13 cm.
"連環圖畫"
$0.29

1. China—History—Civil War, 1945-1949—Fiction. I. Ch'ang-ch'un tien ying chih p'ien ch'ang. II. Jen min mei shu ch'u pan she, Peking.

PL2832.5.C465 78-843178

Fen, Yung.
(Niu shih ti chan tou)
牛市的战斗 / 奋永编；李昌柏绘. — 第1版. — [广州]：广东人民出版社, 1976.
79 p. : ill. ; 11 x 13 cm.
"广东民兵革命斗争故事连环画"
$0.11

1. China—History—Civil War, 1945-1949—Fiction. I. Li, Ch'ang-po. II. Title.

PL2857.E45N5 78-843185

Hai, T'ao.
(Hsiao yen)
硝烟 / 海涛. — 第1版. — [合肥]：安徽人民出版社：安徽省新華書店, 1978.
3, 652 p. : ill. ; 19 cm.
Fiction.
$1.22

1. China—History—Civil War, 1945-1949—Fiction. I. Title.

PL2861.A57H74 79-841094

K'o-yang.
(To tao)
夺刀 / 克扬. — 第1版. — 太原：山西人民出版社, 1977, 1978 printing.
685 p., [3] leaves of plates : ill. ; 19 cm.
$1.60

1. China—History—Civil War, 1945-1949—Fiction. I. Title.

PL2872.Z8T6 79-835938

Liang, Hsing.
(Lung hu feng yün chi)
龙虎风云记 / 梁信. — 北京第1版. — 北京：人民文学出版社, 1978.
578 p. ; 19 cm.
Fiction.
$1.10

1. China—History—Civil War, 1945-1949—Fiction. I. Title.

PL2878.A62L8 79-836922

Lu, Hung-tao.
(San yüeh san)
三月三 / 原著詹炎；改编陆鸿道；绘画郑波. — 第1版. — 上海：上海人民出版社, 1976.
86 p. : ill. ; 94 x 125 mm.
"連環圖畫"
$0.09

1. China—History—Civil War, 1945-1949—Fiction. I. Chan, Yen. San yüeh san. II. Cheng, Po. III. Title.

PL2880.U1533S2 78-843186

Sun, Wu.
(Lin-fen lü)
临汾旅 / 孙吴编著. — 第2版. — 北京：解放军文艺社, 1978.
189 p. ; 19 cm.
$0.45

1. China—History—Civil War, 1945-1949—Fiction. I. Title.

PL2905.N278L5 1978 79-840098

Ting, Lü-chin.
(O kuo)
惡果 / 丁履進著. — 台北：長歌出版社, 民國67 [1978]
379 p. ; 19 cm. — (長歌傳記文學叢刊)
Edition for 1953 published under title: Ku tu meng nan chi.
Series romanized: Ch'ang ko chuan chi wen hsüeh ts'ung k'an.
NT$80.00 ($3.00 U.S.)

1. China—History—Civil War, 1945-1949—Fiction. I. Title.

PL2909.I565K8 1978 78-843615

T'ung, Meng-hou.
(Hsüeh jan chan ch'i hung)
血染战旗红 / 编文童孟侯；绘画金奎, 赵仁年. — 第1版. — 上海：上海人民出版社, 1977.
114 p. : ill. ; 94 x 127 mm. — (硬骨头六连之歌 ; 1)
"連環圖畫"
$0.12

1. Sino-Japanese Conflict, 1937-1945—Fiction. 2. China—History—Civil War, 1945-1949—Fiction. I. Chin, K'uei. II. Chao, Jen-nien. III. Title. IV. Series: Ying ku t'ou liu lien chih ko ; 1.

PL2914.N547H78 79-836682

Jmemoto, Sutezō, 1903-
(Sansei ni kaimetsushita Motoizumi Ryodan)
山西に潰滅した元泉旅団 / [著者 楳本捨三]. — 東京：秀英書房, 昭和53 [1978]
341 p. : ill. ; 20 cm. — (楳本捨三著作集 ; 第3巻) Ja 78-26618
¥1400

1. Motoizumi, Kaoru, 1893-1948—Fiction. 2. China—History—Civil War, 1945-1949—Fiction. I. Title. II. Series: Umemoto, Sutezō, 1903- Umemoto Sutezō chosaku shū ; dai 3-kan.

PL840.M38S2 79-804553

Wu, Yu-heng.
(Pei shan chi)
北山记 / 吴有恒. — 第1版. — [广州]：广东人民出版社, 1978.
2, 453 p. ; 21 cm. — (粤海文丛)
$1.00

1. China—History—Civil War, 1945-1949—Fiction. I. Title. II. Series: Yüeh hai wen ts'ung.

PL2921.Y68P4 79-836858

79 C

Wu, Yu-heng.
(Shan hsiang feng yün lu)
山乡风云录 / 吴有恒. — [香港?: s. n., 1978?]
2, 347 p. ; 21 cm.
Reprint of the 1962 (1978 printing) ed. published by Kuang-tung jen min ch'u pan she, in series: Yüeh hai wen ts'ung.

1. China—History—Civil War, 1945-1949—Fiction. I. Title. II. Series: Yüeh hai wen ts'ung.

PL2921.Y68S5 1978 79-838115

CHINA (Continued)

—HISTORY—CIVIL WAR, 1945-1949 —PERSONAL NARRATIVES

Feng, I-lu.
⌜Hsü Pang chan i chien wen lu⌝
國共徐蚌大會戰 / ⌜馮亦魯⌝. — 九龍：現代出版公司, ⌜1978?⌝
32, 2, 248 p. : ill. ; 19 cm.
Cover title.
Running title: Hsü Pang chan i chien wen lu
Reprint of the 1963 ed.
1. China—History—Civil War, 1945-1949—Personal narratives. 2. Feng, I-lu. I. Title. II. Title: Hsü Pang chan i chien wen lu.
Title romanized: Kuo kung Hsü Pang ta hui chan.
DS777.54.F4 1978 79-836119

⌜**Kuo, I-tung**⌝, 1920-
⌜Hsüeh chan i yü shih i nien⌝
異域 / 馬克騰著. — 臺北：迅雷出版社：經銷者星光書報社, 民國 -65 ⌜ -1976⌝
v. ; 19 cm.
Reprint of Kuo I-tung's Hsüeh chan i yü shih i nien, originally published by Tzu li wan pao, Taipei.
1. China—History—Civil War, 1945-1949—Personal narratives. 2. Kuo, I-tung, 1920- I. Title.
Title romanized: I yü.
DS777.54.K84 1976 78-842946

Lo, Shih-pu.
(Yüeh ma i yü)
躍馬異域 / 羅石補著. — 台北：南京出版公司, 民國68 ⌜1979⌝
2, 2, 190 p. ; 20 cm.
"滇, 緬, 寮, 印, 棉邊區游擊二十年"
NT$45.00
1. China—History—Civil War, 1945-1949—Personal narratives. 2. Lo, Shih-pu. I. Title.
DS777.54.L58 79-838717

Wang, P'ei-yüan.
(K'ang chan k'an luan hui i lu)
抗戰戡亂回憶錄：共匪殺戮人民實況 / 王培驥撰記. — 臺北：王培驥：總經銷正中書局, 民國67 ⌜1978⌝
2, 2, 60, 158 p., ⌜6⌝ leaves of plates : ports. ; 20 cm.
"附退休抒感詩二首及慶祝蔣經國先生當選第六任總統詩八道": p. 155-157.
NT$60.00
1. Sino-Japanese Conflict, 1937-1945—Personal narratives, Chinese. 2. China—History—Civil War, 1945-1949—Personal narratives. 3. Wang, P'ei-yüan. I. Title.
DS777.53.W338 79-835479

—HISTORY—CIVIL WAR, 1945-1949 —PERSONAL NARRATIVES —ADDRESSES, ESSAYS, LECTURES

P'an, Yu-hsin.
(Hsien shih ti yeh yü)
現實的揶揄 / 潘酉辛作. — ⌜香港?：s. n., 1978?⌝
vi, ii, 188 p., ⌜14⌝ leaves of plates : ill. ; 22 cm.
Reprint of the 1948 ed. published by Ta kuang pao she, Kuang-chou.
1. China—Civil War, 1945-1949—Personal narratives—Addresses, essays, lectures. 2. China—Social conditions—Addresses, essays, lectures. 3. P'an, Yu-hsin—Addresses, essays, lectures. 4. Journalists—China—Biography—Addresses, essays, lectures. I. Title.
DS777.54.P32 1978 79-835451

—HISTORY—1949-1976

Andō, Hikotarō, 1917-
(Chūka Jimmin Kyōwakoku no ayumi)
中華人民共和国の歩み / 安藤彦太郎. — 東京：講談社, 昭和52 ⌜1977⌝
204 p. : ill. (some col.) ; 27 cm. — (図説中国の歴史 ; 10) Ja***
Bibliography: p. 190-191.
Includes index.
¥2400
1. China—History—1945-1976. 2. China—Social life and customs—1949- I. Title. II. Series: Zusetsu Chūgoku no rekishi ; 10.
DS735.Z86 vol. 10 78-812159
[DS777.55]

Chesneaux, Jean.
China : the People's Republic, 1949-1976 / Jean Chesneaux, with the assistance of J. Bellassen ... ⌜et al.⌝ ; translated from the French by Paul Auster and Lydia Davis. — 1st American ed. — New York : Pantheon Books, c1979.
xii, 255 p. : maps ; 25 cm. — (The Pantheon Asia library, new approaches to the new Asia)
Translation of La Chine.
Includes bibliographical references and index.
ISBN 0-394-42873-0. ISBN 0-394-73623-0 pbk.
1. China—History—1949-1976. I. Title.
DS777.55.C44572313 1979 951.05 78-51797
78 MARC

(Chūgoku kakumei no shōri)
中国革命の勝利 / ⌜執筆者 吉沢南 ... et al.⌝. — 東京：東京大学出版会, 1978.
v, 305, 6 p. ; 21 cm. — (講座中国近現代史 ; 7) Ja 78-27051
Includes bibliographical references.
¥2400
1. China—History—Civil War, 1945-1949. 2. China—History—1949-1976. I. Yoshizawa, Minami. II. Series: Kōza Chūgoku kingendai shi ; 7.
DS755.K69 vol. 7 79-801655
[DS777.54]

Furui, Yoshimi, 1903-
(Nitchū jūhachinen)
日中十八年：一政治家の軌跡と展望 / 古井喜實. — 東京：牧野出版, 1978.
243 p., ⌜2⌝ leaves of plates : port. ; 19 cm. Ja 78-16736
¥1600
1. Japan—Foreign relations—China. 2. China—Foreign relations—Japan. 3. China—History—1949-1976. I. Title.
DS849.C6F78 78-814478

Great Britain. Schools Council. History 13-16 Project.
The rise of Communist China / ⌜Project team David Sylvester et al.⌝. — ⌜Student's book⌝. — Edinburgh : Holmes McDougall, 1977.
88 p. : ill., col. maps, ports. ; 25 cm. — (Modern world studies) GB77-21638
ISBN 0-7157-1558-5 : £2.05
1. China—History—1949-1976. 2. China—History—Republic, 1912-1949. I. Title. II. Series.
DS777.55.G685 1977 951.04 78-314102
78 MARC

Guillermaz, Jacques.
La Chine populaire / Jacques Guillermaz, ... — 7e éd. mise à jour. — ⌜Paris⌝ : PUF, 1977.
126, ⌜2⌝ p. : maps ; 18 cm. — (Que sais-je? ; 840) F78-1300
Bibliography: p. ⌜127⌝
9.00F
1. China—History—1949-1976. 2. China—History—Republic, 1912-1949. I. Title.
DS777.55.G833 1977 320.9'51'04 79-337042
79 MARC

Harris, Nigel.
The mandate of heaven : Marx and Mao in modern China / ⌜by⌝ Nigel Harris. — London ; New York : Quartet Books, 1978.
xi, 307 p. ; 23 cm. GB78-38125
Bibliography: p. ⌜x⌝-⌜xi⌝
Includes index.
ISBN 0-7043-2191-2 : £6.95
1. China—History—1949-1976. 2. Chung-kuo kung ch'n tang—History. 3. Chung-kuo kuo min tang—History. I. Title.
DS777.55.H317 951.05 79-309845
79 MARC

Hutheesing, Gunottam Purushottam.
⌜The great peace. Japanese⌝
毛沢東の国 / ラージャ・フーシーシン著；朝川清訳. — 東京：共同出版社, 昭和29 ⌜1954⌝
254 p. ; 19 cm.
Colophon inserted.
1. China—History—1949-1976. 2. China—Description and travel—1949-1976. 3. Hutheesing, Gunottam Purushottam. I. Title.
Title romanized: Mō Takutō no kuni.
DS777.55.H8516 79-801569

Kobayashi, Fumio, 1933-
(Chūgoku gendai shi no shūhen)
中国現代史の周辺 / 小林文男著. — 東京：アジア経済研究所；アジア経済出版会発売, 1976.
vi, 179 p. ; 18 cm. — (アジアを見る眼 ; 50) Ja***
1. China—History—1949-1976. I. Title. II. Series: Ajia o miru me ; 50.
DS777.55.K578 78-811401

Meisner, Maurice J 1931-
Mao's China : a history of the People's Republic / Maurice Meisner. — New York : Free Press, c1977.
xiv, 416 p. ; 24 cm. — (The Transformation of modern China series)
Includes bibliographical references and index.
ISBN 0-02-920820-3 : $17.95
1. China—History—1949-1976. I. Title.
DS777.55.M455 1977 951.05 76-51566
77⌜r79⌝rev MARC

Peyrefitte, Alain, 1925-
Quand la Chine s'éveillera ... le monde tremblera; regards sur la voie chinoise, par Alain Peyrefitte. ⌜Paris⌝ Fayard ⌜1973⌝
xx, 475 p. illus. 24 cm. 40.00F F***
Bibliography: p. ⌜461⌝-462.
1. China—History—1949-1976. I. Title.
DS777.55.P45 1973 73-177158
73⌜r79⌝rev MARC

Peyrefitte, Alain, 1925-
⌜Quand la Chine s'éveillera ... le monde tremblera. Japanese⌝
中国が目ざめるとき世界は震撼する / アラン・ペールフィット著；杉辺利英訳. — 東京：白水社, 1974.
2 v. : ill. ; 20 cm. Ja 74-20657 (v. 1)
¥1300 (v. 1) ¥1200 (v. 2)
1. China—History—1949-1976. I. Title.
Title romanized: Chūgoku ga mezameru toki sekai wa shinkansuru.
DS777.55.P4516 76-804446

Peyrefitte, Alain, 1925-
⌜Quand la Chine s'éveillera ... le monde tremblera. Chinese. Selections⌝
法國人看中國大陸 / 著者亞蘭·皮爾裴特；譯者楊仲男. — 台北：五洲出版社, 民國65 ⌜1976⌝
122 p. ; 20 cm.
Colophon title.
CONTENTS: 皮爾裴特·亞蘭 法國人看大陸(原名當中國大陸甦醒時)—楊仲男 看「中共大鬧巴黎」 他從大陸來 「毛澤東囚徒」作者包若望訪問記「法國人看中國大陸」在巴黎引起激烈辯論
1. China—History—1949-1976. I. Yang, Chung-nan. II. Title.
Title romanized: Fa-kuo jen k'an Chung-kuo ta lu.
DS777.55.P45213 77-838447

—HISTORY—1949-1976 —ADDRESSES, ESSAYS, LECTURES

Asakawa, Kenji, 1910-1975.
(Kokuun o hiraki seiten o miru)
披黒雲睹青天：浅川謙次追悼遺稿集. — 東京：浅川謙次追悼遺稿集刊行委員会, 1977.
357 p. : ill. ; 22 cm. Ja 78-20117
List of the author's works: p. 346-355.
1. China—History—1949-1976—Addresses, essays, lectures. 2. China—Study and teaching—Japan—Addresses, essays, lectures. I. Title.
DS777.55.A8144 1977 78-814611

China : the impact of the cultural revolution / edited by Bill Brugger. — London : Croom Helm ; New York : Barnes & Noble Books, c1978.
300 p. ; 23 cm. GB***
Bibliography: p. 280-290.
Includes index.
ISBN 0-06-490760-0 (Barnes & Noble)
1. China—History—1949-1976—Addresses, essays, lectures. I. Brugger, Bill.
DS777.55.C4486 1978b 951.05 78-322657
79 MARC

China, new Soviet studies. — Moscow : Social Sciences Today Editorial Board, USSR Academy of Sciences, 1975.
176 p. ; 22 cm. — (Problems of the contemporary world ; 34) USSR***
Includes bibliographical references.
1. China—History—1949-1976—Addresses, essays, lectures. I. Series.
DS777.55.C44689 951.9'043 79-303729
79 MARC

(Chūkyō kōwa)
中共講話 / 日本外政学会編. — 東京：同学会, 昭和32 ⌜1957⌝
326 p. ; 18 cm.
Includes bibliographical references.
CONTENTS: 植田捷雄 本書の読者のために—リッグ·ロバート·B. 中共軍の実力はどの程度か—パーマ·ノーマン·D. 外交政策は何を狙うか—韋楚 真の指導権は誰が握るか—呉元立 中共の経済は欧米諸国を追越すか—趙国鈞 農民の生活は果して向上しつつあるか—戴聖與 毛沢東主義とソ連共産主義の相異は何か—陳錫恩 教育は中国人を変形させうるか—スタイナー·H. アーサー 労仂者の地位は向上したか—チャン·C. M. 中共の謎と実体—ヒントン·ハロルド·C. 中国共産党第八回全国代表大会
1. China—History—1949-1976—Addresses, essays, lectures. 2. Communism—China—Addresses, essays, lectures. I. Nihon Gaisei Gakkai.
DS777.55.C465749 78-812760

CHINA

—HISTORY—1949-1976
—ADDRESSES, ESSAYS, LECTURES
(Continued)

Strong, Anna Louise, 1885-1970.
(Chung-kuo ti chü pien)
中國的巨變 / 安娜·路易斯·斯特朗著. — 重排本. — 香港：南粵出版社, 1977.
2, 140 p. ; 21 cm.
"附錄...路易·艾黎著:《我所知道的斯特朗》."
$4.60
1. China—History—1949-1976—Addresses, essays, lectures. I. Alley, Rewi, 1897- II. Title.
DS777.55.S8256 1977 79-836797

—HISTORY—1949-1976
—ANECDOTES, FACETIAE, SATIRE, ETC.

Ku, Cheng-chien.
(Hui yin pi)
回音壁 / 作者古錚劍. — 臺北：白雲文化事業有限公司, 民國67 [1978]
4, 4, 206 p. ; 22 cm. — (白雲叢書 ; 8)
Series romanized: Pai yün ts'ung shu.
NT$45.00
1. China—History—1949-1976—Anecdotes, facetiae, satire, etc. 2. China—History—1976- —Anecdotes, facetiae, satire, etc. I. Title.
DS777.55.K7158 79-835890

(Nung ts'un hsin shih)
农村新事. — [广州]：广东人民出版社, 1975.
79 p. ; 19 cm.
CONTENTS: 人满七十粮过三千—"穷棒子"精神传后代—他当了第二把手以后—藏族老阿爸的儿子回来了—做雷锋那样的人—送牛记——次移风易俗的结婚喜诗会—科研花开又一枝—农民女画家—独有英雄驱虎豹—北京来的"庄稼汉"
$0.17
1. China—History—1949-1976—Anecdotes, facetiae, satire, etc.
DS777.55.N84 78-843156

—HISTORY—1949-1976—BIOGRAPHY

(Kuo kung feng yün ming jen lu)
國共風雲名人錄 / 李約等著. — 香港：廣角鏡出版社, 19
v. : ill. ; 19 cm. — (廣角鏡人物叢書)
Series romanized: Kuang chiao ching jen wu ts'ung shu.
$5.50 (v. 3)
1. China—History—Republic, 1912-1949—Biography. 2. China—History—1949-1976—Biography. 3. China—Biography. I. Li, Yüeh.
DS778.A1K87 78-842847
C

Mamo, David, 1948-
Dizionario storico-biografico della Cina moderna / [di] David Mamo, Laurie Upson. — Firenze : Vallecchi, 1977.
xxvii, 458 p. ; 20 cm. — (Il Pellicano : Storia) It78-Sept
Bibliography: p. [449]-455.
Includes index.
L8000
1. China—History—Republic, 1912-1949—Biography. 2. China—History—1949-1976—Biography. 3. China—Biography. I. Title.
DS778.A1M35 920'.051 78-396318
*79 MARC

(Mo kuei hsing hsiang)
魔鬼形像. — 台北：大陸資料出版社：總經銷新亜出版社, 民國56 [1967]
9, 257 p. : ill. ; 19 cm.
Colophon title.
CONTENTS: 鄧小平—陶鑄—羅瑞卿—陳雲—王光美—朱德—譚—波—陳毅
1. China—History—1949-1976—Biography. 2. China—Biography. 3. Communists—China—Biography. I. Ta lu tzu liao ch'u pan she.
DS778.A1M58 78-843817

—HISTORY—1949-1976—BIOGRAPHY
—ANECDOTES, FACETIAE, SATIRE, ETC.

Ch'en, K'o-wei.
[Chung kung t'e ch'üan chieh chi sheng huo cheng hsiang]
中共高級幹部的生活 / 陳克煒編. — 九龍：阿爾泰出版社, 1977.
2, 212 p., [2] leaves of plates : ill. ; 19 cm.
Reprint of the ed. published by Li ming wen hua shih yeh kung ssu, Taipei under title: Chung kung t'e ch'üan chieh chi sheng huo cheng hsiang; with intro. deleted.
$6.50 ($1.40 U.S.)
1. China—History—1949-1976—Biography—Anecdotes, facetiae, satire, etc. 2. China—Biography—Anecdotes, facetiae, satire, etc. I. Title.
Title romanized: Chung kung kao chi kan pu ti sheng huo.
DS778.A1C464 1977 78-843450

—HISTORY—1949-1976
—CARICATURES AND CARTOONS

(Chieh fa kung fei pao hsing man hua chuan chi)
揭發共匪暴行漫畫專輯 / 編纂王安慶；著作人黃書文等；[臺灣省立彰化社會教育館編印]. — 彰化：該館, 民國65 [1976]
4, 44 p. : chiefly ill. (some col.) ; 27 cm. — (研究叢書 - 臺灣省立彰化社會教育館 ; 7)
Colophon title.
"非賣品"
1. China—History—1949-1976—Caricatures and cartoons. I. Wang, An-ch'ing. II. Wang, Shu-wen. III. T'ai-wan-sheng li Chang-hua she hui chiao yü kuan. IV. Series: T'ai-wan sheng li Chang-hua she hui chiao yü kuan. Yen chiu ts'ung shu - T'ai-wan sheng li Chang-hua she hui chiao yü kuan ; 7.
DS777.55.C44637 78-842428

—HISTORY—1949-1976—HANDBOOKS, MANUALS, ETC.

Guía de la nueva China. — Pekin : Ediciones en Lenguas Extranjeras, 1958.
299 p., [3] leaves of plates : ill. ; 19 cm.
1. China—History—1949-1976—Handbooks, manuals, etc.
DS777.55.G77 951.05 79-110226
79 MARC

—HISTORY—1949-1976
—PERIODICALS—INDEXES

Index to Survey of People's Republic of China press, Selections from People's Republic of China magazines, and Current background. Jan./Mar. 1977–
Hong Kong, American Consulate General.
v. 28 cm. quarterly.
Continues: United States. Consulate General. Hongkong. Index to Survey of China mainland press, Extracts from China mainland magazines, and Current background.
Key title: Index to Survey of People's Republic of China press, Selections from People's Republic of China magazines, and Current background, ISSN 0148-6365
1. China—History—1949-1976—Periodicals—Indexes. 2. China—History—1976- —Periodicals—Indexes. I. United States. Consulate General. Hongkong. II. Survey of People's Republic of China press. III. Selections from People's Republic of China magazines. IV. United States. Consulate General. Hongkong. Current background.
DS777.55.U518 951.05 77-643182
MARC-S

Japan. Naikaku Kambō. Naikaku Chōsashitsu.
("Chūkyō kankei shiryō" jikōbetsu sakuin)
「中共関係資料」事項別索引：自1957年12月7日至1970年4月28日. — [東京]：内閣官房内閣調査室, [1970]
516 p. ; 25 cm.
Cover title.
1. China—History—1949-1976—Periodicals—Indexes. 2. China—Periodicals—Indexes. I. Title.
DS777.55.J3 1970 78-812867

—HISTORY—1949-1976—SOURCES

The People's Republic of China : a documentary history of revolutionary change / edited by Mark Seldon, with Patti Eggleston. — New York : Monthly Review Press, c1979.
xviii, 718 p. ; 22 cm.
Bibliography: p. 716-718.
ISBN 0-85345-466-3 : $20.00
1. China—History—1949-1976—Sources. I. Selden, Mark. II. Eggleston, Patti.
DS777.55.P423 951.05 78-13916
78 MARC

—HISTORY—1976-

Evans, Les.
China after Mao / Leslie Evans. — New York : Monad Press : distributed by Pathfinder Press, 1978.
191 p. ; 22 cm.
Includes bibliographical references and index.
ISBN 0-913460-61-3 : $16.00
1. China—History—1976- . I. Title.
DS779.2.E93 951.05 78-59264
79 MARC

Macrae, Norman.
La Cina verso il 2000 / a cura di Norman Macrae, Brian Beedham ed Emily MacFarquhar. — Milano : Editoriale del Corriere della sera, 1978.
100 p. ; 18 cm. — (I Libri de Il Mondo) It79-Apr
Supplement to n. 15 de Il Mondo.
1. China—History—1976- . I. Beedham, Brian, joint author. II. MacFarquhar, Emily, joint author. III. Il Mondo. IV. Title.
DS779.2.M32 951.05 79-363097
*79 MARC

Munthe-Kaas, Harald.
Kinas nye ansikt / Harald Munthe-Kaas. — [Oslo] : Tiden Norsk Forlag, c1979.
315 p. : ill. ; 22 cm. N***
ISBN 8210017802 : kr94.00
1. China—History—1976- . I. Title.
DS779.2.M86 79-375018
79 MARC

Rongen, Ole Bjørn.
Kina—den nye lange marsjen / Ole Bjørn Rongen. — Oslo : PaxForlag, 1979.
215 p. : ill. ; 19 cm. — (En Pax-bok ; 625) N***
Includes bibliographical references.
ISBN 8253010060
1. China—History—1976- . I. Title.
DS779.2.R66 79-363158
79 MARC

Ruge, Gerd.
Begegnung mit China : eine Weltmacht im Aufbruch / Gerd Ruge ; [aus dem Englischen übers. und bearb. von Hans Thomas unter Beteiligung von Michaela Charlé und Claudia Richarz]. — 26.-35. Tausend. — Düsseldorf ; Wien : Econ Verlag, 1979.
518 p. ; 22 cm. GFR***
Includes index.
ISBN 3-430-17837-1 : DM39.80
1. China—History—1976- . I. Title.
DS779.2.R83 951.05 79-367274
79 MARC

Tames, Richard L
China today / [by] Richard Tames. — London : Kaye and Ward [etc.], 1978.
96 p. : ill., 2 maps, ports. ; 23 cm. GB78-30297
Bibliography: p. 96.
ISBN 0-7182-0458-1 : £2.95
1. China—History—1976- . I. Title.
DS779.2.T35 951.05 79-311831
79 MARC

—HISTORY—1976- —ADDRESSES, ESSAYS, LECTURES

Materialien zur China-Diskussion : Texte einer interdisziplinären Wissenschaftlergruppe über ihre Studienreise im September 1978 in die Volksrepublik China / von Günter Abraham ... [et al.] ; hrsg. von Sigmar Groeneveld. — Saarbrücken : Breitenbach, 1979.
vi, 308 p. : ill. ; 21 cm. — (Occasional papers ; Nr. 16 ISSN 0342-0701) GFR***
Bibliography: p. 302-308.
ISBN 3-88156-111-0
1. China—History—1976- —Addresses, essays, lectures. I. Abraham, Günter, 1929- . II. Groeneveld, Sigmar.
DS779.2.M37 951.05 79-670211
79 MARC

—HISTORY—1976- —ANECDOTES, FACETIAE, SATIRE, ETC.

Ku, Cheng-chien.
(Hui yin pi)
回音壁 / 作者古錚劍. — 臺北：白雲文化事業有限公司, 民國67 [1978]
4, 4, 206 p. ; 22 cm. — (白雲叢書 ; 8)
Series romanized: Pai yün ts'ung shu.
NT$45.00
1. China—History—1949-1976—Anecdotes, facetiae, satire, etc. 2. China—History—1976- —Anecdotes, facetiae, satire, etc. I. Title.
DS777.55.K7158 79-835890

CHINA (Continued)

—HISTORY—1976- —BIOGRAPHY

Ting, Wang.
(Hua Kuo-feng Chi Teng-k'uei ho hsin ch'i ti i tai)
華國鋒紀登奎和新起的一代 / 丁望著. — 香港：明報月刊社, 1977.
10, iv, 191 p. : ill. ; 21 cm. — (現代中國研究叢刊：傳記類；1) (明報月刊叢書；10)
Title in English in colophon and on cover Hua Kuo-feng, Chi Teng-kuei and a new generation.
Series 2 romanized: Ming pao yüeh k'an ts'ung shu.
Includes index.
Bibliography: p. 181-184.
$18.00 ($4.50 U. S.)

1. China—History—1976- —Biography. 2. China—Biography. 3. Communists—China—Biography. 4. Statesmen—China—Biography. I. Title. II. Title: Hua Kuo-feng, Chi Teng-kuei and a new generation. III. Series: Hsien tai Chung-kuo yen chiu ts'ung k'an : Chuan chi lei ; 1.
DS779.28.T56 78-842374

—HISTORY—1976- —COLLECTED WORKS

Chung-kuo ta lu yen chiu ch'u pan she.
(Chung-kuo ta lu chin k'uang chien pao)
中國大陸近況簡報 / 中國大陸研究出版社編. — [台北]：該社, 民國67- [1978-
v. ; 22 cm.
"編著者: 中國大陸問題研究中心"
NT$20.00

1. China—History—1976- —Collected works. I. Chung-kuo ta lu wen t'i yen chiu chung hsin. II. Title.
DS779.2.C48 1978 78-841577

—HISTORY—1976- —CONGRESSES

Sino-American Conference on Mainland China, 7th, T'ai-pei, 1978.
Proceedings of the Seventh Sino-American Conference on Mainland China, June 6-9, 1978. — Taipei : Institute of International Relations, 1978.
ca. 450 p. in various pagings ; 27 cm.
Includes bibliographies.
1. China—History—1976- —Congresses. I. Chung-hua min kuo chi kuan hsi yen chiu so, T'ai-pei.
DS779.16.S56 1978 951.05 79-112703
79 MARC

—HISTORY—1976- —PERIODICALS—INDEXES

Index to Survey of People's Republic of China press, Selections from People's Republic of China magazines, and Current background. Jan./Mar. 1977-
Hong Kong, American Consulate General.
v. 28 cm. quarterly.
Continues: United States. Consulate General. Hongkong. Index to Survey of China mainland press, Extracts from China mainland magazines, and Current background.
Key title: Index to Survey of People's Republic of China press, Selections from People's Republic of China magazines, and Current background, ISSN 0148-6365
1. China—History—1949-1976—Periodicals—Indexes. 2. China—History—1976- —Periodicals—Indexes. I. United States. Consulate General. Hongkong. II. Survey of People's Republic of China press. III. Selections from People's Republic of China magazines. IV. United States. Consulate General. Hongkong. Current background.
DS777.55.U518 951.05 77-643182
MARC-S

—HISTORY—1976- —YEARBOOKS

China facts & figures annual. v. 1
1978-
[Gulf Breeze, Fla.] Academic International Press.
v. ill. 24 cm. (Academic International reference series.)
Editor: 1978- J. L. Scherer.
Key title: China facts & figures annual, ISSN 0190-602X.
1. China—History—1976- —Yearbooks. I. Scherer, John L.
DS779.15.C48 951'.005 79-640300
MARC-S

—HISTORY—ADDRESSES, ESSAYS, LECTURES

Boodberg, Peter Alexis, 1903-
Selected works of Peter A. Boodberg / compiled by Alvin P. Cohen. — Berkeley : University of California Press, c1979.
xix, 501 p. ; 27 cm.
Bibliography: p. 496-501.
ISBN 0-520-03314-0
1. China—History—Addresses, essays, lectures. 2. Chinese philology—Addresses, essays, lectures. I. Cohen, Alvin P. II. Title.
DS736.B66 951 76-24580
79 MARC

Katz, Albert, 1858-1923.
Die Juden in China, von Albert Katz. Berlin, A. Katz, 1900.
31, [1] p. 22 cm.
1. Jews in China—History—Addresses, essays, lectures. 2. China—History—Addresses, essays, lectures. I. Title.
DS135.C5K36 951'.004'924 a 12-1490
[r79]rev MARC

(Kitaĭ)
Китай : история, культура и историография : [Сборник статей] / АН СССР, Ин-т востоковедения ; [Отв. ред. Н. Ц. Мункуев]. — Москва : Наука, 1977.
248 p. ; 22 cm. USSR 77
Includes bibliographical references.
2.10rub
1. China—History—Addresses, essays, lectures. I. Munkuev, Nikolaĭ T͡Syrendorzhievich. II. Akademii͡a nauk SSSR. Institut vostokovedenii͡a.
DS736.K57 79-408694

Miao, Yüeh.
(Tu shih ts'un kao)
讀史存稿 / 缪钺著. — 香港第1版. — 香港生活, 讀書, 新知三联书店, 1978.
229 p. ; 21 cm.
Reprint of the 1963 ed. published in Peking.
$8.00

1. China—History—Addresses, essays, lectures. I. Title.
DS736.M47 1978 79-839137

Miyazaki, Ichisada, 1901-
(Ajia shi kenkyū)
アジア史研究 / 宮崎市定. — 京都：同朋舎, 19
v. ; 22 cm. — (東洋史研究叢刊 ; 4の) Ja 78-21210 (v. 5)
Added t. p.: Asiatica : studies in the Oriental history.
Includes indexes.
¥7000 (v. 5)

1. China—History—Addresses, essays, lectures. I. Title. II. Series: Tōyōshi kenkyū sōkan ; 4, V [etc.]
DS735.M578 78-814848

(Rekishi ronshū)
歴史論集：生江義男先生還暦記念. — 調布：生江義男先生還暦記念歴史論集刊行委員会, 1978.
11, 2, 374 p., [1] leaf of plates : ill. ; 22 cm. Ja 79-4262
Table of contents also in English with title: A collection of treatises on history in commemoration of the sixtieth anniversary of Prof. Namae's birth.
Includes bibliographical references.
"生江義男先生著作目録": p. 7-11 (1st group)
1. Japan—History—Addresses, essays, lectures. 2. China—History—Addresses, essays, lectures. 3. Europe—History—Addresses, essays, lectures. 4. History—Study and teaching—Addresses, essays, lectures. 5. Korea—Religious life and customs—Addresses, essays, lectures. 6. Namae, Yoshio, 1917- I. Namae, Yoshio, 1917-
DS835.R418 79-808368

(Rekishigaku rombun shū)
歴史学論文集：日本大学史学科五十周年記念 / [編集 日本大学]史学科創立五十周年記念事業実行委員会. — 東京：同委員会, 1978.
681 p. : ill. ; 22 cm. Ja 78-20036
Includes bibliographies.
Not in trade.
1. Japan—History—Addresses, essays, lectures. 2. China—History—Addresses, essays, lectures. 3. Nihon Daigaku, Tokyo. Shigakka. I. Nihon Daigaku, Tokyo. Shigakka. II. Nihon Daigaku, Tokyo. Shigakka. Sōritsu Gojisshūnen Kinen Jigyō Jikkō Iinkai.
DS837.R49 78-817106

Shih, Nien-hai.
(Ho shan chi)
河山集 / 史念海著. — 香港：廣角鏡出版社：華風書局發行, 1977.
3, 2, 302 p., [4] leaves of plates : ill. ; 21 cm.
Includes bibliographical references.
$10.00

1. China—History—Addresses, essays, lectures. I. Title.
D735.S48 1977 79-838670

(Wen shih hsüan chi)
文史選集. — 臺北：中央日報社, 民國68- [1979-
v. ; 19 cm. — (中央日報專刊叢書)
Series romanized: Chung yang jih pao chuan k'an ts'ung shu.
NT$50.00 (v. 1)

1. China—History—Addresses, essays, lectures. 2. Chinese literature—History and criticism—Addresses, essays, lectures. 3. Philosophy, Chinese—Addresses, essays, lectures. 4. Taiwan—History—Addresses, essays, lectures.
DS736.W45 79-840512

Wu, Yü-chang, 1878-1966.
(Li shih wen chi)
历史文集 / 吳玉章. — 香港：生活·讀書·新知三联书店香港分店, 1978.
6, 138 p., [2] leaves of plates : ill. ; 22 cm.
Reprint ed.
Includes bibliographical references.
$4.00

1. China—History—Addresses, essays, lectures. 2. China—History—Methodology—Addresses, essays, lectures. I. Title.
DS736.W87 1978 79-839569

Yeh, Shih, 1150-1223.
(Hsi hsüeh chi yen hsü mu)
习学记言序目 / 叶适著. — 第1版. — 北京：中华书局, 1977.
2 v. (3, 11, 771 p.) ; 21 cm.
Reprint ed.; with new introd.
"叶适起初只是「辑录经史百氏条目」没有加评论, 这时名为《习学记言》后来「研玩群书」撰写了论述, 乃以「序目」名书"
$2.30

1. Chinese literature—History and criticism—Addresses, essays, lectures. 2. China—History—Addresses, essays, lectures. I. Title.
PL2261.Y44 1977 78-841010

78 C

—HISTORY—ANECDOTES, FACETIAE, SATIRE, ETC.

Chao, T'ing-chün.
(Shih shih jen wu t'an)
史事人物談 / 趙廷俊著. — 臺北：臺灣學生書局, 民國47 [1958], 民國64 [1975] printing.
7, 211 p. ; 20 cm. — (學生書苑 ; 38)
Series romanized: Hsüeh sheng shu yüan.
NT$60.00

1. China—History—Anecdotes, facetiae, satire, etc. I. Title.
DS736.C4715 78-842051

(Ch'eng yü ku shih)
成語故事：附插圖：中國歷史故事集成. — 台南：利大出版社：總經銷第一書店, 民國67 [1978]
7, 357 p. : ill. ; 19 cm.
Cover title.
NT$35.00

1. China—History—Anecdotes, facetiae, satire, etc. 2. Chinese language—Terms and phrases.
DS736.C49 78-842138

78 C

Chiang, Yung, 1681-1762.
(Ssu shu ku jen tien lin)
四書古人典林 / 江永撰. — 台北：廣文書局, 民國67 [1978]
260 leaves in various foliations ; 22 cm. — (漢學彙編)
Reprint ed.
NT$390.00

1. China—History—Anecdotes, facetiae, satire, etc. 2. Ssu shu. I. Title. II. Series: Han hsüeh hui pien.
DS736.C5138 1978 79-841440

CHINA

—HISTORY—ANECDOTES, FACETIAE, SATIRE, ETC. (Continued)

Ch'ing kuo ming hua)
傾國名花：中國名女人故事 / [河洛出版社編印]. — 台北：該社, 民國67 [1978]
6, 200 p. ; 22 cm.
Reprint ed.?
NT$50.00

1. China—History—Anecdotes, facetiae, satire, etc. 2. Women—China—Anecdotes, facetiae, satire, etc. I. Ho lo t'u shu ch'u pan she. II. Title: Chung-kuo ming nü jen ku shih.
DS736.C5164 1978 79-837257

Chuang, Lien.
(K'an tien shih shuo chang ku)
看電視說掌故 / 莊練著. — 台北：求精出版社：總經銷精益書局, 民國66 [1977]
3, 244 p. ; 19 cm. — (精益叢書 ; 9)
Series romanized: Ching i ts'ung shu.
NT$45.00

1. China—History—Anecdotes, facetiae, satire, etc. I. Title.
DS736.C5196 78-841892

(Chung-kuo jen ti chih hui)
中國人的智慧. — [s. l.]：蘇宿出版社；高雄：心影書店, [1977?]
6, 184 p. ; 19 cm.
NT$30.00

1. China—History—Anecdotes, facetiae, satire, etc. 2. China—Biography—Anecdotes, facetiae, satire, etc.
DS736.C539 78-843641

Hsü, Chih-ching.
(Li shih kan ying t'ung chi)
歷史感應統紀：[4卷] / 王震著 [i. e. 許止淨編纂]. — 台北：銀河出版社：總經銷新音書局, 民國65 [1976]
558 p. in various pagings ; 22 cm. — (銀河文庫 ; 105)
Real authorship statement from captions.
Series romanized: Yin ho wen k'u.
Reprint of the 1933 ed. published by Hung hua she, Soochow.
NT$120.00
(Continued on next card)
1. China—History—Anecdotes, facetiae, satire, etc. I. Wang, Chen, writer on I ching. II. Title.
DS736.H757 1976 79-835416

Hsü, Tung-che.
(Chih lüeh yü chi mou)
智略與計謀 / 徐東哲編著. — 香港：新力書局, [1978?]
8, 169 p. ; 19 cm.
Cover title.
$4.50

1. Strategy. 2. China—History—Anecdotes, facetiae, satire, etc. I. Title.
U162.H77 78-843951

Li, I-ting.
(Chung-kuo hua hui shih hua)
中國花卉史話 / 李邁[李奕定]著. — 台北：總經銷中國花卉雜誌社, 民國66 [1977]
176 p. : ill. ; 22 cm. — (中國花卉叢書)
NT$120.00

1. Chinese poetry—History and criticism—Addresses, essays, lectures. 2. Flowers in literature—Addresses, essays, lectures. 3. China—History—Anecdotes, facetiae, satire, etc. I. Title. II. Series: Chung-kuo hua hui ts'ung shu.
PL2308.L49 79-837063

Li, Yü-chang, 1871-1923.
(Ching-t'ao an tu shih lu)
景陶庵讀史錄 / [編著者黎豫樟；校訂者黎驥]. — 廣州：綜合出版社：發行者廣東省立圖書館, 民國37 [1948]
69 double leaves : port. ; 23 cm.
In case.

1. China—History—Anecdotes, facetiae, satire, etc. I. Title.
DS736.L512 1948 78-842160

Liu, Hsiang, 77?-6? B. C.
(Shuo yüan)
說苑 / [劉向撰]. — [漢口?]：湖北崇文書局, 光緒紀元 [1875]
2 v. (on double leaves) ; 27 cm.
Printed in 20 chüan.
Block print.

1. China—History—Anecdotes, facetiae, satire, etc. I. Title.
DS736.L53477 1875 78-843773

Lu, Chia-chi.
(Tui lien hsin yü)
對聯新語 / 陸家驥編著. — 臺北：臺灣商務印書館, 民國67 [1978]
2, 2, 4, 300 p. ; 18 cm. — (人人文庫 ; 特563)
Series romanized: Jen jen wen k'u.
NT$30.00

1. Couplets. 2. Chinese poetry. 3. China—History—Anecdotes, facetiae, satire, etc. I. Title.
PL2519.C7L7 78-842609

Meng, Chung-jen.
(Chung-kuo yeh shih ku shih)
中國野史故事 / 孟仲仁編著. — 臺北：出版家文化事業有限公司出版部, 民國67 [1978]
4, 200 p. ; 20 cm. — (出版家叢書 ; 54)
Series romanized: Ch'u pan chia ts'ung shu.
NT$40.00

1. China—History—Anecdotes, facetiae, satire, etc. I. Title.
DS736.M425 78-843618

Pai, Wei-liang.
(Shih shuo hsin yü ~~hsin~~ shih)
世說新語新釋 / 劉義慶著；白惟良譯註. — 高雄：大衆書局, 67 [1978]
5, 430 p. : ill. ; 21 cm. — (古典文學 ; 第9輯)
NT$125.00

1. Liu, I-ch'ing, 403-444. Shih shuo hsin yü. 2. China—History—Anecdotes, facetiae, satire, etc. I. Liu, I-ch'ing, 403-444. Shih shuo hsin yü. 1978. II. Title. III. Series: Ku tien wen hsüeh ; ti 9 chi.
DS736.L5363P34 79-837646

Su, T'ung-ping.
(Yen lo, Pao lao yu Chang Lung, Chao Hu)
閻羅·包老與張龍·趙虎 / 莊練[蘇同炳]著. — 台北：求精出版社：總經銷精益書局, 民國66 [1977]
13, 1, 223 p. ; 19 cm. — (精益叢書 ; 3)
In colophon: King of death, Judgi [i. e. Judge] Pao & his colleagues, by Chuang-lien.
Series romanized: Ching i ts'ung shu.
NT$40.00
1. China—History—Anecdotes, facetiae, satire, etc. I. Title.
DS736.S93 78-843780

Tso, Sung-ch'ao.
(Shuo yüan chi cheng)
說苑集證 [左松超撰 台北 文史哲出版社 民國62 i. e. 1973]
2 v. (40, 1761 p.) 22 cm. NT$500.00
國立臺灣師範大學博士論文
Bibliography: v. 2, p. 1739-1761.

1. Liu, Hsiang, 77?-6? B. C. Shuo yüan. 2. China—History—Anecdotes, facetiae, satire, etc. I. Liu, Hsiang, 77?-6? B. C. Shuo yüan. 1973. II. Title.
DS736.L534773T78 73-898138

Tung, Chung-shu, 2d cent. B. C.
(Ch'un ch'iu fan lu)
春秋繁露 / [董仲舒撰；孫鑛等評]. 新序 / [劉向撰]. — [台北?]：中國子學名著集成編印基金會, [1978?]
2, 3, 4, 710 p. ; 22 cm. — (中國子學名著集成 ; 027 : 珍本初編. 儒家子部)
Reprint of 2 works.
Each work has separate t. p.

1. Confucius. Ch'un ch'iu. 2. China—History—Anecdotes, facetiae, satire, etc. I. Sun, K'uang, 1542-1613. II. Liu, Hsiang, 77?-6? B. C. Hsin hsü. 1978. III. Title. IV. Series: Chung-kuo tzu hsüeh ming chu chi ch'eng ; 027.
PL2470.Z7T8 1978 78-842839

Wu, Chia-ch'ing, 1901-
(Shih ch'ü lu)
拾趣錄 / 伍稼青輯[述]. — 臺北：臺灣學生書局, 民國67 [1978] printing.
3, 19, 391 p. ; 20 cm. — (學生書苑 ; 60)
Series romanized: Hsüeh sheng shu yüan.
"本書輯述者其他重要作品簡目": p. 391.
NT$80.00

1. China—History—Anecdotes, facetiae, satire, etc. I. Title.
DS736.W76 79-835137

—HISTORY—BIBLIOGRAPHY

Phillips, Leona Rasmussen.
Chinese history : a bibliography / by Leona Rasmussen Phillips. — New York : Gordon Press, 1978.
119 p. ; 24 cm. — (Gordon Press bibliographies for librarians series)
ISBN 0-8490-1391-7
1. China—History—Bibliography. I. Title.
Z3106.P48 016.951 78-5136
[DS706] 78 MARC

—HISTORY—CHRONOLOGY

Ch'en, Yüan, 1880-1971.
(Erh shih shih shuo jun piao)
二十史朔閏表：附西曆回曆 / 陳垣著. — 新1版. — 北京：中華書局, 1962, 1978 printing.
242 p. ; 27 cm.
$3.00

1. Calendar, Chinese. 2. China—History—Chronology. I. Title.
CE15.C453 1978 79-838045

(Chung-kuo li shih chi nien piao)
中國歷史紀年表 / 本社編訂. — 台北：華世出版社, 1978.
163 p. ; 22 cm.
Includes index.
NT$60.00

1. Calendar, Chinese. 2. China—History—Chronology. I. Hua shih ch'u pan she.
CE15.C485 1978 79-839136

Ho, San-yüan.
(Chung-kuo li tai ti wang tao t'ung)
中國歷代帝王道統 / [何三源撰；編輯王春茂；中華道德進修協會編印]. — 台中：該會：發行所新學出版社, 民國65 [1976]
2, 62 p., [2] leaves of plates : ill. ; 21 cm. — (歷史叢書)
Cover title.
Series romanized: Li shih ts'ung shu.
"非賣品"
"太上黃庭外景玉經": p. 59-62.
1. China—History—Chronology. I. Wang, Ch'un-mao. II. Chung-hua tao te chin hsiu hsieh hui. III. Huang t'ing ching. 1976.
DS733.H625 78-843322

Hsü, Shih-li.
(Chung-kuo li shih ta shih chi nien)
中国历史大事纪年：远古至一九四九年 / 徐师历编. — 第1版. — 澳門：爾雅社, 1979.
287 p. ; 27 cm.
Includes index.
$30.00

1. China—History—Chronology. I. Title.
DS733.H68 79-840533

Kuo, I-tung, 1920-
(Chung-kuo jen shih kang)
中國人史綱 / 柏楊[郭衣洞]著. — 臺北：星光出版社, 民國68 [1979]
2 v. (12, 994 p.) : ill. ; 22 cm. — (柏楊歷史研究叢書 ; 第1部)
NT$480.00 ($15.00 U.S.)

1. China—History—Chronology. I. Title. II. Series: Kuo, I-tung, 1920- Pai Yang li shih yen chiu ts'ung shu ; ti 1 pu.
DS733.K77 79-837199

CHINA

—HISTORY—CHRONOLOGY (Continued)

Liu, Shih-hsin.
(Shang ku ti wang ta shih nien piao)
上古帝王大事年表 / 劉士心編. — [香港]：用舍堂書屋, 1973.
2, 8, 169 p. ; 22 cm.
"研讀文史必備參考書"

1. China—History—Chronology. 2. China—Kings and rulers—Chronology. I. Title.
DS733.L455 78-840917

—HISTORY—COLLECTED WORKS

(Li shih hsüeh pao)
歷史學報
[台南] 國立成功大學歷史學系
no. ill. 27 cm.
Spine title. : Kuo li Ch'eng-kung ta hsüeh li shih hsüeh hsi li shih hsüeh pao.
Other title. : Bulletin of History Department. National Ch'eng Kung University.
Chinese or English.

1. China—History—Collected works. 2. Taiwan—History—Collected works. I. Kuo li Ch'eng-kung ta hsüeh. Li shih hsüeh hsi. II. Kuo li Ch'eng-kung ta hsüeh. Li shih hsüeh hsi. Kuo li Ch'eng-kung ta hsüeh li shih hsüeh hsi li shih hsüeh pao. III. Kuo li Ch'eng-kung ta hsüeh. Li shih hsüeh hsi. Bulletin of History Department. National Ch'eng Kung University.
DS735.L475 79-642211 (MARC-S)

Miyazaki, Ichisada, 1901–
(Miyazaki Ichisada Ajia shi ron kō)
宮崎市定アジア史論考. — 東京：朝日新聞社, 昭和51 [1976]
3 v. : ill. ; 22 cm. Ja 76-3753 (v. 1)
Includes bibliographical references and indexes.
CONTENTS: 1巻 概説編—2巻 古代中世編—3巻 近世編
¥4500 per vol.

1. Asia—History—Collected works. 2. China—History—Collected works. I. Title. II. Title: Ajia shi ron kō.
DS33.M566 76-806603

—HISTORY—DICTIONARIES

Dillon, Michael, Dr.
Dictionary of Chinese history / Michael Dillon. — London : F. Cass, 1979.
viii, 239 p. : map ; 22 cm. GB***
ISBN 0-7146-3107-8 : £9.95 ($22.50 U.S.)
1. China—History—Dictionaries. I. Title.
DS733.D54 1979 951'.003 79-310371
79 MARC

Yen, T'ai-pai.
(Chung-kuo li shih tz'u hui)
中國歷史辭彙 / [編作者嚴太白]. — 臺北：希代書版有限公司, 民國67 [1978]
191 p. ; 19 cm. — (希代文叢 ; 3)
Series romanized: Hsi tai wen ts'ung.
NT$45.00

1. China—History—Dictionaries. I. Title.
DS733.Y46 78-843481

—HISTORY—ERRORS, INVENTIONS, ETC.

Ch'ien, Ta-hsin, 1728–1804.
(Chu shih shih i)
諸史拾遺 / 錢大昕撰. — 台北：廣文書局, 民國67 [1978]
219 p. ; 22 cm. — (史料七編)
Reprint of the ed. published by Ch'ang-sha Lung shih chia shu.
NT$180.00

1. China—History—Errors, inventions, etc. I. Title. II. Series: Shih liao ch'i pien.
DS736.C515 1978 79-838699

(Ssu shih pien i)
四史辨疑 / 楊家駱主編. — 臺北：鼎文書局, 民國66 [1977]
2, 21, 572 p. ; 31 cm. — (中國學術類編)
CONTENTS: 梁玉繩 史記志疑.—錢大昭 漢書辨疑. 後漢書辨疑. 續漢書辨疑. 三國志辨疑.

1. China—History—Errors, inventions, etc. I. Yang, Chia-lo. II. Liang, Yü-sheng, 1744–1819. Shih chi chih i. 1977. III. Ch'ien, Ta-chao, 1744–1813. IV. Series: Chung-kuo hsüeh shu lei pien.
DS736.S76 79-838607

—HISTORY—ADDRESSES, ESSAYS, LECTURES

Wu, Yü-chang, 1878–1966.
(Li shih wen chi)
历史文集 / 吳玉章. — 香港：生活·讀書·新知三联书店香港分店, 1978.
6, 138 p., [2] leaves of plates : ill. ; 22 cm.
Reprint ed.
Includes bibliographical references.
$4.00

1. China—History—Addresses, essays, lectures. 2. China—History—Methodology—Addresses, essays, lectures. I. Title.
DS736.W87 1978 79-839569

—HISTORY—JUVENILE LITERATURE

Shichiri, Jūkei.
(Seishōnen no tame no yūhō Shina rekishi monogatari)
青少年のための友邦支那歴史物語 / 七理重惠著. — 東京：フタバ書院成光館, 昭和17 [1942]
3, 1, 16, 224 p., [1] leaf of plates : ill. ; 22 cm.

1. China—History—Juvenile literature. I. Title. II. Title: Shina rekishi monogatari.
DS735.S4784 78-814781

—HISTORY—MISCELLANEA

Chang, Wen-ts'an.
(Chung-kuo shih wen t'i hsiang hsi)
中國史問題詳析 / 張文粲編著. — 九龍：漢文出版社, 1970.
4, 129 p. ; 19 cm.

1. China—History—Miscellanea. I. Title.
DS736.C4577 78-842463

(Chung-kuo shih hsüeh ming chu hsüan tu)
中國史學名著選讀 / 中國文化學院史學系編印. — [陽明山]：該系, [1971?]
254 p. ; 21 cm.
Cover title.

1. China—History—Miscellanea. I. Chung-kuo wen hua hsüeh yüan. Shih hsüeh hsi.
DS736.C555 79-840628
C

Hsiang-kang ch'ing nien ch'u pan she.
(Chung wai li shih wen t'i hsüan chiang)
中外歷史問題選講 / [本社編]. — 香港：香港青年出版社, 1977.
6, 184 p. ; 19 cm. — (基本知識叢刊)
Series romanized: Chi pen chih shih ts'ung k'an.
$5.00

1. China—History—Miscellanea. 2. History—Miscellanea. I. Title.
DS736.H72 1977 79-840788

—HISTORY—PICTORIAL WORKS

(Chung-hua min kuo shih hua)
中華民國史畫 / [編輯者中華民國史畫編纂小組]. — 臺北：近代中國出版社：經銷中央文物供應社, 民國67 [1978]
3 v. : ill. (some col.) ; 27 x 39 cm.
Includes bibliographical references.
NT$3000.00 ($100.00 U.S.)

1. China—History—Pictorial works. 2. Taiwan—History—Pictorial works. I. Chung-hua min kuo shih hua pien tsuan hsiao tsu.
DS736.C537 78-843008

Hsieh, Min-ts'ung.
(Chung-hua li shih t'u chien)
中華歷史圖鑑 / 編著者謝敏聰. — 台北：聯經出版事業公司, [民國67 i. e. 1978序]
492 p. : ill. (some col.) ; 31 cm.
Bibliography: p. 491-492.
NT$1,200.00

1. China—History—Pictorial works. I. Title.
DS736.H744 79-836818

—HISTORY—PROPHECIES

Yasui, Kōzan, 1921–
(Yogen to kakumei)
予言と革命：ここにも大変革の危機がある / 安居香山. — 京都：探究社, 昭和51 [1976]
250 p. : ill. ; 19 cm. Ja 77-9281
¥850

1. Japan—History—Prophecies. 2. China—History—Prophecies. I. Title.
DS835.Y34 78-812750

—HISTORY—STUDY AND TEACHING

(Chung-kuo ku shih yen chiu fa)
中國古史研究法 / 本社編. — 台北：漢苑出版社；板橋：總經銷竹林圖書屋, 民國66 [1977]
6, 302 p. ; 22 cm. — (東亞文學叢書)
At head of title: Ta hsüeh yung shu.
NT$120.00

1. China—History—Study and teaching. I. Han yüan ch'u pan she. II. Series: Tung Ya wen hsüeh ts'ung shu.
DS734.95.C547 79-838708

—HISTORY, LOCAL—BIBLIOGRAPHY

Byon, Jae-Hyon.
Local gazeteers of Southwest China : a handbook / by Jae-Hyon Byon ; [calligraphy by Eva Chung]. — Seattle : School of International Studies, University of Washington, 1979.
xii, 140 p. ; 27 cm. — (Parerga ; 5)
Bibliography: p. 139-140.
1. China—History, Local—Bibliography. I. Title. II. Series.
Z3107.A2B96 016.9151'003 79-118036
[DS737] 79 MARC

—HISTORY, LOCAL—BIBLIOGRAPHY—CATALOGS

Tōyō Bunko, Tokyo.
(Tōyō Bunko chihōshi mokuroku)
東洋文庫地方志目錄：支那·滿洲·臺灣. — 東京：東洋文庫, 昭和10 [1935]
2, 4, 2, 237, 36 p. ; 27 cm.
Includes index.

1. China—History, Local—Bibliography—Catalogs. 2. Manchuria—History, Local—Bibliography—Catalogs. 3. Taiwan—History, Local—Bibliography—Catalogs. 4. Tōyō Bunko, Tokyo—Catalogs. I. Title.
Z3106.T68 1935 78-813185
[DS735]

CHINA (Continued)

—HISTORY, MILITARY

Ch'en, Sen-fu.
(Hsi shuo Hsi-pei chün)
細說西北軍 / 陳森甫著. — 臺南：德華出版社, 民國66 [1977]
2, 3, 538 p., [12] leaves of plates : ill., ports. ; 20 cm. — (愛書人文庫 ; 055)
Series romanized: Ai shu jen wen k'u.
"馮玉祥年譜": p. 517–537.
Bibliography: p. 538.
NT$150.00
1. China—History, Military. 2. Feng, Yü-hsiang, 1882–1948. 3. Kuo min chün—History. 4. Generals—China—Biography. I. Title. II. Title: Hsi-pei chün.
UA837.C439 79–836454

China (People's Republic of China, 1949–). Chung-kuo jen min chieh fang chün. 52843 pu tui.
(Ku tai chan cheng shih li)
古代战争十例 / [中国人民解放軍]五二八四三部队, 山西师范学院中文系. — [香港?：s. n., 1977?]
2, 191 p. : maps ; 22 cm.
Reprint of the 1975 ed. published by Shan-hsi jen min ch'u pan she, T'ai-yüan.
1. China—History, Military. 2. Battles—China. I. Shan-hsi shih fan hsüeh yüan. Chung wen hsi. II. Title.
DS738.C524 1977 78–841265

Feng, Tso-min.
(Chung-kuo ming chiang lieh chuan)
中國名將列傳 / 馮作民編著. — 臺北：星光出版社, 民國67 [1978]
271 p. ; 20 cm. — (雙子星叢書 ; 66號)
"本書是以日人守屋洋的「中國名將列傳」為藍本改寫而成."
Series romanized: Shuang tzu hsing ts'ung shu.
NT$45.00
1. Generals—China—Biography. 2. China—History, Military. I. Moriya, Hiroshi, 1932– Chūgoku saishō retsuden. II. Title.
DS734.F45 79–835419

Ku, Tsu-yü, 1624–1680.
[Ku shih tu shih fang yü chi yao ching lüeh. Selections]
顧氏讀史方輿紀要精略 / 顧祖禹撰. — 台北：老古出版社, 民國67 [1978]
310, 318, [13] p. ; 22 cm. — (正統謀略學彙編：初輯 ; 第18册)
"精略之京省每章序文, 取疏達通甫先生註釋為準, 京省內容則取辛鍾靈先生輯本"
"附: 顧炎武 天下郡國利病書論略" p. 1–318 (2d group)
1. China—Historical geography. 2. China—History, Military. 3. China—Description and travel. I. Ku, Yen-wu, 1613–1682. T'ien hsia chün kuo li ping shu. Selections. 1978. II. Title. III. Series: Cheng t'ung mou lüeh hsüeh hui pien : Ch'u chi ; ti 18 ts'e.
Title romanized: Ku shih tu shih fang yü chi yao ching lüeh.
DS706.5.K822 1978 79–840130

Li, T'ien-ming.
(Sung Yüan Hsiang Fan chan i chih yen chiu)
宋元襄樊戰役之研究 / 李天鳴撰. — [台北?：李天鳴], 民國62 [1973]
202 double leaves : ill., maps ; 26 cm.
Cover title.
Reproduced from typescript.
"私立中國文化學院史學研究所碩士論文"
Bibliography: leaves 191–202.
1. China—History, Military. 2. Battles—China. 3. China—History—Sung dynasty, 960–1279. 4. China—History—Yüan dynasty, 1260–1368. I. Title.
DS738.L45 1973b 79–835350

Liu, Yü-chang, 1903 or 4–
(Jung ma wu shih nien)
戎馬五十年：劉玉章回憶錄. — [s. l. : s. n.], 民國64 [1975] (臺北：海天印刷廠)
16, 8, 228 p., [9] leaves of plates (1 fold.) : ill. ; 22 cm.
Autobiographical.
1. China—History, Military. 2. Liu, Yü-chang, 1903 or 4– I. Title.
DS738.L53A34 78–840797

Watanabe, Ryūsaku, 1903–
(Banri no chōjō)
万里の長城：攻防戰史 / 渡辺龍策著. — 東京：秀英書房, 1978.
306, [3] p. : ill. ; 20 cm. Ja 78–26443
Bibliography: p. [308]–[309]
¥1400
1. Great Wall of China. 2. China—History, Military. I. Title.
DS793.G67W34 79–805614

—HISTORY, MILITARY—ADDRESSES, ESSAYS, LECTURES

(Hsing huo liao yüan)
星火燎原：选编之一. — 第1版. — 北京：中国人民解放军战士出版社, 1977.
233 p. ; 21 cm.
CONTENTS: 朱德 人民军队人民战争.—聂荣臻 南昌起义的历史意义和经验教训.—罗荣桓 秋收起义与我军初创时期.—陈毅 学习毛主席的马克思列宁主义的创造作风.—叶剑英 大革命失败与广州起义.—徐向前奔向海陆丰.—贺龙 湘鄂西初期的革命斗争.—徐向前 鄂豫皖红军的反围攻斗争.—刘伯承 回顾长征.—聂荣臻 结束第二次国内革命战争的最后一仗.—刘伯承 我们在太行山上.—陈毅 江南抗战之春.—聂荣臻 中国人民怎样战胜了日本法西斯侵略者.—刘伯承千里跃进大别山.—叶剑英 伟大的战略决战.
$0.57
1. China—History, Military—Addresses, essays, lectures. 2. Chung-kuo jen min chieh fang chün—History—Addresses, essays, lectures.
DS738.H724 78–842667
C

—IMPRINTS

Chang, Chih-tung, 1837–1909.
(Shu mo ta wen)
書目答問 / 張文襄[之洞]著. — 上海：掃葉山房, 宣統元年 [1909]
2 v. (on double leaves) ; 20 cm.
Lithoprinted.
1. China—Imprints. I. Title.
Z3101.C42 1909 78–843739

Kobase, Keikichi, 1865–1944.
(Nihon-koku genzaisho mokuroku kaisetsu kō)
日本國見在書目錄解說稿 附·同書目錄同書索引 [藤原佐世撰] 小長谷惠吉著 [再刊] 東京 小宮山書店 昭和31 [1956]
88, 22, 28, 10 p. illus. 22 cm.
300部刊行の第83番
1. Fujiwara, Sukeyo, d. 898. Nihon-koku genzaisho mokuroku. 2. China—Imprints. I. Fujiwara, Sukeyo, d. 898. Nihon-koku genzaisho mokuroku, 1956. II. Title.
Z3102.K6 1956 J 67–4016

K'ung, Chien-kuo.
(Wen hsien t'ung k'ao ching chi k'ao yen chiu)
文獻通考經籍考研究 / 孔建國撰. — [台北?：s. n., 1978?]
137 p. (on double leaves) ; 27 cm.
Reproduced from typescript.
Thesis (M.A.)—Kuo li cheng chih ta hsüeh, 1975.
Bibliography: p. 131–137.
1. China—Imprints. 2. Ma, Tuan-lin, fl. 1273. Wen hsien t'ung k'ao. I. Title.
Z3102.K84 1978 79–838248

P'an, Mei-yüeh.
(San ch'ao pen k'ao)
三朝本考 / 潘美月著. — [台北?：潘美月], 民國53 [1964]
74 double leaves ; 27 cm.
Cover title.
"國立台灣大學中國文學研究所碩士論文"
Reproduced from ms. copy.
Bibliography: leaves 72–74.
1. China—Imprints. I. Title.
Z3101.P36 79–835893

(Pi chi hsiao shuo ta kuan tsung so yin)
筆記小說大觀總索引. — [台北：新興書局, 1978?]
123, 14, 4, 234 p. : 22 cm.
Appendices: 新書簡報.—范希曾 書目答問補正.
1. Chinese essays—Bibliography. 2. Chinese fiction—Bibliography. 3. China—Imprints. I. Fan, Hsi-tseng, 1900–1930. Shu mu ta wen pu cheng. 1978.
Z3108.L5P5 79–836830
[PL2395]

A Sung bibliography = Bibliographie des Sung / initiated by Etienne Balazs ; edited by Yves Hervouet. — Hong Kong : Chinese University Press, c1978.
xxxiv, 598 p. ; 28 cm.
Introduction and instructions to the reader in English and French. Notices in English or French.
Errata slip inserted.
ISBN 9622011586 : $60.00
1. China—Imprints. 2. China—History—Sung dynasty, 960-1279—Bibliography. I. Balazs, Étienne, 1905-1963. II. Hervouet, Yves. III. Title: Bibliographie des Sung.
Z3102.S77 016'.951'02 79-106852
79 MARC

(Wen lu t'ang shu ying)
文祿堂書影 / [編輯者王文進]. — 臺北縣板橋：藝文印書館, 民國65 [1976]
10, [101] p. (on double leaves) ; 38 cm.
In case.
Reprint ed.
1. China—Imprints. 2. Bibliography—Rare books. I. Wang, Wen-chin, fl. 1937–
Z3101.W45 1976 78–842336

Yajima, Genryō.
(Wa-Kan kosho mokuroku hō no chishiki)
和漢古書目録法の知識 / 矢島玄亮著. — 仙台：万葉堂書店, 昭和51 [1976]
291 p. ; 22 cm. Ja 77–5245
¥2800
1. Bibliography—Theory, methods, etc. 2. Bibliography—Early printed books. 3. China—Imprints. 4. Japan—Imprints. I. Title.
Z1001.Y33 79–800219

[Yeh, Ch'ang-ch'ing] 1898 or 9–
(Han shu i wen chih wen ta)
漢書藝文志問答 正中書局編審委員會編著 [臺北] 正中書局 [民國58 i. e. 1969]
2, 204 p. 22 cm. (國學萃編)
Series romanized: Kuo hsüeh ts'ui pien.
Originally published in 1940 under the author's name 葉長青
1. Pan, Ku, 32–92. Han shu i wen chih. 2. China—Imprints. I. Cheng chung shu chü, T'ai-pei. II. Title.
Z3102.P292Y43 1969 70–840534

—IMPRINTS—CATALOGS

Koryŏ Taehakkyo, Seoul, Korea. Chungang Tosŏgwan.
(Hwasan mun'go Hanjŏk mongnok)
華山文庫漢籍目錄 = Catalogue of the Hwa-san collection. — [서울]：高麗大學校中央圖書館, 1976.
203 p., [1] leaf of plates : port. ; 26 cm. — (高麗大學校藏書目錄 ; 第13輯)
In Korean.
Includes index.
1. Chinese imprints—Korea—Catalogs. 2. China—Imprints—Catalogs. 3. Bibliography—Rare books—Catalogs. 4. Yi, Sŏng-ŭi—Library—Catalogs. 5. Koryŏ Taehakkyo, Seoul, Korea. Chungang Tosŏgwan—Catalogs. I. Title. II. Title: Catalogue of the Hwa-san collection. III. Series: Koryŏ Taehakkyo, Seoul, Korea. Koryŏ Taehakkyo changsŏ mongnok ; che 13-chip.
Z3316.K69 1976 78–826504

Koryŏ Taehakkyo, Seoul, Korea. Chungang Tosŏgwan.
(Kyŏnghwadang mun'go mongnok)
景和堂文庫目録 = Catalogue of the Gyeong-hwa-dang collection. — [서울]：高麗大學校中央圖書館, 1975.
138 p., [1] leaf of plates : port. ; 26 cm. — (高麗大學校藏書目録 ; 第12輯)
Includes index.
1. Chinese imprints—Catalogs. 2. China—Imprints—Catalogs. 3. Bibliography—Rare books—Catalogs. 4. Pak, Hyŏng-yun—Library—Catalogs. 5. Koryŏ Taehakkyo, Seoul, Korea. Chungang Tosŏgwan—Catalogs. I. Title. II. Title: Catalogue of the Gyeong-hwa-dang collection. III. Series: Koryŏ Taehakkyo, Seoul, Korea. Koryŏ Taehakkyo changsŏ mongnok ; che 12-chip.
Z3316.K69 1975 78–826800

CHINA

—IMPRINTS—CATALOGS (Continued)

Temmagū, Osaka.
(Ōsaka Temmangū gobunko Kanseki bunrui mokuroku)
大阪天満宮御文庫漢籍分類目録 / [長澤規矩也, ちづ編]. — 大阪：大阪天満宮, 1977.
4, 81, 14 p., [2] leaves of plates : ill. ; 26 cm. Ja 78-4494
Includes index.

1. China—Imprints—Catalogs. 2. Chinese imprints—Japan—Catalogs. 3. Japan—Imprints—Catalogs. 4. Temmagū, Osaka—Library—Catalogs. I. Nagasawa, Kikuya, 1902- II. Nagasawa, Chizu. III. Title.
Z3301.T44 78-812537

—IMPRINTS—PERIODICALS

(T'u shu kuan nien chien)
圖書年鑑
[Washington, Center for Chinese Research Materials, Association of Research Libraries,
v. 22 x 28 cm.
Reprint of the 1933 yearbook published by Chung-kuo t'u shu ta tz'u tien pien chi kuan, Nanking.
Vol. for issued in 2 or more parts.

1. Libraries—China—Periodicals. 2. Publishers and publishing—China—Periodicals. 3. China—Imprints—Periodicals.
Z845.C5T82 79-642137
MARC-S

—INDUSTRIES

Hubbard, Gilbert Ernest, 1885- comp.
Eastern industrialization and its effect on the West, with special reference to Great Britain and Japan / by G. E. Hubbard, assisted by Denzil Baring ; with a conclusion by T. E. Gregory. — Wesport, Conn. : Greenwood Press, 1979.
xxii, 395 p. ; 23 cm.
Reprint of the 1935 ed. published by Oxford University Press, London.
Bibliography: p.
Includes index.
ISBN 0-8371-5726-9
1. Japan—Industries. 2. China—Industries. 3. India—Industries. 4. Great Britain—Commerce. 5. Great Britain—Economic conditions—1918-1945. 6. Competition, International. I. Baring, Denzil. II. Title.
HC462.H87 1979 338'.095 77-138614
79 MARC

Nearing, Scott, 1883-
Whither China? : An economic interpretation of recent events in the Far East / by Scott Nearing. — Westport, Conn. : Hyperion Press, 1977, c1927.
225 p. ; 23 cm. — (China studies)
Reprint of the ed. published by International Publishers, New York.
Bibliography: p. 211-214.
Includes index.
ISBN 0-88355-385-6 : $16.50
1. China—Economic conditions—1912-1949. 2. China—Industries. 3. China—Social conditions. 4. China—History—1900- . 5. China—Foreign relations. I. Title. II. Series: China studies (Westport, Conn.)
HC427.N38 1977 330.9'51'05 75-39029
77 MARC

United States. Central Intelligence Agency. National Foreign Assessment Center.
China: gross value of industrial output, 1965-77 : a research paper / National Foreign Assessment Center. — [Washington] : The Center, 1978.
iv, 69 p. ; 27 cm.
Cover title.
"ER78-10355."
1. Gross national product—China. 2. China—Industries. I. Title.
HC430.I5U53 1978 339.351 78-602573
79 MARC

Yang, Ta-chin, ed.
(Hsien tai Chung-kuo shih yeh chih)
現代中國實業誌 / 楊大金編. — 台1版. — 台北：華世出版社, 民國67 [1978]
3 v. ; 22 cm. — (中國經濟史料叢書：第一輯；第5種)
CONTENTS: 第1-2冊. 製造業.—第3冊. 礦冶業.
NT$1400.00

1. China—Industries. I. Title. II. Series: Chung-kuo ching chi shih liao ts'ung shu : Ti i chi ; ti 5 chung.
HC427.8.Y3 1978 79-841395

—INDUSTRIES—1949-1976

Ching, Ch'ih.
(Tzu pen chu i kung shang yeh ti she hui chu i kai tsao)
资本主义工商业的社会主义改造 / 景池. — 第1版. — 上海：上海人民出版社, 1976.
80 p. ; 19 cm.
Includes bibliographical references.
$0.16
1. China—Economic conditions—1949- 2. China—Industries—1949- I. Title.
HC427.9.C56548 79-835450

Christiansen, Eigil, 1945-
Kina—industri, plan, menneske / [af] Eigil Christiansen og Kjeld A. Larsen ; [Kort og diagrammer er udført af Mogens Skjoldager]. — [København] : Gyldendal, 1978.
103 p. : ill. ; 30 cm. — (Regionalgeografiske analyser) D78-21
Bibliography: p. 92-93.
Includes index.
ISBN 8701588311 : kr68.50
1. China—Industries—1949-1976. 2. China—Economic policy—1949- . I. Larsen, Kjeld Allan, 1942- joint author. II. Title. III. Series.
HC427.9.C58116 78-380758
79 MARC

Ting, Kuo-chen, fl. 1950- ed.
(Hsin kung shang ts'an k'ao tzu liao)
新工商參考資料 丁國楨編 北京 五十年代出版社 1950.
2, 5, 238 p. tables. 18 cm.

1. China—Economic policy—1949-1976. 2. China—Industries—1949-1976. 3. China—Commerce. I. Title.
HC427.9.T57 C 64-1039

Ting, Kuo-chen, fl. 1950- ed.
(Hsin kung shang ts'an k'ao tzŭ liao)
新工商参考資料 丁國楨編 增訂本 北京 五十年代出版社 1952.
2, 6, 313 p. tables. 19 cm.

1. China—Economic policy—1949-1976. 2. China—Industries—1949-1976. 3. China—Commerce. I. Title.
HC427.9.T57 1952 C 60-5309

—INDUSTRIES—1949-1976
—ADDRESSES, ESSAYS, LECTURES

Hua ch'iao kung shang chieh hui kuo kuan kuang t'uan.
(Hsin Chung-kuo kung shang yeh chia ti tao lu)
新中國工商業家的道路 / 華僑工商界回國觀光團編印. — [香港：s. n., 1978?]
285 p. : ill. ; 27 cm.
Reprint of the 1950 ed. published by Hua ch'iao ching chi ch'u pan she.

1. China—Industries—1949-1976—Addresses, essays, lectures. I. Title.
HC427.9.H833 1978 78-843945

(Kung yeh hsüeh Ta-ch'ing)
工业学大庆. — [s. l.]：越秀区工业局翻印, 1977.
3, 64 p. ; 18 cm.
Cover title.
Reprint ed.

1. China—Industries—1949-1976—Addresses, essays, lectures.
HC427.9.K856 1977 79-838269

—INDUSTRIES—1976-

(Chūgoku no keikōgyō no genjō)
中国の軽工業の現状. — 東京：日中経済協会, 1978.
167, 5 p. ; 25 cm. — (中国経済関係調査報告書) (日中経報 ; no. 109)
Ja 78-19935
Cover title.
At head of title: 52-chō-16.
Includes bibliographical references.
¥630
1. China—Industries—1976- I. Nitchū Keizai Kyōkai. II. Series: Chūgoku keizai kankei chōsa hōkokusho. III. Series: Nitchū Keizai Kyōkai. Nitchūkei hō ; no. 109.
HC427.92.C5 78-814312

(Chūgoku shuyō sangyō no genjō)
中国主要産業の現状. — 東京：日中経済協会, 1978.
200 p. : ill. ; 25 cm. — (日中経報 ; no. 101) Ja 78-19893
Cover title.
"52-調-8"

1. China—Industries—1976- 2. Industry and state—China. I. Nitchū Keizai Kyōkai. II. Series: Nitchū Keizai Kyōkai. Nitchūkei hō ; no. 101.
HC427.92.C52 78-813974

*78 J

—INDUSTRIES—1976-
—ADDRESSES, ESSAYS, LECTURES

Ch'üan kuo kung yeh hsüeh Ta-ch'ing hui i, Ta-ch'ing, China and Peking, China, 1977.
(Chien ch'ih tang ti chi pen lu hsien tsai tou cheng chung hsüeh Ta-ch'ing)
坚持党的基本路线在斗争中学大庆：全國工业学大庆会议发言汇编. — 第1版. — 北京：人民出版社, 1977.
4, 395 p. ; 21 cm.
$0.82
1. China—Industries—1976- —Addresses, essays, lectures. I. Title.
HC427.92.C48 1977 78-841460

C

(Jen chen hsüeh hsi Ta-ch'ing ching yen, chien ch'ih ho li kuei chang chih tu)
认真学习大庆经验, 坚持合理规章制度. — [香港? : s. n., 1978?]
59 p. ; 21 cm.
Reprint of the 1977 ed. published by Kuang-tung jen min ch'u pan she.
CONTENTS: 中国共产党大庆委员会 大庆油田政治工作要点.—政治工作要结合经济工作一道去做.—中国共产党大庆委员会 领导班子革命化的措施.—彻底的革命精神.—中共大庆机关委员会 发扬党的优良作风, 搞好机关革命化.—大庆油田政治部 加強基层建设的六条要求.—石油化学工业部大批判组 加強基层建设, 搞好企业大治.—大庆革命委员会 岗位责任制度.—在革命的岗位上.

1. China—Industries—1976- —Addresses, essays, lectures. 2. Industry and state—China—Addresses, essays, lectures. 3. Chung-kuo kung ch'an tang—Party work—Addresses, essays, lectures.
HC427.92.J46 1978 79-836867

—INDUSTRIES—1976- —
CONGRESSES

Ch'üan kuo kung yeh hsüeh Ta-ch'ing hui i, Ta-ch'ing, China and Peking, China, 1977.
The National Conference on Learning from Taching in Industry, selected documents. — 1st ed. — Peking : Foreign Languages Press, 1977.
89 p. ; 19 cm.
Translation of Ch'üan kuo kung yeh hsüeh Ta-ch'ing hui i wen chien hsüan pien.
1. China—Industries—1976- —Congresses. 2. Industry and state—China—Congresses. I. Title.
HC427.92.C48 1977a 338'.0951 77-155805
79 MARC

—INDUSTRIES—HANDBOOKS, MANUALS, ETC.

Terry, Edith.
China's foreign trade organizations, domestic corporations, and ministries / compiled by Edith Terry. — [Washington] : National Council for US-China Trade, 1979.
68 p. : ill. ; 23 x 11 cm.
Cover title: Pocket guide to China's foreign trade organizations, domestic corporations, and ministries.
1. China—Commerce—Handbooks, manuals, etc. 2. China—Industries—Handbooks, manuals, etc. 3. Foreign trade promotion—China—Handbooks, manuals, etc. I. National Council for United States-China Trade. II. Title. III. Title: Pocket guide to China's foreign trade organizations, domestic corporations, and ministries.
HF3837.T47 382'.025'51 79-88564
79 MARC

CHINA (Continued)

—INDUSTRIES—HISTORY

Kung, Chün.
(Chung-kuo hsin kung yeh fa chan shih ta kang)
中國新工業發展史大綱 / 龔俊 [i. e. 駿] 編. — 台1版. — 台北 : 華世出版社, 民國67 [1978]
2, 2, 13, 302 p. ; 22 cm.
Bibliography: p. 292-302.
NT$80.00

1. China—Industries—History. I. Title.

HC427.7.K85 79-838234

Shima, Ichirō, 1937-
(Chūgoku minzoku kōgyō no tenkai)
中国民族工業の展開 / 島一郎著. — 京都 : ミネルヴァ書房, 1978.
x, 324 p. ; 22 cm. Ja 78-17419
Bibliography: p. viii-ix.
¥2600

1. China—Industries—History. 2. Textile industry—China—History. I. Title.

HC427.8.S529 78-813634

—INDUSTRIES—HISTORY—SOURCES

Sun, Yü-t'ang, 1910-
[Chung-kuo chin tai kung yeh shih tzu liao]
中國近代史工業史資料 / 孫毓棠編. — [台北縣永和鎮] : 文海出版社, [1979?-
v. in : ill. ; 22 cm. — (近代中國史料叢刊續編 [i. e. 輯] ; 611- : 第62輯)
Reprint of the 1957 ed. published by K'o hsüeh ch'u pan she, Peking, which was issued as Chung-kuo chin tai ching chi shih ts'an k'ao tzu liao ts'ung k'an, ti 2 chung.
1. China—Industries—History—Sources. I. Title. II. Series: Shen, Yü-lung, comp. Chin tai Chung-kuo shih liao ts'ung k'an hsü chi ; 611. III. Series: Chung-kuo chin tai ching chi shih ts'an k'ao tzu liao ts'ung k'an ; ti 2 chung.
Title romanized: Chung-kuo chin tai shih kung yeh shih tzu liao.

HC427.7.S79 1979 79-840605

—INTELLECTUAL LIFE

see also May fourth movement

(Gendai Shina shisō no shomondai)
現代支那思想の諸問題 / 神谷正男譯編. — 東京 : 生活社, 昭和15 [1940]
5, 12, 271, 16, 21 p. : 22 cm.
Bibliography: p. 257-271.
Includes indexes.
"郭湛波氏の「近五十年中國思想史」...綵尚志氏の論文「近三十年來の中國思想界」を加えて本文となし、外に二・三の参考論文を附録として編纂飜譯した"
1. China—Intellectual life. 2. China—Civilization—Occidental influences. 3. Learning and scholarship—China. I. Kamiya, Masao, 1910-1972. II. Kuo, Chan-po. Chin wu shih nien Chung-kuo ssu hsiang shih. Japanese. Selections. 1940. III. Ts'ai, Shang-ssu, 1905- Kin sanjūnenrai no Chūgoku shisōkai. 1940.

DS721.G36 78-815255

Jaeger, Georgette.
Les Lettrés chinois : poètes T'ang et leur milieu / Georgette Jaeger. — Neuchâtel : Éditions de la Baconnière, cop. 1977.
347 p. ; 21 cm. Sw78-978
Bibliography: p. 339-342.
Includes index.
ISBN 2-8252-0950-3 : 39.00F
1. Chinese poetry—T'ang dynasty, 618-907—History and criticism. 2. China—Intellectual life. 3. China—History—T'ang dynasty, 618-907. I. Title.
PL2321.J3 895.1'3'09 78-368780
78 MARC

Kan, Takayuki, 1939-
(Takeuchi Yoshimi ron)
竹内好論 : 亜細亜への反歌 / 菅孝行. — 東京 : 三一書房, 1976.
273 p. ; 20 cm. Ja 76-6359
List of Y. Takeuchi's works: p. 272.
Bibliography: p. 273.
¥1600
1. Takeuchi, Yoshimi, 1910-1977. 2. Philologists—China—Biography. 3. Philologists—Japan—Biography. 4. Japan—Intellectual life. 5. China—Intellectual life. I. Title.

PL1064.T3K3 78-815397

Lin, Yü-sheng, 1934-
The crisis of Chinese consciousness : radical antitraditionalism in the May Fourth era / Lin Yü-sheng. — Madison : University of Wisconsin Press, 1979.
xiv, 201 p. ; 24 cm.
Added Chinese title romanized: Chung-kuo i shih ti wei chi.
Bibliography: p. 165-179.
Includes index.
ISBN 0-299-07410-2 : $20.00
1. China—Intellectual life. 2. May fourth movement. I. Title.
DS721.L567 951.05 77-91057
78 MARC

Okuzaki, Hiroshi, 1935-
(Chūgoku kyōshin jinushi no kenkyū)
中國郷紳地主の研究 / 奥崎裕司著. — 東京 : 汲古書院, 1978.
15, 599, 9, 57 p., [1] fold. leaf of plates : ill. ; 22 cm. Ja***
Includes bibliographical references and index.
¥8000

1. Upper classes—China. 2. China—Gentry. 3. Land tenure—China. 4. China—Intellectual life. 5. Yüan, Huang, chin shih 1586. 6. China—Biography. I. Title.

HT653.C6O48 79-800227

(Sud'by kul'tury KNR, 1949–1974)
Судьбы культуры КНР, 1949–1974 / [отв. редактор В. А. Кривцов]. — Москва : Наука, 1978.
380, [1] p. ; 21 cm. USSR***
At head of title: Akademiâ nauk SSSR. Institut Dal'nego Vostoka.
Bibliography : p. 372-[381]
1.40rub
1. China—Intellectual life. I. Krivtsov, Vladimir Alekseevich. II. Akademiâ nauk SSSR. Institut Dal'nego Vostoka.
DS777.55.S8284 79-412032

(T'a men chao ch'i p'eng po)
他们朝气蓬勃 / 广州部队生产建设兵团《兵团战士报》供稿. — 第1版. — [广州] : 广东人民出版社, 1973.
38 p. : chiefly ill. ; 22 cm.

1. China—Intellectual life. 2. Youth—China. I. Ping t'uan chan shih pao.

DS724.T28 78-843152

—INTELLECTUAL LIFE—ADDRESSES, ESSAYS, LECTURES

Ai, Ssu-ch'i, 1905-1966.
(Lun Chung-kuo t'e shu hsing chi ch'i ta)
論中國特殊性及其他 / 艾思奇著. — [香港? : s. n.] ; Hong Kong : 經銷處神州圖書公司, [1977?]
3, 274 p. ; 19 cm.
Reprint of the 1941 ed. published by Ch'en kuang shu tien, Hongkong.

1. China—Intellectual life—Addresses, essays, lectures. 2. Dialectical materialism—Addresses, essays, lectures. I. Title.

DS721.A37 1977 78-843831

Cheng pao ch'u pan she, Hongkong.
(Kuan yü chih shih fen tzu ti kai tsao)
関於知識分子的改造 正報出版社編 香港 1949.
45 p. 19 cm.

1. China—Intellectual life—Addresses, essays, lectures. 2. Communism—China—Addresses, essays, lectures. I. Title.

DS721.C48123 1949 C 62-3016 ‡
rev
[r78]

(Chih shih fen tzu ti ssu hsiang kai tsao wen t'i)
知識分子的思想改造問題 / 中國民主同盟南方總支部宣傳委員會編. — [香港? : s. n., 1978?]
2, 2, 204 p. ; 21 cm.
Reprint of the 1952 ed. published by Jen chien shu wu, Canton.

1. China—Intellectual life—Addresses, essays, lectures. 2. Intellectuals—China—Addresses, essays, lectures. I. Chung-kuo min chu t'ung meng. Nan fang tsung chih pu. Hsüan ch'uan wei yüan hui.

DS723.C426 1978 78-843976

Fei, Hai-chi.
(Han hsüeh fan pu chi)
漢學反哺集 費海璣編者 [臺北] 臺灣商務印書館 [1966]
2, 3, 2, 326 p. 18 cm. (人人文庫080-081)

1. China—Intellectual life—Addresses, essays, lectures. 2. China—Study and teaching—France—Addresses, essays, lectures. I. Title.

DS721.F36 C 67-2463

(Feng hua cheng mao)
风华正茂 : 上山下乡知识青年先进事迹选 / 福建人民出版社编辑. — [香港? : s. n., 1979?]
2, 108 p. ; 21 cm.
Reprint of the 1973 ed. published by Fu-chien jen min ch'u pan she.
CONTENTS: "她给俺们留下的是共产主义精神".—火红的青春.—让青春在广阔天地里大放光彩.—农村也是大学.—前进在金光大道上.—年轻的女书记.—扎根农村 茁壮成长.—青春献给北大荒.—她热爱祖国的农村.—阳光雨露育新苗.—我们这一代青年人.—铁手绘新图.—"奋斗"之歌.
1. China—Intellectual life—Addresses, essays, lectures. 2. Youth—China—Addresses, essays, lectures. I. Fu-chien jen min ch'u pan she, Foochow, China.
DS724.F46 1979 .79-840268

Kyōto Daigaku Shina Tetsugakushi Kenkyūkai.
(Tōyō bunka no mondai)
東洋文化の問題 / 京都大学支那哲学史研究會編, [代表者 重澤俊郎]. 第1號. — 京都 : 甲文社, 昭和24 [1949]
4, 311 p. ; 21 cm.
Colophon inserted.
No more published.

1. China—Intellectual life—Addresses, essays, lectures. I. Shigezawa, Toshio, 1906- II. Title.

DS721.K96 1949 78-813570

(Nung k'en hsin chün)
农垦新军 / 农林部农垦局編. — 第1版. — [北京] : 农业出版社, 1975.
2, 102 p. ; 19 cm.
$0.20

1. China—Intellectual life—Addresses, essays, lectures. 2. Youth—China—Addresses, essays, lectures. I. China (People's Republic of China, 1949-). Nung lin pu. Nung k'en chü.

DS724.N83 79-840055

Shan ho nung ch'ang. Chih shih ch'ing nien pien ch'uang tsu.
(Chih tsai pien chiang)
志在边疆 / 山河农场知识青年编创组. — 第1版. — 北京 : 农业出版社 : 新华书店北京发行所发行, 1976.
2, 104 p. : ill. ; 19 cm.

1. China—Intellectual life—Addresses, essays, lectures. 2. Youth—China—Addresses, essays, lectures. I. Title.

DS724.S48 1976 79-837733

(Ta yu hsi wang ti shih yeh)
大有希望的事业 : 推荐株洲市厂社挂钩, 集体安置知识青年到社队农·林·茶场的经验. — 本社编. — 第1版. — [長沙?] : 湖南人民出版社, 1974.
29 p. ; 19 cm.
Colophon title.

1. China—Intellectual life—Addresses, essays, lectures. 2. Youth—China—Addresses, essays, lectures. I. Hu-nan jen min ch'u pan she.

DS724.T3 1974 79-837154

Yen-an ti ch'ü, China. Ko ming wei yüan hui. Cheng kung tsu.
(Yen-an ching shen yü hsin jen)
延安精神育新人 : 北京插队知识青年在延安 / 陕西省延安地区革命委员会政工组编. — [香港? : s. n., 1978?]
4, 145 p. ; 21 cm.
Reprint of the 1972 ed. published by Jen min ch'u pan she, Peking.
1. China—Intellectual life—Addresses, essays, lectures. 2. Yenan, China—Intellectual life—Addresses, essays, lectures. I. Title.
DS724.Y46 1978 78-843460

CHINA

—INTELLECTUAL LIFE—ADDRESSES, ESSAYS, LECTURES (Continued)

(Ying feng chan ch'ih)
迎风展翅 / 湖南省株洲市委宣传部组织编写. — 第1版. — 北京：人民体育出版社, 1975.
116 p. : ill. ; 19 cm.
"报告文学集"
CONTENTS: 朱欣, 舒青 锻炼身体 坚持乡村.—王文成 工农呕心血 根实苗更壮.—春岭 茶场雏燕斗风雨.—张良善, 刘建辉 迎风展翅.—龚龙淦 山乡开遍映山红.—吳荣福 欲与天公试比高.—陈有源 春満山村.
1. China—Intellectual life—Addresses, essays, lectures. 2. Youth—China—Addresses, essays, lectures. 3. Physical fitness—Addresses, essays, lectures. I. Chu-chou, China. Shih wei. Hsüan ch'uan pu.
DS724.Y56 1975 79-836049

—INTELLECTUAL LIFE —BIBLIOGRAPHY

Sakade, Yoshinobu, 1934-
(Shin Kan shisō kenkyū bunken mokuroku)
秦漢思想研究文献目録 / 坂出祥伸編. — 吹田：関西大学出版広報部, 1978.
8, 137 p. ; 26 cm. Ja 78-17058
¥4800
1. China—Intellectual life—Bibliography. 2. China—History—Ch'in dynasty, 221-207 B. C.—Bibliography. 3. China—History—Han dynasty, 202 B. C.-220 A. D.—Bibliography. I. Title.
Z3108.C55S24 78-813240
[DS721]
[pt. 1] 本多龍成 支那經學史 I. [pt. 2] 山岸德平 日本漢文學史 [pt. 3] 松枝茂夫 支那現代文 II. [pt. 4] 長澤規矩也 漢籍解題 I. [pt. 5] 小柳司氣太 漢文學研究導言 阿部吉雄 近世儒學史上に於ける闇齋學の地位—第3卷 [pt. 1] 高田眞治 儒教倫理 I. [pt. 2] 内田泉之助 支那文學史 II. [pt. 3] 原富男 諸子學 II. [pt. 4] 杖下隆之 漢文典 I. [pt. 5] 藤原楚水 書道及金石—第4卷 [pt. 1] 内田泉之助 支那文學史 III. [pt. 2] 長澤規矩也 支那小說 I. [pt. 3] 松枝茂夫 支那現代文 III. [pt. 4] 林秀一 本邦に於ける鄭註孝經の刊行について—第5卷 [pt. 1] 高田眞治 儒教倫理 II. [pt. 2] 青木正兒 支那戲曲史 I.

—INTELLECTUAL LIFE —COLLECTED WORKS

(Kambungaku kōza)
漢文學講座 / [編輯代表者 長澤規矩也]. — 東京：共立社, 昭和8-9 [1933-1934]
7 v. : ill. ; 23 cm.
Cover title.
L. C. set: Parts of each vol. bound together.
Includes bibliographies.
CONTENTS: 第1卷 [pt. 1] 内田泉之助 支那文學史 I. [pt. 2] 原富男 諸子學 I. [pt. 3] 松枝茂夫 支那現代文 I. [pt. 4] 加藤常賢 支那家族制度に於ける主要問題 後藤俊瑞 朱子の倫理思想—第2卷 [pt. 3] 山口察常 漢文教授法 I. [pt. 4] 目加田誠 現代文藝 橋本成文 清朝尙書學—第6卷 [pt. 1] 原富男 諸子學 III. [pt. 2] 杖下隆之 漢文典 II. [pt. 3] 山口察常 漢文教授法 II. [pt. 4] 藤島亥治郎 建築 [pt. 5] 山田準 陽明學概論—第7卷 [pt. 1] 内田泉之助 支那文學史 IV. [pt. 2] 原富男 諸子學 IV. [pt. 3] 青木正兒 支那戲曲史 II. [pt. 4] 島田鈞一 清朝儒學史概說
1. Chinese philology—Collected works. 2. China—Intellectual life—Collected works. I. Nagasawa, Kikuya, 1902-
PL1025.K3 1933 78-815878

—INTELLECTUAL LIFE —SOURCES

(Genten Chūgoku kindai shisō shi)
原典中国近代思想史 / 西順蔵編. — 東京：岩波書店, 1976-1977.
6 v. : ill. ; 19 cm. Ja 77-798 (v. 1)
Includes bibliographical references.
CONTENTS: 1. アヘン戦争から太平天国まで—2. 洋務運動と変法運動—3. 辛亥革命—4. 五四運動から国民革命まで—5. 毛澤東思想の形成と発展—6. 国共分裂から解放戦争まで
¥10600
1. China—Intellectual life—Sources. I. Nishi, Junzō, 1914-
DS721.G368 77-812627

—JUVENILE LITERATURE

Rau, Margaret.
Our world : the People's Republic of China / by Margaret Rau. — New York : J. Messner, c1978.
127 p. : ill. ; 22 cm.
Includes index.
SUMMARY: An introduction to the third largest country in the world emphasizing the people's way of life.
ISBN 0-671-32951-0 lib. bdg. : $7.29.
1. China—Juvenile literature. [1. China] I. Title. II. Title: The People's Republic of China.
DS706.R38 1978 951.05 78-107425
79 MARC

—KINGS AND RULERS

Ming Ch'eng-tsu, Emperor of China, 1360-1424,
(Sheng hsüeh hsin fa)
聖學心法 / 明成祖撰. — [臺北?]：中國子學名著集成編印基金會, [1977?]
3, 3, 596 p. ; 22 cm. — (中國子學名著集成 ; 038 : 珍本初編, 儒家子部)
"四卷."
"明永樂七年内府刊本."
1. Kings and rulers—Duties. 2. China—Kings and rulers. I. Title. II. Series: Chung-kuo tzu hsüeh ming chu chi ch'eng ; 038.
JC381.M65 1977 79-837192

—KINGS AND RULERS —ANECDOTES, FACETIAE, SATIRE ETC.

(Chiu ch'ao t'an tsuan)
九朝談纂 / 明·不著撰人. — 臺北：偉文圖書出版社, 民國66 [1977]
3 v. (1502 p.) ; 22 cm. — (清代禁燬書叢刊 : 第一輯)
Photoreprint ed.
NT$1000.00 ($26.40 U. S.)
1. China—Kings and rulers—Anecdotes, facetiae, satire, etc. 2. China—History—Ming dynasty, 1368-1644—Anecdotes, facetiae, satire, etc. I. Series: Ch'ing tai chin hui shu ts'ung k'an ; ti i chi.
DS753.5.C46 1977 78-841445

—KINGS AND RULERS—BIOGRAPHY

Ch'ing Hsüan-t'ung, Emperor of China, 1906-1967.
From Emperor to citizen : the autobiography of Aisin-Gioro Pu Yi / [translated by W. J. F. Jenner]. — 2d ed. — Peking : Foreign Languages Press, 1979-
v. : ill. ; 22 cm.
Translation of Wo ti ch'ien pan sheng.
1. Ch'ing Hsüan-t'ung, Emperor of China, 1906-1967. 2. China—Kings and rulers—Biography. I. Title.
DS773.C51513 1979 951'.03'0924 79-112550
79 MARC

Der Ling, Princess.
Two years in the forbidden city / by the Princess Der Ling. — San Francisco : Chinese Materials Center, 1977 [c1911]
ix, 383 p., [17] leaves of plates : ill. ; 22 cm. — (Reprint series - Chinese Materials Center ; no. 68)
Reprint of the 1912 ed. published by Moffat, Yard, New York.
1. Tz'u-hsi, Empress dowager of China, 1835-1908. 2. China—Kings and rulers—Biography. 3. China—Court and courtiers. I. Title. II. Series: Chinese Materials Center. Reprint series - Chinese Materials Center ; no. 68.
DS763.T8D43 1977 951'.03 78-306763
78 MARC

Hsü, Liang-chih.
(I tai hsiung ts'ai Li Shih-min)
一代雄才李世民 / [著作者徐亮之]. — 臺北：河洛圖書出版社, 民國67 [1978]
2, 2, 239 p., [2] leaves of plates : ill. ; 22 cm. — (一代叢書 ; 4)
Reprint ed.
NT$50.00
1. T'ang T'ai-tsung, Emperor of China, 597-649. 2. China—History—T'ang dynasty, 618-907. 3. China—Kings and rulers—Biography. I. Title. II. Series: I tai ts'ung shu ; 4.
DS749.42.T35H76 1978 79-841853

Hsü, Liang-chih.
(T'ang T'ai-tsung hsin chuan)
唐太宗新傳 / 徐亮之著. — 九龍：現代出版公司, [1978?]
2, 2, 239 p., [2] leaves of plates : ill. ; 19 cm.
Cover title.
Reprint of the 1955 ed. published by Tzu yu ch'u pan she, Kowloon.
$5.00
1. T'ang T'ai-tsung, Emperor of China, 597-649. 2. China—Kings and rulers—Biography. I. Title.
DS749.42.T35H77 1978 79-837791

Kao, Hung-chün.
(Ying hsiung hao han)
英雄好漢 / 高宏俊編著. — 台北縣新店鎮：常春樹書坊, 民國67 [1978]
321 p. (p. 317-321 advertisements) : ill. ; 19 cm.
NT$50.00
1. Statesmen—China—Biography. 2. China—Kings and rulers—Biography. I. Title.
DS734.K37 79-840332

Moriya, Hiroshi, 1932-
(Hasha no senryaku to ketsudan)
覇者の戦略と決断 / 守屋洋著. — 東京：産業能率短期大学出版部, 1978.
218 p. : ill. ; 19 cm. Ja 78-23422
¥980
1. China—History—Spring and autumn period, 722-481 B. C.—Biography. 2. China—Kings and rulers—Biography. 3. China—Biography. I. Title.
DS747.15.M67 79-809183

—KING AND RULERS—CHRONOLOGY

Liu, Shih-hsin.
(Shang ku ti wang ta shih nien piao)
上古帝王大事年表 / 劉士心編. — [香港]：用舍堂書屋, 1973.
2, 8, 169 p. ; 22 cm.
"研讀文史必備參考書"
1. China—History—Chronology. 2. China—Kings and rulers—Chronology. I. Title.
DS733.L455 78-840917

—LAW see Law—China

—MAPS

Heibonsha.
(Chūgoku chizuchō)
中国地図帳 平凡社版 東京 平凡社 1973.
80 p. 29 cm. Ja 73-13801
付：中国地名表記便覧
1. China—Maps. 2. China—Gazetteers. 3. Names, Geographical—China. I. Title.
G2305.H35 1973 73-808398

Ti t'u ch'u pan she.
(Chung-kuo ti t'u ts'e)
中国地图册 = Zhongguo dituce / 地图出版社编制]. — 普及本第2版. — 北京：該社, 1977.
33 p. : all col. maps : 19 x 27 cm.
Cover title.
"本图上中国国界线系按照我社1971年出版的《中华人民共和国地图》绘制·本图行政区划资料截止期1976年12月"
$0.45
1. China—Maps. I. Title. II. Title: Zhongguo dituce.
G2305.T445 1977 78-842597
C

The Times atlas of China / [editors and chief contributors, P. J. M. Geelan, D. C. Twitchett ; cartography by Kyobunkaku Limited, John Bartholomew & Son Limited, and Hunting Surveys Limited]. — London : Times Books, 1974.
xl, 145, 27 p. : chiefly col. maps ; 39 cm. GB75-00895
Includes index.
ISBN 0-7230-0118-9 : £25.00
1. China—Maps. 2. China—Description and travel. 3. Names, Geographical—China. I. Geelan, P. J. M. II. Twitchett, Denis Crispin. III. Kyobunkaku. IV. Bartholomew (John) and Son, ltd. V. Hunting Surveys, ltd. VI. Title: Atlas of China.
G2305.T47 1974 912'.51 75-313716
75[r79]rev2 MARC

—MILITARY POLICY

Kang, Chong-Sook.
China in Waffen : d. Rüstungs- u. Abrüstungspolitik d. Volksrepublik China von 1969 bis zum Tod Mao-Tse-Tungs / Chong-Sook Kang. — Frankfurt/Main : Haag und Herchen, 1979.
143 p. ; 21 cm. — (Internationale Beziehungen ; Bd. 2 ISSN 0171-4414)
GFR79-A
Summary in English.
Bibliography: p. 125-143.
ISBN 3-88129-208-X : DM19.80
1. China—Military policy. 2. China—Foreign relations—1949-1976. 3. China—Foreign relations—1976- . 4. Munitions—China. 5. Arms control. I. Title. II. Series.
UA835.K36 79-376036
79 MARC

CHINA

—MILITARY POLICY (Continued)

Korea (Republic). Kukt'o T'ongirwŏn. Chosa Yŏn'gusil.

(Chuyo kukka ŭi kunsa chŏllyak)

主要國家의軍事戰略 / 刊行責任 李範赫. — ［서울］: 國土統一院調査研究室, 1976.

167 p. : ill. ; 26 cm. — (周辺勢力調査研究)

"국통조 76-12-100."

1. Military policy. 2. United States—Military policy. 3. Russia—Military policy. 4. China—Military policy. 5. Japan—Military policy. I. Yi, Pŏm-hyŏk, 1928– II. Title. III. Series: Chubyŏn seryŏk chosa yŏn'gu.

UA11.K67 1976 78-826293

—MISSIONS see Missions—China

—NOBILITY—GENEALOGY

Sun, Yao.

(Ch'un ch'iu shih tai chih shih tsu)

春秋時代之世族 / 孫曜撰. — 台1版. — ［台北］: 華世出版社, 民國65 ［1976］

2, 192 p. ; 21 cm.

Reprint of the 1931 ed.
NT$50.00

1. China—Nobility—Genealogy. 2. China—History—Spring and Autumn period, 722–481 B.C. I. Title.

CS1164.S9 1976 79-838276

Wang, I-t'ung, 1914–

(Wu ch'ao men ti)

五朝門第 / 王伊同著. — 重刊第1版. — 香港: 中文大學出版社, 1978.

2 v. (viii, 331 p., ［43］ leaves of plates (39 fold.)) : 75 Geneal. tables ; 24 cm.

Bibliography: p. ［290］-296.
"附高門權門世系婚姻表": v. 2.
ISBN 962-201-147-0
ISBN 962-201-150-0 pbk.

1. China—Nobility—Genealogy. 2. China—History—Chin dynasty, 265–419. 3. China—History—Southern and northern dynasties, 317–589. I. Title.

CS1164.W36 79-840333

—OFFICIALS AND EMPLOYEES

Sun, Kuo-tung.

(T'ang tai chung yang chung yao wen kuan ch'ien chuan t'u ching yen chiu)

唐代中央重要文官遷轉途徑研究 / 孫國棟. — 香港: 龍門書店, 1978.

2, 6, 565 p. : ill. ; 25 cm. — (新亞研究所叢書)

$70.00

1. China—Officials and employees. 2. China—History—T'ang dynasty, 618–907—Sources. I. Title. II. Series: Hongkong. Chinese University. Institute of Advanced Chinese Studies and Research. Hsin ya yen chiu so ts'ung shu.

JQ1512.S88 79-837757

T'ang liu tien.

唐六典: ［30卷 / 唐元宗御撰; 李林甫等奉敕注］. — ［日本東都?］: 官板, ［天保7 i.e. 1836?］

8 v. (on double leaves) in case ; 26 cm.

Caption title.
Block print.

1. China—Officials and employees. 2. China—History—T'ang dynasty, 618–907—Sources. I. T'ang Hsüan-tsung, Emperor of China, 685–762. II. Li, Lin-fu, d. 752.

Title romanized: T'ang liu tien.

JQ1512.T3 1836 77-836510

T'ang liu tien.

大唐六典 ［30卷］ 唐玄宗御撰 李林甫奉敕註 台北 文海出版社 民國51 ［1962］

522 p. 20 cm.

Facsimile ed.
Each page represents 2 leaves of the original.

1. China—Officials and employees. 2. China—History—T'ang dynasty, 618–907—Sources. I. T'ang Hsüan-tsung, Emperor of China, 685–762. II. Li, Lin-fu, d. 752. III. Title.

Title romanized: Ta T'ang liu tien.

JQ1512.T3 1962 C 63-2101

(T'ang liu tien)

唐六典 / ［張九齡等撰］. — ［台北: 台灣］商務［印書館, 1976?］

3 v. ; 20 cm. — (四庫全書珍本六集; 117-119)

Cover title.
Reproduced from Wen yüan ko copy.

1. China—Officials and employees. 2. China—History—T'ang dynasty, 618–907—Sources. I. Chang, Chiu-ling, 678–740. II. Series: Ssu k'u ch'üan shu. Ssu k'u ch'üan shu chen pen liu chi ; 117–119.

AC149.S7 1976, vol. 117–119 78-843860
[JQ1512]

Wang, Ya-nan.

(Chung-kuo kuan liao cheng chih yen chiu)

中國官僚政治研究: 中國官僚政治之經濟的歷史的解析 / 王亞南著. — ［香港?］: 習之出版社; 香港: 經銷處一山書屋, ［1979?］

2, 8, 321 p. ; 19 cm.

Reprint of the 1948 ed. published by Shih tai wen hua ch'u pan she. Shanghai.

1. China—Officials and employees. 2. Bureaucracy. I. Title.

JQ1512.W3 1979 79-839156

—OFFICIALS AND EMPLOYEES —BIOGRAPHY

Kuwabara, Jitsuzō, 1870–1931.

(Ho Jukō no jiseki)

蒲壽庚の事蹟: 唐宋時代に於けるアラブ人の支那通商の概況殊に宋末の提擧市舶西域人 / ［著者桑原騭藏］. — 東京: 岩波書店, 昭和10 ［1935］, 昭和23 ［1948］ printing.

5, 11, 228, 18 p., ［1］ leaf of plates : ill. ; 26 cm.

Colophon inserted.
Includes bibliographies and index.

1. P'u, Shou-keng, fl. 1245–1284. 2. Muslims in China—Biography. 3. China—Officials and employees—Biography. 4. China—Relations (general) with Arab countries. 5. Arab countries—Relations (general) with China. 6. Customs administration—China. I. Title.

DS751.6.P8K88 1935 79-801279

—OFFICIALS AND EMPLOYEES —CORRESPONDENCE

Chiang, Ch'ao-yüeh, 1898–

(Ting ssu sheng ch'i chi)

丁巳聲氣集 / 江山異生［姜超嶽］. — ［台北］: 三民書局, 民國66 ［1977］

3, 70 p. : ill. ; 22 cm.

1. Chiang, Ch'ao-yüeh, 1898– 2. China—Officials and employees—Correspondence. I. Title.

DS778.C5373A4 1977 79-841415

—OFFICIALS AND EMPLOYEES —HISTORY

Makino, Shūji, 1931–

(Gendai kōtōkan no taikeiteki kenkyū)

元代勾當官の體系的研究 / 牧野修二著. — 東京: 大明堂, 1979.

v, 204 p. ; 22 cm. Ja 79-8100

Originally presented as the author's thesis, Kyūshū Daigaku, 1972.
Includes bibliographical references.
Errata slip inserted.
¥3500

1. China—Officials and employees—History. 2. China—History—Yüan dynasty, 1260–1368. I. Title.

JQ1512.Z1M34 1979 79-810184

—PERIODICALS

China and us.

New York, U. S.—China Peoples Friendship Association.

v. ill. 28 cm. bimonthly.

"Publication of the U. S.—China Peoples Friendship Association."
Key title: China and us, ISSN 0191-3166.

1. United States—Relations (general) with China—Periodicals. 2. China—Relations (general) with the United States—Periodicals. 3. China—Periodicals. I. U. S.—China Peoples Friendship Association.

E183.8.C5C464 301.29'73'051 76-648679
MARC-S

(Kitaĭskaîa Narodnaîa Respublika)

Китайская Народная Республика.

Москва, Изд-во Наука, Глав. ред. восточной лит-ры ［etc.］

v. ill. 23 cm. annual.

"Политика, экономика, идеология" (varies slightly).
Vols. for issued by Akademiîa nauk SSSR, Institut Dal'nego Vostoka.

1. China—Periodicals. I. Akademiîa nauk SSSR. Institut Dal'nego Vostoka.

DS701.K54 79-641408
(MARC-S)

(Ṣīn al-muṣawwarah)

الصين المصورة. 1–

［بكين، الصين، s. n.］ 1970–

no. ill. 37 cm. monthly.

1. China—Periodicals.

DS701.S467 79-645917
(MARC-S)

—PERIODICALS—HISTORY

(Senzen no Chūgoku jironshi kenkyū)

戦前の中国時論誌研究 / 小島麗逸編. — 東京: アジア経済研究所: 発売 アジア経済出版会, 1978.

241 p. ; 26 cm. — (中国関係新聞雑誌解題; 2) (文献解題; 26) Ja***

Includes bibliographical references.
¥5000

1. China—Periodicals—History. 2. Japanese periodicals—History. 3. Japanese periodicals—Indexes. I. Kojima, Reiitsu, 1934– II. Series: Chūgoku kankei shimbun zasshi kaidai ; 2. III. Series: Bunken kaidai ; 26.

DS701.S37 78-817224

—PERIODICALS—INDEXES

Japan. Naikaku Kambō. Naikaku Chōsashitsu.

("Chūkyō kankei shiryō" jikōbetsu sakuin)

「中共関係資料」事項別索引: 自1957年12月7日至1970年4月28日. — ［東京］: 内閣官房内閣調査室, ［1970］

516 p. ; 25 cm.

Cover title.

1. China—History—1949–1976—Periodicals—Indexes. 2. China—Periodicals—Indexes. I. Title.

DS777.55.J3 1970 78-812867

—POETRY

Alley, Rewi, 1897-

Snow over the pines : poems / by Rewi Alley ; with a foreword by George Hatem (Ma Hai-teh). — Auckland : New Zealand-China Friendship Society, 1977.

142 p., ［2］ leaves of plates : ill. ; 22 cm. NZ***

1. China—Poetry. I. Title.

PR9639.3.A5S6 821 79-316975
79 MARC

—POLITICS AND GOVERNMENT

Ma, Ch'i-hua.

(Chen-kuan cheng lun)

貞觀政論 / 馬起華著. — 台北板橋: 漢苑出版社, 民國66 ［1977］

143, ［9］ p. ; 22 cm. — (政大政治叢書; 1)

At head of title: Ta hsüeh yung shu.
Includes bibliographical references.

1. China—Politics and government. 2. China—History—T'ang dynasty, 618–907. 3. T'ang T'ai-tsung, Emperor of China, 597–649. I. Title. II. Series: Kuo li cheng chih ta hsüeh. Cheng ta cheng chih ts'ung shu ; 1.

JQ1508.M29 79-841071

Menzel, Ulrich.

Wirtschaft und Politik im modernen China : e. Sozial- u. Wirtschaftsgeschichte von 1842 bis nach Maos Tod / Ulrich Menzel. — Opladen : Westdeutscher Verlag, 1978.

230 p. ; 19 cm. GFR79-A

Bibliography: p. 222-230.
ISBN 3-531-11460-3 : DM19.80

1. China—Economic conditions. 2. China—Polities and government. I. Title.

HC427.7.M46 79-355874
79 MARC

CHINA

—POLITICS AND GOVERNMENT (Continued)

Pai, Hsiao-sheng.
(Pei-ching to ch'üan mi hsin)
北京奪權秘辛 / 百笑生著. — 台中：新企業世界出版社, 民國66 [1977]
227 p. ; 20 cm.
Reprint ed.
NT$40.00

1. China—Politics and government I. Title.
PL2892.A34P4 1977b — 79-835565

Sanyal, Narayan.
(Cīna-Bhārata laṅ mārca)
চীন-ভারত লঙ্ মার্চ / নারায়ণ সান্যাল. — কলিকাতা : মণ্ডল বুক হাউস, 1977.
319 p., [2] leaves of plates : ill., maps (on lining papers) ; 23 cm.
In Bengali.
Includes bibliographical references and index.
Rs20.00

1. China—Politics and government. 2. China—Relations (general) with India. 3. India—Relations (general) with China. I. Title.
DS740.S36 — 78-911112

Sarkesian, Sam Charles.
Comparative politics / Sam C. Sarkesian, James H. Buck. — Sherman Oaks, Calif. : Alfred Pub. Co., c1979.
xii, 359 p. : ill. ; 24 cm.
Includes bibliographical references and index.
ISBN 0-88284-067-2 : $13.95
1. Comparative government. 2. United States—Politics and government. 3. Russia—Politics and government. 4. China—Politics and government. 5. Japan—Politics and government. I. Buck, James Harold., joint author. II. Title.
JF51.S28 — 320.3 / 78 — 78-12083 MARC

—POLITICS AND GOVERNMENT—TO 1643

Chang, Yang-hao, 1269-1329.
(San shih chung kao)
三事忠告 / [張養浩撰]. 七國考 / [董說撰]. — [臺北：臺灣]商務[印書館, 1979]
4 v. ; 22 cm. — (四庫全書珍本九集 ; 165-168)
Cover title.
"三事忠告四卷…七國攷十四卷"
Photoreprint ed. of the Wen Yüan Ko copy.
1. Local government—China. 2. Civil service ethics. 3. China—Politics and government—To 1643. 4. China—History—Warring States, 403-221 B. C. I. Tung, Shuo, 1620-1686. Ch'i kuo k'ao. 1979. II. Title. III. Series: Ssu k'u ch'üan shu. Ssu k'u ch'üan shu chen pen chiu chi ; 165-168.
AC149.S7 1979, vol. 165-168 — 79-840282
[JS7353.A3]

Forte, Antonio.
Political propaganda and ideology in China at the end of the seventh century : inquiry into the nature, authors and function of the Tunhuang document S.6502, followed by an annotated translation / Antonino Forte. — Napoli : Istituto universitario orientale, Seminario di studi asiatici, 1976.
xii, 312 p., [18] leaves of plates : facsims. ; 25 cm. — It78-May
Bibliography: p. [285]-296.
Includes index.
1. China—Politics and government—To 1643. 2. Buddhism and state—China. 3. Tun-huang manuscripts. 4. China—History—T'ang dynasty, 618-907. I. Title.
DS749.3.F63 — 951'.01 / *79 — 79-309494 MARC

Harada, Taneshige, 1911-
(Jōgan seiyō)
貞観政要 / 原田種成著. — 東京：明治書院, 1978-
v. ; 22 cm. — (新釈漢文大系 ; 95-) — Ja 78-14185 (v. 1)
Colophon inserted.
"原文は拙編の「貞観政要定本」を底本とした"
¥5200 (v. 1)
1. China—Politics and government—To 1643. 2. Political ethics. 3. T'ang T'ai-tsung, Emperor of China, 597-649. I. Wu, Ching, 670-749. Chen-kuan cheng yao. Japanese & Chinese. 1978. II. Title. III. Series: Shinshaku Kambun taikei ; 95 [etc]
JQ1500.A5 1978 — 79-801891

Hsieh, Kuo-chen, 1901-
(Ming Ch'ing tang she yün tung k'ao)
明清黨社運動考 / 著者謝國楨. — 臺1版. — 板橋：漢苑出版社, 民國64 [1975]
2, 2, 270 p. : 20 cm.
Colophon title.

1. Political parties—China. 2. China—Politics and government—To 1643. 3. China—Politics and government—1644-1911. I. Title.
JQ1519.A45H72 — 79-835913

P'an, Sheng-te.
(Pei Sung yü lun yü tang cheng chih yen chiu)
北宋輿論與黨爭之研究 / 潘昇德. — [台北? : s. n., 1978?]
79 double leaves ; 27 cm.
Reproduced from ms. copy.
Originally presented as the author's thesis (M.A.), Kuo li cheng chih ta hsüeh, 1961.
Bibliography: leaves 77-79.
1. China—History—Sung dynasty, 960-1279. 2. Public opinion—China. 3. China—Politics and government—To 1643. I. Title.
DS751.P29 1978 — 79-835700

Yang, Shu-fan.
(Liao Chin chung yang cheng chih chih tu)
遼金中央政治制度 / 楊樹藩著. — 臺北：臺灣商務印書館, 民國67 [1978]
207, [17] p. ; 19 cm. — (岫廬文庫 ; 034)
Series romanized: Hsiu-lu wen k'u.
NT$36.00

1. China—Politics and government—To 1643. 2. Khitan Mongols—History. 3. Ju-chen (Tribe)—History. I. Title.
JQ1509 1978.Y36 — 79-835787

Yang, Shu-fan.
(Ming tai chung yang cheng chih chih tu)
明代中央政治制度 / 楊樹藩著. — 臺北：臺灣商務印書館, 民國67 [1978]
3, 246, [13] p. ; 20 cm. — (岫廬文庫 ; 36)
Series romanized: Hsiu lu wen k'u.
Includes bibliographical references.
NT$36.00
1. China—Politics and government—To 1643. 2. China—History—Ming dynasty, 1368-1644. I. Title.
JQ1509 1368.Y36 — 78-841510

Yang, Shu-fan.
(Yüan tai chung yang cheng chih chih tu)
元代中央政治制度 / 楊樹藩著. — 臺北：臺灣商務印書館, 民國67 [1978]
[16], 212 p., [1] fold. leaf ; 19 cm. — (岫廬文庫 ; 035)
Series romanized: Hsiu-lu wen k'u.
NT$36.00

1. China—Politics and government—To 1643. 2. China—History—Yüan dynasty, 1260-1368. I. Title.
JQ1509 1260.Y36 — 78-842346

—POLITICS AND GOVERNMENT —1644-1911

Fu, Tsung-mao, 1927-
(Ch'ing tai chün chi ch'u tsu chih chi chih chang chih yen chiu)
清代軍機處組織及職掌之研究 傅宗懋著 [臺北 嘉新水泥公司文化基金會 1967]
14, 3, 694 p. 21 cm. (嘉新水泥公司文化基金會研究論文第69種) (嘉新水泥公司文化基金會叢書)
Bibliography: p. 522-527.
1. China. Chün chi ch'u. 2. China—Politics and government—1644-1911. I. Title.
JQ1512.Z4F8 — 72-836696

Hsieh, Kuo-chen, 1901-
(Ming Ch'ing tang she yün tung k'ao)
明清黨社運動考 / 著者謝國楨. — 臺1版. — 板橋：漢苑出版社, 民國64 [1975]
2, 2, 270 p. : 20 cm.
Colophon title.

1. Political parties—China. 2. China—Politics and government—To 1643. 3. China—Politics and government—1644-1911. I. Title.
JQ1519.A45H72 — 79-835913

Hu, Chien-kuo.
(Ch'ing tai wen kuan k'ao k'o chih tu chih yen chiu)
清代文官考課制度之研究 / 胡健國撰. — [台北? : s. n., 1978?]
2 v. (467 p. on double leaves) ; 27 cm.
Reproduced from typescript.
Originally presented as the author's thesis (M.A.), Kuo li cheng chih ta hsüeh, 1971.
Bibliography: p. 461-467.
1. Civil service—China. 2. Employees, Rating of—China. 3. China—Politics and government—1644-1911. I. Title.
JQ1512.Z13R33 1978 — 79-835055

Huang, Kuang-liang.
(Ch'ing tai k'o chü chih tu chih yen chiu)
清代科舉制度之研究 / 黃光亮著. — Taipei：嘉新水泥公司文化基金會, 民國65 [1976]
4, 2, 434 p. ; 22 cm. — (嘉新水泥公司文化基金會研究論文 ; 第311種) (嘉新水泥公司文化基金會叢書)
Series 1 romanized: Chia hsin shui ni kung ssu wen hua chi chin hui yen chiu lun wen. Series 2 romanized: Chia hsin shui ni kung ssu wen hua chi chin hui ts'ung shu.
Bibliography: p. 426-434.
1. Civil service—China—Examinations. 2. China—Politics and government—1644-1911. I. Title.
JQ1512.Z1H84 — 78-842398

Tuan, Ch'ang-kuo.
(Kung ch'in wang I-hsin yü Hsien T'ung chih chi ti wai chiao yü cheng chih chiu fen)
恭親王奕訢與咸同之際的外交與政治糾紛：1858-1865 / 段昌國. — [台北? : s. n., 1972?]
70 double leaves ; 27 cm.
"國立臺灣大學歷史學研究所(近代史組)六十一學年度碩士畢業論文"
Reproduced from the typescript.
Bibliography: leaves 67-70.
1. China—History—Hsien-feng, 1850-1861. 2. China—History—T'ung-chih, 1861-1875. 3. China—Politics and government—1644-1911. 4. I-hsin, 1833-1898. I. Title.
DS758.T8 — 79-841141

Yang, Shu-fan.
(Ch'ing tai chung yang cheng chih chih tu)
清代中央政治制度 / 楊樹藩著. — 臺北：臺灣商務印書館, 民國67 [1978]
3, 222, [14] p. ; 20 cm. — (岫廬文庫 ; 37)
Series romanized: Hsiu lu wen k'u.
Includes bibliographical references.
NT$36.00

1. China—Politics and government—1644-1911. I. Title.
JQ1509 1644.Y36 — 78-841079

—POLITICS AND GOVERNMENT —1644-1911—ADDRESSES, ESSAYS, LECTURES

Liu, Chia-chü.
(Ch'ing ch'u cheng chih fa chan shih lun chi)
清初政治發展史論集 / 劉家駒著. — 臺北：臺灣商務印書館, 民國67 [1978]
220, [16] p. ; 20 cm. — (岫廬文庫 ; 048)
Series romanized: Hsiu lu wen k'u.
Includes bibliographical references
NT$36.00
1. China—Politics and government—1644-1912—Addresses, essays, lectures. I. Title.
DS754.2.L57 — 79-837868

—POLITICS AND GOVERNMENT—19TH CENTURY

Li, Chien-nung.
The political history of China, 1840-1928. Translated and edited by Ssu-yu Teng and Jeremy Ingalls. Stanford, Calif., Stanford University Press [1967, c1956]
xii, 545 p. maps. 22 cm.
"SP60."
Translation of Chung-kuo chin pai nien chêng chih shih.
Bibliography: p. 515-528.
1. China—Politics and government—19th century. 2. China—Politics and government—1900-1949. 3. China—Foreign relations. I. Têng, Ssū-yü, 1906- ed. II. Ingalls, Jeremy, 1911- ed. III. Title.
DS755.L4313 1967 — 951'.03 — 67-26648 [r78]rev2 MARC

CHINA (Continued)

—POLITICS AND GOVERNMENT—1900-

(Bokkōseru Shina no kokkashugi shisō)
勃興せる支那の國家主義思想 / ［編輯 南滿洲鐵道株式會社北京公所研究室］. — 北京：同研究室，大正15 ［1926］
166 p. ; 22 cm. — (北京滿鐵月報特刊；第5)
Cover title.

1. Nationalism—China. 2. China—Politics and government—1900- I. Minami Manshū Tetsudō Kabushiki Kaisha. Pekin Kōsho. Kenkyūshitsu. II. Series: Minami Manshū Tetsudō Kabushiki Kaisha. Pekin Kōsho. Kenkyūshitsu. Pekin Mantetsu geppō tokkan ; dai 5.
DS740.2.B64 78-811110

Chen, Percy, 1901-
China called me : my life inside the Chinese Revolution / by Percy Chen. — 1st ed. — Boston : Little, Brown, c1979.
xxi, 423 p., [8] leaves of plates : ill., map (on lining papers ; 25 cm.
Includes index.
ISBN 0-316-13849-5
1. Chen, Percy, 1901- . 2. Statesmen—China—Biography. 3. China—Politics and government—1900- . I. Title.
DS778.C5267A33 951.04'092'4 79-13945
79 MARC

Lam, Sai-chun.
Modern China in struggle : an ideological study of the late Ch'ing / Lam Sai-chun. — Hong Kong : Cosmos Books, 1978.
128 p. ; 22 cm.
Includes bibliographies.
CONTENTS: A political polemic.—The Three-people's principles.
$7.00
1. China—Politics and government—1900- . 2. Sun, Yat-sen, 1866-1925. San min chu i. I. Title.
DS764.L35 320.9'51'03 78-112050
79 MARC

Takagi, Keizō, 1941-
(Shōdan Tō Shōhei)
正伝 鄧小平 / 高木桂蔵著. — 東京：秀英書房，昭和53 ［1978］
14, 301 p., [1] leaf of plates : ill. ; 20 cm. Ja***
Includes bibliographical references.
¥1300

1. Teng, Hsiao-p'ing, 1900- 2. Statesmen—China—Biography. 3. China—Politics and government—1900- I. Title.
DS778.T39T34 79-800431

Tyau, Min-ch'ien T Z 1888-
China's new constitution and international problems / by Min-ch'ien T. Z. Tyau. — Reprint ed. — Arlington, Va. : University Publications of America, 1976.
xv, 286 p. ; 24 cm. — (China studies) (Studies in Chinese government and law)
Reprint of the 1918 ed. published by Commercial Press, Shanghai.
ISBN 0-89093-064-3
1. China—Constitutional law. 2. China—Politics and government—1900-1949. 3. China—Foreign relations. 4. European War, 1914-1918—China. I. Title.
342'.51'029 76-55090
78 MARC

—POLITICS AND GOVERNMENT—1900-1949

Chan Lau, Kit-ching.
Anglo-Chinese diplomacy in the careers of Sir John Jordan and Yüan Shih-kai / Chan Lau Kit-ching. — [Hong Kong] : Hong Kong University Press, 1978.
viii, 168 p. ; 23 cm.
Bibliography: p. [144]-156.
Includes index.
ISBN (invalid) 0-9622090-1-09
1. Great Britain—Foreign relations—China. 2. China—Foreign relations—Great Britain. 3. Great Britain—Foreign relations—1901-1910. 4. Great Britain—Foreign relations—1910-1936. 5. China—Politics and government—1900-1949. 6. China—Foreign relations—To 1912. 7. China—Foreign relations—1912-1949. 8. Yüan, Shih-k'ai, 1859-1916. 9. Jordan, John Newell, Sir, 1852- . I. Title.
DA47.9.C6C45 327.41'051 79-106848
79 MARC

Li, Chien-nung.
The political history of China, 1840-1928. Translated and edited by Ssu-yu Teng and Jeremy Ingalls. Stanford, Calif., Stanford University Press [1967, c1956]
xii, 545 p. maps. 22 cm.
"SP60."
Translation of Chung-kuo chin pai nien chêng chih shih.
Bibliography: p. 515-528.
1. China—Politics and government—19th century. 2. China—Politics and government—1900-1949. 3. China—Foreign relations. I. Têng, Ssŭ-yü, 1906- ed. II. Ingalls, Jeremy, 1911- ed. III. Title.
DS755.L4313 1967 951.03 67-26648
[r78]rev2 MARC

—POLITICS AND GOVERNMENT—1911-1949—SOURCES

Melis, Giorgio, 1925-
La rivoluzione cinese / Giorgio Melis. — Firenze : Le Monnier, 1978.
vi, 121 p. ; 19 cm. — (Le Contraddizioni irrisolte del mondo contemporaneo) (Storia parallela ; Sezione 5) It78-Dec
"Documenti": p. [17]-94.
Bibliography: p. 108-110.
L2200
1. China—Politics and government—1911-1949—Sources. 2. China—Politics and government—1949-1976—Sources. I. Title. II. Series.
DS775.M43 320.9'51'042 79-347754
*79 MARC

—POLITICS AND GOVERNMENT—1912-1937

Ch'en, Wei-lun, 1897-
(Huang K'o-ch'iang hsien sheng chuan chi)
黃克強先生傳記 / 陳維綸撰；［編輯者中國國民黨中央委員會黨史委員會］. — 臺北：經銷處中央文物供應社，民國62 ［1973］
2, 4, 318 p., [4] leaves of plates : ill. ; 22 cm.
Includes bibliographical references.
1. Huang, Hsing, 1874-1916. 2. Statesmen—China—Biography. 3. China—History—Revolution, 1911-1912. 4. China—Politics and government—1912-1937. I. Chung-kuo kuo min tang. Chung yang wei yüan hui. Tang shih wei yüan hui. II. Title.
DS778.H85C47 78-842983

Holubnychy, Lydia.
Michael Borodin and the Chinese revolution, 1923-1925 / by Lydia Holubnychy. — Ann Arbor, Mich. : Published for East Asian Institute, Columbia University by University Microfilms International, 1979.
ix, 503 p. : map ; 25 cm. — (Monograph publishing : Sponsor series)
Bibliography: p. [479]-503.
ISBN 0-8357-0386-X : $24.25
1. China—Politics and government—1912-1937. 2. Communism—China. 3. Borodin, Mikhail Markovich, 1884-1953. I. Title.
DS777.44.H64 320.9'51'041 78-31689
78 MARC

Mizuno, Baigyō.
(Shina no henkyoku)
支那の變局 / 水野梅曉著. — 東京：東方通信社調查部，大正10 ［1921］
1, 6, 277, 2 p. : map ; 19 cm.
"支那變局ノ現勢圖" (fold. leaf) inserted at end.

1. China—Politics and government—1912-1937. I. Title.
DS777.44.M59 79-805327

Nakayama, Kyūshirō, 1874-1961.
(Saikin Shina no seiji oyobi bunka)
最近支那の政治及文化 / 中村［山］久四郎著. — 東京：三省堂，大正11 ［1922］
2, 5, 285 p. ; 18 cm. — (日本百科叢書)
Series romanized: Nihon hyakka sōsho.

1. China—Politics and government—1912-1937. 2. China—Civilization. I. Title.
DS777.44.N37 79-803670

(Saikin ni okeru Ō-Beijin no Shina-kan)
最近ニ於ケル歐米人ノ支那觀. — ［東京］：外務省情報部，大正2-3 ［1913-1914］
6 v. in 1 ; 23 cm.

1. China—Politics and government—1912-1937. 2. China—Foreign economic relations. 3. China—Foreign opinion. I. Japan. Gaimushō. Jōhōbu.
DS774.S3 79-802817

Ōtaka, Jirō.
(Shina jihen to rekkyō no dōkō)
支那事變と列強の動向 / 大高二郎著. — 東京：日本青年外交協會，昭和13 ［1938］
11, 107 p., [3] leaves of plates : ill. ; 18 cm.

1. China—Politics and government—1912-1937. 2. Sino-Japanese Conflict, 1937-1945. 3. World politics, 1933-1945. I. Title.
DS774.O79 79-808033

—POLITICS AND GOVERNMENT—1912-1937—ADDRESSES, ESSAYS, LECTURES

Fang, Chih-min, 1900-1935.
(K'o ai ti Chung-kuo)
可爱的中国 / 方志敏著. — 香港：生活·讀書·新知三联书店香港分店，1975.
2, 98 p., [1] leaf of plates : ill. ; 19 cm.
$1.50

1. China—Politics and government—1912-1937—Addresses, essays, lectures. I. Title.
DS777.47.F33 1975 75-840645

(Nei kuo wen t'i t'ao lun ta kang)
內國問題討論大綱 = China's internal problems / 編輯者陳伯華；審訂者青年協會書報部. — 上海：該部，民國15-16 ［1926-1927］
2 v. ; 19 cm. — (公民教育叢刊；第13-14種)
Colophon title.
In Chinese.
Series romanized: Kung min chiao yü ts'ung k'an.
L. C. set incomplete: v. 2 wanting.
1. China—Politics and government—1912-1937—Addresses, essays, lectures. I. Ch'en, Po-hua.
JQ1502.N44 78-841349

—POLITICS AND GOVERNMENT—1912-1937—CONGRESSES

Kuo nan hui i, Lo-yang, China, 1932.
(Kuo nan hui i chi lu)
國難會議紀錄 / 國難會議秘書處編. — ［台北縣永和鎮］：文海出版社，［1977?］
10, 304 p., [15] leaves of plates : ill. ; 22 cm. — (近代中國史料叢刊續編 [i. e. 輯]；484：第四十九輯)
Reprint ed.

1. China—Politics and government—1912-1937—Congresses. I. Kuo nan hui i mi shu ch'u. II. Series: Shen, Yün-lung, comp. Chin tai Chung-kuo shih liao ts'ung k'an hsü chi ; 484.
DS755.S532 vol. 484 78-841435
[DS777.47]

—POLITICS AND GOVERNMENT—1912-1949

see also May fourth movement

Ch'eng, T'ien-ku, 1889-1974.
(Ch'eng T'ien-ku hui i lu)
程天固回憶錄. — 香港：龍門書店，1978.
2, 3, 3, 539 p. : port. ; 22 cm.
In colophon: Ch'u pan che Ch'eng Ta-min.
$80.00

1. Ch'eng, T'ien-ku, 1889-1974. 2. China—Politics and government—1912-1949. 3. Statesmen—China—Biography. I. Title.
DS778.C5363A38 1978 79-837615

Chiang, Mei-Ling Sung, 1897-
Conversations with Mikhail Borodin / by Madame Chiang Kai-shek. — London : [Free Chinese Center], [1978?]
[1], 96, 63 p. ; 21 cm. GB78-18929
Added t.p.: Yü Pao-lo-t'ing t'an hua ti hui i.
1. Borodin, Mikhail Markovich, 1884-1953. 2. Communism—China. 3. China—Politics and government—1912-1949. I. Title. II. Title: Yü Pao-lo-t'ing t'an hua ti hui i.
DK268.B65C48 335.43'01 78-388982
78 MARC

Chou, Ching-wen, 1909-
(Lun Chung-kuo to shu jen ti cheng chih lu hsien)
論中國多數人的政治路線 周鯨文著 香港 時代批評社 民國36 ［1947］
27 p. 19 cm. (時代叢書政治組第五種)
Series romanized: Shih tai ts'ung shu : Cheng chih tsu.

1. China—Politics and government—1912-1949. 2. Communism—China. I. Title.
[DS777.55.C4575] C 61-4299
Yale Univ. Library rev

CHINA

—POLITICS AND GOVERNMENT
—1912-1949 (Continued)

Lo, Kang.

[Wo men tui Jih Wang hsieh ting ying yu chih jen shih. Japanese]

日汪協定に對する吾人の認識 / [羅剛]. — [上海]：大陸調査會上海調査室, 昭和15 [1940]

26 p. (on double leaves) ; 26 cm. — (大調政治資料 ; 第3輯)

Cover title.

Translation of article published in Shang-hai chou pao, v. 2, no. 26, Dec. 7, 1940.

1. China—Foreign relations—Japan—Foreign opinion, Chinese. 2. Japan—Foreign relations—China—Foreign opinion, Chinese. 3. Public opinion—China. 4. China—Politics and government—1912-1949. I. Title. II. Series: Tairiku Chōsakai. Shanhai Chōsashitsu. Taichō seiji shiryō ; dai 3-shū.

Title romanized: Nichi-Ō kyōtei ni taisuru gojin no ninshiki.

DS740.5.J3L616 78-812948

Mao, Tse-tung, 1893-1976.

[Chung-kuo ko ming yü Chung-kuo kung ch'an tang]

中國革命和中國共產黨　毛澤東著　北京　人民出版社　1953.

14 p.　19 cm.

1. China—Politics and government—1912-1949. 2. Communism—China. I. Title.

Title romanized: Chung-kuo ko ming ho Chung-kuo kung ch'an tang.

DS775.M355 1953 C 60-3222

Ōmori, Minoru, 1922-

(Mō Takutō)

毛沢東 = Mao Zedong：不世出の巨星 / 大森実. — 東京：講談社, 1979.

394 p. : ill. ; 19 cm. — (His 人物現代史 ; 9) Ja***

In Japanese.

Bibliography: p. 383-386.

Includes indexes.

1. Mao, Tse-tung, 1893-1976. 2. China—Politics and government—1912-1949. 3. China—Politics and government—1949-1976. 4. Heads of state—China—Biography. I. Title. II. Title: Mao Zedong. III. Series: Ōmori, Minoru, 1922- Jimbutsu gendai shi ; 9.

DS778.M3O45 79-809034

Takagi, Rikurō, 1880-

(Nikka kōyūroku)

日華交友錄 / 高木陸郎著. — 東京：救護會出版部, 昭和18 [1943]

6, 6, 236 p. ; 18 cm.

1. China—Politics and government—1912-1949. 2. China—Foreign relations—Japan. 3. Japan—Foreign relations—China. I. Title.

DS774.T26 79-801739

Takagi, Shōnosuke, 1900-

(Kitō kara Chūka shinseiken e)

冀東から中華新政權へ / 高木翔之助著. — 東京：北支那社, 昭和13 [1938]

4, 4, 184 p., [1] leaf of plates : ill. ; 19 cm.

Cover title.

1. China—Politics and government—1912-1949. 2. Hopei, China (Province)—Politics and government. 3. Chi-tung fang kung tzŭ chih chêng fu. I. Title.

DS777.47.T267 79-800467

Takata, Fusao, 1923-

(Chūgoku Kyōsantō)

中国共産党 / 高田富佐雄. — [東村山]：教育社：東京：販売　教育社出版サービス, [1978]

170 p. : ports. ; 18 cm. — (時事問題解説 ; 107) (入門新書) Ja 78-25753

Cover title.

Series 2 romanized: Nyūmon shinsho.

¥400

1. Chung-kuo kung ch'an tang—History. 2. China—Politics and government—1912-1949. I. Title. II. Series: Jiji mondai kaisetsu ; 107.

JQ1519.C553T34 79-806882

[T'ang, Liang-li] 1901- ed.

Suppressing communist-banditry in China. — Hyperion reprint ed. — Westport, Conn. : Hyperion Press, 1978.

vii, 110 p., [14] leaves of plates : ill. ; 23 cm. — (China studies)

Reprint of the 1934 ed. published by China United Press, Shanghai, in series: "China to-day" series.

ISBN 0-88355-395-3 : $16.50

1. China—Politics and government—1912-1949. 2. Communism—China. I. Title. II. Series. III. Series: China studies (Westport, Conn.) IV. Series: "China to-day" series.

DS777.47.T32 1978 951.04 75-39039
78 MARC

Tsuda, Motonori, b. 1871.

(Shina jihen hishi)

支那事變秘史 / 津田元德著. — 大連：大阪屋號書店, 昭和16 [1941]

4, 8, 312 p., [7] leaves of plates : ill. ; 19 cm.

Front cover mutilated.

1. China—Politics and government—1912-1949. 2. Sino-Japanese Conflict, 1937-1945. I. Title.

DS774.T69 79-804384

Vogau, Boris Andreevich, 1894-1937.

Une femme russe en Chine : [Journal de Xénia Mikhaïlovna] / [par] B. Pilniak [i.e. B. A. Vogau] ; [collab.] A. Rogozina ; trad., préf. et notes de Jocelyne de Maack. — Lausanne : L'Age d'homme, 1976.

127 p. ; 21 cm. — (Contemporains) Sw77-4338

19.00F

1. China—Politics and government—1912-1949. 2. Biespalykh, Xénia Mikhaïlovna. 3. China—Biography. 4. Russia—Biography. I. Rogozina, A., joint author. II. Title.

DS775.V6414 320.9'51'04 78-400604
*79 MARC

Wang, Fan-Hsi, 1907-

(Shuang-shan hui i lu)

雙山回憶錄 / 王凡西著. — 九龍：周記行, 1977.

4, 322 p. ; 20 cm.

$8.00

1. Wang, Fan-Hsi, 1907- 2. Communists—China—Biography. 3. Communism—China. 4. China—Politics and government—1912-1949. I. Title.

DS778.W3445A37 78-842364

Wang, Sheng.

(Tui kung fei ho t'an ts'e lüeh ti fen hsi)

對共匪和談策略的分析 / [王昇著]. — [台北]：世界反共聯盟中華民國分會, 民國67 [1978]

58 p. ; 19 cm.

Cover title.

1. China—Politics and government—1912-1949. 2. China—Politics and government—1949-1976. I. Title.

DS775.W339 79-835236

—POLITICS AND GOVERNMENT—
1912-1949—ADDRESSES, ESSAYS,
LECTURES

Chung, Shu-nan.

(Kai tsao hsien chuang)

改造現狀：又名一瓣香 / 著作者鍾樹楠；校訂者雷渝齊. — 臺北：中國報導·亞洲世紀台北總社, 民國67 [1978]

256 p. : ill. ; 20 cm.

Colophon title.

NT$80.00

1. China—Politics and government—1912-1949—Addresses, essays, lectures. 2. China—Politics and government—1949-1976—Addresses, essays, lectures. 3. Taiwan—Politics and government—1945- —Addresses, essays, lectures. I. Title. II. Title: I pan hsiang.

DS775.C53876 79-838633

(Chung yang jih pao wu shih nien lai she lun hsüan chi)

中央日報五十年來社論選集. — 台北：中央日報社, 民國67 [1978]

2, 16, 372 p. ; 22 cm. — (中央日報創刊五十週年紀念叢書 ; 1)

Cover title.

NT$120.00

1. China—Politics and government—1912-1949—Addresses, essays, lectures. 2. Taiwan—Politics and government—1945- —Addresses, essays, lectures. I. Chung yang jih pao, T'ai-pei. II. Chung yang jih pao she. III. Series: Chung yang jih pao ch'uang k'an wu shih chou nien chi nien ts'ung shu ; 1.

DS777.47.C566 79-841419

Hung, Tsun-yuan.

(Tsen yang chiu kuo)

怎樣救國? / 洪尊元著. — [香港? : s. n., 1978?]

37 p. ; 21 cm.

Reprint of the 1940 ed. published by Hsin chien ti she, Chungking.

1. China—Politics and government—1912-1949—Addresses, essays, lectures. I. Title.

DS775.H88 1978 79-837883

(Kuang jung ti chan ti)

光荣的战地 / [题图陈亦逊, 肖映川]. — [香港? : s. n., 1978?]

158 p. : ill. ; 21 cm.

Reprint of the 1977 ed. published by Chieh fang chün wen i she, Peking.

$60.00

1. China—Politics and government—1912-1949—Addresses, essays, lectures. 2. Chung-kuo jen min chieh fang chün—Addresses, essays, lectures.

DS775.K767 1978 78-842566

Ōtani, Kōzui, 1876-1948.

(Shina ron)

支那論 / 大谷光瑞著. — 東京：民友社, 大正12 [1923]

2, 148 p. ; 19 cm.

Cover title.

1. China—Politics and government—1912-1949—Addresses, essays, lectures. I. Title.

DS774.O83 78-816569

Umwälzung einer Gesellschaft : zur Sozialgeschichte der chinesischen Revolution (1911-1949) / hrsg. von Richard Lorenz. — 1. Aufl., Erstausg. — Frankfurt am Main : Suhrkamp Verlag, 1977.

511 p. ; 18 cm. — (Edition Suhrkamp ; 870) GFR***

Bibliography: p. 495-501.

1. China—Economic conditions—1912-1949—Addresses, essays, lectures. 2. China—Politics and government—1912-1949—Addresses, essays, lectures. 3. Labor and laboring classes—China—Addresses, essays, lectures. 4. Agriculture—China—Addresses, essays, lectures. I. Lorenz, Richard.

HC427.9.U44 330.9'51'04 78-393231
79 MARC

—POLITICS AND GOVERNMENT—
1912-1949—CARICATURES AND
CARTOONS

Erling, Johnny, 1952-

Tigermaske und Knochengespenst : d. neue chines. Karikatur / Erling, v. Graeve. — Köln : Prometh-Verlag, 1978.

140 p. : numerous ill. (some col.) ; 20 x 24 cm. GFR79-A

Includes bibliographical references and index.

ISBN (invalid) 3-922009-6 -: DM26.00

1. China—Politics and government—1912-1949—Caricatures and cartoons. 2. China—Politics and government—1949-1976—Caricatures and cartoons. I. Graeve, Detlev von, 1944- joint author. II. Title.

DS775.E62 320.9'51'05 79-364907
79 MARC

—POLITICS AND GOVERNMENT
—1912-1949—YEARBOOKS

(Chūgoku nenkan)

中國年鑑　宣統3-　年 [1911-

[台北　天一出版社　民國62-　i. e. 1973-

v.　27 cm.

Reprint of a yearbook with original title: Shina nenkan, published 1911-14 in Tokyo by Tōa Dōbunkai Chōsa Hensanbu.

Title in Chinese: Chung-kuo nien chien.

1. China—Economic conditions—1912-1949—Yearbooks. 2. China—Politics and government—1912-1949—Yearbooks. I. Tōa Dōbunkai. II. Title: Shina nenkan. III. Title: Chung-kuo nien chien.

HC426.C5192 78-643208
(MARC-S)

—POLITICS AND GOVERNMENT—
1937-1945

Akaike, Nō, b. 1879.

(Shina jihen to Yudayajin)

支那事変と猶太人 / 赤池濃著. — 東京：政経書房, 昭和14 [1939]

4, 104 p. ; 19 cm. — (国際秘密力研究叢書 ; 第4册)

Series romanized: Kokusai himitsu ryoku kenkyū sōsho.

1. China—Politics and government—1937-1945. 2. China—Economic conditions—1912-1949. 3. Sino-Japanese Conflict, 1937-1945. 4. Jews in China. I. Title.

DS777.53.A7366 78-815789

CHINA
—HISTORY—1937-1945 (Continued)

Arai, Takeo, 1919-
(Ka Kokuhō den)
華国鋒伝 / 新井宝雄著. — 東京：青年出版社, 1978.
341 p., [2] leaves of plates : ill. ; 20 cm. Ja 78-15779
¥2800

1. Hua, Kuo-feng. 2. Heads of state—China—Biography. 3. China—History—1937-1945. 4. China—History—1945- I. Title.
DS779.29.H8A8 78-812874

(Kōnichi sensō)
抗日戦争 / [執筆者 吉沢南 ..., et al.]. — 東京：東京大学出版会, 1978.
v, 303 p. ; 21 cm. — (講座中国近現代史 ; 6) Ja***
Includes bibliographical references.

1. Sino-Japanese Conflict, 1937-1945. 2. China—Politics and government—1937-1945. I. Yoshizawa, Minami. II. Series: Kōza Chūgoku kingendai shi ; 6.
DS755.K69 vol. 6 79-804425
[DS777.53]

Kuan, Chung, 1940-
(Ho t'an ti ch'ien chien)
和談的前鑑：抗戰時期國共和談的再認識 / 關中著. — 臺北：中央日報社, 民國68 [1979]
71 p. : maps ; 19 cm.
Cover title.
NT$15.00

1. China—Politics and government—1937-1945. I. Title.
DS777.53.K765 79-841067

Liang, Chin-tung, 1890-
Patrick Hurley, the China mediator / by Chin-tung Liang. — Jamaica, N.Y. : Center of Asian Studies, St. John's University, [1977?]
37 leaves ; 28 cm. — (St. John's papers in Asian studies ; no. 23)
Includes bibliographical references.
1. China—Politics and government—1937-1945. 2. United States—Foreign relations—China. 3. China—Foreign relations—United States. 4. Hurley, Patrick Jay, 1883-1963. I. Title. II. Series.
DS777.53.L465 320.9'51'042 79-311171
79 MARC

Lin, Wen-lung.
(Sansen to kensetsu Kahoku)
参戰と建設華北 / 林文龍著. — 東京：聯合出版社, 昭和18 [1943]
2, 2, 148 p., [1] leaf of plates : port. ; 19 cm.
Cover title.

1. World War, 1939-1945—China. 2. China—Politics and government—1937-1945. 3. China—Foreign relations—Japan. 4. Japan—Foreign relations—China. I. Title.
D767.3.L55 79-804166

Schaller, Michael, 1947-
The U.S. crusade in China, 1938-1945 / Michael Schaller. — New York : Columbia University Press, 1979.
xiii, 364 p. ; 24 cm.
Bibliography: p. 345-355.
Includes index.
ISBN 0-231-04454-2
1. United States—Foreign relations—China. 2. China—Foreign relations—United States. 3. China—Politics and government—1937-1945. 4. World War, 1939-1945—China. I. Title.
E183.8.C5S325 940.53'22'73 78-15032
78 MARC

Taylor, George Edward, 1905-
The struggle for north China / by George E. Taylor. — 1st AMS ed. — New York : AMS Press, 1978.
xiv, 247 p. : map ; 22 cm.
Reprint of the 1940 ed. published by International Secretariat, Institute of Pacific Relations, New York, in the I.P.R. inquiry series.
Includes bibliographical references and index.
ISBN 0-404-59566-9
1. China—Politics and government—1937-1945. 2. Sino-Japanese Conflict, 1937-1945. I. Title.
DS777.53.T39 1978 320.9'51'042 75-30100
78 MARC

Wang, Sheng.
An analysis of the Chinese Communist tactics of peace talks / by Wang Sheng. — [Taipei?] : World Anti-Communist League, China Chapter, 1978.
76 p. ; 19 cm. — ([English publications of the APACL/ROC] : Pamphlets ; 210)
Cover title.
$0.70
1. China—Politics and government—1937-1945. 2. Communist strategy. I. World Anti-Communist League. China Chapter. II. Title.
DS777.53.W347 320.9'51'042 78-109092
78 MARC

Wu, Ch'eng-yü.
(Shinsei Chūgoku no seiji shihyō)
新生中國の政治指標 / 伍澄宇著；宮坂政治譯. — 東京：泉書房, 昭和19 [1944]
10, 209 p. ; 18 cm.

1. China—Politics and government—1937-1945. I. Title.
DS777.53.W76816 79-805653

—POLITICS AND GOVERNMENT
—1937-1945—ADDRESSES, ESSAYS, LECTURES

(Kōa kenkoku undō no riron to shuchō)
興亞建國運動の理論と主張 / 上海日本總領事館特別調査班譯. — [s. l. : s. n.], 昭和17 [1942]
8, 6, 564 p. ; 21 cm.
"民國二十八年(昭和十四年)秋より約一ヶ年の間に興亞建國運動の幹部同志によって發表された論文を邦譯編輯したもの"
1. China—Politics and government—1937-1945—Addresses, essays, lectures. 2. Sino-Japanese Conflict, 1937-1945—Collaborationists—Addresses, essays, lectures. 3. China—Foreign relations—Japan—Addresses, essays, lectures. 4. Japan—Foreign relations—China—Addresses, essays, lectures. I. Japan. Sōryōjikan, Shanghai.
DS777.53.K54 78-815769

Mao, Tse-tung, 1893-1976.
[Yü shih chieh hsüeh lien tai piao t'uan K'o-le-man hsien sheng, Ya-te hsien sheng, Fu-lu-te hsien sheng, Lei-k'o-nan hsien sheng chih t'an hua. Japanese]
自由獨立ノ支那建設ヲ目指シテ：其ノ他. — [大連]：滿鐵調査部, 昭和13 [1938]
1, 16 leaves ; 27 cm. — (ソ聯紙情報 ; 第8號)
Cover title.
CONTENTS: 毛澤東 自由獨立ノ支那建設ヲ目指シテ—王明 國共兩黨提携ノ發展ト鞏化
1. China—Politics and government—1937-1945—Addresses, essays, lectures. 2. Communism—China—Addresses, essays, lectures. I. Ch'en, Shao-yü, 1907-1974. Kuan yü yung hu kuo min cheng fu shih shih k'ang chan chien kuo kang ling, t'i an ti shuo ming. Japanese. 1938. II. Title. III. Series: Soren shi jōhō ; dai 8-gō.
Title romanized: Jiyū dokuritsu no Shina kensetsu o mezashite.
DS777.53.M25717 78-815761

—POLITICS AND GOVERNMENT
—1937-1949

Chou, Ching-wen.
(Lun hsien chieh tuan Chung-kuo cheng chih ti t'ui ch'ao)
論現階段中國政治的退潮 周鯨文著 香港 時代批評社 [時代書店總經售] 民國30 [1941]
6, 50 p. 19 cm.

1. China—Politics and government—1937-1949. I. Title.
DS777.53.C555 C 67-1825

Chou, Ching-wen.
(Min chu chu i ti tou cheng)
民主主義的鬥爭 / 周鯨文著. — 香港：時代批評社, 民國29 [1940]
2, 2, 178 p. ; 22 cm. — (時代叢書 : 政治組 ; 第3種)
Series romanized: Shih tai ts'ung shu.
Includes bibliographical references.

1. China—Politics and government—1937-1949. I. Title.
DS777.53.C5538 78-843149

Daini Kokuminkai. Henshūbu.
(Ugoku Hokushi)
動く北支 / [編者 第二國民會編輯部；校閲 高橋坦]. — 東京：第二國民會出版部：發賣所 栗田書店, 昭和12 [1937]
2, 4, 277 p., [9] leaves of plates : ill. ; 19 cm.

1. China—Politics and government—1937-1949. 2. China—Foreign relations—Japan. 3. Japan—Foreign relations—China. 4. China—Armed Forces. 5. Marco Polo Bridge Incident, 1937. I. Title.
DS777.53.D27 1939 78-815478

Hasegawa, Satoru, 1896-1969.
(Meirō shin Shina no kensetsu)
明朗新支那の建設 / 長谷川了著. — 東京：第二国民会出版部, 昭和13 [1938]
2, 4, 299 p. ; 19 cm.

1. China—Politics and government—1937-1949. 2. China—Foreign relations—Japan. 3. Japan—Foreign relations—China. 4. Russia—Foreign relations. 5. Communist strategy. I. Title.
DS777.53.H337 78-816935

Hiramatsu, Takashi.
(Shina no jissai o mite chōki kensetsu o kataru)
支那の實際を見て長期建設を語る / 平松卓著. — 東京：國際同盟通信社, 昭和13 [1938]
140 p., [1] leaf of plates : ill. ; 19 cm.

1. China—Description and travel—1901-1948. 2. China—Politics and government—1937-1949. 3. Hiramatsu, Takashi. I. Title. II. Title: Chōki kensetsu o kataru.
DS710.H69 79-804115

Japan. Dai Tōashō. Sōmukyoku. Sōmuka.
(Chūkyō gaisetsu)
中共概說. — [東京]：大東亞省總務局總務課, [1944]
8, 219 p. ; 20 cm. — (大東亞資料 ; 第11號)
Cover title.
"極秘"

1. China—Politics and government—1937-1949. 2. Communism—China. I. Title. II. Series: Dai Tōa shiryō ; dai 11-gō.
DS777.53.J3217 1944 79-802260

Ssŭ-ma, Lu, 1919-
鬥爭十八年 司馬璐著 香港 亞洲出版社 民國41 [1952]
288 p. 19 cm.

1. China—Politics and government—1937-1949. 2. Ssu-ma, Lu, 1919- 3. China—Biography. I. Title.
Title romanized: Tou chêng shih pa nien.
DS777.47.S75 C 62-2134 ‡

Ssu-ma, Lu, 1919-
[Tou cheng shih pa nien. Japanese]
聖地延安：失われし祖国よ・青春よ / 司馬璐著；矢內茂訳. — 東京：生活社, 1955.
262 p. : ill. ; 19 cm.
On verso of t. p.: My strife in eighteen years.

1. China—Politics and government—1937-1949. 2. Ssu-ma, Lu, 1919- 3. China—Biography. I. Title.
Title romanized: Seichi En'an.
DS777.47.S7516 79-800646

—POLITICS AND GOVERNMENT
—1937-1949—ADDRESSES, ESSAYS, LECTURES

Ch'ang, Yen-sheng, 1898-1947.
(Ch'ing ning chi)
清寧集 / 常乃悳[燕生]著. — [香港?：s. n., 1977?]
1, 2, 144 p. ; 21 cm.
Reprint of the 1947 ed. published by Chung liu ch'u pan she, Shanghai.
"本集...所搜有關政治性論文...三十二篇"
1. China—Politics and government—1937-1949—Addresses, essays, lectures. I. Title.
DS777.53.C3654 1977 78-842570

(Ch'ü nien kuo kung t'an p'an wei shen mo shih pai)
去年國共談判為什広失敗. — [s. l.]：遼東建國書社, [194-?]
2, 47 p. ; 19 cm. — (時論選輯 ; 3)
Cover title.
Series romanized: Shih lun hsüan chi.
CONTENTS: 林祖涵關於國共談判的報告.—國共談判雙方來往文件.—延安權威人士評國共談判.—周恩來闡明國共談判眞像.—如何解決(周恩來雙十節報告).—林祖涵再致國民政府代表的信.—蔣介石演說具有危險性.
1. China—Politics and government, 1937-1949—Addresses, essays, lectures.
DS777.53.C568 79-839953

CHINA

—POLITICS AND GOVERNMENT —1937-1949—ADDRESSES, ESSAYS, LECTURES (Continued)

(Chung-kuo yu chi)
中國遊記. — ［s. l.］：遼東建國書社, 1946 printing.
62 p. ; 19 cm. — (時論選集 ; 14)
Series romanized: Shih lun hsüan chi.

1. China—Politics and government—1937-1949—Addresses, essays, lectures.
DS777.53.C5838 78-843153

Fukuoka, Jun'yū, 1896-
(Ō Seiei ni atō)
汪精衛に與ふ / 福岡醇祐. — 東京：亞洲青年聯盟出版部：發賣所 万岳樓書房, 昭和15 ［1940］
248 p. ; 19 cm.

1. China—Politics and government—1937-1949—Addresses, essays, lectures. I. Wang, Chao-ming, 1883-1944. II. Title.
DS777.53.F857 79-801742

(K'ang chan san nien)
抗戰三年：香港國民日報周年紀念册 / 編輯者國民日報社. — ［香港?：s. n., 1978?］
100 p. ; 26 cm.
Colophon title.
Reprint of the 1940 ed. published by Kuo min jih pao she, Hongkong.

1. Sino-Japanese Conflict, 1937-1945—Chronology. 2. China—Politics and government—1937-1949—Addresses, essays, lectures. 3. Kuo min jih pao she—Addresses, essays, lectures. I. Kuo min jih pao she.
DS777.53.K2726 78-842050

Lin, T'ien-hsing.
(Chung-kuo nei mu)
中國内幕 / 編輯者林天行. 第1輯. — 福州：大達圖書公司, 民國37 ［1948］ printing.
4, 190 p. ; 18 cm.
No more published?

1. China—Politics and government—1937-1949—Addresses, essays, lectures. I. Title.
DS777.53.L548 79-837399

Lu, Pi.
(Lun Chang Hsüeh-liang)
論張學良 / 魯泌著. — Hong Kong：神州圖書公司, ［1978?］
7, 132 p. ; 19 cm.
Reprint of the 1948 ed. published by Shih tai p'i p'an she, Hong Kong, in series: Shih tai ts'ung shu : Jen wen tsu.
1. Chang, Hsüeh-liang, 1898- —Addresses, essays, lectures. 2. China—History—Sian Incident, 1936—Addresses, essays, lectures. 3. China—Politics and government—1937-1949—Addresses, essays, lectures. I. Title.
DS778.C495L8 1978 78-842246

(Lun chieh chi li ch'ang yü min tsu li ch'ang ti i chih)
論階級立場與民族立場的一致 / 洛甫著；［編緝新華日報館］. — ［香港?：s. n., 1978?］
2, 132 p. ; 18 cm.
Reprint of the 1939 ed. published by Hsin hua jih pao kuan, Chungking.
CONTENTS: 洛甫 論共產黨的階級立場與民族立場的一致.—洛甫 在民族自衛戰最前綫的崗位上.—彭德懷 彭德懷談堅持河北抗戰與鞏固團結.—呂正操 呂正操談冀中博野事件真相.—蕭伺榮 八路軍新四軍抗戰兩年來的戰績.—平江慘案特輯.—毛澤東談話.
1. Sino-Japanese Conflict, 1937-1945—Addresses, essays, lectures. 2. China—Politics and government—1937-1949—Addresses, essays, lectures. I. Chang, Wen-t'ien, 1900- II. Hsin hua jih pao kuan, Chungking, China.
DS777.53.L866 1978 78-843459

—POLITICS AND GOVERNMENT—1949-

Chiang, Ching-kuo, 1906-
(Chao huan)
召喚：蔣主席告大陸同胞書 = A clarion call / by Chiang Ching-kuo. — 臺北：黎明文化事業股份有限公司, 民國64 ［1975］
16 p. ; 21 cm.
"中英對照"
NT$6.00
1. China—Politics and government—1949- I. Title. II. Title: A clarion call.
DS777.55.C445846 78-843643

Ch'oe, Yŏng.
(Pukpang samgak kwan'gye chŏnmang kwa kŭgŏt i Hanbando e mich'inŭn yŏnghyang)
北方三角關係展望과그것이韓半島에미치는影響 / 研究執筆責任 崔榮；刊行責任 金炳璘. — ［서울］：國土統一院政策企劃室, 1977.
84 p. ; 26 cm. — (周邊環境與件)
"국통정77-1-1119."
Includes bibliographical references.
1. China—Politics and government—1949- 2. China—Relations (general) with Russia. 3. Russia—Relations (general) with China. 4. Korea (Democratic People's Republic)—Foreign relations. I. Kim, Pyŏng-nin. II. Title. III. Series: Chubyŏn hwan'gyŏng yŏkŏn.
DS777.55.C4539 79-825194

Hŏ, Kyŏng-gu.
(Pukhan ŭi ch'ejejŏk pyŏnsu ka kwŏllyŏk kujo pyŏnhwa e mich'inŭn yŏnghyang)
北韓의体制的變數가權力構造變化에미치는影響：中共과의比較 / 研究執筆責任 許景九；刊行責任 李啓熙. — ［서울］：國土統一院調査研究室, 1977.
69 p. ; 26 cm. — (北韓의政治)
"국통조77-12-1317."
Includes bibliographical references.

1. China—Politics and government—1949- 2. Korea (Democratic People's Republic)—Politics and government. 3. Power (Social sciences) I. Yi, Kye-hŭi. II. Title. III. Series: Pukhan ŭi chŏngch'i.
DS777.55.H5575 79-825027

Mathur, Girish, 1920-
China in Africa / Girish Mathur. — New Delhi : Kalamkar Prakashan, 1978.
40 p. ; 22 cm.
Includes bibliographical references.
Rs5.00
1. Africa—Relations (general) with China. 2. China—Relations (general) with Africa. 3. National liberation movements—Africa. 4. Africa—Politics and government—1960- . 5. China—Politics and government—1949- . I. Title.
DT38.9.C5M35 301.29'51'06 79-902567
79 MARC

—POLITICS AND GOVERNMENT —1949—PERIODICALS

(Chū-So jijō)
中ソ事情 Socialist world review.

東京 ジャパン・プレス・サービス
no. 18 x 25 cm.
"編集 國際事情研究会"
Published every other day.
In Japanese.
1. China—Politics and government—1949- —Periodicals. 2. Russia—Politics and government—1945- —Periodicals. 3. World politics—1945- —Periodicals. I. Kokusai Jijō Kenkyūkai. II. Title: Socialist world review.
DS777.55.C46553354 78-642524 (MARC-S)

—POLITICS AND GOVERNMENT— 1949-1976

Abu Imran.
China's policies : other side of the picture / Abu Imran. — Lahore : People's Pub. House, 1976.
65 p. ; 16 cm.
Rs1.50
1. China—Politics and government—1949-1976. 2. China—Foreign relations—1949-1976. I. Title.
DS777.55.A667 941.05 76-938702
78 MARC

Adjitorop, Jusuf.
Integrasi kekuasaan politik dan sistim hukum dgn revolusi di Tiongkok Rakjat / Jusuf Adjitorop. — ［Djakarta］ : Jajasan Pendidikan dan Kebudajaan Baperki Djakarta, 1964.
52 p. : ill. ; 22 cm.
Includes bibliographical references.
1. China—Politics and government—1949-1976. 2. China—Constitutional law. I. Title.
JQ1508.A35 76-949952
79 MARC

Chang, Parris H 1936-
Power and policy in China / Parris H. Chang. — 2d and enl. ed. — University Park : Pennsylvania State University Press, c1978.
331 p. : map ; 23 cm.
Bibliography: p. ［310］-322.
Includes index.
ISBN 0-271-00543-2. ISBN 0-271-00544-0 pbk.
1. China—Politics and government—1949-1976. I. Title.
DS777.55.C375 1978 320.9'51'05 78-50773
79 MARC

Ch'en, Yü-ch'ing, 1918-
(Fen sui kung fei t'ung chan yin mou k'ai ch'uang fan kung fu kuo hsin chü)
粉碎共匪統戰陰謀開創反共復國新局 / 陳裕清著；海外出版社編印. — 1st ed. — 臺北：該社, 1973.
18, 28 p. ; 18 cm.
English and Chinese.
Added t. p.: Chinese communist united front scheme vs. ROC's mainland recovery mission.

1. China—Foreign relations—1949-1976. 2. China—Politics and government—1949-1976. 3. Taiwan—Foreign relations. 4. Taiwan—Politics and government—1945- I. Title. II. Title: Chinese communist united front scheme vs. ROC's mainland recovery mission.
DS777.55.C44517 78-841699

Chin, Hsiung-pai, 1903-
(Chūkyō no jūdai mondai)
中共の十大問題 / 金雄白；本郷賀一訳. — 東京：時事通信社：発売所 同社出版局, 昭和38 ［1963］
443 p. ; 19 cm.

1. China—Politics and government—1949-1976. 2. China—Economic conditions—1949-1976. 3. Communism—China. I. Title.
DS777.55.C446638 78-814380

Chin, Hsiung-pai, 1903-
(Chūkyō no uchimaku)
中共の内幕 / 金雄白；本郷賀一訳. — 東京：時事通信社：発売所 同社出版局, 昭和37 ［1962］
364 p. ; 19 cm.

1. China—Politics and government—1949-1976. 2. Socialism in China. I. Title.
DS777.55.C44664 78-809442

Chin, Hsiung-pai, 1903-
(Kono nazo no kyojin Chūgoku)
この謎の巨人、中国：新聞が書かなかったその実像 / 金雄白著；伊達政之訳. — 東京：蒼洋社：発売 英潮社, 1978.
271 p., ［9］ leaves of plates : ill. ; 19 cm. — (蒼洋社選書 ; 1)
Ja 78-25403
Series romanized: Sōyōsha sensho.
¥1200
1. China—Politics and government—1949-1976. 2. China—Social life and customs—1949- I. Title. II. Title: Chūgoku.
DS777.55.C446645 79-801519

China (Republic of China). Kuang fu ta lu she chi yen chiu wei yüan hui.
(Kung fei pien chiang tzu chih ch'ü hsien k'uang chih yen chiu)
共匪邊疆自治區現況之研究 / 光復大陸設計研究委員會研擬. — ［台北］：該會, 民國65 ［1976］
6, 476 p. ; 22 cm. — (邊疆研究叢書)
Cover title.
1. Minorities—China. 2. China—Race relations. 3. China—Politics and government—1949-1976. I. Title. II. Series: Pien chiang yen chiu ts'ung shu.
DS730.C4482 1976a 78-841893

Ch'iu, K'ung-yüan.
Mainland situation after the Fourth National People's Congress / ［by Chiu Kung-yuan］. — Taipei : Chung Hwa Information Service, ［1978?］
29 p. ; 20 cm.
On cover: 3.
Includes bibliographical references.
1. China—Politics and government—1949-1976. 2. China (People's Republic of China, 1949-). Ch'üan kuo jen min tai piao ta hui. I. Title.
DS777.55.C4528 320.9'51'05 78-103830
78 MARC

Chŏng, Yŏng.
(Chunggong ŭi tang, chŏng, kun ŭi yŏkhak kwan'gye wa Hwa Kuk-pong ch'eje ŭi chŏnmang)
中共의党·政·軍의力学関係와華國鋒体制의展望 / 研究執筆責任 鄭永；刊行責任 朴鏞德. — ［서울］：國土統一院調査研究室, 1977.
45 p. ; 26 cm. — (周邊環境與件：中共)
"국통조 77-1-1103."
1. China—Politics and government—1949-1976. 2. China—Politics and government—1977- I. Pak, Yong-dŏk. II. Title. III. Series: Chubyŏn hwan'gyŏng yŏkŏn.
DS777.55.C4543 78-826292

CHINA
—POLITICS AND GOVERNMENT
—1949-1976 (Continued)

Chou, Ching-wen, 1909-
(Feng pao shih nien)
風暴十年—中國紅色政權的眞面貌　周鯨文著
香港　時代批評社　民國48 [1959]
588 p. 20 cm.

1. China—Politics and government—1949-1976. I. Title.
DS777.55.C457 C 59-1429

Chou, Ching-wen, 1909-
Ten years of storm; the true story of the Communist regime in China, by Chow Ching-wen. Foreword by Lin Yutang. Translated and edited by Lai Ming. Westport, Conn., Greenwood Press [1973, c1960]
xxii, 323 p. 22 cm.
Translation of Feng pao shih nien.
Includes bibliographical references.
ISBN 0-8371-6685-3
1. China—Politics and government—1949-1976. I. Title.
[DS777.55.C4573 1973] 951.05 72-12632
00[r79]rev MARC

Chou, Ching-wen, 1909-
(Wo wei li shih tso cheng)
我爲歷史作証　周鯨文著　香港　時代批評社
民國48 [1959]
149 p. 19 cm.

1. China—Politics and government—1949-1976. 2. Communism—China. I. Title.
DS777.55.C458 C 60-314
rev

(Chūkyō no jittai to Nitchū shinkankei no arikata)
中共の実態と日中新関係の在り方：中国総合研究会・第二十一次研究報告. — 東京：中国総合研究会, 昭和43 [1968]
376 p. ; 18 x 25 cm.
Cover title.
Not in trade.

1. China—Politics and government—1949-1976. 2. China—Foreign relations—Asia. 3. Asia—Foreign relations—China. I. Chūgoku Sōgō Kenkyūkai.
DS777.55.C465762 78-813750

Chung-kuo kuo min tang. Chung yang wei yüan hui. Ti liu tsu.
(Kuo min ko ming tsai Chung-kuo ta lu ti yen hsü)
國民革命在中國大陸的延續 [秦孝儀序 臺北 中國國民黨中央委員會第六組編印 民國57 [1968]
2, 4, 66 p. 21 cm.
Cover title.
1. China—Politics and government—1949-1976. I. Ch'in, Hsiao-i. II. Title.
DS777.55.C47584 73-836987

Enseñanza y revolución en China / [por] Mao Tse Tung y otros autores ; edición a cargo de Joan Senent-Josa. — Barcelona : Anagrama, D.L. 1977.
334 p., 1 leaf ; 21 cm. — (Colección Documentos ; 30) Sp77-Oct
Bibliography: p. 331-334.
ISBN 8433901303 : 350ptas
1. Mao Tse Tung, 1893-1976. 2. Education—Philosophy. 3. China—Politics and government—1949-1976. I. Mao Tse-Tung, 1893-1976. II. Senent, Juan.
LB880.M332E57 78-379932
78 MARC

Fan, Yüan-yen.
(Wei pa i k'u nan t'ung pao tso cheng)
為八億苦難同胞作證 / 范園焱口述；劉本炎執筆；中央日報編印. — 臺北：該報社, 民國66 [1977]
186 p. : ports. ; 19 cm.
NT$20.00 ($0.50 U. S.)
1. China—Politics and government—1949-1976. 2. China—Social conditions. 3. Fan, Yüan-yen. 4. Refugees, Political—Taiwan—Biography. 5. Refugees, Political—China—Biography. I. Liu, Pen-yen. II. Chung yang jih pao, T'ai-pei. III. Title.
DS777.55.F3377 78-843502

Fan, Yüan-yen.
(Wo chia Mi-ko shih chiu fei lai)
我駕米格十九飛來 / 范園焱著. — 台北：文藝月刊社：經銷者遠東書報社, 民國67 [1978]
264 p. : col. ill. ; 20 cm. — (文藝月刊叢書 ; 11)
Series romanized: Wen i yüeh k'an ts'ung shu.
NT$70.00

1. China—Politics and government—1949-1976. 2. China—Social conditions. 3. Fan, Yüan-yen. 4. Refugees, Political—Taiwan—Biography. 5. Refugees, Political—China—Biography. I. Title.
DS777.55.F338 78-843497

Głabiński, Stanisław.
[Mao i inni. Bulgarian]
Мао и другите : Чжоу Ен-Лай, Лю Шао-Ци, Лин Бяо, Пън Дъ-Хуай / Станислав Гломбински ; [превел от полски Красимир Иванджийски]. — София : Военно изд-во, 1978.
295 p. : ill. ; 20 cm. — (Контрасти) Bu***
Series romanized: Kontrasti.
"Биографични бележки": p. 285-[294]
Includes bibliographical references.
0.62 lv
1. China—Politics and government—1949-1976. I. Title.
Title romanized: Mao i drugite.
DS777.55.G53812 79-407297

Han, Suyin, pseud.
Wind in the tower : Mao Tsetung and the Chinese Revolution, 1949-1976 / [by] Han Suyin ; with a foreword by Malcolm MacDonald. — St. Albans : Triad, 1978.
476 p. ; 18 cm. GB78-31861
Sequel to The Morning deluge.
Includes bibliographical references and index.
ISBN 0-586-04505-8 : £1.75
1. China—Politics and government—1949-1976. 2. Mao, Tse-tung, 1893-1976. I. Title.
DS777.55.H2975 1978 951.05'092'4 79-308822
79 MARC

Hoffmann, Rainer.
Kampf zweier Linien : zur polit. Geschichte d. chines. Volksrepublik 1949-1977 / Rainer Hoffmann. — 1. Aufl. — Stuttgart : Klett-Cotta, 1977.
141 p. ; 24 cm. GFR78-A
Includes bibliographical references and index.
ISBN 3-12-910180-2 : DM38.00
1. China—Politics and government—1949-1976. I. Title.
DS777.55.H5735 320.9'51'05 78-376436
78 MARC

Hoffmann, Rainer.
Maos Rebellen : Sozialgeschichte d. chines. Kulturrevolution / Rainer Hoffmann. — 1. Aufl. — Hamburg : Hoffmann und Campe, 1977.
227 p. ; 21 cm. — (Historische Perspektiven ; 8) GFR78-A
Summary in English.
Bibliography: p. 216-222.
Includes index.
ISBN 3-455-09220-9 : DM32.00
1. China—Social conditions. 2. China—Politics and government—1949-1976. I. Title.
HN737.H63 309.1'51'05 78-389857
79 MARC

Hsin min chu ch'u pan she, Hongkong.
(Chung-hua jen min kung ho kuo k'ai kuo wen hsien)
中華人民共和國開國文獻. — 重版. — 香港：文化資料供應社, 1978.
10, 311 p. ; 19 cm.
Reprint of the 1949 ed. published by Hsin min chu ch'u pan she, Hongkong.
$10.00
1. China—Politics and government—1949-1976. I. Title.
JQ1503 1978.H77 1978 79-835150

(Hsüeh hsi Mao chu hsi ti kuang hui wen hsien Lun shih ta kuan hsi)
学习毛主席的光辉文献《论十大关系》. — [香港?：s.n., 1978?]
4, 137 p. ; 21 cm.
Reprint of the 1977 ed. published by Chiang-su jen min ch'u pan she.

1. Mao, Tse-tung, 1893-1976. Lun shih ta kuan hsi. 2. China—Politics and government—1949-1976.
DS777.55.M32173H78 1978 79-840652

(Hsüeh hsi shih liu t'iao shou ts'e)
学习十六条手册. — 增订本. — 第1版. — 北京：人民出版社：新华书店发行, 1966.
3, 83 p., [1] leaf of plates : ports. ; 13 cm.
CONTENTS: 中国共产党中央委员会关于无产阶级文化大革命的决定.—林彪同志在庆祝无产阶级文化大革命群众大会上的讲话.—林彪同志在接见外地来京革命师生大会上的讲话.—林彪同志在接见全国各地来京革命师生大会上的讲话.—周恩来同志在接见外地来京革命师生大会上的讲话.—周恩来同志在接见全国各地来京革命师生大会上的讲话.—学习十六条, 熟悉十六条, 运用十六条.—革命青少年要向解放军学习.—用文斗, 不用武斗.—掌握斗争的大方向.—工农群众和革命学生在毛泽东思想旗帜下团结起来.—向工农兵致敬, 向工农兵学习.

1. Chung-kuo kung ch'an tang. Chung yang wei yüan hui. Chung-kuo kung ch'an tang chung yang wei yüan hui kuan yü wu ch'an chieh chi. 2. Chung-kuo kung ch'an tang—Purges. 3. China—Politics and government—1949-1976.
JQ1519.A5H8526 79-837859

Jiji Tsūshinsha. Gaishinbu.
(Pekin Taiwan Kokusai Rengō)
北京·台湾·国際連合 / 時事通信社外信部編. — 東京：時事通信社, 昭和36 [1961]
239 p. ; 18 cm. — (時事新書)
Series romanized: Jiji shinsho.

1. China—Politics and government—1949-1976. 2. Taiwan—Politics and government—1945- 3. United Nations—Membership. 4. World politics—1945- I. Title.
DS777.55.J52 1961 78-813116

Jōno, Hiroshi.
(Chūgokugaku genri)
中国学原理 / 城野宏. — 東京：原書房, 1978.
214 p. ; 20 cm. Ja 78-18115
¥1300

1. China—Study and teaching. 2. China—Politics and government—1949-1976. 3. Japan—Foreign relations—China. 4. China—Foreign relations—Japan. I. Title.
DS734.95.J66 78-813203

Kao, Tsu-hsing.
(T'ien t'ang yü ti yü)
天堂與地獄 / 高祖興著. — 臺北：中國大陸問題研究所, 民國67 [1978]
68 p. : ill. ; 22 cm. — (中國大陸問題叢書 ; 11)
NT$30.00

1. Taiwan—Politics and government—1945- 2. China—Politics and government—1949-1976. I. Title. II. Series: Chung-kuo ta lu wen t'i ts'ung shu ; 11.
DS799.8.K38 79-839949

Kaushik, Devendra.
China, an ideological puzzle / by Devendra Kaushik. — New Delhi : Sopan Pub. House, c1979.
vii, 96 p. ; 23 cm.
Bibliography: p. [89]-91.
Includes index.
Rs30.00 ($6.00 U.S.)
1. China—Politics and government—1949-1976. I. Title.
DS777.55.K33654 320.9'51'05 79-902328
79 MARC

Kikuchi, Masanori, 1930-
(Pekin Shinkyō kikō)
北京·新疆紀行：革命と民衆を訪ねて / 菊地昌典, 山本満. — 東京：筑摩書房, 1978.
iv, 206 p., [2] leaves of plates : ill. ; 20 cm. Ja 78-16389
¥1300

1. China—Politics and government—1949-1976. 2. Sinkiang—Description and travel. I. Yamamoto, Mitsuru, 1927- joint author. II. Title.
DS777.55.K475 78-812957

Kokusai Jōsei Kenkyūkai.
(Chūkyō naigai seisaku no kihon hōshin)
中共内外政策の基本方針. — 東京：国際情勢研究会, 昭和41 [1966]
37 p. ; 25 cm.
Cover title.

1. China—Foreign relations—1949- 2. China—Politics and government—1949-1976. I. Title.
DS777.55.K615 1966a 79-803468

Li, T'ien-min, 1910-
Chang Chun-chiao, a noticeable likely successor to the Chinese communist leadership / [by Li T'ien-min]. — Taipei : Chung Hwa Information Service, [1978?].
12 p. ; 20 cm.
On cover: 8.
Includes bibliographical references.
1. China—Politics and government—1949-1976. 2. Chang, Ch'un-ch'iao. I. Title.
DS777.55.L434313 320.9'51'05 78-103821
79 MARC

CHINA

—POLITICS AND GOVERNMENT
—1949-1976 (Continued)

Ma, Jui-hsüeh.
(Shu ssu chi)
抒思集 / 馬瑞雪著. — 臺北：中華日報：總經銷星光書報社, 民國67 [1978]
4, 205 p. ; 20 cm. — (中華日報甲種叢書 ; 60)
In colophon: Ch'u pan che Chung-hua jih pao she.
Series romanized: Chung-hua jih pao chia chung ts'ung shu.
NT$50.00
1. Ma, Jui-hsüeh. 2. Authors, Chinese—20th century—Biography. 3. China—Politics and government—1949-1976. I. Title.
PL2881.J8Z474 79-839124

Mao, Tse-tung, 1893-1976.
(Lun shih ta kuan hsi)
论十大关系 = Lun shi da guanxi / 毛泽乐. — 注音本. — 北京：文字改革出版社, 1978.
55 p. ; 19 cm.
In Chinese characters and Pinyin.
$0.16
1. China—Politics and government—1949-1976. I. Title. II. Title: Lun shi da guanxi.
DS777.55.M3217 1978 79-837977

(Mei-kuo t'e wu chien tieh yin mou wu chuang pao tung an)
美國特務間諜陰謀武裝暴動案 / [編輯者羣衆書店]. — Washington : Center for Chinese Research Materials, Association of Research Libraries, [1979?]
50 p. on [25] leaves : ill. ; 22 x 28 cm.
Imprint stamped on cover.
Reprint of the 1951 ed. published by Ch'ün chung shu tien, Peking. "M788"
1. Espionage, American—China. 2. Trials (Espionage)—China. 3. China—Politics and government—1949-1976. I. Ch'ün chung shu tien.
DS777.55.M4544 1979 79-841045

Ōmori, Minoru, 1922-
(Mō Takutō)
毛沢東 = Mao Zedong：不世出の巨星 / 大森実. — 東京：講談社, 1979.
394 p. : ill. ; 19 cm. — (His 人物現代史 ; 9) Ja***
In Japanese.
Bibliography: p. 383-386.
Includes indexes.
1. Mao, Tse-tung, 1893-1976. 2. China—Politics and government—1912-1949. 3. China—Politics and government—1949-1976. 4. Heads of state—China—Biography. I. Title. II. Title: Mao Zedong. III. Series: Ōmori, Minoru, 1922- Jimbutsu gendai shi ; 9.
DS778.M3O45 79-809034
79 J

Pei-ching nei jan chi tsung ch'ang. Kung jen li lun tsu.
(Kuang hui ti wen hsien, Ch'iang ta ti wu ch'i)
光辉的文献, 强大的武器：学习毛主席《论十大关系》讲座 / 北京内燃机总厂工人理论组, 中国社会科学院经济研究所计划经济组. — 第1版. — 北京：人民出版社, 1977.
2, 106 p. ; 19 cm. — (农村政治理论丛书)
$0.19
1. Mao, Tse-tung, 1893-1976. Lun shih ta kuan hsi. 2. China—Politics and government—1949-1976. I. Chung-kuo she hui k'o hsüeh yüan. Chi hua ching chi tsu. II. Title. III. Series: Nung ts'un cheng chih li lun ts'ung shu.
DS777.55.M32173P44 1977 79-840651
79 C

Sha, K'o-fu.
(Chung kung nei mu)
中共内幕 / 沙可夫編著. — [香港?：s. n., 1978?]
16 p. ; 21 cm.
Reprint of a combined ed. completed in two parts called "ch'ien chi" and "hou chi", issued in series as Hu sheng pao ts'ung shu chih i.
1. China—Politics and government—1949-1976. I. Title.
DS777.55.S4536 1978 78-843142

(Ts'uan tang to ch'üan ti i ko ta yin mou)
篡党夺权的一个大阴谋：批判四人帮对《论总纲》,《汇报提纲》,《二十条》的批判. — [北京]：人民出版社, 1977.
70 p. ; 19 cm.
CONTENTS: 向群 打着反复辟的旗号搞复辟.—中国科学院理论组 要知松高洁 待到雪化时.—国家计委大批判组 一场篡党夺权的反革命丑剧.
$0.14
1. China—Politics and government—1949-1976.
DS777.55.T796 78-842703
C

Wang, Sheng.
(Tui kung fei ho t'an ts'e lüeh ti fen hsi)
對共匪和談策略的分析 / [王昇著]. — [台北]：世界反共聯盟中華民國分會, 民國67 [1978]
58 p. ; 19 cm.
Cover title.
1. China—Politics and government—1912-1949. 2. China—Politics and government—1949-1976. I. Title.
DS775.W339 79-835236

Wang, Ssu-ch'eng.
(Mao Tse-tung yü hung huo)
毛澤東與紅禍 / 王思誠著. — [再版]. — 台北：黎明文化事業股份有限公司, 民國66 [1977]
10, 9, 714 p. ; 20 cm.
NT$120.00
1. Mao, Tse-tung, 1893-1976. 2. Heads of state—China—Biography. 3. Chung-kuo kung ch'an tang—History. 4. China—Politics and government—1949-1976. I. Title.
DS778.M3W3 1977 78-842252
C

Ward, Doug.
China-the sick dragon / by Doug Ward. — [s.l. : s.n. ; San Francisco : distributor, Sundowner Services, c1978]
vii, 187 p. : ill. ; 22 cm.
Includes bibliographical references.
ISBN 0-932410-06-5 : $8.00
1. China—Politics and government—1949-1976. 2. China—Politics and government—1976- I. Title.
DS777.55.W396 320.9'51'05 78-65281
79 MARC

Whiting, Allen Suess, 1926-
Chinese domestic politics and foreign policy in the 1970's / by Allen S. Whiting. — Ann Arbor : Center for Chinese Studies, University of Michigan, 1979.
v, 85 p. ; 23 cm. — (Michigan papers in Chinese studies ; no. 36)
CONTENTS: Introduction.—PRC media images of the U.S. and Taiwan, January-April 1976 and 1977.—Domestic politics and foreign trade in the PRC, 1971-1976.
ISBN 0-89264-036-7 : $4.00
1. China—Politics and government—1949-1976. 2. Mass media—Political aspects—China. I. Title. II. Series.
DS777.55.W443 320.9'51'05 78-31865
79 MARC

Yin, Fang.
(Chung kung ch'ün chien jan hsi lu)
中共羣奸燃犀錄 / 尹方輯. — [台北]：台北市海州同鄉會, 民國66 [1977]
400 p. : ill. ; 22 cm.
NT$60.00 ($2.00 U.S.)
1. China—Politics and government—1949-1976. 2. Chung-kuo kung ch'an tang. I. Title.
DS777.55.Y47 78-841172
C

Yuasa, Makoto, 1941-
(Tō Shōhei)
鄧小平 / 湯浅誠編著. — 東京：日本文芸社, [1978]
260 p. : ill. ; 19 cm. Ja 78-14507
¥900
1. Teng, Hsiao-p'ing, 1900- 2. Statesmen—China—Biography. 3. China—Politics and government—1949-1976. 4. China—Politics and government—1976- I. Title.
DS778.T39Y8 78-812746

—POLITICS AND GOVERNMENT
—1949-1976—ADDRESSES, ESSAYS, LECTURES

(Ch'en Tsai-tao tzu liao chi)
陳再道資料集：原名「武漢事件資料專輯」. — [香港?：s. n., 1970]
52 leaves : ill. ; 23 x 34 cm.
Cover title.
Publication date stamped on p. [4] of cover.
A collection of articles and newspaper clippings reprinted from various sources.
1. Ch'en, Tsai-tao—Addresses, essays, lectures. 2. Generals—China—Biography—Addresses, essays, lectures. 3. China—Politics and government—1949-1976—Addresses, essays, lectures. I. Title: Wu-han shih chien tzu liao chuan chi.
DS778.C53497C46 79-837920

(Ch'ou chien chi)
抽剑集. — 第1版. — [广州]：广东人民出版社, 1977.
99 p. ; 19 cm.
$0.21
1. China—Politics and government—1949-1976—Addresses, essays, lectures.
DS777.55.C4563 1977b 79-838634

(Chūgoku bunka daikakumei no saikentō)
中国文化大革命の再検討 / 東亜文化研究所紀要編集委員会編. — 東京：霞山会, 1978-
v. ; 22 cm. — (東亜文化叢書 ; 4-) Ja 79-787 (v. 1)
¥1500 (v. 1)
1. China—Politics and government—1949-1976—Addresses, essays, lectures.
DS777.55.C465558 79-806332

Chung-shan ta hsüeh, Canton, China. Li shih hsüeh hsi. Ch'i erh chi. Pien hsieh tsu.
(Lao tung jen min fan K'ung tou cheng shih hua)
劳动人民反孔斗争史话 / 中山大学历史[学]系七二级编写组编. — [广州]：广东人民出版社, 1974.
3, 92 p., [7] leaves of plates : ill. ; 19 cm.
$0.23
1. Confucius—Addresses, essays, lectures. 2. China—Politics and government—1949-1976—Addresses, essays, lectures. I. Title.
B128.C8C555 1974 75-838164

Chung, Shu-nan.
(Kai tsao hsien chuang)
改造現狀：又名一瓣香 / 著作者鍾樹楠；校訂者雷渝齊. — 臺北：中國報導·亞洲世紀台北總社, 民國67 [1978]
256 p. : ill. ; 20 cm.
Colophon title.
NT$80.00
1. China—Politics and government—1912-1949—Addresses, essays, lectures. 2. China—Politics and government—1949-1976—Addresses, essays, lectures. 3. Taiwan—Politics and government—1945- —Addresses, essays, lectures. I. Title. II. Title: I pan hsiang.
DS775.C53876 79-838633

(Hua Kuo-feng yü Teng Hsiao-p'ing)
華國鋒與鄧小平 / 王建[譯]；[編者毛國倫]. — 香港：明報有限公司出版部, 1977.
6, 270 p. : ports. ; 19 cm. — (中共近人近事譯叢)
Includes bibliographical references.
CONTENTS: 米徹爾·奧辛伯 華國鋒的湖南政績.—斯托厄·舒爾霖 華國鋒主持編輯的「論十大關係」.—約翰·吉廷斯 鄧小平的三份重大文件.—愛德華·賴爾斯 鄧小平的再起和再落.—杜勉 動盪中的過渡時期.—天安門的雷霆風暴.—米徹爾·奧辛伯 千秋功罪細評說.
$10.00 ($2.50 U.S.)
1. China—Politics and government—1976- —Addresses, essays, lectures. 2. China—Politics and government—1949-1976—Addresses, essays, lectures. 3. Hua, Kuo-feng—Addresses, essays, lectures. 4. Teng, Hsiao-p'ing, 1900- —Addresses, essays, lectures. I. Wang, Chien. II. Mao, Kuo-lun. III. Series: Chung kung chin jen chin shih i ts'ung.
DS779.26.H84 78-842944

(Kabe shimbun ni yoru Chūkyō「bunka daikakumei」kanren shiryō shū)
壁新聞による中共「文化大革命」関連資料集. — [s. l. : s. n., 1967?-
v. ; 25 cm.
Cover title.
1. China—Politics and government—1949-1976—Addresses, essays, lectures. I. Title: Chūkyō「bunka daikakumei」kanren shiryō shū.
DS777.55.K18 78-809453

CHINA

—POLITICS AND GOVERNMENT —1949-1976—ADDRESSES, ESSAYS, LECTURES (Continued)

(Kōeihei shimbun ni yoru Chūkyō「bunka daikakumei」kanren shiryō shū)
紅衛兵新聞による中共「文化大革命」関連資料集. — [s. l. : s. n., 1967?-
v. : ill. ; 25 cm.
Cover title.

1. China—Politics and government—1949-1976—Addresses, essays, lectures. I. Title: Chūkyō bunka daikakumei kanren shiryō shū.
DS777.55.K595 78-810964

(Kuo ch'ing hsüan ch'uan tzu liao)
國慶宣傳資料. — [香港] : 香港同胞慶祝中華人民共和國成立二十五周年大會籌備委員會宣傳部, 1974.
59 p. ; 19 cm.
Cover title.
1. China—Politics and government—1949-1976—Addresses, essays, lectures. 2. World politics—1965-1975—Addresses, essays, lectures. I. Hsiang-kang t'ung pao ch'ing chu Chung-hua jen min kung ho kuo ch'eng li erh shih wu chou nien ta hui ch'ou pei wei yüan hui. Hsüan ch'uan pu.
DS777.55.K7518 78-841492

Li, Cheng-chung.
(Fan kung ko ming ch'ing shih chih fen hsi)
反共革命情勢之分析 / 李正中講 ; 教育部社會教育司編印. — [台北] : 該司, 民國65 [1976]
18 p. ; 19 cm. — (文化講座專集 ; 71)
"中華民國六十五年六月廿一日講於臺南高雄"

1. China—Politics and government—1949-1976—Addresses, essays, lectures. 2. Communism—Addresses, essays, lectures. I. China (Republic of China). Chiao yü pu. She hui chiao yü ssu. II. Title. III. Series: Wen hua chiang tso chuan chi ; 71.
DS777.55.L4283 78-841469

Li, Chung-tao, fl. 1974-
(Mao ch'iu mi wen)
毛酋秘聞 / 李忠道著. — 台北 : 李忠道 : 總經銷黎明文化事業股份有限公司, 民國64 [1975]
5, 206 p. : ill. ; 20 cm.
"曾自六十三年八月二十日起在香港華僑日報連載三個月"

1. China—Politics and government—1949-1976—Addresses, essays, lectures. 2. Mao, Tse-tung, 1893-1976—Addresses, essays, lectures. I. Title.
DS777.55.L4284 76-836587

(Li shih ch'e lun pu jung tao chuan)
历史车轮不容倒转 : 上海工人批林批孔文选. — 第1版. — [上海人民出版社编辑]. — 上海 : 該社, 1974.
4, 2, 77 p. ; 19 cm.

1. China—Politics and government—1949-1976—Addresses, essays, lectures. 2. Confucius—Addresses, essays, lectures. 3. Lin, Piao, 1908-1971—Addresses, essays, lectures. I. Shang-hai jen min ch'u pan she.
DS777.55.L4358 78-843823

Lin, Mu-yen.
(Kuang fu ta lu)
光復大陸 : 匪共必降論 / 林穆嚴著. — 臺北 : 正中書局, 民國66 [1977]
38, 35 p., [1] leaf of plates : port. ; 22 cm.
Title in English on p. [4] of cover: On national recovery : the certainty of Chinese Communists' surrender.
English and/or Chinese.
1. China—Politics and government—1949-1976—Addresses, essays, lectures. I. Title. II. Title: On national recovery.
DS777.55.L4584 78-842328

Liu, Chia-chü.
(Li shih yü hsien shih)
歷史與現實 / 劉家駒著. — [香港] : 中國學人出版社, [1978?]
165 p. ; 22 cm.
$8.00

1. China—Politics and government—1949-1976—Addresses, essays, lectures. 2. Taiwan—Politics and government—1945- —Addresses, essays, lectures. I. Title.
DS777.55.L4948 79-837277

Mao, Tse-tung, 1893-1976.
(Tsai k'uo ta ti chung yang kung tso hui i shang ti chiang hua)
在扩大的中央工作会议上的讲话 : [一九六二年一月三十日] / 毛泽东. — 香港 : 生活讀書新知三联书店, 1978.
30 p. ; 21 cm.
$0.70
1. China—Politics and government—1949-1976—Addresses, essays, lectures. I. Title.
DS777.55.M339 1978 79-836122

Pei-ching yu hsien tien ch'ang. Kung jen li lun tsu.
(Chi feng pao yü ti tou cheng)
急风暴雨的斗争 / [北京有线电厂工人理论组编写]. — 第1版. — 北京 : 中国少年兒童出版社, 1977.
2, 2, 81 p. : ill. ; 19 cm.
$0.19

1. China—Politics and government—1949-1976—Addresses, essays, lectures.
DS777.55.P3738 1977 79-835728

(P'i K'ung ts'an k'ao ts'ai liao)
批孔参考材料. — [香港? : s. n., 1979?]
67 p. ; 21 cm.
Reprint ed.

1. China—Politics and government—1949-1976—Addresses, essays, lectures. 2. Confucius—Addresses, essays, lectures.
DS777.55.P51225 1979 79-837835

The Revolution is dead, long live the revolution : readings on the great proletarian cultural revolution from an ultra-left perspective / compiled and edited by the 70's. — Hong Kong : The 70's, 1976.
v. 291 p. ; 20 cm.
CONTENTS: Progressive Labour Party. The great proletarian cultural revolution and the reversal of workers' power in China.—Haris, N. China.—Wang, F. H. On "the great proletarian cultural revolution".—Dunayevskaya, R. Mao's China and the "proletarian cultural Revolution".—Situationist International. The explosion point of ideology in China.—Brendel, C. Theses on the Chinese revolution.—Kwok, K. C. "Everything remains the same after so much ado".—Sheng-Wu-Lien. Whither China?.—Shuet, Y. The dusk of rationality.—Man, Wu. Introduction to revelations that move the earth to tears.—An interview with an ultra-leftist.—Li, I. C. Concerning socialist democracy and legal system.—See, L. Y. and Che, W. Some thoughts on the Chinese revolution.

1. China—Politics and government—1949-1976—Addresses, essays, lectures. I. Ch'i shih nien tai.
DS777.55.R46 320.9'51'05 78-102598
79 MARC

Ssu-ma, Ch'ang-feng.
(Wen ko hou ti Chung kung)
文革後的中共 / 司馬長風著. — 台北 : 時報文化出版事業有限公司, 民國66 [1977]
3, 5, 549 p., [1] fold. leaf of plates : graph ; 20 cm. — (時報書系 ; 89)
Series romanized: Shih pao shu hsi.
NT$100.00
1. China—Politics and government—1949-1976—Addresses, essays, lectures. 2. China—Politics and government—1976- —Addresses, essays, lectures. I. Title.
DS777.55.S7278 79-835196

79 C

(Tsai sheng li ti ta tao shang fen yung ch'ien chin)
在勝利的大道上奮勇前進 : 熱烈慶祝中華人民共和國成立十九周年. — 香港 : 香港三聯書店, 1968.
17 p. ; 19 cm. — (學習叢書 ; 28)
Cover title.
Series romanized: Hsüeh hsi ts'ung shu.
CONTENTS: 林彪副主席在中華人民共和國成立十九周年慶祝大會上的講話—《人民日報》,《紅旗》雜誌,《解放軍報》社論 在勝利的大道上奮勇前進—周恩來總理在慶祝中華人民共和國成立十九周年招待會上的講話
1. China—Politics and government—1949-1976—Addresses, essays, lectures. I. Lin, Piao, 1908-1971. II. Chou, En-lai, 1898-1976.
DS777.55.T7236 79-839618

Tso, Shun-sheng, 1893?-1969.
(Tso Shun-sheng tzu hsüan chi)
左舜生自選集 / 左舜生著. — [臺北縣永和鎮] : 文海出版社, [1978]
2 v. : ill. ; 22 cm. — (近代中國史料叢刊續編 [i. e. 輯] ; 528-529 : 第五十三輯)
CONTENTS: [1] 政論, 時評.—[2] 近作, 雜文
1. China—Politics and government—1949-1976—Addresses, essays, lectures. 2. World politics—1945-1955—Addresses, essays, lectures. I. Title. II. Series: Shen, Yün-lung, comp. Chin tai Chung-kuo shih liao ts'ung k'an hsü chi ; 528-529.
DS755.S532 vol. 528-529 78-843363
[DS777.55]

(Wu ch'an chieh chi chuan cheng hsia chin hsing ko ming ti li lun wu ch'i)
無產階級專政下進行革命的理論武器 : 紀念關於正確處理人民内部矛盾的問題發表十周年. — 香港 : 香港三聯書店, 1967.
28 p. ; 19 cm. — (學習叢書 ; 第26輯)
Cover title.
Series romanized: Hsüeh hsi ts'ung shu.
CONTENTS: 無產階級專政下進行革命的理論武器 《紅旗》雜誌社論—正確區分兩類社會矛盾 團結·批評·團結 《人民日報》社論—解放軍總政號召全軍認眞學習《關於正確處理人民内部矛盾的問題》

1. Mao, Tse-tung, 1893-1976. Kuan yü cheng ch'üeh ch'u li jen min nei pu mao tun ti wen t'i—Addresses, essays, lectures. 2. China—Politics and government—1949-1976—Addresses, essays, lectures. 3. Communism—China—Addresses, essays, lectures. 4. Dictatorship of the proletariat—Addresses, essays, lectures. I. Hung ch'i.
DS777.55.M26W8 79-840076

(Wu ch'an chieh chi wen hua ta ko ming hao te hen)
无产阶级文化大革命好得很. — [香港? : s. n., 1978?]
2, 108 p. ; 20 cm.
Colophon title.
Reprint of the 1974 ed. published by Ho-pei jen min ch'u pan she, Tientsin.

1. China—Politics and government—1949-1976—Addresses, essays, lectures.
DS777.55.W8232 78-843211

—POLITICS AND GOVERNMENT —1949-1976—BIBLIOGRAPHY

Documents on contemporary China, 1949-1975 : a selected research collection in microfiche / consulting editor, George Lowy. — Greenwich, Conn. : JAI Press, c1976- c1977.
2 v. ; 29 cm.
CONTENTS: [1] Sorich, R., ed. Cultural revolution: Red Guard translations, bibliography, index.—[2] Shera, E. B., compilor. Bibliography, index for sections: Bibliography/reference and leadership information ...
ISBN 0-89232-062-1 (v. 1)
1. China—Politics and government—1949-1976—Bibliography. 2. Chung-kuo kung ch'an tang—Purges—Bibliography. 3. Hung wei ping—Bibliography. I. Lowy, George. II. Sorich, Richard. III. Shera, Elizenija B.
Z3108.A5D62 016.3209'51'05 76-150965
[DS777.55] 00 MARC

—POLITICS AND GOVERNMENT— 1949-1976—CARICATURES AND CARTOONS

Erling, Johnny, 1952-
Tigermaske und Knochengespenst : d. neue chines. Karikatur / Erling, v. Graeve. — Köln : Prometh-Verlag, 1978.
140 p. : numerous ill. (some col.) ; 20 x 24 cm. GFR79-A
Includes bibliographical references and index.
ISBN (invalid) 3-922009-6 : DM26.00
1. China—Politics and government—1912-1949—Caricatures and cartoons. 2. China—Politics and government—1949-1976—Caricatures and cartoons. I. Graeve, Detlev von, 1944- joint author. II. Title.
DS775.E62 320.9'51'05 79-364907
79 MARC

—POLITICS AND GOVERNMENT —1949-1976—CONGRESSES

China. Kuo fang pu. Ch'ing pao chü.
(I chiu wu ch'i nien fei wei chung yao hui i)
一九五七年匪僞重要會議彙編 [臺北] 國防部情報局編印 民國47 [1958]
2, 130 p. 19 cm.
Cover title.

1. China—Politics and government—1949-1976—Congresses. 2. Congresses and conventions—China. I. Title. II. Title: Fei wei chung yao hui i.
DS777.55.C446723 1958 70-841251

CHINA

—POLITICS AND GOVERNMENT —1949-1976—CONGRESSES (Continued)

China (Republic of China). Kuang fu ta lu she chi yen chiu wei yüan hui.
(Kuang fu ta lu she chi yen chiu wei yüan hui ch'üan t'i wei yüan ti erh shih i tz'u hui i)
光復大陸設計研究委員會全體委員第二十一次會議錄 / 光復大陸設計研究委員會秘書處彙編. — [台北：該處], 民國64 [1975]
4, 372 p. ; 20 cm.
Cover title.
1. China—Politics and government—1949-1976—Congresses. 2. Taiwan—Politics and government—1945- —Congresses. I. China (Republic of China). Kuang fu ta lu she chi yen chiu wei yüan hui. Mi shu ch'u. II. Title.
DS777.55.C4467236 1975 — 79-835669

International Conference on World Politics, 7th, Hague, 1969.
Conference papers / Seventh International Conference on World Politics. — Hague : Secretariat, [1969-
v. ; 30 cm.
Cover title.
Includes bibliographical references.
1. China—Politics and government—1949-1976—Congresses.
DS777.55.I59 1969 — 320.9'51'05 — 79-301952
79 — MARC

—POLITICS AND GOVERNMENT—1949-1976—SOURCES

And Mao makes 5 : Mao Tsetung's last great battle / edited with an introd. by Raymond Lotta. — Chicago : Banner press, 1978.
vi, 552 p. : ill. ; 22 cm.
Includes bibliographical references.
ISBN 0-916650-08-1 : $5.95
1. China—Politics and government—1949-1976—Sources. 2. China—Politics and government—1976- —Sources. I. Lotta, Raymond.
DS777.55.A745 — 320.9'51'05 — 78-70431
79 — MARC

Melis, Giorgio, 1925-
La rivoluzione cinese / Giorgio Melis. — Firenze : Le Monnier, 1978.
vi, 121 p. ; 19 cm. — (Le Contraddizioni irrisolte del mondo contemporaneo) (Storia parallela ; Sezione 5) — It78-Dec
"Documenti": p. [17]-94.
Bibliography: p. 108-110.
L2200
1. China—Politics and government—1911-1949—Sources. 2. China—Politics and government—1949-1976—Sources. I. Title. II. Series.
DS775.M43 — 320.9'51'042 — 79-347754
*79 — MARC

—POLITICS AND GOVERNMENT—1949-1976—DICTIONARIES—KOREAN

Kukhoe Tosŏgwan, Seoul, Korea. Haeoe Charyoguk.
(Chunggong sajŏn)
中共事典 / [編輯人孫允睦]. — [서울]：國會圖書館海外資料局, 1978.
352 p. ; 23 cm. — (海外資料 ; 第38輯)
Not in trade.
1. China—Politics and government—1949-1976—Dictionaries—Korean. I. Son, Yun-mok. II. Title. III. Series: Haeoe charyo ; che 38-chip.
DS777.55.K7456 1978 — 78-826877

—POLITICS AND GOVERNMENT —1949-1976—SOURCES

(Chung kung chi mi wen chien hui pien)
中共機密文件彙編 / 國立政治大學國際關係研究中心輯印. — 臺北：該中心, 民國67 [1978]
5, 8, 579 p. ; 24 cm.
NT$600 ($15.00 U.S.)
1. China—Politics and government—1949-1976—Sources. I. Kuo li cheng chih ta hsüeh. Kuo chi kuan hsi yen chiu chung hsin.
DS777.55.C468613 — 78-841991

—POLITICS AND GOVERNMENT —1949-1976—SOURCES—INDEXES

Ajia Keizai Kenkyūjo, Tokyo.
(Kōeihei hō sōmoku)
红卫兵报总目 / [矢吹晋, 加々美光行編]. — 東京：アジア経済研究所, 昭和49- [1974-
v. ; 25 cm.
Cover title.
"所内資料 調査研究部 no. 49-7 ; 中国文革研 no. 2"
1. Red Guard publications—Indexes. 2. China—Politics and government—1949-1976—Sources—Indexes. I. Yabuki, Susumu, 1938- II. Kagami, Mitsuyuki. III. Title.
DS777.55.A676 1974 — 78-813933

—POLITICS AND GOVERNMENT —1949-1976—TERMINOLOGY

Kuo li cheng chih ta hsüeh. Kuo chi kuan hsi yen chiu chung hsin.
A comprehensive glossary of Chinese communist terminology / edited by Warren Kuo. — Taipei : Institute of International Relations, National Chengchi University, Republic of China, 1978.
907 p., [7] fold. leaves of plates : ill., maps ; 27 cm.
Includes indexes.
1. China—Politics and government—1949-1976—Terminology. I. Kuo, Warren. II. Title.
DS777.55.K756 1978 — 320'9'51'05 — 78-112849
79 — MARC

—POLITICS AND GOVERNMENT—1976-

Asano, Yūzō, 1912-
(Mō Takutō naki ato no Chūgoku)
毛沢東なきあとの中国：現状とその分析 / 浅野雄三. — 東京：中国事情研究会, 1978.
78 p. ; 19 cm. — (中国事情研究資料 ; 2) — Ja 78-22066
¥300
1. China—Politics and government—1976- I. Title. II. Series: Chūgoku jijō kenkyū shiryō ; 2.
DS779.26.A8 — 78-816001

Bettelheim, Charles.
Questions sur la Chine après la mort de Mao Tsé-toung / Charles Bettelheim. — Paris : F. Maspero, 1978.
153 p. ; 20 cm. — (Économie et socialisme ; 35) — F***
Includes bibliographical references.
ISBN 2-7071-1026-4 : 25.00F
1. China—Economic conditions—1976- . 2. China—Politics and government—1976- . 3. Socialism in China. I. Title.
HC427.92.B47 — 335.43'0951 — 78-387336
79 — MARC

Chi Hsin (Research group)
The case of the Gang of Four : with first translation of Teng Hsiao Ping's "Three poisonous weeds" / by Chi Hsin. — Hong Kong : Cosmos Books, 1977.
295 p., [12] leaves of plates : ill. ; 19 cm.
$12.00
1. China—Politics and government—1976- . I. Teng, Hsiao-p'ing, 1900- . II. Title.
DS779.26.C43183 1977 — 320.9'51'05 — 78-105738
79 — MARC

Chin, Steve S. K.
(Jen wang cheng hsi hu)
人亡政息乎 = China after Mao：毛澤東身後的中國 / 金思愷. — 香港：工貿一周：發行者歐美書業香港有限公司, 1976.
30 p. : ill. ; 22 cm.
"原載工貿一週"
In Chinese.
$2.00
1. China—Politics and government—1976- 2. China—Registers. I. Title.
DS779.26.C47 — 77-836677

China (People's Republic of China, 1949-). Ch'üan kuo jen min tai piao ta hui.
(Chung-hua jen min kung ho kuo ti wu chieh ch'üan kuo jen min tai piao ta hui)
中华人民共和国第五届全国人民代表大会第一次会议文件. — 第1版. — 北京：人民出版社, 1978.
2, 145 p. ; 19 cm.
$0.28
1. China—Politics and government—1976- 2. China—Registers. I. Title.
J671.H64 1978 — 79-837844

Chŏng, Yŏng.
(Chunggong ŭi tang, chŏng, kun ŭi yŏkhak kwan'gye wa Hwa Kuk-pong ch'eje ŭi chŏnmang)
中共의党·政·軍의力学関係와華國鋒体制의展望 / 研究執筆責任 鄭永 ; 刊行責任 朴鏞德. — [서울]：國土統一院調査研究室, 1977.
45 p. ; 26 cm. — (周邊環境與件 : 中共)
"국통조 77-1-1103."
1. China—Politics and government—1949-1976. 2. China—Politics and government—1977- I. Pak, Yong-dŏk. II. Title. III. Series: Chubyŏn hwan'gyŏng yŏkŏn.
DS777.55.C4543 — 78-826292

(Chūgoku Zenjindai, Seikyō Zenkoku Kaigi tokushūgō)
中国全人代·政協全国会議特集号 / [編集者 ラヂオプレス]. — 東京：同プレス, 1978.
183, 10 p. ; 25 cm. — Ja***
Cover title.
"中華人民共和国憲法": p. 76-92.
"政協会議規約": p. 120-127.
Includes index.
1. China—Politics and government—1976- 2. China—Constitutional law. 3. China—Registers. I. China (People's Republic of China, 1949-). Ch'üan kuo jen min tai piao ta hui. II. Chung-kuo jen min cheng chih hsieh shang hui i. III. Rajio Puresu.
JQ1503 1978.C47 — 79-805288

Daubier, Jean.
Les nouveaux maîtres de la Chine / Jean Daubier. — Paris : B. Grasset, c1979.
287 p. ; 21 cm. — F***
Includes bibliographical references.
ISBN 2-246-00720-8 : 48.00F
1. China—Politics and government—1976- . 2. Hua, Kuo-feng. 3. Teng, Hsiao-p'ing, 1904- . I. Title.
DS779.26.D38 — 320.9'51'05 — 79-366082
79 — MARC

Digan, Parig.
The Christian China-watchers : a post-Mao perspective / Parig Digan. — Brussels : Pro Mundi Vita, 1978.
60 p. ; 21 cm. — Be***
Bibliography: p. 56-60.
1. Christianity—China. 2. China—Politics and government—1976- . I. Title.
BR1288.D53 — 275.1 — 79-317120
79 — MARC

Glebov, Vladimir.
Maoism, words & deeds / Vladimir Glebov. — New Delhi : Sterling, c1979.
115 p. ; 22 cm.
Includes bibliographical references.
Rs25.00
1. China—Politics and government—1976- . I. Title.
DS779.26.G53 — 335.43'4 — 79-902637
79 — MARC

Gudoshnikov, Leonid Moiseevich.
China after Mao / Leonid Gudoshinikov, Rostislav Neronov. — Moscow : Novosti Press Agency Pub. House, 1979.
85 p. ; 17 cm. — USSR***
0.25rub
1. China—Politics and government—1976- . I. Neronov, Rostislav Mikhaĭlovich, joint author. II. Title.
DS779.26.G82 — 79-317388
79 — MARC

(Hsien ch'i hsüeh hsi Mao chu hsi chu tso ti hsin kao ch'ao)
掀起学习毛主席著作的新高潮：热烈欢呼《毛泽东选集》第五卷出版. — 第1版. — [广州]：广东人民出版社, 1977.
46 p., [1] leaf of plates : port. ; 19 cm.
"《毛泽东选集》第五卷是一九四九年九月到一九五七年的重要著作"
$0.12
1. China—Politics and government—1976- 2. Mao, Tse-tung, 1893-1976.
DS779.26.H72 1977b — 78-841694

Hua, Kuo-feng.
Speech at the second National Conference on Learning from Tachai in Agriculture : December 25, 1976 / Hua Kuo-Feng. — 1st ed. — Peking : Foreign Languages Press, 1977.
44 p. ; 19 cm.
Translation of Chung-kuo kung ch'an tang chung yang wei yüan hui chu hsi Hua Kuo-feng t'ung chih tsai ti erh tz'u ch'üan kuo nung yeh hsüeh Ta-chai hui i shang ti chiang hua.
1. China—Politics and government—1976- . I. Title.
DS779.26.H8213 — 320.9'51'05 — 79-301029
79[r79]rev — MARC

Kindermann, Gottfried Karl, 1926-
Pekings chinesische Gegenspieler : Theorie und Praxis nationalchinesischen Widerstandes auf Taiwan / Gottfired-Karl Kindermann. — Düsseldorf : Droste, c1977.
290 p. ; 21 cm. — GFR***
Bibliography: p. 280-290.
ISBN 3-7700-0325-7
1. China—Politics and government—1976- . 2. Taiwan—Politics and government—1945- . I. Title.
DS779.26.K56 — 320.9'51'05 — 79-359414
79 — MARC

CHINA
—POLITICS AND GOVERNMENT —1976- (Continued)

Li, Hsing-k'o.
The fall of Madam Mao / by Ly Singko. — 1st ed. — New York : Vantage Press, c1979.
136 p. ; 21 cm.
ISBN 0-533-03678-X : $6.50
1. China—Politics and government—1976- . 2. Chiang, Ch'ing. I. Title.
DS779.26.L5173 320.9'51'05 78-54378
79 MARC

Li, Ming-hua.
(Ts'ung p'i Lin p'i K'ung tao Pei-p'ing hsin feng pao)
從「批林批孔」到北平新風暴：導向毀滅的三年中共內部鬪爭 / ［黎明華著］. — 台北：中國大陸問題資料研究中心，民國66 ［1977］
4, 2, 116 p. ; 22 cm. — (中國大陸問題資料研究叢書 ; 第2種)
Includes bibliographical references.
NT$60.00
1. China—Politics and government—1976- I. Title. II. Series: Chung-kuo ta lu wen t'i tzu liao yen chiu ts'ung shu ; ti 2 chung.
DS779.26.L519 79-835461

Mooneyham, Walter Stanley, 1926-
China, a new day / by W. Stanley Mooneyham. — Plainfield, N.J. : Logos International, c1979.
241 p. : ill. ; 18 cm.
ISBN 0-88270-364-1 : $2.50
1. Christianity—China. 2. China—Civilization—1949- . 3. China—Politics and government—1976- . I. Title.
BR1285.M55 275.1 79-83789
79 MARC

Myrdal, Jan.
China notebook, 1975-1978 / Jan Myrdal ; photos. by Gun Kessle ; translated from the Swedish by Rolf vom Dorp. — Chicago : Liberator Press, [c1979]
146 p. : ill. ; 22 cm.
Translation of 2 works: Kinesiska frågor från Liu Ling 1975 and Kina efter Mao Tsetung.
ISBN 0-930720-58-X : $3.95
1. China—Social life and customs—1949- . 2. Lieou-lin, China. 3. China—Politics and government—1976- . I. Myrdal, Jan. Kina efter Mao Tsetung. English. 1979. II. Title.
DS777.55.M8913 951.05 79-88412
79 MARC

Myrdal, Jan.
Kina efter Mao Tsetung : anteckningar kring up görelsen med "de fyra" / Jan Myrdal. — Stockholm : Ordfront, 1977.
104 p., [4] leaves of plates : 4 ill. ; 19 cm. S77-32/33
ISBN 9173240478 : kr19.00
1. China—Politics and government—1976- . I. Title.
DS779.26.M97 77-571886
78[r79]rev MARC

Nitchū Keizai Kyōkai Kaihō Henshūsha Yūkō Hōchū-dan.
(Chūgoku no seiji to keizai, sen-kyūhyaku-shichijūshichi-nen hachigatsu)
中国の政治と経済—1977年8月：日中経済協会会報編集者友好訪中団報告書. — 東京：同協会, 1978.
119 p. ; 25 cm. — (日中経報 ; no. 104) Ja 78-19892
Cover title.
1. China—Economic policy—1976- 2. China—Politics and government—1976- I. Title. II. Series: Nitchū Keizai Kyōkai. Nitchūkei hō ; no. 104.
HC427.92.N58 1978 78-814261

Shibata, Minoru, 1930-
(Mō Shū igo no Chūgoku)
毛·周以後の中国 / 柴田穂. — ［東村山］：教育社；東京：販売 教育社出版サービス, ［1978］
195 p. : ill. ; 18 cm. — (時事問題解説 ; 74) (入門新書) Ja 78-24005
Cover title.
Series 2 romanized: Nyūmon shinsho.
¥400
1. China—Politics and government—1976- I. Title. II. Series: Jiji mondai kaisetsu ; 74.
DS779.26.S54 79-801466

Shibata, Minoru, 1930-
(Shin Chūgoku sangokushi)
新中国三国志：毛王朝秘史 / 柴田穂著. — 東京：日本工業新聞社, 1978.
327 p., [1] leaf of plates : ill. ; 20 cm. Ja 78-26444
¥2200
1. China—Politics and government—1976- I. Title.
DS779.26.S55 79-802823

Smith, Bruce.
Mao's last battle : the next stage / by Bruce Smith. — London : China Policy Study Group, [1978]
32 p. ; 21 cm. GB78-27920
ISBN 0-9502015-4-5 : £0.40
1. China—Politics and government—1976- . I. Title.
DS779.26.S6 320.9'51'05 79-303343
79 MARC

Teng, Hsiao-p'ing, 1900-
Programm für Chinas Zukunft : Dèng Xiǎopíngs Dokumente zur Lage der Nation auf dem Höhepunkt des Machtkampfes 1975 : Übersetzung und Kommentar / Martin Krott. — Hamburg : Institut für Asienkunde, 1978.
126 p. : ill. ; 21 cm. — (Mitteilungen des Instituts für Asienkunde Hamburg ; Nr. 92) GFR***
Summary in English.
Includes bibliographical references.
ISBN 3-921469-41-4 : DM12.00
1. China—Politics and government—1976- . 2. China—Economic policy—1976- . 3. Science and state—China. I. Krott, Martin, 1953- . II. Title. III. Series: Institut für Asienkunde. Mitteilungen ; Nr. 92.
DS1.I55 Nr. 92 78-376858
[DS779.26] 78 MARC

Teng, Hsiao-p'ing, 1904-
Programm für Chinas Zukunft : Dèng Xiǎopíngs Dokumente zur Lage der Nation auf dem Höhepunkt des Machtkampfes 1975 : Übersetzung und Kommentar / Martin Krott. — 2. erw. Aufl. — Hamburg : Institut für Asienkunde, 1978.
143 p. ; 21 cm. — (Mitteilungen des Instituts für Asienkunde Hamburg ; Nr. 95) GFR***
Summary in English.
Includes bibliographical references.
ISBN 3-921469-45-7
1. China—Politics and government—1976- . 2. China—Economic policy—1976- . 3. Science and state—China. I. Krott, Martin, 1953- . II. Title. III. Series: Institut für Asienkunde. Mitteilungen ; Nr. 95.
DS1.I55 Nr. 95 950 s 79-372086
[DS799.26] 79 MARC

(Teng Hsiao-p'ing)
鄧小平：復出，受辱內幕經過. — 九龍：星光公司, ［1977?］
[36] p. : ill. ; 26 cm.
Cover title.
$2.00
1. Teng, Hsiao-p'ing, 1904- 2. China—Politics and government—1976-
DS778.T39T456 79-839095

Ward, Doug.
China-the sick dragon / by Doug Ward. — [s.l. : s.n. ; San Francisco : distributor, Sundowner Services, c1978]
vii, 187 p. : ill. ; 22 cm.
Includes bibliographical references.
ISBN 0-932410-06-5 : $8.00
1. China—Politics and government—1949-1976. 2. China—Politics and government—1976- . I. Title.
DS777.55.W396 320.9'51'05 78-65281
79 MARC

Yuasa, Makoto, 1941-
(Tō Shōhei)
鄧小平 / 湯浅誠編著. — 東京：日本文芸社, ［1978］
260 p. : ill. ; 19 cm. Ja 78-14507
¥900
1. Teng, Hsiao-p'ing, 1900- 2. Statesmen—China—Biography. 3. China—Politics and government—1949-1976. 4. China—Politics and government—1976- I. Title.
DS778.T39Y8 78-812746

—POLITICS AND GOVERNMENT—1976- —ADDRESSES, ESSAYS, LECTURES

Avakian, Bob.
Mao Tsetung's immortal contributions / by Bob Avakian. — Chicago : RCP Publications, [1979?]
342 p. : ill. ; 22 cm.
Articles originally published in Revolution.
Includes bibliographical references.
ISBN 0-89851-045-7
1. Mao, Tse-tung, 1893-1976—Addresses, essays, lectures. 2. China—Politics and government—1976- —Addresses, essays, lectures. I. Title.
DS778.M3A94 951.05'092'4 79-88177
00 MARC

(Ch'e ti po tiao "ssu jen pang" chia tso p'ai ti wei chuang).
彻底剥掉"四人幫"假左派的伪装 / 稭思等著. — ［香港? : s. n., 1979?］
2, 120 p. ; 22 cm.
Reprint of the 1978 ed. published by Ch'ün chung ch'u panshe, Peking.
Includes bibliographical references.
CONTENTS: 稭思 彻底剥掉四人幫假左派的伪装.—孙旭培 四人幫的假左派伪装和极右实质.—王澈 假左真右—从阶级斗争问题看四人幫的反革命修正主义.—石竹 批修只能批右不能批左吗?—华泽 假左真右和宁左勿右.—人民日报理论部 围绕批判极左思潮的一场激烈斗争.—李洪林 四人幫路线的实质和倾向.—胡汉平 彻底剥开四人幫左的假面具.—金汶 四人幫的假左伪装是什么货色?.—第一机械工业部理论组 驳四人幫几个假左真右的口号.—斉平 四人幫为什么反对批极左?
1. China—Politics and government—1976- —Addresses, essays, lectures. I. Chi, Ssu.
DS779.26.C395 1979 79-841075

(Ch'eng sheng ch'ien chin)
乘胜前进：《人民日报》，《红旗》杂志，《解放军报》一九七七年元旦社论. — 第1版. — ［广州］：广东人民出版社, 1977.
6 p. ; 19 cm.
Cover title.
$0.03
1. China—Politics and government—1976- —Addresses, essays, lectures. I. Jen min jih pao, Peking. II. Hung ch'i. III. Chieh fang chün pao, Peking.
DS779.26.C42 1977 78-841043

(Ch'eng sheng ch'ien chin)
乘胜前进：人民日报·红旗杂志·解放军报一九七七年元旦社论. — 第1版. — ［北京］：人民出版社, 1977.
5 p., [2] leaves of plates : ports. ; 21 cm.
$0.10
1. China—Politics and government—1976- —Addresses, essays, lectures. I. Jen min jih pao, Peking. II. Hung ch'i. III. Chieh fang chün pao, Peking.
DS779.26.C42 1977b 79-836093

(Chien ch'ih she hui chu i ti min chu yüan tse)
坚持社会主义的民主原则 / 《人民日报》特约评论员. — 第1版. — 北京：人民出版社：新華書店发行, 1978.
39 p. ; 21 cm.
$0.21
1. China—Politics and government—1976- —Addresses, essays, lectures. I. Jen min jih pao, Peking.
DS779.26.C463 79-840638

China im Spiegel der Weltpresse : Dokumentation ausgew. Beitr. u. Daten zu polit. u. wirtschaftl. Entwicklungen 1977 u. 1978 / Wilhelm G. Franken, unter Mitarb. von Detlev Passarge, Marlies Prigge, Hubert-Günter Striefler. — Hamburg : Verlag Weltarchiv, 1978.
iv, 820 p. ; 29 cm. — (Veröffentlichungen des HWWA-Institut für Wirtschaftsforschung-Hamburg : Aktuelle Fachinformation) GFR79-A
"Abgeschlossen: 31.10.1978."
English, French, German, or Italian.
Research carried out by HWWA-Institut für Wirtschaftsforschung-Hamburg, Hauptabteilung Dokumentation, Bibliothek und Archive.
Bibliography: p. 803-817.
ISBN 3-87895-179-5 : DM68.00
1. China—Politics and government—1976- —Addresses, essays, lectures. 2. China—Economic conditions—1976- —Addresses, essays, lectures. 3. China—Foreign opinion—Addresses, essays, lectures. I. Franken, Wilhelm G. II. HWWA-Institut für Wirtschaftsforschung-Hamburg. Hauptabteilung Dokumentation, Bibliothek und Archive.
DS779.26.C4735 320.9'51'05 79-369082
79 MARC

Entwicklungsprobleme und Lösungsversuche in der Volksrepublik China : Maos Erbe im Widerspruch / Werner Pfennig, Klaus Voll. — Berlin : Dialogus mundi, 1977.
320 p. : ill., map ; 21 cm. — (Berliner Studien zur internationalen Politik ; Bd. 4) GFR78-A50
Filmography: p. 306.
Bibliography: p. 307-318.
DM19.80
1. China—Politics and government—1976- —Addresses, essays, lectures. I. Pfennig, Werner, 1944- . II. Voll, Klaus. III. Series.
DS779.26.E57 320.9'51'05 79-372099
79 MARC

The Gang of Four : first essays after the fall / edited by Steve S. K. Chin. — [Hong Kong] : Centre of Asian Studies, University of Hong Kong, 1977.
191 p. ; 27 cm. — (Selected seminar papers on contemporary China ; 2) (Centre of Asian Studies occasional papers and monographs ; no. 23 ISSN 0378-2689)
Includes bibliographical references.
HK$25.00
1. China—Politics and government—1976- —Addresses, essays, lectures. I. Chin, Steve S. K. II. Series. III. Series: Hongkong. University. Centre of Asian Studies. Occasional papers and monographs ; no. 23.
DS777.55.S434 vol. 2 951.05'08 s 78-318825
[DS779.26] 78 MARC

CHINA

—POLITICS AND GOVERNMENT —1976- —ADDRESSES, ESSAYS, LECTURES (Continued)

Han, Shou-shih.
(Tui kung fei wen hua t'ung chan yin mou ti fen hsi)
對共匪文化統戰陰謀的分析 / 韓守湜. — ［台北］：警備總司令部政戰部，［1978?］
1, 22 p. ; 22 cm.
Cover title.
At head of title: Chuan t'i pao kao.

1. Propaganda, Chinese—Addresses, essays, lectures. 2. China—Politics and government—1976- —Addresses, essays, lectures. I. Title.

DS777.57.H3 79-835617

(Hsi k'an ta hao hsing shih)
喜看大好形势：彻底砸烂四人幫，夺取革命，生产双胜利. — 第1版. — 上海：上海人民出版社，1976.
2, 95 p. ; 19 cm.
"选自一九七六年十月二十四日至十一月七日的人民日报，红旗杂志(一九七六年第十一期)和解放军报.
$0.19

1. China—Politics and government—1976- —Addresses, essays, lectures.

DS779.26.H7 78-842587

(Hsin shih ch'i tsung jen wu chiang hua)
新时期总任务讲话 / 谢立等编写. — 第1版. — 香港：生活·讀書·新知三联书店，1978.
83 p. ; 19 cm.
$1.50

1. China—Politics and government—1976- —Addresses, essays, lectures. I. Hsieh, Li.

DS779.26.H74 79-839522

Hua, Kuo-feng.
Hua Kuo feng on Teng Hsiao-ping : a special collection. — Kowloon : Altai Books, 1979.
28 p. ; 21 cm.
Cover title.
Added t.p.: Hua Kuo-feng p'i Teng chuan chi.
English and Chinese.
1. Teng, Hsiao-p'ing, 1904- —Addresses, essays, lectures. 2. China—Politics and government—1976- —Addresses, essays, lectures. I. Title. II. Title: Hua Kuo-feng p'i Teng chuan chi.
DS778.T39H813 951.05'092'4 79-103954
79 MARC

Hua, Kuo-feng.
(T'uan chieh ch'i lai, wei chien she she hui chu i ti hsien tai hua ch'iang kuo erh fen tou)
团结起来，为建設社会主义的现代化强国而奋斗 / 华国锋. — 第1版. — 北京：人民出版社，1978.
68 p. ; 21 cm.
"一九七八年二月二十六日在第五届全國人民代表大会第一次会议上的政府工作报告"
1. China—Politics and government—1976- —Addresses, essays, lectures. I. Title.
JQ1503 1978.H8 79-835621

(Hua chu hsi, Yeh fu chu hsi, Teng fu chu hsi tsai ch'üan chün cheng chih kung tso hui i shang ti)
华主席，叶副主席，邓副主席在全军政治工作会议上的讲话. — 第1版. — ［北京］：人民出版社，1978.
43 p. ; 21 cm.
$0.13

1. China—Politics and government—1976- —Addresses, essays, lectures. I. Hua, Kuo-feng. II. Yeh, Chien-ying. III. Teng, Hsiao-p'ing, 1900-

DS779.26.H828 79-835125

79 C

(Hua Kuo-feng yü Teng Hsiao-p'ing)
華國鋒與鄧小平 / 王建［譯］；［編者毛國倫］. — 香港：明報有限公司出版部，1977.
6, 270 p. : ports. ; 19 cm. — (中共近人近事譯叢)
Includes bibliographical references.
CONTENTS: 米徹爾·奥辛伯 華國鋒的湖南政績.—斯托厄·舒爾霖 華國鋒主持編輯的「論十大關係」.—約翰·吉廷斯 鄧小平的三份重大文件.—愛德華·賴爾斯 鄧小平的再起和再落.—杜勉 動盪中的過渡時期.—天安門的雷霆風暴.—米徹爾·奥辛伯 千秋功罪細評説.
$10.00 ($2.50 U.S.)

1. China—Politics and government—1976- —Addresses, essays, lectures. 2. China—Politics and government—1949-1976—Addresses, essays, lectures. 3. Hua, Kuo-feng—Addresses, essays, lectures. 4. Teng, Hsiao-p'ing, 1900- —Addresses, essays, lectures. I. Wang, Chien. II. Mao, Kuo-lun. III. Series: Chung kung chin jen chin shih i ts'ung.

DS779.26.H84 78-842944

(Jen min ti ying ming ling hsiu)
人民的英明领袖：歌颂华主席的光辉实践 / 本社编. — 第1版. — 上海：上海人民出版社：新華書店上海发行所发行，1977.
6, 442 p., ［1］ leaf of plates : ill. ; 21 cm.
$1.00

1. Hua, Kuo-feng—Addresses, essays, lectures. 2. China—Politics and government—1976- —Addresses, essays, lectures. I. Shang-hai jen min ch'u pan she.

DS779.29.H8J46 79-838595

(Kang Ao ai kuo t'ung pao chien chüeh yung hu Chung kung chung yang liang hsiang chüeh i)
港澳愛國同胞堅决擁護中共中央兩項决議熱烈慶祝文化大革命十周年 / ［港九工會聯合會宣教委員會編印］. — ［香港?］：該會，1976.
4, 84 p., ［1］ leaf of plates : ill. ; 19 cm.

1. China—Politics and government—1976- —Addresses, essays, lectures. I. Kang Chiu kung hui lien ho hui. Hsüan chiao wei yüan hui. II. Title: Chien chüeh yung hu Chung kung chung yang liang hsiang chüeh i.

DS779.26.K29 79-835022

(Mao chu hsi ti chiao yü fang chen ch'i jung ts'uan kai)
毛主席的教育方針岂容篡改. — ［香港? : s. n., 1978?］
2, 104 p. ; 21 cm.
Reprint of the 1977 ed. published by Chiang-su jen min ch'u pan she.

1. Education and state—China—Addresses, essays, lectures. 2. China—Politics and government—1976- —Addresses, essays, lectures.

LC94.C5M36 1978 79-837968

(Pei-ching cheng t'an feng yün)
北京政壇風雲：華國鋒，鄧小平，王洪文，江青，張春橋，姚文元事蹟落齊. — ［香港? : s. n., 1977?］
32 p. : ill. ; 27 cm.
Cover title.

1. China—Politics and government—1976- —Addresses, essays, lectures.

DS779.26.P45 79-839101

(Pi hsü chung shih ch'iao wu kung tso)
必須重视侨务工作. — 第1版. — 北京：人民出版社，1978.
26 p. ; 21 cm.
Contains newspaper articles from Jen min jih pao.
CONTENTS: 必须重视侨务工作 《人民日报》社论.—廖承志 批判"四人帮"所谓"海外关系"问题的反动谬论.—在英明领袖华主席、党中央的亲切关怀下全国侨务会议预备会在京举行.
$0.10
1. China—Politics and government—1976- —Addresses, essays, lectures. 2. Chinese in foreign countries—Addresses, essays, lectures. I. Jen min jih pao, Peking.

DS779.22.P5 1978b 79-838628

(Pi hsü chung shih ch'iao wu kung tso)
必须重视侨务工作. — 第2版. — ［北京?］：人民出版社：新華書店发行，1978.
42 p. ; 21 cm.
CONTENTS: 必须重视侨务工作.—国务院侨务办公室负责人就我国侨务政策问题答记者问.—廖承志 批判四人幫所谓海外关係问题的反动谬论.—连贯 华侨的历史及其光荣传统.—在英明领袖华主席，党中央的亲切关怀下全国侨务会议预备会在京举行.
$0.15
1. China—Politics and government—1976- —Addresses, essays, lectures. 2. Chinese in foreign countries—Addresses, essays, lectures.

DS779.22.P5 1978c 79-840664

(P'i p'an ssu jen pang ch'üan mien ts'uan kai Ma-k'o-ssu chu i li lun wen chi)
批判"四人帮"全面篡改马克思主义理论文集. — 第1版. — 北京：人民出版社：新華書店发行，1978.
318 p. ; 21 cm.
Includes bibliographical references.
$0.75

1. Communism—China—Addresses, essays, lectures. 2. China—Politics and government—1976- —Addresses, essays, lectures.

HX419.P5 79-837752

Ssu-ma, Ch'ang-feng.
(Wen ko hou ti Chung kung)
文革後的中共 / 司馬長風著. — 台北：時報文化出版事業有限公司，民國66 ［1977］
3, 5, 549 p., ［1］ fold. leaf of plates : graph ; 20 cm. — (時報書系 ; 89)
Series romanized: Shih pao shu hsi.
NT$100.00

1. China—Politics and government—1949-1976—Addresses, essays, lectures. 2. China—Politics and government—1976- —Addresses, essays, lectures. I. Title.

DS777.55.S7278 79-835196

Sung, Chen-t'ing.
(P'ing "ssu jen pang" ti fan tung shih chieh kuan)
评"四人帮"的反动世界观 / 宋振庭著. — 第1版. — 北京：中國青年出版社：新华书店北京发行所发行，1978.
1, 43 p. ; 19 cm.
Includes bibliographical references.
$0.12
1. China—Politics and government—1976- —Addresses, essays, lectures. I. Title.
DS779.26.S92 79-837882

(Ta cho fan fu pi ti ch'i hao kao fu pi)
打着反復辟的旗号搞复辟. — 第1版. — ［广州］：广东人民出版社，1977.
53 p. ; 19 cm.
Contains 3 newspaper articles.
CONTENTS: 向群 打着反复辟的旗号搞复辟.—中國科學院理论组 要知松高洁 待到雪化时.—国家计委大批判组 一杨篡党夺权的反革命丑剧.
$0.11
1. China—Politics and government—1976- —Addresses, essays, lectures.
DS779.26.T345 78-841466

(Tong-Puga ŭi chŏngch'i wa sahoe)
東北亞의政治와社會 / 李聖根編. — ［서울］：東北亞研究所，1978.
335 p. ; 21 cm.
English title on p. ［4］ of cover: The dynamics of political societies in East Asia.
Includes bibliographical references.
Not in trade.
1. East Asia—Politics and government—Addresses, essays, lectures. 2. China—Politics and government—1976- —Addresses, essays, lectures. 3. Russia—Politics and government—1953- —Addresses, essays, lectures. 4. Korea (Democratic People's Republic)—Politics and government—Addresses, essays, lectures. I. Rhee, Seung-Keun. II. Tong-Puga Yŏn'guso. III. Title: Dynamics of political societies in East Asia.
DS518.1.T67 79-826435

Wang, Sheng.
(Hua fei Kuo-feng ti nan t'i)
華匪國鋒的難題 / 王昇著. — 臺北：黎明文化事業公司，民國66 ［1977］
44 p. ; 16 cm.
NT$5.00

1. China—Politics and government—1976- —Addresses, essays, lectures. 2. Hua, Kuo-feng—Addresses, essays, lectures. I. Title.
DS779.26.W37 78-842987

—POLITICS AND GOVERNMENT —1976- —PICTORIAL WORKS

(Jen hsing ti chao huan)
人性的召喚 = An appeal to human nature. — 台北：瞻望出版社，1977.
92 p. : ill. ; 23 cm.
English and Chinese.

1. Taiwan—Politics and government—1945- —Pictorial works. 2. China—Politics and government—1976- —Pictorial works. I. Title: An appeal to human nature.

DS799.8.J44 78-842004

—POLITICS AND GOVERNMENT—1976- —SOURCES

And Mao makes 5 : Mao Tsetung's last great battle / edited with an introd. by Raymond Lotta. — Chicago : Banner press, 1978.
vi, 552 p. : ill. ; 22 cm.
Includes bibliographical references.
ISBN 0-916650-08-1 : $5.95
1. China—Politics and government—1949-1976—Sources. 2. China—Politics and government—1976- —Sources. I. Lotta, Raymond.
DS777.55.A745 320.9'51'05 78-70431
79 MARC

CHINA

—POLITICS AND GOVERNMENT
—1976- —SOURCES (Continued)

China (People's Republic of China, 1949-). Ch'üan kuo jen min tai piao ta hui.
(Chung-hua jen min kung ho kuo ti wu chieh ch'üan kuo jen min tai piao ta hui)
中华人民共和国第五届全国人民代表大会第一次会议文件. — 香港：生活·讀書·新知三联书店香港分店, 1978.
2, 140 p. : music ; 21 cm.
$2.00
1. China—Politics and government—1976- —Sources. I. Title.
DS779.26.C473 1978 78-843257

China (People's Republic of China, 1949-). Ch'üan kuo jen min tai piao ta hui.
(Chung-hua jen min kung ho kuo ti wu chieh ch'üan kuo jen min tai piao ta hui ti i tz'u...)
中华人民共和国第五届全国人民代表大会第一次会议文件. — 北京：人民出版社, 1978.
2, 197 p., [4] leaves of plates : ill., ports. : 21 cm.
$0.67
1. China—Politics and government—1976- —Sources. I. Title.
DS779.26.C473 1978b 78-843260

Chung-kuo jen min cheng chih hsieh shang hui i. Ch'üan kuo wei yüan hui, 5th, Peking, 1978.
(Chung-kuo jen min cheng chih hsieh shang hui i ti wu chieh ch'üan kuo wei yüan hui ti i tz'u)
中国人民政治协商会议第五届全国委员会第一次会议文件. — 第1版. — 北京：人民出版社, 1978.
2, 94 p., [4] leaves of plates : ill. ; 21 cm.
$0.41
1. China—Politics and government—1976- —Sources. I. Title.
JQ1503 1978.C48 1978 79-836447

—POLITICS AND GOVERNMENT
—PERIODICALS

Minami Manshū Tetsudō Kabushiki Kaisha. Sōmubu. Chōsaka.
(Chōsa shiryō)
調査資料 第10- 輯
[大連] 南満洲鐵道株式會社總務部調査課 [大正8- i. e. 1919-
v. 22 cm.
Earlier vols. classified separately in L. C.
Superseded by: Minami Manshū Tetsudō Kabushiki Kaisha. Shomubu. Chōsaka. Chōsa shiryō.

1. China—Economic conditions—Periodicals. 2. China Politics and government—Periodicals. 3. East Asia—Periodicals.
HC426.M562a 75-791578
(MARC-S)

—POPULAR CULTURE

Popular media in China : shaping new cultural patterns / edited by Godwin C. Chu ; foreword by A. Doak Barnett. Honolulu : Published for the East-West Center by the University Press of Hawaii, c1978.
x, 263 p. : ill. ; 22 cm.
Bibliography: p. [249]-254.
Includes index.
ISBN 0-8248-0622-0
1. Mass media—China. 2. China—Popular culture. I. Chu, Godwin C., 1927-
P92.C5P6 301.16'1'0951 78-13282
78 MARC

—POPULATION

Ch'ai, Sung-lin, 1934-
Population and population policy in Mainland China / by Chai Sunglin. — [Taipei] : Asia and the World Forum, 1977.
90 p. : graphs ; 23 cm. — (Asia and world monographs ; 6)
Includes bibliographical references.
$1.50
1. China—Population. 2. China—Population policy. I. Title. II. Series: Asia and the World Forum. Monograph series - Asia & the World Forum ; 6.
HB3654.A3C47 301.32'9'51 78-111742
79 MARC

Fischer, Harald, 1949-
China : d. neue Selbstbewusstsein : Gesellschaft, Erziehung, Bevölkerung / Harald Fischer, Christian Graf von Krockow, Hermann Schubnell. — [1.-5. Tsd.]. — München ; Zürich : Piper, 1978.
320 p. ; 19 cm. GFR78-A
Bibliography: p. 316-320.
ISBN 3-492-02313-4 : DM32.00
1. China—Social conditions. 2. China—Population. 3. Education—China—1949- . 4. Birth control—China. I. Krockow, Christian, Graf von, joint author. II. Schubneil, Hermann, joint author. III. Title.
HN737.F57 309.1'51'05 78-384499
79 MARC

—POPULATION POLICY

Ch'ai, Sung-lin, 1934-
Population and population policy in Mainland China / by Chai Sunglin. — [Taipei] : Asia and the World Forum, 1977.
90 p. : graphs ; 23 cm. — (Asia and world monographs ; 6)
Includes bibliographical references.
$1.50
1. China—Population. 2. China—Population policy. I. Title. II. Series: Asia and the World Forum. Monograph series - Asia & the World Forum ; 6.
HB3654.A3C47 301.32'9'51 78-111742
79 MARC

Chen, Pi-chao.
China's population program at the grassroots level : report on a field trip, summer, 1972 / Pi-chao Chen. — [Pasadena : California Institute of Technology, c1973.
24 p. ; 23 cm. — (Occasional papers - Caltech Population Program ; ser. 1, no. 3)
Cover title.
Includes bibliographical references.
1. Birth control—China. 2. China—Population policy. I. Title. II. Series: Caltech Population Program. Occasional papers - Caltech Population Program ; ser. 1, no. 3.
HQ766.5.C6C43 362.8'2 79-303189
79 MARC

—PRESIDENTS—BIOGRAPHY

Ch'en, Chien-fu.
[Kuo fu ch'üan chuan]
孫文傳 / [陳健夫著]. — 九龍：理代出版公司, [1978?]
9, 534 p., 112 leaves of plates : ill., ports. ; 19 cm.
Cover title.
Originally published in 1964 by Tzu yu T'ai p'ing yang wen hua shih yeh kung ssu, T'ai-pei, under title: Kuo fu ch'üan chuan.
$9.50
1. Sun, Yat-sen, 1866-1925. 2. China—Presidents—Biography. I. Title. II. Title: Kuo fu ch'üan chuan.
Title romanized: Sun Wen chuan.
DS777.C528 1978 78-842086
C

Ch'en, Pin-ho, fl. 1926-
(Lun Chiang Chieh-shih)
論蔣介石 陳彬龢著 上海 太平書局 民國31 [1942]
62 p. 26 cm.
Colophon title.

1. Chiang, Kai-shek, 1887-1975. 2. China—Presidents—Biography. I. Title.
DS778.C55C4355 74-840266
rev
70 [r78] C

Chook, Shiu Heng.
Chiang Kai-shek close-up : a personal view / Shiu Heng Chook ; edited by Edward K. Chook. — Oakland, CA. : United California University Press, [1978?]
161 p., [8] leaves of plates : ill. ; 22 cm.
1. Chiang, Kai-shek, 1887-1975. 2. China—Presidents—Biography. I. Chook, Edward K. II. Title.
DS778.C55C496 951.04'2'0924 78-109073
78 MARC

Ikeguchi, Ekan, 1936-
(Shō Kaiseki)
蔣介石：その偉大なる生涯 / 池口恵観. — 東京：論稿社, 昭和51 [1976]
215 p. : ill. ; 20 cm. Ja***
¥1200

1. Chiang, Kai-shek, 1887-1975. 2. China—Presidents—Biography. 3. China—History—Republic, 1912-1949. 4. Japan—Foreign relations—China. 5. China—Foreign relations—Japan. I. Title.
DS778.C55I33 78-814965

Matsushima, Shūe, 1871-1935.
(Shinchō matsuro hishi)
清朝末路秘史 / 松島宗衞. — 長崎村(東京市外)：大星社, 大正14 [1925]
6, 271 p., [6] leaves of plates : ports. ; 22 cm.
L.C. copy imperfect: all before pref. wanting; title from caption.

1. Yuan, Shih-k'ai, 1859-1916. 2. China—Presidents—Biography. 3. China—History—19th century. I. Title.
DS778.Y84M37 78-814187

Sakamaki, Teiichirō.
(Shina bunkatsu ron)
支那分割論：附袁世凱 / [著者 酒卷貞一郎]. — 東京：啓成社, 大正2 [1913]
584, 204, [35] p., [5] leaves of plates : maps ; 23 cm.

1. China—History—1912-1937. 2. China—Foreign relations—1912-1949. 3. China—History—Revolution, 1911-1912. 4. Revolutions—China. 4. Yüan, Shih-k'ai, 1859-1916. 6. China—Presidents—Biography. I. Title. II. Sakamaki, Teiichirō. En Seigai. 1913.
DS774.S332 79-809704

—PRESIDENTS—BIOGRAPHY
—ADDRESSES, ESSAYS, LECTURES

(Tsung t'ung Chiang kung hsing i chuan chi)
總統蔣公行誼專集 / 教育部訓育委員會編. — 台北：該會, 民國66 [1977]
2, 2, 166 p., [1] leaf of plates : port. ; 22 cm.
Cover title.
CONTENTS: 史振鼎 總統蔣公永恆不朽的生命.—李正中 總統蔣公的革命勳業.—李雲漢 總統蔣公早年的讀書生活.—沈若雲 總統蔣公的睿知遠見.—周何 總統蔣公是真正不朽的偉人.—周道濟 總統蔣公逝世的歷史意義.—馬起華 總統蔣公的偉大出於平凡.—張京育 總統蔣公的大智大仁大勇.—鄔昆如 總統蔣公的信仰.—楊崇森 總統蔣公是偉大的愛國者.—趙際良 總統蔣公是現代的完人.—蔣一安 總統蔣公對人類歷史的貢獻.—蔣永敬 總統蔣公怎樣領導抗戰.—儲鴻舉 總統蔣公的立德立功立言

1. Chiang, Kai-shek, 1887-1975—Addresses, essays, lectures. 2. China—Presidents—Biography—Addresses, essays, lectures. I. China (Republic of China). Chiao yü pu. Hsün yü wei yüan hui.
DS778.C55T774 79-841413

(Wo men ti kuo fu)
我們的國父 / 吳稚暉等著. — 臺北：中央文物供應社, 民國67 [1978]
2, 222 p. ; 22 cm. — (大衆文庫)
Series romanized: Ta chung wen k'u.
Includes bibliographical references.
NT$40.00

1. Sun, Yat-sen, 1866-1925—Addresses, essays, lectures. 2. China—Presidents—Biography—Addresses, essays, lectures. I. Wu, Ching-heng, 1864-1953.
DS777.A595W6 79-836507

(Yüan Shih-k'ai shih liao)
袁世凱史料. — [香港?] : s. n., 1977-1978]
2 v. : ill. ; 31-22 x 31 cm.
Cover title.
Date of publication stamped on p. [4] of cover.
A collection of articles and newspaper clippings reprinted from various sources, including one in English.

1. Yüan, Shih-k'ai, 1859-1916—Addresses, essays, lectures. 2. China—Presidents—Biography—Addresses, essays, lectures.
DS778.Y84Y84 78-842188
78 C

—PRINCES AND PRINCESSES—
BIOGRAPHY

Shu-Chiung.
Yang Kuei-fei, the most famous beauty of China / by Shu Chiung (Mrs. Wu Lien teh). — Taipei : Ch'eng Wen Pub. Co., 1974.
xvii, 102 p., [5] leaves of plates : ill. ; 20 cm.
Reprint of the 1923 ed. published by Commercial Press, 1923.
1. Yang Kuei-fei. 2. China—Princes and princesses—Biography. I. Title.
DS749.42.Y36S58 1974 951'.01'0924 79-311934
79 MARC

CHINA (Continued)

—RACE RELATIONS

China (Republic of China). Kuang fu ta lu she chi yen chiu wei yüan hui.
(Kung fei pien chiang hsing cheng ching chi wen chiao ts'o shih chih yen chiu)
共匪邊疆行政·經濟·文教措施之研究 / 光復大陸設計研究委員會研擬. — ［台北］：該會，民國65 ［1976］
2, 204 p. ; 22 cm. — (邊疆研究叢書)
Cover title.
1. China—Boundaries. 2. China—Race relations. I. Title. II. Series: Pien chiang yen chiu ts'ung shu.
DS737.C4884 1976 78–841058

China (Republic of China). Kuang fu ta lu she chi yen chiu wei yüan hui.
(Kung fei pien chiang tzu chih ch'ü hsien k'uang chih yen chiu)
共匪邊疆自治區現況之研究 / 光復大陸設計研究委員會研擬. — ［台北］：該會，民國65 ［1976］
6, 476 p. ; 22 cm. — (邊疆研究叢書)
Cover title.
1. Minorities—China. 2. China—Race relations. 3. China—Politics and government—1949-1976. I. Title. II. Series: Pien chiang yen chiu ts'ung shu.
DS730.C4482 1976a 78–841893

O, Hai.
(Fu Miao lu)
撫苗錄 / 鄂海撰. — 台北：廣文書局，民國67 ［1978］
372 p. ; 22 cm. — (史料七編)
Reprint ed.
NT$300.00
1. Miao people. 2. China—History—Ch'ing dynasty, 1644-1912—Sources. 3. China—Race relations. I. Title. II. Series: Shih liao ch'i pien.
DS731.M5O12 1978 78–842307

—REGISTERS

Chin, Steve S. K.
(Jen wang cheng hsi hu)
人亡政息乎 = China after Mao：毛澤東身後的中國 / 金思愷. — 香港：工貿一周：發行者歐美書業香港有限公司，1976.
30 p. : ill. ; 22 cm.
"原載工貿一週"
In Chinese.
$2.00
1. China—Politics and government—1976- 2. China—Registers. I. Title.
DS779.26.C47 77–836677

China (People's Republic of China, 1949-). Ch'üan kuo jen min tai piao ta hui.
(Chung-hua jen min kung ho kuo ti wu chieh ch'üan kuo jen min tai piao ta hui)
中华人民共和国第五届全国人民代表大会第一次会议文件. — 第1版. — 北京：人民出版社，1978.
2, 145 p. ; 19 cm.
$0.28
1. China—Politics and government—1976- 2. China—Registers. I. Title.
J671.H64 1978 79–837844

(Chūgoku oyobi Manshūkoku jimmei chimei benran)
中國及滿洲國人名地名便覽：英文對照 / ［編纂者タイムス出版社編輯部］. — 東京：タイムス出版社，昭和8 ［1933］
277 p. in various pagings ; 19 cm.
1. China—Registers. 2. Manchuria—Registers. 3. China—Gazetteers. 4. Manchuria—Gazetteers. I. Taimusu Shuppansha. Henshubu.
DS705.C5253 79–804239

(Chūgoku Zenjindai, Seikyō Zenkoku Kaigi tokushūgō)
中国全人代·政協全国会議特集号 / ［編集者 ラヂオプレス］. — 東京：同プレス，1978.
183, 10 p. ; 25 cm. Ja***
Cover title.
"中華人民共和国憲法": p. 76-92.
"政協会議規約": p. 120-127.
Includes index.
1. China—Politics and government—1976- 2. China—Constitutional law. 3. China—Registers. I. China (People's Republic of China, 1949-). Ch'üan kuo jen min tai piao ta hui. II. Chung-kuo jen min cheng chih hsieh shang hui i. III. Rajio Puresu.
JQ1503 1978.C47 79–805288

(Hsin Chung-kuo ming jen ying chi)
新中國名人影集. — ［香港］：週末報社，［1978?］
25 leaves : ports. ; 15 x 21 cm. — (週末報叢書；2)
Series romanized: Chou mo pao ts'ung shu.
1. China—Registers.
JQ1507.H74 79–837669

K'an, Tao-ming.
(Chung-kuo li tai han lin hsüeh shih tsuan yao)
中國歷代翰林學士纂要 / 著作者闞道明. — 新竹：三人出版社，民國61 ［1972］
410 p. ; 21 cm.
Colophon title.
1. China—Registers. 2. Civil service—China. 3. Scholars—China. I. Title.
JQ1507.K35 79–835713

Lamb, Malcolm.
Directory of Chinese officials and organizations, 1968-1978 / by Malcolm Lamb. — Rev. — Canberra : Contemporary China Centre, Research School of Pacific Studies, Australian National University, 1978.
xxiii, 285 p. ; 23 cm. — (Contemporary China papers ; no. 10) Aus***
Cover title: Directory of officials in the People's Republic of China, 1968-1978.
Previous ed. published in 1976 under title: Directory of Central officials in the People's Republic of China, 1968-1975.
ISBN 0-909596-25-5 : $6.00 ($7.00 U.S.)
1. China—Registers. I. Title. II. Title: Directory of officials in the People's Republic of China, 1968-1978.
JQ1507.L35 1978 951'.0025 79–308192
79 MARC

Schultz, Jeffrey.
China's foreign trade corporations ... organization and personnel / by Jeffrey Schultz ; executive editor, Nicholas H. Ludlow. — Washington : National Council for US-China Trade, 1979.
339 p. : ill. ; 29 cm.
Bibliography: p. 31-32.
Includes index.
1. China—Commerce. 2. Foreign trade promotion—China. 3. China—Registers. I. Ludlow, Nicholas H. II. National Council for United States-China Trade. III. Title.
HF3836.5.S38 354'.51'00827 78-71596
00 MARC

(Shao-hsing shih pa nien t'ung nien hsiao lu)
紹興十八年同年小錄. 中州人物考 / ［孫奇逢撰］. — ［台北：臺灣］商務［印書館，1978?］
2 v. ; 20 cm. — (四庫全書珍本八集；82-83)
Cover title.
Photoreprint ed. of two works, both of the Wen Yüan Ko copy.
1. China—Registers. 2. Civil service—China. 3. China—History—Sung dynasty, 960-1279—Biography. 4. China—Biography. 5. China—History—Ming dynasty, 1368-1644—Biography. 6. Honan, China (Province)—Biography. I. Sun, Ch'i-feng, 1585-1675. Chung chou jen wu k'ao. 1978?. II. Title: Chung chou jen wu k'ao. III. Series: Ssu k'u ch'uan shu. Ssu k'u ch'uan shu chen pen pa chi ; 82–83.
AC149.S7 1978, vol. 82–83 78–841087
[JQ1507]
78 C

United States. Central Intelligence Agency. National Foreign Assessment Center.
Appearances and activities of leading Chinese officials. 1978–
［Washington］ National Foreign Assessment Center.
v. 27 cm. (A Reference aid)
Continues: United States. Central Intelligence Agency. National Foreign Assessment Center. Appearances and activities of leading personalities of the Peoples' Republic of China.
Key title: Appearances and activities of leading Chinese officials, ISSN 0198-8681.
1. China—Registers. I. Title.
JQ1507.U452a 354'.51'002 79–643426
MARC-S

—REGISTERS—YEARBOOKS

(Chūgoku soshikibetsu jimmeibo)
China directory. 中国組織別人名簿 1979–
［東京］ ラヂオプレス
v. 27 cm.
In English and Japanese.
Title romanized also in Chinese: Chung-kuo tsu chih pieh jen ming pu.
1. China—Registers—Yearbooks. I. Title: Chūgoku soshikibetsu jimmeibo. II. Title: Chung-kuo tsu chih pieh jen ming pu.
JQ1507.C527 951'.0025 79–642283
(MARC-S)

—RELATIONS (GENERAL) WITH AFRICA

Bogoslovskiĭ, Viktor Vasil'evich.
(Proiski Pekina v Afrike)
Происки Пекина в Африке / В. В. Богословский. — Киев : Политиздат Украины, 1978.
174 p. ; 17 cm. USSR***
Includes bibliographical references.
0.25rub
1. Africa—Relations (general) with China. 2. China—Relations (general) with Africa. I. Title.
DT38.9.C5B63 79–413386

China in Africa. — ［s.l. : s.n., 1972］
46 p. ; 24 cm.
1. Africa—Relations (general) with China. 2. China—Relations (general) with Africa.
DT38.9.C5C5 301.29'6'051 78-109327
79 MARC

Mathur, Girish, 1920-
China in Africa / Girish Mathur. — New Delhi : Kalamkar Prakashan, 1978.
40 p. ; 22 cm.
Includes bibliographical references.
Rs5.00
1. Africa—Relations (general) with China. 2. China—Relations (general) with Africa. 3. National liberation movements—Africa. 4. Africa—Politics and government—1960- 5. China—Politics and government—1949- I. Title.
DT38.9.C5M35 301.29'51'06 79-902567
79 MARC

—RELATIONS (GENERAL) WITH ARAB COUNTRIES

Kuwabara, Jitsuzō, 1870-1931.
(Ho Jukō no jiseki)
蒲壽庚の事蹟：唐宋時代に於けるアラブ人の支那通商の概況殊に宋末の提擧市舶西域人 / ［著者桑原隲藏］. — 東京：岩波書店，昭和10 ［1935］，昭和23 ［1948］ printing.
5, 11, 228, 18 p., ［1］ leaf of plates : ill. ; 26 cm.
Colophon inserted.
Includes bibliographies and index.
1. P'u, Shou-keng, fl. 1245-1284. 2. Muslims in China—Biography. 3. China—Officials and employees—Biography. 4. China—Relations (general) with Arab countries. 5. Arab countries—Relations (general) with China. 6. Customs administration—China. I. Title.
DS751.6.P8K88 1935 79–801279

—RELATIONS (GENERAL) WITH FOREIGN COUNTRIES

Hsiang, Ta, 1900-
(Chung wai chiao t'ung hsiao shih)
中外交通小史 / 向達著. — ［香港?：s. n., 1978?］
2, 2, 112 p. ; 19 cm.
Cover title.
Reprint of the ed. published by Shang wu yin shu kuan, Shanghai, in series: Wan yu wen k'u : Ti i chi, i ch'ien chung.
1. China—Relations (general) with foreign countries. 2. China—Civilization. I. Title.
DS740.4.H716 1978 79–835981

—RELATIONS (GENERAL) WITH FRANCE

Bernard, Henri, S.J.
La Chine en France durant la renaissance, par Henri Bernard ... Shanghai, Imprimerie de T'ou-sè-wè, 1940.
cover-title, 14 p. 23 cm.
"Extrait du Bulletin de l'Université l'Aurore déc. 1940."
"Liste des impressions françaises sur la Chine avant 1610": p. 13-14.
1. China—Relations (general) with France. 2. France—Relations (general) with China. 3. China in literature. I. Title.
DS740.5.F8B4 301.29'51'044 47-37465
［r79］rev MARC

CHINA (Continued)

—RELATIONS (GENERAL) WITH GHANA

Chung, Kuan.
A case study of Chinese Communist subversive activity in Africa : relations between Red China and Ghana, 1964-1966 / by Kuan Chung. — ꞁTaipeiꞁ : World Anti-communist League, China Chapter, 1977.
37 p. ; 19 cm. — (Pamphlet - Asian Peoples' Anti-communist League, Republic of China ; no. 195)
Cover title.
Includes bibliographical references.
$0.50 (U.S.)
1. Ghana—Relations (general) with China. 2. China—Relations (general) with Ghana. 3. Ghana—Politics and government—1957- . 4. Nkrumah, Kwame, Pres. Ghana, 1909-1972. I. Title.
DT512.C48 301.29'51'0667 79-311056
79 MARC

—RELATIONS (GENERAL) WITH GREAT BRITAIN

Appleton, William Worthen.
A cycle of Cathay : the Chinese vogue in England during the seventeenth and eighteenth centuries / by William W. Appleton. — New York : Octagon Books, 1979, c1951.
xii, 182 p., ꞁ4ꞁ leaves of plates : ill. ; 22 cm.
Originally presented as the author's thesis, Columbia University.
Reprint of the ed. published by Columbia University Press, New York.
Includes bibliographical references and index.
ISBN 0-374-90277-1 : $12.00
1. China—Relations (general) with Great Britain. 2. Great Britain—Relations (general) with China. I. Title.
ꞁDS740.5.G5A7 1979ꞁ 301.29'51'042 78-15422
78 MARC

—RELATIONS (GENERAL) WITH INDIA

Sanyal, Narayan.
(Cīna-Bhārata laṅ mārca)
চীন-ভারত লঙ্ মার্চ / নারায়ণ সান্যাল. — কলিকাতা : মণ্ডল বুক হাউস, 1977.
319 p., ꞁ2ꞁ leaves of plates : ill., maps (on lining papers) ; 23 cm.
In Bengali.
Includes bibliographical references and index.
Rs20.00

1. China—Politics and government. 2. China—Relations (general) with India. 3. India—Relations (general) with China. I. Title.

DS740.S36 78-911112
SA

—RELATIONS (GENERAL) WITH IRAN

Tashakkurī, ʻAbbās.
(Īrān bi-rivāyat-i Chīn-i bāstān)
ایران بروایت چین باستان : نگاهی به روابط سیاسی ، تجارتی و فرهنگی ایران و چین تا اسلام / عباس تشکری . — ꞁتهرانꞁ : مؤسسهٔ روابط بین المللی ، ꞁ1977 or 1978ꞁ 2536.
4, 187 p., ꞁ1ꞁ leaf of plates : ill. ; 24 cm.
Chinese or Persian.
Bibliography : p. 172–187.
1. Iran — Relations (general) with China. 2. China — Relations (general) with Iran. I. Title.
DS274.2.C6T37 78-970926

—RELATIONS (GENERAL) WITH JAPAN

Kojima, Naoki, 1919-
(Yōjō no ten)
洋上の点：森恪という存在 / 小島直記. — 東京：中央公論社，1978.
413 p. ; 20 cm. Ja 78-12323
Originally published in Chūō kōron, Jan. 1976-Dec. 1977.
Bibliography: p. 412.
¥1200
1. Mori, Kaku, 1882-1932. 2. Politicians—Japan—Biography. 3. Japan—Politics and government—1912-1945. 4. Japan—Relations (general) with China. 5. China—Relations (general) with Japan. I. Title.
DS885.5.M67K64 78-812502

Nitchū Keizai Kyōkai.
(Nitchū jinji kōryū jōkyō)
日中人事交流状況. — 東京：日中経済協会，197
v. ; 19 x 26 cm. — (日中経報 ; no.) Ja***
Cover title.
"51-調-6"
All tables.
CONTENTS: —3. 1976年1月-12月
1. Japan—Relations (general) with China. 2. China—Relations (general) with Japan. 3. Foreign trade promotion—Japan. 4. Foreign trade promotion—China. I. Title. II. Series: Nitchū Keizai Kyōkai. Nitchūkei hō ; no.
DS849.C6N5623 1977 79-804220

Pak, Yŏng-sŏk.
(Manbosan sakŏn yŏn'gu)
萬寶山事件研究：日帝大陸侵略政策의一環으로서의 / 朴永錫著. — 서울：亜細亜文化社，1978.
vi, 288 p. ; 23 cm.
"萬寶山事件과朝鮮에서의中國人排斥事件關係日誌": p. 221-251.
Bibliography: p. 252-265.
Includes index.
₩3200
1. Koreans in Manchuria. 2. Chinese in Korea. 3. Korea—Relations (general) with China. 4. China—Relations (general) with Korea. 5. Japan—Relations (general) with China. 6. China—Relations (general) with Japan. I. Title.
DS731.K6P34 79-825462

Saeki, Arikiyo.
(Saigo no kentōshi)
最後の遣唐使 / 佐伯有清. — 東京：講談社，1978.
203 p. : ill. ; 18 cm. — (講談社現代新書) Ja 78-27008
Series romanized: Kōdansha gendai shinsho.
¥390

1. Japan—Relations (general) with China. 2. China—Relations (general) with Japan. I. Title.
DS849.C6S24 79-805647

Wang, Chi-wu, 1899-
(Chung-kuo Jih-pen chiao tʻung shih)
中國日本交通史　王輯五著　主編者：王雲五 傳緯平　上海　商務　民國26 ꞁ1937ꞁ
1, 4, 224 p. tables. 19 cm. (中國文化史叢書第2輯)
Includes bibliographical references.

1. China—Relations (general) with Japan. 2. Japan—Relations (general) with China. 3. Japan—Civilization—Chinese influences. I. Title. II. Series: Chung-kuo wen hua shih tsʻung shu, ti 2 chi.
DS740.5.J3W27 C 59-5091
rev

Yoshino, Sakuzō, 1878-1933.
(Taishi mondai)
對支問題 / 吉野作造著. — 東京：日本評論社，昭和5 ꞁ1930ꞁ
4, 154 p. ; 19 cm. — (時事問題講座 ; 7)

1. Japan—Relations (general) with China. 2. China—Relations (general) with Japan. I. Title. II. Series: Jiji mondai kōza ; 7.
DS849.C6Y58 78-816242
78 J

—RELATIONS (GENERAL) WITH JAPAN—ADDRESSES, ESSAYS, LECTURES

(Hsü Fu yü Jih-pen)
徐福與日本 / ꞁ主編林建同，朱振聲ꞁ. — 九龍：香港徐福會；香港：經銷南天書業公司，1976.
5, 258 p., ꞁ23ꞁ leaves of plates : ill. ; 21 cm.
Colophon inserted.
$15.00 ($3.00 U. S.)

1. Hsü, Fu, 3d cent. B. C.—Addresses, essays, lectures. 2. China—Relations (general) with Japan—Addresses, essays, lectures. 3. Japan—Relations (general) with China—Addresses, essays, lectures. I. Lin, Jen Tung, 1911-
DS747.5.H78H77 78-843499

Sei, Iku-do, 1911-
(Shin Kan minzoku kara Yamato minzoku e)
新・漢民族から大和民族へ：春風吹イテ又生ズ / 盛毓度. — 東京：東洋経済新報社，1978.
219 p. : ill. ; 19 cm. — (V books) Ja 78-22761
¥880
1. Japan—Civilization—Addresses, essays, lectures. 2. National characteristics, Japanese—Addresses, essays, lectures. 3. Japan—Relations (general) with China—Addresses, essays, lectures. 4. China—Relations (general) with Japan—Addresses, essays, lectures. I. Title.
DS821.S329 78-816786

Tsʻai, Pʻei-huo.
(Tōa no ko·kaku omou)
東亜の子かく思ふ / 蔡培火著. — 東京：岩波書店，昭和12 ꞁ1937ꞁ
9, 2, 232 p. ; 20 cm.

1. China—Relations (general) with Japan—Addresses, essays, lectures. 2. Japan—Relations (general) with China—Addresses, essays, lectures. I. Title.
DS740.5.J3T77 78-816955

Tsuji, Masanobu, 1902-
(Zairyū Hōjin ni uttō)
在留邦人に愬ふ / 辻政信講演. — 上海：上海興亞報國會，昭和19 ꞁ1944ꞁ
2, 53 p. ; 20 cm.

1. World War, 1939-1945—Campaigns—Solomon Islands—Guadalcanal Island—Addresses, essays, lectures. 2. World War, 1939-1945—Addresses, essays, lectures. 3. Japan—Relations (general) with China—Addresses, essays, lectures. 4. China—Relations (general) with Japan—Addresses, essays, lectures. I. Title.
D767.98.T76 78-816247

—RELATIONS (GENERAL) WITH JAPAN—CHRONOLOGY

(Sen-kyūhyaku-shichijūshichinen no Nitchū kankei)
1977年の日中関係. — ꞁ東京ꞁ：日中経済協会，1978.
28 p. ; 26 x 37 cm. — (日中経報 ; no. 97) Ja***
Cover title.

1. Japan—Relations (general) with China—Chronology. 2. China—Relations (general) with Japan—Chronology. I. Nitchū Keizai Kyōkai. II. Series: Nitchū Keizai Kyōkai. Nitchūkei hō ; no. 97.
DS849.C6S476 79-802020

—RELATION (GENERAL) WITH JAPAN—CONGRESSES

Tōa Bunka Shinkō Kyōgikai, Tokyo, Japan, 1937.
(Tōa Bunka Shinkō Kyōgikai sokkiroku)
東亞文化振興協議會速記錄：日支文化提携 / 大東文化協會東亞文教國策委員會編. — 東京：大東文化協會東亞文教國策委員會事務局，昭和12 ꞁ1937ꞁ
10, 215 p., ꞁ21ꞁ leaves of plates : ill. ; 23 cm.
On spine: Nisshi bunka teikei kaigi sokkiroku.
1. Japan—Relations (general) with China—Congresses. 2. China—Relations (general) with Japan—Congresses. I. Daitō Bunka Kyōkai. Tōa Bunkyō Kokusaku Iinkai. II. Title. III. Title: Nisshi bunka teikei kaigi sokkiroku.
DS849.C5T6 1937 78-812088

—RELATIONS (GENERAL) WITH JAPAN—PERIODICALS

(Koyū)
滬友
ꞁ東京　滬友会ꞁ
no. ill. 21 cm.

1. Shanghai—Periodicals. 2. Japan—Relations (general) with China—Periodicals. 3. China—Relations (general) with Japan—Periodicals. I. Koyūkai.
DS796.S2K68 78-641832
(MARC-S)

CHINA (Continued)

—RELATIONS (GENERAL) WITH KOREA

Pak, Yŏng-sŏk.
(Manbosan sakŏn yŏn'gu)
萬寶山事件研究：日帝大陸侵略政策의一環으로서의 / 朴永錫著. — 서울：亜細亜文化社，1978.
vi, 288 p. ; 23 cm.
"萬寶山事件과朝鮮에서의中國人排斥事件關係日誌": p. 221-251.
Bibliography: p. 252-265.
Includes index.
₩3200
1. Koreans in Manchuria. 2. Chinese in Korea. 3. Korea—Relations (general) with China. 4. China—Relations (general) with Korea. 5. Japan—Relations (general) with China. 6. China—Relations (general) with Japan. I. Title.
DS731.K6P34 79-825462

—RELATIONS (GENERAL) WITH ROMANIA

Sub semnul prieteniei frăţeşti româno-chineze : vizita oficială de prietenie a tovarăşului Nicolae Ceauşescu, secretar general al Partidului Comunist Român, preşedintele Republicii Socialiste România, în Republica Populară Chineză [la invitaţia tovarăşului Hua Kuo-fen] 14-20 mai 1978. — Bucureşti : Editura Politică, 1978.
220 p., [33] leaves of plates : ill. (some col.) ; 20 cm. R78-3871
lei25.00
1. Romania—Relations (general) with China. 2. China—Relations (general) with Romania. 3. Visits of state—China. 4. Ceauşescu, Nicolae—Journeys—China. I. Ceauşescu, Nicolae.
DR229.C5S9 79-349439
*79 MARC

—RELATIONS (GENERAL) WITH RUSSIA

Ch'oe, Yŏng.
(Pukpang samgak kwan'gye chŏnmang kwa kŭgŏt i Hanbando e mich'inŭn yŏnghyang)
北方三角關係展望과그것이韓半島에미치는影響 / 研究執筆責任 崔榮；刊行責任 金炳璘. — [서울]：國土統一院政策企劃室，1977.
84 p. ; 26 cm. — (周邊環境與件)
"국통정77-1-1119."
Includes bibliographical references.
1. China—Politics and government—1949- 2. China—Relations (general) with Russia. 3. Russia—Relations (general) with China. 4. Korea (Democratic People's Republic)—Foreign relations. I. Kim, Pyŏng-nin. II. Title. III. Series: Chubyŏn hwan'gyŏng yŏkŏn.
DS777.55.C4539 79-825194

Min, Pyŏng-ch'ŏn.
(Chung-So taerip i Hanbando ŭi Nam-Pukhan kwan'gye e mich'inun yŏnghyang)
中蘇對立이韓半島의南北韓關係에미치는影響 / 研究執筆責任 閔丙天；刊行責任 朴鍾徹. — [서울]：國土統一院調査研究室，1976.
91 p. ; 26 cm. — (周邊勢力調査研究)
"국통조76-12-129."
1. China—Relations (general) with Russia. 2. Russia—Relations (general) with China. 3. Korea (Democratic People's Republic)—Foreign relations. 4. Korea—Foreign relations. I. Pak, Chong-ch'ŏl. II. Title. III. Series: Chubyŏn seryŏk chosa yŏn'gu.
DS740.5.R9M56 79-825033
79 K

—RELATIONS (GENERAL) WITH SINGAPORE

Lee, Lai-to.
China's changing attitude towards Singapore, 1965-1975 / by Lee Lai-to. — [Singapore] : Dept. of Political Science, University of Singapore, 1975.
48 p. ; 30 cm. — (Occasional paper series - Department of Political Science, University of Singapore ; no. 22)
Includes bibliographical references.
$2.30
1. Singapore—Relations (general) with China. 2. China—Relations (general) with Singapore. I. Title. II. Series: Singapore (City). University. Dept. of Political Science. Occasional paper series - Department of Political Science, University of Singapore ; no. 22.
DS598.S7L44 301.29'51'05952 76-940114
79 MARC

—RELATIONS (GENERAL) WITH SOUTHEASTERN ASIA—ADDRESSES, ESSAYS, LECTURES

Mengapa 600 [i.e. enam ratus] djuta rakjat R.R.T. bergerak keselatan? : untuk tanah, beras, timah dan kekajaan alam. — Djakarta : New Nusantara Pub. Coy, [1959?]
58 p. : map ; 23 cm. — (Seri kewaspadaan nasional)
1. Chinese in Southeastern Asia—Addresses, essays, lectures. 2. China—Relations (general) with Southeastern Asia—Addresses, essays, lectures. 3. Asia Southeastern—Relations (general) with China—Addresses, essays, lectures.
DS509.5.C5M46 77-986859
79 MARC

—RELATIONS (GENERAL) WITH THE NEAR EAST

Tseng, Wen-wu.
(Chung-kuo ching ying Hsi-yü shih)
中國經營西域史 / 曾問吾著. — [台北縣永和鎮]：文海出版社，[1978?]
3 v. (713, [24] p., [31] leaves of plates (5 fold.)) : ill., maps ; 22 cm. — (近代中國史料叢刊讀編 [i. e.輯] ; 513 : 第五十二輯)
Reprint of the 1936 ed. published by Shang-wu yin shu kuan, Shanghai.
Includes bibliographies.
1. Sinkiang—History. 2. China—Relations (general) with the Near East. 3. Near East—Relations (general) with China. I. Title. II. Series: Shen, Yün-lung, comp. Chin tai Chung-kuo shih liao ts'ung k'an hsü chi ; 513.
DS755.S532 vol. 513 79-835323
[DS793.S62]

—RELATIONS (GENERAL) WITH THE UNITED STATES

Han-fu.
(Chung-kuo yü Mei-kuo)
中國与美國 漢夫著 上海 引擎出版社 民國26 [1937]
2, 128 p. 17 cm. (中國与世界叢書第2種)
1. United States—Relations (general) with China. 2. China—Relations (general) with the United States. I. Title.
E183.8.C5H2 C 66-2640

Harding, Harry, 1946-
China and the U.S. : normalization and beyond / by Harry Harding, Jr. — [New York] : China Council of the Asia Society, c1979.
32, [1] p : ill. ; 28 cm.
Cover title.
Page [33] is p. [3] of cover.
Bibliography: p. [33]
$1.50
1. United States—Relations (general) with China. 2. China—Relations (general) with the United States. I. Title.
E183.8.C5H33 301.29'73'051 79-50107
79 MARC

Pugach, Noel H
Paul S. Reinsch, open door diplomat in action / Noel H. Pugach. — Millwood, N.Y. : KTO Press, 1979.
xii, 310 p. : ill. ; 24 cm. — (KTO studies in American history)
Bibliography: p. [294]-298.
Includes index.
ISBN 0-527-73050-5
1. Reinsch, Paul Samuel, 1869-1923. 2. United States—Relations (general) with China. 3. China—Relations (general) with the United States. 4. Eastern question (Far East) 5. Diplomats—United States—Biography. I. Title. II. Series.
E748.R33P83 327.2'092'4 79-1503
79 MARC

Sergeĭchuk, S
USA a Čína : politika USA vůči Číně v letech 1949-1972 / S. Sergejčuk ; [z ruského originálu SŠA i Kitaj přel. Vlasta Lassu]. — 1. vyd. — Praha : Academia, 1975.
250 p. ; 21 cm. Cz***
Includes bibliographical references.
Kčs23.00
1. United States—Relations (general) with China. 2. China—Relations (general) with the United States. I. Title.
E183.8.C5S4312 78-420895
79 MARC

United States. Congress. House. Committee on International Relations. Subcommittee on Asian and Pacific Affairs.
A new realism : factfinding mission to the People's Republic of China, July 3-13, 1978 : report / by the Subcommittee on Asian and Pacific Affairs to the Committee on International Relations, U.S. House of Representatives. — Washington : U.S. Govt. Print. Off., 1978.
vii, 132 p. ; 24 cm.
At head of title: 95th Congress, 2d session. Committee print.
Includes bibliographical references.
1. United States—Relations (general) with China. 2. China—Relations (general) with the United States. 3. China—Economic conditions—1976- I. United States. Congress. House. Committee on International Relations. II. Title.
E183.8.C5U48 1978 301.29'73'051 79-600630
79 MARC

—RELATIONS (GENERAL) WITH THE UNITED STATES—ADDRESSES, ESSAYS, LECTURES

Teng, Hsiao-p'ing, 1904-
Remarks of His Excellency Deng Xiaoping, Vice Premier of the State Council of the People's Republic of China, and President Carter at the signing of the agreement on scientific and technological cooperation and the cultural agreement between China and the United States, January 31, 1979. — [Washington : s.n., 1979]
[2] p. ; 23 cm.
Seal of the President on cover.
1. United States—Relations (general) with China—Addresses, essays, lectures. 2. China—Relations (general) with the United States—Addresses, essays, lectures. I. Carter, Jimmy, 1924- . II. Title.
E183.8.C5T4 301.29'51'073 79-311102
79 MARC

—RELATIONS (GENERAL) WITH THE UNITED STATES—PERIODICALS

China and us.
New York, U. S.—China Peoples Friendship Association.
v. ill. 28 cm. bimonthly.
"Publication of the U. S.—China Peoples Friendship Association."
Key title: China and us, ISSN 0191-3166.
1. United States—Relations (general) with China—Periodicals. 2. China—Relations (general) with the United States—Periodicals. 3. China—Periodicals. I. U. S.—China Peoples Friendship Association.
E183.8.C5C464 301.29'73'051 76-648679
MARC-S

—RELATIONS (GENERAL) WITH TIBET

Nag-dbaṅ-chos-ldan.
Subud erikẹ = ein Rosenkranz aus Perlen : die Biographie des 1. Pekinger lČan skya Khutukhtu Ṅag dbaṅ blo bzaṅ č'os ldan / verfasst von Ṅag dbaṅ č'os ldan alias Šes rab dar rgyas ; hrsg., übers. und kommentiert von Klaus Sagaster. — Wiesbaden : Harrassowitz, 1967.
430, 164 p. : facsims. ; 25 cm. — (Asiatische Forschungen ; Bd. 20)
A revision of the editor's thesis, Bonn, 1959.
Includes texts in English, Mongolian, and Tibetan.
Translation of the Subud erike, originally written in Mongolian.
Bibliography of works by Lcaṅ-skya Ṅag-dbaṅ -blo-bzaṅ-chos-ldan: p. 351-373.
Bibliography: p. [374]-382.
Includes indexes.
1. Lcaṅ-skya Ṅag-dbaṅ-blo-bzaṅ-chos-ldan, 1642-1717. 2. Lamas—Tibet—Biography. 3. China—Relations (general) with Tibet. 4. Tibet—Relations (general) with China. I. Sagaster, Klaus. II. Title. III. Title: Ein Rosenkranz aus Perlen. IV. Series.
BQ970.C347N33 1967 79-342753
79 MARC

—RELATIONS (GENERAL) WITH VIETNAM

Hu, Lien.
(Ch'u shih Yüeh-nan chi)
出使越南記 / 胡璉著. — 台北：中央日報社，民國67 [1978]
162, [10] p., [3] leaves of plates : ill. ; 20 cm.
NT$50.00
1. Vietnam—History—1945-1975. 2. China—Relations (general) with Vietnam. 3. Vietnam—Relations (general) with China. I. Title.
DS556.9.H8 78-843091

—RELIGION

see also Christianity and other religions—Chinese

Bernard, Henri, S.J.
Sagesse chinoise et philosophie chrétienne; essai sur leurs relations historiques par Henri Bernard ... Hautes études ... Tientsin. Tientsin, En vente à la procure de la Mission de Sienshien, 1935.
2 p.l., 277, vi p. 24 cm.
1. Philosophy, Chinese. 2. Christianity and other religions—Chinese. 3. Jesuits—China—History. 4. China—Religion. I. Tien-chin kung shang hsüeh yüan. II. Title.
B126.B4 1935 181'.11 39-23523
[r79]rev MARC

Bernard, Henri, S. J.
Sagesse chinoise et philosophie chrétienne, par Henri Bernard-Maitre. Paris, Cathasia [1950]
277, v p. 24 cm. (Les Humanités d'Extrême-Orient) (Série culturelle des Hautes Études de Tientsin)
Bibliographical footnotes.
1. Philosophy, Chinese. 2. Christianity and other religions—Chinese. 3. Jesuits—China—History. 4. China—Religion. I. Title. II. Series: Tientsin. Institut des hautes études industrielles et commerciales. Série culturelle.
B126.B4 1950 181'.11 52-67676
[r79]rev MARC

CHINA

—RELIGION (Continued)

Bush, Richard Clarence, 1923-
Religion in China / by Richard C. Bush ; [map, Homer Grooman]. — Niles, Ill. : Argus Communications, c1977.
xiii, 82 p. : ill. ; 23 cm. — (Major world religions series)
Bibliography: p. 81-82.
ISBN 0-913592-98-6 : $2.95
1. China—Religion. I. Title. II. Series.
BL1810.B87 299'.51 77-82796
79 MARC

Creel, Herrlee Glessner, 1905-
Sinism : a study of the evolution of the Chinese world-view / by Herrlee Glessner Creel. — Westport, Conn. : Hyperion Press, 1975, c1929.
vii, 127 p. ; 23 cm.
Reprint of the ed. published by Open Court Pub. Co., Chicago.
Includes bibliographical references.
ISBN 0-88355-165-9
1. Philosophy, Chinese. 2. China—Religion. I. Title.
B126.C7 1975 181'.11 74-2904
74[r79]rev MARC

Groot, Jan Jakob Maria de, 1854-1921.
[Jaarlijksche Feesten en Gebruiken van de Emoy-Chineezen. French]
Les fêtes annuellement célébrées à Émoui (Amoy) : étude concernant la religion populaire des chinois / par J. J. M. de Groot ; traduite du hollandais, avec le concours de l'auteur par C. G. Chavannes ; ill. par Félix Régamey ; with a new introd. by Inez de Beauclair and Harvey Molé. — San Francisco : Chinese Materials Center, 1977.
2 v. (xxiv, xxv, 832 p., [24] leaves of plates) : ill. ; 27 cm. — (Reprint series - Chinese Materials Center ; no. 77)
Translation of Jaarlijksche Feesten en Gebruiken van de Emoy-Chineezen.
Reprint of the 1886 ed. published by E. Leroux, Paris, which was issued as v. 10-11 of Annales du Musée Guimet.
Includes bibliographical references and indexes.
ISBN 0-89644-514-3
1. Amoy, China—Festivals, etc. 2. Amoy, China—Religious life and customs. 3. China—Religion. I. Title. II. Series: Chinese Materials Center. Reprint series - Chinese Materials Center ; no. 77. III. Series: Paris. Musée Guimet. Annales ; t. 11-12.
BL1802.G7614 299'.51 78-112310
79 MARC

Howard, Leslie G
Can there be a Chinese church / Leslie G. Howard ; Chinese translation by Fung Yue Ling. — Hong Kong : International Book Services, [1978]
54, 65 p., [2] leaves of plates : ill. ; 21 cm.
Added t.p.: Neng fou yu i ko chen cheng ti Chung-kuo chiao hui.
English and Chinese.
Errata slip inserted.
1. Missions—China. 2. China—Religion. 3. China—Civilization. I. Title. II. Title: Neng fou yu i ko chen cheng ti Chung-kuo chiao hui.
BV3415.2.H68 266'.00951 79-374213
79 MARC

Li, Shih-yü, fl. 1948–
(Hsien tsai Hua-pei mi mi tsung chiao)
現在華北秘密宗教 / 李世瑜著. — 台北 : 古亭書屋, 民國64 [1975]
2, viii, 176 p., [2] leaves of plates : ill. ; 22 cm. — (古亭集刊 ; 17)
Series romanized: Ku t'ing chi k'an.
Reprint ed.

1. China—Religion. 2. Sects—China. I. Title.
BL1801.L5 1975 78-842055

Mission catholique chinoise.
Les pratiques, rites et croyances de la réligion populaire chinoise. — [Port Louis] : Mission catholique chinoise, [1975]
56 p. : ill. ; 22 cm.
Cover title.
1. China—Religion. 2. China—Religious life and customs. I. Title.
BL1810.M5 1975 200'.951 78-101784
79 MARC

Sawada, Mizuho.
(Hōkan no kenkyū)
寳巻の研究 / 沢田瑞穂. — 増補. — 東京 : 国書刊行会, 昭和50 [1975]
461, 12 p. ; 22 cm. Ja 75-16694
Bibliography: p. 447-461.
Includes index.
¥4300

1. Religious literature, Chinese—History and criticism. 2. China—Religion. I. Title.
BL1802.S28 79-800286

Shen, Ping-wen.
Chinese communist criminal acts in persecution of religions / by Shen Ping-wen. — [Taipei] : World Anti-Communist League, China Chapter, Asian Peoples' Anti-Communist League, Republic of China, 1978.
133, xii p. : ill. ; 19 cm. — (Pamphlet - Asian Peoples' Anti-Communist League, Republic of China ; 208)
Cover title.
Translation of Kung fei p'o hai tsung chiao ti tsui hsing.
$1.45
1. China—Religion. 2. Persecution—China. 3. Communism and religion. I. Title.
BL1802.S5313 322'.1'0951 78-106333
79 MARC

Thompson, Laurence G
Chinese religion : an introduction / Laurence G. Thompson. — 3d ed. — Belmont, Calif. : Wadsworth, c1979.
xvii, 163 p. : ill. ; 24 cm. — (The Religious life of man)
Includes bibliographical references and index.
ISBN 0-87872-224-6
1. China—Religion. I. Title.
BL1802.T5 1979 299'.51 79-4683
79 MARC

—RELIGION—ADDRESSES, ESSAYS, LECTURES

Legge, James, 1815-1897.
The religions of China : Confucianism and Tâoism described and compared with Christianity / by James Legge. — Philadelphia : R. West, 1978.
ix, 310 p. ; 23 cm.
Reprint of the 1880 ed. published by Hodder and Stoughton, London.
Includes bibliographical references.
ISBN 0-8492-1545-5 lib. bdg. : $30.00
1. China—Religion—Addresses, essays, lectures. 2. Confucianism—Relations—Christianity—Addresses, essays, lectures. 3. Taoism—Relations—Christianity—Addresses, essays, lectures. 4. Christianity and other religions—Confucianism—Addresses, essays, lectures. 5. Christianity and other religions—Taoism—Addresses, essays, lectures. I. Title.
[BL1801.L4 1978] 299'.51 78-2685
78 MARC

—RELIGION—CONGRESSES

China and Christianity : historical and future encounters / James D. Whitehead, Yu-ming Shaw, N.J. Girardot, editors. — Notre Dame, Ind. : Center for Pastoral and Social Ministry, University of Notre Dame ; distributed by University of Notre Dame Press, c1979.
xv, 293 p. ; 23 cm.
Selected proceedings of a conference on "China, the religious dimensions," held Je. 29-July 2, 1977 at the University of Notre Dame.
Includes bibliographical references.
ISBN 0-268-00730-6 : $8.95
1. Christianity—China—Congresses. 2. China—Religion—Congresses. I. Whitehead, James D. II. Shaw, Yu-ming. III. Girardot, N. J. IV. Notre Dame, Ind. University. Center for Pastoral and Social Ministry. V. Title: China, the religious dimensions.
BR1285.C525 275.1 79-52165
79 MARC

International Conference on Taoist Studies, 2d, Chino, Japan, 1972.
Facets of Taoism : essays in Chinese religion / edited by Holmes Welch, Anna Seidel. — New Haven : Yale University Press, 1979.
301 p. ; 24 cm.
Includes bibliographical references and index.
ISBN 0-300-01695-6
1. Taoism—Congresses. 2. China—Religion—Congresses. I. Welch, Holmes. II. Seidel, Anna K. III. Title.
BL1899.5.I57 1972 299'.514 77-28034
77 MARC

—RELIGIOUS LIFE AND CUSTOMS

Masuda, Fukutarō, 1903–
(Tōa hōchitsujo josetsu)
東亞法秩序序説 : 民族信仰を中心として / 増田福太郎著. — 東京 : ダイヤモンド社, 昭和17 [1942]
3, 6, 309, 10 p. : ill. ; 22 cm.
Includes index.
"附錄　南島寺廟探訪記": p. [189]-309.

1. China—Religious life and customs. I. Title.
BL1801.M38 79-804285

Mission catholique chinoise.
Les pratiques, rites et croyances de la réligion populaire chinoise. — [Port Louis] : Mission catholique chinoise, [1975]
56 p. : ill. ; 22 cm.
Cover title.
1. China—Religion. 2. China—Religious life and customs. I. Title.
BL1810.M5 1975 200'.951 78-101784
79 MARC

—RURAL CONDITIONS

Chung-kuo kung ch'an tang. Chung yang pan kung t'ing.
Socialist upsurge in China's countryside / edited by the General Office of the Central Committee of the Communist Party of China. — 1st ed. — Peking : Foreign Languages Press, 1978.
v, 547 p. ; 19 cm.
"Originally published in three volumes in Chinese .... The present English translation is from an abridged Chinese edition published by the People's Publishing House, Peking, in January 1956."
1. Collective farms—China. 2. China—Rural conditions. I. Title.
HD1492.C5C59 1978 79-100155
79 MARC

(Hsüeh hsi Chung-kuo nung ts'un ti she hui chu i kao ch'ao ti hsü yen ho an yü)
学习《中国农村的社会主义高潮》的序言和按语 / 中共南宁地委宣传部主編. — 第1版. — [南宁?] : 广西人民出版社, 1977.
2 v. ; 19 cm.
Vol. 2 has compiler statement: Chung kung Yü-lin ti wei hsüan ch'uan pu chu pien; Kuang-hsi jen min ch'u pan she pien chi.
$0.42
1. Chung-kuo kung ch'an tang. Chung yang pan kung t'ing. Chung-kuo nung ts'un ti she hui chu i kao ch'ao. 2. Agriculture, Cooperative—China. 3. China—Rural conditions. I. Chung-kuo kung ch'an tang. Nan-ning ti wei. Hsüan ch'uan pu. II. Chung-kuo kung ch'an tang. Yü-lin ti wei. Hsüan ch'uan pu. III. Kuang-hsi jen min ch'u pan she.
HD1491.C6H69 79-841136

Institute of Pacific Relations.
Agrarian China : selected source materials from Chinese authors / compiled and translated by the Research Staff of the Secretariat, Institute of Pacific Relations ; with an introd. by R. H. Tawney. — 1st AMS ed. — New York : AMS Press, 1978.
xviii, 258 p. ; 23 cm.
Reprint of the 1938 ed. published by University of Chicago Press, Chicago, issued as a report in the International research series of the Institute of Pacific Relations.
ISBN 0-404-59532-4
1. Land tenure—China. 2. Agriculture—Economic aspects—China. 3. Agriculture—China. 4. China—Rural conditions. I. Title. II. Series: Institute of Pacific Relations. International research series.
HD925.I57 1978 338.1'0951 75-30060
78 MARC

Kulp, Daniel Harrison, 1888–
[Country life in South China. Japanese]
南支那の村落生活 : 家族主義の社會學 / D·H·カルプ著 ; 喜多野清一, 及川宏譯. — 東京 : 生活社, 昭和15 [1940], 昭和17 [1942] printing.
[44], 467, 28 p., [9] leaves of plates : ill. ; 23 cm.
Bibliography: p. 451–453.
Includes index.
1. China—Social life and customs. 2. China—Rural conditions. 3. Family—China. I. Title.
Title romanized: Minami Shina no sonraku seikatsu.
DS721.K8517 78-816556

Parish, William L
Village and family in contemporary China / William L. Parish, Martin King Whyte. — Chicago : University of Chicago Press, 1978.
xiii, 419 p. : ill. ; 24 cm.
Includes bibliographical references and index.
ISBN 0-226-64590-8
1. Rural families—China. 2. Villages—China—Case studies. 3. China—Rural conditions. I. Whyte, Martin King, joint author. II. Title.
HQ684.P37 301.42'0951 78-3411
78 MARC

Roll, Charles Robert.
Incentives and motivation in China : the "reality" of rural inequality / Charles R. Roll. — Santa Monica, Calif. : Rand Corp., 1976.
36 p. ; 28 cm. — (The Rand paper series ; P-5563)
Cover title.
Includes bibliographical references.
1. Income distribution—China. 2. Land reform—China. 3. China—Rural conditions. I. Title. II. Series: Rand Corporation. Paper ; P-5563.
AS36.R28 no. 5563 081 s 79-311601
[HC427.9] 79 MARC

—RURAL CONDITIONS—ADDRESSES, ESSAYS, LECTURES

(Chūkyō nōson tembyō)
中共·農村点描. — 東京 : 綜合研究所, 昭和40- [1965-
v. ; 19 cm.
"昭和40年8月　綜研資料　10"

1. Agriculture—Economic aspects—China—Addresses, essays, lectures. 2. China—Rural conditions—Addresses, essays, lectures. I. Sōgō Kenkyūjo.
HD2097.C48 79-804132

Huang, Philip C 1940-
Chinese Communists and rural society, 1927-1934 / Philip C. C. Huang, Lynda Schaefer Bell, Kathy Lemons Walker. — Berkeley : Center for Chinese Studies, University of California, c1978.
121 p. : maps ; 23 cm. — (China research monographs ; no. 13)
Bibliography: p. 99-115.
ISBN 0-912966-18-1 : $5.00
1. Communism—China—Addresses, essays, lectures. 2. China—Rural conditions—Addresses, essays, lectures. I. Bell, Lynda Schaefer, joint author. II. Walker, Kathy Lemons, joint author. III. Title. IV. Series.
HX417.H8 335.43'4 78-620018
79 MARC

CHINA

—RURAL CONDITIONS—ADDRESSES, ESSAYS, LECTURES (Continued)

(Yung she hui chu i chan ling nung ts'un ssu hsiang wen hua chen ti)
用社会主义占领农村思想文化阵地 / 农村版图书编选小组选编. — [香港? : s. n., 1979?]
82 p. ; 21 cm.
Reprint of the 1st (1974) ed. published by Nung ts'un tu wu ch'u pan she, Peking, in series: Nung ts'un pan t'u shu.
1. Communism—China—Addresses, essays, lectures. 2. China—Rural conditions—Addresses, essays, lectures. I. Nung ts'un pan t'u shu pien hsüan hsiao tsu.
HX419.Y86 1979 79-838605

—RURAL CONDITIONS—DICTIONARIES—CHINESE

Broadbent, K P
A Chinese/English dictionary of China's rural economy / by Kieran Broadbent. — Farnham Royal : Commonwealth Agricultural Bureaux, 1978.
xiii, 406 p. : ill., col. map (on lining papers) ; 26 cm. GB78-3198
Bibliography: p. 345-363.
Includes indexes.
ISBN 0-85198-381-2 : £30.00
1. Agriculture—Economic aspects—China—Dictionaries—Chinese. 2. China—Rural conditions—Dictionaries—Chinese. 3. Chinese language—Dictionaries—English. I. Title.
HD2097.B76 330.9'51'05 79-306829
79 MARC

—SOCIAL CONDITIONS

Crow, Carl, 1883–1945.
[Four hundred million customers. Japanese]
支那で成功する道 / カール・クロウ著 ; 柴田賢一譯. — 東京 : 高山書院, 昭和15 [1940]
7, 8, 289 p. ; 19 cm.
1. China—Social life and customs. 2. China—Social conditions. 3. China—Economic conditions. 4. Crow, Carl, 1883–1945. I. Title.
Title romanized: Shina de seikōsuru michi.
DS721.C9316 79-800152

Fan, Yüan-yen.
(Wei pa i k'u nan t'ung pao tso cheng)
為八億苦難同胞作證 / 范園焱口述 ; 劉本炎執筆 ; 中央日報編印. — 臺北 : 該報社, 民國66 [1977]
186 p. : ports. ; 19 cm.
NT$20.00 ($0.50 U. S.)
1. China—Politics and government—1949–1976. 2. China—Social conditions. 3. Fan, Yüan-yen. 4. Refugees, Political—Taiwan—Biography. 5. Refugees, Political—China—Biography. I. Liu, Pen-yen. II. Chung yang jih pao, T'ai-pei. III. Title.
DS777.55.F3377 78-843502

Fan, Yüan-yen.
(Wo chia Mi-ko shih chiu fei lai)
我駕米格十九飛來 / 范園焱著. — 台北 : 文藝月刊社 : 經銷者遠東書報社, 民國67 [1978]
264 p. : col. ill. ; 20 cm. — (文藝月刊叢書 ; 11)
Series romanized: Wen i yüeh k'an ts'ung shu.
NT$70.00
1. China—Politics and government—1949–1976. 2. China—Social conditions. 3. Fan, Yüan-yen. 4. Refugees, Political—Taiwan—Biography. 5. Refugees, Political—China—Biography. I. Title.
DS777.55.F338 78-843497

Fischer, Harald, 1949-
China : d. neue Selbstbewusstsein : Gesellschaft, Erziehung, Bevölkerung / Harald Fischer, Christian Graf von Krockow, Hermann Schubnell. — [1.-5. Tsd.]. — München ; Zürich : Piper, 1978.
320 p. ; 19 cm. GFR78-A
Bibliography: p. 316-320.
ISBN 3-492-02313-4 : DM32.00
1. China—Social conditions. 2. China—Population. 3. Education—China—1949- . 4. Birth control—China. I. Krockow, Christian, Graf von, joint author. II. Schubnell, Hermann, joint author. III. Title.
HN737.F57 309.1'51'05 78-384499
79 MARC

Freedman, Maurice.
The study of Chinese society : essays / by Maurice Freedman ; selected and introduced by G. William Skinner. — Stanford, Calif. : Stanford University Press, 1979.
xxiv, 491 p. ; 24 cm.
Bibliography: p. [445]-475.
ISBN 0-8047-0964-5 : $28.50
1. China—Social conditions. 2. Chinese in Southeastern Asia. I. Skinner, George William, 1925- . II. Title.
HN733.F73 309.1'51 78-65395
00 MARC

Hoffmann, Rainer.
Maos Rebellen : Sozialgeschichte d. chines. Kulturrevolution / Rainer Hoffmann. — 1. Aufl. — Hamburg : Hoffmann und Campe, 1977.
227 p. ; 21 cm. — (Historische Perspektiven ; 8) GFR78-A
Summary in English.
Bibliography: p. 216-222.
Includes index.
ISBN 3-455-09220-9 : DM32.00
1. China—Social conditions. 2. China—Politics and government—1949-1976. I. Title.
HN737.H63 309.1'51'05 78-389857
79 MARC

Kuo, Mo-jo, 1892–1978.
(Chung-kuo ku tai she hui yen chiu)
中國古代社会研究 / 郭沫若著. — 第2版. — [北京] : 人民出版社, 1964, 1977 printing.
x, 281 p. ; 21 cm.
$0.71
1. China—Social conditions. I. Title.
HN733.5.K86 1964 79-835751

Kuo, Mo-jo, 1892–1978.
(Chung-kuo ku tai she hui yen chiu)
中國古代社會研究 / 郭沫若著. — 香港 : 生活·讀書·新知三联书店, 1978.
x, 281 p. ; 22 cm.
"根据人民出版社一九六四年十月第二版重印"
Includes bibliographical references.
$8.00
1. China—Social conditions. I. Title.
HN733.K86 1978 79-838270

Mader, Jean.
Mille questions, mille réponses sur la République populaire de Chine / Jean Mader ; [publié par l'Association des élèves de l'Institut supérieur de gestion]. — [Paris] (B, rue de Lota, 75116) : [Association des élèves de l'Institut supérieur de gestion], 1977.
126 p., [20] p. of col. plates ; 21 cm. F78-1824
30.00F
1. China—Economic conditions—1949-1976. 2. China—Social conditions. I. Title.
HC427.9.M15 330.9'51'05 79-343311
79 MARC

Nearing, Scott, 1883-
Whither China? : An economic interpretation of recent events in the Far East / by Scott Nearing. — Westport, Conn. : Hyperion Press, 1977, c1927.
225 p. ; 23 cm. — (China studies)
Reprint of the ed. published by International Publishers, New York.
Bibliography: p. 211-214.
Includes index.
ISBN 0-88355-385-6 : $16.50
1. China—Economic conditions—1912-1949. 2. China—Industries. 3. China—Social conditions. 4. China—History—1900- . 5. China—Foreign relations. I. Title. II. Series: China studies (Westport, Conn.)
HC427.N38 1977 330.9'51'05 75-39029
77 MARC

Wu, Po-hsiao.
(Lu-an feng wu)
潞安風物 / 吳伯簫著. — [香港? : s. n., 1978?]
2, 145 p. ; 19 cm.
Reprint of the 1947 ed. published by Hai yang shu wu, Hongkong, which was issued as ti 2 chi of Pei fang wen ts'ung.
1. Sino-Japanese Conflict, 1937–1945—Personal narratives, Chinese. 2. China—Social conditions. 3. Wu, Po-hsiao. 4. Authors, Chinese—20th century—Biography. I. Title.
DS777.53.W828 1978 79-835139

—SOCIAL CONDITIONS—1949-1960

Peregalli, Arturo.
Introduzione alla storia della Cina : analisi critica della società cinese / Arturo Peregalli. — Pistoia : CEIDEM, 1976.
242 p. ; 17 cm. — (Collana diretta da Umberto Melotti) It78-Aug
Bibliography: p. 237-240.
L3000
1. China—Social conditions—1949-1960. I. Title.
HN737.P47 309.1'51'05 78-398335
•79 MARC

—SOCIAL CONDITIONS—1949-1976

Chen, Ting Kai.
Die Volksrepublik China : Nord u. Süd in d. Entwicklung / Ting Kai Chen, mit d. Beitr., Ausgewählte Kommunen, regionale Gegensätze und Differenzierungen (aus, K. Buchanan, The transformation of the Chinese earth) u. China nach Mao von Harry Hamm. — 1. Aufl. — Stuttgart : Klett, 1977.
219 p. : ill., maps ; 23 cm. — (Länderprofile, geographische Strukturen, Daten, Entwicklungen) GFR78-A2
Edition for 1974 published under title: Nord- und Südchina.
Bibliography: p. 215-219.
ISBN 3-12-928050-2 : DM28.00
1. China—Economic conditions—1949-1976. 2. China—Social conditions—1949-1976. 3. Physical geography—China. I. Buchanan, Keith M., joint author. II. Hamm, Harry, joint author. III. Title. IV. Series.
HC427.9.C52172 1977 309.1'51'05 79-355023
79 MARC

—SOCIAL CONDITIONS—1976-

Tschudi, Aadel Brun, 1909-
Gjensyn med Kina / Aadel Brun Tschudi. — Oslo : Grøndahl, 1978 (Lillestrøm : Lillestrøm industritrykk)
189 p. : ill. ; 26 cm. N78-Nov.
ISBN 8250403320 : kr88.00
1. China—Social conditions—1976- . 2. China—Description and travel—1976- . 3. Missions—Norway—History. 4. Missions—China—History. I. Title.
HN733.5.T7 79-341862
79 MARC

—SOCIAL CONDITIONS—ADDRESSES, ESSAYS, LECTURES

Abe, Kazunari, 1927-
(Gendai no chōryū)
現代の潮流 : 安部一成エッセイ集. — 東京 : 東洋図書出版, 1978.
355 p. ; 20 cm. Ja 78-16798
Includes bibliographical references.
¥1600
1. Japan—Economic policy—1945- —Addresses, essays, lectures. 2. Economics—Addresses, essays, lectures. 3. Peace—Addresses, essays, lectures. 4. China—Social conditions—Addresses, essays, lectures. I. Title.
HC462.9.A62 78-812263

P'an, Yu-hsin.
(Hsien shih ti yeh yü)
現實的揶揄 / 潘酉辛作. — [香港? : s. n., 1978?]
vi, ii, 188 p., [14] leaves of plates : ill. ; 22 cm.
Reprint of the 1948 ed. published by Ta kuang pao she, Kuang-chou.
1. China—Civil War, 1945–1949—Personal narratives—Addresses, essays, lectures. 2. China—Social conditions—Addresses, essays, lectures. 3. P'an, Yu-hsin—Addresses, essays, lectures. 4. Journalists—China—Biography—Addresses, essays, lectures. I. Title.
DS777.54.P32 1978 79-835451

(Tzu yu ti chen han)
自由的震撼 : 范園焱義士駕機來歸的深遠意義 / [著作者本公司編譯部編]. — 臺北 : 黎明文化事業公司, 民國66 [1977]
454, [7] p., [11] leaves of plates : ill., group ports. ; 19 cm.
CONTENTS: 邵希彥 立志東飛—高佑宗 覆巢之下—劉承司 飛向自由
NT$50.00
1. Fan, Yüan-yen—Addresses, essays, lectures. 2. Refugees, Political—China—Biography—Addresses, essays, lectures. 3. Refugees, Political—Taiwan—Biography—Addresses, essays, lectures. 4. China—Social conditions—Addresses, essays, lectures. I. Shao, Hsi-yen. Li chih tung fei. 1977. II. Kao, Yu-tsung. Fu ch'ao chih hsia. 1977. III. Liu, Ch'eng-ssu. Fei hsiang tzu yu. 1977. IV. Li ming wen hua kung ssu. Pien i pu.
DS778.F26T97 78-843442

—SOCIAL CONDITIONS—CASE STUDIES

Erasmus Universiteit. China Studiegroep.
Hoe ver is China? : protocollen en artikelen naar aanleiding van de studiereis naar de Volksrepubliek China in Oktober 1977 van de China Studiegroep van de Erasmus Universiteit te Rotterdam / organisatie van de reis De China Studiegroep, in samenwerking met het Bureau Buitenlandse Betrekkingen van de Erasmus Universiteit Rotterdam ; comité van aanbeveling voor de reis, H. C. Bos ... [et al.]. — [3000 DR] Rotterdam, (Postbus 1738) : [Bureau Buitenlandse Betrekkingen], 1978.
x, 619 p. : with ill. ; 21 cm. Ne78-34
Includes bibliographies.
fl 17.50
1. China—Social conditions—Case studies. 2. China—Economic conditions—1976- —Case studies. 3. Medical care—China—Case studies. I. Bos, Hendricus Cornelis, 1926- . II. Title.
HN737.E72 1978 79-342561
MARC

CHINA (Continued)

—SOCIAL CONDITIONS—CONGRESSES

Studies in Chinese society / edited by Arthur P. Wolf ; contributors, Emily M. Ahern ... [et al.]. — Stanford, Calif. : Stanford University Press, 1978.
xii, 372 p. : ill. ; 23 cm.
Papers, from conferences held between 1966 and 1973, which were first published in the seven volume series: Studies in Chinese society.
Includes bibliographical references and index.
ISBN 0-8047-1006-6
1. China—Social conditions—Congresses. 2. China—Economic conditions—Congresses. I. Wolf, Arthur P. II. Ahern, Emily M.
HN733.S8 309.1'51 78-62272
00 MARC

—SOCIAL CONDITIONS—CHRONOLOGY

Kimura, Ikujirō, 1924-
(Chūgoku rōdō undō shi nempyō)
中国労動運動史年表：1557-1949 / 木村郁二郎編. — 東京：[木村]：発売所 汲古書院, 1978.
451, 7 p. ; 27 cm. Ja 78-23915
Bibliography: p. 1-7 (2d group)
¥9000
1. Labor and laboring classes—China—Political activity—Chronology. 2. Trade-unions—China—Chronology. 3. China—Social conditions—Chronology. I. Title.
HD8738.K55 79-803746

—SOCIAL LIFE AND CUSTOMS

Angus, Fay.
The white pagoda / Fay Angus. — Wheaton, Ill. : Tyndale House, c1978.
192 p. : ill. ; 22 cm.
ISBN 0-8423-8213-5 : $3.95
1. Angus, Fay. 2. China—Biography. 3. China—Social life and customs. I. Title.
CT1828.A53A38 951.04'2'0924 77-83555
00 MARC

Bernard, Henri, S.J.
Le père Matthieu Ricci et la société chinoise de son temps (1552-1610) Tientsin, Hautes études, 1937.
2 v. 24 cm.
"Bibliographie": v. 1, p. xiii-xvi.
1. Ricci, Matteo, 1552-1610. 2. Missionaries—China. 3. Jesuits—China. 4. China—Social life and customs. I. T'ien-chin kung shang hsüeh yüan. II. Title.
BV3427.R46B475 266'.2'0924 38-33085
[r79]rev MARC

Ch'ên, Ta, 1896-
Emigrant communities in south China : a study of overseas migration and its influence on standards of living and social change / by Ta Chen ; English version edited by Bruno Lasker. — New York : AMS Press, 1978.
xvi, 287 p. ; 23 cm.
Reprint of the 1940 ed. published by the Institute of Pacific Relations, New York and issued under the auspices of the China Institute of Pacific Relations International research series.
Includes bibliographical references and index.
ISBN 0-404-59515-4
1. China—Social life and customs. 2. China—Emigration and immigration. I. Lasker, Bruno, 1880-1965. II. Title. III. Series: Institute of Pacific Relations. International research series.
DS721.C476 1978 301.32'9'51 75-30052
78 MARC

(Ch'in ting ta Ch'ing t'ung li)
欽定大清通禮 / [來保等編纂]. — [台北：台灣]商務[印書館, 1978?]
6 v. ; 20 cm. — (四庫全書珍本八集 ; 125-130)
Cover title.
'Photoreprint ed. of the Wen Yüan Ko copy.
1. China—Social life and customs. 2. Rites and ceremonies—China. 3. China—History—Ch'ing dynasty, 1644-1912—Sources. I. Lai, Pao. II. Title: Ta Ch'ing t'ung li. III. Series: Ssu k'u ch'üan shu. Ssu k'u ch'üan shu chen pen pa chi ; 125-130.
AC149.S7 1978, vol. 125-130 78-841106
[DS721]

Chou, Ching-wen, 1909-
[Chung-hua min tsu tsai i ch'ieh min tsu ko ming tou cheng chung ti ling tao ti wei]
中華民族奮鬥的方向 周鯨文著 [香港] 時代批評社 [1959]
97 p. 19 cm.
First ed. publishhed in 1939 under title: Chung-hua min tsu tsai i ch'ieh min tsu ko ming tou cheng chung ti ling tao ti wei.
I. Title.
Title romanized: Chung-hua min tsu fen tou ti fang hsiang.
DS721.C54 1959 C 60-586

Chow, Ching Lie, 1936-
Journey in tears : memory of a girlhood in China / by Chow Ching-li ; translated from the French by Abby Israel. — New York : McGraw-Hill, c1978.
262 p., [4] leaves of pplates : ill. ; 24 cm.
Translation of Le palanquin des larmes.
ISBN 0-07-010818-8 : $11.95
1. Chow, Ching Lie, 1936- . 2. China—Social life and customs. 3. China—Biography. I. Title.
CT1828.C544A3413 951.05'092'4 78-18171
78 MARC

Chow, Ching Lie, 1936-
Le Palanquin des larmes / Chow Ching Lie ; récit recueilli par Georges Walter .... — Paris : France loisirs, 1976.
371 p., [8] p. of plates ; 25 cm. F76-18049
30.00F
1. Chow, Ching Lie, 1936- . 2. China—Social life and customs. I. Walter, Georges. II. Title.
CT1828.C544A34 1976 951.05'092'4 79-372055
79 MARC

Chu, Yü, 1899 or 1900.
(T'ai-wan yü chung yüan wen hua hsüeh t'ung ti kuan hsi)
臺灣與中原文化血統的關係 / 祝毓編著. — [臺北：s. n., 1977?]
36, 362 p. ; 22 cm.
Cover title.
Reprint ed.?
1. Taiwan—Social life and customs. 2. China—Social life and customs. I. Title.
DS799.4.C45 1977 78-841576

Cormack, Annie.
Chinese birthday, weddings, funeral, and other customs / by Mrs. J. G. Cormack. — 3d ed. — Taipei : Ch'eng Wen Pub. Co., 1974.
220 p., [8] leaves of plates : ill. ; 20 cm.
Reprint of the 1927 ed. published by China Booksellers, Peking; 4th ed. published in 1935 under title: Every day customs in China.
1. China—Social life and customs. I. Title.
DS721.C75 1974 390'.0951 78-322807
79 MARC

Crow, Carl, 1883-1945.
[Four hundred million customers. Japanese]
支那で成功する道 / カール・クロウ著 ; 柴田賢一譯. — 東京：高山書院, 昭和15 [1940]
7, 8, 289 p. ; 19 cm.
1. China—Social life and customs. 2. China—Social conditions. 3. China—Economic conditions. 4. Crow, Carl, 1883-1945. I. Title.
Title romanized: Shina de seikōsuru michi.
DS721.C9316 79-800152

Gotō, Asatarō, 1881-1945.
(Shina minzoku no tembō)
支那民俗の展望 / 後藤朝太郎著. — 東京：冨山房, 昭和11 [1936]
6, 18, 752, 2 p., [18] leaves of plates : ill. ; 22 cm.
1. China—Social life and customs. I. Title.
DS721.G765 78-813444

Hu, P'u-an, 1878- comp.
[Chung-hua ch'üan kuo feng su chih]
中華風俗志 / 胡樸安[編輯]. — Taipei：東方文化書局複刊, [1975?]
4 v. ; 21 cm. — (國立北京大學, 中國民俗學會民俗叢書 ; 145-148 : 第八輯) (東方文叢)
Caption title: Chung-hua ch'üan kuo feng su chih.
Added t. p.: Modern & ancient Chinese folkways.
Series 2 romanized: Tung fang wen ts'ung.
Reprint ed.
1. China—Social life and customs. I. Title. II. Title: Modern & ancient Chinese folkways. III. Title: Chung-hua ch'üan kuo feng su chih. IV. Series: Kuo li Pei-ching ta hsüeh, Chung-kuo min su hsüeh hui min su ts'ung shu ; 145-148.
Title romanized: Chung-hua feng su chih.
GR335.K87 vol. 145-148 78-841518
[DS721]

Kulp, Daniel Harrison, 1888-
[Country life in South China. Japanese]
南支那の村落生活：家族主義の社會學 / D・H・カルプ著 ; 喜多野清一, 及川宏譯. — 東京：生活社, 昭和15 [1940], 昭和17 [1942] printing.
[44], 467, 28 p., [9] leaves of plates : ill. ; 23 cm.
Bibliography: p. 451-453.
Includes index.
1. China—Social life and customs. 2. China—Rural conditions. 3. Family—China. I. Title.
Title romanized: Minami Shina no sonraku seikatsu.
DS721.K8517 78-816556

(Ming chi li)
明集禮 / [徐一夔 ... et al.]. — [台北：台灣商務印書館, 1978?]
12 v. : ill. ; 20 cm. — (四庫全書珍本八集 ; 113-124)
Cover title.
Reprint ed.; originally issued in 53 chüan.
1. China—Social life and customs. 2. Rites and ceremonies—China. 3. China—History—Ming dynasty, 1368-1644—Sources. I. Hsü, I-k'uei, 14th cent. II. Series: Ssu k'u ch'üan shu. Ssu k'u ch'üan shu chen pen pa chi ; 113-124.
AC149.S7 1978, vol. 113-124 78-841089
[DS721]

Pruitt, Ida.
A China childhood / by Ida Pruitt ; with a foreword by John K. Fairbank. — San Francisco : Chinese Materials Center, 1978.
xiii, 205 p., [4] leaves of plates : ill., maps (on lining papers) ; 22 cm. — (Asian library series ; no. 10)
ISBN 0-89644-523-2
1. China—Social life and customs. 2. Pruitt, Ida. 3. Missionaries—China—Biography. 4. Missionaries—United States—Biography. I. Title. II. Series.
DS721.P77 951.04'092'4 79-100810
79 MARC

(Ta chin chi li)
大金集禮. — [台北：臺灣]商務[印書館, 1978?]
4 v. ; 20 cm. — (四庫全書珍本八集 ; 109-112)
Cover title.
Photoreprint ed. of the Wen Yüan Ko copy.
1. China—Social life and customs. 2. Rites and ceremonies—China. 3. Ju-chen (Tribe)—History—Sources. I. Series: Ssu k'u ch'uan shu. Ssu k'u ch'uan shu chen pen pa chi ; 109-112.
AC149.S7 1978, vol. 109-112 78-841505
[DS721]

Uchiyama, Kiyoshi, 1887-
(Bōekijō yori mitaru Shina fūzoku no kenkyū)
貿易上ヨリ見タル支那風俗之研究 / 內山清著. — 上海：日々新聞社：賣捌所 日本堂, 大正4 [1915]
8, 17, 538 p., [16] leaves of plates : ill. ; 23 cm.
Originally published in Nichinichi shimbun, Jan.-July 1915.
1. China—Social life and customs. 2. China—Commerce. I. Title. II. Title: Shina fūzoku no kenkyū.
DS721.U29 78-808535

Wilkinson, Hiram Parkes, 1866-
The family in classical China / by H. P. Wilkinson. — Taipei : Ch'eng Wen Pub. Co., 1974.
239 p., [1] fold. leaf of plates : diagr. ; 20 cm.
Reprint of the 1926 ed. published by Macmillan, London.
Includes bibliographical references and index.
1. Family—China—History. 2. Marriage—China—History. 3. China—Social life and customs. I. Title.
HQ684.W54 1974 301.42'0951 79-308756
79 MARC

—SOCIAL LIFE AND CUSTOMS—1949-

Andō, Hikotarō, 1917-
(Chūka Jimmin Kyōwakoku no ayumi)
中華人民共和国の歩み / 安藤彦太郎. — 東京：講談社, 昭和52 [1977]
204 p. : ill. (some col.) ; 27 cm. — (図説中国の歴史 ; 10) Ja***
Bibliography: p. 190-191.
Includes index.
¥2400
1. China—History—1945-1976. 2. China—Social life and customs—1949- I. Title. II. Series: Zusetsu Chūgoku no rekishi ; 10.
DS735.Z86 vol. 10 78-812159
[DS777.55]

Myrdal, Jan.
China notebook, 1975-1978 / Jan Myrdal ; photos. by Gun Kessle ; translated from the Swedish by Rolf vom Dorp. — Chicago : Liberator Press, [c1979]
146 p. : ill. ; 22 cm.
Translation of 2 works: Kinesiska frågor från Liu Ling 1975 and Kina efter Mao Tsetung.
ISBN 0-930720-58-X : $3.95
1. China—Social life and customs—1949- . 2. Lieou-lin, China. 3. China—Politics and government—1976- . I. Myrdal, Jan. Kina efter Mao Tsetung. English. 1979. II. Title.
DS777.55.M8913 951.05 79-88412
79 MARC

Myrdal, Jan.
Kinesiska frågor från Liu Ling 1975 / Jan Myrdal. — Stockholm : PAN/Norstedt, 1975.
52 p. ; 19 cm. — (En PANbok) S76-6/7
ISBN 9117515211 : kr19.00
1. China—Social life and customs—1949- . 2. Lieou-lin, China. I. Title.
DS777.55.M89 78-412248
*78 MARC

CHINA
—SOCIAL LIFE AND CUSTOMS —1949- (Continued)

(Shin Chūgoku shuzaiki)
新中国取材記 / NHK取材班. — 東京：日本放送出版協会, 1978-
v. : ill. ; 19 cm. Ja 78-12310 (v. 1)
CONTENTS: 1. 民衆生活の素顔—2. 文明と風土
¥900 (v. 1)

1. China—Social life and customs—1949- I. Nippon Hōsō Kyōkai. Shuzaihan.
DS724.S55 78-814298

—SOCIAL LIFE AND CUSTOMS —1949-1976

Chin, Hsiung-pai, 1903-
(Kono nazo no kyojin Chūgoku)
この謎の巨人、中国：新聞が書かなかったその実像 / 金雄白著；伊達政之訳. — 東京：蒼洋社：発売 英潮社, 1978.
271 p., ₍9₎ leaves of plates : ill. ; 19 cm. — (蒼洋社選書 ; 1)
Ja 78-25403
Series romanized: Sōyōsha sensho.
¥1200
1. China—Politics and government—1949-1976. 2. China—Social life and customs—1949- I. Title. II. Title: Chūgoku.
DS777.55.C446645 79-801519

—SOCIAL LIFE AND CUSTOMS —ADDRESSES, ESSAYS, LECTURES

Inoue, Kōbai.
(Shina kakuchi fūzoku sōdan)
支那各地風俗叢談 / 井上紅梅著. — 上海：日本堂書店, 大正13 ₍1924₎
10, 379 p., ₍2₎ leaves of plates : ill. ; 20 cm.
Spine title.

1. China—Social life and customs—Addresses, essays, lectures. I. Title.
DS721.I63 79-804385

Kao, Teng-wei.
(Pa pai chung Chung-kuo feng ch'ing)
八〇〇種中國風情 / 著者高登偉. — ₍台北₎：金文圖書有限公司, 民國66 ₍1977₎
313 p. ; 19 cm. — (黄金文庫 ; 13)
Series romanized: Huang chih wen k'u.
NT$55.00

1. China—Social life and customs—Addresses, essays, lectures. 2. Ethnology—China—Addresses, essays, lectures. I. Title.
DS721.K176 78-842113

Kumagai, Yasushi.
(Shina gōchin zatsuwa)
支那郷鎮雜話：長江宣撫記錄 / 熊谷康著. — 大連：大連日日新聞社, 昭和18 ₍1943₎
264 p. ; 19 cm.

1. China—Social life and customs—Addresses, essays, lectures. 2. Kumagai, Yasushi—Addresses, essays, lectures. I. Title.
DS721.K8519 78-817166
78 J

Kuo, Li-ch'eng.
(Min su hsieh ch'ü)
民俗擷趣 / 郭立誠著. — 臺北：出版家文化事業有限公司出版部, 民國66 ₍1977₎
212 p. : ill. ; 19 cm.— (出版家叢書 ; 38)
Series romanized: Ch'u pan chia ts'ung shu.
NT$40.00

1. China—Social life and customs.—Addresses, essays, lectures. I. Title.
DS721.K867 78-843373

Uchiyama, Kanzō, 1885-1959.
(Shanhai fūgo)
上海風語 / 内山完造. — 東京：改造社, 昭和16 ₍1941₎, 昭和18 ₍1943₎ printing.
2, 7, 348 p. ; 19 cm.

1. Shanghai—Social life and customs—Anecdotes, facetiae, satire, etc. 2. China—Social life and customs—Anecdotes, facetiae, satire, etc. 3. Uchiyama, Kanzō, 1885-1959. I. Title.
DS796.S25U24 78-815787

—STATISTICS

United States. Central Intelligence Agency. National Foreign Assessment Center.
China, economic indicators / National Foreign Assessment Center. — Washington : The Center : may purchase from Photoduplication Service, Library of Congress, 1978.
iii, 50 p., ₍1₎ fold. leaf of plates : ill. ; 27 cm. — (A Reference aid - National Foreign Assessment Center)
"ER 78-10750."
Chiefly tables.
1. Economic indicators—China. 2. China—Economic conditions—1949-1976. 3. China—Economic conditions—1976- . 4. China—Statistics. I. Title.
HC427.9.U462 1978 330.9'51'05 79-602325
79 MARC

United States. Central Intelligence Agency. National Foreign Assessment Center.
China, a statistical compendium / National Foreign Assessment Center. — Washington : The Center, 1979.
ii, 13 p. : ill. ; 27 cm. — (A Reference aid)
"ER 79-10374."
1. China—Statistics. I. Title.
HA1706.U53 1979 315.1 79-603267
79 MARC

—STATISTICS—COLLECTED WORKS

China. Nung kung shang pu.
(Nung kung shang pu t'ung chi piao)
農工商部統計表
₍Washington, Center for Chinese Research Materials, Association of Research Libraries,
v. 28 cm.
Reprint.

1. China—Statistics—Collected works. I. Title.
HA1701.N86a 315.1 79-642154
(MARC-S)

—STATISTICS—YEARBOOKS

China. Nung shang pu. Tsung wu t'ing. T'ung chi k'o.
(Nung shang t'ung chi piao)
農商統計表 第1-10次 民國1-10年 ₍1912-1921₎
₍Washington, Center for Chinese Research Materials, Association of Research Libraries, 1978?₎
10 v. ill. 28 cm.
Reprint of the annual published by Shang-hai Chung-hua shu chü. Shanghai and Nung shang pu tsung wu t'ing t'ung chi k'o, Peking; with a new introd. in English.
Some vols. issued in parts.
Vols. for 1920-21 issued together.
Vols. for 1912-16 published in Shang-hai by Shang-hai Chung-hua shu chü.

1. China—Statistics—Yearbooks. I. Title.
HA1704.C47a 79-642040
(MARC-S)
79 C

—STATISTICS, VITAL

China. Chu chi ch'u. T'ung chi chü.
(Chung-kuo t'u ti jen k'ou, tsu tien chih tu chih t'ung chi fen hsi)
中國土地人口, 租佃制度之統計分析 / 本社輯. — 台1版. — ₍台北₎：華世出版社, 民國67 ₍1978₎
112, 109, 168, ₍13₎ p., ₍2₎ fold. leaves ; 22 cm. — (中國經濟史料叢書：第一輯；第1種)
"本書為國民政府主計處統計局所出版的...「中國土地問題之統計分析」(民國三十年七月),「中國人口問題之統計分析」(民國三十三年一月),「中國租佃制度之統計分析」(民國三十一年八月)三種之合刊."
NT$300.00

1. Land use—China—Statistics. 2. Land tenure—China—Statistics. 3. China—Statistics, Vital. I. Hua shih ch'u pan she. II. Title. III. Series: Chung-kuo ching chi shih liao ts'ung shu : Ti i chi ; ti i chung.
HD925.C47 1978 79-837764

—STUDY AND TEACHING

see also Sinologists

Jōno, Hiroshi.
(Chūgokugaku genri)
中国学原理 / 城野宏. — 東京：原書房, 1978.
214 p. ; 20 cm. Ja 78-18115
¥1300

1. China—Study and teaching. 2. China—Politics and government—1949-1976. 3. Japan—Foreign relations—China. 4. China—Foreign relations—Japan. I. Title.
DS734.95.J66 78-813203

—STUDY AND TEACHING —FRANCE—ADDRESSES, ESSAYS, LECTURES

Fei, Hai-chi.
(Han hsüeh fan pu chi)
漢學反哺集 費海璣編著 ₍臺北₎ 臺灣商務印書館 ₍1966₎
2, 3, 2, 326 p. 18 cm. (人人文庫080-081)

1. China—Intellectual life—Addresses, essays, lectures. 2. China—Study and teaching—France—Addresses, essays, lectures. I. Title.
DS721.F36 C 67-2463

—STUDY AND TEACHING —GREAT BRITAIN

Akademiia nauk SSSR. Institut Dal'nego Vostoka.
(Kitaevedenie Anglii)
Китаеведение Англии / ₍В. Н. Никифоров, Л. Ф. Конькова, Е. П. Синицын ... et al. ; Отв. ред. В. Н. Никифоров₎ ; АН СССР, Ин-т Дальнего Востока. — Москва : Наука, 1977.
189 p. ; 21 cm. — (Очерки зарубежной историографии Китая)
USSR 77
Summary in English.
Bibliography : p. 159–₍181₎
Includes index.
1.40rub
1. China—Study and teaching—Great Britain. I. Nikiforov, V. N. II. Series: Ocherki zarubezhnoĭ istoriografii Kitaia.
DS734.97.G7A37 1977 78-421021

—STUDY AND TEACHING—INDIA—BIBLIOGRAPHY

Kumar, Brij Kishore, 1938-
China through Indian eyes : a select bibliography, 1911-1977 / B. K. Kumar. — Delhi : Concept Pub. Co., 1978.
140 p. ; 22 cm.
Includes indexes.
Rs60.00 ($12.00 U.S.)
1. China—Bibliography. 2. China—Study and teaching—India—Bibliography. I. Title.
Z3106.K84 016.951 78-904380
₍DS706₎ 78 MARC

—STUDY AND TEACHING —JAPAN—ADDRESSES, ESSAYS, LECTURES

Asakawa, Kenji, 1910-1975.
(Kokuun o hiraki seiten o miru)
披黒雲睹青天：浅川謙次追悼遺稿集. — 東京：浅川謙次追悼遺稿集刊行委員会, 1977.
357 p. : ill. ; 22 cm. Ja 78-20117
List of the author's works: p. 346-355.

1. China—History—1949-1976—Addresses, essays, lectures. 2. China—Study and teaching—Japan—Addresses, essays, lectures. I. Title.
DS777.55.A8144 1977 78-814611

CHINA
—STUDY AND TEACHING—JAPAN
—ADDRESSES, ESSAYS, LECTURES (Continued)

Kanda, Kiichirō, 1897-
(Bokurin kanwa)
墨林閒話 / 神田喜一郎. — 東京：岩波書店，1977.
vi, 236 p., [10] leaves of plates : ill. ; 19 cm. Ja 77-26881
¥1500
1. Chinese literature—Japanese authors—History and criticism—Addresses, essays, lectures. 2. China—Study and teaching—Japan—Addresses, essays, lectures. 3. Calligraphy, Japanese—Addresses, essays, lectures. 4. Calligraphy, Chinese—Addresses, essays, lectures. I. Title.
PL3043.K3 79-801651

—STUDY AND TEACHING—UNITED STATES—CONGRESSES

An evaluation of Chinese studies in American universities and colleges, 1958-1975 / edited with an introd. and pref. by Paul K. T. Sih. — New York : Center of Asian Studies, St. John's University, c1978.
xx, 121 p. : ill. ; 22 cm.
Papers presented before the 17th annual conference of the American Association of Teachers of Chinese Language and Culture, Oct. 24 and 25th, 1975, Center of Asian Studies, St. John's University, New York.
Includes bibliographies.
1. China—Study and teaching—United States—Congresses. I. Sih, Paul Kwang Tsien, 1909- II. American Association of Teachers of Chinese Language and Culture.
DS734.97.U6E92 951'.007'1173 78-56779
78 MARC

CHINA. CHÜN CHI CH'U

Fu, Tsung-mao, 1927-
(Ch'ing tai chün chi ch'u tsu chih chi chih chang chih yen chiu)
清代軍機處組織及職掌之研究 傅宗懋著 [臺北 嘉新水泥公司文化基金會 1967]
14, 3, 694 p. 21 cm. (嘉新水泥公司文化基金會研究論文第69種)(嘉新水泥公司文化基金會叢書)
Bibliography: p. 522-527.
1. China. Chün chi ch'u. 2. China—Politics and government—1644-1911. I. Title.
JQ1512.Z4F8 72-836696

CHINA. K'UNG CHÜN. AMERICAN VOLUNTEER GROUP

Nalty, Bernard C
Tigers over Asia / Bernard C. Nalty. — 1st ed. — New York : Elsevier-Dutton, c1978.
vii, 182 p. : ill. ; 24 cm. — (Men and battle)
"A Talisman/Parrish book."
Bibliography: p. 175-176.
Includes index.
ISBN 0-525-93007-8
1. World War, 1939-1945—Aerial operations, American. 2. China. K'ung chün. American Volunteer Group. 3. United States. Army Air Forces. 14th Air Force. 4. Chennault, Claire Lee, 1890-1958. I. Title. II. Series.
D790.N33 1978 940.54'49'73 78-56711
78 MARC

—HISTORY—PICTORIAL WORKS

Rosholt, Malcolm Leviatt, 1907-
Days of the ching pao : a photographic record of the Flying Tigers-14th Air Force in China in World War II / by Malcolm Rosholt. — [s.l. : s.n.], c1978 (Amherst, Wis. : Palmer Publications)
192 p. : chiefly ill. ; 32 cm.
1. World War, 1939-1945—Aerial operations, American—Pictorial works. 2. China. K'ung chün. American Volunteer Group—History—Pictorial works. 3. United States. Air Force. 14th Air Force—History—Pictorial works. I. Title.
D790.R67 940.54'49'73 78-52589
79 MARC

CHINA. LU CHÜN. CH'ANG SHENG CHÜN

Smith, Richard Joseph, 1944-
Mercenaries and Mandarins : the Ever-Victorious Army in nineteenth century China / Richard J. Smith ; foreword by John K. Fairbank. — Millwood, N.Y. : KTO Press, c1978.
xxiii, 271 p. : ill. ; 24 cm. — (KTO studies in American history)
Based on the author's thesis, University of California at Davis.
Bibliography: p. [245]-253.
Includes index.
ISBN 0-527-83950-7
1. China. Lu chün. Ch'ang sheng chün. 2. Taiping Rebellion, 1850-1864. I. Title. II. Series.
UA838.C4S54 951'.03 78-6394
78 MARC

CHINA. LU CHUN. HSIN CHIEN LU CHUN

Liu, Feng-han, 1930-
(Hsin chien lu chün)
新建陸軍 / 劉鳳翰著；[編印者中央研究院近代史研究所]. — 臺北縣南港鎮：該所，1967.
5, 15, 434 p. : ill. ; 21 cm. — (His 新軍志；第1篇)(中央研究院近代史研究所專刊；20)
Title on p. [4] of cover: The New-Founded Army.
Includes index.
Bibliography: p. 414-420.
1. China. Lu chün. Hsin chien lu chün. 2. Yüan, Shih-k'ai, 1859-1916. I. Chung yang yen chiu yüan. Chin tai shih yen chiu so. II. Title. III. Title: The New-Founded Army. IV. Series: Chung yang yen chiu yüan. Chin tai shih yen chiu so. Chung yang yen chiu yüan chin tai shih yen chiu so chuan k'an ; 20.
UA837.L52 vol. 1 78-840926

CHINA. LU CHUN. TI 8 LU CHÜN

Ishinryō.
(Man-Shi ni okeru kyōsan higun no katsudō)
滿支に於ける共產匪軍の活動. — 東京：維新寮，昭和13 [1938]
64 p. (on double leaves) ; 24 cm.
Cover title.
1. China. Lu chün. Ti 8 lu chün. 2. Manchuria—Politics and government. 3. Communism—China—Manchuria—History. I. Title.
UA837.I78 1938 79-804700

CHINA. LU CHÜN. TI 8 LU CHÜN. 120 SHIH. 7 T'UAN. 6 LIEN

Lu, Yen-shih.
(Hsüeh jan chan ch'i tai tai hung)
血染战旗代代红：硬骨头六连的故事 / 陆岩石. — [香港?：s. n., 1978?]
2, 2, 158 p., [2] leaves of plates : ill. ; 21 cm.
Reprint of the 1977 ed. published by Shang-hai jen min ch'u pan she, Shanghai.
1. China. Lu chün. Ti 8 lu chün. 120 shih. 7 t'uan. 6 lien—Fiction. I. Title.
PL2880.U22H78 1978 78-840925

CHINA (PEOPLE'S REPUBLIC OF CHINA, 1949-)

Chou, Ching-wen, 1909-
Ten years of storm; the true story of the Communist regime in China, by Chow Ching-wen. Foreword by Lin Yutang. Translated and edited by Lai Ming. [1st ed.] New York, Holt, Rinehart and Winston [1960]
xxii, 323 p. 22 cm.
Translation of Feng pao shih nien.
1. China (People's Republic of China, 1949-). I. Title.
DS777.55.C4573 1960 951.05 60-12858
[r78]rev MARC

CHINA (PEOPLE'S REPUBLIC OF CHINA. 1949-). CH'ÜAN KUO JEN MIN TAI PIAO TA HUI

Ch'iu, K'ung-yüan.
Mainland situation after the Fourth National People's Congress / [by Chiu Kung-yuan]. — Taipei : Chung Hwa Information Service, [1978?]
29 p. ; 20 cm.
On cover: 3.
Includes bibliographical references.
1. China—Politics and government—1949-1976. 2. China (People's Republic of China. 1949-). Ch'üan kuo jen min tai piao ta hui. I. Title.
DS777.55.C4528 320.9'51'05 78-103830
78 MARC

CHINA (PEOPLE'S REPUBLIC OF CHINA, 1949-). CHUNG-KUO JEN MIN CHIH FANG CHUN

Chou, Ching-wen, 1909-
(Mao-Tse-tung ti chün tui)
毛澤東的軍隊 周鯨文著 香港 時代批評社 1964.
6, 4, 252 p. illus. 21 cm.
Includes bibliographical references.
1. China (People's Republic of China, 1949-). Chung-kuo jen min chieh fang chün. I. Title.
UA835.C5 C 66-2014

Furukawa, Mantarō, 1933-
(Mō Takutō no guntai)
毛沢東の軍隊：中国人民解放軍の50年 / 古川万太郎著. — 東京：講談社，1978.
355 p. : ill. ; 20 cm. Ja 78-26786
Includes bibliographical references.
¥1200
1. China (People's Republic of China, 1949-) Chung-kuo jen min chieh fang chün. 2. Chung kuo jen min chieh fang chün. I. Title.
UA837.F85 79-803396

Kawashima, Kōzō.
(Chūgoku Jimmin Kaihōgun)
中国人民解放軍 / 川島弘三. — [東村山]：教育社；東京：販売 教育社出版サービス，[1978]
209 p., [1] leaf of plates : ill. ; 18 cm. — (時事問題解説；no. 75)(入門新書) Ja 78-23757
Cover title.
Series 2 romanized: Nyūmon shinsho.
Includes bibliographical references.
¥400
1. China (People's Republic of China, 1949-). Chung-kuo jen min chieh fang chün. I. Title. II. Series: Jiji mondai kaisetsu ; 75.
UA837.K39 79-800063

(Lin Piao shih liao chi)
林彪史料集. — [香港?：s. n., 1972]
2 v. : ill. ; 22 x 31-29 cm.
Cover title.
Publication date stamped on p. [4] of cover.
A collection of articles reprinted from various sources.
1. Lin, Piao, 1908-1971—Addresses, essays, lectures. 2. Generals—China—Biography—Addresses, essays, lectures. 3. China (People's Republic of China, 1949-). Chung-kuo jen min chieh fang chün—Biography—Addresses, essays, lectures.
DS778.L4725L523 78-842242

(Wei ta ti chün tui, kuang hui ti li ch'eng)
伟大的军队，光辉的历程：庆祝中国人民解放军建军五十周年. — [香港?：s. n., 1978?]
3, 219 p. ; 21 cm.
Reprint of the 1977 ed. published by Chiang-su jen min ch'u pan she.
1. China (People's Republic of China, 1949-). Chung-kuo jen min chieh fang chün. I. Title: Ch'ing chu Chung-kuo jen min chieh fang chün chien chün wu shih chou nien.
UA835.W4 1978 79-836376
79 C

—ADDRESSES, ESSAYS, LECTURES

(Kuang hui ti wu shih nien)
光輝的五十年：庆祝中国人民解放军五十周年. — [香港?：s. n., 1978?]
2, 166 p., [4] leaves of plates : ill., ports. ; 21 cm.
Reprint of the 1977 ed. published by Jen min ch'u pan she, Peking.
1. China (People's Republic of China, 1949-). Chung-kuo jen min chieh fang chün—Addresses, essays, lectures.
UA837.K78 1978 79-835088
79 C

—BIOGRAPHY

Chieh fang chün hua pao she, Peking.
(Wei ta ti kung ch'an chu i chan shih)
伟大的共产主义战士：雷锋 / 解放军画报社编. — 第1版. — 上海：上海人民出版社，1977.
40 p., [3] leaves of plates : ill. ; 26 cm.
$1.35
1. Lei, Feng, 1939-1962. 2. Soldiers—China—Biography. 3. China (People's Republic of China, 1949-). Chung-kuo jen min chieh fang chün—Biography. I. Title.
DS778.L3C53 1977 78-842141

CHINA (PEOPLE'S REPUBLIC, 1949-). CHUNG-KUO JEN MIN CHIEH FANG CHUN

—BIOGRAPHY (Continued)

(Jen min ti hung ssu ling)
人民的红司令：深切怀念敬爱的朱德同志. — [香港? : s. n., 1978]
3, 348 p., [2] leaves of plates : ports. ; 22 cm.
Reprint of the 1978 ed. published by Shang-hai jen min ch'u pan she, Shanghai.

1. Chu, Te, 1886-1976—Addresses, essays, lectures. 2. Generals—China—Biography—Addresses, essays, lectures. 3. China (People's Republic, 1949-). Chung-kuo jen min chien fang chün—Biography.

DS778.C6J45 1978 — 79-840331

Lin, Tan.
(Lei Feng ti ku shih)
雷鋒的故事 / 改编林丹；原著陈广生，崔家骏；绘画毛震耀. — 第1版. — 上海：上海人民出版社，1977.
124 p. : ill. ; 10 x 13 cm.
"連環圖畫"
$0.13

1. Lei, Feng, 1939-1962. 2. Soldiers—China—Biography. 3. China (People's Republic of China, 1949-). Chung-kuo jen min chieh fang chün—Biography. I. Ch'en, Kuang-sheng. II. Ts'ui, Chia-chün. III. Title.

DS778.L3L56 — 78-841092

(P'eng Te-hui)
彭德懷. — [香港?]：勁草書房，[1978?]
3, 200 p. : ill. ; 19 cm.
On cover: P'eng Te-huai tzu liao chi.
Reprint of 1969 ed. published under the compiler: Chung-kuo wen t'i yen chiu chung hsin by Tzu lien ch'u pan she, Kowloon.
$9.00

1. P'eng, Te-hui, 1900- 2. Generals—China—Biography. 3. China (People's Republic of China, 1949-). Chung-kuo jen min chieh fang chün—Biography. I. Chung-kuo wen t'i yen chiu chung hsin. II. Title: P'eng Te-huai tzu liao chi.

DS778.P6P4 1978 — 79-838731

Yang, T'ung-fang, 1905-
(Chi Ho Lung)
记贺龙 / 沙汀 [楊同芳]. — 第2版. — 北京：中国青年出版社，1978.
131 p. ; 19 cm.
$0.28

1. Ho, Lung, 1896-1969. 2. China (People's Republic of China, 1949-). Chung-kuo jen min chieh fang chün—Biography. 3. Generals—China—Biography. I. Title.

DS778.H6Y3 1978 — 79-836133

C

(Yung ts'un ti yeh chi, wei ta ti i sheng)
永存的业绩，伟大的一生：缅怀我军几位杰出的将领 / [广西人民出版社编辑]. — [香港? : s. n., 1978?]
2, 113 p. ; 21 cm.
Reprint of the 1977 ed. published by Kuang-hsi jen min ch'u pan she.

1. Generals—China—Biography. 2. China (People's Republic of China, 1949-). Chung-kuo jen min chieh fang chün—Biography. I. Kuang-hsi jen min ch'u pan she.

DS778.A1Y86 1978 — 79-835437

—BIOGRAPHY—ADDRESSES. ESSAYS, LECTURES

(Ho Lung)
賀龍 / 沙汀，王震等著. — 香港：廣角鏡出版社，1977.
202 p., [6] leaves of plates : ill. ; 19 cm. — (廣角鏡人物叢書)
Series romanized: Kuang chiao ching jen wu ts'ung shu.
$5.50

1. Ho, Lung, 1896-1969—Addresses, essays, lectures. 2. Generals—China—Biography—Addresses, essays, lectures. 3. China (People's Republic of China, 1949-). Chung-kuo jen min chieh fang chün—Biography—Addresses, essays, lectures. I. Yang, T'ung-fang, 1905- II. Wang, Chen, 1908 or 9-

DS778.H6H6 — 78-841178

(Hsü Hsiang-ch'ien tzu liao chi)
徐向前資料集. — [修訂]. — [香港? : s. n., 1978]
27 leaves ; 22 x 34 cm.
Cover title.
A collection of articles and newspaper clippings reprinted from various sources.

1. Hsü, Hsiang-ch'ien, 1902- —Addresses, essays, lectures. 2. Generals—China—Biography—Addresses, essays, lectures. 3. China (People's Republic of China, 1949-). Chung-kuo jen min chieh fang chün—Biography—Addresses, essays, lectures.

DS778.H8154H7 — 78-841668

—BIOGRAPHY —JUVENILE LITERATURE

Ch'i, Jung, fl. 1977-
(Li Wen-chung)
李文忠 / 齐榕. — 第1版. — 上海：上海人民出版社，1977.
159 p. : ill. ; 19 cm. — (青年英雄故事丛书) (少年儿童文艺读物)
Series 1 romanized: Ch'ing nien ying hsiung ku shih ts'ung shu. Series 2 romanized: Shao nien erh t'ung wen i tu wu.
$0.29

1. Li, Wen-chung, d. 1967—Juvenile literature. 2. Soldiers—China—Biography—Juvenile literature. 3. China (People's Republic of China, 1949-). Chung-kuo jen min chieh fang chün—Biography—Juvenile literature. I. Title.

DS778.L4193C47 — 78-842907

Shen, Chi-feng.
(Lei Feng ching shen tai tai ch'uan)
雷锋精神代代传：雷锋班的故事 / 沈继锋；[插图陈剑英，颜志强]. — 第1版. — 上海：少年儿童出版社，1978.
3, 2, 127 p. : ill. ; 19 cm.
Juvenile literature.
$0.27

1. Lei, Feng, 1939-1962—Juvenile literature. 2. Soldiers—China—Biography—Juvenile literature. 3. China (People's Republic of China, 1949-). Chung-kuo jen min chieh fang chün—Biography—Juvenile literature. I. Title.

DS778.L3S53 — 78-842811

Tu, Wen-yung.
(Chu tsung ssu ling ti ku shih)
朱总司令的故事 / 杜文勇編；[绘者林岗等]. — 第1版. — 北京：人民美術出版社，1977.
[56] p. : ill. ; 16 cm.
"連環圖畫"
$0.18
1. Chu, Te, 1886-1976—Juvenile literature. 2. Generals—China—Biography—Juvenile literature. 3. China (People's Republic of China, 1949-). Chung-kuo jen min chieh fang chün—Biography—Juvenile literature. I. Lin, Kang. II. Title.

DS778.C6T82 — 78-842906

78 C

—CIVIC WORK—ADDRESSES, ESSAYS, LECTURES

(Feng huo sui yüeh)
烽火岁月 / 彭寿生，舒兴等口述. — [香港? : s. n., 1977?]
157 p., [6] leaves of plates : ill. ; 21 cm.
Reprint of the 1976 ed. published by Chiang-hsi jen min ch'u pan she, Nan-ch'ang.

1. China (People's Republic of China, 1949-). Chung-kuo jen min chieh fang chün—Fiction. 2. Chinese fiction—20th century. I. P'eng, Shou-sheng. II. Shu, Hsing.

PL2653.F437 1977 — 79-835691

79 C

(Ko ming jung lu)
革命熔炉. — [香港? : s. n., 1979?]
116 p. ; 21 cm.
Colophon title.
Reprint of the 1971 ed. published by Kuang-tung jen min ch'u pan she, Canton.

1. China (People's Republic of China, 1949-). Chung-kuo jen min chieh fang chün—Civic work—Addresses, essays, lectures.

UA837.K6 1979 — 79-840634

—HISTORY

(Chung-kuo hung chün)
中國紅軍. — 香港：海星出版印刷中心，[1979?]
34 p. : ill. ; 26 cm.
Cover title.
$2.00

1. China (People's Republic of China, 1949-). Chung-kuo jen min chieh fang chün—History.

UA837.C498 1979 — 79-840009

(Hsüeh hsi Mao chu hsi chien chün lu hsien)
学习毛主席建军路线. — 第1版. — 北京：人民出版社，1977.
3, 120 p. ; 19 cm.
$0.23

1. China (People's Republic of China, 1949-). Chung-kuo jen-min chieh fang chün—History.

UA837.H77 — 78-843869

—HISTORY—ADDRESSES, ESSAYS, LECTURES

(Kuang hui ti chieh jih, shen ch'ieh ti huai nien)
光辉的节日，深切的怀念：纪念中国人民解放军建军五十周年文集. — [香港：s. n., 1977?]
3, 216 p. ; 21 cm.
Reprint of the 1977 ed. published by Kuang-tung jen min ch'u pan she, Canton.

1. China (People's Republic of China, 1949-) Chung-kuo jen min chieh fang chün—History—Addresses, essays, lectures.

UA837.K77 — 78-842507

—HISTORY—PICTORIAL WORKS

(Pa i Nan-ch'ang ch'i i)
八一南昌起义 /《八一南昌起义》画册编辑组编. — 第1版. — [南昌]：江西人民出版社，1977.
77 p. : chiefly col. ill. ; 27 cm.
"纪念中国人民解放军建军五十周年, 1927-1977"

1. Nan-ch'ang, China (City)—History—Pictorial works. 2. Communism—China—History—Pictorial works. 3, China (People's Republic of China, 1949-). Chung-kuo jen min chieh fang chün—History—Pictorial works. I. Pa i Nan-ch'ang ch'i i hua ts'e pien chi tsu.

DS796.N15P29 — 78-841003

—POLITICAL ACTIVITY

China (People's Republic of China, 1949-). Chung-kuo jên min chieh fang chün. Tsung chêng chih pu.
工作通訊 机密發到团 第1- 期 1961-
[北京]
no. in v. 29 cm. irregular.
Supersedes 八一杂志
Photocopy.
1. China (People's Republic of China, 1949-). Chung-kuo jên min chieh fang chün—Political activity. 2. Chung-kuo kung ch'an tang—Party work. I. Title.
Title romanized: Kung tso t'ung hsün.

UA839.3.A3 — C 64-668

CHINA (PEOPLE'S REPUBLIC OF CHINA, 1949-). CHUNG-KUO JEN MIN CHIEH FANG CHUN. HAI CHÜN

T'ai-wan tsao ch'uan ku fen yu hsien kung ssu.
(Kung fei hai chün tsao ch'uan neng li yü chien t'ing chien li chih yen chiu)
共匪海軍造船能力與艦艇建立之研究 / 臺灣造船股份有限公司編. — [基隆]：該公司，民國61 [1972]
1, 50 p. ; 21 cm. — (專題研究)
Cover title.
Series romanized: Chuan t'i yen chiu.

1. China (People's Republic of China, 1949-). Chung-kuo jen min chieh fang chün. Hai Chün. 2. Warships. I. Title.

VA633.T34 1972 — 78-843874

CHINA (PEOPLE'S REPUBLIC OF CHINA, 1949-). CHUNG-KUO JEN MIN CHIEH FANG CHÜN. K'UNG CHÜN

—SURGEONS—BIOGRAPHY—ADDRESSES, ESSAYS, LECTURES

(Chih hsing Mao chu hsi ko ming lu hsien feng ch'ui lang ta pu hui t'ou)

执行毛主席革命路线风吹浪打不回头：驻上海空军某部军医韩玉芬的先进事迹 ［上海人民出版社编辑］. — ［香港?：s. n., 1978?］

2, 65 p. : 20 cm.

Reprint of the 1971 ed. published by Shang-hai jen min ch'u pan she, Shanghai.

1. Han, Yü-fen—Addresses, essays, lectures. 2. China (People's Republic of China, 1949-). Chung-kuo jen min chieh fang chün. K'ung chün—Surgeons—Biography—Addresses, essays, lectures. 3. Surgeons—China—Biography—Addresses, essays, lectures. I. Shang-hai jen min ch'u pan she.

UH347.H36C48 1978 79-835865

CHINA (REPUBLIC OF CHINA). CH'IAO WU WEI YÜAN HUI

—ADDRESSES, ESSAYS, LECTURES

Ch'iao wu erh shih wu nien pien chi wei yüan hui.

(Ch'iao wu erh shih wu nien)

僑務二十五年 / 僑務二十五年編輯委員會編印. — 台北：海外出版社, 民國46 ［1957］

6, 5, 261 p., ［9］ leaves of plates (1 fold.) : ill. ; 24 cm.

1. China (Republic of China). Ch'iao wu wei yüan hui—Addresses, essays, lectures. 2. Chinese in foreign countries—Addresses, essays, lectures. I. Title.

JQ1532.Z314C47 1957 79-840679

CHINA (REPUBLIC OF CHINA). CHIAO YÜ PU. TI YU SSŬ

China (Republic of China). Chiao yü pu. T'i yü ssu.

(T'i yü ssu ch'eng li wu chou nien kung tso pao kao)

體育司成立五週年工作報告 / 教育部體育司編印. — ［台北］：該司, 民國67 ［1978］

2, 302 p. ; 27 cm.

Cover title.

1. China (Republic of China). Chiao yü pu. Ti yü ssu. I. Title.

GV303.T28C46 1978 79-838285

CHINA (REPUBLIC OF CHINA). CHIEN CH'A YÜAN

(Chung-hua min kuo chien ch'a yüan chih yen chiu)

中華民國監察院之研究 / 傅啓學 ... ［et al.］. — Washington : Center for Chinese Research Materials, Association of Research Libraries, ［1978］

3, 23, 1080 p. on ［277］ leaves ; 22 x 29 cm.

Reprint of the 1967 ed. issued by the authors, with added introduction in English.
Each leaf represents four pages of the original.
Includes bibliographical references.
$48.00

1. China (Republic of China). Chien ch'a yüan. I. Fu, Ch'i-hsüeh, 1902-

JQ1530.5.C49 1978 79-840087

Lei, Chen, 1897-

(Chien ch'a yüan chih chiang lai)

監察院之將來 雷震著 台北 自由中國出版社 民國42 ［1953］

5, 6, 173 p. 19 cm. (自由中國社叢書之 23)

Includes bibliographical references.

1. China (Republic of China). Chien ch'a yüan. 2. China—Constitutional law. I. Title.

JQ1530.5.L44 C 66-1055

CHINA (REPUBLIC OF CHINA). HSING CHENG YÜAN. CH'ING NING FU TAO WEI YÜAN HUI

China (Republic of China). Hsing cheng yüan. Ch'ing nien fu tao wei yüan hui.

(Ch'ing nien fu tao kung tso chien chieh)

青年輔導工作簡介 = Briefing of youth guidance service / 行政院青年輔導委員會編印. — ［台北］：該會, 1977.

［44］ p. : ill. ; 18 x 20 cm.

Cover title.
English and Chinese.

1. China (Republic of China). Hsing cheng yüan. Ch'ing ning fu tao wei yüan hui. I. Title. II. Title: Briefing of youth guidance service.

HF5382.5.T3C48 1977 331.3 79-837566

CHINA (REPUBLIC OF CHINA) HSING CHENG YÜAN. KUO CHÜN T'UI CHUI KUAN PING FUTAO WEI YÜAN HOIC

—ADDRESSES, ESSAYS, LECTURES

(Fu tao kung tso erh shih nien chi nien chi)

輔導工作二十年紀念集 / ［編輯者輔導工作二十年紀念集編輯小組；行政院國軍退除役官兵輔導委員會編印］. — 台北：該會, 民國63 ［1974］

2 v. (914 p.) : ill. ; 22 cm.

Colophon title: Kuo chün t'ui ch'u i kuan ping fu tao kung tso erh shih nien chi nien chi.

1. Veterans—Taiwan—Addresses, essays, lectures. 2. China (Republic of China). Hsing cheng yüan. Kuo chün t'ui ch'u i kuan ping fu tao wei yüan hui—Addresses, essays, lectures. I. Fu tao kung erh shih nien chi nien chi pien chi hsiao tsu. II. China (Republic of China). Hsing cheng yüan. Kuo chün t'ui ch'u i kuan ping fu tao wei yüan hui. III. Title: Kuo chün t'ui ch'u i kuan ping fu tao kung tso erh shih nien chi nien chi.

UB359.T28F8 79-835151

CHINA (REPUBLIC OF CHINA). K'AO SHIH YÜAN

China (Republic of China). K'ao shih yüan. Mi shu ch'u.

(K'ao shih yüan kai k'uang)

考試院概況 / 考試院秘書處編印. — ［台北］：該處, 民國66 ［1977］

2, 43 p., ［2］ leaves of plates : ill. ; ports. ; 21 cm.

1. China (Republic of China). K'ao shih yüan. I. Title.

JQ1532.Z32C47 1977 78-843389

CHINA (REPUBLIC OF CHINA). LU CHÜN

—ANECDOTES, FACETIAE, SATIRE, ETC.

Chung, Yao.

(Yang Sen chiang chün ti ch'uan ch'i ku shih)

楊森將軍的傳奇故事：紀念惠公逝世週年 / 著者鍾堯. — 台北：鍾堯, 民國67 ［1978］

101 p. : ill. ; 20 cm.

Cover title.
NT$50.00

1. Yang, Sen, 1883-1977—Anecdotes. 2. Generals—China—Anecdotes, facetiae, satire, etc. 3. China (Republic of China). Lu chün—Anecdotes, facetiae, satire, etc. I. Title.

DS778.Y28C48 79-839565

—BIOGRAPHY

［Hu Tsung-nan shang chiang nien p'u pien tsuan wei yüan hui］

［Hu Tsung-nan shang chiang nien p'u］

胡上將宗南年譜 / ［胡宗南上將年譜］編纂委員會編. — ［台北縣永和鎮］：文海出版社, ［1978?］

2, 2, 2, 8, 310 p., ［13］ leaves of plates : ill. ; 22 cm. — (近代中國史料叢刊續編 ［i. e. 輯］ ; 488)

1. Hu, Tsung-nan, 1896-1962—Chronology. 2. Generals—China—Biography. 3. Generals—Taiwan—Biography. 4. China (Republic of China). Lu chün—Biography. I. Title. II. Series: Shen, Yün-lung, comp. Chin tai Chung-kuo shih liao ts'ung k'an hsü chi ; 488.
Title romanized: Hu shang chiang Tsung-nan nien p'u.

DS755.S532 vol. 488 78-840945
［DS778.H843］

C

Wan, Yao-huang, 1890-

(Wan Yao-huang chiang chün jih chi)

萬耀煌將軍日記 / 著作人萬耀煌. — 台北：湖北文獻社, 民國67 ［1978］

2 v. : ill. ; 22 cm. — (湖北文獻社叢書之一)

Colophon title.
Series romanized: Hu-pei wen hsien she ts'ung shu.
NT$120.00 per vol.

1. Wan, Yao-huang, 1890- 2. Generals—China—Biography. 3. China (Republic of China). Lu chün—Biography. I. Title.

DS778.W28A38 79-836099

CHINA (REPUBLIC OF CHINA). TS'AI CHENG PU. T'U SHU SHIH

China (Republic of China). Ts'ai cheng pu. T'u shu shih.

(Ts'ai cheng pu t'u shu shih t'u shu ch'i k'an mu lu)

財政部圖書室圖書期刊目錄 / 財政部圖書室編印. — ［台北］：該室, 民國 ［19

v. ; 26 cm.

1. China (Republic of China). Ts'ai cheng pu. T'u shu shih—Catalogs. I. Title.

Z955.T3332C48 1976 78-843006

CHINA (REPUBLIC OF CHINA). YU CHENG TSUNG CHÜ

—ADDRESSES, ESSAYS, LECTURES

(Yao ch'uan "yu chü chi t'i wu pi" an chen hsiang)

謠傳「郵局集體舞弊」案真相 / 今日郵政月刊社編印. — ［台北］：該社, 民國50 ［1961］

47 p. ; 21 cm.

1. China (Republic of China). Yu cheng tsung chü—Addresses, essays, lectures. I. Chin jih yu cheng yüeh k'an she.

HE7299.Y36 79-840307

—DESCRIPTION AND TRAVEL

Benningsen, A. P., Count.

(Gendai Mōko)

現代蒙古 / 農商務省商工局譯 ; ［ア・ペ・ベンニングセン伯著］. — 東京：時事彙存社, 大正3 ［1914］

2, 2, 2, 217 p. ; 19 cm.

1. China, Northwest—Description and travel. 2. Mongolia (Inner Mongolia)—Description and travel. 3. Benningsen, A. P., Count. I. Title.

DS793.N6B46 79-804133

CHINA (PORCELAIN) see Porcelain

CHINA-CLAY see Kaolin

CHINA FAIRINGS, VICTORIAN

—CATALOGS

Victorian fairings and their values / ［edited by Margaret Anderson］. — 2d ed. — Worthing ［Eng.］ : Apollo Press, 1978.

126 p. : chiefly ill. ; 23 cm. GB***
Includes index.
ISBN 0-902921-81-9 : £2.80 ($5.95 U.S.)
1. China fairings, Victorian—Catalogs. I. Anderson, Margaret, fl. 1975-

NK4660.V52 1978 738.8 78-323394
79 MARC

CHINA IN ART

Hirayama, Ikuo, 1930–
(Taiga)
大河：第一中国画集 / 平山郁夫. — 東京：講談社, 1978.
107 p. : chiefly col. ill. ; 30 x 31 cm. Ja 78-25488
¥4800

1. Hirayama, Ikuo, 1930– 2. China in art. 3. Tibet in art. 4. Silk Road in art. I. Title.
ND1059.H56A4 1978b 79-801749

CHINA IN LITERATURE

Bernard, Henri, S.J.
La Chine en France durant la renaissance, par Henri Bernard ... Shanghai, Imprimerie de T'ou-sè-wè, 1940.
cover-title, 14 p. 23 cm.
"Extrait du Bulletin de l'Université l'Aurore déc. 1940."
"Liste des impressions françaises sur la Chine avant 1610": p. 13-14.
1. China—Relations (general) with France. 2. France—Relations (general) with China. 3. China in literature. I. Title.
DS740.5.F8B4 301.29'51'044 ₍r79₎rev 47-37465 MARC

Chen, Cheng-siang.
(Shih ti ti li)
詩的地理 / 陳正祥著. — 香港：商務印書館, 1978.
2, 89 p., ₍3₎ fold. leaves of plates : maps ; 19 cm.
Includes bibliographical references.
$5.00

1. Chinese poetry—History and criticism. 2. China in literature. 3. Poetry of places—China—History and criticism. I. Title.
PL2307.C454 78-841332

CHINA INLAND MISSION

—HISTORY

Broomhall, Marshall, 1866-
The jubilee story of the China Inland Mission / by Marshall Broomhall. — San Francisco : Chinese Materials Center, 1977.
xvi, 386 p., ₍16₎ leaves of plates (1 fold.) : ill. ; 22 cm. — (Reprint series - Chinese Materials Center ; no. 76)
Reprint of the 1915 ed. published by Morgan & Scott, London.
Includes bibliographical references and index.
1. China Inland Mission—History. 2. Missions—China—History. I. Title. II. Series: Chinese Materials Center. Reprint series - Chinese Materials Center ; no. 76.
BV3410.C56B67 1977 266'.00951 79 79-100478 MARC

CHINA-JAPAN WAR, 1894-1895 see Chinese-Japanese War, 1894-1895

CHINA LANGUAGE (AFRICA) see Shona language

CHINA PAINTING

—EXHIBITIONS

Chicago, Judy, 1939-
The dinner party : a symbol of our heritage / written and illustrated by Judy Chicago ; designed by Sheila Levrant de Bretteville. — 1st ed. — Garden City, N.Y. : Anchor Press/Doubleday, 1979.
255 p., ₍20₎ leaves of plates : ill. (some col.) ; 29 cm.
Includes index.
ISBN 0-385-14566-7 : $24.95. ISBN 0-385-14567-5 pbk. : 12.95
1. China painting—Exhibitions. 2. Needlework—Exhibitions. 3. Women in art—Exhibitions. I. Title.
NK4605.C45 738 78 78-69653 MARC

—TECHNIQUE

Reboulleau, M E F
Nouveau manuel complet de la peinture sur verre, sur porcelaine et sur émail, des émaillages industriels et de la fabrication des émaux et des couleurs vitrifiables ... / par M. Reboulleau, ... — Paris : L. Laget, 1977.
419 p., ₍1₎ fold. leaf of plates ; 18 cm. F78-6019
Reprint of the 1866 ed. published by Roret, Paris, in series: Manuels Roret.
"Nouvelle édition entièrement refondue et considérablement augmentée par M. M.-D. Magnier."
On spine: M.-D. Magnier, Peinture sur verre.
Five hundred copies printed.
ISBN 2-85204-025-5 : 135F
1. Glass painting and staining—Technique. 2. China painting—Technique. 3. Enamel and enameling—Technique. I. Magnier, M. Désiré. II. Title. III. Title: Peinture sur verre.
NK5304.R4 1977 738 79 79-350488 MARC

—ASIA—EXHIBITIONS

Tenri Gyararī.
(Saitō)
彩陶. — 東京：天理ギャラリー, 1977.
₍38₎ p. : ill. ; 26 cm. — (天理ギャラリー；第47回展) Ja***
Cover title.
On p. ₍4₎ of cover: Painted pottery, in the collection of Tenri Sankōkan Museum.
Catalog of an exhibition held at Tenri Gallery, July 15–Oct. 22, 1977.

1. China painting—Asia—Exhibitions. 2. Pottery, Prehistoric—Asia—Exhibitions. 3. Pottery, Oriental—Exhibitions. I. Title. II. Title: Painted pottery. III. Series: Tenri Gyarari. Tenri Gyarari ; dai 47-kaiten.
NK4605.5.A78T46 1977 78-814127

—AUSTRIA—VIENNA

Kothgasser, Anton, 1769-1851.
Die Einschreibebüchlein des Wiener Glas- und Porzellanmalers Anton Kothgasser : (1769-1851) / Rudolf von Strasser. Anh., "Die Geschichte der geheimen Aufzeichnungen des Glasmalers Gottlob S. Mohn, Laxenburg" / ₍K. H. Heine₎. — Karlsruhe : Corona-Verlag Heine, ₍1977?₎
264 p. : ill. ; 22 cm. — (₍Schätze der Glaskunst ; Bd. 4₎) GFR78-A
Includes bibliographical references.
ISBN (invalid) 3-921711-03-X : DM48.00
1. Kothgasser, 1769-1851. 2. Mohn, Gottlob Samuel, 1789-1825. 3. China painting—Austria—Vienna. 4. Glass painting and staining—Austria—Vienna. I. Strasser, Rudolf von. II. Heine, Karl-Heinrich. Die Geschichte der geheimen Aufzeichnungen des Glasmalers Gottlob S. Mohn, Laxenburg. ₍1977?₎ III. Title. IV. Series.
NK5398.K67A4 1977 79 79-383760 MARC

—FLORIDA

Florida State Federation of China Painters.
Florida flora and fauna in fine art : Florida Federation of China Painters tenth anniversary convention, April 13, 14, 15, 1978 / editor, Evelyn Spillan. — St. Petersburg, Fla. : Valkyrie Press, c1978.
97 p. : ill. ; 28 cm.
Includes index.
1. China painting—Florida. 2. Florida in art. I. Spillan, Evelyn. II. Title.
NK4605.5.U62F63 738.1'5 78₍r78₎rev 78-55554 MARC

—JAPAN

Kitaōji, Rosanjin, 1883–1959.
(Rosanjin sobyō shū)
魯山人素描集 / 北大路魯山人；秦秀雄編. — 東京：五月書房, 昭和52 ₍1977₎
₍72₎ p. : chiefly ill. (some col.) ; 26 cm. Ja 77-26213
¥3600

1. Kitaōji, Rosanjin, 1883–1959. 2. China painting—Japan. 3. Artists' preparatory studies—Japan. I. Hata, Hideo. II. Title.
ND1059.K533A4 1977 78-809529

—SPAIN—ARAGON

Alvaro Zamora, María Isabel.
Cerámica aragonesa decorada : desde la expulsión de los moriscos a la extinción de los alfares (siglos XVII-comienzos del XX) / ₍por₎ Ma. Isabel Alvaro Zamora. — Zaragoza : Libros Pórtico, 1978.
254 p. : ill. ; 23 cm. — (Estudios ; 2) Sp78-Dec
Bibliography: p. 247-249.
ISBN 8485264282
1. Pottery, Spanish—Spain—Aragon. 2. China painting—Spain—Aragon. I. Title.
NK4124.A7A483 79 79-380150 MARC

CHINA TRADE ART

—CATALOGS

Howard, David Sanctuary.
China for the West : Chinese porcelain & other decorative arts for export illustrated from the Mottahedeh Collection / David Howard and John Ayers ; foreword by Nelson A. Rockefeller. — London ; New York : Sotheby Parke Bernet, 1978.
2 v. (698 p.) : ill. (some col.) ; 34 cm. GB***
Bibliography: p. 673-678.
Includes index.
ISBN 0-85667-035-9
1. China trade porcelain—Catalogs. 2. China trade art—Catalogs. 3. Mottahedeh, Rafi—Art collections—Catalogs. 4. Mottahedeh, Mildred—Art collections—Catalogs. I. Ayers, John, joint author. II. Title.
NK4565.5.H67 738.2'0951 78 78-319971 MARC

CHINA TRADE PORCELAIN

—CATALOGS

Howard, David Sanctuary.
China for the West : Chinese porcelain & other decorative arts for export illustrated from the Mottahedeh Collection / David Howard and John Ayers ; foreword by Nelson A. Rockefeller. — London ; New York : Sotheby Parke Bernet, 1978.
2 v. (698 p.) : ill. (some col.) ; 34 cm. GB***
Bibliography: p. 673-678.
Includes index.
ISBN 0-85667-035-9
1. China trade porcelain—Catalogs. 2. China trade art—Catalogs. 3. Mottahedeh, Rafi—Art collections—Catalogs. 4. Mottahedeh, Mildred—Art collections—Catalogs. I. Ayers, John, joint author. II. Title.
NK4565.5.H67 738.2'0951 78 78-319971 MARC

—EXHIBITIONS

Oriental Ceramic Society of Hong Kong.
South-east Asian and Chinese trade pottery : an exhibition catalogue, January 26 to April 2, 1979, Hong Kong Museum of Art / presented by the Oriental Ceramic Society of Hong Kong and the Urban Council, Hong Kong. — ₍Hong Kong₎ : The Society, 1979.
vii, 264 p. (p. 264 blank) : ill. (some col.) ; 27 cm.
English and Chinese.
Title also in Chinese: Tung-nan Ya Tz'u yü Chung-kuo ch'u k'ou tz'u.
Bibliography: p. 156-157.
1. China trade porcelain—Exhibitions. 2. Porcelain, Chinese—Ming dynasty, 1368-1644—Exhibitions. 3. Porcelain, Chinese—Sung-Yüan dynasties, 960-1368—Exhibitions. 4. Porcelain, Southeast Asian—Exhibitions. I. Hongkong. Urban Council. II. Hong Kong Museum of Art. III. Title. IV. Title: Tung-nan Ya tz'u yü Chung-kuo ch'u k'ou tz'u.
NK4565.5.O74 1979 738'.095'074095125 79 79-104670 MARC

—UNITED STATES—EXHIBITIONS

Cairns-Malatt, Linda.
Yang khi, vases of the sea : an exhibition of American China trade porcelain, July 1, 1979—January 1, 1980, Decatur House, National Trust for Historic Preservation, 748 Jackson Place, N.W., Washington, D.C. / by Linda Cairns-Malatt ; photography by Wm. Edmund Barrett. — Washington : Preservation Press, 1979.
29, ₍2₎ p. : ill. ; 22 cm.
Bibliography: p. ₍30₎-₍31₎
1. China trade porcelain—United States—Exhibitions. 2. Porcelain, Chinese—Ming-Ch'ing dynasties, 1368-1912—Exhibitions. I. National Trust for Historic Preservation in the United States. II. Title.
NK4565.5.C34 738.2'0951'0740153 79 79-115995 MARC

CHINA WOOD-OIL TREE see Tung tree

CHINAWARE see Porcelain; Pottery

CHINCOTEAGUE PONY

—JUVENILE LITERATURE

Ford, Barbara.
The island ponies : an environmental study of their life on Assateague / by Barbara Ford & Ronald R. Keiper ; photos. by Ronald R. Keiper. — 1st ed. — New York : Morrow, 1979.
95 p. : ill. ; 25 cm.
Includes index.
SUMMARY: Describes the scientific study being undertaken on the wild ponies of Assateague Island which explores their role in the ecology of the barrier island.
ISBN 0-688-22179-3. ISBN 0-688-32179-8 lib. bdg.
1. Chincoteague pony—Juvenile literature. 2. Animal ecology—Assateague Island—Juvenile literature. 3. Assateague Island—Juvenile literature. ₍1. Chincoteague pony. 2. Assateague Island. 3. Animal ecology. 4. Ecology₎ I. Keiper, Ronald R., joint author. II. Title.
SF315.2.C4F67 599'.725 79 79-11026 MARC

CHONCOTEAGUE PONY
—JUVENILE LITERATURE (Continued)

Scott, Jack Denton, 1915-
Island of wild horses / words by Jack Denton Scott ; photos. by Ozzie Sweet. — New York : Putnam, c1978.
60 p. : ill. ; 28 cm.
SUMMARY: Pesents theories about the history of the wild horses of Assateague Island and describes their present life and habits.
ISBN 0-399-20648-5 : $7.95
1. Chincoteague pony—Juvenile literature. 2. Assateague Island—Juvenile literature. [1. Chincoteague pony. 2. Ponies] I. Sweet, Ozzie. II. Title.
SF315.2.C4S36 1978 636.1'6 78-17380
78 MARC

CHINESE ALMANACS see Almanacs, Chinese

CHINESE-AMERICAN NEWSPAPERS

—DIRECTORIES

Lo, Karl, 1935-
Chinese newspapers published in North America, 1854-1975 / compiled by Karl Lo and H. M. Lai. — Washington : Center for Chinese Research Materials, Association of Research Libraries, c1977.
xiii, 138 p. ; 28 cm. — (Bibliographical series - Center for Chinese Research Materials, Association of Research Libraries ; no. 16)
Includes index.
1. Chinese-American newspapers—Directories. 2. Chinese newspapers—Canada—Directories. I. Lai, H. Mark, joint author. II. Title. III. Series: Association of Research Libraries. Center for Chinese Research Materials. Bibliographical series ; no. 16.
Z6953.5.C45L6 071'.3 77-89070
[PN4885.C5] 79 MARC

CHINESE AMERICANS

Gibson, Otis.
The Chinese in America / Otis Gibson. — New York : Arno Press, 1978 [c1877]
403 p. ; 21 cm. — (The Asian experience in North America : Chinese and Japanese)
Reprint of the ed. published by Hitchock & Walden, Cincinnati.
ISBN 0-405-11272-6
1. Chinese Americans. I. Title. II. Series: Asian experience in North America : Chinese and Japanese.
E184.C5G5 1978 973'.04'951 78-54816
78 MARC

—ECONOMIC CONDITIONS—STATISTICS

Sung, Betty Lee.
Statistical profile of the Chinese in the United States : 1970 census / Betty Lee Sung. — New York : Arno Press, 1978.
ix, 128 p. ; 29 cm. — (The Asian experience in North America : Chinese and Japanese)
Reprint of the 1975 ed. published by the U.S. Dept. of Labor, Manpower Administration, Washington.
"Prepared for the Manpower Administration, U.S. Department of Labor."
ISBN 0-405-11289-0
1. Chinese Americans—Statistics. 2. Chinese Americans—Economic conditions—Statistics. 3. Chinese Americans—Social conditions—Statistics. I. United States. Dept. of Labor. Manpower Administration. II. Title. III. Series.
E184.C5S96 1978 301.45'19'51073 78-54832
78 MARC

—HISTORY

Steiner, Stanley.
Fusang, the Chinese who built America / by Stan Steiner. — 1st ed. — New York : Harper & Row, c1979.
xii, 259 p. ; 22 cm.
Bibliography: p. [239]-251.
Includes index.
ISBN 0-06-014087-9 : $11.95
1. Chinese Americans—History. 2. America—Discovery and exploration—Chinese. I. Title.
E184.C5S76 1979 970'.004'951 78-2172
78 MARC

—HISTORY—ADDRESSES, ESSAYS, LECTURES

Anti-Chinese violence in North America / edited by Roger Daniels. — New York : Arno Press, 1978.
204 p. in various paging, [2] leaves of plates : ill. ; 24 cm. — (The Asian experience in North America : Chinese and Japanese)
Includes bibliographical references.
Reprinted from various journals.
CONTENTS: Wilcox, W. P. Anti-Chinese riots in Washington.—Crane, P. and Larson, A. The Chinese massacre.—Karlin, J. A. The indemnification of aliens injured by mob violence.—Rudolph, F. Chinamen in Yankeedom.—Karlin, J. A. The anti-Chinese outbreaks in Seattle, 1885-1886.—Karlin, J. A. The anti-Chinese outbreak in Tacoma, 1885.—Locklear, W. R. The celestials and the angels.—Carranco. L. Chinese expulsion from Humboldt County.—Wortman, R. T. Denver's anti-Chinese riot, 1880.—Wynne, R. E. American labor leaders and the Vancouver anti-oriental riot.—Jacques, L. M. D. The Chinese massacre in Torreon (Coahuila) in 1911.
ISBN 0-405-11263-7
1. Chinese Americans—History—Addresses, essays, lectures. 2. Violence—United States—History—Addresses, essays, lectures. I. Daniels, Roger. II. Series: Asian experience in North America : Chinese and Japanese.
E184.C5A57 1978 973'.04'951 78-54807
78 MARC

—HISTORY—JUVENILE LITERATURE

McCunn, Ruthanne Lum.
An illustrated history of the Chinese in America / Ruthanne Lum McCunn. — San Francisco : Design Enterprises of San Francisco, c1979.
133 p. : ill. ; 29 cm.
Bibliography: p. 130.
Includes index.
SUMMARY: Traces the history of the Chinese in the United States focusing on their struggle for acceptance by the white population and their contributions to the development of their new country.
ISBN 0-932538-01-0 : $11.95
1. Chinese Americans—History—Juvenile literature. [1. Chinese Americans—History] I. Title.
E184.C5M2 973'.04'951 79-50114
00 MARC

—SOCIAL CONDITIONS—STATISTICS

Sung, Betty Lee.
Statistical profile of the Chinese in the United States : 1970 census / Betty Lee Sung. — New York : Arno Press, 1978.
ix, 128 p. ; 29 cm. — (The Asian experience in North America : Chinese and Japanese)
Reprint of the 1975 ed. published by the U.S. Dept. of Labor, Manpower Administration, Washington.
"Prepared for the Manpower Administration, U.S. Department of Labor."
ISBN 0-405-11289-0
1. Chinese Americans—Statistics. 2. Chinese Americans—Economic conditions—Statistics. 3. Chinese Americans—Social conditions—Statistics. I. United States. Dept. of Labor. Manpower Administration. II. Title. III. Series.
E184.C5S96 1978 301.45'19'51073 78-54832
78 MARC

—STATISTICS

Sung, Betty Lee.
Statistical profile of the Chinese in the United States : 1970 census / Betty Lee Sung. — New York : Arno Press, 1978.
ix, 128 p. ; 29 cm. — (The Asian experience in North America : Chinese and Japanese)
Reprint of the 1975 ed. published by the U.S. Dept. of Labor, Manpower Administration, Washington.
"Prepared for the Manpower Administration, U.S. Department of Labor."
ISBN 0-405-11289-0
1. Chinese Americans—Statistics. 2. Chinese Americans—Economic conditions—Statistics. 3. Chinese Americans—Social conditions—Statistics. I. United States. Dept. of Labor. Manpower Administration. II. Title. III. Series.
E184.C5S96 1978 301.45'19'51073 78-54832
78 MARC

—CALIFORNIA—SAN FRANCISCO

Salter, Christopher L
San Francisco's Chinatown : how Chinese a town? / Christopher L. Salter. — San Francisco : R & E Research Associates, c1978.
xii, 82 p. : ill. ; 28 cm.
Bibliography: p. 79-82.
ISBN 0-88247-500-2 : $8.00
1. Chinese Americans—California—San Francisco. 2. San Francisco—Buildings. 3. Architecture—California—San Francisco. I. Title.
F869.S39C57 979.4'61'004951 77-91458
79 MARC

—CALIFORNIA—SIERRA NEVADA MOUNTAINS—JUVENILE FICTION

Chang, Kathleen.
The Iron Moonhunter / written & illus. by Kathleen Chang ; translated into Chinese by Chiu-Chung Liao & Florence Lau. — San Francisco : Children's Book Press/Imprenta de Libros Infantiles, c1977.
24 p. : col. ill. ; 25 cm. — (Fifth world tales)
Title also in Chinese: Chui yüeh hao.
English and Chinese.
SUMMARY: While they work on the Central Pacific Railroad in the Sierra Nevada mountains, the Chinese workers build an unusual railroad of their own.
ISBN 0-89239-011-5 : $2.95
1. Chinese Americans—California—Sierra Nevada Mountains—Juvenile fiction. 2. Railroad construction workers—California—Sierra Nevada Mountains—Juvenile fiction. 3. Central Pacific Railroad—History—Juvenile fiction. [1. Chinese Americans—Fiction. 2. Railroads—Trains—Fiction. 3. Chinese language—Readers] I. Title. II. Title: Chui yüeh hao.
PZ10.831.C4624 I 7 813'.5'4 77-73783
79 MARC

—MONTANA—BUTTE—SOCIAL LIFE AND CUSTOMS

Lee, Rose Hum.
The growth and decline of Chinese communites in the Rocky Mountain region / Rose Hum Lee. — New York : Arno Press, 1978.
vi, 371 p. ; 24 cm. — (The Asian experience in North America : Chinese and Japanese)
Bibliography: p. 364-271.
Reprint of the author's thesis, University of Chicago, 1947.
ISBN 0-405-11279-3
1. Chinese Americans—Montana—Butte—Social life and customs. 2. Chinese Americans—Rocky Mountain region—Social life and customs. 3. Butte, Mont.—Social life and customs. 4. Rocky Mountain region—Social life and customs. I. Title. II. Series: Asian experience in North America : Chinese and Japanese.
F739.B8L43 1978 978.6'68 78-54823
78 MARC

—NORTHWEST, PACIFIC

Wynne, Robert Edward.
Reaction to the Chinese in the Pacific Northwest and British Columbia, 1850-1910 / Robert Edward Wynne. — New York : Arno Press, 1978.
vi, 511 p. ; 21 cm. — (The Asian experience in North America : Chinese and Japanese)
Originally presented as the author's thesis, University of Washington, 1964.
Bibliography: p. 500-511.
ISBN 0-405-11298-X
1. Chinese Americans—Northwest, Pacific. 2. Chinese in British Columbia. 3. Northwest, Pacific—Race relations. 4. British Columbia—Race relations. I. Title. II. Series: Asian experience in North America : Chinese and Japanese.
F855.2.C5W96 1978 979.5'004'951 78-54847
78 MARC

—ROCKY MOUNTAIN REGION—SOCIAL LIFE AND CUSTOMS

Lee, Rose Hum.
The growth and decline of Chinese communites in the Rocky Mountain region / Rose Hum Lee. — New York : Arno Press, 1978.
vi, 371 p. ; 24 cm. — (The Asian experience in North America : Chinese and Japanese)
Bibliography: p. 364-271.
Reprint of the author's thesis, University of Chicago, 1947.
ISBN 0-405-11279-3
1. Chinese Americans—Montana—Butte—Social life and customs. 2. Chinese Americans—Rocky Mountain region—Social life and customs. 3. Butte, Mont.—Social life and customs. 4. Rocky Mountain region—Social life and customs. I. Title. II. Series: Asian experience in North America : Chinese and Japanese.
F739.B8L43 1978 978.6'68 78-54823
78 MARC

—TEXAS—JUVENILE LITERATURE

Martinello, Marian L
Who are the Chinese Texans? / Marian L. Martinello, William T. Field, Jr. — San Antonio : University of Texas, Institute of Texan Cultures at San Antonio, c1979.
80 p. : ill. ; 27 cm. — (Stories for young readers)
SUMMARY: Questions and answers survey the history and way of life of Chinese Americans who live in Texas.
ISBN 0-933164-46-7
1. Chinese Americans—Texas—Juvenile literature. 2. Texas—Juvenile literature. [1. Chinese Americans. 2. Texas—History] I. Field, William T., joint author. II. Title. III. Series.
F395.C5M37 976.4'004'951 78-57796
00 MARC

—WASHINGTON (STATE)—HISTORY

Hildebrand, Lorraine
Straw hats, sandals, and steel : the Chinese in Washington State / Lorraine Barker Hildebrand. — Tacoma : Washington State American Revolution Bicentennial Commission, 1977.
xiv, 97 p. : ill. ; 23 cm. — (Ethnic history series ; 2)
Bibliography: p. 87-90.
ISBN 0-917048-05-9
1. Chinese Americans—Washington (State)—History. 2. Washington (State)—Race relations. 3. Washington (State)—History—To 1889. I. Title. II. Series.
F900.C5H54 979.7'004'51 77-78792
00 MARC

CHINESE ANTROLOGY see Antrology, Chinese

CHINESE ARTS see Arts, Chinese

CHINESE ASTROLOGY see Astrology, Chinese

CHINESE BALLADS AND SONGS

—HISTORY AND CRITICISM

(Yüeh fu shih yen chiu lun wên chi)
乐府詩研究論文集　作家出版社編輯部編　北京　作家出版社　1957-[70.　香港?　1970]
2 v.　22 cm.　(中國古典文學研究論文集滙編第[1], 3輯)　$11.50 (U.S.)
Vol. 1, without series statement, forms one of the 第1輯 of the series.
Vol. 2 contains articles selected and edited by the 中國語文學社 from the 文學遺產增刊 and other sources.
1. Chinese poetry—History and criticism.　2. Chinese ballads and songs—History and criticism.　I. Tso chia ch'u pan shê, Peking. II. Chung-kuo yü wên hsüeh shê.　III. Wên hsüeh i ch'an.　Wên hsüeh i ch'an tsêng k'an.　Selections. 1970.　IV. Series: Chung-kuo ku tien wên hsüeh yen chiu lun wên chi hui pien.
PL2307.Y8　78-843679

CHINESE BRONZES see Bronzes, Chinese

CHINESE CALENDAR see Calendar, Chinese

CHINESE CALLIGRAPHY see Calligraphy, Chinese

CHINESE CHARACTERS

Đào Duy Anh.
Chữ Nôm : nguồn gốc, cấu tạo, diễn biến / Đào Duy Anh.— Hà nội : Khoa Học Xã Hội, 1975.
223 p. ; 19 cm.
Includes bibliographical references.
1. Vietnamese language—Writing.　2. Chinese characters.　3. Vietnamese poetry.　I. Title.
PL4372.D36　79-984025
79　MARC

—ADDRESSES, ESSAYS, LECTURES

(Nippongo no gemba)
日本語の現場 / 読売新聞社会部編. — 東京：読売新聞社, 昭和50-52 [1975-1977]
4 v. : ill. ; 19 cm.　Ja 75-23271 (v. 1)
Originally published in Yomiuri shimbun.
¥750 per vol.
1. Japanese language—Addresses, essays, lectures.　2. Chinese characters—Addresses, essays, lectures.　I. Yomiuri Shimbun Sha. Shakaibu.　II. Yomiuri shimbun, Tokyo.　III. Title: Nihongo no gemba.
PL527.N56　1975　76-812316

—DICTIONARIES—JAPANESE

(Gakken Kan-Wa daijiten)
学研 漢和大字典 / 藤堂明保編. — 東京：学習研究社, 昭和53 [1978]
4, 1740, 90 p. : ill. ; 22 cm.　Ja***
Includes indexes.
¥3800
1. Chinese characters—Dictionaries—Japanese.　I. Tōdō, Akiyasu, 1915-　II. Title: Kan-Wa daijiten.
PL677.5.G3　78-812683
J

Gakushū jambo kanji hyakka jiten.
三省堂学習漢字図解辞典 / 石井勲主幹 ; 三省堂編修所編. — 東京：三省堂, 昭和52 [1977]
996, 35 p. : col. ill. ; 22 cm.　Ja***
First ed. published in 1974 under title: Gakushū jambo kanji hyakka jiten.
Includes index.
¥3000
1. Chinese characters—Dictionaries—Japanese.　2. Japanese language—Dictionaries.　I. Ishii, Isao, 1919-　II. Sanseidō (Founded 1915)　III. Title.　IV. Title: Gakushū kanji zukai jiten.
Title romanized: Sanseidō gakushū kanji zukai jiten.
PL677.5.G34　1977　78-813238

Kanno, Dōmei, 1865-1938.
(Jigen)
字源 / 簡野道明著. — 東京：北辰館：發賣所 明治書院, 大正12 [1923]
2765 p. in various pagings : ill. ; 23 cm.
Includes indexes.
1. Chinese characters—Dictionaries—Japanese.　I. Title.
PL677.5.K36　78-809341

Kojima, Yukie.
(Yasokai-ban Rakuyōshū sōsakuin)
耶蘇会板落葉集総索引：字訓索引·字音索引 / 小島幸枝編. — 東京：笠間書院, 1978.
338, 264 p. ; 22 cm. — (笠間索引叢刊 ; 55)　Ja 78-20782
Series romanized: Kasama sakuin sōkan.
"本文篇": p. [1]-220 (2d group)
¥12000
1. Rakuyōshū—Indexes.　2. Chinese characters—Dictionaries—Japanese—Indexes.　3. Rakuyōshū.　4. Chinese characters—Dictionaries—Japanese.　5. Japanese language—Dictionaries—Early works to 1800. I. Rakuyōshū. 1978.　II. Title.　III. Title: Rakuyōshū sōsakuin.
PL674.5.R32K64　78-815966

Kokugo Kenkyūkai.
(Shin kokugo jiten)
新国語辞典：新制当用漢字：現代かなづかい / [編者　国語研究会]. — 増補. — 東京：一歩社, 昭和24 [1949]
12, 242, 8 p. ; 15 cm.
Includes indexes.
1. Chinese characters—Dictionaries—Japanese.　2. Japanese language—Dictionaries.　I. Title.
PL677.5.K6　79-800870

Shin'inshū.
新韻集：[2巻] / 正宗敦夫編纂校訂. — 東京：現代思潮社, 1978.
24, 192 p. ; 16 cm. — (覆刻日本古典全集)　Ja 79-895
"釋萬里撰すといふ"
Photoreprint of the 1944 ed. (reproduction of Awa no Kuni Bunko bon) published by Nihon Koten zenshū Kankōkai, Tokyo.
1. Japanese language—Dictionaries—Early works to 1800.　2. Chinese characters—Dictionaries—Japanese.　I. Masamune, Atsuo, 1881-1958.　II. Shūku, b. 1428.　Shin'inshū.　III. Series: Nihon koten zenshū.
Title romanized: Shin'inshū.
PL674.5.S45　1978　79-803841

Wagokuhen.
倭玉篇：[3巻] / 正宗敦夫編纂校訂. — 東京：現代思潮社, 昭和52 [1977]
11, [484] p. ; 16 cm. — (覆刻日本古典全集)　Ja 78-1202
Photoreprint of the 1936 ed. published by Nihon Koten Zenshū Kankōkai, Tokyo, which is photoreprint of Yasuda Bunko copy published during the Keichō period.
1. Japanese language—Dictionaries—Early works to 1800.　2. Chinese characters—Dictionaries—Japanese.　I. Masamune, Atsuo, 1881-1958.　II. Series: Nihon koten zenshū.
Title romanized: Wagokuhen.
PL674.5.W3　1977　79-802969

—DICTIONARIES—JAPANESE —INDEXES

Kojima, Yukie.
(Yasokai-ban Rakuyōshū sōsakuin)
耶蘇会板落葉集総索引：字訓索引·字音索引 / 小島幸枝編. — 東京：笠間書院, 1978.
338, 264 p. ; 22 cm. — (笠間索引叢刊 ; 55)　Ja 78-20782
Series romanized: Kasama sakuin sōkan.
"本文篇": p. [1]-220 (2d group)
¥12000
1. Rakuyōshū—Indexes.　2. Chinese characters—Dictionaries—Japanese—Indexes.　3. Rakuyōshū.　4. Chinese characters—Dictionaries—Japanese.　5. Japanese language—Dictionaries—Early works to 1800. I. Rakuyōshū. 1978.　II. Title.　III. Title: Rakuyōshū sōsakuin.
PL674.5.R32K64　78-815966

—DICTIONARIES—KOREAN

Yu, Chŏng-gi.
(Sŏlmun chajŏn)
說文字典 / 柳正基著. — 大田：農經出版社 ; 서울：發賣元　通文館, 1973.
14, 300, 91, [8] p. ; 22 cm.
"附　啓蒙源典": p. 1-91, 1-8 (3d and 4th groups)
1. Chinese characters—Dictionaries—Korean.　2. Chinese language—Readers.　I. Title.　II. Title: Kyemong wŏnjŏn.
PL935.Y85　78-826828

—STUDY AND TEACHING —JAPAN

Ishii, Isao, 1919-
(Kanji ni yoru sainō kaihatsu)
漢字による才能開発　三歳からの漢字教育　石井勲著　東京　講談社　昭和45 (1970)
245 p.　20 cm.　Ja 70-15469
¥560
1. Language arts (Preschool)—Japan.　2. Language arts (Elementary)—Japan.　3. Chinese characters—Study and teaching—Japan. I. Title.
LB1140.5.L3I74　78-812467

—STUDY AND TEACHING (ELEMENTARY)—JAPAN

Japan. Mombushō.
(Jidō seito no kanji o kaku nōryoku to sono kijun)
児童生徒の漢字を書く能力とその基準 / [著作文部省]. — 東京：明治図書出版, 昭和27 [1952]
2, 3, 327 p., [3] leaves of plates : ill. ; 21 cm. — (初等教育研究資料：第1集)
1. Japanese language—Study and teaching (Elementary)—Japan.　2. Chinese characters—Study and teaching (Elementary)—Japan.　3. Japanese language—Study and teaching (Secondary)—Japan.　4. Chinese characters—Study and teaching (Secondary)—Japan.　I. Title.　II. Series: Shotō kyōiku kenkyū shiryō ; dai 1-shū.
LB1577.J3J295　1952　78-817037

—STUDY AND TEACHING (SECONDARY)—JAPAN

Japan. Mombushō.
(Jidō seito no kanji o kaku nōryoku to sono kijun)
児童生徒の漢字を書く能力とその基準 / [著作文部省]. — 東京：明治図書出版, 昭和27 [1952]
2, 3, 327 p., [3] leaves of plates : ill. ; 21 cm. — (初等教育研究資料：第1集)
1. Japanese language—Study and teaching (Elementary)—Japan.　2. Chinese characters—Study and teaching (Elementary)—Japan.　3. Japanese language—Study and teaching (Secondary)—Japan.　4. Chinese characters—Study and teaching (Secondary)—Japan.　I. Title.　II. Series: Shotō kyōiku kenkyū shiryō ; dai 1-shū.
LB1577.J3J295　1952　78-817037

—JAPAN

Amagaya, Fumio.
(Toyo kanjihyo seikai)
当用漢字表精解：よみかた·かきかた·送りがなのしかた / 天ケ谷文男著. — 東京：中文書院, 昭和25 [1950]
1, 210 p. ; 19 cm.
Includes indexes.
1. Chinese characters—Japan.　I. Title.
PL528.A46　79-805000

Hasegawa, Motoi, 1884-
(Tōyō kanji shinjitai no kaisetsu)
当用漢字新字体の解説 / 長谷川基著. — 東京：学習研究社, 昭和30 [1955]
7, 251 p. ; 22 cm.
Colophon inserted.
Includes index.
1. Chinese characters—Japan.　2. Japanese language—Writing.　3. Japanese language—Reform.　I. Title.
PL528.H28　79-804999

Matsuzaka, Tadanori, 1902-
(Tōyō kanji handobukku)
当用漢字ハンドブック / 松坂忠則著. — 東京：共和出版社, 1950.
11, 204 p. ; 19 cm.
1. Chinese characters—Japan.　I. Title.
PL528.M37　79-804998

CHINESE CHARACTERS —JAPAN (Continued)

Shiraishi, Daiji, 1912–
(Tōyō kanjihyō jimmei kanji gakushū kanji, kaitei tōyō kanji onkunhyō, gendai kanazukai, kaitei okurigana no tsukekata)
当用漢字表 人名漢字 学習漢字・改定当用漢字音訓表・現代かなづかい・改定送り仮名の付け方 / 白石大二編. — 増補改訂版. — 東京 : 大蔵省印刷局, 1977.
511 p. : 19 cm. Ja 78-11314
¥1000

1. Chinese characters—Japan. I. Title.
PL528.S51 79-801943

Sugimoto, Tsutomu, 1927–
(Itaiji to wa nani ka)
異体字とは何か / [著者 杉本つとむ]. — 東京 : 桜楓社, 1978.
270 p., [2] leaves of plates : ill. ; 19 cm. — (杉本つとむ日本語講座 ; 1) Ja 79-2970
Includes bibliographical references and indexes.
Errata slip inserted.
¥1500

1. Chinese language—Writing. 2. Chinese characters—Japan. 3. Chinese characters—Korea. I. Title. II. Series: Sugimoto, Tsutomu, 1927– Sugimoto Tsutomu Nihongo kōza ; 1.
PL1171.S77 79-808462

—JAPAN—DICTIONARIES —JAPANESE

(Tōyō kanji hyōki jiten)
当用漢字表記辞典 / 佐藤喜代治監修 ; 加藤彰彦, 飛田良文編. — 増訂版. — 東京 : 桜楓社, 1978.
594 p. ; 19 cm. Ja 78-10922
Includes index.
¥680

1. Chinese characters—Japan—Dictionaries—Japanese. 2. Japanese language—Dictionaries. 3. Japanese language—Orthography and spelling. I. Satō, Kiyoji, 1912– II. Katō, Akihiko. III. Hida, Yoshifumi, 1933–
PL677.5.T68 1978 79-804941

—KOREA

Sugimoto, Tsutomu, 1927–
(Itaiji to wa nani ka)
異体字とは何か / [著者 杉本つとむ]. — 東京 : 桜楓社, 1978.
270 p., [2] leaves of plates : ill. ; 19 cm. — (杉本つとむ日本語講座 ; 1) Ja 79-2970
Includes bibliographical references and indexes.
Errata slip inserted.
¥1500
1. Chinese language—Writing. 2. Chinese characters—Japan. 3. Chinese characters—Korea. I. Title. II. Series: Sugimoto, Tsutomu, 1927– Sugimoto Tsutomu Nihongo kōza ; 1.
PL1171.S77 79-808462

J

CHINESE CHESS

(Chung-kuo hsiang ch'i chüeh sai p'ing shu)
中国象棋决赛评述. — 第1版. — [北京] : 人民体育出版社 : 新华书店北京发行所发行, 1978.
160 p. : ill. ; 19 cm.
At head of title: Chung-hua jen min kung ho kuo ti san chieh yün tung hui.
$0.42

1. Chinese chess. I. Chung-hua jen min kung ho kuo yün tung hui. 3d, Peking, 1975.
GV1458.C5C558 79-837411

Huang, Chin-huan.
(Nan-yang hsiang ch'i p'u)
南洋象棋譜 : 殘局篇 / 黃金環著. — 香港 : 新力書局, [1976?] printing.
2 v. : ill. ; 19 cm.
Cover title.
$6.00

1. Chinese chess. I. Title.
GV1458.C5H8 78-843161

Leventhal, Dennis A
The chess of China / by Dennis A. Leventhal. — A Mei Ya international ed. — Taipei : Mei Ya Publications, 1978.
ii, 165 p. : ill. ; 20 cm. & chess set (32 pieces) in box.
Bibliography: p. 161-165.
$6.00
1. Chinese chess. I. Title.
GV1458.C5L43 794.1'8 79-105274
79 MARC

Spindler, Pierre Éric.
Traité du jeu d'échecs chinois, Xiang-Qi / Pierre-Éric Spindler ; avec le concours de Pierre Rival. — [Paris] : Flammarion, 1977.
268 p. : ill. ; 25 cm. — (Ludothèque de l'impensé radical) F79-561
On cover: Traité du jeu de Xiang-Qi, échecs chinois.
ISBN 2-08-202204-8 : 48.00F
1. Chinese chess. I. Rival, Pierre, joint author. II. Title. III. Title: Traité du jeu de Xiang-Qi, échecs chinois. IV. Title: Xiang-Qi.
GV1458.C5S65 794.1'8 79-380882
79 MARC

CHINESE CHILDREN'S SONGS see Children's songs, Chinese

CHINESE CHRISTIAN POETRY see Christian poetry, Chinese

CHINESE CLASSICS

Lo, Chen-yü, 1866–1940.
(Han shih ching ts'an tzu chi lu hsü pien)
漢石經殘字集錄續編 : 一卷, 補遺一卷 / [羅振玉寫]. — [s. l. : s. n., 1929?]
72 double leaves in various foliations : ill. ; 27 cm. — (松翁居遼後所著書)
Issued in case with the author's Han Hsi-p'ing shih ching ts'an tzu chi lu. [s. l., 1929]
Caption title: Han Hsi-p'ing shih ching ts'an tzu chi lu hsü pien.
Lithoprinted.
"漢石經殘字集錄三編 : 一卷, 補遺一卷": at end of work.

1. Chinese classics. 2. Inscriptions, Chinese. I. Title. II. Series: Lo, Chen-yü, 1866–1940. Sung-weng chü Liao hou so chu shu.
PL2459.A49 79-836439
79 C

Lo, Chen-yü, 1866–1940.
(Han shih ching ts'an tzu chi lu ssu pien)
漢石經殘字集錄四編 : 一卷, 補遺一卷 / [羅振玉寫]. — [s. l. : s. n., 庚午 i. e. 1930序]
85 double leaves in various foliations : ill. ; 27 cm. — (松翁居遼後所著書)
Issued in case with the author's Han Hsi-p'ing shih ching ts'an tzu chi lu. [s. l., 1929]
Lithoprinted.
1. Chinese classics. 2. Inscriptions, Chinese. I. Title. II. Series: Lo, Chen-yü, 1866–1940. Sung-weng chü Liao hou so chu shu.
PL2459.A49 79-836440
79 C

T'ieh, Hsia-sheng.
(Tso chuan Kuo yü yin ching shuo ching chih yen chiu)
左傳國語引經說經之研究 / 鐵夏生. — [台北? : s. n.], 民國55 [1966]
2 v. (373 p. on double leaves) ; 27 cm.
Reproduced from ms. copy.
"國立台灣大學中文研究所碩士論文"
Bibliography: p. 371-373.

1. Tso-ch'iu, Ming. Tso chuan. 2. Tso-ch'iu, Ming. Kuo yü. 3. Chinese classics. I. Title.
PL2470.Z7T5 1966 79-835176

—HISTORY AND CRITICISM

Ch'ien, Mu, 1895–
(Liang Han ching hsüeh chin ku wen p'ing i)
兩漢經學今古文平議 / 錢穆著. — 再版. — 臺北 : 東大圖書有限公司 : 總經銷三民書局, 1978.
5, 7, 434, 2 p. ; 22 cm. — (滄海叢刊 : 國學)
Series romanized: Ts'ang hai ts'ung k'an.

1. Chinese classics—History and criticism. I. Title.
PL2461.Z7C56 1978 79-837963

Kan, P'eng-yün, b. 1861.
(Kuo hsüeh pi t'an)
國學筆談 / 甘鵬雲撰. — 台1版. — [台北縣永和鎮] : 華世出版社, 民國66 [1977]
35, 15 leaves ; 22 cm.
Reprint of the ed. published by Ch'ung ya t'ang, in series: Kan shih chia ts'ang ts'ung kao.
NT$30.00

1. Chinese classics—History and criticism. I. Title.
PL2461.Z7K27 1977 79-840539

Li, Chen-hsing.
(Wang Su chih ching hsüeh)
王肅之經學 / 李振興. — [s. l. : s. n., 1978?]
3 v. (708 p. on double leaves) ; 27 cm.
Spine title.
Reproduced from ms. copy.
Bibliography: p. 694–708.
"聖證論": p. 676–693.
1. Chinese classics—History and criticism. 2. Wang, Su, 195–256. I. Wang, Su, 195–256. II. Wang, Su, 195–256. Sheng cheng lun. 1978. III. Shu ching. IV. Title.
PL2461.Z6L47 78-843521

Lin, Cheng-hua, 1946–
(Huang Chen chih ching hsüeh)
黃震之經學 / 林政華. — [台北? : s. n., 1978?]
3 v. (4 double leaves, 12, 516 p. on double leaves) ; 27 cm.
Spine title.
Reproduced from typescript, with pref. dated 1977.
Bibliography: p. 491–516.

1. Chinese classics—History and criticism. 2. Huang, Chen, chin shih 1256. I. Title.
PL2461.Z6L5 78-842901

Wan, Ching-yü.
(Chou li hsin chuan)
周禮新傳 : 經國政略大典 / 萬靖宇著. — 蘇澳鎮(宜蘭縣) : 萬靖宇, 民國66 [1977]
10, 8, 502 p. ; 22 cm.
"著者著作簡介": p. 501–502.
Includes bibliographies.
NT$180.00

1. State, The. 2. Utopias. 3. Chinese classics—History and criticism. 4. Chou li. I. Title.
JC327.W27 79-836525

—TRANSLATIONS INTO JAPANESE

(Sentetsu icho Kanseki kokujikai zensho)
先哲遺著漢籍國字解全書 / [編輯 早稻田大學出版部]. — 東京 : 同部, 昭和23- [1948-
v. ; 22 cm.
Reprint of the 1909 ed. published by Waseda Daigaku Shuppanbu.

1. Chinese classics—Translations into Japanese. 2. Japanese literature—Translations from Chinese. I. Waseda Daigaku, Tokyo. II. Title: Kanseki kokujikai zensho.
PL2488.A7 1948 79-800924

CHINESE COOKERY see Cookery, Chinese

CHINESE DETECTIVE STORIES see Detective and mystery stories, Chinese

CHINESE DRAMA

Under this heading, without subdivision, are entered individual plays and collections ofplays by one author irrespective of whether they are comedies of tragedies

see also One-act plays, Chinese

Chang, Tao-fan, 1897–1968.
(Liu hsüeh sheng chih lien)
留學生之戀 / 張道藩著. — 臺北 : 道藩文藝圖書館, 民國67 [1978]
103 p., [2] leaves of plates : ill. ; 22 cm.
Play.

I. Title.
PL2740.T3L5 1978 79–835566

Chao, Huan.
(Nan hai ch'ang ch'eng)
南海长城 : 五場話剧 / 赵寰著 ; 解放軍文艺丛书編輯部編. — [香港? : s. n., 1978?]
99 p., [4] leaves of plates : ill. ; 21 cm.
Play.
Reprint of the 1964 ed. published by Chung-kuo hsi chü ch'u pan she, Peking.

I. Title.
PL2838.H83N3 1978 79–836379

Chiang-hsi sheng hua chü t'uan.
(Pa i feng pao)
八一风暴 : 九场革命历史话剧 / 江西省话剧团集体创作. — 第1版. — 北京 : 人民文学出版社, 1977.
119 p., [2] leaves of plates : ill. ; 19 cm.
"江西省话剧团一九七七年一月演出本"
$0.30

1. Nan-ch'ang, China (District)—History—Drama. I. Title.
PL2844.H78P3 78–843295

Ch'ü, Chün-chieh.
(T'ai-wan erh nü)
台湾儿女 : 六幕话剧 / 编剧瞿俊杰执笔, 赵志华. — 第1版. — 北京 : 解放军文艺社, 1978.
87 p., [1] leaf of plates : ill. ; 21 cm.
"一九七八年三月演出本"
$0.34

I. Chao, Chih-hua. II. Title.
PL2852.C83T34 79–838190

Ch'un-ch'ao.
(Chüeh lieh)
决裂 : 电影文学剧本 / 春潮, 周杰. — 第1版. — 上海 : 上海人民出版社, 1976.
86 p., [3] leaves of plates : ill. ; 19 cm.
Play.
$0.23

I. Chou, Chieh, joint author. II. Title.
PL2853.C47C5 1976b 78–843610

Chung, I-ch'un.
(Huan niu)
还牛 : 花鼓戏 / 钟宜淳, 徐叔华编剧. — [香港? : s. n., 1978?]
48 p. ; 22 cm.
Play.
In number notation.
Reprint of the 1978 ed. published by Hu-nan jen min ch'u pan she, in series: Hu-nan ti fang hsiao hsi hsüan.
I. Hsü, Shu-hua, joint author. II. Title. III. Series: Hu-nan ti fang hsiao hsi hsüan.
PL2854.I2H8 1978 79–835696

Chung-kuo jen min chieh fang chün hai chün cheng chih pu wen kung t'uan. Hua chü t'uan.
(Chia wu hai chan)
甲午海战 八場历史剧 中国人民解放軍海軍政治部文工团話剧团改編 朱祖貽 李恍执笔 北京 中国戏剧出版社 1961.
102 p. illus. 19 cm.
Adaptation of 希侬等著电影剧本甲午风云
1. Chinese-Japanese War, 1894–1895—Drama. I. Chu, Tsu-i, fl. 1961– II. Hsi-nung, pseud. Chia wu feng yün. III. Title.
C 64–549

Chungking, China. Wen hua chü. Hsi ch'ü kung tso wei yüan hui.
(Shen Chi-p'ing)
審吉平 / 重慶市文化局戲曲工作委員会編. — [香港? : s. n., 1978?]
55 p. ; 18 cm.
Colophon title.
Play.
Reprint of the 1956 ed. published by Ch'ung-ch'ing jen min ch'u pan she, Chungking.
I. Title.
PL2854.Z52S5 1978 79–840011

(Hsiang sheng ch'uan t'ung tso p'in hsüan)
相声傳統作品选 / 中国曲艺研究会主編. — [香港? : s. n., 1979?]
2, 260 p. ; 20 cm.
Reprint of the 1957 ed. published by Tso chia ch'u pan she, Peking.

1. Chinese drama. 2. Mime. I. Chung-kuo ch'ü i yen chiu hui.
PL2579.M5H69 1979 79–840273

Hsiao, K'o-ou.
(Ou chih ko)
鷗之歌 / [蕭]可鷗. — 九龍 : 高原出版社, 1974.
73 p. : ports. ; 18 cm.
Poems.

I. Title.
PL2862.I3171409 78–840923

(Hsiu ting P'ing chü hsüan)
修訂平劇選 / 國立編譯館編訂 ; [修訂者姜作棟, 林柏年, 李效厂]. — 臺1版. — 台北 : 該館 : 發行所正中書局, 民國47– [19
v. ; 19 cm.

1. Chinese drama. I. Kuo li pien i kuan, T'ai-pei. II. Chiang, Tso-tung. III. Lin, Po-nien. IV. Li, Hsiao-an. V. Title: P'ing chü hsüan.
PL2568.H75 79–838719

Hsü, Hsi-che.
(Fei ts'ui cho)
翡翠鐲 / 著者許希哲. — 台北 : 照明出版社, 1977.
110 p. ; 19 cm.
Colophon title.
"三幕六場話劇"
Play.
NT$30.00
I. Title.
PL2862.U156F4 78–841330

Hsüeh, Shou-hsien.
(Ch'e lun kun kun)
车轮滚滚 : 电影文学剧本 / 薛寿先. — 第1版. — 北京 : 人民文学出版社 : 新华书店北京发行所发行, 1978.
71 p., [2] leaves of plates : ill. ; 19 cm.
$0.21

I. Title.
PL2862.U33774C5 79–836871

Hu, Ch'i, 1918–
(Pao kung tan)
報功單 胡奇著 [北京] 天下圖書公司 [1949]
67 p. 17 cm. (大衆文藝叢書)
Plays.
Series romanized: Ta chung wen i ts'ung shu.

I. Title.
PL2863.C52P3 C 68–837

Kao, Ming, fl. 1345.
(P'i p'a chi)
琵琶記 : 附劄記 / 高明撰. — 臺1版. — 臺北 : 臺灣商務印書館, 民國67 [1978]
2, 2, 216 p. ; 19 cm. — (人人文庫 ; 特561)
Series romanized: Jen jen wen k'u.
NT$30.00

I. Title.
PL2694.K36P5 1978 78–843288

Kuo, Mo-jo, 1892–1978.
(Mo-jo chü tso hsüan)
沫若剧作选. — 第1版. — 北京 : 人民文学出版社 : 新華書店北京发行所发行, 1978.
404 p., [2] leaves of plates : ill. ; 21 cm.
Plays.
"《沫若剧作选》选收了《棠棣之花》、《屈原》、《虎符》和《蔡文姬》四个剧本。本选集是根据本社一九五九年四月出版的《沫若选集》第二卷重印的。"
$1.55

I. Title.
PL2778.O2A19 1978 79–838625

Li, Chien-wu, 1908?–
(Che pu kuo shih ch'un t'ien)
這不過是春天 / 李健吾. — [香港? : s. n., 1978?]
119, vi p. ; 18 cm.
Play.
Reprint of the 1940 ed. published by Wen hua sheng huo ch'u pan she, Shanghai, which was issued as the 11th of Wen chi ts'ung shu.

I. Title.
PL2780.C5C45 1978 78–840887

Li, Chien-wu, 1908?–
(Li Chien-wu hsi chü chi)
李健吾戲劇集 / [李健吾]. — [香港? : s. n., 19
v. ; 21 cm.
Cover title.
Plays.
Reprint of the 19 editions published by Wen hua sheng huo ch'u pan she, Shanghai.

$30.00 (v. 5)

I. Title.
PL2780.C5 1978 79–836256

Li, Chien-wu, 1908?–
(Huang hua)
黄花 / 李健吾. — [香港? : s. n., 1978]
114 p. ; 21 cm.
Play.
Reprint of the 1947 ed. published by Wen hua sheng huo ch'u pan she, Shanghai, which was issued as the fifth of Li Chien-wu hsi chü chi.
$30.00

I. Title.
PL2780.C5 1978, vol. 5 79–836257

Li, Chun, 1928–
(Tsou hsiang chi)
走乡集 李准[著] 北京 中国电影出版社 1963.
317 p. 21 cm.
Plays.
CONTENTS: 老兵新传—小康人家—耕云播雨—一个女儿

I. Title.
PL2877.C8T7 C 64–1269

CHINESE DRAMA (Continued)

Li, Yün-liang.
(Ch'ing ch'un)
青春 / 李云良, 王炼. — 第1版. — 上海 : 上海文艺出版社 : 新華書店上海发行所发行, 1978.
88, [2] leaves of plates : ill. ; 19 cm. — (电影文学剧本)
Series romanized: Tien ying wen hsüeh chü pen.
$0.21

I. Wang, Lien, joint author. II. Title.
PL2877.Y818C47 79-837269

Liu, Fei-lieh, 1922-1958.
(Ch'ü chung meng hui)
曲終夢回 : 廣播劇 / 劉非烈著 ; [中國廣播公司編輯]. — 臺北 : 中國新聞出版公司, 民國45 [1956]
38 p. ; 20 cm.
Play.

I. Chung-kuo kuang po kung ssu. II. Title.
PL2879.F4C5 76-839224

(Liu k'ou tui chang)
流寇隊長 : 三幕劇 / 王震之執筆 ; [著作者魯迅藝術學院戲劇系集體創作]. — [香港 : s.n., 1977?]
79 p. ; 21 cm.
Cover title.
Reprint of the 1938 ed. published by Hsin-hua jih pao, Canton, issued as ti 2 chung in series: Yen-an Lu Hsün i shu hsüeh yüan hsi ch'ü ts'ung k'an.
I. Wang, Chen-chih. II. Yen-an Lu Hsün i shu hsüeh yüan. Hsi chü hsi. III. Series: Yen-an Lu Hsün i shu hsüeh yüan. Yen-an Lu Hsün i shu hsüeh yüan hsi ch'ü ts'ung k'an ; ti 2 chung.
PL2879.Z7 1977 78-843662

Lu, Ya-nung.
(Pai chan pai sheng)
百戰百勝 : 歌劇 / 魯亞農著 ; 止怡曲 ; 東北文工二團編. — [香港? : s.n., 1977?]
60 p. ; 21 cm.
Play.
Includes songs in number notation.
Reprint of the 1949 ed. published by Tung-pei shu tien, Mukden.

I. Chih, I. II. Tung-pei wen kung erh t'uan. III. Title.
PL2880.U194P34 1977 79-837749

Ma, Chi-hsing.
(Pao-tzu-wan chan tou)
豹子湾战斗 : 四幕话剧 / 马吉星编剧. — 第2版. — 北京 : 人民文学出版社, 1965, t.p. 1978.
110 p., [2] leaves of plates : ill. ; 19 cm.
"一九七六年十二月中国人民解放军武汉军区政治部话剧团演出本"
$0.29

1. Sino-Japanese Conflict, 1937-1945—Drama. I. Title.
PL2881.C48P3 1978 79-835415

Ma, Shao-po.
(Ch'ien nien ping ho k'ai le tung)
千年冰河開了凍 / 馬少波, 辛大明編. — [香港 : s.n., 1977?]
169 p., [2] leaves of plates : ill., music ; 21 cm.
Play.
Reprint of the 1950 ed. published by Hsin hua shu tien Hua tung fen tien, Shanghai.

1. Prostitution—China—Drama. I. Hsin, Ta-ming, joint author. II. Title.
PL2881.S5C5 1977 79-837376

Meng, Hsien-chang, 1894-
(Nan-hai chi chan)
南海激战 : 电影文学剧本 / 孟宪章. — [香港? : s.n., 1978]
88 p. ; 21 cm.
Reprint of the 1978 ed. published by Hu-nan jen min ch'u pan she, Ch'ang-sha.

I. Title.
PL2883.N36N51554 1978 79-838189

Na, Sha.
(Jang chan mo fa tou pa)
讓戰魔發抖吧 : 四幕話劇 / 那沙著. — [香港? : s.n., 1978?]
126 p., [4] leaves of plates : ill. ; 21 cm.
Play.
Reprint of the 1950 ed. published by Hsin hua shu tien, issued in series: Tung pei wen i ts'ung shu.

I. Title.
PL2887.A2J3 1978 79-835743

Pao, Ch'i-ch'üan.
(Shuang t'ien hsiao chiao)
霜天曉角 : 青年守則劇本 / 包起權編著. — [香港? : s.n., 1978?]
160 p. ; 21 cm.
Play.
Reprint of the 1947 ed. published by Cheng chung shu chü, Shanghai.

I. Title.
PL2892.A6317S5 1978 78-840871

Pei-ching hua chü t'uan.
(Pao p'o chih ch'ien)
爆破之前 : 独幕话剧 / 北京话剧团集体创作. — 第1版. — 北京 : 人民文学出版社, 1976.
28 p. ; 19 cm.
"群众演唱"
$0.09

I. Title.
PL2892.E512P3 1976 79-840002

(Pien chiang hsin miao)
边疆新苗 : 六场话剧 / 编剧沙叶新等. — 第1版. — 上海 : 上海人民出版社, 1977.
105 p., [4] leaves of plates : ill. ; 19 cm.
Play.
$0.28

I. Title.
PL2895.Y38P5 79-835954

Shen, Hsi-meng.
(Ni hung teng hsia ti shao ping)
霓虹灯下的哨兵 : 九場話劇 / 沈西蒙, 漠雁, 呂兴臣編劇 ; 沈西蒙执笔. — [香港? : s.n., 1977?]
119 p., [5] leaves of plates : ill. ; 18 cm.
Play.
Reprint of the 1964 ed. published by Chung-kuo hsi chü ch'u pan she, Peking.
I. Mo-yen, joint author. II. Lü, Hsing-ch'en, fl. 1963- joint author. III. Title.
PL2898.N16N5 1977 78-841317

Shen, Hsi-meng.
(Ni hung teng hsia ti shao ping)
霓虹灯下的哨兵 : 九场话剧 / 沈西蒙, 漠雁, 吕兴臣编剧. — 第1版. — 北京 : 人民文学出版社, 1964 t.p. 1978.
119 p., [3] leaves of plates : ill. ; 19 cm.
Play.
$0.32
I. Mo-yen, joint author. II. Lü, Hsing-ch'en, joint author. III. Title.
PL2898.N16N5 1978 78-843611

78 C

Shu, Ch'ing-ch'un, 1898-1966.
(Hsi wang Ch'ang-an)
西望長安 : 五幕六場話劇 / 老舍[舒慶春]. — [香港? : s.n., 1978?]
36 p. ; 26 cm.
Drama.
Reprint ed.

I. Title.
PL2804.C5H76 1978 78-843895

Sun, Wei-shih.
(Ch'u sheng ti t'ai yang)
初升的太阳 : 六场話劇 / 孙维世编剧. — [香港? : s.n., 1978?]
2, 92 p., [4] leaves of plates : ill. ; 21 cm.
Play.
Reprint of the 1977 ed. published by Jen min wen hsüeh ch'u pan she, Peking.
Includes a song in number notation.
I. Title.
PL2905.N34C48 1978 79-838080

Sung, Chih-ti, 1914-1956.
(Ts'ao mu chieh ping)
草木皆兵 : 三幕劇 / 作者宋之的, 夏衍, 于伶. — [香港? : s.n., 1978?]
125 p. ; 21 cm.
Three acts play.
Reprint of the 1946 ed. published by Mei hsüeh ch'u pan she, Chungking.

I. Shen, Tuan-hsien, 1900- joint author. II. Yü, Ling, pseud., joint author. III. Title.
PL2905.N43T7 1978 78-843730

Sung, Kuang-tsu.
(Shui shang chiao t'ung chan)
水上交通站 : 沪剧小戏 / 宋光祖编剧 ; 頤迪雄, 朱洪泽, 谈敬德作曲 ; 陆国源舞台设计. — [香港? : s.n., 1977?]
62 p. : ill. ; 21 cm.
Play.
"此剧根据《解放军文艺》一九七二年十二期杨钟的同名故事编剧"
Includes songs in number notation.
Reprint of the 1976 ed. published by Shang-hai jen min ch'u pan she.
I. Ku, Ti-hsiung. II. Yang, Chung. Shui shang chiao t'ung chan. III. Title.
PL2905.N444S5 1977 79-835254

T'ang, Hsüeh-ch'u.
(Shih fen chung)
十分鈡 : 独幕越剧 / 湯学楚作. — [香港 : s.n., 1978?]
17 p. ; 19 cm.
Reprint of the 1958 ed. published by Tung hai wen i ch'u pan she, Hang-chow; issued in series: Che-chiang sheng wen i ta yüeh chin ta hui ch'uang tso hsüan.
I. Title. II. Series: Che-chiang sheng wen i ta yüeh chin ta hui, 1958. Che-chiang sheng wen i ta yüeh chin ta hui ch'uang tso hsüan.
PL2908.N35914S5 1978 79-836391

T'ao, Hsiung.
(Fan chien tieh)
反間諜 : 三幕劇 / 陶熊著. — [香港? : s.n., 1978?]
4, 146 p. ; 20 cm.
Cover title.
Play.
Reprint of the ed. published by Wen chiang t'u shu kung ssu.

I. Title.
PL2811.O18F3 1978 78-842637

Ts'ao, Yu.
[Plays]
曹禺选集. — 第1版. — 北京 : 人民文学出版社, 1961, t.p. 1978.
426 p., [1] leaf of plates : port. ; 21 cm.
$1.05

I. Title. Title romanized: Ts'ao Yü hsüan chi.
PL2815.A8A6 1978 79-835947

79 C

Ts'ui, Wei, 1912-
(Tung p'ing fu)
東平府 崔嵬著 中南文聯籌委會編 上海 上海 雜誌公司 1951.
128 p. 19 cm. (新戲曲叢書第1輯)
Play.
Series romanized: Hsin hsi ch'ü ts'ung shu.

I. Title.
PL2912.U34T8 C 66-705

CHINESE DRAMA (Continued)

Tung, Mei-k'an.
(Ch'in-huai hsing huo)
秦淮星火 / 董每戡著. — [香港?: s.n., 1977?]
4, 176 p. ; 22 cm.
Play.
Reprint of the 1948 ed. published by Wen t'ung shu chü, Shanghai, in series: Wen i ts'ung shu.
Includes music in number notation.
I. Title.
PL2914.N545C5 1977 — 78-841038

Wang, Sheng-shan.
(Ch'en hsi)
晨曦 / 王生善著. — 臺北：正中書局；九龍：海外總經銷集成圖書公司, 民國67 [1978]
222 p., [1] leaf of plates : ill. ; 20 cm. — (正中文藝叢書)
Play.
Series romanized: Cheng chung wen i ts'ung shu.
I. Title.
PL2919.S487C53 — 79-841877

Wang, Shih-fu, fl. 1295-1307.
[Hsi hsiang chi]
西廂 / [編輯者泰東圖書局]. — [香港 : s.n., 1978?]
2, 11, 306 p. ; 21 cm.
"王實父作"
"凡唱白全依實獲齋藏板...關漢卿所續四齣概行刪去"
Reprint of the 1921 ed. published by T'ai tung t'u shu chü, Shanghai: in series: Ming ch'ü ts'ung k'an; ti 1 chung.
I. T'ai tung t'u shu chü. II. Title. III. Series: Ming ch'ü ts'ung k'an ; ti 1 chung.
Title romanized: Hsi hsiang.
PL2693.H75 1978 — 78-841983

Wang, Wei-ch'eng, fl. 1969-
(Yüan yüan liu ch'ang)
源遠流長 一景四幕社會教育喜劇 王慰誠編劇 [臺北?] 改造出版社 民國58 [1969]
2, 78 p. 19 cm.
Cover title.
I. Title.
PL2919.W35Y8 — 70-838597

Wu-hsi, China. Wen hua chü. Hsi ch'ü ch'uang tso yen chiu shih.
(Hung hua ch'ü)
紅花曲：錫剧 / 无锡市文化局戏曲創作研究室集体創作. — 第1版. — 上海：上海文化出版社, 1965.
106 p., [1] leaf of plates : ill. ; 19 cm. — (华东戏剧丛刊)
Series romanized: Hua-tung hsi chü ts'ung k'an.
Includes songs in number notations.
I. Title.
PL2921.Z73H8 1965 — 78-842290

Yao, Hsin-nung, 1905-
(Ch'ing kung yüan)
清宮怨 / [著者姚克i.e.姚莘農]. — 臺北：聯經出版事業公司, 民國67 [1978]
3, 212 p. ; 20 cm. — (姚克戲劇選集)
Play.
Series romanized: Yao K'o hsi chü hsüan chi.
NT$30.00
I. Title.
PL2923.H73C5 1978 — 79-835601

Yü, Ping-k'un.
(Ta lang t'ao sha)
大浪淘沙 / 于炳坤, 朱道南, 伊琳；于炳坤执笔. — 第1版. — 北京：人民文学出版社, 1977.
88 p., [2] leaves of plates : ill. ; 19 cm. — (电影文学剧本)
Play.
Series romanized: Tien ying wen hsüeh chü pen.
$0.25
I. Chu, Tao-nan, joint author. II. I, Lin, joint author. III. Title.
PL2928.P49T3 — 78-842102

Yü, Ta-kang, 1908-1977.
(Liao yin ko chü tso)
寥音閣劇作 / 俞大綱著. — 臺北：俞大綱基金會：發行者河洛圖書出版社, 民國66 [1977]
127 p. ; 21 cm. — (俞大綱全集 ; 第1集)
CONTENTS: 國劇簡介.—李亞仙.—王魁負桂英.—楊八妹
NT$60.00
I. Title.
PL2831.T35A6 1977, vol. 1 — 79-838660

—YÜAN DYNASTY, 1260-1368

(Ku pen Yüan Ming tsa chü)
孤本元明雜劇 北京 中國戲劇出版社 1958.
4 v. (various pagings) 22 cm.
涵芬樓藏版
據商務印書館原紙型重印
1. Chinese drama—Yüan dynasty, 1260-1368. 2. Chinese drama—Ming dynasty, 1368-1644. I. Title: Yüan Ming tsa chü.
PL2594.K84 1958 — C 59-585

(Ku pen Yüan Ming tsa chü)
孤本元明雜劇 / 王實甫等撰. — 臺1版. — 臺北：臺灣商務印書館, 民國66 [1977]
10 v. ; 20 cm.
"涵芬樓藏版"
NT$1,500.00
1. Chinese drama—Yüan dynasty, 1260-1368. 2. Chinese drama—Ming dynasty, 1368-1644. I. Wang, Shih-fu, fl. 1295-1307. II. Title: Yüan Ming tsa chü.
PL2594.K84 1977 — 78-842997

[**Sui, Shu-sen**], 1906- ed.
(Ch'üan Yüan san ch'ü)
全元散曲 / 本局編輯. — 台北：啓業書局, 民國66 [1977]
2 v. (34, 1924 p.) ; 22 cm. — (中國學術名著 : 詩詞類)
Series romanized: Chung-kuo hsüeh shu ming chu : Shih tz'u lei.
Reprint of the 1964 ed. published by Chung-hua shu chü, Peking, under the editor: Sui Shu-sen; with pref. reset and partially reworded.
Bibliography: p. 17-25 (1st group)
Includes indexes.
NT$300.00
(Continued on next card)
1. Chinese drama—Yüan dynasty, 1260-1368. I. Ch'i yeh shu chü. II. Title.
PL2594.S88 1977 — 79-835916

(Yüan jen tsa chü hsüan chu)
元人雜劇選注 / [主編者楊家駱]. — 5版. — 臺北：世界書局, 民國64 [1975]
4, 548 p. : ill. ; 20 cm. — (曲學叢書 : 第一集 ; 第4冊) (中國學術名著 : 第二輯)
Series 1 romanized: Chung-kuo hsüeh shu ming chu : Ti erh chi.
1. Chinese drama—Yüan dynasty, 1260-1368. I. Yang, Chia-lo. II. Series: Ch'ü hsüeh ts'ung shu : Ti i chi ; ti 4 ts'e.
PL2594.Y823 1975 — 79-838680

(Yüan jen tsa chü hsüan chu)
元人雜劇選注 / 本社編輯. — 台北：成偉出版社, [1978?]
4, 548 p. : ill. ; 22 cm. — (中國學術名著叢刊)
Series romanized: Chung-kuo hsüeh shu ming chu ts'ung k'an.
Reprint of the 1976 ed. published by Shih chieh shu chü, T'ai-pei, in series: Shih chieh wen k'u : Ssu pu k'an yao.
1. Chinese drama—Yüan dynasty, 1260-1368. I. Ch'eng wei ch'u pan she.
PL2594.Y823 1978 — 78-843438

—YÜAN DYNASTY, 1260-1368
—HISTORY AND CRITICISM

Ch'i, Hsiao-feng.
(Yüan tai kung an chü yen chiu)
元代公案劇研究 / 齊曉楓撰. — [台北?: s.n., 1978?]
3, 207, [5] p. (on double leaves) ; 27 cm.
Reproduced from typescript.
Originally presented as the author's thesis (M.A.), Fu jen ta hsüeh, 1975.
Bibliography: p. [209]-[212]
1. Chinese drama—Yüan dynasty, 1260-1368—History and criticism. I. Title.
PL2384.C4 1978 — 78-843913

Perng, Ching-Hsi.
Double jeopardy : a critique of seven Yüan courtroom dramas / by Ching-Hsi Perng. — Ann Arbor : Center for Chinese Studies, University of Michigan, 1978.
xi, 178 p. ; 23 cm. — (Michigan papers in Chinese studies ; no. 35)
Bibliography: p. 173-178.
ISBN 0-89264-035-9 : $4.00
1. Chinese drama—Yüan dynasty, 1260-1368—History and criticism. 2. Trials—China—Drama. I. Title. II. Series.
PL2384.P4 — 895.1'2'4093 — 78-13029
78 — MARC

Ts'ung, Ching-wen, 1927-
(Yüan tai lien ai chü shih chung chi ch'iao yen chiu)
元代戀愛劇十種技巧研究 / 叢靜文著. — 臺北：臺灣商務印書館, 民國67 [1978]
2, 2, 1, 133 p. ; 19 cm. — (人人文庫 ; 2403-2404)
Series romanized: Jen jen wen k'u.
Bibliography: p. 133.
NT$18.00
1. Chinese drama—Yüan dynasty, 1260-1368—History and criticism. 2. Love in literature. I. Title.
PL2384.T77 — 79-841106

—MING DYNASTY, 1368-1644

(Ku pen Yüan Ming tsa chü)
孤本元明雜劇 北京 中國戲劇出版社 1958.
4 v. (various pagings) 22 cm.
涵芬樓藏版
據商務印書館原紙型重印
1. Chinese drama—Yüan dynasty, 1260-1368. 2. Chinese drama—Ming dynasty, 1368-1644. I. Title: Yüan Ming tsa chü.
PL2594.K84 1958 — C 59-585

(Ku pen Yüan Ming tsa chü)
孤本元明雜劇 / 王實甫等撰. — 臺1版. — 臺北：臺灣商務印書館, 民國66 [1977]
10 v. ; 20 cm.
"涵芬樓藏版"
NT$1,500.00
1. Chinese drama—Yüan dynasty, 1260-1368. 2. Chinese drama—Ming dynasty, 1368-1644. I. Wang, Shih-fu, fl. 1295-1307. II. Title: Yüan Ming tsa chü.
PL2594.K84 1977 — 78-842997

(Ming Ch'ing Min-nan hsi ch'ü ssu chung)
明清閩南戲曲四種 / 吳守禮, 林宗毅同輯. 清乾隆間刊同窓琴書記校理；清光緒間刊荔枝記校理 / 吳守禮著. — [台北?]：林宗毅, 1978.
3 v. in case : ill. ; 26 cm. — (定靜堂叢書 ; 第3函)
Colophon title.
Series romanized: Ting ching t'ang ts'ung shu.
"函裝壹百部之第柒陸號"
CONTENTS: [1] 明清閩南戲曲四種 : 嘉靖刊本「荔鏡記」戲文. 萬曆刊本「荔枝記」戲文. 順治刊本「荔枝記」戲文. 萬曆刊本「金花女」戲文.—[2]-[3] 吳守禮 清乾隆間刊「同窓琴書記」校理. 清光緒間刊荔枝記校理.
1. Chinese drama—China—Fukien (Province) 2. Chinese drama—Ming dynasty, 1368-1644. 3. Chinese drama—Ch'ing dynasty, 1644-1912. I. Wu, Shou-li. II. Lin, Tsung-i. III. Wu, Shou-li. Ch'ing Ch'ien-lung chien k'an T'ung ch'uang ch'in shu chi chiao li. 1978. IV. Wu, Shou-li. Ch'ing Kuang-hsü chien k'an Li chih chi chiao li. 1978.
PL3031.F82M5 — 78-842452

—MING DYNASTY, 1368-1644
—HISTORY AND CRITICISM

Tseng, Yung-i.
(Ming tsa chü kai lun)
明雜劇概論 / 曾永義著；鄭騫, 張敬指導. — [台北]：嘉新水泥公司文化基金會, 民國67 [1978]
3, 3, 405 p. ; 22 cm. — (嘉新水泥公司文化基金會研究論文 ; 第302種) (嘉新水泥公司文化基金會叢書)
Series 1 romanized: Chia hsin shui ni kung ssu wen hua chi chin hui yen chiu lun wen. Series 2 romanized: Chia hsin shui ni kung ssu wen hua chi chin hui ts'ung shu.
"國立台灣大學中國文學研究所論文"
Appendices (p. 380-405): 1. 明代雜劇年表.—2. 參考書目.
1. Chinese drama—Ming dynasty, 1368-1644—History and criticism. I. Title.
PL2386.T73 — 79-836544

CHINESE DRAMA
—MING DYNASTY, 1368-1644
—HISTORY AND CRITICISM
(Continued)

Tseng, Yung-i.
(Ming tsa chü yen chiu)
明雜劇研究 / 曾永義. — [台北?]：曾永義, [1978?]
5 v. (on double leaves) : 27 cm.
Spine title.
Reproduced from ms. copy.
Appendices (at end of v. 5): 1. 明代雜劇體製提要.—2. 明代雜劇年表.—3. 參考書目.
1. Chinese drama—Ming dynasty, 1368–1644—History and criticism. I. Title.
PL2386.T74 1978 78–842996

—CH'ING DYNASTY, 1644-1912

(Hung lou meng hsi ch'ü chi)
紅樓夢戲曲集 / 阿英編. — 第1版. — 北京：中華書局, 1978.
2 v. (1, 4, 820 p.) ; 21 cm.
$2.45
1. Chinese drama—Ch'ing dynasty, 1644–1912. 2. Ts'ao, Chan, ca. 1717–1763. Hung lou meng. I. Ch'ien, Hsing-ts'un, 1905– II. Ts'ao, Chan, ca. 1717–1763. Hung lou meng.
PL2597.H8 78–843597

(Ming Ch'ing Min-nan hsi ch'ü ssu chung)
明清閩南戲曲四種 / 吳守禮, 林宗毅同輯. 清乾隆間刊同窓琴書記校理；清光緒間刊荔枝記校理 / 吳守禮著. — [台北?]：林宗毅, 1978.
3 v. in case : ill. ; 26 cm. — (定静堂叢書；第3函)
Colophon title.
Series romanized: Ting ching t'ang ts'ung shu.
"函裝壹百部之第柴陸號"
CONTENTS: [1] 明清閩南戲曲四種：嘉靖刊本「荔鏡記」戲文. 萬曆刊本「荔枝記」戲文. 順治刊本「荔枝記」戲文. 萬曆刊本「金花女」戲文.—[2]–[3] 吳守禮 清乾隆間刊「同窓琴書記」校理. 清光緒間刊荔枝記校理.
1. Chinese drama—China—Fukien (Province) 2. Chinese drama—Ming dynasty, 1368–1644. 3. Chinese drama—Ch'ing dynasty, 1644–1912. I. Wu, Shou-li. II. Lin, Tsung-i. III. Wu, Shou-li. Ch'ing Ch'ien-lung chien k'an T'ung ch'uang ch'in shu chi chiao li. 1978. IV. Wu, Shou-li. Ch'ing Kuang-hsü chien k'an Li chih chi chiao li. 1978.
PL3031.F82M5 78–842452

—20TH CENTURY

(Cheng to)
争夺：曲艺专辑. — [香港?：s. n., 1977?]
33 p. ; 21 cm.
Cover title.
At head of title: Ch'ün chung yen ch'ang.
Reprint of the 1975 ed. published by Jen min wen hsüeh ch'u pan she, Peking.
1. Chinese drama—20th century.
PL2603.C367 1977 78–840775

(Hai chün ch'ü i tso p'in hsüan)
海军曲艺作品选 / 中国人民解放军海军政治部文化部編. — [香港?：s. n., 1978?]
2, 2, 347 p. ; 21 cm.
Reprint of the 1978 ed. published by Jen min wen hsüeh ch'u pan she, Peking.
1. Chinese drama—20th century. I. China (People's Republic of China, 1949–). Chung-kuo jen min chieh fang chün. Hai chün. Cheng chih pu. Wen hua pu.
PL2603.H3 1978 79–838130

(Hua chu hsi tao kuo tsan shao so)
华主席到过咱哨所：文艺演唱集 / [福建人民出版社编辑出版]. — [香港?：s. n., 1978?]
123 p. ; 21 cm.
Includes songs in number notation.
Reprint of the 1977 ed. published by Fu-chien jen min ch'u pan she.
1. Chinese drama—20th century. I. Fu-chien jen min ch'u pan she, Foochow, China.
PL2603.H825 1978 79–835931

(Huan hu sheng li ti shih yüeh)
欢呼胜利的十月：工农兵演唱集. — 第1版. — [西安]：陕西人民出版社, 19
v. ; 19 cm.
Includes songs in number notation.
$0.23 (v.7)
1. Chinese drama—20th century.
PL2603.H83 79–836262

(Huo, t'ung hung ti huo)
火·通红的火. — 上海：上海人民出版社, 1976.
376 p. : ill. ; 21 cm. — (朝霞丛刊)
Series romanized: Chao hsia ts'ung k'an.
CONTENTS: 中篇小说 朱敏慎 初战—段荃法, 张有德, 樊俊智 青砖歌—话剧剧本 陆天明 火·通红的火四幕话剧·征文选刊
$0.82
1. Chinese fiction—20th century. 2. Chinese drama—20th century. I. Lu, T'ien-ming. II. Chu, Min-shen. Ch'u chan. 1976. III. Tuan, Ch'üan-fa. Ch'ing chuan ko. 1976.
PL2653.H88 79–835741

(I chiu san liu nien Chung-kuo tsui chia tu mu chü chi)
一九三六年中國最佳獨幕劇集 / 阿英編. — [香港?：s. n., 1977?]
278 p. ; 22 cm.
Reprint of the 1937 ed. published by Hsi chü shih tai ch'u pan she, Shanghai.
1. Chinese drama—20th century. 2. One-act plays, Chinese. I. Ch'ien, Hsing-ts'un, 1905– II. Title: Chung-kuo tsui chia tu mu chü chi.
PL2602.I18 1977 78–841001

(K'ang chan hsi chü hsüan)
抗戰戲劇選：集體創作 / 胡春冰編. — [香港：s. n., 1978?]
4, 84 p., [3] leaves of plates : ill. ; 21 cm.
Reprint of the 1938 ed. published by Nu hou ch'u pan she.
1. Chinese drama—20th century. 2. Sino-Japanese Conflict, 1937–1945—Drama. I. Hu, Ch'un-ping.
PL2602.K27 1978 79–835175

(Mu ch'in ti hsin)
母親的心 / 劉滄浪等著. — [香港 ?：s. n., 1978?]
11, 125 p., [2] leaves of plates : ill. ; 19 cm.
Reprint of the 1951 ed. published by Sheng huo, tu shu, hsin chih san lien shu tien, Peking, in series: Hsi chü ch'uang tso ts'ung shu.
CONTENTS: 劉滄浪 母親的心.—李東絲 吃驚病.—石慰慈 滾出中國去.
1. Chinese drama—20th century. 2. One-act plays, Chinese. I. Liu, Ts'ang-lang. II. Li, Shu-ssu. Ch'ih ching ping. 1978. III. Shih, Wei-tz'u. Kun ch'u Chung-kuo ch'ü. 1978. IV. Series: Hsi chü ch'uang tso ts'ung shu.
PL2603.M76 1978 78–843210
78 C

(Shang-hai shuo ch'ang chi)
上海说唱集 / 本社编. — [香港?：s. n., 1978]
ii, 2, 184 p. ; 22 cm.
Includes songs in number notation.
Reprint of the 1978 ed. published by Shang-hai wen i ch'u pan she, Shanghai, in series: Ch'ü i ch'uang tso ts'ung shu.
1. Chinese drama—China—Shanghai. 2. Chinese drama—20th century. I. Shang-hai wen i ch'u pan she. II. Series: Ch'ü i ch'uang tso ts'ung shu.
PL3032.S482S48 1978 79–836857
79 C

(Shen ch'ieh huai nien ching ai ti Chou tsung li)
深切怀念敬爱的周总理：小戏集 / 人民文学出版社编辑部编. — 第1版. — 北京：該社：新华书店北京发行所发行, 1977.
2, 217 p. ; 19 cm.
$0.44
1. Chinese drama—20th century. 2. Chou, En-lai, 1898–1976—Drama. I. Jen min wen hsüeh ch'ü pan she, Peking.
PL2603.S465 79–837439

(Shen-yang pu tui ch'ü i tso p'in hsüan)
沈阳部队曲艺作品选 / 中国人民解放军沈阳部队政治部文化部编. — 北京：人民文学出版社, 1977.
2, 3, 501 p. ; 21 cm.
$1.00
1. Soldiers' writings, Chinese. 2. Chinese drama—20th century. I. China (People's Republic of China, 1949–). Chung-kuo jen min chieh fang chün. Shen-yang pu tui. Cheng chih pu. Wen hua pu. II. Title: Ch'ü i tso p'in hsüan.
PL2515.5.S6S5 78–843456

(Ting Hsiu-ch'in shen i tzu)
丁秀芹审椅子. — 第1版. — 北京：人民出版社：新华书店发行, 1976.
47 p. ; 19 cm.
Colophon title.
"曲艺专辑"
CONTENTS: 丁秀芹审椅子.—梨园哨卡.—阵地哨兵.
1. Chinese drama—20th century. I. Ch'en, Tseng-chih. Ting Hsiu-ch'in shen i tzu. 1976. II. Chu, Chen-yüeh. Li yüan shao k'a. 1976. III. Liu, Ssu-ch'ang. Chen ti shao ping. 1976.
PL2603.T59 79–840790

(Ts'ung nu li tao chiang chün)
从奴隶到将军 / [本社编]. — 第1版. — 上海：上海文艺出版社：新華書店上海发行所发行, 1978.
414 p., [5] leaves of plates : ill. ; 21 cm. — (电影与戏剧丛刊)
Series romanized: Tien ying yü hsi chü ts'ung k'an.
$1.10
1. Chinese drama—20th century. I. Shang-hai wen i ch'u pan she.
PL2603.T78 79–839614

(T'uan chieh ch'ien chin)
團結前進 / 丁毅 ... [et al.] 編劇；四野政治部文工團改編；[編輯者中南文學術藝 i. e. 藝術界聯合會籌委會. — 漢口：新華書店中南總分店, 1950.
155 p. ; 18 cm. — (長江文藝叢書)
Includes songs in number notation.
Series romanized: Ch'ang-chiang wen i ts'ung shu.
CONTENTS: 團結前進.—胡果剛 沃老大娘瞅「後兜.—荒草 買紅薯.—寧森 找嚮導.—丁洪, 唐克 好班長.
1. Chinese drama—20th century. I. Ting, I, 1921– II. Ssu yeh cheng chih pu wen kung t'uan. III. Chung nan wen hsüeh i shu chieh lien ho hui ch'ou wei hui.
PL2603.T82 79–839166

—20TH CENTURY
—HISTORY AND CRITICISM

(Chung-kuo hua chü yün tung wu shih nien shih liao chi)
中國話劇運動五十年史料集：1907-1957 / 「中國話劇運動五十年史料集」編委會；[田漢 ... et al.]. — 香港：文化資料供應社, 1978 printing.
2, 316 p., [9] leaves of plates : ill. ; 21 cm.
"一九五八年初由中國劇劇出版社出版了第一輯...可惜其後就未見有第二輯...本書係根據一九五八年出版的第一輯重印的"
1. Chinese drama—20th century—History and criticism. I. T'ien, Han, 1890– II. Chung-kuo hua chü yün tung wu shih nien shih liao chi pien wei hui.
PN2874.C55 1958b 78–842893

—20TH CENTURY—HISTORY AND CRITICISM—ADDRESSES, ESSAYS, LECTURES

T'ang, Wen-piao.
(T'ien kuo pu shih wo men ti)
天國不是我們的 / 著者唐文標. — 香港：文化出版事業公司, [1978?]
287 p., [2] leaves of plates : ill. ; 19 cm.
Colophon title.
$7.00
1. Chinese poetry—20th century—History and criticism—Addresses, essays, lectures. 2. Chinese drama—History and criticism—Addresses, essays, lectures. 3. Moving-picture plays—History and criticism—Addresses, essays, lectures. I. Title.
PL2333.T32 1978 79–835519

CHINESE DRAMA (Continued)

—BIBLIOGRAPHY

Cheng, Chen-to, 1897–1958.
(Chung-kuo hsi ch'ü ti hsüan pen)
中國戲曲的選本 / ⌈鄭振鐸著⌉. — 香港：龍門書店，1969.
42 p. ; 21 cm. — (中國文學研究叢編 ; 第1輯)

1. Chinese drama—Bibliography. I. Title. II. Series: Chung-kuo wen hsüeh yen chiu ts'ung pien ; ti 1 chi.

Z3108.L5C4653 1969 79–838106
[PL2357]

Chung, Ssu-ch'eng, fl. 1321.
(Lu kuei pu)
錄鬼薄（外四種）鍾嗣成等著 上海 古典文学出版社 1957.
7, 2, 374, 37 p. facsim. 21 cm.
錄鬼薄補校 孟稱舜刊本錄鬼薄: p. 352–374.
PARTIAL CONTENTS: 錄鬼薄續編 明·闕名撰—太和正音譜 2卷 朱權撰—曲品 2卷 呂天成撰—傳奇品 2卷 高奕撰

1. Chinese drama—Bibliography. I. Title.

Z3108.L5C54 1957 C 60–2390
[PL2357]
rev

Chung, Ssu-ch'eng, fl. 1321.
(Lu kuei pu)
錄鬼簿，外四種 / 鍾嗣成等著. — 新1版. — 上海：上海古籍出版社：新华書店上海发行所发行，1978.
2, 2, 374, 37 p., ⌈1⌉ leaf of plates ; 21 cm.
CONTENTS: 鍾嗣成 錄鬼簿 2卷.—錄鬼簿續編 1卷.—朱權 太和正音譜 2卷.—呂天成 曲品 2卷.—高奕 傳奇品 2卷.—錄鬼簿補校.—孟稱舜刊本錄鬼簿.
1. Chinese drama—Bibliography. I. Title.

Z3108.L5C54 1978 79–837249
[PL2357]

—BIOGRAPHY see Dramatists, Chinese

—HISTORY AND CRITICISM

⌈**Huang, Hsü-wu**⌉
(Shih tz'u ch'ü yen chiu)
詩詞曲研究 / ⌈編輯者本社編輯部⌉. — 臺北：莊嚴出版社，民國67 ⌈1978⌉
4, 243 p. ; 22 cm. — (古典大學叢書 ; 2)
Reprint of an earlier ed. by Huang Hsü-wu.
NT$60.00
1. Chinese poetry—History and criticism. 2. Chinese drama—History and criticism. 3. Tz'u—History and criticism. I. Chuang yen ch'u pan she. Pien chi pu. II. Title. III. Series: Ku tien ta hsüeh ts'ung shu ; 2.

PL2307.H795 1978 78–840909

Scott, Adolphe Clarence, 1909-
The classical theatre of China / by A. C. Scott ; with ill. by the author. — Westport, Conn. : Greenwood Press, 1978.
250, ⌈6⌉ p., ⌈3⌉ leaves of plates : ill. ; 24 cm.
Reprint of the 1957 ed. published by Macmillan, New York.
Bibliography: p. ⌈9⌉
Includes indexes.
ISBN 0-313-20022-X
1. Theater—China. 2. Chinese drama—History and criticism. I. Title.
PN2871.S4 1978 792'.0951 77-16447
77⌈r79⌉rev MARC

Yang, Tsung-chen, 1919–
(Chung-kuo hsi ch'ü shih)
中國戲曲史 / ⌈著作者孟瑤 i. e. 楊宗珍 ; 編輯者傳記文學雜誌社⌉. — 再版. — 臺北：傳記文學出版社，民國65 ⌈1976⌉
4 v. (7, 23, 910 p.) ; 19 cm. — (傳記文學叢書 ; 66)
Vol. 4 consists principally of excerpts from Chinese opera in number notation.
Series romanized: Chuan chi wen hsüeh ts'ung shu.
Bibliography: p. 905–910.

NT$30.00 per vol.

1. Chinese drama—History and criticism. I. Chuan chi wen hsüeh tsa chih she. II. Title.
PL2357.Y3 1976 78–843462

—HISTORY AND CRITICISM
—ADDRESSES, ESSAYS, LECTURES

T'ang, Wen-piao.
(T'ien kuo pu shih wo men te)
天國不是我們的 / 唐文標. — 臺北：聯經出版事業公司，民國65 ⌈1976⌉
288 p., ⌈3⌉ leaves of plates : ill. ; 20 cm. — (文化叢刊)
Series romanized: Wen hua ts'ung k'an.
Includes bibliographical references.
CONTENTS: 電影·戲劇的盲視.—現代詩的狂想.—大地的事.
NT$10.00
1. Chinese poetry—20th century—History and criticism—Addresses, essays, lectures. 2. Chinese drama—History and criticism—Addresses, essays, lectures. 3. Moving-picture plays—History and criticism—Addresses, essays, lectures. I. Title.
PL2333.T32 1976 76–840657

—TRANSLATIONS INTO ENGLISH

Eight Chinese plays from the thirteenth century to the present / translated ⌈from the Chinese⌉ with an introduction by William Dolby. — London : Elek, 1978.
164 p. 23 cm. GB78-13133
Includes bibliographical references.
CONTENTS: The battling doctors, excerpt from Cai Shun shares the mulberries (Cai Shun fen-shen), Yuanben play attributed to Liu Tangqing (late thirteenth or early fourteenth century).—Grandee's son takes the wrong career (Huan-men zi-di cuo li-shen), Nanxi play (anon., late thirteenth or early fourteenth century).—Qiu Hu tries to seduce his own wife (Qiu Hu xi-qi), Yuan zaju play by Shi Junbao (1192-1276).—Secret liaison with Chancellor Bo Pi', Act VII from Washing silk (Wan-sha ji), Chuanqi play by Liang Chenyu (1520-c. 1580).—Wolf of Mount Zhong (Zhong-sham lang), Brief zaju play by Wang Jiusi (1468-1551).—Buying rouge (Mai yan-zhi), Early Qing farce (anon., eighteenth century).—Hegemon King says farewell to his queen (Ba-wany bie-ji), Peking Opera, version by Mei Lanfang (1894-1961).—Identifying footprints in the snow (Ping-xue bian-zong), Chuanju drama, version (anon.) of the mid-1950's.
1. Chinese drama—Translations into English. 2. English drama—Translations from Chinese. I. Dolby, A. W. E.
PL2658.E5E37 1978b 895.1'2'008 78-324894
79 MARC

—CHINA—FURIEN (PROVINCE)

(Ming Ch'ing Min-nan hsi ch'ü ssu chung)
明清閩南戲曲四種 / 吳守禮，林宗毅同輯. 清乾隆間刊同窓琴書記校理；清光緒間刊荔枝記校理 / 吳守禮著. — ⌈台北?⌉：林宗毅，1978.
3 v. in case : ill. ; 26 cm. — (定静堂叢書 ; 第3函)
Colophon title.
Series romanized: Ting ching t'ang ts'ung shu.
"函裝壹百部之第柒陸號"
CONTENTS: ⌈1⌉ 明清閩南戲曲四種：嘉靖刊本「荔鏡記」戲文. 萬曆刊本「荔枝記」戲文. 順治刊本「荔枝記」戲文. 萬曆刊本「金花女」戲文.—⌈2⌉-⌈3⌉ 吳守禮 清乾隆間刊「同窓琴書記」校理. 清光緒間刊荔枝記校理.

1. Chinese drama—China—Fukien (Province) 2. Chinese drama—Ming dynasty, 1368–1644. 3. Chinese drama—Ch'ing dynasty, 1644–1912. I. Wu, Shou-li. II. Lin, Tsung-i. III. Wu, Shou-li. Ch'ing Ch'ien-lung chien k'an T'ung ch'uang ch'in shu chi chiao li. 1978. IV. Wu, Shou-li. Ch'ing Kuang-hsü chien k'an Li chih chi chiao li. 1978.
PL3031.F82M5 78–842452

78 C

—CHINA—KIANGSU (PROVINCE)

(Hsi chü hsiao hsi k'ao)
錫剧小戲考 / 言前編. — 第1版. — 上海：上海文化出版社，1058.
2, 118 p., ⌈3⌉ leaves of plates : ill. ; 15 cm.

1. Chinese drama—China—Kiangsu (Province) I. Yen, Ch'ien.
PL3031.K552H7 78–840599

—CHINA—SHANGHAI

(Shang-hai shuo ch'ang chi)
上海说唱集 / 本社编. — ⌈香港?：s. n., 1978⌉
ii, 2, 184 p. ; 22 cm.
Includes songs in number notation.
Reprint of the 1978 ed. published by Shang-hai wen i ch'u pan she, Shanghai, in series: Ch'ü i ch'uang tso ts'ung shu.

1. Chinese drama—China—Shanghai. 2. Chinese drama—20th century. I. Shang-hai wen i ch'u pan she. II. Series: Ch'ü i ch'uang tso ts'ung shu.
PL3032.S482S48 1978 79–836857

CHINESE DRAMATISTS see Dramatists, Chinese

CHINESE DRAWING see Drawing, Chinese

CHINESE EASTERN RAILWAY

China. Treaties, etc., 1875-1908 (Ch'ing Te-tsung)
⌈Chung O ting li Chi-lin mu chih ho t'ung. Japanese⌉
伐木ニ關スル露(東支鐵道會社)支(吉林省)間協定. — ⌈新京：滿鐵經濟調査會，1934⌉
8 leaves : 27 cm. — (北鐵林區接收及經營ニ關スル對策案參考資料 : 第2)
Cover title.
At head of title: Gokuhi (rubber stamped)
1. Lumber trade—China—Heilungkiang (Province) 2. Chinese Eastern Railway. 3. China—Commercial treaties. 4. Russia—Commercial treaties. I. Russia. Treaties, etc., 1894-1917 (Nicholas II) II. Chinese Eastern Railway. III. Heilungkiang, China (Province) IV. Title. V. Series: Hokutetsu rinku sesshū oyobi keiei ni kansuru taisakuan sankō shiryō : dai 3.
Title romanized: Batsuboku ni kansuru Ro-Shi, Kokuryūkō-shō, kan no kyōtei.
K4611.L86C57416 79–803803

China. Treaties, etc., 1875-1908 (Ch'ing Te-tsung)
⌈Hei-lung chiang t'ieh lu kung ssu fa mu ho t'ung. Japanese⌉
伐木ニ關スル露(東支鐵道會社)支(黒龍江省)間ノ協定. — ⌈新京：滿鐵經濟調査會，1934?⌉
8 leaves : 27 cm. — (北鐵林區接收及經營ニ關スル對策案參考資料 : 第3)
Cover title.
At head of title: Gokuhi (rubber stamped)
1. Lumber trade—China—Kirin (Province) 2. Chinese Eastern Railway. 3. China—Commercial treaties. 4. Russia—Commercial treaties. I. Russia. Treaties, etc., 1894-1917 (Nicholas II) II. Chinese Eastern Railway. III. Kirin, China (Province) IV. Title. V. Series: Hokutetsu rinku sesshū oyobi keiei ni kansuru taisakuan sankō shiryō : dai 2.
Title romanized: Batsuboku ni kansuru Ro-Shi, Kitsurin-shō, kan no kyōtei.
K4611.L86C57316 79–803802

—ADDRESSES, ESSAYS, LECTURES

Ning, Hua-t'ing.
(Jih O tui chih chung chih Chung-tung t'ieh lu)
日俄對峙中之中東鐵路 / 甯華庭著. — ⌈香港?：s. n., 1978?⌉
58 p. ; 18 cm.
Reprint of the 1931 ed. published by Liang yu t'u shu yin shua kung ssu, Shanghai, in series: I chiao ts'ung shu; ti 16 chung.

1. Chinese Eastern Railway—Addresses, essays, lectures. 2. Russians in Manchuria—Addresses, essays, lectures. 3. Japanese in Manchuria—Addresses, essays, lectures. I. Title.
HE3290.C45N56 1978 78–841633

—FINANCE—STATISTICS

Minami Manshū Tetsudō Kabushiki Kaisha. Sōmubu. Shiryōka.
(Hokutetsu soshiki oyobi zaisan chōsho)
北鐵組織及財產調書. — ⌈大連：南滿洲鐵道株式會社⌉總務部資料課，昭和9 ⌈1934⌉
155 leaves ; 27 cm. — (綜合情報 ; 9, 第16號)
Cover title.
At head of title: Gokuhi (rubber stamped)
1. Chinese Eastern Railway—Management—Statistics. 2. Chinese Eastern Railway—Finance—Statistics. I. Title. II. Series: Minami Manshū Tetsudō Kabushiki Kaisha. Sōmubu. Shiryōka. Sōgō jōhō ; 9, dai 16-gō.
HE3290.C45M58 1934 78–816155

CHINESE EASTERN RAILWAY (Continued)

—MANAGEMENT STATISTICS

Minami Manshū Tetsudō Kabushiki Kaisha. Sōmubu. Shiryōka.
(Hokutetsu soshiki oyobi zaisan chōsho)
北鐵組織及財産調書. — ⌊大連：南滿洲鐵道株式會社⌋總務部資料課, 昭和9 ⌊1934⌋
155 leaves ; 27 cm. — (綜合情報 ; 9, 第16號)

Cover title.
At head of title: Gokuhi (rubber stamped)
1. Chinese Eastern Railway—Management—Statistics. 2. Chinese Eastern Railway—Finance—Statistics. I. Title. II. Series: Minami Manshū Tetsudō Kabushiki Kaisha. Sōmubu. Shiryōka. Sōgō jōhō ; 9, dai 16-gō.

HE3290.C45M58 1934 78-816155

CHINESE EASTERN RAILWAY. ECONOMIC BUREAU

Minami Manshū Tetsudō Kabushiki Kaisha. Sōmubu. Shiryōka.
(Kyū Hokutetsu Keizai Chōsakyoku no gaibō)
舊北鐵經濟調査局ノ概貌. — ⌊大連：南滿洲鐵道株式會社⌋總務部資料課, 昭和10 ⌊1935⌋
a-b, 27 leaves ; 27 cm.

Cover title.
At head of title: Hi (rubber stamped)
"舊北鐵經濟調査局發行圖書及地圖一覽": leaves 19-23.
1. Chinese Eastern Railway. Economic Bureau. I. Title.

H62.5.C5M59 1935 78-816491

CHINESE ESSAYS

Hsü, Pao-kuang, d. 1723.
⌊Chung-shan ch'uan hsin lu. Japanese & Chinese⌋
中山傳信錄：⌊6卷 / 徐葆光撰⌋；長澤規矩也解題. — ⌊東京⌋：古典研究會：發賣所 汲古書院, 昭和52 ⌊1977⌋
388 p. : ill. ; 25 cm. — (和刻本漢籍隨筆集 ; 15) Ja 77-22330

In Chinese with Japanese reading marks.
Photoreprint ed.
CONTENTS: 徐葆光撰 中山傳信錄—高濂撰 遵生寶訓—姜紹書 韻石齋筆談—葛洪著 涉史隨筆—孟棨著 本事詩—風密著 浩然齋雅談—范攄撰 雲溪友議—袾宏著 竹窓隨筆
¥4500

1. Chinese essays. I. Nagasawa, Kikuya, 1902- II. Title. III. Series: Wakokubon Kanseki zuihitsu shū ; 15.
Title romanized: Chūzan denshinroku.

PL2606.W3 vol. 15 79-809118

J

Li, Lung.
(Chüeh lien ch'i wen hsü chi)
絕聯奇文續集 / 李隆編著. — 臺北：星光出版社, 民國67 ⌊1978⌋
216 p. ; 19 cm. — (雙子星叢書 ; 59)
Series romanized: Shuang tzu hsing ts'ung shu.
"文章·詩歌·對聯·謎語"
NT$38.00

1. Couplets. 2. Chinese poetry. 3. Chinese essays. 4. Riddles, Chinese. I. Title.

PL2519.C7L464 79-836266

79 C

(Li tai san wen hsüan)
歷代散文選 / ⌊編著者王熙元 ... et al.⌋. — 臺北：南嶽出版社, 民國65 ⌊1976⌋
2, 10, 716 p. ; 22 cm.
NT$180.00

1. Chinese essays. I. Wang, Hsi-yüan.

PL2607.L45 79-835371

Pan, Ku, 32-92.
⌊Pai hu t'ung. Japanese & Chinese⌋
白虎通德論 / ⌊班固撰 ; 郎壁金校⌋ ; 長澤規矩也解題. — ⌊東京⌋：古典研究會：汲古書院発売, 昭和47 ⌊1972⌋, 昭和54 ⌊1979⌋ printing.
333 p. ; 25 cm. — (和刻本漢籍隨筆集 ; 10) Ja***

In Chinese, with Japanese reading marks.
Photoreprint ed.
CONTENTS: 班固 白虎通德論—應劭 風俗通義—顏之推 顏氏家訓—蔡邕 獨斷—崔豹 古今注—范成大 桂海虞衡志—嵇含 南方草木狀—王嘉 拾遺記
¥4500

1. Chinese essays. I. Lang, Pi-chin. II. Nagasawa, Kikuya, 1902- III. Title. IV. Series: Wakokubon Kanseki zuihitsu shū ; 10.
Title romanized: Byakko tsūtoku ron.

PL2606.W3 vol. 10 79-811830

Tso, Hsüan, fl. 1811.
⌊San yü ou pi. Japanese & Chinese⌋
三餘偶筆 / ⌊左暄著⌋ ; 長澤規矩也解題. — 東京：古典研究会：汲古書院発売, 昭和49 ⌊1974⌋, 昭和51 ⌊1976⌋ printing.
310 p. ; 25 cm. — (和刻本漢籍隨筆集 ; 9) Ja 74-4761

In Chinese, with Japanese reading marks.
Photoreprint ed.
CONTENTS: 左暄 三餘偶筆—趙翼 簷曝雜記—陳星瑞撰 ; 曹夢資編 集古偶錄—張英撰 聰訓齋語—陳繼儒撰 讀書十六觀—陳鑑撰 捶鍼十六觀—林逋撰 省心錄—邵雍 漁樵問答

1. Chinese essays. I. Nagasawa, Kikuya, 1902- II. Title. III. Series: Wakokubon Kanseki zuihitsu shū ; 9.
Title romanized: San'yo gūhitsu.

PL2606.W3 vol. 9 79-809820

Tsung, Lin, ca. 500-ca. 563.
⌊Ching Ch'u sui shih chi. Japanese & Chinese⌋
荊楚歲時記 / ⌊宗懍著⌋ ; 長澤規矩也解題. — ⌊東京⌋：古典研究会：汲古書院発売, 昭和49 ⌊1974⌋, 昭和54 ⌊1979⌋ printing.
305 p. ; 25 cm. — (和刻本漢籍隨筆集 ; 11) Ja***

Photoreprint ed.
In Chinese, with Japanese reading marks.
CONTENTS: 宗懍 荊楚歲時記—小窓別紀—田汝成 熙朝樂事—茶磨山人 清嘉録
¥4500
1. Chinese essays. I. Nagasawa, Kikuya, 1902- II. Title. III. Series: Wakokubon Kanseki zuihitsu shū ; 11.
Title romanized: Keiso saijiki.

PL2606.W3 vol. 11 79-811844

(Wen pien)
文編 / ⌊唐順之編⌋. — ⌊臺北：臺灣⌋商務⌊印書館, 1978⌋
16 v. ; 20 cm. — (四庫全書珍本八集 ; 227-242)
Cover title.
Photoreprint ed. of the Wen Yüan Ko copy.

1. Chinese essays. I. T'ang, Shun-chih, 1507-1560. II. Series: Ssu k'u ch'üan shu. Ssu k'u ch'üan shu chen pen pa chi ; 227-242.

AC149.S7 1978, vol. 227-242 78-843775
[PL2606]

(Wen yen san wen ti p'u t'ung hua fan i)
文言散文的普通话翻译 / 于在春翻译. — 上海：上海教育出版社：新華書店上海发行所发行, 1978.
4, 3, 250 p. ; 19 cm.
Title in Pinyin on cover: Wenyan sanwen de putonghua fanyi.
$0.56

1. Chinese essays. I. Yü, Tsai-ch'un. II. Title: Wenyan sanwen de putonghua fanyi.

PL2607.W46 79-839470

—20TH CENTURY

(Ai ch'ing yü jen sheng)
愛情與人生 / 陳銀櫃編著. — 台南：大府城出版有限公司：總經銷中央書局, 民國67 ⌊1978⌋
190 p. ; 20 cm. — (大府城書系 ; 20)
Series romanized: Ta fu ch'eng shu hsi.
NT$40.00

1. Chinese essays—20th century. I. Ch'en, Yin-kuei.

PL2623.A38 79-837032

—BIBLIOGRAPHY

(Pi chi hsiao shuo ta kuan tsung so yin)
筆記小說大觀總索引. — ⌊台北：新興書局, 1978?⌋
123, 14, 4, 234 p. ; 22 cm.
Appendices: 新書簡報.—范希曾 書目答問補正.

1. Chinese essays—Bibliography. 2. Chinese fiction—Bibliography. 3. China—Imprints. I. Fan, Hsi-tseng, 1900-1930. Shu mu ta wen pu cheng. 1978.

Z3108.L5P5 79-836830
[PL2395]

79 C

—CH'IN AND HAN DYNASTIES, 221 B. C. -220 A. D.

(Hsi Han wen chi)
西漢文紀 / ⌊梅鼎祚編⌋. — ⌊台北：台灣⌋商務⌊印書館, 1978?⌋
6 v. ; 20 cm. — (四庫全書珍本八集 ; 327-332)
Cover title.
Reproduced from Wen yüan ko copy.

1. Chinese essays—Ch'in and Han dynasties, 221 B. C.-220 A. D. I. Mei, Ting-tso, 1553-1619. II. Series: Ssu k'u ch'üan shu. Ssu k'u ch'üan shu chen pen pa chi ; 327-332.

AC149.S7 1978, vol. 327-332 79-837131
[PL2619]

—CHIN DYNASTY, 265-419

(Hsi Chin wen chi)
西晋文紀 / ⌊梅鼎祚編⌋. — ⌊台北：台灣⌋商務⌊印書館, 1978?⌋
6 v. ; 20 cm. — (四庫全書珍本八集 ; 333-338)
Cover title.
Reproduced from Wen yüan ko copy.

1. Chinese essays—Chin dynasty, 265-419. I. Mei, Ting-tso, 1553-1619. II. Series: Ssu k'u ch'üan shu. Ssu k'u ch'üan shu chen pen pa chi ; 333-338.

AC149.S7 1978, vol. 333-338 79-837130
[PL2619]

—T'ANG DYNASTY, 618-907

Feng, Ming-chih⌋
⌊T'ang Sung pa ta chia ku wen hsüan chu⌋
唐宋八大家古文今讀 ⌊臺北⌋ 江南出版社 ⌊1967⌋
2, 2, 134 p. 19 cm. 7.00
At head of title: 古典文學基礎讀物
In colophon: 出版者 大明書局
Reprint of the ed. with pref. dated 1962 published under title: T'ang Sung pa ta chia ku wen hsüan chu, by Feng Ming-chih.
1. Chinese essays—T'ang dynasty, 618-907. 2. Chinese essays—Sung dynasty, 960-1279. I. Title.
Title romanized: T'ang Sung pa ta chia ku wen chin tu.

PL2619.F4 1967 70-835922

⌊**Feng, Ming-chih**⌋
(T'ang Sung pa ta chia ku wen hsüan chu)
唐宋八大家古文選注 / ⌊編者本社編輯部⌋. — 臺北：學海出版社, 民國63 ⌊1974⌋
2, 2, 134 p. ; 22 cm.

Reprint of the ed. with pref. dated 1962 published under the author: Feng Ming-chih.

1. Chinese essays—T'ang dynasty, 618-907. 2. Chinese essays—Sung dynasty, 960-1279. I. Hsüeh hai ch'u pan she, T'ai-pei. Pien chi pu. II. Title.

PL2619.F4 1974 75-838117

rev

Feng, Ming-chih.
(T'ang Sung pa ta chia ku wen hsüan chu)
唐宋八大家古文選注 / 馮式馮明之注釋. — 香港：萬里書店, 1978.
2, 2, 134 p. ; 19 cm. — (古典文學初步讀物)
Reprint ed. with pref. dated 1962.
$2.00

1. Chinese essays—T'ang dynasty, 618-907. 2. Chinese essays—Sung dynasty, 960-1279. I. Title. II. Series: Ku tien wen hsüeh ch'u pu tu wu.

PL2619.F4 1978 79-835476

CHINESE ESSAYS

—TANG DYNASTY, 618-907 (Continued)

Li, Chao, fl. 806–820.

[T'ang kuo shih pu. Japanese & Chinese]
唐國史補 / [李肇撰] ; 長澤規矩也解題. — 東京 : 出版 古典研究会 : 発行所 汲古書院, 昭和48 [1973], 昭和53 [1978] printing.
297 p. ; 25 cm. — (和刻本漢籍随筆集 ; 6) Ja***

In Chinese, with Japanese reading marks.
Photoreprint ed.
CONTENTS: 李肇 唐國史補—段成式 酉陽雜俎—王仁裕 開元天寶遺事
¥4500

1. China—History—T'ang dynasty, 618–907—Anecdotes, facetiae, satire, etc. 2. Chinese essays—T'ang dynasty, 618–907. I. Nagasawa, Kikuya, 1902– II. Title. III. Series: Wakokubon Kanseki zuihitsu shū ; 6.
Title romanized: Tō kokushi ho.

PL2606.W3 vol. 6 — 79–809165
[DS749.3]

(T'ang Sung pa ta chia wen hsüan)
唐宋八大家文選 / 主編國語日報古今文選編輯室. — 第2版. — 臺北 : 國語日報附設出版部, 民國67 [1978]
ca. 450 p. in various pagings ; 27 cm.

Colophon title.
Text accompanied with Kuo yü chu yin.
NT$145.00

1. Chinese essays—T'ang dynasty, 618–907. I. Ku chin wen hsüan.

PL2619.T3 — 79–840001

(Tseng chu T'ang ts'e)
增注唐策. — [台北 : 臺灣] 商務 [印書館, 1978?]
2 v. : 20 cm. — (四庫全書珍本八集 ; 216–217)

Cover title.
Photoreprint ed. of the Wen Yüan Ko copy.

1. Chinese essays—T'ang dynasty, 618–907. I. Title: T'ang ts'e. II. Series: Ssu k'u ch'üan shu. Ssu k'u ch'üan shu chen pen pa chi ; 216–217.

AC149.S7 1978, vol. 216–217 — 78–843744
[PL2619]

—SUNG DYNASTY, 960-1279

[Feng, Ming-chih]

[T'ang Sung pa ta chia ku wen hsüan chu]
唐宋八大家古文今讀 [臺北] 江南出版社 [1967]
2, 2, 134 p. 19 cm. 7.00

At head of title: 古典文學基礎讀物
In colophon: 出版者 大明書局
Reprint of the ed. with pref. dated 1962 published under title: T'ang Sung pa ta chia ku wen hsüan chu, by Feng Ming-chih.

1. Chinese essays—T'ang dynasty, 618–907. 2. Chinese essays—Sung dynasty, 960–1279. I. Title.
Title romanized: T'ang Sung pa ta chia ku wen chin tu.

PL2619.F4 1967 — 70–835922

[Feng, Ming-chih]

(T'ang Sung pa ta chia ku wen hsüan chu)
唐宋八大家古文選注 / [編者本社編輯部]. — 臺北 : 學海出版社, 民國63 [1974]
2, 2, 134 p. ; 22 cm.

Reprint of the ed. with pref. dated 1962 published under the author: Feng Ming-chih.

1. Chinese essays—T'ang dynasty, 618–907. 2. Chinese essays—Sung dynasty, 960–1279. I. Hsüeh hai ch'u pan she, T'ai-pei. Pien chi pu. II. Title.

PL2619.F4 1974 — 75–838117

—MING DYNASTY, 1368-1644

(Hsien ch'ing i ch'ü)
閒情逸趣 : [明清小品賞析] / 選註邱琇環, 陳幸蕙 ; 主編校訂吳宏一. — 台北 : 長橋出版社, 民國65 [1976]
231 p. : ill. ; 19 cm. — (中國古典文學精選叢書 : 第1輯)

"作者小傳": p. 224–230.
NT$42.00
1. Chinese essays—Ming dynasty, 1368–1644. 2. Chinese essays—Ch'ing dynasty, 1644–1912. I. Ch'iu, Hsiu-huan. II. Ch'en, Hsing-hui. III. Wu, Hung-i, 1943– IV. Title: Ming Ch'ing hsiao p'in shang hsi. V. Series: Chung-kuo ku tien wen hsüeh ching hsüan ts'ung shu.
PL2620.H7 — 78–843838

(Hsien ch'ing i ch'ü)
閒情逸趣 / 主編吳宏一 ; 選註邱琇環, 陳幸蕙. — 九龍 : 新光出版公司, [1977?]
230 p. : ill. ; 19 cm.
"明清小品賞析"

1. Chinese essays—Ming dynasty, 1368–1644. 2. Chinese essays—Ch'ing dynasty, 1644–1912. I. Wu, Hung-i, 1943– II. Ch'iu, Hsiu-huan. III. Ch'en, Hsing-hui.

PL2620.H7 1977 — 79–836021

—CH'ING DYNASTY, 1644-1912

(Hsien ch'ing i ch'ü)
閒情逸趣 : [明清小品賞析] / 選註邱琇環, 陳幸蕙 ; 主編校訂吳宏一. — 台北 : 長橋出版社, 民國65 [1976]
231 p. : ill. ; 19 cm. — (中國古典文學精選叢書 : 第1輯)

"作者小傳": p. 224–230.
NT$42.00
1. Chinese essays—Ming dynasty, 1368–1644. 2. Chinese essays—Ch'ing dynasty, 1644–1912. I. Ch'iu, Hsiu-huan. II. Ch'en, Hsing-hui. III. Wu, Hung-i, 1943– IV. Title: Ming Ch'ing hsiao p'in shang hsi. V. Series: Chung-kuo ku tien wen hsüeh ching hsüan ts'ung shu.
PL2620.H7 — 78–843838

(Hsien ch'ing i ch'ü)
閒情逸趣 / 主編吳宏一 ; 選註邱琇環, 陳幸蕙. — 九龍 : 新光出版公司, [1977?]
230 p. : ill. ; 19 cm.
"明清小品賞析"

1. Chinese essays—Ming dynasty, 1368–1644. 2. Chinese essays—Ch'ing dynasty, 1644–1912. I. Wu, Hung-i, 1943– II. Ch'iu, Hsiu-huan. III. Ch'en, Hsing-hui.

PL2620.H7 1977 — 79–836021

(Wan Ming erh shih chia hsiao p'in)
晚明二十家小品 / 施蟄存編. — 台北 : 新文豐出版公司, 民國66 [1977]
3, 15, 379 p. ; 20 cm.

Reprint ed.
Appendices (p. 345–379): 甲 諸家小傳—乙 本書採輯書目
NT$150.00

1. Chinese essays—Ming dynasty, 1368–1644. I. Shih, Chih-ts'un.

PL2620.W3 1977 — 78–841521

—20TH CENTURY

Ch'en, Chih-hao.

(Lun shuo wen chiang hua)
論說文精華 / 陳志豪著. — 台南 : 光田出版社, 民國67 [1978]
9, 338 p. ; 21 cm.

Caption title: Kuo wen shih yung ta ch'üan.
"歷屆高普檢特考試試題"
NT$80.00
1. Chinese language—Style. 2. Chinese essays—20th century. 3. Government correspondence. I. Title. II. Title: Kuo wen shih yung ta ch'üan.

PL1271.C4253 — 78–840805

(Chi feng ching ts'ao)
疾风劲草 : 报告文学集 / [福建人民出版社编辑]. — [香港? : s. n., 1979?]
96 p. ; 22 cm.

Reprint of the 1978 ed. published by Fu-chien jen min ch'u pan she.
CONTENTS: 赵敬伯, 王芸亭 疾风劲草.—卢秀启, 吴惠聪 当逆流滚滚的时候.—李玉光 吴拢炳 三个共产党员.—中共厦门大学委员会宣传部 坚持真理的人.

1. Chinese essays—20th century. I. Fu-chien jen min ch'u pan she, Foochow, China.

PL2623.C4825 1979 — 79–840549

(Chi mo chih ko)
寂寞之歌 / 張秀亞等著. — 台南 : 青溪出版社 : 總經銷東風圖書供應社, 民國66 [1977]
149 p. ; 20 cm. — (東風文叢 ; 1)

Series romanized: Tung feng wen ts'ung.
NT$35.00

1. Chinese essays—20th century. I. Chang, Hsiu-ya, 1918–

PL2623.C484 — 78–843286

(Chiang shang ts'ai hung)
江上彩虹 / [編者彩虹青年叢書編輯部]. — 臺北 : 彩虹出版社, 民國66 [1977]
224 p. ; 20 cm. — (青年散文 ; 第4輯) (彩虹叢書 ; 95)

Series 1 romanized: Ch'ing nien san wen. Series 2 romanized: Ts'ai hung ts'ung shu.
NT$25.00

1. Chinese essays—20th century. I. Ts'ai hung ch'ing nien ts'ung shu pien chi pu.

PL2623.C487 — 79–835933

(Chin hsiu t'ien lun)
錦繡天倫 : 明道文藝思親徵文選集 / [編輯明道文藝雜誌社]. — 臺中縣烏日鄉 : 明道文藝雜誌社, 民國67 [1978]
8, 163 p. ; 20 cm. — (明道文藝叢書 ; 2)

Series romanized: Ming tao wen i ts'ung shu.
NT$30.00

1. Chinese essays—20th century. 2. Parent and child. I. Ming tao wen i tsa chih she.

PL2623.C4944 — 78–842871

(Ch'ing kan ti huo hua)
情感的火花 / 羅蘭等著. — 臺北 : 中華日報社 : 總經銷星光書報社, 民國67– [1978–
v. ; 19 cm. — (中華日報甲種叢書 ; 50)

Series romanized: Chung-hua jih pao chia chung ts'ung shu.
NT$50.00 (v. 1)

1. Chinese essays—20th century. I. Chin, P'ei-fen.

PL2623.C512 — 78–842733

(Ch'un hua ch'iu yüeh chi)
春花秋月集 / 韓立人編著. — 台中 : 立誼出版社, 民國67 [1978]
6, 176 p. ; 20 cm.

NT$35.00

1. Chinese essays—20th century. I. Han, Li-jen.

PL2623.C5195 — 79–838191

(Chung-kuo hsien tai san wen hsüan)
中國現代散文選 編輯者余中生 臺北 文馨出版社 民國62 [1973]
2, 10, 446 p. 19 cm. NT$45.00

Colophon title.
Hongkong ed. published under title: Ming chia san wen hsüan tu.

1. Chinese essays—20th century. I. Yü, Chung-sheng. II. Title: Ming chia san wen hsüan tu.

PL2622.M53 1973 — 73–896578

rev
73 [r79] C

(Chung-kuo san wen chan)
中國散文展 : 六十六年期 / 張力, 單德興, 周素鳳編著. — 臺南 : 長河出版社, 民國67 [1978]
2 v. (10, 626 p.) ; 20 cm. — (長河叢書 ; 15)

Series romanized: Ch'ang ho ts'ung shu.
NT$110.00

1. Chinese essays—20th century. I. Chang, Li. II. Chou, Su-feng. III. Shan, Te-hsing.

PL2623.C537 — 79–836026

79 C

(Fei hsiang huang hun)
飛向黃昏 : 青年雜感輯 / [編者彩虹青年叢書編輯部]. — 臺北 : 彩虹出版社, 民國67 [1978]
224 p. ; 20 cm. — (彩虹叢書 ; 134)

Series romanized: Ts'ai hung ts'ung shu.
NT$33.00

1. Chinese essays—20th century. 2. Chinese essays—Taiwan. I. Ts'ai hung ch'ing nien ts'ung shu pien chi pu.

PL3031.T32F35 — 79–839460

CHINESE ESSAYS
—20TH CENTURY (Continued)

(Fei hsiang lan t'ien)
飛向藍天：青年散文輯. — 臺北：彩虹出版社, 民國67 [1978]
224 p. ; 20 cm. — (彩虹叢書 ; 132)
Series romanized: Ts'ai hung ts'ung shu.
NT$30.00

1. Chinese essays—20th century. 2. Chinese essays—Taiwan.
PL3031.T32F36 79-840798

(Fei yang ti hsüan lü)
飛揚的旋律：青年散文輯. — 台北：彩虹出版社, 民國66 [1977]
224 p. ; 19 cm. — (彩虹叢書 ; 114)
Series romanized: Ts'ai hung ts'ung shu.
NT$25.00

1. Chinese essays—20th century.
PL2623.F36 79-835246

(Feng yeh chuan hung shih)
楓葉轉紅時 / 黃隆標等著. — 台南：西北出版社, 民國66 [1977]
127, 252 p. ; 19 cm.
NT$45.00

1. Chinese essays—20th century. I. Huang, Lung-piao.
PL2623.F47 79-836741

(Hsien tai ming chia san wen ching hsüan)
現代名家散文精選 / 胡適等著 ; [編輯者王同世]. — 合訂本. — 台南：慈暉出版社 ; 台東：人文書局總經銷, 民國64 [1975]
200, 132 p. ; 20 cm.

1. Chinese essays—20th century. I. Hu, Shih, 1891-1962. II. Wang, T'ung-shih.
PL2607.H72 78-841616

(Hsien tai wen chang hsüan tu)
现代文章选读：记叙文 / 贵阳师范学院中文系编写. — [香港?：s. n., 1978?]
1, 238 p. ; 21 cm.
Reprint of the 1976 ed. published by Kuei-chou jen min ch'u pan she.
CONTENTS: 中國工人阶级的先锋战士：铁人王进喜.—县委书记的榜样：焦裕禄.—"拉革命车不松套，一直拉到共产主义"：记无产阶级优秀战士王国福.—谁是最可爱的人.—人民的好医生李月华.—生命线.—英雄修建成昆路 万水千山只等闲.—驯水记.—万杆红旗舞东风.—要使红旗飘万代 重在教育后一代.——块银元.
1. Chinese essays—20th century. I. Kuei-yang shih fan hsüeh yüan. Chung wen hsi.
PL2623.H737 1978 78-842865

(Hsin yü hsiao p'in)
心語小品 / 黃玉娥編著. — 台南：龍輝出版社, 民國67 [1978]
186 p. : ill. ; 19 cm. — (散文叢書)
Series romanized: San wen ts'ung shu.
NT$30.00

1. Chinese essays—20th century. 2. Chinese poetry—20th century. I. Huang, Yü-o.
PL2623.H743 79-836055

(Hsin yüeh san wen hsüan)
新月散文選 / [編輯者梁實秋, 葉公超]. — 臺北：雕龍出版社, 民國67 [1978]
2, 4, 453 p. ; 22 cm. — (彫龍文學叢書 ; 1)
Series romanized: Tiao lung wen hsüeh ts'ung shu.
Includes bibliographical references.
NT$95.00

1. Chinese essays—20th century. I. Liang, Shih-ch'iu, 1901- II. Yeh, Kung-ch'ao, 1904-
PL2622.H795 78-841972

(Kao chi lun shuo wen fan)
高級論説文範 / 林貴珠編選. — 修訂3版. — 台南：利大出版社：總經銷第一書店, 民國67 [1978]
2, 7, 326 p. ; 22 cm.
Cover title.
NT$45.00

1. Chinese language—Style. 2. Chinese essays—20th century. I. Lin, Kuei-chu.
PL1271.K343 79-835523

(Liu lang ti mai i jen)
流浪的賣藝人：學生創作選第二集 / 古梅等著；中國學生週報編. — 九龍：該報, 民國42 [1953]
51 p. ; 18 cm. — (中國學生叢書)
In colophon: Pien yin che Chung-kuo hsüeh sheng chou pao she.
Series romanized: Chung-kuo hsüeh sheng ts'ung shu.

1. Chinese essays—20th century. 2. School prose. I. Ku, Mei. II. Chung-kuo hsüeh sheng chou pao. III. Chung-kuo hsüeh sheng chou pao she, Kowloon.
PL2623.L55 78-841733

(Mei hua ti jen sheng)
美化的人生. — 台南：綜合出版社, 民國66 [1977]
4, 219 p. ; 20 cm.
Reprint ed.?
NT$25.00

1. Chinese essays—20th century.
PL2623.M36 1977 79-835080
79 C

(Ming chia san wen ching hua)
名家散文精華 / 朱自清等著；彩虹出版社編印. — 香港：該社, [1978?]
5, 326 p. ; 19 cm.
$6.00

1. Chinese essays—20th century. I. Chu, Tzu-ch'ing, 1898-1948. II. Ts'ai hung ch'u pan she.
PL2623.M49 79-835112

(Ming chia san wen hsüan tu)
名家散文選讀 / [余中生編]. — 增訂本. — 香港：文正出版公司, [1978?]
2, 10, 446 p. ; 19 cm.
Running title: Chung-kuo hsien tai san wen hsüan.
Reprint of the 1973 ed. published by Wen hsin ch'u pan she, T'ai-pei, under title: Chung-kuo hsien tai san wen hsüan.
$7.50

1. Chinese essays—20th century. I. Yü, Chung-sheng. II. Title: Chung-kuo hsien tai san wen hsüan.
PL2622.M53 1978 78-843570

(Nan fang san wen kuang ch'ang)
南方散文廣場：當代青年作家散文選 / 林仙龍主編. — 台中：文林出版社；台南總經銷處大山書店, 民國66 [1977]
2 v. ; 20 cm.
"本書共分兩輯, 收輯四十七家作品"
CONTENTS: [1] 青春行.—[2] 青草地.

1. Chinese essays—20th century. I. Lin, Hsien-lung. II. Title: Ch'ing ch'un hsing. III. Title: Ch'ing ts'ao ti.
PL2623.N345 79-836841
79 C

(Pen t'eng ti yin fu)
奔騰的音符：青年雜感輯. — 台北：彩虹出版社, 民國66 [1977]
224 p. ; 19 cm. — (彩虹叢書 ; 116)
Series romanized: Ts'ai hung ts'ung shu.
NT$25.00

1. Chinese essays—20th century.
PL2623.P44 78-842070

(P'iao i hsiao p'in)
飄逸小品 / 劉載福編. — [台南]：綜合[出版社, 1978?]
8, 196 p. ; 19 cm.

1. Chinese essays—20th century. I. Liu, Tsai-fu.
PL2623.P48 79-836244

(Sheng huo ch'ing ch'ü)
生活情趣 林語堂等著 [邵逸青編輯 嘉義] 文友書局總經銷 [1971?]
260 p. 18 cm. (中外名家散文集粹1)(名家散文集粹1) NT$15.00
In colophon: 出版者 袁平文化出版社

1. Chinese essays—20th century. I. Shao, I-ch'ing, comp. II. Lin, Yutang, 1895-
PL2607.S517 74-898117

(Sheng ming chih ko)
生命之歌 / 陳錫書編著. — 台南：大府城出版有限公司：總經銷中央書局, 民國67 [1978]
186, 188 p. : ill. ; 20 cm. — (大府城書系)
Series romanized: Ta fu ch'eng shu hsi.
NT$60.00

1. Chinese essays—20th century. I. Ch'en, Hsi-shu.
PL2623.S5116 79-840361

(Sheng ming ti hua kuo)
生命的花果 / 葉慶炳等著. — 臺北：中華日報社：總經銷星光書報社, 民國67- [1978-
v. ; 19 cm. — (中華日報甲種叢書 ; 51)
Series romanized: Chung-hua jih pao chia chung ts'ung shu.
NT$50.00 (v. 1)

1. Chinese essays—20th century. I. Yeh, Ch'ing-ping.
PL2623.S512 78-843103

(Sui tao wai ti fang ts'ao ti)
隧道外的芳草地 / 艾靈編. — 臺中：藍燈文化事業股份有限公司, 民國66 [1977]
289 p. ; 20 cm. — (藍燈青年文庫)
Series romanized: Lan teng ch'ing nien wen k'u.
NT$50.00

1. Chinese essays—20th century. I. Ai, Ling.
PL2623.S85 78-841910
78 C

(Tang yang ti hsin hsien)
盪漾的心弦：青年抒情散文輯. — 台北：彩虹出版社, 民國66 [1977]
224 p. ; 20 cm. — (彩虹叢書 ; 117)
Series romanized: Ts'ai hung ts'ung shu.
NT$25.00

1. Chinese essays—20th century.
PL2623.T34 79-837909
79 C

(T'ien lai, hsin ch'ü)
天籟, 心曲 / 韓立人編著. — 台中：立誼出版社, 民國67 [1978]
5, 176 p. ; 20 cm.
NT$35.00

1. Chinese essays—20th century. I. Han, Li-jen.
PL2623.T5 79-837723

CHINESE ESSAYS
—20TH CENTURY (Continued)

(Wan hsiang hsiao p'in)
萬象小品 / 本社編輯部編. — 臺北：源成文化圖書供應社：香港：總經銷平價書屋, 民國66 [1977]
8, 202 p. ; 19 cm.
NT$30.00

1. Chinese essays—20th century. I. Yüan ch'eng wen hua t'u shu kung ying she. Pien chi pu.
PL2623.W3 79-836287

(Wo hsü yao i fen ai)
我需要一份愛 / [編輯委員張家華, 雪柔, 陳美卿]. — 台北：龍風實業有限公司圖書出版部, 民國66 [1977]
237, 6 p., [1] leaf of plates : ports. ; 22 cm. — (龍風叢書 ; 2)
"當代散文, 詩歌, 小説選輯"
Series romanized: Lung feng ts'ung shu.
NT$60.00
1. Chinese essays—20th century. 2. Chinese poetry—20th century. 3. Chinese fiction—20th century. I. Chang, Chia-hua. II. Hsüeh-jou. III. Ch'en, Mei-ch'ing.
PL2623.W58 78-843124

(Wo ti hsin sheng)
我的心聲 / [中華世界資料供應出版社編印]. — 台北：該社, [1978?]
8, 280 p. ; 20 cm.
NT$60.00

1. Chinese essays—20th century. I. Chung-hua shih chieh tzu liao kung ying ch'u pan she.
PL2623.W63 78-843755

(Wo ti ling i pan)
我的另一半 / 言曦等著. — 合訂版. — 臺北：中華日報社：總經銷星光書報社, 民國67 [1978]
3 v. in 1 ; 20 cm. — (中華日報甲種叢書 ; 27, 32, 40)
On t.p., v. 1: 5th ed.; v. 2: 4th ed. ; v. 3: 3d ed.
Series romanized: Chung-hua jih pao chia chung ts'ung shu.
NT$150.00

1. Chinese essays—20th century. I. Ch'iu, Yen-hsi.
PL2623.W64 1978 78-841915

(Wo ti mu ch'in)
我的母親 / [主編者幼獅文化事業公司]. — [臺北]：中國青年反共救國團, 民國66 [1977]
5, 4, 168 p. ; 18 cm.
In colophon: Ch'u pan che Yu shih wen hua shih yeh kung ssu.
"本書為全國高初中學生散文比賽甄選出的最優作品."
NT$15.00
1. Chinese essays—20th century. 2. School prose, Chinese—Taiwan. 3. Mothers—Taiwan—Addresses, essays, lectures. I. Yu shih wen hua shih yeh kung ssu.
PL2607.W6 79-841430

Wu, Po-hsiao.
(Pei chieh hsing)
北极星 / 吴伯萧. — 增订再版. — 北京：人民文学出版社, 1978.
2, 140 p. : ill. ; 19 cm.
$0.30

1. Chinese essays—20th century. I. Title.
PL2623.W78 79-835456

(Wu-i ch'en ch'ü)
武夷晨曲：散文集. — [香港?：s. n., 1978?]
2, 96 p. : ill. ; 21 cm.
Reprint of the 1972 ed. published by Fu-chien jen min ch'u pan she.

1. Chinese essays—20th century.
PL2623.W795 1978 79-838129

—20TH CENTURY
—HISTORY AND CRITICISM
—ADDRESSES, ESSAYS, LECTURES

Cheng, Ming-li, 1950-
(Hsien tai san wen hsin shang)
現代散文欣賞 / 鄭明娳著. — 臺北：東大圖書公司：總經銷三民書局, 1978.
2, 212 p. ; 22 cm. — (滄海叢刊 : 文學)
Essays.
Series romanized: Ts'ang hai ts'ung k'an : Wen hsüeh.

1. Chinese essays—20th century—History and criticism—Addresses, essays, lectures. I. Title.
PL2395.C474 78-842523

—WOMEN AUTHORS

(Chi yü liu ssu)
寄語柳絲：少女抒情散文 / 江秀雲編. — 臺北：彩虹出版社, 民國67 [1978]
208 p. ; 20 cm. — (彩虹叢書 ; 131)
Series romanized: Ts'ai hung ts'ung shu.
NT$35.00

1. Chinese essays—Women authors. I. Chiang, Hsiu-yün.
PL2515.C4555 79-836608

—TAIWAN

(Fei hsiang huang hun)
飛向黃昏：青年雜感輯 / [編者彩虹青年叢書編輯部]. — 臺北：彩虹出版社, 民國67 [1978]
224 p. ; 20 cm. — (彩虹叢書 ; 134)
Series romanized: Ts'ai hung ts'ung shu.
NT$33.00

1. Chinese essays—20th century. 2. Chinese essays—Taiwan. I. Ts'ai hung ch'ing nien ts'ung shu pien chi pu.
PL3031.T32F35 79-839460

(Fei hsiang lan t'ien)
飛向藍天：青年散文輯. — 臺北：彩虹出版社, 民國67 [1978]
224 p. ; 20 cm. — (彩虹叢書 ; 132)
Series romanized: Ts'ai hung ts'ung shu.
NT$30.00

1. Chinese essays—20th century. 2. Chinese essays—Taiwan.
PL3031.T32F36 79-840798

CHINESE ETHICS see ETHICS, CHINESE

CHINESE EXPORT ART see China trade art

CHINESE EXPORT PORCELAIN see China trade porcelain

CHINESE FAN PAINTING see Fan painting, Chinese

CHINESE FICTION

Under this heading, without subdivision, are entered individual works and collections by one author

see also Detective and mystery stories, Chinese; Ghost stories, Chinese

Ai, Hsüan.
(Shan yü yü lai)
山雨欲来 / 艾煊. — 第1版. — 上海：上海文艺出版社：新華書店上海发行所发所, 1978.
309 p. ; 21 cm.
Fiction.
$0.78

I. Title.
PL2833.I23S5 79-836852

Ai, Wu.
(Nan hsing chi)
南行記 / 艾蕪. — [香港?：s. n., 1978?]
vii, 170 p. ; 22 cm.
Short stories.
Reprint of the 1946 ed. published by Wen hua sheng huo ch'u pan she, Shanghai, which was issued as no. 1 of Wen hsüeh ts'ung k'an.
CONTENTS: 人生哲學的一課.—山峽中.—松嶺上.—在茅草地.—洋官與雞.—我詛咒你那麼一笑.—我們的友人.—我的愛人.

I. Title.
PL2736.I27N3 1978 79-835624

Ai, Yün.
(Ho T'ien-t'ien)
何田田 / 艾雲著. — 香港：天地圖書有限公司, 1977.
344 p. ; 19 cm. — (天地叢書 ; 8)
Fiction.
Series romanized: T'ien ti ts'ung shu.
$8.50

I. Title.
PL2833.I38H6 78-841122

Ao-te-ssu-erh.
(Yao yüan ti ko pi)
遥远的戈壁 / 敖德斯尔. — 第2版. — 北京：人民文学出版社, 1977.
2, 444 p. ; 19 cm.
Short stories.
$0.89

I. Title.
PL2833.O2Y3 1977 78-842693

Chang, Eileen.
(Chang Ai-ling tuan p'ien hsiao shuo chi)
張愛玲短篇小説集. — 本版. — 臺北：皇冠出版社, 民國66 [1977]
507 p. ; 19 cm.
Short stories.
NT$70.00

I. Title.
PL2837.E35A6 1977 79-835636

79 C

Chang, Fang.
(Sha-ho ts'un)
沙河村 / 張放著. — 台北：文豪出版社：總代理學生圖書供應社, 民國66 [1977]
293 p. : 20 cm.
Short stories.
CONTENTS: 山.—暖流.—風土情.—沙河村.
NT$45.00 ($4.00 U.S.)

I. Title.
PL2837.F25S5 79-836630

79 C

Chang, Fang-hsü.
(Chin I hsiao shuo chi)
靳以小說集. — 香港：中流出版社, 1977.
2, 348 p. ; 19 cm.
Short stories.
$8.00

I. Title.
PL2740.F3C485 79-836075

CHINESE FICTION (Continued)

Chang, Hai-p'ing.
(Pai se ti yu yü)
白色的憂鬱：長篇寫實文藝小說 / 張海屏著. — 再版. — 彰化縣鹿港鎮：雄鷄之聲雜誌社, 民國62 [1973]
122 p. ; 20 cm. — (雄鷄之聲叢書 ; 4)
Fiction.
Series romanized: Hsiung chi chih sheng ts'ung shu.

I. Title.
PL2837.H28P3 79-841454

Chang, Hsi-kuo.
(P'i mu shih cheng chuan)
皮牧師正傳 / 張系國. — 臺北：洪範書店, 民國67 [1978]
220 p. ; 19 cm. — (洪範文學叢書 ; 35)
Fiction.
Series romanized: Hung fan wen hsüeh ts'ung shu.
NT$50.00

I. Title.
PL2837.H68P5 79-837981

Chang, Hsi-kuo.
(Tso jih chih nu)
昨日之怒 / 張系國. — 臺北：洪範書店：總經銷爾雅出版社, 民國67 [1978]
20, 300 p. ; 19 cm. — (洪範文學叢書 ; 25)
Fiction.
Series romanized: Hung fan wen hsüeh ts'ung shu.
NT$55.00

I. Title.
PL2837.H68T75 78-841002

Chang, Hsiao-feng, 1941-
(K'u ch'iang)
哭牆 / 張曉風著. — 九龍：華聯出版社, [1978?]
3, 1, 1, 179 p. ; 19 cm.
Short stories.
CONTENTS: 哭牆.—鐘.—訴.—嗯, 很甜.—樹.—潘渡娜
$6.00

I. Title.
PL2837.H753K9 1978 79-837366

Chang, Hsiao-feng, 1941-
(Wu-ling jen)
武陵人 / 張曉風著. — 臺北：靈聲雜誌社, 1973.
96, 56, [9] p. ; 18 cm. — (藝契叢書)
Added colophon title: Return from Utopia / Esther Chang.
Play.
Includes music.
Series romanized: I ch'i ts'ung shu.

I. Title.
PL2837.H753W8 79-838126

Chang, Hsing.
(Wu-ling shan hsia)
武陵山下 / 张行. — 第3版. — [长沙]：湖南人民出版社, 1978.
872 p., [6] leaves of plates : ill. ; 21 cm.
Fiction.
$1.95

I. Title.
PL2837.H764W8 1978 78-842384

Chang, I, 1941-
(Yüan)
源 / 張毅著. — 台北：台灣新生報, 民國67 [1978]
2 v. (4, 612 p., [4] leaves of plates) : ill. ; 20 cm.
Fiction.
NT$50.00

I. Title.
PL2837.I15Y8 79-838060

Chang, Ju-i.
(Ch'ing ch'un jih chi)
青春日記 / 章如意著. — 香港：上海書局, 1978.
2, 451 p. ; 21 cm. — (Her 青春三部曲 ; 2)
Cover title.
Series romanized: Ch'ing ch'un san pu ch'ü.
$11.00

I. Title.
PL2837.J77C47 vol. 2 78-843508

Chang, Ju-i.
(Ch'ing ch'un san pu ch'ü)
青春三部曲 / 章如意著. — 香港：上海書局, 1977-
v. ; 21 cm.
Cover title.
CONTENTS: 1. 少女日記.—2. 青春日記.
$8.70 (v. 1)
$11.00 (v. 2)

I. Title.
PL2837.J77C47 78-843509

Chang, Ju-i.
(Shao nü jih chi)
少女日記 / 章如意著. — 香港：上海書局, 1977.
356 p. ; 21 cm. — (Her 青春三部曲 ; 1)
Fiction.
Series romanized: Ch'ing ch'un san pu ch'ü.
$8.70

I. Title.
PL2837.J77S5 78-838864
rev

Chang, Liang-tse, 1939-
(Chang Liang-tse tuan p'ien ch'uang tso chi)
張良澤短篇創作集. — 台南：大行出版社, 民國65- [1976-
v. ; 20 cm.
Short stories.
CONTENTS: 1. 生存的條件
NT$35.00 (v. 1)

I. Title.
PL2837.L49A15 1976 79-836363

Chang, Liang-tse, 1939-
(Sheng ts'un ti t'iao chien)
生存的條件 / 著作者張良澤. — 台南：大行出版社, 民國65 [1976]
221 p., [4] leaves of plates : ports. ; 20 cm. — (張良澤短篇創作集 ; 1)
Short stories.
NT$35.00

I. Title.
PL2837.L49A15 1976, vol. 1 79-836362

Chang, Lin.
(Huo yen hua)
火焰花：列车长的日记 / 张林著. — 第1版. — 上海：上海文艺出版社, 1978.
153 p. ; 19 cm.
Fiction.
$0.37

I. Title.
PL2837.L498H8 79-839117

Chang, Shu-han.
(Fei meng t'ien ya)
飛夢天涯 / 張漱菡著. — 臺北：皇冠出版社, 民國67 [1978]
343 p. ; 19 cm.
Fiction.
NT$55.00

I. Title.
PL2837.S5F344 79-838124

Chang, Shu-han.
(I nan wang)
意難忘 / 張漱菡著. — 臺北：皇冠出版社, 民國59 [1970], 民國67 [1978] printing.
272 p. ; 19 cm.
Fiction.
NT$50.00

I. Title.
PL2837.S5I2 79-837070

Chang, T'ien-hsin, 1924-
(T'ien ya yu chih yin)
天涯有知音 = The collection of short stories / by Timothy Chang. — 臺北：大地出版社 ; N.Y.：海外經銷處 K. Chang, 1978.
2, 2, 199 p. ; 19 cm. — (萬卷文庫 ; 66)
In Chinese.
Series romanized: Wan chüan wen k'u.
NT$75.00
I. Title. II. Title: The collection of short stories.
PL2837.T48T5 79-839476

Chang, T'ien-i, 1906-
(Chi pei yü nai tzu)
脊背與奶子 / 張天翼著. — [香港?：s. n., 1977?]
62 p. ; 18 cm.
Fiction.
Reprint of the 1933 ed. published by Shang-hai Liang yu t'u shu yin shua kung ssu, issued as ti 58 chung in series: I chiao ts'ung shu.

I. Title.
PL2740.T5C54 1977 78-841008

Chang, Ting-ya.
(Fang niu wa chi chih sha ti)
放牛娃机智杀敌 / 张定亚编文 ; 王西京绘画. — [西安]：陕西人民出版社, 1973.
[40] p. : ill. ; 15 cm.
CONTENTS: 张定亚编文 王西京绘画 放牛娃机智杀敌—张定亚编文 延波绘画 深入虎穴炸敌车—西安市人防展览会供稿 王西京绘画 越南南方女英雄
1. Vietnamese Conflict, 1961-1975—Fiction. I. Wang, Hsi-ching. II. Yen, Po. III. Chang, Ting-ya. Shen ju hu hsüeh cha ti ch'e. 1973. IV. Title. V. Title: Yüeh-nan nan fang nü ying hsiung.
PL2837.T56F3 78-838159
rev

Chang, Tung-p'ing.
(Hsiao hsiang tao)
小向导 / 原著栗谦 ; 改编张东平 ; 绘画冯越, 冯远. — 第1版. — 上海：上海人民出版社, 1976.
70 p. : ill. ; 10 x 13 cm.
Title in romanization on cover: Xiao xiang dao.
"連環圖畫"
$0.07

1. China—History—Civil War, 1945-1949—Fiction. I. Su, Ch'ien. II. Feng, Yüeh. III. Title. IV. Title: Hiao siang dao.
PL2837.T8H7 78-843189

Chao, Mu-sung.
(Chiao hsin hsia ti sheng ming)
脚心下的生命 / 作者趙慕嵩. — [高雄?]：德馨室出版社, 民國66 [1977]
247 p. ; 20 cm.
Colophon title.
Fiction.
NT$40.00

I. Title.
PL2838.M8C47 78-842387

Chao, P'ing-fu, 1901-1931.
(Jou-shih hsüan chi)
柔石选集. — 第1版. — 北京：人民文学出版社：新华书店北京发行所发行, 1978.
230 p., [1] leaf of plates : ill. ; 21 cm.
Short stories.
$0.59

I. Title.
PL2741.P5J6 1978 79-837679

CHINESE FICTION (Continued)

Chao, Shu-hsia.
(Hsi ch'uang i yeh yü)
西窗一夜雨 / 趙淑俠著. — 臺北：道聲出版社：總經銷聯合圖書供應中心，民國67 [1978]
5, 246 p. ; 20 cm. — (道聲百合文庫)
Short stories.
Series romanized: Tao sheng pai ho wen k'u.
CONTENTS: 王博士的巴黎假期.—龐提老爹的新屋.—影與鏡.—寂寞之舟.—西窗一夜雨.—賽納河之王.—爬山的兒子.—異國之夜.—韓森太太的一天.
NT$50.00

I. Title.
PL2838.S47H74 79-837840

Chao, Tzu-fan.
(Pan hsia liu she hui)
半下流社會 / 趙滋蕃著. — 台北：大漢出版社，民國67 [1978]
4, 6, 326 p., [8] leaves of plates : ill. ; 20 cm. — (大漢叢書 ; 12)
Fiction.
Series romanized: Ta han ts'ung shu.
NT$55.00

I. Title.
PL2838.T93P28 1978 78-841975

Chao, Tzu-fan.
(Pan shang liu she hui)
半上流社會 / 趙滋蕃著. — 台北：大漢出版社：總經銷忠佑實業有限公司，民國67 [1978]
399 p., [1] leaf of plates : port. ; 20 cm. — (大漢叢書 ; 13)
Title in English on p. [4] of cover: Semi-high society.
Fiction.
Series romanized: Ta han ts'ung shu.
NT$59.00

I. Title. II. Title: Semi-high society.
PL2838.T93P3 1978 79-837672

Chao, Tzu-fan.
(Tzu wu hsien shang)
子午線上 / 趙滋蕃著. — 台北：長歌出版社，民國65 [1976]
960 p. ; 19 cm. — (長歌當代創作叢刊 ; 2)
Fiction.
Series romanized: Ch'ang ko tang tai ch'uang tso ts'ung k'an.
"本書作者作品書目": p. [959]-960.
NT$150.00 ($5.00 U.S.)

I. Title.
PL2838.T93T9 1976 78-842339

Ch'en, Che.
(I k'o hung tou)
一顆紅豆 / 瓊瑤 [陳喆] 著. — 臺北：皇冠雜誌社，民國67 [1978]
319 p. ; 20 cm.
Fiction.
NT$55.00

I. Title.
PL2840.C35H8 79-838678

9 C

Ch'en, Ch'en, b. ca. 1608.
(Shui hu hou chuan)
水滸後傳：白話本 / [原著者陳忱]. — 臺初版. — 臺北：河洛圖書出版社，民國67 [1978]
3, 370 p. ; 22 cm. — (白話本中國古典小說叢刊 ; 18)
Series romanized: Pai hua pen Chung-kuo ku tien hsiao shuo ts'ung k'an.
Reprint ed.
NT$120.00

I. Shui hu chuan. II. Title.
PL2698.C38S5 1978 79-836968

Ch'en, Chien-yao.
(Yao yüan ti Chin-chu chai)
遥远的金竹寨 / 陈见尧. — 第1版. — 北京：人民文学出版社：新华书店北京发行所发行，1978.
304 p. ; 19 cm.
Fiction.
$0.59

I. Title.
PL2840.C557Y3 79-838210

Ch'en, Ch'ing-yün.
(Chien ch'u kuei shen ching)
劍出鬼神驚 / 陳青雲著. — 香港：春秋書店，[1978?]
5 v. (1671 p.) ; 19 cm.
Fiction.
$30.00

I. Title.
PL2840.C56C5 78-840867

Ch'en, Ch'ing-yün.
(Chin chü ssu chüeh)
金菊四絕 / 陳青雲著. — 澳門：毅力出版社；香港：總代理陳湘記書局，[1978?]
6 v. ; 18 cm.
Fiction.
$39.60 (H.K.)

I. Title.
PL2840.C56C55 78-842163

Ch'en, Ch'ing-yün.
(Yün chü i hsiu)
雲菊一秀 / 陳青雲著. — 香港：春秋書店，[1978?]
6 v. ; 18 cm. — (新派長篇武俠奇情小說)
Fiction.
Series romanized: Hsin p'ai ch'ang p'ien wu hsia ch'i ch'ing hsiao shuo.
$33.60

I. Title.
PL2840.C56Y8 78-841128

Ch'en, Fei, 1936–
(San lang an)
三狼案：二十年目擊奇案精選 / 陳非著. — 九龍：文藝書屋，1978.
2, 204 p. ; 19 cm. — (文星叢刊 ; 323)
Title in English in colophon: The murder case.
Fiction.
Series romanized: Wen hsing ts'ung k'an.
$6.00

I. Title. II. Title: The murder case.
PL2840.F4S2 78-841485

Ch'en, Hui-t'ung.
(Chiang hai erh nü)
江海儿女 / 陈惠彤. — 第1版. — 北京：人民文学出版社：新华书店北京发行所发行，1978.
2, 568 p. ; 19 cm.
Fiction.
$1.10

I. Title.
PL2840.H824C5 79-840513

Ch'en, Jung, fl. 1975–
(Kuang ming yü hei an)
光明与黑暗 / 谌容. — 第1版. — 北京：人民文学出版社，1978.
469 p. ; 19 cm.
Fiction.
$0.89

I. Title.
PL2840.J8K8 79-838103

Ch'en, K'o-huan.
(Feng lou yeh t'an)
風樓夜談 / 陳克環著. — 台北：環球書社：總發行乾隆圖書無限公司，民國67 [1978]
9, 318 p. ; 19 cm. — (女作家叢書 ; 2 : 散文類 ; 2)
NT$55.00

I. Title. II. Series: San wen lei ; 2. III. Series: Nü tso chia ts'ung shu ; 2 : San wen lei ; 2.
PL2840.K77F46 79-837770

Ch'en, Kuei-chu.
(Tai yüeh hsi hsiang)
待月西廂 / [編著者陳桂珠]. — 臺北：莊嚴出版社，民國67 [1978]
1, 226 p. ; 19 cm. — (中國古典愛情故事 ; 2) (古典新刊 ; 18)
Short stories.
CONTENTS: 霍小玉.—金玉奴.—李亞仙.—西廂記.—玉堂春
NT$45.00

I. Title. II. Series: Chung-kuo ku tien ai ch'ing ku shih ; 2. III. Series: Ku tien hsin k'an ; 18.
PL2840.K818T34 78-842556

[Ch'en, Kuei-fen]
(Hsi-hu ku shih)
西湖故事. — 九龍：光華書局，[1978?]
1, 235 p. ; 19 cm.
Reprint ed. originally published under Ch'en Kuei-fen.

I. Title.
PL2840.K82H76 1978 78-843015

Ch'en, Li-te, fl. 1977–
(Ch'ien ch'ü)
前驱 / 陈立德. — 第1版. — 北京：人民文学出版社，1964, 1978 printing.
2 v. (734 p.) ; 19 cm. — (His 大革命三部曲)
Fiction.
$1.45

I. Title. II. Series: Ch'en, Li-te, fl. 1977– Ta ko ming san pu ch'ü.
PL2840.L55C5 79-835017

Ch'en, Ming-p'an.
(Pu lo, ssu k'a yeh ta)
部落·斯卡也答 / 陳銘磻著. — 臺南：德華出版社，民國66 [1977]
216 p. ; 20 cm. — (愛書人文庫 ; 56)
Fiction.
Series romanized: Ai shu jen wen k'u.
NT$40.00

I. Title.
PL2840.M45P8 79-835303

Ch'en, Ming-p'an.
(Shih-fang li ti ku shih)
石坊里的故事 / 陳銘磻著；陳銘松圖. — 臺北：出版家文化事業有限公司出版部，民國68 [1979]
167 p. : ill. ; 19 cm. — (愛書人叢書 ; 1)
Short stories.
Series romanized: Ai shu jen ts'ung shu.
NT$50.00
I. Title.
PL2840.M45S5 79-839098

Ch'en, Sen, fl. 1823–1849.
(P'in hua pao chien)
品花寶鑑 / [石函氏 i. e. 陳森書]. — [s. l. : s. n., 19--?]
24 v. (on double leaves) ; 20 cm.
Caption title.
Block print.

I. Title.
PL2705.E5448P5 1900z 78-841721

CHINESE FICTION (Continued)

Ch'en, Sen-fu.
(Ch'un yü han)
春雨寒 / 陳森甫著. — 臺南：德華出版社, 民國67 [1978]
1, 550 p. ; 19 cm. — (愛書人文庫 ; 81)
Fiction.
Series romanized: Ai shu jen wen k'u.
CONTENTS: 上卷 理學的家庭.—中卷 離亂的人間.—下卷 伏之歌.
NT$100.00
I. Title.
PL2840.S37C47 79-839189

Ch'en, She.
(Hsiang yang ti jen)
像樣的人 / 陳涉作. — [香港?：s. n., 1978?]
241 p. ; 21 cm.
Fiction.
Reprint of the 1937 ed. published by Shang-hai liang yu t'u shu yin shua kung ssu, Shanghai.
I. Title.
PL2840.S43H8 1978 78-843095

Ch'en, Shih-ho, b. 1887.
(Shuo ku tao chin)
說古道今 / [陳士和講述]. — [台]北：河洛圖書出版社, 民國67 [1978]
15, 129 p. : ill. ; 21 cm.
"中國說書欣賞"
"陳士和的說書, 以蒲松齡的原作為基礎"
NT$50.00
1. P'u, Sung-ling, 1640-1715. Liao chai chih i. II. Title.
PL2705.E5449S5 1978 79-836129

Ch'en, Ts'an-yün.
(Hsiang p'iao ssu chi)
香飘四季 / 陈残云. — [香港?：s. n., 1978?]
4, 506 p. ; 22 cm.
Fiction.
Reprint of the 1973 ed. published by Kuang-tung jen min ch'u pan she, Canton, in series: Yüeh hai wen ts'ung.
I. Title. II. Series: Yüeh hai wen ts'ung.
PL2840.T8H8 1978 79-836795

Ch'en, Yen, fl. 1974-
(Hsin wo ti wei hsiao)
心渦的微笑 / 陳彥著. — 臺北：水芙蓉出版社：總經銷星光書報社, 民國63 [1974]
4, 2, 234 p. ; 20 cm. — (水芙蓉書庫 ; 21)
Short stories.
Series romanized: Shui fu jung shu k'u.
NT$36.00
I. Title.
PL2840.Y4H7 75-836476

Ch'en, Yü-fu.
(Yü ko tzu)
漁歌子 / 作者陳郁夫. — 臺北：白雲文化事業有限公司, 民國66 [1977]
1, 1, 250 p. ; 22 cm. — (白雲叢書 ; 1)
Short stories.
Series romanized: Pai yün ts'ung shu.
CONTENTS: 孺子.—奈何天. (附)蕭南 天涯長憶舊時情.—阿福伯.—蝙蝠與飛象.—青原與關關的故事.—夜訪.—漁歌子.—浴在火光中的觀音.—鷹未揚.—大橋下.—心燈.—山中.
NT$60.00
I. Title.
PL2840.Y738Y8 78-842804

Cheng, Chen-to, 1897-1958.
(Ch'ü huo che ti tai pu)
取火者的逮捕 / 郭源新 [鄭振鐸]. — [香港?：s. n., 1977?]
20, 230 p. ; 19 cm.
Short stories.
Reprint of the 1934 ed. published by Sheng huo shu tien, Shanghai, which was issued as the 8th of Ch'uang tso wen k'u.
CONTENTS: 取火者的逮捕.—亞凱諾的誘惑.—埃娥.—神的滅亡.
I. Title.
PL2744.C4C5 1977 79-836432

Cheng, Chia-chen, fl. 1964-
(Chiang p'an ch'ao yang)
江畔朝阳 郑加真著 [上海] 上海人民出版社 [1972]
695 p. illus. 19 cm. $1.25
Fiction.
I. Title.
PL2841.C45C54 73-899917

Cheng, Pao-chih.
(Pi ch'üan chih chan)
碧泉之战 / 郑保志. — 第1版. — 太原：山西人民出版社, 1977-
v. : ill. ; 19 cm.
Fiction.
$1.20 (v. 1)
1. Korean War, 1950-1953—Fiction. I. Title.
PL2841.P3P5 79-836912

Cheng, Po-ch'i.
(K'uan-ch'eng-tzu ta chiang)
寬城子大將 / 鄭伯奇創作. — [香港?：s. n., 1977?]
60 p. ; 18 cm.
Fiction.
Reprint of the 1932 ed. published by Liang yu t'u shu yin shua kung ssu, Shanghai, in series: I chiao ts'ung shu.
I. Title.
PL2744.P6K8 1977 78-843081

Chennault, Anna.
(Ch'en Hsiang-mei tuan p'ien hsiao shuo hsüan)
陳香梅短篇小說選 / 陳香梅著. — 台北：傳記文學出版社, 民國67 [1978]
2, 170 p., [1] leaf of plates : port. ; 20 cm.
Short stories.
CONTENTS: 窄門.—矛盾之事.—秋之花.—女秘書及其他.—灰色的吻.—無聲的祝福.—聖誕的故事.—永恒的春天.—金縷衣.—離婚.—愛.—茶花怨.—死戀.—愛神的嘆息.
NT$40.00
I. Title.
PL2841.5.N3C47 79-838122

Chi, Kang.
[Kun kun Liao-ho. Japanese. Selections]
滾滾遼河：「満州国」治下・中国知識人の獄中闘争 / 紀剛著 ; 加藤豊隆訳. — 松山：元在外公務員援護会, 1978.
266 p. : ill. ; 22 cm. Ja 79-2158
Fiction.
¥3000
I. Title.
Title romanized: Konkon Ryōga.
PL2842.K3K8216 1978 79-808981

Ch'i, Ko.
(Chi che hsün fang pi chi)
記者尋芳筆記 / 齊歌著. — 台北：聯亞出版社, 民國67 [1978]
211 p. : ill. ; 20 cm. — (聯亞叢刊)
Short stories.
Series romanized: Lien ya ts'ung k'an.
NT$40.00
I. Title.
PL2842.K6C5 79-840103

Ch'i, Mien.
(Pi k'ung hsiung ying)
碧空雄鷹 / 齐勉著. — [香港?：s. n., 1978?]
360 p. ; 21 cm.
Fiction.
Reprint of the 1974 ed. published by Shan-tung jen min ch'u pan she, Chi-nan.
I. Title.
PL2842.M5P5 1978 78-841486

Ch'i, Wen, fl. 1972-
(Lien)
煉 / 起文著. — 臺北：中華日報社：總經銷星光書報社, 民國66 [1977] printing.
602 p. ; 20 cm. — (中華日報甲種叢書 ; 48)
Fiction.
Series romanized: Chung hua jih pao chia chung ts'ung shu.
NT$80.00
I. Title.
PL2842.W45L5 1977 79-836611

(Chi kung chuan)
濟公傳：宋代傳奇事蹟. — 香港：明亮書局, [197-?]
3 v. (16, 1192 p.) ; 19 cm.
At head of title: Ta tzu tsu pen ku tien wen hsüeh.
"四卷二八〇回"
1. Tao-chi, Shih, 1148-1209—Fiction.
PL2832.5.C4743 78-840994

Ch'i-teng-sheng.
(San pu ch'ü hei ch'iao)
散步去黑橋 / 七等生著. — 台北：遠景出版社, 民國67 [1978]
6, 253 p. : port. ; 20 cm. — (遠景叢刊 ; 90)
Chiefly short stories.
Series romanized: Yüan ching ts'ung k'an.
CONTENTS: 復職.—小林阿達.—回鄉印象.—迷失的蝶.—散步去黑橋.—夜湖.—寓言.—書簡.—我年輕的時候
NT$55.00
I. Title.
PL2842.Z85S3 1978 79-836609

Chiang, Hsiao-yün.
(Sui yüan)
隨緣：蔣曉雲短篇小說集之一 / [蔣曉雲著]. — 臺北：皇冠出版社, 民國66 [1977]
216 p. ; 19 cm.
Short stories.
CONTENTS: 隨緣.—宜室宜家.—驚喜.—掉傘天.—口角春風.—姻：三部曲之一.—姻：三部曲之二.—姻：三部曲之三.—快樂頭家緣.
NT$40.00
I. Title.
PL2844.H82S9 79-838121

Chiang, Hsing-yü, fl. 1973-
(T'an huang tao hsia)
彈簧刀下 / 江杏雨著. — 香港：香港宏業書局, 1977.
2, 335 p. ; 21 cm.
Fiction.
$8.30
I. Title.
PL2844.H83T3 78-841129

Chiang, Kuei.
(Yün Han yu yu)
雲漢悠悠 / 姜貴著. — 台北：時報文化出版公司, 民國67 [1978]
132 p. ; 19 cm. — (時報書系 ; 99)
Fiction.
Series romanized: Shih pao shu hsi.
NT$35.00
I. Title.
PL2844.K8Y8 79-836610

79

Chiang, Pao-hsüan.
(Tieh hua chieh)
蝶花劫 / 著作者蔣箸超 [抱玄]. — [香港?：s. n., 1977?]
2, 2, 174 p. ; 22 cm.
Colophon title.
Fiction.
Reprint of the 1914 ed. published by Min ch'üan ch'u pan pu, Shanghai.
I. Title.
PL2747.P28T5 79-836690

CHINESE FICTION (Continued)

Chiao, Ch'in.
(Hsin liao chai)
新聊齋 / 焦琴著；唐賢翔編. — 臺北：偵探雜誌社，民國57 [1968]
3 v. : ill. ; 20 cm.

I. T'ang, Hsien-hsiang. II. Title.
PL2845.C48H74 78–843007

Ch'iao, Mu.
(Ch'ing ch'un hua ch'e)
青春花車 / 喬木著；中央日報編印. — 臺北：中央日報社，民國67 [1978]
262 p. ; 19 cm.
Fiction.
NT$45.00

I. Chung yang jih pao, T'ai-pei. II. Title.
PL2845.M8C44 79–841093

Ch'iao, T'ing.
(Ai ti ch'eng no)
愛的承諾 / 喬婷著. — 臺北：藍帶雜誌社；Silver Spring, Md.：美國總代理 Tang's Corporation, 1976.
352 p. ; 20 cm. — (小說創作叢書)
Fiction.
Series romanized: Hsiao shuo ch'uang tso ts'ung shu.
NT$40.00
I. Title.
PL2845.T5A74 78–840928

Chiao, Tsu-yao.
(Tsung kung ch'eng shih ho t'a ti nü erh)
总工程师和他的女儿 / 焦祖尧. — 第1版. — 北京：人民文学出版社：新华书店北京发行所发行，1978.
579 p. ; 19 cm.
$1.10

I. Title.
PL2845.T75T78 79–840112

Chieh, Yüeh.
(Ling tui ti ta yen)
领队的大雁 / 解越. — [香港?：s. n., 1979?]
2, 282 p. ; 21 cm.
Fiction.
Reprint of the 1978 ed. published by Pei-ching jen min ch'u pan she.

I. Title.
PL2846.H28L5 1979 79–838209

Chih, Hsia, 1917–
(T'ieh tao yu chi tui)
铁道游击队 / 知侠. — 新1版. — 上海：上海人民出版社，1977.
2, 610 p., [1] leaf of plates : ill. ; 19 cm.
Fiction.
$1.10

1. Sino-Japanese Conflict, 1937–1945—Fiction. I. Title.
PL2847.H75T55 1977 79–835305

Chih, Yen.
(Hsing yu ti yü chi)
醒遊地獄記：小本小說 / [編纂者指嚴]. — [香港?：s. n., 1978?]
132 p. ; 21 cm.
Fiction.
Reprint of the 1914 ed. published by Shang wu yin shu kuan, Shanghai.

I. Title.
PL2705.I4H7 1978 79–837507

Chin, I, fl. 1970–
(Ch'u ju shih t'u)
初入世途 / 金依著；阿立挿圖. — 香港：文教出版社，1978.
217 p., [5] leaves of plates : ill. ; 19 cm.
Fiction.
$5.50

I. Title.
PL2848.I2C54 78–842875

Ch'in, Mu.
(Huang chin hai an)
黄金海岸 / 秦牧. — [香港?：s. n., 1978?]
144 p. ; 21 cm.
Fiction.
Reprint of the 1955 ed. published by Kuang-tung jen min ch'u pan she, Canton, China, in series: Yüeh hai wen ts'ung.

I. Title. II. Series: Yüeh hai wen ts'ung.
PL2848.M8H84 1978 79–839560

Chin, Yung.
(Hsüeh shan fei hu)
雪山飛狐 / 金庸著. — 第2次修訂. — 香港：明河社，1977.
425 p., [8] leaves of plates : ill. ; 21 cm. — (金庸作品集；第13冊)
Title in English in colophon: Flying fox of the snowy mountains.
Short stories.
CONTENTS: 雪山飛狐.—鴛鴦刀.—白馬嘯西風
$14.00 ($3.00 U. S.)
I. Title. II. Title: Flying fox of the snowy mountains. III. Series: Chin, Yung. Chin Yung tso p'in chi ; ti 13 ts'e.
PL2848.Y8H75 78–841971

Chin-hsiao-chieh.
(Hei shih fu jen)
黑市夫人 / 金小姐著. — 台北：哲志出版社，民國66 [1977]
222 p. ; 20 cm.
Fiction.
NT$35.00

I. Title.
PL2848.Z7H4 78–841452

Ch'in-yün.
(Tien wang huo hua)
電網火花 / 沁韻著. — 台北：新星出版社：仁愛書店經銷，民國64 [1975]
149 p. ; 20 cm.
Fiction.

I. Title.
PL2848.Z9T55 78–841130

(Ching-kang shan tou cheng ku shih)
井冈山斗争故事 / 井冈山地区文教局. — [香港?：s. n., 1978?]
202 p. : ill. ; 21 cm.
Reprint of the 1974 ed. published by Chiang-hsi jen min ch'u pan she, Nan-ch'ang.

1. Ching-kang Shan region—History—Fiction. I. Ching-kang shan ti ch'ü, China. Wen chiao chü.
PL3031.C552C5 1978 78–841101

Ch'iu, Heng-ts'ung.
(K'uang piao)
狂飆 / 邱恒聪著. — 第1版. — 南昌：江西人民出版社，1977.
513 p. ; 19 cm.
Fiction.
$0.94

I. Title.
PL2849.U398K8 78–841111

Ch'iu, Meng-hen.
(Hsüeh mien sha)
血面煞 / 秋夢痕著. — 香港：春秋書店，[1978?]
6 v. ; 18 cm. — (新派長篇武俠奇情小說)
Fiction.
Series romanized: Hsin p'ai ch'ang p'ien wu hsia ch'i ch'ing hsiao shuo.
$30.00
I. Title.
PL2849.U54H7 78–841146

Ch'iu, Meng-hen.
(Li chieh ch'i t'ung)
歷劫奇童 / 秋夢痕著. — 香港：春秋書店，[1978?]
6 v. ; 18 cm. — (新派長篇武俠奇情小說)
Fiction.
Series romanized: Hsin p'ai ch'ang p'ien wu hsia ch'i ch'ing hsiao shuo.
$5.60 per vol.
I. Title.
PL2849.U54L5 78–842555

Ch'iu, Meng-hen.
(Shen nü chien)
神女劍 / 秋夢痕著. — 香港：春秋書店，[1978?]
6 v. ; 18 cm. — (新派長篇武俠奇情小說)
Fiction.
Series romanized: Hsin p'ai ch'ang p'ien wu hsia ch'i ch'ing hsiao shuo.
$30.00
I. Title.
PL2849.U54S48 78–842707

Ch'iu, Yang, fl. 1959–
(Tan chien p'ien)
胆劍篇 根据曹禺等創作剧本改編 邱揚改編 陈長明填曲 程十髮繪圖 北京 人民美术出版社 1963.
100 p. illus. 19 cm.

1. Kou Chien, fl. 496 B. C.—Fiction. I. Ch'eng, Shih-fa. II. Ts'ao, Yü. Tan chien p'ien. III. Title.
PL2849.U77T3 1963 C 64–1323

Chou, Chieh-ch'en.
(Chou Chieh-ch'en tzu hsüan chi)
周介塵自選集 / [著作者周介塵]. — 臺北：黎明文化事業公司，民國67 [1978]
2, 276 p., [2] leaves of plates : ill. ; 20 cm. — (新中國文學叢刊；56)
Short stories.
Series romanized: Hsin Chung-kuo wen hsüeh ts'ung k'an.
CONTENTS: 老鴇和鳳凰.—蛙聲停止的時候.—龍三叔.—第七號審判庭.—狄師傅.—遺囑.—借刀.—賊.—抉擇.—白龍峯.
NT$70.00
I. Title.
PL2851.C49A15 1978 79–839911

Chou, Chieh-ch'en.
(Huo yü hsüeh ti nien tai)
火與血的年代 / 周介塵著. — 臺北：名人出版社，民國66 [1977]
265 p. ; 19 cm.
Fiction.
NT$45.00

I. Title.
PL2851.C49H8 78–843567

Chou, Ch'üan-p'ing.
(Meng li ti wei hsiao)
夢裏的微笑：小說集 / [周]全平作；葉靈鳳畫. — [香港?：s. n., 1977?]
iv, 241 p., [7] leaves of plates : ill. ; 22 cm.
Short stories.
Reprint of the 1925 (1926 printing) ed. published by Kuang hua shu chü, Shanghai; in series Ch'uang tsao she ts'ung shu.
CONTENTS: 林中.—聖誕之夜.—愛與血的交流.—舊夢.
I. Title.
PL2851.C78M4 1977 78–840842

CHINESE FICTION (Continued)

Chou, Lao-fu.
(K'ai hsin k'ai hsin chi)
開心開心集 / 周老夫著. — 臺北：星光出版社：經銷者星光書報社, 民國66 [1977]
201 p. ; 20 cm. — (雙子星叢書 ; 38)
Essays.
Series romanized: Shuang tzu hsing ts'ung shu.
NT$30.00

I. Title.
PL2851.L3K3 79-835599

Chou, Pai-p'in.
(Chung-kuo sha jen wang ku shih)
中國殺人王故事 / 作者周白蘋. — 九龍：馬錦記書局, [1977?]
36, 37, 107 p. ; 19 cm.
Colophon title.
Fiction.
CONTENTS: 中國殺人王大鬧紐約.—中國殺人王大戰芝加哥.

I. Title.
PL2754.P34C46 79-837271

Chou, Pai-p'in.
(Chung-kuo sha jen wang ta nao Shang-hai)
中國殺人王大鬧上海 / 周白蘋著. — 九龍：馬錦記書局, [1977?]
187 p. in various pagings ; 19 cm.
Cover title.
Colophon title: Chung-kuo sha jen wang ku shih.
Short stories.
CONTENTS: 中國殺人王登陸上海上下集.—殺人王大戰三K黨.—機器英雄橫掃安南.—機器英雄大鬧廣西.—機器英雄大鬧重慶.
$2.80 (Continued on next card)
I. Title. II. Title: Ta nao Shang-hai.
PL2754.P34C47 79-836181

Chou, Po-nai.
(Wen p'ing la chi tui)
文憑垃圾堆 / 帆隱 [周伯乃] 著. — 台北：黄河出版社：總經銷聯亞出版社, 民國67 [1978]
5, 244 p. ; 20 cm. — (黄河文萃 ; 1)
Fiction.
Series romanized: Huang-ho wen ts'ui.
NT$50.00

I. Title.
PL2851.P6W4 79-837220

Chou, Shu-p'in.
(Shih lo ti shu hsin)
失落的書信 / 周叔蘋著. — 臺北：正中書局, 民國65 [1976]
2, 162 p. ; 20 cm. — (正中文藝叢書)
Two stories.
Series romanized: Cheng chung wen i ts'ung shu.
CONTENTS: 第一封信. 彩虹.—第二封信. 親情.

I. Title.
PL2851.S57S5 78-842139

Chou, Shu-p'in.
(Wu meng meng)
霧濛濛 / 周叔蘋著. — 臺北：正中書局, 民國66 [1977]
215 p. ; 20 cm. — (正中文藝叢書)
Fiction.
Series romanized: Cheng chung wen i ts'ung shu.

I. Title.
PL2851.S57W8 78-841943

Chou, Wen.
(Tsai Pai-sen chen)
在白森鎮 / 周文創作. — [香港?：s. n., 1976?]
227, 2 p. ; 18 cm.
Fiction.
Reprint of the 1937 ed. published by Liang yu t'u shu kung ssu, Shanghai; issued in series: Chung p'ien ch'uang tso hsin chi ; ti ch'i chung.

I. Title.
PL2754.W4T75 1976 78-841339

Chu, Hsing-ho, 1936–
(Hai pien ti nü lang)
海邊的女郎 / 朱星鶴著. — 台南：鳳凰城圖書公司, 民國67 [1978]
257 p. ; 20 cm. — (鳳凰新書 ; 406)
Short stories.
Series romanized: Feng huang hsin shu.
CONTENTS: 海邊的女郎.—鵝情.—閒事.—車上車下.—一隻火腿.—我要她活下去.—含淚的微笑.—玉碎.—明天·天晴.—未完成的夢.—從哀痛中站起.—安魂曲.—醒在朦朧的睡裡.—皈依.—終站.
NT$49.00

I. Title.
PL2852.H74H3 79-837953

Chu, Huan-wen, fl. 1962–
(Pai lang t'ao t'ien)
白浪滔天 / 朱煥文著. — 台南：鳳凰城圖書公司, 民國67 [1978]
264 p. ; 20 cm. — (鳳凰新書 ; 403)
Short stories.
Series romanized: Feng huang hsin shu.
CONTENTS: 白浪滔天.—生命的悲歌.—乳燕飛.—召喚.—結.—勇者的畫像.—石上花.—五根鋼筋.——夥的.—炎陽下
NT$49.00
I. Title.
PL2852.H85P3 79-836958

Chu, Hui-chieh, 1916–
(Huo chi)
火祭 / 朱慧潔著. — 台南：長河出版社：總經銷建立書局, 民國67 [1978]
4, 2, 6, 200 p. ; 19 cm. — (長河叢書 ; 11)
Fiction.
Series romanized: Ch'ang ho ts'ung shu.
NT$45.00

I. Title.
PL2755.H85H8 78-843287

Chu, Meng-p'ing, 1955–
(Ai ti chui tsung)
愛的追踪 / 朱夢萍著. — 台南：青溪出版社 ; 總經銷東風圖書供應社, 民國67 [1978]
191 p. : ill. ; 19 cm. — (東風文叢 ; 9)
Fiction.
Series romanized: Tung feng wen ts'ung.
NT$40.00

I. Title.
PL2852.M4A74 79-835485

Chu, Yü, fl. 1971–
(Ch'üan t'ao)
圈套 / 朱羽著. — 台北縣土城：漢麟出版社, 民國65 [1976]
301 p. ; 20 cm.
Fiction.
NT$50.00

I. Title.
PL2852.Y8C567 78-841369

Chu, Yü, fl. 1971–
(Feng, yü, lei)
風, 雨, 雷 / 朱羽著. — 台北：漢麟出版社, 民國66 [1977]
332 p. ; 20 cm.
Fiction.
NT$50.00

I. Title.
PL2852.Y8F45 78-842649

Chu, Yü, fl. 1971–
(Huang se hsiung ch'e)
黃色凶車 / 朱羽著. — 台北縣土城：漢麟出版社, 民國67 [1978]
324 p. ; 20 cm.
Two short stories.
"危險人物": p. 163–324.
NT$50.00

I. Chu, Yü, fl. 1971– Wei hsien jen wu. 1978. II. Title.
PL2852.Y8H85 79-836488

Chu, Yü, fl. 1971–
(Hsüeh jou mo fang)
血肉磨坊 / 朱羽著. — 台北縣土城：漢麟出版社, 民國67 [1978]
482 p. ; 19 cm.
Fiction.
NT$70.00

I. Title.
PL2852.Y8H76 79-838722

Chu, Yü, fl. 1971–
(Jen t'ou yen)
人頭宴 / 朱羽著. — 台北縣土城：漢麟出版社, 民國66 [1977]
266 p. ; 20 cm.
Fiction.
NT$50.00

I. Title.
PL2852.Y8J4 78-841597

Chu, Yü, fl. 1971–
(Ti ch'i hao chih ling)
第七號指令 / 朱羽著. — 台北：漢麟出版社, 民國66 [1977]
296 p. ; 19 cm. — (His 美人局故事 ; 7–8)
Two stories.
CONTENTS: 第七號指令—橫掃美人局
NT$50.00

I. Title. II. Series: Chu, Yü, fl. 1971– Mei jen chü ku shih ; 7–8.
PL2852.Y8T48 78-843207

Chu, Yü, fl. 1971–
(T'ieh hsüeh ling)
鐵血嶺 / 朱羽著. — 台北：漢麟出版社, 民國67 [1978]
298 p. ; 19 cm.
Fiction.
NT$50.00

I. Title.
PL2852.Y8T49 79-835241

Chu, Yü, fl. 1971–
(T'ien lo ti wang)
天羅地網 / 朱羽著. — 台北縣土城：漢麟出版社, 民國67 [1978]
327 p. ; 19 cm.
Fiction.
NT$50.00

I. Title.
PL2852.Y8T53 79-838711

Chu, Yü, fl. 1971–
(Tsui o chih ch'iao)
罪惡之橋 / 朱羽著. — 台北：漢麟出版社, 民國67 [1978]
406 p. ; 20 cm.
Fiction.
NT$70.00

I. Title.
PL2852.Y8T78 79-836222

Chu, Yü, fl. 1971–
(Yung ch'uang hei t'ien t'ang)
勇闖黑天堂 / 朱羽著. — 台北：漢麟出版社, 民國67 [1978]
312 p. ; 20 cm.
Fiction.
NT$50.00

I. Title.
PL2852.Y8Y8 79-835638

CHINESE FICTION (Continued)

Chu-ko, Ch'ing-yün.
(Chu ch'un ling)
朱唇令 / 諸葛青雲著. — 台北：春秋出版社：總經銷呂氏書店, 民國67 [1978]
2 v. (628 p.) ; 20 cm.
Fiction.
NT$80.00

I. Title.
PL2852.Z45C5 79-836066

Chu-ko, Ch'ing-yün.
(Fei ts'ui ch'uan)
翡翠船 / 諸葛青雲著. — 台北：春秋出版社：總經銷呂氏書店, 民國67 [1978]
2 v. (968 p.) ; 20 cm.
Fiction.
NT$60.00 per vol.

I. Title.
PL2852.K59F4 79-836921

Chu-ko, Ch'ing-yün.
(I chien kuang han shih ssu chou)
一劍光寒十四州 / 諸葛青雲著. — 台北：春秋出版社：總經銷呂氏書店, 民國67 [1978]
2 v. (631 p.) ; 20 cm.
At head of title: Wu hsia ch'i ch'ing ming chu.
Fiction.
NT$80.00

I. Title.
PL2852.Z45I24 79-836039

Chu-ko, Ch'ing-yün.
(Pi hsüeh feng huang)
碧血鳳凰：長篇武俠名著 / 諸葛青雲著. — [台北]：春秋出版社：總經銷呂氏書店, 民國66 [1977]
2 v. (579 p.) ; 18 cm.
Fiction.
NT$35.00 per vol.

I. Title.
PL2852.K59P5 79-835590

Chün-ch'ing, pseud.
(Li ming ti ho pien)
黎明的河边 / 峻青. — 北京第2版. — 北京：人民文学出版社：新华书店北京发行所发行, 1978.
2, 527 p. ; 19 cm.
Short stories.
$1.00

I. Title.
PL2853.C5L5 1978 79-837719

Chün-jan.
(Ch'ang ch'ang ti Wu-la yin ho)
长长的乌拉银河 / 俊然. — 第1版. — 北京：解放军文艺社, 1977.
521 p. ; 19 cm.
Fiction.
$1.10

I. Title.
PL2853.J3C5 78-841484

Chung-hui.
(Hsüeh lei huang yüan huan hsin t'ien)
血泪荒原换新天 / 众慧編写. — [香港?：s. n., 1978?]
127 p. ; 21 cm.
Fiction.
Reprint of the 1966 ed. published by Nei Meng-ku jen min ch'u pan she, Kuei-sui.

I. Title.
PL2854.Z5H7 1978 79-837175

(Chung-kuo ch'i t'an)
中國奇談 / 林耀川編譯. — 臺北永和：常春樹書坊；[台北]：總經銷星光書報社, 民國66 [1977]
319 p. ; 20 cm. — (中國人的書)
Series romanized: Chung-kuo jen ti shu.
NT$5.00

1. Fantastic fiction, Chinese. I. Lin, Yao-ch'uan.
PL2629.F35C5 79-836347

Chung-kuo k'o hsüeh yüan. Wen hsüeh yen chiu so, Peking.
(Pu p'a kuei ti ku shih)
不怕鬼的故事 / 中国科学院文學研究所編. — 香港：港青出版社, 1978.
5, 105 p. ; 21 cm.
Reprint ed.
$4.00

1. Ghost stories, Chinese. I. Title.
PL2629.G5C57 1978 79-835961

(Chung-kuo ku tien wen hsüeh hui pien)
中國古典文學彙編 / 主編博愛編譯中心. — 第3版. — 台北：博愛圖書公司：發行者博愛圖書中心, [1978?]
3 v. ; 20 cm.
"節著改編"
CONTENTS: 第1輯 三國演義. 封神演義. 西遊記. 水滸傳.—第2輯 紅樓夢. 金瓶梅. 儒林外史. 聊齋誌異.—第3輯 白蛇傳. 西廂記. 桃花扇. 鏡花緣.

1. Chinese fiction. I. Po ai pien i chung hsin.
PL2625.C74 79-835894

(Chung-kuo ku tien wu hsia ta chan)
中國古典武俠大展 / [著i. e.編者石蒲雨]. — 臺北：武陵出版社, 民國66 [1977] printing.
4, 225 p. ; 20 cm. — (愛書人文庫 ; A27)
Series romanized: Ai shu jen wen k'u.
NT$40.00

1. Chinese fiction. I. Shih, P'u-yü.
PL2625.C75 78-843251

(Chung-kuo ku tien wu hsia ta chan)
中國古典武俠大展 / 石蒲雨譯[i. e. 編]. — 澳門：武陵出版社, [1978?]
4, 225 p. ; 19 cm.
$6.00

1. Chinese fiction. I. Shih, P'u-yü.
PL2625.C75 1978 78-843249

(Chung-kuo min chien t'ung su hsiao shuo)
中國民間通俗小說. — 台南：利大出版社：總經銷第一書店, 民國66 [1977]
108, 137, 114, 263 p. ; 20 cm. — (民間通俗小說) (中國古典文學)
Series 1 romanized: Min chien t'ung su hsiao shuo. Series 2 romanized: Chung-kuo ku tien wen hsüeh.
"古典文學大字足本"
CONTENTS: 七世夫妻.—觀世音.—花木蘭.—三戲白牡丹.
NT$80.00

1. Folk literature, Chinese. 2. Chinese fiction.
PL2446.C56 78-842957

78 C

Chung-shan hsien wen hua chü Yeh ch'uang Chiu-ch'ü sha ch'uang tso tsu.
(Yeh ch'uang Chiu-ch'ü sha)
夜闯九曲沙 / 中山县文化局《夜闯九曲沙》创作组 ; 余松岩编 ; 黄剑培, 黄国宏绘. — 第1版. — [广州]：广东人民出版社, 1975.
92 p. : ill. ; 11 x 13 cm.
"广东民兵革命斗争故事连环画"
1. Sino-Japanese Conflict, 1937-1945—Fiction. I. Yü, Sung-yen. II. Huang, Chien-p'ei. III. Huang, Kuo-hung. IV. Title.
PL2854.Z76Y4 1975 79-836669

Fan-lu, pseud.
(Fan-lu tzu hsüan chi)
繁露自選集 / [著作者繁露]. — 臺北：黎明文化事業公司, 民國67 [1978]
2, 268 p., [2] leaves of plates : ill. ; 20 cm. — (中國新文學叢刊 ; 63)
Short stories.
Series romanized: Chung-kuo hsin wen hsüeh ts'ung k'an.
CONTENTS: 山的故事.—泛泛之交.—春桃姑娘.—徵婚啓事.—小姨.—潭畔思仇.—蕩婦.—瓜瓞緜緜.—忘憂草.—洪流.—鑽石戒指.
NT$70.00
I. Title.
PL2857.A56A6 1978 79-837848

Fan-lu, pseud.
(K'uang k'uang chung ti jen)
框框中的人 / [繁露著]. — 臺中：學人文化事業有限公司, 民國67 [1978]
391 p. ; 20 cm.
Fiction.
NT$60.00

I. Title.
PL2857.A56K8 79-837476

Fei, Meng.
(Chih yeh hsiung shou)
職業兇手 / 費蒙著. — 臺北：名人出版社, 民國67 [1978]
2 v. (1107 p.) ; 19 cm.
Fiction.
NT$90.00 per vol.

I. Title.
PL2857.E39C46 79-836558

Feng, Chü-chih.
(Liu lei ti yün)
流淚的雲 / 馮菊枝著. — 台北：皇冠雜誌社, 民國67 [1978]
208 p. ; 19 cm.
Short stories.
CONTENTS: 蝶飛.—暗流.—倦鳥.—流淚的雲.—昨夜.—湖底.—風在廻盪.
NT$40.00
I. Title.
PL2857.E525L5 79-837320

Feng, Feng, 1935–
(K'o-p'iao hu)
柯飄湖 馮馮[著 台北 皇冠雜誌社 民國61 i. e. 1972]
349 p. 19 cm. (皇冠叢書第329種) NT$28.00
• Short stories.
Series romanized: Huang kuan ts'ung shu.

I. Title.
PL2857.E535K6 73-897207

rev

Feng, Feng, 1935–
(Meng yen ti nü shen)
蒙眼的女神 / 馮馮著. — 台北：皇冠雜誌社, [1974?]
199 p. ; 19 cm. — (皇冠叢書 ; 第375種)
Short stories.
Series romanized: Huang kuan ts'ung shu.
NT$30.00

I. Title.
PL2857.E535M4 75-835537

Feng, Feng, 1935–
(Ping yai)
冰崖 / 馮馮著. — 臺北：皇冠雜誌社, 民國66 [1977]
378 p. ; 19 cm.
Fiction.
NT$60.00

I. Title.
PL2857.E535P5 78-838459

CHINESE FICTION (Continued)

Feng, Feng, 1935–
(Wei hsiao)
微笑 馮馮著 ［臺北 皇冠出版社 民國61 i. e. 1972］
210 p. 19 cm. (皇冠叢書第323種) NT$18.00
Cover title.
Short stories.
Series romanized: Huang kuan ts'ung shu.

I. Title.
PL2857.E535W4 73–896481

Feng, Meng-lung, 1574?–1645?
(Ching shih t'ung yen)
警世通言 / 馮夢龍編. — 九龍：中華書局, 1978.
2 v. (5, 3, 647 p.) ; 21 cm.
Reprint of the 1965 ed.
$20.00

1. Short stories, Chinese. I. Title.
PL2627.F38 1978 79–836546

Feng, Meng-lung, 1574?–1645? — ［Hsin lieh kuo chih］...
［1978］
NT$120.00

1. China—History—Chou dynasty, 1122–221 B. C.—Fiction. I. Ts'ai, Yüan-fang, fl. 1736–1770. II. Ho lo t'u shu ch'u pan she. III. Title.
Title romanized: Tung Chou lieh kuo chih.
PL2698.F4H8 1978 79–837316

Feng, Meng-lung, 1574–1645? comp.
(Hsing shih heng yen)
醒世恆言 / 馮夢龍編；顧學頡校注. — 港1版. — 九龍：中華書局香港分局, 1965, 1978 printing.
2 v. (4, 3, 864 p.) ; 21 cm.
"以「世界文庫」覆排明葉敬池刻本醒世恆言為底本, 並參校了衍慶堂刻本和今古奇觀等書."
$26.00
1. Short stories, Chinese. I. Ku, Hsüeh-chieh. II. Title.
PL2627.F43 1965 79–836952

Feng, Meng-lung, 1574?–1645? comp.
(Yü shih ming yen)
喻世明言 / 馮夢龍編. — 港1版. — 九龍：中華書局香港分局, 1965, 1978 printing.
2 v. (15, 2, 3, 642 p.) : ill. ; 21 cm.
In 40 chüan.
"喻世明言初版本名古今小說...當古今小說一刻增補再版的時候,書名即改成了喻世明言"
$22.00
1. Short stories, Chinese. I. Feng, Meng-lung, 1574?–1645? comp. Ku chin hsiao shuo. II. Title.
PL2627.F45 1965 79–837150

Fêng, Yao.
(Wo wei ch'ing k'uang)
我爲卿狂 鳳瑤著 台南 王家出版社 民國59 ［1970］
381 p. 19 cm. NT$22.00
Fiction.

1. Title.
PL2857.E594W6 72–841982

Feng, Yü-ch'i.
(Jen yü i p'ien hsin)
人語一片心：長篇社會言情小說 / 馮玉奇著. — ［香港］：香港互助圖書社, ［1957?–
v. ; 18 cm.
Fiction.

I. Title.
PL2857.E6J4 78–842293

Feng shen yen i.
封神演義 / 許仲琳編. — 九龍：中華書局, 1970, 1976 printing.
2 v. (7, 1017 p.) ; 19 cm.
"一百回"
$14.00

I. Hsü, Chung-lin, 16th cent.
Title romanized: Feng shen yen i.
PL2698.F428 1970 78–841127

Fu, Chi-mei.
(Ch'i meng lu)
奇夢錄 / ［編著者符濟梅］. — 臺北：莊嚴出版社, 民國66 ［1977］
7, 185 p. ; 19 cm. — (古典新刊 ; 16)
Short stories.
NT$40.00

1. Dreams—Fiction. I. Title. II. Series: Ku tien hsin k'an ; 16.
PL2858.C415C5 79–835524

Hai, Hsin.
(Ch'ing ch'un ti hsin shih)
青春的心事 / 海辛著. — 香港：中流出版社, 1978.
2, 263 p. ; 19 cm.
Fiction.
$7.50

I. Title.
PL2861.A47C54 78–841960

Hai, Wen.
(Yüeh ts'an hsing hui)
月殘星輝 / 海汶著. — 台北：文豪出版社：總代理學生圖書供應社, 民國67 ［1978］
242 p. ; 20 cm. — (文豪叢書)
Fiction.
Series romanized: Wen hao ts'ung shu.
NT$45.00 ($5.00 U.S.)

I. Title.
PL2861.A575Y9 79–835955

Hai, Ying.
(Ch'en yü ti hsiao wu)
沉鬱的下午 / 海瑩著. — 台北：環球書社：總發行乾隆圖書無限公司, 民國68 ［1979］
2, 3, 179 p. ; 20 cm. — (女作家叢書 ; 6 : 小說類 ; 1)
Short stories.
CONTENTS: 惑.—農家新曲.—死繭.—沉鬱的下午.—惡夢.—瑪芳.—馥玲.—迴流.—凝聚的霧.—礁.—提親.—迷失.—半圓.—琦眞的週末.—嬗遞季.
NT$50.00
I. Title. II. Series: Nü tso chia ts'ung shu ; 6 : Hsiao shuo lei ; 1.
PL2861.A5813C5 79–840552

9 C

Han, Chiang-hung.
(Hi! Chung-kuo jen?)
Hi! 中國人? / 韓江鴻著. — 九龍：創作出版社, 1978–
v. ; 19 cm.
Fiction.
$6.00 (v. 1)

I. Title.
PL2861.A582H5 78–843314

Han, Chiang-hung.
(K'u yü ch'i feng)
苦雨淒風 / 韓江鴻著. — 九龍：創作出版社, 1978.
7, 167 p. ; 19 cm.
Fiction.
$6.00

I. Title.
PL2861.A582K8 78–843877

Han, Chüeh, 1917–
(Hsin yen pu mei chi)
信言不美集 / 寒爵著. — 台北：水芙蓉出版社, 民國67 ［1978］
2, 5, 256 p. ; 19 cm. — (水芙蓉書庫 ; 117)
Series romanized: Shui fu jung shu k'u.
NT$48.00

I. Title.
PL2861.A586H74 79–838120

Han, Hsin.
(Ch'un kuei meng hen)
春閨夢痕 / ［寒心著］. — 台北：德昌出版社, 民國66 ［1977］
194 p. ; 19 cm. — (奇異鬼孤集錦故事)
"奇異鬼孤小說"
CONTENTS: 花魂.—愁海情癡.—意難忘.—蠱香.
NT$30.00

1. Ghost stories, Chinese. I. Title. II. Series: Ch'i i kuei hu chi chin ku shih.
PL2861.A62C4 79–836056

Han, Hsin.
(Han yüeh ch'iao ch'iao)
寒月俏俏 / 寒心著. — 台北：德昌出版社, 民國66 ［1977］
189 p. ; 19 cm. — (奇異鬼狐集錦故事)
Cover title.
"奇異鬼狐小說"
CONTENTS: 説鬼者.—死神約會.—鏡中幽靈.—此情不渝.—石頭奇案.—惡有惡報.
NT$30.00
1. Ghost stories, Chinese. I. Title. II. Series: Ch'i i kuei hu chi chin ku shih.
PL2861.A62H3 79–836551

Han, Hsin.
(Pi po p'iao hun)
碧波飄魂 / 寒心著. — 台北：德昌出版社, 民國66 ［1977］
189 p. ; 19 cm. — (奇異鬼狐集錦故事)
Cover title.
"奇異鬼狐小說"
CONTENTS: 碧波飄魂.—鬼母恨.—情兮悠悠.—冤債.
NT$30.00
1. Ghost stories, Chinese. I. Title. II. Series: Ch'i i kuei hu chi chin ku shih.
PL2861.A62P5 79–835392

Han, Hsin.
(Yeh niang)
夜娘 / ［寒心著］. — 台北：德昌出版社, 民國66 ［1977］
184 p. ; 19 cm. — (奇異鬼狐集錦故事)
"奇異鬼狐小說"
CONTENTS: 蠱香.—誰是兇手.—血屍疑雲.—憶難忘.
NT$30.00

1. Ghost stories, Chinese. I. Title. II. Series: Ch'i i kuei hu chi chin ku shih.
PL2861.A62Y4 79–837364

(Han hsiu ts'ao)
含羞草. — 香港：香港海鷗出版公司, ［1975?］
51 p. : ill. ; 15 x 17 cm. — (民間故事連環圖)
Fiction.
$1.20

I. Series: Min chien ku shih lien huan t'u.
PL2832.5.H33 78–842273

78 C

Hao-shang-k'o.
(Li tai ti wang yü nü jen)
歷代帝王與女人 / 濠上客編著. — 高雄：勝夫書局, 民國66 ［1977］
3 v. ; 20 cm.
NT$120.00

1. China—Court and courtiers—Fiction. I. Title.
PL2861.A78L5 79–838199

CHINESE FICTION (Continued)

Ho, Mei-ling.
(Ssu tao lu hsiang)
四道爐香 / [編著者何美玲]. — 臺北：莊嚴出版社，民國67 [1978]
1, 220 p. ; 19 cm. — (中國古典愛情故事 ; 3) (古典新刊 ; 19)
Short stories.
CONTENTS: 又見珍珠衫.—傳奇.—雙妹記.—花魁女.
NT$45.00

I. Title. II. Series: Chung-kuo ku tien ai ch'ing ku shih ; 3. III. Series: Ku tien hsin k'an ; 19.
PL2861.O315S68 78-841112

Ho, Yü-po, 1906–
(T'a ti hsiao hsi)
她的消息 / 賀玉波著. — [香港? : s. n., 1977?]
6, 147 p. ; 22 cm.
Short stories.
Reprint of the 1937 ed.
CONTENTS: 同命鳥.—離京.—營長太太.—破產.—離婚後.—她的消息.—自沈.—選擧.

I. Title.
PL2764.O27T3 1977 78-841572

Hsi, Ch'ing.
(Chu Lei)
朱蕾 = Zhu Lei / 奚青著. — 第1版. — 上海：上海文艺出版社，1978.
386 p. ; 19 cm.
Fiction.
$0.80

I. Title. II. Title: Zhu Lei.
PL2862.I12C47 78-843196

Hsia, Chih-yen.
(Pei-ching tsui han leng ti tung t'ien)
北京最寒冷的冬天 / 夏之炎著；丁祖威譯. — 臺北：聯合報社：總經銷聯經出版事業公司，民國66 [1977]
3, 322 p. ; 19 cm. — (聯合報叢書)
Fiction.
Series romanized: Lien ho pao ts'ung shu.
NT$40.00
I. Ting, Tsu-wei. II. Title.
PL2862.I179P4 1977b 79-841868

Hsiao, Ch'ao-ch'ün.
(Hsiao-sa hsiao shuo hsüan)
瀟颯小說選 / [著作者蕭超羣]. — 臺北縣永和：廣城出版社，民國66 [1977]
6, 2, 320 p. ; 22 cm.
Short stories.
NT$80.00

I. Title.
PL2862.I262H75 78-840946

Hsiao, Hai.
(Hung chün ch'iang tu Ta-tu ho)
红军强渡大渡河 / 啸海改编；汪国新，田期松绘画. — 第1版. — [武汉]：湖北人民出版社，1976.
56 p. : ill. ; 10 x 13 cm.
"根据一九七五年十月十九日《湖北日报》杨得志同志《强渡大渡河》一文改编"
1. Long March, 1934–1935—Fiction. I. Wang, Kuo-hsin. II. T'ien, Ch'i-sung. III. Yang, Te-chih, 1910– Ch'iang tu Ta-tu ho. IV. Title.
PL2862.I3143H8 79-835250

Hsiao, I, fl. 1978–
(Hsüeh yü chien hua hung)
血雨濺花紅 / 蕭逸著. — [台北]：春秋出版社：總經銷呂氏書店，民國67 [1978]
462 p. ; 20 cm. — (長篇武俠名著)
Fiction.
Series romanized: Ch'ang p'ien wu hsia ming chu.
NT$60.00

I. Title.
PL2862.I3155H7 79-835945

Hsiao, I.
(Na ts'ai yün)
那彩雲 / 蕭逸著. — 香港：春秋書店，[1978?]
5 v. (1389 p.) ; 19 cm. — (新派長篇武俠奇情小說)
Fiction.
Series romanized: Hsin p'ai ch'ang p'ien wu hsia ch'i ch'ing hsiao shuo.
$5.00 per vol.

I. Title.
PL2862.I3148N3 78-842524

Hsiao, Pai, 1925–
(Ta-lien-lo hsi)
大澰洛溪 / 蕭白著. — 台北：環球書社：總發行乾隆圖書無限公司，民國67 [1978]
249 p. ; 20 cm. — (當代文學叢書 : 散文類 ; 1)
NT$50.00

I. Title. II. Series: Tang tai wen hsüeh ts'ung shu : San wen lei ; 1.
PL2862.I319T3 79-837769

Hsiao, Sa.
(Erh tu mi yüeh)
二度蜜月 / 蕭颯著. — 臺北：聯經出版事業公司，民國67 [1978]
2, 261 p. ; 20 cm. — (現代小說叢刊)
Short stories.
Series romanized: Hsien tai hsiao shuo ts'ung k'an.
CONTENTS: 郭老太的困擾.—姿美的一日.—臨時演員.—浮光鏡影.—鬧鐘吵醒的早上.—最後的桂圓.—二度蜜月.—尷尬.—酒宴.—意外.—老師！吃餅.
NT$45.00

I. Title.
PL2862.I3227E7 79-838068

C

Hsiao, T'ung, 1928–
(Yin hai)
銀海 / 蕭銅著. — 香港：中流出版社，1978.
336 p. ; 19 cm.
Fiction.
$9.00

I. Title.
PL2862.I325Y5 78-841961

C

Hsiao-ch'an.
(Ch'ien fan chih wai)
千帆之外 / 小嬋著. — 台北：黎明文化事業公司，民國67 [1978]
172 p. ; 19 cm.
NT$40.00

I. Title.
PL2862.I34424C5 79-841115

Hsiao-ch'iu.
(Lang li hsing)
浪里行 / 晓丘，东旋. — 第1版. — 北京：人民文学出版社：新华书店北京发行所发行，1978.
240 p. ; 19 cm.
Fiction.
$0.47

I. Tung, Hsüan, joint author. II. Title.
PL2862.I34426L36 79-839123

Hsiao-ho, 1955–
(Ch'i chiao shou)
祈教授 / 小赫著. — 臺北：林白出版社，民國67 [1978]
256 p. ; 20 cm. — (林白叢書 ; 111)
Short stories.
Series romanized: Lin pai ts'ung shu.
CONTENTS: 祈教授.—功在杏林.—黃昏大決鬪.—凱達法警長之夢.—安安的葬禮.—最後的球賽.—風箏.
NT$55.00
I. Title.
PL2862.I3444C5 79-839116

Hsiao-hsiao-sheng.
[Chin P'ing Mei tz'u hua]
金瓶梅 / [笑笑生]. — 星洲：星洲世界書局，1978.
3 v. ([62], 1110 p.) ; 21 cm.
"附錄「金瓶梅與王世貞」—吳晗": v. 1, prelim. p. 1–47.
$22.00

I. Wu, Han, 1909– Chin P'ing Mei yü Wang shih-chen. 1978. II. Title.
Title romanized: Chin P'ing Mei.
PL2698.H73C5 1978 79-836764

Hsieh, Yu-ch'un.
(Keng Hsiao-liang kan ch'e)
耿小梁赶车 / 谢友纯. — [香港? : s. n., 1979?]
103 p. : ill. ; 21 cm.
Fiction.
Title in Pinyin on cover: Geng Xiao Liang gan che.
Reprint of the 1978 ed. published by Shao nien erh t'ung ch'u pan she, Shanghai.

I. Title.
PL2862.I4216K4 1979 79-838214

Hsin, Yü, 1933–
(Wo kei na pai ch'ih i k'uai ch'ien)
我給那白痴一塊錢 / 辛鬱著. — 臺北：天華出版事業股份有限公司，民國67 [1978]
2, 235 p. ; 19 cm. — (天華文學叢刊 ; 6)
Short stories.
Series romanized: T'ien hua wen hsüeh ts'ung k'an.
Pbk : NT$45.00

I. Title.
PL2862.I45219W6 79-836341

Hsin-tai.
(Pu hsieh ti A-chih)
補鞋的阿枝 / 心岱著. — 台北：皇冠出版社，民國68 [1979]
222 p. ; 20 cm.
Short stories.
CONTENTS: 情景.—獨木舟.—故事.—鏡子.—黑貓.—白雪之戀.—無花果.—阿羅漢.
NT$50.00
I. Title.
PL2862.I455P8 79-841139

Hsü, Chen-ya, fl. 1916–
(Pai hua Yü li hun)
白話玉梨魂 徐枕亞著 [惜紅館主譯文] 台南 一鳴書局 民國51 [1962]
310 p. 19 cm.

I. Title. II. Title: Yü li hun.
PL2765.U173P3 C 63-2484 ‡

rev [r79]

Hsü, Chia-shih, 1944–
(Shang sheng ti hai yang)
上昇的海洋 / 許家石著. — 臺北：聯經出版事業公司，民國65 [1976]
1, 10, 303 p. ; 20 cm. — (許家石短篇小說集)
Short stories.
CONTENTS: 採石船的故事.—上昇的海洋.—費茘蒙與島.—水.—三十五歲.—我的朋友蕭木茲.—長門賦.—朝露.—林阿枝的舞蹈事業.
NT$50.00
I. Title. II. Series: Hsü Chia-shih, 1944– Hsü Chia-shih tuan p'ien hsiao shuo chi.
PL2862.U1419S45 78-841451

CHINESE FICTION (Continued)

Hsü, Chih-mo, 1896–1931.
[Selected works]
想飛 / 徐志摩. — 台北：偉文圖書出版社，民國67 [1978]
102 p. ; 19 cm. — (民初散文作家選集)
CONTENTS: 我所知道的康橋.—翡冷翠山居閒話.—北戴河海濱的幻想.—天目山中筆記.—迎上前去.—悼沉叔薇.—我的彼得.—弔劉叔和.—自剖.—再剖.—求醫.—我的祖母之死.—傷雙栝老人.—想飛.—羅曼羅蘭.
NT$25.00
I. Title. II. Series: Min ch'u san wen tso chia hsüan chi.
Title romanized: Hsiang fei.
PL2765.U2A16 1978 79–839483

Hsü, En-mei, 1934–
[Feng tsai lin shao]
春歸路 / 徐薏藍[恩楣]著. — 台北：皇冠出版社，民國66 [1977]
395 p. ; 19 cm.
Fiction.
Previous ed. published by Li chih ch'u pan she, under title: Feng tsai lin shao.
NT$60.00
I. Title.
Title romanized: Ch'un kuei lu.
PL2862.U1495F4 1977 78–842535

Hsü, En-mei, 1934–
(Ho shang ti yüeh kuang)
河上的月光 / 徐薏藍[恩楣]著. — 臺北：皇冠出版社，民國66 [1977]
349 p. ; 19 cm.
Fiction.
NT$55.00
I. Title.
PL2862.U1495H6 1977 79–835232

Hsü, En-mei, 1934–
(Hsi yü ssu ssu)
細雨絲絲 / 徐薏藍[恩楣]著. — 第1版. — 臺北：四季出版事業有限公司，民國67 [1978]
1, 270 p. ; 19 cm. — (四季書庫 ; 30)
Short stories.
Series romanized: Ssu chi shu k'u.
CONTENTS: 細雨絲絲.—那年雨季.—永恒的愛.—綺年悲夢.
NT$45.00
I. Title.
PL2862.U1495H68 79–835260

Hsü, En-mei, 1934–
(Hsüan ko)
弦歌 / 徐薏藍[恩楣]著. — 臺北：皇冠出版社，民國66 [1977]
409 p. ; 19 cm.
Fiction.
NT$60.00
I. Title.
PL2862.U1495H7 78–840999

Hsü, En-mei, 1934–
(Liu yün)
流雲 / 徐薏藍[恩楣]著. — 臺北：皇冠出版社，民國56 [1967]，民國67 [1978] printing.
316 p. ; 19 cm.
Fiction.
NT$55.00
I. Title.
PL2862.U1495L5 79–835997

Hsü, En-mei, 1934–
(Mei kuei yüan)
玫瑰園 / 徐薏藍[恩楣]著. — 臺北：星光出版社，民國67 [1978]
2, 259 p. ; 19 cm. — (雙子星叢書 ; 51號)
Short stories.
Series romanized: Shuang tzu hsing ts'ung shu.
CONTENTS: 玫瑰園.—千里共嬋娟.—椽.—往事.—春恨.—泡影.—青春之歌.—愛與死.—悲歌.—擇侶.—春霧.—迷惘.—踩着冷雨.—她的故事.—冷月
NT$38.00
I. Title.
PL2862.U1495M4 79–837518

Hsü, Hao-chiung.
(Liu Hua)
刘华 / 编文许豪炯；绘画罗希贤，钱生发，苏民. — 第1版. — 上海：上海人民美術出版社：新華書店上海发行所发行，1978.
158 p. : ill. ; 96 x 128 mm.
"連環圖畫."
$0.17
1. Liu, Hua, 1899–1925—Fiction. I. Lo, Hsi-hsien. II. Ch'ien, Sheng-fa. III. Su, Min, fl. 1965–
PL2862.U1545L5 79–836872

Hsü, Hsü-kung.
(Shuang ch'eng nü tzu)
雙城女子 / 徐吁公著；[評者四懺詞人]. — [香港? : s. n., 1978?]
114, [34] p. ; 21 cm.
"奇情小說"
Reprint of the 1916 ed. published by Hsiao shuo ts'ung pao she, Shanghai.
I. Ssu-ch'an-tz'u-jen. II. Title.
PL2765.U244S5 1978 79–835534

Hsü, Tan-lin, 1949–
(Wo yao tsou chin jen hai)
我要走進人海 / 田間[許丹林]著. — 臺北：浩瀚出版社，民國64 [1975]
2, 168 p. ; 20 cm. — (浩瀚文庫 ; 14 : 創作類 ; 8)
Short stories.
Series romanized: Hao han wen k'u.
NT$30.00
I. Title.
PL2862.U256W6 76–838291

Hsü, Ying-chieh.
(Hai-yen chan ch'ih)
海燕展翅 / 编文徐英杰；绘画王双贵. — [石家庄?]：河北人民出版社，1977.
73 p. : ill. ; 10 x 13 cm.
"連環圖畫"
$0.10
I. Wang, Shuang-kuei. II. Title.
PL2862.U3225H3 79–835251

Hsu, Yu, 1908–
(Hua shen)
花神 / [著作者徐訏]. — 臺北：黎明文化事業公司，民國66 [1977]
323 p. ; 20 cm.
Short stories.
CONTENTS: 父親.—鳥叫.—來高陞路的一個女人.—自殺.—投海.—蓋棺論定.—花神.—新寡.
NT$50.00
I. Title.
PL2765.U32H78 78–840992

Hsüan, Chi.
(Pu tse shou tuan)
不擇手段 / 玄機著. — [台北]：金蘭文化出版社，民國67 [1978]
240 p. ; 20 cm.
Fiction.
NT$40.00
I. Title.
PL2862.U327P8 79–838277

79 C

Hsüan, Chi.
(To ming hsieh tzu)
奪命蠍子 / 玄機著. — [台北]：金蘭文化出版社，民國67 [1978]
263 p. ; 20 cm.
Fiction.
NT$40.00
I. Title.
PL2862.U327T58 79–839480

Hsüan, Chi.
(T'ou t'ien huan jih)
偷天換日 / 玄機著. — [台北]：金蘭文化出版社，民國68 [1979]
264 p. ; 20 cm.
Fiction.
NT$40.00
I. Title.
PL2862.U327T6 79–838279

Hsüan, Chien-jen.
(Ch'in ts'ao ch'ü)
琴操曲 / 宣建人著. — 臺北：臺灣學生書局，民國67 [1978]
3, 202 p. ; 20 cm. — (學生書苑 ; 65)
Series romanized: Hsüeh sheng shu yüan.
NT$70.00
I. Title.
PL2862.U328C49 78–836492

Hsüan, Hsiao-fo.
(Chui chui hsing ch'en)
綴綴星辰 / 玄小佛著. — 臺北縣土城：漢麟出版社，民國67 [1978]
347 p. ; 20 cm.
Fiction.
Reprint of the 1975? ed. published by Nan ch'i ch'u pan she, Taipei, in series: Nan ch'i wen i ts'ung shu.
NT$60.00
I. Title.
PL2862.U333C5 1978 79–839925

Hsüan, Hsiao-fo.
(Hua jui pin fen)
花蕊繽紛 / 玄小佛著. — 臺北：小說創作社；Wheaton, MD.：美國總代理 D. Tang, 1978.
311 p. ; 20 cm. — (小說創作叢書)
Fiction.
Series romanized: Hsiao shuo ch'uang tso ts'ung shu.
NT$40.00
I. Title.
PL2862.U333H8 78–842968

Hsüan, Hsiao-fo.
(Ts'ai se ti meng)
彩色的夢 / 玄小佛著. — 台北：漢麟出版社，民國67 [1978]
311 p. ; 19 cm.
Two short stories.
"流浪的雲": p. 191–311.
NT$55.00
I. Hsüan, Hsiao-fo. Liu lang ti yün. 1978. II. Title.
PL2862.U333T74 79–837360

Hsüan, Hsiao-fo.
(Tso jih yü hsiao hsiao)
昨日雨瀟瀟 / 玄小佛著. — 台北：漢麟圖書有限公司，民國67 [1978]
275 p. ; 20 cm.
Fiction.
NT$55.00
I. Title.
PL2862.U333T745 79–837478

Hsüan, Hsiao-fo.
(T'ung hsing lien tzu shu)
同性戀自述 / 玄小佛著. — [香港]：文娛出版社：發行者吳興記書報社，[1978?]
334 p. ; 19 cm.
Fiction.
$6.00
I. Title.
PL2862.U333T8 78–840844

CHINESE FICTION (Continued)

Huang, T'ien-shih.
(Wo chih mi yüeh)
我之蜜月 / [黃天石著]. — [香港：s. n., 1978?]
74 p. ; 21 cm.
Cover title.
At head of caption title: Hsieh shih hsiao shuo.
Reprint ed., with original preface dated 1922.

I. Title.
PL2768.T54W6 1978 78–840838

(Hsüeh Jen-kuei cheng Liao shih lüeh)
薛仁貴征遼事略 / 著作者趙萬里. — 臺景印版. — 台北：河洛圖書出版社, 民國66 [1977]
76 p., 1 leaf of plates : 21 cm. — (小說彙編)
Reprint of the 1957 ed. published by Ku tien wen hsüeh ch'u pan she, Shanghai.
NT$30.00

1. Hsüeh, Jen-kuei, 614–683—Fiction. I. Chao, Wan-li, 1905– II. Series: Hsiao shuo hui pien.
PL2687.H67 1977 79–837407

Hsüeh-k'o, pseud.
(Chan tou ti ch'ing ch'un)
战斗的青春 / 雪克著. — 新3版. — 上海：上海文艺出版社, 1962, 1978 printing.
vi, 603 p. ; 21 cm.
Fiction.
$1.40

I. Title.
PL2862.U34C5 1978 79–835045

Hsüeh-yen.
(Hei mo)
黑魔 / [雪雁著]. — 香港：春秋書店, [1978?]
6 v. ; 18 cm. — (新派長篇武俠奇情小說)
Fiction.
Series romanized: Hsin p'ai ch'ang p'ien wu hsia ch'i ch'ing hsiao shuo.
$5.00 per vol.

I. Title.
PL2862.U46H44 78–842356

Hsüeh-yen.
(Pai chien ling ma)
白劍鈴馬 / 雪雁著. — 香港：春秋書店, [1978?]
5 v. ; 18 cm. — (新派長篇武俠奇情小說)
Fiction.
Series romanized: Hsin p'ai ch'ang p'ien wu hsia ch'i ch'ing hsiao shuo.
$29.00
I. Title.
PL2862.U46P3 78–841114

Hsüeh-yen.
(Tzu tien, han ch'üan)
紫電·寒泉 / 雪雁著. — 香港：春秋書店, [1978?]
3 v. (1046 p.) ; 19 cm. — (新派長篇武俠奇情小說)
Fiction.
Series romanized: Hsin p'ai ch'ang p'ien wu hsia ch'i ch'ing hsiao shuo.
$19.50
I. Title.
PL2862.U46T9 78–842527

Hu, Ch'i, 1918–
(Nü shui shou)
女水手 胡奇著 [上海] 上雜出版社 [1953]
1, 108 p. 19 cm.
Short stories.

I. Title.
PL2863.C52N8 C 68–832

Hu, Hsi-shan.
(Ch'ung hsiao ch'ü)
冲霄曲 / 胡锡山. — [s. l. : s. n., 1978?]
528 p., [8] leaves of plates : ill. ; 21 cm.
Fiction.
Reprint of the 1977 ed. published by Chiang-su jen min ch'u pan she,

I. Title.
PL2863.H75C4 1978 79–838570

Hu, Hsiao, 1922–
(Li ming ch'ien)
黎明前 / [呼嘯著]. — 臺初版. — 臺北：正中書局, 民國56 [1967]
132 p. ; 20 cm. — (正中文叢叢書)
Fiction.
Series romanized: Cheng chung wen i ts'ung shu.

I. Title.
PL2863.H76L5 78–841463

Hu, Hsiao, 1922–
(Shan meng)
山盟 / 呼嘯著. — 臺北：求精出版社：總經銷精益書局, 民國67 [1978]
2, 251 p. ; 19 cm. — (精益叢書 ; 15)
Short stories.
Series romanized: Ching i ts'ung shu.
CONTENTS: 山盟.—勳章.—最後一封信.—會前插曲.—一幅畫.—滄海遺珠.—釣魚.—路.—候診室.—畫的滄桑.—迷惑.—流水不回頭.—勝利者.
NT$45.00
I. Title.
PL2863.H76S48 78–842007

Hu, Hsüeh-fang.
(Chiao yin chi)
脚印集 / 胡学方. — [香港? : s. n., 1978?]
2, 134 p. : ill. ; 21 cm.
Short stories.
Reprint of the 1977 ed. published by Kuang-hsi jen min ch'u pan she.

I. Title.
PL2863.H84C4 1978 78–843881

Hu, K'o.
(Ying-hua ho p'an)
映花河畔 / 胡柯. — [香港? : s. n., 1978?]
263 p. ; 21 cm.
Fiction.
Reprint of the 1977 ed. published by Hu-nan jen min ch'u pan she, Ch'ang-sha.

I. Title.
PL2863.K59Y5 1978 78–842938

Hu, K'un-chung, 1948–
(Tsou tsai t'ien keng shang)
走在田埂上 / 胡坤仲著. — 臺中：新企業世界出版社, 民國66 [1977]
3, 206 p. ; 20 cm.
Short stories.
NT$40.00

I. Title.
PL2863.K85T7 78–841117

78 C

Hu, Lun-ch'ing, ed.
(Ch'uan ch'i hsiao shuo hsüan)
傳奇小說選 / [編註者胡倫淸]. — 臺5版. — 臺北：正中書局, 民國66 [1977]
2, 3, 238 p. ; 20 cm. — (國文精選叢書)
Series romanized: Kuo wen ching hsüan ts'ung shu.

1. Chinese fiction. I. Title.
PL2625.H8 1977 78–843759

Hua, Chia.
(Tung ch'ü ch'un lai)
冬去春来 / 华嘉. — [香港? : s. n., 1979?]
142 p. ; 21 cm.
Fiction.
Reprint of the 1st (1954) ed. published by Kuang-tung jen min ch'u pan she, Canton, China, in series: Yüeh hai wen ts'ung.

I. Title. II. Series: Yüeh hai wen ts'ung.
PL2863.5.C5T8 1979 79–838254

Huang, Ch'un-ming.
(Wo ai Ma-li)
我愛瑪莉 / 黃春明. — 台北：遠景出版社, 民國68 [1979]
16, 200 p. ; 20 cm. — (遠景叢刊 ; 115)
Short stories.
Series romanized: Yüan ching ts'ung k'an.
CONTENTS: 我愛瑪莉.—蘋果的滋味.—莎喲娜啦·再見.—一個作者的卑鄙心靈.
NT$55.00
I. Title.
PL2865.C56W6 79–840561

Huang, Chung-su.
(Ch'ou ch'ang)
惆悵 / 著者黃仲蘇. — [香港? : s. n., 1977?]
2, 90 p. ; 22 cm.
Colophon title.
Short stories.
Reprint of the 1934 ed. published by Chung-hua shu chü, Shanghai, in series: Wen i hsiao ts'ung shu ; ti 1 chung.
CONTENTS: 惆悵.—了解.—童心.—樓外樓.—夜半呼聲.—雁語與回憶.—此時此地.—批也爾·羅地之死.
I. Title.
PL2865.C57C5 1977 78–840919

Huang, Ku-liu.
(Hsia-ch'iu chuan)
蝦球傳 / 黃谷柳著. — Hong Kong：宏文圖書公司, 1978.
3, 334 p., [8] leaves of plates : ill. ; 21 cm.
Fiction.
Reprint of the 1957 ed. published by T'ung su wen i ch'u pan she, Peking.
$9.50
I. Title.
PL2865.K8 1978 79–838754

Huang, I-yün.
(Hou pu ju kuei)
後不如歸：言情小說 / [著作人黃翼雲]. — [香港? : s. n., 1978?]
2, 114 p. ; 21 cm.
Fiction.
Reprint of the 1915 ed. published by Shang wu yin shu kuan, Shang-hai, issued in series: Shuo pu ts'ung shu, erh chi, ti 97 pien.

I. Title.
PL2865.I25H68 1978 78–842559

Huang, Ying.
(T'ien tao)
天刀：沈勝衣傳奇 / 黃鷹著. — 台北：春秋出版社：總經銷呂氏書店, 民國68 [1979]
156, 166 p. ; 19 cm.
Fiction.
"鬼簫": p. 1–166 (2d group)
NT$50.00

I. Huang, Ying. Kuei hsiao. 1979. II. Title.
PL2865.Y49T5 79–841097

(Huang-ho hsing)
黃河星. — 香港：香港海鷗出版公司, [1975?]
55 p. : ill. ; 15 x 17 cm. — (民間故事連環圖)
Fiction.
$1.00

I. Series: Min chien ku shih lien huan t'u.
PL2832.5.H8 78–841714

CHINESE FICTION (Continued)

Hung, Hsing-fu, 1949–
(Hei mien Ch'ing-tzu)
黑面慶仔 / 洪醒夫著. — 臺北：爾雅出版社, 民國67 [1978]
243 p. ; 19 cm. — (爾雅叢書 ; 48)
Short stories.
Series romanized: Erh ya ts'ung shu.
"洪醒夫寫作年表": p. 241–243.
CONTENTS: 散戲.—素芬出嫁這日.—僵局.—神轎.—跛脚天助和他的牛.—金樹坐在灶坑前.—扛.—黑面慶仔.—人鬼遊戲.—吾土.
NT$50.00
I. Title.
PL2866.N49H4 79–840514

Hung, Huang.
(Pa hai hsiung feng)
霸海雄風 / 洪荒著. — 台北：春秋出版社：總經銷呂氏書店, 民國66 [1977]
3 v. (972 p.) ; 18 cm. — (長篇武俠名著)
Fiction.
Series romanized: Ch'ang p'ien wu hsia ming chu.
NT$115.00
I. Title.
PL2866.N532P3 79–836013

I, K'o.
(Ti chai ying sheng)
邸宅嬰聲 / 異客著. — 台北：金蘭文化出版社, 民國67 [1978]
109 p. ; 19 cm. — (靈異小說情選)
Short stories.
Series romanized: Ling i hsiao shuo ching hsüan.
CONTENTS: 猛鬼陳天彩.—邸宅嬰聲.—情人谷.
NT$30.00
1. Ghost stories, Chinese. I. Title.
PL2867.K6T5 79–835775

I, Ta.
(Jan erh yu t'a)
然而有她 / 依達 ; [培新圖 ; 編者環球文藝編輯委員會]. — 香港：環球圖書雜誌出版社, 1978.
48 p. : ill. ; 19 cm. — (環球文藝 ; 第524期)
Cover title.
Fiction.
$0.60
I. Title. II. Series: Huan ch'iu wen i ; ti 524 ch'i.
PL2867.T3J3 79–835483

I, Ta.
(Kuei)
歸 / 依達著. — 香港：環球圖書雜誌出版社, 1975.
309 p. ; 17 cm. — (文藝創作小說)
Short stories.
Series romanized: Wen i ch'uang tso hsiao shuo.
I. Title.
PL2867.T3K8 79–836390

I, Ta.
(Shan-shan, pieh k'u!)
珊珊·別哭!＝又名桔色果 / 依達著. — [臺北?]：張昭琮 ; 臺北：總經銷光大圖書雜誌服務中心, 民國64 [1975]
224 p. ; 20 cm.
Fiction.
NT$32.00
I. Title.
PL2867.T3S47 78–841367

I, Ta.
(T'ang-mi pieh k'u)
湯米別哭 / 依達著. — [台北?]：環怡出版社 ; 台北：總經銷和風書報社, 民國64 [1975]
300 p. ; 20 cm.
Colophon and caption title: Pieh k'u T'ang-mi.
Short stories.
CONTENTS: 別哭湯美—愛的故事—冬戀—花花公子
NT$40.00
I. Title. II. Title: Pieh k'u T'ang-mi.
PL2867.T3T36 78–841366

I-hsün.
(Hsiang-kang ming jen ying hsiung chuan)
香港名人英雄傳 / 著者宜迅. — 九龍：百靈出版社, [1978?]
201 p. ; 19 cm.
Colophon title.
Short stories.
$5.00
I. Title.
PL2867.Z58H7 79–835428

I-shu.
(Wo ch'ou)
我醜 / 作者亦舒. — 香港：明窗出版社：發行明報有限公司出版部, 1979.
332 p. ; 19 cm.
Colophon title.
Short stories.
CONTENTS: 我醜.—表哥.—圓舞.—舊時.
$6.00 ($2.00 U.S.)
I. Title.
PL2867.Z75W6 79–840622

I-wen, fl. 1975–
(Chin-ling shih erh ch'ai)
金陵十二釵 / 憶文著. — 香港：春秋書店, [1978?]
5 v. ; 18 cm.
Fiction.
$24.00
I. Title.
PL2867.W43C48 78–842092

I-wen, fl. 1975–
(Ch'ing nu)
情奴 / 憶文著. — 香港：春秋書店, [1978?]
5 v. (1679 p.) ; 19 cm.
Fiction.
$32.00
I. Title.
PL2867.W43C5 78–841574

I-wen, fl. 1975–
(Pai yang t'u)
白陽圖 / 憶文著. — 香港：春秋書店, [1978?]
4 v. (1154 p.) ; 19 cm. — (新派長篇武俠奇情小說)
Fiction.
Series romanized: Hsin p'ai ch'ang p'ien wu hsia ch'i ch'ing hsiao shuo.
$22.00
I. Title.
PL2867.W43P28 78–842536

I-wen, fl. 1975–
(P'an lung chen yü nei)
蟠龍震宇内 / 憶文著. — 香港：春秋書店, [1978?]
3 v. ; 19 cm. — (新派長篇武俠奇情小說)
Fiction.
Series romanized: Hsin p'ai ch'ang p'ien wu hsia ch'i ch'ing hsiao shuo.
$13.50
I. Title.
PL2867.W43P3 78–842250

I-ying.
(Ch'ien yüan nü kung)
千元女工：工廠生活故事集 / 亦影著. — 香港：萬源圖書公司, 1977.
2, 161 p. : ill. ; 19 cm.
Fiction.
$4.50
I. Title.
PL2867.Z9C45 79–836078

Inoue, Yasushi, 1907–
[Inoue Yasushi shōsetsu zenshū. Chinese. Selections]
井上靖小说选 / 唐月梅译. — 第1版. — 北京：人民文学出版社, 1977.
162 p. ; 19 cm.
"根据《井上靖小说全集》(第3, 4, 10, 11卷)新潮1975年版译出"
$0.36
I. Title.
Title romanized: Ching-shang Ch'ing hsiao shuo hsüan.
PL830.N63A54 1977 78–841057

Jen-wei.
(Ku mu han ying)
古墓寒影 / [人畏著]. — 台北：德昌出版社, 民國66 [1977]
187 p. ; 19 cm. — (奇異鬼狐集錦故事)
Short stories.
"奇異鬼狐小說"
CONTENTS: 離魂曲.—愁海魔女.—墓底墊石.—影形替身.
NT$30.00
1. Ghost stories, Chinese. I. Title. II. Series: Ch'i i kuei hu chi chin ku shih.
PL2868.E75K8 79–836512

Jen-wei.
(Meng hun chi yü)
夢魂寄語 / [人畏著]. — 台北：德昌出版社, 民國66 [1977]
195 p. ; 19 cm. — (奇異鬼狐集錦故事)
Short stories.
"奇異鬼狐小說"
CONTENTS: 犬魂.—悠悠惧.—情海魅影.—血影.
NT$30.00
1. Ghost stories, Chinese. I. Title. II. Series: Ch'i i kuei hu chi chin ku shih.
PL2868.E75M4 79–837607

Jen-wei.
(Tzu yeh fang hun)
子夜芳魂 / [人畏著]. — 台北：德昌出版社, 民國66 [1977]
189 p. ; 19 cm. — (奇異鬼狐集錦故事)
Short stories.
"奇異鬼狐小說"
NT$30.00
1. Fantastic fiction, Chinese. I. Title. II. Series: Ch'i i kuei hu chi chin ku shih.
PL2868.E75T9 79–836489

Jo-meng.
(Ch'i nan miao nü)
奇男妙女：社會傳眞·名家小說 / 若夢著. — 台北：哲志出版社, 民國66 [1977]
207 p. ; 20 cm.
Short stories.
CONTENTS: 花花和尚.—風流胚.—老妻少夫.—吃女人的男人.—文夫出售.—淫術.
NT$35.00
I. Title.
PL2868.O28C48 78–842837

Jo-ping.
(Lang tzu p'o)
浪子魄 / 若冰著. — [台北]：志明[出版社, 1978?]
304 p. ; 20 cm. — (志明小說選)
Cover title.
Fiction.
Series romanized: Chih ming hsiao shuo hsüan.
I. Title.
PL2868.O3L3 79–837987
79 C

Jo-ping.
(Sha shou)
殺手 / 若冰著. — 台北：志明出版社, 民國66 [1977]
294 p. ; 20 cm. — (志明小說選輯)
Fiction.
"民初遊俠鬥智小說"
Series romanized: Chih ming hsiao shuo hsüan chi.
NT$50.00
I. Title.
PL2868.O3S5 78–841944

CHINESE FICTION (Continued)

Juan, Lang.
(Hai chiao ch'un hui)
海角春回 / 阮朗著. — 香港：上海書局, 1978.
142 p. : ill. ; 19 cm.
Fiction.
$3.70

I. Title.
PL2868.U35H3 78-841964

K'ang, Pai.
(Fan-lung shan)
蟠龍山 / 康白著. — 台北：震旦出版社：總經銷平陽傳播廣告事業有限公司, 民國65 [1976]
4, 190 p. ; 19 cm.
Fiction.

I. Title.
PL2869.N554F3 79-841057

K'ang, Pai.
(Yüan ch'i yüan mieh)
緣起緣滅 / 康白著. — 臺北：聯經出版事業公司, 民國67 [1978]
207 p. ; 20 cm.
Fiction.
NT$40.00

I. Title.
PL2869.N554Y8 78-843531

Kao, Feng.
(Tung shih chang)
董事長 / 高峯著. — 台北：哲志出版社, 民國67 [1978]
290 p. : ill. ; 20 cm.
At head of title: She hui hsiao shuo shuo, p'ien shu ta kuan.
Fiction.
NT$50.00

I. Title.
PL2869.O1175T8 79-838036

Kao, Kao, fl. 1969–
(Ch'ien jen chung)
千人塚 新派武俠小說 高皐著 香港 武林出版社 1970.
2 v. (652 p.) 17 cm. $7.80
Fiction.

I. Title.
PL2869.O135C5 75-843461

Kao, Kao, fl. 1969–
(Feng po t'ing)
風波亭 新派武俠小說 高皐著 [香港?] 武林出版社 1970.
338 p. 18 cm. $3.50
Fiction.

I. Title.
PL2869.O135F4 77-843739

Kao, Kao, fl. 1969–
(Lo jih hui ko)
落日揮戈—新派武俠小說 高皐著 [香港?] 武林出版社 1969.
260 p. 18 cm. $3.00
Fiction.

I. Title.
PL2869.O135L6 70-843397

Kao, T'ao, 1902–1950.
(Hsin sheng tai)
新生代 / [著者齊同 i.e. 高滔]. — 香港：文教出版社, 1978 printing.
300 p. ; 19 cm. — (中國現代文學叢書)
Fiction.
Series romanized: Chung-kuo hsien tai wen hsüeh ts'ung shu.
$7.50

I. Title.
PL2869.O164H7 1978 78-841976

Kao, Yang.
(Hsiao feng hsien)
小鳳仙 / 高陽著. — 臺北：南京出版公司, 民國67 [1978]
2 v. (748 p.) ; 20 cm.
Fiction.
NT$45.00 per vol.

I. Title.
PL2869.O167H74 78-842732

Kao, Yang.
(Hua k'uei)
花魁 / 高陽著. — 臺北：時報文化出版事業有限公司, 民國68 [1979]
205 p. ; 19 cm. — (時報書系 ; 171)
Fiction.
Series romanized: Shih pao shu hsi.
Based on Mai yu lang tu chan hua k'uei.
NT$50.00

I. Mai yu lang tu chan hua k'uei. II. Title.
PL2869.O167H83 79-841163

Kao, Yang.
(Mu-tzu chün ch'en)
母子君臣 高陽著 [臺北 皇冠出版社 1973?]
381 p. 19 cm. His 慈禧全傳之1 皇冠叢書第313種 NT$40.00

1. Tz'u-hsi, Empress dowager of China, 1835–1908—Fiction. I. Title.
PL2869.O167T9 vol. 1 74-896319

Kao, Yang.
(P'i p'a yüan)
琵琶怨 / 高陽著. — 台北：南京出版社, 民國67 [1978]
432 p. ; 20 cm.
Fiction.
NT$68.00

I. Title.
PL2869.O167P55 79-840370

Kao, Yang.
(Ts'ui-p'ing shan)
翠屏山 / 高陽著. — 台北：求精出版社, 民國67 [1978]
311 p. ; 20 cm. — (精益叢書 ; 12)
Fiction.
Series romanized: Ching i ts'ung shu.
NT$60.00

I. Title.
PL2869.O167T7 79-835810

Kao, Ying.
(Yün yai ch'u nuan)
云崖初暖：《奴隶们起来》第一部 / 高缨. — 第1版. — 北京：人民文学出版社：新华书店北京发行所发行, 1978.
656 p. ; 19 cm.
Fiction.
$1.25

I. Title.
PL2869.O169Y8 79-837687

Kao, Yung.
(Feng liu sha shou)
風流殺手 / 高庸著. — 台北：春秋出版社：總經銷呂氏書店, 民國67 [1978]
2 v. (1006 p.) ; 20 cm. — (長篇武俠名著)
Fiction.
Series romanized: Ch'ang p'ien wu hsia ming chu.
NT$70.00 per vol.

I. Title.
PL2869.O28F4 79-839481

Kao, Yung.
(Hsüan feng shih pa ch'i)
旋風十八騎 / 高庸著. — 台北：春秋出版社：香港：總經銷陳湘記書局, 民國67 [1978]
2 v. (717 p.) ; 19 cm. — (長篇武俠名著)
Fiction.
Series romanized: Ch'ang p'ien wu hsia ming chu.
NT$50.00 per vol.

I. Title.
PL2869.O28H7 79-835915

Keng-hsin.
(Hsüeh shan feng yün)
雪山風雲 / 耕心著. — 台中：弘業文化實業股份有限公司, 民國66 [1977]
266 p. ; 19 cm.
Short stories.
CONTENTS: 我們是孤寂的.—煙斗.—老相士.—雪山風雲.—在故鄉的表演.—鍾老伯與阿龍.—臘鼓聲中.—離婚.—藍先生.—新奇的把戲.—新年.—後記.
NT$40.00
I. Title.
PL2870.N45H78 78-842154

Kim, Sŏn-o.
(Hansi ch'oyŏn)
漢詩抄演 / 金善晤著. — 서울：景仁文化社, 1976.
176 p. : ill. ; 18 cm.
₩700

I. Title.
PL2871.M5H3 79-825097

K'o, Ling.
(Hui ming)
晦明 / 柯靈著. — [香港?]：創作書社, [1978?]
ii, 117 p. ; 19 cm.
Cover title.
Reprint of the 1946 ed. published by Wen hua sheng huo ch'u pan she, Shanghai, in series: Wen hsüeh ts'ung k'an : Ti ch'i chi.

I. Title.
PL2872.L5H8 1978 79-835243

K'o-yang.
(To tao)
奇刀 / 克扬. — 第1版. — 太原：山西人民出版社, 1977, 1978 printing.
685 p., [3] leaves of plates : ill. ; 19 cm.
$1.60

1. China—History—Civil War, 1945–1949—Fiction. I. Title.
PL2872.Z8T6 79-835938

79 C

Ku, Chün-ch'iao.
(Ch'ing shao nien hu po shao)
青少年护泊哨 / 顾骏翘. — [香港?]：s. n., 1978?]
230 p., [3] leaves of plates : ill. ; 21 cm.
Fiction.
Reprint of the 1976 ed. published by Chiang-su jen min ch'u pan she.

I. Title.
PL2874.C75C5 1978 79-836854

CHINESE FICTION (Continued)

Ku, Lung.
(Ch'i hsing lung wang)
七星龍王 / 古龍著. — 臺北：春秋出版社：發行所呂氏書店，民國67 [1978]
414 p. ; 22 cm. — (古龍小說特輯) (寶龍系列 ; 2)
Fiction.
Series 2 romanized: Pao lung hsi lieh.
NT$140.00

I. Title. II. Series: Ku, Lung. Ku Lung hsiao shuo t'e chi.
PL2874.L8C38 79-839558

Ku, Lung.
(Chien hua yen yü chiang nan)
劍花煙雨江南 / [古龍著]. — 台北：漢鱗 [i. e. 麟] 出版社，民國67 [1978]
190 p. ; 22 cm. — (古龍小說專輯 ; 11) (漢麟系列)
Fiction.
Series 2 romanized: Han lin hsi lieh.
NT$80.00

I. Title. II. Series: Ku, Lung. Ku Lung hsiao shuo chuan chi ; 11.
PL2874.L8C42 79-840615

Ku, Lung.
(Chiu yüeh ying fei)
九月鷹飛 / [古龍著]. — 台北：漢麟出版社，民國67 [1978]
2 v. (655 p.) ; 22 cm. — (古龍小說專輯 ; 5)
Fiction.
NT$90.00 per vol.

I. Title. II. Series: Ku, Lung. Ku Lung hsiao shuo chuan chi ; 5.
PL2874.L8C45 78-843757

Ku, Lung.
(Chüeh pu ti t'ou)
絕不低頭：古龍著. — 台北縣：漢麟出版社，民國67 [1978]
314 p. ; 20 cm.
Fiction.
NT$60.00

I. Title.
PL2874.L8C49 1978 79-835556

Ku, Lung.
(Hsüeh ying wu)
血鸚鵡 / [古龍著]. — 台北：漢麟出版社，民國67 [1978]
2 v. (3, 656 p.) ; 21 cm. — (古龍小說專輯 ; 10) (漢麟系列)
Fiction.
Series 2 romanized: Han lin hsi lieh.
NT$200.00

I. Title. II. Series: Ku, Lung. Ku Lung hsiao shuo chuan chi ; 10.
PL2874.L8H78 79-836037

Ku, Lung.
(Hu hua ling)
護花鈴 / 古龍著. — 臺北：春秋出版社，民國67 [1978]
2 v. (792 p.) ; 22 cm.
Fiction.
NT$100.00 per vol.

I. Title.
PL2874.L8H79 79-836057

Ku, Lung.
(Huan le ying hsiung)
歡樂英雄 / [古龍著]. — 台北：漢麟出版社，民國67 [1978]
2 v. (762 p.) ; 22 cm. — (古龍小說專輯 ; 8) (漢麟系列)
Fiction.
Series 2 romanized: Han lin hsi lieh.
NT$100.00 per vol.

I. Title. II. Series: Ku, Lung. Ku Lung hsiao shuo chuan chi ; 8.
PL2874.L8H795 79-83735

Ku, Lung.
(Huo ping Hsiao-shih-i-lang)
火併蕭十一郎 / [古龍著]. — 台北：漢麟出版社，民國67 [1978]
2 v. (658 p.) : 22 cm. — (古龍小說專輯 ; 9)
Fiction.
NT$180.00

I. Title. II. Series: Ku Lung. Ku Lung hsiao shuo chuan chi ; 9.
PL2874.L8H84 79-837438

Ku, Lung.
(Ming chien feng liu)
名劍風流 / 古龍著. — 台北：漢麟出版社，民國67 [1978]
3 v. (1203 p.) ; 22 cm. — (古龍小說專輯 ; 14)
Fiction.
NT$300.00

I. Title. II. Series: Ku, Lung. Ku Lung hsiao shuo chuan chi ; 14.
PL2874.L8M5 79-836054

Ku, Lung.
(Pa wang ch'iang)
霸王槍：七種武器之五. — 台北：漢鱗 [i. e. 麟] 出版社，民國67 [1978]
322 p. ; 22 cm. — (古龍小說專輯 ; 13) (漢麟系列)
Fiction.
Series 2 romanized: Han lin hsi lieh.
NT$90.00

I. Title. II. Title: Ch'i chung wu ch'i. III. Series: Ku, Lung. Ku Lung hsiao shuo chuan chi ; 13
PL2874.L8P27 79-839545

Ku, Lung.
(Pi yü tao)
碧玉刀：七種武器之二 / [古龍著]. — 台北：漢麟出版社，民國67 [1978]
225 p. ; 22 cm. — (古龍小說專輯 ; 13) (漢麟系列)
Fiction.
Series 2 romanized: Han lin hsi lieh.
NT$80.00

I. Title. II. Title: Ch'i chung wu ch'i. III. Series: Ku Lung. Ku Lung hsiao shuo chuan chi ; 13.
PL2874.L8P45 79-839547

Ku, Lung.
(P'iao hsiang chien yü)
飄香劍雨 / 古龍著. — 香港：毅力出版社，[1978?]
3 v. (751 p.) ; 18 cm.
Fiction.
"新派武俠小說"
$15.00

I. Title.
PL2874.L8P46 79-836459

Ku, Lung.
(P'iao hsiang chien yü hsü chi)
飄香劍雨續集 / 古龍著. — 澳門：毅力出版社，[1978?]
3 v. (864 p.) ; 18 cm.
Cover title.
Fiction.
"新派武俠小說"
$15.00 (H.K.)

I. Title.
PL2874.L8P47 79-836460

Ku, Lung.
(P'iao po ying hsiung chuan)
飄泊英雄傳 / 古龍著. — 臺北：南琪出版社：海外總代理新代出版社，民國67 [1978]
2 v. (653 p.) ; 22 cm. — (古龍長篇武俠小說) (南琪武俠叢書)
Fiction.
Series 2 romanized: Nan ch'i wu hsia ts'ung shu.
NT$100.00 per vol.

I. Title. II. Series: Ku, Lung. Ku Lung ch'ang p'ien wu hsia hsiao shuo.
PL2874.L8P48 79-838236

Ku, Lung.
(Pei tou han)
北斗寒 / [作者古龍]. — 臺北：桂冠圖書有限公司：發行四季，桂冠聯合發行部，民國66 [1977]
5 v. : ill. ; 19 cm. — (古龍小說專輯 ; 21) (桂冠系列)
Fiction.
Series 2 romanized: Kuei kuan hsi lieh.
NT$40.00 per vol.

I. Title. II. Series: Ku, Lung. Ku Lung hsiao shuo chuan chi ; 21.
PL2874.L8P4 79-835923

Ku, Lung.
(Pien ch'eng lang tzu)
邊城浪子 / 古龍著 — [香港?]：武林出版社，1976.
2 v. (615 p.) ; 17 cm.
Fiction.
"新派武俠小說"

I. Title.
PL2874.L8P5 1976 79-836961

Ku, Lung.
(Shen chien ch'i hsia)
神劍奇俠：正宗長篇武俠名著 / 古龍著. — 澳門：武藝出版社，[1978?]
4 v. (1148 p.) : ill. ; 20 cm.
Fiction.
"武俠長篇奇情小說"
$4.80 per vol.

I. Title.
PL2874.L8S5 79-837437

Ku, Lung.
(T'ien ya ming yüeh tao)
天涯明月刀：新派武俠奇情故事 / 古龍. — 臺北：華新出版有限公司：總經銷桂冠圖書有限公司，民國66 [1977]
2 v. (517 p.) ; 19 cm. — (古龍小說專輯) (長篇武俠奇情小說)
Fiction.
Series 2 romanized: Ch'ang p'ien wu hsia ch'i ch'ing hsiao shuo.
Pbk. : NT$200.00
I. Title. II. Series: Ku, Lung. Ku Lung hsiao shuo chuan chi.
PL2874.L8T5 1977 78-843899

Ku, Lung.
(Wu lin wai shih)
武林外史 / [古龍著]. — 台北縣土城：漢林出版社，民國67 [1978]
4 v. (1404 p.) ; 22 cm. — (古龍小說專輯 ; 6) (漢麟系列)
Fiction.
Series 2 romanized: Han lin hsi lieh.
NT$360.00

I. Title. II. Ku, Lung. Ku Lung hsiao shuo chuan chi ; 6.
PL2874.L8W8 78-843107

Ku, Lung.
(Yüan yüeh, wan tao)
圓月，彎刀 / [古龍著]. — 台北：漢麟出版社，民國67 [1978]
2 v. (735 p.) ; 22 cm. — (古龍小說專輯 ; 7)
Fiction.
NT$100.00 per vol.

I. Title. II. Series: Ku, Lung. Ku Lung hsiao shuo chuan chi ; 7.
PL2874.L8Y84 78-843756

C

Ku, Ying.
(Hsüeh shih i yün)
血屍疑雲 / [谷影著]. — [台北]：德昌出版社，[民國66 i. e. 1977]
187 p. ; 19 cm. — (奇異鬼狐集錦故事)
"奇異鬼狐小說"
CONTENTS: 檢來的夢.—巫山驚夢.—影像之謎.—鬼父托孤
NT$30.00

1. Ghost stories, Chinese. I. Title. II. Series: Ch'i i kuei hu chi chin ku shih (T'ai-pei ed.)
PL2874.Y56H7 79-837127

CHINESE FICTION (Continued)

Ku, Ying.
(Lun hui ch'ien hsiang)
輪迴千相 / [谷影著]. — [台北] : 德昌出版社, [民國66 i. e. 1977]
193 p. ; 19 cm. — (奇異鬼狐集錦故事)
"奇異鬼狐小說"
CONTENTS: 情塚埋恨.—托孤.—屍影.—擋花悲魔.—果報不爽
NT$30.00

1. Ghost stories, Chinese. I. Title. II. Series: Ch'i i kuei hu chi chin ku shih (T'ai-pei ed.)
PL2874.Y56L86 79-837107

Ku, Ying.
(Nieh hai hun fei)
孽海魂飛 / 谷影著. — [台北 : 德昌出版社, 民國66 i. e. 1977]
184 p. ; 19 cm. — (奇異鬼狐集錦故事)
"奇異鬼狐小說"
At head of title: Tuan p'ien shu i wen i hsiao shuo.
CONTENTS: 雷劫.—棄屍悲鳴.—天網.—夢中戀痕
NT$30.00
1. Ghost stories, Chinese. I. Title. II. Series: Ch'i i kuei hu chi chin ku shih (T'ai-pei ed.)
PL2874.Y56N5 79-836511

Ku, Ying.
(Ying hai shih i)
瀛海拾異 / 谷影著. — [台北 : 德昌出版社, 民國66 i. e. 1977]
184 p. ; 19 cm.
"奇異鬼狐小說"
At head of title: Tuan p'ien shu i wen i hsiao shuo.
CONTENTS: 叠巫斷魂.—恨海葬情.—幽明兩廂憶.—墓中人
NT$30.00

1. Ghost stories, Chinese. I. Title.
PL2874.Y56Y5 79-837129

(Ku chin yao kuai ta kuan)
古今妖怪大觀. — 台北 : 華貿出版社 : 總經銷精益書局, 民國65 [1976]
5, 4, 75, 68 p. : ill. ; 19 cm.
Colophon title.
Reprint ed.
NT$40.00

1. Fantastic fiction, Chinese.
PL2629.F35K8 1976 79-835490

(Ku tai tuan p'ien hsiao shuo)
古代短篇小說. — 香港 : 海鷗出版公司, 1978.
235 p. : ill. ; 21 cm.
CONTENTS: 碾玉观音.—錯斬崔寧.—金玉奴棒打薄情郎.—汪信之一死救全家.—沈小霞相会出师表.—白娘子永鎮雷峯塔.—杜十娘怒沉百宝箱.—賣油郎独占花魁.—灌園叟晚逢仙女.—錢秀才錯占鳳凰儔.
$6.00

1. Short stories, Chinese.
PL2625.K87 79-835369

Ku-yü.
(Po shih yü wo)
博士與我 / 穀雨著. — 第1次版. — 新加坡 : 黃氏方外居, 1975.
7, 280, 2 p. ; 19 cm.
Fiction.

I. Title.
PL2874.Z9P6 78-940159

Kuan, Wang-hsing.
(K'uang ai ch'ien ts'ai ti fu jen)
狂愛錢財的婦人 / 觀望星著. — 台北 : 金星出版社, 民國67 [1978]
128 p. ; 20 cm.
NT$35.00

I. Title.
PL2875.A57K8 79-839479

Kuang, T'ai.
(Chi mo ti hsi yang)
寂寞的夕陽 / 光泰著. — 臺北 : 時報文化出版事業有限公司, 民國66 [1977]
211 p. ; 19 cm. — (時報書系 ; 61)
Fiction.
Series romanized: Shih pao shu hsi.
NT$38.00

I. Title.
PL2875.A64C48 79-835148

Kuang, T'ai.
(Ning meng shu)
檸檬樹 / 光泰著. — 臺北 : 小說創作社 ; Silver Spring, Md. : 美國總代理 Tang's Corporation, 民國65 [1976]
319 p. ; 20 cm. — (小說創作叢書)
Fiction.
Series romanized: Hsiao shuo ch'uang tso ts'ung shu.
NT$40.00
I. Title.
PL2875.A64N5 78-841468

Kuei, Chien-ch'ou.
(Hsiang chung i yün)
香塚疑雲 / 鬼見愁著. — 台北 : 金蘭文化出版社, 民國67 [1978]
198 p. ; 20 cm. — (靈異小說精選)
Two novels.
Series romanized: Ling i hsiao shuo ching hsüan.
CONTENTS: 香塚疑雲.—琴挑文君.
NT$30.00

1. Ghost stories, Chinese. I. Title.
PL2875.E38H74 79-835905

Kuei, Chien-ch'ou.
(Lang tsung mei ying)
狼踪魅影 / 鬼見愁著. — 台北 : 金蘭文化出版社, 民國67 [1978]
192 p. ; 20 cm. — (靈異小說精選)
Fiction.
Series romanized: Ling i hsiao shuo ching hsüan.
NT$30.00

I. Title.
PL2875.E38L3 78-843541

Kuei, Chien-ch'ou.
(Shih ying)
屍影 / 鬼見愁著. — 台北 : 金蘭文化出版社, 民國67 [1978]
192 p. ; 19 cm. — (靈異小說精選)
Short stories.
Series romanized: Ling i hsiao shuo ching hsüan.
CONTENTS: 屍影.—舞台命案.—烟鬼.
NT$30.00

I. Title.
PL2875.E38S54 78-841783

Kuei-ku-tzu.
(Lu pien yu hun)
路邊遊魂 / 鬼谷子著. — 台北 : 金蘭文化出版社, 民國67 [1978]
193 p. ; 20 cm. — (靈異小說精選)
Fiction.
Series romanized: Ling i hsiao shuo ching hsüan.
NT$30.00

I. Title.
PL2875.E415L8 79-835233

C

Kuei-ku-tzu.
(Pi po ch'ih hsing)
碧波魑形 / 鬼谷子著. — 台北 : 金蘭文化出版社, 民國67 [1978]
180 p. ; 20 cm. — (靈異小說精選)
Fiction.
Series romanized: Ling i hsiao shuo ching hsüan.
NT$30.00

I. Title.
PL2875.E415P5 79-835526

Kuei-ku-tzu.
(Tao shih)
盗屍 / 鬼谷子著. — 台北 : 德昌出版社, 民國66 [1977]
188 p. ; 19 cm. — (奇異鬼狐集錦故事)
Cover title.
Short stories.
"奇異鬼狐小說"
CONTENTS: 情劫.—大榕樹下.—盗屍.—模特兒之謎.
NT$30.00

1. Ghost stories, Chinese. I. Title. II. Series: Ch'i i kuei hu chi chin ku shih.
PL2875.E45T36 79-836513

Kuo, I-tung, 1920-
(Cheng cha)
掙扎 / 著者郭衣洞. — 九龍 : 天風圖書公司, [1978?]
6, 161 p. ; 19 cm. — (郭衣洞小說全集 ; 第4集)
Short stories.
CONTENTS: 兀鷹.—相思樹.—一葉.—火車上.—進酒.—歸巢.—辭行.—窄路.—路碑.—平衡.—客人.—朋友.
$5.50
I. Title.
PL2875.O17A15 1978, vol. 4 79-835990

Kuo, I-tung, 1920-
(Kuang yeh)
曠野 / 著者郭衣洞. — 九龍 : 天風圖書公司, [1978?]
6, 3, 371 p. ; 19 cm. — (郭衣洞小說全集 ; 第3集)
Fiction.
$8.50

I. Title.
PL2875.O17A15 1978, vol. 3 79-835989

Kuo, I-tung, 1920-
(Kuang yeh)
曠野 / 著者郭衣洞. — [臺北] : 星光出版社, 民國66 [1977]
6, 3, 371 p. ; 20 cm. — (郭衣洞小說全集 ; 第3集) (雙子星叢書 ; 51)
Series 1 romanized: Kuo I-tung hsiao shuo ch'üan chi. Series 2 romanized: Shuang tzu hsing ts'ung shu.
NT$45.00

I. Title.
PL2875.O17A15 1977, vol. 3 78-836822
[PL2875.O17]

Kuo, I-tung, 1920-
(Kuo I-tung hsiao shuo ch'üan chi)
郭衣洞小說全集 / 著者郭衣洞. — 九龍 : 天風圖書公司, [1978?-
v. ; 19 cm.
CONTENTS: 第1集 秘密.—第2集 莎羅冷.—第3集 曠野.—第4集 掙扎.—第5集 怒航.
$6.00 (v. 1) varies.

I. Title.
PL2875.O17A15 1978 79-835985

Kuo, I-tung, 1920-
(Mi mi)
秘密 / 著者郭衣洞. — 九龍 : 天風圖書公司, [1978?]
6, 8, 2, 217 p. ; 19 cm. — (郭衣洞小說全集 ; 第1集)
Short stories.
CONTENTS: 峽谷.—秘密.—龍眼粥.—强水街.—沉船.—窗前.—結.—蓮.—拱橋.—塑像.
$6.00
I. Title.
PL2875.O17A15 1978, vol. 1 79-835987

Kuo, I-tung, 1920-
(Nu hang)
怒航 / 著者郭衣洞. — [臺北] : 星光出版社 ; 臺北 : 經銷者星光書報社, 民國66 [1977]
6, 3, 231 p. ; 19 cm. — (郭衣洞小說全集 ; 第5集) (雙子星叢書 ; 53)
Series 1 romanized: Kuo I-tung hsiao shuo ch'üan chi. Series 2 romanized: Shuang tzu hsing ts'ung shu.
CONTENTS: 七星山.—隆格.—周琴.—晚霞.—鋼鞭.—微笑.—夜歸.—時代.—屈膝.—閒步港.—鬼屋.—出獄.
NT$35.00

I. Title.
PL2875.O17A15 1977, vol. 5 78-836821
[PL2875.O17]

CHINESE FICTION (Continued)

Kuo, I-tung, 1920–
(Nu hang)
怒航 / 著者郭衣洞. — 九龍：天風圖書公司, [1978?]
6, 3, 231 p. ; 19 cm. — (郭衣洞小說全集 ; 第5集)
Short stories.
CONTENTS: 七星山.—隆格.—周琴.—晚霞.—鋼鞭.—微笑.—夜歸.—時代.—屈膝.—閒步港.—鬼屋.—重逢.
$6.50
I. Title.
PL2875.O17A15 1978, vol. 5 79–835991

Kuo, I-tung, 1920–
(Sha-lo-leng)
莎羅冷 / 著者郭衣洞. — 九龍：天風圖書公司 [1978?]
6, 3, 165 p. ; 19 cm. — (郭衣洞小說全集 ; 第2集)
Fiction.
$5.50
I. Title.
PL2875.O17A15 1978, vol. 2 79–835988

Kuo, Liang-hui.
[Ch'ing ch'ing ts'ao]
青草青青 / 郭良蕙著. — 台北縣土城：漢麟出版社, 民國67 [1978]
277 p. ; 20 cm.
Previous ed. published under title: Ch'ing ch'ing ts'ao.
Fiction.
NT$50.00
I. Title.
Title romanized: Ch'ing ts'ao ch'ing ch'ing.
PL2875.O2C47 1978 79–835771

Kuo, Liang-hui.
(Hao ko ch'iu)
好個秋 / 作者郭良蕙. — 臺北：天下圖書公司：總經銷文藝月刊社業務部, 民國65 [1976]
397 p. ; 20 cm. — (天下小說 ; 6)
Colophon title.
Fiction.
Series romanized: T'ien hsia hsiao shuo.
NT$50.00
I. Title.
PL2875.O2H36 78–842389

Kuo, Liang-hui.
(Liang chung i wai ti)
兩種以外的 / 郭良蕙著. — 台北：漢麟出版社, 民國67 [1978]
270 p. ; 19 cm.
Fiction.
NT$60.00
I. Title.
PL2875.O2L47 79–840356

Kuo, Liang-hui.
(T'a men ti kuo shih)
他們的故事 / 郭良蕙著. — 台北：漢麟出版社, 民國67 [1978]
395 p. ; 20 cm.
NT$60.00
I. Title.
PL2875.O2T35 78–843436

Kuo, Ssu-fen.
(Hua chieh tzu)
花街子 / 郭嗣汾著；中央日報編印. — 臺北：中央日報社, 民國67 [1978]
386 p. ; 19 cm.
Fiction.
NT$50.00
I. Title.
PL2875.O25H85 79–835949

Kuo, Ting-t'ang.
(T'a)
塔 / 郭鼎堂著. — [香港?：s.n., 1976?]
325 p., [1] leaf of plates : ill. ; 19 cm.
Short stories.
Reprint of the 1930 ed. published by Shang wu yin shu kuan, Shanghai, issued as no. 1 in series: Hsüeh i ts'ung shu.
I. Title.
PL2778.O25T3 1976 79–836665

Lang, Lan.
(T'ieh tao ch'ien shao)
鐵道前哨 / [郎瀾著]. — [香港?：s.n., 1977?]
2, 446 p. ; 22 cm.
Fiction.
Reprint of the 1977 ed. published by Shang-hai jen min ch'u pan she, Shanghai.
I. Title.
PL2876.A49T5 1977 78–841487

Lao, Hsiang.
(Ch'üan chia ts'un)
全家村 / 老向著；[主編者周黎庵]. — [香港?：s.n., 1977?]
182 p. ; 19 cm.
Fiction.
Reprint of the 1940 ed. published by Yü chou feng she, Shanghai, issued as ti 5 ts'e in series: Yü chou feng she yüeh shu.
I. Title.
PL2779.A58C48 1977 79–838598

Lei, Hua.
(Yü mien sha shou)
玉面殺手 / 雷華著. — 台北：金蘭文化出版社, 民國67 [1978]
233 p. ; 20 cm.
Fiction.
NT$40.00
I. Title.
PL2876.E46Y8 79–838216

Lei, Yin.
(K'ung pu mo hua)
恐怖魔畫 / 雷吟著. — [台北?]：金蘭文化出版社, 民國68 [1979]
250 p. ; 20 cm. — (長篇奇情偵探小說)
Fiction.
NT$40.00
I. Title. II. Series: Ch'ang p'ien ch'i ch'ing chen t'an hsiao shuo.
PL2876.E49K8 79–841450

Li, Ch'iao, 1934–
(Ch'ing ch'ing hsiao shu)
青青校樹 / 李喬著. — 臺中：臺灣省政府新聞處, 民國67 [1978]
206 p. ; 19 cm. — (省政文藝叢書 ; 71)
Cover title.
Fiction.
Series romanized: Sheng cheng wen i ts'ung shu.
I. Title.
PL2877.C519C5 79–841698

Li, Chien-wu, 1908?–
(Shih ming)
使命 / 李健吾. — [香港?：s.n., 1977?]
167, iii p. ; 21 cm.
Short stories.
Reprint of the ed. published by Wen hua sheng huo ch'u pan she, in series: Wen hsüeh ts'ung k'an, with postscript dated 1939.
CONTENTS: 使命.—結束.—死的影子.—田原上.—私情.—買賣.
I. Title.
PL2780.C5S5 1977 78–841041

Li, Chun.
(Li Shuang-shuang hsiao chuan)
李双双小传 / 李准. — 第1版. — 北京：人民文学出版社, 1977.
2, 477 p. ; 19 cm.
Short stories.
$0.92
I. Title.
PL2877.C8L5 78–842480

Li, Chun, 1928–
(Ch'e lun ti che yin)
車輪的轍印 李准著 北京 人民文学出版社 1959.
183 p. 21 cm.
I. Title.
PL2877.C8C45 C 60–618

Li, Chun, 1928–
(Li Shuang-shuang hsiao chuan)
李双双小傳 李准著 北京 作家出版社 1961.
170 p. illus. 18 cm.
Short stories.
I. Title.
C 62–4386
Harvard Univ. Chinese-Japanese Library 5768

Li, Chun, 1928–
(Lu hua fang pai ti shih hou)
蘆花放白的时候 李准著 北京 作家出版社 1957.
226 p. 19 cm.
I. Title.
PL2877.C8L8 C 59–2199

Li, Chun, 1928–
(Ping hua hsüeh hsiao)
冰化雪消 李准著 北京 通俗讀物出版社 1956.
77 p. illus. 19 cm.
I. Title.
PL2877.C8P5 C 59–2241

Li, Chun, 1928–
(Pu neng tsou na t'iao lu)
不能走那條路 李準 [i. e.准] 著 北京 中國青年出版社 1955.
85 p. 19 cm.
Short stories.
I. Title.
PL2877.C8P8 C 63–1320

Li, Chun, 1928–
(Ta ho pen liu)
大河奔流：电影新作集 / [李准]. — 第1版. — 北京：人民文学出版社, 1977.
324 p. ; 19 cm.
Short stories.
CONTENTS: 李准 大河奔流.—杨时文 斯民三 周泱 普通党员.
$0.67
I. Yang, Shih-wen. P'u t'ung tang yüan. 1977. II. Ta ho pen liu. [Motion picture]
PL2877.C8T3 78–843534

CHINESE FICTION (Continued)

Li, Chun, 1928–
(Yeh ku niang)
野姑娘　李准著　北京　中國青年出版社 1956.
60 p. illus. 19 cm.

I. Title.
PL2877.C8Y1 C 59-2200

Li, Ch'ung-k'o.
(Shou k'ao ho yin tzu)
手銬和銀子 / 李崇科著. — 台中：弘業文化實業股份有限公司；發行所當代文化出版社, 民國66 [1977]
258 p. ; 19 cm.
Short stories.
CONTENTS: 手銬和銀子.—虎稱.—阿義.—夜歸人.—早來的聖誕節.—龍騰虎躍.—爸爸不賭多好.—繭.—水邊唇.—鄉下來的學生.
NT$40.00
I. Title.
PL2877.C825S5 78-843072

Li, Fei-kan, 1905–
(Ch'u lien)
初戀：巴金短篇抒情小說選. — 香港：駱駝出版社, 1977.
207 p. ; 19 cm.
Short stories.
CONTENTS: 初戀.—洛伯爾先生.—不幸的人.—愛的摧殘.—父與女.—老年.—愛的十字架.—好人.——封信.—墮落的路.—愛.—白鳥之歌
$6.50
I. Title.
PL2780.F4C57 78-842321

Li, Fei-kan, 1905–
(Ti ssu ping shih)
第四病室 / 作者巴金 [李芾甘]. — Hong Kong：大方出版社, 1966.
4, 210 p. ; 19 cm.
Fiction.
Reprint of the ed. published by Ch'en kuang ch'u pan kung ssu.

I. Title.
PL2780.F4T5 1966 79-837996

Li, Fei-meng.
(Hsün hsiung chi)
尋兇記 / 李費蒙著. — 臺北：聯亞出版社, 民國66 [1977]
359 p. ; 20 cm. — (聯亞叢刊 ; 31)
Fiction.
Series romanized: Lien ya ts'ung k'an.
NT$55.00

I. Title.
PL2877.F385H78 79-837204

Li, Fei-meng.
(Lo t'o ch'i an)
駱駝奇案 / [李] 費蒙著. — [修訂版]. — 臺北：平凡出版社；經銷名人出版社, 民國68 [1979]
2 v. (976 p.) ; 20 cm.
Fiction.
NT$200.00

I. Title.
PL2877.F385L6 79-841455

Li, Fei-meng.
(Tu kuo ch'ou ch'eng)
睹國仇城 / [李] 費蒙著. — 修訂版. — 臺北：名人出版社, 民國67 [1978]
2 v. (648 p.) : 20 cm.
Fiction.
NT$130.00

I. Title.
PL2877.F385T83 79-835629

Li, Hua-lan.
(Yü hua shih)
雨花石 / 李华岚；[陳达林插图]. — [香港?：s. n., 1978]
4, 146 p. : ill. ; 22 cm.
Title in Pinyin on cover: Yu hua shi.
Juvenile literature.
Reprint of the ed. published in the same year by Shao nien erh t'ung ch'u pan she, Shanghai.
I. Title.
PL2877.H7773Y8 1978 79-836633

Li, K'o, fl. 1953–
(Ti tao chan)
地道戰　李克　李微含著　上海　新文藝出版社 1953.
351 p. 18 cm.

1. Sino-Japanese Conflict, 1937-1945—Fiction. I. Li, Wei-han, joint author. II. Title.
PL2877.K6T5 C 58-6227 ‡

Li, Kuang-t'ien, 1907–
(Yin li)
引力 / 李廣田作. — [香港?]：創作書社, [1978?]
240, 7 p., [9] leaves of plates : ill. ; 19 cm.
Cover title.
Fiction.
Reprint of the 1948 (1949 printing) ed. published by Ch'en kuang ch'u pan kung ssu, Shanghai; issued in series: Ch'en kuang wen hsüeh ts'ung shu ; ti 25 chung.
I. Title.
PL2780.K8Y556 1978 78-841479

Li, P'eng.
(Ch'eng-chi-ssu-han chih pao tsang)
成吉思汗之寶藏 / 李芃著. — Hong Kong：中外出版社, 1977.
134 p. : ill. ; 18 cm. — (驚險小說叢書)
Series romanized: Ching hsien hsiao shuo ts'ung shu.
$4.60

I. Title.
PL2877.P4C5 78-840813

Li, Po-t'ung.
(Hsi-t'ai-hou mi shih yen i)
西太后秘史演義 / [李伯通編]. — 台北：新文豐出版股份有限公司, 民國67 [1978]
2 v. : ill. : 19 cm. — (零玉碎金集刊 : 28)
At head of title: Li shih hsiao shuo.
Series romanized: Ling yü sui chin chi k'an.
Reprint ed.
NT$110.00
1. Tz'u-hsi, Empress dowager of China, 1835-1908—Fiction. I. Title.
PL2780.P6H7 79-837482

Li, Ti.
(Yao yüan ti Pin-lang chai)
遥远的槟榔寨 / 李迪. — [香港?：s. n., 1978?]
382 p. : ill. ; 21 cm.
Fiction.
Reprint of the 1977 ed. published by Yüan-nan jen min ch'u pan she, K'un-ming.
$150.00

I. Title.
PL2877.T48Y3 1978 78-843540
C

Li, Yen-sheng.
(Li Shih-shih wai chuan)
李師師外傳 / 李餘生著. — [香港?]：古今出版社, [pref. 癸卯 i. e. 1963]
2, 98 p. : ill. ; 19 cm.
"此書在澳門綠村電台轉播"

1. Li, Shih-shih—Fiction. I. Title.
PL2877.Y38L5 79-836649

Li, Yün-te.
(Fei t'eng ti ch'ün shan)
沸騰的群山　李云德 [著]　北京　人民文学出版社 1966–
v. in 19 cm.
Fiction.

I. Title.
PL2877.Y82F4 C 66-2991

Li, Yün-te.
(Fei t'eng ti ch'ün shan)
沸騰的群山　李云德 [著　修改]　北京　人民文学出版社 1972–
v. ; 21 cm.
Fiction.
$0.90 (v. 1) various.

I. Title.
PL2877.Y82F4 1972 72-836508

Liang, Hsing.
(Lung hu feng yün chi)
龙虎风云记 / 梁信. — 北京第1版. — 北京：人民文学出版社, 1978.
578 p. ; 19 cm.
Fiction.
$1.10

1. China—History—Civil War, 1945-1949—Fiction. I. Title.
PL2878.A62L8 79-836922

Liang, Pin, 1914–
(Fan shen chi shih)
翻身记事 / 梁斌. — 第1版. — 北京：人民文学出版社, 1978.
507 p. ; 19 cm.
Fiction.
$0.97

I. Title.
PL2878.A66F3 78-842400

Liang, Pin, 1914–
(Hung ch'i p'u)
紅旗譜 / 梁斌. — [再版]. — 北京：中国青年出版社, 1978.
559 p. ; 19 cm.
Fiction.
$1.15

I. Title.
PL2878.A66H8 1978 79-837689

Liang, Te-so, 1905–
(Lao Mao ti jih chi)
老毛的日記 / 梁得所編. — [香港?：s. n., 1978?]
60 p. ; 19 cm.
Fiction.
Reprint of the 1933 ed. published by Liang yu t'u shu yin shua kung ssu, Shanghai, in series: I chiao ts'ung shu; ti 19 chung.

I. Title.
PL2781.A545L3 1978 78-841160

Liang, Yü-sheng.
(Lien chien feng yün lu)
聯險風雲錄 / 梁羽生著；插圖雲君. — 九龍：偉青書店, [1978?]
5 v. : ill. ; 19 cm.
Fiction.
"散花女俠姉妹作"
$15.00

I. Title.
PL2878.A76L5 79-836325

CHINESE FICTION (Continued)

Liao, T'ing.
(Pan t'ou ming ti yeh)
半透明的夜 / 廖汀著. — 臺北：東英出版社：總經銷雨辰書報社, 民國66 [1977]
233 p. ; 20 cm.
Short stories.
NT$35.00

I. Title.
PL2878.A78P3 78-842782

Lin, Chien.
(Kuei hsiang)
歸鄉 / 林間著. — 台中：新企業世界出版社, 民國66 [1977]
2, 168 p. ; 20 cm.
Short stories.
CONTENTS: 鷲鷲山風雲.—誰說去年春天沒有天牛.—外線牛之歌.—擁抱今日的吶喊.—車上的女乘客.—狗戲.—回鄉.—雲那麼高的爆竹.—上路.—守候.—風雨之夜.—線民.—揮碎整街的笑.—黑暗到黎明.—孤獨的蛇.
NT$35.00

I. Title.
PL2878.N12138K8 1977 78-843437

Lin, Hai-yin.
(Tung ch'ing shu)
冬青樹 / 林海音著. — 臺北：重光文藝出版社：總經銷新亞出版社, 民國44 [1955], 民國60 [1971] printing.
3, 3, 137 p. ; 19 cm.
Short stories.

I. Title.
PL2878.N13T86 78-841365

Lin, Nu-ch'ao.
(Ho ti kuang lin)
闔第光臨 / 林怒潮著. — 臺北：星光出版社, 民國67 [1978]
2, 221 p. ; 19 cm. — (雙子星叢書 ; 60)
Short stories.
Series romanized: Shuang tzu hsing ts'ung shu.
NT$35.00

I. Title.
PL2879.N1468H6 79-835390

Lin, Nu-ch'ao.
(Lo hu chi)
落戶記 / 林怒潮著. — 台北：星光出版社：經銷星光書報社, 民國67 [1978]
165 p. ; 19 cm. — (雙子星叢書 ; 68)
Short stories.
Series romanized: Shuang tzu hsing ts'ung shu.
NT$30.00

I. Title.
PL2878.N1468L6 79-835393

Lin, P'ei-fen, 1957-
(I chiu ch'i pa nien ch'un)
一九七八年春 / 林佩芬著. — 臺北：爾雅出版社, 民國67 [1978]
97 p. ; 19 cm. — (爾雅叢書 ; 46)
Fiction.
Series romanized: Erh ya ts'ung shu.
NT$35.00

I. Title.
PL2878.N1469I17 79-838019

Lin, Tan.
(Feng huo Ch'i-p'an-yai)
烽火棋盘崖 / 原著刘丹 ; 改编林丹 ; 绘画汪大伟. — [第1版]. — [上海]：上海人民出版社, [1977]
[92] p. : ill. ; 95 x 128 mm.
"連環圖畫"
$0.09
1. Ching-kang Shan region—History—Fiction. I. Liu, Tan, fl. 1977- II. Wang, Ta-wei. III. Title.
PL2878.N1489F4 78-843188

Lin, Ts'ang-yü, 1955-
(Yüeh kuang pien chao)
月光遍照 / [林蒼鬱著]. — 臺北：龍田出版社, 民國67 [1978]
13, 377 p. ; 20 cm. — (年輪書系 ; 1)
Short stories.
Series romanized: Nien lun shu hsi.
CONTENTS: 萬般郴聲響不去.—蛇與青蛙.—曹晁男與歐陽司陰.—悲儀朱蘿哀.—蓋.—跛族的愛神.—離訣.—月光遍照.—後記.—作品年表.
NT$100.00 ($3.00 U. S.)
I. Title.
PL2878.N1495Y8 79-836361

Lin, Wei-wen.
(Yung heng ti ssu nien)
永恒的思念 / 林蔚文編著. — 台南：大府城出版有限公司：總經銷中央書局, 民國67 [1978]
147 p. ; 20 cm. — (大府城書系)
Short stories.
Series romanized: Ta fu ch'eng shu hsi.
CONTENTS: 獻禮—評「永恒的思念」.—我讀「永恒的思念」.—永恒的思念.—疑雲.—房東的女兒.—除夕舞會.—歸宿.—敲詐者.—窮女婿.—後記.
NT$30.00
I. Title.
PL2878.N155Y8 1978 79-838119

Lin, Wen-lieh.
(Kuei ch'iao erh nü)
归侨儿女 / 林文烈. — 第1版. — [广州]：广东人民出版社, 1978.
2, 424 p. ; 19 cm.
Fiction.
$0.80

I. Title.
PL2878.N16K8 79-840166

Lin, Yutang, 1895-1976.
[Chinatown family. Chinese]
唐人街 / 林語堂著 ; 唐強譯. — 臺南：德華出版社, 民國66 [1977], 民國67 [1978] printing.
315 p. ; 19 cm. — (愛書人文庫 ; 053)
Series romanized: Ai shu jen wen k'u.
NT$55.00

I. Title.
Title romanized: T'ang jen chieh.
PR9470.9.L5C513 1977b 79-837726

Ling, Ts'ao.
(Yu Kuei-shan)
遊龜山：戲曲故事 / 淩操改編 ; 董天野繪圖. — 新1版. — 上海：上海文化出版社：上海圖書發行公司發行, 1955.
60 p. : ill. ; 18 cm.
Fiction.
"根據舊劇「蝴蝶盃」的前半部改編而成."

I. Hu tieh pei. II. Title.
PL2878.N52Y8 79-837750

Liu, Ch'i, fl. 1962-
(Chan cheng ch'i kuan)
战争奇观 / 柳杞. — 第1版. — [北京]：解放军文艺社：新华书店北京发行所发行, 1978.
385 p. ; 19 cm.
Fiction.
$0.87

I. Title.
PL2879.C4C48 79-839172

C

Liu, Chih-kuei.
(Hsi-shuang-pan-na ti ku shih)
西双版纳的故事 / 刘治贵, 白德松编 ; 白德松, 葛明艺绘. — 第1版. — [成都]：四川人民出版社, 1978.
142 p. : ill. ; 97 x 127 mm.
"連環圖畫"
$0.16

I. Pai, Te-sung, joint author. II. Ko, Ming-i. III. Title.
PL2879.C485H7 79-836651

Liu, Ch'ing, 1916-
(Hen t'ou t'ieh)
狠透铁：1957年纪事 / 柳青著. — [香港?：s. n., 1979?]
78 p., [1] leaf of plates : ill. ; 21 cm.
Title in Pinyin on cover: Hen tou tie.
"短篇小说"
Fiction.
Reprint of the 1978 ed. published by Shan-hsi jen min ch'u pan she.

I. Title. II. Title: Hen tou tie.
PL2879.C5H4 1979 79-840564

Liu, Ch'uan, fl. 1976-
(Ti erh ko ch'un t'ien)
第二个春天 / 刘川, 贺宝贤执笔. — 第1版. — 北京：人民文学出版社, 1976.
79 p., [2] leaves of plates : ill. ; 19 cm. — (电影文学剧本)
Fiction.
"根据同名话剧改编"
Series romanized: Tien ying wen hsüeh chü pen.
$0.23

I. Ho, Pao-hsien, joint author. II. Title.
PL2879.C8T5 77-838653

Liu, Fu, fl. 1073.
(Ch'ing so kao i)
青瑣高議 / [著作者劉斧]. — 臺景印初版. — 臺北：河洛圖書出版社, 民國66 [1977]
1, 232 p. ; 21 cm. — (小說彙編)
Reprint of the 1958 ed. published by Ku tien wen hsüeh ch'u pan she, Shanghai, with part of the prefatory matter deleted.
CONTENTS: 青瑣高議前集 10卷—青瑣高議後集 10卷—青瑣高議別集 7卷
NT$70.00

1. Fantastic fiction, Chinese. I. Title. II. Series: Hsiao shuo hui pien.
PL2687.L55C4 1977 79-836914

Liu, Ts'an-yang.
(Tao chün tzu)
盜君子 / 柳殘陽著. — 台北：春秋出版社：總經銷呂氏書店, 民國67 [1978]
206 p. ; 19 cm. — (長篇武俠名著)
Fiction.
Series romanized: Ch'ang p'ien wu hsia ming chu.
"神手無相故事之三"
NT$30.00
I. Title.
PL2879.T66T36 79-838118

Liu, Ts'an-yang.
(Ti sha t'ien hsin)
地煞天心 / 柳殘陽著. — 台北：春秋出版社：總經銷呂氏書店, 民國67 [1978]
3 v. (1200 p.) : 20 cm.
Fiction.
NT$165.00

I. Title.
PL2879.T66T5 79-838290

Liu, Ts'an-yang.
(Tuan jen)
斷刃：新派武俠小說 / 柳殘陽. — 台北：春秋出版社：總經銷呂氏書店, 民國67 [1978]
2 v. (769 p.) ; 20 cm.
"長篇武俠名著"
Fiction.
NT$50.00 per vol.

I. Title.
PL2879.T66T8 79-836542

Liu, Ts'an-yang.
(Wu lin ssu ta chia)
武林四大家 / 柳殘陽著. — 香港：春秋書店, [1978?]
8 v. ; 18 cm.
Fiction.
NT$40.00

I. Title.
PL2879.T66W8 78-842334

CHINESE FICTION (Continued)

Liu, Ts'an-yang.
(Yin chiao chen pa huang)
銀角震八荒 / 柳殘陽著. — 台北：春秋出版社：總經銷呂氏書店, 民國67 [1978]
3 v. (1154 p.) ; 19 cm. — (長篇武俠名著)
Fiction.
Series romanized: Ch'ang p'ien wu hsia ming chu.
NT$180.00

I. Title.
PL2879.T66Y56 79-840725

Lo, Hsüan.
(Nan kuo feng yen)
南國烽烟 / [罗旋]. — [香港?：s.n., 1978?]
480 p., [6] leaves of plates : ill. ; 21 cm.
Reprint of the 1977 ed. published by Chiang-hsi jen min ch'u pan she, Nan-ch'ang.

I. Title.
PL2880.O2195N3 1978 78-843290

Lo, K'o.
(Hei se t'ien t'ang)
黑色天堂 / 洛克著. — 臺北：小說創作社；Silver Spring, Md.：美國總代理 T'ang's Corp., 民國66 [1977]
192 p. ; 20 cm. — (小說創作叢書)
Fiction.
Series romanized: Hsiao shuo ch'uang tso ts'ung shu.
NT$30.00
I. Title.
PL2880.O237H4 78-841115

Lo, Kuan-chung, ca. 1330–ca. 1400.
[P'ing yao chuan. Japanese]
平妖傳 / [羅貫中・馮夢龍著]；佐藤春夫譯. — 東京：世紀書房, 昭和26- [1951-
v. ; 18 cm.
Fiction.
Colophon inserted.

I. Feng, Meng-lung, 1574?–1645?, joint author. II. Title.
Title romanized: Heiyōden.
PL2690.P5J3 1951 79-807243

[**Lo, Kuan-chung**] ca. 1330–ca. 1400.
[San-kuo chih yen i]
繡像第一才子書 / 毛聲山評. — [s.l.]：經國堂發兌, [順治甲申 i.e. 1644序]
20 v. (on double leaves) in 2 cases : ill. ; 25 cm. — (四大奇書 : 第1種)
Running title: Ti i ts'ai tzu shu.
In 51 chüan, 120 hui.
Later editions published under the author: Lo Kuan-chung and the title: San-kuo chih yen i.
"聖歎外書"
Block print.
1. China—History—Three kingdoms, 220–265—Fiction. I. Mao, Sheng-shan. II. Title. III. Title: Ti i ts'ai tzu shu. IV. Series: Ssu ta ch'i shu : ti 1 chung.
Title romanized: Hsiu hsiang ti i ts'ai tzu shu.
PL2690.S3 1644 79-836164

Lo, Kuan-chung, ca. 1330–ca. 1400.
[San kuo chih yen i]
三國演義：大字足本古典文學 / 羅貫中著. — 台南：第一書店, [1978?]
805 p. ; 20 cm.
In colophon: Fa hsing so, Li ta ch'u pan she.
Originally published under title: San kuo chih yen i.
NT$55.00
1. China—History—Three kingdoms, 220–265—Fiction. I. Title. II. Title: San kuo chih yen i.
Title romanized: San kuo yen i.
PL2690.S3 1978 78-841316

Lo, Kuang-pin.
(Hung yen)
红岩 / 罗广斌, 杨益言著. — 香港：生活·讀書·新知三联书店, 1977.
597 p., [12] leaves of plates : ill. ; 22 cm.
Fiction.
Reprint of the 1961 ed. published by Chung-kuo ch'ing nien ch'u pan she, Peking.
$13.50
I. Yang, I-yen, joint author. II. Title.
PL2880.O25H8 1977 78-841081

Lü, Chiang, 1953-
(Fei hsiang chih ying)
飛翔之鷹 / 履彊著. — 臺北：皇冠出版社, 民國67 [1978]
293 p. ; 19 cm.
Short stories.
NT$50.00

I. Title.
PL2880.U1195F4 79-835420

Lu, Jen.
(Lun Wen-hsü san hsi Liu Hsien-k'ai)
倫文叙三戲柳先開：廣東故事鬥智小説 / 魯人撰. — [香港]：創基出版社：經售紅文書局, [1978?]
211 p. in various pagings ; 19 cm.
Fiction.
$4.00

I. Title.
PL2783.U213L8 79-835264

Lu, Li.
(Wo shih tiao ch'a yüan)
我是調查員 / 陸勵著. — 臺北：小說創作社：美國總代理 Silver Spring, Md. : Tang's Corporation, 民國65 [1976]
255 p. ; 19 cm. — (小說創作叢書)
Fiction.
Series romanized: Hsiao shuo ch'uang tso ts'ung shu.

I. Title.
PL2880.U1577W6 79-837490

Lu, Mo, fl. 1977-
(Che san ko nü hai)
這三個女孩 / 盧沫著. — [台北]：德昌出版社, [1978?]
216 p. ; 20 cm. — (德昌新刊 : 53)
Short stories.
"感情探討系列小說"
Series romanized: Te ch'ang hsin k'an.
CONTENTS: 兩行淚.——個欺騙愛情的故事.—漂泊的小車.—出賣幸福的女人.—隨風而逝.—泥涼.—這三個女孩.
I. Title.
PL2880.U16423C4 79-835555

Lu, Mo, fl. 1977-
(Jou ch'ing chih yeh)
柔情之夜 / 盧沫著. — [台北]：德昌出版社, [1979?]
238 p. ; 19 cm.
Cover title.
Short stories.

I. Title.
PL2880.U16423J6 79-840704

Lu, Mo, fl. 1977-
(T'ai-pei yeh t'an)
台北夜譚 / 盧沫著. — 臺北：聯亞出版社, 民國66 [1977]
2, 223 p. ; 20 cm. — (聯亞叢刊 : 15)
Short stories.
Series romanized: Lien ya ts'ung k'an.
CONTENTS: 夜話—遲春—贖罪—電話戰—單身的告白—兔子事件—漩渦追踪—五個男人—毛毛的命運—舊事
NT$40.00
I. Title.
PL2880.U16423T27 78-835368

rev

Lu, Mo, fl. 1977-
(Tan shen ti kao pai)
單身的告白 / 盧沫著. — 臺北：聯亞出版社, 民國66 [1977]
2, 201 p. ; 19 cm. — (聯亞叢刊 ; 32)
Short stories.
Series romanized: Lien ya ts'ung k'an.
CONTENTS: 丹丹的禮物.—銀子的選擇.—房客.—雞飛狗跳.—單身的告白.—漩渦追踪.—慾望之沼.
NT$38.00
I. Title.
PL2880.U16423T3 78-841153

Lü, Ta-ming.
(Ta ti sung)
大地頌 / 呂大明著. — 臺中：光啓出版社, 民國66 [1977]
167 p. ; 19 cm. — (文藝叢刊 ; 56)
Series romanized: Wen i ts'ung k'an.

I. Title.
PL2880.U16496T3 79-837362

Lu, Ti, fl. 1968-
(Hsiao tsu tsung)
小祖宗 / 魯帝著. — 台北：哲志出版社；香港：總經銷陳湘記書局, 民國67 [1978]
307 p. ; 19 cm. — (鄉野俠情粗獷小說)
Fiction.
Series romanized: Hsiang yeh hsia ch'ing ts'u kuang hsiao shuo.
NT$50.00

I. Title.
PL2880.U165H74 79-835711

Lu, Ti, fl. 1968-
(Min ch'u hsiang yeh ts'u kuang hsiao shuo)
民初鄉野粗獷小說 / 魯帝著. — 台北：哲志出版社, 民國6 [19
v. ; 19 cm. — (鄉野俠情粗獷小說)
Series romanized: Hsiang yeh hsia ch'ing ts'u kuang hsiao shuo.
CONTENTS:
—第6部. 無敵九段.
NT$50.00 (v. 6)
I. Title.
PL2880.U165M5 79-839126

Lu, Ti, fl. 1968-
(Nan tien yüan jih chi)
男店員日記 / 魯帝著. — 台北：哲志出版社, 民國66 [1977]
191 p. ; 19 cm. — (哲志叢書 ; 27)
Cover title varies slightly.
Fiction.
Series romanized: Che chih ts'ung shu.
NT$40.00
I. Title.
PL2880.U165N36 78-842648

Lu, Ti, fl. 1968-
(Wu ti chiu tuan)
無敵九段 / 魯帝著. — 台北：哲志出版社, 民國67 [1978]
294 p. ; 19 cm. — (民初鄉野粗獷小說 ; 第6部) (鄉野俠情粗獷小說)
Series 1 romanized: Min ch'u hsiang yeh ts'u kuang hsiao shuo. Series 2 romanized: Hsiang yeh hsia ch'ing ts'u kuang hsiao shuo.
NT$50.00
I. Title.
PL2880.U165M5 vol. 6 79-839125

Lu, Ti, fl. 1968-
(Yao ch'ien shu)
搖錢樹 / 魯帝著. — 台北：哲志出版社, 民國67 [1978]
263 p. ; 20 cm.
At head of title: Min ch'u hsing yeh hsia ch'ing ts'u k'uang hsiao shuo.
Fiction.

I. Title.
PL2880.U165Y3 79-840358

Lu, Yen.
(Hua chu hsi tsai chan huo fen fei ti nien tai)
华主席在战火纷飞的年代 / 陆岩, 保国. — 第1版. — 北京：解放军文艺社, 1978.
150 p. ; 19 cm.
$0.34

1. Hua, Kuo-feng—Fiction. I. Pao, Kuo. II. Title.
PL2880.U1995H8 79-836957

CHINESE FICTION (Continued)

Lu, Yen, fl. 1933-
(Lu Yen hsüan chi)
魯彥選集. — 香港：中流出版社, 1977.
2, 291 p. ; 19 cm.
Short stories.
Reprint of the ed. with postscript dated 1933.
$7.00
I. Title.
PL2880.U1995A6 1977 78-841040

Lu, Yen-shih.
(Hsüeh jan chan ch'i tai tai hung)
血染战旗代代红：硬骨头六连的故事 / 陆岩石. — ［香港?：s.n., 1978?］
2, 2, 158 p., ［2］ leaves of plates : ill. ; 21 cm.
Reprint of the 1977 ed. published by Shang-hai jen min ch'u pan she, Shanghai.
1. China. Lu chün. Ti 8 lu chün. 120 shih. 7 t'uan. 6 lien—Fiction. I. Title.
PL2880.U22H78 1978 78-840925

(Lü yeh hsien tsung)
綠野仙蹤. — 香港：藝美圖書公司, 1978.
65 p. : ill. ; 19 cm. — (世界名著兒童故事精選)
Bound with Pai-hsüeh kung chu. Hongkong. 1978.
$2.00
I. Series: Shih chieh ming chu erh t'ung ku shih ching hsüan.
PZ10.831.P32 79-837265

Lung, Chi, fl. 1969-
(Chiu ming kuai mao)
九命怪貓 / 龍驥著. — 台北：哲志出版社, 民國67 ［1978］
187 p. ; 20 cm.
Fiction.
NT$40.00
I. Title.
PL2880.U48C56 79-835778

Lung, Chi, fl. 1969-
(Hu tieh yüan)
蝴蝶怨：大社會中的小騙局 / 龍驥著. — 台北：哲志出版社, 民國67 ［1978］
207 p. ; 20 cm. — (小騙局 ; 2)
Short stories.
Series romanized: Hsiao p'ien chü.
NT$40.00
I. Title.
PL2880.U48H84 79-835770

Lung, Chi, fl. 1969-
(Shen ch'iang shuang chieh ku shih)
神槍雙傑故事　龍驥著　香港　環球圖書雜誌出版社　1969-70.
4 v. 18 cm.
Fiction.
CONTENTS: 第1集　紅眼貓—第2集　櫻花血—第3集　挑戰女人島—第4集　死亡單行道
I. Title.
PL2880.U48S5 72-835121

Lung, Chi, fl. 1969-
(Ssu ch'eng)
死城 / 龍驥著. — 台北：哲志出版社, 民國67 ［1978］
194 p. : 20 cm. — (紙上劇場 : 3)
Short stories.
Series romanized: Chih shang chü ch'ang.
CONTENTS: 死城.—巴大叔.—酣海遊魂.—劊神的化身.—倩女幽魂.—復活.—白日夢.
NT$40.00
I. Title.
PL2880.U48S6 79-835637

Lung, Chi, fl. 1969-
(T'iao chan ssu wang chiao)
挑戰死亡角 / ［作者龍驥］. — 台北：哲志出版社；香港：海外總經銷陳湘記書局, 民國67 ［1978］
233 p. ; 19 cm. — (龍驥長篇鬪智小說專輯)
Fiction.
NT$45.00
I. Title. II. Series: Lung, Chi, fl. 1969- Lung Chi ch'ang p'ien tou chih hsiao shuo chuan chi.
PL2880.U48T67 79-836142

Lung, Chi, fl. 1969-
(Wang p'iao)
玩票：［法網專案］ / 龍驥著. — 臺北：聯亞出版社, 民國66 ［1977］
2, 188 p. ; 20 cm. — (聯亞叢刊 ; 27)
Short stories.
Series romanized: Lien ya ts'ung k'an.
CONTENTS: 玩票.—偷樑計.—媚倫夢.—放長線釣大魚
NT$40.00
I. Title.
PL2880.U48W3 78-841201

78 C

Lung, Ching.
(Kuei niang tzu)
鬼娘子 / 龍井著. — 台北：德昌出版社, 民國62 ［1973］
204 p. ; 20 cm. — (奇異鬼狐集錦故事)
"怪異小說"
Short stories.
CONTENTS: 碧潭鬼戀.—鬼新娘.—尤十娘.—沙漠女鬼.—棺中飛煞.—人?鬼?
I. Title. II. Series: Ch'i i kuei hu chi chin ku shih (T'ai-pei ed.)
PL2880.U485K8 79-836135

Ia, Feng.
(Liu Hu-lan chuan)
刘胡兰传 / 马烽. — 第1版. — 北京：中國青年出版社, 1978.
2, 2, 409 p. ; 19 cm.
$0.85
1. Liu, Hu-lan, 1932-1947—Fiction. I. Title.
PL2881.F4L58 78-842835

78 C

Ma, Ling-shu.
(T'ao li chieh)
桃李劫 / 馬靈殊著. — 香港：上海書局, 1978.
229 p. ; 21 cm.
Fiction.
$7.40
I. Title.
PL2881.L5T36 79-840167

C

Ma, Pin.
(Hsin lu)
新路 / 南宫搏著. — 臺北：九歌出版社：總經銷爾雅出版社, 民國67 ［1978］
347 p. : ill. ; 19 cm. — (九歌文庫 ; 6)
Title in English on verso of t. p.: From Williamsburg to Washington.
Fiction.
Series romanized: Chiu ko wen k'u.
NT$60.00
I. Title. II. Title: From Williamsburg to Washington.
PL2881.P5H7 79-835412

C

Ma, Sen.
(Sheng huo tsai p'ing chung)
生活在瓶中 / 馬森著. — 第1版. — 臺北：四季出版事業有限公司：發行四季, 桂冠聯合發行部, 民國67 ［1978］
2, 296, 4 p., ［1］ leaf of plates : ports. ; 20 cm. — (四季文萃 ; 22)
Short stories.
Series romanized: Ssu chi wen ts'ui.
CONTENTS: 懷念在巴黎的那段日子(代序).—法國的小農生活.—老于的故事.—安娜的夢.—加特琳的婚禮.—碧姬的迷惘.—娜娜奶.—野鴨.—生活在瓶中
I. Title.
PL2881.S38S5 78-842635

Ma, Sen-liang.
(Chu-chiang shui yu han)
珠江水猶寒 / 馬森亮著. — 臺北：正中書局；New York：美國總經銷華强圖書公司, 民國68 ［1979］
8, 4, 407 p. ; 20 cm. — (正中文藝叢書)
Fiction.
Series romanized: Cheng chung wen i ts'ung shu.
I. Title.
PL2881.S4C5 79-840598

Ma, Te, fl. 1978-
(San ta pai ku ching)
三打白骨精 / 马得编绘. — 第1版. — 北京：人民美術出版社, 1978.
69 p. : col. ill. ; 10 x 14 cm.
"連環圖畫"
$0.22
I. Title.
PL2881.T43S2 79-836516

Man-sheng.
(Yü lung ti yen ching)
玉龙的眼睛 / 曼生. — ［香港?：s.n., 1978?］
281 p., ［8］ leaves of plates : ill. ; 22 cm.
Fiction.
Reprint of the 1977 ed. published by Chiang-su jen min ch'u pan she, Nanking.
I. Title.
PL2882.N53Y8 1978 79-835186

Man-t'ien-hsüeh.
(Yin-hao ti shih chieh)
銀好的世界 / 漫天雪著. — 香港：大光出版社, 1978.
131 p. : ill. ; 19 cm.
Short stories.
CONTENTS: 這不過是死神.—苗.—手.—月明她需要控訴.—仇恨的延續.—銀好的世界.—世界是美麗的.
$4.00
I. Title.
PL2882.N55Y5 78-843032

Matsumoto, Seichō, 1909-
［Kiiroi fūdo. Chinese］
黃色風土：長篇推理小說 / 松本清張著；公洲譯. — 香港：天地圖書有限公司, 1978.
342 p. ; 19 cm. — (天地叢書)
Fiction.
Series romanized: T'ien ti ts'ung shu.
$11.00
I. Title.
Title romanized: Huang se feng t'u.
PL856.A8K4813 79-837706

Meng, Ch'i.
(She hui ching tzu)
社會鏡子：在西門町釣凱子的一個女孩自述 / 孟琪. — ［台北］：德昌出版社, ［1978?］
171 p. ; 19 cm.
"民族晚報專欄連載"
Short stories.
I. Title.
PL2883.N38S5 79-840114

Meng, Chün.
(Chen ai)
真愛 / 孟君著. — 香港：海濱圖書公司, ［1978?］
438 p. ; 19 cm.
Fiction.
$7.00
I. Title.
PL2883.N5C52 79-835924

CHINESE FICTION (Continued)

Meng, Chün.
(San ko nan jen)
三個男人 / 孟君著. — 香港：海濱圖書公司, [1978?]
455 p. ; 19 cm.
Fiction.
"曾在「婦女與家庭」畫報連載"

I. Title.
PL2883.N5S2 78-843897

Meng, Hua.
(Hua chung hsin yin)
畫中心音 / 夢華著. — 台北：金蘭文化出版社, 民國67 [1978]
196 p. ; 19 cm. — (靈異小說精選)
Short stories.
Series romanized: Ling i hsiao shuo ching hsüan.
CONTENTS: 畫中心音.—鬼屋疑雲.—狐姦.—鬼踪.—艷鬼探春.
NT$30.00
1. Ghost stories, Chinese. I. Title.
PL2883.N5156H76 79-835849

Meng, Hua.
(Hun hsi p'iao p'iao)
魂兮飄飄 / 夢華著. — 台北：金蘭文化出版社, 民國67 [1978]
204 p. ; 19 cm. — (靈異小說精選)
Fiction.
Series romanized: Ling i hsiao shuo ching hsüan.
NT$30.00

I. Title.
PL2883.N5156H8 78-843609

Meng, Hua.
(Kuei ch'i)
鬼妻 / 夢華著. — 台北：金蘭文化出版社, 民國67 [1978]
194 p. ; 19 cm. — (靈異小說精選)
Three novels.
Series romanized: Ling i hsiao shuo ching hsüan.
CONTENTS: 魂離情歸.—鬼妻.—現世報.
NT$30.00
1. Ghost stories, Chinese. I. Title.
PL2883.N5156K8 79-837011

Meng, Hua.
(Yu hun chui ch'ou)
幽魂追仇 / 夢華著. — 台北：金蘭文化出版社, 民國67 [1978]
192 p. ; 20 cm. — (靈異小說精選)
Fiction.
Series romanized: Ling i hsiao shuo ching hsüan.
NT$30.00

I. Title.
PL2883.N5156Y8 78-843060

Mo-jen, pseud.
(Shih jen ko ming chia)
詩人革命家：胡漢民傳 / [著作者墨人]. — 臺北：近代中國出版社：總經銷中央文物供應社, 民國67 [1978]
4, 346 p., [4] leaves of plates : ill. ; 22 cm. — (先烈先賢傳記叢刊) (近代中國叢書)
Bibliography: p. 344-345.
NT$45.00
1. Hu, Han-min, 1879-1936—Fiction. I. Title. II. Series: Hsien lieh hsien hsien chuan chi ts'ung k'an. III. Series: Chin tai Chung-kuo ts'ung shu.
PL2886.O18S5 79-840362

Mo-shang-sang.
(T'a mei yu hsiang ch'ou)
他沒有鄉愁 / 陌上桑著. — 台北：楓樺出版社：總經銷杏文出版社, 民國67 [1978]
2, 341 p. : 20 cm.
Short stories.
NT$70.00

I. Title.
PL2886.O68T3 79-835304

Mu, Chung-nan.
(Mu Chung-nan tzu hsüan chi)
穆中南自選集. — 臺北：黎明文化事業公司, 民國67 [1978]
2, 254 p., [2] leaves of plates : ill. ; 20 cm. — (中國新文學叢刊 ; 64)
Short stories.
Series romanized: Chung-kuo hsin wen hsüeh ts'ung k'an.
CONTENTS: 第十九個.—苦難中成長.—楊賓樓與小白龍.—一張支票.—毒火燃燒的女人.—荒唐的回憶.
NT$70.00
I. Title.
PL2886.U143M8 79-841881

Mu, Hsiang.
(Chin-yang ch'iu)
晋阳秋：新波旧澜第一部 / 慕湘. — 第2版. — 北京：解放军文艺社：新华书店北京发行所發行, 1978.
592 p. ; 21 cm.
Fiction.
$1.40

1. Sino-Japanese Conflict, 1937-1945—Fiction. I. Title.
PL2886.U17C5 1978 79-837941

Mu, Hsüan.
(Chiang ch'ao lieh huo)
江潮烈火：江苏民兵斗争故事 / 穆烜, 理陶编文；徐晓平, 马明仲, 王忠山绘. — 第1版. — [南京]：江苏人民出版社, 1975, 1978 printing.
86 p. : ill. ; 11 x 13 cm.
"南通市创作办公室供稿"
"連環圖畫"
$0.12
1. Sino-Japanese Conflict, 1937-1945—Fiction. I. Li, T'ao, joint author. II. Hsü, Hsiao-p'ing. III. Ma, ming-chung. IV. Wang, Chung-shan. V. Title.
PL2886.U16C4 79-836487

Mu-jung, Mei.
(Lung hu sha shou)
龍虎殺手 / 慕容美著. — 台北：春秋出版社：總經銷呂氏書店, 民國68 [1979]
2 v. (884 p.) ; 19 cm. — (長篇武俠名著)
Fiction.
Series romanized: Ch'ang p'ien wu hsia ming chu.
NT$130.00

I. Title.
PL2886.U196L8 79-840518

Mu-jung, Mei.
(Pei hung che i)
悲鴻折翼 / 慕容美著. — 香港：春秋書店, [1978?]
8 v. ; 18 cm. — (新派長篇武俠奇情小說)
Fiction.
Series romanized: Hsin p'ai ch'ang p'ien wu hsia ch'i ch'ing hsiao shuo.
$46.40
I. Title.
PL2886.U196P4 78-841575

Ni-lo, pseud.
(Chin hsiang ch'ing ch'ieh)
近鄉情怯 / 尼洛著. — 台北：世系出版社：經銷瑞泰文物供應社, 民國67 [1978]
312 p. ; 20 cm. — (世系文學叢刊 ; 2)
Two stories.
Series romanized: Shih hsi wen hsüeh ts'ung k'an.
CONTENTS: 近鄉情怯.—咆哮荒塚.
NT$65.00
I. Ni-lo, pseud. P'ao hsiao huang chung. 1978. II. Title.
PL2888.L6C5 79-839927

Ni-lo, pseud.
(Ni-lo tzu hsüan chi)
尼洛自選集. — 臺北：黎明文化事業公司, 民國67 [1978]
2, 276 p., [2] leaves of plates : ill. ; 20 cm. — (中國新文學叢刊 ; 65)
Short stories.
Series romanized: Chung-kuo hsin wen hsüeh ts'ung k'an.
CONTENTS: 三代.—潞潞與我.—島外的島.—十年.—陌生的重逢.—悼意.—戰南麻.—倫青.—蜕
NT$70.00
I. Title.
PL2888.L6A6 1978 79-837071

Ni-lo, pseud.
(Shan ch'a yü lu)
山茶與露 / 尼洛著. — 台北：世系出版社：經銷瑞泰文物供應社, 民國67 [1978]
483 p. ; 20 cm. — (世系文學叢刊 ; 3)
Fiction.
Series romanized: Shih hsi wen hsüeh ts'ung k'an.
NT$90.00
I. Title.
PL2888.L6S5 79-838677

Pai, Hsiao-sheng.
(Pei-ching to ch'üan mi hsin)
北京奪權秘辛 / 百笑生著. — 台中：新企業世界出版社, 民國66 [1977]
227 p. ; 20 cm.
Reprint ed.
NT$40.00
1. China—Politics and government—1976- —Fiction. I. Title.
PL2892.A34P4 1977b 79-835565

Pai, Hsien-yung, 1937-
(Yu yüan ching meng)
遊園驚夢 / [著者白先勇]. — 臺北：仙人掌出版社, 民國57 [1968]
155, [23] p. ; 18 cm. — (仙人掌文庫 ; 6)
Title in English in colophon: Wake in the garden.
Short stories.
Series romanized: Hsien jen chang wen k'u.

I. Title.
PL2892.A345Y8 78-840845

Pai, Hsü-chih.
(Lao ch'u nü)
老處女 / 白序之著. — [香港? : s. n. 1978?]
4, 184 p. ; 21 cm.
Short stories.
Reprint of the 1934 ed. published by Ta tung shu chü, Shanghai; in series: Hsin wen hsüeh ts'ung shu ; 1.
CONTENTS: 老處女.—矛盾.—孝女.—一個女兵.—混沌.—一個人力車夫.—我的國文教師.—一夕話.—乞丐.
I. Title.
PL2892.A3458L3 1978 78-842006

Pai, Ling.
(Ch'ien mien chih jen)
千面紙人 / 白翎譯著. — [台北?]：哲志出版社, [1977?]
212 p. ; 20 cm. — (西洋怪談故事 ; 22)
Short stories.
CONTENTS: 紙人—幽靈票友—店員帽—弄假成眞—絕不低頭—預兆—凶宅七小時—石道士—陌生女客—舐犢情深—陰陽的問題飛女—候補判官
NT$30.00
I. Title. II. Series: Hsi yang kuai t'an ku shih ; 22.
PL2892.A375C5 77-842509

Pai, Ling.
(Kuei hsüeh)
鬼穴 / 白翎譯著. — [台北]：哲志出版社, [1978?]
219 p. ; 19 cm. — (西洋怪談故事 ; 23)
Short stories.
CONTENTS: 魔襪.—五分鐘的百萬富翁.—現場表演.—追到陰間.—冰屍.—斷手.—潑辣幽魂.—午夜情人.—海邊艷屍.—雛妓.
NT$30.00
1. Ghost stories, Chinese. I. Title. II. Series: Hsi yang kuai t'an ku shih ; 23.
PL2892.A375K79 79-836417

Pai, Ling.
(Mo kuei kou hun)
魔鬼勾魂 / 白翎譯著. — [臺北?]：哲志出版社, [1977?]
212 p. ; 20 cm. — (西洋怪談故事 ; 24)
Short stories.
CONTENTS: 蜜月飛魂.—色狼遇鬼.—與鬼決鬥.—模特兒之戀.—吸血的白蘭黛.—穿馬褂的人.—幽冥兄弟.—申一刀.—模特兒.—我吻過她.—馬小蓉
NT$30.00
1. Title. II. Series: Hsi yang kuai t'an ku shih ; 24.
PL2892.A375M6 79-836175

Pai, Tieh-hun.
(Fei ying chieh)
飛英刼：中國文學名著 / [白蝶魂著]. — 台北：新文豐出版公司, 民國67 [1978]
2, 144 p. ; 20 cm. — (零玉碎金集刊 ; 6)
Fiction.
Series romanized: Ling yü sui chin chi k'an.
Reprint ed.
NT$30.00
I. Title.
PL2795.A44F44 1978 79-837839

CHINESE FICTION (Continued)

P'an, Lei.
(Ching ching ti Hung-ho)
靜靜的紅河：長篇小說 / [著者潘壘]. — 臺北：聯經出版事業公司, 民國67 [1978]
5, 2, 610 p. ; 20 cm. — (潘壘作品集 ; 3)
Fiction.
Series romanized: P'an Lei tso p'in chi.
NT$100.00

I. Title.
PL2892.A6C53 79-837707

P'an, Lei.
(Mo kuei shu)
魔鬼樹：孽子三部曲之一 / [潘壘]. — 臺北：聯經出版事業公司, 民國67 [1978]
2 v. (1240 p.) ; 20 cm. — (潘壘作品集 ; 18)
Fiction.
Series romanized: P'an Lei tso p'in chi.
NT$240.00

I. Title. II. Title: Nieh tzu.
PL2892.A6A6 1977, vol. 18 79-838180

P'an, P'ei-yüan.
(Yung pu hsiao shih ti tien po)
永不消逝的电波 / 原著林金；改编潘培元；绘画华三川. — 第2版. — 上海：上海人民出版社, 1978.
118 p. : ill. ; 10 x 13 cm.
"連環圖畫"
$0.18

I. Lin, Chin. II. Hua, San-ch'uan. III. Title.
PL2892.A613Y8 1978 79-839469

P'an, Shou-k'ang.
(K'u kuo k'uang yeh)
哭過曠野：中國武俠欣賞 / 潘壽康編. — 臺排印初版. — 臺北：河洛圖書出版社, 民國67 [1978]
2, 304 p. ; 22 cm.
Short stories.
CONTENTS: 霹靂閃.—刺客.—鐵傘俠.—荒江野俠.—兩代梟雄.—初試.—斷雁殘鴻.
NT$50.00
I. Title.
PL2892.A614K8 78-842636

Pan-hsia-lou-chu.
(Yen wang ling)
閻王令 / 伴霞樓主著. — [台]北：有志圖書公司：羅氏書局印行, 民國67 [1978]
3 v. (1296 p.) ; 20 cm.
Fiction.
NT$60.00
I. Title.
PL2892.A615Y4 1978 79-835807

Pao, Hsiao-hui.
(Ai tao shen ch'u wu yüan yu)
愛到深處無怨尤 / 鮑曉輝. — 台北：龍風實業有限公司圖書出版部, 民國66 [1977]
218 p. ; 20 cm. — (龍風叢書 ; 4)
Short stories.
Series romanized: Lung feng ts'ung shu.
NT$45.00

I. Title.
PL2892.A637A73 78-841782

P'eng, Hsüan-hsien.
(Ai hsiao ti ta niu)
愛笑的大妞 / 彭選賢著. — 高雄：三信出版社, 民國65 [1976]
2, 241 p. ; 20 cm.
Short stories.
NT$35.00
I. Title.
PL2892.E58A73 78-842798

P'eng, Hsüan-hsien.
(Mien mien ch'ing shan)
綿綿青山 / 彭選賢著. — 台中：新企業世界出版社, 民國66 [1977]
2, 1, 214 p. ; 20 cm.
Short stories.
Reprint of the 1976 ed. published by Ta sheng ch'u pan she, Yüan-lin, which was issued as ti 6 chung of Ta sheng shu k'u.
CONTENTS: 少年男女.—懷念曲.—跑在林間.—嫁.—綿綿青山.—我一個會唱歌的太太.—憂鬱的莎岡.—落霧.—書蟲世界.—遇.—白鶴與鼓槌.—北市李.—輕塵.—兩個人看電影.—打魚殺家.
NT$35.00
I. Title.
PL2892.E58M5 1977 78-843595

P'eng, P'ei-p'ei.
(Wang jih ch'ing huai)
往日情懷 / 彭珮珮. — 台北：遠行出版社：總經銷遠景出版社, 民國67 [1978]
5, 208 p. ; 19 cm. — (小草叢刊 ; 37)
Fiction.
Series romanized: Hsiao ts'ao ts'ung k'an.
NT$40.00

I. Title.
PL2892.E63W3 79-835394

Pi, Chen, 1926–
(Ai ch'ing ti hao chiao)
愛情的號角 / [著作者畢珍]. — 台北：新文豐出版公司, 民國67 [1978]
358 p. ; 20 cm. — (畢珍專集 ; 7)
Short stories.
Series romanized: Pi Chen chuan chi.
CONTENTS: 姐姐的男朋友.—朱門假日.—疑雲.—陞官圖.—白老哥.—罪人群.—想起那眼.—愛情的號角.
NT$80.00

I. Title.
PL2892.I14A6 1978, vol. 7 79-840257

Pi, Chen, 1926–
(Lung hu hsiung ti)
龍虎兄弟 / [著作者畢珍]. — 臺北：新文豐出版公司, 民國67 [1978]
3 v. (3, 1012 p.) ; 20 cm. — (畢珍專集 ; 2)
Fiction.
Series romanized: Pi Chen chuan chi.
NT$210.00

I. Title.
PL2892.I14A6 1978, vol. 2 79-837539

Pi, Chen, 1926–
(Nü ch'ien sui)
女千歲 / 畢珍著. — 臺北：國家出版社, 民國67 [1978]
2 v. (859 p.) ; 20 cm. — (His 中國奇女子列傳 ; 7)
Fiction.
Series romanized: Chung-kuo ch'i nü tzu lieh chuan.
NT$60.00

I. Title.
PL2892.I14A6 1975, vol. 7 79-836913

Pi, Chen, 1926–
(Nü tang chia)
女當家 / 畢珍著. — 臺北：國家書店, 民國67 [1978]
2 v. (822 p.) ; 19 cm. — (His 中國奇女子列傳 6)
Fiction.
Series romanized: Chung-kuo ch'i nü tzu lieh chuan.
NT$60.00 per vol.

I. Title.
PL2892.I14A6 1975, vol. 6 79-836228

Pi, Chen, 1926–
(Pi Chen chuan chi)
畢珍專集. — 臺北：新文豐出版公司, 民國19
v. : ill. ; 20 cm.
CONTENTS: —2. 龍虎兄弟.—3. 縱橫長江一書生.—4. 發棺.
NT$210.00 per vol.

I. Title.
PL2892.I14A6 1978 79-837540

Pi, Chen, 1926–
(Tsung heng Ch'ang-chiang i shu sheng)
縱橫長江一書生 / [著作者畢珍]. — 臺北：新文豐出版公司, 民國67 [1978]
3 v. (2, 1008 p.) : ill. ; 19 cm. — (畢珍專集 ; 3)
Fiction.
Series romanized: Pi Chen chuan chi.
NT$210.00

I. Title.
PL2892.I14T7 79-837537

Pi, Chen, 1926–
(Wo chu feng wu ti shih hou)
我住風屋的時候 / 畢珍著. — 臺北：水芙蓉出版社：經銷者星光書報社, 64 [1975]
1, 257 p. ; 19 cm. — (水星文庫 ; 17)
Short stories.
Series romanized: Shui hsing wen k'u.
Reprint of the 1968 ed. published by Li chih ch'u pan she, T'ai-pei, in series: Li chih wen ts'ung.
CONTENTS: 啼地.—槍親.—呼喚.—響鞭.—歸去來兮.—萬里長麵.—收穫的日子.—大房·槽坊·磨坊.—我住風屋的時候.—武破鑼獅過火焰山.
NT$38.00

I. Title.
PL2892.I14W6 1975 78-841973

Pi, Fan.
(Tso yeh p'ai hui)
昨夜徘徊 / 畢凡. — 香港：環球圖書雜誌出版社, 1978.
48 p. : ill. ; 19 cm. — (環球文藝 ; 546)
Cover title.
Fiction.
$0.70

I. Title. II. Series: Huan ch'iu wen i ; 546.
PL2892.I1417T7 79-837745

Pi, P'u, 1923–
(Hsi t'ou yüeh)
溪頭月 / 畢璞著. — 臺中：學人文化事業有限公司, 民國67 [1978]
2, 214 p. ; 20 cm.
Short stories.
CONTENTS: 阿梅的故事.—獨腳戲.—女為悅己者容.—大學女生與小丑.—洋夢.—妒夫.—凝香, 這可愛的女孩.—贖罪的丈夫.—黑色的雪.—心魔.—溪頭月.—他的聖母瑪利亞.

I. Title.
PL2892.I144H66 79-837938

Pi-chu.
(Po ko tien hua hei wo)
撥個電話給我 / 碧竹著. — 台北：遠景出版社, 民國67 [1978]
2, 2, 377 p. ; 19 cm. — (遠景叢刊 ; 91)
Short stories.
Series romanized: Yüan ching ts'ung k'an.
NT$70.00

I. Title.
PL2892.I16P6 79-837375

Pi-lu-shan-ch'iao.
(Hsiang hsi kan shih chi)
湘西趕屍記 / 碧廬山樵著. — 台北：中華藝林文物出版有限公司, 民國66 [1977]
148 p. ; 20 cm.
Short stories.
CONTENTS: 湘西趕屍記.—大江東去.
NT$50.00

I. Title.
PL2892.I215H8 79-835043

Ping-tu-shan.
(Chui ch'iu feng man)
追求豐滿：又名一個隆乳的少女 / 冰度山著作. — 台北：三佑出版社, 民國67 [1978]
212 p., [4] leaves of plates : ill. ; 20 cm. — (三佑書系 ; 8)
Fiction.
Series romanized: San yu shu hsi.
NT$45.00

I. Title.
PL2892.I54C48 79-838211

P'u, Sung-ling, 1640–1715.
[Liao chai chih i]
增圖補校但刻聊齋志異 / 劉階平編校. — 臺北：臺灣學生書局, 民國67 [1978]
4 v. : ill. ; 22 cm.
"「同文」石印「詳註聊齋志異圖詠」本, 共繪四百四十四圖. 本編每卷各擷取八圖, 十六卷, 計收有一百二十八圖, 分列各卷首."
NT$1000.00
I. Tan, Ming-lun. II. Liu, Chieh-p'ing. III. Lü, Chan-en. Hsiang chu Liao chai chih i t'u yung. Selections. 1978. IV. Title. V. Title: Tan k'o Liao chai chih i.
Title romanized: Tseng t'u pu chiao Tan k'o Liao chai chih i.
PL2722.U2L5 1978b 79-838056

CHINESE FICTION (Continued)

Sai-weng.
(Sheng chien chien ch'ing)
聖劍殲情 / 塞翁著. — 台北：春秋出版社：總經銷呂氏書店, 民國67 [1978]
474 p. ; 20 cm.
Fiction.
"香港眞報連載"
NT$65.00

I. Title.
PL2894.A37S5 79-838567

Sang, P'in-tsai, 1939–
(Yü ch'ing)
餘情 / 桑品載著. — 臺北：世界文物出版社, 民國67 [1978]
203 p. ; 19 cm.
Short stories.
NT$35.00

I. Title.
PL2894.A5Y8 78-843383

Sen-sen.
(Huang chung han shih)
荒塚寒屍 / 森森著. — 台北：金蘭文化出版社. 民國67 [1978]
200 p. ; 19 cm. — (靈異小說精選)
Fiction.
Series romanized: Ling i hsiao shuo ching hsüan.
NT$30.00

I. Title.
PL2894.E54H8 78-843690

Shan, Hsüeh-p'eng.
(Yen ling feng yün)
燕岭风云 / 単学鵬. — 北京第1版. — 北京：人民文学出版社, 1977.
3, 669 p. ; 19 cm.
Fiction.
$1.20

I. Title.
PL2896.N135Y4 78-841152

Shang-kuan, Yung.
(T'an wu ta chi t'uan)
貪汚大集團 / 上官庸著. — 香港：武林出版社, [1976?]
146 p. ; 17 cm. — (小鬼子傳奇 ; 第29集)
Series romanized: Hsiao kuei tzu ch'uan ch'i.
Fiction.
$2.00

I. Title.
PL2896.N627T36 78-841364

Shao, Chien, 1934–
(Shao Chien tzu hsüan chi)
邵們自選集. — 臺北：黎明文化事業公司, 民國67 [1978]
3, 4, 406 p., [2] leaves of plates : ill. ; 20 cm. — (新中國文學叢刊 ; 50)
Short stories.
Series romanized: Hsin Chung-kuo wen hsüeh ts'ung k'an.
Series title in colophon: Chung-kuo hsin wen hsüeh ts'ung k'an.
NT$80.00
I. Title.
PL2896.O12A15 1978 79-836960

Shen, Ts'ung-wen, 1902–
(Hei yeh)
黑夜 / 沈從文著. — 香港：香港滙通書店, 1977.
133 p. ; 19 cm. — (**His** 從文創作 ; 7)
Short stories.
Series romanized: Ts'ung-wen ch'uang tso.
CONTENTS: 會明.—黑夜.—泥塗.—燈.—若墨醫生.
$4.20

I. Title.
PL2801.N18H4 1977 79-835633

Shen, Ts'ung-wen, 1902–
(K'ang k'ai ti wang tzu)
慷慨的王子 / 沈從文作. — [香港? : s. n., 1978?]
57 p. ; 18 cm.
Fiction.
Reprint of the 1933 ed. published by Liang yu t'u shu yin shua kung ssu, Shanghai, which was issued as ti 63 chung of I chiao ts'ung shu.
$12.00

I. Title.
PL2801.N18K3 1978 79-835851

Shen, Ts'ung-wen, 1902–
(Shen wu chih ai)
神巫之愛 / 沈從文著. — 香港：滙通書店, 1977.
1, 91 p. ; 19 cm. — (從文創作 ; 6)
Fiction.
Series romanized: Ts'ung-wen ch'uang tso.
$3.50

I. Title.
PL2801.N18T77 vol. 6 79-835372

Shen, Ts'ung-wen, 1902–
(Ts'ung-wen ch'uang tso)
從文創作 / 沈從文著. — 香港：滙通書店, 19
v. : 19 cm.
Fiction.
CONTENTS:
—6. 神巫之愛
$3.50 (v. 6)

I. Title.
PL2801.N18T77 79-835373

Shen, Yen-ping, 1896–
[Selected works]
茅盾短篇小說選. — 香港：文教出版社, 1978.
2, 314 p. ; 19 cm. — (中國現代文學叢書)
Reprint of the 1955 ed. published by Jen min wen hsüeh ch'u pan she, Peking under title: Mao Tun tuan p'ien hsiao shuo hsüan chi.
Series romanized: Chung-kuo hsien tai wen hsüeh ts'ung shu.
$7.50
I. Title.
Title romanized: Mao Tun tuan p'ien hsiao shuo hsüan.
PL2801.N2A6 1978 79-835049

Shen, Yen-ping, 1896–
(Tzu yeh)
子夜 / 茅盾 [沈雁冰]. — 第3版. — 北京：人民文学出版社, 1960, 1977 printing.
576 p. ; 21 cm.
$1.35

I. Title.
PL2801.N2T9 78-843082

Shih, Chün-feng.
(Ssu sheng tzu)
私生子 / [著作者石峻峰]. — [台北?] : 樂群出版社, 民國67 [1978]
111 p. ; 19 cm.
Cover title.
Fiction.
NT$30.00

I. Title.
PL2899.H2177S75 78-841941

Shih, Kuo.
(Ch'in Hsiang-lien)
秦香蓮 / 史果編著 ; 徐進, 仁年揷圖. — 澳門：右文出版社, 1978.
3, 84 p. : ill. ; 19 cm.
Fiction.
$3.20

I. Title.
PL2899.H265C48 1978 79-836345

Shu, Ch'ing-ch'un, 1898–1966.
(Lao-she ming chu hsüan)
老舍名著選 / 侃如編選. — 香港：滙通書店, 1978.
76 p. ; 19 cm.
Short stories.
$3.20

I. K'an-ju. pueud. II. Title.
PL2804.C5L35 1978 78-843244

Shu, Ch'ing-ch'un, 1898–1966.
(Lao-she tuan p'ien hsiao shuo hsüan)
老舍短篇小說選. — 重版. — 香港：文教出版社, 1978.
1, 206 p. ; 19 cm. — (中國現代文學叢書)
Series romanized: Chung-kuo hsien tai wen hsüeh ts'ung shu.
Reprint of the 1956 ed. published by Jen min wen hsüeh ch'u pan she, Peking.
$6.00

I. Title.
PL2804.C5L37 1978 79-836284

Shu, Ch'ing-ch'un, 1898–1966.
(Pu ch'eng wen t'i ti wen t'i)
不成問題的問題 / 老舍 [舒慶春] 著. — 3版. — 香港：上海書局, 1977.
142 p. ; 18 cm. — (中國文學名著小叢書)
Series romanized: Chung-kuo wen hsüeh ming chu hsiao ts'ung shu.
$3.60

I. Title.
PL2804.C5P8 1977 79-835262

Shu, Ch'ing-ch'un, 1898–1966.
(T'ing lai ti ku shih)
聽來的故事 / 老舍 [舒慶春] 著. — 香港：南國出版社, 1977 printing.
1, 202 p. ; 19 cm.
Short stories.
Reprint ed.
CONTENTS: 老字號.—斷魂槍.—聽來的故事.—新時代的舊悲劇.—且說屋裏.—新韓穆烈德.—哀啓.
$5.00
I. Title.
PL2804.C5T5 1977 79-837066

Shu, Fen.
(Ku-su ch'un)
姑苏春 / 树棻. — 第1版. — 上海：上海文艺出版社：新華書店上海发行所发行, 1978.
450 p. : 19 cm
$0.95

1. Sino-Japanese Conflict, 1937–1945—Fiction. I. Title.
PL2901.F4K8 79-837688

Shu, Ying.
(Jen tsai ch'un shan wai)
人在春山外 / 疏影著. — 臺北：小說創作社, 民國64 [1975]
399 p. ; 20 cm. — (小說創作長篇文藝叢書)
Fiction.
Series romanized: Hsiao shuo ch'uang tso ch'ang p'ien wen i ts'ung shu.

I. Title.
PL2901.Y5J45 78-842641

Shui hu chuan.
水滸傳 / 施耐庵著. — [台南?] : 利大出版社 ; 台南：總經銷第一書店, [1977?]
661 p. ; 20 cm.
"大字足本·古典文學"
NT$55.00

I. Title.
Title romanized: Shui hu chuan.
PL2694.S5 1977b 78-842580

CHINESE FICTION (Continued)

Ssu, Ch'in, 1948-
(Ch'ang t'i lüeh ying)
長堤掠影 / 思芹. — [台北]：文豪出版社：總代理學生圖書供應社, 民國67 [1978]
187 p. ; 19 cm. — (文豪叢書)
Short stories.
Series romanized: Wen hao ts'ung shu.
CONTENTS: 長堤掠影.—異域蒼涼.—古架疑情.
NT$45.00 ($4.00 U.S.)

I. Title.
PL2903.S7C5 78-670205

Ssu, Ma.
(Chien k'o lieh chuan)
劍客列傳 / 司馬著. — 台北：漢麟出版社, 民國67 [1978]
422 p. ; 19 cm.
Fiction.
NT$70.00

I. Title.
PL2903.S74C4 79-836258

Ssu, Ma.
(Hsiao hsiang yüeh)
瀟湘月 / 司馬著. — 台北縣土城鄉：漢麟出版社, 民國67 [1978]
2 v. (847 p.) ; 19 cm.
Fiction.
NT$140.00

I. Title.
PL2903.S74H7 79-836568

Ssu, Ma.
(Jih lo tzu chin ch'eng)
日落紫禁城 / 司馬著. — 台北縣土城：漢麟出版社, 民國67 [1978]
330 p. ; 20 cm.
NT$50.00

1. Ch'ing Shih-tsung, Emperor of China, 1677-1735—Fiction. I. Title.
PL2903.S74J5 79-835498

Ssu, Ma.
(Ku tu feng yü)
故都風雨 / 司馬著. — 台北縣土城：漢麟出版社, 民國67 [1978]
2 v. (543 p.) ; 19 cm.
Fiction.
NT$45.00 per vol.

I. Title.
PL2903.S74K8 79-836904

Ssu-ma, Ch'ang-feng.
(Piao k'o)
鏢客 / 司馬長風著. — 台北：春秋出版社：總經銷呂氏書店, 民國67 [1978]
271 p. ; 20 cm. — (長篇武俠名著)
Cover title.
Fiction.
Series romanized: Ch'ang p'ien wu hsia ming chu.
NT$35.00

I. Title.
PL2903.S825P5 79-837075

Ssu-ma, Tzu-yen.
(Yu hsia lieh chuan)
遊俠列傳 / 司馬紫烟著. — 台北：春秋出版社：總經銷呂氏書店, 民國 [19
v. ; 20 cm.
Cover title.
Short stories.

I. Title.
PL2903.S847Y8 79-836134

Su, Hsüan-ying, 1884-1918.
[Selected works]
蘇曼殊選集. — 香港：香港文學研究社, [1978?]
4, 141 p. ; 21 cm. — (現代文學叢書) (中國現代文學作家選集叢書)
Series 1 romanized: Hsien tai wen hsüeh ts'ung shu. Series 2 romanized: Chung-kuo hsien tai wen hsüeh tso chia hsüan chi ts'ung shu.
$6.00

I. Title.
Title romanized: Su Man-shu hsüan chi.
PL2724.U2A6 1978 79-835263

Su, P'ei-yüan.
(I sheng ti ai yü chi mo)
醫生的愛與寂寞 / 作者蘇培源. — 臺北：遠流出版社, 民國67 [1978]
4, 199 p. : ill. ; 20 cm.
Colophon title.
NT$30.00

I. Title.
PL2904.P4I2 79-840562

Sun, Shou-k'uang.
(Je hsüeh hen shuo chi)
熱血痕說集 / [著作者孫受匡]. — [香港?：s. n., 1978?]
111 p. ; 18 cm.
Reprint of the 1923 ed. published by Yü ch'u hsiao shuo she.

I. Title.
PL2808.N215J4 1978 79-838127

Sun, Yü-hsin.
(Fu ch'ou ku)
復仇谷 / 孫玉鑫著. — 台北：春秋出版社：總經銷呂氏書店, 民國67 [1978]
2 v. (1075 p.) ; 19 cm. — (長篇武俠名著)
Fiction.
Series romanized: Ch'ang p'ien wu hsia ming chu.
NT$160.00

I. Title.
PL2905.N284F8 79-838280

Sun, Yü-hsin.
(Hsüeh chien)
血劍 / 孫玉鑫著. — 台北：春秋出版社：總經銷呂氏書店, 民國67 [1978]
2 v. ; 19 cm.
"長篇武俠名著"
Fiction.
NT$60.00 per vol.

I. Title.
PL2905.N284H7 79-839923

Sung, Sheng-fa.
(Ken sui Ch'en I t'ung chih ta yu chi)
跟随陈毅同志打游击 / 宋生发. — 第1版. — 北京：解放军文艺社：新华书店北京发行所发行, 1960, 1978 printing.
93 p., [4] leaves of plates : ill. ; 19 cm.
$0.28

1. Ch'en, I, 1901-1972—Fiction. I. Title.
PL2905.N457K4 79-839482

Sung, Tse-lai.
(Ta-niu-nan ts'un)
打牛湳村 / 宋澤萊著. — 台北：遠景出版社, 民國67 [1978]
27, 1, 318 p. ; 20 cm. — (遠景叢刊 ; 94)
Fiction.
Series romanized: Yüan ching ts'ung k'an.
NT$70.00

I. Title.
PL2905.N46T3 79-836771

Tai, Yen.
(Hsiao-tou-tou ti jih chi)
小豆豆的日記 / [著者戴嚴]. — 台北：代言出版社, 民國67 [1978]
146 p. : ill. ; 20 cm. — (代言叢書 ; 5)
Series romanized: Tai yen ts'ung shu.
NT$35.00

I. Title.
PL2908.I27H8 79-840560

T'an, Ch'i.
(Chiang hu lang tzu)
江湖浪子 / 譚奇著. — [台北]：志明, [1978?]
240 p. ; 20 cm.
Fiction.
"現代粗獷鬪智小說."

I. Title.
PL2908.N14C5 79-838037

T'an, T'an.
(Chang Fei miao)
張飛廟 / 譚談著. — 台北：志明出版社, 民國66 [1977]
260 p. ; 20 cm. — (譚談小說專輯 ; 26)
Fiction.
"鄉野粗獷傳奇小說"
Series romanized: T'an T'an hsiao shuo chuan chi.
NT$50.00
I. Title.
PL2908.N3589 1977, vol. 26 79-840663

T'an, T'an.
(Chin feng huang)
金鳳凰 / 譚談著. — 台北：志明出版社, 民國66 [1977]
293 p. ; 20 cm. — (譚談小說專輯 ; 25)
"鄉野粗獷鬥智小說"
Fiction.
Series romanized: T'an T'an hsiao shuo chuan chi.
NT$50.00
I. Title.
PL2908.N3589 1977, vol. 25 79-837655

T'an, T'an.
(Hsieh niang tzu)
蠍娘子 / 譚談著. — 台北：志明出版社, [1978?]
260 p. ; 20 cm. — (譚談小說專輯)
Fiction.
Colophon stamped at end of work.
"鄉野傳奇粗獷小說."

I. Title. II. Series: T'an T'an. T'an T'an hsiao shuo chuan chi.
PL2908.N3589H75 79-839495

T'an, T'an.
(Liang hsin p'ing)
良心秤 / 譚談著. — [台北?：志明出版社, 1979?]
300 p. ; 20 cm. — (志明文藝叢書)
Fiction.
Series romanized: Chih ming wen i ts'ung shu.

I. Title.
PL2908.N3589L5 79-839146

T'an, T'an.
(P'i li wu hao)
霹靂五號 / 譚談著. — [台北]：志明, [1978?]
464 p. ; 22 cm. — (譚談諜報專輯 ; 1)
Cover title.
Fiction.
Series romanized: T'an T'an tieh pao chuan chi.
Sequel: Yü tien chi hua.

I. Title.
PL2908.N3589 1978, vol. 1 79-838207

CHINESE FICTION (Continued)

T'an, T'an.
(T'an T'an hsiao shuo chuan chi)
譚談小説專輯. — 台北：志明出版社, 民國 [19
v. ; 20 cm.
"鄉野粗獷鬥智小說"
Fiction.
CONTENTS:
25. 金鳳凰
NT$25.00 (v. 25)
I. Title.
PL2908.N3589 1977 79-837654

T'an, T'an.
(T'an T'an tieh pao chuan chi)
譚談諜報專輯 / [譚談著]. — [台北]：志明, [1978?-
v. : 22 cm.
Running title.
Fiction.
CONTENTS: 1. 霹靂五號.

I. Title.
PL2908.N3589 1978 79-838206

T'an, T'an.
(T'an T'an tou chih hsiao shuo chuan chi)
譚談鬥智小說專輯. — [台北]：志明, [1978?]
v. ; 22 cm.
Cataloged from running title.

I. Title.
PL2908.N3589 1978a 79-838198

T'an, T'an.
(To ming chin p'ai)
奪命金牌 / 譚談著. — [台北]：志明, [1978?]
360 p. ; 22 cm. — (譚談鬥智小說專輯 ; 1)
Cover title.
Fiction.
Series romanized: T'an T'an tou chih hsiao shuo chuan chi.

I. Title.
PL2908.N3589 1978a, vol. 1 79-838197

T'an, T'an.
(Tung-t'ing-lei)
洞庭雷 / [譚談著]. — [台北]：志明, [1978?]
293 p. ; 20 cm. — (志明文藝叢書)
Fiction.
Series romanized: Chih ming wen i ts'ung shu.

I. Title.
PL2908.N3589T8 78-841812

T'an, T'an.
(Wan ma ku)
萬馬谷 / 譚談著. — 台北：志明出版社, 民國 67 [1978]
258 p. ; 20 cm. — (譚談小說專輯 ; 27)
Fiction.
Series romanized: T'an T'an hsiao shuo chuan chi.
"鄉野粗獷傳奇小說"
NT$50.00
I. Title.
PL2908.N3589 1977, vol. 27 79-839497

T'ang, Mei-sheng.
(Ting cho ch'iang feng ti han tzu)
頂着强風的漢子 / 唐梅生著. — 臺中：新企業世界出版社, 民國66 [1977]
196 p. ; 20 cm.
Short stories.
CONTENTS: 頂着强風的漢子.—希望.—奮鬥.—明天的太陽.—媽媽, 您在那裡.—傳家之寶.—明星夢醒.—守望相助.—神童.—姻緣.—迷失
NT$40.00
I. Title.
PL2908.N445T5 78-841757

Teng, Ai-mei.
(I p'ien feng hung)
一片楓紅 / 鄧藹梅著. — 臺北：小說創作社, 民國64 [1975]
374 p. ; 20 cm. — (小說創作長篇文藝叢書)
Fiction.
Series romanized: Hsiao shuo ch'uang tso ch'ang p'ien wen i ts'ung shu.

I. Title.
PL2909.E37I16 78-841200

Teng, Ai-mei.
(Wan yün)
晚雲 / 鄧藹梅著. — 臺北：小說創作社, 民國 63 [1974]
390 p. ; 20 cm. — (小說創作長篇文藝叢書)
Fiction.
Series romanized: Hsiao shuo ch'uang tso ch'ang p'ien wen i ts'ung shu.

I. Title.
PL2909.E37W3 78-840937

T'eng, Kang, fl. 1929.
(Mo jih)
末日 / 勝剛著. — [香港 : s. n., 1978?]
2, 72 p. ; 21 cm.
Four stories.
CONTENTS: 回家以前.—末日.—妻底職業.—夜未央.
Reprint of the 1929 ed. published by Chin wu shu tien, Shanghai.

I. Title.
PL2812.E55M6 1978 79-836143

Teng, Wen-lai, 1932-
(Ta mo ts'ang ts'ang)
大漠蒼蒼 / 鄧文來著. — 臺北：名人出版社, 民國66 [1977]
317 p. ; 20 cm.
Fiction.
NT$55.00

I. Title.
PL2909.E45T3 78-840952

T'ien, Ch'en.
(Yu chien chao ch'ing)
又見朝晴 / 田塵. — 香港：環球圖書雜誌出版社, 1978.
48 p. : ill. ; 19 cm. — (環球文藝 ; 549)
Cover title.
Fiction.
"小小青蘋果 姬小苔.": p. 39-48.
$0.70
I. Chi, Hsiao-t'ai, Hsiao hsiao ch'ing p'ing kuo. 1978. II. Title. III. Series: Huan ch'iu wen i ; 549.
PL2909.I36Y8 79-837746

T'ien, Yüan, 1927-
(Ta ti chih ko)
大地之歌 / 田原著. — 再版. — 臺北：水芙蓉出版社, 民國67 [1978]
2, 319 p. ; 19 cm. — (水芙蓉書庫 ; 128)
Fiction.
Series romanized: Shui fu jung shu k'u.
NT$50.00

I. Title.
PL2909.I47T35 79-836909

Ting, Yin-sheng.
(K'ung-tzu yen i)
孔子演義 / 丁寅生著. — 台北：廣文書局, 民國67 [1978]
2 v. ; 22 cm. — (漢學彙編)
Reprint ed.
NT$480.00

1. Confucius—Fiction. I. Title. II. Series: Han hsüeh hui pien.
PL2909.I68K8 1978 79-835486

Ting, Ying, 1930-
(Pai se ti jih chi)
白色的日記 / 丁穎著. — 臺中：藍燈文化事業股份有限公司, 民國66 [1977]
209 p. ; 20 cm. — (藍燈叢書)
Fiction.
Series romanized: Lan teng ts'ung shu.
NT$40.00

I. Title.
PL2909.I683P3 79-835822

Ting, Ying-ch'uan.
("Ling ssu" hao ch'an p'in)
"04"号产品 / 丁盈川. — 第1版. — [西安]：陕西人民出版社：陕西省新華書店发行, 1975, 1977 printing.
383 p., [6] leaves of plates : ill. ; 19 cm.
Fiction.
$0.77

I. Title.
PL2909.I69L5 79-840517

Ts'ai, Tung-fan.
[Yüan shih t'ung su yen i]
元世祖 / [著作者李唐]. — 台北：河洛圖書出版社, 民國67 [1978]
376, [8] p. ; 22 cm. — (中國歷史小叢書 ; 23) (歷史小叢書)
"元世祖我們採用蔡東藩的元史通俗演義"
Series 1 romanized: Chung-kuo li shih hsiao ts'ung shu. Series 2 romanized: Li shih hsiao ts'ung shu.
Pbk. : NT$35.00

1. China—History—Yüan dynasty, 1260-1368—Fiction. I. Li, T'ang. II. Title.
Title romanized: Yüan Shih-tsu.
PL2912.A5Y8 79-835710

(Ts'ai kuei chi)
才鬼記 / 梅鼎祚輯 ; [編者本公司編輯部]. — 臺北：偉文圖書出版社, 民國66 [1977]
247 p. ; 22 cm. — (秘笈叢編 ; R007)
NT$200.00 ($5.30 U.S.)

1. Ghost stories, Chinese. I. Mei, Ting-tso, 1553-1619. II. Wei wen t'u shu ch'u pan she. Pien chi pu. III. Series: Mi chi ts'ung pien ; R007.
PL2629.G5T75 79-837519

Ts'ao, Chan, ca. 1717-1763.
[Hung lou meng]
戚蓼生序鈔本石頭記：[8卷：80回 / 曹霑]. — 台北：廣文書局, 民國66 [1977]
4 v. (20, 3168 p.) ; 22 cm. — (紅樓夢叢書)
Reprint ed.
Series romanized: Hung lou meng ts'ung shu.

I. Title. II. Title: Shih t'ou chi.
Title romanized: Ch'i Liao-sheng hsü ch'ao pen Shih t'ou chi.
PL2727.S2A1 1977b 78-838497

rev

Ts'ao, Yu-fang.
(Feng ch'en li)
風塵裏 / 曹又方著. — 臺北：皇冠出版社, 民國67 [1978]
202 p. ; 19 cm. — (鴛鴦譜系列小說 ; 2)
Short stories.
Series romanized: Yüan yang p'u hsi lieh hsiao shuo.
CONTENTS: 人在風塵裏.—出走.—岸.—再生緣.—寒樓.—露.
NT$40.00

I. Title.
PL2912.A59F4 79-836238

Ts'ao, Yu-fang.
(Shih shih ti ch'un)
溼溼的春 / 曹又方著. — 臺北：皇冠出版社, 民國67 [1978]
230 p. ; 19 cm. — (鴛鴦譜系列小說 ; 3)
Short stories.
Series romanized: Yüan yang p'u hsi lieh hsiao shuo.
CONTENTS: 愛情遊記.—育亡.—純情的獨白.—群鶯會.—溼溼的春.—這樣就叫幸福.
NT$40.00
I. Title.
PL2912.A59S5 79-835473

CHINESE FICTION (Continued)

(Ts'ao mang hao hsia chuan)
草莽豪俠傳. — ［台］北：河洛圖書出版社，民國66 ［1977］
4, 332 p. : ill. ; 22 cm. — (中國草莽人物故事)
Pbk. : NT$50.00

1. Chinese fiction. I. Series: Chung-kuo ts'ao mang jen wu ku shih.
PL2625.T75 79-835505

Ts'en, K'ai-lun.
(Hsing fu hua)
幸福花：文藝創作小説 / 岑凱倫著. — 香港：環球圖書雜誌出版社，1978.
384 p. ; 17 cm.
Fiction.
$5.50

I. Title.
PL2912.E38H75 79-837373

Tseng, Hsin-i.
(Ts'ai-feng ti hsin yüan)
彩鳳的心願 / 曾心儀著. — 台北：遠景出版社，民國67 ［1978］
8, 1, 194 p. ; 19 cm. — (遠景叢刊 ; 92)
Short stories.
Series romanized: Yüan ching ts'ung k'an.
CONTENTS: 彩鳳的心願.—一個作家的畫像.—郛淪. 我的朱麗亞.—李萌的三個尷尬時期.—杜鵑花開.—支票.
NT$40.00
I. Title.
PL2912.E555T7 79-836259

Tsou, Lang.
(Ch'ieh kuo che)
竊國者 / 鄒郎. — 第1版. — 臺北：四季出版事業有限公司，民國67 ［1978］
202, ［22］ p., ［1］ leaf of plates : ill. ; 22 cm. — (鄒郎小説專輯 ; 1)
Series romanized: Tsou Lang hsiao shuo chuan chi.
Edition for 1957 published under title: Chan tou ch'eng.
"鄒郎先生著作年表": at end of text.
NT$65.00
1. T'ai-yüan, China—History—Fiction. I. Title.
PL2912.O35A15 1978, vol. 1 78-842881

Tsou, Lang.
(Ti hsia ssu ling)
地下司令 / ［作者鄒郎；插圖呂游銘］. — 第1版. — 臺北：四季出版事業有限公司，民國67 ［1978］
422, ［22］ p. : ill. ; 22 cm. — (鄒郎小説專輯 ; 4)
Fiction.
NT$120.00
I. Title.
PL2912.O35A15 1978, vol. 4 78-842885

Tsou, Lang.
(Tsou Lang hsiao shuo chuan chi)
鄒郎小説專輯. — 臺北：四季出版事業有限公司，民國 ［19
v. ; 22 cm.
Fiction.
"鄒郎先生著作年表": v. 3, p. 1-7 (4th group)
CONTENTS:
2-3. 陰謀者
NT$130.00 per vol.
I. Title.
PL2912.O35A15 1978 78-841102

Tsou, Lang.
(Yin mou che)
陰謀者 / 鄒郎. — 臺北：四季出版事業有限公司，民國67 ［1978］
2 v. ; 22 cm. — (鄒朗小説專輯 ; 2-3)
Vol. 2 has subtitle: P'an tieh.
Fiction.
Series romanized: Tsou Lang hsiao shuo chuan chi.
"鄒郎先生著作年表": v. 2, p. 1-7 (4th group)
NT$260.00
I. Title.
PL2912.O35A15 1978, vol. 2-3 78-841103

Tu, Ai.
(Pu hsiu ti ch'eng)
不朽的城 / 杜埃. — ［香港?：s. n., 1979］
2, 285 p. ; 22 cm.
Short stories.
Reprint of the 1979 ed. published by Kuang-tung jen min ch'u pan she; issued in series: Yüeh hai wen ts'ung.

I. Title. II. Series: Yüeh hai wen ts'ung.
PL2913.A47P8 1979 79-840099

Tu-ku, Hung.
(Lung cheng hu tou)
龍爭虎鬭 / 獨孤紅著. — 台北：漢麟出版社，民國67 ［1978］
2 v. (825 p.) ; 20 cm.
Fiction.
NT$70.00 per vol.

I. Title.
PL2913.Z47L9 79-837369

Tu-ku, Hung.
(Mo nü Shih-san niang)
魔女十三娘 / 獨孤紅著. — 香港：春秋書店，［1978?］
5 v. (1357 p.) ; 19 cm. — (新派長篇武俠奇情小説)
Fiction.
Series romanized: Hsin p'ai ch'ang p'ien wu hsia ch'i ch'ing hsiao shuo.
$25.00
I. Title.
PL2913.Z47M6 78-843078

Tu-ku, Hung.
(Pao en chien)
報恩劍 / 獨孤紅著. — 台北：春秋出版社：總經銷呂氏書店，民國66 ［1977］
2 v. ; 19 cm. — (長篇武俠名著)
Fiction.
Series romanized: Ch'ang p'ien wu hsia ming chu.
NT$65.00

I. Title.
PL2913.Z47P3 79-835774

Tu-ku, Hung.
(Yung Ch'ien fei lung chuan)
雍乾飛龍傳 / 獨孤紅著. — 台北：春秋出版社：總經銷呂氏書店. 民國67 ［1978］
2 v. (972 p.) : 20 cm.
"長篇武俠名著"
NT$130.00

I. Title.
PL2913.Z48Y8 79-836059

Tuan, K'o-ch'ing.
(T'ieh chih)
鐵汁 / 段可情著. — ［香港?：s. n., 1977?］
197 p. ; 22 cm.
Short stories.
Reprint of the 1935 ed. published by Ch'i chih shu chü, Shanghai.
CONTENTS: 一封退回的信.—鐵汁.—查票員.—一封英兵遺落的信.—梆票匪的供狀

I. Title.
PL2914.A426T5 1977 78-840872

Tuan, Ts'ai-hua, 1933-
(Liu lang ch'üan wang)
流浪拳王 / 段彩華著. — 臺北：天華出版事業股份有限公司，民國67 ［1978］
1, 275 p. ; 20 cm. — (天華文學叢刊 ; 8)
Short stories.
Series romanized: T'ien hua wen hsüeh ts'ung k'an.
CONTENTS: 孩子, 小鳥, 蜂窩.—淇河渡口.—老漁人.—野棉花.—嬰兒.—掃映的片子.—流浪拳王.—心病.—櫻花恨.

I. Title.
PL2914.A45L5 79-836998

Tuan, Ts'ai-hua, 1933-
(Lung p'ao chieh)
龍袍刼 / 段彩華. — 臺北：名人出版社，民國66 ［1977］
346 p. ; 20 cm.
Fiction.
NT$60.00

I. Title.
PL2914.A45L8 78-841271

Tung, Nien.
(Lo yü ti hsiao chen)
落雨的小鎮 / 東年著. — 臺北：聯經出版事業公司，民國66 ［1977］
1, 219 p. ; 20 cm. — (現代小説叢刊)
Short stories.
Series romanized: Hsien tai hsiao shuo ts'ung k'an.
NT$45.00

I. Title.
PL2914.N554L6 78-840829

T'ung, Shih-chang.
(Jen sheng t'an so)
人生探索 / 童世璋著. — 台中：弘業文化實業股份有限公司，民國66 ［1977］
4, 240 p. ; 19 cm.
NT$40.00

I. Title.
PL2817.N24J4 79-835239

Tung, Tzu-wei.
(T'ieh tao yu chi tui)
铁道游击队 / 原著刘知侠；改编董子畏；绘画丁斌曾，韩和平. — 第2版. — 上海：上海人民美術出版社，1978.
10 v. : ill. ; 99 x 125 mm.
"連環圖"
CONTENTS: 1. 血染洋行.—2. 飞车夺枪.—3. 夜袭临城.—4. 杨集除奸.—5. 巧打冈村.—6. 苗庄血战.—7. 两雄遇难.—8. 湖上神兵.—9. 三路出击.—10. 胜利路上.
$1.43

1. Sino-Japanese Conflict, 1937-1945—Fiction. I. Liu, Chih-hsia. T'ieh tao yu chi tui. II. Ting, Pin-tseng. III. Han, Ho-p'ing. IV. Title.
PL2914.N572T5 1978 79-836559

Tung-fang, Pai.
(Lu i hu)
露意湖 / 東方白著. — 臺北：爾雅出版社，民國67 ［1978］
16, 450 p. ; 19 cm. — (爾雅叢書 ; 39)
Fiction.
Series romanized: Erh ya ts'ung shu.
NT$75.00

I. Title.
PL2914.N63L8 79-837015

c

Tung-fang, Yü.
(Feng ch'en san ch'ih chien)
風塵三尺劍 / 東方玉. — ［台北］：金蘭文化出版社，民國68 ［1979］
368 p. ; 22 cm. — (東方玉小説專輯 ; 1)
"臺灣新生報連載"
Fiction.
Series romanized: Tung-fang Yü hsiao shuo chuan chi.
"金蘭：武俠系列"
NT$120.00

I. Title.
PL2914.N684T84 vol. 1 79-841086

CHINESE FICTION (Continued)

Tung-fang, Yü.
(San che chien)
三折劍 / 東方玉著. — 臺北：南琪出版社，民國68 [1979]
2 v. (830 p.) ; 22 cm. — (南琪武俠叢書)
Fiction.
Series romanized: Nan ch'i wu hsia ts'ung shu.
NT$200.00

I. Title.
PL2914.N684S2 79-840367

Tung-fang, Yü.
(T'ung-sha tao)
銅沙島 / 東方玉著. — 香港：春秋書店，[1978?]
8 v. ; 18 cm. — (新派長篇武俠奇情小說)
Fiction.
Series romanized: Hsin p'ai ch'ang p'ien wu hsia ch'i ch'ing hsiao shuo.
$40.00
I. Title.
PL2914.N684T86 78-841784

Wang, Chia-yü.
(Fu hsiao chieh)
芙小姐 / 王家棫創作. — [香港?：s.n., 1978?]
60 p. ; 19 cm.
Short stories.
Reprint of 1932 ed. published by Liang yu t'u shu yin shua kung ssu, Shanghai, issued as ti 23 chung in series: I chiao ts'ung shu.
CONTENTS: 芙小姐.—玫瑰花叢裏的一隻蝴蝶.

I. Wang, Chia-yü. Mei kuei hua ts'ung li ti i chih hu tieh. 1978? II. Title.
PL2919.C438F8 1978 78-843729

Wang, Ching-hsi, 1932-
(Ch'i hsing liao)
七星寮—小說集 王敬羲著 高雄 大業書店 民國44 [1955]
104 p. 18 cm. (今日文叢第3輯之1)
Fiction.

I. Title.
C 64-2203

Wang, Ching-hsi, 1932-
(Hsüan shou)
選手 王敬羲著 九龍 友聯出版社 1955.
97 p. 18 cm.
Fiction.

I. Title.
PL2919.C5H75 C 60-2025 ‡

Wang, Ching-hsi, 1932-
(K'ang T'ung ti kuei lai)
康同的歸來 小說集 王敬羲著 九龍 香港正文出版社 1967.
2, 1, 143 p. 18 cm.
Colophon title.
Fiction.

I. Title.
PL2919.C5K3 C 68-2159

Wang, Ching-hsi, 1932-
(Pao yü tsou lai)
暴雨驟來 王敬羲著 臺北 文星書店 1966.
2, 2, 194 p. 19 cm. (文星叢刊187)
Short stories.

I. Title.
PL2919.C5P3 C 67-204

Wang, Ching-hsi, 1932-
(Pen ch'ao shan chuang)
奔潮山莊 王敬羲著 九龍 高原出版社 1962.
160 p. 19 cm. (中國當代文藝叢書)
Fiction.

I. Title.
PL2919.C5P4 C 63-110

Wang, Ching-hsi, 1932-
(Sheng tan li wu)
聖誕禮物 王敬羲著 臺北 明華書局 民國44 [1955]
132 p. 19 cm.

I. Title.
PL2919.C5S5 C 60-856‡

Wang, Ching-hsi, 1932-
(Sui yüeh chih ko)
歲月之歌—長篇小說 王敬羲著 九龍 正文出版社 1963.
168 p. 19 cm
Fiction.

I. Title.
PL2919.C5S8 C 64-936

Wang, Ching-hsi, 1932-
(To ts'ai ti huang hun)
多彩的黃昏 王敬羲著 [九龍] 友聯出版社 [1954]
92 p. 18 cm. (中國學生叢書)

I. Title.
PL2919.C5T6 C 68-2475
rev

Wang, Chung-chang.
(T'ien ya p'ing tsung)
天涯萍踪 / 王仲章著. — 臺中：青山出版社，民國66 [1977]
200 p. ; 19 cm.
NT$40.00

I. Title.
PL2919.C88T5 78-841945

Wang, Liang-ying.
(Hai shang chen ch'a ping)
海上侦察兵 / 原著章明 ; 改编王良莹 ; 绘画上海市美术学校第六期工农兵美术创作学习班. — 第1版. — 上海：上海人民出版社，1977.
110 p. : ill. ; 95 x 126 mm.
"連環圖畫"
$0.12
I. Chang, Ming, fl. 1965- II. Shang-hai shih mei shu hsüeh hsiao. Ti liu ch'i kung nung ping mei shu ch'uang tso hsüeh hsi pan. III. Title.
PL2919.L39H3 79-836689

Wang, Lin-t'ai.
(Lan se ti yüeh liang)
藍色的月亮 / 著作者王臨泰. — 台中：亞洲文學出版社，民國66 [1977]
250 p. ; 19 cm.
Colophon title.
Fiction.

I. Title.
PL2919.L53L3 1977 79-840040

Wang, Lin-t'ai.
(Wu hsi ho p'an)
烏溪河畔 / 著作者王臨泰. — 臺中：亞洲大學出版社，民國66 [1977]
278 p. : ill. ; 19 cm.
Colophon title.
Fiction.
NT$50.00

I. Title.
PL2919.L53W8 1977 79-840063

Wang, Shang-cheng.
(Hai hsia li ming)
海峽黎明 / 王尚政. — [香港?：s.n., 1979?]
240 p. ; 22 cm.
Fiction.
Reprint of the 1978 ed. published by Fu-chien jen min ch'u pan she.

I. Title.
PL2919.S43H3 1979 79-840115

Wang, Su, fl. 1977-
(Ch'ing nien chin wei chün)
青年近卫军 / 王素改编. — 第1版. — 北京：人民美術出版社：发行者新华书店北京发行所，1977.
3 v. : ill. ; 11 x 13 cm.
"连环画"
"根据[苏联]法捷耶夫原著"
$0.60
1. World War, 1939-1945—Fiction. I. Fadeev, Aleksandr Aleksandrovich, 1901-1956. Molodaîa gvardiîa. II. Title.
PL2919.S75C47 79-836917

Wang, Te-ming.
(Pen t'eng ti Ya-lu-tsang-pu chiang)
奔腾的雅鲁藏布江 / 王德明，赵志立. — 第1版. — 北京：人民文学出版社，1978.
507 p. ; 19 cm.
Fiction.
$0.97

I. Chao, Chih-li. II. Title.
PL2919.T43P4 79-839171

Wang, T'o, 1944-
(Wang chün tsao kuei)
望君早歸 / 王拓著. — 台北：遠景出版社，民國66 [1977]
13, 246 p. ; 20 cm. — (遠景叢刊 ; 78)
Short stories.
Series romanized: Yüan ching ts'ung k'an.
CONTENTS: 春牛圖.—獎金二〇〇〇元.—一個年輕的中學教員.—車站.—望君早歸.
NT$45.00
I. Title.
PL2919.T6W36 78-842730

Wang, Tung-man.
(Chang-ho ch'un)
漳河春 / 王东满. — [香港?：s.n., 1978?]
2, 725 p. ; 21 cm.
Fiction.
Reprint of the 1976 (1977 printing) ed. published by Shan-hsi jen min ch'u pan she, T'ai-yüan.

I. Title.
PL2919.T84C5 1978 78-840918

Wang, Yang, 1926-
(Shih chi mo ti ch'ing tiao)
世紀末的情調 / 汪洋著. — 台南：青溪出版社：總經銷東風圖書供應社，民國67 [1978]
219 p. : ill. ; 20 cm. — (東風文叢 ; 6)
Short stories.
Series romanized: Tung feng wen ts'ung.
CONTENTS: 玫瑰幽蘭.—舊愛新歡.—灰色的週末.—怪老人.—愛情的滋味.—說謊的人.—舊夢.—哀愁的背影.—世紀末的情調.—春雨迷濛.—失去的春天.—夕陽西下.—輪回.
I. Title.
PL2919.Y29S5 79-837674

CHINESE FICTION (Continued)

Wang, Ying-hsiang.
(I to liu liu ti yün)
一朵溜溜地雲 / 王映湘著. — 台南：鳳凰城圖書公司, 民國67 ⌊1978⌋
217 p. ; 20 cm. — (鳳凰新書 ; 405)
Short stories.
Series romanized: Feng huang hsin shu.
CONTENTS: 一朵溜溜地雲.—意外.—放榜之後.—龍子.—無根的愛.—當仁不讓.—刮鬍刀.—啞戰.—芳鄰.—歸來.—詩人會.
NT$45.00
I. Title.
PL2919.Y47I2 79-837355

Wang, Ying-hsiang.
(Jui-chen)
瑞珍 / 王映湘著. — 台中：弘業文化實業股份有限公司, 民國66 ⌊1977⌋
245 p. ; 19 cm.
Short stories.
CONTENTS: 愛心.—弄巧成拙.—重負.—大姐.—瑞珍.—酒女.—回頭.—疚.—當仁不讓.—吉安村.——廂情願.
NT$40.00
I. Title.
PL2919.Y47J8 78-843094

Wei, Chen-chung.
(Ch'iao na kang lou)
巧拿岗楼 / 韦振中画. — 第1版. — ⌊广州⌋：广东人民出版社, 1972.
⌊24⌋ p. : chiefly col. ill. ; 16 cm.
I. Title.
PL2920.E32C5 78-842310

Wei, Ssu-li.
(Ku huo)
蠱惑 / 衛斯理著. — 香港：明窗出版社：發行明報有限公司出版部, 1978.
198 p. ; 19 cm. — (科學幻想小說 ; 第5集)
Two short stories.
Series romanized: K'o hsüeh huan hsiang hsiao shuo.
"再来一次.": p. 109–198.
$6.00 ($2.00 U.S.)
I. Wei, Ssu-li. Tsai lai i tz'u. 1978. II. Title.
PL2920.E526K8 79-838034

Wei, Ssu-li.
(Lan hsüeh jen)
藍血人 / 衛斯理著. — 香港：明窗出版社：發行明報有限公司出版部, 1978.
358 p. ; 19 cm. — (科學幻想小說 ; 第2集)
Series romanized: K'o hsüeh huan hsiang hsiao shuo.
$6.00 ($2.00 U. S.)
I. Title.
PL2920.E526L29 79-836589

Wei, Ssu-li.
(Lao mao)
老貓 / 衛斯理著. — ⌊香港⌋：明窗出版社, 發行明報有限公司出版部, 1978.
357 p. ; 19 cm. — (科學幻想小說 ; 第1集)
Short stories.
Series romanized: K'o hsüeh huan hsiang hsiao shuo.
CONTENTS: 老貓.—聚寶盆.—影子
$6.00 ($2.00 U.S.)
I. Title.
PL2920.E526L3 79-835956

Wen, Chu.
(Yen yü hsieh yang)
煙雨斜陽 / 文竺著. — 臺北：小說創作社, 1977.
380 p. ; 20 cm. — (小說創作叢書)
Fiction.
Series romanized: Hsiao shuo ch'uang tso ts'ung shu.
NT$40.00
I. Title.
PL2920.E62Y4 78-841080

Wen, Hsiao-p'ing, 1951–
(Cheng yüeh hsin liang)
正月新娘 / 溫小平著. — 臺北：小說創作社, 民國64 ⌊1975⌋
350 p. ; 20 cm.
Fiction.
I. Title.
PL2920.E654C4 79-835421

Wen, Hsin, 1930–
(Hei yü yü feng liu kua fu)
黑獄與風流寡婦 / 文心, 龍驥合著. — 台南：青溪出版社：總經銷東風圖書供應社, 民國67 ⌊1978⌋
191 p. ; 19 cm. — (東風文叢 ; 7)
Series romanized: Tung feng wen ts'ung.
CONTENTS: 黑獄.—狼吻.—風流寡婦.—血剪刀.—妖鳥.—龍舟.
NT$40.00
I. Lung, Chi, fl. 1969– joint author. II. Title.
PL2920.E658H4 79-836930

Wen, I-ch'i.
(Ch'ien ying mei ying)
倩影魅影 / 文亦奇著. — 臺北：皇冠出版社, 民國67 ⌊1978⌋
234 p. ; 20 cm. — (His 古今奇譚 ; 2)
Two short stories.
Series romanized: Ku chin ch'i t'an.
CONTENTS: 倩影魅影.—麗莎的微笑.
NT$40.00
I. Title.
PL2920.E66A6 1978, vol. 2 79-839972

Wen, I-ch'i.
(Yü shou hsüeh shou)
玉手血手 / 文亦奇著. — 臺北：皇冠出版社, 民國67 ⌊1978⌋
226 p. ; 20 cm. — (古今奇譚 ; 1)
Short stories.
Series romanized: Ku chin ch'i t'an.
CONTENTS: 玉手血手.—死亡之吻
NT$40.00
I. Wen, I-ch'i. Ssu wang chih wen. 1978. II. Title.
PL2920.E66A6 1978, vol. 1 79-836149

Wen, Jui-an.
(Pai i Fang Chen-mei)
白衣方振眉 / ⌊温瑞安⌋. — 台南：長河出版社：總經銷建立書局, 民國67 ⌊1978⌋
305 p. ; 20 cm. — (温瑞安新武俠專書 ; 1) (長河叢書 ; 9) (長河文庫 ; 9)
Short stories.
Series 1 romanized: Wen Jui-an hsin wu hsia chuan shu. Series 2 romanized: Ch'ang ho ts'ung shu. Series 3 romanized: Ch'ang ho wen k'u.
NT$55.00
I. Title.
PL2920.E6626 1978, vol, 1 79-836157

Wen, Jui-an.
(Wen Jui-an hsin wu hsia chuan shu)
温瑞安新武俠專書. — 台南：長河出版社：總經銷建立書局, 民國67– ⌊1978–
v. ; 20 cm. — (長河叢書 ; 9) (長河文庫 ; 9)
Cover title: Hsin wu hsia chuan shu.
Short stories.
Series 1 romanized: Ch'ang ho ts'ung shu. Series 2 romanized: Ch'ang ho wen k'u.
CONTENTS: 1. 白衣方振眉.
NT$55.00 (v. 1)
I. Title. II. Title: Hsin wu hsia chuan shu.
PL2920.E6626 1978 79-836158

Wo-lung-sheng.
(Chien ch'i she ch'ang hung)
劍氣射長虹 / 卧龍生著. — 香港：春秋書店, ⌊1978?⌋
5 v. (1771 p.) ; 19 cm. — (新派長篇武俠奇情小說)
Fiction.
Series romanized: Hsin p'ai ch'ang p'ien wu hsia ch'i ch'ing hsiao shuo.
$32.50
I. Title.
PL2920.O2C45 78-843602

Wo-lung-sheng.
(Chin yen shen chi)
金眼神姬 / 卧龍生著. — 香港：春秋書店, ⌊1978?⌋
5 v. (1320 p.) ; 19 cm. — (新派長篇武俠奇情小說)
Fiction.
Series romanized: Hsin p'ai ch'ang p'ien wu hsia ch'i ch'ing hsiao shuo.
$25.00
I. Title.
PL2920.O2C47 78-842706

Wo-lung-sheng.
(Ch'un ch'iu pi)
春秋筆 / 臥龍生著. — 台北：春秋出版社, 民國67 ⌊1978⌋
3 v. (1618 p.) ; 19 cm. — (長篇武俠名著)
Fiction.
Series romanized: Ch'ang p'ien wu hsia ming chu.
"武俠奇情名著"
NT$180.00
I. Title.
PL2920.O2C5 79-835812

Wo-lung-sheng.
(Hsüan yüan shen chien)
軒轅神劍 / 卧龍生著. — 香港：春秋書店, ⌊1978?⌋
5 v. (1629 p.) ; 19 cm. — (新派長篇武俠奇情小說)
Fiction.
Series romanized: Hsin p'ai ch'ang p'ien wu hsia ch'i ch'ing hsiao shuo.
$25.00
I. Title.
PL2920.O2H7 78-842610

Wo-lung-sheng.
(Lung feng ch'i hsing)
龍鳳七星 / 卧龍生著. — 香港：春秋書店, ⌊1978?⌋
4 v. (1127 p.) ; 19 cm. — (新派長篇武俠奇情小說)
Fiction.
Series romanized: Hsin p'ai ch'ang p'ien wu hsia ch'i ch'ing hsiao shuo.
$20.00
I. Title.
PL2920.O2L85 78-842532

Wo-lung-sheng.
(Pa chien hui)
八劍會 / 卧龍生著. — 台北：文康印刷出版公司, ⌊1978?⌋.
5 v. (1528 p.) ; 18 cm.
Fiction.
"新派長篇武俠名著"
NT$45.00 per vol.
I. Title.
PL2920.O2P3 78-841573

Wu, Ch'eng-en, ca. 1500–ca. 1582.
(Hsi yu chi)
西遊記 / 吳承恩撰. — 台南：第一書店, ⌊1977?⌋
1055 p. ; 20 cm.
"大字足本古典文學"
"一百回"
NT$70.00
I. Title.
PL2697.H75 1977 78-842550

CHINESE FICTION (Continued)

Wu, Ch'iang, 1910-
(Hung jih)
红日 / 吴强著. — [2次修订本]. — 北京 : 中国青年出版社 : 新华书店北京发行所发行, 1978.
12, 563 p. ; 19 cm.
Fiction.
$1.15

I. Title.
PL2921.C5H8 1978 79-837717

Wu, Chien-ch'ing.
(Yün Fei-chang)
云飞嶂 / 武剑青. — 第1版. — [广州] : 广东人民出版社 : 广东省新華書店, 1978.
645 p. ; 21 cm.
Fiction.
$1.42

I. Title.
PL2921.C534Y8 79-838568

Wu, Ching-tzu, 1701-1754.
(Ju lin wai shih)
儒林外史 / 吳敬梓著. — 臺北 : 聯經出版事業公司, 民國67 [1978]
4, 4, 550 p., 4 leaves ; 22 cm. — (中國古典小說新刊)
Reprint ed.
NT$100.00

I. Title. II. Series: Chung-kuo ku tien hsiao shuo hsin k'an.
PL2732.U22J8 1978b 79-835816

Wu, Ching-tzu, 1701-1754.
(Ju lin wai shih)
儒林外史 / 吳敬梓著: — 香港 : 廣昌書局, [1978?]
302, [36] p., [17] leaves of plates : ill. ; 19 cm.
Fiction.
"五十五回."
$5.00

I. Title.
PL2732.U22J8 1978 78-843990

Wu, Ch'u, fl. 1959-
(Hui sheng ku)
迴聲谷 / 吳楚著. — 台北 : 南京出版公司, 民國67 [1978]
556 p. ; 20 cm.
Fiction.
NT$90.00

I. Title.
PL2921.C58H84 79-837115

Wu, Ch'u.
(Lu-lu)
露露 / 吳楚著. — 台北 : 南京出版公司, 民國67 [1978]
527 p. ; 20 cm.
Fiction.
NT$90.00

I. Title.
PL2921.C58L8 79-837365

Wu, Ch'ung-lan.
(Tu jen)
渡人 / 吳崇蘭著. — 臺初版. — 臺北 : 正中書局, 民國66 [1977]
171 p. ; 20 cm. — (正中文藝叢書)
Fiction.
Series romanized: Cheng chung wen i ts'ung shu.

I. Title.
PL2921.C8T8 78-842442

Wu, Ch'ung-lan.
(Wei na ch'ing ch'ing i wen)
爲那輕輕一吻 / 吳崇蘭著. — 台北 : 皇冠出版社, 民國68 [1979]
335 p. ; 19 cm.
Short stories.
NT$60.00

I. Title.
PL2921.C8W4 79-839549

Wu, Hsiang-hui.
(Tuan chih shao nien)
斷指少年 / [作者吳祥輝]. — 臺北 : 遠流出版社, 民國67 [1978]
225 p. ; 19 cm.
Fiction.
NT$40.00

I. Title.
PL2921.H756T8 79-841096

Wu, Tsu-hsiang.
(Shan hung)
山洪 / 吳組緗. — [香港?] : 創作書社, [1978?]
226 p. ; 21 cm.
Fiction.
Reprint of the 1946 ed. published by Hsing ch'ün ch'u pan kung ssu, Shanghai.

I. Title.
PL2921.T72S5 1978 79-835632

Wu, Yen-mei, 1933-
(T'iao teng lien tan)
挑燈練膽 : 秉燭夜譚之六 / 司馬中原 [吳延玫] 著. — 臺北 : 皇冠雜誌社, 民國67 [1978]
221 p. ; 19 cm.
Short stories.
NT$45.00

I. Title.
PL2921.Y4T52 79-836923

Wu, Yu-heng.
(Pei shan chi)
北山记 / 吴有恒. — 第1版. — [广州] : 广东人民出版社, 1978.
2, 453 p. ; 21 cm. — (粤海文丛)
$1.00

1. China—History—Civil War, 1945-1949—Fiction. I. Title. II. Series: Yüeh hai wen ts'ung.
PL2921.Y68P4 79-836858

Wu, Yu-heng.
(Shan hsiang feng yün lu)
山乡风云录 / 吴有恒. — [香港? : s. n., 1978?]
2, 347 p. ; 21 cm.
Reprint of the 1962 (1978 printing) ed. published by Kuang-tung jen min ch'u pan she, in series: Yüeh hai wen ts'ung.

1. China—History—Civil War, 1945-1949—Fiction. I. Title. II. Series: Yüeh hai wen ts'ung.
PL2921.Y68S5 1978 79-838115

(Wu Hsün chuan)
武訓傳. — [香港? : s. n., 1977?]
34 p. : ill. ; 19 cm.
"電影小說"
Reprint of the ed. published by Shih chieh ch'u pan she, Hongkong.

1. Wu, Hsün, 1838-1897—Fiction.
PL2832.5.W8 1977 78-841009

Wu-ling-ch'iao-tzu.
(Chin shih meng)
金石盟 / 武陵樵子著. — 香港 : 春秋書店, [1978?]
3 v. (810 p.) ; 19 cm. — (新派長篇武俠奇情小說)
Fiction.
Series romanized: Hsin p'ai ch'ang p'ien wu hsia ch'i ch'ing hsiao shuo.
$16.50

I. Title.
PL2921.Z78C47 78-842152

Wu-ling-ch'iao-tzu.
(San hua chui hun)
三花追魂 / 武陵樵子著. — 香港 : 春秋書店, [1978?]
3 v. (791 p.) ; 19 cm.
Fiction.
$15.00

I. Title.
PL2921.L52S2 78-840868

Wu-ming-shih, pseud.
(Chin se ti she yeh)
金色的蛇夜 / 無名氏著. — 香港 : 新聞天地社 ; 台北 : 總經銷遠景出版社, 民國66 [1977]
478 p. ; 19 cm. — (無名氏全書)
Fiction.
$12.00

I. Title. II. Series: Wu-ming-shih, pseud. Wu-ming-shih ch'üan shu.
PL2921.Z83C5 1977 79-836177

Wu-ming-shih, pseud.
(T'a li ti nü jen)
塔裏的女人 / 無名氏著. — 香港 : 新聞天地社, 民國65 [1976]
182 p. ; 19 cm.
Fiction.
"無名氏全書"
$5.00

I. Title.
PL2921.Z83T3 1976 79-835492

Yang, I, 1907-
(Sheng ti fan jao)
生底煩擾 / 歐陽山 [楊儀]. — [香港? : s. n., 1979?]
vii, 2, 209 p. ; 18 cm.
Short stories.
Reprint of the 1936 ed. published by Wen hua sheng huo ch'u pan she, Shanghai, in series: Wen hsüeh ts'ung k'an : Ti erh chi.
CONTENTS: 新客.—人底界限.—痞棍世界.—消遣.—第一夢.—銀狗仔.—禁令.—姊妹.—生底煩擾.—邱家兄弟.—憂心者.
I. Title.
PL2922.I3S5 1979 79-839552

Yang, Nien-tz'u.
(Yang Nien-tz'u tzu hsüan chi)
楊念慈自選集 / [著作者楊念慈]. — 臺北 : 黎明文化事業公司, 民國66 [1977]
2, 286 p., [2] leaves of plates : ill. ; 20 cm. — (中國新文學叢刊 : 46)
Short stories.
Series romanized: Chung-kuo hsin wen hsüeh ts'ung k'an.
CONTENTS: 小傳.—恩愛.—貓的故事.—金龜背的風水.—半城王和他的兒子.—鴨的悲喜劇.—枯楊生稊.—師道.—交棒.—作品書目.
NT$70.00
I. Title.
PL2922.N5A15 1977 78-843267

Yang, Tsung-chen, 1919-
(Li ming ch'ien)
黎明前 / [孟瑤 i. e. 楊宗珍著]. — 臺中 : 學人文化事業有限公司, 民國67 [1978]
780 p. ; 19 cm.
Fiction.
NT$150.00

I. Title.
PL2922.T78L48 79-838589

CHINESE FICTION (Continued)

Yang, Ts'un-jen.
(K'u men)
苦悶 / 著者楊邨人. — [香港?: s. n., 1977?]
4, 155 p. ; 22 cm.
Short stories.
Reprint of the 1934 ed. published by Ch'i chih shu chü, Shanghai.
CONTENTS: 殘忍.—博愛.—賣稿.—人生的陰影.—驚喜.—等待.

I. Title.
PL2922.T76K8 1977 78-841320

Yang, Tzu-wei.
(Shih nien hsin lien)
十年心戀 / 楊子崴著. — 台北: 哲志出版社, 民國66 [1977]
209 p. ; 20 cm.
Short stories.
CONTENTS: 十年心戀.—紅鸞禧.—女教員戀曲.—反徒的畫像.
NT$35.00

I. Title.
PL2922.T95S57 78-841969

Yang, Tzu-wei.
(Wu niang p'iao hsiang meng)
舞孃飄香夢 / 著作者楊子崴. — 台北縣永和鎮: 集成出版社, 民國67 [1978]
200 p. ; 19 cm. — (集成文叢)
Colophon title.
Fiction.
Series romanized: Chi ch'eng wen ts'ung.
NT$40.00

I. Title.
PL2922.T95W8 79-839167

Yang, Wei, fl. 1973-
(Ai chih meng)
愛之夢 / 何索[楊蔚]著. — 台北: 南京出版公司, 民國68 [1979]
191 p. ; 20 cm.
Fiction.
NT$45.00

I. Title.
PL2922.W373A7 79-839924

Yang, Wei, fl. 1973-
(Ch'ing ch'un ch'ü)
青春曲 / 何索[楊蔚]著. — 臺北: 出版家文化事業有限公司出版部, 民國67 [1978]
156 p. ; 20 cm. — (出版家叢書; 48)
Short stories.
Series romanized: Ch'u pan chia ts'ung shu.
NT$40.00

I. Title.
PL2922.W373C53 78-842445

Yang, Wei, fl. 1973-
(Kei lang man)
給浪漫 / 何索[楊蔚]著. — 臺北: 聯亞出版社, 民國67 [1978]
187 p. ; 20 cm. — (聯亞叢刊; 47)
Fiction.
Series romanized: Lien ya ts'ung k'an.
NT$40.00

I. Title.
PL2922.W373K4 79-838009

Yang, Wei, fl. 1973-
(Lao shu yü hu tieh)
老鼠與蝴蝶 / 何索[楊蔚]著. — 臺北: 聯亞出版社, 民國67 [1978]
207 p. ; 19 cm. — (聯亞叢刊; 43)
Fiction.
Series romanized: Lien Ya ts'ung k'an.

I. Title.
PL2922.W373L36 79-838100

Yang, Wei, fl. 1973-
(Nü jen k'uang hsiang ch'ü)
女人狂想曲 / 何索[楊蔚]著. — 臺北: 林白出版社, 民國67 [1978]
243 p. ; 20 cm. — (林白叢書; 104)
Short stories.
Series romanized: Lin pai ts'ung shu.
CONTENTS: 女人狂想曲.—兩面夏娃.—落花怨.—三個愛.—那個空虛的夏天.—哭泣的娃娃.
NT$45.00
I. Title.
PL2922.W373N8 79-837363

Yang, Wei, fl. 1973-
(Pu chü)
佈局 / 何索著. — [台北]: 德昌出版社, [67 i. e. 1978序]
2 v. ; 19 cm. — (何索社會諷刺幽默小說)(德昌新刊; 80)
Fiction.
Series 2 romanized: Te ch'ang hsin k'an.

I. Title. II. Series: Yang, Wei, fl. 1973- Ho So she hui feng tz'u yu mo hsiao shuo.
PL2922.W373P8 79-840553

Yang, Yü-lung, 1929-
(Ch'üeh le chiao ti huang yüeh liang)
缺了角的黃月亮 / 楊御龍著. — 台中: 弘業文化實業股份有限公司: 發行所當代文化出版社, 民國66 [1977]
2, 222 p. ; 19 cm.
Short stories.
NT$40.00

I. Title.
PL2922.Y75C4 79-835240

Yang, Yü-lung, 1929-
(Hei tu)
黑渡 / 楊御龍著. — 台南: 鳳凰城圖書公司, 民國67 [1978]
237 p. ; 20 cm. — (鳳凰新書; 404)
Short stories.
Series romanized: Feng huang hsin shu.
NT$49.00

I. Title.
PL2922.Y75H4 79-836240

Yao, Hsin-nung, 1905-
(Ch'u Pa-wang)
楚霸王 / [著者姚克 i. e. 姚莘農]. — 臺北: 聯經出版事業公司, 民國67 [1978]
3, 4, 165 p. ; 20 cm. — (姚克戲劇選集)
Play.
Series romanized: Yao K'o hsi chü hsüan chi.
NT$30.00

1. Hsiang, Chi, 232-202 B. C.—Fiction. I. Title.
PL2923.H73C54 79-835600

Yao, Hsin-nung, 1905-
(Mei jen chi)
美人計 / [著者姚克 i. e. 姚莘農]. — 臺北: 聯經出版事業公司, 民國67 [1978]
6, 176 p. ; 20 cm. — (姚克戲劇選集)
Drama.
Series romanized: Yao K'o hsi chü hsüan chi.
NT$30.00

1. China—History—Three kingdoms, 220-265—Fiction. I. Title.
PL2923.H73M4 79-836216

Yao, Hsüeh-yin.
(Li Tzu-ch'eng)
李自成 / 姚雪垠著. — 第2版. — 北京: 中國青年出版社, 1977-
v. in : ill. ; 21 cm.
$2.40 (v. 1)

1. Li, Tzu-ch'eng, 1606-1645—Fiction. I. Title.
PL2923.H8L5 1977 78-841091

Yao, Ting-sheng.
(T'u ti shih p'ien)
土地诗篇 / 姚鼎生. — 第1版. — 福州: 福建人民出版社: 新華書店发行, 1978-
v. ; 19 cm.
Fiction.
$1.50 (v. 1)

I. Title.
PL2923.T5T8 79-839484

Yao, Yüan-ch'u.
(Ch'un lien yen ying)
春奩豔影 / [著作者姚鵷雛]. — 台北: 新文豐出版公司, 民國67 [1978]
122 p. ; 19 cm. — (中國文學名著)(零玉碎金集刊; 8)
Fiction.
Series 2 romanized: Ling yü sui chin chi k'an.
Reprint ed.
Pbk. : NT$20.00

I. Title. II. Series: Chung-kuo wen hsüeh ming chu.
PL2826.Y8C5 1978 79-839178

Yao, Yüan-ch'u.
(Yen ts'u cheng hsien lu)
燕蹴箏絃錄 / [著作者姚鵷雛]. — 台北: 新文豐出版公司, 民國67 [1978]
2, 4, 167, 2 p. ; 19 cm. — (零玉碎金集刊; 7)
At head of title: Chung-kuo wen hsüeh ming chu.
Fiction.
Series romanized: Ling yü sui chin chi k'an.
Reprint ed.
Pbk : NT$30.00
I, Title.
PL2826.Y8Y4 1978 79-838208

Yeh, Fei.
(Lao-nung hsi ti wu yen)
荖濃溪的嗚咽 / 葉菲著. — 臺中: 光啓出版社, 民國65 [1976]
2, 225 p. ; 19 cm. — (小說叢刊; 44)
Short stories.
Series romanized: Hsiao shuo ts'ung k'an.
CONTENTS: 荖濃溪的嗚咽.—山中的老人.—一種喜悅.—分家.—利慾薰心.—老師的心意.—老駱.—選戰.—醒悟.—割稻.—耕耘與收穫.—結.—我是傻公.—抉擇
NT$30.00

I. Title.
PL2924.F35L3 78-840869

Yeh, Hsin, fl. 1977-
(Yen ying)
岩鷹 / 叶辛, 忻豹. — [香港?: s. n., 1978?]
2, 342 p. ; 22 cm.
Fiction.
Reprint of the 1978 ed. published by Shang-hai wen i ch'u pan she, Shanghai.

I. Hsin-pao, joint author. II. Title.
PL2924.H855Y4 1978 79-837954

Yeh, Shan.
(Yang Mu shih chi)
楊牧詩集. — 臺北: 洪範書店, 民國67- [1978-
v. ; 20 cm. — (洪範文學叢書; 31)
Series romanized: Hung fan wen hsüeh ts'ung shu.
Includes indexes.
CONTENTS: 1. 1956-1974.
NT180.00 (v. 1)

I. Title.
PL2924.S47A6 1978 79-838113

Yeh, Shao-chün, 1893-
(Ch'eng chung)
城中 / 著者葉紹鈞. — 九龍: 百靈出版社, [1978?]
157 p. ; 18 cm.
Colophon title.
Short stories.
Reprint of the 1934 ed. published by K'ai ming shu tien.
CONTENTS: 病夫.—前途.—演講.—城中.—雙影.—在民間.—晨.—微波.—搭班子.
$5.80

I. Title.
PL2827.S4C48 1978 78-841181

CHINESE FICTION (Continued)

Yeh, Shih-t'ao, 1925–
(Ts'ai liu chi)
採硫記 / 葉石濤著. — 台北：龍田出版社, 民國68 [1979]
220 p. ; 19 cm. — (葉石濤小說集)
Short stories.
CONTENTS: 採硫記.—決鬥.—斷層.—等待.—玫瑰項圈.—墮胎.—叛國者.—敗戰者.—晴天和陰天.
NT$60.00
I. Title. II. Series: Yeh, Shih-t'ao, 1925– Yeh Shih-t'ao hsiao shuo chi.
PL2924.S55T7 79-841427

Yeh-feng.
(Chih i i yüan ti t'ien o)
值一億圓的天鵝 / 野楓著. — 台南：震平文化出版社：總經銷南海書報社, 民國68 [1979]
236 p. ; 19 cm.
Fiction.
"社會傳眞·五光十色"
NT$40.00
I. Title.
PL2924.Z42C5 79-840520

Yen, Ch'in.
(Ai shen ti ying tzu)
愛神的影子 / 嚴沁著. — 臺北：小說創作社, 民國65 [1976]
2, 350 p. ; 20 cm. — (小說創作叢書)
L.C. copy imperfect: First two pages duplicated.
Series romanized: Hsiao shuo ch'uang tso ts'ung shu.
NT$40.00 ($1.50 U.S.)
I. Title.
PL2925.C444A757 78-841147

Yen, Ch'in.
(Ch'ien fan chin ch'u)
千帆盡處 / 嚴沁著. — 臺北：小說創作社；Silver Spring, Md.：美國總代理 Tang's Corp., 1977.
352 p. ; 20 cm. — (小說創作長篇文藝叢書)
Fiction.
Series romanized: Hsiao shuo ch'uang tso ch'ang p'ien wen i ts'ung shu.
NT$40.00
I. Title.
PL2925.C444C526 78-842909

Yen, Ch'in.
(Ch'in yün hsin sheng)
琴韻心聲 / 嚴沁著. — 臺北：小說創作社；Silver Spring, Md.：美國總代理 T'ang's Corp., 民國66 [1977]
320 p. ; 20 cm. — (小說創作叢書)
Fiction.
Series romanized: Hsiao shuo ch'uang tso ts'ung shu.
Series title varies slightly.
NT$40.00
I. Title.
PL2925.C444C546 78-841462

Yen, Ch'in.
(Ching t'ing chi mo)
靜聽寂寞 / 嚴沁著. — 台北：環怡出版社：發行所和風書報社, 民國66 [1977]
303 p. ; 20 cm.
Short stories.
CONTENTS: 靜聽寂寞—白色的幽靈—夢幻情天—烽火情淚—飛躍浪端.
NT$45.00
I. Title.
PL2925.C444C554 78-842762

Yen, Ch'in.
(Hsia hsien yüeh)
下弦月 / 嚴沁著. — 臺北：小說創作社, 1978.
318 p. ; 19 cm. — (小說創作叢書)
Short stories.
Series romanized: hsiao shuo ch'uang tso ts'ung shu.
CONTENTS: 無情荒地有情天.—世俗以外的人.—下弦月.—草淚戀歌.
NT$45.00
I. Title.
PL2925.C444H67 78-842401

Yen, Ch'in.
(Hsin yüan)
心願 / 嚴沁著. — 臺北：小說創作社；Silver Spring, Md.：美國總代理 T'ang's Corporation, 民國64 [1975]
348 p. ; 20 cm. — (小說創作長篇文藝叢書)
Short stories.
Series romanized: Hsiao shuo ch'uang tso ch'ang p'ien wen i ts'ung shu.
CONTENTS: 心願.—抓不住的情.—青色田園.—黑夜足音.
I. Title.
PL2925.C444H73 78-840915

Yen, Ch'in.
(Huai nien)
懷念 / 嚴沁著. — 臺北：小說創作社：發行者藍帶雜誌社, 1978.
296 p. ; 20 cm. — (小說創作叢書)
Series romanized: Hsiao shuo ch'uang tso ts'ung shu.
NT$45.00
I. Title.
PL2925.C444H78 79-835640

Yen, Ch'in.
(K'ung ku yu lan)
空谷幽蘭 / 嚴沁著. — 臺北：小說創作社：發行者藍帶雜誌社, 1976.
336 p. ; 20 cm. — (小說創作長篇文藝叢書) (小說創作叢書)
Fiction.
Series 1 romanized: Hsiao shuo ch'uang tso ch'ang p'ien wên i ts'ung shu. Series 2 romanized: Hsiao shuo ch'uang tso ts'ung shu.
NT$40.00
I. Title.
PL2925.C444K86 78-840840

Yen, Ch'in.
(Lan se ti yu yü)
藍色的憂鬱 / 嚴沁著. — 臺北：小說創作社；Wheaton, MD.：美國總代理 Donald Tang, 民國64 [1975]
348 p. ; 19 cm. — (小說創作長篇文藝叢書)
Short stories.
Series romanized: Hsiao shuo ch'uang tso ch'ang p'ien wen i ts'ung shu.
CONTENTS: 雨夜.—愛與同情.—藍色的憂鬱.—餘情未了.
I. Title.
PL2925.C444L37 78-842964

Yen, Ch'in.
(Liang hsin hsiang hsü)
兩心相許 / 嚴沁著. — 臺北：小說創作社；Wheaton, Md.：美國總代理 D. Tang, 1976.
356 p. ; 20 cm. — (小說創作長篇文藝叢書)
Fiction.
Series romanized: Hsiao shuo ch'uang tso ch'ang p'ien wen i ts'ung shu.
NT$50.00
I. Title.
PL2925.C444L48 78-842643

C

Yen, Ch'in.
(Pan to yeh chü)
半朵野菊 / 嚴泌著. — 香港：環球圖書雜誌出版社, 1973.
296 p. ; 17 cm. — (文藝創作小說)
Short stories.
Series romanized: Wen i ch'uang tso hsiao shuo.
CONTENTS: 半朵野菊.—笑在淚中.—投影.—永遠不再.
I. Title.
PL2925.C444P3 79-836394

Yen, Feng-le.
(Hsien wei shu chi)
县委书记 / 阎丰乐. — [香港?：s. n., 1978?]
v. ; 22 cm.
Cover title.
Fiction.
Reprint of the 1976 ed. published by Jen min ch'u pan she, Peking.
I. Title.
PL2925.F4H7 1978 79-835188

Yen, Fu, 1853–1921.
(Yü yeh t'ang shih chi)
瘉壄堂詩集 / 嚴復著. — 臺北：臺灣商務印書館, 民國66 [1977]
22, 39, 1 leaves, [10] p. : 19 cm. — (人人文庫 ; 2339-2340)
Series romanized: Jen jen wen k'u.
NT$18.00
I. Title.
PL2828.F8Y8 1977 79-835484

Yen, T'ing-yün.
(Ching hu yüeh)
鏡湖月 / 華嚴 [嚴停雲] 著. — 臺北：皇冠出版社, 民國67 [1978]
309 p. ; 19 cm.
Fiction.
NT$60.00
I. Title.
PL2925.T5C57 79-836223

Yen-hsiang.
(T'iao ts'ao ti hsüeh sheng)
跳槽的學生 / 雁翔編著. — 臺北：聯亞出版社, 民國67 [1978]
3, 219 p. ; 19 cm. — (聯亞叢刊 ; 33)
Short stories.
Series romanized: Lien ya ts'ung k'an.
NT$40.00
I. Title.
PL2925.Z67T5 79-836224

Yü, Ch'i-min.
(I tao ts'ai hung)
一道彩虹 / 禹其民著. — 台北：哲志出版社；香港：海外總經銷陳湘記書局, 民國67 [1978]
276 p. ; 19 cm.
At head of title: Ch'ang p'ien wen i ch'uang tso hsiao shuo.
Fiction.
NT$50.00
I. Title.
PL2928.C53I28 79-839464

Yü, Ch'i-min.
(K'ung liu ch'ing yüan)
空留情緣 / 禹其民著. — 台北：哲志出版社, 民國67 [1978]
276 p. ; 20 cm.
At head of title: Ch'ang pien wen i ch'uang tso hsiao shuo.
Fiction.
NT$45.00
I. Title.
PL2928.C53K78 78-843548

Yü, Ch'i-min.
(Li ch'ao niao)
離巢鳥 / 禹其民著. — 台北：哲志出版社：香港：海外總經銷陳湘記書局, 民國67 [1978]
276 p. ; 19 cm.
At head of title: Ch'ang p'ien wen i ch'uang tso hsiao shuo.
Fiction.
NT$50.00
I. Title.
PL2928.C53L49 79-836910

Yü, Ch'i-min.
(Pien se ti yüeh liang)
變色的月亮 / 禹其民著. — 台北：哲志出版社, 民國67 [1978]
277 p. ; 19 cm.
At head of title: Ch'ang p'ien wen i ch'uang tso hsiao shuo.
Fiction.
I. Title.
PL2928.C53P47 79-836241

CHINESE FICTION (Continued)

Yü, Ch'i-min.
(Wo shih ni ti chih hsin jen)
我是妳的知心人 / 著作者禹其民. — 台北：哲志出版社，民國66 [1977]
332 p. ; 19 cm.
Colophon title.
Fiction.
NT$40.00

I. Title.
PL2928.C53W59 78-842644

Yü, Feng, 1917-
(Wu ch'an che)
无产者 / 于逢. — [香港? : s. n., 1979?]
607 p. ; 21 cm.
Fiction.
Reprint of the 1978 ed. published by Kuang-tung jen min ch'u pan she, Canton.

I. Title.
PL2928.F43W8 1979 79-838577

Yü, Hsüeh.
(Pei ai ti ling hun)
悲哀的靈魂：在中國大陸的土地上 / 歲雪著. — 香港：紫蘭書屋，1978.
188 p. ; 21 cm.
Short stories.
CONTENTS: 青面獠.—魏阿大.—老楊嫂.—靈活.—母與子.—罪與罰.—海棠姑娘.—悲哀的靈魂.—玉蘭.
$8.00

I. Title.
PL2928.H745P4 79-837986

Yü, Huan-su.
(Ch'ing ch'uan li li)
晴川歷歷 / 于還素著. — 臺北：天華出版事業股份有限公司，民國67 [1978]
4, 4, 247 p. ; 19 cm. — (天華文學叢刊 ; 4)
Series romanized: T'ien hua wen hsüeh ts'ung k'an.
Pbk. : NT$55.00

I. Title.
PL2928.H8C47 79-835204

Yü, I, fl. 1958-
(Chin-men chih lü)
金門之旅 余依著 香港 長興書局 [1958]
132 p. 18 cm.
Sequel: 人生的果實

I. Title.
PL2928.I2C5 C 59-120

Yü, I, fl. 1958-
(Jen sheng ti kuo shih)
人生的果實 余依著 香港 長興書局 [1959?]
133-275 p. 18 cm.
Sequel to: 金門之旅

I. Title.
?928.I2J4 C 59-121

rev

Yü, Li-hua, 1931-
(Fu chia ti erh nü men)
傅家的兒女們 / 於梨華著. — 香港：天地圖書有限公司，1978.
4, 524 p. ; 19 cm.
Fiction.
$12.00

I. Title.
PL2928.L43F8 78-841984

Yü, Min, essayist.
(Ti i ko hui ho)
第一个回合 / 于敏. — 第1版. — 北京：中國青年出版社，1978.
2 v. (1119 p.) ; 19 cm.
Fiction.
$2.25

I. Title.
PL2928.M5T5 79-836010

Yü, Sung-yen.
(Hai-hua)
海花 / 余松岩. — 修订本. — [香港? : s. n., 1978?]
139 p. : ill. ; 21 cm.
Fiction.
Reprint of the 1977 ed. published by Kuang-tung jen min ch'u pan she, in series: Shao nien wen i ts'ung shu.

I. Title.
PL2928.S8H3 1978 78-843550

Yü, Ta-fu, 1896-1945.
[Selected works]
達夫自選集. — 香港：聯合圖書公司，[1978?]
3, 366 p., [2] leaves of plates : ill. ; 19 cm.
Cover title.
Chiefly short stories.
Reprint of the ed. published by T'ien ma shu tien, Shanghai.
$8.00

I. Title. Title romanized: Ta-fu tzu hsüan chi.
PL2831.T3A6 1978 78-841906

Yu, Tseng-hui.
(Chang sheng i wai)
掌聲以外 / 尤增輝著. — 臺中：藍燈文化事業股份有限公司，民國65 [1976]
2, 237 p. ; 19 cm. — (青年文庫 ; 5)
Short stories.
Series romanized: Ch'ing nien wen k'u
NT$40.00

I. Title.
PL2928.T73C4 78-841347

Yu, Tseng-hui.
(Jung chen ch'un hsing)
榕鎮春醒 / [著作人尤增輝]. — 臺中：臺灣省政府新聞處，民國66 [1977]
210 p. : port. ; 19 cm. — (省政文藝叢書 ; 68)
Fiction.
Series romanized: Sheng cheng wen i ts'ung shu.

I. Title.
PL2928.T73J8 78-843289

Yü-wen, Yao-chi.
(Ssu lin)
死林 / [宇文瑤璣著]. — 台北：春秋出版社，民國61 [1972]
2 v. (751 p.) ; 19 cm.
Cover title.
Fiction.

I. Title.
PL2928.Z9S68 79-837858

Yün, Ching.
(Feng ling ti hu huan)
風鈴的呼喚 / 雲菁著. — 臺北：皇冠雜誌社，民國68 [1979]
330 p. ; 20 cm.
Fiction.
NT$60.00

I. Title.
PL2929.N13F38 79-840169

Yün, Ching.
(Yüeh erh wan wan)
月兒彎彎 / 雲菁著. — 臺北：皇冠出版社，民國66 [1977]
753 p. ; 20 cm.
Fiction.
NT$100.00

I. Title.
PL2929.N13Y8 79-835501

—T'ANG DYNASTY, 618-907

(Tangdae chŏn'gi sosŏlsŏn)
唐代傳奇小說選 / 丁範鎭譯 ; [韓國中國學會編]. — 서울：汎學圖書，1975.
246 p. ; 23 cm.
Chinese and Korean.

1. Chinese fiction—T'ang dynasty, 618-907. 2. Chinese fiction—Translations into Korean. 3. Korean fiction—Translations from Chinese. I. Chŏng, Pŏm-jin, 1935- II. Han'guk Chunggŭk Hakhoe.
PL2641.T365 78-826784

—T'ANG DYNASTY, 618-907
—HISTORY AND CRITICISM

Kondō, Haruo, 1914-
(Tōdai shōsetsu no kenkyū)
唐代小説の研究 / 近藤春雄著. — 東京：笠間書院，1978.
457, 9 p. ; 22 cm. — (笠間叢書 ; 111) Ja 79-5156
Series romanized: Kasama sōsho.
Includes bibliographical references and index.
¥10000

1. Chinese fiction—T'ang dynasty, 618-907—History and criticism.
I. Title.
PL2431.K6 79-808864

Yi, Tong-hyang.
(T'ang tai ch'uan ch'i hsiao shuo ts'ung k'ao)
唐代傳奇小說叢考 / 李東鄉. — [台北? : s. n.], 民國59 [1970]
105 double leaves ; 27 cm.
"國立臺灣大學中文研究所碩士論文"
Cover title.
Reproduced from typescript copy.
Bibliography: leaves 102-105.
1. Chinese fiction—T'ang dynasty, 618-907—History and criticism.
I. Title.
PL2431.Y5 78-843914

—MING DYNASTY, 1368-1644
—HISTORY AND CRITICISM

Li, Kuei-chu.
(Ming ch'uan ch'i so chien ti Chung-kuo nü hsing)
明傳奇所見的中國女性 / 李桂柱. — [台北? : s. n., 1978?]
96 double leaves ; 27 cm.
Reproduced from ms. copy.
Thesis (M.A.)—Kuo li T'ai-wan ta hsüeh, 1970.
Bibliography: leaves 94-96.

1. Chinese fiction—Ming dynasty, 1368-1644—History and criticism. 2. Women in literature. I. Title.
PL2436.L5 79-835005

Shuang, I.
(San yen hsüan t'an)
「三言」選談 / 雙翼編著. — 香港：上海書局，1977.
5, 273 p. ; 19 cm.
$7.30

1. Feng, Meng-lung, 1574?-1645? Ching shih t'ung yen. 2. Feng, Meng-lung, 1574?-1645? Hsing shih heng yen. 3. Feng, Meng-lung, 1574?-1645? Yü shih ming yen. 4. Chinese fiction—Ming dynasty, 1368-1644—History and criticism. I. Title.
PL2436.S5 79-835455

CHINESE FICTION (Continued)

—CH'ING DYNASTY, 1644-1912
—HISTORY AND CRITICISM

Lin, Jui-ming.
(Wan Ch'ing ch'ien tse hsiao shuo ti li shih i i)
晚淸譴責小說的歷史意義 / 林瑞明. — ［臺北 : s. n.］, 民國65 ［1976］
2 v. (283 p.) on double leaves ; 27 cm.
Reproduced from typescript.
"國立臺灣大學歷史學研究所碩士論文"
Bibliography: p. 271-283.

1. Chinese fiction—Ch'ing dynasty, 1644-1912—History and criticism. I. Title.
PL2437.L5 79-841859

—20TH CENTURY

(Chan tou ying hsiung ku shih hsüan)
战斗英雄故事选 / 中国人民解放军总政治部, 人民出版社编. — 第1版. — 北京 : 人民出版社 : 新華書店发行, 1978.
3, 537 p. : ill. ; 19 cm.
$0.98

1. Short stories, Chinese. 2. Chinese fiction—20th century. I. China (People's Republic of China, 1949-). Chung-kuo jen min chieh fang chün. Tsung cheng chih pu. II. Jen min ch'u pan she, Peking.
PL2653.C4194 79-839473

(Chan tou ying hsiung ku shih hsüan chi)
戰鬥英雄故事選集 / ［編輯者山東人民出版社］. — ［香港? : s. n., 1978?］
50 p. ; 18 cm.
Reprint of the 1952 ed. published by Shan-tung jen min ch'u pan she, Tsinan.
CONTENTS: 青年英雄營長劉奎基.—黎明 射擊英雄魏來國.—智勇雙全的郭繼勝.—嚴寒 記青年戰鬭英雄林茂成同志.—張明 你的事蹟永遠留在我們心裏.—青年指揮員張明.—戰鬥英雄劉從信.—邵荫 戰鬥英雄王國田.
1. Short stories, Chinese. 2. Chinese fiction—20th century. I. Shan-tung jen min ch'u pan she, Tsinan, China.
PL2653.C4195 1978 79-839604

(Ch'i lo t'o ti jen)
騎骆驼的人 : 短篇小说选. — ［香港? : s. n., 1978?］
378 p. ; 21 cm.
Short stories.
Reprint of the 1977 ed. published by Nei Meng-ku jen min ch'u pan she, Hu-ho-hao-t'e.

1. Short stories, Chinese. 2. Chinese fiction—20th century.
PL2653.C51261 1978 79-837111

(Chin tai Chung-kuo chien tieh ku shih)
近代中國間諜故事 : ［1921-1971 / 名家聯合執筆］. — ［台北］: 新萬象月刊社, 民國67- ［1978-
v. ; 20 cm. — (新萬象月刊叢書 ; 3)
Series romanized: Hsin wan hsiang yüeh k'an ts'ung shu.
NT$50.00 (v. 1)

1. Spy stories, Chinese. 2. Chinese fiction—20th century. I. Title: Chung-kuo chien tieh ku shih.
PL2653.C514775 79-839550

(Chin tai ming chia chieh tso hsüan)
近代名家傑作選 / 巴金等著. — 香港 : 南國出版社, 1977.
2, 244 p. ; 19 cm.
Short stories.
$3.80

1. Short stories, Chinese. 2. Chinese fiction—20th century. I. Li, Fei-kan, 1905-
PL2652.C4 79-835231

(Ch'ing chu t'u ts'ui)
青竹吐翠 : 短篇小说集 / 厦门大学中文系七二级工农兵学员编. — ［香港? : s. n., 1977?］
274 p. : ill. ; 22 cm.
Reprint of the 1975 ed. published by Fu-chien jen min ch'u pan she.
CONTENTS: 刘胥 新的里程.—张永和 开春时节.—王敏 机播风云.—陈乙森 铁道哨兵.—方叶 新主任.—何端端 青竹吐翠.—林正让 主人.—汤滔 新书记上任.—林荔 春风吹放校园花.—陈乙森 列车飞驰.—郑以灵 迎风挺拔.—曾宪沧 灿烂征途第一步.—李军 移山造田的风波.—李启宇 这里也是战场

1. Short stories, Chinese. 2. Chinese fiction—20th century. I. Hsia-men ta hsüeh Chung wen hsi ch'i erh chi kung nung ping hsüeh yüan.
PL2653.C514867 1977 78-841033

(Ching-kang feng huo)
井冈烽火 / 井冈山地区文教局编. — ［香港? : s. n., 1978?］
2, 286 p., ［12］ leaves of plates : ill. ; 21 cm.
At head of title: Ching-kang shan tou cheng ku shih.
Reprint of the 1977 ed. published by Chiang-hsi jen min ch'u pan she, Nan-ch'ang.

1. Short stories, Chinese. 2. Chinese fiction—20th century. 3. Ching-kang Shan region—History—Fiction. I. Ching-kang shan ti ch'ü, China. Wen chiao chü.
PL2653.C51487 1978 79-836067

(Ch'üan kuo tso chia hsüan chi)
全國作家選集 / 編著者符兆祥. — 台北 : 全國出版社, 民國66 ［1977］
2 v. : ports. ; 20 cm.
Colophon title.
NT$35.00 per vol.

1. Short stories, Chinese—Taiwan. 2. Chinese fiction—20th century. I. Fu, Chao-hsiang.
PL3031.T32C524 79-837117

(Ch'uang yeh chin hsing ch'ü)
创业进行曲 / 本社编. — ［香港 : s. n., 1978?］
2, 214 p. ; 21 cm.
Reprint of the 1977 ed. published by Shang-hai jen min ch'u pan she, Shanghai.

1. Short stories, Chinese—China—Ta-ch'ing. 2. Chinese fiction—20th century. I. Shang-hai jen min ch'u pan she.
PL3031.T242C5 1978 78-843133

(Chun hsing)
准星 : 广州民兵革命斗争故事集 / 中国人民解放军广州警备区政治部編. — ［香港 : s. n., 1978?］
158 p. : ill. ; 21 cm.
Reprint of the 1977 ed. published by Kuang-tung jem nin ch'u pan she.
CONTENTS: 邓华棠 神奇的武器.—陈国凯, 江夏 老民兵智捉大老鼠.—钟闻远 铁流.—彭尔清 红霞满天.—戴胜德 准星.—孫吴远, 陈国凯, 彭尔清 攻关.—区荣光 伶仃洋上捉"蟹"记.—肖敏刚 出车之前.—陈柏坚, 志彤 节日的喜剧.—丁培强 特殊的岗位.—韦庆忠, 吴永顺 大桥头的故事.
1. Canton, China—History—Fiction. 2. Short stories, Chinese—China—Canton. 3. Chinese fiction—20th century. I. China (People's Republic of China, 1949-). Chung-kuo jen min chieh fang chün. Kuang-chou ching pei ch'ü. Cheng chih pu.
PL3031.C362C47 1978 79-836024

(Chung-kuo hsien tai hsiao shuo hsüan chiang)
中国現代小説選講. — 香港 : 港青出版社, 1978.
252 p. ; 19 cm.
CONTENTS: 魯迅 故鄉 ; 祝福 ; 傷逝.—郭沫若 岐路.—蔣光慈 鴨綠江上.—葉聖陶 抗爭.—郁達夫 薄奠.—王統照 湖畔兒語.—許欽文 石宕.—鄭伯奇 最初之課.—共庵 一個貞烈的女孩子.—利民 三天勞工底自述.—高世華 沉自己的船.—王思玷 偏枯
$7.00

1. Short stories, Chinese. 2. Chinese fiction—20th century.
PL2653.C5765 78-843594

(Fei ping ch'ih shui)
飞兵赤水 : 短篇小説集 / ［《飞兵赤水》编辑组编］. — ［香港 : s. n., 1978?］
317 p. : ill. ; 21 cm.
Reprint of the 1977 ed. published by Ssu-ch'uan jen min ch'u pan she, Ch'eng-tu.
CONTENTS: 陈国光 号令.—玉川 歌声.—李天雄 奇袭赤水镇.—钟明冰 当红军去.—谭远身 渔家女.—李相友 根深叶茂.—王均道 苗家兄弟.—王旋 铁锤连长.—张方正 苗民支队.—卢家贵 "盼"大爹.—王均道, 谭远身 风雪路上.—钟明冰 夺舟記.—李天雄 赤河铺路.—王川告示.
1. Short stories, Chinese. 2. Chinese fiction—20th century. I. Fei ping ch'ih shui pien chi tsu.
PL2653.F354 1978 78-840912

(Feng huo sui yüeh)
烽火岁月 / 彭寿生, 舒兴等口述. — ［香港? : s. n., 1977?］
157 p., ［6］ leaves of plates : ill. ; 21 cm.
Reprint of the 1976 ed. published by Chiang-hsi jen min ch'u pan she, Nan-ch'ang.

1. China (People's Republic of China, 1949-). Chung-kuo jen min chieh fang chün—Fiction. 2. Chinese fiction—20th century. I. P'eng, Shou-sheng. II. Shu, Hsing.
PL2653.F437 1977 79-835691

79 C

(Feng yü t'ung chou chan k'uang t'ao)
风雨同舟战狂涛 : 抗洪斗争英雄录 / 本社編. — ［香港? : s. n., 1978?］
3, 2, 200 p. : ill. ; 21 cm.
Reprint of the 1964 ed. published by Ho-pei jen min ch'u pan she, Tientsin.

1. Short stories, Chinese. 2. Chinese fiction—20th century. I. Ho-pei jen min ch'u pan she.
PL2653.F48 1978 79-840013

79 C

Hsia-men ta hsüeh Chung wen hsi ch'i erh chi kung nung ping hsüeh yüan.
(T'e pieh hang ch'eng)
特别航程 : 前线民兵斗争故事 / 厦门大学中文系七二级工农兵学员编写. — ［香港? : s. n., 1978?］
165 p. : ill. ; 21 cm.
Reprint of the 1975 ed. published by Fu-chien jen min ch'u pan she, Foochow.
CONTENTS: 打坦克.—阿龙嫂.—特别航程.—惊堂木的故事.—海岛红鹃.—代理班长.—三闯狼牙滩.—崇山常青水长绿.—深山炮响.—熊熊的火焰

1. Short stories, Chinese. 2. Chinese fiction—20th century. I. Title.
PL2653.H698 1978 78-842665

78 C

(Hsiao shuo hsüan chi)
小説選集 / 李南衡主編, ［助理編輯林邊, 梁景峯］. — 台北 : 明潭出版社, 民國68 ［1979］
2 v. ; 22 cm. — (日據下台灣新文學 : 明集 ; 2-3)
NT$250.00 ($10.00 U.S.) per vol.

1. Short stories, Chinese—Taiwan. 2. Chinese fiction—20th century. I. Li, Nan-heng. II. Lin, Pien. III. Liang, Ching-feng. IV. Series: Jih chü hsia T'ai-wan hsin wen hsüeh : Ming chi ; 2-3.
PL3031.T32H68 79-840154

78 C

(Hsin jen hsiao shuo hsüan)
新人小説選 / 亦舒等著. — 九龍 : 友聯出版社, 1971 printing.
4, 2, 198 p. ; 16 cm. — (友聯文庫 ; 10)
Series romanized: Yu lien wen k'u.

1. Short stories, Chinese. 2. Chinese fiction—20th century. I. I, Shu.
PL2653.H7834 79-835896

79 C

(Hsin ti chan tou)
新的战斗. — 第1版. — 北京 : 北京人民出版社, 1976.
265 p. ; 19 cm.
CONTENTS: 北师大中文系七三级三班 新的战斗.—窦孝鹏 人民的战士.—阎丰乐, 曹秀清 父亲.—陈建功 算账.—赵巨鹏 新来的副书记.—张建军, 张树杞 清积雪.—李国馥 草原的女儿.—毋国政 北疆风雪.—北师大中文系七三级三班 心愿.—于文香 大虎伏"虎".—李存葆 蜜桃花开.—梁向东 一丝不苟.—马霄 扣子.—陈昌本 龙爪松下.
$0.51
1. Short stories, Chinese. 2. Chinese fiction—20th century.
PL2653.H78695 78-843125

CHINESE FICTION —20TH CENTURY (Continued)

(Hsin yüeh hsiao shuo hsüan)
新月小說選 / ［編輯者梁實秋，葉公超］. — 臺北：雕龍出版社，民國67 ［1978］
4, 3, 450 p. ; 22 cm. — (雕龍文學叢書 ; 2)
Series romanized: Tiao lung wen hsüeh ts'ung shu.
NT$95.00
1. Short stories, Chinese. 2. Chinese fiction—20th century. I. Liang, Shih-ch'iu, 1901- II. Yeh, Kung-ch'ao, 1904-
PL2652.H828 79-836388

Hu-t'ou shan hsia tou Chiang yao pien hsieh tsu.
(Hu-t'ou shan hsia tou Chiang yao)
虎头山下斗江妖 / 《虎头山下斗江妖》编写组编. — 第1版. — 上海：上海人民出版社，1977.
2, 33 p. ; 18 cm.
Colophon title.
$0.10
1. Short stories, Chinese. 2. Chinese fiction—20th century. 3. Chiang, Ch'ing—Fiction. I. Title.
PL2653.H79693 1977c 78-841506

(Hua man shan ch'eng)
花滿山城 / ［著作人陳正之等］. — 臺中：臺灣省政府新聞處，民國67 ［1978］
248 p. : ill. ; 19 cm. — (省政文藝叢書 ; 70)
Series romanized: Sheng chen wen i ts'ung shu.
1. Short stories, Chinese—Taiwan. 2. Chinese fiction—20th century. I. Ch'en, Cheng-chih, fl. 1978-
PL3031.T32H77 79-836999

(Hung ch'i ju hua)
红旗如画：学大寨的故事. — 第1版. — ［广州］：广东人民出版社，1977.
156 p. : ill. ; 19 cm.
CONTENTS: 龙奇, 黄向农, 余松岩 华主席率领我们勇登攀.—徐一行 贡献.—陈堪进 新寮新天地.—郑文贤 飞沙走石.—吴茂信, 郑文贤 英雄兄妹.—黄循洛 宏图.—黄森涛, 许本立, 张奇杰 奋战狮子山.—陈泽, 朱飞 水寨春秋.—赵澄卓, 阳秀 白沙湾凯歌.—阳秀, 朱飞 开辟新渔场.—程贤章, 凌霄 水神爷爷.
$0.33
1. Short stories, Chinese. 2. Chinese fiction—20th century.
PL2653.H826 78-842941

(Huo, t'ung hung ti huo)
火·通红的火. — 上海：上海人民出版社，1976.
376 p. : ill. ; 21 cm. — (朝霞丛刊)
Series romanized: Chao hsia ts'ung k'an.
CONTENTS: 中篇小说 朱敏慎 初战—段荃法, 张有德, 樊俊智 青砖歌—话剧剧本 陆天明 火·通红的火四幕话剧·征文选刊
$0.82
1. Chinese fiction—20th century. 2. Chinese drama—20th century. I. Lu, T'ien-ming. II. Chu, Min-shen. Ch'u chan. 1976. III. Tuan, Ch'üan-fa. Ch'ing chuan ko. 1976.
PL2653.H88 79-835741

(K'ai lu hsien feng)
开路先锋：报告文学 / 江西九〇九地质队工人写作组. — ［香港?：s.n., 1977?］
197 p., ［3］ leaves of plates : ill. ; 19 cm.
Reprint of the 1972 ed. published by Chiang-hsi jen min ch'u pan she, Nan-ch'ang.
CONTENTS: 地下宝藏侦察兵—红鹰展翅—钻机高歌—踏遍青山—突破—忠诚的战士—向红层进军—争气矿—并肩战斗—龙腾虎跃—光辉的道路—后記
1. Short stories, Chinese—China—Kiangsi (Province) 2. Chinese fiction—20th century. I. Chiang-hsi chiu ling chiu ti chih tui. Kung jen hsieh tso tsu.
PL3031.K52K34 1977 79-835424

Ku, Ying.
(Ch'i yüan kuo pao)
奇冤果報 / 谷影著. — ［台北：德昌出版社，民國66 i.e. 1977］
185 p. ; 19 cm. — (奇異鬼狐集錦故事)
"奇異鬼狐小說"
Cover title.
CONTENTS: 鬼宅.—再世鴛盟.—鬼兒子.—魔影
NT$30.00
1. Ghost stories, Chinese. 2. Chinese fiction—20th century. I. Title. II. Series: Ch'i i kuei hu chi chin ku shih (T'ai-pei ed.)
PL2871.Y56C5 79-837957

(Li shih fu tai t'u ti)
李师傅带徒弟：革命故事集 / 肇庆地区文化局编. — ［香港?：s.n., 1978?］
84 p. ; 21 cm.
Reprint of the Kuang-tung jen min ch'u pan she, Canton.
CONTENTS: 关夕芝 小球迷和大西瓜.—范卫宁 哑炮的秘密.—严瑞清 黎晓 养鸡风波.—林镒 大傻瓜的故事.—陈锦润 李师傅带徒弟.—胡思源 巧换孙儿说家公.—李三元 猪栏栅的故事.—郭添全 伏虎记.—梁驹 火眼金睛
1. Short stories, Chinese. 2. Chinese fiction—20th century. I. Chao-ch'ing ti ch'ü, China. Wen hua chü.
PL2653.L5317 1978 79-835255

(Liao-ho feng huo)
辽河烽火：辽宁民兵革命斗争故事集 / 辽宁省军区政治部编. — ［香港：s.n., 1977?］
138 p. : ill. ; 22 cm.
Reprint of the 1973 ed. published by Liao-ning jen min ch'u pan she, Mukden.
1. Liaoning, China (Province)—History—Fiction. 2. Short stories, Chinese—China—Liaoning. 3. Chinese fiction—20th century. I. China (People's Republic of China, 1949-). Chung-kuo jen min chieh fang chün. Liao-ning sheng chün ch'ü. Cheng chih pu.
PL3031.L52L5 1977 78-841198

(Lien ho pao liu ch'i nien tu hsiao shuo chiang tso p'in chi)
聯合報六七年度小說獎作品集 / ［張子樟等著］. — 臺北：聯合報社：總經銷聯經出版事業公司，民國67 ［1978］
2, 32, 369 p. ; 22 cm. — (聯合報叢書)
Series romanized: Lien ho pao ts'ung shu.
NT$80.00
1. Short stories, Chinese. 2. Chinese fiction—20th century. I. Chang, Tzu-chang. II. Title: Hsiao shuo chiang tso p'in chi.
PL2653.L5372 79-841424

(Ma-an ti pao kao)
马安的報告. — ［香港?：s.n., 1979?］
179 p., ［6］ leaves of plates : ill. ; 21 cm.
Reprint of the 1975 ed. published by Kuang-tung jen min ch'u pan she.
CONTENTS: 藍放 反修防修的坚强战士.—王为, 肖玲 红旗飘揚在马安山.—陈甲优 矿山尖兵.—杜峻 火红的旗帆.—王旺, 胡天佑, 王振生 闪亮的乌金.—曾德中 金色的道路.—姚涌, 谢作尧, 廉海生 汽笛欢歌.—王稔 红色管家人.—欧阳斌, 林仕权 不卷刃的尖刀.—杨昭科, 王稔 矿工的贴心人.—欧阳斌 攻关记.—龚夫, 陈佑 煤海浪花.
1. Ma-an mei k'uang—Fiction. 2. Chinese fiction—20th century.
PL2653.M23 1979 79-841695

(Man t'ien fei hsia)
满天飞霞 = Mantianfeixia / 大港油田政治部编. — 第1版. — 天津：天津人民出版社，1975.
234 p., ［3］ leaves of plates : ill. ; 19 cm. — (天津工人创作丛书)
CONTENTS: 宋新英 油田的早晨.—许国英 特别护士.—金安福 满天飞霞.—单士航 油田参谋.—韩效闵 心向大海.—刘元章, 高飞 关书记.—许鹏銮 战士.—赵凤珍, 晓山 带班.—曾望贤 油田人家.—欣英, 世航 油田春潮.—赵慧敏, 张宝生 任务面前.
1. Short stories, Chinese. 2. Chinese fiction—20th century. I. Ta-kang yu t'ien cheng chih pu. II. Title: Mantianfeixia.
PL2653.M34 1975 79-840119

(Min hai chi lang)
闽海激浪：福建民兵革命斗争故事集 / 福建省军区政治部编. — 第1版. — ［福州?］：福建人民出版社，1977.
2, 507 p. : ill. ; 19 cm.
$0.92
1. Short stories, Chinese. 2. Chinese fiction—20th century. 3. Fukien, China (Province)—Militia—Fiction. I. China (People's Republic of China, 1949-). Chung-kuo jen min chieh fang chün. Fu-chien sheng chün ch'ü. Cheng chih pu.
PL2653.M5 79-837789

(San ma ju yü)
三馬入峪 / 段彩華等著. — 再版. — 臺北：黎明文化事業公司，民國66 ［1977］
110 p. ; 20 cm. — (金像獎叢書)
Series romanized: Chin hsiang chiang ts'ung-shu.
CONTENTS: 臧冠華 硬漢.—何祖武 兩伙伴.—曹抄 沉舟.—段彩華 三馬入峪.—楊御龍 小材大用.
NT$20.00
1. Short stories, Chinese. 2. Chinese fiction—20th century. I. Tuan, Ts'ai-hua, 1933-
PL2653.S319 78-842956

(Shen sheng ti shih ming)
神圣的使命 / 本社编. — 第1版. — 上海：上海文艺出版社：新華書店上海发行所发行，1979.
348 p. ; 21 cm.
$0.91
1. Short stories, Chinese. 2. Chinese fiction—20th century. I. Shang-hai wen i ch'u pan she.
PL2653.S482 79-841875

(Shuang ch'ien niu)
双牵牛：河北遵化农业机械化故事集 / 河北遵化县文化馆编. — ［香港：s.n., 1978?］
179 p. : ill. ; 21 cm.
Reprint of the 1977 ed. published by Jen min wen hsüeh ch'u pan she, Peking.
CONTENTS: 化宣 迎着斗争的风浪, 前进!.—高集夫 双牵牛.—胡天启 农机姓农.—王树丰 小厂飞出金凤凰.—姚学忠 铁师傅夺钢.—赵宗金 火线重逢.—李树州 开快车的人.—李国 红管家新传.—徐浦森 取經送宝.—姚学忠 雪地遥控.—高夫集 新来的女机手.—姚学忠 孔雀开屏.—高夫集 雹灾之后.—张连瑞 一架土播机的誕生.
1. Short stories, Chinese. 2. Chinese fiction—20th century. I. Ho-pei Tsun-hua hsien wen hua kuan. II. Title: Ho-pei Tsun-hua nung yeh chi hsieh hua ku shih chi.
PL2653.S54 1978 79-836179

Ssu-ch'uan ta hsüeh, Chengtu, China. Chung wen hsi.
(Hsüeh hai sheng ch'ou shou tsu yüan)
血海深仇收租院：四川省大邑贫下中农家史 / 四川大学中文系，四川省大邑阶级教育展览馆编. — ［香港?：s.n., 1978?］
4, 129 p. : ill. ; 20 cm.
Reprint of the 1975 ed. published by Ssu-ch'uan jen min ch'u pan she, Chen-tu.
$25.00
1. Szechwan, China—History—Fiction. 2. Short stories, Chinese—China—Szechwan. 3. Chinese fiction—20th century. I. Ssu-ch'uan sheng Ta-i chieh chi chiao yü chan lan kuan. II. Title.
PL3031.S92S78 1978 79-835859

(T'ien yüan ch'un hsiao)
田園春曉 / ［著作人楊御龍等］. — 臺中：臺灣省政府新聞處，民國67 ［1978］
238 p. : ports. ; 20 cm. — (省政文藝叢書 ; 69)
Series romanized: Sheng cheng wen i ts'ung shu.
CONTENTS: 楊御龍 田園春曉.—羅枝土 創業.—朱成才 重整家園.—唐震寰 鬼妻.—黃中全 吾愛吾村.—廖妙麗 失足幽蘭.—左堯安 阿秀.—王映湘 上山村的龍脈.—常劍飛 正義村的春天.—呂燕珠 雙連渡.—劉華齋 開路風波.—周一之 源遠流長.—余伯牙 雙喜臨門.—翁麗華 綠夏.
1. Short stories, Chinese. 2. Chinese fiction—20th century. I. Yang, Yü-lung, 1929-
PL2653.T545 79-836268

(T'ing-chiang ch'ih wei tui)
汀江赤卫队 / 龙岩地区文化局，龙岩军分区政治部编. — ［香港?：s.n., 1978］
204 p. : ill. ; 21 cm.
Reprint of the ed. published by Fu-chien jen min ch'u pan she, Foochow.
1. Short stories, Chinese. 2. Chinese fiction—20th century. I. Lung-yen ti ch'ü, China. Wen hua chü. II. China (People's Republic of China, 1949-). Chung-kuo jen min chieh fang chün. Lung-yen chün fen ch'ü. Cheng chih pu.
PL2653.T55 1978 79-835709

(Tsai chi liu chung)
在激流中 / 天津市化工局政治部編. — 第1版. — 天津：天津人民出版社，1976.
255 p. : ill. ; 19 cm. — (天津工人创作丛书)
$0.48
1. Short stories, Chinese. 2. Chinese fiction—20th century. I. Tientsin. Hua kung chü. Cheng chih pu. II. Series: T'ien-chin kung jen ch'uang tso ts'ung shu.
PL2653.T573 79-837319

(Ts'ui ho chi)
淬火集 / 中共第一重型机器厂委员会政治部［編］. — 第1版. — 北京：人民文学出版社，1977.
5, 220 p. ; 19 cm.
CONTENTS: 程树榛 期望.—王修亚 搬山记.—陶世才 三访巴特热.—马小平 师傅.—边祥林 路.—张连儒 探伤.—刘学秀 擎天柱.—王忠侠 张春生 闯炉长.—陈沫 加油.—聂振亚 嘱托.—王治新 酒的故事.—兼良 金钥匙.—王治新 新的阵地.
$0.43
1. Short stories, Chinese. 2. Chinese fiction—20th century. I. Chung kung ti i chung hsing chi ch'i ch'ang. Cheng chih pu.
PL2653.T837 78-842807

CHINESE FICTION
—20TH CENTURY (Continued)

(Tuan p'ien hsiao shuo hsüan)
短篇小說選 / 劉守宜主編. — 臺北：聯經出版事業公司, 民國67 [1978]
4 v. ; 22 cm.
At head of title: Wen hsüeh tsa chih tso p'in chi.
NT$375.00

1. Short stories, Chinese. 2. Chinese fiction—20th century. I. Liu, Shou-i. II. Title: Wen hsüeh tsa chih tso p'in chi.
PL2653.T845 79-835999

(Tung wu yü wo)
動物與我：[丹華文學獎第二輯] / 曾炤等著. — 九龍：高原出版社, 1978.
4, 5, 229 p. ; 19 cm. — (種子文藝叢書)
Series romanized: Chung tzu wen i ts'ung shu.
$5.00

1. Short stories, Chinese. 2. Chinese fiction—20th century. I. Tseng, Yen.
PL2653.T89 79-836082

(Wan ch'üan ho p'an)
万泉河畔 / 中国人民解放军工程兵政治部编. — 天津：天津人民出版社, 1976.
384 p. ; 19 cm.
Title also in Pinyin on cover: Wan quan he pan.
$0.72

1. Short stories, Chinese. 2. Chinese fiction—20th century. I. China (People's Republic of China, 1949–). Chung-kuo jen min chieh fang chün. Kung ch'eng ping cheng chih pu. II. Title: Wan quan he pan.
PL2653.W28 1976 79-835752

(Wei ta ti shih yeh)
伟大的事业：短篇小说集 / 山东文艺社编. — [香港?：s. n., 1979?]
285 p. ; 22 cm.
Short stories.
Reprint of the 1978 ed. published by Shan-tung jen min ch'u pan she, Chi-nan.

1. Short stories, Chinese. 2. Chinese fiction—20th century. I. Shan-tung wen i she.
PL2653.W347 1979 79-841871

(Wo hsü yao i fen ai)
我需要一份愛 / [編輯委員張家華, 雪柔, 陳美卿]. — 台北：龍風實業有限公司圖書出版部, 民國66 [1977]
237, 6 p., [1] leaf of plates : ports. ; 22 cm. — (龍風叢書 ; 2)
"當代散文, 詩歌, 小說選輯"
Series romanized: Lung feng ts'ung shu.
NT$60.00
1. Chinese essays—20th century. 2. Chinese poetry—20th century. 3. Chinese fiction—20th century. I. Chang, Chia-hua. II. Hsüeh-jou. III. Ch'en, Mei-ch'ing.
PL2623.W58 78-843124

(Ya-lü chiang nu t'ao)
鸭绿江怒涛 / 辽宁省军区政治部编. — [香港?：s. n., 1978?]
2, 258 p. : ill. ; 21 cm.
Reprint of the 1977 ed. published by Liao-ning jen min ch'u pan she, Mukden, in series: Liao-ning min ping ko ming tou cheng ku shih chi.

1. Short stories, Chinese. 2. Chinese fiction—20th century. 3. Korean War, 1950–1953—Fiction. I. China (People's Republic of China, 1949–). Chung-kuo jen min chieh fang chün. Liao-ning sheng chün ch'ü. Cheng chih pu. II. Series: Liao-ning min ping ko ming tou cheng ku shih chi.
PL2653.Y22 1978 79-836842

—20TH CENTURY—BIBLIOGRAPHY

Tsai, Meishi, 1942-
Contemporary Chinese novels and short stories, 1949-1974 : an annotated bibliography / by Meishi Tsai, with the assistance of I-mei Tsai. — Cambridge, Mass. : Council on East Asian Studies, Harvard University : distributed by Harvard University Press, 1979, c1978.
xvi, 408 p. ; 29 cm. — (Harvard East Asian monographs ; 78)
Includes indexes.
ISBN 0-674-16681-7 : $25.00
1. Chinese fiction—20th century—Bibliography. I. Title. II. Series.
Z3108.L5T78 1979 016.8951'3'09 78-15643
[PL2415] 78 MARC

—20TH CENTURY—HISTORY AND CRITICISM

Ch'un-hua.
(Hsiao ku shih hsieh tso tsa t'an)
小故事写作杂谈 / 春华, 晓阳编写. — [香港?：s. n., 1979?]
82 p. (on double leaves) ; 21 cm.
Reprint of the 1976 ed. published by Shang-hai jen min ch'u pan she, Shanghai.

1. Short stories, Chinese—History and criticism. 2. Chinese fiction—20th century—History and criticism. I. Hsiao-yang, joint author. II. Title.
PL2443.C48 1979 79-838724

Ch'un-hua.
(Hsiao ku shih hsieh tso tsa t'an)
小故事写作杂谈 / 春华, 晓阳编写. — [香港?：s. n., 1979?]
82 p. (on double leaves) ; 21 cm.
Reprint of the 1976 ed. published by Shang-hai jen min ch'u pan she, Shanghai.

1. Short stories, Chinese—History and criticism. 2. Chinese fiction—20th century—History and criticism. I. Hsiao-yang, joint author. II. Title.
PL2443.C48 1979 79-838724

Hsia, Chih-tsing, 1921-
A history of modern Chinese fiction / by C. T. Hsia. — 2d ed. — New Haven : Yale University Press, 1971.
xvi, 701 p. ; 21 cm.
First ed. published in 1961 under title: A history of modern Chinese fiction, 1917-1957.
Bibliography: p. 643-684.
Includes index.
ISBN 0-300-01462-7
1. Chinese fiction—20th century—History and criticism. I. Title.
PL2442.H8 1971 895.1'3'509 79-108225
79 MARC

—BIBLIOGRAPHY

(Pi chi hsiao shuo ta kuan tsung so yin)
筆記小說大觀總索引. — [台北：新興書局, 1978?]
123, 14, 4, 234 p. ; 22 cm.
Appendices: 新書簡報.—范希曾　書目答問補正.

1. Chinese essays—Bibliography. 2. Chinese fiction—Bibliography. 3. China—Imprints. I. Fan, Hsi-tseng, 1900–1930. Shu mu ta wen pu cheng. 1978.
Z3108.L5P5 79-836830
[PL2395]

Wang, I-chao, fl. 1972-
(Wang I-chao so ts'un Chung-kuo chiu t'ung su hsiao shuo shu mu)
王以昭所存中國通俗小說書目. — [臺北?：s. n., 1974?]
102 p. (on double leaves) ; 19 cm.
Cover title.

1. Chinese fiction—Bibliography. I. Title. II. Title: Chung-kuo t'ung su hsiao shuo shu mu.
Z3108.L5W28 1974 77-841365
[PL2415]
rev

—HISTORY AND CRITICISM
—ADDRESSES, ESSAYS, LECTURES

Hu, Chü-jen.
(Hung lou Shui hu yü hsiao shuo i shu)
紅樓水滸與小說藝術 / 胡菊人著. — 港1版. — 香港：百葉書舍, 1977.
253 p. ; 19 cm. — (百葉文叢)
Series romanized: Pai yeh wen ts'ung.
$8.00

1. Ts'ao, Chan, ca. 1717–1763. Hung lou meng. 2. Shui hu chuan. 3. Chinese fiction—History and criticism—Addresses, essays, lectures. I. Title.
PL2727.S2H747 78-841885

Sun, K'ai-ti, 1902-
(Hsiao shuo p'ang cheng)
小說旁证 / 孙楷第. — [香港?：s. n., 1978?]
30 p. ; 26 cm.
Photoreprint ed.

1. Chinese fiction—History and criticism—Addresses, essays, lectures. I. Title.
PL2416.S9 1978 78-843668

Wang, T'o, 1944-
(Chang Ai-ling yü Sung Chiang)
張愛玲與宋江 / 王拓著. — 臺中：藍燈文化事業股份有限公司, 民國65 [1976]
1, 3, 227 p. ; 19 cm. — (藍燈新刊 ; 2) (藍燈叢書)
Series 1 romanized: Lan teng hsin k'an. Series 2 romanized: Lan teng ts'ung shu.
NT$40.00

1. Chinese fiction—History and criticism—Addresses, essays, lectures. I. Title.
PL2416.W37 76-839874

—TRANSLATIONS FROM SPANISH

Cervantes Saavedra, Miguel de, 1547-1616.
[Don Quixote. Chinese]
堂吉河德 / 塞万提斯著 ; 杨绛译. — 第1版. — 北京：人民文学出版社, 1978.
2 v. : 21 cm.
Fiction.
Includes bibliographical references.
$2.55

I. Yang, Chiang. II. Title.
Title romanized: T'ang Chi-ho-te.
PQ6332.9.C5A2 1978 79-835032

—TRANSLATIONS INTO ENGLISH

Chang, Nai-ying, 1911-1942.
The field of life and death and tales of Hulan River : two novels / by Hsiao Hung [i.e. Chang Nai-ying]. — Bloomington : Indiana University Press, c1979.
xxvi, 291 p. ; 22 cm. — (Chinese literature in translation)
Translation of Shen ssu ch'ang translated by Howard Goldblatt and Ellen Yeung, and Hu-lan ho chuan translated by Howard Goldblatt.
Includes bibliographical references.
ISBN 0-253-15821-4
I. Chang, Nai-ying, 1911-1942. Hu-lan ho chuan. English. 1979. II. Title. III. Series.
PZ3.C36415 Fi 1979 895.1'3'5 78-19549
[PL2740.N3] 78 MARC

Ch'en, Jo-hsi.
The execution of Mayor Yin, and other stories from the great proletarian cultural revolution / Chen Jo-hsi ; translated from the Chinese by Nancy Ing and Howard Goldblatt. — Bloomington : Indiana University Press, c1978.
xxviii, 220 p. ; 22 cm. — (Chinese literature in translation)
CONTENTS: The execution of Mayor Yin.—Chairman Mao is a rotten egg.—Night duty.—Residency check.—Jen Hsiu-lan.—The big fish.—Keng Erh in Peking.—Nixon's press corps.
ISBN 0-253-12475-1
I. Title. II. Series.
PZ4.C5178 Ex 895.1'3'5 78-1956
[PL2840.J6] 78 MARC

[Hsiao-hsiao-sheng]
Cin p'ing mei; the adventurous history of Hsi Men and his six wives. With an introd. by Arthur Waley. London, Bodley Head [1959]
xxii, 852 p. 23 cm.
Translation of Chin P'ing Mei tz'ŭ hua.
"English translation by Bernard Miall from the abridged version by Franz Kuhn."
Attributed to Hsiao-hsiao-sheng. Cf. C. T. Hsia. The classic Chinese novel, 1968, p. 165-202.
Bibliographical references included in "Notes to the introduction": (p. xvii-xviii)
I. Title.
[PZ3.H85923 Ch 1959] 895.1'3'4 66-4000
[PL2698.H73] [r79]rev MARC

Li, Fei-kan, 1905-
Cold nights : a novel / by Pa Chin [i.e. F. Li] ; translated by Nathan K. Mao & Liu Ts'un-yan. — Hong Kong : Chinese University Press ; Seattle : University of Washington Press, 1978.
xxxiv, 181 p. : ill. ; 24 cm.
Translation of Han yeh.
Bibliography: p. 177-181.
I. Title.
PZ3.L612 Co 895.1'3'5 78-21758
[PL2780.F4] 79 MARC

CHINESE FICTION
—TRANSLATION INTO ENGLISH
(Continued)

Li, Fei-kan, 1905-
The family / by Pa Chin [i.e. F. Li] ; [translated by Sidney Shapiro ; ill. by Liu Tan-Chai]. — 3d ed. — Peking : Foreign Languages Press, 1978.
284 p., [7] leaves of plates : ill. ; 22 cm.
Translation of Chia.
The 1st vol. of the trilogy, Turbulent stream; the 2d vol. is Spring; the 3d, Autumn.
I. Title.
PZ3.L612 Fam 1978 895.1'3'5 79-105565
[PL2780.F4] 79 MARC

Ma, Pin.
The river flows east / by Nan-Kung Po [i.e. P. Ma]. — 1st ed. — Hong Kong : Rainbow Press, 1957.
186 p., [8] leaves of plates : ill. ; 19 cm.
Translation of Shui tung liu.
I. Title.
PZ4.M11125 Ri 895.1'3'5 78-109183
[PL2881.P5] 79 MARC

Shên, Yen-ping, 1896-
Midnight / Mao Tun [Shen Yen-ping ; ill. by Yeh Chien-yu]. — [1st AMS ed.] — New York : AMS Press, [1979]
524 p., [16] leavse of plates : ill. ; 22 cm.
Translation of Tzu yeh.
Reprint of the 1957 ed. published by Foreign Languages Press, Peking.
ISBN 0-404-14485-3
I. Title.
[PZ3.S54723 Mi 1979] 895.1'3'5 75-36237
[PL2801.N2] 79 MARC

Shu, Ch'ing-ch'un, 1898-1966.
Rickshaw = the novel Lo-t'o Hsiang Tzu / by Lao She [i.e. Shu Sh'ing-ch'un] ; translated by Jean M. James ; [maps by William C. Stanley]. — Honolulu : University Press of Hawaii, c1979.
xi, 249 p. : maps ; 22 cm.
ISBN 0-8248-0616-6
I. Title. II. Title: Lo-to Hsiang Tzu.
PZ3.S5619 Rh 895.1'3'5 79-10658
[PL2804.C5] 79 MARC

Ts'ao, Chan, ca. 1717-1763.
A dream of red mansions / Tsao Hsueh-chin [i.e. C. Ts'ao] and Kao Ngo ; [translated by Yang Hsien-yi and Gladys Yang ; illustrated by Tai Tun-pang]. — 1st ed. — Peking : Foreign Languages Press, 1978-
v. : ill. ; 22 cm.
Title also in Chinese on cover: Hung lou meng.
I. Kao, E., fl. 1792, joint author. II. Title.
PZ3.T783 Dq 1978 895.1'3'4 78-106814
[PL2727.S2] 79[r79]rev MARC

Ts'ao, Chan, 1717 (ca.)-1763.
The story of the stone : a Chinese novel in five volumes / by Cao Xueqin ; translated by David Hawkes. — Bloomington : Indiana University Press, [1979-
v. ; 24 cm. — (Chinese literature in translation)
Translation of Hung lou meng.
CONTENTS: v. 1. The golden days.
ISBN 0-253-19266-8 (set)
I. Hawkes, David. II. Title. III. Series.
PZ3.T783 St 1979 895.1'3'4 78-20279
[PL2727.S2] 78 MARC

Yen ling tui ti ku shih. English
Wild-goose guerrillas / by Liu Fu-hai and others ; [ill. by Hsin Ho-chiang]. — 1st ed. — Peking : Foreign Language Press, 1978.
142 p., [8] leaves of plates : ill. ; 19 cm.
Translation of Yen ling tui ti ku shih.
1. Sino-Japanese Conflict, 1937-1945—Fiction. I. Liu, Fu-hai.
PZ1.W654 1978 823 78-109631
[PL2925] 79 MARC

—TRANSLATIONS INTO FRENCH

Hao-jan, 1932-
Les enfants de Xisha : roman / Haoran ; traduit du chinois par Liang Paitchin ; avec une préf. par Michelle Loi. — Lausanne [7, rue de Genève] : Alfred Eibel, 1976.
310 p. ; 18 cm. — (La Chine d'aujourd'hui ; 1) Sw76-6779
Translation of Hsi-sha erh nü.
Bibliography: p. 293-309.
ISBN 2-8274-0002-2 : 15.10F
I. Title. II. Series.
PL2861.A76H7514 1976 895.1'3'5 78-377021
*78 MARC

Lo, Kuang-pin.
[Hung yen. Japanese]
紅岩 / 羅広斌, 楊益言著 ; 三好一訳. — 東京 : 講談社, 1978.
2 v. : 20 cm. Ja 78-13039
Fiction.
¥1200 per vol.

I. Yang, I-yen, joint author. II. Title.
Title romanized: Kōgan.
PL2880.O25H816 79-801289

—TRANSLATIONS INTO KOREAN

(Tangdae chŏn'gi sosŏlsŏn)
唐代傳奇小說選 / 丁範鎭譯 ; [韓國中國學會編]. — 서울 : 汎學圖書, 1975.
246 p. : 23 cm.
Chinese and Korean.

1. Chinese fiction—T'ang dynasty, 618-907. 2. Chinese fiction—Translations into Korean. 3. Korean fiction—Translations from Chinese. I. Chŏng, Pŏm-jin, 1935- II. Han'guk Chungguk Hakhoe.
PL2641.T365 78-826784

—TRANSLATIONS INTO RUSSIAN

Chin P'ing Mei tz'ŭ hua. Russian.
[Hsiao-hsiao-shêng]
T͡Svety slivy v zolotoĭ vaze : ili T͡Szin', Pin, Meĭ : Roman / Perevod s kit. V. Manukhina ; [Podgot. teksta L. Sycheva ; Komment. B. Riftina ; Khudozh. A. Lepiatskiĭ]. — Moskva : Khudozh. lit., 1977.
2 v. : ill. ; 21 cm. USSR78
Translation of Chin P'ing Mei tz'ŭ hua.
Attributed to Hsiao-hsiao-shêng. Cf. C. T. Hsia. The classic Chinese novel, 1968, p. 165-202.
2.10rub (v. 1) 2.40rub (v. 2)
I. Title.
PL2698.H73C518 79-364027
79 MARC

CHINESE FOLK LITERATURE see Folk literature, Chinese

CHINESE FOLK-LORE see Folk-lore, Chinese

CHINESE GARDENS see Gardens, Chinese

CHINESE GHOST STORIES see Ghost stories, Chinese

CHINESE HUMOROUS POETRY see Humorous poetry, Chinese

CHINESE IMPRINTS

—CATALOGS

Cornell University. Libraries.
The catalog of the Wason Collection on China and the Chinese, Cornell University Libraries / compiled and edited by Paul P. W. Cheng. — Washington : Center for Chinese Research Materials, Association of Research Libraries, [1978-
v. : port. ; 39 cm. — (Bibliographical series - Center for Chinese Research Materials, Association of Research Libraries ; no. 17a)
Includes index.
CONTENTS: pt. 1. Serials catalog.
1. China—Bibliography—Catalogs. 2. Chinese imprints—Catalogs. 3. Wason, Charles William, 1854-1918—Library—Catalogs. I. Cheng, Paul P. W. II. Title. III. Series: Association of Research Libraries. Center for Chinese Research Materials. Bibliographical series ; no. 17a-.
Z3109.C67 1978 015.51 79-113498
[DS706] 00 MARC

Hsing cheng yüan chu chi ch'u tien tzu ch'u li tzu liao chung hsin.
(Hsing cheng yüan chu chi ch'u tien tzu ch'u li tzu liao chung hsin Chung wen t'u shu mu lu)
行政院主計處電子處理資料中心中文圖書目錄. — [台北] : 該中心, 民國66 [1977]
134 p. : 27 cm.

1. Chinese imprints—Catalogs. 2. Hsing cheng yüan chu chi ch'u tien tzu ch'u li tzu liao chung hsin. Library—Catalogs. I. Title: Chung wen t'u shu mu lu.
Z955.T33325H84 1977 78-843005

Koryŏ Taehakkyo, Seoul, Korea. Chungang Tosŏgwan.
(Kyŏnghwadang mun'go mongnok)
景和堂文庫目録 = Catalogue of the Gyeong-hwa-dang collection. — [서울] : 高麗大學校中央圖書館, 1975.
138 p., [1] leaf of plates : port. : 26 cm. — (高麗大學校藏書目録 ; 第12輯)
Includes index.
1. Chinese imprints—Catalogs. 2. China—Imprints—Catalogs. 3. Bibliography—Rare books—Catalogs. 4. Pak, Hyŏng-yun—Library—Catalogs. 5. Koryŏ Taehakkyo, Seoul, Korea. Chungang Tosŏgwan—Catalogs. I. Title. II. Title: Catalogue of the Gyeong-hwa-dang collection. III. Series: Koryŏ Taehakkyo, Seoul, Korea. Koryŏ Taehakkyo changsŏ mongnok ; che 12-chip.
Z3316.K69 1975 78-826800

T'ai-wan sheng li T'ai-chung t'u shu kuan.
(T'ai-wan sheng li T'ai-chung t'u shu kuan t'u shu mu lu)
臺灣省立臺中圖書館圖書目錄. — [台中] : 該館, 民國60- [1971-
v. ; 27 cm.
Cover title.

1. Chinese imprints—Catalogs. 2. T'ai-wan sheng li T'ai-chung t'u shu kuan—Catalogs. I. Title.
Z955.T3338T35 1971 78-842950

—JAPAN—CATALOGS

Kotohiragū, Kotohira-chō, Japan. Toshokan.
(Kotohiragū Kanseki jun Kansekibunrui mokuroku kō)
金刀比羅宮漢籍準漢籍分類目錄稿 / [長澤規矩也, 松原秀明, 長澤ちづ共編]. — [琴平町(香川県)] : 金刀比羅宮圖書館, [1977]
41 p. : 21 cm. Ja 78B-3014

1. Chinese imprints—Japan—Catalogs. 2. Kotohiragū, Kotohira-cho, Japan—Catalogs. 3. Japan—Imprints—Catalogs. I. Nagasawa, Kikuya, 1902- II. Matsubara, Hideaki. III. Nagasawa, Chizu. IV. Title.
Z3101.K645 1977 78-814769

Temmagū, Osaka.
(Ōsaka Temmangū gobunko Kanseki bunrui mokuroku)
大阪天満宮御文庫漢籍分類目録 / [長澤規矩也, ちづ編]. — 大阪 : 大阪天満宮, 1977.
4, 81, 14 p., [2] leaves of plates : ill. ; 26 cm. Ja 78-4494
Includes index.

1. China—Imprints—Catalogs. 2. Chinese imprints—Japan—Catalogs. 3. Japan—Imprints—Catalogs. 4. Temmagū, Osaka—Library—Catalogs. I. Nagasawa, Kikuya, 1902- II. Nagasawa, Chizu. III. Title.
Z3301.T44 78-812537

—KOREA—CATALOGS

Koryŏ Taehakkyo, Seoul, Korea. Chungang Tosŏgwan.
(Hwasan mun'go Hanjŏk mongnok)
華山文庫漢籍目錄 = Catalogue of the Hwa-san collection. — [서울] : 高麗大學校中央圖書館, 1976.
203 p., [1] leaf of plates : port. ; 26 cm. — (高麗大學校藏書目錄 ; 第13輯)
In Korean.
Includes index.
1. Chinese imprints—Korea—Catalogs. 2. China—Imprints—Catalogs. 3. Bibliography—Rare books—Catalogs. 4. Yi, Sŏng-ŭi—Library—Catalogs. 5. Koryŏ Taehakkyo, Seoul, Korea. Chungang Tosŏgwan—Catalogs. I. Title. II. Title: Catalogue of the Hwa-san collection. III. Series: Koryŏ Taehakkyo, Seoul, Korea. Koryŏ Taehakkyo changsŏ mongnok ; che 13-chip.
Z3316.K69 1976 78-826504

--TAIWAN--T'AI-PEI (CITY)
--UNION LISTS

T'ai-wan ta hsüeh, T'ai-pei. T'u shu kuan.
(Kuo li T'ai-wan ta hsüeh p'u t'ung pen hsien chuang shu mu)
國立臺灣大學普通本線裝書目 / 國立臺灣大學圖書館編印. — [台北] ; 該館, 民國60 [1971] 民國67 [1978]序.
3, 2, 12, 497 p. ; 21 cm.

1. Chinese imprints—Taiwan—T'ai-pei (City)—Union lists. 2. Catalogs, Union—Taiwan—T'ai-pei (City) I. Title.
Z955.T3339T33 1978 79-837235

CHINESE IMPRINTS (Continued)

—UNITED STATES—UNION LISTS

Chinese cooperative catalog. Jan./Feb. 1978–
Washington, Library of Congress.
sheets 11 x 15 cm. bimonthly. (Library of Congress catalogs)
Microfiche.
Each issue cummulative from the beginning of the calendar year.
Continues: Chinese cooperative catalog, ISSN 0095–1072. In printed form.
Key title: Chinese cooperative catalog, ISSN 0095–1072.
1. Chinese imprints—United States—Union lists. 2. Catalogs, Union—United States—Periodicals. I. United States. Library of Congress. II. Series: United States. Library of Congress catalogs.
Microfiche 5166 Z 79–641787
[Z881.U49] [Z663.7.A73] MARC-S

CHINESE IN AUSTRALIA

—ECONOMIC CONDITIONS

Liu, Ta-jen, fl. 1936–
(Ao-chou Hua ch'iao ching chi)
澳洲華僑經濟 劉達人 田心源編著 台北 海外出版社 民國47 [1958]
172 p. illus. 19 cm. (華僑經濟叢書)
Series romanized: Hua ch'iao ching chi ts'ung shu.
1. Chinese in Australia—Economic conditions. 2. Chinese in New Zealand—Economic conditions. I. T'ien, Hsin-yüan, joint author. II. Title.
HC605.L5 C 59–5361

—HISTORY

Yong, C F
The new gold mountain : the Chinese in Australia, 1901-1921 / C. F. Yong. — Richmond, Australia : Raphael Arts, 1977.
xii, 301 p. : ill., map, ports.; 24 cm. Aus***
Bibliography: p. 276-290.
Includes index.
ISBN 0-908046-05-7
1. Chinese in Australia—History. 2. Australia—History—20 century. I. Title.
DU122.C5Y66 994'.004'951 78-326207
79 MARC

CHINESE IN BANDUNG, INDONESIA

(Bandon han Kakyō bōdō)
バンドン反華僑暴動：(八·五事件)：ナショナリズムの爆発. — 東京：インドネシアセンター, 1973.
72 p. ; 26 cm. Ja 74–17718
Cover title.
1. Chinese in Indonesia. 2. Chinese in Bandung, Indonesia. 3. Indonesia—Politics and government—1966– 4. Nationalism—Indonesia. I. Indoneshia Sentā.
DS632.3.C5B36 78–815486

CHINESE IN BRITISH COLUMBIA

Wynne, Robert Edward.
Reaction to the Chinese in the Pacific Northwest and British Columbia, 1850-1910 / Robert Edward Wynne. — New York : Arno Press, 1978.
vi, 511 p. ; 21 cm. — (The Asian experience in North America : Chinese and Japanese)
Originally presented as the author's thesis, University of Washington, 1964.
Bibliography: p. 500-511.
ISBN 0-405-11298-X
1. Chinese Americans—Northwest, Pacific. 2. Chinese in British Columbia. 3. Northwest, Pacific—Race relations. 4. British Columbia—Race relations. I. Title. II. Series: Asian experience in North America : Chinese and Japanese.
F855.2.C5W96 1978 979.5'004'951 78-54847
78 MARC

CHINESE IN BURMA

Li, Li-kuo, 1954–
(Ts'ung i yü tao T'ai-wan)
從異域到台灣 / 李利國編著. — 台南：長河出版社：總經銷建立書局, 民國67 [1978]
4, 237 p., [4] leaves of plates : ill. ; 20 cm. — (長河叢書 ; 8)
Series romanized: Ch'ang ho ts'ung shu.
NT$50.00
1. Chinese in Burma. 2. Anti-communist movements—Burma. 3. Anti-communist movements—China. 4. Guerrillas—Burma. 5. Guerrillas—China. I. Title.
DS530.4.L5 79–840763

Lo, Shih-pu.
(Man huang hsing chiao)
蠻荒行脚 / 羅石補著. — 臺北：南京出版公司, 民國67 [1978]
4, 2, 1, 443 p. ; 20 cm.
NT$70.00
1. Chinese in Burma. 2. Guerrillas—Burma. 3. Guerrillas—China. 4. Anti-communist movements—Burma. 5. Anti-communist movements—China. I. Title.
DS530.4.L6 79–840642

CHINESE IN CALIFORNIA

California. State Board of Control.
California and the Oriental / State Board of Control of California. — New York : Arno Press, 1978.
231 p. : ill. ; 24 cm. — (The Asian experience in North America : Chinese and Japanese)
Reprint of the 1920 ed. published by California State Printing Office, Sacramento.
Includes index.
ISBN 0-405-11265-3
1. Japanese in California. 2. Chinese in California. 3. East Indians in California. I. Title. II. Series: Asian experience in North America : Chinese and Japanese.
JV6888.C2A2 1978 301.45'19'50794 78-54809
78 MARC

Canada. Royal Commission on Chinese Immigration.
Report of the Royal Commission on Chinese Immigration. — New York : Arno Press, 1978.
xii, 487 p. ; 24 cm. — (The Asian experience in North America : Chinese and Japanese)
Reprint of 1885 ed. printed by order of the Commission, Ottawa, which was issued as no. 54a of the 1884-1885 Canadian Parliament's Sessional papers.
ISBN 0-405-11267-X
1. Chinese in Canada. 2. Chinese in California. I. Title. II. Series: Asian experience in North America : Chinese and Japanese. III. Series: Canada. Parliament. Sessional papers ; no. 54a.
JV7285.C5C353 1978 301.45'19'51073 78-54810
78 MARC

CHINESE IN CANADA

Canada. Royal Commission on Chinese Immigration.
Report of the Royal Commission on Chinese Immigration. — New York : Arno Press, 1978.
xii, 487 p. ; 24 cm. — (The Asian experience in North America : Chinese and Japanese)
Reprint of 1885 ed. printed by order of the Commission, Ottawa, which was issued as no. 54a of the 1884-1885 Canadian Parliament's Sessional papers.
ISBN 0-405-11267-X
1. Chinese in Canada. 2. Chinese in California. I. Title. II. Series: Asian experience in North America : Chinese and Japanese. III. Series: Canada. Parliament. Sessional papers ; no. 54a.
JV7285.C5C353 1978 301.45'19'51073 78-54810
78 MARC

Canada. Royal Commission on Chinese and Japanese Immigration.
Report of the Royal Commission on Chinese and Japanese Immigration. — New York : Arno Press, 1978.
xiv, 430 p. ; 24 cm. — (The Asian experience in North America : Chinese and Japanese)
Reprint of the 1902 ed. printed by S. E. Dawson, Ottawa, which was issued as Sessional paper no. 54 of the 1902 Parliament of Canada.
ISBN 0-405-11268-8
1. Chinese in Canada. 2. Japanese in Canada. 3. Canada—Emigration and immigration. I. Series: Asian experience in North America : Chinese and Japanese. II. Series: Canada. Parliament, 1902-1903. Sessional paper ; no. 54.
JV7285.C5C35 1978 325'.251'0971 78-54812
78 MARC

—CONGRESSES

Intercultural seminars, Chinese. — [Toronto] : Ontario Ministry of Culture and Recreation, Multicultural Development Branch, [1978?]
48 p. ; 21 cm. C***
Bibliography: p. 27.
1. Chinese in Canada—Congresses. I. Ontario. Multicultural Development Branch.
F1035.C5I55 301.45'19'51071 79-302016
79 MARC

CHINESE IN CUBA

Pérez de la Riva, Juan.
El barracón : esclavitud y capitalismo en Cuba / Juan Pérez de la Riva ; prólogo de Jordi Maluquer de Motes. — Barcelona : Crítica, 1978.
185 p. ; 20 cm. — (Crítica/historia ; 6) Sp***
Includes bibliographical references.
ISBN 8474230551
1. Cuba—Economic conditions. 2. Slavery in Cuba. 3. Chinese in Cuba. 4. Labor and laboring classes—Dwellings—Cuba. I. Title.
HC157.C9P46 79-341018
79 MARC

—ECONOMIC CONDITIONS

Helly, Denise.
Idéologie et ethnicité : les Chinois Macao à Cuba, 1847-1886 / Denise Helly. — Montréal : Presses de l'Université de Montréal, 1979.
345 p. : ill. ; 22 cm. — (Collection Recherches caraïbes) C***
Bibliography: p. [333]-342.
ISBN 0-8405-0383-0
1. Chinese in Cuba—Economic conditions. 2. Chinese in Cuba—Social conditions. 3. Sugar growing—Cuba. 4. Slavery in Cuba. 5. Cuba—Economic conditions. 6. Cuba—Social conditions. I. Title.
F1789.C53H44 301.45'19'5107291 79-382571
79 MARC

—SOCIAL CONDITIONS

Helly, Denise.
Idéologie et ethnicité : les Chinois Macao à Cuba, 1847-1886 / Denise Helly. — Montréal : Presses de l'Université de Montréal, 1979.
345 p. : ill. ; 22 cm. — (Collection Recherches caraïbes) C***
Bibliography: p. [333]-342.
ISBN 0-8405-0383-0
1. Chinese in Cuba—Economic conditions. 2. Chinese in Cuba—Social conditions. 3. Sugar growing—Cuba. 4. Slavery in Cuba. 5. Cuba—Economic conditions. 6. Cuba—Social conditions. I. Title.
F1789.C53H44 301.45'19'5107291 79-382571
79 MARC

CHINESE IN FOREIGN COUNTRIES

China (Republic of China). Ch'iao wu wei yüan hui.
(Chin jih Hua ch'iao)
今日華僑 = Overseas Chinese today, 1977–1978 / 僑務委員會編印. — [台北]：該會, 民國66 [1977]
139 p. : chiefly col. ill. ; 22 cm.
English and Chinese.
1. Chinese in foreign countries. I. Title. II. Title: Overseas Chinese today, 1977–1978.
DS732.C543 1977 78–843927

(Hua ch'iao ts'an cheng ch'üan ch'üan an)
華僑參政權全案 / [編輯者劉士木；校正者藍欣禾]. — [上海]：上海華僑聯合會, 民國2 [1913]
224 p. in various pagings, [3] leaves of plates : ports. ; 22 cm.
1. Chinese in foreign countries. 2. Repatriation—China. 3. Suffrage—China. I. Liu, Shih-mu.
DS732.H7865 1913 78–843825

Kuang-tung chien she hsieh chin hui ch'ou pei wei yüan hui.
(Kao ch'iao pao shu)
告僑胞書 / [廣東建設協進會籌備委員會]. — [香港? : s. n., 1978?]
22 p. ; 13 x 18 cm.
Reprint ed.
1. Kwangtung, China (Province)—Economic conditions. 2. Chinese in foreign countries. 3. Repatriation—China—Kwangtung (Province) I. Title.
HC428.K9K8 1978 79–835846

—ADDRESSES, ESSAYS, LECTURES

Ch'iao wu erh shih wu nien pien chi wei yüan hui.
(Ch'iao wu erh shih wu nien)
僑務二十五年 / 僑務二十五年編輯委員會編印. — 台北：海外出版社, 民國46 [1957]
6, 5, 261 p., [9] leaves of plates (1 fold.) : ill. ; 24 cm.
1. China (Republic of China). Ch'iao wu wei yüan hui—Addresses, essays, lectures. 2. Chinese in foreign countries—Addresses, essays, lectures. I. Title.
JQ1532.Z314C47 1957 79–840679

CHINESE IN FOREIGN COUNTRIES —ADDRESSES, ESSAYS, LECTURES (Continued)

(Pi hsü chung shih ch'iao wu kung tso)
必须重视侨务工作. — 第1版. — 北京：人民出版社，1978.
26 p. ; 21 cm.
Contains newspaper articles from Jen min jih pao.
CONTENTS: 必须重视侨务工作 《人民日报》社论.—廖承志 批判"四人帮"所谓"海外关系"问题的反动谬论.—在英明领袖华主席、党中央的亲切关怀下全国侨务会议预备会在京举行.
$0.10
1. China—Politics and government—1976- —Addresses, essays, lectures. 2. Chinese in foreign countries—Addresses, essays, lectures. I. Jen min jih pao, Peking.
DS779.22.P5 1978b 79-838628

(Pi hsü chung shih ch'iao wu kung tso)
必须重视侨务工作. — 第2版. — ［北京?］：人民出版社：新華書店发行，1978.
42 p. ; 21 cm.
CONTENTS: 必须重视侨务工作.—国务院侨务办公室负责人就我国侨务政策问题答记者问.—廖承志 批判四人幫所谓海外关係问题的反动谬论.—连贯 华侨的历史及其光荣传统.—在英明领袖华主席，党中央的亲切关怀下全国侨务会议预备会在京举行.
$0.15
1. China—Politics and government—1976- —Addresses, essays, lectures. 2. Chinese in foreign countries—Addresses, essays, lectures.
DS779.22.P5 1978c 79-840664
79 C

--ECONOMIC CONDITIONS --CONGRESSES

Hua ch'iao ching chi chien t'ao hui i, T'ai-pei, 1956.
(Hua ch'iao ching chi chien t'ao hui i shih lu)
華僑經濟檢討會議實錄 / 編輯者華僑經濟檢討會議秘書處. — ［台北］：該處，民國46 ［1957］
6, 120 p., [2] leaves of plates : ill. ; 21 cm.
Colophon title.
1. Chinese in foreign countries—Economic conditions—Congresses. I. Hua ch'iao ching chi chien t'ao hui i, T'ai-pei, 1956. Mi shu ch'u.
DS732.H7815 1956 79-840671
79 C

—LEGAL STATUS, LAWS, ETC.

China. Laws, statutes, etc.
(Hsien hsing ch'iao wu fa kuei chi yao)
現行僑務法規輯要 / 中央僑務委員會編印. — ［南京］：該會，民國20 ［1931］
ca. 120 p. in various pagings, [27] fold. leaves ; 19 cm.
1. Chinese in foreign countries—Legal status, laws, etc. 2. Repatriation—Law and legislation—China. I. China. Ch'iao wu wei yüan hui. II. Title.
78-842891
78 C

CHINESE IN HAWAII

Coman, Katharine, 1857-1915.
The history of contract labor in the Hawaiian Islands / Katharine Coman and Hawaii's Japanese / Andrew W. Lind. — New York : Arno Press, 1978.
68, viii, 264 p. : ill. ; 23 cm. — (The Asian experience in North America : Chinese and Japanese)
Reprint of 2 works, the 1st published in 1903 for the American Economic Association by Macmillan, New York, which was issued as 3d ser., v. 4, no. 3 of the Association's Publications; the 2d published in 1946 by Princeton University Press, Princeton, N.J.
Includes bibliography and index.
ISBN 0-405-11209-6
1. Japanese Americans—Hawaii. 2. World War, 1939-1945—Hawaii. 3. Contract labor—Hawaii. 4. Chinese in Hawaii. I. Lind, Andrew William, 1901- Hawaii's Japanese. 1978. II. Title. III. Series: Asian experience in North Aamerica : Chinese and Japanese. IV. Series: American Economic Association. Publications ; 3d ser., v. 4, no. 3.
DU624.7.J3C65 1978 996.9'004'956 78-54813
78 MARC

CHINESE IN INDONESIA

(Bandon han Kakyō bōdō)
バンドン反華僑暴動：(八・五事件)：ナショナリズムの爆発. — 東京：インドネシアセンター，1973.
Ja 74-17718
72 p. ; 26 cm.
Cover title.
1. Chinese in Indonesia. 2. Chinese in Bandung, Indonesia. 3. Indonesia—Politics and government—1966- 4. Nationalism—Indonesia. I. Indoneshia Sentā.
DS632.3.C5B36 78-815486

Garna, Yudistira K
Masyarakat Cina dan pembangunan Indonesia : suatu studi latar belakang masalah Cina di Indonesia / oleh Yudistira K. Garna. — Bandung : Fakultas Sastra, Universitas Negeri Padjadjaran, 1972.
15 leaves ; 29 cm.
"Etn. 9/72."
Includes bibliographical references.
1. Chinese in Indonesia. 2. Indonesia—Race relations. I. Title.
DS632.3.C5G37 73-940729
79 MARC

Hidajat Z M.
Masyarakat dan kebudayaan Cina Indonesia / Hidajat Z. M. — Bandung : Tarsito, 1977.
195 p. ; 21 cm.
Includes bibliographical references.
1. Chinese in Indonesia. 2. Indonesia—Civilization. I. Title.
DS632.3.C5H52 79-940444
79 MARC

Ma, Shu-li, 1910-
(Yin-ni ti pien yü luan)
印尼的變與亂 馬樹禮著 Indonesia in turmoil, by S. L. Mah. ［再版 台北］ 海外出版社 ［1965］
2, 6, 2, 180, 22 p. illus. 19 cm.
In Chinese.
1. Indonesia—Politics and government—1950-1966. 2. Chinese in Indonesia. I. Title.
DS644.M19 72-837547

—POLITICS AND GOVERNMENT

Badan Permusjawaratan Kewarganegaraan Indonesia. Pengurus Pusat Harian.
Pedoman kampanje perdjoangan Baperki dalam pemilihan umum. — Djakarta : Pengurus Harian Baperki Pusat, [1955]
78 p. : ill. ; 22 cm.
1. Elections—Indonesia. 2. Chinese in Indonesia—Politics and government. I. Title.
JQ778.B3 1955 77-986942
79 MARC

Matsuda, Tokiko, 1905-
(Chitei no hitobito)
地底の人々 / 松田解子. — 東京：世界文化社，昭和28 ［1953］
274 p. : ill. ; 19 cm.
Colophon inserted.
1. World War, 1939-1945—Conscript labor—Japan—Fiction. 2. Chinese in Japan—Fiction. I. Title.
PL833.A8C54 79-800832

CHINESE IN JAVA

—POPULATION—STATISTICS —COLLECTED WORKS

Indonesia. Biro Pusat Statistik.
Penduduk Cina luar Jawa: hasil registrasi penduduk. Chinese population of outer Java: results of population registration. 1973/75-
Jakarta, Biro Pusat Statistik.
v. 21 x 33 cm. (Statistik penduduk & [i. e. dan] tenaga kerja)
English and Indonesian.
1. Java—Population—Statistics—Collected works. 2. Chinese in Java—Population—Statistics—Collected works. I. Indonesia. Biro Pusat Statistik. Chinese population of outer Java: results of population registration. II. Title. III. Series.
HA1817.J3 I 53d 312'.93 78-942135
MARC-S

CHINESE IN KOREA

Pak, Yŏng-sŏk.
(Manbosan sakŏn yŏn'gu)
萬寶山事件硏究：日帝大陸侵略政策의一環으로서의 / 朴永錫著. — 서울：亜細亜文化社，1978.
vi, 288 p. ; 23 cm.
"萬寶山事件과朝鮮에서의中國人排斥事件關係日誌": p. 221-251.
Bibliography: p. 252-265.
Includes index.
₩3200
1. Koreans in Manchuria. 2. Chinese in Korea. 3. Korea—Relations (general) with China. 4. China—Relations (general) with Korea. 5. Japan—Relations (general) with China. 6. China—Relations (general) with Japan. I. Title.
DS731.K6P34 79-825462

—ECONOMIC CONDITIONS

Lu, Kuan-ch'ün.
(Han-kuo Hua ch'ia ching chi)
韓國華僑經濟 盧冠群著 臺北 海外出版社 民國45 ［1956］
2, 2, 4, 95 p. tables. 19 cm. (華僑經濟叢書)
Bibliography: p. 95.
1. Chinese in Korea—Economic conditions.
DS904.5.C5L8 C 60-5170
rev

CHINESE IN MALAYA

—ECONOMIC CONDITIONS

T'ang, Shih-ch'ing.
(Hsin-chia-p'o Malai-ya Hua ch'iao ching chi)
新加坡馬來亞華僑經濟 唐史青著 臺北 海外出版社 民國45 ［1956］
2, 6, 3, 178 p. tables. 19 cm. (華僑經濟叢書)
Series romanized: Hua ch'iao ching chi ts'ung shu.
1. Chinese in Singapore—Economic conditions. 2. Chinese in Malaya—Economic conditions. I. Title
HC445.8.T36 C 60-5179

—POLITICS AND GOVERNMENT

Ku, Hung-ting, 1941-
Kuomintang's mass movement and the Kreta Ayer incident (1927) in Malaya / by Ku Hung-ting. — [Singapore] : Institute of Humanities and Social Sciences, College of Graduate Studies, Nanyang University, 1976.
37 leaves ; 28 cm. — (Occasional paper series - Institute of Humanities and Social Sciences, Nanyang University ; no. 13)
Summary in Chinese.
Includes bibliographical references.
1. Chinese in Malaya—Politics and government. 2. Chinese in Singapore—Politics and government. 3. Malaya—Politics and government. 4. Singapore—Politics and government. 5. Chung-kuo kuo min tang. I. Title. II. Series: Nanyang University. Institute of Humanities and Social Sciences. Occasional paper series - Institute of Humanities and Social Sciences, Nanyang University ; no. 13.
DS595.2.C5K8 959.5'004'951 78-940442
79 MARC

CHINESE IN MALAYSIA

Chui, Kwei-chiang.
Late Ch'ing's modern enterprises and the Chinese in Singapore and Malaya, 1904-1911 / by Chui Kwei-chiang. — Singapore : Institute of Humanities and Social Sciences, College of Graduate Studies, Nanyang University, 1976.
36 ; 28 cm. — (Occasional paper series - Institute of Humanities and Social Sciences, Nanyang University ; no. 17)
Summary in Chinese.
Includes bibliographical references.
1. Investments—China. 2. Chinese in Singapore. 3. Chinese in Malaysia. I. Title. II. Series: Nanyang University. Institute of Humanities and Social Sciences. Occasional paper series - Institute of Humanities and Social sciences, Nanyang University ; no. 17.
HG5782.C48 332.6'7351'0595 78-940362
79 MARC

CHINESE IN MALAYSIA (Continued)

Goh, Cheng Teik.
Integration in a plural society : the Chinese in Malaysia / [Goh Cheng Teik]. — [Kuala Lumpur] : Goh, c1978.
39 p. : port. ; 27 cm.
Includes bibliographical references.
$3.00
1. Chinese in Malaysia. 2. Malaysia—Race relations. I. Title.
DS595.2.C5G64 301.45'19'510595 78-942331
79 MARC

Tham, Seong Chee.
Malays and modernization : a sociological interpretation / Tham Seong Chee. — [Singapore] : Singapore University Press, c1977.
xiii, 319 p. ; 23 cm.
Bibliography: p. [301]-309.
Includes index.
1. Malaysia—Economic conditions. 2. Social classes—Malaysia. 3. Occupational prestige—Malaysia. 4. Chinese in Malaysia. 5. Education—Malaysia. 6. Social values. I. Title.
HC445.5.T53 309.1'595'05 77-941949
79 MARC

—ADDRESSES, ESSAYS, LECTURES

25 [i.e. Duapuluh lima] tahun, 1949-1974 MCA. — [Kuala Lumpur] : MCA, [1974]
70, 100 p. : ill. ; 29 cm.
Cover title.
Added t.p.: Ma Hua kung hui nien wu chou nien chi nien t'e k'an, 1949-1974.
Chinese and English.
Includes bibliographical references.
CONTENTS: 25th anniversary of the MCA.—Wong, Kuan Lee. Constitutional developments and the Chinese.—Thong, Saw Pak. The changing world power pattern and our future.—Foo, Peng Sang & Lee, Ker Foon. The open university of United Kingdom: a great innovation in education.—Teh, Hock Heng. Mutual cultural acceptance as basis of national solidarity.—Chan, Choe Thor. The development of KSM: past, present and the future.—Yip, Yat Hoong. Higher education in Asia in the next decade.—Tan, Chang Soong. A brief on the development of Tunku Abdul Rahman College.—Teng, Sin Wu. Population projections for peninsular Malaysia (1967-1997), and the geographical distribution of community groups up to 1970 in Malaysia.—Ho, R. Ung Hun. The part-time politician.—Loh, W. Population growth and employment.—Restructuring of the Malaysian society.
1. Malaysia—Addresses, essays, lectures. 2. Chinese in Malaysia—Addresses, essays, lectures. I. Malaysian Chinese Association. II. Title: Ma Hua kung hui nien wu chou nien chi nien t'e k'an, 1949-1974.
DS592.D82 301.45'19'510595 76-941886
78 MARC

—ECONOMIC CONDITIONS

(Ma-lai-hsi-ya Hua jen ching chi wen t'i chuan chi)
馬來西亞華人經濟問題專輯 = Malaysian Chinese economic problems = Masaalah ekonomi keturunan China Malaysia / 雪蘭莪中華總商會編. — Kuala Lumpur : 該會, 1978.
109 p. ; 27 cm.
Cover title.
First article in Chinese, English, and Malay; others in Chinese or English.
Bibliography: p. 98.
CONTENTS: 雪蘭莪中華總商會提呈予馬華工商聯合會 關于我國華人經濟問題之備忘錄.—馬來亞汽車鋸業公會提呈予雪華總商會 關于我國華人經濟問題之備忘錄.—冷眼 在國家經濟發展中華人經濟地位問題檢討.—Chin, P. An appraisal of the economic status and problems of the Malaysian Chinese in the context of national economic development.—傅係中 在國家經濟發展中華人經濟地位問題檢討
$3.00
1. Chinese in Malaysia—Economic conditions. I. Selangor Chinese Chamber of Commerce. II. Title: Malaysian Chinese economic problems. III. Title: Masaalah ekonomi keturunan China Malaysia.
HC445.5.M32 78-942940

CHINESE IN MANILA

Amyot, Jacques, 1920–
[The Manila Chinese. Japanese]
マニラの中国人 = The Manila Chinese / Jacques Amyot ; 監修 加藤秀俊 ; 訳 田村愛理. — 東京 : 学習院大学東洋文化研究所, 1977.
57 p. : ill. ; 26 cm. — (調査研究報告 - 学習院大学東洋文化研究所 ; no. 2) Ja***
Bibliography: p. 50–57.
1. Chinese in Manila. I. Title. II. Series: Gakushūin Daigaku. Tōyō Bunka Kenkyūjo. Chōsa kenkyū hōkoku - Gakushūin Daigaku, Tōyō Bunka Kenkyūjo ; no. 2.
Title romanized: Manira no Chūgokujin.
DS689.M2A6616 959.9'1'004951 78–816997

CHINESE IN NEW YORK (CITY)

—FICTION

Chu, Louis, 1915-
Eat a bowl of tea / by Louis Chu ; introd. by Jeffery Chan. — Seattle : University of Washington Press, 1979, c1961.
250 p. ; 21 cm.
Reprint, with a new introd., of the ed. published by L. Stuart, New York.
ISBN 0-295-95607-0 : $6.95
1. Chinese in New York (City)—Fiction. I. Title.
PZ4.C558 Eat 1979 813'.5'4 78-21209
[PS3553.H776] 79 MARC

CHINESE IN NEW ZEALAND

—ECONOMIC CONDITIONS

Liu, Ta-jen, fl. 1936–
(Ao-chou Hua ch'iao ching chi)
澳洲華僑經濟 劉達人 田心源編著 台北 海外出版社 民國47 [1958]
172 p. illus. 19 cm. (華僑經濟叢書)
Series romanized: Hua ch'iao ching chi ts'ung shu.
1. Chinese in Australia—Economic conditions. 2. Chinese in New Zealand—Economic conditions. I. T'ien, Hsin-yüan, joint author. II. Title.
HC605.L5 C 59-5361

CHINESE IN PERU

—ADDRESSES, ESSAYS, LECTURES

Rodríguez Pastor, Humberto.
Los trabajadores chinos culíes en el Perú : artículos históricos / Humberto Rodríguez P. — Lima : [s.n.], 1977.
ii, 53 p. : ill. ; 30 cm.
Cover title.
Includes bibliographies.
1. Chinese in Peru—Addresses, essays, lectures. I. Title.
F3619.C5R62 985'.004'951 78-110015
79 MARC

CHINESE IN ROBE, AUSTRALIA

—HISTORY

Cawthorne, Ellen Mary.
The long journey : the story of the Chinese landings at Robe during the gold rush era, 1852-63 / [by] Ellen Mary Cawthorne. — [Robe, S.A.] : [E. M. Cawthorne], [1974]
[16] p. : ill. ; 13 x 21 cm. Aus
Cover title.
ISBN 0-9597128-0-1
1. Chinese in Robe, Australia—History. 2. Robe, Australia—History. I. Title.
DU330.R6C38 325'.251'099423 79-300089
79 MARC

CHINESE IN SINGAPORE

Chui, Kwei-chiang.
Late Ch'ing's modern enterprises and the Chinese in Singapore and Malaya, 1904-1911 / by Chui Kwei-chiang. — Singapore : Institute of Humanities and Social Sciences, College of Graduate Studies, Nanyang University, 1976.
36 ; 28 cm. — (Occasional paper series - Institute of Humanities and Social Sciences, Nanyang University ; no. 17)
Summary in Chinese.
Includes bibliographical references.
1. Investments—China. 2. Chinese in Singapore. 3. Chinese in Malaysia. I. Title. II. Series: Nanyang University. Institute of Humanities and Social Sciences. Occasional paper series - Institute of Humanities and Social sciences, Nanyang University ; no. 17.
HG5782.C48 332.6'7351'0595 78-940362
79 MARC

Lee, Poh Ping.
Chinese society in nineteenth century Singapore / Lee Poh Ping. — Kuala Lumpur ; New York : Oxford University Press, 1978.
xi, 139 p. : maps ; 23 cm. — (East Asian historical monographs)
Bibliography: p. [120]-131.
Includes index.
ISBN 0-19-580384-1
1. Chinese in Singapore. 2. Singapore—Economic conditions. 3. Singapore—Social conditions. I. Title.
DS598.S7L45 301.45'19'5105952 78-941542
79 MARC

—ECONOMIC CONDITIONS

T'ang, Shih-ch'ing.
(Hsin-chia-p'o Malai-ya Hua ch'iao ching chi)
新加坡馬來亞華僑經濟 唐史青著 臺北 海外出版社 民國45 [1956]
2, 6, 3, 178 p. tables. 19 cm. (華僑經濟叢書)
Series romanized: Hua ch'iao ching chi ts'ung shu.
1. Chinese in Singapore—Economic conditions. 2. Chinese in Malaya—Economic conditions. I. Title
HC445.8.T36 C 60–5179

—POLITICS AND GOVERNMENT

Ku, Hung-ting, 1941-
Kuomintang's mass movement and the Kreta Ayer incident (1927) in Malaya / by Ku Hung-ting. — [Singapore] : Institute of Humanities and Social Sciences, College of Graduate Studies, Nanyang University, 1976.
37 leaves ; 28 cm. — (Occasional paper series - Institute of Humanities and Social Sciences, Nanyang University ; no. 13)
Summary in Chinese.
Includes bibliographical references.
1. Chinese in Malaya—Politics and government. 2. Chinese in Singapore—Politics and government. 3. Malaya—Politics and government. 4. Singapore—Politics and government. 5. Chung-kuo kuo min tang. I. Title. II. Series: Nanyang University. Institute of Humanities and Social Sciences. Occasional paper series - Institute of Humanities and Social Sciences, Nanyang University ; no. 13.
DS595.2.C5K8 959.5'004'951 78-940442
79 MARC

CHINESE IN SOUTH AFRICA

—ATTITUDES

Smedley, Linda N
The Chinese community in South Africa / L. N. Smedley, D. C. Groenewald. — Pretoria : Institute for Sociological, Demographic and Criminological Research, 1976-
v. ; 21 cm. — (Report - Institute for Sociological, Demographic, and Criminological Research ; no. S-44-) SA***
Summary in Afrikaans.
Bibliography: v. 1, p. 101-103.
CONTENTS: phase 1. Background and attitudes of the white population group towards the Chinese minority group.
lis, joint author. II. Title. III. Series: Human Sciences Research Council. Institute for Sociological, Demographic, and Criminological Research. Report - Institute for Sociological, Demographic, and Criminological Research ; no. S-44-.
DT764.C45S63 78-319898
301.15'43'968004951
79 MARC

—PUBLIC OPINION

Smedley, Linda N
The Chinese community in South Africa / L. N. Smedley, D. C. Groenewald. — Pretoria : Institute for Sociological, Demographic and Criminological Research, 1976-
v. ; 21 cm. — (Report - Institute for Sociological, Demographic, and Criminological Research ; no. S-44-) SA***
Summary in Afrikaans.
Bibliography: v. 1, p. 101-103.
CONTENTS: phase 1. Background and attitudes of the white population group towards the Chinese minority group.
ISBN 0-86965-347-4 (v. 1) : R2.50 (v. 1)
1. Chinese in South Africa—Public opinion. 2. Public opinion—South Africa. 3. Chinese in South Africa—Attitudes. I. Groenewald, Dirk Cornelis, joint author. II. Title. III. Series: Human Sciences Research Council. Institute for Sociological, Demographic, and Criminological Research. Report - Institute for Sociological, Demographic, and Criminological Research ; no. S-44-.
DT764.C45S63 78-319898
301.15'43'968004951
79 MARC

CHINESE IN SOUTHEASTERN ASIA

China (Republic of China). Ch'iao wu wei yüan hui. Ti san ch'u.
(Hua ch'iao yü k'uo chan Tung-nan Ya mao i wen t'i)
華僑與擴展東南亞貿易問題 / 僑務委員會第三處編印. — 台北：該處, 民國46 [1957]
4, 115 p. ; 19 cm. — (華僑經濟專題研究叢書 ; 3) (華僑經濟參考叢刊)
Series 1 romanized: Hua ch'iao ching chi chuan t'i yen chiu ts'ung shu. Series 2 romanized: Hua ch'iao ching chi ts'an k'ao ts'ung k'an.
1. Taiwan—Foreign economic relations—Asia, Southeastern. 2. Asia, Southeastern—Foreign economic relations—Taiwan. 3. Chinese in Southeastern Asia. I. Title.
HF1606.Z4A783 79-840442

Freedman, Maurice.
The study of Chinese society : essays / by Maurice Freedman ; selected and introduced by G. William Skinner. — Stanford, Calif. : Stanford University Press, 1979.
xxiv, 491 p. ; 24 cm.
Bibliography: p. [445]-475.
ISBN 0-8047-0964-5 : $28.50
1. China—Social conditions. 2. Chinese in Southeastern Asia. I. Skinner, George William, 1925- II. Title.
HN733.F73 309.1'51 78-65395
00 MARC

Gotō, Asatarō, 1881-1945.
(Nan'yō no kakyō)
南洋の華僑 / 後藤朝太郎著. — 東京：高山書院, 昭和17 [1942]
6, 3, 5, 393 p. ; 19 cm.

1. Chinese in Southeastern Asia. I. Title.
DS509.5.C5G67 78-808566

Sung, Ming-shun.
(Tung-nan Ya Hua jen chi ch'i ch'ien t'u)
东南亞華人及其前途：民族主义及社會主义的衝击 / 宋明順. — Singapore：南洋大學研究院人文與社會科學研究所, 1976.
46 leaves ; 30 cm. — (研究論文 - 南洋大學研究院人文與社會科學研究所 ; no. 58)
Includes bibliographical references.
1. Chinese in Southeastern Asia. I. Title. II. Series: Nanyang University. Institute of Humanities and Social Sciences. Occasional paper series - Institute of Humanities and Social Sciences, Nanyang University ; no. 58.
DS509.5.C5S87 78-940416

Taiwan Ginkō. Chōsaka.
(Nan'yō ni okeru Kakyō)
南洋ニ於ケル華僑(支那移住民) 附·爲替關係 臺灣銀行總務部調查課 [臺北] 臺灣銀行 大正3 [1914]
172 p. illus. 22 cm.

1. Chinese in Southeastern Asia. 2. China—Emigration and immigration. I. Title.
DS732.T25 1914 J 58-4065 ‡

Taiwan Takushoku Kabushiki Kaisha. Chōsaka.
(Nan'yō kakyō to sono taisaku)
南洋華僑と其の對策. — 臺北：臺湾拓殖株式会社調査課, 昭和17 [1942]
2, 2, 169 p. ; 22 cm. — (臺調資 ; A第27號)
Cover title.
Not in trade.

1. Chinese in Southeastern Asia. I. Title. II. Series: Taiwan Takushoku Kabushiki Kaisha. Chōsaka. Taichō shi ; A dai 27-gō.
DS509.5.C5T3 1942 78-816606

—ADDRESSES, ESSAYS, LECTURES

Mengapa 600 [i.e. enam ratus] djuta rakjat R.R.T. bergerak keselatan? : untuk tanah, beras, timah dan kekajaan alam. — Djakarta : New Nusantara Pub. Coy, [1959?]
58 p. : map ; 23 cm. — (Seri kewaspadaan nasional)
1. Chinese in Southeastern Asia—Addresses, essays, lectures. 2. China—Relations (general) with Southeastern Asia—Addresses, essays, lectures. 3. Asia Southeastern—Relations (general) with China—Addresses, essays, lectures.
DS509.5.C5M46 77-986859
79 MARC

Shee, Poon Kim.
The roots of Sinophobia in the ASEAN countries : a comparative perspective / by Shee Poon Kim. — [Singapore] : Institute of Humanities and Social Sciences, College of Graduate Studies, Nanyang University, 1977.
27, 2 leaves ; 30 cm. — (Occasional paper series - Institute of Humanities and Social Sciences, Nanyang University ; no. 74)
Includes bibliographical references.
1. Chinese in Southeastern Asia—Addresses, essays, lectures. 2. Asia, Southeastern—Race relations—Addresses, essays, lectures. 3. Association of South East Asian Nations. I. Title. II. Series: Nanyang University. Institute of Humanities and Social Sciences. Occasional paper series - Institute of Humanities and Social Sciences, Nanyang University ; no. 74.
DS509.5.C5S47 301.45'19'51059 78-940460
79 MARC

—ECONOMIC CONDITIONS

Nan'yō Kyōkai.
(Nan'yō no kakyō)
南洋の華僑 / 南洋協會編纂. — 再版. — 東京：同協會, 昭和15 [1940]
2, 6, 194 p. : maps ; 23 cm.

1. Chinese in Southeastern Asia—Economic conditions. I. Title.
HC412.N27 1940b 79-801351

CHINESE IN THE UNITED STATES

see also Chinese Americans

Condit, Ira M
The Chinaman as we see him / Ira M. Condit. — New York : Arno Press, 1978.
233 p. : ill. ; 21 cm. — (The Asian experience in North America : Chinese and Japanese)
Reprint of the 1900 ed. published by F. H. Revell Co., Chicago under title: The Chinaman as we see him, and fifty years of work for him.
ISBN 0-404-11305-6
1. Missions to Chinese—United States. 2. Chinese in the United States. I. Title. II. Series.
BV2787.C66 1978 266'.022 78-54839
78 MARC

Gulick, Sidney Lewis, 1860-1945.
American democracy and Asiatic citizenship / Sidney Lewis Gulick. — New York : Arno Press, 1978 [c1918]
xii, 257 p., [1] leaf of plates : graph ; 21 cm. — (The Asian Experience in North America : Chinese and Japanese)
Reprint of the ed. published by Scribner, New York.
Includes bibliographical references.
ISBN 0-405-11307-2
1. Japanese in the United States. 2. Chinese in the United States. 3. United States—Emigration and immigration. I. Title. II. Series: Asian experience in North America : Chinese and Japanese.
JV6884.G78 1978 301.32'8'51073 78-54841
78 MARC

Li, Tien-Lu.
Congressional policy of Chinese immigration / Tien-Lu Li. — New York : Arno Press, 1978 [c1916]
132 p. ; 24 cm. — (The Asian experience in North America : Chinese and japanese)
Reprint of the ed. printed for the author by the pub. house of the Methodist Episcopal Church, South Nashville, Tenn.
Originally presented as the author's thesis, Vanderbilt University, 1916.
Bibliography: p. [126]-132.
ISBN 0-405-11280-7
1. United States—Emigration and immigration. 2. China—Emigration and immigration. 3. Chinese in the United States. I. Title. II. Series: Asian experience in North America, : Chinese and Japanese.
JV6874.L5 1978 325'.251'0973 78-54824
78 MARC

Liang, Chi Shad.
The changing size and changing character of Chinese immigration to the United States / by Liang Chi Shad. — [Singapore] : Institute of Humanities and Social Sciences, College of Graduate Studies, Nanyang University, 1976.
54 leaves ; 28 cm. — (Occasional paper series - Institute of Humanities and Social Sciences, Nanyang University ; no. 15)
Summary in Chinese.
Includes bibliographical references.
1. Chinese in the United States. 2. United States—Emigration and immigration. I. Title. II. Series: Nanyang University. Institute of Humanities and Social Sciences. Occasional paper series - Institute of Humanities and Social Sciences, Nanyang University ; no. 15.
JV6874.L52 325'.251'0973 78-322572
79 MARC

McKee, Delber L 1923-
Chinese exclusion versus the open door policy, 1900-1906 : clashes over China policy in the Roosevelt era / Delber L. McKee. — Detroit : Wayne State University Press, 1977.
905/1(d
292 p. : ill. ; 24 cm.
Bibliography: p. 277-283.
Includes index.
ISBN 0-8143-1565-8
1. United States—Foreign relations—China. 2. China—Foreign relations—United States. 3. Chinese in the United States. 4. United States—Emigration and immigration. 5. United States—Foreign relations—1901-1909. 6. Roosevelt, Theodore, Pres. U.S., 1858-1919. I. Title.
E183.8.C5M29 327.73'051 76-47024
76[r79]rev MARC

United States. Congress. Joint Special Committee to Investigate Chinese Immigration.
Report of the Joint Special Committee to Investigate Chinese Immigration, United States Senate, Forty-fourth Congress. — New York : Arno Press, 1978.
viii, 1281 p. ; 24 cm. — (The Asian experience in North America : Chinese and Japanese)
Reprint of the 1877 ed. published by U.S. Govt. Print. Off., Washington, which was issued as Report no. 689 of the 44th Congress, 2d session, Senate.
ISBN 0-405-11294-7
1. United States—Emigration and immigration. 2. Chinese in the United States. I. Series: Asian experience in North America : Chinese and Japanese. II. Series: United States. Congress. Senate. Report ; no. 689.
JV6874.U54 1978 301.32'9'73 78-54836
78 MARC

—ADDRESSES, ESSAYS, LECTURES

Chou, Ming-ch'i.
(I chiu ling wu nien ti fan Mei ai kuo yün tung)
1905年的反美爱国运动 / 编写者周明綺. — [香港?：s. n., 1978?]
40 p. : ill. ; 19 cm.
Colophon title.
Title in romanization on cover: 1905 nian de fan Mei ai guo yun dong.
Reprint of the 1962 ed. published by Chung-hua shu chü, Peking, issued in series: Chung-kuo li shih hsiao ts'ung shu.
1. Chinese in the United States—Addresses, essays, lectures. I. Title. II. Title: Yi jiu ling wu nian de fan Mei ai guo yun dong.
E184.C5C53 1978 78-841489
C

—EDUCATION

Yung, Wing, 1828-1912.
My life in China and America / Yung Wing. — New York : Arno Press, 1978 [c1909]
vi, 286 p. : port. ; 24 cm. — (The Asian experience in North America : Chinese and Japanese)
Reprint of the ed. published by H. Holt, New York.
Includes index.
ISBN 0-405-11301-3
1. Yung, Wing, 1828-1912. 2. China. 3. Education—China. 4. Chinese in the United States—Education. I. Title. II. Series: Asian experience in North America : Chinese and Japanese.
DS763.Y8A3 1978 951'.03'0924 78-54850
78 MARC

—HISTORY

Zo, Kil Young.
Chinese emigration into the United States, 1850-1880 / Kil Young Zo. — New York : Arno Press, 1978, c1971.
226 p. : map ; 24 cm. — (The Asian experience in North America : Chinese and Japanese)
Reprint of the author's thesis, Columbia.
Bibliography: p. 207-224.
ISBN 0-405-11302-1
1. United States—Emigration and immigration—History. 2. China—Emigration and immigration—History. 3. Chinese in the United States—History. I. Title. II. Series: Asian experience in North America : Chinese and Japanese.
JV6874.Z6 1978 301.32'8'51073 78-54851
78 MARC

CHINESE IN VIETNAM

(Yüeh-nan ch'ü kan Hua ch'iao hui kuo chen hsiang)
越南驱赶华侨回国真相. — 香港：生活, 讀書, 新知三联书店香港分店, 1978.
35 p. ; 21 cm.

1. Chinese in Vietnam. 2. China—Foreign relations—Vietnam. 3. Vietnam—Foreign relations—China.
DS556.45.C55Y83 79-835445

CHINESE INSCRIPTIONS see Inscriptions, Chinese

CHINESE-JAPANESE CONFLICT, 1937-1945 see Sino-Japanese Conflict, 1937-1945

CHINESE-JAPANESE WAR, 1894-1895

Honda, Jimpachi, 1871-1909.
[Nisshin senshi kōkyūroku. Chinese]
日清戰史講授錄 / 譽田甚八著 ; [訓練總監部]
軍學編譯處譯 ; 沈雲龍主編. — 臺北縣永和鎮 :
文海出版社, 民國65 [1976]
2, 6, 348, 76 p. ; 22 cm. — (近代中國史料叢刊續編 [i. e. 輯] ; 380 : 第38輯)
Reprint of the 1936 ed.
1. Chinese-Japanese War, 1894-1895. I. Title. II. Series: Shen, Yün-lung, Comp. Chin tai Chung-kuo shih liao ts'ung k'an hsü chi ; 380 : ti 38 chi.
Title romanized: Jih Ch'ing chan shih chiang shou lu.
DS755.S532 vol. 380 77-842732
[DS765]
rev
78 [r79] C

—DIPLOMATIC HISTORY

Itō, Hirobumi, 1841-1909.
(T'ai-wan ko jang Chung Jih t'an p'an mi hua lu)
台灣割讓中日談判秘話錄 : 伊藤博文·李鴻章一問一答. — 再版. — 臺北 : 西南書局, 民國64 [1975]
162 p., [2] leaves of plates : ill. ; 22 cm. — (學術叢書 ; 43)
"李鴻章手記"
Chinese and Japanese.
Series romanized: Hsüeh shu ts'ung shu.
1. Chinese-Japanese War, 1894-1895—Diplomatic history. 2. Japan—Foreign relations—China. 3. China—Foreign relations—Japan. 4. Taiwan—History—To 1895. I. Li, Hung-chang, 1823-1901. II. Title. III. Title: Chung Jih t'an p'an mi hua.
DS765.I93 1975 79-836789
79 C

—DRAMA

Chung-kuo jen min chieh fang chün hai chün cheng chih pu wen kung t'uan. Hua chü t'uan.
(Chia wu hai chan)
甲午海战　八場历史剧　中国人民解放軍海軍政治部文工团話剧团改編　朱祖貽　李恍执笔　北京　中国戏剧出版社　1961.
102 p. illus. 19 cm.
Adaptation of 希依等著电影剧本甲午风伝
1. Chinese-Japanese War, 1894-1895—Drama. I. Chu, Tsu-i, fl. 1961- II. Hsi-nung, pseud. Chia wu feng yün. III. Title.
C 64-549
rev 2

—FICTION

Lin, K'ai.
(Chia wu hai chan)
甲午海战 : 根据朱祖贻, 李恍同名话剧改编 / 林锴编绘. — 第1版. — 北京 : 人民美術出版社, 1977.
136 p. : ill. ; 11 x 13 cm.
"連環圖畫."
$0.17
1. Chinese-Japanese War, 1894-1895—Fiction. I. Chu, Tsu-i. II. China (People's Republic of China, 1949-). Chung-kuo jen min chieh fang chün. Hai chün cheng chih pu. Chia wu hai chan. III. Title.
PL2878.N1432C5 78-842801
78 C

—PERSONAL NARRATIVES, JAPANESE

Teikoku Zaigō Gunjinkai. Kamikamo Bunkai.
(Kaikoroku)
回顧錄 : 日清日露戰役上賀茂村出身者 / 帝國在鄉軍人會上賀茂分會編纂. — 京都 : 同分會, 昭和11 [1936]
276, 4 p., [5] leaves of plates : ill. ; 23 cm.
On spine: Nisshin Nichi-Ro sen'eki kaikoroku.
1. Chinese-Japanese War, 1894-1895—Personal narratives, Japanese. 2. Russo-Japanese War, 1904-1905—Personal narratives, Japanese. 3. Japan. Rikugun—Biography. 4. Kyoto—Biography. I. Title: Nisshin Nichi-Ro sen'eki kaikoroku.
DS765.T44 1936 78-811848

—REGISTERS OF DEAD—JAPAN —AKITA (PREFECTURE)

(Boshin no eki Seinan no eki Nisshin sensō Nichi-Ro sensō Akita-ken sembotsusha jimmeiroku)
戊辰の役·西南の役·日清戦争·日露戦争秋田県戦歿者人名録 : 稿 / [編集　三島亮吉, 渋谷鉄五郎, 桝谷健夫]. — [秋田] : 誠忠舘設立委員会 : ツバサ広業, 昭和49 [1974]
167 p. : ill. ; 26 cm. Ja 75-6205
Cover title.
Includes bibliographies.
1. Akita, Japan (Prefecture)—Biography. 2. Japan—History—Civil War, 1868—Registers of dead. 3. Satsuma Rebellion, 1877—Registers of dead. 4. Chinese-Japanese War, 1894-1895—Registers of dead—Japan—Akita (Prefecture) 5. Russo-Japanese War, 1904-1905—Registers of dead—Japan—Akita (Prefecture) I. Mishima, Ryōkichi. II. Shibuya, Tetsugorō. III. Masutani, Takeo. IV. Title: Akita-ken sembotsusha jimmeiroku.
DS894.39.A396B67 78-812979

CHINESE LANGUAGE

Gotō, Asatarō, 1881-1945.
(Gendai Shinagogaku)
現代支那語學 / 後藤朝太郎著. — 東京 : 博文館, 明治41 [1908]
3, 6, 286 p. ; 23 cm. — (帝國百科全書 ; 第177編)
Series romanized: Teikoku hyakka zensho.
Includes bibliographies.
1. Chinese language. I. Title.
PL1071.G6 79-803422

Humboldt, Wilhelm, Freiherr von, 1767-1835.
Brief an M. Abel-Rémusat über die Natur grammatischer Formen im allgemeinen und über den Geist der chinesischen Sprache im besonderen / Wilhelm von Humboldt ; nach d. Ausg. Paris 1827 ins Dt. übertr. u. mit e. Einf. versehen von Christoph Harbsmeier. Zur philosophischen Grammatik des Altchinesischen im Anschluss an Humboldts Brief an Abel-Rémusat / Christoph Harbsmeier. — Stuttgart-Bad Cannstatt : Frommann-Holzboog, 1979.
297 p. ; 19 cm. — (Grammatica universalis ; Bd. 17) GFR79-A
Cover title: Wilhelm von Humboldts Brief an Abel Rémusat und die philosophische Grammatik des Altchinesischen.
Translation of Lettre à M. Abel-Rémusat.
Reprint of the ed. published by Dondey-Dupré.
Bibliography: p. 280-293.
Includes index.
ISBN 3-7728-0673-2 : DM148.00
1. Chinese language. I. Rémusat, Jean Pierre Abel, 1788-1832. II. Harbsmeier, Christoph. Zur philosophischen Grammatik des Altchinesischen im Anschluss an Humboldts Brief an Abel Rémusat. 1979. III. Title. IV. Title: Wilhelm van Humboldts Brief an Abel Rémusat und die philosophische Grammatik des Altchinesischen.
PL1073.H815 1979 495.1 79-376225
79 MARC

—ACCENTS AND ACCENTUATION

Chang, Hsiao-yü.
(Ch'ing sheng pien i chü li)
輕聲辨義舉例 / 張孝裕著. — 臺北 : 學海出版社, 民國66 [1977]
2, 11, 304 p. ; 21 cm.
Bibliography: p. 302-304.
Includes indexes.
NT$80.00
1. Chinese language—Intonation. 2. Chinese language—Accents and accentuation. I. Title.
PL1213.C4 79-841879

—ADDRESSES, ESSAYS, LECTURES

Chien, Tsung-wu.
(Jen shih wo men ti yü yen wen tzu)
語識我們的語言文字 / 簡宗梧著. — 台北 : 偉文圖書出版社, 民國67 [1978]
124 p. ; 20 cm. — (中國語文叢書)
NT$40.00
1. Chinese language—Addresses, essays, lectures. I. Title.
PL1095.C47 79-838059

Chinese language use / edited by Beverly Hong. — Canberra : Contemporary China Centre, Research School of Pacific Studies, Australian National University, 1978.
xx, 138 p. ; 25 cm. — (Contemporary China papers ; 13) Aus***
Errata slip inserted.
Includes bibliographies.
1. Chinese language—Addresses, essays, lectures. I. Hong, Beverly.
PL1095.C54 495.1'5 79-308188
79 MARC

Hong Kong language papers / edited by Robert Lord. — Hong Kong : Hong Kong University Press, c1979.
ix, 246 p. ; 22 cm.
Includes bibliographies.
ISBN 9622090125
1. Bilingualism—Hongkong—Addresses, essays, lectures. 2. Chinese language—Addresses, essays, lectures. 3. Chinese language—Bibliography—Addresses, essays, lectures. I. Lord, Robert, 1931-
P115.5.H85H6 401 79-106833
79 MARC

Hui-chün.
(Han yü chi pen chih shih)
漢語基本知識 / 卉君編. — 港初版. — 香港 : 商務印書館, 1977.
11, 170 p. : ill. ; 19 cm. — (中學生文庫)
Series romanized: Chung hsüeh sheng wen k'u.
$5.00
1. Chinese language—Addresses, essays, lectures. I. Title.
PL1095.H84 78-841909

Kuang-tung shih fan hsüeh yüan. Han tzu ch'ang shih pien hsieh tsu.
(Han tzu ch'ang shih)
汉字常识 / 广东师范学院中文系《汉字常识》编写组. — [香港? : s. n., 1978?]
74 p. : 21 cm.
Reprint of the 1977 ed. published by Kuang-tung jen min ch'u pan she, Canton.
"常见错别字的辨认": p. 50-74.
1. Chinese language—Addresses, essays, lectures. I. Title.
PL1095.K8 1978 78-843614

Ogaeri, Yoshio, 1910-1966.
(Tairiku no gengo to bungaku)
大陸の言語と文學 / 魚返善雄著. — 東京 : 三省堂, 昭和15 [1940]
2, 9, 260 p., [6] leaves of plates : ill. ; 19 cm.
1. Chinese language—Addresses, essays, lectures. 2. Chinese literature—20th century—History and criticism—Addresses, essays, lectures. I. Title.
PL1071.O33 78-816297

Strindberg, August, 1849-1912.
Kina och Japan : studier / af August Strindberg. — [Staffanstorp : ABISAG, 1977]
86, [1] p. : ill. ; 21 cm. — (Abisags asienserie ; nr. 2) S77-34
Reprint of the 1911 ed. published by Björck & Börjesson, Stockholm; with a new introduction by L. Utterström.
ISBN 9185438022 : kr25.00
1. Chinese language—Addresses, essays, lectures. 2. Japanese language—Addresses, essays, lectures. I. Title. II. Series.
PL1073.S78 1977 79-349178
79 MARC

—ADVERB

Chen, Chung-Yu.
Reducibility of preverbal and postverbal locatives / by Chen Chung-Yu. — [Singapore] : Institute of Humanities and Social Sciences, College of Graduate Studies, Nanyang University, 1976.
37, [1] leaves : ill. ; 30 cm. — (Occasional paper series - Institute of Humanities and Social Sciences, Nanyang University ; no. 52)
Bibliography: leaf [38]
1. Chinese language—Adverb. I. Title. II. Series: Nanyang University. Institute of Humanities and Social Sciences. Occasional paper series - Institute of Humanities and Social Sciences, Nanyang University ; no. 52.
PL1235.C49 495.1'5 78-940411
79 MARC

—ALPHABET see Chinese chracters

CHINESE LANGUAGE (Continued)

—ARCHAIC CHINESE

Hashimoto, Mantaro J
Phonology of ancient Chinese / Mantaro J. Hashimoto. — Tokyo : Institute for the Study of Languages & Cultures of Asia & Africa, 1978-
v. ; 26 cm. — (Study of languages & cultures of Asia & Africa, monograph series, ; no. 10) Ja***
1. Chinese language—Archaic Chinese. 2. Chinese language—Phonology. I. Title. II. Series.
PL2245.H3 495.1'1'5 79 78-324893 MARC

P'an, Wen-kuang.
(Ku Han yü yü yin yen chiu tzu liao ch'u t'an)
古汉语语音研究资料初探 / 潘文光. — Singapore：南洋大學研究院人文與社會科學研究所, 1977.
40 leaves ; 29 cm. — (研究論文 - 南洋大學研究院人文與社會科學研究所 : no. 62)
Reproduced from typescript.
Includes bibliographical references.
1. Chinese language—Archaic Chinese. 2. Chinese language—Phonology. I. Title. II. Series: Nanyang University. Institute of Humanities and Social Sciences. Occasional paper series - Institute of Humanities and Social Sciences, Nanyang University ; no. 62.
PL2245.P3 78-940458

Wang, Chia-ping.
(Ku shu nan tzu)
古書難字 / 王嘉炳編註. — 臺北縣新店鎮：常春樹書坊, 民國67 [1978]
257 p. ; 20 cm.
NT$45.00
1. Chinese language—Vocabulary. 2. Chinese language—Archaic Chinese. I. Title.
PL1495.W24 79-835165

—BIBLIOGRAPHY—ADDRESSES, ESSAYS, LECTURES

Hong Kong language papers / edited by Robert Lord. — Hong Kong : Hong Kong University Press, c1979.
ix, 246 p. ; 22 cm.
Includes bibliographies.
ISBN 9622090125
1. Bilingualism—Hongkong—Addresses, essays, lectures. 2. Chinese lanaguage—Addresses, essays, lectures. 3. Chinese language—Bibliography—Addresses, essays, lectures. I. Lord, Robert, 1931-
P115.5.H85H6 401 79 79-106833 MARC

—COMPOSITION AND EXERCISES

Cheng, Fa-ming.
(Tsen yang hsieh i lun wen)
怎樣寫議論文 / 張希文校訂；鄭發明著；國語日報語文中心主編. — 第1版. — 臺北：國語日報出版部, 民國66 [1977]
2, 2, 3, 248 p. ; 22 cm.
Text accompanied with Kuo yü chu yin.
NT$60.00
1. Chinese language—Composition and exercises. I. Kuo yü jih pao yü wen chung hsin. II. Title.
PL1271.C435 79-838291

Chia, Ming, fl. 1973-
(Hsiao p'eng yu hsüeh ying yung wen)
小朋友學應用文 / 賈銘著. — 再版. — 香港：宏業書局, 1978.
2, 125 p. ; 19 cm.
$2.80
1. Letter-writing, Chinese. 2. Chinese language—Composition and exercises. I. Title.
PL1275.C4295 78-843607

(Chung hsüeh sheng tso wen chiang p'ing)
中学生作文讲评 / [本社编]. — 第1版. — 上海：上海人民出版社, 1973.
65 p. : ill. ; 19 cm.
1. Chinese language—Composition and exercises. I. Shang-hai jen min ch'u pan she.
PL1271.C593 79-836859

Lai, Ch'ing-hsiung.
(Tso wen p'ing yü shih li)
作文評語示例 / 賴慶雄著. — 香港：語文中心 [1978?]
2, 3, 139 p. ; 21 cm.
$5.50
1. Chinese language—Composition and exercises. 2. Chinese language—Study and teaching. I. Title.
PL1271.L28 1978 79-837855

T'ai-pei (City). Chiao yü chü. Pien chi hsiao tsu.
(Ju ho chih tao erh t'ung hsieh tso)
如何指導兒童寫作 / 主編台北市政府教育局；編著者[該局]編輯小組. — 台北：該局, 民國66 [1977]
2, 154 p. : ill. ; 21 cm. — (國民小學輔導叢書 ; 3)
Colophon title.
1. Chinese language—Composition and exercises. 2. Chinese language—Study and teaching (Elementary)—Taiwan. I. T'ai-pei (City). Chiao yü chü. II. Title. III. Series: Kuo min hsiao hsüeh fu tao ts'ung shu ; 3.
LB1577.C48T34 1977 78-841234

Tseng, Hsin-hsiung, 1945-
(T'i tsao hsieh tso yin tao)
提早寫作引導 / 曾信雄著. — 第7版. — [台北?]：新兒童出版社, 民國66 [1977]
112 p. : ill., port. ; 19 cm.
Cover title.
Text accompanied by kuo yü p'in yin.
"附錄 兒童優秀作品": p. 8-11.
NT$30.00
1. Creative writing (Elementary education) 2. Chinese language—Composition and exercises. I. Title.
LB1576.T79 78-843063

Wang, Chin-hsiang.
(Tso wen p'ing yü chü li)
作文評語舉例 / 王進祥著；[總編輯林洋慈]. — 台北：國家出版社, 民國67 [1978]
3, 134 p. ; 19 cm.
Appendices (p. 93-134): 作文命題參考.—幾本有益作文的書.
NT$50.00
1. Chinese language—Composition and exercises. 2. Chinese language—Study and teaching. I. Lin, Yang-tz'u. II. Title.
PL1271.W254 78-842477

—CONVERSATION

Teng, Shou-hsin.
A basic course in Chinese grammar : a graded approach through conversational Chinese / Teng Shou-hsin. — San Francisco : Chinese Materials Center, 1977.
xii, 135 p. ; 22 cm. — (Asian library series ; no. 5)
1. Chinese language—Grammar. 2. Chinese language—Conversation. I. Title. II. Series.
PL1111.T4 495.1'8'2421 79 79-312822 MARC

—CONVERSATION AND PHRASE BOOKS

Chang, Tan, fl. 1977-
(Shih yung p'u t'ung hua)
實用普通話 / [張丹]. — [香港]：香港中文大學校外進修部, 1977.
174 p. : ill. ; 21 cm.
Text accompanied with romanization.
1. Chinese language—Conversation and phrase books. I. Title.
PL1121.C5C35 79-837148

DeFrancis, John Francis, 1911-
Character text for Beginning Chinese / by John De Francis, with the assistance of Yung Teng Chia-yee. — 2d ed. — New Haven : Yale University Press, 1976.
519 p. : ill. ; 26 cm. — (Yale linguistic series)
Prefatory matter in English, text in Chinese.
Includes index.
ISBN 0-300-02055-4. ISBN 0-300-02059-7 pbk.
1. Chinese language—Grammar. 2. Chinese language—Conversation and phrase books. I. De Francis, John Francis, 1911- Beginning Chinese. II. Yung Teng, Chia-yee, joint author. III. Title. IV. Series.
PL1111.D37 1976 76-5105

Liu, Ch'iu-sheng.
(Chien ming shih yung kuo yü hui hua)
簡明實用國語會話 / 劉秋生編著. — 香港：陽光文化服務公司；九龍：經銷處利通圖書公司, 1977.
2, 133 p. ; 19 cm. & cassette tapes (2-track. mono.) in case.
Romanized title on label of cassette box: Jian ming shi yong guo yu hui hua.
"大衆國語初級篇中級篇合訂本"
$30.00
1. Chinese language—Conversation and phrase books. I. Title. II. Title: Jian ming shi yong guo yu hui hua.
PL1121.C5L5 78-840977
C

—CONVERSATION AND PHRASE BOOK—DANISH

Fønss, Nina.
Kina-parlør : sproglære, rejseparlør og mini-ordbog / [af] Nina Fønss. — [København] : Rhodos, 1978.
167 p. : ill. ; 22 cm. D78-24
Bibliography: p. 124.
ISBN 8774965824 : kr58.00
1. Chinese language—Conversation and phrase book—Danish. 2. Danish language—Terms and phrases—Chinese. I. Title.
PL1125.D3F6 78 78-384432 MARC

—CONVERSATION AND PHRASE BOOKS—JAPANESE

Iwasaki, Keitarō.
(Shinsen Nittai gengo shū)
新撰日臺言語集 / 岩崎敬太郎著. — 臺北：新撰日臺言語集發行所, 大正2 [1913], 大正5 [1916] printing.
2, 14, 696 p. ; 23 cm.
1. Chinese language—Dialects—Taiwan. 2. Chinese language—Conversation and phrase books—Japanese. I. Title. II. Title: Nittai gengo shū.
PL1834.I9 78-815845

Tanaka, Seiichirō, 1906-
(Chūgokugo jukugo jiten)
中国語熟語辞典 / 田中清一郎編. — 東京：白水社, 1977.
31, 266 p. ; 18 cm. Ja 77-22330
Includes index.
¥2500
1. Chinese language—Conversation and phrase books—Japanese. I. Title.
PL1121.J3T3 78-813915

Tsou, Chia-yen.
(Lao ch'i ta yen chieh tan tzu so yin)
老乞大諺解単字索引 / 鄒嘉彦. — Tokyo : 国立共同利用研アジア・アフリカ言語文化研究所言語情報機械処理プロジェクト, 1976.
245 p. ; 26 cm. — (CAAAL 単刊シリーズ : 1) (アジア・アフリカ語の計数研究 ; 第4号) Ja***
Cover title.
Title on p. [4] of cover: A single character index to the No-geol-dae eon-hae.
In colophon: Hakkō, Tōkyō Gaikokugo Daigaku Ajia Afurika Gengo Bunka Kenkyūjo.
"問題となる言語形式を含む文を網羅的にリストしたもので…機械力を利用し…「老乞大諺解」の中国文の部分を全部入力し上述の目的に合うように印出したもの"
Includes index.
1. Nogŏltae ŏnhae—Indexes. 2. Chinese language—Conversation and phrase books—Korean—Indexes. I. Title. II. Title: Rō kitsudai genkai tanji sakuin. III. Series: CAAAL tankan shirīzu ; 1. IV. Series: Ajia-Afurikago no keisū kenkyū ; dai 4-gō.
P381.A75A38 vol. 4 79-800201
[PL1121.K6]

CHINESE LANGUAGE (Continued)

—DATA PROCESSING

Chu, Fu, 1937-
(Chung wen shu ju yen chiu)
中文輸入研究 / 編著朱邦復[朱復]；編輯沈紅蓮，馬毓德，劉亞玲. — [台北]：大周建設公司，民國67 [1978]
219 p. ; 27 cm.

1. Chinese language—Data processing. I. Shen, Hung-lien. II. Ma, Yü-te. III. Liu, Ya-ling. IV. Title.
PL1074.5.C46 79-83747

—DIALECTS—CHINA—AMOY—GLOSSARIES, VOCABULARIES, ETC.

Nakajima, Motoki.
(Bingo Tōzantō hōgen kiso goi shū)
閩語東山島方言基礎語彙集 / 中嶋幹起著. — 東京：東京外国語大学アジア・アフリカ言語文化研究所，1977.
276 p. : ill. ; 26 cm. Ja 78-4299
Added t. p.: A study of the basic vocabulary of the Min dialect in the Tung Shan Island.
English and Japanese.
Accompanied by "日本語索引" (52 p.)
Includes indexes.
1. Chinese language—Dialects—China—Amoy—Glossaries, vocabularies, etc. 2. Chinese language—Dialects—China—Tung-shan Island—Glossaries, vocabularies, etc. I. Title. II. Title: A study of the basic vocabulary of the Min dialect in the Tung Shan Island.
PL1704.N35 495.1'7 79-801436
*79 J

—DIALECTS—CHINA—CANTON

Chuang, Shih-kuang.
(Kuang-tung hua chih nan)
廣東話指南 / 莊世光著. — 臺北：文史哲出版社，民國66 [1977]
7, 208 p. ; 22 cm.
Bibliography: p. 207-208.
NT$120.00

1. Chinese language—Dialects—China—Canton. I. Title.
PL1734.C48 78-842068

Phoon, Mun-Kwong.
A synchronic phonology of modern Cantonese dialect / by Phoon Mun-Kwong. — [Singapore] : Institute of Humanities and Social Sciences, College of Graduate Studies, Nanyang University, 1976.
39 leaves ; 30 cm. — (Occasional paper series - Institute of Humanities and Social Sciences, Nanyang University ; no. 39)
Includes bibliographical references.
1. Chinese language—Dialects—China—Canton. 2. Chinese language—Phonology. I. Title. II. Series: Nanyang University. Institute of Humanities and Social Sciences. Occasional paper series - Institute of Humanities and Social Sciences, Nanyang University ; no. 39.
PL1739.P47 495.1'7 78-940457
79 MARC

Wang, Tzu-ming.
(Shih yung Kuang-chou hua)
實用廣州話 = Practical Cantonese / 王子明編著；會槇冰校訂；羅國維，殷巧兒講錄. — 香港：萬里書店，1976.
3, 198 p. ; 22 cm. & cassette. — (萬里有聲叢書)
In container.
English and Chinese.
Series romanized: Wan li yu sheng ts'ung shu.
"本書附錄音帶一卷"
1. Chinese language—Dialects—China—Canton. I. Lo, Kuo-wei. II. Yin, Ch'iao-erh. III. Title. IV. Title: Practical Cantonese.
PL1735.W3 495.1'7 79-840091

—DIALECTS—CHINA—CANTON—DICTIONARIES

Fok, Bo Choi.
(Yin hsing chieh chien tzu tien)
音形捷檢字典：國粵注音・漢英合解 = Chinese-English Dictionary / 霍寶材撰著. — 香港：霍氏出版公司，1977.
134, 221, 357-376, 79, [5] p. ; 20 x 27 cm.
English and/or Chinese.
$50.00
1. Chinese language—Dictionaries. 2. Chinese language—Dialects—China—Canton—Dictionaries. 3. Chinese language—Dictionaries—English. I. Title. II. Title: Chinese-English Dictionary.
PL1420.F6 78-840804

—DIALECTS—CHINA—CANTON—DICTIONARIES—ENGLISH

Choy, Rita Mei-Wah.
Read and write Chinese : a simplified guide to the Chinese characters / Rita Mei-Wah Choy. — 2d ed. — San Francisco, Calif. : Choy, c1979.
xxix, 310 p. ; 18 cm.
English and Chinese.
Includes index.
$5.95
1. Chinese language—Dictionaries—English. 2. Chinese language—Dialects—China—Canton—Dictionaries—English. I. Title.
PL1455.C62 1979 495.1'8'2421 79-119535
79 MARC

—DIALECTS—CHINA—JU-KAO HSIEN

Ting, Pang-hsin.
(Ju-kao fang yen ti yin yün)
如皐方言的音韻 / 丁邦新. — [台北? : s. n., 1978?]
84 p. (on double leaves), [5] fold. leaves ; 27 cm.
Spine title.
Reproduced from ms. copy with original date 1963 at end of work.
Bibliography: leave 84.

1. Chinese language—Dialects—China—Ju-kao hsien. 2. Chinese language—Phonology. I. Title.
PL1901.T54 1978 79-835068

—DIALECTS—CHINA—KANSU

Zav'i͡alova, Ol'ga Isaakovna.
Dialekty Gan'su / O. I. Zav'i͡alova. — Moskva : Nauka, 1979.
119 p. : ill. ; 21 cm. USSR79
At head of title: Akademii͡a nauk SSSR. Institut Dal'nego Vostoka.
Bibliography: p. 66-[68]
1.00rub
1. Chinese language—Dialects—China—Kansu. I. Title.
PL1905.Z3 79-380476
79 MARC

—DIALECTS—CHINA—KWEICHOW (PROVINCE)

Li, Fang-kuei.
(Shui hua yen chiu)
水話研究 / 李方桂著；[編輯者中央研究院歷史語言研究所出版品編輯委員會]. — 臺北：該所，民國66- [1977-
v. : 26 cm. — (專刊 - 中央研究院歷史語言研究所 ; 73)
Added t. p.: A study of the Sui language, texts and translations.
Pbk. : NT$200.00
1. Chinese language—Dialects—China—Kweichow (Province) I. Title. II. Title: A study of the Sui language, texts and translations.
PL1621.K8L5 79-836713

—DIALECTS—CHINA—NANKING

Wang, Su-ts'un.
(Nan-ching hua)
南京話 / 王素存編纂. — [台北? : s. n., 1977?]
2 v. ; 27 cm.
Caption title.
Reproduced from ms. copy on double leaves.

1. Proverbs, Chinese—China—Nanking. 2. Children's poetry, Chinese—China—Nanking. 3. Chinese language—Dialects—China—Nanking. I. Title.
PN6519.C5W36 78-840812

—DIALECTS—CHINA—TUNG-SHAN ISLAND—GLOSSARIES, VOCABULARIES, ETC.

Nakajima, Motoki.
(Bingo Tōzantō hōgen kiso goi shū)
閩語東山島方言基礎語彙集 / 中嶋幹起著. — 東京：東京外国語大学アジア・アフリカ言語文化研究所，1977.
276 p. : ill. ; 26 cm. Ja 78-4299
Added t. p.: A study of the basic vocabulary of the Min dialect in the Tung Shan Island.
English and Japanese.
Accompanied by "日本語索引" (52 p.)
Includes indexes.
1. Chinese language—Dialects—China—Amoy—Glossaries, vocabularies, etc. 2. Chinese language—Dialects—China—Tung-shan Island—Glossaries, vocabularies, etc. I. Title. II. Title: A study of the basic vocabulary of the Min dialect in the Tung Shan Island.
PL1704.N35 495.1'7 79-801436

—DIALECTS—TAIWAN

Iwasaki, Keitarō.
(Shinsen Nittai gengo shū)
新撰日臺言語集 / 岩崎敬太郎著. — 臺北：新撰日臺言語集發行所，大正2 [1913]，大正5 [1916] printing.
2, 14, 696 p. ; 23 cm.

1. Chinese language—Dialects—Taiwan. 2. Chinese language—Conversation and phrase books—Japanese. I. Title. II. Title: Nittai gengo shū.
PL1834.I9 78-815845

Lien, Heng, 1878-1936.
(T'ai-wan yü tien)
臺灣語典；雅言 / 連橫撰. — [臺北縣永和鎮]：文海出版社，[1978]
3, 2, 108, 130 p. ; 22 cm. — (近代中國史料叢刊續編[i. e.輯] ; 507 : 第五十一輯)
Reprint of two works.
1. Chinese language—Dialects—Taiwan. 2. Taiwan—Description and travel. 3. Taiwan—Social life and customs. I. Lien, Heng, 1878-1936. Ya yen. 1978. II. Title. III. Title: Ya yen. IV. Series: Shen, Yün-lung, comp. Chin tai Chung-kuo shih liao ts'ung k'an hsü chi ; 507.
DS755.S532 vol. 507 78-843948
[PL1837.L5]

Lin, Shuang-fu.
The grammar of disjunctive questions in Taiwanese / by Shuang-fu Lin. — Taipei : Taiwan Students Book Co., [c1975]
iii, 243 p. : diagrs. ; 22 cm.
English and Taiwanese.
Bibliography: p. 240-243.
1. Chinese language—Dialects—Taiwan. 2. Chinese language—Interrogative. 3. Chinese language—Syntax. I. Title.
PL1833.L5 495.1'7 78-106798
79 MARC

Lu, Shu-mei.
(T'ai-wan Min-nan yü yin yün yen chiu)
臺灣閩南語音韻研究 / 盧淑美著. — 臺北：文史哲出版社，民國66 [1977]
122 p. ; 22 cm.
Bibliography: p. 122.
NT$50.00

1. Chinese language—Dialects—Taiwan. 2. Chinese language—Phonology. I. Title.
PL1839.L8 78-841594

—DICTIONARIES

Ch'en, Kuo-hung.
(Tz'u ch'üan)
辭全 / 編著者陳國弘. — 台北：誠正出版社：總經銷誠正書局，民國64 [1975]
2046, [150] p. : ill. ; 16 cm.
Colophon title.

1. Chinese language—Dictionaries. I. Title.
PL1420.C47 78-842995

CHINESE LANGUAGE
—DICTIONARIES (Continued)

Ch'iu, Ch'un-lin.
(Ch'iu shih tzu tien)
丘氏字典 / 編著者丘春林；編輯助理葉秀馨. — 台北：道光文化事業有限公司, 民國67 [1978]
1102 p. ; 20 cm.
Colophon title.
At head of title: Erh san su chien fa.
NT$180.00
1. Chinese language—Dictionaries. I. Yeh, Hsiu-hsin. II. Title.
PL1420.C518 78-842000

Chou, Fa-kao.
(Kuang ya shu cheng yin shu so yin)
廣雅疏證引書索引 / 主編周法高；編纂范國 .. [et al.]. — [香港]：中文大學出版社, 1978.
18, 679 p. ; 21 cm.
ISBN 9-622-01164-0 : $40.00
1. Wang, Nien-sun, 1744-1832. Kuang ya shu cheng—Concordances. 2. Chinese language—Dictionaries. I. Fan, Kuo. II. Title.
PL1465.C53W332 78-843302

Chou, T'ien-hsiang.
(Ch'en kuang hsüeh sheng tz'u tien)
晨光學生詞典 / [編著者周天翔]. — 再版. — 台南：晨光出版社, 民國67 [1978]
6, 916, 83 p. : ill. ; 14 cm.
Includes index.
"附注音符號檢字表": p. 1-83 (last group)
NT$50.00
1. Chinese language—Dictionaries. I. Title. II. Title: Hsüeh sheng tz'u tien.
PL1420.C5215 78-842486

Fok, Bo Choi.
(Yin hsing chieh chien tzu tien)
音形捷檢字典：國粵注音·漢英合解 = Chinese-English Dictionary / 霍寶材撰著. — 香港：霍氏出版公司, 1977.
134, 221, 357-376, 79, [5] p. ; 20 x 27 cm.
English and/or Chinese.
$50.00
1. Chinese language—Dictionaries. 2. Chinese language—Dialects—China—Canton—Dictionaries. 3. Chinese language—Dictionaries—English. I. Title. II. Title: Chinese-English Dictionary.
PL1420.F6 78-840804

(Hsin hsiu K'ang-hsi tzu tien)
新修康熙字典 / [張玉書等總閱；凌紹雯等纂修；高樹藩重修]. — 台北：啓業書局, 民國68 [1979]
2 v. (71, 2424, 276 p.) ; 22 cm.
NT$600.00
1. Chinese language—Dictionaries. I. Chang, Yü-shu, 1642-1711. II. Kao, Shu-fan. III. Title: K'ang-hsi tzu tien.
PL1420.K3 1979 79-840599

(Hsin tz'u tien)
新辭典：附四角號碼索引. — 星洲：世界書局, [1957?]
5, 6, 2107, 118 p. ; 15 cm.
Cover title.
Includes index.
1. Chinese language—Dictionaries.
PL1420.H7978 1957b 78-843301

Hsing-chün, Shih, fl. 997.
[Lung k'an shou chien]
龍龕手鑑 / [行均]；正宗敦夫編纂校訂. — 東京：現代思潮社, 昭和52 [1977]
10, 414, 2 p. ; 16 cm. — (覆刻日本古典全集) Ja 78-1266
Photoreprint of the 1934 ed. (reproduction of Sung pen hsin hsiu Lung k'an shou chien with new footnotes) published by Nihon Koten Zenshū Kankōkai, Tokyo; with explanatory text in Japanese.
1. Chinese language—Dictionaries. I. Masamune, Atsuo, 1881-1958. II. Title. III. Title: Lung k'an shou chien. IV. Series: Nihon koten zenshū.
Title romanized: Ryōgan shukan.
PL1465.H75 1977 79-804552

Hsü, Ch'ing-t'i.
(Tz'u yüan)
辭淵 / 許清梯主編. — 臺北：大新書局, 民國67 [1978]
2, 77, 2152, 197 p. : ill. ; 17 cm.
NT$200.00
1. Chinese language—Dictionaries. I. Title.
PL1420.H843 79-836127

K'ang, Che-mao.
(Hsin pien kuo yü tzu tien)
新編國語字典：附注音符號檢字表 / [著作者康哲茂]. — 台南：光田出版社, 民國67 [1978]
6, 305, 65 p. : ill. ; 16 cm.
Includes indexes.
NT$20.00
1. Chinese language—Dictionaries. I. Title.
PL1420.K266 79-839151

K'ang, Che-mao.
[Kuo yü ta tz'u tien]
國語新辭典：附注音符號檢字表 / [編輯者康哲茂]. — 台南：綜合出版社, 民國63 [1974]
6, 899, 65 p. : ill. ; 20 cm.
1. Chinese language—Dictionaries. I. Title.
Title romanized: Kuo yü hsin tz'u tien.
PL1420.K27 1974b 78-842003

K'ang, Che-mao.
[Kuo yü ta tz'u tien]
綜合國語辭典 / 康哲茂編著. — 台南：臺灣綜合出版社, 民國63 [1974]
6, 899, 65 p. : ill. ; 15 cm.
"注音符號檢字表": p. 1-65 (last group)
1. Chinese language—Dictionaries. I. Title. II. Title: Kuo yü tz'u tien.
Title romanized: Tsung ho kuo yü tz'u tien.
PL1420.K27 1974c 78-842002

K'ang, Che-mao.
[Kuo yü ta tz'u tien]
國語新辭典：附注音符號檢字表 / [著作人康哲茂]. — 第1版. — 嘉義：中國晉學出版社, 民國63 [1974]
6, 786, 65 p. : ill. ; 15 cm.
Cover title: Piao chun kuo yü hsin tz'u tien.
1. Chinese language—Dictionaries. I. Title. II. Title: Piao chun kuo yü hsin tz'u tien.
Title romanized: Kuo yü hsin tz'u tien.
PL1420.K27 1974d 78-841996

K'ang, Che-mao.
[Kuo yü ta tz'u tien]
新國語詞典：附注音符號檢字表 / [編者康哲茂. — 臺南：啓仁書局, 民國63 [1974]
6, 899, 65 p. : ill. ; 14 cm.
1. Chinese language—Dictionaries. I. Title.
Title romanized: Hsin kuo yü tz'u tien.
PL1420.K27 1974e 78-841994

K'ang, Che-mao.
[Kuo yü ta tz'u tien]
模範學生辭典：附注音符號檢字表 / [編輯者康哲茂]. — 臺南：臺南新世紀出版社, 民國63 [1974]
6, 786, 65 p. : ill. ; 14 cm.
1. Chinese language—Dictionaries. I. Title. II. Title: Hsüeh sheng tz'u tien.
Title romanized: Mo fan hsüeh sheng tz'u tien.
PL1420.K27 1974f 78-841979

K'ang, Che-mao.
[Kuo yü ta tz'u tien]
標準國音學生詞典：附注音符號檢字法 [i. e. 表 / 編者康哲茂]. — 臺南：綜合出版社, 民國63 [1974]
6, 786, 65 p. : ill. ; 14 cm.
1. Chinese language—Dictionaries. I. Title. II. Title: Hsüeh sheng tz'u tien.
Title romanized: Piao chun kuo yin hsüeh sheng tz'u tien.
PL1420.K27 1974g 78-841980

Kao, Ching-nan.
(Kuo yü ta tz'u tien)
國語大辭典 / 主編人高靜男. — 台南：正業書局, 民國64 [1975]
8, 955, 166, 64 p. ; 16 cm.
Colophon title.
On spine: Tsui hsin kuo yü chu yin.
"注音符號檢字表": p. 1-64 (last group)
1. Chinese language—Dictionaries. I. Title.
PL1420.K344 79-835147

Kung nung ping tzu tien pien hsieh tsu.
(Kung nung ping tzu tien)
工农兵字典 = Gong nong bing zi dian / 《工农兵字典》编写组. — [香港? : s. n., 1978]
886 p. ; 18 cm.
Reprint of the 1978 ed. (rev. ed.) published by Shang-hai chiao yü ch'u pan she, Shanghai.
1. Chinese language—Dictionaries. I. Title. II. Title: Gong nong bing zi dian.
PL1420.K827 1978 79-835720

(Mo fan hsüeh sheng tzu tien)
模範學生字典：詞性分類 / [編輯者啓明編譯部；校訂者齊鐵恨]. — 再版. — 九龍：香港啓明書局, 1977.
1, 7, 327, 56 p.
$3.00
1. Chinese language—Dictionaries. I. Ch'i, T'ieh-heng. II. Ch'i ming pien i pu.
PL1420.M6 79-835329

(Piao chun hsüeh sheng tzu tien)
標準學生字典：最新國語注音 / [編著者正業書局]. — 台南：該局, 民國64 [1975]
8, 438, 64 p. ; 14 cm.
Based on Kuo yü ta tz'u tien, by Kao Ching-nan.
"注音符號檢字表": p. 1-64 (last group)
1. Chinese language—Dictionaries. I. Kao, Ching-nan. Kuo yü ta tz'u tien. II. Cheng yeh shu chü. III. Title: Hsüeh sheng tzu tien.
PL1420.P487 78-841762

(Piao chun kuo yü chu yin tz'u hai)
標準國語注音辭海 / 主編人高靜男. — 台南：正業書局, 民國64 [1975]
1909 p. in various pagings ; 16 cm.
Colophon title.
Title on cover and spine: Tz'u hai.
CONTENTS: 辭海—中國成語辭典—六体字鑑—篆字體
NT$110.00
1. Chinese language—Dictionaries. 2. Chinese language—Writing—Dictionaries. 3. Chinese language—Terms and phrases. I. Kao, Ching-nan. II. Title: Tz'u hai.
PL1420.P49 76-836693

Piao chun kuo yü chu yin tz'u hai.
辭典 / 主編者高靜男. — 台南：西北出版社, 民國67 [1978]
1909 p. in various pagings ; 16 cm.
Colophon title.
Title on spine: Piao chun kuo yü chu yin tz'u tien.
Reprint of the 1975 ed. published by Cheng yeh shu chü, Tainan.
CONTENTS: 辭海.—中國成語辭典.—六体字鑑.—篆字體.
NT$120.00
1. Chinese language—Dictionaries. 2. Chinese language—Writing—Dictionaries. 3. Chinese language—Terms and phrases. I. Kao, Ching-nan. II. Title.
Title romanized: Tz'u tien.
PL1420.P49 1978 79-835448

CHINESE LANGUAGE
—DICTIONARIES (Continued)

(Piao chun kuo yü chu yin Tz'u hai)
標準國語注音辭海 / 主編者高靜男；編輯者正業書局. — 台南：該局, 民國67 [1978]
7, 1811, 200, 64 p. ; 20 cm.
Cover title.
"大字足本"
"附錄": p. 1-200 (3d group).
"注音符號檢字表": p. 1-64 (last group)
NT$220.00
1. Chinese language—Dictionaries. 2. Chinese language—Writing—Dictionaries. 3. Chinese language—Terms and phrases. I. Cheng yeh shu chü. II. Kao, Ching-nan. III. Title: Tz'u hai.
PL1420.P49 1978b 79-839150

Shih, Shu-ling.
(Hsüeh sheng kuo yü tz'u tien)
學生國語辭典：附注音符號檢字表 / [編著者施淑玲]. — 台南：龍門圖書股份有限公司, 民國66 [1977]
6, 916, 83 p. ; 14 cm.
Includes index.
NT$50.00
1. Chinese language—Dictionaries. I. Title. II. Title: Kuo yü tz'u tien.
PL1420.S584 79-839148

Shih, Shu-ling.
(Piao chun kuo yü tz'u tien)
標準國語詞典：附注音符號檢字表 / [編著者施淑玲]. — 台南：龍門圖書股份有限公司, 民國66 [1977]
6, 1015, 83 p. ; 16 cm.
"部首檢字法"
NT$90.00
1. Chinese language—Dictionaries. I. Title. II. Title: Kuo yü tz'u tien.
PL1420.S5844 79-839147

T'ao, Ch'eng-chiu.
(Chu yin kuo yü hsin tz'u tien)
注音國語新辭典 / 陶承九主編. — 台北：力行書局, 民國65 [1976]
40, 1242, 46 p. ; 22 cm.
1. Chinese language—Dictionaries. I. Title. II. Title: Kuo yü hsin tz'u tien.
PL1420.T34 78-842090

T'ao, Ch'eng-chiu.
(Hsüeh sheng kuo yin tz'u tien)
學生國音辭典 / 陶承九主編. — 台北：力行書局, 民國67 [1978]
50, 875 p. ; 16 cm.
NT$80.00
1. Chinese language—Dictionaries. I. Title. II. Title: Kuo yin tz'u tien.
PL1420.T346 79-841162

(Tz'u hai)
辞海 = Ci hai. — 修订本. — 上海：上海辞书出版社, 19
v. : ill. ; 21 cm.
Includes index.
$0.45 (v. 13)
1. Chinese language—Dictionaries. I. Title: Ci hai.
PL1420.T86 1978 78-843468

(Tz'u hai)
辭海：一九六五年新編本 / [辭海編輯委員會編]. — 九龍：中華書局香港分局, 1965, 1979 printing.
2 v. (32, 4257 p.) : ill. ; 28 cm.
1. Chinese language—Dictionaries. I. Tz'u hai pien chi wei yüan hui.
PL1420.T86 1965 79-838574

Wang, Fu-hua.
(Tsui hsin piao chun kuo yü tz'u tien)
最新標準國語辭典 / 編著者王復華. — 台北：哲志出版社, 民國64 [1975]
10, 771 p. ; 16 cm.
Colophon title.
"注音符號檢字表": p. 721-771.
1. Chinese language—Dictionaries. I. Title. II. Title: Kuo yü tz'u tien.
PL1420.W276 78-841988

Wang, Hsing-hua.
(Kuo yin tzu tien)
國音字典：國音號碼檢字法 / 王星華編著. — 臺8版. — 臺北：正中書局, 民國65 [1976]
565 p. ; 18 cm.
Cover title.
1. Chinese language—Dictionaries. I. Title.
PL1420.W29 79-841070

Weng, Ch'uan-yü.
(Kuo min hsin tz'u tien)
國民新詞典 / 編著人翁傳鈺. — 高雄：和記百成書局, 民國64 [1975]
971, [91] p. ; 16 cm.
Colophon title.
On cover: Pai ch'eng shu tien yin hsing.
1. Chinese language—Dictionaries. I. Title.
PL1420.W47 78-841760

Weng, Ch'uan-yü.
[Kuo min hsin tz'u tien]
國中新辭典 / 編著人翁傳鈺. — 高雄：和記百成書局, 民國64 [1975]
971, [91] p. ; 16 cm.
Colophon title.
On cover: Pai ch'eng shu tien yin hsing.
1. Chinese language—Dictionaries. I. Title. II. Title: Kuo min hsin tz'u tien.
Title romanized: Kuo chung hsin tz'u tien.
PL1420.W47 1975b 78-842089

—DICTIONARIES—ENGLISH

Anderson, Olov Bertil, 1920-
A companion volume to R. H. Mathews' Chinese-English dictionary / Olov Bertil Anderson. — [2d ed.]. — [Lund, Sweden] : Studentlitteratur ; London : Curzon Press, 1978.
335 p. ; 23 cm. — (His An investigation into the present state of standard Chinese pronounciation ; pt.1a) GB78-17372
ISBN 9144152213. ISBN 0-7007-0081-1 : £7.50 (Curzon Press)
1. Mathews, Robert Henry. A Chinese-English dictionary. 2. Chinese language—Dictionaries—English. I. Mathews, Robert Henry. A Chinese-English dictionary. II. Title.
PL1209.A5 1978 pt. 1a 495.1'152 s 78-320299
[PL1455.M343A5] 79 MARC

Broadbent, K P
A Chinese/English dictionary of China's rural economy / by Kieran Broadbent. — Farnham Royal : Commonwealth Agricultural Bureaux, 1978.
xiii, 406 p. : ill., col. map (on lining papers) ; 26 cm. GB78-3198
Bibliography: p. 345-363.
Includes indexes.
ISBN 0-85198-381-2 : £30.00
1. Agriculture—Economic aspects—China—Dictionaries—Chinese. 2. China—Rural conditions—Dictionaries—Chinese. 3. Chinese language—Dictionaries—English. I. Title.
HD2097.B76 330.9'51'05 79-306829
79 MARC

China (Republic of China). Kuo fang pu.
(Hua Mei chün yü tz'u tien)
華美軍語詞典 / 國防部印頒. — [臺北]：該部, 民國59- [1970-
v. in ; 24 cm.
Edition of 1964 published under title: Mei Hua chün yü tz'u tien.
CONTENTS: [1] 陸軍之部 上·下册.—[2] 海軍之部 上·下册.—[3] 空軍之部 上·下册.
—[5] 聯戰之部.—[6] 政戰之部.
1. Military art and science—Dictionaries—Chinese. 2. Chinese language—Dictionaries—English. I. Title.
U25.C47 1970 78-842027

Choy, Rita Mei-Wah.
Read and write Chinese : a simplified guide to the Chinese characters / Rita Mei-Wah Choy. — 2d ed. — San Francisco, Calif. : Choy, c1979.
xxix, 310 p. ; 18 cm.
English and Chinese.
Includes index.
$5.95
1. Chinese language—Dictionaries—English. 2. Chinese language—Dialects—China—Canton—Dictionaries—English. I. Title.
PL1455.C62 1979 495.1'8'2421 79-119535
79 MARC

Devloo, Edmond.
An etymological Chinese-English dictionary : a handbook for the systematical study of the most useful 8000 Chinese characters with the etymological explanation of the 200 primitives (stemwords) and of the 800 phonetics heading the different groups of ideograms / compiled by Edmond Devloo. — Tai-pei : Hua Ming Press, 1969.
917 p. ; 22 cm.
At head of title: A.M.D.G.
Title also in Chinese: Han tzu tzu yüan yen chiu tzu tien.
Errata slip inserted.
Includes indexes.
1. Chinese language—Dictionaries—English. I. Title. II. Title: Han tzu tzu yüan yen chiu tzu tien.
PL1455.D4 495.1'2 78-378501
78 MARC

(Fen hsi chi shu tz'u tien)
分析技术词典. — 第1版. — 北京：科学出版社, 1977-
v. : ill. ; 19 cm.
CONTENTS: [1] 色谱分析
$0.42 (v. 1)
1. Chemistry, Analytic—Dictionaries—Chinese. 2. Chinese language—Dictionaries—English.
QD131.F46 78-842970

Fok, Bo Choi.
(Yin hsing chieh chien tzu tien)
音形捷檢字典：國粵注音·漢英合解 = Chinese-English Dictionary / 霍寶材撰著. — 香港：霍氏出版公司, 1977.
134, 221, 357-376, 79, [5] p. ; 20 x 27 cm.
English and/or Chinese.
$50.00
1. Chinese language—Dictionaries. 2. Chinese language—Dialects—China—Canton—Dictionaries. 3. Chinese language—Dictionaries—English. I. Title. II. Title: Chinese-English Dictionary.
PL1420.F6 78-840804

(Hsin Han Ying tz'u tien)
新漢英詞典 = A new Chinese-English dictionary. — 香港：香港青年出版社, 1979.
6, 718 p. ; 19 cm.
Includes indexes.
$26.00
1. Chinese language—Dictionaries—English. I. Title: A new Chinese-English dictionary. II. Title: Han Ying tz'u tien.
PL1455.H785 495.1'3'21 79-839099

(Hsiu chen Han Ying tz'u tien)
袖珍漢英詞典 = A Pocket Chinese-English dictionary. — 九龍：中華書局, 1978.
620 p. ; 14 cm.
Includes indexes.
$5.50
1. Chinese language—Dictionaries—English. I. Title: A Pocket Chinese-English dictionary.
PL1455.H792 78-841189

Liu, Ta-jen, fl. 1936-
(Liu shih Han Ying tz'u tien)
劉氏漢英辭典 = Liu's Chinese-English dictionary / 劉達人編. — 台北：華英出版社：經銷商全台書城, 民國67 [1978]
ixxx, [i. e. xxxi], 1554 p. : ill. ; 27 cm.
Colophon on label inserted.
NT$1200.00
1. Chinese language—Dictionaries—English. I. Title. II. Title: Liu's Chinese-English dictionary.
PL1455.L68 1978b 78-841592

CHINESE LANGUAGE
—DICTIONARIES—ENGLISH
(Continued)

(She hui kung tso tz'u tien)
社會工作辭典 / [主編人社會工作辭典編印委員會] ; 主編蔡漢賢. — [台北] : 中華民國社區發展研究訓練中心, 民國66 [1977]
6, 170, 961 p. ; 19 cm.
Includes indexes.
1. Social service—Dictionaries—Chinese. 2. Chinese language—Dictionaries—English. 3. Social service—Dictionaries. 4. English language—Dictionaries—Chinese. 5. Social service—Bibliography. I. Ts'ai, Han-hsien. II. She hui kung tso tz'u tien pien yin wei yüan hui.
HV12.S53 78-841809

Shih, Te-chung.
(Ying Han nung yao tz'u tien)
英汉农药辞典 / 石得中, 胡笑形, 曹为超编. 第1版. — 北京 : 石油化学工业出版社, 1978.
[x], 699 p. ; 20 cm.
Bibliography: p. [iii]-[iv]
Appendices (p. 448-699): 日英汉对照农药名称.—汉英日对照农药名称.—农药分子式索引.—英拉汉对照常见病, 虫, 草害名称.—日拉汉对照常见病虫, 草害名称
1. Agricultural chemicals—Dictionaries. 2. English language—Dictionaries—Chinese. 3. Agricultural chemicals—Dictionaries—Chinese 4. Chinese language—Dictionaries—English. I. Hu, Hsiao-hsing joint author. II. Ts'ao, Wei-ch'ao, joint author. III. Title.
S585.S5514 79-837461

(Tsui hsin Han Ying tz'u tien)
最新汉英词典 = A new Chinese-English dictionary. — [s. l.] : 云南书局, 1978.
36, 744 p. ; 19 cm.
1. Chinese language—Dictionaries—English. I. Title: A new Chinese-English dictionary.
PL1455.T79 495.1'3'21 78-843762

United States. Air Force. 7602nd Air Intelligence Group.
Chinese-English rocketry dictionary. — [San Francisco] : 7602nd Air Intelligence Group, 1977-
v. ; 27 cm.
Includes index.
1. Rocketry—Dictionaries—Chinese. 2. Chinese language—Dictionaries—English.
TL780.5.U54 1977 629.47'5'03 79-600811
79 MARC

(Wan yung Ying wen shou ts'e)
萬用英文手册 = Modern English handbook / [編著中美出版中心]. — 香港 : 該中心, [1978?]
ca. 550 p. in various pagings : ill. ; 15 cm.
"特大號"
$6.00
1. Chinese language—Terms and phrases. 2. Chinese language—Dictionaries—English. I. Chung Mei ch'u pan chung hsin. II. Title: Modern English handbook.
PL1497.W27 1978 78-842093

Wang, Ting-ho.
(Han Ying shih i tsao chü ta tzu tien)
漢英釋意造句大字典 = A dictionary to the use of the 2000 basic Chinese characters / 主編王定和 ; 協編 s. Contanceau, J. Pach. — 臺北 : 成文出版社, 民國67 [1978]
xxxiii, 1280 p. ; 22 cm.
NT$380.00
1. Chinese language—Dictionaries—English. 2. Chinese language—Sentences. I. Title. II. Title: A dictionary to the use of the 2000 basic Chinese characters.
PL1455.W19 495.1'3'21 79-838273

—DICTIONARIES—GERMAN

Hoffmann, Alfred, 1911-
Glossar der wichtigsten Säugetiere Chinas : e. lexikograph. Beitr. zur modernen chines. Sprache : chines.-dt. u. dt.-chines. : mit e systemat. Übersicht über d. wichtigsten Säugetiere Chinas / von Alfred Hoffmann. — Wiesbaden : Harrassowitz, 1978.
xi, 103 p. ; 24 cm. — (Veröffentlichungen des Ostasien-Instituts der Ruhr-Universität Bochum ; Bd. 24) GFR79-A
Includes bibliographical references.
ISBN 3-447-02001-6
1. Mammals—Nomenclature (Popular)—Dictionaries—Chinese. 2. Chinese language—Dictionaries—German. 3. Mammals—Nomenclature (Popular)—Dictionaries—German. 4. German language—Dictionaries—Chinese. 5. Mammals—China. I. Title. II. Series: Ruhr-Universität Bochum. Ostasien-Institut. Veröffentlichungen ; Bd. 24.
QL708.H63 79-361044
79 MARC

Hoffmann, Alfred, 1911-
Neue chinesische Vogelnamen : Nachträge zum Glossar der heute gültigen chinesischen Vogelnamen : nebst einer neuen systematischen Übersicht über die Vögel Chinas / von Alfred Hoffmann. — Wiesbaden : Harrassowitz, 1978.
x, 113 p. ; 24 cm. — (Veröffentlichungen des Ostasien-Instituts der Ruhr-Universität Bochum ; Bd. 23) GFR***
ISBN 3-447-01980-8
1. Birds—Nomenclature (Popular)—Dictionaries—Chinese. 2. Chinese language—Dictionaries—German. 3. Birds—Nomenclature (Popular)—Dictionaries—German. 4. German language—Dictionaries—Chinese. 5. Birds—China. I. Hoffmann, Alfred, 1911- Glossar der heute gültigen chinesischen Vogelnamen. II. Title. III. Series: Ruhr-Universität Bochum. Ostasien-Institut. Veröffentlichungen ; Bd. 23.
QL677.H69 79-361040
79 MARC

—DICTIONARIES—INDONESIAN

Lie, Joek Koey.
(Chung-hua Yin-ni ta tz'u tien)
中華印尼大辭典 = Kamus baharu bahasa Tionghoa-Indonesia / 李毓愷編. — Tjet. 32. — [Djakarta : Warga, 1972]
330 p. ; 19 cm.
Chinese and Indonesian.
Includes indexes.
1. Chinese language—Dictionaries—Indonesian. I. Title. II. Title: Kamus baharu bahasa Tionghoa-Indonesia.
PL5077.L48 1972 74-940820

—DICTIONARIES—JAPANESE

(Hsien tai Chung Jih tz'u tien)
現代中日辭典 = Modern Chinese Japanese dictionary / 總編輯鈕先銘. — 台北 : 名山出版社, 民國67 [1978]
943, 60 p. ; 17 cm.
Colophon title: Ming shan hsien tai Chung Jih tz'u tien.
Includes index.
NT$240.00 ($8.00 U.S.)
1. Chinese language—Dictionaries—Japanese. I. Niu, Hsien-ming, 1912- II. Title: Modern Chinese Japanese dictionary. III. Title: Ming shan hsien tai Chung Jih tz'u tien.
PL681.C5H79 79-841437

—DICTIONARIES—KOREAN

(Hanŭihak yongŏ sajŏn)
漢醫學用語辭典 / 編輯部編. — 서울 : 杏林出版社, 1978.
283 p. ; 22 cm.
₩4000
1. Medicine, Korean—Dictionaries—Korean. 2. Medicine, Chinese—Dictionaries—Chinese. 3. Chinese language—Dictionaries—Korean. I. Haengnim Ch'ulp'ansa. P'yŏnjippu.
R121.H27 78-826394

—DICTIONARIES—POLYGLOT

Takenouchi, Yasumi.
(Kan-Nichi-Ō taishō sekai jimmei jiten)
漢日欧対照世界人名辞典 / 竹之内安巳著. — 鹿児島 : 鹿児島短期大学南日本文化研究所, 1976.
267 p. ; 25 cm. — (南日本文化研究所研究叢書 ; 2) Ja 76-14315
Cover title.
First ed. published in 1964 under title: Sekai jimmei Chūgokugo jiten.
Bibliography: 11th-12th prelim. pages.
1. Names, Personal—Dictionaries—Polyglot. 2. Chinese language—Dictionaries—Polyglot. 3. Biography. I. Title. II. Title: Sekai jimmei jiten. III. Series: Kagoshima Tanki Daigaku. Minami Nihon Bunka Kenkyūjo. Minami Nihon Bunka Kenkyūjo kenkyū sōsho ; 2.
CS2305.T34 1976 79-804587

—DICTIONARIES—SPANISH

(Hua Hsi tsung ho ta tz'u tien)
華西綜合大辭典 = Diccionario español de la lengua china / [主編者新陸書局編輯部]. — 台北 : 該書局, 民國67 [1978]
xxxviii, 1140, 173 p. ; 27 cm.
Spanish and/or Chinese.
Includes indexes.
NT$350.00
1. Chinese language—Dictionaries—Spanish. I. Hsin lu shu chü. T'ai-pei. Pien chi pu. II. Title: Diccionario español de la lengua china.
PL1459.S6H8 495.1'3'61 79-838225

—DICTIONARIES—VIETNAMESE

Hoàng Thúc Trâm.
Hán-Việt tân từ-điển / Hoàng Thúc-Trâm. — [Bản thứ 2]. — Saigon : Vĩnh Bảo, c1951.
1505 p. ; 20 cm.
1. Chinese language—Dictionaries—Vietnamese. I. Title.
PL4377.H55 1951 79-984029
79 MARC

—ETYMOLOGY

Ch'en, Chien-hou, 1824-1887.
(Shuo wen pu shou t'i yao yü chin tu)
說文部首提要與今讀 / 陳建信[i. e.侯], 錢玄同撰. — 臺北縣板橋 : 藝文印書館, 民國66 [1977]
2, 102 p. ; 20 cm.
Reprint of 2 works, the 1st one selected from Shuo wen chieh tzu chu, with author's pref. dated 1872; and the 2d from the author's ms. dated 1935.
CONTENTS: 陳建侯 說文提要.—錢玄同 說文部首今讀(聲次)
1. Hsü, Shen, d. 120? Shuo wen chieh tzu. 2. Chinese language—Etymology. I. Hsü, Shen, d. 120? Shuo wen chieh tzu. II. Ch'en, Chien-hou, 1824-1887. Shuo wen t'i yao. 1977. III. Ch'ien, Hsüan-t'ung, 1887-1939. Shuo wen pu shou chin tu. 1977. IV. Title.
PL1281.H83C3 79-841423

h'en, Kuang-cheng.
(Hui i yen chiu)
會意研究 / [著者陳光政]. — [台北] : 文史哲出版社 ; [高雄 : 總經銷復文書局], 民國65 [1976]
2, 248 p. ; 22 cm.
Distributor's name stamped on verso of t. p.
Includes bibliographical references.
1. Chinese language—Etymology. I. Title.
PL1281.C5165 79-838616

Chiang, Shan-kuo, 1898-
(Hsing sheng tzu ti fen hsi)
形聲字的分析 / 蔣善國著. — [香港? : s. n., 1978?]
35 p. ; 27 cm.
Reprint of the 1959 ed. published by Wen tzu kai ko ch'u pan she, Peking.
1. Chinese language—Etymology. I. Title.
PL1281.C544 1978 78-840841

Chou, Hsiao-p'ing.
(Shuo wen hsing sheng tzu sheng mu chia chieh shuo fa fan)
說文形聲字聲母假借說發凡 / 周小萍. — [陽明山? : s. n.], 民國63 [1974]
221 p. (on double leaves) ; 27 cm.
Reproduced from ms. copy.
"中國文化學院中國文學研究所碩士論文"
Bibliography: p. 215-221.
1. Hsü, Shen, d. 120? Shuo wen chieh tzu. 2. Chinese language—Etymology. I. Title.
PL1281.H83C56 79-841833

Fang, Chün-chi.
(Shih ming k'ao shih)
釋名考釋 / [著者方俊吉]. — 臺北 : 文史哲出版社, 民國67 [1978]
2, 1, 159 p. ; 22 cm.
Bibliography: p. 157-159.
NT$80.00
1. Liu, Hsi, fl. 200. Shih ming. 2. Chinese language—Etymology. I. Title.
PL1281.L5563F3 79-835491

Huang, K'an, 1886-1936.
(Huang K'an lun hsüeh tsa chu)
黃侃論學雜著 中華書局上海編輯所編輯 北京 中華書局 1964.
14, 4, 2, 488 p. 21 cm.
1. Chinese language—Phonology. 2. Hsü, Shen, d. 120? Shuo wen chieh tzu. 3, Chinese language—Etymology. 4. Erh ya. 5. Rites and ceremonies—China. I. Chung-hua shu chü. Shang-hai pien chi so.
PL1201.H89 1964 C 66-298
rev

CHINESE LANGUAGE
—ETYMOLOGY (Continued)

Huang, K'an, 1886-1936.
(Huang K'an lun hsüeh tsa chu)
黃侃論學雜著 ［臺1版］ 臺北 : 臺灣中華書局 ［1969］
11, 1, 2, 488 p. 22 cm.
Originally published in 1964 by 北京中華書局 with editor statement: 中華書局上海編輯所編輯
1. Chinese language—Phonology. 2. Hsü, Shen, d. 120? Shuo wen chieh tzu. 3. Chinese language—Etymology. 4. Erh ya. 5. Rites and ceremonies—China. I. Chung-hua shu chü. Shang-hai pien chi so.
PL1201.H89 1969 70-837916

Huang, K'an, 1886-1936.
[Huang K'an lun hsüeh tsa chu
黃季剛先生論學名著 / 著作人黃侃. — 台1版. — 台北 : 九思出版社, 民國66 [1977]
4, 2, 488 p. ; 22 cm. — (九思叢書 ; 13)
Colophon title.
Series romanized: Chiu ssu ts'ung shu.
Originally published under title: Huang K'an lun hsüeh tsa chu.
NT$180.00
1. Chinese language—Phonology. 2. Hsü, Shen, d. 120? Shuo wen chieh tzu. 3. Chinese language—Etymology. 4. Erh ya. 5. Rites and ceremonies—China. I. Title.
Title romanized: Huang Chi-kang hsien sheng lun hsüeh ming chu.
PL1201.H89 1977 78-841131

K'o, Shu-ling.
(Shuo wen shang sheng tzu ken yen chiu)
說文上聲字根研究 / 柯淑齡. — [台北? : s. n., 1978?]
2 v. (4, 177 double leaves) ; 27 cm.
Spine title.
Reproduced from ms. copy.
Bibliography: leaves 175-177.
1. Hsü, Shen, d. 120? Shuo wen chieh tzu. 2. Chinese language—Etymology. I. Title.
PL1281.H83K6 1978 79-835090

Li, Hsiao-ting.
(Lun wen tzu hsüeh ti shih yung hsing)
论文字学的实用性 / 李孝定. — Singapore : 南洋大學研究院人文與社會科學研究所, 1976.
11 leaves ; 28 cm. — (研究論文 - 南洋大學研究院人文與社會科學研究所 ; no. 11)
1. Chinese language—Etymology. I. Title. II. Series: Nanyang University. Institute of Humanities and Social Sciences. Occasional paper series - Institute of Humanities and Social sciences, Nanyang University ; no. 11.
PL1281.L516 78-940248

Liu, Kuo-hsiang.
(Chung-kuo wen tzu hsüeh)
中國文字學 / 作者劉國祥. — 台北 : 天工書局, 民國67 [1978]
497 p. : ill. ; 22 cm.
Colophon title.
Errata slip inserted.
NT$190.00
1. Chinese language—Etymology. I. Title.
PL1281.L57 78-843866

Lü, Fo-t'ing.
(Wen tzu hua yen chiu)
文字畫研究 / 呂佛庭. — 臺1版. — 臺北 : 華正書局, 民國64 [1975]
5, 3, 51, 96 p. : chiefly ill. (some col.) ; 39 cm.
Colophon and spine title: Chung-kuo wen tzu hua yen chiu.
1. Lü, Fo-t'ing. 2. Painting, Chinese. 3. Chinese language—Etymology. I. Title. II. Title: Chung-kuo wen tzu hua yen chiu.
ND1049.L7798A4 1975 78-842468

Murakami, Kyūkichi, 1898-
(Jigen o saguru)
字原を探る / 村上久吉. — 東京 : 叢文社, 1978.
378 p. : ill. ; 22 cm. Ja 78-21281
Includes indexes.
¥2000
1. Chinese language—Etymology. I. Title.
PL1281.M8 78-816321

Nakazawa, Mareo, 1907-
(Kanji Kango gaisetsu)
漢字·漢語概説 / 中沢希男著. — 東京 : 教育出版, 1978.
234, 4 p., [1] leaf of plates : ill. ; 21 cm. Ja 78-6407
Bibliography: p. 233-234.
Includes index.
¥1700
1. Chinese language—Etymology. 2. Chinese language—Pronunciation. 3. Kundoku—History. I. Title.
PL1281.N34 79-800080

(Shuo wen chieh tzu ku lin cheng pu ho pien)
說文解字詁林正補合編 / 合編者鼎文書局編輯部. — 臺北 : 鼎文書局, 民國66 [1977]
12 v. ; 24 cm. — (中國學術類編)
Colophon title.
Reprint ed.
Vol. 12 is index.
1. Hsü, Shen, d. 120? Shuo wen chieh tzu. 2. Chinese language—Etymology. I. Ting wen shu chü. Pien chi pu. II. Series: Chung-kuo hsüeh shu lei pien.
PL1281.H83S55 1977 78-843906

Strindberg, August, 1849-1912.
Kinesiska språkets herkomst : studier / af August Strindberg ; [förord: Lennart Utterström]. — [Staffanstorp : Abisag, 1977]
2 v. ; 21 cm. — (Abisags asienserie ; v. 3-4) S77-40
Reprint of the 1912 ed. published by Björck & Börjesson, Stockholm; with a new introduction by L. Utterström.
"Bilaga till Kina och Japan."
ISBN 9185438049 : kr40.00
1. Chinese language—Etymology. I. Strindberg, August, 1849-1912. Kina och Japan. II. Title. III. Series.
PL1281.S74 1977 79-346694
79 MARC

Tai, T'ung, 13th cent.
(Liu shu ku)
六書故 / [戴侗撰]. — [台北 : 台灣]商務[印書館, 1976?]
5 v. ; 20 cm. — (四庫全書珍本六集 ; 52-56)
Cover title.
Reproduced from Wen yüan ko copy.
1. Chinese language—Etymology. I. Title. II. Series: Ssu k'u ch'üan shu. Ssu k'u ch'üan shu chen pen liu chi ; 52-56.
AC149.S7 1976, vol. 52-56 78-843851
[PL1281]

Tai, T'ung, 13th cent.
The six scripts; or, The principles of Chinese writing. A translation by L. C. Hopkins. With a memoir of the translator by W. Perceval Yetts. Cambridge, [Eng.] University Press, 1954.
xxvii, 84 p. port., facsim. 22 cm.
Facsimile of the translation published in Amoy, 1881: p. 1-83.
1. Chinese language—Etymology. I. Hopkins, Lionel Charles, 1854-1952, tr. II. Yetts, Walter Perceval, 1878- . III. Title.
PL1281.T26513 1954 495.1'1'1 54-2892
[r79]rev2 MARC

T'ang, Lan, 1900-
(Ku wen tzu hsüeh tao lun)
古文字學導論 / 唐蘭著. — 香港 : 太平書局, 1965, 1978 printing.
52, 85, [13] leaves, [4] leaves of plates : ill. ; 21 cm.
$10.00
1. Chinese language—Etymology. I. Title.
PL1281.T338 79-837671

Teng, Ho.
(Chung-kuo wen tzu chieh kou hsüan chieh)
中國文字結構選解 / 鄧和著. — 臺北 : 正中書局 ; 九龍 : 總經銷集成圖書公司, 民國67 [1978]
279, [10] p. ; 22 cm.
1. Chinese language—Etymology. I. Title.
PL1281.T46 79-836765

Tung-lai-hsien-sheng.
(Wo yao jen shih Chung-kuo tzu)
我要認識中國字 / 東萊先生著. — 臺北 : 名人出版社, 民國67 [1978]
401 p. : ill. ; 19 cm.
NT$90.00
1. Chinese language—Etymology. I. Title.
PL1281.T95 1978 79-836570

(Wen tzu hsüeh)
文字學. — 香港 : 港青出版社, 1978.
2, 86 p. : ill. ; 19 cm.
$3.00
1. Chinese language—Etymology.
PL1281.W344 78-842172

Yen, K'o-chün, 1762-1843.
(Shuo wen sheng lei)
說文聲類 / 嚴可均撰. — 台北 : 廣文書局, 民國55 [1966]
71, 54, 1 leaves ; 20 cm. — (音韻學叢書)
Includes original t. p.: Shuo wen sheng lei erh chüan.
Reprint of the Ssu lu t'ang pen of Wei-nan Yen shih printed by Shih kuo hsüeh chai, Ch'eng-tu, in 1924.
1. Hsü, Shen, d. 120? Shuo wen chieh tzu. 2. Chinese language—Etymology. 3. Chinese language—Phonology. I. Title. II. Series: Yin yün hsüeh ts'ung shu.
PL1281.H83Y4 1966 79-839145

—ETYMOLOGY—ADDRESSES, ESSAYS, LECTURES

Kao, Ming, 1909-
(Kao Ming hsiao hsüeh lun ts'ung)
高明小學論叢 / 高明著. — 臺北 : 黎明文化事業公司, 民國67 [1978]
4, 2, 516 p. ; 22 cm. — (國學叢書)
Running title: Kao Ming wen chi.
Series romanized: Kuo hsüeh ts'ung shu.
NT$100.00
1. Chinese language—Phonology—Addresses, essays, lectures. 2. Chinese language—Etymology—Addresses, essays, lectures. I. Title. II. Title: Kao Ming wen chi.
PL1201.K29 79-837670

—ETYMOLOGY
—DICTIONARIES

Yeh, Ch'ang-ch'ing, 1898 or 9-
(Wen tzu hsüeh ming tz'u ch'üan shih)
文字學名詞詮釋 / 著作者葉長青. — [香港? : s. n., 1978?]
8, 224 p. : ill. ; 20 cm.
Colophon title.
Reprint of the 1927 ed. published by Ch'ün chung t'u shu kung ssu, Shanghai, in series: Kuo hsüeh chuan k'an she ts'ung shu.
1. Chinese language—Etymology—Dictionaries. I. Title. II. Series: Kuo hsüeh chuan k'an she. Kuo hsüeh chuan k'an she ts'ung shu.
PL1281.Y4 1978 78-843067

—EXAMINATIONS, QUESTIONS, ETC.

Hua-nan shih yüan fu chung.
(Kao k'ao yü wen fu hsi t'i kang)
高考語文覆習提綱 / 华南师院附中编 [广州 : 該院], 1978.
134 p. ; 19 cm.
Cover title.
1. Chinese language—Examinations, questions, etc. I. Title.
PL1119.H82 1978 79-840348

—GLOSSARIES, VOCABULARIES, ETC.

Chin, Shou-shen.
(Ssu yung shou ts'e)
四用手册 : 念·講·寫·用 / 金受申編. — 北京 : 寶文堂書店, 1952.
20 p. ; 18 cm. — (速成識字基本讀物)
Cover title.
At head of title: Su ch'eng shih tzu kuai kun pang chu.
With kuo yü chu yin.
1. Chinese language—Glossaries, vocabularies, etc. I. Title.
PL1489.C496 79-839468

CHINESE LANGUAGE
—GLOSSARIES, VOCABULARIES, ETC.
(Continued)

Feng, Ch'uan-huang.
(Hua yü ti ch'ing sheng tz'u hui)
华语的轻声词汇 = Huayu de qingsheng cihui / 冯传璜. — [新加坡]：南洋大学华语研究中心, 1975.
viii, 38 leaves ; 28 cm.
Cover title.

1. Chinese language—Intonation. 2. Chinese language—Glossaries, vocabularies, etc. I. Title. II. Title: Huayu de qingsheng cihui.
PL1213.F4 78–940562

Hsiao, Ts'ung-fang.
(Nung ts'un ssu yen ch'ang yung tzu)
农村四言常用字 / 萧从方, 刘毅然編. — [香港?：s. n., 1978?]
62 p. : ill. ; 21 cm.
Reprint of the 1964 ed. published by Nung ts'un tu wu ch'u pan she, Peking.

1. Chinese language—Glossaries, vocabularies, etc. I. Liu, I-jan, joint author. II. Title.
PL1498.H75 1978 78–843013
C

Hsieh, Ping, 1885–
(Ch'ang yung tzu tzu hsing fen pien piao)
常用字字形分辨表 謝冰編 太原 山西人民出版社 1958.
2, 39 p. 19 cm.

1. Chinese language—Glossaries, vocabularies, etc. 2. Chinese language—Homonyms. I. Title.
PL1489.H7 C 66–16

(Kosa ŏwŏn sajŏn)
古事語源事典 / [編纂古事熟語編纂委員會, 國典刊行會]. — [서울?]：國典刊行會出版部, 1977.
1220 p. : ill. ; 16 cm.
₩3800

1. Korean language—Dictionaries. 2. Korean language—Glossaries, vocabularies, etc. 3. Chinese language—Glossaries, vocabularies, etc. I. Kosa Sugŏ P'yŏnch'an Wiwŏnhoe. II. Kukchŏn Kanhaenghoe.
PL935.K67 79–826444

Liu, Ta-chieh, 1904–
(Chung-kuo wei hsüeh fa ta shih tzu hui)
中國文學發達史字彙 = Glossary for The history of Chinese literature development / 劉大杰原著 ; 劉崇純編注. — [台北?]：美國各大學中國語文聯合研習所, [1978?]
ca. 350 p. ; 27 cm.
English, Chinese, and Kuo yü romanization.
"This is a glossary of the first twelve chapters of the Chung-kuo wen-hsüeh fa-ta shih ... The edition used is that printed by the Taiwan Chung-hua shu-chü, 1962 (6th printing)."
Reproduced from typescript.

1. Chinese literature—History and criticism. 2. Chinese language—Glossaries, vocabularies, etc. I. Liu, Charles A. II. Title. III. Title: Glossary for The history of Chinese literature development.
PL2264.L539 78–841985
C

McNaughton, William, 1933-
Reading and writing Chinese : a guide to the Chinese writing system, the student's 1,020 list, the official 2,000 list / by William McNaughton. — Rutland, Vt. : Tuttle, c1979.
376 p. ; 19 cm.
ISBN 0-8048-1188-1 : $17.50
1. Chinese language—Glossaries, vocabularies, etc. 2. Chinese language—Writing. I. Title.
PL1495.M3 495.1'1'1 77-77699
79 MARC

Taiwan. Mi shu ch'u.
(Ch'ang yung ming tz'u hui pien)
常用名詞彙編 / 臺灣省政府秘書處編. — 南投縣中興新村：該處, 民國51 [1962]
4, 402 p. ; 19 cm.
Some terminology also in English.

1. Chinese language—Glossaries, vocabularies, etc. I. Title.
PL1489.T3 1962 78–842067
78 C

—GRAMMAR

DeFrancis, John Francis, 1911–
Character text for Beginning Chinese / by John De Francis, with the assistance of Yung Teng Chia-yee. — 2d ed. — New Haven : Yale University Press, 1976.
519 p. : ill. ; 26 cm. — (Yale linguistic series)
Prefatory matter in English, text in Chinese.
Includes index.
ISBN 0-300-02055-4. ISBN 0-300-02059-7 pbk.
1. Chinese language—Grammar. 2. Chinese language—Conversation and phrase books. I. De Francis, John Francis, 1911– Beginning Chinese. II. Yung Teng, Chia-yee, joint author. III. Title. IV. Series.
PL1111.D37 1976 76–5105

Dougherty, Ching-yi.
Chinese for Americans / by Ching-yi Dougherty. — [s.l. : s.n.], 1975-1978.
2 v. (487 p.) ; 28 cm.
Cover title.
1. Chinese language—Grammar. I. Title.
PL1111.D68 495.1'8'2421 79-105811
79 MARC

Goldenthal, Allan B
Think Chinese, speak Chinese / Allan B. Goldenthal. — New York : Regents Pub. Co., c1978.
431 p. : ill. ; 27 cm.
Title also in Chinese: Hsiang Chung wen, shuo Chung wen.
Bibliography: p. 423.
Includes index.
1. Chinese language—Grammar. I. Title. II. Title: Hsiang Chung wen, shuo Chung wen.
PL1111.G57 495.1'8'2421 79-101041
79 MARC

(Hsien tai Han yü yü fa chiang hua)
現代汉语语法讲话 / 丁声树 ... [et al.] ; 中國語文杂志社編. — 北京：商务印書館, 1963.
x, 228 p. ; 22 cm. — (中國語文丛書)
Series romanized: Chung-kuo yü wen ts'ung shu.

1. Chinese language—Grammar. I. Ting, Sheng-shu. II. Chung-kuo yü wen tsa chih she, Peking.
PL1103.H7 78–843613

Li, Chin-hsi, 1890–
(Chung-kuo yü fa chiao ts'ai)
中國語法教材 / 黎錦熙, 刘世儒著. — [香港]：嵩華出版事業公司：發行波文書局, [19
v. ; 20 cm.
A teaching manual for Tsen yang chiao hsüeh Chung-kuo yü fa.
CONTENTS:
第3編 复式句和篇章結构
1. Chinese language—Grammar. I. Liu, Shih-ju, joint author. II. Li, Chin-hsi, 1890– Tsen yang chiao hsüeh Chung-kuo yü fa. III. Title.
PL1103.L446 1978 78–842171
78 C

Liu, Shih-ju.
(Hsien tai Han yü yü fa chiang i)
現代汉语语法讲义 / 刘世儒編著. — [香港?：s. n., 1978?]
xviii, 212 p. : ill. ; 19 cm.
Reprint of the 1964 ed. published by Shang wu yin shu kuan, Peking
Includes bibliographical references.

1. Chinese language—Grammar. I. Title.
PL1103.L58 1978 79–836472

Ma, Chu-ts'un.
(Hsien tai Han yü yü fa)
現代汉語語法 / 馬竹邨編著. — 香港：神州圖書公司, [1978?]
3, 94 p. ; 18 cm.
Reprint of the 1960 ed. published by Shan-tung jen min ch'u pan she, Tsinan, in series: Yü yen hsüeh chi ch'u chih shih.

1. Chinese language—Grammar. I. Title. II. Series: Yü yen hsüeh chi ch'u chih shih.
PL1103.M287 1978 79–836080

Marney, John.
A handbook of modern Chinese grammar / John Marney. — San Francisco : Chinese Materials Center, 1977.
viii, 78 p. ; 22 cm. — (Asian library series ; no. 3)
$3.75
1. Chinese language—Grammar. I. Title. II. Series.
PL1111.M23 495.1'8'2421 79-312821
79 MARC

T'an, Ch'üan-chi.
(Ku tai Han yü chi ch'u)
古代漢語基礎 / 譚全基編. — 九龍：中華書局香港分局, 1978.
6, 234 p. ; 19 cm.
$6.50

1. Chinese language—Grammar. I. Title.
PL1109.T3 79–838721

Teng, Shou-hsin.
A basic course in Chinese grammar : a graded approach through conversational Chinese / Teng Shou-hsin. — San Francisco : Chinese Materials Center, 1977.
xii, 135 p. ; 22 cm. — (Asian library series ; no. 5)
1. Chinese language—Grammar. 2. Chinese language—Conversation. I. Title. II. Series.
PL1111.T4 495.1'8'2421 79-312822
79 MARC

Yanagimachi, Tatsuya.
(Kambun dokkai jiten)
漢文読解辞典 / 柳町達也. — 東京：角川書店, 1978.
543 p. ; 19 cm. — (角川小辞典 ; 18) Ja 78–11914
Series romanized: Kadokawa shōjiten.
Bibliography: p. 540–543.
¥1800

1. Kunten. 2. Chinese language—Grammar. I. Title.
PL627.Y25 78–814326

—GRAMMAR—ADDRESSES, ESSAYS, LECTURES

(Yü wen hsüeh hsi ts'an k'ao tzu liao)
語文學習參考資料：内部學習材料 / 東北教育社編印. — [香港?：s. n., 1978?]
2, 112 p. ; 21 cm.
Reprint of the 1951? ed.

1. Chinese language—Grammar—Addresses, essays, lectures. I. Tung pei chiao yü she, Mukden.
PL1103.Y85 1978 79–838131

—GRAMMAR—THEORY, ETC.

Cheng, Tien-mu.
Ansätze und Methoden einer "kommunikativen Grammatik" des Chinesischen / Cheng Tien-mu. — Bochum : Studienverlag Brockmeyer, 1978-
v. ; 21 cm. GFR78-A
Includes bibliographical references and index.
ISBN 3-88339-020-8 (v. 1) : DM3.80 (v. 1)
1. Chinese language—Grammar—Theory, etc. I. Title.
PL1107.C46 78-392610
79 MARC

CHINESE LANGUAGE (Continued)

—GRAMMAR, HISTORICAL

Chan, Hsiu-hui.
(Nan pei ch'ao chu i shu ssu chung yü fa yen chiu)
南北朝著譯書四種語法研究 / 詹秀惠撰. — [台北] : 詹秀惠, 民國64 [1975]
252, [22] p. ; 27 cm. & abstract.
Cover title.
"國立臺灣大學中國文學研究所博士論文"
Bibliography: p. 244–252.

1. Chinese language—Grammar, Historical. I. Title.
PL1101.C37 79–836396

—HETERONYMS

Kawaguchi, Eiichi.
(P'o yin tzu yen chiu)
破音字研究 / 川口榮一. — 民國59 [1970]
134 double leaves ; 27 cm.
Cover title.
Reproduced from typescript.
Thesis (M. A.)—Kuo li T'ai-wan ta hsüeh.
Bibliography: leaves 133–134.

1. Chinese language—Heteronyms. I. Title.
PL1313.K3 79–841876

Tseng, Kuo-t'ai.
(P'o yin tzu ta ch'üan)
破音字大全 / 曾國泰著. — 臺北 : 名人出版社, 民國67 [1978]
405 p. ; 19 cm.
Text accompanied by kuo yü p'in yin.
NT$90.00

1. Chinese language—Heteronyms. I. Title.
PL1313.T75 79–835114

—HISTORY

[**Wang, Li**], 1901–
(Chung-kuo yü yen hsüeh shih)
中國語言學史 / [編著者莊嚴出版社]. — 臺北 : 該社, 民國67 [1978]
2, 228 p. ; 22 cm. — (古典大學叢書 ; 3)
Reprint of the 1967 ed. published in Hongkong, under the author's name: Wang Li, with some of the bibliographical references deleted.
Includes bibliographical references.
NT$80.00
1. Chinese language—History. I. Chuang yen ch'u pan she. II. Title. III. Series: Ku tien ta hsüeh ts'ung shu ; 3.
PL1075.W268 1978 78–843651

—HOMONYMS

Chu, Hsüeh-ch'iung.
(Lun yü chia chieh tzu k'ao)
論語假借字考 / 朱學瓊. — [台北? : s. n., 1978?]
3 v. (291 double leaves) ; 27 cm.
Reproduced from ms. copy, which was originally presented as the author's thesis (M. A.), Kuo li T'ai-wan ta hsüeh, 1968; with handwritten t. p.
Bibliography: leaves 288–291.
1. Confucius. Lun yü. 2. Chinese language—Homonyms. I. Confucius. Lun yü. Selections. 1978. II. Title.
PL2471.Z9H63 1978 78–842894

Hsieh, Ping, 1885–
(Ch'ang yung tzu t[..] hsing fen pien piao)
常用字字形分辨表 謝冰編 太原 山西人民出版社 1958.
2, 39 p. 19 cm.

1. Chinese language—Glossaries, vocabularies, etc. 2. Chinese language—Homonyms. I. Title.
PL1489.H7 C 66–16

Liu, Ju.
(Kuo yü t'ung yin tz'u lei pien)
國語同音詞類辨 / 編輯者劉儒 ; 校訂者黎錦熙, 方毅. — [香港? : s. n., 1978?]
7, 46 p. ; 21 cm.
Cover title.
Reprint of the ed. published by Shang wu yin shu kuan, Shanghai.

1. Chinese language—Homonyms. I. Title.
PL1311.L5 1978 79–837001

(Piao chun kuo yin t'ung yin tzu hui)
標準國音同音字彙. — 臺中 : 藍燈文化事業股份有限公司, 民國66 [1977]
3, 157 p. : ill. ; 19 cm. — (藍燈叢書) (應用叢書)
Cover title.
Series 1 romanized: Lan teng ts'ung shu. Series 2 romanized: Ying yung ts'ung shu.
NT$20.00

1. Chinese language—Homonyms. I. Title: Kuo yin t'ung yin tzu hui.
PL1311.P5 79–835425

—IDIOMS, CORRECTIONS, ERRORS

Hung, Wen-hu.
(Tzu pien tz'u tien)
字別辭典 / 洪文虎編著. — 台北縣新店鎮 : 常春樹書坊, 民國67 [1978]
ca. 300 p. in various pagings ; 20 cm. — (中國人的書)
Series romanized: Chung-kuo jen ti shu.
Includes index.
NT$50.00

1. Chinese language—Idioms, corrections, errors. I. Title.
PL1271.H98 79–835193

Li, Chih-ch'üan.
(Pieh tzu pien cheng)
別字辨正 / 李植泉編著. — 臺北 : 正中書局 ; 九龍 : 總經銷集成圖書公司, 民國67 [1978]
2, 2, 210 p. ; 21 cm.

1. Chinese language—Idioms, corrections, errors. I. Title.
PL1271.L45 78–843941

Tz'u tien pien chi tsu.
(Ch'ang yung tz'u pien hsi tz'u tien)
常用詞辨析詞典 / 《辭典》編輯組編. — 香港 : 駱駝出版社, 1977.
230 p., [1] fold. leaf ; 15 cm.
$4.50

1. Chinese language—Idioms, corrections, errors. I. Title.
PL1271.T95 78–841011

Yang, Wan-kuei.
(Tzu pien)
字辨 / [編者楊莞畂]. — 臺北 : 武陵出版社, 民國67 [1978]
260 p. ; 19 cm. — (愛書人文庫 ; A36)
Cover title: Ch'ang yung tzu pien shou ts'e.
Series romanized: Ai shu jen wen k'u.
NT$55.00

1. Chinese language—Idioms, corrections, errors. I. Title. II. Title: Ch'ang yung tzu pien shou ts'e.
PL1271.Y34 79–837675

—IDIOMS, CORRECTIONS, ERRORS —DICTIONARIES

(Hsin pien ch'eng yü tz'u tien)
新編成語詞典 / [編印香港萬源圖書公司]. — 第1版. — 香港 : 該公司, 1979.
88, 889 p. ; 14 cm.
Title in English on cover: A dictionary of Chinese idioms.
Text accompanied with Pinyin.
$10.00

1. Chinese language—Idioms, corrections, errors—Dictionaries. I. Hsiang-kang wan yüan t'u shu kung ssu. II. Title: Dictionary of Chinese idioms.
PL1271.H81433 79–840122

—IDIOMS, CORRECTIONS, ERRORS —DICTIONARIES—ENGLISH

Wei, S S
A practical dictionary of Chinese idioms, English idioms, English synonyms / S. S. Wei. — Hong Kong : Practical English Press, 1978.
ix, 26, 392 p. ; 20 cm.
Includes indexes.
$35.00 ($7.00 U.S.)
1. Chinese language—Idioms, corrections, errors—Dictionaries—English. 2. English language—Idioms, corrections, errors—Dictionaries. 3. English language—Synonyms and antonyms—Dictionaries. I. Title.
PL1271.W38 495.1'3'21 79–108083
00 MARC

—INTERROGATIVE

Lin, Shuang-fu.
The grammar of disjunctive questions in Taiwanese / by Shuang-fu Lin. — Taipei : Taiwan Students Book Co., [c1975]
iii, 243 p. : diagrs. ; 22 cm.
English and Taiwanese.
Bibliography: p. 240-243.
1. Chinese language—Dialects—Taiwan. 2. Chinese language—Interrogative. 3. Chinese language—Syntax. I. Title.
PL1833.L5 495.1'7 78–106798
79 MARC

—INTONATION

Chang, Hsiao-yü.
(Ch'ing sheng pien i chü li)
輕聲辨義舉例 / 張孝裕著. — 臺北 : 學海出版社, 民國66 [1977]
2, 11, 304 p. ; 21 cm.
Bibliography: p. 302–304.
Includes indexes.
NT$80.00

1. Chinese language—Intonation. 2. Chinese language—Accents and accentuation. I. Title.
PL1213.C4 79–841879

Ch'i, Kung.
(Shih wen sheng lü lun kao)
詩文聲律論稿 / 啓功著. — 九龍 : 中華書局, 1978.
2, 128 p. ; 21 cm.
At head of colophon title: I lin hsin lu.
$5.50

1. Chinese language—Intonation. 2. Chinese language—Versification. I. Title.
PL1213.C447 78–841455

Feng, Ch'uan-huang.
(Hua yü ti ch'ing sheng tz'u hui)
华语的轻声词汇 = Huayu de qingsheng cihui / 冯传璜. — [新加坡] : 南洋大学华语研究中心, 1975.
viii, 38 leaves ; 28 cm.
Cover title.

1. Chinese language—Intonation. 2. Chinese language—Glossaries, vocabularies, etc. I. Title. II. Title: Huayu de qingsheng cihui.
PL1213.F4 78–940562

Tai, Huai-ch'ing.
(Chung-kuo yü yin chuan hua)
中國語音轉化 / 戴淮清. — 臺北 : 東方文化書局, 1974.
7, 193 p. ; 20 cm. — (國立北京大學中國民俗學會民俗叢書 ; 164 : 第9輯) (東方文叢)
Added t. p.: The changes of Chinese enunciation.
Spine title: Chung-kuo tzu yin chuan hua.
1. Chinese language—Intonation. I. Title. II. Title: The changes of Chinese enunciation. III. Series: Kuo li Pei-ching ta hsüeh, Chung-kuo min su hsüeh hui min su ts'ung shu ; 164.
GR335.K87 vol. 164 78–842834
[PL1213]

Tai, James H-Y
Phonological changes in modern standard Chinese in the People's Republic of China since 1949 / by James H-Y. Tai ; prepared for Office of Research, United States Information Agency, Washington, D.C. — Washington : The Office, [1978?].
iv, 122 p. ; 27 cm.
1. Chinese language—Pronunciation. 2. Chinese language—Intonation. I. United States. Information Agency. Office of Research. II. Title.
PL1209.T3 495.1'1'5 78–603573
79 MARC

CHINESE LANGUAGE (Continued)

—LEXICOGRAPHY

Ch'en, Li-fu, 1900–
[Wu pi chien tzu hsüeh sheng tzu tien. Selections]
五筆檢字法 / 陳立夫著. — 臺北：中國語文研究中心, 民國60 [1971]
18, 90, [20] p. ; 19 cm. — (中國語文研究叢書)
Cover title.
"除序言外均取材自…五筆檢字學生字典"
1. Chinese language—Lexicography. I. Title. II. Series: Chung-kuo yü wen yen chiu ts'ung shu.
Title romanized: Wu pi chien tzu fa.
PL1411.C515 1971 78–841942

Das Eckzahlsystem : Einf. u. Anwendung : e. neues Nachschlagsystem für d. chines.-dt. Wörterbuch / Cheng Tien-mu ... [et al.]. — Bochum : Studienverlag Brockmeyer, 1979.
106 p. ; 21 cm. — (Programm zur Modernisierung des sinologischen Studiums, Reihe 1 ; Bd. 1) GFR79-A
"Ergänzungsband zu dem Chinesisch-deutschen-Wörterbuch."
Bibliography: p. 101-104.
ISBN 3-88339-040-2 : DM9.80
1. Chinese language—Lexicography. 2. Chinese language—Writing. 3. Chinese language—Phonetics. 4. Chinese language—Transliteration into German. I. Cheng, Tien-mu. II. Series.
PL1411.E25 495.1 79-368516
79 MARC

—PARTICLES

Huang, Ch'i-yüan.
(Tu Fu shih hsü tzu yen chiu)
杜甫詩虛字研究 / 黃啓原著. — 台北：洙泗出版社, 民國66 [1977]
6, 8, 331, [1] p. ; 25 cm.
Bibliography: p. [332]
NT$80.00
1. Tu, Fu, 712–770—Language. 2. Chinese language—Particles. I. Title.
PL2675.H853 78–841000

Pierson, Jan Lodewijk, 1893–
The characters者之 … 也and所 : their part in the structure of Japanese and Chinese sentences. Leiden, E. J. Brill, 1960.
170 p. 28 cm.
"Reprint ... [of the author's articles in English] published in Monumenta Nipponica, vol. XI, no. 1, 1955, and vol. XI, no. 4, 1956, with M. Mullie's criticism [in French] Mon. Nipp. vol. XII, nos. 3/4, 1958, and vol. XIV, nos. 1/2, 1958, and ... [the author's] answer to these in full."
1. Chinese language—Particles. 2. Japanese language—Particles.
PL1237.P5 495.1'59 61–4947 ‡
(MARC)

Tseng, I-fu, 1903–
(Ti tzu ti yung fa yü fen hua)
"的"字底用法与分化 / 曾毅夫編著；黎錦熙校訂. — [香港?：s. n., 1979?]
6, 74 p. ; 20 cm.
Cover title.
Reprint of the 1957 ed. published by Ho-pei jen min ch'u pan she, Pao-ting.
Includes bibliographical references.
1. Chinese language—Particles. I. Title.
PL1237.T75 1979 79–841402
79 C

—PHILOSOPHY

Jullien, François, 1951-
Encre de Chine : la Révolution et sa lettre / François Jullien. — Lausanne : A. Eibel : diffusion Éditions L'Age d'homme, 1978.
122 p. ; 18 cm. — (Asie/Asie du Sud-Est ; 8) Sw78-6621
ISBN 2-8274-0037-5 : 30.00F
1. Chinese language—Philosophy. I. Title.
PL1035.J8 495.1'01 79-344256
79 MARC

—PHONETICS

Das Eckzahlsystem : Einf. u. Anwendung : e. neues Nachschlagsystem für d. chines.-dt. Wörterbuch / Cheng Tien-mu ... [et al.]. — Bochum : Studienverlag Brockmeyer, 1979.
106 p. ; 21 cm. — (Programm zur Modernisierung des sinologischen Studiums, Reihe 1 ; Bd. 1) GFR79-A
"Ergänzungsband zu dem Chinesisch-deutschen-Wörterbuch."
Bibliography: p. 101-104.
ISBN 3-88339-040-2 : DM9.80
1. Chinese language—Lexicography. 2. Chinese language—Writing. 3. Chinese language—Phonetics. 4. Chinese language—Transliteration into German. I. Cheng, Tien-mu. II. Series.
PL1411.E25 495.1 79-368516
79 MARC

Wang, I, 1877–
(Kuo yü fa yin hsüeh)
國語發音學 = Chinese phonetics / 汪怡編. — [香港?：s. n., 1977?]
325, [33] p., [1] fold. leaf : ill. ; 20 cm.
Cover title.
At head of title: Hsin chu.
In Chinese.
Reprint of the 1924 ed. published by Shang wu yin shu kuan, Shanghai.
1. Chinese language—Phonetics. I. Title. II. Title: Chinese phonetics.
PL1205.W334 1977 78–841755

—PHONOLOGY

Chang, Cheng-t'i.
(Chung-hua yün hsüeh)
中華韻學 / 張正體, 張婷婷著. — 臺北：臺灣商務印書館, 民國67 [1978]
539, [26] p. ; 20 cm.
NT$118.00
1. Chinese language—Phonology. I. Chang, T'ing-t'ing, joint author. II. Title.
PL1201.C3 79–840526

Chuang, Chia-t'ing.
(Ch'ieh yün chih chang t'u kuang p'ien)
切韻指掌圖廣篇 / 作者莊嘉廷. — [臺北縣永和鎮]：文津出版社, 民國66 [1977]
2, 150 p. ; 22 cm.
Cover title.
NT$40.00
1. Chinese language—Phonology. I. Title.
PL1201.C45 79–836182

Hashimoto, Mantaro J
Phonology of ancient Chinese / Mantaro J. Hashimoto. — Tokyo : Institute for the Study of Languages & Cultures of Asia & Africa, 1978-
v. ; 26 cm. — (Study of languages & cultures of Asia & Africa, monograph series ; no. 10) Ja***
1. Chinese language—Archaic Chinese. 2. Chinese language—Phonology. I. Title. II. Series.
PL2245.H3 495.1'1'5 78-324893
79 MARC

Hu, Chia-yang.
(Chung yüan yin yün yü kuo yü yin hsi chih yin chih chu tzu pi chiao yen chiu)
中原音韻與國語音系之音質逐字比較研究 / 胡嘉陽. — 民國54 [1965]
2 v. (155 double leaves) ; 27 cm.
Cover title.
Reproduced from ms. copy.
Thesis (M. A.)—Kuo li T'ai-wan ta hsüeh.
1. Chou, Te-ch'ing, fl. 1314–1324. Chung yüan yin yün. 2. Chinese language—Phonology. I. Title.
PL1201.C3853H8 79–841846

Huang, K'an, 1886–1936.
(Huang K'an lun hsüeh tsa chu)
黃侃論學雜著 中華書局上海編輯所編輯 北京 中華書局 1964.
14, 4, 2, 488 p. 21 cm.
1. Chinese language—Phonology. 2. Hsü, Shen, d. 120? Shuo wen chieh tzu. 3. Chinese language—Etymology. 4. Erh ya. 5. Rites and ceremonies—China. I. Chung-hua shu chü. Shang-hai pien chi so.
PL1201.H89 1964 C 66–298

Huang, K'an, 1886–1936.
(Huang K'an lun hsüeh tsa chu)
黃侃論學雜著 [臺1版] 臺北：臺灣中華書局 [1969]
14, 4, 2, 488 p. 22 cm.
Originally published in 1964 by 北京中華書局 with editor statement: 中華書局上海編輯所編輯
1. Chinese language—Phonology. 2. Hsü, Shen, d. 120? Shuo wen chieh tzu. 3. Chinese language—Etymology. 4. Erh ya. 5. Rites and ceremonies—China. I. Chung-hua shu chü. Shang-hai pien chi so.
PL1201.H89 1969 70–837916

Huang, K'an, 1886–1936.
[Huang K'an lun hsüeh tsa chu]
黃季剛先生論學名著 / 著作人黃侃. — 台1版. — 台北：九思出版社, 民國66 [1977]
4, 2, 488 p. ; 22 cm. — (九思叢書 ; 13)
Colophon title.
Series romanized: Chiu ssu ts'ung shu.
Originally published under title: Huang K'an lun hsüeh tsa chu.
NT$180.00
1. Chinese language—Phonology. 2. Hsü, Shen, d. 120? Shuo wen chieh tzu. 3. Chinese language—Etymology. 4. Erh ya. 5. Rites and ceremonies—China. I. Title.
Title romanized: Huang Chi-kang hsien sheng lun hsüeh ming chu.
PL1201.H89 1977 78-841131

Lu, Shu-mei.
(T'ai-wan Min-nan yü yin yün yen chiu)
臺灣閩南語音韻研究 / 盧淑美著. — 臺北：文史哲出版社, 民國66 [1977]
122 p. ; 22 cm.
Bibliography: p. 122.
NT$50.00
1. Chinese language—Dialects—Taiwan. 2. Chinese language—Phonology. I. Title.
PL1839.L8 78–841594

P'an, Ch'ung-kuei, 1908–
(Chung-kuo sheng yün hsüeh)
中國聲韻學 / 潘重規, 陳紹棠著. — 臺北：東大圖書有限公司：總經銷三民書局, 1978.
3, 290 p. ; 22 cm. — (滄海叢刊：語言)
Series romanized: Ts'ang hai ts'ung k'an : Yü yen.
1. Chinese language—Phonology. I. Ch'en, Shao-t'ang, joint author. II. Title.
PL1201.P3 79–835173

P'an, Wen-kuang.
(Ku Han yü yü yin yen chiu tzu liao ch'u t'an)
古汉语语音研究资料初探 / 潘文光. — Singapore：南洋大學研究院人文與社會科學研究所, 1977.
40 leaves ; 29 cm. — (研究論文 - 南洋大學研究院人文與社會科學研究所 ; no. 62)
Reproduced from typescript.
Includes bibliographical references.
1. Chinese language—Archaic Chinese. 2. Chinese language—Phonology. I. Title. II. Series: Nanyang University. Institute of Humanities and Social Sciences. Occasional paper series - Institute of Humanities and Social Sciences, Nanyang University ; no. 62.
PL2245.P3 78–940458

Phoon, Mun-Kwong.
A synchronic phonology of modern Cantonese dialect / by Phoon Mun-Kwong. — [Singapore] : Institute of Humanities and Social Sciences, College of Graduate Studies, Nanyang University, 1976.
39 leaves ; 30 cm. — (Occasional paper series - Institute of Humanities and Social Sciences, Nanyang University ; no. 39)
Includes bibliographical references.
1. Chinese language—Dialects—China—Canton. 2. Chinese language—Phonology. I. Title. II. Series: Nanyang University. Institute of Humanities and Social Sciences. Occasional paper series - Institute of Humanities and Social Sciences, Nanyang University ; no. 39.
PL1739.P47 495.1'7 78-940457
79 MARC

Ting, Pang-hsin.
(Ju-kao fang yen ti yin yün)
如皐方言的音韻 / 丁邦新. — [台北?：s. n., 1978?]
84 p. (on double leaves), [5] fold. leaves ; 27 cm.
Spine title.
Reproduced from ms. copy with original date 1963 at end of work.
Bibliography: leave 84.
1. Chinese language—Dialects—China—Ju-kao hsien. 2. Chinese language—Phonology. I. Title.
PL1901.T54 1978 79–835068

CHINESE LANGUAGE
—PHONOLOGY (Continued)

Yen, K'o-chün, 1762–1843.
(Shuo wen sheng lei)
說文聲類 / 嚴可均撰. — 台北：廣文書局, 民國55 ⌜1966⌝
71, 54, 1 leaves ; 20 cm. — (音韻學叢書)
Includes original t. p.: Shuo wen sheng lei erh chüan.
Reprint of the Ssu lu t'ang pen of Wei-nan Yen shih printed by Shih kuo hsüeh chai, Ch'eng-tu, in 1924.
1. Hsü, Shen, d. 120? Shuo wen chieh tzu. 2. Chinese language—Etymology. 3. Chinese language—Phonology. I. Title. II. Series: Yin yün hsüeh ts'ung shu.
PL1281.H83Y4 1966 79-839145

—PHONOLOGY—ADDRESSES, ESSAYS, LECTURES

Kao, Ming, 1909–
(Kao Ming hsiao hsüeh lun ts'ung)
高明小學論叢 / 高明著. — 臺北：黎明文化事業公司, 民國67 ⌜1978⌝
4, 2, 516 p. ; 22 cm. — (國學叢書)
Running title: Kao Ming wen chi.
Series romanized: Kuo hsüeh ts'ung shu.
NT$100.00
1. Chinese language—Phonology—Addresses, essays, lectures. 2. Chinese language—Etymology—Addresses, essays, lectures. I. Title. II. Title: Kao Ming wen chi.
PL1201.K29 79-837670

Lo, Ch'ang-p'ei, 1899–1958.
(Lo Ch'ang-p'ei yü yen hsüeh lun wen hsüan chi)
羅常培語言學論文選集. — 台1版. — 台北：九思出版社, 民國67 ⌜1978⌝
3, 235 p., ⌜4⌝ fold. leaves : ill. ; 27 cm. — (九思叢書 ; 28)
Series romanized: Chiu ssu ts'ung shu.
Reprint ed.
Includes bibliographical references.
NT$300.00
1. Chinese language—Phonology—Addresses, essays, lectures. I. Title.
PL1201.L656 1978 78-842109

—PREPOSITIONS

Hagège, Claude, 1936-
Le problème linguistique des prépositions et la solution chinoise : avec un essai de typologie à travers plusieurs groupes de langues / par Claude Hagège. — ⌜Paris⌝ : Société de linguistique de Paris ; Louvain : dépositaire, Peeters, 1975.
429 p., ⌜2⌝ fold. leaves of plates : ill. ; 25 cm. — (Collection linguistique - Société de linguistique de Paris ; 71) F***
Bibliography: p. ⌜413⌝-421.
Includes indexes.
ISBN 2-8017-0023-1
1. Grammar, Comparative and general—Prepositions. 2. Chinese language—Prepositions. I. Title. II. Series: Société de linguistique de Paris. Collection linguistique ; 71.
P285.H3 415 78-349965
78⌜r79⌝rev MARC

--PRIMERS see Primers, Chinese

—PRONOUN

Han, Yao-lung.
(Chia Chin wen chung ch'eng tai tz'u yung fa yen chiu)
甲金文中稱代詞用法研究 / 韓耀隆. — ⌜臺北? : s. n.⌝, 民國56 ⌜1967⌝
2, 6, 192 p. (on double leaves) ; 27 cm.
"國立臺灣大學中文研究所碩士論文"
Reproduced from ms. copy.
Bibliography: p. 184–192.
1. Chinese language—Pronoun. 2. Inscriptions, Chinese. 3. Oracle bones. 4. Bronzes, Chinese. I. Title.
PL1234.H3 78-842937

Kim, Sang-ju.
(Meng-tzu san shen tz'u yen chiu)
孟子三身詞研究 / 金相姝撰. — ⌜台北? : s. n., 1978?⌝
138, ⌜2⌝ p. (on double leaves) ; 27 cm.
Reproduced from ms. copy.
Thesis (M.A.)—Kuo li T'ai-wan ta hsüeh, 1972.
Bibliography: p. ⌜139⌝-⌜140⌝
1. Mencius. Meng-tzu. 2. Chinese language—Pronoun. I. Title.
PL2474.Z9P74 79-835050

—PRONUNCIATION

Anderson, Olov Bertil, 1920-
An investigation into the present state of standard Chinese pronounciation / Olov Bertil Anderson. — ⌜2d ed.⌝. — ⌜Lund, Sweden⌝ : Studentlitteratur ; London : Curzon Press, 1978-
v. ; 23 cm. GB78-17372
CONTENTS: pt.1a. A companion volume to R. H. Mathews' Chinese-English dictionary.
ISBN 9144152213. ISBN 0-7007-0081-1 : £7.50 (Curzon Press)
1. Chinese language—Pronunciation. I. Title.
PL1209.A5 1978 495.1'152 78-320298
79 MARC

Chang, Pu-hsiu.
(Tzu mu yin pien hui pien)
子母音變彙編 / 張卜庥纂. — 臺北：臺灣商務印書館, 民國67 ⌜1978⌝
3, 151 p. : 27 cm.
Reproduced from ms. copy.
Includes index.
NT$50.00
1. Chinese language—Pronunciation. I. Title.
PL1209.C386 79-835560

China (Republic of China). T'ai-wan tien hsin kuan li chü.
(Tien hua ch'uan pao hsing ming chu chih chieh tzu yung yü)
電話傳報姓名住址解字用語. — ⌜台北⌝：台灣電信管理局, 民國67 ⌜1978⌝
96 p. ; 27 cm.
With kuo yu chu yin.
Includes index.
1. Chinese language—Spoken Chinese. 2. Chinese language—Pronunciation. 3. Telephone etiquette. I. Title.
PL1137.S6C56 1978 79-840366

Lin, Wu-hsien, 1944–
(Yü wen pien cheng shou ts'e)
語文辨正手册 / 林武憲. — 台北：遠流出版社, 民國67 ⌜1978⌝
209 p. : ill. : 22 cm. — (遠流青少年百科全書 : 11 : 第3輯. 讀書與寫作的藝術)
NT$40.00
1. Chinese language—Pronunciation. 2. Chinese language—Writing. I. Title. II. Series: Yüan liu ch'ing shao nien pai k'o ch'üan shu : 11.
PL1209.L56 79-837506

(Ming pen p'ai tzu Chiu ching chih yin)
明本排字九經直音；經咫 / ⌜陳祖范撰⌝. — ⌜臺北：臺灣⌝商務⌜印書館, 1979⌝
100, 135, 61, ⌜7⌝ leaves ; 20 cm. — (四庫全書珍本九集 ; 60)
Cover title.
Photoreprint ed. of the Wen Yüan Ko copy.
1. Wu ching. 2. Chinese language—Pronunciation. I. Ch'en, Tsu-fan, 1676–1754. Ching chih. 1979. II. Title: Chiu ching chih yin. III. Series: Ssu k'u ch'üan shu. Ssu k'u ch'üan shu chen pen chiu chi ; 60.
AC149.S7 1979, vol. 60 79-840776
[PL2462.Z9P76]

Nakazawa, Mareo, 1907–
(Kanji Kango gaisetsu)
漢字·漢語概説 / 中沢希男著. — 東京：教育出版, 1978.
234, 4 p., ⌜1⌝ leaf of plates : ill. ; 21 cm. Ja 78-6407
Bibliography: p. 233–234.
Includes index.
¥1700
1. Chinese language—Etymology. 2. Chinese language—Pronunciation. 3. Kundoku—History. I. Title.
PL1281.N34 79-800080

Tai, James H-Y
Phonological changes in modern standard Chinese in the People's Republic of China since 1949 / by James H-Y. Tai ; prepared for Office of Research, United States Information Agency, Washington, D.C. — Washington : The Office, ⌜1978?⌝.
iv, 122 p. ; 27 cm.
1. Chinese language—Pronunciation. 2. Chinese language—Intonation. I. United States. Information Agency. Office of Research. II. Title.
PL1209.T3 495.1'1'5 78-603573
79 MARC

—PUNCTUATION

Liu, Yü-ch'en, 1921–
(Piao tien fu hao yung fa)
標點符號用法 / 劉玉琛著. — 第2版. — 臺北：國語日報出版部, 民國67 ⌜1978⌝
2, 95 p. : ill. ; 22 cm.
NT$30.00
1. Chinese language—Punctuation. I. Title.
PL1272.L5 79-837725
C

—READERS

Asano, Michiari, 1930–
(Kanshi yōkai)
漢詩要解 / 浅野通有著. — 東京：有精堂, 昭和40 ⌜1965⌝
6, 158 p. : map (on lining paper) ; 19 cm. — (文法解明叢書 ; 44)
Colophon inserted.
Series romanized: Bumpō kaimei sōsho.
Includes index.
1. Chinese language—Readers. 2. Chinese poetry—Translations into Japanese. 3. Japanese poetry—Translations from Chinese. 4. Chinese language—Text-books for foreigners—Japanese. I. Title.
PL1117.A8 79-800630

DeFrancis, John Francis, 1911-
Beginning Chinese reader / by John DeFrancis, with the assistance of Yung Teng Chia-yee and Yung Chih-sheng. — 2d ed. — New Haven : Yale University Press, 1977.
2 v. (xxxiv, 1012 p.) ; 25 cm. — (Yale linguistics series)
Includes index.
ISBN 0-300-02056-2 (v. 1) : $32.50 (v. 1). ISBN 0-300-02060-0 pbk. (v. 1) : $6.95 (v. 1)
1. Chinese language—Readers. I. Yung Teng, Chia-yee, joint author. II. Yung, Chih-sheng, joint author. III. Title.
PL1115.D4 1977 495.1'86'421 76-5103
77⌜r79⌝rev MARC

(Erh, san nien chih chuan k'o hsüeh hsiao kuo wen hsüan)
二, 三年制專科學校國文選 / 編輯者國文編輯委員. — 再版. — 臺北：幼獅文化事業公司, 民國64 ⌜1975⌝
1, 3, 192 p. ; 22 cm.
Colophon title.
1. Chinese language—Readers. I. Kuo wen pien chi wei yüan. II. Title: Kuo wen hsüan.
PL1117.E7 78-843942

(Han yü tu pen)
汉语读本 Chinese reader. 北京 商务印书馆 1972.
2 v. illus. 24 cm.
"英文译释"
1. Chinese language—Readers. I. Title: Chinese reader.
PL1117.H3 73-835519

⌜**Ho, Hsing-ssu**⌝
⌜San tzu ching chu chieh pei yao⌝
三字經：白話註解. — 台南：利大出版社：總經銷第一書店, 民國66 ⌜1977⌝
294 p. ; 20 cm.
Cover title.
Spine title: San tzu ching pai hua chü chieh.
Previously published under title: San tzu ching chu chieh pei yao and under author: Ho Hsing-ssu.
Appendices (p. 129–294): 二十四孝.—百家姓.—人生必讀.—弟子規.—千金文.—千金譜.—菜根譚.—昔時賢文.—朱子治家格言.
NT$30.00
1. Wang, Ying-lin, 1223–1296. San tzu ching. 2. Chinese language—Readers. I. Wang, Ying-lin, 1223–1296. San tzu ching. 1977. II. Title. III. Title: San tzu ching pai hua chü chieh.
Title romanized: San tzu ching.
PL1115.W393H6 1977 79-836556

CHINESE LANGUAGE
—READERS (Continued)

(Hsiao P'a-ti ti sheng jih)
小帕蒂的生日：小学课外阅读文选 = Xiao Padi de shengri / 上海市虹口区教师进修学校编. — 第1版. — 上海：上海人民出版社, 1976.
153 p. : ill. ; 19 cm.
CONTENTS: 李心田"王防".—继路, 梦林 拳顶山上一支枪.—贾平凹 弹弓和南瓜的故事.—李风琪 捉蟹.—逢春 小海星的故事.—玉锡维, 张德民 小帕蒂的生日.—鞠宇东 召唤.—张发良 边防狩猪.—杨钟 水上交通站.
$0.28
1. Chinese language—Readers. I. Wang, Hsi-wei. II. Shang-hai shih Hung-k'ou ch'ü chiao shih chin hsiu hsüeh hsiao. III. Title: Xiao Padi de shengri.
PL1117.H69 79-840350

(Kao chung kuo wen yüeh tu wen hsüan)
高中國文閱讀文選 / 趙庚颺主編；何容校訂. — 臺北：臺灣開明書店, 民國66- [1977- 民國67- [1978- printing.
v. ; 22 cm.
1. Chinese language—Readers. I. Chao, Keng-yang. II. Ho, Jung, 1903-
PL1117.K26 79-840568

(Ming ch'uan kuo wen hsüan)
銘傳國文選 / 私立銘傳女子商業專科學校國文教學研究會編纂. — [台北]：該校, 民國59-66 [1970-77]
4 v. ; 22 cm.
1. Chinese language—Readers. I. Ming-ch'uan nü tzu shang yeh chuan k'o hsüeh hsiao. Kuo wen chiao hsüeh yen chiu hui.
PL1117.M52 79-838143

Miyawaki, Kennosuke.
(Shintai Shina jibun kōyō)
新體支那時文綱要 / 宮脇賢之介著. — 東京：晾文堂, 昭和2 [1927]
3, 2, 8, 307, 34 p. ; 20 cm.
Added t. p.: Einfuehrung in die moderne chinesische Schriftsprache.
Includes index.
1. Chinese language—Readers. 2. Chinese language—Text-books for foreigners—Japanese. I. Title. II. Title: Shina jibun kōyō. III. Title: Einfuehrung in die moderne chinesische Schriftsprache.
PL1117.M554 79-801366

Ogata, Tsurukichi, 1897-
(Kindai Nihon ni okeru senjimongata kyōkasho no kenkyū)
近代日本における千字文型教科書の研究 / 尾形裕康[鶴吉]著. — 東京：早稲田大学出版部, 昭和53 [1978]
18, 1078 p., [3] leaves of plates : ill. ; 27 cm. Ja***
Includes bibliographical references and indexes.
¥48000
1. Chinese language—Readers. 2. Chinese language—Text-books for foreigners—Japanese. I. Title. II. Title: Senjimongata kyōkasho no kenkyū.
PL1117.O36 78-815939
J

(San tzu ching chu chieh)
三字經註解：國語注音 / [編者本社編輯部]. — [台北?]：五福出版社；台北永和鎮：總經銷銘祥書局, 民國67 [1978]
217 p. ; 20 cm.
On spine: San tzu ching chu chieh: chu yin hsiang chieh.
NT$30.00
1. Wang, Ying-lin, 1223-1296. San tzu ching. 2. Chinese language—Readers. I. Wang, Ying-lin, 1223-1296. San tzu ching. 1978. II. Wu fu ch'u pan she. Pien chi pu.
PL1115.W393S24 79-840271

(Taehak Hanmun)
大學漢文 / [編註李丙疇]. — 서울：研修社, 1977.
280, 35 p., [4] leaves of plates : maps ; 23 cm.
Colophon inserted.
₩3500
1. Chinese language—Readers. 2. Chinese language—Text-books for foreigners—Korean. I. Yi, Pyong-ju, 1921-
PL1117.T32 78-82653

Wang, Ying-lin, 1223-1296.
(San tzu ching)
三字經 / 何子文編註. — 台中：立誼出版社, 民國66 [1977]
89 p. ; 19 cm.
Colophon title.
"王應麟撰"
"注音詳註"
NT$10.00
1. Chinese language—Readers. 2. Primers, Chinese. I. Ho, Tzu-wen. II. Title.
PL1115.W39 1977 79-836225

Yu, Chŏng-gi.
(Sŏlmun chajŏn)
說文字典 / 柳正基著. — 大田：農經出版社；서울：發賣元 通文館, 1973.
14, 300, 91, [8] p. ; 22 cm.
"附 啓蒙源典": p. 1-91, 1-8 (3d and 4th groups)
1. Chinese characters—Dictionaries—Korean. 2. Chinese language—Readers. I. Title. II. Title: Kyemong wŏnjŏn.
PL935.Y85 78-826828

—REFORM

(Chi ch'eng Mao chu hsi i chih tso hao wen tzu kai ko kung tso)
继承毛主席遗志做好文字改革工作 = Jicheng Mao zhuxi yizhi zuohao wenzi gaige gongzuo / [本社编]. — 第1版. — 北京：文字改革出版社, 1977.
2, 42 p. ; 19 cm.
$0.11
1. Chinese language—Reform. I. Wen tzu kai ko ch'u pan she, Peking. II. Title: Jicheng Mao zhuxi yizhi zuohao wenzi gaige gongzuo.
PL1175.C517 79-836008

Chia, Shih-Yar.
(Lun chien t'i tzu ti hun luan chi ch'i chiao hsüeh)
论简体字的混乱及其教学 / 谢世涯. — Singapore：南洋大學研究院人文與社會科學研究所, 1977.
28 leaves ; 30 cm. — (研究論文 - 南洋大學研究院人文與社會科學研究所 ; no. 72)
Reproduced from typescript.
Includes bibliographical references.
1. Chinese language—Reform. 2. Chinese language—Study and teaching—Singapore. I. Title. II. Series: Nanyang University. Institute of Humanities and Social Sciences. Occasional paper series - Institute of Humanities and Social Sciences, Nanyang University ; no. 72.
PL1175.C518 78-940285

Chiu, Rosaline Kwan-wai.
The three-fold objective of the language reform in Mainland China in the last two decades / Rosaline Kwan-wai Chiu. — [Quebec] : International Center for Research on Bilingualism, 1969.
10, [1] leaves ; 28 cm. — (Publication - International Center for Research on Bilingualism ; B-7)
"Paper prepared for the symposium organised under the auspices of the Comité québécois des études de la Chine et des autres pays d'Asie, Université Laval, Québec, April 14, 1969."
Bibliography: leaf [11]
1. Chinese language—Reform. I. Université Laval. Centre international de recherches sur le bilinguisme. II. Title.
PL1175.C527 495'.1'1 79-307877
79 MARC

Chou, Yu-kuang.
(Han tzu kai ko kai lun)
漢字改革概論 / 周有光著. — 修訂本. — 澳門：爾雅社, 1978.
3, 2, 358 p. ; 21 cm.
Includes bibliographical references.
$13.00
1. Chinese language—Reform. I. Title.
PL1175.C545 1978 78-840911

Chung-kuo wen tzu kai ko wei yüan hui.
(Chien hua tzu tsung piao)
简化字总表 = Jianhuazi zongbiao / 中国文字改革委员会编. — 第2版. — 北京：文字改革出版社, 1977.
36 p. ; 19 cm.
"附《第二次汉字简化方案(草案)》第一表"
$0.10
1. Chinese language—Reform. I. Title. II. Title: Jianhuazi zongbiao.
PL1175.C5582 1977 78-843648

—REFORM—ADDRESSES, ESSAYS, LECTURES

Language reform in China : documents and commentary / edited with an introd. by Peter J. Seybolt and Gregory Kuei-Ke Chiang. — White Plains, N.Y. : M. E. Sharpe, 1979.
xiii, 410 p. ; 24 cm.
ISBN 0-87332-081-6
1. Chinese language—Reform—Addresses, essays, lectures. I. Seybolt, Peter J. II. Chiang, Gregory Kuei-Ke.
PL1175.L36 495.1'1 76-4302
79 MARC

—REFORM—DICTIONARIES

(Chien t'i tzu tsung piao)
简体字总表. — 再版. — Singapore：教育出版社, 1977.
82 p. ; 19 cm.
Appendices: (p. 68-82): 异体字表.—新旧字形对照表.
$1.00
1. Chinese language—Reform—Dictionaries.
PL1489.C495 78-943647

—RHETORIC

Ch'en, I-tseng, 14th cent.
(Wen shuo)
文說 藝圃擷餘 唐音癸籤 [台北 臺灣商務印書館 1972]
2 v. 19 cm. (四庫全書珍本三集 399-400)
Cover title.
Reprint of 文淵閣手抄本
CONTENTS: 1. 陳繹曾 文說 王世懋 藝圃擷餘 胡震亨 唐音癸籤 卷1-13.—2. 胡震亨 唐音癸籤 卷14-33.
1. Chinese language—Rhetoric. 2. Chinese poetry—History and criticism. I. Wang, Shih-mao, 1536-1588. I p'u hsieh yü. 1972. II. Hu, Chen-heng, fl. 1630. T'ang yin kuei ch'ien. 1972. III. Title. IV. Title: I p'u hsieh yü. V. Title: T'ang yin kuei ch'ien. VI. Series: Ssu k'u ch'üan shu. Ssu k'u ch'üan shu chen pen san chi 399-400.
AC149.S7 1972, vol. 399-400 73-898048
[PL1271]

Hsü, Ch'in-t'ing, 1941-
(Wen chang p'o t'i chi ch'iao yü hsiu tz'u chih yen chiu)
文章破題技巧與修辭之研究 / 著者徐芹庭. — 台北：成文出版社, 民國65 [1976]
3, 494 p. ; 22 cm.
Colophon title.
Cover title: Wen chang p'o t'i chi ch'iao chi hsiu tz'u fang fa chih yen chiu.
Bibliography: p. 494.
NT$150.00
1. Chinese language—Style. 2. Chinese language—Rhetoric. I. Title.
PL1271.H81435 79-835310

Lin, Wan.
(Hsien tai tso chia hsieh tso fan li)
現代作家寫作範例 / 林宛編. — 香港：復興書社, [1978?]
4, 305 p. ; 19 cm.
$9.00
1. Chinese language—Rhetoric. 2. Description (Rhetoric) I. Title.
PL1271.L528 79-836146

CHINESE LANGUAGE
—RHETORIC (Continued)

Lo chi yü fa hsiu tz'u man t'an pien hsieh tsu.
(Lo chi yü fa hsiu tz'u man t'an)
逻辑语法修辞漫谈 / 《逻辑语法修辞漫谈》编写组. — 第1版. — 上海：上海教育出版社, 1978.
3, 3, 187 p. ; 19 cm. — (青年自学丛书)
Series romanized: Ch'ing nien tzu hsüeh ts'ung shu.
$0.37

1. Chinese language—Syntax. 2. Chinese language—Rhetoric. I. Title.
PL1241.L6 1978 78-843201

Ts'ai, Tsung-yang.
(Wen teng)
文燈：文章作法講話 / 蔡宗陽著. — 第2版. — 臺北：國語日報出版部, 民國66 ₍1977₎
5, 3, 4, 263 p. ; 21 cm.
Bibliography: p. 260-263.
NT$65.00

1. Chinese language—Rhetoric. I. Title.
PL1271.T74 79-838617

C

—RHYME

Fu, Hsi-jen.
(Ch'u tz'u ku yün k'ao shih)
楚辭古韻考釋 / ₍著述者傅錫壬₎. — 臺北：淡江文理學院出版指導委員會, 民國62 ₍1973₎
200, ₍4₎ p. ; 22 cm. — (淡江文史叢書)
Series romanized: Tan-chiang wen shih ts'ung shu.
Bibliography: p. 199-200.

1. Ch'u tz'u. 2. Chinese language—Rime. I. Title.
PL2521.C54F8 79-836470

Hsiung, Chung.
₍Ku chin yün hui chü yao₎
古今韻會擧要. — 서울：亞細亞文化社, 1975.
vii, 499 p. ; 26 cm. — (韓國學古辭典시리즈)
"Hsiung Chung published a condensed version of Ku-chin yün-hui ₍by Huang Kung-shao₎, called Ku-chin yün-hui chü-yao...in 1297."
Series romanized: Han'gukhak ko sajŏn ssirijŭ.
Photoreprint of the 17th cent.(?) Korean ed. with new introd. in English, with caption title: A bibliographical note on Ku-chin yün-hui chü-yao.
1. Chinese language—Rhyme. I. Huang, Kung-shao, chin shih 1925. Kun chin yün hui. II. Title.
Title romanized: Kogŭm unhoe kŏyo.
PL1279.H74 1975 79-825651

Ōshima, Masatake, 1859-1938.
(Inkyō on'in kō)
韻鏡音韻考 / ₍大島正健著₎. — 東京：啓成社, 明治45 ₍1912₎
6, 173 p. ; 23 cm.
Errata slip inserted.

1. Yün ching. 2. Chinese language—Rime. I. Title.
PL1279.O8 78-816983

Ōya, Tōru, 1851-1928.
(Inkyō kō)
韻鏡考 / ₍著作者 大矢透₎. — 東京：大矢, 大正13 ₍1924₎
2, 4, 2, 220 p. : ill. ; 26 cm.

1. Yün ching. 2. Chinese language—Rime. I. Title.
PL1279.Y85309 1924 79-801551

Sung, Shu-p'ing.
(Pai Chü-i ku t'i shih yü yüeh fu shih chih yün li chi yung yün k'ao)
白居易古體詩與樂府詩之韻例及用韻考 / 宋淑萍. — ₍台北?：s. n., 1978?₎
2 v. (326 p. on double leaves) ; 27 cm.
Title from spine.
Reproduced from typescript.
Bibliography: p. 325-326.
1. Pai, Chü-i, 772-846—Criticism and interpretation. 2. Chinese language—Rime. I. Title.
PL2674.S85 1978 78-842867

(Ts'ung Chu-tzu Shih chi chuan hsieh yün chung k'an tang shih ti yü yin hsien hsiang)
從朱子詩集傳叶韻中看當時的語音現象. — ₍台北?：s. n., 1978?₎
3 v. (345 double leaves) ; 27 cm.
Title from spine.
Reproduced from ms. copy.
Bibliography: leaves 345.

1. Chu, Hsi, 1130-1200. Shih chi chuan. 2. Chinese language—Rime.
PL2466.R3313T78 78-843884

C

—RHYME—DICTIONARIES

(Shih tz'u yün chi)
詩詞韻輯 / 姚詩雅輯. — 台北：廣文書局, 民國67 ₍1978₎
249 leaves in various foliations ; 22 cm. — (漢學彙編)
Reprint of the Ching shih chai ts'ang pan printed by Hua-t'ai kuan she, 1865.
CONTENTS: 佩文詩韻五卷.—戈載 詞林正韻發凡一卷.—戈載 詞林正韻三卷
NT$350.00
1. Chinese language—Rhyme—Dictionaries. I. Yao, Shih-ya. II. Ko, Tsai. Tz'u lin cheng yün fa fan. 1978. III. Ko, Tsai. Tz'u lin cheng yün. 1978. IV. P'ei wen yün fu. Selections. 1978. V. Series: Han hsüeh hui pien.
PL1279.S55 1978 79-837425

Wang, Hsi-yüan.
(Tz'u lin yün tsao)
詞林韻藻 / 王熙元, 陳滿銘, 陳弘治合編, ₍助編者胡慧怡 ... et al.₎. — 臺北：臺灣學生書局, 民國67 ₍1978₎
5, 12, 438 p. ; 22 cm. — (師大國文系文學叢書 ; 1)
NT$160.00

1. Chinese language—Rhyme—Dictionaries. I. Ch'en, Man-ming, joint author. II. Ch'en, Hung-chih, joint author. III. Title.
PL1279.W3 79-840569

Yü, Chao, fl. 1800.
(Shih yün chi ch'eng)
詩韻集成 / 編者余照. — 臺中：曾文出版社, 民國67 ₍1978₎
4, 264 p. ; 22 cm.
Colophon title.
Title on cover: Tseng kuang shih yün chi ch'eng.
"詞林典腋": p. 1-248.
NT$50.00
1. Chinese language—Rhyme—Dictionaries. I. Title. II. Title: Tseng kuang shih yün chi ch'eng. III. Title: Tz'u lin tien i.
PL1279.Y78 1978 79-837206

C

—RHYTHM

Hsieh, Yün-fei.
(Wen hsüeh yü yin lü)
文學與音律 / 謝雲飛著. — 臺北：東大圖書公司：總經銷三民書局, 1978.
8, 2, 121 p. ; 22 cm. — (滄海叢刊 : 語言)
Series romanized: Ts'ang hai ts'ung k'an : Yü yen.
Includes bibliographical references.

1. Chinese language—Rhythm. I. Title.
PL1279.H73 79-839174

—SEMANTICS

T'an, Ch'üan.
(Shih yung tz'u hui hsin chih shih)
實用詞彙新知識 / 譚全編著. — 香港：商務印書館, 1978.
3, 4, 160 p. ; 19 cm. — (中學生文庫)
Series romanized: Chung hsüeh sheng wen k'u.
$4.50

1. Chinese language—Semantics. I. Title.
PL1291.T3 79-836128

(Tz'u i pien hsi)
詞义辨析 / ₍語文学习杂志社編₎. — ₍香港?：s. n., 1979?₎
3 v. ; 21 cm.
Title in Pinyin on cover: Ciyi bianxi.
Reprint of the 1962 (1963 printing) ed. published by Jen min chiao yü ch'u pan she, Peking, in series: Yü wen hsüeh hsi ts'ung shu.

1. Chinese language—Semantics. I. Title: Ciyi bianxi. II. Yü wen hsüeh hsi tsa chih she.
PL1291.T9 1979 79-841456

—SENTENCES

Lü, Shu-hsiang.
(Ts'ung chu yü pin yü ti fen pieh t'an kuo yü chü tzu ti fen hsi)
從主語賓語的分別談國語句子的分析 / ₍呂叔湘著₎. — 香港：龍門書店, 1969.
55 p. ; 21 cm. — (中國文學研究叢編 ; 第1輯)

1. Chinese language—Sentences. I. Title. II. Series: Chung-kuo wen hsüeh yen chiu ts'ung pien : Ti 1 chi.
PL1241.L84 79-840319

Wang, Ting-ho.
(Han Ying shih i tsao chü ta tzu tien)
漢英釋意造句大字典 = A dictionary to the use of the 2000 basic Chinese characters / 主編王定和；協編 s. Contanceau, J. Pach. — 臺北：成文出版社, 民國67 ₍1978₎
xxxiii, 1280 p. ; 22 cm.
NT$380.00
1. Chinese language—Dictionaries—English. 2. Chinese language—Sentences. I. Title. II. Title: A dictionary to the use of the 2000 basic Chinese characters.
PL1455.W19 495.1'3'21 79-838273

—SPOKEN CHINESE

China (Republic of China). T'ai-wan tien hsin kuan li chü.
(Tien hua ch'uan pao hsing ming chu chih chieh tzu yung yü)
電話傳報姓名住址解字用語. — ₍台北₎：台灣電信管理局, 民國67 ₍1978₎
96 p. ; 27 cm.
With kuo yu chu yin.
Includes index.
1. Chinese language—Spoken Chinese. 2. Chinese language—Pronunciation. 3. Telephone etiquette. I. Title.
PL1137.S6C56 1978 79-840366

—STUDY AND TEACHING

Lai, Ch'ing-hsiung.
(Tso wen p'ing yü shih li)
作文評語示例 / 賴慶雄著. — 香港：語文中心, ₍1978?₎
2, 3, 139 p. ; 21 cm.
$5.50

1. Chinese language—Composition and exercises. 2. Chinese language—Study and teaching. I. Title.
PL1271.L28 1978 79-83785?

CHINESE LANGUAGE
—STUDY AND TEACHING (Continued)

Wang, Chin-hsiang.
(Tso wen p'ing yü chü li)
作文評語舉例 / 王進祥著 ; ［總編輯林洋慈］. — 台北 : 國家出版社, 民國67 ［1978］
3, 134 p. ; 19 cm.
Appendices (p. 93–134): 作文命題參考.—幾本有益作文的書.
NT$50.00

1. Chinese language—Composition and exercises. 2. Chinese language—Study and teaching. I. Lin, Yang-tz'u. II. Title.
PL1271.W254 78–842477

—STUDY AND TEACHING
—KOREA—HISTORY

Kang, Sin-hang, 1930–
(Yijo sidae ŭi yŏkhak chŏngch'aek kwa yŏkhakcha)
李朝時代의譯學政策과譯學者 / 姜信沆著. — 서울 : 塔出版社, 1978.
114 p. ; 21 cm. — (國語學研究選書 ; 4)
Includes bibliographical references.
₩1500

1. Translating and interpreting—Korea—History. 2. Translators—Korea—Biography. 3. Chinese language—Study and teaching—Korea—History. I. Title. II. Series: Kugŏhak yŏn'gu sŏnsŏ ; 4.
P306.8.K6K35 79–825627

—STUDY AND TEACHING
—SINGAPORE

Chia, Shih-Yar.
(Lun chien t'i tzu ti hun luan chi ch'i chiao hsüeh)
论简体字的混乱及其教学 / 谢世涯. — Singapore : 南洋大學研究院人文與社會科學研究所, 1977.
28 leaves ; 30 cm. — (研究論文 - 南洋大學研究院人文與社會科學研究所 ; no. 72)
Reproduced from typescript.
Includes bibliographical references.
1. Chinese language—Reform. 2. Chinese language—Study and teaching—Singapore. I. Title. II. Series: Nanyang University. Institute of Humanities and Social Sciences. Occasional paper series - Institute of Humanities and Social Sciences, Nanyang University ; no. 72.
PL1175.C518 78–940285

—STUDY AND TEACHING (ELEMENTARY)
—CHINA

Shang-hai shih ti liu shih fan hsüeh hsiao. Yü wen chiao yen tsu.
(Hsiao hsüeh chi hsü wen chiao hsüeh tzu liao)
小学记叙文教学資料 / 上海市第六师范学校语文教研组. — ［香港? : s. n., 1979?］
1, 77 p. : map ; 21 cm.
Title in Pinyin on cover: Xiaoxue jixuwen jiaoxue ziliao.
Reprint of the 1978 ed. published by Shang-hai chiao yü ch'u pan she, Shanghai.
1. Chinese language—Study and teaching (Elementary)—China. 2. Reading (Elementary)—China. I. Title. II. Title: Xiaoxue jixuwen jiaoxue ziliao.
LB1525.S48 1979 79–839974

—STUDY AND TEACHING
(ELEMENTARY)—TAIWAN

T'ai-pei (City). Chiao yü chü. Pien chi hsiao tsu.
(Ju ho chih tao erh t'ung hsieh tso)
如何指導兒童寫作 / 主編台北市政府教育局 ; 編著者［該局］編輯小組. — 台北 : 該局, 民國66 ［1977］
2, 154 p. : ill. ; 21 cm. — (國民小學輔導叢書 ; 3)
Colophon title.
1. Chinese language—Composition and exercises. 2. Chinese language—Study and teaching (Elementary)—Taiwan. I. T'ai-pei (City). Chiao yü chü. II. Title. III. Series: Kuo min hsiao hsüeh fu tao ts'ung shu ; 3.
LB1577.C48T34 1977 78–841234

Wang, Pu-hua.
(Ch'i chih pan kuo yü k'o chiao hsüeh ti yen chiu)
啓智班國語科教學的研究 / ［著者王步華］. — 臺北 : 文史哲出版社, 民國67 ［1978］
4, 237 p. : ill. ; 22 cm.
Bibliography: p. 231–237.
NT$80.00

1. Slow learning children—Taiwan. 2. Chinese language—Study and teaching (Elementary)—Taiwan. I. Title.
LC4637.T28W36 79–839536

C

—STUDY AND TEACHING (SECONDARY)

Shen, Hui.
(Chung hsüeh tso wen chih tao yü p'i kai)
中学作文指导与批改 / 沈辉. — 第1版. — 上海 : 上海教育出版社 : 新華書店上海发行所发行, 1978.
2, 132 p. ; 19 cm.
Title in Pinyin on cover: Zhongxue zuowen zhidao yu pigai.
Includes bibliographical references.
$0.27
1. Chinese language—Style. 2. Chinese language—Study and teaching (Secondary) I. Title. II. Title: Zhongxue zuowen zhidao yu pigai.
PL1271.S475 79–839459

C

—STYLE

Ch'en, Chih-hao.
(Kuo wen shih yung ta ch'üan)
國文實用大全 / 陳志豪著. — 台南 : 光田出版社, 民國67 ［1978］
9, 338 p. ; 21 cm.
At head of title: Li chieh kao p'u chien t'e k'ao shih shih t'i hsin pan.
NT$80.00

1. Letter-writing, Chinese. 2. Chinese language—Style. I. Title.
PL1275.C419 78–843608

Ch'en, Chih-hao.
(Lun shuo wen chiang hua)
論說文精華 / 陳志豪著. — 台南 : 光田出版社, 民國67 ［1978］
9, 338 p. ; 21 cm.
Caption title: Kuo wen shih yung ta ch'üan.
"歷屆高普檢特考試試題"
NT$80.00
1. Chinese language—Style. 2. Chinese essays—20th century. 3. Government correspondence. I. Title. II. Title: Kuo wen shih yung ta ch'üan.
PL1271.C4253 78–840805

Chia, Shih-Yar.
(Lun hsin chiu ying yung wen ti yung yü)
论新旧应用文的用语 / 谢世涯. — Singapore : 南洋大學研究院人文與社會科學研究所, 1977.
20, 4 leaves ; 30 cm. — (研究論文 - 南洋大學研究院人文與社會科學研究所 ; no. 80)
Reproduced from typescript.
Includes bibliographical references.
1. Letter-writing, Chinese. 2. Chinese language—Style. I. Title. II. Series: Nanyang University. Institute of Humanities and Social Sciences. Occasional paper series - Institute of Humanities and Social Sciences, Nanyang University ; no. 80.
PL1271.C4464 78–940291

Hsü, Ch'in-t'ing, 1941–
(Wen chang p'o t'i chi ch'iao yü hsiu tz'u chih yen chiu)
文章破題技巧與修辭之研究 / 著者徐芹庭. — 台北 : 成文出版社, 民國65 ［1976］
3, 494 p. ; 22 cm.
Colophon title.
Cover title: Wen chang p'o t'i chi ch'iao chi hsiu tz'u fang fa chih yen chiu.
Bibliography: p. 494.
NT$150.00

1. Chinese language—Style. 2. Chinese language—Rhetoric. I. Title.
PL1271.H81435 79–835310

Huang, Chia-huan.
(Hsieh tso tz'u tien)
寫作辭典 / 黃嘉煥, 李紹經編著. — 台南 : 翰林出版社, 民國66 ［1977］
7, 400 p. ; 21 cm.
Cover title.
"青年學生一般讀者閱讀寫作最佳資料"
NT$55.00

1. Chinese language—Style. I. Li, Shao-ching, joint author. II. Title.
PL1271.H846 78–843199

(Kao chi lun shuo wen fan)
高級論說文範 / 林貴珠編選. — 修訂3版. — 台南 : 利大出版社 : 總經銷第一書店, 民國67 ［1978］
2, 7, 326 p. ; 22 cm.
Cover title.
NT$45.00

1. Chinese language—Style. 2. Chinese essays—20th century. I. Lin, Kuei-chu.
PL1271.K343 79–835523

Kuo, Ching-hsing.
(Hsin pien ying yung wen)
新編應用文 / 郭景星編著. — 臺北 : 南京出版公司, 民國67 ［1978］
4, 433 p. ; 21 cm.
"大專院校·高普考試·公務人員·出國留學參考用書"
NT$125.00

1. Letter-writing, Chinese. 2. Chinese language—Style. I. Title.
PL1275.K86 79–835170

Lin, Ch'ing-hsi.
(Tso wen fang fa)
作文方法 / 林清習編著. — 台中 : 林清習 : 總經銷新民書局, 民國66 ［1977］
4, 217 p., ［1］ fold. leaf of plates : ill. ; 21 cm.
Cover title.
Errata slip inserted.
NT$30.00

1. Chinese language—Style. I. Title.
PL1271.L517 79–837678

Shen, Chien-shih.
(Ying yung wen chü yü)
應用文舉隅 / 沈兼士著. — 臺北 : 五南圖書出版公司, 民國67 ［1978］
4, 8, 381 p. ; 22 cm.
"大學用書"
Errata slip inserted.
NT$120.00

1. Letter-writing, Chinese. 2. Chinese language—Style. I. Title.
PL1275.S48 79–837074

Shen, Hui.
(Chung hsüeh tso wen chih tao yü p'i kai)
中学作文指导与批改 / 沈辉. — 第1版. — 上海 : 上海教育出版社 : 新華書店上海发行所发行, 1978.
2, 132 p. ; 19 cm.
Title in Pinyin on cover: Zhongxue zuowen zhidao yu pigai.
Includes bibliographical references.
$0.27
1. Chinese language—Style. 2. Chinese language—Study and teaching (Secondary) I. Title. II. Title: Zhongxue zuowen zhidao yu pigai.
PL1271.S475 79–839459

Sung, Wen-wei.
(P'ing chu wen fa chin liang)
評註文法津梁 / 宋文蔚編 ; 楊游校. — 高雄 : 啓聖圖書公司 : 總經銷復文書局, 民國63 ［1974］
128, 102, 96, ［12］ p. ; 19 cm.
Reprint ed.
Distributor's name stamped in colophon.

1. Chinese language—Style. I. Title. II. Title: Wen fa chin liang.
PL1271.S84 1974 79–838251

CHINESE LANGUAGE
—STYLE (Continued)

Tseng, Ch'i-shan.
(Ko lei wen t'i tso fa)
各類文體作法 / 曾期善著. — 台北：偉文圖書出版社, 民國67 [1978]
130 p. ; 19 cm. — (中國語文叢書)
Series romanized: Chung-kuo yü wen ts'ung shu.
NT$40.00

1. Chinese language—Style. I. Title.
PL1271.T76 79-839551

Wang, Cheng-fu.
(Tso wen chieh t'i)
作文解題 / 王正夫著. — 台北：偉文圖書出版社, 民國67 [1978]
144 p. ; 19 cm. — (中國語文叢書)
Series romanized: Chung-kuo yü wen ts'ung shu.
NT$40.00

1. Chinese language—Style. I. Title.
PL1271.W23 79-840347

Yüan, Chin-shu.
(Hsin pien ying yung wen)
新編應用文 / 袁金書編著. — 修訂4版. — 臺北：袁金書, 民國66 [1977]
494, [15] p. : ill. ; 22 cm.
At head of title: Ta chuan shih yung.
NT$120.00

1. Letter-writing, Chinese. 2. Chinese language—Style. I. Title.
PL1275.Y87 1977 79-835499
C

—STYLE—STUDY AND TEACHING —TAIWAN

Wang, Chung-hua.
(Ying yung wen chiao hsüeh chih shang chüeh)
應用文教學之商榷 / 王忠華著. — 台北：世紀書局, 民國67 [1978]
4, 70 p. ; 22 cm.
Cover title.
Bibliography: p. 69-70.
NT$60.00

1. Chinese language—Style—Study and teaching—Taiwan. 2. Letter-writing, Chinese—Study and teaching—Taiwan. I. Title.
PL1068.T3W3 79-839553
C

—SYNTAX

Ch'en, Yü-jo.
(Hsiao ching yü fa chih yen chiu)
孝經語法之研究 / 陳于若. — [台北? : s. n., 1978?]
2 v. (30, 293 p. on double leaves) ; 27 cm.
Reproduced from typescript.
Originally presented as the author's thesis (M.A.), T'ai-wan ta hsüeh, 1973.
Bibliography: p. 279-293.

1. Hsiao ching. 2. Chinese language—Syntax. I. Title.
PL2476.Z7C58 1978 79-835072

Lin, Shuang-fu.
The grammar of disjunctive questions in Taiwanese / by Shuang-fu Lin. — Taipei : Taiwan Students Book Co., [c1975]
iii, 243 p. : diagrs. ; 22 cm.
English and Taiwanese.
Bibliography: p. 240-243.
1. Chinese language—Dialects—Taiwan. 2. Chinese language—Interrogative. 3. Chinese language—Syntax. I. Title.
PL1833.L5 495.1'7 78-106798
79 MARC

Lo chi yü fa hsiu tz'u man t'an pien hsieh tsu.
(Lo chi yü fa hsiu tz'u man t'an)
逻辑语法修辞漫谈 / 《逻辑语法修辞漫谈》编写组. — 第1版. — 上海：上海教育出版社, 1978.
3, 3, 187 p. ; 19 cm. — (青年自学丛书)
Series romanized: Ch'ing nien tzu hsüeh ts'ung shu.
$0.37

1. Chinese language—Syntax. 2. Chinese language—Rhetoric. I. Title.
PL1241.L6 1978 78-843201

—TERMS AND PHRASES

Ch'ang, Ch'ing, fl. 1952-
(Ch'eng yü kuo shih ching hsüan)
成語故事精選 / 常青編 ; 陳文等繪. — 香港：新雅兒童教育出版社, 1978.
2 v. : col. ill. ; 20 cm.
$6.00 per vol.

1. Anecdotes—China. 2. Chinese language—Terms and phrases. I. Ch'en, Wen, fl. 1975- II. Title.
PN6267.C5C44 79-83735

Chang, Hsün-ch'i.
(Wen hsüeh chang ku)
文學掌故 / 張迅齊, 張忠江編著. — 臺北永和：常春樹書坊 ; 臺北：總經銷星光書報社, 民國66 [1977]
479 p. ; 19 cm. — (中國人的書)
Cover title: Chung-kuo wen hsüeh ti chang ku.
Series romanized: Chung-kuo jen ti shu.
NT$80.00
1. Anecdotes—China. 2. Chinese language—Terms and phrases. I. Chang, Chung-chiang, joint author. II. Title. III. Title: Chung-kuo wen hsüeh ti chang ku.
PN6267.C5C464 78-841620

(Ch'eng yü ku shih)
成語故事：附插圖：中國歷史故事集成. — 台南：利大出版社：總經銷第一書店, 民國67 [1978]
7, 357 p. : ill. ; 19 cm.
Cover title.
NT$35.00

1. China—History—Anecdotes, facetiae, satire, etc. 2. Chinese language—Terms and phrases.
DS736.C49 78-842138

(Chung-kuo ch'eng yü ku shih)
中國成語故事. — [九龍：百靈出版社, 1978?]
163 p. ; 19 cm.
Cover title.
$4.80

1. Anecdotes—China. 2. Chinese language—Terms and phrases.
PN6267.C5C585 78-841296

Hsü, Yeh-chi.
(Shih yung ch'eng yü tz'u tien)
實用成語辭典 / 許業基編著. — 九龍：時代圖書有限公司, 1978.
3, 66, 443 p. ; 18 cm.
$9.80

1. Chinese language—Terms and phrases. I. Title.
PL1497.H776 8-843031

Kan-su shih fan ta hsüeh. Han yü ch'eng yü tz'u tien pien hsieh tsu.
(Han yü ch'eng yü tz'u tien)
汉语成语词典 = Hanyu chengyu cidian / 甘肃师范大学中文系《汉语成语词典》编写组. — 第1版. — 上海：上海教育出版社：新華书店上海发行所发行, 1978.
88, 889 p. ; 14 cm.
Includes index.
$1.75

1. Chinese language—Terms and phrases. I. Title. II. Title: Hanyu chengyu cidian.
PL1497.K28 1978 79-838684

Kao, Kuo-shu.
(Chung-kuo ch'eng yü tz'u tien)
中國成語辭典 / [著作者高國書]. — 台北：五福出版社：總經銷銘祥書局, 民國67 [1978]
19, 61, 239, 6, 319 p. : ill. ; 20 cm.
"成語故事": last two groups.
NT$100.00

1. Chinese language—Terms and phrases. I. Kao, Kuo-shu. Chung-kuo min chien ku shih. 1978. II. Title.
PL1497.K32 1978 79-838230

Kao, Kuo-shu.
(Chung-kuo ch'eng yü tz'u tien)
中國成語辭典 / 高國書編著. — 台南：西北出版社, 民國68 [1979]
19, 61, 239, 166 p. ; 20 cm.
NT$45.00

1. Chinese language—Terms and phrases. I. Title.
PL1497.K32 1979 79-841165

(Ku tien fu yin tz'u hui chi lin)
古典複音詞彙輯林. — 臺北：鼎文書局, 民國67 [1978]
8 v. ; 24 cm. — (中國學術類編)

1. Chinese language—Terms and phrases. I. Series: Chung-kuo hsüeh shu lei pien.
PL1497.K78 79-836062

Kung, Hsien-nan.
(Chung-kuo yü yen ching hua)
中國語言精華 / 龔顯男編著. — 雲林縣虎尾鎮：龔氏出版社, 民國68- [1979-
v. ; 19 cm.
NT$125.00 (v. 1)

1. Chinese language—Terms and phrases. 2. Chinese wit and humor. 3. Maxims, Chinese. I. Title.
PL1497.K84 79-839555

Liu, Hsing-chai.
(Chung-kuo ch'eng yü ta tz'u tien)
中國成語大辭典 / 劉省齋編. — 再版. — 台南：綜合出版社, 民國67 [1978]
2, 97, 421 p. ; 21 cm.
Includes index.
NT$55.00

1. Chinese language—Terms and phrases. 2. Proverbs, Chinese. I. Title.
PL1497.L585 79-837852

Liu, Li-yeh.
(Chung-kuo ch'eng yü ta tz'u tien)
中國成語大辭典：書信·作文·演講·求解·措詞·寫作 / 劉麗葉編著. — 台南：大府城出版有限公司, 民國68 [1979]
103, 375 p. ; 20 cm. — (大府城書系)
Caption title: Chung-kuo ch'eng yü tz'u tien.
Series romanized: Ta fu ch'eng shu hsi.
Includes index.
NT$50.00
1. Chinese language—Terms and phrases. I. Title.
PL1497.L587 79-838246

CHINESE LANGUAGE
—TERMS AND PHRASES (Continued)

Liu, Li-yeh.

[Chung-kuo ch'eng yü ta tz'u tien]
最新成語大辭典：書信·作文·演講·求解·措詞·寫作：附插圖 / 劉麗葉編著. — 台南：大府城出版有限公司, 民國68 [1979]
103, 375, 7, 357 p. : ill. ; 20 cm. — (大府城書系)

Caption title: Chung-kuo ch'eng yü tz'u tien.
A collection of two works; one work was published under title: Chung-kuo ch'eng yü ta tz'u tien, the other work includes anecdotes.
Series romanized: Ta fu ch'eng shu hsi.
Includes index.
NT$85.00

1. Chinese language—Terms and phrases. 2. Anecdotes—China. I. Title.
Title romanized: Tsui hsin ch'eng yü ta tz'u tien.

PL1497.L587 1979b 79-838229

(Ming Ch'ing su yü tz'u shu chi ch'eng)
明清俗語辭書集成 / 長澤規矩也編. — 東京：汲古書院, 昭和49- [1974-
v. ; 27 cm.

Prefatory matter in Japanese.
明清俗語辭書集成 不鮮明箇所解讀表: (16 p.) in booklet issued with the works.

1. Chinese language—Terms and phrases. I. Nagasawa, Kikuya, 1902-

PL1497.M5 75-836321

rev

(Piao chun kuo yü chu yin tz'u hai)
標準國語注音辭海 / 主編人高靜男. — 台南：正業書局, 民國64 [1975]
1909 p. in various pagings ; 16 cm.

Colophon title.
Title on cover and spine: Tz'u hai.
CONTENTS: 辭海—中國成語辭典—六体字鑑—篆字體
NT$110.00

1. Chinese language—Dictionaries. 2. Chinese language—Writing—Dictionaries. 3. Chinese language—Terms and phrases. I. Kao, Ching-nan. II. Title: Tz'u hai.

PL1120.P49 76-836693

Piao chun kuo yü chu yin tz'u hai.
辭典 / 主編者高靜男. — 台南：西北出版社, 民國67 [1978]
1909 p. in various pagings ; 16 cm.

Colophon title.
Title on spine: Piao chun kuo yü chu yin tz'u tien.
Reprint of the 1975 ed. published by Cheng yeh shu chü, Tainan.
CONTENTS: 辭海.—中國成語辭典.—六体字鑑.—篆字體.
NT$120.00

1. Chinese language—Dictionaries. 2. Chinese language—Writing—Dictionaries. 3. Chinese language—Terms and phrases. I. Kao, Ching-nan. II. Title.
Title romanized: Tz'u tien.

PL1420.P49 1978 79-835448

(Piao chun kuo yü chu yin Tz'u hai)
標準國語注音辭海 / 主編者高靜男；編輯者正業書局. — 台南：該局, 民國67 [1978]
7, 1811, 200, 64 p. ; 20 cm.

Cover title.
"大字足本"
"附錄": p. 1-200 (3d group).
"注音符號檢字表": p. 1-64 (last group)
NT$220.00

1. Chinese language—Dictionaries. 2. Chinese language—Writing—Dictionaries. 3. Chinese language—Terms and phrases. I. Cheng yeh shu chü. II. Kao, Ching-nan. III. Title: Tz'u hai.
PL1420.P49 1978b 79-839150

Shan-tung shih fan hsüeh yüan. Liao-ch'eng fen yüan. Han yü ch'eng yü tz'u tien pien hsieh tsu.
(Mao Tse-tung hsüan chi li ti ch'eng yü ku shih)
毛泽东选集里的成語故事 / 山东师范学院聊城分院中文系汉语成语词典编写组. — 第1版. — 北京：中國少年兒童出版社, 1978.
6, 189 p. : ill. ; 19 cm.

$0.36
1. Mao, Tse-tung, 1893-1976—Addresses, essays, lectures. 2. Chinese language—Terms and phrases. I. Title.
DS778.M3S48 1978 79-835441

(Wan yung Ying wen shou ts'e)
萬用英文手册 = Modern English handbook / [編著中美出版中心]. — 香港：該中心, [1978?]
ca. 550 p. in various pagings : ill. ; 15 cm.
"特大號"
$6.00

1. Chinese language—Terms and phrases. 2. Chinese language—Dictionaries—English. I. Chung Mei ch'u pan chung hsin. II. Title: Modern English handbook.
PL1497.W27 1978 78-842093

Wang, Tsung-chieh.
(Chung-kuo ko yen yen yü ssu yung ta tz'u tien)
中國格言諺語四用大辭典 / 著者王宗傑, 洪慧貞. — 台北：金文圖書有限公司, 民國68 [1979]
540 p. ; 20 cm.
NT$160.00

1. Chinese language—Terms and phrases. I. Hung, Hui-chen, joint author. II. Title.
PL1497.W33 79-840592

Wu, Jui-shu
(Tso wen ch'eng yü tz'u tien)
作文成語辭典 / 陳斯良編著. — 台北：祥生出版社, 民國64 [1975]
31, 28, 274 p. ; 20 cm.
Formerly published under author: Wu Jui-shu.
NT$40.00

1. Chinese language—Terms and phrases. I. Ch'en, Ssu-liang. II. Title.
PL1497.W82 1975 77-839557

—TEXT-BOOKS FOR FOREIGNERS
—ENGLISH

Pei-ching yü yen hsüeh yüan.
(Han yü k'o pen)
汉语课本 = Modern Chinese readers / 北京语言学院编. — 北京：商务印书馆：中国国际书店发行, 19
v. : ill. ; 21 cm.
In Chinese.

1. Chinese language—Text-books for foreigners—English. I. Title. II. Title: Modern Chinese readers.
PL1109.P425 79-839616

—TEXT-BOOKS FOR FOREIGNERS
—FRENCH

Kao, Tching.
Le chinois quotidien : par des textes diversifiés et des exercices essentiellement basés sur des locutions usuelles / Tching Kao, Makoto Kanehisa. — Paris : Librairie d'Amérique et d'Orient Maisonneuve, 1976.
192 p. ; 25 cm. — (Langues d'Amérique et d'Orient) F***
French or Chinese.
1. Chinese language—Text-books for foreigners—French. I. Kanehisa, Makoto, joint author. II. Title. III. Series.
PL1129.F7K3 495'.1'82441 78-368813
78 MARC

—TEXT-BOOKS FOR FOREIGNERS
—JAPANESE

Asano, Michiari, 1930-
(Kanshi yōkai)
漢詩要解 / 浅野通有著. — 東京：有精堂, 昭和40 [1965]
6, 158 p. : map (on lining paper) ; 19 cm. — (文法解明叢書 ; 44)
Colophon inserted.
Series romanized: Bumpō kaimei sōsho.
Includes index.
1. Chinese language—Readers. 2. Chinese poetry—Translations into Japanese. 3. Japanese poetry—Translations from Chinese. 4. Chinese language—Text-books for foreigners—Japanese. I. Title.
PL1117.A8 79-800630

Baba, Harukichi.
(Shinago tsūkai)
支那語通解：發音は注音符號で / 岩村成允先生校閲；馬場春吉著. — 東京：岡崎屋書店, 昭和14 [1939]
5, 326, 26 p. ; 19 cm.

1. Chinese language—Text-books for foreigners—Japanese. I. Title.
PL1129.J3B3 79-801860

Kōsaka, Jun'ichi, 1915-
(Chūgoku hyōjun bumpō)
中国標準文法 / 香坂順一, 宮田一郎著. — 東京：江南書院, 1958.
152 p. ; 21 cm.

1. Chinese language—Text-books for foreigners—Japanese. I. Miyata, Ichirō, 1923- joint author. II. Title.
PL1129.J3K578 79-801050

Miyawaki, Kennosuke.
(Shintai Shina jibun kōyō)
新體支那時文綱要 / 宮脇賢之介著. — 東京：瞭文堂, 昭和2 [1927]
3, 2, 8, 307, 34 p. ; 20 cm.
Added t. p.: Einfuehrung in die moderne chinesische Schriftsprache.
Includes index.

1. Chinese language—Readers. 2. Chinese language—Text-books for foreigners—Japanese. I. Title. II. Title: Shina jibun kōyō. III. Title: Einfuehrung in die moderne chinesische Schriftsprache.
PL1117.M554 79-801366

Ogata, Tsurukichi, 1897-
(Kindai Nihon ni okeru senjimongata kyōkasho no kenkyū)
近代日本における千字文型教科書の研究 / 尾形裕康[鶴吉]著. — 東京：早稲田大学出版部, 昭和53 [1978]
18, 1078 p., [3] leaves of plates : ill. ; 27 cm. Ja***
Includes bibliographical references and indexes.
¥48000
1. Chinese language—Readers. 2. Chinese language—Text-books for foreigners—Japanese. I. Title. II. Title: Senjimongata kyōkasho no kenkyū.
PL1117.O36 78-815939

Tōdō, Akiyasu, 1915-
(Chūgoku koten no yomikata)
中國古典の讀みかた：漢文の文法 / 藤堂明保, 近藤光男著. — 東京：江南書院, 1956.
370 p., [1] leaf of plates : ill. ; 19 cm.

1. Chinese language—Text-books for foreigners—Japanese. I. Kondō, Mitsuo, 1921- joint author. II. Title.
PL1129.J3T64 79-802259

—TEXT-BOOKS FOR FOREIGNERS
—KOREAN

Ch'oe, Sin-ho.
(Hanmun kanghwa)
漢文講話 / 崔信浩著. — 서울：玄岩社, 1977.
340 p. ; 19 cm. — (玄岩新書 ; 48)
Series romanized: Hyŏnam sinsŏ.
₩2400

1. Chinese language—Text-books for foreigners—Korean. I. Title.
PL1129.K6C55 78-826350

Kim, Kwi-dal.
(Mobŏm Chungguŏ)
模範中國語：讀本·文法 / 金貴達著. — 서울：進明文化社, 1976.
170 p. ; 23 cm.
Includes index.
₩2300

1. Chinese language—Text-books for foreigners—Korea. I. Title.
PL1109.K48 79-826356

CHINESE LANGUAGE
—TEXT-BOOKS FOR FOREIGNERS
—KOREAN (Continued)

(Taehak Hanmun)
大學漢文 / [編註李丙疇]. — 서울：研修社, 1977.
280, 35 p., [4] leaves of plates : maps ; 23 cm.
Colophon inserted.
₩3500

1. Chinese language—Readers. 2. Chinese language—Text-books for foreigners—Korean. I. Yi, Pyong-ju, 1921–
PL1117.T32 78–826535

Yi, Pyŏng-hun.
(Algi swiun Hanmun kanghwa)
알기쉬운漢文講話 / 李炳薰著. — 서울：創造社, 1975.
288 p. ; 18 cm. — (百科新書 : 31)
Series romanized: Paekkwa sinsŏ.

1. Chinese language—Text-books for foreigners—Korean. I. Title.
PL1109.Y48 79–825220

—TRANSLATING INTO ENGLISH

Chang, Ta-ts'ung.
(Fan i chih yüan li yü chi ch'iao)
翻譯之原理與技巧 = The principles and techniques of translation / 張達聰著. — 台北：國家書店, 民國68 [1979]
iii, vii, 645 p. ; 22 cm.
English and/or Chinese.
In colophon: Ch'u pan che Kuo chia ch'u pan she.
Includes index.
NT$280.00
1. Chinese language—Translating into English. 2. English language—Translating into Chinese. I. Title. II. Title: Principles and techniques of translation.
PL1277.C543 495.1'8'02 79–839973

Huang, Hsüan-fan.
(Chung Ying fan i)
中英翻譯：理論與實踐 / 黃宣範編著. — 台北：文鶴出版有限公司：總經銷衆文圖書公司, 民國67 [1978]
14, 413 p. ; 22 cm.
NT$135.00

1. Chinese language—Translating into English. 2. English language—Translating into Chinese. I. Title.
PL1277.H84 79–840797

—TRANSLATING INTO JAPANESE

see also Kundoku

—TRANSLITERATION

China (People's Republic of China, 1949–). Kuo wu yüan. Han yü p'in yin fang an shen ting wei yüan hui.
(Han yü p'in yin fang an)
汉语拼音方案 = Hanyu pinyin fang'an. — 第1版. — 北京：文字改革出版社, 1978.
20 p. ; 19 cm.
"汉语拼音方案草案, 经中国文字改革委员会提出后…由国务院组织汉语拼音方案审订委员会加以审核修订…提请全国人民代表大会讨论和批准…"
"关于当前文字改革工作和汉语拼音方案的报告 吴玉章": p. 5–20.

1. Chinese language—Transliteration. I. Wu, Yü-chang, 1878–1966. Kuan yü tang ch'ien wen tzu kai ko kung tso ho Han yü p'in yin. II. China (People's Republic of China, 1949–). Ch'üan kuo jen min tai piao ta hui. III. Chung-kuo wen tzu kai ko wei yüan hui. Han yü p'in yin fang an. IV. Title. V. Title: Hanyu pinyin fang'an.
PL1181.C54 1978 79–837233

(Han yü p'in yin kuang po chiang tso)
汉语拼音广播讲座 Hanyu pinyin guangbo jiangzuo. 中國文字改革委员会推广处編 北京 Wenzi gaige chubanshe. 文字改革出版社 1965.
22 p. 13 cm.
Cover title.
1965年中央人民广播电台与中国文字改革委员会应各方面人士的要求联合举办汉语拼音广播讲座这本书就是这次广播使用的教材1.
1. Chinese language—Transliteration. I. Chung-kuo wen tzu kai ko wei yüan hui. T'ui kuang ch'u. II. Chung yang jen min kuang po tien t'ai. Peking.
PL1185.H32 1965 C 68–1293

(Han yü p'in yin kuang po chiang tso)
汉语拼音广播讲座 = Hanyu pinyin guangbo jiangzuo. — 北京：文字改革出版社, 1974.
32 p. ; 19 cm.
Cover title.
"中国文字改革委员会和中央人民广播电台联合举办"汉语拼音广播讲座"这本书就是广播讲座的课本"

1. Chinese language—Transliteration. I. Chung-kuo wen tzu kai ko wei yüan hui. II. Chung yang jen min kuang po tien t'ai, Peking.
PL1185.H32 1974 78–841053

—TRANSLITERATION INTO GERMAN

Das Eckzahlsystem : Einf. u. Anwendung : e. neues Nachschlagsystem für d. chines.-dt. Wörterbuch / Cheng Tien-mu ... [et al.]. — Bochum : Studienverlag Brockmeyer, 1979.
106 p. ; 21 cm. — (Programm zur Modernisierung des sinologischen Studiums, Reihe 1 ; Bd. 1) GFR79-A
"Ergänzungsband zu dem Chinesisch-deutschen-Wörterbuch."
Bibliography: p. 101-104.
ISBN 3-88339-040-2 : DM9.80
1. Chinese language—Lexicography. 2. Chinese language—Writing. 3. Chinese language—Phonetics. 4. Chinese language—Transliteration into German. I. Cheng, Tien-mu. II. Series.
PL1411.E25 495.1 79-368516
79 MARC

—VERSIFICATION

Ch'i, Kung.
(Shih wen sheng lü lun kao)
詩文聲律論稿 / 啓功著. — 九龍：中華書局, 1978.
2, 128 p. ; 21 cm.
At head of colophon title: I lin hsin lu.
$5.50

1. Chinese language—Intonation. 2. Chinese language—Versification. I. Title.
PL1213.C447 78–841455

Ch'ü, Hsüan-ying, 1892–
[Hsüeh shih ch'ien shuo]
詩學義海：中國古典詩詞入門. — 九龍：百靈出版社, [1978?]
4, 216 p. ; 19 cm.
Originally published in 1961 by Shang-hai shu chü, Hongkong, under the title: Hsüeh shih ch'ien shuo and the authors' name: Ch'ü T'o-yüan and Chou Tzu-i, with preface deleted.
$7.00
1. Chinese poetry—History and criticism. 2. Chinese language—Versification. I. Chou, Tzu-i, joint author. II. Title. III. Title: Chung-kuo ku tien shih tz'u ju men.
Title romanized: Shih hsüeh i hai.
PL2307.C5737 1978 79–836515

(Kuan yü shih ko ko lü wen t'i t'ao lun chuan chi)
關於詩歌格律問題討論專輯 / 編者辛墾；主編者中國新文學資料室. — [香港]：神州圖書公司, 1978.
87 p. ; 27 cm.
Colophon title.
Reprint ed.
Includes bibliographical references.
$20.00
1. Chinese language—Versification. I. Hsin, K'en.
PL1279.K76 1978 79–837029

Kūkai, 774–835.
[Bunkyo hifu ron]
文鏡秘府論 / 弘法大師(空海)撰. — 臺景印初版. — 台北：河洛圖書出版社, 民國65 [1976]
6, 3, 266 p. ; 22 cm. — (夏學叢書)
Text in Chinese characters.
Series romanized: Hsia hsüeh ts'ung shu.
NT$120.00
1. Chinese language—Versification. 2. Chinese poetry—History and criticism. I. Title.
Title romanized: Wen ching mi fu lun.
PL1279.K84 1976 78–842101

Lung, Mu-hsün, 1902–1966.
(T'ang Sung tz'u ko lü)
唐宋詞格律 / 龍榆生[沐勛]編撰. — 第1版. — 上海：上海古籍出版社：新華書店上海发行所发行, 1978.
2, 2, 9, 203 p. ; 19 cm.
"本书是…在大学讲授唐宋词的讲义, 原名《唐宋词定格》…张珍怀同志并编辑《词韵简编》附于书后"
$0.55
1. Chinese language—Versification. 2. Tz'u—History and criticism. I. Title.
PL1279.L85 1978 79–838012

Saitō, Shō, 1898–
(Kanshi nyūmon)
漢詩入門 / 齋藤晌著. — 東京：元々社, 昭和29 [1954]
3, 287 p., [1] leaf of plates : port. ; 18 cm. — (民族教養新書 ; 1)
Series romanized: Minzoku kyōyō shinsho.

1. Chinese poetry—History and criticism. 2. Chinese language—Versification. 3. Poetics. I. Title.
PL2307.S27 78–816927

—VOCABULARY

Wang, Chia-ping.
(Ku shu nan tzu)
古書難字 / 王嘉炳編註. — 臺北縣新店鎮：常春樹書坊, 民國67 [1978]
257 p. ; 20 cm.
NT$45.00

1. Chinese language—Vocabulary. 2. Chinese language—Archaic Chinese. I. Title.
PL1495.W24 79–835165

—WORD FREQUENCY

Kuo li T'ai-wan shih fan ta hsüeh. Kuo wen yen chiu so.
(Ch'ang yung kuo tzu piao chun tzu t'i piao)
常用國字標準字體表：初稿名國民常用字表. — [台北]：教育部, 民國67 [1978]
68, 481, 3, 8 p. ; 21 cm.
Cover title.
"教育部社會教育司委託國立臺灣師範大學國文研究所研訂"
1. Chinese language—Word frequency. 2. Chinese language—Writing. I. China (Republic of China). Chiao yü pu. She hui chiao yü ssu. II. Title.
PL1498.K8 1978 79–836915

—WRITING

see also Chinese characters

Chou, Chieh-ch'en.
(Wo pu tsai hsieh ts'o tzu)
我不再寫錯字 / 周介塵. — 臺北：名人出版社, 民國67 [1978]
231 p. ; 19 cm.
Includes index.
NT$70.00

1. Chinese language—Writing. I. Title.
PL1171.C59 79–836906

Das Eckzahlsystem : Einf. u. Anwendung : e. neues Nachschlagsystem für d. chines.-dt. Wörterbuch / Cheng Tien-mu ... [et al.]. — Bochum : Studienverlag Brockmeyer, 1979.
106 p. ; 21 cm. — (Programm zur Modernisierung des sinologischen Studiums, Reihe 1 ; Bd. 1) GFR79-A
"Ergänzungsband zu dem Chinesisch-deutschen-Wörterbuch."
Bibliography: p. 101-104.
ISBN 3-88339-040-2 : DM9.80
1. Chinese language—Lexicography. 2. Chinese language—Writing. 3. Chinese language—Phonetics. 4. Chinese language—Transliteration into German. I. Cheng, Tien-mu. II. Series.
PL1411.E25 495.1 79-368516
79 MARC

CHINESE LANGUAGE —WRITING (Continued)

Kao, T'ang.
(K'ai shu, hsing shu, ts'ao shu, san t'i tzu fan)
楷書·行書·草書三體字範 / 高堂書. — 九龍：江南書局：發行利通圖書公司，[1978?]
270 p. : all facsims. ; 22 cm.
$6.50

1. Penmanship, Chinese. 2. Chinese language—Writing, Cursive. 3. Chinese language—Writing. I. Title. II. Title: San t'i tzu fan.
Z44.K386 79-839959

Kuo li T'ai-wan shih fan ta hsüeh. Kuo wen yen chiu so.
(Ch'ang yung kuo tzu piao chun tzu t'i piao)
常用國字標準字體表：初稿名國民常用字表. — [台北]：教育部，民國67 [1978]
68, 481, 3, 8 p. : 21 cm.
Cover title.
"教育部社會教育司委託國立臺灣師範大學國文研究所研訂"
1. Chinese language—Word frequency. 2. Chinese language—Writing. I. China (Republic of China). Chiao yü pu. She hui chiao yü ssu. II. Title.
PL1498.K8 1978 79-836915

Lai, Fang-pin.
(Shu fa t'ung lun)
書法通論 / 賴芳彬編著. — [s. l.]：震平文化出版社；嘉義：文友書局，民國60 [1971]
2 v. (2, 10, 1157 p.) : ill. ; 27 cm.

1. Calligraphy, Chinese. 2. Chinese language—Writing. I. Title.
NK3634.A2L32 78-840877

Ledderose, Lothar.
Die Siegelschrift (Chuan-shu) in der Ch'ing-Zeit. Ein Beitr. z. Geschichte d. chines. Schriftkunst. Mit 141 Abb. Wiesbaden, F. Steiner, 1970.
267 p., 30 l. with illus. 24 cm. (Studien zur ostasiatischen Schriftkunst, Bd. 1) DM28.00 GDB70-A49-15
Originally presented as the author's thesis, Heidelberg.
Summary in English.
Bibliography: p. 228-237.
1. Seals (Numismatics)—China. 2. Chinese language—Writing. I. Title. II. Series.
CD5385.L4 1970 737'.0951 79-564163
79[r79]rev MARC

Lin, Wu-hsien, 1944–
(Yü wen pien cheng shou ts'e)
語文辨正手册 / 林武憲. — 台北：遠流出版社，民國67 [1978]
209 p. : ill. ; 22 cm. — (遠流青少年百科全書 ; 11 : 第3輯. 讀書與寫作的藝術)
NT$40.00

1. Chinese language—Pronunciation. 2. Chinese language—Writing. I. Title. II. Series: Yüan liu ch'ing shao nien pai k'o ch'üan shu ; 11.
PL1209.L56 79-837506

McNaughton, William, 1933-
Reading and writing Chinese : a guide to the Chinese writing system, the student's 1,020 list, the official 2,000 list / by William McNaughton. — Rutland, Vt. : Tuttle, c1979.
376 p. ; 19 cm.
ISBN 0-8048-1188-1 : $17.50
1. Chinese language—Glossaries, vocabularies, etc. 2. Chinese language—Writing. I. Title.
PL1495.M3 495.1'1'1 77-77699
79 MARC

Ming jen ch'u pan she.
(Yu ch'ü ti Chung-kuo tzu)
有趣的中國字 / 名人出版社編著. — 台北：該社，民國67 [1978]
160 p. : ill. ; 19 cm.
NT$58.00

1. Word games. 2. Chinese language—Writing. I. Title.
GV1507.W8M56 1978 79-837121

P'an, Fu-min.
(Hsin tseng o tzu yüeh wei)
新增訛字閱微 / [潘阜民編著]. — Hong Kong：偉志印務公司，[1977?]
[28] p. ; 19 cm.
Cover title.

1. Chinese language—Writing. I. Title. II. Title: O tzu yüeh wei.
PL1171.P27 79-835328

Sugimoto, Tsutomu, 1927–
(Itaiji to wa nani ka)
異体字とは何か / [著者 杉本つとむ]. — 東京：桜楓社，1978.
270 p., [2] leaves of plates : ill. ; 19 cm. — (杉本つとむ日本語講座 ; 1) Ja 79-2970
Includes bibliographical references and indexes.
Errata slip inserted.
¥1500
1. Chinese language—Writing. 2. Chinese characters—Japan. 3. Chinese characters—Korea. I. Title. II. Series: Sugimoto, Tsutomu, 1927– Sugimoto Tsutomu Nihongo kōza ; 1.
PL1171.S77 79-808462

—WRITING—ADDRESSES, ESSAYS, LECTURES

Cooper, Arthur D
The creation of the Chinese script / by Arthur Cooper. — London : China Society, 1978.
40, [1] p. : 2 ill. ; 21 cm. — (China Society occasional papers ; no. 20 ISSN 0306-6665) GB78-16191
"Based on a lecture delivered to the China Society on 2nd March, 1977."
£2.00
1. Chinese language—Writing—Addresses, essays, lectures. I. Title. II. Series: China Society, London. China Society occasional papers : New series ; no. 20.
PL1171.C66 495.1'1'1 78-320360
78 MARC

—WRITING—DICTIONARIES

(Chuan li hsing ts'ao shu fa p'u)
篆隸行草書法譜 / 主編者徐心涵，黃玉書. — 台中：先知出版社：經銷中華書畫中心，民國66 [1977] printing.
653 p. ; 22 cm. — (學術巨著 ; 第1輯)
Colophon title.
Series romanized: Hsüeh shu chü chu.
NT$200.00

1. Chinese language—Writing—Dictionaries. I. Hsü, Hsin-han. II. Title: Shu fa p'u.
PL1469.C52 78-842022

(Piao chun kuo yü chu yin tz'u hai)
標準國語注音辭海 / 主編人高静男. — 台南：正業書局，民國64 [1975]
1909 p. in various pagings : 16 cm.
Colophon title.
Title on cover and spine: Tz'u hai.
CONTENTS: 辭海—中國成語辭典—八体字彙—篆字體
NT$110.00
1. Chinese language—Dictionaries. 2. Chinese language—Writing—Dictionaries. 3. Chinese language—Terms and phrases. I. Kao, Ching-nan. II. Title: Tz'u hai.
PL1420.P49 76-836693

Piao chun kuo yü chu yin tz'u hai.
辭典 / 主編者高静男. — 台南：西北出版社，民國67 [1978]
1909 p. in various pagings : 16 cm.
Colophon title.
Title on spine: Piao chun kuo yü chu yin tz'u tien.
Reprint of the 1975 ed. published by Cheng yeh shu chü, Tainan.
CONTENTS: 辭海.—中國成語辭典.—八体字彙.—篆字體.
NT$120.00
1. Chinese language—Dictionaries. 2. Chinese language—Writing—Dictionaries. 3. Chinese language—Terms and phrases. I. Kao, Ching-nan. II. Title.
Title romanized: Tz'u tien.
PL1420.P49 1978 79-835448

(Piao chun kuo yü chu yin Tz'u hai)
標準國語注音辭海 / 主編者高静男；編輯者正業書局. — 台南：該局，民國67 [1978]
7, 1811, 200, 64 p. : 20 cm.
Cover title.
"大字足本"
"附錄": p. 1-200 (3d group).
"注音符號檢字表": p. 1-64 (last group)
NT$220.00
1. Chinese language—Dictionaries. 2. Chinese language—Writing—Dictionaries. 3. Chinese language—Terms and phrases. I. Cheng yeh shu chü. II. Kao, Ching-nan. III. Title: Tz'u hai.
PL1420.P49 1978b 79-839150

—WRITING—HISTORY

(Seiiki shutsudo mokkan)
西域出土木簡：展大 / 編集 飯島稲太郎 [春敬]. — 東京：書藝文化新社，昭和45 [1970]，昭和49 [1974] printing.
[34] p. (on double leaves) ; 34 cm. & explanatory text. — (和漢墨寶選集 ; 第6巻) Ja***
Cataloged from colophon (inserted).
Folded book, issued in case.
¥2700
1. China—Antiquities. 2. Inscriptions, Chinese. 3. Inscriptions—China. 4. Chinese language—Writing—History. I. Iijima, Shunkei, 1906– II. Series: Wa-Kan bokuhō senshū ; dai 6-kan.
ND1457.J33W3 vol. 6 78-814353
[DS715]

—WRITING, CURSIVE

Ch'en, Ch'un, 1483–1544.
(Ming Ch'en Tao-fu Ku shih shih chiu shou)
明陳道復古詩十九首 / [陳淳]. — 第1版. — 北京：文物出版社，1977.
[39] p. (on double leaves) : facsims. ; 42 cm.
In case.
$12.50

1. Ch'en, Ch'un, 1483–1544. 2. Calligraphy, Chinese. 3. Chinese language—Writing, Cursive. I. Ku shih shih chiu shou. 1977. II. Title.
NK3634.C395A4 1977 79-835626

Chu, Yün-ming, 1460–1526.
(Ming Chu Yün-ming ts'ao shu Ch'ien Hou Ch'ih-pi fu)
明祝允明草書前後赤壁賦：上海博物館藏. — 香港：書譜社，[1978?]
[59] p. ; 41 cm. — (書譜叢帖 : 第3輯 ; 17)
Series romanized: Shu p'u ts'ung t'ieh.
Reprint ed.
$22.00
1. Chu, Yün-ming, 1460–1526. 2. Calligraphy, Chinese. 3. Chinese language—Writing, Cursive. I. Su, Shih, 1036–1101. Ch'ien Ch'ih-pi fu. 1978? II. Su, Shih, 1036–1101. Hou Ch'ih-pi fu. 1978? III. Shang-hai po wu kuan. IV. Title. V. Title: Ch'ih-pi fu.
NK3634.C48A4 1978 78-843140

Hsü, Chih-ch'ien.
(Mao chu hsi shih tz'u san shih chiu shou ts'ao shu t'ieh)
毛主席诗词三十九首草书帖 / 编辑出版荣宝斋；书写者徐之谦. — 第1版. — 北京：該斋：发行者新华书店北京发行所，1977.
75 p. : all facsims. ; 26 cm.
Colophon title.
$0.45

1. Hsü, Chih-ch'ien. 2. Calligraphy, Chinese. 3. Chinese language—Writing, Cursive. I. Mao, Tse-tung, 1893–1976. Poems. 1977. II. Jung pao chai, Peking. III. Title.
NK3634.H795A4 1977 79-837105

Huai-su, 725–785.
[K'u sun t'ieh]
唐釋懷素苦筍帖：上海博物館藏. — 香港：書譜社，[1977]
[22] p. : chiefly facsims. ; 41 cm. — (書譜叢帖 : 第三輯 ; 2)
Series romanized: Shu p'u ts'ung t'ieh : Ti san chi.
Reprint ed.
$8.00

1. Huai-su, 725–785. 2. Calligraphy, Chinese. 3. Chinese language—Writing, Cursive. I. Shang-hai po wu kuan. II. Title. III. Title: K'u sun t'ieh.
Title romanized: T'ang shih Huai-su k'u sun t'ieh.
NK3634.H825A67 1977 78-842279

Huang, T'ing-chien, 1045–1105.
(Sung Huang T'ing-chien Chu shang tso)
宋黃庭堅諸上座 / [编者故宫博物院]. — 北京：故宫博物院，1977.
[20] double leaves : all facsims. ; 43 cm.
In case.
$12.00
1. Huang, T'ing-chien, 1045–1105. 2. Calligraphy, Chinese. 3. Chinese language—Writing, Cursive. I. Ku kung po wu yüan, Peking. II. Chu shang tso. 1977. III. Title.
NK3634.H83A735 1977 78-840836

CHINESE LANGUAGE
—WRITING, CURSIVE (Continued)

Kao, T'ang.
(K'ai shu, hsing shu, ts'ao shu, san t'i tzu fan)
楷書·行書·草書三體字範 / 高堂書. — 九龍：江南書局：發行利通圖書公司, [1978?]
270 p. : all facsims. ; 22 cm.
$6.50

1. Penmanship, Chinese. 2. Chinese language—Writing, Cursive. 3. Chinese language—Writing. I. Title. II. Title: San t'i tzu fan.
Z44.K386 79-839959

C

Kao-hsien, 9th cent.
(T'ang shih Kao-hsien ts'ao shu Ch'ien tzu wen)
唐釋高閑草書千字文：上海博物館藏. — 香港：書譜社, [1977]
[30] p. : chiefly facsims. ; 41 cm. — (書譜叢帖：第三輯 ; 3)
Series romanized: Shu p'u ts'ung t'ieh : Ti san chi.
Reprint ed.
$15.00
1. Kao-hsien, 9th cent. 2. Calligraphy, Chinese. 3. Chinese language—Writing, Cursive. I. Chou, Hsing-ssu, d. 521. Ch'ien tzu wen. 1977. II. Shang-hai po wu kuan. III. Title.
NK3634.K3A75 1977 78-842795

C

(Kuei-chi Wang shih shu han chi)
會稽王氏書翰集. — 台北：學海出版社, 民國65 [1976]
[22] p. ; 32 cm.
Reprint ed.
NT$60.00

1. Calligraphy, Chinese. 2. Chinese language—Writing, Cursive. 3. Wang, Hsi-chih, 321–379.
NK3634.A4C874 1976 78-842379

C

Lu, Chü-jen, d. ca. 1377.
(Yüan Lu Chü-jen T'iao chih shui shih)
元陸居仁苕之水詩. — 第1版. — 北京：文物出版社, 1977.
[15] p. (on double leaves) : chiefly facsims. ; 43 cm.
Caption title: T'iao chih shui.
In case.
"故宮博物院藏"
$5.00
1. Lu, Chü-jen, d. ca. 1377. 2. Calligraphy, Chinese. 3. Chinese language—Writing, Cursive. I. Ku kung po wu yüan, Peking. II. Title. III. Title: T'iao chih shui.
NK3634.L82A78 1977 78-842678

C

Lu, Yu, 1125–1210.
[Sung Lu Yu tzu shu shih]
陸游自書詩. — 臺北：學海出版社, 民國65 [1976]
[38] p. : all facsims. ; 31 cm.
Cover title: Sung Lu Yu tzu shu shih.
Reprint ed.

1. Lu, Yu, 1125–1210. 2. Calligraphy, Chinese. 3. Chinese language—Writing, Cursive. I. Title.
Title romanized: Lu Yu tzu shu shih.
NK3634.L84A4 1976 78-840954

Shen, Ts'an, 15th cent.
(Ming Shen Ts'an Ch'ien tzu wen)
明沈粲千字文. — 第1版. — 北京：文物出版社, 1977.
[33] p. (on double leaves) : facsims ; 42 cm.
"故宮博物院藏"
$10.00
1. Shen, Ts'an, 15th cent. 2. Calligraphy, Chinese. 3. Chinese language—Writing, Cursive. I. Chou, Hsing-ssu, d. 521. Ch'ien tzu wen. 1977. II. Ku kung po wu yüan, Peking. III. Title. IV. Title: Ch'ien tzu wen.
NK3634.S486A4 1977 79-835788

Shih, Tzu-ch'en.
(Ts'ao shu i shu)
草書藝術 / 史紫忱. — 臺北：華正書局, 民國66 [1977]
3, 75 p. : ill., facsims. ; 21 cm.
NT$50.00

1. Calligraphy, Chinese. 2. Chinese language—Writing, Cursive. I. Title.
NK3634.A4C877 78-842263

Sich'ŏnggak Kyoyuk Sa. Sŏye Kyoyukpu.
(Sipch'ilch'ŏp)
十七帖 / 草書王羲之 ; [編著者視聴覺教育社書藝教育部]. — 서울：同社, 1974, 1975 printing.
96 p. : chiefly ill. ; 33 cm. — (書藝技法시리즈 ; 6)
Colophon inserted.

1. Wang, Hsi-chih, 321–379. Shih ch'i t'ieh. 2. Calligraphy, Chinese. 3. Chinese language—Writing, Cursive. I. Wang, Hsi-chih, 321–379. Shih ch'i t'ieh. 1975. II. Title. III. Series: Sŏye kipŏp sirijŭ ; 6.
NK3634.W3A7537 79-826530

Sun, Kuo-t'ing, 648–703.
[Shu p'u. Japanese & Chinese]
書譜 / 孫過庭. — 東京：書藝文化新社, 昭和45 [1970], 昭和49 [1974] printing.
[102] p. (on double leaves) ; 34 cm. & explanatory text. — (和漢墨寶選集 ; 第5巻)
Ja***
Cataloged from colophon (inserted).
Folded book, issued in case.
Reproduced from ms. copy, with printed transcript.
¥3800
1. Calligraphy, Chinese. 2. Chinese language—Writing, Cursive. 3. Sun, Kuo-t'ing, 648–703. I. Title. II. Title: Son Katei Shofu. III. Series: Wa-Kan bokuhō senshū ; dai 5-kan.
Title romanized: Shofu.
ND1457.J33W3 vol. 5 79-800371
[NK3634.A4C87]

Sun, Kuo-t'ing, 648–703.
(T'ang Sun Kuo-t'ing shu p'u chen chi)
唐孫過庭書譜真跡. — Kowloon : Grand Books & Art Co., [1978?]
[71] p. : chiefly facsims. ; 32 cm. — (天下叢書 ; 9)
Series romanized: T'ien hsia ts'ung shu.
$20.00

1. Sun, Kuo-t'ing, 648–703. 2. Calligraphy, Chinese. 3. Chinese language—Writing, Cursive. I. Title.
NK3634.S92A4 1978 78-841902

Takahashi, Kanjō.
(Mōhitsu pen santai kanji no kuzushikata)
毛筆ペン三體漢字のくづし方 / 高橋観城編著. — 東京：松榮堂, 昭和7 [1932]
198 p. ; 19 cm.

1. Calligraphy, Japanese. 2. Calligraphy, Chinese. 3. Japanese language—Writing, Cursive. 4. Chinese language—Writing, Cursive. I. Title. II. Title: Santai kanji no kuzushikata.
Z44.T2363 78-817000

Tu, Mu, 803–853?
(T'ang Tu Mu Chang Hao-hao shih)
唐杜牧張好好詩. — 第1版. — 北京：文物出版社, 1977.
[8] double leaves in case : ill. ; 43 cm.
"故宮博物院藏"
$5.50

1. Tu, Mu, 803–853? 2. Calligraphy, Chinese. 3. Chinese language—Writing, Cursive. I. Title.
NK3634.T78A75 1977 78-842696

Tung, Ch'i-ch'ang, 1555–1636.
(Tung Ch'i-ch'ang P'an-ku hsü)
董其昌盤谷序. — [台1版]. — [台北：書法文物館：總經銷精益書局, 民國66 i. e. 1977]
[20] p. (on double leaves) : all facsims. ; 30 cm. — (書法文物叢刊)
Title from label on cover.
Edition statement and imprint from label mounted at end of the text.
"版本收藏周燕謀"
In case.
1. Tung, Ch'i-ch'ang, 1555–1636. 2. Calligraphy, Chinese. 3. Chinese language—Writing, Cursive. I. Chou, Yen-mou. II. Title. III. Series: Shu fa wen wu ts'ung k'an.
NK3634.T84A74 1977 78-842350

Tung, Ch'i-ch'ang, 1555–1636.
(Tung Ch'i-ch'ang sung Chang sheng tz'u)
董其昌送章生詞. — 台1版. — [台北]：書法文物館：總經銷精益書局, 民國66 [1977]
[26] p. (on double leaves) : all facsims. ; 30 cm. — (書法文物叢刊)
Title from label on cover.
Colophon on label inserted.
"版本收藏 周燕謀"
In case.
1. Tung, Ch'i-ch'ang, 1555–1636. 2. Calligraphy, Chinese. 3. Chinese language—Writing, Cursive. I. Title. II. Series: Shu fa wen wu ts'ung k'an.
NK3634.T84A745 1977 78-842348

Wang, Ch'ung, 1494–1533.
(Ming Wang Ch'ung tzu shu shih)
明王寵自書詩：上海博物館藏. — 香港：書譜社, [1977]
[28] p. : facsims. ; 41 cm. — (書譜叢帖 : 第三輯 ; 19)
Series romanized: Shu p'u ts'ung t'ieh : Ti san chi.
"釋文": p. [25]–[27]
$15.00
1. Wang, Ch'ung, 1494–1533. 2. Calligraphy, Chinese. 3. Chinese language—Writing, Cursive. I. Shang-hai po wu kuan. II. Title.
NK3634.W28A4 1977 78-841554

Wu, K'uan, 1435–1504.
(Ming Wu K'uan Shen Chou tz'u yün tung yüan chü hua shih)
明吳寬沈周次韻東園菊花詩. — 第1版. — 北京：文物出版社, 1977.
[21] p. (on double leaves) : facsims. ; 42 cm.
"故宮博物院藏"
$7.00
1. Wu, K'uan, 1435–1504. 2. Shen, Chou, 1427–1509. 3. Calligraphy, Chinese. 4. Chinese language—Writing, Cursive. I. Shen, Chou, 1427–1509. II. Ku kung po wu yüan, Peking. III. Title.
NK3634.W85A4 1977a 79-835531

Yü, Yu-jen, 1878–1964.
(Yü Yu-jen hsien sheng mo pao)
于右任先生墨寶 / 編輯者于右任先生百年誕辰紀念籌備委員會. — [台北]：國史館, 民國67 [1978]
[107] p. : facsims., port. ; 31 cm.
Colophon title.
In case.
NT$150.00 ($4.00 U. S.)
1. Yü, Yu-jen, 1878–1964. 2. Calligraphy, Chinese. 3. Chinese language—Writing, Cursive. I. Yü Yu-jen hsien sheng pai nien tan ch'en chi nien ch'ou pei wei yüan hui. II. Title.
NK3634.Y8A4 1978 78-842697

—WRITING, LI STYLE

(Han Lou Shou pei)
漢婁壽碑. — 台北：華正書局, 民國65 [1976]
[14] double leaves : facsims. ; 38 cm.
Colophon title.
Reprint ed.
In case.
NT$150.00

1. Calligraphy, Chinese. 2. Chinese language—Writing, Li style. I. Lou, Shou.
NK3634.A4L533 1976 78-841171

Ku kung po wu yüan, Peking.
(Ming t'a Han Ts'ao Ch'üan pei)
明拓漢曹全碑 故宮博物院藏 [北京] 文物出版社 1964.
[53] p. (on double leaves) 37 cm. (故宮博物院藏歷代善本碑帖之一)
Reproduced from rubbings.

1. Calligraphy, Chinese. 2. Chinese language—Writing, Li style. I. Title.
NK3634.Z6K8 1964 C 65-1736

rev

CHINESE LANGUAGE
—WRITING, LI STYLE (Continued)

(Li shu tzu t'ieh)
隸書字帖 / 編選者萬應軒. — 台北：藝術圖書公司, 民國67 [1978]
112 p. ; 22 cm.
CONTENTS: 怎樣學寫隸書.—隸法行筆示意圖.—隸書臨摹字帖.—曹全碑隸書.—孔宙碑.—孔宙碑選字.—婁壽碑選字.—唐人隸書小字帖.

1. Penmanship, Chinese. 2. Chinese language—Writing, Li style. I. Wan, Ying-hsüan.
Z44.L553 79–836686

C

(Ming ch'u t'a Shih Ch'en ch'ien hou pei)
明初拓史晨前後碑. — 台北：藝文印書館, 民國60 [1971]
[65] p. (on double leaves) ; 34 cm.
Colophon title.
Cover title on label: Ming t'a Shih Ch'en ch'ien hou pei.
In case.

1. Calligraphy, Chinese. 2. Chinese language—Writing, Li style. I. Title: Shih Ch'en ch'ien hou pei. II. Title: Ch'ien hou pei. III. Title: Ming t'a Shih Ch'en ch'ien hou pei.
NK3634.A4L55 78–841170

C

(Ming t'a Ts'ao Ch'üan pei)
明拓曹全碑 / 主編者李光業. — [台北]：中國書法學會, 民國60 [1971]
68 p. (on double leaves) : facsims. ; 35 cm.
Colophon title.
In case.
Reprint ed.
"曹全碑釋文": p. 67–68.

1. Calligraphy, Chinese. 2. Chinese language—Writing, Li style. I. Ts'ao, Ch'üan. II. Li, Kuang-yeh. III. Title: Ts'ao Ch'üan pei.
NK3634.A4L557 1971 79–835134

C

Shih, Ch'en, fl. 168–169.
(Han Shih Ch'en ch'ien pei)
漢史晨前碑：普及本. — 香港：書譜出版社, [1978?]
[40] p. : chiefly facsims. ; 29 cm.
Reprint ed.
$4.00

1. Shih, Ch'en, fl. 168–169. 2. Calligraphy, Chinese. 3. Chinese language—Writing, Li style. I. Title.
NK3634.S49A67 1978 78–843379

C

Sich'ŏnggak Kyoyuk Sa. Sŏye Kyoyukpu.
(Yegibi)
礼器碑 / [編著者視聴覺教育社書藝教育部]. — 서울：同社, 1975.
96 p. : chiefly ill. ; 33 cm. — (書藝技法시리즈 ; 8)
Title on colophon (inserted): Yesŏ Han yegibi.

1. Calligraphy, Chinese. 2. Chinese language—Writing, Li style. I. Han li ch'i pei. 1975. II. Title. III. Title: Yesŏ Han yegibi. IV. Series: Sŏye kipŏp sirijŭ ; 8.
NK3634.A4L558 79–826249

K

Wu, Jui, 1298–1355.
(Yüan Wu Jui li shu Li sao chuan shu Ch'ien tzu wen)
元吳叡隸書離騷篆書千字文：上海博物館藏. — 香港：書譜社, [1977]
[35] p. ; 41 cm. — (書譜叢帖：第三輯 ; 14)
Series romanized: Shu p'u ts'ung t'ieh : Ti san chi.
$15.00
1. Wu, Jui, 1298–1355. 2. Calligraphy, Chinese. 3. Chinese language—Writing, Li style. I. Ch'ü, Yüan, ca. 343–ca. 277 B.C. Li sao. 1977. II. Chou, Hsing-ssu, d. 521. Ch'ien tzu wen. 1977. III. Title.
NK3634.W844A4 1977 78–842262

—WRITING, SEAL STYLE

Chang, Ping-lin, 1868–1936.
(Chang Ping-lin chuan shu t'ieh)
章炳麟篆書帖. — [台北：書法文物館, 1979?]
[30] p. (on double leaves) ; 26 cm.
Title from label on cover.
"中華民國書法碑帖"
Imprint from label on p. [3] of cover.
In case.
Reprint ed.
1. Chang, Ping-lin, 1868–1936. 2. Calligraphy, Chinese. 3. Chinese language—Writing, Seal style. I. Title.
NK3634.C3726A63 1979 79–841836

(Chou Yü ting ming)
周盂鼎銘：精拓放大本. — [台北：書法文物館, 1978?]
[26] p. (on double leaves) : all facsims ; 30 cm.
Publisher stamped on label mounted on cover.
In case.

1. Calligraphy, Chinese. 2. Chinese language—Writing, Seal style. 3. Inscriptions, Chinese.
NK3634.A4S417 78–842327

Chuang, Po-ho.
(Ch'ing jen chuan shu san chia)
清人篆書三家：鄧完白·吳讓之·趙之謙 / 莊伯和編譯. — 台北：藝術圖書公司, 民國67 [1978]
96 p. : ill. ; 22 cm.
NT$50.00

1. Calligraphy, Chinese. 2. Chinese language—Writing, Seal style. 3. Teng, Shih-ju, 1743–1805. 4. Wu, Jang-chih, 1799–1870. 5. Chao, Chih-ch'ien, 1829–1884. I. Title.
NK3634.A2C4863 79–838282

Lo, Fu-i, 1905–
(Han yin wen tzu cheng)
汉印文字征 / 罗福颐编. — 第1版. — 北京：文物出版社, 1978.
ca. 350 leaves of various foliations ; 27 cm.
Errata slip inserted.
Reprint ed.
Includes bibliographies.
$8.00
1. Chinese language—Writing, Seal style. 2. Seals (Numismatics)—China. I. Title.
PL1469.L59 1978 79–837516

—WRITING, SEAL STYLE
—DICTIONARIES

Taiho, Kiei.
(Tensho jiten)
篆書字典 / 太甫熙永編. — 東京：国書刊行会, 1978.
560, 88 p. ; 23 cm. **Ja 78–25603**
¥8000

1. Chinese language—Writing, Seal style—Dictionaries. I. Title.
PL1469.T295 **79–805220**

CHINESE LEGENDS see Legends, Chinese

CHINESE LETTERS

(Ch'ing tai pa chia ch'ih tu)
清代八家尺牘. — 台北：學海出版社, 民國65 [1976]
48 p. ; 32 cm.
"嚴繩孫, 惲壽平, 張英, 汪士鋐, 何焯, 張照, 姚鼐, 陳介祺"
NT$130.00

1. Chinese letters.
PL2610.C438 78–840931

(Ming hsien shou cha)
名賢手札 / 珍藏者周樹聲. — [台北]：周樹聲, 民國65 [1976]
[65] p. : ill. ; 27 cm.
Colophon title.

1. Chinese letters. 2. Chou, Shu-sheng. I. Chou, Shu-sheng.
PL2610.M516 78–842443

CHINESE LITERATURE

Under this heading, without subdivision, are entered individual works of creative writing not clearly indicated to be dramatic works, fiction or poetry, and collections of works of mixed genre by individual authors

see also Classification—Books—Chinese literature

A-feng.
(San wen erh shih liu p'ien)
散文二十六篇 / 阿鳳著. — [香港?：s. n., 1978?]
132 p. ; 18 cm.
Essays.
Reprint of the 1957 ed. published by Chung-kuo ch'ing nien ch'u pan she, Peking.

I. Title.
PL2833.A4S3 1978 78–840993

A-shui.
(Feng, yü, yün, wu)
風, 雨, 雲, 霧 / 編著者阿水. — [台北?]：五福出版社, 民國67 [1978]
7, 211 p. ; 19 cm.
Colophon title.
Literature.
NT$40.00

I. Title.
PL2833.A8F4 79–835495

Abe, Kōbō, 1924–
[Selected works. Chinese]
砂丘之女及其他 / 安部公房著. — 九龍：新天出版社, [1978?]
10, 251 p. ; 19 cm.
CONTENTES: 安部公房與砂丘之女.—砂丘之女.—紅色的繭.—棍子.—魔筆.—無關的死.
$7.00

I. Title.
Title romanized: Sha ch'iu chih nü chi ch'i t'a.
PL845.B4A53 1978 78–840815

(Ai ch'ing yü jen sheng)
愛情與人生 / 陳銀櫃編著. — 台南：大府城出版有限公司：總經銷中央書局, 民國67 [1978]
190 p. ; 20 cm. — (大府城書系 ; 20)
Series romanized: Ta fu ch'eng shu hsi.
NT$40.00

1. Chinese essays—20th century. I. Ch'en, Yin-kuei.
PL2623.A38 79–837032

Chan, Jui-lin, 1949–
(Hsin yüan san ch'ü)
心園散曲 / 詹瑞麟著. — 臺北：國家書店, 民國67 [1978]
2, 2, 3, 202 p. ; 20 cm. — (讀書人叢刊 ; 65)
In colophon: Ch'u pan che Kuo chia ch'u pan she.
Series romanized: Tu shu jen ts'ung k'an.
NT$50.00

I. Title.
PL2836.J84H74 79–838576

CHINESE LITERATURE (Continued)

Chang, Ch'e.
(Chang Ch'e tsa wen)
張徹雜文. — 臺北：皇冠出版社，民國67 [1978]
265 p. ; 19 cm.
NT$40.00

I. Title.
PL2837.C45A6 1978 — 79-837685

Ch'ang, Cheng, fl. 1976–
(Wang shih, Chung-kuo jen)
往事，中國人 / 長争著. — 香港：朝陽出版社，1978.
2, 169 p. ; 21 cm.
$4.80

I. Title.
PL2837.C465W34 — 78-840948

Chang, Chih-chieh.
(Tsou ch'u shih yen shih)
走出實驗室：散文·雜文集 / 張之傑著. — 台北：自然科學文化事業公司出版部，民國67 [1978]
230 p. : ill. ; 22 cm.
NT$50.00

I. Title.
PL2837.C525T7 — 79-840620

Chang, Fang-hsü.
(Chin I san wen chi)
靳以散文集. — 香港：中流出版社，1977.
3, 202 p. ; 19 cm.
$4.70

I. Title.
PL2740.F3C49 — 79-835897

Chang, Fang-hsü.
(Chin-i wen chi)
靳以文集. — [香港?；s. n., 1978]
31 leaves : ill. ; 31 cm.
Cover title.
A collection of articles reprinted from various sources.
Date of publication stamped on p. [4] of cover.
CONTENTS: 我在上海遇到了長春汽車厂工人.—創造奇迹的時代.—老孟泰来到了上海.—想起在朝鮮的日子.—光輝的苏联文学引着我前進.—一只美丽的酒瓶.—献給向荒地進攻的青年們.—葛琴　靳以.—孫陵　幽冥永隔寬恕靳以.—麥默　痴情的悲哀.

I. Ko, Ch'in. Chin-i. 1978. II. Sun, Ling. Yu ming yung ko k'uan shu Chin-i. 1978. III. Mai, Mo. Ch'ih ch'ing ti pei ai. 1978. IV. Title.
PL2740.F3C53 — 79-836251

Chang, Hsi-kuo.
(K'ung-tzu chih ssu)
孔子之死 / 張系國. — 臺北：洪範書店，民國67 [1978]
222 p. ; 19 cm. — (洪範文學叢書 ; 36)
Literature.
Series romanized: Hung fan wen hsüeh ts'ung shu.
Includes bibliographical references.
NT$50.00

I. Title.
PL2837.H68K84 — 79-840359

Chang, Hsiu-ya, 1918–
(Chang Hsiu-ya hsüan chi)
張秀亞選集. — 香港：香港文學研究社，[1978?]
4, 156 p. ; 21 cm. — (中國現代文選叢書)
$6.00

I. Title. II. Series: Chung-kuo hsien tai wen hsüan ts'ung shu.
PL2837.H765A6 1978 — 79-839613

Chang, Hsiu-ya, 1918–
(Jen sheng hsiao ching)
人生小景 / 張秀亞著. — 台北：水芙蓉出版社，民國67 [1978]
5, 4, 180 p. ; 19 cm. — (水芙蓉書庫 ; 110)
Series romanized: Shui fu jung shu k'u.
NT$40.00

I. Title.
PL2837.H765J4 — 79-836490

Chang, Hsüeh-yin.
(P'iao hu ti yün ying)
飄忽的雲影 / 張雪茵著. — 臺北：臺灣學生書局，民國67 [1978]
2, 248 p. ; 20 cm. — (學生書苑 ; 62)
Series romanized: Hsüeh sheng shu yüan.
CONTENTS: 飄忽的雲影.—兩顆發亮的貝殼.—孽緣.—三表哥.—紅樓.—重逢.—招魂曲.—慈暉.—紫色的愛.—輓歌.—西天的晚霞.
NT$70.00

I. Title.
PL2837.H77P5 — 79-837010

Chang, Kuo, fl. 1968–
(K'en-ting ch'ing ko)
墾丁情歌 / 張過著. — 臺北：臺灣商務印書館，民國67 [1978]
2, 3, 2, 113 p. ; 19 cm. — (人人文庫 ; 2405-2406)
"雜文"
Series romanized: Jen jen wen k'u.
NT$18.00

I. Title.
PL2837.K845K4 — 79-841063

Chang, P'u, 1602–1641.
(Ch'i lu chai lun lüeh)
七錄齋論略 / 張溥撰. — 臺北：偉文圖書出版社，民國66 [1977]
2 v. (842 p.) ; 22 cm. — (清代禁燬書叢刊 : 第一輯)
Reprint ed.
"七錄齋集": p. 171-842.
NT$600.00 ($15.81 U. S.)

I. Title. II. Series: Ch'ing tai chin hui shu ts'ung k'an : Ti i chi.
PL2698.C316C48 1977 — 78-842019

Chang, Shu-p'ing, d. 1970.
(Ch'üan-an i kao)
蜷厂遺稿 / [編輯者曾霽虹]. — [s. l. : s. n.], 民國63 [1974] (台北：新文化彩色印書館)
84, [25] p., [1] leaf of plates : port. ; 22 cm.

I. Tseng, Chi-hung. II. Title.
PL2837.S52C4 1974 — 79-835410

Chang, T'eng-chiao.
(Chang T'eng-chiao tzu hsüan chi)
張騰蛟自選集. — 臺北：黎明文化事業公司，民國67 [1978]
8, 380 p., [2] leaves of plates : ill. ; 20 cm. — (中國新文學叢刊 ; 61)
Literature.
Series romanized: Chung-kuo hsin wen hsüeh ts'ung k'an.
NT$80.00

I. Title.
PL2837.T4C5 — 79-837314

Chang, T'eng-chiao.
(Wo ai shan lin, wo ai yüan yeh)
我愛山林，我愛原野 / 張騰蛟著. — 臺北：聯亞出版社，民國66 [1977]
2, 4, 210 p. ; 20 cm. — (聯亞叢刊 ; 29)
Essays.
Series romanized: Lien ya ts'ung k'an.
NT$40.00

I. Title.
PL2837.T4W6 — 78-841161

Chang, T'o-wu, 1928–
(Tai ma shu tsu yü chi)
代馬輸卒餘記 / 張拓蕪著. — 臺北：爾雅出版社，民國67 [1978]
15, 223 p. ; 19 cm. — (爾雅叢書 ; 41)
Series romanized: Erh ya ts'ung shu.
Appendices (p. 191-223): 三毛　張拓蕪的傳奇.—三毛　我的筆友張拓蕪.—張系國　勇者的畫像.—張默　血淚染織的奇書.—姜穆　張拓蕪的代馬輸卒.—張拓蕪寫作年表.
NT$50.00
I. Title.
PL2837.T6T33 — 79-837446

Chao, Kuo-ts'ai.
(Chien ho yen yü)
劍河煙雨 / 趙國材著. — 台北：台湾新生報，民國67 [1978]
5 v. : ill. ; 20 cm.
Vols. 1-2: 4 ed.
NT$50.00 per vol.

I. Title.
PL2838.K58C5 1978 — 79-838123

Chao, Pao-ch'u, comp.
(T'ai-hu Chao shih chia chi ts'ung k'o)
太湖趙氏家集叢刻　趙寶初輯　[臺北縣永和鎮] 文海出版社　[民國50 i. e. 1970]
2 v. 20 cm. (近代中國史料叢刊 585-589)
Reprint ed.
CONTENTS: 1. 趙文楷　石柏山房詩存9卷—2. 趙昀　逡園詩鈔6卷　趙昀　逡翁自訂年譜　趙繼元　靜觀堂遺集 2卷　趙曾重　閑黑
1. Chinese literature. 2. Chao family. I. Title. II. Series: Shen, Yün-lung, comp. Chin tai Chung-kuo shih liao ts'ung k'an 585-589.
DS755.S53 no. 585-589 — 73-899786
[PL2516.C48]

Chao, Te-chu.
(Ma p'i p'ai pu ch'uan)
馬屁拍不穿 / 趙德鉒著. — [臺北?]：德昌出版社，[1978?]
239 p. ; 20 cm.

I. Title.
PL2838.T4M3 — 79-838710

Chao, Tzu-fan.
(Chi chien chi)
擊劍集 / 趙滋蕃著. — 臺北：彩虹出版社，民國66 [1977]
224 p. ; 19 cm. — (彩虹叢書 ; 119)
Series romanized: Ts'ai hung ts'ung shu.
NT$35.00

I. Title.
PL2838.T93C485 — 78-843126

(Che li hsiao p'in)
哲理小品 / 編譯羅國才. — 香港：弘毅文教社，1978.
144 p. ; 22 cm.
Colophon title.
At head of cover title: Chung-kuo wen hsüeh p'u chi tu wu.
$6.00

1. Chinese literature. I. Lo, Kuo-ts'ai.
PL2455.C39 — 78-842537

CHINESE LITERATURE (Continued)

Ch'en, Ch'ao-tung.
(Hsiao chen ch'ing shen)
小鎮情深 / 陳朝棟著. — 台北：水芙蓉出版社, 民國67 [1978]
2, 6, 240 p. ; 20 cm. — (水芙蓉書庫 ; 106)
Essays.
Series romanized: Shui fu jung shu k'u.
NT$42.00

I. Title.
PL2840.C33H8 78-841483

Ch'en, Chi-t'ai, 1573-1640.
(I wu chi)
已吾集 / 陳際泰撰. — 臺北：偉文圖書出版社, 民國66 [1977]
2 v. (870 p.) ; 22 cm. — (清代禁燬叢刊 : 第一輯)
Reprint of the ed. in the possession of the National Central Library, Taipei.
CONTENTS: 上 陳際泰 已吾集 卷1-10.—下 陳際泰 已吾集 卷11-14. 陳孝威 壺山集 3卷. 陳孝逸 癡山集 6卷.
NT$600.00 ($15.81 U. S.)

I. Ch'en, Hsiao-wei. Hu shan chi. 1977. II. Ch'en, Hsiao-i. Ch'ih shan chi. 1977. III. Title. IV. Series: Ch'ing tai chin hui shu ts'ung k'an : Ti i chi.
PL2698.C385I2 1977 78-842020

Ch'en, Fan.
(Ch'en meng chi)
塵夢集 / 陳凡著. — 香港：廣角鏡出版社, 1978.
367 p. ; 21 cm.
$20.00

I. Title.
PL2840.F3A6 1978 79-836964

Ch'en, Hsi-wen.
(Shuang yeh i hsiao p'in)
雙頁一小品 / 陳錫文編著. — 台南：大府城出版有限公司：總經銷中央書局, 民國67 [1978]
203 p. ; 20 cm. — (大府城書系 ; 39)
Series romanized: Ta fu ch'eng shu hsi.
NT$60.00

I. Title.
PL2840.H72S5 79-838024

Ch'en, Hsiao-lin, 1949-
(Lang mang shao nien hsing)
浪莽少年行 / 陳曉林著. — 第1版. — 臺北：四季出版事業有限公司, 民國67 [1978]
4, 10, 4, 272 p. ; 20 cm. — (四季文萃 ; 23 : 論述)
Series romanized: Ssu chi wen ts'ui.
Appendices (p. 159-272): 1. 丘容 文明是不會没落的.—2. 婁學文 [illegible]學的與非數學的.—3. 李師鄭 結束與開端.—4. 孫瑋芒 前塵與後浪.—5. 王汎森 開拓史界及翻譯界之心胸.—6. 蔣芸 蔣芸「告別文壇」聲明.—7. [illegible]振煜 評蔣芸的抄襲與尹雪曼的掩飾.—8. 金玉書 蔣芸盜作事件的面面觀.—9. 寒爵 陳曉林的「我論瓊瑤」.—10. 芥川龍之介 竹藪中.
NT$45.00

I. Title.
PL2840.H74L3 1978 79-837359

Ch'en, Hsiao-p'in.
(Hsiang yeh ti meng)
鄉野的夢 / 編著者陳曉品. — 台中：天宏出版社, 民國67 [1978]
6, 180 p. ; 20 cm.
Colophon title.
Literature.
NT$35.00

I. Title.
PL2840.H743H75 79-835902

Ch'en, Hsin-ch'uan, 1952-
(Lo hsia man chien)
落霞滿箋 / 陳新川著. — 台北：國家書店, 民國67 [1978]
4, 3, 239 p. ; 20 cm. — (讀書人叢刊 ; 52)
In colophon: Ch'u pan che, Kuo chia ch'u pan she.
Essays.
Series romanized: Tu shu jen ts'ung k'an.
NT$45.00

I. Title.
PL2840.H767L6 78-840853

Ch'en, Jen-t'ai.
(Teng tai hua k'ai)
等待花開 / 陳仁泰著. — 苗栗縣竹南鎮：七燈出版社, 民國67 [1978]
216 p. ; 20 cm. — (七燈藝術文庫 ; 6)
Series romanized: Ch'i teng i shu wen k'u.
NT$40.00

I. Title.
PL2840.J4T4 79-838025

Ch'en, K'o-huan.
(Chin hsiu nien hua)
錦繡年華：給成長中的女孩 / 陳克環著. — 臺北：九歌出版社, 民國68 [1979]
188 p. ; 19 cm. — (九歌文庫 ; 19)
Series romanized: Chiu ko wen k'u.
NT$45.00

I. Title.
PL2840.K77C5 79-840100

Ch'en, K'o-huan.
(Yu yu wo ssu)
悠悠我思 / 陳克環著. — 第1版. — 台北：彥博出版社, 民國67 [1978]
3, 210 p. ; 20 cm. — (彥博文藝叢書 ; 3)
Series romanized: Yen po wen i ts'ung shu.
NT$45.00

I. Title.
PL2840.K77Y8 79-837846

Ch'en, Meng-chia, 1911-
(Pu k'ai hua ti ch'un t'ien)
不開花的春天 / 陳夢家創作. — [香港? : s.n., 1977?]
63 p. ; 18 cm.
Essays.
Reprint of the 1932 ed. published by Liang yu t'u shu yin shua kung ssu, Shanghai; in series: I chiao ts'ung shu ; ti 3 chung.

I. Title.
PL2743.M39P8 1977 78-843543

Ch'en, Shao-t'ing.
(O o chi)
諤諤集 / 陳少廷著. — 臺中：藍燈文化事業股份有限公司, 民國67 [1978]
2, 268 p. ; 19 cm. — (藍燈新刊 ; 18)
Essays.
Series romanized: Lan teng hsin k'an.
NT$45.00

I. Title.
PL2840.S425O15 79-837505

Ch'en, T'ien-lan, 1939-
(Hui kuei tzu jan)
回歸自然 / 陳天嵐著. — 臺北：星光出版社, 民國66 [1977]
170 p. ; 20 cm. — (雙子星叢書 ; 46)
Prose works.
Series romanized: Shuang tzu hsing ts'ung shu.
NT$35.00

I. Title.
PL2840.T47H84 78-841247

Ch'en, Ying-feng.
(K'ai fang ti jen sheng)
開放的人生 / 陳英鳳編著. — 台南：大府城出版有限公司：總經銷中央書局, 民國67 [1978]
152 p. ; 20 cm. — (大府城書系 ; 24)
Series romanized: Ta fu ch'eng shu hsi.
NT$40.00

I. Title.
PL2840.Y535K3 79-839462

Cheng, Chen-to, 1897-1958.
(Ch'ü huo che ti tai pu)
取火者的逮捕 鄭振鐸著 上海 新文藝出版社 1956.
98 p. 19 cm.

I. Title.
PL2744.C4C5 1956 C 61-331 ‡

Cheng, Chia-chen, fl. 1964-
(Chan tou tsai Pei-ta-huang)
战斗在北大荒 牡丹江青年垦荒队的故事 郑加真 萧沉等著 上海 少年儿童出版社, 1964.
2, 158 p. 19 cm.

I. Hsiao, Ch'en, joint author. II. Title.
PL2841.C45C5 C 66-3155

Cheng, Ch'ing-ho.
(Ch'u lien jih chi)
初戀日記：難忘的初戀, 永恒的記憶 / 鄭清和著. — 台中：新企業世界出版社, 民國66 [1977]
3, 213 p. ; 20 cm.
NT$40.00

I. Title.
PL2841.C49C5 79-837686

Cheng, I-mei.
(I-mei hsiao p'in)
逸梅小品 / [著作者鄭逸梅]. — 台北：新文豐出版公司, 民國67 [1978]
6, 4, 6, 280 p. ; 19 cm. — (零玉碎金集刊 ; 10)
Series romanized: Ling yü sui chin chi k'an.
Reprint ed.
NT$70.00

I. Title.
PL2841.I17I2 1978 79-838249

Cheng, I-mei.
(I-mei ts'ung t'an)
逸梅叢談 / 著作者鄭逸梅. — 台北：新文豐出版公司, 民國67 [1978]
10, 440 p. ; 19 cm. — (零玉碎金集刊 ; 12)
Colophon title.
Series romanized: Ling yü ts'ui chin chi k'an.
Pbk.: NT$70.00
Essays.

I. Title.
PL2841.I17A16 1978 79-838712

Cheng, Shu-min.
(Tzu tsai chi)
自在集 / 鄭淑敏. — 臺北：洪建全教育文化基金會, 民國67 [1978]
13, 241 p. ; 20 cm.
Cover title.
Essays.
NT$45.00

I. Title.
PL2841.S565T9 78-842238

CHINESE LITERATURE (Continued)

(Chi yü liu ssu)
寄語柳絲：少女抒情散文 / 江秀雲編. — 臺北：彩虹出版社, 民國67 [1978]
208 p. ; 20 cm. — (彩虹叢書 ; 131)
Series romanized: Ts'ai hung ts'ung shu.
NT$35.00

1. Chinese essays—Women authors. I. Chiang, Hsiu-yün.
PL2515.C4555 79-836608

Chiang, Hai-chou, 1932-
(Chih hu shui yüeh)
知乎水月 / 碧果 [姜海洲] 著. — 臺北：源成文化圖書供應社, 民國66 [1977]
4, 222 p. ; 20 cm.
Essays.
NT$40.00

I. Title.
PL2844.H3C5 77-841855

Chiang, I-chih.
(Ni nung wo nung shih shih ch'ing)
妳儂我儂世世情 / 姜藝之著. — 永和鎮：石室出版社 ; 台北：總經銷博大書局, 民國66 [1977]
222 p. ; 20 cm.
Literature.
NT$40.00

I. Title.
PL2844.I2N5 79-838213

Chiao, Chü-yin.
(Yeh k'u)
夜哭：[散文詩集] / 焦菊隱著. — 增訂4版. — [香港? : s. n., 1978?]
20, 61 p. ; 21 cm.
Reprint of the 1933 ed. published by Pei hsin shu chü.

I. Title.
PL2845.C49Y4 1978 78-843604

Ch'iao, Hung, 1951-
(P'iao hsiang hai mei)
飄向海湄 / 喬洪著. — 臺北：彩虹出版社, 民國67 [1978]
172 p. ; 20 cm. — (青年最佳散文) (彩虹叢書 ; 129)
Essays.
Series 1 romanized: Ch'ing nien tsui chia san wen. Series 2 romanized: Ts'ai hung ts'ung shu.
NT$30.00
I. Title.
PL2845.H8P5 79-840363

Ch'ien, Chung-shu.
(Ch'ien Chung-shu hsüan chi)
錢鍾書選集. — 香港：香港文學研究社, [1978?]
5, 1, 175 p. ; 21 cm. — (現代文學叢書)
Prose.
Series romanized: Hsien tai wen hsüeh ts'ung shu.
Includes bibliographical references.
$5.50
I. Title.
PL2749.C8A15 1978 78-843266

Ch'ien, Ko-ch'uan.
(Li hsia pi t'an)
籬下筆談 / 錢歌川著. — 臺北：臺灣開明書店, 民國67 [1978]
4, 323 p. ; 19 cm. — (開明文學叢刊)
Essays.
Series romanized: K'ai ming wen hsüeh ts'ung k'an.

I. Title.
PL2846.N25L5 78-843203

Ch'in, Mu.
(I hai shih pei)
艺海拾贝 / 秦牧. — 第2版. — 上海：上海文艺出版社, 1978.
3, 240 p. ; 21 cm.
$0.62

I. Title.
PL2848.M8I2 1978 79-837420

Chin, P'ei-fen.
(Lo Lan hsiao yü)
羅蘭小語 / [靳佩芬] 著. — 香港：崑崙出版公司, [19
v. ; 19 cm.
CONTENTS:
第3輯 成功的兩翼
$5.00 (v. 3)

I. Title.
PL2848.P4L59 1978 78-842034

Chin, P'ei-fen.
(Lo Lan san wen)
羅蘭散文 / 羅蘭 [靳佩芬] 著. — 香港：宇宙出版社, [19
v. ; 19 cm.
CONTENTS: —第6輯
淡煙疏雨.
$6.50 (v. 6)

I. Title.
PL2848.P4A16 1978 79-837527

Chin, P'ei-fen.
(Ch'eng kung ti liang i)
成功的兩翼 / 羅蘭 [靳佩芬] 著. — 香港：崑崙出版公司, [1978?]
5, 221 p. ; 19 cm. — (羅蘭小語 ; 第3輯)
Series romanized: Lo Lan hsiao yü.
$5.00

I. Title.
PL2848.P4L59 1978, vol. 3 78-842035

Chin, P'ei-fen.
(Tan yen shu yü)
淡煙疏雨 / 羅蘭 [靳佩芬] 著. — 香港：宇宙出版社, [1978?]
4, 205 p. ; 19 cm. — (羅蘭散文 ; 第6輯)
Series romanized: Lo Lan san wen.
$6.50

I. Title.
PL2848.P4A16 1978, vol. 6 79-837526

(Ch'ing ching fan)
青精飯. — 香港：海鷗出版公司, [1975?]
45 p. : ill. ; 15 x 17 cm. — (民間故事連環圖)
$1.00

I. Series: Min chien ku shih lien huan t'u.
PL2832.5.C478 78-841874

Chou, Ch'i-tzu, 1912-
(Wei mai an tuan shu)
未埋庵短書 / [著作者周棄子]. — 改板. — 台北：領導出版社, 民國67 [1978]
4, 6, 4, 272 p. : ill. ; 20 cm.
Essays.
NT$50.00 ($1.40 U.S.)

I. Title.
PL2851.C47W4 1978 79-835521

Chou, Ching-wen, 1909-
(Chi chien chi)
擊劍集 周鯨文著 香港 時代批評社 1964.
263 p. 21 cm.
Essays.

I. Title.
PL2851.C58C5 C 65-978
rev

Chou, Ch'un-t'i.
(Chiang-nan i)
江南憶 / 周春堤著. — 臺北：天華出版事業股份有限公司, 民國67 [1978]
2, 5, 223 p. ; 19 cm. — (天華文學叢刊 ; 7)
Chiefly essays.
Series romanized: T'ien hua wen hsüeh ts'ung k'an.
NT$45.00

I. Title.
PL2851.C85C47 78-843204

Chou, Eric, 1915-
(Shih wei kuan tsa chi)
侍衛官雜記 宋喬著 香港 學文書店 1957.
2 v. (761 p.) in 1. 19 cm.

I. Title.
PL2851.E73S5 C 60-5045
rev

Chou, Mi, 1232-1308.
(Ts'ao ch'uang yün yü)
草窗韻語 / [周密撰]. — 臺北：臺灣學生書局, 民國62 [1973]
10, 242, 180 p. ; 22 cm. — (歷代畫家詩文集 ; 24 : 第三輯)
"私立東海大學藏本；民國初年蔣氏密韻樓景宋刊本；兩種合刊景印本"
Appendices (p. 1-180 (3d group)): 周密 雲煙過眼錄.—湯允謨 雲煙過眼錄續集
I. Chou, Mi, 1232-1308. Yün yen kuo yen lu. 1973. II. T'ang, Yün-mo. Yün yen kuo yen lu hsü chi. 1973. III. Title. IV. Series: Li tai hua chia shih wen chi ; 24.
PL2687.C635T7 1973 78-841609

Library of Congress 78 C

Chou, Shu-jen, 1881-1936.
(K'ung I-chi)
孔乙己 / 節錄魯迅 [周樹人] 原著 ; 程十髮繪畫. — 九龍：香港江南書局, [1978?]
24 leaves : col. ill. ; 22 cm.
Originally published in 1958 by Jen min mei shu ch'u pan she, Peking.
$6.50

I. Ch'eng, Shih-fa, illus. II. Title.
PL2754.S5K86 1978 79-836144

Library of Congress C

Chu, Ch'en-tung, 1933-
(Ch'uang ti mei hsüeh)
窗的美學 朱沉冬著 高雄 光明出版社 民國61 [1972]
153 p. 19 cm. (光明文庫 46) NT$15.00
Essays.

I. Title.
PL2852.C39C5 73-897271
rev

CHINESE LITERATURE (Continued)

Chu, Hsi-lin.
(Chu Hsi-lin wen chi)
朱錫林文集 / 著作朱錫林. — 臺北：臺灣新興出版社, 民國 ₍19
v. ; 19 cm.
Colophon title.
CONTENTS:
—3. 心靈的呼聲.
NT$25.00 (v. 3)
I. Title.
PL2852.H67 1978 78–842343

Chu, Hsi-lin.
(Hsin ling ti hu sheng)
心靈的呼聲 / 朱錫林著. — 臺北：台灣新興出版社, 民國67 ₍1978₎
5, 3, 121 p. ; 19 cm. — (朱錫林文集 ; 3)
Series romanized: Chu Hsi-lin wen chi.
NT$25.00
I. Title.
PL2852.H67 1978, vol. 3 78–842342

Chu, Hui-chieh, 1916–
(T'i lan hua chiu)
提籃話舊 / 朱慧潔著. — 台南：長河出版社：總經銷建立書局, 民國67 ₍1978₎
4, 200 p. ; 20 cm. — (長河叢書 ; 17)
Series romanized: Ch'ang ho ts'ung shu.
NT$45.00
I. Title.
PL2755.H85T5 79–836950

Chu, Kuang-ch'ien, 1898–
(Chu Kuang-ch'ien hsüan chi)
朱光潛選集. — 香港：香港文學研究社, ₍1978?₎
5, 1, 182 p. ; 21 cm. — (現代文學叢書)
Series romanized: Hsien tai wen hsüeh ts'ung shu.
$5.50
I. Title.
PL2755.K8A6 1978 78–842876

Ch'ü, Ta-chün, 1630–1696.
(Sao hsieh)
騷屑 / 屈大均撰. — ₍香港?₎：何氏至樂樓, ₍1978?₎
₍146₎ p. (on double leaves) : ill. ; 27 cm. — (何氏至樂樓叢書 ; 第17)
Cover title on label: Sao hsieh tz'u.
In case.
Series romanized: Ho shih chih le lou ts'ung shu.
"景印清末國學扶輪會本"
I. Title. II. Title: Sao hsieh tz'u.
PL2698.C793S36 1978 78–842388

Chu, Te-jun, 1294–1365.
(Ts'un fu chai wen chi)
存復齋文集 / ₍朱德潤撰₎. — 臺北：臺灣學生書局, 民國62 ₍1973₎
10, 336, 4 p. ; 22 cm. — (歷代畫家詩文集 ; 28)
"國立中央圖書館藏本"
Reprint of the ed. published in 10 chüan and fu lu 1 chüan.
I. Title. II. Series: Li tai hua chia shih wen chi ; 28.
PL2694.C78T75 1973 78–840814

Chu, Tzu-ch'ing, 1898–1948.
(Chu Tzu-ch'ing san wen, shih hsüan chi)
朱自清散文·詩選集. — 台北：華聯出版社：總經銷啓源書局, 民國67 ₍1978₎
4, 159 p. ; 19 cm.
NT$30.00
I. Title.
PL2755.T9A6 1978 79–837690

Chu, Tzu-ch'ing, 1898–1945.
(Twit mosŭp)
뒷모습 / 朱自清外 ; ₍許世旭譯₎. — 서울：汎友社, 1976.
116 p. ; 19 cm. — (汎友에세이選 ; 12)
Series romanized: Pŏmu esei sŏn.
PARTIAL CONTENTS: 朱自清 뒷모습.—荷塘月色.—月朦朧鳥朦朧簾捲海棠紅.—春春.—零落.—沈默.—亡妻에게.—郁達夫 立秋날밤.—불나방葬送記.—어느文學青年에게보내는公開狀.—故事.—나그네혼자서.—日記零拾.
W280
I. Hŏ, Se-uk. II. Yü, Ta-fu, 1896–1945. III. Title.
PL2755.T9T86 1976 79–826353

Chu-ch'iao, 1948–
(Lü ti hsin yüan)
綠的心園 / 竹樵著. — 臺中：新企業出版社, 民國66 ₍1977₎
190 p. ; 19 cm.
Essays.
NT$40.00
I. Title.
PL2852.Z3L8 78–843011

Chuang, K'o-ch'ang.
(Pi keng yü t'an)
筆耕餘談 / 莊克昌著. — 臺北：維新書局, 民國60 ₍1971₎
2, 4, 154 p. ; 20 cm. — (His 散文集 ; 7)
Series romanized: San wen chi.
NT$20.00
I. Title.
PL2852.5.A56P5 78–842100

Chuang, Yin, 1933–
(Hua pen hsieh tzu hui shuo)
話本楔子彙說 / ₍著作者莊因₎. — ₍臺北₎：國立臺灣大學文學院, 1965.
1, 180 p. ; 21 cm. — (國立臺灣大學文史叢刊 ; 16)
Summary in English, with added title: The evolution of the prologue in the hua-pen of the Sung, Yuan, and Ming dynasties.
Originally presented as the author's thesis (M.A.), Kuo li T'ai-wan ta hsüeh, 1964.
Bibliography: p. 173–177.
I. Title. II. Series: T'ai-wan ta hsüeh, T'ai-pei. Kuo li T'ai-wan ta hsüeh wen shih ts'ung k'an ; 16.
PL2445.C58 1965 79–839554

(Ch'un hua ch'iu yüeh chi)
春花秋月集 / 韓立人編著. — 台中：立誼出版社, 民國67 ₍1978₎
6, 176 p. ; 20 cm.
NT$35.00
1. Chinese essays—20th century. I. Han, Li-jen.
PL2623.C5195 79–838191

(Chung-kuo san wen chan)
中國散文展：六十六年期 / 張力, 單德興, 周素鳳編著. — 臺南：長河出版社, 民國67 ₍1978₎
2 v. (10, 626 p.) ; 20 cm. — (長河叢書 ; 15)
Series romanized: Ch'ang ho ts'ung shu.
NT$110.00
1. Chinese essays—20th century. I. Chang, Li. II. Chou, Su-feng. III. Shan, Te-hsing.
PL2623.C537 79–836026

Chung yung.
中庸章句大全 / ₍胡廣等撰₎. 中庸諸註糾正 / ₍胡渭撰₎. — ₍臺北₎：中國子學名著集成編印基金會, ₍1977?₎
2, 2, 657 p. ; 22 cm. — (中國子學名著集成 ; 017 : 珍本初編, 儒家子部)
"中庸章句大全一卷…為明内府朱絲闌寫本…中庸諸注糾正三卷…清人鈔本."
I. Hu, Kuang, 1370–1418. II. Hu, Wei, 1633–1714. Chung yung chu chu chiu cheng. 1977. III. Title. IV. Series: Chung-kuo tzu hsüeh ming chu chi ch'eng ; 017.
Title romanized: Chung yung chang chü ta ch'üan.
PL2473.D5 78–841998

Confucius.
₍Lun yü₎
論語註疏解經 / 何晏集解 ; 邢昺疏. — ₍台北?₎：中國子學名著集成編印基金會, ₍1977?₎
12, 3, 2, 604 p. ; 22 cm. — (中國子學名著集成 ; 002 : 珍本初編, 儒家子部)
"二十卷"
"明熊九岳等校刊本 墨校近人阮閣手書題記"
I. Ho, Yen, d. 249. II. Hsing, Ping, 932–1010. III. Title. IV. Series: Chung-kuo tzu hsüeh ming chu chi ch'eng ; 002.
Title romanized: Lun yü chu shu chieh ching.
PL2471.D4 1977 78–841786

Fei, Sheng-fu.
(Feng pao)
风暴 / 费声福, 吴兆修改编 : 费声福绘. — 第2版. — 北京：人民美術出版社, 1977.
168 p. : ill. ; 11 x 13 cm.
"連環圖畫"
"根据金山原著同名电影剧本"
$0.26
I. Wu, chao-hsiu, joint author. II. Chin, Shan. Feng pao. III. Title.
PL2857.E395F4 79–836550

Feng, Meng-lung, 1574?–1645?
(Ts'ai tzu yü chia jen)
才子與佳人 / 馮夢龍原著. — 台南：大夏出版社, 民國67 ₍1978₎
231 p. ; 20 cm. — (明代話本小說精選集)
NT$35.00
I. Title. II. Series: Ming tai hua pen hsiao shuo ching hsüan chi.
PL2698.F4T75 1978 79–836796

Feng, Sha-yen.
(Mao tun chi)
矛盾集 / 風沙雁著. — Singapore：洪炉文化企业公司, 1976.
74 p. ; 19 cm.
Essays.
$1.00
I. Title.
PL2857.E578M3 78–940009

Feng, Sha-yen.
(Sha li chi)
砂砾集 / 风沙雁. — Singapore：万里文化企业公司, 1974.
83 p. ; 19 cm.
"詩歌·散文"
$1.20
I. Title.
PL2857.E578S5 75–940168

Feng, Tzu-k'ai, 1898–1975.
(Feng Tzu-k'ai hsüan chi)
豐子愷選集. — 香港：香港文學研究社, ₍1978?₎
6, 3, 179 p. ; 21 cm. — (現代文學叢書)
Essays.
Series romanized: Hsien tai wen hsüeh ts'ung shu.
I. Title.
PL2760.E55A16 1978 78–842868

CHINESE LITERATURE (Continued)

Feng-ying, 1958–
(Shen ch'iu feng yu hung)
深秋楓又紅 / 作者楓影. — [台南?]：莊家出版社；台南：總經銷世一書局, 民國66 [1977], 民國67 [1978] printing.
3, 200 p. ; 19 cm.
Colophon title.
Literature.
NT$35.00
I. Title.
PL2857.E77S53 79–835429

Fu, Ch'ih.
(P'ing ts'ung)
萍蕟 / 傅遲著. — 臺北：智燕出版社, 民國67 [1978]
2, 2, 248 p. ; 19 cm.
NT$50.00
I. Title.
PL2858.C57P5 79–841056

Fu, Hsiao-hsien.
(Wu hua ti yüan ti)
無花的園地 / 傅孝先著. — 臺北：九歌出版社：總經銷爾雅出版社, 民國67 [1978]
201 p. ; 19 cm. — (九歌文庫 ; 3)
Essays.
Series romanized: Chiu ko wen k'u.
NT$45.00
I. Title.
PL2858.H74W8 79–835356

Hai, Mo.
(Mei mei yü wo hsü chi)
妹妹與我續集 / 海默著. — 臺北：星光出版社, 民國65 [1976]
218 p. ; 20 cm.
Essays.
NT$30.00
I. Title.
PL2861.A5M4 78–842796

Han, Chiang-hung.
(Jen cha chih ko).
人渣之歌 / 韓江鴻著. — 九龍：創作出版社, 1978.
6, 106 p. ; 19 cm.
Essays.
$4.00
I. Title.
PL2861.A582J4 78–841322

Han, Yü, 768–824.
(Han Ch'ang-li chüan chi)
韓昌黎全集 / 韓愈著. — 台北：新文豐出版公司, 民國66 [1977]
126, 166, 174, 174, [34] p. ; 22 cm.
Reprint ed.
CONTENTS: 詩文集 40卷.—外集 10卷.—遺文.—附朱子校昌黎先生集傳目錄.—附韓集校勘4卷.
NT$200.00
I. Title.
PL2670.A1 1977 78–843987

Ho, Ch'i-fang, 1911–1977.
(Ho Ch'i-fang hsüan chi)
何其芳選集. — 香港：香港文學研究社, [1978?]
5, 2, 147 p. ; 21 cm. — (現代文學叢書)
Chiefly essays.
Series romanized: Hsien tai wen hsüeh ts'ung shu.
I. Title.
PL2861.O16A6 1978 78–841297

Ho, Ch'i-fang, 1911–1977.
[San wen hsüan chi]
何其芳散文選集. — 香港：文教出版社, 1978.
2, 3, 142 p. ; 19 cm. — (中國現代文學叢書)
Series romanized: Chung-kuo hsien tai wen hsüeh ts'ung shu.
Reprint of the 1957 ed. published by Jen min wen hsüeh ch'u pan she, Peking, under title: San wen hsüan chi.
$3.80
I. Title.
Title romanized: Ho Ch'i-fang san wen hsüan chi.
PL2861.O16S2 1978 79–835768

Hsia, Ch'eng-ying, 1910–
(Pu an p'ai li ch'u p'ai)
不按牌理出牌 / 何凡[夏承楹]著. — 臺北：純文學出版社, 民國66 [1977]
3, 4, 194 p. ; 19 cm. — (純文學叢書 ; 78)
Essays.
Series romanized: Ch'un wen hsüeh ts'ung shu.
NT$40.00
I. Title.
PL2862.I178P8 1977 79–840044

Hsia, Yüan-yü.
(Liu hsing yü)
流星雨 / 夏元瑜著. — 臺北：九歌出版社, 民國67 [1978]
275 p. : ill. ; 20 cm. — (九歌文庫 ; 11)
Series romanized: Chiu ko wen k'u.
NT$55.00
I. Title.
PL2862.I228L5 79–836543

Hsiao, Pai, 1925–
(Hsiao Pai san wen ching hsüan chi)
蕭白散文精選集 / 蕭白著. — 第1版. — 臺北：源成文化圖書供應社, 民國67 [1978]
5, 244 p. ; 20 cm.
NT$45.00
I. Title.
PL2862.I319H69 78–842634

Hsiao, Pai, 1925–
(I chiang teng ying)
一槳燈影 / 蕭白著. — 臺北：黎明文化事業公司, 民國67 [1978]
2, 4, 232 p. ; 20 cm.
Essays.
NT$50.00
I. Title.
PL2862.I319I12 79–836420

Hsiao ching.
郭煌本孝經類纂 / [編纂者陳鐵凡]. — 台北：燕京文化事業股份有限公司, 民國66 [1977]
173 p. : facsims. ; 27 cm.
I. Ch'en, T'ieh-fan. II. Tun-huang manuscripts. III. Title.
Title romanized: Tun-huang pen Hsiao ching lei tsuan.
PL2476.A47 78–842399

Hsiao-hsiao-sheng.
(Chin P'ing Mei tz'u hua)
金瓶梅詞話：明萬曆丁巳刻本. — 臺北：聯經出版事業公司, 民國67 [1978]
20 v. (on double leaves) in 2 cases : ill. ; 29 cm.
"笑笑生作"
"景印限定版三百部"
NT$3800.00
I. Title.
PL2698.H73C5 1978b 79–838624

Hsiao-min.
(Hui-i ch'ü)
回憶曲 / 小民著. — 臺北：林白出版社：門市部中國書城, 民國67 [1978]
194 p. ; 20 cm. — (林白叢書 ; 95)
Literature.
Series romanized: Lin pai ts'ung shu.
NT$40.00
I. Title.
PL2862.I3178H8 78–842447

Hsiao-ming.
(Su-su! ni ti jih tzu ni?)
素素! 你的日子呢? / 文筱明；圖小虎. — 台北：代言出版社, 民國66 [1977]
125 p. (p. 124–125 advertisements) : ill. ; 18 cm. — (代言叢書 ; 3)
Series romanized: Tai yen ts'ung shu.
I. Hsiao-hu. II. Title.
PL2862.I3179S9 78–841606

Hsieh, Hsiu-tsung.
(Feng wu hsiao yü)
風屋小語 / 謝秀宗著. — 台中：新企業世界出版社, 民國66 [1977]
7, 170 p. ; 20 cm.
Essays.
NT$35.00
I. Title.
PL2862.I359F3 1977 78–843596

Hsieh, Kuei-hsiang, 1954–
(P'iao tsai feng chung ti ko)
飄在風中的歌 / 謝桂香著. — 台中：新企業世界出版社, 民國66 [1977]
201 p. ; 20 cm.
Essays.
NT$40.00
I. Title.
PL2862.I365P5 78–841583

Hsin, Sheng.
(Ken tsung chui chi)
跟踪追击 / 原著安忠民, 朱向群；改编辛生；绘画高适. — 第2版. — 上海：上海人民美術出版社, 1978.
133 p. : ill. ; 10 x 13 cm.
"連環圖畫."
$0.15
I. Kao, Shih. II. An, Chung-min. Ken tsung chui chi. III. Title.
PL2862.I438K4 78–841584

Hsin-hsin.
(Hua ching)
花景 / 芯心著. — 臺北：源成文化圖書供應社, 民國65 [1976]
6, 216 p. ; 20 cm.
Essays.
NT$40.00
I. Title.
PL2862.I4525H9 78–841145

Hsü, Chih-mo, 1896–1931.
[Selected works]
徐志摩詩輯, 附小說集 / 徐志摩編著. — 高雄：莊家出版社, 民國64 [1975]
106, 1, 92 p. ; 20 cm.
Cover title.
NT$25.00
I. Title.
Title romanized: Hsü Chih-mo shih chi, fu hsiao shuo chi.
PL2765.U2A6 1974b 75–837076

CHINESE LITERATURE (Continued)

Hsü, Chih-mo, 1896-1931.
[Selected works]
徐志摩散文，日記選集：附新詩代表作．— 台中：立誼出版社，民國66 [1977]
2, 2, 207 p. ; 19 cm. — (讀者叢書)
Cover title.
Series romanized: Tu che ts'ung shu.
NT$26.00
I. Title.
Title romanized: Hsü Chih-mo san wen, jih chi hsüan chi.
PL2765.U2A6 1977c 79-835554

Hsü, Chih-mo, 1896-1931.
[Selected works]
徐志摩選集．— 香港：香港文學研究社，[1978?]
6, 5, 224 p. ; 21 cm. — (現代文學叢書)
Series romanized: Hsien tai wen hsüeh ts'ung shu.

I. Title.
Title romanized: Hsü Chih-mo hsüan chi.
PL2765.U2A6 1978 78-841571

Hsü, Chih-mo, 1896-1931.
[Works]
徐志摩全集；朱自清全集．— [台南?]：第一書店，[1977]
528, 380 p. ; 20 cm.
Cover title.
NT$85.00

1. Chu, Tzu-ch'ing, 1898-1948. Works. 1977. II. Title. III. Title: Chu Tzu-ch'ing ch'üan chi.
Title romanized: Hsü Chih-mo ch'üan chi.
PL2765.U2 1977 78-841453

Hsü, Chih-mo, 1896-1931.
[Works]
徐志摩全集：詩輯，日記，散文，小說．— 再版．— 台南：利大出版社：總經銷第一書店，民國66 [1977]，cover 民國67 [1978]
12, 380 p. : 20 cm.
Cover title.
NT$35.00
I. Title.
Title romanized: Hsü Chih-mo ch'üan chi.
PL2765.U2 1978 78-843761

Hsü, Ch'iu, 1636-1708.
(Nan chou ts'ao t'ang chi)
南州草堂集 / [徐釚撰]．— 景印初版．— 臺北：臺灣學生書局，民國62 [1973]
2 v. (10, 684, 4 p.) ; 22 cm. — (歷代畫家詩文集；33：第三輯)
"南州草堂續集": v. 2, p. 555-684.

I. Title. II. Series: Li tai hua chia shih wen chi ; 33.
PL2710.S824N3 1973 78-841610

Hsü, Fang, 1622-1694.
(Chü i t'ang chi)
居易堂集 / 徐枋撰．— 臺北：臺灣學生書局，民國62 [1973]
10, 555, 4 p., [1] leaf of plates : port. ; 22 cm. — (歷代畫家詩文集；31)
"民國八年羅振玉排印明季三孝廉集本景印本"

I. Title. II. Series: Li tai hua chia shih wen chi ; 31.
PL2698.H7716C48 1973 78-840830

Hsü, Hsi-che.
(Hsü Hsi-che tzu hsüan chi)
許希哲自選集．— 臺北：黎明文化事業公司，民國67 [1978]
347 p., [2] leaves of plates : ill. ; 20 cm. — (中國新文學叢刊；60 新中國文學叢刊；60)
Prose.
Series 1 romanized: Chung-kuo hsin wen hsüeh ts'ung k'an. Series 2 romanized: Hsin Chung-kuo wen hsüeh ts'ung k'an.
NT$80.00
I. Title.
PL2862.U156A16 1978 79-837691

Hsü, Ju-lin.
(Ku ying hsing)
孤鷹行：徐如林的散文集．— [台]北：伙伴出版公司：總經銷長橋出版社，民國67 [1978]
212 p. : ill. ; 20 cm.
NT$40.00

I. Title.
PL2862.U216K8 79-837027

Hsü, Po-hua.
(Chang fu yung yüan shih nan jen)
丈夫永遠是男人 / [許伯華著]．— 台北：許伯華；新店[台北縣]：總經銷巨鷹文化企業有限公司，民國66 [1977]
215, [12] p., [2] leaves of plates : col. ill. ; 20 cm.
Essays.
"合歡山滑雪十日記 龔剴": p. 199-212.
NT$45.00
I. Kung, K'ai. Ho-huan shan hua hsüeh shih jih chi. 1977. II. Title.
PL2862.U235C5 78-841113

Hsü, Ta-jan,
(Yüan fang)
遠方 / 許達然．— 臺北：遠行出版社，民國67 [1978]
111 p. ; 19 cm. — (小草叢刊；32)
Series romanized: Hsiao ts'ao ts'ung k'an.
NT$10.00

I. Title.
PL2862.U254Y8 1978 79-837578

Hsü, Ti-shan, 1893-1941.
(Ch'un ti lin yeh)
春底林野 / 許地山．— 台北：偉文圖書出版社，民國67 [1978]
121 p. ; 20 cm. — (民初散文作家選集)
NT$25.00

I. Title. II. Series: Min ch'u san wen tso chia hsüan chi.
PL2765.U28C513 1978 79-838612

Hsü, Ti-shan, 1893-1941.
(Hsü Ti-shan hsüan chi)
許地山選集．— 香港：香港文學研究社，[1978?]
5, 157 p. ; 21 cm. — (現代文學叢書)
Essays.
Series romanized: Hsien tai wen hsüeh ts'ung shu.
$5.00

I. Title.
PL2765.U28A16 1978 78-842878

Hsü, Yu.
(Fang k'uai san pai p'ien)
方塊三百篇 / 許由著．— 高雄：台灣時報社；台北：總經銷名人出版社，民國66 [1977]
4, 8, 302 p. ; 19 cm. — (台灣時報叢書)
Essays.
Series romanized: T'ai-wan shih pao ts'ung shu.
Distributor's name stamped in colophon.
N$100.00
I. Title.
PL2862.U323F3 78-841532

Hsüan, Chien-jen.
(Sheng ming ti hua to)
生命的花朵：名作家散文 / 宣建人著．— 臺北：彩虹出版社，民國67 [1978]
176 p. ; 20 cm. — (彩虹叢書；125)
Series romanized: Ts'ai hung ts'ung shu.
NT$30.00

I. Title.
PL2862.U328S45 78-843071

Hsün, Hsü, d. 289.
(Hsün Kung-tseng chi)
荀公曾集 / [荀勗撰]．梁貞白先生陶隱居集 / [陶弘景撰]．宣和御製宮詞 / [宋徽宗撰]．— 臺北：臺灣學生書局，民國62 [1973]
10, 68, 140, 63 p. ; 22 cm. — (歷代畫家詩文集；22：第三輯)
"三種合刊景印本，國立中央圖書館藏本"
NT$250.00

1. Chinese literature. I. T'ao, Hung-ching, 452-536. Liang Chen pai hsien sheng T'ao Yin-chü chi. 1973. II. Sung Hui-tsung, Emperor of China, 1082-1135. Hsuan-ho yü chih kung tz'u. 1973. III. Title. IV. Series: Li tai hua chia shih wen chi ; 22.
PL2455.H83 1973 79-835690

Hu, P'in-ch'ing, 1920-
(Shui ching ch'iu)
水晶球 / 胡品清著．— 台北：水芙蓉出版社，民國66 [1977]
240, [8] p. ; 20 cm. — (水芙蓉書庫；111)
Essays.
Series romanized: Shui fu jung shu k'u.
NT$42.00

I. Title.
PL2863.P5S47 78-841963

(Huai nien Kuo Mo-jo)
怀念郭沫若：诗文集．— 香港：生活·讀書·新知三联书店，1978.
4, 285 p., [3] leaves of plates : ill. ; 21 cm.

1. Kuo, Mo-jo, 1892-1978—Criticism and interpretation. 2. Kuo, Mo-jo, 1892-1978, in fiction, drama, poetry, etc.
PL2778.O2Z68 79-836972

Huang, Che-yung, 1953-
(Ssu hsü to to)
思絮朵朵 / 黃哲永著．— 再版．— 臺中：新企業出版社，民國66 [1977]
6, 198 p. ; 19 cm.
NT$40.00

I. Title.
PL2865.C35S7 78-842513

Huang, Ch'ing-hui, 1951-
(Han hsing hsiao yü)
寒星小語 / 黃晴輝編著．— 台南：大府城出版有限公司：總經銷中央書局，民國67 [1978]
254 p. ; 20 cm. — (大府城書系)
Literature.
Series romanized: Ta fu ch'eng shu hsi.
NT$50.00

I. Title.
PL2865.C463H26 79-838021

Huang, Ch'ing-hui, 1951-
(Han hsing ti t'iao yüeh)
寒星的跳躍 / 黃晴輝編著．— 台南：大府城出版有限公司：總經銷中央書局，民國67 [1978]
254 p. ; 20 cm. — (大府城書系)
Literature.
Series romanized: Ta fu ch'eng shu hsi.
NT$50.00

I. Title.
PL2865.C463H3 79-838020

(Huang pa wen chi)
皇霸文紀 / [梅鼎祚編]．— [臺北：臺灣]商務[印書館，1976]
2 v. ; 20 cm. — (四庫全書珍本六集；343-344)
Cover title.
"十三卷."
Photoreprint ed. of the Wen Yüan Ko copy.

1. Chinese literature. I. Mei, Ting-tso, 1553-1619. II. Series: Ssu k'u ch'üan shu. Ssu k'u ch'üan shu chen pen liu chi ; 343-344.
AC149.S7 1976, vol. 343-344 79-835380
[PL2450]

CHINESE LITERATURE (Continued)

Hung, I-yung, 1951–
(Chang chien hsing)
仗劍行 / ［著者洪宜勇］. — 高雄：德馨室出版社，民國67 ［1978］
154 p. : ill. ; 20 cm. — (德馨論著 ; 1)
Series romanized: Te hsin lun chu.
NT$40.00

I. Title.
PL2866.N536C5 79–835693

I-yeh, fl. 1969–
(Hua yeh hsü yü)
花葉絮語 / 一葉著. — 九龍：海洋文藝社，1978.
3, 240 p. ; 19 cm. — (海洋文藝叢書)
Essays.
Series romanized: Hai yang wen i ts'ung shu.
$6.20

I. Title.
PL2867.Y4H8 78–841065

Jen, Chen.
(Hsiang ch'ing)
鄉情 / 任眞著. — 台南：鳳凰城圖書公司，民國67 ［1978］
247 p. ; 20 cm. — (鳳凰新書 ; 203)
Literature.
Series romanized: Feng huang hsin shu.
NT$50.00

I. Title.
PL2868.E53H7 79–837072

Jen, Mei, fl. 1975–
(I t'iao mao t'an)
一条毛毯：根据林志常文 / 任梅改编；沈尧伊绘. — 第1版. — 北京：人民美術出版社：新华书店北京发行所發行，1978.
19 p. : col. ill. ; 14 x 16 cm.
"連環圖畫"
Title in Pinyin on cover: Yitiao maotan.
$0.11
I. Lin, Chih-ch'ang. I t'iao mao t'an. II. Shen, Yao-i. III. Title. IV. Title: Yitiao maotan.
PL2868.E64I2 79–838107

K'ang, Yüan.
(Wu ku san chi)
霧谷散記 / 康原著. — 台中：新企業世界出版社，民國66 ［1977］
4, 5, 175 p. : 20 cm.
Literature.
NT$35.00

I. Title.
PL2869.N557W8 1977 78–843382

Kao, Chin-ch'iang.
(Na ko jen)
那個人 / 高金鏘著. — 台南：綜合出版社，民國67 ［1978］
4, 250 p. ; 20 cm. — (綜合文叢 ; 12)
Literature.
Series romanized: Tsung ho wen ts'ung.
NT$45.00

I. Title.
PL2869.O1125N3 79–838023

Kim, Si-sŭp, 1435–1493.
［Works］
국역매월당집. — 서울：세종대왕기념사업회，1977–
v. : ill. ; 22 cm.
Includes original text.
Includes bibliographical references.
₩8000 per vol.

I. Sejong Taewang Kinyŏm Saŏphoe. II. Title.
Title romanized: Kugyŏk Maewŏltang chip.
PL2698.K5 1977 79–825162

Ku, Neng-hao.
(Yeh lü mo shang ti ch'un)
葉綠脈上的純 / 古能豪著. — 高雄：心影出版社，民國67 ［1978］
297 p. ; 20 cm. — (心影文庫 ; 18)
Series romanized: Hsin ying wen k'u.
NT$50.00

I. Title.
PL2874.N4Y4 79–838026

Ku-yin.
(Lien hung erh ch'ih)
臉紅耳赤 / 姑隱著. — 再版. — 台南：青溪出版社：總經銷東風圖書供應社，民國67 ［1978］
7, 177 p. ; 20 cm. — (東風文叢 ; 3)
Series romanized: Tung feng wen ts'ung.
NT$40.00

I. Title.
PL2874.Z8L5 79–837028

Kuan, Kuan, 1929–
(Ch'ing tso yüeh liang ch'ing tso)
請坐月亮請坐 管管著 高雄 大業書店 民國58 ［1969］
315 p. 18 cm. (大業現代文學叢書4) 20.00
Colophon title.
Essays.

I. Title.
PL2875.A52C5 70–837793

rev

Kuan, Kuan, 1929–
(Kuan Kuan san wen chi)
管管散文集. — 台北：中華文藝月刊社，民國65 1976］
256 p. : 19 cm. — (中華文藝叢書 ; 7)
Essays.
Series romanized: Chung-hua wen i ts'ung shu.
"散文詩": p. 196–254.
NT$50.00 ($2.00 U.S.)

I. Title
PL2875.A52A6 1976 76–840295

rev

Kuan, Shih-ming, 1738–1798.
(Yün-shan t'ang wen chi)
韞山堂文集 / 管世銘撰. — 臺北：石門圖書公司，民國65 ［1976］
ca. 130 leaves in various foliations ; 22 cm. — (清人文集 ; 3)
Reprint ed.

I. Title. II. Series: Ch'ing jen wen chi ; 3.
PL2717.K33Y8 1976 78–842363

Kuan, Tzu.
(Hsin k'ou tz'u huang)
信口雌黃 / 關子著. — ［台南］：莊家出版社：世一書局總經銷，民國65 ［1976］，民國66 ［1977］ printing.
5, 148 p. : 20 cm.
Essays.
NT$30.00

I. Title.
PL2875.A56H7 78–842797

Kuei, Jen, 1927–
(Kuei Jen tzu hsüan chi)
歸人自選集. — 臺北：黎明文化事業公司，民國67 ［1978］
6, 326 p., ［2］ leaves of plates : ill. ; 20 cm. — (中國新文學叢刊 ; 47)
Series romanized: Chung-kuo hsin wen hsüeh ts'ung k'an.
NT$80.00

I. Title.
PL2875.E414A16 1978 79–841880

Kuo, Mo-jo, 1892–1978.
［Selected works］
郭沫若文集. — ［香港? : s. n., 1978?］
443 p. ; 21 cm.
Reprint of the 1937 ed. published by Hu lung shu tien, Shanghai.

I. Title.
Title romanized: Kuo Mo-jo wen chi.
PL2778.O2A6 1978 78–842032

Lai, Ho, 1894–1943.
(Lai Ho hsien sheng ch'üan chi)
賴和先生全集 / 李南衡主編；［編輯助理林邊，梁景峯］. — 台北：明潭出版社，民國68 ［1979］
12, 6, 510 p., ［4］ leaves of plates : ill. ; 22 cm. — (日據下台灣新文學 : 明集 ; 1)
Literature.
NT$250.00 ($10.00 U.S.)

I. Li, Nan-heng. II. Lin, Pien. III. Liang, Ching-feng. IV. Title. V. Series: Jih chü hsia T'ai-wan hsin wen hsüeh : Ming chi ; 1.
PL2779.A4 1979 79–840084

Lei, Chia, 1915–
(Ts'ung ping tou tao ta ch'uan)
从冰斗到大川 / 雷加. — 第1版. — 上海：上海文艺出版社：新華書店上海发行所发行，1978.
3, 196 p. : ill. ; 19 cm.
$0.45

I. Title.
PL2876.E43T78 79–840165

Li, Ao, 1935–
(Li Ao kao pieh wen t'an shih shu)
李敖告别文壇十書 / 李敖著. — 九龍：文藝書屋，19
v. ; 19 cm.
CONTENTS: —3. 媽離不了你

I. Title.
PL2877.A6L45 79–835039

Li, Chi-yeh, 1904–
(Kei shao nan shao nü)
給少男少女 李霽野著 香港 僑益書局 1958.
113 p. 18 cm.

I. Title.
PL2877.C513K4 C 59–1421 ‡

rev

Li, Chia-mien, 1901–
(I yüeh lou yin ts'ao)
倚月樓吟草 / ［著作者李加勉］. — ［臺北?］：區天佑，［1977?］
18, 227 p., ［1］ leaf of plates : ports. : 20 cm.

I. Title.
PL2780.C485I16 79–837473

Li, Chun, 1928–
(Yeh tsou Lo-t'o ling)
夜走駱駝岭 李準［著］ 北京 作家出版社 1959.
152 p. 19 cm.

I. Title.
C 61–4537

CHINESE LITERATURE (Continued)

Li, Feng, fl. 1959-
(Kei nü p'eng yu ti hsin)
給女朋友的信　李風著　九龍　文偉書店 1959.
212 p. 18 cm.

I. Title.
PL2877.F416K44 1959 C 59-3114 ‡

Li, Feng, fl. 1959-
(Kei nü p'eng yu ti hsin)
給女朋友的信　李風著　九龍　康美書店　1961
212 p. 18 cm.

I. Title.
PL2877.F416K44 1961 C 62-241 ‡

Li, Fu, fl. 1965-
(Shu ch'ing min yao)
抒情民謠 / 黎父. — 臺中：藍燈文化事業股份有限公司, 民國67 [1978]
3, 6, 155 p. : ill. ; 20 cm. — (藍燈叢書)
Title in English in colophon: Lyrics and ballads.
Series romanized: Lan teng ts'ung shu.
NT$40.00

I. Title. II. Title: Lyrics and ballads.
PL2877.F8S5 79-837673

Li, Hsiu.
(Tung jih i shih)
冬日已逝 / 李秀著. — 台南：鳳凰城圖書公司, 民國67 [1978]
257 p. ; 20 cm. — (鳳凰新書 ; 204)
Literature.
Series romanized: Feng huang hsin shu.
NT$49.00

I. Title.
PL2877.H7734T8 79-835903

Li, Hua-chang, 1955-
(T'o lo ti hui i)
陀螺的回憶 / 李化彰著. — 台南：綜合出版社, 民國68 [1979]
5, 3, 242 p. ; 19 cm. — (綜合文叢 ; 25)
"散文選集"
Series romanized: Tsung ho wen ts'ung.
NT$48.00

I. Title.
PL2877.H7772T6 79-840536

Li, Kuang-ling, 1952-
(Liang ko shih chieh)
兩個世界 / 李光鈴著. — 台南：綜合出版社, 民國67 [1978]
8, 2, 236 p. ; 20 cm. — (綜合文叢 ; 18)
Literature.
Series romanized: Tsung ho wen ts'ung.
NT$48.00

I. Title.
PL2877.K82L5 79-837825

Li, Kuang-t'ien, 1907-
(Li Kuang-t'ien hsüan chi)
李廣田選集. — 香港：香港文學研究社, [1978?]
6, 156 p. ; 21 cm. — (中國現代文學作家選集叢書)(現代文學叢書)
Literature.
Series 1 romanized: Chung-kuo hsien tai wen hsüeh tso chia hsüan chi ts'ung shu. Series 2 romanized: Hsien tai wen hsüeh ts'ung shu.
$6.00

I. Title.
PL2780.K8A6 1978 79-837604

Li, Li-kuo, 1954-
(Shou ma hsing)
瘦馬行 / 李利國著. — 高雄：德馨室, 民國67 [1978]
184 p. : port. ; 20 cm. — (德馨論著 ; 2)
Series romanized: Te hsin lun chu.
NT$40.00

I. Title.
PL2877.L525S5 79-837317

Li, Li-kuo, 1954-
(Yao yüan ti lu sheng)
遙遠的櫓聲 / 李利國著. — 高雄：德馨室, 民國67 [1978]
2, 202 p., [1] leaf of plates : port. ; 20 cm. — (德馨文選 ; 1)
Series romanized: Te hsin wen hsüan.
NT$40.00

I. Title.
PL2877.L525Y3 79-836733

Li, Meng-lung.
(Chün chu shou chi)
捃燭手記 / 李夢龍著. — 臺北：南京出版公司, 民國66 [1977]
283 p. : port. ; 19 cm.
The edition published in the same year by Ming jen ch'u pan she, T'ai-pei under title: Sheng ming ti chiao ao.
NT$50.00

I. Title.
PL2877.M4C48 1977b 78-842731

Li, Meng-lung.
(Shih nien hsin shih shih nien teng)
十年心事十年燈 / 李夢龍著 ; [編者劉蘋華]. — 臺北：水牛出版社, 民國67 [1978]
2, 274 p., [8] leaves of plates : ill. ; 19 cm. — (水牛文庫 ; 208)
Title in English in colophon: Ten years: a pilgrimage of a writer.
Series romanized: Shui niu wen k'u.
NT$50.00

I. Liu, P'in-hua, 1951- II. Title.
PL2877.M4S5 79-836569

Li, Mu-hua.
(Shu ti lien hsiang)
書的聯想 / 李牧華著. — 臺北：名人出版社, 民國67 [1978]
201 p. ; 19 cm.
Essays.
NT$39.00

I. Title.
PL2877.M8S5 79-837026

Li, P'ei-cheng, 1920-
(Hsiao ch'uan chih ko)
小船之歌 / 李佩徵著. — 臺北：世界文物出版社, 1977.
193 p., [1] leaf of plates : col. ports. ; 22 cm.
Added t. p.: Songs from a small boat, by Li Pei-cheng ; with English translation by John M. McLellan.
NT$100.00 ($4.00 U. S.)

I. Title. II. Title: Songs from a small boat.
PL2877.P38H74 78-842392

Li, Po, 705?-762.
[Works]
李太白全集 / 王琦注. — 第1版. — 北京：中華書局：新華書店北京發行所發行, 1977.
3 v. (10, 32, 1694, 22 p.) ; 21 cm.
Includes index.
$5.10

I. Wang, Ch'i, fl. 1758. II. Title.
Title romanized: Li T'ai-pai ch'üan chi.
PL2671.A1 1977 79-836834

Li, Po, 705?-762.
[Works]
李太白集註 [36卷 李白撰 王琦輯註 台北 新興書局 民國57 i. e. 1968]
592 p. 22 cm.
Original t. p. reads: 李太白文集輯註
Caption title: 李太白文集
Each page represents 2 leaves of the original.
影印臺灣大學圖書館藏清乾隆戊寅(1758)聚錦堂原刻本
I. Wang, Ch'i, fl. 1758, ed. II. Title.
Title romanized: Li T'ai-pai chi chu.
PL2671.A1 1968 72-835707

Li, Shuang-tse, 1949-1977.
(Tsai chien, shang kuo)
再見, 上國：李雙澤作品集 / 梁景峯編輯. — 臺北：總經銷長橋出版社, 1978.
4, 2, 290 p., [4] leaves of plates : col. ill. ; 21 cm.
Cover title.
Includes songs in number notation.
NT$80.00

I. Liang, Ching-feng. II. Title.
PL2877.S56T7 1978 79-839546

Li, Ya-nan, 1954-
(Li Ya-nan tzu hsüan chi)
李亞南自選集. — 高雄：德馨室出版社, 民國67 [1978]
6, 285 p. : ill. ; 20 cm.
Literature.
NT$40.00

I. Title.
PL2877.Y26A16 1978 78-843253

Li, Ya-nan, 1954-
(Ou ko, ou ko)
謳歌, 謳歌 / 李亞南著. — 臺北：水芙蓉出版社：總經銷星光書報社, 民國63 [1974]
2, 4, 212 p. ; 20 cm. — (水芙蓉書庫 ; 22)
Prose.
Series romanized: Shui fu jung shu k'u.
NT$35.00

I. Title.
PL2877.Y26O8 75-836478

Li, Ya-nan, 1954-
(Sheng ming ti ch'ü tzu)
生命的曲子 / 李亞南著. — 臺北：水芙蓉出版社：總經銷星光書報社, 民國66 [1977]
2, 4, 230 p. ; 19 cm. — (水芙蓉書庫 ; 92)
Essays.
Series romanized: Shui fu jung shu k'u.
NT$40.00

I. Title.
PL2877.Y26S5 78-842292

Li, Ya-nan, 1954-
(Ya-nan hsiao yü)
亞南小語　李亞南著　[臺北]　彩虹出版社 [民國62 i. e. 1973]
168 p. 19 cm. (彩虹叢書15) NT$20.00
Essays.
Series romanized: Ts'ai hung ts'ung shu.

I. Title.
PL2877.Y26Y3 73-898117

Li-ni.
(Chiang chih ko)
江之歌 / 麗尼著. — [香港]：創作書社, [1978]
62 p. ; 19 cm. — (天馬叢書 ; 29)
Cover title.
Series romanized: T'ien ma ts'ung shu.
$3.00

I. Title.
PL2877.Z4C5 78-843539

CHINESE LITERATURE (Continued)

Liang, Ch'i-ch'ao, 1873–1929.
(Liang Ch'i-ch'ao hsüan chi)
梁啓超選集. — 香港：香港文學研究社, [1978?]
5, 161 p. ; 21 cm. — (現代文學叢書)
Series romanized: Hsien tai wen hsüeh ts'ung shu.
$6.00

I. Title.
PL2781.A5A6 1978 79–837146

Liang, Ch'i-ch'ao, 1873–1929.
(Tsui k'u yü tsui le)
最苦與最樂 / 梁啓超. — 台北：偉文圖書出版社, 民國67 [1978]
112 p. ; 20 cm. — (民初散文作家選集)
Title in English on cover: Joys and sorrows.
"梁啓超散文選"
NT$25.00

I. Title. II. Title: Joys and sorrows. III. Series: Min ch'u san wen tso chia hsüan chi.
PL2781.A5T7 1978 79–838076

Liang, Chu-chu.
(Ch'ing ch'un shih che yang ti mei hao)
青春是這樣的美好 / 梁珠珠, 張麗娥, 徐萍生合著. — [台北?]：德昌出版社, [1977?]
241 p. ; 19 cm. — (德昌新刊 ; 37)
"大學女生日記選"
Series romanized: Te ch'ang hsin k'an.

I. Chang, Li-o. II. Hsü, P'ing-sheng. III. Title.
PL2878.A58C47 78–841246

Liang, Shih-ch'iu, 1901–
(Liang Shih-ch'iu cha chi)
梁實秋札記 / 梁實秋著. — 台北：時報文化出版事業有限公司, 民國67 [1978]
276 p. ; 20 cm. — (時報書系 ; 135)
Series romanized: Shih pao shu hsi.
NT$95.00

I. Title.
PL2781.A54A16 1978b 79–837983

Liang, Shih-ch'iu, 1901–
(Liang Shih-ch'iu hsüan chi)
梁實秋選集. — 香港：香港文學研究社, [1978?]
5, 3, 191 p. ; 21 cm. — (現代文學叢書)
Essays.
Series romanized: Hsien tai wen hsüeh ts'ung shu.
$5.50

I. Title.
PL2781.A54A16 1978 78–842522

Liang-hsüan.
(Hsi p'in ch'ih chung wei)
細品癡中味 / [亮軒著]. — 臺北：皇冠雜誌社, 民國67 [1978]
201 p. ; 20 cm. — (亮軒專欄集 ; 2)
Series romanized: Liang-hsüan chuan lan chi.
NT$40.00

I. Title.
PL2878.A763A6 1978, vol. 2 79–839186

Liang-hsüan.
(Liang-hsüan chuan lan chi)
亮軒專欄集 / [亮軒著]. — 臺北：皇冠雜誌社, 民國6 [19
v. ; 20 cm.
CONTENTS: —2. 細品癡中味.
NT$40.00 (v. 2)

I. Title.
PL2878.A763A6 1978 79–839185

Lien, Shih-sheng, 1907–1973.
(Ch'iu shui chi)
秋水集：連士升文集. — [s. l. : s. n.], 1974 (新加坡：日昇美術印務有限公司)
192 p. ; 21 cm. — (海濱寄簡 ; 第7集)

I. Title. II. Series: Lien, Shih-sheng, 1907–1973. Hai pin chi chien ; ti 7 chi.
PL2781.E55A6 1974, vol. 7 78–940162

Lin, Ch'ing-hsüan.
(Hu tieh wu hsü)
蝴蝶無鬚 / 林清玄著. — 臺北：皇冠出版社, 民國67 [1978]
259 p. ; 19 cm.
Prose.
NT$50.00

I. Title.
PL2878.N1237H8 79–836613

Lin, Chu-ying.
(Ni shuo tsen mo pan?)
你說怎麼辦? / 林珠英著. — 三重：世新出版社, 民國65 [1976]
9, 342 p. ; 20 cm. — (眞善美書系 ; 4)
Series romanized: Chen shan mei shu hsi.
NT$40.00

I. Title.
PL2878.N1245N5 78–841124

Lin, Hsi-chia.
(Shu yü shan ti jih tzu)
屬於山的日子 / 林錫嘉著. — 台北：水芙蓉出版社, 民國67 [1978]
3, 5, 204 p. ; 20 cm. — (水芙蓉書庫 ; 105)
Series romanized: Shui fu jung shu k'u.
NT$40.00

I. Title.
PL2878.N1385S5 78–842322

Lin, Huan-chang.
(Tso hsieh hsiao meng)
做些小夢：林煥彰抒情散文集. — 中和[臺北縣]：再興出版社 ; [台北]：總經銷國立臺灣大學員工福利社圖書部, 民國64 [1975]
8, 119 p. ; 19 cm. — (朶朶叢書 ; 2)
Essays.
Series romanized: To to ts'ung shu.
NT$30.00
I. Title.
PL2878.N14237T7 79–840123

Lin, Wen-i.
(Ch'eng en men)
承恩門 / 林文義著. — 台北：水芙蓉出版社, 民國67 [1978]
10, 2, 162 p. ; 20 cm. — (水芙蓉書庫 ; 118)
Essays.
Series romanized: Shui fu jung shu k'u.
NT$35.00

I. Title.
PL2878.N157C5 79–835171

Lin, Yün, 1955–
(Lin Yün san wen chi)
林芸散文集. — 再版. — 台北：水芙蓉出版社, 民國68 [1979]
3, 4, 220 p. ; 20 cm. — (水芙蓉書庫 ; 134)
Series romanized: Shui fu jung shu k'u.
NT$45.00

I. Title.
PL2878.N27A6 1979 79–837474

Lin, Yutang, 1895–1976.
(Jen sheng ti chiao pu)
人生的脚步 / 林語堂著. — 臺中：曾文出版社, 民國67 [1978] printing.
265 p. ; 20 cm.
Cover title.
NT$35.00

I. Title.
PL2781.N2J44 1978 79–837693

Lin, Yutang, 1895–1976.
(Lin Yü-t'ang sheng huo yü ssu hsiang)
林語堂生活與思想. — 高雄：莊家出版社 ; [臺北?]：總經銷世一書局, 民國64 [1975]
5, 208 p. ; 19 cm. — (當代名家文選)
Cover title.
Essays.
Series romanized: Tang tai ming chia wen hsüan.
NT$30.00
I. Title.
PL2781.N2A16 1975b 78–843303

Lin, Yutang, 1895–1976.
(P'i ching chi)
披荆集 / 林語堂著. — 臺北：南京出版公司, 民國67 [1978]
3, 129 p. ; 19 cm.

I. Title.
PL2781.N2P5 1978 79–835110

Lin, Yutang, 1895–1976.
[Selected works]
哲人的沉思：林語堂散文精選集. — 第1版. — 臺北：源成文化圖書供應社, 民國66 [1977]
4, 4, 268 p. : 20 cm.
Essays.
Appendices (p. 249–268): 薛光前 愛國反共的林語堂先生.—陳鼎環 突破湘山幾片雲.—傅一勤 「開明英文讀本」介論.—畢璞 風趣灑脫的文人.
NT$45.00
I. Title.
Title romanized: Che jen ti ch'en ssu.
PL2781.N2A6 1977d 79–835553

Lin, Yutang, 1895–1976.
[Selected works]
林語堂選集. — 香港：香港文學研究社, [1978?]
6, 3, 216 p. ; 21 cm. — (現代文學叢書)
Series romanized: Hsien tai wen hsüeh ts'ung shu.
$5.00

I. Title.
Title romanized: Lin Yü-t'ang hsüan chi.
PL2781.N2A6 1978 78–842476

Liu, Chin-te.
(Hsin ling hsiao yü)
心靈小語 / 劉錦得著. — 臺中：新企業世界出版社, 民國66 [1977]
6, 234 p. ; 20 cm.
NT$40.00

I. Title.
PL2879.C49H7 79–835649

Liu, Fei-lieh, 1922–1958.
(Liu Fei-lieh hsüan chi)
劉非烈選集 / 朱白水選輯. — 台北：黎明文化事業公司, 民國67 [1978]
2, 260 p. ; 19 cm. — (中國新文學叢刊 ; 13)
Series romanized: Chung-kuo hsin wen hsüeh ts'ung k'an.
NT$70.00

I. Chu, Pai-shui. II. Title.
PL2879.F41.5 1978 79–836929

CHINESE LITERATURE (Continued)

Liu, Jui-lin, 1936–
(I lin wai shih)
藝林外史 / 劉亞倫[瑞林]著. — 高雄：德馨室出版社, 民國67 [1978]
6, 4, 232 p. ; 20 cm. — (德馨萬象 ; 01)
Series romanized: Te hsin wan hsiang.
NT$50.00

I. Title.
PL2879.J8I17 79–839548

Liu, Kuo, 1154–1206.
(Lung-chou chi)
龙洲集 / 刘过著. — 第1版. — 上海：上海古籍出版社：新華書店上海发行所发行, 1978.
6, 15, 148 p. ; 21 cm.
$0.64

I. Title.
PL2687.L562L8 1978 79–840149

Liu, Li-yeh.
(Shan liang ti ming teng)
閃亮的明燈 / 劉麗葉編著. — 台南：大府城出版有限公司：總經銷中央書局, 民國67 [1978]
203 p. ; 20 cm. — (大府城書系 ; 19)
Series romanized: Ta fu ch'eng shu hsi.
NT$40.00

I. Title.
PL2879.L48S5 79–837050

Liu, Pai-yü.
(Liu Pai-yü san wen hsüan)
刘白羽散文选. — 第1版. — 北京：人民文学出版社, 1978.
2, 300 p. ; 21 cm.
Essays.
$0.72

I. Title.
PL2879.P3S36 78–842869

Liu, Tsai-fu.
(Nan yeh hsing hui)
南夜星輝 / 劉載福著. — 台南：綜合出版社, 民國66 [1977]
3, 189 p. ; 20 cm.
NT$25.00

I. Title.
PL2879.T65N3 1977 79–835850

Liu, Yung, fl. 1973–
(Ying ch'uang hsiao yü)
螢窗小語 / 劉墉著. — 臺北：劉墉：總經銷星光書報社, [19
v. ; 19 cm.
NT$50.00 (v. 5)

I. Title.
PL2879.Y85Y5 1978 79–835588

Liu, Yung, fl. 1973–
(Ying ch'uang sui pi)
螢窗隨筆 / 劉墉著 ; [攝影柯作平]. — 臺北：劉墉：總經銷星光書報社, [1977?]
208 p. : ill. ; 21 cm.
"劉墉散文詩畫集"
Essays.
$110.00

I. Title.
PL2879.Y85Y55 78–838022

Lo, Chen-yü, 1866–1940.
(Liao chü i kao)
遼居乙稿 / [羅振玉]. — [s. l. : s. n., 辛未 i. e. 1931序]
2, 81 double leaves ; 27 cm. — (松翁居遼後所著書)
Issued in case with the author's Han Hsi-p'ing shih ching ts'an tzu chi lu. [s. l., 1929]
Lithoprinted.

I. Title. II. Lo, Chen-yü, 1866–1940. Sung-weng chü Liao hou so chu shu.
PL2459.A49 79–836442
[AC150]

Lo, Chen-yü, 1866–1940.
(Liao chü tsa chu)
遼居雜著 / [羅振玉]. — [s. l. : s. n., 己巳 i. e. 1929序]
2 v. (on double leaves) ; 27 cm. — (松翁居遼後所著書)
Issued in case with the author's Han Hsi-p'ing shih ching ts'an tzu chi lu. [s. l., 1929]
Lithoprinted.
CONTENTS: [1] 矢彝考釋 璽印姓氏徵補正 漢兩京以來鏡銘集錄 鏡話 蒿里遺文目錄續編 補遺—[2] 李文田 和林金石錄 敦煌古寫本毛詩校記 帝範校補 宋槧文苑英華殘本校記

I. Li, Wen-t'ien, 1834–1895. Ho-lin chin shih lu. 1929. II. Title. III. Series: Lo, Chen-yü, 1866–1940. Sung-weng chü Liao hou so chu shu.
PL2459.A49 79–836444
[AC150] C

Lo, Yüeh-ling.
(Hsia jih ti hui i)
夏日的回憶 / 著作者羅悦玲 ; 插圖劉振偉 ; 總編輯顧俊 ; 編輯章英華, 王鳳生, 蘇永安. — 台北：經銷者領導出版社, 民國66 [1977]
41 p. : col. ill. ; 22 cm.
Colophon title.
Juvenile literature.
Text with kuo yü chu yin.
NT$30.00

I. Liu, Chen-wei, fl. 1977– II. Ku, Chün. III. Chang, Ying-hua. IV. Title.
PL2880.O3675H7 79–840522

Lu, Sheng-yen, 1945–
(Ch'ing shan chih wai)
青山之外 / 盧勝彥著. — [台中]：青山出版社；總經銷學海書局, 民國64 [1975]
2, 195 p. ; 20 cm. — (盧勝彥文集 ; 第18冊)
Essays.
Series romanized: Lu Sheng-yen wen chi.
NT$35.00

I. Title.
PL2880.U16477C55 78–841450

Lu, Sheng-yen, 1945–
(Ni p'u sa ti huo ch'i)
泥菩薩的火氣：盧勝彥的評論專作 / 盧勝彥著. — 臺中：新企業世界出版社, 民國67 [1978]
3, 191 p. ; 20 cm. — (盧勝彥文集 ; 第35冊)
Essays.
Series romanized: Lu Sheng-yen wen chi.
NT$45.00

I. Title.
PL2880.U16477N5 79–839463

Lu, Sheng-yen, 1945–
(Tsai cho ling ssu ti hsiao chou)
載着靈思的小舟：靈感的片羽 / 盧勝彥著. — 臺中：新企業世界出版社, 民國67 [1978]
6, 196 p. ; 20 cm. — (盧勝彥文集 ; 第32冊)
Essays.
Series romanized: Lu Sheng-yen wen chi.
NT$40.00

I. Title.
PL2880.U16477T7 78–841448

Lu, Sheng-yen, 1945–
(Tung fang ti fei chan)
東方的飛氈：不可思議的發現 / 盧勝彥著. — 九龍：清山出版社, [1977?]
5, 222 p. ; 19 cm. — (盧勝彥文集 ; 第31冊)
Literature.
Series romanized: Lu Sheng-yen wen chi.
$6.00

I. Title.
PL2880.U16477T8 78–842879

Lu, Yao-tung, 1933–
(Le ma ch'ang ch'eng)
勒馬長城 / 逯耀東著. — 3版. — 台北：時報出版公司, 民國67 [1978]
2, 289 p. ; 20 cm. — (時報書系 ; 117)
Essays.
Series romanized: Shih pao shu hsi.
NT$55.00

I. Title.
PL2880.U197L4 1978 79–836966

Lü, Yüan.
(Hsiao tung mu)
小冬木 = Xiao dong mu / 吕运著. — 第1版. — [北京]：中国少年兒童出版社：新华书店北京发行所, 1962, 1978 printing.
62 p. : ill. ; 19 cm.
Juvenile literature.
$0.15

I. Title. II. Title: Xiao dong mu.
PZ10.843.L8H74 79–838104

Lü shih ch'un ch'iu.
呂氏春秋 / 呂不韋撰. — 台北：老古出版社, 民國67 [1978]
2 v. (892, [56] p.) ; 22 cm. — (正統謀略學彙編 : 初輯 ; 第4–5冊)
Reprint of the block print ed.

I. Lü, Pu-wei, d. 235 B. C. II. Title. III. Series: Cheng t'ung mou lüeh hsüeh hui pien : Ch'u chi ; ti 4–5 ts'e.
Title romanized: Lü shih ch'un ch'iu.
PL2663.L8 1978 79–840102

Ma, Jui-hsüeh.
(Sung kei ku hsiang ti ko)
送給故鄉的歌 / 馬瑞雪. — 臺北：夏林含英：發行者純文學出版社, 民國66 [1977]
3, 131 p. ; 20 cm. — (純文學叢書 ; 80)
Series romanized: Ch'un wen hsüeh ts'ung shu.
CONTENTS: 幻滅.—黑鰻.—冰山下的戀歌.—送給故鄉的歌.
NT$40.00

I. Title.
PL2881.J8S9 79–836343

Mencius.
[Meng-tzu. Japanese & Chinese]
孟子：[14卷] / 武内義雄, 小林勝人譯註. — 東京：岩波書店, 昭和11 [1936], 昭和30 [1955] printing.
331 p. ; 15 cm. — (岩波文庫 ; 1309–1310a)
Series romanized: Iwanami bunko.
"原文は續古逸叢書と四部叢刊とに収められてゐる影清內府藏宋刊大字本にもとづいた"
I. Takeuchi, Yoshio, 1886– II. Kobayashi, Katsundo.
Title romanized: Mōshi.
PL2488.P8 79–801624

Mencius.
(Meng-tzu i chu)
孟子譯注 / 本社編審. — 台北：河洛圖書出版社, 民國66 [1977]
12, 483 p. ; 22 cm. — (夏學叢書)
Series romanized: Hsia hsüeh ts'ung shu.
Reprint of the 1960 ed. published by Chung-hua shu chü, Peking under the real author: Lan-chou ta hsüeh Chung wen hsi Meng-tzu i chu hsiao tsu.
NT$140.00

I. Ho lo t'u shu ch'u pan she. II. Lan-chou ta hsüeh. Chung-kuo yü yen wen hsüeh hsi. Meng-tzu i chu hsiao tsu. III. Title.
PL2474.R47 1977 78–840976

CHINESE LITERATURE (Continued)

Meng, Hsüan.
(Hsieh tsai shan chü ti jih tzu li)
寫在山居的日子裏 / 孟旋著. — 台南：綜合出版社, 民國67 [1978]
5, 2, 214 p. ; 20 cm. — (綜合文叢 ; 16)
Literature.
Series romanized: Tsung ho wen ts'ung.
NT$45.00

I. Title.
PL2883.N51557H7 79-837824

(Miao-shan ch'ing lü)
苗山情侶. — Hong kong：海歐出版公司, [1975?]
82, 40 p. : chiefly ill. ; 13 x 18 cm.
$2.00

PL2832.5.M5 78-841716

Mo-shang-ch'en.
(Yü hsiang chi)
玉香集 / 作者陌上塵. — 高雄：德馨室出版社, 民國67 [1978]
10, 207 p. ; 20 cm. — (德馨詩集 ; 7)
Colophon title.
Literature.
Series romanized: Te hsin shih chi.
NT$50.00
I. Title.
PL2886.O67Y8 79-838022

Mo-sheng-jen, 1948–
(Chi na ch'ang fa ti nü hai)
給那長髮的女孩 / 陌生人編著. — 台南：大府城出版有限公司：總經銷中央書局, 民國67 [1978]
6, 200 p. ; 20 cm. — (大府城書系 ; 8)
Series romanized: Ta fu ch'eng shu hsi.
NT$35.00

I. Title.
PL2886.O69C5 79-836434

(Nan hai hsüeh shu)
南海血書 / 譯著人朱桂等 ; [中央日報編印]. — 臺北：中央日報社, 民國68 [1979] printing.
95 p. : ill. ; 19 cm.
Colophon title.
CONTENTS: 阮天仇 南海血書.—薛翔川 故鄉行.—涂靜怡 從苦難中成長.—温良恭 商青.
NT$15.00

1. Chinese literature—20th century. I. Chu, Kuei, fl. 1976– II. Chung yang jih pao, T'ai-pei.
PL2513.N27 79-841429

Nung, Fu.
(Shui ch'e chi)
水車集 / [作者農婦]. — 香港：明窗出版社：發行明報有限公司出版部, 1978.
8, 337 p. : ill. ; 19 cm.
$6.00 ($2.00 U.S.)

I. Title.
PL2890.U5S54 79-838587

Ou-yang, Chan.
(Oú-yang Hsing-chou wen chi)
歐陽行周文集 / 歐陽詹著. — [臺北]：臺北市閩南同鄉會, [postscript 民國66 i. e. 1977]
156 leaves in various foliations, 7 p., [1] leaf of plates : port. ; 21 cm.
"國立中央圖書館特藏善本"
Photoreprint ed.

I. Title.
PL2677.O89O9 1977 79-835631

Ou-yang, Hsiu, 1007–1072.
(Kuei t'ien lu)
歸田錄 / [歐陽修撰]. — 據元刻本校涵芬樓版. — 上海：商務印書館, 民國8 [1919]
13, 12, [3] double leaves ; 27 cm.

I. Title.
PL2683.A6 1919 78-841013

C

Ou-yang, Hsiu, 1007–1072.
(Ou-yang Hsiu san wen hsüan)
歐陽修散文選 / 余民天編. — 臺北：河洛圖書出版社, 民國64 [1975]
4, 14, 184 p. ; 22 cm. — (河洛文庫 ; 52)
Series romanized: Ho lo wen k'u.
Reprint of the ed. with prelim. paging dated 1934.

I. Yü, Min-t'ien. II. Title.
PL2683.A6 1975 79-840594

C

Ou-yang, Tzu, 1939–
(I chih ti ying hua)
移植的櫻花 / 歐陽子著. — 台北：爾雅出版社, 民國67 [1978]
207 p. ; 19 cm. — (爾雅叢書 ; 37)
Essays.
Series romanized: Erh ya ts'ung shu.
CONTENTS: 入院記.—美國人的價值觀念.—美國人的處世態度.—美國人的人際關係.—美國人的精神信仰.—移植的櫻花.—週末午後.—我與美國少棒.—農耕之樂.—我兒世松.—附錄 關於我自已.
NT$40.00

I. Title.
PL2891.U25I17 79-836253

C

P'an, Han-nien, 1901–
(Li hun)
離婚 / 潘漢年作. — [香港?：s. n., 1978?]
123 p. ; 21 cm.
Literature.
Reprint of the 1930 ed. published by Kuang hua shu chü, Shanghai.
CONTENTS: 離婚.—情人.—苦杯.—她和她.—求愛.—無聊人的半天.—白皮鞋.—混沌中

I. Title.
PL2795.A582L5 1978 78-842806

C

P'an, Hsi-chen.
(Ch'ien li huai jen yüeh tsai feng)
千里懷人月在峯 / 琦君[潘希真]著. — 3版. — 臺北：爾雅出版社, 民國67 [1978]
214 p. ; 19 cm. — (爾雅叢書 ; 42)
Series romanized: Erh ya ts'ung shu.
NT$45.00

I. Title.
PL2892.A54C57 79-837937

C

P'eng, Chen-ch'iu.
(Hui chia)
回家 / 彭震球著. — 修訂再版. — 臺北：水牛出版社, 民國66 [1977]
3, 2, 3, 199 p. ; 20 cm. — (水牛文庫 ; 142)
Title in English in colophon: To return home.
Series romanized: Shui niu wen k'u.
NT$35.00

I. Title. II. Title: To return home.
PL2892.E527H84 79-838116

P'eng, P'in-kuang, 1926–
(P'eng-p'ai ch'en ssu chi)
澎湃沉思集 / 彭品光著. — 臺北：地球出版社：總代理熱門書店, 民國67 [1978]
6, 4, 241 p. ; 19 cm.
Essays.
NT$50.00

I. Title.
PL2892.E64P37 79-836011

P'eng, P'in-kuang, 1926–
(P'eng-p'ai hsin sheng chi)
澎湃心聲集 / 彭品光著. — 臺北：大林出版社, 民國66 [1977]
4, 10, 408 p. ; 19 cm. — (大林文庫 ; 135)
Essays.
Series romanized: Ta lin wen k'u.
NT$55.00

I. Title.
PL2892.E64P378 79-836631

P'eng, P'in-kuang, 1926–
(P'eng-p'ai nu ch'ao chi)
澎湃怒潮集 / 澎湃[彭品光]著. — 臺北：星光出版社, 民國67 [1978]
2, 6, 272 p. ; 19 cm. — (雙子星叢書 ; 73號)
Series romanized: Shuang tzu hsing ts'ung shu.
NT$50.00

I. Title.
PL2892.E64P39 79-839465

P'eng, Shao-sheng, 1740–1796.
(Erh lin chü chi)
二林居集 / 彭紹升撰. — 臺北：石門圖書公司, 民國65 [1976]
2 v. ; 22 cm. — (清人文集 ; 2)
Reprint ed.; originally issued in 24 chüan.

I. Title. II. Series: Ch'ing jen wen chi ; 2.
PL2722.E464E7 1976 78-841758

Pi-chu.
(Pi-chu san wen tzu hsüan chi)
碧竹散文自選集 / 黃燕德[碧竹]著. — 台北：水芙蓉出版社, 民國67 [1978]
26, 4, 432 p. ; 20 cm. — (水芙蓉書庫 ; 126)
Series romanized: Shui fu jung shu k'u.
NT$80.00

I. Title.
PL2892.I16A6 1978 79-839599

(P'iao i hsiao p'in)
飄逸小品 / 劉栽福編. — [台南]：綜合[出版社, 1978?]
8, 196 p. ; 19 cm.

1. Chinese essays—20th century. I. Liu, Tsai-fu.
PL2623.P48 79-836244

P'u, Sung-ling, 1640–1715.
(Hsing shih yin yüan)
醒世姻緣 / 西周生[蒲松齡]著. — 九龍：明亮書局, [1978?]
3 v. (2, 5, 6, 1070 p.) ; 19 cm.
Colophon title.
"大字足本古典文學"

I. Title.
PL2722.U2H8 1978 79-836545

CHINESE LITERATURE (Continued)

P'u, Sung-ling, 1640–1715.
(Liao chai chih i)
聊齋志異：[蒲松齡手稿本]. — 香港：南天書業公司, 1969.
17, 798, 25 p. : port. ; 22 cm.
Reprint of the 1955 ed. published by Wen hsüeh ku chi k'an hsing she, Peking; with prefatory matters deleted.
"聊齋誌異附册": p. 1–25 (last group)

I. Title.
PL2722.U2L5 1969 77-836952

P'u, Sung-ling, 1640–1715.
[Liao chai chih i]
鑄雪齋抄本聊齋志異 / [蒲松齡著]. — 第1版. — 上海：上海人民出版社, 1974.
12 v. (on double leaves) in case ; 26 cm.
"《鑄雪齋抄本聊齋志異》共十二卷…这次影印版框也照原樣"

I. Title. II. Title: Liao chai chih i.
Title romanized: Chu hsüeh chai ch'ao pen Liao chai chih i.
PL2722.U2L5 1974b 78-843569

P'ü, Sung-ling, 1640–1715.
(Liao chai chih i)
聊齋志異：會校會注會評本 / [蒲松齡著] ; 張友鶴輯校. — 新1版. — 上海：上海古籍出版社：新華書店上海发行所发行, 1978.
4 v. (1735 p.) : facsims. ; 21 cm.
$6.85

I. Chang, Yu-ho. II. Title.
PL2722.U2L5 1978 79-836971

Pu, Sung-ling, 1640–1715.
[Liao chai chih i. Selections]
聊斋故事选译 = Liaozhai gushi xuanyi / 于在春选译. — 第1版. — 上海：上海古籍出版社：新華書店上海发行所发行, 1978.
2, 3, 160 p. : ill. ; 19 cm. — (中国古典文学作品选读)
$0.37
I. Yü, Tsai-ch'un. II. Title. III. Title: Liaozhai gushi xuanyi. IV. Series: Chung-kuo ku tien wen hsüeh tso p'in hsüan tu.
Title romanized: Liao chai ku shih hsüan i.
PL2722.U2L5215 1978b 79-839539

P'u-t'i, 1931–
(Chih feng ts'ao)
知風草 / 菩提著. — 台北：水芙蓉出版社, 民國65 [1976]
2, 3, 196 p. ; 19 cm. — (水芙蓉書庫 ; 59)
Series romanized: Shui fu jung shu k'u.
Essays.
NT$35.00

I. Title.
PL2892.U28C5 76-840397

Sang, Jou.
(Shao nien tao)
少年島 / 桑柔著. — 臺北：天華出版事業股份有限公司, 民國67 [1978]
4, 215 p. ; 20 cm. — (天華文學叢刊 ; 5)
Literature.
Series romanized: T'ien hua wen hsüeh ts'ung k'an.
Pbk. : NT$45.00

I. Title.
PL2894.A493S5 78-842009

Sang, P'in-tsai, 1939–
(Ssu-yang wen ts'ui)
司陽文粹 / 桑品載著. — 台北：水芙蓉出版社, 民國67 [1978]
6, 186 p. ; 20 cm. — (水芙蓉書庫 ; 101)
Series romanized: Shui fu jung shu k'u.
NT$40.00

I. Title.
PL2894.A5S7 78-842484

Sha, Sui, 1948–
(Yen-chi)
燕姬 / [著者沙穗]. — 高雄：心影出版社, 民國68 [1979]
3, 211 p. ; 19 cm. — (心影文庫 ; 25)
Literature.
Series romanized: Hsin ying wen k'u.
NT$50.00

I. Title.
PL2895.S9Y46 79-840176

Shao, T'ing-ts'ai, 1648–1711.
(Ssu fu t'ang wen chi)
思復堂文集 / 邵廷采撰. — 臺北：華世出版社, 民國66 [1977]
2 v. (12, 1086 p.) ; 22 cm.
"本書根據清光緒十九年(一八九三)會稽徐氏(友蘭) 鑄學齋刊本影印, 原收在紹興先正遺書中"
NT$380.00

I. Title.
PL2724.H3S7 1977 78-842838

Shen, Jen-k'ang.
(Huo pa)
火把 / 沈仁康. — [香港?: s. n., 1978?]
2, 214 p. ; 21 cm.
Essays.
Reprint of the 1977 ed. published by Kuang-hsi jen min ch'u pan she, Nan-ning.

I. Title.
PL2898.N19H86 1978 78-842619

Shen, K'uo, 1031?–1095?
(Meng-hsi pi t'an)
夢溪筆談 / 沈括撰 ; 彭元瑞手書. — [台北?] : 中國子學名著集成編印基金會, [1978?]
3, 4, 648 p. ; 22 cm. — (中國子學名著集成 ; 096 : 珍本初編, 雜家子部)
"明覆刊宋乾道二年本"
"子略 高似孫撰": p. 483–648.
I. P'eng, Yüan-jui, 1731–1803. II. Kao, Ssu-sun, chin shih 1184. Tzu lüeh. 1978? III. Title. IV. Series: Chung-kuo tzu hsüeh ming chu chi ch'eng ; 096.
PL2687.S5M4 1978 78-842705

Shen, K'uo, 1031?–1095?
[**Meng-hsi pi t'an. Japanese**]
夢溪筆談 / 沈括 ; 梅原郁訳注. — 東京：平凡社, 1978–
v. : ill. ; 18 cm. — (東洋文庫 ; 344–) Ja 79-3689 (v. 1)
Series romanized: Tōyō bunko.
Includes bibliographical references.
¥1000 (v. 1)
I. Umehara, Kaoru, 1934– II. Title.
Title romanized: Mukei hitsudan.
PL2687.S5M416 79-808136

Shen, Ts'ung-wen, 1902–
(Fei yu ts'un ti)
廢郵存底 / 沈從文著. — 香港：滙通書店, 1976.
iv, 131 p. ; 19 cm.
"答辭, [by] 蕭乾": p. 65–131.
Reprint of the 1949 ed. published by K'ai ming shu tien, Shanghai.
$3.80

I. Hsiao, Ch'ien, 1911– Ta tz'u. 1976. II. Title.
PL2801.N18F4 1976 78-842898

Shen, Ts'ung-wen, 1902–
(Hei feng chi)
黑鳳集 / 沈從文著. — 香港：滙通書店, 1977.
1, 156 p. ; 19 cm.
At head of title: Ts'ung-wen ch'uang tso chih ssu.
Reprint of the 1949 ed. published by K'ai ming shu tien, Peking, in series: Shen ts'ung-wen chu tso chi chih i.

I. Title.
PL2801.N18H38 1977 79-835488

Shen, Yen-ping, 1896–
(Pai yang li tsan)
白楊禮讚 / 茅盾 [沈雁冰]. — [s. l. : s. n., 1978?]
ii, 131 p. ; 19 cm. — (龍山文化叢書)
Cover title.
Series romanized: Lung shan wen hua ts'ung shu.
Reprint of the 1949 ed. published by Wen hua sheng huo ch'u pan she, Shanghai, in series: Wen hsüeh ts'ung k'an.

I. Title.
PL2801.N2P3 1978 78-843529

Shen, Yen-ping, 1896–
[Selected works]
茅盾散文集 / 茅盾 [沈雁冰] 著. — 香港：滙通書店, 1976.
2, 4, 119 p. ; 19 cm.

I. Title.
Title romanized: Mao Tun san wen chi.
PL2801.N2A6 1976 79-836176

Shih ching.
詩經譯注 / 陳慎初著. — 台北：洪氏出版社：總經銷樂天書局, 民國66 [1977] printing.
2, 2, 100 p. ; 19 cm.
Reprint of the 1959 ed. published by Hsiang-kang wan li shu tien, Hongkong, under the author's name: Yü Kuan-ying and issued in series: Ku tien wen hsüeh ch'u pu tu wu.

I. Ch'en, Shen-ch'u. II. Yü, Kuan-ying, tr.
Title romanized: Shih ching i chu.
PL2466.R8 1977 79-841069

Shih ching. Selections.
詩經續本：注音詳解 / 朱令譽編註. — 台南：大夏出版社, 民國67 [1978]
4, 7, 137 p. ; 21 cm.
NT$30.00

I. Chu, Ling-yü. II. Title.
Title romanized Shih ching tu pen.
PL2466.R8 1978 79-836275

Shu, Ko.
(Hsiao hsiao p'ao mo)
小小泡沫 / 書戈著. — 臺南：聞道出版社, 民國65 [1976]
75 p. ; 19 cm. — (文藝小叢書 ; 020)
Literature.
Series romanized: Wen i hsiao ts'ung shu.
CONTENTS: 耕夜的人.—龍二爺.—天平山上的猴子.—狗屎爹.—永順行.—小小泡沫.—也算渡假.
I. Title.
PL2901.K6H7 79-840597

Shui fu jung ch'u pan she pien chi tsu chi mei kung tsu t'ung jen.
(Chia wen li t'u)
佳文麗圖 / 本社編輯組及美工組同仁合著. — 台北：水芙蓉出版社, 民國67 [1978]
6, 150 p. : ill. ; 20 cm. — (水芙蓉書庫 ; 113)
Series romanized: Shui fu jung shu k'u.
NT$40.00

I. Title.
PL2902.I26C4 1978 79-835853

Shui hu chuan. Selections.
黑旋風李逵 / 著者施耐庵. — 台北：遠流出版社, 民國67 [1978]
132 p. : ill. ; 22 cm. — (遠流青少年百科全書 ; 15 : 第4輯, 中國古典文學名著)
NT$40.00

I. Shih, Nai-an, ca. 1290–ca. 1365. II. Title. III. Series: Yüan liu ch'ing shao nien pai k'o ch'üan shu ; 15.
Title romanized: Hei hsüan feng Li K'uei.
PL2694.S52 1978c 79-839474

CHINESE LITERATURE (Continued)

Ssu shu.
四書集註 / [朱熹章句]. — 서울 : 曺龍承, 1977.
409 p. in various pagings ; 23 cm.
Colophon inserted.
CONTENTS: 大學章句一卷.—中庸章句一卷.—論語集註十卷.—孟子集註七卷.
₩4000

I. Chu, Hsi, 1130–1200. II. Cho, Yong-sŭng. III. Title.
Title romanized: Sasŏ chipchu.
PL2463.B3 78–826357

Ssu shu.
新譯四書讀本 : 國語注音, 註解, 語譯, 章旨 / [編著者陳基政]. — 台南 : 新世紀出版社, 民國66 [1977]
6, 625 p. ; 21 cm.
Cover title: Ssu shu tu pen.
NT$80.00

I. Ch'en, Chi-cheng. II. Title. III. Title: Ssu shu tu pen.
Title romanized: Hsin i Ssu shu tu pen.
PL2463.R26 79–839928

Ssu shu.
四書析義 / [編輯部華星出版社 ; 校訂者呂茗青]. — 新板. — 板橋 : 該社, 民國66 [1977]
16, 5, 384 p. ; 22 cm.
At head of title: Hsüeh shuo yen chiu, p'ien chang fen hsi.
Colophon title: Hsin pien Ssu shu hsi i.
NT$100.00

I. Hua hsing ch'u pan she. II. Title. III. Title: Hsin pien Ssu shu hsi i.
Title romanized: Ssu shu hsi i.
PL2463.R52 79–836962

Sun, Ju-ling.
(Fang k'uai pai p'ien)
方塊百篇 / 孫如陵著. — [台北?] : 中國文選社, 民國67 [1978]
9, 319 p. ; 19 cm.
NT$50.00

I. Title.
PL2905.N234F3 78–841056

Sung, Ching-i.
(Na i chen yü)
那一陣雨 / 宋晶宜著. — 臺北 : 名人出版社, 民國66 [1977]
247 p. ; 19 cm.
Essays.
NT$44.00

I. Title.
PL2905.N433N3 79–835494

Sung, Tse-lai.
(T'iao ku jih chi)
糶穀日記 / 宋澤萊著. — 台北 : 遠景出 社 : 門市部出版家書城, 民國68 [1979]
246 p. ; 20 cm. — (遠景叢刊 ; 101)
Short stories.
Series romanized: Yüan ching ts'ung k'an.
CONTENTS: 花鼠仔立志的故事.—糶穀日記.—漁仔寮案 櫻城之衰.
NT$55.00
I. Title.
PL2905.N46T5 79–840589

Ta hsüeh. Korean & Chinese.
大學. 中庸 / 權泰益譯註. — 서울 : 瑞文堂, 1974.
285 p. ; 18 cm. — (瑞文文庫 ; 125)
Series romanized: Sŏmun mun'go.

I. Kwŏn, T'ae-ik. II. Chungyung. Korean & Chinese. 1974. III. Title.
Title romanized: Taehak.
PL2489.M77 79–826355

Ta-huang.
(Tsai wu tien ti hsiao chan)
在誤點的小站 / 大荒著. — 臺北 : 天華出版事業股份有限公司, 民國67 [1978]
4, 4, 208 p. ; 20 cm. — (天華文學叢刊 ; 3)
Series romanized: T'ien hua wen hsüeh ts'ung k'an.
NT$45.00

I. Title.
PL2907.Z4T3 78–842967

Tan, Fei.
(Jan chi chi)
燃箕集 / 丹扉著. — 大甲鎮[台中縣] : 慧龍文化有限公司 : 總經銷文星書店, 民國67 [1978]
4, 221 p. ; 19 cm. — (慧龍書系 ; 317)
Series romanized: Hui lung shu hsi.
NT$45.00

I. Title.
PL2908.N2J3 1978 79–837947

T'ang, I-cheng.
(K'o Lan wen chi)
柯藍文集 / [唐 一正]. — [香港? : s. n., 1978]
49 leaves ; 31 cm.
Cover title.
A collection of articles reprinted from various sources.
Publication date stamped on p. [4] of cover.
CONTENTS: 鋼匙.—魚鷹.—鐵窗烈火.

I. Title.
PL2908.N4K6 1978 79–835513

T'ang, Jen, pseud.
(Hua hua jen sheng)
話畫人生 / [唐人著]. — 台南 : 西北出版社, 民國65 [1976]
6, 208 p. : ill. ; 20 cm.
NT$35.00
Literature.

I. Title.
PL2908.N422H8 79–835093

T'ang, Shih-sheng, 1551–1636.
(San i chi)
三易集 / 唐時升撰. — 臺北 : 偉文圖書出版社, 民國66 [1977]
2 v. (997 p.) ; 22 cm. — (明代論著叢刊 : 第三輯)
Reprint ed.
NT$600.00

I. Title. II. Series: Ming tai lun chu ts'ung k'an : Ti san chi.
PL2698.T27S2 1977 78–842091

T'ang, Wei-po.
(Hsün ch'iu)
尋秋 : 散文精選 / 湯為伯著. — 台南 : 綜合出版社, 民國67 [1978]
6, 221 p. ; 20 cm.
NT$30.00

I. Title.
PL2908.N53H7 79–838008

T'ang, Wei-po.
(Ku hsiang ti wu yüeh)
故鄉的五月 / 湯爲伯著. — 台南 : 綜合出版社, 民國67 [1978]
5, 224 p. ; 20 cm. — (綜合文叢 ; 11)
Series romanized: Tsung ho wen ts'ung.
NT$45.00

I. Title.
PL2908.N53K8 79–839467

Teng, Wen-lai, 1932–
(Teng Wen-lai tzu hsüan chi)
鄧文來自選集. — 臺北 : 黎明文化事業公司, 民國66 [1977]
316 p., [2] leaves of plates : ill. ; 20 cm. — (中國新文學叢刊 ; 49)
Series romanized: Chung-kuo hsin wen hsüeh ts'ung k'an.
NT$80.00

I. Title.
PL2909.E45A6 1977 79–838611

T'ieh, Ying.
(Feng huang shu chuan lan)
鳳凰樹專欄 / 鐵英著. — 台北 : 遠景出版社, 民國68 [1979]
6, 151 p. ; 20 cm. — (遠景叢刊 ; 113)
Essays.
Series romanized: Yüan ching ts'ung k'an.
NT$40.00

I. Title.
PL2909.I29F46 79–840168

T'ien, Han, 1899–
(T'ien Han ch'uang tso hsüan)
田漢創作選 / 編者筱梅. — 香港 : 經銷處神州圖書公司, [1978?]
2, 2, 200 p. ; 19 cm.
Colophon title.
Reprint of the 1936 ed. published by Fang ku shu tien, Shanghai.
CONTENTS: 戲劇: 南歸. 一九三二年的月光曲. 古潭的聲音. 回春之曲. 洪水.—散文: 離鄉的滋味. 夢裏的故鄉. 不堪回首的那一夜.—詩: 七夕. 春月的下面. 東都春雨曲. 暴風雨後的春朝. 黃昏.
I. Ch'en, Hsiao-mei. II. Title.
PL2812.I4T48 1978 79–835868

Ting, Wang.
(Yeh tu sui pi)
夜讀隨筆 / 丁望著. — 台北 : 時報文化出版事業有限公司, 民國66 [1977]
3, 187 p. : ill. ; 20 cm. — (時報書系 ; 86)
Series romanized: Shih pao shu hsi.
NT$40.00

I. Title.
PL2909.I62Y4 78–842251

Ts'ai, I-ch'iu.
(Hsiao yüan che hsüeh)
校園哲學 / 蔡益秋著. — 台南 : 綜合出版社, 民國68 [1979]
11, 237 p. ; 19 cm. — (綜合文叢 ; 20)
Series romanized: Tsung ho wen ts'ung.
NT$48.00

I. Title.
PL2912.A453H7 79–840519

Ts'ai, Ying-hu.
(Hu shih)
壺史 / [著作者蔡瀛壺]. — 台北 : 新文豐出版公司, 民國66 [1977]
2 v. (on double leaves) in case : ill. ; 22 cm.
Reprint ed.
NT$240.00

I. Title.
PL2815.A52H8 1977 79–838250

Ts'ai, Yüan-p'ei, 1867–1940.
[Selected works]
蔡元培選集. — 香港 : 香港文學研究社, [1977?]
6, 149 p. ; 20 cm. — (中國現代文學作家選集叢書) (現代文學叢書)
Series 1 romanized: Chung-kuo hsien tai wen hsüeh tso chia hsüan chi ts'ung shu. Series 2 romanized: Hsien tai wen hsüeh ts'ung shu.
$6.00
I. Title.
Title romanized: Ts'ai Yüan-p'ei hsüan chi.
PL2815.A55A6 1977 79–836535

CHINESE LITERATURE (Continued)

(Ts'ai lou chi)
彩楼記 / 無名氏著；黃裳校注. — 第1版. — 上海：上海古典文學出版社, 1956.
1, 86 p. : ill. ; 19 cm.

I. Huang, Shang.
PL2698.T74 1956 78-841821

Ts'ao, Chan, 1717 (ca.)-1763.
[Hung lou meng]
乾隆甲戌脂硯齋重評石頭記. — 3版. — 台北：胡適紀念館, 民國64 [1975]
246, 9 leaves, [2] fold. leaves of plates : ill. ; 27 cm.
Running title: Shih t'ou chi.
"…曹雪芹…寫"
"劉福銓舊藏的…抄本四大册, 共有十六回: 第一到第八回, 第十三到第十六回, 第二十五到第二十八回…這個鈔本後來就稱為「甲戌本」"
"跋乾隆甲戌脂硯齋重評石頭記影印本 胡適": leaves 1-9 (last group)
I. Hu, Shih, 1891-1962. Pa Ch'ien-lung chia hsü Chih yen chai ch'ung p'ing Shih t'ou chi ying yin pen. 1975. II. Title. III. Title: Shih t'ou chi.
Title romanized: Ch'ien-lung chia hsü Chih yen chai ch'ung p'ing Shih t'ou chi.
PL2727.S2A1 1975 78-842962

Ts'ao, Chan, 1717 (ca.)-1763.
[Hung lou meng]
校本紅樓夢 / [曹雪芹 i. e. 曹霑著]. — 臺北：華正書局, 民國66 [1977]
[15], 929, 419 p. ; 22 cm.
"一百二十回"
NT$200.00

I. Title.
Title romanized: Chiao pen Hung lou meng.
PL2727.S2A1 1977i 78-843603

Ts'ao, Chih, 192-232.
[Works]
曹子建全集 / 曹植著；[丁晏箋]. — 臺北：清流出版社, 民國65 [1976]
222 leaves in various foliations ; 22 cm. — (國學珍本叢書 ; 5)
Includes original t. p.: Ts'ao Tzu-chien chi.
Running title: Ts'ao chi ch'üan p'ing.
Reprint ed.
NT$200.00

I. Ting, Yen, 1794-1875. II. Title. III. Title: Ts'ao Tzu-chien chi. IV. Series: Kuo hsüeh chen pen ts'ung shu ; 5.
Title romanized: Ts'ao Tzu-chien ch'üan chi.
PL2664.T75 1976 79-835222

Tseng, Ching-wei.
(Kei shao nien p'eng yu ti hsin)
給少年朋友的信 曾勁韋著 星洲 星洲世界書局 [1960?]
55 p. 19 cm.
Letters.

I. Title.
PL2912.E4Z53 1960 C 62-161 ‡
rev

Tu, Chao-t'ang.
(Wo hsien chi)
蝸涎集 杜召棠著 臺北 民智出版社 民國52 [1963]
2 v. (2, 8, 576 p.) in 1. 19 cm.
——續集 [台北 德志出版社 1969]
2 v. (1, 8, 481, 4 p.) 19 cm. 80.00
PL2913.C3W6 1963 Suppl.

I. Title.
PL2913.C3W6 1963 C 63-2394
rev 2

Tu, Fu, 712-770.
(Tu si ŏnhae)
杜詩諺解 / 杜甫；李炳疇註解；高銀譯. — 서울：民音社, 1976.
151 p. : ill. ; 20 cm. — (民音社世界詩人選 ; 49)
W350

I. Yi, Pyŏng-ju, 1921- II. Ko, Ŭn, 1933- III. Title. IV. Series: Minŭmsa segye siinsŏn ; 49.
PL2675.T764 1976 78-826646

T'u, Lung, 1542-1605.
(Pai yü chi)
白榆集 / 屠隆撰. — 臺北：偉文圖書出版社, 民國66 [1977]
2 v. (1034 p.) ; 22 cm. — (明代論著叢刊 ; 第三輯)
Reprint ed.
NT$700.00 ($18.44 U.S.)

I. Title. II. Series: Ming tai lun chu ts'ung k'an : Ti san chi.
PL2698.T812P3 1977 78-841612

T'u, Lung, 1542-1605.
(Yu-ch'üan chi)
由拳集 / 屠隆撰. — 臺北：偉文圖書出版社, 民國66 [1977]
3 v. (1214 p.) ; 22 cm. — (明代論著叢刊 : 第三輯)
Photoreprint ed.
NT$900.00 ($23.71 U.S.)

I. Title. II. Series: Ming tai lun chu ts'ung k'an : ti san chi.
PL2698.T812Y8 1977 78-842008

Tu, Pen, 1276-1350.
(Ch'ing-chiang pi chang chi)
清江碧嶂集 / [撰者杜本]. — 臺北：臺灣學生書局, 民國62 [1973]
10, 74, 4 p. ; 22 cm. — (歷代畫家詩文集 ; 27 ; 第三輯)
"國立中央圖書館藏本明末汲古閣刊本景印"

I. Title. II. Series: Li tai hua chia shih wen chi ; 27.
PL2694.T86C5 1973 79-836711

Tu-ku, Hung.
(Tao shen)
刀神 / 獨孤紅著. — 台北：春秋出版社：總經銷呂氏書店, 民國67 [1978]
2 v. (708 p.) ; 20 cm.
NT$90.00

I. Title.
PL2913.Z47T36 79-836273

T'ung, Shih-chang, 1918-
(Lo ch'ih sui feng ts'ui)
落翅隨風摧：幽默諷刺集 / 童世璋著. — 臺中：新企業世界出版社, 民國67 [1978]
206 p. ; 19 cm. — (諾貝爾文庫)
NT$40.00

I. Title. II. Series: No-pei-erh wen k'u.
PL2817.N24L6 79-836554

Tung, Shu-fan, 1933-
(Hsin ling ti hsi yüeh)
心靈的喜悅 / 董樹藩著. — 臺北：國家書店, 民國67 [1978]
7, 200 p., [4] leaves of plates : ill. ; 20 cm. — (讀書人叢刊 ; 54)
Series romanized: Tu shu jen ts'ung k'an.
NT$45.00

I. Title.
PL2914.N565H74 78-842646

Tung, Pao-chung.
(Meng-ku k'ao jou pa lei wu)
蒙古烤肉芭蕾舞 / 董保中著. — 臺北：九歌出版社：總經銷爾雅出版社, 民國67 [1978]
209 p. ; 19 cm. — (九歌文庫 ; 10)
Essays.
Series romanized: Chiu ko wen k'u.
NT$45.00

I. Title.
PL2914.N558M4 79-835104

Tung-fang, Pai.
(Huang chin meng)
黃金夢 / 東方白著. — 臺北：臺灣學生書局, 民國66 [1977]
6, 2, 196 p. ; 19 cm. — (學生書苑 ; 59)
Literature.
Series romanized: Hsüeh sheng shu yüan.
CONTENTS: 黃金夢.—莎河與我.—白溪與我.—鯤.—草原上.—坊子.—熊的兒子.—鵝.—學生不老.—復活.—上帝知道一切、等待吧.—大文豪與小方塊.
NT$40.00

I. Title.
PL2914.N63H8 78-843677

(Wan hsiang hsiao p'in)
萬象小品 / 本社編輯部編. — 臺北：源成文化圖書供應社：香港：總經銷平價書屋, 民國66 [1977]
8, 202 p. ; 19 cm.
NT$30.00

1. Chinese essays—20th century. I. Yüan ch'eng wen hua t'u shu kung ying she. Pien chi pu.
PL2623.W3 79-836287

Wang, Chi-ts'ung.
(Wang Chi-ts'ung tzu hsüan chi)
王集叢自選集. — 臺北：黎明文化事業公司, 民國67 [1978]
8, 420 p., [2] leaves of plates : ill. ; 20 cm. — (中國新文學叢刊 ; 55)
Essays.
Series romanized: Chung-kuo hsin wen hsüeh ts'ung k'an.
Includes bibliographical references.
Appendices (p. 363-420): 任卓宣 方開始的王集叢先生.—吳曼君 我所認識的王集叢先生.—馬璧 值得敬仰的文藝理論家王集叢先生.—周伯乃 論王集叢.—鄭學稼 讀晨霧.—鳳兮 撥開晨霧.—東郭牙 走入晨霧.—白圭 我看晨霧.—五大雜文 三十年代的面目.—王集叢 中央, 掃蕩, 新華
NT$80.00

I. Title.
PL2919.C433A16 1978 79-836970

Wang, Chien-yü, 1953-
(Hsin hsiang pan pan)
心香瓣瓣 / 王建裕, 綠沙合著. — 台南：龍輝出版社, 民國67 [1978]
10, 199 p. ; 19 cm.
NT$40.00

I. Lü-sha, 1952- joint author. II. Title.
PL2919.C474H73 79-837926

Wang, Chien-yü.
(Wo ssu wo yü)
我思我語 / 王建裕著. — 苗栗縣竹南鎮：七燈出版社, 民國66 [1977]
186 p. ; 20 cm. — (七燈惠衆文庫 ; 6)
Essays.
Series romanized: Ch'i teng hui chung wen k'u.
NT$35.00

I. Title.
PL2919.C474W6 78-842482

Wang, Chien-yü, 1953-
(Yao chi ch'ien li)
遙寄千里 / 王建裕著. — 臺中：曾文出版社, 民國67 [1978]
235 p. ; 20 cm.
Cover title.
Prose.
NT$40.00

I. Title.

CHINESE LITERATURE (Continued)

Wang, Ching-hsi, 1897–1968.
(Hsüeh yeh)
雪夜 / 汪敬熙著. — [香港? : s. n., 1978?]
4, 2, 96 p. ; 21 cm.
Essays.
Reprint of the 1925 ed. published by Ya tung t'u shu kuan, Shanghai.

I. Title.
PL2822.C485H7 1978 78–843568

Wang, Ching-hsi, 1932–
(Kuan t'ien chi)
觀天集 王敬羲著 [九龍 文藝書屋 1969]
2, 4, 151 p. 18 cm. (文星叢刊275) 2.50
Essays.

I. Title.
PL2919.C5K8 74–839492

Wang, Erh-min.
(Shui shou ti hua)
水手的話 / 王爾敏著. — 臺北 : 世界文物出版社, 民國66 [1977]
183 p. ; 19 cm. — (世界文叢 ; 35)
Essays.
Series romanized: Shih chieh wen ts'ung.
NT$30.00

I. Title.
PL2919.E7S5 78–842976

Wang, Hsing-hua.
(Chü yüan shih wen chi)
莒園詩文集 / 王星華編印. — [台北?] : 王星華, [1975?]
2, 193 p. ; 22 cm.
Cover title.
Literature.

I. Title.
PL2919.H8285C5 79–841114

Wang, Hsiu-nan.
(Nan kuo pi keng)
南國筆耕 : 王秀南教授散文集. — Hong Kong : 香港东南亚研究所 ; Singapore : 总経销新加坡东南亚教育研究中心, 1978.
293 p. : ill. ; 22 cm.
Cover title.
Colophon title: Wang Hsiu-nan chiao shou san wen chi : Nan kuo pi keng.
$16.00
I. Title. II. Title: Wang Hsiu-nan chiao shou san wen chi.
PL2919.H835N3 79–837721

Wang, Jo-wang.
(Chih lao hu)
紙老虎 : 新型京劇 / 王若望, 江悬天, 王震伯作. — [香港? : s. n., 1978?]
55 p., [2] leaves of plates : ill. ; 21 cm.
Reprint of the 1951 ed. published by Lao tung ch'u pan she, Shanghai, in series: Lao tung hsi ch'ü ts'ung shu.

I. Chiang, Yü-t'ien, joint author. II. Wang, Chen-po, joint author. III. Title.
PL2919.J6C4 1978 79–835242

Wang, Lu-sung.
(Sheng ming ti t'ou ying)
生命的投影 / 王祿松著. — 台北 : 水芙蓉出版社, 民國66 [1977]
2, 2, 6, 232 p. ; 20 cm. — (水芙蓉書庫 ; 112)
Essays.
Series romanized: Shui fu jung shu k'u.
"青年戰士報「新文藝」副刊專欄"
NT$42.00

I. Title.
PL2919.L8S5 78–841334

Wang, Shih-chen.
(Tsun yen yü pei ai)
尊嚴與悲哀 / [王世禎著]. — 台南 : 西北出版社, 民國65 [1976]
5, 199 p. ; 19 cm.
NT$35.00

I. Title.
PL2919.S488T7 78–842870

Wang, Shu-ch'uan, 1920–
(Wang Shu-ch'uan tzu hsüan chi)
王書川自選集. — 臺北 : 黎明文化事業公司, 民國67 [1978]
8, 426 p., [2] leaves of plates : ill. ; 20 cm. — (中國新文學叢刊 ; 62)
Literature.
Series romanized: Chung-kuo hsin wen hsüeh ts'ung k'an.
NT$80.00

I. Title.
PL2919.S497W3 79–839118

Wang, Ssu-jen, 1575–1646.
(Wang Chi-ch'ung tsa chu)
王季重雜著 / 王思任撰. — 臺北 : 偉文圖書出版社, 民國66 [1977]
2 v. (874 p.) ; 22 cm. — (明代論著叢刊 : 第三輯)
Reproduced from the ed. in the possession of Chung yang t'u shu kuan.
NT$600.00 ($15.81 U.S.)

I. Title. II. Series: Ming tai lun chu ts'ung k'an : Ti san chi.
PL2698.W322A6 1977 78–843537

Wang, Ting-chün.
(Ling kan)
靈感 / 王鼎鈞. — [台北?] : 王鼎鈞 ; 台北 : 總經銷四季出版公司, 民國67 [1978]
189 p. : ill. ; 20 cm.
Literature.
NT$45.00

I. Title.
PL2919.T5L5 79–840360

Wang, Ting-ho.
(Wang kung tzu ta k'o wen)
王公子答客問 / 王定和著. — 台北 : 王定和 : 總經銷遠東書報社, 民國67 [1978]
158 p. : ill. ; 20 cm.
NT$40.00

I. Title.
PL2919.T52W3 79–835105

Wang, Wei, 701–761.
(Wang yu ch'eng chi chien chu)
王右丞集箋註 / 趙松谷箋註. — 台北 : 廣文書局, 民國66 [1977]
2 v. (1076, [230] p.) ; 22 cm. — (筆記六編)
Reprint ed.
NT$780.00

I. Chao, Tien-ch'eng, 1683–1756. II. Title. III. Series: Pi chi liu pien.
PL2676.A1 1977 78–843645

Wang, Wei, 701–761.
(Wang Yu sisŏn)
王維詩選 / 李炳漢譯註. — 서울 : 民音社, 1976.
98 p. : ill. ; 19 cm. — (民音社世界詩人選 ; 67)
W350

I. Yi, Pyŏng-han. II. Title. III. Series: Minŭmsa segye siinsŏn ; 67.
PL2676.A6 1976 78–826645

Wei, Tzu-yün, 1918–
(Ching wa chi)
井蛙集 / 魏子雲著. — 臺北 : 大林出版社, 民國67 [1978]
5, 159 p. ; 20 cm. — (大林文庫 ; 137)
Series romanized: Ta lin wen k'u.
NT$35.00

I. Title.
PL2823.E4C5 78–843535

Wei, Tzu-yün, 1918–
(Wu shu chi)
梧鼠集 / 魏子雲著. — 臺北 : 大林出版社, [1978?]
4, 162 p. ; 19 cm. — (大林文庫 ; 138)
Series romanized: Ta lin wen k'u.
NT$35.00

I. Title.
PL2823.E4W8 79–835047

Wen, K'o-shih.
(K'uai le ti chiao pu)
快樂的脚步 / 文可式編. — 台北 : 哲志出版社, 民國67 [1978]
162 p. : ill. ; 19 cm.
Literature.
NT$35.00

I. Title.
PL2920.E666K8 79–835388

Wen, T'ung, 1018–1079.
[Tan yüan chi]
新刻石室先生丹淵集 / [文同撰]. — 臺北 : 臺灣學生書局, 民國62 [1973]
3 v. (10, 986, 4 p.) ; 22 cm. — (歷代畫家詩文集 ; 23)
Running title: Tan yüan chi.
"國立中央圖書館藏本"
Reprint of the 1612 ed. published in 40 chüan, shih i 2 chüan, and fulu 1 chüan.
I. Title. II. Title: Tan yüan chi. III. Series: Li tai hua chia shih wen chi ; 23.
Title romanized: Hsin k'o Shih-shih hsien sheng Tan yüan chi.
PL2687.W43 1973 78–841202

(Wen yen san wen ti p'u t'ung hua fan i)
文言散文的普通话翻译 / 于在春翻译. — 上海 : 上海教育出版社 : 新華書店上海发行所发行, 1978.
4, 3, 250 p. ; 19 cm.
Title in Pinyin on cover: Wenyan sanwen de putonghua fanyi.
$0.56

1. Chinese essays. I. Yü, Tsai-ch'un. II. Title: Wenyan sanwen de putonghua fanyi.
PL2607.W46 79–839470

Weng, Fu-lai.
(Wo hsin shen ch'u)
我心深處 / 翁福來著. — 嘉義 : 南北出版社, 民國67 [1978]
190 p. ; 22 cm.
Cover title.
"一九七七年集紀念"
NT$60.00

I. Title.
PL2920.E6767W6 79–841405

Wu, Cheng-jen, 1956–
(Pai yün p'iao p'iao)
白雲飄飄 / 吳正任著. — 臺北 : 彩虹出版社, 民國67 [1978]
192 p. ; 20 cm. — (彩虹叢書 ; 124)
"青年最佳散文"
Series romanized: Ts'ai hung ts'ung shu.
NT$30.00

I. Title.
PL2921.C43P3 78–843542

CHINESE LITERATURE (Continued)

Wu, Chia-ch'ing, 1901-
(Chia-ch'ing ts'ung kao)
稼青叢稿 [伍受真(稼青)著 臺北 臺灣中華書局經售 1968]
208 p. 19 cm.

I. Title.
PL2921.C46C5 C 68-237
rev

Wu, Chih-hsiung.
(Ch'iao jan hui shou hua i hsieh)
悄然回首花已謝 / 吳志雄著. — 台南：綜合出版社, 民國67 [1978]
6, 3, 192 p. : port. ; 20 cm. — (綜合文叢 ; 13)
Literature.
Series romanized: Tsung ho wen ts'ung.
NT$40.00

I. Title.
PL2921.C553C48 79-838039

Wu, Chih-hsiung.
(Hsing hsing ti hsin yüan)
星星的心願 / 吳志雄著. — 台南：綜合出版社, 民國67 [1978]
4, 6, 203 p. ; 20 cm. — (綜合文叢 ; 14)
Literature.
Series romanized: Tsung ho wen ts'ung.
NT$40.00

I. Title.
PL2921.C553H7 79-839456

Wu, Ching-tzu, 1701-1754.
(Ju lin wai shih)
儒林外史：大字足本古典文學 / 吳敬梓著. — [台南]：利大出版社：總經銷第一書店, [1977?]
4, 496 p. ; 20 cm.
"五十六回"
NT$42.00

I. Title.
PL2732.U22J8 1977b 78-842519

Wu, Ching-tzu, 1701-1754.
(Ju lin wai shih)
儒林外史 / 吳敬梓著. — 臺北：聯經出版事業公司, 民國67 [1978]
4, 4, 550 p., 4 leaves ; 22 cm. — (中國古典小說新刊)
Reprint ed.
NT$100.00

I. Title. II. Series: Chung-kuo ku tien hsiao shuo hsin k'an.
PL2732.U22J8 1978b 79-835816

Wu, Cho-liu, 1900-1976.
[Yoakemae no Taiwan. Chinese]
黎明前的台灣 / [著者吳濁流；編者張良澤]. — 台北：遠行出版社, 民國66 [1977]
43, 214 p. ; 19 cm. — (吳濁流作品集 ; 5) (遠行叢刊 ; 29)
Series 2 romanized: Yüan hsing ts'ung k'an.
CONTENTS: 歷史很多漏洞—我最景仰的偉人—蘿蝠星的詩與人—北埔事件抗日烈士蔡清琳—回顧日據時代的臺灣文學—黎明前的臺灣—日本應往何處去—臺灣文學的現狀—新文學運動的氛圍氣—有關文化的雜感—二—漫談文化沙漠的文化—光復二十年來的感想—忘卻歌唱的金絲雀—素富貴, 行乎富貴—川端康成的絃外之音—別人無份的世界猶之乎熄火山
NT$45.00

I. Chang, Liang-tse, 1939- II. Title. III. Series: Wu, Cho-liu, 1900-1976. Wu Cho-liu tso p'in chi ; 5.
Title romanized: Li ming ch'ien ti T'ai-wan.
PL841.5.C5Y612 1977 78-842033

Wu, Po-hsiao.
(Yü shu)
羽書 / 吳伯蕭著. — [香港?]：創作書社, [1978?]
ii, 118 p. ; 19 cm.
Cover title.
Essays.
Reprint of the 1941 ed. published by Wen hua sheng huo ch'u pan she, Shanghai, in series: Wen hsüeh ts'ung k'an : Ti ch'i chi.

I. Title.
PL2921.P6Y8 1978 79-835854

Wu, Tung-ch'üan.
(Jen yen hsiao p'in)
人言小品 / 吳東權著. — 臺北：水芙蓉出版社, 民國67 [1978]
7, 236 p. ; 19 cm. — (水芙蓉書庫 ; 121)
Series romanized: Shui fu jung shu k'u.
NT$45.00

I. Title.
PL2921.T83J4 79-836239

Wu, Yen-mei, 1933-
(Yün shang ti sheng yin)
雲上的聲音 / 司馬中原[吳延玫]著. — 再版. — 臺北：源成文化圖書供應社, 民國65 [1976]
3, 216 p. : 20 cm.
Essays.
Includes bibliographical references.
NT$40.00
I. Title.
PL2921.Y4Y84 78-842386

Yang, Ch'ing-ch'u.
(Pi sheng ti hui hsiang)
筆聲的迴響 / 楊青矗著. — 高雄：敦理出版社, 民國67 [1978]
238 p. ; 19 cm. — (敦理叢刊 ; 7)
Series romanized: Tun li ts'ung k'an.
NT$45.00 ($2.00 U.S.)

I. Title.
PL2922.C52P5 79-836547

Yang, Hung-nung, 1919-
(Ch'ang feng wan li ch'ing)
長風萬里情 / 著者楊俊[弘農]. — 台北：燕京文化事業股份有限公司, 民國66 [1977]
2, 182 p. ; 19 cm.
Colophon title.
Literature.
NT$40.00

I. Title.
PL2825.H8C5 78-843291

Yang, I, 1907-
(Ch'ien t'u ssu chin)
前途似锦 / 欧阳山[楊儀]. — [香港?：s.n., 1978]
76 p. ; 21 cm.
Reprint of the 1955 (1978 printing) ed. published by Kuang-tung jen min ch'u pan she.

I. Title.
PL2922.I3C5 1978 78-842642

Yang, Ling-yeh.
(Chien shan chien shui chi)
見山見水集 / 羊令野著. — 臺北：大林出版社, 民國67 [1978]
5, 205 p. ; 19 cm. — (大林文庫 ; 139)
Essays.
Series romanized: Ta lin wen k'u.
NT$40.00

I. Title.
PL2922.L48C5 79-835070

Yang, Ta-heng.
(T'i-yüan t'an hui)
梯園談薈 / [楊大亨編]. — 台北：新文豐出版公司, 民國67 [1978]
4, 120 p. ; 19 cm. — (零玉碎金集刊 ; 9)
At head of title: Chung-kuo wen hsüeh ming chu.
Literature.
Series romanized: Ling yü sui chin chi k'an.
Reprint ed.
NT$20.00
I. Title.
PL2733.A56T5 1978 79-838215

Yang, Wei, fl. 1973-
(Ho So k'uang hsiang)
何索狂想 / 何索[楊蔚]著. — 臺北：九歌出版社：總經銷爾雅出版社, 民國67 [1978]
224 p. ; 19 cm. — (九歌叢刊 ; 2)
Series romanized: Ch'u ko ts'ung k'an.
NT$45.00

I. Title.
PL2922.W373H62 79-838101

(Yang chia chiang lien huan t'u hua)
楊家將連環圖畫. — [香港]：海鷗出版公司, [1978-
v. : ill. ; 21 cm.
Spine title.
CONTENTS: 第1輯 楊業歸宋 楊七郎打擂 李陵碑 孟良搬兵 穆桂英招親 南北和 楊金花奪帥印.

PL2832.5.Y3 79-840150

Yao, Shou, 1423-1495.
(Ku-an chi hsüan)
穀菴集選 / [姚綬著]. — [台北]：台灣學生書局, [1973]
10, 372, 4 p., [1] fold. leaf of plates : ill. ; 22 cm. — (歷代畫家詩文集 ; 29 : 第三輯)
"十卷附錄二卷...茲據重刻本景印"

I. Title. II. Series: Li tai hua chia shih wen chi ; 29.
PL2698.Y32K8 1973 78-841608

Yeh, Ch'ing-ping.
(I t'ung tien hua)
一通電話 / 葉慶炳著. — 臺北：九歌出版社：總經銷爾雅出版社, 民國67 [1978]
226 p. ; 19 cm. — (His 晚鳴軒散文集 ; 第4集) (九歌文庫 ; 8)
Series 1 romanized: Wan ming hsüan san wen chi. Series 2 romanized: Chiu ko wen k'u.
NT$50.00
I. Title.
PL2924.C54A6 1978, vol. 4 79-835163

Yeh, Ch'ing-ping.
(Shui lai k'an wo)
誰來看我 / 葉慶炳著. — 臺北：九歌出版社：總經銷爾雅出版社, 民國67 [1978]
227 p. ; 19 cm. — (His 晚鳴軒散文集 ; 3) (九歌文庫 ; 4)
Series 1 romanized: Wan ming hsüan san wen chi. Series 2 romanized: Chiu ko wen k'u.
NT$50.00
I. Title.
PL2924.C54A6 1978, vol. 3 78-842514

Yeh, Ch'ing-ping.
(Wan ming hsüan san wen chi)
晚鳴軒散文集 / 葉慶炳著. — 臺北：九歌出版社：總經銷爾雅出版社, 民國 [19
v. ; 19 cm.
CONTENTS: —3. 誰來看我.

I. Title.
PL2924.C54A6 1978 78-842512

CHINESE LITERATURE (Continued)

Yeh-ssu, 1948–
(Shen hua wu ts'an)
神話午餐 / 也斯. — 臺北：洪範書店：總經銷爾雅出版社, 民國67 [1978]
7, 230 p. ; 19 cm. — (洪範文學叢書 ; 26)
Essays.
Series romanized: Hung fan wen hsüeh ts'ung shu.
NT$45.00

I. Title.
PL2924.Z7S5 78–842378

Yen, Chen-kuo.
(Liu-hsia Chih t'ung ch'ih K'ung lao erh)
柳下跖痛斥孔老二 / 改编严振国；绘画沈阳市苏家屯区文化馆宣传组. — 沈阳：辽宁人民出版社, 1974.
[30] p. : col. ill. ; 15 cm.
$0.14

I. Shen-yang shih Su-chia-tun ch'ü wen hua kuan. Hsüan ch'uan tsu. II. Title.
PL2925.C42L5 78–841842

Yen, Yüan-shu, 1933–
(She hui hsieh shih wen hsüeh chi ch'i t'a)
社會寫實文學及其他 / 顏元叔著. — 1版. — 臺北：巨流圖書公司, 民國67 [1978]
2, 282 p. ; 20 cm. — (巨流叢書 ; 21)
Series romanized: Chü liu ts'ung shu.
NT$60.00

I. Title.
PL2925.Y84S5 79–839177

Yen-hsiang.
(Ta hsüeh sheng ti p'ou ch'ing chi)
大學生的剖情記 / 雁翔著. — 臺北：聯亞出版社, 民國66 [1977]
213 p. ; 19 cm. — (聯亞叢刊 ; 24)
Essays.
Series romanized: Lien ya ts'ung k'an.
NT$40.00

I. Title.
PL2925.H76T3 78–841062

Yin, Hsüeh-man.
(Li shih ti ching tzu)
歷史的鏡子 / 尹雪曼著. — 臺北：大林出版社, [1976?]
10, 317 p. ; 19 cm. — (大林文庫 ; 136)
Essays.
Series romanized: Ta lin wen k'u.
NT$50.00

I. Title.
PL2926.N25L5 79–835230

Yü, Ho-fang.
(Yü Ho-fang hsüan chi)
虞和芳選集 / 虞和芳著. — 臺北：星光出版社, 民國68 [1979]
2, 6, 2, 233 p. : 20 cm. — (雙子星叢書 ; 65號)
Literature.
Series romanized: Shuang tzu hsing ts'ung shu.
NT$45.00

I. Title.
PL2928.H58Y8 79–839113

Yü, Jui-lin.
(Ta Han chih sheng)
大漢之聲 / 余瑞霖著. — 台南：鳳凰城圖書公司, 民國67 [1978]
199 p. : 20 cm. — (鳳凰新書 : 202)
Essays.
Series romanized: Feng huang hsin shu.
NT$40.00

I. Title.
PL2928.J86T3 79–835740

Yü, Ta-kang, 1908–1977.
(Yü Ta-kang ch'üan chi)
俞大綱全集. — 臺北：俞大綱基金會：河洛圖書出版社印行, 民國66- [1977-
v. : 21 cm.
CONTENTS: 第1集 寥音閣劇作

I. Title.
PL2831.T35A6 1977 79–838659

Yüan-yüan.
(Hsien hsien chi pi)
閒閒幾筆 / 圓圓. — [香港?]：明窗出版社, 1978.
13, 293 p. ; 19 cm.
Literature.
$6.00 ($2.00 U.S.)

I. Title.
PL2929.A68H75 79–835256

—TO 221 B. C.

see also Chinese classics

Ch'en, Ching-shan.
(Hsien Ch'in wen hsüeh hsüan shih)
先秦文學選釋 / 編註者陳靜珊. — 台北：天山出版社, 民國67- [1978-
v. : 22 cm. — (大專國學參考用書)
Colophon title.
NT$100.00 (v. 1)

1. Chinese literature—To 221 B. C. I. Title. II. Series: Ta chuan kuo hsüeh ts'an k'ao yung shu.
PL2492.C5 79–836393

—TO 221 B. C.—HISTORY AND CRITICISM

Pei-ching ta hsüeh. Chung-kuo yü yen wen hsüeh hsi. Chung-kuo wen hsüeh shih chiao yen shih.
(Hsien Ch'in wen hsüeh shih ts'an k'ao tzu liao)
先秦文學史參考資料 / 北京大學中國文學史教研室選注. — 新1版. — 北京：中華書局：新中華書店北京發行所發行, 1962, 1978 printing.
4, 554 p. : 21 cm.
$1.70
1. Chinese literature—To 221 B. C.—History and criticism. I. Title.
PL2281.P4 1962 79–838720

—CH'IN AND HAN DYNASTIES, 221 B.C.-220 A.D.

Pei-ching ta hsüeh. Chung-kuo yü yen wen hsüeh hsi. Chung-kuo wen hsüeh shih chiao yen shih.
(Liang Han wen hsüeh shih ts'an k'ao tzu liao)
兩漢文學史參考資料 / 北京大學中國文學史教研室選注. — 新1版. — 北京：中華書局：新華書店北京發行所發行, 1962, 1978 printing.
2, 612 p. : 21 cm.
$1.90
1. Chinese literature—Ch'in and Han dynasties, 221 B. C.–220 A. D. I. Title.
PL2281.P43 1962 79–836816

—CH'IN AND HAN DYNASTIES, 221 B. C.—220 A. D.—HISTORY AND CRITICISM

Liu, Chi-hua, 1943–
(Han Wei chih chi wen hsüeh ti hsing shih yü nei jung)
漢魏之際文學的形式與內容 / 劉紀華著. — 台北：世紀書局; 民國67 [1978]
142 p. ; 22 cm.
Bibliography: p. 140–142.
NT$60.00

1. Chinese literature—Ch'in and Han dynasties, 221 B. C.—220 A. D.—History and criticism. 2. Chinese literature—Three kingdoms, 220–265—History and criticism. I. Title.
PL2284.L49 79–840730

—THREE KINGDOMS, 220-265 —HISTORY AND CRITICISM

Liu, Chi-hua, 1943–
(Han Wei chih chi wen hsüeh ti hsing shih yü nei jung)
漢魏之際文學的形式與內容 / 劉紀華著. — 台北：世紀書局; 民國67 [1978]
142 p. ; 22 cm.
Bibliography: p. 140–142.
NT$60.00

1. Chinese literature—Ch'in and Han dynasties, 221 B. C.—220 A. D.—History and criticism. 2. Chinese literature—Three kingdoms, 220–265—History and criticism. I. Title.
PL2284.L49 79–840730

C

—SUI DYNASTY, 581-618 —HISTORY AND CRITICISM

(Sui T'ang Wu tai wen hsüeh p'i p'ing tzu liao hui pien)
隋唐五代文學批評資料彙編 / 羅聯添編輯；國立編譯館主編. — 臺北：成文出版社, 民國67 [1978]
2, 11, 39, 298 p. : 22 cm. — (中國文學批評資料彙編 : 2)
Includes bibliographies.
NT$105.00

1. Chinese literature—Sui dynasty, 581–618—History and criticism. 2. Chinese literature—T'ang dynasty, 618–907—History and criticism.; 3. Chinese literature—Five dynasty, 907–960—History and criticism. I. Lo, Lien-t'ien. II. Series: Chung-kuo wen hsüeh p'i p'ing tzu liao hui pien : 2.
PL2290.S9 79–838689

—T'ANG DYNASTY, 618-907 —HISTORY AND CRITICISM

Liu, Chung-ho.
(T'ang tai wen hsüeh ch'üan chi)
唐代文學全集 / 劉中和著. — 臺北：世界文物出版社：經銷者世界文物供應社, 民國68- [1979-
v. ; 22 cm.
Colophon title: T'ang tai wen hsüeh.
NT$150.00 (v. 1)

1. Chinese literature—T'ang dynasty, 618–907—History and criticism. I. Title.
PL2291.L5 79–838257

(Sui T'ang Wu tai wen hsüeh p'i p'ing tzu liao hui pien)
隋唐五代文學批評資料彙編 / 羅聯添編輯；國立編譯館主編. — 臺北：成文出版社, 民國67 [1978]
2, 11, 39, 298 p. : 22 cm. — (中國文學批評資料彙編 : 2)
Includes bibliographies.
NT$105.00

1. Chinese literature—Sui dynasty, 581–618—History and criticism. 2. Chinese literature—T'ang dynasty, 618–907—History and criticism.; 3. Chinese literature—Five dynasty, 907–960—History and criticism. I. Lo, Lien-t'ien. II. Series: Chung-kuo wen hsüeh p'i p'ing tzu liao hui pien : 2.
PL2290.S9 79–838689

—FIVE DYNASTY, 907-960 —HISTORY AND CRITICISM

(Sui T'ang Wu tai wen hsüeh p'i p'ing tzu liao hui pien)
隋唐五代文學批評資料彙編 / 羅聯添編輯；國立編譯館主編. — 臺北：成文出版社, 民國67 [1978]
2, 11, 39, 298 p. : 22 cm. — (中國文學批評資料彙編 : 2)
Includes bibliographies.
NT$105.00

1. Chinese literature—Sui dynasty, 581–618—History and criticism. 2. Chinese literature—T'ang dynasty, 618–907—History and criticism.; 3. Chinese literature—Five dynasty, 907–960—History and criticism. I. Lo, Lien-t'ien. II. Series: Chung-kuo wen hsüeh p'i p'ing tzu liao hui pien : 2.
PL2290.S9 79–838689

CHINESE LITERATURE (Continued)

—SUNG DYNASTY, 960-1279
—HISTORY AND CRITICISM

(Chin tai wen hsüeh p'i p'ing tzu liao hui pien)
金代文學批評資料彙編 / 林明德編輯；國立編譯館主編. — 臺北：成文出版社，民國68 [1979]
2, 20, 219 p. ; 22 cm. — (中國文學批評資料彙編 ; 5)
"部編大學用書"
NT$95.00

1. Chinese literature—Sung dynasty, 960-1279—History and criticism. I. Lin, Ming-te, 1946- II. Kuo li pien i kuan, T'ai-pei. III. Series: Chung-kuo wen hsüeh p'i p'ing tzu liao hui pien ; 5.
PL2293.C55 79-840540

(Nan Sung wen hsüeh p'i p'ing tzu liao hui pien)
南宋文學批評資料彙編. / 張健編輯. — 台北：成立出版社，民國67 [1978]
2, 11, 580 p. ; 21 cm. — (中國文學批評資料彙編 ; 4)
NT$205.00

1. Chinese literature—Sung dynasty, 960-1279—History and criticism. I. Chang, Chien, 1939- II. Series: Chung-kuo wen hsüeh p'i p'ing tzu liao hui pien ; 4.
PL2293.N36 79-838743

—YÜAN DYNASTY, 1368-1644
—BIBLIOGRAPHY

Liu, Chao-yu, 1937-
(Ssu k'u chu lu Yüan jen pieh chi t'i yao pu cheng)
四庫著錄元人別集提要補正 / 劉兆祐著. — 臺北：私立東吳大學中國學術著作獎助委員會：總經銷臺灣商務印書館，民國67 [1978]
2, 2, 8, 284 p. : ill. ; 22 cm. — (私立東吳大學中國學術著作獎助委員會叢書 ; 78)
Series romanized: Ssu li Tung wu ta hsüeh, Chung-kuo hsüeh shu chu tso chiang chu wei yüan hui ts'ung shu.
Bibliography: p. 206-227.
NT$170.00 ($4.00 U. S.)

1. Chinese literature—Yüan dynasty, 1368-1644—Bibliography. 2. Ssu k'u ch'üan shu—Bibliography. I. Ssu k'u ch'üan shu. II. Title.
Z3108.L5L57 79-838643
[PL2264]

—MING DYNASTY, 1368-1644
—HISTORY AND CRITICISM

Ch'en, Wan-i, fl. 1976-
(Wan Ming hsing ling wen hsüeh ssu hsiang yen chiu)
晚明性靈文學思想研究 / 陳萬益. — [台北：s. n., 1978?]
250 p. (on double leaves) ; 27 cm.
Spine title.
Reproduced from ms. copy.
Bibliography: p. 234-250.

1. Chinese literature—Ming dynasty, 1368-1644—History and criticism. I. Title.
PL2296.C46 1978 78-842958

—MING DYNASTY, 1368-1644
—HISTORY AND CRITICISM
—ADDRESSES, ESSAYS, LECTURES

Iriya, Yoshitaka, 1910-
(Mindai shibun)
明代詩文 / 入矢義高. — 東京：筑摩書房，1978.
268 p., [1] leaf of plates : facsims. ; 20 cm. — (中国詩文選 ; 23)
Ja 78-10554
Bibliography: p. 266-268.
¥1500

1. Chinese literature—Ming dynasty, 1368-1644—History and criticism—Addresses, essays, lectures. I. Title. II. Series: Chūgoku shibun sen ; 23.
PL2296.I7 78-814622

Yokota, Terutoshi, 1927-
(Mindai bungaku ron no tenkai)
明代文学論の展開 / 横田輝俊. — [広島：広島大学文学部]，1977.
v. ; 21 cm.
Ja***
cover title.
Errata slip inserted.
Includes bibliographical references.
"広島大学文学部紀要第三七巻特輯号"
1. Chinese literature—Ming dynasty, 1368-1644—History and criticism—Addresses, essays, lectures. I. Hiroshima Daigaku. Bungakubu. Hiroshima Daigaku Bungakubu kiyō. II. Title.
PL2296.Y6 78-815458

—CH'ING DYNASTY, 1644-1912

(Huang Ch'ing wen ying)
皇清文穎 / 張廷玉等編輯]. — [台北：台灣]商務[印書館，1978?]
24 v. ; 20 cm. — (四庫全書珍本八集 ; 351-374)
Cover title.
Photoreprint ed. of the Wen Yüan Ko copy.

1. Chinese literature—Ch'ing dynasty, 1644-1912. I. Chang, T'ing-yü, 1672-1755. II. Series: Ssu k'u ch'üan shu. Ssu k'u ch'üan shu chen pen pa chi ; 351-374.
AC149.S7 1978, vol. 351-374 78-843776
[PL2507]

—CH'ING DYNASTY, 1644-1912
—HISTORY AND CRITICISM

Chang, Jung-hui.
(Ch'ing tai T'ung-ch'eng p'ai wen hsüeh chih yen chiu)
清代桐城派文學之研究 / 張榮輝. — [台北?：s. n., 1978?]
3 v. (3 p. on double leaves, 294 double leaves) ; 27 cm.
Reproduced from typescript; with t. p. in ms.
Originally presented as the author's thesis (M.A.), Kuo li cheng chih ta hsüeh, 1966.

1. Chinese literature—Ch'ing dynasty, 1644-1912—History and criticism. I. Title.
PL2297.C47 1978 78-843933

Fu tan ta hsüeh, Shanghai. Chung-kuo yü yen wen hsüeh hsi. I chiu wu liu chi. Chung-kuo chin tai wen hsüeh shih pien hsieh hsiao tsu.
(Chung-kuo chin tai wen hsüeh shih kao)
中國近代文學史稿 / 復旦大學中文系1956級中國近代文學史編寫小組編著. — 第1版. — 香港：達文社，1978.
2, 400 p. ; 21 cm.
Reprint of the 1960 ed. published by Chung-hua shu chü, Shanghai.
1. Chinese literature—Ch'ing dynasty, 1644-1912—History and criticism. 2. Chinese literature—20th century—History and criticism. I. Title.
PL2297.F8 1978 79-839472

Yang, Shih-chi.
(Wen yüan t'an wang)
文苑談往 / 楊世驥撰. — 台1版. — 台北：華世出版社，民國67 [1978]
3, 4, 170 p. ; 22 cm.
"附錄 晚清文學史話": p. 165-170.
NT$60.00

1. Chinese literature—Ch'ing dynasty, 1644-1912—History and criticism. I. Title.
PL2297.Y3 79-837912

—CH'ING DYNASTY, 1644-1912
—HISTORY AND CRITICISM
—ADDRESSES, ESSAYS, LECTURES

Chao, Yu-p'ei.
(Chung-kuo chin pai nien lai ti wen hsüeh)
中國近百年來的文學 / 趙友培講；教育部社會教育司編印. — [台北]：該司，民國65 [1976]
24 p. : port. ; 20 cm. — (文化講座專集 ; 77)

1. Chinese literature—Ch'ing dynasty, 1644-1912—History and criticism—Addresses, essays, lectures. 2. Chinese literature—20th century—History and criticism—Addresses, essays, lectures. I. Title. II. Series: Wen hua chiang tso chuan chi ; 77.
PL2297.C52 78-841149

—19TH CENTURY

I ching.
易經精華 / 著作者薛悟邨. — 台北：傳統書局，民國67 [1978]
240 leaves in various foliations ; 27 cm.
Colophon title.
Reprint of the 1821 ed. published in 6 chüan and chüan mo 1 chüan.
NT$240.00

1. Hsüeh, Chia-ying, 19th cent. II. Title.
Title romanized: I ching ching hua.
PL2464.K5 1978 79-835423

—20TH CENTURY

(Ch'i yüeh liu huo)
七月流火 / 李磬 ... [et al.；仙枝 ... et al. 主編]. — 臺北：皇冠雜誌社，民國67 [1978]
256 p. : ill. ; 20 cm. — (三三集刊 ; 第16輯)
Series romanized: San san chi k'an.
NT$40.00

1. Chinese literature—20th century. 2. Chinese literature—20th century—History and criticism. I. Li, Ch'ing, 1906- II. Hsien, Chih, 1953-
PL2513.C519 79-838745

(Chien men)
劍門 / 朱西寧 ... [et al.；馬叔禮 ... et al. 主編]. — 臺北：皇冠雜誌社，民國66 [1977]
253 p. ; 20 cm. — (三三集刊 ; 第8輯)
Series romanized: San san chi k'an.
NT$40.00

1. Chinese literature—20th century. I. Chu, Hsi-ning, 1927- II. Ma, Shu-li.
PL2513.C523 79-836520

(Ch'uang tsao jih hui k'an)
創造日彙刊 / 創造社編. — [香港?：s. n., 1978?]
6, 1, 2, 507 p. ; 19 cm.
Reprint of the 1929 ed. published by Kuang hua shu chü, Shanghai.

1. Chinese literature—20th century. I. Ch'uang tsao she.
PL2512.C45 1978 79-836180

(Chung-kuo hsin wen i ta hsi)
中國新文藝大系. — 九龍：百靈出版社，[1978?-
v. ; 19 cm.
Cover title.
Reprint of the 1976 ed. published by Ta han ch'u pan she, T'ai-pei, which was issued as no. 1- of Ta han hsin k'an, with preface by Hsing Kuang-tsu deleted.
CONTENTS: 1-2集 散文 郁達夫主編.—

1. Chinese literature—20th century. I. Yü, Ta-fu, 1896-1945.
PL2512.C48 1978 78-841613

(Chung-kuo ti nu hou)
中國的怒吼 / 周錦選編. — 臺北：智燕出版社：總經銷衆文圖書股份有限公司，民國68 [1979]
250 p. : ill. ; 19 cm. — (中國新文學創作叢刊 ; 2)
A collection of articles selected from various newspapers and periodicals published before 1949.
NT$50.00

1. Chinese literature—20th century. 2. Sino-Japanese Conflict, 1937-1945—Literary collections. I. Chou, Chin, fl. 1973- II. Series: Chung-kuo hsin wen hsüeh ch'uang tso ts'ung k'an ; 2.
PL2512.C492 79-840563

(I po)
衣缽 / 履彊 ... [et al.；仙枝 ... et al. 主編]. — 臺北：皇冠雜誌社，民國67 [1978]
255 p. ; 19 cm. — (三三集刊 ; 第11輯)
Series romanized: San san chi k'an.
NT$40.00

1. Chinese literature—20th century. 2. Chinese literature—20th century—History and criticism. I. Lü, Chiang, 1953- II. Hsien, Chih, 1953-
PL2513.I25 79-836077

CHINESE LITERATURE
—20TH CENTURY (Continued)

(Jih ch'u hsi shan yü)
日出西山雨 / 楊日旭 ... [et al.]. — 臺北：皇冠雜誌社，民國67 [1978]
253 p. : ill. ; 20 cm. — (三三集刊；第15輯)
Series romanized: San san chi k'an.
NT$40.00

1. Chinese literature—20th century. 2. Chinese literature—20th century—History and criticism. I. Yang, Jih-hsü.
PL2513.J5 79-837990

(Lo chiang ch'ien ti shou shih)
落江前的手勢 / 朱西寧 ... [et al.]. — 臺北：皇冠雜誌社，民國67 [1978]
258 p. : ill. ; 20 cm. — (三三集刊；第9輯)
Series romanized: San san chi k'an.
NT$40.00

1. Chinese literature—20th century. I. Chu, Hsi-ning, 1927-
PL2513.L6 78-842940

(Lu hsiao)
盧笑 / 朱西寧 ... [et al. ; 馬叔禮 ... et al. 主編]. — 臺北：皇冠雜誌社，民國66 [1977]
251 p. : ill. ; 19 cm. — (三三集刊；第7輯)
Series romanized: San san chi k'an.
NT$40.00

1. Chinese literature—20th century. I. Chu, Hsi-ning, 1927- II. Ma, Shu-li, 1910-
PL2513.L8 78-843590

(Man tso i kuan ssu hsüeh)
滿座衣冠似雪 / 温瑞安等著. — 臺北：皇冠出版社，民國67 [1978]
268 p. ; 19 cm. — (神州文集；1)
Series romanized: Shen chou wen chi.
NT$40.00

1. Chinese literature—20th century. I. Wen, Jui-an. II. Series: Shen chou wen chi ; 1.
PL2513.M27 79-835405

(Shui shih)
水勢 / 郭為藩 ... [et al. ; 仙枝 ... et al. 主編]. — 臺北：皇冠雜誌社，民國68 [1979]
247 p. : ill. ; 20 cm. — (三三集刊；第18輯)
Series romanized: San san chi k'an.
NT$40.00

1. Chinese literature—20th century. 2. Chinese literature—20th century—History and criticism—Addresses, essays, lectures. I. Kuo, Wei-fan. II. Hsien, Chih, 1953-
PL2513.S55 79-840159

(Tung feng wan li)
东风万里 / 山东省革委文化局创作办公室. — [香港?：s. n., 1978?]
4, 336 p. ; 22 cm.
Reprint of the 1977 ed. published by Shan-tung jen min ch'u pan she, Tsinan.

1. Chinese literature—20th century. I. Shantung, China. Ko ming wei yüan hui. Wen hua chü. Ch'uang tso pan kung shih.
PL2513.T85 1978 79-835708

(Wei hsi ti i hsien hsiang)
微細的一線香 / 主編張德本，許素蘭. — 臺北：鴻蒙文學出版有限公司：總經銷故鄉出版社，民國67 [1978]
341 p. : ill. ; 22 cm. — (前衛文學叢刊；第1輯)
Colophon title.
NT$60.00

1. Chinese literature—Taiwan. 2. Chinese literature—20th century. I. Chang, Te-pen. II. Hsü, Su-lan. III. Series: Ch'ien wei wen hsüeh ts'ung k'an ; ti 1 chi.
PL3031.T32W4 79-836069

Wu-han shih ti san mien fang chih ch'ang. Kung jen hsieh tso tsu.
(Sha ting nu hou)
纱锭怒吼 / 武汉市第三棉纺织厂工人写作组. — 第1版. — [漢口]：湖北人民出版社，1977.
316 p. : ill. ; 19 cm. — (厂史故事集)
Fiction.
Series romanized: Li shih ku shih chi.
$0.60

1. Short stories, Chinese. 2. Chinese fiction—20th century. 3. Textile workers—China—Fiction. I. Title.
PL2653.W74 1977 79-835946

—20TH CENTURY—BIBLIOGRAPHY

Iida, Kichirō, 1922- ed.
(Gendai Chūgoku bungaku kenkyū bunken mokuroku)
現代中国文学研究文献目録：1919-1945 / 飯田吉郎編. — 東京：中国文化研究会：発売所「大安」，1961?
87, [7] p. ; 26 cm.
On cover: A bibliography for the study of modern Chinese literature in Japan.
Reprint of the 1959 ed.; with new suppl.
Includes indexes.
1. Chinese literature—20th century—Bibliography. 2. Chinese literature—20th century—Translations into Japanese—Bibliography. 3. Japanese literature—20th century—Translations from Chinese—Bibliography. I. Title. II. Title: A bibliography for the study of modern Chinese literature in Japan.
Z3108.L5I4 1961 78-814744
[PL2602]

—20TH CENTURY—BIO-BIBLIOGRAPHY
—INDEXES

Li, Li-ming.
(Chung-kuo hsien tai liu pai tso chia hsiao chuan tzu liao so yin)
中國現代六百作家小傳資料索引 / 李立明編著. — 香港：波文書局，1978.
3, 21, 319 p., [1] leaf of plates : port. ; 20 cm.

1. Li, Li-Ming. Chung-kuo hsien tai liu pai tso chia hsiao chuan—Indexes. 2. Chinese literature—20th century—Bio-bibliography—Indexes. I. Li, Li-ming. Chung-kuo hsien tai liu pai tso chia hsiao chuan. II. Title.
Z3108.L5L382 1978 79-835674
[PL2264]

—20TH CENTURY—HISTORY AND CRITICISM

(Ch'i yüeh liu huo)
七月流火 / 李磬 ... [et al. ; 仙枝 ... et al. 主編]. — 臺北：皇冠雜誌社，民國67 [1978]
256 p. : ill. ; 20 cm. — (三三集刊；第16輯)
Series romanized: San san chi k'an.
NT$40.00

1. Chinese literature—20th century. 2. Chinese literature—20th century—History and criticism. I. Li, Ch'ing, 1906- II. Hsien, Chih, 1953-
PL2513.C519 79-838745

Chia, Yang.
(Hsüeh yüan, t'u ti, ch'uan t'ung)
血緣，土地，傳統 / 之華[嘉揚]著. — 台北：求精出版社：總經銷精益書局，1977.
4, 3, 344 p., [1] leaf of plates : port. ; 19 cm. — (精益叢書；5)
In colophon: Blood, land, tradition, by Chi-hua.
Series romanized: Ching i ts'ung shu.
NT$60.00
1. Chinese literature—Taiwan—History and criticism. 2. Chinese literature—20th century—History and criticism. 3. Authors, Chinese—Anecdotes, facetiae, satire, etc. I. Title. II. Title: Blood, land, tradition.
PL3031.T3C46 78-841244

Chu, Hsi-ning, 1927-
(Ch'ü li p'ien)
曲理篇 / 朱西寧著. — 大甲鎮：慈龍文化有限公司：總經銷文星書店，民國67 [1978]
253 p. ; 20 cm. — (慈龍書系；316)
Essays.
Series romanized: Hui lung shu hsi.
NT$48.00

1. Chinese literature—20th century—History and criticism. I. Title.
PL2303.C438 79-837943

Fang, Hsiu.
Notes on the history of Malayan Chinese new literature, 1920-1942 / [by] Fang Xiu ; translated by Angus W. McDonald ; [edited by Kazuo Enoki]. — Tokyo : Centre for East Asian Cultural Studies, [1977]
xiv, 352 p. ; 20 cm. — (East Asian cultural studies series ; no. 18)
Ja78-39
A translation with revisions of Ma Hua hsin wen hsüeh shih kao.
Notes, indexes, glossary, maps, and errata (71 p.) tipped in.
1. Chinese literature—Malaysia—History and criticism. 2. Chinese literature—20th century—History and criticism. I. Enoki, Kazuo, 1913- II. Title. III. Series.
PL3097.M3F313 895.1'099595 79-347498
79 MARC

Fu tan ta hsüeh, Shanghai. Chung-kuo yü yen wen hsüeh hsi. Hsien tai wen hsüeh tsu.
(Chung-kuo hsien tai wen hsüeh shih)
中国现代文学史：1919-1942 / 复旦大学中文系現代文学組編著. — 香港：神州圖書公司，[1978?]
2 v. (vii, 600 p.) ; 19 cm.
Cover title.
Reprint of the ed. published by Shang-hai wen i ch'u pan she.
Includes bibliographical references.

1. Chinese literature—20th century—History and criticism. I. Title.
PL2302.F86 1978 79-836022

Fu tan ta hsüeh, Shanghai. Chung-kuo yü yen wen hsüeh hsi. I chiu wu liu chi. Chung-kuo chin tai wen hsüeh shih pien hsieh hsiao tsu.
(Chung-kuo chin tai wen hsüeh shih kao)
中國近代文學史稿 / 復旦大學中文系1956級中國近代文學史編寫小組編著. — 第1版. — 香港：達文社，1978.
2, 400 p. ; 21 cm.
Reprint of the 1960 ed. published by Chung-hua shu chü, Shanghai.
1. Chinese literature—Ch'ing dynasty, 1644-1912—History and criticism. 2. Chinese literature—20th century—History and criticism. I. Title.
PL2297.F8 1978 79-839472

(I po)
衣缽 / 履彊 ... [et al. ; 仙枝 ... et al. 主編]. — 臺北：皇冠雜誌社，民國67 [1978]
255 p. ; 19 cm. — (三三集刊；第11輯)
Series romanized: San san chi k'an.
NT$40.00

1. Chinese literature—20th century. 2. Chinese literature—20th century—History and criticism. I. Lü, Chiang, 1953- II. Hsien, Chih, 1953-
PL2513.I25 79-836077

(Jih ch'u hsi shan yü)
日出西山雨 / 楊日旭 ... [et al.]. — 臺北：皇冠雜誌社，民國67 [1978]
253 p. : ill. ; 20 cm. — (三三集刊；第15輯)
Series romanized: San san chi k'an.
NT$40.00

1. Chinese literature—20th century. 2. Chinese literature—20th century—History and criticism. I. Yang, Jih-hsü.
PL2513.J5 79-837990

Pai, Hsien-yung, 1937-
(Mo jan hui shou)
驀然回首 / 白先勇著. — 臺北：爾雅出版社，民國67 [1978]
184 p. : port. ; 20 cm. — (爾雅叢書；43)
Series romanized: Erh ya ts'ung shu.
Appendices (p. 165-184): 林懷民 白先勇回家.—白先勇寫作年表
NT$45.00.

1. Chinese literature—20th century—History and criticism. I. Title.
PL2303.P3 79-838157

Ragvald, Lars, 1942-
Yao Wenyuan as a literary critic and theorist : the emergence of Chinese Zhdanovism / by Lars Ragvald. — Stockholm : Univ., 1978.
ix, 225 p. ; 24 cm. S78-13
Thesis—Stockholm.
Extra t.p. with thesis statement inserted.
Bibliography: p. 214-225.
ISBN 9171460098
1. Chinese literature—20th century—History and criticism. 2. Yao, Wen-yüan. 3. Literature and state—China. I. Title.
PL2303.R34 895'.1'09005 78-398757
79 MARC

CHINESE LITERATURE —20TH CENTURY—HISTORY AND CRITICISM (Continued)

Shih, Tzu-ch'en.
(Wen hsüeh jen)
文學人 / 史紫忱著. — 臺北：星光出版社, 民國67 [1978]
306 p. ; 19 cm. — (雙子星叢書 ; 69號)
Series romanized: Shuang tzu hsing ts'ung shu.
NT$50.00

1. Chinese literature—History and criticism. 2. Literature. I. Title.

PL2264.S55 79-837147

Ssu-ma, Ch'ang-feng.
(Chung-kuo hsin wen hsüeh shih)
中國新文學史 / 司馬長風著. — 香港：昭明出版社, 1975-1978.
3 v. ; 23 cm.
Includes bibliographical references.
$20.00 (v. 1) varies.

1. Chinese literature—20th century—History and criticism. I. Title.

PL2302.S8 75-837124

Ts'ai, I-chung.
[Ts'ung Ch'en Tu-hsiu ti wen hsüeh ko ming tao Li Chin-fa ti hsiang cheng p'ai hsin shih]
從陳獨秀到李金髮 / 蔡義忠著. — 台北：清流出版社, 民國66 [1977]
206 p. ; 22 cm.
"包括大陸作家鄧拓等燕山夜話"
Previously published in 1973 under title: Ts'ung Ch'en Tu-hsiu ti wen hsüeh ko ming tao Li Chin-fa ti hsiang cheng p'ai hsin shih.
NT$35.00
1. Chinese literature—20th century—History and criticism. 2. Authors, Chinese—20th century. I. Title.
Title romanized: Ts'ung Ch'en Tu-hsiu tao Li Chin-fa.

PL2302.T8 1977 78-840927

(Wen i tso p'in fen hsi)
文藝作品分析 / 姚重等著. — [香港?]：半島書樓, [1978?]
ix, 173 p. ; 19 cm. — (雋思文化叢書)
Cover title.
Series romanized: Chün ssu wen hua ts'ung shu.
Reprint of the 1953 ed. published by Hsiang-kang hsüeh sheng shu tien, Hongkong.

1. Chinese literature—20th century—History and criticism. 2. Chinese literature—Study and teaching. I. Yao, Chung.

PL2303.W47 1978 79-835906

Yü, Lei.
(Hsien tai Chung-kuo wen hsüeh ming chu p'ing hsi)
現代中國文學名著評析 / 于蕾編. — 香港：萬源圖書公司, 1978.
2, 208 p. ; 19 cm. — (自學知識文庫)
Series romanized: Tzu hsüeh chih shih wen k'u.
$6.00

1. Chinese literature—20th century—History and criticism. I. Title.

PL2303.Y75 79-836074

—20TH CENTURY —HISTORY AND CRITICISM —ADDRESSES, ESSAYS, LECTURES

Chao, Yu-p'ei.
(Chung-kuo chin pai nien lai ti wen hsüeh)
中國近百年來的文學 / 趙友培講 ; 教育部社會教育司編印. — [台北]：該司, 民國65 [1976]
24 p. : port. ; 20 cm. — (文化講座專集 ; 77)

1. Chinese literature—Ch'ing dynasty, 1644-1912—History and criticism—Addresses, essays, lectures. 2. Chinese literature—20th century—History and criticism—Addresses, essays, lectures. I. Title. II. Series: Wen hua chiang tso chuan chi ; 77.

PL2297.C52 78-841149

(Chungguk sin munhak p'yŏngnon sŏn)
中國新文學評論選 / 車柱環註釋. — 서울：新雅社, 1976.
131 p. ; 22 cm.
₩1200

1. Chinese literature—20th century—History and criticism—Addresses, essays, lectures. I. Ch'a, Chu-hwan.

PL2262.C64 78-826421

(Hsien tai wen hsüeh tso p'in hsüan chiang)
現代文学作品选讲 / 吉林师范大学中文系文选写作教研室. — 长春：吉林人民出版社, 1978.
317 p. ; 19 cm.
CONTENTS: 毛泽东 蝶恋花·答李淑一.—賀敬之 中国的十月.—鲁迅 祝福.—蒋子龙 三个起重工.—袁鹰 青山翠竹.—陈洪山 塞北鱼米乡.—大庆油田工人写作组 铁人的故事.—尚华, 徐太国 昔阳人民胜天図.—魏巍 谁是最可爱的人.—恪牛 柿子红了.
$0.61
1. Chinese literature—20th century—History and criticism—Addresses, essays, lectures. I. Chi-lin shih fan ta hsüeh, Ch'ang-ch'un, China. Wen hsüan hsieh tso chiao yen shih.
PL2303.H73 79-835192

Hsüeh-wei, pseud.
(Liang chien chi)
兩間集 / 雪葦著. — 香港：神州圖書公司, [1978?]
227 p. ; 19 cm.
"文藝理論"
Reprint of the 1951 ed. published by Hsin wen i ch'u pan she, Shanghai.
1. Chinese literature—20th century—History and criticism—Addresses, essays, lectures. 2. Creative writing—Addresses, essays, lectures. I. Title.

PL2302.H78 1978 79-836023

I-ch'ün, pseud.
(Wu ssu wen hsüeh ko ming yün tung ti chen mien mu)
「五四」文學革命運動的真面目 / 以群. — 九龍：富壤書房, 1973.
117 p. ; 21 cm.
"根據1963年作家出版社「文學問題漫論」一書第一部份重印"
Includes bibliographical references.
1. Chinese literature—20th century—History and criticism—Addresses, essays, lectures. 2. May fourth movement—Addresses, essays, lectures. I. I-ch'ün, pseud. Wen hsüeh wen t'i man lun. II. Title.

PL2302.I2 78-841596

Juan, Wu-ming.
(Chung-kuo hsin wen t'an mi lu)
中國新文壇秘錄 / 阮无名編. — Washington : Center for Chinese Research Materials, Association of Research Libraries, [1978?]
2, 3, 289 p. on [148] leaves ; 22 x 28 cm.
Reprint of the 1933 ed. published by Nan ch'iang shu chü, Shanghai.

1. Authors, Chinese—20th century—Anecdotes, facetiae, satire, etc. 2. Chinese literature—20th century—History and criticism—Addresses, essays, lectures. I. Title.

PL2277.J84 1978 79-840796

(K'ang chan wen i t'ung hsün)
抗戰文藝通訊 / 張葉舟編. — 香港：神州圖書公司, [1978?]
4, 9, 401 p. ; 19 cm.
Reprint of the 1939 ed. published by Kuo chi ch'u pan she, Shanghai.

1. Sino-Japanese Conflict, 1937-1945—Literary collections. 2. Chinese literature—20th century—History and criticism—Addresses, essays, lectures. I. Chang, Yeh-chou.

DS777.53.K2733 1978 78-842179

Li, Chien-wu, 1908?-
(Chu hua chi)
咀華集 ; 咀華二集 / 劉西渭 [李健吾]. — [香港?]：創作書社, [1978?]
375 p. in various pagings ; 19 cm.
Cover title.
Reprint of the 1947-1948 eds. published by Wen hua sheng huo ch'u pan she, Shanghai; issued in series: Wen hsüeh ts'ung k'an; ti 3, 7 chi.
1. Chinese literature—20th century—History and criticism—Addresses, essays, lectures. I. Li, Chien-wu, 1908?- Chu hua erh chi. 1978. II. Title.

PL2302.L454 78-841974

Li, Chien-wu, 1908?-
(Chü hua erh chi)
咀華二集 劉西渭 [李健吾] 作 上海 文化生活出版社 民國36 [1947]
163p. 17 cm. (文學叢刊)

1. Chinese literature—20th century—History and criticism—Addresses, essays, lectures. I. Title.

PL2302.L455 C 60-3180 ‡

Li, Ho-lin.
(Kuan yü Chung-kuo hsien tai wen hsüeh)
关于中國現代文学 / 李何林著. — Washington : Center for Chinese Research Materials, Association of Research Libraries, [1978]
v, 124 p. on [65] leaves ; 22 x 28 cm.
Reprint of the 1956 ed. published by Hsin wen i ch'u pan she, Shanghai, with an English annotation at head of work.
$7.60
1. Chinese literature—20th century—History and criticism—Addresses, essays, lectures. I. Title.
PL2302.L465 1978 895.1'09'005 79-840623

Li, Kuang-t'ien, 1907-
(Lun wen hsüeh chiao yü)
論文学教育 / 李廣田. — 港1版. — [香港]：創作書社, 1977.
2, 2, 199 p. ; 19 cm.
"據一九五〇年文化工作社版重印"
$8.00

1. Chinese literature—20th century—History and criticism—Addresses, essays, lectures. 2. Literature—Addresses, essays, lectures. I. Title.

PL2303.L44 1977 79-835309

Ogaeri, Yoshio, 1910-1966.
(Tairiku no gengo to bungaku)
大陸の言語と文學 / 魚返善雄著. — 東京：三省堂, 昭和15 [1940]
2, 9, 260 p., [6] leaves of plates : ill. ; 19 cm.

1. Chinese language—Addresses, essays, lectures. 2. Chinese literature—20th century—History and criticism—Addresses, essays, lectures. I. Title.

PL1071.O33 78-816297

Shao, Ch'üan-lin.
(Chung-kuo hsien tai wen hsüeh tso p'in hsüan chiang)
中國現代文學作品選講 / [邵]荃麟, 葛琴編講. — 香港：創作書社, 1978.
365 p. ; 19 cm.
$12.00

1. Chinese literature—20th century—History and criticism—Addresses, essays, lectures. I. Ko, Ch'in, 1910- joint author. II. Title.

PL2302.S52 79-840120

(Shui shih)
水勢 / 郭為藩 ... [et al. ; 仙枝 ... et al. 主編]. — 臺北：皇冠雜誌社, 民國68 [1979]
247 p. : ill. ; 20 cm. — (三三集刊 ; 第18輯)
Series romanized: San san chi k'an.
NT$40.00

1. Chinese literature—20th century. 2. Chinese literature—20th century—History and criticism—Addresses, essays, lectures. I. Kuo, Wei-fan. II. Hsien, Chih, 1953-

PL2513.S55 79-840159

(T'an shu chi)
談書集 / 著者文壽, 胡品清等. — 台北：大漢出版社, 民國66 [1977]
3, 126 p. : 19 cm. — (大漢文庫 ; 8)
Colophon title.
Series romanized: Ta Han wen k'u.
Includes bibliographical references.
NT$25.00
1. Chinese literature—20th century—History and criticism—Addresses, essays, lectures. I. Chao, Tzu-fan. II. Hu, P'in-ch'ing, 1920-

PL2302.T34 79-835634

CHINESE LITERATURE —20TH CENTURY—HISTORY AND CRITICISM—ADDRESSES, ESSAYS, LECTURES (Continued)

Ts'ai, Yüan-huang.
(Chi mo ti chieh)
寂寞的結 / 蔡源煌著. — 臺北：聯經出版事業公司, 民國67 [1978]
2, 6, 272 p. ; 20 cm. — (文化叢刊)
NT$50.00

1. Chinese literature—20th century—History and criticism—Addresses, essays, lectures. 2. Literature, Modern—20th century—History and criticism—Addresses, essays, lectures. I. Title. II. Series: Wen hua ts'ung k'an.
PL2303.T66 79–838682

Wen, Jui-an.
(Hui shou mu yün yüan)
回首暮雲遠 / 温瑞安著. — 第1版. — 台北：四季出版事業有限公司, 民國66 [1977]
2, 251 p. ; 20 cm. — (四季文萃 ; 15)
Series romanized: Ssu chi wen ts'ui.
NT$45.00

1. Chinese literature—20th century—History and criticism—Addresses, essays, lectures. I. Title.
PL2303.W44 78–843536

—20TH CENTURY—PERIODICALS

Modern Chinese literature newsletter. v. 1– Feb. 1975–
[Los Angeles, etc., University of California, Oriental Languages Dept.]
v. ill. 22 cm. semiannual.
Vols. for 1975–spring 1976 issued by California University, Center for Chinese Studies, fall 1976– by University of California, Oriental Languages Dept.
Key title: Modern Chinese literature newsletter, ISSN 0190–2369.
1. Chinese literature—20th century—Periodicals. I. California. University. Center for Chinese Studies. II. California. University. University at Los Angeles. Oriental Languages Dept.
PL2303.M64 895.1'08'005 79–640255
MARC-S

(Ya-lü chiang)
鸭绿江 Yalujiang.
[沈阳 鸭绿江文艺月刊社]
no. ill. 26 cm. monthly.

1. Chinese literature—20th century—periodicals. I. Title: Yalujiang.
PL2303.Y27 78–648579
(MARC-S)

—20TH CENTURY—PERIODICALS —BIBLIOGRAPHY

Hsien tai wen hsüeh ch'i k'an lien ho tiao ch'a hsiao tsu.
(Chung-kuo hsien tai wen hsüeh ch'i k'an mu lu)
中国現代文学期刊目錄：初稿 / 現代文学期刊联合調査小組. — [香港? : s. n., 1978?]
vii, 110 p. ; 18 x 26 cm.
Reprint of the 1961 ed. published by Shang-hai wen i ch'u pan she, in series: Chung-kuo hsien tai wen hsüeh shih tz'u liao ts'ung shu : Chia chung.
Includes index.
1. Chinese literature—20th century—Periodicals—Bibliography. I. Title.
Z3108.L5H68 1978 78–842952
[PL2303]
C

—20TH CENTURY—TRANSLATIONS INTO JAPANESE—BIBLIOGRAPHY

Iida, Kichirō, 1922– ed.
(Gendai Chūgoku bungaku kenkyū bunken mokuroku)
現代中国文学研究文献目錄：1919–1945 / 飯田吉郎編. — 東京：中国文化研究会：発売所「大安」, 1961?
87, [7] p. : 26 cm.
On cover: A bibliography for the study of modern Chinese literature in Japan.
Reprint of the 1959 ed.: with new suppl.
Includes indexes.

1. Chinese literature—20th century—Bibliography. 2. Chinese literature—20th century—Translations into Japanese—Bibliography. 3. Japanese literature—20th century—Translations from Chinese—Bibliography. I. Title. II. Title: A bibliography for the study of modern Chinese literature in Japan.
Z3108.L5I4 1961 78–814744
[PL2602]

—BIBLIOGRAPHY

Shen chou t'u shu kung ssu.
(Lu Hsün yen chiu yü hsin wen hsüeh ts'an k'ao shu mu)
魯迅研究與新文學參考書目 / 神州圖書公司. — 香港：經銷處該公司, [1975?]
80 p., [2] leaves of plates : ill. ; 21 cm.
Cover title.

1. Chou, Shu-jen, 1881–1936—Bibliography. 2. Chinese literature—Bibliography. I. Title.
Z8168.2.S47 1975 79–835009
[PL2754.S5]

—CHRONOLOGY

Cheng, Hui-wen.
(Chung-kuo wen hsüeh chia ku shih)
中國文學家故事 / [編著者鄭惠文]. — 臺北：莊嚴出版社, 民國66 [1977]
4, 296 p., [2] leaves of plates : ports. ; 20 cm. — (古典新刊 ; 9)
"附錄 中國文學年表 西諦": p. 221–296.
NT$55.00

1. Authors, Chinese—Biography. 2. Chinese literature—Chronology. I. Cheng, Chen-to, 1897–1958. Chung-kuo wen hsüeh nien piao. 1977. II. Title. III. Series: Ku tien hsin k'an ; 9.
PL2277.C425 78–842258

—DICTIONARIES—CHINESE

Liu, Ta-kang.
(Tz'u hai Ying Hua)
辭海英華 / 劉大剛著. — 南投鎮[南投縣]：九陽雜誌社, 民國68 [1979]
35, 1119 p. : ill. ; 22 cm.
NT$800.00

1. Chinese literature—Dictionaries—Chinese. I. Title.
PL2257.L58 79–840160

—HISTORY AND CRITICISM

Chang, Jen-ch'ing.
(Wei Chin nan pei ch'ao wen hsüeh ssu hsiang shih)
魏晉南北朝文學思想史 / 張仁青著. — 臺北：文史哲出版社, 民國67 [1978]
6, 13, 6, 828 p. ; 22 cm. — (文史哲學集成)
Series romanized: Wen shih che chi ch'eng.
Bibliography: p. 813–828.
NT$360.00

1. Chinese literature—History and criticism. I. Title.
PL2264.C424 79–838287

Ch'ien, Chi-po, 1887–
(Chung-kuo wen hsüeh shih)
中國文學史 錢基博著 [臺北] 海國書局 [民國59–60 i. e. 1970–1971 (v. 1, 1971)]
2 v. (6, 758 p.) 22 cm.
CONTENTS: 上冊 上古文學(先秦) 中古文學(漢·魏·六朝·唐)—下冊 近古文學(宋·金)

1. Chinese literature—History and criticism. I. Title.
PL2264.C4489 70–843142

Chu, Wei-chih.
(Chung-kuo wen i ssu ch'ao shih lüeh)
中國文藝思潮史略 朱維之著 [澳門] 合作出版社 [1966]
2, 6, 174 p. 19 cm.
Originally published in 1939 by 上海合作出版社

1. Chinese literature—History and criticism. I. Title.

Chu, Wei-chih.
(Chung-kuo wen i ssu ch'ao shih lüeh)
中國文藝思潮史略 / 朱維之著. — 香港：港青出版社, 1978.
160 p. ; 19 cm.
Reprint ed.
$5.80

1. Chinese literature—History and criticism. I. Title.
PL2264.C475 1978 79–836145

(Chung-kuo wen hsüeh shih kai shu)
中國文學史概述. — 3版. — 香港：上海書局, 1977.
85 p. ; 21 cm.
This work is the v. 1 of the 1957 ed.

1. Chinese literature—History and criticism.
PL2264.C54 1977 79–836072

Debon, Günther, 1921-
Grundbegriffe der chinesischen Schrifttheorie und ihre Verbindung zu Dichtung und Malerei / von Günther Debon. — Wiesbaden : Steiner, 1978.
x, 103 p. : 24 cm. — (Studien zur ostasiatischen Schriftkunst ; Bd. 3)
GFR78-A
Summaries in English.
Includes bibliographical references and indexes.
ISBN 3-515-02769-6
1. Calligraphy, Chinese. 2. Chinese literature—History and criticism. 3. Painting, Chinese. I. Title. II. Series.
NK3634.A2D4 745.6'199'51 79-337030
79 MARC

Deeney, John J
(Pi chiao wen hsüeh yen chiu chih hsin fang hsiang)
比較文學研究之新方向 / 李達三著. — 臺北：聯經出版事業公司, 民國67 [1978]
2, 6, 393 p. ; 22 cm.
Title in English on p. [4] of cover: New orientations for comparative literature.
Includes bibliographical references and indexes.
NT$140.00
1. Literature, Comparative—Chinese and European. 2. Literature, Comparative—European and Chinese. 3. Literature, Comparative. 4. Chinese literature—History and criticism. 5. European literature—History and criticism. I. Title. II. Title: New orientations for comparative literature.
PL2274.D4 78–843014

Hsiang-kang ch'ing nien ch'u pan she.
(Hsin pien Chung-kuo wen hsüeh fa chan shih)
新編中國文學發展史 / [本社編]. — 香港：香港青年出版社, 1978.
2 v. ; 19 cm. — (基本知識叢刊)
Series romanized: Chi pen chih shih ts'ung k'an.
$12.00

1. Chinese literature—History and criticism. I. Title. II. Title: Chung-kuo wen hsüeh fa chan shih.
PL2264.H69 1978 78–841959

Huang, Ch'un-kuei.
(Wen hsin tiao lung chih ch'uang tso lun)
文心雕龍之創作論 / 黃春貴著. — 臺北：文史哲出版社, 民國67 [1978]
10, 6, 207 p. ; 22 cm. — (文史哲學集成)
Series romanized: Wen shih che hsüeh chi ch'eng.
Bibliography: p. 199–207.
NT$80.00

1. Liu, Hsieh, ca. 465–ca. 522. Wen hsin taio lung. 2. Chinese literature—History and criticism. I. Title.
PL2263.L54H78 79–835422

(Ku tien wen hsüeh hsin shang)
古典文學欣賞：注音詳解 / 周益津編註. — 台南：大夏出版社, 民國66 [1977]
444 p. in various pagings ; 22 cm.
Text accompanied with Chu yin fu hao.
CONTENTS: 詩經.—古詩.—唐詩.—宋詞.—散文.
NT$75.00

1. Chinese literature—History and criticism. I. Chou, I-chin.
PL2264.K855 79–837851

CHINESE LITERATURE
—HISTORY AND CRITICISM
(Continued)

Li, Tan-lang.
(Wen hsüeh yü hsieh tso)
文學與寫作 / 李丹郎著. — 臺北：新文豐出版股份有限公司, 民國67 [1978]
2, 3, 209 p ; 19 cm. — (創作叢刊：第1輯；4)
NT$68.00

1. Chinese literature—History and criticism. 2. Creative writing. I. Title. II. Series: Ch'uang tso ts'ung k'an : Ti 1 chi ; 4.

PL2264.L513 79-841392

Liu, Hsi-tsai, 1813-1881.
(I kai)
藝概 / 劉熙載撰. — 第1版. — 上海：上海古籍出版社：新華書店上海发行所发行, 1978.
6, 1, 1, 184 p. ; 19 cm.
$0.58

1. Chinese literature—History and criticism. I. Title.

PL2263.L45 1978 79-840524

Liu, Hsieh, ca. 465-ca. 522.
[Wen hsin tiao lung]
文心雕龍校注 / [劉勰著；黄叔琳注；李詳補注]；楊明照校注[拾遺]. — 臺北：河洛圖書出版社, 民國65 [1976]
10, 5, 469 p. ; 22 cm. — (夏學叢書)
"十卷"
Series romanized: Hsia hsüeh ts'ung shu.
Reprint of the 1958 ed. published by Ku tien wen hsüeh ch'u pan she, Shanghai; with original postscript deleted.
Bibliography: p. 454-469.
NT$130.00

1. Chinese literature—History and criticism. I. Huang, Shu-lin, 1672-1756. II. Li, Hsiang, 1859-1931. III. Yang, Ming-chao. IV. Title.
Title romanized: Wen hsin tiao lung chiao chu.

PL2263.L5 1976b 78-842910

Liu, Hsieh, ca. 465-522.
[Wen hsin tiao lung. Japanese & Chinese]
文心雕龍 / [劉勰撰]；長澤規矩也解題. — 東京：古典研究会：發賣所 汲古書院, 1977.
383 p. ; 25 cm. — (和刻本漢籍隨筆集；16) Ja 77-24213
In Chinese, with Japanese reading marks.
Photoreprint ed.
CONTENTS: 劉勰 文心雕龍—二家詩品—皎然 詩式—齊己 風騷旨格—宋三家詩話—尤袤 全唐詩話—陳繹曾 文章歐冶—懐悦 詩法源流—都穆 南濠詩話—高琦 文章一貫
1. Chinese literature—History and criticism. I. Nagasawa, Kikuya, 1902- II. Title. III. Series: Wakokubon Kanseki zuihitsu shū ; 16.
Title romanized: Bunshin chōryū.

PL2606.W3 vol. 16 79-810427
[PL2263]

Liu, James J Y
Essentials of Chinese literary art / James J. Y. Liu. — North Scituate, Mass. : Duxbury Press, c1979.
xvi, 150 p. ; 24 cm.
Includes bibliographical references and index.
ISBN 0-87872-191-6
1. Chinese literature—History and criticism. I. Title.
PL2265.L48 815.1'009 78-12907
78 MARC

Liu, Lin-sheng, 1894-
(Chung-kuo wen hsüeh shih)
中國文學史 / 劉麟生著. — 臺北：中新書局：總經銷道明書局, 民國66 [1977]
444 p. ; 22 cm.
NT$180.00

1. Chinese literature—History and criticism. I. Title.

PL2264.L534 79-835354

Liu, Ta-chieh, 1904-
(Chung-kuo wei hsüeh fa ta shih tzu hui)
中國文學發達史字彙 = Glossary for The history of Chinese literature development / 劉大杰原著；劉崇純編注. — [台北?]：美國各大學中國語文聯合研習所, [1978?]
ca. 350 p. ; 27 cm.
English, Chinese, and Kuo yü romanization.
"This is a glossary of the first twelve chapters of the Chung-kuo wen-hsüeh fa-ta shih ... The edition used is that printed by the Taiwan Chung-hua shu-chü, 1962 (6th printing)."
Reproduced from typescript.

1. Chinese literature—History and criticism. 2. Chinese language—Glossaries, vocabularies, etc. I. Liu, Charles A. II. Title. III. Title: Glossary for The history of Chinese literature development.

PL2264.L539 78-841985

Lu, K'an-ju, 1903-
(Liu Hsieh ho Wen hsin tiao lung)
劉勰和文心雕龍 / 陆侃如, 牟世金. — 第1版. — 上海：上海古籍出版社：新華書店上海发行所发行, 1978.
77 p. ; 19 cm. — (中国古典文学基本知识丛书)
Includes bibliographical references.
$0.20
1. Liu, Hsieh, ca. 465-ca. 522. Wen hsin tiao lung. 2. Chinese literature—History and criticism. I. Mou, Shih-chin, joint author. II. Title. III. Series: Chung-kuo ku tien wen hsüeh chi pen chih shih ts'ung shu.

PL2263.L54L78 79-836814

Oka, Fuhō.
(Koten sōmoku zakkō)
古典草木雜考 / 岡不崩著. — 東京：第一書房, 昭和51 [1976]
2, 1, 12, 365 p., [10] leaves of plates : ill. ; 22 cm. Ja 77-25630
Photoreprint of the 1935 ed. published by Ōokayama Shoten, Tokyo.
Includes index.
¥4800
1. Materia medica—East Asia. 2. Plants in literature. 3. Japanese literature—To 1600—History and criticism. 4. Chinese literature—History and criticism. I. Title.

RS179.O38 1976 79-804211

Pei-ching ta hsüeh. Chung-kuo yü yen wen hsüeh hsi. Chung-kuo wen hsüeh shih chiao yen shih.
(Wei Chin Nan-pei ch'ao wen hsüeh shih ts'an k'ao tzu liao)
魏晉南北朝文學史參考資料 / 北京大學中國文學史教研室選注. — 北京：中華書局：新華書店北京發行所發行, 1962, 1978 printing.
2 v. (13, 734 p.) ; 21 cm.
$2.30
1. Chinese literature—History and criticism. I. Title.
PL2283.P4 1962b 79-836815

Shih, Leei.
(Wen hsin tiao lung yü Fo Ju erh chiao i li lun chi)
文心雕龍與佛儒二教義理論集 / 石壘著. — 香港：雲在書屋：經銷處龍門書店, 1977.
4, 2, 187, 4 p. ; 22 cm.
On p. [4] of cover: A collection of treatises on the theory of the Wen hsin tiao lung derived from Buddhism and Confucianism.
Includes bibliographical references.
$50.00
1. Liu, Hsieh, ca. 465-ca. 522. Wen hsin tiao lung. 2. Chinese literature—History and criticism. 3. Buddhism. 4. Confucianism. I. Title. II. Title: A collection of treatises on the theory of the Wen hsin tiao lung derived from Buddhism and Confucianism.

PL2263.L54S5 79-839144

[T'an, Cheng-pi]
[Chung-kuo nü hsing ti wen hsüeh sheng huo]
中國婦女文學史. — 九龍：百靈出版社, [1978?]
8, 4, 466, 64 p. ; 19 cm.
Colophon title.
Originally published in 1931 by Kuang ming shu chü, Shanghai, under title: Chung-kuo nü hsing ti wen hsüeh sheng huo and the author's name: T'an Cheng-pi.
Includes bibliographies.
1. Women authors—China. 2. Chinese literature—History and criticism. I. Title.
Title romanized: Chung-kuo fu nü wen hsüeh shih.

PL2278.T29 1978 79-835116

Toda, Kōgyō, 1910-
(Bunshin chōryū)
文心雕龍 / 戸田浩暁著. — 東京：明治書院, 昭和49-53 [1974-1978]
2 v. : ill. ; 22 cm. — (新釈漢文大系；64-65) Ja 75-504 (v. 1)
Includes index.
¥4800 per vol.
1. Liu, Hsieh, ca. 465-ca. 522. Wen hsin tiao lung. 2. Chinese literature—History and criticism. I. Liu, Hsieh, ca. 465-ca. 522. Wen hsin tiao lung. Japanese & Chinese. Selections. 1974. II. Title. III. Series: Shinshaku Kambun taikei ; 64-65.

PL2263.L54T56 1974 75-802485

Ts'ai, I.
(Wen hsüeh ch'ang shih)
文学常识 / 蔡仪. — 第1版. — 北京：人民文学出版社：新华书店北京发行所发行, 1978.
89 p. ; 19 cm.
"本书根据原作家出版社一九六二年版旧型重印"
$0.20

1. Chinese literature—History and criticism. I. Title.

PL2264.T74 1978 79-840739

Tseng, I.
(Chung-kuo wen hsüeh shih)
中國文學史 / 曾毅撰著. — 台1版. — 臺北：文史哲出版社, 民國66 [1977]
152, 254, [28] p. ; 22 cm. — (中外學術名著叢刊；第1種)
Reprint of the 1929 ed.
NT$120.00

1. Chinese literature—History and criticism. I. Title. II. Series: Chung wai hsüeh shu ming chu ts'ung k'an ; ti 1 chung.

PL2264.T77 1977 78-841182

Tu, Jo.
(Wen jen yü wen hsüeh)
文人與文學 / 杜若著. — 臺北：水芙蓉出版社：總經銷星光書報社, 民國63- [1974-
v. ; 19 cm. — (水芙蓉書庫；11, 124)
NT$35.00 (v. 1) varies

1. Chinese literature—History and criticism. I. Title.
PL2264.T797 75-836402

Wang, Ying-lin, 1223-1296.
[K'un hsüeh chi wen. Japanese & Chinese]
困學紀聞：[20卷 / 王應麟撰]. 攷古質疑：[6卷 / 葉大慶撰]. 肎綮錄 / [趙叔向撰]；長澤規矩也解題. — 東京：古典研究會, 昭和49 [1974], 昭和53 [1978] printing.
343 p. ; 25 cm. — (和刻本漢籍隨筆集；12)
In Chinese, with Japanese reading marks.
Photoreprint of the 1661 ed. published in 15 v. by D. Nakano, Kyoto, of the 1803 ed. published in 2 v. by Horinoya Nihē, Tokyo, and of the 1829 ed. published by Shōheizaka Gakumonjo, Tokyo.
¥4500

1. Chinese literature—History and criticism. 2. Errors and blunders, Literary. I. Yeh, Ta-ch'ing, chin shih 1205. K'ao ku chih i. Japanese & Chinese. 1974. II. Chao, Shu-hsiang. K'en ch'i lu. Japanese & Chinese. 1974. III. Nagasawa, Kikuya, 1902- IV. Title. V. Series: Wakokubon Kanseki zuihitsu shū ; 12.
Title romanized: Kongaku kibun.

PL2606.W3 vol. 12 79-809143

Wei, Ch'ing-chih, fl. 1240-1244.
[Shih jen yü hsieh. Japanese & Chinese]
詩人玉屑 / [魏慶之撰]；長澤規矩也解題. — 東京：古典研究會：汲古書院発売, 昭和52 [1977]
369 p. ; 25 cm. — (和刻本漢籍隨筆集；17) Ja 77-27740
In Chinese, with Japanese reading marks.
Photoreprint ed.
CONTENTS: 魏慶之 詩人玉屑—梁橋 氷川詩式—王世貞 文章九命—王守謙 古今文評
¥4500

1. Chinese literature—History and criticism. I. Nagasawa, Kikuya, 1902- II. Title. III. Series: Wakokubon Kanseki zuihitsu shū ; 17.
Title romanized: Shijin gyokusetsu.

PL2606.W3 vol. 17 79-809545
[PL2263]

CHINESE LITERATURE
—HISTORY AND CRITICISM (Continued)

(Wen hsin tiao lung yü shih p'in yen chiu)
文心雕龍與詩品研究 / [編輯者本社編輯部]. — 臺北：莊嚴出版社, 民國67 [1978]
210 p. ; 22 cm. — (古典大學叢書 ; 4)
NT$70.00

1. Liu, Hsieh, ca. 465–ca. 522. Wen hsin tiao lung. 2. Chinese literature—History and criticism. 3. Ssu-k'ung, T'u, 837–908. Shih p'in. 4. Chinese poetry—History and criticism. I. Chuang yen ch'u pan she. Pien chi pu. II. Series: Ku tien ta hsüeh ts'ung shu ; 4.
PL2263.L54W45 79–840734

Wu, Na, 1372–1457.
[Wen chang pien t'i]
文章辨體序說 / 吳納著 ; 于北山校點. 文體明辨序說 / 徐師曾著 ; 羅根澤校點. — 香港：太平書局, 1965, 1977 printing.
177 p. ; 21 cm.
$6.00

1. Chinese literature—History and criticism. 2. Literary form. I. Yü, Pei-shan. II. Hsü, Shih-tseng, 1517–1580. Wen t'i ming pien. 1977. III. Lo, Ken-tse, 1903– IV. Title. V. Title: Wen t'i ming pien.
Title romanized: Wen chang pien t'i hsü shuo.
PL2264.W79 1965 79–836070

—HISTORY AND CRITICISM
—ADDRESSES, ESSAYS, LECTURES

Aoki, Masaru, 1887–1964.
(Shina bungei ronsō)
支那文藝論藪 / 青木正兒著. — 東京：弘文堂, 昭和2 [1927]
9, 8, 563 p. ; 23 cm.
Includes bibliographical references.

1. Chinese literature—History and criticism—Addresses, essays, lectures. 2. Arts, Chinese—Addresses, essays, lectures. I. Title.
PL2272.5.A5 79–802743

(Bungaku to tetsugaku no aida)
文学と哲学のあいだ / 中国古典文学研究会編. — 東京：笠間書院, 1978.
245 p. ; 19 cm. — (中国文学の世界 ; 3) (笠間選書 ; 107) Ja 79–409
Series 2 romanized: Kasama sensho.
Includes bibliographical references.
¥1300

1. Chinese literature—History and criticism—Addresses, essays, lectures. I. Chūgoku Koten Bungaku Kenkyūkai. II. Series: Chūgoku bungaku no sekai ; 3.
PL2272.5.B86 79–803524

Ch'i, I-chün, 1930–
(Nü hsing mei t'an)
女性美譚 / 戚宜君著. — 台北：水芙蓉出版社, 民國67 [1978]
4, 4, 230 p. : ill. ; 20 cm. — (水芙蓉書庫 ; 125)
Series romanized: Shui fu jung shu k'u.
NT$45.00

1. Chinese literature—History and criticism—Addresses, essays, lectures. 2. Women—Literary collections—Addresses, essays, lectures. I. Title.
PL2275.W6C5 79–837371

Chou, Yüeh-jan.
(Pan pen yü shu chi)
版本與書籍 / 周越然著. — 上海：知行出版社, 民國34 [1945]
140 p. ; 18 cm.

1. Chinese literature—History and criticism—Addresses, essays, lectures. 2. Books—Addresses, essays, lectures. I. Title.
PL2272.5.C5355 78–843241

(Chung-kuo ku tien wen hsüeh yen chiu ts'ung k'an)
中國古典文學研究叢刊 / [主編柯慶明, 林明德]. — 臺北：巨流圖書公司, 民國66–67 [1977–1978]
3 v. in 6 ; 22 cm.
CONTENTS: [1] 詩歌之部 1–2.—[2] 小說之部 1–3.—[3] 散文與論評之部.
NT$480.00

1. Chinese literature—History and criticism—Addresses, essays, lectures. I. K'o, Ch'ing-ming. II. Lin, Ming-te, 1932 or 3–
PL2272.5.C5394 78–842184

(Chung-kuo li tai wen lun hsüan)
中國歷代文論選 / 郭紹虞主編. — 第1版. — 香港：達文社, 1978.
15, 3, 4, 489 p. ; 21 cm.
$20.00

1. Chinese literature—History and criticism—Addresses, essays, lectures. I. Kuo, Shao-yü, 1893–
PL2272.5.C53943 78–843243

(Chung-kuo wen hsüeh shih lun wen hsüan chi)
中國文學史論文選集 / 羅聯添編. — 臺北：臺灣學生書局, 民國67– [1978–
v. ; 22 cm.
Includes bibliographical references.
NT$260.00 (v. 1–2)

1. Chinese literature—History and criticism—Addresses, essays, lectures. I. Lo, Lien-t'ien.
PL2272.5.C549 79–836007

Hayashi, Hideichi, 1902–
(Hayashi Hideichi Hakushi sonkō)
林秀一博士存稿. — 岡山：林秀一先生古稀記念出版会, 昭和49 [1974]
15, 238 p. : ports. ; 22 cm. Ja 75–8059
Includes bibliographical references.
List of the author's works: p. 232–236.
Not in trade.
1. Philosophy, Chinese—Addresses, essays, lectures. 2. Chinese literature—History and criticism—Addresses, essays, lectures. 3. Hayashi, Hideichi, 1902– I. Title.
B126.H36 79–801445

Hou, Chien, 1926–
(Wen hsüeh, ssu hsiang, shu)
文學·思想·書 / 侯健著. — 台北：皇冠出版社, 民國67 [1978]
262 p. ; 19 cm. — (文學瞭望 ; 1)
Includes bibliographical references.
NT$50.00

1. Chinese literature—History and criticism—Addresses, essays, lectures. I. Title. II. Series: Wen hsüeh liao wang ; 1.
PL2264.H65 79–839164

Hsieh, K'ang.
(Wen shih lun ts'ung)
文史論叢 / 著作者謝康. — 臺北：中外圖書出版社, 民國67 [1978]
2, 4, 354 p., [1] leaf of plates : port. ; 19 cm. — (中外文庫 ; 29)
Colophon title.
Series romanized: Chung wai wen k'u.
NT$60.00

1. Chinese literature—History and criticism—Addresses, essays, lectures. 2. Literature—Addresses, essays, lectures. I. Title.
PL2272.5.H773 1978 78–843650

Huang, Ch'ing-hsüan, 1932–
(Chung-kuo wen hsüeh chien shang chü yü)
中國文學鑑賞舉隅 / 黃慶萱, 許家鸞著. — 臺北：東大圖書有限公司 : 總經銷三民書局, 1979.
2, 2, 336 p. ; 22 cm. — (滄海叢刊 : 中國文學)
Series romanized: Ts'ang hai ts'ung k'an : Chung-kuo wen hsüeh.

1. Chinese literature—History and criticism—Addresses, essays, lectures. I. Hsü, Chia-luan, joint author. II. Title.
PL2272.5.H84 79–841422

Kao, Chün, 1938–
(Wen hsüeh yü she hui kai tsao)
文學與社會改造 / 高準著. — 臺南：德華出版社, 民國67 [1978]
8, 288 p. ; 20 cm. — (愛書人文庫 ; 076)
Series romanized: Ai shu jen wen k'u.
Appendices (p. 271–288): 1. 陳鴻森 另外的聲音.—2. 何懷碩 有其「功能」 絕不「次等」
CONTENTS: 文學與社會.—中國現代文學的主潮.—論中國現代詩的流變與前途方向.—「七十年代詩選」批判.—剷除麻痺人心的「小市民思想」.—論臺灣社會的革新之道.—略論臺灣人口問題.—托爾斯泰思想述略.—李白詩中的戰鬪性與入世精神.—為「詩潮」答辯流言.
NT$75.00

1. Chinese literature—History and criticism—Addresses, essays, lectures. 2. Literature and society—Addresses, essays, lectures. I. Ch'en, Hung-sen, 1950– Ling wai ti sheng yin. 1978. II. Ho, Hwai Shouh, 1941– Yu ch'i kung neng, chüeh pu tz'u teng. 1978. III. Title.
PL2272.5.K33 79–841113

C

Ku, T'ien-hung, 1945–
(T'an so tsai ku tien ti lu shang)
探索在古典的路上 / 古添洪著. — 台中：普天出版社, 民國66 [1977]
4, 2, 232 p. : ill. ; 20 cm. — (普天新刊)
Series romanized: P'u t'ien hsin k'an.
Includes bibliographical references.
NT$45.00

1. Chinese literature—History and criticism—Addresses, essays, lectures. I. Title.
PL2272.5.K7 78–840848

C

(Ku tien wen hsüeh ch'ang shih)
古典文學常識. — 九龍：百靈出版社, [1978?]
1, 177 p. ; 19 cm.
Reprint ed.
Includes bibliographical references.
$6.50

1. Chinese literature—History and criticism—Addresses, essays, lectures.
PL2272.5.K74 1978 78–843606

C

Li, Ch'en-tung.
(Wen hsüeh yen chiu hsin t'u ching)
文學研究新途徑 / 李辰冬著. — 台北：啓德出版社, 民國61 [1972]
3, 2, 344 p. ; 22 cm.

1. Chinese literature—History and criticism—Addresses, essays, lectures. I. Title.
PL2272.5.L443 78–842954

C

Li, I-chih, 1909–
(Pan-t'ang wen liao)
半𡒄文聊 李一之著 [臺北] 臺灣商務印書館 [民國59 i. e. 1970]
2, 2, 2, 76 p. 18 cm. (人人文庫 1403) $8.00
Essays.

1. Chinese literature—History and criticism—Addresses, essays, lectures. I. Title.
PL2272.5.L48 72–839848

rev

P'an, Hsi-chen.
(Tu shu yü sheng huo)
讀書與生活 / 琦君[潘希真]著. — 臺北：東大圖書公司 : 總經銷三民書局, 1978.
3, 201 p. ; 22 cm. — (滄海叢刊 : 文學)
Series romanized: Ts'ang hai ts'ung k'an : Wen hsüeh.
NT$45.00

1. Chinese literature—History and criticism—Addresses, essays, lectures. I. Title.
PL2272.5.P3 78–841499

CHINESE LITERATURE
—HISTORY AND CRITICISM
—ADDRESSES, ESSAYS, LECTURES
(Continued)

(Wen hsüeh lun chi)
文學論集 / ［李殿魁主編］. — ［台北］：中華學術院，民國67 ［1978］
2, 4, 936, 13 p. : ports. ; 22 cm. — (中華學術與現代文化叢書 ; 第2册)
In colophon: Ch'u pan che Hua kang ch'u pan yu hsien kung ssu.
NT$260.00

1. Chinese literature—History and criticism—Addresses, essays, lectures. 2. Literature—History and criticism—Addresses, essays, lectures. I. Li, Tien-k'uei. II. Series: Chung-hua hsüeh shu yü hsien tai wen hua ts'ung shu ; ti 2 ts'e.

PL2272.5.W434 79-837149

Wen hsüeh pai t'i.
文學手册 / 主編傅東華；撰述蔡元培 ...et al.］. — 臺北：大漢出版社，民國66 ［1977］
390 p. ; 20 cm. — (大漢新刊 ; 19)
Series romanized: Ta han hsin k'an.
Previous ed. published in 1935 by Sheng huo shu tien, Shanghai under title: Wen hsüeh pai t'i and editor's statement: Cheng Chen-to, Fu Tung-hua pien.
NT$70.00

1. Literature—History and criticism—Addresses, essays, lectures. 2. Chinese literature—History and criticism—Addresses, essays, lectures. I. Fu, Tung-hua, 1895- II. Ts'ai, Yüan-p'ei, 1867-1940. III. Cheng, Chen-to, 1897-1958. IV. Title.
Title romanized: Wen hsüeh shou ts'e.

PN595.C6W45 1977 78-842549

c

(Wen hsüeh p'ing lun ts'ung k'an)
文学评论丛刊 / 《文学评论》编辑部编. — 第1版. — 北京：中國社会科学出版社：新华书店北京发行所发行，1978-
v. ; 21 cm.
Title in Pinyin on cover of v. 1: Wenxue pinglun congkan.
Includes bibliographical references.
$0.78 (v. 1)
1. Chinese literature—History and criticism—Addresses, essays, lectures. 2. Literature—Addresses, essays, lectures. I. Wen hsüeh p'ing lun pien chi pu. II. Title: Wenxue pinglun congkan.

PL2272.5.W436 79-841079

(Wen hsüeh yen chiu lun ts'ung)
文學研究論叢 / ［編輯者本社編輯部］. — 臺北：莊嚴出版社，民國67 ［1978］
6, 375 p. ; 22 cm. — (古典大學叢書 ; 5)
NT$110.00

1. Chinese literature—History and criticism—Addresses, essays, lectures. I. Chuang yen ch'u pan she. Pien chi pu. II. Series: Ku tien ta hsüeh ts'ung shu ; 5.

PL2272.5.W438 79-837768

(Wen shih hsüan chi)
文史選集. — 臺北：中央日報社，民國68- ［1979-
v. ; 19 cm. — (中央日報專刊叢書)
Series romanized: Chung yang jih pao chuan k'an ts'ung shu.
NT$50.00 (v. 1)

1. China—History—Addresses, essays, lectures. 2. Chinese literature—History and criticism—Addresses, essays, lectures. 3. Philosophy, Chinese—Addresses, essays, lectures. 4. Taiwan—History—Addresses, essays, lectures.

DS736.W45 79-840512

Wong, Yoon-wah.
(Chung hsi wen hsüeh kuan hsi yen chiu)
中西文學關係研究 / 王潤華著. — 臺北：東大圖書公司：總經銷三民書局，1978.
3, 2, 275 p. ; 22 cm. — (滄海叢刊 : 文學)
Series romanized: Ts'ang hai ts'ung k'an : Wen hsüeh.
Includes bibliographies.
NT$55.00

1. Literature, Comparative—Chinese and European—Addresses, essays, lectures. 2. Literature, Comparative—European and Chinese—Addresses, essays, lectures. 3. Chinese literature—History and criticism—Addresses, essays, lectures. 4. European literature—History and criticism—Addresses, essays, lectures. I. Title.

PL2274.W6 78-843044

Yang, Hung-lieh, 1901?-
(Chung-kuo wen hsüeh tsa lun)
中國文學雜論 / 楊鴻烈著. — 臺北：中新書局：總經銷道明書局，民國66 ［1977］
2, 228 p. ; 22 cm.
Reprint of the 1928 ed. published by Ya tung t'u shu kuan, Shanghai.
NT$130.00

1. Chinese literature—History and criticism—Addresses, essays, lectures. I. Title.

PL2272.5.Y35 1977 78-842621

Yeh, Shih, 1150-1223.
(Hsi hsüeh chi yen hsü mu)
习学记言序目 / 叶适著. — 第1版. — 北京：中华书局，1977.
2 v. (3, 11, 771 p.) ; 21 cm.
Reprint ed.; with new introd.
"叶适起初只是「辑录经史百氏条目」没有加评论，这时名为《习学记言》，后来「研玩群书」撰写了论述，乃以「序目」名书"
$2.30
1. Chinese literature—History and criticism—Addresses, essays, lectures. 2. China—History—Addresses, essays, lectures. I. Title.

PL2261.Y44 1977 78-841010

Yoshikawa, Kōjirō, 1904-
(Gakuji shiji)
學事詩事：随筆集 / 吉川幸次郎. — 東京：筑摩書房，昭和35 ［1960］
292 p. ; 20 cm.
Essays.

1. Chinese literature—History and criticism—Addresses, essays, lectures. I. Title.

PL2272.5.Y554 78-816258

—HISTORY AND CRITICISM
—COLLECTED WORKS

Alekseev, Vasiliĭ Mikhaĭlovich, 1881-1951.
(Kitaĭskaia literatura)
Китайская литература : Избр. тр. / В. М. Алексеев ; ［Авт. вступ. статьи Л. З. Эйдлин, с. 6-28］. — Москва : Наука, 1978.
595 p. : ill. ; 26 cm. USSR 78
On leaf preceding t. p.: Akademiia nauk SSSR. Institut vostokovedeniia.
Includes bibliographical references and indexes.
4.20rub
1. Chinese literature—History and criticism—Collected works. I. Eĭdlin, Lev Zalmanovich. II. Title.
PL2270.A4 1978 79-411509

—HISTORY AND CRITICISM—THEORY, ETC.

Lisevich, Igor' Samoĭlovich.
Literaturnaia mysl' Kitaia na rubezhe drevnosti i srednikh vekov / I. S. Lisevich. — Moskva : Nauka, 1979.
266 p. ; 22 cm. USSR79
At head of title: Akademiia nauk SSSR. Institut vostokovedeniia.
Bibliography: p. 240-［250］
Includes indexes.
1.80rub
1. Chinese literature—History and criticism—Theory, etc. I. Title.
PL2273.L5 79-378587
79 MARC

—JAPANESE AUTHORS

(Hiji-han senken bunshū shiryō)
日出藩先賢文集史料 / ［編纂者 佐藤暁］. — 日出町(大分県)：日出藩史料刊行会，1977-1978.
2 v. ; 24 cm. Ja 78-15745 - 78-15746
"大分県日出藩史料(別冊)"
Limited edition: 25 copies printed.
CONTENTS: その1. 甲原漁莊稿 豊筑二觀 関蕉川文集 関星巖著 忘眠夜話 関星巖詩稿—その2. 関星巖著 東郭日記 関星巖文集 関蕉川著 夫婦論 伊藤松軒文集 伊藤松太郎履歴調書
Not in trade.

1. Chinese literature—Japan—Hiji-han. 2. Chinese literature—Japanese authors. 3. Hiji-han, Japan—History—Sources. I. Satō, Akatsuki. II. Ōita-ken Hiji-han shiryō.

PL3057.H5H5 78-816703

—JAPANESE AUTHORS
—HISTORY AND CRITICISM

Kamimura, Kankō, 1873-1926.
(Gozan bungaku shōshi)
五山文學小史 / 上村觀光著；長田偶德校. — 東京：裳華房，明治39 ［1906］
4, 242, 40 p., ［2］ leaves of plates ; 23 cm.
"日本禪林撰述書目": p. 1-10 (3d group)

1. Chinese literature—Japanese authors—History and criticism. 2. Japanese literature—Zen authors—History and criticism. 3. Zen literature—Japan—History and criticism. I. Title.

PL3046.K27 79-805018

—KOREAN AUTHORS

Yi, Il-lo, 1152-1220.
(P'ahanjip)
破閑集 / 李仁老著；柳在泳譯註. — 서울：一志社，1978.
287 p. ; 22 cm.
Includes original text.
₩5000

1. Chinese literature—Korean authors. 2. Chinese literature—Korean authors—History and criticism. I. Yu, Chae-yŏng. II. Title.

PL3073.Y5 1978 79-825755

—KOREAN AUTHORS
—HISTORY AND CRITICISM

Ku, Cha-gyun, 1912-1964.
(Chosŏn p'yŏngmin munhaksa)
朝鮮平民文學史 / 具滋均著. — 서울：民學社，1974.
117 p. ; 18 cm. — (民學叢書 ; 2)
First ed. published in 1955 under title: Han'guk p'yŏngmin munhaksa.
Bibliography: p. 113-117.
₩400
1. Korean literature—1598-1800—History and criticism. 2. Chinese literature—Korean authors—History and criticism. 3. Popular literature—Korea—History and criticism. I. Title.

PL958.25.K8 1974 79-825991

Yi, Il-lo, 1152-1220.
(P'ahanjip)
破閑集 / 李仁老著；柳在泳譯註. — 서울：一志社，1978.
287 p. ; 22 cm.
Includes original text.
₩5000

1. Chinese literature—Korean authors. 2. Chinese literature—Korean authors—History and criticism. I. Yu, Chae-yŏng. II. Title.

PL3073.Y5 1978 79-825755

—KOREAN AUTHORS
—HISTORY AND CRITICISM
—ADDRESSES, ESSAYS, LECTURES

Kanda, Kiichirō, 1897-
(Bokurin kanwa)
墨林閒話 / 神田喜一郎. — 東京：岩波書店，1977.
vi, 236 p., ［10］ leaves of plates : ill. ; 19 cm. Ja 77-26881
¥1500
1. Chinese literature—Japanese authors—History and criticism—Addresses, essays, lectures. 2. China—Study and teaching—Japan—Addresses, essays, lectures. 3. Calligraphy, Japanese—Addresses, essays, lectures. 4. Calligraphy, Chinese—Addresses, essays, lectures. I. Title.

PL3043.K3 79-801651

Yi, Ka-wŏn, 1917-
(Yŏnmin kukhak san'go)
淵民國學散藁 / 李家源著. — 서울：東西文化社，1977.
448 p. ; 21 cm.
₩1990

1. Korea—Civilization—Addresses, essays, lectures. 2. Chinese literature—Korean authors—History and criticism—Addresses, essays, lectures. 3. Books—Reviews. I. Title. II. Title: Kukhak san'go.

DS904.Y485 78-826642

CHINESE LITERATURE (Continued)

—KOREAN AUTHORS —HISTORY AND CRITICISM —PERIODICALS

(Han'guk hanmunhak yŏn'gu)
韓國漢文學研究 第1- 輯
[서울] 韓國漢文學研究會 1976-
no. 21 cm.

Other title, no. 1- : Journal of Korean literature in Chinese.

1. Chinese literature—Korean authors—History and criticism—Periodicals. 2. Chinese literature—Korea—History and criticism—Periodicals. I. Han'guk Hanmunhak Yŏn'guhoe. II. Journal of Korean literature in Chinese.

PL3060.H35 78–647792 (MARC-S)

K

—LIBRARY RESOURCES—WASHINGTON, D.C.

Hu, Shu Chao.
The development of the Chinese collection in the Library of Congress / Shu Chao Hu. — Boulder, Colo. : Westview Press, 1979.
xvi, 259 p. ; 23 cm. — (A Westview replica edition)
Bibliography: p. [227]-242.
Includes index.
ISBN 0-89158-552-4 : $18.00
1. Chinese literature—Library resources—Washington, D.C. 2. United States. Library of Congress. I. Title.
Z3101.H768 026'.951 79-1741
79 MARC

—PERIODICALS

(Chungguk munhak)
中國文學
[서울] 韓國中國語文學會
no. 26 cm.

Other title, : Journal of Chinese literature.
Chinese or Korean.

1. Chinese literature—Periodicals. I. Han'guk Chunggukŏ Munhakhoe. II. Title: Journal of Chinese literature.

PL2250.C59 78–649282 (MARC-S)

K

–STUDY AND TEACHING

Cheng, Chen-to, 1897–1958.
(Yen chiu Chung-kuo wen hsüeh ti hsin t'u ching)
研究中國文學的新途徑 / 鄭振鐸著. — 香港 : 龍門書店, 1969.
26 p. ; 21 cm. — (中國文學研究叢編 : 第1輯)

1. Chinese literature—Study and teaching. I. Title. II. Series: Chung-kuo wen hsüeh yen chiu ts'ung pien : Ti 1 chi.

PL2258.C49 1969 79–840269

(Wen i tso p'in fen hsi)
文藝作品分析 / 姚重等著. — [香港?] : 半島書樓, [1978?]
ix, 173 p. : 19 cm. — (島思文化叢書)

Cover title.
Series romanized: Chün ssu wen hua ts'ung shu.
Reprint of the 1953 ed. published by Hsiang-kang hsüeh sheng shu tien, Hongkong.

1. Chinese literature—20th century—History and criticism. 2. Chinese literature—Study and teaching. I. Yao, Chung.

PL2303.W47 1978 79–835906

—TRANSLATIONS FROM ENGLISH

Shakespeare, William, 1564–1616.
[Works. Chinese]
莎士比亚全集. — 第1版. — 北京 : 人民文学出版社, 1978.
11 v. : port. ; 21 cm.
"主要根据解放前出版的朱生豪译本 ... 这次 ... 对译文作了一次全面校订 ... 补上删节部分 ... 朱生豪未译的六个历史剧和全部诗歌也都译出补全"
Includes bibliographical references.
$19.25
I. Chu, Sheng-hao. II. Title.
Title romanized: Sha-shih-pi-ya ch'üan chi.

PR2796.C5C5 1978 79–841052

—TRANSLATIONS FROM FOREIGN LANGUAGES

(No-pei-erh wen hsüeh chiang tso p'in hsüan chi)
諾貝爾文學獎作品選集. — 九龍 : 百靈出版社, [1977?]
2, 266 p. ; 19 cm.

Reprint of the 1965 ed. translated by Lin Chün-te and published by Hsing kuang ch'u pan she, T'ai-pei, issued as no. 26 in series: Shuang tzu hsing ts'ung shu.
$6.50

1. Chinese literature—Translations from foreign languages. 2. Literature—Translations into Chinese. 3. Nobel prizes. I. Lin, Chün-te.

PL2655.N6 1977 79–840240

Ta-hsüeh. English.
Ta hio : the great learning / newly rendered into the American language by Ezra Pound. — Folcroft, Pa. : Folcroft Library Editions, 1975.
35 p. ; 24 cm.
"The Confucian classics are customarily divided into the Five ching and the Four shu. The first of the Four shu (or books) is the Ta hio, a work of which the first chapter is ascribed to Confucius, and the remainder to one of his disciples, Thseng-tseu (Tsang Tzu)."
Reprint of the 1928 ed. published by the University of Washington Book Store, Seattle, which was issued as no. 14 of the University of Washington chapbooks.
ISBN 0-8414-6705-6 lib. bdg. : $7.50
I. Confucius. II. Tsêng, Shên, 505-437 B.C. III. Pound, Ezra Loomis, 1885-1972. IV. Series: Washington (State). University. University of Washington chapbooks ; no. 14.
PL2478.M6 1975 181'.09'512 75-14162
75[r79]rev MARC

Ta-hsüeh. English.
Ta-hio = The great learning / newly rendered into the American language by Ezra Pound. — Norwood, Pa. : Norwood Editions, 1977 [c1928]
35 p. ; 22 cm.
"The Confucian classics are customarily divided into the Five ching and the Four shu. The first of the Four shu (or books) is the Ta hio, a work of which the first chapter is ascribed to Confucius, and the remainder to one of his disciples, Thseng-tseu (Tsang Tzu)."
Reprint of the ed. published by the University of Washington Book Store, Seattle, which was issued as no. 14 of the University of Washington chapbooks.
ISBN 0-8482-2099-4 : $10.00
I. Confucius. II. Tsêng, Shên, 505-437 B.C. III. Pound, Ezra Loomis, 1885-1972. IV. Series: Washington (State). University. University of Washington chapbooks ; no. 14.
[PL2478.M6 1977] 181'.09'512 77-5658
77[r79]rev MARC

Wei, Chuang, chin shih 894.
The song-poetry of Wei Chuang (836-910 A.D.) / translated with an introd. and supplementary material by John Timothy Wixted ; calligraphy by Eugenia Y. Tu. — [Tempe, Ariz.] : Center for Asian Studies, Arizona State University, 1979.
iii, 144 p. ; 23 cm. — (Occasional paper - Center for Asian Studies, Arizona State University ; no. 12)
English and/or Chinese.
Bibliography: p. 133-144.
I. Wixted, John Timothy. II. Tu, Eugenia Y. III. Title. IV. Series: Arizona. State University, Tempe. Center for Asian Studies. Occasional paper ; no. 12.
PL2677.W4A28 1979 895.1'1'3 79-118498
79 MARC

—TRANSLATIONS INTO JAPANESE

see also Kundoku

—JAPAN—HIJI-HAN

(Hiji-han senken bunshū shiryō)
日出藩先賢文集史料 / [編纂者 佐藤暁]. — 日出町(大分県) : 日出藩史料刊行会, 1977-1978.
2 v. ; 24 cm. Ja 78-15745 - 78-15746
"大分県日出藩史料(別冊)"
Limited edition: 25 copies printed.
CONTENTS: その1. 甲原漁莊稿 豊筑二觀 関蕉川文集 関星巌著 忘眠夜話 関星巌詩稿—その2. 関星巌著 東郭日記 関星巌文集 関蕉川著 夫婦論 伊藤松軒文集 伊藤松太郎履歴調書
Not in trade.

1. Chinese literature—Japan—Hiji-han. 2. Chinese literature—Japanese authors. 3. Hiji-han, Japan—History—Sources. I. Satō, Akatsuki. II. Ōita-ken Hiji-han shiryō.

PL3057.H5H5 78–816703

J

—KOREA—HISTORY AND CRITICISM —PERIODICALS

(Han'guk hanmunhak yŏn'gu)
韓國漢文學研究 第1- 輯
[서울] 韓國漢文學研究會 1976-
no. 21 cm.

Other title, no. 1- : Journal of Korean literature in Chinese.

1. Chinese literature—Korean authors—History and criticism—Periodicals. 2. Chinese literature—Korea—History and criticism—Periodicals. I. Han'guk Hanmunhak Yŏn'guhoe. II. Journal of Korean literature in Chinese.

PL3060.H35 78–647792 (MARC-S)

K

—MALAYSIA—HISTORY AND CRITICISM

Fang, Hsiu.
Notes on the history of Malayan Chinese new literature, 1920-1942 / [by] Fang Xiu ; translated by Angus W. McDonald ; [edited by Kazuo Enoki]. — Tokyo : Centre for East Asian Cultural Studies, [1977]
xiv, 352 p. ; 20 cm. — (East Asian cultural studies series ; no. 18)
Ja78-39
A translation with revisions of Ma Hua hsin wen hsüeh shih kao.
Notes, indexes, glossary, maps, and errata (71 p.) tipped in.
1. Chinese literature—Malaysia—History and criticism. 2. Chinese literature—20th century—History and criticism. I. Enoki, Kazuo, 1913- II. Title. III. Series.
PL3097.M3F313 895.1'099595 79-347498
79 MARC

—TAIWAN

(T'ai-wan shih hui)
臺灣詩薈 : 中華民國六十六年丁巳端午詩集 / [編撰人連横]. — 台北 : 成文出版社, 民國66 [1977]
2 v. : ill. ; 22 cm.

Reprint ed.; with new pref.

1. Chinese poetry—Taiwan. 2. Chinese literature—Taiwan. I. Lien, Heng, 1878–1936.

PL3031.T32T329 1977 78–840776

(Wei hsi ti i hsien hsiang)
微細的一線香 / 主編張德本, 許素蘭. — 臺北 : 鴻蒙文學出版有限公司 : 總經銷故鄉出版社, 民國67 [1978]
341 p. : ill. ; 22 cm. — (前衛文學叢刊 ; 第1輯)

Colophon title.
NT$60.00

1. Chinese literature—Taiwan. 2. Chinese literature—20th century. I. Chang, Te-pen. II. Hsü, Su-lan. III. Series: Ch'ien wei wen hsüeh ts'ung k'an ; ti 1 chi.

PL3031.T32W4 79–836069

C

—TAIWAN —HISTORY AND CRITICISM

Chia, Yang.
(Hsüeh yüan, t'u ti, ch'uan t'ung)
血緣, 土地, 傳統 / 之華 [嘉揚] 著. — 台北 : 求精出版社 : 總經銷精益書局, 1977.
4, 3, 344 p., [1] leaf of plates : port. ; 19 cm. — (精益叢書 ; 5)

In colophon: Blood, land, tradition, by Chi-hua.
Series romanized: Ching i ts'ung shu.
NT$60.00

1. Chinese literature—Taiwan—History and criticism. 2. Chinese literature—20th century—History and criticism. 3. Authors, Chinese—Anecdotes, facetiae, satire, etc. I. Title. II. Title: Blood, land, tradition.

PL3031.T3C46 78–841244

CHINESE LITERATURE (Continued)

—TAIWAN—HISTORY AND CRITICISM—ADDRESSES, ESSAYS, LECTURES

(Che yang ti chiao shou Wang Wen-hsing)
這樣的教授王文興：從鄉土文學論戰到農業問題論戰 / 潘榮禮, 蕭國和編輯. — 高雄：敦理出版社, 民國67 [1978]
230 p. (p. 228-230 advertisements) : 20 cm. — (敦理叢刊; 13)
Series romanized: Tun li ts'ung k'an.
NT$45.00
1. Chinese literature—Taiwan—History and criticism—Addresses, essays, lectures. 2. Farmers—Taiwan—Addresses, essays, lectures. 3. Wang, Wen-hsing, 1939- —Addresses, essays, lectures. I. P'an, Jung-li. II. Hsiao, Kuo-ho.
PL3031.T3C453 79-836708

Hou, Li-ch'ao.
(Hsiang t'u wu ai)
鄉土吾愛：評鄉土文學論爭 / 侯立朝著. — 臺北：博學出版社：總經銷傳眞書店, 民國66 [1977]
8, 246 p., [4] leaves of plates : ill. ; 20 cm.
NT$55.00
1. Chinese literature—Taiwan—History and criticism—Addresses, essays, lectures. I. Title.
PL3031.T3H6 79-835557

(Hsiang t'u wen hsüeh t'ao lun chi)
鄉土文學討論集 / 編著尉天聰. — 臺北：尉天聰：總經銷遠流, 長橋聯合發行部, 民國67 [1978]
5, 6, 44, 850 p. ; 20 cm.
Colophon title.
NT$160.00 ($6.00 U.S.)
1. Chinese literature—Taiwan—History and criticism—Addresses, essays, lectures. I. Wei, T'ien-ts'ung, 1935-
PL3031.T3H74 78-842155

(Wen hsien tzu liao hsüan chi)
文獻資料選集 / 李南衡主編 ; [助理編輯林邊, 梁景峯]. — 台北：明潭出版社, 民國68 [1979]
12, 4, 504 p. ; 22 cm. — (日據下台灣新文學 : 明集 ; 5)
NT$250.00 ($10.00 U.S.)
1. Chinese literature—Taiwan—History and criticism—Addresses, essays, lectures. I. Li, Nan-heng. II. Lin, Pien. III. Liang, Ching-feng. IV. Series: Jih chü hsia T'ai-wan hsin wen hsüeh : Ming chi ; 5.
PL3031.T3W45 79-840170

CHINESE LITERATURE (PEKING)

—INDEXES

Hinrup, Hans J
An index to "Chinese literature" 1951-1976 / Hans J. Hinrup. — London : Curzon Press, 1978.
231 p. ; 23 cm. — (Scandinavian Institute of Asian Studies monograph series ; no. 38 ISSN 0069-1712) GB***
Includes index.
ISBN 0-7007-0122-2 : £6.00
1. Chinese literature (Peking)—Indexes. 2. Arts, Chinese—Periodicals—Indexes. I. Title. II. Series: Centralinstitut for nordisk Asienforskning. Monograph series ; no. 38.
NX583.A1H56 700'.951 79-300218
79 MARC

CHINESE LOGOGRAPHS see Chinese characters

CHINESE MANUSCRIPTS see Manuscripts, Chinese

CHINESE MATHEMATICS see Mathematics, Chinese

CHINESE MEDICINE see Medicine, Chinese

CHINESE MURAL PAINTING AND DECORATION see Mural painting and decoration, Chinese

CHINESE MYSTERY STORIES see Detective and mystery stories, Chinese

CHINESE MYTHOLOGY see Mythology, Chinese

CHINESE NEWSPAPERS

—BIBLIOGRAPHY—UNION LISTS

London. University. Contemporary China Institute.
A bibliography of Chinese newspapers and periodicals in European libraries / The Contemporary China Institute. — Cambridge ; New York : Cambridge University Press, 1975.
68, 1025, 12 p. ; 25 cm. — (Contemporary China Institute publications) GB***
ISBN 0-521-20950-1
1. Chinese periodicals—Bibliography—Union lists. 2. Chinese newspapers—Bibliography—Union lists. 3. Catalogs, Union—Europe. I. Title. II. Series: London. University. Contemporary China Institute. Publications.
Z6958.C5L78 1975 016.05 75-332818
76[r79]rev MARC

—CANADA—DIRECTORIES

Lo, Karl, 1935-
Chinese newspapers published in North America, 1854-1975 / compiled by Karl Lo and H. M. Lai. — Washington : Center for Chinese Research Materials, Association of Research Libraries, c1977.
xiii, 138 p. ; 28 cm. — (Bibliographical series - Center for Chinese Research Materials, Association of Research Libraries ; no. 16)
Includes index.
1. Chinese-American newspapers—Directories. 2. Chinese newspapers—Canada—Directories. I. Lai, H. Mark, joint author. II. Title. III. Series: Association of Research Libraries. Center for Chinese Research Materials. Bibliographical series ; no. 16.
Z6953.5.C45L6 071'.3 77-89070
[PN4885.C5] 79 MARC

—TAIWAN

Hsiang, Hsia-jen.
(Pao chih p'ing lun chih fen hsi)
報紙評論之分析 / 項下人著. — [台北? : s. n., 民國66 i. e. 1977]
87 p. ; 20 cm.
Originally presented as the author's thesis (M. A.), Kuo li cheng chih ta hsüeh, 1957.
Includes bibliographical references.
NT$40.00
1. Newspapers—Sections, columns, etc.—Editorials. 2. Chinese newspapers—Taiwan. I. Title.
PN4778.H7 1977 79-836974

CHINESE OIL TREE see Tung tree

CHINESE ONE-ACT PLAYS see One-act plays, Chinese

CHINESE OPERA see Opera, Chinese

CHINESE PEOPLE'S LIBERATION ARMY see China (People's Republic of China, 1949-) Chung-kuo jên min chieh fang chün

CHINESE PERIODICALS

—BIBLIOGRAPHY—CATALOGS

Chung-kuo kuo ming tang. Chung yang wei yüan hui. T'u shu kuan.
(Chung-kuo kuo min tang chung yang wei yüan hui t'u shu kuan tsa chih mu lu)
中國國民黨中央委員會圖書館雜誌目錄 / 中國國民黨中央委員會編印. — [臺北]：該館, 民國59 [1970]
14, 56, 12 p. ; 27 cm.
Cover title.
Includes index.
1. Chinese periodicals—Bibliography—Catalogs. 2. Chung-kuo kuo ming tang. Chung yang wei yüan hui. T'u shu kuan—Catalogs. I. Title.
Z6958.T34C47 1970 78-842441
[PN5370.2]

United States. Library of Congress.
Chinese periodicals in the Library of Congress / compiled by Han Chu Huang. — Washington : Library of Congress : for sale by the Supt. of Docs., U.S. Govt. Print. Off., 1978.
521 p. ; 27 cm.
Supt. of Docs. no.: LC 17.2:C43/4
ISBN 0-8444-0222-2
1. Chinese periodicals—Bibliography—Catalogs. 2. United States. Library of Congress. I. Huang, Han Chu, 1913- II. Title.
Z6958.C5U53 1978 016.05 76-608330
[PN5362] MARC
—— —— Copy 3. Z663.382.C473
76

—BIBLIOGRAPHY—UNION LISTS

London. University. Contemporary China Institute.
A bibliography of Chinese newspapers and periodicals in European libraries / The Contemporary China Institute. — Cambridge ; New York : Cambridge University Press, 1975.
68, 1025, 12 p. ; 25 cm. — (Contemporary China Institute publications) GB***
ISBN 0-521-20950-1
1. Chinese periodicals—Bibliography—Union lists. 2. Chinese newspapers—Bibliography—Union lists. 3. Catalogs, Union—Europe. I. Title. II. Series: London. University. Contemporary China Institute. Publications.
Z6958.C5L78 1975 016.05 75-332818
76[r79]rev MARC

CHINESE PERIODICALS (GENERAL)

(Che i tai tsa chih)
這一代雜誌 New generation. 第1- 期 民國66年7月- [1977-
台北 這一代雜誌社 總經銷 聯豐書報社]
no. 27 cm. monthly.
I. Title: New generation.
AP95.C4C5346 78-645314
(MARC-S)

(Cheng ming)
爭鳴 第1- 期 1977年11月-
[香港 百家出版社]
no. ill. 27 cm. monthly.
AP95.C4C5366 78-647011
(MARC-S)

CHINESE PERIODICALS (GENERAL) (Continued)

(Kuan ch'a chia)
観察家　第1-　期　1977年11月-
［九龍　觀察家出版社］
no.　ill.　27 cm.　monthly.
Other title, Nov. 1977-　: The Observers monthly.

I. Title: The Observers monthly.
AP95.C4K8367　78-647087 (MARC-S)

(Kuang-tung ch'ing nien)
广东青年
［广州　广东青年杂志社］
no.　ill.　26 cm.

AP95.C4K843　78-647765 (MARC-S)

(Pei tou)
北斗　1-　［1977］6月-
［九龍　北斗出版委員會］
no.　ill.　27 cm.　monthly.

AP95.C4P47　78-647864 (MARC-S)

(Ta Chung-hua)
大中華　The great China.　1-　民國67年3月-［1978-
台北　大中華月刊社］
no.　ill.　26 cm.　monthly.
In Chinese.

I. Title: The great China.
AP95.C4T28　78-646039 (MARC-S)

T'ai-wan tien li kung ssu.
(T'ai tien yüeh k'an)
台電月刊
［台北　台灣電力公司］
v.　ill.　27 cm.
Began with vol. for 1963.

I. Title.
AP95.C4T44a　78-646064 (MARC-S)

CHINESE PHILOLOGY

—ADDRESSES, ESSAYS, LECTURES

Boodberg, Peter Alexis, 1903-
. Selected works of Peter A. Boodberg / compiled by Alvin P. Cohen. — Berkeley : University of California Press, c1979.
xix, 501 p. ; 27 cm.
Bibliography: p. 496-501.
ISBN 0-520-03314-0
1. China—History—Addresses, essays, lectures. 2. Chinese philology—Addresses, essays, lectures. I. Cohen, Alvin P. II. Title.
DS736.B66　951　76-24580
79　MARC

Hsia, P'ing, fl. 1977-
(Chi chiu erh chi)
急就式集 / 夏平著. — 九龍：中華書局, 1978.
1, 4, 215 p. ; 21 cm.
$8.50

1. Chinese philology—Addresses, essays, lectures. I. Title.
PL1051.H7　79-836746

Huang, Tsun-sheng.
(Chung-kuo yü wen hsin lun)
中國語文新論 / ［黃尊生著］. — 香港：弘文出版社, 民國67 ［1978］
203 p. ; 21 cm.
$15.00

1. Chinese philology—Addresses, essays, lectures. I. Title.
PL1051.H84　78-843252

Ni, Hai-shu.
(Yü wen man t'an)
語文漫談 / 倪海曙著. — 香港：神州圖書公司, ［1978?］
2, 2, 109 p. : ill. ; 19 cm.
Reprint of the 1953 ed. published by Tung fang shu tien, Shanghai.

1. Chinese philology—Addresses, essays, lectures. I. Title.
PL1051.N5 1978　78-843727

(Tairiku no gengo to bungaku)
大陸の言語と文學 / 魚返善雄編著. — 東京：三省堂, 昭和15 ［1940］
2, 9, 260 p., ［6］ leaves of plates : ill. ; 20 cm.
CONTENTS: ダントン·G. 新支那の言語—林語堂　支那の方言—ヒューズ·E. R. 新支那の文學—林語堂　飜譯論—魚返善雄　大陸の人と言葉

1. Chinese philology—Addresses, essays, lectures. I. Ogaeri, Yoshio, 1910-1966.
PL1071.T34　78-811714

Yü, Tseng-wei.
(Chung-kuo wen hsüeh ch'ien shuo)
中國文學淺説 / 郁增偉著. — 香港：郁增偉, 民國67 ［1978］
1, 2, 104 p. ; 19 cm.
Errata slip inserted.
$2.00

1. Chinese philology—Addresses, essays, lectures. I. Title.
PL1051.Y84　79-836518

—ANECDOTES, FACETIAE, SATIRE, ETC.

Li, Wen-chung.
(Yü wen ch'ü t'an)
語文趣談 / 李文忠著. — 台中；立誼出版社, 民國67 ［1978］
4, 225 p. ; 20 cm.
NT$40.00

1. Chinese philology—Anecdotes, facetiae, satire, etc. I. Title.
PL1052.L5　79-837989

—COLLECTED WORKS

(Kambungaku kōza)
漢文學講座 / ［編輯代表者　長澤規矩也］. — 東京：共立社, 昭和8-9 ［1933-1934］
7 v. : ill. ; 23 cm.
Cover title.
L. C. set: Parts of each vol. bound together.
Includes bibliographies.
CONTENTS: 第1巻 ［pt. 1］ 内田泉之助　支那文學史 I. ［pt. 2］ 原富男　諸子學 I. ［pt. 3］ 松枝茂夫　支那現代文 I. ［pt. 4］ 加藤常賢　支那家族制度に於ける主要問題　後藤俊瑞　朱子の倫理思想—第2巻 ［pt. 1］ 本多龍成　支那經學史 I. ［pt. 2］ 山岸德平　日本漢文學史 ［pt. 3］ 松枝茂夫　支那現代文 II. ［pt. 4］ 長澤規矩也　漢籍解題 I. ［pt. 5］ 小柳司氣太　漢文學研究導言　阿部吉雄　近世儒學史上に於ける闇齋學の地位—第3巻 ［pt. 1］ 高田眞治　儒教倫理 I. ［pt. 2］ 内田泉之助　支那文學史 II. ［pt. 3］ 原富男　諸子學 II. ［pt. 4］ 枚下隆之　漢文典 I. ［pt. 5］ 藤原楚水　書道及金石—第4巻 ［pt. 1］ 内田泉之助　支那文學史 III. ［pt. 2］ 長澤規矩也　支那小說 I. ［pt. 3］ 松枝茂夫　支那現代文 III. ［pt. 4］ 林秀一　本邦に於ける鄭註孝經の刊行について—第5巻 ［pt. 1］ 高田眞治　儒教倫理 II. ［pt. 2］ 青木正兒　支那戲曲史 I. ［pt. 3］ 山口察常　漢文教授法 I. ［pt. 4］ 目加田誠　現代文藝　橋本成文　清朝尙書學—第6巻 ［pt. 1］ 原富男　諸子學 III. ［pt. 2］ 枚下隆之　漢文典 II. ［pt. 3］ 山口察常　漢文教授法 II. ［pt. 4］ 藤島亥治郎　建築 ［pt. 5］ 山田準　陽明學概論—第7巻 ［pt. 1］ 内田泉之助　支那文學史 IV. ［pt. 2］ 原富男　諸子學 IV. ［pt. 3］ 青木正兒　支那戲曲史 II. ［pt. 4］ 島田鈞一　清朝儒學史概說

1. Chinese philology—Collected works. 2. China—Intellectual life—Collected works. I. Nagasawa, Kikuya, 1902-
PL1025.K3 1933　78-815878
J

—DICTIONARIES—JAPANESE

Kondō, Haruo, 1914-
(Chūgoku gakugei daijiten)
中国学芸大事典 / 近藤春雄. — 東京：大修館書店, 1978.
5, 1000 p. ; 27 cm.　Ja 78-26440
Includes bibliographies.
¥17000

1. Chinese philology—Dictionaries—Japanese. 2. China—Dictionaries and encyclopedias. I. Title.
PL1071.K6　79-801491
J

—PERIODICALS

(Chungmun hakpo)
中文學報　創刊號-
［서울］　淑明女子大學校中國語中國文學科 1977-
no.　21 cm.
Chinese or Korean.

1. Chinese philology—Periodicals. I. Sungmyŏng Yŏja Taehakkyo. Chunggukŏ Chungguk Munhakkwa.
PL1007.C48　79-640699 (MARC-S)
K

—STUDY AND TEACHING—JAPAN —HISTORY

Ogata, Korekiyo.
(Nihon Kambungaku shi kōgi)
日本漢文学史講議 / 緒方惟精著. — 東京：評論社, 昭和36 ［1961］
2, 6, 205 p. ; 22 cm.
Colophon inserted.
Bibliography: p. 205.

1. Chinese philology—Study and teaching—Japan—History. I. Title.
PL1068.J3O35　78-815221
J

—STUDY AND TEACHING —SINGAPORE—PERIODICALS

(Yü wen)
語文　Yuwen.　创刊号-　1976年4月-
新加坡　新加坡华文第二语文教师协会
no.　ill.　27 cm.　semiannual.

1. Chinese philology—Study and teaching—Singapore—Periodicals. I. Hsin-cha-p'o Hua wen ti erh yü wen chiao shih hsieh hui. II. Title: Yuwen.
PL1068.S5Y8　78-647764 (MARC-S)

CHINESE PHILOLOGY (Continued)

—STUDY AND TEACHING (ELEMENTARY)—ADDRESSES, ESSAYS, LECTURES

(Hsiao hsüeh kuo wen cheng ming lun chan)
小學國文正名論戰 / 沙學浚編著. — 再版. — 臺北：大林出版社, 1976.
10, 16, 1017 p., [3] leaves of plates : ill. ; 21 cm.
NT$200.00

1. Chinese philology—Study and teaching (Elementary)—Addresses, essays, lectures. I. Sha, Hsüeh-chün.
PL1065.H723 1976 — 79-837349

—STUDY AND TEACHING (SECONDARY)—TAIWAN

Li, Chin-ch'eng.
(Chung hsüeh kuo wen chiao hsüeh lun heng)
中學國文教學論衡 / 李金城著. — 嘉義：興國出版社；台南：總經銷復文書局, 民國66 [1977]
14, 173 p. ; 22 cm.
Cover title.
At head of title: Ta hsüeh yung shu.
NT$70.00

1. Chinese philology—Study and teaching (Secondary)—Taiwan. I. Title.
PL1068.T3L47 — 79-839170

CHINESE POETRY

Under this heading, without subdivision, are entered individual works and collections by one author

see also Fu; Humorous poetry, Chinese; Tz'u

(A-shih-ma)
阿詩瑪：撒尼人民間叙事長詩. — 香港：海鷗出版公司, [1977?]
6 v. (93 p.) in 1 : col. ill. ; 18 x 19 cm.
"這本連環畫是根據雲南撒尼人民間叙事長詩阿詩瑪改編的"

PL2832.5.A2 — 78-841342

Chang, Chien, 1939-
(Wu li ti hsüeh hua)
屋裏的雪花 / 張健著. — 高雄：德馨室出版社, 民國67 [1978]
2, 204 p. ; 20 cm. — (德馨詩集 ; 06)
Poems.
Series romanized: Te hsin shih chi.
NT$40.00

I. Title.
PL2837.C49W8 — 79-839475

Chang, Ch'in, 1828-1883.
[Lo yeh shih]
張松譜先生落葉詩箋註及題詞 / [張琴, 陳寒光箋註]. — [台北：梅州文獻社, 62 i. e. 1973 hsü]
8, 5, 81 p. : ill. ; 19 cm.
Cover title.
Imprint stamped on p. 1 (first group)

I. Ch'en, Han-kuang. II. Title. III. Title: Lo yeh shih.
Title romanized: Chang Sung-p'u hsien sheng Lo yeh shih chien chu chi t'i tz'u.

PL2705.A46L6 1973 — 78-843988

Chang, Hsiang-hua.
(Pu mien ti ch'ing ch'ing ts'ao)
不眠的青青草 / 著者張香華. — 臺北：星光出版社, 民國67 [1978]
4, 2, 6, 178 p. ; 19 cm. — (星光詩叢)
Poems.
Series romanized: Hsing kuang shih ts'ung.
NT$35.00

I. Title.
PL2837.H727P8 — 78-842357

Chang, Hsiu-ya, 1918-
(Shih jen ti hsiao mu wu)
詩人的小木屋 / 張秀亞著. — 臺中：光啓出版社, 民國67 [1978]
4, 7, 195 p. ; 20 cm. — (文藝叢書 ; 58)
Prose.
Series romanized: Wen i ts'ung shu.
NT$45.00

I. Title.
PL2837.H765S45 — 79-840446

Chang, Mo.
(Chang Mo tzu hsüan chi)
張默自選集. — 臺北：黎明文化事業公司, 民國67 [1978]
8, 300 p., [2] leaves of plates : ill. ; 20 cm. — (中國新文學叢刊 : 48)
Series romanized: Chung-kuo hsin wen hsüeh ts'ung k'an.
NT$70.00

I. Title.
PL2837.M57A6 1978 — 78-843538

Chao, Nai-ting, 1943-
(I chung ti ch'i ch'iu)
異種的企求：趙廼定詩集. — 臺北：笠詩社, 民國64 [1975]
128 p. ; 19 cm. — (笠叢書)
Poems.
Series romanized: Li ts'ung shu.
NT$40.00

I. Title.
PL2838.N34I24 — 79-835487

Chao, P'u-ch'u.
(P'ien shih chi)
片石集 / 赵朴初. — 第1版. — 北京：人民文学出版社, 1978.
10, 15, 372 p. ; 20 cm.
Poems.
$1.30

I. Title.
PL2838.P8P5 — 78-841556

Ch'en, Ch'un, 1483-1544.
(Ch'en Pai-yang chi)
陳白陽集 / [陳淳]. — 臺北：臺灣學生書局, 民國62 [1973]
10, 370, 4 p. ; 22 cm. — (歷代畫家詩文集 ; 30 : 第三輯)
"國立中央圖書館藏本; 明萬曆四十三年陳仁錫刊本景印本"

I. Title. II. Series: Li tai hua chia shih wen chi ; 30.
PL2698.C39A6 1973 — 78-841186

Ch'en, I-chih.
(Lo jih ch'ang yen)
落日長煙 / 陳義芝著. — [高雄]：德馨室出版社；台中縣沙鹿鎮：總經銷詩人季刊社, 民國66 [1977]
209 p. ; 20 cm. — (詩季叢書 ; 2)
Poems.
Series romanized: Shih chi ts'ung shu.
NT$40.00
I. Title.
PL2840.I23L6 — 79-841870

Ch'en, I-k'ang.
(Ta-chai hung ch'i ying feng p'iao)
大寨红旗迎风飘 / 陈镒康, 糜佳乐等写；邵隆海, 余小仪等画. — 第1版. — 上海：上海人民出版社, 1977.
[28] p. : col. ill. ; 13 x 19 cm.
Title also in Pinyin on cover: Dazhai hongqi ying feng piao.
$0.17
1. Ta-chai, China—Poetry. I. Mi, Chia-le, joint author. II. Shao, Lung-hai. III. Yü, Hsiao-i. IV. Title. V. Title: Tazhai hongqi ying feng piao.
PL2840.I26T3 — 78-841245

Ch'en, Meng-chia, 1911-
(Ko chung chih ko)
歌中之歌 / 陳夢家作. — [香港? : s. n., 1978?]
4, 55 p. ; 18 cm.
Reprint of the 1932 ed. published by Liang yu t'u shu yin shua kung ssu, in series: I chiao ts'ung shu ; ti 50 chung.

I. Title.
PL2743.M39K6 1978 — 79-836071

Ch'en, Min-hua, 1933-
(Ch'in ch'uang shih ch'ao)
琴窗詩鈔 / 陳敏華著. — 再版. — 臺北：三民書局, 民國64 [1975]
8, 208 p. ; 20 cm. — (三民文庫 ; 147)
Poems.
Series romanized: San min wen k'u.

I. Title.
PL2840.M43C48 — 79-838588

Ch'en, Tse-p'o.
(Chia yüan)
家園 / [陳澤坡著]. — [香港? : s. n., 1978?]
80 p. ; 21 cm.
Poems.
Reprint of the 1931 ed.

I. Title.
PL2743.T77C5 — 79-837718

Chen, Wen-liang.
(Shih nien kan k'u chi)
十年甘苦集 / 甄文亮. — [香港? : s. n., 1978?]
60 p. (on double leaves) ; 22 cm.
Poems.
Reprint of the 1958 ed.

I. Title.
PL2743.W4S5 1978 — 79-835259

Cheng, Nan.
(Ch'ui chih ko)
锤之歌 / 郑南. — [香港? : s. n., 1978?]
3, 127 p. ; 22 cm.
Colophon title.
Poems.
Reprint of the 1977 ed. published by Kuang-tung jen min ch'u pan she, Canton.

I. Title.
PL2841.N3C5 1978 — 78-840870

Ch'eng, Wen-yen.
(T'ung hsin chih t'an tz'u)
同心梔彈詞 / 程文棪編；王蘊章校. — [香港? : s. n., 1977?]
3, 67 p. ; 22 cm.
Reprint of the 1928 ed. published by Shang wu yin shu kuan, Shanghai.

I. Title.
PL2744.W46T8 1977 — 78-843754

CHINESE POETRY (Continued)

Ch'iao-yen.
(Kao shang ti jen)
高尚的人：白求恩在抗日前线 / 峭岩. — 第1版. — 太原：山西人民出版社：山西省新華書店发行, 1976, 1977 printing.
2, 206 p., ₍5₎ leaves of plates : ill. ; 19 cm.
$0.52

I. Bethune, Norman—Poetry. I. Title.
PL2845.Z85K3 79–841834

Chien, An-liang.
(Lo yeh ti i shu)
落葉的遺書 / 簡安良著. — 高雄：三信出版社, 民國67 ₍1978₎
4, 113 p. ; 20 cm. — (現代詩叢 ; 6)
Poems.
Series romanized: Hsien tai shih ts'ung.
NT$30.00

I. Title.
PL2846.N114L6 79–835386

Ch'ien, Yü.
(Man lu tz'u)
曼廬詞 / ₍錢宇₎. — ₍台北? : s. n., 1978?₎
₍26₎ leaves ; 27 cm.
Cover title.
Poems.
Reproduced from ms. copy.

I. Title.
PL2846.N34M36 1978 79–83810[illegible]

Ch'in, Sung.
(Ch'ang i chih kung t'ung ti ko)
唱一支共同的歌 / 秦松著. — 香港：海洋文藝社, 1978.
3, 112 p. ; 19 cm. — (海洋詩叢)
Poems.
Series romanized: Hai yang shih ts'ung.
$4.00

I. Title.
PL2848.S8C5 79–837272

Chou, Chih-ch'i, 1782–1862.
(Chin liang meng yüeh tz'u)
金梁夢月詞 / 周之琦撰. — 臺1版. — 臺北：臺灣商務印書館, 民國67 ₍1978₎
2, 5, 71 p. ; 18 cm. — (人人文庫 ; 2398)
Poems.
Series romanized: Jen jen wen k'u.
NT$12.00

I. Title.
PL2705.O772C5 1978 79–841862

Chou, Shu-jen, 1881–1936.
(Lu Hsün shih kao)
鲁迅詩稿 / ₍上海鲁迅纪念馆编辑₎. — 北京：文物出版社：新华书店发行, 1976.
6, 89 p. : 27 cm.
Reprint ed.

I. Shang-hai Lu Hsün chi nien kuan. II. Title.
PL2754.S5L79 1976b 79–836418

Chou, Yün-shih.
(Hsiao lo hao ch'ui chan ko)
小螺号吹战歌 / 周云石 ; ₍插图卢宪民 ... et al.₎. — ₍香港? : s. n., 1978?₎
3, 90 p. : ill. ; 21 cm.
Poems.
Reprint of the 1st (1978) ed. published by Fu-chien jen min ch'u pan she, Fu-chou.

I. Lu, Hsien-min. II. Title.
PL2851.Y84H7 1978 79–836855

Chu, Chieh-ju.
(T'ang Sung shih chi lien)
唐宋詩集聯 / 朱介如稿. — ₍香港? : 味天穡室, ₍1971?₎
146 p. ; 21 cm.
Poems.

I. Title.
PL2755.C4T3 79–836850

Chu, Te, 1886–1976.
(Chu Te shih hsüan chi)
朱德詩選集. — 第1版. — 香港：萬源圖書公司, 1977.
3, 73 p., ₍6₎ leaves of plates : ports. ; 19 cm.
Poems.
$4.00

I. Title.
PL2755.T4A6 1977b 78–841121

Chuang, Chin-kuo.
(Hsiang t'u yü ming t'ien)
鄉土與明天 / 莊金國著. — 臺北：大漢出版社, 民國67 ₍1978₎
8, 178 p. ; 20 cm. — (大漢詩叢 ; 1)
Cover title.
Poems.
Series romanized: Ta han shih ts'ung.
NT$50.00

I. Title.
PL2852.5.A535H75 79–836772

(Ch'un hsia ch'iu tung shih hsüan)
春夏秋冬詩選 / 王維 ... 等著 ; ₍編者鐘意出版公司編輯部 ; 插圖華藝選輯₎. — Kowloon：鐘意出版公司, ₍1978?₎
209 p. : ill. ; 19 cm. — (鐘意暢銷叢書 ; 5)
Series romanized: Chung i ch'ang hsiao ts'ung shu.
$5.00

1. Chinese poetry. I. Wang, Wei, 701–761. II. Chung i ch'u pan kung ssu. Pien chi pu.
PL2518.C56 79–836824

(Chung-kuo wan lien)
中國輓聯 / 佚名. — Taipei：東方文化書局複刊, ₍1978?₎
6, 142 p. ; 21 cm. — (國立北京大學, 中國民俗學會民俗叢書 ; 152 : 第八輯) (東方文叢)
Added t. p.: Chinese funeral scrolls.
"陸美英集"
Series 2 romanized: Tung fang wen ts'ung.
Reprint of the 1961 ed.

1. Couplets. 2. Funeral rites and ceremonies—China. 3. Chinese poetry. I. Lu, Mei-ying. II. Title: Chinese funeral scrolls. III. Series: Kuo li Pei-ching ta hsüeh, Chung-kuo min su hsüeh min su ts'ung shu ; 152.
CR335.K87 vol. 152 79–836326
[PL2519.C7]

Fang, Ming, 1954–
(Ping shou ti yüeh)
病瘦的月 / 方明. — 臺北：文津出版社, 民國66 ₍1977₎
6, 172 p. ; 20 cm. — (台大詩社叢書)
Poems.
Series varies slightly.
NT$40.00

I. Title. II. Series: T'ai ta shih she. T'ai ta shih she ts'ung shu.
PL2857.A642P5 78–840865

Fang, O-chen, 1954–
(O mei fu)
蛾眉賦 / 方娥真著. — 第1版. — 臺北：四季出版事業有限公司, 民國66 ₍1977₎
23, 6, 239 p. ; 20 cm. — (四季文萃 ; 16)
Poems.
Series romanized: Ssu chi wen ts'ui.
NT$40.00

I. Title.
PL2857.A644O2 78–840884

Fang, Tung-mei.
(Chien pai ching she shih chi)
堅白精舍詩集 / 方東美 ; ₍編纂者方東美先生全集編纂委員會₎. — 臺北：黎明文化事業股份有限公司, 民國67 ₍1978₎
2, 85, 501 p., ₍2₎ leaves of plates : ill. ; 22 cm. — (東美全集)
Cover title.
"本集根據方東美先生詩詞手稿所編"
NT$140.00
I. Fang Tung-mei hsien sheng ch'üan chi pien tsuan wei yüan hui. II. Title. III. Series: Fang, Tung-mei. Tung-mei ch'üan chi.
PL2760.A55C5 79–836925

Hai, Meng.
₍Pao feng yü ch'ien chin₎
激變 / 海蒙. — ₍香港? : s. n., 1978?₎
48 p. ; 18 cm.
Two poems.
Reprint of the 1948 ed. published by Hsin shih ko she, Hongkong, in series: Hsin shih ko ts'ung shu.
CONTENTS: 暴風雨前進.—飄呀, 解放的旗.

1. Hai, Meng. P'iao ya, chieh fang ti ch'i. 1978. II. Title. III. Series: Hsin shih ko ts'ung shu.
Title romanized: Chi pien.
PL2861.A49P3 1978 79–836851

Ho, Ching-chih, 1924–
(Lei Feng chih ko)
雷鋒之歌 賀敬之著 北京 中國青年出版社 1964.
67 p. 17 cm.

1. Lei, Fêng, 1939–1962—Poetry. I. Title.
PL2861.O23L4 1964 C 64–1618
rev

Hsiang, Yin-Tso.
(Ssu chi hui wen shih chi)
四季迴文詩集 / 熊應祚吟草. — 1st ed. — San Francisco：熊應祚, 1978.
144 p. : col. ill. ; 38 cm.
Title in English on cover and in colophon: Palindrome poems of four seasons.

I. Title. II. Title: Palindrome poems of four seasons.
PL2765.I32S7 78–71437

Hsiang-yün, Shih.
(P'ing tsung shih ts'ao)
萍踪詩草 / 祥雲法師著. — 臺北：傳賢文化事業公司, 民國60 ₍1971₎, 民國66 ₍1977₎ printing.
1, 4, 116 p. ; 19 cm.
Poems.
NT$25.00

I. Title.
PL2862.I346P5 79–835271

Hsiao, Hsiao, 1947–
(Chü mu)
舉目 / 蕭蕭. — 彰化縣社頭鄉：大昇出版社：詩人季刊社發行, 民國67 ₍1978₎
110 p., ₍1₎ leaf of plates : ill. ; 20 cm. — (詩人小集 ; 6)
Poems.
NT$30.00

I. Title. II. Series: Shih jen hsiao chi ; 6.
PL2862.I3144C53 79–837318

Hsiao, Pai, 1925–
(Yeh yen)
野煙 / 蕭白著. — 台北：水芙蓉出版社, 民國68 ₍1979₎
3, 225 p. ; 19 cm. — (水芙蓉書庫 ; 130)
Prose.
Series romanized: Shui fu jung shu k'u.
NT$45.00

I. Title.
PL2862.I319Y43 79–841166

CHINESE POETRY (Continued)

Hsieh, K'ang.
(Shih lien hsin hua)
詩聯新話 / 謝康著. — 臺北：中外圖書出版社, 民國66 [1977]
2, 267 p. : port. ; 19 cm. — (中外文庫 ; 28)
Cover title.
Reprint ed.
Series romanized: Chung wai wen k'u.
NT$40.00
1. Chinese poetry—History and criticism. 2. Couplets. 3. Chinese poetry. 4. Anecdotes—China. I. Title.
PL2307.H78 1977 78-843080

Hsieh, Ssu-yen.
(P'ai men chi)
排悶集 / 謝似顏作. — [台北? : s. n., 1979?]
3, 50 p. ; 22 cm.
Poems.

I. Title.
PL2862.I369P3 79-841164

(Hsin pien k'ai hsin shih hsüan)
新編開心詩選 / 鐘意出版公司編輯部. — Kowloon : 鐘意出版公司, [1978?]
159 p. : ill. ; 19 cm. — (鐘意暢銷叢書 ; 2)
Series romanized: Chung i ch'ang hsiao ts'ung shu.

1. Humorous poetry, Chinese. I. Chung i ch'u pan kung ssu. Pien chi pu.
PL2519.H8H7 79-838030

Hsü, Ch'i-cheng.
(Lü yin shen ch'u)
綠蔭深處 / 許其正著. — 臺中：光啓出版社, 民國67 [1978]
4, 180 p. ; 19 cm. — (文藝叢書 ; 57)
Prose works.
Series romanized: Wen i ts'ung shu.
NT$35.00

I. Title.
PL2862.U14167L7 79-835496

Hsü, Lan-shan.
(Yü hu ping)
玉壺冰：許藍山詩集. — 高雄：德馨室出版社, 民國67 [1978]
158 p. : port. ; 20 cm. — (德馨詩集 ; 4)
Poems.
Series romanized: Te hsin shih chi.
NT$40.00

I. Title.
PL2862.U2244Y8 79-835959

Hsü, Shu.
(I t'uan huo)
一團火 / 徐淑. — 九龍：新思社, 1975.
94 p. ; 19 cm.
Poems.
"思潮二集"
$5.00

I. Title.
PL2862.U247I2 76-838462

Hsu, Yu, 1908-
(Pien hen chi)
鞭痕集 / [著者徐訏]. — 臺北：正中書局 ; 九龍：海外總經銷集成圖書公司, 民國66 [1977]
8, 222 p. ; 19 cm. — (正中文藝叢書)
Poems.
Series romanized: Cheng chung wen i ts'ung shu.

I. Title.
PL2765.U32P53 79-841866

Hu, I-min, fl. 1928-1948.
(Hu lao yin hsiao)
虎牢吟嘯 / 胡逸民著. — 香港：[s. n.], 1978.
5, 94 p. : port. ; 21 cm.
Poems.
Original pref. dated 1948.
$5.00

I. Title.
PL2863.I2H8 1978 79-836754

Hu, P'in-ch'ing, 1920-
(Po li jen)
玻璃人 / 胡品清著. — 臺中：學人文化事業有限公司, 民國67 [1978]
8, 5, 260 p. ; 20 cm.
Colophon title.
Poems.
NT$50.00

I. Title.
PL2863.P5P6 79-838007

Hu, Yung-huai.
(Chien feng hao)
劍鋒号：叙事诗 / 胡永槐, 金瑞华. — [香港? : s. n., 1978?]
2, 106 p. : ill. ; 21 cm.
Reprint of the 1977 ed. published by Shang-hai jen min ch'u pan she, Shanghai.

I. Chin, Jui-hua, joint author. II. Title.
PL2863.Y85C4 1978 79-835814

Huai, Yü.
(Huan le ti shou ku)
欢乐的手鼓 / 怀宇. — [香港 : s. n., 1978?]
3, 128 p. ; 21 cm.
"这是一本反映国际题材的抒情短诗集"
Reprint of the 1977 ed. published by T'ien-chin jen min ch'u pan she, Tientsin.
Includes bibliographical references.

I. Title.
PL2864.I2H8 1978 78-841611

Huang, Hun-hsing.
(Liang an teng huo)
两岸燈火：黃昏星, 周清嘯詩集. — 台北：神州詩社, 民國67 [1978]
6, 16, 230 p. ; 20 cm. — (神州叢書)
Series romanized: Shen-chou ts'ung shu.
NT$45.00

I. Chou, Ch'ing-hsiao, joint author. II. Title.
PL2865.H94L5 79-836755

Huang, Kuo-pin.
(Chih huan)
指環 / 黃國彬著. — 第1版. — 香港：詩風社, 1976.
10, 172 p. ; 19 cm. — (詩風叢書 ; 3)
Poems.
Series romanized: Shih feng ts'ung shu.
$6.00

I. Title.
PL2865.K86C5 79-836073

Huang, Kuo-pin.
(P'an yüeh kuei ti hai tzu)
攀月桂的孩子 / [作者黃國彬]. — 臺北：林白出版社, 1975.
3, 129 p. ; 20 cm. — (詩風叢書 ; 1)
Poems.
Series romanized: Shih feng ts'ung shu.
NT$35.00

I. Title.
PL2865.K86P3 78-841999

Huang, Kuo-pin.
(Ti chieh)
地劫 / [黃國彬著]. — 香港：詩風社, 1977.
2, 135 p. ; 19 cm. — (詩風叢書 ; 7)
Series romanized: Shih feng ts'ung shu.
$6.00

I. Title.
PL2865.K86T5 78-842341

Ju, Chih-chüan.
(Kao kao ti pai yang shu)
高高的白杨树 / 茹志鹃. — 第2版. — 上海：上海文艺出版社, 1978.
159 p. ; 19 cm.
Title in pinyin on cover: Gao gao de bai yang shu.
Prose.
CONTENTS: 关大妈.—妯娌.—在果树园里.—新当选的团支书.—百合花.—高高的白杨树.—澄河边上.—如愿.—鱼圩边.—黎明前的故事.—在社会主义的轨道上.—收获时节.—重印后記
$0.33

I. Title. II. Title: Gao gao de bai yang shu.
PL2868.U25K3 1978 79-840621

Jung-tzu, 1928-
(Hsüeh shih wo ti t'ung nien)
雪是我的童年 / 蓉子著. — 台北：環球書社：總發行乾隆圖書無限公司, 民國67 [1978]
3, 4, 128 p. ; 20 cm. — (女作家叢書：新詩類 : 1)
Poems.
NT$35.00

I. Title. II. Series: Nü tso chia ts'ung shu : Hsin shih lei : 1.
PL2868.U65H7 79-838683

Jung-tzu, 1928-
(Jung-tzu tzu hsüan chi)
蓉子自選集. — 臺北：黎明文化事業公司, 民國67 [1978]
10, 298 p., [2] leaves of plates : ill. ; 20 cm. — (中國新文學叢刊 : 57)
Poems.
Series romanized: Chung-kuo hsin wen hsüeh ts'ung k'an.
Series title on t. p.: Hsin Chung-kuo wen hsüeh ts'ung k'an.
NT$80.00

I. Title.
PL2868.U65A6 1978 79-836965

K'ang-fu.
(Ch'iung t'u)
窮途 / 康夫著. — 香港：羅盤詩雙月刊, 1977.
174 p. ; 20 cm. — (羅盤叢書 ; 第1號)
Title also in English in colophon: Black alley.
Poems.
$10.00

I. Title. II. Title: Black alley.
PL2869.N57C5 78-843989

Ko, Yang, fl. 1948-
(Hsüeh ch'ou)
血仇 / 戈陽. — [香港? : s. n., 1978?]
32 p. : ill. ; 18 cm.
"新詩歌"
Poems.
Reprint of the 1948 ed. published by Hsin shih ko she, Hongkong, in series: Hsin shih ko ts'ung shu.

I. Title. II. Series: Hsin shih ko ts'ung shu.
PL2775.Y3H7 1978 79-838571

K'o, Yüan.
(Lang hua tao)
浪花岛 / 柯原. — 第1版. — [广州]：广东人民出版社, 1977.
4, 122 p. ; 19 cm. — (南方诗丛)
Poems.
Series romanized: Nan fang shih ts'ung.
$0.27

I. Title.
PL2872.Y8L3 78-841232

CHINESE POETRY (Continued)

(Ku shih i jih i shou)
古詩一日一首 / 東萊先生著. — 台北：名人出版社, 民國67 [1978]
4 v. (783 p., [8] leaves of plates) : ill. ; 20 cm.
"本社商請東萊先生…共收錄中國古詩數百首"
NT$40.00 per vol.

1. Chinese poetry. I. Tung-lai-hsien-sheng.
PL2518.K8 79-836835

Kuan, Chao-ch'i.
(Ying hai fu yen tz'u)
瀛海浮煙詞 / 關照祺著. — 臺北：臺灣學生書局, 民國67 [1978]
6, 53 p. ; 21 cm.
Poems.
NT$30.00

I. Title.
PL2778.A5Y5 79-835694

Kuan, Kuan, 1929–
[Huang wu chih lien]
荒蕪之臉 [管管著 臺中 普天出版社 民國 61 i. e. 1972]
6, 204 p. 19 cm. (普天文庫 42) NT$20.00
Poems.
管管和他的詩 辛鬱作—管管及其「三朵色的罌粟花」 張默作—詮釋管管的三首詩 魏子雲作: p. 178–204.

I. Title.
PL2875.A52H8 72-837302

Kuo, Mo-jo, 1892–1978.
(Shu tao ch'i)
蜀道奇 / 郭沫若. — 第1版. — 成都：四川人民出版社, 1978.
75 p., [5] leaves of plates (1 fold.) : ill. ; 19 cm.
Poems.
$0.31

I. Title.
PL2778.O2S55 79-840162

Lei, Pai-wei, 1908–
(Ch'un ni tz'u)
春泥詞 / 雷白韋著. — 台北：雷白韋：經售處臺灣大學法學院事務組, 民國67 [1978]
8, 48 p. ; 21 cm.
Poems.
NT$50.00

I. Title.
PL2779.E57C5 79-841453

Li, Chin-fa.
(Wei hsing fu erh ko)
為幸福而歌 / 李金髮著. — [香港?]：創作書社, [1978?]
xiv, 296 p., [4] leaves of plates : ill. ; 19 cm.
Poems.
Reprint of the 1931 ed. published by Shang wu yin shu kuan, Shanghai, in series: Wen hsüeh yen chiu hui ts'ung shu.
I. Title.
PL2877.C534W4 1978 78-843128

Li, Ch'in-pai.
(Hsieh tsai ch'uan ch'ang ti shih)
写在船厂的诗 / 李槟白. — Singapore：风云出版社, 1978.
50 p. ; 19 cm.
Poems.
$0.90

I. Title.
PL2877.C5344H7 79-940548

Li, Chün, fl. 1977–
(Shan hai ch'ing huai)
山海情怀 / [李钧]. — [香港?：s. n., 1978?]
2, 106 p. : ill. ; 22 cm.
Poems.
Reprint of the 1977 ed. published by T'ien-chin jen min ch'u pan she, Tientsin.

I. Title.
PL2877.C812S5 1978 79-835187

Li, Chung-ch'iu, 1942–
(Lü hen)
履痕 / 著作者李仲秋. — 台北：領導出版社, 民國67 [1978]
177 p. ; 19 cm.
Poems.
"The stream of life, by Lee Chung-chin, translated by John M. McLellan" p. 139–175.
NT$50.00
I. Li, Chung-ch'iu, 1942– Sheng ming ti liu. English & Chinese. 1978. II. Title.
PL2877.C824L8 78-843649

Li, Hsien-sheng, 1948–
(Ming p'ien yü k'a p'ien)
名片與卡片 / 李仙生 ; [策畫者詩人季刊社]. — 彰化縣社頭鄉：大昇出版社：[臺北?]：詩人季刊社發行, 民國67 [1978]
117 p. ; 20 cm. — (詩人小集 ; 2)
Poems.
NT$30.00
I. Shih jen chi k'an she. II. Title. III. Series: Shih jen hsiao chi ; 2.
PL2877.H765M5 79-838569

Li, K'uei-hsien.
(Ch'ih lo ti ch'iang wei)
赤裸的薔薇 / 李魁賢著. — 高雄：三信出版社, 民國65 [1976]
2, 7, 182 p. ; 20 cm.
Poems.
Appendices (p. 149–182): 英譯作品.—趙天儀 孤獨的靈魂.—後記.—作者書目.—評論文獻.—作者年譜.

I. Title.
PL2877.K824C5 79-841869

Li, Lung.
(Chüeh lien ch'i wen hsü chi)
絕聯奇文續集 / 李隆編著. — 臺北：星光出版社, 民國67 [1978]
216 p. ; 19 cm. — (雙子星叢書 ; 59)
Series romanized: Shuang tzu hsing ts'ung shu.
"文章·詩歌·對聯·謎語"
NT$38.00
1. Couplets. 2. Chinese poetry. 3. Chinese essays. 4. Riddles, Chinese. I. Title.
PL2519.C7L464 79-836266

Li, Po, 705?–762.
[Poems]
李白诗选 / 复旦大学中文系古典文学教研组选注. — 第2版. — 北京：人民文学出版社：新华书店北京发行所发行, 1977.
9, 14, 300 p., [1] leaf of plates : ill. ; 19 cm.
$0.72
I. Fu tan ta hsüeh, Shanghai. Chung-kuo yü yen wen hsüeh hsi. Ku tien wen hsüeh chiao yen tsu. II. Title.
Title romanized: Li Pai shih hsüan
PL2671.A6 1977 79-837470

Li, Po, 705?–762.
[Poems]
李白诗选 / 复旦大学中文系古典文学教研组选注. — 第2版. — 北京：人民文学出版社：新华书店北京发行所发行, 1977.
9, 14, 300 p., [1] leaf of plates : ill. ; 19 cm.
$0.72
I. Fu tan ta hsüeh, Shanghai. Chung-kuo yü yen wen hsüeh hsi. Ku tien wen hsüeh chiao yen tsu. II. Title.
Title romanized: Li Pai shih hsüan.
PL2671.A6 1977 79-837470

Li, Shang-yin, 813–858.
[Li I-shan shih chi chi p'ing]
李義山詩集 / 朱鶴齡箋注 ; 沈厚塽輯評. — 港1版. — 九龍：中華書局, 1978.
279 leaves in various foliations ; 21 cm.
"上中下三卷."
Reprint of the 1870 ed. published by Kuang-chou ts'ui shu, under title: Li I-shan shih chi chi p'ing.
$16.00
I. Chu, Ho-ling, 1606–1683. II. Shen, Hou-shuang. III. Title.
Title romanized: Li I-shan shih chi.
PL2672.A1 1978 78-841318

Li, Shang-yin, 813–858.
(Li Shang-yin shih hsüan)
李商隱詩選 / 安徽师范大学中文系古代文学教研组选注. — 第1版. — 北京：人民文学出版社：新华书店北京发行所发行, 1978.
6, 29, 241 p. ; 19 cm. — (中国古典文学读本丛书)
"李商隱年表": p. 207–241.
$0.70
I. An-hui shih fan ta hsüeh. Ku tai wen hsüeh chiao yen tsu. II. Title. III. Series: Chung-kuo ku tien wen hsüeh tu pen ts'ung shu.
PL2672.A6 1978 79-837471

Li, Yü, 937–978.
(Li Hou-chu tz'u ch'üan chi)
李後主詞全集 / [李煜] ; 編輯者倪炳富. — 新竹：太平洋出版有限公司, 民國66 [1977]
111 p. in various pagings ; 20 cm.
Colophon title.
Appendices (p. [87–111]): 年譜.—集評.—填詞五法
NT$45.00

I. Ni, Ping-fu. II. Title.
PL2678.L48L5 1977 79-835591

Li-sha.
(Chin sha yün hsia)
金沙云霞 / 里沙. — [香港?：s. n., 1977?]
3, 104 p. ; 21 cm.
Poems.
Reprint ed.; originally published by Ssu-ch'uan jen min ch'u pan she.

I. Title.
PL2877.Z6C4 1977 78-841589

Lien, Heng, 1878–1936.
(T'ai-wan shih sheng)
臺灣詩乘 / [原著者連橫] ; 臺灣省文獻委員會編. — [再版]. — 臺中：該會, 民國64 [1975]
264, [12] p. ; 20 cm. — (雅堂叢刊 ; 3)
Series romanized: Ya-t'ang ts'ung k'an.

1. Taiwan—Description and travel—Poetry. 2. Poetry of places—Taiwan. I. Title.
PL2781.E5T3 1975 76-838118
rev

Lin, Ch'ing-hsüan.
(Leng yüeh chung ti)
冷月鐘笛 / 林清玄著. — 臺北：皇冠出版社, 民國68 [1979]
218 p. ; 20 cm.
Prose.
NT$50.00

I. Title.
PL2878.N1237L4 79-841426

Lin, Lin, poet.
(A-lai-yeh shan)
阿萊耶山 / 林林. — [香港?：s. n., 1978?]
123 p. ; 18 cm.
Poems.
Reprint of the 1950 ed. published by Jen chien shu wu, Kuang-chou, in series: Jen chien wen ts'ung.

I. Title.
PL2878.N1436A62 1978 79-837908

CHINESE POETRY (Continued)

Lin, Lü.
(Fu hsin)
覆信 / 林綠著. — 台北 : 環球書社 : 總發行乾隆圖書無限公司, 民國67 [1978]
5, 3, 170 p. ; 20 cm. — (當代文學叢書 : 新詩類 ; 1)
Poems.
NT$50.00

I. Title. II. Series: Tang tai wen hsüeh ts'ung shu : Hsin shih lei ; 1.
PL2878.N144F8 79–837845

Lin, T'ing-ch'iao.
(Han ya chi)
寒鴉集 / [林廷橋]. — [臺北? : s. n.], 民國63 [1974]
9, 9, 59 p. ; 18 cm.
Poems.

I. Title.
PL2878.N1494H3 78–841344

Liu, Chang, fl. 1978–
(Yen-shan ch'un)
燕山春 / 刘章. — [香港? : s. n., 1978?]
3, 133 p. : ill. ; 21 cm.
Poems.
Reprint of the 1978 ed. published by T'ien-chin jen min ch'u pan she, Tientsin.

I. Title.
PL2879.C28Y46 1978 79–838275

Liu, Fu, 1891–1934.
(Meng)
夢 劉半農[劉復]著 香港 上海書局 1964.
56 p. 18 cm. (中國文學名著小叢書)
Series romanized: Chung-kuo wen hsüeh ming chu hsiao ts'ung shu.

I. Title.
PL2782.F8M46 1964 C 65–1391

Liu, Hsiao-pin.
(Ching chüeh ti ts'ao yüan)
警觉的草原 : 叙事诗 / 刘晓滨. — 第1版. — 太原 : 山西人民出版社, 1975.
2, 98 p., [6] leaves of plates : ill. ; 19 cm.
Poems.
$0.23

I. Title.
PL2879.H675C5 1975 78–843202

Liu, Lung-min.
(Lung mien lien hua)
龍眠聯話 : 續編 / 劉隆民編著. — [再版]. — 臺北 : 臺灣學生書局, 民國67 [1978]
396, [10] p. : ill. ; 20 cm. — (學生書苑 ; 61)
Series romanized: Hsüeh sheng shu yüan.
NT$100.00

1. Couplets. 2. Chinese poetry. I. Title.
PL2519.C7L5 1978 79–835515

Liu, Tsu-p'ei.
(Lu sheng chan ko)
芦笙战歌 / 刘祖培. — 第1版. — 北京 : 人民文学出版社, 1978.
326 p. ; 19 cm.
Poem.
$0.67

I. Title.
PL2879.T67L8 79–841872

Lo, Chih-ch'eng.
(Kuang chih shu)
光之書 / 著者羅智成. — 台北 : 龍田出版社, 民國68 [1979]
206 p. : ill. ; 19 cm.
Colophon title.
"羅智成詩集"
Poems.
Includes index.
NT$50.00
I. Title.
PL2880.O212K8 79–840516

Lu, Chia-chi.
(Tui lien chang ku ts'ung t'an)
對聯掌故叢談 / 陸家驥編著. — 臺北 : 陸家驥, 民國66 [1977]
2, 168 p. ; 20 cm.
NT$50.00

1. Couplets. 2. Chinese poetry. 3. Anecdotes—China. I. Title.
PL2519.C7L69 78–843998

Lu, Chia-chi.
(Tui lien hsin yü)
對聯新語 / 陸家驥編著. — 臺北 : 臺灣商務印書館, 民國67 [1978]
2, 2, 4, 300 p. ; 18 cm. — (人人文庫 ; 特563)
Series romanized: Jen jen wen k'u.
NT$30.00

1. Couplets. 2. Chinese poetry. 3. China—History—Anecdotes, facetiae, satire, etc. I. Title.
PL2519.C7L7 78–842609

Lu, Chien-hung.
(T'ien chi)
天機 / 陸健鴻著. — 第1版. — 香港 : 詩風社, 1977.
3, 95 p. ; 19 cm. — (詩風叢書 ; 6)
Poems.
Series romanized: Shih feng ts'ung shu.
$5.00

I. Title.
PL2880.U1315T5 79–835106

Ma, Chen.
(Hsia wai shih chi)
霞外詩集 / 撰者元・馬臻. — 景印本. — 臺北 : 臺灣學生書局, 民國62 [1973]
10, 508, 4 p. ; 22 cm. — (歷代畫家詩文集 ; 25)
"明末汲古閣刊元十家詩集本"
NT$150.00

I. Title. II. Series: Li tai hua chia shih wen chi ; 25.
PL2694.M27H7 78–841162

Ma, Te, fl. 1972–
(Ko cho ch'ang t'i)
隔着長堤 馬德[著 新加坡] 萬里文化企業公司 [1972]
69 p. 19 cm. $0.60
Poems.

I. Title.
PL2881.T4K6 73–940378

Mao, Tse-tung, 1893–1976.
[Poems]
毛主席詩詞三十九首. — 上海 : 上海書畫社, 1977.
3, 22 double leaves ; 35 cm.
In case.
I. Title. Title romanized: Mao chu hsi shih tz'u san shih chiu shou.
PL2785.O3M287 1977 78–841333

Meng, Hsiang-sen.
(Wan ch'an chi)
萬蟬集 / 孟祥森著. — 台北 : 遠景出版社, 民國67 [1978]
194 p. ; 19 cm. — (遠景叢刊 ; 86)
Poems.
Series romanized: Yüan ching ts'ung k'an.
NT$45.00

I. Title.
PL2883.N5155W3 79–835904

Mi, Chün.
(Je tai shih ch'ao)
熱帶詩抄 / 米軍著. — [香港? : s. n., 1978?]
40 p. ; 18 cm.
Poems.
Series romanized: Ch'ih tao wen i ts'ung shu.
Reprint of the 1950 ed. published by Ch'ih tao ch'u pan she. in series: Ch'ih tao wen i ts'ung shu.

I. Title.
PL2884.C8J4 1978 79–838575

(Miao lien ch'iao tui)
妙聯巧對 / 鄭嘉善編著. — 臺北 : 星光出版社, 民國65– [1976–
v. ; 20 cm. — (雙子星叢書 ; 35,)
Vol. 2 has title: Ch'iao lien miao tui hsü chi; v. 3– : Ch'iao lien miao tui.
Series romanized: Shuang tzu hsing ts'ung shu.
NT$45.00 per vol.

1. Couplets. 2. Chinese poetry. I. Cheng, Chia-shan. II. Title: Ch'iao lien miao tui.
PL2519.C7M5 77–838344
rev

Miao lien ch'iao tui.
巧聯妙對 / 鄭嘉善編著. — 九龍 : 華聯出版社, [1977?]
355 p. ; 19 cm.
Previous ed. published in 1976 by Hsing kuang ch'u pan she, T'ai-pei under title: Miao lien ch'iao tui.

1. Couplets. 2. Chinese poetry. I. Cheng, Chia-shan. II. Title. Title romanized: Ch'iao lien miao tui.
PL2519.C7M5 1977 78–841321

(Ming chia shih tz'u hsin shang)
名家詩詞欣賞 / [朱自清讀法指導 ; 余雪曼評釋]. — 臺南 : 臺南東海出版社, 民國67 [1978]
232, 133, 130, [13] p. ; 21 cm.
"朱自清著「唐詩三百首」指導大概": p. 1–20 (1st group)
CONTENTS: 唐詩三百首.—千家詩.—三李詩詞集.
NT$55.00

1. Chinese poetry. 2. Tz'u. I. Chu, Tzu-ch'ing, 1898–1948. II. She, Hsüeh-man. III. Sun, Chu, 1711–1778. T'ang shih san pai shou. 1978. IV. Hsieh, Fang-te, 1226–1289. Ch'ien chia shih. 1978. V. San Li shih tz'u chi. 1978.
PL2518.M44 79–835096

Mu, Hung.
(Hu shang ch'ü)
湖上曲 / 沐鴻作. — [香港? : s. n., 1978?]
173 p. ; 21 cm.
Poem.
Reprint of the 1929 ed. published by Nan hua t'u shu chü.

I. Title.
PL2886.U174H8 1978 79–837273

Mu, Yin, 1951–
(Hei lien)
黑臉 / 牧尹. — 彰化縣社頭鄉 : 大昇出版社, 民國67 [1978]
112 p. ; 20 cm. — (詩人小集 ; 3)
Poems.
"作品年表": p. 109–112.
NT$30.00

I. Title. II. Series: Shih jen hsiao chi ; 3.
PL2886.U187H44 79–835958

CHINESE POETRY (Continued)

P'an, Fei-sheng.
(Shuo chien t'ang chi)
說劍堂集 / 潘飛聲撰. — 香港：龍門書店, 1977.
10, 3, 95 p. ; 21 cm.
Reprint of the 1891 ed.
CONTENTS: 海山詞.—花語詞.—珠江低唱.—長相思詞
$30.00

I. Title.
PL2722.A57S55 1977 — 79-838108

Pao, Shao-yu.
(Li ching hua yüan t'i hua shin chi)
麗精畫苑題畫詩集 / 作者鮑少游. — 香港：麗精美術學院, 1966.
66 p. : ill. ; 19 cm.
Reprint ed.?
Errata slip inserted.
$3.00

I. Title.
PL2892.A66L5 1966 — 79-837367

P'eng, Hsü-yüan.
(Su chung ch'ing shih tz'u chi)
訴衷情詩詞集 / 彭須源著. — 台北：學藝出版社, 民國67 [1978]
109 p. : port. ; 22 cm.
Poems.
NT$50.00

I. Title.
PL2892.E57S8 — 79-837370

Pu-lin-pei-ho, Pa.
(Hsing chün)
星群 / 巴·布林贝赫. — 第1版. — 呼和浩特：内蒙古人民出版社, 1976.
117 p. : ill. ; 19 cm.
Poems.
$0.27

I. Title.
PL2892.U25H74 — 78-841781

Shen, Chün-ju.
(Liao liao chi)
寥寥集 / [沈钧儒著：沈叔羊编：画像沈叔羊]. — 第1版. — 北京：生活·读书·新知三联书店：新华书店发行, 1978.
10, 14, 174 p., [1] leaf of plates : port. ; 21 cm.
Poems.
$0.71

I. Shen, Shu-yang. II. Title.
PL2801.N12L5 1978 — 79-837722

Shih, Ch'ing-wen.
(Yang kuang chih wai)
陽光之外 / [作者施清文]. — 台北：萬年青書店：總代理環宇出版社, 67 [1978]
4, 115 p. ; 19 cm.
Poems.
NT$35.00

I. Title.
PL2899.H21736Y3 — 78-843471

Shih, Yung-fu.
(Hsing huo chi)
星火集 / 时永福. — [香港? : s. n., 1978?]
3, 163 p. : ill. ; 21 cm.
Poems.
Reprint of the 1977 ed. published by Chiang-hsi jen min ch'u pan she, Nan-ch'ang.

I. Title.
PL2899.H296H74 1978 — 78-840938

Shih ching. Selections.
詩經國風：夢幻的古典 / 吉川幸次郎選；洪順隆評析譯註. — 台北：林白出版社, 民國68- [1979-
v. ; 20 cm. — (中國歷代詩人選集 ; 1-)
Cover title: Kuo feng.
NT$60.00 (v. 1)

I. Yoshikawa, Kōjirō, 1904- II. Hung, Shun-lung. III. Title. IV. Title: Kuo feng. V. Series: Chung-kuo li tai shih jen hsüan chi ; 1-
Title romanized: Shih ching kuo feng.
PL2466.R795 1979 — 79-839097

Shou-yang hsien Ta-chai ching shen sung ch'uang tso tsu.
(Ta-chai ching shen sung)
大寨精神颂 / 寿阳县《大寨精神颂》创作组. — 第1版. — 北京：人民文学出版社：新华书店北京发行所发行, 1977.
3, 58 p. ; 19 cm. — (农村文艺演唱丛书)
$0.15

I. Title. II. Series: Nung ts'un wen i yen ch'ang ts'ung shu.
PL2900.U85T3 1977 — 79-838205

Shu, Ch'ing-ch'un, 1898-1966.
(Nü tien yüan)
女店員：三幕十三場話劇 / 老舍[舒慶春]. — [香港? : s. n., 1978?]
31 p. : ill. ; 26 cm.
Play.
Reprint ed.

I. Title.
PL2804.C5N8 1978 — 79-835658

Sŏ, Kyŏng-bo.
(Tongyang myŏngsi)
東洋名詩：一鵬禪詩第十二集 / 徐京保著. — 서울：信興出版社, 1977.
116 p. ; 19 cm.
Poems.
Chinese and Korean.
₩500

1. Zen poetry. I. Title.
PL992.74.K95T6 — 79-825207

[Song, Ik-p'il] 1534-1599.
(Kwibong sijip)
龜峰詩集 / [沈宗直纂集]. — 三陟：三陟鄉校, 檀紀4296 [1964]
2, 31, 4 double leaves ; 29 cm.
Title from label on cover.
Poems.
Caption title: Pisŏn Kwibong Sŏnsaeng sijip.

I. Sim, Chong-jik, ed. II. Title.
PL2698.S6K9 1964 — 79-825745

Su, Shao-lien, 1949-
(Mang mang chi)
茫茫集 / 蘇紹連. — 彰化縣社頭鄉：大昇出版社：詩人季刊社發行, 民國67 [1978]
128 p. ; 20 cm. — (詩人小集 ; 5)
Poems.
NT$ 30.00

I. Title. II. Series: Shih jen hsiao chi ; 5.
PL2904.S54M3 — 79-836969

Sun, Ching-hsüan.
(Huang-ho ti erh tzu)
黄河的儿子：长篇叙事诗 / 孙静轩. — 第1版. — [武汉]：湖北人民出版社, 1978.
385 p. : ill. ; 19 cm.
Poem.
$0.86

I. Title.
PL2905.N165H8 — 79-841708

Sun, Chu, 1711-1778, comp.
[T'ang shih san pei shou]
注音唐詩三百首 / 蘅塘退士[孫洙]選輯. 千家詩欣賞. — 再版. — [高雄]：莊家出版社；台南：總經銷世一書局, 民國67 [1978]
251, 162, [17] p. ; 22 cm.
Cover title.
"朱自清著 「唐詩三百首」指導大概": p. 1-20.
Text accompanied with Chu yin fu hao.
NT$45.00

1. Chinese poetry—T'ang dynasty, 618-907. 2. Chinese poetry. I. Hsieh, Fang-te, 1226-1289, comp. Ch'ien chia shih. 1978. II. Title.
Title romanized: Chu yin T'ang shih san pai shou.
PL2531.S842 1978b — 79-840318

Sun, Yüan-heng.
(Ch'ih-ch'ien chi)
赤嵌集 [4卷] 孫元衡著 臺灣銀行經濟研究室編輯 臺北 臺灣銀行 民國47 [1958]
4, 83 p. 19 cm. (臺灣文獻叢刊第10種)
Cover title.
Poems.

I. Title. II. Series: T'ai-wan wen hsien ts'ung k'an, ti 10 chung.
PL2724.U5C4 1958 — C 61-2188

Sung, Fu-sen.
(Ch'e lao pan ti ko)
车老板的歌 / 宋福森. — [香港? : s. n., 1978?]
3, 85 p. : ill. ; 19 cm.
Poems.
Reprint of the 1975 ed. published by Chi-lin jen min ch'u pan she, Ch'ang-ch'un.
"本书收…诗歌四十首"

I. Title.
PL2905.N436C5 1978 — 79-838573

Tai, Wang-shu, 1905-1950.
(Wang-shu ts'ao)
望舒草 / 戴望舒著. — [香港?]：創作書社, [1978?]
15, 5, 116 p. ; 19 cm.
Poems.
Reprint of the 1936 ed. published by Fu hsing shu chü, Shanghai.
"詩論零札": p. 111-115.

I. Title.
PL2811.I3A17 1978 — 78-843475

T'ao, Ch'ien, 372?-427?
(T'ao Yüan-ming)
陶淵明：田園心聲 / 一海知義選；洪順隆析譯注. — 台北：林白出版社, 民國68 [1979]
8, 210 p., [1] fold. leaf of plates : ill. ; 20 cm. — (中國歷代詩人選集 ; 5)
"這裏只選五言詩六十五首, 附四言詩一首"
NT$60.00

I. Ikkai, Tomoyoshi, 1929- II. Hung, Shun-lung. III. Title. IV. Series: Chung-kuo li tai shih jen hsüan chi ; 5.
PL2665.T3A6 1979 — 79-839112

T'ao, Yung.
(T'ao ling shih ling)
陶令詩零 / [陶鏞]. — [香港? : s. n., 1978?]
13 double leaves ; 25 cm.
Poems.
Reprint ed.

I. Title.
PL2811.O3T3 — 79-835185

Te-liang.
(Hua shih)
画室：德亮詩画集. — 高雄：德馨室出版社, 民國67 [1978]
30, 145 p., [10] leaves of plates : ill. ; 20 cm. — (德馨詩集 ; 2)
Poems.
Series romanized: Te hsin shih chi.
NT$40.00

I. Title. II. Title: Te-liang shih hua chi.
PL2909.E34H8 — 79-836911

CHINESE POETRY (Continued)

Ts'ai, Yen-p'ei.
(Hsiao shih san chüan)
小詩三卷 / 蔡炎培. — 香港：明窗出版社：發行明報有限公司出版部, 1978.
182 p. ; 19 cm.
Poems.
$6.00 ($2.00 U.S.)

I. Title.
PL2912.A5153H7 79–841095

Tsang, K'o-chia, 1910–
(Sheng ming ti chiao han)
生命的叫喊 / 臧克家著. — 香港：上海書局, 1977.
2, 68 p. ; 18 cm. — (中國文學名著小叢書)
Poems.
Series romanized: Chung-kuo wen hsüeh ming chu hsiao ts'ung shu.
$2.00

I. Title.
PL2815.A6S48 78–843381

Tsao, Tsen-Cha, 1901–
(Shuang chien lou shih kao)
雙劍樓詩稿 = Poems and translations / 作者趙曾珏, 秦昭華. — [紐約] : 趙曾珏 ; 台北：經銷處台灣中華書局, 民國61 [1972]
15, 16, 140, 60 p. ; 20 cm.
Colophon title.
English and/or Chinese.
I. Tsao, Christina Ching, joint author. II. Title. III. Title: Poems and translations.
PL2815.A73S5 79–841406

Tu, Chih-min.
(Shao so feng hsüeh yeh)
哨所风雪夜 / 杜志民. — 第1版. — 呼和浩特：内蒙古人民出版社, 1977.
2, 124 p. ; 19 cm.
Poems.
$0.24

I. Title.
PL2913.C475S5 78–842551

Tu, Fu, 712–770.
[Poems]
杜詩評鈔 / [杜甫] ; 沈德潛纂. — [香港? : s. n., 1977?]
276 p. ; 22 cm. — (文學叢書)
In 4 chüan.
Series romanized: Wen hsüeh ts'ung shu.
Reprint of the ed. printed by Bunkyūdō, Kyoto.
I. Shen, Te-ch'ien, 1673–1769. II. Title.
Title romanized: Tu shih p'ing ch'ao.
PL2675.A17 1977 78–843999

T'u, T'ung-kuei.
(Yün yün chou shih)
孕雲盩詩 / [著者涂同軌]. — 九龍：涂公遂重印, 民國66 [1977]
3, 2, 25 double leaves, [1] leaf of plates : port. ; 28 cm.
In case.
Reprint of the 1934 ed.; with new postscript.

I. Title.
PL2816.T86Y86 1977 79–837847

Tuan, Chin-yün.
(Chin lu hsiao shih)
錦廬小詩 / 作者段錦雲. — 台南：興文齋書局, 民國66 [1977]
67 p., [1] leaf of plates : port. ; 19 cm.
Colophon title.
Errata slip inserted.

I. Title.
PL2914.A413C5 79–837368

(Tui lien ching hua)
對聯精華 / 艾蓮編. — 香港：駱駝出版社, 1977.
4, 131 p., [1] leaf of plates : ill. ; 19 cm.
$5.00

1. Couplets. 2. Chinese poetry. I. Ai, Lien.
PL2519.C7T83 79–835225

(Tui lien i wan lien)
對聯一萬聯 / 沈一忠編. — 台南：綜合出版社, 民國66 [1977]
12, 277 p. ; 20 cm.
$28.00

1. Couplets. 2. Chinese poetry. I. Shen, I-chung.
PL2519.C7T86 79–837048

T'ung, Chia-t'ung.
(Chin se ti yen ying)
金色的岩鹰：长诗 / 童嘉通. — [香港? : s. n., 1978?]
2, 237 p., [6] leaves of plates : ill. ; 21 cm.
Reprint of the 1977 ed. published by Ssu-ch'uan min tsu ch'u pan she, Chengtu.

I. Title.
PL2914.N4C5 1978 78–843985

T'ung, Chia-t'ung.
(Pien chiang shan yüeh)
边疆山月 / 童嘉通. — 第1版. — 成都：四川人民出版社, 1976.
4, 124 p. : ill. ; 19 cm.
Poems.
"詩丛"
$0.29

I. Title.
PL2914.N49P5 79–836973

Tung-shan, pseud.
(Pi shang Liang-shan)
逼上梁山 / 凍山著. — [香港? : s. n., 1978?]
52 p. : ill. ; 18 cm.
Poems.
Reprint of the 1948 ed. published by Shih ko ch'u pan she, Hongkong.

I. Title.
PL2817.N238P5 1978 79–837595

(Wakokubon Kanshi shūsei)
和刻本漢詩集成 / 長澤規矩也編. — [東京] : 古典研究会 : 汲古書院発売, 昭和50–52 [1975–1977]
20 v. ; 27 cm.
Photoreprint ed.
CONTENTS: 1–10. 唐詩—11–16. 宋詩—17–20. 補篇

1. Chinese poetry. 2. Poets, Chinese—Biography. I. Nagasawa, Kikuya, 1902– II. Title: Kanshi shūsei.
PL2518.W2 1975 76–810794
J

(Wakokubon Kanshi shūsei, sōshūhen)
和刻本漢詩集成·總集篇 / 長澤規矩也編. — [東京] : 出版 古典研究会 : 發行 汲古書院, 1978–
v. ; 27 cm.
Ja 78–25604 (v. 1)
Photoreprint ed.
¥6500 (v. 1)

1. Chinese poetry. I. Nagasawa, Kikuya, 1902– II. Title: Kanshi shūsei.
PL2518.W24 1978 79–802088

(Wan yu tui lien ta ch'üan)
萬有對聯大全 / 沈一忠編. — 台南：綜合出版社, 民國66 [1977]
12, 277, 12, 190 p. ; 19 cm.
NT$50.00

1. Couplets. 2. Chinese poetry. I. Shen, I-chung.
PL2519.C7W29 78–843986

Wang, Chiao-wo.
(T'ao se san san ch'ü)
桃色三三曲 / 王皎我著. — [香港? : s. n., 1978?]
2, 2, 234 p. ; 21 cm.
Poems.
Reprint of the 1931 ed. published by Ta tung shu chü, Shanghai, issued in series: Hsin wen hsüeh ts'ung shu.

I. Title.
PL2822.C35T3 1978 78–843551

Wang, Keng-hsin.
(Feng wu chi)
風物集 / 王耕心著. — 九龍：穫果出版社, 1977.
111 p. ; 19 cm.
Poems.
$6.00

I. Title.
PL2919.K4F4 78–840839

Wang, Lu-sung.
(Feng yü chung ti kuo hun)
風雨中的國魂 / 王祿松著. — 臺北：水芙蓉出版社, 民國67 [1978]
2, 3, 5, 247 p. : ill. ; 19 cm. — (水芙蓉書庫 ; 129)
Poems.
Series romanized: Shui fu jung shu k'u.

I. Title.
PL2919.L8F44 79–837374

Wen, Ch'an.
(Hsien yin chi)
閒吟集 / 文嬋著. — 臺北：啓明書局, 1977.
10, 4, 82 p. (on double leaves) ; 19 cm.
Poems.

I. Title.
PL2920.E555H7 78–841756

Wong, Yoon-wah.
(Nei wai chi)
内外集 / 王潤華著. — 臺北：國家出版社, 民國67 [1978]
3, 6, 167 p. ; 19 cm. — (讀書人叢刊 ; 58)
Poems.
Series romanized: Tu shu jen ts'ung k'an.
NT$45.00

I. Title.
PL2920.O44N4 79–835642

Wu, Wei-yeh, 1609–1672.
[Wu Mei-ts'un shih chi]
吳梅村詩集箋注 / 吳梅村[偉業]撰 ; [箋注者吳翌鳳]. — 臺景印初版. — 臺北：河洛圖書出版社, 民國64 [1975]
532, [26] p., [1] leaf of plates : port. ; 22 cm. — (夏學叢書)
Series romanized: Hsia hsüeh ts'ung shu.
NT$160.00
I. Wu, I-feng, 1742–1819. II. Title.
Title romanized: Wu Mei-ts'un shih chi ch'ien chu.
PL2732.U238W8 1975b 79–836166

CHINESE POETRY (Continued)

Ya-hsien.
(Ya-hsien tzu hsüan chi)
瘂弦自選集 / ⌐著作者瘂弦⌐. — 臺北：黎明文化事業公司, 民國66 ⌐1977⌐
8, 264 p., ⌐2⌐ leaves of plates : ill. ; 20 cm. — (中國新文學叢刊 ; 14)
Poems.
Series romanized: Chung-kuo hsin wen hsüeh ts'ung k'an.
Appendices (p. 227-264): 理論與態度.—有那麼一個人.—作品書目.—作品評論引得.
NT$70.00
I. Title.
PL2824.5.H8A17 1977 — 79-840346

Ya-t'ing.
(Hsiang ta ti)
向大地 / 亞汀著. — 臺北：龍門出版社, 民國46 ⌐1957⌐
ii, 50 p. ; 19 cm. — (新藝叢書 ; 4)
Poems.
Series romanized: Hsin i ts'ung shu.
I. Title.
PL2921.5.Z8H7 — 79-836548

Yang, T'ao-ying.
(Hu pien)
湖邊 / 楊桃英著 ; ⌐挿繪者吳璧人⌐. — 臺北：林白出版社, 民國66 ⌐1977⌐
91 p. : ill. ; 19 cm. — (詩與畫叢書 ; 15)
Poems.
NT$35.00
I. Title. II. Series: Shih yü hua ts'ung shu ; 15.
PL2922.T35H8 — 79-841085

Yang, T'ing, 1956–
(Ching t'ing liu shui)
靜聽流水 / 楊亭. — 彰化縣社頭鄉：大昇出版社：詩人季刊社發行, 民國67 ⌐1978⌐
122 p. ; 20 cm. — (詩人小集 ; 4)
Poems.
NT$30.00
I. Title. II. Series: Shih jen hsiao chi ; 4.
PL2922.T53C5 — 79-835189

Yang, Tzu-chien, 1953–
(Chien-ch'en shih ch'ao)
劍塵詩鈔：一卷 / 楊子澗著. — 嘉義：興國出版社 ; 高雄：總經銷高雄復文書局, 民國66 ⌐1977⌐
102 p. ; 20 cm.
NT$30.00
I. Title.
PL2922.T917C48 — 79-838253

Yeh, Shan.
(Pei tou hsing)
北斗行 / 楊牧⌐葉珊⌐. — 臺北：洪範書店：總經銷爾雅出版社, 民國67 ⌐1978⌐
6, 5, 217 p. ; 19 cm. — (洪範文學叢書 ; 28)
Poems.
Series romanized: Hung fan wen hsüeh ts'ung shu.
NT$50.00
I. Title.
PL2924.S47P4 — 78-841151

Yü, Kuang-chung.
(Yü yung heng pa ho)
與永恒拔河 / 余光中. — 臺北：洪範書店, 民國68 ⌐1979⌐
8, 213 p. ; 19 cm. — (洪範文學叢書 ; 41)
Poems.
Series romanized: Hung fan wen hsüeh ts'ung shu.
NT$55.00
I. Title.
PL2928.K8Y97 — 79-841404

Yü, Ta-kang, 1908–
(Liao yin ko chi)
寥音閣集 / 俞大綱著. — 台北：九思出版社, 民國66 ⌐1977⌐
22, 116 p., ⌐1⌐ leaf of plates : ports. ; 22 cm. — (九思叢書 ; 8)
Series romanized: Chiu ssu ts'ung shu.
CONTENTS: 寥音閣詩—寥音閣詞—寥音閣集外編
NT$80.00
I. Title.
PL2831.T35L5 — 77-842564
rev

Yüan, Hui.
(Ying hsiung ti hua hsiang)
英雄的画像 / 元辉. — 第1版. — 北京：人民文学出版社：新华书店北京发行所发行, 1978.
4, 172 p. ; 19 cm.
Poems.
$0.37
I. Title.
PL2929.A548Y5 — 79-838572

Yüan, Shui-p'o, 1916–
(Ma Fan-t'o ti shan ko)
馬凡陀的山歌 / ⌐著者馬凡陀 i. e. 袁水拍⌐. — ⌐香港? : s. n., 1978?⌐
6, 245 p. ; 19 cm.
Poems.
Reprint of the 1948 ed. published by Sheng huo shu tien, Hongkong.
I. Title.
PL2929.A6M3 1978 — 79-837856

Yüan, Tien.
(Shui liu ch'ien li)
水流千里 / 原甸著. — Hong Kong：新馬文化社, 1978.
67 p. ; 19 cm.
Poem.
$2.60
I. Title.
PL2929.A616S5 — 79-841860

—THREE KINGDOMS, 220-265

(Ts'ao Ts'ao, Ts'ao P'i, Ts'ao chih shih hsüan)
曹操曹丕曹植詩選 / 編輯部編選. — 板橋：廣城出版社, 民國66 ⌐1977⌐
5, 149 p. ; 22 cm.
NT$60.00
1. Chinese poetry—Three kingdoms, 220-265. I. Ts'ao, Ts'ao, 155-220. Poems. 1977. II. Ts'ao, P'i, 187-226. Poems. 1977. III. Ts'ao, Chih, 192-232. Poems. 1977. IV. Kuang ch'eng ch'u pan she. Pien chi pu.
PL2525.T75 — 78-843997

—T'ANG DYNASTY, 618-907

Fujiwara, Kintō, 966-1041, comp.
(Wa-Kan rōei shū)
倭漢朗詠集：⌐2巻 / 藤原公任撰 ; 伝藤原行成筆⌐. — 京都：便利堂, 昭和50 ⌐1975⌐
3 v. ; 21 cm. — Ja 76-12249
Cataloged from title piece mounted on case.
Title in colophon (mounted on case): Gyobutsu Wa-Kan rōei shū.
Issued in a case.
Reproduced from the Imperial Household Agency ms. copy.
CONTENTS: ⌐1⌐-⌐2⌐ 倭漢朗詠集 上下—⌐3⌐ 御物本倭漢朗詠集解説并釋文 吉澤義則監修
1. Japanese poetry—Heian period, 794-1185. 2. Chinese poetry—Japanese authors. 3. Waka. 4. Chinese poetry—T'ang dynasty, 618-907. I. Fujiwara, Yukinari, 971 or 2-1027. II. Yoshizawa, Yoshinori, 1876-1954. III. Title. IV. Title: Gyobutsu Wa-Kan rōei shū.
PL763.2.F8 1975 — 78-812679

Hsü, Cho, 1624-1713.
(Ch'üan T'ang shih lu)
全唐詩錄 / 徐倬編著. — 台北：宏業書局, 民國65 ⌐1976⌐
4 v. (10, 44, 54, 4468 p.) ; 22 cm.
Reprint of the 1706 ed.
NT$900.00
1. Chinese poetry—T'ang dynasty, 618-907. I. Title.
PL2531.H77 1976 — 78-841864

Shen, Te-ch'ien, 1673-1769, comp.
⌐T'ang shih pieh ts'ai chi⌐
唐詩別裁 / 沈德潛選. — 臺1版. — 臺北：臺灣商務印書館, 民國67 ⌐1978⌐
2 v. ; 19 cm. — (人人文庫 ; 552-553)
Originally published under title: T'ang shih pieh ts'ai chi.
Series romanized: Jen jen wen k'u.
NT$30.00 per vol.
1. Chinese poetry—T'ang dynasty, 618-907. I. Title.
Title romanized: T'ang shih pieh ts'ai.
PL2531.S5 1978 — 79-839121

Sun, Chu, 1711-1778, comp.
(T'ang shih san pai shou)
唐詩三百首：國語注音·言文對照·語譯註解·作法小傳 / ⌐蘅塘退士 i. e. 孫洙選輯⌐. — 台南：利大出版社：總經銷第一書店, 民國66 ⌐1977⌐
6, 241 p. ; 20 cm.
"朱自清著「唐詩三百首」指導大概": p. 1-21 (2d group)
Reprint ed.
NT$22.00
1. Chinese poetry—T'ang dynasty, 618-907. I. Chu, Tzu-ch'ing, 1898-1948. II. Title.
PL2531.S842 1977 — 78-842581

Sun, Chu, 1711-1778, comp.
(T'ang shih san pai shou)
唐詩三百首 / 蘅塘退士⌐孫洙⌐編訂. — 九龍：華聯出版社, ⌐1978?⌐
8, 380 p. ; 19 cm.
"言文對照, 考正注釋, 題解欣賞"
$4.00
1. Chinese poetry—T'ang dynasty, 618-907. I. Title.
PL2531.S842 1978 — 79-835042

Sun, Chu, 1711-1778, comp.
⌐T'ang shih san pei shou⌐
注音唐詩三百首 / 蘅塘退士⌐孫洙⌐選輯. 千家詩欣賞. — 再版. — ⌐高雄⌐：莊家出版社 ; 台南：總經銷世一書局, 民國67 ⌐1978⌐
251, 162, ⌐17⌐ p. ; 22 cm.
Cover title.
"朱自清著 「唐詩三百首」指導大概": p. 1-20.
Text accompanied with Chu yin fu hao.
NT$45.00
1. Chinese poetry—T'ang dynasty, 618-907. 2. Chinese poetry. I. Hsieh, Fang-te, 1226-1289, comp. Ch'ien chia shih. 1978. II. Title.
Title romanized: Chu yin T'ang shih san pai shou.
PL2531.S842 1978b — 79-840318

(T'ang shih hsüan)
唐詩选 / 中国社会科学院文学研究所编. — 第1版. — 北京：人民文学出版社, 1978.
2 v. ; 19 cm. — (中国古典文学读本丛书)
Series romanized: Chung-kuo ku tien wen hsüeh tu pen ts'ung shu.
Includes bibliographical references.
$2.10
1. Chinese poetry—T'ang dynasty, 618-907. I. Chung-kuo she hui k'o hsüeh yüan. Wen hsüeh yen chiu so.
PL2531.T277 — 78-842645

(T'ang shih i pai shou)
唐詩一百首 = Tangshi yibaishou / ⌐上海古籍出版社编辑⌐. — 第1版. — 上海：該社, 1978.
2, 4, 6, 154 p. ; 19 cm. — (中国古典文学作品选读)
$0.37
1. Chinese poetry—T'ang dynasty, 618-907. I. Shang-hai ku chi ch'u pan she. II. Title: Tangshi yibaishou. III. Series: Chung-kuo ku tien wen hsüeh tso p'in hsüan tu.
PL2531.T2775 — 79-835016

CHINESE POETRY (Continued)

—T'ANG DYNASTY, 618-907—HISTORY AND CRITICISM

Ch'en, Yin-k'o, 1890–1969.
(Yüan Pai shih chien cheng kao)
元白詩箋證稿 / 陳寅恪著. — 新1版. — 上海 : 上海古籍出版社, 1978.
5, 371 p. ; 21 cm.
$1.20

1. Chinese poetry—T'ang dynasty, 618–907—History and criticism. 2. Yüan, Chen, 779–831—Criticism and interpretation. 3. Pai, Chü-i, 772–846—Criticism and interpretation. I. Title.
PL2321.C53 1978 79–836014

Ho, Chi-p'eng.
(T'ang tai pien sai shih yen chiu)
唐代邊塞詩研究 / 何寄澎. — ⌊台北? : s. n.⌋, 民國62 ⌊1973⌋
166 p. (on double leaves) ; 27 cm.
Reproduced from typescript.
"國立臺灣大學中文研究所碩士論文"
Bibliography: p. 159–166.

1. Chinese poetry—T'ang dynasty, 618–907—History and criticism. I. Title.
PL2321.H54 79–841867

Ho, Chi-p'eng.
(Tsung shih yü kuan ch'ing)
總是玉關情 : 唐代邊塞詩初探 / 何寄澎著. — 臺北 : 聯經出版事業公司, 民國67 ⌊1978⌋
4, 168 p. ; 19 cm. — (文化叢刊)
Series romanized: Wen hua ts'ung k'an.
NT$40.00

1. Chinese poetry—T'ang dynasty, 618–907—History and criticism. I. Title.
PL2321.H55 79–837392

Jaeger, Georgette.
Les Lettrés chinois : poètes T'ang et leur milieu / Georgette Jaeger. — Neuchâtel : Éditions de la Baconnière, cop. 1977. Sw78-978
347 p. ; 21 cm.
Bibliography: p. 339-342.
Includes index.
ISBN 2-8252-0950-3 : 39.00F
1. Chinese poetry—T'ang dynasty, 618-907—History and criticism. 2. China—Intellectual life. 3. China—History—T'ang dynasty, 618-907. I. Title.
PL2321.J3 895.1'3'09 78-368780
78 MARC

Jeng, Lillian Leung.
Lectures on Chinese poetry : Ts'u poets in T'ang and Sung dynasties / Lillian Leung Jeng. — Taipei : Hua Hsia Cultural & Publication Center, c1978.
98, ⌊14⌋ p. : ill. ; 19 cm.
Title in Chinese on p. ⌊4⌋ of cover: T'ang Sung tz'u chia chü yao.
English and Chinese.
Bibliography: p. ⌊101⌋-⌊103⌋
1. Tz'u—History and criticism. 2. Poets, Chinese. 3. Chinese poetry—T'ang dynasty, 618-907—History and criticism. 4. Chinese poetry—Sung dynasty, 960-1279—History and criticism. I. Title. II. Title: Tz'u poets in T'ang und Sung dynasties. III. Title: T'ang Sung tz'u chia chü yao.
PL2343.J4 895.1'1'309 78-670204
79 MARC

(Li Pai Tu Fu Pai Chü-i)
李白杜甫白居易 / 歐陽彬 ; 許愼知編註. — 台南 : 大夏出版社, 民國67 ⌊1978⌋
148, 66, 67, ⌊34⌋ p. ; 21 cm.
"注音·詳註·白話"
CONTENTS: 白居易詩今釋.—李白詩今釋.—杜甫詩今釋.
NT$70.00
1. Chinese poetry—T'ang dynasty, 618–907—History and criticism. 2. Li, Po, 705?–762—Criticism and interpretation. 3. Tu, Fu, 712–770—Criticism and interpretation. 4. Pai, Chü-i, 772–846—Criticism and interpretation. I. Ou-yang, Pin. II. Hsü, Shen-chih. III. Pai, Chü-i, 772–846. IV. Li, Po, 705?–762. V. Tu, Fu, 712–770.
PL2321.L48 79–835117

(Li Tu shih chin shih)
李杜詩今釋 : 注音詳解 / 歐陽彬編注. — 台南 : 大夏出版社, 民國67 ⌊1978⌋
66, 67, ⌊23⌋ p. ; 21 cm.
Cover title: Chu yin pai hua Li Tu shih chin shih.
CONTENTS: 李白詩今釋.—杜甫詩今釋.
NT$30.00
1. Chinese poetry—T'ang dynasty, 618–907—History and criticism. 2. Li, Po, 705?–762—Criticism and interpretation. 3. Tu, Fu, 712–770—Criticism and interpretation. I. Ou-yang, Pin. II. Li, Po, 705?–762. Poems. 1978. III. Tu, Fu, 712–770. Poems. 1978. IV. Title: Chu yin pai hua Li Tu shih chin shih.
PL2321.L49 79–837677

Lü, Cheng-hui.
(Yüan Pai pi chiao yen chiu)
元白比較研究 / 呂正惠. — ⌊台北? : s. n.⌋, 民國63 ⌊1974⌋
119 p. (on double leaves) ; 27 cm.
"國立臺灣大學中文研究所碩士論文"
Cover title.
Reproduced from typescript copy.
Bibliography: p. 113–119.
1. Chinese poetry—T'ang dynasty, 618–907—History and criticism. 2. Yüan, Chen, 779–831. 3. Pai, Chü-i, 772–846. 4. Poets, Chinese—Biography. I. Title.
PL2321.L8 78–843893

Murakami, Tetsumi, 1930–
(Santaishi)
三体詩 / 村上哲見⌊著⌋. — 東京 : 朝日新聞社, 1978–
v. ; 15 cm. — (中国古典選 ; 29–) Ja 78–20828 (v. 1)
"増註本=泉南本に訓点を施したものを底本とする"
Includes original text.
Bibliography: v. 1, p. 27–38.
¥380 (v. 1)

1. Chinese poetry—T'ang dynasty, 618–907—History and criticism. 2. Chinese poetry—T'ang dynasty, 618–907—Translations into Japanese. 3. Japanese poetry—Translations from Chinese. I. San t'i T'ang shih. 1978. II. Title. III. Series: Chūgoku koten sen ; 29 ⌊etc.⌋
PL2321.M77 1978 78–816839

Soin, d. 1626.
(Santaishi Soin shō)
三体詩素隠抄 / 中田祝夫編. — 東京 : 勉誠社, 昭和52 ⌊1977⌋
2 v. : ill. ; 22 cm. — (抄物大系) Ja 77–26253
"影印底本には木版本の寛永14年丁丑版(国会図書館亀田文庫所蔵本)をとった"
¥20000
1. Chinese poetry—T'ang dynasty, 618–907—History and criticism. I. Nakada, Norio, 1915– II. San t'i T'ang shih. Japanese & Chinese. 1977. III. Title. IV. Series: Shōmono taikei.
PL2321.S6 1977 79–800098

Yokoyama, Iseo, 1935–
(Tōshi no kanshō)
唐詩の鑑賞 : 珠玉の百首選 / 横山伊勢雄. — 東京 : ぎょうせい, 1978.
3, 10, 429 p., ⌊5⌋ leaves of plates : ill. ; 22 cm. Ja 78–20827
¥2500

1. Chinese poetry—T'ang dynasty, 618–907—History and criticism. 2. Chinese poetry—T'ang dynasty, 618–907—Translations into Japanese. 3. Japanese poetry—Translations from Chinese. I. Title.
PL2321.Y57 78–815963

J

—T'ANG DYNASTY, 618-907—TRANSLATIONS INTO JAPANESE

Murakami, Tetsumi, 1930–
(Santaishi)
三体詩 / 村上哲見⌊著⌋. — 東京 : 朝日新聞社, 1978–
v. ; 15 cm. — (中国古典選 ; 29–) Ja 78–20828 (v. 1)
"増註本=泉南本に訓点を施したものを底本とする"
Includes original text.
Bibliography: v. 1, p. 27–38.
¥380 (v. 1)

1. Chinese poetry—T'ang dynasty, 618–907—History and criticism. 2. Chinese poetry—T'ang dynasty, 618–907—Translations into Japanese. 3. Japanese poetry—Translations from Chinese. I. San t'i T'ang shih. 1978. II. Title. III. Series: Chūgoku koten sen ; 29 ⌊etc.⌋
PL2321.M77 1978 78–816839

Takagi, Masakazu, 1912–
(Tōshisen)
唐詩選 高木正一⌊著 ⌊東京⌋ 朝日新聞社 ⌊1955⌋
12, 302, 20, 4 p. col. map. 19 cm. (中国古典選)
Series romanized: Chūgoku koten sen.
Includes original text.

1. Chinese poetry—T'ang dynasty, 618–907—Translations into Japanese. 2. Japanese poetry—Translations from Chinese. I. Li, P'an-lung, 1514–1570, comp. T'ang shih hsüan. Japanese & Chinese. 1955.
PL779.C3T3 J 67–3406

Takagi, Masakazu, 1912–
(Tōshisen)
唐詩選 高木正一⌊著 東京⌋ 朝日新聞社 ⌊1965–66⌋
2 v. illus. 19 cm. (中国古典選 第14–15巻)
Series romanized: Chūgoku koten sen.
Includes original text.
1. Chinese poetry—T'ang dynasty, 618–907—Translations into Japanese. 2. Japanese poetry—Translations from Chinese. I. Li, P'an-lung, 1514–1570, comp. T'ang shih hsüan. Japanese & Chinese. 1965. II. Title.
PL779.C3T3 1965 76–806433
rev

Yokoyama, Iseo, 1935–
(Tōshi no kanshō)
唐詩の鑑賞 : 珠玉の百首選 / 横山伊勢雄. — 東京 : ぎょうせい, 1978.
3, 10, 429 p., ⌊5⌋ leaves of plates : ill. ; 22 cm. Ja 78–20827
¥2500

1. Chinese poetry—T'ang dynasty, 618–907—History and criticism. 2. Chinese poetry—T'ang dynasty, 618–907—Translations into Japanese. 3. Japanese poetry—Translations from Chinese. I. Title.
PL2321.Y57 78–815963

—FIVE DYNASTIES, 907-960—HISTORY AND CRITICISM

Ho, Chin-lan.
(Wu tai shih jen chi ch'i shih)
五代詩人及其詩 / 何金蘭. — ⌊台北 : s. n.⌋, 民國66 ⌊1977⌋
2 v. (432 p. on double leaves) ; 26 cm.
"國立臺灣大學中國文學研究所博士論文"
Bibliography: v. 2, p. 421–432.

1. Chinese poetry—Five dynasties, 907–960—History and criticism. 2. Poets, Chinese—Biography. I. Title.
PL2322.H6 78–842912

(Sung shih i pai shou)
宋詩一百首 : Songshi yibaishou / ⌊上海古籍出版社编辑出版⌋. — 新1版. — 上海 : 該社, 1978.
2, 4, 7, 141 p. : 19 cm. — (中国古典文学作品选读
"本書係將原中华书局上海编辑所编辑出版... 补充, 修订出版"
$0.36

1. Chinese poetry—Sung dynasty, 960–1279. I. Shang-hai ku chi ch'u pan she. II. Chung-hua shu chü. Shang-hai pien chi so. Sung shih i pai shou. III. Title; Songshi yibaishou. IV. Series: Chung-kuo ku tien wen hsüeh tso p'in hsüan tu.
PL2533.S89 79–837907

Sung shih pai i ch'ao.
宋詩別裁 / 張景星選. — 臺1版. — 臺北 : 臺灣商務印書館, 民國67 ⌊1978⌋
2, 9, 121 p. : 19 cm. — (人人文庫 : 特554)
Series romanized: Jen jen wen k'u.
Originally published in 1761 under title: Sung shih pai i ch'ao.
NT$30.00

1. Chinese poetry—Sung dynasty, 960–1279. I. Chang, Ching-hsing, fl. 1761. II. Title.
Title romanized: Sung shih pieh ts'ai.
PL2533.S9 1978 79–837851

—SUNG DYNASTY, 960-1279—HISTORY AND CRITICISM

Hu, Yu-feng, 1951–
(Chin shih yen chiu)
金詩研究 / 胡幼峯著. — Taipei : 嘉新水泥公司文化基金會, 民國67 ⌊1978⌋
2, 129 p. : 22 cm. — (嘉新水泥公司文化基金會研究論文 ; 第410種) (嘉新水泥公司文化基金會叢書)
Series 1 romanized: Chia-hsin shui ni kung ssu wen hua chi chin hui yen chiu lun wen. Series 2 romanized: Chia-hsin shui ni kung ssu wen hua chi chin hui ts'ung shu.
Bibliography: p. 126–129.

1. Chinese poetry—Sung dynasty, 960–1279—History and criticism. I. Title.
PL2323.H78 79–838627

CHINESE POETRY
—SUNG DYNASTY, 960-1279
—HISTORY AND CRITICISM
(Continued)

Iritani, Sensuke, 1933-
(Sōshisen)
宋詩選 / 入谷仙介. — 東京：朝日新聞社, 1979.
2 v. ; 15 cm. — (中国古典選 ; 33-34) Ja 79-4722 (v. 1)
"本書の選択の基礎となったのは…「宋詩鈔」である"
Includes original text.
¥360 (v. 1) ¥320 (v. 2)
1. Chinese poetry—Sung dynasty, 960-1279—History and criticism. 2. Chinese poetry—Sung dynasty, 960-1279—Translations into Japanese. 3. Japanese poetry—Translations from Chinese. I. Title. II. Series: Chūgoku koten sen ; 33-34.
PL2323.I7 1979 79-809001

Jeng, Lillian Leung.
Lectures on Chinese poetry : Ts'u poets in T'ang and Sung dynasties / Lillian Leung Jeng. — Taipei : Hua Hsia Cultural & Publication Center, c1978.
98, [14] p. : ill. ; 19 cm.
Title in Chinese on p. [4] of cover: T'ang Sung tz'u chia chü yao.
English and Chinese.
Bibliography: p. [101]-[103]
1. Tz'u—History and criticism. 2. Poets, Chinese. 3. Chinese poetry—T'ang dynasty, 618-907—History and criticism. 4. Chinese poetry—Sung dynasty, 960-1279—History and criticism. I. Title. II. Title: Tz'u poets in T'ang und Sung dynasties. III. Title: T'ang Sung tz'u chia chü yao.
PL2343.J4 895.1'1'309 78-670204
79 MARC

(Sōshi)
宋詩：附·金 / 著者 佐藤保. — 東京：明治書院, 1978.
244 p. ; 19 cm. — (中国の名詩鑑賞 ; 8) Ja 78-21936
Colophon title.
¥1500
1. Chinese poetry—Sung dynasty, 960-1279—History and criticism. 2. Chinese poetry—Sung dynasty, 960-1279—Translations into Japanese. 3. Japanese poetry—Translations from Chinese. I. Satō, Tamotsu, 1934- II. Series: Chūgoku no meishi kanshō ; 8.
PL779.C3C465 vol. 8 78-815395
[PL2323]

Ts'ui, Jui-yü.
(Liu Yung yü Chou Pang-yen)
柳永與周邦彥 / 崔瑞郁. — [台北? : s.n.], 民國65 [1976]
57 double leaves ; 27 cm.
"國立台灣大學中文研究所碩士論文"
Reproduced from the typescript.
Bibliography: leaves 56-57.
1. Chinese poetry—Sung dynasty, 960-1279—History and criticism. 2. Liu, Yung, chin shih 1034—Criticism and interpretation. 3. Chou, Pang-yen, 1056-1121—Criticism and interpretation. I. Title.
PL2323.T78 78-843915

—SUNG DYNASTY, 960-1279
—TRANSLATIONS INTO JAPANESE

Iritani, Sensuke, 1933-
(Sōshisen)
宋詩選 / 入谷仙介. — 東京：朝日新聞社, 1979.
2 v. ; 15 cm. — (中国古典選 ; 33-34) Ja 79-4722 (v. 1)
"本書の選択の基礎となったのは…「宋詩鈔」である"
Includes original text.
¥360 (v. 1) ¥320 (v. 2)
1. Chinese poetry—Sung dynasty, 960-1279—History and criticism. 2. Chinese poetry—Sung dynasty, 960-1279—Translations into Japanese. 3. Japanese poetry—Translations from Chinese. I. Title. II. Series: Chūgoku koten sen ; 33-34.
PL2323.I7 1979 79-809001

(Sōshi)
宋詩：附·金 / 著者 佐藤保. — 東京：明治書院, 1978.
244 p. ; 19 cm. — (中国の名詩鑑賞 ; 8) Ja 78-21936
Colophon title.
¥1500
1. Chinese poetry—Sung dynasty, 960-1279—History and criticism. 2. Chinese poetry—Sung dynasty, 960-1279—Translations into Japanese. 3. Japanese poetry—Translations from Chinese. I. Satō, Tamotsu, 1934- II. Series: Chūgoku no meishi kanshō ; 8.
PL779.C3C465 vol. 8 78-815395
[PL2323]

—YÜAN DYNASTY, 1260-1368—
TRANSLATIONS INTO ENGLISH

Sui, Shu-sen, 1906- ed.
The wine of endless life : Taoist drinking songs from the Yuan dynasty / edited and translated by Jerome B. Seaton. — Ann Arbor, Mich. : Ardis, c1978.
viii, 59 p. ; 22 cm.
"All poems were taken from Ch'uan Yuan San-ch'u, edited by Sui Shu-sen."
Includes index.
1. Chinese poetry—Yüan dynasty, 1260-1368—Translations into English. 2. English poetry—Translations from Chinese. I. Seaton, Jerome B. II. Title.
PL2658.E3S77 895.1'1'408 78-110701
79 MARC

—MING DYNASTY, 1368-1644

Shen, Te-ch'ien, ed.
(Ming shih pieh ts'ai)
明詩別裁 / 沈德潛, 周準選. — 臺1版. — 臺北：臺灣商務印書館, 民國67 [1978]
2, 4, 13, 118, 131 p. ; 18 cm. — (人人文庫 ; 特556)
Series romanized: Jen jen wen k'u.
NT$30.00
1. Chinese poetry—Ming dynasty, 1368-1644. I. Chou, Chun, d. 1756. II. Title.
PL2536.S5 1978 79-838727

—MING DYNASTY, 1368-1644
—HISTORY AND CRITICISM

Matsushita, Tadashi, 1908-
(Min Shin no sanshisetsu)
明·清の三詩説 / 松下忠著. — 東京：明治書院, 1978.
9, 220 p. ; 22 cm. Ja 79-4723
Based on the author's Edo jidai no shifū shiron, part 2, published in 1969.
¥2500
1. Chinese poetry—Ming dynasty, 1368-1644—History and criticism. 2. Chinese poetry—Ch'ing dynasty, 1644-1912—History and criticism. I. Matsushita, Tadashi, 1908- Edo jidai no shifū shiron. II. Title.
PL2326.M3 79-808519

—MING DYNASTY, 1368-1644
—SOCIETIES, ETC.

Huang, Chih-min.
(Ming jen shih she chih yen chiu)
明人詩社之研究 / 黃志民. — [台北? : s.n.], 民國61 [1972]
134 double leaves ; 27 cm.
"國立政治大學中文研究所碩士論文"
Reproduced from typewritten copy.
Bibliography: leaves 129-134.
1. Chinese poetry—Ming dynasty, 1368-1644—Societies, etc. I. Kuo li cheng chih ta hsüeh. Chung wen yen chiu so. II. Title.
PL2252.H8 79-836130

—CH'ING DYNASTY, 1644-1912

(Chin tai shih hsüan)
近代詩選 / 北京大學中文系文學專門化一九五五級《近代詩選》小組選註. — 北京：人民文學出版社, 1963.
17, 494 p. ; 19 cm.
1. Chinese poetry—Ch'ing dynasty, 1644-1912. 2. Chinese poetry—Ch'ing dynasty, 1644-1912—History and criticism. 3. Chinese poetry—20th century. 4. Chinese poetry—20th century—History and criticism. I. Pei-ching ta hsüeh. Chin tai shih hsüan hsiao tsu.
PL2537.C47 78-840960

Kuo ch'ao shih pieh ts'ai chi.
清詩別裁 / 沈德潛選. — 臺1版. — 臺北：臺灣商務印書館, 民國67 [1978]
4 v. ; 19 cm. — (人人文庫 ; 特557-560)
Series romanized: Jen jen wen k'u.
Reprint ed.
NT$30.00 per vol.
1. Chinese poetry—Ch'ing dynasty, 1644-1912. I. Shen, Te-ch'ien, 1673-1769. II. Title.
Title romanized: Ch'ing shih pieh ts'ai.
PL2537.K84 1978 79-838679

(T'ai-p'ing-t'ien-kuo shih ko hsüan)
太平天国诗歌选 / 太平天国历史博物馆编. — 第1版. — 上海：上海人民出版社：新華書店上海发行所发行, 1978.
3, 8, 212 p. : ill. ; 19 cm.
Edition for 1961 published under title: T'ai-p'ing-t'ien-kuo ko yao.
$0.46
1. Taiping Rebellion, 1850-1864—Poetry. 2. Chinese poetry—Ch'ing dynasty, 1644-1912. I. T'ai-p'ing-t'ien-kuo li shih po wu kuan.
PL2537.T3 79-840161

—CH'ING DYNASTY, 1644-1912
—HISTORY AND CRITICISM

(Chin tai shih hsüan)
近代詩選 / 北京大學中文系文學專門化一九五五級《近代詩選》小組選註. — 北京：人民文學出版社, 1963.
17, 494 p. ; 19 cm.
1. Chinese poetry—Ch'ing dynasty, 1644-1912. 2. Chinese poetry—Ch'ing dynasty, 1644-1912—History and criticism. 3. Chinese poetry—20th century. 4. Chinese poetry—20th century—History and criticism. I. Pei-ching ta hsüeh. Chin tai shih hsüan hsiao tsu.
PL2537.C47 78-840960

Kao, Pai-shih.
(Kuang Hsüan shih t'an tien chiang lu chiao chu)
光宣詩壇點將錄斠註 / 高拜石撰 ; [原著者汪辟疆]. — [台]北：河洛圖書出版社, 民國65 [1976]
178 p. ; 21 cm. — (夏學新編 ; 005)
Series romanized: Hsia hsüeh hsin pien.
NT$100.00
1. Wang, Kuo-yüan, 1886- Kuang Hsüan shih t'an tien chiang lu. 2. Chinese poetry—Ch'ing dynasty, 1644-1912—History and criticism. I. Wang, Kuo-yüan, 1886- Kuang Hsüan shih t'an tien chiang lu. 1976. II. Title.
PL2327.W363K36 79-836038

Matsushita, Tadashi, 1908-
(Min Shin no sanshisetsu)
明·清の三詩説 / 松下忠著. — 東京：明治書院, 1978.
9, 220 p. ; 22 cm. Ja 79-4723
Based on the author's Edo jidai no shifū shiron, part 2, published in 1969.
¥2500
1. Chinese poetry—Ming dynasty, 1368-1644—History and criticism. 2. Chinese poetry—Ch'ing dynasty, 1644-1912—History and criticism. I. Matsushita, Tadashi, 1908- Edo jidai no shifū shiron. II. Title.
PL2326.M3 79-808519

T'ien-chin shih tsao chih kung ssu. Cheng chih ch'u. Li lun tsu.
(T'ai p'ing t'ien kuo shih ko ch'ien t'an)
太平天国诗歌浅谈 / 天津市造纸公司政治处理论组, 天津市历史研究所文学研究室. — [香港? : s. n., 1978?]
60 p., [1] leaf of plates : ill. ; 22 cm.
At head of title: Chung-kuo wen hsüeh shih chih shih tu wu.
Reprint of the 1978 ed. published by Chung-hua shu chü, Peking.
Includes bibliographical references.
1. Chinese poetry—Ch'ing dynasty, 1644-1912—History and criticism. 2. Taiping Rebellion, 1850-1864—Poetry. I. T'ien-chin shih li shih yen chiu so. Wen hsüeh yen chiu shih. II. Title.
PL2537.T5 1978 79-838193

—CH'ING DYNASTY, 1644-1912—HISTORY AND CRITICISM—ADDRESSES, ESSAYS, LECTURES

Cheng, Tzu-yu.
(Shih lun yü shih chi)
詩論與詩紀 / [鄭子瑜著]. — 九龍：中華書局, 1978.
4, 2, 156 p., [2] leaves of plates : facsims. ; 21 cm.
$7.00
1. Chinese poetry—20th century—History and criticism—Addresses, essays, lectures. 2. Chinese poetry—Ch'ing dynasty, 1644-1912—History and criticism—Addresses, essays, lectures. I. Title.
PL2332.C46 78-842153

CHINESE POETRY (Continued)

—20TH CENTURY

(Chin tai shih hsüan)
近代詩選 / 北京大學中文系文學專門化一九五五級《近代詩選》小組選註. — 北京：人民文學出版社, 1963.
17, 494 p. ; 19 cm.

1. Chinese poetry—Ch'ing dynasty, 1644–1912. 2. Chinese poetry—Ch'ing dynasty, 1644–1912—History and criticism. 3. Chinese poetry—20th century. 4. Chinese poetry—20th century—History and criticism. I. Pei-ching ta hsüeh. Chin tai shih hsüan hsiao tsu.
PL2537.C47 78–840960

(Ch'ing fa huo che hua lien)
青髮或者花臉 / [編著者詩人畫會]. — 臺北：香草山出版有限公司, 民國65 [1976]
345 p. : ill. ; 21 cm.
Errata slip inserted.
NT$80.00

1. Chinese poetry—Taiwan. 2. Chinese poetry—20th century. 3. Painting, Modern—20th century—Taiwan. I. Shih jen hua hui.
PL3031.T32C485 79–836342

(Chung-kuo hsien tai shu ch'ing shih i pai shou)
中國現代抒情詩一百首 / 璧華編著. — 香港：天地圖書有限公司, 1978.
3, 4, 247 p. ; 19 cm.
$8.00

1. Chinese poetry—20th century. I. Pi-hua.
PL2542.C55 79–835948

(Erh ch'i kuang ch'ang shih ch'ao)
二七广场诗抄. — [香港?：s. n., 1979?]
2, 81 p., [1] leaf of plates : ill. ; 22 cm.
Reprint of the 1978 ed. published by Ho-nan jen min ch'u pan she.

1. Chou, En-lai, 1898–1976—Poetry. 2. Chinese poetry—20th century.
PL2543.E7 1979 79–840097

(Fei yüeh yü ch'ao yüeh)
飛躍與超越 / [編輯者中國現代詩獎基金會]. — 臺北：該會, 民國63 [1974]
106 p. ; 22 cm.
"我國旅越名詩人吳望堯設置第一屆中國現代詩獎紀念特輯"
CONTENTS: 紀弦詩選.—羅青得獎作品.

1. Chinese poetry—20th century. I. Chi, Hsüan, 1913– II. Lo, Ch'ing, 1948– III. Chung-kuo hsien tai shih chiang chi chin hui.
PL2543.F34 79–835094

Fu-chien sheng chün ch'ü cheng chih pu, Lung-yen ti ch'ü ko ming wei yüan hui Ku t'ien sung pien hsieh tsu.
(Ku t'ien sung)
古田颂 / 福建省军区政治部, 龙岩地区革命委员会《古田颂》编写组. — [香港?：s. n., 1978?]
4, 256 p. ; 21 cm.
Reprint of the 1977 ed. published by Fu-chien jen min ch'u pan she.
$90.00

1. Chinese poetry—20th century. 2. Chung-kuo jen min chieh fan chün—Poetry. I. Title.
PL2543.F8 1978 79–835290

(Hsien tai ch'ing nien san wen hsüan)
現代青年散文選 / 盧勝彥編. — 8版. — [台中]：青山出版社：總經銷學海書局, 民國63 [1974]
148, 140 p. ; 19 cm.
NT$35.00

1. Chinese prose literature—20th century. 2. Chinese poetry—20th century. I. Lu, Sheng-yen, 1945–
PL2623.H734 1974 78–841968

(Hsin chung ti ko)
心中的歌：周总理在人民心中永生 = Xin zhong de ge / [本社编]. — 第1版. — 上海：上海人民出版社, 1977.
32 p. : ill. ; 26 cm.
"汉语注音读物"
$0.18

1. Chou, En-lai, 1898–1976—Poetry. 2. Chinese poetry—20th century. I. Shang-hai jen min ch'u pan she. II. Title: Xin zhong de ge.
PL2543.H7817 79–835922

(Hsin chung ti ko)
心中的歌：怀念敬爱的周总理. — [香港?：s. n., 1977]
3, 156 p. ; 21 cm.
Reprint of the 1977 ed. published by Kuang-tung jen min ch'u pan she.

1. Chinese poetry—20th century. 2. Chou, En-lai, 1898–1976—Poetry.
PL2543.H7816 1977 78–843871

(Hsin yü hsiao p'in)
心語小品 / 黃玉娥編著. — 台南：龍輝出版社, 民國67 [1978]
186 p. : ill. ; 19 cm. — (散文叢書)
Series romanized: San wen ts'ung shu.
NT$30.00

1. Chinese essays—20th century. 2. Chinese poetry—20th century. I. Huang, Yü-o.
PL2623.H743 79–836055

(Hsüeh hua ti yüeh hui)
雪花的約會 / 羊令野等著. — 臺北：天華出版事業股份有限公司, 民國67 [1978]
2, 8, 272 p. : ports. ; 19 cm. — (天華文學叢刊 ; 2)
Series romanized: T'ien hua wen hsüeh ts'ung k'an.
NT$60.00

1. Korea—Description and travel—1953– 2. Korea—Description and travel—1953– —Poetry. 3. Chinese poetry—20th century. 4. Chinese poetry—Translations from Korean. 5. Korean poetry—Translations into Chinese. I. Yang, Ling-yeh.
DS902.4.H77 78–843461

(Hua chu hsi ti tsu chi)
华主席足迹 / [诗刊社]. — [香港?：s. n., 1978?]
4, 164 p. ; 22 cm.
Reprint of the 1977 ed. published by Jen min ch'u pan she, Peking.

1. Chinese poetry—20th century. 2. Hua, Kuo-feng—Poetry. I. Shih k'an she.
PL2543.H83 1978 78–843894

(Huan ch'ing ti ku sheng)
欢庆的鼓声 / [福建人民出版社编辑]. — [香港?：s. n., 1978?]
138 p. ; 21 cm.
Reprint of the 1977 ed. published by Fu-chien jen min ch'u pan she, Foochow.

1. Chinese poetry—20th century. 2. Hua, Kuo-feng—Poetry. I. Fu-chien jen min ch'u pan she, Foochow, China.
PL2543.H85 1978 78–843245

(Hung mien hua k'ai)
红棉花开：广西叙事民歌选 / 广西文艺创作办公室编. — [香港?：s. n., 1979?]
233 p. : ill. ; 21 cm.
Reprint of the 1978 ed. published by Jen min wen hsüeh ch'u pan she, Peking.

1. Chinese poetry—China—Kwangsi (Province) 2. Chinese poetry—20th century. I. Kuang-hsi Chuang tsu tzu chih ch'ü. Ko ming wei yüan hui. Wen i ch'uang tso pan kung shih.
PL3031.K882H79 1979 79–840733

(I yeh hsiao shih)
一葉小詩 / 編著者黎亞衡. — 台南：龍輝出版社, 民國67 [1978]
9, 187 p., [1] leaf of plates : ill. ; 19 cm.
Colophon title.
NT$30.00

1. Chinese poetry—20th century. I. Li, Ya-heng.
PL2543.I25 79–837069

(Ko ming lieh shih shih ch'ao)
革命烈士诗抄 / 本社编. — 第2版. — [北京]：中國青年出版社：新华书店北京发行所, 1962, 1978 printing.
4, 256 p., [4] leaves of plates : ill. ; 19 cm.
$0.57

1. Revolutionary poetry, Chinese. 2. Chinese poetry—20th century. 3. Communism—Poetry. I. Chung-kuo ch'ing nien ch'u pan she. Peking.
PL2519.R48K6 1962 79–838194

(Kung she hsin ch'ü)
公社新曲 / 東鹿县文化馆编 ; [東鹿县美术创作组插图]. — 第1版. — 天津：天津人民出版社：天津市新华书店发行, 1974.
4, 130 p. : ill. ; 19 cm.
Title in Pinyin on cover: Gongshe xin qu.

1. Chinese poetry—20th century. I. Shu-lu hsien wen hua kuan. II. Shu-lu hsien mei shu ch'uang tso tsu. III. Title: Gongshe xin qu.
PL2543.K85 79–837232

(Mao chu hsi ch'i chih p'iao wan tai)
毛主席旗帜飘万代 / [甘肃人民出版社编辑, 出版]. — [香港?：s. n., 1978?]
4, 328 p. : ill. ; 21 cm.
Reprint of the 1977 ed. published by Kan-su jen min ch'u pan she, Lan-chou.
$120.00

1. Chinese poetry—20th century. 2. Mao, Tse-tung, 1893–1976, in fiction, drama, poetry, etc. I. Kan-su jen min ch'u pan she.
PL2543.M33 1978 78–842478

(Mao chu hsi yung yüan ho wo men tsai i ch'i)
毛主席永远和我们在一起 / [甘肃人民出版社编辑]. — [香港?：s. n., 1978?]
6, 243 p., [1] leaf of plates : port. ; 22 cm.
Reprint of the 1977 ed. published by Kan-su jen min ch'u pan she, Lan-chou.

1. Chinese poetry—20th century. 2. Mao, Tse-tung, 1893–1976, in fiction, drama, poetry, etc. I. Kan-su jen min ch'u pan she.
PL2543.M337 1978 79–837389

(Mo-t'e-erh i shu ying chi)
模特兒藝術影集 / 翁文苑主編. — 佳里鎮(台南縣)：全一出版社：欣文書局總經銷, 民國66 [1977]
ca. 150 p. : ill. ; 22 cm. — (欣文書摘 ; 1)
Cover title.
Series romanized: Hsin wen shu che.
NT$50.00

1. Chinese poetry—20th century. 2. Ballets—Pictorial works. I. Weng, Wen-yüan.
PL2543.M6 78–840580

CHINESE POETRY
—20TH CENTURY (Continued)

(Shih hsüan chi)
詩選集 / 李南衡主編 ; ₍助理編輯林邊, 梁景峯₎. — 台北 : 明潭出版社, 民國68 ₍1979₎
12, 9, 446 p. ; 22 cm. — (日據下台灣新文學 : 明集 ; 4)
NT$250.00 ($10.00 U.S.)

1. Chinese poetry—Taiwan. 2. Chinese poetry—20th century. I. Li, Nan-heng. II. Lin, Pien. III. Liang, Ching-feng. IV. Series: Jih chü hsia T'ai-wan hsin wen hsüeh : Ming chi ; 4.
PL3031.T32S52 79–840153

(Shou-feng shih she hsü chi)
壽峰詩社續集 / 著作各社員 ; 編校王天賞. — 高雄 : 壽峰詩社, 民國65 ₍1976₎
4, 295 p., ₍3₎ leaves of plates : ports. ; 22 cm.
Colophon title.
"非賣品"

1. Chinese poetry—Taiwan. 2. Chinese poetry—20th century. 3. Shou-feng shih she. I. Wang, T'ien-shang, 1903–
PL3031.T32S55 79–835013

(Shu chi ti tou li)
书记的斗笠 : 敍事诗集 / 浙江省纪念毛主席《在延安文艺座谈会上的讲话》发表三十周年征文办公室编. — ₍香港? : s. n., 1979?₎
103 p. ; 22 cm.
"纪念毛主席《在延安文艺座谈会上的讲话》发表卅周年征文选"
Reprint of the 1972 ed. published by Che-chiang jen min ch'u pan she, Hangchow.
CONTENTS: 崔銘先 书记的斗笠.—王河 采石歌.—卢祥耀 团结渠.

1. Chinese poetry—20th century. I. Che-chiang sheng chi nien Mao chu hsi tsai Yen-an wen i tso t'an hui shang ti chiang hua fa piao san shih chou nien cheng wen pan kung shih.
PL2543.S58 1979 79–840132

C

(Sung ko hsien kei Hua chu hsi)
頌歌献给华主席 / 本社编. — 第1版. — 太原 : 山西人民出版社, 1977.
3, 233 p. ; 21 cm.
"我们…编辑了诗集《頌歌献给华主席》, 以小三十二开本及时出版. 后来…我们又加以搜集整理, 在原基础上重新编辑, 出版大三十二开本"
$0.73

1. Chinese poetry—20th century. 2. Hua, Kuo-feng—Poetry. I. Shan-hsi jen min ch'u pan she, T'ai-yüan, China.
PL2543.S823 79–838042

C

Ta-chai ta tui. Yeh yü ch'uang tso tsu.
(Ta-chai hsin min ko hsüan)
大寨新民歌选 / 大寨大队业余创作组 ; ₍尾花孙平, 蒋松涛₎. — ₍香港? : s. n., 1978?₎
vii, 179 p., ₍4₎ leaves of plates : ill. ; 21 cm.
Reprint of the 1977 ed. published by Shang-hai jen min ch'u pan she, Shanghai.
$70.00

1. Chinese poetry—China—Ta-chai. 2. Chinese poetry—20th century. I. Title.
PL3031.T22T3 1978 78–842402

C

(Tieh lo tsai yü chung ti chi i)
跌落在雨中的記憶. — 台中 : 立誼出版社, 民國67 ₍1978₎
6, 153 p. ; 19 cm.
NT$35.00

1. Chinese poetry—20th century.
PL2543.T475 79–840012

(T'ien-an-men ko ming shih wen hsüan)
天安門革命詩文选 / 童怀周编. — ₍香港? : s. n., 1979?₎
5, 10, 584 p., ₍16₎ leaves of plates : ill. ; 21 cm.
"纪念敬爱的周总理逝世两周年"
Reprint ed. with postscript dated 1977 published by Fu-chien jen min ch'u pan she.

1. Chou, En-lai, 1898–1976—Poetry. 2. Chinese poetry—20th century. I. T'ung, Huai-chou.
PL2543.T483 1979 79–838686

(T'ien-an-men ko ming shih wen hsüan, hsü chi)
天安門革命詩文选 : 续集 / ₍童怀周编₎. — ₍香港? : s. n., 1979?₎
2, 11, 448 p., ₍10₎ leaves of plates : ill. ;
Cover title.
Reprint ed. with pref. dated 1978.

1. Chou, En-lai, 1898–1976—Poetry. 2. Chinese poetry—20th century. I. T'ung, Huai-chou.
PL2543.T4834 1979 79–838755

(T'ien-an-men shih wen chi)
天安门诗文集 / 童怀周编. — 第1版. — ₍北京₎ : 北京出版社 : 北京市新华书店发行, 1979.
18, 4, 488 p., ₍8₎ leaves of plates : ill. ; 19 cm.
Title also in Pinyin on cover: Tianmen shiwenji.
$1.15

1. Chou, En-lai, 1898–1976—Poetry. 2. Chinese poetry—20th century. I. T'ung, Huai-chou. II. Title: Tianmen shiwenji.
PL2543.T484 79–838685

(Tsai cho ko ti ch'uan)
載着歌的船 / 劉載福編. — 台南 : 綜合出版社, 民國66 ₍1977₎
9, 187 p. : ill. ; 19 cm.
NT$25.00

1. Chinese poetry—20th century. I. Liu, Tsai-fu.
PL2543.T72 79–836243

(Wan k'o chen chu sa Liang-shan)
万颗珍珠撒凉山 / 《万颗珍珠撒凉山》编辑组编. — ₍香港? : s. n., 1978?₎
5, 199 p. : ill. ; 21 cm.
Reprint of the 1977 ed. published by Ssu-ch'uan min tsu ch'u pan she, Chengtu.

1. Chinese poetry—20th century. I. Wan k'o chen chu sa Liang-shan pien chi tsu.
PL2543.W29 1978 78–842552

(Wang chu hsü chi)
網珠續集 ₍郭亦園編輯₎. — ₍香港 : 郭亦園, 己酉 i. e. 1969序₎
176 p. in various pagings ; 27 cm.
Cover title.

1. Chinese poetry—20th century. I. Kuo, I-yüan. II. Kuo, I-yüan. Wang chu chi.
PL2543.W36 79–837031

Wen, Jen-p'ing.
(Ching chih ti ting)
精緻的鼎 / 温任平. — 台南 : 長河出版社, 民國67 ₍1978₎
3, 2, 261 p. ; 20 cm. — (長河叢書 ; 20)
Essays.
Series romanized: Ch'ang ho ts'ung shu.
NT$60.00

1. Chinese poetry—20th century—History and criticism. I. Title.
PL2333.W4 79–836887

(Wo hsü yao i fen ai)
我需要一份愛 / ₍編輯委員張家華, 雪柔, 陳美卿₎. — 台北 : 龍風實業有限公司圖書出版部, 民國66 ₍1977₎
237, 6 p., ₍1₎ leaf of plates : ports. ; 22 cm. — (龍風叢書 ; 2)
"當代散文, 詩歌, 小說選輯"
Series romanized: Lung feng ts'ung shu.
NT$60.00
1. Chinese essays—20th century. 2. Chinese poetry—20th century. 3. Chinese fiction—20th century. I. Chang, Chia-hua. II. Hsüeh-jou. III. Ch'en, Mei-ch'ing.
PL2623.W58 78–843124

(Wo men ti tui wu hsiang t'ai yang)
我们的队伍向太阳 / 本社编. — ₍香港? : s. n., 1978₎
11, 222 p. : ill. ; 21 cm.
Reprint of the 1978 ed. published by Shang-hai wen i ch'u pan she, Shanghai.

1. Chinese poetry—20th century. I. Shang-hai wen i ch'u pan she.
PL2543.W55 1978 79–835811

(Yang kuang sa man wu ch'i lu)
阳光洒满五·七路 : 五·七干校诗选. — ₍香港? : s. n., 1979?₎
3, 137 p. : ill. ; 21 cm.
Poems.
Reprint of the 1977 ed. published by Jen min wen hsüeh ch'u pan she, Peking.

1. Wu ch'i kan hsiao—Poetry. 2. Chinese poetry—20th century.
PL2543.Y28 1979 79–841700

(Yen-an sung)
延安颂 : 诗集 / 延安大学中文系编. — 增订本. — ₍西安₎ : 陕西人民出版社 : 陕西省新華書店发行, 1977.
10, 370 p., ₍9₎ leaves of plates : ill. ; 21 cm.
$2.00

1. Chinese poetry—20th century. 2. Yenan, China—Poetry. I. Yen-an ta hsüeh. Chung wen hsi.
PL2543.Y43 79–837849

(Yin hsü chia ku wen tui lien)
殷墟甲骨文對聯 / ₍羅振玉 ... et al.集₎. — 台南 : 鳴宇出版社, 民國67 ₍1978₎
215 p. ; 20 cm.
NT$60.00

1. Couplet. 2. Chinese poetry—20th century. I. Lo, Chen-yü, 1866–1940.
PL2519.C7Y56 79–840270

(Ying cho chao yang shang shao wei)
迎着朝阳上哨位 / 广州部队政治部宣传部编. — ₍香港? : s. n., 1979?₎
4, 108 p. (on double leaves) ; 21 cm.
Reprint of the 1975 ed. published by Kuang-tung jen min ch'u pan she, Canton.

1. Chinese poetry—20th century. I. China (People's Republic of China, 1949–). Chung-kuo jen min chieh fang chün. Kuang-chou pu tui. Cheng chih pu. Hsüan ch'uan pu.
PL2543.Y496 1979 79–838613

—20TH CENTURY—HISTORY AND CRITICISM

(Chin tai shih hsüan)
近代詩選 / 北京大學中文系文學專門化一九五五級《近代詩選》小組選註. — 北京 : 人民文學出版社, 1963.
17, 494 p. ; 19 cm.

1. Chinese poetry—Ch'ing dynasty, 1644–1912. 2. Chinese poetry—Ch'ing dynasty, 1644–1912—History and criticism. 3. Chinese poetry—20th century. 4. Chinese poetry—20th century—History and criticism. I. Pei-ching ta hsüeh. Chin tai shih hsüan hsiao tsu.
PL2537.C47 78–840960

CHINESE POETRY
—20TH CENTURY
—HISTORY AND CRITICISM
(Continued)

Li, Yu, fl. 1973–
(Hung ping lou shih hua)
紅拉樓詩話 / 李猷著. — 臺北：臺灣商務印書館, 民國67 [1978]
[8], 164 p. ; 19 cm. — (人人文庫 ; 特590)
Series romanized: Jen jen wen k'u.
NT$30.00

1. Chinese poetry—20th century—History and criticism. I. Title.

PL2333.L45 79–836437

Lo, Ch'ing, 1948–
(Tsung Hsü Chih-mo tao Yü Kuang-chung)
從徐志摩到余光中 / 羅青著. — 臺北：爾雅出版社, 民國67– [1978–
v. ; 20 cm. — (爾雅叢書 ; 45–)
Series romanized: Erh ya ts'ung shu.
NT$60.00 (v. 1)

1. Chinese poetry—20th century—History and criticism. I. Title.

PL2333.L6 79–838615

Mu, Yü.
(Tang tai ming chia hsien tai shih hsi chieh)
當代名家現代詩析解：附篇擊轅吟 / 作者牧宇. — 九龍：自由報, 民國68 [1979]
170 p. ; 20 cm.
Colophon title.
$1.50

1. Chinese poetry—20th century—History and criticism. I. Title.

PL2333.M8 79–841709

Yang, Ch'ang-nien.
(Hsin shih p'in shang)
新詩品賞 / 楊昌年著. — [再版]. — 臺北：牧童出版社, 民國67 [1978]
3, 10, 612 p. ; 19 cm. — (牧原叢刊 ; 39)
Title in English in colophon: Appreciation of modern poetry.
Edition for 1975 published under title: Hsin shih yen chiu.
Series romanized: Mu yüan ts'ung k'an.
Bibliography: p. 605–612.
NT$125.00
1. Chinese poetry—20th century—History and criticism. 2. Poetry—History and criticism. 3. Poetics. I. Title. II. Title: Appreciation of modern poetry.

PL2333.Y3 1978 79–841452

—20TH CENTURY
—HISTORY AND CRITICISM
—ADDRESSES, ESSAYS, LECTURES

Cheng, Tzu-yu.
(Shih lun yü shih chi)
詩論與詩紀 / [鄭子瑜著]. — 九龍：中華書局, 1978.
4, 2, 156 p., [2] leaves of plates : facsims. ; 21 cm.
$7.00

1. Chinese poetry—20th century—History and criticism—Addresses, essays, lectures. 2. Chinese poetry—Ch'ing dynasty, 1644–1912—History and criticism—Addresses, essays, lectures. I. Title.

PL2332.C46 78–842153

Kuo, Hsiao-ch'uan.
(T'an shih)
谈诗 / 郭小川. — 第1版. — 上海：上海文艺出版社, 1978.
2, 131 p., [1] leaf of plates : ill. ; 19 cm.
$0.34

1. Chinese poetry—20th century—History and criticism—Addresses, essays, lectures. I. Title.

PL2333.K8 79–840096

T'ang, Wen-piao.
(T'ien kuo pu shih wo men te)
天國不是我們的 / 唐文標. — 臺北：聯經出版事業公司, 民國65 [1976]
288 p., [3] leaves of plates : ill. ; 20 cm. — (文化叢刊)
Series romanized: Wen hua ts'ung k'an.
Includes bibliographical references.
CONTENTS: 電影·戲劇的盲視.—現代詩的狂想.—大地的事.
NT$10.00
1. Chinese poetry—20th century—History and criticism—Addresses, essays, lectures. 2. Chinese drama—History and criticism—Addresses, essays, lectures. 3. Moving-picture plays—History and criticism—Addresses, essays, lectures. I. Title.

PL2333.T32 1976 76–840657

T'ang, Wen-piao.
(T'ien kuo pu shih wo men ti)
天國不是我們的 / 著者唐文標. — 香港：文化出版事業公司, [1978?]
287 p., [2] leaves of plates : ill. ; 19 cm.
Colophon title.
$7.00
1. Chinese poetry—20th century—History and criticism—Addresses, essays, lectures. 2. Chinese drama—History and criticism—Addresses, essays, lectures. 3. Moving-picture plays—History and criticism—Addresses, essays, lectures. I. Title.

PL2333.T32 1978 79–835519

—BIOGRAPHY see Poets, Chinese

—HISTORY AND CRITICISM

Chen, Cheng-siang.
(Shih ti ti li)
詩的地理 / 陳正祥著. — 香港：商務印書館, 1978.
2, 89 p., [3] fold. leaves of plates : maps ; 19 cm.
Includes bibliographical references.
$5.00

1. Chinese poetry—History and criticism. 2. China in literature. 3. Poetry of places—China—History and criticism. I. Title.

PL2307.C454 78–841332

Ch'en, Hsiang.
(Ying lien ku chin t'an)
楹聯古今談 / 陳香編著. — 臺北：國家書店, 民國67 [1978]
2 v. (2, 913 p.) ; 20 cm. — (讀書人叢刊 ; 69)
Poems.
Series romanized: Tu shu jen ts'ung k'an.
NT$80.00 per vol.

1. Couplets—History and criticism. 2. Chinese poetry—History and criticism. I. Title.

PL2309.C68C4 79–839457

Ch'en, I-tseng, 14th cent.
(Wen shuo)
文説 藝圃擷餘 唐音癸籤 [台北 臺灣商務印書館 1972]
2 v. 19 cm. (四庫全書珍本三集 399–400)
Cover title.
Reprint of 文淵閣手抄本
CONTENTS: 1. 陳繹曾 文説 王世懋 藝圃擷餘 胡震亨 唐音癸籤 卷1-13.—2. 胡震亨 唐音癸籤 卷14–33.
1. Chinese language—Rhetoric. 2. Chinese poetry—History and criticism. I. Wang, Shih-mao, 1536–1588. I p'u hsieh yü. 1972. II. Hu, Chen-heng, fl. 1630. T'ang yin kuei ch'ien. 1972. III. Title. IV. Title: I p'u hsieh yü. V. Title: T'ang yin kuei ch'ien. VI. Series: Ssu k'u ch'üan shu. Ssu k'u ch'üan shu chen pen san chi 399–400.

AC149.S7 1972, vol. 399–400 73–898048
[PL1271]

Ch'en, Yüan-fu.
[Ch'en-shan lou k'o erh shih hua. Japanese & Chinese]
枕山樓課児詩話 / [陳元輔撰] ; 長澤規矩也解題. — 東京：古典研究会：汲古書院発売, 1978.
375 p. ; 25 cm. — (和刻本漢籍随筆集 ; 20) Ja 78–10057
In Chinese, with Japanese reading marks.
Photoreprint ed.
CONTENTS: 陳元輔 枕山樓課児詩話—査為仁 蓮坡詩話—袁枚 隨園詩話—袁枚 續詩品三十二首—趙翼 甌北詩話—詩話三種: 滄浪詩話·談藝錄·秋圃擷餘
¥4500
1. Chinese poetry—History and criticism. I. Nagasawa, Kikuya, 1902– II. Title. III. Series: Wakokubon Kanseki zuihitsu shū ; 20.
Title romanized: Chinzanrō kaji shiwa.

PL2606.W3 vol. 20 79–809396
[PL2306]

Ch'eng, Ming-cheng.
(Sui-yüan shih hua shu p'ing)
隨園詩話述評 / 研究生程明琤. — [台北? : s.n.], 民國52 [1963]
17, 119 double leaves ; 27 cm.
Reproduced from ms. copy.
"國立台灣大學中文研究所碩士論文"

1. Yüan, Mei, 1716–1798. Sui-yüan shih hua. 2. Chinese poetry—History and criticism. I. Title.

PL2306.Y83C47 1963 78–842808

Chi, Che.
(Chung-kuo shih tz'u yen chin shih)
中國詩詞演進史 / [嵇哲著 : 編輯者本社編輯部]. — 臺北：莊嚴出版社, 民國67 [1978]
9, 316 p. ; 22 cm. — (古典大學叢書 ; 6)
Reprint of the 1971 ed. published by Ta hsüeh t'u shu ch'u pan she, T'ai-pei, in series: Ta hsüeh wen hsüeh ts'ung shu.
NT$90.00
1. Chinese poetry—History and criticism. 2. Tz'u—History and criticism. I. Chuang yen ch'u pan she. P'ien chi pu. II. Title. III. Series: Ku tien ta hsüeh ts'ung shu ; 6.

PL2307.C54 1978 79–838245

Chou, I-chin.
(Ming chia shih hsüan hsi)
名家詩選析：注音詳解 / 周益津編注. — 台南：大夏出版社, 民國67 [1978]
6, 161 p. ; 21 cm.
"注音白話"
NT$30.00

1. Chinese poetry—History and criticism. I. Title.

PL2307.C572 79–836575

Ch'ü, Hsüan-ying, 1892–
[Hsüeh shih ch'ien shuo]
詩學養海：中國古典詩詞入門 / [編印者莊嚴出版社]. — 臺北：該社：發行者草葉出版社, 民國66 [1977]
4, 216 p. ; 20 cm. — (古典新刊 ; 2)
Reprint of the 1961 ed. published by Shang-hai shu chü, Hongkong, under the title: Hsüeh shih ch'ien shuo and the authors' name: Ch'ü T'o-yüan [i. e. Hsüan-ying] and Chou Tzu-i.

NT$50.00

1. Chinese poetry—History and criticism. 2. Poetics. I. Chuang yen ch'u pan she. II. Chou, Tzu-i, joint author. III. Title. IV. Series: Ku tien hsin k'an ; 2.
Title romanized: Shih hsüeh i hai.

PL2307.C58 1977 77–843501

Ch'ü, Hsüan-ying, 1892–
[Hsüeh shih ch'ien shuo]
詩學養海：中國古典詩詞入門. — 九龍：百靈出版社, [1978?]
4, 216 p. ; 19 cm.
Originally published in 1961 by Shang-hai shu chü, Hongkong, under the title: Hsüeh shih ch'ien shuo and the authors' name: Ch'ü T'o-yüan and Chou Tzu-i, with preface deleted.
$7.00

1. Chinese poetry—History and criticism. 2. Chinese language—Versification. I. Chou, Tzu-i, joint author. II. Title. III. Title: Chung-kuo ku tien shih tz'u ju men.
Title romanized: Shih hsüeh i hai.

PL2307.C5737 1978 79–836515

C

(Chūgoku shibun sen)
中國詩文選 = Zhungguo shi wen hyan / 藤堂明保, 牛島徳次訳註. — 東京：江南書院, 1957.
170 p. ; 21 cm.

1. Chinese poetry—History and criticism. 2. Chinese poetry—Translations into Japanese. 3. Japanese poetry—Translations from Chinese. I. Tōdō, Akiyasu, 1915– II. Ushijima, Tokuji, 1918– III. Title: Zhungguo shi wen hyan.

PL2307.C597 78–816142

CHINESE POETRY
—HISTORY AND CRITICISM (Continued)

Fang, Tzu-tan.
(Chung-kuo li tai shih hsüeh t'ung lun)
中國歷代詩學通論 / 方子丹著. — [臺北?] : 大海文化事業股份有限公司, 民國67 [1978]
1398, [20] p. ; 22 cm.
Errata slip inserted.
NT$460.00

1. Chinese poetry—History and criticism. I. Title.
PL2307.F36 79-837525

Ho, Ch'i-fang, 1911-1977.
(Shih ko hsin shang)
诗歌欣赏 / 何其芳. — 第1版. — 北京 : 人民文学出版社 ; 新华书店北京发行所发行, 1962, t. p. 1978.
115 p. ; 19 cm.
$0.25

1. Chinese poetry—History and criticism. I. Title.
PL2307.H65 1962 79-840729

Hsieh, K'ang.
(Shih lien hsin hua)
詩聯新話 / 謝康著. — 臺北 : 中外圖書出版社, 民國66 [1977]
2, 267 p. : port. ; 19 cm. — (中外文庫 ; 28)
Cover title.
Reprint ed.
Series romanized: Chung wai wen k'u.
NT$40.00
1. Chinese poetry—History and criticism. 2. Couplets. 3. Chinese poetry. 4. Anecdotes—China. I. Title.
PL2307.H78 1977 78-843080

Huang, Chia-huan.
(Nan yüan lien hua)
楠園聯話 / 黃嘉煥著. — 台南 : 鳳凰城圖書公司, 民國67 [1978]
9, 258 p. : 20 cm. — (鳳凰新書 ; 002)
Series romanized: Feng huang hsin shu.
NT$49.00

1. Couplets—History and criticism. 2. Chinese poetry—History and criticism. I. Title.
PL2309.C68H8 79-836951

Huang, Chia-huan.
(Shih tz'u ku shih)
詩詞故事 / 黃嘉煥著. — 台南 : 鳳凰城圖書公司, 民國67 [1978]
16, 351 p. ; 20 cm. — (鳳凰新書 ; 1)
Series romanized: Feng huang hsin shu.
NT$59.00

1. Poets, Chinese—Anecdotes, facetiae, satire, etc. 2. Chinese poetry—History and criticism. I. Title.
PL2277.H78 79-837378

[**Huang, Hsü-wu**]
(Shih tz'u ch'ü yen chiu)
詩詞曲研究 / [編輯者本社編輯部]. — 臺北 : 莊嚴出版社, 民國67 [1978]
4, 243 p. ; 22 cm. — (古典大學叢書 ; 2)
Reprint of an earlier ed. by Huang Hsü-wu.
NT$60.00
1. Chinese poetry—History and criticism. 2. Chinese drama—History and criticism. 3. Tz'u—History and criticism. I. Chuang yen ch'u pan she. Pien chi pu. II. Title. III. Series: Ku tien ta hsüeh ts'ung shu ; 2.
PL2307.H795 1978 78-840909

Iritani, Sensuke, 1933-
(Koshisen)
古詩選 / 入谷仙介 ; 監修 吉川幸次郎. — 東京 : 朝日新聞社, 1978.
2 v. ; 15 cm. — (中国古典選 ; 23-24) Ja 78-15938 (v. 1)
¥360 (v. 1) ¥320 (v. 2)

1. Chinese poetry—History and criticism. 2. Chinese poetry—Translations into Japanese. 3. Japanese poetry—Translations from Chinese. I. Title. II. Series: Chūgoku koten sen ; 23-24.
PL2307.I74 78-814813

Kūkai, 774-835.
[Bunkyo hifu ron]
文鏡秘府論 / 弘法大師(空海)撰. — 臺景印初版. — 台北 : 河洛圖書出版社, 民國65 [1976]
6, 3, 266 p. ; 22 cm. — (夏學叢書)
Text in Chinese characters.
Series romanized: Hsia hsüeh ts'ung shu.
NT$120.00
1. Chinese language—Versification. 2. Chinese poetry—History and criticism. I. Title.
Title romanized: Wen ching mi fu lun.
PL1279.K84 1976 78-842101

Maeno, Naoaki, 1920-
(Kanshi no kaishaku to kanshō jiten)
漢詩の解釈と鑑賞事典 / 前野直彬, 石川忠久編. — 東京 : 旺文社, 1979.
686 p. ; 18 cm. Ja 79-8977
Includes indexes.
¥2000

1. Chinese poetry—History and criticism. 2. Chinese poetry—Translations into Japanese. 3. Japanese poetry—Translations from Chinese. I. Ishikawa, Tadahisa, 1932- joint author. II. Title.
PL2307.M3 79-810449

Nyogetsu, fl. 1517-1524.
(Chūka jakuboku shi shō)
中華若木詩抄 [東山如月和尚註] 十七行古活字版 [東京] 東洋文庫 [昭和45 i.e. 1970]
1 v. (various pagings) 22 cm. N.T. Ja***
岩崎文庫所蔵の古板本のうちから複製

1. Chinese poetry—History and criticism. 2. Chinese poetry—Japanese authors—History and criticism. I. Title.
PL2306.N9 1970b 74-801292

Nyogetsu, fl. 1517-1524.
[Chūka jakuboku shi shō]
語学資料としての中華若木詩抄 : [4巻] : 校本 / [東山如月註 ; 編者] 亀井孝. — 大阪 : 清文堂, 昭和52 [1977]
5, 529 p. ; 22 cm. Ja 77-22888
CONTENTS: 影印の部(古活字版・十八行本)—校異の部
¥9800
1. Chinese poetry—History and criticism. 2. Chinese poetry—Japanese authors—History and criticism. I. Kamei, Takashi, 1912- II. Title.
Title romanized: Gogaku shiryō to shite no Chūka jakuboku shi shō.
PL2306.N9 1977 78-816368

Nyogetsu, fl. 1517-1524.
(Chūka jakuboku shi shō)
中華若木詩抄 : [3巻 / 如月著 ; 中田祝夫編・解説]. — [再版]. — 東京 : 勉誠社, 昭和52 [1977]
3, 314, 57 p. ; 21 cm. — (勉誠社文庫 ; 20) Ja 77-26252
Series romanized: Benseisha bunko.
Reproduced from the Kokuritsu Kokkai Toshokan Kameda Bunko ms. copy dated 1633.
"抄物研究図書論文目録": p. [23]-57 (2d group)
¥2500
1. Chinese poetry—History and criticism. 2. Chinese poetry—Japanese authors—History and criticism. I. Nakada, Norio, 1915- II. Title.
PL2306.N9 1977b 79-800066

P'ei, P'u-hsien.
(Chi chü shih yen chiu hsü chi)
集句詩研究續集 / 裴普賢著. — 臺北 : 臺灣學生書局, 民國68 [1979]
3, 7, 309 p. ; 20 cm. — (學生書苑 ; 68)
Series romanized: Hsüeh sheng shu yüan.
NT$105.00

1. Chinese poetry—History and criticism. I. Title.
PL2307.P344 79-840591

Saitō, Shō, 1898-
(Kanshi nyūmon)
漢詩入門 / 齋藤晌著. — 東京 : 元々社, 昭和29 [1954]
3, 287 p., [1] leaf of plates : port. ; 18 cm. — (民族教養新書 ; 1)
Series romanized: Minzoku kyōyō shinsho.

1. Chinese poetry—History and criticism. 2. Chinese language—Versification. 3. Poetics. I. Title.
PL2307.S27 78-816927

Tai, Li-chu, 1946-
(Shih yü hua)
詩與畫 / 戴麗珠著. — 臺北 : 聯經出版事業公司, 民國67 [1978]
2, 2, 143 p., [8] leaves of plates : ill. ; 20 cm. — (文化叢刊)
Series romanized: Wen hua ts'ung k'an.
Bibliography: p. 135-143.
NT$40.00

1. Chinese poetry—History and criticism. 2. Painting, Chinese. 3. Art and literature. I. Title.
PL2307.T27 79-839162

Tanaka, Shū, 1912-
(Chūgokugo de manabu Kanshi)
中国語で学ぶ漢詩 : 発音注記言文対照 / 田中秀. — 東京 : 永和語学社 : 発売元 白馬出版, 昭和52 [1977]
230 p. : ill. ; 20 cm. Ja 77-8076
Colophon inserted.
¥1500
1. Chinese poetry—History and criticism. 2. Poetics. 3. Chinese language—Text-books for foreigners—Japanese. I. Title.
PL2307.T29 79-802316

Ts'ai, Cheng-sun, 13th cent.
[Ching hsüan ku chin ming hsien ts'ung hua shih lin kuang chi. Japanese & Chinese]
詩林廣記 : [20巻 / 蔡正孫撰]. 藝苑巵言 : [8巻 / 王世貞撰]. 名賢詩評 : [6巻 / 兪允文編] ; 長澤規矩也解題. — 東京 : 古典研究会 : 汲古書院, 1977.
343 p. ; 25 cm. — (和刻本漢籍随筆集 ; 18) Ja 78-5074
In Chinese, with Japanese reading marks.
Photoreprint ed.
Errata slip inserted.
¥4500
1. Chinese poetry—History and criticism. I. Nagasawa, Kikuya, 1902- II. Wang, Shih-chen, 1526-1590. I yüan chih yen. Japanese & Chinese. 1977. III. Yü, Yün-wen, 1512-1579. Ming hsien shih p'ing. Japanese & Chinese. 1977. IV. Title. V. Series: Wakokubon Kanseki zuihitsu shū ; 18.
Title romanized: Shirin kōki.
PL2606.W3 vol. 18 79-809160
[PL2306]

(Tz'u lin tien ku)
詞林典故 / [張廷玉 ... et al. 修輯]. — [台北 : 台灣]商務[印書館, 1976?]
2 v. ; 20 cm. — (四庫全書珍本六集 ; 120-121)
Cover title.
Photoreprint ed. of the Wen Yüan Ko copy.

1. Chinese poetry—History and criticism. I. Chang, T'ing-yü, 1672-1755. II. Series: Ssu k'u ch'üan shu. Ssu k'u ch'üan shu chen pen liu chi ; 120-121.
AC149.S7 1976, vol. 120-121 78-843918
[PL2306]

Wang, Pao-chen.
(Yü erh t'ung hsiao wan ku ch'ou)
與爾同銷萬古愁 / 王保珍著. — 臺北 : 長安出版社, 民國67 [1978]
2, 4, 6, 206 p. ; 19 cm.
"中國古典詩詞欣賞"
NT$45.00

1. Chinese poetry—History and criticism. 2. Tz'u—History and criticism. I. Title.
PL2307.W28 79-836840

Wang, Shih-chen, 1526-1590.
[Ming shih p'ing. Japanese & Chinese]
明詩評 / [王世貞撰] ; 長澤規矩也解題. — 東京 : 古典研究會, 1978.
395 p. ; 25 cm. — (和刻本漢籍随筆集 ; 19) Ja 78-8101
In Chinese, with Japanese reading marks.
Photoreprint ed.
CONTENTS: 王世貞 明詩評—胡應麟 詩藪—李沂 秋星閣詩話—蔣一葵 詩評集解—左培 書文式 馮班 鈍吟老人雑録—徐增 而菴詩話—王士禎 漁洋詩話—陳廷敬 杜律詩話—沈德潛 說詩晬語
¥4500

1. Chinese poetry—History and criticism. I. Nagasawa, Kikuya, 1902- II. Title. III. Series: Wakokubon Kanseki zuihitsu shū ; 19.
Title romanized: Minshihyō.
PL2606.W3 vol. 19 79-809147

CHINESE POETRY
—HISTORY AND CRITICISM (Continued)

Wei, Ch'ing-chih, fl. 1240–1244.
(Shih jen yü hsieh)
詩人玉屑 / 魏慶之編. — 新1版. — 上海：上海古籍出版社, 1978.
2 v. (2, 24, 630 p.) ; 19 cm.
$1.60

1. Chinese poetry—History and criticism. 2. Poetics. I. Title.
PL2306.W4 1978 79-835228

Yamada, Katsumi, 1909–
(Chūgoku meishi kanshō jiten)
中国名詩鑑賞辞典 / 山田勝美. — 東京：角川書店, 1978.
567 p. : ill. ; 19 cm. — (角川小辞典 ; 19) Ja 78-19278
Chinese and Japanese.
Series romanized: Kadokawa shōjiten.
Includes index.
¥1600
1. Chinese poetry—History and criticism. 2. Chinese poetry—Translations into Japanese. 3. Japanese poetry—Translations from Chinese. I. Title.
PL2307.Y26 78-814501

Yang, Sung-nien.
(Chung-kuo shih lun tso p'in ch'ien ch'üeh hsi t'ung ti wen t'i)
中国詩論作品欠鉠系統的問題 / 楊松年. — Singapore：南洋大學研究院人文與社會科學研究所, 1976.
17 leaves, [1] p. ; 28 cm. — (研究論文 - 南洋大學研究院人文與社會科學研究所 ; no. 27)
Reproduced from typescript.
Includes bibliographical references.

1. Chinese poetry—History and criticism. I. Title. II. Series: Nanyang University. Institute of Humanities and Social Sciences. Occasional paper series - Institute of Humanities and Social Sciences, Nanyang University ; no. 27.
PL2307.Y28 78-940252

(Yüeh fu shih yen chiu lun wên chi)
乐府詩研究論文集 作家出版社編輯部編 北京 作家出版社 1957-[70. 香港? 1970]
2 v. 22 cm. (中國古典文學研究論文集滙編第[1], 3輯) $11.50 (U.S.)
Vol. 1, without series statement, forms one of the 第1輯 of the series.
Vol. 2 contains articles selected and edited by the 中國語文學社 from the 文學遺產增刊 and other sources.
1. Chinese poetry—History and criticism. 2. Chinese ballads and songs—History and criticism. I. Tso chia ch'u pan shê, Peking. II. Chung-kuo yü wên hsüeh shê. III. Wên hsüeh i ch'an. Wên hsüeh i ch'an tsêng k'an. Selections. 1970. IV. Series: Chung-kuo ku tien wên hsüeh yen chiu lun wên chi hui pien.
PL2307.Y8 78-843679

—HISTORY AND CRITICISM —ADDRESSES, ESSAYS, LECTURES

Abe, Shōjirō, 1923–
(Chūgoku no shushiwa)
中国の酒詩話 / 阿部正次郎著. — 東京：高文堂出版社, 1978.
108 p. ; 18 cm. — (高文堂新書 ; 24) Ja 78-22508
Series romanized: Kōbundō shinsho.
¥480
1. Chinese poetry—History and criticism—Addresses, essays, lectures. 2. Drinking in literature—Addresses, essays, lectures. I. Title.
PL2308.A2 78-816309

Ch'ien, Chung-shu, 1911–
(Chung-kuo shih yü Chung-kuo hua)
中國詩與中國畫 / [錢鍾書著]. — 香港：龍門書店, 1969.
18, 7 p. ; 21 cm. — (中國文學研究叢編 ; 第1輯)
"原刊開明書店二十周年紀念文集, 因輯出單行·書次並刊入潘力山著從學理上論中國詩一文, 則采自小説月報第十七卷號外中國文學研究專刊."
Includes bibliographical references.
1. Chinese poetry—History and criticism—Addresses, essays, lectures. 2. Painting, Chinese—Addresses, essays, lectures. I. P'an, Li-shan. Ts'ung hsüeh li shang lun Chung-kuo shih. 1969. II. Title. III. Series: Chung-kuo wen hsüeh yen chiu ts'ung pien ; ti 1 chi.
PL2307.C554 79-840233

Chūgoku Koten Bungaku Kenkyūkai.
(Dokushojin no bungaku to shisō)
読書人の文学と思想：中国文学の世界 / 中国古典文学研究会編. — 東京：笠間書院, 昭和51 [1976]
269 p. ; 19 cm. — (笠間選書 ; 55) Ja 76-11799
Series romanized: Kasama sensho.
Includes bibliographical references.
¥1300
1. Chinese poetry—History and criticism—Addresses, essays, lectures. I. Title.
PL2307.C595 1976 78-814263

Ho, Ch'i-fang, 1911–1977.
(Hsin shih hua)
新詩話 / 何其芳. — 香港：神州圖書公司, 1977.
84 p., [1] leaf of plates : ill. ; 19 cm.
Reprint ed.
$6.50

1. Chinese poetry—History and criticism—Addresses, essays, lectures. I. Title.
PL2307.H58 1977 79-835920

Li, I-ting.
(Chung-kuo hua hui shih hua)
中國花卉史話 / 李邁[李奕定]著. — 台北：總經銷中國花卉雜誌社, 民國66 [1977]
176 p. : ill. ; 22 cm. — (中國花卉叢書)
NT$120.00

1. Chinese poetry—History and criticism—Addresses, essays, lectures. 2. Flowers in literature—Addresses, essays, lectures. 3. China—History—Anecdotes, facetiae, satire, etc. I. Title. II. Series: Chung-kuo hua hui ts'ung shu.
PL2308.L49 79-837063

Lin, Lin, poet.
(Shih ko tsa lun)
詩歌雜論 / 林林著. — [香港：神州圖書公司, 1978?]
147 p. ; 19 cm.
Reprint of the 1949 ed. published by Jen chien shu wu, Hongkong, in series: Jen chien wen ts'ung.

1. Chinese poetry—History and criticism—Addresses, essays, lectures. I. Title.
PL2307.L563 79-837150

Liu, Ta-pai, 1881–1932.
(Chiu shih hsin hua)
舊詩新話 / [劉大白]. — 台北：莊嚴出版社, 民國66 [1977]
6, 4, 230, 2 p. ; 20 cm. — (古典新刊 ; 5)
CONTENTS: 白屋詩話.—舊詩新話.—白屋說詩.
Pbk. : NT$50.00

1. Chinese poetry—History and criticism—Addresses, essays, lectures. I. Title. II. Series: Ku tien hsin k'an ; 5.
PL2307.L594 1977 78-840929

Liu, Ta-pai, 1881–1932.
(Chiu shih hsin hua)
舊詩新話 / [劉大白著]. — 九龍：百靈出版社, [1978?]
6, 230 p. ; 19 cm.
$7.50

1. Chinese poetry—History and criticism—Addresses, essays, lectures. I. Title.
PL2307.L594 1978 79-835285

Nyogetsu, fl. 1517–1524.
(Chūka jakuboku shi shō)
中華若木詩抄 [東山如月和尚註] 十七行古活字版 [東京] 東洋文庫 [昭和45 i.e. 1970]
1 v. (various pagings) 22 cm. N.T. Ja***
岩崎文庫所蔵の古板本のうちから複製

1. Chinese poetry—History and criticism. 2. Chinese poetry—Japanese authors—History and criticism. I. Title.
PL2306.N9 1970b 74-801292

(Sui-yüan shih fa ts'ung hua)
隨園詩法叢話 / 袁枚纂輯. — 台北：廣文書局, 民國67 [1978]
ca. 150 leaves in various foliations ; 22 cm. — (漢學彙編)
Reprint ed.
NT$250.00

1. Chinese poetry—History and criticism—Addresses, essays, lectures. I. Yüan, Mei, 1716–1798. II. Series: Han hsüeh hui pien.
PL2306.S9 1978 79-836058

Yang, Ling-yeh.
(Ch'ien shou ch'ien yen chi)
千手千眼集 / 羊令野著. — 台北：大林出版社, [1978?]
4, 197 p. ; 19 cm. — (大林文庫 ; 133)
Essays.
Series romanized: Ta lin wen k'u.
NT$40.00

1. Chinese poetry—History and criticism—Addresses, essays, lectures. 2. Arts, Chinese—Addresses, essays, lectures. I. Title.
PL2307.Y277 78-843938

—INDEXES

Wong, Kai-chee.
A research guide to English translation of Chinese verse : Han Dynasty to T'ang Dynasty / Kai-chee Wong, Pung Ho and Shu-leung Dang. — Hong Kong : Chinese University Press, 1977.
xii, 368 p. ; 27 cm.
Title on verso of t.p.: Chung shih Ying i so yin; Han chih T'ang.
"For each entry, the name of the poet and the title of the poem are given, both as the Chinese original and the English translation. Sequence of the poets and poems follows that of the Ch'uan Han San-kuo Chin Nan-Pei-chao shih ... by Ting Fu-pao ..., 1959 edition, and the Ch'uan T'ang shih ..., 1960 edition."
Bibliography: p. ix-xi.
ISBN 9622011411 : $60.00
1. Chinese poetry—Translations into English—Indexes. 2. English poetry—Translations from Chinese—Indexes. 3. Ting, Fu-pao, 1874-1952, ed. Ch'üan Han San-kuo Chin Nan-pei-ch'ao shih—Indexes. 4. Ch'üan T'ang shih—Indexes. 5. Chinese poetry—Indexes. I. Ho, P'eng, joint author. II. Teng, Shih-liang, joint author. III. Title. IV. Title: Chung shih Ying i so yin.
PL2658.E3W576 016.8951'1'008 73-92412
79 MARC

—JAPANESE AUTHORS

Fujiwara, Kintō, 966–1041, comp.
(Wa-Kan rōei shū)
倭漢朗詠集：[2巻 / 藤原公任撰；伝藤原行成筆]. — 京都：便利堂, 昭和50 [1975]
3 v. ; 21 cm. Ja 76-12249
Cataloged from title piece mounted on case.
Title in colophon (mounted on case): Gyobutsu Wa-Kan rōei shū.
Issued in a case.
Reproduced from the Imperial Household Agency ms. copy.
CONTENTS: [1]-[2] 倭漢朗詠集 上下—[3] 御物本倭漢朗詠集解說并釋文 吉澤義則監修

1. Japanese poetry—Heian period, 794–1185. 2. Chinese poetry—Japanese authors. 3. Waka. 4. Chinese poetry—T'ang dynasty, 618–907. I. Fujiwara, Yukinari, 971 or 2–1027. II. Yoshizawa, Yoshinori, 1876–1954. III. Title. IV. Title: Gyobutsu Wa-Kan rōei shū.
PL763.2.F8 1975 78-812679

(Gendai Nihon shishū, gendai Nihon Kanshi shū)
現代日本詩集·現代日本漢詩集 / [著者代表 島崎藤村]. — 東京：改造社, 昭和4 [1929]
595 p. : ports. ; 23 cm. — (現代日本文學全集 ; 37)

1. Japanese poetry—Meiji period, 1868–1912. 2. Japanese poetry—Taishō period, 1912–1926. 3. Chinese poetry—Japanese authors. 4. Poets, Japanese—Biography. I. Shimazaki, Tōson, 1872–1943. II. Title: Gendai Nihon Kanshi shū. III. Series: Gendai Nihon bungaku zenshū (Tokyo, Kaizō Sha) ; 37.
PL755.6.G42 vol. 37 79-802756
[PL763.6]

Hayashi, Kokei, 1875–1947.
(Kaifūsō shinchū)
懷風藻新註 / 林古溪[著；編者 林大]. — 東京：明治書院, 昭和33 [1958]
2, 2, 2, 336 p., [1] leaf of plates : facsims. ; 22 cm.
"本文と序との注釋で…流布の板本を以て底本とし、まま群書類従本等を參看"
Includes indexes.
1. Kaifūsō. 2. Chinese poetry—Japanese authors. 3. Chinese poetry—Translations into Japanese. 4. Japanese poetry—Translations from Chinese. I. Hayashi, Ōki, 1913– II. Kaifūsō. 1958. III. Title.
PL3054.A2K334 1958 79-808508

CHINESE POETRY
—JAPANESE AUTHORS (Continued)

Inokuchi, Atsushi, 1915–
(Josei to Kanshi)
女性と漢詩：和漢女流詩史 / 猪口篤志著. — 東京：笠間書院, 1978.
351 p., [1] leaf of plates : ill. ; 19 cm. — (笠間選書 ; 103)
Ja 78-19699
Series romanized: Kasama sensho.
¥1800
1. Chinese poetry—Women authors. 2. Chinese poetry—Japanese authors. 3. Chinese poetry—Translations into Japanese. 4. Japanese poetry—Translations from Chinese. I. Title.
PL2515.I5 78-816813

(Nijūhachihon narabini kuhon shīka)
二十八品并九品詩歌；現存三十六人詩歌屛風詩哥 / [編集 岡山大学池田家文庫等刊行会 責任編集 赤羽学]. — 岡山：福武書店, 昭和50 [1975]
145 p., [4] leaves of plates : ill. ; 22 cm. — (岡山大学国文学資料叢書 ; 3)
Ja 75-19786
Includes indexes.
¥2500
1. Waka. 2. Japanese poetry—1185-1600. 3. Saddharmapuṇḍarika—Poetry. 4. Chinese poetry—Japanese authors. I. Akahane, Manabu, 1928– II. Genson sanjūrokunin shika byōbu shika. 1975. III. Series: Okayama Daigaku. Okayama Daigaku kokubungaku shiryō sōsho ; 3.
PL758.N55 78-811648

Sugimoto, Yukio, 1907–
(Kaifūsō)
懐風藻 / 杉本行夫註釋. — 東京：弘文堂書房, 昭和18 [1943]
7, 2, 13, 410 p. ; 19 cm.
"寛政五年本を底本とした"
1. Kaifūsō. 2. Chinese poetry—Japanese authors. 3. Chinese poetry—Translations into Japanese. 4. Japanese poetry—Translations from Chinese. I. Kaifūsō. 1943. II. Title.
PL3054.A2K338 79-806172

—JAPANESE AUTHORS—HISTORY AND CRITICISM

(Gozan shisō)
五山詩僧 / 玉村竹二[解説]. — 東京：講談社, 1978.
488 p., [4] leaves of plates : ill. ; 20 cm. — (日本の禅語録 ; 8)
Ja 78-17596
Bibliography: p. 441-444.
CONTENTS: 現代語訳—原文：義堂周信 空華集—絶海中津 蕉堅藁—中巌円月 東海一漚集・文明軒雑談—虎関師錬 済北集—雪村友梅 岷峨集・宝覚真空禅師録
¥1800
1. Chinese poetry—Japanese authors—History and criticism. 2. Zen poetry—Japan—History and criticism. I. Tamamura, Takeji, 1911– II. Series: Nihon no Zen goroku ; 8.
PL3050.G68 79-800730

Nyogetsu, fl. 1517-1524.
[Chūka jakuboku shi shō]
語学資料としての中華若木詩抄：[4巻]：校本 / [東山如月註；編者] 亀井孝. — 大阪：清文堂, 昭和52 [1977]
5, 529 p. ; 22 cm. Ja 77-22888
CONTENTS: 影印の部(古活字版・十八行本)—校異の部
¥9800
1. Chinese poetry—History and criticism. 2. Chinese poetry—Japanese authors—History and criticism. I. Kamei, Takashi, 1912– II. Title.
Title romanized: Gogaku shiryō to shite no Chūka jakuboku shi shō.
PL2306.N9 1977 78-816368

Nyogetsu, fl. 1517-1524.
(Chūka jakuboku shi shō)
中華若木詩抄：[3巻 / 如月著；中田祝夫編・解説]. — [再版]. — 東京：勉誠社, 昭和52 [1977]
3, 314, 57 p. ; 21 cm. — (勉誠社文庫 ; 20) Ja 77-26252
Series romanized: Benseisha bunko.
Reproduced from the Kokuritsu Kokkai Toshokan Kameda Bunko ms. copy dated 1633.
"抄物研究図書論文目録": p. [23]-57 (2d group)
¥2500
1. Chinese poetry—History and criticism. 2. Chinese poetry—Japanese authors—History and criticism. I. Nakada, Norio, 1915– II. Title.
PL2306.N9 1977b 79-800066

—JAPANESE AUTHORS—INDEXES

Tatsumi, Masaaki.
(Kaifūsō kanji sakuin)
懐風藻漢字索引：付 本文篇 / 辰巳正明編. — 東京：新典社, 1978.
246 p. ; 22 cm. Ja 78-27148
"底本は「本朝儷風詩集」(寛政五年竹笣楼再校系版本)を使用"
¥3000
1. Kaifūsō—Indexes. 2. Chinese poetry—Japanese authors—Indexes. 3. Chinese poetry—Translations into Japanese—Indexes. 4. Japanese poetry—Translations from Chinese—Indexes. I. Kaifūsō. 1978. II. Title.
PL3054.A2K32 1978 79-804691

—KOREAN AUTHORS

(Chŏngdangmae sijip)
政堂梅詩集 / [編輯 姜大寬]. — 서울：晋州姜氏中央宗會, 1975.
255 p., [4] leaves of plates : ill. ; 22 cm.
Errata slip inserted.
"姜通亭·通溪兄弟兩先生…紀念事業으로詩集을發刊"
1. Chinese poetry—Korean authors. 2. Kang, Hoe-baek, 1357-1402—Anecdotes. 3. Kang, Hoe-jung, 1360-1421—Anecdotes. I. Kang, Hoe-baek, 1357-1402. II. Kang, Hoe-jung, 1360-1421. III. Kang, Tae-gon.
PL3074.C54 78-826854

(Han'guk yŏktae aejŏng Hansi sŏn)
韓國歷代愛情漢詩選 / 韓喆熙編譯. — 서울：益文社, 1976.
252 p. ; 18 cm.
Includes original texts.
W800
1. Chinese poetry—Korean authors. 2. Korean poetry—Translations from Chinese. 3. Chinese poetry—Translations into Korean. 4. Love poetry, Chinese. 5. Love poetry, Korean. I. Han, Ch'ŏr-hŭi.
PL982.C3H27 79-825734

(Hŏnsusong)
獻壽頌. — [서울]：公報室, [1955?]
44 double leaves, [4] leaves of plates : ill. ; 27 cm.
1. Rhee, Syngman, 1875-1965—Anniversaries, etc. 2. Rhee, Syngman, 1875-1965—Poetry. 3. Chinese poetry—Korean authors. I. Rhee, Syngman, 1875-1965. II. Korea (Republic). Kongbosil.
DS916.5.R5H66 78-826792

(Sijo, Hansi)
時調·漢詩 / 譯註金鼎烈. — 서울：書榮出版社, 1978.
562 p. : ill. ; 23 cm. — (精選韓國古典文學全集 ; 4)
Colophon inserted.
W4500
1. Sijo. 2. Chinese poetry—Korean authors. 3. Korean poetry—To 1900. I. Kim, Chŏng-nyŏl, fl. 1978– II. Series: Chŏngsŏn Han'guk kojŏn munhak chŏnjip ; 4.
PL975.6.S49 79-825899

(Taedong sisŏn)
大東詩選 / [張志淵, 權純九編輯；洪弼周, 鄭萬朝, 尹喜求校閱]. — 서울：曹龍承, 1978.
1379 p. in various pagings ; 22 cm.
Limited edition: 100 copies printed.
Colophon inserted.
Photoreproduction of the 1918 ed. in 12 kwŏn.
W15000
1. Chinese poetry—Korean authors. I. Chang, Chi-yŏn, 1864-1921. II. Kwŏn, Sun-gu.
PL3074.T3 1978 78-826553

—THEMES, MOTIVES

Lee, Peter H 1929-
Celebration of continuity : themes in classic East Asian poetry / Peter H. Lee. — Cambridge : Harvard University Press, 1979.
viii, 264 p. ; 24 cm.
Includes bibliographical references and index.
ISBN 0-674-10457-9
1. Chinese poetry—Themes, motives. 2. Korean poetry—Themes, motives. 3. Japanese poetry—Themes, motives. I. Title.
PL2307.L4 895 78-26145
78 MARC

—TRANSLATIONS FROM ENGLISH

Byron, George Gordon Noël Byron, Baron, 1788-1824.
[Works. Chinese]
拜倫詩選 / 梁貞譯. — 再版. — 香港：上海書局, 1977.
106 p. ; 18 cm.
$2.40
I. Title.
Title romanized: Pai-lun shih hsüan.
PR4354.Z5C5 1977 79-837196

C

—TRANSLATIONS FROM FOREIGN LANGUAGES

Ch'in, Tzu-hao, 1912-1963.
(Shih chieh ming shih hsin shang)
世界名詩欣賞 / 覃子豪著. — 台中：新企業世界出版社, 民國66 [1977]
3, 215 p. ; 20 cm. — (詩歌文庫)
Reprint ed.
NT$35.00
1. Poetry—History and criticism. 2. Chinese poetry—Translations from foreign languages. 3. Poetry—Translations into Chinese. I. Title. II. Series: Shih ko wen k'u.
PN1119.C5C57 1977 78-843667

C

—TRANSLATIONS FROM KOREAN

(Hsüeh hua ti yüeh hui)
雪花的約會 / 羊令野等著. — 臺北：天華出版事業股份有限公司, 民國67 [1978]
2, 8, 272 p. : ports. ; 19 cm. — (天華文學叢刊 ; 2)
Series romanized: T'ien hua wen hsüeh ts'ung k'an.
NT$60.00
1. Korea—Description and travel—1953– 2. Korea—Description and travel—1953– —Poetry. 3. Chinese poetry—20th century. 4. Chinese poetry—Translations from Korean. 5. Korean poetry—Translations into Chinese. I. Yang, Ling-yeh.
DS902.4.H77 78-843461

Tu, Fu, 712-770.
[Poems. Korean and Chinese]
杜甫詩選 / 杜甫著；李元燮譯. — 서울：正音社, 1976.
289 p. ; 18 cm. — (正音文庫 ; 115)
Series romanized: Chŏngŭm mun'go.
Colophon inserted.
W350
I. Yi, Wŏn-sŏp. II. Title.
Title romanized: Tu Po sisŏn.
PL2675.A55 1976 79-826354

—TRANSLATIONS INTO ENGLISH

Ch'en, Mai-tzu, 1914–
(Mai-tzu Chung Ying wen ho k'an shih hsüan)
邁子中英文合刊詩選 = Selection of Chinese poems by Prof. Chen Mai-tze with English translation. — [臺北? : s. n., 1978?]
75, 2 p., [8] leaves of plates : ill. ; 22 cm.
Cover title.
Chinese and English.
I. Title. II. Title: Selection of Chinese poems by Prof. Chen Mai-tze with English translation.
PL2840.M28A25 79-835773

Chou, En-lai, 1398-1976.
In quest ; poems of Chou En-lai / translated by Nancy T. Lin. — Hongkong : Joint Pub. Co., 1979.
iii, 51 p., [3] leaves of plates : ill. ; 22 cm.
English and Chinese.
Title also in Chinese: Chou En-lai shih hsüan.
ISBN 9620400011 : $10.00 ($3.45 U.S.)
I. Title. II. Title: Chou En-lai shih hsüan.
PL2754.E5A229 895.1'1'5 79-104222
79 MARC

CHINESE POETRY
—TRANSLATIONS INTO ENGLISH
(Continued)

Han, Yü, 768-824.
Growing old alive : poems / by Han Yü ; in versions by Kenneth O. Hanson. — Port Townsend, Wash. : Copper Canyon Press, c1978.
46 p. : ill. ; 23 cm.
ISBN 0-914742-33-7 : $3.50
I. Hanson, Kenneth O., 1922- . II. Title.
PL2670.A24 1978 895.1'1'3 78-55419
78 MARC

Han-shan, fl. 627-649.
Le Clodo du Dharma : 25 poèmes / de Han-shan ... ; présentés ⌊et traduits⌋ par Jacques Pimpaneau ; ⌊publié par le⌋ Centre de publication Asie orientale .... — Paris (Tour centrale, 2, place Jussieu, 75221, cedex 05) : Centre de publication Asie orientale, 1975.
66 p. : ill. ; 30 cm. — (Bibliothèque asiatique ; 18) F76-8698
French translations and Chinese text of Han-shan shih chi.
"Calligraphies de Li Kwok-wing."
I. Pimpaneau, Jacques. II. Li, Kuo-jung, 1929- . III. Title.
PL2677.H3A36 895.1'1'2 79-368512
79 MARC

Hong Kong : images on shifting waters : an anthology / compiled by Joyce Hsia, T. C. Lai. — Hong Kong : Kelly & Walsh, c1977.
150 p. : ill. ; 23 cm.
$25.00
1. English poetry—20th century. 2. English poetry—Hong Kong. 3. Hong Kong—Poetry. 4. English poetry—Translations from Chinese. 5. Chinese poetry—Translations into English. I. Hsia, Joyce. II. Lai, T'ien-ch'ang.
PR1195.H59H66 821 78-102569
79 MARC

Ikkyū, 1394-1481.
Ikkyū Sōjun: a Zen monk and his poetry. ⌊Translation and commentary by⌋ Sonja Arntzen. ⌊Bellingham⌋ Program in East Asian Studies, Western Washington State College ⌊1973⌋
x, 171 p. illus. 28 cm. (Western Washington State College. Program in East Asian Studies. Occasional paper no. 4) $3.75
Contains 56 poems from the Kyōunshū with text in Japanese (in Chinese characters) with romanization, and translations and commentary in English.
Bibliography: p. 167-170.
I. Arntzen, Sonja, 1945- tr. II. Series: Washington (State). Western Washington State College, Bellingham. Program in East Asian Studies. Occasional paper no. 4.
PL2698.I35K913 1973 895.6'1'2 73-620051
73⌊r78⌋rev MARC

Ku shih shih chiu shou.
The nineteen ancient poems / translated and with an introd. by Kenneth P. H. Ho. — Hong Kong : Kelly & Walsh, 1977.
158 p. : ill. ; 22 cm.
Title also in Chinese: Ku shih shih chiu shou.
English and/or Chinese.
Errata slip inserted.
Bibliography: p. 155-158.
1. Chinese poetry—Translations into English. 2. English poetry—Translations from Chinese. I. Ho, P'ei-hsiung. II. Title.
PL2658.E3K8 1977 895.1'1'2 79-100429
79 MARC

Lai, T'ien-ch'ang.
Nobel fragrance : Chinese flowers & trees / T. C. Lai. — Kowloon, Hong Kong : Swindon Book Co., c1977.
228 p. : ill. ; 23 cm.
1. Chinese poetry—Translations into English. 2. English poetry—Translations from Chinese. 3. Flowers in literature. 4. Trees in literature. I. Title.
PL2658.E5L3 895.1'1'5 78-106334
79 MARC

Mei, Lan-fang, 1894-1961.
Homecoming of a Phoenix / translation and verse adaptation by Sophia Shangkuan Min. — ⌊Taipei⌋ : Yen Huan Wu, c1976
60 p. ; 27 cm.
"This adaptation is based on ... the revised version performed by the famous actor Mei Lan-fang ..."
Title in Chinese on cover: Feng huan ch'ao.
I. Min, Sophia Shangkuan. II. Title. III. Title: Feng huan ch'ao.
PL2883.I244F413 1976 782.1'2 79-348021
79 MARC

The Penguin book of Zen poetry / edited and translated by Lucien Stryk and Takashi Ikemoto ; with an introd. by Lucien Stryk. — Chicago : Swallow Press, 1977.
160 p. : ill. ; 21 cm.
ISBN 0-8040-0792-6 : $8.95
1. Zen poetry—Translations into English. 2. Haiku—Translations into English. 3. Japanese poetry—Translations into English. 4. Chinese poetry—Translations into English. 5. English poetry—Translations from Japanese. 6. English poetry—Translations from Chinese. I. Stryk, Lucien. II. Ikemoto, Takashi, 1906-
PL782.E3P4 1977b 895.6'1'008 77-83237
78 MARC

The Penguin book of Zen poetry / edited and translated ⌊from the Japanese and Chinese⌋ by Lucien Stryk and Takashi Ikemoto ; with an introduction by Lucien Stryk. — London : Allen Lane, 1977.
160 p. : ill., ports. ; 21 cm. GB78-06538
ISBN 0-7139-1041-0 : £3.95
1. Zen poetry—Translations into English. 2. Haiku—Translations into English. 3. Japanese poetry—Translations into English. 4. Chinese poetry—Translations into English. 5. English poetry—Translations from Japanese. 6. English poetry—Translations from Chinese. I. Stryk, Lucien. II. Ikemoto, Takashi, 1906-
PL782.E3P4 1977 895.6'1'008 78-319502
78 MARC

Wang Wei, 701-761.
Six poems from Wang Wei / ⌊translations⌋ by David Young. — ⌊Oberlin, Ohio⌋ : Triskelion Press, 1969.
⌊8⌋ leaves ; 22 cm.
85 copies, no. 66
I. Title.
PL2676.A17 1969 895.1'1'3 79-303544
79 MARC

—TRANSLATIONS INTO ENGLISH
—INDEXES

Wong, Kai-chee.
A research guide to English translation of Chinese verse : Han Dynasty to T'ang Dynasty / Kai-chee Wong, Pung Ho and Shu-leung Dang. — Hong Kong : Chinese University Press, 1977.
xii, 368 p. ; 27 cm.
Title on verso of t.p.: Chung shih Ying i so yin; Han chih T'ang.
"For each entry, the name of the poet and the title of the poem are given, both as the Chinese original and the English translation. Sequence of the poets and poems follows that of the Ch'uan Han San-kuo Chin Nan-Pei-chao shih ... by Ting Fu-pao ..., 1959 edition, and the Ch'uan T'ang shih ..., 1960 edition."
Bibliography: p. ix-xi.
ISBN 9622011411 : $60.00
1. Chinese poetry—Translations into English—Indexes. 2. English poetry—Translations from Chinese—Indexes. 3. Ting, Fu-pao, 1874-1952, ed. Ch'üan Han San-kuo Chin Nan-pei-ch'ao shih—Indexes. 4. Ch'üan T'ang shih—Indexes. 5. Chinese poetry—Indexes. I. Ho, P'eng, joint author. II. Teng, Shih-liang, joint author. III. Title. IV. Title: Chung shih Ying i so yin.
PL2658.E3W576 016.8951'1'008 73-92412
79 MARC

—TRANSLATIONS INTO FRENCH

Li, Ch'ing-chao, 1081-ca. 1141.
Œuvres poétiques complètes / de Li Qingzhao ... ; traduit du chinois par Liang Paitchin. — ⌊Paris⌋ : Gallimard, 1977.
177 p. : ill. ; 23 cm. — (Connaissance de l'Orient ; 45) F78-4225
Title also in Chinese: Li Ch'ing-chao shih tz'u fu ch'üan chi.
55.00F
I. Liang, Paitchin. II. Title: Li Ch'ing-chao shih tz'u fu ch'üan chi. III. Series: Connaissance de l'Orient : Collection UNESCO d'œuvres représentatives ; 45.
PL2682.A26 1977 895.1'1'4 79-347345
79 MARC

—TRANSLATIONS INTO JAPANESE

Asano, Michiari, 1930–
(Kanshi yōkai)
漢詩要解 / 浅野通有著. — 東京：有精堂，昭和40 ⌊1965⌋
6, 158 p. : map (on lining paper) ; 19 cm. — (文法解明叢書 ; 44)
Colophon inserted.
Series romanized: Bumpō kaimei sōsho.
Includes index.
1. Chinese language—Readers. 2. Chinese poetry—Translations into Japanese. 3. Japanese poetry—Translations from Chinese. 4. Chinese language—Text-books for foreigners—Japanese. I. Title.
PL1117.A8 79–800630

Chou, Shu-jen, 1881–1936.
⌊Poems. Japanese & Chinese⌋
魯迅詩選 / ⌊周樹人著⌋. — 北京：外文出版社，1974.
1 portfolio (⌊15⌋ leaves) ; 38 cm.

I. Title: Lu Hsün shih hsüan. Title romanized: Ro Jin shisen.
PL2754.S5A57 1974 78–816286

(Chūgoku shibun sen)
中國詩文選 = Zhungguo shi wen hyan / 藤堂明保，牛島徳次訳註. — 東京：江南書院，1957.
170 p. ; 21 cm.

1. Chinese poetry—History and criticism. 2. Chinese poetry—Translations into Japanese. 3. Japanese poetry—Translations from Chinese. I. Tōdō, Akiyasu, 1915– II. Ushijima, Tokuji, 1918– III. Title: Zhungguo shi wen hyan.
PL2307.C597 78–816142

Fujiwara, Sosui.
·(Chūkai Kanshi kayō gafu rōei sen)
註解漢詩歌謡樂府朗詠選 / 藤原楚水著. — 東京：省心書房，1977–1978.
Ja 79–984
2 v. : ill. ; 27 cm.
Includes index.
¥7500 (v. 1) ¥8500 (v. 2)
1. Chinese poetry—Translations into Japanese. 2. Japanese poetry—Translations into Chinese. I. Title. II. Title: Kanshi kayō gafu rōei sen.
PL779.C3F84 79–805891

Hayashi, Kokei, 1875–1947.
(Kaifūsō shinchū)
懐風藻新註 / 林古溪⌊著；編者 林大⌋. — 東京：明治書院，昭和33 ⌊1958⌋
2, 2, 2, 336 p., ⌊1⌋ leaf of plates : facsims. ; 22 cm.
"本文と序との注釋で…流布の板本を以て底本とし、まま群書類従本等を参看"
Includes indexes.
1. Kaifūsō. 2. Chinese poetry—Japanese authors. 3. Chinese poetry—Translations into Japanese. 4. Japanese poetry—Translations from Chinese. I. Hayashi, Ōki, 1913– II. Kaifūsō. 1958. III. Title.
PL3054.A2K334 1958 79–808508

Inokuchi, Atsushi, 1915–
(Josei to Kanshi)
女性と漢詩：和漢女流詩史 / 猪口篤志著. — 東京：笠間書院，1978.
351 p., ⌊1⌋ leaf of plates : ill. ; 19 cm. — (笠間選書 ; 103)
Ja 78–19699
Series romanized: Kasama sensho.
¥1800
1. Chinese poetry—Women authors. 2. Chinese poetry—Japanese authors. 3. Chinese poetry—Translations into Japanese. 4. Japanese poetry—Translations from Chinese. I. Title.
PL2515.I5 78–816813

Iritani, Sensuke, 1933–
(Koshisen)
古詩選 / 入谷仙介；監修 吉川幸次郎. — 東京：朝日新聞社，1978.
2 v. ; 15 cm. — (中国古典選 ; 23–24) Ja 78–15938 (v. 1)
¥360 (v. 1) ¥320 (v. 2)

1. Chinese poetry—History and criticism. 2. Chinese poetry—Translations into Japanese. 3. Japanese poetry—Translations from Chinese. I. Title. II. Series: Chūgoku koten sen ; 23–24.
PL2307.I74 78–814813

Maeno, Naoaki, 1920–
(Kanshi no kaishaku to kanshō jiten)
漢詩の解釈と鑑賞事典 / 前野直彬，石川忠久編. — 東京：旺文社，1979.
686 p. ; 18 cm. Ja 79–8977
Includes indexes.
¥2000

1. Chinese poetry—History and criticism. 2. Chinese poetry—Translations into Japanese. 3. Japanese poetry—Translations from Chinese. I. Ishikawa, Tadahisa, 1932– joint author. II. Title.
PL2307.M3 79–810449

·**Sugimoto, Yukio,** 1907–
(Kaifūsō)
懐風藻 / 杉本行夫註釋. — 東京：弘文堂書房，昭和18 ⌊1943⌋
7, 2, 13, 410 p. ; 19 cm.
"寛政五年本を底本とした"

1. Kaifūsō. 2. Chinese poetry—Japanese authors. 3. Chinese poetry—Translations into Japanese. 4. Japanese poetry—Translations from Chinese. I. Kaifūsō. 1943. II. Title.
PL3054.A2K338 79–806172

Yamada, Katsumi, 1909–
(Chūgoku meishi kanshō jiten)
中国名詩鑑賞辞典 / 山田勝美. — 東京：角川書店，1978.
567 p. : ill. ; 19 cm. — (角川小辞典 ; 19) Ja 78–19278
Chinese and Japanese.
Series romanized: Kadokawa shōjiten.
Includes index.
¥1600
1. Chinese poetry—History and criticism. 2. Chinese poetry—Translations into Japanese. 3. Japanese poetry—Translations from Chinese. I. Title.
PL2307.Y26 78–814501

—TRANSLATIONS INTO JAPANESE
—INDEXES

Tatsumi, Masaaki.
(Kaifūsō kanji sakuin)
懐風藻漢字索引：付 本文篇 / 辰巳正明編. — 東京：新典社，1978.
246 p. ; 22 cm. Ja 78–27148
"底本は「本朝懐風詩集」(寛政五年竹笆楼再校系版本)を使用"
¥3000
1. Kaifūsō—Indexes. 2. Chinese poetry—Japanese authors—Indexes. 3. Chinese poetry—Translations into Japanese—Indexes. 4. Japanese poetry—Translations from Chinese—Indexes. I. Kaifūsō. 1978. II. Title.
PL3054.A2K32 1978 79–804691

CHINESE POETRY (Continued)

—TRANSLATIONS INTO KOREAN

(Han'guk yŏktae aejŏng Hansi sŏn)
韓國歷代愛情漢詩選 / 韓喆熙編譯. — 서울 : 益文社, 1976.
252 p. ; 18 cm.
Includes original texts.
W800

1. Chinese poetry—Korean authors. 2. Korean poetry—Translations from Chinese. 3. Chinese poetry—Translations into Korean. 4. Love poetry, Chinese. 5. Love poetry, Korean. I. Han, Ch'ŏr-hŭi.
PL982.C3H27 79-825734

—WOMEN AUTHORS

(Ch'ien ku chüeh ch'ang)
千古絕唱：歷代才女詩詞 / 季靈編著. — 臺北：星光出版社：經銷者星光書報社, 民國67 ⌊1978⌋
16, 236 p. ; 19 cm. — (雙子星叢書 ; 49號)
Series romanized: Shuang tzu hsing ts'ung shu.
NT$38.00

1. Chinese poetry—Women authors. 2. Tz'u—Women authors. I. Chi, Ling.
PL2515.C458 78-843070

Inokuchi, Atsushi, 1915–
(Josei to Kanshi)
女性と漢詩：和漢女流詩史 / 猪口篤志著. — 東京：笠間書院, 1978.
351 p., ⌊1⌋ leaf of plates : ill. ; 19 cm. — (笠間選書 ; 103)
Ja 78-19699
Series romanized: Kasama sensho.
¥1800
1. Chinese poetry—Women authors. 2. Chinese poetry—Japanese authors. 3. Chinese poetry—Translations into Japanese. 4. Japanese poetry—Translations from Chinese. I. Title.
PL2515.I5 78-816813

(Li tai ts'ai nü shih tz'u hsüan)
歷代才女詩詞選 / 季靈編著. — 香港：志明出版社, ⌊1978?⌋
16, 236 p. ; 19 cm.
$6.00

1. Chinese poetry—Women authors. I. Chi, Ling.
PL2515.L5 79-835018

(Su shou yin)
素手吟 / 乃鼎⌊編⌋著. — 臺北：南京出版公司：總經銷華一書局, 民國66 ⌊1977⌋
7, 210 p. ; 19 cm.
"歷代才女詩詞欣賞"
NT$40.00

1. Chinese poetry—Women authors. I. Nai-ting.
PL2515.S88 78-843666

—CHINA—KIANSI (PROVINCE)

Fu, Hsüan-ts'ung.
(Huang T'ing-chien ho Chiang-hsi shih p'ai chüan)
黃庭堅和江西詩派卷 / 傅璇琮. — 第1版. — 北京：中華書局：新華書店北京發行所發行, 1978.
2 v. (84, 966 p.) ; 21 cm. — (古典文學研究資料彙編)
Series romanized: Ku tien wen hsüeh yen chiu tzu liao hui pien.
$3.65
1. Chinese poetry—China—Kiangsi (Province) 2. Huang, T'ing-chien, 1045–1105, in fiction, drama, poetry, etc. I. Title.
PL3031.K552F8 79-836395

(Hung mien hua k'ai)
红棉花开：广西叙事民歌选 / 广西文艺创作办公室编. — ⌊香港? : s. n., 1979?⌋
233 p. : ill. ; 21 cm.
Reprint of the 1978 ed. published by Jen min wen hsüeh ch'u pan she, Peking.

1. Chinese poetry—China—Kwangsi (Province) 2. Chinese poetry—20th century. I. Kuang-hsi Chuang tsu tzu chih ch'ü. Ko ming wei yüan hui. Wen i ch'uang tso pan kung shih.
PL3031.K882H79 1979 79-840733

—CHINA—TA-CHAI

Ta-chai ta tui. Yeh yü ch'uang tso tsu.
(Ta-chai hsin min ko hsüan)
大寨新民歌选 / 大寨大队业余创作组 ; ⌊尾花孙平, 蒋松涛⌋. — ⌊香港? : s. n., 1978?⌋
vii, 179 p., ⌊4⌋ leaves of plates : ill. ; 21 cm.
Reprint of the 1977 ed. published by Shang-hai jen min ch'u pan she, Shanghai.
$70.00

1. Chinese poetry—China—Ta-chai. 2. Chinese poetry—20th century. I. Title.
PL3031.T22T3 1978 78-842402

—TAIWAN

(Ch'ing fa huo che hua lien)
青髮或者花臉 / ⌊編著者詩人畫會⌋. — 臺北：香草山出版有限公司, 民國65 ⌊1976⌋
345 p. : ill. ; 21 cm.
Errata slip inserted.
NT$80.00

1. Chinese poetry—Taiwan. 2. Chinese poetry—20th century. 3. Painting, Modern—20th century—Taiwan. I. Shih jen hua hui.
PL3031.T32C485 79-836342

(Ping ch'en tuan yang ch'üan sheng shih jen lien yin ta hui chuan chi)
丙辰端陽全省詩人聯吟大會專輯 / 省立台中圖書館編印. — ⌊台中⌋：該館, 民國65 ⌊1976⌋
2 leaves, 3–33 p, ⌊3⌋ leaves of plates : ill. ; 27 cm.
Cover title.

1. Chinese poetry—Taiwan. I. T'ai-wan sheng li T'ai-chung t'u shu kuan. II. Title: Ch'üan sheng shih jen lien yin ta hui chuan chi.
PL3031.T32P5 78-843593

(Shih hsüan chi)
詩選集 / 李南衡主編 ; ⌊助理編輯林邊, 梁景峯⌋. — 台北：明潭出版社, 民國68 ⌊1979⌋
12, 9, 446 p. ; 22 cm. — (日據下台灣新文學 : 明集 ; 4)
NT$250.00 ($10.00 U.S.)

1. Chinese poetry—Taiwan. 2. Chinese poetry—20th century. I. Li, Nan-heng. II. Lin, Pien. III. Liang, Ching-feng. IV. Series: Jih chü hsia T'ai-wan hsin wen hsüeh : Ming chi ; 4.
PL3031.T32S52 79-840153

(Shou-feng shih she hsü chi)
壽峰詩社續集 / 著作各社員 ; 編校王天賞. — 高雄：壽峰詩社, 民國65 ⌊1976⌋
4, 295 p., ⌊3⌋ leaves of plates : ports. ; 22 cm.
Colophon title.
"非賣品"

1. Chinese poetry—Taiwan. 2. Chinese poetry—20th century. 3. Shou-feng shih she. I. Wang, T'ien-shang, 1903–
PL3031.T32S55 79-835013

(T'ai-wan shih hui)
臺灣詩薈：中華民國六十六年丁巳端午詩集 / ⌊編撰人連橫⌋. — 台北：成文出版社, 民國66 ⌊1977⌋
2 v. : ill. ; 22 cm.
Reprint ed.; with new pref.

1. Chinese poetry—Taiwan. 2. Chinese literature—Taiwan. I. Lien, Heng, 1878–1936.
PL3031.T32T329 1977 78-840776

—TAIWAN—HISTORY AND CRITICISM

Lien, Heng, 1878–1936.
(T'ai-wan shih sheng)
臺灣詩乘 / 連橫撰. — ⌊臺北縣永和鎮⌋：文海出版社, ⌊1978⌋
5, 262, 4, 88 p. ; 22 cm. — (近代中國史料叢刊續編⌊i. e.輯⌋ ; 506 : 第五十一輯)
"臺灣詩薈雜文鈔 連橫," at end of work.
1. Chinese poetry—Taiwan—History and criticism. I. Lien, Heng, 1878–1936. T'ai-wan shih hui tsa wen ch'ao. 1978. II. Title. III. Series: Shen, Yün-lung, comp. Chin tai Chung-kuo shih liao ts'ung k'an hsü chi ; 506.
DS755.S532 vol. 506 79-835346
[PL3031.T3]

—TAIWAN—PERIODICALS

(Shih ch'ao)
詩潮 第1- 集 民國66年5月- ⌊1977-
台中⌋ 藍燈文化事業公司
v. ill. 21 cm.
Vols. for May 1977- prepared by Shih ch'ao she.

1. Chinese poetry—Taiwan—Periodicals.
PL3031.T3S53 78-641963
(MARC-S)

CHINESE POETS see Poets, Chinese

CHINESE POLITICAL BALLADS AND SONGS see Political ballads and songs, Chinese

CHINESE PORCELAIN see Porcelain, Chinese

CHINESE PORTRAIT PAINTING see Portrait painting, Chinese

CHINESE POTTERY see Pottery, Chinese

CHINESE PROPAGANDA see Proaganda, Chinese

CHINESE PROSE LITERATURE

see also Fu

Chou, Ya-p'ing.
(K'ao cheng Ku wen kuan chih)
攷正古文觀止 / 周亞平註譯. — 台南：龍門圖書股份有限公司, 民國66 ⌊1977⌋
2, 10, 723 p. ; 21 cm.
"國語注音, 爾注語譯"
NT$85.00

1. Wu, Ch'eng-ch'üan, fl. 1711, ed. Ku wen kuan chih. 2. Chinese prose literature. I. Wu, Ch'eng-ch'üan, fl. 1711, ed. Ku wen kuan chih. 1977. II. Title. III. Title: Ku wen kuan chih.
PL2606.W83C68 79-838228

CHINESE PROSE LITERATURE (Continued)

(Wakokubon Kanseki bunshū)
和刻本漢籍文集 / 長澤規矩也編. — ₍東京₎ : 出版古典研究會 : 發行所 汲古書院, 昭和52-
₍1977-
v. ; 27 cm. Ja 77-27565 v. 1
Photoreprint ed.
¥7000 (v. 1)

1. Chinese prose literature. I. Nagasawa, Kikuya, 1902- II. Title: Kanseki bunshū.
PL2607.W3 1977 79-804946

—20TH CENTURY

Chu, Ch'en-tung, 1933- , comp.
(Ch'ing nien san wen hsüan)
青年散文選 朱沉冬主編 高雄 光明出版社
民國62 ₍1973₎
3, 8, 208, 2 p. 20 cm. NT$18.00

1. Chinese prose literature—20th century. I. Title.
PL2623.C513 73-898845
rev

Chu, Ch'en-tung, 1933- , comp.
(Kao ch'ing san wen hsüan)
高青散文選 朱沉冬主編 高雄 光明出版社
民國62 ₍1973₎
5, 16, 462, 2 p. 20 cm. (高青叢書) NT$30.00

1. Chinese prose literature—20th century. I. Title.
PL2623.C515 73-898453
rev

Chu, Ch'en-tung, 1933- , comp.
(Kao ch'ing san wen hsüan chi)
高青散文選集 朱沉冬 林義雄主編 ₍高雄 殷雷出版社 民國60 i. e. 1971₎
11, 149 p. 22 cm. (現代青年文學叢書) NT$25.00
在高青文粹發表過的作品中選出一部份優秀作品整理出版

1. Chinese prose literature—20th century. I. Lin, I-hsiung, joint comp. II. Kao ch'ing wen ts'ui. III. Title.
PL2623.C517 73-899493
rev

Chu, Ch'en-tung, 1933- , comp.
(San wen ming tso hsüan)
散文名作選 朱沉冬主編 高雄 光明出版社
民國62 ₍1973₎
3, 9, 253, 2 p. 20 cm. (高青叢書) NT$18.00

1. Chinese prose literature—20th century. I. Title.
PL2623.C518 73-898811
rev

(Chung-kuo san wen ta chan)
中國散文大展 / 主編高登偉. — 台北 : 金陵圖書有限公司, 民國67- ₍1978-
v. ; 19 cm. — (中國當代散文傑作大特輯 ; 1)
NT$55.00 (v. 1)

1. Chinese prose literature—20th century. I. Kao, Teng-wei. II. Series: Chung-kuo tang tai san wen chieh tso ta t'e chi ; 1.
PL2623.C538 79-841421

(Chung-kuo ti k'u nan)
中國的苦難 / 周錦選編 ; 孫陵 ... ₍et al.₎ 校訂. — 臺北 : 智燕出版社 : 總經銷衆文圖書股份有限公司, 民國68 ₍1979₎
250 p. ; 19 cm. — (中國新文學創作叢刊 ; 1)
NT$50.00

1. Chinese prose literature—20th century. I. Chou, Chin, fl. 1973- II. Sun, Ling. III. Series: Chung-kuo hsin wen hsüeh ch'uang tso ts'ung k'an ; 1.
PL2622.C474 79-839103

Fu-hsüan-ko-chu, comp.
(I yeh i hsiao p'in)
一頁一小品 / 撫萱閣主主編. — 九龍 : 光華書局, ₍19
v. ; 19 cm.
$6.50 (v. 3)

1. Chinese prose literature—20th century. I. Title.
PL2622.F8 1978 78-842618

(Hou ch'in chan shih)
后勤战士 / 武汉部队后勤部征文小组编. — ₍香港? : s. n., 1978₎
140 p. ; 22 cm.
Reprint of the 1978 ed. published by Hu-pei jen min ch'u pan she.

1. Chinese prose literature—20th century. I. China (People's Republic of China, 1949-). Chung kuo jen min chieh fang chün. Wu-han pu tui. Hou ch'in pu. Cheng wen hsiao tsu.
PL2623.H65 1978b 79-836267

(Hsien tai ch'ing nien san wen hsüan)
現代青年散文選 / 盧勝彥編. — 8版. — ₍台中₎ : 青山出版社 : 總經銷學海書局, 民國63 ₍1974₎
148, 140 p. ; 19 cm.
NT$35.00

1. Chinese prose literature—20th century. 2. Chinese poetry—20th century. I. Lu, Sheng-yen, 1945-
PL2623.H734 1974 78-841968

(Jen sheng ch'ien wan hsü)
人生千萬絮 / 著作者胡適等. — ₍台南₎ : 慈暉出版社 : 龍輝出版社, 民國66 ₍1977₎
200, 132 p. ; 19 cm.
Issued also under the title: Hsien tai ming chia san wen ching hsüan.
NT$40.00

1. Chinese prose literature—20th century. I. Hu, Shih, 1891-1962.
PL2622.J46 79-837468

(Ma jen ti wen i)
罵人的文藝 / 白牧編輯. — 台南 : 王家出版社, 民國66 ₍1977₎
201 p. ; 19 cm.
Reprint of the 1971 ed. published by Lo hua sheng ch'u pan she, Taipei, which was issued as ti 17 chung of Mei kuei ts'ung shu.
NT$30.00

1. Invective—Anecdotes, facetiae, satire, etc. 2. Chinese prose literature—20th century. I. Title.
PN6231.I65P3 1977 79-835302

—HISTORY AND CRITICISM

Fang, Hsüan-ch'en.
(Fan wen shang hsi)
範文賞析 / 方玄琛著. — 台北 : 偉文圖書出版社, 民國67 ₍1978₎
3, 151 p. ; 20 cm. — (中國語文叢書)
Series romanized: Chung-kuo yü wen ts'ung shu.
NT$40.00

1. Chinese prose literature—History and criticism. I. Title.
PL2395.F29 79-839477

Inokuchi, Atsushi, 1915-
(Zoku bunshō kihan)
続文章軌範 / 猪口篤志著. — 東京 : 明治書院, 昭和52 ₍1977₎
2 v. : ill. ; 22 cm. — (新釈漢文大系 ; 56-57) Ja***
Bibliography: v. 1, p. 12-14.
Includes index.
¥5200 per vol.

1. Hsü wen chang kuei fan. 2. Chinese prose literature—History and criticism. 3. Chinese prose literature—Translations into Japanese. 4. Japanese prose literature—Translations from Chinese. I. Hsü wen chang kuei fan. Japanese & Chinese. 1977. II. Title. III. Series: Shinshaku Kambun taikei ; 56-57.
PL2395.H73I5 77-818406
rev

Shimizu, Shigeru, 1925-
(Tō Sō hakka bun)
唐宋八家文 / 清水茂. — 東京 : 朝日新聞社, 1978-
v. ; 15 cm. — (中国古典選 ; 35) Ja 79-985 (v. 1)
Series romanized: Chūgoku koten sen.
¥400 (v. 1)

1. Chinese prose literature—History and criticism. I. Title.
PL2409.S5 1978 79-805330

—TRANSLATIONS INTO JAPANESE

Inokuchi, Atsushi, 1915-
(Zoku bunshō kihan)
続文章軌範 / 猪口篤志著. — 東京 : 明治書院, 昭和52 ₍1977₎
2 v. : ill. ; 22 cm. — (新釈漢文大系 ; 56-57) Ja***
Bibliography: v. 1, p. 12-14.
Includes index.
¥5200 per vol.
1. Hsü wen chang kuei fan. 2. Chinese prose literature—History and criticism. 3. Chinese prose literature—Translations into Japanese. 4. Japanese prose literature—Translations from Chinese. I. Hsü wen chang kuei fan. Japanese & Chinese. 1977. II. Title. III. Series: Shinshaku Kambun taikei ; 56-57.
PL2395.H73I5 77-818406
rev

CHINESE PROVERBS see **Proverbs, Chinese**

CHINESE QUESTION see **China—History; Eastern question (Far East)**

CHINESE RED ARMY see **China (People's Republic of China, 1949-) Chung-kuo jên min chieh fang chün**

CHINESE RIDDLES see **Riddles, Chinese**

CHINESE SCULPTURE see **Sculpture, Chinese**

CHINESE SHADOWS see Shadow pantomimes and plays

CHINESE SNUFF BOXES AND BOTTLES see **Snuff boxes and bottles, Chinese**

CHINESE STUDENTS IN FRANCE

Ho, Ch'ang-kung, 1900-
⌐Ch'in kung chien hsüeh sheng huo hui i. Japanese⌐
フランス勤工倹学の回想：中国共産党の一源流 / 何長工著；河田悌一，森時彦訳. — 東京：岩波書店，1976.
xii, 228, 4 p. : ill. ; 18 cm. — (岩波新書 : 956) Ja 76-9600
Series romanized: Iwanami shinsho.
¥230
1. Chinese students in France. I. Kawata, Teiichi, 1945- II. Mori, Tokihiko, 1947- III. Title.
Title romanized: Furansu kinkō kengaku no kaisō.
LC3085.F7H616 79-800175

Sheng, Ch'eng, 1899-
(Hai wai kung tu shih nien chi shih)
海外工讀十年紀實 / 盛成著. — ⌐台北⌐：文海出版社，⌐1978?⌐
2, 356 p., ⌐6⌐ leaves of plates : ill. ; 22 cm. — (近代中國史料叢刊續編 ⌐i. e. 輯⌐ ; 498 : 第50輯)
Caption title: Mon Odyssée en Europe.
Photoreprint ed.
1. Chinese students in France. 2. Europe—Description and travel—1919- 3. Sheng, Ch'eng, 1899- 4. Intellectuals—China—Biography. I. Title. II. Title: Mon Odyssée en Europe. III Series: Shen, Yün-lung, Comp. Chin tai Chung-kuo shih liao ts'ung k'an hsü chi ; 498.
DS755.S532 vol. 498 79-83755(
[LC3085.F7]

CHINESE STUDENTS IN JAPAN

Yeh, Lung-yen, 1940-
(Ch'ing mo min ch'u chih fa cheng hsüeh t'ang)
清末民初之法政學堂 / 葉龍彥. — ⌐台北? : s. n.⌐，民國63 ⌐1974⌐
2 v. (417 double leaves) ; 27 cm.
"私立中國文化學院三民主義研究所博士論文"
Reproduced from typescript.
Bibliography: v. 1, leaves 407-417.
1. Education, Higher—China—History. 2. Law—Study and teaching (Higher)—China. 3. Political science—Study and teaching (Higher)—China. 4. Chinese students in Japan. I. Title.
LA1133.Y4 79-835399

—HISTORY

Sanetō, Keishū, 1896-
(Chūgokujin Nihon ryūgaku shikō)
中國人日本留學史稿 / 實藤惠秀著. — 東京：日華學會，昭和14 ⌐1939⌐
8, 368 p., ⌐9⌐ leaves of plates : ill. ; 22 cm.
Cover title.
Originally published in Nikka gakuhō, Nov. 25, 1936-Dec. 13, 1938.
Bibliography: p. 360-365.
1. Chinese students in Japan—History. I. Title.
LC3089.J3S29 79-807517

—REGISTERS

Ko sheng kuan fei tzu fei pi yeh hsüeh sheng hsing ming piao.
清末各省官費自費留日學生姓名表. — ⌐台北⌐：文海出版社，⌐1978?⌐
395 p. ; 22 cm. — (近代中國史料叢刊續編 ⌐i. e. 輯⌐ ; 494 : 第五十輯)
Original t. p. reads: Ko sheng kuan fei tzu fei pi yeh hsüeh sheng hsing ming piao.
Reprint ed.
1. Chinese students in Japan—Registers. I. Title. II. Series: Shen, Yün-lung, comp. Chin tai Chung-kuo shih liao ts'ung k'an hsü chi ; 494.
Title romanized: Ch'ing mo ko sheng kuan fei tzu fei liu Jih hsüeh sheng hsing ming piao.
DS755.S532 vol. 494 78-841434
[LC3089.J3]
78 (

CHINESE STUDIES

see also Sinologists

CHINESE SUGAR-CANE see Sorgo

CHINESE TALES see Tales, Chinese

CHINESE TRADE PORCELAIN see Porcelain, Chinese

CHINESE WATER-COLOR PAINTING see Water-color painting, Chinese

CHINESE WIT AND HUMOR

I, Ho.
(Hsin hsiao tien)
新笑典 / 頤和編. — 臺北：花孩兒出版有限公司，民國67 ⌐1978⌐
2, 263 p. ; 20 cm.
NT$50.00
1. Chinese wit and humor. I. Title.
PN6222.C5I2 79-835257

Jo-hung.
(Hsiao ti i shu)
笑的藝術 / 若虹著. — 台北：哲志出版社，民國67 ⌐1978⌐
10, 209 p. ; 20 cm. — (哲志叢書 ; 69)
Series romanized: Che chih ts'ung shu.
NT$40.00
1. Laughter. 2. Chinese wit and humor. I. Title.
PN6149.P5J6 1978 79-835520

Kuan, Kuan.
(I yü shuang kuan)
一語雙關 / 關關著. — 台北縣永和鎮：雲水出版社，⌐66 i. e. 1977序⌐
189 p. ; 20 cm.
NT$45.00
1. Puns and punning. 2. Chinese wit and humor. I. Title.
PN6231.P8K8 78-843106

Kung, Hsien-nan.
(Chung-kuo yü yen ching hua)
中國語言精華 / 龔顯男編著. — 雲林縣虎尾鎮：龔氏出版社，民國68- ⌐1979-
v. ; 19 cm.
NT$125.00 (v. 1)
1. Chinese language—Terms and phrases. 2. Chinese wit and humor. 3. Maxims, Chinese. I. Title.
PL1497.K84 79-839555

Lung, Chi, fl. 1969-
(Liang hsing ti yu mo)
兩性的幽默 / ⌐著者龍驥⌐. — 台南：青溪出版社：總經銷東風圖書供應社，民國67 ⌐1978⌐
24, 223, 2 p. ; 19 cm. — (東風益智書系 ; 6)
Colophon and spine title: Liang hsing yu mo.
Series romanized: Tung feng i chih shu hsi.
NT$40.00
1. Chinese wit and humor. I. Title.
PN6222.C5L8 79-837003

Meng, Chung-jen.
(Chung-kuo hsiao hua ku shih)
中國笑話故事 / 孟仲仁編著. — 臺北：出版家文化事業有限公司，民國67 ⌐1978⌐
166 p. ; 19 cm. — (出版家叢書 ; 53)
Series romanized: Ch'u pan chia ts'ung shu.
NT$40.00
1. Chinese wit and humor. I. Title.
PN6222.C5M4 79-835502

Wang, Shih-hsin, comp.
(Ma jen ti i shu yü wen hsüeh)
罵人的藝術與文學 王士新編著 ⌐臺北⌐ 常春樹書坊 ⌐星光書報社總經銷 民國61 i. e. 1972⌐
182 p. 18 cm. NT$20.00
At head of title: 智慧的書
1. Chinese wit and humor. I. Title.
PN6222.C5W34 1972 73-897263
rev

CHINESE WIT AND HUMOR, PICTORIAL

Feng, Tzu-k'ai, 1898-1975.
(Hu sheng hua chi)
護生畫集 / 豐子愷畫；弘一法師書. — 香港：波文書局，⌐1978?⌐
5 v. : ill. ; 21 cm.
Reprint ed.
$50.00
1. Chinese wit and humor, Pictorial. I. Hung-i, 1880-1942. II. Title.
NC1699.F4A4 1978 78-842704

CHINESE WOOD-ENGRAVING see Wood-engraving, Chinese

CHINESE WOOD-OIL TREE see Tung tree

CHING, FANG, 77-37 B. C. CHING SHIH I CHUAN

Liang, Hsiang-jun.
(Ching Fang I)
京房易 / 梁湘潤著. — 台北：行卯出版社，民國65 ⌐1976⌐
300 p. : ill. ; 20 cm.
NT$450.00
1. Ching, Fang, 77-37 B. C. Ching shih I chuan. 2. I ching. I. Title.
PL2464.Z6L4925 77-837392

CH'ING, MA CH'IA

—FICTION

Ponthier, François.
Chère Madame Ch'ing / François Ponthier. — ⌐Paris⌐ : Stock, c1978.
320 p. ; 23 cm. F***
ISBN 2-234-00889-1 : 42.00F
1. Ch'ing, Ma Ch'ia—Fiction. I. Title.
PQ2631.O6476C5 843'.9'14 78-387669
79 MARC

CHING, RAYMOND, 1939-

Ching, Raymond, 1939-
Raymond Ching, the bird paintings : water colours and pencil drawings, 1969-1975 / text by David Snow, A. H. Chisholm & M. F. Soper. — London : Collins, c1978.
142 p. : ill. (some col.) ; 48 cm. GB***
ISBN 0-00-219685-9
1. Birds—Pictorial works. 2. Birds in art. 3. Ching, Raymond, 1939- . I. Snow, David William. II. Chisholm, Alexander Hugh, 1890- . III. Soper, M. F. IV. Title. V. Title: The bird paintings.
QL674.C44 598.2'02'22 79-300554
79 MARC

CHING, T'ING-PIN, 1861-1902

Chia, I-chün, 1902-
(Ching T'ing-pin sao Ch'ing mieh yang ch'i i)
景廷宾扫清灭洋起义 / 編写者賈逸君；責任編委刘桂五. — [香港?：s. n., 1978?]
36 p. : ill. ; 19 cm.
Colophon title.
Title in pinyin on cover: Jing Ting Bin sao Qing mie yang qiyi.
Reprint of the 1962 ed. published by Chung-hua shu chü, Peking, in series: Chung-kuo li shih hsiao ts'ung shu.
1. Ching, T'ing-pin, 1861-1902. 2. Revolutionists—China—Biography. I. Liu, Kuei-wu. II. Title. III. Title: Jing Ting Bin sao Qing mie yang qiyi.
DS763.C49C46 1978 78-841464

CH'ING DYNASTY, 1644-1912 see China—History—Ch'ing dynasty, 1644-1912

CHING-HSIEN, CHINA (HOPEI)

—HISTORY—SOURCES

(Ching-hsien chih)
景縣志：河北省 / 耿兆棟監修；張汝漪總纂. — 臺1版. — 台北：成文出版社，民國65 [1976]
6 v. (2136 p., [1] fold. leaf of plates) : ill. ; 27 cm. — (中國方志叢書：華北地方；第500號)
"據民國二十一年鉛印本影印"
1. Ching-hsien, China (Hopei)—History—Sources. I. Keng, Chao-tung. II. Chang, Ju-i, chü jen 1891. III. Series: Chung-kuo fang chih ts'ung shu : Hua-pei ti fang ; ti 500 hao.
DS793.C4729C46 78-842023

CHING-HSING MEI K'UANG KU FEN YU HSIEN KUNG SSU

Minami Manshū Tetsudō Kabushiki Kaisha. Hokushi Keizai Chōsajo, Peking.
(Seikei. Seihō ryōtankō rōdō gaiyō chōsa hōkoku)
井陘·正豐兩炭礦勞働概要調査報告 / [執筆者關戸嘉明]. — [北京]：滿鐵·北支經濟調査所，昭和16 [1941]
4, 85 leaves : 26 cm.
Cover title.
"第一次調査報告書(第六班)"
"極秘" (rubber stamped)
1. Coal miners—China—Hopei (Province) 2. Ching-hsing mei k'uang ku fen yu hsien kung ssu. 3. Cheng-feng mei k'uang kung ssu. I. Sekido, Yoshiaki, 1917- II. Title.
HD8039.M62C639 79-806639

CH'ING HSÜAN-T'UNG, EMPEROR OF CHINA, 1906-1967

Ch'ing Hsüan-t'ung, Emperor of China, 1906-1967.
From Emperor to citizen : the autobiography of Aisin-Gioro Pu Yi / [translated by W. J. F. Jenner]. — 2d ed. — Peking : Foreign Languages Press, 1979-
v. : ill. ; 22 cm.
Translation of Wo ti ch'ien pan sheng.
1. Ch'ing Hsüan-t'ung, Emperor of China, 1906-1967. 2. China—Kings and rulers—Biography. I. Title.
DS773.C51513 1979 951'.03'0924 79-112550
79 MARC

CH'ING JEN-TSUNG, EMPEROR OF CHINA, 1760-1820

—FICTION

(Chia-ch'ing chün yu T'ai-wan)
嘉慶君遊台灣. — 三重 [台北縣]：世新出版社，民國65 [1976]
51 p. : ill. ; 19 cm.
"中國歷史通俗小說"
NT$7.00
1. Ch'ing Jen-tsung, Emperor of China, 1760-1820—Fiction.
PL2832.5.C475 78-841370
C

CHING KANG SHAN REGION

—DESCRIPTION AND TRAVEL

Chiang-hsi shih fan hsüeh yüan. Ti li pien hsieh tsu.
(Ching-kang shan ti li)
井冈山地理 / 江西师范学院地理编写组. — [香港?：s. n., 1978?]
4, 81 p., [1] fold. leaf of plates : ill. : 21 cm.
Reprint of the 1976 ed. published by Chiang-hsi jen min ch'u pan she, Nan-ch'ang, in series: Ti li hsiao ts'ung shu.
1. Ching Kang Shan region—Description and travel. I. Title.
DS793.C53C47 1978 78-843331
C

—DESCRIPTION AND TRAVEL—VIEWS

(Ching-kang shan)
井岡山：纪念毛主席创建井冈山革命根据地五十周年，1927-1977 / [《井冈山》画册编辑组编]. — 第1版. — [南昌]：江西人民出版社，1977.
135 p., [4] leaves of plates : chiefly col. ill. ; 26 cm.
$12.50
1. Ching-kang Shan region—Description and travel—Views. I. Ching-kang shan hua ts'e pien chi tsu.
DS793.C533C47 78-841546
C

—HISTORY

Chung-kuo kung ch'an tang. Ching-kang shan ti ch'ü wei yüan hui. Hsüan ch'uan pu.
(Ching-kang shan ko ming wen wu)
井冈山革命文物 / 中共井冈山地区委员会宣传部主编. — [香港?：s. n., 1978?]
112 p., [17] leaves of plates : ill. ; 21 cm.
Reprint of the 1977 ed. published by Chiang-hsi jen min ch'u pan she, Nan-ch'ang.
1. Ching-kang Shan region—History. I. Title.
DS793.C533C48 1978 79-836337
C

—HISTORY—ADDRESSES, ESSAYS, LECTURES

(Mao wei yüan tsai Ching-kang shan)
毛委员在井冈山. — [香港?：s. n., 1978?]
4, 312 p., [2] leaves of plates : ill. ; 21 cm.
Reprint of the 1977 ed. published by Chiang-hsi jen min ch'u pan she, Nan-ch'ang.
1. Mao, Tse-tung, 1893-1976—Addresses, essays, lectures. 2. Ching-kang Shan region—History—Addresses, essays, lectures.
DS778.M3M344 1978 78-841693

—HISTORY—FICTION

(Ching-kang feng huo)
井冈烽火 / 井冈山地区文教局编. — [香港?：s. n., 1978?]
2, 286 p., [12] leaves of plates : ill. ; 21 cm.
At head of title: Ching-kang shan tou cheng ku shih.
Reprint of the 1977 ed. published by Chiang-hsi jen min ch'u pan she, Nan-ch'ang.
1. Short stories, Chinese. 2. Chinese fiction—20th century. 3. Ching-kang Shan region—History—Fiction. I. Ching-kang shan ti ch'ü, China. Wen chiao chü.
PL2653.C51487 1978 79-836067

(Ching-kang shan tou cheng ku shih)
井冈山斗争故事 / 井冈山地区文教局. — [香港?：s. n., 1978?]
202 p. : ill. ; 21 cm.
Reprint of the 1974 ed. published by Chiang-hsi jen min ch'u pan she, Nan-ch'ang.
1. Ching-kang Shan region—History—Fiction. I. Ching-kang shan ti ch'ü, China. Wen chiao chü.
PL3031.C552C5 1978 78-841101

Lin, Tan.
(Feng huo Ch'i-p'an-yai)
烽火棋盘崖 / 原著刘丹；改编林丹；绘画汪大伟. — [第1版]. — [上海]：上海人民出版社，[1977]
[92] p. : ill. ; 95 x 128 mm.
"連環圖畫"
$0.09
1. Ching-kang Shan region—History—Fiction. I. Liu, Tan, fl. 1977- II. Wang, Ta-wei. III. Title.
PL2878.N1489F4 78-843188

—POETRY

Ch'iao-yen.
(Fang ko Ching-kang shan)
放歌井冈山 / 峭岩. — 第1版. — 南昌：江西人民出版社，1976.
3, 155 p. ; 19 cm.
Poems.
$0.30
1. Ching-kang Shan region—Poetry. I. Title.
PL2845.Y4F3 1976 79-838713

CH'ING MEN

Shuai, Hsüeh-fu.
[Ch'ing Hung shu yüan]
中國幫會史 / [帥學富著]. — 九龍：現代出版公司，[1978?]
3, 3, 302 p., [4] leaves of plates : ill. ; 19 cm.
Cover title.
"洪門搜秘": p. 213-302.
$4.00
1. Secret societies—China—History. 2. Ch'ing men. 3. Hung men. I. Fei-lieh, pseud. Hung men sou mi. 1978. II. Title.
Title romanized: Chung-kuo pang hui shih.
HS310.S56 1978 79-836160

CH'ING NIEN CHAN SHIH PAO, T'AI-PEI

(Feng yü sheng hsin hsin)
風雨生信心：青年戰士報創刊廿五週年紀念專輯. — [臺北：青年戰士報社]，民國66 [1977]
124 p. : ill. ; 39 cm.
Cover title.
1. Ch'ing nien chan shih pao, T'ai-pei. I. Ch'ing nien chan shih pao, T'ai-pei.
PN5370.5.C5F4 78-842368

CH'ING SHENG-TSU, EMPEROR OF CHINA, 1654-1722

Wu, Silas H L 1929-
Passage to power : K'ang-hsi and his heir apparent, 1661-1722 / Silas H. L. Wu. — Cambridge : Harvard University Press, 1979.
xv, 252 p. : maps ; 24 cm. — (Harvard East Asian series ; 91)
Bibliography: p. 227-240.
Includes index.
ISBN 0-674-65625-3
1. Ch'ing Sheng-tsu, Emperor of China, 1654-1722. 2. China—History—K'ang-hsi, 1662-1722. I. Title. II. Series.
DS754.4.C53W8 951'.03'0924 79-4191
79 MARC

—QUOTATIONS

Ch'ing Sheng-tsu, Emperor of China, 1654–1722.
[T'ing hsün ko yen]
聖祖庭訓格言 / [清聖祖]. 御製日知薈説 / [乾隆]. — [台北：台灣]商務[印書館, 1978]
ca. 250 leaves in various foliations ; 20 cm. — (四庫全書珍本八集 ; 132)
Cover title.
Caption title: Sheng-tsu Jen huang ti T'ing hsün ko yen.
"庭訓格言一卷. 御製日知薈説四卷."
Photoreprint ed. of the Wen Yüan Ko copy.
1. Ch'ing Sheng-tsu, Emperor of China, 1654–1722—Quotations. I. Ch'ing Kao-tsung, Emperor of China, 1711–1799. Yü chih jih chih hui shuo. 1978. II. Title. III. Title: Sheng-tsu Jen huang ti T'ing hsün ko yen. IV. Title: T'ing hsün ko yen. V. Title: Yü chih jih chih hui shuo. VI. Series: Ssu k'u ch'üan shu. Ssu k'u ch'üan shu chen pen pa chi ; 132.
Title romanized: Sheng-tsu T'ing hsün ko yen.
AC149.S7 1978, vol. 132 78–841503
[DS754.4]

Ssu, Ma.
(Jih lo tzu chin ch'eng)
日落紫禁城 / 司馬著. — 台北縣土城：漢麟出版社, 民國67 [1978]
330 p. ; 20 cm.
NT$50.00
1. Ch'ing Shih-tsung, Emperor of China, 1677-1735—Fiction. I. Title.
PL2903.S74J5 79–835498

CH'ING-T'IEN HSIEN, CHINA

—PERIODICALS

(Ch'ing-t'ien chi k'an)
青田季刊 第1- 期 民國63年6月- [1974- 台北 台北市青田同鄉會]
no. 19 cm. quarterly.
1. Ch'ing-t'ien hsien, China—Periodicals. I. T'ai-pei shih Ch'ing-t'ien t'ung hsiang hui.
DS793.C5427C47 78–646954
(MARC-S)

CHING-YUAN HSIEN, CHINA

–HISTORY–SOURCES

(Ching-yüan hsien chih)
靖遠縣志：甘肅省 / 陳之驥編次. — 臺1版. — 臺北：成文出版社, 民國65 [1976]
2 v. (817 p.) ; 27 cm. — (中國方志叢書：華北地方 ; 第551號)
"據清道光十三年刊本 民國十四年鉛字重印本影印"
1. Ching-yüan hsien, China—History—Sources. I. Ch'en, Chih-chi, chin shih 1826. II. Series: Chung-kuo fang chih ts'ung shu : Hua-pei ti fang ; ti 551 hao.
DS793.C545C47 1976 79–838218

CHINGLEPUT, INDIA (DISTRICT)

—RURAL CONDITIONS

Mencher, Joan P 1930-
Agriculture and social structure in Tamil Nadu : past origins, present transformations, and future prospects / by Joan P. Mencher. — Durham, N.C. : Carolina Academic Press, c1978.
xix, 314 p., [7] leaves of plates (1 fold.) : ill. ; 22 cm.
Bibliography: p. [301]-307.
Includes index.
ISBN 0-89089-101-X
1. Agriculture and state—India—Chingleput (District) 2. Agriculture and state—India—Tamil Nadu. 3. Land tenure—India—Chingleput (District) 4. Chingleput, India (District)—Rural conditions. I. Title.
HD2075.C47M46 1978 338.1'0954'82 77-80768
79 MARC

CHINGPAW LANGUAGE see Kachin language

CHINGPAWS see Kachin tribes

CHINHAE HWAHAK CHUSIK HOESA

Chinhae Hwahak Simnyŏnsa P'yŏnch'an Wiwŏnhoe.
(Simnyŏnsa)
十年史 / [鎭海化學十年史編纂委員會]. — [鎭海]：鎭海化學株式會社, 1975.
555 p. : ill. ; 27 cm.
On cover: 10 year history.
Summary in English.
Bibliography: p. 478–479.
1. Chinhae Hwahak Chusik Hoesa.
HD9483.K64C473 79–825662

CHINJU CHONGNYON HOEUISO

Chinju Ch'ŏngnyŏn Hoeŭiso. Hongbo Punkwa Wiwŏnhoe.
(Chinju Ch'ŏngnyŏn Hoeŭiso)
晋州青年会議所 = The Jin Ju Jaycees guide book. — 晋州：晋州青年會議所弘報分科委員會, 1976.
115 p. : ill. ; 27 cm.
Cover title: Chinju Ch'ŏngnyŏn Hoeŭiso simnyŏnsa.
In Korean.
"非賣品"
1. Chinju Ch'ŏngnyŏn Hoeŭiso. I. Title: Chinju Ch'ŏngnyŏn Hoeŭiso simnyŏnsa.
HF331.K65C4523 78–826664

CHINO-JAPANESE WAR, 1894-1895 see Chinese-Japanese War, 1894-1895

CHINOOK SALMON

Kissner, Paul D
Annual performance report for a study of Chinook salmon in southeast Alaska / Paul D. Kissner. — [Juneau] : Alaska Dept. of Fish and Game, Sport Fish Division, [1976]
37 p. : ill. ; 29 cm. — ([Anadromous fish studies] ; study AFS 41)
At head of title: Volume 17. Study AFS 41.
Includes bibliographies.
1. Chinook salmon. 2. Fish-culture—Alaska. 3. Fishery management—Alaska. I. Alaska. Division of Sport Fish. II. Title. III. Title: A study of Chinook salmon in southeast Alaska. IV. Series.
QL638.S2K56 597'.55 77-620581
77[r79]rev MARC

Kissner, Paul D
Annual progress report for a study of chinook salmon in southeast Alaska / by Paul D. Kissner, Jr. — [Juneau] : Alaska Dept. of Fish and Game, Division of Sport Fish, 1973.
i, 24 leaves ; 29 cm. — (Anadromous fish studies ; study AFS-41-1)
Cover title: A study of chinook salmon in southeast Alaska.
At head of title: Volume 14. Study AFS-41-1.
Bibliography: leaves 23-24.
1. Chinook salmon. 2. Fishes—Racial analysis. 3. Fishes—Alaska. 4. Scales (Fishes) I. Title. II. Title: A study of chinook salmon in southeast Alaska. III. Series.
QL638.S2K57 597'.55 76-358762
76[r79]rev MARC

CHINY (COMTÉ)

—HISTORY

Jeantin, Jean François Louis.
Les Chroniques de l'Ardenne et des Woëpvres : ou, Revue et examen des traditions locales antérieures au onzième siècle pour servir a l'histoire de l'ancien Comté de Chiny / Jean-F.-L. Jeantin. — Genève : Mégariotis, 1978.
2 v. (xxix, 594 p., [9] leaves of plates, xiii, 623 p., [8] leaves of plates) ; 23 cm. Sw78-11518
Reprint of the 1851 ed. published by L. Maison, Paris.
Includes bibliographical references.
250F
1. Chiny (Comté)—History. I. Title.
DH801.C44J4 1978 949.3'5 79-340718
79 MARC

CHINZEI GAKUIN

Long, Caroll Summerfield, 1850–1890.
(C. S. Rongu Nihon senkyōki)
C. S. ロング日本宣教記 鮫島盛隆著 東京 キリスト新聞社 昭和49 (1974)
163 p. ports. 19 cm. (鎮西学院研究叢書 1) Ja 74–7607
¥600
1. Long, Caroll Summerfield, 1850–1890. 2. Missionaries—Japan—Biography. 3. Missionaries—United States—Biography. 4. Educators—Japan—Biography. 5. Chinzei Gakuin. I. Samejima, Moritaka, ed. II. Title. III. Series: Chinzei Gakuin. Chinzei Gakuin kenkyū sōsho, 1.
BV3457.L59A33 78–812814

CHION'IN, KYOTO

Hōnen Shōnin gyōjō ezu.
法然上人絵伝 = Hōnen Shōnin e-den / [編集担当 塚本善隆；監修 田中一松]. — 東京：角川書店, 昭和52 [1977]
132, 171, 19 p., [22] leaves of plates : ill. (some col.) ; 37 cm. — (新修日本絵巻物全集 ; 14) Ja 77–19832
In Japanese with summary and legends in English.
"法然上人繪傳解説": p. 1–57 (2d group)
Bibliography: p. 162–163.
¥20000
1. Hōnen Shōnin eden (Scrolls) 2. Painting, Japanese—Kamakura-Momoyama periods, 1185–1600. 3. Chion'in, Kyoto. I. Tsukamoto, Zenryū, 1898- II. Title: Hōnen Shōnin e-den. III. Series: Shinshū Nihon emakimono zenshū ; 14.
Title romanized: Hōnen Shōnin eden.
ND1053.N5 1975, vol. 14 78–814867
[ND1053.4]

CHIPANGALI WILDLIFE ORPHANAGE

Rayner, Richard, 1914-
Who cares? : Chipangali Wildlife Orphanage / by Richard Rayner. — Salisbury : Regal Publishers, 1977.
88 p. : ill. ; 21 cm.
ISBN 0-7974-0297-7
1. Chipangali Wildlife Orphanage. 2. Wildlife rescue—Rhodesia, Southern. I. Title.
QL83.2.R39 639'.9'096891 78-317327
78 MARC

Wilson, Vivian J
Orphans of the wild : the story behind Chipangali / Vivian J. Wilson ; line-drawings by Paddy Wilson. — Bulawayo : Books of Rhodesia, 1977.
254 p., [1] leaf of plates : ill. ; 23 cm.
ISBN 0-86920-162-X
1. Chipangali Wildlife Orphanage. 2. Wild animal collecting—Africa, Southern. 3. Wilson, Vivian J. 4. Zoologists—Rhodesia, Southern—Biography. I. Title.
QL83.2.W54 596'.09689'1 79-306170
79 MARC

CHIPAYA LANGUAGE

—READERS FOR NEW LITERATES

Quintunaca liyaquicha. Vamos a leer cuentos. Cochabamba, Bolivia, Instituto Lingüístico de Verano, 1966-
v. illus. 21 cm.
In Chipaya and Spanish.
Introd. signed: Ronaldo Olson K. and Francisca Bundy de Olson.
1. Chipaya language—Readers for new literates. 2. Spanish language—Readers for new literates. I. Olson K., Ronaldo. II. Olson, Francisca Bundy de.
PM5814.C52Q5 78-396121
79 MARC

CHIPBOARD see Particle board

CHIPHYŎNJŎN

Yi, Chae-ch'ŏl.
(Chiphyŏnjŏn ko)
集賢殿考 / 李載喆著. — 서울 : 韓國圖書館協會, 1978.
v, 76 p. ; 25 cm.
Summary in English.
Bibliography: p. 64-69.
₩4000

1. Chiphyŏnjŏn. I. Title.
Z675.N2Y47 79-825486

CHIPP, DONALD LESLIE, 1925-

Chipp, Donald Leslie, 1925-
Don Chipp : the third man / Don Chipp, John Larkin. — Adelaide : Rigby, [1978]
ix, 254 p., [4] leaves of plates : ill. ; 23 cm. Aus***
Includes index.
ISBN 0-7270-0827-7 : $9.95
1. Chipp, Donald Leslie, 1925- . 2. Australia—Politics and government—1945- . 3. Statesmen—Australia—Biography. I. Larkin, John, joint author.
DU117.2.C44A33 329'.0092'4 79-302440
79 MARC

Hewat, Timothy, 1928-
Don Chipp / [by] Tim Hewat [and] David Wilson. — Camberwell, Vic. : Widescope, 1978.
154 p., [4] leaves of plates : ill. ; 22 cm. Aus
Includes index.
ISBN 0-86932-042-4 : $8.95
1. Chipp, Donald Leslie, 1925- . 2. Australia—Politics and government—1945- . 3. Politicians—Australia—Biography. I. Wilson, David, joint author.
DU117.2.C44H48 329'.0092'4 79-300915
79 MARC

CHIPPENDALE, THOMAS, 1718-1779

Gilbert, Christopher.
The life and work of Thomas Chippendale / Christopher Gilbert. — London : Studio Vista, 1978.
2 v. : ill. (some col.) ; 28 cm. GB***
Bibliography: v. 1, p. 310-315.
Includes index.
ISBN 0-289-70839-7 : £49.00
1. Chippendale, Thomas, 1718-1779. 2. Cabinet-workers—England—Biography. 3. Furniture—England—History—18th century. I. Title.
NK2542.C5G5 749.2'2 79-315221
79 MARC

Loughlin, David.
The case of Major Fanshawe's chairs / David Loughlin. — New York : Universe Books, 1978.
160 p. : ill. ; 22 cm.
Includes bibliographical references and index.
ISBN 0-87663-317-3 : $10.95
1. Chairs—Ireland—Collectors and collecting. 2. Decoration and ornament—Chippendale style. 3. Fanshawe, R. G. 4. Chippendale, Thomas, 1718-1779. I. Title. II. Title: Fanshawe's chairs.
NK2715.L63 749'.3 78-52196
79 MARC

CHIPPENDALE STYLE see Decoration and ornament--Chippendale style

CHIPPEWA CO., MINN.

—DIRECTORIES

Midland Atlas Company, Milbank, S.D.
Atlas of Chippewa County, Minnesota / compiled by the Midland Atlas Company, inc., Milbank, South Dakota. — Milbank, S.D. : The Company, c1978.
19, [36] leaves : ill., maps, ports. ; 36 cm.
Cadastral maps of townships, scale ca. 1:46,000.
Includes rural directories.
$42.50
1. Chippewa Co., Minn.—Maps. 2. Real property—Minnesota—Chippewa Co.—Maps. 3. Chippewa Co., Minn.—Directories. I. Title.
G1428.C4M5 1978 912'.776'39 79-311764
00 MARC

—MAPS

Midland Atlas Company, Milbank, S.D.
Atlas of Chippewa County, Minnesota / compiled by the Midland Atlas Company, inc., Milbank, South Dakota. — Milbank, S.D. : The Company, c1978.
19, [36] leaves : ill., maps, ports. ; 36 cm.
Cadastral maps of townships, scale ca. 1:46,000.
Includes rural directories.
$42.50
1. Chippewa Co., Minn.—Maps. 2. Real property—Minnesota—Chippewa Co.—Maps. 3. Chippewa Co., Minn.—Directories. I. Title.
G1428.C4M5 1978 912'.776'39 79-311764
00 MARC

CHIPPEWA INDIANS

Copway, George, Chippewa chief, 1818-1863?
Indian life and Indian history by an Indian author : embracing the traditions of the North American Indians regarding themselves, particularly of that most important of all the tribes, the Ojibways / by Kah-ge-ga-gah-bowh, known also by the English name of George Copway. — 1st AMS ed. — New York : AMS Press, 1978.
266 p. : ill. ; 18 cm.
Reprint of the 1860 ed. published by A. Colby, Boston.
Originally published under title: The traditional history and characteristic sketches of the Ojibway nation.
ISBN 0-404-15517-0
1. Chippewa Indians. I. Title.
E99.C6C8 1978 970'.004'97 76-43684
78 MARC

—BIBLIOGRAPHY—UNION LISTS

Scott, Patricia.
Chippewa and Cree : a bibliography of books, newspaper articles, government documents, and other printed and written matter in various libraries of the United States and Canada. — Rocky Boy, Mont. : Rocky Boy School, 1976, c1977.
xv, 277 p., [40] leaves of plates : ill. (some col.) ; 28 cm.
Compiled by Patricia Scott.
Includes index.
1. Chippewa Indians—Bibliography—Union lists. 2. Cree Indians—Bibliography—Union lists. 3. Catalogs, Union—United States. 4. Catalogs, Union—Canada. I. Title.
Z1210.C5S37 016.97'0004'97 78-109129
[E99.C6] 79 MARC

—CAPTIVITIES

Mrs. Huggins, the Minnesota captive. — New York : Garland Pub., 1978.
17 p. ; 23 cm. — (The Garland library of narratives of North American Indian captivities ; v. 86)
Reprinted from the Boys' books of romance and adventure, no. 10, 1864.
Issued with the reprint of the 1871 ed. of Reary, J. Sketch of the life of John Reary. New York, 1978.
ISBN 0-8240-1710-2 : $29.50
1. Chippewa Indians—Wars—1862-1865. 2. Chippewa Indians—Captivities. 3. Huggins, Sophia Josephine. 4. Indians of North America—Minnesota—Captivities. 5. Minnesota—Biography. I. Series.
E85.G2 vol. 86 973'.04'97 s 78-21477
[E99.C6] 78 MARC

—CHILDREN

Hilger, Inez.
Chippewa child life and its cultural background / by M. Inez Hilger. — 1st AMS ed. — New York : AMS Press, 1979.
xiv, 204 p., [17] leaves of plates : ill. ; 23 cm.
Reprint of the 1951 ed. published by the U.S. Govt. Print. Off., Washington, which was issued as no. 146 of the U.S. Bureau of American Ethnology bulletin.
Bibliography: p. 173-187.
Includes index.
ISBN 0-404-15581-2
1. Chippewa Indians—Children. 2. Indians of North America—Minnesota—Children. I. Title. II. Series: United States. Bureau of American Ethnology. Bulletin ; 146.
[E99.C6H46 1979] 970'.004'97 76-43740
79 MARC

—DANCES

Barrett, Samuel, Alfred, 1879-1965.
The dream dance of the Chippewa and Menominee Indians of northern Wisconsin / S. A. Barrett. — New York : Garland Pub., 1979.
p. 252-415 : ill. ; 23 cm. — (Classics of anthropology ; 14)
Reprint of the 1911 ed published by the Public Museum, Milwaukee, and issued as v. 1, article 4 of the Bulletin of the Public Museum of the City of Milwaukee.
Includes index.
ISBN 0-8240-9634-7 : $20.00
1. Chippewa Indians—Dances. 2. Menominee Indians—Dances. 3. Indians of North America—Wisconsin—Dances. I. Title. II. Series. III. Series: Milwaukee. Public Museum. Bulletin ; v. 1, article 4.
E99.C6B35 1979 793.3'19775 78-66483
78 MARC

Barrett, Samuel Alfred, 1879-1965.
The dream dance of the Chippewa and Menominee Indians of northern Wisconsin / by S. A. Barrett. — 1st AMS ed. — New York : AMS Press, 1979.
p. 252-406 : ill. ; 23 cm.
Reprint of the 1911 ed. published by the Public Museum, Milwaukee, and issued as v. 1, article 4, of the Bulletin of the Public Museum of the City of Milwaukee.
Includes index.
ISBN 0-404-15482-4
1. Chippewa Indians—Dances. 2. Menominee Indians—Dances. 3. Indians of North America—Wisconsin—Dances. I. Title. II. Series: Milwaukee. Public Museum. Bulletin ; v. 1, article 4.
[E99.C6B35 1979b] 793.3'19775 76-43647
79 MARC

—FICTION

Johnston, Basil.
Moose meat and wild rice / Basil H. Johnston. — Toronto : McClelland and Stewart, 1978.
188 p. ; 22 cm. C78-1252-6
ISBN 0-7710-4443-7 : $12.95
1. Chippewa Indians—Fiction. 2. Indians of North America—Canada—Fiction. I. Title.
PZ4.J7216 Mo 813'.5'4 79-302600
[PR9199.3.J597] 79[r79]rev MARC

—HISTORY

Danziger, Edmund Jefferson, 1938-
The Chippewas of Lake Superior / by Edmund Jefferson Danziger, Jr. — 1st ed. — Norman : University of Oklahoma Press, c1978.
xiv, 263 p., [3] leaves of plates : ill. ; 24 cm. — (The Civilization of the American Indian series ; v. 148)
Bibliography: p. 245-254.
Includes index.
ISBN 0-8061-1487-8
1. Chippewa Indians—History. I. Title. II. Series.
E99.C6D25 970'.004'97 78-58130
78 MARC

—LEGENDS

Morriseau, Norval.
Legends of my people, the great Ojibway / illustrated and told by Norval Morriseau ; edited by Selwyn Dewdney. — Toronto ; New York : McGraw-Hill Ryerson, 1977, c1965.
xxii, 130 p. : ill. ; 24 cm. C77-10887-3
ISBN 0-07-077714-4
1. Chippewa Indians—Legends. 2. Indians of North America—Ontario—Legends. I. Dewdney, Selwyn H., ed. II. Title.
E99.C6M78 1977 398.2'097 78-319823
78 MARC

—SOCIAL CONDITIONS

Lithman, Yngve Georg.
The community apart : a case study of a Canadian Indian reserve community / Yngve Georg Lithman. — Stockholm : Dept. of Social Anthropology, Univ. of Stockholm, 1978.
vii, 239 p. : ill. ; 21 cm. — (Stockholm studies in social anthropology ; 6 ISSN 0347-0830) S78-24
Thesis—Stockholms universitet.
Bibliography: p. 233-239.
ISBN 9185284106
1. Chippewa Indians—Social conditions. 2. Indians of North America—Manitoba—Social conditions. 3. Fort Alexander Indian Reservation, Man. I. Title. II. Series.
E99.C6L53 301.45'19'707127 79-304704
79 MARC

CHIPPEWA INDIANS (Continued)

—SOCIAL LIFE AND CUSTOMS

Densmore, Frances, 1867-1957.
Chippewa customs / by Frances Densmore. — Reprint ed. / with an introd. by Nina Marchatti Archabal. — St. Paul : Minnesota Historical Society Press, 1979.
xii, 204 p., ⌐45⌐ leaves of plates : ill. ; 23 cm. — (Publications of the Minnesota Historical Society)
Reprint of the 1929 ed. published by the U.S. Govt. Print. Off., Washington, which was issued as Bulletin 86 of Smithsonian Institution's Bureau of American Ethnology.
Bibliography: p. 195-196.
Includes index.
ISBN 0-87351-142-5
1. Chippewa Indians—Social life and customs. 2. Indians of North America—Minnesota—Social life and customs. I. Title. II. Series: Minnesota Historical Society. Publications. III. Series: United States. Bureau of American Ethnology. Bulletin ; 86.
E99.C6D38 1979 970'.004'97 79-15400
79 MARC

—WARS—1862-1865

Mrs. Huggins, the Minnesota captive. — New York : Garland Pub., 1978.
17 p. ; 23 cm. — (The Garland library of narratives of North American Indian captivities ; v. 86)
Reprinted from the Boys' books of romance and adventure, no. 10, 1864.
Issued with the reprint of the 1871 ed. of Reary, J. Sketch of the life of John Reary. New York, 1978.
ISBN 0-8240-1710-2 : $29.50
1. Chippewa Indians—Wars—1862-1865. 2. Chippewa Indians—Captivities. 3. Huggins, Sophia Josephine. 4. Indians of North America—Minnesota—Captivities. 5. Minnesota—Biography. I. Series.
E85.G2 vol. 86 973'.04'97 s 78-21477
⌐E99.C6⌐ 78 MARC

CHIPPEWA LANGUAGE

see also Algonquian languages

—GLOSSARIES, VOCABULARIES, ETC.

Johnston, Basil.
Ojibway language lexicon for beginners / by Basil Johnston. — ⌐Ottawa⌐ : Indian and Northern Affairs, 1978.
134 p. ; 28 cm. C***
ISBN 0-662-10145-6
1. Chippewa language—Glossaries, vocabularies, etc. I. Title.
PM853.J6 497'.3 79-312099
79 MARC

—TEXTS

Katolik anamie-masinaigan. A Catholic prayerbook and catechism in the Otchipwe-Indian language by Rt. Rev. Bishop F. Baraga, D.D., with an appendix of the mass and vespers in Latin; and prayers in the Ottawa-Indian language, by Rev. John B. Weikamp, TERT. O.S.F. New York, Cincinnati, Chicago, Benziger brothers ⌐ca. 1880⌐
346 p. front., illus. 16 cm.
1. Catholic Church—Prayer-books and devotions—Chippewa. 2. Catholic Church—Catechisms and creeds—Chippewa. 3. Catholic Church—Prayer-books, and devotions—Ottawa. 4. Chippewa language—Texts. 5. Ottawa language—Texts. I. Baraga, Friedrich, Bp., 1797-1868. II. Weikamp, John B.
PM854.C3 26-17206
⌐r79⌐rev MARC

CHIPROVT͡SI REGION, BULGARIA

Milanov, Milan Vasilev.
(Chiprovt͡si)
Чипровци : ⌐краеведски очерк с маршрути⌐ / Милан Василев Миланов. — 2. прераб. и доп. изд. — София : Медицина и физкултура, 1977 (Пловдив : печ. Д. Благоев)
60, ⌐4⌐ p. : ill. ; 19 cm. — (Малка туристическа библиотека)
Bu 77-2089
First ed. published in 1970 under title: Chiprovt͡si i Chiprovskata planina.
Series romanized: Malka turisticheska biblioteka.
Bibliography: p. ⌐62⌐
0.25 lv
1. Chiprovt͡si region, Bulgaria. I. Title.
DR95.C46M54 1977 79-411723

CHIPS, WOOD see Wood chips

CHIQUITOS, BOLIVIA (PROVINCE)

—HISTORY

Ovando, Jorge.
La invasión brasileña a Bolivia en 1825 : una de las causas del Congreso de Panamá / Jorge Alejandro Ovando Sanz. — 1. ed. — La Paz, Bolivia : Ediciones Isla, 1977.
182 p. ; 20 cm.
"Publicaciones de Antonio Paredes-Candia."
Limited edition of 500 copies.
Bibliography: p. 177-178.
Includes index.
1. Bolivia—History—Wars of Independence, 1809-1825. 2. Chiquitos, Bolivia (Province)—History. 3. Bolivia—Foreign relations—Brazil. 4. Brazil—Foreign relations—Bolivia. I. Title.
F3323.O93 984'.041 78-102468
79 MARC

CHIRAC, JACQUES, 1932-

Vivian, Pierre.
Réflexions d'un Français sur le mystère Chirac / Pierre Vivian. — ⌐Paris⌐ : Librairie Notre temps, c1978.
190 p. ; 21 cm. F***
Bibliography: p. ⌐189⌐-190.
28.00F
1. France—Politics and government—1969- . 2. Chirac, Jacques, 1932- . I. Title.
DC417.V58 320.9'44'083 78-399060
79 MARC

CHIRADZULU, MALAWI (DISTRICT)

—CENSUS

Malawi, National Statistical Office.
Chiradzulu population pilot census, August 1973 / National Statistical Office. — Zomba : The Office, 1974.
ii, 39 p. ; 33 cm.
Cover title.
1. Chiradzulu, Malawi (District)—Census. I. Title.
HA1977.M32C465 316.89'7 76-980219
79 MARC

CHIRALITY

—CONGRESSES

Nato Advanced Study Institute, University of Sussex, 1978.
Optical activity and chiral discrimination : proceedings of the NATO Advanced Study Institute, held at the University of Sussex, Falmer, England, September 10-22, 1978 / edited by Stephen F. Mason. — Dordrecht, Holland ; Boston : D. Reidel Pub. Co., c1979.
xi, 372 p. : graphs ; 25 cm. — (NATO advanced study institutes series : Series C, Mathematical and physical sciences ; v. 48)
Includes bibliographical references and index.
ISBN 9027709823
1. Optical rotatory dispersion—Congresses. 2. Circular dichroism—Congresses. 3. Chirality—Congresses. I. Mason, Stephen Finney, 1923- II. Title. III. Series.
QD473.N37 1978 541'.7 79-10532
79 MARC

CHIRGOTIS, WILLIAM G., 1910-

Chirgotis, William G 1910-
250 home plans for today's living / from the drawing board of William G. Chirgotis. — ⌐Passaic, N.J.⌐ : Federal Marketing Corp. ; New York : distributed exclusively by Sterling Pub. Co., c1979.
239 p. : ill. ; 28 cm.
ISBN 0-932944-01-9
1. Chirgotis, William G., 1910- . 2. Architecture, Domestic—United States—Designs and plans. 3. Architecture, Modern—20th century—United States—Designs and plans. I. Title.
NA737.C4A4 1979 728.3 78-65787
79 MARC

CHIRIACKA, ERNEST, 1920-

—EXHIBITIONS

Chiriacka, Ernest, 1920-
Ernest Chiriacka, recent paintings : May 15 to June 5, 1979. — New York : Kennedy Galleries, ⌐c1979⌐
⌐16⌐ p. : ill. (some col.) ; 28 cm.
1. Chiriacka, Ernest, 1920- —Exhibitions. 2. Indians of North America—Pictorial works—Exhibitions. 3. The West in art—Exhibitions. I. Kennedy Galleries, inc., New York. II. Title.
ND237.C4924A4 1979 759.13 79-109870
79 MARC

CHIRICAHUA INDIANS see Apache Indians

CHIRICO, ANDREA DE, 1891-1952

—EXHIBITIONS

Alberto Savinio : Palazzo delle esposizioni, Roma, 18 maggio-18 luglio 1978 / scritti di Maurizio Fagiolo, Daniela Fonti, Pia Vivarelli ; ⌐catalogo a cura di Pia Vivarelli⌐. — Roma : De Luca, ⌐1978⌐
295 p. : ill. ; 24 cm. It78-Sept
At head of title: S.P.Q.R. Assessorato belle arti e problemi della cultura.
Bibliography: p. 291-295.
1. Chirico, Andrea de, 1891-1952—Exhibitions. I. Chirico, Andrea de, 1891- . II. Rome (City). Palazzo delle esposizioni.
N6923.C497A4 1978 709'.2'4 78-396755
*79 MARC

CHIRICO, GIORGIO DE, 1888-

Chirico, Giorgio de, 1888-
Pictor optimus pinxit : Giorgio De Chirico (1888-1978) / Giorgio Ruggeri. — Bologna : Galleria Marescalchi, 1979.
103 p. : ill. (some col.) ; 24 cm. — (I Maestri ; 16) It79-July
Bibliography: p. 98-103.
1. Chirico, Giorgio de, 1888- . I. Ruggeri, Giorgio. II. Title.
N6923.C5A4 1979 79-384058
*79 MARC

Costantini, Costanzo.
Il pittore glorioso / Costanzo Costantini. — Milano : SugarCo, ⌐1978⌐
159 p. : ill. ; 22 cm. It***
L8500
1. Chirico, Giorgio de, 1888- . 2. Painters—Italy—Biography. I. Title.
ND623.C56C67 79-383999
79 MARC

CHIRIGUANO INDIANS

Nordenskiöld, Erland, friherre, 1877-1932.
The changes in the material culture of two Indian tribes under the influence of new surroundings / by Erland Nordenskiöld. — 1st AMS ed. — New York : AMS Press, 1979.
xvi, 245 p. : ill. ; 23 cm.
Reprint of the 1920 ed. published by Pehrssons förlag, Göteborg, Sweden, which was issued as v. 2 of Comparative ethnographical studies.
Bibliography: p. 214-236.
ISBN 0-404-15142-6
1. Chiriguano Indians. 2. Chané Indians. I. Title. II. Series: Nordenskiöld, Erland, friherre, 1877-1932. Comparative ethnographical studies ; 2.
F3320.2.C4N67 1979 984'.004'98 75-46055
78 MARC

—BIOGRAPHY

Dos mujeres indígenas. — 1. ed. — México : Instituto Indigenista Interamericano, 1976.
xii, 184 p., ⌐2⌐ leaves of plates : ill. ; 19 cm. — (Serie Antropología social ; 14)
Includes a translation from English of Basilia by J. Nash.
CONTENTS: Nash, J. Basilia.—Rocca, M. M. Facundina.
1. Basilia, 1912- . 2. Miranda de Méndez, Facundina. 3. Kechua Indians—Biography. 4. Chiriguano Indians—Biography. I. Nash, June C., 1927- Basilia. Spanish. 1976. II. Rocca, Manuel María. Facundina. 1976. III. Series: Inter-American Indian Institute. Serie Antropología social; 14.
F2230.2.K4B373 980'.004'98 79-375530
79 MARC

CHIROGNOMY see Palmistry

CHIROGRAPHY see Penmanship; Writing

CHIROL, VALENTINE, SIR, 1852-1929

Frey, Werner, 1947-
Sir Valentine Chirol : die britische Position und Politik in Asien 1895-1925 / Werner Frey. — Zürich : Juris-Verlag, 1976.
viii, 256 p. ; 23 cm. Sw76-6891
Thesis Zürich.
Vita.
Bibliography: p. 230-255.
ISBN 3-260-04080-3 : 48.00F
1. Asia—Foreign relations—Great Britain. 2. Great Britain—Foreign relations—Asia. 3. India—Politics and government—20th century. 4. Chirol, Valentine, Sir, 1852-1929.
DS33.4.G7F73 327.41'05 78-372116
*78 MARC

CHIROMANCY see Palmistry

CHIRONOMIDAE

see also Chironomus

Platzer-Schultz, Ilse.
Unsere Zuckmücken, Chironomidae / von Ilse Platzer-Schultz. — Wittenberg Lutherstadt : A. Ziemsen, 1974.
104 p. : ill. ; 21 cm. — (Die Neue Brehm-Bücherei ; 477) GDR
Bibliography: p. 98-101.
9.10M
1. Chironomidae. 2. Chironomus. I. Title.
QL537.C456P5 75-569407
79 MARC

CHIRONOMUS

Platzer-Schultz, Ilse.
Unsere Zuckmücken, Chironomidae / von Ilse Platzer-Schultz. — Wittenberg Lutherstadt : A. Ziemsen, 1974.
104 p. : ill. ; 21 cm. — (Die Neue Brehm-Bücherei ; 477) GDR
Bibliography: p. 98-101.
9.10M
1. Chironomidae. 2. Chironomus. I. Title.
QL537.C456P5 75-569407
79 MARC

CHIROPODISTS see Podiatrists

CHIROPODY see Podiatry

CHIROPRACTIC

see also Naturopathy

Barge, Fred H
Tortipelvis / by Fred H. Barge. — 2d ed. — Davenport, Iowa : Bawden Bros., 1978.
vii, 128 p. : ill. ; 23 cm. — (His Chiropractic technic ; v. 1)
Includes indexes.
1. Tortipelvis. 2. Spinal adjustment. 3. Chiropractic. I. Title. II. Series.
RZ242.B37 1978 615'.534 79-101936
79 MARC

DeGiacomo, F P
Man's greatest gift to man—chiropractic / F. P. DeGiacomo. — Old Bethpage, N.Y. : LSR Learning Associates, c1978.
xiii, 273, [1] p. : ill. ; 23 cm.
Bibliography: p. [274]
ISBN 0-932544-00-2 : $14.00
1. Chiropractic. 2. Chiropractic—History. I. Title. II. Title: Chiropractic.
RZ242.D43 615'.534 78-63055
79 MARC

—HISTORY

DeGiacomo, F P
Man's greatest gift to man—chiropractic / F. P. DeGiacomo. — Old Bethpage, N.Y. : LSR Learning Associates, c1978.
xiii, 273, [1] p. : ill. ; 23 cm.
Bibliography: p. [274]
ISBN 0-932544-00-2 : $14.00
1. Chiropractic. 2. Chiropractic—History. I. Title. II. Title: Chiropractic.
RZ242.D43 615'.534 78-63055
79 MARC

—POPULAR WORKS

Chiropractic health care : a conservative approach to health restoration, maintenance, and disease resistance / [edited by] R. C. Schafer. — 3d ed. — Des Moines : Foundation for Chiropractic Education and Research, 1979, c1978.
viii, 120 p. : ill. ; 23 cm.
Bibliography: p. 115-119.
1. Chiropractic—Popular works. I. Schafer, R. C.
RZ244.C48 1979 615'.534 79-103686
79 MARC

—PRACTICE

Basic chiropractic paraprofessional manual : practice administration and management / R. C. Schafer, author/editor. — 1st ed. — Des Moines : American Chiropractic Association, c1978.
ca. 500 p. in various pagings : ill. ; 29 cm.
Includes bibliographies and index.
1. Chiropractic—Practice. 2. Chiropractic clinics—Administration. I. Schafer, R. C.
RZ232.2.B37 615'.534 78-11974
78 MARC

—TEXAS—HISTORY

Rhodes, Walter R
The official history of chiropractic in Texas / by Walter R. Rhodes. — Austin : Texas Chiropractic Association, c1978.
188 p. : ill. ; 24 cm.
Includes bibliographical references.
1. Chiropractic—Texas—History. I. Texas Chiropractic Association. II. Title.
RZ225.U6R48 615'.534'09764 78-110420
78 MARC

CHIROPRACTIC CLINICS

—ADMINISTRATION

Basic chiropractic paraprofessional manual : practice administration and management / R. C. Schafer, author/editor. — 1st ed. — Des Moines : American Chiropractic Association, c1978.
ca. 500 p. in various pagings : ill. ; 29 cm.
Includes bibliographies and index.
1. Chiropractic—Practice. 2. Chiropractic clinics—Administration. I. Schafer, R. C.
RZ232.2.B37 615'.534 78-11974
78 MARC

CHIROPRACTORS

—LICENSES—MINNESOTA
—PERIODICALS

Minnesota. Chiropractic Board of Examiners.
Biennial report of examining and licensing boards. Minneapolis, Chiropractic Board of Examiners.
v. 28 cm.
Period covered by reports ends June 30.
1. Minnesota. Chiropractic Board of Examiners—Periodicals. 2. Chiropractors—Licenses—Minnesota—Periodicals.
RZ205.U52M63 353.9'776'008243 78-645357
MARC-S

—FRANCE—PARIS—BIOGRAPHY

Hanoteau, Guillaume.
Les doigts du miracle : [Lucien Vaimber] / Guillaume Hanoteau. — Paris : La Table ronde, [1975]
285 p. ; 22 cm. F***
39.00F
1. Vaimber, Lucien. 2. Chiropractors—France—Paris—Biography. 3. Paris—Biography. I. Title.
RZ232.V35H36 615'.534'0924 78-388377
79 MARC

CHIROSOPHY see Palmistry

CHIROTHERAPY see Massage

CHISEL-TOOTHED KANGAROO RAT see Dipodomys microps

CHISHIMA ISLANDS see Kuril Islands

CHISHIRE, ENG.

—HISTORY—SOURCES—BIBLIOGRAPHY
—CATALOGS

Cheshire, Eng. Record Office.
Cheshire Record Office and Chester Diocesan Record Office. — [Chester] : [Cheshire County Council], [1975]
38 leaves in various pagings ; 30 cm. GB76-22549
Cover title.
"Compiling a family tree" (sheet ([2] sides)) as insert.
ISBN 0-9501590-3-4 : £0.40
1. Cheshire, Eng. Record Office—Catalogs. 2. Cheshire, Eng. (Diocese)—Record Office—Catalogs. 3. Chishire, Eng.—History—Sources—Bibliography—Catalogs. 4. Cheshire, Eng.—Church history—Sources—Bibliography—Catalogs. I. Chester, Eng. (Diocese) Record Office. II. Title.
CD1065.C47C47 1975 016.9427'1 78-317709
78 MARC

CHISHITSU CHŌSAJO

—PERIODICALS

Chishitsu Chōsajo.
(Chishitsu Chōsajo nempō)
地質調査所年報

[川崎] 工業技術院地質調査所
v. ill. 26 cm.
Other title, : Annual report of Geological Survey of Japan.
In Japanese.
1. Geology—Periodicals. 2. Geology, Economic—Periodicals. 3. Chishitsu Chōsajo—Periodicals. I. Chishitsu Chōsajo. Annual report of Geological Survey of Japan.
QE1.C43a 78-644940
(MARC-S)
J

CHIT MAUNG, U, D. 1946

Ma Ma Lay, 1916–
(Thu lo lu)
သူလိုလူ / မမလေး. — 7. အကြိမ်နှိပ်ခြင်း. — ရန်ကုန် : ဂျာနယ်ကျော်မမလေးစာပေ, 1976.
794 p. ; 18 cm.
In Burmese; appendices in English.
"ဂျာနယ်ကျော်ဦးချစ်မောင်၏နေ့စဉ်မှတ်တမ်း": p. [589]–746.
Appendices: Anti-Fascist People's Freedom League statement.—The real origin and causes of the Burma rebellion, 1930–32.
K20.00

1. Chit Maung, U, d. 1946. I. Title.

PN5449.B8Z746 1976 77–986487
S A

CHITA, SIBERIA. OBLASTNAIA BIBLIOTEKA

–CATALOGS

Sapozhnikova, K L
Zemli͡a rodnai͡a / [sost. K. L. Sapozhnikova]. — [s. l. : s. n., 196–?]
9 p.
In colophon: Chitinskai͡a oblastnai͡a biblioteka im. A. S. Pushkina. Spravochno-bibliograficheskiĭ otdel.
L.C. copy on microfilm only.
An annotated bibliography.
——— Microfilm. On 1 reel with other items ; 35 mm.
Microfilm Slavic 12982 Z
1. Chitinskaya oblast', Russia—Bibliography—Catalogs. 2. Chita, Siberia. Oblastnai͡a biblioteka—Catalogs. I. Chita, Siberia. Oblastnai͡a biblioteka. Spravochno-bibliograficheskiĭ otdel. II. Title.
[Z2514.C5Z44] 79–367359
[DK771.C35]

CHITAL see Axis axis

CHITALISHTE VŬZRAZHDANE

—ADDRESSES, ESSAYS, LECTURES

50 [i.e. Petdeset] godini Chitalishte Vŭzrazhdane, kvartal Podui͡ane, Sofii͡a, 1928-1978 : i͡ubileen sb. / redaktsionna kolegii͡a, Atanas Dushkov ... [et al.]. — Sofii͡a : OF, 1978.
109, [19] p. : ill. ; 29 cm. Bu78-1654
0.96 lv
1. Chitalishte Vŭzrazhdane—Addresses, essays, lectures. I. Dushkov, Atanas, 1909-
Z839.C47P47 79-379458
*79 MARC

CHITIN

Subramanian, V 1919-
Studies on the formation of chitin-metal complexes / by V. Subramanian, T. Yoshinari and Bruno d'Anglejan. — Montreal : McGill University, Marine Sciences Centre, 1974.
75 leaves : graphs ; 28 cm. — (M.S.C. manuscript report - McGill University, Marine Sciences Centre ; no. 27) C***
Bibliography: leaves 71-75.
1. Chitin. 2. Organometallic compounds. 3. Complex compounds. I. Yoshinari, Takeshi, joint author. II. D'Anglejan, Bruno F., joint author. III. Title. IV. Series: McGill University. Marine Sciences Centre. Marine Sciences Centre manuscript report ; no. 27.
QD321.S924 547'.782 78-323350
79 MARC

CHITINSKAYA OBLAST', RUSSIA

—BIBLIOGRAPHY—CATALOGS

Sapozhnikova, K L
Zemli͡a rodnai͡a / [sost. K. L. Sapozhnikova]. — [s. l. : s. n., 196-?]
9 p.
In colophon: Chitinskai͡a oblastnai͡a biblioteka im. A. S. Pushkina. Spravochno-bibliograficheskiĭ otdel.
L.C. copy on microfilm only.
An annotated bibliography.
——— Microfilm. On 1 reel with other items ; 35 mm.
Microfilm Slavic 12962 Z
1. Chitinskaya oblast', Russia—Bibliography—Catalogs. 2. Chita, Siberia. Oblastnai͡a biblioteka—Catalogs. I. Chita, Siberia. Oblastnai͡a biblioteka. Spravochno-bibliograficheskiĭ otdel. II. Title.
[Z2514.C5Z44] 79-367859
[DK771.C35]

—HISTORY, LOCAL

Balabanov, Viktor Fedorovich.
(V debri͡akh nazvaniĭ)
В дебрях названий : [Геогр. названия Забайкалья : Справочник] / Виктор Балабанов. — Иркутск : Вост.-Сиб. кн. изд-во, 1977.
78 p. ; 20 cm. USSR 77
0.22rub
1. Names, Geographical — Russian Republic — Chitinskaya oblast', Russia. 2. Chitinskaya oblast', Russia—History, Local. I. Title.
DK771.C35B34 79-409975

CHITTAGONG, BANGLADESH (DISTRICT)

—HISTORY

Māhabuba-ula Ālama, 1898–
(Caṭṭagrāmera itihāsa)
চট্টগ্রামের ইতিহাস. [লেখক] মাহবুব-উল্ আলম. চট্টগ্রাম, নয়া-লোক প্রকাশনী [1965–66]
3 v. 19 cm.
In Bengali.
CONTENTS: [1] পুরানা আমল.—[2] নবাবী আমল.—[3] কোম্পানী আমল.
1. Chittagong, Bangladesh (District)—History. I. Title.
DS396.8.C43M33 S A 68-797
PL 480: EP-B-1103

CHITTAGONG HILL TRACTS, BANGLADESH (DISTRICT)

Hutchinson, Robert Henry Sneyd, d. 1930.
Chittagong Hill Tracts / by R. H. Sneyd Hutchinson. — Delhi : Vivek Pub. Co., 1978.
vii, 108 p. ; 23 cm.
Caption title: Gazetteer of the Chittagong Hill Tracts' District.
Reprint of the 1909 ed. printed at the Pioneer Press, Allahabad.
Includes index.
Rs36.00
1. Chittagong Hill Tracts, Bangladesh (District) I. Title. II. Title: Gazetteer of the Chittagong Hill Tracts' District.
DS396.8.C432H87 1978 954.9'23 78-912262
79 MARC

CHIU, CHANG-CHUN, 1148-1227

Wang, Kuo-wei, 1877–1927.
[Ch'ang-ch'un-chen-jen hsi yu chi chu]
長春眞人西遊記校注 [上·下卷] 校注者王國維 蒙古律例 [12卷] 臺北 廣文書局 民國61 [1972]
4, 146, 129 p. 22 cm. (史料四編) NT$200.00
Colophon title.
Caption title: Ch'ang-ch'un-chen-jen hsi yu chi chu.
Reprint ed.
1. Li, Chih-ch'ang, 1193–1278. Ch'ang-chun-chen-jen hsi yu chi. 2. China—Description and travel. 3. Mongolia—Description and travel. 4. Asia, Central—Description and travel. 5. Ch'iu, Ch'ang-ch'un, 1148–1227. 6. Law—Mongolia—Sources. I. Li, Chih-ch'ang, 1193–1278. Ch'ang-ch'un-chen-jen hsi yu chi. 1972. II. Title. III. Title: Meng-ku lü li. IV. Series: Shih liao ssu pien.
Title romanized: Ch'ang-ch'un-chen-jen hsi yu chi chiao chu.
DS707.L53W36 1972 73-897702
rev C

CH'IU, CHIN, 1877-1907

(Ch'iu Chin)
秋瑾. — 香港：海鷗出版社, [1976?]
80 p. : ill. ; 18 x 19 cm.
$3.00
1. Ch'iu, Chin, 1877–1907. 2. Revolutionists—China—Biography.
DS763.C5C44 79–840054
79 C

—ANECDOTES

(Ch'iu Chin yü Chou Yün-hsien chih ch'i jen ch'i shih)
秋瑾與周韻僊之奇人奇事 / [朱琬珍 ... et al.]. — [台北 ? : s. n., 1978?]
[18] p. : facsims. ; 27 cm.
Cover title.
1. Ch'iu, Chin, 1877–1907—Anecdotes. 2. Revolutionists—China—Anecdotes, facetiae, satire, etc. 3. Chou, Yü-hsien—Addresses, essays, lectures. I. Chu, Wan-chen.
DS763.C5C46 78–841920
C

CH'IU, YING, 1522-1560. SPRING MORNING IN THE HAN PALACE

Ch'iu, Ying, 1522–1560.
[Han kung ch'un hsiao]
明仇英漢宮春曉. — [台北]：國立故宮博物院, [197-]
1 scroll (8 ft. 2 in.) in case : col. ill. ; 13 cm.
Title from label mounted on case.
Explanatory note ([2] p.) in Chinese and English inserted.
1. Ch'iu, Ying, 1522–1560. Spring morning in the Han Palace. I. Kuo li ku kung po wu yüan. II. Title.
Title romanized: Ming Ch'iu Ying Han kung ch'un hsiao.
ND1049.C518A7 1970z 78–843676

CHIU-CHIANG, CHINA (KIANGSI PROVINCE)

—HISTORY—SOURCES

(Chiu-chiang fu chih)
九江府志：江西省 / 達春布修；黃鳳樓纂. — 臺1版. — 台北：成文出版社, 民國64 [1975]
2 v. (1170 p.) : ill. ; 27 cm. — (中國方志叢書華中地方；第267號)
Reprint of the 1874 ed.
1. Chiu-chiang, China (Kiangsi Province)—History—Sources. I. Ta-ch'un-pu, 19th cent. II. Huang, Feng-lou, chin shih 1832. III. Series: Chung-kuo fang chih ts'ung shu Hua-chung ti fang ; ti 267 hao.
DS796.C58C48 1975 78–841472

CHIVALRY

see also Bushidō; Civilization, Medieval; Courtly love; Feudalism; Heraldry; Honor; Hwarangdo

Bulfinch, Thomas, 1796-1867.
Bulfinch's Mythology. — New York : Avenel Books : distributed by Crown Publishers, [1978]
xvii, 957 p., [25] leaves of plates : ill. ; 23 cm.
The three works, popularly known as Bulfinch's mythology, were originally written and published separately.
Includes index.
CONTENTS: The age of fable.—The age of chivalry.—Legends of Charlemagne.
ISBN 0-517-26277-0
1. Mythology. 2. Chivalry. 3. Folk-lore—Europe. 4. Charlemagne, 742-814—Romances. I. Title. II. Title: Mythology.
BL310.B82 1978 398.2 78-13042
78 MARC

Jacq, Christian.
De sable et d'or : symbolique héraldique, l'honneur du nom / Christian Jacq, Patrice Delaperrière. — Paris (15, rue Racine, 75006) : Éditions des Trois mondes, 1976.
271 p. : ill. ; 22 cm. — (Chemins des symboles ; 2) F76-12930
Bibliography: p. 264-271.
ISBN 2-901037-05-4 : 45.00F
1. Heraldry. 2. Chivalry. 3. Symbolism. I. Delaperrière, Patrice, joint author. II. Title.
CR21.J32 929.6 78-400691
79 MARC

—ROMANCES see Romances

CHIVALRY IN LITERATURE

Ruggieri, Ruggero M
L'umanesimo cavalleresco italiano : da Dante all'Ariosto / Ruggero M. Ruggieri. — Nuova ed. riveduta e ampliata. — Napoli : Fratelli Conte, 1977.
xii, 349 p., [4] leaves of plates : ill. ; 23 cm. — (Riscontri ; 4) It78-Nov
Includes bibliographical references and indexes.
L9600
1. Italian literature—To 1400—History and criticism. 2. Italian literature—15th century—History and criticism. 3. Chivalry in literature. 4. Humanism in literature. I. Title.
PQ4067.R79 1977 851'.1'09353 79-352966
*79 MARC

CHIVERS, THOMAS HOLLEY, 1809-1858

—AUTHORSHIP

Bell, Landon Covington, 1880-1960.
Poe and Chivers. [Folcroft, Pa.] Folcroft Library Editions, 1973 [c1931]
101 p. 24 cm.
Reprint of the ed. published by C. A. Trowbridge, Columbus, Ohio.
ISBN 0-8414-3133-7 (lib. bdg.)
1. Poe, Edgar Allan, 1809-1849—Authorship. 2. Chivers, Thomas Holley, 1809-1858—Authorship. 3. Plagiarism. I. Title.
PS2636.B45 1973 818'.3'09 73-9500
73[r79]rev2 MARC

Bell, Landon Covington, 1880-1960.
Poe and Chivers / by Landon C. Bell. — Norwood, Pa. : Norwood Editions, 1976 [c1931]
101 p. ; 24 cm.
Reprint of the ed. published by C. A. Trowbridge Co., Columbus, Ohio.
ISBN 0-8482-0298-8 : $12.50
1. Poe, Edgar Allan, 1809-1849—Authorship. 2. Chivers, Thomas Holley, 1809-1858—Authorship. 3. Plagiarism. I. Title.
PS2636.B45 1976 818'.3'09 76-13159
76[r79]rev MARC

CHIVERS, THOMAS HOLLEY, 1809-1858 (Continued)

—CRITICISM AND INTERPRETATION

Lombard, Charles M
Thomas Holley Chivers / by Charles M. Lombard. — Boston : Twayne, 1979.
148 p. : port. ; 21 cm. — (Twayne's United States authors series ; TUSAS 325)
Bibliography: p. 146-147.
Includes index.
ISBN 0-8057-7258-8
1. Chivers, Thomas Holley, 1809-1858—Criticism and interpretation.
PS1294.C4Z78 811'.3 78-12109
78 MARC

CHIVILIKHIN, VLADIMIR ALEKSEEVICH

—CRITICISM AND INTERPRETATION

Sergovantšev, Nikolaĭ Mikhaĭlovich.
(Vladimir Chivilikhin)
Владимир Чивилихин / Н. Сергованцев. — Москва : Сов. Россия, 1978.
77 p. ; 16 cm. — (Писатели Советской России) USSR 78
Series romanized: Pisateli Sovetskoĭ Rossii.
0.15rub
1. Chivilikhin, Vladimir Alekseevich—Criticism and interpretation.
PG3479.5.H55Z88 79-410214

CHIYO JINJA, HIKONE, JAPAN

Shiga, Japan. Kyōiku Iinkai.
(Jūyō bunkazai Chiyo Jinja honden shūri kōji hōkokusho)
重要文化財千代神社本殿修理工事報告書 / ［編集 滋賀県教育委員会］. — ［大津］：同委員会，昭和41 ［1966］
25 p., 69 leaves of plates : 79 ill. ; 30 cm.
1. Chiyo Jinja, Hikone, Japan. 2. Shinto shrines—Japan—Hikone—Conservation and restoration. 3. Hikone, Japan—Shrines—Conservation and restoration. I. Title. II. Title: Chiyo Jinja honden shūri kōji hōkokusho.
NA6057.H423S53 1966 79-800678

CHIYODA-KU, TOKYO

—DESCRIPTION

Suzuki, Masao, 1926–
(Chiyoda-ku no rekishi)
千代田区の歴史 / 鈴木理生・文. — 東京：名著出版，1978.
260, 37 p. : ill. ; 19 cm. — (東京ふるさと文庫 ; 5) Ja 78-27027
Bibliography: p. 37 (2d group)
¥980
1. Chiyoda-ku, Tokyo—History. 2. Tokyo—History. 3. Chiyoda-ku, Tokyo—Description. 4. Tokyo—Description. I. Title. II. Series: Tōkyō furusato bunko ; 5.
DS896.7.C5S97 79-803635

—HISTORY

Suzuki, Masao, 1926–
(Chiyoda-ku no rekishi)
千代田区の歴史 / 鈴木理生・文. — 東京：名著出版，1978.
260, 37 p. : ill. ; 19 cm. — (東京ふるさと文庫 ; 5) Ja 78-27027
Bibliography: p. 37 (2d group)
¥980
1. Chiyoda-ku, Tokyo—History. 2. Tokyo—History. 3. Chiyoda-ku, Tokyo—Description. 4. Tokyo—Description. I. Title. II. Series: Tōkyō furusato bunko ; 5.
DS896.7.C5S97 79-803635

Suzuki, Masao, 1926–
(Meiji umare no machi Kanda Misaki-chō)
明治生れの町神田三崎町 / 鈴木理生著. — 東京：青蛙房，1978.
447 p. : ill. ; 22 cm. Ja 78-18099
Y4500
1. Chiyoda-ku, Tokyo—History. 2. Kanda-ku, Tokyo—History. 3. Tokyo—History. I. Title. II. Title: Kanda Misaki-chō.
DS896.7.C5S98 78-814816

—JUVENILE LITERATURE

(Iidabashi mukashi mukashi)
飯田橋むかしむかし / ［編集 飯田橋ディベロッパーズクラブ］. — 東京：同クラブ，昭和51 ［1976］
2 v. : ill. ; 30 cm. Ja 76-36
Cover title.
Vol. 2 has subtitle: Ohanashi.
Bibliography: v. 2, p. 1.
¥800
1. Chiyoda-ku, Tokyo—Juvenile literature. 2. Tokyo—Historical geography—Juvenile literature. 3. Tokyo—Historical geography—Maps—Juvenile literature. I. Iidabashi Diberoppāzu Kurabu.
DS896.7.C5I33 78-809399

CHLAMYDOMONAS

Ettl, Hanuš, 1931-
Die Gattung Chlamydomonas Ehrenberg / von H. Ettl. — Vaduz : Cramer, 1976.
1122 p. : numerous ill. ; 24 cm. — (His Chlamydomonas und die nächstverwandten Gattungen ; 2) (Beihefte zur Nova Hedwigia ; Heft 49) GFR***
Label mounted on verso of t.p.: Exclusive Distributor: ISBS, Inc., Forest Grove.
Bibliography: p. 744-774.
Includes index.
ISBN 3-7682-5449-6 : DM100.00
1. Chlamydomonas. I. Title. II. Series. III. Series: Nova Hedwigia : Beihefte ; Heft 49.
QK504.N62 Heft 49 586'.008 s 76-483801
［QL368.V6］ 76［r79］rev MARC

CHLAMYS

Heald, D I
Pecten scallop survey in southern Geographe Bay, September 1976 / by D. I. Heald. — Perth : Dept. of Fisheries and Wildlife, 1977.
43 p. : ill. ; 30 cm. — (Report - Department of Fisheries and Wildlife ; no. 31) Aus***
Bibliography: p. 16-17.
1. Scallop fisheries—Australia—Geographe Bay. 2. Fishery resources—Australia—Western Australia. 3. Scallops. 4. Chlamys. 5. Geographe Bay, Australia. I. Title. II. Series: Western Australia. Dept. of Fisheries and Wildlife. Report - Department of Fisheries and Wildlife ; no. 31.
SH372.2.A8H43 79-316105
79 MARC

CHLIDONIAS NIGER see Black tern

CHLORAL

Mžinić, Radosna.
On the use of anaesthetics in the transportation of sardines. Rome, Food and Agriculture Organization of the United Nations, 1970.
23 p. 28 cm. (General Fisheries Council for the Mediterranean. Studies and reviews, no. 47) It***
Summary in English and French; text in English.
Bibliography: p. 23.
1. Animal anesthesia. 2. Fishes—Transportation. 3. Sardines. 4. Chloral. 5. Tricaine. I. Title. II. Series.
SH1.G317 no. 47 597'.0922 78-316823
［SH156.9］ 78 MARC

CHLORAL-HYDRATE see Chloral

CHLORAMPHENICOL see Chloromycetin

CHLORAOACETIC ACIDS

see also Dichlorophenoxyacetic acid

CHLORETHYLENE see Vinyl chloride

CHLORIDES

see also Vinyl chloride

Pavlov, Dmitriĭ Igorevich.
(Ėkzogennye khloridnye vody i ėndogennoe rudoobrazovanie)
Экзогенные хлоридные воды и эндогенное рудообразование / Д. И. Павлов. — Москва : Недра, 1975.
102 p. : ill. ; 22 cm. USSR 75
Bibliography : p. 90–［101］
0.71rub
1. Hydrothermal deposits. 2. Chlorides. I. Title.
QE364.P38 78-420639

CHLORINATED PARAFFIN see Paraffin, Chlorinated

CHLORINATED RUBBER see Rubber, Chlorinated

CHLORINATION

see also Rubber, Chlorinated

CHLORINE

Suidan, Makram T
Reduction of aqueous free chlorine with granular activated carbon : final report, project no. A-066-ILL / by Makram T. Suidan and Vernon L. Snoeyink. — Urbana : University of Illinois, Water Resources Center, 1975.
xiii, 108 p. : graphs ; 28 cm. — (WRC research report ; no. 103)
Originally presented as M. T. Suidan's thesis, University of Illinois at Urbana-Champaign.
Bibliography: p. 88-89.
1. Chlorine. 2. Reduction, Chemical. 3. Carbon, Activated. I. Snoeyink, Vernon L., joint author. II. Title. III. Series: Illinois. University at Urbana-Champaign. Water Resources Center. WRC research report ; noo. 103.
HD1694.A136 no. 103 333.9'1'008 s 78-620572
［QD181.C5］ 78 MARC

—CONGRESSES

Diaphragm cells for chlorine production : proceeding of a symposium held at the City University, London, England, 16-17 June 1976. — London : Society of Chemical Industry, 1977.
vi, 115 p. : ill. ; 21 cm. GB78-20382
Organized by the Electrochemical Technology Group of the Society of Chemical Industry.
Includes bibliographical references.
ISBN 0-901001-50-3 : £10.00
1. Electrolytic cells—Congresses. 2. Chlorine—Congresses. I. Society of Chemical Industry, London. Electrochemical Technology Group.
TP159.E4D5 665'.83 78-324747
79 MARC

—PATENTS

Powell, Robert, 1918-
Chlorine and caustic soda manufacture; recent developments, 1969. Park Ridge, N.J., Noyes Development Corp. ［1969］
265 p. illus. 28 cm. (Chemical process review no. 33) 35.00
Bibliography: p. 263-264.
ISBN 0-8155-0264-8
1. Chlorine—Patents. 2. Sodium hydroxide—Patents. I. Title. II. Series.
TP245.C5P67 661'.07'320272 70-78755
69［r79］rev MARC

—TOXICOLOGY

Martens, D W
Dechlorination of municipal sewage using sulfur dioxide / D. W. Martens and J. A. Servizi. — New Westminster, B.C. : International Pacific Salmon Fisheries Commission, 1975.
24 leaves : diagr. ; 28 cm. — (Progress report - International Pacific Salmon Fisheries Commission ; no. 32) C***
Bibliography: leaves 23-24.
1. Fishes, Effect of water pollution on. 2. Water quality bioassay. 3. Sewage—Purification—Chlorination—Environmental aspects. 4. Chlorine—Toxicology. 5. Sewage—Toxicology. 6. Sulphur dioxide—Toxicology. 7. Pacific salmon—Mortality. I. Servizi, J. A., joint author. II. Title. III. Series: International Pacific Salmon Fisheries Commission (U.S. and Canada). Progress report ; no. 32.
SH346.I62 no. 32 79-316640
［SH177.C43］ 79 MARC

CHLORINE (Continued)

—TOXICOLOGY—ILLINOIS

Roseboom, Donald P
Acute toxicity of residual chlorine and ammonia to some native Illinois fishes / by Donald P. Roseboom and Dorothy L. Richey. — Urbana : Illinois State Water Survey, 1977.
42 p. : ill. ; 28 cm. — (Report of investigation - Illinois State Water Survey ; 85)
Includes bibliographical references.
1. Fishes, Effect of water pollution on—Illinois. 2. Chlorine—Toxicology—Illinois. 3. Ammonia—Toxicology—Illinois. 4. Water quality bioassay—Illinois. 5. Bluegill—Mortality. 6. Channel catfish—Mortality. I. Richey, Dorothy L., joint author. II. Title. III. Series: Illinois. Water Survey. Report of investigation ; 85.
GB705.I3A3 no. 85 553'.7'09773 s 78-622477
[SH177.C43] 79 MARC

—TOXICOLOGY—MICHIGAN

Basch, R E
Calculated residual chlorine concentrations safe for fish / R. E. Basch and J. G. Truchan. — [Lansing] : Michigan Water Resources Commission, Bureau of Water Management, Water Quality Appraisal Section, 1974.
iii, 29 leaves ; 29 cm. — (Technical bulletin - Bureau of Water Management ; 74-2)
Bibliography: leaves 26-29.
1. Fishes, Effect of water pollution on—Michigan. 2. Chlorine—Toxicology—Michigan. 3. Water quality bioassay—Michigan. 4. Electric power-plants—Environmental aspects—Michigan. I. Truchan, J. G., joint author. II. Title. III. Series: Michigan. Bureau of Water Management. Technical bulletin - Bureau of Water Management ; 74-2.
SH177.C43B37 597'.05'222 79-315386
79 MARC

CHLORINE DIOXIDE

Chao, Shun-chung.
(Erh yang hua lü chih tsao fang fa chih yen chiu)
二氧化氯製造方法之研究 = Study on the processes of manufacturing chlorine dioxide / 趙順中, 潘登灶. — 臺北：臺灣省林業試驗所, 1971.
10 p. : ill. ; 27 cm. — (臺灣省林業試驗所報告；第201號)
Chiefly Chinese.
Includes bibliographical references.
1. Chlorine dioxide. I. P'an, Teng-tsao, joint author. II. Title. III. Title: Study on the processes of manufacturing chlorine dioxide. IV. Series: T'ai-wan sheng lin yeh shih yen so, T'ai-pei. T'ai-wan sheng lin yeh shih yen so pao kao : ti 201 hao.
SD86.5.T328a no. 201 79-837709
[TP245.C5]

CHLORINE INDUSTRY

—BRAZIL—ALAGOAS (STATE)

Clan S.A.
Polo cloroquímico para o Estado de Alagoas / Clan. — [Maceió] : Companhia de Desenvolvimento do Estado de Alagoas, [between 1974 and 1978]
182 leaves : ill. ; 31 cm.
At head of title: Governo do Estado de Alagoas, Secretaria de Planejamento, Secretaria da Indústria e Comércio.
1. Chlorine industry—Brazil—Alagoas (State) I. Companhia de Desenvolvimento de Alagoas. II. Title.
TP245.C5C54 1974 78-393886
79 MARC

—CHINA

China (Republic of China). Ta lu shih yu chi hua hsüeh kung yeh yen chiu tsu.
(Ta lu fei ch'ü lü chien kung yeh chih yen chiu)
大陸匪區氯鹼工業之研究 / [大陸石油及化學工業研究組]；經濟部編印. — [台北]：該部, 民國67 [1978]
30 p. ; 22 cm. — (專題研究)
Series romanized: Chuan t'i yen chiu.
1. Chlorine industry—China. 2. Alkali industry and trade—China. I. Title.
HD9660.C53C427 79-840771

CHLORINE OXIDES

Masschelein, W
Chlorine dioxide : chemistry and environmental impact of oxychlorine compounds / by W. J. Masschelein ; edited for publication by Rip G. Rice. — Ann Arbor, Mich. : Ann Arbor Science, c1979.
ix, 190 p. : ill. ; 24 cm.
Revised and updated translation of the author's Les Oxydes de chlore et le chlorite de sodium.
Includes bibliographical references and index.
ISBN 0-250-40224-6
1. Chlorine oxides. 2. Sodium chlorite. I. Rice, Rip G. II. Title.
QD181.C5M3413 546'.732'2 77-92595
79 MARC

CHLOROALKANE see Paraffin, Chlorinated

CHLOROCOCCALES

Hindák, František.
Studies on the chlorococcal algae, Chlorophyceae / František Hindák. — Bratislava : VEDA, Pub. House of the Slovak Academy of Sciences, 1977-
v. : ill. ; 24 cm. — (Treatise on biology ; XXIII/4) CzS***
Summaries also in Slovak and Russian.
Bibliography: v. 1, p. 179-183.
Includes index.
Kcs30.00 (v. 1)
1. Chlorococcales. 2. Algae—Czechoslovakia. I. Title. II. Series: Biologické práce ; XXIII/4 .
QH7.B47 vol. 23/4, etc. 79-339718
[QK569.C496] 79 MARC

CHLORODIHYDROMETHYL-PHENYLBENZODIAZEPINONE see Diazepam

CHLOROFORM

Błaszkiewicz, Barbara.
Badanie wiązań wodorowych chloroformu i metanolu z cyklicznymi akceptorami protonu metodą przesunięć chemicznych / Barbara Błaszkiewicz. — Wyd. 1. — Poznań : Wydawn. Naukowe Uniwersytetu im. Adama Mickiewicza, 1976.
34 p. : ill. ; 24 cm. — (Seria Fizyka ; Nr 24) P***
Summary in English.
Bibliography: p. [32]-34.
zł8.00
1. Hydrogen bonds. 2. Chloroform. 3. Methanol. I. Title. II. Series.
QD461.B58 79-339081
79 MARC

CHLOROHYDROCARBONS

see also Paraffin, Chlorinated; Tetrachlorodibenzodioxin

CHLOROIDODOQUINOLINOL see Iodochlorhydroxyquin

CHLOROMONAS

see also Chlamydomonas

CHLOROMYCETIN

United States. Library of Congress. Science Policy Research Division.
Competitive problems in the drug industry : chloramphenicol (chloromycetin) : summary and analysis / Select Committee on Small Business, United States Senate, Subcommittee on Monopoly, Economic Concentration, and Anticompetitive Activities ; prepared by Congressional Research Service, Library of Congress. — Washington : U.S. Govt. Print. Off., 1979.
vii, 97 p. : graphs ; 24 cm.
At head of title: 96th Congress, 1st session. Committee print.
"A summary analysis and discussion of issues highlighted during the 1968 and 1969 hearings on chloramphenicol (Competitive problems in the drug industry—volumes 6 and 11), with a review of current findings and subsequent government actions relating to these drugs."
Summary by Irene Jillson, private consultant, and edited by Vikki Zegel of CRS Science Policy Research Division.
Bibliography: p. 77-81.
1. Drug trade—United States. 2. Competition—United States. 3. Chloromycetin. I. Jillson, Irene. II. United States. Congress. Senate. Select Committee on Small Business. Subcommittee on Monopoly, Economic Concentration, and Anticompetitive Activities. III. Title.
HD9666.4.U56 1979 338.8 79-601965
79 MARC

CHLOROPHENOLS

see also Pentachlorophenol

—METABOLISM

Ahlborg, Ulf G 1938-
Metabolism of chlorophenols : studies on dechlorination in mammals / by Ulf G. Ahlborg. — [Stockholm : Statens naturvårdsverk, Forskningssekretariatet], 1977.
134 p. : ill. ; 24 cm. — (Rapport från Statens naturvårdsverk) S***
"SNV PM 895."
Includes articles by the author and others originally published in various journals.
Includes bibliographies.
1. Chlorophenols—Metabolism. 2. Chlorophenols—Toxicology. 3. Pentachlorophenol—Metabolism. 4. Mammals—Physiology. I. Title. II. Series: Sweden. Statens naturvårdsverk. Rapport från Statens naturvårdsverk.
QP801.C48A34 615.9'51'2 78-318494
78 MARC

—TOXICOLOGY

Ahlborg, Ulf G 1938-
Metabolism of chlorophenols : studies on dechlorination in mammals / by Ulf G. Ahlborg. — [Stockholm : Statens naturvårdsverk, Forskningssekretariatet], 1977.
134 p. : ill. ; 24 cm. — (Rapport från Statens naturvårdsverk) S***
"SNV PM 895."
Includes articles by the author and others originally published in various journals.
Includes bibliographies.
1. Chlorophenols—Metabolism. 2. Chlorophenols—Toxicology. 3. Pentachlorophenol—Metabolism. 4. Mammals—Physiology. I. Title. II. Series: Sweden. Statens naturvårdsverk. Rapport från Statens naturvårdsverk.
QP801.C48A34 615.9'51'2 78-318494
78 MARC

see also Desmidiales

CHLOROPHYLL

see also Porphyrin and porphyrin compounds

—CONGRESSES

Symposium on Chlorophyll Organization and Energy Transfer in Photosynthesis, Ciba Foundation, 1978.
Chlorophyll organization and energy transfer in photosynthesis. — Amsterdam ; New York : Excerpta Medica, 1979.
xi, 374 p. : ill. ; 25 cm. — (Ciba Foundation symposium ; 61 (new ser.))
Proceedings of the symposium held Feb. 7-9, 1978 in London.
Includes bibliographies and indexes.
ISBN 0-444-90044-6
1. Photosynthesis—Congresses. 2. Chlorophyll—Congresses. 3. Energy transfer—Congresses. I. Title. II. Series: Ciba Foundation. Symposium ; new ser., 61.
QK882.S946 1978 581.1'3342 78-12368
78 MARC

CHLOROPLASTS

see also Chlorophyll

Gillham, Nicholas W
Organelle heredity / Nicholas W. Gillham. — New York : Raven Press, c1978.
xv, 602 p. : ill. ; 27 cm.
Includes bibliographical references and index.
ISBN 0-89004-102-4
1. Cytoplasmic inheritance. 2. Mitochondria. 3. Chloroplasts. 4. Cytogenetics. I. Title.
[DNLM: 1. Genetics. 2. Chloroplasts. 3. Mitochondria. QH452.G479o]
QH452.G54 575.2'1 75-43195
78 MARC

CHLOROPLASTS (Continued)

The Intact chloroplast / edited by J. Barber. — Amsterdam ; New York : Elsevier Scientific Pub. Co. ; New York : distributors for the U. S. and Canada, Elsevier/North-Holland, 1976.
xi, 476 p. : ill. ; 25 cm. — (Topics in photosynthesis ; v. 1)
Includes bibliographies and indexes.
ISBN 0-444-41451-7
1. Chloroplasts. I. Barber, James, 1940-
QK882.I53 581.8'733 76-23261
76 MARC

—CONGRESSES

International Symposium on Chloroplast Development, Island of Spetsai, Greece, 1978.
Chloroplast development : proceedings of the International Symposium on Chloroplast Development held on the Island of Spetsai, Greece, July 9-15, 1978 / editors, G. Akoyunoglou and J. H. Argyroudi-Akoyunoglou. — Amsterdam ; New York : Elsevier/North-Holland Biomedical Press ; New York : sole distributors for the USA and Canada, Elsevier North-Holland, 1978.
xvi, 888 p. : ill. ; 25 cm. — (Developments in plant biology ; v. 2)
Includes bibliographical references and index.
ISBN 0-444-80084-0
1. Chloroplasts—Congresses. 2. Akoyunoglou, G. I. Argyroudi-Akoyunoglou, J. H. II. Title. III. Series.
QK725.I496 1978 581.8'733 78-23489
78 MARC

CHLUP, OTOKAR, 1875-1965

Chlup, Otokar, 1875-1965.
Otokar Chlup a perspektivy výchovy : [život, dílo a ukázky z díla O. Chlupa / výběr uspoř. a pův. texty naps. Vlastimil Pařízek]. — 1. vyd. — Praha : SPN, 1977.
217, [1] p., [1] leaf of plates : port. ; 20 cm. — (Z dějin pedagogiky)
Cz78
Includes bibliographical references.
Kčs15.00
1. Chlup, Otokar, 1875-1965. 2. Education—Philosophy. I. Pařízek, Vlastimil. II. Title.
LB880.C4256O87 78-388920
*79 MARC

CHMARA, GREGORY, D. 1970

Volmane, Véra.
Gregory Chmara, l'homme expressif / Véra Volmane et Charles Ford. — Paris : la Table ronde, 1979.
159 p., [16] p. of plates ; 24 cm. F79-3950
Filmography: p. 153-154.
Includes index.
42.00F
1. Chmara, Gregory, d. 1970. 2. Actors—Russia—Biography. I. Ford, Charles, 1908- joint author. II. Title.
PN2728.C56V64 79-381032
79 MARC

CHMIELOWSKI, JOACHIM BENEDYKT, 1700-1763. NOWE ATENY

Rybicka-Nowacka, Halina.
Nowe Ateny Benedykta Chmielowskiego : metoda, styl, język / Helena Rybicka-Nowacka. — Wyd. 1. — Warszawa : Państwowe Wydawn. Naukowe, 1974.
159 p. : ill. ; 24 cm. — (Rozprawy Uniwersytetu Warszawskiego ; 75)
Summary in French.
Bibliography: p. 158-159.
zł36.00
1. Chmielowski, Joachim Benedykt, 1700-1763. Nowe Ateny. I. Title. II. Series: Warsaw. Uniwersytet. Rozprawy ; 75.
PG7157.C55N6837 79-351557
79 MARC

CHNUMHOTEP

—TOMB

Moussa, Ahmed M
Das Grab des Nianchchnum und Chnumhotep / von Ahmed M. Moussa und Hartwig Altenmüller ; Photographien von D. Johannes ; Zeichnungen von W. Ruhm. — Mainz am Rhein : P. v. Zabern, c1977.
178 p., [58] leaves of plates : ill. (some col.) ; 36 cm. — (Old Kingdom tombs at the causeway of King Unas at Saqqara) (Archäologische Veröffentlichungen ; 21) GFR***
Includes bibliographical references and indexes.
ISBN 3-8053-0050-6 : DM198.00
1. Sakkara—Tombs. 2. Chnumhotep—Tomb. 3. Nianchchnum—Tomb. 4. Egypt—Antiquities. I. Altenmüller, Hartwig, joint author. II. Title. III. Series. IV. Series: Deutsches Archäologisches Institut. Abteilung Kairo. Archäologische Veröffentlichungen ; 21.
DT57.D47 no. 21 78-384091
[DT73.S3] MARC

CHO, AE-SIL

Cho, Ae-sil.
(Ch'arari t'onggok igirŭl)
차라리痛哭이기를 : 趙愛實隨想集. — 서울 : 傳獒苑, 1977.
269 p. ; 21 cm.
Essays.
₩1200
1. Cho, Ae-sil. 2. Women—Korea—Biography. 3. Christian biography—Korea. I. Title.
CT1848.C45A32 78-826860
K

CHO, YONG-GI, 1936-

Cho, Yong-gi, 1936-
The fourth dimension / by Paul Yonggi Cho ; [with a foreword by Robert Schuller]. — Plainfield, N.J. : Logos International, c1979.
186 p. ; 21 cm.
ISBN 0-88270-380-3 : $3.95
1. Christian life—1960- . 2. Cho, Yong-gi, 1936- . I. Title.
BV4501.2.C49 248'.4 79-65588
79 MARC

CHO, YUN-JE, 1904-1976

(Tonam Cho Yun-je Paksa kohŭi kinyŏm nonch'ong)
陶南趙潤濟博士古稀紀念論叢 / [編輯陶南趙潤濟博士古稀紀念論叢刊行委員會]. — 서울 : 螢雪出版社, 1976.
512 p. : ill. ; 23 cm.
"著作要目": p. 24-30.
Includes bibliographical references.
1. Korean literature—History and criticism—Addresses, essays, lectures. 2. Cho, Yun-je, 1904-1976—Bibliography. 3. Cho, Yun-je, 1904-1976. I. Tonam Cho Yun-je Paksa Kohŭi Kinyŏm Nonch'ong Kanhaeng Wiwŏnhoe.
PL957.1.T66 79-825495
K

—BIBLIOGRAPHY

(Tonam Cho Yun-je Paksa kohŭi kinyŏm nonch'ong)
陶南趙潤濟博士古稀紀念論叢 / [編輯陶南趙潤濟博士古稀紀念論叢刊行委員會]. — 서울 : 螢雪出版社, 1976.
512 p. : ill. ; 23 cm.
"著作要目": p. 24-30.
Includes bibliographical references.
1. Korean literature—History and criticism—Addresses, essays, lectures. 2. Cho, Yun-je, 1904-1976—Bibliography. 3. Cho, Yun-je, 1904-1976. I. Tonam Cho Yun-je Paksa Kohŭi Kinyŏm Nonch'ong Kanhaeng Wiwŏnhoe.
PL957.1.T66 79-825495
K

CHŌ FAMILY

—ARCHIVES—CATALOGS

Chō Ke Shiryō Chōsa Iinkai.
(Chō Ke shiryō mokuroku)
長家史料目録 / [編集 長家史料調査委員会]. — 穴水町(石川県) : 穴水町教育委員会, 1978.
15, 137 p., [8] leaves of plates : ill. ; 26 cm. — (古文書等緊急調査報告書 ; 昭和52年度)
Ja 78-12845
1. Chō family—Archives—Catalogs. 2. Ishikawa, Japan (Prefecture)—History—Sources—Bibliography—Catalogs. I. Title. II. Series: Komonjo tō kinkyū chōsa hōkokusho ; Shōwa 52-nendo.
CD2189.5.C46C46 1978 78-816141

CHO-PYŎNG-HWA, 1921-

—CORRESPONDENCE

Cho, Pyŏng-hwa, 1921-
(Siin ŭi p'yŏnji)
詩人의便紙 : 詩人金大圭와의文學論書簡集 / 趙炳華. — 서울 : 靑潮社, 1977.
365 p. ; 21 cm.
Letters.
₩1200
1. Cho, Pyŏng-hwa, 1921- —Correspondence. 2. Kim, Tae-gyu, 1942- —Correspondence. I. Kim, Tae-gyu, 1942- II. Title.
PL980.4.C54 78-826314

CH'OBALSIM CHAGYONGMUN

Sim, Chae-yŏl.
(Ch'obalsim chagyŏngmun kangŭi)
初發心自警文講義 / 沈載烈講義 ; 李鍾益校閲. — [서울] : 經林出版社 : 供給處良友堂, 1976.
331 p. ; 23 cm.
Includes original text and indexes.
₩900
1. Ch'obalsim chagyŏngmun. 2. Monastic and religious life (Buddhism) 3. Buddhist novices—Discipline. I. Title.
BQ6115.S55 79-825701

CHOCH'ŎN-NI, KOREA

Kim, Po-hyŏn, fl. 1976-
(Choch'ŏn chi)
朝天誌 / 金寶鉉著. — [朝天?] : 金, 1976.
182 p., [2] leaves of plates (1 fold.) : ill. ; 22 cm.
Errata slip inserted.
Bibliography: p. 181.
Not in trade.
1. Choch'ŏn-ni, Korea. I. Title.
DS725.C47K55 78-826632

CHOCIŁOWSKI, JERZY

Chociłowski, Jerzy.
Indyjska szarada / Jerzy Chociłowski. — Wyd. 1. — Warszawa : Krajowa Agencja Wydawnicza, 1977 (Katow : RSW "PKR")
148, [4] p., [16] leaves of plates : ill. ; 17 cm. — (Z reporterskich wedrówek)
P78-705
zł35.00
1. India—Description and travel—1947- . 2. Chociłowski, Jerzy. I. Title.
DS414.C539 78-382366
78 MARC

CHOCOLATE

see also Cocoa; Cookery (Chocolate)

Kolpas, Norman.
The chocolate lovers' companion / by Norman Kolpas. — New York : Quick Fox, 1977.
155 p. : ill. ; 24 cm.
On cover: A delicious sampler of history, poetry, stories, recipes, and tips about the world's favorite food.
Originally published under title: The chocolate (and coca) connoisseur's companion.
ISBN 0-8256-3126-2 : $6.95
1. Cookery (Chocolate) 2. Chocolate. I. Title.
TX767.C5K64 641.3'37'4 78-56219
78 MARC

Marcus, Adrianne.
The chocolate bible / Adrianne Marcus. — New York : Putnam, c1979.
279 p., [4] leaves of plates : ill. ; 24 cm.
ISBN 0-399-14042-4 : 12.95
1. Chocolate. 2. Chocolate industry. 3. Cookery (Chocolate) I. Title.
TX791.M36 1979 641.3'37'4 78-25784
78 MARC

CHOCOLATE (Continued)

Pelletier, Irène Dubé.
Les secrets d'une chocolatière : ou, Comment ouvrir une fabrique de chocolat dans sa cuisine / Irène Dubé Pelletier. — Montréal : Éditions du jour, c1977.
88 p. : ill. ; 22 cm. C***
"C-177."
ISBN 0-7760-0754-8 : $5.00
1. Confectionery. 2. Chocolate. I. Title.
TX791.P44 641.8'53 78-377509
78 MARC

—EARLY WORKS TO 1800

Lavedán, Antonio.
Tratado de los usos, abusos, propiedades y virtudes del tabaco, café, té y chocolate; extractado de los mejores autores que han tratado de esta materia, á fin de que su uso no perjudique á la salud, antes bien pueda servir de alivio y curación de muchos males. Madrid, En la Imprenta Real, 1796.
237 p. 20 cm.
On spine: Usos del tabaco.
1. Health—Early works to 1800. 2. Tobacco—Early works to 1800. 3. Coffee—Early works to 1800. 4. Tea—Early works to 1800. 5. Chocolate—Early works to 1800. I. Title. II. Title: Usos del tabaco.
RA775.L68 78-411600
78 MARC

—JUVENILE LITERATURE

Allamand, Pascale.
Cocoa beans and daisies : how Swiss chocolate is made / Pascale Allamand. — New York : F. Warne, c1978.
[47] p. : ill. ; 24 cm.
SUMMARY: Explains how cocoa beans from Africa come together with milk of Switzerland to make chocolate. Includes two recipes.
ISBN 0-7232-6156-3
1. Confectionery—Juvenile literature. 2. Chocolate—Juvenile literature. [1. Chocolate] I. Title. II. Title: Swiss chocolate is made.
TX792.A44 641.3'37'4 78-7043
00 MARC

CHOCOLATE INDUSTRY

Marcus, Adrianne.
The chocolate bible / Adrianne Marcus. — New York : Putnam, c1979.
279 p., [4] leaves of plates : ill. ; 24 cm.
ISBN 0-399-14042-4 : 12.95
1. Chocolate. 2. Chocolate industry. 3. Cookery (Chocolate) I. Title.
TX791.M36 1979 641.3'37'4 78-25784
78 MARC

CHOCOLATE-TREE see Cacao

CHOCTAW INDIANS

—ANTIQUITIES

Lintz, Christopher.
An archaeological survey of the Kerr-Mcgee Choctaw coal mine facility, Haskell County, Oklahoma / by Christopher Lintz. — [Norman] : Archaeological Research and Management Center, University of Oklahoma, 1978.
vi, 191 p. : ill. ; 28 cm. — (Project report series ; no. 2 ISSN 0160-3078)
Bibliography: p. 187-191.
1. Haskell Co., Okla—Antiquities. 2. Choctaw Indians—Antiquities. 3. Excavations (Archaeology)—Oklahoma—Haskell Co. 4. Coal mines and mining—Oklahoma—Haskell Co. 5. Oklahoma—Antiquities. 6. Kerr-McGee Corporation. I. Oklahoma. University. Archaeological Research and Management Center. II. Title. III. Series.
F702.H4L56 976.6'77 79-620759
79 MARC

—LAW see Law, Choctaw

—TREATIES

Choctaw Nation. Constitution.
Constitution and laws of the Choctaw Nation : together with the treaties of 1837, 1855, 1865, and 1866 / published by authority of the General Council by A. R. Durant, and Davis Homer, and Ben Watkins, assistant compilers. — Wilmington, Del. : Scholarly Resources, 1973.
352 p. ; 23 cm. — (The Constitutions and laws of the American Indian tribes ; v. 12)
Reprint of the 1894 ed. published by J. F. Worley, Dallas.
Includes index.
ISBN 0-8420-1706-2
1. Law, Choctaw. 2. Choctaw Indians—Treaties. I. Durant, A. R. II. Homer, Davis A. III. Watkins, Ben. IV. Choctaw Nation. Laws, statutes, etc. V. Choctaw Nation. Treaties, etc. VI. Title. VII. Series.
KF8228.C8A5 1973 340.5'2 73-88765
77[r79]rev MARC

CHOCTAW JARGON see Mobilian trade language

CHOCTAW LANGUAGE

see also Mobilian trade language

CHOCTAW LAW see Law, Choctaw

CHODHRIS

Shah, Ghanshyam.
Socio-economic condition of Chodhras : a restudy / Ghanshyam Shah. — Surat : Centre for Social Studies, [1977?]
207 leaves [1] leaf of plates : map ; 29 cm.
Mimeographed.
Includes bibliographical references.
Rs40.00
1. Chodhris. I. Title.
DS432.C53S5 309.1'54'75 78-907799
79 MARC

CHODOWIECKI, DANIEL NICOLAUS, 1726-1801

—EXHIBITIONS

Maerker, Peter, 1937-
Bürgerliches Leben im 18. Jahrhundert : Daniel Chodowiecki 1726-1801 : Zeichnungen und Druckgraphik : [Ausstellung], Städelsches Kunstinstitut und Städtische Galerie Frankfurt am Main, 8. Juni 1978 bis 20. August 1978 : [Katalog / Peter Märker]. — Frankfurt am Main : Städel, c1978.
175 p. : ill. ; 24 cm. GFR***
Bibliography: p. 173-175.
1. Chodowiecki, Daniel Nicolaus, 1726-1801—Exhibitions. 2. Germany—Social life and customs—Pictorial works—Exhibitions. 3. Germany—History—18th century—Pictorial works—Exhibitions. 4. Germany in art—Exhibitions. I. Chodowiecki, Daniel Nicolaus, 1726-1801. II. Städelsches Kunstinstitut und Städtische Galerie Frankfurt am Main. III. Title.
N6888.C48M33 78-393269
79 MARC

CHODRAS see Chodhris

CH'OE, CHAE-HUI, 1914-

—ADDRESSES, ESSAYS, LECTURES

Ch'oe, Chae-hŭi, 1914-
(Pawi e taech'im ŭl)
바위에대침을 / 崔載喜著. — 서울 : 徽文出版社, 1978.
441 p. ; 22 cm. — (나의人生觀)
₩2670
1. Ch'oe, Chae-hui, 1914- —Addresses, essays, lectures. 2. Philosophers—Korea—Biography—Addresses, essays, lectures. 3. Philosophy, Modern—Addresses, essays, lectures. I. Title. II. Series: Na ŭi insaenggwan.
B5254.C492A36 78-826904

CH'OE, HAK-SONG, 1901-1933

—CRITICISM AND INTERPRETATION

Cho, Nam-hyŏn.
(Ilche ha ŭi chisigin munhak)
일제하의지식인문학 / [조남현지음]. — 서울 : 평민사, 1978.
117 p. ; 21 cm. — (평민서당 ; 10)
Series romanized: P'yŏngmin sŏdang.
₩500
1. Korean literature—20th century—History and criticism. 2. Yu, Chin-o, 1906- —Criticism and interpretation. 3. Ch'oe, Hak-song, 1901-1933—Criticism and interpretation. I. Title.
PL958.6.C49 79-825959

CH'OE, HŬNG-JO, 1918-

Ch'oe, Hŭng-jo, 1918-
(Nae ka ch'ajŭn Hananim)
내가찾은하나님 : 논·픽션 / 崔興朝著. — 서울 : 自由社出版部, 1976.
305 p. ; 19 cm.
₩900
1. Ch'oe, Hŭng-jo, 1918- 2. Converts—Korea—Biography. I. Title.
BV4935.C45A33 78-826794

CH'OE, TŎK-HYU, 1922-

Ch'oe, Tŏk-hyu, 1922-
(Ch'oe Tŏk-hyu)
崔德休 = Choi Duk-hyu : 作家와作品 / 金仁煥. — 서울 : 金星出版社, 1977.
[32] p. : chiefly ill. (some col.) ; 33 cm. — (韓國現代美術代表作家100人選集 : 56)
The introductory review of the artist and the legend also in English.
₩1500
1. Ch'oe, Tŏk-hyu, 1922- I. Kim, In-hwan. II. Title: Choi Duk-hyu. III. Series: Han'guk hyŏndae misul taep'yo chakka paegin sŏnjip : 56.
ND1069.C485A4 1977 78-826409

CHOICE, SOCIAL see Social choice

CHOICE (ECONOMIC THEORY) see Consumer's preferences

CHOICE (PSYCHOLOGY)

see also Decision-making; Dissonance (Psychology); Risk-taking (Psychology); Social choice

Choice and perceived control / edited by Lawrence C. Perlmutter, Richard A. Monty ; sponsored by the U.S. Army Human Engineering Laboratory. — Hillsdale, N.J. : Lawrence Erlbaum Associates ; New York : distributed by Halsted Press, 1979.
xiv, 384 p. ; 24 cm.
Includes bibliographies an indexes.
Papers presented at a conference held at Virginia Polytechnic Institute and State University, Blacksbury, Va., Feb. 13-15, 1978.
ISBN 0-470-26852-2
1. Choice (Psychology) 2. Control (Psychology) I. Perlmuter, Lawrence C. II. Monty, Richard A. III. United States. Army Human Engineering Laboratories.
BF441.C38 153.8 79-19042
79 MARC

Crouch, Robert L
Human behavior : an economic approach / Robert Crouch. — North Scituate, Mass. : Duxbury Press, c1979.
xvii, 204 p. : ill. ; 24 cm.
Includes bibliographical references and index.
ISBN 0-87872-205-X
1. Economics—Psychological aspects. 2. Human behavior. 3. Choice (Psychology) I. Title.
HB74.P8C76 330'.01'9 78-10391
78 MARC

CHOICE (PSYCHOLOGY) (Continued)

Luce, Robert Duncan.
Individual choice behavior : a theoretical analysis / R. Duncan Luce. — Westport, Conn. : Greenwood Press, ₍1979₎ c1959.
xii, 153 p. ; 24 cm.
Reprint of the ed. published by Wiley, New York.
Bibliography: p. 147-149.
Includes index.
ISBN 0-313-20778-X
1. Psychometrics. 2. Choice (Psychology) I. Title.
₍BF39.L8 1979₎ 153.8'3 78-25881
78 MARC

Meyners, Robert, 1922-
Sexual style : facing and making choices about sex / Robert Meyners and Claire Wooster. — 1st ed. — New York : Harcourt Brace Jovanovich, c1979.
186 p. : ill. ; 25 cm.
Bibliography: p. ₍180₎-186.
ISBN 0-15-181399-X
1. Sex customs. 2. Choice (Psychology) I. Wooster, Claire, 1942- joint author. II. Title.
HQ12.M48 301.41'7 78-22266
79 MARC

Musaph, Herman, 1915-
Seksualiteit en partnerkeuze. Psychologische fundering. Door H. Musaph. Haarlem, Bohn, 1970.
vii, 199 p. 22 cm. 20.00
Bibliography: p. 197-199.
ISBN (invalid) 906051022X
1. Sex (Psychology). 2. Choice (Psychology). I. Title.
₍DNLM: 1. Sex Behavior. HQ21 M985s 1970₎
BF692.M8 72-500141
Shared Cataloging with 79₍r79₎rev MARC
DNLM

Nelson, Richard C
Choosing : a better way to live / Richard C. Nelson. — North Palm Beach, Fla. : Guidelines Press, c1977.
xv, 240 p. : ill. ; 21 cm.
ISBN 0-932570-00-3 : $5.95
1. Success. 2. Choice (Psychology) I. Title.
BF637.S8N43 158'.1 78-67188
79 MARC

Pessemier, Edgar A 1922-
Understanding and analyzing contingency data / by Edgar A. Pessemier. — West Lafayette, Ind. : Institute for Research in the Behavioral, Economic, and Management Sciences, Krannert Graduate School of Management, Purdue University, 1978.
25, ₍2₎, 8 p. : ill. ; 28 cm. — (Paper - Institute for Research in the Behavioral, Economic, and Management Sciences, Krannert Graduate School of Management No. 654)
Bibliography: p. ₍26₎-₍27₎
1. Consumers' preferences. 2. Choice (Psychology) I. Title. II. Series: Krannert Graduate School of Management. Institute for Research in the Behavioral, Economic and Management Sciences. Paper - Institute for Research in the Behavioral, Economic, and Management Sciences, Krannert Graduate School of Management.
HD6483.P8 no. 654 658'.001'9 s 78-622302
₍HF5415.3₎ 79 MARC

Schelling, Thomas C 1921-
Micromotives and macrobehavior / Thomas Schelling. — 1st ed. — New York : Norton, c1978.
252 p. : ill. ; 22 cm. — (Fels lectures on public policy analysis)
Includes index.
ISBN 0-393-05701-1. ISBN 0-393-09009-4 pbk.
1. Policy sciences. 2. Choice (Psychology) 3. Social choice. 4. Collective behavior. 5. Social problems. I. Title. II. Series.
H61.S355 1978 301.11'3 78-17119
78 MARC

Wagenaar, W A
De beste stuurlui dempen de put : hoe mensen beslissen, toegelicht aan de hand van Nederlandse spreekwoorden / door Willem Albert Wagenaar. — Baarn : Ambo, 1977.
117 p. ; 21 cm. — (Amboboeken) Ne77-20
Bibliography: p. 114-115.
Includes index.
ISBN 9026303777 : fl 12.50
1. Choice (Psychology) 2. Advertising—Psychological aspects. 3. Proverbs, Dutch—Psychological aspects. I. Title.
BF441.W28 79-364124
*79 MARC

CHOICE BY LOT

see also Divination; Oracles

CHOICE OF BOOKS see Book selection; Books and reading

CHOICE OF COLLEGE see College, Choice of

CHOICE OF PERIODICALS see Periodical selection

CHOICE OF PROFESSION see Vocational guidance

CHOICE OF SCHOOL see School, Choice of

CHOICE OF TRANSPORTATION

Allis-Chalmers Corporation. Advanced Technology Center.
Milwaukee County dual-mode systems study / prepared by program manager, Allis-Chalmers Corp., Advanced Technology Center. — ₍Washington₎ : United States Dept. of Transportation, Urban Mass Transportation Administration, 1971-
v. ; 28 cm.
Cover title.
Includes bibliographies.
CONTENTS: v. 1. Summary report.
1. Local transit—Wisconsin—Milwaukee Co. 2. Transportation planning—Wisconsin—Milwaukee Co. 3. Choice of transportation. I. United States. Urban Mass Transit Administration. II. Title.
HE309.W6A44 1971 388.4'09775'94 78-318844
78 MARC

Fajans, Michael.
An alternative to the automobile : public and campus operated transit services at the University of California / prepared by Michael Fajans and Ira Stephen Fink. — Berkeley : Regents of the University of California, Systemwide Administration, Assistant Vice President—Physical Planning, Construction and Operations, 1977.
ix, 137 p. : ill. ; 24 cm.
Bibliography: p. 136-139.
$4.00
1. California. University—Students. 2. Choice of transportation. 3. Transportation—California. I. Fink, Ira Stephen, joint author. II. Title.
LD766.F34 378.1'98'709794 77-71339
79 MARC

Hensher, David A 1947-
Behavioral intentions as predictors of very specific behavior / David A. Hensher and Jordan Louviere. — ₍Sydney₎ : School of Economic & Financial Studies, Macquarie University, 1977.
34 p. : graphs ; 30 cm. — (Research paper - Macquarie University, School of Economic & Financial Studies ; no. 146) Aus***
Bibliography: p. 31-34.
ISBN 0-85937-261-4
1. Choice of transportation. 2. Urban transportation—Australia. I. Louviere, Jordan J., joint author. II. Title. III. Series: Macquarie University. School of Economic and Financial Studies. Research paper - School of Economic and Financial Studies, Macquarie University ; no. 146.
HE336.C5H44 301.1 78-321693
78 MARC

Sullivan, Brian E
The timed transfer focal point : a refinement in public transport service design / by Brian E. Sullivan. — Vancouver, B.C. : Bureau of Transit Services, Dept. of Municipal Affairs, Province of British Columbia, 1975, 1976 printing.
20, ₍2₎ p., ₍2₎ leaves of plates : ill. ; 28 cm. — (Technical report - Transit Services) C***
Includes bibliographical references.
1. Local transit. 2. Choice of transportation. 3. Motor bus lines. I. Title. II. Series: British Columbia. Bureau of Transit Services. Technical report - Transit Services.
HE305.S86 388.4'1322 79-311677
79 MARC

Sur l'analyse de la répartition modale : besoins et ressources : rapport final / étude réalisée pour le Comité des transports de la région de Montréal ₍par un₎ groupe de travail, R. Chapleau ... ₍et al.₎. — Montréal : Université de Montréal, Centre de recherche sur les transports, 1977.
11 leaves ; 28 cm. — (Publication - Université de Montréal, Centre de recherche sur les transports ; 78) C78-5975-1
Bibliography: leaf 11.
$1.25
1. Urban transportation—Québec (Province)—Montréal metropolitan area. 2. Montréal metropolitan area, Québec—Transit systems. 3. Choice of transportation. I. Chapleau, Robert, 1945- II. Comité des transports de la région de Montréal. III. Title.
HE311.C32M657 388.4'131'0971428 79-367933
79 MARC

—CONGRESSES

National Research Council. Transportation Research Board.
Preferences, perceptions and market segments in travel behavior / Transportation Research Board, Commission on Sociotechnical Systems, National Research Council. — Washington : National Academy of Sciences, 1977.
iii, 48 p. : graphs ; 28 cm. — (Transportation research record ; 649)
Includes bibliographical references.
ISBN 0-309-02678-4
1. Choice of transportation—Congresses. I. Title. II. Series.
TE7.H5 no. 649 380.5'08 s 78-16514
₍HE336.C5₎ 78 MARC

—MATHEMATICAL MODELS

Banister, David.
Car availability and usage : a modal split model based on these concepts / ₍by₎ David Banister. — Reading : Department of Geography, University of Reading, 1977.
₍2₎, ii, 36 p. : ill. ; 21 cm. — (Geographical papers ; no. 58 ISSN 0305-5914) GB78-07463
Includes bibliographical references.
ISBN 0-7049-0468-3
1. Choice of transportation—Mathematical models. 2. Transportation, Automotive—Mathematical models. I. Title. II. Series.
HE336.C5B36 388.3'21 78-323750
79 MARC

Geraud, Patrice.
Les modèles de demande en transport interurbain de voyageurs / Patrice Geraud. — Arcueil : Institut de recherche des transports, 1977.
60 p. ; 30 cm. — (Note d'information I.R.T. ; no 10 ISSN 0339-8676) F***
Bibliography: p. 57-58.
ISBN 2-85782-025-9
1. Traffic surveys—France—Matematical models. 2. Choice of transportation—Mathematical models. I. Title. II. Series: Institut de recherche des transports. Note d'information - I.R.T. ; no 10.
HE373.F6G47 388.3'2 78-398537
79 MARC

Hensher, David A 1947-
Interdependencies of travel decisions, other activities, and household members in the frequency, timing, and mode of travel to the shops : empirical evidence / David A. Hensher. — North Ryde, Australia : School of Economic and Financial Studies, Macquarie University, 1978.
24, ₍2₎ p. ; 30 cm. — (Research paper - School of Economic and Financial Studies, Macquarie University ; no. 168) Aus***
Bibliography: p. ₍25₎-₍26₎
ISBN 0-85837-337-8
1. Travel time (Traffic engineering)—Mathematical models. 2. Choice of transportation—Mathematical models. I. Title. II. Series: Macquarie University. School of Economic and Financial Studies. Research paper - School of Economic and Financial Studies, Macquarie University ; no. 168.
HE336.T76H443 388.4 79-307408
79 MARC

Westin, Richard B
Parking location and transit demand : a case study of endogenous attributes in disaggregate mode choice models / by Richard B. Westin and David W. Gillen. — Toronto : Institute for Policy Analysis, University of Toronto, 1977.
28, ₍10₎ p. : ill. ; 28 cm. — (Working paper series - Institute for Policy Analysis, University of Toronto ; no. 7704 ISSN 0082-5301) C***
Bibliography: p. ₍32₎-₍33₎
1. Choice of transportation—Mathematical models. 2. Urban transportation—Mathematical models. 3. Automobile parking—Mathematical models. I. Gillen, David W., joint author. II. Title. III. Series: University of Toronto. Institute for the Quantitative Analysis of Social and Economic Policy. Working paper series - Institute for the Quantitative Analysis of Social and Economic Policy, University of Toronto ; no. 7704.
HE336.C5W48 388.4 78-318805
78 MARC

—MATHEMATICAL MODELS—CONGRESSES

National Research Council. Transportation Research Board.
Transportation forecasting and travel behavior. — Washington : Transportation Research Board, Commission on Sociotechnical Systems, National Research Council, National Academy of Sciences, 1978.
v, 211 p. : ill. ; 28 cm. — (Transportation research record ; 673)
Includes bibliographical references.
ISBN 0-309-02813-2
1. Transportation planning—Congresses. 2. Choice of transportation—Mathematical models—Congresses. 3. Urban transportation—Congresses. I. Title. II. Series.
TE7.H5 no. 673 380.5'08 s 79-12520
₍HE193₎ 79 MARC

—UNITED STATES—ADDRESSES, ESSAYS, LECTURES

National Research Council. Transportation Research Board.
Transit marketing. — Washington : Transportation Research Board, Commission on Sociotechnical Systems, National Research Council, National Academy of Sciences, 1976.
iii, 30 p. : ill. ; 28 cm. — (Transportation research record ; 590)
Includes bibliographical references.
ISBN 0-309-02558-3
1. Local transit—United States—Addresses, essays, lectures. 2. Choice of transportation—United States—Addresses, essays, lectures. 3. Urban transportation—United States—Public opinion—Addresses, essays, lectures. 4. Public opinion—United States—Addresses, essays, lectures. I. Title. II. Series.
TE7.H5 no. 590 380.5'08 s 76-47615
₍HE308₎ 76 MARC

CHOIR-STALLS

—NETHERLANDS

Verspaandonk, J A J M
1918–
Haarlem. De Grote of Sint Bavokerk. ⌜Door⌝ J. A. J. M. Verspaandonk. ⌜Amsterdam, Buijten & Schipperheijn, 1973⌝.
160 p. with illus. 19 cm. (Misericordereeks, 1) Ne 73-8
Dutch and English.
Bibliography: p. 160.
ISSN 90-6064-202-3 : fl 25.00
1. Wood-carving—Netherlands—Haarlem. Wood-carving, Medieval—Haarlem. Choir-stalls—Netherlands—Netherlands. 4. Haarlem. Sint-Bavokerk.
NK9754.H32V47 74-312551

CHOIR-STALLS, BAROQUE

—SILESIA

Heyer, Karl Johannes.
Das barocke Chorgestühl in Schlesien : e. Darst. d. Chorgestühle u.e. Beitr. zur Geschichte von Kunst u. Kunsthandwerk im Barock / von Karl Johannes Heyer ; mit e. musikgeschichtl. Beitr. von Johannes Aengenvoort. — Frankfurt am Main : Weidlich, 1977.
xvi, 247, ⌜106⌝ p. : 286 ill. ; 27 cm. — (Bau- und Kunstdenkmäler des deutschen Ostens : Reihe C, Schlesien ; Bd. 6 ISSN 0522-490X) GFR78-A37
A revision of the author's thesis, Breslau, 1929.
Bibliography: p. 226-231.
Includes indexes.
ISBN 3-8035-8609-7 : DM98.00
1. Wood-carving, Baroque—Silesia. 2. Choir-stalls, Baroque—Silesia. 3. Church decoration and ornament, Baroque—Silesia. I. Aengenvoort, Johannes. II. Title. III. Series.
NK9771.P6H49 1977 79-357549
79 MARC

CHOIRS (MUSIC)

see also Choral music; Choral societies; Conducting, Choral

Dakers, Lionel.
Making church music work / by Lionel Dakers. — London : Mowbrays, 1978.
157, ⌜60⌝ p. : music ; 18 cm. — (Addington Press series ; no. 1) GB**
"Check list of anthems": p. ⌜1⌝-⌜60⌝
ISBN 0-264-66470-1 : £4.50
1. Church music. 2. Choirs (Music) 3. Conducting, Choral. 4. Church music—Bibliography. I. Title. II. Series.
MT88.D27 783'.026 79-302784
79 MARC

Kalugina, Nonna Vasil'evna.
(Metodika raboty s russkim narodnym khorom)
Методика работы с русским народным хором : ⌜Учебник для муз. вузов⌝ / Н. Калугина ; Гос. муз.-пед. ин-т им. Гнесиных, Кафедра хорового дирижирования. — 2-е изд., перераб. и доп. — Москва : Музыка, 1977.
255 p. : music ; 27 cm. USSR 77
First ed. published under title: Osnovy metodiki raboty s russkim narodnym khorom.
Bibliography: p. 181–⌜186⌝
1.49rub
1. Choirs (Music) I. Title.
MT88.K35 1977 79-407608

Lamb, Gordon H
Choral techniques / Gordon H. Lamb. — 2d ed. — Dubuque, Iowa : W. C. Brown Co., c1979.
xii, 292 p. : ill., music ; 24 cm.
"Lists of choral music": p. 269-279; bibliography: p. 285-287.
ISBN 0-697-03417-8
1. Conducting, Choral. 2. Choirs (Music) I. Title.
MT85.L3 1979 784.9'6 78-51478
79 MARC

Soboleva, Galina Gennadievna.
(Sovremennyĭ russkiĭ narodnyĭ khor)
Современный русский народный хор / Г. Г. Соболева. — Москва : Знание, 1978.
88 p., ⌜8⌝ leaves of plates : ill. ; 20 cm. — (Народный университет : Факультет литературы и искусства) USSR 78
Series romanized: Narodnyĭ universitet : Fakul'tet literatury i iskusstva.
Bibliography: p. ⌜87⌝
0.40rub
1. Choral societies—Russia. 2. Choirs (Music) I. Title.
ML1537.S66 79-408641

—PRAYER-BOOKS AND DEVOTIONS —ENGLISH

Patt, Richard W
Psallite, series B : devotions and prayers for church choir singers : based on the Gospels, series B, from The church year calendar and lectionary / by Richard W. Patt. — St. Louis : Concordia Pub. House, c1978.
79 p. ; 21 cm.
ISBN 0-570-03784-0 : $2.25
1. Choirs (Music)—Prayer-books and devotions—English. I. Title.
BV4596.C48P369 242'.6'9 78-9025
78 MARC

CHŌJŪ GIGA

Chōjū giga.
鳥獣人物戯画 / ⌜編者 小松茂美 ; 執筆者 小松茂美, 上野憲示⌝. — 東京 : 中央公論社, 昭和52 ⌜1977⌝
179 p. : ill. (some col.) ; 36 cm. — (日本絵巻大成 ; 6) Ja***
"作者に鳥羽僧正 ...を当てはめる説が根強いが ...複数の作家を想定すべきであ⌜る⌝"
Bibliography: p. 178–179.
¥9800
1. Chōjū giga. I. Komatsu, Shigemi, 1925– II. Ueno, Kenji, 1948– III. Toba Sōjō, 1053–1140. IV. Series: Nihon emaki taisei ; 6.
Title romanized: Chōjū jimbutsu giga.
ND1059.6.C46C46 1977 78-812925

CHOL LANGUAGE see Cholti language

CHOLA ARCHITECTURE see Architecture, Chola

CHOLAM see Durra

CHOLAS

Sethuraman, N 1930-
The Cholas : mathematics reconstructs the chronology / N. Sethuraman. — Kumbakonam : Sethuraman, 1977.
193 p., ⌜2⌝ leaves of plates : ill., geneal. tables ; 26 cm.
Includes bibliographical references.
Rs30.00
1. Cholas. 2. South India—History. I. Title.
DS484.S48 954'.8 78-902478
79 MARC

—KINGS AND RULERS—BIOGRAPHY

Sadasiva Pandarathar, T V 1892-1960.
Mutar̲ kulōttuṅkacōl̲an̲ / T. V. Catācivapaṇṭārattār el̲utiyatu. — 3. patippu — ⌜Cen̲n̲ai⌝ : Jūpiṭṭar Accukkūṭattil Patippikkappeṟṟatu, 1941.
9, 106 p. ; 19 cm.
In Tamil.
Includes bibliographical references.
1. Kulottunga I, Chola King, fl. 1070-1118. 2. Tamil Nadu—History. 3. Cholas—Kings and rulers—Biography. I. Title.
DS485.M28K867 1941 78-912892
79 MARC

CHOLERA, ASIATIC

—DIAGNOSIS

World Health Organization. Bacterial Diseases Unit.
Guidelines for the laboratory diagnosis of cholera / prepared by the WHO Bacterial Diseases Unit. — Geneva : World Health Organization ; ⌜Albany : available from Q Corporation⌝, 1974.
23 b. : ill. ; 22 cm. Sw75-A-2485
Corrigenda: leaf inserted.
Includes bibliographical references.
ISBN 9241540346 : 5.00F
1. Cholera, Asiatic—Diagnosis. 2. Diagnosis, Laboratory. I. Title.
QR201.C5W67 1974 616.9'32'0756 79-304325
79 MARC

—EGYPT—AT TŪR

Hermann, Wolfgang, Dr.
Die El-Tor-Cholera-Pandemie 1961 bis 1968. Mit einem Nachwort von R.-E. Bader. Leipzig, J. A. Barth, 1973.
76 p. illus., maps. 24 cm. (Beiträge zur Hygiene und Epidemiologie, Heft 21) GDR***
Bibliography: p. ⌜71⌝-74.
1. Cholera, Asiatic—Egypt—at Tūr. I. Title. II. Series.
RC134.E3T874 73-340076
79⌜r79⌝rev MARC

—INDIA—MISCELLANEA

Pilgrimages, cholera, &c. = उरूस व यात्रेसंबंधी जरीमरी यांविषयीं संभाषणें. — 3d ed. — ⌜Bombay⌝ : Bombay Tract and Book Society, 1860.
52 p. ; 13 cm. — (Bombay Tract and Book Society ; no. 45)
In Marathi.
1. Cholera, Asiatic—India—Miscellanea. 2. Pilgrims and pilgrimages—India—Miscellanea. I. Bombay Tract and Book Society. II. Title: Urūsa va yātresambandhī jarīmarī yāvishayī sambhāshaṇẽ.
RC134.I3P54 1860 78-984337

—ITALY—NAPLES, 1836

Forti Messina, Annalucia.
Società ed epidemia : il colera a Napoli nel 1836 / Annalucia Forti Messina. — Milano : F. Angeli, c1979.
174 p. : ill. ; 22 cm. — (Studi e richerche storiche ; 4) It***
Includes bibliographical references and index.
ISBN 8820414848 : L5000
1. Cholera, Asiatic—Italy—Naples, 1836. I. Title. II. Series.
RC133.I9N384 79-361766
79 MARC

CHOLESTEROL

see also Blood cholesterol

CHOLESTEROL DIETARY see Low-cholesterol diet

CHOLESTEROL METABOLISM

see also Hypercholesteremia

CHOLIC ACID

see also Ursodeoxycholic acid

CHOLINE

see also Acetylcholine

—THERAPEUTIC USE

Brain acetylcholine and neuropsychiatric disease / edited by Kenneth L. Davis and Philip A. Berger. — New York : Plenum Press, c1979.
xvi, 597 p. : ill. ; 26 cm.
Includes bibliographies and index.
ISBN 0-306-40157-6
1. Neuropsychiatry. 2. Acetylcholine—Physiological effect. 3. Choline—Therapeutic use. 4. Brain cemistry. 5. Parasympathomimetic agents—Testing. I. Davis, Kenneth L. II. Berger, Philip A.
RC343.B664 615'.78 79-862
79 MARC

CHOLINERGIC BLOCKING AGENTS see Parasympatholytic agents

CHOLINERGIC RECEPTORS

Cholinergic-monoaminergic interactions in the brain / edited by Larry L. Butcher. — New York : Academic Press, 1978.
xiii, 402 p. : ill. ; 24 cm. — (Behavioral biology series)
Includes bibliographies and index.
ISBN 0-12-147850-5
1. Brain chemistry. 2. Adrenergic mechanisms. 3. Cholinergic receptors. 4. Neural transmission. I. Butcher, Larry L. II. Series.
[DNLM: 1. Acetylcholine—Metabolism. 2. Biogenic amines—Metabolism. 3. Brain—Metabolism. 4. Brain—Physiology. 5. Brain chemistry. WL300.3 C547]
QP376.C488 599'.01'88 78-14964
78 MARC

Persson, Henry.
Studies on regulatory mechanisms of the peripheral autonomic nervous system, with special regard to adrenoceptor functions. Lund, Zoophysiological Institute, University of Lund [1971]
49 p. illus. 24 cm. S***
Originally presented as the author's thesis, Lund.
Bibliography: p. 42-49.
1. Nervous system, Autonomic. 2. Adrenergic receptors. 3. Cholinergic receptors. 4. Biological control systems. I. Title.
QP368.P47 1971 599'.01'88 78-321843
79 MARC

—CONGRESSES

The cholinergic synapse / edited by Stanislav Tůcek ; co-editors: Eric A. Barnard ... [et al.]. — Amsterdam ; New York : Elsevier Scientific Pub. Co. ; New York : sole distributors for the U.S.A. and Canada, Elsevier/North-Holland, 1979.
xx, 511 p. : ill. ; 27 cm. — (Progress in brain research ; vol. 49)
Includes bibliographies and indexes.
ISBN 0-444-80105-7 (v. 49)
1. Synapses—Congresses. 2. Cholinergic receptors—Congresses. 3. Neural transmission—Congresses. 4. Myasthenia gravis—Congresses. I. Tůcek, S. II. Barnard, Eric A., 1927- . III. Series.
[DNLM: 1. Synapses—Congresses. 2. Acetylcholine—Congresses. 3. Receptors, Cholinergic—Congresses. 4. Neuromuscular junction—Congresses. W1 PR667J v. 49 / WL102.8 I61c 1978]
QP376.P7 vol. 49 612'.82'08 s 79-577
[QP364] 79 MARC

Federation of European Biochemical Societies.
Properties of purified cholinergic and adrenergic receptors / editor, Maria Wollemann. — [Amsterdam] : North-Holland ; New York : American Elsevier, 1975.
vii, 144 p. : ill. ; 23 cm. — (Its Proceedings of the ninth FEBS meeting ; [7]) ([FEBS meeting] ; v. 37) Ne75-45
Includes bibliographies and index.
ISBN 0-444-10937-4 (American Elsevier)
1. Cholinergic receptors—Congresses. 2. Adrenergic mechanisms—Congresses. I. Wollemann, Maria. II. Title. III. Series: Federation of European Biochemical Societies. FEBS meeting ; v. 37.
QH345.F43 1975 vol. 7 574.1'92'08 s 76-350323
[QP363] *76[r79]rev MARC

CHOLINESTERASE

Drews, Ulrich.
Cholinesterase in embryonic development / Ulrich Drews. — Stuttgart : Fischer, 1975.
52 p. : 38 ill. ; 24 cm. — (Progress in histochemistry and cytochemistry ; v. 7, no. 3) GFR75-A
Bibliography: p. 46-49.
Includes index.
ISBN 3-437-10403-9 : DM38.00
1. Developmental cytology. 2. Cholinesterase. I. Title. II. Series.
QL963.5.D73 591.3'33 76-479701
76[r79]rev MARC

CHOLINOLYTIC AGENTS see Parasympatholytic agents

CHŎLLA-NAMDO, KOREA

—BIOGRAPHY

Chŏlla-namdo Kyoyuk Yŏn'guwŏn.
(Uri kojang ŭi ch'unghyoyŏl)
우리고장의忠孝烈：國民精神教育資料解説書. — [光州]：全羅南道教育研究院, 1978-
v. : ill. ; 21 cm. — (교육연구 ; 제171호)
Not in trade.
1. Filial piety—Korea—Chŏlla-namdo. 2. Allegiance—Korea—Chŏlla-namdo. 3. Chŏlla-namdo, Korea—Biography. 4. Monuments—Korea—Chŏlla-namdo. I. Title. II. Series: Kyoyuk yŏn'gu ; che 171-ho.
BJ1533.F5C45 1978 79-825055

—HISTORIC HOUSES, ETC.

(Wang In Paksa yujŏkchi chonghap chosa pogosŏ)
王仁博士遺蹟址綜合調査報告書. — [光州]：全羅南道：王仁博士顯彰協會, 1974.
93 p., [16] leaves of plates : ill. ; 27 cm.
Includes one article with summary in English.
"500部限定版"
"非賣品"
1. Wani, 4th cent. 2. Chŏlla-namdo, Korea—Historic houses, etc. 3. Korea—Relations (general) with Japan. 4. Japan—Relations (general) with Korea. I. Chŏlla-namdo, Korea. II. Wang In Paksa Hyŏnch'ang Hyŏphoe.
DS852.W34W36 79-825554

CHŎLLA-PUKTO, KOREA

—DIRECTORIES

(Chŏlla-pukto myŏnggam)
全羅北道名鑑
[全州] 全羅北道名鑑社
v. ill. 27 cm. annual.

1. Chŏlla-pukto, Korea—Directories.
DS924.C46C45 78-647885
(MARC-S)

CHOLLA-PUKTO, KOREA (PROVINCE)

—STATISTICS—YEARBOOKS

Chŏlla-pukto, Korea.
(T'onggye yŏnbo - Chŏlla Pukto)
통계연보 - 전라북도
[전주] 전라북도
v. 27 cm.
Other title, : Statistical year book of Jeon Buk.

1. Chŏlla-pukto, Korea (Province)—Statistics—Yearbooks. I. Title: Statistical year book of Jeon Buk.
HA1857.C52C46a 79-641917
(MARC-S)

CHOLTÍ LANGUAGE

—DICTIONARIES—SPANISH

Schumann G , Otto.
La lengua chol, de Tila (Chiapas) [por] Otto Schumann G. [1. ed.] México, UNAM, Coordinación de Humanidades, 1973.
113 p. 22 cm. (Universidad Nacional Autónoma de México. Centro de Estudios Mayas. Cuaderno 8)
Bibliography: p. 113.
1. Choltí language—Grammar. 2. Choltí language—Dictionaries—Spanish. 3. Spanish language—Dictionaries—Choltí. I. Title. II. Series: Mexico (City). Universidad Nacional. Centro de Estudios Mayas. Cuaderno 8.
PM3649.S37 78-399378
79 MARC

—GRAMMAR

Schumann G , Otto.
La lengua chol, de Tila (Chiapas) [por] Otto Schumann G. [1. ed.] México, UNAM, Coordinación de Humanidades, 1973.
113 p. 22 cm. (Universidad Nacional Autónoma de México. Centro de Estudios Mayas. Cuaderno 8)
Bibliography: p. 113.
1. Choltí language—Grammar. 2. Choltí language—Dictionaries—Spanish. 3. Spanish language—Dictionaries—Choltí. I. Title. II. Series: Mexico (City). Universidad Nacional. Centro de Estudios Mayas. Cuaderno 8.
PM3649.S37 78-399378
79 MARC

CHOLULA, MEXICO

—ANTIQUITIES

Lagunas Rodríguez, Zaíd.
Enterramientos humanos de la zona arqueológica de Cholula, Puebla / Zaíd Lagunas Rodríguez, Carlos Serrano Sánchez, Sergio López Alonso. — México : Instituto Nacional de Antropología e Historia, SEP, Departamento de Antropología Física, 1976.
130 p. : 56 ill. ; 27 cm. — (Colección científica - Instituto Nacional de Antropología e Historia ; 44 : Antropología física)
Summary in English, French, and Spanish.
Bibliography: p. 119-122.
1. Indians of Mexico—Cholula—Antiquities. 2. Mounds—Mexico—Cholula. 3. Cholula, Mexico—Antiquities. 4. Mexico—Antiquities. I. Serrano Sánchez, Carlos, joint author. II. López Alonso, Sergio, joint author. III. Title. IV. Series: Mexico. Instituto Nacional de Antropología e Historia. Colección científica - Instituto Nacional de Antropología e Historia ; 44.
F1219.1.P9L28 972'.4 78-104182
79 MARC

CHOLUS BREVISCAPUS

—CLASSIFICATION

Vaurie, Patricia.
Revision of Cholus (Aphyoramphus) / Patricia Vaurie. — New York : American Museum of Natural History, 1978-
v. : ill. ; 26 cm. — (American Museum novitates ; no. 2657 ISSN 0003/0082)
Cover title.
CONTENTS: —Pt. 2. New species of breviscapus Group (Coleoptera, Curculionidae, Cholinae)
$0.60
1. Cholus howdenae—Classification. 2. Cholus guttatus—Classification. 3. Cholus breviscapus—Classification. 4. Insects—Classification. I. Title. II. Series.
QL1.A436 no. 2657 500.9'08 s 78-109395
[QL596.C9] 79 MARC

CHOLUS GUTTATUS

—CLASSIFICATION

Vaurie, Patricia.
Revision of Cholus (Aphyoramphus) / Patricia Vaurie. — New York : American Museum of Natural History, 1978-
v. : ill. ; 26 cm. — (American Museum novitates ; no. 2657 ISSN 0003/0082)
Cover title.
CONTENTS: —Pt. 2. New species of breviscapus Group (Coleoptera, Curculionidae, Cholinae)
$0.60
1. Cholus howdenae—Classification. 2. Cholus guttatus—Classification. 3. Cholus breviscapus—Classification. 4. Insects—Classification. I. Title. II. Series.
QL1.A436 no. 2657 500.9'08 s 78-109395
[QL596.C9] 79 MARC

CHOLUS HOWDENAE

—CLASSIFICATION

Vaurie, Patricia.
Revision of Cholus (Aphyoramphus) / Patricia Vaurie. — New York : American Museum of Natural History, 1978-
v. : ill. ; 26 cm. — (American Museum novitates ; no. 2657 ISSN 0003/0082)
Cover title.
CONTENTS: —Pt. 2. New species of breviscapus Group (Coleoptera, Curculionidae, Cholinae)
$0.60
1. Cholus howdenae—Classification. 2. Cholus guttatus—Classification. 3. Cholus breviscapus—Classification. 4. Insects—Classification. I. Title. II. Series.
QL1.A436 no. 2657 500.9'08 s 78-109395
[QL596.C9] 79 MARC

CHOMEAUX, ROGER, 1907-

Chomeaux, Roger, 1907-
Chomo : un pavé dans la vase intellectuelle / [Roger Chomeaux] ; propos recueillis par Laurent Danchin. — [Paris] : J. C. Simoën, c1978.
297 p., [4] leaves of plates : ill. ; 24 cm. F***
46.00F
1. Chomeaux, Roger, 1907- . 2. Artists—France—Biography. I. Danchin, Laurent, 1946- . II. Title.
NX549.Z9C482 1978 700'.92'4 79-371487
79 MARC

CHOMOLUNGNA see Everest, Mount

CHOMSKY, NOAM

England, Barbara R
Glossarized charts of Noam Chomsky's grammar / by Barbara R. England. — [Bloomington, Ind.] : Physsardt : distributed by Bloomington Distribution Group, 1978.
vii, 161 p. : ill. ; 23 cm.
Bibliography: p. 121-123.
Includes indexes.
ISBN 0-916062-03-1
1. Chomsky, Noam . 2. Generative grammar—Charts, diagrams, etc. I. Title.
P85.C47E54 415 78-56641
79 MARC

Matthews, Peter H
Generative grammar and linguistic competence / P. H. Matthews. — London ; Boston : G. Allen & Unwin, 1979.
112 p. ; 23 cm. GB***
Includes bibliographical references and index.
ISBN 0-04-410002-7 : £6.95
1. Generative grammar. 2. Competence and performance (Linguistics) 3. Chomsky, Noam. I. Title.
P158.M37 415 79-317280
79 MARC

Sampson, Geoffrey.
Liberty and language / Geoffrey Sampson. — Oxford ; New York : Oxford University Press, 1979.
viii, 251 p. ; 23 cm. GB79
Includes bibliographical references and index.
ISBN 0-19-215951-8 : £5.75 ($15.95)
1. Chomsky, Noam. 2. Languages—Political aspects. 3. Liberalism. I. Title.
P85.C47S26 410'.92'4 78-40941
78 MARC

Smith, Neil.
Modern linguistics : the results of Chomsky's revolution / Neil Smith and Deirdre Wilson. — Bloomington : Indiana University Press, c1979.
334 p. ; 22 cm.
Bibliography: p. [315]-325.
Includes index.
ISBN 0-253-19457-1
1. Linguistics. 2. Language and languages. 3. Chomsky, Noam. I. Wilson, Deirdre, joint author. II. Title.
P121.S58 410 78-20405
78 MARC

CHON, SOK-CHAE, 1916-

(Sinbu Chŏn Sŏk-chae Paksa hwannyŏk kinyŏm nonmunjip)
神父全碩在博士還曆紀念論文集 / [編輯神父全碩在博士還曆紀念論文集刊行委員會]. — [大邱] : 曉星女子大學同窓會, 1976.
782 p., [11] leaves of plates : ill. ; 26 cm.
English, German or Korean; pref. also in Latin.
Includes bibliographies.
1. Chŏn, Sŏk-chae, 1916- I. Chŏn, Sŏk-chae, 1916- II. Sinbu Chŏn Sŏk-chae Paksa Hwannyŏk Kinyŏm Nonmunjip Kanhaeng Wiwŏnhoe.
AC147.S57 79-826543

CHON, T'AE-IL, 1948-1970

Kim, Yong-gi.
(Honoo yo, watashi o tsutsume)
炎よ、わたしをつつめ : ある韓国青年労働者の生と死 / 金英琪著 ; 李浩培訳. — 東京 : たいまつ社, 1978.
380 p., [2] leaves of plates : ill. ; 19 cm. Ja 79-797
¥1300
1. Chŏn, T'ae-il, 1948-1970. 2. Labor and laboring classes—Korea—Biography. 3. Labor and laboring classes—Korea. 4. Korea—Social conditions. I. Title.
HD8730.5.K534 79-808774

CHŌNAN-MACHI, JAPAN

—TEMPLES—CONSERVATION AND RESTORATION

Jūyō Bunkazai Kasamoridera Kannondō Shūri Iinkai.
(Jūyō bunkazai Kasamoridera Kannondō shūri kōji hōkokusho)
重要文化財笠森寺観音堂修理工事報告書 / [編集 重要文化財笠森寺観音堂修理委員会]. — [長南町(千葉県)] : 同委員会, 昭和35 [1960]
57 p., [45] leaves of plates : ill. ; 30 cm.
1. Kasamoridera, Chōnan-machi, Japan. 2. Temples, Buddhist—Japan—Chōnan-machi—Conservation and restoration. 3. Chōnan-machi, Japan—Temples—Conservation and restoration. I. Title. II. Title: Kasamoridera Kannondō shūri kōji hōkokusho.
NA6057.C484J89 1960 78-815091

CHONBUK TREHAKKYO

Chŏnbuk Taehakkyo Isibonyŏnsa P'yŏnch'an Wiwŏnhoe.
(Chŏnbuk Taehakkyo isibonyŏnsa)
全北大學校二十五年史, 1952-1977 / [編輯人 全北大學校25年史編纂委員會]. — [全州] : 全北大學校, 1978.
824 p. : ill. ; 26 cm.
On verso of t. p.: The twenty-five-year history of Jeonbug National University.
Errata slip inserted.
Not in trade.
1. Chŏnbuk Taehakkyo. I. Title. II. Title: The twenty-five-year history of Jeonbug National University.
LG281.C554C47 1978 79-825816

Nam, Sa-go, 1506-1571.
(Kyŏgam yurok)
格庵遺錄 / [南師古]. 三易大經 / [崔濟愚]. 大巡典經 / [姜一淳] ; 註解者趙聖紀. — 서울 : 太宗出版社, 1977.
656 p., [5] leaves of plates : ill. ; 25 cm. — (韓國著名 [i. e. 名著] 大全集 ; 1)
Colophon inserted.
₩8500
1. Religious literature, Korean. 2. Ch'ŏndogyo—Sacred books. 3. Taesun Chillihoe—Sacred books. I. Ch'oe, Che-u, 1824-1864. Samyŏk taegyŏng. 1977. II. Kang, Il-sun, 1871-1909. Taesun chŏn'gyŏng. 1977. III. Cho, Sŏng-gi. IV. Title. V. Series: Han'guk myŏngjŏ taejŏnjip (Sŏul) ; 1.
B5250.N35 1977 79-826532

CH'ŎNDOGYO

—SACRED BOOKS

CHONDRIOSOMES see Mitochondria

CHŎNDŬNGSA, KANGHWAGUN, KOREA

An, Chinho.
(Chŏndŭng ponmal saji)
傳燈本末寺誌 ; 奉先本末寺誌 / [安震湖]. — 서울 : 亜細亜文化社, 1978.
viii, 417, 246 p. : ill. ; 27 cm. — (韓國寺誌叢書 ; 第4輯)
Colophon inserted.
₩10000
1. Temples, Buddhist—Korea—Kyŏnggi. 2. Chŏndŭngsa, Kanghwa-gun, Korea. 3. Pongsŏnsa, Yangju-gun, Korea. I. Title. II. Title: Pongsŏn ponmal saji. III. Series: Han'guk saji ch'ongsŏ ; che 4-chip.
BQ6351.K9A5 79-825046

Han'guk Pulgyo Yŏn'guwŏn.
(Chŏndŭngsa)
傳燈寺 / 韓國佛教研究院著 ; 監修李箕永, 黃壽永 ; 執筆李箕永 ... [et al.] ; 寫眞에드워드B·아담스. — 서울 : 一志社, 1978.
95 p. : ill. ; 22 cm. — (韓國의寺刹 ; 14)
Series romanized: Han'guk ŭi sach'al.
Bibliography: p. 91.
₩1000
1. Chŏndŭngsa, Kanghwa-gun, Korea. I. Yi, Ki-yŏng, 1922- II. Title.
BQ6351.K352C483 1978 79-825045

CHŌNEN, D. 1016

Satō, Haruo, 1892-1964.
(Shakadō monogatari)
釋迦堂物語 / 佐藤春夫著. — 東京 : 平凡社, 昭和32 [1957]
147 p., [4] leaves of plates : ill. ; 23 cm.
Originally published in Kokoro, Jan.-Dec. 1956.
Limited edition: 500 copies printed. No. 489.
1. Chōnen, d. 1016. 2. Priests, Buddhist—Japan—Biography. 3. Buddhist pilgrims and pilgrimages—China—Wu-t'ai shan. 4. Seiryōji, Kyoto. I. Title.
BQ946.O537S27 78-813607
78 J

CHŎNG, CH'ANG-SŎP. 1927-

Chŏng, Ch'ang-sŏp, 1927-
(Chŏng Ch'ang-sŏp)
丁昌燮 = Joung Chang-sop : 作家와作品 / 李逸. — 서울 : 金星出版社, 1977.
[32] p. : chiefly ill. (some col.) ; 33 cm. — (韓國現代美術代表作家100人選集 ; 59)
The introductory review of the artist and the legend also in English.
₩1500
1. Chŏng, Ch'ang-sŏp, 1927- I. Yi, Il, 1932- II. Title: Joung Chang-sop. III. Series: Han'guk hyŏndae misul taep'yo chakka paegin sŏnjip ; 59.
ND1069.C514A4 1977 78-826303
K

CHŎNG, CHŎM-SIK, 1917-

Chŏng, Chŏm-sik, 1917-
(Chŏng Chŏm-sik)
鄭點植 = Joung Jum-sik : 作家와作品 / 吳光洙. — 서울 : 金星出版社, 1977.
[32] p. : chiefly ill. (some col.) ; 33 cm. — (韓國現代美術代表作家100人選集 ; 57)
The introductory review of the artist and the legend also in English.
₩1500
1. Chŏng, Chŏm-sik, 1917- I. O, Kwang-su. II. Title: Joung Jum-sik. III. Series: Han'guk hyŏndae misul taep'yo chakka paegin sŏnjip ; 57.
ND1069.C516A4 1977 78-826410
78 K

CHONG, KI-RYONG, 1562-1622

Chŏng, Ŭl-byŏng.
(Myŏngjang Chŏng Ki-ryong)
名將鄭起龍 / 鄭乙炳著. — 서울 : 智炤林, 1978.
363 p. ; 19 cm.
₩1300
1. Chŏng, Ki-ryong, 1562-1622. 2. Korea—History—Japanese invasions, 1592-1598. 3. Generals—Korea—Biography. I. Title.
DS913.45.C47C48 79-825788

CHŎNG, MYŎN-SŎP, 1892-1944

Chŏng, Myŏn-sŏp, 1892–1944.
(Chŏng Sosŏk Sŏnsaeng yumuk)
鄭紹石先生遺墨 / [鄭冕燮]. — 錦山：鄭堯臣, 1978.
ca. 250 p. : port. : 27 cm.

1. Chŏng, Myŏn-sŏp, 1892–1944. 2. Calligraphy, Korean. 3. Calligraphy, Chinese. I. Title.
ND1457.K66C472 1978 79–825438

CHŎNG, P'YŎNG-HAE

Chŏng, P'yŏng-hae.
(Samch'ŏn-yukpaek-osip ŭi kijŏk)
3650의기적：信仰手記 / 丁平海著. — 서울：惠宣文化社, 1977, 1978 printing.
286 p. ; 19 cm.
₩1200

1. Chŏng, P'yŏng-hae. 2. Christian biography—Korea. 3. Women—Korea—Biography. I. Title.
BR1337.C48A36 78–826861
K

CHŎNG, SŎN, 1676-1759

Chŏng, Sŏn, 1676–1759.
(Kyŏmjae Chŏng Sŏn)
謙齋鄭敾 / [編輯人 洪思重]. — 서울：中央日報·東洋放送, 1977,
186 p. : chiefly ill. (some col.) ; 27 cm. — (韓國의美 ; 1)
Includes bibliographical references.
₩5800

1. Chŏng, Sŏn, 1676–1759. 2. Landscape painters, Korean. I. Hong, Sa-jung. II. Title. III. Series; Han'guk ŭi mi (Sŏul) ; 1.
ND1069.C518A4 1977 78–826858

CHŎNG, YŎNG-NYŎL, 1934-

Chŏng, Yŏng-nyŏl, 1934–
(Chŏng Yŏng-nyŏl)
鄭永烈 = Joung Yung-yul : 作家와作品 / 金仁煥. — 서울：金星出版社, 1977.
[32] p. : chiefly ill. (some col.) ; 33 cm. — (韓國現代美術代表作家100人選集 ; 70)
The introductory review of the artist and the legend also in English.
₩1500
1. Chŏng, Yŏng-nyŏl, 1934– I. Kim, In-hwan. II. Title: Joung Yung-yul. III. Series: Han'guk hyŏndae misul taep'yo chakka paegin sŏnjip ; 70.
ND1069.C52A4 1977 78–826387

CHON'GUK CHOLLYOK NODONG CHOHAP

—PERIODICALS

Chŏn'guk Chŏllyŏk Nodong Chohap.
(Hwaltong pogo - Chŏn'guk Chŏllyŏk Nodong Chohap)
活動報告 - 全國電力勞動組合
[s. l.] 全國電力勞動組合
v. ill. 26 cm.

1. Chŏn'guk Chŏllyŏk Nodong Chohap—Periodicals.
HD8039.E3C46 79–641441
(MARC-S)

CHŎN'GUK CH'ŎLTO NODONG CHOHAP. SŎUL CHIBU

Chŏn'guk Ch'ŏlto Nodong Chohap. Sŏul Chibu.
(Hwaltong pogo)
活動報告, 1977. 4. 1—1978. 3. 31：1978年度代議員大會, 1978. 4. 7. — [서울]：全國鐵道勞動組合서울支部, [1978]
199 p., [2] leaves of plates : ill. ; 26 cm.

1. Chŏn'guk Ch'ŏlto Nodong Chohap. Sŏul Chibu. I. Title.
HD6835.5.Z7R23 78–826623
K

CHON'GUK KUMYUNG NODONG CHOHAP. HAN'GUK SANGOP UNHAENG CHIBU

Chŏn'guk Kŭmyung Nodong Chohap. Han'guk Sangŏp Ŭnhaeng Chibu.
(Sangŭn Nojo sibonyŏnsa)
商銀勞組十五年史 / [編輯人朴鍾武]. — [서울]：全國金融勞動組合韓國商業銀行支部, 1978.
329 p., [10] leaves of plates : ill. ; 21 cm.
Not in trade.

1. Chŏn'guk Kŭmyung Nodong Chohap. Han'guk Sangŏp Ŭnhaeng Chibu. I. Pak, Chong-mu. II. Title.
HD6835.5.Z7B264 79–825947
K

CH'ŎNJU see Rice wines

CH'ŎNJUGYO HYEHWADONG KYOHOE, SEOUL. KOREA

—HISTORY

(Paektong pansegi)
柏洞半世紀：惠化洞聖堂50年史. — 서울：天主教惠化洞教會, 1977.
427 p. : ill. ; 24 cm.

1. Ch'ŏnjugyo Hyehwadong Kyohoe, Seoul, Korea—History. I. Title: Hyehwadong Sŏngdang osimnyŏnsa.
BX1670.7.S46P33 78–826621
K

CH'ONT'AE (SECT) see Tendai (Sect)

CHOPIN, FRYDERYK FRANCISZEK, 1810-1849

Chopin in Silesia : [lectures / from the Symposium Chopin in Silesia ; translated by Czesław Boniakowski. — Katowice : Music College Library Katowice, 1974.
24 p. ; 24 cm. — (Fascicles of the Archives Music Culture in Silesia ; no. 1)
1. Chopin, Fryderyk Franciszek, 1810-1849. 2. Silesia—History—Addresses, essays, lectures. I. Boniakowski, Czesław. II. Series: Katowice, Poland (City). Panstwowa Wyzsza Szkoła Muzyczna. Biblioteka. Archiwum Slaskiej Kultury Muzycznej. Prace Archiwum Slaskiej Kultury Muzycznej ; Nr. 1.
ML410.C54C4673 786.1'092'4 76-360956
76[r79]rev MARC

Jordan, Ruth.
Nocturne : a life of Chopin / [by] Ruth Jordan. — London : Constable, 1978.
286 p., leaf of plate, 12 p. of plates : ill., facsims, ports. ; 23 cm.
Bibliography: p. [277]-280.
Includes index.
ISBN 0-09-462330-9 : £6.95
1. Chopin, Fryderyk Franciszek, 1810-1849. 2. Composers—Biography. I. Title.
ML410.C54J85 786.1'092'4 78-323027
79 MARC

Marek, George Richard, 1902-
Chopin / George R. Marek, Maria Gordon-Smith. — 1st ed. — New York : Harper & Row, c1978.
ix, 289 p. [4] leaves of plates : ill. ; 24 cm.
Bibliography: p. 277-282.
Includes index.
ISBN 0-06-012843-7 : $12.95
1. Chopin, Fryderyk Franciszek, 1810-1849. 2. Composers—Biography. I. Gordon-Smith, Maria, joint author.
ML410.C54M37 786.1'092'4 76-57880
78 MARC

Wierzyński, Kazimierz, 1894-1969.
Życie Chopina / Kazimierz Wierzyński. — Wyd. 1. — Kraków : Wydaw. Literackie, 1978 (Krak. : DNar)
379, [1] p. ; 21 cm. P79-1466
Based on the author's Życie Chopina published in 1953 by Roy, New York.
zł90.00
1. Chopin, Fryderyk Franciszek, 1810-1849. 2. Composers—Biography. I. Title.
ML410.C54W57 79-381181
79 MARC

—DISCOGRAPHY—CATALOGS

Grudziński, Albert.
Chopin na płytach : katalog wystawy, Zamek Ostrogskich, marzec - październik 1978 / oprac. Albert Grudziński. — [Warszawa] : Towarzystwo im. Fryderyka Chopina, [1978] (Warsz. : WDA)
[54] p. : ill. (some col.) ; 21 cm. P78-2675
Introduction also in English.
Includes indexes.
1. Chopin, Fryderyk Franciszek, 1810-1849—Exhibitions. 2. Chopin, Fryderyk Franciszek, 1810-1849—Discography—Catalogs. I. Towarzystwo imienia Fryderyka Chopina, Warsaw. II. Title.
ML141.W2C45 78-395091
79 MARC

Hirai, Minako.
(Gakusei monogatari)
樂聖物語 / 平井美奈子. — 東京：實業之日本社, 昭和23 [1948], 昭和24 [1949] printing.
4, 256 p. ; 19 cm.

1. Beethoven, Ludwig van, 1770–1827—Juvenile fiction. 2. Schubert, franz Peter, 1797–1828—Juvenile fiction. 3. Chopin, Fryderyk Franciszek, 1810–1849—Juvenile fiction. 4. Schumann, Robert Alexander, 1810–1856—Juvenile fiction. I. Title.
PZ49.31.H515 79–803518

—EXHIBITIONS

Grudziński, Albert.
Chopin na płytach : katalog wystawy, Zamek Ostrogskich, marzec - październik 1978 / oprac. Albert Grudziński. — [Warszawa] : Towarzystwo im. Fryderyka Chopina, [1978] (Warsz. : WDA)
[54] p. : ill. (some col.) ; 21 cm. P78-2675
Introduction also in English.
Includes indexes.
1. Chopin, Fryderyk Franciszek, 1810-1849—Exhibitions. 2. Chopin, Fryderyk Franciszek, 1810-1849—Discography—Catalogs. I. Towarzystwo imienia Fryderyka Chopina, Warsaw. II. Title.
ML141.W2C45 78-395091
79 MARC

—JUVENILE POETRY

Chotomska, Wanda.
Muzyka pana Chopina / Wanda Chotomska ; ilustrował Wiesław Majchrzak. — Wyd. 1. — Warszawa : Nasza Księgarnia, 1978.
36 p. : col. ill. ; 25 cm. p***
zł27.00
1. Chopin, Fryderyk Franciszek, 1810-1849—Juvenile poetry. I. Majchrzak, Wiesław. II. Title.
PZ69.C5M8 79-350934
79 MARC

CHOPIN, FRYDERYK FRANCISZEK, 1810-1849. WORKS

Jonson, George Charles Ashton, 1861-
A handbook to Chopin's works / by G. C. Ashton Jonson. — Boston : Longwood Press, 1978.
287 p. ; 22 cm.
Reprint of the 2d ed., rev., 1912, published by Scribner, New York.
Includes bibliographical references.
ISBN 0-89341-078-0 : $20.00
1. Chopin, Fryderyk Franciszek, 1810-1849. Works.
ML410.C54J8 1978 786.1'092'4 77-8735
78 MARC

CHOPPERS see Forage harvesting machinery

CHOPUNNISH INDIANS see Nez Percé Indians

CHOQUETTE, ADRIENNE

—CRITICISM AND INTERPRETATION

Paradis, Suzanne.
Adrienne Choquette lue par Suzanne Paradis : une analyse de l'œuvre littéraire d'Adrienne Choquette / Suzanne Paradis. — Notre-Dame-des-Laurentides, Québec : Presses laurentiennes, 1978.
220 p. : ill. ; 19 cm. C***
Includes selections from the writings of Adrienne Choquette (p. ₍143₎-220)
ISBN 0-919705-01-5 : $7.75
1. Choquette, Adrienne—Criticism and interpretation. I. Choquette, Adrienne. II. Title.
PQ3919.C538Z84 843 78-382782
78 MARC

CHORAL CONDUCTING see Conducting, Choral

CHORAL JOURNAL

—INDEXES

Paine, Gordon.
The Choral journal : an index to volumes 1-18 / by Gordon Paine. — Lawton, Okla. : American Choral Directors Association, 1978.
xv, 170 p. ; 28 cm. — (Monograph - American Choral Directors Association ; no. 3)
1. Choral journal—Indexes. I. Title. II. Series: American Choral Directors Association. Monograph - American Choral Directors Association ; no. 3.
ML1.C6563P3 784.1'05 79-109624
00 MARC

CHORAL MUSIC

see also Choral societies; Conducting, Choral; Madrigal; Motet

—BIBLIOGRAPHY

Jacobs, Arthur.
Choral music : a symposium / edited by Arthur Jacobs. — Middlesex, Eng. ; New York : Penguin Books, 1963, 1978 printing.
448 p. : music ; 18 cm.
Bibliography: p. ₍405₎-413.
"Recommended musical editions": p. ₍415₎-426.
"Recommended gramophone records": p. ₍427₎-440.
ISBN 0-14-020533-0 : £1.25 ($3.95 US)
1. Choral music—History and criticism. 2. Choral music—Bibliography. 3. Choral music—Discography. I. Title.
ML1500.J3 1978 784'.1 79-111282
79 MARC

Valentin, Erich, 1906-
Handbuch der Chormusik / Erich Valentin ; hrsg. im Auftrag der Arbeitsgemeinschaft Deutscher Chorverbände. — Berichtigte Neuaufl. — Regensburg : G. Bosse, 1968-
v. : ill. ; 17 cm. GFR***
Vol. 2, "unter Mitarbeit von Hermann Handerer."
Includes some text in Latin.
Includes indexes.
1. Choral music—History and criticism. 2. Choral music—Bibliography. I. Handerer, Hermann. II. Title.
ML128.V7V3 1968 784'.1 78-380889
79 MARC

—BIBLIOGRAPHY—CATALOGS

Canadian Music Centre.
Catalogue of Canadian choral music, available for perusal from the library of the Canadian Music Centre. Toronto ₍1966₎
195 l. 28 cm. $1.00 Can. C67-1429
1. Choral music—Bibliography—Catalogs. 2. Music, Canadian—Bibliography—Catalogs. 3. Canadian Music Centre—Catalogs. I. Title.
ML128.V7C25 016.784'106 68-84122
₍r79₎rev MARC

—DISCOGRAPHY

Jacobs, Arthur.
Choral music : a symposium / edited by Arthur Jacobs. — Middlesex, Eng. ; New York : Penguin Books, 1963, 1978 printing.
448 p. : music ; 18 cm.
Bibliography: p. ₍405₎-413.
"Recommended musical editions": p. ₍415₎-426.
"Recommended gramophone records": p. ₍427₎-440.
ISBN 0-14-020533-0 : £1.25 ($3.95 US)
1. Choral music—History and criticism. 2. Choral music—Bibliography. 3. Choral music—Discography. I. Title.
ML1500.J3 1978 784'.1 79-111282
79 MARC

—HISTORY AND CRITICISM

Jacobs, Arthur.
Choral music : a symposium / edited by Arthur Jacobs. — Middlesex, Eng. ; New York : Penguin Books, 1963, 1978 printing.
448 p. : music ; 18 cm.
Bibliography: p. ₍405₎-413.
"Recommended musical editions": p. ₍415₎-426.
"Recommended gramophone records": p. ₍427₎-440.
ISBN 0-14-020533-0 : £1.25 ($3.95 US)
1. Choral music—History and criticism. 2. Choral music—Bibliography. 3. Choral music—Discography. I. Title.
ML1500.J3 1978 784'.1 79-111282
79 MARC

Valentin, Erich, 1906-
Handbuch der Chormusik / Erich Valentin ; hrsg. im Auftrag der Arbeitsgemeinschaft Deutscher Chorverbände. — Berichtigte Neuaufl. — Regensburg : G. Bosse, 1968-
v. : ill. ; 17 cm. GFR***
Vol. 2, "unter Mitarbeit von Hermann Handerer."
Includes some text in Latin.
Includes indexes.
1. Choral music—History and criticism. 2. Choral music—Bibliography. I. Handerer, Hermann. II. Title.
ML128.V7V3 1968 784'.1 78-380889
79 MARC

—HISTORY AND CRITICISM —BULGARIA—VARNA

Dimitrov, Milko.
(Sedemdeset i pet godini samodeĭno khorovo izkustvo v grad Varna)
75 години самодейно хорово изкуство в град Варна / Милко Димитров. — Варна : Окръжен съвет за изкуство и култура, 1975.
29 p. ; 21 cm. Bu***
At head of title: Okrŭzhen t͡sentŭr za khudozhestvena samodeĭnost.
1. Choral music—History and criticism—Bulgaria—Varna. I. Title.
ML1551.B8D5 79-412414

—GERMANY—HISTORY AND CRITICISM

Künneke, Burkhard.
Der Deutsche Sängerbund : Entstehung, Entwicklung u. Stellung in d. heutigen Gesellschaft / Burkhard Künneke. — 1. Aufl. — Frankfurt/Main : Haag und Herchen, 1978.
117 p. : diagrs. ; 21 cm. GFR78-A22
Bibliography: p. 114-117.
ISBN 3-88129-104-0 : DM16.80
1. Deutscher Sängerbund. 2. Choral music—Germany—History and criticism. I. Title.
ML2660.K83 79-343929
79 MARC

—HUNGARY

A Magyar kórusművészet 30 éve / Baross Gabor ... ₍et al.₎ tanulmányai ; szerk. Nagy Olivér. — Budapest : Zeneműkiadó, 1975.
232 p. ; 19 cm. — (Zeneélet)
Includes bibliographical references.
ISBN 9633301084 : 16.00Ft
1. Choral music—Hungary. I. Nagy, Olivér.
ML1551.H85M2 784'.1'09439 79-344918
79 MARC

—JAPAN—TOKUSHIMA

(Awa no kotanuki Amerika ni utau)
阿波の子タヌキアメリカに歌う / ₍編集 四国放送株式会社₎. — 徳島 : 出版, 昭和52 ₍1977₎
95 p. : ill. ; 20 cm. Ja 78-3056
Cover title: Tokushima Children's Choir in U. S. A.
¥700

1. Tokushima Shōnen Shōjo Gasshōdan. 2. Choral music—Japan—Tokushima. I. Shikoku Hōsō Kabushiki Kaisha. II. Title: Tokushima Children's Choir in U. S. A.
ML340.8.T592T6 79-802038
Library of Congress *79 J MN

—POLAND—KRAKOW

Życzkowski, Józef.
Gaudeamus igitur ... : dzieje Krakowskiego Chóru Akademickiego / Józef Życzkowski ; ostatni rozdział napisał, całość opracował i przypisami opatrzył Stanisław Has. — Wyd. 1. — Kraków : Wydawn. Literackie, 1977.
317 p. : ill. ; 21 cm. — (Cracoviana : Seria 2, Ludzie i wydarzenia) P***
Includes index.
zł70.00
1. Krakowski Chór Akademicki. 2. Choral music—Poland—Krakow. I. Has, Stanisław. II. Title.
ML297.8.K72K79 79-342371
79 MARC

—UZBEKISTAN

(Maktab programmalari sakkiz ĭillik maktablar uchun)
Мактаб программалари саккиз йиллик мактаблар учун : ашула : 1966–1967 ўқув йили / ₍редактор И. Аҳмедов₎. — Тошкент : Ўқитувчи, 1966.
21 p. ; 20 cm.
At head of title: Ŭzbekiston SSR Maorif Ministrligi. Pedagogika fanlari ilmiĭ tekshirish instituti.
1. Choral music—Uzbekistan. I. Akhmedov, I. II. Nauchno-issledovatel'skiĭ institut pedagogicheskikh nauk, Tashkend.
ML1537.M25 79-408494

CHORAL READING see Choral speaking

CHORAL RECITATIONS

Pour une poésie spectacle / ₍textes choisis par₎ Jan Delignac ; gravures originales de Monique Martin. — Avanton : Didascol, 1978.
1 portfolio (6 pieces) : ill. ; 28 cm. F***
Includes bibliography.
90.00F
1. French poetry. 2. Choral recitations. I. Delignac, Jane.
PQ1165.P79 841'.008 79-345795
79 MARC

CHORAL SINGING

see also Choirs (Music); Choral societies; Conducting, Choral

—UKRAINE

Herasymova-Persyds'ka, Nina Oleksandrivna.
Khorovyĭ kontsert na Ukraïni v XVII-XVIII st. / N. Herasymova-Persyds'ka ; peredmova D. S. Lykhachova. — Kyïv : Muzychna Ukraïna, 1978.
179 p., ₍5₎ leaves of plates : ill. ; 21 cm. USSR***
Includes music.
Summary in Russian.
Bibliography: p. 172-₍173₎
1.10rub
1. Choral singing—Ukraine. I. Title.
ML1551.U4H5 79-373976
79 MARC

CHORAL SOCIETIES

see also Choirs (Music); Community music

—BANAT

Ilić, Sava.
Prilog o horskoj delatnosti Srba u Temišvaru i Banatu : 1868-1944 / Sava Ilić ; ₍naslovna stranica Mita Čonka₎. — Bukurešt : Kriterion, 1978.
301 p. : ill. ; 21 cm. R78-965
Half-title: Srpski horovi u Banatu.
Bibliography: p. 203-₍204₎
Includes index.
lei14.50
1. Choral societies—Timişoara, Romania. 2. Choral societies—Banat. I. Title. II. Title: Srpski horovi u Banatu.
ML27.R6I4 79-369934
79 MARC

—ROMANIA—TIMIŞOARA

Ilić, Sava.
Prilog o horskoj delatnosti Srba u Temišvaru i Banatu : 1868-1944 / Sava Ilić ; ₍naslovna stranica Mita Čonka₎. — Bukurešt : Kriterion, 1978.
301 p. : ill. ; 21 cm. R78-965
Half-title: Srpski horovi u Banatu.
Bibliography: p. 203-₍204₎
Includes index.
lei14.50
1. Choral societies—Timişoara, Romania. 2. Choral societies—Banat. I. Title. II. Title: Srpski horovi u Banatu.
ML27.R6I4 79-369934
79 MARC

—RUSSIA

Soboleva, Galina Gennadievna.
(Sovremennyĭ russkiĭ narodnyĭ khor)
Современный русский народный хор / Г. Г. Соболева. — Москва : Знание, 1978.
88 p., ₍8₎ leaves of plates : ill. ; 20 cm. — (Народный университет : Факультет литературы и искусства) USSR 78
Series romanized: Narodnyĭ universitet : Fakul'tet literatury i iskusstva.
Bibliography: p. ₍87₎
0.40rub
1. Choral societies—Russia. 2. Choirs (Music) I. Title.
ML1537.S66 79-408641

CHORAL SPEAKING

see also Choral recitations

CHORALE

Krummacher, Friedhelm.
Die Choralbearbeitung in der protestantischen Figuralmusik zwischen Praetorius und Bach / Friedhelm Krummacher. — Kassel ; Basel ; Tours ; London : Bärenreiter, 1978.
xvi, 546 p. : music ; 24 cm. — (Kieler Schriften zur Musikwissenschaft ; Bd. 22)
Includes bibliographical references.
ISBN 3-7618-0526-8
1. Chorale. 2. Church music—Protestant churches. I. Title. II. Series.
ML3184.K76 78-388503
00 MARC

Liggett, Clayton E
Concert theatre / Clayton E. Liggett ; ill. by Martha Merrill. — Rev. ed. — New York : Richards Rosen Press, c1970.
185 p. : ill. ; 29 cm. — (The Theatre student)
Bibliography: p. 183-185.
SUMMARY: Defines Concert Theatre and describes its three basic types—Reader's Theatre, Chamber Theatre, and Choric Theatre.
ISBN 0-8239-0194-7
1. Readers' theater. 2. Choral speaking. 3. Amateur theatricals. ₍1. Oral interpretation. 2. Choral speaking. 3. Amateur theatricals₎ I. Merrill, Martha. II. Title.
PN2081.R4L54 1970b 815'.05 78-111991
79 MARC

CHORDATA

see also Vertebrates

CHOREOGRAPHERS

—BIOGRAPHY

Coton, A V
The new ballet : Kurt Jooss and his work / by A. V. Coton. — London : D. Dobson, 1946.
156 p. : ill. ; 28 cm.
"The ballets of Kurt Jooss and other ballets existing in the repertory of the Ballets Jooss": p. 154.
Includes index.
1. Jooss, Kurt, 1901- . 2. Choreographers—Biography. 3. Ballet—History. 4. Ballets—Stories, plots, etc. I. Title.
GV1785.J57C67 1946 793.3'2'0924 79-109903
79 MARC

Leshkov, D I
Marius Petipa / by D. I. Leshkov ; adapted from the Russian ; edited by Cyril Beaumont. — London : C. W. Beaumont, 1971.
32 p., ₍1₎ leaf of plates : ports. ; 20 cm.
This adaptation from the Russian of Leshkov's Marius Petipa, Petrograd, 1922, first appeared in Dance Journal, 1937-1938.
1. Petipa, Marius Ivanovich, 1822-1910. 2. Choreographers—Biography. I. Beaumont, Cyril William, 1891- . II. Title.
GV1785.P42L4713 792.8'2'0924 79-108578
79 MARC

—DENMARK—BIOGRAPHY

Bournonville, August, 1805-1879.
Breve til barndomshjemmet. Lettres à la maison de son enfance. Udg. af Det Danske Sprog-og Litteraturselskab ved Nils Schiørring og Svend Kragh-Jacobsen. København, Munksgaard, 1969-78.
3 v. 23 cm. 56.25 (v. 1) varies D69-49
Vol. 3 published by Reitzel.
CONTENTS: 1. 1824-30.—2. Journal 1820.—3. Kommentarer og registre.
1. Bournonville, August, 1805-1879. 2. Choreographers—Denmark—Biography. I. Schiørring, Nils, ed. II. Kragh-Jacobsen, Svend, ed. III. Title. IV. Title: Lettres à la maison de son enfance.
GV1785.B64A4 1969 76-562475
79₍7908r79₎rev MARC

Bournonville, Auguste, 1805-1879.
My theatre life / August Bournonville ; translated from the Danish by Patricia N. McAndrew ; introd. by Svend Kragh-Jacobsen. — Middletown, Conn. : Wesleyan University Press, c1979.
xx, 709 p., ₍16₎ leaves of plates : ill. ; 26 cm.
Translation of Mit theaterliv.
Includes index.
ISBN 0-8195-5035-3 : $37.50
1. Bournonville, Auguste, 1805-1879. 2. Choreographers—Denmark—Biography. 3. Kongelige Teater og kapel. I. Title.
GV1785.B64A3313 792.8'092'4 78-27349
78 MARC

Gruen, John.
Erik Bruhn, danseur noble / John Gruen. — New York : Viking Press, 1979.
xii, 252 p., ₍19₎ leaves of plates : ill. ; 25 cm.
Filmography: p. 239-240.
Bibliography: p. 241-244.
Includes index.
ISBN 0-670-29771-2 : $14.95
1. Bruhn, Erik. 2. Dancers—Denmark—Biography. 3. Choreographers—Denmark—Biography. I. Title.
GV1785.B78G78 792.8'092'4 79-12719
79 MARC

—SWEDEN—BIOGRAPHY

Näslund, Erik, 1948-
Birgit Cullberg / Erik Näslund. — Stockholm : Norstedt, 1978.
369, ₍4₎ p. : ill. ; 22 cm. S78-32
Bibliography of works by and about Birgit Cullberg: p. 334-338.
"Verkförteckning": p. 339-360.
Includes indexes.
ISBN 9117832926
1. Cullberg, Birgit. 2. Choreographers—Sweden—Biography.
GV1785.C83N33 79-351781
79 MARC

—UNITED STATES—BIOGRAPHY

Beckford, Ruth.
Katherine Dunham, a biography / Ruth Beckford ; foreword by Arthur Mitchell. — New York : M. Dekker, c1979.
ix, 146 p., ₍11₎ leaves of plates : ill. ; 24 cm. — (The Dance program ; v. 14)
Includes index.
ISBN 0-8247-6828-0
1. Dunham, Katherine. 2. Dancers—United States—Biography. 3. Choreographers—United States—Biography. I. Series.
GV1785.D82B4 793.3'2'0924 79-4577
79 MARC

McDonagh, Don.
Martha Graham : a biography / Don McDonagh. — Newton Abbot ₍Eng.₎ : David & Charles, 1974, c1973.
x, 341 p., ₍16₎ leaves of plates : ill. ; 24 cm.
Bibliography: p. 333-334.
Includes index.
ISBN 0-7153-6716-1
1. Graham, Martha. 2. Dancers—United States—Biography. 3. Choreographers—United States—Biography.
GV1785.G7M32 1974 793.3'092'4 79-104841
00 MARC

Page, Ruth.
Page by page / by Ruth Page ;edited and with an introd. by Andrew Mark Wentink. — Brooklyn, N.Y. : Dance Horizons, c1978.
xvi, 224 p. : ill. ; 25 cm.
ISBN 0-87127-102-8 : $14.95
1. Page, Ruth. 2. Dancers—United States—Biography. 3. Choreographers—United States—Biography. I. Wentink, Andre Mark. II. Title.
GV1785.P26A36 792.8'092'4 78-65648
00 MARC

Shawn, Ted, 1891-1972.
One thousand and one night stands / by Ted Shawn, with Gray Poole ; new introd. by Walter Terry. — 1st paperback ed. — New York : Da Capo Press, 1979, c1960.
288 p., ₍8₎ leaves of plates : ill. ; 21 cm. — (A Da Capo paperback)
Reprint of the ed. published by Doubleday, Garden City, N.Y.
ISBN 0-306-80095-0 : $6.95
1. Shawn, ed, 1891-1972. 2. Choreographers—United States—Biography. 3. Denishawn school of dancing, New York. I. Poole, Gray Johnson, joint author. II. Title.
₍GV1785.S5A3 1979₎ 792.8'092'4 78-20818
78 MARC

CHOREOGRAPHY

see also Ballet; Choreographers; Dance notation

Allen, Judith.
That all American game / choreography, Judith Allen, February, 1976 ; music, Dvorak, Humoresque, Strauss, The fire-bell polka ; notator, Judith Allen, August, 1977 ; autographer, Patricia Hanick. — ₍Iowa City?₎ : Allen, c1978.
x leaves, 25 p. : ill. ; 22 x 37 cm.
1. Dance notation. 2. Allen, Judith. 3. Choreography. 4. Modern dance. I. Hanick, Patricia. II. Title.
GV1587.A44 793.3'2 79-101122
79 MARC

Humphrey, Doris, 1895-1958.
Doris Humphrey, the collected works. — New York : Dance Notation Bureau Press, 1978-
v. : ill. ; 29 cm.
Notations by Odette Blum and others.
CONTENTS: v. 1. Water study. Shakers. Partita V.

ISBN 0-932582-00-1
1. Humphrey, Doris, 1895-1958. 2. Choreography. 3. Dance notation. I. Blum, Odette. II. Title.
GV1782.5.H85 1978 793.3'2 78-67320
79 MARC

Van Praagh, Peggy.
The choreographic art; an outline of its principles and craft, by Peggy Van Praagh and Peter Brinson. Wtih a foreword by Cyril Beaumont. Photos. and drawings by Robert Bruce Church London, A. & C. Black ₍1963₎
xxiv, 387 p. illus. 26 cm.
Bibliography: p. 372-375.
1. Choreography. I. Brinson, Peter, joint author.
GV1782.V3 1963a 63-25424
MARC

—ADDRESSES, ESSAYS, LECTURES

Siegel, Marcia B
Two American dance classics / Marcia B. Siegel. — ₍Bronxville, N.Y.₎ : Sarah Lawrence College, 1978.
24 p. ; 18 cm. — (Essays from the faculty - Sarah Lawrence College ; v. 4., no. 1)
Cover title.
1. Choreography—Addresses, essays, lectures. 2. Balanchine, George—Addresses, essays, lectures. 3. Humphrey, Doris, 1895-1958—Addresses, essays, lectures. 4. Dancing—United States—Addresses, essays, lectures. I. Title. II. Series: Essays from Sarah Lawrence faculty ; v. 4, no. 1.
GV1782.5.S53 792.8'4 79-107394
79 MARC

CHORIONIC GONADOTROPINS

—THERAPEUTIC USE

Colpron, Guy.
Un petit livre pour un gros problème / par Guy Colpron. — Montréal : Éditions Héritage, [c1976]
159 p. : ill. ; 18 cm. C***
On cover: Docteur Colpron et l'obésité.
"Distributions Éclair ... Anjou, Québec."—Canadiana.
Bibliography: p. 156.
ISBN 0-7773-3901-3 : $3.95
1. Obesity. 2. Reducing diets. 3. Chorionic gonadotropins—Therapeutic use. I. Title. II. Title: Docteur Colpron et l'obésité.
RC628.C58 616.3'98 79-341807
79 MARC

CHORIORETINITIS

see also Toxoplasmosis, Ocular

CHORISTONEURA OCCIDENTALIS see Western spruce budworm

CHORNY, KUZ'MA, PSEUD.

—ICONOGRAPHY

Kudraŭtsaŭ, Ivan.
Kuz'ma Chorny : zhytstsīo i tvorchasts' : matėryīaly dlīa vystaŭki ŭ shkole / skladal'nik I. F. Kudraŭtsaŭ. — Vydanne 2., dapoŭnenae. — Minsk : Narodnaīa asveta, 1979.
150 p. : ill. ; 29 cm. USSR***
At head of title: Navukova-dasledchy instytut pedahohiki Ministerstva asvety BSSR.
1.30rub
1. Chorny, Kuz'ma, pseud.—Iconography. 2. Chorny, Kuz'ma, pseud.—Illustrations. I. Minsk. Navukova-dasledchy instytut pedagogiki.
PG2835.C5Z744 1979 79-372845
79 MARC

—ILLUSTRATIONS

Kudraŭtsaŭ, Ivan.
Kuz'ma Chorny : zhytstsīo i tvorchasts' : matėryīaly dlīa vystaŭki ŭ shkole / skladal'nik I. F. Kudraŭtsaŭ. — Vydanne 2., dapoŭnenae. — Minsk : Narodnaīa asveta, 1979.
150 p. : ill. ; 29 cm. USSR***
At head of title: Navukova-dasledchy instytut pedahohiki Ministerstva asvety BSSR.
1.30rub
1. Chorny, Kuz'ma, pseud.—Iconography. 2. Chorny, Kuz'ma, pseud.—Illustrations. I. Minsk. Navukova-dasledchy instytut pedagogiki.
PG2835.C5Z744 1979 79-372845
79 MARC

CHOROID

—DISEASES

Morse, Peter H 1935-
Vitreoretinal disease : a manual for diagnosis and treatment / Peter H. Morse. — Chicago : Year Book Medical Publishers, c1979.
xii, 625 p. : ill. ; 26 cm.
Includes bibliographies and index.
ISBN 0-8151-5962-5 : $39.95
1. Retina—Diseases. 2. Vitreous body—Diseases. 3. Choroid—Diseases. I. Title.
[DNLM: 1. Vitreous body. 2. Retinal diseases. 3. Choroid. WW250 M886v]
RE551.M67 617.7'3 78-50441
Shared Cataloging with 78 MARC
DNLM

CHOROID PLEXUS

see also Blood-brain barrier

CHOROTI INDIANS

Karsten, Rafael, 1879-1956.
Indians tribes of the Argentine and Bolivian Chaco : ethnological studies / by Rafael Karsten. — 1st AMS ed. — New York : AMS Press, [1979]
x, 236, [1] p. : ill. ; 24 cm.
Reprint of the 1932 ed. published by Societas Scientiarum Fennica, Helsinki, which was issued as v. 4, no. 1 of its series: Commentationes humanarum litterarum.
Includes bibliographical references and index.
ISBN 0-404-15942-7
1. Indians of South America—Chaco, El Gran. 2. Choroti Indians. 3. Toba Indians. 4. Mataco Indians. 5. Toba language (Indian)—Glossaries, vocabularies, etc. 6. Choroti language—Glossaries, vocabularies, etc. I. Title. II. Series: Suomen Tiedeseura. Commentationes humanarum litterarum ; v. 4, no. 1.
F2821.1.C47K37 1979 982'.3 76-44743
78 MARC

Nordenskiöld, Erland, friherre, 1877-1932.
An ethno-geographical analysis of the material culture of two Indian tribes in the Gran Chaco / by Erland Nordenskiöld. — [1st AMS ed.]. — New York : AMS Press, 1979.
xi, 295 p. : ill. ; 23 cm.
Translation of Eine geopgraphische und ethnographische Analyse der materiellen Kultur zweier Indianerstämme in El Gran Chaco (Südamerika).
Reprint of the 1919 ed. published by the author, Göteborg, Sweden, which was issued as v. 1 of Comparative ethnographical studies.
Bibliography: p. 270-293.
ISBN 0-404-15141-8
1. Choroti Indians. 2. Ashluslay Indians. 3. Indians of South America—Chaco, El Gran. I. Title. II. Series: Nordenskiöld, Erland, friherre, 1877-1932. Comparative ethnographical studies ; 1.
F3320.2.C5N6713 1979 982'.3'004'98 75-46054
78 MARC

CHOROTI LANGUAGE

—GLOSSARIES, VOCABULARIES, ETC.

Karsten, Rafael, 1879-1956.
Indians tribes of the Argentine and Bolivian Chaco : ethnological studies / by Rafael Karsten. — 1st AMS ed. — New York : AMS Press, [1979]
x, 236, [1] p. : ill. ; 24 cm.
Reprint of the 1932 ed. published by Societas Scientiarum Fennica, Helsinki, which was issued as v. 4, no. 1 of its series: Commentationes humanarum litterarum.
Includes bibliographical references and index.
ISBN 0-404-15942-7
1. Indians of South America—Chaco, El Gran. 2. Choroti Indians. 3. Toba Indians. 4. Mataco Indians. 5. Toba language (Indian)—Glossaries, vocabularies, etc. 6. Choroti language—Glossaries, vocabularies, etc. I. Title. II. Series: Suomen Tiedeseura. Commentationes humanarum litterarum ; v. 4, no. 1.
F2821.1.C47K37 1979 982'.3 76-44743
78 MARC

CHORUS (DRAMA) see Drama—Chorus (Greek drama)

CHŌSEN MUENTAN KABUSHIKI KAISHA

—HISTORY—ADDRESSES, ESSAYS, LECTURES

(Chōsen Muentan Kabushiki Kaisha kaikoroku)
朝鮮無煙炭株式会社回顧録：西鮮中央鉄道株式会社・元山北港株式会社・城山農業株式会社：汗と油・血と涙の記録 / [編集者 朝無社社友会回顧録編集委員会]. — 福山：朝無社社友会事務局, 1978.
268 p. : ill. ; 27 cm. Ja 78-21620
"朝鮮無煙炭(株) 回顧錄說明書" (4 leaves) in pocket.
1. Chōsen Muentan Kabushiki Kaisha—History—Addresses, essays, lectures. I. Chōmusha Shayūkai. Kaikoroku Henshū Iinkai.
HD9556.K84C563 78-817080

CHOSEN PEOPLE (JEWS) see Jews—Election, Doctrine of

CHOSEN YESUGYO CHANGNOHOE see Taehan Yesugyo Changnohoe

CHOSES see Personal property

CHOSES IN ACTION

see also Assignments; Negotiable instruments

CHŌSHŪ, JAPAN

—HISTORY

Yoshimura, Tōshū.
(Geishūguchi senki)
藝州口戰記：附 茸狩日記 / 吉村藤舟著. — 下関：防長史料出版社, 昭和51 [1976]
178 p., [2] leaves of plates : ill. ; 19 cm. — (His 四境戰爭 ; 其ノ4)
Ja 76-22498
Reprint of the 1937 ed. published by Kyōdoshi Kenkyūkai, Shimonoseki, which was issued as v. 19 of Kyōdo monogatari.
1. Chōshū, Japan—History. 2. Hiroshima-han, Japan—History. 3. Japan—History—Restoration, 1853-1870. I. Itō, Tsunetaru, 1774-1858. Kinokogari nikki. 1976. II. Title. III. Series: Yoshimura, Tōshu. Shikyō sensō ; sono 4. IV. Series: Kyōdo monogatari ; dai 19-shū.
DS894.79.Y349C4678 1976 78-811008

—HISTORY—PERIODICALS

Bō-Chō Shidankai.
(Bō-Chō Shidankai zasshi)
防長史談會雜誌 第1-4卷(第1-38號)；明治42年9月-大正3年9月 [1909-14. 東京] 国書刊行会 [昭和51 i. e. 1976]
4 v. ill. 23 cm.
Reprint of a monthly periodical issued by Bō-Chō Shidankai, published in Tokyo.
1. Chōshū, Japan—History—Periodicals. 2. Japan—History—Restoration, 1853-1870—Periodicals.
DS894.79.Y349C462 78-646229
(MARC-S)

—HISTORY, MILITARY —SOURCES

(Chōshū-han Daini Kiheitai dattai bōdō shiryō shū)
長州藩第二奇兵隊脱隊暴動史料集 / 国広哲也編. — [光]：光地方史研究会, 昭和52 [1977]
144 p., [4] leaves of plates : ill. ; 22 cm. Ja 78-1034
Errata slip inserted.
1. Chōshū, Japan—History, Military—Sources. 2. Kiheitai—History—Sources. 3. Japan—History—Restoration, 1853-1870—Sources. I. Kunihiro, Tetsuya.
DS894.79.Y349C463 79-805671

CHOSŎN HYŎNGMYŎNG PANGMULGWAN, P'YŎNGYANG, KOREA

Chosŏn Hyŏngmyŏng Pangmulgwan, P'yŏngyang, Korea.
Korean Revolutionary Museum. — Pyongyang : Foreign Languages Pub. House, 1963.
[126] p. : chiefly ill., maps (1 fold.), ports ; 23 cm.
1. Chosŏn Hyŏngmyŏng Pangmulgwan, P'yŏngyang, Korea. 2. Korean resistance movements, 1905-1945.
DS930.3.P9C48 1963 79-312481
951.9'3'03074095193
79 MARC

CHOSŎN MINSOK PANGMULGWAN

—PICTORIAL WORKS

Korea (Democratic People's Republic). Munmul Pojon Chidoguk. Hwach aek P'yŏnjipsil.
(Ch'ao-hsien min su po wu kuan)
朝鲜民俗博物馆 / [编辑朝鲜民主主义人民共和国文物保存指导局画册编辑室]. — [平壤?]：朝鲜民俗博物馆, 1977.
[36] p. : chiefly col. ill. ; 18 x 24 cm.
1. Chosŏn Minsok Pangmulgwan—Pictorial works. I. Title,
GN36.K72P934 79-835011

CHOSŎN NODONGDANG

Chosŏn Minjujuŭi Inmin Konghwaguk Sahoe Kwahagwŏn.
(“Kkosŭttarikka Sahoedang wiwŏnjang imyŏ Kkosŭttarikka Kija Tongmaeng wiwŏnjang in Kkosŭttarikka Chosŏn Ch‘insŏn Munhwa Hyŏphoe wiwŏnjang i chegi han chilmun e taehan taedap” e taehayŏ)
《꼬스따리까사회당위원장이며꼬스따리까기자동맹위원장인꼬스따리까조선친선문화협회위원장이제기한질문에대한대답》에대하여 / [해설 사회과학원 ; 편집 김일성동지로작해설편집부]. — [평양] : 사회과학출판사, 1976.
65 p. ; 19 cm. — (위대한수령김일성동지의로작해설문고)
₩0.20
1. Kim, Il-sŏng, 1912- Kkosŭttarikka Sahoedang wiwŏnjang imyŏ Kkosŭttarikka Kija Tongmaeng wiwŏnjang in Kkosŭttarikka Chosŏn Ch‘insŏn Munhwa Hyŏphoe wiwŏnjang i chegi han chilmun e taehan taedap. 2. Korean reunification question (1945-) 3. Chosŏn Nodongdang. 4. World politics—1965-1975. I. Kim Il-sŏng Tongji Nojak Haesŏl P‘yŏnjippu. II. Title. III. Series: Widae han suryŏng Kim Il-sŏng Tongji ŭi nojak haesŏl mun'go.
DS917.25.K543C48 1976 78-826737

—ADDRESSES, ESSAYS, LECTURES

Kim, Il-song, 1912-
(Kuan yü Ch‘ao-hsien lao tung tang ti chien she)
关于朝鲜劳动党的建设 / 金日成. — 平壤 : 外国文出版社, 1977-
v. : port. ; 22 cm.
1. Chosŏn Nodongdang—Addresses, essays, lectures. I. Title.
JQ1729.5.A98C4279685 79-840364

—PARTY WORK

Kim, Il-sŏng, 1912-
(Uri tang tanch‘edŭl ŭi kwaŏp e taehayŏ)
우리당단체들의과업에대하여 : 북조선로동당평안남도순천군당대표회의에서한연설 1948년1월24일 / 김일성. — [평양] : 조선로동당출판사, 1977.
22 p., [1] leaf of plates : port. ; 19 cm.
₩0.45
1. Chosŏn Nodongdang—Party work. I. Title.
JQ1729.5.A98C4338 79-826005

—PARTY WORK—ADDRESSES, ESSAYS, LECTURES

Kim, Il-sŏng, 1912-
On the building of the Workers' Party of Korea / Kim Il Sung. — Pyongyang, Korea : Foreign Languages Pub. House, 197
v. : port. ; 22 cm.
1. Chosŏn Nodongdeng—Party work—Addresses, essays, lectures. I. Title.
JQ1729.5.A98C4283 329.9'519'3 79 79-347807 MARC

“Man is the most precious in the world”. — Pyongyang : Foreign Languages Pub. House, 1977.
67 p., [1] leaf of plates : port. ; 19 cm. — (Among the people)
At head of title: The Party History Institute of the C.C. of the Workers' Party of Korea.
1. Chosŏn Nodongdang—Party work—Addresses, essays, lectures. I. Chosŏn Nodongdang. Chungang Wiwŏnhoe. Tang Yŏksa Yŏn'guso. II. Series.
JQ1729.5.A98C4387 329.9'519'3 78 78-318711 MARC

—PURGES

Chayu P‘yŏngnonsa.
(Pukkoe ŭi sukch‘ŏng kwa panhang undong)
北傀의肅清과反抗運動 : 成分分類에依한彈壓과그에따른抗爭. — 서울 : 自由評論社, 1977.
123 p. ; 22 cm.
Not in trade.
1. Political purges—Korea (Democratic People's Republic) 2. Chosŏn Nodongdang—Purges. 3. Anti-communist movements—Korea (Democratic People's Republic) I. Title.
DS935.55.C48 1977 79-825599

CHOSON SAHOEJUUI NODONG CH'ONGNYON TONGMAENG

—CONGRESSES

Chosŏn Sahoejuŭi Nodong Ch‘ŏngnyŏn Tongmaeng. Taehoe. 5th, P‘yŏngyang, Korea, 1964.
(Chosŏn Minju Ch‘ŏngnyŏn Tongmaeng Che Och‘a Taehoe munhŏnjip)
조선민주청년동맹제5차대회문헌집. — [평양] : 조선사회주의로동청년동맹출판사, 1964.
263 p. ; 21 cm.
1. Chosŏn Sahoejuŭi Nodong Ch‘ŏngnyŏn Tongmaeng—Congresses. I. Title.
HQ799.K62C46 1964aa 79-825766

CHOTA NAGPUR, INDIA

—HISTORY

Sa, Fidelis de, 1935-1970.
Crisis in Chota Nagpur : with special reference to the judicial conflict between Jesuit missionaries and British government officials, November 1889-March 1890 / Fidelis de Sa ; [edited by Alvino Noronha, with the assistance of Achilles Meersman]. — Bangalore : Redemptorist Publications, 1975.
lx, 357 p. : maps ; 23 cm.
Bibliography: p. [xxxi]-xlix.
Includes index.
1. Missions—India—Chota Nagpur. 2. Jesuits—Missions. 3. Mundas. 4. Oraons. 5. Chota Nagpur, India—History. I. Noronha, Alvino. II. Meersman, Achilles. III. Title.
BV3312.C47S2 1975 266'.00954'12 79 78-908957 MARC

CHOU, CHIH-TAO

Chou, Chih-tao.
(Pa shih i wang)
八十憶往 / 周志道. — [台北? : s. n., 1979?]
2, 355 p., [6] leaves of plates : ill. ; 27 cm.
Cover title.
NT$100.00
1. Chou, Chih-tao. 2. Generals—China—Biography. I. Title.
DS778.C5929A35 79-840720

CHOU, EN-LAI, 1898-1976

Ajia Seikei Gakkai.
(Kōgun sōkenki no Mō Takutō to Shū Onrai)
紅軍創建期の毛沢東と周恩来 : 立三路線論考 / [編集責任者 蜂屋亮子]. — 東京 : アジア政経学会 : 発売所 巌南堂書店, 1978.
6, 190 p. ; 21 cm. — (現代中国研究叢書 ; 15) Ja 78-7624
Includes bibliographical references.
¥1000
1. China—History—1912-1937. 2. Chung-kuo kung ch'an tang—History. 3. Chung-kuo kung nung hung chün—History. 4. Mao, Tse-tung, 1893-1976. 5. Chou, En-lai, 1898-1976. I. Hachiya, Akiko. II. Title. III. Series: Gendai Chūgoku kenkyū sōsho ; 15.
DS777.55.G43 vol. 15 78-816766
[DS777.47]

—ADDRESSES, ESSAYS, LECTURES

(Chou En-lai tsung li pa shih tan ch‘en chi nien shih wen hsüan)
周恩来总理八十诞辰纪念诗文选 / [人民出版社编輯·出版]. — 第1版. — 北京 : 該社 : 新華書店发行, 1978.
2, 339 p., [8] leaves of plates : ill. ; 21 cm.
$1.40
1. Chou, En-lai, 1898-1976—Addresses, essays, lectures. 2. Prime ministers—China—Biography—Addresses, essays, lectures. I. Jen min ch‘u pan she, Peking.
DS778.C593C5319 79-836809

(Chou tsung li ti ying ming yü shan ho kung ts‘un)
周总理的英名与山河共存 / [水利电力部政治部宣传处编]. — [香港? : s. n., 1978?]
100 p., [5] leaves of plates : ill. ; 21 cm.
Reprint of the 1977 (1978 printing) ed. published by Shui li tien li ch‘u pan she, Peking.
1. Chou, En-lai, 1898-1976—Addresses, essays, lectures. I. China (People's Republic of China, 1949-). Shui li tien li pu. Cheng chih pu. Hsüan ch‘uan ch‘u.
DS778.C593C537 1978 79-836897

(Chü kung chin ts‘ui wei jen min)
鞠躬尽瘁为人民 : 周总理革命实践片断. — 第1版. — 上海 : 上海人民出版社, 1977.
3, 149 p. : ill. ; 19 cm.
$0.28
1. Chou, En-lai, 1898-1976—Addresses, essays, lectures.
DS778.C593C54 78-842219

(Chü kung chin ts‘ui wei jen min)
鞠躬尽瘁为人民 : 周总理革命实践片断. — 第2版. — 上海 : 上海人民出版社, 1977.
4, 210 p. : ill. ; 19 cm.
$0.41
1. Chou, En-lai, 1898-1976—Addresses, essays, lectures.
DS778.C593C54 1977b 79-837878

Chung-kuo li shih po wu kuan, Peking.
(Chi nien Chou En-lai tsung li wen wu hsüan pien)
紀念周恩來总理文物选编 / 中国历史博物馆编. — 第1版. — 北京 : 文物出版社, 1977.
2, 245 p., [2] leaves of plates : ill. ; 21 cm.
$0.75
1. Chou, En-lai, 1898-1976—Addresses, essays, lectures. 2. Prime ministers—China—Biography—Addresses, essays, lectures. I. Title.
DS778.C593C548 1977 79-835180

(Hung se ti shih yüeh)
红色的十月 = Hong se de shi yue : 学生课外阅读文选 / 上海市虹口区教师进修学校编. — 第1版. — 上海 : 上海人民出版社, 1978.
98 p. : 19 cm.
$0.21
1. Hua, Kuo-feng—Addresses, essays, lectures. 2. Mao, Tse-tung, 1893-1976—Addresses, essays, lectures. 3. Chou, En-lai, 1895-1976—Addresses, essays, lectures. I. Shang-hai shih Hung-k‘ou ch‘ü chiao shih chin hsiu hsüeh hsiao. II. Title: Hong se de shi yue.
DS779.29.H8H88 79-835140

(Jen min ti hao tsung li Chou En-lai t‘ung chih yung yüan huo tsai wo men hsing chung hsü pien)
人民的好总理周恩来同志永远活在我们心中续编. — 第1版. — 上海 : 上海人民出版社, 1977.
7, 552 p., [1] leaf of plates : port. ; 19 cm.
$0.98
1. Chou, En-lai, 1898-1976—Addresses, essays, lectures.
DS778.C593J46 78-842243

CHOU, EN-LAI, 1898-1976
—ADDRESSES, ESSAYS, LECTURES (Continued)

Liao, Ch'i-k'ang.
(Sui wei ching ai ti Chou fu chu hsi)
随卫敬爱的周付主席 / 廖其康. — [香港? : s. n., 1978?]
2, 63 p., [2] leaves of plates : ill. ; 22 cm.
Title also in Pinyin on cover: Suiwei jingaide zhoufuzhuxi.
Reprint of the 1978 ed. published by Ssu-ch'uan jen min ch'u pan she, Ch'eng-tu.
1. Chou, En-lai, 1898-1976—Addresses, essays, lectures. I. Title. II. Title: Suiwei jingaide Zhoufuzhuxi.
DS778.C593L53 1978 79-835794

Lo, Jui-ch'ing, 1906-
(Hsi-an shih pien ho Chou En-lai t'ung chih)
西安事变和周恩来同志 / 罗瑞卿, 吕正操, 王炳南. — 第1版. — 北京 : 人民出版社 : 新華書店发行, 1978.
2, 101 p., [3] leaves of plates : ill. ; 21 cm.
Errata slip inserted.
$0.31
1. China—History—Sian Incident, 1936—Addresses, essays, lectures. 2. Chou, En-lai, 1898-1976.—Addresses, essays, lectures. I. Lü, Cheng-ts'ao. II. Wang, Ping-nan. III. Title.
DS777.47.L6 79-839135

—ANECDOTES

Chou, T'i.
(Chou tsung li ti ku shih)
周总理的故事 = Zhouzonglidegushi / [周倜, 洪子诚编写]. — [香港? : s. n., 1978?]
3, 2, 119 p., [5] leaves of plates : ill. ; 22 cm.
Reprint of the 1977 ed. published by T'ien-chin jen min ch'u pan she, Tientsin.
1. Chou, En-lai, 1898-1976—Anecdotes. 2. Prime ministers—China—Anecdotes, facetiae, satire, etc. I. Hung, Tzu-ch'eng, joint author. II. Title. III. Title: Zhouzonglidegushi.
DS778.C593C5345 1978 79-835025

—DEATH AND BURIAL
—PICTORIAL WORKS

(Jen min ti tao nien)
人民的悼念 / 《人民的悼念》联合编辑组编 ; [特約执行编辑吳鹏 ; 特约编辑高强 ... et al.]. — 第1版. — [北京] : 北京出版社 : 新华书店北京发行所发行, 1979.
[236] p. : chiefly ill. (some col.) ; 26 cm.
$5.00
1. Chou, En-lai, 1898-1976—Death and burial—Pictorial works. I. Wu, P'eng. II. Kao, Ch'iang. III. Jen min ti tao nien lien ho pien chi tsu.
DS778.C593J474 79-838748

(Shen ch'ieh huai nien ching ai ti Chou tsung li)
深切怀念敬爱的周总理 : 小戏集 / 人民文学出版社编辑部编. — 第1版. — 北京 : 該社 : 新华书店北京发行所发行, 1977.
2, 217 p. ; 19 cm.
$0.44
1. Chinese drama—20th century. 2. Chou, En-lai, 1898-1976—Drama. I. Jen min wen hsüeh ch'ü pan she, Peking.
PL2603.S465 79-837439

—JUVENILE LITERATURE

(Hsing fu ti hui i)
幸福的回忆 / [福建人民出版社编辑]. — [香港? : s. n., 1978]
2, 82 p., [7] leaves of plates : ill. ; 21 cm.
Reprint of the 1978 ed. published by Fu-chien jen min ch'u pan she.
1. Chou, En-lai, 1898-1976—Juvenile literature. I. Fu-chien jen min ch'u pan she, Foochow, China.
DS778.C593H79 1978 79-837932

—POETIC WORKS

Chou, Hung-hsing.
(Chou tsung li yü shih ko)
周总理与诗歌 / 周红兴, 李如鸾. — [香港? : s. n., 1978]
3, 3, 212 p., [4] leaves of plates : ill. ; 21 cm.
Title in Pinyin on cover: Zhouzongli yu shige.
Reprint of the 1978 ed. published by Ssu-ch'uan jen min ch'u pan she, Ch'eng-tu.
1. Chou, En-lai, 1898-1976—Poetic works. I. Li, Ju-luan, joint author. II. Title. III. Title: Zhouzongli yu shige.
PL2754.E5Z6 1978 79-837341

—POETRY

(Erh ch'i kuang ch'ang shih ch'ao)
二七广场诗抄. — [香港? : s. n., 1979?]
2, 81 p., [1] leaf of plates : ill. ; 22 cm.
Reprint of the 1978 ed. published by Ho-nan jen min ch'u pan she.
1. Chou, En-lai, 1898-1976—Poetry. 2. Chinese poetry—20th century.
PL2543.E7 1979 79-840097

(Hsin chung ti ko)
心中的歌 : 怀念敬爱的周总理. — [香港? : s. n., 1977]
3, 156 p. ; 21 cm.
Reprint of the 1977 ed. published by Kuang-tung jen min ch'u pan she.
1. Chinese poetry—20th century. 2. Chou, En-lai, 1898-1976—Poetry.
PL2543.H7816 1977 78-843871

(T'ien-an-men ko ming shih wen hsüan)
天安門革命詩文选 / 童怀周编. — [香港? : s. n., 1979?]
5, 10, 584 p., [16] leaves of plates : ill. ; 21 cm.
"纪念敬爱的周总理逝世两周年"
Reprint ed. with postscript dated 1977 published by Fu-chien jen min ch'u pan she.
1. Chou, En-lai, 1898-1976—Poetry. 2. Chinese poetry—20th century. I. T'ung, Huai-chou.
PL2543.T483 1979 79-838686

(T'ien-an-men ko ming shih wen hsüan, hsü chi)
天安門革命詩文选 : 续集 / [童怀周编]. — [香港? : s. n., 1979?]
2, 11, 448 p., [10] leaves of plates : ill.
Cover title.
Reprint ed. with pref. dated 1978.
1. Chou, En-lai, 1898-1976—Poetry. 2. Chinese poetry—20th century. I. T'ung, Huai-chou.
PL2543.T4834 1979 79-838755

(T'ien-an-men shih wen chi)
天安门诗文集 / 童怀周编. — 第1版. — [北京] : 北京出版社 : 北京市新华书店发行, 1979.
18, 4, 488 p., [8] leaves of plates : ill. ; 19 cm.
Title also in Pinyin on cover: Tianmen shiwenji.
$1.15
1. Chou, En-lai, 1898-1976—Poetry. 2. Chinese poetry—20th century. I. T'ung, Huai-chou. II. Title: Tianmen shiwenji.
PL2543.T484 79-838685

—PORTRAITS, CARICATURES, ETC.

(Chou En-lai t'ung chih wei kung ch'an chu i shih yeh kuang hui chan tou ti i sheng)
周恩来同志为共产主义事业光辉战斗的一生 / [广东人民出版社编]. — 第1版. — [广州] : 該社, 1977.
60 p. : chiefly ill., group ports. ; 27 cm.
$1.00
1. Chou, En-lai, 1898-1976—Portraits, caricatures, etc.
DS778.C593C532 1977 78-841686

CHOU, FU, 1837-1921
—CHRONOLOGY

Chou, Fu, 1837-1921.
(Min kuo Chou Yü-shan hsien sheng Fu tzu ting nien p'u)
民國周玉山先生馥自訂年譜 / [周]學淵 ... [et al.]校字. — 臺北 : 臺灣商務印書館, 民國67 [1978]
168 p. ; 20 cm. — (新編中國名人年譜集成)
Reprint ed.
NT$34.00
1. Chou, Fu, 1837-1921—Chronology. 2. Statesmen—China—Biography. I. Chou, Hsüeh-yüan. II. Title. III. Series: Hsin pien Chung-kuo ming jen nien p'u chi ch'eng.
DS763.C53A35 1978 79-840608

CHOU, PANG-YEN, 1056-1121
—CRITICISM AND INTERPRETATION

Ts'ui, Jui-yü.
(Liu Yung yü Chou Pang-yen)
柳永與周邦彥 / 崔瑞郁. — [台北? : s. n.], 民國65 [1976]
57 double leaves ; 27 cm.
"國立台灣大學中文研究所碩士論文"
Reproduced from the typescript.
Bibliography: leaves 56-57.
1. Chinese poetry—Sung dynasty, 960-1279—History and criticism. 2. Liu, Yung, chin shih 1034—Criticism and interpretation. 3. Chou, Pang-yen, 1056-1121—Criticism and interpretation. I. Title.
PL2323.T78 78-843915

CHOU, PANG-YEN, 1056-1121. CH'ING CHEN TZ'U

Yü, P'ing-po, 1899-
(Ch'ing chen tz'u shih)
清真詞釋 俞平伯著 上海 開明書店 民國38 [1949]
xviii, 76 p. 18 cm.
1. Chou, Pang-yen, 1056-1121. Ch'ing chen tz'u. I. Chou, Pang-yen, 1056-1121. Ch'ing chen tz'u. 1949. II. Title.
PL2687.C64C539 1949 C 59-1582

Yü, P'ing-po, 1899-
(Ch'ing chen tz'u shih)
清真詞釋 / [俞平伯著]. — [香港?] : 一山書屋, [1978?]
ii, 74 p. ; 19 cm.
Reprint ed.
1. Chou, Pang-yen, 1056-1121. Ch'ing chen tz'u. I. Chou, Pang-yen, 1056-1121. Ch'ing chen tz'u. 1978. II. Title.
PL2687.C64C539 1978 79-837600

CHOU, SHU-JEN, 1881-1936 YEH T'SAO

Li, Ho-lin.
(Lu Hsün Yeh ts'ao chu chieh)
鲁迅《野草》注解 / 李何林. — 修订本第2版. — [西安] : 陕西人民出版社, 1975, 1977 printing.
2, 226 p., [1] leaf of plates : port. ; 19 cm.
$0.47

1. Chou, Shu-jen, 1881-1936. Yeh ts'ao. I. Chou, Shu-jen, 1881-1936. Yeh ts'ao. 1975. II. Title.

PL2754.S5Y4579 1975 79-835108

CHOU, SHU-JEN, 1881-1936

Chang, Ching.
(Lu Hsün tsai Kuang-chou)
鲁迅在广州 / 张竞. — [香港? : s. n., 1978?]
121 p. : ill. ; 21 cm.
Reprint of the 1977 ed. published by Kuang-tung jen min ch'u pan she, Canton.
Includes bibliographical references.

1. Chou, Shu-jen, 1881-1936. 2. Chou, Shu-jen, 1881-1936—Political and social views. 3. Authors, Chinese—20th century—Biography. I. Title.

PL2754.S5Z565 1978 79-838128

T'ang, T'ao, 1913-
[Lu Hsün hsien sheng ti ku shih. Japanese]
魯迅先生 / 唐弢著 ; 三島孚滋雄訳. — 東京 : 満江紅, 1978.
109 p. : ports. ; 20 cm. Ja 79-1608
¥800

1. Chou, Shu-jen, 1881-1936. 2. Authors, Chinese—20th century—Biography. I. Title.
Title romanized: Ro Jin Sensei.

PL2754.S5Z86316 79-804878

—ANECDOTES

(Lu Hsün hui i lu)
鲁迅回憶錄 / 宋庆龄, 周建人, 茅盾等 ; [本社編]. — 第1版. — 上海 : 上海文艺出版社, 1978-
v. : ill. ; 21 cm.
$0.77 (v. 1)

1. Chou, Shu-jen, 1881-1936—Anecdotes. 2. Authors, Chinese—20th century—Biography—Anecdotes, facetiae, satire, etc. I. Sun, Ch'ing-ling Sung, 1890- II. Chou, Chien-jen. III. Shen, Yen-ping, 1896- IV. Shang-hai wen i ch'u pan she.

PL2754.S5Z7536 78-842475

—BIBLIOGRAPHY

Shen chou t'u shu kung ssu.
(Lu Hsün yen chiu yü hsin wen hsüeh ts'an k'ao shu mu)
魯迅研究與新文學參考書目 / 神州圖書公司. — 香港 : 經銷處該公司, [1975?]
80 p., [2] leaves of plates : ill. ; 21 cm.
Cover title.

1. Chou, Shu-jen, 1881-1936—Bibliography. 2. Chinese literature—Bibliography. I. Title.

Z8168.2.S47 1975 79-835009
[PL2754.S5]

—BIOGRAPHY

Oda, Takeo, 1900-
[Ro Jin no shōgai. Chinese]
魯迅傳 / 小田嶽夫著 ; 范泉譯. — 第1版. — 澳門 : 爾雅社, 1978.
3, 118 p., [1] leaf of plates : port. ; 19 cm. — (魯迅研究資料叢書)
"魯迅著作年表": p. 106-115.

1. Chou, Shu-jen, 1881-1936—Biography. 2. Authors, Chinese—20th century—Biography. I. Title. II. Series: Lu Hsün yen chiu tzu liao ts'ung shu.
Title romanized: Lu Hsün chuan.

PL2754.S5Z77713 79-835852

—CHARACTERS

Chou, Tso-jen, 1885-1966.
(Lu Hsün hsiao shuo li ti jen wu)
魯迅小說裏的人物 周遐壽著 上海 上海出版公司 1954.
17, 3, 326 p. 19 cm. (魯迅研究資料之2)
本書是作者根據魯迅先生著的吶喊彷徨兩部小說以及朝花夕拾(部分)裏的人物
Series romanized: Lu Hsün yen chiu tzu liao.

1. Chou, Shu-jen, 1881-1936—Characters.

PL2754.S5Z64 C 63-284

rev

—CORRESPONDENCE

Chung-shan ta hsüeh, Canton, China. Chung wen hsi.
(Lu Hsün tsai Kuang-chou)
鲁迅在廣州 : 資料專輯 / 中山大學中文系編. — 第1版. — 澳門 : 爾雅社, 1978.
2, 285 p., [3] leaves of plates : ill. ; 19 cm. — (魯迅研究資料叢書)
Includes bibliographical references.
$12.00 (H.K.)
1. Chou, Shu-jen, 1881-1936—Homes and haunts—China—Canton. 2. Chou, Shu-jen, 1881-1936—Correspondence. 3. Authors, Chinese—20th century—Correspondence. I. Title. II. Series: Lu Hsün yen chiu tzu liao ts'ung shu.

PL2754.S5Z655 78-843528

—CRITICISM AND INTERPRETATION

Wu, Chung-chieh.
(Lun Lu Hsün ti hsiao shuo ch'uang tso)
论鲁迅的小说创作 / 吴中杰, 高云. — 第1版. — 上海 : 上海文艺出版社 : 新華書店上海发行所发行, 1978.
193 p. ; 21 cm. — (中国现代文学研究丛书)
Series romanized: Chung-kuo hsien tai wen hsüeh yen chiu ts'ung shu.
Includes bibliographical references.
$0.55
1. Chou, Shu-jen, 1881-1936—Criticism and interpretation. I. Kao, Yün, joint author. II. Title.
PL2754.S5Z963 79-841078

—CRITICISM AND INTERPRETATION —ADDRESSES, ESSAYS, LECTURES

Chang, Hsiang-t'ien.
(Lu Hsün shih wen sheng huo tsa t'an)
魯迅詩文生活雜談 / 張向天著. — 香港 : 上海書局, 1977.
5, 250 p. : ill. ; 21 cm.
$6.50

1. Chou, Shu-jen, 1881-1936—Criticism and interpretation—Addresses, essays, lectures. I. Title.

PL2754.S5Z573 79-836740

Ch'iao, Feng.
(Lüeh chiang kuan yü Lu Hsün ti shih ch'ing)
略講關於魯迅的事情 / 喬峯著. — 第1版. — 澳門 : 爾雅社, 1978.
53 p. ; 19 cm. — (魯迅研究資料叢書)
Reprint of the 1954 ed. published by Jen min wen hsüeh ch'u pan she, Peking.
$2.50 (H.K.)
1. Chou, Shu-jen, 1881-1936—Criticism and interpretation—Addresses, essays, lectures. I. Title. II. Series: Lu Hsün yen chiu tzu liao ts'ung shu.

PL2754.S5Z63 1978 79-838737

Chou, Shu-jen, 1881-1936.
[Selected works]
鲁迅批判"狄克" / [周樹人] ; 北京汽车制造厂工人理论研究所, 中国科学院文学研究所近代文学组选注. — 第1版. — 北京 : 中华书局, 1976.
34 p. ; 19 cm.
CONTENTS: 任平 一个地地道道的老投降派—鲁迅 三月的租界—鲁迅 出关—鲁迅致王冶秋信—鲁迅 《八月的乡村》序

1. Chou, Shu-jen, 1881-1936—Criticism and interpretation—Addresses, essays, lectures. 2. Chang, Ch'un-ch'iao—Criticism and interpretation—Addresses, essays, lectures. I. Pei-ching ch'i ch'e chih tsao ch'ang. Kung jen li lun yen chiu so. II. Chung-kuo k'o hsüeh yüan. Wen hsüeh yen chiu so, Peking. Chin tai wen hsüeh tsu. III. Title.
Title romanized: Lu Hsün p'i p'an Ti-k'o.

PL2754.S5A6 1976e 78-841667

Han, Ch'ang-ching.
(Hsüeh hsi Lu Hsün)
學習魯迅 / 韓長經著. — 第1版. — 澳門 : 爾雅社, 1978.
iii, 81 p. ; 19 cm. — (魯迅研究資料叢書)
$3.20 (H. K.)

1. Chou, Shu-jen, 1881-1936—criticism and interpretation—Addresses, essays, lectures. I. Title. II. Series: Lu Hsün yen chiu tzu liao ts'ung shu.

PL2754.S5Z662 1978 79-835548

Hsü, Kuang-p'ing, 1907-1968.
(Kuan yü Lu Hsün ti sheng huo)
關於魯迅的生活 / 許廣平著. — 第1版. — 澳門 : 爾雅社, 1978.
69 p. ; 19 cm. — (魯迅研究資料叢書)
$2.80

1. Chou, Shu-jen, 1881-1936—Criticism and interpretation—Addresses, essays, lectures. I. Title. II. Series: Lu Hsün yen chiu tzu liao ts'ung shu.

PL2754.S5Z6738 1978 79-835337

Hu, Ping.
(Lu Hsün yen chiu cha chi)
魯迅研究札記 / 胡冰著. — 第1版. — 澳門 : 爾雅社, 1978.
1, 134 p. ; 19 cm. — (魯迅研究資料叢書)
Includes bibliographical references.
$8.00 (H.K.)

1. Chou, Shu-jen, 1881-1936—Criticism and interpretation—Addresses, essays, lectures. I. Title. II. Series: Lu Hsün yen chiu tzu liao ts'ung shu.

PL2754.S5Z682 1978 78-841550

C

(Lu Hsün yen chiu tzu liao)
鲁迅研究资料 / 鲁迅研究资料编辑部编. — 第1版. — 澳門 : 爾雅社, 1979-
v. : ill. ; 20 cm.
Title in Pinyin on cover of v. 1: Luxun yanjiu ziliao.
$28.00 (v. 1)

1. Chou, Shu-jen, 1881-1936—Criticism and interpretation—Addresses, essays, lectures. I. Lu Hsün yen chiu tzu liao pien chi pu. II. Title: Luxun yanjiu ziliao.

PL2754.S5Z75377 79-841707

Pei-ching ta hsüeh. Chung-kuo yü yen wen hsüeh hsi. Hsieh tso chiao hsüeh hsiao tsu.
(Lu Hsün tso p'in hsüan chiang)
魯迅作品選講 / 北京大學中[國語言]文[學]系寫作教學小組編. — 第1版. — 澳門 : 爾雅社, 1978.
15 v. in 2 ; 19 cm. — (魯迅研究資料叢書)
Includes bibliographical references.
$30.00

1. Chou, Shu-jen, 1881-1936—Criticism and interpretation. I. Chou, Shu-jen, 1881-1936. Selections. 1978. II. Title. III. Series: Lu Hsün yen chiu tzu liao ts'ung shu.

PL2754.S5Z814 1978 78-843200

—HOMES AND HAUNTS

(Lu Hsün chi nien kuan)
魯迅紀念館 / 晨光出版公司編輯. — [香港? : s. n., 1978?]
[44] p. : chiefly ill. ; 21 cm.
Reprint of the 1952 ed. published by Ch'en kuang ch'u pan she, Shanghai; in series: Hsin Chung-kuo hua k'u ; 32.

1. Chou, Shu-jen, 1881-1936—Museums, relics, etc. 2. Chou, Shu-jen, 1881-1936—Homes and haunts. I. Ch'en kuang ch'u pan kung ssu.

PL2754.S5Z7498 1978 78-842474

CHOU, SHU-JEN, 1881-1936 (Continued)

—HOMES AND HAUNTS—CHINA —CANTON

Chung-shan ta hsüeh, Canton, China. Chung wen hsi.
(Lu Hsün tsai Kuang-chou)
魯迅在廣州：資料專輯 / 中山大學中文系編. — 第1版. — 澳門：爾雅社, 1978.
2, 285 p., [3] leaves of plates : ill. ; 19 cm. — (魯迅研究資料叢書)
Includes bibliographical references.
$12.00 (H.K.)
1. Chou, Shu-jen, 1881-1936—Homes and haunts—China—Canton. 2. Chou, Shu-jen, 1881-1936—Correspondence. 3. Authors, Chinese—20th century—Correspondence. I. Title. II. Series: Lu Hsün yen chiu tzu liao ts'ung shu.
PL2754.S5Z655 78-843528

Tseng, Min-chih.
(Lu Hsün tsai Kuang-chou ti jih tzu)
魯迅在廣州的日子 / 曾敏之著. — 第1版. — 澳門：爾雅社, 1978.
96 p., [5] leaves of plates : ill. ; 19 cm. — (魯迅研究資料叢書)
Reprint of the 1956 ed. published by Kuang-tung jen min ch'u pan she, Canton.
Includes bibliographical references.
$4.00
1. Chou, Shu-jen, 1881-1936—Homes and haunts—China—Canton. 2. Authors, Chinese—20th century—Biography. I. Title. II. Series: Lu Hsün yen chiu tzu liao ts'ung shu.
PL2754.S5Z88 1978 79-836519

—ILLUSTRATIONS

Ting, Ts'ung.
(Lu Hsün hsiao shuo ch'a t'u)
魯迅小說插圖 / 丁聰. — 第1版. — 北京：人民美術出版社：新华书店北京发行所发行, 1978.
33 p. : ill. ; 19 cm.
$0.52
1. Chou, Shu-jen, 1881-1936—Illustrations. I. Chou, Shu-jen, 1881-1936. Selections. 1978. II. Title.
PL2754.S5Z867 79-840538

—KNOWLEDGE—ART

Huang, Meng-t'ien.
(Lu Hsün yü mei shu)
魯迅與美術 黃蒙田著 [香港] 大光出版社 [1972-
v. illus. 19 cm. $2.90
1. Chou, Shu-jen, 1881-1936—Knowledge—Art. I. Title.
PL2754.S5Z683 73-898132
rev

—MUSEUMS, RELICS, ETC.

(Lu Hsün chi nien kuan)
魯迅紀念館 / 晨光出版公司編輯. — [香港?：s. n., 1978?]
[44] p. : chiefly ill. ; 21 cm.
Reprint of the 1952 ed. published by Ch'en kuang ch'u pan she, Shanghai; in series: Hsin Chung-kuo hua k'u ; 32.
1. Chou, Shu-jen, 1881-1936—Museums, relics, etc. 2. Chou, Shu-jen, 1881-1936—Homes and haunts. I. Ch'en kuang ch'u pan kung ssu.
PL2754.S5Z7498 1978 78-842474

—PHILOSOPHY

Lin, Fei.
(Lu Hsün ch'ien ch'i ssu hsiang fa chan shih lüeh)
鲁迅前期思想发展史略 / 林非. — 第1版. — 上海：上海文艺出版社：新華書店上海发行所发行, 1978.
148 p. ; 21 cm. — (中国现代文学研究丛书)
Series romanized: Chung-kuo hsien tai wen hsüeh yen chiu ts'ung shu.
$0.43
1. Chou, Shu-jen, 1881-1936—Philosophy. I. Title.
PL2754.S5Z724 79-841706

—POLITICAL AND SOCIAL VIEWS

Chang, Ching.
(Lu Hsün tsai Kuang-chou)
鲁迅在广州 / 张竞. — [香港?：s. n., 1978?]
121 p. : ill. ; 21 cm.
Reprint of the 1977 ed. published by Kuang-tung jen min ch'u pan she, Canton.
Includes bibliographical references.
1. Chou, Shu-jen, 1881-1936. 2. Chou, Shu-jen, 1881-1936—Political and social views. 3. Authors, Chinese—20th century—Biography. I. Title.
PL2754.S5Z565 1978 79-838128

Ch'en, Sou-yü.
(Lu Hsün yü nü shih ta hsüeh sheng yün tung)
鲁迅与女师大学生运动 / 陈漱渝. — [香港?：s. n., 1978]
145 p. : ill. ; 21 cm.
Reprint of the 1978 ed. published by Pei-ching jen min ch'u pan she, Peking.
Includes bibliographical references.
1. Chou, Shu-jen, 1881-1936—Political and social views. 2. Student movements—China—Peking. I. Title.
PL2754.S5Z589 1978 79-839615

—PORTRAITS, ETC.

Chung-shan ta hsüeh, Canton, China. Chung wen hsi.
(Lu Hsün tsai Kuang-chou)
鲁迅在广州 / 中山大学中文系编；汤小铭绘. — 第1版. — [广州]：广东人民出版社, 1977.
[42] p. : col. ill. ; 21 cm.
"連環圖畫"
$1.20
1. Chou, Shu-jen, 1881-1936—Portraits, etc. I. T'ang, Hsiao-ming. II. Title.
PL2754.S5Z6552 79-837061

CHOU, SHU-SHENG

(Ming hsien shou cha)
名賢手札 / 珍藏者周樹聲. — [台北]：周樹聲, 民國65 [1976]
[65] p. : ill. ; 27 cm.
Colophon title.
1. Chinese letters. 2. Chou, Shu-sheng. I. Chou, Shu-sheng.
PL2610.M516 78-842443

CHOU, TE-CH'ING, FL. 1314-1324. CHUNG YUAN YIN YUN

Hu, Chia-yang.
(Chung yüan yin yün yü kuo yü yin hsi chih yin chih chu tzu pi chiao yen chiu)
中原音韻與國語音系之音質逐字比較研究 / 胡嘉陽. — 民國54 [1965]
2 v. (155 double leaves) ; 27 cm.
Cover title.
Reproduced from ms. copy.
Thesis (M. A.)—Kuo li T'ai-wan ta hsüeh.
1. Chou, Te-ch'ing, fl. 1314-1324. Chung yüan yin yün. 2. Chinese language—Phonology. I. Title.
PL1201.C3853H8 79-841846

CHOU, TUN-I, 1017-1073

Chang, Po-hsing, 1652-1725.
(Sung Chou Lien-hsi hsien sheng Tun-i nien p'u)
宋周濂溪先生惇頤年譜 / 張伯行編. — 臺北：臺灣商務印書館, 民國67 [1978]
6, 154 p. ; 20 cm. — (新編中國名人年譜集成；第一輯)
Caption title: Chou Lien-hsi hsien sheng ch'üan chi.
Running title: Chou Lien-hsi chi.
Reprint of the 1866 ed. published by Cheng i shu chü, Foochow.
CONTENTS: 周濂溪先生全集.—年譜.
NT$32.00
1. Chou, Tun-i, 1017-1073. 2. Philosophers—China—Biography. I. Chou, Tun-i, 1017-1073. Chou Lien-hsi hsien sheng ch'üan chi. Selections. 1978. II. Title. III. Title: Chou Lien-hsi hsien sheng ch'üan chi. IV. Title: Chou Lien-hsi chi. V. Series: Hsin pien Chung-kuo ming jen nien p'u chi ch'eng.
B128.C44C45 1978 79-841077

CHOU, YÜ-HSIEN

—ADDRESSES, ESSAYS, LECTURES

(Ch'iu Chin yü Chou Yün-hsien chih ch'i jen ch'i shih)
秋瑾與周韻僊之奇人奇事 / [朱琬珍 ... et al.]. — [台北?：s. n., 1978?]
[18] p. : facsims. ; 27 cm.
Cover title.
1. Ch'iu, Chin, 1877-1907—Anecdotes. 2. Revolutionists—China—Anecdotes, facetiae, satire, etc. 3. Chou, Yü-hsien—Addresses, essays, lectures. I. Chu, Wan-chen.
DS763.C5C46 78-841920

CHOU KING see Shu ching

CHOU LI

Wan, Ching-yü.
(Chou li hsin chuan)
周禮新傳：經國政略大典 / 萬靖宇著. — 蘇澳鎮(宜蘭縣)：萬靖宇, 民國66 [1977]
10, 8, 502 p. ; 22 cm.
"著者著作簡介": p. 501-502.
Includes bibliographies.
NT$180.00
1. State, The. 2. Utopias. 3. Chinese classics—History and criticism. 4. Chou li. I. Title.
JC327.W27 79-836525

CHOU-SHAN ISLANDS, CHINA

—HISTORY

(Chou-shan ch'e t'ui nien ch'i chou nien chi nien t'e k'an)
舟山撤退廿七週年紀念特刊 / 編輯人張行周. — 臺北：舟山同鄉會, 民國66 [1977]
3, 170 p. : ill. ; 22 cm.
Colophon title.
1. Chou-shan Islands, China—History. 2. China—History—Civil War, 1945-1949. I. Chang, Hsing-chou.
DS793.C555C47 1977 79-836121

CHOU WU-WANG, EMPEROR OF CHINA, 1169-1116 B.C.

Li, T'ang.
(Chou Wu-wang)
周武王 / 李唐. — 臺景印初版. — 臺北：河洛圖書出版社, 民國67 [1978]
3, 2, 83 p. ; 21 cm. — (中國歷史小叢書；002)
Series romanized: Chung-kuo li shih hsiao ts'ung shu.
Reprint of the 1961 ed. published by Hsiang-kang hung yeh shu chü, Hong Kong, in series: Li shih hsiao ts'ung shu.
Pbk.: NT$35.00
1. China—History—Chou dynasty, 1122-221 B. C. 2. Chou Wu-wang, Emperor of China, 1169-1116 B. C.
DS747.L48 1978 79-836048

CHOUANS

see also Vendean War, 1793-1800

Orain, Adolphe, 1834-1918.
La chouannerie en pays Gallo, suivi de Mes souvenirs : textes inédits / Adolphe Orain. — Rennes : Armor, [c1977]
239 p. : port. ; 20 cm. F***
40.00F
1. Ille-et-Vilaine, France (Dept.)—History. 2. France—History—Revolution, 1793-1799. 3. Chouans. 4. Orain, Adolphe, 1834-1918. 5. Ille-et-Vilaine, France (Dept.)—Biography. I. Orain, Adolphe, 1834-1918. Mes souvenirs. 1977. II. Title.
DC611.I3O7 1977 944'.15 78-398134
79 MARC

Roincé, Job de.
Le drame de Quiberon, 1795 / Job de Roincé. — [Rennes : J. de Roincé, 1976]
205 p. ; 21 cm. F***
Bibliography: p. 203-204.
30.00F
1. Quiberon, France—History. 2. France—History—Revolution, 1795. 3. Chouans. I. Title.
DC195.Q5R64 944'.13 79-354033
79 MARC

Roincé, Job de.
Histoires de chouans / Job de Roincé. — Paris : F. Lanore, 1978-
v. ; 20 cm. F78-12318
Bibliography: v. 1, p. [193]-194.
30.00F (v. 1)
1. Chouans. I. Title.
DC218.5.R64 944.04 79-371476
79 MARC

—BIBLIOGRAPHY

Lemière, Edmond.
Bibliographie de la contre-révolution dans les provinces de l'ouest ou des guerres de la Vendée & de la chouannerie (1793, 1815, 1832) / par Edmond Lemière ; continuée par son fils Yves Lemière. — Nantes : Librairie nantaise, 1976.
583 p. ; 23 cm. F***
Reprint of the 1904 ed. published by F. Guyon, Saint-Brieuc.
1. Vendean War, 1793-1800—Bibliography. 2. Chouans—Bibliography. 3. France—History—Revolution, 1789-1799—Bibliography. I. Lemière, Yves. II. Title.
Z2178.L45 1976 016.944'04 79-367932
[DC218] 79 MARC

—BIOGRAPHY

Dupont, Étienne, 1864-1928.
La chouannerie dans l'Ouest : Mesdames Fleurdelys / Étienne Dupont. — Coutances : OCEP, c1977.
155 p. ; 19 cm. F***
Bibliography: p. 151-153.
ISBN 2-7134-0019-8
1. Chouans—Biography. 2. Women—France—Biography. I. Title.
DC218.1.D86 1977 920.72'0944 78-401044
79 MARC

—FICTION

Duvivier de Fortemps de Loneux, Jean Luc.
Ramibuchaille, chouan ardennais / Jean-Luc Duvivier de Fortemps de Loneux ; préf. de Jean-Pierre Lambot. — Bruxelles [r. de Brabant 91] : p. de Méyère, [1977]
184 p. : ill. ; 21 cm. Be78-5
285F
1. Ardennes—History—Fiction. 2. Chouans—Fiction. I. Title.
PQ2664.U92R3 843'.9'14 78-388436
•78 MARC

CHOUDHURY, PRASANNALAL, 1901-

—ADDRESSES, ESSAYS, LECTURES

(Kabi Prasannalāla)
কবি প্রসন্নলাল : অভিনন্দন গ্রন্থ / সম্পাদক অক্ষয় কুমার মিশ্র. — 1. সংস্করণ. — বৰপেটা : কমল বুক ষ্টল, 1978.
119 p. ; 23 cm.
In Assamese
List of works of P. Choudhury: p. 119
Rs6.00
1. Choudhury, Prasannalal, 1901- —Addresses, essays, lectures. I. Choudhury, Prasannalal, 1901- II. Miśra, Akshaya Kumāra, 1944-
PK1569.C58Z74 78-906357

CHOUDHURY, SASANKAMOHAN

—ANECDOTES

Choudhury, Sasankamohan.
(Bārabelā baiṭhaka)
বারবেলা বৈঠক. [লেখক] শশাঙ্কমোহন চৌধুরী. কলিকাতা, জিজ্ঞাসা [1972]
155 p. 23 cm. Rs6.00
In Bengali.
1. Choudhury, Sasankamohan—Anecdotes. 2. Authors, Bengali—20th century—Biography. I. Title.
PK1730.17.O865Z462 73-901903

CHOURAQUI, ANDRÉ, 1917-

—ADDRESSES, ESSAYS, LECTURES

Chouraqui, André, 1917-
Ce que je crois / André Chouraqui. — Paris : B. Grasset, c1979.
345 p. ; 21 cm. F***
ISBN 2-246-00726-7 : 55.00F
1. Chouraqui, André, 1917- —Addresses, essays, lectures. 2. Judaism—Addresses, essays, lectures. 3. Israel—Biography—Addresses, essays, lectures. I. Title.
DS126.6.C47A33 956.94'05'0924 79-367943
79 MARC

CHOUTEAU FAMILY

Cunningham, Mary B 1914-
The founding family of St. Louis / Mary B. Cunningham, Jeanne C. Blythe. — St. Louis : Midwest Technical Publications, c1977.
x, 416 p. : ill. ; 24 cm.
Includes index.
ISBN 0-913656-16-X
1. Chouteau family. I. Blythe, Jeanne C., 1908- joint author. II. Title.
CS71.C552 1977 929'.2'0973 77-94385
79 MARC

CHOW, CHING LIE, 1936-

Chow, Ching Lie, 1936-
Journey in tears : memory of a girlhood in China / by Chow Ching-li ; translated from the French by Abby Israel. — New York : McGraw-Hill, c1978.
262 p., [4] leaves of pplates : ill. ; 24 cm.
Translation of Le palanquin des larmes.
ISBN 0-07-010818-8 : $11.95
1. Chow, Ching Lie, 1936- . 2. China—Social life and customs. 3. China—Biography. I. Title.
CT1828.C544A3413 951.05'092'4 78-18171
78 MARC

Chow, Ching Lie, 1936-
Le Palanquin des larmes / Chow Ching Lie ; récit recueilli par Georges Walter .... — Paris : France loisirs, 1976.
371 p., [8] p. of plates ; 25 cm. F76-18049
30.00F
1. Chow, Ching Lie, 1936- . 2. China—Social life and customs. I. Walter, Georges. II. Title.
CT1828.C544A34 1976 951.05'092'4 79-372055
79 MARC

CHOW, HUNG-CHING, 1902-1957

Chow, Hung-ching, 1902-1957.
The collected papers of Hung-ching Chow. — Taipei, Taiwan : Institute of Mathematics, Academia Sinica, 1959.
ix, 181 p., [1] leaf of plates : port. ; 26 cm.
Includes title in Chinese.
Includes bibliographies.
1. Mathematics—Collected works. 2. Power series—Collected works. 3. Summability theory—Collected works. 4. Chow, Hung-ching, 1902-1957. I. Chung yang yen chiu yüan. Shu hsüeh chiu so. II. Title.
QA3.C495 1959 510'.8 78-322947
79 MARC

CHOW CHOWS (DOGS)

Green, Honor.
The chow chow / by Honor Green. — Leicester : K and R Books, 1975.
100 p., [6] p. of plates : ill. (some col.), ports. ; 20 cm. — (The Dog owner's library) GB78-09636
Bibliography: p. 96-97.
Includes index.
ISBN 0-903264-03-X : £2.50
1. Chow chows (Dogs) I. Title.
SF429.C5G73 1975 636.7'2 78-317616
78 MARC

CHOZENKAN

—HISTORY—SOURCES

(Chōzenkan gakujuku shiryō)
長善館学塾史料. — [新潟] : 新潟県教育委員会, 昭和49 [1974]
2 v. : ill. ; 26 cm. — (新潟県文化財調査報告書 ; 第14) Ja 75-2133
In colophon: Niigata-ken bunkazai nempō ; dai 14.
CONTENTS: 1. 漢詩集·漢詩文·書輸集—2. 日記·学規·門人姓名簿
1. Chōzenkan—History—Sources. I. Niigata, Japan (Prefecture). Kyōiku Iinkai. II. Series: Niigata-ken bunkazai chōsa hōkokusho ; dai 14.
LG275.Y6993C47 79-807235

—HISTORY—SOURCES —BIBLIOGRAPHY—CATALOGS

Niigata Toshokan.
(Chōzenkan gakujuku shiryō mokuroku)
長善館学塾資料目録 : 新潟県立新潟図書館所蔵. — 新潟 : 新潟県立新潟図書館, 昭和49 [1974]
46 p. ; 26 cm. Ja 75-2128
Cover title.
1. Chōzenkan—History—Sources—Bibliography—Catalogs. 2. Niigata Toshokan—Catalogs. I. Chōzenkan. II. Title.
Z5816.C48N5 1974 79-807516
[LG275.Y6993]

CHRAPOWIECKI, JAN ANTONI, 1612-1685

Chrapowicki, Jan Antoni, 1612-1685.
Diariusz / Jan Antoni Chrapowicki ; oprac. i wstępem poprzedził Tadeusz Wasilewski. — Wyd. 1. — Warszawa : Pax, 1978-
(Łódź : ŁZG)
v. : ill. ; 21 cm.
Includes bibliographical references and indexes.
CONTENTS: cz. 1. Lata 1656-1664.
zł120.00 (v. 1)
1. Chrapowiecki, Jan Antoni, 1612-1685. 2. Politicians—Poland—Biography. 3. Poland—Gentry—Biography. I. Wasilewski, Tadeusz. II. Title.
DK4307.5.C48A33 79-371788
79 MARC

CHRESTIEN DE TROYES, 12TH CENT. CHEVALIER AU LYON

Lonigan, Paul R 1935-
Chretien's Yvain : a study of meaning through style / by Paul R. Lonigan. — Ann Arbor : Published for the Graduate School and University Center, City University of New York by University Microfilms International, c1978.
vii, 152 p. ; 26 cm. — (Monograph publishing : Sponsor series)
Bibliography: p. 149-152.
ISBN 0-8357-0369-1 : $12.50
1. Chrestien de Troyes, 12th cent. Chevalier au lyon. 2. Ywain—Romances—History and criticism. I. New York (City). City University of New York. Graduate School/University Center. II. Title.
PQ1445.Y83L66 841'.1 78-27707
78 MARC

CHRISTIEN DE TROYES, 12TH CENT.
CHEVALIER AU LYON (Continued)

Wigger, Marianne, Dr. phil.
Tempora in Chrétiens "Yvain" : e. textlinguist. Unters. / Marianne Wigger. — Frankfurt am Main ; Bern ; Las Vegas : Lang, 1978.
217 p. ; 21 cm. — (Studia romanica et linguistica ; 6) GFR78-A
Bibliography: p. 203-217.
ISBN 3-261-02610-3
1. Chrestien de Troyes, 12th cent. Chevalier au lyon. 2. Chrestien de Troyes, 12th cent.—Language—Grammar. 3. French language—To 1500—Tense. I. Title. II. Series.
PQ1445.Y83W48 841'.1 78-395795
79 MARC

CHRESTIEN DE TROYES, 12TH CENT.
CLIGÉS

Freeman, Michelle A
The poetics of translatio studii and conjointure : Chrétien de Troyes's Cligés / by Michelle A. Freeman. — Lexington, Ky. : French Forum, c1979.
199 p. ; 23 cm. — (French forum monographs ; 12)
A revision of the author's thesis, Princeton, 1976, published under the title: Translatio studii and romance composition.
Bibliography: p. ₍193₎-199.
1. Chrestien de Troyes, 12th cent. Cligés. I. Title. II. Series.
PQ1445.C6F7 1979 841'.1 78-54262
00 MARC

CHRESTIEN DE TRYOES, 12TH CENTURY
PERCEVAL LE GALLOIS

Le Rider, Paule.
Le chevalier dans le conte du Graal de Chrétien de Troyes / Paule Le Rider. — Paris : Société d'édition d'enseignement supérieur, c1978.
389 p. ; 18 cm. — (Bibliothèque du Moyen Age) (Collection Littérature) F***
French or Latin.
Includes bibliographical references and indexes.
ISBN 2-7181-5015-7 : 46.00F
1. Chréstien de Troyes, 12th cent. Perceval le Gallois. 2. Grail—Legends—History and criticism. I. Title.
PQ1445.P7L4 841'.1 78-367558
78 MARC

CHRESTIEN DE TROYES, 12TH CENT.

—LANGUAGE—GRAMMAR

Wigger, Marianne, Dr. phil.
Tempora in Chrétiens "Yvain" : e. textlinguist. Unters. / Marianne Wigger. — Frankfurt am Main ; Bern ; Las Vegas : Lang, 1978.
217 p. ; 21 cm. — (Studia romanica et linguistica ; 6) GFR78-A
Bibliography: p. 203-217.
ISBN 3-261-02610-3
1. Chrestien de Troyes, 12th cent. Chevalier au lyon. 2. Chrestien de Troyes, 12th cent.—Language—Grammar. 3. French language—To 1500—Tense. I. Title. II. Series.
PQ1445.Y83W48 841'.1 78-395795
79 MARC

—STYLE

Hilka, Alfons, 1877-1939.
Die direkte Rede als stilistisches Kunstmittel in den Romanen des Kristian von Troyes / Alfons Hilka. — Genève : Slatkine Reprints, 1979.
177 p. ; 23 cm. Sw***
Some text in Old French.
Originally presented as part of the author's thesis, Breslau, 1902.
Reprint of the 1903 ed. published by M. Niemeyer, Halle.
Includes bibliographical references.
1. Chrestien de Troyes, 12th cent.—Style. 2. Dialogue. I. Title.
PQ1451.H5 1979 841'.1 79-370250
79 MARC

CHRESTOMATHIES see Anthologies; Readers; and subdivision Readers under names of the other languages

CHRISM

Dionysius Areopagita, Pseudo-
Das Sakrament der Myron-Weihe in der Schrift De ecclesiastica hierarchia des Pseudo-Dionysios Areopagita : in syr. Übers. u. Kommentaren / von Werner Strothmann. — 1. Aufl. — Wiesbaden : Harrassowitz, 1977-1978.
2 v. ; 24 cm. — (Göttinger Orientforschungen : Reihe 1, Syriaca ; Bd. 15) GFR77-A
Includes bibliographical references with indexes in Syriac and Greek.
CONTENTS: T. 1. Syrischer Text mit Wortverzeichnissen.—T. 2. Einführung, Übersetzung.
ISBN 3-447-01906-9 (v. 1) : DM38.00 (v. 1)
1. Chrism. I. Strothmann, Werner, 1907- . II. Title. III. Series.
BR65.D63D4215 1977 78-391787
79 MARC

Syrische Hymnen zur Myron-Weihe / hrsg., übers. u. mit e. vollst. Wortverz. vers. von Werner Strothmann. — 1. Aufl. — Wiesbaden : Harrassowitz, 1978.
xxv, 145 p. ; 24 cm. — (Göttinger Orientforschungen : Reihe 1, Syriaca ; Bd. 16) GFR78-A41
German and Syriac.
From Jacobite and Catholic sources.
Includes bibliographical references and indexes.
ISBN 3-447-01957-3 : DM32.00
1. Chrism. 2. Hymns, Syriac. 3. Hymns, German—Translations from Syriac. I. Strothmann, Werner, 1907- . II. Jacobite Church. Liturgy and ritual. 'Al kudash murun. German & Syriac. Selections. 1978. III. Catholic Church. Liturgy and ritual. IV. Series.
BV196.O5S97 79-340934
79 MARC

CHRISMAN FAMILY

Miletich, Ruby Pearl Crissman, 1924-
Our McKaughans/McCoins and Crissmans in 1978 / by Ruby Pearl Crissman Miletich, Gladys Moverene Crissman Guydes, Minnie Othello Crissman Culler. — Fayetteville, NC : Fayetteville Office Supply Co., 1978.
x, 85, ₍48₎ p. : ill. ; 28 cm.
Bibliography: p. ₍48₎ (3d group)
1. McKaughan family. 2. Chrisman family. I. Guydes, Gladys Moverene Crissman, 1914- joint author. II. Culler, Minnie Othello Crissman, 1911-1977, joint author. III. Title.
CS71.M15271978 929'.2'0973 79-63031
79 MARC

CHRIST see Jesus Christ

CHRIST CHURCH, ALBEMARLE COUNTY, VA.

A History of Christ Church, Glendower, and the early history of St. Anne's Parish, Albemarle County, Virginia : one hundred and twenty fifth anniversary, June 30, 1957. — ₍s.l. : s.n., 1957₎ (₍s.l. : King Lindsay Print. Corp.₎)
67 p., ₍2₎ leaves of plates : ill. ; 24 cm.
Cover title.
CONTENTS: Langhorne, E. C. A history of Christ Church, Glendower.—Brydon, G. M. A sketch of the early history of Saint Anne's Parish in Albemarle County, Virginia.
1. Christ Church, Albemarle County, Va. 2. St. Anne's Parish, Albemarle County, Va. I. Langhorne, Elizabeth Coles. A history of Christ Church, Glendower. 1957. II. Brydon, George MacLaren, 1875- A sketch of the early history of Saint Anne's Parish in Albemarle County, Virginia. 1957.
BX5980.G48C483 283'.755'482 79-311458
79 MARC

CHRIST COMMUNAL ORGANIZATION

Richardson, James T 1941-
Organized miracles : a study of contemporary, youth, communal, fundamentalist organization / James T. Richardson, Mary White Stewart, Robert B. Simmonds. — New Brunswick, N.J. : Transaction Books, c1979.
xxviii, 368 p. : ill. ; 24 cm.
Bibliography: p. 349-361.
Includes index.
ISBN 0-87855-284-7 : $16.95
1. Christ Communal Organization. I. Stewart, Mary White, 1945- joint author. II. Simmonds, Robert B., 1945- joint author. III. Title.
BV4407.3.R52 301.5'8 78-55937
78 MARC

CHRISTADELPHIANS

—GREAT BRITAIN

Wilson, Bryan R
Sects and society : a sociological study of three religious groups in Britain / Bryan R. Wilson. — Westport, Conn. : Greenwood Press, 1978, c1961.
397 p. ; 23 cm.
Reprint of the ed. published by W. Heinemann, London, in series: Heinemann books on sociology.
Bibliography: p. 374-388.
Includes index.
ISBN 0-313-20439-X
1. Elim Foursquare Gospel Alliance. 2. Christian Science—Great Britain. 3. Christadelphians—Great Britain. 4. Church and social problems—Great Britain. I. Title.
₍BX7435.E54W5 1978₎ 289 78-5993
78 MARC

CHRISTCHURCH HARMONIC SOCIETY

Pritchard, Brian W
Words and music : a jubilee history of the Christchurch Harmonic Society / by Brian Pritchard. — Christchurch, N.Z. : Pegasus, 1977.
142 p., ₍8₎ leaves of plates : ill. ; 22 cm. NZ***
Includes index.
1. Christchurch Harmonic Society. I. Title.
ML28.C45C56 784'.06'293155 79-302065
79 MARC

CHRISTCHURCH TEACHERS COLLEGE

Christchurch Teachers College centennial 1977 / editor, Michael Harlow. — Christchurch : Christchurch Teachers College Centennial Committee, 1977.
141 p. : ill., facsim., music, ports. ; 21 cm. NZ77-9
Cover title: Centennial publication, Christchurch Teachers College.
$3.95
1. Christchurch Teachers College. I. Harlow, Michael. II. Title: Centennial publication, Christchurch Teachers College.
LB2139.C483C48 370'.73'09931 78-323793
79 MARC

CHRISTELIJK-HISTORISCHE UNIE

—ADDRESSES, ESSAYS, LECTURES

Dulst, A J van, 1929-
Varen we wèl in de welvaart? ₍Door₎ A. J. van Dulst. Kampen, Kok, 1971.
135 p. 21 cm. fl 9.75 Ne71-21
ISBN 9024222733
1. Netherlands—Social conditions—Addresses, essays, lectures. 2. Church and social probems—Netherlands—Addresses, essays, lectures. 3. Christelijk-Historische Unie—Addresses, essays, lectures. I. Title.
HN517.D84 78-355005
78 MARC

CHRISTENBERRY, WILLIAM, 1936-

Livingston, Jane.
William Christenberry, color photographs : December 21, 1978-February 11, 1979, The Corcoran Gallery of Art, Washington, D.C. — Washington : The Gallery, ₍1979₎
₍16₎ p. : col. ill. ; 21 cm.
Bibliography: p. ₍15₎
1. Photography, Artistic—Exhibitions. 2. Christenberry, William, 1936- . I. Christenberry, William, 1936- . II. Corcoran Gallery of Art, Washington, D.C. III. Title.
TR647.C47 1979 779'.092'4 78-65911
00 MARC

CHRISTENING see Baptism

CHRISTENSEN, AKSEL ERHARDT, 1906-

Christensen, Aksel Erhardt, 1906-
Danmark, Norden og Østersøen : udvalgte afhandlinger / Aksel E. Christensen. — København : Danske historiske forening, 1976.
282 p., ₍1₎ leaf of plates : ill. ; 24 cm. D***
"Udgivet på halvfjerdsårsdagen den 11. september 1976."
Danish, English, French or German; summaries in English.
"Aksel E. Christensens forfatterskab 1928-1976": p. ₍277₎-282.
Includes bibliographical references.
ISBN 8787462052
1. Denmark—History—Addresses, essays, lectures. 2. Scandinavia—History—Addresses, essays, lectures. 3. Baltic Sea—Addresses, essays, lectures. 4. Christensen, Aksel Erhardt, 1906- —Addresses, essays, lectures. I. Title.
DL148.C47 948 78-399908
79 MARC

CHRISTENSEN FAMILY

Dansk slægtsforskning.
Slægtsbog for efterkommere efter Peder Christensen, gårdejer i Midtgård i Videbæk, Vorgod sogn, født 1799. — Fredericia : Dansk slægtsforskning, 1975.
ca. 250 leaves : ill. ; 28 cm. D***
"Bibliotekseksemplar."
1. Christensen family. 2. Denmark—Genealogy. I. Title.
CS909.C56 1975 78-397876
79 MARC

CHRISTENSEN FAMILY (Continued)

Dansk slægtsforskning.
Slægtsbog for efterkommere efter Peder Christensen, gårdejer i Ø. Mårupgård, Fjaltring Sogn, født 1807 / [Dansk Slægtsforskning]. — Fredericia : Dansk Slægtsforskning, 1975.
[96] leaves, [14] leaves of plates : ill. ; 28 cm. D78-25
"Biblioteksekseemplar."
Not in trade
1. Christensen family. 2. Denmark—Genealogy. I. Title: Slægtsbog gor efterkommere efter Peder Christensen.
CS909.C56 1975a 78-393098
79 MARC

Nordisk slægtsforskning.
En slægt Christensen fra Bejstrup sogn (Ø. Han herred) : om Christen Michelsen, født 1824, og hans hustru, deres forfædre og efterkommere. — Skals : Nordisk Slægtsforskning, 1976.
ca. 150 leaves : ill. ; 29 cm. D76-44
One fold. geneal. table inserted.
Not in trade.
1. Christensen family. 2. Denmark—Genealogy. I. Title.
CS909.C56 1976 79-338005
79 MARC

Nordisk slægtsforskning.
En slægt Christiansen fra Bastholm, Vrejlev sogn : om Christian Pedersen, født 1918, og hustru, deres forfædre og efterkommere. — Skals : Nordisk Slægtsforskning, 1976.
ca.100 leaves : ill. ; 29 cm. D76-44
Fold. geneal. table inserted.
Not in trade.
1. Christensen family. 2. Denmark—Genealogy. I. Title.
CS909.C56 1976a 79-338780
79 MARC

Slægtsarkivet, Viborg, Denmark.
Slægtsbog familien Christiansen : omhandler skolelærer i Svingelbjerg i Vesterbølle sogn og indremissionær i Viborg Niels Christiansen, 1852-1935 og hustru Mette Kirstine Christensen, 1858-1940, deres forfædre og efterkommere / [Slægtsarkivet]. — Viborg (Falkevej 19, 8800) : Slægtsarkivet, [1978].
[100] leaves, [5] leaves of plates : ill. ; 30 cm. D78-35
Cover title.
Photoreprint of Bet'e Niels, Stu'e Niels and Gammel Niels by N. Christiansen, published 1927-29 by O. Lohse, Copenhagen.
Geneal. table inserted.
1. Christensen family. 2. Denmark—Genealogy. I. Christiansen, Niels, 1852-1935. II. Title.
CS909.C56 1978 78-391164
78 MARC

Slægtsarkivet, Viborg, Denmark.
Slægtsbog familien Kristensen : omhandler brødrene Morten Kristensen, f. 18. september 1849, og Niels Christian Kristensen, f. 17. april 1859, begge gårdsgårdmænd i Svingelbjerg, Vesterbølle sogn : deres forfædre og efterkommere / [Slægtsarkivet]. — Viborg (Falkevej 19, 8800) : Slægtsarkivet, [1978]
172 leaves, [25] leaves of plates : ill. ; 30 cm. D78-35
Cover title.
Four geneal. tables inserted.
kr300.00
1. Christensen family. 2. Denmark—Genealogy. I. Title.
CS909.C56 1978a 79-350696
79 MARC

Slægtsarkivet, Viborg, Denmark.
Slægtsbog for familien Bønlykke (Bønnelykke) : anetavle for samt efterkommere efter gårdejer i Hals, senere husmand i Gandrup, Vester Hassing sogn, Niels Christian Christiansen, f. 1805 i Brovst sogn, død 1885 i Gandrup, og hustru Mariane Olesdatter, f. 1813 i Gøttrup sogn, død 1882 i Gandrup / [Slægtsarkivet]. — Viborg (Falkevej 19, 8800) : Slægtsarkivet, [1978]
186 leaves, 13 leaves of plates : ill. ; 30 cm. D78-45
Cover title.
3 genealogical tables inserted.
Includes index.
1. Bønlykke family. 2. Christensen family. 3. Denmark—Genealogy. I. Title.
CS909.B7335 1978 79-380576
79 MARC

CHRISTIAN II, KING OF DENMARK, 1481-1559

Huitfeldt, Arild, 1546-1609.
Christian II's [i.e. den andens] historie / Arild Huitfeldt. — København : Rosenkilde og Bagger, 1976.
340 p. ; 22 cm. — (His Danmarks riges krønike) D***
Reprint of the 1596 ed. printed in Copenhagen by M. Vingaard, with original t.p.: Historiske beskriffuelse om hues sig haffuer tildragit under Kong Christiern den anden.
Includes index.
1. Denmark—History—1448-1660. 2. Christian II, King of Denmark, 1481-1559. I. Title. II. Series.
DL183.8.H83 1976 78-371551
78 MARC

Venge, Mikael.
Når vinden føjer sig : spillet om magten i Danmark marts-december 1523 / Mikael Venge. — [Odense] : Odense universitetsforlag, 1977.
179 p. ; 23 cm. — (Odense University studies in history and social sciences ; v. 47) D***
Thesis—Odense.
Summary in German.
Bibliography: p. 165-[168]
Includes index.
ISBN 8774921916 : kr80.00
1. Denmark—History—1448-1660. 2. Christian II, King of Denmark, 1481-1559. I. Title. II. Series: Odense universitet. Studies in history and social sciences ; v. 47.
DL183.8.V47 78-394113
79 MARC

CHRISTIAN III, KING OF DENMARK, 1503-1559

Huitfeldt, Arild, 1546-1609.
Christian III's [i.e. den tredies] historie / Arild Huitfeldt. — København : Rosenkilde og Bagger, 1976.
ca 300 p. ; 22 cm. — (His Danmarks riges krønike) D***
Reprint of the 1595 ed. printed in Copenhagen by M. Vingaard, with original t.p.: En kaart historiske beskriffuelse paa hues merckeligt som sig aarlige under Kong Christian den tredie ...
Includes index.
1. Denmark—History—Christian III, 1534-1559. 2. Christian III, King of Denmark, 1503-1559. I. Title. II. Series.
DL187.H84 1976 78-371552
78 MARC

Lausten, Martin Schwarz.
Religion og politik : studier i Christian IIIs forhold til det tyske rige i tiden 1544-1559 / Martin Schwarz Lausten. — København : Akademisk Forlag : [eksp. DBK], 1977.
374 p. ; 24 cm. — (Kirkehistoriske studier) D77-49
Thesis—Copenhagen.
Summary in German.
Bibliography: p. 349-365.
Includes index.
ISBN 8750017721 : kr118.00
1. Denmark—History—Christian III, 1534-1559. 2. Denmark—Foreign relations—Germany. 3. Germany—Foreign relations—Denmark. 4. Germany—History—Charles V, 1519-1556. 5. Christian III, King of Denmark, 1503-1559. I. Title. II. Series: Kirkehistoriske studier (Copenhagen, 1974-)
DL187.L38 79-346779
79 MARC

CHRISTIAN VII, KING OF DENMARK, 1749-1808

—FICTION

Hebsgaard, Mark.
Det eneste hoved / [af] Mark Hebsgaard. — København : Rhodos, c1978.
185 p. ; 22 cm. D78-47
ISBN 8774966596 : kr78.00
1. Frederik V, King of Denmark, 1723-1766—Fiction. 2. Christian VII, King of Denmark, 1749-1808—Fiction. 3. Denmark—History—1660-1814—Fiction. I. Title.
PT8176.18.E2E5 839.8'1'374 79-342682
79 MARC

—PERSONALITY

Christiansen, Viggo, 1867-1939.
Christian den VII's sindssygdom / [af] Viggo Christiansen. — Genoptrykt efter udgaven 1906 / med forord af Johan Schioldann Nielsen og en efterskrift af Niels Juel-Nielsen. — [Odense] : Odense Universitetsforlag, 1978.
73 p. ; 19 cm. D78-43
Reprint of the 1906 ed. published by Gyldendal, Copenhagen.
Includes bibliographical references.
ISBN 8774922599 : kr53.10
1. Christian VII, King of Denmark, 1749-1808—Personality. 2. Schizophrenia—Biography. 3. Denmark—Kings and rulers—Biography. I. Schioldann Nielsen, Johan. II. Juel-Nielsen, Niels. III. Title.
DL199.3.C48 1978 79-381834
79 MARC

CHRISTIAN, ANTON, 1940-

—EXHIBITIONS

Christian, Anton, 1940-
Anton Christian : Zeichnungen und Objekte : [Ausstellung] 12.1.-26.2.1978, Städtisches Museum Leverkusen, Schloss Morsbroich : [Katalog]. — Leverkusen : Städtisches Museum, [1978]
[44] p. : chiefly ill. (some col.) ; 20 x 21 cm. GFR***
1. Christian, Anton, 1940- —Exhibitions. I. Leverkusen. Städtisches Museum.
N6811.5.C48A4 1978 78-394253
79 MARC

CHRISTIAN, DÉSIRÉ, 1846-1907

Traub, Jules S
The glass of Désiré Christian, ghost for Gallé / Jules S. Traub. — Chicago : Art Glass Exchange, c1978.
143 p. : ill. (some col.) ; 31 cm.
Bibliography: p. 137-138.
ISBN 0-932988-01-6 : $39.50
1. Christian, Désiré, 1846-1907. 2. Gallé, Charles Martin Emile, 1846-1904. 3. Glassware—France—History—19th century. I. Title.
NK5198.C44T72 748.2'94 78-73256
79 MARC

CHRISTIAN, ROY, 1914-

Christian, Roy, 1914-
Derbyshire / Roy Christian. — London : Batsford, 1978.
191 p., [8] leaves of plates : ill., map ; 25 cm. GB***
Includes index.
ISBN 0-7134-1295-X : £5.95
1. Derbyshire, Eng.—Description and travel—Guide-books. 2. Christian, Roy, 1914- I. Title.
DA670.D43C42 942.5'1 78-321094
79 MARC

CHRISTIAN AND MISSIONARY ALLIANCE

—GOVERNMENT

Bailey, Keith M
Servants in charge : a training manual for elders and deacons / Keith M. Bailey. — Harrisburg, PA : Christian Publications, c1979.
123 p. ; 19 cm. — (Christian life & ministry series)
Includes bibliographical references.
1. Elders (Church officers) 2. Deacons. 3. Christian and Missionary Alliance—Government. I. Title.
BV680.B34 253 79-64968
00 MARC

CHRISTIAN ANTIQUITIES

see also Architecture, Gothic; Catacombs; Church furniture; Church vestments; Crosses; Fasts and feasts; Heraldry, Sacred; Sepulchral monuments

Bagin, Anton.
Kresťanská archeológia. Úvod do staroslovienčiny, časť historická / Anton Bagin. — Bratislava : Spolok sv. Vojtecha v CN, 1977.
175 p. ; 29 cm. CzS77
Cover title.
At head of title: Rímskokatolícka Cyrilometodejská bohoslovecká fakulta v Bratislave.
Includes bibliographies.
Kčs58.00
1. Christian antiquities. 2. Church Slavic language. I. Bagin, Anton. Úvod do staroslovienčiny, časť historická. 1977. II. Title.
BR131.B26 79-379916
*79 MARC

—GEORGIA (TRANCAUCASIA)

Mepisashvili, Rusudan.
The arts of ancient Georgia / Rusudan Mepisashvili and Vakhtang Tsintsadze ; photographs by Rolf Schrade ; foreword by David M. Lang. — London : Thames and Hudson, 1979.
310 p. : ill. (some col.) ; 28 cm. GB***
Translation of Die Kunst des Alten Georgien.
Bibliography: p. 299-[304]
Includes index.
ISBN 0-500-23289-X : £12.00
1. Art, Georgian. 2. Art—Georgia (Trancaucasia) 3. Georgia (Trancaucasia)—Antiquities. 4. Christian antiquities—Georgia (Trancaucasia) I. Tsintsadze, Vakhtang, joint author. II. Schrade, Rolf. III. Title.
N7292.9.M4613 709'.47'95 79-309720
79 MARC

CHRISTIAN ANTIQUITIES (Continued)

—GREECE, MODERN

Pallas, D
Les monuments paléochrétiens de Grèce découverts de 1959 à 1973 / D. Pallas. — Città del Vaticano : Pontificio Istituto di archeologia cristiana, 1977.
327 p., [2] fold. leaves of plates : ill. ; 24 cm. — (Sussidi allo studio delle antichità cristiane ; 5) VC***
Date in title printed as 1958, corrected by label.
Includes bibliographical references and indexes.
Errata slip inserted.
1. Greece, Modern—Antiquities. 2. Christian antiquities—Greece, Modern. 3. Excavations (Archaeology)—Greece, Modern. 4. Church history—Primitive and early church, ca. 30-600. I. Title. II. Series.
DF129.P34 937 78-394663
79 MARC

—IRELAND

Streit, Jakob.
Sonne und Kreuz : Irland zwischen Megalithkultur u. frühem Christentum / Jakob Streit. — Stuttgart : Verlag Freies Geistesleben, 1977.
208 p. : ill., maps ; 24 cm. GFR78-A
Includes bibliographical references.
ISBN 3-7725-0681-X : DM42.00
1. Ireland—Church history—To 1172. 2. Ireland—Civilization. 3. Missions—Europe. 4. Megalithic monuments—Ireland. 5. Crosses—Ireland. 6. Christian antiquities—Ireland. 7. Ireland—Antiquities. I. Title.
BR794.S85 78-382545
78 MARC

—ITALY—AQUILEIA

Fink, Josef, 1912-
Der Ursprung der ältesten Kirchen am Domplatz von Aquileja. Münster, Böhlau-Verlag, 1954.
89 p. plates, plans. 23 cm. (Münstersche Forschungen, Heft 7)
Habilitationsschrift—Münster.
Bibliographical footnotes.
1. Aquileia—Churches. 2. Christian antiquities—Italy—Aquileia. I. Title.
NA5621.A6F5 55-25982
[r79]rev MARC

—ITALY—ROME (CITY)

Reekmans, Louis.
Die Situation der Katakombenforschung in Rom / Louis Reekmans. — Opladen : Westdeutscher Verlag, 1979.
65 p. : 30 ill., map ; 24 cm. — (Geisteswissenschaften, Vorträge - Rheinisch-Westfälische Akademie der Wissenschaften ; G233) GFR79-A
"231. Sitzung am 21. Juni 1978 in Düsseldorf."
Includes bibliographical references.
ISBN 3-531-07233-1
1. Catacombs—Italy—Rome (City) 2. Christian antiquities—Italy—Rome (City) 3. Excavations (Archaeology)—Italy—Rome (City) 4. Italy—Antiquities. I. Title. II. Series: Rheinisch-Westfälische Akademie der Wissenschaften. Geisteswissenschaften, Vorträge ; G 233.
DG807.4.R43 937'.6 79-365052
79 MARC

—JAPAN—TOKYO

Yamamoto, Hideteru.
(Edo Kirishitan yashiki no shiseki)
江戸切支丹屋敷の史蹟 / 山本秀煌著. — 東京：イデア書院：發賣所 慶文堂, 大正13 [1924]
7, 2, 163 p., [2] leaves of plates : ill. ; 20 cm. — (イデア叢書 ; 第1篇)
Series romanized: Idea sōsho.
1. Catholic Church in Japan. 2. Christian antiquities—Japan—Tokyo. 3. Japan—Church history—To 1868. I. Title.
BX1668.Y34 79-800395

—OHRID LAKE REGION, MACEDONIA AND ALBANIA

Bitrakova-Grozdanova, Vera.
Monuments paléochrétiens de la région d'Ohrid = Starohristijanski spomenjci vo Ohridsko / Vera Bitrakova Grozdanova ; grafička skica Gorgi Krstevski ; fotograf Petar Alčev ; [prevod na francuski Klimentina Hadži-Lega ; likovna oprema Niko Tozi]. — Ohrid : Zavod za zaštita na spomenicite na kulturata : Naroden muzej, 1975.
108 p., [23] leaves of plates : col. ill. ; 26 cm. Yu76-5765
Macedonian and French on opposite pages.
Part of illustrative matter in pocket.
Includes bibliographical references.
1. Christian antiquities—Ohrid Lake region, Macedonia and Albania. 2. Ohrid Lake region, Macedonia and Albania—Antiquities. I. Title. II. Title: Starohristijanski spomenjci vo Ohridsko.
BR133.036B5714 726.5'0938'1 79-343059
79 MARC

—PALESTINE--GAZA STRIP

Chachanidze, Vasiliĭ Illarionovich.
(Petr Iver i arkheologicheskie raskopki Gruzinskogo monastyri͡a v Ierusalime)
Петр Ивер и археологические раскопки Грузинского монастыря в Иерусалиме / В. И. Чачанадзе. — Тбилиси : Мецниереба, 1977.
235 p., [1] leaf of plates : ill. ; 22 cm. USSR 77
On leaf preceding t. p.: Akademii͡a nauk Gruzinskoĭ SSR. Institut filosofii.
Added t. p. in English and Georgian; English title: Peter the Iberian and archaeological excavations of the Georgian monastery in Jerusalem; Georgian title: Petre Iberieli da K'art'uli monastris ark'eologiuri gat'xrebi Ierusalimši.
Text in Russian with contributions in Italian, French, and English; summary in English.
Errata slip inserted.
Includes bibliographical references.
1.75rub
1. Petrus Iberus, Bp. of Majuma, 411?-491? 2. Bishop—Palestine—Gaza Strip—Biography. 3. Christian antiquities—Palestine—Gaza Strip. 4. Excavations (Archaeology)—Palestine—Gaza Strip. 5. Gaza Strip — Antiquities. 6. Palestine — Antiquities. 7. Gaza Strip—Biography. I. Title. II. Title: Peter the Iberian and archaeological excavations of the Georgian monastery in Jerusalem. III. Title: Petre Iberieli da K'art'uli monastris ark'eologiuri gat'xrebi Ierusalimši.
BR1720.P43C45 79-410910

—RUSSIA—EXHIBITIONS

Redin, Egor Kuz'mich, 1863-1908.
(Katalog dvi͡enadt͡satago Arkheologicheskago s"i͡ezda v g. Khar'kovi͡e)
Каталогъ XII Археологическаго съѣзда въ г. Харьковѣ : Отдѣлъ церковныхъ древностей / составилъ Е. К. Рѣдинъ. — Харьковъ : Тип. Губ. Правленія, 1902.
iii, 162, 28 p. ; 24 cm.
Includes index.
Exhibition catalog.
1. Church decoration and ornament—Russia—Exhibitions. 2. Orthodox Eastern Church and art—Exhibitions. 3. Christian antiquities—Russia—Exhibitions. I. Arkheologicheskiĭ s"ezd, 12th, Kharkov, 1903. II. Kharkov. Universitet. Istoriko-filologicheskoe obshchestvo. Otdel t͡serkovnykh drevnosteĭ. III. Title.
NK1653.R9R42 79-414620

—SPAIN—MOTA DEL MARQUÉS

Parrado del Olmo, Jesús María.
Antiguo Partido Judicial de Mota del Marqués / por Jesús María Parrado del Olmo. — [Valladolid] : Diputación Provincial de Valladolid, 1976.
350 p., [102] leaves of plates : ill. ; 24 cm. — (Catálogo monumental de la Provincia de Valladolid ; t. 9) Sp***
Bibliography: p. [329]-333.
Includes index.
ISBN 8450014182
1. Art, Spanish—Spain—Mota del Marqués. 2. Architecture—Spain—Mota del Marqués. 3. Christian antiquities—Spain—Mota del Marqués. 4. Mota del Marqués, Spain—Antiquities. I. Title. II. Series: García Chico, Esteban. Catálogo monumental de la Provincia de Valladolid ; t. 9.
N7109.V25G3 t. 9 78-399731
[N7109.M67] 79 MARC

—SWITZERLAND—GENEVA

Bonnet, Charles.
Les Premiers édifices chrétiens de la Madeleine à Genève / Charles Bonnet. — Genève : A. Jullien, 1977.
x, 224 p., [40] p. of plates : ill. ; 32 cm. — (Mémoires et documents publiés par la Société d'histoire et d'archéologie de Genève ; t. 8) Sw78-1177
Originally presented as the author's thesis, Université de Lyon II.
Bibliography: p. [199]-206.
Includes indexes.
75.00F
1. Église Sainte-Marie-Madeleine, Geneva. 2. Christian antiquities—Switzerland—Geneva. I. Title. II. Series: Société d'histoire et d'archéologie de Genève. Mémoires et documents ; t. 8.
DQ441.S68 t. 8 949.4'5'008 s 79-346540
[NA5851.G39] 79 MARC

—TURKEY—ANI

Cuneo, Paolo.
L'architettura della scuola regionale di Ani nell'Armenia medievale : relazione svolta nella seduta del 1° marzo 1976 / Paolo Cuneo. — Roma : Accademia Nazionale dei Lincei, 1977.
80 p., [35] leaves of plates (3 fold.) : ill. ; 27 cm. — (Problemi attuali di scienza e di cultura ; quaderno n. 235) It***
Includes bibliographical references.
L9500
1. Church architecture—Turkey—Ani. 2. Architecture, Medieval—Turkey—Ani. 3. Christian antiquities—Turkey—Ani. I. Title. II. Series: Accademia nazionale dei Lincei, Rome. Problemi attuali di scienza e di cultura ; quaderno n. 235.
AS222.R53 n. 235 085 s 78-378233
[NA5871.A54] 78 MARC

—TURKEY—PHRYGIA

Gibson, Elsa.
The "Christians for Christians" inscriptions of Phrygia : Greek texts, translation and commentary / by Elsa Gibson. — Missoula, Mont. : Scholars Press, 1978.
xiii, 161 p., [18] leaves of plates : ill. ; 24 cm. — (Harvard theological studies ; 32 ISSN 0073-0726)
Includes bibliographical references and index.
ISBN 0-89130-262-X pbk.
1. Inscriptions—Turkey—Phrygia. 2. Christian antiquities—Turkey—Phrygia. 3. Phrygia—Antiquities. 4. Turkey—Antiquities. I. Title. II. Series.
CN415.P4G52 939.2'6 78-12688
78 MARC

—WALES

Smith, J Beverley.
Abaty Ystrad Fflur, Dyfed / [gan] J. Beverley Smith, W. G. Thomas. — Llundain [i.e. London] : H.M.S.O., 1977.
32 p., fold. plate : ill., plan (tipped in) ; 21 cm. — (Adran yr Amgylchedd llawlyfr swyddogol) GB77-32966
At head of title: Paratowyd gan Adran yr Amgylchedd ar ran Ysgrifennydd Gwladol Cymru.
Bibliography: p. 32.
ISBN 0-11-670341-5 : £0.50
1. Strata Florida Abbey, Wales—History. 2. Christian antiquities—Wales. 3. Wales—Antiquities. I. Thomas, William Gwyn, joint author. II. Great Britain. Dept. of the Environment. III. Great Britain. Welsh Office. IV. Title. V. Series: Great Britain. Dept. of the Environment. Department of the Environment official handbook.
BR133.G72W347 78-390370
79 MARC

—YUGOSLAVIA—SERBIA
—HISTORIOGRAPHY
—EXHIBITIONS

(Izlozi Srpskog učenog društva)
Излози Српског ученог друштва : истраживања српске средњовековне уметности 1871-1884 / [текст Соња Богдановић, Љиљана Мишковић-Прелевић ; одговорни уредник Недељко Гвозденовић ; превод на француски Матилда Станојевић ; фотографије Миодраг Ђорђевић. — Београд : Српска академија наука и уметности, 1978.
212 p., [9] leaves of plates : ill. (some col.), fold. map ; 23 cm. — ([Каталог] - Галерија Српске академије наука и уметности ; 34) Yu***
Exhibition of works by M. Valtrović and D. Milutinović organized by Galerija Srpske akademije nauka i umetnosti and Istorijski muzej Srbije.
Summaries in French; legends also in French.
Includes bibliographical references and index.
1. Art, Medieval — Yugoslavia — Serbia — Historiography — Exhibitions. 2. Christian antiquities—Yugoslavia—Serbia—Historiography—Exhibitions. 3. Valtrović, Mihailo, 1839-1915—Exhibitions. 4. Milutinović, Dragutin, 1840-1900—Exhibitions. I. Bogdanović, Sonja. II. Mišković-Prelević, Ljiljana. III. Istorijski muzej Srbije. IV. Series: Srpska akademija nauka i umetnosti, Belgrad. Galerija. Katalog ; 34.
N7249.S4 I 95 78-423032

CHRISTIAN ARCHAEOLOGY see Christian antiquities

CHRISTIAN ART AND SYMBOLISM

see also Altarpieces; Art, Byzantine; Art, Early Christian; Art, Gothic; Art, Medieval; Art, Romanesque; Art and religion; Bible—Illustrations; Catacombs; Cathedrals; Christian antiquities; Church architecture; Church decoration and ornament; Church furniture; Church vestments; Crosses; Ecclesiastical embroidery; Emblems; Heraldry, Sacred; Fish (in religion, folk-lore, etc.); Holy cards; Icons; Icons, Byzantine; Image (Theology); Jewish art and symbolism; Liturgical objects; Mosaics; Orthodox Eastern Church and art; Reformation and art; Relics and reliquaries; Religious articles; Symbolism of numbers; and also phrase headings of the type [subject] in art, e.g. Christian saints in art; and subdivision Art under names of Saints, living before 1400, biblical chracters, etc., e.g. Mary, Virgin—Art

CHRISTIAN ART AND SYMBOLISM (Continued)

Cowen, Painton.
Rose windows / Painton Cowen. — San Francisco : Chronicle Books, c1979.
144 p. : ill. ; 29 cm. — (Art and imagination)
"A Prism edition."
Bibliography: p. 142.
Includes index.
1. Rose windows. 2. Christian art and symbolism. 3. Glass painting and staining, Medieval. 4. Cathedrals. I. Purce, Jill. II. Title.
NK5308.C64 748.5'9 78-23286
78 MARC

Esmeijer, Anna C
Divina Quaternitas : a preliminary study in the method and application of visual exegesis / Anna C. Esmeijer. — Assen : Gorcum, 1978.
XI, 263 p. : with ill. ; 25 cm. Ne78-40
Bibliography: p. 224-238.
Includes index.
ISBN 9023215672 : fl 97.50
1. Bible—Illustrations. 2. Art, Medieval. 3. Christian art and symbolism. 4. Visual perception. I. Title.
N8023.E8 704.948'2 79-301022
*79 MARC

Fink, Josef, 1912-
Noe der Gerechte in der frühchristlichen Kunst. Münster, Böhlau, 1955.
123 p. 63 plates. 23 cm. (Archiv für Kulturgeschichte. Beihefte, Heft 4)
1. Noah—Art. 2. Christian art and symbolism.
N8110.F5 a 56-4254
Cornell Univ. Library [r79]rev MARC

Fulcanelli, pseud.
Les Demeures philosophales : et le symbolisme hermétique dans ses rapports avec l'art sacré et l'ésotérisme du grand œuvre / Fulcanelli ; avec 3 préfaces de Eugène Canseliet, ... ; dessins de Julien Champagne .... — [Nouvelle éd.]. — Paris : J.-J. Pauvert, 1977.
2 v. (468 p., 384 p. (xlv leaves of plates)) ; 22 cm. — (Collection curieuse d'alchimie traditionnelle) F78-18941
Includes bibliographical references and index.
ISBN 2-7202-0100-6 : 144F
1. Alchemy. 2. Christian art and symbolism. 3. Architecture, Medieval. 4. Decoration and ornament, Architectural. I. Title.
QD26.F84 1977 540'.1 79-376389
79 MARC

Guerra, Manuel.
Simbología románica : el cristianismo y otras religiones en el arte románico / Manuel Guerra. — Madrid : Fundación Universitaria Española, 1978.
411 p., [36] leaves of plates : ill. ; 24 cm. — (Publicaciones de la Fundación Universitaria Española : Bellas Artes ; 1) Sp***
Includes bibliographical references and indexes.
ISBN 8473920430
1. Art, Romanesque. 2. Christian art and symbolism. 3. Symbolism in art. 4. Art and religion. I. Title. II. Series: Fundación Universitaria Española. Publicaciones.
N7853.G83 704.948 79-352854
79 MARC

Hauschildt, Karl, 1920-
Bejahtes Leben und Sterben : Gedanken zu d. 7 Steintaf. auf. d. Friedhof Neumünster / Karl Hauschildt ; mit Zeichn. u. Aufn. von Hermann Pohl ; [hrsg. vom Ev.-luth. Kirchengemeindeverb. Neumünster]. — Neumünster : Wachholtz, 1975.
96 p. : ill. ; 24 cm. GFR76-A
ISBN 3-529-06156-5 : DM17.50
1. Christian art and symbolism. 2. Pohl, Hermann, 1917- . 3. Neumünster, Ger. Friedhof. I. Pohl, Hermann, 1917- . II. Title.
BV153.G3H38 79-358626
79 MARC

Kapner, Gerhardt.
Barocker Heiligenkult in Wien und seine Träger / Gerhardt Kapner. — Wien : Verl. f. Geschichte u. Politik, 1978.
147 p. ; 21 cm. — (Österreich Archiv) Au78-20-311
Includes bibliographical references.
ISBN 3-7028-0107-3 : S176.00
1. Christian saints—Cult—Austria—Vienna. 2. Vienna—Church history. 3. Art, Baroque—Austria—Vienna. 4. Christian art and symbolism. I. Title. II. Series.
BX2333.K36 79-358628
*79 MARC

King, Bucky.
Ecclesiastical crafts / Bucky King, Jude Martin. — New York : Van Nostrand Reinhold, 1978.
132 p. : ill. (some col.) ; 24 cm.
Bibliography: p. 128-129.
Includes index.
ISBN 0-442-22966-6
1. Liturgical objects. 2. Church vestments. 3. Christian art and symbolism. 4. Church decoration and ornament. I. Martin, Jude, joint author. II. Title.
NK1650.K47 745.5 77-25033
77 MARC

Kretzenbacher, Leopold.
Das verletzte Kultbild : Voraussetzungen, Zeitschichten und Aussagewandel eines abendländischen Legendentypus / Leopold Kretzenbacher. — München : Verlag der Bayerischen Akademie der Wissenschaften ; in Kommission bei C. H. Beck, 1977.
123 p., [4] leaves of plates : ill. ; 22 cm. — (Sitzungsberichte - Bayerische Akademie der Wissenschaften, Philosophisch-Historische Klasse ; Jahrg. 1977, Heft 1 ISSN 0342-5991) GFR***
Includes bibliographical references and index.
1. Christian art and symbolism. 2. Idols and images—Worship. 3. Legends, Christian—History and criticism. I. Title. II. Series: Akademie der Wissenschaften, Munich. Philosophisch-Historische Klasse. Sitzungsberichte ; Jahrg. 1977, Heft 1.
BV150.K69 78-386879
79 MARC

Lafontaine-Dosogne, Jacqueline.
Iconographie de l'enfance de la Vierge dans l'Empire byzantin et en Occident. Bruxelles, 1964-65.
2 v. plates. 30 cm. (Académie royale de Belgique. Classe des beaux-arts. Mémoires. Collection in-4°., 2. sér., t. 11, fasc. 3-3b)
Includes bibliographies.
1. Mary, Virgin—Art. 2. Christian art and symbolism. I. Title. II. Series: Académie royale des sciences, des lettres et des beaux-arts de Belgique. Classe des beaux-arts. Mémoires. Collection in-4°, 2. sér., t. 11, fasc. 3-3b.
AS242.B3415 ser. 2, t. 11, fasc. 3 a 66-192
[N8070] [r79]rev2 MARC

Loukotka, Jiří.
O náboženství a umění / Jiří Loukotka. — 1. vyd. — Praha : Horizont, 1974.
128 p., [8] leaves of plates : ill. ; 20 cm. — (Malá moderní encyklopedie ; sv. 79) Cz74
Includes bibliographical references and index.
Kčs11.00
1. Arts and religion. 2. Christian art and symbolism. 3. Březina, Otokar, pseud., 1868-1929—Religion and ethics. I. Title.
NX660.L68 75-533215
*78[r79]rev MARC

Matsumoto, Fujio, 1930–
(Iesu no genfūkei)
イエスの原風景：聖の図像学 / 松本富士男. — 東京：新泉社, 1975.
242, iv p. : ill. ; 22 cm. Ja 76–2269
Bibliography: p. i–iv.
¥3000
1. Christian art and symbolism. 2. Jesus Christ—Iconography. I. Title.
N7831.M28 79–808965

Maximus Tyrius.
Maximus of Tyre. [Conceived, lettered, and illustrated by Ben Shahn. Meriden? Conn., 1963]
[32] p. illus. 18 x 24 cm.
1. Christian art and symbolism. I. Shahn, Ben, 1898-1969. II. Title.
BV150.M38 Rare Bk Coll 64-5761
——— Copy 3. Rosenwald Coll. MARC
[r79]rev

Meer, Frédéric van der, 1904-
L'Apocalypse dans l'art / Frédéric van der Meer ; préf. de Jean Delumeau. — [Paris] : Chêne, [1978]
368 p. : ill. (some col.) ; 34 cm. F***
Includes indexes.
ISBN 2-85108-193-4 : 370F
1. Apocalyptic art. 2. Christian art and symbolism. I. Title.
N8038.5.M43 704.948'2 79-344863
79 MARC

Meer, Frederik van der, 1904-
Apokalypse : die Visionen des Johannes in der europäischen Kunst / Frits van der Meer ; [ins Deutsche übers. von Frans Stoks]. — Freiburg : Herder, 1978.
368 p. : ill. (chiefly col.) ; 34 cm. GFR***
Translation of Apocalypse.
Includes text of Revelation.
Published simultaneously in English, French, and Dutch.
ISBN 3-451-18223-8
1. Apocalyptic art. 2. Christian art and symbolism. I. Bible. N.T. Revelation. German. II. Title.
N8038.5.M4315 79-381694
79 MARC

Meer, Frederick van der, 1904-
Apocalypse : visions from the Book of Revelation in western art / [by] Frederick van der Meer. — [English ed.]. — London : Thames and Hudson, 1978.
3-367, [1] p. : ill., (some col.), facsims., (some col.), col. map ; 34 cm. GB79-17730
Issued also under title: L'Apocalypse dans l'art.
Ill. on lining papers.
Includes the text of the Book of Revelation.
ISBN 0-500-23294-6 : £40.00
1. Apocalyptic art. 2. Christian art and symbolism. I. Bible. N.T. Revelation. English. 1978. II. Title.
N8038.5.M4313 704.948'2 79-315968
79 MARC

Raynor, Louise A
Church needlepoint : patterns and instructions / Louise A. Raynor and Carolyn H. Kerr. — Wilton, Conn. : Morehouse-Barlow Co., c1976.
75 p. : ill. (some col.) ; 19 x 24 cm.
Bibliography: p. 71.
ISBN 0-8192-1214-8
1. Canvas embroidery—Patterns. 2. Christian art and symbolism. I. Kerr, Carolyn H., joint author. II. Title.
TT778.C3R39 746.9'5 78-324530
79 MARC

Schell, Herman, 1850-1906.
Herman Schell als Wegbereiter zum II. [i.e. Zweiten] Vatikanischen Konzil : sein Briefwechsel mit Franz Brentano u. Nachschr. seiner Vorlesungen über Friedrich Nietzsche, über christl. Kunst u. über Fundamentaltheologie / eingel., kommentiert u. hrsg. von Josef Hasenfuss. — München ; Paderborn ; Wien : Schöningh, 1978.
221 p. ; 22 cm. — (Abhandlungen zur Philosophie, Psychologie, Soziologie der Religion und Ökumenik ; n.F., Heft 38) GFR78-A
ISBN 3-506-70188-6 : DM9.80
1. Schell, Herman, 1850-1906. 2. Brentano, Franz Clemens, 1838-1917. 3. Nietzsche, Frederick Wilhelm, 1844-1900. 4. Christian art and symbolism. 5. Theology, Doctrinal. I. Brentano, Franz Clemens, 1838-1917. II. Hasenfuss, Josef, 1901- . III. Title. IV. Series.
BX4705.S5A25 1978 79-359642
79 MARC

Seppälä, Hilkka, 1943-
Johannes Edelläkävijän ja Kastajan ikonit / [Hilkka Seppälä ; piirrokset, Petros Sasaki ; valokuvat, Vesa-P. Takala]. — Kuopio : Ortodoksinen kirkkomuseo, 1976.
75 p. : ill. ; 21 cm. Fi76-Aug.
Bibliography: p. 73-74.
ISBN 9519396004
1. John the Baptist—Art. 2. Icon painting. 3. Christian art and symbolism. I. Title.
N8187.5.S46 79-357274
*79 MARC

Stevenson, J
The catacombs : rediscovered monuments of early Christianity / [by] J. Stevenson. — London : Thames and Hudson, 1978.
179 p. : ill., maps, plans, ; 25 cm. — (Ancient peoples and places) GB78-38564
Bibliography: p. 169-170.
Includes index.
ISBN 0-500-02091-4 : £7.50 ($16.95 U.S.)
1. Catacombs. 2. Christian art and symbolism. 3. Church history—Primitive and early church, ca. 30-600. I. Title.
BR132.S74 270'.1 79-317142
79 MARC

Stillwell, Margaret Bingham, 1887-
Noah's ark, in early woodcuts and modern rhymes, by Margaret Bingham Stillwell. New York, E. B. Hackett, the Brick row bookshop, inc. [1942]
49 p., 1 l. illus. 24 cm.
"First edition."
"Three hundred copies have been printed ... at the Walpole printing office in Mount Vernon, New York."
1. Noah's ark. 2. Wood-engravings. 3. Illustration of books. 4. Illustrated books—15th and 16th centuries. 5. Christian art and symbolism. I. Title.
N8180.S75 769'.4'84 42-4891
——— Copy 3. Rosenwald Coll. MARC
[r79]rev

Toyama, Usaburō.
(Genshi Kirisutokyō no bijutsu)
原始キリスト教の美術 / 外山卯三郎著. — 東京：造形美術協会出版局：発売元 日貿出版社, 昭和51 [1976]
370 p. : ill. ; 21 cm. Ja 77–7436
Includes bibliographical references.
¥3600
1. Art, Early Christian. 2. Christian art and symbolism. I. Title.
N7832.T68 79–804740

Verdier, Rafael.
Cinco bolas y un misterio. Málaga, 1973.
60 p. plates. 23 cm. 100ptas Sp***
ISBN 8440058845
1. Iglesia de San Juan, Málaga, Spain. 2. Málaga, Spain (City)—Religious life and customs. 3. Christian art and symbolism. I. Title.
BX4636.M34I348 79-408243
79 MARC

Vorreux, Damien.
Un Symbole franciscain, le Tau : histoire, théologie et iconographie / Damien Vorreux. — Paris : Éditions franciscaines, 1977.
109 p., [14] p. of plates ; 18 cm. — (Présence de saint François ; 30) F78-13112
Includes bibliographical references.
1. Crosses. 2. Tau (The Greek letter) 3. Christian art and symbolism. 4. Franciscans. I. Title.
BV160.V67 246'.558 79-369128
79 MARC

CHRISTIAN ART AND SYMBOLISM (Continued)

Vranos, Iōannēs Charilaos.
(Theōria hagiographias)
Θεωρία ἁγιογραφίας / Ἰω-Χαριλάου Βράνου. — Θεσσαλονίκη : Ἐκδόσεις Π. Πουρνάρα, 1977.
402 p., ⌈7⌉ leaves of plates : ill. (some col.) ; 25 cm.
Bibliography: p. ⌈403⌉–⌈405⌉
1. Orthodox Eastern Church and art. 2. Christian art and symbolism. 3. Icon painting. I. Title.
N72.R4V72 79–413756

Weitzmann, Kurt, 1904-
The icon : holy images, sixth to fourteenth century / ⌈by⌉ Kurt Weitzmann. — London : Chatto and Windus, 1978.
135 p. : ill. (chiefly col.) ; 29 cm. GB79-00500
Bibliography: p. 34.
ISBN 0-7011-2336-2 : £12.95
1. Icons. 2. Icon painting. 3. Votive offerings in art. 4. Christian art and symbolism. I. Title.
N8187.W44 1978b 704.948′2 79-306923
79 MARC

Weitzmann, Kurt, 1904-
The icon : holy images—sixth to fourteenth century / Kurt Weitzmann. — New York : G. Braziller, 1978.
134 p. : ill. (some col.) ; 29 cm.
Bibliography: p. 34.
ISBN 0-8076-0892-0 : $24.95. ISBN 0-8076-0893-9 pbk. : $11.95
1. Icons. 2. Icon painting. 3. Votive offerings in art. 4. Christian art and symbolism. I. Title.
N8187.W44 1978 704.948′2 78-6495
78 MARC

—MEDIEVAL, 500-1500

Lambert, Élie, 1888-
L'architecture des templiers / Élie Lambert. — Paris : A. et J. Picard, 1955, 1978 printing.
104 p. : ill. ; 23 cm. F***
Includes bibliographical references.
ISBN 2-7084-0025-8 : 48.00F
1. Church architecture—Philosophy. 2. Christian art and symbolism—Medieval, 500-1500. 3. Templars. I. Title.
NA4828.L35 729′.58′2 79-379107
79 MARC

Piltz, Elisabeth.
Trois sakkoi byzantins : analyse iconographique / Elisabeth Piltz. — Stockholm : Almqvist & Wiksell, c1976.
82 p., ⌈31⌉ leaves of plates : ill. ; 27 cm. — (Figura ; nova ser. 17) S76-51/52
"Acta Universitatis Upsaliensis."
Errata slip inserted.
Bibliography: p. 73-76.
Includes indexes.
ISBN 9155403514
1. Sakkos. 2. Ecclesiastical embroidery, Byzantine. 3. Christian art and symbolism—Medieval, 500-1500. I. Title. II. Series.
N25.F5 n.s., no. 17 79-373153
⌈NK9310⌉ *79 MARC

Scharfe, Siegfried.
Am Rande des Heiligen : fröhl. Bildkunst in alten Kirchen / Siegfried Scharfe. — 1. Aufl. — Berlin : Evangelische Verlagsanstalt, 1977.
142 p. : ill. ; 21 cm. GDR78-A
7.80M
1. Christian art and symbolism—Medieval, 500-1500. 2. Art, Medieval—Themes, motives. 3. Wit and humor, Pictorial. 4. Wit and humor, Medieval. I. Title.
N7850.S3 79-362968
79 MARC

—MEDIEVAL, 500-1500—CONGRESSES

Aktuelle Kunst in Spiegelungen des Mittelalters : ⌈Tagung d. Arbeitskreises Bildende Kunst v. 5.-8. Mai 1975 auf d. Liebfrauenberg im Elsass / Evang. Akad. Arnoldshain⌉ ; mit Beitr. von Hans R. Blankensteijn ... ⌈et al.⌉. — Als Ms. gedr. — Arnoldshain/Taunus : Evang. Akad. Arnoldshain, 1975.
77 p. ; 30 cm. — (Arnoldshainer Protokolle ; 1975, 3)
Cover title.
DM4.00
1. Art, Medieval—Congresses. 2. Art, Modern—20th century—Congresses. 3. Christian art and symbolism—Medieval, 500-1500—Congresses. 4. Christian art and symbolism—Modern, 1500- —Congresses. I. Blankensteijn, Hans R. II. Arbeitskreis Bildende Kunst. III. Evangelische Akademie in Hessen und Nassau. IV. Series.
N5940.A38 79-344177
79 MARC

—MEDIEVAL, 500-1500—EARLY WORKS TO 1800

Theophilus, called also Rugerus.
On divers arts : the foremost medieval treatise on painting, glassmaking, and metalwork / Theophilus ; translated from the Latin with introd. and notes by John G. Hawthorne and Cyril Stanley Smith. — New York : Dover Publications, 1979.
xxxv, 216 p., ⌈9⌉ leaves of plates : ill. ; 24 cm.
Translation of Diversarum artium schedula.
"An unabridged and corrected republication of the work originally published in 1963 by the University of Chicago Press."
Bibliography: p. 195-200.
ISBN 0-486-23784-2 : $5.00
1. Art—Early works to 1800. 2. Christian art and symbolism—Medieval, 500-1500—Early works to 1800. 3. Technology—Early works to 1800. I. Hawthorne, John G. II. Smith, Cyril Stanley, 1903- . III. Title.
N7420.T383 1979 709′.02 78-74298
79 MARC

—MEDIEVAL, 500-1500—AUSTRIA—CARINTHIA

Fritz, Anton.
Kärntens Flügelaltäre / Anton Fritz. — Klagenfurt : Verl. Carinthia, 1975.
184 p. : ill. (some col.) ; 27 cm. Au76-7-195
Bibliography: p. 181-183.
Includes indexes.
ISBN 3-85378-012-1 : S850.00
1. Christian art and symbolism—Medieval, 500-1500—Austria—Carinthia. 2. Altarpieces, Austrian—Austria—Carinthia. 3. Altarpieces, Gothic—Austria—Carinthia. I. Title.
N7948.A3C374 79-362756
*79 MARC

—EXHIBITIONS

—MEDIEVAL, 500-1500—AUSTRIA—VIENNA

Perger, Richard, 1928-
Die mittelalterlichen Kirchen und Klöster Wiens / Richard Perger, Walther Brauneis. — Wien ; Hamburg : P. Zsolnay, 1977.
335 p. : ill. ; 22 cm. — (Wiener Geschichtsbücher ; Bd. 19/20) Au***
Includes bibliographical references.
ISBN 3-552-02913-3
1. Vienna—Churches. 2. Vienna—Monasteries. 3. Christian art and symbolism—Medieval, 500-1500—Austria—Vienna. 4. Vienna—History. I. Brauneis, Walther, joint author. II. Title. III. Series.
NA5510.V54P47 79-382440
79 MARC

—MEDIEVAL, 500-1500—BYZANTINE EMPIRE

Bank, Alisa Vladimirovna.
(Prikladnoe iskusstvo Vizantii devi͡atogo-dvenadt͡satogo vv.)
Прикладное искусство Византии IX–XII вв. : Очерки / А. В. Банк. — Москва : Наука, 1978.
203, ⌈108⌉ p. : ill. ; 21 cm. — (Культура народов Востока) USSR 78
At head of title: Akademii͡a nauk SSSR. Otdelenie istorii, and other organizations.
Summary in English.
Series romanized: Kul'tura narodov Vostoka.
Bibliography: p. 167–⌈183⌉
Includes index.
1.70rub
1. Art objects, Byzantine. 2. Christian art and symbolism—Medieval, 500–1500—Byzantine Empire. I. Title.
NK715.B36 79–407206

Lazarev, Viktor Nikitich, 1897-1976.
Vizantiĭskoe i drevnerusskoe iskusstvo : Stat'i i materialy / V. N. Lazarev. — Moskva : Nauka, 1978.
335 p. : ill. ; 27 cm. USSR79
On leaf preceding t.p.: Akademii͡a nauk SSSR. Otdelenie istorii.
Includes bibliographical references and indexes.
5.60rub
1. Art, Byzantine. 2. Christian art and symbolism—Medieval, 500-1500—Byzantine Empire. 3. Art, Russian. 4. Christian art and symbolism—Medieval, 500-1500—Russian Republic. I. Title.
N7852.5.L39 1978 79-377961
79 MARC

—MEDIEVAL, 500-1500—CZECHOSLOVAKIA

Homolka, Jaromír.
Studie k počátkům umění krásného slohu v Čechách : k problematice společenské funkce výtvarného umění v předhusitských Čechách / Jaromír Homolka. — Vyd. 1. — Praha : Universita Karlova, 1976.
121 p., ⌈10⌉ leaves of plates : ill. ; 24 cm. — (Acta Universitatis Carolinae : Philosophica et historica : Monographia ; 55/1974) Cz***
Summary in Russian and German.
Bibliography: p. 115-121.
Includes index.
Kčs20.00
1. Art, Gothic—Czechoslovakia. 2. Art, Czech. 3. Christian art and symbolism—Medieval, 500-1500—Czechoslovakia. I. Title. II. Series: Prague. Universita Karlova. Acta : Philosophica et historica : Monographia ; 55/1974.
AS141.A522 no. 55 1974 79-349561
⌈N6826⌉ 79 MARC

—MEDIEVAL, 500-1500—FRANCE—MORLAAS

Lacoste, Jacques.
Sainte-Foy de Morlàas / Jacques Lacoste ; photos, Jacques Lacoste. — Jurançon : Amis des églises anciennes du Béarn, ⌈1976⌉
43 p., ⌈4⌉ leaves of plates : ill. ; 21 cm. F***
Includes bibliographical references.
1. Sainte-Foy de Morlaas (Church) 2. Christian art and symbolism, medieval, 500-1500—France—Morlaas. I. Title.
NA5551.S234L32 914.4′79 79-384853
79 MARC

—MEDIEVAL, 500-1500—GERMANY—CATALOGS

Braun, Joseph, 1857-
Meisterwerke der deutschen Goldschmiedekunst der vorgotischen Zeit / Joseph Braun. — New York : Garland, 1979.
2 v. in 1 : ill. ; 27 cm. — (A Dealers' and collectors' bookshelf : Metalwork ; 4)
Reprint of the 1922 ed., published in Munich by Riehn & Reusch in 2 vols.
ISBN 0-8240-3354-X : $40.00
1. Goldwork—Germany—Catalogs. 2. Goldwork, Medieval—Germany—Catalogs. 3. Christian art and symbolism—Medieval, 500-1500—Germany—Catalogs. I. Title. II. Series: Dealers' and collectors' booklist : Metalwork ; 4.
NK7150.A1B7 1979 739.2′282′0943 78-50268
79 MARC

—MEDIEVAL, 500-1500—GERMANY, WEST—BADEN-WÜRTTEMBERG—EXHIBITIONS

Mystik am Oberrhein und in benachbarten Gebieten : ⌈Ausstellung⌉ Augustinermuseum Freiburg im Breisgau 10. September bis 22. Oktober 1978 : ⌈Katalog / Gesamtleitung, Hans H. Hofstätter, Dietmar Lüdke⌉. — Freiburg i. Br. : Städtische Museen, ⌈1978⌉
217 p. : numerous ill. (some col.) ; 20 x 21 cm. GFR***
"Die Leihgeber": leaf inserted.
Bibliography: p. 207-214.
1. Sculpture, German—Germany, West—Baden-Württemberg—Exhibitions. 2. Sculpture, Gothic—Germany, West—Baden-Württemberg—Exhibitions. 3. Mysticism—Germany, West—Baden-Württemberg—Exhibitions. 4. Christian art and symbolism—Medieval, 500-1500—Germany, West—Baden-Württemberg—Exhibitions. 5. Illumination of books and manuscripts, German—Germany, West—Baden-Württemberg—Exhibitions. 6. Illumination of books and manuscripts, Gothic—Germany, West—Baden-Württemberg—Exhibitions. 7. Glass painting and staining—Germany, West—Baden-Württemberg—Exhibitions. 8. Glass painting and staining, Gothic—Germany, West—Baden-Württemberg—Exhibitions. I. Hofstätter, Hans Hellmut, 1928- . II. Lüdke, Dietmar. III. Augustinermuseum, Freiburg i.B.
NB572.M9 79-355118
79 MARC

—MEDIEVAL, 500-1500—GERMANY, WEST—KLEVE (NORTH RHINE-WESTPHALIA)—EXHIBITIONS

Kunstschätze aus dem Klevischen : Ausstellung anlässl. d. 800-Jahrfeier d. Klever Pfarrkirche 1174-1974 : Städt. Museum Haus Koekkoek, Kleve, 27. Okt. - 24. Nov. 1974 : ⌈Katalog⌉. — Kleve : Städt. Museum Haus Koekkoek, 1974.
⌈63⌉ p. : numerous ill. ; 21 cm. GFR75-B5
1. Christian art and symbolism—Medieval, 500-1500—Germany, West—Kleve (North Rhine-Westphalia)—Exhibitions. 2. Art, German—Germany, West—Kleve (North Rhine-Westphalia)—Exhibitions. 3. Pfarrkirche Sankt Mariä Himmelfahrt, Kleve (North Rhine-Westphalia)—Exhibitions. I. Städtisches Museum Haus Koekkoek.
N7950.K53K86 79-369125
79 MARC

—MEDIEVAL, 500-1500—GERMANY, WEST—PALATINATE

Mittelalterliche Wandmalereien in der Pfalz / hrsg. von d. Pfalzgalerie d. Bezirksverb. Pfalz, Kaiserslautern. — Kaiserslautern : Pfalzgalerie d. Bezirksverb. Pfalz, 1976.
⌈154⌉ p. : numerous ill. ; 25 cm. GFR77-B4
1. Mural painting and decoration, German—Germany, West—Palatinate. 2. Mural painting and decoration, medieval—Germany, West—Palatinate. 3. Christian art and symbolism, Medieval, 500-1500—Germany, West—Palatinate. I. Pfalzgalerie.
ND2750.P34M57 79-381234
79 MARC

—MEDIEVAL, 500-1500—ITALY

Mutti, Claudio.
Simbolismo e arte sacra : il linguaggio segreto dell'Antelami / Claudio Mutti. — 1. ed. — Parma : Edizioni All'insegna del Veltro, c1978.
67 p., ⌈4⌉ leaves of plates : ill. ; 17 cm. It***
Includes bibliographical references.
L2000
1. Antelami, Benedetto, fl. 1177-1233. 2. Christian art and symbolism—Medieval, 500-1500—Italy. I. Title.
NB623.A47M87 730′.92′4 79-382002
79 MARC

CHRISTIAN ART AND SYMBOLISM (Continued)

—MEDIEVAL, 500-1500—ITALY—LAZIO —CONGRESSES

I Cistercensi e il Lazio : atti delle giornate di studio dell'Istituto di storia dell'arte dell'Università di Roma, 17-21 maggio 1977 / Istituto nazionale di archeologia e storia dell'arte, Istituto di storia dell'arte dell'Università di Roma. — 1. ed. — Roma : Multigrafica, 1978.
x, 302 p. : ill. ; 28 cm. It78-Sept
Italian or German.
Includes bibliographical references and indexes.
1. Art, Cistercian—Italy—Lazio—Congresses. 2. Christian art and symbolism—Medieval, 500-1500—Italy—Lazio—Congresses. 3. Cistercians in Italy—Congresses. I. Rome (City). Università. Istituto di storia dell'arte. II. Istituto nazionale di archeologia e storia dell'arte.
N7952.A3L392 79-364090
*79 MARC

—MEDIEVAL, 500-1500—RUSSIAN REPUBLIC

Högy, Tatjana.
Russische Ikonen : Einf. in Bildtypen, Malweise u. Sinn russ. Ikonen / von Tatjana Högy. — Giessen : Schmitz, 1978.
146 p. : 70 ill. (some col.) ; 21 cm. GFR78-A
Bibliography: p. 146.
Includes index.
DM16.80
1. Icon painting—Russian Republic. 2. Christian art and symbolism—Medieval, 500-1500—Russian Republic. I. Title.
N8189.R9H63 755'.2 79-351959
79 MARC

Lazarev, Viktor Nikitich, 1897-1976.
Vizantiĭskoe i drevnerusskoe iskusstvo : Stat'i i materialy / V. N. Lazarev. — Moskva : Nauka, 1978.
335 p. : ill. ; 27 cm. USSR79
On leaf preceding t.p.: Akademiiā nauk SSSR. Otdelenie istorii.
Includes bibliographical references and indexes.
5.60rub
1. Art, Byzantine. 2. Christian art and symbolism—Medieval, 500-1500—Byzantine Empire. 3. Art, Russian. 4. Christian art and symbolism—Medieval, 500-1500—Russian Republic. I. Title.
N7852.5.L39 1978 79-377961
79 MARC

Ryndina, Anna Vadimovna.
(Drevnerusskaiā melkaiā plastika)
Древнерусская мелкая пластика : Новгород и Центральная Русь XIV–XV веков / А. В. Рындина. — Москва : Наука, 1978.
189 p., [2] leaves of plates : ill. ; 22 cm. — (Памятники древнерусского искусства) USSR***
On leaf preceding t. p.: Akademiiā nauk SSSR. Vsesoiūznyĭ nauchno-issledovatel'skiĭ institut iskusstvoznaniiā Ministerstva kul'tury SSSR.
Series romanized: Pamiātniki drevnerusskogo iskusstva.
Includes bibliographical references.
1.40rub
1. Art, Decorative—Russian Republic. 2. Christian art and symbolism—Medieval, 500–1500—Russian Republic. I. Title.
NK1653.R9R93 79–412289

—MEDIEVAL, 500-1500—RUSSIAN REPUBLIC—SUZDAL, RUSSIA

Ovchinnikov, Adolf' Nikolaevich.
(Suzdal'skie zlatye vrata)
Суздальские златые врата = Golden gates in Suzdal / А. Н. Овчинников ; [фотографии А. А. Александрова]. — Москва : Искусство, 1978.
35 p., [64] leaves of plates : ill. ; 27 cm. USSR***
Russian and English.
Includes bibliographical references.
4.10rub
1. Goldwork, Medieval—Russian Republic—Suzdal, Russia. 2. Metal doors—Russian Republic—Suzdal, Russia. 3. Christian art and symbolism—Medieval, 500–1500—Russian Republic—Suzdal, Russia. 4. Sobor Rozhdestva bogorodiīsy v Suzdale. I. Aleksandrov, Aleksei Alekseevich. II. Title. III. Title: Golden gates in Suzdal.
NK7156.S88O9 739.2'2747'31 79–413811

—MEDIEVAL, 500-1500—SPAIN—PALENCIA (PROVINCE)

Sancho Campo, Angel.
Santa María y Santiago en el arte palentino / Angel Sancho Campo. — Palencia : [s.n.], 1975.
650 p. : chiefly ill. ; 23 cm. — (His El arte sacro en Palencia ; v. 4) Sp***
"Número extraordinario del Boletín oficial del Obispado de Palencia."
Includes bibliographical references.
1. Mary, Virgin—Art. 2. James, Saint, apostle—Art. 3. Christian art and symbolism—Medieval, 500-1500—Spain—Palencia (Province) 4. Christian art and symbolism—Modern, 1500- —Palencia (Province) I. Title.
N7962.A1S27 vol. 4 79-368052
[N7962.A3] 79 MARC

—MEDIEVAL, 500-1500—SPAIN—RIPOLL

Rico, Francisco.
Signos e indicios en la portada de Ripoll ... / Francisco Rico. — Barcelona : Fundación Juan March, 1976.
63 p. : ill. ; 24 cm. Sp***
ISBN 8470750283
1. Santa Maria de Ripoll (Basilica). Portada. 2. Sculpture, Romanesque—Spain—Ripoll. 3. Christian art and symbolism—Medieval, 500-1500—Spain—Ripoll. I. Title.
NB811.R56R52 79-377513
79 MARC

—MODERN, 1500- —CONGRESSES

Aktuelle Kunst in Spiegelungen des Mittelalters : [Tagung d. Arbeitskreises Bildende Kunst v. 5.-8. Mai 1975 auf d. Liebfrauenberg im Elsass / Evang. Akad. Arnoldshain] ; mit Beitr. von Hans R. Blankensteijn ... [et al.]. — Als Ms. gedr. — Arnoldshain/Taunus : Evang. Akad. Arnoldshain, 1975.
77 p. ; 30 cm. — (Arnoldshainer Protokolle ; 1975, 3)
Cover title.
DM4.00
1. Art, Medieval—Congresses. 2. Art, Modern—20th century—Congresses. 3. Christian art and symbolism—Medieval, 500-1500—Congresses. 4. Christian art and symbolism—Modern, 1500- —Congresses. I. Blankensteijn, Hans R. II. Arbeitskreis Bildende Kunst. III. Evangelische Akademie in Hessen und Nassau. IV. Series.
N5940.A38 79-344177
79 MARC

—MODERN PERIOD, 1500- —CALIFORNIA

Nolan, James Lee, 1930-
Discovery of the lost art treasures of California's first mission / by James L. Nolan ; commissioned by Helen K. Copley ; edited by Richard F. Pourade. — [La Jolla, Calif. : Copley Books, c1978]
ix, 118 p. : ill. (some col.) ; 29 cm.
Includes bibliographical references and index.
ISBN 0-913938-20-3 : $20.00
1. Christian art and symbolism—Modern period, 1500- —California. 2. San Diego mission. I. Pourade, Richard F. II. Title.
N7910.C2N64 979.4'98 78-73173
79 MARC

—MODERN PERIOD, 1500- —ITALY—EXHIBITIONS

Arte sacra di Pino Casarini. — Pordenone : Concordia 7, 1977.
iv, 77 p., [36] leaves of plates : ill. (some col.) ; 27 cm. — (Edizioni d'arte ; nuova ser., 9) It79-Aug
Exhibition and catalog sponsored by the Centro iniziative culturali Pordenone.
"Catalogo ... a cura di Mario Possamai."
"Mostra d'arte alla Galleria sagittaria ... 1977."
Bibliography: p. 75-77.
1. Casarini, Pino, 1897-1972—Exhibitions. 2. Christian art and symbolism—Modern, 1500- —Italy—Exhibitions. I. Casarini, Pino, 1897-1972. II. Possamai, Mario. III. Centro iniziative culturali Pordenone. IV. Galleria d'arte sagittaria.
N6923.C358A4 1977 79-382621
*79 MARC

—MODERN PERIOD, 1500- —ITALY—TRIESTE—EXHIBITIONS

Le Benedettine e il Monastero di San Cipriano / a cura dei Civici musei di storia e arte di Trieste ; direzione della Mostra e del catalogo Laura Ruaro Loseri ; selezione delle opere L. Ruaro e M. Walcher ; schede e testi di Luisa Crusvar, Laura Ruaro Loseri, Maria Walcher. — Udine : Istituto per l'Enciclopedia del Friuli-Venezia Giulia, 1978.
59 p. : ill. (some col.) ; 23 cm. It79-Apr
Exhibition held in Trieste in 1978.
Bibliography: p. 59.
1. Christian art and symbolism—Modern period, 1500- —Italy—Trieste—Exhibitions. 2. Monastero di San Cipriano—Exhibitions. I. Ruaro Loseri, Laura. II. Civici musei di storia ed arte di Trieste.
N7952.T74B46 79-359328
*79 MARC

—MODERN PERIOD, 1500- —MEXICO

Haufe, Hans.
Funktion und Wandel christlicher Themen in der mexikanischen Malerei des 20. Jahrhunderts / Hans Haufe. — Berlin : Colloquium-Verlag, 1978.
227 p. [68] leaves of plates : ill. ; 21 cm. — (Bibliotheca Ibero-Americana ; Bd. 25) GFR78-A
Includes bibliographical references and index.
ISBN 3-7678-0449-2 : DM39.00
1. Mural painting and decoration—20th century—Mexico. 2. Mural painting and decoration, Mexican—Mexico. 3. Christian art and symbolism—Modern period, 1500- —Mexico. I. Title. II. Series.
ND2644.H38 79-348464
79[r79]rev MARC

—MODERN PERIOD, 1500- —NEW ENGLAND—EXHIBITIONS

Benes, Peter.
New England meeting house and church, 1630-1850 : a loan exhibition at the Currier Gallery of Art, Manchester, New Hampshire / Peter Benes & Philip D. Zimmerman. — Boston : Published by Boston University for the Dublin Seminar for New England Folklife, c1979.
xi, 177 p. : ill. ; 22 cm.
Bibliography: p. 110.
ISBN 0-87270-050-X : $7.95
1. Christian art and symbolism—Modern, 1500- —New England—Exhibitions. 2. Church decoration and ornament—New England—Exhibitions. 3. Churches, Friend—New England—Exhibitions. 4. Churches, Protestant—New England—Exhibitions. I. Zimmerman, Philip D., joint author. II. Currier Gallery of Art, Manchester, N.H. III. Dublin Seminar for New England Folklife. IV. Title.
N7905.B46 79-53115
709.948'2'097407401428
00 MARC

MODERN PERIOD, 1500- —PALENCIA (PROVINCE

Sancho Campo, Angel.
Santa María y Santiago en el arte palentino / Angel Sancho Campo. — Palencia : [s.n.], 1975.
650 p. : chiefly ill. ; 23 cm. — (His El arte sacro en Palencia ; v. 4) Sp***
"Número extraordinario del Boletín oficial del Obispado de Palencia."
Includes bibliographical references.
1. Mary, Virgin—Art. 2. James, Saint, apostle—Art. 3. Christian art and symbolism—Medieval, 500-1500—Spain—Palencia (Province) 4. Christian art and symbolism—Modern, 1500- —Palencia (Province) I. Title.
N7962.A1S27 vol. 4 79-368052
[N7962.A3] 79 MARC

—MODERN PERIOD, 1500- —RUSSIAN REPUBLIC —EXHIBITIONS

(Russkaiā zhivopis' semnadtsatogo–vosemnadtsatogo vekov)
Русская живопись XVII–XVIII веков : Из собраний Гос. Рус. музея, Гос. Третьяковской галереи, Музея древнерус. искусства им. Андрея Рублева : Каталог выставки / [Предисл. Т. Ананьевой ; редактор-составитель С. В. Ямщиков]. — Ленинград : Гос. рус. музей, 1977.
130 p. : ill. ; 22 cm. USSR 77
0.90rub
1. Icons, Russian—Exhibitions. 2. Christian art and symbolism—Modern period, 1500- —Russian Republic—Exhibitions. I. Iāmshchikov, Saveliĭ Vasil'evich. II. Leningrad. Gosudarstvennyĭ russkiĭ muzeĭ. III. Moscow. Gosudarstvennaiā Tret'iākovskaiā galleriā. IV. Muzeĭ drevnerusskogo iskusstva.
N8189.R9R87 79-408950

—MODERN PERIOD, 1500- —SWEDEN—NORRLAND

Telhammer, Ingrid.
Predikstolsmakare och predikstolskonst före 1777 i norra Hälsingland, Medelpad och Ångermanland / Ingrid Telhammer. — Umeå : Skytteanska samf., 1978.
248 p. : ill. ; 24 cm. — (Kungl. Skytteanska samfundets handlingar ; no. 20 ISSN 1560-2416) S78-34
Thesis—Umeå.
Summaries in English and German.
Bibliography: p. 236-241.
Includes indexes.
kr60.00
1. Pulpits—Sweden—Norrland. 2. Wood-carving, Baroque—Sweden—Norrland. 3. Wood-carving—Sweden—Norrland—History—18th century. 4. Church decoration and ornament, Baroque—Sweden—Norrland. 5. Church decoration and ornament—Sweden—Norrland—History—18th century. 6. Christian art and symbolism—Modern period, 1500- —Sweden—Norrland. 7. Wood-carvers—Sweden—Biography. I. Title. II. Series: Skytteanska samfundet. Handlingar ; no. 20.
DL601.S54 no. 20 748.5 s 79-362532
[NK9761.A3] 79 MARC

—ADDRESSES, ESSAYS, LECTURES

Spätantike und frühes Christentum / [hrsg.] von Beat Brenk ; mit Beitr. von Hugo Brandenburg ... [et al.] ; Übers. fremdsprach. Texte, Liliane Châtelet-Lange ... et al. ; Literaturverz., Jürg Andreas Bossardt ; Namen- u. Sachreg., Marianne Eaton-Krauss ; Strichzeichn., Dieter Götze ... et al.]. — Frankfurt am Main ; Berlin ; Wien : Propyläen-Verlag, 1977.
351, 400 p. : numerous ill. (some col.) ; 29 cm. — (Propyläen Kunstgeschichte : Supplementbände ; 1) GFR78-A18
One col. ill. inserted.
Bibliography: p. 333-343 (1st group)
Includes index.
ISBN 3-549-05085-2 : DM260.00. ISBN 3-549-15095-4 Hldr. : DM290.00
1. Art, Ancient—Addresses, essays, lectures. 2. Art, Early Christian—Addresses, essays, lectures. 3. Christian art and symbolis—Medieval, 500-1500—Addresses, essays, lectures. 4. Jewish art and symbolism—Addresses, essays, lectures. I. Brenk, Beat. II. Brandenburg, Hugo. III. Series: Propyläen Kunstgeschichte (Berlin, 1966-) : Supplementbände ; 1.
N5340.S68 78-394561
79 MARC

CHRISTIAN ART AND SYMBOLISM (Continued)

—CATALOGS

Artisti per l'Anno santo 1975 / testo di Giovanni Fallani. — ⌈Città del Vaticano⌉ : Libreria editrice vaticana, 1976.
168 p. : ill. ; 32 cm. It78-July
1. Art, Modern—20th century—Catalogs. 2. Christian art and symbolism—Catalogs. 3. Holy Year, 1975—Catalogs. I. Fallani, Giovanni.
N7880.A77 78-393439
*79 MARC

—DICTIONARIES

Bottomley, Frank.
The church explorer's guide to symbols and their meaning / ⌈by⌉ Frank Bottomley ; illustrated by Connie Green. — London : Kaye and Ward, 1978.
xvi, 176 p. : ill., coats of arms ; 20 cm. GB78-22465
Bibliography: p. 161.
Includes index.
ISBN 0-7182-1182-0. ISBN 0-7182-1187-1 pbk. : £1.95
1. Christian art and symbolism—Dictionaries. I. Green, Connie. II. Title.
BV150.B63 1978 246'.03 78-324908
79 MARC

Richter, Gert, 1929-
Lexikon der Kunstmotive : antike u. christl. Welt / Gert Richter, Gerhard Ulrich ; mit Zeichn. von Gerhard Ulrich. — München : Bertelsmann-Lexikon-Verlag, 1978.
319 p. : ill. ; 21 cm. GFR78-A
Includes bibliographical references and index.
ISBN 3-570-01568-8 : DM28.00
1. Mythology, Classical, in art—Dictionaries. 2. Christian art and symbolism—Dictionaries. I. Ulrich, Gerhard, joint author. II. Title.
N7760.R52 704.948 78-378826
78 MARC

—EXHIBITIONS

Die Hl. ⌈i.e. Heilige⌉ Ursula und ihre elftausend Jungfrauen : ⌈Ausstellung⌉, Wallraf-Richartz-Museum, Köln 6. Juli bis 3. September 1978 / ⌈Katalogbearbeitung, Frank Günter Zehnder⌉. — Köln : Wallraf-Richartz-Museum, 1978.
74 p. : numerous ill. ; 27 cm. GFR***
Bibliography: p. 74.
1. Ursula, Saint—Art—Exhibitions. 2. Christian art and symbolism—Exhibitions. I. Zehnder, Frank Günter. II. Cologne. Wallraf-Richartz-Museum.
N8080.U77H44 704.948'6 78-401159
79 MARC

—ALPS

Beitl, Klaus.
Volksglaube : Zeugnisse religiöser Volkskunst / Klaus Beitl ; mit 48 Farbtaf. ⌈nach Farbaufn. v. Josef Dapra⌉. — Salzburg ; Wien : Residenz Verl., c1978.
159 p. : ill. ; 24 cm. Au79-4
Bibliography: p. 159.
ISBN 3-7017-0201-2 : S298.00
1. Christian art and symbolism—Alps. 2. Folk art—Alps. I. Title.
N7949.6.B44 79-360297
*79 MARC

—AUSTRIA

Brauneck, Manfred.
Religiöse Volkskunst : Votivgaben, Andachtsbilder, Hinterglas, Rosenkranz, Amulette / Manfred Brauneck, unter Mitarb. von Hildegard Brauneck ; u. mit Fotos von Wulf Brackrock. — Köln : DuMont, 1978.
387 p. : ill. (some col.) ; 21 cm. — (DuMont Dokumente) GFR78-A
Bibliography: p. 371-387
ISBN 3-7701-0967-8
1. Religious articles—Austria. 2. Religious articles—Germany, West—Bavaria. 3. Folk art—Austria. 4. Folk art—Germany, West—Bavaria. 5. Christian art and symbolism—Austria. 6. Christian art and symbolism—Germany, West—Bavaria. I. Brauneck, Hildegard, joint author. II. Brackrock, Wulf. III. Title.
NK1653.A9B72 704.948'2'094 78-369753
78 MARC

—AUSTRIA—ALTAUSSEE

Amon, Karl.
Altaussee, Stmk. : ⌈die Pfarrkirche St. Ägid zu Altaussee, Salzkammergut, Steiermark / Karl Amon ; Fotos, Albert Rastl⌉. — Salzburg : Verl. St. Peter, 1975.
11 p. : ill. ; 17 cm. — (Christliche Kunststätten Österreichs ; Nr. 109)
Au78-16-231
Bibliography: p. 11.
S11.00
1. Pfarrkirche Sankt Ägid, Altaussee, Austria. 2. Christian art and symbolism—Austria—Altaussee. I. Title. II. Series.
N6801.C5 Nr. 109 914.36'04'508 s 78-392135
⌈NA5510.A54⌉ *78 MARC

—AUSTRIA—STEIN AN DER DONAU

Lechner, Gregor Martin.
Stift Göttweig und seine Kunstschätze / Gregor M. Lechner ; Bilddokumentation, Herbert Fasching. — 1. Aufl. — St. Pölten ; Wien : Verl. Niederösterr. Pressehaus, 1977.
104 p., ⌈32⌉ leaves of plates : ill. (some col.) ; 28 cm. Au77-13-232
Bibliography: p. 102-104.
ISBN 3-85326-437-9 : S390.00
1. Göttweig (Benedictine abbey) 2. Christian art and symbolism—Austria—Stein an der Donau. I. Fasching, Herbert. II. Title.
NA5510.G643L42 78-395120
*79 MARC

—BELGIUM—SAINT-TROND—EXHIBITIONS

Een Keuze uit het kerkelijk kunstbezit van de parochies van Gott-Sint-Truiden : Tentoonstelling, ingericht in het Stadhuis te Sint-Truiden, 17 september-1 oktober 1977. — 3800 Sint-Truiden : Provinciaal Museum voor Religieuze Kunst, 1977.
58 p. : ill. ; 30 cm. — (Kunst en oudheden in Limburg ; 19) Be78-2
"Ingericht ter gelegenheid vaan de zevenjaarlijkse Sint-Trudofeesten..."
Errata slip inserted.
Bibliography: p. 18-22.
100F
1. Christian art and symbolism—Belgium—Saint-Trond—Exhibitions. I. Series.
N7955.S27K48 78-398643
*79 MARC

—ENGLAND—EALING

Ealing Museum, Art, and History Society.
Art, architecture and history of Ealing churches : a survey of Christian churches within the London Borough of Ealing / ⌈Ealing Museum, Art and History Society⌉. — ⌈Ealing⌉ : Ealing Museum, Art and History Society, Publications Committee, ⌈1978?⌉
⌈71⌉ p. : ill. ; 21 cm. GB78-12070
£1.50
1. Christian art and symbolism—England—Ealing. 2. Ealing, Eng.—Churches. I. Title.
N7944.E24E24 1978 726'.5'0942184 78-323612
79 MARC

—ETHIOPIA

see also Ethiopian magic scrolls

—EUROPE, EASTERN

Kádár, Zoltán, fl. 1958-
Középkori művészet Kelet-Európában. Budapest, Gondolat Kiadó, 1961.
53 p. illus., map, plans. 17 cm. (Művészettörténet, 16)
Bibliography: p. ⌈55⌉
1. Art, Medieval—Europe, Eastern. 2. Art—Europe, Eastern. 3. Christian art and symbolism—Europe, Eastern. I. Title.
N7850.K3 75-292836
77⌈r79⌉rev MARC

—FRANCE

L'Art religieux : abbayes, églises et cathédrales. — Paris : Larousse, 1978.
480 p. : col. ill. ; 30 cm. — (Beautés de la France) (Découvrir)
F78-20420
Includes index.
ISBN (invalid) 2-03-013275-9 : 197F
1. Christian art and symbolism—France. 2. Church architecture—France. 3. France—Civilization.
N7949.A1A77 914.4'04'83 79-361629
79 MARC

—FRANCE—ALSACE—EXHIBITIONS

Coutumes et rites de protection en Alsace : exposition au Musée alsacien, Strasbourg, 8 mai-2 octobre 1977. — Strasbourg : Le Musée, 1977.
67 p., ⌈12⌉ leaves of plates : ill. (some col.) ; 24 cm. F***
At head of title: Ville de Strasbourg.
Bibliography: p. 65-67.
18.00F
1. Votive offerings—France—Alsace—Exhibitions. 2. Votive offerings in art—Exhibitions. 3. Rites and ceremonies—France—Alsace—Exhibitions. 4. Christian art and symbolism—France—Alsace—Exhibitions. I. Strasbourg. Musée alsacien.
N7949.A3A552 78-369021
704.948'2'0944383074043835
78 MARC

—GERMANY, EAST—NAUMBURG AN DER SAALE

Schubert, Ernst, Dr. phil.
Naumburg : Dom u. Altstadt / Ernst Schubert ; Aufnahmen von Fritz Hege. — 1. Aufl. — Berlin : Union-Verlag, 1978.
243 p. : numerous ill. (some col.) ; 28 cm. GFR79-A11
Bibliography: p. 239-243.
52.00M
1. Naumburg an der Saale. Dom. 2. Architecture—Germany, East—Naumberg an der Saale. 3. Christian art and symbolism—Germany, East—Naumburg an der Saale. 4. Naumburg an der Saale—Description. I. Hege, Fritz. II. Title.
NA5586.N3S283 79-382617
79 MARC

—GERMANY, WEST—ABENBERG

Kornbacher, Franz.
Abenberg : Stilla-Klosterkirche / ⌈Franz Kornbacher ; Fotos, Hans Gundel ... et al.⌉. — 1. Aufl. — München ; Zürich : Schnell und Steiner, 1978.
18, ⌈1⌉ p. : numerous ill. ; 17 cm. — (Kunstführer ; Nr. 1116) GFR78-A
Caption title.
Cover title: Stilla-Klosterkirche, Abenberg.
Bibliography: p. ⌈19⌉
1. Stilla-Klosterkirche, Abenberg, Ger. 2. Christian art and symbolism—Germany, West—Abenberg. I. Title. II. Title: Stilla-Klosterkirche, Abenberg. III. Series: Kleine deutsche Kirchenführer; Nr. 1116.
NA5564.K53 Nr. 1116 914.3'04'8708 s 78-398029
⌈NA5586.A33⌉ 78⌈r79⌉rev MARC

—GERMANY, WEST—BAVARIA

Brauneck, Manfred.
Religiöse Volkskunst : Votivgaben, Andachtsbilder, Hinterglas, Rosenkranz, Amulette / Manfred Brauneck, unter Mitarb. von Hildegard Brauneck ; u. mit Fotos von Wulf Brackrock. — Köln : DuMont, 1978.
387 p. : ill. (some col.) ; 21 cm. — (DuMont Dokumente) GFR78-A
Bibliography: p. 371-387
ISBN 3-7701-0967-8
1. Religious articles—Austria. 2. Religious articles—Germany, West—Bavaria. 3. Folk art—Austria. 4. Folk art—Germany, West—Bavaria. 5. Christian art and symbolism—Austria. 6. Christian art and symbolism—Germany, West—Bavaria. I. Brauneck, Hildegard, joint author. II. Brackrock, Wulf. III. Title.
NK1653.A9B72 704.948'2'094 78-369753
78 MARC

—GERMANY, WEST—COLOGNE—CATALOGS

Schulten, Walter, 1924-
Kostbarkeiten in Köln : Erzbischöfl. Diözesan-Museum, Katalog / Walter Schulten. — Köln : Greven, 1978.
126 p. : ill. (some col.) ; 23 cm. GFR78-A
Includes bibliographical references.
ISBN 3-7743-0161-1
1. Christian art and symbolism—Germany, West—Cologne—Catalogs. 2. Erzbischöfliches Diözesan-Museum—Catalogs. I. Erzbischöfliches Diözesan-Museum Köln. II. Title.
N7823.G3C647 79-354736
79 MARC

—GERMANY, WEST—DETTELBACH

Schnell, Hugo, 1904-
Wallfahrtskirche Dettelbach, Main / ⌈Hugo Schnell⌉. — 3., (überarb.) Aufl. — München ; Zürich : Schnell und Steiner, 1975.
14 p. : numerous ill. (some col.) ; 17 cm. — (Kunstführer ; Nr. 679)
GFR75-A
Bibliography: p. 14.
1. Wallfahrtskirche Maria im Sande, Dettelbach, Ger. 2. Christian art and symbolism—Germany, West—Dettelbach. I. Title. II. Series: Kleine deutsche Kirchenführer ; Nr. 679.
NA5564.K53 Nr. 679 79-381619
⌈NA5586.W268⌉ 79 MARC

CHRISTIAN ART AND SYMBOLISM (Continued)

—GERMANY, WEST—EAST FRIESLAND

Rödiger, Hans-Bernd.
Friesische Kirchen im Jeverland und Harlingerland / Hans-Bernd Rödiger, Klaus Wilkens. — Jever : Mettcker, 1978.
87 p. : chiefly ill. ; 27 cm. GFR79-A4
Includes bibliographical references and index.
DM26.80
1. Churches—Germany, West—East Friesland—Guide-books. 2. Christian art and symbolism—Germany, West—East Friesland. I. Wilkens, Klaus, joint author. II. Title.
NA5574.E23R63 79-361947
79 MARC

—GERMANY, WEST—ESSEN

Küppers, Leonhard, 1903-
Essen : Dom u. Domschatz / Leonhard Küppers ; Aufn. von Peter Happel. — Königstein im Taunus : Köster, 1975.
80 p. : chiefly ill. (some col.) ; 27 cm. — (Die Blauen Bücher) GFR76-A
Bibliography: p. ₍2₎
ISBN 3-7845-4410-X : DM16.80
1. Essen. Cathedral. 2. Christian art and symbolism—Germany, West—Essen. I. Title.
NA5586.E7K79 726'.6'094355 79-348449
79 MARC

—GERMANY, WEST—FRANCONIA

Schemmel, Bernhard.
Figuren und Reliefs an Haus und Hof in Franken / von Bernhard Schemmel. — Würzburg : Kommissionsverlag F. Schöningh, 1978.
68 p., ₍7₎ leaves of plates : ill. ; 23 cm. — (Quellen und Forschungen zur Geschichte des Bistums und Hochstifts Würzburg ; Bd. 31) GFR***
Includes bibliographical references and indexes.
ISBN 3-87717-032-3
1. Sculpture, German—Germany, West—Franconia. 2. Relief (Sculpture)—Germany, West—Franconia. 3. Decoration and ornament, Architectural—Germany, West—Franconia. 4. Shrines—Germany, West—Franconia. 5. Christian art and symbolism—Germany, West—Franconia. I. Title. II. Series.
BX1538.W8Q4 Bd. 31 79-348230
₍NB574.F7₎ 79 MARC

—GERMANY, WEST—GÖPPINGEN (LANDKREIS)

Hummel, Heribert.
Wandmalereien im Kreis Göppingen / Heribert Hummel ; Aufnahmen von Traute Uhland-Clauss. — Weissenhorn : Konrad, 1978.
132, ₍82₎ p. : numerous ill. (some col.) ; 25 cm. — (Veröffentlichungen des Kreisarchivs Göppingen ; Bd. 6) (Kunst in Bayern und Schwaben ; Bd. 2) GFR79-A
Bibliography: p. 124-126.
Includes index.
ISBN 3-87437-150-6 : DM39.00
1. Mural painting and decoration—Germany, West—Göppingen (Landkreis) 2. Mural painting and decoration, German—Germany, West—Göppingen (Landkreis) 3. Christian art and symbolism—Germany, West—Göppingen (Landkreis) 4. Göppingen, Ger. (Landkreis)—History. I. Title. II. Series: Kreisarchiv Göppingen. Veröffentlichungen ; Bd. 6. III. Series: Kunst in Bayern und Schwaben ; Bd. 2.
ND2750.G63H85 79-367438
79₍r79₎rev MARC

—GERMANY, WEST—IPHOFEN

Groll, Max.
Katholische Stadtpfarrkirche St. Veit, Iphofen : Kreis Kitzingen a. Main, seit 1810 Erzdiözese Bamberg, Patron St. Vitus (Fest 15. Juni), Fest der Kirchweihe 22. Sept. / ₍Max Groll₎. — 4., erw. Aufl. — München : Schnell & Steiner, 1976.
18 p. : ill. (some col.) ; 17 cm. — (Kunstführer ; Nr. 333) GFR***
Caption title.
On cover: Iphofen.
Includes bibliographical references.
1. Sankt Veitkirche, Iphofen, Ger. 2. Christian art and symbolism—Germany, West—Iphofen. I. Title. II. Series: Kleine deutsche Kirchenführer ; Nr. 333.
NA5564.K53 Nr. 333 79-381618
₍NA5586.S2596₎ 79 MARC

—GERMANY, WEST—MEMMINGEN

Schnell, Hugo, 1904-
Sankt Josef Memmingen : Katholische Stadtpfarrkirche / ₍Hugo Schnell, Otto Lutz₎. — 2., neubearb. Aufl. — München : Schnell & Steiner, 1975.
13 p. : ill. ; 17 cm. — (Kunstführer ; Nr. 202) GFR***
Caption title.
Includes bibliographical references.
1. Sankt Josef (Church), Memmingen, Ger. 2. Christian art and symbolism—Germany, West—Memmingen. I. Lutz, Otto, Stadtpfarrer, joint author. II. Title. III. Series: Kleine deutsche Kirchenführer ; Nr. 202.
NA5564.K53 Nr. 202 79-381621
₍NA5586.S257₎ 79 MARC

—GERMANY, WEST—MUNICH

Altmann, Lothar.
Katholische Stadtpfarrkirche St.₍i.e. Sankt₎ Maximilian : Dekanat München-Mitte, Patron Hl. Maximilian, (12. Okt.) / Lothar Altmann, Hugo Schnell. — 2., neubearb. Aufl. — München : Schnell & Steiner, 1976.
14 p. : ill. ; 17 cm. — (Kunstführer ; Nr. 284) GFR***
Caption title.
On cover: St. Maximilian/München.
Includes bibliographical references.
1. Sankt Maximilian (Church), Munich. 2. Christian art and symbolism—Germany, West—Munich. I. Schnell, Hugo, 1904- joint author. II. Title. III. Title: Sankt Maximilian, München. IV. Series: Kleine deutsche Kirchenführer ; Bd. 284.
NA5564.K53 Nr.284 79-381620
₍NA5586.S2585₎ 79 MARC

—GERMANY, WEST—OSNABRÜCK

Schewe, Josef.
Der Dom zu Osnabrück / ₍Josef Schewe₎. — 1. Aufl. — München ; Zürich : Schnell und Steiner, 1977.
22 p. : ill. (some col.) ; 17 cm. — (Kunstführer ; Nr. 1119) GFR78-A
Caption title.
Bibliography: p. 22.
1. Osnabrück. Cathedral. 2. Christian art and symbolism—Germany, West—Osnabrück. I. Title. II. Series: Kleine deutsche Kirchenführer ; Nr. 1119.
NA5564.K53 Nr. 1119 914.3'04'8708 s 78-395791
₍NA5586.O83₎ 79 MARC

—GERMANY, WEST—PASSAU

Schäffer, Gottfried.
Wallfahrten im Passauer Land / Gottfried Schäffer ; Gregor Peda. — Freilassing : Pannonia, 1978.
48 p. : 25 ill. ; 17 cm. — (Kleine Pannonia-Reihe ; 69) GFR78-A
Cover title.
ISBN 3-7897-0069-X : DM3.80
1. Church architecture—Germany, West—Passau. 2. Christian art and symbolism—Germany, West—Passau. I. Peda, Gregor. II. Title.
NA5576.P35S26 914.3'35 78-388781
79 MARC

—GERMANY, WEST—RATISBON

Altmann, Lothar.
Die Stiftskirche St. ₍i.e. Sankt₎ Johann in Regensburg / ₍Lothar Altmann ; Fotos, Wilkin Spitta₎. — 1. Aufl. — München ; Zürich : Schnell und Steiner, 1977.
15 p. : numerous ill. ; 17 cm. — (Kunstführer ; Nr. 1114) GFR78-A
Caption title.
Includes bibliographical references.
1. Stiftskirche Sankt Johann, Ratisbon. 2. Christian art and symbolism—Germany, West—Ratisbon. I. Title. II. Series: Kleine deutsche Kirchenführer ; Nr. 1114.
NA5564.K53 Nr. 1114 914.3'04'8708 s 78-395794
₍NA5586.R3₎ 79 MARC

—GERMANY, WEST—SCHLESWIG

Støtt, Henrik.
Kirkekunst i Sydslesvig / Henrik Støtt ; ₍fotos, Elke Prien₎. — ₍København₎ : Grænseforeningen : ₍eksp.₎, Idag, ₍1976₎
40 p. : ill. ; 21 cm. — (Grænseforeningens årbog ; 1976) D76-Feb.
kr15.75
1. Christian art and symbolism—Germany, West—Schleswig. I. Prien, Elke. II. Title. III. Series: Grænseforeningen. Årbog ; 1976.
N7950.A3S347 78-383696
78 MARC

—GERMANY, WEST—TRAUNSTEIN

Schierghofer, Georg.
Pfarrkirche St. ₍i.e. Sankt₎ Oswald, Traunstein : Erzdiözese München-Freising, früher Archidiakonat Baumburg a. Alz / ₍Georg Schierghofer ; überarb. u. erg., Wilhelm Schels₎. — 2., neubearb. Aufl. — München ; Zürich : Schnell und Steiner, 1976.
14 p. : numerous ill. ; 17 cm. — (Kunstführer ; Nr. 536) GFR76-A
Caption title.
Cover title: St. Oswald, Traunstein.
Bibliography: p. 14.
1. Pfarrkirche Sankt Oswald, Traunstein, Ger. 2. Christian art and symbolism—Germany, West—Traunstein. I. Schels, Wilhelm. II. Title. III. Series: Kleine deutsche Kirchenführer ; Nr. 536.
NA5564.K53 Nr. 536 79-381617
₍NA5586.P336₎ 79 MARC

—GERMANY, WEST—WALDKIRCH

Birth, Adalbert, 1927-
St. ₍i.e. Sankt₎ Margaretha Waldkirch : Ldkr. Emmendingen : Erzbistum Frieburg i. B. / ₍Adalbert Birth₎. — 1. Aufl. — München ; Zürich : Schnell und Steiner, 1977.
18, ₍1₎ p. : numerous ill. (some col.) ; 17 cm. — (Kunstführer ; Nr. 1117)
Caption title.
Cover title: Waldkirch, St. ₍i.e. Sankt₎ Margaretha.
Bibliography: p. ₍19₎
GFR78-A
1. Sankt Margaretha (Church) Waldkirch, Ger. 2. Christian art and symbolism—Germany, West—Waldkirch. I. Title. II. Title: Waldkirch, Sant Margaretha. III. Series: Kleine deutsche Kirchenführer ; Nr. 1117.
NA5564.K53 Nr. 1117 914.3'04'8708 s 78-398051
₍NA5586.W24₎ 78₍r79₎rev MARC

—GERMANY, WEST—WITTLAER

Scheiermann, Friedrich.
Düsseldorf-Wittlaer / von Friedrich Scheiermann. — 1. Aufl. — Neuss : Gesellschaft für Buchdruckerei, 1976.
15 p. : ill. (1 col.) ; 21 cm. — (Rheinische Kunststätten ; Heft 185 (8/1976))
Caption title.
Bibliography: p. 12.
ISBN 3-88094-161-0
1. Pfarrkirche Sankt Remigius, Wittlaer. 2. Christian art and symbolism—Germany, West—Wittlaer. I. Title. II. Series: Rheinische Kunststätten ; Heft 185.
N3.R45 Heft 185 914.3'42 s 79-362585
₍NA5586.P34₎ 79 MARC

—GREECE, MODERN—LEVKAS ISLAND

Rontogiannēs, Panos G
(Hē Christianikē technē stē Leukada)
'Η Χριστιανικὴ τέχνη στὴ Λευκάδα / Πάνου Γ. Ροντογιάννη. — 'Αθῆναι : ₍'Εταιρεία Λευκαδικῶν Μελετῶν₎ : κεντρικὴ πώλησις Βιβλιοπωλεῖον Γρηγόρη, 1974.
25, 628 p. : ill. ; 24 cm. — ('Επετηρίς - 'Εταιρεία Λευκαδικῶν Μελετῶν ; τ. 3–1973)
Includes bibliographical references.
1. Christian art and symbolism—Greece, Modern—Levkas Island. I. Title. II. Series: Hetaireia Leukadikōn Meletōn. Epetēris - Hetaireia Leukadikōn Meletōn ; 1973/t. 3.
DF901.L4H47a 1973 t. 3 77–517195
[N7951.A3]

—ITALY

Schumacher, Walter Nikolaus.
Hirt und "Guter Hirt" : Studien zum Hirtenbild in d. röm. Kunst vom 2. bis zum Anfang d. 4. Jh. unter bes. Berücks. d. Mosaiken in d. Südhalle von Aquileja / Walter Nikolaus Schumacher. — Rom ; Freiburg im Breisgau ; Wien : Herder, 1977.
334, 64 p. : 328 ill. ; 24 cm. — (Römische Quartalschrift für christliche Altertumskunde und Kirchengeschichte : Supplementheft ; 34) GFR78-A17
Habilitationsschrift—Freiburg im Breisgau, 1968.
Bibliography: p. 7-18.
Includes indexes.
ISBN 3-451-17920-2 : DM173.00
1. Art, Roman—Italy. 2. Art, Early Christian—Italy. 3. Shepherds in art. 4. Christian art and symbolism—Italy. 5. Sepulchral monuments—Italy. 6. Mural painting and decoration, Early Christian—Italy. 7. Mosaics, Early Christian—Italy. 8. Aquileia. Cattedrale. I. Title. II. Series.
N5740.S38 79-357899
79 MARC

—ITALY—ADDRESSES, ESSAYS, LECTURES

Chiese e cattedrali / ₍a cura di Maria Rossella Bigi, Maria Raffaella Ceccopieri₎ ; premessa storica di Giuseppe Alberigo ; saggio critico di Marcello Fagiolo. — Milano : Touring club italiano, ₍1978₎
207 p. : ill. ; 30 cm. — (Italia meravigliosa ; 8) It78-Dec
Bibliography: p. 206.
Includes index.
1. Christian art and symbolism—Italy—Addresses, essays, lectures. I. Touring club italiano.
N7952.A1C44 79-350611
*79 MARC

—ITALY—ABRUZZI—EXHIBITIONS

1. ₍i.e. Primo₎ Premio Marsica : Mostra regionale d'arte sacra : pittura : Scuola media C. Corradini, 16-30 luglio 1978, Avezzano. — Avezzano : Comitato ricerca espressione artistica sacra, 1978.
159 p. : ill. ; 21 cm. It79-Feb
1. Painting, Italian—Italy—Abruzzi—Exhibitions. 2. Painting, Modern—20th century—Italy—Abruzzi—Exhibitions. 3. Christian art and symbolism—Italy—Abruzzi—Exhibitions. 4. Premio Marsica—Exhibitions. I. Comitato ricerca espressione artistica sacra. II. Title: Mostra regionale d'arte sacra.
ND1432.I8P74 79-364548
*79 MARC

CHRISITAN ART AND SYMBOLISM (Continued)

—ITALY—ASSISI—CATALOGS

Museo francescano.
Il Museo francescano : catalogo / a cura di P. Gerlach, S. Gieben, M. D'Alatri. — Roma : Istituto storico dei cappuccini, 1973.
104 p., [13] leaves of plates : ill. ; 22 cm. It***
Includes bibliographical references and index.
1. Christian art and symbolism—Italy—Assisi—Catalogs. 2. Museo francescano—Catalogs. I. Gerlach, P. II. Gieben, S. III. D'Alatri, Mariano.
N7823.I8R655 78-387865
79 MARC

—ITALY—NAPLES

La Real cappella del Tesoro di S. Gennaro : documenti inediti / [a cura di] Franco Strazzullo ; introduzione di Raffaello Causa. — Napoli : Società editrice napoletana, [1978]
xv, 229 p., [30] leaves of plates : ill. ; 24 cm. — (Studi e testi di storia e critica dell'arte ; 5) It78-Nov
Bibliography: p. 197-198.
Includes indexes.
L15000
1. Naples. Duomo. Cappella del Tesoro. 2. Christian art and symbolism—Italy—Naples. I. Strazzullo, Franco.
BX4634.N3D867 79-338860
•79 MARC

—ITALY—SANTORSO

Cuman, Fiorenzo.
Capitelli a Santorso : saggio sulla pietà popolare nell'Alto Vicentino / Fiorenzo Cuman, Tarcisio Pirocca ; presentazione di Giovanni Mantese. — [s.l.] : Ascledum, 1976.
66 p. : ill. (some col.) ; 23 cm. It***
Includes bibliographical references.
1. Santorso, Italy—Shrines. 2. Christian art and symbolism—Italy—Santorso. I. Pirocca, Tarcisio, joint author. II. Title.
NB1793.I8C85 79-380153
79 MARC

—ITALY—TRIESTE—EXHIBITIONS

Tesori delle comunità religiose di Trieste : Mostra promossa dal Comune di Trieste e dall'Azienda autonoma di soggiorno e turismo di Trieste : Museo del Castello di San Giusto, luglio-settembre 1978 / a cura dei Civici musei di storia ed arte di Trieste ; direzione della Mostra e del catalogo Laura Ruaro Loseri ; schede e testi di Marisa Bianco Fiorin ... [et al.]. — Udine : Istituto per l'Enciclopedia del Friuli-Venezia Giulia, 1978.
118 p. : ill. (some col.) ; 23 cm. It79-Apr
1. Christian art and symbolism—Italy—Trieste—Exhibitions. I. Ruaro Loseri, Laura. II. Civici musei di storia ed arte di Trieste.
N7952.T74T47 79-361961
•79 MARC

—JAPAN—NAGASAKI—EXHIBITIONS

Tōkyō Kokuritsu Hakubutsukan.
(Kirishitan kankei ihin ten)
キリシタン関係遺品展：東京国立博物館所蔵長崎奉行所収納品を中心とする：[昭和48年10月20日-11月11日：長崎県立美術博物館] / 東京国立博物館，長崎県教育委員会，長崎県立美術博物館. — [東京]：東京国立博物館，昭和48 [1973]
32 p. : ill. ; 26 cm.
Catalog of an exhibition.
1. Christian art and symbolism—Japan—Nagasaki—Exhibitions. 2. Christian art and symbolism—Japan—Tokyo—Exhibitions. 3. Nagasaki Bugyōsho. I. Nagasaki Kenritsu Bijutsu Hakubutsukan. II. Nagasaki (Prefecture). Kyōiku Iinkai. III. Title.
N7824.5.J3N337 79-802990

—JAPAN—TOKYO—EXHIBITIONS

Tōkyō Kokuritsu Hakubutsukan.
(Kirishitan kankei ihin ten)
キリシタン関係遺品展：東京国立博物館所蔵長崎奉行所収納品を中心とする：[昭和48年10月20日-11月11日：長崎県立美術博物館] / 東京国立博物館，長崎県教育委員会，長崎県立美術博物館. — [東京]：東京国立博物館，昭和48 [1973]
32 p. : ill. ; 26 cm.
Catalog of an exhibition.
1. Christian art and symbolism—Japan—Nagasaki—Exhibitions. 2. Christian art and symbolism—Japan—Tokyo—Exhibitions. 3. Nagasaki Bugyōsho. I. Nagasaki Kenritsu Bijutsu Hakubutsukan. II. Nagasaki (Prefecture). Kyōiku Iinkai. III. Title.
N7824.5.J3N337 79-802990

—POLAND

Bohdziewicz, Piotr.
Studia z dziejów sztuki polskiej w okresie baroku i rokoka : kościół i klasztor PP. Wizytek w Warszawie, kościół fundacyjny w Kobyłce pod Warszawą i jego dekoracja malarska, kościół i klasztor OO. Bernardynów w Dubnie / Piotr Bohdziewicz. — Wyd. 1. — Lublin : Wydawn. Towarzystwa Naukowego Katolickiego Uniwersytetu Lubelskiego, 1973.
159 p., [56] leaves of plates : ill. ; 24 cm. — (Towarzystwo Naukowe Katolickiego Uniwersytetu Lubelskiego. Rozprawy Wydziału Historyczno-Filologicznego ; 37)
Includes bibliographical references.
1. Art, Polish. 2. Art, Baroque—Poland. 3. Art, Rococo—Poland. 4. Christian art and symbolism—Poland. 5. Churches—Poland. 6. Monasteries—Poland. I. Title.
N7255.P6B63 79-413686
79 MARC

Sztuka sakralna w Polsce na Ziemiach Zachodnich i Północnych / oprac. Tadeusz Dobrzeniecki. — Warszawa : Ars Christiana, 1976 (Warsz. : Tamka)
52, [4] p., [128] leaves of plates : ill. ; 30 cm. P77-9334
Errata slip inserted.
Bibliography: p. [54]
zł350.00
1. Christian art and symbolism—Poland. 2. Oder-Neisse area. I. Dobrzeniecki, Tadeusz.
N7971.P6S9 78-382361
78 MARC

—POLAND—CZĘSTOCHOWA

Die Kunstschätze des Klosters Jasna Góra / Janusz St. Pasierb, Jan Samek, Kazimierz Szafraniec ; [Fotos, Jan Michlewski, Janusz Rosikon ; übertr. aus d. Poln., Josef Adamiak u. Siegfried Schmidt]. — 1. Aufl. — Leipzig : Seemann, VEB, 1977.
148, [1] p. : chiefly col. ill. ; 32 cm. GFR78-A4
Bibliography: p. 148-[149]
60.00M
1. Christian art and symbolism—Poland—Częstochowa. 2. Częstochowa, Poland. Klasztor Paulinów. I. Pasierb, Janusz St. II. Samek, Jan. III. Szafraniec, Kazimierz.
N7971.P62C974 704.948'2'094385 78-383522
78 MARC

—POLAND—GLATZ (GRAFSCHAFT)—PICTORIAL WORKS

Grafschafter Kirchen in heutiger Zeit : Aufnahmen 1976-1978 / Bildausw. u. Zs.-Stellung Jörg Marx. — Leimen/Heidelberg : Marx, 1978.
340 p. : chiefly ill. ; 29 cm. GFR79-A8
Includes bibliographical references and index.
DM69.00
1. Churches—Poland—Glatz (Grafschaft)—Pictorial works. 2. Christian art and symbolism—Poland—Glatz (Grafschaft)—Pictorial works. I. Marx, Jörg.
NA5955.P62G454 79-379398
79 MARC

—POLAND—RZESZÓW (VOIVODESHIP)

Carmazinu Cacovschi, V
(Mystetstvo Lemkivs'koï tserkvy)
Мистецтво Лемківської церкви = De arte sacra ecclesiarum Lemcoviensium / В. Кармазин-Каковський. — Рим : [s. n.], 1975.
306 p. : ill. ; 24 cm. — (Видання Українського католицького університету ім. св. Климента Папи) (Праці Філософічно-гуманістичного факультету ; т. 12) It***
Summary in English, French, and Latin.
Bibliography: p. 185-187.
1. Christian art and symbolism—Poland—Rzeszów (Voivodeship) 2. Stave churches—Poland—Rzeszów (Voivodeship) I. Title. II. Title: De arte sacra ecclesiarum Lemcoviensium. III. Series: Rome (City). Ukraïns'kyĭ katolyts'kyĭ universytet. Vydannia. IV. Series: Rome (City). Ukraïns'kyĭ katolyts'kyĭ universytet. Filosofichno-humanistychnyĭ fakul'tet. Pratsi ; t. 12.
N7971.P6C37 76-530615

—RUSSIAN REPUBLIC—ZVENIGOROD

Nikolaeva, Tat'iana Vasil'evna.
(Drevniĭ Zvenigorod)
Древний Звенигород = Old Zvenigorod : Архитектура, искусство : [Альбом] / Т. В. Николаева. — Москва : Искусство, 1978.
207 p. : col. ill. ; 26 cm. USSR 78
Summary in English; list of illustrations also in English.
Bibliography: p. [201]-203.
1. Church architecture—Russian Republic—Zvenigorod. 2. Architecture—Russian Republic—Zvenigorod. 3. Christian art and symbolism—Russian Republic—Zvenigorod. I. Title. II. Title: Old Zvenigorod.
NA5697.Z84N54 79-410972

—SCOTLAND

Willsher, Betty.
Stones : a guide to some remarkable eighteenth century gravestones / by Betty Willsher and Doreen Hunter ; photos. by Betty Willsher. — Edinburgh : Canongate, 1978.
xi, 139 p. : ill. ; 23 cm. GB***
Bibliography: p. 133.
Includes index.
ISBN 0-903937-36-0 : £7.95
1. Sepulchral monuments—Scotland. 2. Christian art and symbolism—Scotland. 3. Emblems. I. Hunter, Doreen, joint author. II. Title.
NB1862.W54 730'.9411 79-304209
79 MARC

Willsher, Betty.
Stones : a guide to some remarkable eighteenth century gravestones / by Betty Willsher and Doreen Hunter ; photos. by Betty Willsher. — New York : Taplinger Pub. Co., 1979, c1978.
xi, 139 p. : ill. ; 23 cm.
Bibliography: p. 133.
Includes index.
ISBN 0-8008-7447-1 : $14.95
1. Sepulchral monuments—Scotland. 2. Christian art and symbolism—Scotland. 3. Emblems. I. Hunter, Doreen, joint author. II. Title.
NB1862.W54 1979 730'.9411 78-62268
79 MARC

—SWEDEN—VÄSTERÅS

Ekström, Gunnar, 1893-
Västerås domkyrkas inventarier genom tiderna / Gunnar Ekström. — Västerås : [Kulturnämnden], 1976.
263 p., [12] leaves of plates : ill. (some col.) ; 31 cm. — (Västerås kulturnämnds skriftserie ; nr. 3) S77-39
Bibliography: p. 263.
Includes indexes.
kr75.00
1. Västerås domkyrka. 2. Christian art and symbolism—Sweden—Västerås. I. Title. II. Series: Västerås kulturnämnd. Västerås kulturnämnds skriftserie ; nr. 3.
NA5791.V26E37 77-579054
78 MARC

—SWITZERLAND

Speich, Klaus.
Kirchen und Klöster in der Schweiz / Klaus Speich, Hans R. Schläpfer ; [hrsg. von Niklaus Flüeler ; an d. Texten arbeiteten mit, Henriette Bon ... [et al.] ; die Grundrisse u. Isometrien zeichnete Andreas Adam ; Kt. Heinz Engel]. — München : Beck, 1978.
344 p. : numerous ill. (some col.), maps ; 30 cm. GFR79-A
Bibliography: p. 340-341.
Includes index.
ISBN 3-406-04111-6 : DM88.00
1. Churches—Switzerland. 2. Monasteries—Switzerland. 3. Convents and nunneries—Switzerland. 4. Christian art and symbolism—Switzerland. I. Schläpfer, Hans Ruedi, 1929- II. Flüeler, Niklaus, 1934- III. Title.
NA5841.S63 726'.5'09494 79-358419
79 MARC

—SWITZERLAND—AARGAU

Weber, Hans, 1948-
Zeichen des Glaubens : religiöse Bildwelt im Alltag / [Photos] Hans Weber ; [einführender Text] Rolf Thalmann. — Lenzburg : E. C. Otz, [1977]
16 p., [25] leaves of plates : ill. (some col.) ; 23 cm. Sw77-10950
19.80F
1. Christian art and symbolism—Switzerland—Aargau. 2. Religious articles—Switzerland—Aargau. 3. Folk art—Switzerland—Aargau. I. Thalmann, Rolf. II. Title.
N7964.A3A188 79-362970
•79 MARC

—SWITZERLAND—RHÄZÜNS

Wyss, Alfred.
Razén/Rhäzüns : Sogn Gieri, Sogn Paul, Nossadunna / Alfred Wyss. — Basel [Münzgässlein 16] : Gesellschaft für Schweizerische Kunstgeschichte, c1977.
19 p. : ill. ; 20 cm. — (Schweizerische Kunstführer) Sw78-3752
Cover title.
Bibliography: p. 19.
1. Christian art and symbolism—Switzerland—Rhäzüns. 2. Rhäzüns, Switzerland—Churches. I. Title. II. Series.
N7964.R45W9 914.94'7 79-367480
79 MARC

CHRISTIAN ART AND SYMBOLISM (Continued)

—UKRAINE

Lavrs'kyĭ muzeĭ kul'tiv i pobutu. Viddil pis'ma ĭ druku.
(Ksylohrafichni doshky Lavrs'koho muzeiu)
Ксилографічні дошки Лаврського музею ... обробив до друку і вступний нарис написав П. М. Попов. Київ, З друкарні поліграфічного факультету Художнього інституту, 1927–
v. illus., plates. 47 cm.
At head of title, v. 1– : Всеукраїнський музейний городок. Лаврський музей культів і побуту. Відділ письма й друку.
Added t.-p., v. 1– : Clichés xylographiques du Musée de Lavra.
Résumé (in v. 1) in French.
1. Wood-engraving, Ukrainian. 2. Christian art and symbolism—Ukraine. I. Popov, Pavlo Mykolaĭovych, 1890– II. Title. III. Title: Clichés xylographiques du Musée de Lavra.
NE1156.4.U37L38 1927 40–38548

CHRISTIAN ATHEISM see Death of God theology

CHRISTIAN BIOGRAPHY

see also Abbots; Apostles; Bishops; Christian hagiography; Christian martyrs; Christian saints; Christian saints, Women; Fathers of the church; Hermits; Missionaries; Monasticism and religious orders; Puritans; Reformers; Saints; Theologians

Chandler, Russell.
The overcomers / Russell Chandler. — Old Tappan, N.J. : Revell, c1978.
158 p. ; 22 cm.
CONTENTS: His will brings her joy: Elisabeth Elliot.—Towering theologian: Carl F. H. Henry.—They just praise the Lord: Bill and Gloria Gaither.—Hang in there, brother: Edward Victor Hill.—Woman of prayer: Vonette Bright.—Changing her world: Ann Kiemel.—Yielded novice: Maria Von Trapp.—Thoughtful Quaker: D. Elton Trueblood.—He tries God: Walter Hoving.—Pastoral partners: Frederick W. and Ruth P. Cropp.—Southern singer: Joanne Cash Yates.—Elder statesman: Paul S. Rees.—Conclusion: through it all.
ISBN 0-8007-0944-6 : $6.95
1. Christian biography. 2. Christian life—1960- I. Title.
BR1700.2.C46 209'.2'2 78-16621
78 MARC

Foertsch, Ursula.
Mein Leben, sein Sieg : Corrie ten Boom / Ursula Foertsch. — Giessen ; Basel : Brunnen-Verlag, 1978.
96 p. ; 18 cm. — (ABCteam ; 3029 : Berichte, Erzählungen, Lebensbilder)
GFR78-A
Includes bibliographical references.
ISBN 3-7655-3029-8 : DM3.95
1. Ten Boom, Corrie. 2. Christian biography. I. Title.
BR1725.T35F67 79-384316
79 MARC

The Guideposts treasury of hope. — Garden City, N.Y. : Doubleday, 1979, c1976.
305 p. ; 22 cm.
ISBN 0-385-14975-1 : $8.95
1. Hope. 2. Christian biography.
BV4638.G83 248'.2 79-114012
00 MARC

Kelsey, Morton T
The age of miracles : seven journeys to faith / Morton Kelsey. — Notre Dame, Ind. : Ave Maria Press, c1979.
77 p. ; 21 cm.
ISBN 0-87793-169-0 : $2.45
1. Faith-cure. 2. Christian biography. I. Title.
[BT732.5.K38] 248'.2 78-74095
79[r79]rev MARC

Knowles, Leo.
Candidates for sainthood / by Leo Knowles. — St. Paul : Carillon Books, 1978.
ix, 138 p. ; 22 cm.
Bibliography: p. 137-138.
CONTENTS: Charles de Foucauld.—John Green Hanning.—Matt Talbot.—Miguel Pro.—Brother Andre.—Edel Mary Quinn.—Luigi Orione.—Vincent McNabb.—Edith Stein.—Titus Brandsma.—Bahkita.—Tom Dooley.
ISBN 0-89310-035-8 : $8.95. ISBN 0-89310-036-6 pbk.
1. Catholics—Biography. 2. Christian biography. I. Title.
BX4651.2.K56 282'.092'2 78-55247
79 MARC

Ndebele, Salu Daka.
Guerrilla for Christ / Salu Daka Ndebele, with Dan Wooding. — Old Tappan, N.J. : F. H. Revell Co., c1978.
125 p. ; 18 cm. — (Spire books)
ISBN 0-8007-8351-4
1. Ndebele, Salu Daka. 2. Christian biography. 3. Communism and Christianity. I. Wooding, Dan, joint author. II. Title.
BR1725.N35A33 248'.2 78-13087
78 MARC

Nigg, Walter, 1903-
Heilige ohne Heiligenschein / Walter Nigg. — Olten [etc.] : Walter, cop. 1978.
254 p. ; 21 cm. Sw78-11200
ISBN 3-530-61208-1 : 26.00F
1. Christian biography. I. Title.
BR1700.2.N483 79-341748
79 MARC

Our struggle to serve : the stories of 15 evangelical women / Virginia Hearn, editor. — Waco, Tex. : Word Books, c1979.
191 p. ; 23 cm.
Includes bibliographical references.
ISBN 0-8499-0089-1 : $7.95
1. Christian biography. 2. Women—Biography. I. Hearn, Virginia.
BR1713.O97 280'.4'0922 78-59434
79[r79]rev MARC

Searle, John Derek.
Twentieth century Christians : twelve mini-biographies / John D. Searle. — Edinburgh : Saint Andrew Press, 1977.
vii, 182 p. ; 20 cm. GB***
Includes bibliographical references.
ISBN 0-7152-0367-3 : £2.10
1. Christian biography. I. Title.
BR1700.2.S36 78-314266
78 MARC

Thompson, Phyllis.
Capturing voices : the story of Joy Ridderhof / by Phyllis Thompson. — London [etc.] : Hodder and Stoughton, 1978.
190 p. ; 18 cm. — (Hodder Christian paperbacks) GB78-17984
American ed. published under title: Count it all Joy!
ISBN 0-340-22447-9 : £0.95
1. Ridderhof, Joy. 2. Christian biography. 3. Phonorecords in missionary work. I. Title.
BR1725.R56T48 1978b 269'.2'0924 79-309584
79 MARC

—JUVENILE LITERATURE

(Hikari o kakagate)
光をかかげて / [編者　上沢謙二, 永井明]. — 東京：ヨルダン社, 1967.
2 v. : ill. ; 21 cm.
CONTENTS: [1] 日本篇—[2] 外国編
1. Christian biography—Juvenile literature. I. Uesawa, Kenji, 1890– II. Nagai, Akira, 1921–
BR1704.H54 79–809455

—CALIFORNIA—LOS ANGELES

Barrett, Ethel.
Barrett : a street cop who cared / as told by Ethel Barrett. — Old Tappan, N.J. : F. H. Revell Co., c1978.
189 p. ill. ; 24 cm.
ISBN 0-8007-0918-7
1. Barrett, Gary, Sgt. 2. Christian biography—California—Los Angeles. 3. Los Angeles—Police—Biography. I. Title.
BR1725.B347A33 248'.2 78-710
78[r79]rev MARC

—CALIFORNIA—ORANGE CO.

Nason, Donna.
Tara, child of hopes & dreams / by Donna Nason. — Wheaton, Ill. : Tyndale House Publishers, c1978.
223 p. : ill. ; 21
ISBN 0-8423-6920-1 (pbk.) : $4.95
1. Brain-damaged children—California—Orange Co.—Biography. 2. Nason, Tara. 3. Christian biography—California—Orange Co. I. Title.
RJ496.B7N37 248'.2 78-54087
00 MARC

—CHINA

Hefley, James C
China : Christian martyrs of the 20th century : an excerpt from By their blood / by James and Marti Hefley. — Milford, MI : Mott Media, c1978.
78 p. ; 23 cm.
ISBN 0-915134-28-4. ISBN 0-915134-16-0 pbk.
1. Christian martyrs—Biography. 2. Missionaries—China—Biography. 3. Christian biography—China. I. Hefley, Marti, joint author. II. Title.
BR1608.C6.H43 272'.092'2 78-6187
78 MARC

—FINLAND

Salo, Jussi, 1928-
Viikon kahdeksas päivä / Jussi Salo. — Lapua : Herättäjäyhdistys, 1976.
139 p. ; 19 cm. Fi
ISBN 9519013474
1. Salo, Jussi, 1928- . 2. Christian biography—Finland. 3. Christian life—1960- . I. Title.
BR1725.S244A38 79-364820
*79 MARC

—FRANCE

Pierre, Jean.
Les mecs de la boîte / Jean Pierre. — Paris : Les Éditions ouvrières, c1979.
115 p. ; 18 cm. — (Collection Jeunes travailleurs) F***
ISBN 2-7082-1991-X : 23.00F
1. Pierre, Jean. 2. Christian biography—France. 3. Christian life—1960- . 4. Trade-unions—France. I. Title.
BR1725.P515A35 248'.2'0924 79-370456
79 MARC

—GEORGIA—ATLANTA

Haggai, John Edmund.
My son Johnny / John Edmund Haggai. — Wheaton, Ill. : Tyndale House, c1978.
238 p., [4] leaves of plates : ill. ; 21 cm.
ISBN 0-8423-4647-3 : $4.95
1. Brain-damaged children—Georgia—Atlanta—Biography. 2. Haggai, John Edmund, 1950-1975. 3. Christian biography—Georgia—Atlanta. I. Title.
RJ496.B7H33 362.7'8'1968588 75-7229
00 MARC

—GERMANY

Voss, Gustav.
(Nihon no chichi e)
日本の父へ / グスタフ・フォス. — 東京：新潮社, 昭和52 [1977]
217 p., [1] leaf of plates : port. ; 20 cm. Ja 77–8332
¥750
1. Voss, Gustav. 2. Educators—Germany—Biography. 3. Christian biography—Germany. 4. Fathers. 5. Domestic education—Japan. I. Title.
LB775.V692A36 79–805972

Wer mir dienen will : 24 Lebensbilder von Männern u. Frauen in Dienst d. Liebe / hrsg. im Auftr. von Innere Mission u. Hilfswerk d. Evang. Kirchen in d. DDR von Gerhard Bosinski u. Paul Toaspern ; [Porträtzeichn. von Werner Juza]. — 1. Aufl. — Berlin : Evangelische Verlagsanstalt, 1978.
440 p. : ports. ; 22 cm. GDR78-A
Includes bibliographical references.
14.80M
1. Christian biography—Germany. 2. Inner missions—Biography. 3. Church charities—Germany—Biography. I. Bosinski, Gerhard. II. Toaspern, Paul.
BR859.W47 79-382239
79 MARC

—GERMANY—ADDRESSES, ESSAYS, LECTURES

Sie riefen zum Leben / hrsg. von Arno Pagel. — 1. Aufl. — Marburg an d. Lahn : Verlag der Francke-Buchh., 1977.
183 p. : ill. ; 21 cm. GFR78-A13
ISBN 3-88224-006-7 : DM13.80
1. Christian biography—Germany—Addresses, essays, lectures. I. Pagel, Arno.
BR1700.2.S53 78-379027
78 MARC

CHRISTIAN BIOGRAPHY (Continued)

—INDIA

Enoch, H
Following the Master : memoirs of a professor / H. Enoch. — 1st ed. — Bombay : Jyoti Pocketbooks, 1977.
108 p., [4] leaves (p. 107-108 advertisements) : ill. ; 20 cm.
Rs4.50
1. Enoch, H. 2. Christian biography—India. I. Title.
BR1725.E57A34 289 78-900026
79 MARC

Kiṟistutās, T A
Rhenius, apostle of Tirunelveli / by D. A. Christadoss. — [டேனிஷ்பேட்டை, சேலம் மாவட்டம்] : Bethel Publication, 1976.
x, 472 p., [2] leaves of plates : ill., map ; 19 cm.
Cover title in Tamil: Tirunelvēli appostalaṉ Rēṉiyas.
In Tamil.
Tamil works by Rhenius: p. 471–472.
Includes bibliographical references and index.
Rs7.00
1. Rhenius, Charles Theophilus Ewald, 1790–1838. 2. Christian biography—India. I. Title: Tirunelvēli appostalaṉ Rēṉiyas.
BR1725.R48K57 78–903376

Kiṟistutās, T A
(Tirunelvēli Tiruccapaiyiṉ mutalām irattaccāṭci Tāvītu Cuntarāṉantam)
திருநெல்வேலி திருச்சபையின் முதலாம் இரத்தச்சாட்சி தாவீது சுந்தரானந்தம் / தே. அ. கிறிஸ்துதாஸ். — [Palayamkottai : கிடைக்குமிடம் Diocesan Book Depot, pref. 1976]
71 p. ; 19 cm.
In Tamil.
Includes bibliographical references.
Rs2.00
1. Cuntarānantam, Tāvītu. 2. Christian biography—India. I. Title.
BR1725.C845K57 77–913297

Kiṟistutās, T A
(Tirunelvēli Tiruccapaiyiṉ pattoṉpatām nūṟṟāṇṭu paricuttavāṭṭikaḷ paṉṉiruvar)
திருநெல்வேலி திருச்சபையின் பத்தொன்பதாம் நூற்றுண்டு பரிசுத்த வாட்டிகள் பன்னிருவர் / ஆக்கியோன் D. A. கிறிஸ்துதாஸ். — [பாளையங் கோட்டை : கிடைக்குமிடம் அத்தியட்சாதீன புத்தகசாலை], 1976.
62 p. ; 20 cm.
In Tamil.
Rs1.75
1. Christian biography—India. I. Title.
BR1700.2.K5 77–913291

—INDIA—TIRUNELVELI (DISTRICT)

Kiristutās, T A
Maraiyaviruntha manikka karkal / by D. A. Chirstadoss. — [Palayamkottai : kiṭaikkumiṭam Diocesan Book Depot], 19
v. ; 20 cm.
In Tamil.
Includes bibliographical references and index.
Rs3.00 (v. 2)
1. Christian biography—India—Tirunelveli (District) 2. Tirunelveli, India (District)—Biography. I. Title.
BR1170.T57K57 78-907083
79 MARC

—INDONESIA—JAVA

Sumanto Wp., I
Kyai Sadrach, seorang pencari kebenaran : sebabak sejarah pekabaran Injil di Jawa Tengah / disusun oleh I. Sumanto Wp. — Jakarta : BPK Gunung Mulia, 1974.
98 p. : ill. ; 18 cm.
1. Radin, Sadrach, Kyai, 1840 or 41-1924. 2. Gereja Kristen Jawa. 3. Christian biography—Indonesia—Java. I. Title.
BX7800.G455S95 76-940674

—ITALY

Bardelli, Francesco, 1946-
David Lazzeretti / Francesco Bardelli. — 2. ed. — [Siena] : Cantagalli, 1978.
129 p. : ill. ; 22 cm. It78-Oct
Includes bibliographies.
L4000
1. Lazzaretti, Davide, 1834-1878. 2. Christian biography—Italy.
BR1725.L26B37 1978 79-358959
*79 MARC

Moscato, Antonio.
Davide Lazzaretti, il messia dell'Amiata : l'ultima delle eresie popolari agli albori del movimento operaio e contadino / Antonio Moscato. — 2. ed. — Roma : Savelli, 1978.
154 p. ; 19 cm. — (Cultura politica ; 213 : Sezione antropolgia ; 3)
It79-Jan
Includes bibliographical references.
L3000
1. Lazzaretti, Davide, 1834-1878. 2. Christian biography—Italy. I. Title.
BR1725.L26M67 1978 79-354717
*79 MARC

O'Donnell, James Joseph, 1950-
Cassiodorus / James J. O'Donnell. — Berkeley : University of California Press, c1979.
xvi, 303 p. ; 23 cm.
Bibliography: p. 273-296.
Includes index.
ISBN 0-520-03646-8
1. Cassiodorus Senator, Flavius Magnus Aurelius, ca. 487-ca. 580. 2. Christian biography—Italy.
BR1720.C4O36 271'.0092'4 77-93470
79 MARC

—JAPAN

Akai, Yonekichi, 1887-1974.
(Kono michi)
この道 : [赤井米吉遺稿集]. — 武蔵野 : 赤井つる, 昭和50 [1975]
213 p. : ill. ; 19 cm. Ja ***
1. Akai, Yonekichi, 1887-1974. 2. Educators—Japan—Biography. 3. Christian biography—Japan. I. Title.
LB775.A4342A34 1975 78-809463

Akiyama, Sōzō, 1916-1942.
(Akiyama Sōzō Guntai nikki shō hoka)
秋山宗三『軍隊日記』(抄)ほか / 藤田若雄編. — 東京 : 秋山彰三, 昭和48 [1973]
83 p., [2] leaves of plates : ports. ; 21 cm. Ja 76-7337
Colophon inserted.
¥400
1. Akiyama, Sōzō, 1916-1942. 2. World War, 1939-1945—Personal narratives, Japanese. 3. World War, 1939-1945—Japan. 4. Christian biography—Japan. I. Fujita, Wakao, 1912- II. Title. III. Guntai nikki shō hoka.
D811.A549 78-813747

Ch'a, Yun-sun, 1933-
(Salmŭl wihan chŏngji)
삶을위한停止 / [著者 車潤順]. — 서울 : 普文出版社 : 普及處 새한문고, 1977.
131 p. : 19 cm.
₩700
1. Yoshinari, Minoru. 2. Yoshinari, Yoshiko. 3. Converts—Japan—Biography. 4. Lepers—Japan—Biography. 5. Blind—Japan—Biography. 6. Christian biography—Japan. I. Title.
BV4930.C43 79-825123

Iino, Jūzō.
(Ai wa akeyuku)
愛は明けゆく / 飯野十造著. — 静岡 : 静岡英枝基督教会, 昭和25 [1950], 昭和26 [1951] printing.
306 p., [1] leaf of plates : ill. ; 19 cm.
1. Iino, Jūzō. 2. Missions to lepers. 3. Christian biography—Japan. I. Title.
BU3457.I36A3 78-810108

Ishihara, Hyōei, 1895-
(Kaishinki)
回心記 / 石原兵永著. — 東京 : 新教出版社, 昭和17 [1942], 昭和26 [1951] printing.
229 p. : 19 cm.
1. Ishihara, Hyōei, 1895- 2. Christian biography—Japan. I. Title.
BR1317.I825A35 79-805794

Kanai, Tameichirō, 1887-1963.
(Kanai Tameichirō chosaku shū)
金井為一郎著作集 金井為一郎著作集刊行委員会編 東京 キリスト新聞社 昭和49-52 (1974-77)
3 v. port. 22 cm. Ja 74-6514 (v. 1)
1. Theology—Collected works—20th century. 2. Nihon Kirisuto Kyōdan—Collected works. 3. Kanai, Tameichirō, 1887-1963. 4. Christian biography—Japan.
BX8750.N54K36 74-811070

Masutani, Fumio.
(Shinkō to jinkaku)
信仰と人格 / 増谷文雄著. — 東京 : 青山書院, 昭和22 [1947]
6, 171 p. ; 19 cm.
1. Christian biography—Japan. 2. Priests, Shin—Japan—Biography. I. Title.
BR1316.M37 78-812741

Matsuda, Kiichi, 1921-
(Ōmura Sumitada den)
大村純忠伝 : 付・日葡交渉小史 / 松田毅一著. — 改訂. — 東京 : 教文館, 1978.
463 p. ; 22 cm. Ja 79-3536
Edition for 1963 published under title: Nippo kōshō shi.
Bibliography: p. 462-463.
¥3000
1. Ōmura, Sumitada, 1533-1587. 2. Christian biography—Japan. 3. Daimyo—Biography. 4. Japan—Church history—To 1868. 5. Japan—Relations (general) with Portugal. 6. Portugal—Relations (general) with Japan. I. Title.
BR1317.O45M36 1978 79-809095

Oketani, Hideaki, 1932-
(Tenshin Kanzō Kafū)
天心 鑑三 荷風 / 桶谷秀昭. — 東京 : 小沢書店, 昭和51 [1976]
291 p. ; 20 cm. Ja 76-13822
¥1500
1. Okakura, Kakuzō, 1862-1913. 2. Art critics—Japan—Biography. 3. Uchimura, Kanzō, 1861-1930. 4. Christian biography—Japan. 5. Nagai, Kafū, 1879-1959—Biography. 6. Authors, Japanese—20th century—Biography. I. Title.
N7483.O37O35 79-805271

Ōta, Aito, 1928-
(Meiji Kirisutokyō no ryūiki)
明治キリスト教の流域 : 静岡バンドと幕臣たち / 太田愛人. — 東京 : 築地書館, 1979.
231 p. ; 20 cm. Ja 79-6672
Bibliography: p. 218-227.
¥1400
1. Protestant churches—Japan. 2. Japan—Church history—1868-1945. 2. Christian biography—Japan. I. Title.
BR1306.5.O8 79-809369

Pacifism in Japan : the Christian and socialist tradition / edited by Nobuya Bamba and John F. Howes. — Kyoto : Minerva Press, [1978]
xiv, 300 p. ; 23 cm. Ja78-42
Includes index.
Bibliography: p. [273]-295.
Errata slip inserted.
ISBN 0-7748-0072-0 : Y6000
1. Japan—History—1868- —Biography. 2. Pacifism. 3. Christian biography—Japan. 4. Socialists—Japan—Biography. 5. Japan—Biography. I. Bamba, Nobuya, 1937- . II. Howes, John F.
DS883.P3 920'.052 79-314423
*79 MARC

Saitō, Takeshi, 1887-
(Natsukashiki hitobito)
なつかしき人々 / [著者 斎藤勇]. — 東京 : 研究社出版, 1978.
528 p., [1] leaf of plates : port. ; 22 cm. — (斎藤勇著作集 ; 別巻)
Ja 78-5080
On verso of t. p.: Old familiar faces.
Series romanized: Saitō Takeshi chosaku shū.
List of the author's works: p. 467-498.
Includes index.
¥7000
1. Biography—20th century. 2. Intellectuals. 3. Christian biography—Japan. 4. Saitō, Takeshi, 1887- I. Title. II. Title: Old familiar faces.
PR99.S224 1975, vol. 8 78-812124
[CT203J]

CHRISTIAN BIOGRAPHY
—JAPAN (Continued)

Sakata, Tasuku, 1878–1969.
(Onchō no shōgai)
恩寵の生涯：新編 / 坂田祐；[編者　洒枝義旗]. — 東京：待晨堂, 昭和51 [1976]
9, 586 p., [4] leaves of plates : ill. ; 19 cm. Ja 77-9087
¥2500

1. Sakata, Tasuku, 1878–1969. 2. Uchimura, Kanzō, 1861–1930. 3. Christian biography—Japan. 4. Educators—Japan—Biography. I. Sakaeda, Yoshitaka, 1898- II. Title.

BR1317.S13A36 78–811753

Shimizu, Yasuzō, 1891-
(Nozomi o ushinawazu)
希望を失わず：續朝陽門外　清水安三手記. — 改訂第五版. — 東京：櫻美林出版部, 昭和26 [1951]
253 p. : ill. ; 19 cm.

1. Shimizu, Yasuzo, 1891- 2. Christian biography—Japan. 3. Educators—Japan—Biography. 4. Ch'ung chen hsüeh yüan. 5. Ōbirin Gakuen. I. Shimizu, Yasuzo, 1891 Choyomongai. II. Title.

BR1317.S5A36 79–806191

Takamizawa, Junko.
(Eien no ashioto)
永遠のあしおと：真実な神に出会った人たち / 高見沢潤子. — 東京：主婦の友社, 昭和51 [1976]
239 p. : ports. ; 19 cm. Ja 76–9177
¥980

1. Christian biography—Japan. I. Title.

BR1316.T34 78–812709

Tazaki, Kensaku.
(Sutemi de ikiru)
捨身で生きる：ある牧師の生活と意見 / 田崎健作著. — 東京：日本YMCA同盟出版部, 1964.
259 p. ; 18 cm. — (キリスト教教養新書)
Series romanized: Kirisutokyō kyōyō shinsho.

1. Tazaki, Kensaku. 2. Christian Biography—Japan. 3. Christian life. I. Title.

BX8750.N58T397 79–801864

(Uemura Masahisa to sono jidai)
植村正久と其の時代 / [編者　佐波亘, 小澤三郎]. — 東京：教文館, 昭和 - [
v. : ports. ; 23 cm. Ja***
Includes bibliographies.
CONTENTS:
[8] 新補遺
¥7000 (v. 8)
1. Uemura, Masahisa, 1857–1925. 2. Christian biography—Japan. 3. Protestantism—Japan—History. 4. Japan—Church history—1868–1945. I. Sawa, Wataru, 1881–1958. II. Ozawa, Saburō, 1909–1969.
BR1317.U27U36 79–805717

—JAPAN—COLLECTED WORKS

Fujita, Wakao, 1912–
(Jūgonen sensō to Mukyōkai nidaime)
十五年戦争と無教会二代目 / 藤田若雄. — 東京：大河原礼三, 1973–1975.
4 v. ; 25 cm. — (内村鑑三記念講演研究：中間報告 ; 1–4)

Vol. 2 and 4 published by Kirisutokyō Shakai Shisō Kenkyūkai.
Includes bibliographical references.
"講演内容分解表" issued as suppl. with v. 1–2 and 4.
CONTENTS: 第1篇　満州事変から日中戦争勃発まで—第2篇　日中戦争から太平洋戦争終結まで—第3篇　敗戦の神義論—第4篇　内村二十年記念講演

1. Mukyōkaishugi Shūkai—Collected works. 2. War and religion—Collected works. 3. World War, 1939–1945—Mukyōkaishugi Shūkai—Collected works. 4. Christian Biography—Japan—Collected works. 5. Uchimura, Kanzō, 1861–1930. I. Title. II. Series: Uchimura Kanzō kinen kōen kenkyū : Chūkan hōkoku ; 1–4.

BX8699.M8F84 78–813398

—KOREA

Chang, Yŏng-ch'ang.
(Sae segye ero)
새世界에로：나의體驗記 / 張泳暢著. — 서울：同志社, 1976.
260 p. : 19 cm.

1. Segye Kidokkyo T'ongil Sillyŏng Hyŏphoe. 2. Chang, Yŏng-ch'ang. 3. Christian biography—Korea. I. Title.

BX9750.S4C464 79–825862

Cho, Ae-sil.
(Ch'arari t'onggok igirŭl)
차라리痛哭이기를：趙愛實隨想集. — 서울：傳藝苑, 1977.
269 p. : 21 cm.
Essays.
W1200

1. Cho, Ae-sil. 2. Women—Korea—Biography. 3. Christian biography—Korea. I. Title.

CT1848.C45A32 78–826860

Chŏng, P'yŏng-hae.
(Samch'ŏn-yukpaek-osip ŭi kijŏk)
3650의기적：信仰手記 / 丁平海著. — 서울：惠宣文化社, 1977, 1978 printing.
286 p. : 19 cm.
₩1200

1. Chŏng, P'yŏng-hae. 2. Christian biography—Korea. 3. Women—Korea—Biography. I. Title.

BR1337.C48A36 78–826861

(Minnŭn cha ege ttarŭnŭn p'yojŏk)
믿는자에게따르는표적 / 조용기 [편] 著. — 서울：영산출판사, 1976.
241 p. : ports. ; 19 cm.

1. Faith-cure. 2. Christian biography—Korea. I. Cho, Yong-gi, 1936–

BT732.5.M56 79–825600

—KOREA—ADDRESSES, ESSAYS, LECTURES

Ch'oe, Sŏk-chu, 1900–
(Nae ka pon insaeng paekkyŏng)
내가본人生百景 / [著者崔錫柱]. — 서울：大韓基督教書會, 1974.
384 p. : ports. ; 20 cm. — (崔錫柱選集 ; 4)
Series romanized: Ch'oe Sŏk-chu sŏnjip.
Includes bibliographies.
1. Christian biography—Korea—Addresses, essays, lectures. 2. Korea—Biography—Addresses, essays, lectures. 3. Biography—Addresses, essays, lectures. I. Title.

BX8915.C623 1972, vol. 4 79–826361
[BR1336]

—LEVANT

Les Formes syriaques de la matière de l'Histoire lausiaque / par René Draguet. — Louvain : Secrétariat du CorpusSCO, 1978.
2 v. in 4 ; 26 cm. — (Corpus scriptorum Christianorum Orientalium ; v. 389-390, 398-399 : Scriptores Syri ; t. 169-170, 173-174) Be***
Introd. and notes in French; Syriac text and French translation.
Includes bibliographical references and indexes.
CONTENTS: I, [1] Les manuscrits. Édition des pièces liminaires et des ch. 1-19.—I, [2] Les recensions. Version des pièces liminaires et des ch. 1-19.—II, [1] Édition des ch. 20-71, Épilogue, (72-73).—II, [2] Version des ch. 20-71, Épilogue. Appendice (72-73).
ISBN 2-8017-0078-9 (v. I, 1)
1. Palladius, successively Bp. of Helenopolis and of Aspona, d. ca. 430. Historia Lausiaca. 2. Christian biography—Levant. 3. Monasticism and religious orders—Levant. I. Draguet, René. II. Palladius, successively Bp. of Helenopolis and of Aspona, d. ca. 430. Historia Lausiaca. III. Series: Corpus scriptorum Christianorum Orientalium ; v. 389-390, [etc.]
BR60.C5S8 t. 169-170, etc. 79-343060
281'.08 s
[BR65.P3] 79 MARC

—NETHERLANDS

Ten Boom, Corrie.
He sets the captive free / Corrie Ten Boom. — Boston : G. K. Hall, 1978, c1977.
viii, 89 p. ; 25 cm. — (Her Jesus is victor)
Large print ed.
ISBN 0-8161-6627-7
1. Ten Boom, Corrie. 2. Christian biography—Netherlands. 3. World War, 1939-1945—Personal narratives, Dutch. 4. Ravensbrück (Concentration camp) 5. Christian life—1960- . 6. Large type books. I. Title. II. Series.
[BR1725.T35A33 1978] 269'.2'0924 78-10616
78 MARC

—PALESTINE

Baldwin, Louis.
Jesus of Galilee : his story in everyday language / Louis Baldwin. — Valley Forge, PA : Judson Press, c1979.
142 p. : map ; 22 cm.
Bibliography: p. 137-139.
Includes index.
ISBN 0-8170-0841-1 : $4.95
1. Jesus Christ—Biography. 2. Christian biography—Palestine. I. Title.
BT301.2.B22 232.9'01 79-11587
79 MARC

Bible. N.T. Gospels. English. Authorized. Selections. 1977.
Story of the life of Christ : a harmony of the four Gospels / by Eldin Ricks. — 1st ed. — Salt Lake City : Deseret Press, 1977.
453 p. : ill. ; 23 cm.
Bibliography: p. 453.
Includes index.
1. Jesus Christ—Biography—Sources, Biblical. 2. Christian biography—Palestine. I. Ricks, Eldin. II. Title.
BT299.2.R53 1977 226'.1 77-73540
79 MARC

Bogolíûbov, Andreĭ.
(Syn chelovecheskiĭ)
Сын человеческий / Андрей Боголюбов. — Bruxelles : Жизнь с Богом, [1968?]
335 p. : ill. ; 19 cm.
Bibliography : p. 323–327.
1. Jesus Christ—Biography. 2. Christian biography—Palestine. I. Title.
BT301.2.B55 79–414250

Borer, Wilbur J
Book of the Lord : reflections on the life of Christ / Wilbur J. Borer ; ill. by Jean Charlot. — Huntington, Ind. : Our Sunday Visitor, c1978.
xi, 493 p. : ill. ; 24 cm.
ISBN 0-87973-852-9
1. Jesus Christ—Biography. 2. Christian biography—Palestine. I. Title.
BT301.2.B57 232.9'01 78-50696
78 MARC

Converse, Gordon N
Come see the place : the Holy Land Jesus knew / photos. by Gordon N. Converse ; text by Robert J. Bull and B. Cobbey Crisler. — Englewood Cliffs, N.J. : Prentice-Hall, c1978.
144 p. : ill. ; 23 X 29 cm.
ISBN 0-13-152538-7
1. Palestine—Description and travel—Views. 2. Jesus Christ—Biography. 3. Christian biography—Palestine. I. Bull, Robert J. II. Crisler, B. Cobbey, 1933- . III. Title.
DS108.5.C65 779'.9'2209 78-7054
00 MARC

Endō, Shūsaku, 1923-
A life of Jesus / by Shusaku Endo ; translated from the Japanese by Richard A. Schuchert. — New York : Paulist Press, c1978.
v, 179 p. : ill. ; 24 cm.
Translation of Iesu no shōgai.
ISBN 0-8091-0269-2 : $9.95
1. Jesus Christ—Biography. 2. Christian biography—Palestine. I. Title.
BT301.2.E5313 232.9'01 78-61721
79 MARC

Fosdick, Harry Emerson, 1878-1969.
The Man from Nazareth as his contemporaries saw Him / by Harry Emerson Fosdick. — Westport, Conn. : Greenwood Press, 1978, c1949.
282 p. ; 23 cm.
Reprint of the ed. published by Harper, New York.
Includes bibliographical references and indexes.
ISBN 0-313-20603-1
1. Jesus Christ—Biography. 2. Christian biography—Palestine. I. Title.
BT303.F62 1978 232.9'03 78-16469
78 MARC

Goodspeed, Edgar Johnson, 1871-1962.
A life of Jesus / by Edgar J. Goodspeed. — Westport, Conn. : Greenwood Press, 1979, c1950.
248 p. ; 22 cm.
Reprint of the ed. published by Harper, New York.
Includes bibliographical references and index.
ISBN 0-313-20728-3
1. Jesus Christ—Biography. 2. Christian biography—Palestine. I. Title.
BT301.G73 1979 232.9'01 78-21540
78 MARC

CHRISTIAN BIOGRAPHY
—PALESTINE (Continued)

Grant, Michael, 1914-
Jesus / ₍by₎ Michael Grant. — Revised ed. — London : Sphere, 1978.
320 p. ; 18 cm. GB78-33535
Bibliography: p. 310.
Includes index.
ISBN 0-7221-3995-0 : £1.95
1. Jesus Christ—Biography. 2. Christian biography—Palestine. I. Title.
BT301.2.G68 1978 232.9'01 79-301118
79 MARC

Hirschmann, Maria Anne.
Follow me / Maria Anne Hirschmann and Betty Pershing. — Huntington Beach, Calif. : S.P.A.R.C. Pub. Co., c1979.
223 p. ; 21 cm. — (Hansi books)
Bibliography: p. 223.
ISBN 0-932878-01-6 : $3.95
1. Jesus Christ—Biography. 2. Christian biography—Palestine. I. Pershing, Betty, joint author. II. Title.
BT301.2.H57 232.9'01 79-84331
00 MARC

Iitsuka, Reiji, 1904-
("Kono hito o miyo")
"この人を見よ" = Ecce home / ₍著者 飯塚羚兒₎. — 東京 : 南北書園, 昭和23 ₍1948₎
341, 6 p., ₍1₎ leaf of plates : ill. ; 19 cm.
In Japanese.
1. Jesus Christ—Biography. 2. Christian biography—Palestine. I. Title.
BT301.I38 79-808353

Ijjas, Antal.
Jézus története / Ijjas Antal. — Budapest : Ecclesia, 1978.
2 v. (815 p.) ; 20 cm. H78
Bibliography: p. ₍777₎-₍813₎
ISBN 9633634067 : 150.00Ft
1. Jesus Christ—Biography. 2. Christian biography—Palestine. I. Title.
BT301.2.I33 1978 79-371191
*79 MARC

Márkus, Mihály.
Jézus / Márkus Mihály. — Budapest : Kiadja a Református Zsinati Iroda Sajtóosztálya, 1977.
64 p. ; 20 cm. H77
ISBN 9637030530 : 11.50Ft
1. Jesus Christ—Biography. 2. Christian biography—Palestine. I. Title.
BT301.2.M29 79-375306
*79 MARC

Oursler, Fulton, 1893-1952.
Fulton Oursler's Greatest : The greatest book ever written, The greatest story ever told, The greatest faith ever known. — 1st ed. — Garden City, N.Y. : Doubleday, 1979.
1052 p. ; 21 cm. — (A Doubleday-Galilee book)
ISBN 0-385-14659-0 : $8.95
1. Bible. O.T.—History of Biblical events. 2. Jesus Christ—Biography. 3. Christian biography—Palestine. 4. Apostles. 5. Bible. N.T.—History of Biblical events. I. Title.
BS1197.O79 220.9'505 78-69663
79 MARC

Rustomjee, Framroz.
An interpretation of the life of Jesus the Christ / by Framroz Rustomjee. — Colombo : Rustomjee, 1976.
171 p., ₍1₎ leaf of plates : ill. ; 22 cm.
1. Jesus Christ—Biography. 2. Christian biography—Palestine. I. Title.
BT301.2.R83 232 78-911698
79 MARC

Skvorfšov, K
(Zhizn' Īisusa Khrista, po Evangelīi͡am i narodnym predanīi͡am)
Жизнь Іисуса Христа, по Евангеліямъ и народнымъ преданіямъ / К. Скворцова. — Кіевъ : Тип. С. В. Кульженко, 1876.
iv, ii, 334 p., ₍1₎ leaf of plates : ill. ; 23 cm.
1. Jesus Christ—Biography. 2. Christian biography—Palestine. I. Title.
BT301.2.S56 Russian Imperial Coll. 78-421757

Van Dyk, Fay Blix, 1951-
His touch is love / Fay Blix Van Dyk. — Nashville : Southern Pub. Association, c1978.
94 p. ; 20 cm.
ISBN 0-8127-0189-5
1. Jesus Christ—Biography. 2. Christian biography—Palestine. I. Title.
BT301.2.V35 232.9'01 78-6523
78 MARC

Wilkinson, John Donald.
Jerusalem as Jesus knew it : archaeology as evidence / John Wilkinson. — London : Thames and Husdon, c1978.
208 p. : ill. ; 20 cm. GB***
Photo. ₍1₎ leaf inserted.
Bibliography: p. 201-203.
Includes index.
ISBN 0-500-05031-7 : $8.95 (U.S.)
1. Jerusalem—Antiquities. 2. Jesus Christ—Biography—Passion Week. 3. Christian biography—Palestine. I. Title.
DS109.W7 933 77-20743
78₍r79₎rev MARC

—PALESTINE—ADDRESSES, ESSAYS, LECTURES

Bogdani, Pjetër, ca. 1625-1689.
Cuneus prophetarum / a Pietro Bogdano. — München : Trofenik, 1977.
436 p. in various pagings : ill. ; 25 cm. — (Beiträge zur Kenntnis Südosteuropas und des Nahen Orients ; Bd. 24) GFR78-A
Introductory matter in Albanian, German, Italian, or Latin: text in Italian and Albanian.
Reprint of the 1685 ed. printed by A. Candianus, Padua.
ISBN 3-87828-107-2 : DM124.00
1. Prophets—Addresses, essays, lectures. 2. Sibyls—Addresses, essays, lectures. 3. Jesus Christ—Biography—Addresses, essays, lectures. 4. Christian biography—Palestine—Addresses, essays, lectures. I. Title. II. Series.
BS1198.B6316 1977 78-384008
78 MARC

—PALESTINE—JUVENILE LITERATURE

Carriger, Sally A
I-can-read-it-myself Bible stories : stories about Jesus / by Sally A. Carriger ; illustrated by Terry McBride. — Washington : Review and Herald Pub. Association, c1978.
79 p. : ill. ; 20 cm.
SUMMARY: Presents seven New Testament stories about Jesus.
$4.25
1. Jesus Christ—Biography—Juvenile literature. 2. Christian biography—Palestine—Juvenile literature. ₍1. Jesus Christ. 2. Bible stories—N.T.₎ I. McBride, Terry, 1950- . II. Title.
BT302.C34 232.9'01 77-83636
78 MARC

Degering, Etta.
Once upon a Bible time / told by Etta B. Degering ; illustrated by Vernon Nye. — Washington : Review and Herald Pub. Association, c1976-
v. : col. ill. ; 19 x 22 cm.
Vol. 2 illustrated by H. W. Munson.
SUMMARY: Retells four New Testament stories: "The Temple Tax Coin", "On Jericho Road", "The Woman Bent-Over Double", and "Small Daughter".
1. Jesus Christ—Biography—Juvenile literature. 2. Christian biography—Palestine—Juvenile literature. ₍1. Bible stories—N.T.₎ I. Nye, Vernon. II. Munson, Harold W. III. Title.
BT302.D33 225.9'505 76-14118
77₍r79₎rev MARC

Griffiths, Kitty Anna.
Come, meet Jesus, the baby : the story of Matthew 1 and Luke 1-2:20 / Kitty Ann Griffiths ; illustrated by "Willy." — 1st zondervan ed. — Grand Rapids, Mich. : Zondervan, 1978, c1976.
111 p. : ill. ; 21 cm. — (Come, meet series)
SUMMARY: Relates the events surrounding Jesus' birth including the genealogy of Joseph, the angels' announcements to Elizabeth and to Mary, the births of John the Baptist and of Jesus, and the visit of the shepherds to the stable in Bethlehem.
ISBN 0-310-25241-5
1. Jesus Christ—Nativity—Juvenile literature. 2. Bible stories, English—N.T. Gospels. 3. Christian biography—Palestine—Juvenile literature. ₍1. Jesus Christ—Nativity. 2. Bible stories—N.T.₎ I. Willy. II. Title.
BT315.2.G74 1978 232.9'21 78-16703
78 MARC

Griffiths, Kitty Anna.
Come, meet Jesus, the boy : the story of Matthew 2, Luke 2:21-52 / Kitty Ann Griffiths ; illustrated by "Willy." — 1st Zonderamin ed. — Grand Rapids : Zondervan Pub. House, 1978, c1976.
96 p. : ill. ; 21 cm. — (Come, meet series)
SUMMARY: Retells the events of Jesus Christ's childhood including the visit from the wise men, his family's flight into Egypt, and his meeting with temple scholars.
ISBN 0-310-25251-2
1. Jesus Christ—Childhood—Juvenile literature. 2. Bible stories, English—N.T. Gospels. 3. Christian biography—Palestine—Juvenile literature. ₍1. Jesus Christ—Childhood. 2. Bible stories—N.T.₎ I. Willy. II. Title.
BT320.G74 1978 232.9'2 78-16680
78 MARC

Villiers, Marjorie Howard, 1903-
Jesus has come / with text by Marjorie Villiers, based on the French original by Martine Douillet and Geneviève Guilhem ; illustrated by Philippe Joudiou. — Valley Forge, Pa. : Judson Press, 1978, c1973.
₍28₎ p. : col. ill. ; 21 cm.
Based on M. Douillet's Jésus est venu.
SUMMARY: Relates the birth and early life of Jesus Christ.
ISBN 0-8170-0810-1 : $2.75
1. Jesus Christ—Biography—Juvenile literature. 2. Christian biography—Palestine—Juvenile literature. ₍1. Jesus Christ—Nativity₎ I. Douillet, Martine. Jésus est venu. II. Joudiou, Philippe. III. Title.
BT302.V63 232.9'01 78-105555
78 MARC

Villiers, Marjorie Howard, 1903-
Jesus with us / with text by Marjorie Villiers ; illustrated by Philippe Joudiou ; based on the French original by Martine Douillet and Geneviève Guilhem. — Valley Forge, Pa. : Judson Press, 1978, c1973.
₍28₎ p. : col. ill. ; 21 cm.
Based on M. Douillet's Jésus avec nous.
SUMMARY: Describes the adult life of Jesus through His Crucifixion and Resurrection.
ISBN 0-8170-0811-X : $2.75
1. Jesus Christ—Biography—Juvenile literature. 2. Christian biography—Palestine—Juvenile literature. ₍1. Jesus Christ₎ I. Douillet, Martine. Jésus avec nous. II. Joudiou, Philippe. III. Title.
BT302.V64 232.9'01 78-105560
78 MARC

—RUSSIA

Hamm, Gerhard.
Du hast uns nie verlassen : Erfahrungen christl. Familien in d. Sowjetunion / Gerhard Hamm. — Wuppertal : Brockhaus, 1978.
175 p. : ill. ; 21 cm. — (ABC-team ; 151 : Berichte, Erzählungen, Lebensbilder) GFR78-A
ISBN 3-417-12151-5 : DM15.80
1. Christian biography—Russia. 2. Hamm Family. 3. Persecution—Russia. I. Title.
BR1700.2.H27 79-351865
79 MARC

—SCOTLAND

Hudson, J Harrison.
Let the fire burn : a study of R. M. McCheyne, Robert Annan, Mary Slessor / by J. Harrison Hudson, Thomas W. Jarvie, Jock Stein. — Dundee : Handsel Publications, 1978.
₍4₎, 67 p. : ports. ; 21 cm. GB78-18852
Second chapter a reprint of T. W. Jarvie's booklet entitled: A broken vessel restored.
Includes bibliographies.
£0.80
1. McCheyne, Robert Murray, d. 1843. 2. Annan, Robert, 1834-1867. 3. Slessor, Mary Mitchell, 1848-1915. 4. Christian biography—Scotland. I. Jarvie, Thomas W., joint author. II. Stein, Jock, joint author. III. Title.
BX9225.M17H8 285'.2'0922 78-323752
79 MARC

—SWEDEN

Fogelklou, Emilia, 1878-1972.
Brev till vännerna / Emilia Fogelklou ; urval och inl. av Gunnel Vallquist. — Stockholm : Natur o. kultur, 1979.
231, ₍1₎ p. ; 21 cm. S79
"Emilia Fogelklous skrifter i urval": p. 18.
Includes index.
ISBN 9127008444
1. Fogelklou, Emilia, 1878-1972. 2. Christian biography—Sweden. I. Vallquist, Gunnel. II. Title.
BR1725.F53A313 79-371617
79 MARC

—UNITED STATES

Bailey, Ney.
Faith is not a feeling / Ney Bailey with Sharon Fischer. — San Bernardino, Ca : Here's Life Publishers, c1978.
ix, 114 p. ; 21 cm.
ISBN 0-918956-45-5 : $2.95
1. Bailey, Ney. 2. Christian biography—United States. I. Fischer, Sharon, joint author. II. Title.
BR1725.B327A33 248'.2 78-60077
79 MARC

Bauer, Clare.
Some through the waters / by Clare Bauer. — Nashville : T. Nelson Publishers, c1979.
160 p., ₍2₎ leaves of plates : ill. ; 20 cm.
Includes index.
ISBN 0-8407-5687-9 : $3.95
1. Bauer, Clare. 2. Christian biography—United States. 3. Christian life—1960- . I. Title.
BR1725.B383A37 248'.2 79-2438
79 MARC

Berven, Ken.
I love being married to a grandma / Ken Berven ; foreword by Billy Graham. — Boston : G. K. Hall, 1978.
xii, 139 p. ; 24 cm.
"Large print ed."
ISBN 0-8161-6632-3
1. Berven, Ken. 2. Christian biography—United States. 3. Grandparents—Religious life. 4. Large type books. I. Title.
₍BR1725.B46A34 1978b₎ 248'.85'0924 78-10547
78 MARC

Boone, Shirley.
The honeymoon is over / Shirley and Pat Boone. — Carol Stream, Ill. : Creation House, c1977, 1978 printing.
185 p., ₍7₎ leaves of plates : ill. ; 23 cm.
ISBN 0-88419-130-3 : $6.95
1. Boone, Shirley. 2. Boone, Charles Eugene. 3. Christian biography—United States. I. Boone, Charles Eugene, joint author. II. Title.
BR1702.B58 248'.2 77-75818
00 MARC

CHRISTIAN BIOGRAPHY
—UNITED STATES (Continued)

Cash, June Carter, 1929-
Among my klediments / June Carter Cash. — Grand Rapids : Zondervan Pub. House, c1979.
152 p., [2] leaves of plates : ill. ; 22 cm.
ISBN 0-310-38170-3
1. Cash, June Carter, 1929- . 2. Country musicians—United States—Biography. 3. Christian biography—United States. I. Title.
ML420.C2653A3 784'.092'4 79-10959
79 MARC

Grace, John Patrick, 1942-
Hearing His voice / John Patrick Grace. — Notre Dame, Ind. : Ave Maria Press, c1979.
157 p. : ill. ; 21 cm.
ISBN 0-87993-187-9 : $3.50
1. God—Will. 2. Pentecostalism—United States. 3. Christian biography—United States. I. Title.
BV4501.2.G69 269 79-54696
79 MARC

Greiner, Claire, 1932-
God, who else? / By Claire and Ruth Greiner. — Bala Cynwyd, Pa. : Family Christian Inspiration, c1979.
120 p., [6] leaves of plates : ill. ; 22 cm.
1. Greiner, Claire, 1932- . 2. Greiner, Ruth, 1931- . 3. Christian biography—United States. I. Greiner, Ruth, 1931- joint author. II. Title.
[BR1725.G863A33] 248'.2 79-52248
00 MARC

Horton, Marilee.
Dear mamma please don't die / Marilee Horton. — Nashville : Thomas Nelson, c1979.
160 p. ; 21 cm.
ISBN 0-8407-5689-5
1. Horton, Marilee. 2. Chritian biography—United States. 3. Multiple sclerosis—Biography. 4. Depression, Mental—Biography. I. Title.
BR1725.H664A33 248'.2 79-13455
79 MARC

Howard, Linda.
The secret life of a housewife / by Linda Howard. — Plainfield, N.J. : Logos International, c1978.
v, 149 p. ; 21 cm.
ISBN 0-88270-296-3 : $2.95
1. Howard, Linda. 2. Christian biography—United States. 3. Wives—Religious life. I. Title.
BR1725.H688A37 248'.2 78-60946
78 MARC

Hughes, Allison.
Love, honor, and frustration / Allison Hughes. — Grand Rapids : Zondervan Pub. House, 1979, c1977.
167 p. ; 21 cm.
Published in 1977 under title: Love, honor, and obesity.
ISBN 0-310-26331-X
1. Hughes, Allison. 2. Christian biography—United States. 3. Depression, Mental. 4. Corpulence—Psychological aspects. I. Title.
BR1725.H77A35 1979 248'.2'0924 79-11909
79 MARC

Hughes, Harold E
The man from Ida Grove : a senator's personal story / Dick Schneider. — Lincoln, Va. : Chosen Books ; Waco Tex. : distributed by Word Books, c1979.
346 p. : ill. ; 24 cm.
ISBN 0-912376-38-4
1. Hughes, Harold E. 2. Legislators—United States—Biography. 3. United States. Congress. Senate—Biography. 4. Christian biography—United States. I. Schneider, Richard, 1922- joint author. II. Title.
E840.8.H83A33 328.73'092'4 78-31152
78 MARC

Jefferson, Ted, 1932-
One bad dude : the miraculous transformation of a four-time loser / by Ted Jefferson with Michael W. Fedo ; [original art by Dean Clark, photos. by Judy Fedo]. — Kalamazoo, Mich. : Master's Press, c1978.
176 p. : ill. ; 23 cm.
ISBN 0-89251-050-1 : $6.95
1. Jefferson, Ted, 1932- . 2. Christian biography—United States. I. Fedo, Michael W., joint author. II. Title.
BR1725.J386A34 248'.24 78-54222
79 MARC

Johnson, Lois Walfrid.
Either way, I win : a guide for growth in the power of prayer / Lois Walfrid Johnson. — Minneapolis : Augsburg Pub. House, c1979.
192 p. ; 18 cm.
ISBN 0-8066-1706-3 : $2.50
1. Johnson, Lois Walfrid. 2. Christian biography—United States. 3. Cancer—Biography. I. Title.
[BR1725.J59A34] 248'.2 79-50078
79 MARC

Jones, Margaret Jean.
The world in my mirror / Margaret Jean Jones. — Nashville : Abingdon, c1979.
160 p. ; 21 cm.
ISBN 0-687-46270-3
1. Jones, Margaret Jean. 2. Christian biography—United States. 3. Paraplegia—Biography. I. Title.
BR1725.J645A38 248'.2 79-17730
79 MARC

Kelley, Frances.
Better than I was / Frances Kelley with Helen Kooiman Hosier. — Nashville : T. Nelson, c1979.
154 p. ; 20 cm.
ISBN 0-8407-5671-2
1. Kelley, Frances. 2. Christian biography—United States. 3. Faith-cure. I. Hosier, Helen Kooiman, joint author. II. Title.
BR1725.K43A34 248'.2 79-1095
79 MARC

Linehan, Kevin, 1948-
Such were some of you : the spiritual odyssey of an ex-gay Christian / Kevin Linehan. — Scottdale, Pa. : Herald Press, 1979.
231 p. ; 20 cm.
ISBN 0-8361-1890-1 : $5.95
1. Linehan, Kevin, 1948- . 2. Christian biography—United States. 3. Homosexuals—United States—Biography. 4. Homosexuality and Christianity. I. Title.
BR1725.L446A35 248'.2 79-12178
79 MARC

Littauer, Florence, 1928-
I could be happy, if only ... / by Florence Littauer. — Scottsdale, Ariz. : Good Life Productions, c1978.
187 p. ; 21 cm.
$3.95
1. Littauer, Florence, 1928- . 2. Christian biography—United States. I. Title.
BR1725.L487A3 248'2. 78-67209
78 MARC

McClary, Clebe.
Living proof / Clebe McClary, [with Diane Barker]. — Atlanta, GA : Cross Roads Publishers, c1978.
150 p. : ill. ; 22 cm.
$3.95
1. McClary, Clebe. 2. Christian biography—United States. 3. Vietnamese Conflict, 1961-1975—Personal narratives, American. 4. Soldiers—United States—Biography. I. Barker, Diane, joint author. II. Title.
BR1725.M3117A34 248'.2 78-66754
78 MARC

Monson, Gabriele.
Say hi to Jesus for me / Gabriele Monson. — Minneapolis : Augsburg Pub. House, c1979.
175 p. ; 20 cm.
ISBN 0-8066-1690-3 : $3.95
1. Monson, Todd. 2. Christian biography—United States. 3. Cancer—Biography. 4. Monson, Gabriele. I. Title.
BR1725.M524M66 248'.2 78-66951
79 MARC

My most memorable encounter with God / [Stephen Olford ... et al.] ; edited by David Enlow. — Wheaton, Ill. : Tyndale House Publishers, c1977.
223 p. : ports. ; 21 cm.
ISBN 0-8423-4655-4 : $3.95
1. Christian biography—United States. 2. Spiritual life. I. Olford, Stephen F. II. Enlow, David R.
BR569.M9 248'.2 77-72442
79 MARC

Nordtvedt, Matilda.
Living beyond depression / by Matilda Nordtvedt. — Minneapolis, Minn. : Bethany Fellowship, c1978.
128 p. ; 18 cm. — (Dimension books)
Includes bibliographical references.
ISBN 0-87123-339-8 : $1.95
1. Nordtvedt, Matilda. 2. Christian biography—United States. 3. Christian life—1960- . 4. Depression, Mental—Biography. I. Title.
BR1725.N54A35 248'.86 78-58082
78 MARC

Payton, Everett J
I won't be crippled when I see Jesus / Everett J. Payton. — Minneapolis, Minn. : Augsburg Pub. House, c1979.
158 p. ; 20 cm.
ISBN 0-8066-1716-0
1. Payton, Everett Lee. 2. Christian biography—United States. 3. Payton, Everett J. 4. Blind—Biography. 5. Hydrocephalus—Biography. 6. Cerebral palsy—United States—Biography. I. Title.
BR1725.P33P39 248'.86 79-50084
00 MARC

Potterbaum, Charlene.
His eye is on the sparrow so ... this is really for the birds / Charlene Potterbaum. — Plainfield, N.J. : Logos International, c1979.
ix, 190 p. ; 21 cm.
On spine: This is really for the birds.
ISBN 0-88270-354-4 : $3.50
1. Potterbaum, Charlene. 2. Christian biography—United States. 3. Christian life—1960- . I. Title. II. Title: This is really for the birds.
BR1725.P63A34 248'.2'0924 79-83788
79 MARC

Pulkingham, Betty.
Little things in the hands of a big God / Betty Pulkingham. — Waco, Tex. : Word Books, c1977.
142 p. : ill. ; 21 cm.
ISBN 0-8499-2855-9 : $4.95
1. Pulkingham, Betty. 2. Christian biography—United States. 3. Christian life—1960- . I. Title.
BR1725.P84A34 283'.092'4 79-63930
00 MARC

Raines, Robert Arnold.
Going home / Robert A. Raines. — 1st ed. — San Francisco : Harper & Row, c1979.
x, 145 p. ; 21 cm.
ISBN 0-06-066768-0
1. Raines, Robert Arnold. 2. Christian biography—United States. I. Title.
BR1725.R25A33 1979 287'632'0924 78-15834
78 MARC

Robertson, Dede.
My God will supply : how the Lord provides in times of shortage / Dede Robertson, with John Sherrill. — Lincoln, Va. : Chosen Books ; Waco, Tex. : distributed by Word Books, c1979.
172 p. ; 22 cm.
ISBN 0-912376-48-1
1. Robertson, Dede. 2. Christian biography—United States. 3. Soybean as food. 4. Cookery (Soybeans) I. Sherrill, John, joint author. II. Title.
BR1725.R616A35 248'.2 79-16126
79 MARC

Roman, LuLu.
LuLu / LuLu Roman. — Old Tappan, N.J. : F. H. Revell, c1978.
173 p. : ill. ; 22 cm.
ISBN 0-8007-0956-X : $6.95
1. Roman, LuLu. 2. Christian biogrphy—United States. 3. Entertainers—United States—Biography. I. Title.
BR1725.R65A34 248'.24 78-15510
78 MARC

Satre, Elizabeth Dahl.
The story of Ellen : how love transforms a troubled child / Elizabeth Dahl Satre. — Minneapolis : Augsburg Pub. House, c1979.
110 p. ; 20 cm.
ISBN 0-8066-1691-1 : $2.95
1. Satre, Ellen. 2. Christian biography—United States. 3. Problem children—United States—Biography. I. Title.
BR1725.S3S27 248'.2 78-66952
79 MARC

Sheetz, Ann Kindig.
Born again, but still wet behind the ears / Ann Kindig Sheets. — 1st ed. — Chappaqua, N.Y. : Christian Herald Books, c1979.
160 p. ; 21 cm.
ISBN 0-915684-43-8 : $6.95
1. Sheetz, Ann Kindig. 2. Christian biography—United States. 3. Christian life—1960- . I. Title.
BR1725.S454A33 248'.2 78-64839
79 MARC

Sherrill, John L
My friend, the Bible / John Sherrill. — Lincoln, Va. : Chosen Books ; Waco, Tex. : distributed by Word Books, c1978.
169 p. ; 23 cm.
"Bible reading programs": p. 167-169.
ISBN 0-912376-37-6
1. Bible—Use. 2. Sherrill, John L. 3. Christian biography—United States. I. Title.
BS538.3.S48 248'.2 78-15895
78 MARC

Stamps, Ellen de Kroon, 1940-
My years with Corrie / Ellen de Kroon Stamps. — Old Tappan, N.J. : F. H. Revell Co., c1978.
128 p. ; 22 cm.
ISBN 0-8007-0957-8 : $5.95
1. Stamps, Ellen de Kroon, 1940- . 2. Christian biography—United States. 3. Ten Boom, Corrie. I. Title.
BR1725.S733A35 269'.2'0922 78-13979
78 MARC

Turner, Dean.
Krinkle nose : a prayer of thanks / by Dean Turner. — Old Greenwich, Conn. : Devin-Adair Co., c1979.
92 p., [2] leaves of plates : ill. ; 24 cm.
ISBN 0-8159-6002-6 : $4.95
1. Turner, Taos Lee, 1972- . 2. Christian biography—United States. 3. Turner, Dean. I. Title.
BR1725.T77T87 242 77-78424
79 MARC

CHRISTIAN BIOLOGY

see also Clergy

CHRISTIAN BROTHERS COLLEGE, ROSTREVOR

Moore, Peter L
Pride of the hills : the story of Rostrevor House / by Peter L. Moore. — Blackwood, S.A. : Lynton Publications, 1975.
187 p. : ill. (1 col.), maps, plans, ports. ; 22 cm.
Aus
ISBN 0-86946-200-8
1. Rostrevor House, South Australia. 2. Christian Brothers College, Rostrevor. I. Title.
DU330.R67M66 994.2'3 78-322325
78[r78]rev MARC

CHRISTIAN CHURCH (DISCIPLES OF CHRIST)

—CLERGY—BIOGRAPHY

Living responsibility. — Glen Irish, Vic. : Vital Publications, 1976.
142 p. : ill. ; 18 cm. Aus
CONTENTS: Williams, E. Lyall. Reflections.—White, A. E. Biography.
ISBN 0-909116-04-0
1. Williams, E. Lyall, 1906- . 2. Christian Church (Disciples of Christ)—Clergy—Biography. 3. Clergy—Australia—Biography. 4. Christian life—Christian Church (Disciples of Christ) authors. I. Williams, E. Lyall, 1906- . II. White, A. E.
BX7343.W53L58 286'.63 79-315513
79 MARC

Tucker, Johnny.
Like a meteor across the horizon : the Jesse B. Ferguson story / by Johnny Tucker. History and true position of the Church of Christ in Nashville / by Tolbert Fanning, et al. — Fayetteville, Tenn. : Tucker Publications, 1978.
iii, 46, 61 p. : ill. ; 21 cm.
Second work is a reprint of the 1854 ed. published by Cameron & Fall, Nashville.
Includes bibliographical references.
1. Ferguson, Jesse Babcock, 1819-1870. 2. Christian Church (Disciples of Christ)—Clergy—Biography. 3. Clergy—Tennessee—Nashville—Biography. 4. Spiritualists—Tennessee—Nashville—Biography. 5. Nashville—Biography. 6. Nashville. Christ Church. I. Fanning, Tolbert, 1810-1874. II. History and true position of the Church of Christ in Nashville. 1978. III. Title.
BX7343.F45T8 286'.6'0924 79-101049
79 MARC

—PERIODICALS

Impact. no. 1-
[Claremont, Calif.] Disciples Seminary Foundation, 1978–
v. 23 cm.
"A journal of thought of Disciples of Christ on the Pacific Slope."
Published by the Disciples Seminary Foundation in association with the School of Theology at Claremont.
Key title: Impact (Claremont), ISSN 0194-0422.
1. Christian Church (Disciples of Christ)—The West—Periodicals. 2. Christian Church (Disciples of Christ)—Periodicals. I. Disciples Seminary Foundation. II. Key title.
BX7301.I 47 286'.679 79-643794
MARC-S

—THE WEST—PERIODICALS

Impact. no. 1-
[Claremont, Calif.] Disciples Seminary Foundation, 1978–
v. 23 cm.
"A journal of thought of Disciples of Christ on the Pacific Slope."
Published by the Disciples Seminary Foundation in association with the School of Theology at Claremont.
Key title: Impact (Claremont), ISSN 0194-0422.
1. Christian Church (Disciples of Christ)—The West—Periodicals. 2. Christian Church (Disciples of Christ)—Periodicals. I. Disciples Seminary Foundation. II. Key title.
BX7301.I 47 286'.679 79-643794
MARC-S

CHRISTIAN CIVILIZATION see Civilization, Chiristian

CHRISTIAN COLLEGES see Church colleges

CHRISTIAN COMMUNICATION see Communication (Theology)

CHRISTIAN COMMUNITIES

see also Monasticism and religious orders; Religious communities

Arnold, Heini, 1913-
Gemeinsames Leben : ein Weg zu wahrer Brüderlichkeit : ein Brief vom Bruderhof (Society of Brothers) in Einheit verbunden mit den Brüdern, die man die Hutterischen nennt / von Heini und Annemarie Arnold. — Rifton, N.Y. : Plough Pub. House, c1977.
30 p. : ill. ; 22 cm.
Translation of Living in community.
ISBN 0-87486-180-2 : $1.00
1. Bruderhof Communities. 2. Christian communities. I. Arnold, Annemarie, joint author. II. Title.
BX8129.B64A8315 289.7'3 77-1417
77 MARC

Gish, Arthur G
Living in Christian community / Arthur G. Gish. — Scottdale, Pa. : Herald Press, 1979.
379 p. ; 22 cm.
Bibliography: p. [371]-373.
Includes index.
ISBN 0-8361-1887-1 : $8.95
1. Christian communities. 2. Church renewal. 3. Christian life—Church of the Brethren authors. 4. Church. I. Title.
BV4405.G57 262 79-11848
79 MARC

Jackson, Dave.
Coming together / Dave Jackson. — Minneapolis : Bethany Fellowship, c1978.
199 p. : ill. ; 21 cm.
Includes index.
ISBN 0-87123-087-9 : $3.95
1. Christian communities. 2. Christian communities—United States. I. Title.
BV4405.J32 301.34 78-16123
78 MARC

Pro Mundi Vita (Society)
Basic communities in the Church / [compiled by J. Kerkhofs, with the cooperation of A. Mendoza and L. Hertsens] — [Brussels] : Pro Mundi Vita, 1976.
32 p. ; 27 cm. — (Pro Mundi Vita bulletin ; 62) Be***
Includes bibliographical references.
1. Christian communities. 2. Sociology, Christian (Catholic) I. Kerkhofs, Jan. II. Mendoza, A. III. Hertsens, L. IV. Title. V. Series: Pro Mundi Vita (Society). Bulletin ; 62.
BV4405.P76 1976 259 78-320325
78 MARC

Tapp, Michael, 1933-
Why ritual? / Michael Tapp. — London : Floris Books, 1978.
41 p. ; 18 cm. GB***
ISBN 0-903540-13-4 : £1.00
1. Christian Community. 2. Rites and ceremonies. I. Title.
BP605.C45T36 265 78-314494
78 MARC

Wagemaker, Herbert.
A special kind of belonging : the Christian community / Herbert Wagemaker, Jr. — Waco, Tex. : Word Books, c1978.
126 p. ; 23 cm.
Bibliography: p. 125-126.
ISBN 0-8499-0072-7 : $5.95
1. Christian communities. I. Title.
BV4405.W33 248 77-92473
78 MARC

—ADDRESSES, ESSAYS, LECTURES

La Comunità : corso di spiritualità : studio teologico S. Bernardino, Verona, 1978. — Verona : L.I.E.F., 1978.
174 p. ; 21 cm. — (Esperienze dello spirito ; 2) It78-Sept
Includes bibliographies.
L3500
1. Christian communities—Addresses, essays, lectures.
BV4405.C65 78-397611
*79 MARC

—FRANCE

Hébrard, Monique.
Les nouveaux disciples : voyage à travers les communautés charismatiques / Monique Hébrard. — Paris : Centurion, c1979.
284 p., [2] leaves of plates : ill. ; 22 cm. F***
Includes bibliographical references.
ISBN (invalid) 2-227-35507-9 : 55.00F
1. Pentecostalism—France. 2. Pentecostalism—Catholic Church. 3. Christian communities—France. I. Title.
BX2350.57.H43 282'.44 79-380549
79 MARC

—GERMANY, WEST—BENSHEIM

Hofmann, Irmela, 1924-
Kein Tag wie jeder andere : [Tagebuch aus 10 Jahren e. Bensheimer Grossfamilie] / Irmela Hofmann. — Wuppertal : Aussaat-Verlag, 1978.
163 p. : ill. ; 20 x 21 cm. GFR78-A38
ISBN 3-7615-4801-X
1. Offensive Junger Christen—History. 2. Bensheim, Ger.—Religions and ecclesiastical institutions. 3. Christian communities—Germany, West—Bensheim. I. Title.
BR1642.G3H63 79-364044
79 MARC

—ITALY

Barbero, Franco.
Maestri di nessuno : cristiani di base contro ogni presunzione magisteriale / Franco Barbero. — Torino : Claudiana, 1978.
183 p. ; 20 cm. — (Collana Nostro tempo ; 26) It78-Oct
Includes bibliographical references.
L3900
1. Sociology, Christian (Catholic) 2. Liberation theology. 3. Christian communities—Italy. I. Title.
BX1753.B34 79-344785
*79 MARC

Sciubba, Roberto, 1946-
Le comunità di base in Italia / Roberto Sciubba, Rossana Sciubba Pace. — 1. ed. — Roma : Coines, 1976.
2 v. ; 19 cm. — (Universale Coines ; 43-44 : Cristianesimo oggi)
It76-Aug
Bibliography: v. 1, p. 107-113.
CONTENTS: v. 1. Storia e cronaca.—v. 2. La mappa del movimento.
L1500 (v. 1); L3000 (v. 2)
1. Christian communities—Italy. I. Sciubba Pace, Rossana, joint author. II. Title.
BV4406.I8S35 77-479132
*77[r79]rev MARC

—ITALY—GUDO GAMBAREDO

Campi, Maretta.
Costruire una casa, costruire la Chiesa. Milano, Jaca book, 1970.
160 p. 23 cm. (Cronache alla prova : Chiesa e società, n. 23) L1500
It71-July
At head of title: Redattore Maretta Campi.
1. Christian communities—Italy—Gudo Gambaredo. 2. Gudo Gambaredo, Italy—Church history. I. Title.
BV4406.I8C35 79-414335
79 MARC

—ITALY—LOPPIANO

Lorit, Sergio C
Die Loppiano-Story : Entstehung einer Stadt / S. C. Lorit. — 2., verb. Aufl. — München : Zürich ; Wien : Verlag Neue Stadt, 1977.
93 p. ; 19 cm. GFR78-A14
ISBN 3-87996-060-7 : DM4.00
1. Christian communities—Italy—Loppiano. 2. Catholics—Italy—Loppiano. 3. Loppiano, Italy—Church history. I. Title.
BX1548.L66L67 1977 78-367936
78 MARC

—ITALY—RIESI

Vinay, Tullio, 1909-
Riesi ou la force de l'agapê / Tullio Vinay et Georges Richard-Molard. — Paris : Buchet-Chastel, 1976.
194 p. ; 21 cm. F77-9533
34.85F
1. Vinay, Tullio, 1909- . 2. Waldenses in Riesi, Italy—Biography. 3. Christian communities—Italy—Riesi. 4. Riesi, Italy—Biography. I. Richard-Molard, Georges. II. Title.
BX4883.V5A33 284'.4 78-388842
79 MARC

—UNITED STATES

Jackson, Dave.
Coming together / Dave Jackson. — Minneapolis : Bethany Fellowship, c1978.
199 p. : ill. ; 21 cm.
Includes index.
ISBN 0-87123-087-9 : $3.95
1. Christian communities. 2. Christian communities—United States. I. Title.
BV4405.J32 301.34 78-16123
78 MARC

CHRISTIAN CONFERENCE OF ASIA

O'Grady, Ron.
Singapore to Penang : the Christian Conference of Asia from 1973 to 1977 / Ron O'Grady. — Singapore : CCA, 1977.
62 p. : ill. ; 21 cm.
Includes index.
1. Christian Conference of Asia. I. Title.
BR1060.C45O33 262'.001 78-940977
79 MARC

CHRISTIAN CONVERTS see Converts

CHRISTIAN DEMOCRACY

see also Catholic action; Center parties; Church and labor; Socialism, Christian; Trade-unions, Catholic

Alzaga Villaamil, Oscar.
Le origini della democrazia cristiana in Spagna / Oscar Alzaga Villaamil. — Roma : Cinque lune, 1978.
365 p. ; 22 cm. — (Collana di storia del movimento cattolico ; 42)
It79-July
Translation of La primera democracia cristiana en España.
Includes bibliographical references.
L6000
1. Partido Social Popular. 2. Christian democracy. I. Title.
JN8395.S58A6216 329.9'45 79-375950
*79 MARC

Dettling, Warnfried, 1943-
Die christliche Demokratie vor den Europawahlen / Warnfried Dettling. — Hannover : Niedersächsische Landeszentrale für Politische Bildung, 1978.
112 p. ; 21 cm. GFR***
Includes bibliographical references.
1. Christlich-Demokratische Union (Germany (Federal Republic)) 2. Christlich-Soziale Union. 3. Center parties—Europe. 4. Christian democracy. I. Title.
JN3971.A98C4238 329.9'43 79-350767
79 MARC

Fogarty, Michael Patrick.
Christian democracy in Western Europe, 1820-1953, by Michael P. Fogarty. Westport, Conn., Greenwood Press [1974]
xxxiii, 448 p. illus. 24 cm.
Reprint of the 1957 ed. published by Routledge & K. Paul, London.
Bibliography: p. 436-448.
ISBN 0-8371-7114-8
1. Christian democracy. 2. Europe—Social conditions. I. Title.
[HN373.F6 1974] 261.8 73-11997
73 MARC

Letamendia, Pierre.
La Démocratie chrétienne / Pierre Letamendia, ... — [Paris] : PUF, 1977.
126 p., [2] ; 18 cm. — (Que sais-je? ; 1692) F77-17342
Bibliography: p. [127]
9.00F
1. Christian democracy. I. Title.
HN37.C3L438 335'.7 78-390070
79 MARC

Niunka, Vladas, 1907-
Klerikalinės partijos ir Vatikanas : Krikščionių demokratų partijos politinėje imperializmo sistemoje / V. Niunka. — Vilnius : Mintis, 1974.
78, [2] p. ; 17 cm. USSR***
Bibliography: p. 76-[79]
0.15rub
1. Catholic Church—Doctrinal and controversial works—Miscellaneous authors. 2. Catholic Church in Lithuania. 3. Christian democracy. I. Title.
BX1767.N49 78-400199
79 MARC

Sopicki, Stanisław, 1903-
Chrzesciańscy demokraci / Stanisław Sopicki. — London : Nakł. Chrzescijańskiego Zjednoczenia Zawodowego Polskiego, [1946?]
27 p. ; 18 cm.
Cover title.
"Odczyt wygłoszony 21 czerwca 1946 r. w Londynie."
1. Christian democracy. I. Title.
HN37.C3S66 79-342347
79 MARC

—ADDRESSES, ESSAYS, LECTURES

Das Elend der Christdemokraten : Ortsbestimmung d. polit. Mitte Europas / hrsg. von Gerd-Klaus Kaltenbrunner ; [Beitr. von Alfons Dalma ... et al.]. — Orig.-Ausg. — Freiburg [im Breisgau] ; Basel ; Wien : Herder, 1977.
191 p. ; 19 cm. — (Herderbücherei ; 9521 : Initiative ; 21) GFR78-A22
Bibliography: p. [172]-187.
Includes bibliographical references.
ISBN 3-451-09521-1 : DM10.90
1. Center parties—Europe—Addresses, essays, lectures. 2. Christian democracy—Addresses, essays, lectures. I. Kaltenbrunner, Gerd, 1939-
JN94.A979E38 329'.02'094 79-358824
79 MARC

—HISTORY—ADDRESSES, ESSAYS, LECTURES

Il Partito cristiano : D.C. e mondo cattolico in Piemonte, 1900-1975. — Torino : Stampatori, [1978]
229 p. ; 20 cm. — (Argomenti di politica e cultura contemporanea)
It78-Mar
L5000
1. Partito della democrazia cristiana—History—Addresses, essays, lectures. 2. Christian democracy—History—Addresses, essays, lectures. 3. Church and social problems—Italy—Piemonte—History—Addresses, essays, lectures. I. Series.
JN5657.D43P37 78-381166
*79 MARC

—PERIODICALS

European digest.
Luxembourg, Christian-Democratic Group, European Parliament.
v. 30 cm.
1. Center parties—Europe—Periodicals. 2. Christian democracy—Periodicals. 3. International economic relations—Periodicals. 4. World politics—1975-1985—Periodicals. I. European Parliament. Christian-Democratic Group.
JN94.A979E847 329'.02'094 79-649713
MARC-S

CHRISTIAN DENOMINATIONS see Christian sects

CHRISTIAN DEVOTIONAL CALENDARS see Devotional calendars

CHRISTIAN DEVOTIONAL LITERATURE see Devotional literature

CHRISTIAN DOCTRINE see Theology, Doctrinal

CHRISTIAN DRAMA

see also Christmas plays

—BIBLIOGRAPHY

Johnson, Albert.
Best church plays : a bibliography of religious drama / Albert Johnson. — Gloucester, Mass. : P. Smith, 1978, c1968.
viii, 180 p. ; 21 cm.
Reprint of the ed. published by Pilgrim Press, Philadelphia.
Includes index.
ISBN 0-8446-2328-8
1. Christian drama—Bibliography. I. Title.
[Z5784.R3J6 1978] 016.822'051 78-112302
[PN1880] 79 MARC

CHRISTIAN DRAMA, AMERICAN

Waddy, Lawrence.
Drama in worship / by Lawrence Waddy. — New York : Paulist Press, c1978.
210 p. ; 23 cm.
ISBN 0-8091-2107-7 : $8.95
1. Christian drama, American. 2. Amateur theatricals. I. Title.
PN6120.R4W22 821'.9'14 78-58952
78 MARC

CHRISTIAN DRAMA, ENGLISH

—HISTORY AND CRITICISM

Craig, Hardin, 1875-1968.
English religious drama of the Middle Ages / by Hardin Craig. — Westport, Conn. : Greenwood Press, 1978, c1960.
vi, 421 p. ; 24 cm.
Reprint of the ed. first published in 1955 by Clarendon Press, Oxford.
Bibliography: p. [390]-401.
Includes index.
ISBN 0-313-20496-9
1. English drama—To 1500—History and criticism. 2. Christian drama, English—History and criticism. 3. Mysteries and miracle plays, English—History and criticism. I. Title.
PR641.C7 1978 822'.1'07 78-6893
78 MARC

CHRISTIAN DRAMA, SPANISH

—HISTORY AND CRITICISM

Fothergill-Payne, Louise.
La alegoría en los autos y farsas anteriores a Calderón / [por] Louise Fothergill-Payne. — London : Tamesis Books, D.L. 1977.
223 p. ; 24 cm. — (Colección Tamesis : Serie A, Monografías ; 66)
Bibliography: p. 212-215.
Includes index.
ISBN 0-7293-0032-3
1. Christian drama, Spanish—History and criticism. 2. Spanish drama—Classical period, 1500-1700—History and criticism. 3. Allegory. I. Title.
PQ6121.R3F6 862'.3 78-394403
79 MARC

CHRISTIAN EDUCATION

see also Bible—Study; Bible colleges; Catechetics; Catechisms; Catechumens; Church and college; Church and education; Church schools; Drama in Christian education; Jesus Christ—Teaching methods; Motivation in Christian education; Theology—Study and teaching; Week-day church schools

Barrena, José.
La libertad de enseñanza / José Barrena. — Madrid : Biblioteca de Autores Cristianos, c1978.
30 p. ; 19 cm. — (Cuadernos BAC ; 1) Sp***
Bibliography: p. 30.
ISBN 8422003894
1. Education—Aims and objectives. 2. Christian education. 3. Education and state. I. Title.
LB41.B282 78-380432
78 MARC

Byrne, Herbert W 1917-
Improving church education / H. W. Byrne. — Birmingham, Ala. : Religious Education Press, c1979.
xiv, 352 p. ; 20 cm.
Bibliography: p. 344-346.
Includes index.
ISBN 0-89135-017-9
1. Christian education. I. Title.
BV1471.2.B93 268 79-10852
79 MARC

Coudreau, François.
Risques et chances de l'enseignement religieux aujourd'hui / par François Coudreau, P.S.S. — Metz : Église de Metz", 1976.
167 p. ; 20 cm. F78-2191
Cover title.
"Supplément 'Église de Metz', septembre 1976."
20.00F
1. Christian education. I. Église de Metz (Periodical) ; 1976, septembre (Supplement) II. Title.
BV1471.2.C6583 268 78-388847
79 MARC

Cove, Mary K
Regarding religious education / Mary K. Cove and Mary Louise Mueller. — Mishawaka, Ind. : Religious Education Press, c1977.
vii, 181 p. : ill. ; 21 cm.
Includes bibliographical references and index.
ISBN 0-89135-011-X
1. Christian education. I. Mueller, Mary Louise, joint author. II. Title.
BV1471.2.C67 268 77-10873
77 MARC

CHRISTIAN EDUCATION (Continued)

Gatti, Gaetano.
Il mistero del catechista : itinerari di fede per la formazione spirituale dei ministri della parola / Gaetano Gatti. — Bologna : EDB, 1978.
266 p. ; 19 cm. — (Educatori e catechisti) It78-Dec
Bibliography: p. 261-262.
L4000
1. Christian education. I. Title. II. Series.
BV1471.2.G39 79-342153
*79 MARC

Giacomelli, Renzo.
Quale catechesi : dossier Sinodo 1977 / Renzo Giacomelli. — Leumann/[Rivoli] : Elle Di Ci, 1978.
191 p. ; 21 cm. — (Problemi e proposte di catechesi ; 6) It78-June
ISBN 8801152256 : L2600
1. Christian education. 2. Episcopal conferences (Catholic) I. Title. II. Series.
BV1471.2.G47 78-377630
*78 MARC

Hughes, Herbert Daniel.
Addysg grefyddol : i be' ac i bwy? / [gan] Herbert Hughes. — Abertawe [i.e. Swansea] : Penry, 1978.
130 p. ; 19 cm. GB79-02329
Bibliography: p. 127-130.
ISBN 0-903701-12-X : £1.25
1. Christian education. I. Title.
BV1471.2.H83 79-378781
79 MARC

Kao, Hsin.
(Chi-tu chiao chiao yü kai lun)
基督教教育概論 / 高炘編著. — 台北：中華福音神學院編譯著述委員會；總經銷校園團契出版社，1977.
2, 2, 176 p. : ill. ; 22 cm.
Cover title.
Includes bibliographies.
1. Christian education. I. Title.
BV1521.K36 78-841060

Kelsey, Morton T
Can Christians be educated? : A proposal for effective communication of our Christian religion / Morton Kelsey ; compiled and edited by Harold William Burgess. — Mishawaka, Ind. : Religious Education Press, c1977.
154 p. ; 23 cm.
Bibliography: p. 151-153.
ISBN 0-89135-008-X
1. Christian education. I. Title.
BV1471.2.K44 207 77-3691
77[r79]rev MARC

Korherr, Edgar Josef, 1928-
Methodik des Religionsunterrichts / Edgar Josef Korherr ; [Illustr., Bruno Wegscheider]. — Wien ; München : Jugend & Volk ; Wien : Österr. Bundesverl., 1977.
93 p. : ill. ; 30 cm. — (Materialien zur Pädagogik) Au78-3-234
Bibliography: p. 84-93.
ISBN 3-7141-5926-6 : S78.00
1. Christian education. I. Title.
BV1471.2.K58 79-371754
*79 MARC

Lagarde, Claude.
Apprendre à dire Dieu : pour une initiation à la symbolique chrétienne / Claude et Jacqueline Lagarde. — Paris : Centurion, c1978.
206 p. ; 21 cm. — (Champs nouveaux) F***
Includes bibliographical references and index.
ISBN 2-227-35701-0 : 38.00F
1. Christian education. I. Lagarde, Jacqueline, joint author. II. Title.
BV1471.2.L33 268'.432 79-353419
79 MARC

Mongoven, Anne Marie.
Signs of catechesis : an overview of the National catechetical directory / by Anne Marie Mongoven. — New York : Paulist Press, c1979.
xii, 140 p. ; 19 cm.
ISBN 0-8091-2152-2 : $1.95
1. Catholic Church. National Conference of Catholic Bishops. Sharing the light of faith. 2. Catechetics. 3. Christian education. I. Title.
BX1961.M66 268'.6 78-58964
79 MARC

Negri, Giancarlo.
Catechesi e mentalità di fede : metodologia catechistica fondamentale / Giancarlo Negri. — Torino Leumann : Elle Di Ci, [1976]
295 p. ; 23 cm. — (Quaderni di pedagogia catechistica ; C, 1) It78-Feb
Includes bibliographical references.
L3500
1. Christian education. I. Title.
BV1471.2.N36 79-382710
*79 MARC

Nichols, Kevin.
Cornerstone / by Kevin Nichols. — Slough : St Paul Publications, 1978.
137 p. ; 24 cm. — (Guidelines for religious education ; 1) GB79-10797
"Commissioned by and published the authority of the Bishops' Conference of England and Wales."
Includes index.
ISBN 0-85439-157-6 : £2.50
1. Christian education. I. Title. II. Series.
BV1471.2.N47 268'.6 79-312426
79 MARC

O'Brien, Robert Yorke.
Clarity in religious education / Robert Yorke O'Brien. — Wellington, Ind. : Religious Education Press, c1978.
131 p. ; 20 cm.
Includes bibliographical references and index.
ISBN 0-89135-013-6
1. Christian education. I. Title.
BV1471.2.O25 268 78-3497
78 MARC

Sharing the light of faith : national catechetical directory for Catholics of the United States. — [Washington] : United States Catholic Conference, Dept. of Education, c1979.
vii, 182 p. : ill. ; 28 cm.
Includes bibliographical references and index.
1. Christian education. I. United States Catholic Conference. Dept. of Education.
BV1471.S52 268'.6 79-108061
79 MARC

Stuchbery, Ian.
Growing in Christ : new patterns of initiation and education in the parish community / Ian Stuchbery. — Toronto : Anglican Book Centre, c1978.
109 p. ; 22 cm. C78-6567-0
Bibliography: p. [105]-107.
ISBN 0-919030-23-8 : £3.25
1. Christian education. 2. Church renewal—Anglican Church of Canada. I. Title.
BV1471.2.S86 262'.001 79-312815
79 MARC

Weber, Hartwig.
Religiöse Erziehung : Informationen u. Anregungen / Hartwig Weber. — Orig.-Ausg. — Gütersloh : Gütersloher Verlagshaus Mohn, 1978.
127 p. ; 19 cm. — (Gütersloher Taschenbücher : Siebenstern ; 279) GFR78-A
ISBN 3-579-03679-3 : DM9.80
1. Christian education. I. Title.
BV1471.2.W42 79-337226
79 MARC

—ADDRESSES ESSAYS, LECTURES

Aesthetic dimensions of religious education / edited by Gloria Durka and Joanmarie Smith. — New York : Paulist Press, c1979.
xiii, 235 p. ; 23 cm.
Includes bibliographical references.
ISBN 0-8091-2164-6 : $8.50
1. Christian education—Addresses, essays, lectures. 2. Aesthetics—Addresses, essays, lectures. I. Durka, Gloria. II. Smith, Joanmarie.
BV1473.A33 207 78-65903
00 MARC

Aspekte christlicher Erziehung in der Schule / hrsg. von d. Landeskonferenz für Schule u. Erziehung ; [hrsg. von Hermann Maier]. — 1. Aufl. — Regensburg : Pustet, 1978.
271 p. ; 19 cm. GFR78-A
Includes bibliographical references.
ISBN 3-7917-0562-8
1. Christian education—Addresses, essays, lectures. I. Maier, Hermann, 1936- II. Landeskonferenz für Schule und Erziehung.
BV1473.A85 78-384113
78 MARC

La Catechesi oggi / [scritti del] card. M. Pellegrino ... [et al.]. — Roma : Città nuova ; [Roma] : Pontificia università lateranense, 1978.
103 p. ; 21 cm. — (Temi di pastorale ; 2) It78-Aug
Includes bibliographical references.
L2000
1. Christian education—Addresses, essays, lectures. I. Pellegrino, Michele.
BV1473.C29 79-386303
*79 MARC

Christian education : an introduction to its scope / edited by Douglas J. Simpson. — Nashville : Randall House Publications, c1978.
252 p. ; 23 cm.
Incudes bibliographical references.
ISBN 0-89265-053-2
1. Christian education—Addresses, essays, lectures. I. Simpson, Douglas J.
BV1471.2.C498 207 79-107271
79 MARC

Homegrown Christian education : planning and programming for Christian education in the local congregation / David W. Perry, editor. — New York : Seabury Press, 1979.
viii, 151 p. ; 21 cm.
"A Crossroad book."
Includes bibliographies.
ISBN 0-8164-2212-5
1. Christian education—Addresses, essays, lectures. I. Perry, David W.
BV1471.2.H64 1979 268 78-21500
78 MARC

Process and relationship : issues in theology, philosophy, and religious education : a festschrift for Randolph Crump Miller / edited by Iris V. Cully and Kendig Brubaker Cully. — Birmingham, Ala. : Religious Education Press, c1978.
xii, 139 p. ; 23 cm.
Bibliography: p. 124-133.
Includes index.
ISBN 0-89135-012-8
1. Christian education—Addresses, essays, lectures. 2. Theology—Addresses, essays, lectures. 3. Philosophy—Addresses, essays, lectures. 4. Miller, Randolph Crump, 1910- —Addresses, essays, lectures. I. Cully, Iris V. II. Cully, Kendig Brubaker. III. Miller, Randolph Crump, 1910-
BV1473.P76 230 77-18760
77 MARC

Totosaus, Josep Maria.
Transmetre la fe, avui / Josep M. Totosaus. — 1. ed. — [Montserrat] : Publicacions de l'Abadia de Montserrat, 1978.
64 p. ; 19 cm. — (Col·lecció L'Espiga ; 7) Sp***
ISBN 8472022870
1. Christian education—Addresses, essays, lectures. I. Title.
BV1473.T67 79-372379
79 MARC

Towards a Christian theology of education / [by] M. Calvert ... [et al.]. — London : Church Literature Association for the editors, 1977.
[2], 5, [1] p. ; 21 cm. — (Dolphin papers ; 7) GB78-05769
Cover title.
£0.15
1. Christian education—Addresses essays, lectures. 2. Church and education—Addresses essays, lectures. I. Calvert, M.
BV1473.T684 261.1 78-320319
78 MARC

—AUDIO-VISUAL AIDS

see also Christian education—Bulletin boards; Puppets and puppet-plays in Christian education

McNulty, Edward N 1936-
Equipping the church / by Edward McNulty. — King of Prussia, Pa. : Religious Pub. Co., c1978.
iii, 67 p. : ill. ; 20 cm. — (His Using AV's in the church ; v. 3)
Bibliography: p. 66-67.
1. Christian education—Audio-visual aids. I. Title.
BV1535.M33 vol. 3 268'.635 s 79-100972
79 MARC

McNulty, Edward N 1936-
Plugging in / by Edward McNulty. — King of Prussia, Pa. : Religious Pub. Co., c1978.
iii, 72 p. : ill. ; 21 cm. — (His Using AV's in the church ; v. 1)
Bibliography: p. 69-72.
1. Christian education—Audio-visual aids. I. Title.
BV1535.M33 vol. 1 268'.635 s 79-101005
79 MARC

Strube, Hans-Heinrich.
Lernen mit Medien : Kurzmodelle, Medienverb. / Hans-Heinrich Strube, Weert Flemmig. — Gütersloh : Gütersloher Verlagshaus Mohn, 1977.
90 p. : ill. ; 30 cm. & slides (2 slides : col. ; 1 3/8 x 1 1/2 in.) unmounted, in pocket. — (Ku-Praxis ; 7) GFR77-A25
ISBN 3-579-04407-9 : DM13.80
1. Confirmation—Instruction and study. 2. Christian education—Audio-visual aids. I. Flemmig, Weert, joint author. II. Title. III. Series.
BV815.S88 78-387079
79 MARC

—BULLETIN BOARDS

Staffeld, Jean, 1928-
38 recipes for bulletin boards and art projects that Christian kids can make / by Jean Staffeld, Carol Ball, Jo Anne Klemp ; Illustrated by Ted Smith. — St. Louis : Concordia Pub. House, c1978.
94 p. : ill. ; 28 cm.
ISBN 0-570-03774-3 : $3.95
1. Christian education—Bulletin boards. 2. Creative activities and seat work. I. Ball, Carol, 1952- joint author. II. Klemp, Jo Anne, 1950- joint author. III. Smith, Ted. IV. Title.
BV1535.25.S73 207 77-29138
78 MARC

CHRISTIAN EDUCATION (Continued)

—CONGRESSES

Convegno liturgico-pastorale, 15th, Vatican City, 1973.
Educazione alla fede e iniziazione cristiana : atti del XV Convegno liturgico-pastorale / relazioni di Luciano Borello ... [et al.] ; introduzione del card. Ferdinando Antonelli. — Milano : Edizioni O.R., 1973.
158 p. ; 21 cm. — (Sussidi liturgico-pastorali ; nuova ser., 5) It***
Bibliography: p. 53-54.
L2000
1. Baptism—Catholic Church—Congresses. 2. Confirmation—Catholic Church—Congresses. 3. First communion—Congresses. 4. Christian education—Congresses. I. Borello, Luciano. II. Title.
BX2205.C5 1973 265 79-337860
79 MARC

—CURRICULA—PERIODICALS

United Methodist Church (United States). Board of Discipleship. Curriculum Resources Committee.
Curriculum plans.
Nashville, Tenn., Curriculum Resources Committee of the United Methodist Church.
v. 23 cm. annual.
Key title: Curriculum plans, ISSN 0160-0885.
1. United Methodist Church (United States)—Education—Periodicals. 2. Christian education—Curricula—Periodicals. I. Title.
BX8225.A1U54a 268'.6 78-649071
MARC-S

—DICTIONARIES—GERMAN

Korherr, Edgar Josef, 1928-
Praktisches Wörterbuch der Religionspädagogik und Katechetik / hrsg. v. Edgar Josef Korherr u. Gottfried Hierzenberger, unter Mitarb. v. P. Anzenberger ... [et al.]. — 2. Aufl. — Wien ; Freiburg ; Basel : P. Herder, 1978.
[12] p., 1138 columns : graphs ; 20 cm. — (Religionsunterricht, Information und Verkündigung) Au78-22-25
Bibliography: columns 1089-1138.
ISBN 3-210-24433-2 : S250.00
1. Christian education—Dictionaries—German. I. Hierzenberger, Gottfried, 1937- joint author. II. Title.
BV1461.K67 1978 79-362962
*79 MARC

—HISTORY

Grassi, Joseph A
Jesus as teacher : a New Testament guide to learning "the way" / by Joseph A. Grassi. — Winona, Minn. : St. Mary's College Press, c1978.
159 p. ; 22 cm. — (A Pace book)
Published in 1973 under title: The teacher in the primitive church and the teacher today.
Includes bibliographical references.
ISBN 0-88489-106-2 : $4.95
1. Christian education—History. 2. Jesus Christ—Teaching methods. 3. Education—History. I. Title.
BV1465.G7 1978 268'.3 78-62676
79 MARC

—HISTORY—EARLY CHURCH, CA. 30-600

Nunes, Ruy Afonso da Costa.
História da educação na antiguidade cristã : o pensamento educacional dos mestres e escritores cristãos no fim do mundo antigo / Ruy Afonso da Costa Nunes. — São Paulo : Editora Pedagógica e Universitária, c1978.
xi, 246 p. ; 21 cm.
Includes bibliographies and indexes.
1. Christian education—History—Early church, ca. 30-600. I. Title.
BV1465.N86 79-105258
79 MARC

Santomauro, Gaetano.
La paideia del monachesimo popolare dei primi secoli / Gaetano Santomauro. — Bari : Adriatica, 1975-
v. ; 17 cm. It77-Dec
Includes bibliographical references.
CONTENTS: 1. Il problema dell'origine storica.

L2500 (v. 1)
1. Christian education—History—Early church, ca. 30-600. 2. Monasticism and religious orders—Early church, ca. 30-600. I. Title.
BV1465.S16 78-385332
*78 MARC

—PHILOSOPHY

Schmidt, Ludwig, 1924-
Glaubensumsetzung in der Erwachsenenpädagogik / Ludwig Schmidt. — 1. Aufl. — Frankfurt/Main : Haag und Herchen, 1978.
vi, 122 p. ; 21 cm. GFR78-A
Bibliography: p. 105-112.
Includes index.
ISBN 3-88129-107-5 : DM14.80
1. Christian education of adults. 2. Christian education—Philosophy. 3. Hermeneutics. I. Title.
BV1488.S35 78-367862
78 MARC

Thompson, Andrew D
Where do we go from here : a survey of U.S. diocesan statements of goals and objectives for religious education and catechesis / by Andrew D. Thompson and Robert P. Stamschror. — Washington : United States Catholic Conference, Dept. of Education, c1979.
ii, 49 p. : ill. ; 22 cm.
Includes bibliographical references.
1. Christian education—Philosophy. 2. Christian education—United States. I. Stamschror, Robert P., joint author. II. United States Catholic Conference. Dept. of Education. III. Title.
BV1464.T47 207 79-120037
79 MARC

United Methodist Chruch (United States). Board of Discipleship. Division of Education.
Foundations for teaching and learning in the United Methodist Church : a statement of the Division of Education, Board of Discipleship, the United Methodist Church. — Nashville : Discipleship Resources, c1979.
82 p. : ill. ; 22 cm.
Includes bibliographical references.
1. United Methodist Church (United States)—Doctrinal and controversial works—Methodist authors. 2. Christian education—Philosophy. I. Title.
BX8377.U54 1979 230'.7'673 79-115597
79 MARC

Werbs, Ulrich.
Die Bedeutung des Hörers für die Verkündigung : pastoraltheolog. Überlegungen zum antropolog. Ansatz der Verkündigung im holländ. Katechismus für Erwachsene / von Ulrich Werbs. — 1. Aufl. — Leipzig : St.-Benno-Verlag, 1978.
xxiii, 187 p. ; 23 cm. — (Erfurter theologische Studien ; 39) GDR79-A
Bibliography: p. xi-xxiii.
19.80M
1. Christian education—Philosophy. 2. Christian education of adults. 3. De Nieuwe katechismus. 4. Catholic Church—Catechisms and creeds—Dutch. I. Title. II. Series.
BV1464.W47 79-360186
79 MARC

Wyler, Alain.
L'éducateur au service de la foi : étude théologique, psychologique, pédagogique et pastorale / Alain Wyler. — Paris : Le Centurion, c1978.
163 p. ; 22 cm. F***
Includes bibliographical references.
ISBN 2-227-31537-7 : 32.00F
1. Christian education—Philosophy. I. Title.
BV1464.W94 268'.433 78-384870
78 MARC

—PHILOSOPHY—ADDRESSES, ESSAYS, LECTURES

Don Peter, W L A
Studies in Christian education / by W. L. A. Don Peter. — Colombo : Wesley Press, 1974.
100 p. ; 23 cm.
Includes bibliographical references.
Rs5.50
1. Christian education—Philosophy—Addresses, essays, lectures. I. Title.
BV1464.D66 207 77-906381
79 MARC

Heath, Douglas H
The peculiar mission of a Quaker school / Douglas H. Heath. — Wallingford, Pa. : Pendle Hill Publications, 1979.
31 p. ; 20 cm. — (Pendle Hill pamphlet ; 225)
Includes bibliographical references.
ISBN (invalid) 0-87574-255-4 : $1.10
1. Friends, Society of—Education—Addresses, essays, lectures. 2. Christian education—Philosophy—Addresses, essays, lectures. I. Title.
BX7748.E4H42 207 79-84919
00 MARC

—PSYCHOLOGY

see also Motivation in christian education

Zwergel, Herbert A
Religiöse Erziehung und Entwicklung der Persönlichkeit : psychologischer Leitfaden für Religionslehrer und praktische Theologen / Herbert A. Zwergel. — Einsiedeln ; Zürich ; Köln : Benziger, 1976.
94 p. ; 20 cm. — (Religionspädagogik : Theorie und Praxis ; 33) Sw76-5297
Includes bibliographical references.
ISBN 3-545-25121-2 : 17.80F
1. Christian education—Psychology. 2. Psychology, Religious. I. Title.
BV1471.2.Z9 268'.01'9 76-487436
*77[r79]rev MARC

—TEACHER TRAINING—ADDRESSES, ESSAYS, LECTURES

Qualifiziert als Lehrer und Erzieher im Religionsunterricht : zur Lehrerfortbildung im Medienverbund / Helmut Hanisch ... [et al. ; im Auftr. d. Religionspädag. Projektentwicklung in Baden u. Württemberg (RPE) hrsg. von Klaus Dessecker u. Gerhard Martin]. — 1. Aufl. — Stuttgart : Calwer Verlag ; München : Kösel, 1978.
210 p. ; 24 cm. GFR78-A
Includes bibliographical references.
ISBN 3-7668-0557-6
1. Christian education—Teacher training—Addresses, essays, lectures. I. Hanisch, Helmut, 1943-
BV1533.Q34 78-379038
78 MARC

—TEACHER TRAINING—HANDBOOKS, MANUALS, ETC.

Adams, Margee.
Developing leadership for the church's teaching ministry / Margee and Jim Adams. — Livermore, CA : Griggs Educational Service, c1978.
88 p. ; 28 cm.
Includes bibliographical references.
1. Christian education—Teacher training—Handbooks, manuals, etc. 2. Christian leadership—Handbooks, manuals, etc. I. Adams, Jim, joint author. II. Title.
BV1533.A3 268'.3 77-92679
78 MARC

—TEACHING METHODS

Grom, Bernhard.
Methoden für Religionsunterricht, Jugendarbeit und Erwachsenenbildung / Bernhard Grom. — 4., erw. Aufl. — Düsseldorf : Patmos-Verlag ; Göttingen : Vandenhoeck und Ruprecht, [1977]
271 p. : ill. ; 21 cm. GFR78-A
Bibliography: p. 259-265.
Includes index.
ISBN 3-491-78364-X : DM22.80
1. Christian education—Teaching methods. I. Title.
BV1534.G74 1977 78-376596
78 MARC

Keithahn, Mary Nelson, 1934-
How to teach Bible stories / Mary Nelson Keithahn and Marilyn H. Dunshee. — Nashville : Abingdon, c1978.
112 p. ; 28 cm.
Cover title: How to teach Bible stories to children.
Bibliography: p. 106-108.
Includes index.
ISBN 0-687-17960-2
1. Bible stories—Study and teaching. 2. Christian education—Teaching methods. I. Dunshee, Marilyn H., 1923- joint author. II. Title. III. Title: How to teach Bible stories to children.
BS546.K44 268'.6 78-8714
78 MARC

Leavitt, Guy P
Teach with success / written by Guy P. Leavitt ; revision and leader's guide by Eleanor Daniel ; illustrated by Pat Karch. — Cincinnati : Standard Pub., [1978?], c1956.
160 p. : ill. ; 28 cm.
Bibliography: p. 160.
ISBN 0-87239-231-7
1. Christian education—Teaching methods. I. Daniel, Eleanor Ann. II. Title.
BV1534.L43 1978 268'.6 78-63285
79 MARC

—TEXT-BOOKS

Harbig, Horst, 1938-
Die Kinderbibelwoche : Erfahrungen u. Materialien für e. kinderfreundl. Gemeindearbeit / Horst Harbig, Wolf-Rüdiger Marsen, Dietrich Petersmann. — 1. Aufl. — München : Claudius-Verlag, 1978.
203 p. : ill. ; 24 cm. GFR78-A
Includes music.
Includes index.
ISBN 3-532-61619-X
1. Bible—Study—Outlines, syllabi, etc. 2. Christian education—Text-books. I. Marsen, Wolf-Rüdiger, 1946- joint author. II. Petersmann, Dietrich, joint author. III. Title.
BS592.H33 78-383702
78 MARC

CHRISTIAN EDUCATION —TEXT-BOOKS (Continued)

Hsiao yüan t'uan ch'i ch'u pan she.
(Chih hui yü shen liang)
智慧與身量：中學生聚會手冊 / 校園團契出版社. — [修訂版]. — 臺北：該社, 1975.
117 p. ; 19 cm. — (校園工作訓練材料 ; 4)
Title in English in colophon: Wisdom and stature; manual for secondary fellowship.
Series romanized: Hsiao yüan kung tso hsün lien ts'ai liao.
Includes bibliographical references.
1. Christian education—Text-books. I. Title.
BV1561.H73 1975 79-840062

—TEXT-BOOKS—CATHOLIC

Aparisi, Antonio.
Plan de pastoral de adolescentes : documentos de base, tercer año / [por] Antonio Aparisi y equipo de educadores. — Madrid : Marova, 1977 printing.
207 p. ; 24 cm. — (His Odres nuevos) Sp78-Jan
ISBN 8426903568 : 300ptas
1. Christian education—Text-books—Catholic. 2. Christian education of adolescents. I. Title.
BX930.A64 78-389075
79 MARC

—TEXT-BOOKS FOR CHILDREN—CATHOLIC

Tuneu Vila, Ramón.
Caminamos con Jesús hacia la casa del Padre : ciclo de iniciación de los niños a la vida cristiana : cuaderno del 2. año de catequesis / [preparado por Ramón Tuneu Vila ; con la colaboración del Secretariado Interdiocesano de Catequesis de Cataluña]. — 43. ed. corregida y aumentada. — [Barcelona : Secretariado Interdiocesano de Catequesis de Cataluña : distribuye, Centro de material Catequístico, 1977.
94 p. : col. ill. ; 22 cm. Sp***
ISBN 8440063555
1. Christian education—Text-books for children—Catholic. I. Title.
BX930.T86 1977 79-371002
79 MARC

—TEXT-BOOKS FOR CHILDREN—SALVATION ARMY

Salvation Army.
Army beliefs and characteristics (a re-written course) : a teaching directory for Young People's Sergeant-Majors and class teachers. — New ed. — London : Salvation Army, 1978-
v. : ill. ; 22 cm. GB***
Includes index.
ISBN 0-85412-320-2 (v. 1)
1. Christian education—Text-books for children—Salvation Army. I. Title.
BX9714.A47 1978 268'.432 79-313052
79 MARC

—TEXT-BOOKS FOR YOUNG PEOPLE—CATHOLIC

McKenna, Megan.
New Testament, understandings of Jesus / by Megan McKenna and Darryl Ducote. — New York : Paulist Press, 1978, c1979.
85 p., [28] leaves : ill. ; 28 cm. — (Followers of the way series ; v. 2)
Contains a section of exercises for students.
Includes bibliographical references.
SUMMARY: Explores the life and teachings of Jesus Christ with discussion questions, enclosed handouts, music, and other suggested activities.
ISBN 0-8091-9543-7
1. Jesus Christ—Person and offices—Juvenile literature. 2. Christian education—Text-books for young people—Catholic. [1. Jesus Christ—Person and offices. 2. Catholic Church—Doctrinal and controversial works. 3. Christian life] I. Ducote, Darryl, joint author. II. Title. III. Series.
BT202.M3117 232 78-71529
00 MARC

Uzima mpya : kitabu cha mazungumuzo ya dini kwa watoto. — Bukavu : Centre interdiocésain de pastorale et de catéchèse, [19]
v. ; 21 cm.
1. Christian education—Text-books for young people—Catholic. I. Centre interdiocésain de pastorale et de catéchèse, Bukavu, Zaire.
BX930.U95 79-362823
79 MARC

—ASIA—ADDRESSES, ESSAYS, LECTURES

Models of christian education in Asia. — Singapore : Christian Conference of Asia, [1977?]
55 p. : ill. ; 22 cm.
Cover title.
1. Christian education—Asia—Addresses, essays, lectures. I. Christian Conference of Asia.
BV1477.M62 207 78-940085
79 MARC

—BRAZIL—SERGIPE (STATE)—HISTORY—SOURCES

A **Catequese/1575 [i.e. mil quinhentos e setenta e cinco]** : documentos para a sua história em Sergipe / seleção e notas de Luiz Antonio Barreto. — [Aracaju] : Governo do Estado de Sergipe, Secretaria da Educação e Cultura, Departamento de Cultura e Patrimônio Histórico, 1975.
31 p. ; 24 cm. — (Coleção Documentos para a história de Sergipe ; v. 1)
CONTENTS: Anchieta, J. de. Primeiros aldeamentos na Bahia (excerto).—Sacchino, F. Historiae Societatis Jesu (excerto).—Tholosa, I. de. Carta provincial (excerto).
1. Christian education—Brazil—Sergipe (State)—History—Sources. 2. Sergipe, Brazil (State)—Church history—Sources. I. Barreto, Luiz Antonio. II. Anchieta, José de, 1534-1597. Primeiros aldeamentos na Bahia. Selections. 1975. III. Sacchini, Francesco, 1570-1625. Historiæ Societatis Jesu (parte tertia, sive Borgia). Selections. 1975. IV. Toloza, Ignacio. V. Series.
BV1517.B72C37 78-400568
79 MARC

—FRANCE—CONTROVERSIAL LITERATURE

Bonnet de Viller, Adrien Marie.
Pour une scolarité catholique / A.-M. Bonnet de Viller. — Vouillé : Éditions de Chire : Diffusion de la Pensée française, c1976.
123 p. ; 18 cm. F***
ISBN 2-85190-025-0 : 15.00F
1. Christian education—France—Controversial literature. 2. Catholic Church—Education—Controversial literature. 3. Education—France—Controversial literature. I. Title.
BV1470.F7B58 377'.8'244 78-393354
79 MARC

Casin, Renée.
Instituteur catholique mon frère. Montsurs, Résiac [1973]
214 p. 19 cm. 18.00F F***
1. Catholic Church—Doctrinal and controversial works. 2. Communism and Christianity—Catholic Church. 3. Catholic Church—Education—Controversial works. 4. Christian education—France—Controversial works. I. Title.
BX1751.2.C3433 230'.2 79-380358
79 MARC

—FRANCE—TEACHING METHODS—CONTROVERSIAL LITERATURE

François, Dominique.
Un Nouveau piège de la subversion : l'expression corporelle / Dominique François ; préface du R. P. Lelong. — Paris (13, rue Mazarine, 75006) : Éditions du Cèdre, 1976.
75 p. ; 21 cm. F77-8721
Includes bibliographical references.
18.00F
1. Catholic Church in France—Controversial literature. 2. Christian education—France—Teaching methods—Controversial literature. I. Title.
BX1530.2.F727 268'.6 78-389005
79 MARC

—GERMANY—HISTORY

Peil, Dietmar.
Zur "angewandten Emblematik" in protestantischen Erbauungsbüchern : Dilherr, Arndt, Francisci, Scriver / Dietmar Peil ; [diese Arbeit ist im Sonderforschungsbereich 7 Mittelalterforschung in Münster entstanden]. — Heidelberg : Winter, 1978.
91 p., [4] leaves of plates : ill. ; 24 cm. — (Beihefte zum Euphorion, Zeitschrift für Literaturgeschichte ; Heft 11 ISSN 0531-2167) GFR78-A
Includes bibliographical references.
ISBN 3-533-02703-1
1. Christian education—Germany—History. 2. Emblem books, German. 3. Dilherr, Johannes Michael, 1604-1669. 4. Arndt, Johann, 1555-1621. 5. Francisci, Erasmus, 1627-1694. 6. Scriver, Christian, 1629-1693. I. Title. II. Series: Euphorion, Zeitschrift für Literaturgeschichte : Beihefte ; Heft 11.
BV1470.G3P44 78-391785
79 MARC

—GERMANY, WEST

Sorge, Helga, 1940-
Religion und Ideologie : Analysen u. Materialien für d. Unterricht / Helga Sorge, Siegfried Vierzig. — 1. Aufl. — Stuttgart ; Berlin ; Köln ; Mainz : Kohlhammer, 1977.
177 p. ; 19 cm. — (Urban-Taschenbücher ; Bd. 634 : T.-Reihe) GFR77-A
Includes bibliographical references and indexes.
ISBN 3-17-004365-X
1. Religion—Philosophy. 2. Ideology. 3. Christian education of young people. 4. Christian education—Germany, West. I. Vierzig, Siegfried, joint author. II. Title.
BL51.S6225 79-382708
79 MARC

—GERMANY, WEST—CASE STUDIES

Schuh, Hans, 1943-
Interaktionsanalyse : eine empirische Untersuchung zur Praxis des Religionsunterrichts / Hans Schuh. — Zürich [etc.] : Benziger, c1978.
374 p. : ill. ; 23 cm. — (Studien zur praktischen Theologie ; 17) Sw78-6023
Originally presented as the author's thesis, Mainz, 1977.
Bibliography: p. 366-374.
ISBN 3-545-21517-2 : 29.80F
1. Christian education—Germany, West—Case studies. 2. Interaction analysis in education—Case studies. I. Title. II. Series: Studien zur praktischen Theologie (Zürich) ; 17.
BV1517.G3S38 1978 78-384274
79 MARC

—GERMANY, WEST—CASE STUDIES—ADDRESSES, ESSAYS, LECTURES

Sozialethischer Unterricht, dokumentiert und analysiert : eine Untersuchung zur Praxis des problemorientierten Religionsunterrichts / Günter Stachel (Hrsg.) ; mit Beitr. von Alberto Bondolfi ... [et al.]. — Zürich : Benziger, c1977.
200 p. ; 23 cm. — (Studien zur praktischen Theologie ; 15) Sw77-6222
Includes bibliographical references.
ISBN 3-545-21515-6 : 28.80F
1. Christian education—Germany, West—Case studies—Addresses, essays, lectures. 2. Christian ethics—Study and teaching—Case studies—Addresses, essays, lectures. 3. Social ethics—Study and teaching—Case studies—Addresses, essays, lectures. I. Stachel, Günter. II. Series: Studien zur praktischen Theologie (Zürich) ; 15.
BV1470.G3S66 78-388595
*79 MARC

—INDONESIA

Indonesia. Proyek Peningkatan Penelitian/Survey Keagamaan.
Laporan survey dan Seminar Pendidikan Agama pada Perguruan Umum. [Djakarta, 1970]
ix, 607 p. illus. 22 cm.
Rp2600
1. Christian education — Indonesia. I. Seminar Pendidikan Agama pada Perguruan Umum, Tjiawi, Indonesia, 1970.
BV1470.I 54 I 52 1970 207'.598 72-940410

—ITALY

Riva, Silvio Antonio.
Il movimento catechistico italiano : dal Vaticano II al sinodo della catechesi / [di] Silvio Riva e Gianni Catti. — Bologna : EDB, 1977.
239 p. ; 19 cm. — (Educatori e catechisti ; 7) It78-Dec
1. Christian education—Italy. I. Catti, Giovanni, joint author. II. Title. III. Series.
BV1470.I8R57 79-352306
*79 MARC

—ITALY—CONGRESSES

Associazione guide e scouts cattolici italiani.
Credo la Chiesa : atti del Convegno catechetico nazionale AGESCI 1978 sulla Chiesa / a cura di Luigi Moro ; [relazioni di] L. Sartori ... [et al.]. — Roma : Borla, 1979.
118 p. ; 19 cm. It79-July
Includes bibliographical references.
L3000
1. Catechetics—Catholic Church—Congresses. 2. Christian education—Italy—Congresses. I. Moro, Luigi. II. Sartori, Luigi. III. Title.
BX1968.A87 1979 79-382231
*79 MARC

CHRISTIAN EDUCATION (Continued)

—PAKISTAN

Hall, Mary, Ph.D.
A quest for the liberated Christian : examined on the basis of a mission, a man, and a movement as agents of liberation / Mary Hall. — Frankfurt am Main ; Bern ; Las Vegas : Lang, 1978.
342 p. ; 21 cm. — (Studies in the intercultural history of Christianity ; Bd. 19) GFR78-A
Bibliography: p. 319-341.
ISBN 3-261-02668-5
1. Christian education—Pakistan. 2. Câmara, Hélder, 1909- . 3. Pentecostalism—Catholic Church. 4. Pentacostalism—United States. I. Title. II. Series.
BV1470.P18H34 282'.09'04 79-306108
79 MARC

—QUÉBEC (PROVINCE)

Québec (Province). Conseil supérieur de l'éducation. Comité catholique.
Voies et impasses / Comité catholique du Conseil supérieur de l'éducation. — Montréal : Fides, [1975]
viii, 286 p. ; 23 cm. C***
Includes bibliographical references.
Includes index.
ISBN 0-7755-0539-0
1. Christian education—Québec (Province) I. Title.
BV1470.C2Q39 1975 377'.8'2714 76-453567
76[r79]rev MARC

—QUÉBEC (PROVINCE)—CHAMBLY

Beaudin, Pierre.
L'enseignement religieux au Québec appliqué à la Régionale de Chambly, 1963-1974 / Pierre Beaudin. — [Montréal] : Fides, c1978.
123 p. ; 24 cm. — (Collection Essais et recherches : Section religion) C79-1290-1
Thesis—Université du Québec, Montréal.
Bibliography: p. 115-119.
ISBN 0-7755-0702-4 : $5.50
1. Christian education—Québec (Province)—Chambly. 2. Commission scolaire régionale de Chambly. I. Title.
BV1470.C2B4 207'.714'37 79-365900
79 MARC

—UNITED STATES

Piveteau, Didier J
Resurgence of religious instruction : conception and practice in a world of change / Didier-Jacques Piveteau and J. T. Dillon. — Notre Dame, Ind. : Religious Education Press, c1977.
viii, 291 p. : ill. ; 23 cm.
Includes bibliographical references and indexes.
ISBN 0-89135-007-1 : $9.50
1. Christian education—United States. I. Dillon, J. T., joint author. II. Title.
BV1467.P58 207 76-55588
78 MARC

Thompson, Andrew D
Where do we go from here : a survey of U.S. diocesan statements of goals and objectives for religious education and catechesis / by Andrew D. Thompson and Robert P. Stamschror. — Washington : United States Catholic Conference, Dept. of Education, c1979.
ii, 49 p. : ill. ; 22 cm.
Includes bibliographical references.
1. Christian education—Philosophy. 2. Christian education—United States. I. Stamschror, Robert P., joint author. II. United States Catholic Conference. Dept. of Education. III. Title.
BV1464.T47 207 79-120037
79 MARC

—UNITED STATES—STATISTICS

United States Catholic Conference.
A national inventory of parish catechetical programs / conducted by the United States Catholic Conference, in cooperation with the Boys Town Center for the Study of Youth Development —the Catholic University of America. — Washington : Publications Office, United States Catholic Conference, 1978.
iv, 70 p. : ill. ; 22 cm.
1. Christian education—United States—Statistics. I. Boys Town for the Study of Youth Development. II. Title.
BV1467.U54 1978 268'.8'2 78-109919
78 MARC

CHRISTIAN EDUCATION (THEOLOGY)

Blanke, Ingrid, 1937-
Sinn und Grenze christlicher Erziehung : Kierkegaard u. d. Problematik d. christl. Erziehung in unserer Zeit / Ingrid Blanke. — Frankfurt am Main ; Bern ; Las Vegas : Lang, 1978.
xix, 286 p. ; 21 cm. — (Europäische Hochschulschriften : Reihe 11, Pädagogik ; Bd. 55) GFR78-A
Bibliography: p. 277-286.
ISBN 3-261-02505-0
1. Kierkegaard, Søren Aabye, 1813-1885. 2. Christian education (Theology) 3. Education—Philosophy. I. Title. II. Series.
BX4827.K5B56 78-387245
78 MARC

Fraas, Hans-Jürgen.
Glauben und Lernen : e. theol.-didakt. Leitf. für d. Elementarerziehung / Hans-Jürgen Fraas. — 1. Aufl. — Göttingen : Vandenhoeck und Ruprecht, 1978.
92 p. ; 19 cm. — (Kleine Vandenhoeck-Reihe ; 1444) GFR78-A
Bibliography: p. 92.
ISBN 3-525-33424-9
1. Christian education (Theology) I. Title.
BV1464.F72 79-382705
79 MARC

—ADDRESSES, ESSAYS, LECTURES

Weiterfahrung und christliche Hoffnung / Erich Feifel (Hrsg.) ; mit Beitr. von Günter Biemer ... [et al.]. — Donauwörth : Auer, 1977.
151 p. : graph ; 20 cm. GFR77-A40
Includes bibliographical references.
ISBN 3-403-00722-7 : DM14.80
1. Christian education (Theology)—Addresses, essays, lectures. 2. Experience (Religion)—Addresses, essays, lectures. I. Feifel, Erich.
BV1464.W44 79-375858
79 MARC

CHRISTIAN EDUCATION OF ADOLESCENTS

Aparisi, Antonio.
Plan de pastoral de adolescentes : documentos de base, tercer año / [por] Antonio Aparisi y equipo de educadores. — Madrid : Marova, 1977 printing.
207 p. ; 24 cm. — (His Odres nuevos) Sp78-Jan
ISBN 8426903568 : 300ptas
1. Christian education—Text-books—Catholic. 2. Christian education of adolescents. I. Title.
BX930.A64 78-389075
79 MARC

CHRISTIAN EDUCATION OF ADULTS

Downs, Tom, 1938-
Parish as learning community : modeling for parish and adult growth / by Tom Downs. — New York : Paulist Press, c1979.
119 p. : ill. ; 29 cm. — (A Deus book)
Includes bibliographical references.
ISBN 0-8091-2172-7 : $2.95
1. Christian education of adults. I. Title.
BV1488.D68 268'.434 78-70816
79 MARC

Eltern- und Familienbildung in evangelischer Trägerschaft : Begründungen, Ansätze, Perspektiven / hrsg. von Horst Exner ... [et al.] ; in Zsarb. mit d. Wiss. Begleitauschuss d. Bildungspolit. Ausschusses d. Evang. Kirche in Deutschland (EKD). — Gelnhausen ; Berlin ; Stein : Burckhardthaus-Laetare Verlag, 1979.
vii, 245 p. ; 21 cm. — (Grundlagen evangelischer Erwachsenenbildung) GFR79-A
Bibliography: p. 227-232.
ISBN 3-7664-3070-X
1. Socialization—Study and teaching—Germany, West. 2. Parenting—Study and teaching—Germany, West. 3. Family—Germany, West—Religious life. 4. Christian education of adults. I. Exner, Horst. II. Evangelische Kirche in Deutschland. Bildungspolitischer Ausschuss. Wissenschaftlicher Begleitausschuss. III. Series.
HQ783.E44 301.15'7'071043 79-382915
79 MARC

Evans, Gary T
Equipping God's people : basic concepts for adult education / prepared by Gary T. Evans and Richard E. Hayes ; with the assistance of a group of editorial advisors under the direction of the Church's teaching series use guide and leader development committee. — New York : Seabury Press, 1979.
vi, 73 p. ; 21 cm. — (The Church's teaching series ; v. 7)
Bibliography: p. 52-73.
ISBN 0-8164-2239-9
1. Christian education of adults. I. Hayes, Richard E., joint author. II. Title. III. Series.
[BV1488.E87] 268'.434 79-14252
79 MARC

Growing persons : the report of the adult education working party [presented to] the British Council of Churches, Division of Community Affairs. — London : British Council of Churches, [1977]
vi, 59, [1] p. ; 21 cm. — (Britain today & tomorrow ; no. 2) GB78-15372
£0.95
1. Adult education—England. 2. Christian education of adults. I. British Council of Churches. Division of Community Affairs. II. Series.
LC5256.G7G78 261.1 78-320225
78 MARC

Hungs, Franz-Josef, 1932-
Altenbildung, Altenpastoral : Erfahrungen in d. theol. Erwachsenenbildung mit älteren Menschen / Franz-Josef Hungs. — München : Kösel, 1978.
96 p. ; 22 cm. GFR78-A
Bibliogrpahy: p. 88-90.
ISBN 3-466-36032-3
1. Church work with the aged. 2. Aged. 3. Christian education of adults. I. Title.
BV4435.H86 79-377618
79 MARC

Moran, Gabriel.
Education toward adulthood / Gabriel Moran. — New York : Paulist Press, c1979.
v, 151 p. : ill. ; 21 cm.
Includes bibliographical references.
ISBN 0-8091-2194-8 : $5.95
1. Christian education of adults. I. Title.
BV1488.M67 207 78-65897
79 MARC

Netzer, Johannes.
Das Taufgespräch in der Gemeinde : pastoraltheologische Untersuchung zum rechten Vollzug der erneuerten Taufpraxis / Johannes Netzer. — 2. Aufl. — Freiburg ; Basel ; Wien : Herder, 1977.
238, [1] p. ; 21 cm. — (Untersuchungen zur praktischen Theologie ; Bd. 4) GFR77-A23
Bibliography: p. 213-[239]
ISBN 3-451-17536-3 : DM32.00
1. Infant baptism. 2. Christian education of adults. 3. Evangelistic work—Germany, West. I. Title. II. Series.
BV813.2.N47 1977 234'.1612 78-377074
78 MARC

Schmidt, Ludwig, 1924-
Glaubensumsetzung in der Erwachsenenpädagogik / Ludwig Schmidt. — 1. Aufl. — Frankfurt/Main : Haag und Herchen, 1978.
vi, 122 p. ; 21 cm. GFR78-A
Bibliography: p. 105-112.
Includes index.
ISBN 3-88129-107-5 : DM14.80
1. Christian education of adults. 2. Christian education—Philosophy. 3. Hermeneutics. I. Title.
BV1488.S35 78-367862
78 MARC

Werbs, Ulrich.
Die Bedeutung des Hörers für die Verkündigung : pastoraltheolog. Überlegungen zum antropolog. Ansatz der Verkündigung im holländ. Katechismus für Erwachsene / von Ulrich Werbs. — 1. Aufl. — Leipzig : St.-Benno-Verlag, 1978.
xxiii, 187 p. ; 23 cm. — (Erfurter theologische Studien ; 39) GDR79-A
Bibliography: p. xi-xxiii.
19.80M
1. Christian education—Philosophy. 2. Christian education of adults. 3. De Nienwe katechismus. 4. Catholic Church—Catechisms and creeds—Dutch. I. Title. II. Series.
BV1464.W47 79-360186
79 MARC

—ADDRESSES, ESSAYS, LECTURES

Erwachsenenbildung in der pluralen Gesellschaft / hrsg. von Franz Henrich ; mit Beitr. von Wolfgang Böhme ... [et al.]. — 1. Aufl. — Düsseldorf : Patmos-Verlag, 1978.
171 p. ; 21 cm. — (Patmos-Paperbacks) (Schriften der Katholischen Akademie in Bayern ; Bd. 84) GFR78-A
Revised papers originally presented at two meetings sponsored by the Katholische Akademie in Bayern, 15-16 Mar. 1975, and 8-9 Oct. 1977.
Includes bibliographical references.
ISBN 3-491-77589-2
1. Adult education—Addresses, essays, lectures. 2. Christian education of adults—Addresses, essays, lectures. I. Henrich, Franz. II. Böhme, Wolfgang, 1919- . III. Katholische Akademie in Bayern.
LC5215.E714 79-378539
79 MARC

Kirchliche Erwachsenenarbeit / Jürgen Lott (Hrsg.). — 1. Aufl. — Stuttgart ; Berlin ; Köln ; Mainz : Kohlhammer, 1977.
138 p. ; 19 cm. — (Urban-Taschenbücher ; Bd. 633 : T-Reihe) GFR77-A46
Bibliography: p. 131-137.
ISBN 3-17-004430-3 : DM12.00
1. Christian education of adults—Addresses, essays, lectures. 2. Church work—Addresses, essays, lectures. I. Lott, Jürgen, 1943-
BV1488.K57 79-366276
79 MARC

CHRISTIAN EDUCATION OF BOYS

—COLLECTED WORKS

Bosco, Giovanni, Saint, 1815-1888.
Obras fundamentales / ₍por₎ San Juan Bosco ; edición dirigida por Juan Canals Pujol y Antonio Martínez Azcona ; estudio introductorio de Pedro Braido. — Madrid : Edica, 1978.
xxxii, 831 p. : port. ; 20 cm. — (Biblioteca de autores cristianos ; 402) Sp***
Bibliography: p. ₍11₎-15.
Includes index.
ISBN 8422008785 : 1000ptas
1. Catholic Church—Doctrinal and controversial works—Catholic authors—Collected works. 2. Christian education of boys—Collected works. 3. Christian education of young people—Collected works. 4. Salesians—Collected works. I. Canals Pujol, Juan. II. Martínez Azcona, Antonio.
BX1751.B64 1978 79-111094
79 MARC

CHRISTIAN EDUCATION OF CHILDREN

see also Christian education of boys; Christian education of mentally handicapped children; Christian education of socially handicapped children; Christian education of young people

Balmes, Jaime.
La relijion demostrada al alcance de los niños / por Jaime Balmes. — Quito : Reimpreso por M. Rivadeneira, 1853.
116 p. ; 16 cm.
1. Christian education of children. 2. Catholic Church—Doctrinal and controversial works, Popular. I. Title.
BV1475.B234 79-369536
79 MARC

Ban, Arline J
Teaching and learning with older elementary children : a manual for workers in the church / Arline J. Ban ; ill. by Linda Weller. — Valley Forge, PA : Judson Press, c1979.
64 p. : ill. ; 22 cm. — (Teaching and learning series)
Bibliography: p. 62-63.
Includes index.
ISBN 0-8170-0799-7 : $2.50
1. Christian education of children. I. Title. II. Title: Older elementary children. III. Series.
BV1475.2.B33 268'.432 78-25808
78 MARC

Cully, Iris V
Christian child development / Iris V. Cully. — 1st ed. — San Francisco : Harper & Row, c1979.
x, 166 p. ; 21 cm.
Includes bibliographical references and index.
ISBN 0-06-061648-2
1. Christian education of children. 2. Children—Religious life. 3. Child development. I. Title.
BV1475.2.C814 1979 268'.432 78-19507
78 MARC

Davis, Cos H
Children and the Christian faith / Cos H. Davis, Jr. — Nashville : Broadman Press, c1979.
120 p. ; 21 cm.
ISBN 0-8054-6221-X
1. Christian education of children. 2. Family—Religious life. I. Title.
BV1475.2.D38 249 79-50339
79 MARC

Harger, Grace B
Teaching and learning with young children : a manual for workers in the church / Grace B. Harger and Arline J. Ban ; ill. by Linda Weller. — Valley Forge, PA : Judson Press, c1979.
63 p. : ill. ; 22 cm. — (Teaching and learning series)
Bibliography: p. 60-62.
Includes index.
ISBN 0-8170-0803-9 : $2.50
1. Christian education of children. I. Ban, Arline J., joint author. II. Title. III. Title: Young children. IV. Series.
BV1475.2.H35 268'.432 78-26020
78 MARC

Kuzma, Kay.
Building character / by Kay and Jan Kuzma. — Mountain View, Calif. : Pacific Press Pub. Association, c1979.
173 p. : ill. ; 19 cm. — (Harvest series ; 108)
Includes bibliographical references.
1. Christian education of children. I. Kuzma, Jan, joint author. II. Title.
BV1475.2.K89 649'.7 79-83592
79 MARC

Larsen, Earnest.
Will religion make sense to your child? / Earnest Larsen and Patricia Galvin. — Rev. ed. — Liguori, Mo. : Liguori Publications, 1979, c1970.
128 p. (p. 127-128 publisher's list) : ill. ; 18 cm.
ISBN 0-89243-109-1 : $1.95
1. Christian education of children. 2. Theology, Doctrinal—Popular works. I. Galvin, Patricia, joint author. II. Title.
BV1475.2.L3 1979 649'.7 79-88234
00 MARC

Marranzini, Alfredo, 1920-
I giovani al V Sinodo dei vescovi / Alfredo Marranzini. — Roma : A.V.E., 1978.
174 p. ; 21 cm. — (Pastorale oggi ; 12) It***
Includes bibliographical references.
L3500
1. Catholic Church. Synodus Episcoporum, 5th, 1977. 2. Christian education of young people. 3. Christian education of children. I. Title.
BX831 1977.M37 79-361769
79 MARC

Mendía Gallardo, Rafael.
De la catequesis a la fiesta : la expresión religiosa del niño y preadolescente en el tiempo libre / ₍por₎ Rafael Mendía, José Garayalde. — Salamanca ; Tejares : Instituto Pontificio San Pío X, 1976.
219 p. ; 22 cm. — (Colección Fe-experiencia ; 11) Sp76-June
Includes bibliographical references.
ISBN 8472212351
1. Christian education of children. I. Garayalde, José, joint author. II. Title.
BV1475.2.M44 79-367391
79 MARC

Riva, Silvio Antonio.
L'educazione religiosa del fanciullo : liberazione dell'uomo, promozione del credente / Silvio Riva. — Bologna : EDB, 1977.
279 p. ; 19 cm. — (Educatori e catechisti ; 4) It78-Sept
Includes bibliographical references.
L4300
1. Christian education of children. I. Title. II. Series.
BV1475.2.R55 78-400302
*79 MARC

Schneider, Dirk, 1942-
Ich werden, Partner werden : Sexualerziehung im Religionsunterricht : Analysen, Methoden, Modelle / Dirk Schneider. — Zürich ₍etc.₎ : Benziger, c1978.
343 p. ; 21 cm. Sw78-6022
Originally presented as the author's thesis, Universität Erlangen-Nürnberg, 1974, under the title: Sexualpädagogik im Religionsunterricht.
Bibliography: p. 328-343.
ISBN 3-545-26138-7 : 25.00F
1. Christian education of children. 2. Christian education of young people. 3. Sex instruction for children. I. Title.
BV1475.2.S36 1978 79-384326
79 MARC

Ward, Ted Warren, 1930-
Values begin at home / Ted Ward. — Wheaton, Ill. : Victor Books, c1979.
120 p. ; 21 cm. — (The Family concern series)
Bibliography: p. 119-120.
ISBN 0-88207-637-X : $3.50
1. Christian education of children. 2. Family—Religious life. 3. Moral education. 4. Values. I. Title. II. Series: Family concern series.
BV1475.2.W37 649'.7 76-68855
79 MARC

CHRISTIAN EDUCATION OF MENTALLY HANDICAPPED CHILDREN

Bissonnier, Henri, 1911-
The pedagogy of resurrection : the Christian formation of the handicapped / Henri Bissionnier ₍i.e. Bissonnier₎ ; translated by Carolyn Frederick. — New York : Paulist Press, c1979.
225 p. ; 21 cm.
Translation of Pédagogie de résurrection.
Includes bibliographical references.
ISBN 0-8091-2214-6 : $6.95
1. Christian education of socially handicapped children. 2. Christian education of mentally handicapped children. I. Title.
BV1615.S6B5713 248'.86 79-52104
79 MARC

CHRISTIAN EDUCATION OF SOCIALLY HANDICAPPED CHILDREN

Bissonnier, Henri, 1911-
The pedagogy of resurrection : the Christian formation of the handicapped / Henri Bissionnier ₍i.e. Bissonnier₎ ; translated by Carolyn Frederick. — New York : Paulist Press, c1979.
225 p. ; 21 cm.
Translation of Pédagogie de résurrection.
Includes bibliographical references.
ISBN 0-8091-2214-6 : $6.95
1. Christian education of socially handicapped children. 2. Christian education of mentally handicapped children. I. Title.
BV1615.S6B5713 248'.86 79-52104
79 MARC

CHRISTIAN EDUCATION OF YOUNG PEOPLE

see also Christian education—Text-books for young people

Marranzini, Alfredo, 1920-
I giovani al V Sinodo dei vescovi / Alfredo Marranzini. — Roma : A.V.E., 1978.
174 p. ; 21 cm. — (Pastorale oggi ; 12) It***
Includes bibliographical references.
L3500
1. Catholic Church. Synodus Episcoporum, 5th, 1977. 2. Christian education of young people. 3. Christian education of children. I. Title.
BX831 1977.M37 79-361769
79 MARC

Martinelli, Raffaello.
Rinnovamento psicopedagogico della pastorale giovanile : ricerche e orientamenti / Raffaello Martinelli. — Roma : Pontificia Universitas lateranensis, 1977.
190 p. ; 25 cm. It***
Bibliography: p. 17-36.
Abstract of thesis—Pontificia Universitas lateranensis.
1. Church work with youth. 2. Christian education of young people. I. Title.
BV4447.M33 79-356993
79 MARC

Schneider, Dirk, 1942-
Ich werden, Partner werden : Sexualerziehung im Religionsunterricht : Analysen, Methoden, Modelle / Dirk Schneider. — Zürich ₍etc.₎ : Benziger, c1978.
343 p. ; 21 cm. Sw78-6022
Originally presented as the author's thesis, Universität Erlangen-Nürnberg, 1974, under the title: Sexualpädagogik im Religionsunterricht.
Bibliography: p. 328-343.
ISBN 3-545-26138-7 : 25.00F
1. Christian education of children. 2. Christian education of young people. 3. Sex instruction for children. I. Title.
BV1475.2.S36 1978 79-384326
79 MARC

Sorge, Helga, 1940-
Religion und Ideologie : Analysen u. Materialien für d. Unterricht / Helga Sorge, Siegfried Vierzig. — 1. Aufl. — Stuttgart ; Berlin ; Köln ; Mainz : Kohlhammer, 1977.
177 p. ; 19 cm. — (Urban-Taschenbücher ; Bd. 634 : T.-Reihe) GFR77-A
Includes bibliographical references and indexes.
ISBN 3-17-004365-X
1. Religion—Philosophy. 2. Ideology. 3. Christian education of young people. 4. Christian education—Germany, West. I. Vierzig, Siegfried, joint author. II. Title.
BL51.S6225 79-382708
79 MARC

—COLLECTED WORKS

Bosco, Giovanni, Saint, 1815-1888.
Obras fundamentales / ₍por₎ San Juan Bosco ; edición dirigida por Juan Canals Pujol y Antonio Martínez Azcona ; estudio introductorio de Pedro Braido. — Madrid : Edica, 1978.
xxxii, 831 p. : port. ; 20 cm. — (Biblioteca de autores cristianos ; 402) Sp***
Bibliography: p. ₍11₎-15.
Includes index.
ISBN 8422008785 : 1000ptas
1. Catholic Church—Doctrinal and controversial works—Catholic authors—Collected works. 2. Christian education of boys—Collected works. 3. Christian education of young people—Collected works. 4. Salesians—Collected works. I. Canals Pujol, Juan. II. Martínez Azcona, Antonio.
BX1751.B64 1978 79-111094
79 MARC

CHRISTIAN ETHICS

see also Bible—Ethics; Bible. N.T.—Ethics; Cardinal virtures; Casuistry; Christian life; Christianity and economics; Commandments, Ten; Fundamental options (Ethics); Golden rule; Guilt; Law and gospel; Love (Theology); Merit (Christianity);

Pastoral medicine; Pastoral psychology; Sin; Sins; Social ethics; Vices; Virtues; and subdivision Ethics under names of Christian theologians; and subdivision Moral and religious aspects under specific subjects, e.g. Property—Moral and religious aspects

Birch, Bruce C
The predicament of the prosperous / Bruce C. Birch and Larry L. Rasmussen. — 1st ed. — Philadelphia : Westminster Press, c1978.
212 p. ; 20 cm. — (Biblical perspectives on current issues)
Bibliography: p. ₍207₎-210.
Includes index.
ISBN 0-664-24211-1
1. Sociology, Christian. 2. Wealth—Biblical teaching. 3. Human ecology—Moral and religious aspects. 4. Christian ethics. 5. Christianity and economics. I. Rasmussen, Larry L., joint author. II. Title. III. Series.
BT738.B52 261.8 78-18412
78 MARC

CHRISTIAN ETHICS (Continued)

Chiavacci, Enrico.
Morale della vita fisica / Enrico Chiavacci. — Bologna : EDB, 1976.
179 p. ; 24 cm. — (Corso di teologia morale : Morale speciale) It78-Jun
Includes bibliographical references.
L2800
1. Church and social problems—Catholic Church. 2. Christian ethics. 3. Bioethics. I. Title.
HN37.C3C43 78-392243
*79 MARC

Cudworth, Ralph, 1617-1688.
A treatise concerning eternal and immutable morality / Ralph Cudworth. — New York : Garland Pub., 1976.
xii, 303 p., [1] leaf of plates : port. ; 23 cm. — (British philosophers and theologians of the 17th and 18th centuries)
Reprint of the 1731 ed. printed for J. and J. Knapton, London.
Includes bibliographical references.
ISBN 0-8240-1768-4 : $25.00
1. Christian ethics. I. Title. II. Series.
BJ1241.C8 1976 241 75-11214
76[r78]rev MARC

Cumberland, Richard, Bp. of Peterborough, 1631-1718.
A treatise of the laws of nature, 1727 / Richard Cumberland. — New York : Garland Pub., 1978.
740 p. in various pagings, [2] fold. leaves of plates : ill. ; 26 cm. — (British philosophers and theologians of the 17th and 18th centuries)
Translation of De legibus naturae.
Reprint of the ed. printed by R. Phillips, London.
Includes bibliographical references and indexes.
ISBN 0-8240-1770-6 : $29.50
1. Ethics. 2. Christian ethics. I. Title. II. Series.
B1201.C83D423 1978 170 75-11216
79 MARC

Gaffney, James.
Newness of life : a modern introduction to Catholic ethics / James Gaffney. — New York : Paulist Press, c1979.
viii, 350 p. : ill. ; 21 cm.
Bibliography: p. 338-349.
ISBN 0-8091-2202-2 : $4.95
1. Christian ethics. I. Title.
BX1758.2.G33 241'.04'2 79-84404
79 MARC

Gyr, Constantin.
Gewissensentscheid in der klinischen Krankenpflege : ein moraltheologischer Beitrag zur Frage der Berufsethik des Krankenpflegepersonals / Constantin Gyr. — Bern ; Frankfurt a.M. ; Las Vegas : Peter Lang, c1977.
353 p. ; 21 cm. — (Europäische Hochschulschriften : Reihe 23, Theologie ; Bd. 98) Sw77-12099
Originally presented as the author's thesis, Theologische Fakultät Luzern.
Three hundred copies printed.
Bibliography: p. 296-305.
ISBN 3-261-00375-X : 62.00F
1. Medical ethics. 2. Nursing ethics. 3. Medicine, Clinical—Decision making. 4. Conscience. 5. Christian ethics. 6. Medicine and Christianity. I. Title. II. Series.
R724.G95 1977 174'.2 78-385432
*79 MARC

Honecker, Martin.
Das Recht des Menschen : Einf. in d. evang. Sozialethik / Martin Honecker. — Orig.-Ausg., 1. Aufl. — Gütersloh : Gütersloher Verlagshaus Mohn, 1978.
217 p. ; 19 cm. — (Gütersloher Taschenbücher ; 290 : Siebenstern) GFR78-A
"Eine Veröffentlichung des Comenius-Instituts, Münster."
Includes bibliographical references.
ISBN 3-579-03690-4
1. Church and civil rights. 2. Christian ethics. 3. Social ethics. I. Title.
BT738.15.H66 79-341747
79

Medisinsk etikk : grunnlagsproblemer i kristent lys / Arne Emil Schjøth, med bidrag av Willey Olsson ... [et al.]. — Oslo : Luther, 1978 (Drammen : Tangen-trykk)
181 p. ; 22 cm. N78-May
Bibliography: p. 181.
ISBN 8253172141 : kr65.00
1. Medical ethics. 2. Christian ethics. 3. Medicine and Christianity. I. Schjøth, Arne Emil.
R724.M32 79-337369
79 MARC

Miss, Werner, 1945-
Sigmund Freuds Kritik am christlichen Liebesgebot als Herausforderung an den christlichen Glauben / vorgelegt von Werner Miss. — Zürich : Juris-Verlag, 1976.
ii, 265 p. ; 23 cm. Sw76-11996
Thesis—Zürich.
Vita.
Bibliography: p. 257-265.
ISBN 3-260-04156-7 : 48.00F
1. Love (Theology)—History of doctrines. 2. Summary of the law (Theology) 3. Freud, Sigmund, 1856-1939. 4. Psychoanalysis and religion. 5. Christian ethics. I. Title.
BV4639.M55 79-362885
*79 MARC

Rauschenbusch, Walter, 1861-1918.
The social principles of Jesus / by Walter Rauschenbusch ; written under the direction of Sub-committee on College Courses, Sunday School Council of Evangelical Denominations and Committee on Voluntary Study, Council of North American Student Movements. — Philadelphia : R. West, 1978 [c1916]
198 p. ; 26 cm.
Reprint of the ed. published by Association Press, New York, which was issued as 4th year, pt. 1, of College voluntary study courses.
ISBN 0-8492-2258-3 : $20.00
1. Jesus Christ—Teachings. 2. Sociology, Christian. 3. Christian ethics. I. Title. II. Series: College voluntary study courses ; 4th year, pt. 1.
[BS2417.S7R3 1978] 261 78-4230
78 MARC

Rüttgardt Jan Olaf.
Heiliges Leben in der Welt : Grundzüge christl. Sittlichkeit nach Philipp Jakob Spener / von Olaf Rüttgardt. — Bielefeld : Luther-Verlag, 1978.
202 p. ; 24 cm. — (Arbeiten zur Geschichte des Pietismus : Bd. 16) GFR78-A
A revision of the author's thesis, Göttingen, 1972.
Bibliography: p. 191-200.
Includes indexes.
ISBN 3-7858-0243-9 : DM56.00
1. Spener, Philipp Jakob, 1635-1705. 2. Christian ethics. I. Title. II. Series.
BR1653.S6R83 1978 78-396514
79 MARC

Schulze, Hans, 1912-
Theologische Sozialethik : Grundlegung, Methodik, Programmatik / Hans Schulze. — Gütersloh : Gütersloher Verlagshaus Mohn, 1979.
392 p. ; 23 cm. GFR79-A
Bibliography: p. 363-379.
Includes indexes.
ISBN 3-579-04079-0 : DM38.00
1. Christian ethics. 2. Social ethics. I. Title.
BJ1253.S389 79-378812
79 MARC

Scorer, Charles Gordon.
Life in our hands : a study of human values / [by] C. G. Scorer. — Leicester : Inter-Varsity Press, 1978.
160 p. ; 20 cm. GB78-14073
Includes bibliographical references.
ISBN 0-85110-585-8 : £1.95
1. Life and death, Power over. 2. Christian ethics. 3. Medical ethics. I. Title.
BJ1469.S36 241 78-320320
78 MARC

Siegmund, Georg, 1903-
Sexus und Kultur / Georg Siegmund. — 1. Aufl. — Leutesdorf am Rhein : Johannes-Verlag, 1977.
44 p. ; 18 cm. — (Aktuelle Schriften) GFR78-A
Includes bibliographical references.
ISBN 3-7794-0670-5 : DM2.00
1. Sexual ethics. 2. Christian ethics. I. Title.
HQ32.S49 176 78-391353
79 MARC

Weber, Max, 1864-1920.
Die protestantische Ethik : eine Aufsatzsammlung / Max Weber ; [hrsg. von Johannes Winckelmann] . — 5., erneut überarb. u. um e. Nachw. vers. Aufl. (31.-36. Tsd.). — Gütersloh : Gütersloher Verlagshaus Mohn, 1979-
v. ; 19 cm. — (Gütersloher Taschenbücher Siebenstern ; 53-) GFR79-A(v.
Includes bibliographical references.
ISBN 3-579-03798-6 (v. 1) : DM14.80 (v. 1)
1. Christianity and economics. 2. Christian ethics. 3. Protestantism. 4. Capitalism. 5. Religion and sociology. I. Winckelmann, Johannes. II. Title.
BR115.E3W37 1979 79-371571
79 MARC

White, Jerry E 1937-
Honesty, morality & conscience / Jerry White. — Colorado Springs, Colo. : Navpress, c1978.
240 p. : ill. ; 21 cm. — (A Navigator book)
Includes bibliographical references.
ISBN 0-89109-439-3. ISBN 0-89109-431-8 pbk. : $3.95
1. Christian ethics. 2. Honesty. I. Title.
BJ1251.W54 241 78-61619
79 MARC

Williams, Harri, 1913-
Duw a phob daioni : llawlyfr ar foeseg Gristnogol / [gan] Harri Williams. — Llandysul : Gwasg Gomer, 1978.
152 p. ; 22 cm. GB78-27130
Bibliography: p. 149.
Includes index.
ISBN 0-85088-590-6 : £1.50
1. Christian ethics. I. Title.
BJ1185.W44W54 78-398111
79 MARC

—EARLY CHURCH, CA. 30-600

Teaching of the Twelve Apostles. French & Greek.
La doctrine des Douze Apôtres = Didachè / introd., traduction, notes, appendice et index par Willy Rordorf et André Tuilier. — Paris : Éditions du Cerf, 1978.
228 p. ; 20 cm. — (Sources chrétiennes ; no 248) F***
Bibliography: p. [129]-135.
Includes indexes.
ISBN 2-204-01278-5 : 169F
1. Christian ethics—Early church, ca. 30-600. 2. Church polity—Early church, ca. 30-600. I. Rordorf, Willy. II. Tuilier, André. III. Title.
BS2940.T4F8 1978 230'.1'1 79-374963
79 MARC

—ADDRESSES, ESSAYS, LECTURES

Catholic Church. Irish Episcopal Conference.
The work of justice : Irish bishops' pastoral. — Dublin : Veritas Publications, 1977.
81 p. ; 19 cm. GB78-35846
"Full text of the Pastoral Letter issued by the Irish Catholic Bishops' Conference on September 11th 1977."
Imprint covered by label: Sole distributors in U.K., Fowler Wright Books Limited, Tenbury Wells, Worcs., Eng.
"Cat. no. 3362."
Bibliography: p. 78-[81]
ISBN 0-905092-24-4 : £0.66
1. Christianity and justice—Addresses, essays, lectures. 2. Christian ethics—Addresses, essays, lectures. 3. Social ethics—Addresses, essays, lectures. 4. Ireland—Economic conditions—Addresses, essays, lectures. 5. Catholic Church—Pastoral letters and changes. I. Title.
BR115.J8C37 1977 261.8 79-316882
79 MARC

Christian social ethics / Perry C. Cotham, editor. — Grand Rapids : Baker Book House, c1979.
265 p. ; 22 cm.
Includes bibliographical references.
ISBN 0-8010-2424-2 : $4.95
1. Christian ethics—Addresses essays, lectures. 2. Social ethics—Addresses essays, lectures. I. Cotham, Perry C.
BJ1251.C52 261 78-68118
79 MARC

Doing evil to achieve good : moral choice in conflict situations / edited by Richard A. McCormick and Paul Ramsey. — Chicago : Loyola University Press, c1978.
267 p. ; 24 cm.
Includes bibliographical references.
ISBN 0-8294-0285-3 : $12.95
1. Christian ethics—Addresses, essays, lectures. I. McCormick, Richard A., 1922- . II. Ramsey, Paul.
BJ1251.D64 241 78-11316
78 MARC

Ethik im Kontext des Glaubens : Probleme, Grundsätze, Methoden / Hrsg., Dietmar Mieth, Francesco Compagnoni ; [Beitr. von Klaus Demmer ... et al.]. — Freiburg/Schweiz : Universitätsverlag ; Freiburg i.Br. [etc.] : Herder, c1978.
186 p. ; 23 cm. — (Studien zur theologischen Ethik ; 3) Sw78-6507
Includes bibliographies.
ISBN (invalid) 3-7278-0184-X : 32.60F
1. Christian ethics—Addresses, essays, lectures. I. Mieth, Dietmar. II. Compagnoni, Francesco. III. Series: Études d'éthique chrétienne ; 3.
BJ1253.E84 79-380112
79 MARC

Jongeneel, J A B
Dialog Kristen dengan azas-azas etik non-Kristen di Indonesia : orasi yang diucapkan pada upacara dies natalis XIV pada Fakultas Theologia, Universitas Kristen Indonesia di Tomohon, pada hari Kamis 7 Oktober 1976 / oleh J. A. B. Jongeneel ; [diterjemahkan oleh M. B. Jongeneel-Touw; dengan perbaikan redaksi oleh F. S. Lalamentik dan P. S. Naipospos]. — Jakarta : BPK Gunung Mulia, [1977]
48 p. ; 22 cm.
1. Christian ethics—Addresses, essays, lectures. 2. Religious ethics—Addresses, essays, lectures. 3. Indonesia—Religion—Addresses, essays, lectures. I. Title.
BJ1275.J66 78-940623
79 MARC

Law, morality, and the Bible : a symposium / edited by Bruce Kaye & Gordon Wenham ; [pref. by J. I. Packer]. — Downers Grove, Ill. : InterVarsity Press, c1978.
ix, 252 p. ; 21 cm.
Includes bibliographical references and indexes.
ISBN 0-87784-795-9 : $5.95
1. Law (Theology)—Biblical teaching—Addresses, essays, lectures. 2. Christian ethics—Addresses, essays, lectures. I. Kaye, Bruce. II. Wenham, Gordon J.
BS680.L33L38 241 78-18549
79 MARC

McDonagh, Enda.
Social ethics and the Christian : towards freedom in communion / Enda McDonagh. — Manchester : Manchester University Press, 1979.
89 p. ; 20 cm. GB***
Based on the Ferguson lectures delivered at the University of Manchester, 1978.
ISBN 0-7190-0739-9 : £3.95
1. Christian ethics—Addresses, essays, lectures. 2. Social ethics—Addresses, essays, lectures. I. Title.
BJ1275.M27 261 79-315193
79 MARC

Oppenheimer, Helen, Lady, 1926-
The character of Christian ethics / by Helen Oppenheimer. — London : Athlone Press, 1977.
28 p. ; 22 cm. — (The John Coffin memorial lecture ; 1977 ISSN 0449-0789) GB78-23962
Distributed in the U.S.A. by Humanities Press.
"Delivered before the University of London on 15 March 1977."
Includes bibliographical references.
ISBN 0-485-16211-3 : £0.75
1. Christian ethics—Addresses, essays, lectures. I. Title. II. Series.
BJ1275.O66 241 78-326808
79 MARC

CHRISTIAN ETHICS
—ADDRESSES, ESSAYS, LECTURES
(Continued)

Ricerca morale e scienze umane / ₍di₎ Autori vari ; a cura di Aniceto Molinaro. — Bologna : EDB, 1979.
143 p. ; 19 cm. — (Collana Studi e ricerche ; 28) It79-Aug
Includes bibliographical references.
L2500
1. Science and ethics—Addresses, essays, lectures. 2. Christian ethics—Addresses, essays, lectures. I. Molinaro, Aniceto, 1936-
BJ57.R53 79-382214
*79 MARC

Schöpferische Nachfolge : Festschr. für Heinz Eduard Tödt / hrsg. von Christofer Frey u. Wolfgang Huber, in Zsarb. mit Wolfgang Lienemann ... ₍et al.₎. — Heidelerg : Forschungsstätte d. Evang. Studiengemeinschaft, 1978.
582 p. : port. ; 21 cm. — (Texte und Materialien der Forschungsstätte der Evangelischen Studiengemeinschaft : Reihe A ; Nr. 5) GFR78-B14
Bibliography of H. E. Tödt's works: p. 573-581.
Includes bibliographical references.
ISBN 3-88257-004-0 : DM18.00
1. Christian ethics—Addresses, essays, lectures. 2. Theology—Addresses, essays, lectures. 3. Tödt, Heinz Eduard—Addresses essays, lectures. 4. Tödt, Heinz Eduard—Bibliography. I. Tödt, Heinz Eduard. II. Frey, Christofer. III. Huber, Wolfgang, 1931- . IV. Series: Evangelische Studiengemeinschaft. Forschungsstätte. Texte und Materialien der Forschungsstätte der Evangelischen Studiengemeinschaft : Reihe A ; Nr. 5.
BJ1253.S387 79-365270
79 MARC

Was sind Grundwerte? : zum Problem ihrer Inhalte u. ihrer Begründung / hrsg. von Otto Kimminich ; mit Beitr. von Wilhelm Dreier ... ₍et al.₎. — 1. Aufl. — Düsseldorf : Patmos Verlag, 1977.
134 p. ; 21 cm. — (Patmos-Paperback) (Schriften der Katholischen Akademie in Bayern ; Bd. 80) GFR78-A16
Includes bibliographical references.
ISBN 3-491-77579-5 : DM13.80
1. Ethics—Addresses, essays, lectures. 2. Christian ethics—Addresses, essays, lectures. 3. Values—Addresses, essays, lectures. I. Kimminich, Otto.
BJ1114.W3 78-398039
79 MARC

—ANGLICAN AUTHORS

Brill, Earl H
The Christian moral vision / written by Earl H. Brill with the assistance of a group of editorial advisors under the direction of the Church's Teaching Series Committee. — New York : Seabury Press, 1979.
xviii, 236 p. ; 22 cm. — (The Church's teaching series ; 6)
Bibliography: p. 226-229.
Includes index.
ISBN 0-8164-0423-2
1. Christian ethics—Anglican authors. I. Title. II. Series.
BJ1251.B74 241'.04'3 79-15499
79 MARC

Pittenger, William Norman, 1905-
Loving says it all / by Norman Pittenger. — New York : Pilgrim Press, c1978.
126 p. ; 22 cm.
ISBN 0-8298-0352-1 : $6.95
1. Christian ethics—Anglican authors. 2. Process theology. I. Title.
BJ1251.P57 241'.03 78-5671
78 MARC

—BAPTIST AUTHORS

McLaughlin, Raymond W
The ethics of persuasive preaching / Raymond W. McLaughlin. — Grand Rapids : Baker Book House, c1979.
215 p. ; 23 cm.
Bibliography: p. 195-206.
Includes indexes.
ISBN 0-8010-6051-6 : $9.95
1. Preaching. 2. Christian ethics—Baptist authors. I. Title.
BV4235.E75M23 241'.6'41 78-59311
79 MARC

—CATHOLIC AUTHORS

Cultrera, Francesco.
Mutabilità e immutabilità della legge naturale / Francesco Cultrera. — Napoli : Edizioni dehoniane, 1977.
146 p. ; 21 cm. It79-Feb
Bibliography: p. 139-144.
L2500
1. Christian ethics—Catholic authors. 2. Bible—Ethics. 3. Natural law. I. Title.
BJ1249.C77 79-382606
*79 MARC

González de Cardedal, Olegario, 1934-
Etica y religión : la conciencia española entre el dogmatismo y la desmoralización / ₍por₎ Olegario González de Cardedal. — Madrid : Cristiandad, D.L. 1977.
345 p. ; 18 cm. — (Epifanía) (El Libro de bolsillo Cristiandad ; 41) Sp78-Feb
Bibliography: p. ₍331₎-339.
Includes index.
ISBN 8470572156
1. Christian ethics—Catholic authors. 2. Religion. I. Title.
BJ1249.G68 78-383705
78 MARC

Hall, Theodore, 1921-
That mysterious fundamental option / by Theodore Hall. — Chicago : Franciscan Herald Press, c1978.
70 p. ; 18 cm. — (Synthesis series)
Includes bibliographical references.
ISBN 0-8199-0746-4 : $0.75
1. Christian ethics—Catholic authors. I. Title.
BJ1249.H28 241'.04'2 78-15095
78 MARC

Hammes, John A
Human destiny : exploring today's value-systems / John A. Hammes. — Huntington, Ind. : Our Sunday Visitor, c1978.
x, 292 p. ; 21 cm.
Bibliography: p. 273-288.
Includes index.
ISBN 0-87973-752-2 : $5.95
1. Christian ethics—Catholic authors. 2. Meaning (Philosophy) 3. Values. I. Title.
BJ1249.H33 241'.04'2 78-62241
78 MARC

Häring, Bernhard, 1912-
Free and faithful in Christ : moral, theology for clergy and laity / Bernard Häring. — New York : Seabury Press, 1978-
v. ; 22 cm.
"A Crossroad book."
Includes bibliographical references and index.
CONTENTS: v. 1. General moral theology.
ISBN 0-8164-0398-8 (v. 1) : $17.50
1. Christian ethics—Catholic authors. I. Title.
BX1758.2.H27 241'.04'2 78-12253
78 MARC

Häring, Bernhard, 1912-
Free and faithful in Christ : moral theology for priests and laity / ₍by₎ Bernard Häring. — Slough : St Paul Publications, 1978-
v. ; 23 cm. GB78-37450
Includes bibliographical references and index.
CONTENTS: v. 1. General moral theology.
ISBN 0-85439-147-9 (v. 1) : £6.50 (v. 1)
1. Christian ethics—Catholic authors. I. Title.
BX1758.2.H27 1978b 241'.04'2 79-302735
79 MARC

Hughes, Gerard J
Authority in morals : an essay in Christian ethics / by Gerard J. Hughes. — London : Heythrop College, 1978.
₍6₎, ix, 136 p. ; 21 cm. — (Heythrop monographs ; 3 ISSN 0309-4723) GB79-02338
Text on inside back cover.
Bibliography: p. 133-136.
ISBN 0-905764-02-1 : £4.75 ($10.00 U.S.)
1. Christian ethics—Catholic authors. 2. Authority. I. Title. II. Series.
BJ1249.H78 241 79-307798
79 MARC

Innocentius III, Pope, 1160 or 61-1216.
De miseria condicionis humane / Lotario dei Segni (Pope Innocent III) ; edited by Robert E. Lewis. — Athens : University of Georgia Press, c1978.
xiv, 303 p. ; 25 cm. — (The Chaucer library)
English and Latin on opposite pages.
First published under title: De contemptu mundi.
ISBN 0-8203-0395-X : $25.00
1. Man (Christian theology) 2. Christian ethics—Catholic authors. I. Lewis, Robert E. II. Title. III. Series.
BT700.I513 1978 233 75-26119
79 MARC

John Paul II, Pope, 1920-
Educazione all'amore / Karol Wojtyla ; traduzione di Elzbieta Cywiak e Vladyslaw Kujawski. — 2. ed. — Roma : Logos, 1978.
157 p. ; 21 cm. — (Biblioteca universale cristiana ; 19) It***
Translation of Elementarz etyczny.
L3600
1. Christian ethics—Catholic authors. I. Title.
BJ1249.J5716 1978 79-358932
79 MARC

Lio, Hermenegildus.
Morale perenne e morale nuova nella formazione ed educazione della coscienza / Ermenegildo Lio. — ₍Roma₎ : Pontificia università lateranense ; Roma : Città nuova, 1979.
354 p. ; 20 cm. — (Studi e testi di teologia morale ; 2) It79-June
Includes bibliographical references and indexes.
L7000
1. Christian ethics—Catholic authors. I. Title. II. Series.
BX1758.2.L56 79-378079
*79 MARC

Macrelli, Ciro.
Panorama del mondo morale : (sinossi di teologia morale) / Ciro Macrelli. — Pinerolo : Alzani, 1976.
314 p. ; 21 cm. It77-Dec
Bibliography: p. 299-306.
L4000
1. Christian ethics—Catholic authors. 2. Social ethics. I. Title.
BJ1249.M163 78-397935
*79 MARC

O'Connell, Timothy E
Principles for a Catholic morality / Timothy E. O'Connell. — New York : Seabury Press, 1978.
xv, 233 p. ; 24 cm.
"A Crossroad book."
Bibliography: p. ₍228₎-233.
ISBN 0-8164-0404-6 : $11.95
1. Christian ethics—Catholic authors. I. Title.
BJ1249.O26 241'.04'2 78-17884
78 MARC

Pennock, Michael
Moral problems : what does a Christian do? / Michael Pennock. — Notre Dame, Ind. : Ave Maria Press, c1979.
239 p. : ill. ; 23 cm.
Includes bibliographies.
SUMMARY: Presents a systematic method, based in Catholic morality, for resolving problems concerning such issues as drugs, alcohol, premarital sex, abortion, euthanasia, poverty, prejudice, and the arms race.
ISBN 0-87793-177-1 : $3.95
1. Youth—Conduct of life. 2. Christian ethics—Catholic authors. ₍1. Ethics. 2. Conduct of life. 3. Catholic Church₎ I. Title.
BJ1661.P4 241'.04'2 79-51015
00 MARC

Reidy, Maurice, 1943-
Foundations for a medical ethic : a personal and theological exploration of the ethical issues in medicine today / Maurice Reidy. — New York : Paulist Press, 1979, c1978.
104 p. ; 19 cm. — (A Deus book)
Bibliography: p. 103-104.
ISBN 0-8091-2216-2 : $2.45
1. Medical ethics. 2. Christian ethics—Catholic authors. 3. Medicine and Christianity. I. Title.
R724.R44 174'.2 79-65621
79 MARC

Reverendo padre / a cura di Carmelita Tognetti ; prefazione di Domenico Porzio. — 1. ed. — ₍Milano₎ : Libreria della famiglia, 1978-
v. ; 20 cm. It78-June
"Raccolta di trent'anni di corrispondenza con i lettori di Famiglia Cristiana."
CONTENTS: v. 1. Amore, famiglia e sesso in Italia.
L4000
1. Sexual ethics. 2. Christian ethics—Catholic authors. I. Tognetti, Carmelita. II. Famiglia cristiana.
HQ32.R48 241'.6'6 79-340616
*79 MARC

Riedl, Alfons.
Die kirchliche Lehrautorität in Fragen der Moral nach den Aussagen des Ersten Vatikanischen Konzils / Alfons Riedl. — Freiburg : Herder, 1979.
x, 534 p. ; 23 cm. — (Freiburger theologische Studien ; Bd. 117) GFR***
Originally presented as the author's thesis, Bonn.
Bibliography: p. 507-532.
ISBN 3-451-18449-4
1. Catholic Church—Teaching office. 2. Church—Authority—History of doctrines. 3. Christian ethics—Catholic authors. 4. Vatican Council, 1869-1870. I. Title. II. Series.
BX1746.R515 1979 79-375422
79 MARC

Romić, Jakov Rafael.
U vremenu s Kristom : uvod u kršćansku etiku / Jakov Rafael Romić ; ₍omot Roko Bolanča₎. — Zagreb : Kršćanska sadašnjost, 1976 (Pula : "Otokar Keršovani")
208 p. ; 21 cm. — (Biblioteka Centra za koncilska istraživanja, dokumentaciju i informacije Kršćanska sadašnjost : Priručnici ; sv. 16) Yu77-27
Includes bibliographical references.
1. Christian ethics—Catholic authors. I. Title. II. Series: Centar za koncilska istraživanja, dokumentaciju i informacije Kršćanska sadašnjost. Biblioteka : Priručnici ; sv. 16.
BJ1249.R77 78-357608
78₍r79₎rev MARC

Vallery, J
L'identité de la morale chrétienne : points de vue de quelques théologiens contemporains de langue allemande / J. Vallery. — Louvain-la-Neuve : Centre Cerfaux-Lefort, 1976.
xxxii, 195 p. ; 24 cm. — (Lex Spiritus Vitae ; 2) Be***
Includes bibliographical references.
1. Christian ethics—Catholic authors. 2. Rotter, Hans, 1932- . 3. Schüller, Bruno. I. Title. II. Series.
BX1758.2.V32 241'.04'2 78-390015
79 MARC

CHRISTIAN ETHICS (Continued)

—CATHOLIC AUTHORS—ADDRESSES, ESSAYS, LECTURES

Curran, Charles E
Transition and tradition in moral theology / Charles E. Curran. — Notre Dame, Ind. : University of Notre Dame Press, c1979.
xiv, 255 p. ; 21 cm.
Includes bibliographical references and index.
ISBN 0-268-01837-5 : $11.95
1. Christian ethics—Catholic authors—Addresses, essays, lectures. 2. Social ethics—Addresses, essays, lectures. 3. Medical ethics—Addresses, essays, lectures. I. Title.
BJ1249.C843 241'.04'2 78-20877
78 MARC

Gilbert, Maurice.
Morale et ancien Testament / M. Gilbert, J. L'Hour, J. Scharbert ; introd. par Ph. Delhaye. — Louvain-la-Neuve : Université catholique de Louvain : diffusion, Centre Cerfaux-Lefort, 1976.
184 p. ; 24 cm. — (Lex spiritus vitae ; 1) Be***
Includes bibliographical references and indexes.
1. Bible. O.T.—Ethics—Addresses, essays, lectures. 2. Christian ethics—Catholic authors—Addresses, essays, lectures. I. L'Hour, Jean, joint author. II. Scharbert, Josef, joint author. III. Title. IV. Series.
BS1199.E8G54 221.6 78-388825
78 MARC

Moral norms and Catholic tradition / edited by Charles E. Curran and Richard A. McCormick. — New York : Paulist Press, c1979.
viii, 363 p. ; 21 cm. — (Readings in moral theology ; no. 1)
Includes bibliographical references.
ISBN 0-8091-2203-0 : $5.95
1. Christian ethics—Catholic authors—Addresses, essays, lectures. 2. Ethics—Addresses, essays, lectures. I. Curran, Charles E. II. McCormick, Richard A., 1922- . III. Series.
BX1758.2.M62 241'.04'2 79-84237
79 MARC

—CATHOLIC AUTHORS—CONGRESSES

Congresso nazionale dei teologi moralisti, 7th, Catania, 1977.
La fondazione della norma morale nella riflessione teologica e marxista contemporanea / VII Congresso nazionale dei teologi moralisti ; ₍di₎ Autori vari. — Bologna : EDB, 1979.
303 p. ; 19 cm. — (Collana Studi e ricerche ; 29) It79-June
Includes bibliographical references.
L5000
1. Christian ethics—Catholic authors—Congresses. 2. Christian ethics—Controversial literature—Congresses. 3. Communist ethics—Congresses. 4. Communism and Christianity—Catholic Church—Congresses. I. Title.
BX1758.2.C62 1977 79-371396
*79 MARC

—COLLECTED WORKS

Handbuch der christlichen Ethik / hrsg. von Anselm Hertz ... ₍et al.₎ ; mit Beitr. von G. Altner ... ₍et al.₎. — Freiburg ; Basel ; Wien : Herder ; Gütersloh : Gütersloher Verlagshaus G. Mohn, 1978.
2 v. ; 25 cm. GFR78-A
Includes bibliographical references and indexes.
ISBN 3-451-17321-2
1. Christian ethics—Collected works. I. Hertz, Anselm.
BJ1253.H28 78-394443
79 MARC

—CONGRESSES

American Society of Christian Ethics.
Selected papers. Annual meeting - the American Society of Christian Ethics.
₍Newton Centre, MA, etc.₎ American Society of Christian Ethics ₍etc.₎; Waterloo, Ont., etc., distributed by the Council on the Study of Religion, etc.₎
v. 23 cm. annual.
Key title: Selected papers. Annual meeting - American Society of Christian Ethics, ISSN 0146-5821
1. Christian ethics—Congresses. I. Title.
BJ1189.A4a 241 78-647150
MARC-S

—CONTROVERSIAL LITERATURE

Zuev, IÛ P
(Psevdogumanizm khristianskoĭ propovedi lîubvi k bliznemu)
Псевдогуманизм христианской проповеди любви к ближнему / Ю. П. Зуев. — Москва : Знание, 1978.
64 p. : 20 cm. — (Новое в жизни, науке, технике : Серия Научный атеизм ; № 9, 1978) USSR 78
Includes bibliographical references.
0.11rub
1. Christian ethics—Controversial literature. 2. Love (Theology) I. Title. II. Series: Novoe v zhizni, nauke, tekhnike : Serifa Nauchnyĭ ateizm ; 1978, no. 9.
BJ1256.Z83 79-408295

—CONTROVERSIAL LITERATURE—CONGRESSES

Congresso nazionale dei teologi moralisti, 7th, Catania, 1977.
La fondazione della norma morale nella riflessione teologica e marxista contemporanea / VII Congresso nazionale dei teologi moralisti ; ₍di₎ Autori vari. — Bologna : EDB, 1979.
303 p. ; 19 cm. — (Collana Studi e ricerche ; 29) It79-June
Includes bibliographical references.
L5000
1. Christian ethics—Catholic authors—Congresses. 2. Christian ethics—Controversial literature—Congresses. 3. Communist ethics—Congresses. 4. Communism and Christianity—Catholic Church—Congresses. I. Title.
BX1758.2.C62 1977 79-371396
*79 MARC

—DICTIONARIES

Concise dictionary of Christian ethics / edited by Bernhard Stoeckle ; with contributions by J. Dominian, Anton Vögtle, Laurence Bright, and others. — New York : Seabury Press, c1979.
x, 285 p. ; 24 cm.
"A Crossroad book."
Includes bibliographies.
ISBN 0-8164-0357-0 : $19.50
1. Christian ethics—Dictionaries. I. Stoeckle, Bernhard, 1927- . II. Dominian, Jacob. III. Vögtle, Anton. IV. Bright, Laurence.
BJ1199.C66 241'.03 79-50666
79 MARC

—HISTORY

Hoffmann, Paul, 1933-
Jesus von Nazareth und eine christliche Moral : sittliche Perspektiven der Verkündigung Jesu / Paul Hoffmann, Volker Eid. — Freiburg : Herder, 1975.
259 p. ; 22 cm. — (Quaestiones disputatae ; 66) GFR***
Bibliography: p. 257-259.
ISBN 3-451-02066-1
1. Jesus Christ—Ethics. 2. Christian ethics—History. I. Eid, Volker, joint author. II. Title.
BS2417.E8H63 241 75-511835
75₍r79₎rev MARC

Hoffmann, Paul, 1933-
Jesus von Nazareth und eine christliche Moral : sittl. Perspektiven d. Verkündigung Jesu / Paul Hoffmann, Volker Eid. — 2., durchges. u. erg. Aufl. — Freiburg im Breisgau ; Basel ; Wien : Herder, 1976.
259 p. ; 22 cm. — (Quaestiones disputatae ; 66) GFR76-A
Bibliography: p. 257-259.
ISBN 3-451-02066-1 : DM35.00
1. Jesus Christ—Ethics. 2. Christian ethics—History. I. Eid, Volker, joint author. II. Title.
BS2417.E8H63 1976 241 77-571689
78₍r79₎rev MARC

Reiter, Johannes, 1944-
Der Moraltheologe Ferdinand Probst : (1816-1899) : e. Studie zur Geschichte d. Moraltheologie im Übergang von d. Romantik zur Neuscholastik / Johannes Reiter. — 1. Aufl. — Düsseldorf : Patmos-Verlag, 1978.
270 p. ; 22 cm. — (Moraltheologische Studien : Historische Abt. ; Bd. 4) GFR78-A
Bibliography: p. 247-259.
Includes indexes.
ISBN 3-491-78443-3
1. Christian ethics—History. 2. Probst, Ferdinand, 1816-1899. I. Title.
BJ1227.R44 79-340244
79 MARC

—HISTORY OF DOCTRINES

Tödt, Heinz Eduard.
Rudolf Bultmanns Ethik der Existenztheologie / Heinz Eduard Tödt. — Orig. -Ausg. — Gütersloh : Gütersloher Verlagshaus Mohn, 1978.
128 p. ; 19 cm. — (Ethiker des Protestantismus ; Bd. 1) (Gütersloher Taschenbücher Siebenstern ; 440) GFR78-A47
Bibliography of R. K. Bultmann's works: p. 15-16.
Bibliography: p. 16-17.
Includes index.
ISBN 3-579-03917-2 : DM9.80
1. Christian ethics—History of doctrines. 2. Bultmann, Rudolf Karl, 1884-1976. I. Title. II. Series.
BJ1253.T6 79-362224
79 MARC

—MORMON AUTHORS

Bennion, Lowell Lindsay, 1908-
The things that matter most / Lowell L. Bennion. — Salt Lake City : Bookcraft, c1978.
63 p. ; 22 cm.
ISBN 0-88494-346-1 : $3.95
1. Christian ethics—Mormon authors. I. Title.
BJ1275.B45 241 78-62211
79 MARC

—ORTHODOX EASTERN AUTHORS

Gregorios, Paulos.
The human presence : an orthodox view of nature / Paulos Gregorios. — Geneva : World Council of Churches, c1978.
103 p. ; 22 cm. Sw***
Includes bibliographical references.
ISBN 2-8254-0575-2
1. Nature (Theology)—History of doctrines. 2. Human ecology. 3. Christian ethics—Orthodox Eastern authors. I. Title.
BT695.5.G73 291.2'12 79-311714
79 MARC

—STUDY AND TEACHING—CASE STUDIES—ADDRESSES, ESSAYS, LECTURES

Sozialethischer Unterricht, dokumentiert und analysiert : eine Untersuchung zur Praxis des problemorientierten Religionsunterrichts / Günter Stachel (Hrsg.) ; mit Beitr. von Alberto Bondolfi ... ₍et al.₎. — Zürich : Benziger, c1977.
200 p. ; 23 cm. — (Studien zur praktischen Theologie ; 15) Sw77-622
Includes bibliographical references.
ISBN 3-545-21515-6 : 28.80F
1. Christian education—Germany, West—Case studies—Addresses, essays, lectures. 2. Christian ethics—Study and teaching—Case studies—Addresses, essays, lectures. 3. Social ethics—Study and teaching—Case studies—Addresses, essays, lectures. I. Stachel, Günter. II. Series: Studien zur praktischen Theologie (Zürich) ; 15.
BV1470.G3S66 78-388595
*79 MARC

—STUDY AND TEACHING—SPAIN—SALAMANCA

Teodoro del Santísimo Sacramento, padre.
El curso moral salamanticense : estudio histórico y valoración crítica / Teodoro del Santísimo Sacramento. — ₍Salamanca₎ : Universidad Pontificia de Salamanca, Facultad de Derecho Canónico, 1968.
xxiii, 227 p. ; 25 cm.
Bibliography: p. ₍xiii₎-xxiii.
1. Christian ethics—Study and teaching—Spain—Salamanca. 2. Salamanca. Pontificia Universidad Eclesiástica—History. I. Title.
BJ1200.T46 79-367374
79 MARC

CHRISTIAN EVIDENCES see Apologetics

CHRISTIAN EXISTENTIALISM
see Christianity and existentialism

CHRISTIAN FICTION

Fickett, Harold L
Mrs. Sunday's problem, and other stories / Harold Fickett. — Old Tappan, N.J. : Revell, c1979.
158 p. ; 22 cm.
CONTENTS: A sycamore tree.—Christbirth.—The missionary conference.—Persecuted for Christ's sake.—Palm Sunday.—Mrs. Sunday's problem.
ISBN 0-8007-0996-9
1. Christian fiction. I. Title.
PZ4.F449 Mi 813'.5'4 78-31469
₍PS3556.I33₎ 78 MARC

CHRISTIAN GIVING

see also Stewardship, Christian

Cunningham, Richard B
Creative stewardship / Richard B. Cunningham. — Nashville : Abingdon, c1979.
128 p. ; 20 cm. — (Creative leadership series)
Includes bibliographical references.
ISBN 0-687-09844-0
1. Stewardship, Christian. 2. Christian giving. 3. Church finance. I. Title. II. Series.
BV772.C779 248'.6 79-973
79 MARC

CHRISTIAN GIVING (Continued)

Fisher, Wallace E
All the good gifts : on doing Biblical stewardship / Wallace E. Fisher. — Minneapolis : Augsburg Pub. House, c1979.
111 p. ; 20 cm.
Bibliography: p. 109-111.
ISBN 0-8066-1702-0 : $2.95
1. Stewardship, Christian. 2. Christian giving. I. Title.
BV772.F48 248'.6 79-50077
79 MARC

MacArthur, John F
Giving—God's way / John F. MacArthur, Jr. — Wheaton, Ill. : Tyndale House Publishers, 1978.
104 p. ; 18 cm.
ISBN 0-8423-1034-7 : $1.95
1. Christian giving. I. Title.
BV772.M25 248'.6 77-93748
79 MARC

—ADDRESSES, ESSAYS, LECTURES

Eguchi, Shin'ichi, 1914-
(Chi no shio no hako)
地の塩の箱：ある幸福論 / 江口榛一. — 東京：新潮社, 昭和49 [1974]
251 p. ; 20 cm. Ja 74-18399
¥780

1. Christian giving—Addresses, essays, lectures. 2. Christian life—Addresses, essays, lectures. 3. Eguchi, Shin'ichi, 1914- —Addresses, essays, lectures. I. Title.
BV772.E42 78-812453

CHRISTIAN HAGIOGRAPHY

Johnson, Phyllis, 1937-
Le vain siecle guerpir : a literary approach to sainthood through Old French hagiography of the twelfth century / by Phyllis Johnson and Brigitte Cazelles. — Chapel Hill : U.N.C. Dept. of Romance Languages : [distributed by University of North Carolina Press], 1979.
321 p. ; 24 cm. — (North Carolina studies in the Romance languages and literatures ; no. 205)
Bibliography: p. [311]-321.
ISBN 0-8078-9205-X
1. French poetry—To 1500—History and criticism. 2. Christian hagiography. 3. Christian saints in literature. I. Cazelles, Brigitte, joint author. II. Title. III. Series.
PQ178.J64 841'.1'09352 79-10678
79 MARC

CHRISTIAN HERESIES AND HERETICS

Rochfort, Luke.
The genealogie of Protestants / Luke Rochfort. The reformed Protestant / John Brereley [i.e. L. Anderton]. — Ilkley [Eng.] : Scolar Press, 1978.
199, 162 p. ; 20 cm. — (English recusant literature, 1558-1640 ; v. 383)
GB***
Rochfort's work is a reprint of the 1621 ed. from a copy in Edinburgh University Library. Reference: Allison and Rogers 731; not in STC. Anderton's work is a reprint of the 1621 ed. from a copy in the British Library. Not in Allison and Rogers; not in STC.
ISBN 0-85967-480-0
1. Catholic Church—Doctrinal and controversial works—Catholic authors. 2. Protestantism—Controversial literature. 3. Christian heresies and heretics. I. Anderton, Lawrence. The reformed Protestant. 1978. II. Title. III. Series: Rogers, David Morrison, comp. English recusant literature, 1558-1640 ; v. 383.
BX1750.A1E5 vol. 383 282'.41 s 79-304200
79 MARC

CHRISTIAN LABOR UNIONS see Trade-unions, Catholic

CHRISTIAN LEADERSHIP

see also Church consultation; Peer counseling in the church

Armerding, Hudson T
Leadership / Hudson T. Armerding. — Wheaton, Ill. : Tyndale House Publishers, c1978.
170 p. ; 21 cm.
ISBN 0-8423-2126-8
1. Christian leadership. I. Title.
BV652.1.A75 262'.1 78-64673
79 MARC

Dayton, Edward R
Strategy for leadership / Edward R. Dayton, Ted. W. Engstrom. — Old Tappan, N.J. : F. H. Revell Co., c1979.
240 p. : ill. ; 24 cm.
Bibliography: p. 233-235.
Includes index.
ISBN 0-8007-0994-2 : $8.95
1. Church management. 2. Christian leadership. I. Engstrom, Theodore Wilhelm, 1916- joint author. II. Title.
BV652.D38 658'.91'25 79-11180
79 MARC

Engstrom, Theodore Wilhelm, 1916-
The Christian executive : a practical reference for Christians in management positions, leaders of Christian organizations, Christian educators, pastors, and other Christian workers / Ted W. Engstrom & Edward R. Dayton. — Waco, Tex. : Word Books, c1979.
216 p. ; 23 cm.
Bibliography: p. 212-213.
Includes index.
ISBN 0-8499-0136-7 : $9.95
1. Christian leadership. I. Dayton, Edward R., joint author. II. Title.
BV652.1.E525 658'.91'2 78-65812
79 MARC

Harper, Michael, 1931-
Let my people grow! : ministry and leadership in the Church / by Michael Harper. — London : Hodder and Stoughton, 1977.
254 p. ; 22 cm. — (Ecclesia books) GB77-19209
Bibliography: p. [247]-248.
Includes index.
ISBN 0-340-21573-9 : £2.95
1. Church renewal—Church of England. 2. Church polity. 3. Christian leadership. I. Title.
BV600.2.H35 1977b 262'.001 77-379727
78[r79]rev MARC

Harper, Michael, 1931-
Let my people grow : ministry and leadership in the church / Michael Harper. — Plainfield, N.J. : Logos International, c1977.
254 p. ; 21 cm.
Bibliography: p. [247]-248.
Includes index.
ISBN 0-88270-236-X : $3.50
1. Church renewal—Church of England. 2. Church polity. 3. Christian leadership.
BV600.2.H35 1977 262'.001 77-73840
77[r79]rev MARC

McDonough, Reginald M
Keys to effective motivation / Reginald M. McDonough. — Nashville : Broadman Press, c1979.
140 p. : ill. ; 20 cm.
Includes bibliographical references.
ISBN 0-8054-3226-4
1. Christian leadership. 2. Motivation (Psychology) 3. Man (Christian theology) I. Title.
BV652.1.M29 254'.6 77-26532
79 MARC

Miller, Herb.
Tools for active Christians / Herb Miller. — St. Louis : Bethany Press, c1979.
127 p. ; 23 cm.
Bibliography: p. 127.
ISBN 0-8272-3624-7 : $5.95
1. Christian leadership. I. Title.
BV652.1.M54 254 79-14795
79 MARC

Powers, Bruce P
Christian leadership / Bruce P. Powers. — Nashville, Tenn. : Broadman Press, c1979.
133 p. : ill. ; 20 cm.
Bibliography: p. 131-133.
ISBN 0-8054-3227-2
1. Christian leadership. I. Title.
BV652.1.P65 353 78-72841
79 MARC

Sill, Sterling W
Leadership. Salt Lake City, Bookcraft [1958]-
v. ports. 24 cm. $5.95 (v. 3)
Includes indexes.
1. Christian leadership. 2. Leadership. I. Title.
BV652.1.S57 62-6243
[r79]rev MARC

—ADDRESSES, ESSAYS, LECTURES

Leadership profiles from Bible personalities / compiled by Ernest E. Mosley. — Nashville : Broadman Press, c1979.
130 p. ; 20 cm.
Bibliography: p. 129-130.
CONTENTS: Mosley, E. E. Introduction.—Bennett, W. A. Moses, leadership that overcomes obstacles.—Faulkner, B. R. Joshua, leader in his own right.—Ishee, J. A. Elijah, can a prophet be a leader?—Mosley, E. E. Nehemiah, builder of walls and people.—McDonough, R. M. Paul, a self-actualized leader.—Dale, R. D. Barnabas, hidden leader of the New Testament.—Grubbs, B. Simon Peter, shaped a powerful model.
ISBN 0-8054-2532-2
1. Christian leadership—Addresses, essays, lectures. 2. Bible—Biography—Addresses, essays, lectures. I. Mosley, Ernest E.
BV652.1.L4 262'.1 78-74503
79 MARC

—CASE STUDIES

Winter, Uwe-Helmut, 1943-
Der Gemeindekirchenrat als Entscheidungsgremium in seiner Bedeutung für die Beteiligung der Gemeinde am gemeindeleitenden Handeln : eine empirische Analyse am Beispiel der Evangelischen Kirche in Berlin-Brandenburg / vorgelegt von Uwe-Helmut Winter. — Berlin : [s.n.], 1975.
324, 21 p. ; 21 cm. GFR***
Thesis—Kirchliche Hochschule, Berlin.
Vita.
Bibliography: p. 308-324.
1. Parish councils. 2. Evangelische kirche in Berlin-Brandenburg. 3. Christian leadership—Case studies. I. Title: Der Gemeindekirchenrat als Entscheidungsgremium ...
BV652.9.W55 78-376940
78 MARC

—HANDBOOKS, MANUALS, ETC.

Adams, Margee.
Developing leadership for the church's teaching ministry / Margee and Jim Adams. — Livermore, CA : Griggs Educational Service, c1978.
88 p. ; 28 cm.
Includes bibliographical references.
1. Christian education—Teacher training—Handbooks, manuals, etc. 2. Christian leadership—Handbooks, manuals, etc. I. Adams, Jim, joint author. II. Title.
BV1533.A3 268'.3 77-92679
78 MARC

CHRISTIAN LEGENDS see Legends, Christian

CHRISTIAN LIFE

see also Asceticism; Character; Christian communities; Christian education; Conduct of life; Conversion; Devotional exercises; Faith; Monastic and religious life of women; Religiousness; Revivals; Spiritual life; Stewardship, Christian; and subdivision Religious life under classes of persons and institutions, e.g. Children—Religious life

Abe, Fujio, 1897-
(Onchō no fukuin)
恩寵の福音 / 安部藤夫著. — 東京：ホーリネス教會出版部, 昭和3 [1928]
4, 188 p. ; 19 cm.

1. Christian life. I. Title.
BV4509.J3A23 79-805811

Arslanian, Gēorg, 1869-1951.
(Klank'i dasě molut'iwnneru handēp)
Կեանքի դասը մոլութիւններու հանդէպ / գրեց Գէորգ Արսլանեան. — Կ. Պոլիս : Տպագրութիւն "Կիւթէմպէրկ," 1927.
109 p. : port. ; 19 cm.
1. Vices. 2. Christian life. I. Title.
BV4625.A77 79-971011

Beschi, Costantino Giuseppe, 1630-ca. 1746.
வேதியரொழுக்கம்; or, Instructions to catechists, by C. J. Beschi. Madras, Printed and sold at the American Mission Press, 1849.
232 p. 16 cm.
In Tamil.

1. Religious education. 2. Evangelistic work. 3. Christian life. I. Title. II. Title: Instructions to catechists.
Title romanized: Vētiyaroḻukkam.

BV1471.B38 1849 72-209110

Clarkson, Edith Margaret, 1915-
So you're single! / Margaret Clarkson. — Wheaton, Ill. : H. Shaw Publishers, c1978.
167 p. ; 22 cm.
ISBN 0-87788-772-1 : $2.95
1. Single women. 2. Chastity. 3. Christian life.
HQ800.C55 301.4 78-53012
79 MARC

CHRISTIAN LIFE (Continued)

[Gossner, Johannes Evangelist] 1773-1858.
Das Herz des Menschen, ein Tempel Gottes, oder eine Werkstätte des Satans : in zehn Figuren sinnbildlich dargestellt zur Erweckung und Beförderung des christlichen Sinnes : nach der 5. verb. Augsburger Aufl., 1815. — 2. deutsche americanische Ausg. — Reading : H. B. Sage, 1823.
58 p. : ill. ; 17 cm.
"Aus dem Französichen übersetzt und im Jahre 1732 zu Würzburg von dem Universitätskupferstecher unter dem Titel: Geistlicher Sittenspiegel ... herausgegeben." However, according to the Neue deutsche Biographie the author is G. E. Gossner, according to Holzmann-Bohatta, G. E. Gossner and P. F. Pöschel. Attributed to J. N. J. Schmidt in the 1891 edition.
1. Christian life. I. Pöschel. Philipp Friedrich. II. Schmidt, Johannes Nikolaus Jürgens. III. Title. IV. Title: Geistlicher Sittenspiegel.
BV4515.G67 1823 78-378663
78 MARC

[Gossner, Johannes Evangelist] 1773-1858.
Der Spiegel der Seele : das Herz des Menschen, ein Tempel Gottes, oder auch eine Behausung des Satans : in Bildern versinnbildlicht, dargestellt, erläutert und beschrieben / hrsg. von Ernest E. Pates. — New York : [s.n.], 1891.
116 p. : ill. ; 16 cm.
Published in 1823 under title: Das Herz des Menschen, ein Tempel Gottes oder eine Werkstätte des Satans.
Attributed to J. N. J. Schmidt in the preface. However according to the Neue deutsche Biographie the author is J. E. Gossner, according to Holzmann-Bohatta, J. E. Gossner und P. F. Pöschel.
1. Christian life. I. Pates, Ernst E. II. Schmidt, Johannes Nikolaus Jürgens. III. Pöschel, Philipp Friedrich. IV. Title.
BV4515.G67 1891 78-378662
78 MARC

Howell, John C
Senior adult family life : a source book / John C. Howell. — Nashville : Broadman Press, c1979.
137 p. ; 21 cm.
Includes bibliographical references.
ISBN 0-8054-5423-3
1. Aged—United States—Family relationships. 2. Aged—United States—Social conditions. 3. Christian life. I. Title.
HQ1064.U5H68 301.43'5 79-51139
79 MARC

Ishikawa, Takeyoshi, 1887-1961.
(Shinkō no michi)
信仰の道 / 石川武美著. — 東京：主婦之友社, 昭和21 [1946], 昭和23 [1948] printing.
13, 200 p. ; 19 cm.

1. Christian life. 2. Faith. I. Title.
BV4509.J3I79 79-809048

Janney, John R
Husbandmanship : the art of being a good husband / John R. Janney. — 1st ed. — Bridgeton, N.J. : Christian Family Crusades, c1979.
184 p. ; 22 cm.
Includes bibliographical references.
1. Husbands. 2. Christian life. I. Title.
[HQ756.J36] 301.42'7 79-65564
79 MARC

Law, William, 1686-1761.
A serious call to a devout and holy life ; The spirit of love / William Law ; edited from the first editions by Paul G. Stanwood ; introd. by Austin Warren and Paul G. Stanwood ; pref. by John Booty. — New York : Paulist Press, c1978.
x, 526 p. ; 23 cm. — (The Classic of Western spirituality)
Errata slip inserted.
Bibliography: p. 499-500
Includes indexes.
ISBN 0-8091-0265-X. ISBN 0-8091-2144-1 pbk. : $6.95
1. Christian life. I. Stanwood, P. G. II. Law, William, 1686-1761. The spirit of love. 1978. III. Title. IV. Series.
BV4500.L3 1978 248'.4 78-61418
78 MARC

Newton, David F b. 1796.
A holiday present : or, Educating little Mary for the heavenly kingdom / by D. F. Newton. — New York : Newton, 1871.
540 p., [1] leaf of plates : ill. ; 19 cm.
1. Christian life. I. Title.
BV4501.N444 649'.7 78-320371
78 MARC

Ōtsuka, Noyuri, 1924-
(Shinjitsu o otte)
真実を追って：女教師のノート / 大塚野百合著. — 東京：日本基督教団出版部, 昭和36 [1961]
170 p. ; 18 cm. — (女性と生活シリーズ)
Series romanized: Josei to seikatsu shirizu.

1. Christian life. 2. Women in Christianity. 3. Education—Philosophy. I. Title.
BV4501.2.O8 79-806154

Sako, Jun'ichirō, 1919-
(Kirisutosha no tame no jinseiron)
キリスト者のための人生論 / 佐古純一郎編. — 東京：ぐろりあ：発売元 教文館, 昭和34 [1959]
184 p. ; 18 cm. — (ぐろりあ選書)
Series romanized: Guroria sensho.

1. Christian life. I. Title.
BV4509.J3S245 79-807533

Satō, Teikichi, 1887-
(Shi ni chokumenseru taiken)
死に直面せる体験 / 佐藤定吉. — 東京：イエスの僕會, 大正14 [1925], 昭和9 [1934] printing.
2, 334 p., [1] leaf of plates : port. ; 19 cm.

1. Death—Meditations. 2. Christian life. I. Title.
BT825.S27 78-813793

Stevens, Abel, 1815-1897.
Christian work and consolation : the problem of an effective and happy life / by Abel Stevens. — New York : Phillips & Hunt, 1882.
206 p. ; 20 cm.
1. Christian life. 2. Consolation. I. Title.
BV4501.S79446 248'.4 79-312365
79 MARC

Tazaki, Kensaku.
(Sutemi de ikiru)
捨身で生きる：ある牧師の生活と意見 / 田崎健作著. — 東京：日本YMCA同盟出版部, 1964.
259 p. ; 18 cm. — (キリスト教教養新書)
Series romanized: Kirisutokyō kyōyō shinsho.

1. Tazaki, Kensaku. 2. Christian Biography—Japan. 3. Christian life. I. Title.
BX8750.N58T397 79-801864

Thavōritēs, M
(Tetralogia me stochasmo kai pistē)
Τετραλογία μὲ στοχασμὸ καὶ πίστη / Μ. Θαβωρίτη. — Ἀθήνα : [s. n.], 1976.
96 p. ; 25 cm.
1. Logos. 2. Christian life. I. Title.
BT210.T46 79-414221

Uchimura, Kanzō, 1861-1930.
(Kirisuto shinto no nagusame)
基督信徒のなぐさめ / 内村鑑三著. — 改版. — 東京：岩波書店, 1976.
133 p. ; 15 cm. — (岩波文庫) Ja 77-4994
Series romanized: Iwanami bunko.
¥100

1. Christian life. I. Title.
BV4501.2.U23 1976 79-806177

Uemura, Masahisa, 1857-1925.
(Inori no seikatsu)
祈の生活 / [著作者 植村正久]. — 東京：原成吉：發賣所 伊藤書店, 大正8 [1919]
2, 2, 313 p. ; 20 cm.
Errata slip inserted.

1. Christian life. 2. Prayer. I. Title.
BV4509.J3U35 79-808461

--DISCIPLES OF CHRIST AUTHORS
see Christian life--Christian Church (Disciples of Christ) authors

—EARLY CHURCH, CA. 30-600

Chrēstou, Panagiōtēs, 1917-
(Nēptika kai ēsychastika)
Νηπτικὰ καὶ ἡσυχαστικά / Παναγιώτου Κ. Χρήστου. — Θεσσαλονίκη : Πατριαρχικὸν Ἵδρυμα Πατερικῶν Μελετῶν, 1977.
262 p. ; 25 cm. — (His Θεολογικὰ μελετήματα ; 3)
Series romanized: Theologika meletēmata.
Includes bibliographical references.
1. Christian life—Early church, ca. 30-600. 2. Diadochus, Bp. of Photice, 5th cent. 3. Palamus, Gregorius, Saint, Abp. of Thessalonica, ca. 1296-ca. 1359. 4. Hesychasm. I. Title.
BR45.C48 vol. 3 78-421768
[BR195.C5]

—MIDDLE AGES, 600-1500

Albertus Magnus, Saint, Bp. of Ratisbon, 1193?-1280. Spurious and doubtful works.
Das Anhangen an Gott; ein Unterricht des Albertus Magnus, gewesenen Bischoffs zu Regensburg. Germantown, Gedruckt bey C. Saur, 1764.
[7], 56 p. 12 cm.
Translation of De adhaerendo Deo.
Previously published in Germany in 1737 under title: Albertus Magni gewesenen Bischoff zu Regenspurg Unterricht vom Anhangen an Gott.
Includes an introd. and two poems by the translator, believed to be G. Tersteegen.
Evans 9850.
1. Mysticism—Middle Ages, 600-1500. 2. Christian life—Middle Ages, 600-1500. 3. Hymns, German. I. Tersteegen, Gerhard, 1697-1769. II. Title.
BV5080.A415 78-368405
78 MARC

—1910-

Buckingham, Jamie.
Coping with criticism / by Jamie Buckingham. — Plainfield, N.J. : Logos International, c1978.
158 p. ; 22 cm.
ISBN 0-88270-327-7 : $6.95
1. Christian life—1910- . 2. Psychology—Popular works. I. Title.
BV4501.2.B819 248'.4 78-60994
79 MARC

—1960-

Aldrich, Joe.
Secrets to inner beauty / [Joe Aldrich]. — Santa Ana, Calif. : Vision House Publishers, c1977.
142 p. ; 21 cm.
ISBN 0-88449-071-8 : $2.98
1. Christian life—1960- . I. Title.
BV4501.2.A374 248'.4 77-87103
78 MARC

Allen, Charlotte Hale.
Full-time living / Charlotte Hale Allen. — Old Tappan, N.J. : F. H. Revell Co., c1978.
159 p. ; 22 cm.
ISBN 0-8007-0939-X
1. Christian life—1960- . I. Title.
BV4501.2.A423 248'.4 78-14867
78 MARC

Armbruster, Wally.
It's still lion vs Christian in the corporate arena / Wally Armbruster. — St. Louis : Concordia Pub. House, c1979.
116 p. : ill. ; 24 cm.
ISBN 0-570-03791-3 : $1.95
1. Christian life—1960- . 2. Business ethics. I. Title.
BV4501.2.A65 248'.88 78-31394
78 MARC

Baker, Cozy, 1927-
Holiday frame of mind / by Cozy Baker ; photos. by Barbi Baker ; [back cover port. by Migs Elliot]. — Washington : Acropolis Books, c1978.
63 p. : ill. (some col.) ; 23 cm.
1. Christian life—1960- . I. Baker, Barbi. II. Title.
BV4501.2.B376 248'.4 78-78351
79 MARC

Basansky, Bill.
How to know the Father's voice / Bill Basansky. — [Tulsa, Okla. : Bill Basansky Evangelistic Association, c1978]
103 p. ; 22 cm.
$2.95
1. Christian life—1960- . I. Title.
BV4501.2.B38418 248'.4 78-57651
78 MARC

CHRISTIAN LIFE
—1960- (Continued)

Bauer, Clare.
Some through the waters / by Clare Bauer. — Nashville : T. Nelson Publishers, c1979.
160 p., [2] leaves of plates : ill. ; 20 cm.
Includes index.
ISBN 0-8407-5687-9 : $3.95
1. Bauer, Clare. 2. Christian biography—United States. 3. Christian life—1960- . I. Title.
BR1725.B383A37 248'.2 79-2438
79 MARC

Baughen, Michael A
The Moses principle : leadership and the venture of faith / Micheal Baughen. — Wheaton, Ill : H. Shaw, c1978.
118 p. ; 21 cm.
ISBN 0-87788-558-3
1. Moses. 2. Bible. O.T.—Biography. 3. Christian life—1960- . I. Title.
BS580.M6B29 1978 253 78-27498
78 MARC

Baumann, Dan.
Extraordinary living for ordinary people / Dan Baumann. — Irvine, Calif. : Harvest House Publishers, c1978.
127 p. ; 21 cm.
ISBN 0-89081-149-0 : $2.95
1. Christian life—1960- . 2. Virtues. I. Title.
BV4501.2.B38453 248'.4 78-58432
79 MARC

Beall, James Lee.
How to achieve security, confidence, and peace / James Lee Beall. — Plainfield, N.J. : Logos International, c1978.
417 p. ; 21 cm.
ISBN 0-88270-268-8 : $4.95
1. Christian life—1960- . I. Title.
BV4501.2.B3867 248'.4 78-50752
78 MARC

Bednarczyk, Krzysztof.
Formuła mocy / Krzysztof Bednarczyk. — Wyd. 1. — Warszawa : Słowo Prawdy, 1977 (Pozn. : Pallottinum)
220, [4] p. : ill. ; 17 cm. P78-929
Bibliography: p. 220.
1. Christianity—20th century. 2. Church history. 3. Christian life—1960- . I. Title.
BR121.2.B36 79-341547
79 MARC

Beerman-De Roos, C L
Een waagstuk / C. L. Beerman-De Roos. — Blaricum : Bigot & Van Rossum, [1963]
140 p. ; 20 cm. — (Op weg)
Includes bibliographical references.
1. Christian life—1960- . I. Title.
BV4509.D8B4 79-365037
79 MARC

Bowman, George M 1920-
Clock wise / George M. Bowman. — Old Tappan, N.J. : F. H. Revell, c1979.
127 p. ; 22 cm.
ISBN 0-8007-0987-X
1. Christian life—1960- . I. Title.
BV4501.2.B683 248'.4 78-26844
78 MARC

Brandt, Henry R
I want happiness now! / Henry Brandt, with Phil Landrum. — Grand Rapids : Zondervan Pub. House, c1978.
175 p. : ill. ; 21 cm.
ISBN 0-310-21640-0. ISBN 0-310-21641-9 pbk.
1. Christian life—1960- . I. Landrum, Phil, joint author. II. Title.
BV4501.2.B686 248'.4 78-14510
78 MARC

Brandt, Priscilla.
Two-way prayer / Priscilla Brandt ; foreword by Joyce Landorf. — Waco, Tex. : Word Books, c1979.
151 p. ; 23 cm.
Bibliography: p. 147-151.
ISBN 0-8499-0022-0 : $6.95
1. Prayer. 2. Christian life—1960- . I. Title.
BV210.2.B685 248'.3 78-65802
79 MARC

Braun, Jon E
It ain't gonna reign no more / Jon E. Braun. — 1st ed. — Nashville : T. Nelson, c1978.
x, 154 p. ; 21 cm.
ISBN 0-8407-5649-6 : $3.95
1. Christian life—1960- . I. Title.
BV4501.2.B688 248'.4 78-1375
78[r79]rev MARC

Brooks, Pat.
Healing of the mind / Pat Brooks. — Skyland, N.C. : New Puritan Library, c1978.
151 p. ; 18 cm.
ISBN 0-932050-00-X : $1.95
1. Christian life—1960- . 2. Christianity and occult sciences. 3. Brooks, Pat. I. Title.
[BV4501.2.B766] 248'.4 78-59366
79 MARC

Butler, Marie.
Ron's story : a legacy of love / by Marie Butler. — Westchester, Ill. : Good News Publishers, c1978.
143 p. ; 21 cm.
ISBN 0-89107-155-5 : $3.50
1. Lymphatics—Cancer—Biography. 2. Christian life—1960- . 3. Butler, Ron. I. Title.
RC280.L9B87 362.1'9'6994420924 78-52686
79 MARC

Chandler, Russell.
The overcomers / Russell Chandler. — Old Tappan, N.J. : Revell, c1978.
158 p. ; 22 cm.
CONTENTS: His will brings her joy: Elisabeth Elliot.—Towering theologian: Carl F. H. Henry.—They just praise the Lord: Bill and Gloria Gaither.—Hang in there, brother: Edward Victor Hill.—Woman of prayer: Vonette Bright.—Changing her world: Ann Kiemel.—Yielded novice: Maria Von Trapp.—Thoughtful Quaker: D. Elton Trueblood.—He tries God: Walter Hoving.—Pastoral partners: Frederick W. and Ruth P. Cropp.—Southern singer: Joanne Cash Yates.—Elder statesman: Paul S. Rees.—Conclusion: through it all.
ISBN 0-8007-0944-6 : $6.95
1. Christian biography. 2. Christian life—1960- . I. Title.
BR1700.2.C46 209'.2'2 78-16621
78 MARC

Chapian, Marie.
Free to be thin / Marie Chapian. — Minneapolis : Bethany Fellowship, 1979.
179, [3] p. : ports. ; 21 cm.
Bibliography: p. 182.
ISBN 0-87123-560-9 : $3.95
1. Reducing—Moral and religious aspects. 2. Christian life—1960- . I. Title.
RM222.2.C44 613.2'5 79-15656
79 MARC

Cho, Yong-gi, 1936-
The fourth dimension / by Paul Yonggi Cho ; [with a foreword by Robert Schuller]. — Plainfield, N.J. : Logos International, c1979.
186 p. ; 21 cm.
ISBN 0-88270-380-3 : $3.95
1. Christian life—1960- . 2. Cho, Yong-gi, 1936- . I. Title.
BV4501.2.C49 248'.4 79-65588
79 MARC

Cobb, Mary Ann.
Lorie : a story of hope / Mary Ann Cobb. — Nashville : T. Nelson, c1979.
181 p. ; 21 cm.
Bibliography: p. 179-181.
ISBN 0-8407-5677-1
1. Down's syndrome—Biography. 2. Christian life—1960- . 3. Cobb, Lorie Lynn, 1974- .
RJ506.D68C6 362.7'8'30926 78-20951
78 MARC

Cook, Robert A 1912-
Walk with the King today / Robert A. Cook. — 1st ed. — Chappaqua, N.Y. : Christian Herald Books, c1978.
253 p. ; 22 cm.
ISBN 0-915684-42-X : $4.95
1. Christian life—1960- . I. Title.
BV4501.2.C673 248'.4 78-56977
78 MARC

Copley, Derek.
Building with bananas : people problems in the local church / [by] Derek & Nancy Copley. — Exeter : Paternoster Press, 1978.
160 p. : ill. ; 18 cm. GB78-38908
ISBN 0-85364-223-0 : £1.50
1. Christian life—1960- . 2. Interpersonal relations. I. Copley, Nancy, joint author. II. Title.
BV4501.2.C6775 254 79-315833
79 MARC

Cranor, Phoebe.
How am I supposed to love myself? / Phoebe Cranor. — Minneapolis : Bethany Fellowship, c1979.
144 p. ; 21 cm.
Includes index.
ISBN 0-87123-236-7 : $2.95
1. Wives—Religious life. 2. Christian life—1960- . 3. Self-love (Theology) I. Title.
BV4527.C72 248'.843 79-14229
79 MARC

Day, Afton.
From fat to fit in four grueling months / Afton Day. — Salt Lake City : Bookcraft, c1979.
135 p. ; 24 cm.
Includes bibliographical references.
ISBN 0-88494-367-4 : $4.95
1. Reducing—Psychological aspects. 2. Reducing diets. 3. Christian life—1960- . I. Title.
RM222.2.D36 613.2'5 78-75367
79 MARC

Dobbert, John.
Give yourself a chance / John Dobbert. — Old Tappan, N.J. : Revell, c1979.
160 p. ; 22 cm.
ISBN 0-8007-0988-8
1. Success. 2. Christian life—1960- . I. Title.
BJ1611.2.D6 248'.4 78-21106
78 MARC

Donnelly, Doris.
Learning to forgive / Doris Donnelly. — New York : Macmillan, c1979.
xiv, 127 p. ; 22 cm.
Includes bibliographical references and index.
ISBN 0-02-532140-4
1. Forgiveness of sin. 2. Christian life—1960- . I. Title.
BT795.D66 248'.4 79-15964
79 MARC

Dresselhaus, Richard L
The joy of belonging : a study in church membership / Richard L. Dresselhaus. — Springfield, Miss. : Gospel Pub. House, c1978.
128 p. ; 18 cm. — (Radiant books)
ISBN 0-88243-526-4 : $1.50
1. Church membership. 2. Christian life—1960- . I. Title.
BV820.D73 260 78-66868
00 MARC

Douglass, Stephen B
Managing yourself : practical help for Christians in personal planning, time scheduling, and self-control / Stephen B. Douglass ; foreword by Bill Bright. — San Bernardino, CA : Here's Life Publishers, c1978.
223 p. : ill. ; 21 cm.
Includes bibliographies.
ISBN 0-918956-49-8 : $3.95
1. Christian life—1960- . 2. Conduct of life. I. Title.
BV4501.2.D65 248'.4 78-70647
79 MARC

Dubois, Paul André.
Vie chrétienne et authenticité / Paul-André Dubois. — Genève ; Paris : La Maison de la Bible, 1977.
111 p. ; 18 cm. Sw77-12084
ISBN 2-8260-3208-9 : 7.80F
1. Christian life—1960- . I. Title.
BV4501.2.D73 248'.4 79-377313
*79 MARC

Dugan, LeRoy.
The uncomplicated Christian : a simple guidebook on living up to God's potential for your life / LeRoy Dugan. — Minneapolis : Bethany Fellowship, c1978.
123 p. ; 18 cm. — (Dimension books)
ISBN 0-87123-572-2 : $1.95
1. Christian life—1960- . I. Title.
BV4501.2.D74 248'.4 78-66886
79 MARC

Eastman, Addison J
A handful of pearls : the Epistle of James / by Addison J. Eastman. — 1st ed. — Philadelphia : Westminster Press, c1978.
106 p. ; 19 cm.
Includes bibliographical references.
ISBN 0-664-24202-2
1. Bible. N.T. James—Criticism, interpretation, etc. 2. Christian life—1960- . I. Title.
BS2785.2.E17 227'.91'077 78-5797
78 MARC

Father Divine.
A treatise on overpopulation taken from interviews, sermons, and lectures by Father Divine. Philadelphia, New Day Pub. Co., 1967.
v, 41 p. 28 cm.
1. Christian life—1960- . 2. Population. I. Title.
BX7350.Z6F377 261.8'3 73-172243
73[r79]rev MARC

Ferguson, Ben.
The shaping of a man of faith / Ben E. Ferguson. — Wheaton, Ill. : Victor Books, c1979.
95 p. ; 18 cm.
ISBN 0-88207-516-0 : $1.75
1. Abraham, the patriarch. 2. Patriarchs (Bible)—Biography. 3. Bible. O.T.—Biography. 4. Christian life—1960- . I. Title.
[BS580.A3F47] 222'.11'0924 78-55178
00 MARC

Föller, Oskar, 1948-
Verbindliches Leben nicht nur im Mitarb.-Kreis! : Anstösse für d. Praxis / Oskar Föller. — 1. Aufl. — Neuhausen/Stuttgart : Hänssler, 1978.
120 p. ; 18 cm. — (Telos-Taschenbuch ; 212) GFR78-A27
ISBN 3-7751-0344-9 : DM4.80
1. Christian life—1960- . I. Title.
BV4503.F58 79-384253
79 MARC

CHRISITAN LIFE
—1960- (Continued)

Galloway, Dale E
How to feel like a somebody again / Dale Galloway. — Irvine, Calif. : Harvest House Publishers, c1978.
144 p. ; 21 cm.
ISBN 0-89081-130-X : $2.95
1. Christian life—1960- . 2. Self-acceptance. I. Title.
BV4501.2.G29 248'.4 78-55587
79 MARC

Getz, Gene A
A profile for a Christian life style : a study of Titus / Gene A. Getz. — Grand Rapids : Zondervan Pub. House, c1973.
200 p. ; 18 cm.
ISBN 0-310-25092-7 : $1.95
1. Bible. N.T. Titus—Criticism, interpretation, etc. 2. Christian life—1960- . I. Title.
BS2755.2.G47 248'.4 78-5814
78 MARC

Gregory, Jacob.
(Kĕ hawatam hawatats'ealin hawitenakan apahovowt'ean, bayts' ov ē hawatats'ealĕ?)
Կը հաւատամ հաւատացեալին յաւիտենական ապահովութեան, բայց ով է հաւատացեալը? / Ճէյքըպ Կրէկըրի. — Պէյրութ : ⌈ի դիմաց Հայ Աւետարանական Եղբայրներու Եկեղեցիներուն, ընդհ. քարտուղարուհի եւ գանձապահուհի Կ. Աճէմեան, 196–?⌉
471, 3 p. ; 21 cm.
Translated from English.
1. Apologetics—20th century. 2. Christian life—1960- I. Title.
BT1102.G74712 79-971141

Griffith, Harry C
Adventure in discipleship : a study book / by Harry C. Griffith. — Grand Rapids : Zondervan Pub. House, c1978.
141 p. ; 21 cm.
Includes bibliographical references.
ISBN 0-310-37501-0
1. Christian life—1960- . I. Title.
BV4501.2.G758 248'.4 78-23756
78⌈r79⌉rev MARC

Hadidian, Allen, 1950-
Successful discipling / by Allen Hadidian. — Chicago : Moody Press, c1979.
176 p. ; 22 cm.
Includes bibliographical references.
ISBN 0-8024-8408-5 : $3.95
1. Evangelistic work. 2. Christian life—1960- . I. Title.
BV3790.H23 269'.2 79-98
79 MARC

Hansel, Tim.
When I relax I feel guilty / Tim Hansel. — Elgin, Ill. : D. C. Cook Pub. Co., c1979.
150 p. : ill. ; 21 cm.
Includes bibliographical references.
ISBN 0-89191-137-5 : $3.95
1. Christian life—1960- . 2. Recreation—Moral and religious aspects. I. Title.
BV4501.2.H335 248'.4 78-73460
79 MARC

Harper, Michael, 1931-
Spiritual warfare. London, Hodder & Stoughton, 1970.
128 p. 21 cm. 21/- B70-08722
ISBN 0-340-12913-1
1. Devil. 2. Christian life—1960- . I. Title.
BT981.H35 1970 235'.4 70-513848
70⌈r79⌉rev MARC

⌈Harris, Clifton⌉
From medicine to miracles / Kathryn Kuhlman. — Minneapolis, Minn. : Bethany Fellowship, c1978.
96 p. : ill. ; 18 cm. — (Dimension books)
"The story of Clifton Harris, M.D., as told to a reporter of the Kathryn Kuhlman Foundation."
ISBN 0-87123-383-5 : $1.50
1. Osteoarthritis—Biography. 2. Faith-cure. 3. Christian life—1960- . 4. Harris, Clifton. 5. Physicians (General practice)—Louisiana—Pineville—Biography. I. Kuhlman, Kathryn. II. Title.
RC931.O67H37 615'.852'0924 78-67100
79 MARC

Hart, Archibald D
Feeling free / Archibald D. Hart. — Old Tappan, N.J. : Revell, c1979.
191 p. : ill. ; 22 cm.
Includes bibliographical references and index.
ISBN 0-8007-0973-X : $7.95
1. Emotions. 2. Christian life—1960- . I. Title.
BF561.H38 152.4 78-26474
78 MARC

Herngren, Leif, 1936-
Att vara människa / Leif Herngren. — Stockholm : Gummesson, 1978.
39, ⌈1⌉ p. ; 19 cm. S78
Includes bibliographical references.
ISBN 9170705380
1. Christian life—1960- . I. Title.
BV4501.2.H377 79-380515
79 MARC

Herron, Orley R
A Christian executive in a secular world / Orley R. Herron. — Nashville : Nelson, c1979.
143 p. ; 21 cm.
ISBN 0-8407-5151-6
1. Christian life—1960- . 2. Executives—Religious life. I. Title.
BV4596.E93H47 248'.88 79-4228
79 MARC

Hirschmann, Maria Anne.
I am but a child in Christ / Maria Anne Hirschmann. — Huntington Beach, Calif. : S.P.A.R.C. Pub. Co., c1977.
189 p. ; 21 cm. — (Hansi books)
ISBN 0-932878-00-8
1. Christian life—1960- . I. Title.
BV4501.2.H523 248'.4 77-89331
79 MARC

Hitt, Russell T
How Christians grow / Russell T. Hitt. — New York : Oxford University Press, 1979.
viii, 153 p. ; 22 cm.
Includes bibliographical references.
ISBN 0-19-502558-X
1. Christian life—1960- . I. Title.
BV4501.2.H524 248'.4 79-2261
79 MARC

Horban, Michael.
Hanging loose in an uptight world / Michael Horban. — Toronto : Full Gospel Pub. House ; Springfield, Mo. : Gospel Pub. House, c1978.
123 p. : ill. ; 18 cm. — (Radiant books) C***
ISBN 0-88243-733-X : $1.50
1. Christian life—1960- . I. Title.
BV4501.2.H59 248'.4 78-67444
79 MARC

Hurnard, Hannah.
Walking among the unseen / Hannah Hurnard. — Wheaton, Ill. : Tyndale House Publishers, c1977.
207 p. ; 18 cm.
ISBN 0-8423-7805-7 : $1.50
1. Christian life—1960- . I. Title.
BV4501.2.H86 248'.4 77-72435
00 MARC

Hyder, O Quentin, 1930-
Shape up / O. Quentin Hyder. — Old Tappan, N.J. : F. H. Revell Co., c1979.
158 p. : ill. ; 21 cm.
Bibliography: p. 157-158.
ISBN 0-8007-0975-6
1. Health. 2. Christian life—1960- . I. Title.
RA776.5.H93 613 78-11765
78 MARC

Jackson, Catherine.
The Christian's secret of a happy life for today : a paraphrase of Hannah Whitall Smith's classic / Catherine Jackson. — Old Tappan, N.J. : Revell, c1979.
323 p. ; 22 cm.
ISBN 0-8007-0976-4
1. Christian life—1960- . 2. Faith. I. Smith, Hannah Whitall, 1832-1911. The Christian's secret of a happy life. II. Title.
BV4501.2.J3125 248'.4 78-9896
78 MARC

Johnson, Keith.
Life's priorities / by Keith Johnson ; edited by Kandy Klein. — Carol Stream, Ill. : Zoé Publications, ⌈1979⌉
128 p. ; 22 cm.
ISBN 0-89841-000-2 : $2.95
1. Christian life—1960- . I. Klein, Kandy. II. Title.
BV4501.2.J55 248'.4 79-63739
00 MARC

Judd, Wayne, 1941-
How to wait for Jesus / Wayne Judd. — Nashville : Southern Pub. Association, c1979.
31 p. ; 14 cm.
ISBN 0-8127-0206-9
1. Christian life—1960- . I. Title.
⌈BV4501.2.J82⌉ 248'.4 78-13975
78 MARC

Kaiser, Richard.
The mending heart / ⌈Richard Kaiser⌉. — Irvine, Calif. : Harvest House Publishers, ⌈c1979⌉
125 p. ; 21 cm.
Includes bibliographical references.
ISBN 0-89081-201-2 : $3.95
1. Christian life—1960- . I. Title.
BV4501.2.K26 248'.86 79-84746
79 MARC

Keller, Weldon Phillip, 1920-
A gardener looks at the fruits of the spirit / W. Phillip Keller. — Waco, Tex. : Word Books, c1979.
187 p. ; 23 cm.
ISBN 0-8499-0144-8 : $7.95
1. Christian life—1960- . 2. Gifts, Spiritual. I. Title.
BV4501.2.K425 248'.4 79-63125
79 MARC

Kerr, Graham.
The love feast : how good, natural, wholesome food can create a warm and lasting Christian family / Graham Kerr. — New York : Simon and Schuster, c1978.
256 p. : ill. ; 24 cm.
Includes index.
ISBN 0-671-24052-8
1. Cookery. 2. Nutrition. 3. Christian life—1960- . I. Title.
TX652.K43 641.5'63 78-7221
78 MARC

Kiemel, Ann.
I love the word impossible / Ann Kiemel. — Wheaton, Ill. : Tyndale House Publishers, 1976.
155 p. : ill. ; 22 cm.
ISBN 0-8423-1575-6 : $5.95
1. Christian life—1960- . 2. Witness bearing (Christianity) 3. Kiemel, Ann. I. Title.
BV4501.2.K486 248'.5 75-42908
77⌈r79⌉rev MARC

Lange, Birgit Elisabeth, 1901-
Mosaisk dröm / Birgit Lange. — Lund ⌈Pedellg. 42 : Förf.⌉, 1977.
304 p., ⌈4⌉ leaves of plates : ill. ; 22 cm. S77-35
English or Swedish.
Includes bibliographical references.
1. Christian life—1960- . I. Title.
BV4501.2.L31843 78-397059
79 MARC

Leonard, Donna.
Lord, I'm listening / Donna Leonard. — Wheaton, Ill. : Victor Books, c1979.
132 p. : ill. ; 21 cm.
ISBN 0-88207-745-7 : $1.95
1. Christian life—1960- . I. Title.
BV4501.2.L447 248'.4 78-63637
79 MARC

Lighterness, Tony.
Living in the kingdom here and now / by Tony Lighterness. — Plainfield, N.J. : Logos International, c1979.
ix, 189 p. ; 21 cm.
ISBN 0-88270-357-9 : $3.50
1. Christian life—1960- . I. Title.
BV4501.2.L493 248'.4 79-63858
00 MARC

Ligon, Bill.
Discipleship : the Jesus view (an alternative to extremism) / by Bill Ligon, with Robert Paul Lamb. — Plainfield, N.J. : Logos International, c1979.
xix, 250 p. ; 20 cm.
ISBN 0-88270-348-X : $3.95
1. Christian life—1960- . I. Lamb, Robert Paul, joint author. II. Title.
BV4501.2.L495 248'.4 79-87506
79 MARC

(Ling hun ti chi tung)
靈魂的悸動 / 章陵等著. — 臺北 : 宇宙光出版社, 1978.
269 p. : ill. ; 20 cm. — (宇宙光真實故事集 ; 3)
Series romanized: Yü chou kuang chen shih ku shih chi.
NT$48.00

1. Christian life—1960- . I. Chang, Ling.
BV4510.2.L56 79-835871

Littauer, Florence, 1928-
Blow away the black clouds / Florence Littauer. — Scottsdale, Ariz. : Good Life Productions, c1979.
136 p. ; 21 cm.
Bibliography: p. 135-136.
$3.95
1. Christian life—1960- . 2. Depression, Mental. I. Title.
BV4501.2.L55 248'.86 79-50380
79 MARC

Little, Paul E
How to give away your life / by Paul E. Little. — Santa Ana, Calif. : Vision House, c1978.
139 p. ; 22 cm.
ISBN 0-88449-063-7 : $3.95
1. Christian life—1960- . I. Title.
⌈BV4501.2.L555⌉ 248'.4 78-71261
00 MARC

Lockyer, Herbert.
A cure for troubled hearts / ⌈Herbert Lockyer⌉. — Irvine, Calif. : Harvest House Publishers, ⌈c1979⌉
125 p. ; 21 cm.
$2.95
1. Christian life—1960- . I. Title.
⌈BV4501.2.L63⌉ 248'.4 79-87770
79 MARC

CHRISTIAN LIFE
—1960- (Continued)

Long, James, 1949-
Life, Jesus-style : scaling the sermon on the mount / by James Long. — Wheaton, Ill. : Victor Books, c1978.
160 p. : ill. ; 18 cm. — (SonPower youth publication)
ISBN 0-88207-575-6 : $1.95
1. Sermon on the mount. 2. Christian life—1960- . I. Title.
BT380.2.L6 248'.48 78-58692
79 MARC

Lundstrom, Lowell.
How to enjoy supernatural prosperity / Lowell Lundstrom. — Irvine, Calif. : Harvest House Publishers, c1979.
112 p. ; 21 cm.
ISBN 0-89081-204-7 : $2.95
1. Success. 2. Christian life—1960- . I. Title.
[BJ1611.2.L87] 248'.4 79-87771
79 MARC

Lutzer, Erwin W
How to say no to a stubborn habit, even when you feel like saying yes / Erwin W. Lutzer. — Wheaton, Ill. : Victor Books, c1979.
143 p. ; 21 cm.
ISBN 0-88207-787-2 : $2.95
1. Christian life—1960- . I. Title.
BV4501.2.L88 248'.4 79-64039
00 MARC

Mackay, Malcolm G 1919-
More than coincidence / Malcolm G. Mackay. — Edinburgh : Saint Andrew Press, 1979.
242 p. ; 23 cm. GB***
Includes bibliographical references.
ISBN 0-7152-0416-5 : £6.95
1. Christian life—1960- . 2. Civilization, Modern—1950- . I. Title.
BV4501.2.M24145 261 79-315306
79 MARC

Mains, Karen Burton.
Karen! Karen! : One woman's response to the whispers of God / by Karen Burton Mains. — Wheaton, Ill. : Tyndale House Publishers, c1979.
191 p. ; 22 cm.
ISBN 0-8423-2025-3 : $6.95
1. Christian life—1960- . 2. Mains, Karen Burton. I. Title.
BV4501.2.M327 248'.2 78-64675
79 MARC

Mains, Karen Burton.
The key to a loving heart / Karen Burton Mains. — Elgin, Ill. : D.C. Cook Pub. Co., c1979.
168 p. ; 22 cm.
ISBN 0-89191-184-7 : $6.95
1. Christian life—1960- . I. Title.
BV4501.2.M328 248'.4 79-51746
00 MARC

Marshall, Catherine Wood, 1914-
The helper / Catherine Marshall. — Boston : G. K. Hall, 1979, c1978.
xxv, 318 p. ; 24 cm.
"Published in large Print."
ISBN 0-8161-6695-1
1. Holy Spirit—Meditations. 2. Christian life—1960- . 3. Large type books. I. Title.
[BT122.M35 1979] 231'.3 79-10016
79 MARC

McDill, Wayne.
The power to become / Wayne McDill. — Old Tappan, N.J. : F. H. Revell Co., c1979.
128 p. ; 22 cm.
ISBN 0-8007-1053-3
1. Christian life—1960- . 2. Self-actualization. I. Title.
BV4501.2.M224 248'.4 79-13034
79 MARC

McKee, Bill.
Go for the gold : how to win the crowns God gives at the end of the Christian race / Bill McKee. — Scottsdale, Ariz. : Good life Productions, c1979.
169 p. ; 21 cm.
$3.95
1. Christian life—1960- . I. Title.
[BV4501.2.M2417] 248'.4 79-50378
79 MARC

Mīnā, Mūsá Wahbah.
(Shajarat al-dam)
شجرة الدم / بقلم موسى وهبه مينا. — [القاهرة] : التربية الكنسية بكنيسة الشهيد العظيم مار جرجس، خمارويةـ شبرا ـ مصر، [1974?]
55 p. : ill. ; 17 cm.
1. Christian life—1960- I. Title.
BV4515.2.M56 77-960059

Moore, Marvin, 1937-
How to handle your imagination / Marvin Moore. — Nashville : Southern Pub. Association, c1979.
46 p. ; 19 cm. — (Better living series)
ISBN 0-8127-0195-X
1. Success. 2. Imagination. 3. Christian life—1960- . I. Title. II. Series.
[BF637.S8M62] 158 78-13660
78 MARC

Mori, Arimasa, 1911-1976.
(Ikani ikiru ka)
いかに生きるか / 森有正. — 東京 : 講談社, 昭和51 [1976]
204 p. : 18 cm. — (講談社現代新書) Ja 76-21528
Series romanized: Kōdansha gendai shinsho.
¥370
1. Christian life—1960- I. Title.
BV4509.J3M67 77-802174

Moster, Mary Beth.
Living with cancer / Mary Beth Moster. — Chicago : Moody Press, c1979.
179 p. ; 22 cm.
Includes bibliographical references.
ISBN 0-8024-4947-6 : $5.95
1. Cancer—Psychological aspects. 2. Cancer—Social aspects. 3. Christian life—1960- . I. Title.
RC263.M65 362.1'9'6994 78-23665
78 MARC

Muggeridge, Malcolm, 1903-
A twentieth century testimony / Malcolm Muggeridge ; introd. by Thomas Howard. — Nashville : T. Nelson, c1978.
[77] p. : ill. ; 23 x 29 cm.
Includes index.
ISBN 0-8407-5143-5 : $12.50
1. Christian life—1960- . 2. Muggeridge, Malcolm, 1903- . I. Title.
BV4501.2.M74 248'.4 78-15925
78 MARC

Murchison, Anne Ferrell.
Milk for babes : using the Bible to find the answers to life's questions / Anne Ferrell Murchison. — Waco, Tex. : Word Books, c1979.
255 p. ; 22 cm.
ISBN 0-8499-2868-0 : $6.95
1. Christian life—1960- . 2. Murchison, Anne Ferrell. 3. Theology, Doctrinal—Popular works. I. Title.
BV4501.2.M776 230 78-63745
00 MARC

Nordtvedt, Matilda.
Living beyond depression / by Matilda Nordtvedt. — Minneapolis, Minn. : Bethany Fellowship, c1978.
128 p. ; 18 cm. — (Dimension books)
Includes bibliographical references.
ISBN 0-87123-339-8 : $1.95
1. Nordtvedt, Matilda. 2. Christian biography—United States. 3. Christian life—1960- . 4. Depression, Mental—Biography. I. Title.
BR1725.N54A35 248'.86 78-58082
78 MARC

Olsson, Karl A
When the road bends : a book about the pain and joy of passage / Karl A. Olsson. — Minneapolis : Augsburg Pub. House, c1979.
234 p. ; 20 cm.
Includes bibliographical references.
ISBN 0-8066-1695-4
1. Rites and ceremonies. 2. Christian life—1960- . I. Title.
BV178.O47 248'.4 78-66948
79 MARC

Ortlund, Anne.
Love me with tough love : disciplines for living together in the Body of Christ / Anne Ortlund. — Waco, Tex. : Word Books, c1979.
185 p. ; 23 cm.
Includes bibliographical references.
ISBN 0-8499-0145-6 : $7.95
1. Christian life—1960- . I. Title.
BV4501.2.O727 248'.4 79-63937
79 MARC

Osgood, Don.
Pressure points : the Christian's response to stress / Don Osgood. — 1st ed. — Chappaqua, N.Y. : Christian Herald Books, c1978.
219 p. : ill. ; 21 cm.
Bibliography: p. 219.
ISBN 0-915684-40-3 : $6.95
1. Christian life—1960- . 2. Stress (Psychology) I. Title.
BV4501.2.O76 248'.4 78-56975
78 MARC

Oyarzabal, Antonio.
Gudari : soldaduz-karako argibideak / Antonio Oyarzabal. — Arrate-Eibar : [s.n.], 1974.
144 p. ; 18 cm. Sp74-Dec
ISBN 8440077998
1. Christian life—1960- . I. Title.
BV4509.B3O9 78-383808
78 MARC

Perry, Wilma I
Awake the sleeping giant within you / Wilma I. Perry. — Old Tappan, N.J. : F. H. Revell Co., c1979.
142 p. ; 22 cm.
ISBN 0-8007-1035-5
1. Christian life—1960- . 2. Creative ability. I. Title.
BV4501.2.P424 248'.4 78-27690
78 MARC

Pfürtner, Stephan H 1922-
Christ sein, Mensch sein [von] Stephan H. Pfürtner. [Würzburg] Echter Verlag [c1972]
62, [1] p. 23 cm. GDB***
Bibliography: p. 61-[63]
ISBN 3-429-00230-3
1. Christian life—1960- . I. Title.
BV4503.P47 72-330837
79[r79]rev MARC

Phillips, Mike, 1946-
A survival guide for tough times / Mike Phillips. — Minneapolis : Bethany Fellowship, 1979.
176 p. ; 21 cm.
Includes bibliographical references.
ISBN 0-87123-498-X : $2.95
1. Christian life—1960- . 2. Family—Religious life. 3. Human ecology—Moral and religious aspects. 4. End of the world. I. Title.
BV4501.2.P53 248'.4 79-4261
79 MARC

Pierre, Jean.
Les mecs de la boîte / Jean Pierre. — Paris : Les Éditions ouvrières, c1979.
115 p. ; 18 cm. — (Collection Jeunes travailleurs) F***
ISBN 2-7082-1991-X : 23.00F
1. Pierre, Jean. 2. Christian biography—France. 3. Christian life—1960- . 4. Trade-unions—France. I. Title.
BR1725.P515A35 248'.2'0924 79-370456
79 MARC

Poland, Larry W
Rise to conquer : a call for Christian action / Larry W. Poland. — 1st ed. — Chappaqua, N.Y. : Christian Herald Books, c1979.
131 p. ; 21 cm.
ISBN 0-915684-48-9 : $6.95
1. Christian life—1960- . 2. Christianity—20th century. 3. Kingdom of God. I. Title.
BV4501.2.P55547 248'.4 78-64843
78 MARC

Poland, Larry W
Spirit power : all you need when you need it! / Larry W. Poland. — San Bernardino, Calif. : Campus Crusade for Christ International, c1978.
125 p. : ill. ; 21 cm.
ISBN 0-918956-39-0 : $2.95
1. Christian life—1960- . 2. Holy Spirit—Miscellanea. I. Title.
BV4501.2.P5555 248'.4 77-91763
78 MARC

Potterbaum, Charlene.
His eye is on the sparrow so ... this is really for the birds / Charlene Potterbaum. — Plainfield, N.J. : Logos International, c1979.
ix, 190 p. ; 21 cm.
On spine: This is really for the birds.
ISBN 0-88270-354-4 : $3.50
1. Potterbaum, Charlene. 2. Christian biography—United States. 3. Christian life—1960- . I. Title. II. Title: This is really for the birds.
BR1725.P63A34 248'.2'0924 79-83788
79 MARC

Powell, Terry.
Nobody's perfect / by Terry Powell. — Wheaton, Ill. : Victor Books, c1979.
116 p. : ill. ; 18 cm. — (SonPower youth publication)
ISBN 0-88207-577-2 : $1.75
1. Bible. N.T.—Biography. 2. Christian life—1960- . I. Title.
[BS2430.P68] 248'.4 78-65556
79 MARC

Pulkingham, Betty.
Little things in the hands of a big God / Betty Pulkingham. — Waco, Tex. : Word Books, c1977.
142 p. : ill. ; 21 cm.
ISBN 0-8499-2855-9 : $4.95
1. Pulkingham, Betty. 2. Christian biography—United States. 3. Christian life—1960- . I. Title.
BR1725.P84A34 283'.092'4 79-63930
00 MARC

CHRISTIAN LIFE
—1960- (Continued)

Rommel, Kurt.
Ich möchte heute fromm sein : Erfahrungen u. Erlebnisse, Erkenntnisse u. Ergebnisse / Kurt Rommel. — Stuttgart : Steinkopf, 1977.
144 p. ; 21 cm. GFR78-A
ISBN (invalid) 3-7984-0335-8 : DM14.80
1. Christian life—1960- . 2. Rommel, Kurt. I. Title.
BV4501.2.R66 79-372638
79 MARC

Russell, Letty M
The fuuture of partnership / by Letty M. Russell. — 1st ed. — Philadelphia : Westminster Press, c1979.
200 p. ; 20 cm.
Includes bibliographical references.
ISBN 0-664-24240-5
1. Christian life—1960- . 2. Man (Christian theology) 3. God. 4. Freedom (Theology) I. Title.
BV4501.2.R85 261.8'34 78-20805
78 MARC

Sall, Millard J
The emotions of God's people : helping those who are hurting / Millard J. Sall. — Grand Rapids : Zondervan Pub. House, c1978.
155 p. ; 21 cm.
ISBN 0-310-32441-6
1. Bible—Psychology. 2. Christianity—Psychology. 3. Psychology, Religious. 4. Psychology—Popular works. 5. Christian life—1960- . I. Title.
BS645.S24 155.2 78-15717
78 MARC

Salo, Jussi, 1928-
Viikon kahdeksas päivä / Jussi Salo. — Lapua : Herättäjäyhdistys, 1976.
139 p. ; 19 cm. Fi
ISBN 9519013474
1. Salo, Jussi, 1928- . 2. Christian biography—Finland. 3. Christian life—1960- . I. Title.
BR1725.S244A38 79-364820
*79 MARC

Scanzoni, John H 1935-
Love and negotiate : creative conflict in marriage / John Scanzoni. — Waco, Tex. : Word Books, c1979.
148 p. ; 21 cm.
Includes bibliographical references.
ISBN 0-8499-0100-6 : $6.95
1. Marriage—United States. 2. Family—United States. 3. Christian life—1960- . 4. Interpersonal relations. I. Title.
HQ734.S3758 301.42 78-65816
79 MARC

Schneider, Hans J 1935-
Masters of legalized confusion and their puppets / by Hans J. Schneider. — Ashland, OR : World Wide Pub. Corp., 1979.
iv, 28 p. : ill. ; 22 cm.
Includes index.
ISBN 0-930294-11-4 : $1.45
1. Christianity—20th century. 2. Christian life—1960- . I. Title.
[BR124.S36 1979] 209'.04 78-12609
78 MARC

Sheetz, Ann Kindig.
Born again, but still wet behind the ears / Ann Kindig Sheets. — 1st ed. — Chappaqua, N.Y. : Christian Herald Books, c1979.
160 p. ; 21 cm.
ISBN 0-915684-43-8 : $6.95
1. Sheetz, Ann Kindig. 2. Christian biography—United States. 3. Christian life—1960- . I. Title.
BR1725.S454A33 248'.2 78-64839
79 MARC

Shelly, Judith Allen.
Caring in crisis : Bible studies for helping people / Judith Allen Shelly. — Downers Grove, Ill. : InterVarsity Press, c1979.
94 p. ; 21 cm.
ISBN 0-87784-563-8 : $2.95
1. Christian life—1960- . 2. Sympathy. 3. Bible—Study—Textbooks. I. Title.
[BV4501.2.S438] 248.4 78-13878
79 MARC

Shostrom, Everett L 1921-
Healing love : how God works within the personality / Everett L. Shostrom & Dan Montogomery. — Nashville : Abingdon, c1978.
141 p. : ill. ; 23 cm.
Bibliography: p. 135-136.
Includes index.
ISBN 0-687-16739-6
1. Christian life—1960- . 2. Love. 3. Christianity—Psychology. I. Montgomery, Dan, 1946- joint author. II. Title.
BV4501.2.S452 248'.4 78-8294
78 MARC

Skoglund, Elizabeth.
Loving begins with me / Elizabeth Skoglund. — 1st ed. — San Francisco : Harper & Row, c1979.
x, 98 p. ; 21 cm.
Includes bibliographical references.
ISBN 0-06-067393-1
1. Christian life—1960- . 2. Self-respect. I. Title.
BV4501.2.S4757 1979 248'.4 78-3364
79 MARC

Smith, Jerry L
The searcher : how to satisfy a searching soul / by Jerry L. Smith ; introd. by W. A. Criswell. — Melbourne, Ark. : Christo-Centric Ministries, c1978.
79 p. : ill. ; 22 cm.
Includes bibliographical references.
ISBN 0-9602136-0-0
1. Christian life—1960- . I. Title.
BV4501.2.S53415 248'.4 78-65150
00 MARC

Smith, Paul B
The church on the brink / Paul B. Smith. — Wheaton, Ill. : Tyndale House Publishers, c1977.
223 p. ; 21 cm.
ISBN 0-8423-0278-6 : $3.95
1. Pentecostalism. 2. Love (Theology) 3. Revivals. 4. Christian life—1960- . I. Title.
BR1644.S55 269 78-80733
79 MARC

Staton, Knofel.
Check your lifestyle : putting proverbs into practice / by Knofel Staton. — Cincinnati, : New Life Books, c1979.
112 p. ; 18 cm.
ISBN 0-87239-233-3
1. Christian life—1960- . 2. Bible. O.T. Proverbs—Criticism, interpretation, etc. I. Title.
[BV4501.2.S734] 248'.4 78-66436
79[r79]rev MARC

Stearns, Bill.
From rock bottom to mountaintop : making sense out of your ups and downs / Bill Stearns. — Wheaton, Ill. : Victor Books, c1979.
160 p. : ill. ; 18 cm. — (SonPower youth publication)
ISBN 0-88207-580-2 : $2.25
1. Christian life—1960- . 2. Emotions. I. Title.
BV4501.2.S737 248'.4 79-65040
00 MARC

Ten Boom, Corrie.
Don't wrestle, just nestle / Corrie ten Boom. — Old Tappan, N.J. : Revell, c1978.
95 p. ; 20 cm. — (Her Jesus is victor)
ISBN 0-8007-0848-2
1. Christian life—1960- . I. Title. II. Series.
BV4501.2.T38 248'.4 78-14214
78 MARC

Ten Boom, Corrie.
He sets the captive free / Corrie Ten Boom. — Boston : G. K. Hall, 1978, c1977.
viii, 89 p. ; 25 cm. — (Her Jesus is victor)
Large print ed.
ISBN 0-8161-6627-7
1. Ten Boom, Corrie. 2. Christian biography—Netherlands. 3. World War, 1939-1945—Personal narratives, Dutch. 4. Ravensbrück (Concentration camp) 5. Christian life—1960- . 6. Large type books. I. Title. II. Series.
[BR1725.T35A33 1978] 269'.2'0924 78-10616
78 MARC

Tenney, Merrill Chapin, 1904-
Roads : a Christian must travel : fresh insights into the principles of Christian experience / Merrill C. Tenney. — Wheaton, Ill. : Tyndale House Publishers, c1979.
108 p. ; 21 cm.
ISBN 0-8423-5675-4 : $2.95
1. Christian life—1960- . 2. Bible. N.T. Luke and Acts—Criticism, interpretation, etc. I. Title.
BV4501.2.T422 248'.4 78-64672
79 MARC

Thompson, James, 1942-
Our life together / James Thompson. — Austin, Tex. : SPC, c1977, 1978 printing.
144 p. ; 19 cm. — (Journey books)
ISBN 0-8344-0095-2 : $1.95
1. Fellowship. 2. Christian life—1960- . I. Title.
BV4517.5.T48 262'.001 77-79338
79 MARC

Timmons, Tim.
The ultimate lifestyle / [Tim Timmons]. — Santa Ana, Calif. : Vision House Publishers, c1977.
xii, 212 p., [1] leaf of plates : ill. ; 21 cm.
Bibliography: p. [211]-212.
ISBN 0-88449-027-0 : $3.98
1. Christian life—1960- . I. Title.
BV4501.2.T54 248'.4 77-20594
78 MARC

'Uwayṭ, Mīshāl.
(Milḥ al-arḍ)
ملح الأرض / [ميشال عويط]. — [s. l. : s. n., 1977?] ([جونية : مطابع الكريم الحديثة])
205 p. ; 20 cm.
1. Christian life—1960- I. Title.
BV4501.2.U9 79-970787

Vaughn, Ruth.
To be a graduate / by Ruth Vaughn. — Nashville : T. Nelson, c1979.
95 p. ; 21 cm.
ISBN 0-8407-4073-5 : $4.95
1. Christian life—1960- . 2. Vaughn, Ruth. I. Title.
BV4501.2.V37 248'.2 79-1404
79 MARC

Wallace, Joanne.
The image of loveliness / Joanne Wallace ; ill. by Jan Noble. — Old Tappan, N.J. : F. H. Revell Co., c1978.
158 p. : ill. ; 22 cm.
ISBN 0-8007-0963-2
1. Beauty, Personal. 2. Charm. 3. Etiquette for women. 4. Christian life—1960- . I. Title.
RA778.W215 646.7 78-15887
78 MARC

Warner, Paul L
Feeling good about feeling bad : a look at irrational guilt, its causes and solutions / Paul L. Warner, with Sam Crabtree, Jr. — Waco, Tex. : Word Books, c1979.
87 p. : ill. ; 22 cm.
ISBN 0-8499-0104-9 : $4.95
1. Christian life—1960- . 2. Guilt. 3. Mental health. I. Title.
BV4501.2.W375 248'.4 78-59431
79 MARC

White, John Wesley.
The man from Krypton : the gospel according to Superman / John Wesley White ; foreword by Billy Graham. — Minneapolis : Bethany Fellowship, c1978.
175 p. ; 18 cm.
ISBN 0-87123-384-3 : $2.25
1. Apologetics—20th century. 2. Christian life—1960- . 3. Superman (Comic strip)—Moral and religious aspects. I. Title.
BT1102.W477 248'.4 78-73455
79 MARC

Whitehead, Evelyn Eaton.
Christian life patterns : the psychological challenges and religious invitations of adult life / Evelyn Eaton Whitehead and James D. Whitehead. — 1st ed. — Garden City, N.Y. : Doubleday, 1979.
216 p. ; 22 cm.
Bibliography: p. [205]-209.
Includes index.
ISBN 0-385-15130-6 : $8.95
1. Christian life—1960- . 2. Adulthood. I. Whitehead, James D., joint author. II. Title.
BV4501.2.W47 248'.84 78-22543
79 MARC

Wiersbe, Warren W
5 secrets of living / by Warren W. Wiersbe. — Wheaton, Ill. : Tyndale House Publishers, 1978, c1977.
88 p. : ill. ; 18 cm.
ISBN 0-8423-0870-9 : $1.95
1. Christian life—1960- . I. Title.
BV4501.2.W5183 248'.4 77-83549
79 MARC

Wiersbe, Warren W
Be mature : an expository study of the Epistle of James / Warren W. Wiersbe. — Wheaton, Ill. : Victor Books, c1978.
176 p. ; 18 cm.
Includes text of the Epistle of James.
ISBN 0-88207-771-6 : $2.25
1. Bible. N.T. James—Commentaries. 2. Christian life—1960- . I. Bible. N.T. James. English. Authorized. 1978. II. Title.
BS2785.3.W5 227'.91'06 78-52558
79 MARC

Williams, Laura Wineburg, 1928-
Our runaway / L. Weinberg Williams. — Valley Forge, PA : Judson Press, c1978.
64 p. ; 22 cm.
ISBN 0-8170-0813-6 : $2.95
1. Williams, Mark. 2. Juvenile delinquents—Texas—Biography. 3. Runaway youth—Texas—Biography. 4. Christian life—1960- . I. Title.
HV9105.T44W54 362.8'2'0926 78-15132
78 MARC

Wise, Karen.
God knows I won't be fat again / Karen Wise. — Nashville : T. Nelson, c1978.
159 p., [4] leaves of plates : ill. ; 21 cm.
ISBN 0-8407-5665-8
1. Obesity. 2. Reducing. 3. Christian life—1960- . I. Title.
RC628.W53 613.2'5 78-10517
78 MARC

Wood, George O
You can't beat the beatitudes / George O. Wood & William J. Krutza. — Springfield, Mo. : Gospel Pub. House, c1978.
88 p. ; 18 cm. — (Radiant books)
ISBN 0-88243-719-4 : $1.25
1. Beatitudes. 2. Christian life—1960- . I. Krutza, William J., joint author. II. Title.
[BT382.W66] 226'.93'077 78-58721
79 MARC

CHRISTIAN LIFE (Continued)

—1960—ADDRESSES, ESSAYS, LECTURES

Berti Arnoaldi-Veli, Francesco.
Coi miei compagni io devo restare / Francesco Berti Arnoaldi ; prefazione di Ferruccio Parri. — 1. ed. — Venezia ; Padova : Marsilio, 1974.
113 p. ; 22 cm. — (Testimonianze ; n. 5) It76-Feb
Includes bibliographical references.
L2500
1. Christian life—1960— —Addresses, essays, lectures. 2. Christianity and politics—Addresses, essays, lectures. 3. Radicalism—Addresses, essays, lectures. I. Title.
BV4501.2.B42 79-355335
*79 MARC

Chang, Hung-ch'ing.
(Ling kuang tien ti)
靈光點滴 / 張宏卿著. — [台北?] : 張宏卿, 1977-
v. ; 20 cm.
NT$30.00 (v. 1)
1. Christian life—1960- —Addresses, essays, lectures. I. Title.
BV4510.2.C48 79-839523

(Hsin ling chih lü)
心靈之旅. — 臺北 : 校園團契出版社, 1978.
3, 110 p. ; 19 cm. — (校園文選 ; 9)
Title in English in colophon: Journey of the soul.
Series romanized: Hsiao yüan wen hsüan.
1. Christian life—1960- —Addresses, essays, lectures.
BV4510.2.H74 79-836445

Ich habe einen Traum : Visionen u. Wirklichkeiten / hrsg. von Meinold Krauss. — 1. Aufl., (1.-8. Tsd.). — Stuttgart ; Berlin : Kreuz-Verlag, 1978.
221 p. ; 21 cm. GFR78-A
Includes bibliographical references.
ISBN 3-7831-0544-7 : DM19.80
1. Christianity—20th century—Addresses, essays, lectures. 2. Christian life —1960- —Addresses, essays, lectures. I. Krauss, Meinold, 1944-
BR123.I25 78-381925
78 MARC

(Wo chao tao le)
我找到了 / 韓偉等著. — 台北 : 中國信徒佈道會, 民國66 [1977] 民國67 [1978] printing.
131, v, 54, [4] p. ; 20 cm.
Includes bibliographical references.
1. Christian life—1960- —Addresses, essays, lectures. I. Han, Wei, 1928-
BV4510.2.W6 79-840264

—1960- —COLLECTED WORKS

Edvardsen, Aril, 1938-
Fram til urkristant menighetsliv / Aril Edvardsen. — Kvinesdal : Logos, c197
v. ; 18 cm. N77-Apr.
CONTENTS:
—bd. 2. Åndens frukt-kjærligheten.
ISBN 8270670464 (v. 2) : kr15.00 (v. 2)
1. Christian life—1960- —Collected works. I. Title.
BV4495.E38 79-347462
79 MARC

—1960- —MISCELLANEA

Blaiklock, E M
The answer's in the Bible : one hundred and one answers to familiar questions / by E. M. Blaiklock. — London : Hodder and Stoughton, 1978.
186 p. ; 18 cm. GB***
ISBN 0-340-23260-9 : £1.00
1. Theology—Miscellanea. 2. Christian life—1960- —Miscellanea. I. Title.
BR96.B58 230 79-302347
79 MARC

—ADDRESSES, ESSAYS, LECTURES

Chi, Wŏn-sang.
(Mal hanŭn kaltae)
말하는갈대 : 지원상수필집 / [저자池元祥]. — 서울 : 컨콜디아사, 1977.
229 p. ; 19 cm.
On verso of t. p.: Speaking reed.
W700
1. Christian life—Addresses, essays, lectures. 2. Christianity—Korea—Addresses, essays, lectures. I. Title. II. Title: Speaking reed.
BV4509.K6C45 79-825811

Eguchi, Shin'ichi, 1914-
(Chi no shio no hako)
地の塩の箱 : ある幸福論 / 江口榛一. — 東京 : 新潮社, 昭和49 [1974]
251 p. ; 20 cm. Ja 74-18399
¥780
1. Christian giving—Addresses, essays, lectures. 2. Christian life—Addresses, essays, lectures. 3. Eguchi, Shin'ichi, 1914- —Addresses, essays, lectures. I. Title.
BV772.E42 78-812453

Fujibayashi, Ekizō, 1907-
(Ichihōritsuka no seikatsu to shinkō)
一法律家の生活と信仰 / 藤林益三著. — 東京 : 東京布井出版, 1978.
ix, 289 p. ; 20 cm. Ja 78-20118
¥2000
1. Fujibayashi, Ekizō, 1907- —Addresses, essays, lectures. 2. Lawyers—Japan—Biography—Addresses, essays, lectures. 3. Christian life—Addresses, essays, lectures. I. Title.
79-808459

Hong, Harold S 1911-
(Millim sok e t'ŭin kil)
密林속에트인길 : 知性人을위한기독교엣세이 / 洪顯卨著. — 서울 : 혜선문화사, 1976.
129 p. ; 22 cm.
Essays.
₩500
1. Christian life—Addresses, essays, lectures. I. Title.
BV4509.K6H65 78-826302

Trueblood, David Elton, 1900-
The encourager : periodical essays / by D. Elton Trueblood. — Nashville : Broadman Press, c1978.
144 p. ; 22 cm.
ISBN 0-8054-1950-0
1. Christian life—Addresses, essays, lectures. I. Title.
BV4501.2.T75 248'.4 78-62527
78 MARC

Yancey, Philip.
Unhappy secrets of the Christian life / Philip Yancey & Tim Stafford. — Grand Rapids : Zondervan Pub. House, c1979.
157 p. : ill. ; 21 cm.
ISBN 0-310-35420-X. ISBN 0-310-35421-8 pbk.
1. Christian life—Addresses, essays, lectures. I. Stafford, Tim, joint author. II. Title.
BV4501.2.Y32 248'.4 78-26460
78 MARC

—ANECDOTE, FACETIAE, SATIRE, ETC.

De Rosa, Peter.
Prayers for pagans and hypocrites / by Peter De Rosa ; ill. by Haro. — 1st U.S. ed. — New York : Morrow, 1979, c1977.
128 p. : ill. ; 22 cm.
ISBN 0-688-03349-7
1. Christian life—Anecdote, facetiae, satire, etc. I. Title.
BV4501.2.D48 1979 248 78-27802
78 MARC

—ANGLICAN AUTHORS

Holloway, Richard.
A new heaven / by Richard Holloway. — 1st American ed. — Grand Rapids : Eerdmans, 1979, c1978.
125 p. ; 21 cm.
ISBN 0-8028-1811-0 : $2.95
1. Jesus Christ—Crucifixion—Meditations. 2. Christian life—Anglican authors. I. Title.
BT450.H67 1979 232 79-118505
79 MARC

Reid, Gavin.
Living the new life / Gavin Reid. — Nashville : Abingdon, 1979, c1977.
128 p. ; 20 cm.
ISBN 0-687-22370-9
1. Theology, Doctrinal—Introductions. 2. Christian life—Anglican authors. I. Title.
BT65.R43 1979 201'.1 78-11980
78 MARC

Tucker, Graham.
It's your life : create a Christian lifestyle / Graham Tucker. — Toronto : Anglican Book Centre, c1977.
176 p. : ill. ; 22 cm. C78-7194-8
Includes bibliographical rferences.
ISBN 0-919030-22-X : $4.95
1. Christian life—Anglican authors. I. Title.
BV4501.2.T78 248'.48'3 79-315836
79 MARC

—ANGLICAN AUTHORS —ADDRESSES, ESSAYS, LECTURES

Packer, James Innell.
For man's sake! / J. I. Packer. — Exeter [Eng.] : Paternoster Press, c1978.
64 p. ; 18 cm. — (Paternoster punchline ; no. 9) GB***
Bibliography: p. 64.
ISBN 0-85364-217-6 : £0.70
1. Christian life—Anglican authors—Addresses, essays, lectures. 2. Secularism—Great Britain—Addresses, essays, lectures. I. Title.
BV4510.2.P3 261.2 78-314897
78 MARC

—ANGLICAN AUTHORS—COLLECTED WORKS

Lewis, Clive Staples, 1898-1963.
The joyful Christian : 127 readings from C. S. Lewis. — Boston : G. K. Hall, 1978, c1977.
xvii, 421 p. ; 24 cm.
Large print ed.
Bibliography: p. 417-421.
ISBN 0-8161-6634-X
1. Theology—Collected works—20th century. 2. Church of England—Collected works. 3. Christian life—Anglican authors—Collected works. 4. Large type books. I. Title.
[BX5037.L4 1978] 230'.092'4 78-10549
78 MARC

—ASSEMBLIES OF GOD AUTHORS

Strader, Karl.
Mountain-moving motivation : how to experience the power of God! / By Karl Strader, with Stephen Strang. — Orlando, Fla. : Joy Books, c1978.
127 p. : ill. ; 21 cm.
ISBN 0-933102-00-3
1. Christian life—Assemblies of God authors. I. Strang, Stephen, joint author. II. Title.
BV4501.2.S797 248'.48'99 78-61497
79 MARC

—BAPTIST AUTHORS

Blevins, William L
Care and maintenance for the Christain life / William L. Blevins. — Waco, Texas : Word Books, c1979.
116 p. ; 22 cm.
Includes bibliographical references.
ISBN 0-8499-0139-1 : $6.95
1. Christian life—Baptist authors. I. Title.
BV4501.2.B557 248'.4 78-65818
00 MARC

Burchett, Harold Ewing.
People helping people / by Harold Ewing Burchett. — Chicago : Moody Press, c1979.
127 p. (p. 123-127 blank for "Notes") : ill. ; 22 cm.
ISBN 0-8024-6457-2 : $6.95
1. Peer counseling in the church. 2. Christian life—Baptist authors. I. Title.
BV4409.B87 248'.07 78-21546
78 MARC

Carter, Evelyn.
No ground / Evelyn Carter with Leona Choy. — Plainfield, N.J. : Logos International, c1978.
xi, 200 p. ; 21 cm.
ISBN 0-88270-291-2 : $6.95
1. Christian life—Baptist authors. I. Choy, Leona, joint author. II. Title.
BV4501.2.C333 248'.48'61 78-73008
79 MARC

CHRISTIAN LIFE
—BAPTIST AUTHORS (Continued)

Flynn, Leslie B
God's will : you can know it / Leslie and Bernice Flynn. — Wheaton, Ill. : Victor Books, c1979.
144 p. ; 18 cm.
ISBN 0-88207-779-1 : 2.25
1. God—Will. 2. Christian life—Baptist authors. I. Flynn, Bernice, joint author. II. Title.
[BV4501.2.F578] 248'.4 78-64361
79 MARC

Lavender, John Allan.
Hang tough in a hostile world : based on the second half of Hebrews / John Allan Lavender. — Denver : Accent Books, c1978.
192 p. : ill. ; 21 cm.
Bibliography: p. 191-192.
ISBN 0-916406-94-6 : $3.95
1. Bible. N.T. Hebrews VI-XIII—Criticism, interpretation, etc. 2. Christian life—Baptist authors. I. Bible. O.T. Hebrews. English. New American standard. Selections. 1978. II. Title.
BS2775.2.L38 227'.87'06 77-93249
79 MARC

Lifsey, Harvey M
How to receive from God / by Harvey M. Lifsey. — Atlanta : Peachtree Publishers ; [Oceanside, Calif. : order from Christian Dynamics], c1978.
79 p. ; 22 cm.
ISBN 0-931948-01-0
1. Christian life—Baptist authors. 2. Faith. I. Title.
BV4501.2.L49 248'.48'61 78-65640
79 MARC

Mann, Gerald.
The seven deadly virtues / Gerald Mann. — Waco, Tex. : Word Books, c1979.
115 p. ; 21 cm.
Includes bibliographical references.
ISBN 0-8499-2853-2 : $4.95
1. Christian life—Baptist authors. I. Title.
BV4501.2.M337 248'.48'61 78-65809
79 MARC

McCray, Walter Arthur.
Black folks and Christian liberty : Black ... Christian ... and free to be cultural and social / [Walter Arthur McCray]. — 1st ed. — Chicago : Black Light Fellowship, c1979.
viii, 194 p. ; 22 cm.
"Black Light Fellowship series."
Bibliography: p. 190-193.
$4.95
1. Christian life—Baptist authors. 2. Afro-Americans—Religion. I. Title.
BV4501.2.M223 248'.48'6133 78-71258
00 MARC

McIntire, Russell M
Live your faith! / By Russell M. McIntire. — Gretna, La. : Pelican Pub. Co., 1979.
167 p. ; 23 cm.
ISBN (invalid) 0-88289-216-9
1. Christian life—Baptist authors. I. Title.
BV4501.2.M236 248'.48'61 78-25579
78 MARC

Miller, Lew, 1917-
You can beat those spiritual blahs / Lew Miller. — Denver : Accent Books, c1979.
128 p. ; 21 cm.
Commentary on Second Peter.
ISBN 0-89636-015-6 : $2.95
1. Bible. N.T. 2 Peter—Commentaries. 2. Christian life—Baptist authors. I. Bible. New Testament. 2 Peter. 1979. II. Title.
BS2795.3.M54 248'.4 78-72866
79 MARC

Saul, Martha Roper.
God sees a beautiful you / Martha Roper Saul. — Nashville, Tenn. : Broadman Press, c1979.
152 p. ; 20 cm.
Includes bibliographies.
ISBN 0-8054-5267-2
1. Christian life—Baptist authors. 2. Bible. N.T. Romans—Criticism, interpretation, etc. I. Title.
BV4501.2.S274 248'.48'61 78-72275
79 MARC

Tassell, Paul, 1934-
Secrets of the blessed man / Paul Tassell. — Schaumburg, Ill. : Regular Baptist Press, c1966, 1979 printing.
123p. ; 21 cm.
ISBN 0-87227-033-5
1. Christian life—Baptist authors. I. Title.
BV4501.2.T265 248'.48'61 78-22065
78 MARC

Taylor, Judson S
Freeing the captives : an allegory illustrating the acts of sin and grace, exposing Satan's false reign, disclosing man's base servitudes, manifesting Jehovah's dominion, and revealing saving mercy, with the plan of its operation / by Judson S. Taylor. — St. Louis, Mo. : National Baptist Pub. Co., c1886.
239 p. ; 18 cm.
1. Christian life—Baptist authors. I. Title.
BV4515.T38 230'.6'1 78-324419
79 MARC

Warlick, Harold C
Conquering loneliness / Harold C. Warlick, Jr. — Waco, Tex. : Word Books, c1979.
125 p. ; 21 cm.
Includes bibliographical references.
ISBN 0-8499-0169-3 : $5.95
1. Loneliness. 2. Christian life—Baptist authors. I. Title.
BV4911.W34 248'.48'61 79-65189
79 MARC

Welch, Bobby H
This comes first! / Bobby H. Welch. — Nashville : Broadman, c1979.
137 p. ; 19 cm.
ISBN 0-8054-5504-3
1. Christian life—Baptist authors. I. Title.
BV4501.2.W417 248'.48'61 79-84786
79 MARC

—BIBLICAL TEACHING

Calian, Carnegie Samuel.
For all your seasons : Biblical direction through life's passages / by Carnegie Samuel Calian. — Atlanta : John Knox Press, c1979.
135 p. ; 21 cm.
ISBN 0-8042-2084-0
1. Christian life—Presbyterian authors. 2. Christian life—Biblical teaching. I. Title.
BV4501.2.C243 248'.4 78-71050
79 MARC

Hill, Harold, 1905-
Instant answers for King's kids in training / by Harold Hill ; compiled by Gretchen Zimmer Black, with Irene Burk Harrell. — Plainfield, N.J. : Logos International, c1978.
vi, 206 p. ; 21 cm.
ISBN 0-88270-277-7 : $3.50
1. Christian life—Biblical teaching. 2. Christian life—Quotations, maxims, etc. I. Black, Gretchen Zimmer, joint author. II. Harrell, Irene Burk, joint author. III. Title.
BS2545.C48H54 248'.4 78-58862
78 MARC

Sono, Ayako, 1931-
(Watakushi o kaeta Seisho no kotoba)
私を変えた聖書の言葉 / 曽野綾子. — 東京 : 講談社, 1978.
227 p. ; 20 cm. Ja 79-3619
¥750
1. Christian life—Biblical teaching. I. Title.
BV4509.J3S66 79-805940

—BIOGRAPHY see Christian biography

—CATHOLIC AUTHORS

Albertinus, Aegidius, 1560-1620.
Der Welt Thurnierplatz : darinn erstl. d. geistl. Manns- u. Weibs Personen in ihren Zierden und Eigenschafften, folgends die weltlichen ... auffziehen ... / durch Aegidium Albertinum. — Fotomechan. Neudr. d. Orig.-Ausg., München, N. Henricus, 1615. — Leipzig : Zentralantiquariat der DDR, 1975.
356 p. ; 21 cm. GDR76-A
Running title: Der Welt Schau- und Thurnierplatz.
90.00M
1. Christian life—Catholic authors. I. Title. II. Title: Der Welt Schau- und Thurnierplatz.
BX2349.A58 1975 79-349718
79 MARC

Armstrong, April Oursler, 1926-
Cry Babel : the nightmare of aphasia and a courageous woman's struggle to rebuild her life / April Oursler Armstrong. — 1st ed. — Garden City, N.Y. : Doubleday, 1979.
ix, 252 p. ; 22 cm.
ISBN 0-385-13529-7 : $8.95
1. Aphasia—Biography. 2. Christian life—Catholic authors. 3. Armstrong, April Oursler, 1926- 4. Cerebrovascular disease—Biography. 5. Authors—United States—Biography. I. Title.
RC425.A75 616.8'552'09 77-26511
78 MARC

Barros, Neimar de.
Os fantoches / Neimar de Barros. — 1. ed. — São Paulo : Sorvil, c1978.
90 p. : ill. ; 21 cm.
Cr$50.00
1. Christian life—Catholic authors. 2. Meditations. I. Title.
BX2350.2.B3247 79-338358
79 MARC

Basadonna, Giorgio.
Nel cuore della resurrezione / Giorgio Basadonna. — 1. ed. — Milano : Rusconi, 1979.
231 p. ; 22 cm. — (Dimensione religiosa) It79-Mar
L7500
1. Christian life—Catholic authors. I. Title.
BX2350.2.B329 79-357825
*79 MARC

Bessière, Gérard.
Dieu est bien jeune : essai sur le goût chrétien de vivre / Gérard Bessière. — Paris : Éditions du Cerf, 1976.
203 p. ; 20 cm. — (Épiphanie) F77-1855
Includes bibliographical references.
ISBN 2-204-01074-X : 30.00F
1. Meditations. 2. Christian life—Catholic authors. I. Title.
BX2183.B473 248'.48'2 79-375845
79 MARC

Bona, Giovanni, Cardinal, 1609-1674.
Guida al cielo / Giovanni Bona ; a cura di M. C. Borgogno. — Roma : Istituto missionario pia società San Paolo, 1944.
191 p. ; 17 cm. — (Il Fiore dei santi padri, dottori e scrittori ecclesiastici)
Translation of Manuductio ad cœlum.
1. Christian life—Catholic authors. I. Borgogno, C. II. Title. III. Series.
BX2349.B5716 79-378399
79 MARC

Boros, Ladislaus, 1927-
Being a Christian today / Ladislaus Boros ; translated by M. Benedict Davies. — New York : Seabury Press, 1979.
124 p. ; 22 cm.
Translation of Heute Christ sein.
"A Crossroad book."
ISBN 0-8164-0440-2
1. Christian life—Catholic authors. I. Title.
BX2350.5.B6713 248'.48'2 79-13607
79 MARC

Brasó, Gabriel M
Servir en l'amor / Gabriel M. Brasó ; edició a cura de Josep M. Soler i Canals. — 1. ed. — [Montserrat] : Abadia De Montserrat, 1978.
192 p. ; 19 cm. — (Col·lecció Saurí ; 41) Sp***
ISBN 847202282X
1. Christian life—Catholic authors. I. Title.
BX2350.2.B68 78-380960
78 MARC

Catholic Church. Vicariate Apostolic of Juba, Sudan. Bishop, 1961-1967 (Mazzoldi)
Farewell pastoral letter of His Lordship, Bishop Sisto Mazzoldi, Vicar Apostolic of Juba (Sudan). — [s.l. : s.n.], 1964.
11 p. ; 15 cm.
1. Catholic Church—Pastoral letters and charges. 2. Missions—Sudan—Juba. 3. Christian life—Catholic authors. I. Title: Farewell pastoral letter of His Lordship, Bishop Sisto Mazzoldi ...
BX874.M43F37 78-305152
79 MARC

Chasseurs d'aube : livre des commencements / Claude Cesbron ... [et al.]. — Paris : Le Centurion, c1979.
205 p. ; 25 cm. F***
Includes bibliographical references.
ISBN 2-227-62000-5 : 42.00F
1. Christian life—Catholic authors. I. Cesbron, Claude.
BX2350.2.C46 248 79-378321
79 MARC

Costa, Rovílio.
Primado da pessoa na vida cristã / Rovílio Costa. — Porto Alegre : Escola Superior de Teologia São Lourenço de Brindes : distribuidora, Livraria Sulina, 1974.
96 p. ; 21 cm.
1. Christian life—Catholic authors. 2. Prayer. I. Title.
BX2350.2.C64 78-361074
78 MARC

Couto, Gabriel P Bueno.
O homem ea sua realização; o roteiro da felicidade [por] Gabriel P. Bueno Couto. São Paulo, Ed. Paulinas, 1968.
99 p. 18 cm. (Coleção Caminho, verdade e vida, 6) NCr$3.80
BBM69-3
LACAP69-0388
Bibliographical footnotes.
1. Man (Christian theology) 2. Christian life—Catholic authors. I. Title.
BT701.2.C64 70-445179
79[r79]rev2 MARC

Dobraczyński, Jan, 1910-
Sprawy ludzi mocnych. Wyd. 1. W co wierzę? Wyd. 4. / Jan Dobraczyński. — Warszawa : Pax, 1978 (Łódź : ŁZG)
262, [2] p. ; 21 cm. — (His Dzieła) P78-8775
zł75.00
1. Christian life—Catholic authors. 2. Mary, Virgin. 3. Catholics—Poland. I. Dobraczyński, Jan, 1910- W co wierzę? 1978. II. Title. III. Series.
BX2350.2.D573 1978 79-359072
79 MARC

CHRISTIAN LIFE
—CATHOLIC AUTHORS (Continued)

Doherty, Catherine de Hueck, 1900-
I live on an island / Catherine de Hueck Doherty. — Notre Dame, Ind. : Ave Maria Press, c1979.
126 p. : ill. ; 21 cm.
ISBN 0-87793-170-4. ISBN 0-87793-171-2 pbk. : $2.75
1. Christian life—Catholic authors. 2. Doherty, Catherine de Hueck, 1900- . 3. Catholics—Ontario—Biography. 4. Ontario—Biography. I. Title.
BX2350.2.D623 248'.48'2 78-74433
79 MARC

Doughty, Thomas.
IHS Maria Joseph, the second part, 1619, by Thomas Doughty. Menston, Scolar P., 1970.
[40], 479 p. 16 cm. (English recusant literature, 1558-1640, v. 17)
B70-12558
Reprinted from a copy in the library of Heythrop College. References: Allison and Rogers 281; not in STC.
ISBN 0-85417-253-X
1. Christian life—Catholic authors. I. Title. II. Series: Rogers, David Morrison, comp. English recusant literature, 1558-1640, v. 17.
BX1750.A1E5 vol. 17 282'.41 s 70-599473
[BX2349] 71[r79]rev MARC

Ellard, Gerald, 1894-1963.
Christian life and worship / Gerald Ellard ; with an introd. by Leo Klein. — Reprint ed. — New York : Arno Press, 1978 [c1940].
xxi, 420 p. : ill. ; 24 cm. — (The American Catholic tradition)
Reprint of the 1948 rev. and enl. ed. published by Bruce Pub. Co., Milwaukee, issued in Science and culture series.
Includes bibliographies and index.
ISBN 0-405-10819-2
1. Sacraments (Liturgy) 2. Catholic Church. Liturgy and ritual. 3. Christian life—Catholic authors. I. Title. II. Series.
BX2200.E6 1978 248'.3 77-11282
77 MARC

Fernández Ramos, Felipe.
Interpelado por la palabra / Felipe Fernández Ramos. — Madrid : Narcea, [1975]
156 p. ; 21 cm. Sp75-Nov
ISBN 8427702469
1. Bible—Criticism, interpretation, etc. 2. Christian life—Catholic authors. I. Title.
BS517.2.F47 78-386482
*79 MARC

Forrester, David.
Listening with the heart / David Forrester ; introd. by Michael Hollings. — New York : Paulist Press, c1978.
95 p. ; 21 cm.
Bibliography: p. 95.
ISBN 0-8091-2183-2 : $2.95
1. Prayer. 2. Christian life—Catholic authors. I. Title.
BV210.2.F67 1978 248'.3 78-70822
79 MARC

Fuchs, Konstantin, 1910-
Sag ja zu dir : Selbstverwirklichung als Aufgabe d. Christen / Konstantin Fuchs. — Freiburg im Breisgau ; Basel ; Wien : Herder, 1977.
160 p. ; 20 cm. GFR78-A
Bibliography: p. 160.
ISBN 3-451-17967-9 : DM16.80
1. Self-realization. 2. Christian life—Catholic authors. I. Title.
BJ1470.F82 248'.48'2 78-399244
79 MARC

Ha, Chae-byŏl.
(Chasin ŭl k'iunŭn saramdŭl)
자신을키우는사람들 / 하재별저. — 서울 : 한국천주교중앙협의회, 1977.
253 p. ; 20 cm.
₩800

1. Christian life—Catholic authors. I. Title.
BX2350.2.H26 78-826672

Herman, Nicolas, 1611-1691.
The practice of the presence of God / by Brother Lawrence of the Resurrection [i.e. N. Herman] ; translated by Sister Mary David. — Rev. ed. — New York : Paulist Press, 1978.
127 p. ; 19 cm. — (The Spiritual masters series)
Translation of La pratique de la présence de Dieu.
ISBN 0-8091-2137-9 : $2.25
1. Christian life—Catholic authors. I. Title.
BX2349.H42 1978 248'.2 78-61665
78 MARC

Labarrière, Pierre Jean.
L'existence réconciliée. Paris-[Bruges], Desclée De Brouwer, (1967).
206 p. 20 cm. (Collection Christus, no 26. Essais) bfr 165.- Be68-3:
Bibliographical footnotes.
1. Christian life—Catholic authors. I. Title.
BX2350.2.L2 72-358087
79[r79]rev MARC

Liguori, Alfonso Maria de', Saint, 1696-1787.
The mission-book of the Congregation of the Most Holy Redeemer. — New York : Arno Press, 1978 [c1862].
502 p. ; 21 cm. — (The American Catholic tradition)
Reprint of the new, rev. and enl. ed. published by Kelly & Piet, Baltimore.
ISBN 0-405-10843-5
1. Catholic Church—Prayer-book and devotions—English. 2. Christian life—Catholic authors. I. Redemptorists. II. Title. III. Series.
BX2110.L6 1978 264'.02 77-11300
77 MARC

Luis de Granada, 1504-1588.
[Guia de pecadores. Japanese]
ぎやどぺかどる / [Luis de Granada 著]. 妙貞問答 ; 破提宇子 / [ハビアン著]. 顯僞錄 / [澤野忠庵著 ; 與謝野寛, 正宗敦夫, 與謝野晶子編纂校訂]. — 東京 : 現代思潮社, 1978.
[44], 186, 134, 73, 30, 25 p. ; 16 cm. — (覆刻日本古典全集)
Ja 79-825
Spine title.
Photoreprint of the 1927 ed. published in 2 v. by Nihon Koten Zenshū Kankōkai, Tokyo.

1. Christian life—Catholic authors. 2. Christianity and other religions. 3. Christianity—Controversial literature. I. Fabian, Fucan, fl. 1583-1607. Myōtei mondō. 1978. II. Fabian, Fucan, fl. 1583-1607. Ha Deusu. 1978. III. Ferreira, Christovão, 1580-ca. 1652. Kengiroku. 1978. IV. Title. V. Series: Nihon koten zenshū.
Title romanized: Giya do pekadoru.
BX2349.L716 1978 79-805675

Maliński, Mieczysław.
Nasz chleb powszedni / Mieczysław Maliński. — London : Veritas, 1977.
207 p. ; 19 cm. — (Biblioteka "Gazety Niedzielnej") GB78-02988
£1.50
1. Christian life—Catholic authors. 2. Meditations. I. Title.
BX2350.2.M3147 79-384242
79 MARC

Marstin, Ronald.
Beyond our tribal gods : the maturing of faith / Ronald Marstin. — Maryknoll, N.Y. : Orbis Books, c1979.
viii, 150 p. ; 21 cm.
Bibliography: p. 146-150.
ISBN 0-88344-030-X
1. Faith. 2. Faith—Psychology. 3. Christian life—Catholic authors. 4. Sociology, Christian (Catholic) I. Title.
BT771.2.M28 261.8'5 79-4354
79 MARC

Mirewicz, Jerzy.
Człowiek—miasto oblężone / Jerzy Mirewicz. — Londyn : Księża Jezuici, 1978.
124 p. ; 22 cm. GB***
1. Christian life—Catholic authors. 2. Temptation. I. Title.
BX2350.2.M49 79-379021
79 MARC

O'Shea, Kevin.
The way of tenderness / by Kevin O'Shea. — New York : Paulist Press, c1978.
76 p. ; 19 cm. — (A Deus book)
ISBN 0-8091-2166-2 : $1.95
1. Christian life—Catholic authors. I. Title.
[BX2350.2.O79] 248'.48'2 78-61728
79 MARC

Preston, Geoffrey, 1936-1977.
God's way to be human : meditations on following Christ through Scripture and sacrament / Geoffrey Preston ; posthumous texts prepared, with a memoir, by Aidan Nichols. — New York : Paulist Press, c1978.
105 p. ; 23 cm.
Includes bibliographical references.
ISBN 0-8091-0280-3 : $4.95
1. Jesus Christ—Meditations. 2. Christian life—Catholic authors. I. Nichols, Aidan. II. Title.
BT306.4.P73 1978 232 78-65902
79 MARC

Rey, Alfonso, fl. 1973-
En la fuente de la alegría / Alfonso Rey. — 3. ed. — Madrid : Ediciones Palabra, 1973.
229 p. ; 18 cm. — (Cuadernos Palabra ; 24) Sp***
Includes bibliographical references.
ISBN 8471180669
1. Christian life—Catholic authors. I. Title.
BX2186.R43 1973 78-385741
79 MARC

Rivera, José, fl. 1975-
Libres del mundo : está es la victoria que ha vencido al mundo, nuestra fe (I Jn 5, 4) / José Rivera, José María Iraburu. — Burgos : Aldecoa, [1975?]
48 p. ; 19 cm. — (Cuadernos de espiritualidad ; 6)
Cover title.
Includes bibliographical references.
ISBN 8470091751 : 150ptas
1. Christian life—Catholic authors. I. Iraburu, José María, joint author. II. Title.
BX2350.2.R58 77-568580
*78[r79]rev MARC

Santaner, Marie Abdon.
L'Avenir de la communauté religieuse : vivre ensemble l'Alliance / M.-A. Santaner. — Paris : Éditions ouvrières, 1978.
183 p. ; 18 cm. — (Collection Lumière des hommes) F78-13106
ISBN 2-7082-1954-5 : 32.00F
1. Christian life—Catholic authors. 2. Church. I. Title.
BX2350.2.S254 248'.48'2 79-349262
79 MARC

Serra Estellés, Francisco.
Al tener que pensar en voz alta / [por] Francisco Serra Estellés. — Madrid : Sociedad de Educación Atenas, 1977.
221 p. ; 19 cm. — (Temas vivos ; 45) Sp77-May
Includes bibliographical references.
ISBN 8470200801
1. Christian life—Catholic authors. I. Title.
BX2186.S43 78-379660
78 MARC

Sheed, Francis Joseph, 1897-
The instructed heart : soundings at four depths / F. J. Sheed. — Huntington, Ind. : Our Sunday Visitor, inc., c1979.
128 p. ; 22 cm.
ISBN 0-87973-739-5 : $5.95
1. Ward, Maisie, 1889- . 2. Heart—Biblical teaching. 3. Christian life—Catholic authors. 4. Mary, Virgin. I. Title.
BX4705.W287S52 248 79-87925
79 MARC

Smith, Herbert F
How to get what you want from God / Herbert F. Smith. — Huntington, Ind. : Our Sunday Visitor, c1979.
170 p. : ill. ; 20 cm.
ISBN 0-87973-632-1 : $3.95
1. Christian life—Catholic authors. I. Title.
BV4501.2.S5333 248'.4 79-87927
79 MARC

Tanaka, Eikichi, 1902-
(Shiawase wa doko ni)
しあわせはどこに / 田中英吉. — 東京 : 中央出版社, 昭和50 [1975]
252 p. ; 18 cm. Ja 76-977
¥800

1. Christian life—Catholic authors. I. Title.
BX2350.2.T34 79-806153

Urteaga Loidi, Jesús, 1921-
Los defectos de los cantos / Jesús Urteaga Loidi. — Madrid : Ediciones Rialp, c1978.
413 p. ; 20 cm. — (Patmos ; 172) Sp***
Includes bibliographical references and index.
ISBN (invalid) 0-8432-1196-6
1. Christian life Catholic authors. 2. Bible Biography. 3. Christian saints—Biography. I. Title.
BX2350.2.U68 79-357652
79 MARC

Wolff, Pierre, 1929-
May I hate God? / By Pierre Wolff. — New York : Paulist Press, c1979.
76 p. ; 19 cm. — (A Deus book)
ISBN 0-8091-2180-8 : $1.95
1. Christian life—Catholic authors. I. Title.
BX2350.2.W6 248'.48'2 78-70815
79 MARC

—CATHOLIC AUTHORS—ADDRESSES, ESSAYS, LECTURES

Lohfink, Norbert.
Unsere grossen Wörter : d. Alte Testament zu Themen dieser Jahre / Norbert Lohfink. — Freiburg im Breisgau ; Basel ; Wien : Herder, 1977.
253 p. ; 22 cm. GFR78-A
Contains revised lectures, some previously published.
Includes bibliographical references.
ISBN 3-451-17976-8 : DM29.80
1. Bible. O.T.—Criticism, interpretation, etc.—Addresses, essays, lectures. 2. Christian life—Catholic authors—Addresses, essays, lectures. I. Title.
BS1174.2.L63 79-369798
79 MARC

McNulty, Frank J
Should you ever feel guilty? / By Frank J. McNulty and Edward Wakin. — New York : Paulist Press, c1978.
91 p. ; 19 cm. — (A Deus book)
Includes bibliographical references.
ISBN 0-8091-2149-2 : $2.95
1. Guilt—Addresses, essays, lectures. 2. Christian life—Catholic authors—Addresses, essays, lectures. I. Wakin, Edward, joint author. II. Title.
BJ1471.5.M35 241'.1 78-70627
79 MARC

CHRISTIAN LIFE (Continued)

—CHRISTIAN CHURCH (DISCIPLES OF CHRIST) AUTHORS

Living responsibility. — Glen Irish, Vic. : Vital Publications, 1976.
142 p. : ill. ; 18 cm. Aus
CONTENTS: Williams, E. Lyall. Reflections.—White, A. E. Biography.
ISBN 0-909116-04-0
1. Williams, E. Lyall, 1906- . 2. Christian Church (Disciples of Christ)—Clergy—Biography. 3. Clergy—Australia—Biography. 4. Christian life—Christian Church (Disciples of Christ) authors. I. Williams, E. Lyall, 1906- . II. White, A. E.
BX7343.W53L58 286'.63 79-315513
79 MARC

—CHURCH OF THE BRETHREN AUTHORS

Gish, Arthur G
Living in Christian community / Arthur G. Gish. — Scottdale, Pa. : Herald Press, 1979.
379 p. ; 22 cm.
Bibliography: p. [371]-373.
Includes index.
ISBN 0-8361-1887-1 : $8.95
1. Christian communities. 2. Church renewal. 3. Christian life—Church of the Brethren authors. 4. Church. I. Title.
BV4405.G57 262 79-11848
79 MARC

—CONGREGATIONALIST AUTHORS

Mathews, William Barnes.
Life's abiding resources / by William Barnes Mathews. — North Quincy, Mass. : Christopher Pub. House, 1978.
303 p. ; 22 cm.
ISBN 0-8158-0371-0 : $6.95
1. Christian life—Congregationalist authors. I. Title.
BV4501.2.M379 248'.48'58 78-66872
79 MARC

Ortlund, Raymond C
Intersections : with Christ at the crossroads of life / Raymond C. Ortlund. — Waco, Tex. : Word Books, c1979.
151 p. ; 23 cm.
ISBN 0-8499-0127-8 : $6.95
1. Christian life—Congregational authors. 2. Jesus Christ—Person and offices. I. Title.
BV4501.2.O728 248'.48'58 78-65798
79 MARC

—CONGRESSES

Heiligungskonferenz auf Sankt Chrischona, 28th, 1977.
Lernet von mir / Bericht d. 28. Heiligungskonferenz auf St. Chrischona bei Basel vom 1.-8. August 1977 ; Redner, Fritz Aeschlimann ... [et al.]. — Als Ms. gedr. — Giessen ; Basel : Brunnen-Verlag, 1977.
96 p. : 21 cm. GFR78-A
ISBN 3-7655-0463-7 : DM9.80
1. Sermons, German—Congresses. 2. Christian life—Congresses. I. Aeschlimann, Fritz. II. Title.
BV4244.H43 1977 252'.00943 78-368509
78 MARC

—FICTION

see also Exempla

—FRIEND AUTHORS

Steere, Douglas Van, 1901-
Prayer and worship / Douglas V. Steere. — Richmond, Ind. : Friends United Press, [foreword 1978]
ix, 88 p. ; 18 cm.
Bibliography: p. 87-88.
ISBN 0-913408-44-1 : $1.95
1. Christian life—Friend authors. 2. Prayer. 3. Worship. I. Title.
BV4501.S794 1978 248'.3 78-70480
79 MARC

—HUTTERITE BRETHREN AUTHORS

Arnold, Heini, 1913-
Printed letters from Heini and Annemarie Arnold and others. — [Rifton, NY : Plough Pub. House], c19
v. : ill. ; 23 cm.
Includes index.
CONTENTS:
—v. 8. 131-150.
1. Arnold, Heini, 1913- . 2. Arnold, Annamarie, 1909- . 3. Christian life—Hutterite Brethren authors. I. Arnold, Annemarie, 1909- joint author. II. Title.
BX8141.A76 289.7'3 79-114838
00 MARC

—JUVENILE FICTION

The Entertaining, moral, and religious repository; containing, upwards of three score separate performances, all of which are written in a simple yet pleasing stile ... for the amusement and instruction of the youth of both sexes. Published by a society in Great-Britain, instituted for the beneficent purpose of aiding the intention of His Majesty, as expressed in his royal proclamation for the suppression of vice and immorality. Elizabeth-Town, Printed by S. Kollock, for C. Davis, 1799.
324 p. 17 cm.
Evans 35298. Probably a companion vol. to Evans 35297 (same title; designated v. 1)
Previously published in 1798, with "In two volumes. Vol. II" on t. p. (Evans 35296. v. 2) In the present ed. the same sheets are used for p. [9]-324.
CONTENTS: Contents originally published in England as Cheap repository tracts.
1. Christian life—Juvenile fiction. 2. Children's stories, English. I. Cheap repository.
PZ6.E63 1799 78-317619
78 MARC

—JUVENILE LITERATURE

Kremer, Eva Maria, 1928-
Jesus hat viele Gesichter : wie Kinder in fernen Ländern Jesus erleben / Eva-Maria Kremer, Werner Zurfluh ; [Ill., Mona Helle-Ineichen]. — Luzern ; München : Rex-Verlag, c1975.
100 p. : ill. ; 22 cm. Sw75-A-7776
ISBN 3-7252-0279-6 : 18.00F
1. Children's stories, German. 2. Christian life—Juvenile literature. I. Zurfluh, Werner, joint author. II. Title.
PZ33.K6446 79-383696
*79 MARC

—LITERARY COLLECTIONS

The Peanut butter hamster, and other animal tails / edited by Grace Fox Anderson ; illustrated by Richard Johnson. — Wheaton, Ill. : Victor Books, c1979.
132 p. : ill. ; 20 cm. — (A Winner book)
SUMMARY: Presents short stories and poems featuring animals in a religious setting.
ISBN 0-88207-484-9 : $1.75
1. Animals—Literary collections. 2. Christian life—Literary collections. 3. Children's literature, American. [1. Animals—Literary collections. 2. Christian life—Literary collections. 3. American literature—Collections] I. Fox, Grace. II. Johnson, Richard, 1953-
PZ5.P37 [Fic] 78-65204
79 MARC

—LUTHERAN AUTHORS

Arndt, Johann, 1555-1621.
True Christianity / Johann Arndt ; translation and intod. by Peter Erb ; pref. by Heiko A. Oberman. — New York : Paulist Press, c1979.
xvii, 301 p. ; 23 cm. — (The Classics of Western Spitituality)
Translation of book 1 and parts of books 2-6 of Wahres Christenthum.
Includes bibliographical references and indexes.
ISBN 0-8091-0281-1
1. Christian life—Lutheran authors. 2. Mysticism. I. Title. II. Series.
BV4503.A76213 1979 248'.48'41 78-72046
00 MARC

Christenson, Laurence.
Back to square one / Larry Christenson. — Minneapolis : Bethany Fellowship, c1979.
144 p. ; 20 cm.
ISBN 0-87123-025-9 : $3.50
1. Christian life—Lutheran authors. I. Title.
BV4501.2.C497 248'.48'41 79-16413
79 MARC

Hong, Edna Hatlestead, 1913-
The downward ascent / Edna Hong. — Minneapolis : Augsburg Pub. House, c1979.
93 p. ; 20 cm.
ISBN 0-8066-1679-2 : $3.50
1. Man (Christian theology) 2. Christian life—Lutheran authors. I. Title.
BT701.2.H66 233 78-66942
79 MARC

Schmidt, Ernst G 1931-
Choose to win! / Ernst G. Schmidt. — Nashville : Abingdon, c1979.
127 p. ; 21 cm.
ISBN 0-687-06700-6
1. Christian life—Lutheran authors. I. Title.
BV4501.2.S295 248'.48'41 79-30
79 MARC

Thielicke, Helmet, 1908-
Being a Christian when the chips are down / Helmut Thielicke ; translated by H. George Anderson. — Philadelphia : Fortress Press, c1979.
125 p. ; 23 cm.
Translation of Der Christ im Ernstfall.
ISBN 0-8006-0541-1 : $5.95
1. Christian life—Lutheran authors. 2. Church year—Meditations. I. Title.
BV4503.T5213 1979 248'.48'41 78-54562
78 MARC

Timmons, Tim.
Maximum living in a pressure-cooker world / Tim Timmons ; research and editorial assistance by Martha Greene. — Waco, Tex. : Word Books, c1979.
237 p. ; 23 cm.
Includes bibliographical references.
ISBN 0-8499-0117-0 : $8.95
1. Christian life—Lutheran authors. I. Greene, Martha. II. Title.
BV4501.2.T53 248'.4 78-65807
00 MARC

Wentz, Frederick K
Getting into the act : opening up lay ministry in the weekday world / Frederick K. Wentz. — Nashville : Abingdon, c1978.
144 p. ; 20 cm.
Includes bibliographical references.
ISBN 0-687-14125-7
1. Christian life—Lutheran authors. 2. Laity. 3. Liberation theology. I. Title.
BV4501.2.W4198 248'.48'41 78-17179
78 MARC

Z biblią na co dzień.
[Warszawa, Zwiastun]
v. 17 cm.
Vols. for issued as a supplement to: Kalendarz ewangelicki.
1. Christian life—Lutheran authors. 2. Devotional calendars—Lutheran Church. I. Kalendarz ewangelicki.
BV4509.P6Z18 78-646049
MARC-S

—MEDITATIONS

Indahl, Pam.
Lessons in living / [Pam Indahl]. — Spokane : Indahl, c1978.
xi, 272 p. ; 21 cm.
Includes index.
1. Christian life—Meditations. I. Title.
BV4501.2.I48 248'.4 77-95160
79 MARC

Thielicke, Helmut, 1908-
Der Christ im Ernstfall : d. kleine Buch d. Hoffnung : Meditationen, Reflexionen / Helmut Thielicke. — Originalausg. — Freiburg im Breisgau ; Basel ; Wien : Herder, 1977.
154 p. ; 18 cm. — (Herderbücherei ; Bd. 600) GFR77-A
ISBN 3-451-07600-4 : DM5.90
1. Christian life—Meditations. I. Title.
BV4503.T52 78-388451
79 MARC

—METHODIST AUTHORS

Conners, Kenneth Wray.
Lord, have You got a minute? / By Kenneth Wray Conners. — Valley Forge, PA : Judson Press, c1979.
127 p. : ill. ; 22 cm.
ISBN 0-8170-0816-0 : $4.95
1. Meditations. 2. Christian life—Methodist authors. I. Title.
BV4832.2.C627 248'.48'7 78-15297
78 MARC

Langford, Thomas A
Christian wholeness / Thomas A. Langford. — Nashville : Upper Room, c1978.
141 p. ; 21 cm.
Includes bibliographical references.
ISBN 0-8358-0383-X
1. Christian life—Methodist authors. I. Title.
[BV4501.2.L3185] 248'.48'76 78-58011
79 MARC

McClain, Charles Allen, 1923-
First came the wings / Charles A. McClain. — Nashville : Upper Room, c1979.
144 p. ; 22 cm.
Includes bibliographical references.
ISBN 0-8358-0387-2
1. Christian life—Methodist authors. I. Title.
BV4501.2.M19 248'.48'7 79-63893
79 MARC

CHRISTIAN LIFE
—METHODIST AUTHORS (Continued)

Oden, Marilyn Brown.
The courage to care / Marilyn Brown Oden. — Nashville : Abingdon, c1979.
95 p. ; 21 cm.
Includes bibliographical references.
ISBN 0-687-09760-6
1. Christian life—Methodist authors. 2. Courage. I. Title.
BV4501.2.O324 248'.48'7 78-13008
78 MARC

—MORMON AUTHORS

Ashton, Marvin J 1915-
What is your destination? / Marvin J. Ashton. — Salt Lake City : Deseret Book Co., 1978.
x, 197 p. ; 24 cm.
Includes index.
ISBN 0-87747-719-1 : $5.95
1. Christian life—Mormon authors. I. Title.
BX8656.A83 248'.49'3 78-14982
78 MARC

Packham, E Richard.
Born of the spirit / E. Richard Packham. — Salt Lake City : Bookcraft, c1979.
x, 80 p. ; 24 cm.
Includes index.
ISBN 0-88494-361-5 : $4.50
1. Christian life—Mormon authors. I. Title.
BX8656.P3 248'.48'933 79-83680
79 MARC

Yorgason, Blaine M 1942-
Others / Blaine M. Yorgason and Brenton G. Yorgason. — Salt Lake City : Bookcraft, c1978..
xiv, 140 p. : ill. ; 24 cm.
Includes index.
ISBN 0-88494-349-6 : $4.00
1. Christian life—Mormon authors. I. Yorgason, Brenton G., joint author. II. Title.
BX8656.Y67 248'.48'33 78-67218
78 MARC

—ORTHODOX EASTERN AUTHORS

Dudko, Dimitriĭ.
Ein ungeschriebenes Buch : Aufzeichnungen eines russ. Priesters / Dimitrij Dudko ; ⌜übers. von Erwin Lohneisen⌝ ; mit einem Vorw. v. Kardinal Joseph Ratzinger. — Graz ; Wien ; Köln : Verl. Styria, 1978.
175 p. ; 21 cm. Au78-11-33
Translation of Verīū, Gospodi.
ISBN 3-222-11081-6 : S148.00
1. Christian life—Orthodox Eastern authors. 2. Orthodox Eastern Church—Doctrinal and controversial works—Orthodox Eastern authors. I. Title.
BX382.D815 79-347442
•79 MARC

Neplīūev, Nikolaĭ Nikolaevich, 1851–1908.
(Partīīa mirnago progressa)
Партія мирнаго прогресса : ея идеальныя основы и жизненная программа / Н. Неплюевъ. — Глуховъ : Тип. А. К. Нестерова, 1906.
42 p. ; 27 cm.
Bound with the author's Golos vīeruīūshchago mirīanina po povodu predstoīashchago Sobora. Kiev, 1906.
1. Political parties—Russia. 2. Krestovozdvizhenskoe pravoslavnoe trudovoe bratstvo. 3. Christian life—Orthodox Eastern authors. I. Title.
BX510.N46 Russian Imperial Coll. 78-422083
[DK263.16]

Neplīūev, Nikolaĭ Nikolaevich, 1851–1908.
(Pis'mo k dukhovenstvu)
Письмо къ духовенству / Н. Неплюевъ. — Кіевъ : Лито-тип. И. И. Чоколова, 1905.
20 p. ; 27 cm.
Bound with the author's Golos vīeruīūshchago mirīanina po povodu predstoīashchago Sobora. Kiev, 1906.
1. Christian life—Orthodox Eastern authors. I. Title.
BX510.N46 Russian Imperial Coll. 78-422088
[BX382]

Neplīūev, Nikolaĭ Nikolaevich, 1851–1908.
(Pis'mo k sel'skomu naselenīīū)
Письмо къ сельскому населенію / Н. Неплюевъ. — Кіевъ : Лито-тип. И. И. Чоколова, 1905.
12 p. ; 27 cm.
Bound with the author's Golos vīeruīūshchago mirīanina po povodu predstoīashchago Sobora. Kiev, 1906.
1. Orthodox Eastern Church, Russian—Sermons. 2. Sermons, Russian. 3. Christian life—Orthodox Eastern authors. I. Title.
BX510.N46 Russian Imperial Coll. 78-422087
[BX513]

(Troitskīe listki)
Троицкіе листки : духовно-нравственное чтеніе для народа. — Вып. 1. и 2. — ⌜Сергіевскій посадъ⌝ : Изд. Св. Троицкія Сергіевы лавры, 1880–⌜188⌝1.
70 pamphlets : ill. ; 22 cm.
Cover title.
Pamphlet no. 1–8, 12, 15–16, and 30: "Изданіе второе."
1. Christian life—Orthodox Eastern authors. I. Troitskaīa Sergieva lavra.
BX382.T76 Russian Imperial Coll. 79–409351

—ORTHODOX EASTERN AUTHORS—
ADDRESSES, ESSAYS, LECTURES

Galaction, Gala, 1879-1961.
Ziua Domnului. București, Editura Institutului Biblic și de Misiune Ortodoxă, 1958.
vii, 341 p. 21 cm.
At head of title: Pagini creștine.
1. Christian life—Orthodox Eastern authors—Addresses, essays, lectures. 2. Orthodox Eastern Church, Romanian—Addresses, essays, lectures. I. Title. II. Title: Pagini creștine.
BX382.G34 79-384566
79 MARC

—ORTHODOX EASTERN AUTHORS
—PERIODICALS

(al-Nūr)
النور.
⌜بيروت⌝ حركة الشبيبة الارثوذكسية.
v. ill. 24 cm. 10 no. a year.
1. Christian life—Orthodox Eastern authors—Periodicals. 2. Theology—Periodicals. I. Ḥarakat al-Shabībah al-Urthūdhuksīyah (Lebanon)
BX382.N87 79–642270
(MARC–S)

—PRESBYTERIAN AUTHORS

Boice, James Montgomery, 1938-
Awakening to God / James Montgomery Boice. — Downers Grove, Ill. : Inter-Varsity Press, c1979.
233 p. ; 21 cm. — (His Foundations of the Christian faith ; v. 3)
Includes bibliographical references and index.
ISBN 0-87784-745-2 : $5.95
1. Holy Spirit. 2. Salvation. 3. Christian life—Presbyterian authors. 4. Election (Theology) I. Title. II. Series.
BT121.2.B64 234 78-25014
79 MARC

Calian, Carnegie Samuel.
For all your seasons : Biblical direction through life's passages / by Carnegie Samuel Calian. — Atlanta : John Knox Press, c1979.
135 p. ; 21 cm.
ISBN 0-8042-2084-0
1. Christian life—Presbyterian authors. 2. Christian life—Biblical teaching. I. Title.
BV4501.2.C243 248'.4 78-71050
79 MARC

Hudnut, Robert K
The bootstrap fallacy : what the self-help books don't tell you / Robert K. Hudnut. — Cleveland : Collins, 1978.
146 p. ; 22 cm.
Includes bibliographical references.
ISBN 0-529-05495-7 : $7.95
1. Christian life—Presbyterian authors. 2. Psychology—Popular works—Controversial literature. I. Title.
BV4501.2.H73 1978 248'.48'51 78-69974
79 MARC

Ogilvie, Lloyd John.
When God first thought of you : the full measure of love as found in 1, 2, 3 John / Lloyd John Ogilvie. — Waco, Tex. : Word Books, c1978.
214 p. ; 23 cm.
ISBN 0-8499-0102-2 : $6.95
1. Bible. N.T. Epistles of John—Meditations. 2. Christian life—Presbyterian authors. I. Title.
BS2805.4.O35 242'.2 78-64793
79 MARC

Smith, M Blaine.
Knowing God's will : Biblical principles of guidance / M. Blaine Smith ; foreword by Richard C. Halverson. — Downers Grove, Ill. : InterVarsity Press, c1979.
141 p. ; 21 cm.
Includes bibliographical references.
ISBN 0-87784-610-3 : $3.95
1. Christian life—Presbyterian authors. 2. God—Will. I. Title.
BV4501.2.S53437 248'.48'51 78-24756
79 MARC

Sueltz, Arthur Fay.
If I should die before I live : with study guide for personal reflection or group interaction / Arthur Fay Sueltz. — Waco, Tex. : Word Books, c1979.
144 p. ; 22 cm.
ISBN 0-8499-0090-5 : $6.95
1. Christian life—Presbyterian authors. I. Title.
BV4501.2.S828 248'.48'51 78-65800
79 MARC

—QUOTATIONS, MAXIMS, ETC.

Hill, Harold, 1905-
Instant answers for King's kids in training / by Harold Hill ; compiled by Gretchen Zimmer Black, with Irene Burk Harrell. — Plainfield, N.J. : Logos International, c1978.
vi, 206 p. ; 21 cm.
ISBN 0-88270-277-7 : $3.50
1. Christian life—Biblical teaching. 2. Christian life—Quotations, maxims, etc. I. Black, Gretchen Zimmer, joint author. II. Harrell, Irene Burk, joint author. III. Title.
BS2545.C48H54 248'.4 78-58862
78 MARC

Macdonald, George, 1824-1905.
The world of George Macdonald : selections from his works of fiction / edited by Rolland Hein. — Wheaton, Ill. : Harold Shaw Publishers, c1978.
199 p. ; 21 cm. — (The Wheaton literary series)
On spine: George Macdonald.
ISBN 0-87788-922-8 : $4.95
1. Christian life—Quotations, maxims, etc. I. Hein, Rolland. II. Title.
PR4966.H4 1978 823'.8 78-62193
79 MARC

Petty, Jo.
Life is for living / Jo Petty. — Old Tappan, N.J. : F. H. Revell Co., c1979.
158 p. ; 21 cm.
ISBN 0-8007-0978-0 : $5.95
1. Christian life—Quotations, maxims, etc. I. Title.
BV4501.2.P433 248'.4 78-10803
78 MARC

Wilson, James Bright.
Spiritual potpourri / James Bright Wilson. — Anaheim, Calif. : KNI Inc., c1979.
206 p. ; 22 cm.
1. Christian life—Quotations, maxims, etc. I. Title.
BV4513.W47 242 79-84567
00 MARC

—REFORMED AUTHORS

Calvin, Jean, 1509-1564.
Concerning scandals / by John Calvin ; translated by John W. Fraser. — Grand Rapids : Eerdmans, c1978.
xi, 119 p. ; 23 cm.
Translation of De scandalis.
ISBN 0-8028-3511-2 : $6.95
1. Reformed Church—Doctrinal and controversial works. 2. Christian life—Reformed authors. I. Title.
BX9420.D4213 248'.48'42 78-8675
78 MARC

Jabay, Earl.
Precisely how to take care of your self / by Earl Jabay. — Plainfield, N.J. : Logos International, c1979.
ix, 98 p. ; 21 cm.
ISBN 0-88270-346-3 : $2.95
1. Christian life—Reformed authors. I. Title.
BV4501.2.J29 248'.48'57 78-71526
79 MARC

Ridderbos, Simon Jan.
Het lege midden : over geloof en psychische moeilijkheden / S. J. Ridderbos. — Baarn : Ten Have, ⌜1977⌝
101 p. ; 21 cm. Ne78-2
ISBN 9025941249 : fl 12.50
1. Apologetics—20th century. 2. Christian life—Reformed authors. I. Title.
BT1102.R5137 79-377533
•79 MARC

—SEVENTH-DAY ADVENTIST
AUTHORS

Anderson, Roy Allan.
God's unique love for you : the secret of joy, peace, and power / by Roy Allan Anderson. — Mountain View, Calif. : Pacific Press Pub. Association, c1979.
158 p. ; 22 cm. — (Dimension ; 136)
"Previously published under the title You can be free."
1. Christian life—Seventh-Day Adventist authors. I. Title.
⌜BV4501.2.A475 1979⌝ 248'.48'673 79-115756
79 MARC

Chaij, Fernando.
The impending drama / by Fernando Chaij. — Nashville : Southern Pub. Association, c1979.
128 p. ; 20 cm.
ISBN 0-8127-0208-5
1. End of the world. 2. Christian life—Seventh-Day Adventist authors. I. Title.
BT876.C46 248'.48'673 78-22061
78 MARC

CHRISTIAN LIFE
—SEVENTH-DAY ADVENTIST AUTHORS

Chaij, Fernando.
The key to victory : companion volume to the impending drama / by Fernando Chaij. — Nashville : Southern Pub. Association, c1979.
126 p. ; 20 cm. — (Horizon)
ISBN 0-8127-0224-7 : $3.95
1. Christian life—Seventh-Day Adventist authors. 2. Holy Spirit. 3. Eschatology. I. Title.
BV4501.2.C45 248'.48'673 79-548
79 MARC

Davis, Thomas A
Was Jesus really like us? / Thomas A. David. — Washington : Review and Herald Pub. Association, c1979.
189 p. ; 21 cm.
Includes bibliographical references.
1. Jesus Christ—Humanity. 2. Christian life—Seventh-Day Adventist authors. I. Title.
BT218.D38 232.9'03 78-26842
78 MARC

Hackett, Willis J
Molding the Christian mind / Willis J. Hackett. — Washington : Review and Herald Pub. Association, c1978.
160 p. : ill. ; 21 cm.
1. Christian life—Seventh-Day Adventist authors. 2. Body, Human—Moral and religious aspects. I. Title.
BV4501.2.H23 613 78-23292
78 MARC

Hilde, Reuben.
You can live radiantly every day / Reuben Hilde. — Washington : Review and Herald Pub. Association, c1978.
128 p. ; 18 cm.
$4.50
1. Christian life—Seventh-Day Adventist authors. I. Title.
BV4501.2.H448 248'48'673 78-25731
78 MARC

Keough, G Arthur, 1909-
Infinitely happy / by G. Arthur Keough. — Nashville : Southern Pub. Association, c1978.
128 p. ; 20 cm. — (Horizon)
ISBN 0-8127-0213-1
1. Christian life—Seventh-Day Adventist authors. I. Title.
BV4501.2.K435 248'.48'673 78-21952
78 MARC

Knoche, Keith.
Lord, give me patience—but hurry up about it! / By Keith Knoche. — Mountain View, Calif. : Pacific Press Pub. Association, c1978.
126 p. : ill. ; 18 cm. — (Agapē ; 107)
1. Christian life—Seventh-Day Adventists. I. Title.
BV4501.2.K57 248'.48'673 77-93362
79 MARC

Schurch, Maylan.
Pilgrimage : a young man's odyssey to do God's will / Maylan Schurch. — Washington : Review and Herald Pub. Association, c1979.
127 p. ; 21 cm.
1. Christian life—Seventh-Day Adventist authors. I. Title.
BV4501.2.S324 286'.73 79-505
79 MARC

Strong, June.
Where are we running? / By June Strong. — Nashville : Southern Pub. Association, c1979.
128 p. ; 17 cm.
ISBN 0-8127-0207-7
1. Christian life—Seventh-Day Adventist authors. I. Title.
[BV4501.2.S818] 248'.4 78-26271
78 MARC

—UNITED CHURCH OF CANADA AUTHORS

MacDougall, Bruce.
Rejoice in the Lord / Bruce MacDougall. — Nashville : Abingdon, c1979.
128 p. ; 20 cm.
ISBN 0-687-35933-3
1. Christian life—United Church of Canada authors. 2. MacDougall, Bruce. I. Title.
BV4501.2.M228 248'.2 78-10610
78 MARC

CHRISTIAN LIFE IN LITERATURE

Cazelles, Brigitte.
La faiblesse chez Gautier de Coinci / Brigitte Cazelles. — Saratoga, Calif. : Anma Libri, 1978.
180 p. ; 24 cm. — (Stanford French and Italian studies ; v. 14)
Bibliography: p. 172-180.
ISBN 0-915838-27-3
1. Gautier de Coincy, 1177?-1236. Les miracles de Nostre Dame. 2. Mary, Virgin, in fiction, drama, poetry, etc. 3. Christian life in literature. I. Title. II. Series.
PQ1463.G44M533 841'.1 79-100570
79 MARC

CHRISTIAN LITERATURE

see also Christian drama; Christianity in literature

Philipp, Wolfgang, 1915-1969, comp.
Der Protestantismus im 19. und 20. Jahrhundert. Bremen, Schünemann (1965).
cxxxv, 479 p. 18 cm. (Klassiker des Protestantismus, Bd. 8) DM19.80 (subscription price DM17.80) GDNB66-A10-62
Sammlung Dieterich, Bd. 273.
Includes bibliographies.
1. Christian literature. I. Title. II. Series.
BR53.P45 67-98087
[r79]rev MARC

—AUTHORSHIP

Writing and winning articles / Glenn F. Arnold, editor. — Nashville : T. Nelson, c1979.
247 p. ; 23 cm.
Bibliography: p. 223-225.
Includes index.
ISBN 0-8407-5682-8
1. Christian literature—Authorship. I. Arnold, Glenn F.
BR117.W7 1979 808'.066'2 79-10863
79 MARC

—DICTIONARIES—GERMAN

Kranz, Gisbert.
Lexikon der christlichen Weltliteratur / Gisbert Kranz. — Freiburg i. Br. ; Basel ; Wien : Herder, 1978.
1094 columns ; 25 cm. GFR78-A
Includes indexes.
ISBN 3-451-17949-0 : DM98.00
1. Christian literature—Dictionaries—German. I. Title.
BR117.3.K7 78-375743
78 MARC

—HISTORY AND CRITICISM

Schrey, Helmut.
Leben im Licht und Was ich sagen kann : Hören u. Antworten in d. Erbauungsliteratur d. Quäker : Versuch auf d. Grenze von Literatur u. Literaturwiss., Konfessionskunde u. Glaubenswirklichkeit : Antworten auf offene Fragen, Woodbrooker Ergebnisse Sommer 1976 / Helmut Schrey. — Kastellaun : Henn [in Komm.], 1978.
129 p. ; 21 cm. — (Hochschulschriften : wissenschaftliche Grenzgänge) GFR79-A
Includes texts in English.
Bibliography of the author's works: p. 128-129.
Includes bibliographical references.
ISBN 3-450-19939-0 : DM12.80
1. Friends, Society of—Doctrinal and controversial works—Friend authors. 2. Christian literature—History and criticism. I. Title.
BX7731.2.S37 79-368569
79 MARC

CHRISTIAN LITERATURE, EARLY

see also Fathers of the church; Jesus Christ—Biography—Apocryphal and legendary literature; Latin literature, Medieval and modern; Literature, Medieval; Liturgies, Early Christian; Theology—Collected works—Early church ; Theology —Early Church, Ca. 30-600

Apostolic Fathers. English.
The Apostolic Fathers / edited by Jack N. Sparks. — Nashville : T. Nelson, c1978.
xi, 319 p. ; 20 cm.
CONTENTS: The letter of Clement of Rome to Corinth (I Clement).—An ancient homily by an unknown author (II Clement).—The letters of Ignatius of Antioch.—Polycarp of Smyrna.—The Shepherd of Hermas.—The letter of Barnabas.—The teaching of the Lord to the Gentiles by the twelve apostles (the Didache).
ISBN 0-8407-5661-5 pbk. : $4.95
1. Christian literature, Early. I. Sparks, Jack N.
BR60.A62 1978 230 78-14870
78 MARC

Giles, Edward, ed.
Documents illustrating papal authority, A.D. 96-454 / edited and introduced by E. Giles. — Westport, Conn. : Hyperion Press, 1979.
xxi, 344 p. ; 22 cm.
Reprint of the 1952 ed. published by S.P.C.K., London.
Includes bibliographical references and indexes.
ISBN 0-88355-696-0 : $25.00
1. Papacy—History—Sources. 2. Church history—Primitive and early church, ca. 30-600—Sources. 3. Christian literature, Early. I. Title.
[BX953.G5 1979] 262'.13 78-59023
78 MARC

Harkness, Georgia Elma, 1891-
A devotional treasury from the early church, compiled by Georgia Harkness. Nashville, Abingdon Press [1968]
160 p. 21 cm.
Bibliography: p. 157-160.
1. Devotional literature. 2. Christian literature, Early. I. Title.
BV4801.H27 242'.1'09015 68-17436
[r79]rev MARC

Leonetti, Alfonso, 1909- comp.
Dieu ou Mammon; choix de textes de l'Écriture et des premiers Pères de l'Église, autour de la question sociale. Paris, Éditions ouvrières [1954]
141 p. 19 cm.
1. Wealth. 2. Poverty. 3. Christian literature, Early (Selections: Extracts, etc.) I. Bible. French. Selections. 1954. II. Title.
BR115.W4L4 55-18709
[r79]rev2 MARC

—COLLECTED WORKS

Philokalia tōn hierōn nēptikōn. Italian.
Φιλοκαλία : testi di ascetica e mistica della Chiesa orientale / a cura di Giovanni Vannucci. — Firenze : Libreria editrice fiorentina, 1978–
v. ; 21 cm. — (Ricerca del Graal ; 2)
It 78–Aug (v. 1)
"In the eighteenth century, Macarius of Corinth and Nicodemus of the Holy Mountain compiled ... the Philokalia ... [which was] published in Venice in ... 1792," under title: Philokalia tōn hierōn nēptikōn." Cf. Early Fathers from the Philokalia, 1954.
L2500 (v. 1)
1. Spiritual life—Orthodox Eastern authors. 2. Christian literature, Early—Collected works. I. Makarios, Saint, Metropolitan of Corinth, d. 1805. II. Nikodēmos Hagioreitēs, 1748 or 9–1809. III. Vannucci, Giovanni. IV. Title.
BX382.P4516 79–412093

—GREEK AUTHORS—BIBLIOGRAPHY

Geerard, Maurice.
Clavis Patrum Graecorum / cura et studio Mauritii Geerard. — Turnhout : Brepols, 1974-
v. ; 26 cm. — (Corpus Christianorum)
Added t.p.: Clavis Patrum Graecorum : qua optimae quaeque scriptorum Patrum Graecorum recensiones a primaevis saeculis usque ad Octavum commode recluduntur.
Bibliography: v. 2, p. [xiii]-xxii.
Includes index.
CONTENTS: v. 2. Ab Athanasio ad Chrysostomum.
1. Christian literature, Early—Greek authors—Bibliography. 2. Theology—Early church, ca. 30-600—Bibliography. I. Title. II. Series.
Z7791.G43 016.23'01'1 79-114761
[BR60] 79 MARC

—GREEK AUTHORS—HISTORY AND CRITICISM

Pellegrino, Michele.
Letteratura greca cristiana / Michele Pellegrino. — 3. ed. riveduta e aggiornata. — Roma : Studium, 1978.
230 p. ; 17 cm. — (Nuova universale Studium ; 29) It78-Nov
Bibliography: [209]-230.
L4000
1. Christian literature, Early—Greek authors—History and criticism. I. Title.
BR67.P43 1978 79-341704
*79 MARC

—HISTORY AND CRITICISM

Altaner, Berthold, 1885-1964.
Patrologie : Leben, Schriften u. Lehre d. Kirchenväter / Berthold Altaner, Alfred Stuiber. — 8., durchges. u. erw. Aufl. — Freiburg im Breisgau ; Basel ; Wien : Herder, 1978.
xxii, 672 p. ; 22 cm. GFR79-A
Includes bibliographical references and index.
ISBN 3-451-13580-9 : DM78.00
1. Christian literature, Early—History and criticism. 2. Fathers of the church—Bio-bibliography. 3. Church history—Primitive and early church, ca. 30-600—Bibliography. I. Stuiber, Alfred, joint author. II. Title.
BR67.A37 1978 79-353776
79 MARC

Dibelius, Martin, 1883-1947.
A fresh approach to the New Testament and early Christian literature / by Martin Dibelius. — Westport, Conn. : Greenwood Press, 1979.
280 p. ; 23 cm.
Reprint of the 1936 ed. published by I. Nicholson and Watson, London, in series: The International library of Christian Knowledge.
Bibliography: p. 274.
Includes index.
ISBN 0-8371-4219-9
1. Bible. N.T.—Introductions. 2. Christian literature, Early—History and criticism. I. Title. II. Series: The International library of Christian knowledge.
BS2330.D5 1979 225.6 78-32096
79 MARC

CHRISTIAN LITERATURE, EARLY —HISTORY AND CRITICISM (Continued)

Leone, Luigi.
Cristianesimo e cultura antica in alcuni scrittori cristiani / Luigi Leone. — ⌐Lecce⌐ : Milella, ⌐1976?⌐
102 p. ; 21 cm. It78-June
Bibliography: p. ⌐93⌐-102.
L3000
1. Christianity and culture—History. 2. Christian literature, Early—History and criticism. I. Title.
BR67.L46 78-398374
*79 MARC

Ogilvie, Robert Maxwell.
The library of Lactantius / by R. M. Ogilvie. — Oxford ⌐Eng.⌐ : Clarendon Press, 1978.
viii, 121 p. ; 23 cm. GB***
Includes bibliographical references and index.
ISBN 0-19-826645-6 : £7.50
1. Lactantius, Lucius Caecilius Firmianus. 2. Classical literature—History and criticism. 3. Christian literature, Early—History and criticism. I. Title.
BR65.L26O35 270.1'092'4 78-40087
78 MARC

Papadopoulos, Stylianos G
(Patrologia)
Πατρολογία / Στυλιανοῦ Γ. Παπαδοπούλου. — 'Αθήνα : Παπαδόπουλος, 1977.
v. ; 25 cm.
L. C. copy imperfect : v. 1, p. 133–134 mutilated.
Includes bibliographies and index.
CONTENTS : τ. 1. Εἰσαγωγή, Β' καὶ Γ' αἰώνας.
1. Christian literature, Early—History and criticism. I. Title.
BR67.P36 78–420782

Plutarch's ethical writings and early Christian literature / ed. by Hans Dieter Betz. — Leiden : Brill, 1978.
xi, 584 p. ; 25 cm. — (Studia ad corpus Hellenisticum Novi Testamenti ; v. 4) Ne78-41
Includes bibliographical references and indexes.
ISBN 9004056599 : fl 280.00
1. Plutarchus—Criticism and interpretation. 2. Plutarchus—Religion and ethics. 3. Christian literature, Early—History and criticism. I. Betz, Hans Dieter. II. Series.
PA4383.P58 888'.01'07 79-303632
*79 MARC

White, Helen Constance, 1896-1967.
The Tudor books of private devotion / by Helen C. White. — Westport, Conn. : Greenwood Press, 1979, c1951.
284 p., ⌐2⌐ leaves of plates : ill. ; 24 cm.
Reprint of the ed. published by University of Wisconsin Press, Madison.
Bibliography: p. ⌐251⌐-255.
Includes index.
ISBN 0-313-21063-2
1. Devotional literature—History and criticism. 2. Christian literature, English—History and criticism. I. Title.
⌐BV4818.W48 1979⌐ 242 78-21661
78 MARC

—HISTORY AND CRITICISM— ADDRESSES, ESSAYS, LECTURES

Latinität und alte Kirche : Festschrift f. Rudolf Hanslik zum 70. Geburtstag / ⌐die Red. besorgten Herbert Bannert u. Johannes Divjak⌐. — Wien ; Köln ; Graz : Böhlau, 1977.
372 p. ; 24 cm. — (Wiener Studien : Beiheft ; 8) Au77-15-165
Includes texts in English and Latin.
Includes bibliographical references.
Bibliography of R. Hanslik's works: p. ⌐363⌐-372.
ISBN 3-205-07026-0 : S824.00
1. Latin philology—History—Addresses, essays, lectures. 2. Christian literature, Early—History and criticism—Addresses, essays, lectures. 3. Hanslik, Rudolf. I. Hanslik, Rudolf. II. Bannert, Herbert. III. Divjak, Johannes. IV. Series.
PA2026.H28L3 78-395289
*79 MARC

—HISTORY AND CRITICISM— CONGRESSES

Christianisme et formes littéraires de l'Antiquité tardive en Occ dent : huit exposés suivis de discussions / par Alan Cameron ⌐et al.⌐ ; avec la participation de Helena Junod-Ammerbauer François Paschoud ; entretiens préparés et présidés par Manfr Fuhrmann. — Genève : Fondation Hardt ; Berne : dépositai pour la Suisse, Francke, cop. 1977.
500 p. ; 23 cm. — (Entretiens sur l'antiquité classique ; t. 23) Sw78-27
French, English, or German.
Includes bibliographical references and index.
70.00F
1. Christian literature, Early—History and criticism—Congresses. 2. Lat literature—History and criticism—Congresses. I. Cameron, Alan. II. Fuh mann, Manfred. III. Series.
BR67.C5 281'.1 78-3849
78 MAR

CHRISTIAN LITERATURE, ENGLISH

—COLLECTED WORKS

Bunyan, John, 1623-1688..
Upon a penny loaf : the wisdom of John Bunyan, author of Pilgrim's progress / selected and edited by Roger C. Palms. — Minneapolis : Bethany Fellowship, c1978.
152 p. ; 21 cm.
Includes index.
ISBN 0-87123-573-0 : $3.50
1. Christian literature, English—Collected works. I. Palms, Roger C. II. Title.
BR75.B7 1978 230 78-12239
78 MARC

—HISTORY AND CRITICISM

Hill, John Spencer, 1943-
John Milton, poet, priest and prophet : a study of divine vocation in Milton's poetry and prose / John Spencer Hill. — London : Macmillan, 1979.
xiv, 233 p. ; 23 cm. GB***
Includes bibliographical references and index.
ISBN 0-333-23768-4 : £10.00
1. Milton, John, 1608-1674—Criticism and interpretation. 2. Vocation in literature. 3. Christian literature, English—History and criticism. I. Title.
PR3588.H48 1979b 821'.4 79-310696
79 MARC

Hill, John Spencer, 1943-
John Milton, poet, priest, and prophet : a study of divine vocation in Milton's poetry and prose / John Spencer Hill. — Totowa, N.J. : Rowman and Littlefield, 1979.
xiv, 233 p. ; 23 cm.
Includes bibliographical references and index.
ISBN 0-8476-6124-5
1. Milton, John, 1608-1674—Criticism and interpretation. 2. Vocation in literature. 3. Christian literature, English—History and criticism. I. Title.
PR3588.H48 1979 821'.4 78-9941
78 MARC

—WALES

Jenkins, Geraint H
Literature, religion and society in Wales, 1660-1730 / by Geraint H. Jenkins. — Cardiff : University of Wales Press ⌐for⌐ the History and Law Committee of the Board of Celtic Studies, 1978.
⌐10⌐, 351 p., plate : facsim. ; 23 cm. — (Studies in Welsh history ; 2) GB78-14068
Bibliography: p. ⌐310⌐-332.
Includes index.
ISBN 0-7083-0669-1 : £9.50
1. Christian literature, English—Wales. 2. Christian literature, Welsh—Wales. 3. Wales—Church history. 4. Wales—Religious life. 5. Wales—Social life and customs. I. Wales. University. Board of Celtic Studies. History and Law Committee.. II. Title. III. Series.
BR117.J46 274.29 79-301119
79 MARC

CHRISTIAN LITERATURE, FRENCH

Vie de nostre benoît Sauveur IhesusCrist.
La vie de nostre benoît Sauveur IhesusCrist and La Saincte vie de Nostre Dame : translatee a la requeste de tres hault et Puissant Prince Iehan, duc de Berry / ⌐edited by⌐ Millard Meiss and Elizabeth H. Beatson. — New York : New York University Press for the College Art Association of America, 1977.
xxxvi, 186 p. : ill. ; 29 cm. — (Monographs on archaeology and the fine arts ; 32)
Text based chiefly on Mss. fr. 992, in the Bibliothèque nationale, Paris.
The first part of La Vie de nostre benoît Sauveur IhesusCrist is attributed to J. Gerson.
ISBN 0-8147-5410-4
1. Christian literature, French. 2. French literature—To 1500. I. Meiss, Millard. II. Beatson, Elizabeth Home. III. Gerson, Joannes, 1363-1429. La vie de nostre benoît Sauveur IhesusCrist. IV. Saincte vie de Nostre Dame. 1977. V. Series.
PQ1391.V5 843'.1 76-16657
76⌐r78⌐rev MARC

CHRISTIAN LITERATURE, GERMAN

—HISTORY AND CRITICISM

Hoffmann, Rüdiger.
Religiöse Jugendliteratur : eine Analyse d. Weltbildes kirchl. Verteilblätter / Rüdiger Hoffmann. — Hildesheim : Gerstenberg, 1974.
xiii, 234 p. : ill. ; 21 cm. — (Studia irenica ; 16) GFR75-A
"Eine überarbeitete Fassung der Arbeit, die ich unter dem Titel 'Analyse des Weltbildes religiöser Jugendliteratur ...' in Frankfurt am Main zur ersten Staatsprüfung vorgelegt habe."
Bibliography: p. 227-234.
ISBN 3-8067-0216-0 : DM28.00
1. Christian literature for children—History and criticism. 2. Christian literature, German—History and criticism. I. Title. II. Series.
BR117.5.H63 79-376085
79 MARC

Kuschel, Karl-Josef, 1948-
Jesus in der deutschsprachigen Gegenwartsliteratur / Karl-Josef Kuschel ; mit einem Vorw. von Walter Jens. — 2. Aufl., 3. Taus. — Zürich ⌐etc.⌐ : Benziger ; Gütersloh : Gütersloher Verlagshaus G. Mohn, cop. 1978.
xviii, 383 p. ; 23 cm. — (Ökumenische Theologie ; Bd. 1) Sw79-572
Originally presented as the author's thesis, Tübingen.
Bibliography: p. 356-375.
Includes index.
ISBN 3-545-24057-6 (Benziger) : 39.00F
1. German literature—20th century—History and criticism. 2. Christian literature, German—History and criticism. 3. Jesus Christ in fiction, drama, poetry, etc. I. Title. II. Series.
PT405.K79 1978 830'.9'351 79-368183
79 MARC

CHRISTIAN LITERATURE, JAPANESE

(Kirishitanbanshū)
きりしたん版集. — 天理：天理大学出版部 ; 東京：発売　八木書店, 昭和51-　⌐1976-
v. in ; 22 cm. — (天理図書館善本叢書 : 和書之部 ; 第38　巻) Ja***
Photoreprint of ms. copies.
CONTENTS: 1. pt. 1 ばうちずもの授けやう・ぎやどぺかるど(上巻)・おらしよの翻譯・こんてむつすむん地—1. pt. 2附録　ぎやどぺかるど(下巻)
Bibliography: v. 1, pt. 1, p. 4–7 (2d group)
¥ 108000 (v. 1)

1. Christian literature, Japanese. I. Tenri Toshokan. Tenri Toshokan zempon sōsho ; Washo no bu ; dai 38-kan etc.
BR53.K635 1976 79–805688

CHRISTIAN LITERATURE, LATIN

—HISTORY AND CRITICISM— ADDRESSES, ESSAYS, LECTURES

Saggi e documenti di letteratura religiosa medioevale / a cura dell'Istituto di storia e dell'Istituto di filologia moderna della Facoltà di magistero, Università degli studi di Trieste. — San Daniele del Fr⌐i⌐uli⌐ : Grillo, 1975.
152 p. : ill. ; 24 cm. — (Collana Saggi e documenti ; 1) It77-Jan
Includes bibliographical references.
L4000
1. Christian literature, Latin—History and criticism—Addresses, essays, lectures. 2. Latin literature, Medieval and modern—Addresses essays, lectures. I. Trieste. Università. Facoltà di magistero. Istituto di storia. II. Trieste. Università. Facoltà di magistero. Istituto di filologia moderna.
BR67.S26 78-379546
*78 MARC

CHRISTIAN LITERATURE, WELSH

—WALES

Jenkins, Geraint H
Literature, religion and society in Wales, 1660-1730 / by Geraint H. Jenkins. — Cardiff : University of Wales Press ⌐for⌐ the History and Law Committee of the Board of Celtic Studies, 1978.
⌐10⌐, 351 p., plate : facsim. ; 23 cm. — (Studies in Welsh history ; 2) GB78-14068
Bibliography: p. ⌐310⌐-332.
Includes index.
ISBN 0-7083-0669-1 : £9.50
1. Christian literature, English—Wales. 2. Christian literature, Welsh—Wales. 3. Wales—Church history. 4. Wales—Religious life. 5. Wales—Social life and customs. I. Wales. University. Board of Celtic Studies. History and Law Committee.. II. Title. III. Series.
BR117.J46 274.29 79-301119
79 MARC

CHRISTIAN LITERATURE FOR CHILDREN

—ADDRESSES, ESSAYS, LECTURES

Kinderbuch und Religion / hrsg. von Reinhold Jacobi ; mit Beitr. von Anton Baumeister ... ⌐et al.⌐. — Regensburg : Pustet, 1979.
136 p. ; 19 cm. GFR79-A
Includes bibliographical references.
ISBN 3-7917-0594-6 : DM16.80
1. Christian literature for children—Addresses, essays, lectures. I. Jacobi, Reinhold, 1941-
BR117.5.K56 79-381287
79 MARC

CHRISTIAN LITERATURE FOR CHILDREN (Continued)

—HISTORY AND CRITICISM

Hoffmann, Rüdiger.
Religiöse Jugendliteratur : eine Analyse d. Weltbildes kirchl. Verteilblätter / Rüdiger Hoffmann. — Hildesheim : Gerstenberg, 1974.
xiii, 234 p. : ill. ; 21 cm. — (Studia irenica ; 16) GFR75-A
"Eine überarbeitete Fassung der Arbeit, die ich unter dem Titel 'Analyse des Weltbildes religiöser Jugendliteratur ...' in Frankfurt am Main zur ersten Staatsprüfung vorgelegt habe."
Bibliography: p. 227-234.
ISBN 3-8067-0216-0 : DM28.00
1. Christian literature for children—History and criticism. 2. Christian literature, German—History and criticism. I. Title. II. Series.
BR117.5.H63 79-376085
79 MARC

CHRISTIAN MARTYRS

see also Martyrologies

Torres de Castilla, Alfonso.
Historia de las persecuciones / A. Torres de Castilla. — Barcelona : Petronio, c1978.
2 v. : ill. ; 20 cm. Sp***
Published in 1863-66 under title: Historia de las persecuciones políticas y religiosas.
ISBN 847250591X : 300ptas per vol.
1. Persecution. 2. Christian martyrs. 3. Religion and science—1860-1899. 4. Inquisition. 5. Europe—History—1492- I. Title.
BR1601.T7 1978 78-380546
78 MARC

—BIOGRAPHY

Augsburger, Myron S
Faithful unto death : fifteen young people who were not afraid to die for their faith / Myron S. Augsburger. ; [etchings by Jan Luyken]. — Waco, Tex. : Word Books, c1978.
125 p. : ill. ; 22 cm.
SUMMARY: Presents protraits drawn from court testimony, letters, and other historical records of 16th-century Anabaptist youths, primarily from Holland, Belgium, and Germany, who were persecuted for their religious beliefs.
ISBN 0-8499-0067-0 : $5.95
1. Anabaptists—Biography. 2. Christian martyrs—Biography. 3. Youth—Religious life. [1. Anabaptists. 2. Martyrs. 3. Christian life] I. Title.
BX4940.A9 284'.3 77-92462
79 MARC

Hefley, James C
China : Christian martyrs of the 20th century : an excerpt from By their blood / by James and Marti Hefley. — Milford, MI : Mott Media, c1978.
78 p. ; 23 cm.
ISBN 0-915134-28-4. ISBN 0-915134-16-0 pbk.
1. Christian martyrs—Biography. 2. Missionaries—China—Biography. 3. Christian biography—China. I. Hefley, Marti, joint author. II. Title.
BR1608.C6.H43 272'.092'2 78-6187
78 MARC

Märtyrbuch; denckwürdige Reden vnnd Thaten vieler H. Märtyrer, welche nach den Zeitten der Aposteln biss auffs Jar Christi 1597 in Teutschland, Franckreich, Engelland, Schottland, Niderland, Italien, Hispanien c. (etc.) vmb der Euangelischen Warheit willen verfolget vnnd jämmerlich gemartert worden. Auss den grossen französischen Actis martyrum fleissig aussgezogen vnd verteutschet vnd mit dem eylfften vnd zwelfften Buch so vormaln nicht darbey gewesen, gemehret. Basel [L. König] 1597.
[12], 834, [15] p. 18 cm.
Abridged version in German (originally made in ca. 1590 by C. Rab?) of J. Crespin's Le livre des martyrs, published by S. Goulart in updated editions under title Histoire des martyrs; the present German text revised and enlarged by L. König on the basis of the French ed. published in Geneva in 1597 which included two new chapters by Goulart.
Includes index.
1. Christian martyrs—Biography. 2. Persecution. I. Rab, Christoph, 1552-1620. II. Crespin, Jean, d. 1572. Histoire des martyrs. III. Goulart, Simon, 1543-1628. IV. König, Ludwig, 1572-1641.
BR1600.M28 78-375849
78 MARC

—ANGOLA—BIOGRAPHY

Fernando da Riese Pio X, Father.
Un martire triestino : P. Piergiovanni Filippi / Fernando da Riese Pio X. — 2. ed. — Montuzza-Trieste : Convento dei frati minori cappuccini, 1962.
88 p., [8] leaves of plates : ill. ; 17 cm.
1. Filippi, Piergiovanni, 1927-1961. 2. Missionaries—Angola—Biography. 3. Missionaries—Italy—Biography. 4. Christian martyrs—Angola—Biography. I. Title.
BV3625.A62F543 1962 79-376732
79 MARC

—BULGARIA—SOFIA

Sluzhba s zhitiem i stradaniem svetago velikomuchenika Georgia Novago : [otpech. i izd. ot Nikola Karastoĭanov] Samokov, 1855 / [predgovor Petŭr Dinekov ; redaktor Panaĭot Penev]. — Fototip. izd. — Sofiia : Septemvri, 1978.
74 p. : ill. ; 22 cm. Bu79-134
Reprint of the 1855 ed. published by Nikola Karastoĭanov, Samokov.
Accompanying text containing foreword by Petŭr Dinekov, issued as suppl. (16 p.) and inserted at end.
2.00lv
1. Novi, Georgi, d. 1515. 2. Christian martyrs—Bulgaria—Sofia. 3. Orthodox Eastern Church—Biography. 4. Sofia—Biography. I. Penev, Panaĭot.
BX395.N63S55 1978 79-369843
*79 MARC

—ENGLAND

Martin, Gregory, d. 1582.
The love of the soule / Gregory Martin. A catalogue of martyrs in England / Thomas Worthington. M. Antonius ... declares the cause of his returne / Marco Antonio de Dominis. — Ilkley [Eng.] : Scolar Press, 1978.
ca. 400 p. in various pagings ; 19 cm. — (English recusant literature, 1558-1640 ; v.363) GB***
Martin's work is a reprint of the 1619 ed. from a copy in the library of St. Edmund's College, Ware. References: Allison and Rogers 530; STC 17506/17134. Worthington's work is a reprint of the 1608 ed. from a copy in the library of Oscott College. Reference: Allison and Rogers 916; not in STC. Dominis' work is a reprint of the 1623 ed. from a copy in the library of Trinity College, Cambridge. References: Allison and Rogers 272; STC 7000.
ISBN 0-85967-448-7 : £10.00
1. Catholic Church—Doctrinal and controversial works—Catholic authors. 2. Protestantism—Controversial literature. 3. Christian martyrs—England. I. Worthington, Thomas, 1549-1627. A catalogue of martyrs in England. 1978. II. Dominis, Marcoantonio de, Abp., 1566-1624. M. Antonius ... declares the cause of his returne. 1978. III. Title. IV. Series: Rogers, David Morrison, comp. English recusant literature, 1558-1640 ; v. 363.
BX1750.A1E5 vol. 363 282.41 s 78-320571
78 MARC

—FRANCE—AVRANCHES—BIOGRAPHY

Toussaint, Joseph, chanoine.
Le Chanoine de Gouvets et les martyrs de l'Avranchin / J. Toussaint. — Avranches (20, rue de la Constitution, 50300) : Éditions l'Avranchin, [1977]
176 p. : ill. ; 24 cm. F78-4904
Includes bibliographies.
45.00F
1. Gouvets, Louis Georges de, d. 1796. 2. Catholic Church—Clergy—Biography. 3. Clergy—France—Biography. 4. Christian martyrs—France—Avranches—Biography. 5. Avranches, France—Biography. I. Title.
BX4705.G61752T68 282'.092'2 78-381708
78 MARC

—JAPAN

Rebellato, Leone.
Pellegrino in Giappone : nel centenario della scoperta dei cristiani / Leone Rebellato. — Padova : Segretariato missioni francescane, 1977.
251 p., [10] leaves of plates : ill. ; 24 cm. It78-Oct
Bibliography: p. 245-246.
1. Catholic Church in Japan—History. 2. Missions—Japan—History. 3. Christian martyrs—Japan. 4. Japan—Description and travel—1945- 5. Rebellato, Leone. I. Title.
BX1668.R4 79-358947
*79 MARC

—POLAND—BIOGRAPHY—ADDRESSES, ESSAYS, LECTURES

Saint Stanisław, Bishop of Kraków, in commemoration of the 900th anniversary of his martyrdom in 1079 / with a foreword by John Cardinal Krol ; [edited by Doyce B. Nunis, Jr.]. — Santa Barbara, Calif. : Saint Stanisław Publications Committee, 1979.
83 p. : ill. ; 27 cm.
ISBN 0-9602944-0-6
1. Stanisław, Saint, Bp. of Krakow, 1030?-1079—Addresses, essays, lectures. 2. Christian martyrs—Poland—Biography—Addresses, essays, lectures. 3. Krakow—History—Addresses, essays, lectures. I. Nunis, Doyce Blackman.
BX4700.S69S24 282'.092'4 79-65592
79 MARC

—SPAIN

Martínez, Emilio, 1849-1919.
Recuerdos de antaño : los mártires españoles de la Reforma del siglo XVI y la Inquisición / [por] Emilio Martínez. — Tarrasa, Barcelona : Clie, D.L. 1977.
411 p., [7] leaves of plates : ill. ; 18 cm. Sp77-Dec
Published in 1949 under title: Los mártires de España.
Includes bibliographical references.
ISBN 8472283178 : 270ptas.
1. Christian martyrs—Spain. 2. Persecution—Spain. 3. Protestants—Spain. 4. Reformation—Spain. 5. Inquisition—Spain. I. Title.
BR1608.S8M3 1977 78-387621
79 MARC

—UGANDA

Tarcisio, Brother.
The blood of the martyrs : drama (outline) and brief account of the Uganda martyrs / by Bro. Tarcisio. — [Kiteredde? Kyotera, Masaka, Uganda : s.n., 1969]
49 p. ; 21 cm.
Cover title.
1. Christian martyrs—Uganda. 2. Tarcisio, Brother. The blood of the martyrs. I. Title.
BR1608.U45T33 272 76-983584
79 MARC

CHRISTIAN METHODIST EPISCOPAL CHURCH

—DOCTRINAL AND CONTROVERSIAL WORKS

Johnson, Joseph Andrew, 1914-
Basic Christian Methodist beliefs / Joseph A. Johnson, Jr. — Shreveport, La. : Fourth Episcopal District Press, c1978.
iii, 268 p. ; 24 cm.
Bibliography: p. [257]-264.
Includes index.
1. Christian Methodist Episcopal Church—Doctrinal and controversial works. I. Title.
BX8467.J63 230'.7'83 78-111646
79 MARC

CHRISTIAN MINISTRY see Clergy—Office

CHRISTIAN NAMES see Names, Personal

CHRISTIAN PATRON SAINTS

Trüb, Carl Ludwig Paul, 1894-
Heilige und Krankheit / von C. L. Paul Trüb. — 1. Aufl. — Stuttgart : Klett-Cotta, 1978.
307 p. ; 23 cm. — (Geschichte und Gesellschaft ; Bd. 19) GFR78-A
Bibliography: p. 220-247.
Includes index.
ISBN 3-12-913150-7 : DM74.00
1. Sick. 2. Medicine and religion. 3. Faith-cure. 4. Christian patron saints. I. Title.
BT732.5.T78 79-339062
79 MARC

CHRISTIAN PERFECTION see Perfection

CHRISTIAN PILGRIMS AND PILGRIMAGES

see also Christian saints; Christian shrines

Baumer, Iso.
Wallfahrt heute / Iso Baumer, Walter Heim. — Freiburg/Schweiz [etc.] : Kanisius-Verlag, c1978.
62 p. ; 20 cm. Sw78-5982
ISBN 3-85764-057-X : 5.80F
1. Christian pilgrims and pilgrimages. I. Heim, Walter, 1922- . II. Title.
BX2323.B36 79-378410
79 MARC

Besnard, Albert Marie.
Par un long chemin vers Toi : le pèlerinage chrétien / A.-M. Besnard. — Paris : Éditions du Cerf, 1978, c1959.
124 p. ; 18 cm. — (Foi vivante ; 184) F***
Includes bibliographical references.
ISBN 2-204-01237-8 : 13.00F
1. Spiritual life—Catholic authors. 2. Christian pilgrims and pilgrimages. I. Title.
BX2350.2.B435 1978 248'.29 79-350826
79 MARC

—BIOGRAPHY

Récits d'un pèlerin russe / traduits et présentés par Jean Laloy. — Paris : Seuil, 1978.
184 p. ; 18 cm. — (Points) (Sagesses ; 14) F78-10904
Translation of Otkrovennye rasskazy strannika dukhovnomu svoemu ottsu, published in 1948 in French under title: Récits d'un pèlerin à son père spirituel.
Reprint of the "nouvelle édition revue et mise à jour," published by Seuil, 1966.
ISBN 2-02-004775-6 : 9.50F
1. Christian pilgrims and pilgrimages—Biography. I. Laloy, Jean.
BX597.Z9O74 1978 248'.29 79-375625
00 MARC

CHRISTIAN PILGRIMS AND PILGRIMAGES —BIOGRAPHY (Continued)

Relazioni di un pellegrino : da un manoscritto del Monte Athos / a cura di Divo Barsotti. — 2. ed. — Firenze : Libreria editrice fiorentina, 1978.
127 p. ; 21 cm. — (Ricerca del Graal ; 5) It78-Oct
Translation of Otkrovennye rasskazy strannika dukhovnomu svoemu otts͡u.
L1500
1. Christian pilgrims and pilgrimages—Biography. I. Barsotti, Divo, 1914-
BX597.Z9O743 1978 79-361754
•79 MARC

—BELGIUM—DIKSMUIDE

Ghein, Rik de.
Een halve eeuw Vlaams idealisme : de IJzerbedevaarten in de archieven van het IJzerbedevaartcomité en De standaard en haar lezers / Rik de Ghein, Clem de Ridder. — Tielt : Lannoo, 1977.
128 p. : ill. ; 26 cm. Ne77-26
ISBN 902090681X : fl 25.00
1. Flemish movement. 2. Demonstrations—Belgium—Diksmuide. 3. Dixmude, Battle of, 1904. 4. Christian pilgrims and pilgrimages—Belgium—Diksmuide. I. Ridder, Clem de, joint author. II. IJzerbedevaartcomité. III. De Standaard. IV. Title.
DH491.G46 78-394047
•79 MARC

—ENGLAND

Finucane, Ronald C
Miracles and pilgrims : popular beliefs in medieval England / Ronald C. Finucane. — Totowa, N.J. : Rowman and Littlefield, 1977.
248 p., ₍8₎ leaves of plates : ill. ; 24 cm.
Bibliography: p. ₍242₎-244.
Includes index.
ISBN 0-87471-831-7
1. England—Church history—Medieval period, 1066-1485. 2. Miracles. 3. Christian pilgrims and pilgrimages—England. 4. Christian saints—Cult. 5. Christian shrines—England. I. Title.
BR747.F56 1977 274.2 77-21771
77₍r79₎rev MARC

—FRANCE

Belot, Victor R
La France des pèlerinages / Victor H. ₍sic₎ Belot. — Verviers, r. de Limbourg 65 : Marabout, ₍1977₎
255 p. : ill. ; 18 cm. — (Guide Marabout ; 16) Be77-7
Includes indexes.
130F
1. Christian shrines—France. 2. Christian pilgrims and pilgrimages—France. I. Title.
BX2320.B4 914.4'04'83 78-382523
•79 MARC

—FRANCE—LOURDES

Zoffoli, Egle.
Un lungo cammino : diario di un'esperienza / Egle Zoffoli. — Bologna : EDB, 1978.
384 p. : ill. ; 19 cm. — (Testimoni ; 2) It78-Dec
L4500
1. Christian pilgrims and pilgrimages—France—Lourdes. 2. Zoffoli, Egle. 3. Italy, Northern—Description and travel. 4. France, Southern—Description and travel. I. Title.
BT653.Z63 79-342776
•79 MARC

—FRANCE—MONTLIGEON

Dutot, Louis.
Le pèlerinage Notre-Dame de Montligeon : son fondateur, son histoire / Louis Dutot. — ₍s.l. : s.n., 1977₎ (La Chapelle-Montligeon : Impr. de Montligeon)
161 p., ₍6₎ leaves of plates : ill. ; 23 cm. F***
Includes bibliographical references.
1. Buguet, Paul Joseph, 1843-1918. 2. Catholic Church—Clergy—Biography. 3. Clergy—France—Biography. 4. Basilique Notre-Dame, La Chapelle-Montligeon, France—History. 5. Christian pilgrims and pilgrimages—France—Montligeon. I. Title.
BX4705.B8947D87 282'.092'4 79-350514
79 MARC

—FRANCE—VÉZELAY

Lebailly, Monique.
Vézelay ; ou, les Quatre vents de l'Esprit / Monique Lebailly. — Paris : Éditions S.O.S., 1978.
208 p., ₍8₎ p. of plates ; 21 cm. — (Collection Hauts lieux de spiritualité) F79-5607
Bibliography: p. ₍207₎-208.
ISBN 2-7185-0848-5 : 45.00F
1. Christian pilgrims and pilgrimages—France—Vézelay. 2. Vézelay, France (Benedictine abbey) 3. Vézelay, France—Religious life and customs. 4. Lebailly, Monique. I. Title.
BX2321.V49L4 263'.042'4441 79-378320
79 MARC

—GREAT BRITAIN

Adair, John Eric, 1934-
The pilgrims' way : shrines and saints in Britain and Ireland / John Adair ; photos. by Peter Chèze-Brown. — London : Thames and Hudson, 1978.
208 p. : ill. (some col.) ; 27 cm. GB***
Bibliography: p. 205.
Includes index.
ISBN 0-500-25061-8 : £5.95
1. Christian shrines—Great Britain. 2. Christian shrines—Ireland. 3. Christian pilgrims and pilgrimages—Great Britain. 4. Christian pilgrims and pilgrimages—Ireland. I. Title.
BX2320.A3 914.1'04'857 78-320821
79 MARC

—IRELAND

Adair, John Eric, 1934-
The pilgrims' way : shrines and saints in Britain and Ireland / John Adair ; photos. by Peter Chèze-Brown. — London : Thames and Hudson, 1978.
208 p. : ill. (some col.) ; 27 cm. GB***
Bibliography: p. 205.
Includes index.
ISBN 0-500-25061-8 : £5.95
1. Christian shrines—Great Britain. 2. Christian shrines—Ireland. 3. Christian pilgrims and pilgrimages—Great Britain. 4. Christian pilgrims and pilgrimages—Ireland. I. Title.
BX2320.A3 914.1'04'857 78-320821
79 MARC

—NETHERLANDS

Buijtenen, M P van.
Langs de heiligenweg : perspectief van enige vroegmiddeleeuwse verbindingen met Noord-Nederland / M. P. van Buijtenen. — Amsterdam : Noord-Hollandsche U.M., 1977.
152 p., ₍6₎ leaves of plates (1 fold.) : ill. ; 26 cm. — (Verhandelingen - Koninklijke Nederlandse Akademie van Wetenschappen, Afd. Letterkunde ; nieuwe reeks, deel 94) Ne77-11
Includes bibliographical references.
ISBN 0-7204-8341-7 : fl 55.00
1. Christian pilgrims and pilgrimages—Netherlands. 2. Netherlands—Church history. 3. Netherlands—History—To 1384. I. Title. II. Series: Akademie van Wetenschappen, Amsterdam. Afdeling Letterkunde. Verhandelingen ; nieuwe reeks, deel 94.
BX2323.B83 78-377753
•78 MARC

—PALESTINE

Martin, James, 1921-
A pilgrim's guide to the Holy Land / James Martin. — Philadelphia : Westminster Press, ₍1979₎ c1978.
viii, 100 p., ₍8₎ leaves of plates : ill. ; 21 cm.
British ed. published in 1978 under title: A plain man in the Holy Land.
ISBN 0-664-24276-6
1. Palestine—Description and travel. 2. Christian pilgrims and pilgrimages—Palestine. 3. Martin, James, 1921- I. Title.
DS107.4.M3454 1979 915.694'04'5 79-14506
79 MARC

Martin, James, 1921-
A plain man in the Holy Land / by James Martin. — Edinburgh : St Andrew Press, 1978.
ix, 100 p., ₍18₎ p. of plates : ill. ; 20 cm. GB78-16516
ISBN 0-7152-0373-8 : £1.50
1. Palestine—Description and travel. 2. Christian pilgrims and pilgrimages—Palestine. 3. Martin, James, 1921- I. Title.
DS107.4.M3454 915.694'04'50924 78-319946
78 MARC

Smith, Alfred Charles, 1822-1898.
Narrative of a modern pilgrimage through Palestine on horseback, and with tents / by Alfred Charles Smith. — London : Society for Promoting Christian Knowledge, ₍1873?₎
xxiv, 416 p. : ill. ; 19 cm.
Includes bibliographical references and indexes.
1. Palestine—Description and travel. 2. Christian pilgrims and pilgrimages—Palestine. 3. Smith, Alfred Charles, 1822-1898. I. Title.
DS107.S58 915.694'04'3 79-311596
79 MARC

Tägt, Nils, 1913-
Från Siljan till Gennesaret : glimtar från en pilgrimsfärd till det Heliga landet / Nils Tägt ; ₍fotogr. av Nils-Olof Tägt₎. — Stockholm : Gummesson, 1979.
208 p., ₍10₎ leaves of plates : ill. ; 19 cm. S79
ISBN 9170705542
1. Israel—Description and travel. 2. Christian pilgrims and pilgrimages—Palestine. 3. Tägt, Nils, 1913- . I. Title.
DS107.4.T27 79-374316
79 MARC

—SILESIA

Peus, Busso, 1942-
Schlesische Wallfahrtsmedaillen vor 1945 ₍i.e. neunzehnhundertfünfundvierzig₎ / Busso Peus. — Hildesheim : Lax, 1977.
p. 161-225, ₍4₎ leaves of plates : ill. ; 21 cm. GFR78-A7
Cover title.
Bibliography: p. 161-162.
DM6.00
1. Medals, Devotional—Silesia. 2. Christian pilgrims and pilgrimages—Silesia. I. Title.
CJ5793.R34P48 78-382370
78 MARC

—SPAIN—SANTIAGO DE COMPOSTELA

Stokstad, Marilyn Jane, 1929-
Santiago de Compostela : in the age of great pilgrimages / by Marilyn Stokstad. — Norman : University of Oklahoma Press, c1978.
xiii, 177 p. : ill. ; 20 cm. — (The Centers of civilization series)
Bibliography: p. 165-172.
Includes index.
ISBN 0-8061-1454-1 : $5.95.
1. Santiago de Compostela—History. 2. Christian pilgrims and pilgrimages—Spain—Santiago de Compostela. I. Title. II. Series.
DP402.S23S77 946'.11 77-18612
79 MARC

—SPAIN—SANTIAGO DE COMPOSTELA —EXHIBITIONS

Vidal, Marguerite.
Moissac et le chemin de Compostelle : ₍exposition / notices du catalogue, Marguerite Vidal₎. — ₍Moissac₎ : Musées de Moissac, 1976.
vi, 25 p. : ill. ; 18 x 21 cm. F***
1. Moissac, France (Tarn et Garonne)—Exhibitions. 2. Quercy—Exhibitions. 3. Christian pilgrims and pilgrimages—Spain—Santiago de Compostela—Exhibitions. I. Title.
DC801.M69V523 704.948'2'0740475 77-550547
77₍r79₎rev MARC

—SPAIN—SEVILLE

Martínez Morellá, F
Romero en La Peregrina : singladuras del cuaderno de bitácora / F. Martínez Morellá. — ₍2. ed.₎. — Alicante : ₍s.n.₎, 1974.
28 p. ; 22 cm. Sp***
1. Christian pilgrims and pilgrimages—Spain—Seville. 2. Cano, Juan Sebastián del, d. 1526. I. Title.
BX2321.S48M37 1974 78-387749
78 MARC

CHRISTIAN POETRY

see also Jesus Christ—Poetry

Firth, Peter J
Lord of the seasons / by Peter Firth ; illustrations by Noeline Kelly. — Guildford ₍etc.₎ : Lutterwork Press, 1978.
80 p. : ill. ; 20 cm.
ISBN 0-7188-2355-9 : £1.20
1. Christian poetry. 2. Seasons—Poetry. I. Title.
PR6056.I525L6 821'.9'14 78-322890
79 MARC

CHRISTIAN POETRY, AMERICAN

Calkin, Ruth Harms.
Lord, I keep running back to you / Ruth Harms Calkin. — Wheaton, Ill. : Tyndale House Publishers, c1979.
112 p. : ill. ; 18 cm.
ISBN 0-8423-3820-9 : $1.95
1. Christian poetry, American. I. Title.
₍PS3553.A417L57₎ 811'.5'4 78-58748
00 MARC

CHRISTIAN POETRY, AMERICAN (Continued)

The Country of the risen King : an anthology of Christian poetry / Merle Meeter, compiler. — Grand Rapids : Baker Book House, c1978.
xii, 446 p. ; 25 cm.
Includes indexes.
ISBN 0-8010-6042-7 : $12.95
1. Christian poetry, American. 2. Christian poetry, English. I. Meeter, Merle.
PS595.C47C6 821'.008'031 77-87993
78[r78]rev MARC

Fey, Mark D
Water song : a photo and musical essay / [photography and verse, Mark D. Fey ; editors, Linda Merz, Gail Denham, Leona Fey]. — Tacoma, WA : Christographics, c1979.
40 p. : col. ill. ; 21 x 26 cm. & disc (33 1/3 rpm. stereo. 7 in.)
Cover title.
Disc: Water song, music and lyrics by Gary Baller.
1. Christian poetry, American. I. Merz, Linda. II. Denham, Gail. III. Fey, Leona. IV. Title.
PS3556.E95W3 811'.5'4 79-54764
79 MARC

Hutchens, James B
Trusting Jesus : poems / by James B. Hutchens, Jr. — [Morgantown, N.C.] : Hutchens, c1977.
[28] leaves ; 22 cm.
1. Jesus Christ—Poetry. 2. Christian poetry, American. I. Title.
PS3558.U815T7 811'.5'4 79-107277
79 MARC

Ideas on wings : a collection of poems from the Christian Science periodicals / [edited by Carol Chapin Lindsey ; photos. by Gordon N. Converse]. — Boston : Christian Science Pub. Society, c1978.
[124] p. : ill. ; 23 cm.
Includes index.
ISBN 0-87510-120-8
1. Christian poetry, American. 2. American poetry—Christian Science authors. I. Lindsey, Carol Chapin.
PS595.C47I3 811'.5'408031 78-110066
79 MARC

Kiemel, Ann.
I'm celebrating / Ann Kiemel ; [photos. by Fred Kiemel]. — Old Tappan, N.J. : F. H. Revell, c1979.
96 p. : ill. ; 24 cm.
ISBN 0-8007-6010-7
1. Christian poetry, American. I. Title.
PS3561.I345I4 811'.5'4 79-10400
79 MARC

Miller, Ursula, 1882-1972.
Prairie praises / by Ursula Miller. — Protection, Kan. : [L. M. Selzer], c1978.
viii, 130 p. : port. ; 23 cm.
1. Christian poetry, American. I. Title.
PS3563.I42146P7 1978 811'.5'4 78-61903
79 MARC

Mulliken, Frances Hartman, 1924-
Poetic voices of the Restoration. Independence, Mo., Herald Pub. House [1960]
227 p. 20 cm.
1. Christian poetry, American. 2. Mormons and Mormonism—Poetry. 3. Reorganized Church of Jesus Christ of Latter-Day Saints. I. Title.
PS595.R4M84 811'.5'208 60-14175
[r79]rev MARC

Poems of inspiration from the masters / compiled by James R. Mills. — Old Tappan, N.J. : Revell, c1979.
173 p. ; 24 cm.
ISBN 0-8007-1054-1
1. Christian poetry, English. 2. Christian poetry, American. I. Mills, James R., 1927-
PR1195.C48P6 821'.008'031 79-14266
79 MARC

Tanksley, Perry.
Come share the joy / Perry Tanksley. — Old Tappan, N.J. : Revell, c1979.
80 p. ; 27 cm.
ISBN 0-8007-0990-X
1. Christian poetry, American. I. Title.
PS3570.A53C6 811'.5'4 79-84153
79 MARC

—HISTORY AND CRITICISM

Hatfield, Edwin Francis, 1807-1883.
The poets of the church : a series of biographical sketches of hymn-writers with notes on their hymns / by Edwin F. Hatfield. — Boston : Longwood Press, 1977.
vii, 719 p. ; 22 cm.
First published in 1884 by A. D. F. Randolph, New York.
Includes indexes.
ISBN 0-89341-195-7
1. Christian poetry, English—History and criticism. 2. Christian poetry, American—History and criticism. 3. Hymns, English—History and criticism. 4. Hymn writers—Biography. I. Title.
PR508.C65H3 1977 264'.2'0922 77-91533
78 MARC

CHRISTIAN POETRY, ARGENTINE

Brunero, María Alicia.
Tiempo de morir / María Alicia Brunero. — Buenos Aires : Editorial Guadalupe, c1976.
119 p. ; 20 cm. — (Colección Contemplación y acción ; 16)
1. Christian poetry, Argentine. 2. Death—Poetry. 3. Meditations. 4. Death—Meditations. I. Title.
PQ7798.12.R794T5 861 78-109431
78 MARC

Güiraldes, Ricardo, 1886-1927.
Poemas místicos / Ricardo Güiraldes. — 2. ed. — [Buenos Aires] : Editorial Ricardo Güiraldes, 1977.
26 p. ; 30 cm.
1. Christian poetry, Argentine. I. Title.
PQ7797.G75A17 1977 861 78-108431
79 MARC

CHRISTIAN POETRY, ARMENIAN

Hovhannēsian, Awetis.
(Erku osokhner)
Երկու ոսոխներ : մարմինն ու հոգին իրարու դէմ / գրեց Աւետիս Յովհաննէսեան. — 2. տպագրութիւն, 1. տպեալէն նորոգ բարեփոխեալ յաւելուածով մը. — Երուսաղէմ : ի Տպարանի Առաքելական Ս. Աթոռոյ, Սրբոց Յակոբեանց, 1911.
63 p. ; 19 cm.
A dialog in verse.
1. Christian poetry, Armenian. I. Title. II. Title: Marminn u hogin iraru dēm.
PK8548.H647E7 78-970776

Mkhit'ar Sebastats'i, 1676-1749.
Sei inni sacri di Mechitar di Sebaste in onore della SS. Vergine e di S. Antonio abate / traduzione italiana a cura di Giuseppe Frasson in occasione del terzo centenario della nascita di Mechitar, 1676-1976. — Venezia : S. Lazzaro, 1977.
51 p. ; 21 cm. It***
Added t.p. in Armenian.
Texts in Armenian and Italian.
1. Christian poetry, Armenian. 2. Christian poetry, Italian. I. Title.
PK8547.M53A56 1977 891'.992'14 79-351378
79 MARC

CHRISTIAN POETRY, BRAZILIAN

Lima, Heber Salvador de.
Reflexos / Heber Salvador de Lima. — São Paulo : Edições Loyola, [1978?]
143 p. : ill. ; 20 cm.
Cr$70.00
1. Christian poetry, Brazilian. I. Title.
PQ9698.22.I5234R4 869'.1 79-363832
79 MARC

CHRISTIAN POETRY, CANADIAN

Pallen, Anna Belle.
One Canada! : a poetic narration / by Anna Belle Pallen. — Toronto : Spiritual Press, c1977.
31 p. ; 18 cm. C78-9418-2
ISBN 0-910970-79-X : $2.00
1. Christian poetry, Canadian. 2. Canada—Poetry. I. Title.
PR9199.3.P32O5 811'.5'4 79-302662
79 MARC

CHRISTIAN POETRY, CHINESE

Ya-i.
(Ching yeh)
靜夜 / 雅億著. — 台北：中國主日學協會, [1976?]
[24] p. : ill. ; 19 cm. — (沙崙文選)
Cover title.
Poems.
Series romanized: Sha-lun wen hsüan.
1. Christian poetry, Chinese. I. Title.
PL2921.5.I19C5 78-841341

CHRISTIAN POETRY, DALMATIAN

Kašić, Bartol, 1575-1650.
Pjesni duhovne / Bartol Kašić. — Nachdruck / besorgt und eingeleitet von Hans-Joachim Härtel. — München : R. Trofenik, 1978.
[32], 167, [8] p. ; 23 cm. — (Beiträge zur Kenntnis Südosteuropas und des Nahen Orients ; Bd. 27) GFR***
Reprint of the 1617 ed., pri Bartolomeu Zannèttu u Rimu.
Introduction in German.
Bibliography of works by and about B. Kašić. : p. [13]-[16] (1st group)
ISBN 3-87828-126-9
1. Christian poetry, Dalmatian. I. Härtel, Hans-Joachim. II. Title. III. Series.
PG1658.K32P57 1978 79-357636
79 MARC

CHRISTIAN POETRY, DANISH

Christiansen, Lauritz U
Fra en vintræsgren : salmer og religiøse digte / af Lauritz U. Christiansen. — Esbjerg : L. U. Christiansen, 1977.
55 p. ; 21 cm. D77-8
ISBN 8798053604
1. Christian poetry, Danish. I. Title.
PT8176.13.H76F7 78-397990
79 MARC

CHRISTIAN POETRY, EARLY

—HISTORY AND CRITICISM

Gamber, Stanislas.
Le Livre de la Genèse dans la poésie latine au 5e siècle : versions françaises et provençales de la Genèse au Moyen Age / traduction de la Bible en langue d'oc, Stanislas Gamber. — Genève : Slatkine, 1977.
xvi, 263 p. ; 23 cm. Sw78-106
Reprint of the 1899 ed. published by A. Fontemoing, Paris.
Includes bibliographical references.
60.00F
1. Christian poetry, Latin—History and criticism. 2. Christian poetry, Early—History and criticism. 3. Bible in literature. I. Title.
PA6053.G35 1977 871'.01 79-344678
79 MARC

CHRISTIAN POETRY, ENGLISH

Colman, Henry.
Divine meditations : 1640 / Henry Colman ; edited with introd. and commentary by Karen E. Steanson. — New Haven : Published for the Elizabethan Club [by] Yale University Press, 1979.
xii, 206 p. : facsim. ; 21 cm. — (The Elizabethan Club seris ; 6)
Errata slip inserted.
Bibliography: p. 199-201.
Includes indexes.
ISBN 0-300-02305-7
1. Christian poetry, English. I. Steanson, Karen E. II. Title. III. Series: Elizabethan Club series ; 6.
PR3359.C27D5 1979 821'.4 78-25908
78 MARC

Firth, Peter J
Lord of the seasons / by Peter Firth ; ill. by Noeline Kelly. — Atlanta : John Knox Press, c1978.
80 p. : ill. ; 20 cm.
ISBN 0-8042-2408-0
1. Christian poetry, English. 2. Seasons—Poetry. I. Title.
PR6056.I525L6 1978b 823'.9'14 78-71047
78 MARC

George Herbert and the seventeenth-century religious poets : authoritative texts criticism / selected and edited by Mario A. Di Cesare. — 1st ed. — New York : Norton, c1978.
xvi, 401 p. ; 22 cm. — (A Norton critical edition)
Bibliography: p. 387-401.
ISBN 0-393-04398-3. ISBN 0-393-09254-2 pbk.
1. English poetry—Early modern, 1500-1700. 2. Christian poetry, English. 3. English poetry—Early modern, 1500-1700—History and criticism—Addresses, essays, lectures. 4. Religion in literature—Addresses, essays, lectures. I. Herbert, George, 1593-1633. II. Di Cesare, Mario A..
PR1209.G4 1978 821'.7'0931 77-28074
77 MARC

Kleinman, Hyman H
The religious sonnets of Dylan Thomas : a study in imagery and meaning / H. H. Kleinman. — New York : Octagon Books, 1979, c1963.
xii, 153 p. ; 23 cm.
Reprint of the ed. published by University of California Press, Berkeley, which was issued as no. 13 of Perspectives in criticism.
Bibliography: p. 147.
Includes index.
ISBN 0-374-94589-6 : $12.00
1. Thomas, Dylan, 1914-1953—Criticism and interpretation. 2. Sonnets, English—History and criticism. 3. Christian poetry, English—History and criticism. I. Title. II. Series: Perspectives in criticism ; 13.
[PR6039.H52Z74 1979] 821'.9'12 78-26190
78 MARC

CHRISTIAN POETRY, ENGLISH (Continued)

Lanier, Emilia.
The poems of Shakespeare's dark lady : Salve Deus Re Judaeorum / by Emilia Lanier ; introd. by A. L. Rowse. — London : Cape, 1978.
144 p., [6] leaves of plates : ill. ; 22 cm.
Text of the poems based on copies in the Bodelian Library and the British Library, which were originally published in 1611.
Includes bibliographical references.
ISBN 0-224-01631-8 : £4.95
1. Christian poetry, English. 2. Women—Poetry. I. Rowse, Alfred Leslie, 1903- . II. Title. III. Title: Shakespeare's dark lady.
PR2296.L27S2 1978 821'.3 79-300269
79 MARC

Lanier, Emilia.
The poems of Shakespeare's Dark Lady = Salve deus rex judaeorum / by Emilia Lanier ; introd. by A. L. Rowse. — New York : C. N. Potter : distributed by Crown Publishers, 1979.
xiii, 144 p., [6] leaves of plates : ports. ; 22 cm.
ISBN 0-517-53745-1
1. Christian poetry, English. 2. Women—Poetry. I. Rowse, Alfred Leslie, 1903- II. Title. III. Title: Shakespeare's Dark Lady. IV. Title: Salve deus rex judaeorum.
PR2296.L27S2 1979 821'.3 78-31141
79 MARC

Poems of inspiration from the masters / compiled by James R. Mills. — Old Tappan, N.J. : Revell, c1979.
173 p. ; 24 cm.
ISBN 0-8007-1054-1
1. Christian poetry, English. 2. Christian poetry, American. I. Mills, James R., 1927-
PR1195.C48P6 821'.008'031 79-14266
79 MARC

—HISTORY AND CRITICISM

Hatfield, Edwin Francis, 1807-1883.
The poets of the church : a series of biographical sketches of hymn-writers with notes on their hymns / by Edwin F. Hatfield. — Boston : Longwood Press, 1977.
vii, 719 p. ; 22 cm.
First published in 1884 by A. D. F. Randolph, New York.
Includes indexes.
ISBN 0-89341-195-7
1. Christian poetry, English—History and criticism. 2. Christian poetry, American—History and criticism. 3. Hymns, English—History and criticism. 4. Hymn writers—Biography. I. Title.
PR508.C65H3 1977 264'.2'0922 77-91533
78 MARC

Lewalski, Barbara Kiefer, 1931-
Protestant poetics and the seventeenth-century religious lyric / Barbara Kiefer Lewalski. — Princeton, N.J. : Princeton University Press, c1979.
xiv, 536 p., [6] leaves of plates : ill. ; 25 cm.
Includes bibliographical references and index.
ISBN 0-691-06395-8 : $22.50
1. English poetry—Early modern, 1500-1700—History and criticism. 2. Christian poetry, English—History and criticism. 3. Protestantism in literature. I. Title.
PR545.R4L48 821'.4'0931 78-70305
79 MARC

Osmond, Percy Herbert, 1872-
The mystical poets of the English church / by Percy H. Osmond. — Norwood, Pa. : Norwood Editions, 1978.
xi, 436 p. ; 23 cm.
Reprint of the 1919 ed. published by the Society for Promoting Christian Knowledge, London.
Bibliography: p. 420-431.
Includes index.
ISBN 0-8482-2019-6 lib. bdg. : $40.00
1. English poetry—History and criticism. 2. Mysticism in literature. 3. Christian poetry, English—History and criticism.
[PR508.M8O8 1978] 821'.009 78-2470
78 MARC

CHRISTIAN POETRY, FINNISH

Tiililä, Osmo.
Syvällä kulkee väkevä virta : valitut runot / Osmo Tiililä (Morosus"). — [Helsinki] : Alea-kirja, 1978 (Saarijärvi : Ari-p.).
176 p. ; 25 cm. Fi78-6
ISBN 9519272445
1. Christian poetry, Finnish. I. Title.
PH355.T477S95 79-372452
79 MARC

Turtiainen, Eino M
Tunnustan sinulle, meri / Eino M. Turtiainen. — Turku : E. Turtiainen, 1977 (Turku : Pika).
56 p. : ill. ; 21 cm. Fi78-5
1. Christian poetry, Finnish. I. Title.
PH355.T835T8 79-356050
79 MARC

Viikilä, Marjatta.
Avoimella kämmenellä : runoja / Marjatta Viikilä. — Tikkurila : RV, 1975.
67 p. ; 21 cm. Fi77Dec.
ISBN 9516052851
1. Christian poetry, Finnish. I. Title.
PH355.V436A96 79-356630
79 MARC

Viljamaa, Ritva.
Kultaomenia hopeamaljassa : runoja / Ritva Viljamaa. — Hämeenlinna : Päivä, 1977 (Saarijärvi : Ari-p.).
114, [2] p. ; 18 cm. — (Päivän runosarja ; 5) Fi78-4
ISBN 951622170X
1. Christian poetry, Finnish. I. Title.
PH355.V497K8 79-339855
79 MARC

CHRISTIAN POETRY, FRENCH

Matossian, André.
L'Enjeu / André Matossian. — Paris : Éditions Saint-Germain-des-Prés, 1977.
73 p. ; 19 cm. F78-254
ISBN 2-243-00569-7 : 32.00F
1. Christian poetry, French. I. Title.
PQ2673.A7844E5 841'.9'14 79-339454
79 MARC

Murard, François Régis de.
Symphonie en noir et blanc : florilège / François-Régis de Murard ; [publié par l'Abbaye Saint-Michel de Frigolet]. — [Tarascon] (Abbaye de Frigolet, 13150) : F.-R. de Murard, 1976.
60 p. ; 21 cm. F77-1070
1. Christian poetry, French. I. Title.
PQ2673.U577S9 841'.9'14 78-398623
78 MARC

Vuaillat, Jean.
Mariales, suivi de Signets pour Noël / Jean Vuaillat. — [s.l. : s.n., 1978] (Aurillac : Impr. Gerbert)
62 p. ; 20 cm. F***
1. Christian poetry, French. I. Vuaillat, Jean. Signets pour Noël. 1978. II. Title: Mariales.
PQ2643.U4M3 841'.9'14 79-362513
79 MARC

CHRISTIAN POETRY, GASCON

Lesbats, Julien.
Chantons en gascon! / par J. Lesbats. — Labrit (40420) : J. Lesbats, 1975.
48 p. ; 24 cm. F77-5820
In Gascon dialect and French.
Cover title: Cantam en gascoum!
5.00F
1. Christian poetry, Gascon. I. Title. II. Title: Cantam en gascoum!
PC3402.L47C47 849'.1'54 78-397067
78 MARC

CHRISTIAN POETRY, GERMAN

Busch, Ernst.
Stiltypen der deutschen freirhythmischen Hymne aus dem religiösen Erleben / Ernst Busch. — Reprograph. Nachdr. d. Ausg. Frankfurt am Main, 1934. — Hildesheim : Gerstenberg, 1975.
131 p. ; 22 cm. GFR75-A31
Reprint of the ed. published by M. Diesterweg, which was issued as v. 6 of Frankfurter Quellen und Forschungen zur germanischen und romanischen Philologie.
ISBN 3-8067-0296-9 : DM24.00
1. German poetry—18th century—History and criticism. 2. Christian poetry, German—History and criticism. 3. German language—Versification. I. Title. II. Series: Frankfurter Quellen und Forschungen zur germanischen und romanischen Philologie ; Heft 6.
PT535.B8 1975 831'.5'09 78-390226
79 MARC

—HISTORY AND CRITICISM

Psalmen vom Expressionismus bis zur Gegenwart / hrsg. von Paul Konrad Kurz. — Freiburg im Breisgau ; Basel ; Wien : Herder, 1978.
325 p. ; 22 cm. GFR79-A3
ISBN 3-451-18258-0 : DM34.00
1. German poetry—20th century—History and criticism. 2. Christian poetry, German—History and criticism. I. Kurz, Paul Konrad.
PT553.P7 831'.9'10931 79-362004
79 MARC

CHRISTIAN POETRY, ITALIAN

Le Laude dei Bianchi contenute nel Codice vaticano chigiano L. VII 266 : edizione critica / a cura di Bernard Toscani. — [Firenze] : Libreria editrice fiorentina, 1979.
213 p. : ill. ; 20 cm. It79-Mar
Includes bibliographical references.
L5000
1. Italian poetry—To 1400. 2. Christian poetry, Italian. I. Toscani, Bernard. II. Vatican. Biblioteca vaticana. MSS. (Chigi L. VII 266)
PQ4212.A2L3 79-381081
*79 MARC

Mkhit'ar Sebastats'i, 1676-1749.
Sei inni sacri di Mechitar di Sebaste in onore della SS. Vergine e di S. Antonio abate / traduzione italiana a cura di Giuseppe Frasson in occasione del terzo centenario della nascita di Mechitar, 1676-1976. — Venezia : S. Lazzaro, 1977.
51 p. ; 21 cm. It***
Added t.p. in Armenian.
Texts in Armenian and Italian.
1. Christian poetry, Armenian. 2. Christian poetry, Italian. I. Title.
PK8547.M53A56 1977 891'.992'14 79-351378
79 MARC

Rossetti, Bartolomeo.
Er Vangelo seconno noantri / Bartolomeo Rossetti ; prefazione di Raffaello Lavagna. — [4. ed.] — Roma : Città nuova, 1978.
xx, 343 p. ; 20 cm. It78-Aug
Poems in romanesco.
L5000
1. Jesus Christ—Poetry. 2. Christian poetry, Italian. I. Title.
PQ4878.O85E7 1978 78-385491
*78 MARC

—ITALY—BISACQUINO, SICILY

Canti popolari religiosi bisacquinesi / a cura del decano don Calogero Di Vincenti. — Palermo : Scuola grafica salesiana, [1977?]
39 p. ; 22 cm. It78-Apr
At head of title: Associazione turistica Pro-Bisacquino.
1. Folk-songs—Italy—Bisacquino, Sicily. 2. Folk-songs, Sicilian. 3. Christian poetry, Italian—Italy—Bisacquino, Sicily. I. Di Vincenti, Calogero. Bisacquino.
PQ4218.B48C36 784.4'9458 79-366286
*79 MARC

—ITALY—POZZUOLI—HISTORY AND CRITICISM

Mugione, Antonio.
I poeti della Madonna di Campiglione. [Di] Antonio Mugione, Gaetano Capasso. Napoli, Athena mediterranea, 1971.
200 p. 20 cm. L1000 ItSuppl-6
Includes bibliographical references.
1. Christian poetry, Italian—Italy—Pozzuoli—History and criticism. 2. Mary, Virgin, in literature. 3. Santuario di Campiglione in literature. I. Capasso, Gaetano, joint author. II. Title.
PQ4126.R4M84 851'.009'352 78-422924
79 MARC

CHRISTIAN POETRY, LATIN

—HISTORY AND CRITICISM

Gamber, Stanislas.
Le Livre de la Genèse dans la poésie latine au 5e siècle : versions françaises et provençales de la Genèse au Moyen Age / traduction de la Bible en langue d'oc, Stanislas Gamber. — Genève : Slatkine, 1977.
xvi, 263 p. ; 23 cm. Sw78-106
Reprint of the 1899 ed. published by A. Fontemoing, Paris.
Includes bibliographical references.
60.00F
1. Christian poetry, Latin—History and criticism. 2. Christian poetry, Early—History and criticism. 3. Bible in literature. I. Title.
PA6053.G35 1977 871'.01 79-344678
79 MARC

Liberatore Fiorani, Francesca.
Lirica mediolatina : sequenza e tropo / Francesca Liberatore Fiorani. — Roma : Edizioni dell'Ateneo & Bizzarri, 1977.
123 p. ; 22 cm. — (Nuovi saggi ; n. 68) It78-Mar
Includes bibliographical references.
L4000
1. Christian poetry, Latin—History and criticism. 2. Latin poetry, Medieval and modern—History and criticism. I. Title.
PA8056.L5 871'.03'09 78-401075
*79 MARC

CHRISTIAN POETRY, MEXICAN

Muñoz, Esaú P 1884-1967.
Lámpara es a mis pies : poesías de Esaú P. Muñoz ; editores, Rosalío F. Muñoz y Josefina Muñoz de Rodríguez. — Las [sic] Angeles : [s.n.], 1978.
59 p., [2] leaves of plates : ill. ; 22 cm. — (Obras de Esaú P. Muñoz)
1. Christian poetry, Mexican. I. Muñoz, Rosalío F. II. Muñoz de Rodríguez, Josefina. III. Title. IV. Series: Muñoz, Esaú P., 1884-1967. Works. 1978.
PQ7297.M746L3 1978 861 79-103264
79 MARC

CHRISTIAN POETRY, NORWEGIAN

Sandnes, Asbjørn, 1911-
Morgon og kveld : dikt, salmar og viser / Asbjørn Sandnes. — Oslo : Luther, 1978.
57 p. ; 22 cm. N78-Dec.
ISBN 8153132174 : kr35.00
1. Christian poetry, Norwegian. I. Title.
PT8951.29.A52M6 79-358126
79 MARC

Vibe-Müller, Ingrid, 1921-
Bare som glimt / Ingrid Vibe-Müller. — Oslo : Lunde, 1978 (Drammen : Tangen-trykk)
59 p. ; 21 cm. N78-July
Poems.
ISBN 8252030351 : kr32.00
1. Christian poetry, Norwegian. I. Title.
PT8951.32.I2B3 78-395195
79 MARC

CHRISTIAN POETRY, SPANISH

Cantores de la Virgen de la Montaña / prólogo, selección y notas de Valeriano Gutiérrez Macías. — Cáceres : [Valeriano Gutiérrez Macías], 1975.
361 p. ; 21 cm. Sp76-Feb
ISBN 8440090854 : 400ptas
1. Christian poetry, Spanish. 2. Montaña, Virgen de la—Poetry. 3. Mary, Virgin—Poetry. 4. Cáceres, Spain (City)—Religious life and customs—Poetry. I. Gutiérrez Macías, Valeriano.
PQ6208.R4C27 861'.008'0351 79-361318
79 MARC

Correa Marín, Gil.
La soledad acompañada / Gil Correa Marín. — Barcelona : [s.n., 1977?]
64 p. ; 20 cm. Sp***
Poems.
ISBN 8440032242
1. Christian poetry, Spanish. I. Title.
PQ6653.O7135S6 79-122398
00 MARC

Dicastillo, Miguel de, 17th cent.
Aula de Dios, Cartuxa real de Zaragoza / Miguel de Dicastillo ; con estudio preliminár de Aurora Egido. — Zaragoza : Libros Pórtico, 1978.
69, [31], 139 p. ; 23 cm. — (Estudios ; 4) Sp***
Reprint of the 1637 edition published by Diego Dormer, Zaragoza.
Includes bibliographical references.
ISBN 8485264290
1. Aula Dei (Monastery), Saragossa—Poetry. 2. Christian poetry, Spanish. I. Egido, Aurora. II. Title.
PQ6388.D535A94 1978 861'.3 79-366779
79 MARC

Mendoza, Iñigo de, ca. 1424–ca. 1507.
Vita Christi fecho por coplas. [Burgos, Friedrich Biel, ca. 1490–93]
xcix+ l., all before leaf v, leaf x, leaf xci, and probably 1 leaf after xcix wanting. f°. 27.2 cm.
Signatures: a^{10}? (L. C. has only 5 leaves numbered v–ix, the 1st signed a_4), b–k^{10} ($k_{1, 10}$ wanting) ; 38 lines ; 99G.
Unrecorded. Probably a reprint of the undated ed. printed at Zamora ca. 1483 by Antonio de Centenera (edition C in Rodríguez-Puértolas. Fray Iñigo de Mendoza y sus "Coplas de Vita Christi" [1968]) L. C. copy mistakenly registered under M–489 in Goff. Third census.
CONTENTS: Vita Christi.—Sermon trobado.—Coplas ... en vituperio delas malas hembras.—Coplas ... en que declara como ... es reparada nuestra castilla.—Dechado.—Question y diferencia ... entre la razon y sensualidad.—Los gozos de nuesta señora.—La cena.—Coplas ... ala veronica.—Coplas al spiritu sancto.—Manrique, Jorge. Coplas ... ala muerte ... de ... su padre.—Lamentacion ala quinta angustia.—Mena, Juan de. Coplas ... contra los peccados mortales (with continuation by Gómez Manrique)—Pregunta de sancho de rojas aun aragones (with Respuesta del aragones)—Manrigue, Jorge. Coplas ... sobre que es amor.
1. Spanish poetry—To 1500. 2. Christian poetry, Spanish. 3. Jesus Christ—Poetry. I. Manrique, Jorge, 1440?–1470. II. Mena, Juan de, 1411–1456.
Incun. X.M52 77–551379
[PQ6180]

Nuño, Daniel.
Enciclopedia de poesía evangélica : monólogos y diálogos para fiestas cristianas / Daniel Nuño. — 2. ed. española. — Tarrasa [España] : Clie, 1972.
365 p. ; 18 cm.
1. Christian poetry, Spanish. I. Title.
PQ6664.U57E5 1972 861'.6'4 78-112889
79 MARC

Serrano Segovia, Sebastián, comp.
Antología de poesías a la Virgen de la Caridad. Cartagena, 1973.
240 p. 24 cm. 200ptas Sp***
1. Christian poetry, Spanish. 2. Caridad, Virgen de la—Poetry. 3. Mary, Virgin—Poetry. 4. Cartagena, Spain—Religious life and customs—Poetry. I. Title.
PQ6208.R4S47 861'.008'0351 79-364788
79 MARC

—HISTORY AND CRITICISM

Keller, John Esten.
Pious brief narrative in medieval Castilian & Galican verse : from Berceo to Alfonso X / John Esten Keller. — Lexington : University Press of Kentucky, c1978.
x, 142 p. ; 22 cm. — (Studies in Romance languages ; 21)
Bibliography: p. [133]-135.
Includes index.
ISBN 0-8131-1381-4 : $15.00
1. Spanish poetry—To 1500—History and criticism. 2. Narrative poetry, Spanish—History and criticism. 3. Christian poetry, Spanish—History and criticism. 4. Gonzalo de Berceo, 13th cent.—Criticism and interpretation. 5. Alfonso X, el Sabio, King of Castile and Leon, 1221-1284. Cantigas de Santa María. I. Title. II. Series.
PQ6060.K45 861'.1'09 77-84064
78 MARC

CHRISTIAN POETRY, SWEDISH

Edhardt, Carl-Gustaf, 1942-
Ögonblick av evighet : dikter och bilder / Carl-Gustaf Edhardt. — Stockholm : Norman, 1978.
95, [1] p. : ill. ; 25 cm. S79
ISBN 9153636600
1. Christian poetry, Swedish. I. Title.
PT9876.15.D45O34 79-367403
79 MARC

Olivecrona, Eva, 1934-
Om jag började tala med dig : dikter / av Eva Olivecrona ; [ill. och omsl., Ann-Kristin Simonsson]. — Stockholm : Gummesson, 1979.
68 p. : ill. ; 19 cm.
ISBN 9170705437
1. Christian poetry, Swedish. I. Title.
PT9876.25.L5O4 79-383709
79 MARC

Olsson, Helga, 1908-
Buketten : dikter / av Helga Olsson. — [Klintehamn (Skolkvior 34) : Förf.], 1977.
28, [1] p. : ill. ; 21 cm. S78-4
kr18.00
1. Christian poetry, Swedish. I. Title.
PT9876.25.L693B8 839.7'1'74 78-396193
79 MARC

CHRISTIAN POETRY, UKRAINIAN

Domashovets', H
(Vybrani tvory)
Вибрані твори : [збірка віршів] / Григорій Домашовець. — Ювілейне вид. — Гар[т]форд : Домашовець, 1960.
263 p. : ill. ; 22 cm.
1. Christian poetry, Ukrainian.
PG3969.D57A6 1960 78–422979

CHRISTIAN POETRY, VENEZUELAN

Las Celestiales / compilación y notas de Iñaki de Errandonea ; pref. de Miguel Otero Silva ; ilus. de Joseba Escucarreta. — 2. ed. — Caracas : J. A. Catala, 1974.
[118] p. : ill. ; 33 cm.
Bibliography: p. [116]-[117]
1. Folk poetry, Venezuelan. 2. Christian poetry, Venezuelan. 3. Humorous poetry, Venezuelan. I. Errandonea, Iñaki de. II. Escucarreta, Joseba. III. Title.
PQ8545.C44 1974 398.2 79-360081
79 MARC

CHRISTIAN POETRY, WELSH

Thomas, Madathethu Abraham.
Codi pontydd : myfyrdodau ar genhadaeth a neges i'n hoes / [gan] M. A. Thomas ; cyfieithiwyd [o'r Saesneg] gan Cyril G. Williams. — [Tregaron : Gwasg Tŷ Gwyn, 1978]
77 p. ; 19 cm. GB78-24493
Translation of About you and me.
Poems.
£1.00
1. Christian poetry, Welsh. I. Williams, Cyril Glyndwr. II. Title.
PB2298.T55A6418 79-369592
79 MARC

CHRISTIAN SAINTS

see also Apostles; Christian hagiography; Christian martyrs; Christian patron saints; Church calendar; Fathers of the church; Martyrologies

Viola sanctorum.
Viola sanctorum. Augsburg, Johann Froschauer, 23 Dec. (x Kal. Ianuarii) 1496.
[10], cxxxiiii l. 4°. 23.3 cm.
Brit. Mus. Cat. (XV cent.) II, p. 395 (IA. 6493) ; Goff. Third census, M–338.
Signatures: 10 leaves unsigned, a–d^8, e–x^4.
Contemporary calfskin binding, stamped in blind.
Ex libris Jacobi P. R. Lyell.
Bound with Passio Sancti Meynradi. Basel, 20 Sept. 1496 ; and John, Prester. De ritu et moribus Indorum. [Deventer, ca. 1490]
1. Christian saints.
Incun. 1496.V5 Rosenwald Coll. 55–3633
——— Copy 2. 21 cm. Inscriptions: Cart. Buxiae; Canoniae Rothensis 1772; A. A. Smets, Savannah, June 2, 1834. Stamp: dem Kloster Roth. Batchelder Coll.
——— Copy 3. 19 cm. Rare Bk. Coll.
[BX4654]

—ATTRIBUTES

Cruz, Joan Carroll, 1931-
The incorruptibles : a study of the incorruption of the bodies of various Catholic saints and beati / by Joan Carroll Cruz. — Rockford, Ill. : Tan Books and Publishers, 1977.
310 p. : ill. ; 21 cm.
Bibliography: p. 303-310.
ISBN 0-89555-066-0 : $5.00
1. Christian saints—Biography. 2. Christian sanits—Attributes. I. Incorruption. II. Title.
BX4669.C78 282'.092'2 77-93992
00 MARC

Misch, Jürgen.
Die Elite Gottes : Heilige zwischen Wahn u. Heldentum / Jürgen Misch. — Pfaffenhofen : Ludwig, 1978.
232, [12] p. : 12 ill. ; 22 cm. GFR79-A
Bibliography: p. 231-232.
ISBN 3-7787-2027-9 : DM24.00
1. Christian saints—Biography. 2. Christian saints—Attributes. I. Title.
BX4661.M5 79-361669
79 MARC

Mullin, Redmond.
Miracles and magic : the miracles and spells of saints and witches / by Redmond Mullin ; ill. by Fiona Cairns. — Oxford : Mowbray, 1979, c1978.
299 p. : ill. ; 23 cm. GB***
Bibliography: p. 273-283.
Includes index.
ISBN 0-264-66434-5 : £6.95
1. Miracles. 2. Christian saints—Attributes. I. Title.
BT97.2.M84 1979 133 79-309370
79 MARC

Negro, Franco.
Il santo nella concezione cattolica / Franco Negro ; prefazione di Ernest Zmarzlý ... ; con la collaborazione di Mario Duni ... [et al.]. — 2. ed. con 42 figure fuori testo. — Modena : S.T.E.M.-Mucchi, 1977.
[24], 266 p., [42] leaves of plates : ill. ; 24 cm. It78-Dec
"Edizione in 200 esemplari numerati e firmati dall'autore. N. 173."
Includes bibliographical references.
L30000
1. Christian saints—Attributes. I. Title.
BX4661.N43 1977 79-360282
*79 MARC

—BIOGRAPHY

Agapit, Archimandrite.
(Zhizn' svíàtago Grigoríià Bogoslova, arkhiepiskopa konstantinopol'skago, i ego pastyrskaià díeiàtel'nost')
Жизнь святаго Григорія Богослова, архіепископа константинопольскаго, и его пастырская дѣятельность / оршанскихъ Покровскаго и Богоявленскаго монастырей настоятеля, архимандрита Агапита. — Санктпетербургъ : Тип. духовнаго журнала Странникъ, 1869.
iv, 288 p. ; 24 cm.
On cover: M. Θ.
Includes bibliographical references.
1. Gregorius Nazianzenus, Saint, Patriarch of Constantinople. 2. Christian saints—Biography. I. Title.
BR1720.G7A45 78–422796
Russian Imperial Coll.

Aretino, Pietro, 1492-1556.
Le vite dei santi : Santa Caterina vergine, San Tommaso d'Aquino, 1540-1543 / Pietro Aretino ; testo con introduzione e commento di Flavia Santin. — Roma : Bonacci, 1978.
345 p. ; 21 cm. — (L'Ippogrifo ; 13) It78-Sept
Bibliography: p. 15-16.
Includes index.
L12000
1. Catharina, Saint, of Alexandria. 2. Thomas Aquinas, Saint, 1225?-1274. 3. Christian saints—Biography. I. Santin, Flavia. II. Title. III. Series: L'Ippogrifo (Rome) ; 13.
BX4700.C45A73 1978 78-388449
*79 MARC

CHRISTIAN SAINTS
—BIOGRAPHY (Continued)

Badurina, Teodoro.
Vjerski i prosvetni rad ss. Ćirila i Metoda među Slavenima / Teodor Badurina. — Washington, D.C. : Badurina, 1966.
ix, 99 p. : facsims. ; 22 cm.
Includes bibliographical references.
1. Cyrillus, Saint, of Thessalonica, 827 (ca.)-869. 2. Methodius, Saint, Abp. of Moravia, d. 885. 3. Christian saints—Biography. I. Title.
BX4700.C9B23 1966 79-348882
79 MARC

Badurina, Teodoro.
Vjerski i prosvjetni rad ss. Ćiril i Metoda među Slavenima. 2. izd. Teodor Badurina. Washington, D.C., 1967.
114 p. facsims. 22 cm.
Latin or Serbo-Croatian.
Includes bibliographical references.
1. Cyrillus, Saint, of Thessalonica, 827 (ca.)-869. 2. Methodius, Saint, Abp. of Moravia, d. 885. 3. Christian saints—Biography. I. Title.
BX4700.C9B23 1967 70-859960
79[r79]rev MARC

Conzelmann, Hans.
Bemerkungen zum Martyrium Polykarps / Hans Conzelmann. — Göttingen : Vandenhoeck & Ruprecht, 1978.
20 p. ; 25 cm. — (Nachrichten der Akademie der Wissenschaften in Göttingen, Philologisch-Historische Klasse ; Jahrg. 1978, Nr. 2) GFR78-A40
Pages also numbered 42-58 continuing the paging of the preceding number.
Includes bibliographical references.
ISBN 3-525-85116-2 : DM4.00
1. Martyrdom of Polycarp. 2. Polycarpus, Saint, Bp. of Smyrna. 3. Christian saints—Biography. I. Title. II. Series: Akademie der Wissenschaften, Göttingen. Philologisch-Historische Klasse. Nachrichten ; Jahrg. 1978, Nr. 2.
AS182.G823 Jahrg. 1978, Nr. 2 79-357634
[BR1720.P7] 79 MARC

Cruz, Joan Carroll, 1931-
The incorruptibles : a study of the incorruption of the bodies of various Catholic saints and beati / by Joan Carroll Cruz. — Rockford, Ill. : Tan Books and Publishers, 1977.
310 p. : ill. ; 21 cm.
Bibliography: p. 303-310.
ISBN 0-89555-066-0 : $5.00
1. Christian saints—Biography. 2. Christian sanits—Attributes. I. Incorruption. II. Title.
BX4669.C78 282'.092'2 77-93992
00 MARC

Don Peter, W L A
Xavier as educator / by W. L. A. Don Peter. — Delhi : Jesuit Educational Association of India, [foreword 1974]
x, 191 p. ; 22 cm.
Bibliography: p. [181]-184.
Includes index.
1. Francisco Xavier, Saint, 1506-1552. 2. Christian saints—Biography. 3. Educators—Biography. I. Title.
BX4700.F8D6 266'.2'0924 78-911553
78 MARC

Douze récits byzantins sur Saint Jean Chrysostome / publiés par François Halkin. — Bruxelles : Société des bollandistes, 1977.
562 p. ; 25 cm. — (Subsidia hagiographica ; no 60) Be***
Texts in Greek with headings and notes in French.
Includes bibliographical references and indexes.
CONTENTS: Vie, par Théodore de Trimithonte.—Item, abrégé anonyme.—Vie, par Georges d'Alexandrie.—Fragment d'une vie perdue.—Vie abrégée.—Vie anonyme du Vaticanus 1669.—Vie, par Cosme Vestitor.—Vie attribuée à Hésychius.—Translation par Syméon Métaphraste.—Translation du Sabaiticus 242.—Récit oratoire de la translation.—Translation du Mosquensis 26.
1. Chrysostomus, Joannes, Saint, Patriarch of Constantinople, d. 407. 2. Christian saints—Biography. I. Halkin, François. II. Series.
BR1720.C5D68 270.2'092'4 78-388823
78 MARC

Dreyer, Petrus Secundus.
Die wankelende rots : 'n lewensgeskiedenis van die apostel Petrus / P. S. Dreyer. — Kaapstad : H.A.U.M., 1977.
133 p. ; 22 cm. SA
Bibliography: p. 133.
ISBN 0-7986-0333-X : R4.20
1. Peter, Saint, Apostle. 2. Bible. N.T.—Biography. 3. Christian saints—Biography. I. Title.
BS2515.D73 79-384535
79 MARC

Un Éloge de Jacques, le frère du Seigneur = [A eulogy of James, the brother of the Lord] / par un pseudo-André de Crète, avec une paraphrase ancienne de l'épître catholique de Saint Jacques ; éd., traduction et notes critiques par Jacques Noret, avec la collab. de Herman Gaspart. — Toronto : Institut pontifical d'études médiévales, 1978.
112 p. ; 26 cm. — (Studies and texts - Institut pontifical d'études médiévales ; 44) C78-1140-6
Includes French translation and Greek text of Vios kai martyrion tou hagiou Iakōvou tou apostolou kai adelphou tou Kyriou.
Bibliography: p. [9]-11.
Includes indexes.
ISBN 0-88844-044-8 : $8.00
1. James, Saint, apostle. 2. Christian saints—Biography. 3. Bible. N.T.—Biography. 4. Bible. N.T. James—Paraphrases, Greek. I. Andreas, Saint, Abp. of Crete. Vios kai martyrion tou hagiou Iakōvou tou apostolou kai adelphou tou Kyriou. French & Greek. 1978. II. Noret, Jacques, 1939- III. Gaspart, Herman. IV. Title: A eulogy of James, the brother of the Lord. V. Series: Pontifical Institute of Mediaeval Studies. Studies and texts ; 44.
BS2454.J3E43 225.9'24 79-359061
79 MARC

Fremantle, Anne Jackson, 1909-
Saints alive : The lives of thirteen heroic saints / Anne Fremantle. — 1st ed. — Garden City, N.Y. : Doubleday, 1978.
182 p. ; 22 cm.
ISBN 0-385-12441-4 : $7.95
1. Christian saints—Biography. I. Title.
BX4655.2.F66 209'.2'2 78-4789
78 MARC

Grauwen, Wilfried Marcel.
Norbertus aartsbisschop van Maagdenburg (1126-1134) / door Wilfried Marcel Grauwen. — Brussel : Paleis der Academiën, 1978.
lxii, 689 p., [1] fold. leaf of plates : map ; 27 cm. — (Verhandelingen van de Koninklijke Academie voor Wetenschappen, Letteren en Schone Kunsten van België, Klasse der Letteren ; jaarg. 40, nr. 86) Be***
Summary in English.
Bibliography: p. xi-lix.
Includes index.
1. Norbert, Saint, Abp. of Magdeburg, d. 1134. 2. Christian saints—Biography. 3. Premonstratensians—History. I. Title. II. Series: Vlaamse Academie voor Wetenschappen, Letteren en Schone Kunsten van België. Klasse der Letteren. Verhandelingen, jaarg. 40, nr. 86.
AS242.B624 jaarg. 40, nr. 86 78-392370
[BX4700.N8] 79 MARC

Hanley, Boniface.
Ten Christians : by their deeds you shall know them / Boniface Hanley. — Notre Dame, Ind. : Ave Maria Press, c1979.
269 p. : ill. ; 22 cm.
ISBN 0-87793-183-6 : $5.95
1. Christian saints—Biography. 2. Catholics—Biography. I. Title.
BX4655.2.H27 282'.092'2 79-53836
79 MARC

Hiyane, Antei, 1892-
(Sei Zabieru den)
聖ザビエル傳 / 比屋根安定. — 東京 : 朝日新聞社, 昭和24 [1949]
250 p., [1] leaf of plates : ill. ; 19 cm.
"青少年のために"

1. Francisco Xavier, Saint, 1506-1552. 2. Christian saints—Biography. I. Title.
BX4700.F8H57 79-807715

Manns, Peter.
Testvéreink, a szentek / [a magyar kiadásért felel, Marosi László]. — Eisenstadt : Prugg, 1977.
488 p. ; 21 cm. Au***
Translated selections from Reformer der Kirche which was originally published under title: Die Heiligen in ihrer Zeit.
Includes index.
1. Christian saints—Biography. I. Marosi, László. II. Title.
BX4655.2.M3215 78-384142
78 MARC

Misch, Jürgen.
Die Elite Gottes : Heilige zwischen Wahn u. Heldentum / Jürgen Misch. — Pfaffenhofen : Ludwig, 1978.
232, [12] p. : 12 ill. ; 22 cm. GFR79-A
Bibliography: p. 231-232.
ISBN 3-7787-2027-9 : DM24.00
1. Christian saints—Biography. 2. Christian saints—Attributes. I. Title.
BX4661.M5 79-361669
79 MARC

Reed, Olwen.
An illustrated history of saints and symbols / Olwen Reed. — Bourne End [Eng.] : Spurbooks, c1978.
159 p. : ill. ; 23 cm. GB***
Half title: Saints and symbols.
Includes index.
ISBN (invalid) 9904978311 : £5.00
1. Christian saints—Biography. 2. Christian saints in art. I. Title. II. Title: Saints and symbols.
BX4655.2.R43 270 79-306287
79 MARC

Roig, Rosendo.
Yñigo de Loyola : vida en sociedad, soledad y Compañía / [por] Rosendo Roig. — Bilbao : Mensajero, D.L. 1978.
399 p. ; 18 cm. — (Colección Bolsillo ; 53) Sp78-Sept
ISBN 8427111193
1. Loyola, Ignacio de, Saint, 1491-1556. 2. Christian saints—Biography. I. Title.
BX4700.L7R57 79-351202
79 MARC

Sgarbossa, Mario.
Il santo del giorno / [di] Mario Sgarbossa, Luigi Giovannini. — Roma : Edizioni paoline, 1978.
639 p. : ill. ; 25 cm. It78-Nov
Includes bibliographical references and index.
L9000
1. Christian saints—Biography. I. Giovannini, Luigi, joint author. II. Title.
BX4655.2.S48 270'.092'2 79-337862
*79 MARC

Stables, Pierre.
Deux clefs initiatiques de la "Légende dorée", la kabbale et le "Yi-king" / Pierre Stables. — [Paris] : Dervy, 1975.
307 p. : ill. ; 22 cm. — (Histoire et tradition) F76-4229
Bibliography: p. [300]-303.
ISBN 2-85076-010-2 : 40.00F
1. Jacobus de Varagine. Legenda aurea. 2. Christian saints—Biography. 3. Cabala. 4. I ching. I. Title.
BX4654.S67 282'.092'2 79-350539
79 MARC

Urteaga Loidi, Jesús, 1921-
Los defectos de los cantos / Jesús Urteaga Loidi. — Madrid : Ediciones Rialp, c1978.
413 p. ; 20 cm. — (Patmos ; 172) Sp***
Includes bibliographical references and index.
ISBN (invalid) 0-8432-1196-6
1. Christian life—Catholic authors. 2. Bible—Biography. 3. Christian saints—Biography. I. Title.
BX2350.2.U68 79-357652
79 MARC

Widdowson, Gregory.
An outline of lay sanctity / Gregory Widdowson. — Huntington, Ind. : Our Sunday Visitor, c1979.
160 p. ; 21 cm.
ISBN 0-87973-737-9 : $2.95
1. Christian saints—Biography. 2. Laity. I. Title.
BX4661.W47 282'.092'2 79-84532
00 MARC

—BIOGRAPHY—ADDRESSES, ESSAYS, LECTURES

Ignatius von Loyola / mit e. Essay von Karl Rahner, e. histor. Einf. von Paul Imhof, 52 Farbbildern von Helmuth Nils Loose u. 42 Stichen d. Ignatius-Biographie von 1609. — Freiburg im Breisgau ; Basel ; Wien : Herder, 1978.
111, [8] p. : numerous ill. (some col.) ; 25 cm. GFR78-A41
Erratum slip inserted.
Bibliography: p. [118]-[119]
ISBN 3-451-18006-5 : DM29.50
1. Loyola, Ignacio de, Saint, 1491-1556—Addresses, essays, lectures. 2. Christian saints—Biography—Addresses, essays, lectures. I. Rahner, Karl, 1904- II. Imhof, Paul. III. Loose, Helmuth Nils.
BX4700.L7I34 79-370693
79 MARC

—CORRESPONDENCE

Bocciolini Palagi, Laura.
Il carteggio apocrifo di Seneca e san Paolo : introduzione, testo, commento / Laura Bocciolini Palagi. — Firenze : L. S. Olschki, 1978.
222 p. ; 25 cm. — (Studi Accademia toscana di scienze e lettere La Colombaria ; 46) It78-Apr
Text in Latin with Italian commentary.
Bibliography: p. 207-210.
Includes indexes.
L8500
1. Seneca, Lucius Annaeus. Spurious and doubtful works. Epistolae Senecae ad Paulum et Pauli ad Senecam. 2. Paul, Saint, Apostle. 3. Authors, Latin—Correspondence. 4. Christian saints—Correspondence. I. Seneca, Lucius Annaeus. Spurious and doubtful works. Epistolae Senecae ad Paulum et Pauli ad Senecam. 1978. II. Paul, Saint, apostle. III. Title. IV. Series: Accademia toscana di scienze e lettere La Colombaria. Studi ; 46.
PA6663.E7B6 78-393325
*79 MARC

—CULT

see also Christian shrines

Finucane, Ronald C
Miracles and pilgrims : popular beliefs in medieval England / Ronald C. Finucane. — Totowa, N.J. : Rowman and Littlefield, 1977.
248 p., [8] leaves of plates : ill. ; 24 cm.
Bibliography: p. [242]-244.
Includes index.
ISBN 0-87471-831-7
1. England—Church history—Medieval period, 1066-1485. 2. Miracles. 3. Christian pilgrims and pilgrimages—England. 4. Christian saints—Cult. 5. Christian shrines—England. I. Title.
BR747.F56 1977 274.2 77-21771
77[r79]rev MARC

Trankvilion-Stavrovets'kyĭ, Kyrylo, d. 1646.
[Pērlo mnogōt͡sēnnoē]
Сіа книга названнаѧ Пѣрло многѡцѣнноє. Съставлєнна трѹдолюбіємъ презъ Кирілла Транквѣліона. [Въ Чернигов], 1646.
1 v. (unpaged)
L. C. copy on microfilm only.
—— —— Microfilm. 1 reel ; 35 mm.
Microfilm Slavic 12121 BT
1. Mary, Virgin—Cultus. 2. Christian saints—Cult. 3. Angels—Cultus. I. Title.
Title romanized: Sīa kniga nazvannai͡a Pērlo mnogōt͡sēnnoē.
[BT645.T73] 78-421178

CHRISTIAN SAINTS (Continued)

—CULT—ADDRESSES, ESSAYS, LECTURES

Ermutigung zum Gebet / Johann Baptist Metz, Karl Rahner. — Freiburg im Breisgau ; Basel ; Wien : Herder, 1977.
110 p. ; 20 cm. GFR78-A
Erratum slip inserted.
CONTENTS: Metz, J. B. Ermutigung zum Gebet.—Rahner, K. Gebet zu den Heiligen.
ISBN (invalid) 3-451-17935-7 : DM10.80
1. Prayer—Addresses, essays, lectures. 2. Christian saints—Cult—Addresses, essays, lectures. I. Metz, Johannes Baptist, 1928- Ermutigung zum Gebet. 1977. II. Rahner, Karl, 1904- Gebet zu den Heiligen. 1977.
BV213.E75 79-363290
79 MARC

—CULT—HISTORY

Gödl, Siegfried.
Melanchthons Stellung zur Heiligenanrufung : Anrufung Gottes u. Anrufung d. Heiligen / Siegfried Gödl. — ₍Für den Druck abgeänd.₎. — Wien : VWGÖ, Verb. d. Wissenschaftl. Gesellschaften Österreichs, 1977.
181 p. ; 21 cm. — (Dissertationen der Universität Graz ; 39) Au78-8-32
Originally presented as the author's thesis, Graz, 1975.
Bibliography: p. 9-26.
Includes index.
S155.00
1. Melanchthon, Philipp, 1497-1560. 2. Christian saints—Cult—History. I. Title. II. Series: Graz. Universität. Dissertationen ; 39.
BR335.G63 1977 79-375861
79 MARC

—CULT—AUSTRIA—VIENNA

Kapner, Gerhardt.
Barocker Heiligenkult in Wien und seine Träger / Gerhardt Kapner. — 1. Aufl. — München : Oldenbourg, 1978.
147 p. ; 21 cm. — (Österreich Archiv) GFR78-A
Includes bibliographical references.
ISBN 3-486-48701-9
1. Christian saints—Cult—Austria—Vienna. 2. Vienna—Church history. I. Title. II. Series.
BX4680.K36 78-392637
79 MARC

Kapner, Gerhardt.
Barocker Heiligenkult in Wien und seine Träger / Gerhardt Kapner. — Wien : Verl. f. Geschichte u. Politik, 1978.
147 p. ; 21 cm. — (Österreich Archiv) Au78-20-311
Includes bibliographical references.
ISBN 3-7028-0107-3 : S176.00
1. Christian saints—Cult—Austria—Vienna. 2. Vienna—Church history. 3. Art, Baroque—Austria—Vienna. 4. Christian art and symbolism. I. Title. II. Series.
BX2333.K36 79-358628
*79 MARC

—CULT—FRANCE—ALPES-MARITIMES (DEPT.)

Canestrier, Paul.
Fête populaire et tradition religieuse en pays niçois / Paul Canestrier. — ₍Nice₎ : Serre, 1978.
208 p., ₍7₎ leaves of plates : ill. ; 24 cm. — (Collection Vida) F***
Bibliography: p. ₍199₎-206.
ISBN 2-901175-06-6 : 54.00F
1. Mary, Virgin—Cult—France—Alpes—Maritimes (Dept.) 2. Mary, Virgin—Cult—Monaco. 3. Christian saints—Cult—France—Alpes—Maritimes (Dept.) 4. Christian saints—Cult—Monaco. 5. Alpes—Maritimes, France (Dept.)—Religious life and customs. 6. Monaco—Religious life and customs. I. Title.
BT652.F7C36 282'.44'94 79-352638
79 MARC

—CULT—GERMANY

Beissel, Stephan, 1841-1915.
Die Verehrung der Heiligen und ihrer Reliquien in Deutschland im Mittelalter / Stephan Beissel. — Unveränd. reprograf. Nachdr. der im Verl. Herder erschienenen Originalausg., 1890 u. 1892 / mit e. Vorw. zum Nachdr. von Horst Appuhn. — Darmstadt : Wissenschaftliche Buchgesellschaft, ₍Abt. Verl.₎, 1976.
334 p. in various pagings ; 23 cm. GFR77-A9
Reprint of the author's Die Verehrung der Heiligen und ihrer Reliquien in Deutschland bis zum Beginne des 13. Jahrhunderts (1890) and of his Verehrung der Heiligen und ihrer Reliquien in Deutschland während der zweiten Hälfte des Mittelalters (1892).
Bibliography: p. 1-10 (3d group)
Includes indexes.
ISBN 3-534-06765-7 : DM80.00
1. Relics and reliquaries—Germany. 2. Christian saints—Cult—Germany. 3. Germany—Church history—To 843. 4. Germany—Church history—Middle Ages, 843-1517. I. Appuhn, Horst. II. Title.
BX2333.B44 1976 79-365975
79 MARC

—CULT—ITALY—ROME (CITY)

Jounel, Pierre.
Le culte des saints dans les basiliques du Latran et du Vatican au douzième siècle / Pierre Jounel. — Rome : École française de Rome, 1977.
460 p. : ill. ; 25 cm. — (Collection de l'École française de Rome ; 26) It78-June
Errate slip inserted.
Bibliography: p. ₍427₎-434.
Includes index.
1. Christian saints—Cult—Italy—Rome (City). 2. Rome (City)—Church history. 3. Rome (City). Lateran—History. 4. Vatican—History. I. Title. II. Series: École française de Rome. Collection ; 26.
BX2333.J68 78-388132
*78 MARC

—CULT—MONACO

Canestrier, Paul.
Fête populaire et tradition religieuse en pays niçois / Paul Canestrier. — ₍Nice₎ : Serre, 1978.
208 p., ₍7₎ leaves of plates : ill. ; 24 cm. — (Collection Vida) F***
Bibliography: p. ₍199₎-206.
ISBN 2-901175-06-6 : 54.00F
1. Mary, Virgin—Cult—France—Alpes—Maritimes (Dept.) 2. Mary, Virgin—Cult—Monaco. 3. Christian saints—Cult—France—Alpes—Maritimes (Dept.) 4. Christian saints—Cult—Monaco. 5. Alpes—Maritimes, France (Dept.)—Religious life and customs. 6. Monaco—Religious life and customs. I. Title.
BT652.F7C36 282'.44'94 79-352638
79 MARC

—CULTUS

Bacigalupa, Drew, 1923-
Santos and saints' days. ₍1st ed.₎ Santa Fe, N.M., Sunstone Press, 1972.
32 p. illus. 22 cm. $2.95
ISBN 0-913270-09-1
1. Christian saints—Cultus. 2. Mary, Virgin—Cultus. 3. Santos (Art) I. Title.
BX2333.B3 759.13 73-154725
73₍r79₎rev MARC

—INVOCATION see Christian saints —Cult

—LEGENDS

Veľkomoravské legendy a povesti, ₍prel. zo staroslovien. orig.₎ / ₍zost., prel., štúdiu a pozn. nap. Peter Ratkoš ; úvod nap. Eugen Pauliny ; prebal a väzbu vyprac. Igor Imro ; graf. upr. Zlata Maderová₎. — 1. vyd. — Bratislava : Tatran, 1977.
172 p., ₍4₎ leaves of plates : ill. ; 21 cm. — (Hviezdoslavová knižnica ; zv. 247) CzS77
Kčs18.00
1. Legends—Czechoslovakia—Moravia. 2. Legends, Church-Slavic. 3. Christian saints—Legends. I. Ratkoš, Peter.
GR154.2.M67V44 79-381256
*79 MARC

—POETRY

Vaz, Manuel Inácio dos Santos.
O caminho dos santos / Manuel Inácio dos Santos Vaz. — ₍São Paulo? : Vaz, 1979?₎
152 p. ; 21 cm.
Cr$70.00
1. Christian saints—Poetry. I. Title.
PQ9698.32.A95C3 869'.1 79-114471
79 MARC

—PRAYER-BOOKS AND DEVOTIONS—SWEDISH

Broomé, Catharina B 1923-
Efterföljare : en helgonkalender enligt det nya romerska kalendariet jämte svenska helgondagar / av Catharina Broomé ; med ill. av Kaj Beckman. — Stockholm : Proprius, 1978.
245 p. : ill. ; 22 cm. S78-16
Includes bibliographical references and index.
ISBN 9171182578
1. Christian saints—Prayer-books and devotions—Swedish. I. Title.
BX2166.B76 79-384548
79 MARC

—VENERATION see Christian saints —Cult

—WORSHIP see Christian saints —Cult

—ALGERIA—HIPPO—BIOGRAPHY

Sandvoss, Ernst.
Aurelius Augustinus : e. Mensch auf d. Suche nach Sinn / Ernst Sandvoss. — Orig.-Ausg. — Freiburg im Breisgau ; Basel ; Wien : Herder, 1978.
141 p. ; 18 cm. — (Herderbücherei ; Bd. 674) GFR78-A
ISBN 3-451-07674-8 : DM5.90
1. Augustinus, Aurelius, Saint, Bp. of Hippo. 2. Christian saints—Algeria—Hippo—Biography. 3. Hippo, Algeria—Biography.
BR1720.A9S14 79-353837
79 MARC

—CZECHOSLOVAKIA—BIOGRAPHY

Veľkomoravské legendy a povesti / ₍prel. zo staroslovien. orig., zost., prel. štúdiu a pozn. nap. Peter Ratkoš ; úvod nap. Eugen Pauliny₎. — 1. vyd. — Bratislava : Tatran, 1977.
175 p., ₍3₎ leaves of plates : ill. (some col.) ; 21 cm. — (Hviezdoslavova knižnica ; zv. 247) CzS77
Kčs18.00
1. Cyrillus, Saint, of Thessalonica, 827 (ca.-869)—Legends. 2. Methodius, Saint, Abp. of Moravia, d. 885—Legends. 3. Christian saints—Czechoslovakia—Legends. 4. Christian saints—Czechoslovakia—Biography. I. Ratkoš, Peter.
BX4700.C9V44 79-341240
*79 MARC

—CZECHOSLOVAKIA—LEGENDS

Veľkomoravské legendy a povesti / ₍prel. zo staroslovien. orig., zost., prel. štúdiu a pozn. nap. Peter Ratkoš ; úvod nap. Eugen Pauliny₎. — 1. vyd. — Bratislava : Tatran, 1977.
175 p., ₍3₎ leaves of plates : ill. (some col.) ; 21 cm. — (Hviezdoslavova knižnica ; zv. 247) CzS77
Kčs18.00
1. Cyrillus, Saint, of Thessalonica, 827 (ca.-869)—Legends. 2. Methodius, Saint, Abp. of Moravia, d. 885—Legends. 3. Christian saints—Czechoslovakia—Legends. 4. Christian saints—Czechoslovakia—Biography. I. Ratkoš, Peter.
BX4700.C9V44 79-341240
*79 MARC

—ENGLAND—BIOGRAPHY

Aelfric, Abbot of Eynsham.
Lives of three English saints / ₍by₎ Aelfric ; edited by G. I. Needham. — Revised ed. — ₍Exeter₎ : University of Exeter, 1976.
viii, 119 p. ; 19 cm. — (Exeter medieval English texts) GB78-19642
Text in Anglo-Saxon; introduction and notes in English.
Bibliography: p. 82-85.
CONTENTS: St. Oswald, King and martyr.—St. Edmund, King and martyr.—St. Swithin, Bishop.
ISBN 0-85989-076-7 : £1.00
1. Oswald, Saint, King of Northumbria, 605?-642. 2. Edmund, Saint, King of East Anglia, 841-870. 3. Swithun, Saint, Bp. of Winchester, d. 862. 4. Christian saints—England—Biography. I. Needham, Geoffrey Ivor. II. Title. III. Series.
BX4659.G7A65 1976 78-320126
78 MARC

Berglar-Schröer, Hans Peter, 1919-
Die Stunde des Thomas Morus : einer gegen die Macht / Peter Berglar. — 1. Aufl. — Olten ₍etc.₎ : Walter, 1978.
397 p., ₍4₎ leaves of plates : ill. ; 23 cm. Sw78-7443
Bibliography: p. 389-392.
Includes index.
ISBN 3-530-06150-6 : 39.00F
1. More, Thomas, Sir, Saint, 1478-1535. 2. Statesmen—Great Britain—Biography. 3. Christian saints—England—Biography. 4. Great Britain—History—Henry VIII, 1509-1547. I. Title.
DA334.M8B47 942'.052'0924 78-390805
79 MARC

Reynolds, Ernest Edwin, 1894-
The life and death of St. Thomas More : the field is won / by E. E. Reynolds. — London : Burns & Oates ; New York : Barnes & Noble Books, 1978, c1968.
xiii, 396 p. : ill. ; 23 cm. GB***
Originally published under title: The field is won.
Includes bibliographical references and index.
ISBN 0-06-495854-X
1. More, Thomas, Sir, Saint, 1478-1535. 2. Statesmen—Great Britain—Biography. 3. Christian saints—England—Biography. 4. Great Britain—History—Henry VIII, 1509-1547. I. Title.
DA334.M8R38 1978 942.05'2'0924 78-322932
79 MARC

CHRISTIAN SAINTS
—ENGLAND—BIOGRAPHY (Continued)

Rupp, Ernest Gordon.
Thomas More : the King's good servant / ₍by₎ Gordon Rupp ; ₍photography by Helmuth Nils Loose₎. — London ; New York : Collins, 1978.
63 p., ₍56₎ p. of plates : ill. (chiefly col.), 2 coats of arms (1 col.), facsims. (1 col.), ports. (chiefly col.) ; 24 cm. GB78-24748
Photos. by H. N. Loose are identical with those in Thomas Morus by W. Nigg, published in Freiburg by Herder in 1978.
Bibliography: p. 63.
ISBN 0-529-05494-9 : £4.95 ($14.95 U.S.)
1. More, Thomas, Sir, Saint, 1478-1535. 2. Statesmen—Great Britain—Biography. 3. Christian saints—England—Biography. 4. Great Britain—History—Henry VIII, 1509-1547. I. Loose, Helmuth Nils. II. Title.
DA334.M8R86 1978 942.05'2'0924 78-56261
78 MARC

Walter Daniel, fl. 1170.
The life of Ailred of Rievaulx / by Walter Daniel ; translated from the Latin with introd. and notes by Maurice Powicke. — Oxford : Clarendon Press ; New York : Oxford University Press, 1978.
cii, 81, 81 p., p. ₍82₎-88 ; 23 cm. — (Oxford medieval texts) GB***
Translation of Vita Ailredi.
Introd. and notes in English; text in Latin, with English translation.
Opposite pages numbered in duplicate.
"Based upon ... ₍the translator's₎ earlier work, Ailred of Rievaulx and his biographer Walter Daniel."
Reprint of the 1950 ed. published by Nelson, London and New York, in series Medieval classics.
"Notes on Ailred's writings": p. xcv-cii.
Includes index.
ISBN 0-19-822256-4 : $22.00
1. Ethelred, Saint, 1109?-1166. 2. Christian saints—England—Biography. I. Powicke, Frederick Maurice, Sir, 1879-1963. II. Powicke, Frederick Maurice, Sir, 1879-1963. Ailred of Rievalux and his biographer Walter Daniel. III. Title. IV. Series. V. Series: Medieval classics (London)
BX4700.E7W313 1978 271'.12'024 79-313360
79 MARC

—ETHIOPIA—BIOGRAPHY

Lucatello, Enrico.
L'abuna Yaqob Mariàm (s. Giustino De Jacobis) / ₍di₎ Enrico Lucatello, Luigi Betta. — Roma : a cura della Postulazione generale C.M., 1975.
252 p. : ill. ; 24 cm. It77-Aug
Bibliography: p. 242-247.
1. Jacobis, Giustino de, Saint, 1800-1860. 2. Christian saints—Ethiopia—Biography. I. Betta, Luigi. II. Title.
BX4700.J18L82 266'.2'0924 79-359178
*79 MARC

—EUROPE—BIOGRAPHY—ADDRESSES, ESSAYS, LECTURES

Gaiffier, Baudouin de, 1897-
Recueil d'hagiographie / par Baudouin de Gaiffier ; avant-propos par Jean Leclercq. — Bruxelles : Société des Bollandistes, 1977.
ca. 200 p. in various pagings : ill. ; 25 cm. — (Subsidia hagiographica ; no. 61) Be***
Articles previously published in various journals, 1967-1977.
"Publié à l'occasion du 80me anniversaire de l'auteur."
Includes bibliographical references and indexes.
1. Christian saints—Europe—Biography—Addresses, essays, lectures. 2. Hagiography—Addresses, essays, lectures. I. Title. II. Series.
BX4655.2.G32 282'.092'2 78-388893
79 MARC

—FRANCE—BIOGRAPHY

Bourassin, Emmanuel.
Jeanne d'Arc / Emmanuel Bourassin. — Paris : Perrin, c1977.
349 p., ₍8₎ leaves of plates : ill. ; 21 cm. F***
Bibliography: p. 337-₍343₎
ISBN 2-262-00080-8 : 60.00F
1. Jeanne d'Arc, Saint, 1412-1431. 2. Christian saints—France—Biography.
DC103.B79 944'.026'0924 78-386021
79 MARC

Fabre, Lucien, 1889-1952.
Jeanne d'Arc / Lucien Fabre ; ₍préface de Régine Pernoud₎. — Paris : J. Tallandier, 1978.
541 p., ₍16₎ p. of plates ; 23 cm. — (Figures de proue de l'histoire de France ISSN 0153-2227) F78-16985
ISBN 2-235-00500-4 : 70.00F
1. Jeanne d'Arc, Saint, 1412-1431. 2. Christian saints—France—Biography. I. Series.
DC103.F315 1978 944'.026'0924 79-377655
79 MARC

Lavater-Sloman, Mary, 1891-
Jeanne d'Arc : d. Heilige in Waffen / Mary Lavater-Sloman ; ₍Ergänzungsbibliographie, Zeittafel u. Stammtafeln wurden erarb. von Hubert Fritz₎. — Genehmigte, erw. Taschenbuchausg. — München : Heyne, 1977.
463 p. : ill. ; 18 cm. — (Heyne Biographien ; 37) GFR77-A24
Bibliography: p. 430-₍436₎
Includes index.
ISBN 3-453-55037-4 : DM6.80
1. Jeanne d'Arc, Saint, 1412-1431. 2. Christian saints—France—Biography.
DC103.L37 1977 944'.026'0924 78-385422
79 MARC

Machniewicz, Milton.
O homem da capa preta / pesquisadores, Milton Machniewicz, Xisto P. Bobato. — ₍Curitiba : Editora Gráfica Vicentina, c1975₎
104 p. : ill. ; 23 cm.
Cr$15.00
1. Vincent de Paul, Saint, 1581-1660. 2. Christian saints—France—Biography. I. Bobato, Xisto P., joint author. II. Title.
BX4700.V6M2 78-381213
79 MARC

Passiones vitaeque sanctorum aevi Merovingici et antiquiorum aliquot / ed. Bruno Krusch. — Unveränd. Nachdr. ₍d. Ausg.₎ 1896- — Hannover : Hahn, 1977-
v. ; 31 cm.
Reprint of the ed. published by Hahn, Hannover, which was issued at T. 3 of Monumenta Germaniae historica, Scriptorum rerum Merovingicarum.
Includes bibliographical references and index.
ISBN 3-7752-5296-7 : DM168.00(v. 1)
1. Christian saints—Germany—Biography. 2. Germany—Church history—To 843. 3. Christian saints—France—Biography. 4. France—Church history—To 987. 5. Christian saints—Netherlands—Biography. 6. Netherlands—Church history. I. Krusch, Bruno, 1857-1940. II. Series: Monumenta Germaniae historica inde ab anno Christi quingentesimo usque ad annum millesimum et quingentesimum. Scriptorum rerum Merovingicarum ; T. 3.
BX4683.P37 1977 79-376366
79 MARC

Spens, Willy de, 1911-
Saint Vincent de Paul : ou, le Trésor des pauvres / Willy de Spens. — Paris : Éditions S.O.S., 1978.
216 p., ₍4₎ p. of plates ; 21 cm. — (Collection Hauts lieux de spiritualité) F78-14244
ISBN 2-7185-0840-X : 40.00F
1. Vincent de Paul, Saint, 1581-1660. 2. Christian saints—France—Biography.
BX4700.V6S66 271'.77'024 79-354122
79 MARC

Stevens, Courtenay Edward.
Sidonius Apollinaris and his age / by C. E. Stevens. — Westport, Conn. : Greenwood Press, 1979, c1933.
xiv, 224 p., ₍2₎ leaves of plates : ill. ; 22 cm.
Reprint of the ed. published by the Clarendon Press, Oxford.
Bibliography: p. ₍213₎-220.
Includes index.
ISBN 0-313-20850-6
1. Sidonius Apollinaris, Saint, 431 or 2-ca. 487. 2. Christian saints—France—Biography. I. Title.
BR1720.S5S73 1979 270.2'092'4 78-21112
78 MARC

—FRANCE—BIOGRAPHY—ADDRESSES, ESSAYS, LECTURES

Tyler, Sidney F
Resurrection at Rouen / by Sidney F. Tyler. — 1st ed. — Flagstaff : Northland Press, 1979.
25 p. ; 23 cm.
"An address delivered to the Graduate Club, Tucson, Arizona, November 8, 1978."
Two hundred copies specially bound.
1. Jeanne d'Arc, Saint, 1412-1431—Addresses, essays, lectures. 2. Christian saints—France—Biography—Addresses, essays, lectures. I. Title.
DC103.T96 944'.026'0924 79-109093
00 MARC

—FRANCE—BRIE—BIOGRAPHY

Fontaine, Louis.
Saints et chevaliers oubliés de la Brie / Louis Fontaine ; ill. origionales d'Eugène Collilieux. — Vincennes : Les Éditions de l'Orme rond, c1978.
205 p. : ill. ; 24 cm. F***
Bibliography: p. 205.
45.00F
1. Brie—Church history. 2. Christian saints—France—Brie—Biography. 3. Brie—Biography. I. Title.
BR847.B74F66 282'.44'37 79-378317
79 MARC

—FRANCE—LOURDES—BIOGRAPHY

Guynot, Ernest, 1879-
Sainte Bernadette d'après ses contemporains / Ernest Guynot. — 4. éd. — Paris : Cerf, 1978.
230 p. ; 20 cm. F***
ISBN 2-204-01212-2 : 22.00F
1. Soubirous, Bernadette, Saint, 1844-1879. 2. Christian saints—France—Lourdes—Biography. 3. Lourdes—Biography. I. Title.
BX4700.S65G88 1978 271'.97 78-391712
79 MARC

Laurentin, René.
Vie de Bernadette / René Laurentin. — ₍Paris₎ : Desclée De Brouwer, c1978.
251, ₍5₎ p. : ill. ; 18 cm. F***
On cover: Le livre du centenaire.
Bibliography: p. ₍253₎-₍254₎
ISBN (invalid) 2-220-20157-2 : 98.00F
1. Soubirous, Bernadette, Saint, 1844-1879. 2. Christian saints—France—Lourdes—Biography. 3. Lourdes—Biography. I. Title.
BX4700.S65L38 271'.97 79-377315
79 MARC

Ravier, André.
Bernadette / ₍by₎ André Revier ; with an introduction by Patrick O'Donovan ; translated ₍from the German₎ by Barbara Wall ; photographs by Helmuth Nils Loose. — London ; Cleveland : Collins, 1979.
63 p., ₍40₎ p. of plates : ill. (some col.), facsims., ports. (some col.) ; 24 cm. GB79-22156
Bibliography: p. 61.
ISBN 0-529-05672-0 : $£4.95 ($14.95)
1. Soubirous, Bernadette, Saint, 1844-1879. 2. Christian saints—France—Lourdes—Biography. 3. Lourdes—Biography.
BX470.S65R3813 1979 271'.91'024 79-51569
79 MARC

Saint Pierre, Michel de, 1916-
Bernadette / Michel de Saint Pierre. — Paris : la Table ronde, 1979.
203 p., ₍16₎ p. of plates ; 21 cm. F79-8101
Includes bibliographical references.
42.00F
1. Soubirous, Bernadette, Saint, 1844-1879. 2. Christian saints—France—Lourdes—Biography. 3. Lourdes—Biography. I. Title.
BX4700.S65S32 271'.97 79-385475
79 MARC

—FRANCE—LOURDES—BIOGRAPHY—JUVENILE LITERATURE

Saint-Pierre, Michel de, 1916-
Sainte Bernadette / Michel de Saint-Pierre ; illustrations de Michel Gérard. — ₍Limoges₎ : Droguet-Ardant, 1977.
40 p. : ill. (some col.) ; 28 cm. — (La Vie de nos saints) F78-5781
ISBN 2-7041-0580-4 : 27.00F
1. Soubirous, Bernadette, Saint, 1844-1879—Juvenile literature. 2. Christian saints—France—Lourdes—Biography—Juvenile literature. 3. Lourdes—Biography—Juvenile literature. I. Title.
BX4700.S65S35 271'.97 79-343066
79 MARC

—GERMANY—BIOGRAPHY

Passiones vitaeque sanctorum aevi Merovingici et antiquiorum aliquot / ed. Bruno Krusch. — Unveränd. Nachdr. ₍d. Ausg.₎ 1896- — Hannover : Hahn, 1977-
v. ; 31 cm.
Reprint of the ed. published by Hahn, Hannover, which was issued at T. 3 of Monumenta Germaniae historica, Scriptorum rerum Merovingicarum.
Includes bibliographical references and index.
ISBN 3-7752-5296-7 : DM168.00(v. 1)
1. Christian saints—Germany—Biography. 2. Germany—Church history—To 843. 3. Christian saints—France—Biography. 4. France—Church history—To 987. 5. Christian saints—Netherlands—Biography. 6. Netherlands—Church history. I. Krusch, Bruno, 1857-1940. II. Series: Monumenta Germaniae historica inde ab anno Christi quingentesimo usque ad annum millesimum et quingentesimum. Scriptorum rerum Merovingicarum ; T. 3.
BX4683.P37 1977 79-376366
79 MARC

Saur, Robert Maria.
Gertrud die Grosse : die Wunderblume von Helfta : das Leben einer vergessenen grossen Heiligen / von Robert Maria Saur. — Bern : Bargezzi, cop. 1977.
64 p. ; 17 cm. — (Grosse Marienverehrer ; Bd. 25) Sw78-889
2.00F
1. Gertrude, Saint, surnamed the Great, 1256-1302? 2. Christian saints—Germany—Biography.
BX4700.G6S28 79-372631
79 MARC

—GERMANY, WEST—BIOGRAPHY

Das Leben des heiligen Wolfgang : nach d. Holzschnittbuch d. Johann Weyssenburger aus d. Jahr 1515 / hrsg. von Hans Bleibrunner. — Landshut : Isar-Post, 1976.
₍128₎ p. : ill. ; 23 cm. GFR78-A
Adapted from the original ed. printed by J. Weysenburger in 1515 in Landshut, which was based on an anonymous work written at the Benedictine abbey of Mondsee.
Includes facsimile of the t.p., the first page, and the colophon of the 1515 ed.
Bibliography: p. ₍125₎
1. Wolfgang, Saint, Bp. of Ratisbon, d. 994. 2. Christian saints—Germany, West—Biography. I. Weyssenburger, Johann. II. Bleibrunner, Hans, 1927-
BX4700.W64L42 78-397060
79 MARC

CHRISTIAN SAINTS (Continued)

—GREAT BRITAIN

Vince, John Norman Thatcher.
Discovering saints in Britain. — 2nd ed. / [by] John Vince. — Aylesbury : Shire Publications, 1979.
64 p. ; 18 cm. — (Discovering series ; no. 64) GB79-10016
First ed. (1969) by J. Fidler.
ISBN 0-85263-449-8 : £0.75
1. Christian saints—Great Britain. I. Fidler, John. Discovering saints in Britain. II. Title.
BX4659.G7F5 1979 274.1 79-312690
79 MARC

—GREAT BRITAIN—BIOGRAPHY

Farmer, David Hugh.
The Oxford dictionary of saints / David Hugh Farmer. — Oxford : Clarendon Press, 1978.
xxiv, 435 p. ; 23 cm. GB***
Includes bibliographical references and index.
ISBN 0-19-869120-3 : £7.50
1. Christian saints—Great Britain—Biography. 2. Christian saints—Ireland—Biography. I. Title.
BX4659.G7F37 270 78-326714
79 MARC

—GREECE, MODERN—BIOGRAPHY

Oikonomos, Phōtios G
(Ho Kosmas ho Aitōlos kai hē hierapostolikē drasis autou, 1714–1779)
Ὁ Κοσμᾶς ὁ Αἰτωλὸς καὶ ἡ ἱεραποστολικὴ δρᾶσις αὐτοῦ, 1714–1779 / Θωτίου Γ. Οἰκονόμου. — Ἀθῆναι : [Οἰκονόμος], 1976.
51 p. ; 24 cm.
Includes bibliographical references.
1. Kosmas ho Aitōlos, Saint, 1714–1779. 2. Christian saints—Greece, Modern—Biography. I. Title.
BX395.K67O35 78–422499

—HUNGARY—BIOGRAPHY—ADDRESSES, ESSAYS, LECTURES

Die Heiligen Könige / übers., eingel. u. erkl. v. Thomas von Bogyay, János Bak, Gabriel Silagi. — Graz ; Wien ; Köln : Verl. Styria, 1976.
195 p. ; 20 cm. — (Ungarns Geschichtsschreiber ; Bd. 1) Au76-18-352
Bibliography: p. 13-16.
ISBN 3-222-10900-1 : S180.00
1. Christian saints—Hungary—Biography—Addresses, essays, lectures. 2. Hungary—Kings and rulers—Biography—Addresses, essays, lectures. I. Bogyay, Thomas von. II. Bak, János M. III. Silagi, Gabriel, 1942-
BX4690.H9H43 79-371569
*79 MARC

—INDIA—BIOGRAPHY

Thomas, Sunny, 1945-
Behold a saint : the life & times of Parumala Mar Gregorios / by Sunny Thomas. — [s.l. : s.n., pref. 1977] (New Delhi : Printaid).
x, 49 p. ; 23 cm.
Includes bibliographical references.
Rs10.00
1. Mar Gregorios, Saint, 1848-1902. 2. Christian saints—India—Biography. I. Title.
BX395.M34T46 281.9'092'4 78-904609
79 MARC

—IRELAND—BIOGRAPHY

Bernard de Clairvaux, Saint, 1091?-1153.
The life and death of Saint Malachy, the Irishman / Bernard of Clairvaux ; translated and annotated by Robert T. Meyer. — Kalamazoo, Mich. : Cistercian Publications, 1978.
170 p. ; 22 cm. — (Cistercian Fathers series ; no. 10)
Includes bibliographical references and index.
CONTENTS: The life of Saint Malachy.—Sermon on the passing of Saint Malachy the Bishop.—Homily on the anniversary of the death of Saint Malachy.—Saint Bernard's Epitaph on Saint Malachy.—Saint Bernard's Hymn in honor of Saint Malachy.
ISBN 0-87907-110-0
1. Malachy O'Morgain, Saint, 1094?-1148. 2. Catholic Church—Bishops—Biography. 3. Bishops—Ireland—Biography. 4. Christian saints—Ireland—Biography. I. Title.
BX4700.M23B4613 1978 78-768
282'.092'4
78 MARC

Farmer, David Hugh.
The Oxford dictionary of saints / David Hugh Farmer. — Oxford : Clarendon Press, 1978.
xxiv, 435 p. ; 23 cm. GB***
Includes bibliographical references and index.
ISBN 0-19-869120-3 : £7.50
1. Christian saints—Great Britain—Biography. 2. Christian saints—Ireland—Biography. I. Title.
BX4659.G7F37 270 78-326714
79 MARC

Patrick, Saint, 373?-463?
Confession ; (et) Lettre à Coroticus / saint Patrick ; introduction, texte critique, traduction et notes par Richard P. C. Hanson, ... avec la collaboration de Cécile Blanc. — Paris : Éditions du Cerf, 1978.
198 p. ; 20 cm. — (Sources chrétiennes ; no 249) F79-474
Latin text and French translation of Confessio and Epistola ad milites Corotici.
Bibliography: p. [66]-67.
ISBN 2-204-01277-7 : 109F
1. Patrick, Saint, 373?-463? 2. Christian saints—Ireland—Biography. I. Patrick, Saint, 373?-463? Epistola ad milites Corotici. French & Latin. 1978. II. Hanson, Richard Patrick Crosland. III. Blanc, Cécile. IV. Title.
BR1720.P26A3314 270.2'092'4 79-367273
79 MARC

Patrick, Saint, 373?-463?
St. Patrick, his writings and Muirchu's Life / edited and translated by A. B. E. Hood. — London : Phillimore ; Totowa, N.J. : Rowman and Littlefield, 1978.
101 p. ; 22 cm. — ([Arthurian period sources] ; v. 9) (History from the sources) GB***
English or Latin.
Includes bibliographical references.
ISBN 0-8476-6080-X (Rowman and Littlefield) : £3.00 ($7.50)
1. Patrick, Saint, 373?-463? 2. Christian saints—Ireland—Biography. I. Hood, Allan B. E., 1944- . II. Muirchu maccu Machtheni, fl. 697. Life of St. Patrick. 1978. III. Title. IV. Series.
BX4700.P3A213 1978 270.2'092'4 79-301620
79 MARC

Vita Brigitae. English & Irish.
Bethu Brigte / edited by Donncha Ó hAodha. — [Dublin] : Dublin Institute for Advanced Studies, 1978.
xxx, 91 p. ; 23 cm. Ir***
In Old Irish or Latin, with English translation; introd. and notes in English.
Includes bibliographical references and indexes.
1. Brigid, Saint, of Ireland, ca. 453-ca. 524. 2. Christian saints—Ireland—Biography. I. Ó hAodha, Donncha. II. Title.
BX4700.B85.V5713 1978 79-312013
282'.092'4
79 MARC

Vita Brigitae. Irish.
(Beatha Bhrighde)
Beaṫa Ḃriġde ar Lr. 23L33, R. I. A. Maġnus O Doṁnaill do ċuir i n-eagar. Baile Aṫa Cliaṫ, Oifig an tSoláṫair [1940]
72 p. 19 cm.
A modern version of the life of St. Brigid of Ireland.
1. Brigid, Saint, of Ireland, ca. 453–ca. 524. 2. Christian saints—Ireland—Biography. I. O'Domhnaill, Maoghnas, ed. II. Title.
BX4700.B85V5715 53—51360

—ITALY—BIOGRAPHY

Agresti, Giuliano.
Ritratto della espropriata : (S. Gemma Galgani) / Giuliano Agresti. — Lucca : M. Pacini Fazzi, 1978.
167 p. ; 19 cm. It78-Oct
L3500
1. Galgani, Gemma, Saint, 1878-1903. 2. Christian saints—Italy—Biography. I. Title.
BX4700.G22A615 282'.092'4 78-399096
*79 MARC

Anderson, Robin, 1913-
St. Pius V, a brief account of his life, times, virtues & miracles / by Robin Anderson ; foreword by Alfredo Ottaviani. — Rockford, Ill. : Tan Books and Publishers, 1978.
111 p. : ill. ; 18 cm.
ISBN 0-89555-068-7 : $2.00
1. Pius V, Saint, Pope, 1504-1572. 2. Popes—Biography. 3. Christian saints—Italy—Biography. I. Title.
BX1323.A52 262'.13'0924 78-55637
00 MARC

Basile, Giacinta.
Forever yes : the story of Lucy Filippini / by Giacinta Basile and Geraldine Calabrese. — Philadelphia : Dorrance, c1979.
xvi, 139 p. ; 22 cm.
Bibliography: p. 139.
ISBN 0-8059-2602-X : $5.95
1. Lucia Filippini, Saint, 1672-1732. 2. Christian saints—Italy—Biography. I. Calabrese, Geraldine, joint author. II. Title.
BX4700.L73B37 271'.97 78-72582
79 MARC

Bargellini, Piero, 1897-
Il santo del lavoro / Piero Bargellini. — Torino Leumann : LDC, [1976?]
79 p., [8] leaves of plates : ill. ; 21 cm. It***
L1000
1. Bosco, Giovanni, Saint, 1815-1888. 2. Christian saints—Italy—Biography. I. Title.
BX4700.B75B37 78-387036
79 MARC

Blanc, Pierre, S J
Saint Louis de Gonzague : un saint de 23 ans / Pierre Blanc. — Paris : Téqui, [1977]
149 p., [4] leaves of plates : ill. ; 21 cm. F***
Includes bibliographical references.
ISBN 2-85244-283-3 : 18.00F
1. Luigi Gonzaga, Saint, 1568-1591. 2. Christian saints—Italy—Biography.
BX4700.L75B55 271'.5'0924 78-361309
78 MARC

Bo, Dino del.
Caterina da Genova : l'amore e il purgatorio / Dino Del Bo. — Milano : All'insegna del pesce d'oro, 1978.
178 p. ; 18 cm. It79-Jan
L5500
1. Caterina da Genova, Saint, 1447-1510. 2. Christian saints—Italy—Biography. I. Title.
BX4700.C36B58 79-349392
*79 MARC

Bosco, Giovanni, Saint, 1815-1888.
St. Dominic Savio / by Saint John Bosco ; translated, with notes by Paul Aronica ; foreword by Humberto Cardinal Medeiros. — 2d ed. — New Rochelle, N.Y. : Don Bosco Publications, c1979.
ix, 169 p. : ill. ; 19 cm. — (A Patron book)
Translation of Vita del giovanetto Dom. Savio.
First American ed. published in 1955 under title: The life of Saint Dominic Savio.
1. Savio, Domenico, Saint, 1842-1857. 2. Christian saints—Italy—Biography.
BX4700.S35B63 1979 282'.092'4 78-67221
79 MARC

Bosco, Teresio.
Don Bosco, una biografia nuova / Teresio Bosco ; presentazione di don Giovanni Raineri. — 2. ed. — Torino/Leumann : Elle Di Ci, 1979.
447 p., [16] leaves of plates : ill. ; 21 cm. It***
Bibliography: p. 438.
ISBN 8801116039 : L4500
1. Bosco, Giovanni, Saint, 1815-1888. 2. Christian saints—Italy—Biography. I. Title.
BX4700.B75B57 1979 79-363429
79 MARC

Galgani, Gemma, Saint, 1878-1903.
Gesù solo / Gemma Galgani ; testi scelti a cura di Gabriele Pollice ; introduzione di Giannetto Casoli. — Roma : Città nuova, 1978.
278 p. ; 19 cm. It78-May
Bibliography: p. 40-42.
L4500
1. Galgani, Gemma, Saint, 1878-1903. 2. Christian saints—Italy—Biography. 3. Christian saints—Italy—Correspondence. I. Pollice, Gabriele. II. Title.
BX4700.G22G34 1978 79-383071
*79 MARC

Poage, Godfrey Robert, 1920-
Son of the Passion : the story of Gabriel Francis Possenti, the new patron of Catholic youth for the universal church / by Godfrey Poage. — [Boston] : St. Paul Editions, c1977.
ix, 120 p. [8] leaves of plates : ill. ; 22 cm.
Bibliography: p. 119-120.
$3.50
1. Gabriele, Saint, 1838-1862. 2. Christian saints—Italy—Biography. I. Title.
BX4700.G2P6 1977 270'.62'0924 79-103691
79 MARC

Rinaldi, Peter M
Man with a dream : the story of St. John Bosco / by Peter Rinaldi. — New Rochelle, N.Y. : Don Bosco Publications, c1978.
v, 161 p. : ill. ; 19 cm. — (Patron books)
$2.50
1. Bosco, Giovanni, Saint, 1815-1888. 2. Christian saints—Italy—Biography. I. Title.
BX4700.B75R56 271'.7 78-112476
79 MARC

Sanz de Orodea, Carmen.
Dos grandes santos : Don Bosco, Ignacio de Loyola / Carmen Sanz de Orodea. — Madrid : Sociedad General Española de Librería, c1977.
128 p. : ill. ; 18 cm. — (Colección Textos en español fácil : Nivel elemental) Sp***
ISBN 8471431386 : 125ptas
1. Bosco, Giovanni, Saint, 1815-1888. 2. Loyola, Ignacio de, Saint, 1491-1556. 3. Christian saints—Italy—Biography. I. Title.
BX4700.B75S35 79-375859
79 MARC

Simi, Gino J 1902-
Saint Francis of Paola : God's miracle worker supreme, 1416-1507 / by Gino J. Simi and Mario M. Segreti. — Rockford, Ill. : Tan Books and Publishers, 1977.
209 p. ; 18 cm.
ISBN 0-89555-065-2 : $2.50
1. Franciscus de Paula, Saint, 1416-1507. 2. Christian saints—Italy—Biography. I. Segreti, Mario M., joint author. II. Title.
BX4700.F83S57 271'.06'024 77-78097
79 MARC

CHRISTIAN SAINTS
—ITALY—BIOGRAPHY (Continued)

Stowarzyszenie Apostolstwa Katolickiego : jubileusz wspólnoty ołtarzewskiej 1927-1977 / [aut. Lucjan Balter ... et al.]. — Wyd. 1. — Poznań ; Warszawa : Pallottinum, 1977 dr. ukończ. 1978 (Pozn. : Pallottinum)
184 p., [18] leaves of plates : ill. ; 25 cm. P78-5148
Summary in French.
Includes bibliographies.
zł80.00
1. Pallottines. 2. Pallottines in Poland. 3. Pallotti, Vincenzo, Saint, 1795-1850. 4. Christian saints—Italy—Biography. I. Balter, Lucjan.
BX3866.S76 79-352231
79 MARC

Villepelée, Jean François.
La folie de la Croix : Sainte Gemma Galgani, 1878-1903 / Jean-François Villepelée. — Hauteville : Éditions du Parvis, [1977-
v. : ill. ; 21 cm. Sw***
Includes bibliographical references.
CONTENTS: [1] L'ascension d'une âme.—[2] La contemplation du mystère.

ISBN 2-88022-041-6 (v. 1)
1. Galgani, Gemma, Saint, 1878-1903. 2. Christian saints—Italy—Biography. I. Title.
BX4700.G22V54 282'.092'4 78-382471
79 MARC

Vizzari, Domenico.
San Alfonso Maria de' Liguori, san Paolo della Croce, mons. Emilio Giacomo Cavalieri / [di] Domenico Vizzari, Gioacchino A. De Sanctis. — Napoli : Edizioni dehoniane, 1976.
61 p. ; 24 cm. — (Storia, letteratura e arte del Mezzogiorno ; 5) It78-Apr
Includes bibliographical references and index.
L2000
1. Liguori, Alfonso Maria de', Saint, 1696-1787. 2. Paolo della Croce, Saint, 1694-1775. 3. Cavalieri, Emilio, 1663-1726. 4. Christian saints—Italy—Biography. I. De Sanctis, Gioacchino A., joint author. II. Title.
BX4684.V58 78-378142
*78 MARC

—ITALY—CORRESPONDENCE

Galgani, Gemma, Saint, 1878-1903.
Gesù solo / Gemma Galgani ; testi scelti a cura di Gabriele Pollice ; introduzione di Giannetto Casoli. — Roma : Città nuova, 1978.
278 p. ; 19 cm. It78-May
Bibliography: p. 40-42.
L4500
1. Galgani, Gemma, Saint, 1878-1903. 2. Christian saints—Italy—Biography. 3. Christian saints—Italy—Correspondence. I. Pollice, Gabriele. II. Title.
BX4700.G22G34 1978 79-383071
*79 MARC

—ITALY—ASSISI—BIOGRAPHY

Beobide, Krispin, 1848-1891.
Asis'ko loria / Beobide'ko Aita Fr. Krispin. — [Tolosa] : Auspoa Liburutegia, 1966.
414 p. ; 19 cm. — (Auspoa liburutegia ; 57-58)
1. Francesco d'Assisi, Saint, 1182-1226. 2. Christian saints—Italy—Assisi—Biography. 3. Assisi—Biography. I. Title.
BX4700.F6B47 1966 79-365327
79 MARC

Brogliato, Bortolo, 1926-
Assisi, incontro vivo con frate Francesco e sorella Chiara / Bortolo Brogliato. — Vicenza : L.I.E.F., 1976.
246 p., [14] leaves of plates : ill. ; 20 cm. It77-Oct
Bibliography: p. 236-240.
L3900
1. Francesco d'Assisi, Saint, 1182-1226. 2. Clara, of Assisi, Saint, d. 1253. 3. Christian saints—Italy—Assisi—Biography. 4. Assisi—Biography. 5. Art—Italy—Assisi. I. Title.
BX4700.F6B74 78-388702
*79 MARC

Egan, Maurice Francis, 1852-1924.
The life of St. Francis and the soul of modern man / Maurice Francis Egan. — Albuquerque, N.M. : American Classical College Press, [1979]
60 leaves, [4] leaves of plates : ill. ; 28 cm.
Cover title.
Reprint of the 1912 ed. published by the Century Co., New York, under title: Everybody's St. Francis.
ISBN 0-89266-149-6 : $27.75
1. Francesco d'Assisi, Saint, 1182-1226. 2. Christian saints—Italy—Assisi—Biography. 3. Assisi—Biography. I. Title.
BX4700.F6E5 1979 271'.3'024 79-329
79 MARC

Elcid, Daniel.
El hermano Francisco : el santo que no muere / por Daniel Elcid. — Madrid : Biblioteca de Autores Cristianos, 1978.
xvi, 225 p. ; 19 cm. — (BAC popular ; 15) Sp***
Bibliography: p. [224]-225.
ISBN 8422008750
1. Francesco d'Assisi, Saint, 1182-1226. 2. Christian saints—Italy—Assisi—Biography. 3. Assisi—Biography. I. Title.
BX4700.F6E53 79-377375
79 MARC

Longford, Frank Pakenham, 7th Earl of, 1905-
Francis of Assisi : a life for all seasons / Lord Longford. — London : Weidenfeld and Nicolson, c1978.
xvi, 190 p., [4] leaves of plates : ill. ; 23 cm. GB***
Bibliography: p. [179]-180.
Includes index.
ISBN 0-297-77520-0 : £5.95
1. Francesco d'Assisi, Saint, 1182-1226. 2. Christian saints—Italy—Assisi—Biography. 3. Assisi—Biography. I. Title.
BX4700.F6L64 271'.3'024 78-325804
79 MARC

San Francisco de Asís : escritos, biografías, documentos de la epoca / edición preparada por José Antonio Guerra. — Madrid : Editorial Católica, 1978.
xxiii, 1092 p. : ill. ; 20 cm. — (Biblioteca de autores cristianos ; 399) Sp***
Includes bibliographical references and indexes.
ISBN 8422008742
1. Francesco d'Assisi, Saint, 1182-1226. 2. Christian saints—Italy—Assisi—Biography. 3. Assisi—Biography. 4. Theology—Collected works—Middle Ages, 600-1500. 5. Catholic Church—Collected works. I. Guerra, José Antonio.
BX4700.F6S2717 79-338424
79 MARC

—ITALY—SIENA—BIOGRAPHY

Raymundus de Vineis, 1330-1399.
The life of ... Sainct Catharine of Siena / Raymundus de Vineis. — Ilkley [Eng.] : Scolar Press, 1978.
455 p. ; 19 cm. — (English recusant literature, 1558-1640 ; v. 373) GB***
Translation of Vita Sanctae Catherinae Senensis.
Reprint of the 1609 ed. from a copy in the library of Ushaw College. References: Allison and Rogers 864; STC 4830.
ISBN 0-85967-445-2 : £10.00
1. Caterina da Siena, Saint, 1347-1380. 2. Christian saints—Italy—Siena—Biography. 3. Siena—Biography. I. Title. II. Series: Rogers, David Morrison, comp. English recusant literature, 1558-1640 ; v. 373.
BX1750.A1E5 vol. 373 282'.41 s 78-375621
[BX4700.C4] 78 MARC

—LEBANON—BIOGRAPHY

Benedict, Claire M
Saint Sharbel, mystic of the East, 1828-1898 / Claire M. Benedict. — Cambridge [Mass.] : Ravengate Press, c1977.
xvii, 103, [3] p. : ill. ; 21 cm.
Bibliography: p. [105]-[106]
ISBN 0-911218-11-4 : $6.95. ISBN 0-911218-12-2 pbk.
1. Sharbal Makhlūf, Saint, 1828-1898. 2. Christian saints—Lebanon—Biography. I. Title.
BX4700.S47B46 271'.8 78-107944
78 MARC

Salemi, Elias.
San Charbel Majluf : monje libanés maronita : proyección sobre el país y la vida / [proyección y realizacion, Elias Salemi]. — Montevideo, Uruguay : Misión Maronita, 1977.
90 p. : port. ; 22 cm.
Cover title.
1. Sharbal Makhlūf, Saint, 1828-1898. 2. Christian saints—Lebanon—Biography. I. Title.
BX4700.S47S24 79-382703
79 MARC

—NETHERLANDS—BIOGRAPHY

Passiones vitaeque sanctorum aevi Merovingici et antiquiorum aliquot / ed. Bruno Krusch. — Unveränd. Nachdr. [d. Ausg.] 1896- — Hannover : Hahn, 1977-
v. ; 31 cm.
Reprint of the ed. published by Hahn, Hannover, which was issued at T. 3 of Monumenta Germaniae historica, Scriptorum rerum Merovingicarum.
Includes bibliographical references and index.
ISBN 3-7752-5296-7 : DM168.00(v. 1)
1. Christian saints—Germany—Biography. 2. Germany—Church history—To 843. 3. Christian saints—France—Biography. 4. France—Church history—To 987. 5. Christian saints—Netherlands—Biography. 6. Netherlands—Church history. I. Krusch, Bruno, 1857-1940. II. Series: Monumenta Germaniae historica ab anno Christi quingentesimo usque ad annum millesimum et quingentesimum. Scriptorum rerum Merovingicarum ; T. 3.
BX4683.P37 1977 79-376366
79 MARC

—PALESTINE—BIOGRAPHY

Saba, Bonaventure.
The sinful, the intimate & the mysterious life of Mary Magdalene / Bonaventure Saba. — Albuquerque, N.M. : American Classical College Press, [1978]
21 leaves : port. ; 28 cm. — (The Most meaningful classics in world culture series)
Cover title.
ISBN 0-89266-.33-X
1. Mary Magdalene, Saint. 2. Christian saints—Palestine—Biography. 3. Bible. N.T.—Biography. I. Title. II. Series.
[BS2485.S23] 225.9'2'4 78-11166
78 MARC

—RUSSIA—BIOGRAPHY

Arkhangelov, Serafim Aleksandrovich.
(Zhitie i chudesa svi͡atoĭ Blagovi͡ernoĭ velikoĭ kni͡agini Anny Kashinskoĭ)
Житіе и чудеса святой благовѣрной великой княгини Анны Кашинской / составилъ С. Архангеловъ. — С.-Петербургъ : Тип. Министерства путей сообщенія, Т-ва И. Н. Кушнеревъ, 1909.
178 p. : ill. ; 25 cm.
At head of title: Diven Bog vo svi͡atykh svoikh.
"Посланіе Святѣйшаго Сvнода ... о возстановленіи церковнаго почитанія благовѣрной великой княгини Анны Кашинской" (issued as suppl. no. 21 of the T͡Serkovnyi͡a Vi͡edomosti for 1909) : vi p. laid in.
Includes bibliographical references.

1. Anna Kashinskai͡a, Saint, velikai͡a kni͡aginі͡a, d. 1368. 2. Christian saints—Russia—Biography. I. Title.
BX597.A645A74 Russian Imperial Coll. 78–421755

Nikon, Abp. of Vologda and Tot'ma.
(Zhitie i podvigi Prepodobnago i Bogonosnago ot͡tsa nashego Sergii͡a igumena Radonezhskago i vsei͡a Rossii Chudotvort͡sa)
Житіе и подвиги Преподобнаго и Богоноснаго отца нашего Сергія игумена Радонежскаго и всея Россіи Чудотворца / составлено С. Іеромонахомъ, нынѣ Архимандритомъ Никономъ. — Изд. 3., испр. и доп. многими рисунками. — Сергіевъ посадъ, Моск. губ. : Свято-Троицкая Сергіева Лавра, Ред. Троицкихъ листковъ, Собственная тип., 1898.
266 p., [19] leaves of plates : ill. (1 col.) ; 28 cm.
Cover title varies slightly.
Description of illustrations: p. 259–263.

Includes bibliographical references.
1. Sergiĭ Radonezhskiĭ, Saint, 1314?–1392. 2. Orthodox Eastern Church, Russian—Clergy—Biography. 3. Christian saints—Russia—Biography. I. Title.
BX597.S45N54 1898 Russian Imperial Coll. 78–421104

Orthodox Eastern Church, Russian. Belgorodskai͡a eparkhii͡a. Bishop, 1748–1754 (Ioasaf Gorlenko)
(Dokumenty i di͡ela, otnosi͡ashchiesi͡a k zhizni i di͡ei͡atel'nosti svi͡atiteli͡a Ioasafa (Gorlenko), Episkopa Bi͡elgorodskago i Oboi͡anskago)
Документы и дѣла, относящіеся къ жизни и дѣятельности святителя Іоасафа (Горленко), Епископа Бѣлгородскаго и Обоянскаго : (1748–1754 г.) : (Изъ архива Курской Духовной Консисторіи). Выпускъ первый. — Курскъ : Изд. Курской церковной историко-археологической комиссіи, Епархіальная тип., 1910.
vi, 112, 14, 11 p., [1] leaf of plates : port. ; 30 cm.
Cover title varies slightly.

No more published?
Dedication to Czarina Aleksandra Fedorovna: 1 fold. sheet inserted.
Includes bibliographical references.
1. Ioasaf, Saint, Bp. of Belgorod, 1705–1754. 2. Orthodox Eastern Church, Russian—Bishops—Biography. 3. Bishops—Russia—Biography. 4. Christian saints—Russia—Biography. I. Title.
BX597.I 58O77 1910 Russian Imperial Coll. 78–421033

(Skazanie o zhizni, podvigakh i chudesakh Svi͡atiteli͡a i Chudotvort͡sa Ioasafa, Episkopa Bi͡elogradskago i Oboi͡anskago)
Сказаніе о жизни, подвигахъ и чудесахъ Святителя и Чудотворца Іоасафа, Епископа Бѣлоградскаго и Обоянскаго / предлагаемое благочестивому вниманію курской паствы Соборомъ курскихъ епископовъ. — Курскъ : Электр. тип. Епархіальнаго вѣдомства, 1911–

v. : ill. ; 31 cm. — (Курскій патерикъ ; вып. 1–)
Fold. sheet with dedication to Czarina Aleksandra Fedorovna inserted.
1. Ioasaf, Saint, Bp. of Belgorod, 1705–1754. 2. Christian saints—Russia—Biography. I. Series: Kurskiĭ paterik, v. 1 [etc.]
BX597.I 58S57 78–421728
Russian Imperial Coll.

(Slava svi͡atiteli͡a i chudotvort͡sa Dimitrii͡a, Mitropolita Rostovskago)
Слава святителя и чудотворца Димитрія, Митрополита Ростовскаго. — Сегіевъ посадъ : Тип. Св.-Тр. Сергіевой лавры, 1913.
320 p. : ill. ; 24 cm.
A collection of reports dealing with miracles attributed to St. Dimitriĭ, based on documents from the Spaso-I͡Akovlevskiĭ monastery, the I͡Aroslavskie eparkhial'nyi͡a vi͡edomosti, and other sources.
1. Dimitriĭ, Saint, Metropolitan of Rostov, 1651–1709. 2. Christian saints—Russia—Biography. 3. Miracles.
BX597.D52S57 78–422943

CHRISTIAN SAINTS (Continued)

—SCOTLAND—BIOGRAPHY

Towill, Edwin Sprott.
The saints of Scotland / Edwin Sprott Towill. — Edinburgh : Saint Andrew Press, 1978.
ix, 148 p. ; 20 cm. GB***
Includes bibliographical references.
ISBN 0-7152-0380-0 : £1.95
1. Christian saints—Scotland—Biography. 2. Scotland—Biography. I. Title.
BX4659.S3T68 282'.092'2 78-320898
79 MARC

—SPAIN—BIOGRAPHY

Giner Guerri, Severino, 1930-
El proceso de beatificación de San José de Calasanz. Madrid, ICCE, 1973.
421 p. 25 cm. (Colección Estudios. Sección historia, 2) 250ptas
Sp73
Bibliography: p. 399-407.
ISBN 8472780104
1. José de Calasanz, Saint, 1556-1648. 2. Christian saints—Spain—Biography. I. Title.
BX4700.J58G56 79-376803
79 MARC

Johannes av Korset (Juan de la Cruz) : några anteckningar om hans liv och verk / av Gustaf Renwik. — Nyhamnsläge : Sv. humanistiska förb. ; Lund, Sölveg. 2 : H. Hansson, 1976.
198 p. : ill. ; 23 cm. — (₍Skrifter₎ - Svenska humanistiska förbundet ; 87)
Swedish and/or Spanish.
Bibliography: p. 196-198.
ISBN 9185158046 : kr50.65.
1. Juan de la Cruz, Saint, 1542-1591. 2. Christian saints—Spain—Biography. I. Title. II. Series: Svenska humanistiska förbundet. Skrifter ; 87.
BX4700.J7R46 78-370808
78 MARC

Renwik, Gustaf, 1894-
Johannes av Korset (Juan de la Cruz) : några anteckningar om hans liv och verk / av Gustaf Renwik. — Nyhamnsläge : Sv. humanistiska förb. ; Lund, Sölveg. 2 : H. Hansson, 1976.
198 p. : ill. ; 23 cm. — (₍Skrifter₎ - Svenska humanistiska förbundet ; 87)
Swedish and/or Spanish.
Bibliography: p. 196-198.
ISBN 9185158046 : kr50.65.
1. Juan de la Cruz, Saint, 1542-1591. 2. Christian saints—Spain—Biography. I. Title. II. Series: Svenska humanistiska förbundet. Skrifter ; 87.
BX4700.J7R46 78-370808
78₍r79₎rev MARC

Rubert Candau, Bernardino María, Father, 1903-
San Pascual ?qué hizo por Villarreal? : algunas de las razones que debieron decidir que San Pascual fuera nombrado patrono de Villarreal / Bernardino Rubert Candau. — Villarreal : Impr. Miralles, 1977.
107 p. ; 21 cm. Sp***
Bibliography: p. ₍3₎-₍4₎
150ptas
1. Baylón, Pascual, Saint, 1540-1592. 2. Christian saints—Spain—Biography. 3. Villarreal, Spain—Religious life and customs. I. Title.
BX4700.B2R78 79-377429
79 MARC

Valls Taberner, Fernando, 1888-1942.
San Ramón de Penyafort / ₍por₎ Fernando Valls y Taberner. — Barcelona : Labor, D.L. 1979.
177 p. : ill. ; 20 cm. — (Nueva colección Labor ; 181) Sp***
Bibliography: p. 173-₍178₎
Includes index.
ISBN 843359205X : 400ptas
1. Raymundus de Penyaforte, Saint, 1175?-1275. 2. Christian saints—Spain—Biography. 3. Dominicans—Spain—Biography.
BX4700.R33V3 1979 79-116367
79 MARC

—SPAIN—CORRESPONDENCE

Loyola, Ignacio de, Saint, 1491-1556.
Trost und Weisung : geistliche Briefe / Ignatius von Loyola ; hrsg. von Hugo Rahner ; neu bearb. von Paul Imhof. — Zürich ₍etc.₎ : Benziger, cop. 1979.
217 p. ; 19 cm. — (Klassiker der Meditation) Sw79-7356
Includes bibliographical references.
ISBN 3-545-20514-2 : 11.80F
1. Loyola, Ignacio de, Saint, 1491-1556. 2. Christian saints—Spain—Correspondence. 3. Spiritual life—Catholic authors. 4. Spiritual direction. 5. Jesuits—History—Sources. I. Rahner, Hugo, 1900-1968. II. Imhof, Paul. III. Title.
BX4700.L7A4 1979 79-380495
79 MARC

—SPAIN—AVILA—BIOGRAPHY

Clissold, Stephen.
St. Teresa of Avila / Stephen Clissold. — London : Sheldon Press, 1979.
xv, 272 p., ₍2₎ leaves of plates : map, ports. ; 22 cm. GB***
Includes bibliographical references and index.
ISBN 0-85969-148-9 : £8.95
1. Teresa, Saint, 1515-1582. 2. Christian Saints—Spain—Avila—Biography. 3. Avila, Spain—Biography.
BX4700.T4C55 1979 271'.971'024 79-306067
79 MARC

Efrén de la Madre de Dios, Father.
Tiempo y vida de Santa Teresa / por Efrén de la Madre de Dios, O.C.D., Otger Steggink, O. Carm. — 2. ed. rev. y aum. — Madrid : La Editorial Católica, 1977.
XXIV, 1023 p. ; 20 cm. — (Biblioteca de autores cristianos ; 283)
Sp77-Nov
Bibliography: p. ₍xiii₎-xxiv
ISBN 8422002329 : 990ptas
1. Teresa, Saint, 1515-1582. 2. Christian Saints—Spain—Avila—Biography. 3. Avila, Spain—Biography. I. Steggink, Otger, joint author. II. Title.
BX4700.T4E34 1977 79-351203
79 MARC

Les Fioretti de sainte Thérèse d'Avila ... / ₍textes réunis et présentés par₎ Joseph Gicquel, O.C.D. — Paris : Éditions du Cerf, 1977.
173 p., ₍8₎ p. of plates ; 20 cm. F78-3089
"Recueil de témoignages extraits pour la plupart des procès de béatification et de canonisation de sainte Thérèse d'Avila."—Bibl. de la Fr.
Bibliography: p. 167-₍171₎
ISBN 2-204-01153-3 : 28.00F
1. Teresa, Saint, 1515-1582. 2. Christian saints—Spain—Avila—Biography. 3. Avila, Spain—Biography. I. Gicquel, Joseph.
BX4700.T4F56 271'.971'024 78-398953
79 MARC

—SPAIN—CATALONIA—BIOGRAPHY

Fort y Cogul, Eufemia.
Relacions de Sant Bernat Calbó, com a abat de Santes Creus, amb Vilafranca del Penedès i la seva comarca / per Eufemià Fort i Cogul. — Vilafranca del Penedès : Museu de Vilafranca, 1975.
41 p. ; 22 cm. Sp75-Nov
"Regesta de documents": p. 23-41.
1. Bernardus Calvonius, Saint, Bp. of Vich, 1180-1243. 2. Christian saints—Spain—Catalonia—Biography. 3. Catalonia—Biography. I. Title: Relacions de Sant Bernat Calbó, com a abat de Santes Creus ...
BX4700.B554F67 79-367390
*79 MARC

—TURKEY—LYSIA—BIOGRAPHY

Jones, Charles Williams, 1905-
Saint Nicholas of Myra, Bari, and Manhattan : biography of a legend / Charles W. Jones. — Chicago : University of Chicago Press, 1978.
ix, 558 p. ; 24 cm.
Bibliography: p. 463-492.
Includes indexes.
ISBN 0-226-40699-7
1. Nicholas, Saint, Bp. of Myra. 2. Christian saints—Turkey—Lysia—Biography. 3. Lysia—Biography. I. Title.
BX4700.N55J63 282'.092'4 77-15487
77 MARC

—TURKEY—TARSUS—BIOGRAPHY

Grollenberg, Lucas Hendricus, 1916-
Paul / Lucas Grollenberg. — Philadelphia : Westminster Press, c1978.
179 p. ; 20 cm.
Translation of Die moeilijk Paulus.
ISBN 0-664-24234-0
1. Paul, Saint, Apostle. 2. Bible. N.T.—Biography. 3. Christian saints—Turkey—Tarsus—Biography. 4. Tarsus, Turkey—Biography. 5. Bible. N.T. Epistles of Paul—Theology.
BS2506.G7713 225.9'2'4 78-14372
78 MARC

—TURKEY—TARSUS—BIOGRAPHY—JUVENILE LITERATURE

Belski Lagazzi, Ines.
San Paolo / Ines Belski Lagazzi. — Roma : Edizioni paoline, 1978.
139 p. : ill. ; 27 cm. It78-Nov
L5000
1. Paul, Saint, apostle—Juvenile literature. 2. Christian saints—Turkey—Tarsus—Biography—Juvenile literature. 3. Tarsus, Turkey—Biography—Juvenile literature.
BS2506.B44 79-338135
*79 MARC

—TURKEY—TARSUS—JUVENILE LITERATURE

Schraff, Francis.
The adventures of Peter and Paul : Acts of the Apostles for the young / Francis and Anne Schraff ; with 17 playlets by Suzanne Hockel ; ₍ill., Linda Harris₎. — Liguori, Mo. : Liguori Publications, c1978.
79 p. : col. ill. ; 28 cm.
SUMMARY: Brief stories and plays recount the deeds of Christ's apostles.
ISBN 0-89243-094-X : $2.95
1. Peter, Saint, apostle—Juvenile literature. 2. Apostles—Biography—Juvenile literature. 3. Paul, Saint, Apostle—Juvenile literature. 4. Christian saints—Turkey—Tarsus—Juvenile literature. 5. Tarsus, Turkey—Biography—Juvenile literature. ₍1. Apostles. 2. Bible stories—N.T. 3. Bible plays. 4. Plays₎ I. Schraff, Anne E., joint author. II. Hockel, Suzanne. III. Harris, Linda. IV. Title.
BS2515.S3 225.9'22 78-64755
79 MARC

—UNITED STATES—BIOGRAPHY—JUVENILE LITERATURE

Douglas, Philip, 1929-
Saint of Philadelphia : the life of Bishop John Neumann, 1811-1860 / Philip Douglas. — Cambridge ₍Mass.₎ : Ravengate Press, c1977.
ix, 180 p. : ill. ; 22 cm.
ISBN 0-911218-07-6 : $7.95. ISBN 0-911218-08-4 pbk.
1. Neumann, John Nepomucene, Saint, 1811-1860—Juvenile literature. 2. Christian saints—United States—Biography—Juvenile literature. I. Title.
BX4700.N4D68 271'.64'024 78-107169
79 MARC

—YUGOSLAVIA—SERBIA—BIOGRAPHY

Baranin, Dušan, 1902–1978.
(Stevan Nemanja).
Стеван Немања / Душан Баранин. — Београд : Слово љубве, 1978.
227 p. ; 21 cm. — (Библиотека Историјски роман) Yu***
Series romanized: Biblioteka Istorijski roman.
1. Simeon, Saint, monk, 1114–1200. 2. Serbia—Kings and rulers—Biography. 3. Christian saints — Yugoslavia — Serbia — Biography. I. Title.
DR338.5.S55B37 79–410060

CHRISTIAN SAINTS, PATRON see Christian patron saints

CHRISTIAN SAINTS IN ART

see also Icons; Santos (Art)

Bonnet, Claude.
Les Faïences patronymiques / par Claude Bonnet. — ₍Paris₎ (8, rue de Berri, 75008) : Bibliothèque de l'antiquaire, 1978.
1 v. (unpaged) : ill. (some col.) ; 23 cm. F79-2127
180F
1. Faience. 2. Commemorative plates. 3. Christian saints in art. I. Title.
NK4305.B66 738.3'7 79-370379
79 MARC

Boullet, François.
Ex-voto marins / F. et C. Boullet. — ₍Paris₎ : Éditions maritimes et d'outre-mer, c1978.
189 p. : chiefly ill. (some col.) ; 29 cm. F***
Bibliography: p. 184-188.
195F
1. Votive offerings in art—Europe. 2. Votive offerings—Europe. 3. Primitivism in art—Europe. 4. Ships in art. 5. Christian saints in art. I. Boullet, Colette, joint author. II. Title.
N7942.B68 704.948'2'094 79-378326
79 MARC

Calzada Oliveras, José.
Las claves de bóveda de la Catedral de Gerona / texto, José Calzada Oliveras ; fotografías, compaginación, reproducción, impresión y encuadernación, realizados por los equipos técnicos de F.I.S.A. — 1. ed. — Barcelona : Escudo de Oro, 1975.
61, ₍3₎ p. : ill. (some col.) ; 25 cm. — (Colección Arte en España)
Sp75-Oct
Bibliography: p. ₍63₎
ISBN 8437802504 : 400ptas
1. Bosses (Architecture)—Spain—Gerona. 2. Decoration and ornament, Architectural—Spain—Gerona. 3. Christian saints in art. 4. Gerona, Spain. Catedral. I. Title.
NA3562.G47C34 79-368372
*79 MARC

CHRISTIAN SAINTS IN ART (Continued)

Chionna, Antonio.
Gli insediamenti rupestri del brindisino : un patrimonio da salvare / A. Chionna, G. Lodolo. — Brindisi : ⌈Amministrazione provinciale⌉, 1972.
39 p., ⌈6⌉ leaves of plates : ports. ; 24 cm. — (Quaderni dell'Amministrazione provinciale di Brindisi ; n. 6) It***
Issued on the occasion of an exhibition.
Includes bibliographical references.
1. Mural painting and decoration, Italian—Italy—Brindisi (Province) 2. Mural painting and decoration, Medieval—Italy—Brindisi (Province) 3. Christian saints in art. 4. Crypts—Italy—Brindisi (Province) I. Lodolo, Gabriella, joint author. II. Title. III. Series: Brindisi, Italy (Province). Quaderni ; n. 6.
ND2756.B74C45 79-378224
79 MARC

Museum Narodowe we Wrocławiu.
Śląska sztuka średniowieczna, ⌈ze zbiorów Muzeum Śląskiego. Opracowała Anna Fedorowicz. Wrocław, 1964-
v. illus. 20 cm.
CONTENTS: 1. Malarstwo xiv i pocz. xv w.—2. Madonny na lwach.—3. Figury apostołow z wrocławskiego kościoła Marii Magdaleny.—4. Opłakiwanie z kościoła Panny Marii na Piasku.—5. Rzeźba cechowa pierwszej poł. xv w.—6. Krag mistrza ołtarza św. Barbary.—9. Śląskie pracownie ná przełomie xv-xvi w.—Mistrz lat 1486-87. Mistrz ołtarza z Gościszowic.
—11.—Pejzaż niderlandzki w epitafium z Głogowa.
1. Art, Medieval—Silesia. 2. Art—Silesia. 3. Christian saints in art. 4. Muzeum Narodowe we Wrocławiu. I. Fedorowicz, Anna, ed. II. Title.
N7971.P6M89 1964 68-35672
⌈r79⌉rev2 MARC

Pénicaut, Michel, 1904-
L'Église de Magnac-Bourg; ses vitraux. ⌈Paris, Nouvelles Éditions latines, 1974?⌉
30 p. illus. 19 cm. (Art et tourisme) 6.00F F***
1. Glass painting and staining, Gothic—France—Magnac-Bourg. 2. Glass painting and staining—France—Magnac-Bourg. 3. Christian saints in art. 4. Église de Magnac-Bourg. I. Title.
NK5349.M25P46 914.4'66 75-503400
75⌈r79⌉rev MARC

Reed, Olwen.
An illustrated history of saints and symbols / Olwen Reed. — Bourne End ⌈Eng.⌉ : Spurbooks, c1978.
159 p. : ill. ; 23 cm. GB***
Half title: Saints and symbols.
Includes index.
ISBN (invalid) 9904978311 : £5.00
1. Christian saints—Biography. 2. Christian saints in art. I. Title. II. Title: Saints and symbols.
BX4655.2.R43 270 79-306287
79 MARC

—CATALOGS

Lisbon. Biblioteca Nacional.
Inventário da colecção de registos de santos / organizado e prefaciado por Ernesto Soares. — Lisboa : Biblioteca Nacional, 1955.
xxxvi, 491 p., ⌈12⌉ leaves of plates : ill. ; 29 cm.
1. Prints—Portugal—Lisbon—Catalogs. 2. Christian saints in art—Catalogs. 3. Lisbon. Biblioteca Nacional—Catalogs. I. Soares, Ernesto. II. Title.
NE958.L52 1955 79-376721
79 MARC

CHRISTIAN SAINTS IN LITERATURE

Johnson, Phyllis, 1937-
Le vain siecle guerpir : a literary approach to sainthood through Old French hagiography of the twelfth century / by Phyllis Johnson and Brigitte Cazelles. — Chapel Hill : U.N.C. Dept. of Romance Languages : ⌈distributed by University of North Carolina Press⌉, 1979.
321 p. ; 24 cm. — (North Carolina studies in the Romance languages and literatures ; no. 205)
Bibliography: p. ⌈311⌉-321.
ISBN 0-8078-9205-X
1. French poetry—To 1500—History and criticism. 2. Christian hagiography. 3. Christian saints in literature. I. Cazelles, Brigitte, joint author. II. Title. III. Series.
PQ178.J64 841'.1'09352 79-10678
79 MARC

CHRISTIAN SCIENCE

see also Bible—Medicine, hygiene, etc.; Faith-cure; Mental healing

Kappeler, Max, 1910-
Introduction to the science of Christian Science / ⌈by⌉ Max Kappeler ; ⌈translated from the German by Kathleen Lee⌉. — London : Foundational Book Co., 1978.
171 p. ; 22 cm. GB79-10785
Translation of Einführung in die Wissenschaft der Christlichen Wissenschaft.
Includes bibliographical references.
£6.00
1. Christian Science. I. Title.
BX6943.K2713 230'.9'5 79-313991
00 MARC

—GREAT BRITAIN

Wilson, Bryan R
Sects and society : a sociological study of three religious groups in Britain / Bryan R. Wilson. — Westport, Conn. : Greenwood Press, 1978, c1961.
397 p. ; 23 cm.
Reprint of the ed. published by W. Heinemann, London, in series: Heinemann books on sociology.
Bibliography: p. 374-388.
Includes index.
ISBN 0-313-20439-X
1. Elim Foursquare Gospel Alliance. 2. Christian Science—Great Britain. 3. Christadelphians—Great Britain. 4. Church and social problems—Great Britain. I. Title.
⌈BX7435.E54W5 1978⌉ 289 78-5993
78 MARC

CHRISTIAN SECTS

see also Alumbrados; Camisards; Christian union; Church of the Brethren; Dancers; Friends, Society of; Huguenots; Jansenists; Mormons and Mormonism; Orthodox Eastern Church; Pentecostal churches; Protestant churches

World Council of Churches.
For all God's people : ecumenical prayer cycle. — Geneva : World Council of Churches, c1978.
234, ⌈2⌉ p. : maps ; 24 cm. Sw***
Bibliography: p. ⌈235⌉-⌈236⌉
Includes indexes.
ISBN (invalid) 0-8254-0552-3
1. Devotional calendars. 2. Ecumenical movement. 3. Christian sects. I. Title.
BV4810.W67 1978 264'.1 79-311587
79 MARC

—CONTROVERSIAL LITERATURE

MacCollam, Joel A 1946-
Carnival of souls : religious cults and young people / by Joel A. MacCollam. — New York : Seabury Press, 1979.
188 p. ; 21 cm.
"A Crossroad book."
Bibliography: p. 185-188.
ISBN 0-8164-2211-7
1. Church work with young adults—Case studies. 2. Church work with youth—Case studies. 3. Cults—Controversial literature. 4. Christian sects—Controversial literature. 5. Pastoral theology—Case studies. I. Title.
BV4446.M26 259 78-25976
78 MARC

Vernette, Jean.
Sectes et réveil-religieux : quand l'Occident s'éveille / Jean Vernette. — Mulhouse : Éditions Salvator, 1976.
199 p. ; 19 cm. F77-8662
Includes bibliographical references.
ISBN 2-7067-0021-1 : 35.00F
1. Christianity and other religions. 2. Cults—Controversial literature. 3. Christian sects—Controversial literature. I. Title.
BR127.V4 261.2 78-378234
78 MARC

—CARIBBEAN AREA—DIRECTORIES

Handbook of churches in the Caribbean : specially prepared for the inaugural assembly of the Caribbean Conference of Churches, November 13th-16th, 1973 / edited by Joan A. Brathwaite. — 1st ed. — ⌈Bridgetown, Barbados : Christian Action for Development in the Caribbean⌉, 1973.
234 p. : ill. ; 28 cm.
1. Caribbean Conference of Churches—Directories. 2. Christian Sects—Caribbean area—Directories. I. Brathwaite, Joan A. II. Caribbean Conference of Churches. III. Christian Action for Development in the Caribbean.
BR640.H36 277.29 79-313147
79 MARC

—ENGLAND

see also Peculiar People (Sect)

—FINLAND

see also Laestadians

—GERMANY

see also Evangelische Kirche in Deutschland

—INDIA—KERALA

Pōḷaykkal, Mātyu.
(Knānāya samudāyam)
ക്നാനായ സമുദായം : ഒരു ലഘുചരിത്രം / മാത്യു പോളയ്ക്കൽ. — ⌈Alwaye?⌉ : Pro-manuscripto, 1976.
iv, 79 p. ; 19 cm.
In Malayalam.
Bibliography: p. ⌈75⌉-79.
Rs1.50
1. Christian sects—India—Kerala. 2. Kerala—Church history. I. Title.
BR1170.K4P65 78-902725

Thomas, M A
An outline history of Christian churches and denominations in Kerala / M. A. Thomas. — Trivandrum : Thomas, 1977.
177, ⌈4⌉ p. ; 19 cm.
Bibliography: p. ⌈18⌉
Rs7.00
1. Christian sects—India—Kerala. 2. Kerala—Church history. I. Title.
BR1170.K4T48 275.4'83 78-902343
79 MARC

—RUSSIA

Belov, Anatoliĭ Vasil'evich.
Sekty, sektantstvo, sektanty / A. V. Belov. — Moskva : Nauka, 1978.
151 p. ; 20 cm. — (Nauchno-ateisticheskaia seriia) USSR79
At head of title: Akademiia nauk SSSR.
Includes bibliographical references.
0.50rub
1. Christian sects—Russia. I. Title.
BR935.5.B44 79-372232
79 MARC

Filimonov, Éduard Gennad'evich.
(Novye tendentsii v ideologii i deiatel'nosti sovremennogo sektantstva)
Новые тенденции в идеологии и деятельности современного сектантства / Э. Г. Филимонов, канд. филос. наук. — Москва : Знание, 1977.
63 p. ; 20 cm. — (Новое в жизни, науке, технике : Серия Научный атеизм ; № 11, 1977) USSR 77
Includes bibliographical references.
0.11rub
1. Christian sects—Russia. I. Title. II. Series: Novoe v zhizni, nauke, tekhnike : Seriia Nauchnyĭ ateizm ; 1977, no. 11.
BR936.F48 79-413241

—THAILAND—DIRECTORIES

Christian directory.
Bangkok, Suthep Chaviwan.
v. ill. 27 cm. annual.
Other title, : Christian Thailand directory.
1. Christian sects—Thailand—Directories. I. Title: Christian Thailand directory.
BR1195.C45 275.93'025 78-919022
MARC-S

—UNITED STATES

see also Assemblies of God, General Council; Christian Science; Church of God; Church of Jesus Christ of Latter-Day Saints; Hutterite Brethren; Methodist Church (United States); Reorganized Church of Jesus Christ of Latter-Day Saints

Robertson, Irvine.
What the cults believe / by Irvine Robertson. — 2d ed., rev. — Chicago : Moody Press, c1979.
154 p. ; 22 cm.
Bibliography: p. 149-154.
ISBN 0-8024-9408-0 : $6.95
1. Cults—United States. 2. Christian sects—United States. I. Title.
BL2530.U6R62 1979 291.2 79-18219
79 MARC

CHRISTIAN SECTS
—UNITED STATES (Continued)

Whalen, William Joseph.
Separated brethren : a survey of Protestant, Anglican, Eastern Orthodox, and other denominations in the United States / William J. Whalen. — 3d rev. ed. — Huntington, Ind. : Our Sunday Visitor, c1979.
252 p. ; 23 cm.
Includes index.
Bibliography: p. 247-248.
ISBN 0-87973-829-4 : $5.95
1. Christian sects—United States. 2. Cults—United States. I. Title.
BR510.5.W44 1979 280'.0973 79-83874
00 MARC

—UNITED STATES
—PICTORIAL WORKS

Whitman, Alan.
Christian occasions / photos. and text by Alan Whitman. — 1st ed. — Garden City, N.Y : Dolphin Books, 1978.
144 p. : ill. ; 26 cm.
On cover: A photographic study of unusual styles of religion in American life.
"A Country Music Magazine Press book."
ISBN 0-385-12597-6 : $6.95
1. Christian sects—United States—Pictorial works. 2. Christianity—United States—Pictorial works. I. Title.
BR517.W46 277.3 77-76286
78 MARC

CHRISTIAN SECTS, MEDIEVAL

see also Albigenses; Bogomiles

CHRISTIAN SECTS, PROTESTANT see Protestant churches

CHRISTIAN SERVICE SOCIETY

Christian Service Society, Dacope / M. Abdul Hamid ... [et al.]. — Rajshahi : Dept. of Economics, Rajshahi University, 1976.
xi, 66 p. ; 29 cm. — (Rural development studies series ; 5)
Christian Service Society.
Tk10.00
1. Rural development—Bangladesh—Dacope. 2. Christian Service Society. I. Hamid, M. Abdul. II. Series.
HN690.6.Z9C633 309.2'63'0954922 76-904115
79 MARC

CHRISTIAN SHRINES

—ENGLAND

Finucane, Ronald C
Miracles and pilgrims : popular beliefs in medieval England / Ronald C. Finucane. — Totowa, N.J. : Rowman and Littlefield, 1977.
248 p., [8] leaves of plates : ill. ; 24 cm.
Bibliography: p. [242]-244.
Includes index.
ISBN 0-87471-831-7
1. England—Church history—Medieval period, 1066-1485. 2. Miracles. 3. Christian pilgrims and pilgrimages—England. 4. Christian saints—Cult. 5. Christian shrines—England. I. Title.
BR747.F56 1977 274.2 77-21771
77[r79]rev MARC

—FRANCE

Belot, Victor R
La France des pèlerinages / Victor H. [sic] Belot. — Verviers, r. de Limbourg 65 : Marabout, [1977]
255 p. : ill. ; 18 cm. — (Guide Marabout ; 16) Be77-7
Includes indexes.
130F
1. Christian shrines—France. 2. Christian pilgrims and pilgrimages—France. I. Title.
BX2320.B4 914.4'04'83 78-382523
*79 MARC

—GERMANY, WEST—ALTÖTTING

Brugger, Walter.
Altötting und Neuötting / Walter Brugger [(Text)] ; Adolf Kunzmann [(25 Bilder)]. — Freilassing : Pannonia, 1978.
48 p. : 25 ill. ; 17 cm. — (Kleine Pannonia-Reihe ; 72) GFR78-A31
Cover title.
ISBN 3-7897-0072-X : DM3.80
1. Christian shrines—Germany, West—Altötting. 2. Christian shrines—Germany, West—Neuötting. I. Kunzmann, Adolf. II. Title.
BX2321.A515B78 79-376335
79 MARC

—GERMANY, WEST—NEUÖTTING

Brugger, Walter.
Altötting und Neuötting / Walter Brugger [(Text)] ; Adolf Kunzmann [(25 Bilder)]. — Freilassing : Pannonia, 1978.
48 p. : 25 ill. ; 17 cm. — (Kleine Pannonia-Reihe ; 72) GFR78-A31
Cover title.
ISBN 3-7897-0072-X : DM3.80
1. Christian shrines—Germany, West—Altötting. 2. Christian shrines—Germany, West—Neuötting. I. Kunzmann, Adolf. II. Title.
BX2321.A515B78 79-376335
79 MARC

—GREAT BRITAIN

Adair, John Eric, 1934-
The pilgrims' way : shrines and saints in Britain and Ireland / John Adair ; photos. by Peter Chèze-Brown. — London : Thames and Hudson, 1978.
208 p. : ill. (some col.) ; 27 cm. GB***
Bibliography: p. 205.
Includes index.
ISBN 0-500-25061-8 : £5.95
1. Christian shrines—Great Britain. 2. Christian shrines—Ireland. 3. Christian pilgrims and pilgrimages—Great Britain. 4. Christian pilgrims and pilgrimages—Ireland. I. Title.
BX2320.A3 914.1'04'857 78-320821
79 MARC

—IRELAND

Adair, John Eric, 1934-
The pilgrims' way : shrines and saints in Britain and Ireland / John Adair ; photos. by Peter Chèze-Brown. — London : Thames and Hudson, 1978.
208 p. : ill. (some col.) ; 27 cm. GB***
Bibliography: p. 205.
Includes index.
ISBN 0-500-25061-8 : £5.95
1. Christian shrines—Great Britain. 2. Christian shrines—Ireland. 3. Christian pilgrims and pilgrimages—Great Britain. 4. Christian pilgrims and pilgrimages—Ireland. I. Title.
BX2320.A3 914.1'04'857 78-320821
79 MARC

—PALESTINE—MEDITATIONS

Hyvönen, Väinö E
Astelin pyhillä kivillä / Väinö E. Hyvönen. — [Saarijärvi] : Ari-kustannus, 1977.
224 p. ; 17 cm. Fi77
ISBN 9517860676
1. Palestine—Description and travel. 2. Christian shrines—Palestine—Meditations. I. Title.
DS107.4.H94 78-375873
78 MARC

—POLAND—MASOVIA

Bojanowska-Ziemska, Barbara.
Kapliczki i krzyże przydrożne na Mazowszu z XIX wieku : (na przykładzie kilkunastu osiedli) / Barbara Bojanowska-Ziemska. — Warszawa : Wydaw-a PW, 1978 (Warsz. : PW)
45 p. : ill. ; 24 cm. — (Prace naukowe, budownictwo ; z. 58 ISSN 0137-2297) P78-11104
Cover title.
At head of title: Politechnika Warszawska.
On p. [2] of cover: English title, Nineteenth century wayside shrine and crosses in Mazowsze; Russian title, Pridorozhnye chasovenki i kresty Mazovsha XIX veka.
Summary in English and Russian.
Bibliography: p. 42-43.
zł3.00
1. Christian shrines—Poland—Masovia. 2. Crosses—Poland—Masovia. I. Title. II. Title: Nineteenth century wayside shrine and crosses in Mazowsze. III. Series: Warsaw. Politechnika. Prace naukowe, budownictwo ; z. 58.
TH4.W3715 z. 58 79-363184
[BX2320] 79 MARC

CHRISTIAN SOCIALISM see Socialism, Christian

CHRISTIAN SOCIOLOGY see Sociology, Christian

CHRISTIAN STEWARDSHIP see Stewardship, Christian

CHRISTIAN SYMBOLISM see Christian art and symbolism

CHRISTIAN SYMBOLISM IN ART

Kirigin, Martin.
La mano divina nell'iconografia cristiana / Martin Kirigin. — Città del Vaticano : Pontificio Istituto di archeologia cristiana, 1976.
246 p., [5] leaves of plates : ill. ; 27 cm. — (Studi di antichità cristiana ; 31) VC***
Includes bibliographical references and indexes.
1. Hand in art. 2. Bible—Illustrations. 3. Christian symbolism in art. I. Title. II. Series.
BR130.P6 vol. 31 930 s 78-392151
[N8012.H35] 79 MARC

CHRISTIAN THEOLOGIANS see Theologians

CHRISTIAN THEOLOGY see Theology

CHRISTIAN UNION

see also Disputations, Religious; Ecumenical movement; Pentecostalism and Christian union

Boyer, Charles, 1884-
Le mouvement oecuménique : les faits, le dialogue / Charles Boyer. — Rome : Université grégorienne, 1976.
262 p. ; 22 cm. It78-Jan
Bibliography: p. 251-253.
Includes index.
1. Ecumenical movement—History. 2. Christian union. I. Title.
BX6.5.B65 78-394725
*79 MARC

Gill, Joseph, 1901-
Byzantium and the papacy, 1198-1400 / Joseph Gill. — New Brunswick, N.J. : Rutgers University Press, c1979.
xii, 342 p. ; 24 cm.
Bibliography: p. 311-332.
Includes index.
ISBN 0-8135-0864-9 : $2.50
1. Orthodox Eastern Church—Relations—Catholic Church. 2. Catholic Church—Relations—Orthodox Eastern Church. 3. Byzantine Empire—History. 4. Christian union. I. Title.
DF548.G5 949.5 78-23971
78 MARC

Leenhardt, Franz J 1902-
L'Église : questions aux protestants et aux catholiques / Franz-J. Leenhardt. — Genève : Labor et fides ; Paris : diffusion pour la France Librairie protestante, cop. 1978.
239 p. ; 21 cm. — (Lieux théologiques ; no 1) Sw79-1874
Includes bibliographical references.
38.00F
1. Church. 2. Clergy—Office. 3. Christian union. I. Title.
BV600.2.L42 262'.001 79-368994
79 MARC

Magdalēnē, Monachē.
(Pseudosynodos tōn oikoumenistōn)
Ψευδοσύνοδος τῶν οἰκουμενιστῶν / Μαγδαληνῆς μοναχῆς. — [Κοζάνη] : Ἐκδόσεις Ἱ. Γυν. Ἡσυχαστηρίου Ἀναλήψεως Κοζάνης, 1976.
223 p. : ill. ; 22 cm.
1. Orthodox Eastern Church—Doctrinal and controversial works. 2. Christian union. 3. Councils and synods, Ecumenical. I. Title.
BX324.M33 79-413441

CHRISTIAN UNION (Continued)

[Thompson, Hugh Miller] Bp., 1830-1902.
Unity and its restoration. Addressed to all Christians who desire to hold the faith in unity of the spirit, and in the bond of peace. By a presbyter of the diocese of Illinois. New York, D. Dana, jr., 1860.
109 p. 20 cm.
Edition of 1873: by Hugh Miller Thompson.
1. Christian union. 2. Protestant Episcopal Church in the U.S.A.—Doctrinal and controversial works. I. A presbyter of the diocese of Illinois. II. Title.
BX8.T48 280 34-8379
[r79]rev MARC

Twitchell, Thomas Donn.
"That they may be one" : convincing a puzzled world that God sent His Son / by Thomas Donn Twitchell. — Plainfield, N.J. : Logos International, c1979.
xii, 216 p. ; 21 cm.
ISBN 0-88270-360-9 : $3.95
1. Twitchell, Thomas Donn. 2. Presbyterian Church—Clergy—Biography. 3. Clergy—United States—Biography. 4. Christian union. I. Title.
BX9225.T84A36 285'.1'0924 79-83792
79 MARC

Van der Bent, Ans Joachim.
God so loves the world : the immaturity of world Christianity / A. J. van der Bent ; foreword by David E. Jenkins. — Madras : Christian Literature Society, 1977.
xvi, 143 p. ; 23 cm.
Includes bibliographical references.
Rs10.00
1. Christianity—20th century. 2. Christian union. 3. Christianity and other religions. I. Title.
BR121.2.V324 209'.04 78-900901
78 MARC

World Council of Churches. Commission on Faith and Order.
One baptism, one eucharist, and a mutually recognized ministry : three agreed statements. — Geneva : World Council of Churches, [c1975], 1978 printing.
60 p. : ill. ; 22 cm. — (Faith and order paper ; no. 73) Sw***
Includes bibliographical references.
ISBN 2-8254-0491-8
1. Baptism. 2. Lord's Supper. 3. Pastoral theology. 4. Christian union. I. Title. II. Series.
BV812.W67 1978 265 78-325749
79 MARC

—ADDRESSES, ESSAYS, LECTURES

Die Konziliare Gemeinschaft : eine europäische theologische Diskussion über den Weg und das Endziel der ökumenischen Bewegung / hrsg. im Auftrag der KEK von Gyula Nagy ; mit einer Einleitung von Hanfried Krüger ; Konferenz Europäischer Kirchen. — Frankfurt am Main : In Kommission bei O. Lembeck, 1978.
128 p. ; 24 cm. — (Konferenz Europäischer Kirchen. Studienheft ; 10) GFR***
Bibliography: p. 125.
1. Christian union—Addresses, essays, lectures. I. Nagy, Gyula. II. Conference of European Churches. III. Series: Conference of European Churches. Studienhefte ; 10.
BX9.K77 78-380617
79 MARC

Lausanne 77, 50 years of faith and order, [1927-1977] / [contributors: W. A. Visser't Hooft et al.]. — Geneva : World Council of Churches, [1977]
iv, 82 p. ; 22 cm. — (Faith and order paper ; 82) Sw77-12108
Includes bibliographical references.
ISBN 2-8254-0542-6 : 6.90F
1. Christian union—Addresses, essays, lectures. 2. World Council of Churches. Commission on Faith and Order—Addresses, essays, lectures. I. Visser't Hooft, Willem Adolph, 1900- . II. World Council of Churches.
BX9.L38 270.8'2 79-302574
*79 MARC

Verbindliches Lehren der Kirche heute : Arbeitsbericht aus d. Dt. Ökumen. Studienausschuss u. Texte d. Faith and Order-Konsultation Odessa 1977 / hrsg. vom Dt. Ökumen. Studienausschuss. — Frankfurt am Main : Lembeck, 1978.
227 p. ; 22 cm. — (Beiheft zur Ökumenischen Rundschau ; Nr. 33)
Includes bibliographical references.
ISBN 3-87476-117-7 : DM28.00
1. Theology—20th century—Addresses, essays, lectures. 2. Christian union—Addresses, essays, lectures. I. Deutscher Ökumenischer Studienausschuss. II. World Council of Chruches. Commission on Faith and Order.
BT28.V47 79-339060
79 MARC

Welche Ökumene meinen wir? : eine Bilanz d. Ökumene seit Nairobi / mit Beitr. von Richard Boeckler ... [et al.] u.e. Vorw. von Hanfried Krüger ; hrsg. von Richard Boeckler. — Frankfurt am Main : Lembeck, 1978.
120 p. ; 22 cm. — (Beiheft zur ökumenische Rundschau ; 32) GFR78-A
Includes bibliographical references.
ISBN 3-87476-110-X : DM13.80
1. Christian union—Addresses, essays, lectures. I. Boeckler, Richard.
BX9.W42 78-380165
78 MARC

World Council of Churches. Commission on Faith and Order.
Towards an ecumenical consensus on baptism, the eucharist and the ministry : a response to the Churches / Faith and Order Standing Commission. — Geneva : World Council of Churches, 1977.
30 p. ; 21 cm. — (Faith and order paper ; no. 84) Sw78-876
Includes response of the Commission to replies made by the churches to the statement "One baptism, one Eucharist and a mutually recognized ministry"; an address by the director of the Faith and Order Secretariat upon presentation of the report to the Central Committee of the World Council of Churches; and the text of the covering letter addressed with the document to member churches of the World Council of Churches.
ISBN 2-8254-0560-4 : 2.90F
1. One baptism, one Eucharist and a mutually recognized ministry—Addresses, essays, lectures. 2. Christian union—Addresses, essays, lectures. 3. Baptism and Christian union—Addresses, essays, lectures. 4. Lord's Supper and Christian union—Addresses, essays, lectures. 5. Pastoral theology and Christian union—Addresses, essays, lectures. I. Title. II. Series.
BX9.W64 1977 265 78-326837
79 MARC

—CATHOLIC CHURCH

Catholic Church. Bishops' Conference of England and Wales.
Official response to the ten propositions of the Churches' Unity Commission / the Catholic Bishops' Conference of England and Wales. — Abbots Langley : Catholic Information Services (Bishops' Conference of England and Wales), 1977.
[1], iii, 9, [1] p. ; 21 cm. GB78-07314
£0.20
1. Christian union—England. 2. Churches' Unity Commission. Visible unity. 3. Christian union—Catholic Church. 4. Catholic Church in England. I. Title.
BR759.C37 1977 262'.001 78-317292
78 MARC

Catholic Church. Ecumenical Commission for England and Wales.
The Roman Catholic Church in England and the Ten Propositions of the Churches' Unity Commission : a paper / by the Ecumenical Commission of England and Wales. — Abbots Langley : Catholic Information Office, 1976.
[1], 8 p. ; 21 cm. GB77-223
Cover title.
A provisional response to the report of the Churches' Unity Commission entitled: Visible unity: ten propositions.
ISBN 0-905241-04-5 : £0.15
1. Christian union—England. 2. Churches' Unity Commission. Visible unity. 3. Christian union—Catholic Church. 4. Catholic Church in England. I. Title: The Roman Catholic Church in England and the Ten Propositions ...
BR759.C38 1976 262'.001 77-359592
77[r78]rev MARC

Kantzenbach, Friedrich Wilhelm.
Aktion und Reaktion : Katholizismus d. Gegenwart, evang. gesehen / Friedrich Wilhelm Kantzenbach. — Stuttgart : Steinkopf, 1978.
199 p. ; 21 cm. GFR78-A
Includes bibliographical references.
ISBN 3-7984-0353-8 : DM19.80
1. Catholic Church—Doctrinal and controversial works—Protestant authors. 2. Christian union—Catholic Church. I. Title.
BX1751.2.K3 79-341693
79 MARC

Suenens, Léon Joseph, Cardinal, 1904-
Œcuménisme et renouveau charismatique : orientations théologiques et pastorales / Léon-Joseph Suenens. — Paris : Le Centurion, c1978.
159 p. ; 22 cm. — (Document de Malines ; 2) F***
"Suite à l'étude intitulée 'Orientations théologiques et pastorales sur le renouveau charismatique catholique' (1974), connue sous le nom de Document de Malines."
Includes bibliographical references.
ISBN 2-227-34512-8 : 24.00F
1. Christian union—Catholic Church. 2. Pentecostalism—Catholic Church. I. Title. II. Series.
BX1784.S83 262'.001 79-367899
79 MARC

—CATHOLIC CHURCH—HISTORY

Tavard, Georges Henri, 1922-
Two centuries of ecumenism / by George H. Tavard ; translated by Royce W. Hughes. — Westport, Conn. : Greenwood Press, 1978, c1960.
239 p. ; 22 cm.
Translation of Petite histoire du mouvement oecuménique.
Reprint of the ed. published by Fides Publishers Association, Notre Dame, Ind.
Includes bibliographical references.
ISBN 0-313-20490-X
1. Christian union—Catholic Church—History. 2. Ecumenical movement—History. I. Title.
[BX1784.T3813 1978] 282'.09 78-6449
78 MARC

—CONGRESSES

In each place : towards a fellowship of local churches truly united : [report of a consultation in Geneva, December 10-13, 1976]. — Geneva : World Council of Churches, [1977]
iv, 92 p. ; 22 cm. Sw77-9459
Includes bibliographical references.
ISBN 2-8254-0551-5 : 7.90F
1. Christian union—Congresses. 2. Church—Congresses. I. World Council of Churches.
BX2.I5 262'.001 78-324905
79 MARC

Kirchenrecht und Ökumene : Bericht von e. Tagung / im Auftr. d. Sekretariats d. Bundes d. Evang. Kirchen in d. Dt. Demokrat. Republik hrsg. von Martin Kramer, Manfred Stolpe, Barbara Küntscher. — Berlin : Evangelische Verlagsanstalt, 1978.
61 p. ; 21 cm. GFR79-A9
Meeting organized by the Rechtsausschuss des Bundes der Evangelischen Kirchen in der DDR, May 1974.
Includes bibliographical references.
3.50M
1. Ecclesiastical law—Congresses. 2. Christian union—Congresses. 3. Ecumenical movement—Congresses. 4. Ecclesiastical law—Germany, East—Congresses. I. Kramer, Martin. II. Stolpe, Manfred. III. Küntscher, Barbara. IV. Bund der Evangelischen Kirchen in der DDR. Rechtsausschuss.
262.9 79-372787
79 MARC

Sessione di formazione ecumenica, 13th, Mendola, Italy, 1975.
Ecumenismo oggi : bilancio e prospettive : atti della XIII Sessione di formazione ecumenica organizzata dal Segretariato attività ecumeniche (S.A.E.) : La Mendola (Trento), 29 luglio-6 agosto 1975. — Leumann/[Rivoli] : Elle Di Ci, [1977?]
416 p. ; 21 cm. It79-Feb
Includes bibliographical references.
L5400
1. Christian union—Congresses. I. Segretariato attività ecumeniche. II. Title.
BX8.2.A1S47 1975 79-381285
*79 MARC

World Council of Churches. 3d Assembly, New Delhi, India, 1961.
Work book for the assembly committees. Prepared for the 3rd assembly of the World Council of Churches ... Geneva (1961)
188 p. 21 cm. sfr2.- Sw66-A4-605
1. Christian union—Congresses. I. Title.
BX6.W77 1961f 66-75086
[r79]rev MARC

World Council of Churches. Commission on Faith and Order.
Uniting in hope : reports and documents from the meeting of the Faith and Order Commission, 23 July-5 August, 1974, University of Ghana, Legon. — Geneva : Commission on Faith and Order, World Council of Churches, 1975.
vii, 144 p. ; 22 cm. — (Paper - Commission on Faith and Order, World Council of Churches ; no. 72)
At head of title: Accra 1974.
ISBN 2-8254-0492-6
1. Christian union—Congresses. 2. World Council of Churches. Commission on Faith and Order. I. Title. II. Series: World Council of Churches. Commission on Faith and Order. Paper - Commission on Faith and Order, World Council of Churches ; no. 72.
BX6.W775A2 1974 262'.001 79-302495
79 MARC

—HISTORY

Baier, Klaus Alois.
Unitas ex auditu : die Einheit der Kirche im Rahmen der Theologie Karl Barths / Klaus Alois Baier. — Bern : P. Lang, c1978.
284 p. ; 23 cm. — (Basler und Berner Studien zur historischen und systematischen Theologie ; Bd. 35) Sw***
Originally presented as the author's thesis, Universität Erlangen-Nürnberg, Erlangen, 1974.
Summary in English.
"Schriften Karl Barths": p. 255-261.
Bibliography: p. 262-279.
ISBN 3-261-03046-1
1. Church—Unity—History of doctrines. 2. Christian union—History. 3. Barth, Karl, 1886-1968. I. Title. II. Series.
BV601.5.B34 1978 78-390081
79 MARC

Davies, Rupert Eric, 1909-
The Church in our times : an ecumenical history from a British perspective / [by] Rupert E. Davies. — London : Epworth Press, 1979.
132 p. ; 22 cm. GB79-11438
Bibliography p. 127-128.
Includes index.
ISBN 0-7162-0317-0 : £1.95
1. Christian union—History. I. Title.
BX6.5.D38 274.1 79-314859
79 MARC

CHRISTIAN UNION (Continued)

—ENGLAND

Catholic Church. Ecumenical Commission for England and Wales.
The Roman Catholic Church in England and the Ten Propositions of the Churches' Unity Commission : a paper / by the Ecumenical Commission of England and Wales. — Abbots Langley : Catholic Information Office, 1976.
[1], 8 p. ; 21 cm. GB77-223
Cover title.
A provisional response to the report of the Churches' Unity Commission entitled: Visible unity: ten propositions.
ISBN 0-905241-04-5 : £0.15
1. Christian union—England. 2. Churches' Unity Commission. Visible unity. 3. Christian union—Catholic Church. 4. Catholic Church in England. I. Title: The Roman Catholic Church in England and the Ten Propositions ...
BR759.C38 1976 262'.001 77-359592
77[r78]rev MARC

Leonard, Graham.
The Church of England and the ten propositions / by the Bishop of Truro and Peter Boulton. — London : Church Literature Association for the Church Union, 1977.
[2], 10 p. ; 21 cm. GB78-08069
ISBN 0-85191-105-6 : £0.12
1. Christian union—England. 2. Church of England—Relations. I. Boulton, Peter, joint author. II. The Church Union. III. Title.
BR759.L37 274.1 78-315581
78 MARC

—FRANCE—FRANCHE-COMTÉ
—HISTORY—CONGRESSES

Œcuménisme en Franche-Comté : recherches sur les origines / [avant-propos par Gaston Bordet]. — [Besançon] : Université de Franche-Comté, Faculté des lettres et sciences humaines, 1977.
ix, 404 p. ; 30 cm. — (Recherches comtoises ; no 1-2) F***
Cover title.
At head of title: Institut d'histoire du christianisme (Université de Lyon III); Institut d'études comtoises et jurassiennes; Centre de recherches d'histoire sociale et culturelle de la France de l'Est et de la Confédération helvétique (Université de Franche-Comté).
"Sources et bibliographie": p. 171-180.
Includes indexes.
CONTENTS: Actes du Colloque d'Arc-et-Senens.—Grallet, J. P. Les relations œcuméniques entre catholiques et luthériens dans le pays de Montbéliard et le Territoire de Belfort de 1930 à 1960.
20.00F
1. Christian union—France—Franche-Comté—History—Congresses. 2. Catholic Church—Relations—Lutheran Church—Congresses. 3. Lutheran Church—Relations—Catholic Church—Congresses. I. Grallet, Jean Pierre. Les relations œcuméniques entre catholiques et luthériens dans le pays de Montbéliard et le Territoire de Belfort de 1930 à 1960. 1977. II. Series.
BX843.O4 274.4'45 78-391619
79 MARC

—GERMANY

May, Georg, 1926-
Interkonfessionalismus in der deutschen Militärseelsorge von 1933 bis 1945 / von Georg May. — Amsterdam : B. R. Grüner, 1978.
lxxiii, 529 p. ; 23 cm. — (Kanonistische Studien und Texte ; Bd. 30) Ne78-41
Bibliography: p. xxix-lxxii.
Includes indexes.
ISBN 9060321030
1. Church and state in Germany—1933-1945. 2. Chaplains, Military—Germany. 3. Chaplains, Military—Catholic Church. 4. Catholic Church in Germany—History. 5. Germany—Church history—1933-1945. 6. Christian union—Germany. I. Title. II. Series.
BR856.M284 78-394596
*79 MARC

—GREAT BRITAIN

Catholic Church. Ecumenical Commission for England and Wales.
Local covenants : a working paper of the Ecumenical Commission for England and Wales. — Abbots Langley : Information for Christian Formation, 1978.
14, [2] p. ; 21 cm. GB79-13327
Cover title.
Bibliography: p. 14.
£0.18
1. Christian union—Great Britain. I. Title.
BR759.C38 1978 262'.001 79-315837
79 MARC

—MADAGASCAR

Rabemanahaka, Jean Williberton, 1911-
Hivoatra sy ho faty? / Rabemanahaka Jean Williberton. — [s.l. : s.n., 1977] ([Antanimena-Tananarive : Impr. catholique])
58 p. ; 16 cm. — (His Velona ho an'ny hafa ; laharana 9)
1. Christian union—Madagascar. I. Title.
BR1470.M2R3 78-980218
79 MARC

—NIGERIA—HISTORY

Kalu, Ogbu.
Divided people of God : church union movement in Nigeria, 1875-1966 / by O. U. Kalu ; foreword by James I. McCord. — New York : NOK Publishers, c1978.
xiv, 128 p. : maps ; 24 cm.
Includes bibliographical references and index.
ISBN 0-88357-048-3
1. Christian union—Nigeria—History. 2. Nigeria—Church history. I. Title.
BR1463.N5K34 1978 262'.001 74-81853
00 MARC

—NORWAY

Thorn, Finn.
Katolsk og luthersk kirkefornyelse : en økumens tilbakeblikk og håp / Finn Thorn. — Oslo : Aschehoug, 1979 (Oslo : Norbok)
124 p. ; 19 cm. N79-June
Includes index.
ISBN 8203099203 : kr45.00
1. Catholic Church—Relations—Lutheran Church. 2. Lutheran Church—Relations—Catholic Church. 3. Christian union—Norway. I. Title.
BX8063.7.C3T47 79-378383
79 MARC

—SCOTLAND

Multilateral Church Conversation in Scotland.
The Faith of the Church / [Multilateral Church Conversation in Scotland]. — Edinburgh : St. Andrew Press, 1978.
16, [2] p. ; 22 cm. GB78-24849
ISBN 0-7152-0395-9 : £0.25
1. Theology, Doctrinal. 2. Christian union—Scotland. I. Title.
BT80.M84 230 78-324907
79 MARC

CHRISTIAN UNION AND PASTORAL THEOLOGY see Pastoral theology and Christian union

CHRISTIAN UNION AND THE LORD'S SUPPER see Lord's Supper and Christian union

CHRISTIAN YEAR see Church year

CHRISTIANITY

see also Civilization, Christian; Deism; Ecumenical movement; Homosexuality and Christianity; Jesus Christ; Miracles; Socialism, Christian; Sociology, Christian; Theology; Women in Christianity; and headings beginning with the words Christian and Church; and names of Christian churches and sects, e.g. Catholic Church

Akaiwa, Sakae, 1903-
(Atarashii ningen tanjō)
新しい人間誕生：これからのキリスト者 / 赤岩栄. — 東京：理論社, 1953.
170 p. ; 19 cm.
Colophon inserted.

1. Christianity. I. Title.
BR121.2.A45 79-806713

Allen, John Catling.
The way of the Christian / [by] John Catling Allen. — Amersham : Hulton, 1978.
88 p. : ill. ; 21 cm. GB78-09986
Includes index.
ISBN 0-7175-0782-3 : £1.00
1. Christianity. I. Title.
BR121.2.A46 200 79-313902
79 MARC

Jaikung, H C 1936-
Inpumpekna / by H. C. Jaikung. — 1st ed. — Imphal : Jaikung, 1971.
ii, 125 p. ; 19 cm.
In Hmar.
1. Christianity. I. Title.
BR121.2.J28 78-908650
79 MARC

Maudoodi, Syed Abul 'Ala, Maulana, 1903-
(Yahūdiyyat va Naṣrāniyyat)
یہودیت و نصرانیت / مصنف سید ابوالاعلیٰ مودودی ; مرتب نعیم صدیقی. — کویت : اسلامک بک پبلشرز ; لاہور : ملنے کا پتہ، البدر پبلیکیشنز, 1976.
573 p. ; 25 cm.
In Urdu.
Rs48.00

1. Christianity. 2. Judaism. I. Siddīqī, Na'īm. II. Title.
BR121.2.M37 78-930109

Muramatsu, Takeshi, 1929-
(Kyōyō to shite no Kirisutokyō)
教養としてのキリスト教 / 村松剛著. — 東京：講談社, 1965.
245 p. : ill. ; 18 cm. — (講談社現代新書)
Series romanized: Kōdansha gendai shinsho.
Includes index.

1. Christianity. I. Title.
BR121.2.M86 79-801294

(O religii)
О религіи. — Санктпетербургъ : Въ Синодальной тип., 1862.
81 p. ; 21 cm.
Includes bibliographical references.
CONTENTS: 1. О религіи естественной.—2. О недостаточности ея.—3. Объ откровеніи.—4. О Христіанской религіи.
1. Natural theology. 2. Christianity.
BL190.O2 Russian Imperial Coll. 78-420682

Smart, Ninian, 1927-
The phenomenon of Christianity / [by] Ninian Smart. — London : Collins, 1979.
320 p. ; 23 cm. GB***
Bibliography: p. 313.
Includes index.
ISBN 0-00-215115-4 : £6.95
1. Christianity. I. Title.
BR121.2.S57 270 79-310869
79 MARC

—EARLY CHURCH, CA. 30-600

Fowler, Henry Thatcher, 1867-1948.
The history and literature of the New Testament / by Henry Thatcher Fowler. — Westport, Conn. : Greenwood Press, 1979, c1925.
443 p., [1] leaf of plates : map ; 23 cm.
Reprint of the ed. published by Macmillan, New York which was issued in Religious science and literature series.
Includes bibliographical references and index.
ISBN 0-313-21188-4
1. Bible. N.T.—History of Biblical events. 2. Bible. N.T.—History of contemporary events. 3. Christianity—Early church, ca. 30-600. I. Title. II. Series: Religious science and literature series.
BS2410.F6 1979 225.6 78-12516
78 MARC

Mary Bernard, Sister, 1917-
Agony : can the churches survive without Jesus? / By Sister Mary Bernard. — Harrison, Ark. : New Leaf Press, c1979.
112 p. ; 20 cm.
Includes bibliographical references.
ISBN 0-89221-059-1 : $2.95
1. Jesus Christ—Person and offices. 2. Christianity—Early church, ca. 30-600. 3. Christianity—20th century. I. Title.
BT202.M3763 262'.001 79-84343
00 MARC

CHRISTIANITY
—EARLY CHURCH, CA. 30-600 (Continued)

Miki, Toshiharu, 1920-
(Kirisutokyōteki kodai no kenkyū)
キリスト教的古代の研究：アンブロシウスとアウグスティヌス / 三木利英著. — 東京：風間書房, 昭和52 ⌈1977⌉
3, 2, 390, 5 p. ; 22 cm. Ja 77-7922
Bibliography: p. 389-390.
Includes index.
¥8000
1. Christianity—Early church, ca. 30-600. 2. Ambrosius, Saint, Bp. of Milan. 3. Augustinus, Aurelius, Saint, Bp. of Hipoo. I. Title.
BR165.M64 79-809528

Trimingham, John Spencer.
Christianity among the Arabs in pre-Islamic times / J. Spencer Trimingham. — London ; New York : Longman, 1979.
xiv, 342 p. : maps ; 23 cm. — (Arab background series) GB78
Bibliography: p. 317-329.
Includes index.
ISBN 0-582-78081-0 : £10.00
1. Arabs—Church history. 2. Christianity—Early church, ca. 30-600. 3. Near East—Church history. I. Title.
BR1070.T73 209'.56 78-40599
78 MARC

—EARLY CHURCH, CA. 30-600—ADDRESSES, ESSAYS, LECTURES

Paganisme, judaïsme, christianisme : influences et affrontements dans le monde antique : mélanges offerts à Marcel Simon. — Paris : E. de Boccard, 1978.
390 p., ⌈1⌉ leaf of plates : port. ; 25 cm. F***
French, English, or German.
"Publications de Marcel Simon": p. ⌈371⌉-387.
Includes bibliographical references and indexes.
250F
1. Judaism—History—Post-exilic period, 586 B.C.-210 A.D.—Addresses, essays, lectures. 2. Paganism—Addresses, essays, lectures. 3. Judaism—Relations—Christianity—Addresses, essays, lectures. 4. Christianity and other religions—Judaism—Addresses, essays, lectures. 5. Christianity—Early church, ca. 30-600—Addresses, essays, lectures. 6. Simon, Marcel, 1907- —Addresses, essays, lectures. I. Simon, Marcel, 1907-
BM177.P33 291 78-388850
79 MARC

Stümcke, Ida.
Der Impuls von Lérins : Spuren e. johanneischen Christentums im 4. Jh. / Ida Stümcke, Gérard Klockenbring. — Stuttgart : Verlag Freies Geistesleben, 1977.
65 p. : 20 cm. — (Studien und Versuche ; 14) GFR78-A
Bibliography: p. 63-65.
ISBN 3-7725-0044-7 : DM7.00
1. Christianity—Early church, ca. 30-600—Addresses essays, lectures. 2. Gaul—Church history—Addresses essays, lectures. 3. Lérins (Benedictine abbey)—Addresses essays, lectures. I. Klockenbring, Gérard, joint author. II. Title.
BR205.S8 78-377291
78 MARC

—MIDDLE AGES, 600-1500

Ishihara, Ken, 1882-
(Chūsei Kirisutokyō kenkyū)
中世キリスト教研究 / 石原謙著. — 東京：岩波書店, 昭和27 ⌈1952⌉
5, 2, 340, 10 p. ; 22 cm.
Includes bibliographical references and index.

1. Christianity—Middle Ages, 600-1500. I. Title.
BR252.I83 79-800938

—MIDDLE AGES, 600-1500—COLLECTED WORKS

Constable, Giles
Religious life and thought (11th-12th centuries) / Giles Constable. — London : Variorum Reprints, 1979.
iii, ⌈380⌉ p. ; 24 cm. — (Variorum reprint ; CS 89) GB***
English or French.
Contains 16 articles originally published in 1953-77 in various journals.
Includes bibliographical references and index.
ISBN 0-86078-034-1 : £20.00
1. Christianity—Middle Ages, 600-1500—Collected works. 2. Europe—Church history—Collected works. I. Title.
BR270.C66 282'.09'021 79-312251
00 MARC

—19TH CENTURY

Anderson, Robert, Sir, 1841-1918.
The silence of God / by Robert Anderson. — Grand Rapids, Mich. : Kregel Publications, c1978.
215 p. ; 20 cm.
"Sir Robert Anderson library series."
Reprint of the 8th ed. published in 1907 by Hodder and Stoughton, London.
ISBN 0-8254-2128-4 : $3.50
1. Christianity—19th century. I. Title.
⌈BR121.A57 1978⌉ 231 78-9528
78⌈r79⌉rev MARC

Chateaubriand, François Auguste René, vicomte de, 1768-1848.
Essai sur les révolutions ; Génie du christianisme / Chateaubriand ; texte établi, présenté et annoté par Maurice Regard. — ⌈Paris⌉ : Gallimard, c1978.
xv, 2089 p. ; 18 cm. — (Bibliothèque de la Pléiade ; 272) F***
Bibliography: p. ⌈1969⌉-1994.
Includes index.
155F
1. Revolutions—History. 2. Christianity—19th century. I. Chateaubriand, François Auguste René, vicomte de, 1768-1848. Génie du christianisme. 1978. II. Regard, Maurice. III. Title.
D25.5.C4 1978 909 78-397822
79 MARC

—20TH CENTURY

see also Death of God theology

Altizer, Thomas J J
The descent into hell : a study of the radical reversal of the Christian consciousness / Thomas J. J. Altizer. — New York : Seabury Press, 1979.
217 p. ; 23 cm. — (The Seabury library of contemporary theology)
Reprint of the 1970 ed. published by Lippincott, Philadelphia.
"A Crossroad book."
Includes index.
ISBN 0-8164-1194-8 : $6.95
1. Christianity—20th century. I. Title. II. Series: Seabury library of contemporary theology.
⌈BR121.2.A48 1979⌉ 230'.09'04 79-88310
79 MARC

Anne Françoise, sœur.
Les Religieuses et le petit écran, réflexions d'une téléspectatrice ... Paris, Éditions Fleurus, 1970.
168 p. 20 cm. (La Religieuse dans la pastorale d'aujourd'hui, 22)
F71-4733
Bibliography: p. 163-⌈168⌉
1. Communication (Theology) 2. Mass media—Moral and religious aspects. 3. Christianity—20th century. I. Title.
BV4319.A54 209'.04 79-384768
79 MARC

Bednarczyk, Krzysztof.
Formuła mocy / Krzysztof Bednarczyk. — Wyd. 1. — Warszawa : Słowo Prawdy, 1977 (Pozn. : Pallottinum)
220, ⌈4⌉ p. : ill. ; 17 cm. P78-929
Bibliography: p. 220.
1. Christianity—20th century. 2. Church history. 3. Christian life—1960- . I. Title.
BR121.2.B36 79-341547
79 MARC

Blanco, Guillermo.
El evangelio de Judas / ⌈por⌉ Guillermo Blanco. — Barcelona ⌈etc⌉ : Pomaire, D.L. 1976.
141 p. ; 22 cm. Sp77-Mar
Includes bibliographical references.
ISBN 8428601194 : 300ptas
1. Jesus Christ—Person and offices. 2. Christianity—20th century. I. Title.
BT202.B47 79-381748
79 MARC

Bosanquet, Bernard, 1848-1923.
What religion is / by Bernard Bosanquet. — Westport, Conn. : Greenwood Press, 1979.
xii, 81 p. ; 23 cm.
Reprint of the 1920 ed. published by Macmillan, London.
ISBN 0-313-21202-3
1. Christianity—20th century. I. Title.
BR125.B66 1979 200 78-12709
78 MARC

Bowden, John Stephen.
Voices in the wilderness / John Bowden. — London : SCM Press, 1977, 1978 printing.
85 p. ; 20 cm. GB***
Includes bibliographical references.
ISBN 0-334-01746-7 : £1.10
1. Christianity—20th century. I. Title.
BR479.B68 270.8'2 79-310983
79 MARC

Brown, David, 1922-
God's tomorrow / ⌈by⌉ David Brown. — London : S.C.M., 1977.
vii, 88 p. ; 20 cm. GB78-20829
Includes bibliographical references.
ISBN 0-334-00590-6 : £0.90
1. Christianity—20th century. I. Title.
BR121.2.B73 201'.1 78-322937
79 MARC

Delumeau, Jean.
Le christianisme va-t-il mourir? / Jean Delumeau. — ⌈Paris⌉ : Hachette, c1977.
211 p. ; 23 cm. — (Hachette littérature & ⌈i.e. et⌉ sciences humaines) (Hachette essais) F***
Includes bibliographical references.
ISBN 2-01-003920-3 : 38.00F
1. Christianity—20th century. I. Title.
BR479.D39 270.8'2 78-383218
78 MARC

Dudley, Carl S 1932-
Where have all our people gone? : New choices for old churches / Carl S. Dudley. — New York : Pilgrim Press, c1979.
xiv, 136 p. : ill. ; 21 cm.
Includes bibliographies.
ISBN 0-8298-0359-9 : $5.95
1. Church membership. 2. Church work. 3. Christianity—20th century. I. Title.
BV820.D82 254'.5 79-525
79 MARC

Edwards, David Lawrence.
A reason to hope / ⌈by⌉ David L. Edwards. — London ; New York : Collins, 1978.
252 p. ; 22 cm. GB78-29007
Includes index.
ISBN 0-529-05620-8 : £4.95 ($8.95)
1. Civilization, Modern, 1950- . 2. Christianity—20th century. 3. Christianity and culture. I. Title.
CB430.E34 1978 909 78-56996
78 MARC

Elchinger, Léon Arthur, Bp., 1908-
Je plaide pour l'homme / Léon Arthur Elchinger. — Paris : Fayard, c1976.
301 p. ; 22 cm. F***
Includes bibliographical references.
ISBN 2-213-00357-2 : 37.00F
1. Civilization, Modern—1950- . 2. Christianity—20th century. I. Title.
CB430.E42 909.82 79-341070
79 MARC

Fry, John R
The great apostolic blunder machine : a contemporary attack upon Christendom / John R. Fry. — 1st ed. — San Francisco : Harper & Row, c1978.
183 p. ; 21 cm.
Includes bibliographical references.
ISBN 0-06-063072-8
1. Christianity—20th century. 2. Christianity—Controversial literature. I. Title.
BR121.2.F69 1978 262'.009'04 78-3137
78 MARC

Glen, John Stanley, 1907-
Justification by success : the invisible captivity of the church / J. Stanley Glen. — Atlanta : John Knox Press, c1979.
128 p. ; 21 cm.
Includes bibliographical references.
ISBN 0-8042-0835-2
1. Christianity—20th century. I. Title.
BR121.2.G53 209'.04 78-52446
78 MARC

Gill, Robin.
Faith in Christ : Christian claims in a changing world / by Robin Gill. — London : Mowbrays, 1978.
v, 90 p. ; 18 cm. — (Mowbrays popular Christian paperbacks) GB***
Bibliography: p. 86-90.
ISBN 0-264-66482-5 : £0.80
1. Jesus Christ—Person and offices. 2. Christianity—20th century. 3. Faith. I. Title.
BT202.G5 232 78-319507
78 MARC

Granero, Jesús María.
La aventura de hoy / ⌈por⌉ Jesús M. Granero. — 1. ed. — Madrid : Eapsa, 1976.
xv, 529 p., ⌈10⌉ leaves of plates : ill. ; 21 cm. Sp77-Feb
ISBN 8440019734 : 600ptas
1. Christianity—20th century. 2. Europe—Civilization—1945- I. Title.
BR121.2.G694 79-347933
79 MARC

Greeley, Andrew M 1928-
Crisis in the church : a study of religion in America / by Andrew M. Greeley. — Chicago : Thomas More Press, c1979.
264 p. : diagrs. ; 24 cm.
Includes bibliographical references.
ISBN 0-88347-106-X : $14.95
1. Catholic Church in the United States. 2. Christianity—20th century. 3. Catholic Church—Doctrinal and controversial works—Catholic authors. I. Title.
BX1406.2.G696 301.5'8 79-101825
79 MARC

CHRISTIANITY
—20TH CENTURY (Continued)

Hollenweger, Walter J 1927-
Erfahrungen der Leibhaftigkeit / Walter J. Hollenweger. — München : Kaiser, 1979.
381 p. ; 22 cm. — (His Interkulturelle Theologie ; 1) GFR79-A
Bibliography: p. 356-373.
Includes indexes.
ISBN 3-459-01197-1
1. Christianity—20th century. 2. Theology—20th century. 3. Hollenweger, Walter J., 1927- I. Title.
BR83.H64 vol. 1 79-380806
[BR121.2] 79 MARC

Iammarrone, Giovanni, 1941-
Proposte sulla fede oggi / Giovanni Iammarrone. — Roma : Borla, [1978]
224 p. ; 19 cm. — (Frecce) It***
Includes bibliographical references.
L4000
1. Christianity—20th century. I. Title.
BR121.2.I24 79-352191
79 MARC

John Paul, I, Pope, 1912-1978.
Illustrissimi : lettere del patriarca / Albino Luciani. — Padova : Messaggero, 1976.
350 p. ; 19 cm. — (Collana Messaggero ; 1) It78-Feb
1. Christianity—20th century. 2. Imaginary letters. I. Title.
BR123.J56 78-366908
*78[r79]rev MARC

John Paul I, Pope, 1912-1978.
Illustrissimi : the letters of Pope John Paul I / Albino Luciani ; preface by Cardinal Hume ; translated [from the Italian] by Isabel Quigly ; illustrated by Papas. — London : Collins, 1978.
286 p. : ill. ; 25 cm. GB***
ISBN 0-00-211418-6 : £5.95
1. Christianity—20th century. 2. Imaginary letters. I. Title.
BR123.J5613 230'.2 79-309519
79 MARC

Lackner, Wolfram.
Kirche ohne Religion : Ende d. Protestantismus? / Wolfram Lackner. — Saterland : Saterland-Verlag, 1979.
152 p. : 23 cm. GFR79-A
Bibliography: p. 151-152.
ISBN (invalid) 3-88399-001-0 : DM24.00
1. Christianity—20th century. 2. Protestantism. I. Title.
BR121.2.L22 79-376033
79 MARC

Mancini, Italo.
Con quale cristianesimo / Italo Mancini. — 1. ed. — Roma : Coines, 1978.
237 p. ; 21 cm. — (Nuova biblioteca ; 10) It***
Includes bibliographical references.
L4000
1. Christianity—20th century. I. Title.
BR121.2.M32 78-386967
79 MARC

Marshall, Michael, 1936-
Glory under your feet : the challenge of catholic renewal today / [by] Michael Marshall. — London : Darton, Longman, and Todd, 1978.
137 p. ; 22 cm. GB78-29210
Includes bibliographical references.
ISBN 0-232-51306-5 : £2.95
1. Church renewal. 2. Christianity—20th century. 3. Church. I. Title.
BV600.2.M345 1978 260 79-301452
79 MARC

Mary Bernard, Sister, 1917-
Agony : can the churches survive without Jesus? / By Sister Mary Bernard. — Harrison, Ark. : New Leaf Press, c1979.
112 p. ; 20 cm.
Includes bibliographical references.
ISBN 0-89221-059-1 : $2.95
1. Jesus Christ—Person and offices. 2. Christianity—Early church, ca. 30-600. 3. Christianity—20th century. I. Title.
BT202.M3763 262'.001 79-84343
00 MARC

McBirnie, William Steuart, 1920-
The search for the early church : New Testament principles for today's church / William Steuart McBirnie. — Wheaton, Ill. : Tyndale House Publishers, c1978.
165 p. : ill. ; 21 cm.
ISBN 0-8423-5834-X : $3.95
1. Church history—Primitive and early church, ca. 30-600. 2. Church. 3. Christianity—20th century. I. Title.
BR165.M18 281'.2 78-55982
79 MARC

Meyer, Rudolf, 1896-
Les formes futures du christianisme / Rudolf Meyer. — [Narbonne] : Société du souvenir et des études cathares, [1977]
46 p. ; 22 cm. F***
Translated from the German.
1. Christianity—20th century. I. Title.
BR125.M5714 226'.06 79-370207
79 MARC

Middleton, Robert G
Charting a course for the church : snug harbor or open sea / Robert G. Middleton. — Valley Forge, PA : Judson Press, c1979.
125 p. ; 22 cm.
Includes bibliographical references.
ISBN 0-8170-0844-6 : $5.95
1. Christianity—20th century. I. Title.
BR121.2.M47 261.8 79-9949
79 MARC

Orlandis, José.
Historia y espíritu / José Orlandis. — Pamplona : EUNSA, [1975]
191 p. : 18 cm. — (Temas NT ; 1) (Colección cultural de bolsillo) Sp***
Includes bibliographical references.
ISBN 843130376X
1. History—Philosophy. 2. Christianity—20th century. I. Title.
D16.8.O66 78-381368
78 MARC

Poland, Larry W
Rise to conquer : a call for Christian action / Larry W. Poland. — 1st ed. — Chappaqua, N.Y. : Christian Herald Books, c1979.
131 p. ; 21 cm.
ISBN 0-915684-48-9 : $6.95
1. Christian life—1960- . 2. Christianity—20th century. 3. Kingdom of God. I. Title.
BV4501.2.P55547 248'.4 78-64843
78 MARC

Sarale, Nicolino.
Il cristiano nella diaspora del XX secolo / Nicolino Sarale. — Milano : Pan, 1978.
158 p. ; 21 cm. — (Collana Nuovi saggi ; n. 19) It78-Nov
L4000
1. Catholic Church—Apologetic works. 2. Apologetics—20th century. 3. History (Theology) 4. Christianity—20th century. I. Title.
BX1752.S36 78-400295
*79 MARC

Schneider, Hans J 1935-
Masters of legalized confusion and their puppets / by Hans J. Schneider. — Ashland, OR : World Wide Pub. Corp., 1979.
iv, 28 p. : ill. ; 22 cm.
Includes index.
ISBN 0-930294-11-4 : $1.45
1. Christianity—20th century. 2. Christian life—1960- I. Title.
[BR124.S36 1979] 209'.04 78-12609
78 MARC

Seibel, Alexander, 1943-
Gemeinde Jesu, endzeitlich unterwandert? / Alexander Seibel. — 1. Aufl. — Wuppertal-Elberfeld : Verlag und Schriftenmission d. Evang. Ges. für Deutschland, 1977.
235 p. ; 18 cm. — (Aktuelle Fragen ; Bd. 25) GFR77-A
Includes bibliographical references.
DM9.80
1. Christianity—20th century. 2. Gifts, Spiritual. 3. Bible. N.T. 1 Corinthians—Criticism, interpretation, etc. 4. Imposition of hands. I. Title.
BR121.2.S44 78-398584
79 MARC

Senior, John, 1923-
The death of Christian culture / John Senior. — New Rochelle, N.Y. : Arlington House, c1978.
185 p. ; 24 cm.
Includes bibliographical references and index.
ISBN 0-87000-416-6 : $10.00
1. Civilization, Christian. 2. Christianity—20th century. I. Title.
BR115.C5S46 261.5 78-16154
78 MARC

Six, Jean François.
L'incroyance et la foi ne sont pas ce qu'on croit / Jean-François Six. — Paris : Le Centurion, c1979.
197 p. ; 21 cm. F***
Includes bibliographical references.
ISBN 2-227-31024-3 : 38.00F
1. Christianity—20th century. 2. Non-church affiated people. I. Title.
BR121.2.S56 209'.047 79-378313
79 MARC

Van der Bent, Ans Joachim.
God so loves the world : the immaturity of world Christianity / A. J. van der Bent ; foreword by David E. Jenkins. — Madras : Christian Literature Society, 1977.
xvi, 143 p. ; 23 cm.
Includes bibliographical references.
Rs10.00
1. Christianity—20th century. 2. Christian union. 3. Christianity and other religions. I. Title.
BR121.2.V324 209'.04 78-900901
78 MARC

Van der Bent, Ans Joachim.
God so loves the world : the immaturity of world Christianity / A. J. van der Bent. — U.S. ed. — Maryknoll, N.Y. : Orbis Books, 1979 c1977.
x, 150 p. ; 21 cm.
Includes bibliographical references.
ISBN 0-88344-159-4
1. Christianity—20th century. I. Title.
BR121.2.V333 1979 270.8'2 79-4470
79 MARC

White, John, 1924 (Mar. 5)-
The golden cow : materialism in the twentieth-century church / John White. — Downers Grove, Ill. : Inter-Varsity Press, c1979.
175 p. ; 21 cm.
ISBN 0-87784-490-9 : $3.50
1. Christianity—20th century. 2. Church finance. I. Title.
BR121.2.W45 270.8'2 78-13884
79 MARC

—20TH CENTURY—ADDRESSES, ESSAYS, LECTURES

Arvoja etsimässä : ajankohtaisia puheenvuoroja kristinuskon ja kulttuurin suhteista / [toim., Eino Murtorinne ja T. P. Virkkunen ; kirj., Antti Alhonsaari ... et al.]. — [Helsinki] : Krijapaja, 1978 (Hämeenlinna : Karisto).
164, [2] p. ; 21 cm. Fi78-4
Includes bibliographies.
ISBN (invalid) 9516211977
1. Christianity—20th century—Addresses, essays, lectures. I. Murtorinne, Eino. II. Virkkunen, Timo Päiviö.
BR121.2.A78 79-361409
79 MARC

Dietzfelbinger, Hermann.
Jesus Christus und die Weisen aus dem Abendland / Hermann Dietzfelbinger. — Metzingen : Brunnquell-Verlag der Bibel- u. Missions-Stiftung, 1978.
240 p. ; 20 cm. GFR78-A
Includes bibliographical references.
ISBN 3-7656-0213-2 : DM13.80
1. Christianity—20th century—Addresses, essays, lectures. 2. Christianity and culture—Addresses, essays, lectures. I. Title.
BR123.D495 79-365956
79 MARC

Ich habe einen Traum : Visionen u. Wirklichkeiten / hrsg. von Meinold Krauss. — 1. Aufl., (1.-8. Tsd.). — Stuttgart ; Berlin : Kreuz-Verlag, 1978.
221 p. ; 21 cm. GFR78-A
Includes bibliographical references.
ISBN 3-7831-0544-7 : DM19.80
1. Christianity—20th century—Addresses, essays, lectures. 2. Christian life—1960- —Addresses, essays, lectures. I. Krauss, Meinold, 1944-
BR123.I25 78-381925
78 MARC

(Isip: egi Kidokkyo wa yesul)
20世紀基督教와藝術 / 金文煥譯編. — 서울 : 大韓基督教書會, 1974.
201 p. ; 18 cm. — (現代新書 ; 55)
Series romanized: Hyŏndae sinsŏ.
Includes bibliographical references.
1. Art and religion—Addresses, essays, lectures. 2. Christianity—20th century—Addresses, essays, lectures. I. Kim, Mun-hwan, 1944-
BR115.A8I84 79-825785

Morris, Colin M
Bugles in the afternoon / Colin Morris. — Philadelphia : Westminster Press, [1979], c1977.
117 p. ; 19 cm.
ISBN 0-664-24260-X
1. Sermons, English. 2. Methodist Church—Sermons. 3. Christianity—20th century—Addresses, essays, lectures. I. Title.
BX8333.M584B83 1979 230'.7 78-23273
78 MARC

Norman, Edward R
Christianity and the world order / Edward Norman. — Oxford ; New York : Oxford University Press, 1979.
105 p. ; 20 cm. — (B.B.C. Reith lectures ; 1978) GB78
Includes bibliographical references and index.
ISBN 0-19-215510-5 : £3.25. ISBN 0-19-283019-8 : £1.50
1. Christianity and politics—Addresses, essays, lectures. 2. Christianity—20th century—Addresses, essays, lectures. 3. Secularization (Theology)—Addresses, essays, lectures. I. Title. II. Series: Keith lectures ; 1978.
BR115.P7N597 261.7 78-40857
78 MARC

Paoli, Arturo.
Camminando s'apre cammino / Arturo Paoli. — 3. ed. — Torino : Gribaudi, 1977.
371 p. ; 19 cm. — (La Piscina di Siloe) It79-Feb
L6500
1. Christianity—20th century—Addresses, essays, lectures. 2. Liberation theology—Addresses, essays, lectures.
BR121.2.P22 1977 79-381284
*79 MARC

Poll, G J van de.
Christendom in crisissituatie / G. J. van de Poll. — 's-Gravenhage : Boekencentrum, [1978]
120 p. ; 20 cm. Ne78-51
Includes bibliographical references.
ISBN 9023912179 : fl 16.90
1. Christianity—20th century—Addresses, essays, lectures. I. Title.
BR479.P64 79-360806
*79 MARC

CHRISTIANITY
—20TH CENTURY—ADDRESSES, ESSAYS, LECTURES (Continued)

Schmidt, Hans P
Der Geist und die Geister : 3 Beitr. von Hans P. Schmidt, Walter J. Hollenweger u. Horst Bürkle zu d. Vielfalt religiöser Formen in unserer Zeit. — 1. Aufl. — Konstanz : Bahn, 1976.
79 p. ; 21 cm. GFR77-A
"Aus einer Würdigung von Hans Weissgerber im 'Allgemeinen Deutschen Sonntagsblatt' zu den Vorträgen, die auf dem Heidelberger Pfarrertag 1974 gehalten wurden und hier in überarbeiteter Form vorliegen."
ISBN 3-7621-3056-6 : DM9.80
1. Discernment of spirits—Addresses, essays, lectures. 2. Holy spirit—Addresses, essays, lectures. 3. Experience (Religion)—Addresses, essays, lectures. 4. Christianity—20th century—Addresses, essays, lectures. I. Hollenweger, Walter J., 1927- joint author. II. Bürkle, Horst, 1925- joint author. III. Heidelberger Pfarrertag, 1974. IV. Title.
BV5083.S35 78-398587
79 MARC

Studi sulla produzione sociale del sacro / [di] Ferrarotti, De Lutiis, Macioti, Catucci. — 1. ed. — Napoli : Liguori, 1978-
v. : ill. ; 20 cm. — (Contributi di sociologia ; 44) It79-Mar
Bibliography: v. 1, p. [469]-520.
CONTENTS: v. 1. Forme del sacro in un'epoca di crisi.
ISBN 8820707683 (v. 1) : 12500
1. Christianity—20th century—Addresses, essays, lectures. 2. Secularization (Theology)—Addresses, essays, lectures. I. Ferrarotti, Franco.
BR479.S77 79-349924
*79 MARC

Tendencias y movimientos actuales / J. Ma. González Ruiz ... [et al.] ; dirección y presentación José Antonio Merino. — Madrid : Cisneros, 1978.
172 p. ; 22 cm. Sp78-Mar
"Artículos aparecidos en la revista Verdad y vida, correspondientes a los núms. 139 y 140 de 1977."
Includes bibliographical references.
ISBN 8470470183
1. Christianity—20th century—Addresses, essays, lectures. 2. Civilization, Modern—20th century—Addresses, essays, lectures. I. González Ruiz, José María, 1916- II. Merino, José Antonio.
BR481.T45 78-388705
79 MARC

Wilson, Bryan R
Contemporary transformations of religion / by Bryan Wilson. — London ; New York : Oxford University Press, 1976.
ix, 116 p. ; 19 cm. — (Riddell memorial lectures ; 45th ser)
"The Riddell memorial lectures, forty-fifth series, delivered at the University of Newcastle upon Tyne on 2, 3, 4 December 1974."
Includes bibliographical references.
ISBN 0-19-713914-0 : £2.95
1. Christianity—20th century—Addresses, essays, lectures. 2. Sects—Addresses, essays, lectures. 3. Religion and sociology—Addresses, essays, lectures. I. Title. II. Series.
BR123.W58 209'.04 76-378405
76[r79]rev MARC

Yamamoto, Kanō, 1909-
(Gendai bummei no higeki)
現代文明の悲劇 / 山本和著. — 東京 : 警醒社書店, 昭和24 [1949]
372 p. ; 19 cm.
Colophon inserted.
Includes bibliographical references.
1. Christianity—20th century—Addresses, essays, lectures. I. Title.
BR479.Y35 78-809497

—20TH CENTURY —CONGRESSES

Christianity in independent Africa / edited by Edward Fasholé-Luke ... [et al.]. — London : R. Collings ; Bloomington, IN : Indiana University Press, c1978.
630 p. ; 24 cm.
Includes bibliographical references and index.
ISBN 0-253-37506-1
1. Christianity—Africa—Congresses. 2. Christianity—20th century—Congresses. I. Fasholé-Luke, Edward W.
BR1360.C48 209'.6 77-26351
78 MARC

—ADDRESSES, ESSAYS, LECTURES

Akaiwa, Sakae, 1903-
(Ningen, kono gyakusetsuteki naru mono)
人間・この逆説的なるもの　赤岩榮著　東京　沙羅書房　昭和23 [1948]
7, 275 p. port. 18 cm.
Colophon inserted.
1. Christianity—Addresses, essays, lectures. I. Title.
BR123.A47 J 59-2122

Christians and Jews / edited by Hans Küng and Walter Kasper. — New York : Seabury Press, c1975.
93 p. ; 23 cm. — (Concilium ; new ser., v. 8, no. 10(98), Ecumenism))
"A Crossroad book."
Includes bibliographical references.
ISBN 0-8164-2095-5 : $4.95
1. Christianity and other religious—Judaism—Addresses, essays, lectures. 2. Judaism—Relations—Christianity—Addresses, essays, lectures. 3. Judaism—Addresses, essays, lectures. 4. Christianity—Addresses, essays, lectures. I. Küng, Hans, 1928- II. Kasper, Walter. III. Series: Concilium (New York) ; 98.
BM535.C577 261.2 76-10050
00 MARC

The Christian tradition / [compiled by] World Religions Curriculum Development Center ; project co-directors, Lee Smith and Wes Bodin, project assistants, Joan Voigt and Pat Noyes. — Niles, Ill. : Argus Communications, c1978.
vi, 186 p. : ill. ; 23 cm. — (Religion in human culture)
ISBN 0-89505-013-7
1. Christianity—Addresses, essays, lectures. I. World Religions Curriculum Development Center. II. Series.
BR123.C614 201 78-53804
79 MARC

Dkhar, L 1943-
Ka jingshem mynsiem u Khristan / da L. Dkhar. — [Shillong : Khasi Jaintia Presbyterian Press, pref. 1976]
82 p., [1] leaf of plates : port ; 19 cm.
In Khasi.
Rs2.50
1. Christianity—Addresses, essays, lectures. I. Title.
BR123.D57 78-903874
79 MARC

Fujio, Masahito, 1925-
(Mune ga atsukunaru yō na)
胸が熱くなるような / 藤尾正人著. — 東京 : 同信社, 昭和52 [1977]
xii, 330 p. ; 20 cm. Ja 77-12526
¥2000
1. Christianity—Addresses, essays, lectures. I. Title.
BR123.F92 78-812066

Maxwell, Charles Herbert, 1885-
Adventures of the White girl in her search for God / by Charles Herbert Maxwell. — Norwood, Pa. : Norwood Editions, 1977.
23 p. ; 23 cm.
A reply to G. B. Shaw's The adventures of the Black girl in her search for God.
Reprint of the 1933 ed. published by Lutterworth Press, London.
ISBN 0-8482-1803-5 : $7.50
1. Shaw, George Bernard, 1856-1950. The adventures of the Black girl in her search for God. 2. Christianity—Addresses, essays, lectures. I. Shaw, George Bernard, 1856-1950. The adventures of the Black girl in her search for God. II. Title.
[PR5365.A42M3 1977] 823'.9'12 77-19315
77 MARC

Nakazawa, Kaoru.
(Koshū)
孤愁 / 中澤薫. — 東京 : 長崎次郎, 昭和24 [1949]
181 p., [1] leaf of plates : port. ; 19 cm.
1. Christianity—Addresses, essays, lectures. I. Title.
BR85.N258 79-804695

Troeltsch, Ernst, 1865-1923.
Christian thought, its history and application : lectures written for delivery in England during March 1923 / by Ernst Troeltsch ; translated into English by various hands and edited with an introd. and index by Baron F. von Hügel. — Westport, Conn. : Hyperion Press, 1979.
xxxi, 179 p. ; 22 cm.
Translation of Der Historismus und seine Überwindung.
Reprint of the 1923 ed. published by University of London Press.
Includes index.
ISBN 0-88355-719-3 : $16.50
1. Christianity—Addresses, essays, lectures. I. Hügel, Friedrich, Freiherr von, 1852-1925. II. Title.
BR83.T7313 1979 170 78-59047
78 MARC

—APOLOGETIC WORKS
see Apologetics

—BIOGRAPHY see Christian biography

—CONTROVERSAL LITERATURE

see also Atheism; Christianity and religious humanism; Secularism

Abelard, Miles R
Physicians of no value : the repressed story of ecclesiastical flummery / by Miles R. Abelard. — Winter Park, Fla. : Reality Publications, c1979.
xiii, 146 p. : ill. ; 24 cm.
Includes bibliographical references and index.
ISBN 0-932726-00-3
1. Christianity—Controversial literature. I. Title.
BL2775.2.A23 270 78-63304
79 MARC

Chao, Fang-jen.
(Tsung chiao hsin yang yü wen hua fu hsing)
宗教信仰與文化復興 / 著作者趙芳仁. — [台北] : 趙芳仁, 民國65- [1976-
v. ; 19 cm.
Colophon title.
NT$40.00 (v. 1)
1. Christianity—Controversial literature. 2. Religious thought—Taiwan. I. Title.
BL2780.C49 78-841315

Collins, Anthony, 1676-1729.
A discourse of the grounds and reasons of the Christian religion / Anthony Collins. — New York : Garland Pub., 1976.
xlii [i.e. lxii], 284 p. ; 19 cm. — (British philosophers and theologians of the 17th & 18th centuries)
Reprint of the 1724 ed.
Includes bibliographical references.
ISBN 0-8240-1766-8
1. Christianity—Controversial literature. 2. Bible. O.T.—Quotations in the New Testament. 3. Messiah—Prophecies. 4. Bible. O.T.—Prophecies. 5. Whiston, William, 1667-1752. An essay towards restoring the true text of the Old Testament. 6. Liberty. I. Title. II. Series.
BL2773.C63 1976 230 75-11212
75[r79]rev MARC

De Carli, Franco.
Lettere a Teofilo / Franco De Carli. — Torino : MEB, c1976.
231 p. ; 21 cm. It76-Oct
L3000
1. Christianity—Controversial literature. 2. Stoicism. I. Title.
BL2776.D42 79-380765
*79 MARC

Fabro, Cornelio.
Ludwig Feuerbach, la esencia del cristianismo / Cornelio Fabro. — Madrid : E.M.E.S.A., c1977.
244 p. ; 19 cm. — (Colección Crítica filosófica ; 12) Sp***
Translation of L. Feuerbach, l'essenza del cristianesimo.
Bibliography: p. [233]-238.
Includes index.
ISBN 8426553117
1. Feuerbach, Ludwig Andreas, 1804-1872. Das Wesen des Christentums. 2. Religion—Philosophy. 3. Christianity—Controversial literature.
B2971.W43F3218 78-384141
78 MARC

Feuerbach, Ludwig Andreas, 1804-1872.
A kereszténység lényege / Feuerbach ; fordította Timár Ilona. — 2. változatlan kiad. — Budapest : Akadémiai Kiadó, 1978.
518 p. ; 21 cm. — (Filozófiai írók tára : Új folyam ; 21. köt) H***
Translation of Das Wesen des Christentums.
Bibliography: p. [511]-518.
Includes indexes.
ISBN 9630514486 : 70.00Ft
1. Religion—Philosophy. 2. Christianity—Controversial literature. I. Title. II. Series.
B2971.W42H8 1978 78-391144
79 MARC

Fry, John R
The great apostolic blunder machine : a contemporary attack upon Christendom / John R. Fry. — 1st ed. — San Francisco : Harper & Row, c1978.
183 p. ; 21 cm.
Includes bibliographical references.
ISBN 0-06-063072-8
1. Christianity—20th century. 2. Christianity—Controversial literature. I. Title.
BR121.2.F69 1978 262'.009'04 78-3137
78 MARC

CHRISTIANITY
—CONTROVERSIAL LITERATURE (Continued)

Goldenberg, Naomi R
Changing of the gods : feminism and the end of traditional religions / Naomi R. Goldenberg. — Boston : Beacon Press, c1979.
viii, 152 p. ; 21 cm.
Includes bibliographical references and index.
ISBN 0-8070-1110-X : $9.95
1. Women and religion. 2. Religions (Proposed, universal, etc.) 3. Psychology, Religious. 4. Christianity—Controversial literature. 5. Judaism—Controversial literature. 6. Feminism—Moral and religious aspects. I. Title.
BL458.G64 1979 291.1'7834'12 78-19602
78 MARC

Graham, Lloyd.
Deceptions and myths of the Bible / by Lloyd Graham. — New York : Bell Pub. co., 1979, c1975.
vii, 484 p. : ill. ; 21 cm.
Reprint of the ed. published by University Books, Secaucus, N.J.
Includes index.
ISBN 0-517-27834-0
1. Bible—Criticism, interpretation, etc. 2. Christianity—Controversial literature. I. Title.
[BS533.G68 1979] 220.6 79-103
79 MARC

Ibarreta, Rogelio H de.
La religión al alcance de todos / Rogelio H. de Ibarr eta. — Barcelona : Ediciones Petronio, c1978.
391 p. ; 20 cm. Sp***
ISBN 8472505782
1. Christianity—Controversial literature. I. Title.
BL2747.I22 78-380400
78 MARC

Ibarreta, Rogelio H de.
La religión al alcance de todos / R. H. de Ibarreta. — 1. ed. — Madrid : Ediciones Júcar, 1978.
224 p. ; 18 cm. — (Religión) (Biblioteca Júcar de ciencias humanas ; 56) Sp***
ISBN 8433410563
1. Christianity—Controversial literature. I. Title. II. Series: Biblioteca Júcar de ciencias humanas ; 56.
BL2775.2.I23 79-357897
79 MARC

Islam : the first & final religion. — Karachi : Begum Aisha Bawani waqf, [1977?]
xxv, 238 p. ; 18 cm.
1. Islam—Apologetic works. 2. Christianity—Controversal literature.
BP172.I817 297 77-939225
79 MARC

(Ispoved' pered lûd'mi)
Исповедь перед людьми : [Сборник / Сост. А. А. Круглов, Д. М. Матяс]. — Минск : Беларусь, 1978.
159 p. ; 17 cm. USSR 78
0.25rub
1. Atheists—White Russia—Biography. 2. Christianity—Controversial literature. I. Kruglov, Anatoliĭ Agopeevich. II. Matsfâs, Dzmitryĭ Matsveevich.
BL2785.I 86 78-423084

(Kuo nei chin shih nien lai chih tsung chiao ssu ch'ao)
國内近十年來之宗教思潮 / [選輯者張欽士]. 一 台北：天一出版社, [1977?]
6, 6, 476 p. ; 20 cm.
Colophon title.
Reprint of the 1927 ed. published by Yen ching Hua wen hsüeh hsiao, Peking, with added t. p.: Religious thought movements in China during the last decade, a source book.
NT$520.00 ($13.00 U. S.)
1. Christianity—Controversial literature. 2. Christianity—China. 3. Missions—China—Controversial literature. I. Chang, Ch'in-shih.
BL2775.K86 1977 79-841159

Luis de Granada, 1504-1588.
[Guia de pecadores. Japanese]
ぎやどぺかどる / [Luis de Granada 著]. 妙貞問答 ; 破提宇子 / [ハビアン著]. 顯僞錄 / [澤野忠庵著 ; 與謝野寛, 正宗敦夫, 與謝野晶子編纂校訂]. 一 東京：現代思潮社, 1978.
[44], 186, 134, 73, 30, 25 p. ; 16 cm. — (覆刻日本古典全集)
Ja 79-825
Spine title.
Photoreprint of the 1927 ed. published in 2 v. by Nihon Koten Zenshū Kankōkai, Tokyo.
1. Christian life—Catholic authors. 2. Christianity and other religions. 3. Christianity—Controversial literature. I. Fabian, Fucan, fl. 1583-1607. Myōtei mondō. 1978. II. Fabian, Fucan, fl. 1583-1607. Ha Deusu. 1978. III. Ferreira, Christovão, 1580-ca. 1652. Kengiroku. 1978. IV. Title. V. Series: Nihon koten zenshū.
Title romanized: Giya do pekadoru.
BX2349.L716 1978 79-805675

Malley, William J
Hellenism and Christianity : the conflict between Hellenic and Christian wisdom in the Contra Galilaeos of Julian the Apostate and the Contra Julianum of St. Cyril of Alexandria / William J. Malley. — Roma : Università gregoriana, 1978.
466 p. ; 24 cm. — (Analecta Gregoriana ; v. 210 : Series Facultatis Theologicae, sectio B ; n. 68) It79-May
English, with notes in Greek.
Bibliography: p. [424]-430.
Includes indexes.
1. Julianus, Apostata, Emperor of Rome, 331-363. Kata Galilaiōn. 2. Christianity—Controversial literature. 3. Cyrillus, Saint, Patriarch of Alexandria, 370 (ca.)-444. Contra Julianum. 4. Apologetics—Early church, ca 30-600. I. Title. II. Series: Analecta Gregoriana ; v. 210.
BR160.3.J843M34 239'.4 79-311915
*79 MARC

Marti, Kurt.
Der Mensch ist nicht für das Christentum da : e. Streitgespräch über Gott u.d. Welt zwischen e. Christen u.e. Agnostiker / Kurt Marti, Robert Mächler. — Hamburg : Lutherisches Verlagshaus, 1977.
119 p. ; 19 cm. GFR78-A
Includes bibliographical references.
ISBN (invalid) 3-7859-0425-0 : DM9.80
1. Christianity—Controversial literature. 2. Apologetics—20th century. 3. Agnosticism—Controversial literature. I. Mächler, Robert, 1909- joint author. II. Title.
BL2778.M34 79-371503
79 MARC

Maryam Jameela, 1934-
Islam versus Ahl al-Kitab : past and present / Maryam Jameelah. — 2d rev. ed. — Lahore : Mohammad Yusuf Khan, 1978.
xxvii, 417 p., [1] leaf of plates : port. ; 22 cm.
"A refutation of Judaism and Christianity to combat the menace of Zionism and Christian missionary activity in Muslim lands."
Bibliography: p. [413]-417.
Rs40.00
1. Islam—Apologetic works. 2. Judaism—Controversial literature. 3. Christianity—Controversial literature. I. Title.
BP170.M32 1978 297'.292 78-931208
79 MARC

Netti, A G Hadzarmawit.
Kristen dalam sastra Indonesia / A. G. Hadzarmawit Netti. — Jakarta : BPK Gunung Mulia, 1977.
92 p. ; 18 cm.
1. Indonesian literature—History and criticism. 2. Christianity—Controversial literature. I. Title.
PL5080.5.N47 78-941830
79 MARC

Nistor ha-komer.
ספר נסתור הכומר / נעתק מכ"י אשר בוואטיקאנא ברומא ; ויצא עתה ראשונה לאור עם פתח דבר בל"א ע"י אברהם בהרב צבי הירש ברלינער. — תל-אביב : [י. רובינזון מוכרי ספרים, 1976 or 1977] 737.
15 p. ; 21 cm.
"בת 100 עותקים ... עותק זה מספרו 41."
Reprint of the 635 (1874 or 1875) ed. printed by Gebrüder Bonn, Altona.
1. Christianity—Controversial literature. 2. Judaism—Apologetic works. I. Berliner, Abraham, 1833-1915. II. Title.
Title romanized : Sefer Nistor ha-komer.
BM590.N57 1976 78-952014

Oak, Purushottam Nagesh, 1917-
Christianity is Chrisn-nity / by P. N. Oak. — 1st ed. — New Delhi : Institute for Rewriting Indian History, 1978.
x, 216 p., [3] leaves of plates : ill. ; 23 cm.
Corrections in ms.
Bibliography: p. [208]-210.
Includes index.
Rs15.00 ($4.00 U.S.)
1. Christianity and other religions—Hinduism. 2. Hinduism—Relations—Christianity. 3. Christianity—Controversial literature. 4. Jesus Christ—Historicity. 5. Krishna. 6. Jesus Christ—Hindu interpretations. I. Title.
BR128.H5O22 209 78-908025
79 MARC

Religiîa v vek nauchno-tekhnicheskoĭ revolîutsii : [Sb. stateĭ] / E. M. Babosov ... [et al. ; Pod red. V. I. Garadzhi (SSSR) i O. Klor (GDR)]. — Moskva : Politizdat, 1979.
85 p. ; 20 cm. USSR79
Includes bibliographical references.
0.50rub
1. Religion and science—1946- 2. Christianity—Controversial literature. I. Babosov, Evgeniĭ Mikhaĭlovich. II. Garadzha, Viktor Ivanovich. III. Klohr, Olof.
BL240.2.R42 79-369007
79 MARC

(Soprotivlenie religioznym presledovaniîam)
Сопротивление религиозным преследованиям. — Frankfurt/Main : Посев, 1976.
111 p. ; 17 cm. — (Вольное слово ; вып. 24) GFR***
Includes bibliographical references.
1. Russia—Religion—1917- 2. Orthodox Eastern Church, Russian. 3. Christianity—Controversial literature. 4. World Council of Churches, 5th Assembly, Nairobi, 1975. I. Series: Vol'noe slovo ; vyp. 24.
JC591.V64 vyp. 24 79-412872
[BL980]

Turmeda, Anselmo, 1352-1432?
Autobiografia i atac als partidaris de la creu / per Anselm Turmeda ; introducció de Mikel de Epalza ; i traducció de Mikel de Epalza i Ignasi Riera. — Barcelona : Curial, 1978.
175 p., [1] leaf of plates ; 18 cm. — (Clàssics Curial ; 11) Sp***
Bibliography: p. 95-98.
ISBN 8472561534
1. Islam—Apologetic works. 2. Christianity—Controversial literature. I. Epalza, Miguel de. II. Title.
BP172.T8712 79-359366
79 MARC

—CONTROVERSIAL LITERATURE—ADDRESSES, ESSAYS, LECTURES

Wollschläger, Hans, 1935-
Die Gegenwart einer Illusion : Reden gegen ein Monstrum / Hans Wollschläger. — Zürich : Diogenes-Verlag, cop. 1978.
225 p. ; 18 cm. — (Diogenes-Taschenbuch ; 113) Sw79-2693
Includes bibliographical references.
ISBN 3-257-20576-7 : 7.80F
1. Christianity—Controversial literature—Addresses, essays, lectures. I. Title.
BL2775.2.W64 79-381675
79 MARC

—CONTROVERSIAL LITERATURE—COLLECTED WORKS

Christentum im Spiegel der Weltreligionen : krit. Texte u. Kommentare / hrsg. von Heinz-Jürgen Loth, Michael Mildenberger, Udo Tworuschka. — 1. Aufl. — Stuttgart : Quell-Verlag, 1978.
374 p. ; 20 cm. — (Studienbücher im Gespräch mit der Zeit : Religion) GFR79-A
Includes bibliographical references and indexes.
ISBN 3-7918-6004-6 : DM18.00
1. Christianity and other religions—Collected works. 2. Christianity—Controversial literature—Collected works. I. Loth, Heinz-Jürgen, 1942- II. Mildenberger, Michael, 1934- . III. Tworuschka, Udo.
BR127.C45 79-351008
79 MARC

—DICTIONARIES

Douglas, James Dixon.
The new international dictionary of the Christian church / general editor J. D. Douglas, consulting editor Earle E. Cairns, assistant editor James E. Ruark. — 2d ed. — Exeter [Eng.] : Paternoster Press, 1978.
xiv, 1074 p. ; 25 cm. GB***
Companion volume: The new international dictionary of new testament theology.
ISBN 0-85364-221-4 : £17.00
1. Christianity—Dictionaries. I. Title.
BR95.D68 1978 203 78-320825
79 MARC

Pieterse, Liberius, 1905-1973.
English-Urdu dictionary of Christian terminology / by Liberius Pieterse ; edited by Jan Slomp. — Rawalpindi : Christian Study Centre, 1976.
xxviii, 108 p. ; 25 cm.
Bibliography: p. xxv-xxviii.
1. Christianity—Dictionaries. 2. English language—Dictionaries—Urdu. I. Slomp, Jan, 1932- . II. Title.
BR95.P5 203 77-930423
78 MARC

—ESSENCE, GENIUS, NATURE

see also Demythologization

Corlett, William.
The Christ story / [by] William Corlett & John Moore. — London : Hamilton, 1978.
[5], 154 p. : ill. ; 23 cm. — (Questions) GB78-19838
ISBN 0-241-89771-8 : £3.95
1. Christianity—Essence, genius, nature. I. Moore, John, 1932- joint author. II. Title. III. Series: Questions (London)
BR121.2.C635 1978 201 78-321826
79 MARC

Donovan, Vincent J 1926-
Christianity rediscovered : an epistle from the Masai / Vincent J. Donovan. — Notre Dame, Ind. : Fides/Claretian, c1978.
200 p. ; 20 cm.
ISBN 0-8190-0627-0
1. Christianity—Essence, genius, nature. 2. Missions. 3. Missions to Masai. I. Title.
BT60.D63 230 78-11711
78 MARC

CHRISTIANITY
—ESSENCE, GENIUS, NATURE
(Continued)

Kim, Chun-sik.
(Kidokkyo e taehan kwahakchŏk sasaek)
基督教에對한科學的思索 / 金俊植著. — 서울 : 東明社, 1978.
137 p. ; 18 cm.
₩500

1. Christianity—Essence, genius, nature. I. Title.
BR121.2.K497 79-826274

Kim, Young Pil, 1923-
(Yŏngwŏn kwa ŭi mannam)
永遠과의만남 : 基督敎의理解 / 金永弼著. — 서울 : 現代思想社, 1977.
194 p. ; 19 cm.
On verso of t. p.: On the way to eternity.
₩800

1. Christianity—Essence, genius, nature. I. Title. II. Title: On the way to eternity.
BR121.2.K515 78-826624

Lafon, Guy.
Esquisses pour un christianisme / Guy Lafon. — Paris : Éditions du Cerf, 1979.
228 p. ; 22 cm. — (Cogitatio fidei ; 96) (Théologie et sciences religieuses) F***
Includes bibliographical references.
ISBN 2-204-01267-X : 72.00F
1. Christianity—Essence, genius, nature. I. Title.
BR121.2.L25 201 79-362192
79 MARC

Vilquin, Michel.
De Gaulle, cet inconnu méconnu / M. Vilquin. — Paris : la Pensée universelle, 1972.
119 p. ; 18 cm. F73-2054
15.00F
1. Christianity—Essence, genius, nature. 2. Gaulle, Charles de, Pres. France, 1890-1970—Religion. I. Title.
BR124.V54 230 78-378945
78 MARC

Waterman, Leroy, 1875-
The religion of Jesus : Christianity's unclaimed heritage of prophetic religion / by Leroy Waterman. — Westport, Conn. : Greenwood Press, 1978, c1952.
251 p. ; 23 cm.
Reprint of the 1st ed. published by Harper, New York.
Includes bibliographical references and index.
ISBN 0-313-20586-8
1. Christianity—Essence, genius, nature. 2. Jesus Christ—Teachings. 3. Jesus Christ—History of doctrines—Early church, ca. 30-600. I. Title.
[BR121.W265 1978] 230 78-16405
78 MARC

—EVIDENCES see Apologetics

—HISTORY see Church history

—INFLUENCE—INDIA

Khonglah, J
Ki khanatang jong ka balang Kristan / [J. Khonglah]. — [Shillong : Don Bosco Press], 1971.
175 p. ; 20 cm.
In Khasi.
Rs2.30
1. Church history. 2. Christianity—Influence—India. I. Title.
BR145.2.K48 78-903869
79 MARC

—INFLUENCE—INDIA—SOUTH INDIA—ADDRESSES, ESSAYS, LECTURES

Hirudayam, Ignatius, 1910-
Christianity and Tamil culture / by Ignatius Hirudayam. — [Madras] : Dr. S. Radhakrishnan Institute for Advanced Study in Philosophy, University of Madras, 1977.
vi, 85 p. ; 25 cm.
Includes quotations in Tamil.
"A golden jubilee publication."
Bibliography: p. [84]-85.
Rs8.00
1. Christianity—Influence—India—South India—Addresses, essays, lectures. 2. Tamils—Religion—Addresses, essays, lectures. I. Title.
BR1170.S68H57 261.2 78-905965
79 MARC

—INFLUENCE—SIBERIA—ADDRESSES, ESSAYS, LECTURES

Khristianstvo i lamaizm u korennogo naseleniia Sibiri : Vtoraia polovina XIX—nach. XX v. : [Sb. stateĭ] / AN SSSR, In-t ėtnografii im. N. N. Miklukho-Maklaia ; [Otv. red. i avt. predisl. I. S. Vdovin]. — Leningrad : Nauka, Leningr. otd-nie, 1979.
227 p. : ill. ; 22 cm. USSR79
Includes bibliographical references.
1.90rub
1. Siberia—Religion—Addresses, essays, lectures. 2. Siberia—Native races—Addresses, essays, lectures. 3. Christianity—Influence—Siberia—Addresses, essays, lectures. 4. Lamaisim—Influence—Siberia—Addresses, essays, lectures. I. Vdovin, I. S. II. Akademiia nauk SSSR. Institut ėtnografii.
BL2300.K47 79-382096
79 MARC

—MISCELLANEA

Cremona, Carlo.
La speranza che è in noi : domande e risposte sul messaggio cristiano / Carlo Cremona ; prefazione di Arturo Carlo Jemolo. — 1. ed. — Milano : Rusconi, 1978.
274 p. ; 22 cm. — (Dimensione religiosa) It78-Nov
Includes index.
L7500
1. Christianity—Miscellanea. I. Title.
BR96.C74 78-400305
*79 MARC

—ORIGIN

see also Dead Sea scrolls

Hatano, Seiichi, 1877-1959.
(Kirisutokyō no kigen)
基督教の起源 / 波多野精一著. — 東京 : 創元社, 昭和26 [1951]
198 p. : port. ; 15 cm. — (創元文庫 ; D6)
Series romanized: Sōgen bunko.
Reprint of the 1941 ed.; with explanatory text by S. Yamatani.

1. Christianity—Origin. 2. Theology—Early church. I. Title.
BR129.H29 1951 79-801286

Newsom, Carroll Vincent, 1904-
The roots of Christianity / Carroll V. Newsom. — Englewood Cliffs, N.J. : Prentice-Hall, c1979.
ix, 263 p. ; 24 cm.
Includes index.
ISBN 0-13-783050-5
1. Christianity—Origin. I. Title.
BR129.N44 209 79-9246
79 MARC

—ORIGIN—ADDRESSES, ESSAYS, LECTURES

Flusser, David Gustav.
(Yahadut u-meḳorot ha-Natsrut)
יהדות ומקורות הנצרות : מחקרים ומסות / דוד פלוסר. —
[תל־אביב] : ספרית פועלים, [1979]
486 p. ; 22 cm. — (ספרי דעת־זמננו)
Series romanized: Sifre da'at-zemanenu.
On verso of t. p.: Jewish sources in early Christianity.
Includes bibliographical references and indexes.
1. Bible. N. T.—Criticism, interpretation, etc.—Addresses, essays, lectures. 2. Christianity—Origin—Addresses, essays, lectures. 3. Jesus Christ—Parables—Addresses, essays, lectures. 4. Judaism—Relations—Christianity—Addresses, essays, lectures. 5. Christianity and other religions—Judaism—Addresses, essays, lectures. I. Title.
BS2368.H4F58 79-951639

—PHILOSOPHY

see also Philosophical theology

Beck, Heinrich, 1929 (Apr. 27)-
Anthropologischer Zugang zum Glauben : eine rationale Meditation / Heinrich Beck. — Salzburg ; München : Pustet, c1979.
86 p. ; 21 cm. Au79-10-18
Includes bibliographical references.
ISBN 3-7025-0164-9 : S98.00
1. Christianity—Philosophy. I. Title.
BR100.B38 79-375411
*79 MARC

Boisramé, Maurice.
Propriétaires de la vérité? / Maurice Boisramé, ... René Breton, ... Jean Lemière, ... — Angers (7, rue du Parvis Saint-Maurice, 49000) : C.R.E.R. [Coopérative régionale de l'enseignement religieux], [1975]
55 p. ; 21 cm. — (Foi à l'épreuve ; no 2) F76-8509
12.00F
1. Christianity—Philosophy. 2. Truth. 3. Values. I. Breton, René, joint author. II. Lemière, Jean, joint author. III. Title.
BR100.B538 201'.1 78-388822
78 MARC

Dumitriu, Petru.
Au dieu inconnu / Petru Dumitriu. — Paris : Seuil, c1979.
218 p. ; 21 cm. F***
ISBN 2-02-005210-5 : 48.00F
1. Dumitriu, Petru. 2. Orthodox Eastern Church, Romanian—Biography. 3. Christianity—Philosophy. I. Title.
BX719.D85A32 281.9'3 79-379587
79 MARC

Ebert, Klaus, 1945-
Identität und Metapher : Überlegungen zu e. wissenschaftstheoret. Begründung von Theologie / Klaus Ebert. — Köln ; Wien : Böhlau [in Komm.], 1978.
216 p. ; 20 cm. — (Böhlau Philosophica ; 3) GFR78-A
Includes bibliographical references.
ISBN 3-412-03278-6
1. Identification (Religion) 2. Christianity—Philosophy. I. Title.
BV4509.5.E2 78-396277
79 MARC

Hamer, Colin, 1934-
Voice in the darkness : (an essay in contemporary Catholic existentialism) / [by] Colin Hamer. — Zennor : United Writers Publications, 1978.
143 p. ; 22 cm. GB78-24852
Includes bibliographical references.
ISBN 0-901976-45-8 : £4.95
1. Catholic Church—Apologetic works. 2. Christianity—Philosophy. 3. Hamer, Colin, 1934- . I. Title.
BX1752.H25 230'.2 78-326233
79 MARC

Hazelton, Roger, 1909-
New accents in contemporary theology / by Roger Hazelton. — Westport, Conn. : Greenwood Press, 1979, c1960.
144 p. ; 22 cm.
Reprint of the ed. published by Harper, New York.
Bibliography: p. 141-142.
Includes index.
ISBN 0-313-21181-7
1. Christianity—Philosophy. I. Title.
[BT40.H35 1979] 230 78-12237
78 MARC

Klein, Alessandro.
Antirazionalismo di Kierkegaard / Alessandro Klein. — Milano : Mursia, [1979]
187 p. ; 23 cm. — (Studi di filosofia ; 17) It79-Feb
Bibliography: p. 177-178.
Includes index.
L10000
1. Kierkegaard, Søren Aabye, 1813-1855—Ontology. 2. Ontology. 3. Christianity—Philosophy. I. Title.
B4378.O5K58 79-350837
*79 MARC

Masih, Y
Christian faith and philosophy / Y. Masih. — 1st ed. — Calcutta : Firma KLM, 1978.
vi, 178 p. ; 23 cm.
Includes bibliographical references and index.
Rs28.00
1. Christianity—Philosophy. I. Title.
BR100.M294 201 78-911845
79 MARC

Mendonça, Eduardo Prado de.
O socratismo cristão e as origens da metafísica moderna / Eduardo Prado de Mendonça. — São Paulo : Editora Convívio, 1975.
249 p. ; 22 cm.
1. Metaphysics—History. 2. Socrates—Influence. 3. Christianity—Philosophy. 4. Descartes, René, 1596-1650. I. Title.
BD115.M38 79-370300
79 MARC

Nilson, Jon.
Hegel's phenomenology and Lonergan's insight : a comparison of two ways to Christianity / Jon Nilson. — Meisenheim am Glan : Hain, 1979.
124 p. ; 23 cm. — (Monographien zur philosophischen Forschung ; Bd. 163) GFR79-A
Originally presented as the author's thesis.
Bibliography: p. [122]-124.
ISBN 3-445-01745-X : DM36.00
1. Christianity—Philosophy. 2. Philosophical theology. 3. Hegel, Georg Wilhelm Friedrich, 1770-1831. Phaenomenologie des Geistes. 4. Lonergan, Bernard J. F. Insight. 5. Knowldge, Theory of. I. Title. II. Series.
BR100.N56 1979 201 79-315830
79 MARC

CHRISTIANITY
—PHILOSOPHY (Continued)

Sabetti, Alfredo.
John Toland, un irregolare della società e della cultura inglese tra Seicento e Settecento / Alfredo Sabetti ; in appendice John Toland, Il cristianesimo senza misteri. — 1. ed. — Napoli : Liguori, 1976.
393 p. ; 24 cm. It78-Aug
Includes bibliographical references and index.
L6000
1. Toland, John, 1670-1722. 2. Christianity—Philosophy. 3. England—Intellectual life—17th century. 4. England—Intellectual life—18th century. I. Toland, John, 1670-1722. Christianity not mysterious. 1976. II. Title.
B1393.Z7S2 79-347521
*79 MARC

Sanseverino, Gaetano, 1811-1865.
Un'opera inedita di Gaetano Sanseverino : introduzione e testo / Pasquale Orlando. — Città del Vaticano : Pontificia Accademia teologica romana : Libreria editrice vaticana, 1974.
270 p. ; 24 cm. — (Biblioteca per la storia del tomismo ; 7) VC***
Introductory material and notes in Italian, text in Latin.
Includes bibliographical references and index.
1. Neo-Scholasticism. 2. Philosophy. 3. Christianity—Philosophy. I. Orlando, Pasquale. II. Title.
B839.S26 1974 79-340942
79 MARC

Sousa, João Machado de.
Existe uma filosofia cristã / João Machado de Souza. — Recife : ₍s.n.₎, 1978.
162 p. ; 20 cm.
Bibliography: p. 159-162.
1. Christianity—Philosophy. I. Title.
BR100.S63 78-380102
78 MARC

Trethowan, Illtyd, 1907-
An essay in Christian philosophy. London, New York, Longmans, Green ₍1954₎
186 p. 22 cm.
1. Christianity—Philosophy. I. Title.
BR100.T688 230 54-4910
₍r79₎rev MARC

—PHILOSOPHY—ADDRESSES ESSAYS, LECTURES

La Confession de la foi chrétienne / Philippe d'Harcourt, Michel Sales, Gaston Fessard, Philippe Nemo ... ₍etc.₎ ; textes présentés par Claude Bruaire .... — Paris : Fayard, 1977.
341 p. ; 22 cm. — (Communio ISSN 0339-8471) F77-15997
Includes bibliographical references.
ISBN 2-213-00530-3 : 49.00F
1. Theology—Addresses essays, lectures. 2. Christianity—Philosophy—Addresses essays, lectures. I. Bruaire, Claude, 1932- . II. Harcourt, Philippe d', 1929- . III. Series: Communio (Paris)
BR50.C595 230'.2 78-383038
78 MARC

Severino, Emanuele.
Gli abitatori del tempo : cristianesimo, marxismo, tecnica / Emanuele Severino. — Roma : A. Armando, 1978.
183 p. ; 22 cm. — (Filosofia e problemi d'oggi ; 69) It78-Nov
Bibliography: p. 178.
Includes index.
L5500
1. Dialectical materialism—Addresses, essays, lectures. 2. Alienation (Philosophy)—Addresses, essays, lectures. 3. Christianity—Philosophy—Addresses, essays, lectures. 4. Nihilism (Philosophy)—Addresses, essays, lectures. I. Title.
B809.8.S418 79-341901
*79 MARC

Teilhard de Chardin, Pierre.
The heart of matter / Pierre Teilhard de Chardin ; translated by René Hague. — 1st American ed. — New York : Harcourt Brace Jovanovich, c1979.
276 p. ; 21 cm.
Translation of Le cœur de la matière.
"A Helen and Kurt Wolff book."
"Chronological list of works": p. 241-251.
Bibliography: p. 253-261.
Includes index.
1. Matter—Addresses, essays, lectures. 2. Cosmology—Addresses, essays, lectures. 3. Christianity—Philosophy—Addresses, essays, lectures. 4. Paleontology—Addresses, essays, lectures. I. Title.
B2430.T373C5813 113 78-21707
78 MARC

—PHILOSOPHY—CONGRESSES

Hacia un nuevo humanismo : ₍primer seminario interdisciplinar, 9-16 de setiembre de 1973, Embalse de Río Tercero, Córdoba, Argentina₎ / Estanislao Karlic ... ₍et al.₎. — 1. ed. — Buenos Aires : Editorial BONUM, c1974.
300 p. : ill. ; 24 cm. — (Enfoques latinoamericanos ; no. 11 : Sección Filosofía y teología : Serie Diálogo interdisciplinar)
Includes bibliographical references.
1. Humanism—Congresses. 2. Liberation theology—Congresses. 3. Latin America—Intellectual life—Congresses. 4. Christianity—Philosophy—Congresses. I. Karlic, Estanislao.
B821.H23 76-464053
76₍r79₎rev MARC

—POETRY see Christian poetry

—PSYCHOLOGY

Hétu, Jean Luc.
Quelle foi? : Une rencontre entre l'Évangile et la psychologie / Jean-Luc Hétu. — ₍Montréal₎ : Leméac, ₍c1978₎
310 p. : ill. ; 20 cm. — (Collection Quelle?) C***
Includes bibliographical references.
ISBN 0-7761-0202-8 : $13.25
1. Christianity—Psychology. I. Title.
BL53.H44 201'.9 79-369383
79 MARC

Myers, David G
The human puzzle : psychological research and Christian belief / David G. Myers. — 1st ed. — San Francisco : Harper & Row, c1978.
xxi, 278 p. : ill. ; 20 cm. — (Christian perspectives on counseling and the behavioral sciences)
"A Harper/CAPS book."
Includes bibliographical references and index.
ISBN 0-06-065558-5
1. Christianity—Psychology. I. Title. II. Series.
BR100.M93 1978 201'.9 77-15873
78 MARC

Pfister, Oskar Robert, 1873-1956.
Das Christentum und die Angst / Oskar Pfister ; mit einem Vorw. von Thomas Bonhoeffer. — 2. Aufl. — Olten ; Freiburg i.Br. : Walter, 1975.
xlii, 530 p. ; 23 cm. Sw75-A-3697
Reprint of the 1944 ed. which was published by Artemis Verlag, Zürich.
Bibliography of works about O. R. Pfister: p. xxvii.
Bibliography: p. ₍523₎-530.
Includes indexes.
ISBN 3-530-64010-7 : 63.00F
1. Fear. 2. Christianity—Psychology. I. Bonhoeffer, Thomas, 1931- . II. Title.
BR110.P45 1975 79-374495
79 MARC

Sall, Millard J
The emotions of God's people : helping those who are hurting / Millard J. Sall. — Grand Rapids : Zondervan Pub. House, c1978.
155 p. ; 21 cm.
ISBN 0-310-32441-6
1. Bible—Psychology. 2. Christianity—Psychology. 3. Psychology, Religious. 4. Psychology—Popular works. 5. Christian life—1960- . I. Title.
BS645.S24 155.2 78-15717
78 MARC

Shostrom, Everett L 1921-
Healing love : how God works within the personality / Everett L. Shostrom & Dan Montogomery. — Nashville : Abingdon, c1978.
141 p. : ill. ; 23 cm.
Bibliography: p. 135-136.
Includes index.
ISBN 0-687-16739-6
1. Christian life—1960- . 2. Love. 3. Christianity—Psychology. I. Montgomery, Dan, 1946- joint author. II. Title.
BV4501.2.S452 248'.4 78-8294
78 MARC

Solignac, Pierre.
La névrose chrétienne / Pierre Solignac. — Paris : Éditions de Trévise, c1976.
252 p. ; 22 cm. — (Collection Polémique) F***
ISBN 2-7112-0266-6 : 35.00F
1. Christianity—Psychology. I. Title.
BR110.S66 201'.9 78-378235
78 MARC

Vergote, Antoine.
Dette et désir : deux axes chrétiens et la dérive pathologique / Antoine Vergote. — Paris : Seuil, c1978.
316 p. ; 21 cm. F***
Includes bibliographical references.
ISBN 2-02-005002-1 : 60.00F
1. Christianity—Psychology. 2. Guilt. 3. Desire. 4. Desire for God. I. Title.
BR110.V42 201'.9 79-349708
79 MARC

—RELATIONS see Christianity and other religions

—RENEWAL see Church renewal

—TERMINOLOGY

Sekine, Bunnosuke, 1912-
(Kirisutokyō yōgo jiten)
キリスト教用語辞典 / 関根文之助編. — 大阪 : 新元社, 1959.
116 p. : ill. ; 18 cm.
Includes index.
1. Christianity—Terminology. I. Title.
BR96.5.S45 79-801339

—UNION BETWEEN CHRUCHES see Christian union

—AFRICA

Bürki, Bruno.
La case des chrétiens : essai de théologie pratique sur le lieu de culte en Afrique / Bruno Bürki. — Yaoundé : Éditions Clé, 1973.
127 p. : ill. ; 21 cm. — (Collection théologique Clé)
Includes bibliographical references.
1. Worship. 2. Christianity—Africa. I. Title. II. Series.
BV10.2.B83 79-370218
79 MARC

Ékra, Vangha Mathieu, 1917-
L'échelle sans fin / V. Mathieu Ékra. — Paris : France-Impressions, c1977.
207 p. ; 18 cm. F***
1. Apologetics—20th century. 2. Christianity—Africa. I. Title.
BT1102.E38 230'.7 79-350533
79 MARC

—AFRICA—ADDRESSES, ESSAYS, LECTURES

African Christian spirituality / edited with an introd. by Aylward Shorter. — London : G. Chapman, c1978.
xii, 160 p. ; 22 cm. GB***
Includes bibliographical references.
ISBN 0-225-66238-8 : £2.95
1. Christianity—Africa—Addresses, essays, lectures. 2. Spirituality—Addresses, essays, lectures. I. Shorter, Aylward.
BR1360.A524 230 79-310855
79 MARC

Hastings, Adrian.
A history of African Christianity, 1950-1975 / Adrian Hastings. — Cambridge ; New York : Cambridge University Press, 1979.
xi, 336 p. ; 24 cm. — (African studies series ; ₍26₎)
Bibliography: p. 303-321.
Includes index.
ISBN 0-521-22212-5. ISBN 0-521-29397-9 pbk.
1. Christianity—Africa—Addresses, essays, lectures. I. Title. II. Series.
BR1360.H33 276 78-16599
78 MARC

—AFRICA—CONGRESSES

All Africa Conference of Churches.
Engagement : la deuxième assemblée de la CETA "Abidjan 69" = The second AACC assembly "Abidjan 69". — Nairobi : Conférence des églises de toute l'Afrique, ₍1970?₎
149 p., ₍2₎ leaves of plates : ill. ; 25 cm.
French and English.
1. All Africa Conference of Churches. 2. Christianity—Africa—Congresses. I. Title.
BR1359.A43 1970 262'.001 77-980800
79 MARC

All Africa Conference of Churches. Executive Committee.
Documents from the Executive Committee meeting of the All African ₍sic₎ Conference of Churches, Douala, 5/24-29/77. — ₍s.l. : s.n., 1977?₎
ca. 100 leaves in various pagings ; 34 cm.
English or French.
1. Christianity—Africa—Congresses.
BR1360.A55 1977 262'.001 78-110259
79 MARC

CHRISTIANITY
—AFRICA—CONGRESSES (Continued)

Christianity in independent Africa / edited by Edward Fasholé-Luke ... [et al.]. — London : R. Collings ; Bloomington, IN : Indiana University Press, c1978.
630 p. ; 24 cm.
Includes bibliographical references and index.
ISBN 0-253-37506-1
1. Christianity—Africa—Congresses. 2. Christianity—20th century—Congresses. I. Fasholé-Luke, Edward W.
BR1360.C48 209'.6 77-26351
78 MARC

—AFRICA, SOUTHERN—ADDRESSES, ESSAYS, LECTURES

Dachs, Anthony J
Christianity south of the Zambezi / edited by Anthony J. Dachs. — Gwelo, Rhodesia : Mambo Press, 1973-1977.
2 v. : ill. ; 22 cm.
Vol. 2 by M. F. C. Bourdillon.
Includes bibliographical references.
$2.90 (v. 1). $3.00 (v. 2)
1. Christianity—Africa, Southern—Addresses, essays, lectures. 2. Missions—Africa, Southern—Addresses, essays, lectures. 3. Cults—Africa, Southern—Addresses, essays, lectures. 4. Mashona—Religion—Addresses, essays, lectures. I. Bourdillon, M. F. C. II. Title.
BR1446.D3 209'.689'1 75-303544
75[7812r78]rev MARC

—AFRICA, SUB-SAHARAN—CONGRESSES

Symposium on the Church in Africa, Milligan College, Tenn., 1977.
The church in Africa, 1977 : papers presented at a symposium at Milligan College, March 31-April 3, 1977 / Charles R. Taber,
xii, 212 p. ; 22 cm.
Bibliography: p. 204-212.
ISBN 0-87808-161-5
1. Christianity—Africa, Sub-Saharan—Congresses. I. Taber, Charles Russell. II. Title.
BR1360.S95 1977 276.7 78-14923
78 MARC

—ASIA

Lee, Won Sul.
Beyond ideology : a Christian response to sociopolitical conflict in Asia / Won Sul Lee. — Westchester, Ill. : Cornerstone Books, c1979.
199 p. ; 21 cm.
Includes bibliographical references.
ISBN 0-89107-163-6 : $4.95
1. Christianity—Asia. 2. Asia—Civilization. I. Title.
BR1065.L43 261.8 78-71944
00 MARC

Ruokanen, Tapani.
Tehtävä Aasiassa : Raportti Kaukoidästä / Tapani Ruokanen. — Helsinki : Kirjaneliö, c1978.
355 p. : ill. ; 20 cm. Fi***
ISBN 9516004350
1. Missions—Asia. 2. Lutheran Church—Missions. 3. Missions, Finnish. 4. Christianity—Asia. 5. Asia—Description and travel—1951- . 6. Ruokanen, Tapani. I. Title.
BV3151.R86 79-359163
79 MARC

—BELGIUM—CUESMES

Buslin, Hector.
Histoire des religions pratiquées à Cuesmes à travers les siècles / par Hector Buslin et Raymond G. W. Mahieu. — [Cuesmes : R. G. W. Mahieu], 1977.
93 p. : ill. ; 25 cm. Be***
Bibliography: p. 91.
1. Christianity—Belgium—Cuesmes. 2. Cuesmes, Belgium—Church history. 3. Cuesmes, Belgium—Religion. I. Mahieu, Raymond G. W., joint author. II. Title.
BR828.C83B87 200'.9493'4 78-378458
78 MARC

—CARIBBEAN AREA—CONGRESSES

Moving into freedom : the second CCC assembly, Guyana, 1977 / K. Davis, ed. — Bridgetown, Barbados, W.I. : Cedar Press, c1977.
vi, 98 p. ; 21 cm. — (Working together with Christ series ; no. 2)
Includes bibliographical references.
1. Christianity—Caribbean area—Congresses. I. Davis, Kortright. II. Caribbean Conference of Churches. III. Series.
BR640.M68 277.29 78-324437
79 MARC

—CHINA

Digan, Parig.
The Christian China-watchers : a post-Mao perspective / Parig Digan. — Brussels : Pro Mundi Vita, 1978.
60 p. ; 21 cm. Be***
Bibliography: p. 56-60.
1. Christianity—China. 2. China—Politics and government—1976- I. Title.
BR1288.D53 275.1 79-317120
79 MARC

(Kuo nei chin shih nien lai chih tsung chiao ssu ch'ao)
國内近十年來之宗教思潮 / [選輯者張欽士]. — 台北：天一出版社, [1977?]
6, 6, 476 p. ; 20 cm.
Colophon title.
Reprint of the 1927 ed. published by Yen ching Hua wen hsüeh hsiao, Peking, with added t. p.: Religious thought movements in China during the last decade, a source book.
NT$520.00 ($13.00 U. S.)
1. Christianity—Controversial literature. 2. Christianity—China. 3. Missions—China—Controversial literature. I. Chang, Ch'in-shih.
BL2775.K86 1977 79-841159

Laurentin, René.
Chine et christianisme : après les occasions manquées / René Laurentin. — [Paris] : Desclée De Brouwer, 1977.
370 p. : ill. ; 22 cm. F78-2201
Bibliography: p. [323]-328.
Includes indexes.
ISBN 2-220-02110-6 : 56.00F
1. Christianity—China. 2. Christianity—Taiwan. I. Title.
BR1285.L38 275.1 78-378633
78 MARC

Mooneyham, Walter Stanley, 1926-
China, a new day / by W. Stanley Mooneyham. — Plainfield, N.J. : Logos International, c1979.
241 p. : ill. ; 18 cm.
ISBN 0-88270-364-1 : $2.50
1. Christianity—China. 2. China—Civilization—1949- . 3. China—Politics and government—1976- . I. Title.
BR1285.M55 275.1 79-83789
79 MARC

—CHINA—CONGRESSES

China and Christianity : historical and future encounters / James D. Whitehead, Yu-ming Shaw, N.J. Girardot, editors. — Notre Dame, Ind. : Center for Pastoral and Social Ministry, University of Notre Dame ; distributed by University of Notre Dame Press, c1979.
xv, 293 p. ; 23 cm.
Selected proceedings of a conference on "China, the religious dimensions," held Je. 29-July 2, 1977 at the University of Notre Dame.
Includes bibliographical references.
ISBN 0-268-00730-6 : $8.95
1. Christianity—China—Congresses. 2. China—Religion—Congresses. I. Whitehead, James D. II. Shaw, Yu-ming. III. Girardot, N. J. IV. Notre Dame, Ind. University. Center for Pastoral and Social Ministry. V. Title: China, the religious dimensions.
BR1285.C525 275.1 79-52165
79 MARC

—EL SALVADOR

El Salvador, un pueblo martirizado. — Madrid : Instituto de Estudios Políticos para América Latina y Africa, 1977.
72 p. ; 28 cm. — (América Latina ; 3) Sp***
At head of title: Idoc, España.
1. Christianity—El Salvador. 2. Church and state in El Salvador. 3. El Salvador—Politics and government. I. Idoc. II. Instituto de Estudios Políticos para América Latina y Africa. III. Title. IV. Series: América Latina (Madrid) ; 3.
BR625.E35E4 78-382250
79 MARC

—ESTONIA—COLLECTED WORKS

Vööbus, Arthur.
Studies in the history of the Estonian people : with reference to aspects of social conditions, in particular, the religious and spiritual life and the educational pursuit / by Arthur Vööbus. — Stockholm : ETSE, 1974-
v. ; 26 cm. — (Papers of the Estonian Theological Society in Exile ; 26) S***
Includes bibliographical references and indexes.
1. Estonia—Civilization—Collected works. 2. Christianity—Estonia—Collected works. 3. Christianity and culture—Collected works. I. Title. II. Series: Eesti Usuteadlaste Selts Paguluses. Toimetused ; 26.
DK511.E53V63 947'.41 79-303463
79 MARC

—ETHIOPIA—BIBLIOGRAPHY—CATALOGS

British Library. Dept. of Oriental Manuscripts and Printed Books.
Catalogue of Ethiopian manuscripts in the British Library acquired since the year 1877 / by S. Strelcyn. — London : British Museum Publications, c1978.
xvi, 183 p., [8] leaves of plates : ill. ; 25 cm. GB***
Includes bibliographical references and indexes.
ISBN 0-7141-0649-6 : £35.00
1. Manuscripts, Ethiopic—England—London—Catalogs. 2. British Library. Dept. of Oriental Manuscripts and Printed Books—Catalogs. 3. Christianity—Ethiopia—Bibliography—Catalogs. I. Strelcyn, Stefan. II. Title.
Z6621.B8373E82 016.091 79-300252
[BR1730] 79 MARC

—EUROPE

Clebsch, William A
Christianity in European history / William A. Clebsch. — New York : Oxford University Press, 1979.
xii, 315 p. ; 22 cm.
Includes bibliographical references and index.
ISBN 0-19-502471-0. ISBN 0-19-502472-9 pbk.
1. Christianity—Europe. 2. Europe—History. I. Title.
BR735.C57 209'.4 78-19033
78 MARC

Dunn, Richard S
The age of religious wars, 1559-1715 / Richard S. Dunn. — 2d ed. — New York : Norton, c1978.
ix, 322 p. : ill. ; 21 cm. — (The Norton history of modern Europe)
Includes bibliographical references and index.
ISBN 0-393-05694-5
1. Europe—Politics and government—1517-1648. 2. Europe—Politics and government—1648-1715. 3. Europe—Civilization. 4. Christianity—Europe. I. Title.
D234.D86 1978 940.2'2 78-10525
78 MARC

—EUROPE, EASTERN

Babris, Peter J
Silent churches : persecution of religions in the soviet-dominated areas / by Peter J. Babris. — Arlington Heights, Ill. : Research Publishers, 1978.
531 p. : maps (on lining papers) ; 25 cm.
Bibliography: p. 519-531.
ISBN 0-911252-02-9 : $19.50
1. Christianity—Europe, Eastern. 2. Church and state in Eastern Europe. 3. Persecution—Europe, Eastern. 4. Communism and Christianity. I. Title.
BR738.6.B3 274.7 78-52811
79 MARC

—FRANCE

Vallet, Odon.
Votre commune et l'Église : aspects pratiques, juridiques, financiers, immobiliers / Odon Vallet. — Paris : Éditions du Moniteur, 1978.
179 p. ; 24 cm. — (Collection Vie locale ISSN 0151-0185) F***
Includes bibliographical references.
ISBN 2-86282-029-6 : 68.00F
1. Parishes—France. 2. Christianity—France. 3. Sociology, Christian—France. I. Title.
BR843.V3 254 79-362955
79 MARC

—GERMANY, EAST

Zimmermann, Wolf Dieter, 1911-
Die Welt soll unser Himmel sein : atheistische Propaganda in der DDR / Wolf-Dieter Zimmermann ; [hrsg. von der Evangelischen Zentralstelle für Weltanschauungsfragen]. — 1. Aufl. — Stuttgart : Kreuz-Verlag, c1963.
74 p. ; 19 cm.
1. Atheism—Germany, East. 2. Christianity—Germany, East. 3. Propaganda, Communist—Germany, East. I. Evangelische Zentralstelle für Weltanschauungsfragen. II. Title.
BL2765.G35Z55 79-381010
79 MARC

—GREECE, MODERN

Kakourē, Katerina I
Dionysiaka; aspects of the popular Thracian religion of today, by Katerina J. Kakouri. Translated from the Greek by Helen Colaclides. [1st English ed.] Athens, G. C. Eleftheroudakis [1965]
xix, 174 p. illus., ports. 25 cm.
Translation of Dionysiaka.
Bibliography: p. [149]-159.
1. Christianity—Greece, Modern. 2. Anosténaria. 3. Greeks in Thrace. 4. Folk-lore—Greece, Modern—Thrace. I. Title.
BR136.G8K3413 78-305777
78 MARC

CHRISTIANITY (Continued)

—INDIA

Dubois, Jean Antoine, 1765-1848.
Letters on the state of Christianity in India, in which the conversion of the Hindoos is considered as impracticable, to which is added a vindication of the Hindoos, male and female, in answer to a severe attack made upon both by the Reverend *** / by J. A. Dubois ; edited by Sharda Paul. — New Delhi : Associated Pub. House, c1977.
vii, 122 p. ; 23 cm. — (Associated reprint ; 9)
Running title: The state of Christianity in India.
Reprint of the 1823 ed. published by Longman, London.
Includes index.
Rs35.00
1. Christianity—India. 2. Hinduism—Controversial literature. I. Paul, Sharda, 1942- , II. Title. III. Title: The state of Christianity in India.
BR1155.D8 1977 275.4 77-911368
79 MARC

Godwin, Clement J
Indian religious at the cross-roads / C. J. Godwin. — 1st ed. — Bombay : St. Paul Publications, c1975.
253 p. ; 19 cm.
On spine: Cross-roads.
Rs8.00
1. Christianity—India. 2. Secularism—India. I. Title. II. Title: Cross-roads.
BR1155.G6 209'.54 78-907787
79 MARC

Laloo, Donbok T 1942-
Ka thymmei pyrkhat u Khasi bad ki parom / Donbok T. Laloo. — [Shillong] : Nongiarap A. Laloo bad Marshal R. Susngi, [pref. 1975-
v. ; 20 cm.
In Khasi.
Rs2.00 (v. 1)
1. Christianity—India. 2. Khasis—Religion. I. Title.
BR1155.L34 78-901982
79 MARC

—INDIA—KERALA

Kattackal, J
(Kēraḷasabhā 'sarvē')
കേരളസഭാ 'സർവേ' / ജേക്കബ് കട്ടയ്ക്കൽ, തോമസ് വെള്ളിലാംതടം. — പാലാ : ദീപനാളം പബ്ളിക്കേഷൻസ്, 1978.
vii, 144 p. ; 23 cm.
In Malayalam.
Rs6.00

1. Christianity—India—Kerala. I. Vellilamthadam, Thomas, 1941- joint author. II. Title.
BR1170.K4K37 78-907218

Pulikkunnēl, Jōsaph, 1932-
(Siṃhāsanappōrā)
സിംഹാസനപ്പോരു് / ജോസഫ് പുലിക്കുന്നേൽ. — Palai : ഓശാന പ്രസിദ്ധീകരണം, 1978.
viii, 150 p. ; 19 cm. — (ഉത്തിഷ്ഠത ജാഗ്രത ; 1)
In Malayalam.
Rs6.00

1. Christianity—India—Kerala. 2. Kerala—Church history. I. Title. II. Series: Uttiṣṭhata jāgrata.
BR1170.K4P84 78-907153

—INDONESIA—ADDRESSES, ESSAYS, LECTURES

Sitompul, Adelbert Augustin, 1931-
Masalah modernisasi dewasa ini / A. A. Sitompul. — Jakarta : BPK Gunung Mulia, 1976.
27 p. ; 22 cm.
Includes bibliographical references.
1. Christianity—Indonesia—Addresses, essays, lectures. I. Title.
BR1220.S57 76-941315
78 MARC

—INDONESIA—CONGRESSES

Konsultasi Nasional Pembinaan Warga Gereja, Malang, Indonesia, 1976.
Membina bersama : panggilan untuk pembebasan dan persatuan dalam gereja, masyarakat dan dunia : laporan Konsultasi Nasional Pembinaan Warga Gereja, Malang 16-22 Pebruari 1976. — [Jakarta] : Diterbitkan oleh BPK Gunung Mulia untuk Dewan Gereja-Gereja di Indonesia, 1976.
188 p., [7] leaves of plates : ill. ; 21 cm.
1. Christianity—Indonesia—Congresses. I. Dewan Gereja-Gereja di Indonesia. II. Title.
BR1220.K66 1976 77-940623
79 MARC

—JAPAN

Toda, Yoshio, 1918-
(Nihonjin to sambika)
日本人と讃美歌 / 戸田義雄, 永藤武編著. — 東京 : 桜楓社, 1978.
279 p. : ill. ; 19 cm. Ja 78-19642
Includes bibliographical references and index.
¥980

1. Hymns, Japanese—History and criticism. 2. Christianity—Japan. I. Nagafuji, Takeshi, joint author. II. Title.
BV314.J3T62 79-800699

Usaki, Takesaburō, 1899-
(Kōkoku seishin to Kirisuto no fukuin)
皇國精神と基督の福音 / 宇崎竹三郎著. — 大阪 : 日本橋教會 ; 京都 : 發賣所 聖化社, 昭和15 [1940], 昭和16 [1941] printing.
3, 119 p. ; 19 cm.
"皇紀二千六百年記念出版"

1. Ethics, Japanese. 2. Christianity—Japan. 3. Apologetics—20th century. I. Title.
BJ970.U8 78-812061

—JAPAN—ADDRESSES, ESSAYS, LECTURES

Endō, Yoshimitsu, 1910-
(Nihonjin to Kirisutokyō)
日本人とキリスト教 / 遠藤義光著. — 東京 : キリスト新聞社, 昭和36 [1961]
102 p. ; 18 cm.
Bibliography: p. 101-102.

1. Christianity—Japan—Addresses, essays, lectures. 2. Japan—Civilization—Addresses, essays, lectures. 3. Theology, Doctrinal—Addresses, essays, lectures. I. Title.
BR1305.E52 79-807259

(Nihon ni okeru fukuin to bunka)
日本における福音と文化 / 山本和, 隅谷三喜男編. — 東京 : 教文館, 昭和42 [1967]
122 p. ; 18 cm. — (現代キリスト教双書)
Series romanized: Gendai Kirisutokyō sōsho.

1. Christianity—Japan—Addresses, essays, lectures. I. Yamamoto, Kanō, 1909- II. Sumiya, Mikio, 1916-
BR1305.N52 79-802786

(Nihonjin to fukuin)
日本人と福音 / 松木治三郎 ... [et al.]. — 東京 : 日本基督教団出版部, 昭和37 [1962]
172 p. ; 18 cm.

1. Christianity—Japan—Addresses, essays, lectures. I. Matsuki, Jisaburō, 1906-
BR1305.5.N53 79-806190

Suzuki, Toshirō, 1901-
(Sekaishi to Nihon no mondai)
世界史と日本の問題 : 宗教的視角より見たる / 鈴木俊郎著. — 東京 : 淡路書房, 昭和23 [1948]
188 p. ; 19 cm.

1. Christianity—Japan—Addresses, essays, lectures. I. Title.
BR1305.S89 78-813559

Yuki, Hideo, 1926-
(Kirisutokyō to Nihon)
キリスト教と日本 / [著者 幸日出男]. — 京都 : ルガール社, 1978.
193 p. ; 19 cm. Ja***
Includes bibliographical references.
¥1200

1. Christianity—Japan—Addresses, essays, lectures. I. Title.
BR1309.Y84 79-802447

—JAPAN—HINASE-CHO

Hoshio, Masakazu, 1906-
(Hinase no kankō to Kirishitan)
日生の観光とキリシタン / 星尾正一. — 日生町(岡山県) : 星尾ガラス屋, 1978.
132 p. : ill. ; 15 cm. Ja 78-24487
Cover title.
Bibliography: p. 122.
¥500

1. Hinase-chō, Japan—Description. 2. Christianity—Japan—Hinase-chō. I. Title.
DS897.H466H67 79-802875

—KOREA—ADDRESSES, ESSAYS, LECTURES

Chi, Wŏn-sang.
(Mal hanŭn kaltae)
말하는갈대 : 지원상수필집 / [저자池元祥]. — 서울 : 컨콜디아사, 1977.
229 p. ; 19 cm.
On verso of t. p.: Speaking reed.
W700

1. Christian life—Addresses, essays, lectures. 2. Christianity—Korea—Addresses, essays, lectures. I. Title. II. Title: Speaking reed.
BV4509.K6C45 79-825811

—LATIN AMERICA

Antoine, Charles, 1929-
Le sang et l'espoir : ces chrétiens d'Amérique latine / Charles Antoine. — Paris : Le Centurion, c1978.
144 p. ; 21 cm. F***
ISBN 2-227-34513-6 : 27.00F
1. Christianity—Latin America. I. Title.
BR600.A57 282'.8 79-349710
79 MARC

—LATIN AMERICA—ADDRESSES, ESSAYS, LECTURES

Aufbruch zur Freiheit : christliche Botschaft in Lateinamerika. — Freiburg/Schweiz : Schweizerischer Katholischer Missionsrat ; Basel : Schweizerischer Evangelischer Missionsrat, [1977]
128 p. : ill. ; 23 cm. — (Missionsjahrbuch der Schweiz ; 1977, Jahrg. 44) Sw77-11450
Includes bibliographical references and index.
9.70F
1. Christianity—Latin America—Addresses, essays, lectures. I. Schweizerischer Katholischer Missionsrat. II. Schweizerischer Evangelischer Missionsrat. III. Series.
BR600.A93 78-386680
*79 MARC

Hoffnung der Armen : christl. Leben in Lateinamerika, e. Provokation für Europa / Hermann Gilhaus, Christian Modehn (Hrsg.). — 1. Aufl. — Meitingen ; Freising : Kyrios-Verlag, 1976.
61 p. ; 19 cm. — (Theologie und Leben ; 34) GFR77-A46
Includes bibliographical references.
ISBN 3-7838-0141-9 : DM6.50
1. Christianity—Latin America—Addresses, essays, lectures. 2. Liberation theology—Addresses, essays, lectures. 3. Cardenal, Ernesto—Addresses, essays, lectures. 4. Camara, Helder, 1909- —Addresses, essays, lectures. I. Gilhaus, Hermann. II. Modehn, Christian. III. Series.
BR600.H57 78-367935
78 MARC

CHRISTIANITY (Continued)

—LATIN AMERICA—CONGRESSES

Geschenk der Armen an die Reichen: Zeugnisse aus dem gewaltfreien Kampf d. erneuerten Kirche in Lateinamerika / hrsg. u. kommentiert v. Hildegard Goss-Mayr ; hrsg. v. d. Kath. Sozialakademie Österreichs. — Wien ; München ; Zürich : Europaverl., c1979.
171, ⌊1⌋ p. ; 21 cm. — (Soziale Brennpunkte ; 7) Au79-10-21
Contains papers presented at the 2d and 3d Conferencia General del Episcopado Latinoamericano.
"Quellenverzeichnis": p. 170-⌊172⌋
ISBN 3-203-50707-2 : S118.00
1. Christianity—Latin America—Congresses. 2. Church and social problems—Latin America—Congresses. 3. Nonviolence—Moral and religious aspects—Congresses. I. Goss-Mayr, Hildegard. II. Katholische Sozialakademie Österreichs. III. Conferencia General del Episcopado Latinoamericano.
BR600.G48 79-376032
•79 MARC

—NEAR EAST

Missions Advanced Research and Communication Center.
World Christianity, Middle East / Don M. McCurry, editor. — Monrovia, Calif. : MARC, c1979.
150 p. : ill. ; 28 cm.
ISBN 0-912552-27-1
1. Christianity—Near East. I. McCurry, Don M. II. Title.
BR1070.M48 1979 275.6 79-87790
79 MARC

—NIGERIA

Häselbarth, Hans.
Lebenszeichen aus Afrika : Erfahrungen in Briefen / Hans Häselbarth. — 1. Aufl. — München : Kaiser, 1978.
159 p. ; 21 cm. — (Kaiser Traktate ; 32) GFR78-A
ISBN 3-459-01148-3
1. Christianity—South Africa. 2. Missions—South Africa. 3. Christianity—Nigeria. 4. Missions—Nigeria. I. Title.
BR1450.H33 78-375645
78 MARC

—NIGERIA—ADDRESSES, ESSAYS, LECTURES

Christianity in West Africa : the Nigerian story / edited by Ogbu Kalu. — Ibadan : Daystar Press, 1978.
ix, 394 p., ⌊1⌋ leaves of plates : ill. ; 21 cm.
Includes bibliographical references.
CONTENTS: Crampton, E. P. T. Christianity in Northern Nigeria.—Oduyoye, M. The planting of Christianity in Yorubaland.—Review of F. K. Ekechi, Missionary enterprise and rivalry in Igboland.—Kalu, O. Protestant Christianity in Igboland.—Tasie, G. O. M. Christianity in the Niger Delta.—The United Native African Church.—Ola, C. S. Foundations of the African church in Nigeria.—Ferguson, J. Some Nigerian church founders.—The Christ Apostolic Church.—Ayandele, E. A. The Aladura among the Yoruba.—Akiwowo, A. Christian denominations in Nigeria.
ISBN 9781221372
1. Christianity—Nigeria—Addresses, essays, lectures. 2. Nigeria—Church history—Addresses, essays, lectures. I. Kalu, Ogbu.
BR1463.N5C57 276.69 79-101581
79 MARC

—NIGERIA—NORTHERN NIGERIA

Crampton, Edmund Patrick Thurman.
Christianity in northern Nigeria / Edmund Patrick Thurman Crampton. — ⌊s.l. : s.n.⌋, 1975 (Zaria, Nigeria : Gaskiya Corp.)
ix, 199 p., ⌊8⌋ leaves of plates : ill. ; 22 cm.
Revision of the author's thesis (B.D.), Dublin, 1967.
Bibliography: p. 182-189.
Includes index.
1. Christianity—Nigeria—Northern Nigeria. 2. Missions—Nigeria—Northern Nigeria. 3. Northern Nigeria—Church history. I. Title.
BR1463.N5C7 1975 276.69 77-364796
78 MARC

—NIGERIA—OYO (STATE)—STATISTICS

Oyo, Nigeria (State). Ministry of Finance and Economic Development. Statistics Division.
Report on the survey of religious organisations in the Oyo State. — Ibadan : Ministry of Finance and Economic Development, Statistics Division, 1977.
x, 26 p. ; 25 cm.
Cover title.
Chiefly tables.
N1.00
1. Christianity—Nigeria—Oyo (State)—Statistics. 2. Islam—Nigeria—Oyo (State)—Statistics. I. Title.
BR1463.N5O94 1977 200'.6'26692 79-108572
79 MARC

—OCEANICA—CONGRESSES

Mission, church, and sect in Oceania / edited by James A. Boutilier, Daniel T. Hughes, Sharon W. Tiffany. — Ann Arbor : University of Michigan Press : produced and distributed by University Microfilms International, c1978.
xiv, 500 p. ; 24 cm. — (ASAO monograph ; no. 6) (Monograph publishing : imprint series)
Papers from an ASAO symposium on missionary activity in Oceania held at Stuart, Fla., Mar. 1975.
Bibliography: p. 465-496.
ISBN 0-87991-364-9 : $17.50
1. Missions—Oceanica—Congresses. 2. Christianity—Oceanica—Congresses. I. Boutilier, James A. II. Hughes, Daniel T. III. Tiffany, Sharon W. IV. Association for Social Anthropology in Oceania. V. Series: Association for Social Anthropology in Oceania. ASAO monograph ; no. 6.
BV3670.M49 266'.0099 78-11072
78 MARC

—RUSSIA

Bori, Pier Cesare.
Movimenti religiosi in Russia prima della rivoluzione (1900-1917) / ⌊di⌋ Pier Cesare Bori, Paolo Bettiolo. — Brescia : Queriniana, ⌊1979?⌋
318 p. ; 22 cm. — (Dipartimento di scienze religiose ; 8) It79-Apr
Bibliography: p. 313-316.
L6500
1. Christianity—Russia. 2. Russia—Religion. I. Bettiolo, Paolo, 1947- joint author. II. Title. III. Series.
BR935.5.B67 79-363437
•79 MARC

—SOUTH AFRICA

Etherington, Norman.
Preachers, peasants, and politics in southeast Africa, 1835-1880 : African Christian communities in Natal, Pondoland, and Zululand / Norman Etherington. — London : Royal Historical Society, 1978.
xi, 230 p. : maps ; 23 cm. — (Studies in history ; no. 12) GB***
Bibliography: p. 193-215.
Includes index.
ISBN 0-901050-48-2 : £9.50
1. Christianity—South Africa. 2. Missions—South Africa. 3. Nguni—Religion. I. Title. II. Series: Studies in history (London) ; no. 12.
BR1450.E83 276.8'4 79-317272
79 MARC

Häselbarth, Hans.
Lebenszeichen aus Afrika : Erfahrungen in Briefen / Hans Häselbarth. — 1. Aufl. — München : Kaiser, 1978.
159 p. ; 21 cm. — (Kaiser Traktate ; 32) GFR78-A
ISBN 3-459-01148-3
1. Christianity—South Africa. 2. Missions—South Africa. 3. Christianity—Nigeria. 4. Missions—Nigeria. I. Title.
BR1450.H33 78-375645
78 MARC

—SOUTH AFRICA—ADDRESSES ESSAYS, LECTURES

Delayed action! : an ecumenical witness from the Afrikaans speaking church / by A. S. Geyser ... ⌊et al.⌋. — Pretoria : Distributed for and on behalf of the authors by N. G. Kerkboekhandel, ⌊1960⌋
168 p. ; 22 cm.
Translation of Vertraagde aksie.
1. Christianity—South Africa—Addresses essays, lectures. I. Geyser, A. S.
BR1450.V413 261.8'34'51968 78-395370
79 MARC

—TAIWAN

Laurentin, René.
Chine et christianisme : après les occasions manquées / René Laurentin. — ⌊Paris⌋ : Desclée De Brouwer, 1977.
370 p. : ill. ; 22 cm. F78-2201
Bibliography: p. ⌊323⌋-328.
Includes indexes.
ISBN 2-220-02110-6 : 56.00F
1. Christianity—China. 2. Christianity—Taiwan. I. Title.
BR1285.L38 275.1 78-378633
78 MARC

—TANZANIA

Ujamaa and Christian communities / edited by Christopher Mwoleka and Joseph Healey. — Eldoret, Kenya : Gaba Publications, 1976.
62 p. ; 24 cm. — (Spearhead ; no. 45)
Includes bibliographical references.
1. Collective settlements—Tanzania. 2. Christianity—Tanzania. I. Mwoleka, Christopher. II. Healey, Joseph. III. Series: Spearhead (Eldoret, Kenya) ; no. 45.
HX771.3.A3U37 335'.9'678 77-980736
79 MARC

—UGANDA

Byabazaire, Deogratias M 1941-
The contribution of the Christian chruches to the development of Western Uganda 1894-1974 / Deogratias M. Byabazaire. — Frankfurt am Main ; Bern ; Las Vegas : Lang, 1979.
198 p. : graphs, maps ; 21 cm. — (European university papers : Series 23, Theology ; v. 112) GFR79-A
Bibliography: p. 165-198.
ISBN 3-261-02553-0
1. Christianity—Uganda. 2. Uganda—History. I. Title. II. Series: Europäische Hochschulschriften : Reihe 23, Theologie ; Bd. 112.
BR1443.U33B9 967.6'1 79-317221
79 MARC

Pirouet, Louise.
Black evangelists : the spread of Christianity in Uganda, 1891-1914 / ⌊by⌋ M. Louise Pirouet. — London : Collings, 1978.
xiv, 255 p. : maps ; 23 cm. GB78-15375
A revision of the author's thesis, University of East Africa.
Bibliography: p. ⌊229⌋-246.
Includes index.
ISBN 0-86036-051-2 : £6.00
1. Christianity—Uganda. 2. Evangelistic work—Uganda. I. Title.
BR1443.U33P56 1978 266'.009676'1 78-317867
78 MARC

—UNITED STATES

McHugh, Gerald Austin.
Christian faith and criminal justice : toward a Christian response to crime and punishment / by Gerald Austin McHugh. — New York : Paulist Press, c1978.
x, 234 p. ; 21 cm. — (A Deus book)
Includes bibliographical references.
ISBN 0-8091-2105-0 : $5.95
1. Criminal justice, Administration of—United States. 2. Christianity—United States. I. Title.
HV8138.M29 261.8'33 78-58956
78 MARC

Mizoguchi, Yasuo, 1906-
(Amerika no keisei to Kirisutokyō)
アメリカの形成と基督教 / 溝口靖夫著. — 東京 : 教文館, 昭和26 ⌊1951⌋
6, 260, 7 p., ⌊2⌋ leaves of plates : ill. ; 22 cm.
Includes bibliographical references and index.
1. Mexico—History—Conquest, 1519-1540. 2. South America—History—To 1806. 3. Christianity—United States. 4. Missions—California. I. Title.
F1230.M53 79-800773

—UNITED STATES—ADDRESSES, ESSAYS, LECTURES

Sources of vitality in American church life / edited by Robert L. Moore. — Chicago, Ill. : Exploration Press, c1978.
119 p. ; 23 cm. — (Studies in ministry and parish life)
Includes bibliographical references.
ISBN 0-913552-14-3. ISBN 0-913552-15-1 pbk.
1. Christianity—United States—Addresses, essays, lectures. 2. United States—Religion—1945- —Addresses, essays, lectures. I. Moore, Robert L. II. Series.
BR526.S68 277.3 78-71065
79 MARC

—UNITED STATES—BIBLIOGRAPHY

Sandeen, Ernest Robert, 1931-
American religion and philosophy : a guide to information sources / Ernest R. Sandeen, Frederick Hale. — Detroit : Gale Research Co., 1978.
xvi, 377 p. ; 23 cm. — (American studies informaton guide series ; v. 5) (Gale information guide library)
Includes indexes.
ISBN 0-8103-1262-X
1. Christianity—United States—Bibliography. 2. United States—Religion—Bibliography. 3. Philosophy, American—Bibliography. I. Hale, Frederick, joint author. II. Title.
Z7757.U5S25 016.2'00973 73-17562
⌊BR515⌋ 78 MARC

—UNITED STATES—MISCELLANEA

Ravenhill, Leonard.
America is too young to die / Leonard Ravenhill. — Minneapolis : Bethany Fellowship, c1979.
123 p. : port. ; 21 cm.
ISBN 0-87123-013-5 : $2.95
1. Christianity—United States—Miscellanea. 2. Church renewal—Miscellanea. I. Title.
BR526.R38 262'.001 79-19229
79 MARC

CHRISTIANITY (Continued)

—UNITED STATES
—PICTORIAL WORKS

Whitman, Alan.
Christian occasions / photos. and text by Alan Whitman. — 1st ed. — Garden City, N.Y : Dolphin Books, 1978.
144 p. : ill. ; 26 cm.
On cover: A photographic study of unusual styles of religion in American life.
"A Country Music Magazine Press book."
ISBN 0-385-12597-6 : $6.95
1. Christian sects—United States—Pictorial works. 2. Christianity—United States—Pictorial works. I. Title.
BR517.W46 277.3 77-76286
78 MARC

—WASHINGTON, D.C.

Rebirth in Washington : the Christian impact in the Nation's Capital / Wallace Henley. — Westchester, Ill. : Good News Publishers, 1977.
124 p. ; 21 cm.
ISBN 0-89107-150-4 : $2.95
1. Christianity—Washington, D.C. 2. Washington, D.C.—Religious life and customs. I. Henley, Wallace.
BR560.W35R4 277.53 78-110532
79 MARC

CHRISTIANITY AND ANTISEMITISM

see also Holocaust (Christian theology)

Lazar, Elmer B
The story of the century : "Jews absolved from blame of Jesus Christ's death" / by Elmer B. Lazar. — New York : Philosophical Library, 1978.
vii, 150 p. ; 22 cm.
ISBN 0-8022-2226-9 : $7.50
1. Vatican Council. 2d, 1962-1965. Declaratio de ecclesiae habitudine ad religiones non-Christianas. 2. Catholic Church—Relations—Judaism. 3. Judaism—Relations—Catholic Church. 4. Jesus Christ—Passion—Role of the Jeus. 5. Lazar, Elmer B. 6. Christianity and antisemitism. I. Title.
BX830 1962.A45D4434 262'.5'2 78-50529
79 MARC

Roth, John K
A consuming fire : encounters with Elie Wiesel and the holocaust / John K. Roth ; prologue by Elie Wiesel. — Atlanta : John Knox Press, c1979.
191 p. ; 21 cm.
Bibliography: p. [187]-191.
ISBN 0-8042-0812-3
1. Wiesel, Elie, 1928- —Criticism and interpretation. 2. Holocaust, Jewish (1938-1945) 3. Holocaust (1939-1945)—Moral and religious aspects. 4. Christianity and antisemitism. I. Title.
PQ2683.I32Z86 940'.04'924 78-52442
78 MARC

—ADDRESSES, ESSAYS, LECTURES

Antisemitism and the foundations of Christianity / edited by Alan Davies. — New York : Paulist Press, c1979.
xvii, 258 p. ; 23 cm.
Includes bibliographical references.
ISBN 0-8091-2219-7 : $7.95
1. Christianity and antisemitism—Addresses, essays, lectures. I. Davies, Alan T.
BM535.A64 261.2 79-65620
00 MARC

Bar-Ilan studies in history / [edited by Pinhas Artzi]. — Ramat-Gan, Israel : Bar-Ilan University Press, c1978.
viii, 306 p. ; 25 cm. — (Bar-Ilan departmental researches)
Includes bibliographical references.
1. Near East—History—Addresses, essays, lectures. 2. Christianity and antisemitism—Addresses, essays, lectures. 3. Zionism—History—Addresses, essays, lectures. I. Artzi, P., 1923- . II. Series: Meḥkere Bar-Ilan, pirsumim mahlaktiyim.
DS62.B34 956 79-105198
79 MARC

Baum, Gregory, 1923-
Christian theology after Auschwitz / by Gregory G. Baum. — London : Council of Christians and Jews, [1977?]
3-22 p. ; 22 cm. — (Robert Waley Cohen memorial lecture ; 1976)
GB78-27836
£0.50
1. Holocaust (Christian theology)—Addresses, essays, lectures. 2. Christianity and antisemitism—Addresses, essays, lectures. I. Title. II. Series.
BT93.B38 231'.8 79-314680
79 MARC

Bethge, Eberhard, 1909-
Versuchung des Glaubens : zur Kritik d. christl. Antijudaismus / Eberhard Bethge. — Neukirchen-Vluyn : Neukirchener Verlag, 1976.
16, [2] p. ; 15 cm. — (Erbauliche Reden ; 1) GFR77-A
"Bibelarbeit ... gehalten vor der Landessynode der Evangelischen Kirche im Rheinland am 14. Januar 1976."
Bibliography: p. [18]
ISBN 3-7887-0505-1
1. Christianity and antisemitism—Addresses, essays, lectures. 2. Bible. N.T. 1 Peter II, 5-10—Criticism, interpretation, etc. I. Title.
BM535.B483 78-386218
79 MARC

Erinnern, nicht vergessen : Zugänge zum Holocaust / hrsg. von Martin Stöhr unter Mitarb. von Ulrike Berger ... [et al.] ; im Auftr. d. Arbeitsgemeinschaft Juden u. Christen beim Dt. Evang. Kirchentag. — München : Kaiser, 1979.
178 p. ; 21 cm. — (Kaiser-Traktate ; 43) GFR79-A
Includes bibliographical references.
ISBN 3-459-01210-2
1. Holocaust, Jewish (1939-1945)—Addresses, essays, lectures. 2. Christianity and antisemitism—Addresses, essays, lectures. I. Stöhr, Martin. II. Deutscher Evangelischer Kirchentag.
D810.J4E72 940.53'1503'924 79-370077
79 MARC

Judentum im Mittelalter : 4. Mai-26. Okt. 1978, Ausstellung im Schloss Halbturn : [Katalog] / veranst. v. d. Kulturabt. des Amtes d. Burgenländ. Landesregierung ; [Beitr., Ursula Schubert ... et al. ; Hrsg., Kurt Schubert]. — Eisenstadt : Kulturabt. des Amtes d. Bgld. Landesregierung, 1978.
268 p., [29] leaves of plates : ill. (some col.) ; 24 cm. Au78-17/18-388
At head of title: Schloss Halbturn.
Bibliography: p. 261.
S130.00
1. Jewish art and symbolism—Addresses, essays, lectures. 2. Jews—Rites and ceremonies—Addresses, essays, lectures. 3. Christianity and antisemitism—Addresses, essays, lectures. 4. Jews in Germany—History—Addresses, essays, lectures. 5. Jewish art and symbolism—Exhibitions. I. Schubert, Ursula, Dr. II. Schubert, Kurt, 1923- . III. Halbturn, Austria. Schloss. IV. Burgenland. Kulturabteilung.
N7417.3.J8 79-353872
*79 MARC

—CONGRESSES

Aspects of Jewish culture in the Middle Ages : papers of the eight annual conference of the Center for Medieval and Early Renaissance Studies, State University of New York at Binghamton, 3-5 May, 1974 / edited by Paul E. Szarmach. — Albany : State University of New York Press, 1979.
xxi, 208 p. : ill. ; 23 cm.
Includes bibliographical references.
ISBN 0-87395-165-4
1. Jews—History—70-1789—Historiography—Congresses. 2. Christianity and antisemitism—Congresses. 3. Jews in Spain—History—Congresses. 4. Judaism and art—Congresses. 5. Grail—Congresses. 6. Spain—History—Arab period, 711-1492—Congresses. I. Szarmach, Paul E. II. New York (State). State University at Binghamton. Center for Medieval and Early Renaissance Studies.
DS124.A74 909'.04'924 77-29046
78 MARC

CHRISTIANITY AND ANTISEMITISM IN LITERATURE

Finkelstein, Bluma.
Le monde chrétien vu par des écrivains juifs d'expression française : thèse pour le doctorat de troisième cycle / présentée par / Bluma Finkelstein, sous la direction de Paul Vernois. — [s.l. : s.n.], 1978.
2 v. ; 30 cm.
Thesis (doctorat de troisième cycle)—Université des sciences humaines et sociales, Strasbourg.
Label on t.p.: printed and published in Israel.
Bibliography: v. 2, leaves 237-256.
1. French literature—Jewish authors—History and criticism. 2. Jesus Christ—Jewish interpretations. 3. Jesus Christ, in fiction, drama, poetry, etc. 4. Judaism—Apologetic works. 5. Christianity and antisemitism in literature. I. Title.
PQ150.J4F56 840'.9'352 79-106301
79 MARC

CHRISTIANITY AND ATHEISM

see also Communism and Christianity

Gilardi, Enrico.
La scelta di Dio : ateismo e fede a confronto / Enrico Gilardi. — Leumann/[Rivoli] : Elle Di Ci, [1977]
424 p. ; 24 cm. It79-Jan
Includes bibliographies.
L6800
1. Atheism. 2. Christianity and atheism. I. Title.
BL2747.3.G53 79-382219
*79 MARC

CHRISTIANITY AND CABALA see Cabala and Christianity

CHRISTIANITY AND CAPITALISM

Desmond, Cosmas.
Christians or capitalists? : Christianity and politics in South Africa / Cosmas Desmond. — London : Bowerdean Press, 1978.
160 p. ; 23 cm. GB***
Bibliography: p. [159]-160.
ISBN 0-906097-03-7 : £7.00. ISBN 0-906097-04-5 pbk.
1. Christianity and politics. 2. Church and race relations—South Africa. 3. Christianity and capitalism. 4. Socialism. 5. South Africa—Race relations. 6. South Africa—Politics and government—1961- . I. Title.
BR115.P7D457 261.1'0968 78-325747
79 MARC

Hinkelammert, Franz Josef, 1931-
Las armas ideológicas de la muerte : el discernimiento de los fetiches, capitalismo y cristianismo / Franz J. Hinkelammert. — 1. ed. — [San José] Costa Rica : Editorial Universitaria Centro americana, 1977.
254 p. ; 21 cm. — (Colección DEI)
Includes bibliographical references.
1. Sociology, Christian. 2. Christianity and capitalism. I. Title. II. Series: Universidad Nacional (Costa Rica). Departamento Ecuménico de Investigaciones. Colección DEI.
BT738.H54 78-104471
79 MARC

—CONTROVERSIAL LITERATURE

Fernández Aldabalde, Ramón.
Fetichismo y religión / Ramón Fernández Aldabalde. — 1. ed. — Bilbao : Zero ; Madrid : distribuidor exclusiva, ZYX, 1975.
240 p. ; 20 cm. — (Biblioteca promoción del pueblo ; no. 79 : Serie P)
Sp***
Bibliography: p. 227-236.
ISBN 843170313X
1. Liberation theology. 2. Capitalism—Controversial literature. 3. Christianity and capitalism—Controversial literature. I. Title.
BT83.57.F47 79-376742
79 MARC

CHRISTIANITY AND COMMUNICATION see Communication (Theology)

CHRISTIANITY AND COMMUNISM see Communism and Christianity; Communism and religion

CHRISTIANITY AND CULTURE

Cantalamessa, Raniero.
Cristianesimo e valori terreni : sessualità, politica e cultura / Raniero Cantalamessa. — Milano : Vita e pensiero, 1976.
ix, 169 p. ; 20 cm. — (Fede e mondo moderno ; 1) It78-Nov
Includes bibliographical references.
L2500
1. Christianity and culture. 2. Christianity and politics. I. Title.
BR115.C8C32 79-348990
*79 MARC

Dippel, Cornelis Johannes, 1902-
Verkenning en verwachting : cultuurkritische opstellen / C. J. Dippel. — 's-Gravenhage : Boekencentrum, 1962.
223 p. ; 23 cm.
Includes bibliographical references.
1. Technology and civilization. 2. Christianity and culture. 3. Atomic energy—Social aspects. I. Title.
CB478.D58 79-377527
79 MARC

Edwards, David Lawrence.
A reason to hope / [by] David L. Edwards. — London ; New York : Collins, 1978.
252 p. ; 22 cm. GB78-29007
Includes index.
ISBN 0-529-05620-8 : £4.95 ($8.95)
1. Civilization, Modern, 1950- . 2. Christianity—20th century. 3. Christianity and culture. I. Title.
CB430.E34 1978 909 78-56996
78 MARC

Grunlan, Stephen A
Cultural anthropology : a Christian perspective / Stephen A. Grunlan and Marvin K. Mayers ; with a foreword by Eugene A. Nida. — Grand Rapids : Zondervan, c1979.
309 p. ; 21 cm.
Bibliography: p. 287-294.
Includes indexes.
ISBN 0-310-36321-7
1. Christianity and culture. 2. Anthropology. I. Mayers, Marvin Keene, 1927- joint author. II. Title.
BR115.C8G76 301.2 78-13980
78 MARC

CHRISTIANITY AND CULTURE (Continued)

Kraft, Charles H
Christianity in culture : a study in dynamic Biblical Theologizing in cross-cultural perspective / Charles H. Kraft. — Maryknoll, N.Y. : Orbis Books, c1979.
xviii, 445 p. : ill. ; 24 cm.
Bibliography: p. 405-426.
Includes indexes.
ISBN 0-88344-075-X (pbk.)
1. Christianity and culture. I. Title.
BR115.C8K64 261 78-13736
78 MARC

Lobo Méndez, Gonzalo.
Ideologías y fe cristiana / Gonzalo Lobo Méndez. — Madrid : Editorial Magisterio Español, c1978.
575 p. : ill. ; 25 cm. Sp***
Includes bibliographies and indexes.
ISBN 8426503047
1. Christianity and culture. 2. Theology. 3. Salvation. I. Title.
BR115.C8L59 79-364580
79 MARC

Short, Robert L
Something to believe in : Is Kurt Vonnegut the exorcist of Jesus Christ Superstar? / Robert Short. — 1st ed. — San Francisco : Harper & Row, c1978.
ix, 321 p. : ill. ; 21 cm.
Includes bibliographical references.
ISBN 0-06-067381-8
1. Christianity and culture. 2. Theology—20th century. 3. Vonnegut, Kurt—Religion and ethics. 4. Theology—20th century. 5. United States—Popular culture. 6. Blatty, William Peter. The exorcist. 7. Jesus Christ superstar ₍Motion picture₎ I. Title.
BR115.C8S53 230'.09'04 75-36754
78 MARC

—EARLY CHURCH, CA. 30-600—
ADDRESSES, ESSAYS, LECTURES

Marrou, Henri Irénée.
Christiana tempora : mélanges d'histoire, d'archéologie, d'épigraphie et de patristique / Henri-Irénée Marrou. — Roma : École française de Rome, 1978.
vi, 427 p., ₍1₎ fold. leaf of plates : ill. ; 25 cm. — (Collection de l'École française de Rome ; 35) It78-Oct
In French.
Reprint of previously published articles.
Includes bibliographical references.
1. Civilization, Ancient—Addresses, essays, lectures. 2. Inscriptions, Latin—Addresses, essays, lectures. 3. Inscriptions, Greek—Addresses, essays, lectures. 4. Christianity and culture—Early church, ca. 30-600—Addresses, essays, lectures. 5. Marrou, Henri Irénée—Addresses, essays, lectures. I. Title. II. Series: École française de Rome. Collection ; 35.
CB311.M3 78-393675
*79 MARC

—ADDRESSES, ESSAYS, LECTURES

Dietzfelbinger, Hermann.
Jesus Christus und die Weisen aus dem Abendland / Hermann Dietzfelbinger. — Metzingen : Brunnquell-Verlag der Bibel- u. Missions-Stiftung, 1978.
240 p. ; 20 cm. GFR78-A
Includes bibliographical references.
ISBN 3-7656-0213-2 : DM13.80
1. Christianity—20th century—Addresses, essays, lectures. 2. Christianity and culture—Addresses, essays, lectures. I. Title.
BR123.D495 79-365956
79 MARC

—COLLECTED WORKS

Vööbus, Arthur.
Studies in the history of the Estonian people : with reference to aspects of social conditions, in particular, the religious and spiritual life and the educational pursuit / by Arthur Vööbus. — Stockholm : ETSE, 1974-
v. ; 26 cm. — (Papers of the Estonian Theological Society in Exile ; 26) S***
Includes bibliographical references and indexes.
1. Estonia—Civilization—Collected works. 2. Christianity—Estonia—Collected works. 3. Christianity and culture—Collected works. I. Title. II. Series: Eesti Usuteadlaste Selts Paguluses. Toimetused ; 26.
DK511.E53V63 947'.41 79-303463
79 MARC

—CONGRESSES

Cristianesimo e cultura : atti completi del XLVI corso di aggiornamento culturale dell'Università cattolica : Loreto, 21-26 settembre 1975 : ₍scritti / di₎ Giuseppe Lazzati ... ₍et al.₎. — 2. ed. aumentata. — Milano : Vita e pensiero, 1976.
283 p. ; 22 cm. — (Le Api) (Pubblicazioni dell'Università cattolica del Sacro Cuore) It78-May
"Il materiale del presente volume fa parte di Vita e pensiero, a. LIX, n. 1/2/3 (1976)."
Includes bibliographical references.
L2500
1. Christianity and culture—Congresses. I. Lazzati, Giuseppe, 1909-
BR115.C8C74 1976 79-338444
*79 MARC

—HISTORY

Leone, Luigi.
Cristianesimo e cultura antica in alcuni scrittori cristiani / Luigi Leone. — ₍Lecce₎ : Milella, ₍1976?₎
102 p. ; 21 cm. It78-June
Bibliography: p. ₍93₎-102.
L3000
1. Christianity and culture—History. 2. Christian literature, Early—History and criticism. I. Title.
BR67.L46 78-398374
*79 MARC

CHRISTIANITY AND DEMOCRACY

Dengerink, J D
El cristiano y la democracia moderna : el fundamento bíblico de la democracia / ₍por₎ J. D. Dengerink, J. Grau. — Barcelona : Ediciones Evangélicas Europeas, 1977.
107 p. ; 18 cm. — (Literatura evangélica) Sp77-Dec
Subtitle on cover: Un punto de vista protestante.
Bibliography: p. 105-107.
ISBN 8472850307 : 85ptas
1. Christianity and democracy. I. Grau, José, joint author. II. Title.
BR115.P7D454 78-383690
78 MARC

Llinares, José A
Democracia integral y conciencia cristiana / ₍por₎ José Antonio Llinares Ibáñez. — Salamanca : San Esteban, 1978.
227 p. ; 22 cm. — (Estudio teológico de San Esteban : Glosas ; 2) Sp***
Part 1 previously published in Estudios filosóficos, v. 26, 1977.
Includes bibliographical references.
ISBN 8485045386
1. Christianity and democracy. 2. Communism and Christianity. I. Title.
BR115.P7L54 79-113832
79 MARC

Madiran, Jean.
Les deux démocraties / Jean Madiran. — Paris : Nouvelles Éditions latines, c1977.
199 p. ; 19 cm. F***
ISBN 2-7233-0024-2 : 24.00F
1. Christianity and democracy. 2. Democracy. I. Title.
BR115.P7M3335 261.7 78-388849
79 MARC

—ADDRESSES, ESSAYS, LECTURES

Imai, Saburō.
(Shūkyō to minshu shisō)
宗教と民主思想 / ₍著作 今井三郎₎. — 東京 : ₍今井₎ : 賣捌所 警醒社書店, 大正6 ₍1917₎, 大正7 ₍1918₎ printing.
2, 3, 478 p. ; 19 cm.

1. Christianity and democracy—Addresses, essays, lectures. I. Title.
BR115.P7I46 79-808860

—CONGRESSES

Utz, Arthur Fridolin, 1908-
La concepción cristiana de la democracia pluralista : actas de un simposio internacional en Madrid / Arthur Utz, Heinrich Basilius Streithofen, con la colaboración de numerosos especialistas. — Barcelona : Herder, 1978.
301 p. ; 22 cm. — (Studia Helvetica Friburgensia ; 5) Sp***
Includes bibliographical references and index.
ISBN 8425407486 : 580ptas
1. Christianity and democracy—Congresses. I. Streithofen, Heinrich Basilius, joint author. II. Title.
BR115.P7U89 79-338447
79 MARC

—HISTORY

Uhde, Ute.
Politik und Religion : zum Verhältnis von Demokratie und Christentum bei Alexis de Tocqueville / von Ute Uhde. — Berlin : Duncker & Humblot, c1978.
109 p. ; 24 cm. — (Beiträge zur Politischen Wissenschaft ; Bd. 29) GFR***
Bibliography: p. ₍102₎-109.
ISBN 3-428-04064-3
1. Christianity and democracy—History. 2. Christianity and politics—History. 3. Tocqueville, Alexis Charles Henri Maurice Clérel de, 1805-1859. I. Title. II. Series.
BR115.P7U37 78-375500
78 MARC

—FRANCE—COLLECTED WORKS

Maurras, Charles, 1868-1952.
La Démocratie religieuse / Charles Maurras, ... ; avis au lecteur par Jean Madiran ; notices biographiques de Jacques Vier. — Paris : Nouvelles éditions latines, 1978.
xi, 580 p. ; 23 cm. F78-17971
Includes bibliographical references and index.
CONTENTS: Le dilemme de Marc Sangnier.—La politique religieuse.—L'Action française et la religion catholique.
ISBN 2-7233-0031-5 : 75.00F
1. Christianity and democracy—France—Collected works. 2. Action Française—Collected works. I. Title.
BR115.P7M3432 1978 261.7 79-367399
79 MARC

—SPAIN—HISTORY

Benavides, Domingo.
Democracia y cristianismo en la España de la Restauración, 1875-1931 / Domingo Benavides Gómez. — Madrid : Editora Nacional, ₍c1978₎
391 p. ; 22 cm. — (España en 3 ₍i.e. tres₎ tiempos) Sp***
Bibliography: p. 375-385.
ISBN 8427604513 : 600ptas
1. Catholic Church in Spain—History. 2. Church and state in Spain—History. 3. Christianity and democracy—Spain—History. I. Title.
BX1585.B38 79-376042
79 MARC

CHRISTIANITY AND ECONOMICS

see also Christianity and capitalism; Church and labor; Communism and Christianity; Socialism, Christian; Sociology, Christian; Steawardship, Christian

Antonides, Harry, 1931-
Multinationals and the peaceable kingdom / Harry Antonides. — Toronto : Clarke, Irwin, 1978.
xxi, 248 p. ; 24 cm. C78-1468-5
Bibliography: p. ₍225₎-235.
Includes index.
ISBN 0-7720-1196-6 : $13.95
1. Corporations, American—Canada. 2. Canada—Economic conditions—1945- . 3. Industry and state—Canada. 4. Christianity and economics. I. Title.
HD2810.A57 338.8'8 79-305891
79 MARC

Birch, Bruce C
The predicament of the prosperous / Bruce C. Birch and Larry L. Rasmussen. — 1st ed. — Philadelphia : Westminster Press, c1978.
212 p. ; 20 cm. — (Biblical perspectives on current issues)
Bibliography: p. ₍207₎-210.
Includes index.
ISBN 0-664-24211-1
1. Sociology, Christian. 2. Wealth—Biblical teaching. 3. Human ecology—Moral and religious aspects. 4. Christian ethics. 5. Christianity and economics. I. Rasmussen, Larry L., joint author. II. Title. III. Series.
BT738.B52 261.8 78-18412
78 MARC

Lume, Giovanni.
Il resto per giunta / Giovanni Lume. — Milano : Virgilio, ₍1976?₎
269 p. ; 21 cm. It78-Jan
L4500
1. Christianity and economics. 2. Sociology, Christian. I. Title.
BR115.E3L85 79-374416
*79 MARC

Nankivell, Owen.
All good gifts : a Christian view of the affluent society / ₍by₎ Owen Nankivell. — London : Epworth Press, 1978.
127 p. ; 22 cm. GB78-38207
ISBN 0-7162-0305-7 : £2.50
1. Christianity and economics. I. Title.
BR115.E3N28 261.8'5 79-305753
79 MARC

CHRISTIANITY AND ECONOMICS (Continued)

Tagle Martínez, Hugo.
Visión de la doctrina económica de la Iglesia Católica / Hugo Tagle Martínez. — [Santiago, Chile] : Instituto de Estudios Generales, c1978.
76 p. ; 19 cm.
Includes bibliographical references.
1. Christianity and economics. 2. Sociology, Christian (Catholic) I. Title.
BR115.E3T24 79-107073
79 MARC

Viner, Jacob, 1892-1970.
Religious thought and economic society : four chapters of an unfinished work / by Jacob Viner ; edited by Jacques Melitz and Donald Winch. — Durham, N.C. : Duke University Press, 1978.
211 p. ; 24 cm.
Includes bibliographical references and indexes.
ISBN 0-8223-0398-1 : $11.75
1. Christianity and economics. I. Melitz, Jacques. II. Winch, Donald. III. Title.
BR115.E3V54 1978 261.8'5 77-93857
78 MARC

Weber, Max, 1864-1920.
Die protestantische Ethik : eine Aufsatzsammlung / Max Weber ; [hrsg. von Johannes Winckelmann]. — 5., erneut überarb. u. um e. Nachw. vers. Aufl. (31.-36. Tsd.). — Gütersloh : Gütersloher Verlagshaus Mohn, 1979-
v. ; 19 cm. — (Gütersloher Taschenbücher Siebenstern ; 53-)
GFR79-A(v.
Includes bibliographical references.
ISBN 3-579-03798-6 (v. 1) : DM14.80 (v. 1)
1. Christianity and economics. 2. Christian ethics. 3. Protestantism. 4. Capitalism. 5. Religion and sociology. I. Winckelmann, Johannes. II. Title.
BR115.E3W37 1979 79-371571
79 MARC

—ADDRESSES, ESSAYS, LECTURES

Progresso economico e promozione umana : pagine scelte / [a cura di] Antonino Romeo. — Bresso : Tramontana, 1978.
276 p. ; 18 cm. — (Collana di scienze sociali ; 7A) It79-June
Includes bibliographical references.
L3000
1. Economic development—Addresses, essays, lectures. 2. Christianity and economics—Addresses, essays, lectures. 3. Technology—Social aspects—Addresses, essays, lectures. I. Romeo, Antonino.
HC54.P76 79-378065
*79 MARC

CHRISTIANITY AND EVOLUTION see Evolution and Christianity

CHRISTIANITY AND EXISTENTIALISM

Mistura, Stefano.
Paul Tillich, teologo della nuova psichiatria : tra coraggio di esistere e problema dell'angoscia / Stefano Mistura. — Torino : Claudiana, 1978.
79 p. ; 20 cm. — (Piccola collana moderna ; 33 : Serie teologica)
It78-May
Bibliography: p. 77.
L1700
1. Tillich, Paul, 1886-1965. 2. Psychiatry and religion. 3. Christianity and existentialism. I. Title.
BX4827.T53M55 78-377779
*78 MARC

Rupp, George.
Beyond existentialism and Zen : religion in a pluralistic world / George Rupp. — New York : Oxford University Press, 1979.
xiv, 113 p. ; 22 cm.
Includes bibliographical references and index.
ISBN 0-19-502463-1
1. Religion—Philosophy. 2. Apologetics—20th century. 3. Christianity and existentialism. 4. Zen Buddhism—Philosophy. I. Title.
BL51.R68 200'.1 78-15681
78 MARC

CHRISTIANITY AND EXISTENTIALISM IN LITERATURE

Coles, Robert.
Walker Percy, an American search / by Robert Coles. — 1st ed. — Boston : Little, Brown, c1978.
xx, 250 p. ; 24 cm.
"An Atlantic-Monthly Press book."
Includes index.
ISBN 0-316-15160-2
1. Percy, Walker, 1916- . 2. Christianity and existentialism in literature. 3. Authors, American—20th century—Biography. I. Title.
PS3566.E6912Z6 813'.5'4 78-13629
78 MARC

CHRISTIANITY AND INTERNATIONAL AFFAIRS

see also Catholic Church and world politics; Church and underdeveloped areas

Månson, Thorsten, 1924-
En utmaning. Det internationella ansvaret i blickpunktet. Teckningar av Staffan Lindén. Stockholm, Verbum, 1973.
62 p. illus. 19 cm. (Triangelserien) kr.8.25 S73-42
1. Christianity and international affairs. 2. Church and underdeveloped areas. I. Title.
BR115.I7M28 79-384578
79 MARC

—ADDRESSES, ESSAYS, LECTURES

Frieden, Versöhnung und Menschenrechte. — Gütersloh : Gütersloher Verlagshaus Mohn, 1978-
v. ; 19 cm. — (Die Denkschriften der Evangelischen Kirche in Deutschland ; Bd. 1, T. 1-) (Gütersloher Taschenbücher ; 13- : Siebenstern)
GFR78-A
Includes bibliographical references.
ISBN 3-579-04801-5 (v. 1, pt. 1) : DM12.80 (v. 1, pt. 1)
1. Sociology, Christian—Addresses, essays, lectures. 2. Church work with refugees—Addresses, essays, lectures. 3. Church and race relations—Addresses, essays, lectures. 4. Christianity and international affairs—Addresses, essays, lectures. 5. Peace (Theology)—Addresses, essays, lectures. I. Series.
BT738.F68 78-380301
79 MARC

Wereld in steigers. [n.p.] Interkerkelijk Vredesberaad [1970]
52 p. illus. 28 cm. (Cahiers voor Vredesvraagstukken, 8) Ne***
Cover title.
1. Church and social problems—Netherlands—Addresses, essays, lectures. 2. Christianity and international affairs—Addresses, essays, lectures. I. Interkerkelijk Vredesberaad. II. Series.
HN39.N4W47 79-345428
79 MARC

CHRISTIANITY AND JUSTICE

Casaldáliga, Pedro.
I believe in justice and hope / Pedro Casaldáliga. — Notre Dame, Ind. : Fides/Claretian, c1978.
v, 232 p. ; 20 cm.
Translation of Yo creo en la justicia y en la esperanza!
ISBN 0-8190-0628-9
1. Casaldáliga, Pedro. 2. Catholic Church—Bishops—Biography. 3. Bishops—Brazil—Biography. 4. Church and social problems—Brazil. 5. Church and social problems—Catholic Church. 6. Christianity and justice. I. Title.
BX4705.C33547A3813 282'.092'4 78-21694
78 MARC

Lumbreras Meabe, Juan María.
El cristianismo es amor y exigencia de justicia / [por] Juan María Lumbreras Meabe. — Bilbao : Mensajero, D.L. 1977.
375 p. ; 20 cm. Sp77-Nov
Includes bibliographical references.
ISBN 8427110626
1. God—Love. 2. God—Worship and love. 3. Christianity and justice. I. Title.
BT140.L85 79-385525
79 MARC

—ADDRESSES, ESSAYS, LECTURES

Catholic Church. Irish Episcopal Conference.
The work of justice : Irish bishops' pastoral. — Dublin : Veritas Publications, 1977.
81 p. ; 19 cm. GB78-35846
"Full text of the Pastoral Letter issued by the Irish Catholic Bishops' Conference on September 11th 1977."
Imprint covered by label: Sole distributors in U.K., Fowler Wright Books Limited, Tenbury Wells, Worcs., Eng.
"Cat. no. 3362."
Bibliography: p. 78-[81]
ISBN 0-905092-24-4 : £0.66
1. Christianity and justice—Addresses, essays, lectures. 2. Christian ethics—Addresses, essays, lectures. 3. Social ethics—Addresses, essays, lectures. 4. Ireland—Economic conditions—Addresses, essays, lectures. 5. Catholic Church—Pastoral letters and changes. I. Title.
BR115.J8C37 1977 261.8 79-316882
79 MARC

Rättvisan kan inte vänta. Idé-bok. Ekumeniska u-veckan. Förf.: Bolennart Andersson ... Stockholm, Gummesson; [Solna, Seelig] 1973.
182, (1) p. 19 cm. kr17.65 S73-42
Bibliography: p. 179-182.
ISBN 9170703795
1. Christianity and justice—Addresses, essays, lectures. 2. Church and social problems—Addresses, essays, lectures. I. Andersson, Bolennart, 1939-
BR115.J8R33 78-352848
79 MARC

CHRISTIANITY AND LANGUAGE

Dunning, Stephen Northrup.
The tongues of men / Stephen N. Dunning. — Missoula, Mont. : Scholars Press, c1979.
xi, 260 p. ; 23 cm. — (American Academy of Religion dissertation series ; no. 27 ISSN 0145-272X)
Bibliography: p. 249-260.
ISBN 0-89130-283-2
1. Christianity and language. 2. History (Theology) 3. Hamann, Johann Georg, 1730-1788. 4. Hegel, Georg Wilhelm, Friedrich, 1770-1831—Religion. I. Title. II. Series: American Academy of Religion. Dissertation series - American Academy of Religion ; 27.
BR115.L25D86 901 79-10729
79 MARC

CHRISTIANITY AND LAW

see also Law (Theology)

CHRISTIANITY AND LITERATURE

see also Christianity in literature; Story-telling (Christian theology)

Jones, Robert Maynard.
Llên Cymru a chrefydd ; diben y llenor / [gan] R. M. Jones. — Abertawe [i.e. Swansea] : C. Davies, 1977.
610 p. ; 22 cm. GB78-27001
Includes bibliographical references and index.
ISBN 0-7154-0439-3 : £7.50
1. Welsh literature—History and criticism. 2. Christianity and literature. I. Title.
PB2208.J6 79-359302
79 MARC

Oyola, Eliezer.
Los pecados capitales en la literatura medieval española / Eliezer Oyola. — Barcelona : Puvill, 1979.
270 p. ; 22 cm. — (Biblioteca universitaria Puvill : II, Ensayos ; 3)
Sp***
Bibliography: p. 261-270.
ISBN (invalid) 845202147
1. Spanish literature—To 1500—History and criticism. 2. Deadly sins in literature. 3. Christianity and literature. I. Title.
PQ6060.O96 79-112605
79 MARC

Pym, David.
The religious thought of Samuel Taylor Coleridge / David Pym ; with a foreword by John Coulson. — Gerrards Cross [Eng.] : C. Smythe, 1978.
105 p. ; 23 cm. GB***
Includes bibliographical references and index.
ISBN 0-901072-65-6 : £4.75
1. Coleridge, Samuel Taylor, 1772-1834—Religion and ethics. 2. Christianity and literature. I. Title.
PR4487.R4P9 1978 230'.092'4 79-303215
79 MARC

Pym, David.
The religious thought of Samuel Taylor Coleridge / David Pym ; with a foreword by John Coulson. — New York : Barnes & Noble Books, 1979, c1978.
105 p. : port. ; 22 cm.
Includes bibliographical references and index.
ISBN 0-06-495750-0 : $14.50
1. Coleridge, Samuel Taylor, 1772-1834—Religion and ethics. 2. Christianity and literature. I. Title.
PR4487.R4P9 1978b 821'.7 78-14840
78 MARC

Souffrant, Claude.
Une négritude socialiste : religion et développement chez J. Roumain, J.-S. Alexis, L. Hughes / Claude Souffrant ; préf. de Paul Ricœur. — Paris : Éditions l'Harmattan, [1978].
238 p. ; 22 cm. F***
Bibliography: p. 226-234.
ISBN 2-85802-052-3 : 42.00F
1. Roumain, Jacques, 1907-1944—Criticism and interpretation. 2. Alexis, Jacques Stephen, 1922- —Criticism and interpretation. 3. Hughes, Langston, 1902-1967—Criticism and interpretation. 4. Blacks—Race identity. 5. Race awareness in literature. 6. Socialism and literature. 7. Christianity and literature. I. Title.
PQ3949.R73Z88 78-381466
78 MARC

Thomsen, Allan.
Historical criticism : a reassessment / by Allan Thomsen. — Odense : [Udgivelsesudvalget ved Odense Universitet], 1977.
iii, 5, 100 p. ; 21 cm. — (Udgivelsesudvalgets samling af studenterafhandlinger ; v. 8) D78-4
Thesis Odense.
Summary in Danish.
Bibliograpy: p. [11]-[15]
ISBN 8787482363 : kr19.00
1. Literature, Medieval—History and criticism—Theory, etc. 2. Christianity and literature. 3. Robertson, Durant Waite. I. Title. II. Series: Odense universitet. Konsistorium. Kontaktudvalg med Center-boghandelen. Udgivelsesudvalget. Udgivelsesudvalgets samling af studenterafhandlinger ; v. 8.
PN671.T47 809 78-375326
78 MARC

CHRISTIANITY AND MAGIC

Haack, Friedrich-Wilhelm, 1935-
Aberglaube, Magie, Zauberei / Friedrich-Wilhelm Haack. — 1. Aufl. — München : Evangelischer Presseverband für Bayern, Abt. Schriftenmission, 1977.
43 p. ; 19 cm. — (Münchener Reihe) GFR78-A
Errata slip inserted.
ISBN 3-583-50626-X : DM2.80
1. Christianity and superstition. 2. Superstition—Controversial literature. 3. Christianity and magic. 4. Magic—Controversial literature. I. Title.
BR115.S85H2 79-370220
79 MARC

CHRISTIANITY AND OCCULT SCIENCES

Brooks, Pat.
Healing of the mind / Pat Brooks. — Skyland, N.C. : New Puritan Library, c1978.
151 p. ; 18 cm.
ISBN 0-932050-00-X : $1.95
1. Christian life—1960- 2. Christianity and occult sciences. 3. Brooks, Pat. I. Title.
[BV4501.2.B766] 248'.4 78-59366
79 MARC

Dam, Willem Cornelis van, 1926-
Okkultisme en christelijk geloof / W. C. van Dam. — Den Haag : Voorhoeve, [1978]
120 p. ; 21 cm. — (Pastorale handreiking ; nr. 10) Ne78-18/19
Bibliography: p. 116-117.
ISBN 9029705353 : fl 14.50
1. Christianity and occult sciences. I. Title.
BR115.O3D35 79-385708
*00 MARC

Hoover, David W
How to respond to ... the occult / David W. Hoover. — St. Louis : Concordia Pub. House, c1977.
31 p. ; 23 cm. — (The Response series)
Bibliography: p. 31.
ISBN 0-570-07678-1 : $1.25
1. Christianity and occult sciences. I. Title. II. Title: Occult.
BR115.O3H66 133 77-22055
77 MARC

—CONTROVERSIAL LITERATURE

Inglesby, Eric.
UFO's and the Christian / [by] Eric Inglesby. — London ; New York : Regency Press, 1978.
162 p. ; 23 cm. GB78-30325
Bibliography: p. [8]
£2.40 ($6.00)
1. Christianity and occult sciences—Controversial literature. 2. Catholic Church—Doctrinal and controversial works—Anglican authors. I. Title.
BR115.O3I53 001.9'42 79-305527
79 MARC

CHRISTIANITY AND OTHER RELIGIONS

see also Christianity and religious humanism; Paganism

Bakker, Y W M 1916-
Umat Katolik berdialog dengan umat beragama lain / Y. W. M. Bakker. — Cet. 2. — Yogyakarta : Penerbitan Yayasan Kanisius, 1976.
121 p. ; 21 cm.
First ed. published in 1972 under title: Nostra aetate, zaman kita zaman dialog antar agama.
Includes bibliographies.
1. Christianity and other religions. 2. Catholic Church—Relations. I. Title.
BR127.B28 1976 78-940644
79 MARC

Bühlmann, Walbert.
Alle haben denselben Gott : Begegnung mit d. Menschen u. Religionen Asiens / Walbert Bühlmann. — 1. Aufl. — Frankfurt am Main : Knecht, 1978.
223 p. ; 21 cm. GFR78-A
Bibliography: p. 220-223.
ISBN 3-7820-0414-0
1. Christianity and other religions. 2. Religions. I. Title.
BR127.B76 78-400386
79 MARC

Collet, Paul.
I pagani mi hanno parlato di Dio. Presentazione di Carlo Carretto. Assisi, Cittadella editrice, 1971.
102 p. 19 cm. (Quaderni di spiritualità, 8) L800 It71-July
1. Christianity and other religions. 2. Salvation outside the Catholic Church. 3. Collet, Paul. I. Title.
BR127.C5616 79-384567
79 MARC

Dalmais, Irénée Henri, 1914-
La Foi au Christ parmi les religions des hommes / I. H. Dalmais. — [Paris] : Desclée De Brouwer ; [Montréal] : Bellarmin, 1978.
146 p. ; 20 cm. — (Croire aujourd'hui ISSN 0337-1360) F78-14233
Bibliography: p. 143-146.
ISBN 2-220-02156-4 : 22.00F
1. Christianity and other religions. 2. Religions. I. Title.
BR127.D33 261.2 79-350075
79 MARC

Ferré, Nels Fredrick Solomon, 1908-
The finality of faith and Christianity among the world religions / by Nels F. S. Ferré. — Westport, Conn. : Greenwood Press, 1979, c1963.
x, 115 p. ; 23 cm.
Reprint of the ed. published by Harper & Row, New York.
ISBN 0-313-21182-5
1. Faith. 2. Christianity and other religions. I. Ferré, Nels Fredrick Solomon, 1908- Christianity among the world religions. 1979.
[BT771.2.F4 1979] 234'.2 78-11979
78 MARC

Friedli, Richard.
Fremdheit als Heimat : auf der Suche nach einem Kriterium für den Dialog zwischen den Religionen / Richard Friedli. — Freiburg/Schweiz : Universitätsverlag, c1974.
214 p. ; 23 cm. — (Ökumenische Beihefte zur Freiburger Zeitschrift für Philosophie und Theologie ; 8) Sw74-A
Originally presented as the author's thesis, Fribourg.
Bibliography: p. 14-19.
Includes indexes.
1. Christianity and other religions. 2. Religions. I. Title. II. Series: Freiburger Zeitschrift für Philosophie und Theologie : Ökumenische Beihefte ; 8.
BR127.F74 1974 79-367150
*79 MARC

Giudici, Amilcare.
Religioni e salvezza : un confronto tra la teologia cattolica e la teologia protestante / Amilcare Giudici. — Roma : Borla, 1978.
217 p. ; 21 cm. — (Ricerche teologiche) It78-Oct
Includes bibliographical references.
L5000
1. Salvation outside the church—History of doctrines. 2. Christianity and other religions. I. Title. II. Series.
BT759.G58 78-395546
*79 MARC

Hiyane, Antei, 1892-
(Fukuin to ikyō jiban)
福音と異教地盤 / 比屋根安定著. — 東京：日本基督教団出版部, 昭和36 [1961]
233 p. ; 19 cm.
"附録　基督教対インド宗教に関する参考文献": p. 227-232.
List of the author's works: p. 233.
1. Japan—Religion. 2. Christianity and other religions. I. Title.
BL2202.H55 78-811457

Kharkongngor, Iarington.
La pyndep / da Iarington Kharkongngor. — [Shillong : Ri Khasi Press, pref. 1969]
ii, 52 p. ; 19 cm.
Cover title.
In Khasi.
1. Khasis—Religion. 2. Christianity—Relations. I. Title.
BL2032.K45K46 78-901742
79 MARC

Kim, Kwang-su.
(Kidokkyo wa t'a chonggyo wa ŭi taehwa)
기독교와타종교와의대화 / 김광수. — 서울 : 한국기독교교회협의회, 1977.
146 p. ; 19 cm. — (에큐메니칼문고 ; 5)
Bibliography: p. 143-146.
1. Religions. 2. Christianity and other religions. I. Title. II. Series: Ek'yumenik'al mun'go ; 5.
BL95.K55 79-825738

Luis de Granada, 1504-1588.
[Guia de pecadores. Japanese]
ぎやどぺかどる / [Luis de Granada 著]. 妙貞問答；破提宇子 / [ハビアン著]. 顯僞錄 / [澤野忠庵著；與謝野寛, 正宗敦夫, 與謝野晶子編纂校訂]. — 東京：現代思潮社, 1978.
[44], 186, 134, 73, 30, 25 p. ; 16 cm. — (覆刻日本古典全集)
Ja 79-825
Spine title.
Photoreprint of the 1927 ed. published in 2 v. by Nihon Koten Zenshū Kankōkai, Tokyo.
1. Christian life—Catholic authors. 2. Christianity and other religions. 3. Christianity—Controversial literature. I. Fabian, Fucan, fl. 1583-1607. Myōtei mondō. 1978. II. Fabian, Fucan, fl. 1583-1607. Ha Deusu. 1978. III. Ferreira, Christovão, 1580-ca. 1652. Kengiroku. 1978. IV. Title. V. Series: Nihon koten zenshū.
Title romanized: Giya do pekadoru.
BX2349.L716 1978 79-805675

Romarheim, Arild, 1946-
Krishna, Buddha, Allah eller Kristus / Arild Romarheim. — 2. oppl. — Oslo : Luther, 1978 (Oslo : Furuset boktr.)
135 p. ; 19 cm.
Includes bibliographies and index.
ISBN 8253190670 : kr27.00
1. Christianity and other religions. I. Title.
BR127.R59 1978 79-343647
79 MARC

Schuon, Frithjof, 1907-
De l'unité transcendante des religions / Frithjof Schuon. — Nouv. éd. rev. et augm. — Paris : Seuil, c1979.
185 p. ; 21 cm. F***
ISBN 2-02-005059-5 : 45.00F
1. Religion—Philosophy. 2. Christianity and other religions. 3. Religions. I. Title.
BL51.S464 1979 200'.1 79-353101
79 MARC

Song, Choan-Seng, 1929-
Third-eye theology : theology in formation in Asian settings / Choan-Seng Song. — Maryknoll, N.Y. : Orbis Books, c1979.
xiii, 274 p. ; 24 cm.
Includes bibliographical references.
ISBN 0-88344-474-7
1. Theology, Doctrinal. 2. Christianity and other religions. 3. Asia—Religion. 4. Asia—Civilization. I. Title.
BT78.S66 230 79-4208
79 MARC

Van der Bent, Ans Joachim.
God so loves the world : the immaturity of world Christianity / A. J. van der Bent ; foreword by David E. Jenkins. — Madras : Christian Literature Society, 1977.
xvi, 143 p. ; 23 cm.
Includes bibliographical references.
Rs10.00
1. Christianity—20th century. 2. Christian union. 3. Christianity and other religions. I. Title.
BR121.2.V324 209'.04 78-900901
78 MARC

Vernette, Jean.
Sectes et réveil-religieux : quand l'Occident s'éveille / Jean Vernette. — Mulhouse : Éditions Salvator, 1976.
199 p. ; 19 cm. F77-8662
Includes bibliographical references.
ISBN 2-7067-0021-1 : 35.00F
1. Christianity and other religions. 2. Cults—Controversial literature. 3. Christian sects—Controversial literature. I. Title.
BR127.V4 261.2 78-378234
78 MARC

Wilson, Howard A
Invasion from the East / Howard A. Wilson. — Minneapolis : Augsburg Pub. House, c1978.
160 p. ; 20 cm.
Includes bibliographical references.
ISBN 0-8066-1671-7 : $3.95
1. Religions. 2. Christianity and other religions. I. Title.
BL80.2.W56 261.2 78-52203
79 MARC

Young, Robert Doran, 1928-
Encounter with world religions, by Robert D. Young. Philadelphia, Westminster Press [1970]
223 p. 21 cm. 7.50
Includes bibliographical references.
ISBN 0-664-20876-2
1. Christianity and other religions. I. Title.
BR127.Y6 261.2 75-86531
00[r79]rev MARC

—ADDRESSES, ESSAYS, LECTURES

Panikkar, Raymond, 1918-
The intrareligious dialogue / R. Panikkar. — New York : Paulist Press, c1978.
xxviii, 104 p. : ill. ; 24 cm.
Includes bibliographical references and index.
ISBN 0-8091-0273-0 : $5.95
1. Religions—Addresses, essays, lectures. 2. Christianity and other religions—Addresses, essays, lectures. I. Title.
BL87.P26 291.1'72 78-58962
79 MARC

Schweitzer, Albert, 1875-1965.
Das Christentum und die Weltreligionen : 2 Aufsätze zur Religionsphilosophie / Albert Schweitzer : mit e. Einf. in d. Denken Albert Schweitzers von Ulrich Neuenschwander. — 1. Aufl. — München : Beck, 1978.
124 p. ; 18 cm. — (Beck'sche schwarze Reihe ; Bd. 181) GFR78-A
Includes Das Problem der Ethik by A. Schweitzer.
Includes bibliographical references.
1. Christianity and other religions—Addresses, essays, lectures. 2. Religion—Philosophy—Addresses, essays, lectures. I. Neuenschwander, Ulrich. II. Schweitzer, Albert, 1875-1965. Das Problem der Ethik. 1978. III. Title.
BR127.S32 1978 79-343274
79 MARC

CHRISTIANITY AND OTHER RELIGIONS (Continued)

—BUDDHISM

Amore, Roy C 1942-
Two masters, one message / Roy C. Amore. — Nashville : Abingdon, c1978.
208 p. ; 19 cm.
Bibliography: p. 200-₍201₎
Includes indexes.
ISBN 0-687-42750-9 (cloth)
1. Christianity and other religions—Buddhism. 2. Buddhism—Relations—Christianity. 3. Buddhism—Influence. I. Title.
BR128.B8A47 261.2 77-18062
77 MARC

The Buddhist-Christian encounter. — ₍Brussels₎ : Pro Mundi Vita, 1977.
32 p. ; 27 cm. — (Bulletin - Pro Mundi Vita ; 67) Be***
Caption title.
Bibliography: p. 28-32.
1. Christianity and other religions—Buddhism. 2. Buddhism—Relations—Christianity. I. Series: Pro Mundi Vita (Society). Bulletin ; 67.
BR128.B8B82 261.2 79-305715
79 MARC

De Silva, Lynn A 1919-
The problem of the self in Buddhism and Christianity / Lynn A. de Silva ; foreword by John Hick. — New York : Barnes & Noble, 1979.
xv, 185 p. ; 22 cm. — (Library of philosophy and religion)
Includes bibliographical references and indexes.
ISBN 0-06-491667-7 : $22.50
1. Anātman. 2. Man (Theology) 3. Christianity and other religions—Buddhism. 4. Buddhism—Relations—Christianity. I. Title.
BQ4262.D47 1979 291.2'2 76-19860
78 MARC

Dharmasiri, Gunapala.
A Buddhist critique of the Christian concept of God : a critique of the concept of God in contemporary Christian theology and philosophy of religion from the point of view of early Buddhism / Gunapala Dharmasiri. — Colombo : Lake House Investments, 1974.
xi, 312 p. ; 22 cm.
Originally presented as the author's thesis, University of Lancaster.
Bibliography: p. ₍300₎-306.
Includes index.
1. Christianity and other religions—Buddhism. 2. Buddhism—Relations—Christianity. 3. God. I. Title.
BR128.B8D46 1974 231 75-903604
79 MARC

Mensching, Gustav, 1901-
Buddha und Christus : e. Vergl. / Gustav Mensching. — Stuttgart : Deutsche Verlags-Anstalt, 1978.
287 p. ; 21 cm. GFR78-A
Bibliography: p. ₍279₎-280.
Includes index.
ISBN 3-421-01870-7 : DM32.00
1. Christianity and other religions—Buddhism. 2. Buddhism—Relations—Christianity. I. Title.
BR128.B8M46 78-396279
79 MARC

Nakamura, Hajime, 1912-
Buddhism in comparative light / Hajime Nakamura. — 1st ed. — New Delhi : Islam and the Modern Age Society ; Bombay : distributors, Current Book House, 1975.
xi, 185 p. ; 23 cm.
Includes bibliographical references and index.
Rs40.00
1. Christianity and other religions—Buddhism. 2. Buddhism—Relations—Christianity. I. Title.
BR128.B8N34 294.3 76-900794
78 MARC

Shiotsu, Jun'ichi.
(Kirisuto to Nichiren)
基督と日蓮 / 塩津淳一著. — 東京：巖松堂書店, 昭和23 ₍1948₎
8, 345 p. ; 22 cm.

1. Jesus Christ. 2. Nichiren, 1222-1282. 3. Christianity and other religions—Buddhism. 4. Buddhism—Relations—Christianity. I. Title.
BT202.S52 3·A: 79-807255

—BUDDHISM—ADDRESSES, ESSAYS, LECTURES

Buddhism and Christianity / edited by Claude Geffré and Mariasusai Dhavamony. — New York : Seabury Press, 1979.
x, 126 p. ; 23 cm. — (Concilium ; 116)
"A Crossroad book."
Includes bibliographical references.
ISBN 0-8164-0395-3
1. Christianity and other religions—Buddhism—Addresses, essays, lectures. 2. Buddhism—Relations—Christianity—Addresses, essays, lectures. I. Geffrè, Claude. II. Dhavamony, Mariasusai. III. Series: Concilium (New York) ; 116.
BR128.B8B8 261.2 78-71625
79 MARC

Cannon, Joseph L 1927-
(Bukkyō tai Kirisutokyō)
佛教對キリスト教：討論 = Buddhism vs. Christianity : a debate / Joe Cannon, T. Ishida. — 東京：代々木八幡基督之教会, 昭和28 ₍1953₎
6, 105 p. ; 21 cm.
In Japanese.
Includes bibliographical references.

1. Christianity and other religions—Buddhism—Addresses, essays, lectures. 2. Buddhism—Relations—Christianity—Addresses, essays, lectures. I. Ishida, Tsugio, 1925- II. Title. III. Title: Buddhism vs. Christianity.
BR128.B8C27 78-811111

—BUDDHISM—ADDRESSES, ESSAYS, LECTURES

Phorn Ratanasuwan.
Mental health and religion : lectures delivered at the Thailand Theological Seminary, the Church of Christ in Thailand, Chiengmai, August₍sic₎ 9-11, 1971/2514 / by Phorn Ratanasuwan. — Bangkok : Home of the Psychical Research, ₍1972₎
a-f, 137 p., ₍1₎ leaf of plates : ill. ; 20 cm.
Translated from Thai.
1. Christianity and other religions—Buddhism—Addresses, essays, lectures. 2. Buddhism—Relations—Christianity—Addresses, essays, lectures. 3. Psychology, Religious—Addresses, essays, lectures. I. Title.
BR128.B8P4713 261.2 78-915076
79 MARC

Spae, Joseph John, 1913-
East challenges West : towards a convergence of spiritualities / by Joseph J. Spae. — South Yarra, Australia : Charles Strong Memorial Trust ; Chicago : distribution by Chicago Institute of Theology and Culture, 1979.
81 p. ; 26 cm. — (The Charles Strong Memorial Trust lecture ; 1979)
Includes bibliographical references.
1. Japan—Religion—Addresses, essays, lectures. 2. Christianity and other religions—Buddhism—Addresses, essays, lectures. 3. Buddhism—Relations—Christianity—Addresses, essays, lectures. 4. East and West—Addresses, essays, lectures. I. Title. II. Series: Charles Strong Memorial Trust. Charles Strong Memorial Trust lecture ; 1979.
BL2210.S62 261.2 79-52682
79 MARC

—BUDDHISM—CONGRESSES

(Shūkyō taiken to kotoba)
宗教体験と言葉：仏教とキリスト教との対話 / 南山宗教文化研究所編. — 東京：紀伊国屋書店, 1978.
414 p. ; 20 cm. — (Nanzan symposium ; 1) Ja 78-19627
¥2600

1. Christianity and other religions—Buddhism—Congresses. 2. Buddhism—Relations—Christianity—Congresses. 3. Religion and language—Congresses. I. Nanzan Shūkyō Bunka Kenkyūjo. II. Series.
BR128.B8S46 78-816788

—CHINESE

Bernard, Henri, S.J.
Sagesse chinoise et philosophie chrétienne; essai sur leurs relations historiques par Henri Bernard ... Hautes études ... Tientsin. Tientsin, En vente à la procure de la Mission de Sienshien, 1935.
2 p.l., 277, vi p. 24 cm.
1. Philosophy, Chinese. 2. Christianity and other religions—Chinese. 3. Jesuits—China—History. 4. China—Religion. I. T'ien-chin kung shang hsüeh yüan. II. Title.
B126.B4 1935 181'.11 39-23523
₍r79₎rev MARC

Bernard, Henri, S. J.
Sagesse chinoise et philosophie chrétienne, par Henri Bernard-Maitre. Paris, Cathasia ₍1950₎
277, v p. 24 cm. (Les Humanités d'Extrême-Orient) (Série culturelle des Hautes Ètudes de Tientsin)
Bibliographical footnotes.
1. Philosophy, Chinese. 2. Christianity and other religions—Chinese. 3. Jesuits—China—History. 4. China—Religion. I. Title. II. Series: Tientsin. Institut des hautes études industrielles et commerciales. Série culturelle.
B126.B4 1950 181'.11 52-67676
₍r79₎rev MARC

—COLLECTED WORKS

Christentum im Spiegel der Weltreligionen : krit. Texte u. Kommentare / hrsg. von Heinz-Jürgen Loth, Michael Mildenberger, Udo Tworuschka. — 1. Aufl. — Stuttgart : Quell-Verlag, 1978.
374 p. ; 20 cm. — (Studienbücher im Gespräch mit der Zeit : Religion) GFR79-A
Includes bibliographical references and indexes.
ISBN 3-7918-6004-6 : DM18.00
1. Christianity and other religions—Collected works. 2. Christianity—Controversial literature—Collected works. I. Loth, Heinz-Jürgen, 1942- II. Mildenberger, Michael, 1934- . III. Tworuschka, Udo.
BR127.C45 79-351008
79 MARC

—CONFUCIANISM—ADDRESSES, ESSAYS, LECTURES

Legge, James, 1815-1897.
The religions of China : Confucianism and Tâoism described and compared with Christianity / by James Legge. — Philadelphia : R. West, 1978.
ix, 310 p. ; 23 cm.
Reprint of the 1880 ed. published by Hodder and Stoughton, London.
Includes bibliographical references.
ISBN 0-8492-1545-5 lib. bdg. : $30.00
1. China—Religion—Addresses, essays, lectures. 2. Confucianism—Relations—Christianity—Addresses, essays, lectures. 3. Taoism—Relations—Christianity—Addresses, essays, lectures. 4. Christianity and other religions—Confucianism—Addresses, essays, lectures. 5. Christianity and other religions—Taoism—Addresses, essays, lectures. I. Title.
₍BL1801.L4 1978₎ 299'.51 78-2685
78 MARC

—CONGRESSES

Denkpause im Dialog : Perspektiven d. Begegnung mit anderen Religionen u. Ideologien / Michael Mildenberger ₍Hrsg.₎. — 1. Aufl. — Frankfurt am Main : Lembeck, 1978.
157 p. ; 21 cm. GFR78-A
Most articles translated from English; originally published in Faith in the midst of faiths, Geneva, 1977.
Includes selected material from a conference sponsored by the Sub-unit for Dialogue with People of Living Faiths and Ideologies of the World Council of Churches at Chiang Mai, Thailand, April 18-27, 1977.
ISBN 3-87476-104-5 : DM14.80
1. Christianity and other religions—Congresses. I. Mildenberger, Michael, 1934- . II. World Council of Churches. Sub-unit for Dialogue with People of Living Faiths and Ideologies.
BR127.D46 78-390582
79 MARC

Faith in the midst of faiths : reflections on dialogue in community / ed. by Stanley J. Samartha. — Geneva : World Council of Churches, 1977.
200 p. ; 22 cm. Sw77-9455
Presented to a theological consultation at Chiang Mai, Thailand in April, 1977, organized by the World Council of Churches' sub-unit on dialogue with people of living faiths and ideologies.
Bibliography: p. 199-200.
ISBN 2-8254-0545-0 : 12.50F
1. Christianity and other religions—Congresses. 2. Mission of the church—Congresses. I. Samartha, Stanley J., 1920- . II. World Council of Churches.
BR127.F34 261.2 78-314290
*78 MARC

—HINDUISM

see also Jesus Christ --Hindu interpretations

Hooker, Roger Hardham, 1934-
Uncharted journey ₍by₎ Roger Hooker. London, Church Missionary Society, 1973.
48 p. 22 cm. £0.20 GB73-29114
ISBN 0-900267-06-3
1. Christianity and other religions—Hinduism. 2. Christianity and other religions—Islam. 3. Hinduism—Relations—Christianity. 4. Islam—Relations—Christianity. I. Title.
BR128.H5H674 291.1'72'0954 74-157254
74₍r79₎rev MARC

CHRISTIANITY AND OTHER RELIGIONS —HINDUISM (Continued)

Oak, Purushottam Nagesh, 1917-
Christianity is Chrisn-nity / by P. N. Oak. — 1st ed. — New Delhi : Institute for Rewriting Indian History, 1978.
x, 216 p., ⌜3⌝ leaves of plates : ill. ; 23 cm.
Corrections in ms.
Bibliography: p. ⌜208⌝-210.
Includes index.
Rs15.00 ($4.00 U.S.)
1. Christianity and other religions—Hinduism. 2. Hinduism—Relations—Christianity. 3. Christianity—Controversial literature. 4. Jesus Christ—Historicity. 5. Krishna. 6. Jesus Christ—Hindu interpretations. I. Title.
BR128.H5O22 209 78-908025
79 MARC

—HINDUISM—ADDRESSES, ESSAYS, LECTURES

Mattam, Joseph, 1932-
Land of the trinity : a study of modern Christian approaches to Hinduism / by Joseph Mattam. — Bangalore : Theological Publications in India, ⌜foreword 1975⌝
200 p. ; 23 cm.
"Essays and research."
Bibliography: p. ⌜189⌝-200.
Rs11.00
1. Christianity and other religions—Hinduism—Addresses, essays, lectures. 2. Hinduism—Relations—Christianity—Addresses, essays, lectures. I. Title.
BR128.H5M35 261.2 76-900845
77⌜r79⌝rev MARC

—ISLAM

see also Koran—Relation to the Bible; Missions to Muslims

Assfy, Zaid H
Islam and Christianity : connections and contrasts, together with the stories of the prophets and imams / by Zaid H. Assfy. — York : Sessions, 1977.
x, 122 p. ; 22 cm. GB78-21789
Bibliography: p. ⌜iv⌝
ISBN 0-900657-38-3 : £5.75
1. Shiites. 2. Christianity and other religions—Islam. 3. Islam—Relations—Christionity. 4. Koran stories. I. Title.
BP193.5.A84 297'.122'0922 78-320778
78 MARC

Basetti-Sani, Giulio, 1912-
The Koran in the light of Christ : a Christian interpretation of the sacred book of Islam / by Giulio Basetti-Sani ; translated by W. Russell Carroll and Bede Dauphinee. — Chicago : Franciscan Herald Press, c1977.
xiii, 223 p. ; 21 cm.
First published (c1977) under title: Il Corano nella luce di Cristo.
Includes bibliographical references.
ISBN 0-8199-0713-8
1. Christianity and other religions—Islam. 2. Islam—Relations—Christianity. 3. Koran—Criticism, interpretation, etc. 4. Jesus Christ in the Koran. I. Title.
BP172.B327 1977 297'.1226 76-28786
76 MARC

Dretke, James P 1931-
A Christian approach to Muslims : reflections from West Africa / James P. Dretke. — Pasadena, Calif. : W. Carey Library, c1979.
xviii, 261 p. ; 22 cm. — (The William Carey Library series on Islamic studies)
Bibliography: p. 239-251.
Includes index.
ISBN 0-87808-432-0
1. Islam—Relations—Christianity. 2. Christianity and other religions—Islams. 3. Islam—Africa, West. I. Title.
BP172.D73 261.2 79-11912
79 MARC

Frieling, Rudolf.
Christianity and Islam : a battle for the true image of man / Rudolf Frieling. — Edinburgh : Floris Books, c1978.
144 p. ; 22 cm. GB***
Translation of Christentum und Islam.
Bibliography: p. 138-139.
Includes index.
ISBN 0-903540-11-8 : £5.95
1. Islam—Relations—Christianity. 2. Christianity and other religions—Islam. I. Title.
BP172.F6813 297'.197'2 79-307249
79 MARC

Gibb, Hamilton Alexander Rosskeen, Sir, 1895-1971.
(Maẕhab va siyāsat dar Masīḥīyat va Islām)
مذهب و سياســـت در مسيحيـــت و اسلام / هاميلتون
آ. ر. كيب ⌜i. e.⌝ گيب⌝ ؛ ترجمه م. قائنى. ـــ قم : انتشارات
دار الفكر ، ⌜1972 or 1973⌝ 1351.
96 p. ; 17 cm.
In Persian.
1. Islam and state. 2. Church and state. 3. Islam—Relations—Christianity. 4. Christianity and other religions—Islam. I. Title.
BP173.6.G55 1972 79-971193

Hooker, Roger Hardham, 1934-
Uncharted journey ⌜by⌝ Roger Hooker. London, Church Missionary Society, 1973.
48 p. 22 cm. £0.20 GB73-29114
ISBN 0-900267-06-3
1. Christianity and other religions—Hinduism. 2. Christianity and other religions—Islam. 3. Hinduism—Relations—Christianity. 4. Islam—Relations—Christianity. I. Title.
BR128.H5H674 291.1'72'0954 74-157254
74⌜r79⌝rev MARC

Lubis, M Arsjad Thalib.
Perbandingan agama Kristen dan Islam ⌜oleh⌝ H. M. Arsjad Thalib Lubis. Medan, Islamyah, 1971.
2 v. (494 p.) 22 cm.
Bibliography: v. 1, p. 5-10.
1. Islam—Relations—Christianity. 2. Christianity and other religions—Islam. I. Title.
BP172.L78 73-943340
78 MARC

Lelong, Michel.
Deux fidélités, une espérance : chrétiens et musulmans aujourd'hui / Michel Lelong ; préface de Mgr Huyghe. — Paris : Éditions du Cerf, 1979.
212 p. ; 20 cm. — (Rencontres ; 11 ISSN 0182-2160) F79-8161
Includes bibliographical references.
ISBN 2-204-01312-9 : 46.00F
1. Islam—Relations—Christianity. 2. Christianity and other religions—Islam. I. Title.
BP172.L36 261.2 79-377194
79 MARC

Oster, Kenneth.
Islam reconsidered : a brief historical background to the religion and thought of the Moslem world / Kenneth Oster. — 1st ed. — Hicksville, N.Y. : Exposition Press, c1979.
131 p. : maps ; 22 cm. — (An Exposition-university book)
Bibliography: p. 119-122.
Includes index.
ISBN 0-682-49189-6 : $7.50
1. Islam. 2. Islam—Relations—Christianity. 3. Christianity and other religions—Islam. I. Title.
BP161.2.O86 297 78-62392
79 MARC

Register, Ray G
Dialogue and interfaith witness with Muslims : a guide and sample ministry in the U.S.A. / by Ray G. Register, Jr. — 1st ed. — Kingsport, TN : Distributed in the U.S.A. by Moody Books, c1979.
xii, 99 p. ; 21 cm.
Bibliography: p. 95-99.
ISBN 0-9603018-0-1
1. Missions to Muslims—United States. 2. Christianity and other religions—Islam. 3. Islam—Relations—Christianity. I. Title.
BV2625.R38 261.2 79-6575
79 MARC

Sekali lagi hubungan Islam dan Kristen di Indonesia. — Jakarta : Media Da'wah, 1978.
72 p. ; 18 cm.
Cover title.
1. Missions—Indonesia. 2. Islam—Relations—Christianity. 3. Christianity and other religions—Islam.
BP63.I5S37 79-940097
79 MARC

Shalabī, Ra'ūf.
(Aḍwā' 'alá al-Masīḥīyah)
اضواء على المسيحية : دراسات في أصول المسيحيـــة / بقلم
رؤف شلبي. ـــ صيدا : المكتبة العصرية ، 1975.
176 p. ; 20 cm.
Bibliography : p. 171-172.
1. Islam—Relations—Christianity. 2. Christianity and other religions—Islam. I. Title.
BP172.S38 73-970439

—ISLAM—ADDRESSES ESSAYS, LECTURES

al-'Ālim, Jalāl.
(Qādat al-Gharb yaqūlūn, damirrū al-Islām, abīdū ahlah)
قادة الغرب يقولون ، دمروا الاســلام ، ابيـــدوا اهله / جلال
العالم. ـــ الطبعة 2. ـــ 1975 ,⌜s. l. : s. n.⌝
68 p. ; 20 cm.
Includes bibliographical references.
1. Islam—Relations—Christianity—Addresses, essays, lectures. 2. Christianity and other religions—Islam—Addresses, essays, lectures. 3. Islamic countries—Foreign relations—Addresses, essays, lectures. I. Title.
BP172.A45 1975 79-971502

Aro, Jussi.
Juutalaisuuden ja Islamin peruspiirteitä / Jussi Aro. — Helsinki : Kirjaneliö, 1978 (Pieksämäki : Raamattutalo).
29 p. ; 21 cm. Fi78-7
ISBN 9516004121 : Fmk5.00
1. Islam—Relations—Christianity—Addresses, essays, lectures. 2. Christianity and other religions—Islam—Addresses, essays, lectures. I. Title.
BP172.A74 79-355570
79 MARC

Muslim-Christian conflicts : economic, political, and social origins / edited by Suad Joseph and Barbara L. K. Pillsbury ; foreword by Charle Issawi. — Boulder, Colo. : Westview Press, 1978.
xxi, 245 p. : maps ; 22 cm. — (A Westview/Dawson replica edition)
Includes bibliographies and index.
ISBN 0-89158-256-8 : $17.00
1. Islam—Relations—Christianity—Addresses, essays, lectures. 2. Christianity and other religions—Islam—Addresses essays, lectures. 3. Near East—Politics and government—1945- —Addresses, essays, lectures. 4. Religion and politics—Near East—Addresses, essays, lectures. I. Joseph, Suad. II. Pillsbury, Barbara.
BP172.M797 297'.197'2 78-21772
78 MARC

Seale, Morris S
Qur'an and Bible : studies in interpretation and dialogue / M. S. Seale. — London : Croom Helm, c1978.
124 p. ; 23 cm. GB***
Includes bibliographical references and index.
ISBN 0-85664-818-3 : £8.95
1. Koran—Criticism, interpretation, etc.—Addresses, essays, lectures. 2. Bible—Criticism, inerpretation, etc.—Addresses, essays, lectures. 3. Islam—Relations—Christianity—Addresses, essays, lectures. 4. Christianity and other religions—Islam—Addresses, essays, lectures. 5. Koran—Relation to the Bible—Addresses, essays, lectures. I. Title.
BP130.4.S42 297'.1226 79-303641
79 MARC

World Council of Churches.
Christians meeting Muslims : WCC papers on ten years of Christian-Muslim dialogue. — Geneva : World Council of Churches, c1977.
158 p. ; 24 cm. Sw***
Bibliography: p. 156-158.
ISBN 2-8254-0556-6
1. Islam—Relations—Christianity—Addresses, essays, lectures. 2. Christianity and other religions—Islam—Addresses essays, lectures. I. Title.
BP172.W67 1977 261.2 78-323898
79 MARC

—JAPANESE—ADDRESSES, ESSAYS, LECTURES

Fujiwara, Fujio.
(Nihon seishin to Kirisutokyō)
日本精神と基督教　藤原藤男著　⌜堺⌝　ともしび社　⌜昭和14 i. e. 1939⌝
74 p. 19 cm.
今年の紀元節の日に話した講演の筆記を少しく訂正増補したもの
1. Church and state in Japan—Addresses, essays, lectures. 2. Nationalism and religion—Japan—Addresses, essays, lectures. 3. Kokutai—Addresses, essays, lectures. 4. Christianity and other religions—Japanese—Addresses, essays, lectures. I. Title.
BR1305.5.F83 79-809091

—JUDAISM

see also Cabala and Christianity; Christianity and antisemitism; Jesus Christ—Biography—Sources, Jewish; Judaism (Christian theology); Missions to Jews; Rabbinical literature—Relation to the New Testament

Holmgren, Frederick, 1926-
The God who cares : a Christian looks at Judaism / by Frederick Holmgren. — Atlanta : John Knox Press, c1979.
144 p. ; 21 cm.
Includes bibliographies.
ISBN 0-8042-0588-4 : $4.95
1. Judaism. 2. Christianity and other religions—Judaism. 3. Judaism—Relations—Christianity. I. Title.
BM565.H64 296 78-52445
78 MARC

Weiss-Rosmarin, Trude, 1908-
Judaism and Christianity: the differences, by Trude Weiss-Rosmarin, Ph.D. New York, The Jewish book club ⌜c1943⌝
4 p.l., 7-158 p. 20 cm.
1. Judaism—Relations—Christianity. 2. Christianity and other religions—Judaism. I. Title.
BM535.W43 296.3'87'2 44-43012
⌜r79⌝rev MARC

CHRISTIAN AND OTHER RELIGIONS (Continued)

—JUDAISM—ADDRESSES, ESSAYS, LECTURES

Aron, Robert, 1898-1975.
Où souffle l'esprit : judaïsme et chrétienté / Robert Aron. — Paris : Plon, c1979.
322 p. ; 24 cm. F***
ISBN 2-259-00434-2 : 58.00F
1. Judaism—Relations—Christianity—Addresses, essays, lectures. 2. Christianity and other religions—Judaism—Addresses, essays, lectures. 3. Judaism—Essence, genius, nature—Addresses, essays, lectures. I. Title.
BM535.A765 261.2 79-375235
79 MARC

Christians and Jews / edited by Hans Küng and Walter Kasper. — New York : Seabury Press, c1975.
93 p. ; 23 cm. — (Concilium ; new ser., v. 8, no. 10(98), Ecumenism))
"A Crossroad book."
Includes bibliographical references.
ISBN 0-8164-2095-5 : $4.95
1. Christianity and other religious—Judaism—Addresses, essays, lectures. 2. Judaism—Relations—Christianity—Addresses, essays, lectures. 3. Judaism—Addresses, essays, lectures. 4. Christianity—Addresses, essays, lectures. I. Küng, Hans, 1928- . II. Kasper, Walter. III. Series: Concilium (New York) ; 98.
BM535.C577 261.2 76-10050
00 MARC

Flusser, David Gustav.
(Yahadut u-meḳorot ha-Natsrut)
יהדות ומקורות הנצרות : מחקרים ומסות / דוד פלוסר. — [תל־אביב] : ספרית פועלים, [1979]
486 p. ; 22 cm. — (ספרי דעת־זמננו)
Series romanized: Sifre da'at-zemanenu.
On verso of t. p.: Jewish sources in early Christianity.
Includes bibliographical references and indexes.
1. Bible. N. T.—Criticism, interpretation, etc.—Addresses, essays, lectures. 2. Christianity—Origin—Addresses, essays, lectures. 3. Jesus Christ—Parables—Addresses, essays, lectures. 4. Judaism—Relations—Christianity—Addresses, essays, lectures. 5. Christianity and other religions—Judaism—Addresses, essays, lectures. I. Title.
BS2368.H4F58 79-951639

Gollwitzer, Helmut.
Thema Juden, Christen, Israel : e. Gespräch / Helmut Gollwitzer, Rolf Rendtorff, mit e. Entgegnung von Nathan Peter Levinson ; didaktische Skizze mit einer Projektidee zum Thema Judentum von Helga Sorge. — Stuttgart : Radius-Verlag, 1978.
125 p. ; 19 cm. — (Radius Bücher) GFR79-A4
Includes bibliographical references.
ISBN 3-87173-533-7 : DM14.80
1. Christianity and other religions—Judaism—Addresses, essays, lectures. 2. Judaism—Relations—Christianity—Addresses, essays, lectures. 3. Jewish-Arab relations—Addresses, essays, lectures. I. Rendtorff, Rolf, 1925- joint author. II. Levinson, Nathan Peter, joint author. III. Title.
BM535.G574 79-367965
79 MARC

Hailperin, Herman, 1899-
The three great religions : their theological and cultural affinities / by Herman Hailperin. — Pittsburgh : Duquesne University Library, Herman Hailperin Collection, c1978.
45 p. ; 21 cm.
Includes bibliographical references.
1. Judaism—Relations—Christianity—Addresses, essays, lectures. 2. Christianity and other religions—Judaism—Addresses, essays, lectures. 3. Islam—Origin—Addresses, essays, lectures. I. Title.
BM535.H25 291.1'72 78-56725
78 MARC

Jirjis, Wahīb 'Aṭā Allāh.
(Isrā'īl fī al-mīzān min minẓār Masīḥī)
اسرائيل فى الميزان من منظار مسيحى / الأنبا غريغوريوس [i. e. وهيب عطا الله جرجس]. — [القاهرة] : اسقفية الدراسات [اللاهوتية] العليا والثقافة القبطية والبحث العلمى، 1973.
32 p. ; 17 cm. — (سلسلة المباحث اللاهوتية والعقائدية ؛ 12)
Includes bibliographical references and index.
1. Jewish question—Addresses, essays, lectures. 2. Judaism—Relations—Christianity—Addresses, essays, lectures. 3. Christianity and other religions—Judaism—Addresses, essays, lectures. I. Title. II. Series: Silsilat al-mabāḥith al-lāhūtīyah wa-al-'aqīdīyah ; 12.
DS141.J54 79-971205

Paganisme, judaïsme, christianisme : influences et affrontements dans le monde antique : mélanges offerts à Marcel Simon. — Paris : E. de Boccard, 1978.
390 p., [1] leaf of plates : port. ; 25 cm. F***
French, English, or German.
"Publications de Marcel Simon": p. [371]-387.
Includes bibliographical references and indexes.
250F
1. Judaism—History—Post-exilic period, 586 B.C.-210 A.D.—Addresses, essays, lectures. 2. Paganism—Addresses, essays, lectures. 3. Judaism—Relations—Christianity—Addresses, essays, lectures. 4. Christianity and other religions—Judaism—Addresses, essays, lectures. 5. Christianity—Early church, ca. 30-600—Addresses, essays, lectures. 6. Simon, Marcel, 1907- —Addresses, essays, lectures. I. Simon, Marcel, 1907-
BM177.P33 291 78-388850
79 MARC

—JUDAISM—CONGRESSES

Exodus und Kreuz im ökumenischen Dialog zwischen Juden und Christen : Diskussionsbeiträge für Religionsunterricht u. Erwachsenenbildung / Hans Hermann Henrix, Martin Stöhr (Hrsg.). — 1. Aufl. — Aachen : Einhard-Verlag, 1978.
256 p. ; 21 cm. — (Aachener Beiträge zu Pastoral- und Bildungsfragen ; 8) GFR78-A
Includes bibliographical references and indexes.
1. Christianity and other religions—Judaism—Congresses. 2. Judaism—Relations—Christianity—Congresses. 3. Jesus Christ—Passion—Role of Jews—Congresses. I. Henrix, Hans Hermann, 1941- II. Stöhr, Martin. III. Series.
BM535.E94 78-379451
78 MARC

—JUDAISM—1945- —ADDRESSES, ESSAYS, LECTURES

Jenkins, Graham.
An approach to Christian-Jewish understanding / [by Graham Jenkins, Isaac Levy]. — [London] : [Council of Christians and Jews], [1978?]
[1], 5, 6 p. ; 30 cm. GB78-31181
1. Judaism—Relations—Christianity—1945- —Addresses, essays, lectures. 2. Christianity and other religions—Judaism—1945- —Addresses, essays, lectures. I. Levy, Isaac, 1910- joint author. II. Title.
BM535.J38 261.2 79-305752
79 MARC

—JUDAISM—ADDRESSES, ESSAYS, LECTURES

Lapide, Phinn E 1922-
Jüdischer Monotheismus, christliche Trinitätslehre : e. Gespräch / Pinchas Lapide, Jürgen Moltmann. — München : Kaiser, 1979.
90 p. ; 21 cm. — (Kaiser Traktate ; 39) GFR79-A
Includes bibliographical references.
1. Judaism—Relations—Christianity—Addresses, essays, lectures. 2. Christianity and other religions—Judaism—Addresses, essays, lectures. I. Moltmann, Jürgen, joint author. II. Title.
BM535.L29 79-357396
79 MARC

Torrance, Thomas Forsyth, 1913-
Israel, people of God : God, destiny and suffering / [by] T. F. Torrance. — [London] : [Council of Christians and Jews], [1978]
[1], 15 p. ; 30 cm. GB78-31180
£0.25
1. Judaism—Relations—Christianity—Addresses, essays, lectures. 2. Christianity and other religions—Judaism—Addresses, essays, lectures. 3. Judaism—Israel—Addresses, essays, lectures. I. Title.
BM535.T57 261 79-305024
79 MARC

—TAOISM—ADDRESSES, ESSAYS, LECTURES

Legge, James, 1815-1897.
The religions of China : Confucianism and Tâoism described and compared with Christianity / by James Legge. — Philadelphia : R. West, 1978.
ix, 310 p. ; 23 cm.
Reprint of the 1880 ed. published by Hodder and Stoughton, London.
Includes bibliographical references.
ISBN 0-8492-1545-5 lib. bdg. : $30.00
1. China—Religion—Addresses, essays, lectures. 2. Confucianism—Relations—Christianity—Addresses, essays, lectures. 3. Taoism—Relations—Christianity—Addresses, essays, lectures. 4. Christianity and other religions—Confucianism—Addresses, essays, lectures. 5. Christianity and other religions—Taoism—Addresses, essays, lectures. I. Title.
[BL1801.L4 1978] 299'.51 78-2685
78 MARC

CHRISTIANITY AND PHILOSOPHY see Philosophy and religion

CHRISTIANITY AND POLITICS

see also Christianity and democracy

Boisset, Louis.
Science, idéologies et foi chrétienne / Louis Boisset, Michel Simon. — Lyon : Chronique sociale de France, [1979] F***
177 p. ; 22 cm. — (Collection Débattre)
ISBN 2-85008-004-7 : 39.00F
1. Apologetics—20th century. 2. Religion and science—1946- . 3. Christianity and politics. I. Simon, Michel, joint author. II. Title.
BT1102.B557 239'.8 79-374958
79 MARC

Cantalamessa, Raniero.
Cristianesimo e valori terreni : sessualità, politica e cultura / Raniero Cantalamessa. — Milano : Vita e pensiero, 1976.
ix, 169 p. ; 20 cm. — (Fede e mondo moderno ; 1) It78-Nov
Includes bibliographical references.
L2500
1. Christianity and culture. 2. Christianity and politics. I. Title.
BR115.C8C32 79-348990
*79 MARC

Corsaro, A Maria.
Utopia come speranza : (contributo per una società socialista) / Antonina Maria Corsaro. — Reggio Calabria : Casa del libro, 1976.
98 p. ; 21 cm. It78-Feb
Includes bibliographical references.
L2000
1. Christianity and politics. 2. Communism and Christianity—Italy. 3. Man. I. Title.
BR115.P7C673 78-379061
*78 MARC

Desmond, Cosmas.
Christians or capitalists? : Christianity and politics in South Africa / Cosmas Desmond. — London : Bowerdean Press, 1978.
160 p. ; 23 cm. GB***
Bibliography: p. [159]-160.
ISBN 0-906097-03-7 : £7.00. ISBN 0-906097-04-5 pbk.
1. Christianity and politics. 2. Church and race relations—South Africa. 3. Christianity and capitalism. 4. Socialism. 5. South Africa—Race relations. 6. South Africa—Politics and government—1961- I. Title.
BR115.P7D457 261.1'0968 78-325747
79 MARC

Eells, Robert.
Lonely walk : the life of Senator Mark Hatfield / Robert Eells & Bartell Nyberg. — Chappaqua, N.Y. : Christian Herald Books, c1979.
201 p. ; 24 cm.
Bibliography: p. 195-197.
Includes index.
ISBN 0-915684-49-7 : $8.95
1. Hatfield, Mark O., 1922- . 2. Christianity and politics. 3. United States—Politics and government—1945- . 4. Legislators—United States—Biography. 5. United States. Congress. Senate—Biography. 6. Baptists—United States—Biography. I. Nyberg, Bartell, joint author. II. Title.
E840.8.H3E34 973.92'092'4 79-50942
79 MARC

Fallico, Antonio, 1938-
Gruppi ecclesiali e impegno politico / Antonio Fallico. — Torino : Marietti, [1976]
286 p. ; 20 cm. It78-Mar
Includes bibliographical references.
L4600
1. Christianity and politics. 2. Sociology, Christian. I. Title.
BR115.P7F32 78-367513
*78 MARC

Grassi, Piergiorgio.
La svolta politica della teologia / Piergiorgio Grassi. — Roma : A.V.E., 1976.
173 p. ; 19 cm. — (Minima ; 88) It78-Jan
Includes bibliographical references.
L3000
1. Christianity and politics. 2. Theology—20th century. I. Title.
BR115.P7G73 79-384254
*79 MARC

Innocenti, Ennio.
Dottrina sociale della Chiesa : [breve corso di 20 lezioni] / Ennio Innocenti. — Rovigo : Istituto padano di arti grafiche, c1978-
v. ; 20 cm. It78-Sept
L4000 (v. 1)
1. Sociology, Christian (Catholic) 2. Christianity and politics. I. Title.
BT738.I65 78-394163
*79 MARC

Interdonato, F
La teología en diálogo con la política / Francisco Interdonato. — Bogotá : Ediciones Paulinas, c1978.
181 p. ; 21 cm. — (Colección Liberación pro y contra)
On cover: Reflexión crítica desde América Latina.
Includes bibliographical references.
1. Christianity and politics. 2. Liberation theology—Controversial literature. I. Title.
BR115.P7I55 79-348770
79 MARC

Johnson, John Warren, 1929-
Political Christians : a guide for Christians in public service / John Warren Johnson. — Minneapolis : Augsburg Pub. House, c1979.
158 p. ; 20 cm.
ISBN 0-8066-1689-X : $3.95
1. Christianity and politics. I. Title.
BR115.P7J63 241'.6'22 78-66950
79 MARC

CHRISTIANITY AND POLITICS (Continued)

Kirche, Gesellschaft, Politik : Thesen, Texte / Buchberger...[et al.] ; [hrsg. vom Sozialreferat im Pastoralamt d. Diözese Linz]. — Linz : Oberösterr. Landesverl., 1977.
104 p. ; 21 cm. Au78-1/2-22
Bibliography: p. 101.
ISBN 3-85214-187-7 : S68.00
1. Sociology, Christian (Catholic) 2. Christianity and politics. I. Buchberger, Edeltraud, 1947-
BX1753.K55 78-380451
*79 MARC

Lipschits, Isaac.
De Protestants-christelijke stroming tot 1940 / I. Lipschits. — Deventer : Kluwer, 1977.
123 p. : ill. ; 22 cm. — (Ontstaansgeschiedenis van de Nederlandse politieke partijen ; deel 1) (Cahiers Nederlandse politiek) Ne***
Bibliography: p. 113-117.
Includes indexes.
ISBN 9026705344
1. Political parties—Netherlands. 2. Christianity and politics. 3. Netherlands—Politics and government—1898-1948. I. Title. II. Series. III. Series: Cahiers Nederlandse politiek.
JN5981.L55 79-366029
79 MARC

Ludwig Windthorst, Konrad Adenauer : Porträts zweier christl. Politiker : Niederschr. d. Referate, d. auf e. Akad.-Tagung d. Ludwig Windthorst-Hauses, Holthausen, am 22.-23. Oktober 1976 gehalten wurden / [verantw., S. Guido Dahl]. — Als Ms. hrsg. — Lingen-Holthausen : Ludwig-Windthorst-Haus, 1976.
166 p. ; 21 cm. — (Holthausener Manuskripte ; 77, 1) GFR77-B23
Includes bibliographical references.
1. Windthorst, Ludwig Josef Ferdinand Gustav, 1812-1891. 2. Adenauer, Konrad, 1876-1967. 3. Statesmen—Germany—Biography. 4. Statesmen—Germany, West—Biography. 5. Christianity and politics. I. Dahl, S. Guido. II. Ludwig-Windthorst-Haus. III. Series.
DD205.W4L83 320.9'43'070924 78-377028
78 MARC

Manaranche, André.
Attitudes chrétiennes en politique / André Manaranche. — Paris : Seuil, c1978.
266 p. ; 21 cm. F***
Includes bibliographical references and index.
ISBN 2-02-004814-0 : 45.00F
1. Christianity and politics. I. Title.
BR115.P7M335 261.7 78-378237
78 MARC

Matsuki, Jisaburō, 1906–
(Shin'yaku Seisho ni okeru shūkyō to seiji)
新約聖書に於ける宗教と政治 / 松木治三郎著. — 東京 : 新教出版社, 昭和23 [1948]
231 p. ; 22 cm.
Includes bibliographical references.

1. Christianity and politics. 2. Bible. N. T.—Criticism, interpretation, etc. I. Title. II. Title: Shūkyō to seiji.
BR115.P7M3397 79–802197

Orensanz, Aurelio L
Anarquía y cristianismo / Aurelio L. Orensanz. — Madrid : Mañana Editorial, 1978.
93 p. ; 19 cm. — (Aperos del cristiano ; 16) Sp***
Bibliography: p. 87-93.
ISBN 8474210291
1. Christianity and politics. 2. Revolution (Theology) 3. Eschatology. 4. Anarchism and anarchists. I. Title.
BR115.P7O654 78-376069
78 MARC

Proaño, Leónidas E
Concientización, evangelización, política / Leónidas E. Proaño. — Salamanca : Ediciones Sígueme, 1974.
186 p. ; 18 cm. — (Pedal ; 28) Sp***
ISBN 8430106006 : 120ptas
1. Evangelistic work. 2. Consciousness. 3. Christianity and politics. I. Title.
BV3790.P84 78-417257
78 MARC

Scholl, Hans.
Reformation und Politik : polit. Ethik bei Luther, Calvin u. d. Frühhugenotten / Hans Scholl. — Stuttgart ; Berlin ; Köln ; Mainz : Kohlhammer, 1976.
148 p. ; 19 cm. — (Urban-Taschenbücher ; Bd. 616 : T-Reihe) GFR76-A
Bibliography: p. 141-143.
Includes index.
ISBN 3-17-001962-7
1. Luther, Martin, 1483-1546—Political science. 2. Calvin, Jean, 1509-1564—Political science. 3. Huguenots in France—History. 4. Christianity and politics. 5. France—Church history—16th century. I. Title.
BR333.5.P6S36 79-370579
79 MARC

Seifart, Arnulf, 1936-
Der Gott der "politischen Theologie" : die Entwicklung der Gottesdiskussion vom Kämpfenden Nationalgott bis zur christlich motivierten Strategie des Guerillakrieges / Arnulf Seifart. — Zürich [etc.] : Benziger, cop. 1978.
xv, 377 p. ; 23 cm. Sw79-40
A revision of the author's thesis, Giessen, 1976.
Bibliography: p. 335-350.
Includes indexes.
ISBN 3-545-22087-7 : 36.80F
1. God—History of doctrines. 2. Christianity and politics. 3. Liberation theology. I. Title.
BT102.S388 1978 79-356192
79 MARC

Shiri, G
Karnataka Christians and politics / G. Shiri. — Madras : Published for the Christian Institute for the Study of Religion and Society, Bangalore by the Christian Literature Society, c1978.
95, [1] p. ; 20 cm. — (C.I.S.R.S. social research series ; no. 13)
"This study was undertaken in 1973 as an M.Th. thesis and later revised for publication."
Bibliography: p. [96]
Rs7.00
1. Karnataka, India—Church history. 2. Christianity and politics. I. Title. II. Series: Social research series (Bangalore) ; no. 13.
BR1170.K3S54 301.5'8 78-902993
79 MARC

Taylor, John Vernon, 1914-
Christianity and politics in Africa / John Taylor. — Westport, Conn. : Greenwood Press, 1979.
127 p. ; 23 cm.
Reprint of 1957 ed. published by Penguin Books, London, which was issued as WA9 of Penguin African series.
Bibliography: p. 123-127.
ISBN 0-8371-5951-2
1. Christianity and politics. 2. Africa—Politics and government—1960- I. Title.
BR115.P7T29 1979 261.7 79-32
79 MARC

Wängborg, Sune, 1949-
Evangelisation för samhällsförändring : radikala kristna grupper i Colombia / Sune Wängborg. — [Stockholm] : Verbum, 1976.
58, [1] p., [2] leaves of plates : ill. ; 19 cm. — (Kyrkan i Latinamerika) S77-1/3
Bibliography: p. [59]
kr18.85
1. Evangelistic work—Colombia. 2. Christianity and politics. 3. Colombia—Church history. 4. Catholic Church in Colombia—History. I. Title. II. Series.
BV3777.C7W33 78-354186
78[r79]rev MARC

—ADDRESSES, ESSAYS, LECTURES

100 [i.e. Honderd] jaar partij : bezinning en perspectief : een anti-revolutionaire bijdrage aan christen-democratische politiek / samenstelling en eindred. C. Bremmer en A. Veerman. — Franeker : T. Wever, [1979]
298 p. : ill. ; 22 cm. Ne79-16
"In opdracht van de Anti-Revolutionaire Partij t.g.v. haar honderdjarig bestaan."
Includes bibliographical references.
ISBN 9061352886 : fl 19.90
1. Anti-revolutionaire Partij—Addresses, essays, lectures. 2. Christianity and politics—Addresses, essays, lectures. 3. Netherlands—Economic policy—Addresses, essays, lectures. 4. Netherlands—Social policy—Addresses, essays, lectures. I. Bremmer, C. II. Veerman, A., 1916-
JN5985.A6H65 79-379915
*79 MARC

Bergamín, José, 1895-
El pensamiento perdido : páginas de la guerra y del destierro / José Bergamín. — Madrid : Adra, 1976.
217 p. ; 20 cm. Sp***
ISBN 847418004X
1. Catholic Church in Spain—Addresses, essays, lectures. 2. Christianity and politics—Addresses, essays, lectrues. 3. Spain—History—Civil War, 1936-1939—Moral aspects—Addresses, essays, lectures. 4. Spain—History—Civil War, 1936-1939—Religious aspects—Addresses, essays, lectures. 5. Fascism and the Catholic Church—Addresses, essays, lectures. I. Title.
BX1585.B43 282'.46 78-377071
78 MARC

Berti Arnoaldi-Veli, Francesco.
Coi miei compagni io devo restare / Francesco Berti Arnoaldi ; prefazione di Ferruccio Parri. — 1. ed. — Venezia ; Padova : Marsilio, 1974.
113 p. ; 22 cm. — (Testimonianze ; n. 5) It76-Feb
Includes bibliographical references.
L2500
1. Christian life—1960—Addresses, essays, lectures. 2. Christianity and politics—Addresses, essays, lectures. 3. Radicalism—Addresses, essays, lectures. I. Title.
BV4501.2.B42 79-355335
*79 MARC

Dumas, André.
Political theology and the life of the church / André Dumas. — Philadelphia : Westminster Press, c1978.
x, 148 p. ; 22 cm.
Translation of Théologies politiques et vie de l'église.
Bibliography: p. [145]-148.
ISBN 0-664-24226-X
1. Christianity and politics—Addresses, essays, lectures. I. Title.
BR115.P7D813 261.7 78-16813
78 MARC

Enrique Tarancón, Vicente.
Los cristianos y la política : cartas cristianas / del Cardenal Tarancón. — Madrid : Servicio Editorial del Arzobispado de Madrid Alcalá, D.L. 1977.
134 p. ; 17 cm. Sp78-Feb
Includes bibliographical references.
ISBN 8440039190 : 150ptas
1. Christianity and politics—Addresses, essays, lectures. I. Title.
BR115.P7E64 79-341695
79 MARC

Greinacher, Norbert, 1931-
Vor einem neuen politischen Katholizismus? / Norbert Greinacher, Ferdinand Klostermann. — Frankfurt am Main ; Bern ; Las Vegas : Lang, 1978.
107 p. ; 21 cm. — (Erfahrung und Theologie ; Bd. 1) GFR79-A
Includes bibliographical references.
ISBN 3-261-02628-6
1. Church and state—Catholic Church—Addresses, essays, lectures. 2. Catholic Church—History—1965- —Addresses, essays, lectures. 3. Christianity and politics—Addresses, essays, lectures. I. Klostermann, Ferdinand, joint author. II. Title. III. Series.
BX1791.G73 79-377892
79 MARC

Kompendium des modernen christlich-freiheitlichen Konservatismus / Konrad Bonkosch (Hrsg.), Hans-Georg Roth u. Leo Schütze (Mithrsg.). — Bonn : R-und-S-Verlag, 1978-
v. ; 21 cm. — (Politisch-theoretische Reihe ; Bd. 3-) GFR78-A
CONTENTS: T. 1. Konservative Zeitzeichen. Grundlagen der freiheitlichen Ordnung.

Includes bibliographical references.
ISBN 3-88296-003-5 (v. 1)
1. Conservatism—Germany, West—Addresses, essays, lectures. 2. Christianity and politics—Addresses, essays, lectures. 3. Liberty—Addresses, essays, lectures. 4. Germany, West—Social policy—Addresses, essays, lectures. I. Bonkosch, Konrad. II. Roth, Hans-Georg. III. Schütze, Leo. IV. Series.
JN3971.A91K65 78-378729
78 MARC

La Valle, Raniero.
Fuori dal campo / Raniero La Valle. — 1. ed. — Milano : A. Mondadori, 1978.
315 p. ; 20 cm. — (Saggi ; 102) It***
Includes bibliographical references.
L4500
1. World politics—1945- —Addresses, essays, lectures. 2. Christianity and politics—Addresses, essays, lectures. 3. Catholic Church—History—1965- —Addresses, essays, lectures. 4. La Valle, Raniero—Addresses, essays, lectures. I. Title.
D849.L32 327'.09'045 79-372142
79 MARC

Norman, Edward R
Christianity and the world order / Edward Norman. — Oxford ; New York : Oxford University Press, 1979.
105 p. ; 20 cm. — (B.B.C. Reith lectures ; 1978) GB78
Includes bibliographical references and index.
ISBN 0-19-215510-5 : £3.25. ISBN 0-19-283019-8 : £1.50
1. Christianity and politics—Addresses, essays, lectures. 2. Christianity—20th century—Addresses, essays, lectures. 3. Secularization (Theology)—Addresses, essays, lectures. I. Title. II. Series: Reith lectures ; 1978.
BR115.P7N597 261.7 78-40857
78 MARC

Paoli, Arturo.
Il grido della terra / Arturo Paoli. — Assisi : Cittadella editrice, 1976.
175 p. ; 19 cm. — (Progetti di liberazione) It76-Dec
L2500
1. Catholic Church in Latin America—Addresses, essays, lectures. 2. Christianity and politics—Addresses, essays, lectures. I. Title.
BX1426.2.P28 79-344787
*79 MARC

Personal values in public policy : essays and conversations in government decision-making / edited by John C. Haughey. — New York : Paulist Press, c1979.
vi, 275 p. ; 21 cm.
Includes bibliographical references.
ISBN 0-8091-2201-4 : $5.95
1. Political ethics—Addresses, essays, lectures. 2. Christianity and politics—Addresses, essays, lectures. 3. Social values—Addresses, essays, lectures. 4. United States—Politics and government—Addresses, essays, lectures. I. Haughey, John C.
JA79.P39 172 79-84401
79 MARC

Pluralismo dei cattolici in politica? / [scritti di] G. Andreotti ... [et al.]. — Roma : Borla, 1979.
108 p. ; 19 cm. — (Proposte e dibattiti) It79-Apr
L2500
1. Communism and Christianity—Catholic Church—Italy—Addresses, essays, lectures. 2. Christianity and politics—Addresses, essays, lectures. 3. Sociology, Christian (Catholic)—Addresses, essays, lectures. I. Andreotti, Giulio.
BX1396.4.P55 79-362280
*79 MARC

CHRISTIANITY AND POLITICS (Continued)

—CONGRESSES

Studienkreis Kirche und Befreiung.
Politische Denaturierung von Theologie und Kult / dieser Bd. wurde besorgt von Wilhelm Weber. — Aschaffenburg : Pattloch, 1978.
97 p. ; 19 cm. — (Veröffentlichungen des Studienkreises Kirche und Befreiung) GFR79-A
Papers originally presented at a meeting of the Studienkreis Kirche Und Befreiung, held in Mülheim a. d. Ruhr, June 17-18, 1977.
Includes bibliographical references.
ISBN 3-557-89005-8 : DM14.80
1. Sociology, Christian—Congresses. 2. Christianity and politics—Congresses. I. Weber, Wilhelm, 1925- . II. Title. III. Series: Studienkreis Kirche und Befreiung. Veröffentlichungen des Studienkreises Kirche und Befreiung.
BT738.S79 1978 79-353518
79 MARC

—HISTORY

Bercovitch, Sacvan.
The American jeremiad / Sacvan Bercovitch. — Madison : University of Wisconsin Press, 1978.
xvi, 239 p. ; 23 cm.
Includes Bibliographical references and index.
ISBN 0-299-07350-5 : $15.00
1. American prose literature—History and criticism. 2. Preaching—United States—History. 3. Oratory—United States—History. 4. Puritans—New England. 5. Christianity and politics—History. 6. United States—Civilization. I. Title.
PS362.B43 810'.9'38 78-53283
78 MARC

Caspary, Gerard E
Politics and exegesis : Origen and the two swords / Gerard E. Caspary. — Berkeley : University of California Press, c1979.
xv, 215 p. ; 24 cm.
English and Latin.
Includes bibliographical references and indexes.
ISBN 0-520-03445-7
1. Origenes. 2. Christianity and politics—History. 3. Church and state—History. I. Title.
BR65.O68C37 230'.1'3 77-71058
79 MARC

Hammond, John L
The politics of benevolence : revival religion and American voting behavior / John L. Hammond. — Norwood, N.J. : Ablex Pub. Corp., c1979
xi, 243 p. : graphs ; 24 cm. — (Modern sociology)
Bibliography: p. 221-233.
Includes indexes.
ISBN 0-89391-013-9
1. Revivals—United States—History. 2. Christianity and politics—History. I. Title. II. Series.
BV3773.H35 322.4'4'0973 78-16050
78 MARC

Schjørring, Jens Holger.
Theologische Gewissensethik und politische Wirklichkeit : d. Beispiel Eduard Geismars u. Emanuel Hirsch / von Jens Holger Schjørring ; ⌜übers. aus d. Dän., Eberhard Harbsmeier⌝. — Göttingen : Vandenhoeck und Ruprecht, 1979.
354 p. ; 24 cm. — (Arbeiten zur kirchlichen Zeitgeschichte : Reihe B, Darstellungen ; Bd. 7) GFR79-A
Bibliography: p. ⌜320⌝-338.
Includes index.
ISBN 3-525-55707-8
1. Geismar, Eduard Osvald, 1871- . 2. Hirsch, Emanuel, 1888-1972. 3. Christianity and politics—History. I. Title. II. Series.
BX8080.G37S3415 79-360284
79 MARC

Uhde, Ute.
Politik und Religion : zum Verhältnis von Demokratie und Christentum bei Alexis de Tocqueville / von Ute Uhde. — Berlin : Duncker & Humblot, c1978.
109 p. ; 24 cm. — (Beiträge zur Politischen Wissenschaft ; Bd. 29) GFR***
Bibliography: p. ⌜102⌝-109.
ISBN 3-428-04064-3
1. Christianity and democracy—History. 2. Christianity and politics—History. 3. Tocqueville, Alexis Charles Henri Maurice Clérel de, 1805-1859. I. Title. II. Series.
BR115.P7U37 78-375500
78 MARC

—HISTORY—ADDRESSES, ESSAYS, LECTURES

Evangelische Kirche in Deutschland und die Wiederaufrüstungsdiskussion in der Bundesrepublik : 1950-1955 / hrsg. von Wolf Werner Rausch u. Christian Walther. — Orig.-Ausg. — Gütersloh : Gütersloher Verlagshaus Mohn, 1978.
175 p. ; 19 cm. — (Gütersloher Taschenbücher Siebenstern ; 292) GFR78-A47
On spine: EKD und Wiederaufrüstungsdiskussion.
Bibliography: p. 7-11.
ISBN 3-579-03692-0 : DM7.80
1. Germany, West—Politics and government—Addresses, essays, lectures. 2. Germany, West—Defenses—Addresses, essays, lectures. 3. Evangelische Kirche in Deutschland—Addresses, essays, lectures. 4. Christianity and politics—History—Addresses, essays, lectures. I. Rausch, Wolf Werner, 1948- . II. Walther, Christian. III. Title: EKD und Wiederaufrüstungsdiskussion.
DD259.2.E92 320.9'43'087 79-363958
79 MARC

CHRISTIANITY AND PSYCHICAL RESEARCH

Wilkerson, Ralph.
ESP or HSP? / By Ralph Wilkerson. — Anaheim, Calif. : Melodyland Publishers, c1978.
334 p., ⌜1⌝ leaf of plates : ill. ; 18 cm.
Includes bibliographical references.
ISBN 0-918818-09-5. ISBN 0-918818-04-4 pbk. : $2.45
1. Christianity and psychical research. I. Title.
⌜BR115.P85W45⌝ 231'.74 78-69961
79 MARC

—ADDRESSES, ESSAYS, LECTURES

Savini, Umberto.
Il sesso e la Bibbia. Cristianesimo e reincarnazione. Pescara, Tip. Ferretti, 1972.
201 p. 21 cm. L3000 It73-Mar
1. Sex in the Bible—Addresses, essays, lectures. 2. Reincarnation—Addresses, essays, lectures. 3. Christianity and psychical research—Addresses, essays, lectures. I. Title.
BS680.S5S29 79-374757
79 MARC

CHRISTIANITY AND RELIGIOUS HUMANISM

Aizawa, Genshichi.
(Shūkyō kaikaku to hyūmanizumu)
宗教改革とヒューマニズム：その相関関係の研究 / 相沢源七著. — 仙台：東北教育図書出版, 1977.
341 p., ⌜1⌝ leaf of plates : ill. ; 22 cm. Ja 78-9519
Limited edition: 300 copies printed.
Includes bibliographical references.
¥4500
1. Reformation. 2. Christianity and religious humanism. I. Title.
BR307.A35 78-812141

CHRISTANITY AND REVOLUTION see Revolution (Theology)

CHRISTIANITY AND SCIENCE see Religion and science

CHRISTIANITY AND SUPERSTITION

see also Christianity and magic

Haack, Friedrich-Wilhelm, 1935-
Aberglaube, Magie, Zauberei / Friedrich-Wilhelm Haack. — 1. Aufl. — München : Evangelischer Presseverband für Bayern, Abt. Schriftenmission, 1977.
43 p. ; 19 cm. — (Münchener Reihe) GFR78-A
Errata slip inserted.
ISBN 3-583-50626-X : DM2.80
1. Christianity and superstition. 2. Superstition—Controversial literature. 3. Christianity and magic. 4. Magic—Controversial literature. I. Title.
BR115.S85H2 79-370220
79 MARC

CHRISTIANITY AND THE ARTS

Glaubwürdig : 5 Gespräche über heutige Kunst u. Religion mit Joseph Beuys. Heinrich Böll, Herbert Falken, Kurt Marti, Dieter Wellershoff / Horst Schwebel. — München : Kaiser, 1979.
154 p. ; 21 cm. — (Kaiser Traktate : 40) GFR79-A
ISBN 3-459-01203-X
1. Christianity and the arts. 2. Artists—Germany, West—Interviews. I. Beuys, Joseph. II. Schwebel, Horst.
BR115.A8G56 79-368657
79 MARC

—ADDRESSES, ESSAYS, LECTURES

The Church and the arts / edited by Temple Kingston. — Windsor, Ont. : Canterbury College, University of Windsor, 1977.
vii, 47 p. : ill. ; 22 cm. C***
Includes bibliographical references.
1. Christianity and the arts—Addresses, essays, lectures. I. Kingston, Temple. II. Canterbury College.
BR115.A8C45 261 79-309515
79 MARC

CHRISTIANITY AND THE WORLD see Church and the world

CHRISTIANITY IN LITERATURE

Bolton, Whitney French.
Alcuin and Beowulf : an eighth-century view / W. F. Bolton. — New Brunswick, N.J. : Rutgers University Press, c1978.
xii, 200 p. ; 24 cm.
Bibliography: p. 181-190.
Includes index.
ISBN 0-8135-0865-7
1. Beowulf. 2. Alcuin, 735-804—Criticism and interpretation. 3. Christianity in literature. I. Title.
PR1585.B6 829'.3 78-15569
78 MARC

Hassel, Rudolph Chris, 1939-
Renaissance drama & the English church year / R. Chris Hassel, Jr. — Lincoln : University of Nebraska Press, c1979.
ix, 215 p. : ill. ; 22 cm.
Bibliography: p. 205-210.
Includes index.
ISBN 0-8032-2304-8
1. English drama—Early modern and Elizabethan, 1500-1600—History and criticism. 2. English drama—17th century—History and criticism. 3. Christianity in literature. 4. Church year. 5. Church of England. Book of common prayer. 6. Theater—England—History. I. Title.
PR658.R43H3 822'.3'09 78-24233
78 MARC

Karkainen, Paul A
Narnia explored / Paul A. Karkainen. — Old Tappan, N.J. : F. H. Revell Co., c1979.
192 p. ; 21 cm.
ISBN 0-8007-1068-1 : $8.95
1. Lewis, Clive Staples, 1898-1963. The chronicles of Narnia. 2. Fantastic fiction, English—History and criticism. 3. Christianity in literature. I. Title.
PR6023.E926C535 823'.9'12 79-9243
79 MARC

Kummer, Irène Elisabeth.
Transparenz der Wirklichkeit; Edzard Schaper und die innere Spannung in der christlichen Literatur des 20. Jahrhunderts ⌜von⌝ Irène Sonderegger-Kummer. Berlin, New York, W. de Gruyter, 1971.
xii, 308 p. 23 cm. (Quellen und Forschungen zur Sprach- und Kulturgeschichte der germanischen Völker, n. F., 37 (161)) DM48.00 GDB***
Bibliography: p. ⌜295⌝-306.
1. Schaper, Edzard Hellmuth, 1908- —Religion and ethics. 2. Christianity in literature. I. Title.
PT2638.A68Z75 70-571618
79⌜r79⌝rev MARC

Patrides, C A
Milton and the Christian tradition / by C. A. Patrides. — Hamden, Conn. : Archon Books, 1979, c1966.
xvi, 302 p. ; 23 cm.
Reprint of the ed. published by Clarendon Press, Oxford.
Includes bibliographical references and indexes.
ISBN 0-208-01821-2
1. Milton, John, 1608-1674—Religion and ethics. 2. Christianity in literature. I. Title.
⌜PR3592.R4P3 1979⌝ 821'.4 79-10846
79 MARC

Sammons, Martha C 1949-
A guide through Narnia / Martha C. Sammons. — Wheaton, Ill. : H. Shaw Publishers, c1979.
164 p. : map ; 21 cm. — (Wheaton literary series)
Includes bibliographical references.
ISBN 0-87788-325-4
1. Lewis, Clive Staples, 1898-1963. The chronicles of Narnia. 2. Lewis, Clive Staples, 1898-1963—Criticism and interpretation. 3. Fantastic fiction, English—History and criticism. 4. Christianity in literature. I. Title.
PR6023.E926C537 823'.9'12 78-26476
78 MARC

CHRISTIANITY IN LITERATURE (Continued)

Weber, Antonie.
Die Erzählstruktur von Graham Greenes katholischen Romanen / Antonie Weber. — Bern : Francke Verlag, c1978.
173 p. ; 23 cm. — (Schweizer anglistische Arbeiten ; Bd. 95) Sw***
Summary in English.
Bibliography: p. 163-170.
Includes indexes.
ISBN 3-7720-1401-1
1. Greene, Graham, 1904- —Criticism and interpretation. 2. Christianity in literature. 3. Catholic Church—Doctrinal and controversial works. I. Title. II. Series.
PR6013.R44Z93 823'.9'12 79-355531
79 MARC

—ADDRESSES, ESSAYS, LECTURES

La Littérature contemporaine polonaise et le sacré = Sacrum we współczesnej literaturze polskiej / sous la direction, Irena Sławińska, Stefan Sawicki. — Lublin : Société des lettres et des sciences de l'Université catholique de Lublin, 1978.
179 p. ; 24 cm. Po***.
In French.
At head of title: Université catholique de Lublin, Institut des recherches sur la littérature religieuse, 7.
Includes bibliographical references.
zł110
1. Polish literature—20th century—History and criticism—Addresses, essays, lectures. 2. Polish literature—Catholic authors—History and criticism—Addresses, essays, lectures. 3. Christianity in literature—Addresses, essays, lectures. I. Sławińska, Irena. II. Sawicki, Stefan. III. Lublin (City). Uniwersytet. Zakład Badań nad Literatura Religijna. IV. Title: Sacrum we współczesnej literaturze polskiej.
PG7053.C3L57 79-379382
79 MARC

CHRISTIANS

—BIOGRAPHY see Christian biography

—LEBANON

United States. Congress. Senate. Committee on Foreign Relations. Subcommittee on Near Eastern and South Asian Affairs.
Lebanon : hearing before the Subcommittee on Near Eastern and South Asian Affairs of the Committee on Foreign Relations, United States Senate, Ninety-fifth Congress, second session ... August 16, 1978. — Washington : U.S. Govt. Print. Off., 1978.
iii, 37 p. ; 24 cm.
1. Christians—Lebanon. 2. Lebanon—History—Civil War, 1975- I. Title.
KF26.F662 1978a 301.45'28'095692 78-603301
79 MARC

—TURKEY—İZMIR (CITY)

Missir, Livio Amedeo, 1931-
Appunti familiari : Smirne, mio padre, Ernesto Buonaiuti, introduzione a un epistolario / Livio Missir Reggio Mamachi di Lusignano. — Luxembourg : Euro Editor, 1974.
103 p. ; 21 cm. Lu***
1. Missir family. 2. Italy—Genealogy. 3. Christians—Turkey—İzmir (City) 4. İzmir (City)—History. I. Title.
CS769.M53 1974 78-393128
79 MARC

CHRISTIANS, JEWISH see Jewish Christians

CHRISTIANS (GENERAL CONVENTION OF THE CHRISTIAN CHURCH), AFRO-AMERICAN see Afro-American Christians (General Convention of the Christian Church)

CHRISTIANS AND MOORS, DANCE see Moros y Cristianos (Dance)

CHRISTIANS IN ANTIOCH

—HISTORY

Meeks, Wayne A
Jews and Christians in Antioch in the first four centuries of the common era / by Wayne A. Meeks and Robert L. Wilken. — Missoula, Mont. : Published by Scholars Press for the Society of Biblical Literature, c1978.
x, 127 p. ; 24 cm. — (Sources for Biblical study ; no. 13 ISSN 0145-2762)
Includes translation of 8 letters of Libanius concerning the Jews, Libanius' oration on systems of patronage, and 2 of John Chrysostom's Homilia adversus Judaeos.
Includes bibliographical references.
ISBN 0-89130-229-8
1. Jews in Antioch—History. 2. Christians in Antioch—History. 3. Judaism—Controversial literature. 4. Antioch—History. I. Wilken, Robert Louis, 1936- joint author. II. Libanius. III. Chrysostomus, Joannes, Saint, Patriarch of Constantinople, d. 407. Logo kata ioudaiōn. English. Selections. 1978. IV. Title. V. Series.
DS135.S95M4 209'.564 78-3760
78 MARC

CHRISTIANS IN EUROPE

Brugmans, Hendrik, 1906-
Christen und Europa : d. Beitr. d. christl. Kirchen zur europ. Integration / [Hendrik Brugmans, Bernhard Vogel, Harald Uhl]. — 1. Aufl. — Andernach/Rhein : Pontes-Verlag, 1977.
99 p. ; 21 cm. — (Kleine Europa-Bilbiothek [i.e. -Bibliothek] ; Bd. 10) GFR77-A37
Contribution by H. Brugmans originally issued in French under title: Présence des chrétiens sur le chantier européen.
Includes bibliographical references.
ISBN 3-920061-20-9
1. European federation. 2. Christians in Europe. I. Vogel, Bernhard, joint author. II. Uhl, Harald, joint author. III. Title.
D1060.B758 940.55 78-377043
78 MARC

CHRISTIANS IN INDIA

Vilsan, Kottapalli.
(Dēvuḍi rājakīyaṃ)
దేవుడి రాజకీయం : భారతీయ క్రైస్తవ సాంఘిక విశ్లేషణ / గ్రంథకర్త కొత్తపల్లి విల్సన్. — హైదరాబాదు : భారతీయ సాంస్కృతిక అధ్యయన సంస్థ, 1977.
iv, ii, iv, 84, [2] p. ; 19 cm.
In Telugu.
Bibliography: p. [85]-[86]
Rs2.00
1. Caste—India. 2. Christians in India. I. Title.
DS422.C3V493 77-906186
S A

—ADDRESSES, ESSAYS, LECTURES

Tilak, Ashok Devdatt, 1921-
Mitraho / lekhaka Aśoka Devadatta Ṭiḷaka. — Nāsika : M. A. Ṭiḷaka, 1977.
51 p. ; 20 cm.
In Marathi.
Rs3.00
1. Christians in India—Addresses, essays, lectures. I. Title.
BR1155.T55 79-900584
79 MARC

CHRISTIANS IN KERALA

Mātyu, K J
(Kēraḷasabhācaritram)
കേരളസഭാചരിത്രം / രചയിതാക്കൾ കെ. ജെ. മാത്യു കാവാലം, ജേക്കബ് നെല്ലിക്കുന്നത്ത്. — ചങ്ങനാശ്ശേരി : സന്ദേശനിലയം, 1975.
176 p. ; 19 cm.
In Malayalam.
1. Kerala—Church history. 2. Christians in Kerala. I. Nellikkunattà, Jōkkab, joint author. II. Title.
BR1170.K4M37 78-901553

CHRISTIANS IN NORTH AFRICA

Hamman, Adalbert, 1910-
La vie quotidienne en Afrique du Nord au temps de Saint Augustin / A.-G. Hamman. — [Paris] : Hachette, c1979.
479 p. : map ; 20 cm. F***
Includes bibliographical references.
ISBN 2-01-004214-X : 54.00F
1. Africa, North—Social life and customs. 2. Christians in North Africa. 3. Augustinus, Aurelius, Saint, Bp. of Hippo. I. Title.
DT192.H34 961'.03 79-382384
79 MARC

CHRISTIANS IN RUSSIA

Hartfeld, Hermann, 1942-
Glaube trotz KGB : Christsein in d. UdSSR heute / Hermann Hartfeld ; [Übers. aus d. Russ.]. — 1. Ausg. — Uhldingen : Stephanus, 1976.
416 p. ; 19 cm. GFR77-A
ISBN 3-921213-10-X : DM14.80
1. Christians in Russia. 2. Persecution—Russia. I. Title.
BR936.H2615 78-356731
78 MARC

CHRISTIANS SECTS, MEDIEVAL

see also Circumcellions

CHRISTIANSBORG SLOT

Nørbæk, Jens Fr.
Marmorbroen : en billedbog med tekst / [af] Jens Fr. Nørbæk. — København : Akademisk Forlag : [eksp. DBK], 1978.
144 p. : ill. ; 30 cm. D78-38
Summary in French.
Bibliography: p. 122-131.
Includes indexes.
ISBN 8750018167 : kr70.80
1. Copenhagen—Bridges—Marmorbroen. 2. Christiansborg slot. I. Title.
TG70.C66N63 79-348894
79 MARC

CHRISTIANSEN, GUSTAV, 1919-

Christiansen, Gustav, 1919-
Lerfødderne / fortalt af Gustav. — 1. udg. — København : Medusa, 1978.
295 p. : ill. ; 20 cm. — (Arbejderlitteratur) D78-25
At head of title: Socialistisk kulturfront.
ISBN 8773320358 : kr69.75
1. World War, 1939-1945—Personal narratives, Danish. 2. Christiansen, Gustav, 1919- 3. World War, 1939-1945—Conscript labor—Germany, East—Altenburg. 4. Altenburg, Ger.—Biography. 5. Printers—Denmark—Biography. 6. Printers—Germany, East—Altenburg—Biography. I. Socialistisk kulturfront. II. Title.
D811.5.C45 79-380918
79 MARC

CHRISTIANSEN, VALDEMAR, 1892-1975

Christiansen, Valdemar, 1892-1975.
Vi havde en dejlig barndom : optegnelser / af Valdemar Christiansen. — [Hørsholm : Peter Leth Christiansen], 1977.
220 p. : ill. ; 22 cm. D77-44
Not in trade.
1. Christiansen, Valdemar, 1892-1975. 2. Denmark—Biography. 3. Children—Denmark—Biography. I. Title.
CT1278.C49A38 79-337919
79 MARC

CHRISTIANSHAVN, COPENHAGEN

Copenhagen. Direktoratet for Generalplanlægning.
Christianshavn : dispositionsplan for Christianshavn / Magistratens 4. afd., Direktoratet for Københavns kommunes generalplanlægning. — [Nyt oplag]. — [København] [Østergade 26, 1100 K] : [Direktoratet for Københavns kommunes generalplanlægning], 1976.
115 leaves : ill. ; 30 cm. D77-15/16
kr30.00
1. Copenhagen—City planning. 2. Christianshavn, Copenhagen. 3. City planning—Denmark. I. Title.
HT169.D42C44 78-374779
78 MARC

CHRISTIANSHAVN, COPENHAGEN (Continued)

Grau, Monika.
Christiania : Elendsviertel, soziales Experiment, oder Selbstorganisation Nicht-Angepasster? / Monika Grau & Heiner Gringmuth. — 1. Aufl. — Wetzlar : Neuland-Verlag, 1977.
181 p. : ill. ; 18 x 21 cm. GFR***
Includes bibliographical references.
ISBN 3-921764-00-9
1. Hippies—Denmark—Copenhagen. 2. Collective settlements—Denmark—Copenhagen. 3. Christianshavn, Copenhagen. I. Gringmuth, Heiner, joint author. II. Title.
HN547.G65 78-384312
79 MARC

Langberg, Harald.
Tre huse på Christianshavn / Harald Langberg ; ⌊tysk sammenfatning oversat af A. Frost⌋. — København : Foreningen til gamle Bygningers Bevaring : ⌊eksp. Gyldendal⌋, 1976.
94 p. : ill. ; 26 cm. D77-11
Summary in German.
Includes bibliographical references.
ISBN 8787546035 : kr46.00
1. Christianshavn, Copenhagen. 2. Copenhagen—Dwellings. 3. Architecture, Domestic—Denmark—Copenhagen—Conservation and restoration. I. Title.
NA7375.C6L36 78-387368
79 MARC

—DESCRIPTION

Christianshavn—vandring i tid og rum / ⌊teksten samlet af Henry Fachmann⌋ ; ⌊ill. tegnet af Holger Thorup⌋. — ⌊København⌋ : Fagskolen for Boghaandværk, 1977.
54 p. : ill. ; 24 cm. D78-3
Bibliography: p. ⌊57⌋
Not in trade.
1. Christianshavn, Copenhagen—Description. 2. Copenhagen—Description. I. Fachmann, Henry. II. Thorup, Holger.
DL276.4.C47C47 79-351498
79 MARC

—SOCIAL CONDITIONS

Iversen, Per Løvetand.
Fristaden Christiania som samfundsexperiment. Et forsøg på at belyse strukturerings- og bevidstgørelsesprocessen i Christiania igennem det første år af dets eksistens. København, Eget forlag; Eksp.: Husets Bogcafe, 1972.
91 p. illus. 22 cm. kr10.00 D73-15
"BYT forskningsrapport."
At head of title: Kunstakademiets arkitektskole. Instituttet for byggeteknik.
1. Christianshavn, Copenhagen—Social conditions. I. Title.
HN550.C6I94 78-341748
78 MARC

CHRISTIE, AGATHA MILLER, DAME, 1891-1976

—BIOGRAPHY

Robyns, Gwen.
The mystery of Agatha Christie / Gwen Robyns. — New York : Penguin Books, 1979.
320 p., ⌊16⌋ leaves of plates : ill. ; 18 cm.
Reprint of the 1978 ed. published by Doubleday, New York.
Includes index.
ISBN 0-14-005228-3 (pbk.)
1. Christie, Agatha Miller, Dame, 1891-1976—Biography. 2. Authors, English—20th century—Biography. I. Title.
⌊PS6005.H66Z83 1979⌋ 823'.9'12 79-9781
79 MARC

—CRITICISM AND INTERPRETATION

Ercoli, Emma.
Agatha Christie / di Emma Ercoli. — Firenze : La nuova Italia, 1978.
103 p. ; 17 cm. — (Il Castoro ; 144) It79-Jan
Bibliography: p. 88-96.
L2000
1. Christie, Agatha Miller, Dame, 1891-1976—Criticism and interpretation. 2. Detective and mystery stories, English—History and criticism.
PR6005.H66Z635 823'.9'12 79-347027
*79 MARC

CHRISTIE FAMILY

CHRISTINA, QUEEN OF SWEDEN see Kristina, Queen of Sweden, 1626-1689

CHRISTKATHOLISCHE KIRCHE DER SCHWEIZ

—HISTORY—ADDRESSES, ESSAYS, LECTURES

Christkatholisch : die christkatholische Kirche der Schweiz in Geschichte und Gegenwart : zum 100-Jahr-Jubiläum der Christkatholischen Kirchgemeinde Solothurn / hrsg. von der Christkatholischen Kirchgemeinde Solothurn ; Schriftleitung, Wilhelm Flückiger und Carlo Jenzer. — Einsideln ⌊etc.⌋ : Benziger, cop. 1978.
239 p. ; 22 cm. Sw78-9102
Bibliography: p. 232-236.
ISBN 3-545-25046-6 : 24.00F
1. Christkatholische Kirche der Schweiz—History—Addresses, essays, lectures. 2. Christkatholische Kirchengemeinde Solothurn—Addresses, essays, lectures. I. Flückiger, Wilhelm. II. Jenzer, Carlo. III. Christkatholische Kirchgemeinde Solothurn.
BX4770.S9C47 79-338422
79 MARC

CHRISTKATHOLISCHE KIRCHENGEMEINDE SOLOTHURN

—ADDRESSES, ESSAYS, LECTURES

Christkatholisch : die christkatholische Kirche der Schweiz in Geschichte und Gegenwart : zum 100-Jahr-Jubiläum der Christkatholischen Kirchgemeinde Solothurn / hrsg. von der Christkatholischen Kirchgemeinde Solothurn ; Schriftleitung, Wilhelm Flückiger und Carlo Jenzer. — Einsideln ⌊etc.⌋ : Benziger, cop. 1978.
239 p. ; 22 cm. Sw78-9102
Bibliography: p. 232-236.
ISBN 3-545-25046-6 : 24.00F
1. Christkatholische Kirche der Schweiz—History—Addresses, essays, lectures. 2. Christkatholische Kirchengemeinde Solothurn—Addresses, essays, lectures. I. Flückiger, Wilhelm. II. Jenzer, Carlo. III. Christkatholische Kirchgemeinde Solothurn.
BX4770.S9C47 79-338422
79 MARC

CHRISTKÖNIGSGEMEINDE IN ESCHBORN

Schulz, Heinz-Manfred, 1933-
Ein Jahr in Gottes Werkstatt : e. Gemeinde macht neue Erfahrungen / Heinz-Manfred Schulz. — 1. Aufl. — Mainz : Matthias-Grünewald-Verlag, 1978.
145 p. : 10 ill. ; 21 cm. GFR78-A
ISBN 3-7867-0691-3 : DM15.80
1. Christkönigsgemeinde in Eschborn. 2. Church renewal—Catholic Church. I. Title.
BX4630.E83C487 79-340850
79 MARC

CHRISTLICH-DEMOKRATISCHE UNION (GERMANY (DEMOCRATIC REPUBLIC))

Heyl, Wolfgang.
Verantwortung und Tatbereitschaft : zu Problemen der Leitungstätigkeit / Wolfgang Heyl. — ⌊s.l.⌋ : Sekretariat des Hauptvorstandes der Christlich-Demokratischen Union Deutschlands, 1975.
39 p. ; 19 cm. — (Hefte aus Burgscheidungen ; 197) GDR***
0.50M
1. Socialism in East Germany. 2. Christlich-Demokratische Union (Germany (Democratic Republic)) I. Title. II. Series.
HX536.H38 no. 197 320.9'43 s 78-377576
⌊HX280.5.A6⌋ 78 MARC

—CONGRESSES

Christlich-Demokratische Union (Germany (Democratic Republic)).
Roter Oktober : Wendepunkt d. Menschheitsgeschichte : zum 60. Jahrestag d. Grossen Sozialist. Oktoberrevolution / Pjotr A. Abrassimow, Gerald Götting. — 1. Aufl. — Berlin : Union-Verlag, 1977.
79 p. : ill. ; 20 cm. GDR77-A
Papers of a conference of the Präsidium des Hauptvorstandes of the Christlich-Demokratische Union, held in Burgscheidungen, May 5, 1977.
3.00M
1. Communism—Russia—Congresses. 2. Christlich-Demokratische Union (Germany (Democratic Republic))—Congresses. I. Abrassimov, Petr Andreevich, 1912- . II. Götting, Gerald.
HX313.5.C48 1977 335.43'0947 79-374192
79 MARC

CHRISTLICH-DEMOKRATISCHE UNION (GERMANY (FEDERAL REPUBLIC))

Dettling, Warnfried, 1943-
Die christliche Demokratie vor den Europawahlen / Warnfried Dettling. — Hannover : Niedersächsische Landeszentrale für Politische Bildung, 1978.
112 p. ; 21 cm. GFR***
Includes bibliographical references.
1. Christlich-Demokratische Union (Germany (Federal Republic)) 2. Christlich-Soziale Union. 3. Center parties—Europe. 4. Christian democracy. I. Title.
JN3971.A98C4238 329.9'43 79-350767
79 MARC

Focke, Franz.
Sozialismus aus christlicher Verantwortung : d. Idee e. christl. Sozialismus in d. kath.-sozialen Bewegung u. in d. CDU / Franz Focke. — Wuppertal : Hammer, 1978.
399 p. ; 21 cm. GFR78-A
Bibliography: p. 367-399.
ISBN 3-87294-126-7 : DM36.00
1. Socialism, Christian. 2. Socialism in Germany—History. 3. Socialism in West Germany. 4. Christlich-Demokratische Union (Germany (Federal Republic)) 5. Labor and laboring classes—Germany—History. I. Title.
HX54.F62 335'.7 79-346795
79 MARC

Hackel, Wolfgang, 1942-
Die Auswahl des politischen Nachwuchses in der Bundesrepublik Deutschland : die junge Union als Sprungbrett für politische Karrieren in der CDU / Wolfgang Hackel. — Stuttgart : Verlag Bonn Aktuell, 1978.
351 p. : graphs ; 20 cm. — (Studien zur Politik ; Bd. 2)
Bibliography: p. 337-348.
ISBN 3-87959-098-2
1. Christlich-Demokratische Union (Germany (Federal Republic)) 2. Junge Union Deutschlands. 3. Youth—Germany, West—Political activity. 4. Elite (Social Sciences)—Germany, West. I. Title. II. Series.
JN3971.A98C4347 329.9'43 79-360801
79 MARC

Ley, Richard.
Föderalismusdiskussion innerhalb der CDU/CSU : von d. Parteigründung bis zur Verabschiedung d. Grundgesetzes / von Richard Ley. — Mainz : v. Hase und Koehler, 1978.
194 p. ; 23 cm. — (Beiträge zu Wissenschaft und Politik ; Bd. 17) GFR78-A
A revision of the author's thesis, Mainz, 1976.
Bibliography: p. 180-189.
Includes index.
ISBN 3-7758-0950-3 : DM24.00
1. Federal government—Germany, West. 2. Christlich-Demokratische Union (Germany (Federal Republic)) 3. Christlich-Soziale Union. I. Title. II. Series.
JN3971.A38S8755 1978 79-344879
79 MARC

Das Urteil des Bundesverfassungsgerichts vom 2. März 1977 zur Öffentlichkeitsarbeit von Staatsorganen in Bund und in Ländern : Dokumentation d. Verfahrens u. Materialien / ⌊hrsg. vom Presse- u. Informationsamt d. Bundesverfassungsgericht⌋. — Heidelberg ; Karlsruhe : Müller Juristischer Verl., 1978.
654 p. : ill. ; 24 cm. — (Motive, Texte, Materialien ; Bd. 8) GFR78-A
Materials related to the case of CDU vs. Federal Republic of Germany.
Includes indexes.
ISBN 3-8114-5978-3 : DM78.00
1. Christlich-Demokratische Union (Germany (Federal Republic)). 2. Germany, West. 3. Executive privilege (Government information)—Germany, West. 4. Political parties—Germany, West. I. Germany (Federal Republic, 1949-). Bundesverfassungsgericht. II. Christlich-Demokratische Union (Germany (Federal Republic)) III. Germany (Federal Republic, 1949-). Presse-und Informationsamt.
79-346098
79 MARC

—ADDRESSES, ESSAYS, LECTURES

Grundwerte in der Politik : Analysen u. Beitr. zum Grundsatzprogramm d. Christl.-Demokrat. Union Deutschlands / hrsg. von Heiner Geissler. — Orig.-Ausg. — Frankfurt/M. ; Berlin ; Wien : Ullstein, 1979.
187 p. ; 18 cm. — (Ullstein Buch ; Nr. 34502) GFR79-A
"Grundsatzprogramm der Christlich Demokratischen Union Deutschlands": p. 126-170.
Includes bibliographical references.
ISBN 3-548-34502-6 : DM5.80
1. Christlich-Demokratische Union (Germany (Federal Republic))—Addresses, essays, lectures. 2. Germany, West—Economic policy—Addresses, essays, lectures. 3. Germany, West—Social policy—Addresses, essays, lectures. I. Geissler, Heinrich, 1930- . II. Christlich-Demokratische Union (Germany (Federal Republic))
JN3971.A98C438 329.9'43 79-376699
79 MARC

CHRISTLICH-DEMOKRATISCHE UNION (GERMANY (FEDERAL REPUBLIC)). LANDESVERBAND BADEN-WÜRTTEMBERG

Die CDU in Baden-Württemberg und ihre Geschichte / hrsg. von Paul-Ludwig Weinacht ; mit e. Geleitw. von Hans Filbinger. — Stuttgart ; Berlin ; Köln ; Mainz : Kohlhammer, 1978.
399 p. ; 22 cm. — (Schriften zur politischen Landeskunde Baden-Württembergs ; Bd. 2) GFR78-A
Bibliography: p. 390-392.
Includes index.
ISBN 3-17-004679-9
1. Christlich-Demokratische Union (Germany (Federal Republic)). Landesverband Baden-Württemberg. 2. Baden-Württemberg—Politics and government. I. Weinacht, Paul-Ludwig. II. Series.
JN3971.A98C42289 329.9'43 79-353622
79 MARC

CHRISTLICH-DEUTSCHER TURNVEREIN ÖSTERREICHS

—HISTORY

Hochrainer, Anton.
Die Christlich-deutsche Turnerschaft Österreichs und Mons. Franz Xaver Hofstaetter (obit 1968) / Anton Hochrainer. — Wien : Kirchenhist. Inst., 1972.
ii, 13 leaves ; 30 cm. — (Miscellanea aus dem Kirchenhistorischen Institut der Katholisch-Theologischen Fakultät Wien ; 34) Au73-18-9
Includes bibliographical references.
Not in trade.
1. Religion and sports. 2. Christlich-Deutscher Turnverein Österreichs—History. 3. Hofstaetter, Franz Xaver, 1902-1968. 4. Priests—Austria—Biography. I. Title. II. Series: Vienna. Universität. Katholisch-Theologische Fakultät. Kirchenhistorisches Institut. Miscellanea ; 34.
BR141.V48 no. 34 79-372184
[GV706.8] *79 MARC

CHRISTLICH-SOZIALE UNION

Christlich-Soziale Union. Arbeitskreis Finanz- und Wirtschaftspolitik.
Partner der Zukunft : Leitlinien für e. partnerschaftl. Unternehmensverfassung / von Heinz Broschell ; Vorw. von Franz Josef Strauss ; als Autoren-Team wirkten mit, Hartwig-Ulrich von Both ... [et al.]. — München : Verlag Moderne Industrie, 1979.
168 p. ; 23 cm. GFR79-A
Includes bibliographical references and index.
ISBN 3-478-33790-9 : DM16.80
1. Employee ownership—Germany, West. 2. Industrial relations—Germany, West. 3. Christlich-Soziale Union. I. Broschell, Heinz, 1932- . II. Title.
HD5660.G4C53 1979 79-371777
00 MARC

Dettling, Warnfried, 1943-
Die christliche Demokratie vor den Europawahlen / Warnfried Dettling. — Hannover : Niedersächsische Landeszentrale für Politische Bildung, 1978.
112 p. ; 21 cm. GFR***
Includes bibliographical references.
1. Christlich-Demokratische Union (Germany (Federal Republic)) 2. Christlich-Soziale Union. 3. Center parties—Europe. 4. Christian democracy. I. Title.
JN3971.A98C4238 329.9'43 79-350767
79 MARC

Ley, Richard.
Föderalismusdiskussion innerhalb der CDU/CSU : von d. Parteigründung bis zur Verabschiedung d. Grundgesetzes / von Richard Ley. — Mainz : v. Hase und Koehler, 1978.
194 p. ; 23 cm. — (Beiträge zu Wissenschaft und Politik ; Bd. 17) GFR78-A
A revision of the author's thesis, Mainz, 1976.
Bibliography: p. 180-189.
Includes index.
ISBN 3-7758-0950-3 : DM24.00
1. Federal government—Germany, West. 2. Christlich-Demokratische Union (Germany (Federal Republic)) 3. Christlich-Soziale Union. I. Title. II. Series.
JN3971.A38S8755 1978 79-344879
79 MARC

CHRISTLICHSOZIALE PARTEI

Unger, Günter, 1641-
Die Christlichsoziale Partei im Burgenland. Eisenstadt, Burgenländisches Landesarchiv, 1965.
vi, 201 p. 24 cm. (Burgenländische Forschungen, Heft 49) S90.- GDNB66-B16-289
Bibliography p. v-vi.
1. Christlichsoziale Partei. I. Title.
DB785.B8B78 Heft 49 943.6'15 s 67-100704
[JN2031.C48] [r79]rev MARC

CHRISTMAS

see also Jesus Christ—Nativity; Santa Claus

Christmas / [devised by Terence Goldsmith ; editing by Jacqueline Ridley ; original art work by William Cameron Johnson, David Dowland, and Joyce Smith, original photography by David Couling]. — Huntington, Ind. : Our Sunday Visitor, 1978.
160 p. : ill. ; 27 cm.
ISBN 0-87973-709-3 : $12.95
1. Christmas. I. Goldsmith, Terence. II. Ridley, Jacqueline.
GT4985.C383 1978b 394.2'68282 78-57004
79 MARC

Christmas / [edited by Jacqueline Ridley]. — Poole : Blandford Press, 1978.
160 p. : ill. (chiefly col.), col. facsims. ; 27 cm. GB78-35223
ISBN 0-7137-0846-8 : £5.95
1. Christmas. I. Ridley, Jacqueline.
GT4985.C383 394.2'68282 79-302456
79 MARC

The Christmas book / written by Robert Joseph ; illustrated by Maurice Hughes ; musical arrangements by Donald Siegal ; concept and art direction by Don McAfee. — New York : McAfee Books, c1978.
192 p. : col. ill. ; 32 cm.
Carols for voice and piano: p. 80-131.
ISBN 0-89328-022-4 : $12.95
1. Christmas. 2. Christmas stories. 3. Carols. I. Joseph, Robert.
GT4985.C392 394.2'68282 78-108796
78 MARC

Del Re, Gerard, 1944-
The Christmas almanack / Gerard & Patricia Del Re ; ill. by Doug Jamieson. — 1st ed. — Garden City, N.Y. : Doubleday, 1979.
xvi, 402 p. : ill. ; 26 cm.
Includes index.
ISBN 0-385-13353-7 : $7.95
1. Christmas. I. Del Re, Patricia, 1944- joint author. II. Title.
GT4985.D47 394.2'68282 77-12842
79 MARC

Georgi, Christoph.
Vom Licht der Weihnacht im Erzgebirge / hrsg. und eingeleitet von Günter Hanisch ; Aufnahmen von Christoph Georgi. — 2. Aufl. — Berlin : Evangelische Verlagsanstalt, 1975.
115 p. : chiefly ill. ; 27 cm.
GDR
10.80M
1. Photography, Artistic. 2. Christmas. I. Hanisch, Günter. II. Title.
TR654.G46 1975 78-388022
779'.9'39426282094321
79 MARC

Old fashioned Christmas in stories, poems, pictures / compiled and edited by Edward J. Kutlowski. — Seabrook, N.H. : House of White Birches, c1978.
240 p. : ill. ; 43 cm.
Caption title.
1. Christmas. I. Kutlowski, Edward.
GT4985.O52 394.2'6828 79-100226
79 MARC

Rasband, Ester.
What think you of Christmas? / Ester Rasband ; illustrated by Anna Sturrock. — Salt Lake City : Bookcraft, c1978.
[64] p. : col. ill. ; 24 cm.
ISBN 0-88494-352-6
1. Christmas. I. Sturrock, Anna. II. Title.
BV45.R36 263'.91 78-67241
00 MARC

[Warren, Nathan Boughton], 1815-1898, comp.
Christmas in the olden time, its customs and their origin : the holly and ivy, sports of the eve, Yule log, boar's head, the dinner, mummers, Lord of Misrule, Saturnalia, carols, mysteries and plays, boxes. — Folcroft, Pa. : Folcroft Library Editions, 1976.
47 p. ; 23 cm.
At head of title: The Christmas book.
Reprint of the 1859 ed. published by J. Pattie, London.
ISBN 0-8414-1656-7 lib. bdg. : $8.50
1. Christmas. 2. Carols—History and criticism. I. Title.
GT4985.W4 1976 394.2'68282 76-58002
76[r79]rev MARC

Weber-Kellermann, Ingeborg, 1918-
Das Weihnachtsfest : eine Kultur- und Sozialgeschichte der Weihnachtszeit / Ingeborg Weber-Kellermann. — Luzern [etc.] : Bucher, cop. 1978.
232 p. : numerous ill. (some col.) ; 28 cm. Sw78-12728
Bibliography: p. 223-227.
Includes index.
ISBN 3-7658-0273-5 : 58.00F
1. Christmas. I. Title.
GT4985.W41525 394.2'68282 79-381096
79 MARC

—BIBLIOGRAPHY

Phillips, Leona Rasmussen.
Christmas, an annotated bibliography / by Leona Rasmussen Phillips. — New York : Gordon Press, 1978.
160 p. ; 24 cm. — (Gordon Press bibliographies for librarians series)
ISBN 0-8490-1363-1
1. Christmas—Bibliography. I. Title.
Z5711.C5P58 016.3942'6828 77-964
[GT4985] 77 MARC

—DICTIONARIES

Piø, Iørn, 1927-
Julens hvem hvad hvor : håndbog om julens traditioner / Iørn Piø ; [red., Mogens Rud]. — København : Politiken, 1977.
240 p. : ill. ; 18 cm. D77-49
Bibliography: p. 231-234.
Includes index.
ISBN 8756727852 : kr52.50
1. Christmas—Dictionaries. 2. Christmas—Denmark—Dictionaries. I. Rud, Mogens. II. Title.
GT4985.P47 78-392703
79 MARC

—FICTION see Christmas stories

—JUVENILE POETRY

Desbarats, Peter.
The night the city sang / Peter Desbarats ; ill. by Frank Newfeld. — Toronto : McClelland and Stewart, c1977.
48 p. : ill. (some col.) ; 24 cm. C***
Originally published under title: Halibut York & more.
ISBN 0-7710-2685-4 : $5.95
1. Christmas—Juvenile poetry. I. Newfeld, Frank, 1928- . II. Title.
PR9199.3.D46H3 1977 811'.5'4 79-312534
79 MARC

—LITERARY COLLECTIONS

Christie, Agatha Miller, Dame, 1891-1976.
Star over Bethlehem, and other stories, by Agatha Christie Mallowan. Decorations by Elsie Wrigley. New York, Dodd, Mead [1965]
79 p. col. illus. 21 cm.
1. Christmas—Literary collections. I. Title.
PR6005.H66S8 823'.9'12 65-29638
[r79]rev MARC

Disney (Walt) Productions.
Walt Disney's Christmas treasury / [introd. by Tom Jones]. — New York : Abbeville Press, c1978.
203 p. : col. ill. ; 30 cm.
SUMMARY: An anthology of Christmas stories, illustrations, comic strips, posters, and other miscellanea featuring Disney characters.
ISBN 0-89659-004-6 : $17.95
1. Christmas—Literary collections. [1. Christmas—Literary collections] I. Title. II. Title: Christmas treasury.
PZ5.D64 Wai 1978 [Fic] 78-14209
78 MARC

Fasser, Carl.
Quaida not : gös e versets da Nadal / Carl Fasser. — [Chur] : Lia Rumantscha, 1977.
40 p. : music ; 21 cm. — (La Scena ; n. 50) Sw78-2752
5.50F
1. Christmas—Literary collections. I. Title.
PC959.F34Q3 79-346860
79 MARC

Goepfert, Günter.
Münchner Weihnacht / Günter Goepfert ; [Holzschnitte, Eugen O. Sporer]. — München : Wila Verlag Lampl, 1973.
56 p. : numerous col. ill. ; 27 cm. GFR75-A
DM18.00
1. Christmas—Literary collections. I. Title.
PT2667.O396M8 78-366263
78 MARC

La Navidad en la literature venezolana / compilación e introd. Efraín Subero. — Caracas : Ediciones del Congreso de la República, 1973.
272 p. : ill. ; 24 cm.
"Universidad Católica Andrés Bello, Centro de Investigaciones Literarias, Institutos Humanísticos de Investigación."
Includes bibliographical references.
1. Venezuelan literature. 2. Christmas—Literary collection. I. Subero, Efraín, 1931- . II. Caracas. Universidad Católica Andrés Bello. Centro de Investigaciones Literarias.
PQ8543.N38 860 79-353908
79 MARC

CHRISTMAS
—LITERARY COLLECTIONS (Continued)

Poèmes et contes de Noël. (Précédés de) Le Voyage de Noël / de Marie Noël ; ⌈illustré par Hélène Neveur, Danielle Grassin, Yolande Ardissonne, Jacques Eitel, etc.⌉. — Paris : J. Grassin, 1976.
287 p. : ill. ; 23 cm. — (Collection Grandes anthologies ; 2 ISSN 0395-3076) F77-17630
Three hundred copies printed, No. 272.
ISBN 2-7055-0930-5 : 100F
1. French literature—20th century. 2. Christmas—Literary collections.
PQ1145.C56P6 840'.8'033 79-352584
79 MARC

Pollmer, Karl Hans.
Gloria, Gloria Gott in der Höh' : Advent u. Weihnacht im Erzgebirge / Karl Hans Pollmer ; ⌈Bildtaf., Kurt Eichler⌉. — 1. Aufl. — Berlin : Evangelische Verlagsanstalt, 1976.
143 p., ⌈8⌉ leaves of plates : col. ill. ; 20 cm. GDR76-A
6.80M
1. Christmas—Literary collections. I. Title.
PT2676.O47G57 78-392780
79 MARC

Schrader, Julie, 1882-1939.
Ich fühl' mich wie ein Weihnachtsbaum / Julie Schrader. — München ; Zürich : Piper, 1978.
98 p. : ill. ; 17 cm. — (Piper-Präsent) GFR78-A
ISBN 3-492-02383-5 : DM9.80
1. Christmas—Literary collections. I. Title.
PT2638.R257I34 1978 830'.8'033 79-362145
79 MARC

Seymour, William Kean, 1887- comp.
Happy Christmas / compiled by William Kean Seymour and John Smith ; illustrated by Beryl Sanders. — Philadelphia : Westminster Press, 1979, c1968.
256 p. : ill. ; 26 cm.
Includes index.
SUMMARY: Prose, poetry, carols, plays, and anecdotes, reflecting Christmas, and drawn from the finest writers of the English language from Elizabethan times to the present.
$12.50
1. Christmas—Literary collections. ⌈1. Christmas—Literary collections⌉ I. Smith, John, 1924- . II. Sanders, Beryl. III. Title.
PN6071.C6S4 1979 820'.8'033 79-102624
79 MARC

Weihnachtsveranstaltungen / zsgest., erprobt u. hrsg. von Wolfgang Erk. — Stuttgart : Steinkopf, 1976.
282 p. : col. ill. ; 24 cm. & slides (6 slides : col. ; 2x2 in.) in pockets. GFR77-A
Includes texts of songs in various languages.
Includes bibliographical references.
ISBN 3-7984-0325-2 : DM28.00
1. Christmas—Literary collections. I. Erk, Wolfgang, 1943-
PN6071.C6W38 830'.8'033 79-337322
79 MARC

Zillich, Heinrich, 1898-
Deutsche Weihnachten zwischen Ost und Süd / Heinrich Zillich. — Wien : Österr. Landsmannschaft, 1976.
54 p. ; 18 cm. — (Eckartschriften ; Heft 60) Au77-7-185
Bibliography of H. Zillich's works: p. 48-49.
S35.00
1. Christmas—Literary collections. I. Title.
PT2653.I53D4 833'.9'12 79-365743
*79 MARC

—MEDITATIONS

Luccock, Halford Edwards, 1885-1960.
A sprig of holly / Halford E. Luccock ; Charles S. Hartman, editor. — New York : Pilgrim Press, c1978.
64 p. ; 17 cm.
Includes bibliographical references.
ISBN 0-8298-0354-8 : $2.95
1. Christmas—Meditations. I. Hartman, Charles S., 1913- II. Title.
BV45.L74 1978 242'.33 78-17096
78 MARC

—POETRY

see also Carols

Breda, Emilio Alberto.
Los villancicos porteños / Emilio Breda ; ⌈ilus. de Norah Borges⌉. — 1. ed. — Buenos Aires : Corregidor, c1977.
256 p. : ill. ; 18 cm. — (Serie popular ; no. 1218)
1. Christmas—Poetry. 2. Jesus Christ—Poetry. I. Title.
PQ7798.12.R37V54 861 78-110168
79 MARC

Bryant, Edward Andem, 1869- comp.
Yule-tide cheer : a book of verses for the Christmas season / compiled by Edward A. Bryant. — Great Neck, N.Y. : Granger Book Co., 1978.
xv, 309 p. ; 20 cm. — (Granger poetry library)
Reprint of the 1912 ed. published by Crowell, New York.
Includes index.
ISBN 0-89609-093-0
1. Christmas—Poetry. I. Title. II. Series.
PN6110.C5B7 1978 821'.008'033 78-57854
79 MARC

Derwood, Gene, 1909-1954.
Star. — ⌈New York? : O. Williams?⌉, c1954.
⌈4⌉ p. : facsim., port. ; 18 cm.
Caption title.
On p. ⌈1⌉: Peace on earth, good will toward men.
Commemorative keepsake in the form of a Christmas card with G. Derwood's poem, port., and facsim. signature; copyrighted by Oscar Williams.
L.C. copy has ms. additions by Williams.
1. Christmas—Poetry. I. Williams, Oscar, 1900-1964. II. Title. III. Title: Peace on earth, good will toward men.
PS3507.E716S8 1954 79-309983
79 MARC

Salem, Luis D
Rimas del pesebre / Luis D. Salem. — Miami, Fla. : Editorial Caribe, c1977.
86 p. ; 18 cm.
1. Christmas—Poetry. 2. Jesus Christ—Poetry. I. Title.
PQ7298.29.A366R55 861 77-82265
79 MARC

—PRAYER-BOOKS AND DEVOTIONS

Barnard, Naomi Holt.
The wonder of Christmas / Naomi Holt Barnard. — Nashville : Broadman Press, c1979.
63 p. : ill. ; 20 cm.
Includes bibliographical references.
ISBN 0-8054-5165-X
1. Christmas—Prayer-books and devotions—English. I. Title.
BV45.B28 242'.33 78-67996
79 MARC

—SONGS AND MUSIC see Christmas music

—AUSTRIA, LOWER

Galler, Werner, 1944-
Weihnachten in Niederösterreich / Werner Galler. — 1. Aufl. — St. Pölten ; Wien : Verl. Niederösterr. Pressehaus, 1977.
47 p. : ill. ; 21 cm. — (Wissenschaftliche Schriftenreihe Niederösterreich ; 29/30) Au78-5-225
Includes bibliographical references.
1. Christmas—Austria, Lower. 2. Austria, Lower—Social life and customs. I. Title.
GT4987.47.G34 78-379474
*78 MARC

—BERLIN—LITERARY COLLECTIONS

Weihnachten im alten Berlin : Texte u. Bilder / ges. von Gustav Sichelschmidt. — Berlin : Rembrandt-Verlag, 1978.
161 p. : 26 ill. (some col.) ; 18 cm. GFR79-A
ISBN 3-7925-0260-7 : DM16.80
1. Christmas—Berlin—Literary collections. 2. German literature. 3. Berlin in art. I. Sichelschmidt, Gustav.
PT1110.B4W4 830'.8'033 79-375821
79 MARC

—DENMARK—DICTIONARIES

Piø, Iørn, 1927-
Julens hvem hvad hvor : håndbog om julens traditioner / Iørn Piø ; ⌈red., Mogens Rud⌉. — København : Politiken, 1977.
240 p. : ill. ; 18 cm. D77-49
Bibliography: p. 231-234.
Includes index.
ISBN 8756727852 : kr52.50
1. Christmas—Dictionaries. 2. Christmas—Denmark—Dictionaries. I. Rud, Mogens. II. Title.
GT4985.P47 78-392703
79 MARC

—FINLAND

Jaakkola, Kaisu, 1929-
Muuttuva joulu : kansatieteellinen tutkimus - The changing Christmas : an ethnological study / Kaisu Jaakkola ; kartat ja piirustukset, Topi Valkonen. — Helsinki : Suomen muinaismuistoyhdistys, 1977.
⌈12⌉, 372 p. : ill., maps ; 25 cm. — (Kansatieteellinen arkisto ; 28) Fi78Mar.
Edited by N. Valonen and J. U. E. Lehtonen.
Thesis—Helsinki.
Summary in English.
Bibliography: p. 358-372.
Includes index.
ISBN 9519913491
1. Christmas—Finland. I. Valonen, Niilo. II. Lehtonen, Juhani. III. Title. IV. Title: The changing Christmas. V. Series.
DK450.8.K34 no. 28 79-356729
⌈GT4987.55⌉ 79 MARC

—FINLAND—POHJANMAA

Bregenhøj, Carsten, 1941-
Nuutti : Suomalainen karnevaali = A Finnish carnival / Carsten Bregenhøj, Urpo Vento ⌈valokuvaajat, Annikki Kaivola-Bregenhøj, Ilkka Kolehmainen, Susan Daugthry ; English transl., Leena Anttila⌉. — ⌈Helsinki⌉ : SKS ⌈Suomalaisen Kirjallisuuden Seura⌉, 1975.
48 p. : ill ; 20x22 cm. Fi
English and Finnish.
ISBN 9517170769 : Fmk15.00
1. Christmas—Finland—Pohjanmaa. 2. Festivals—Finland—Pohjanmaa. 3. Carnival—Finland—Pohjanmaa. I. Vento, Urpo, joint author. II. Title.
GT4987.55.B73 79-355733
*79 MARC

—GERMANY

Ruland, Josef.
Christmas in Germany / Josef Ruland ; ⌈with contributions by Emil Barth ... et al. ; transl. by Timothy Nevill⌉. — Bonn : Hohwacht, 1978.
148 p. : numerous ill. (some col.), music ; 24 cm. GFR79-A5
Translation of Weihnachten in Deutschland.
Bibliography: p. 143-146.
ISBN 3-87353-065-1 : DM14.80
1. Christmas—Germany. I. Title.
GT4987.49.R8413 393.2'68282'0943 79-317781
79 MARC

—GERMANY, WEST—FRANCONIA

Worschech, Reinhard.
Fränkische Bräuche zur Weihnachtszeit : von Martini bis Lichtmess / Reinhard Worschech ; Ill., Otto Mayer. — Würzburg : Stürtz, 1978.
168 p. : numerous ill., music ; 20 x 21 cm. GFR79-A
Bibliography: p. 168.
Includes index.
ISBN 3-8003-0125-3 : DM28.00
1. Christmas—Germany, West—Franconia. 2. Franconia—Social life and customs. I. Title.
GT4987.49.W67 394.2'68282'09433 79-369067
79 MARC

—GERMANY, WEST—FRANCONIA—EXHIBITIONS

Nuremberg. Stadtbibliothek. Institut für Fränkische Literatur.
Weihnachten in Franken. Ausstellung des Instituts für Fränkische Literatur der Stadtbibliothek Nürnberg. Aus Anlass der Eröffnung am 5. Dezember 1969 ... hält den Einführungsvortrag Karlheinz Goldmann. ⌈Nürnberg, Fränkische Verlagsanstalt und Buchdruckerei, 1969⌉
16 p. illus. 18 cm. (Ausstellungskatolog der Stadtbibliothek Nürnberg, 67) GDB***
1. Christmas—Germany, West—Franconia—Exhibitions. 2. Franconia—Social life and customs—Exhibitions. I. Goldmann, Karlheinz. II. Title. III. Series: Nuremberg. Stadtbibliothek. Ausstellungs-Katalog 67.
GT4987.49.N87 1969 79-383704
79 MARC

—GREAT BRITAIN—HISTORY

Miall, Antony.
The Victorian Christmas book / Antony and Peter Miall. — London : Dent, 1978.
192 p., ⌈8⌉ leaes of plates : ill. (some col.) ; 25 cm. GB***
ISBN 0-460-12039-5 : £6.95
1. Christmas—Great Britain—History. I. Miall, Peter, joint author. II. Title.
GT4987.43.M52 394.2'68282'0941 78-326305
79 MARC

—NEWFOUNDLAND

Sider, Gerald M
Mumming in outport Newfoundland / Gerald M. Sider. — Toronto : New Hogtown Press, c1977.
31 p. ; 21 cm. C***
Includes bibliographical references.
1. Mumming—Newfoundland. 2. Newfoundland—Social conditions. 3. Christmas—Newfoundland. I. Title.
GT4987.15.S53 393.2'68282 79-317348
79 MARC

CHRISTMAS (Continued)

—PERU—HUÁNUCO (DEPT.)—ADDRESSES, ESSAYS, LECTURES

Antología de "los negritos". — Ciudad de Huánuco ⌈Perú⌉ : Instituto Nacional de Cultura, Filial en Huánuco, 1973.
86 p. ; 22 cm.
CONTENTS: Pulgar Vidal, J. Los negritos.—Pavletich, E. La danza de "los negritos" de Huánuco.—Vizcaya M., N. R. Los Negritos y el mayordomo don Tomás.
1. Negritos (Dance)—Addresses, essays, lectures. 2. Christmas—Peru—Huánuco (Dept.)—Addresses, essays, lectures. I. Pulgar Vidal, Javier, 1913- Los negritos. 1973. II. Pavletich, Esteban, 1906- La danza de "Los negritos" de Huánuco. 1973. III. Vizcaya M., Nicolás R. Los negritos y el mayordomo don Tomás. 1973.
GV1796.N4A57 79-371868
79 MARC

CHRISTMAS BAY, TEX.

McEachron, L W
A fishery survey of Christmas, Drum, and Bastrop Bays, Brazoria County, Texas / by L. W. McEachron, C. R. Shaw, and A. W. Moffett. — ⌈Austin⌉ : Texas Parks & Wildlife Dept., 1977.
83 p. : ill. ; 23 cm. — (Technical series - Texas Parks & Wildlife Department ; no. 20)
Bibliography: p. 43-45.
1. Marine biology—Texas—Brazoria Co. 2. Fishery resources—Texas—Brazoria Co. 3. Oceanography—Texas—Brazoria Co. 4. Christmas Bay, Tex. 5. Drum Bay, Tex. 6. Bastrop Bay, Tex. I. Shaw, C. R., joint author. II. Moffett, Alan W., joint author. III. Title. IV. Series: Texas. Parks and Wildlife Dept. Technical series ; no. 20.
QH105.T4M3 591.9'2'34 78-621507
79 MARC

CHRISTMAS BOOKS see Gift-books (Annuals, etc.); Christmas; Christmas stories

CHRISTMAS CARDS

see also Greeting cards

The Christmas crafts book. — London : Search Press ; New York : Sterling, 1979.
112 p. : ill. (some col.) ; 26 cm. GB***
Bibliography: p. 109.
Includes index.
ISBN 0-8069-5396-9 (Sterling) : $16.95
1. Christmas decorations. 2. Christmas cards. 3. Handicraft.
TT900.C4C45 1979 745.59'41 79-65064
79 MARC

CHRISTMAS CAROLS see Carols

CHRISTMAS COOKERY

see also Gingerbread houses

Another Christmas in the country / ⌈prepared by the staff of Farm wife news. — Milwaukkee : Farm Wife, c1976.
110 p. ; 28 cm.
Cover title.
ISBN 0-89821-015-1
1. Christmas decorations. 2. Christmas cookery. 3. Handicraft. I. Farm wife news. II. Title: Christmas in the country.
TT900.C4A52 745.59'41 79-116425
00 MARC

Christmas in the country / ⌈prepared by the staff of Farm wife news⌉. — Milwaukee : Farm wife news, c1979.
84 p. : ill. ; 28 cm.
Cover title.
ISBN 0-89821-005-4
1. Christmas decoration. 2. Christmas cookery. 3. Handicraft. I. Farm wife news.
TT900.C4C47 745.59'41 79-115737
00 MARC

Domingo, Xavier, 1929-
La cocina de Navidad en 16 recetas / Xavier Domingo. — Madrid : Editorial Cambio 16, ⌈1976?⌉
59 p. ; 18 cm. — (Colección Almanaque ; 1) Sp***
ISBN 848522907X : 60ptas
1. Christmas cookery. 2. Cookery, International. I. Title.
TX739.D65 641.5'68 78-390467
79 MARC

Dutton, June.
A Christmas cook book / recipes by June Dutton ; ill. by Bruce Butte. — San Francisco : Determined Productions, c1978.
126 p., ⌈1⌉ leaf of plates : ill. ; 18 cm.
ISBN 0-915696-06-1
1. Christmas cookery. I. Title.
TX739.D87 641.5'68 77-78295
79 MARC

Farm journal's homespun Christmas / by the editors of Farm journal ; editor, Nancy Steele ... ⌈et al.⌉. — Philadelphia : The Journal ; Garden City, N.Y. : distributed to the trade by Doubleday, c1979.
123 p. : ill. ; 28 cm.
Includes index.
ISBN 0-89795-004-6
1. Christmas cookery. 2. Christmas decorations. I. Steele, Nancy Saunders. II. Farm journal (Philadelphia, 1956-) III. Title: Homespun Christmas.
TX739.F29 394.2'68282 79-12884
79 MARC

Graff, Monika.
Pfeffernuss und Mandelkern : ⌈d. kleine Weihnachtsbackbuch⌉ / Monika Graff. — ⌈1.-10. Tsd.⌉. — Weil der Stadt : Hadecke, 1976.
68 p. : ill. (some col.) ; 21 cm. — (Leckerbissen ; 20) GFR76-A
Includes index.
ISBN 3-7750-0074-7 : DM5.80
1. Christmas cookery. 2. Cookies. 3. Cake. I. Title.
TX739.G68 78-382205
78 MARC

Voth, Norma Jost.
Peppernuts, plain and fancy / by Norma Jost Voth ; illustrated by Ellen Jane Price. — San Jose, Calif. : Voth, c1978.
71 p. : ill. ; 16 cm.
Includes index.
ISBN 0-9601818-1-4
1. Christmas cookery. 2. Cookies. I. Title.
TX739.V67 641.8'659 78-60518
78 MARC

Wittich, Boris.
Weihnachtlicher Küchenzauber / Boris Wittich. — München : Kochbuchverlag Heimeran, 1976.
102 p. : ill. ; 11 x 17 cm. — (Heimerans Kochbuchreihe für Geniesser)
GFR77-A
ISBN 3-8063-1105-6 : DM7.80
1. Christmas cookery. I. Title.
TX739.W58 641.5'68 78-384230
78 MARC

CHRISTMAS DECORATIONS

see also Christmas trees; Gingerbread houses; Glass Christmas decorations

Another Christmas in the country / ⌈prepared by the staff of Farm wife news. — Milwaukkee : Farm Wife, c1976.
110 p. ; 28 cm.
Cover title.
ISBN 0-89821-015-1
1. Christmas decorations. 2. Christmas cookery. 3. Handicraft. I. Farm wife news. II. Title: Christmas in the country.
TT900.C4A52 745.59'41 79-116425
00 MARC

The Christmas crafts book. — London : Search Press ; New York : Sterling, 1979.
112 p. : ill. (some col.) ; 26 cm. GB***
Bibliography: p. 109.
Includes index.
ISBN 0-8069-5396-9 (Sterling) : $16.95
1. Christmas decorations. 2. Christmas cards. 3. Handicraft.
TT900.C4C45 1979 745.59'41 79-65064
79 MARC

Christmas in the country / ⌈prepared by the staff of Farm wife news⌉. — Milwaukee : Farm wife news, c1979.
84 p. : ill. ; 28 cm.
Cover title.
ISBN 0-89821-005-4
1. Christmas decoration. 2. Christmas cookery. 3. Handicraft. I. Farm wife news.
TT900.C4C47 745.59'41 79-115737
00 MARC

Darnell, Jane.
Pods & odd bodikins : ornaments from natural materials / Jane Darnell & Patricia Hawkes ; illustrated by Patricia Hawkes. — 1st ed. — Chester, Conn. : Globe Pequot Press, c1979.
112 p. : ill. ; 23 cm.
Bibliography: p. 110.
Includes index.
ISBN 0-87106-029-9 : $4.95
1. Christmas decorations. 2. Nature craft. I. Hawkes, Patricia, joint author. II. Title.
TT900.C4D37 745.59'41 79-50697
79 MARC

Farm journal's homespun Christmas / by the editors of Farm journal ; editor, Nancy Steele ... ⌈et al.⌉. — Philadelphia : The Journal ; Garden City, N.Y. : distributed to the trade by Doubleday, c1979.
123 p. : ill. ; 28 cm.
Includes index.
ISBN 0-89795-004-6
1. Christmas cookery. 2. Christmas decorations. I. Steele, Nancy Saunders. II. Farm journal (Philadelphia, 1956-) III. Title: Homespun Christmas.
TX739.F29 394.2'68282 79-12884
79 MARC

McCall's Christmas bazaar / by the editors of McCall's needlework & crafts ; editorial director, Rosemary McMurtry ; editor, Charlotte Brem ; art director, Bettan Prichard. — New York : McCall Pattern Co., c1978.
48 p. : ill. (some col.) ; 28 cm.
$1.35
1. Christmas decorations. 2. Handicraft. I. McMurtry, Rosemary. II. Brem, Charlotte. III. McCall's needlework & crafts.
TT900.C4M32 745.59'41 78-110847
79 MARC

Selskabet til haandarbejdets fremme.
Counted cross-stitch designs for Christmas / designs created by the Danish Handcraft Guild. — New York : Scribner, ⌈1978⌉ c1977.
79 p. : ill. ; 21 cm.
First published in Danish and English under title: Sy julekorssting. Cross-stitch designs for Christmas.
ISBN 0-684-15975-9 : $7.95
1. Cross-stitch—Patterns. 2. Christmas decorations. I. Title.
TT778.C76S4413 1978 746.4'4 78-19181
78 MARC

Ströse, Susanne.
Weihnachtliches Gestalten / Susanne Ströse. — 1. Aufl. — München : Don-Bosco-Verlag, 1977.
96 p. : ill. ; 19 cm. GFR78-A8
ISBN 3-7698-0305-1 : DM16.80
1. Christmas decorations. I. Title.
TT900.C4S78 745.59'41 78-399235
79 MARC

—JUVENILE LITERATURE

Shoemaker, Kathryn E
Creative Christmas : simple crafts from many lands / Kathryn E. Shoemaker. — Minneapolis : Winston Press, c1978.
114 p. : ill. ; 22 x 28 cm. — (A Kathryn Shoemaker craft book)
Includes index.
SUMMARY: Presents easy-to-follow instructions for numerous traditional Christmas decorations and gifts from such countries as England, Japan, Italy, Denmark, and Germany.
ISBN 0-03-045716-5
1. Christmas decorations—Juvenile literature. 2. Handicraft—Juvenile literature. ⌈1. Christmas decorations. 2. Handicraft⌉ I. Title. II. Series.
TT900.C4S56 745.59'41 78-59407
79 MARC

CHRISTMAS IN ART

Egger, Hanna.
Weihnachtsbilder im Wandel der Zeit : von d. Spätantike bis zum Barock / Hanna Egger. — Wien ; München : Schroll, c1978.
148 p. : ill. (some col.) ; 24 cm.
Bibliography: p. 146-148.
ISBN 3-7031-0474-0 : S330.00
1. Christmas in art. 2. Art. I. Title.
N8060.E36 704.948'4 79-362725
*79 MARC

Nast, Thomas, 1840-1902.
Thomas Nast's Christmas drawings / by Thomas Nast ; with an introd. by Thomas Nast St. Hill. — New York : Dover Publications, 1978.
viii, 69 p. : chiefly ill. ; 29 cm.
Reprint of the 1890 ed. published by Harper, New York, under title: Thomas Nast's Christmas drawings for the human race: with new introd.
ISBN 0-486-23660-9 pbk. : $3.50
1. Nast, Thomas, 1840-1902. 2. Christmas in art. I. Title. II. Title: Christmas drawings.
⌈NC139.N3A4 1978⌉ 741.9'73 78-52251
78 MARC

—EXHIBITIONS

Klein, Georges, fl. 1970-
Images de Noël : ⌈exposition⌉ Strasbourg, Musée alsacien, 17 décembre 1977-5 février 1978 : ⌈catalogue, textes et notices / Georges Klein⌉. — Strasbourg : Musée alsacien, ⌈1977⌉
58 p., ⌈14⌉ leaves of plates : ill. (some col.) ; 24 cm. F***
Bibliography: p. 58.
1. Folk art—France—Alsace—Exhibitions. 2. Christmas in art—Exhibitions. I. Strasbourg. Musée alsacien. II. Title.
NK948.A47K559 704.948'4 78-383619
78 MARC

CHRISTMAS ISLAND, INDIAN OCEAN

—HISTORY

Bailey, Eric.
The Christmas Island story / ₍by₎ Eric Bailey. — London : Stacey International, 1977.
88 p., xvi p. of plates : ill. (some col.), maps, ports. ; 22 cm. GB78-31870
Maps on lining papers.
Bibliography: p. 80-84.
Includes index.
ISBN 0-905743-08-3 : £2.50
1. Christmas Island, Pacific Ocean—History. I. Title.
DU650.B34 996'.4 79-305425
79 MARC

—PERIODICALS

Australia. Dept. of External Territories.
Annual report on the Territory of Christmas Island. 1967/68–
Canberra, Australian Govt. Pub. Service ₍etc.₎
v. in ill. 34 cm.
Report year ends June 30.
Continues: Australia. Dept. of Territories. Annual report on the Territory of Christmas Island.
Cover title, 1967/68– : Territory of Christmas Island: Report.
1. Christmas Island, Indian Ocean—Periodicals. I. Australia. Dept. of External Territories. Territory of Christmas Island: Report. II. Title.
DS349.9.C45A9a 996'.4 77-642822
MARC-S

Australia. Dept. of Territories.
Annual report on the Territory of Christmas Island. 1963/64–1966/67. Canberra, Dept. of Territories.
4 v. in 1. ill. 34 cm.
Report year ends June 30.
Continues: Australia. Dept. of Territories. Territory of Christmas Island.
Title varies slightly.
Cover title: Territory of Christmas Island: Report.
Continued by: Australia. Dept. of External Territories. Annual report on the Terirtory of Christmas Island.
1. Christmas Island, Indian Ocean—Periodicals. I. Australia. Dept. of Territories. Territory of Christmas Island: Report. II. Title.
DS349.9.C45A9a 996'.4 77-642821
MARC-S

Australia. Dept. of Territories.
Territory of Christmas Island; annual report. 1958/60–1962/63. ₍Canberra, s. n.₎
4 v. in 1. ill. 34 cm.
Report year ends June 30.
Report for 1958/60 covers period Oct. 1958–June 1960.
Continued by: Australia. Dept. of Territories. Annual report on the Territory of Christmas Island.
1. Christmas Island, Indian Ocean—Periodicals. I. Title.
DS349.9.C45A9a 996'.4 77-642820
MARC-S

CHRISTMAS MUSIC

see also Carols

Dancing Christmas carols : a sourcebook for adding movement and gesture to Christmas carols / edited by Doug Adams ; with choreography suggestions by members of the Sacred Dance Guild of the United States and Canada. — Saratoga, CA : Resource Publications, c1978.
135 p. : ill. ; 22 cm.
Bibliography: p. 131-133.
Includes index.
ISBN 0-89390-006-0 : $5.95
1. Movement notation. 2. Dancing. 3. Christmas music. I. Adams, Doug. II. Sacred Dance Guild.
GV1587.D36 793.3 78-63292
79 MARC

—HISTORY AND CRITICISM

Schepping, Wilhelm.
Die Wettener Liederhandschrift und ihre Beziehungen zu den niederländischen Cantiones Natalitiae des 17. Jahrhunderts / Wilhelm Schepping. — Köln : Musikverlage H. Gerig, c1978.
222 p., ₍4₎ leaves of plates : ill., map, facsims., music ; 24 cm. — (Beiträge zur rheinischen Musikgeschichte ; Heft 121) GFR***
Includes the songs of the Wettener Liederhandschrift.
Originally presented as the author's thesis, Cologne, 1977.
"Teil-Auflage: Sonderdruck für die Arbeitsgemeinschaft für rheinische Musikgeschichte, Köln."
Bibliography: p. 208-220.
Includes index.
ISBN 3-87252-087-3
1. Wettener Liederhandschrift. 2. Part-songs—History and criticism. 3. Christmas music—History and criticism. 4. Music—Netherlands—History and criticism. I. Wettener Liederhandschrift. II. Title. III. Series.
ML93.S34 79-343927
79 MARC

CHRISTMAS ORNAMENTS see Christmas decorations

CHRISTMAS PLAYS

Rickenmann, Pius.
Rapperswiler Sternsingerspiel / Pius Rickenmann. — 1. Aufl. — Rapperswil ₍SG₎ : Rapperswiler Sternsinger-Bruderschaft : Auslfg. Arp, 1975.
₍16₎ p. : music ; 17 cm. Sw78-3133
2.00F
1. Christmas plays. I. Title.
PT2678.I3194R3 832'.9'14 78-395929
79 MARC

—HISTORY AND CRITICISM—ADDRESSES, ESSAYS, LECTURES

Steiner, Rudolf, 1861-1925.
Ansprachen zu den Weihnachtspielen aus altem Volkstum : gehalten in Dornach 1915 bis 1924 / Rudolf Steiner ; ₍die Hrg. besorgte Edwin Froböse₎. — Dornach : Rudolf-Steiner-Verlag, 1974.
116 p. ; 23 cm. — (Rudolf Steiner Gesamtausgabe : Vorträge) Sw76-2074
ISBN 3-7274-2740-X : 25.00F
1. Folk-drama,—German—History and criticism—Addresses, essays, lectures. 2. Christmas plays—History and criticism—Addresses, essays, lectures. I. Title.
PT697.C5S7 1974 832'.009'33 79-359877
*79 MARC

CHRISTMAS SERMONS

—HISTORY AND CRITICISM

Egenolf, Hans-Andreas.
Die katholische Weihnachtspredigt nach 1945 / von Hans-Andreas Egenolf. — 1. Aufl. — Leipzig : St.-Benno-Verlag, 1977.
xx, 122 p. ; 21 cm. — (Erfurter theologische Schriften ; 14) GDR78-A
Bibliography: p. xiii-xix.
10.80M
1. Christmas sermons—History and criticism. 2. Preaching—History. I. Title. II. Series.
BV4257.E34 78-388490
78 MARC

CHRISTMAS STORIES

Carroll, Theodus, 1928-
The lost Christmas star / by Theodus C. Carroll ; based on a story by Elizabeth Yates ; illustrated by William Hutchinson. — Champaign, Ill. : Garrard Pub. Co., c1979.
63 p. : ill. ; 23 cm. — (A Mystery book)
SUMMARY: After River Junction's Christmas tree loses its star, two children determine to identify the thief.
ISBN 0-8116-6409-0 (lib. bdg.) : $4.70
1. Mystery and detective stories. 2. Christmas stories. I. Yates, Elizabeth, 1905- II. Hutchinson, William M. III. Title.
PZ7.C2364 Lo ₍E₎ 79-12224
79 MARC

The Christmas book / written by Robert Joseph ; illustrated by Maurice Hughes ; musical arrangements by Donald Siegal ; concept and art direction by Don McAfee. — New York : McAfee Books, c1978.
192 p. : col. ill. ; 32 cm.
Carols for voice and piano: p. 80-131.
ISBN 0-89328-022-4 : $12.95
1. Christmas. 2. Christmas stories. 3. Carols. I. Joseph, Robert.
GT4985.C392 394.2'68282 78-108796
78 MARC

Freeman, James Dillet.
Once upon a Christmas : stories / by James Dillet Freeman ; ₍ill. by Evan Lattimer₎. — 1st ed. — Unity Village, Mo. : Unity Books, c1978.
173 p. : col. ill. ; 26 cm.
CONTENTS: The tale of pig.—Angel with a broken wing.—A gift from Rosa.—Are you sure it's snowing beyond the next corner?—Conversation at an inn.—The King of Heaven is coming tonight.—An event of major importance. —The woman who learned to love Christmas.—My name is Joseph.—Imagination, imagination!
1. Christmas stories. ₍1. Christmas stories. 2. Short stories₎ I. Lattimer, Evan. II. Title.
PZ7.F87483 On ₍Fic₎ 78-53345
79 MARC

Keeping Christmas : stories of the joyous season / selected by Phyllis R. Fenner ; illustrated by Honi Werner. — New York : Morrow, 1979.
221 p. : ill. ; 22 cm.
CONTENTS: Schaefer, J. The riders of Saint Nicholas.—Chamberlain, W. The road to St. Vivien.—Ruben, O. Barren Christmas.—Buck, P. S. Christmas day in the morning.—Dickson, G. R. The Christmas present.—Smith, B. Francie Nolan's Christmas tree.—Chute, B. J. The Christmas doll.—Kampen, I. Greetings from A. Bledsoe Smith.—Buckler, E. Last delivery before Christmas.—Christie, A. The theft of the royal ruby.
ISBN 0-688-22206-4. ISBN 0-688-32206-9 lib. bdg.
1. Christmas stories. ₍1. Christmas stories₎ I. Fenner, Phyllis Reid, 1899- . II. Werner, Honi.
PZ5.K37 ₍Fic₎ 79-15590
79 MARC

Patterson, Lillie.
Christmas trick or treat / by Lillie Patterson ; drawings by Kelly Oechsli. — Champaign, Ill. : Garrard Pub. Co., c1979.
48 p. : col. ill. ; 23 cm.
CONTENTS: Who knocks on Christmas Eve?—Don't forget the nisse.
ISBN 0-8116-7252-2 lib. bdg. : $4.30
1. Christmas stories. ₍1. Christmas stories. 2. Short stories₎ I. Patterson, Lillie. Don't forget the nisse. 1979. II. Oechsli, Kelly. III. Title.
PZ7.P2768 Ch ₍E₎ 78-11308
78 MARC

Die Schönsten Weihnachtsgeschichten der Welt / Alexandra Cordes (Hrsg.) ; mit 30 Federzeichn. von Wilhelm M. Busch. — Orig.-Ausg. — München : Goldmann, 1977.
286 p. : ill. ; 18 cm. — (Goldmann-Taschenbuch ; 3510) GFR78-A
ISBN 3-442-02510-4 : DM7.80
1. Christmas stories. I. Cordes, Alexandra.
PN6071.C6S36 78-364295
78 MARC

Thomas, Leslie, 1931-
Midnight clear / ₍by₎ Leslie Thomas ; illustrated by Shirley Felts. — London : Arlington Books, 1978.
27 p. : col. ill. ; 23 cm. GB78-38045
ISBN 0-85140-291-7 : £2.50
1. Christmas stories. I. Title.
PZ4.T4577 Mi 823'.9'14 79-314105
₍PR6070.H647₎ 79 MARC

Thomas, Vernon.
Suddenly it's Christmas : a collection of yuletide stories / Vernon Thomas. — ₍Calcutta₎ : Writers Workshop ; ₍Thompson, Conn. : agents in U.S., Inter Culture Associates, c1976₎
70 p. ; 22 cm.
"A Writers Workshop greenbird book."
CONTENTS: Suddenly it's Christmas.—The glow of Christmas.—Christmas angel.—Anniversary.—Timothy's Christmas.—Invitations for Christmas.—Another chance.—Christmas roses.—That feeling of Christmas.
Rs15.00
1. Christmas stories. I. Title.
PZ4.T4598 Su 823 78-905461
₍PR9499.3.T55₎ 79 MARC

Weihnachtsgeschichten / Johann Hoffmann-Herreros (Hg.) ; ₍Heinrich Böll ... et al.₎. — 1. Aufl. — Mainz : Matthias-Grünewald-Verlag, 1975.
128 p. ; 18 cm. — (Topos-Taschenbücher ; Bd. 42) GFR75-A
Excerpts from essays by various authors, some translations.
ISBN 3-7867-0531-3 : DM5.80
1. Christmas stories. I. Hoffmann, Hans Peter, 1929-
PN6071.C6W37 78-373054
78 MARC

Weihnachtsgeschichten aus Niedersachsen / hrsg. von Gundel Paulsen. — Husum : Husum-Druck- und -Verl.-Ges., 1977.
135 p. ; 20 cm. — (Kleine HDV-Reihe) GFR78-A03
ISBN 3-88042-043-2
1. German literature—Germany, West—Saxony, Lower. 2. Christmas stories. I. Paulsen, Gundel.
PT3803.S4W44 833'.01 79-366300
79 MARC

CHRISTMAS STORIES, GERMAN

Weihnachtsgeschichten aus Hamburg / hrsg. von Gundel Paulsen. — Husum : Husum Druck- und Verlagsgesellschaft, 1976.
110 p. ; 20 cm. — (Kleine HDV Reihe) GFR77-A
ISBN 3-88042-025-4 : DM7.80
1. Christmas stories, German. 2. German fiction—19th century. 3. German fiction—20th century. 4. German fiction—Germany, West—Hamburg. I. Paulsen, Gundel.
PT1340.C5W38 78-374693
78 MARC

CHRISTMAS TREES

see also Christmas decorations

Chapman, Arthur Glenn, 1895-
Christmas trees for pleasure and profit / Arthur G. Chapman, Robert D. Wray ; ill. by Jim Bernard. — 2d ed. — New Brunswick, N.J. : Rutgers University Press, c1979.
viii, 212 p. : ill. ; 21 cm.
Bibliography: p. 203-205.
Includes index.
ISBN (invalid) 0-8135-0875-X
1. Christmas trees. I. Wray, Robert D., joint author. II. Title.
SB428.C46 1979 635.9'777 79-11878
79 MARC

CHRISTMAS TREES (Continued)

—JUVENILE POETRY

Konopnicka, Maria, 1842-1910.
Choinka / Marji Konopnickiej ; zdobiła graficznie Marja Dziewulska. — Warszawa : W Drukarni i litografji Jan Cotty, 1928.
[10] p. : col. ill. ; 20 cm.
Cover title.
1. Christmas trees—Juvenile poetry. I. Dziewulska, Maria. II. Title.
PZ69.K6C5 1928 79-372385
79 MARC

CHRISTO, 1935- WRAPPED WALK WAYS

Christo, 1935-
Christo, Wrapped walk ways, Loose Park, Kansas City, Missouri, 1977-78 / essay by Ellen R. Goheen ; photos. by Wolfgang Volz. — 1st ed. — New York : H. N. Abrams, 1978.
[96] p. : ill. (some col.) ; 26 x 27 cm.
ISBN 0-8109-0762-3. ISBN 0-8109-2191-X pbk.
1. Christo, 1935- Wrapped walk ways. I. Goheen, Ellen R. II. Volz, Wolfgang. III. Title. IV. Title: Wrapped walk ways.
N7193.C5A78 1978 709'.2'4 78-20619
78 MARC

CHRISTO, CARLOS ALBERTO LIBANIO, 1944-

Christo, Carlos Alberto Libanio, 1944-
Cartas da prisão / Frei Betto [i.e. C. A. Libanio Christo]. — Rio de Janeiro : Civilização Brasileira, 1977.
232 p. ; 21 cm.
Cr$80.00
1. Christo, Carlos Alberto Libanio, 1944- 2. Political prisoners—Brazil—Biography. I. Title.
HV9593.C45 77-568711
77[r79]rev MARC

Christo, Carlos Alberto Libanio, 1944-
Das catacumbas : cartas da prisão, 1969-1971 / Frei Betto [i.e. C. A. Libanio Christo]. — Rio de Janeiro : Civilização Brasileira, 1978.
183 p. ; 21 cm.
Cr$90.00
1. Christo, Carlos Alberto Libanio, 1944- 2. Political prisoners—Brazil—Biography. I. Title.
HV9593.C453 79-339965
79 MARC

Christo, Carlos Alberto Libânio, 1944-
Diário de Puebla / Frei Betto [i.e. C. Christo] ; tradução dos textos castelhanos do Anexo por Glória Rodríguez. — Rio de Janeiro : Civilização Brasileira, 1979.
262 p. ; 21 cm. — (Coleção Documentos da história contemporânea ; v. 73)
Cr$160.00
1. Conferencia General del Episcopado Latinoamericano, 3d, Puebla, Mexico, 1979. 2. Christo, Carlos Alberto Libânia, 1944- I. Title.
BX837.5.C45 79-113102
79 MARC

CHRISTOLOGY see Jesus Christ

CHRISTOPHERSEN, HALFDAN OLAUS, 1902-

Christophersen, Halfdan Olaus, 1902-
Av nød til seir : bilder fra okkupasjonstiden i Norge / H. O. Christophersen. — Oslo : Grøndahl, 1977-
v. : ill. ; 26 cm. N77-Nov.
CONTENTS: bd. 1. Fra 9. april 1940 til kampåret 1942.

ISBN 8250402480 (v. 1) : kr88.00 (v. 1).
1. World War, 1939-1945—Norway—Collected works. 2. Norway—History—German occupation, 1940-1945—Collected works. 3. World War, 1939-1945—Personal narratives, Norwegian. 4. Christophersen, Halfdan Olaus, 1902- . I. Title.
D763.N6C48 79-347868
79 MARC

Christophersen, Halfdan Olaus, 1902-
En vandrer krysser sine spor i Oslo, Drammen- og Ringeriksstraktene / H. O. Christophersen. — 2. utgave. — Oslo : Gyldendal, 1977 (Oslo : Reistad & sønn)
160 p., [12] leaves of plates : ill. ; 24 cm. N77-Nov.
First ed. published in 1961 under title: Vei og varde.
ISBN 8205104336 : kr 98.00. ISBN 8205104328 pbk.
1. Norway—Description and travel—1945- 2. Christophersen, Halfdan Olaus, 1902- . I. Title.
DL419.C47 1977 79-345009
79 MARC

Christophersen, Halfdan Olaus, 1902-
Eventyr i dagslys; nye vandringer. Med fotografier av Jarle Strømme. Oslo, Gyldendal, 1962.
190 p. illus. 23 cm.
1. Norway—Description and travel—1945- . 2. Oslo—Description. 3. Christophersen, Halfdon Olaus, 1902- . I. Title.
DL419.C46 914.81'04'4 64-34030
[r79]rev MARC

CHRISTOPHERSEN, THIES, 1918-

—ADDRESSES, ESSAYS, LECTURES

Anti-Roeder-Arbeitskreis.
NSDAP-Propagandisten unter der Lupe : Dokumentation antisemit., antidemokrat. u. offener NS-Provokationen d. Schönborn-Roeder-Christophersen-Bande u. ihre Deckung seitens staatl. Organe / [Autor, Anti-Roeder-Arbeitskreis]. — 1. Aufl. — Hamburg : Reents, 1978.
203 p. : numerous ill. ; 21 cm. — (Antifaschistische Russell-Reihe ; 3)
GFR78-A
Includes indexes.
ISBN (invalid) 3-88305-003-1 : DM13.50
1. Fascism—Germany, West—Addresses, essays, lectures. 2. Germany, West—Politics and government—Addresses, essays, lectures. 3. Antisemitism—Germany, West—Addresses, essays, lectures. 4. Christophersen, Thies, 1918- —Addresses, essays, lectures. 5. Schönborn, Erwin—Addresses, essays, lectures. 6. Roeder, Manfred, 1929- —Addresses, essays, lectures. I. Title. II. Series.
DD259.2.A69 1978 320.5'33'0943 79-369819
79 MARC

CHRISTOV, SOLVEIG, PSEUD.

—JOURNEYS—EUROPE

Christov, Solveig, pseud.
Reisen for lenge siden / Solveig Christov. — Oslo : Gyldendal, 1978 (Oslo : Reistad & sønn)
196 p., [1] leaf ; 22 cm. N78-Nov.
ISBN 8205114056 : kr89.00. ISBN 8205114048 pbk.
1. Christov, Solveig, pseud.—Journeys—Europe. 2. Europe—Description and travel—1971- 3. Authors, Norwegian—20th century—Biography. I. Title.
PT8950.C6Z476 79-342121
79 MARC

CHRIST'S CHURCH OF THE GOLDEN RULE

Dohrman, H T
California cult; the story of Mankind United. Boston, Beacon Press [1958]
163 p. 22 cm. (Beacon series in the sociology of religion [v. 2])
1. Mankind United. 2. Bell, Arthur L. 3. Christ's Church of the Golden Rule. I. Title.
BP605.M34D63 1958 289.9 58-8395
[r79]rev MARC

Dohrman, H T
California cult : the story of "Mankind United" / H. T. Dohrman. — 1st AMS ed. — New York : AMS Press, 1979, c1958.
xi, 163 p. ; 23 cm.
Reprint of the ed. published by Beacon Press, Boston, which was issued as v. 2 of Beacon series in the sociology of religion.
Includes bibliographical references and index.
Includes index.
ISBN 0-404-60059-X
1. Mankind United. 2. Bell, Arthur L. 3. Christ's Church of the Golden Rule. I. Title.
[BP605.M34D63 1979] 289.9 76-42724
79 MARC

CHROBÁK, DOBROSLAV, 1907-1951

—CRITICISM AND INTERPRETATION

Poézia Chrobákovej prózy / [zost. Ján Števček ; aut.] Albín Bagin ... [et al. ; návrh na prebal a väzbu viprac. Igor Imro]. — 1. vyd. — Bratislava : Tatran, 1977.
276, [3] p. ; 17 cm. — (Interpretácie ; zv. 8) CzS78
Kčs18.00
1. Chrobák, Dobroslav, 1907-1951—Criticism and interpretation. I. Števček, Ján. II. Bagin, Albín.
PG5438.C49Z84 78-385896
*79 MARC

CHROMAFFIN CELLS

see also Catecholamines; Endocrine glands

—TUMORS

see also Carcinoid

CHROMATIC ABERRATION (OPTICS) see Lenses

CHROMATIC VISION see Color vision

CHROMATICS see Color

CHROMATIN

see also Chromosome underreplication; Heterochromatin

—ADDRESSES, ESSAYS, LECTURES

Concepts of the structure and function of DNA, chromatin, and chromosomes / edited by Arnold S. Dion, associate editors, Donald C. Farwell ... [et al.]. — Miami, FL : Symposia Specialists ; Chicago : distributed by Year Book Medical Publishers, c1979.
ix, 318 p. : ill. ; 24 cm.
"Selected lectures from a series sponsored by the Institute for Medical Research, Camden, New Jersey, and held during the academic year 1977-1978."
Title on spine: DNA, chromatin, and chromosomes.
Includes bibliographical references and index.
ISBN 0-8151-2504-6
1. Molecular genetics—Addresses, essays, lectures. 2. Deoxyribonucleic acid—Addresses, essays, lectures. 3. Chromatin—Addresses, essays, lectures. 4. Chromosomes—Addresses, essays, lectures. I. Dion, Arnold S. II. Farwell, Donald C. III. Institute for Medical Research, Camden, N.J. IV. Title: DNA, chromatin, and chromosomes.
QH438.C65 574.8'732 79-65645
79 MARC

—CONGRESSES

Cold Spring Harbor Symposia on Quantitative Biology, 42d, 1978.
Chromatin / Cold Spring Harbor Symposia on Quantitative Biology. — Cold Spring Harbor, N.Y. : Cold Spring Harbor Laboratory, 1978.
2 v. (xi, 1260 p.) : ill. ; 28 cm. — (Cold Spring Harbor Symposia on Quantitative Biology ; v. 42)
Includes bibliographies and indexes.
ISBN 0-87969-041-0
1. Chromatin—Congresses. I. Title. II. Series: Cold Spring Harbor Symposia on Quantitative Biology. Cold Spring Harbor Symposia on Quantitative Biology ; v. 42.
QH301.C6 vol. 42 574'.08 s 79-105187
[QH599] 79 MARC

Nato Advanced Study Institute, Erice, Italy, 1978.
Chromatin structure and function / edited by Claudio A. Nicolini. — New York : Plenum Press, c1979.
2 v. (xix, 880 p.) : ill. ; 26 cm. — (NATO advanced study institutes series : Series A, Life sciences ; v. 21)
Includes bibliographical references and indexes.
CONTENTS: pt. A. Molecular and cellular biophysical methods.—pt. B. Levelsof organization and cell function.
ISBN 0-306-40075-8 (v. 1)
1. Chromatin—Congresses. 2. Carcinogenesis—Congresses. I. Nicolini, Claudio A. II. Title. III. Series.
QH599.N37 1978 574.8'732 78-24268
78 MARC

CHROMATOGRAPHIC ANALYSIS

see also Affinity chromatography; Gas chromatography; Ion exchange chromatography; Liquid chromatography; Paper chromatography; Thin layer chromatography

Chromatography in petroleum analysis / edited by Klaus H. Altgelt, T. H. Gouw. — New York : M. Dekker, c1979.
x, 500 p. : ill. ; 24 cm. — (Chromatographic science ; v. 11)
Includes bibliographical references and indexes.
ISBN 0-8247-6790-X
1. Petroleum—Analysis. 2. Chromatographic analysis. I. Altgelt, Klaus H., 1927- II. Gouw, T. H.
TP691.C57 665'.5'028 79-12150
79 MARC

CHROMATOGRAPHIC ANALYSIS (Continued)

Chromatography of synthetic and biological polymers / compiled and edited by Roger Epton. — Chichester, Eng. : Published for the Chemical Society, Macromolecular Group by Ellis Horwood ; New York : ₍Distributed by₎ Halsted Press, 1978.
2 v. : ill. ; 24 cm. GB***
"Based on lectures given at a Chemical Society International Symposium held at the University of Birmingham ... July 7-9, 1976."
Includes bibliographical references and indexes.
CONTENTS: v. 1. Column packings, GPC, GF, and gradient elution.—v. 2. Hydrophobic, ion exchange, and affinity methods.
ISBN 0-470-99379-0 (v. 1) (Halsted Press) : £17.50 (v. 1)
1. Chromatographic analysis. 2. Polymeters and polymerization—Analysis. I. Epton, Roger. II. Chemical Society, London.
QD79.C4C493 547'.84 78-313092
78₍r79₎rev MARC

Laboratory handbook of chromatographic and allied methods / chief editor, O. Mikeš ; contributors M. Hejtmánek ... ₍et al.₎ ; translator, Procházka ; translation editor, R. A. Chalmers. — Chichester, Eng. : Ellis Horwood ; Chichester, Eng. ; New York : Halsted Press, 1979.
764 p. : ill. ; 24 cm. — (Ellis Horwood series in analytical chemistry) GB***
Originally published in 1966 under title: Laboratory handbook of chromatographic methods, which is a rev. translation of Příručka laboratorních chromatografických metod.
Includes bibliographies and index.
ISBN 0-470-26399-7 (Halsted Press) : £38.00
1. Chromatographic analysis. I. Mikeš, Otakar. II. Hejtmánek, M.
QD79.C4L33 1979 544'.92 78-40595
79 MARC

—ADDRESSES, ESSAYS, LECTURES

Współczesne kierunki w teorii i praktyce chromatograficznej : praca zbiorowa / ₍redaktor naukowy Zdzisław Suprynowicz₎. — Wrocław : Zakład Narodowy im. Ossolińskich, 1976.
310 p. : ill. ; 21 cm. P***
At head of title: Polska Akademia Nauk. Biuro Kadr Naukowych i Spraw Osobowych.
Includes bibliographies.
zł52.00
1. Chromatographic analysis—Addresses, essays, lectures. I. Suprynowicz, Zdzisław. II. Polska Akademia Nauk. Biuro Kadr Naukowych i Spraw Osobowych.
QD79.C4W76 79-410465
79 MARC

—APPARATUS AND SUPPLIES

Woelm, M
Four new Woelm materials for column chromatography : Silica Gel Woelm for adsorption, Silica Gel Woelm for partition, Magnesium Silicate Woelm, Polyamide Woelm. — Eschwege, ₍Ger.₎ : Woelm, 1963.
23 p. ; 21 cm. — (Woelm publication ; AL 11)
Bibliography: p. 21.
Includes index.
1. Chromatographic analysis—Apparatus and supplies. 2. Sorbents. I. Title.
QD79.C4W63 544'.92 78-324227
79 MARC

Woelm, M
Thin-layer chromatography with adsorbents / Woelm. — 2d, enl. ed. — Eschwege, ₍Ger.₎ : Woelm, 1963.
35 p. : ill. ; 21 cm. — (Woelm publication ; AL 10)
Bibliography: p. 31.
Includes index.
1. Thin layer chromatography. 2. Chromatographic analysis—Apparatus and supplies. I. Title.
QD79.C8W63 1963 544'.92 78-324228
79 MARC

—COLLECTED WORKS

Developments in chromatography / edited by C. E. H. Knapman. — London : Applied Science Publishers, c1978-
v. : ill. ; 23 cm. — (Developments series) GB***
Includes bibliographical references and index.
ISBN 0-85334-748-4 (v. 1) : £15.00 (v. 1)
1. Chromatographic analysis—Collected works. I. Knapman, C. E. H.
QD79.C4D48 544'.92 78-319463
79 MARC

(Itogi nauki i tekhniki: Khromatografiía)
Итоги науки и техники: Хроматография. т. 1–
Москва ₍Всес. ин-т науч. и техн. информации₎ 1974–
v. ill. 22 cm.
Other title, v. 1– : Itogi nauki i tekhniki: Serifa Khromatografiía.
1. Chromatographic analysis—Collected works. I. Moscow. Vsesofiuznyĭ institut nauchnoĭ i tekhnicheskoĭ informafsii. II. Title: Khromatografiía. III. Title: Itogi nauki i tekhniki: Serifa Khromatografiía.
QD79.C4 I 87 79-645989
(MARC-S)

—CONGRESSES

International Symposium on Chromatography and Electrophoresis, Riva, Italy, 1978.
Recent developments in chromatography and electrophoresis : proceedings of the 9th International Symposium on Chromatography and Electrophoresis, Riva del Garda, 15-17 May 1978 / edited by Alberto Frigerio, Leïka Renoz. — Amsterdam ; New York : Elsevier Scientific Pub. Co. ; New York : distributors for the U.S. and Canada, Elsevier/North-Holland, 1979.
ix, 357 p. : ill. ; 25 cm. — (Chromatography symposia series ; v. 1)
Includes bibliographical references and index.
ISBN 0-444-41785-0 (v. 1) : fl 120.00 (v. 1)
1. Chromatographic analysis—Congresses. 2. Electrophoresis—Congresses. I. Frigerio, Alberto. II. Renoz, Leïka. III. Title. IV. Series.
QD79.C4I57 1978 544'.92 79-801
79 MARC

—HANDBOOKS, MANUALS, ETC.

Knapp, Daniel R
Handbook of analytical derivatization reactions / Daniel R. Knapp. — New York : Wiley, c1979.
xviii, 741 p. : ill. ; 24 cm.
"A Wiley-Interscience publication."
Includes bibliographical references and indexes.
ISBN 0-471-03469-X
1. Chemistry, Analytic—Technique—Handbooks, manuals, etc. 2. Chromatographic analysis—Handbooks, manuals, etc. 3. Mass spectrometry—Handbooks, manuals, etc. I. Title. II. Title: Analytical derivatization reactions.
QD78.K63 543 78-12944
78 MARC

CHROMATOGRAPHY, THIN LAYER see Thin layer chromatography

CHROMATOPHORES

see also Animal pigments; Chloroplasts

CHROME-NICKEL STEEL

see also Steel, Heat resistant; Steel, Stainless

CHROME STEEL

see also Steel, Stainless

CHROMIUM

Tool and Stainless Steel Industry Committee.
Rhodesian chromium and specialty steel : basic considerations. — 2d ed. — Washington : Tool and Stainless Steel Industry Committee, ₍1973?₎
27 p. : graphs ; 26 cm.
Includes bibliographical references.
1. Chromium. I. Title.
HD9539.C4T66 1973 338.4'7'669734 78-320599
78 MARC

—ENVIRONMENTAL ASPECTS—ILLINOIS

Huff, Linda L
Economic impact of proposed amendments to chromium effluent standards in Illinois, R76-21 / by Linda L. Huff. — ₍Chicago₎ : State of Illinois, Institute for Environmental Quality, 1978.
vii leaves, 57 p. : graphs ; 28 cm. — (IIEQ document ; no. 78/08)
"Project no. 80.103."
Includes bibliographical references.
1. Chromium—Environmental aspects—Illinois. 2. Sewage—Purification—Economic aspects—Illinois. 3. Effluent quality—Standards—Illinois. 4. Water—Pollution—Law and legislation—Illinois. I. Title. II. Series: Illinois Institute for Environmental Quality. IIEQ document ; no. 78/08.
TD427.C57H83 614.7'72 78-623935
79 MARC

CHROMIUM ALLOYS

see also Chromium-iron-nickel alloys

CHROMIUM GROUP

see also Chromium; Molybdenum; Tungsten

—TARIFF see Tariff on chromium-iron alloys

CHROMIUM-IRON-NICKLE ALLOYS

—CONGRESSES

Alloy 800 : proceedings of the Petten international conference / edited by W. Betteridge ... ₍et al.₎. — Amsterdam ; New York : North-Holland Pub. Co. ; New York : distributors for the U.S. and Canada, Elsevier North-Holland, c1978.
xvii, 478 p. : ill. ; 23 cm.
Conference held Mar. 14-16, 1978 and organized by the Directorate-General for Research, Science, and Education of the Commission of the European Communities and by the Petten Establishment of the Joint Nuclear Research Center.
Includes bibliographical references.
ISBN 0-444-85228-X
1. Heat resistant alloys—Congresses. 2. Chromium-iron-nickl alloys—Congresses. I. Betteridge, Walter, 1911- . II. Commission of the European Communities. Directorate-General for Research, Science, and Education. III. Joint Nuclear Research Center, Petten, Netherlands.
TA485.A44 620.1'893 78-11985
78 MARC

CHROMIUM ORES

see also Chromite

CHROMIUM OXIDE

Chu, Wing Fong.
Kinetik der Reduktion von Chromoxid mit Wasserstoff / Wing Fong Chu. — 1. Aufl. — Frankfurt/Main : Haag und Herchen, 1978.
iii, 148 p. : ill. ; 21 cm. GFR78-A
Includes bibliographical references.
ISBN 3-88129-117-2 : DM24.80
1. Reduction, Chemical. 2. Chromium oxide. 3. Hydrogen. I. Title.
QD63.R4C45 541'.393 78-377864
78 MARC

CHROMOLITHOGRAPHY

see also Color-printing; Color prints

—ENGLAND—EXHIBITIONS

Friedman, Joan M
Color printing in England, 1486-1870 : an exhibition, Yale Center for British Art, New Haven, 20 April to 25 June, 1978 / by Joan M. Friedman. — ₍New Haven₎ : The Center, c1978.
ix, 72 p., lxxii p. : ill. (some col.) ; 28 cm.
Bibliography: p. 67-68.
Includes index.
ISBN 0-930606-12-4
1. Color prints—England—Exhibitions. 2. Chromolithography—England—Exhibitions. I. Yale Center for British Art. II. Title.
NE1300.8.G7F74 769'.942'07401468 78-50990
78 MARC

CHROMOLITHOGRAPHY (Continued)

—FRANCE

Cate, Phillip Dennis.
The color revolution : color lithography in France, 1890-1900 / by Phillip Dennis Cate and Sinclair Hitchings ; with a translation by Margaret Needham of André Mellerio's 1898 essay La lithographie originale en couleurs. — Santa Barbara : P. Smith, 1978.
ix, 136 p. : ill. (some col.) ; 31 cm.
Bibliography: p. 130-134.
Includes index.
ISBN 0-87905-032-2 : $9.95
1. Chromolithography—France. I. Hitchings, Sinclair H., joint author. II. Mellerio, André, 1862- La lithographie originale en couleurs. English. 1978. III. Title.
NE2500.C37 764'.2'0944 78-13809
78 MARC

CHROMOPHOTOGRAPHY see Color photography

CHROMOSOME ABNORMALITIES

see also Human chromosome abnormalities; Sex chromosomes abnormalities

CHROMOSOME MAPPING

see also Linkage (Genetics)

CHROMOSOME NUMBERS

see also Aneuploidy; Haploidy

CHROMOSOME REPLICATION

see also Chromosome underreplication

Bianchi, Néstor O
Duplicación cromosómica y heterocromatina a nivel molecular y citológico / por Néstor O. Bianchi. — Washington : Programa Regional de Desarrollo Científico y Tecnológico, Departamento de Asuntos Científicos, Secretaría General de la Organización de los Estados Americanos, 1978.
viii, 98 p. : ill. ; 23 cm. — (Serie de biología ; monografía no. 19)
"78.XXII-B-009-S"
Bibliography: p. 85-93.
$2.00
1. Chromosome replication. 2. Heterochromatin. I. Title. II. Series.
QH600.B53 79-106652
79 MARC

The Structure and replication of genetic material / edited by David M. Prescott, Lester Goldstein. — New York : Academic Press, 1979.
xiii, 491 p. : ill. ; 24 cm. — (Cell biology, a comprehensive treatise ; v. 2)
Includes bibliographies and index.
ISBN 0-12-289502-9
1. Chromosomes. 2. Chromosome replication. 3. Deoxyribonucleic acid repair. I. Prescott, David M., 1926- II. Goldstein, Lester. III. Series.
QH574.C43 vol. 2 574.8'7 s 78-10457
[QH600] 78 MARC

CHROMOSOMES

see also Chromosome replication; Genetic recombination; Genetics; Human chromosomes; Karyotypes; Linkage (Genetics); Plant cytotaxonomy

Faria, Álvaro de.
Mutação e cromossomos / Álvaro de Faria. — Rio de Janeiro : Avenir Editora, 1978.
53 p. ; 22 cm. — (Coleção Depoimentos ; v. 6)
Cr$100.00
1. Mutation (Biology) 2. Chromosomes. I. Title. II. Series.
QH462.A1F37 79-107494
79 MARC

The Structure and replication of genetic material / edited by David M. Prescott, Lester Goldstein. — New York : Academic Press, 1979.
xiii, 491 p. : ill. ; 24 cm. — (Cell biology, a comprehensive treatise ; v. 2)
Includes bibliographies and index.
ISBN 0-12-289502-9
1. Chromosomes. 2. Chromosome replication. 3. Deoxyribonucleic acid repair. I. Prescott, David M., 1926- . II. Goldstein, Lester. III. Series.
QH574.C43 vol. 2 574.8'7 s 78-10457
[QH600] 78 MARC

—ADDRESSES, ESSAYS, LECTURES

Concepts of the structure and function of DNA, chromatin, and chromosomes / edited by Arnold S. Dion, associate editors, Donald C. Farwell ... [et al.]. — Miami, FL : Symposia Specialists ; Chicago : distributed by Year Book Medical Publishers, c1979.
ix, 318 p. : ill. ; 24 cm.
"Selected lectures from a series sponsored by the Institute for Medical Research, Camden, New Jersey, and held during the academic year 1977-1978."
Title on spine: DNA, chromatin, and chromosomes.
Includes bibliographical references and index.
ISBN 0-8151-2504-6
1. Molecular genetics—Addresses, essays, lectures. 2. Deoxyribonucleic acid—Addresses, essays, lectures. 3. Chromatin—Addresses, essays, lectures. 4. Chromosomes—Addresses, essays, lectures. I. Dion, Arnold S. II. Farwell, Donald C. III. Institute for Medical Research, Camden, N.J. IV. Title: DNA, chromatin, and chromosomes.
QH438.C65 574.8'732 79-65645
79 MARC

—ANALYSIS

Granum, Erik.
Om billedbehandling på regnemaskine / af Erik Granum. — Lyngby : Laboratoriet for elektronik, Danmarks tekniske højskole, 1977.
19, 15, 3 leaves : ill. ; 30 cm. D***
Danish or English.
Bibliography: leaves 18-19.
1. Chromosomes—Analysis. 2. Imaging systems in genetics. I. Copenhagen. Polyteknisk læreanstalt. Laboratoriet for elektronik. II. Title.
QH600.G73 78-395627
79 MARC

CHROMOXYLOGRAPHY see Color prints

CHRONIC DISEASES

see also Chronically ill

Tuan, Chen-li.
(Tsen yang chan sheng man hsing chi ping)
怎样战胜慢性疾病 / 段振离. — 第1版. — [郑州] : 河南人民出版社 : 河南省新華書店发行, 1978.
2, 3, 140 p. ; 19 cm. — (卫生知识丛书)
Series romanized: Wei sheng chih shih ts'ung shu.
$0.28
1. Chronic diseases. 2. Medicine, Chinese. I. Title.
RB156.T8 79-837621

—BIOGRAPHY

(Kenkō no yorokobi)
健康のよろこび / 聖教新聞社社会部編. — 東京 : 同社, 昭和52 [1977]
2 v. : ill. ; 18 cm. — (体験談シリーズ ; 4-5) Ja***
Cover title.
Contains collection of articles originally published in Seikyō shimbun, 1974-1977.
CONTENTS: 1. 闘病体験·難病編—2. 闘病体験
¥80 (v. 1) ¥100 (v. 2)
1. Chronic diseases—Biography. 2. Sōka Gakkai Buddhists—Biography. I. Seikyō Shimbun Sha. Shakaibu. II. Seikyō shimbun. III. Series: Taikendan shirizu ; 4-5.
RC108.K46 78-810130

—MISCELLANEA

Hua t'ung pien tsuan wei yüan hui.
(Ch'ien i mo hua ti ping mo)
潛移默化的病魔 / [編著者華通編纂委員會]. — 香港 : 香港華通公司, 1978.
6, 100 p. ; 18 cm.
$3.20
1. Chronic diseases—Miscellanea. I. Title.
RA644.5.H8 1978 79-835286

—PSYCHOLOGICAL ASPECTS—CONGRESSES

New directions in patient compliance / edited by Stuart J. Cohen. — Rev. ed. — Lexington, Mass. : Lexington Books, c1979.
xvii, 170 p. : ill. ; 24 cm.
Papers from a conference held Aug. 11-12, 1978, French Lick, Ind., and sponsored by the Diabetes Research and Training Center, Indiana University School of Medicine.
Includes bibliographies and index.
ISBN 0-669-02721-9
1. Patient compliance—Congresses. 2. Chronic diseases—Psychological aspects—Congresses. I. Cohen, Stuart J. II. Indiana University School of Medicine. Diabetes Research and Training Center.
R726.5.C27 1978 615'.5 78-19707
79 MARC

--STUDY AND TEACHING

Staff manual for teaching patients about chronic obstructive pulmonary diseases. — Chicago : American Hospital Association, c1979.
1 v. ; 28 cm.
Loose-leaf for updating.
"AHA catalog no. 1317."
"Principal author ... Betty Chewning."
"Project of the Center for Health Promotion of the American Hospital Association."
Includes bibliographical references.
ISBN 0-87258-249-3
1. Lungs—Diseases, Obstructive—Study and teaching. 2. Patient education. 3. Chronic diseases—Study and teaching. 4. Medical personnel and patient. I. Chewning, Betty. II. American Hospital Association. Center for Health Promotion.
RC776.O3S73 1979 616.2'4'007 78-27387
78 MARC

—HAWAII—STATISTICS

Wegner, Eldon L
Chronic illness and hospitalization by ethnicity and income in the State of Hawaii, 1969-1971 / by Eldon L. Wegner. — [Honolulu] : Research and Statistics Office, Hawaii State Dept. of Health, 1977.
26 p. ; 28 cm. — (R & S report ; issue no. 19)
Cover title.
Chiefly tables.
1. Hawaii—Statistics, Medical. 2. Chronic diseases—Hawaii—Statistics. 3. Chronically ill—Hawaii—Socioeconomic status—Statistics. 4. Hospital utilization—Hawaii—Statistics. 5. Hospital patients—Hawaii—Socioeconomic status—Statistics. 6. Health and race—Hawaii—Statistics. 7. Income—Hawaii—Statistics. 8. Health surveys—Hawaii. I. Title. II. Series: Hawaii. Dept. of Health. Research and Statistics Office. R & S report ; issue no. 19.
HB3525.H3H33a no. 19 301.32'9'969 s 78-621623
[RA407.4.H3] 79 MARC

—UNITED STATES—STATISTICS

Bonham, Gordon Scott.
Prevalence of chronic skin and musculoskeletal conditions, United States, 1976 / [Gordon Scott Bonham]. — Hyattsville, Md. : U.S. Dept. of Health, Education, and Welfare, Public Health Service, Office of the Assistant Secretary for Health, National Center for Health Statistics, 1978.
iv, 57 p. : ill. ; 26 cm. — (Vital and health statistics : Series 10, Data from the National Health Survey ; no. 124) (DHEW publication ; no. (PHS) 79-1552)
Edition for 1969 by M. H. Wilder.
Includes bibliographical references.
1. Chronic diseases—United States—Statistics. 2. Skin—Diseases—United States—Statistics. 3. Musculoskeletal system—Diseases—United States—Statistics. 4. Health surveys—United States. 5. United States—Statistics, Medical. I. Wilder, Mary H. Prevalence of chronic skin and musculoskeletal conditions, United States, 1969. II. United States. National Center for Health Statistics. III. Title. IV. Series: United States. National Center for Health Statistics. Vital and health statistics : Series 10, Data from the National Health Survey, Data from the health interview survey ; no. 124. V. Series: United States. Dept. of Health, Education, and Welfare. DHEW publication ; no. (PHS) 79-1552.
[DNLM: 1. Skin diseases—Occurrence—United States. 2. Muscular diseases—Occurrence—United States. 3. Bone diseases—Occurrence—United States. W2 A N148vj no. 124]
RA407.3.A346 no. 124 312'.0973 s 78-10839
[RA644.6] 78 MARC

CHRONIC DISEASES
—UNITED STATES—STATISTICS
(Continued)

Ingram, Donald K
Profile of chronic illness in nursing homes, United States : national nursing home survey, August 1973-April 1974 / ⌈Donald K. Ingram⌉. — Hyattsville, Md. : U.S. Dept. of Health, Education, and Welfare, Public Health Service, Health Resources Administration, National Center for Health Statistics, 1977.
iv, 66 p. : ill. ; 26 cm. — (Vital and health statistics : Series 13, Data from the National Health Survey, Data on health resources utilization ; no. 29) (DHEW publication ; no. (PHS) 78-1780)
Includes bibliographical references.
ISBN 0-8496-0096-8
1. Nursing home patients—United States—Statistics. 2. Chronic diseases—United States—Statistics. 3. United States—Statistics, Medical. I. United States. National Center for Health Statistics. II. Title. III. Series: United States. National Center for Health Statistics. Vital and health statistics : Series 13, Data from the National Health Survey, Data on health resources utilization ; no. 29. IV. Series: United States. Dept. of Health, Education, and Welfare. DHEW publication ; no. (HRA) 77-1780.
RA407.3.A349 no. 29 362.1'1'0973 s 77-22564
⌈RA997⌉ 77⌈r79⌉rev MARC

CHRONIC DISEASES IN CHILDREN

—CONGRESSES

Chronic childhood illness : assessment of outcome : a conference sponsored by the John E. Fogarty International Center for Advanced Study in the Health Sciences and the National Institute of Child Health and Human Development, National Institutes of Health, Bethesda, Maryland, May 1-2, 1974 / Gilman D. Grave and I. Barry Pless, editors. — Bethesda, Md. : U.S. Dept. of Health, Education, and Welfare, Public Health Service, National Institutes of Health ; Washington : for sale by the Supt. of Docs., U.S. Govt. Print. Off., 1976.
xiv, 238 p. : ill. ; 27 cm. — (Fogarty International Center series on the teaching of preventive medicine ; v. 3) (DHEW publication ; no. (NIH) 76-877)
Includes bibliographies and index.
Supt. of Docs. no.: HE 20.3714:3
$6.10
1. Chronic diseases in children—Congresses. 2. Chronically ill children—Rehabilitation—Congresses. 3. Child health services—Evaluation—Congresses. I. Grave, Gilman D. II. Pless, Ivan B., 1932- . III. John E. Fogarty International Center for Advanced Study in the Health Sciences. IV. United States. National Institute of Child Health and Human Development. V. Series: John E. Fogarty International Center for Advanced Study in the Health Sciences. Fogarty International Center series on the teaching of preventive medicine ; v. 3. VI. Series: United States. Dept. of Health, Education, and Welfare. DHEW publication ; no. (NIH) 76-877.
RJ380.C58 618.9'2 76-603560
77⌈r79⌉rev MARC

—SCOTLAND

Scotland. Scottish Health Services Council.
The young chronic sick : report of a sub-committee appointed by the council. — Edinburgh : H.M. Stationery Off., 1964.
18 p. ; 25 cm.
At head of title: Scottish Home and Health Department, Scottish Health Services Council.
1. Chronic diseases in children—Scotland. 2. Chronically ill children—Care and treatment—Scotland. I. Title.
RJ380.S39 1964 79-316384
79 MARC

CHRONIC SIMPLE GLAUCOMA see
Glaucoma, Open angle

CHRONICALLY ILL

see also Arthritics; Asthmatics; Diabetics; Epileptics; Nursing home patients

—CARE AND TREATMENT
—NEBRASKA

Nebraska. Bureau of Comprehensive Health Planning.
Long-term care handbook : a description of the alternative modes of health and health-related care for the elderly and disabled in Nebraska / prepared by Bureau of Comprehensive Health Planning, Nebraska Department of Health. — Lincoln, Neb. : The Bureau, 1975.
68 p. ; 28 cm.
1. Long-term care of the sick—Nebraska. 2. Aged—Care and hygiene—Nebraska. 3. Chronically ill—Care and treatment—Nebraska. I. Title.
RA644.7.N2N4 1975 362.6'09782 76-622468
78⌈r79⌉rev MARC

—CARE AND TREATMENT—UNITED STATES—ADDRESSES, ESSAYS, LECTURES

Strauss, Anselm L comp.
Where medicine fails / edited by Anselm L. Strauss. — 3d ed. — New Brunswick, N.J. : Transaction Books, c1979.
296 p. ; 21 cm.
Bibliography: p. 293.
ISBN 0-87855-676-1 : $4.95
1. Medical care—United States—Addresses, essays, lectures. 2. Chronically ill—Care and treatment—United States—Addresses, essays, lectures. I. Title.
⌈DNLM: 1. Chronic disease—Collected works. 2. Economics, Medical—United States—Collected works. 3. Hospitals—United States—Collected works. 4. Medicine—United States—Collected works. 5. Psychiatry—United States—Collected works. WB50 AA1 W5⌉
RA395.A3S86 1979 362.1'0973 78-62893
78 MARC

—INSTITUTIONAL CARE—UNITED STATES

United States. Bureau of the Census.
1976 survey of institutionalized persons : methods and procedures / ⌈prepared by Dennis W. Stoudt⌉. — ⌈Washington⌉ : U.S. Dept. of Commerce, Bureau of the Census : for sale by the Supt. of Docs., U.S. Govt. Print. Off., 1978.
iv, 80 p. ; 28 cm. — (Technical paper - Bureau of the Census : 42)
1. Extended care facilities—United States. 2. Health surveys—United States. 3. Chronically ill—Institutional care—United States. 4. Handicapped—Institutional care—United States. I. Stoudt, Dennis W. II. Title. III. Series: United States. Bureau of the Census. Technical paper ; 42.
RA997.U45 1978 362.1'6'0973 78-606082
78 MARC

—INSTITUTIONAL CARE
—UNITED STATES
—STATISTICS

United States. Bureau of the Census.
1976 survey of institutionalized persons : a study of persons receiving long-term care. — Washington : U.S. Dept. of Commerce, Bureau of the Census : for sale by the Supt. of Docs., U.S. Govt. Print. Off., 1978.
xiii, 440 p. : ill. ; 28 cm. — (Current population reports : Special studies : Series P-23 : no. 69)
Chiefly tables.
Supt. of Docs. no.: Supt. of Docs. No.: C 3.186:23/69
1. Extended care facilities—United States—Statistics. 2. Chronically ill—Institutional care—United States—Statistics. 3. Handicapped—Institutional care—United States—Statistics. 4. Inmates of institutions—United States—Family relationships—Statistics. I. Title. II. Title: Survey of institutionalized persons. III. Series: United States. Bureau of the Census. Current population reports : Special studies : Series PA-23 ; no. 69.
HA203.A218 no. 69 312'.0973 s 78-606054
⌈RA997⌉ 78 MARC

—REHABILITATION

Rehabilitation in chronic renal failure / edited by Samuel B. Chyatte. — Baltimore : Williams & Wilkins, c1979.
xv, 163 p. ; 24 cm. — (Rehabilitation medicine library)
Includes bibliographies and index.
ISBN 0-683-01578-8
1. Renal insufficiency. 2. Chronically ill—Rehabilitation. 3. Renal insufficiency—Psychological aspects. I. Chyatte, Samuel B. II. Series.
⌈DNLM: 1. Kidney failure, Chronic—Rehabilitation. WJ342 R345⌉
RC903.R35 616.6'1 78-13443
78 MARC

—REHABILITATION—SWEDEN

Bergsten, Elisabeth.
A study of patients on chronic haemodialysis : a sociomedical investigation of the experiences of patients on haemodialysis and their quality of life / Elisabeth Bergsten, Hiroyuki Asaba and Jonas Bergström. — Stockholm : Almqvist & Wiksell Periodical Co., c1977.
31 p. : graphs ; 25 cm. — (Scandinavian journal of social medicine : Supplementum ; 11 ISSN 0300-8037) S***
"From the Department of Nephrology, S:t Eriks sjukhus, Stockholm."
Bibliography: p. ⌈28⌉
ISBN 9122001662 : free
1. Hemodialysis—Social aspects—Sweden. 2. Chronically ill—Rehabilitation—Sweden. I. Asaba, Hiroyuki, joint author. II. Bergström, Jonas, 1929- joint author. III. Title. IV. Series.
⌈RC901.7.A7B38⌉ 362.1'9'61 78-318043
78 MARC

—HAWAII—SOCIOECONOMIC STATUS
—STATISTICS

Wegner, Eldon L
Chronic illness and hospitalization by ethnicity and income in the State of Hawaii, 1969-1971 / by Eldon L. Wegner. — ⌈Honolulu⌉ : Research and Statistics Office, Hawaii State Dept. of Health, 1977.
26 p. ; 28 cm. — (R & S report ; issue no. 19)
Cover title.
Chiefly tables.
1. Hawaii—Statistics, Medical. 2. Chronic diseases—Hawaii—Statistics. 3. Chronically ill—Hawaii—Socioeconomic status—Statistics. 4. Hospital utilization—Hawaii—Statistics. 5. Hospital patients—Hawaii—Socioeconomic status—Statistics. 6. Health and race—Hawaii—Statistics. 7. Income—Hawaii—Statistics. 8. Health surveys—Hawaii. I. Title. II. Series: Hawaii. Dept. of Health. Research and Statistics Office. R & S report ; issue no. 19.
HB3525.H3H33a no. 19 301.32'9'969 s 78-621623
⌈RA407.4.H3⌉ 79 MARC

—UNITED STATES—STATES—STATISTICS

United States. National Center for Health Statistics.
State estimates of disability and utilization of medical services : United States, 1974-76 : derived from the health interview survey. — Hyattsville, Md. : U.S. Dept. of Health, Education, and Welfare, Public Health Service, National Center for Health Statistics ; ⌈Washington : for sale by the Supt. of Docs., U.S. Govt. Print. Off.⌉, 1978.
vi, 76 p. : maps ; 27 cm. — (DHEW publication ; no. (PHS) 78-1241)
"A revised and updated version of two earlier reports, Synthetic state estimates of disability derived from the National health survey, 1963-64 and State estimates of disability and utilization of medical services, United States, 1969-71."
Includes bibliographical references.
1. United States—Statistics, Medical. 2. Chronically ill—United States—States—Statistics. 3. Handicapped—United States—States—Statistics. 4. Medical care—United States—States—Utilization—Statistics. 5. Health surveys—United States. I. Title. II. Series: United States. Dept. of Health, Education, and Welfare. DHEW publication ; no. (PHS) 78-1241.
RA407.3.U57 1978 312'.3'0973 79-601474
79 MARC

CHRONICALLY ILL CHILDREN

see also Chronic diseases in children

—CARE AND TREATMENT—SCOTLAND

Scotland. Scottish Health Services Council.
The young chronic sick : report of a sub-committee appointed by the council. — Edinburgh : H.M. Stationery Off., 1964.
18 p. ; 25 cm.
At head of title: Scottish Home and Health Department, Scottish Health Services Council.
1. Chronic diseases in children—Scotland. 2. Chronically ill children—Care and treatment—Scotland. I. Title.
RJ380.S39 1964 79-316384
79 MARC

—EDUCATION

Hertl, Michael.
Kranke und behinderte Kinder in Schule und Kindergarten / Michael Hertl, Renate Hertl. — Stuttgart : Thieme, 1979.
ix, 181 p. : ill. ; 19 cm. — (Flexibles Taschenbuch) GFR79-A
Based on the author's Das chronisch-kranke Kind in der Schule.
Bibliography: p. ⌈173⌉
Includes index.
ISBN 3-13-577401-5 : DM9.80
1. Chronically ill children—Education. 2. Handicapped children—Education. I. Hertl, Renate, joint author. II. Hertl, Michael. Chronisch-kranke Kind in der Schule. III. Title.
LC4580.H42 371.9 79-373933
79 MARC

CHRONICALLY ILL CHILDREN (Continued)

—REHABILITATION—CONGRESSES

Chronic childhood illness : assessment of outcome : a conference sponsored by the John E. Fogarty International Center for Advanced Study in the Health Sciences and the National Institute of Child Health and Human Development, National Institutes of Health, Bethesda, Maryland, May 1-2, 1974 / Gilman D. Grave and I. Barry Pless, editors. — Bethesda, Md. : U.S. Dept. of Health, Education, and Welfare, Public Health Service, National Institutes of Health ; Washington : for sale by the Supt. of Docs., U.S. Govt. Print. Off., 1976.
xiv, 238 p. : ill. ; 27 cm. — (Fogarty International Center series on the teaching of preventive medicine ; v. 3) (DHEW publication ; no. (NIH) 76-877)
Includes bibliographies and index.
Supt. of Docs. no.: HE 20.3714:3
$6.10

1. Chronic diseases in children—Congresses. 2. Chronically ill children—Rehabilitation—Congresses. 3. Child health services—Evaluation—Congresses. I. Grave, Gilman D. II. Pless, Ivan B., 1932- . III. John E. Fogarty International Center for Advanced Study in the Health Sciences. IV. United States. National Institute of Child Health and Human Development. V. Series: John E. Fogarty International Center for Advanced Study in the Health Sciences. Fogarty International Center series on the teaching of preventive medicine ; v. 3. VI. Series: United States. Dept. of Health, Education, and Welfare. DHEW publication ; no. (NIH) 76-877.
RJ380.C58 618.9'2 76-603560
77[r79]rev MARC

CHRONOGRAMS

see also Anagrams

CHRONOLOGY

see also Almanacs; Calendar; Clocks and watches; Day; Geological time; Months; Night; Radioactive dating; Time

Briatore, Luigi.
Cronologia e tecniche della misura del tempo / Luigi Briatore. — [Firenze] : G. Barbèra, 1976.
159 p. : ill. ; 24 cm. — (Pubblicazioni - Centro per la storia della tecnica : Sezione 6, Sussidiari ; v. 2) It78-Feb
Includes bibliographical references.
1. Time. 2. Chronology. 3. Time measurements. I. Title. II. Series: Centro per la storia della tecnica. Pubblicazioni - Centro per la storia della tecnica : Sezione 6, Sussidiari ; v. 2.
QB209.B74 529 78-383562
*78 MARC

Diop, Cheikh Anta.
Physique nucléaire et chronologie absolue / par Cheikh Anta Diop. — Dakar : IFAN, [1974]
155 p. : ill. ; 24 cm. — (Initiations et études africaines ; no 31)
Bibliography: p. [139]-152.
1. Radioactive dating. 2. Nuclear physics. 3. Chronology. I. Title. II. Series.
QC798.D3D56 930'.1'0285 78-398553
79 MARC

(Tarerk' zhamanakagrowt'ean)
Տարերք ժամանակագրութեան : ի պէտս ազգային վարժարանաց . — Ի Վենետիկ : Ի Սուրբն Ղազար, 1857.
46 p. ; 16 cm.
1. Chronology. 2. Calendar. I. Title.
CE11.T37 79–971133

—TABLES

Hull, Terence H
Almanak penanggalan Jawa-Masehi untuk penelitian sosial-ekonomi / [Terence H. Hull]. — Yogyakarta : Lembaga Kependudukan, Universitas Gadjah Mada, 1976.
iii, 12, [55] leaves ; 33 cm. — (Seri metodologi - Universitas Gadjah Mada, Lembaga Kependudukan ; no. 2)
English and Indonesian.
Bibliography: leaf 12 (2d group)
1. Calendar, Javanese. 2. Calendar, Gregorian. 3. Chronology—Tables, etc. I. Title. II. Series: Universitas Gadjah Mada. Lembaga Kependudukan. Seri metodologi - Lembaga Kependudukan, Universitas Gadjah Mada ; no. 2.
CE61.I5H84 76-941449
78 MARC

CHRONOLOGY, BIBLICAL see Bible—Chronology

CHRONOLOGY, BYZANTINE

Bagnall, Roger S
Regnal formulas in Byzantine Egypt / Roger S. Bagnall and K. A. Worp. — Missoula, Mont. : Scholars Press, c1979.
x, 90 p. ; 23 cm. — (Bulletin of the American Society of Papyrologists : Supplements ; no. 2)
Supplements the author's Chronological systems of Byzantine Egypt.
Includes index.
Bibliography: p. [ix]-x.
ISBN 0-89130-280-8 (pbk.)
1. Chronology, Byzantine. 2. Egypt—History—To 640—Chronology. I. Worp, Klaas Anthony, joint author. II. Title. III. Series: American Society of Papyrologists. Bulletin of the American Society of Papyrologists : Supplements ; no. 2.
CE42.B33 932 79-1316
79 MARC

CHRONOLOGY, ECCLESIASTICAL

see also Church calendar

Matthiae, Karl.
Chronologische Übersichten und Karten zur spätjüdischen und urchristlichen Zeit / Karl Matthiae ; [Zeichn. d. Kt. u. Taf., Sigrid Gross-Kaiser]. — 1. Aufl. — Berlin : Evangelische Verlagsanstalt, 1977.
48 p. : ill., col. maps, geneal. tables ; 25 cm. GFR78-A41
Illustrative matter (12 leaves) inserted.
Includes index.
6.80M
1. Chronology, Ecclesiastical. 2. Chronology, Jewish. I. Title.
CE55.M37 79-366895
79 MARC

CHRONOLOGY, EGYPTIAN

Miceli, Jean de.
Nouveaux instruments pour le calcul du temps : chronologie de la XVIIIe dynastie et du temps de l'Exode, Thoutmosis II, pharaon de l'Exode / Jean de Miceli. — Tuile (4, rue Anne-Vialle) : Éditions du livre, 1974.
114 p. : ill. ; 24 cm. F75-11807
On spine: Chronologie de la XVIIIe dynastie, l'Exode.
Bibliography: p. 105-110.
Includes index.
1. Calendar, Egyptian. 2. Chronology, Egyptian. 3. Egypt—History—To 332 B.C.—Chronology. I. Title. II. Title: Thoutmosis II, pharaon de l'Exode. III. Title: Chronologie de la XVIIIe dynastie, l'Exode.
CE29.M5 932'.01 78-386013
79 MARC

CHRONOLOGY, GREEK

Arvanitakēs, G L
(Chronologia tōn archaiōn kai neōterōn Hellēnon)
Χρονολογία τῶν ἀρχαίων καὶ νεωτέρων Ἑλλήνων : σχεδίασμα / ὑπὸ Γ. Λ. Ἀρβανιτάκη. — Ἀθῆναι : Π. Σαββίδη & Ν. Καραγιαννάκη, 1940.
49, [2] p. ; 27 cm.
Bibliography: p. [50]
1. Chronology, Greek. 2. Calendar, Greek. I. Title.
CE42.A78 79–413749

CHRONOLOGY, HEBREW see Chronology, Jewish

CHRONOLOGY, HISTORICAL

see also Calendar; Calendars; History—Yearbooks

Censorinus.
De die natale ("The natal day") by Censorinus (A.D. 238) Life of the Emperor Hadrian by Ælianus Spartianus (circ. A.D. 300) Tr. into English by William Maude. New York, The Cambridge encyclopedia co., 1900.
2 pts. in 1 v. 24 cm.
Part 2 has special t.-p.: Life of the Emperor Hadrian by Ælianus Spartianus ...
——— Microfilm. Microfilm 36506 CE
1. Calendar—Early works to 1800. 2. Chronology, Historical. 3. Hadrianus, emperor of Rome 76-138. 4. Roman emperors—Biography. I. Maude, William, tr.
CE25.C39 1-15408
[r79]rev MARC

Geiss, Imanuel.
Geschichte griffbereit / Imanuel Geiss. — Orig. Ausg. — Reinbek bei Hamburg : Rowohlt, 1979-
v. ; 19 cm. — (Rororo Handbuch) GFR79-A
Bibliography: v. 1, p. [256]
Includes indexes.
CONTENTS: 1. Daten.

ISBN 3-499-16235-0 (v. 1) : DM7.80 (v. 1)
1. Chronology, Historical. I. Title.
D11.G43 902'.02 79-378480
79 MARC

Herman, Jacques.
Mémento d'histoire universelle des origines à nos jours / Jacques Herman ; [cartes par Studiopizz]. — Lausanne : R. Bettex, cop. 1978.
290 p. : maps ; 17 cm. — (Collection Mémento) Sw79-1060
Bibliography: p. [254]
Includes index.
26.00F
1. World history. 2. Chronology, Historical. I. Title.
D20.H514 902'.02 79-371584
79 MARC

Kyōto Daigaku. Bungakubu. Seiyōshi Kenkyūshitsu.
(Sekaishi dainempyō)
世界史大年表 / 京都大學文學部西洋史研究室編. — 東京 : 目黒書店, 昭和22-23 [1947-1948], v. 2, 昭和24 [1949] printing.
2 v. ; 19 x 26 cm.
Bibliography: v. 2, p. 3.

1. Chronology, Historical. I. Title.
D11.K9 1947 78–813122

(Mohan saishin sekai nempyō)
模範最新世界年表 / 喜田貞吉校閲 ; 森田鐵三郎西洋史擔任, 妻木忠太日本史·東洋史擔任 ; 三省堂編輯所編纂. — 新訂改版156版. — 東京 : 三省堂, 昭和10 [1935]
2, 2, 8, 386, 182 p. : geneal. tables ; 19 cm.

1. Chronology, Historical. 2. Kings and rulers—Genealogy. I. Kida, Teikichi. II. Morita, Tetsusaburō. III. Tsumaki, Chūta, 1870– IV. Sanseidō (Founded 1915) V. Title: Saishin sekai nempyō.
D11.M73 78–814415

Sharp, Harold S
Footnotes to world history : a bibliographic source book / by Harold S. Sharp. — Metuchen, N.J. : Scarecrow Press, 1979.
xvii, 734 p. ; 23 cm.
Includes index.
ISBN 0-8108-1185-5
1. World history—Bibliography. 2. Chronology, Historical. I. Title.
Z6201.S52 016.909 78-23417
[D20] 78 MARC

(Shikaku nempyō ni yoru sekaishi)
視覚年表による世界史 / 秀村欣二監修 ; 和田寛編. — 東京 : 蒼風書院, 昭和30 [1955]
90 p. : col. ill. ; 26 cm.
Colophon inserted.
Errata slip inserted.
Includes index.

1. History—Chronology. 2. Geography, Historical—Maps. I. Wada, Hiroshi. II. Title: Sekaishi.
D11.S554 78–811042

Trager, James.
The people's chronology : a year-by-year record of human events from prehistory to the present / edited by James Trager. — 1st ed. — New York : Holt, Rinehart and Winston, c1979.
ix, 1206 p. : ill. ; 24 cm.
Includes index.
ISBN 0-03-017811-8 : $27.95
1. Chronology, Historical. I. Title.
D11.T83 902'.02 78-23808
78 MARC

Weltgeschichte : Daten u. Bilder / [Text, unter Mitarb. von Hiltrud Anlauf, Peter Salz, Gisela Stankiewicz hrsg. von Fritz Winzer]. — Braunschweig : Westermann, 1977.
207 p. : numerous ill. (numerous col.) ; 32 cm. GFR78-A
Includes index.
ISBN 3-14-509032-1 : DM29.80
1. Chronology, Historical. I. Anlauf, Hiltrud. II. Winzer, Fritz.
D11.W42 909 78-400321
79 MARC

CHRONOLOGY, HISTORICAL (Continued)

—JUVENILE LITERATURE

1,000 great events / edited by Lynne Sabel and Philip Steele. — London ; New York : Hamlyn, 1977.
2-352 p. : ill. (some col.), facsims., 2 maps (1 col.), col. plan, ports. (some col.) ; 25 cm. GB77-34270
Includes index.
ISBN 0-600-38743-7 : £2.95
1. Chronology, Historical—Juvenile literature. I. Sabel, Lynne. II. Steele, Philip.
D11.5.O53 902'.02 78-319821
78 MARC

—PERIODICALS

Grundkalendarium.
Berlin, Staatsverlag der Deutschen Demokratischen Republik.
v. 22 cm.
1. Calendar—Periodicals. 2. Chronology, Historical—Periodicals.
CE1.G78 75-647724
MARC-S

—TABLES

Davies, Cyril.
World history in perspective at a glance : from 3000 B.C. to modern times / designed and compiled by Cyril Davies. — [Enlarged ed.]. — [Worthing] : [The author], [1977]
[2], 22 leaves ; 21 x 26 cm. GB78-16480
£1.80
1. Chronology, Historical—Tables. I. Title.
D11.D28 1977 902'.02 78-322753
79 MARC

Freeman-Grenville, Greville Stewart Parker.
Chronology of world history : a calendar of principal events from 3000 BC to AD 1976 / G. S. P. Freeman-Grenville. — 2d ed. — Totowa, N.J. : Rowman and Littlefield, 1978.
746 p. ; 25 cm.
Includes index.
ISBN 0-8476-6040-0 : $40.00
1. Chronology, Historical—Tables. I. Title.
D11.F75 1978 902'.02 78-113042
79 MARC

Las Cases, Emmanuel, comte de, 1766-1842.
Le Sage's historical, genealogical, chronological, and geographical atlas, exhibiting all the royal families in Europe, their origin, descent, marriages &c ... translated from the last and much improved French ed. — London : Printed by R. Juigné, 1813.
2 p., 38 leaves of plates : geneal. tables, col. maps ; 52 cm.
Translation of Atlas historique, généalogique, chronologique et geographique.
1. Geography, Historical—Maps. 2. Chronology, Historical—Tables. 3. Europe—Kings and rulers—Genealogy. I. Title.
G1030.L352 1813 911 78-379102
78 MARC

CHRONOLOGY, INDONESIAN

Casparis, J G de.
Indonesian chronology / by J. G. de Casparis. — Leiden ; Köln : Brill, 1978.
x, 69 p., 4 p. of plates : ill. ; 25 cm. — (Handbuch der Orientalistik : 3. Abt., Indonesien, Malaysia und die Philippinen ; 1. Bd. Geschichte, Lieferung 1) Ne78-49
Bibliography: p. [59]-61.
Includes index.
ISBN 9004057528 : fl 56.00
1. Chronology, Indonesian. I. Title. II. Series.
CE61.I5C37 529'.09598 79-309582
*79 MARC

CHRONOLOGY, ISLAMIC

—PERIODICALS

(al-Taqwīm al-Hijrī wa-mawāqīt al-ṣalāh wa-al-manāzil)
التقويم الهجرى ومواقيت الصلاة والمنازل.
[الخرطوم] وزارة الشئون الدينية والاوقاف، مصلحة الدعوة والتوجيه الدينى.
v. ill. 16 x 21 cm. annual.
1. Calendar, Islamic—Periodicals. 2. Prayer (Islam)—Periodicals. 3. Chronology, Islamic—Periodicals. I. Sudan. Maṣlaḥat al-Da'wah wa-al-Tawjīh al-Dīnī.
CE59.T36 79-649892
(MARC-S)

CHRONOLOGY, JAPANESE

Ueji, Tatsunori, 1932-
(Gengō mondai)
元号問題：その歴史的·現実的意味 / 上地龍典. — [東村山]：教育社；東京：販売 教育社出版サービス, [1979]
152 p. : ill. ; 18 cm. — (時事問題解説 ; 189) (入門新書) Ja 79-5732
Cover title.
Series 2 romanized: Nyūmon shinsho.
Bibliography: p. 152.
¥400
1. Chronology, Japanese. 2. Chronology, Japanese—Public opinion, Japanese. 3. Public opinion—Japan. I. Title. II. Series: Jiji mondai kaisetsu ; 189.
DS836.U35 79-809516

—PUBLIC OPINION, JAPANESE

Ueji, Tatsunori, 1932-
(Gengō mondai)
元号問題：その歴史的·現実的意味 / 上地龍典. — [東村山]：教育社；東京：販売 教育社出版サービス, [1979]
152 p. : ill. ; 18 cm. — (時事問題解説 ; 189) (入門新書) Ja 79-5732
Cover title.
Series 2 romanized: Nyūmon shinsho.
Bibliography: p. 152.
¥400
1. Chronology, Japanese. 2. Chronology, Japanese—Public opinion, Japanese. 3. Public opinion—Japan. I. Title. II. Series: Jiji mondai kaisetsu ; 189.
DS836.U35 79-809516

CHRONOLOGY, JEWISH

see also Calendar, Jewish

Matthiae, Karl.
Chronologische Übersichten und Karten zur spätjüdischen und urchristlichen Zeit / Karl Matthiae ; [Zeichn. d. Kt. u. Taf., Sigrid Gross-Kaiser]. — 1. Aufl. — Berlin : Evangelische Verlagsanstalt, 1977.
48 p. : ill., col. maps, geneal. tables ; 25 cm. GFR78-A41
Illustrative matter (12 leaves) inserted.
Includes index.
6.80M
1. Chronology, Ecclesiastical. 2. Chronology, Jewish. I. Title.
CE55.M37 79-366895
79 MARC

CHRONOLOGY, MAORI

Best, Elsdon, 1856-1931.
The astronomical knowledge of the Maori ; The Maori division of time / [by Elsdon Best]. — New York : AMS Press, 1978.
66, 45 p. ; 24 cm.
Reprint of the 1922 ed. of The astronomical knowledge of the Maori, published by W. A. G. Skinner, Govt. Printer, Wellington, N.Z., which was issued as Dominion Museum monograph no. 3; and of the 1922 ed. of The Maori division of time, published by Skinner, which was issued as Dominion Museum monograph no. 4.
ISBN 0-404-14405-5
1. Astronomy, Maori. 2. Folk-lore, Maori. 3. Chronology, Maori. I. Best, Elsdon, 1856-1931. The Maori division of time. 1978. II. Title. III. Title: The Maori division of time. IV. Series: Wellington, N.Z. Dominion Museum. Dominion Museum monograph ; no. 3-4.
DU423.A85B47 1978 520'.9174'994 75-35226
78 MARC

CHRONOLOGY, ROMAN

Aiôn : le temps chez les Romains / publié par R. Chevallier. — Paris : A. & J. Picard, 1976.
x, 370 p. ; 25 cm. — (Caesarodunum ; 10 bis) F***
Bibliography: p. [363]-370.
ISBN 2-7084-0008-8 : 145F
1. Calendar, Roman. 2. Chronology, Roman. I. Chevallier, Raymond, 1929- II. Series.
CE46.A35 529'.32'2 79-338995
79 MARC

CHRONOPHOTOGRAPHY see Cinematography

CHRYSANTHEMUMS

Machin, Barrie John.
Chrysanthemums, year round growing. — [New ed.] / [by] Barrie Machin and Nigel Scopes. — Poole : Blandford Press, 1978.
x, 233 p., [4] p. of col. plates : ill. (some col.) ; 26 cm. GB78-25246
Third ed. by S. A. Searle and B. J. Machin published in 1968 under title: Chrysanthemums the year round.
Bibliography: p. 219-225.
Includes index.
ISBN 0-7137-0885-9 : £8.75
1. Chrysanthemums. I. Scopes, Nigel Eric Anthony, joint author. II. Searle, Sidney Alexander. Chrysanthemums the year round. III. Title.
SB413.C55S4 1978 635.9'3'355 78-326037
79 MARC

National Chrysanthemum Society (Gt. Brit.)
The chrysanthemum manual of the National Chrysanthemum Society. — 5th revised ed. / revised and edited by S. G. Gosling. — London : The Society, 1976.
391 p. : ill. (some col.) ; 23 cm. GB78-06925
Includes indexes.
1. Chrysanthemums. I. Gosling, Stanley George. II. Title.
SB413.C55N33 1976a 635.9'33'55 78-323011
79 MARC

(Ōgata shisetsu ni okeru kiku no shūnen saibai gijutsu)
大型施設におけるキクの周年栽培技術 / 農林水産技術会議事務局. — 東京：農林統計協会, 1975.
62 p., [1] leaf of plates : ill. ; 21 cm. — (実用化技術レポート ; no. 21) Ja 77-323
Bibliography: p. 50.
1. Chrysanthemums. 2. Greenhouse management—Japan—Inland Sea region. I. Japan. Nōrin Suisan Gijutsu Kaigi. Jimukyoku. II. Title: Kiku no shūnen saibai gijutsu. III. Series: Jitsuyōka gijutsu repōto ; no. 21.
SB413.C55O33 79-806699

Taiwan. Nung lin t'ing.
(Chü hua tsai p'ei)
菊花栽培 / 台灣省政府農林廳編印. — [南投縣]：該廳, [1978?]
36 p. : ill. ; 22 cm.
Cover title.
1. Chrysanthemums. I. Title.
SB413.C55T28 1978 78-843558

—VARIETIES

Miao, P'u-sun.
(Chü p'u)
菊譜 / 繆莆孫繪撰. — 台1版. — 台北：臺灣中華書局, 民國65 [1976]
138 p. in various pagings : chiefly ill. ; 21 cm.
Reprint ed.
1. Chrysanthemums in art. 2. Flower painting and illustration. 3. Painting, Chinese. 4. Chrysanthemums—Varieties. I. Title.
ND1403.C6M5 1976 78-843811

CHRYSANTHEMUMS IN ART

Miao, P'u-sun.
(Chü p'u)
菊譜 / 繆莆孫繪撰. — 台1版. — 台北：臺灣中華書局, 民國65 [1976]
138 p. in various pagings : chiefly ill. ; 21 cm.
Reprint ed.
1. Chrysanthemums in art. 2. Flower painting and illustration. 3. Painting, Chinese. 4. Chrysanthemums—Varieties. I. Title.
ND1403.C6M5 1976 78-843811

CHRYSELEPHANTINE SCULPTURE see Sculpture, Chryselephantine

CHRYSOMELIDAE

—HOST PLANTS

Wilcox, John Avery.
Leaf beetle host plants in northeastern North America (Coleoptera, Chrysomelidae) / by John A. Wilcox. — Kinderhook, NY : Published for Biological Research Institute of America by World Natural History Publications, 1979.
30 p. ; 28 cm.
At head of title: North American Beetle Fauna Project.
Bibliography: p. 27-30.
ISBN 0-916846-09-1
1. Chrysomelidae—Host plants. 2. Host plants—North America. 3. Insects—North America—Host plants. I. Title. II. Title: North American Beetle Fauna Project.
QL596.C5W49 595.7'64 79-111968
79 MARC

CHRYSOSTOMUS, JOANNES, SAINT, PATRIARCH OF CONSTANTINOPLE, D. 407

Douze récits byzantins sur Saint Jean Chrysostome / publiés par François Halkin. — Bruxelles : Société des bollandistes, 1977.
562 p. ; 25 cm. — (Subsidia hagiographics ; no 60) Be***
Texts in Greek with headings and notes in French.
Includes bibliographical references and indexes.
CONTENTS: Vie, par Théodore de Trimithonte.—Item, abrégé anonyme.—Vie, par Georges d'Alexandrie.—Fragment d'une vie perdue.—Vie abrégée.—Vie anonyme du Vaticanus 1669.—Vie, par Cosme Vestitor.—Vie attribuée à Hésychius.—Translation par Syméon Métaphraste.—Translation du Sabaiticus 242.—Récit oratoire de la translation.—Translation du Mosquensis 26.
1. Chrysostomus, Joannes, Saint, Patriarch of Constantinople, d. 407. 2. Christian saints—Biography. I. Halkin, François. II. Series.
BR1720.C5D68 270.2'092'4 78-388823
78 MARC

Goodall, Blake.
The homilies of St. John Chrysostom on the letters of St. Paul to Titus and Philemon : prolegomena to an edition / by Blake Goodall. — Berkeley : University of California Press, c1979.
ix, 102, 4 p. : facsims. ; 26 cm. — (University of California publications : Classical studies ; v. 20)
Bibliography: p. 97-99.
Includes index.
ISBN 0-520-09596-0
1. Bible. N.T. Titus—Criticism, interpretation, etc.—History—Early church, ca. 30-600. 2. Bible. N.T. Philemon—Criticism, interpretation, etc.—History—Early church, ca. 30-600. 3. Chrysostomus, Joannes, Saint, Patriarch of Constantinople, d. 407. I. Title. II. Series: California. University. University of California publications : Classical studies ; v. 20.
BS2755.2.G66 252'.01'4 78-54795
00 MARC

Quattro omelie copte : Vita di Giovanni Crisostomo, Encomi dei 24 vegliardi (ps. Proclo e anonimo), Encomio di Michele arcangelo di Eustazio di Tracia / edizione, traduzione e commento a cura di Antonella Campagnano, Antonella Maresca, Tito Orlandi. — Milano : Cisalpino-Goliardica, 1977.
189 p., 8 leaves of plates : ill. ; 24 cm. — (Testi e documenti per lo studio dell'antichità ; 60) It78-May
Coptic text with translation and commentary in Italian.
Includes bibliographical references and indexes.
L15000
1. Chrysostomus, Joannes, Saint, Patriarch of Constantinople, d. 407. 2. Proclus, Saint, d. 446. 3. Eustathius, Bp. of Trake. 4. Michael, archangel. I. Campagnano, Antonella. II. Maresca, Antonella. III. Orlandi, Tito. IV. Series.
BR1720.C5Q37 79-358113
*79 MARC

—DICTIONARIES, INDEXES, ETC.

Malingrey, Anne Marie.
Indices Chrysostomici / sous la direction de Anne-Marie Malingrey, avec la collaboration de M.-L. Guillaumin. — Hildesheim ; New York : G. Olms, 1978-
v. ; 22 cm. — (Alpha-Omega : Reihe A, Lexika, Indizes, Konkordanzen zur Klassischen philologie ; 31) GFR***
CONTENTS: 1. Ad olympiadem. Ab exilio epistula. De providentia Dei.
Index in Classical Greek with prefatory material in French.
Computer-produced index.
ISBN 3-487-06635-1 (v.1)
1. Chrysostomus, Joannes, Saint, Patriarch of Constantinople, d. 407—Dictionaries, indexes, etc. I. Guillaumin, M. L., joint author. II. Title.
BR65.C455M34 79-344739
79 MARC

CHRYSOTILE

—VERMONT—BELVIDERE MOUNTAIN REGION

Chidester, Alfred Herman, 1914-
Petrology, structure, and genesis of the asbestos-bearing ultramafic rocks of the Belvidere Mountain area in Vermont / by A. H. Chidester, A. L. Albee, and W. M. Cady. — Washington : U.S. Govt. Print. Off., 1978.
v, 95 p. : ill., fold. maps (in pocket) ; 29 cm. — (U.S. Geological Survey professional paper ; 1016)
Bibliography: p. 86-88.
Includes index.
Supt. of Docs. no.: I 19.16:1016
1. Chrysotile—Vermont—Belvidere Mountain region. 2. Petrology—Vermont—Belvidere Mountain region. 3. Asbestos—Vermont—Belvidere Mountain region. I. Albee, Arden Leroy, 1928- joint author. II. Cady, Wallace Martin, 1912- joint author. III. Title. IV. Series: United States. Geological Survey. Professional paper ; 1016.
QE391.C67C48 552'.4 76-608288
76 MARC

CHRZANOWSKI, IGNACY, 1866-1940

—ADDRESSES, ESSAYS, LECTURES

Prace historyczno-literackie : księga zbiorowa ku czci Ignacego Chrzanowskiego. — Kraków : Skł. Gł. w Kasie im. Mianowskiego, 1936.
646 p., [2] leaves of plates : ill. ; 25 cm. — (Prace z historji literatury polskiej ; Nr 1)
"Bibljografja prac Ign. Chrzanowskiego": p. [615]-644.
Includes bibliographical references.
1. Polish literature—History and criticism—Addresses, essays, lectures. 2. Chrzanowski, Ignacy, 1866-1940—Addresses, essays, lectures. I. Chrzanowski, Ignacy, 1866-1940. II. Series.
PG7015.P69 79-355418
79 MARC

CHRZĄSZCZEWSKI, ANTONI, 1770-1851

Chrząszczewski, Antoni, 1770-1851.
Pamiętnik oficjalisty Potockich z Tulczyna / Antoni Chrząszczewski ; wydał, wstępem i komentarzem opatrzył Jerzy Piechowski. — Wrocław : Zakład Narodowy im. Ossolińskich, 1976.
239 p., [4] leaves of plates : ill. ; 21 cm. P***
At head of title: Polska Akademia Nauk. Zakład Narodowy imienia Ossolińskich. Biblioteka.
Includes bibliographical references and index.
zł50.00
1. Potocki family. 2. Poland—Nobility—Biography. 3. Tul'chin—Biography. 4. Chrząszczewski, Antoni, 1770-1851. I. Piechowski, Jerzy. II. Title.
DK4138.P67C48 1976 79-376049
79 MARC

CHTO (THE RUSSIAN WORD)

Kuznetsova, Roza Dmitrievna.
Formirovanie sostavnykh soiûzov v russkom îazyke : Ucheb. posobie / R. D. Kuznetsova. — Kalinin : Kalinin. un-t, 1978.
77 p. ; 20 cm. USSR78
At head of title: Ministerstvo vysshego i srednego spetsial'nogo obrazovaniîa RSFSR. Kalininskiĭ gosudarstvennyĭ universitet.
Bibliography: p. 72-[76]
0.35rub
1. Russian language—Conjunctions. 2. Russian language—Grammar, Historical. 3. Chto (The Russian word) I. Title.
PG2345.K8 79-377467
79 MARC

CHÜ, CH'AO, FL. 1821-1850

Chü, Ch'ao, fl. 1821-1850.
(Chü Ch'ao hua ts'e)
居巢画册. — [上海?: s. n., 1977?]
[13] leaves : chiefly col. ill. ; 36 x 54 cm.
Title from label on spine.
Folded book in case.
1. Chü, Ch'ao, fl. 1821-1850. 2. Fan painting, Chinese. I. Title.
ND1049.C526A4 1977 79-836664

CHU, CHIA. SHU HSUEH KO YEN

Chu, Chia.
(Shu hsüeh ko yen shu cheng)
書學格言疏証 / 祝嘉著. — 九龍：中華書局, 1978.
2, 274 p. ; 21 cm.
$9.50
1. Chu, Chia. Shu hsüeh ko yen. 2. Calligraphy, Chinese—Quotations, Maxims, etc. I. Title.
NK3634.A2C4862 79-839530
C

CHU, HSI, 1130-1200. HSIAO HSUEH

Yi, Ki-sŏk.
(Kyoyuk sedaerŭl wihan Sohak sŏn)
教育世代를위한小學選 / 李基奭編譯. — 서울：培英社, 1977.
224 p. ; 18 cm. — (教育新書 ; 56)
Includes original text.
Series romanized: Kyoyuk sinsŏ.
1. Chu, Hsi, 1130-1200. Hsiao hsüeh. 2. Conduct of life. I. Chu, Hsi, 1130-1200. Hsiao hsüeh. Korean & Chinese. 1977. II. Title.
BJ1558.C5Y5 79-826419
K

CHU, HSI, 1130-1200 SHIH CHI CHUAN

(Ts'ung Chu-tzu Shih chi chuan hsieh yün chung k'an tang shih ti yü yin hsien hsiang)
從朱子詩集傳叶韻中看當時的語音現象. — [台北?: s. n., 1978?]
3 v. (345 double leaves) ; 27 cm.
Title from spine.
Reproduced from ms. copy.
Bibliography: leaves 345.
1. Chu, Hsi, 1130-1200. Shih chi chuan. 2. Chinese language—Rime.
PL2466.R3313T78 78-843884

CHU, HSI. 1130-1200

Chu, Hsi, 1130-1200.
[Chu shu pai hsüan]
御定朱書百選 / [著者朱子 i. e. 朱熹]. — [서울]：培文社, 1978.
4 v. in 1 (on double leaves) ; 24 cm.
Photoreprint ed.
CONTENTS: 朱書百選.—朱門旨訣.
1. Chu, Hsi, 1130-1200. 2. Philosophers—China—Correspondence. I. Chu, Hsi, 1130-1200. Chu men chih chüeh. 1978. II. Title.
Title romanized: Ŏjŏng Chusŏ paeksŏn.
B128.C54A4 1978 78-826651
K

Wang, Mao-hung, 1668-1741.
(Chu tzu nien p'u)
朱子年譜 / 王懋竑撰. — [臺北：臺灣商務印書館, 民國68 i. e. 1979]
3 v. ; 20 cm. — (四庫全書珍本九集 ; 138-140)
Cover title.
Reprint ed.
CONTENTS: 1.-2. 朱子年譜.—3. 朱子年譜考異. 朱子論學切要語.
1. Chu, Hsi, 1130-1200. 2. Philosophers—China—Biography. I. Title. II. Series: Ssu k'u ch'üan shu. Ssu k'u ch'üan shu chen pen chiu chi ; 138-140.
AC149.S7 1979, vol. 138-140 79-840284
[B128.C54]

CHU, HSI, 1130-1200 (Continued)

—ETHICS

Yang, Hui-chieh.
(Chu Hsi lun li hsüeh)
朱熹倫理學 / 楊慧傑著. — 臺北：牧童出版社, 1978.
3. 4, 191 p. : 22 cm. — (牧童文史叢書 ; 23)
English title in colophon: Ethics of Chu-Hsi.
Series romanized: Mu t'ung wen shih ts'ung shu.
Includes bibliographies.
"朱熹年表": p. 169–191.
NT$60.00

1. Chu, Hsi, 1130–1200—Ethics. 2. Ethics, Chinese. I. Title. II. Title: Ethics of Chu Hsi.
B128.C54Y34 78–843003

C

CHU, K'OCHEN, 1890-

Chu, K'o-chen, 1890-
(Chu K'o-chen wen chi)
竺可桢文集. — 第1版. — 北京：科学出版社：新华书店北京发行所发行, 1979.
xvi, 527 p., [2] leaves of plates : ill. ; 27 cm.
Includes bibliographies.
"竺可楨著作目錄": p. 514–525.
$4.85

1. Chu, K'o-chen, 1890- 2. Scientists—China—Biography. 3. Chu, K'o-chen, 1890- —Bibliography. I. Title.
Q143.C56A33 79–840754

C

—BIBLIOGRAPHY

Chu, K'o-chen, 1890-
(Chu K'o-chen wen chi)
竺可桢文集. — 第1版. — 北京：科学出版社：新华书店北京发行所发行, 1979.
xvi, 527 p., [2] leaves of plates : ill. ; 27 cm.
Includes bibliographies.
"竺可楨著作目錄": p. 514–525.
$4.85

1. Chu, K'o-chen, 1890- 2. Scientists—China—Biography. 3. Chu, K'o-chen, 1890- —Bibliography. I. Title.
Q143.C56A33 79–840754

C

CH'U, SUI-LIANG, 596-658. YEN T'A SHENG CHIAO HSU

Sich'ŏnggak Kyoyuk Sa. Sŏye Kyoyukpu.
(Ant'ap sŏnggyosŏ)
雁塔聖教序 / 楷書楮遂良 ; [編著者視聽覺教育社書藝教育部]. — 서울 : 同社, 1974, 1975 printing.
96 p. : chiefly ill. ; 33 cm. — (書藝技法시리즈 ; 4)
Colophon inserted.

1. Ch'u, Sui-liang, 596–658. Yen t'a sheng chiao hsü. 2. Calligraphy, Chinese. I. Ch'u, Sui-liang, 596–658. Yen t'a sheng chiao hsü. 1975. II. Title. III. Series: Sŏye kipŏp sirijŭ ; 4.
NK3634.C47A7937 79–826373

CH'U, SUI-LIANG, 596-658

Ch'u, Sui-liang, 596–658.
(T'ang Ch'u Sui-liang fa shu hsüan)
唐褚遂良法書選. — 普及本. — 香港：書譜出版社, [1978]
[38] p. ; 29 cm.
CONTENTS: 倪寬傳贊.—孟法師碑.
$4.00

1. Ch'u, Sui-liang, 596–658. 2. Calligraphy, Chinese. I. Title.
NK3634.C47A4 1978 79–835132

(Tōjin mosho Rantei jo nishu)
唐人摹書蘭亭序二種 / 編集 飯島稲太郎 [春敬]. — 東京：書藝文化新社, 昭和46 [1971], 昭和49 [1974] printing.
[38] p. (on double leaves) ; 34 cm. & explanatory text. — (和漢墨寶選集 ; 第9巻) Ja***
Cataloged from colophon (inserted)
Label mounted on cover: Tōjin mo Rantei jo.
Folded book, issued in case.
Reproduced from ms. copies, with printed transcript.
CONTENTS: 馮承素 蘭亭叙(故宮博物院蔵)—褚遂良 臨摹蘭亭序(斎藤董盦氏蔵)
¥2800

1. Feng, Ch'eng-su, 7th cent. 2. Ch'u, Sui-liang, 596–658. 3. Calligraphy, Chinese. I. Feng, Ch'eng-su, 7th cent. II. Ch'u, Sui-liang, 596–658. III. Wang, Hsi-chih, 321–379. Lan t'ing chi hsü. 1974. IV. Title: Rantei jo. V. Series: Wa-Kan bokuhō senshū ; dai 9-kan.
ND1457.J33W3 vol. 9 79–800379
[NK3634.F44]

Library of Congress 79 J

CHU, TA, 1626-CA. 1705

Chu, Ta, 1626–ca. 1705.
(Chu Ta shu hua ho ts'e)
朱耷書畫合册. — [上海]：上海博物館, [1956?]
2 v. (on double leaves) : chiefly ill. ; 39 cm.
Title from label on cover.
"珂𭚿版影印"

1. Chu, Ta, 1626–ca. 1705. 2. Calligraphy, Chinese. I. Title.
ND1049.C53A4 1956 78–843839

Chu, Ta, 1626–ca. 1705.
(Pa-ta-shan-jen hsieh sheng ts'e)
[八大山人寫生册]. — [台北 : s. n., 1978]
1 portfolio (12 leaves) : all ill. : 30 x 43 cm.
Title and imprint based on distributor's slip.

1. Chu, Ta, 1626–ca. 1705. I. Title.
ND1049.C53A4 1978 79–835578

CHU, TE, 1886-1976

—ADDRESSES, ESSAYS, LECTURES

(Chan cheng nien tai ti Chu Te t'ung chih)
战争年代的朱德同志. — 第1版. — 北京：人民出版社, 1977.
2, 130 p. ; 19 cm.
"在这本书里 选收了十二篇回忆 ... 的文章"
$0.24

1, Chu, Te, 1886–1976—Addresses, essays, lectures.
DS778.C6C44 1977b 79–837162

Chia, Shou-jen.
(Tsai ken sui Chu Te tsung ssu ling ti jih tzu li)
在跟随朱德总司令的日子里 / [贾守仁]. — [香港? : s. n., 1979?]
3, 2, 144 p., [7] leaves of plates : ill. ; 21 cm.
Reprint of the 1978 ed. published by Shan-tung jen min ch'u pan she, Tsinan.

1. Chu, Te—Addresses, essays, lectures. 2. Long March, 1934-1935—Addresses, essays, lectures. I. Title.
DS778.C6C45 1979 79–839493

(Jen min ti hung ssu ling)
人民的红司令：深切怀念敬爱的朱德同志. — [香港? : s. n., 1978]
3, 348 p., [2] leaves of plates : ports. ; 22 cm.
Reprint of the 1978 ed. published by Shang-hai jen min ch'u pan she, Shanghai.

1. Chu, Te, 1886–1976—Addresses, essays, lectures. 2. Generals—China—Biography—Addresses, essays, lectures. 3. China (People's Republic, 1949-). Chung-kuo jen min chien fang chün—Biography.
DS778.C6J45 1978 79–840331

Library of Congress 79 C

(Shen ch'ieh huai nien ching ai ti Chu Te wei yüan chang)
深切怀念敬爱的朱德委员长. — [香港? : s. n., 1978?]
193 p., [1] leaf of plates : port. ; 21 cm.
Reprint of the 1977 ed. published by Che-chiang jen min ch'u pan she.

1. Chu, Te, 1886–1976—Addresses, essays, lectures.
DS778.C6S42 1978 79–838163

C

—JUVENILE FICTION

Hsieh, Chu-sheng.
(Chu Te t'ung chih ti ku shih ho ch'uan shuo)
朱德同志的故事和传说 / 谢储生 ; [顾炳鑫插图]. — 第1版. — 上海：上海人民出版社, 1977.
2, 116 p., [2] leaves of plates : ill. ; 19 cm. — (少年儿童文艺读物)
Series romanized: Shao nien erh t'ung wen i tu wu.
$0.24

1. Chu, Te, 1886–1976—Juvenile fiction. I. Title.
PZ10.831.H6715C47 78–842531

C

—JUVENILE LITERATURE

Tu, Wen-yung.
(Chu tsung ssu ling ti ku shih)
朱总司令的故事 / 杜文勇編 ; [绘者林岗等]. — 第1版. — 北京：人民美術出版社, 1977.
[56] p. : ill. ; 16 cm.
"連環圖畫"
$0.18

1. Chu, Te, 1886–1976—Juvenile literature. 2. Generals—China—Biography—Juvenile literature. 3. China (People's Republic of China, 1949-). Chung-kuo jen min chieh fang chün—Biography—Juvenile literature. I. Lin, Kang. II. Title.
DS778.C6T82 78–842906

CHU, T'ZU-CH'ING, 1898-1948

Chou, Chin, fl. 1973-
(Chu Tzu-ch'ing yen chiu)
朱自清研究 / 周錦著. — 臺北：智燕出版社, 民國67 [1978]
8, 2, 264 p. ; 20 cm. — (中光文庫 ; 3)
Series romanized: Chung-kuang wen k'u.
NT$80.00

1. Chu, Tzu-ch'ing, 1898–1948. 2. Authors, Chinese—20th century—Biography. I. Title.
PL2755.T9Z6 79–835156

—CRITICISM AND INTERPRETATION —ADDRESSES, ESSAYS, LECTURES

(Chu Tzu-ch'ing tzu liao chi)
朱自清資料集. — [香港?：s. n., 1971]
20 leaves ; 23 x 31 cm.
Cover title.
A collection of articles and newspaper clippings reprinted from various sources.
Date of publication stamped on p. [4] of cover.

1. Chu, Tzu-ch'ing, 1898–1948—Criticism and interpretation—Addresses, essays, lectures.
PL2755.T9Z55 78–841551

CH'U, WEN-CHÜAN

Ch'u, Wen-chüan.
(Hua lo ch'un yu tsai)
花落春猶在 / 編著者褚問鵑. — 臺北：中外圖書出版社, 民國66- [1977-
v. : ill. ; 19 cm. — (中外文庫 ; 26)
Colophon title.
Series romanized: Chung wai wen k'u.
Vol. 2 published by Chung yang t'u shu ch'u pan she, Taipei.
NT$50.00 (v. 1)

1. Ch'u, Wen-chüan. 2. Women—China—Biography. I. Title.
CT1828.C593A34 79–837025

CH'U, YUAN, CA. 343-CA. 277 B. C.

—CRITICISM AND INTERPRETATION

Wang, Hsüeh-lan.
(Ch'ü Yüan chi ch'i tso p'in yen chiu)
屈原及其作品研究 / 王雪蘭. — [台北?：s. n.], 民國60 [1971]
70 double leaves ; 27 cm.
"國立台灣大學中文研究所碩士論文"
Reproduced from typescript.
Bibliography: leaves 68–70.

1. Ch'ü, Yüan, ca. 343–ca. 277 B. C.—Criticism and interpretation. I. Title.
PL2661.Z5W27 79–835020

CHU, YUAN, CA. 343-CA. 277 B. C. LI SAO

—ILLUSTRATIONS

Hsiao, Yün-ts'ung, 1596–1673.
[Ch'in ting pu hui Li sao ch'üan t'u]
欽定補繪蕭雲從離騷全圖. — [台北：台灣]商務[印書館, 1976?]
49, 56, 90, [9] leaves : ill. ; 20 cm. — (四庫全書珍本六集 ; 241)
Cover title.
"欽定補繪離騷全圖三卷"
Photoreprint ed. of the Wen Yüan Ko copy.

1. Ch'u tz'u—Illustrations. 2. Ch'ü, Yüan, ca. 343–ca. 277 B. C. Li sao—Illustrations. I. Ch'u tz'u. Selections. 1976. II. Title. III. Series: Ssu k'u ch'üan shu. Ssu k'u ch'üan shu chen pen liu chi ; 241.
Title romanized: Ch'in ting pu hui Hsiao Yün-ts'ung Li sao ch'üan t'u.
AC149.S7 1976, vol. 241 78–843917
[PL2521.C54]

CHU, YÜN-MING, 1460-1526

Chu, Yün-ming, 1460–1526.
(Ming Chu Yün-ming ts'ao shu Ch'ien Hou Ch'ih-pi fu)
明祝允明草書前後赤壁賦：上海博物館藏. — 香港：書譜社, [1978?]
[59] p. ; 41 cm. — (書譜叢帖：第3輯 ; 17)
Series romanized: Shu p'u ts'ung t'ieh.
Reprint ed.
$22.00

1. Chu, Yün-ming, 1460–1526. 2. Calligraphy, Chinese. 3. Chinese language—Writing, Cursive. I. Su, Shih, 1036–1101. Ch'ien Ch'ih-pi fu. 1978? II. Su, Shih, 1036–1101. Hou Ch'ih-pi fu. 1978? III. Shang-hai po wu kuan. IV. Title. V. Title: Ch'ih-pi fu.
NK3634.C48A4 1978 78–843140

CHÜ CHIA PI YUNG SHIH LEI CH'ÜAN CHI

Nakamura, Shōhachi, 1926-
(Shokkei)
食経 / 中村璋八, 佐藤達全[編著]. — 東京：明徳出版社, 1978.
216 p. ; 20 cm. — (中国古典新書) Ja 78-14480
"飲膳正要": p. 41–141; "居家必用事類全集": p. 142–214.
¥1500

1. Diet—China. 2. Hu-ssu-hui. Yin shan cheng yao. 3. Nutrition—Early works to 1800. 4. Chü chia pi yung shih lei ch'üan chi. 5. Cookery, Chinese—Early works to 1800. I. Satō, Tatsuzen, joint author. II. Hu-ssu-hui. Yin shan cheng yao. Japanese & Chinese. 1978. III. Chü chia pi yung shih lei ch'üan chi. Japanese & Chinese. 1978. IV. Title. V. Series: Chūgoku koten shinsho.
TX360.C6N34 79–804226

CH'Ü-FOU, CHINA

—HISTORY

Shan-tung sheng chan lan kung tso shih.
(Wan o sheng jen chia chan lan tzu liao hsüan pien)
《万恶"圣人家"展览》资料选编 / 山东省展览工作室编. — 第1版. — 济南：山东人民出版社, 1975.
86 p. : ill. ; 27 cm.

1. Ch'ü-fou, China—History. 2. Confucian shrines—China—Ch'ü-fou. I. Title.
DS793.C56143S5 1975 78–843675

CHU HAI SHU YUAN CHIING-KUO WEN SHIH YEN CHIU SO—ALUMNI

Chu hai shu yüan. Chung-kuo wen shih yen chiu so.
(Hsiang-kang Chu hai shu yüan Chung-kuo wen shih yen chiu so ti erh chieh pi yeh t'ung hsüeh lu)
香港珠海書院中國文史研究所第二屆畢業同學錄 / 編輯黃振權, 徐岳寶, 何文華. — [九龍]：該所, 1975.
85 p. : ill. ; 28 cm.
Colophon title.
Chinese and/or English.
1. Chu hai shu yüan. Chung-kuo wen shih yen chiu so—Alumni. I. Wong, Chun Kuen, 1918- II. Tsui, Ngok Po, 1937- III. Ho, Man Wah, 1942- IV. Title.
PL2260.C48C48 1975 79–839152

CH'U-HSIEN, CHINA (CHEKIANG)

—HISTORY—SOURCES

(Ch'ü-chou fu chih)
衢州府志：浙江省 / 楊廷望纂修. — 臺1版. — 台北：成文出版社, 民國64 [1975]
8 v. (2566 p.) : ill. ; 27 cm. — (中國方志叢書：華中地方 ; 第195號)
"據清康熙五十年修 清光緒八年重刊本影印"

1. Ch'ü-hsien, China (Chekiang)—History—Sources. I. Yang, T'ing-wang. II. Series: Chung-kuo fang chih ts'ung shu : Hua chung ti fang ; ti 195 hao.
DS793.C56147C47 1975 79–838258

CHU-HUNG, 1535-1615. SHA MI LÜ ÜEH

Kuang-hua, Shih.
(Sha mi lü i yao lüeh chi chu)
沙彌律儀要略集註 / 廣化述. — 臺北：佛教出版社：流通處大悲精舍, 民國67 [1978]
6, 178 p. ; 21 cm.
NT$40.00

1. Chu-hung, 1535–1615. Sha mi lü üeh. 2. Buddhist novices—Discipline. I. Chu-hung, 1535–1615. Sha mi lü i yao lüeh. 1978. II. Title.
BQ6145.C483K8 79–835603

CHU-JAN, SHIH

(Tōgen, Kyonen)
董源 巨然. — 東京：中央公論社, 昭和52 [1977]
154 p. : ill. (some col.) ; 53 cm. — (文人畫粹編 : 第2卷) Ja***
"中国南宗画史ノート" published as suppl. (16 p.) in case with the work.
Y53000

1. Painting, Chinese—Sung-Yüan dynasties, 960–1368. 2. Tung, Yüan, 10th cent. 3. Chü-jan, Shih. I. Tung, Yüan, 10th cent. II. Chü-jan, Shih. III. Kohara, Hironobu. Chūgoku Nanshūga shi nōto. 1977. IV. Series: Bunjinga suihen : dai 2-kan.
ND1043.4.T63 79–806321

CHU-KO, LIANG, 181-234

Ch'en, Ts'e.
(Shen chi miao suan)
神機妙算：諸葛亮兵法 / 著者陳策. — 臺北：武陵出版社, 民國67 [1978]
244 p. ; 20 cm. — (愛書人文庫)
Series romanized: Ai shu jen wen k'u.
NT$40.00

1. Strategy. 2. Chu-ko, Liang, 181–234. I. Title.
U162.C48 79–836484

Chu-ko, Liang, 181–234.
[Chu-ko Chung-wu hou wen chi]
諸葛亮集 / 段熙仲, 聞旭初編校. — 北京：中華書局, 1960.
30, 244 p., [1] leaf of plates : port. ; 21 cm.
"這次用的是清張澍編的諸葛忠武侯文集的本子...分文集四卷, 附錄二卷, 故事五卷共十一卷...附錄中的卷一是三國時人的文字...卷二則是後人所作關于諸葛亮的論, 贊, 碑, 銘. 故事摭輯各書中的有關材料...卷首原有張澍作的諸葛亮年譜和闕佚文目...沒有採用. 補遺一卷附刻在集後...姚振宗的三國藝文志中輯成諸葛亮著作考附於卷首之末."
——— Another issue. 九龍, 1972.
DS748.2.C52A2 1960 Suppl.

1. Chu-ko, Liang, 181–234. 2. Statesmen—China—Biography. I. Chang, Shu, 1781–1847. II. Tuan, Hsi-chung. III. Wen, Hsü-ch'u.
Title romanized: Chu-ko Liang chi.
DS748.2.C52A2 1960 C 61–1098

Library of Congress rev 2 [r78]

CHU RIVER, CHINA

Kubo, Kenji.
(Kanton Shukō)
廣東珠江 / 久保健二著. — 東京：東亞書林, 昭和17 [1942]
82 p., [17] leaves of plates : ill. ; 19 cm.

1. Chu River, China. I. Title.
DS706.3.K83 79–808127

CHU SAN T'AI-TZU, 1633-1708

Tai, Hsüan-chih.
(Chu San T'ai-tzu an)
朱三太子案 / 戴玄之. — Singapore：南洋大學研究院人文與社會科學研究所, 1977.
15 leaves ; 30 cm. — (研究論文 - 南洋大學研究院人文與社會科學研究所 ; no. 59)
Reproduced from typewritten copy.
Includes bibliographical references.
1. China—History—K'ang-hsi, 1662-1722. 2. Chu San T'ai-tzu, 1633-1708. I. Title. II. Series: Nanyang University. Institute of Humanities and Social Sciences. Occasional paper series - Institute of Humanities and Social Sciences, Nanyang University ; no. 59.
DS754.2.T3 1977 78-940556

CH'U TZ'U

Fu, Hsi-jen.
(Ch'u tz'u ku yün k'ao shih)
楚辭古韻考釋 / [著述者傅錫壬]. — 臺北：淡江文理學院出版指導委員會, 民國62 [1973]
200, [4] p. ; 22 cm. — (淡江文史叢書)
Series romanized: Tan-chiang wen shih ts'ung shu.
Bibliography: p. 199-200.
1. Ch'u tz'u. 2. Chinese language—Rime. I. Title.
PL2521.C54F8 79-836470

—ILLUSTRATIONS

Hsiao, Yün-ts'ung, 1596-1673.
[Ch'in ting pu hui Li sao ch'üan t'u]
欽定補繪蕭雲從離騷全圖. — [台北：台灣]商務[印書館, 1976?]
49, 56, 90, [9] leaves : ill. ; 20 cm. — (四庫全書珍本六集 ; 241)
Cover title.
"欽定補繪離騷全圖三卷"
Photoreprint ed. of the Wen Yüan Ko copy.
1. Ch'u tz'u—Illustrations. 2. Ch'ü, Yüan, ca. 343-ca. 277 B.C. Li sao—Illustrations. I. Ch'u tz'u. Selections. 1976. II. Title. III. Series: Ssu k'u ch'üan shu. Ssu k'u ch'üan shu chen pen liu chi ; 241.
Title romanized: Ch'in ting pu hui Hsiao Yün-ts'ung Li sao ch'üan t'u.
AC149.S7 1976, vol. 241 78-843917
[PL2521.C54]

CH'ÜAN, TE-YÜ, 759-818

Chin, Shih-chün.
(Ch'üan Te-yü yen chiu)
權德輿研究 / 金時俊. — [台北?：金時俊], 民國54 [1965]
123 p. (on double leaves) ; 27 cm.
Cover title.
"國立臺灣大學中文研究所碩士論文"
Reproduced from ms. copy.
Bibliography: p. 119-123.
1. Ch'üan, Te-yü, 759-818. 2. Authors, Chinese—Biography. I. Title.
PL2677.C49Z59 78-843018

CH'UAN-CHIAO HSIEN, CHINA

—HISTORY—SOURCES

(Ch'üan-chiao hsien chih)
全椒縣志：安徽省 / 張其濬修；江克讓纂. — 臺1版. — 台北：成文出版社, 民國63 [1974]
3 v. (1176 p., [4] fold. leaves of plates) : maps ; 27 cm. — (中國方志叢書：華中地方；第225號)
"據民國九年刊本影印"
1. Ch'üan-chiao hsien, China—History—Sources. I. Chang, Ch'i-chün, fl. 1920- II. Chiang, K'o-jang. III. Series: Chung-kuo fang chih ts'ung shu Hua-chung ti fang ; ti 225 hao.
DS793.C563C5 1974 79-836116

CHUAN-SHU (CHINESE WRITING)

see Chinese language—Writing, Seal style

CH'ÜAN T'ANG SHIH

—INDEXES

Wong, Kai-chee.
A research guide to English translation of Chinese verse : Han Dynasty to T'ang Dynasty / Kai-chee Wong, Pung Ho and Shu-leung Dang. — Hong Kong : Chinese University Press, 1977.
xii, 368 p. ; 27 cm.
Title on verso of t.p.: Chung shih Ying i so yin; Han chih T'ang.
"For each entry, the name of the poet and the title of the poem are given, both as the Chinese original and the English translation. Sequence of the poets and poems follows that of the Ch'üan Han San-kuo Chin Nan-Pei-chao shih ... by Ting Fu-pao ..., 1959 edition, and the Ch'üan T'ang shih ..., 1960 edition."
Bibliography: p. ix-xi.
ISBN 9622011411 : $60.00
1. Chinese poetry—Translations into English—Indexes. 2. English poetry—Translations from Chinese—Indexes. 3. Ting, Fu-pao, 1874-1952, ed. Ch'üan Han San-kuo Chin Nan-pei-ch'ao shih—Indexes. 4. Ch'üan T'ang shih—Indexes. 5. Chinese poetry—Indexes. I. Ho, P'eng, joint author. II. Teng, Shih-liang, joint author. III. Title. IV. Title: Chung shih Ying i so yin.
PL2658.E3W576 016.8951'1'008 73-92412
79 MARC

CHUANG FAMILY

(Chuang shih ta tsu p'u)
莊氏大族譜 / 族譜主輯莊吳福；族譜主編莊玉圖. — 中壢：台灣省桃園縣莊和睦营, 民國67 [1978]
487, 32 p., [32] leaves : ill. ; 27 cm.
Colophon title.
1. Chuang family. I. Chuang, Wu-fu. II. Chuang, Wu-t'u.
CS1169.C494 1978 79-840691
C

CHUANG-TZU. NAN HUA CHING

Chang, Pin-ts'un.
(I chuan yü Chuang-tzu ti hsien shih she hui kuan yü li hsiang shih chieh kuan)
易傳與莊子的現實社會觀與理想世界觀 / 張彬村. — [臺北：s.n.], 民國65 [1976]
200, [1] p. (on double leaves) ; 27 cm.
"台灣大學中文研究所碩士論文"
Bibliography: p. 195-[201]
1. I ching. 2. Chuang-tzu. Nan-hua ching. 3. Philosophy, Chinese. I. Title.
PL2464.Z7C39 79-835869
C

Ch'en, Ku-ying.
(Chuang-tzu chin chu chin i)
莊子今註今譯 / 陳鼓應註譯. — 臺北：臺灣商務印書館, 民國64 [1975], 民國66 [1977] printing.
2 v. (4, 2, 2, 995 p.) ; 22 cm.
"古籍今註今譯"
NT$174.00
1. Chuang-tzu. Nan hua ching. I. Chuang-tzu. Nan hua ching. 1977. II. Title.
BL1900.C576C5333 78-841204

Fukunaga, Mitsuji, 1918-
(Sōshi)
莊子 / 福永光司；監修 吉川幸次郎. — 東京：朝日新聞社, 昭和31 [1956]
22, 361, 42 p., [1] leaf of plates : map ; 19 cm. — (中国古典選)
Series romanized: Chūgoku koten sen.
Includes index.
1. Chuang-tzu. Nan-hua ching. I. Chuang-tzu. Nan-hua ching. 1956. II. Title.
BL1900.C576F85 1956 79-806096
J

Fukunaga, Mitsuji, 1918-
(Sōshi)
莊子 / 福永光司. — 東京：朝日新聞社, 1978-
v. ; 15 cm. — (中国古典選 ; 12-) Ja 78-18615 (v. 1)
"本文...統古逸叢書本"
CONTENTS: [1] 內篇—[2]-[4] 外篇
¥440 (v. 1)
1. Chuang-tzu. Nan-hua ching. I. Chuang-tzu. Nan-hua ching. Japanese & Chinese. 1978. II. Title. III. Series: Chūgoku koten sen ; 12 [etc.]
BL1900.C576F85 1978 79-800374
J

Lin, Shu, 1852-1924.
(Chuang-tzu ch'ien shuo)
莊子淺說 / 林紓撰. — 台北：廣文書局, 民國67 [1978]
124 leaves in various foliations ; 22 cm. — (漢學彙編)
NT$230.00
1. Chuang-tzu. Nan-hua-ching. I. Chuang-tzu. Nan-hua ching. 1978. II. Title. III. Series: Han hsüeh hui pien.
BL1900.C576L583 1978 79-841887

Lin, Yün-ming, chin shih 1658.
[Chuang-tzu yin. Japanese & Chinese]
莊子因：標註 / 林雲銘評述；秦鼎補義；東條保標註. — 臺北：蘭臺書局, 中華民國64 [1975]
654 p. ; 27 cm. Ja***
Cover title: Hyōchū hogi Sōji in.
Photoreprint of the 1890 ed.
NT$250.00
1. Chuang-tzu. Nan-hua ching. I. Hata, Kanae, 1761-1831. II. Tōjō, Tamotsu. III. Chuang-tzu. Nan-hua ching. 1975. IV. Title.
Title romanized: Sōji in.
BL1900.C576L58416 1975 78-816404
J

Liu, Shih-p'ei, 1884-1919.
(Chuang-tzu chiao pu)
莊子斠補 / [劉師培]. — [臺北：藝文印書館, 1960?]
22 p. ; 19 cm.
Reprint of the ed. printed by Ning-wu Nan shih, with pref. dated 1911.
1. Chuang-tzu. Nan-hua ching. I. Title.
BL1900.C576L6 1960 79-838005
C

Murobuse, Kōshin, 1889-1970.
(Sōshi)
莊子 / 室伏高信著. — 東京：大東出版社, 昭和14 [1939]
409 p. ; 19 cm.
"莊子內篇本文": p. 367-409.
1. Chuang-tzu. Nan-hua ching. I. Chuang-tzu. Nan-hua ching. Japanese & Chinese. Selections. 1939. II. Title.
BL1900.C576M87 79-805933

CHUANG-TZU. NAN-HUA CHING (Continued)

T'ao, Hung-ch'ing, 1859–1918.
(Lao Chuang cha chi)
老莊札記 / 陶鴻慶撰. — 臺北：廣文書局, 民國64 ⌊1975⌋
2, 4, 72 p. ; 22 cm. — (中國哲學思想要籍叢編)
Reprinted from his work under collective title: Tu chu tzu cha chi, originally published by Tai hsiao lu, with preface dated 1919.
NT$120.00
1. Lao-tzu. Tao te ching. 2. Chuang-tzu. Nan-hua ching. I. Lao-tzu. Tao te ching. Selections. 1975. II. Chuang-tzu. Nan hua ching. Selections. 1975. III. Title. IV. Title: Tu chu tzu cha chi. V. Series: Chung-kuo che hsüeh ssu hsiang yao chi ts'ung pien.
BL1900.L35T328 1975 78–842014

Wang, Shu-min, 1914–
(Han Fei-tzu yü Chuang-tzu)
韓非子與莊子 / 王叔岷. — Singapore：南洋大學研究院人文與社會科學研究所, 1977.
15, 4 leaves ; 29 cm. — (研究論文 - 南洋大學研究院人文與社會科學研究所 ; no. 75)
Reproduced from typescript.
1. Han, Fei, d. 233 B. C. Han Fei-tzu. 2. Chuang-tzu. Nan-hua ching. I. Title. II. Series: Nanyang University. Institute of Humanities and Social Sciences. Occasional paper series – Institute of Humanities and Social Sciences, Nanyang University ; no. 75.
B128.H34W29 78–940246
79 C

CHUANG-TZU

Kuo, Wei, fl. 1977–
(Lao Chuang che hsüeh yü tao hsüeh)
老莊哲學與道學 / 郭為著. — 高雄：復文書局；嘉義：發行者興國出版社, 民國66 ⌊1977⌋
5, 183 p. ; 22 cm.
• NT$50.00
1. Tao. 2. Lao-tzu. 3. Chuang-tzu. 4. Philosophy, Chinese. I. Title.
B127.T3K86 79–838200 •
C

Rajaneesh, Acharya, 1931-
When the shoe fits : talks on the stories of Chuang Tzu / Rajneesh ; compilation, Swami Amrit Pathik ; editing, Ma Prem Veena. — Poona : Rajneesh Foundation, c1976.
370 p. : ill. ; 22 cm.
Rs75.00
1. Chuang-tzŭ. 2. Taoist meditations. I. Title.
BL1900.C576R34 181'.09'514 76-904914
78 MARC

—ADDRESSES, ESSAYS, LECTURES

Cappelletti, Angel J
Lao-tse y Chuang-tse / Angel J. Cappelletti. — ⌊Caracas⌋ : Gobernación del Distrito Federal, ⌊1976⌋
112 p. ; 15 x 16 cm. — (Cuadernos de difusión - Dirección de Cultura de la Gobernación del Distrito Federal ; 8 : Ensayo)
"El presente ensayo es el resultado de la fusión de otros dos trabajos publicados hace ya algunos años en la revista Universidad (Santa Fe, Argentina), incluídos más tarde en mi libro, La filosofía china del período clásico (Puebla-México, 1966)."
Includes bibliographical references.
1. Lao-tzu—Addresses, essays, lectures. 2. Chuang-tzu—Addresses, essays, lectures. I. Title. II. Series: Venezuela (Federal District). Dirección de Cultura. Cuadernos de difusión - Dirección de Cultura de la Gobernación del Distrito Federal ; 8.
BL1900.L35C36 79-347794
79 MARC

Rajaneesh, Acharya, 1931-
The empty boat : talks on the stories of Chuang Tzu / Rajneesh ; compilation, Ma Krishna Priya ; editing, Ma Prem Paras. — Poona : Rajneesh Foundation, c1976.
451 p. (p. 448-451 advertisements) : ill. ; 22 cm.
1. Chuang-tzu—Addresses, essays, lectures. I. Krishna Priya, Ma. II. Prem Paras, Ma. III. Title.
BL1900.C576R33 181'.09'514 78-900916
79 MARC

—BIOGRAPHY—EARLY LIFE

Ch'en, Kuan-hsüeh, 1934–
(Chuang-tzu Sung jen k'ao)
莊子宋人考 / 陳冠學著. — 高雄：三信出版社, 民國66 ⌊1977⌋
5, 124 p. : 20 cm.
NT$30.00
1. Chuang-tzu—Biography—Early life. 2. Chuang-tzu—Name. I. Title.
BL1940.C5C54 79–836140

—NAME

Ch'en, Kuan-hsüeh, 1934–
(Chuang-tzu Sung jen k'ao)
莊子宋人考 / 陳冠學著. — 高雄：三信出版社, 民國66 ⌊1977⌋
5, 124 p. : 20 cm.
NT$30.00
1. Chuang-tzu—Biography—Early life. 2. Chuang-tzu—Name. I. Title.
BL1940.C5C54 79–836140

CHUBA see Ayo (Game)

CHŪBU REGION, JAPAN

—DESCRIPTION AND TRAVEL

(Chūgoku shizen hodō sen-kyūhyakukiro)
中国自然歩道1900キロ：付 広島市周辺の山歩き. — 広島：中国新聞社, 1978.
139 p. : ill. ; 17 x 19 cm. Ja 78–24830
¥800
1. Chūgoku region—Description and travel. 2. Nature trails—Japan—Chūgoku region. I. Chūgoku Shimbun Sha.
DS894.72.C48 79–805287

Tsugita, Kasumi, 1913–
(Utatane)
うたたね / 全訳注 次田香澄. — 東京：講談社, 1978.
156 p. : 15 cm. — (講談社学術文庫) Ja 79–911
"底本は「うたたね」の写本のうち…善本と判断される御物「うたゝね」東山御文庫本"
Series romanized: Kōdansha gakujutsu bunko.
Bibliography: p. 153–154.
¥240
1. Abutsu Ni, ca. 1209-1283. Utatane no ki. 2. Abutsu Ni, ca. 1209-1283—Diaries. 3. Chubu region, Japan—Description and travel. 4. Authors, Japanese—Biography. I. Abutsu Ni, ca. 1209-1283. Utatane no ki. 1978. II. Title.
PL792.A2U837 79–805445

Library of Congress *79 J

—ECONOMIC CONDITIONS

Masaki, Keiji, 1905–
(Tōkai to Ina)
東海と伊那：商品流通と政治・文化交流 / 正木敬二著. — 名古屋：正木, 1978.
xiii, 486 p., ⌊2⌋ leaves of plates : ill. : 22 cm. Ja 79–1942
Includes bibliographies.
Errata slip inserted.
1. Chūbu region, Japan—Economic conditions. 2. Ina region, Japan—Economic conditions. 3. Civil rights—Japan—Ina region—History. 4. Chūbu region, Japan—Politics and government. I. Title.
HC463.C47M37 79–805497

—GOVERNMENT PUBLICATIONS—BIBLIOGRAPHY

(Chūbu hachikenka ni okeru chōsa kankōbutsu shiryō ichiran)
中部8県下における調査・刊行物資料一覧 / ⌊名古屋商工会議所編⌋. — 名古屋：名古屋商工会議所, 昭和49 ⌊1974⌋
104 p. ; 26 cm. Ja 75–14945
Cover title.
1. Social sciences—Japan—Chūbu region—Bibliography. 2. Chūbu region, Japan—Government publications—Bibliography. I. Nagoya Shōkō Kaigisho.
Z7165.J3C48 78–813150
[H53]

—INDUSTRIES

Japan. Sangyō Kōzō Shingikai. Tōkai Hokuriku Chiiki Sangyō Kōzō Bunkakai.
(Tōkai Hokuriku chiiki no sangyō kōzō bijon)
東海北陸地域の産業構造ビジョン：産業構造審議会東海北陸地域産業構造分科会報告 / ⌊編者名古屋通商産業局⌋. — 東京：通商産業調査会, 1978.
344 p. : 21 cm. Ja 78–17897
Errata slip inserted.
1. Chubu region, Japan—Industries. 2. Economic forecasting—Japan—Chubu region. I. Japan. Nagoya Tsūshō Sangyōkyoku. II. Title.
HC463.C47J37 1978 79–805603

—POLITICS AND GOVERNMENT

Masaki, Keiji, 1905–
(Tōkai to Ina)
東海と伊那：商品流通と政治・文化交流 / 正木敬二著. — 名古屋：正木, 1978.
xiii, 486 p., ⌊2⌋ leaves of plates : ill. : 22 cm. Ja 79–1942
Includes bibliographies.
Errata slip inserted.
1. Chūbu region, Japan—Economic conditions. 2. Ina region, Japan—Economic conditions. 3. Civil rights—Japan—Ina region—History. 4. Chūbu region, Japan—Politics and government. I. Title.
HC463.C47M37 79–805497

CHUBUT, ARGENTINE REPUBLIC

—DESCRIPTION AND TRAVEL—CONGRESSES

Gaea, Sociedad Argentina de Estudios Geográficos.
Geografía del Chubut : contribuciones presentadas en la XXXVI semana de geografía / por H. C. Alvarez ... ⌊et al.⌋. — Buenos Aires : Gaea, Sociedad Argentina de Estudios Geográficos, 1978.
209 p. : ill. ; 23 cm. — (Serie especial - Gaea, Sociedad Argentina de Estudios Geográficos ; no. 6)
Includes bibliographies.
1. Chubut, Argentine Republic—Description and travel—Congresses. I. Alvarez, Héctor C. II. Title. III. Series: Gaea, Sociedad Argentina de Estudios Geográficos. Serie especial - Gaea, Sociedad Argentina de Estudios Geográficos ; no. 6.
F2881.G33 1978 918.2'7 79-109242
79 MARC

CHUCUITO, PERU (PROVINCE)

—RURAL CONDITIONS

Palacios Ríos, Félix.
"... Hiwasaha uywa uywataña, uka uywaha hiwasaru uyusitu" : los pastores aymara de Chichillapi / Félix Palacios Ríos. — Lima : Pontificia Universidad Católica del Perú, Programa de Perfeccionamiento en Ciencias Sociales, 1977.
113 p. : ill. ; 30 cm.
Cover title.
Originally presented as the author's thesis (magister en ciencias sociales, Pontificia Universidad Católica del Perú.
Bibliography: p. ⌊111⌋-113.
1. Chucuito, Peru (Province)—Rural conditions. I. Title. II. Title: Los pastores aymara de Chichillapi.
HN350.C45P34 1977 78-102648
79 MARC

CHUDOMIR

—INTERVIEWS

Matev, Georgi.
(Sreshti i razgovori s Chudomir)
Срещи и разговори с Чудомир / Георги Матев. — 1. изд. — Пловдив : Хр. Г. Данов, 1978.
92 p. : ill. ; 21 cm. Bu 78-1008
0.49 lv
1. Chudomir—Interviews. 2. Authors, Bulgarian—20th century—Interviews. I. Chudomir. II. Title.
PG1037.C53Z78 78-422432

CHŪETSU SHOKUSAN KABUSHIKI KAISHA

—HISTORY—SOURCES

(Jinushi dantai shiryō)
地主団体資料：中越殖産株式会社等 / ［編集代表佐藤誠朗，古厩忠夫］. — 見附：見附市史編集委員会，1978.
10, 163 p. ; 21 cm. — (見附市史編集資料；第27集：近・現代編 政治；5) Ja 78-24826
Cover title.
Not in trade.
1. Chūetsu Shokusan Kabushiki Kaisha—History—Sources. 2. Rannan Jinushi Kyōkai—History—Sources. I. Satō, Shigerō. II. Furumaya, Tadao, 1941- III. Series: Mitsuke-shi shi henshū shiryō ; dai 27-shū.
D5897.M5213M57 vol. 27 79-803538
[HD920.M57]
J

CHUFUT-KALE, UKRAINE, IN ART

Shillingovskiĭ, Pavel Aleksandrovich, 1881-1942.
(Chufut-kale)
Чуфут-кале / Павел Шиллинговский. — Ленинград : Художник РСФСР, 1976.
1 portfolio (10 p., [6] leaves of plates : ill.) ; 21 cm. USSR***
"200 нумерованных экземпляров."
9.52rub
1. Shillingovskiĭ, Pavel Aleksandrovich, 1881-1942. 2. Chufut-Kale, Ukraine, in art. I. Title.
NE2056.5.S47A4 1976 79-414061

CHŪGAN SHISŌ see Mādhyamika (Buddhism)

CHŪGOKU KŌGYŌ GIJUTSU SHIKENJO. TOSHOSHITSU

—CATALOGS

Chūgoku Kōgyō Gijutsu Shikenjo. Toshoshitsu.
(Zasshi mokuroku, sen-kyūhyaku-shichijūshichi)
雑誌目録1977 / 中国工業技術試験所図書室；［編集 中国工業技術試験所企画課］. — 呉：同課，昭和52［1977］
25 p. ; 26 cm. Ja***
Cover title.
1. Technology—Periodicals—Bibliography—Catalogs. 2. Sciences—Periodicals—Bibliography—Catalogs. 3. Chūgoku Kōgyō Gijutsu Shikenjo. Toshoshitsu—Catalogs. I. Chūgoku Kōgyō Gijutsu Shikenjo. Kikakuka. II. Title.
Z7916.C48 1977 79-801702
[T45]

CHUGOKU KYOSANTO see Chung-kuo kung ch'an tang

CHŪGOKU REGION

—POLITICS AND GOVERNMENT

(Chūgoku Chihō Sōkanfu shi)
中国地方総監府誌 原爆被災記録 中国地方総監府誌刊行会編 広島 中国地方総監府原爆被災記録刊行会 昭和47 (1972)
162 p. illus., port. 22 cm. Ja 73-9843
1. Japan. Chūgoku Chihō Sōkanfu. 2. Chūgoku region—Politics and government. 3. Hiroshima—Bombardment, 1945.
JQ1699.C524C48 79-808391

—RURAL CONDITIONS

Shichidaigaku Sankanbu Kenkyūkai.
(Chūgoku chihō sankanbu no miryoku)
中国地方山間部の魅力：その現状と課題 / 七大学山間部研究会編. — 東京：東洋図書出版，1978.
271 p. : ill. ; 22 cm. Ja 78-22136
¥2500
1. Chūgoku region—Rural conditions. 2. Regional planning—Japan—Chūgoku region. I. Title.
HN730.C59S54 1978 78-816308
J

CHUIANOV, A.

Chuianov, A
(Na stremnine veka)
На стремнине века : записки секретаря обкома / А. Чуянов ; Литературная обработка Ивана Падерина. — Москва : Политиздат, 1976.
286 p., [8] leaves of plates : ill. ; 21 cm. USSR***
0.90rub
1. Stalingrad, Battle of, 1942-1943—Personal narratives, Russian. 2. Kommunisticheskaia partiia Sovetskogo Soiuza. Volgogradskiĭ oblastnoĭ komitet—Biography. 3. Chuianov, A. 4. Communists—Russia—Biography. I. Paderin, Ivan Grigor'evich, 1918- II. Title.
D764.3.S7C48 1976 76-531913

Chuianov, A
(Na stremnine veka)
На стремнине века : Записки секретаря обкома / А. Чуянов ; Лит. обраб. И. Падерина. — Москва : Политиздат, 1977, c1976.
288 p., [8] leaves of plates : ill. ; 21 cm. USSR 77
0.80rub
1. Stalingrad, Battle of, 1942-1943—Personal narratives, Russian. 2. Kommunicheskaia partiia Sovetskogo Soiuza. Volgogradskiĭ oblastnoĭ komitet—Party work. 3. World War, 1939-1945—Biography. 4. Chuianov, A. 5. Communists—Russia—Biography. I. Paderin, Ivan Grigor'evich, 1918- II. Title.
D764.3.S7C48 1977 78-423073

CHUĬKOV, VASILIĬ IVANOVICH, 1900-

Chuĭkov, Vasiliĭ Ivanovich, 1900-
The end of the Third Reich / V. I. Chuikov ; [translated from the Russian]. — Moscow : Progress, 1978.
263 p. : ill. ; 21 cm. USSR***
Translation of Konets tret'ego reĭkha.
1. World War, 1939-1945—Campaigns—Eastern. 2. Chuĭkov, Vasiliĭ Ivanovich, 1900- . 3. Marshals—Russia—Biography. 4. Russia (1923-U.S.S.R.). Armiia—Biography. 5. World War, 1939-1945—Personal narratives, Russian. I. Title.
D764.C483713 1978 940.54'21'0924 78-396201
79 MARC

CHUKCHI LANGUAGE

—TEXT-BOOKS FOR FOREIGNERS —RUSSIAN

Keul'kut, Viktor Grigor'evich, 1929-1963.
(Russko-chukotskiĭ razgovornik)
Русско-чукотский разговорник. [Магадан] Магаданское книжное изд-во, 1958.
53 p. 10 x 14 cm.
1. Chukchi language—Text-books for foreigners—Russian. I. Title.
PM42.K4 60-36599

CHUKHNOVSKIĬ, BORIS GRIGOR'EVICH

Morozov, Savva Timofeevich.
Oni prinesli kryl'ia v Arktiku : [Dokum. povest' o B. G. Chukhnovskom i A. D. Alekseeve] / Savva Morozov. — Moskva : Mysl', 1979.
173 p., [4] leaves of plates : ill. ; 20 cm. USSR79
0.70rub
1. Chukhnovskiĭ, Boris Grigor'evich. 2. Alekseev, Anatoliĭ Dmitrievich. 3. Aeronautics—Russia—History. 4. Arctic regions—Aerial exploration. 5. Air pilots—Russia—Biography. I. Title.
TL540.C565M67 79-377667
79 MARC

CHUKOVSKAIA, LIDIIA KORNEEVNA

—BIOGRAPHY

Chukovskaia, Lidiia Korneevna.
Protsess iskliucheniia : ocherk literaturnykh nravov / Lidiia Chukovskaia. — [Paris] : YMCA-Press, 1979.
207 p. ; 19 cm. F***
Includes bibliographical references.
1. Chukovskaia, Lidiia Korneevna—Biography. 2. Authors, Russian—20th century—Biography. 3. Soiuz pisateleĭ SSSR. I. Title.
PG3476.C485Z474 79-383784
79 MARC

CHUKOVSKIĬ, KORNEĬ IVANOVICH, 1882-1969

—CRITICISM AND INTERPRETATION —ADDRESSES, ESSAYS, LECTURES

Chukovskiĭ, Korneĭ Ivanovich, 1882-1969.
Les futuristes / [par] Kornei Tchoukovski ; trad. et postface de Gérard Conio. — Lausanne : Éditions L'Age d'homme, [1976]
154 p. ; 21 cm. — (Slavica) Sw77-3748
Translation of Futuristy.
22.00F
1. Russian poetry—20th century—History and criticism. 2. Futurism. 3. Chukovskiĭ, Korneĭ Ivanovich, 1882-1969—Criticism and interpretation—Addresses, essays, lectures. I. Title.
PG3065.F8C4714 891.7'1'409 78-368160
*78 MARC

CHULIKATA-MISHMI LANGUAGE see Idu language

CH'UN CH'IU PERIOD, 722-481 B.C. see China--History--Spring and Autumn period, 722-481 B.C.

CH'UN-HUA HSIEN, CHINA

—HISTORY—SOURCES

(Ch'un-hua hsien chih)
淳化縣志：陝西省 / 萬延樹修；洪亮吉纂. — 臺1版. — 台北：成文出版社, 民國65 [1976]
2 v. (556 p.) ; 27 cm. — (中國方志叢書華北地方；第532號)
Reprint of the 1934 ed.

1. Ch'un-hua hsien, China—History—Sources. I. Wan, T'ing-shu, 18th cent. II. Hung, Liang-chi, 1746–1809. III. Series: Chung-kuo fang chih ts'ung shu Hua-pei ti fang ; ti 532 hao.
DS793.C5653C48 1976 78–841474

CHÜN WARE

—CHINA—LIN-JU HSIEN

Lovell, Hin-cheung.
Report on the investigation of kiln-sites of Ju-type ware and Chün ware in Lin-ju hsien, Honan / by Fêng Hsien-ming ; [translator, Hin-cheung Lovell]. — [London] : Victoria & Albert Museum, c1968.
18 p., [3] leaves of plates : ill. ; 29 cm. — (Chinese translations ; no. 3)
An abstract translated from a report originally published in Wen wu, no. 8, 1964 under title: Ho-nan sheng Lin-ju hsien Sung tai Ju yao i chih tiao ch'a.
1. Lin-ju hsien, China—Antiquities. 2. China—Antiquities. 3. Celadon ware—China—Lin-ju hsien. 4. Chün ware—China—Lin-ju hsien. I. Feng, Hsien-ming. Ho-nan sheng Lin-ju hsien Sung tai Ju yao i chih tiao ch'a. II. Title. III. Series.
DS793.L464L68 931 78-326476
79 MARC

CHUNG, K'UEI

—LEGENDS

[Ch'iao-yün-shan-jen]
Dschung Kuei, Bezwinger der Teufel : altchines. Volksbuch : d. 9. Meisterwerk / [dt. von Clemens i.e. Claude du Bois-Reymond]. — 2. Aufl. — Weimar : Kiepenheuer, 1977.
252 p. : ill. ; 19 cm. GDR77-A
Translation of Chung K'uei cho kuei chuan.
"Dem durchgesehenen und stellenweise überarbeiteten Text liegt die deutsche Erstausgabe, Gustav Kiepenheuer Verlag, Potsdam 1923, zugrunde."
9.00M
1. Chung, K'uei—Legends. I. Du Bois-Reymond, Claude. II. Title.
PL2698.C534C515 1977 895.1'3'4 79-366291
79 MARC

Fu-chien sheng chün ch'ü cheng chih pu, Lung-yen ti ch'ü ko ming wei yüan hui Ku t'ien sung pien hsieh tsu.
(Ku t'ien sung)
古田颂 / 福建省军区政治部，龙岩地区革命委员会《古田颂》编写组. — [香港? : s.n., 1978?]
4, 256 p. ; 21 cm.
Reprint of the 1977 ed. published by Fu-chien jen min ch'u pan she.
$90.00

1. Chinese poetry—20th century. 2. Chung-kuo jen min chieh fan chün—Poetry. I. Title.
PL2543.F8 1978 79–835290

CHUNG, SHIH-P'AN

Chung, Shih-p'an.
(I meng ch'i shih nien)
一夢七十年 / 鍾石磐著. — [s. l. : s. n.] ; [台南縣] 新營：贈送者新營放生服務處, 民國64 [1975]
118 p. : ports. ; 20 cm.
Cover title.
"非賣品"
"附錄": p. 75–118.
1. Chung, Shih-p'an. 2. Taiwan—Biography. I. Title.
CT1828.C63A34 79–837797

CH'UNG CHEN HSUEH YUAN

Shimizu, Yasuzō, 1891-
(Nozomi o ushinawazu)
希望を失わず：續朝陽門外 / 清水安三手記. — 改訂第五版. — 東京：櫻美林出版部, 昭和26 [1951]
253 p. : ill. ; 19 cm.

1. Shimizu, Yasuzo, 1891- 2. Christian biography—Japan. 3. Educators—Japan—Biography. 4. Ch'ung chen hsüeh yüan. 5. Ōbirin Gakuen. I. Shimizu, Yasuzo, 1891- Choyomongai. II. Title.
BR1317.S5A36 79–806191

CHUNG-HO HSIANG, TAIWAN

Chung-ho hsiang, Taiwan.
(Ch'ung hsiu Chung-ho hsiang chih)
重修中和鄉志 / 中和鄉公所編印；主修林德喜；編纂盛清沂，吳基瑞；協纂陳士賓. — 中和鄉[臺北縣]：該公所, 民國66 [1977]
8, 564 p., [4] leaves of plates : ill. ; 27 cm.

1. Chung-ho hsiang, Taiwan. I. Lin, Te-hsi. II. Sheng, Ch'ing-i. III. Wu, Chi-jui. IV. Ch'en, Shih-shih. V. Title. VI. Title: Chung-ho hsiang chih.
DS799.9.C54C49 1977 79–840768

CHUNG HSIN CHEN SO

—ADDRESSES, ESSAYS, LECTURES

Ch'i, Cheng.
(Chieh k'ai kuei tsu i yüan chung hsin chen so ti chen mien mu)
揭開貴族醫院中心診所的真面目：與論會診側記 / 齊正，洛陽編撰. — 台北：齊：四季出版公司總經銷, 民國67 [1978]
265 p. ; 22 cm.
Cover title.
NT$50.00
1. Chung hsin chen so—Addresses, essays, lectures. I. Lo, Yang, joint author. II. Title.
RA990.T284T333 79–838079

CHUNG-HUA MIN KUO CHÜN JEN CHIH YU SHE

—BIOGRAPHY

(Huang ku li shih chang Ch'i-jui ai jung lu)
黃故理事長啓瑞哀榮錄 / 中華民國軍人之友社編印. — [台北]：該社, 民國65 [1976]
2, 177 p. : ill. ; 27 cm.

1. Huang, Ch'i-jui, 1910–1976—Death and burial. 2. Politicians—Taiwan—Biography. 3. Chung-hua min kuo chün jen chih yu she—Biography. I. Chung-hua min kuo chün jen chih yu she.
DS799.82.H85H8 79–836874

CHUNG-HUA MIN KUO HSIN WEN P'ING I WEI YÜAN HUI

Li, Wen-ch'ing.
(Hsin wen p'ing i shih erh nien)
新聞評議十二年：中華民國新聞評議委員會的組織與功能 / 編撰者李文慶. — 臺北：該會, 民國64 [1975]
4, 206 p. ; 19 cm.
Colophon title.
Bibliography: p. 206.
"非賣品"
1. Chung-hua min kuo hsin wen p'ing i wei yüan hui. I. Title.
PN5370.1.C463L5 79–835957

CHUNG-HUA NUNG HSÜEH HUI

—HISTORY

Li, Ch'ung-tao.
(Liu shih nien lai chih Chung-hua nung hsüeh hui)
六十年來之中華農學會 / 李崇道. — [臺北]：中華農學會, 民國65 [1976]
65 p., [1] leaf of plates : ill. ; 22 cm.
Cover title.

1. Chung-hua nung hsüeh hui—History. I. Title.
S322.T28L5 78–842042

CHUNG HUA TSING NIËN HUI

10 [i.e. Sepuluh tahun], 9 Djuni 1946-9 Djuni 1956 Chung Hua Tsing Niën Hui. — Djakarta : Hui, [1956?]
122 p. : ill. ; 23 cm.
1. Chung Hua Tsing Niën Hui. 2. Youth—Indonesia—Jakarta—Societies and clubs. I. Chung Hua Tsing Niën Hui.
HQ799.I612J337 78-950435
79 MARC

CHUNG KUO. [MOTION PICTURE]

A Vicious motive, despicable tricks : a criticism of M. Antonioni's anti-China film China / Renmin Ribao commentator. — Peking : Foreign Languages Press, 1974.
18 p. ; 19 cm.
Title in Chinese in colophon: O tu ti yung hsin, pei lieh ti shou fa.
1. Chung Kuo. [Motion picture] I. Jen min jih pao, Peking. II. Title: O tu ti yung hsin, pei lieh ti shou fa.
PN1997.C46663V5 791.43'7 79-310413
79 MARC

CHUNG-KUO I YAO HSUEH YUAN

—ABSTRACTS

Chung-kuo i yao hsüeh yüan.
(Chung-kuo i yao hsüeh yüan chiao shih chi hsiao yu lun chu che yao)
中國醫藥學院教師暨校友論著摘要：民國47年—民國67年. — 台中：中國醫藥學院圖書館出版組, [1978]
129 p. ; 21 cm.
Cover title.
English and/or Chinese.

1. Medicine—Abstracts. 2. Medicine, Chinese—Abstracts. 3. Chung-kuo i yao hsüeh yüan—Abstracts. I. Title.
R111.C54 79–839936

CHUNG-KUO I YAO HSÜEH YUAN. CHUNG-KUO I HSÜEH YEN CHIU SO

Ch'en, T'ai-hsi.
(Chung-kuo i yao hsüeh yüan Chung-kuo i hsüeh yen chiu so kai k'uang)
中國醫藥學院中國醫學研究所概況 = Graduate Institute of Chinese Medical Sciences, China Medical College : a brief introduction / 陳太羲. — [台中]：該院, [1978?]
13 p. ; 27 cm.
English and Chinese.
"私立中國醫藥學院研究年報 第七期 抽印本 中華民國65年6月發行"

1. Chung-kuo i yao hsüeh yüan. Chung-kuo i hsüeh yen chiu so. I. Title. II. Title: Graduate Institute of Chinese Medical Sciences, China Medical college.
R821.T28C45 78–841582

CHUNG-KUO JEN MIN CHIEH FANG CHÜN

Furukawa, Mantarō, 1933–
(Mō Takutō no guntai)
毛沢東の軍隊：中国人民解放軍の50年 / 古川万太郎著. — 東京：講談社, 1978.
355 p. : ill. ; 20 cm. Ja 78-26786
Includes bibliographical references.
¥1200

1. China (People's Republic of China, 1949–) Chung-kuo jen min chieh fang chün. 2. Chung kuo jen min chieh fang chün. I. Title.
UA837.F85 79-803396

—ADDRESSES, ESSAYS, LECTURES

(Kuang jung ti chan ti)
光荣的战地 / ［题图陈亦逊, 肖映川］. — ［香港?：s. n., 1978?］
158 p. : ill. ; 21 cm.
Reprint of the 1977 ed. published by Chieh fang chün wen i she, Peking.
$60.00

1. China—Politics and government—1912-1949—Addresses, essays, lectures. 2. Chung-kuo jen min chieh fang chün—Addresses, essays, lectures.
DS775.K767 1978 78-842566

—BIOGRAPHY—ADDRESSES, ESSAYS, LECTURES

(Wei ta ti li shih ch'uang chü)
伟大的历史创举：纪念中国人民解放军建军五十周年文选. — ［香港?：s. n., 1978?］
2, 183 p. ; 21 cm.
Reprint of the 1st (1977) ed. published by Shang-hai jen min ch'u pan she, Shanghai.

1. Chung-kuo jen min chieh fang chün—History—Addresses, essays, lectures. 2. Chung-kuo jen min chieh fang chün—Biography—Addresses, essays, lectures. I. China (People's Republic of China, 1949–). Chung-kuo jen min chieh fang chün.
UA837.W44 1978 79-839512

—HISTORY—ADDRESSES, ESSAYS, LECTURES

(Chi nien Chung-kuo jen min chieh fang chün chien chün wu shih chou nien)
纪念中国人民解放军建军五十周年. — ［香港?：s. n., 1978?］
2, 209 p. ; 21 cm.
Reprint of the 1977 ed. published by Jen min ch'u pan she, Peking.

1. Chung-kuo jen min chieh fang chün—History—Addresses, essays, lectures.
UA837.C447 1978 79-837647

(Hsing huo liao yüan)
星火燎原：选编之一. — 第1版. — 北京：中国人民解放军战士出版社, 1977.
233 p. ; 21 cm.
CONTENTS: 朱德 人民军队人民战争.—聂荣臻 南昌起义的历史意义和经验教训.—罗荣桓 秋收起义与我军初创时期.—陈毅 学习毛主席的马克思列宁主义的创造作风.—叶剑英 大革命失败与广州起义.—徐向前 奔向海陆丰.—贺龙 湘鄂西初期的革命斗争.—徐向前 鄂豫皖红军的反围攻斗争.—刘伯承 回顾长征.—聂荣臻 结束第二次国内革命战争的最后一仗.—刘伯承 我们在太行山上.—陈毅 江南抗战之春.—聂荣臻 中国人民怎样战胜了日本法西斯侵略者.—刘伯承 千里跃进大别山.—叶剑英 伟大的战略决战.
$0.57

1. China—History, Military—Addresses, essays, lectures. 2. Chung-kuo jen min chieh fang chün—History—Addresses, essays, lectures.
DS738.H724 78-842667

(Wei ta ti li shih ch'uang chü)
伟大的历史创举：纪念中国人民解放军建军五十周年文选. — ［香港?：s. n., 1978?］
2, 183 p. ; 21 cm.
Reprint of the 1st (1977) ed. published by Shang-hai jen min ch'u pan she, Shanghai.

1. Chung-kuo jen min chieh fang chün—History—Addresses, essays, lectures. 2. Chung-kuo jen min chieh fang chün—Biography—Addresses, essays, lectures. I. China (People's Republic of China, 1949–). Chung-kuo jen min chieh fang chün.
UA837.W44 1978 79-839512

—HISTORY—JUVENILE LITERATURE

(Ho shao nien p'eng yu chiang chieh fang chün ti ch'uan t'ung)
和少年朋友讲解放军的传统. — 北京第1版. — 北京：中國少年兒童出版社, 1977.
2, 2, 128 p. : ill. ; 19 cm.
$0.26

1. Chung-kuo jen min chieh fan chün—History—Juvenile literature. I. Title: Chieh fang chün ti ch'uan t'ung.
UA837.H6 78-841599

CHUNG-KUO JEN MIN CHIH YÜAN CHÜN

—BIOGRAPHY —JUVENILE LITERATURE

Hsiao-yu.
(Lo Sheng-chiao)
罗盛教 / 小友编；沈蓉男, 蒋昌一画. — 第1版. — 上海：上海人民出版社, 1977.
［28］ p. : col. ill. ; 16 cm. — (青年英雄的故事)
"連環圖畫"
Series romanized: Ch'ing nien ying hsiung ti ku shih.
$0.12

1. Lo, Sheng-chiao, 1931-1952—Juvenile literature. 2. Chung-kuo jen min chih yüan chün—Biography—Juvenile literature. 3. Korean War, 1950-1953—Biography—Juvenile literature. I. Shen, Jung-nan. II. Chiang, Ch'ang-i.
DS778.L68H74 78-841358

CHUNG-KUO KUANG PO KUNG SSU

—ADDRESSES, ESSAYS, LECTURES

(Chung kuang wu shih nien chi nien chi)
中廣五十年紀念集 / ［編輯者中國廣播公司研究發展考訓委員會；執行編纂吳疏潭］. — 台北：空中雜誌社, 民國67 ［1978］
6, 390 p. : ill. ; 22 cm. — (中廣叢書)
"1928/1978. 中華民國17年/67年"

1. Radio broadcasting—Taiwan—Addresses, essays, lectures. 2. Radio broadcasting—China—Addresses, essays, lectures. 3. Chung-kuo kuang po kung ssu—Addresses, essays, lectures. I. Wu, Shu-t'an. II. Chung-kuo kuang po kung ssu. Yen chiu fa chan k'ao hsün wei yüan hui. III. Series: Chung-kuo kuang po kung ssu. Chung kuang ts'ung shu.
HE8699.T28C48 79-837210

CHUNG-KUO KUNG CH'AN TANG

Lewis, John Wilson, 1930-
Leadership in Communist China / John Wilson Lewis. — Westport, Conn. : Greenwood Press, 1978, c1963.
xiii, 305 p. ; 23 cm.
Reprint of the ed. published by Cornell University Press, Ithaca, N.Y.
Bibliography: p. 281-297.
Includes index.
ISBN 0-313-20119-6
1. Chung-kuo kung ch'an tang. 2. Communism—China. I. Title.
［JQ1519.A5L4 1978］ 329.9'51 77-25475
77 MARC

Yin, Fang.
(Chung kung ch'ün chien jan hsi lu)
中共羣奸然犀錄 / 尹方輯. — ［台北］：台北市海州同鄉會, 民國66 ［1977］
400 p. : ill. ; 22 cm.
NT$60.00 ($2.00 U.S.)

1. China—Politics and government—1949-1976. 2. Chung-kuo kung ch'an tang. I. Title.
DS777.55.Y47 78-841172

—ADDRESSES, ESSAYS, LECTURES

Ch'ên, Po-ta, 1905–
［Lun Mao Tse-tung ssu hsiang］
毛澤東思想是馬克思列寧主義與中國革命的結合一爲紀念中國共產黨卅週年而作 陳伯達［著 上海］ 新聞日報 1951.
29 p. 21 cm.
Previously published under title: 論毛澤東思想—馬克思列寧主義與中國革命的結合

1. Communism—China—Addresses, essays, lectures. 2. Mao, Tse-tung, 1893-1976—Addresses, essays, lectures. 3. Chung-kuo kung ch'an tang—Addresses, essays, lectures. I. Title.
Title romanized: Mao Tsê-tung ssŭ hsiang shih Ma-k'o-ssŭ Lieh-ning chu i yü Chung-kuo ko ming ti chieh ho.
HX417.C46 1951 C 63-1956 ‡

Ch'ên, Po-ta, 1905–
(Lun Mao Tse-tung ssŭ hsiang)
論毛澤東思想—馬克思列寧主義與中國革命的結合（為紀念中國共產黨的三十週年而作） 陳伯達著北京 人民出版社 1952.
59 p. 18 cm.

1. Communism—China—Addresses, essays, lectures. 2. Mao, Tse-tung, 1893-1976—Addresses, essays, lectures. 3. Chung-kuo kung ch'an tang—Addresses, essays, lectures. I. Title.
HX417.C46 1952 C 58-5412 ‡

Ch'ên, Po-ta, 1905–
［Lun Mao Tse-tung ssu hsiang］
毛澤東論中國革命 （為紀念中國共產黨的三十週年而作） 陳伯達著 北京 山民出版社 1953.
59 p. 18 cm.
Previously published under title: 論毛澤東思想—馬克思列寧主義與中國革命的結合
1. Communism—China—Addresses, essays, lectures. 2. Mao, Tse-tung, 1893-1976—Addresses, essays, lectures. 3. Chung-kuo kung ch'an tang—Addresses, essays, lectures. I. Title.
Title romanized: Mao Tse-tung lun Chung-kuo ko ming.
HX417.C46 1953 C 58-6418 ‡

—BIOGRAPHY—ADDRESSES, ESSAYS, LECTURES

(Cho yüeh ti kung chi, shen ch'ieh ti huai nien)
卓越的功绩, 深切的怀念：缅怀老一辈无产阶级革命家文章集 / ［甘肃人民出版社编辑］. — 第1版. — 兰州：該社, 1977.
3, 255 p. ; 19 cm.
$0.47

1. Chung-kuo kung ch'an tang—Biography—Addresses, essays, lectures. I. Kan-su jen min ch'u pan she.
JQ1519.A5C4787 79-843335

—CONGRESSES

Chung-kuo kung ch'an tang. Ch'üan kuo tai piao ta hui, 11th, Peking, 1977.
(Chung-kuo kung ch'an tang ti shih i tz'u ch'üan kuo tai piao ta hui wen chien hui pien)
中国共产党第十一次全国代表大会文件汇编. — 第1版. — 北京：人民出版社, 1977.
2, 162 p., ［16］ leaves of plates : ill. ; 22 cm.
$0.95

1. Chung-kuo kung ch'an tang—Congresses. I. Title.
JQ1519.A5A79 1977a 78-842682

CHUNG-KUO KUNG CH'AN TANG —CONGRESSES (Continued)

(Hsüeh hsi "shih ta" wen chien ts'an k'ao tzu liao)
学习"十大"文件参考资料. — [香港? : s. n., 1976?-
v. ; 21 cm.
"为了认真学习党的第十次全国代表大会的各项文件 我们从"十大"文献中选出若干个专题...摘编最近一个时期中央报刊的文章和报道中的有关论述"

1. Chung-kuo kung ch'an tang. Ch'üan kuo tai piao ta hui, 10th, Peking, 1973. 2. Chung-kuo kung ch'an tang—Congresses. 3. Communism—China.

JQ1519.A5H8527 77-842803

—HISTORY

Ajia Seikei Gakkai.
(Kōgun sōkenki no Mō Takutō to Shū Onrai)
紅軍創建期の毛沢東と周恩来：立三路線論考 / [編集責任者 蜂屋亮子]. — 東京：アジア政経学会：発売所 巌南堂書店, 1978.
6, 190 p. ; 21 cm. — (現代中国研究叢書 ; 15) Ja 78-7624
Includes bibliographical references.
¥1000
1. China—History—1912-1937. 2. Chung-kuo kung ch'an tang—History. 3. Chung-kuo kung nung hung chün—History. 4. Mao, Tse-tung, 1893-1976. 5. Chou, En-lai, 1898-1976. I. Hachiya, Akiko. II. Title. III. Series: Gendai Chūgoku kenkyū sōsho ; 15.

DS777.55.G43 vol. 15 78-816766
[DS777.47]

Cho, Chae-gwan.
(Chunggong kwŏllyŏk t'ujaeng sarye ŭi yangsang kwa t'ŭksaek punsŏk)
中共權力闘争事例의樣相과特色分析, 1921-48 / 研究執筆責任 趙在瓘 ; 刊行責任 羅昌柱. — [서울] : 國土統一院調査研究室, 1977.
76 p. ; 26 cm. — (周邊環境與件)
"국통조 77-1-1091."
1. Chung-kuo kung ch'an tang—History. I. Na, Ch'ang-ju. II. Title. III. Series: Chubyŏn hwan'gyŏng yŏkŏn.

JQ1519.A5C4786 78-826294

Chou, Ching-wen, 1909-
(Kung ch'an tang ti mo shu)
共產黨的魔術—統戰工作 周鯨文著 [香港] 時代批評社 [1961]
25 p. 19 cm. (時代批評叢書3)
Series romanized: Shih tai p'i p'ing ts'ung shu.

1. Chung-kuo kung ch'an tang—History. I. Title.

JQ1519.A5C483 78-835068

Chu, Tao-nan.
(Tsai ta ko ming ti hung liu chung)
在大革命的洪流中 / 朱道南口述 ; 于炳坤整理. — 新1版. — 上海 : 上海人民出版社, 1977.
231 p., [6] leaves of plates : ill. ; 21 cm.
$0.61

1. Chung-kuo kung ch'an tang—History. I. Yü, Ping-k'un. II. Title.

JQ1519.A5C48514 1977 78-841635

Harris, Nigel.
The mandate of heaven : Marx and Mao in modern China / [by] Nigel Harris. — London ; New York : Quartet Books, 1978.
xi, 307 p. ; 23 cm. GB78-38125
Bibliography: p. [x]-[xi]
Includes index.
ISBN 0-7043-2191-2 : £6.95
1. China—History—1949-1976. 2. Chung-kuo kung ch'n tang—History. 3. Chung-kuo kuo min tang—History. I. Title.
DS777.55.H317 951.05 79-309845
79 MARC

Mao, Tse-tung, 1893-1976.
(Kung ch'an tang jen fa k'an tz'u)
《共产党人》发刊词 / 毛泽东. — [北京] : 人民出版社, 1975, 1976 printing.
19 p. ; 21 cm.
"本版...是根据《毛泽东著作选读(甲种本)》一九六五年四月第二版所载原文重排的"
$0.07

1. Chung-kuo kung ch'an tang—History. 2. Kung ch'an tang jen. I. Title.

JQ1519.A5M333 1975 79-837008

Onate, Andres D 1940-
Chairman Mao and the Chinese Communist Party / Andres D. Onate. — Chicago : Nelson/Hall, c1979.
289 p. : maps ; 22 cm.
Bibliography: p. 273-280.
Includes index.
ISBN 0-88229-250-1. ISBN 0-88229-646-9 pbk.
1. Chung-kuo kung ch'an tang—History. 2. Communism—China—History. 3. Mao, Tse-tung, 1893-1976. I. Title.
JQ1519.A5O52 329.9'51 78-11049
78 MARC

Takata, Fusao, 1923-
(Chūgoku Kyōsantō)
中国共産党 / 高田富佐雄. — [東村山] : 教育社 : 東京 : 販売 教育社出版サービス, [1978]
170 p. : ports. ; 18 cm. — (時事問題解説 ; 107) (入門新書) Ja 78-25753
Cover title.
Series 2 romanized: Nyumon shinsho.
¥400
1. Chung-kuo kung ch'an tang—History. 2. China—Politics and government—1912-1949. I. Title. II. Series: Jiji mondai kaisetsu ; 107.
JQ1519.C553T31 79-806882

Wang, Ssu-ch'eng.
(Mao Tse-tung yü hung huo)
毛澤東與紅禍 / 王思誠著. — [再版]. — 台北 : 黎明文化事業股份有限公司, 民國66 [1977]
10, 9, 714 p. ; 20 cm.
NT$120.00

1. Mao, Tse-tung, 1893-1976. 2. Heads of state—China—Biography. 3. Chung-kuo kung ch'an tang—History. 4. China—Politics and government—1949-1976. I. Title.
DS778.M3W3 1977 78-842252

—MEMBERSHIP—ADDRESSES, ESSAYS, LECTURES

(Chia ch'iang ling tao pan tzu ti chien she)
加强领导班子的建设. — [香港? : s. n., 1977?]
59 p. ; 21 cm.
Reprint of the 1975 ed. published by Chiang-su jen min ch'u pan she, Nanking.

1. Chung-kuo kung ch'an tang—Membership—Addresses, essays, lectures.
JQ1519.A5C475387 1977 79-835054

—PARTY WORK

Mao, Tse-tung, 1893-1976.
(Chung-kuo kung ch'an tang tsai min tsu chan cheng chung ti ti wei)
中国共产党在民族战争中的地位 / 毛泽东. — [北京] : 人民出版社, 1975, 1976 printing.
27 p. ; 21 cm.
"本版...是根据《毛泽东著作选读》甲种本一九六五年四月第二版所载原文重排的"
$0.09
1. Chung-kuo kung ch'an tang—Party work. I. Title.
JQ1519.A5M263 77-838641

—PARTY WORK—ADDRESSES, ESSAYS, LECTURES

An, Tzu-wen.
Training successors for the revolution is the party's strategic task. — Peking : Foreign Languages Press, 1965.
58 p. ; 19 cm.
CONTENTS: An, T. W. Training successors for the revolution is the party's strategic task.—Cultivating and training millions of successors to the proletarian revolution, Renmin Ribao Editorial, August 3, 1964.—Training successors, a long-term plan for the revolutionary cause, Hongqi editorials, no. 14, 1964.
1. Chung-kuo kung ch'an tang—Party work—Addresses, essays, lectures. I. Jen min jih pao, Peking. II. Hung ch'i. III. Title.
JQ1519.A5A8613 329.9'51 78-326973
79 MARC

Chung-kuo kung ch'an tang. Chung yang wei yüan hui.
(Chung kung chung yang kuan yü pan hao ko chi tang hsiao ti chüeh ting)
中共中央关于办好各级党校的决定. 华国锋主席在中共中央党校开学典礼上的讲话. 叶剑英副主席中共中央党校开学典礼上的讲话. — [北京?] : 人民出版社, 1977.
30 p. ; 21 cm.
"中共中央党校副校长胡耀邦在中共中央党校开学典礼上的讲话(摘要)": p. 25-30.

1. Education—China—1949- —Addresses, essays, lectures. 2. Chung-kuo kung ch'an tang—Party work—Addresses, essays, lectures. I. Hua, Kuo-feng. II. Yeh, Chien-ying. III. Title.
LA1131.82.C59 1977 78-841493

(Jen chen tu shu, nu li kai tsao shih chieh kuan)
认真读书努力改造世界观. — [香港? : s. n., 1976?]
34 p. ; 21 cm.
Colophon title.
Reprint of the 1971 ed. published by Kuang-tung jen min ch'u pan she, Canton.
CONTENTS: 红旗杂志评 认真读书·努力改造世界观—金海 学好辩证唯物论的认识论—张福 用革命精神学习革命理论·用革命理论指导革命实践—李华瑞 为巩固无产阶级专政自觉进行两条路线斗争的学习—广州部队某师党委 认真读书·努力提高路线斗争觉悟
1. Chung-kuo kung ch'an tang—Party work—Addresses, essays, lectures. 2. Communism—China—Addresses, essays, lectures.
JQ1519.A5J388 1976 78-840566

(Kung ch'an tang yüan ti chan tou ch'ing ch'un)
共产党员的战斗青春 / 本社编. — [香港? : s. n., 1979?]
127 p. (on double leaves) ; 22 cm.
Reprint of the 1972 ed. published by Chiang-hsi jen min ch'u pan she, Nan-ch'ang.

1. Chung-kuo kung ch'an tang—Party work—Addresses, essays, lectures. I. Chiang-hsi jen min ch'u pan she, Nan-ch'ang, China.
JQ1519.A5K845835 1979 79-839154

(Mao chu hsi ti kan pu cheng ts'e pi hsü jen chen lo shih)
毛主席的干部政策必须认真落实. — 第1版. — 北京 : 人民出版社 : 新華書店发行, 1978.
2, 97 p. ; 19 cm.
$0.21

1. Chung-kuo kung ch'an tang—Party work—Addresses, essays, lectures.
JQ1519.A5M353 79-837756

(Ta-ch'ing ch'ing nien kung tso ching yen)
大庆青年工作经验. — 第1版. — 北京 : 中国青年出版社, 1977.
2, 1, 162 p. ; 19 cm.
CONTENTS: 共青团大庆委员会 高举毛主席树立的大庆红旗培养铁人式的青年工人队伍.——二〇五钻井队团支部 学铁人·做铁人·让铁人精神代代相传.——二〇二钻井队团支部 认真看书学习·尖刀更加锋利. — 采油二部四十九队团支部 带领青年同"四人帮"作斗争.—油建土木二中队团支部 做大干社会主义的突击队.—采油一部中四队团支部 继承会战传统·发扬"三老四严"作风.—科学研究设计院工艺试验室团支部 又红又专是我们的金光大道.—供应指挥部萨尔图仓库配件库团支部 引导青年为革命苦练管库本领.—采油四部女子采油队团支部 认真贯彻"三好"指示·带领青年全面成长.—井下作业二队团支部 努力把铁炼成钢.—采油一部机关小学团支部 用雷锋精神教育新一代.—石油化工总厂仪表车间团支部委员会 朝气蓬勃·发挥一班人的作用.—采油三部团委 主动配合党的中心·开展适合青年特点的活动.—采油二部团委 面向基层抓基层·工作开3新生面.
$0.30

1. Youth—China—Ta-ch'ing—Addresses, essays, lectures. 2. Chung-kuo kung ch'an tang—Party work—Addresses, essays, lectures.
HQ799.C552T37 78-843134

(Ta-ch'ing yu t'ien ti chi hsiang chih tu)
大庆油田的几项制度. — 第1版. — [北京?] : 人民出版社, 1977.
30 p. ; 19 cm.
CONTENTS: 中国共产党大庆委员会 大庆油田政治工作要点—中国共产党大庆委员会 领导班子革命化的措施—大庆油田政治部 加强基层建设的八条要求—大庆革命委员会 岗位责任制度
$0.08
1. Petroleum workers—China—Ta-ch'ing—Addresses, essays, lectures. 2. Petroleum workers—China—Ta-ch'ing—Political activity—Addresses, essays, lectures. 3. Chung-kuo kung ch'an tang—Party work—Addresses, essays, lectures.
HD8039.P42C685 78-842534

(Tsen yang cheng ch'ü tso ko kung ch'an tang yüan)
怎樣争取做個共產党員 / [編者青年團華南工作委員會]. — 三版. — 廣州 : 華南青年出版社 : 新華書店廣東分店總經售, 1952.
62 p. ; 18 cm.
Cover title.

1. Chung-kuo kung ch'an tang—Party work—Addresses, essays, lectures. I. Chung-kuo kung ch'an chu i ch'ing nien t'uan. Hua-nan kung tso wei yüan hui.
JQ1519.A5T716 78-840585

CHUNG-KUO KUNG CH'AN TANG (Continued)

—PURGES

Chén, Yung-kuei.
Thoroughly criticize the "Gang of Four" and bring about a new upsurge in the movement to build Tachai-type counties throughout the country : report at the Second National Conference on Learning from Tachai in Agriculture (December 20, 1976) / Chen Yung-Kuei. — 1st ed. — Peking : Foreign Languages Press, 1977.
47 p. ; 19 cm.
Translation of Ch'e ti p'i p'an ssu jen pang hsien ch'i p'u chi Ta-chai hsien yün tung ti hsin kao ch'ao.
1. Chung-kuo kung ch'an tang—Purges. 2. Agriculture—China.
JQ1519.A5C46913 1977b 78-109525
320.9'51'05
78 MARC

Cheng, Hsüeh-chia, 1906–
(Ts'ung wen ko tao shih i ta)
從文革致十一大：毛澤東扮演的最後悲劇 / 鄭學稼著. — 台北：黎明文化事業公司, 民國67 ₍1978₎
2, 350 p. ; 19 cm.
Includes bibliographical references.
NT$80.00
1. Chung-kuo kung ch'an tang—Purges. 2. Mao, Tse-tung, 1893–1976. I. Title. II. Title: Mao Tse-tung pan yen ti tsui hou pei chü.
JQ1519.A5C47518 78–841776

(Chung kung chung yang ti liang ko chüeh i)
中共中央的两个决议. — 第1版. — 北京：人民出版社：新華書店发行, 1976.
12 p. ; 21 cm.
CONTENTS: 中共中央关于华国锋同志任中共中央第一副主席·国务院总理的决议.—中共中央关于撤销邓小平党内外一切职务的决议. 附录：吴德同志在天安门广场广播讲话. 天安门广场的反革命政治事件. 坚决拥护华国锋同志任中共中央第一副主席和国务院总理. 愤怒批判邓小平搞修正主义妄图复辟资本主义的罪行. 党中央的决议表达了全国人民的心愿.
1. Chung-kuo kung ch'an tang—Purges. 2. Hua, Kuo-feng. 3. Teng, Hsiao-p'ing, 1904–
JQ1519.A5C4866 79–837765

(Hsüeh hsi shih liu t'iao shou ts'e)
学习十六条手册. — 增订本. — 第1版. — 北京：人民出版社：新华书店发行, 1966.
3, 83 p., ₍1₎ leaf of plates : ports. ; 13 cm.
CONTENTS: 中国共产党中央委员会关于无产阶级文化大革命的决定.—林彪同志在庆祝无产阶级文化大革命群众大会上的讲话.—林彪同志在接见外地来京革命师生大会上的讲话.—林彪同志在接见全国各地来京革命师生大会上的讲话.—周恩来同志在接见外地来京革命师生大会上的讲话.—周恩来同志在接见全国各地来京革命师生大会上的讲话.—学习十六条. 熟悉十六条. 运用十六条.—革命青少年要向解放军学习.—用文斗, 不用武斗.—掌握斗争的大方向.—工农群众和革命学生在毛泽东思想旗帜下团结起来.—向工农兵致敬, 向工农兵学习.
1. Chung-kuo kung ch'an tang. Chung yang wei yüan hui. Chung-kuo kung ch'an tang chung yang wei yüan hui kuan yü wu ch'an chieh chi. 2. Chung-kuo kung ch'an tang—Purges. 3. China—Politics and government—1949–1976.
JQ1519.A5H8526 79–837859

(Tsou tzu p'ai chiu shih tang nei ti tzu ch'an chieh chi)
走资派就是党内的资产阶级. — 香港：生活·讀書·新知三联书店香港分店, 1976.
92 p. ; 21 cm. — (批判修正主义路线回击右倾翻案风文章选辑 ; 4)
CONTENTS: 《人民日报》, 《红旗》杂志, 《解放军报》编辑部 文化大革命永放光芒—《人民日报》社论 群众是真正的英雄—程越 加强学习深入批邓—方刚 走资派就是党内的资产阶级—黎章 邓小平全面背叛马克思主义—方海 做批判资产阶级的战士—高路 文化大革命就是搞阶级斗争—新华社通讯员 新华社记者 文化大革命是完全必要的—新华社记者 大庆红旗更鲜艳—谭闻 走同工农结合的道路做反修防修的先锋—《人民日报》社论 天安门广场事件说明了什么
$1.30
1. Chung-kuo kung ch'an tang—Purges. 2. Teng, Hsiao-p'ing, 1900– I. Fang, Kang. II. Series: P'i p'an hsiu cheng chu i lu hsien hui chi yu ch'ing fan an feng wen chang hsüan chi ; 4.
DS777.55.P56 vol. 4 78–842836
[JQ1519.A5]

—PURGES—ADDRESSES, ESSAYS, LECTURES

(She hui chu i yü tzu pen chu i liang t'iao tao lu ti tou cheng)
社会主义与資本主义兩条道路的斗爭：反右派論文集 / ₍山东人民出版社編輯₎. — ₍香港? : s. n., 1978?₎
102 p. ; 19 cm.
Reprint of the 1957 (1958 printing) ed. published by Shan-tung jen min ch'u pan she, Tsinan.
Includes bibliographical references.
1. Chung-kuo kung ch'an tang—Purges—Addresses, essays, lectures. I. Shan-tung jen min ch'u pan she, Tsinan, China. II. Title.
JQ1519.A5S483 1978 79–837754

—PURGES—BIBLIOGRAPHY

Documents on contemporary China, 1949-1975 : a selected research collection in microfiche / consulting editor, George Lowy. — Greenwich, Conn. : JAI Press, c1976- c1977.
2 v. ; 29 cm.
CONTENTS: ₍1₎ Sorich, R., ed. Cultural revolution: Red Guard translations, bibliography, index.—₍2₎ Shera, E. B., compilor. Bibliography, index for sections: Bibliography/reference and leadership information ...
ISBN 0-89232-062-1 (v. 1)
1. China—Politics and government—1949-1976—Bibliography. 2. Chung-kuo kung ch'an tang—Purges—Bibliography. 3. Hung wei ping—Bibliography. I. Lowy, George. II. Sorich, Richard. III. Shera, Elizenija B.
Z3108.A5D62 016.3209'51'05 76-150965
₍DS777.55₎ 00 MARC

CHUNG-KUO KUNG CH'AN TANG. CH'ÜAN KUO TAI PIAO TA HUI, 10TH, PEKING, 1973

(Hsüeh hsi "shih ta" wen chien ts'an k'ao tzu liao)
学习"十大"文件参考资料. — ₍香港? : s. n., 1976?-
v. ; 21 cm.
"为了认真学习党的第十次全国代表大会的各項文件 我们从"十大"文献中选出若干个专题...摘编最近一个时期中央报刊的文章和报道中的有关论述"
1. Chung-kuo kung ch'an tang. Ch'üan kuo tai piao ta hui, 10th, Peking, 1973. 2. Chung-kuo kung ch'an tang—Congresses. 3. Communism—China.
JQ1519.A5H8527 77–842803

CHUNG-KUO KUNG CH'AN TANG. CH'ÜAN KUO TAI PIAO TA HUI, 11TH, PEKING, CHINA, 1977

Rusev, Petko.
(Istinata za edinadesetiiat kongres na KKP)
Истината за XI конгрес на ККП : ₍очерк₎ / Петко Русев. — София : Наука и изкуство, 1978.
49 p. ; 21 cm. Bu 78–1824
Includes bibliographical references.
0.24 lv
1. Chung-kuo Kung Ch'an tang. Ch'üan kuo tai piao ta hui, 11th, Peking, China, 1977. I. Title.
JQ1519.C556R87 79–413184

CHUNG-KUO KUNG CH'AN TANG. CHUNG YANG PANKUNG TING. CHUNG-KUO NUNG TS'UN TI SHE HUI CHU I KAO CH'AO

(Hsüeh hsi Chung-kuo nung ts'un ti she hui chu i kao ch'ao ti hsü yen ho an yü)
学习《中国农村的社会主义高潮》的序言和按语 / 中共南宁地委宣传部主編. — 第1版. — ₍南宁?₎ : 广西人民出版社, 1977.
2 v. ; 19 cm.
Vol. 2 has compiler statement: Chung kung Yü-lin ti wei hsüan ch'uan pu chu pien; Kuang-hsi jen min ch'u pan she pien chi.
$0.42
1. Chung-kuo kung ch'an tang. Chung yang pan kung t'ing. Chung-kuo nung ts'un ti she hui chu i kao ch'ao. 2. Agriculture, Cooperative—China. 3. China—Rural conditions. I. Chung-kuo kung ch'an tang. Nan-ning ti wei. Hsüan ch'uan pu. II. Chung-kuo kung ch'an tang. Yü-lin ti wei. Hsüan ch'uan pu. III. Kuang-hsi jen min ch'u pan she.
HD1491.C6H69 79–841136

CHUNG-KUO KUNG CH'AN TANG. CHUNG YANG WEI YÜAN HUI

United States. Central Intelligence Agency. National Foreign Assessment Center.
China, a look at the 11th Central Committee / National Foreign Assessment Center. — ₍Washington₎ : The Center, 1977.
7 p., ₍1₎ leaf of plates : ill. ; 27 cm. — (A research paper - National Foreign Assessment Center)
Cover title.
"RP 77-10276."
1. Chung-kuo kung ch'an tang. Chung yang wei yüan hui. I. Title. II. Series: United Stats. Central Intelligence Agency. National Foreign Assessment Center. Research paper - Central Intelligence Agency, National Foreign Assessment Center.
JQ1519.A5U55 1977 329.9'51 77-604687
78₍r79₎rev MARC

CHUNG-KUO KUNG CH'AN TANG. CHUNG YANG WEI YÜAN HUI. CHUNG-KUO KUNG CH'AN TANG CHUNG YANG WEI YÜAN HUI KUAN YÜ WU CH'AN CHIEH CHI

(Hsüeh hsi shih liu t'iao shou ts'e)
学习十六条手册. — 增订本. — 第1版. — 北京：人民出版社：新华书店发行, 1966.
3, 83 p., ₍1₎ leaf of plates : ports. ; 13 cm.
CONTENTS: 中国共产党中央委员会关于无产阶级文化大革命的决定.—林彪同志在庆祝无产阶级文化大革命群众大会上的讲话.—林彪同志在接见外地来京革命师生大会上的讲话.—林彪同志在接见全国各地来京革命师生大会上的讲话.—周恩来同志在接见外地来京革命师生大会上的讲话.—周恩来同志在接见全国各地来京革命师生大会上的讲话.—学习十六条. 熟悉十六条. 运用十六条.—革命青少年要向解放军学习.—用文斗, 不用武斗.—掌握斗争的大方向.—工农群众和革命学生在毛泽东思想旗帜下团结起来.—向工农兵致敬, 向工农兵学习.
1. Chung-kuo kung ch'an tang. Chung yang wei yüan hui. Chung-kuo kung ch'an tang chung yang wei yüan hui kuan yü wu ch'an chieh chi. 2. Chung-kuo kung ch'an tang—Purges. 3. China—Politics and government—1949–1976.
JQ1519.A5H8526 79–837859

CHUNG-KUO KUNG NUNG HUNG CHÜN

—ADDRESSES, ESSAYS, LECTURES

(Chin jih ti hung chün)
今日的紅軍 / 朱戈編. — 澳門：文集圖書公司, ₍1978?₎
3, 387 p. ; 19 cm.
Cover title.
Reprint ed.
1. Chung-kuo kung nung hung chün—Addresses, essays, lectures. I. Chu, Ko.
UA837.C4513 1978 79–839968

(Hung chün pu p'a yüan cheng nan)
红军不怕远征难：长征回忆录选编. — 第1版. — 北京：人民出版社, 1977.
198 p., ₍6₎ leaves of plates (2 fold.) ; 21 cm.
$0.58
1. Long March, 1934–1935—Addresses, essays, lectures. 2. Chung-kuo kung nung hung chün—Addresses, essays, lectures.
DS777.47.H8615 1977 79–835318

—BIOGRAPHY

Ch'eng, Fang-wu.
(Ch'ang cheng hui i lu)
长征回忆录 / 成仿吾. — 第1版. — 北京：人民出版社, 1977.
2, 196 p., ₍4₎ leaves of plates (2 fold.) : ill. ; 21 cm.
$0.66
1. Long March, 1934–1935—Personal narratives. 2. Communists—China—Biography. 3. Chung-kuo kung nung hung chün—Biography. I. Title.
DS777.47.C423 78–842031

CHUNG-KUO KUNG NUNG HUNG CHÜN (Continued)

—HISTORY

Ajia Seikei Gakkai.
(Kōgun sōkenki no Mō Takutō to Shū Onrai)
紅軍創建期の毛沢東と周恩来：立三路線論考 / ［編集責任者　蜂屋亮子］. — 東京：アジア政経学会：発売所　巌南堂書店, 1978.
6, 190 p. ; 21 cm. — (現代中国研究叢書 ; 15) Ja 78-7624
Includes bibliographical references.
¥1000

1. China—History—1912-1937. 2. Chung-kuo kung ch'an tang—History. 3. Chung-kuo kung nung hung chün—History. 4. Mao, Tse-tung, 1893-1976. 5. Chou, En-lai, 1898-1976. I. Hachiya, Akiko. II. Title. III. Series: Gendai Chūgoku kenkyū sōsho ; 15.
DS777.55.G43 vol. 15 78-816766
[DS777.47]

CHUNG-KUO KUO MIN TANG

Heinzig, Dieter.
Sowjetische Militärberater bei der Kuomintang : 1923-1927 / Dieter Heinzig. — 1. Aufl. — Baden-Baden : Nomos-Verlagsgesellschaft, 1978.
339 p. ; 23 cm. — (Osteuropa und der internationale Kommunismus ; Bd. 1) GFR78-A
Includes glossary in Chinese with transliteration.
Bibliography: p. 295-309.
Includes index.
ISBN 3-7890-0371-9
1. China—Foreign relations—Russia. 2. Russia—Foreign relations—China. 3. Chung-kuo kuo min tang. I. Title. II. Series.
DS740.5.R8H43 327.47'051 79-341654
79 MARC

Ku, Hung-ting, 1941-
Kuomintang's mass movement and the Kreta Ayer incident (1927) in Malaya / by Ku Hung-ting. — [Singapore] : Institute of Humanities and Social Sciences, College of Graduate Studies, Nanyang University, 1976.
37 leaves ; 28 cm. — (Occasional paper series - Institute of Humanities and Social Sciences, Nanyang University ; no. 13)
Summary in Chinese.
Includes bibliographical references.
1. Chinese in Malaya—Politics and government. 2. Chinese in Singapore—Politics and government. 3. Malaya—Politics and government. 4. Singapore—Politics and government. 5. Chung-kuo kuo min tang. I. Title. II. Series: Nanyang University. Institute of Humanities and Social Sciences. Occasional paper series - Institute of Humanities and Social Sciences, Nanyang University ; no. 13.
DS595.2.C5K8 959.5'004'951 78-940442
79 MARC

Shimizu, Tōzō, 1893-
(Chūgoku Kokumintō)
中國國民黨 / ［淸水董三］. — ［上海］：東亞同文書院支那研究部, 大正15 ［1926］
p. 199-300 ; 22 cm.
Cover title.
"支那研究第十二號別刷"

1. Chung-kuo kuo min tang. I. Tōa Dōbun Shoin Daigaku, Shanghai. Shina Kenkyūbu. II. Title.
JQ1519.A52S54 78-814260

Sun, Cheng-feng.
(Ko ming min chu cheng tang lun)
革命民主政黨論 / 孫正豐著. — 臺北：黎明文化事業公司, 民國67 ［1978］
2, 10, 6, 320 p., [1] fold. leaf of plates : ill. ; 22 cm.
Bibliography: p. 309-320.
NT$100.00

1. Political parties. 2. Chung-kuo kuo min tang. I. Title.
JF2051.S83 79-841702

—ADDRESSES, ESSAYS, LECTURES

Huang, Huang-hsiung.
(Kuo min tang wang ho ch'u ch'ü)
國民黨往何處去? / ［黃煌雄著］. — 臺北：長橋出版社, 民國67 ［1978］
263 p. ; 20 cm.
NT$60.00

1. Chung-kuo kuo ming tang—Addresses, essays, lectures. 2. Taiwan—Politics and government—1945- —Addresses, essays, lectures. I. Title.
JQ1519.A52H93 79-839525

(Li hsing ti nu hou)
理性的怒吼 / 葉扶風輯錄. — 台北：總經銷雨辰書報社, 67 ［1978］
3, 108 p. : ill. ; 22 cm.
NT$70.00 ($7.00 U.S.)

1. Taiwan—Politics and government—1945- —Addresses, essays, lectures. 2. Chung-kuo kuo ming tang—Addresses, essays, lectures. 3. Democracy—Addresses, essays, lectures. 4. Youth—Taiwan—Political activity—Addresses, essays, lectures. I. Yeh, Fu-feng.
DS799.8.L535 79-839930

—CONGRESSES

Chung-kuo kuo min tang. Ch'üan kuo tai piao ta hui, 11th, T'ai-pei, Taiwan, 1976.
(Ti shih i tz'u ch'üan kuo tai piao ta hui chung yao wen hsien)
第十一次全國代表大會重要文獻. — ［台北］：中國國民黨臺北市委員會, 民國65 ［1976］
1, 2, 141 p. ; 19 cm.
Cover title.
1. Chung-kuo kuo min tang—Congresses. I. Chung-kuo kuo min tang. T'ai-pei shih wei yüan hui. II. Title.
JQ1519.A52A58 1976a 78-841391

Chung-kuo kuo min tang. Ch'üan kuo tai piao ta hui, 11th, T'ai-pei, 1976.
(Yü ch'ing fan ying hui pien)
輿情反映彙編 / 十一全大會秘書處編印. — ［臺北］：該處, ［1978?］
2, 4, 132 p. ; 22 cm.
Cover title.
1. Chung-kuo kuo min tang—Congresses. 2. Taiwan—Politics and government—1945- —Congresses. I. Title.
JQ1519.A52A58 1976c 79-838674

—HISTORY

Harris, Nigel.
The mandate of heaven : Marx and Mao in modern China / [by] Nigel Harris. — London ; New York : Quartet Books, 1978.
xi, 307 p. ; 23 cm. GB78-38125
Bibliography: p. [x]-[xi]
Includes index.
ISBN 0-7043-2191-2 : £6.95
1. China—History—1949-1976. 2. Chung-kuo kung ch'n tang—History. 3. Chung-kuo kuo min tang—History. I. Title.
DS777.55.H317 951.05 79-309845
79 MARC

—PARTY WORK

Wu, Li-chün.
(Ch'i jung chien mo?)
豈容緘默?：一個國民黨黨員的自訴 / 武力鈞著. — 台北：當代中國研究所：總經銷中國報導社, 民國67 ［1978］
2, 90 p. ; 20 cm.
NT$30.00

1. Chung-kuo kuo min tang—Party work. 2. Electioneering—Taiwan. I. Title.
JQ1519.A52W8 79-840752
79 C

Chiang, Kai-shek, 1887-1975.
(Huang-p'u hsün lien)
黃埔訓練 / ［講述總裁］; 黨軍社編印. — ［s. l.］：該社；重慶：發行拔提書店, 民國32 ［1943］
4, 184 p. ; 19 cm. — (訓練叢書 - 黨軍社 ; 7)

1. Chung-kuo kuo min tang—Party work. 2. Chung yang lu chün chün kuan hsüeh hsiao. I. Title. II. Series: Tang chün she. Hsün lien ts'ung shu - Tang chün she ; 7.
JQ1519.A52C553 79-836831

Tai, Te-hui.
[Chung-kuo kuo min tang ti tsu chih ho hsün lien]
黨的組織和訓練 / 戴德輝編. — ［香港? : s. n., 1978?］
166 p., [1] fold. leaf of plates ; 20 cm.
Caption title: Chung-kuo kuo min tang ti tsu chih ho hsün lien.
Reprint ed.

1. Chung-kuo kuo min tang—Party work. I. Title.
Title romanized: Tang ti tsu chih ho hsün lien.
JQ1519.A52T3 1978 79-836948

—PARTY WORK—ADDRESSES, ESSAYS, LECTURES

Li, Chung-kuei.
(Kuan ch'e shih i ch'üan ta hui ching shen yü chüeh i)
貫徹十一全大會精神與決議 / 李鍾桂講；教育部社會教育司編印. — ［台北］：該司, 民國66 ［1977］
20 p. ; 19 cm. — (文化講座專集 ; 92)

1. Chung-kuo kuo min tang—Party work—Addresses, essays, lectures. I. Title. II. Series: Wen hua chiang tso chuan chi ; 92.
JQ1519.A52L4 78-841636

CHUNG-KUO KUO MIN TANG. CH'ÜAN KUO TAI PIAO TA HUI, 11TH, T'AI-PEI, 1976

—ADDRESSES, ESSAYS, LECTURES

Chung-kuo kuo min tang. Ch'üan kuo tai piao ta hui, 11th, T'ai-pei, 1976.
(Yü ch'ing fan ying hui pien hsü chi)
輿情反映彙編續集 / 中國國民黨第十一屆全國代表大會秘書處編印. — ［臺北］：該處, ［1978?］
14, 538 p. ; 22 cm.
Cover title.
1. Chung-kuo kuo min tang. Ch'üan kuo tai piao ta hui, 11th, T'ai-pei, 1976—Addresses, essays, lectures. I. Title.
JQ1519.A52A58 1976b 79-835643

CHUNG-KUO KUO MING TANG. CHUNG YANG WEI YÜAN HUI. T'U SHU KUAN

—CATALOGS

Chung-kuo kuo ming tang. Chung yang wei yüan hui. T'u shu kuan.
(Chung-kuo kuo min tang chung yang wei yüan hui t'u shu kuan tsa chih mu lu)
中國國民黨中央委員會圖書館雜誌目錄 / 中國國民黨中央委員會編印. — ［臺北］：該館, 民國59 ［1970］
14, 56, 12 p. ; 27 cm.
Cover title.
Includes index.

1. Chinese periodicals—Bibliography—Catalogs. 2. Chung-kuo kuo ming tang. Chung yang wei yüan hui. T'u shu kuan—Catalogs. I. Title.
Z6958.T34C47 1970 78-842441
[PN5370.2]

CHUNG-KUO NUNG MIN YIN HANG

Chung-kuo nung min yin hang.
Annual report - The Farmers Bank of China. [Taipei] Farmers Bank of China.
v. ill. 27 cm.
1. Chung-kuo nung min yin hang.
HG3349.C48C48a 332.3'1'0951249 78-646181
MARC-S

CHUNG-KUO SHIH YU KU FEN YU HSIEN KUNG SSU

Chung-kuo shih yu chih pien chi hsiao tsu.
(Chung-kuo shih yu chih)
中國石油志 / ⌈編輯中國石油志編輯小組⌉. — ⌈台北⌉ : 中國石油股份有限公司, 民國65 ⌈1976⌉
2 v. (4, 4, 1662 p., ⌈3⌉ fold. leaves of plates) : ill., maps ; 22 cm.
"中國石油股份有限公司三十週年紀念"
Includes index.
1. Petroleum industry and trade—China. 2. Petroleum industry and trade—Taiwan. 3. Chung-kuo shih yu ku fen yu hsien kung ssu. I. Title.
HD9576.C52C54 1976 — 79-835074

CHUNG-KUO TA LU TSAI PAO CHIU CHI TSUNG HUI

Chung-kuo ta lu tsai pao chiu chi tsung hui.
(Erh shih nien ta shih chi)
二十年大事記 / 中國大陸災胞救濟總會. — ⌈台北⌉ : 該會, 民國59 ⌈1970⌉
355 p. ; 22 cm.
1. Chung-kuo ta lu tsai pao chiu chi tsung hui. 2. Refugees—China. 3. Refugees—Taiwan. I. Title.
HV640.5.C5C52 1970 — 79-835209

CUNG-KUO WEN HUA HSÜEH YÜAN

—FACULTY

Wei, Wei-ch'i.
(Hua-kang chiao shou ch'ün chi shu)
華岡教授羣記述 / 魏偉琦著 ; ⌈編輯者中華學術院⌉. — 台北 : 華岡出版有限公司, 民國67- ⌈1978-
v. : ports. ; 22 cm.
Title in English in colophon: A galaxy of Hwakang professors.
NT$75.00 (v. 1)
1. Chung-kuo wen hua hsüeh yüan—Faculty. 2. College teachers—Taiwan. I. Chung-hua hsüeh shu yüan. II. Title. III. Title: A galaxy of Hwakang professors.
LG55.Y35W44 — 79-835294

CHUNG-KUO WEN HUA HSÜEH YÜAN. HSIN WEN HSI

—DIRECTORIES

(She shui chi)
涉水集 / 著作者中國文化學院新聞系 ; 編輯郭俊良 ... ⌈et al.⌉. — ⌈陽明山⌉ : 該系, 民國64 ⌈1975⌉
201 p. : ill. ; 20 cm.
Colophon title.
"中國文化學院新聞系第九屆畢業同學...合力寫成."
1. College prose, Chinese—Taiwan—Chung-kuo wen hua hsüeh yüan. 2. Chung-kuo wen hua hsüeh yüan. Hsin wen hsi—Directories. I. Kuo, Chün-liang. II. Chung-kuo wen hua hsüeh yüan. Hsin wen hsi.
PL3031.T32S488 — 78-841187

CHUNG-SHAN TA HSÜEH, CANTON, CHINA

—ADDRESSES, ESSAYS, LECTURES

(Kuo li Chung-shan ta hsüeh ch'eng li wu shih chou nien t'e k'an)
國立中山大學成立五十周年特刊 / ⌈編輯者國立中山大學成立五十周年特刊編印委員會 ; 國立中山大學校友會編印⌉. — 臺北 : 該校友會, 民國63 ⌈1974⌉
21, 6, 418 p. : ill. ; 28 cm.
Includes bibliographical references.
"非賣品"
1. Chung-shan ta hsüeh, Canton, China—Addresses, essays, lectures. 2. Chung-shan ta hsüeh, Canton, China—Alumni—Addresses, essays, lectures. I. Kuo li Chung-shan ta hsüeh ch'eng li wu shih chou nien t'e k'an pien yin wei yüan hui. II. Kuo li Chung-shan ta hsüeh hsiao yu hui.
LG51.C4K86 — 78-841819

—ALUMNI—ADDRESSES, ESSAYS, LECTURES

(Kuo li Chung-shan ta hsüeh ch'eng li wu shih chou nien t'e k'an)
國立中山大學成立五十周年特刊 / ⌈編輯者國立中山大學成立五十周年特刊編印委員會 ; 國立中山大學校友會編印⌉. — 臺北 : 該校友會, 民國63 ⌈1974⌉
21, 6, 418 p. : ill. ; 28 cm.
Includes bibliographical references.
"非賣品"
1. Chung-shan ta hsüeh, Canton, China—Addresses, essays, lectures. 2. Chung-shan ta hsüeh, Canton, China—Alumni—Addresses, essays, lectures. I. Kuo li Chung-shan ta hsüeh ch'eng li wu shih chou nien t'e k'an pien yin wei yüan hui. II. Kuo li Chung-shan ta hsüeh hsiao yu hui.
LG51.C4K86 — 78-841819

CHUNG-SHAN TA HSÜEH, CANTON, CHINA. T'U SHU KUAN

Chung-shan ta hsüeh, Canton, China. T'u shu kuan.
Sun Yat-sen University Library catalogue : western series. — ⌈Canton⌉ : The University, 1930-
v. ; 27 cm.
Cover title.
Added t.p.: Kuo li Chung-shan ta hsüeh t'u shu kuan hsi wen t'u shu mu lu.
Errata slip inserted.
1. Chung-shan ta hsüeh, Canton, China. T'u shu kuan. I. Title. II. Title: Kuo li Chung-shan ta hsüeh t'u shu kuan hsi wen t'u shu mu lu.
Z955.C536C48 1930 — 017'.1'095127 78 — 78-381696 MARC

CHUNG YANG JIH PAO, T'AI-PEI

—ADDRESSES, ESSAYS, LECTURES

(Chung yang jih pao yü wo)
中央日報與我. — 台北 : 中央日報社, 民國67 ⌈1978⌉
4, 236 p. ; 22 cm. — (中央日報創刊五十週年紀念叢書 ; 2)
NT$100.00
1. Chung yang jih pao, T'ai-pei—Addresses, essays, lectures. I. Series: Chung yang jih pao ch'uang k'an wu shih chou nien chi nien ts'ung shu ; 2.
PN5370.5.C548C5 — 79-841167

CHUNG YANG LU CHÜN CHÜN KUAN HSÜEH HSIAO

Chiang, Kai-shek, 1887-1975.
(Huang-p'u hsün lien)
黃埔訓練 / ⌈講述總裁⌉ ; 黨軍社編印. — ⌈s. l.⌉ : 該社 ; 重慶 : 發行拔提書店, 民國32 ⌈1943⌉
4, 184 p. ; 19 cm. — (訓練叢書 - 黨軍社 ; 7)
1. Chung-kuo kuo min tang—Party work. 2. Chung yang lu chün chün kuan hsüeh hsiao. I. Title. II. Series: Tang chün she. Hsün lien ts'ung shu - Tang chün she ; 7.
JQ1519.A52C553 — 79-836831

CHUNG YANG TIEN YING SHIH YEH KU FEN YU HSIEN KUNG SSU

(Chung yang tien ying shih yeh ku fen yu hsien kung ssu)
中央電影事業股份有限公司 / ⌈編纂者黨營文化事業專輯編纂委員會⌉. — 臺北 : 中國國民黨中央委員會文化工作會, 民國61 ⌈1972⌉
2, 6, 2, 140 p., ⌈7⌉ leaves of plates : ill. ; 21 cm. — (黨營文化事業專輯 ; 6)
Cover title.
1. Chung yang tien ying shih yeh ku fen yu hsien kung ssu. I. Tang ying wen hua shih yeh chuan chi pien tsuan wei yüan hui. II. Series: Tang ying wen hua shih yeh chuan chi ; 6.
PN1999.C47C47 — 79-837270

CHUNG YANG YEN CHIU YÜAN

Chung yang yen chiu yüan.
(Chung yang yen chiu yüan kai k'uang)
中央研究院概況. — ⌈臺北 : s. n.⌉, 民國67 ⌈1978⌉
4, 198 p., ⌈16⌉ leaves of plates : ill. ; 24 cm.
1. Chung yang yen chiu yüan. I. Title.
AS455.C48A54 — 78-842666

(Chung yang yen chiu yüan ch'eng li wu shih chou nien chi nien lun wen chi)
中央研究院成立五十周年紀念論文集 / ⌈中央研究院成立五十周年紀念論文集編輯委員會⌉. — ⌈台北? : 該會⌉, 民國67 ⌈1978⌉
2 v. : ill. ; 27 cm.
English or Chinese.
CONTENTS: 第1輯. 數理科學, 生物科學.—第2輯. 人文社會科學.
1. Chung yang yen chiu yüan. I. Chung yang yen chiu yüan ch'eng li wu shih chou nien chi nien lun wen chi pien chi wei yüan hui. II. Chung yang yen chiu yüan.
AS455.C48C5 — 78-843480

CHUNG YUAN LI KUNG HSUEH YUAN

Chung yüan li kung hsüeh yüan.
(Chung yüan li kung hsüeh yüan kai k'uang)
中原理工學院概況 / 中原理工學院編印. — ⌈中壢⌉ : 該院, 民國67 ⌈1978⌉
170 p., ⌈11⌉ leaves of plates : ill. ; 21 cm.
"一九七八——九七九"
1. Chung yüan li kung hsüeh yüan. I. Title.
T173.C5233A53 — 79-840573

CHUNG YUNG

(Chungyong ŏnhae)
中庸諺解. — 서울 : 漢陽大學校附設國學研究院 : 繇刻大提閣, 1974.
112 p. ; 27 cm.
Photoreprint of Tosan pon, with new explanatory note.
1. Chung yung. I. Chung yung. II. Hanyang Taehakkyo, Seoul, Korea. Kukhak Yŏn'guwŏn.
PL2473.Z6C45 1974 — 78-826778

(Chungyong ŏnhae)
中庸諺解 : 듕용언히. — 서울 : 大提閣, 1976.
112 p. ; 26 cm. — (原本影印韓國古典叢書 : V, 經書言解 ; ⌈7⌉)
Reprint of the 1974 ed. published by Hanyang Taehakkyo Pusŏl Kukhak Yŏn'guwŏn, Seoul, with new explanatory note.
1. Chung yung. I. Chung yung. II. Series: Wŏnbon yŏngin Han'guk kojŏn ch'ongsŏ : V, Kyŏngsŏ ŏnhae ; ⌈7⌉
PL902.W64 V, vol. 7 — 79-825083
[PL2473.Z6]

Lin, Ta-ch'un.
(K'o hsüeh ti hsüeh yung chih yen chiu)
科學的學庸之研究 / 林大椿編著. — 臺初版. — 臺北 : 正中書局, 民國67 ⌈1978⌉
6, 3, 3, 406 p., ⌈1⌉ leaf of plates : facsims. ; 22 cm.
Appendices (p. 341-406): 1. 主要參考書.—2. 次要參考書.—3. 有關學庸之論著.
1. Chiang, Kai-shek, 1887-1975. K'o hsüeh ti hsüeh yung. 2. Ta hsüeh. 3. Chung yung. I. Title.
PL2472.R33 — 79-835169

CHUNG YUNG (Continued)

Ma, Shao-po, 1902 or 3-
(Chung yung hsin i)
中庸新義 / 馬紹伯著. — 修訂版. — 臺北：文史哲出版社, 民國66 [1977]
166 p., [2] leaves of plates : ports. ; 22 cm.
NT$60.00

1. Chung yung. I. Title.
PL2473.Z7M3 79-835932

Shimada, Kenji, 1917-
(Daigaku, Chūyō)
大学·中庸 / 島田虔次著. — 東京：朝日新聞社, 1978.
2 v. ; 15 cm. — (中国古典選 ; 6-7) Ja 78-18614 (v. 1)
"テキストは清の呉志忠が嘉慶16年(1811)校刊した「大学章句」「中庸章句」(東京文求堂影印)を底本とした"
Includes indexes.
¥300 (v. 1) ¥350 (v. 2)
1. Ta hsüeh. 2. Chung yung. I. Ta hsüeh. Japanese & Chinese. 1978. II. Chung yung. Japanese & Chinese. 1978. III. Chu, Hsi, 1130-1200. IV. Title. V. Series: Chūgoku koten sen ; 6-7.
PL2472.Z6S5 1978 78-813858

—COLLECTED WORKS

Yüan, Huan-hsien.
(Wei-mo ching she ts'ung shu)
維摩精舍叢書 [袁煥仙著 台北 永康出版社 民國59 i. e. 1970]
2 v. ports. 21 cm. NT$500.00
Reprint of 民國33(1944)成都維摩精舍刻經處校刻本
CONTENTS: 上 榴窗隨判 1卷 黄葉閒譚 1卷 中庸勝唱 卷上一 下 中庸勝唱 卷下 靈巖語屑 1卷 酬語 1卷
1. Buddhism—Collected works. 2. Chung yung—Collected works. I. Title.
BQ126.Y83 1970 75-837073

CHUNGANG TAEHAKKYO, SEOUL, KOREA

—HISTORY

Chungang Taehakkyo Yuksimnyŏnsa P'yŏnch'an Wiwŏnhoe.
(Chungang Taehakkyo yuksimnyŏnsa)
中央大學校六十年史, 1918-1978 / [編纂中央大學校60年史編纂委員會]. — [서울]：中央大學校, 1978.
789 p. : ill. ; 27 cm.

1. Chungang Taehakkyo, Seoul, Korea—History. I. Title.
LG281.S37C49 1978 79-826243

CH'UNGBUK TAEHAK. TAEHAGWŎN

—DISSERTATIONS—ABSTRACTS
—COLLECTED WORKS

Ch'ungbuk Taehak. Taehagwŏn.
(Taehagwŏn nonmunjip)
大學院論文集
[清州] 忠北大學大學院
no. 26 cm.
Other title, : Thesis collection of the Graduate School, Chung-Buk National University.
Summaries in English.
1. Ch'ungbuk Taehak. Taehagwŏn—Dissertations—Abstracts—Collected works. I. Ch'ungbuk Taehak. Taehagwŏn. Thesis collection of the Graduate School, Chung-Buk National University. II. Title.
LG281.C543C48a 79-641535
(MARC-S)

—STATISTICS

Ch'ungch'ŏng-namdo, Korea. T'onggye Tamdanggwansil.
(Uri ŭi chwap'yo)
우리의좌표：통계로본오늘의충남 = Chungcheongnam-Do in statistics / [편집충청남도통계담당관실]. — [대전]：충청남도, 1978.
206 p., [1] leaf of plates : ill. ; 25 cm.
Cover title.
English and Korean.
Not in trade.

1. Ch'ungch'ŏng-namdo, Korea—Statistics. I. Title. II. Title: Chungcheongnam-Do in statistics.
HA1857.C53C48 1978 315.19 79-825921

CH'UNGCH'ŎNG-PUKTO, KOREA

—DESCRIPTION AND TRAVEL

Ch'ungch'ŏng-pukto, Korea.
(Ch'ungbuk ŭi sanŏp kwa kwan'gwang)
忠北의産業과觀光 = Industry and sight seeing of Chungbuk. — [清州]：忠清北道, [1978]
ca. 150 p. : chiefly col. ill. ; 30 cm.
English and Korean.

1. Ch'ungch'ong-pukto, Korea—Description and travel. 2. Ch'ungch'ong-pukto, Korea—Industries. I. Title. II. Title: Industry and sight seeing of Chungbuk.
DS924.C53C53 1978 78-826412

—INDUSTRIES

Ch'ungch'ŏng-pukto, Korea.
(Ch'ungbuk ŭi sanŏp kwa kwan'gwang)
忠北의産業과觀光 = Industry and sight seeing of Chungbuk. — [清州]：忠清北道, [1978]
ca. 150 p. : chiefly col. ill. ; 30 cm.
English and Korean.

1. Ch'ungch'ong-pukto, Korea—Description and travel. 2. Ch'ungch'ong-pukto, Korea—Industries. I. Title. II. Title: Industry and sight seeing of Chungbuk.
DS924.C53C53 1978 78-826412

CH'UNGCH'ŎNG ŬNHAENG

Ch'ungch'ŏng Ŭnhaeng.
(Ch'ungch'ŏng Ŭnhaeng simnyŏnsa)
忠清銀行十年史. — 大田：충청은행, 1978.
351 p. : ill. ; 27 cm.
Not in trade.

1. Ch'ungch'ŏng Ŭnhaeng. I. Title.
HG3330.5.T33C48 1978 79-825170

CHUNGGUK KONGSANDANG see Chung-kuo kung ch'an tang

CH'UNGMU CH'ŎNGNYŎN HOEŬISO

Ch'ungmu Ch'ŏngnyŏn Hoeŭiso.
(Pulmyŏl ŭi sŏgok)
不滅의序曲：忠武青年会議所10年史 / [編輯人金一信]. — 忠武：忠武青年会議所, 1977.
198 p. : ill. ; 22 cm.
Not in trade.

1. Ch'ungmu Ch'ŏngnyŏn Hoeŭiso. I. Kim, Il-sin. II. Title.
HF331.K65C4727 78-826662

CH'UNGMU, KOREA

—DESCRIPTION

(Kwan'gwang Hallyŏ sudo)
觀光閑麗水道：海上國立公園閑麗水道 = Hanryeo channel Chungmu : National sea park. — [서울]：宇進觀光文化社, [197-]
[40] leaves : all col. ill. ; 14 x 20 cm.
Cover title.
Chiefly in Korean.

1. Ch'ungmu, Korea—Description. I. Ujin Kwan'gwang Munhwasa. II. Title: Hallyŏ sudo. III. Title: Hanryeo channel Chungmu.
DS925.C52K86 79-825679

CH'UNGNYOLSA ALLAK SOWON

—HISTORY

(Ch'ungnyŏlsa chi)
忠烈祠志 / 忠烈祠安樂書院編 ; 丁仲煥, 金錫禧譯. — 서울：民學社, 1978.
243, [108] p. : ill. ; 27 cm.
Includes reprint of the original 1778 ed.

1. Korea—History—Japanese invasions, 1592-1598—Biography. 2. Korea—Biography. 3. Ch'ungnyŏlsa Allak Sŏwŏn—History. I. Ch'ungnyŏlsa Allak sŏwŏn.
DS913.44.C48 1978 79-825499

CHUNGUI, PERU

—ECONOMIC CONDITIONS

Fabián Areas, Glirio.
La semifeudalidad en Anco y Chungui, Provincia La Mar : un caso de economía transicional / Glirio Fabián Areas, Orfilio J. Yupanqui Arteaga. — Lima : Instituto de Investigaciones Económico-Sociales, [between 1975 and 1978]
xi, 215 p., [4] leaves of plates (2 fold.) : ill. ; 29 cm. — (Documentos sobre la realidad nacional ; no. 2)
Cover title.
Bibliography: p. 213-215.
1. Anco, Peru (Ayacucho)—Economic conditions. 2. Chungui, Peru—Economic conditions. 3. Land tenure—Peru—Anco (Ayacucho) 4. Land tenure—Peru—Chungui. I. Yupanqui Arteaga, Orfilio J., joint author. II. Title. III. Series.
HC228.A52F3 333.3'2'09852 78-111060
79 MARC

CHŪŌ-KU, TOKYO, JAPAN

Chūō-ku, Tokyo, Japan
(Chūō-ku shi)
中央区史 東京 東京都中央区役所 昭和33 [1958]
3 v. illus pl tes part fold. col.) p ts maps par fo d col.) di grs., tables. 22 cm
中央区旧日本橋地区 京橋地区各町行政区劃変遷表 (fo d. leaf) in rted at end of v 2.
1. Chūō-ku Tokyo Japan 2. Tokyo—History
DS896.7.C54C49 1958 J 63-429

—DESCRIPTION—ADDRESSES, ESSAYS, LECTURES

Fujiura, Tomitarō, 1885-
(Tōkyō kandan)
東京閑談：京橋·日本橋の思い出 / 藤浦富太郎. — 東京：評論社, 昭和53 [1978]
144 p. : ill. ; 20 cm. — (江戸文化選書 ; 3) Ja***
Colophon inserted.
¥980

1. Tokyo—Description—Addresses, essays, lectures. 2. Chūō-ku, Tokyo—Description—Addresses, essays, lectures. 3. Fujiura, Tomitarō, 1885- I. Title. II. Series: Edo bunka sensho ; 3.
DS896.35.F84 78-812916

CHŪŌ-KU, TOKYO, JAPAN (Continued)

—HISTORY

(Waga machi no ayumi)
わが町の歩み / [編集責任　村上謙重]. — 東京 : 日本橋二丁目通町会商店会, 1978.
376 p., [13] leaves of plates : ill. ; 22 cm. Ja 78-16132
"日本橋二丁目通町会商店会三十年史"

1. Chūō-ku, Tokyo, Japan—History. 2. Tokyo—History. 3. Nihonbashi Nichōmedōri Chōkai Shōtenkai. 4. Retail trades—Japan—Tokyo. I. Murakami, Kaneshige. II. Nihonbashi Nichōmedōri Chōkai Shōtenkai.
DS896.7.C54W33 78-817123

CHUQUICAMATA, CHILE

—SOCIAL LIFE AND CUSTOMS

Chile Exploration Company.
Chuquicamata ... producer of copper in the Atacama Desert of northern Chile. — Anaconda : Chile Exploration Company, [1967?]
32 p. : ill. ; 22 x 28 cm.
Cover title.
1. Copper mines and mining—Chile—Chuquicamata. 2. Chuquicamata, Chile—Social life and customs. I. Title: Chuquicamata ... producer of copper
TN444.C52C48 1967 79-316419
79 MARC

CHUR, SWITZERLAND

—FIRES AND FIRE PREVENTION

Berger, Hans, 1920-
Der Churer Stadtbrand vor 400 Jahren (1574) / Hans und Mathis Berger. — [Chur : Gasser & Eggerling, 1974]
44 p. : ill. ; 18 cm. — (Schriftenreihe der Neuen Bündner Zeitung) Sw75-A-4456
Cover title.
Bibliography: p. 43-44.
3.50F
1. Chur, Switzerland—Fires and fire prevention. I. Berger, Mathis, joint author. II. Title.
TH9571.C49B47 79-371034
*79 MARC

CHURCH

see also Christian communities; Christian union; Christianity; Mission of the church

Corréard, Jacqueline.
A la rencontre du Christ dans son Église / Jacqueline Corréard, Élisabeth Saint-Pierre. — Paris : Éditions Saint-Paul, c1978.
142 p. ; 20 cm. — (A l'écoute de la vie) F***
ISBN (invalid) 2-85049-112-5 : 25.00F
1. Sacraments—Catholic Church. 2. Church. I. Saint-Pierre, Élisabeth, joint author. II. Title.
BX2200.C63 230'.2 79-372795
79 MARC

Egenter, Richard, 1902-
Miteinander umgehen : Pluralismus in d. Kirche, Auftrag u. Chance / Richard Egenter. — 1. Aufl. — München : Don Bosco Verlag, 1978.
134 p. ; 21 cm. GFR
Includes index.
ISBN 3-7698-0330-2 : DM12.80
1. Church. 2. Pluralism (Social sciences) I. Title.
BX1746.E34 79-339066
79 MARC

Florovskiĭ, Georgiĭ Vasil'evich, 1893-
(To sōma tou zōntos Christou)
Τὸ σῶμα τοῦ ζῶντος Χριστοῦ : μία ὀρθόδοξος ἑρμηνεία τῆς Ἐκκλησίας / Γεωργίου Φλωρόφσκυ ; μετάφρασις Ἰ. Κ. Παπαδοπούλου. — Θεσσαλονίκη : Πατριαρχικὸν Ἵδρυμα Πατερικῶν Μελετῶν, 1972.
148 p. ; 21 cm. — (Θεολογικὰ δοκίμια ; 3)
Includes bibliographical references.
1. Church. 2. Jesus Christ—Mystical body. I. Title. II. Series: Theologika dokimia ; 3.
BV600.2.F55 79-408564

Fujii, Takao, 1916-
(Yo ni aru kyōkai no sekinin)
世にある教会の責任 / [執筆者　藤井孝夫, 金井愛明]. — 東京 : 日本基督教団出版部, 1961.
131 p. ; 18 cm.

1. Church. 2. Labor dispute—Japan. I. Kanai, Yoshiaki. II. Title.
BV600.2.F85 79-801319

Gherardini, Brunero.
La Chiesa arca dell'alleanza : la sua genesi, il suo paradosso, i suoi poteri, il suo servizio / Brunero Gherardini. — 2. ed. riveduta e ampliata. — [Roma] : Pontificia università lateranense ; Roma : Città nuova, 1978.
365 p. ; 21 cm. — (Teologia ; 6) It79-Mar
Bibliography: p. [11]-29.
Includes index.
L7500
1. Church. I. Title. II. Series: Teologia (Roma) ; 6.
BX1746.G47 1978 79-367393
*79 MARC

Gish, Arthur G
Living in Christian community / Arthur G. Gish. — Scottdale, Pa. : Herald Press, 1979.
379 p. ; 22 cm.
Bibliography: p. [371]-373.
Includes index.
ISBN 0-8361-1887-1 : $8.95
1. Christian communities. 2. Church renewal. 3. Christian life—Church of the Brethren authors. 4. Church. I. Title.
BV4405.G57 262 79-11848
79 MARC

Hauck, Gary L
Is my church what God meant it to be? / Gary L. Hauck. — Denver : Accent Books, c1979.
160 p. ; 21 cm.
ISBN 0-89636-029-6 : $3.95
1. Theology, Practical. 2. Church. I. Title.
BV3.H35 253 79-50811
79 MARC

Hermanns, Manfred, 1936-
Kirche als soziale Organisation : zwischen Partizipation u. Herrschaft / Manfred Herrmanns. — 1. Aufl. — Düsseldorf : Patmos Verlag, 1979.
160 p. ; 21 cm. — (Patmos-Paperback) GFR79-A
Bibliography: p. 154-160.
ISBN 3-491-77599-X
1. Church. 2. Sociology, Christian. I. Title.
BV600.2.H44 79-365269
79 MARC

Huber, Wolfgang, 1942-
Kirche / Wolfgang Huber. — 1. Aufl., (1.-8. Tsd.). — Stuttgart ; Berlin : Kreuz-Verlag, 1979.
238 p. ; 20 cm. GFR79-A
"Bibliothek Themen der Theologie Ergänzungsband."
Includes bibliographical references.
ISBN 3-7831-0567-6 : DM24.00
1. Church. I. Title. II. Series: Themen der Theologie.
BV600.2.H83 79-370996
79 MARC

Janni, Ugo.
Corpus Domini : la Chiesa, suoi valori eterni, suoi problemi moderni / Ugo Janni. — Roma : A.V.E., 1978.
xxix, 475 p. ; 22 cm. — (Teologia oggi ; 25) It78-June
Includes bibliographical references.
L9000
1. Church. I. Title.
BX1746.J36 78-379035
*78 MARC

Kohler, Léo.
O leigo na Igreja; estudo da Constituição sôbre a Igreja (Lumen gentium) [Pôrto Alegre, Edições Paulinas, 1969]
63 p. 19 cm. (His Cursos de formação, 1. sér) Cr$1.50
1. Vatican Council. 2d, 1962-1965. Constitutio dogmatica de ecclesia. 2. Church. I. Title.
BX830 1962.A45C8757 78-341870
78[r79]rev MARC

Kraus, Clyde Norman.
The authentic witness : credibility and authority / by C. Norman Kraus. — Grand Rapids : W. B. Eerdmans Pub. Co., c1979.
199 p. : ill. ; 21 cm.
ISBN 0-8028-1785-8
1. Church. 2. Sociology, Christian. I. Title.
BV600.2.K67 261.8 78-24012
78 MARC

Leenhardt, Franz J 1902-
L'Église : questions aux protestants et aux catholiques / Franz-J. Leenhardt. — Genève : Labor et fides ; Paris : diffusion pour la France Librairie protestante, cop. 1978.
239 p. ; 21 cm. — (Lieux théologiques ; no 1) Sw79-1874
Includes bibliographical references.
38.00F
1. Church. 2. Clergy—Office. 3. Christian union. I. Title.
BV600.2.L42 262'.001 79-368994
79 MARC

Marshall, Michael, 1936-
Glory under your feet : the challenge of catholic renewal today / [by] Michael Marshall. — London : Darton, Longman, and Todd, 1978.
137 p. ; 22 cm. GB78-29210
Includes bibliographical references.
ISBN 0-232-51306-5 : £2.95
1. Church renewal. 2. Christianity—20th century. 3. Church. I. Title.
BV600.2.M345 1978 260 79-301452
79 MARC

McBirnie, William Steuart, 1920-
The search for the early church : New Testament principles for today's church / William Steuart McBirnie. — Wheaton, Ill. : Tyndale House Publishers, c1978.
165 p. : ill. ; 21 cm.
ISBN 0-8423-5834-X : $3.95
1. Church history—Primitive and early church, ca. 30-600. 2. Church. 3. Christianity—20th century. I. Title.
BR165.M18 281'.2 78-55982
79 MARC

Miller, Hal, 1952-
Christian community, biblical or optional? / Hal Miller. — Ann Arbor, Mich. : Servant Publications, c1979.
144 p. ; 21 cm.
Includes indexes.
ISBN 0-89283-068-9 : $3.50
1. Church. 2. Community. 3. Fellowship. I. Title.
BV601.6.C65M54 262 79-117942
00 MARC

Pagé, Jean Guy.
Foi ou liberté? : Autorité, loi et liberté dans l'Église, mort de Dieu, foi du chrétien, la théologie, ministère ecclésial, place de la femme dans le plan divin / Jean-Guy Pagé ; préf. de Hans Urs von Balthasar. — Montréal : Éditions Bellarmin, 1977.
211 p. ; 21 cm. C77-13477-7
Includes bibliographical references.
ISBN 0-88502-233-5 : $8.95
1. Church. 2. Authority (Religion) 3. Law (Theology) 4. Freedom (Theology) 5. Theology. 6. Women (Christian theology) I. Title.
BX1746.P24 262'.8 78-382490
79 MARC

Philibert, Roger.
Pour quelle Église, quels prêtres / Roger Philibert. — Paris : Éditions ouvrières, [1974]
215 p. ; 18 cm. — (Collection Chercheurs de Dieu) F***
Bibliography: p. [206]-210.
18.00F
1. Church. 2. Pastoral theology—Catholic Church. I. Title.
BX1746.P48 253 79-375270
79 MARC

Riobé, Guy, 1911-
Projet d'Église : une Église libre qui ose / Guy-Marie Riobé. — Paris : Éditions du Cerf, c1979.
104 p. ; 18 cm. — (Dossiers libres) F***
ISBN 2-204-01328-5 : 22.00F
1. Church. I. Title.
BX1746.R54 262'.7 79-371505
79 MARC

Santaner, Marie Abdon.
L'Avenir de la communauté religieuse : vivre ensemble l'Alliance / M.-A. Santaner. — Paris : Éditions ouvrières, 1978.
183 p. ; 18 cm. — (Collection Lumière des hommes) F78-13106
ISBN 2-7082-1954-5 : 32.00F
1. Christian life—Catholic authors. 2. Church. I. Title.
BX2350.2.S254 248'.48'2 79-349262
79 MARC

Schiwy, Günther.
zur Ideologie der Unfehlbarkeitsdiskussion : e. semiot. Strukturanalyse zu Texten von Hans Küng u. d. röm. Glaubenskongregation / Günther Schiwy. — 1. Aufl. — Düsseldorf : Patmos, 1977.
128 p. : ill. ; 21 cm. — (Patmos-paperback) GFR78-A
Bibliography: p. 126-128.
ISBN 3-491-77581-7 : DM23.00
1. Catholic Church—Infallibility. 2. Church. 3. Küng, Hans, 1928- Unfehlbar. 4. Catholic Church. Congretatio pro Doctrina Fidei. Declaratio circa catholicam doctrinam de ecclesia contra nonnullos errores hodiernos tuendam. 5. Semiotics. I. Title.
BV601.6.I5S36 78-369533
79 MARC

Shelley, Bruce Lear, 1927-
The church, God's people / Bruce L. Shelley. — Wheaton, Ill. : Victor Books, c1978.
132 p. ; 21 cm. — (The Victor know & believe series)
Bibliography: p. 132.
ISBN 0-88207-770-8 : $2.95
1. Church. I. Title. II. Series.
BV600.2.S5 260 78-109556
78 MARC

Spillman, James Russell, 1934-
Omega cometh / James R. Spillman. — Old Tappan, N.J. : Revell, c1979.
154 p. ; 22 cm.
ISBN 0-8007-1071-1
1. End of the world. 2. Man (Christian theology) 3. Church. I. Title.
BT876.S68 230 79-9297
79 MARC

CHURCH (Continued)

Thema, Volkskirche : e. Arbeitsbuch für d. Gemeinde / im Auftr. d. Präsidiums d. Synode d. Evang. Kirche in Deutschland hrsg. von d. Kirchenkanzlei ; Bearb., Rüdiger Schloz. — Gelnhausen ; Berlin : Burckhardthaus-Verlag, 1978.
276 p. ; 21 cm. GFR78-A
Bibliography: p. 275-276.
3766400762
1. Church. 2. Evangelische Kirche in Deutschland. I. Schloz, Rüdiger. II. Evangelische Kirche in Deutschland. Kirchenkanzlei.
BV600.2.T484 79-342887
79 MARC

Watson, David C K 1933-
I believe in the church / by David Watson. — 1st American ed. — Grand Rapids : Eerdmans, 1979.
368 p. ; 21 cm.
Includes bibliographical references.
ISBN 0-8028-1788-2
1. Church. I. Title.
BV600.2.W36 1979 262 78-26765
78 MARC

Wegener-Fueter, Hildburg.
Kirche und Ökumene : d. Kirchenbild d. Ökumen. Rates d. Kirchen nach d. Vollversammlungsdokumenten von 1948-1968 / Hildburg Wegener-Fueter. — Göttingen : Vandenhoeck und Ruprecht, 1979.
306 p. ; 25 cm. — (Göttinger theologische Arbeiten ; Bd. 10) GFR79-A
An abridgement of the author's thesis, Göttingen.
Bibliography: p. ⌜295⌝-306.
ISBN 3-525-87363-8
1. Church. 2. World Council of Churches. I. Title. II. Series.
BV600.2.W4252 79-368049
79 MARC

Yi, Chang-sik.
(Hyŏndae kyohoehak)
現代教會學 / 李章植. — 서울 : 大韓基督教書會, 1969, 1974 printing.
420 p. ; 22 cm.
On verso of t. p.: Modern ecclesiology.
Includes bibliographical references.
₩1400
1. Church. I. Title, II. Title: Modern ecclesiology.
BV600.2.Y5 79-825546

—ADDRESSES, ESSAYS, LECTURES

Barr, Browne.
The well church book / Browne Barr. — New York : Seabury Press, c1976.
116 p. : ill. ; 22 cm.
"A Crossroad book."
ISBN 0-8164-0304-X : $7.95
1. Church renewal—Addresses, essays, lectures. 2. Church—Addresses, essays, lectures. 3. Pastoral theology—Addresses, essays, lectures. I. Title.
BV600.2.B36 262'.001 76-18291
76 MARC

Ecclesiologia e cultura moderna : saggi teologici / a cura di Giandomenico Mucci. — ⌜Roma⌝ : Herder, 1979.
xi, 525 p. ; 24 cm. — (Aloisiana ; 12) It79-Mar
Includes bibliographical references and indexes.
1. Theology—Addresses, essays, lectures. 2. Church—Addresses, essays, lectures. I. Mucci, Giandomenico.
BR50.E27 79-359282
•79 MARC

In God's community : essays on the Church and its ministry / edited by David J. Ellis and W. Ward Gasque. — Wheaton, Ill. : H. Shaw Publishers, c1978, 1979 printing.
xiii, 186 p. ; 21 cm.
Bibliography: p. 169-171.
Includes indexes.
ISBN 0-87788-392-0 : $3.95
1. Church—Addresses, essays, lectures. 2. Clergy—Office—Addresses, essays, lectures. 3. Mission of the Church—Addresses, essays, lectures. I. Ellis, David J. II. Gasque, W. Ward.
BV600.2.I5 260 79-12847
79 MARC

Die Kirche : fünfzehn Betrachtungen / hrsg. von Wilhelm Sandfuchs ; ⌜Beitr. von Alfons Auer ... et al.⌝. — Würzburg : Echter, 1978.
208 p. ; 21 cm. GFR79-A
Lectures broadcast in a series by the Bayerischer Rundfunk.
Includes bibliographical references.
DM19.80
1. Church—Addresses, essays, lectures. I. Sandfuchs, Wilhelm. II. München. Bayerischer Rundfunk.
BX1746.K52 79-372929
79 MARC

Kirche und Bibel : Festgabe für Bischof Eduard Schick / hrsg. von d. Professoren d. Phil.-Theol. Hochschule Fulda. — Paderborn ; München ; Wien ; Zürich : Schöningh, 1979.
502 p. : port. ; 24 cm. GFR79-A
Contributions chiefly in German with two in Italian.
Bibliography: p. ⌜501⌝-502.
ISBN 3-506-74450-X : DM58.00
1. Bible—Criticism, interpretation, etc.—Addresses, essays, lectures. 2. Church—Addresses, essays, lectures. 3. Theology—Addresses, essays, lectures. 4. Schick, Eduard, 1906- —Addresses, essays, lectures. I. Schick, Eduard, 1906- . II. Philosophisch-Theologische Hochschule Fulda.
BS540.K53 79-377350
79 MARC

Ratzinger, Joseph.
Living with the Church / Joseph Ratzinger and Karl Lehmann ; translated by Zachary Hayes. — Chicago: Franciscan Herald Press, c1978.
53 p. ; 18 cm. — (Synthesis series)
Translation of Mit der Kirche leben.
Includes bibliographical references.
CONTENTS: Ratzinger, J. Identification with the Church.—Lehmann, K. Is it worthwhile remaining in the Church and living for her?
ISBN 0-8199-0742-1 : $1.50
1. Church—Addresses, essays, lectures. 2. Identification (Religion)—Addresses, essays, lectures. I. Lehmann, Karl, 1936- joint author. II. Title.
BX1746.R3613 262'.7 78-15509
78 MARC

Størst av alt : festskrift til Stephan Tschudi på 70-års dagen 2. januar 1978 / redaksjon, Sverre Inge Apenes ... ⌜et al.⌝ ; illustrasjoner Bjørn Bjørneboe. — Oslo : Land og kirke-Gyldendal, c1978 (Oslo : Reistad)
131 p., ⌜2⌝ leaves : ill. ; 24 cm. N78-March
ISBN 8205108285 : kr68.00
1. Love (Theology)—Addresses, essays, lectures. 2. Church—Addresses, essays, lectures. 3. Tschudi, Stephan—Addresses, essays, lectures. I. Tschudi, Stephan. II. Apenes, Sverre Inge.
BV4639.S75 78-400419
79 MARC

Theological foundations for ministry : selected readings for a theology of the church in ministry / Ray S. Anderson, editor. — Edinburgh : T. & T. Clark ; Grand Rapids, Mich. : Eerdmans, c1978.
xii, 776 p. ; 20 cm.
ISBN 0-8028-1776-9 : $7.50
1. Theology, Practical—Addresses, essays, lectures. 2. Church—Addresses, essays, lectures. I. Anderson, Ray Sherman.
BV3.T44 262'.7 78-13613
78 MARC

—APOSTOLICITY—HISTORY

Maccarrone, Michele.
Apostolicità, episcopato e primato di Pietro : ricerche e testimonianze dal II al V secolo / Michele Maccarrone. — Roma : Lateranum, 1976.
354 p. ; 24 cm. — (Lateranum : Nova series ; a. 42, n. 2) It***
Includes bibliographical references and index.
1. Episcopacy—History of doctrines. 2. Church—Apostolicity—History. 3. Peter, Saint, apostle. I. Title. II. Series.
BV670.2.M3 78-380100
79⌜r79⌝rev MARC

—AUTHORITY

see also Liberty of speech in the church

Morawski, Zdzisław.
Watykan z daleka i z bliska / Zdzisław Morawski. — Wyd. 1. — Warszawa : Książka i Wiedza, 1978.
334 p., ⌜8⌝ leaves of plates : ill. (some col.) ; 21 cm. P***
zł50.00
1. Catholic Church—Government. 2. Church—Authority. 3. Papacy. 4. Vatican. I. Title.
BX1802.M67 78-398256
79 MARC

—AUTHORITY—CONGRESSES

Harker, Aidan.
Commentary on an agreed statement on authority in the Church 1976 / ⌜by⌝ Aidan Harker. — Enfield : Catholic League ; London : Distributed by the Church Literature Association, 1977.
⌜2⌝, 30 p. ; 21 cm. GB78-37460
Includes the text of the statement Authority in the Church.
£0.40
1. Anglican/Roman Catholic International Commission. Agreed statement on authority in the Church. 2. Church—Authority—Congresses. 3. Papacy—Congresses. 4. Pope—Infallibility—Congresses. 5. Catholic Church—Doctrinal and controversial works—Congresses. 6. Anglican Communion—Doctrinal and controversial works—Congresses. I. Anglican/Roman Catholic International Commission. Agreed statement on authority in the Church. 1977. II. Title.
BT91.H37 262'.8 79-305813
79 MARC

—AUTHORITY—HISTORY OF DOCTRINES

Riedl, Alfons.
Die kirchliche Lehrautorität in Fragen der Moral nach den Aussagen des Ersten Vatikanischen Konzils / Alfons Riedl. — Freiburg : Herder, 1979.
x, 534 p. ; 23 cm. — (Freiburger theologische Studien ; Bd. 117) GFR***
Originally presented as the author's thesis, Bonn.
Bibliography: p. 507-532.
ISBN 3-451-18449-4
1. Catholic Church—Teaching office. 2. Church—Authority—History of doctrines. 3. Christian ethics—Catholic authors. 4. Vatican Council, 1869-1870. I. Title. II. Series.
BX1746.R515 1979 79-375422
79 MARC

—BIBLICAL TEACHING

Brown, Raymond Edward.
The community of the beloved disciple / Raymond E. Brown. — New York : Paulist Press, c1979.
204 p. ; 21 cm.
Includes bibliographical references and indexes.
ISBN 0-8091-2174-3 : $3.95
1. Bible. N.T. Johannine literature—Criticism, interpretation, etc. 2. Church—Biblical teaching. I. Title.
BS2601.B76 230 78-65894
79 MARC

Künzel, Georg.
Studien zum Gemeindeverständnis des Matthäus-Evangeliums / Georg Künzel. — 1. Aufl. — Stuttgart : Calwer Verlag, 1978.
295 p. ; 22 cm. — (Calwer theologische Monographien : Reihe A, Bibelwissenschaft ; Bd. 10) GFR78-A
A revision of the author's thesis, Universität Erlangen-Nürnberg, Erlangen, 1975/76.
Bibliography: p. 269-295.
Includes index.
ISBN 3-7668-0583-5
1. Church—Biblical teaching. 2. Theology, Practical—Biblical teaching. 3. Bible. N.T. Matthew—Criticism, interpretation, etc. I. Title.
BS2545.C5K83 1978 78-380579
79 MARC

Stephens, Julius Harold.
The churches and the kingdom / J. Harold Stephens. — Westport, Conn. : Greenwood Press, 1978, c1959.
vi, 119 p. ; 23 cm.
Reprint of the ed. published by Broadman Press, Nashville.
Includes bibliographical references.
ISBN 0-313-20488-8
1. Kingdom of God—Biblical teaching. 2. Church—Biblical teaching. I. Title.
⌜BT94.S78 1978⌝ 231'.7 78-5676
78 MARC

—COMPARATIVE STUDIES see Polity (Religion)

—CONGRESSES

Associazione teologica italiana.
Coscienza e missione di Chiesa : atti del VII Congresso nazionale / Associazione teologica italiana. — Assisi : Cittadella editrice, 1977.
440 p. ; 20 cm. — (Teologia attraverso ..) It78-May
Includes bibliographical references.
L6000
1. Church—Congresses. 2. Mission of the church—Congresses. I. Title.
BX1746.A87 1977 78-368109
•78 MARC

Einheit der Kirche, Einheit der Menschheit : Perspektiven aus Theologie, Ethik u. Völkerrecht / mit Beitr. von Franz Böckle ... ⌜et al.⌝ ; hrsg. von Otto Hermann Pesch. — Freiburg im Breisgau ; Basel ; Wien : Herder, 1978.
175 p. ; 20 cm. GFR79-A
Contributions from a meeting sponsored by the Katholische Akademie Hamburg.
Bibliography: p. 170-175.
ISBN 3-451-18134-7 : DM16.80
1. Church—Congresses. 2. Man (Christian theology)—Congresses. I. Böckle, Franz. II. Pesch, Otto Hermann. III. Katholische Akademie Hamburg.
BX1746.E37 79-357730
79 MARC

In each place : towards a fellowship of local churches truly united : ⌜report of a consultation in Geneva, December 10-13, 1976⌝. — Geneva : World Council of Churches, ⌜1977⌝
iv, 92 p. ; 22 cm. Sw77-9459
Includes bibliographical references.
ISBN 2-8254-0551-5 : 7.90F
1. Christian union—Congresses. 2. Church—Congresses. I. World Council of Churches.
BX2.I5 262'.001 78-324905
79 MARC

CHURCH
—CONGRESSES (Continued)

Joint Lutheran/Roman Catholic Study Commission on the Gospel and the Church.
Evangelium, Welt, Kirche : Schlussbericht und Referate der Römisch-katholisch/evangelisch-lutherischen Studienkommission "Das Evangelium und die Kirche", 1967-1971 / auf Veranlassung des Lutherischen Weltbundes und des Sekretariats für die Einheit der Christen, hrsg. von Harding Meyer. — Frankfurt am Main : O. Lembeck, c1975.
455 p. ; 21 cm. GFR***
English or German.
Includes bibliographical references.
ISBN 3-87476-039-1
1. Church—Congresses. 2. Evangelistic work—Congresses. 3. Catholic Church—Relations—Lutheran Church—Congresses. 4. Luthern Church—Relations—Catholic Church—Congresses. I. Meyer, Harding. II. Lutheran World Federation. III. Catholic Church. Secretariatus ad Christianorum Unitatem Fovendam. IV. Title.
BV600.2.J58 1975 262 76-460030
76[r79]rev MARC

La Pira, Giorgio.
La Chiesa casa di Dio e del popolo / [relazioni di] La Pira, Barsotti, Lorenzi. — Trento : Centro di cultura A. Rosmini, 1976.
87 p. : ill. ; 24 cm. It78-June
Proceedings of a theological meeting held in Levico in 1972.
CONTENTS: La Pira, G. La Chiesa centro della storia.—Barsotti, D. La Chiesa spazio del sacro.—Lorenzi, G. La Chiesa luogo d'incontro e di preghiera.
1. Church—Congresses. I. Barsotti, Divo, 1914- joint author. II. Lorenzi, Guido, 1926- joint author. III. Title.
BX1746.L32 79-367026
*79 MARC

—FOUNDATION

see also Popes—Primacy

—HISTORY OF DOCTRINES

Brunner, Heinz, 1946-
Der organologische Kirchenbegriff in seiner Bedeutung für das ekklesiologische Denken des 19. Jahrhunderts / Heinz Brunner. — Frankfurt am Main ; Bern ; Las Vegas : Lang, 1979.
320 p. ; 21 cm. — (Europäische Hochschulschriften : Reihe 23, Theologie ; Bd. 118) GFR79-A
A revision of the author's thesis, Munich, 1978.
Bibliography: p. 301-318.
ISBN 3-8204-6470-0
1. Church—History of doctrines. I. Title. II. Series.
BX1746.B78 1979 79-379373
79 MARC

Klinger, Elmar.
Ekklesiologie der Neuzeit : Grundlegung bei Melchior Cano u. Entwicklung bis zum 2. Vatikan. Konzil / Elmar Klinger. — Freiburg im Breisgau ; Basel ; Wien : Herder, 1978.
263, [1] p. ; 23 cm. GFR78-A
Habilitationsschrift—Münster.
Bibliography: p. 256-[264]
ISBN 3-451-17435-9 : DM58.00
1. Church—History of doctrines. 2. Cano, Melchor, Bp., 1509?-1560. I. Title.
BV598.K55 79-378409
79 MARC

Wieh, Hermann.
Konzil und Gemeinde : e. systemat.-theol. Unters. zum Gemeindeverständnis d. 2. Vatikan. Konzils in pastoraler Absicht / Hermann Wieh. — 1. Aufl. — Frankfurt am Main : Knecht, 1978.
viii, 248 p. ; 23 cm. — (Frankfurter theologische Studien ; Bd. 25) GFR78-A
Originally presented as the author's thesis, Münster, 1977, under the title: Das Gemeindeverhältnis in den Diskussionen und Dokumenten des Zweiten Vatikanischen Konzils.
Bibliography: p. 229-243.
Includes index.
ISBN 3-7820-0402-7
1. Church—History of doctrines. 2. Vatican Council, 2d, 1962-1965. I. Title. II. Series.
BX1746.W54 1978 79-343926
79 MARC

—HISTORY OF DOCTRINES—MIDDLE AGES, 600-1500

Ryan, John Joseph.
The nature, structure, and function of the Church in William of Ockham / John J. Ryan. — Missoula, Mont. : Scholars Press, c1979.
63 p. ; 24 cm. — (AAR studies in religion ; no. 16 ISSN 0084-6287)
Includes bibliographical references.
ISBN 0-89130-230-1 (pbk.) : $4.50
1. Church—History of doctrines—Middle Ages, 600-1500. 2. Ockham, William, d. ca. 1349. I. Title. II. Series: American Academy of Religion. AAR studies in religion ; no. 16.
BV598.R9 262'.001 78-2891
78 MARC

—HISTORY OF DOCTRINES—20TH CENTURY

Innocenti, Ennio.
La Santa sede nella ecclesiologia del Concilio ecumenico Vaticano II / Ennio Innocenti ; con presentazione di Brunero Gherardini e giudizi di alcuni prelati. — 2. ed. — Rovigo : Istituto padano di arti grafiche, 1977.
317 p. ; 22 cm. It***
Includes bibliographical references and index.
L7500
1. Papacy—History—20th century. 2. Church—History of doctrines—20th century. 3. Vatican Council. 2d, 1962-1965. I. Title.
BX1389.I56 1977 78-395357
79 MARC

Timpe, Nikolaus.
Das kanonistische Kirchenbild vom Codex iuris canonici bis zum Beginn des Vaticanum Secundum : e. histor.-systemat. Unters. / von Nikolaus Timpe. — 1. Aufl. — Leipzig : St.-Benno-Verlag, 1978.
xiv, 293 p. ; 23 cm. — (Erfurter theologische Studien ; Bd. 36) GDR78-A
Bibliography: p. ix-xiv.
Includes index.
23.50M
1. Church—History of doctrines—20th century. 2. Canon law—History. I. Title. II. Series.
BX1746.T54 79-339996
79 MARC

—HISTORY OF DOCTRINES—ADDRESSES, ESSAYS, LECTURES

The Holy Spirit in the life of the church : from Biblical times to the present / edited by Paul D. Opsahl. — Minneapolis : Augsburg Pub. House, c1978.
287, [1] p. ; 22 cm.
Bibliography: p. [288]
ISBN 0-8066-1625-3 : $8.95
1. Holy Spirit—Biblical teaching—Addresses, essays, lectures. 2. Holy Spirit—History of doctrines—Addresses, essays, lectures. 3. Church—History of doctrines—Addresses, essays, lectures. 4. Theology, Lutheran—Addresses, essays, lectures. 5. Pentecostalism—Addresses, essays, lectures. I. Opsahl, Paul D.
BS680.H56H65 231'.3 77-84099
78 MARC

—INFALLIBILITY

see also Catholic Church—Infallibility; Church—Teaching office; Popes—Infallibility

—JUVENILE LITERATURE

Hogan, Benrice.
The church is a who / by Bernice Hogan ; illustrated by John Steiger. — St. Louis : Bethany Press, c1979.
31 p. : ill. ; 27 cm.
SUMMARY: Characterizes the church as people, not buildings.
ISBN 0-8272-0442-6 $5.95
1. Church—Juvenile literature. [1. Church] I. Steiger, John, 1923- II. Title.
BV600.2.H576 261.8 78-24087
78 MARC

—MARKS

see also Church—Unity

—MISSION see Mission of the church

—PURPOSE see Mission of the church

—REFORM see Church renewal

—RENEWAL see Church renewal

—STUDY AND TEACHING—AUDIO-VISUAL AIDS

McNulty, Edward N 1936-
Using AV's in the church / by Edward McNulty. — King of Prussia, Pa. : Religious Pub. Co., c1978-
v. : ill. ; 21 cm.
Vol. 3 also has title: Using audio-visuals in the church.
Bibliography: v. 1, p. 69-72; v. 3, p. 66-67.
CONTENTS: v. 1. Plugging in.—
—v. 3. Equipping the church.
I. Title. II. Title: Using audio-visuals in the church.
BV1535.M33 268'.635 79-102130
79 MARC

—TEACHING OFFICE

see also Catechetics; Christian education; Church and education; Preaching; Prohibited books; Theological seminaries

—UNITY—CONGRESSES

World Council of Churches. 3d Assembly, New Delhi, India, 1961.
New Delhi speaks about Christian witness, service, and unity; the message, appeal, and section reports. Edited by W. A. Visser't Hooft. New York, Association Press [1962]
124 p. 16 cm. (An Association Press reflection book)
1. Ecumenical movement—Congresses. 2. Evangelistic work—Congresses. 3. Church and social problems—Congress. 4. Church—Unity—Congresses. I. Visser 't Hooft, Willem Adolph, 1900- ed. II. Title.
BX6.W77 1961b 280.1 62-11031
[r79]rev MARC

—UNITY—HISTORY OF DOCTRINES

Baier, Klaus Alois.
Unitas ex auditu : die Einheit der Kirche im Rahmen der Theologie Karl Barths / Klaus Alois Baier. — Bern : P. Lang, c1978.
284 p. ; 23 cm. — (Basler und Berner Studien zur historischen und systematischen Theologie ; Bd. 35) Sw***
Originally presented as the author's thesis, Universität Erlangen-Nürnberg, Erlangen, 1974.
Summary in English.
"Schriften Karl Barths": p. 255-261.
Bibliography: p. 262-279.
ISBN 3-261-03046-1
1. Church—Unity—History of doctrines. 2. Christian union—History. 3. Barth, Karl, 1886-1968. I. Title. II. Series.
BV601.5.B34 1978 78-390081
79 MARC

CHURCH, APOSTOLIC see Church history—Primitive and early church, ca. 30-600

CHURCH ADMINISTRATION see Church management

CHURCH ADMINISTRATION, INDIGENOUS see Indigenous Church administration

CHURCH AND CIVIL RIGHTS

Fletcher, Catherine.
Am I free? / Catherine Fletcher. — Niles, Ill. : Argus Communications, c1975.
70 p. : ill. ; 23 cm.
ISBN 0-913592-43-9 : $1.95
1. Freedom (Theology) 2. Civil rights. 3. Church and civil rights. I. Title.
BT810.2.F58 261.8 75-18972
00 MARC

Honecker, Martin.
Das Recht des Menschen : Einf. in d. evang. Sozialethik / Martin Honecker. — Orig.-Ausg., 1. Aufl. — Gütersloh : Gütersloher Verlagshaus Mohn, 1978.
217 p. ; 19 cm. — (Gütersloher Taschenbücher ; 290 : Siebenstern) GFR78-A
"Eine Veröffentlichung des Comenius-Instituts, Münster."
Includes bibliographical references.
ISBN 3-579-03690-4
1. Church and civil rights. 2. Christian ethics. 3. Social ethics. I. Title.
BT738.15.H66 79-341747
79 MARC

—ADDRESSES, ESSAYS, LECTURES

A Christian declaration on human rights : theological studies of the World Alliance of Reformed Churches / edited by Allen O. Miller. — Grand Rapids, MI : Eerdmans, c1977.
190 p. ; 21 cm.
ISBN 0-8028-1717-3
1. Church and civil rights—Addresses, essays, lectures. I. Miller, Allen O. II. World Alliance of Reformed Churches (Presbyterian and Congregational).
BT738.15.C47 261 77-2796
77[r79]rev MARC

CHURCH AND CIVIL RIGHTS (Continued)

—CONGRESSES

Association of Korean Christian Scholars in North America.
Human rights in minority perspectives / edited by Dong Soo Kim and Byong-suh Kim. — Montclair, N.J. : Association of Korean Christian Scholars in North America, c1979.
xiii, 292, xv-xviii p. ; 22 cm. — (Publication series ; The Association of Korean Christian Scholars in North America ; no. 5)
Includes seven articles presented at the 12th annual meeting of the Association of Korean Christian Scholars in North America, held May 25-27, 1978 in Toronto.
Bibliography: p. 269-286.
$8.00
1. Church and civil rights—Congresses. 2. Civil rights—Congresses. 3. Civil rights—Korea—Congresses. I. Kim, Dong Soo. II. Kim, Byong-suh. III. Title. IV. Series: Association of Korean Christian Scholars in North America. Association of Korean Christian Scholars in North America publication series ; no. 5.
BT738.15.A84 1979 261.8 79-88571
79 MARC

CHURCH AND COLLEGE

see also Church and education; Church colleges; Education and state

—CONGRESSES

The Church's ministry in higher education in Pennsylvania : the next ten years : Pennsylvania Commission for United Ministries in Higher Education, tenth anniversary celebration, 1967-1977. — [s.l.] : UMHE, [c1978]
32 p. ; 22 cm.
Cover title.
Two presentations given at the annual meeting of the commission, State College, Pa., Mar. 10-11, 1977.
CONTENTS: Kline, C. M. Pennsylvania higher education.—Palmer, P. The campus and the claims of the Gospel.
1. Church and college—Congresses. 2. Education, Higher—Pennsylvania—Congresses. I. Kline, Caryl M. Pennsylvania higher education. 1978. II. Palmer, Parker J. The campus and the claims of the gospel. 1978.
LC383.C53 377 78-107403
78 MARC

—UNITED STATES

Hesburgh, Teodore Martin, 1917-
The Hesburgh papers : higher values in higher education / Theodore M. Hesburgh. — Kansas City, Kan. : Andrews and McMeel, c1979.
xii, 206 p. ; 24 cm.
ISBN 0-3362-5908-4
1. Catholic universities and colleges—United States. 2. Education, Higher—United States—Aims and objectives. 3. Academic freedom. 4. Church and college—United States. I. Title.
LC501.H44 378'.01'0973 79-10408
79 MARC

—UNITED STATES—PERIODICALS

National Institute for Campus Ministries.
The NICM journal for Jews and Christians in higher education. v. 1-
winter 1976-
[Newton Centre, Mass., National Institute for Campus Ministries]
v. 23 cm. quarterly.
Key title: The NICM journal for Jews and Christians in higher education, ISSN 0362-0794
1. College students—United States—Religious life—Periodicals. 2. Church and college—United States—Periodicals. I. National Institute for Campus Ministries. Journal for Jews and Christians in higher education. II. Title.
LB3609.N33a 378.1'98'0973 79-642695
MARC-S

CHURCH AND EDUCATION

see also Church and college; Church schools; Religion in the public schools; Theology—Study and teaching

Scuola, terra di nessuno : la violenza anticristiana nelle scuole / [a cura del] Comitato per la difesa delle libertà civili ; prefazione di Alberto Crespi. — Torino : Gribaudi, [1978]
116 p., [4] leaves of plates : ill. ; 19 cm. It78-June
"Documenti": p. [77]-116.
L2200
1. Church and education. I. Comitato per la difesa delle libertà civili.
LC371.S38 78-379933
•78 MARC

—ADDRESSES ESSAYS, LECTURES

Kirche und Schule : ÖVP u. SPÖ über ihr Verhältnis zur Kirche, Streitgespräch Cap/Kendol über Abschaffung des Religionsunterrichts ... / [Red., Ilse Arnold, Ursula Pasterk, Walter Sauer ; Zeichn., S. Friedemann]. — Wien ; München : Jugend & Volk, 1978.
104 p. ; 19 cm. — (Schulheft ; 1978, 4) Au79-10-24
Cover title.
Includes bibliographical references.
ISBN 3-7141-5714-X : S48.00
1. Church and education—Addresses, essays, lectures. 2. Österreichische Volkspartei—Addresses, essays, lectures. 3. Sozialistische Partei Österreichs—Addresses, essays, lectures. I. Series.
LC370.K52 79-372766
•79 MARC

Towards a Christian theology of education / [by] M. Calvert ... [et al.]. — London : Church Literature Association for the editors, 1977.
[2], 5, [1] p. ; 21 cm. — (Dolphin papers ; 7) GB78-05769
Cover title.
£0.15
1. Christian education—Addresses essays, lectures. 2. Church and education—Addresses essays, lectures. I. Calvert, M.
BV1473.T684 261.1 78-320319
78 MARC

—CONGRESSES

An Encounter with education for liberation and community. — [Singapore : Christian Conference of Asia], c1975, 1977 printing.
70 p. : ill. ; 22 cm.
1. Education—Asia, Southeastern—Congresses. 2. Church and education—Congresses. I. Christian Conference of Asia.
LA1059.A7842E52 370'.959 78-940027
79 MARC

—HISTORY

Bellinotto, Elena Ofelia.
La educación en el Cristianismo / Elena Ofelia Bellinotto. — Buenos Aires : Editorial Huemul, c1965.
125 p. ; 18 cm. — (Colección Breviarios pedagógicos ; 4)
Bibliography: p. [123]-[124]
1. Church and education—History. 2. Fathers of the church. I. Title.
LA95.B44 78-383812
78 MARC

Olivestam, Carl-Eber, 1942-
Idé och politik : de politiska partierna-skolan och kristendomen : en studie i svensk skolpolitik under 1940-talet = Ideology and politics : the political parties-school and Christianity : a study of educational politics in Sweden during the 1940s' / Carl-Eber Olivestam. — Uppsala : Sv. kyrkohistoriska fören. : Teologiska inst., Uppsala univ., 1977.
352 p. : ill. ; 24 cm. — (Skrifter utg. av Svenska kyrkohistoriska föreningen ; II., ny följd, 27 ISSN 0491-6786)
Extra t.p. with thesis statement inserted.
Thesis Uppsala.
Summary in English.
Bibliography: p. 334-347.
Includes index.
ISBN 9185582018
1. Education and state—Sweden—History. 2. Church and education—History. 3. Politics and education—History. I. Title. II. Title: Ideology and politics. III. Series: Svenska kyrkohistoriska föreningen. Skrifter ; II., ny följd, 27.
LC93.S85O44 379.485 78-373594
78 MARC

—EUROPE—CONGRESSES

Internationales Seminar Europäische Bildung, Herausforderung an die Kirche, Marienberg (Westerwald) Ger., 1977.
Europäische Bildung, Herausforderung an die Kirche : Beitr. zur künftigen Bildungsarbeit in Europa / Christfried Röger, Herbert Schultze (Hg.) ; mit Beitr. von Joachim Dumrese ... [et al.]. — Kastellaun : Henn, 1978.
x, 218 p. ; 20 cm. — (Erziehen heute : Sonderbände ; Sonderheft 6)
GFR78-A
Papers presented at the Internationales Seminar Europäische Bildung, Herausforderung an die Kirche, held Oct. 10-14, 1977, Bad Marienberg.
One paper in English.
Includes bibliographical references.
ISBN 3-450-17925-X : DM12.80
1. Education—Europe—Congresses. 2. Church and education—Europe—Congresses. I. Röger, Christfried. II. Schultze, Herbert. III. Dumrese, Joachim, 1935- . IV. Title. V. Series: Erziehen heute : Sonderheft ; 6.
LA622.I55 1977 79-347964
79 MARC

—FINLAND—ADDRESSES, ESSAYS, LECTURES

Kasvatus ihmisyyteen : Seurakuntaopiston ja Luther-opiston juhlakirja / [toim., Veikko Pöyhönen ja Simo Huhta ; kirj., Erkki Aho ... et al.]. — [Helsinki] : Kirjapaja, 1977.
166 p. : ill. ; 21 cm. FI77Dec.
ISBN 9516211666 : Fmk35.00
1. Lutheran Church in Finland—Education—Addresses, essays, lectures. 2. Church and education—Finland—Addresses, essays, lectures. 3. Seurakuntaopisto—Addresses, essays, lectures. 4. Luther-opisto—Addresses, essays, lectures. I. Pöyhönen, Veikko. II. Huhta, Simo.
BX8029.K38 79-359046
79 MARC

—GERMANY—ADDRESSES, ESSAYS, LECTURES

Leben und Erziehen durch Glauben : Perspektiven bildungspolit. Mitverantwortung d. evang. Kirche / [e. Veröff. d. Comenius-Inst. Münster] ; Red., Hans Kratzert in Zusammenarbeit mit Dieter Aschenbrenner ... [et al.]. — 1. Aufl. — Gütersloh : Gütersloher Verlagshaus Mohn, 1978.
224 p. ; 23 cm. GFR78-A
Includes bibliographical references.
ISBN 3-579-04726-4 : DM24.00
1. Church and education in Germany—Addresses, essays, lectures. 2. Education—Germany, West—Addresses, essays, lectures. I. Kratzert, Hans. II. Comenius-Institut.
LC431.G3L4 78-376909
78 MARC

—PHILIPPINE ISLANDS

Panlasigui, Isidoro.
Religion and education on trial / by Isidoro Panlasigui. — Diliman, Quezon City : [s.n.], 1956.
v, 102 p. ; 19 cm.
1. Church and education—Philippine Islands. I. Title.
LC373.P36 201 78-325160
79 MARC

—SPAIN

Piqué, Antoni.
Escola cristiana, evangelització o tancament / Antoni Piqué. — 1. ed. — [Montserrat] : Publicaciones de l'Abadia de Montserrat, 1976.
138 p. ; 19 cm. — (Col·lecció Saurí ; 31) Sp***
Includes bibliographical references.
ISBN 8472021289
1. Church and education in Spain. 2. Religious education of children. I. Title.
LC431.S7P56 79-345670
79 MARC

—SWEDEN

Sparrman, Karl Johan, 1917-
Folkhögskolor med profil : rörelsekaraktären hos svenska folkhögskolor med anknytning till kristna samfund / Karl Johan Sparrman. — Umeå : Univ. ; Stockholm : Gummesson (distr.), 1978.
[14] leaves, 185, [1] p., [11] leaves : diagr., tables ; 29 cm. S78-25
Thesis—Umeå.
Summary in English.
Includes bibliography.
kr30.05
1. Church and education in Sweden. 2. Education, Secondary—Sweden. I. Title.
LC431.S8S6 1978 79-365818
79 MARC

—SWITZERLAND

Bünter, Willy.
Bildung, Freizeit, Massenmedien / zusgest. und kommentiert von Willy Bünter und Willy Kaufmann. — Zürich ; Einsiedeln ; Köln : Benziger, c1977.
144 p. ; 18 cm. — (Die Synode zum Thema ...) Sw77-4592
Bibliography: p. 82.
ISBN 3-545-21028-6 : 9.80F
1. Church and education—Switzerland. 2. Mass media in religion—Switzerland. 3. Schweizer Synode Zweiundsiebzig. I. Kaufmann, Willy, joint author. II. Title.
LC370.B83 78-400395
•79 MARC

—UNITED STATES

Thayer, Vivian Trow, 1886-
Religion in public education / V. T. Thayer. — Westport, Conn. : Greenwood Press, 1979, c1947.
xi, 212 p. ; 23 cm.
Reprint of the ed. published by Viking Press, New York.
Includes bibliographical references and index.
ISBN 0-313-21212-0
1. Religious education—United States. 2. Church and education in the United States. I. Title.
LC111.T5 1979 377'.1 78-12385
78 MARC

CHURCH AND HOMOSEXUALITY
see Homosexuality and Christianity

CHURCH AND INTERNATIONAL AFFAIRS see Christianity and international affairs

CHURCH AND LABOR

see also Catholic Worker Movement; Christianity and economics; Church and social problems; Trade-unions, Catholic; Work (Theology)

Cantono, Alessandro.
Cristianesimo e lavoro / Alessandro Cantono. — [s.l.] : Società apostolato stampa, [1945]
93 p. ; 19 cm.
Includes bibliographies.
1. Work. 2. Church and labor. I. Title.
HD4904.C253 301.5'5 79-346484
79 MARC

—GERMANY

Khodorkovskiĭ, Leonid Dmitrievich.
(Katolit͡sizm i rabochiĭ klass Germanii, 1871-1933)
Католицизм и рабочий класс Германии, 1871–1933 / Л. Д. Ходорковский. — Москва : Наука, 1978.
334 p. ; 21 cm. USSR 78
At head of title: Akademii͡a nauk SSSR. Institut mezhdunarodnogo rabochego dvizhenii͡a.
Bibliography: p. 313-[328]
Includes index.
2.10rub
1. Church and labor—Germany. 2. Catholic Church in Germany. 3. Deutsche Zentrumpartei. I. Title.
HD6338.2.G3K48 79-412943

—GERMANY—HISTORY

Brandmüller, Walter.
Damals geschehen, heute diskutiert : d. 1. Vatikan. Konzil 1869/70. Kath. Kirche u. Arbeiterschaft im 19. Jh. / Walter Brandmüller. — S[ank]t Ottilien : EOS-Verlag, 1977.
92 p. ; 19 cm. GFR77-A47
Includes bibliographical references.
ISBN 3-88096-028-3 : DM6.80
1. Vatican Council, 1869-1870. 2. Church and labor—Germany—History. I. Title.
BX830 1869.B72 78-386783
79 MARC

—ITALY—FLORENCE

Libro bianco sulle Officine Galileo. — [Firenze] : Libreria editrice fiorentina, [1959]
xi, 209 p. ; 24 cm.
"Raccolta di documenti."
1. Labor disputes—Italy—Florence. 2. Officine Galileo. 3. Church and labor—Italy—Florence.
HD5385.F6L5 79-370914
79 MARC

—KOREA

Hong, Chi-yŏng.
(Sanŏp sŏn'gyo nŭn muŏt ŭl norina?)
산업선교는무엇을노리나? / 홍지영지음. — 서울 : 금란출판사, 1977.
201 p. : 18 cm.
W600
1. Missions—Korea—Industrial work. 2. Chaplains, Industrial—Korea. 3. Church and labor—Korea. I. Title.
BV2695.W6H66 79-825770

Hong, Chi-yŏng.
("Sanŏp sŏn'gyo" wae munjesi toenŭn'ga?)
"産業宣教"왜問題視되는가? : 韓國教界의참된理解를위하여 / 洪志英. — 서울 : 基督教思想社, 1978.
101 p. ; 21 cm.
W300
1. Missions—Korea—Industrial work. 2. Chaplains, Industrial—Korea. 3. Church and labor—Korea. I. Title.
BV2695.W6H67 79-825771

—UTAH—HISTORY

Davies, Joseph Kenneth, 1925-
Deseret's sons of toil : a history of the worker movements of Territorial Utah, 1852-1896 / J. Kenneth Davies. — Salt Lake City : Olympus Pub. Co., c1977.
264 p. : ill. ; 24 cm.
Bibliography: p. 227-252.
Includes indexes.
ISBN 0-913420-64-6 : $12.50
1. Church and labor—Utah—History. 2. Church of Jesus Christ of Latter-Day Saints. 3. Trade-unions—Utah—History. I. Title.
HD6338.2.U52U82 261.8'5 76-26459
76 MARC

CHURCH AND RACE RELATIONS

Adler, Elisabeth.
Oekumene im Kampf gegen Rassismus : ein erster Anfang : Programm d. Oekumen. Rates d. Kirchen z. Bekämpfung d. Rassismus : Bericht über d. ersten 5 Jahre / von Elisabeth Adler ; mit e. Geleitw. von Philip Potter u. e. Beitr. von Gerd-Rainer Fendler z. Diskussionslage in d. Bundesrepublik Deutschland. — Bielefeld ; Frandfurt (Main) : Eckart, 1975.
166 p. ; 21 cm. — (Epd-Dokumentation ; Bd. 14) GFR75-A
An enlargement of the author's A small beginning, published in 1974.
Includes bibliographical references.
ISBN 3-7703-0106-4 : DM12.00
1. World Council of Churches. Programme to Combat Racism. 2. Church and race relations. I. Title. II. Series: Evangelischer Pressedienst. EPD-Pressedienst. EPD-Dokumentation ; Bd. 14.
BT734.2.A3815 76-456182
76[r79]rev MARC

Adler, Elisabeth.
A small beginning. An assessment of the first five years of the Programme to Combat Racism [by] Elisabeth Adler. Geneva, World Council of Churches, (1974).
vi, 102 p. 21 cm. 5.90F Sw74-A-3029
An enl. German ed. was published in 1975 under the title: Oekumene im Kampf gegen Rassismus.
Includes bibliographical references.
1. World Council of Churches. Programme to Combat Racism. 2. Church and race relations. I. Title.
BT734.2.A38 261.8'34'51042 74-182722
74[r79]rev MARC

—ADDRESSES, ESSAYS, LECTURES

Frieden, Versöhnung und Menschenrechte. — Gütersloh : Gütersloher Verlagshaus Mohn, 1978-
v. ; 19 cm. — (Die Denkschriften der Evangelischen Kirche in Deutschland ; Bd. 1, T. 1-) (Gütersloher Taschenbücher ; 13- : Siebenstern)
GFR78-A
Includes bibliographical references.
ISBN 3-579-04801-5 (v. 1, pt. 1) : DM12.80 (v. 1, pt. 1)
1. Sociology, Christian—Addresses, essays, lectures. 2. Church work with refugees—Addresses, essays, lectures. 3. Church and race relations—Addresses, essays, lectures. 4. Christianity and international affairs—Addresses, essays, lectures. 5. Peace (Theology)—Addresses, essays, lectures. I. Series.
BT738.F68 78-380301
79 MARC

—HISTORY—ADDRESSES, ESSAYS, LECTURES

Bassarak, Gerhard, 1918-
Herausforderung des Gewissens : über d. ökumen. Beitrag zum Kampf gegen d. Rassismus / Gerhard Bassarak, Günter Wirth. — 1. Aufl. — Berlin : Union-Verlag, 1977.
77 p. ; 20 cm. — (Fakten, Argumente) GDR78-A
Revised articles that first appeared in the periodical Asien, Afrika, Lateinamerika, v. 4, Heft 3, 1976.
Includes bibliographical references.
4.80M
1. Church and race relations—History—Addresses, essays, lectures. 2. Ecumenical movement—Addresses, essays, lectures. I. Wirth, Günter, joint author. II. Asien, Afrika, Lateinamerika. III. Title.
BT734.2.B35 78-381302
79 MARC

—AFRICA, SOUTHERN

Loth, Heinrich.
Apartheid und Kirchen : südafrikan. Christen im Widerstand : e. histor. Unters. / Heinrich Loth. — 1. Aufl. — Köln : Pahl-Rugenstein, 1977.
237 p. : ill., maps ; 20 cm. GFR78-A
Includes bibliographical references.
ISBN 3-7609-0308-8 : DM14.80
1. Africa, Southern—Race relations. 2. Church and race relations—Africa, Southern. I. Title.
DT737.L67 301.45'1'0968 78-394477
79 MARC

—AFRICA, SOUTHERN—ADDRESSES, ESSAYS, LECTURES

Ecumenical Consultation on Christian Practice and Desirable Action in Social Change and Race Relations in Southern Africa, Kitwe, Zambia, 1964.
Some addresses at an Ecumenical Consultation on Christian Practice and Desirable Action in Social Change and Race Relations in Southern Africa : held at Mindolo Ecumenical Foundation, Kitwe, Northern Rhodesia, May 25-June 2, 1964. — Geneva : Dept. on Church and Society, World Council of Churches ; Sydney : distributed for Ecumenical Book & Documents Club [by] Australian Council of Churches, 1964.
52 p. ; 30 cm.
Cover title: Race relations in southern Africa.
A selection from the addresses originally published as no. 5 in the series Race relations in ecumenical perspective, issued by the Secretariat on Racial and Ethnic Relations of the World Council of Churches.
1. Africa, Southern—Race relations—Addresses, essays, lectures. 2. Church and race relations—Africa, Southern—Addresses, essays, lectures. I. World Council of Churches. Division of Studies. Dept. on Church and Society. II. Title: Some addresses at an Ecumenical Consultation ... III. Title: Race relations in southern Africa.
DT737.E26 1964a 261.8'34'510968 79-373092
79 MARC

—AFRICA, SOUTHERN—CONGRESSES

Ecumenical Consultation on Christian Practice and Desirable Action in Social Change and Race Relations in Southern Africa, Kitwe, Zambia, 1964.
Christians and race relations in Southern Africa : report on an Ecumenical Consultation on Christian Practice and Desirable Action in Social Change and Race Relations in Southern Africa, held at Kitwe, Zambia, May 25-June 2, 1964 .... — Geneva : Dept. on Church and Society, World Council of Churches, 1964.
36 p. ; 21 cm. — (Oikoumene)
"Convened by the Department on Church and Society of the World Council of Churches in co-operation with the Mindolo Ecumenical Foundation and the South African Institute of Race Relations."
1. Africa, Southern—Race relations—Congresses. 2. Church and race relations—Africa, Southern—Congresses. I. World Council of Churches. Dept. on Church and Society. II. Mindolo Ecumenical Foundation. III. South African Institute of Race Relations. IV. Title.
DT737.E26 1964 261.8'34'510968 79-311642
79 MARC

—GERMANY, WEST

Meyers-Herwartz, Christel, 1947-
Die Rezeption des Antirassismus-Programms in der EKD / Christel Meyers-Herwartz. — Stuttgart ; Berlin ; Köln ; Mainz : Kohlhammer, 1979.
378 p. ; 24 cm. GFR79-A
Bibliography: p. 361-378.
ISBN 3-17-004713-2
1. Evangelische Kirche in Deutschland—History. 2. World Council of Churches. Programme to Combat Racism. 3. Church and race relations—Germany, West. 4. Germany, West—Church history. I. Title.
BX8020.M48 79-357963
79 MARC

—ILLINOIS—CHICAGO

Ziegenhals, Walter E 1928-
Urban churches in transition : reflections on selected problems and approaches to churches and communities in racial transition based on the Chicago experience / Walter E. Ziegenhals. — New York : Pilgrim Press, c1978.
201 p. ; 21 cm.
Bibliography: p. 195-201.
ISBN 0-8298-0355-6 : $5.95
1. City churches—Illinois—Chicago. 2. Church and race relations—Illinois—Chicago. 3. Chicago—Church history. 4. Chicago—Race relations. I. Title.
BV637.Z53 78-16422
261.8'34'510420977311
78 MARC

—LOUISIANA

Labbé, Dolores Egger.
Jim Crow comes to church : the establishment of segregated Catholic parishes in South Louisiana / Dolores Egger Labbé. — New York : Arno Press, 1978 [c1971]
104 p. ; 24 cm. — (The American Catholic tradition)
Reprint of the 2d ed. published in 1971 by the University of Southwestern Louisiana, Lafayette, which was issued as no. 4 of USL history series.
Bibliography: p. 96-100.
Includes index.
ISBN 0-405-10838-9
1. Church and race relations—Louisiana. 2. Catholic Church in Louisiana. I. Title. II. Series. III. Series: Louisiana. University of Southwestern Louisiana, Lafayette. The U.S.L. history series, no. 4.
[BX1415.L9L3 1978] 77-11295
261.8'54'51960730763
78 MARC

—SOUTH AFRICA

De Gruchy, John W
The church struggle in South Africa / by John W. De Gruchy. — Grand Rapids : W. B. Eerdmans Pub. Co., c1979.
xv, 267 p. : map ; 21 cm.
Includes bibliographical references and index.
ISBN 0-8028-1786-6
1. South Africa—Race relations. 2. Church and race relations—South Africa. 3. South Africa—Church history. I. Title.
DT763.D397 261.8'34'51968 78-26761
78 MARC

Desmond, Cosmas.
Christians or capitalists? : Christianity and politics in South Africa / Cosmas Desmond. — London : Bowerdean Press, 1978.
160 p. ; 23 cm. GB***
Bibliography: p. [159]-160.
ISBN 0-906097-03-7 : £7.00. ISBN 0-906097-04-5 pbk.
1. Christianity and politics. 2. Church and race relations—South Africa. 3. Christianity and capitalism. 4. Socialism. 5. South Africa—Race relations. 6. South Africa—Politics and government—1961- I. Title.
BR115.P7D457 261.1'0968 78-325747
79 MARC

Kotzé, Jacobus Cornelius Gideon.
Wie beset hierdie bedreigde huis? [Deur] J. C. G. Kotzé. Stellenbosch, Kosmo-Uitgewery, 1964.
214 p. 22 cm.
Includes bibliographical references.
1. South Africa—Race relations. 2. Church and race relations—South Africa. I. Title.
DT763.K64 78-411932
79 MARC

CHURCH AND RACE RELATIONS —SOUTH AFRICA (Continued)

Lückhoff, A H
Cottesloe / A. H. Lückhoff. — Kaapstad : Tafelberg, 1978.
197 p., [2] leaves of plates : ill. ; 22 cm. SA***
Bibliography: p. 187-194.
Includes index.
ISBN 0-624-01178-X
1. Cottesloe Consultation, Johannesburg, 1960. 2. South Africa—Church history. 3. Church and race relations—South Africa. 4. World Council of Churches—History. I. Title.
BR1450.L83 79-353243
79 MARC

Pro Mundi Vita (Society)
Some elements for a better understanding of apartheid in South Africa. — Brussels : Pro Mundi Vita, 1978.
35 p. : ill. ; 27 cm. — (Pro Mundi Vita bulletin ; 72) Be***
Includes bibliographical references.
1. South Africa—Race relations. 2. Church and race relations—South Africa. I. Title. II. Series: Pro Mundi Vita (Society). Bulletin ; 72.
DT763.P78 1978 301.45'1'0968 79-313453
79 MARC

Regehr, Ernie.
Perceptions of apartheid : the churches and political change in South Africa / Ernie Rehehr. — Scottsdale, Pa. : Herald Press, c1979.
309 p. : map ; 23 cm.
Includes bibliographical references and index.
1. South Africa—Race relations. 2. Church and race relations—South Africa. 3. South Africa—Church history. I. Title.
DT763.R387 1979 261.8'34'510968 79-9172
79 MARC

Smith, Nicolaas Johannes.
Die planting van afsonderlike kerke vir nieblanke bevolkingsgroepe deur die Nederduitse Gereformeerde Kerk in Suid-Afrika : 'n histories-missiologiese analise en ekklesiologiese beoordeling / deur Nicolaas Johannes Smith. — Stellenbosch : Universiteit van Stellenbosch, 1972, i.e. 1973.
230 p. ; 24 cm. — (Annale - Universiteit van Stellenbosch ; v. 34, serie B, no. 2) SA***
Thesis—University of Pretoria.
Summary in English.
Bibliography: p. 218-230.
ISBN 0-949954-78-0
1. Nederduitse Gereformeerde Kerk (South Africa)—Membership. 2. Nederduitse Gereformeerde Kerk (South Africa)—Missions. 3. Indigenous church administration. 4. Church and race relations—South Africa. I. Title. II. Series: Stellenbosch, South Africa. University. Annale : Reeks B ; v. 34, no. 2.
AS613.S8 vol. 34, ser. B, no. 2 79-359579
[BX9595.S6] 79 MARC

Villa-Vicencio, Charles M C
The Theology of apartheid / Charles Villa-Vicencio. — Cape Town : Methodist Publishing House, [1977]
38 p. ; 19 cm. SA78
Bibliography: p. 38.
ISBN 0-949942-21-9 : R0.70
1. South Africa—Race relations. 2. Race (Theology) 3. Church and race relations—South Africa. I. Title.
DT763.V49 261.8'34'51968 79-305448
79 MARC

—SOUTH AFRICA—ADDRESSES, ESSAYS, LECTURES

Moorhouse, Fred.
Politics, nonviolence, and social justice : an African perspective / Fred Moorhouse. — [New Malden, Eng.] : Fellowship of Reconciliation, [1977?]
20 p. ; 21 cm. — (Alex Wood memorial lecture ; 1977) GB***
Includes bibliographical references.
£0.25
1. South Africa—Race relations—Addresses, essays, lectures. 2. Church and race relations—South Africa—Addresses, essays, lectures. I. Title. II. Series.
DT763.M586 261.8'34'510968 78-316564
78 MARC

—UNITED STATES

Clark, Mary T 1913-
Discrimination today ; guidelines for civic action, by Mary T. Clark. Foreword by John J. Wright. New York, Hobbs, Dorman [1966]
372 p. 22 cm.
Includes bibliographies.
1. Afro-Americans—Civil rights. 2. Church and race relations—United States. I. Title.
E185.615.C595 301.45'19'6073 65-20271
[r79]rev MARC

Jones, William A 1934-
God in the ghetto / by William A. Jones, Jr. ; [cover photo. by Cheryl Fort]. — Elgin, IL : Progressive Baptist Pub. House, c1979.
164 p. ; 22 cm.
Includes bibliographical references.
ISBN 0-89191-134-0 : $3.00
1. United States—Race relations. 2. Church and race relations—United States. 3. Afro-Americans—Civil rights. 4. Afro-Americans—Economic conditions. 5. Afro-Americans—Religion. 6. Sermons, American. 7. Baptists—Sermons. I. Title.
E185.61.J834 301.45'19'6073 78-71193
79 MARC

—UNITED STATES—CONGRESSES

National Conference on Religion and Race, Chicago, 1963.
Race : challenge to religion : original essays and An appeal to the conscience / edited by Mathew Ahmann. — Westport, Conn. : Greenwood Press, 1979, c1963.
xiv, 178 p. ; 22 cm.
"Convened by agencies of the National Council of Churches, the Synagogue Council of America, and the National Catholic Welfare Conference."
Reprint of the ed. published by H. Regnery Co., Chicago.
Includes bibliographical references.
ISBN 0-313-20796-8
1. Religion and race—Congresses. 2. Church and race relations—United States—Congresses. I. Ahmann, Mathew H. II. National Council of the Churches of Christ in the United States of America. Dept. of Racial and Cultural Relations. III. Title.
[BL65.R3N37 1979] 261.8'34'5196073 78-24276
78 MARC

CHURCH AND REFUGEE PROBLEMS

—AFRICA

All Africa Conference of Churches. Refugee Dept.
Africa's one million refugees : from where they escape, where they find asylum. — [Nairobi : Refugee Dept., All Africa Conference of Churches, 1975?]
22 p. : ill. ; 16 x 24 cm.
Cover title: The African refugee story.
1. Africa—Politics and government—1960- 2. Refugees—Africa. 3. Christianity and refugee problems—Africa. 4. Church and state in Africa. 5. Church work with refugees—Africa. I. Title. II. Title: The African refugee story.
DT30.A467 1975 361.5'3'096 79-314375
79 MARC

CHURCH AND SLAVERY see Slavery and the church

CHURCH AND SOCIAL PROBLEMS

see also Abortion—Religious aspects—Catholic Church; Birth control—Religious aspects; Christianity and international affairs; Church and labor; Church and race relations; Church and refugee problems; Church charities; Civilization, Christian; Liberation theology; Slavery and the church; Socialism, Christian; Sociology, Christian; Sterilization (Birth control)—Moral and religious aspects

Bellenzier Garutti, Maria Teresa.
Donna e famiglia, realtà a confronto / [di] M. T. Garutti Bellenzier. — Roma : A.V.E., 1977.
180 p. ; 22 cm. — (Famiglia e pastorale ; 17) It78-May
Bibliography: p. 175-177.
L3000
1. Family. 2. Feminism. 3. Church and social problems. I. Title.
HQ741.B4 78-378098
*78 MARC

Cormack, Una.
The Church and social work / by Una Cormack. — [Liverpool] : Liverpool Institute of Socio-Religious Studies [for] the Conference of Major Religious Superiors of England and Wales, 1977.
[3], 32 p. ; 21 cm. — (Pastoral investigation of social trends : Working paper ; no. 7) GB77-26954
Bibliography: p. 30-31.
ISBN 0-905052-07-2 : £0.60
1. Church and social problems. 2. Social service. 3. Sociology, Christian (Catholic) I. Title. II. Series.
HV530.C648 361.7'5 78-318211
78 MARC

Esposito, Rosario Francesco, 1921-
La massificazione non esiste : per un'interpretazione e un uso costruttivo dei mass media / [di] Rosario F. Esposito. — Roma : Edizioni paoline, 1978.
352 p. : ill. ; 21 cm. — (Comunicazione : Problemi ; 3) It78-Nov
Bibliography: p. 347-352.
Includes index.
L5000
1. Mass media—Social aspects. 2. Church and social problems. I. Title. II. Series.
HM258.E76 78-396528
*79 MARC

(Kirisutokyō shakai fukushi gaisetsu)
キリスト教社会福祉概説 / 日本基督教社会福祉学会, 日本基督教社会事業同盟編. — 東京 : 日本基督教団出版局, 1978.
335 p. ; 19 cm. Ja 79-4655
Bibliography: p. 317-330.
¥1800
1. Church charities. 2. Church and social problems. 3. Social service—Japan. I. Nihon Kirisutokyō Shakai Fukushi Gakkai. II. Nihon Kirisutokyō Shakai Jigyō Dōmei.
HV530.K57 79-807703

Verdonk, André.
Antwoord op abortus, vivisectie, zelfmoord, drugs. — [De Hoef, Oostzijde 106] : Verdonk, [1977]
94 p. ; 21 cm. Ne78-1
ISBN 9090000569 : fl 7.50
1. Church and social problems. 2. Abortion—Religious aspects. 3. Vivisection—Moral and religious aspects. 4. Suicide—Moral and religious aspects. 5. Drug abuse—Moral and religious aspects. I. Title.
HN31.V42 78-378961
*78 MARC

—ADDRESSES, ESSAYS, LECTURES

Page, Kirby, 1890-1957.
Kirby Page and the social gospel : an anthology / edited, with an introduction by Charles Chatfield and Charles DeBenedetti. — New York : Garland Pub., 1976.
482 p. ; 23 cm. — (The Garland library of war and peace)
Bibliography: p. 37-44.
ISBN 0-8240-0451-5 : $25.00
1. Theology—Addresses, essays, lectures. 2. Church and social problems—Addresses, essays, lectures. 3. Page, Kirby, 1890-1957—Addresses, essays, lectures. I. Title. II. Series.
BR85.P23 1976 230 70-147695
76[r79]rev MARC

Rättvisan kan inte vänta. Idé-bok. Ekumeniska u-veckan. Förf.: Bolennart Andersson ... Stockholm, Gummesson; [Solna, Seelig] 1973.
182, (1) p. 19 cm. kr17.65 S73-42
Bibliography: p. 179-182.
ISBN 9170703795
1. Christianity and justice—Addresses, essays, lectures. 2. Church and social problems—Addresses, essays, lectures. I. Andersson, Bolennart, 1939-
BR115.J8R33 78-352848
79 MARC

Reformasie en revolusie / [onder redaksie van E. J. Smit, J. A. L. Taljaard, P. D. van der Walt ...e.a.]. — Potchefstroom : Instituut vir Bevordering van Calvinisme, Potchefstroomse Universiteit vir Christelike Hoër Onderwys, 1974.
376 p. ; 24 cm. SA
Afrikaans, Dutch, or English.
Includes bibliographical references.
ISBN 0-86990-179-6
1. Church and social problems—Addresses, essays, lectures. 2. Reformation—Addresses, essays, lectures. 3. Revolution (Theology)—Addresses, essays, lectures.
HN31.R439 78-395361
79 MARC

—ANGLICAN COMMUNION

Triolo, Sharon L
Nuclear energy, a Christian concern / [by Sharon L. Triolo]. — London : Catholic Truth Society for Pax Christi and the Commission for International Justice and Peace of the Bishops' Conference of England and Wales, 1977.
[7] p. ; 17 cm. GB78-09998
Cover title.
Bibliography: p. [7]
£0.10
1. Atomic power industry—Social aspects. 2. Church and social problems—Anglican Communion. I. Pax Christi. II. Catholic Church. Bishops' Conference of England and Wales. Commission for International Justice and Peace. III. Title.
HD9698.A2T75 261.8'5 78-321880
79 MARC

—BIBLIOGRAPHY

Centro de Estudios para el Desarrollo e Integración de América Latina.
Complemento a la bibliografía, Desarrollo y revolución, Iglesia y liberación. — Bogotá : Centro de Estudios para el Desarrollo e Integración de América Latina, 1973.
2 v. : 28 cm. — (Boletín bibliográfico - Centro de Estudios para el Desarrollo e Integración de América Latina ; A-1, B-1)
CONTENTS: pt. 1. Desarrollo y revolución.—pt. 2. Iglesia y liberación.
1. Economic development—Bibliography. 2. Church and social problems—Bibliography. 3. Catholic Church in Latin America—Bibliography. I. Centro de Estudios para el Desarrollo e Integración de América Latina. Desarrollo y revolución, Iglesia y liberación. II. Title. III. Series: Centro de Estudios para el Desarrollo e Integración de América Latina. Boletín bibliográfico - Centro de Estudios para el Desarrollo e Integración de América Latina ; A-1, B-1.
Z7164.E15C36 1973 Suppl. 78-376725
[HD85.S7] 78 MARC

Centro de Estudios para el Desarrollo e Integración de América Latina.
Complemento a la bibliografía, Desarrollo y revolución, Iglesia y liberación / CEDIAL, Centro de Estudios para el Desarrollo e integración de América Latina. — Bogotá : CEDIAL, 1974-
2 v. in ; 28 cm. — (Boletín bibliográfico - Centro de Estudios para el Desarrollo e Integración de América Latina ; A-2, B-2, B-3, B-4, B-5)
CONTENTS: pt. 1. Desarrollo y revolución.—pt. 2. Iglesia y liberación.
v.
1. Economic development—Bibliography. 2. Church and social problems—Bibliography. 3. Catholic Church in Latin America—Bibliography. I. Centro de Estudios para el Desarrollo e Integración de América Latina. Desarrollo y revolución, Iglesia y liberación. 1974. II. Title. III. Series: Centro de Estudios para el Desarrollo e Integración de América Latina. Boletín bibliográfico - Centro de Estudios para el Desarrollo e Integración de América Latina ; A-2 [etc.]
Z7164.E15C36 1973 Suppl. 2 78-376726
[HD85.S7] 78 MARC

CHURCH AND SOCIAL PROBLEMS (Continued)

—CATHOLIC CHURCH

see also Catholic action; Catholic Worker Movement; Christian democracy; Socialism and Catholic Church; Sociology, Rural (Catholic); Trade-unions, Catholic

Bender, Ryszard.
Społeczne inicjatywy chrześcijańskie w Królestwie Polskim 1905-1918 / Ryszard Bender. — Wyd. 1. — Lublin : TN KUL, 1978 (Lublin : LZGPKWN)
175, [12] p. ; 24 cm. — (Rozprawy Wydziału Historyczno-Filologicznego, Towarzystwo Naukowe Katolickiego Uniwersytetu Lubelskiego ; 41) P78-3780
Summary in English and French.
Includes bibliographical references and index.
zł95.00
1. Church and social problems—Poland. 2. Church and social problems—Catholic Church. I. Title.
HN39.P7B46 78-390732
79 MARC

Canaletti Gaudenti, Alberto, 1887-
Un corporativista cattolico : Renato de La Tour du Pin : da La Documentation catholique di Parigi / a cura e con prefazione di Alberto Canaletti Gaudenti. — Roma : Editrice Studium, [1935]
148 p. ; 18 cm. — (Quaderni universitari ; 16-17)
1. La Tour du Pin Chambly de la Charce, Charles Humbert René, marquis de, 1834-1924. 2. Corporate state. 3. Church and social problems—Catholic Church. 4. Political scientists—France—Biography. I. La Tour du Pin Chambly de la Charce, Charles Humbert René, marquis de, 1834-1924. II. Title. III. Series.
JC229.L36C36 79-355943
79 MARC

Casaldáliga, Pedro.
I believe in justice and hope / Pedro Casaldáliga. — Notre Dame, Ind. : Fides/Claretian, c1978.
v, 232 p. ; 20 cm.
Translation of Yo creo en la justicia y en la esperanza!
ISBN 0-8190-0628-9
1. Casaldáliga, Pedro. 2. Catholic Church—Bishops—Biography. 3. Bishops—Brazil—Biography. 4. Church and social problems—Brazil. 5. Church and social problems—Catholic Church. 6. Christianity and justice. I. Title.
BX4705.C33547A3813 282'.092'4 78-21694
78 MARC

Chiavacci, Enrico.
Morale della vita fisica / Enrico Chiavacci. — Bologna : EDB, 1976.
179 p. ; 24 cm. — (Corso di teologia morale : Morale speciale) It78-June
Includes bibliographical references.
L2800
1. Church and social problems—Catholic Church. 2. Christian ethics. 3. Bioethics. I. Title.
HN37.C3C43 78-392243
*79 MARC

La Chiesa e la cultura radicale / Gianni Baget Bozzo. — Brescia : Queriniana, [1979?]
291 p. ; 22 cm. — (Dipartimento di scienze religiose ; 11) It79-Apr
Bibliography: p. 275-287.
L6500
1. Church and social problems—Catholic Church. 2. Radicalism. I. Baget Bozzo, Gianni, 1925- II. Series.
HN37.C3C44 261.8'3 79-373175
*79 MARC

Coste, René, 1922-
Les chrétiens et la lutte des classes / René Coste. — Paris : Éditions S.O.S., [1975]
127 p. ; 21 cm. F***
Bibliography: p. 122-126.
ISBN 2-7185-0789-6 : 22.00F
1. Church and social problems—Catholic Church. 2. Social classes. I. Title.
HN37.C3C596 75-519948
76[r79]rev MARC

Fagan, Harry.
Empowerment : skills for parish social action / Harry Fagan. — New York : Paulist Press, c1979.
64 p. : ill. ; 28 cm.
ISBN 0-8091-2210-3 : $3.95
1. Church and social problems—Catholic Church. 2. Social action. I. Title.
HN37.C3F25 261.8'3 79-52016
00 MARC

Fernandes, José.
Problemas sociais : à luz da Igreja / José Fernandes. — Porto [Portugal] : Edições Salesianas, [1976]
153 p. ; 19 cm. — (Colecção Renovar)
1. Church and social problems—Catholic Church. I. Title.
HN37.C3F37 77-155551
79 MARC

Furfey, Paul Hanley, 1896-
Fire on the earth / Paul Hanly Furfey. — Reprinted ed. — New York : Arno Press, 1978, c1936.
ix, 159 p. ; 21 cm. — (The American Catholic tradition)
Reprint of the ed. published by Macmillan, New York, with a new introd.
Includes index.
ISBN 0-405-10830-3
1. Church and social problems—Catholic Church. 2. Catholic Church—Apologetic works. I. Title. II. Series.
HN37.C3F8 1978 261.8'3 77-11287
77 MARC

Gildemeister M , Alfredo.
Apuntes sobre la temática social de la doctrina pontificia / por Alfredo Gildemeister M. — Lima : Universidad del Pacífico, Centro de Investigación, 1977.
23 p. ; 30 cm. — (Trabajos y materiales docentes - Universidad del Pacífico, Centro de Investigación ; no. 1)
1. Church and social problems—Catholic Church. I. Title. II. Series: Universidad del Pacífico. Centro de Investigación. Trabajos y materiales docentes - Universidad del Pacífico, Centro de Investigación ; no. 1.
HN37.C3G465 261.8'3 78-398728
79 MARC

Hollenbach, David.
Claims in conflict : retrieving and renewing the Catholic human rights tradition / by David Hollenbach. — New York : Paulist Press, c1979.
v, 219 p. ; 21 cm. — (Woodstock studies ; 4)
Includes bibliographical references and index.
ISBN 0-8091-0287-0
1. Church and social problems—Catholic Church. 2. Civil rights. I. Title. II. Series.
HN37.C3H584 261.8'3 79-84239
79 MARC

Im, Chin-ch'ang.
(Han'guk ŭi sahoe palchŏn kwa Kat'ollik Kyohoe ŭi yŏkhal)
韓國의社會發展과가톨릭敎會의役割 = Social development of Korea and the role of the Catholic Church / 任珍昌編著. — [서울] : 西江大學校社會問題硏究所, [1976?]
311 p. ; 21 cm.
In Korean; table of contents also in English.
Includes bibliographical references.
1. Church and social problems—Korea. 2. Church and social problems—Catholic Church. I. Title. II. Title: Social development of Korea and the role of the Catholic Church.
HN39.K8I42 78-826442

Keller, Peter.
Die Geschichte der katholischen Arbeitervereine in Unterfranken von 1884-1934 / vorgelegt von Peter Keller. — Würzburg : [Kath. Arbeitnehmer-Bewegung], 1976.
166 p. ; 21 cm. GFR78-B7
Cover title.
Bibliography: p. 159-166.
1. Working-men's clubs—Germany, West—Franconia, Lower. 2. Labor and laboring classes—Germany, West—Franconia, Lower. 3. Church and social problems—Catholic Church. I. Title.
HD6699.F72K44 79-380060
79 MARC

Klüber, Franz.
Katholische Soziallehre und demokratischer Sozialismus / Franz Klüber. — 2., erw. Aufl. — Bonn : Verlag Neue Gesellschaft, 1979.
179 p. ; 22 cm. GFR79-A
Includes bibliographical references.
ISBN 3-87831-296-2 : DM14.00
1. Socialism and Catholic Church. 2. Church and social problems—Catholic Church. I. Title.
HX536.K56 1979 261.7 79-377407
79 MARC

Krasicki, Ignacy.
Watykan na rozdrożu (1960-1970) ; tło i wyniki Soboru Watykańskiego II. [Wyd. 1. Warszawa] Książka i Wiedza, 1971.
445, [1] p. illus. 20 cm. zł50.00
Bibliography: p. 445-[446]
1. Vatican Council. 2d, 1962-1965. 2. Church and social problems—Catholic Church. 3. Church and the world. I. Title.
BX830 1962.K65 77-479384
78 MARC

La Farge, John, 1880-1963.
Interracial justice / John LaFarge. — New York : Arno Press, 1978 [c1937]
xii, 226 p. ; 21 cm. — (The American Catholic tradition)
Reprint of the ed. published by America Press, New York.
Includes index.
Bibliography: p. 195-198.
ISBN 0-405-10839-7
1. United States—Race relations. 2. Afro-Americans. 3. Church and social problems—Catholic Church. I. Title. II. Series.
E185.61.L25 1978 261.8'34'510973 77-11296
77 MARC

Libânio, J B
Formação da consciência crítica / J. B. Libânio. — Petrópolis : Editora Vozes, 197
v. ; 18 cm. — (Coleção Vida religiosa : Temas atuais ; 9/11)
1. Church and social problems—Catholic Church. 2. Subsídios sócio-analíticos. I. Title.
HN37.C3L439 261.8'3 79-114206
79 MARC

López Trujillo, Alfonso.
Hacia una sociedad nueva : socialismo, opción cristiana? / Alfonso López Trujillo. — Bogotá : Ediciones Paulinas, c1978.
146 p. ; 21 cm. — (Colección Estudios sociales ; 8)
Includes bibliographical references.
1. Socialism, Christian. 2. Socialism. 3. Church and social problems—Catholic Church. I. Title.
HX51.L68 335'.7 78-108014
79 MARC

Monaco, Michele.
La vita, le opere, il pensiero di L. A. Muratori e la sua concezione della pubblica felicità / Michele Monaco. — Lecce : Milella, 1977.
142 p. ; 24 cm. — (Collana storica ; 3) It78-Dec
Includes bibliographical references and index.
L4800
1. Muratori, Lodovico Antonio, 1672-1750. 2. Muratori, Lodovico Antonio, 1672-1750. Della pubblica felicità oggetto de' buoni principi. 3. Political science. 4. Church and social problems—Catholic Church. 5. Intellectuals—Italy—Biography. I. Title.
DG545.8.M8M65 907'.2'024 79-353713
*79 MARC

Mysłek, Wiesław.
Socjalizm i katolicyzm : polskie spotkanie w dialogu / Wiesław Mysłek. — Wyd. 1. — Warszawa : Krajowa Agencja Wydawnicza, 1978.
254 p. ; 17 cm. — (Postawy) P***
Includes bibliographical references.
zł22.00
1. Communism and Christianity—Catholic Church—Poland. 2. Church and social problems—Catholic Church. I. Title.
BX1396.4.M96 79-384373
79 MARC

O'Mahony, Patrick J
The fantasy of human rights / by Patrick O'Mahony. — Great Wakering : Mayhew-McCrimmon, 1978.
192 p. : ill. ; 18 cm. GB78-25821
Bibliography: p. 186-188.
Includes index.
ISBN 0-85597-256-4 : £2.00
1. Church and social problems—Great Britain. 2. Church and social problems—Catholic Church. 3. Civil rights. I. Title.
HN39.G7O43 1978 261.7 79-312169
79 MARC

Pimenta, Osvaldo de Melo.
Um grito na imensidão / Osvaldo de Melo Pimenta ; ilus. de Pedro Ribeiro. — São Paulo : Editora Cupolo, 1977.
141 p. : ill. ; 19 cm.
Cr$40.00
1. Church and social problems—Brazil—São Paulo (City) 2. Church and social problems—Catholic Church. I. Title.
HN39.B8P55 261.8'3'09816 79-348750
79 MARC

Reforma agrária interpela a Igreja / Acácio Catarino ... [et al.] — Lisboa : Multinova, [1977]
262 p. ; 21 cm.
1. Land reform—Portugal. 2. Church and social problems—Catholic Church. 3. Church and social problems—Portugal. I. Catarino, Acácio Ferreira.
HD786.R43 79-370117
79 MARC

Salvestrini, Fernando.
La Chiesa e il sesso / Fernando Salvestrini. — 1. ed. — Milano : Massimo, 1976.
160 p. ; 19 cm. — (Collana Sussidi pastorali-liturgici ; n. 43) It78-Mar
Includes bibliographical references.
L1500
1. Sexual ethics. 2. Church and social problems—Catholic Church. I. Title.
HQ32.S24 78-380867
*78 MARC

Sanmiguel, V
Mater et magistra & India / Father Victor — Alwaye North : St. Joseph's Apostolic Seminary, 1963.
x, 306 p., [1] leaf of plates : ports. ; 23 cm.
Bibliography: p. 297-300.
Includes index.
Rs5.50
1. Church and social problems—India. 2. Church and social problems—Catholic Church. 3. Catholic Church. Pope, 1958-1963 (Joannes XXIII) Mater et magistra. I. Title.
HN39.I4S36 261.8'3'0954 77-913540
79 MARC

Segers, Joseph.
Actions pour le développement en République du Zaïre / J. Segers. — Kinshasa-Gombe : Centre d'études pour l'action sociale, 1971.
62 p. ; 27 cm.
Cover title.
1. Economic assistance in Zaire. 2. Church and social problems—Zaire. 3. Church and social problems—Catholic Church. I. Title.
HC591.C6S43 338.91'675'1 79-101843
79 MARC

CHURCH AND SOCIAL PROBLEMS —CATHOLIC CHURCH (Continued)

Terres, propriété et travail des hommes : éléments de réflexion chrétienne sur les problèmes fonciers en espace rural. — ⌜Paris⌝ : Le Centurion, c1978.
117 p. ; 22 cm. F***
At head of title: Conseil national de pastorale rurale, Commission sociale de l'épiscopat.
Includes bibliographical references.
ISBN 2-227-42526-1 : 20.00F
1. Church and social problems—Catholic Church. 2. Sociology, Rural (Catholic) 3. Land use, Rural. I. Conseil national de pastorale rurale. II. Commission sociale de l'épiscopat.
HN37.C3T4 79-341015
79 MARC

Vives Solar, Fernando.
El humanismo de Fernando Vives / Francisco Javier Cid ⌜editor⌝. — 1. ed. — Santiago de Chile : Instituto Chileno de Estudios Humanísticos, ⌜1976⌝
132 p. ; 19 cm.
1. Church and social problems—Catholic Church. 2. Humanism. 3. Vives Solar, Fernando. I. Cid G., Francisco Javier. II. Title.
HN37.C3V499 1976 77-480321
77⌜r79⌝rev MARC

World's Columbian Catholic Congress, Chicago, 1893.
The world's Columbian Catholic congresses. — New York : Arno Press, 1978 ⌜c1893⌝
107 42 p. : ill. ; 24 cm. — (The American Catholic tradition)
Reprint of v. 1 and 3 of the ed. published by J. S. Hyland, Chicago
ISBN 0-405-10871-0
1. Church and social problems—Catholic Church. 2. World's Columbian Catholic Congress, Chicago, 1893. I. Title. II. Series.
HN37.C3W57 1893 261.8'3 77-11323
77 MARC

—CATHOLIC CHURCH—ADDRESSES, ESSAYS, LECTURES

Kirche in der Gesellschaft : der katholische Beitrag 78/79 / hrsg. von Jürgen Wichmann. — München : G. Olzog, c1978.
407 p. ; 18 cm. — (Geschichte und Staat ; Bd. 220-222) GFR***
Includes bibliographies.
ISBN 3-7892-7135-7
1. Church and social problems—Germany, West—Addresses, essays, lectures. 2. Church and social problems—Catholic Church—Addresses, essays, lectures. I. Wichmann, Jürgen, 1930-
HN39.G3K563 261.8'3'0943 79-346831
79 MARC

Kirche und Gesellschaft heute : ⌜Franz Groner zum 65. Geburtstag⌝ / hrsg. von Franz Böckle u. Franz Josef Stegmann. — Paderborn ; München ; Wien ; Zürich : Schöningh, 1979.
297 p. : ill., port. ; 24 cm. GFR79-A
Bibliography of F. Groner's works: p. 293-294.
Includes bibliographical references and index.
ISBN 3-506-74451-8 : DM19.80
1. —Sociology, Christian (Catholic)Addresses, essays, lectures. 2. Church and social problems—Catholic Church—Addresses, essays, lectures. 3. Groner, Franz—Addresses, essays, lectures. I. Groner, Franz. II. Böckle, Franz. III. Stegmann, Franz Josef.
BX1753.K56 79-361102
79 MARC

—CATHOLIC CHURCH—CASE STUDIES

Makil, Perla Q
Dumaguete City urban priorities study : final report / prepared for the Asia Foundation, the National Secretariat for Social Action, and the Redemptorist Fathers of Dumaguete City by the Institute of Philippine Culture on 31 July 1973 ; Perla Q. Makil. — Quezon City : Ateneo de Manila Institute of Philippine Culture, 1973.
62 leaves in various foliations ; 28 cm.
1. Community development, Urban—Philippine Islands—Dumaguete. 2. Church and social problems—Philippine Islands—Dumaguete—Case studies. 3. Church and social problems—Catholic Church—Case studies. I. Ateneo de Manila University. Institute of Philippine Culture. II. Title.
HN730.D84M33 309.2'62'095995 79-305382
79 MARC

Wayland-Smith, Giles, 1935-
The Catholic Church and social change : a research note and some preliminary findings on the Archdiocese of Yucatan / by Giles Wayland-Smith. — ⌜Erie, Pa.⌝ : Institute for Latin American Studies of Northwestern Pennsylvania, Mercyhurst College, c1977.
34 p. : ill. ; 28 cm. — (Latin American monograph series ; no. 5)
Includes bibliographical references.
1. Catholic Church. Archdiocese of Yucatan. 2. Church and social problems—Catholic Church—Case studies. I. Title. II. Series.
BX1430.Y8W39 261 79-101121
79 MARC

—CATHOLIC CHURCH—CONGRESSES

Cattolici e politica in un mondo diviso / ⌜interventi di⌝ Achille Ardigò ... ⌜et al.⌝ ; a cura di Paolo Nepi ; presentazione di Antonio Pavan. — Roma : A.V.E., c1979.
231 p. ; 21 cm. — (Saggi ; 20) It***
"I testi degli interventi al seminario organizzato ... nei giorni 17-19 febbraio 1978 a Montesicuro (Ancona)."
Includes bibliographical references.
L5000
1. Church and social problems—Catholic Church—Congresses. 2. Catholic Church and world politics—Congresses. I. Ardigò, Achille. II. Nepi, Paolo.
HN37.C3C345 261.8'3 79-376967
00 MARC

—CATHOLIC CHURCH—HISTORY—SOURCES

Catholic Church. Pope.
La doctrine sociale de l'Église à travers les siècles : documents pontificaux du XVème au XXème siècle : textes originaux et traductions / publiés et introduits par Arthur F. Utz, avec la collaboration de Médard Boeglin. — 1. éd. — Fribourg, Suisse : Éditions Valores, ⌜1970⌝ 1973 printing.
4 v. (xxxviii, 3091 p.) : 24 cm.
Texts in French or Latin, Italian, English, German, or Spanish, with translations in French.
Includes bibliographical references and indexes.
1. Church and social problems—Catholic Church—History—Sources. 2. Church and social problems—Catholic Church—Papal documents. I. Utz, Arthur Fridolin, 1908- II. Boeglin, Médard. III. Title.
HN37.C3A1 1973 262.9'1 78-375507
78 MARC

—CATHOLIC CHURCH—PAPAL DOCUMENTS

Catholic Church. Pope.
La doctrine sociale de l'Église à travers les siècles : documents pontificaux du XVème au XXème siècle : textes originaux et traductions / publiés et introduits par Arthur F. Utz, avec la collaboration de Médard Boeglin. — 1. éd. — Fribourg, Suisse : Éditions Valores, ⌜1970⌝ 1973 printing.
4 v. (xxxviii, 3091 p.) : 24 cm.
Texts in French or Latin, Italian, English, German, or Spanish, with translations in French.
Includes bibliographical references and indexes.
1. Church and social problems—Catholic Church—History—Sources. 2. Church and social problems—Catholic Church—Papal documents. I. Utz, Arthur Fridolin, 1908- II. Boeglin, Médard. III. Title.
HN37.C3A1 1973 262.9'1 78-375507
78 MARC

—CATHOLIC CHURCH—PAPAL DOCUMENTS—CONGRESSES

Progrès des peuples et solidarité mondiale : table ronde / organisée par l'Institut international d'études sociales à l'occasion du dixième anniversaire de l'encyclique "Populorum progressio" de Sa Sainteté le pape Paul VI, Genève, 24 mars 1977. — Genève : Institut international d'études sociales, 1977.
80 p. : ill. ; 23 cm. Sw***
In French, Spanish, or English.
Includes French text of the encyclical Populorum progressio (p. 55-76)
Includes bibliographical references.
12.00F
1. Church and social problems—Catholic Church—Papal documents—Congresses. I. Catholic Church. Pope, 1963- (Paulus VI). Populorum progressio. French. 1977. II. Geneva. International Institute for Labour Studies.
HN37.C3P77 261.8'3 79-360885
79 MARC

—CATHOLIC CHURCH—PERIODICALS

Religião e sociedade. no. 1-
maio 1977-
⌜São Paulo, Editora de Humanismo, Ciência e Tecnologia⌝
no. ill. 21 cm. semiannual.
Numbers for May 1977- issued by the Centro de Estudos de Religião of the Universidade Estadual de Campinas.
1. Church and social problems — Latin America — Periodicals. 2. Church and social problems—Catholic Church—Periodicals. I. Universidade Estadual de Campinas. Centro de Estudos de Religião.
HN39.L3R44 261.8'3'098 79-645072
MARC-S

—CONGRESSES

Congrès du christianisme social des pays latins, 1st, Geneva, 1931.
Christianisme social et problèmes contemporains, désarmement, chômage, législation sociale, régime industriel : actes du Ier Congrès du christianisme social des pays latins et du 5me Congrès de la Fédération française du christianisme social tenu à Genève du 29 septembre au 2 Octobre 1931. — Saint-Étienne ⌜France⌝ : Le Christianisme social, ⌜1931⌝
249 p. ; 23 cm.
1. Church and social problems—Congresses. I. Fédération française du christianisme social. II. Title.
HN31.C637 1931 261.8'3 79-369189
79 MARC

The Ministry of development in evangelical perspective : a symposium on the social and spiritual mandate / Carl F. H. Henry, convenor ; Robert Lincoln Hancock, editor. — Pasadena, Calif. : William Carey Library, c1979.
viii, 109 p. : ill. ; 22 cm.
Articles presented at the symposium sponsored by Development Assistance Services, inc., held in Nov. 1977, in Colorado Springs.
Includes bibliographical references.
ISBN 0-87808-164-X
1. Church and social problems—Congresses. 2. Missions, Foreign—Congresses. 3. Church and underdeveloped areas—Congresses. I. Henry, Carl Ferdinand Howard, 1913- II. Hancock, Robert Lincoln, 1939- III. Development Assistance Services.
HN31.M64 261.8'3 78-27821
78 MARC

World Council of Churches. 3d Assembly, New Delhi, India, 1961.
New Delhi speaks about Christian witness, service, and unity; the message, appeal, and section reports. Edited by W. A. Visser't Hooft. New York, Association Press ⌜1962⌝
124 p. 16 cm. (An Association Press reflection book)
1. Ecumenical movement—Congresses. 2. Evangelistic work—Congresses. 3. Church and social problems—Congress. 4. Church—Unity—Congresses. I. Visser 't Hooft, Willem Adolph, 1900- ed. II. Title.
BX6.W77 1961b 280.1 62-11031
⌜r79⌝rev MARC

—HISTORY

Foss, Øivind, 1935-
Kirke og klassekamp. Oslo, Gyldendal, 1973.
104 p. 19 cm. (Fakkel-bøker, 257C) kr19.50 N73-18
Bibliography: p. 103-⌜105⌝
ISBN 8205057435
1. Church and social problems—Norway—History. 2. Church and social problems—History. I. Title.
HN39.N6F67 78-414926
78 MARC

—HISTORY—ADDRESSES ESSAYS, LECTURES

Separation without hope? : essays on the relation between the Church and the poor during the industrial revolution and the Western colonial expansion / ⌜authors, Julio A. Barreiro ... et al.⌝ ; ⌜ed., Julio de Santa Ana⌝. — Geneva : World Council of Churches, Commission on the Churches' Participation in Development, c1978.
192 p. ; 22 cm. Sw78-10106
Includes bibliographical references.
ISBN 2-8254-0576-0 : 12.50F
1. Church and the poor—History—Addresses essays, lectures. 2. Church and social problems—History—Addresses essays, lectures. I. Barreiro, Julio. II. Santa Ana, Julio de.
BV639.P6S46 261.8 79-315194
79 MARC

—MEDITATIONS

Pettepiece, Thomas G
Visions of a world hungry : study, prayer & action / Thomas G. Pettepiece. — Nashville : The Upper Room, c1979.
189 p. ; 22 cm.
Bibliography: p. 179-187.
Includes index.
ISBN 0-8358-0371-6
1. Meditations. 2. Food relief—Moral and religious aspects—Meditations. 3. Church and social problems—Meditations. I. Title.
BV4832.2.P485 261.8 79-63097
79 MARC

—METHODIST CHURCH

Jones, Donald G
The sectional crisis and northern Methodism : a study in piety, political ethic, and civil religion, 1864-1876 / by Donald G. Jones. — Metuchen, N.J. : Scarecrow Press, 1979.
viii, 341 p. ; 23 cm. — (Jesse Lee Prize essay of the Commission on Archives and History, the United Methodist Church)
Bibliography: p. 317-337.
Includes index.
ISBN 0-8108-1175-8
1. Methodist Episcopal Church—History. 2. Church and social problems—Methodist Church. I. Title. II. Series: United Methodist Church (United States). Commission on Archives and History. Jesse Lee Prize of Commission on Archives and History, the United Methodist Church .
BX8381.J66 287'.632 78-9978
78 MARC

—ORTHODOX EASTERN CHURCH, RUSSIAN

Nepli͡uev, Nikolaĭ Nikolaevich, 1851-1908.
(Golos vi͡erui͡ushchago miri͡anina po povodu predstoi͡ashchago Sobora)
Голосъ вѣрующаго міряинна по поводу предстоящаго Собора / Н. Неплюевъ. — Кіевъ : Тип. И. И. Горбунова, 1906.
71 p. ; 27 cm.
Reprinted from Trudy of the Kievskai͡a dukhovnai͡a Akademii͡a, nos. 6-7, 1906.
Bound with the author's Pis'mo k sel'skomu naselenii͡u. Kiev, 1905. Pis'mo k dukhovenstvu. Kiev, 1905. Partii͡a mirnago progressa. Glukhov, 1906.
1. Orthodox Eastern Church, Russian—Doctrinal and controversial works. 2. Church and social problems—Orthodox Eastern Church, Russian. 3. Councils and synods—Russia. I. Title.
BX510.N46 78-422081

CHURCH AND SOCIAL PROBLEMS (Continued)

—PERIODICALS

Tradición, familia, propiedad. año 1– (no. 1–) ; feb. 1969–
[Buenos Aires, Sociedad Argentina de Defensa de la Tradición, Familia y Propiedad]
v. ill. 30–48 cm.
Monthly, Feb. 1969–Nov. 1971; bimonthly (irregular), Apr./Mar. 1972–
Supersedes: Cruzada.
1. Church and social problems—Periodicals. 2. Sociology, Christian—Periodicals. I. Sociedad Argentina de Defensa de la Tradición, Familia y Propiedad.
HN30.T72 261.8'3 78–641997 MARC–S

—BRAZIL

Casaldáliga, Pedro.
I believe in justice and hope / Pedro Casaldáliga. — Notre Dame, Ind. : Fides/Claretian, c1978.
v, 232 p. ; 20 cm.
Translation of Yo creo en la justicia y en la esperanza!
ISBN 0-8190-0628-9
1. Casaldáliga, Pedro. 2. Catholic Church—Bishops—Biography. 3. Bishops—Brazil—Biography. 4. Church and social problems—Brazil. 5. Church and social problems—Catholic Church. 6. Christianity and justice. I. Title.
BX4705.C33547A3813 282'.092'4 78-21694
78 MARC

—EAST ASIA—ADDRESSES, ESSAYS, LECTURES

Christian action in the Asian struggle. — Singapore : Christian Conference of Asia, [between 1973 and 1977]
100 p. : ill. ; 24 cm.
Includes bibliographical references.
CONTENTS: Kyaw Than. Introduction.—Thomas, M. M. Christian action in the Asian struggle.—Goh Keng Swee. The Asian struggle.—Won Yong Kang. Asian struggle for justice, liberation, and development.—Nacpil, E. Renewing the church for Christian action.—Castro, E. Forward in ecumenism.—Heuvel, A. H. van den. The conversation of truths.—Visser 't Hooft. Building up the common life.—Coe, S. Across the frontiers.—Dietrich, S. de. The Bible comes alive.—Garett, J. The place of the Bible in an era of technology.—Booth, A. Christian living.
1. Revolution (Theology)—Addresses, essays, lectures. 2. Church and social problems—East Asia—Addresses, essays, lectures. 3. Ecumenical movement—Addresses, essays, lectures. 4. Niles, Daniel Thambyrajah—Addresses, essays, lectures. I. Christian Conference of Asia.
BT738.3.C46 275 77-941957
78 MARC

—EUROPE—CONGRESSES

Conference of European Churches.
Europe after Helsinki and the developing regions : spiritual, moral, and practical responsibilities of our churches : report of a consultation, 7th to 11th March 1977, Gallneukirchen near Linz. — Geneva : Conference of European Churches ; Edinburgh : distributor, Saint Andrew Press, 1977.
100 p. ; 24 cm. — (Occasional paper - Conference of European Churches ; no. 9) Sw***
Includes bibliographical references.
1. Church and social problems—Europe—Congresses. 2. Civil rights—Europe—Congresses. 3. Europe—Economic policy—Congresses. I. Title. II. Series: Conference of European Churches. Occasional papers - Conference of European Churches ; no. 9.
HN39.E9C66 1977 261.8'094 79-303997
79 MARC

—FRANCE

Templier, Joseph, 1921-
Pratique militante et parole d'Église / Joseph Templier. — [Paris] : le Centurion, c1975.
160 p. ; 21 cm. — (Croire et comprendre) F76-1591
Bibliography: p. [157]-158.
ISBN 2-227-30119-8
1. Catholic Church in France. 2. Church and social problems—France. I. Title.
BX1530.2.T45 261.8 79-376751
79 MARC

—GERMANY, WEST—ADDRESSES, ESSAYS, LECTURES

Kirche in der Gesellschaft : der katholische Beitrag 78/79 / hrsg. von Jürgen Wichmann. — München : G. Olzog, c1978.
407 p. ; 18 cm. — (Geschichte und Staat ; Bd. 220-222) GFR***
Includes bibliographies.
ISBN 3-7892-7135-7
1. Church and social problems—Germany, West—Addresses, essays, lectures. 2. Church and social problems—Catholic Church—Addresses, essays, lectures. I. Wichmann, Jürgen, 1930-
HN39.G3K563 261.8'3'0943 79-346831
79 MARC

Kirche in der Gesellschaft : der evangelische Beitrag 78/79 / hrsg. von Hans-Wolfgang Hessler. — München : G. Olzog, c1978.
379 p. ; 18 cm. — (Geschichte und Staat ; Bd. 223-225) GFR***
Includes bibliographical references.
ISBN 3-7892-7136-5
1. Church and social problems—Germany, West—Addresses, essays, lectures. I. Hessler, Hans-Wolfgang, 1928-
HN39.G3K56 261.8'3'0943 79-346713
79 MARC

Soziale Ordnung / mit e. Einf. von Eberhard Müller. — Gütersloh : Gütersloher Verlagshaus Mohn, 1978.
260, [1] p. ; 19 cm. — (Die Denkschriften der Evangelischen Kirche in Deutschland ; Bd. 2) (Gütersloher Taschenbücher ; 415 : Siebenstern) GFR78-A47
Bibliography: p. 255-[261]
ISBN 3-579-04803-1 : DM16.80
1. Church and social problems—Germany, West—Addresses, essays, lectures. I. Müller, Eberhard Johannes. II. Series.
HN39.G3S69 261.8'3'0943 79-346714
79 MARC

—GREAT BRITAIN

Beeson, Trevor.
Britain today and tomorrow / [by] Trevor Beeson. — [London] : Fount Paperbacks, 1978.
284 p. ; 18 cm. GB78-27855
Includes bibliographical references and index.
ISBN 0-00-625126-9 : £1.50
1. Church and social problems—Great Britain. 2. Great Britain—Social conditions—1945- I. Title.
HN39.G7B43 261 79-301988
79 MARC

Nuclear crisis : a question of breeding / edited by David Gosling and Hugh Montefiore. — Dorchester : Prism Press, 1977.
ix, 165 p. ; 23 cm. GB78-09028
ISBN 0-904727-57-2 : £6.00. ISBN 0-904727-58-0 pbk.
1. Atomic power industries—Great Britain. 2. Breeder reactors. 3. Energy policy—Great Britain. 4. Church and social problems—Great Britain. I. Gosling, David L., 1939- II. Montefiore, Hugh.
HD9698.G72N82 621.48'34 78-314092
78 MARC

O'Mahony, Patrick J
The fantasy of human rights / by Patrick O'Mahony. — Great Wakering : Mayhew-McCrimmon, 1978.
192 p. : ill. ; 18 cm. GB78-25821
Bibliography: p. 186-188.
Includes index.
ISBN 0-85597-256-4 : £2.00
1. Church and social problems—Great Britain. 2. Church and social problems—Catholic Church. 3. Civil rights. I. Title.
HN39.G7O43 1978 261.7 79-312169
79 MARC

Wilson, Bryan R
Sects and society : a sociological study of three religious groups in Britain / Bryan R. Wilson. — Westport, Conn. : Greenwood Press, 1978, c1961.
397 p. ; 23 cm.
Reprint of the ed. published by W. Heinemann, London, in series: Heinemann books on sociology.
Bibliography: p. 374-388.
Includes index.
ISBN 0-313-20439-X
1. Elim Foursquare Gospel Alliance. 2. Christian Science—Great Britain. 3. Christadelphians—Great Britain. 4. Church and social problems—Great Britain. I. Title.
[BX7435.E54W5 1978] 289 78-5993
78 MARC

Work or what? : a Christian examination of the employment crisis / introduction by Robin Woods. — London : Church Information Office for the General Synod Board for Social Responsibility, 1977.
[12], 48 p. ; 21 cm. GB78-37457
"An occasional paper of the General Synod Board for Social Responsibility."
Bibliography: p. 44-46.
ISBN 0-7151-6551-8 : £0.70
1. Unemployed—Great Britain. 2. Manpower policy—Great Britain. 3. Church and social problems—Great Britain. I. Church of England. Board for Social Responsibility.
HD5765.A6W67 261.8'5 79-316892
79 MARC

—INDIA

Karokaran, Anto, 1941-
Evangelization and diakonia / Anto Karokaran. — Bangalore : Dharmaram Publications, 1978.
xvii, 285 p. : map ; 23 cm.
"A study in the Indian perspective."
Bibliography: p. [265]-280.
Includes index.
Rs19.00 ($5.00 U.S.)
1. Evangelistic work—India. 2. Missions—India. 3. Church charities—India. 4. Church and social problems—India. 5. Theology, Doctrinal—India—History. I. Title.
BV3777.I4K37 269'.2'0954 78-905126
79 MARC

Sanmiguel, V
Mater et magistra & India / Father Victor — Alwaye North : St. Joseph's Apostolic Seminary, 1963.
x, 306 p., [1] leaf of plates : ports. ; 23 cm.
Bibliography: p. 297-300.
Includes index.
Rs5.50
1. Church and social problems—India. 2. Church and social problems—Catholic Church. 3. Catholic Church. Pope, 1958-1963 (Joannes XXIII) Mater et magistra. I. Title.
HN39.I4S36 261.8'3'0954 77-913540
79 MARC

—INDONESIA—CONGRESSES

Konperensi Gereja dan Masyarakat, 3d, Klender, Indonesia, 1976.
Melihat tanda-tanda jaman : pengamalan Pancasila dalam membangun masa depan : laporan Konperensi Gereja dan Masyarakat, Klender, 15-21 Maret 1976. — [Jakarta] : Diterbitkan oleh BPK Gunung Mulia untuk Dewan Gereja-Gereja di Indonesia, [1976]
295 p. ; 21 cm.
1. Church and social problems—Indonesia—Congresses. 2. Civics, Indonesian—Congresses. I. Dewan Gereja-Gereja di Indonesia. II. Title.
HN39.I53K63 1976 77-941827
79 MARC

—ITALY—CATHOLIC CHURCH—ADDRESSES, ESSAYS, LECTURES

Perché i cattolici a sinistra : rapporto di un'indagine condotta su scala regionale in Emilia-Romagna sullo spostamento politico a sinistra di giovani generazioni cattoliche negli anni 1955-1975 / [di] V. D'Ascenzi ... [et al.]. — Torino : Società editrice internazionale, 1977.
189 p. ; 22 cm. — (Dossier SEI) It78-Apr
Includes bibliographical references.
L3800
1. Socialism and youth—Italy—Emilia-Romagna—Addresses, essays, lectures. 2. Youth—Italy—Emilia-Romagna—Addresses, essays, lectures. 3. Church and social problems—Italy—Catholic Church—Addresses, essays, lectures. I. D'Ascenzi, Vincenzo, 1928-
HX547.P47 301.43'15'09454 78-375575
*78 MARC

—ITALY—HISTORY

Guccione, Eugenio.
Le affittanze collettive nel pensiero politico ed economico dei cattolici tra Otto e Novecento / Eugenio Guccione. — Palermo ; São Paulo : I.L.A. Palma, 1978.
99 p. ; 22 cm. — (Athena) It78-Nov
Includes bibliographical references and index.
L3500
1. Agriculture, Cooperative—Italy, Southern—History. 2. Land tenure—Italy, Southern—History. 3. Church and social problems—Italy—History. I. Title.
HD1491.I8G8 79-340229
*79 MARC

—ITALY—PIEMONTE—HISTORY—ADDRESSES, ESSAYS, LECTURES

Il Partito cristiano : D.C. e mondo cattolico in Piemonte, 1900-1975. — Torino : Stampatori, [1978]
229 p. ; 20 cm. — (Argomenti di politica e cultura contemporanea) It78-Mar
L5000
1. Partito della democrazia cristiana—History—Addresses, essays, lectures. 2. Christian democracy—History—Addresses, essays, lectures. 3. Church and social problems—Italy—Piemonte—History—Addresses, essays, lectures. I. Series.
JN5657.D43P37 78-381166
*79 MARC

—JAPAN

Matsuda, Keiichi.
(Buraku mondai to Kirisutokyō)
部落問題とキリスト教 / [執筆者 松田慶一, 益谷寿, 工藤英一 ; 編集者 日本基督教団宣教研究所]. — 東京 : 日本基督教団宣教研究所第四分科, 1963.
159 p. : map ; 19 cm.
Includes bibliographical references.
1. Buraku people. 2. Church and social problems—Japan. I. Masutani, Hisashi, joint author. II. Kudō, Eiichi, 1922- joint author. III. Nihon Kirisuto Kyōdan. Senkyō Kenkyūjo. IV. Title.
HT725.J3M357 79-805068

CHURCH AND SOCIAL PROBLEMS (Continued)

—KOREA

Im, Chin-ch'ang.
(Han'guk ŭi sahoe palchŏn kwa Kat'ollik Kyohoe ŭi yŏkhal)
韓國의社會發展과가톨릭教會의役割 = Social development of Korea and the role of the Catholic Church / 任珍昌編著. — [서울] : 西江大學校社會問題研究所, [1976?]
311 p. ; 21 cm.
In Korean; table of contents also in English.
Includes bibliographical references.

1. Church and social problems—Korea. 2. Church and social problems—Catholic Church. I. Title. II. Title: Social development of Korean and the role of the Catholic Church.
HN39.K8I42 78-826442

—LATIN AMERICA—ADDRESSES, ESSAYS, LECTURES

Iglesia y seguridad nacional en América Latina. — Madrid : Instituto de Estudios Políticos para América Latina y Africa, 1977.
95 p. ; 28 cm. — (América Latina ; 2) Sp***
At head of title: Idoc, España.
Bibliography: p. 83-95.
1. Church and social problems—Latin America—Addresses, essays, lectures. 2. Latin America—National security—Addresses, essays, lectures. I. Idoc. II. Instituto de Estudios Políticos para América Latina y Africa. III. Title. IV. Series: América Latina (Madrid) ; 2.
HN39.L3I38 78-384311
78 MARC

—LATIN AMERICA—CONGRESSES

Geschenk der Armen an die Reichen: Zeugnisse aus dem gewaltfreien Kampf d. erneuerten Kirche in Lateinamerika / hrsg. u. kommentiert v. Hildegard Goss-Mayr ; hrsg. v. d. Kath. Sozialakademie Österreichs. — Wien ; München ; Zürich : Europaverl., c1979.
171, [1] p. ; 21 cm. — (Soziale Brennpunkte ; 7) Au79-10-21
Contains papers presented at the 2d and 3d Conferencia General del Episcopado Latinoamericano.
"Quellenverzeichnis": p. 170-[172]
ISBN 3-203-50707-2 : $118.00
1. Christianity—Latin America—Congresses. 2. Church and social problems—Latin America—Congresses. 3. Nonviolence—Moral and religious aspects—Congresses. I. Goss-Mayr, Hildegard. II. Katholische Sozialakademie Österreichs. III. Conferencia General del Episcopado Latinoamericano.
BR600.G48 79-376032
*79 MARC

—LATIN AMERICA—PERIODICALS

Religião e sociedade. no. 1–
maio 1977–
[São Paulo, Editora de Humanismo, Ciência e Tecnologia]
no. ill. 21 cm. semiannual.
Numbers for May 1977– issued by the Centro de Estudos de Religião of the Universidade Estadual de Campinas.
1. Church and social problems — Latin America — Periodicals. 2. Church and social problems—Catholic Church—Periodicals. I. Universidade Estadual de Campinas. Centro de Estudos de Religião.
HN39.L3R44 261.8'3'098 79-645072
MARC-S

—NETHERLANDS

Jager, Okke.
Wij zijn niet machteloos : een tijdsbeeld met tegenwicht / Okke Jager. — Baarn : Ten Have, [1978]
123 p. ; 21 cm. Ne78-50
ISBN 9025941435 : fl 13.50
1. Church and social problems—Netherlands. I. Title.
HN39.N4J34 79-370842
*79 MARC

—NETHERLANDS—ADDRESSES, ESSAYS, LECTURES

Dulst, A J van, 1929-
Varen we wèl in de welvaart? [Door] A. J. van Dulst. Kampen, Kok, 1971.
135 p. 21 cm. fl 9.75 Ne71-21
ISBN 9024222733
1. Netherlands—Social conditions—Addresses, essays, lectures. 2. Church and social probems—Netherlands—Addresses, essays, lectures. 3. Christelijk-Historische Unie—Addresses, essays, lectures. I. Title.
HN517.D84 78-355005
78 MARC

Wereld in steigers. [n.p.] Interkerkelijk Vredesberaad [1970]
52 p. illus. 28 cm. (Cahiers voor Vredesvraagstukken, 8) Ne***
Cover title.
1. Church and social problems—Netherlands—Addresses, essays, lectures. 2. Christianity and international affairs—Addresses, essays, lectures. I. Interkerkelijk Vredesberaad. II. Series.
HN39.N4W47 79-345428
79 MARC

—NORWAY—ADDRESSES, ESSAYS, LECTURES

Oppgjør med kirken / Haftor Viestad (red.). — Oslo : Pax, c1976.
286 p. : ill. ; 19 cm. — (En Pax-bok ; 487) N77-Apr.
Includes bibliographical references.
ISBN 8253007779 : kr37.00
1. Church and social problems—Norway—Addresses, essays, lectures. 2. Church and state—Norway—Addresses, essays, lectures. 3. Norske Kirke—Controversial literature—Addresses, essays, lectures. I. Viestad, Haftor, 1941-
HN39.N6O66 78-368185
78 MARC

—NORWAY—HISTORY

Foss, Øivind, 1935-
Kirke og klassekamp. Oslo, Gyldendal, 1973.
104 p. 19 cm. (Fakkel-bøker, 257C) kr19.50 N73-18
Bibliography: p. 103-[105]
ISBN 8205057435
1. Church and social problems—Norway—History. 2. Church and social problems—History. I. Title.
HN39.N6F67 78-414926
78 MARC

—PHILIPPINE ISLANDS—DUMAGUETE —CASE STUDIES

Makil, Perla Q
Dumaguete City urban priorities study : final report / prepared for the Asia Foundation, the National Secretariat for Social Action, and the Redemptorist Fathers of Dumaguete City by the Institute of Philippine Culture on 31 July 1973 ; Perla Q. Makil. — Quezon City : Ateneo de Manila Institute of Philippine Culture, 1973.
62 leaves in various foliations ; 28 cm.
1. Community development, Urban—Philippine Islands—Dumaguete. 2. Church and social problems—Philippine Islands—Dumaguete—Case studies. 3. Church and social problems—Catholic Church—Case studies. I. Ateneo de Manila University. Institute of Philippine Culture. II. Title.
HN730.D84M33 309.2'62'095995 79-305382
79 MARC

—POLAND

Bender, Ryszard.
Społeczne inicjatywy chrześcijańskie w Królestwie Polskim 1905-1918 / Ryszard Bender. — Wyd. 1. — Lublin : TN KUL, 1978 (Lublin : LZGPKWN)
175, [12] p. ; 24 cm. — (Rozprawy Wydziału Historyczno-Filologicznego, Towarzystwo Naukowe Katolickiego Uniwersytetu Lubelskiego ; 41)
P78-3780
Summary in English and French.
Includes bibliographical references and index.
zł95.00
1. Church and social problems—Poland. 2. Church and social problems—Catholic Church. I. Title.
HN39.P7B46 78-390732
79 MARC

Wieśniak, Józef.
Wici - Agraryzm - Siew : ruch młodowiejski w świetle zasad katolickich / Józef Wieśniak. — Kraków : Nakł. Chrześcijańsko-Społecznego Instytutu Oświatowego, 1937.
255, iv p. ; 20 cm.
Bibliography: p. i-iv.
1. Rural youth—Poland. 2. Catholic Church—Doctrinal and controversial works. 3. Church and social problems—Poland. 4. Związek Młodzieży Wiejskiej Rzeczypospolitej Polskiej "Wici". I. Title.
HQ799.P6W56 79-359777
79 MARC

—PORTUGAL

Reforma agrária interpela a Igreja / Acácio Catarino ... [et al.] — Lisboa : Multinova, [1977]
262 p. ; 21 cm.
1. Land reform—Portugal. 2. Church and social problems—Catholic Church. 3. Church and social problems—Portugal. I. Catarino, Acácio Ferreira.
HD786.R43 79-370117
79 MARC

—QUÉBEC (PROVINCE)

Bélanger, André J 1935-
Ruptures et constantes : quatre idéologies du Québec en éclatement : La Relève, la JEC, Cité libre, Parti pris / André-J. Bélanger. — [Montréal] : Hurtubise HMH, [c1977]
219 p. ; 21 cm. — (Sciences de l'homme et humanisme ; 8) C***
Bibliography: p. [203]-219.
ISBN 0-7758-0104-6 : $7.50
1. Political science—Québec (Province)—History. 2. Church and social problems—Québec (Province) 3. Québec (Province)—Politics and government. I. Title.
JA84.C3B44 320.5'09714 79-365380
79 MARC

—UNITED CHURCH OF CHRIST —PERIODICALS

Journal of current social issues. v. 9, no. 5–
spring 1971–
[New York, Division of Higher Education and the American Missionary Association, United Church Board for Homeland Ministries, United Church of Christ.
v. ill. 28 cm. quarterly.
Continues: United Church Board for Homeland Ministries. Division of Higher Education and the American Missionary Association. Journal - [Division of Higher Education and the American Missionary Association, United Church Board for Homeland Ministries]
Published with other similar organizations.
ISSN 0041-7211
1. Church and social problems—United Church of Christ—Periodicals. I. United Church Board for Homeland Ministries. Division of Higher Education and the American Missionary Association.
BX9884.A1U48a 261.8'3'05 75-642289
MARC-S

United Church Board for Homeland Ministries. Division of Higher Education and the American Missionary Association.
Journal - [Division of Higher Education and the American Missionary Association, United Church Board for Homeland Ministries] v.
-9, no. 4; -Jan./Feb. 1971.
[New York, Division of Higher Education and the American Missionary Association, United Church Board for Homeland Ministries, United Church of Christ, etc.]
v. ill. 28 cm.
Monthly (except July)
bimonthly -Jan./Feb. 1971.
Published with other similar organizations.
Continued by: Journal of current social issues, ISSN 0508-3373.
1. Church and social problems—United Church of Christ—Periodicals.
BX9884.A1U48a 261.1 75-642318
MARC-S

—ZAIRE

Segers, Joseph.
Actions pour le développement en République du Zaïre / J. Segers. — Kinshasa-Gombe : Centre d'études pour l'action sociale, 1971.
62 p. ; 27 cm.
Cover title.
1. Economic assistance in Zaire. 2. Church and social problems—Zaire. 3. Church and social problems—Catholic Church. I. Title.
HC591.C6S43 338.91'675'1 79-101843
79 MARC

—ZAMBIA—CONGRESSES

Conference, Churches in National Development, Lusaka, Zambia, 1972.
Conference, Churches in National Development, Lusaka, March 20-23, 1972 : [report]. — [s.l. : s.n., 1972?]
94 p. ; 34 cm.
Cover title.
1. Church and social problems—Zambia—Congresses. 2. Community development—Zambia—Congresses.
HN39.Z3C66 1972 261.8'3'096894 78-317316
78 MARC

CHURCH AND SOCIETY see Church and the world

CHURCH AND STATE

see also Anti-clericalism; Asylum, Right of; Caesaropapism; Christianity and politics; Church and education; Church lands; Church polity; Investiture; Liberty of conscience; Nationalism and religion; Religion and state; Religious liberty; Secularization

Cavallaro, Giovanna.
La pace nella filosofia politica di Marsilio da Padova / Giovanna Cavallaro. — Ferrara : De Salvia, 1973.
45, viii p. ; 24 cm. It76-May
Bibliography: p. iii-viii.
1. Marsilius, of Padua, d. 1342? Defensor pacis. 2. State, The. 3. Church and state. 4. Political science—Early works to 1700. 5. Peace. I. Title.
JC121.M35C38 79-371040
*79 MARC

CHURCH AND STATE (Continued)

Dialogus inter clericum et militem.
Dialogus inter clericum et militem. Cologne, Heinrich Quentell [ca. 1492]
[10] l., the last blank. woodcut: 1 illus. 4°. 19.4 cm.
Leaf ([1a] t. p.): Dialogus inter clericu[m] et militem super dignitate papali et regia. De natiuitate et morib[us] Antichristi.
The occasional ascription of Dialogus to William Ockham is erroneous; the probable author is Pierre Dubois. Cf. Gesamtkat. d. Wiegendr., v. 7, column 378.
Hain. Repertorium (with Reichling's Appendices) 6116; Gesamtkat. d. Wiegendr., 8270; Goff. Third census, D-154.
Initials, paragraph-marks, initial strokes, and underlines supplied in red.
1. Church and state. 2. Antichrist. I. Ockham, William, d. ca. 1349. II. Dubois, Pierre, fl. 1300. III. Compendium de vita antichristi. ca. 1492.
Incun. X.D5 74-227486
[BV629]

Gibb, Hamilton Alexander Rosskeen, Sir, 1895-1971.
(Maẕhab va siyāsat dar Masīḥīyat va Islām)
مذهب و سیاســـت در مسیحیـــت و اسلام / هامیلتون آ. ر. کیب [i. e. گیب] ؛ ترجمه م. قائنی. — قم : انتشارات دار الفکر ، [1972 or 1973] 1351.
96 p. ; 17 cm.
In Persian.
1. Islam and state. 2. Church and state. 3. Islam—Relations—Christianity. 4. Christianity and other religions—Islam. I. Title.
BP173.6.G55 1972 79-971193

Inoue, Yoshihiko, 1928-
(Kyōkai to seiji no mondai)
教会と政治の問題 / [執筆者 井上良彦, 村上伸, 加藤邦雄 ; 編集者 日本基督教団宣教研究所]. — 東京 : 日本基督教団出版部, 1961.
195 p. ; 19 cm.
Bibliography: p. 100-101.
1. Church and state. I. Murakami, Hiroshi, 1930- joint author. II. Katō, Kunio, 1906- joint author. III. Nihon Kirisuto Kyōdan. Senkyō Kenkyūbu. IV. Title.
BV630.2.I53 79-800581

Marsilius, of Padua, d. 1342?
Defensor pacis / Marsilio da Padova ; a cura di Gaetano Garofalo. — 1. ed. — Bergamo : Minerva italica, 1976.
142 p. ; 22 cm. — (Collana di testi filosofici ; 16) It77-May
Bibliography: p. 12-23.
L2000
1. Church and state. 2. State, The. 3. Political science—History. I. Garofalo, Gaetano. II. Title.
JC121.M326 1976 322'.1 77-562694
*77[r79]rev MARC

Pietrzak, Michał.
Prawo wyznaniowe : [skrypt dla studentów wydziałów prawa i administracji] / Michał Pietrzak. — Wyd. 1. — Warszawa : Państ. Wydaw. Naukowe, 1978 (Łódź : ZGWN)
133, [2] p. ; 24 cm. P78-10719
Bibliography: p. 130-[132]
zł22.00
1. Ecclesiastical law. 2. Church and state. 3. Church and state in Poland. I. Title.
K3280.P54 79-365647
79 MARC

Spinelli, Lorenzo, 1915-
Prospettive canonistiche nella Chiesa di oggi. Modena, S.T.E.M., 1973.
179 p. 24 cm. L5000 It73-June
Bibliography: p. [159]-176.
1. Canon law. 2. Religious liberty. 3. Freedom of association (Canon law) 4. Church and state. I. Title.
76-507587
79[r79]rev MARC

—ADDRESSES, ESSAYS, LECTURES

Georvin, Thomas.
Le Monstrueux paradoxe, politique et religion / Thomas Georvin. — Paris : Pensée universelle, [c1975]
93 p. ; 18 cm. F76-18763
21.40F
1. Church and state—Addresses, essays, lectures. I. Title.
BV630.2.G46 261 78-387366
79 MARC

—CATHOLIC CHURCH

see also Church and state in France; Church and state in Italy; Fascism and the Catholic Church

Andreev, Mikhail Vasil'evich.
(Klerikal'nyĭ antikommunizm)
Клерикальный антикоммунизм / М. В. Андреев, канд. филос. наук. — Москва : Знание, 1977.
63 p. ; 21 cm. — (Новое в жизни, науке, технике : Серия Научный атеизм ; № 12, 1977) USSR 77
Includes bibliographical references.
0.11rub
1. Religion—Controversial literature. 2. Propaganda, Anti-communist. 3. Church and state—Catholic Church. I. Title. II. Series: Novoe v zhizni, nauke, tekhnike : Serii͡a Nauchnyĭ ateizm ; 1977, no. 12.
BL2780.A675 79-413635

O'Neill, James Milton, 1881-
Catholicism and American freedom / by James M. O'Neill. — Westport, Conn. : Greenwood Press, 1979, c1952.
xii, 287 p. ; 24 cm.
Reprint of the ed. published by Harper, New York.
Bibliography: p. 273-281.
Includes index.
ISBN 0-313-21153-1
1. Catholic Church in the United States. 2. Church and state in the United States. 3. Church and state—Catholic Church. 4. Blanshard, Paul, 1892- —American freedom and Catholic power. 5. Blanshard, Paul, 1892- —Communism, democracy, and Catholic power. I. Title.
[BX1406.2.O53 1979] 282'.73 78-21495
78 MARC

—CATHOLIC CHURCH—ADDRESSES, ESSAYS, LECTURES

Greinacher, Norbert, 1931-
Vor einem neuen politischen Katholizismus? / Norbert Greinacher, Ferdinand Klostermann. — Frankfurt am Main ; Bern ; Las Vegas : Lang, 1978.
107 p. ; 21 cm. — (Erfahrung und Theologie ; Bd. 1) GFR79-A
Includes bibliographical references.
ISBN 3-261-02628-6
1. Church and state—Catholic Church—Addresses, essays, lectures. 2. Catholic Church—History—1965- —Addresses, essays, lectures. 3. Christianity and politics—Addresses, essays, lectures. I. Klostermann, Ferdinand, joint author. II. Title. III. Series.
BX1791.G73 79-377892
79 MARC

—CATHOLIC CHURCH—CONGRESSES

Die Führung der Kirche in den sozialistischen Staaten Europas / hrsg. von Gabriel Adriányi. — München : Johannes-Berchmans-Verlag, 1979.
166 p. ; 21 cm. — (Sammlung Wissenschaft und Gegenwart) GFR79-A
Revised papers of a conference organized by the Historische Sektion of the Arbeitskreis Wissenschaft und Gegenwart, held in Bonn, Oct. 6-8, 1976.
Includes bibliographies and index.
ISBN 3-87056-019-3 : DM24.00
1. Communism and Christianity—Catholic Church—Congresses. 2. Church and state—Catholic Church—Congresses. 3. Europe, Eastern—Church history—Congresses. 4. Catholic Church in Eastern Europe—History—Congresses. I. Adriányi, Gabriel. II. Arbeitskreis Wissenschaft und Gegenwart. Historische Sektion.
BX1396.4.F83 79-362369
79[r79]rev MARC

—CHURCH OF ENGLAND

see also Church and state in Great Britain

—CONGRESSES

Christ and Caesar in Christian missions / Edwin L. Frizen, Jr. and Wade T. Coggins, editors. — Pasadena, Calif. : William Carey Library, c1979.
viii, 149 p. ; 22 cm.
Addresses and discussions of the AEPM/EFMA/IFMA study conference held in Overland Park, Kan., Sept. 25-29, 1978.
ISBN 0-87808-156-9
1. Missions—Congresses. 2. Church and state—Congresses. I. Frizen, E. L. II. Coggins, Wade T. III. Association of Evangelical Professors of Missions. IV. Evangelical Foreign Missions Association. V. Interdenominational Foreign Mission Association.
BV2390.C46 266 79-17124
79 MARC

Church and state, opening a new ecumenical discussion. — Geneva : World Council of Churches, 1978.
183 p. ; 22 cm. — (Faith and order paper ; no. 85) Sw78-9700
Includes bibliographical references.
ISBN 2-8254-0574-4
1. Church and state—Congresses. I. World Council of Churches. II. Series.
BV630.2.C58 261.7 79-313820
00 MARC

Le Intese tra Stato e confessioni religiose ; problemi e prospettive / a cura di C. Mirabelli. — Milano : A. Giuffrè, 1978.
xi, 239 p. ; 25 cm. — (Pubblicazioni della Facoltà di giurisprudenza, Università di Parma ; 39) It78-July
Includes bibliographical references.
L5500
1. Ecclesiastical law—Congresses. 2. Church and state—Congresses. 3. Religions—Congresses. I. Mirabelli, Cesare. II. Series: Parma. Università. Facoltà di giurisprudenza. Pubblicazioni ; 39.
342.087 79-358401
*79 MARC

—HISTORY

Caspary, Gerard E
Politics and exegesis : Origen and the two swords / Gerard E. Caspary. — Berkeley : University of California Press, c1979.
xv, 215 p. ; 24 cm.
English and Latin.
Includes bibliographical references and indexes.
ISBN 0-520-03445-7
1. Origenes. 2. Christianity and politics—History. 3. Church and state—History. I. Title.
BR65.O68C37 230'.1'3 77-71058
79 MARC

Geiger, Max.
Kirche, Staat, Widerstand : historische Durchgänge und aktuelle Standortbestimmung / Max Geiger. — Zürich : Theologischer Verlag, cop. 1978.
48 p. ; 20 cm. — (Theologische Studien ; 124) Sw79-21
Includes bibliographical references.
ISBN 3-290-17124-8 : 6.00F
1. Church and state—History. I. Title.
BV631.G44 79-351861
79 MARC

Minninger, Monika.
Von Clermont zum Wormser Konkordat : d. Auseinandersetzungen um d. Lehnsnexus zwischen König u. Episkopat / von Monika Minninger. — Köln ; Wien : Böhlau, 1978.
x, 313 p. ; 24 cm. — (Forschungen zur Kaiser- und Papstgeschichte des Mittelalters ; 2) GFR78-A
Some text in Latin.
Originally presented as the author's thesis, Marburg, 1972.
Bibliography: p. [285]-306.
Includes index.
ISBN 3-412-02377-9
1. Investiture—History. 2. Church and state—History. I. Title. II. Series.
BX1198.M56 1978 79-349672
79 MARC

Struve, Tilman.
Die Entwicklung der organologischen Staatsauffassung im Mittelalter / von Tilman Struve. — 1. Aufl. — Stuttgart : Hiersemann, 1978.
viii, 349 p. ; 24 cm. — (Monographien zur Geschichte des Mittelalters ; Bd. 16 ISSN 0026-9832) GFR78-A
A revision of the author's Habilitationsschrift, Stuttgart, 1976.
Bibliography: p. 326-338.
Includes index.
ISBN 3-7772-7805-X : DM154.00
1. State, The—History. 2. Political science—History. 3. Church and state—History. I. Title.
JA82.S77 1978 78-386343
78 MARC

Werner, Ernst, 1920-
Konstantinopel und Canossa : lateinisches Selbstverständnis im 11. Jahrhundert / Ernst Werner. — Berlin : Akademie-Verlag, 1977.
35 p. ; 22 cm. — (Sitzungsberichte der Akademie der Wissenschaften der DDR ; Jahrg. 1977, Nr. 4 G : Gesellschaftswissenschaften) GDR***
Includes bibliographical references.
2.50M
1. Church history—Middle Ages, 600-1500. 2. Church and state—History. 3. Heinrich IV, Emperor of Germany, 1050-1106. 4. Gregorius VII, Saint, Pope, 1015 (ca.)-1085. I. Title. II. Series: Akademie der Wissenschaften der DDR. Sitzungsberichte der Akademie der Wissenschaften der DDR ; Jahrg. 1977, Nr. 4 G.
BR270.W49 79-349244
79 MARC

—HISTORY—ADDRESSES, ESSAYS, LECTURES

Staat und Gesellschaft im politischen Wandel : Beitr. zur Geschichte d. modernen Welt / hrsg. von Werner Pöls. — Stuttgart : Klett-Cotta, 1979.
x, 554 p. ; 24 cm. GFR79-A
Festschrift published on the occasion of the 65th birthday of W. Bussmann.
"Tabula gratulatoria" ([4] p.) inserted.
Includes bibliographical references.
ISBN 3-12-911900-0 : DM138.00
1. Political science—History—Addresses, essays, lectures. 2. Church and state—History—Addresses, essays, lectures. 3. World politics—19th century—Addresses, essays, lectures. 4. World politics—20th century—Addresses, essays, lectures. I. Bussmann, Walter. II. Pöls, Werner.
JA78.S72 320 79-365845
79 MARC

—NORWAY—ADDRESSES, ESSAYS, LECTURES

Oppgjør med kirken / Haftor Viestad (red.). — Oslo : Pax, c1976.
286 p. : ill. ; 19 cm. — (En Pax-bok ; 487) N77-Apr.
Includes bibliographical references.
ISBN 8253007779 : kr37.00
1. Church and social problems—Norway—Addresses, essays, lectures. 2. Church and state—Norway—Addresses, essays, lectures. 3. Norske Kirke—Controversial literature—Addresses, essays, lectures. I. Viestad, Haftor, 1941-
HN39.N6O66 78-368185
78 MARC

CHURCH AND STATE (Continued)

—UNITED STATES

Rice, Charles E
Beyond abortion : the theory and practice of the secular state / by Charles E. Rice. — Chicago : Franciscan Herald Press, c1979.
vii, 159 p. ; 21 cm.
Includes bibliographical references and index.
ISBN 0-8199-0696-4 : $8.95
1. Abortion—United States. 2. Abortion—Law and legislation—United States. 3. Secularism—United States. 4. Church and state—United States. I. Title.
HQ767.5.U5R5 301 78-16248
78 MARC

CHURCH AND STATE IN AFRICA

All Africa Conference of Churches. Refugee Dept.
Africa's one million refugees : from where they escape, where they find asylum. — ₍Nairobi : Refugee Dept., All Africa Conference of Churches, 1975?₎
22 p. : ill. ; 16 x 24 cm.
Cover title: The African refugee story.
1. Africa—Politics and government—1960- 2. Refugees—Africa. 3. Church and refugee problems—Africa. 4. Church and state in Africa. 5. Church work with refugees—Africa. I. Title. II. Title: The African refugee story.
DT30.A467 1975 361.5'3'096 79-314375
79 MARC

CHURCH AND STATE IN ALSACE

—HISTORY—ADDRESSES, ESSAYS, LECTURES

Églises et État en Alsace et en Moselle : changement ou fixité? / Études publiées sous la direction de Jean Schlick. — Strasbourg : Cerdic-publications, 1979.
354 p. ; 22 cm. — (Hommes et Église ; 9) F***
Includes bibliographical references.
ISBN 2-85097-009-3 : 85.00F
1. Church and state in Alsace—History—Addresses, essays, lectures. 2. Church and state in Moselle, France (Dept.)—History—Addresses, essays, lectures. 3. Alsace—Politics and government—Addresses, essays, lectures. 4. Moselle, France (Dept.)—Politics and government—Addresses, essays, lectures. I. Schlick, Jean. II. Series.
BR847.A6E34 261.7'0944'383 79-376035
79 MARC

CHURCH AND STATE IN AUSTRIA

see also Josephinism

—HISTORY

Vocelka, Karl.
Verfassung oder Konkordat? : der publizistische u. polit. Kampf d. österr. Liberalen um die Religionsgesetze des Jahres 1868 / von Karl Vocelka. — Wien : Verl. d. Österr. Akad. d. Wiss., 1978.
238 p., ₍4₎ leaves of plates : ill. ; 24 cm. — (Studien zur Geschichte der Österreichisch-Ungarischen Monarchie ; Bd. 17) (Schriften des DDr. Franz Josef-Mayer-Gunthof-Fonds ; Nr. 12) Au78-16-312
Bibliography: p. ₍221₎-234.
Includes index.
ISBN 3-7001-0241-0 : S350.00
1. Church and state in Austria—History. 2. Catholic Church in Austria—History. 3. Austria—Church history. I. Title. II. Series. III. Series: Franz Josef Mayer-Gunthof-Fond, Vienna. Schriften ; Nr. 12.
DB85.S8 Bd. 17 79-353735
₍BR816₎ *79 MARC

CHURCH AND STATE IN BAHIA, BRAZIL (STATE)

Azevedo, Thales de.
Igreja e Estado em tensão e crise : a conquista espiritual e o padroado na Bahia / Thales de Azevedo. — São Paulo : Editora Atica, 1978.
179 p. ; 21 cm. — (Ensaios ; 51)
Bibliography: p. ₍167₎-179.
Cr$110.00
1. Catholic Church in Bahia, Brazil (State)—History. 2. Church and state in Bahia, Brazil (State). 3. Bahia, Brazil (State)—Church history. I. Title.
BX1467.B3A95 78-395545
79 MARC

CHURCH AND STATE IN BRAZIL

Alves, Marcio Moreira.
A Igreja e a política no Brasil / Marcio Moreira Alves. — 1. ed. — Lisboa : Livraria Sá da Costa Editora, 1978.
313 p. ; 21 cm. — (Terceiro mundo)
Revised translation of L'Église et la politique au Brésil.
1. Catholic Church in Brazil. 2. Church and state in Brazil. I. Title.
BX1466.2.A62417 79-109032
79 MARC

Alves, Márcio Moreira.
A Igreja e a política no Brasil / Márcio Moreira Alves ; pref., Frei Betto. — São Paulo : Editora Brasiliense, 1979.
268 p. ; 21 cm.
Bibliography: p. ₍261₎-268.
Cr$190.00
1. Catholic Church in Brazil. 2. Church and state in Brazil. I. Title.
BX1466.2.A62417 1979 79-114991
79 MARC

O Clero no Parlamento brasileiro / ₍colaboraram na preparação deste trabalho, organização geral do plano, Fernando Bastos de Ávila ; coordinação, Américo Jacobina Lacombe₎. — Brasília : Câmara dos Deputados, Centro de Documentação e Informação, Coordenação de Publicações ; Rio de Janeiro : Centro João XXIII (IBRADES) : Fundação Casa de Rui Barbosa, Centro de Estudos Históricos, 1978-
v. : ill. ; 25 cm. — (Documentos parlamentares - Brasil, Congresso Nacional, Câmara dos Deputados ; 124-)
Includes bibliographical references.
CONTENTS: v. 1. Assembléia Geral Constituinte e Legislativa (1823).
1. Brazil. Congresso—History. 2. Church and state in Brazil. I. Ávila, Fernando Bastos de. II. Lacombe, Américo Jacobina, 1909- . III. Brazil. Congresso. Câmara dos Deputados. Centro de Documentação e Informação. IV. Centro João Vinte e Três. V. Fundação Casa de Rui Barbosa. Centro de Estudos Históricos. VI. Series: Brazil. Congresso. Câmara dos Deputados. Documentos parlamentares ; 124-.
JL2454.C53 328.81'09 79-104666
79 MARC

—HISTORY

Gerson, Brasil.
O regalismo brasileiro / Brasil Gérson. — Rio de Janeiro : Livraria Editora Cátedra, 1978.
282 p. ; 21 cm.
"Em convênio com o Instituto Nacional do Livro."
Includes bibliographical references.
Cr$70.00
1. Church and state in Brazil—History. 2. Brazil—Church history. 3. Brazil—History—1822-1889. I. Title.
BR675.G46 79-109969
79 MARC

CHURCH AND STATE IN CHILE

—HISTORY

Hinkelammert, Franz Josef, 1931-
Ideología de sometimiento : la Iglesia Católica Chilena frente al golpe, 1973-1974 / Franz J. Hinkelammert. — 1. ed. — Ciudad Universitaria Rodrigo Facio, Costa Rica : Editorial Universitaria Centroamericana, 1977.
176 p. ; 21 cm. — (Colección DEI - Departamento Ecuménico de Investigaciones)
1. Church and state in Chile—History. 2. Catholic Church in Chile—History. I. Title. II. Series: Universidad Nacional (Costa Rica). Departamento Ecuménico de Investigaciones. Colección DEI.
BX1468.2.H56 78-104473
78 MARC

CHURCH AND STATE IN COLOMBIA

Identidad cristiana en la acción por la justicia : una visión alternativa : respuesta / elaborada con los aportes de numerosos grupos cristianos del País. — ₍s.l. : s.n., 1977 or 1978₎
55 p., ₍2₎ leaves of plates : ill. ; 21 cm.
On cover: Grupos cristianos responden a los obispos de Colombia.
Includes bibliographical references.
1. Catholic Church—Doctrinal and controversial works—Catholic authors. 2. Catholic Church in Colombia. 3. Church and state in Colombia. 4. Communism and Christianity.
BX1755.I33 79-103365
79 MARC

CHURCH AND STATE IN CYPRUS

Mpilalēs, Spyridōn Sp
(Ho papokaisarismos eis tēn Ekklēsian tēs Kyprou)
'Ο παποκαισαρισμὸς εἰς τὴν Ἐκκλησίαν τῆς Κύπρου : ἱστορικοκανονικὴ θεώρησις τοῦ παποκαισαρισμοῦ τοῦ Μακαρίου / Σπυρίδωνος Σπ. Μπιλαλῆ. — Ἀθῆναι : Ἐκδόσεις Ὀρθοδόξου Τύπου, 1974.
183 p. ; 25 cm.
Includes bibliographies.
1. Church and state in Cyprus. 2. Makarios III, Archbishop of Cyprus, 1913–1977. 3. Cyprus (Archdiocese) 4. Caesaropapism. I. Title.
BX450.M64 79–413150

CHURCH AND STATE IN EASTERN EUROPE

Babris, Peter J
Silent churches : persecution of religions in the soviet-dominated areas / by Peter J. Babris. — Arlington Heights, Ill. : Research Publishers, 1978.
531 p. : maps (on lining papers) ; 25 cm.
Bibliography: p. 519-531.
ISBN 0-911252-02-9 : $19.50
1. Chrisitianity—Europe, Eastern. 2. Church and state in Eastern Europe. 3. Persecution—Europe, Eastern. 4. Communism and Christianity. I. Title.
BR738.6.B3 274.7 78-52811
79 MARC

—ADDRESSES, ESSAYS, LECTURES

Il Nuovo Principe alla prova / ₍a cura di₎ Erre. — 1. ed. — Bologna : CSEO biblioteca, 1978.
254 p. ; 18 cm. — (Paperbacks/Verifiche ; 1) It78-Nov
Articles previously published in CSEO documentazione, 1973-1977.
Includes bibliographical references.
L3000
1. Church and state in Eastern Europe—Addresses, essays, lectures. 2. Europe, Eastern—Church history—Addresses, essays, lectures. I. Erre. II. Centro studi Europa orientale.
BR738.6.N86 78-396278
*79 MARC

CHURCH AND STATE IN ECUADOR

Roncagliolo, Rafael.
Iglesia, prensa y militares : el caso Riobamba y los obispos latinoamericanos / Rafael Roncagliolo, Fernando Reyes Matta. — México : Instituto Latinoamericano de Estudios Transnacionales, c1978.
225 p. ; 21 cm.
"Hechos y documentos": p. ₍87₎-225.
Includes bibliographical references.
1. Catholic Church in Latin America. 2. Catholic Church—Bishops. 3. Bishops—Latin America. 4. Persecution—Ecuador—Riobamba. 5. Church and state in Ecuador. I. Reyes Matta, Fernando, joint author. II. Title.
BX1426.2.R66 78-108426
79 MARC

CHURCH AND STATE IN EL SALVADOR

El Salvador, un pueblo martirizado. — Madrid : Instituto de Estudios Políticos para América Latina y Africa, 1977.
72 p. ; 28 cm. — (América Latina ; 3) Sp***
At head of title: Idoc, España.
1. Christianity—El Salvador. 2. Church and state in El Salvador. 3. El Salvador—Politics and government. I. Idoc. II. Instituto de Estudios Políticos para América Latina y Africa. III. Title. IV. Series: América Latina (Madrid) ; 3.
BR625.E35E4 78-382250
79 MARC

CHURCH AND STATE IN ENGLAND

—ADDRESSES, ESSAYS, LECTURES

Porter, Joshua Roy.
The Crown & the Church / ₍by₎ Roy Porter. — London : Church Literature Association, 1978.
₍2₎, 18 p. ; 21 cm. GB78-32755
Includes bibliographical references.
£0.70
1. Church and state in England—Addresses, essays, lectures. I. Title.
BX5157.P64 322'.1'0941 79-302733
79 MARC

—HISTORY

Pritchard, Arnold, 1949-
Catholic loyalism in Elizabethan England / by Arnold Pritchard. — Chapel Hill : University of North Carolina Press, c1979.
xii, 143 p. ; 24 cm.
Bibliography: p. 231-235.
Includes index.
ISBN 0-8078-1345-1
1. Catholics—England. 2. Archpriest controversy, 1598-1602. 3. Church and state in England—History. I. Title.
BX1492.P78 261.7'0942 78-10208
78 MARC

CHURCH AND STATE IN ESTONIA

Salo, Vello.
Riik ja kirikud 1940-1974 / Vello Salo. — Rooma : Maarjamaa, 1974.
82 p. ; 17 cm. — (Maarjamaa taskuraamat ; n:o 5) It***
Bibliography: p. 67-74.
Includes index.
1. Church and state in Estonia. 2. Estonia—Church history. I. Title.
BR937.E8S24 79-343424
79 MARC

CHURCH AND STATE IN EUROPE

—HISTORY—ADDRESSES, ESSAYS, LECTURES

Church and state in Europe, 1864-1914 / edited with an introd. by Ernst C. Helmreich. — St. Louis : Forum Press, c1979.
114 p., ‹1› leaf of plates : port. ; 24 cm. — (Problems in civilization)
Bibliography: p. 112-114.
ISBN 0-88273-405-9
1. Church and state in Europe—History—Addresses, essays, lectures. I. Helmreich, Ernst Christian. II. Series.
BR735.C5 261.7 78-68021
79 MARC

CHURCH AND STATE IN FINLAND

Arkkila, Reijo.
Tunnustuskirkollisuus / Reijo Arkkila. — ‹Helsinki› : Suomen luterilainen evankeliumiyhdistys, 1976.
48 p. ; 18 cm. Fi77-Apr.
ISBN 9516173713 : Fmk9.85
1. Lutheran Church in Finland—Doctrinal and controversial works—Lutheran authors. 2. Church and state in Finland. 3. Finland—Church history. I. Title.
BX8029.A765 79-357173
79 MARC

Finland. Kirkko ja valtio -komitea.
Kirkko ja valtio -komitean mietintö. — Helsinki : ‹Valtioneuvosto›, 1977 (Tampere : Tampereen keskusp. ; ‹Helsinki› : Painatuskeskus).
281 p. : ill. ; 25 cm. — (Komiteanmietintö ; 1977:21) Fi77-Oct.
ISBN 9514622464
1. Church and state in Finland. I. Series: Finland. Komiteanmietintö ; 1977:21.
J398.R13 1977:21 79-356399
‹BR943› 79 MARC

—HISTORY

Silfverhuth, Voitto.
Kirkon ja keisarin sensuuri : uskonnollisen kirjallisuuden valvonta Suomessa 1809-1865 / Voitto Silfverhuth. — Helsinki : Suomen Kirkkohistoriallinen Seura, 1977.
346 p. ; 21 cm. — (Suomen Kirkkohistoriallisen Seuran toimituksia ; 104 ISSN 0356-0759) Fi***
Summary in English.
Bibliography: p. 314-334.
Includes index.
ISBN 9519021221
1. Censorship—Finland—History. 2. Church and state in Finland—History. 3. Religious literature—History. I. Title. II. Series: Suomen Kirkkohistoriallinen Seura. Toimituksia ; 104.
79-367716
79 MARC

CHURCH AND STATE IN FRANCE

Lecourt, Robert.
Concorde sans concordat : 1952-1957 / Robert Lecourt. — Paris : Hachette, 1978.
187 p. ; 23 cm. F78-11097
At head of title: Entre l'Église et l'État.
Includes bibliographical references.
ISBN 2-01-005313-3 : 35.00F
1. Church and state in France. 2. France—Church history—1945- 3. France—Politics and government—1945-1958. I. Title. II. Title: Entre l'Église et l'État.
BR846.3.L4 282'.44 78-388852
79 MARC

Perrens, François Tommy, 1822-1901.
L'Église et l'État en France sous le règne de Henri IV et la régence de Marie de Médicis / François-Tommy Perrens. — Genève : Mégariotis, ‹1978›
2 v. (xv, 534 ; 508 p.) ; 23 cm. Sw78-3785
Reprint of the 1872 ed. published by A. Durand, Paris.
Includes bibliographical references.
200F
1. Church and state in France. 2. France—History—Henry IV, 1589-1610. 3. France—History—Louis XIII, 1610-1643. I. Title.
DC121.3.P45 1978 322'.1'0944 79-340889
79 MARC

—HISTORY

Kreiser, B Robert, 1943-
Miracles, convulsions, and ecclesiastical politics in early eighteenth-century Paris / B. Robert Kreiser. — Princeton, N.J. : Princeton University Press, 1978.
xvii, 485 p. ; 25 cm.
Bibliography: p. 403-467.
Includes index.
ISBN 0-691-05262-X
1. Jansenists. 2. Miracles. 3. Paris, François de, 1690-1727. 4. Church and state in France—History. 5. Paris—Church history. 6. Catholic Church. Pope, 1700-1721 (Clemens XI). Unigenitus (8 Sept. 1713) 7. Paris. Saint-Médard (Church) I. Title.
BX4722.K73 282'.44'36 77-85546
77‹r79›rev MARC

Latreille, André, 1901-
De Gaulle, la Libération et l'Église catholique / André Latreille ; préf. par Hubert Beuve-Méry. — Paris : Éditions du Cerf, 1978.
146 p. ; 20 cm. — (Rencontres ; nouv. sér., 7) F***
ISBN 2-204-01203-3 : 37.00F
1. France—Politics and government—1945-1958. 2. Gaulle, Charles de, Pres. France, 1890-1970—Religion. 3. Church and state in France—History. 4. Catholic Church in France—History. I. Title.
DC404.L25 320.9'44'0816 78-394424
79 MARC

Rothrock, George A
The Huguenots : a biography of a minority / G. A. Rothrock. — Chicago : Nelson-Hall, c1979.
xxv, 201 p., ‹24› leaves of plates : ill. ; 23 cm.
Bibliography: p. 191-196.
Includes index.
ISBN 0-88229-277-3
1. Huguenots—History. 2. Church and state in France—History. 3. France—History—16th century. 4. France—History—17th century. I. Title.
BX9454.2.R67 272'.4 78-23476
78 MARC

CHURCH AND STATE IN GERMANY

Listl, Joseph.
Kirche und Staat in der neueren katholischen Kirchenrechtswissenschaft / von Joseph Listl. — Berlin : Duncker und Humblot, 1978.
xvi, 279 p. ; 24 cm. — (Staatskirchenrechtliche Abhandlungen ; Bd. 7) GFR78-A
A revision and enlargement of the author's Habilitationsschrift, Ruhr-Universität Bochum, 1977, presented under the title: Die Grundvorstellungen der römischen Kirchenrechtswissenschaft zum Verhältnis von Kirche und Staat vom Ende der Aufklärung bis zum Zweiten Vatikanischen Konzil.
"Theses ex jure publico ecclesiastico": p. ‹236›-244.
Bibliography: p. ‹247›-267.
Includes indexes.
ISBN 3-428-04212-3 : DM78.00
1. Canon law. 2. Catholic Church in Germany. 3. Ecclesiastical law—Germany, West. 4. Church and state in Germany. I. Title.
262.9'0943 79-377128
79 MARC

Volkmann, Klaus J
Die Rechtsprechung staatlicher Gerichte in Kirchensachen : 1933-1945 / Klaus J. Volkmann. — Mainz : Matthias-Grünewald-Verlag, 1978.
xxxix, 241 p. ; 25 cm. — (Veröffentlichungen der Kommission für Zeitgeschichte : Reihe B, Forschungen ; Bd. 24) GFR78-A
Originally presented as the author's thesis, Freiburg i. B., 1976.
Bibliography: p. ‹xii›-xxxv.
ISBN 3-7867-0685-9
1. Ecclesiastical law—Germany. 2. Church and state in Germany. 3. Ecclesiastical law—Germany—Cases—Indexes. I. Title. II. Series: Kommission für Zeitgeschichte. Veröffentlichungen der Kommission für Zeitgeschichte : Reihe B, Forschungen ; Bd. 24.
262.9'0943 78-392573
79 MARC

—19TH CENTURY—ADDRESSES, ESSAYS, LECTURES

Bergsträsser, Ludwig, 1883-1960.
Der politische Katholizismus / Ludwig Bergsträsser. — Hildesheim ; New York : G. Olms, 1976.
2 v. in 1 : ports. ; 20 cm. GFR***
Reprint of the 1921-23 ed. published by Drei Masken Verlag, Munich, which was issued as Reihe 2, Bd. 3 of Der Deutsche Staatsgedanke.
ISBN 3-487-05897-9
1. Church and state in Germany—19th century—Addresses, essays, lectures. 2. Catholic Church in Germany—Addresses, essays, lectures. 3. Deutsche Zentrumspartei—Addresses, essays, lectures. 4. Germany—Politics and government—1789-1900—Addresses, essays, lectures. 5. Germany—Politics and government—1871-1918—Addresses, essays, lectures. I. Title.
BX1536.B4 1976 79-388948
79 MARC

—20TH CENTURY

Aretz, Jürgen.
Katholische Arbeiterbewegung und Nationalsozialismus : d. Verb. Kath. Arbeiter- u. Knappenvereine Westdeutschlands 1923-1945 / von Jürgen Aretz ; ‹dieser Bd. wurde seitens d. Komm. für Zeitgeschichte betreut von Konrad Repgen›. — Mainz : Matthias-Grünewald-Verlag, 1978.
xxix, 252 p. : map, ports. ; 25 cm. — (Veröffentlichungen der Kommission für Zeitgeschichte : Reihe B, Forschungen ; Bd. 25) GFR79-A
Originally presented as the author's thesis, Bonn, 1977.
Bibliography: p. ‹x›-xxvii.
Includes index.
ISBN 3-7867-0715-4
1. Verband Katholischer Arbeiter- und Knappenvereine Westdeutschlands—History. 2. Church and state in Germany—20th century. I. Title. II. Series: Kommission für Zeitgeschichte. Veröffentlichungen der Kommission für Zeitgeschichte : Reihe B, Forschungen ; Bd. 25.
HD6338.2.G3A73 1978 79-374801
79 MARC

Helmreich, Ernst Christian.
The German churches under Hitler : background, struggle, and epilogue / Ernst Christian Helmreich. — Detroit : Wayne State University Press, 1979.
616 p. ; 26 cm.
Bibliography: p. 577-600.
Includes index.
ISBN 0-8143-1603-4 : $30.00
1. Church and state in Germany—20th century. 2. Germany—Church history—20th century. I. Title.
BR856.H443 261.7'0943 78-17737
78 MARC

—1933-1945

Domarus, Max.
Bischof Matthias Ehrenfried und das Dritte Reich / Max Domarus. — Würzburg : M. Domarus, 1975.
95 p. (p. 95 advertisement), ‹4› leaves of plates : ill. ; 21 cm. GFR***
Bibliography: p. 91-92.
Includes indexes.
1. Ehrenfried, Matthias, 1871-1948. 2. Church and state in Germany—1933-1945. 3. Germany—History—1933-1945. I. Title.
BX4705.E343D65 79-354530
79 MARC

Kessemeier, Siegfried, 1930-
Katholische Publizistik im NS-Staat : 1933-1938 : Grundzüge u. Entwicklung / Siegfried Kessemeier. — Münster : Regensberg ‹in Komm.›, 1973.
472 p. : ill. ; 21 cm. — (Arbeiten aus dem Institut für Publizistik der Universität Münster ; Bd. 9) GFR77-A
Originally presented as the author's thesis, Münster, 1969, under the title: Grundzüge und Entwicklung katholischer Publizistik im nationalsozialistischen Staat, 1933-1938.
Bibliography: p. 423-472.
1. Catholic Church in Germany—History. 2. Church and state in Germany—1933-45. 3. Press, Catholic—Germany—History. 4. Public relations and politics. 5. Germany—Church history. I. Title. II. Series: Münster. Universität. Institut für Publizistik. Arbeiten ; Bd. 9.
BX1536.K38 1973 78-371000
78 MARC

May, Georg, 1926-
Interkonfessionalismus in der deutschen Militärseelsorge von 1933 bis 1945 / von Georg May. — Amsterdam : B. R. Grüner, 1978.
lxxiii, 529 p. ; 23 cm. — (Kanonistische Studien und Texte ; Bd. 30) Ne78-41
Bibliography: p. xxix-lxxii.
Includes indexes.
ISBN 9060321030
1. Church and state in Germany—1933-1945. 2. Chaplains, Military—Germany. 3. Chaplains, Military—Catholic Church. 4. Catholic Church in Germany—History. 5. Germany—Church history—1933-1945. 6. Christian union—Germany. I. Title. II. Series.
BR856.M284 78-394596
•79 MARC

Missalla, Heinrich, 1926-
Für Volk und Vaterland : d. kirchl. Kriegshilfe im Zweiten Weltkrieg / Heinrich Missalla. — 1. Aufl. — Königstein/Ts. : Athenäum Verlag, 1978.
xxvi, 215 p. ; 21 cm. GFR78-A
Bibliography: p. ix-xix.
ISBN 3-7610-8024-7
1. World War, 1939-1945—Catholic Church. 2. World War, 1939-1945—War work—Churches. 3. World War, 1939-1945—Germany. 4. Church and state in Germany—1933-1945. I. Title.
D810.C6M57 940.53'1522 78-387521
78 MARC

Niesel, Wilhelm.
Kirche unter dem Wort : d. Kampf d. Bekennenden Kirche d. altpreuss. Union 1933-1945 / Wilhelm Niesel. — Göttingen : Vandenhoeck und Ruprecht, 1978.
xiii, 340 p. ; 23 cm. — (Arbeiten zur Geschichte des Kirchenkampfes : Ergänzungsreihe ; Bd. 11) GFR78-A
Bibliography: ‹317›-320.
Includes index.
ISBN 3-525-55556-3
1. Evangelische Kirche der altpreussischen Union—History. 2. Bekennende Kirche—History. 3. Germany—Church history—1933-1945. 4. Church and state in Germany—1933-1945. I. Title. II. Series.
BX8022.P7N53 78-378688
78 MARC

—ADDRESSES, ESSAYS, LECTURES

Barraclough, Geoffrey, 1908- tr.
Mediaeval Germany, 911-1250 : essays by German historians / translated with an introd. by Geoffrey Barraclough. — 1st AMS ed. — New York : AMS Press, 1979.
2 v. ; 22 cm.
Reprint of the 1938 ed. published by B. Blackwell, Oxford, which was issued as v. 1-2 of Studies in mediaeval history.
CONTENTS: v. 1. Introduction.—v. 2 Essays.
Includes bibliographical references and indexes.
ISBN 0-404-14800-X
1. Germany—History—843-1273—Addresses, essays, lectures. 2. Germany—Constitutional history—Addresses, essays, lectures. 3. Investiture—Addresses, essays, lectures. 4. Church and state in Germany—Addresses, essays, lectures. I. Title. II. Series: Studies in mediaeval history ; v. 1-2.
DD126.B35 1979 943'.02 75-41019
78 MARC

—HISTORY

Weber, Werner, 1904-1976.
Staat und Kirche in der Gegenwart : rechtswissenschaftl. Beitr. aus 4 Jahrzehnten / von Werner Weber. — Tübingen : Mohr, 1978.
viii, 404 p. ; 24 cm. — (Jus ecclesiasticum ; Bd. 25 ISSN 0449-4393) GFR78-A
Includes bibliographical references.
ISBN 3-16-640732-4
1. Ecclesiastical law—Germany, West—History. 2. Church and state in Germany—History. I. Title. II. Series.
262.9'0943 79-364145
79 MARC

CHURCH AND STATE IN GERMANY (Continued)

—HISTORY—ADDRESSES, ESSAYS, LECTURES

Verein für Reformationsgeschichte.
Stadt und Kirche im 16. Jahrhundert / hrsg. von Bernd Moeller. — 1. Aufl. — Gütersloh : Gütersloher Verlagshaus Mohn, 1978.
191 p. ; 23 cm. — (Schriften des Vereins für Reformationsgeschichte ; Bd. 190) GFR78-A
Papers of the second symposium of the Verein für Reformationsgeschichte held in Reinhausen, March 24-26, 1977.
Includes bibliographical references and index.
ISBN 3-579-04334-X : DM48.00
1. Church and state in Germany—History—Addresses, essays, lectures. 2. Church and state in Switzerland—History—Addresses, essays, lectures. 3. Reformation—Germany—Addresses, essays, lectures. 4. Reformation—Switzerland—Addresses, essays, lectures. I. Moeller, Bernd, 1931- . II. Series: Verein für Reformationsgeschichte. Schriften ; Bd. 190.
BR300.V5 Bd. 190 78-379478
[BR309] 78 MARC

CHURCH AND STATE IN GREAT BRITAIN

Crowther-Hunt, Norman Crowther-Hunt, Baron, 1920-
Two early political associations : the Quakers and the dissenting deputies in the age of Sir Robert Walpole / by N. C. Hunt. — Westport, Conn. : Greenwood Press, 1979, c1961.
xvi, 231 p. ; 23 cm.
Reprint of the ed. published at the Clarendon Press, Oxford, Eng.
Bibliography: p. [211]-223.
Includes index.
ISBN 0-313-21036-5
1. Great Britain—Politics and government—1660-1714. 2. Great Britain—Politics and government—1714-1760. 3. Church and state in Great Britain. 4. Friends, Society of—Great Britain. 5. Dissenters, Religious—Great Britain. I. Title.
[DA435.C76 1979] 320.9'41'071 78-23805
78 MARC

Società e rivoluzione in Inghilterra (1640-1689) / Gianpaolo Garavaglia. — Torino : Loescher, [1978]
276 p. ; 22 cm. — (Documenti della storia ; 25 : Sezione di storia moderna) It79-Mar
Includes bibliographies.
4600
1. Great Britain—History—Puritan Revolution, 1642-1660. 2. Great Britain—Politics and government—1660-1688. 3. England—Social conditions—17th century. 4. England—Economic conditions—17th century. 5. Church and state in Great Britain. 6. England—Church history—17th century. I. Garavaglia, Gianpaolo, 1946-
DA405.S62 941.6 79-371541
*00 MARC

CHURCH AND STATE IN INDONESIA

—ADDRESSES, ESSAYS, LECTURES

Gultom, R M S
Tanggung jawab orang Kristen dalam pengamalan Pancasila dan pembangunan / oleh R. M. S. Gultom. — [Salatiga? : s.n., 1976?]
36 p. ; 21 cm.
"Brosur."
Includes bibliographical references.
1. Church and state in Indonesia—Addresses, essays, lectures. 2. Civics—Indonesia—Addresses, essays, lectures. I. Title.
BR1220.G84 78-941055
79 MARC

CHURCH AND STATE IN ITALY

Castelli, Mario.
Nazione Italia e Chiesa in Italia / a cura di Mario Castelli. — Leumann/[Rivoli] : Elle Di Ci, 1978.
90 p. ; 21 cm. — (Problemi e proposte di catechesi ; 8) It79-Apr
Translation of L'Église en Italie.
Originally published in Pro mundi vita bulletin, n. 66-1977.
Bibliography: p. 87-90.
ISBN 8801142269 : L1400
1. Church and state in Italy. 2. Catholic Church in Italy. I. Title. II. Series.
BX1545.2.C3216 79-363428
*79 MARC

Dante politico : individuo e istituzioni nell'autunno del Medioevo / a cura di Gabriele Muresu. — 1. ed. — Torino : Paravia, 1979.
206 p. ; 22 cm. — (Nuovi classici Paravia ; 16) It***
"Antologia": p. [55]-195.
Bibliography: p. [50]-53.
L4200
1. Dante Alighieri, 1265-1321—Political and social views. 2. Dante Alighieri, 1265-1321—Contemporary Italy. 3. Florence—History—To 1421. 4. Italy—History—1268-1492. 5. Church and state in Italy. I. Dante Alighieri, 1265-1321. II. Muresu, Gabriele.
PQ4422.D28 79-382342
*79 MARC

Gioberti, Vincenzo, 1801-1852.
I frammenti Della riforma cattolica e Della libertà cattolica / Vincenzo Gioberti ; a cura di Claudio Vasale. — Padova : CEDAM, 1977.
lxii, 255 p. ; 24 cm. — (His Edizione nazionale della opere edite e inedite ; 31) It78-Dec
Bibliography: p. [lxi]-lxii.
L7000
1. Catholic Church—Doctrinal and controversial works—Catholic authors. 2. Catholic Church in Italy. 3. Church and state in Italy. I. Vasale, Claudio. II. Title.
BX1751.G69 1977 79-356822
*79 MARC

Gli Estremisti di centro : il neo-integralismo cattolico degli anni '70 : comunione e liberazione / a cura di Sandro Bianchi e Angelo Turchini ; con una presentazione di David Maria Turoldo. — Rimini ; Firenze : Guaraldi, c1975.
xx, 172 p. : ill. ; 22 cm. — (Presente imperfetto ; 22) It76-Jan
Includes bibliographical references.
L3500
1. Italy—Politics and government—1945-1976. 2. Church and state in Italy. 3. Catholics—Italy—Political activity. I. Bianchi, Sandro, 1926- II. Turchini, Angelo.
DG577.5.E84 322'.1'0945 78-393715
*79 MARC

Pergameno, Silvio, 1926-
La roba clericale / Silvio Pergameno ; prefazione di Angiolo Bandinelli. — Roma : Savelli, c1975.
139 p. ; 19 cm. — (Battaglie radicali ; 3) It76-Jan
Includes bibliographical references.
L1900
1. Italy—Politics and government—1945-1976. 2. Italy—Social conditions—1945- 3. Church and state in Italy. 4. Catholic Church—Finance. 5. Catholic Church in Italy—Education. I. Title.
DG577.5.P47 322'.1'0945 79-407737
*79 MARC

Rodelli, Luigi.
Patti lateranensi: quanto costano alla democrazia. A cura di Luigi Rodelli e Lorenzo Strik Lievers. [Milano], Dall'Oglio, 1970.
31 p. 22 cm. (Quaderni dell'ALRI, n 2) 500 It70-Feb
On cover: Abolire il concordato.
1. Concordat of 1929 (Italy) 2. Church and state in Italy. I. Strik Lievers, Lorenzo, joint author. II. Title. III. Series: Associazione per la libertà religiosa in Italia. Quaderni dell'ALRI ; 2.
BX1545.R58 78-505049
79[r79]rev MARC

Tomsich, Gustavo.
Comunione & [i.e. e] miliardi / Gustavo Tomsich. — Roma : Sarmi, [1977?]
191 p. ; 20 cm. — (Documenti ; n. 4) It78-Mar
L3000
1. Catholic Church—Finance. 2. Vatican City—Finance. 3. Church and state in Italy. I. Title.
BX1950.T65 79-369662
*79 MARC

Vitale, Antonio.
Ordinamento giuridico e interessi religiosi : corso di diritto ecclesiastico / Antonio Vitale. — Milano : A. Giuffrè, 1979.
134 p. ; 25 cm. It79-June
Includes bibliographical references.
L3600
1. Ecclesiastical law—Italy. 2. Church and state in Italy. I. Title.
262.9'0945 79-377193
*79 MARC

—ADDRESSES, ESSAYS, LECTURES

Boato, Alessandro.
Sinistra e questione cattolica in Italia e nel Trentino / a cura di Alessandro Boato, Marco Boato. — 1. ed. — Trento : U.C.T., 1978.
163 p. ; 21 cm. — (Movimento e istituzione ; 2) It78-Oct
Supplement to Uomo città territorio, 1978, n. 28.9.
Bibliography: p. 156-163.
L2500
1. Italy—Politics and government—1945- —Addresses, essays, lectures. 2. Italy—Politics and government—1914-1945—Addresses, essays, lectures. 3. Catholic Church in Italy—Addresses, essays, lectures. 4. Church and state in Italy—Addresses, essays, lectures. 5. Trent (Province)—Politics and government—Addresses, essays, lectures. 6. Communism and Christianity—Catholic Church—Italy—Addresses, essays, lectures. I. Boato, Marco, joint author. II. Uomo città territorio. III. Title.
DG577.B64 320.9'45'092 78-394326
*79 MARC

Dio in pubblico : il dibattito su fede e politica in Italia / a cura di Giovanni Bianchi ; [scritti di] Ernesto Balducci ... [et al.] ; bibliografia a cura di Renzo Salvi. — Brescia : Queriniana, [1978]
366 p. ; 20 cm. — (Giornale di teologia ; 107) It78-May
Bibliography: p. 331-366.
L5000
1. Church and state in Italy—Addresses, essays, lectures. 2. Catholic Church in Italy—Addresses, essays, lectures. I. Bianchi, Giovanni. II. Balducci, Ernesto, 1922-
BX1545.2.D56 78-390342
*79 MARC

Jemolo, Arturo Carlo, 1891-
Gli uomini e la storia / Arturo Carlo Jemolo ; prefazione di Leopoldo Elia. — Roma : Studium, 1978.
xx, 207 p. ; 19 cm. — (Coscienza del tempo ; 1) It79-July
Previously published in La stampa.
L4200
1. Italy—History—19th century—Addresses, essays, lectures. 2. Italy—History—20th century—Addresses, essays, lectures. 3. Church and state in Italy—Addresses, essays, lectures. I. Title. II. Series.
DG551.J45 945.09 79-377661
79 MARC

Jervolino, Domenico.
Questione cattolica e politica di classe / Domenico Jervolino ; prefazione di Giulio Girardi ; CENDES, Sezione ricerca e informazione. — 1. ed. — Torino : Rosenberg & Sellier, 1979.
137 p. ; 18 cm. — (Quaderni di Fabbrica e stato ; 10) It79-Mar
ISBN 8870110788 : L3200
1. Italy—Politics and government—1945-1976—Addresses, essays, lectures. 2. Italy—Politics and government—1976- —Addresses, essays, lectures. 3. Church and state in Italy—Addresses, essays, lectures. 4. Catholics in Italy—Political activity—Addresses, essays, lectures. 5. Italy—Moral conditions—Addresses, essays, lectures. I. Title. II. Series.
DG577.5.J47 261.7 79-376969
00 MARC

Lai, Benny.
La seconda Conciliazione / Benny Lai ; introduzione di Alberto Sensini. — Firenze : Vallecchi, 1978.
252 p. ; 20 cm. — (Documenti e interventi) It78-June
1. Italy—Politics and government—1976- —Addresses, essays, lectures. 2. Church and state in Italy—Addresses, essays, lectures. 3. Communism and Christianity—Catholic Church—Addresses, essays, lectures. 4. Partito comunista italiano—Addresses, essays, lectures. 5. Catholic Church in Italy—Addresses, essays, lectures. I. Title.
DG581.L34 78-379800
*78 MARC

La Politica dell'ideologia / a cura di Gustavo Guizzardi ; [scritti di] F. Bimbi ... [et al.] ; contributi del Seminario di sociologia, Università di Padova. — 1. ed. — Padova : Cleup, 1978.
178 p. ; 23 cm. It78-Dec
Includes bibliographical references.
L4000
1. Catholic Church in Italy—Addresses, essays, lectures. 2. Church and state in Italy—Addresses, essays, lectures. 3. Italy—Church history—Addresses, essays, lectures. I. Bimbi, Franca. II. Guizzardi, Gustavo.
BX1543.P58 79-354472
*79 MARC

—CONGRESSES

Congresso di storia del Risorgimento italiano, 45th, Rome, 1970.
La fine del potere temporale e il ricongiungimento di Roma all'Italia : atti del XLV Congresso di storia del Risorgimento italiano (Roma, 21-25 settembre 1970). — Roma : Istituto per la storia del Risorgimento italiano, 1972.
xix, 677 p. ; 25 cm. — (Atti del Congresso di storia del Risorgimento italiano ; 45) (Istituto per la storia del Risorgimento italiano. Biblioteca scientifica : Atti dei congressi, 13) It73-Mar
1. Rome (City)—History—Annexation to Italy, 1870—Congresses. 2. Church and state in Italy—Congresses. 3. Italy—History—1849-1870—Congresses. I. Title. II. Series: Istituto per la storia del Risorgimento italiano. Atti del Congresso per la storia del Risorgimento italiano ; 45. III. Series: Biblioteca scientifica : Atti dei congressi ; v. 13.
DG552.A2I7 no. 45 79-370557
[DG812.9] 79 MARC

La Questione cattolica / Baget Bozzo ... [et al.]. — Roma : Nuove edizioni operaie, 1978.
191 p. ; 19 cm. — (Collana Sapere ; 25) It78-Aug
Proceedings of the meeting held in Gorgonzola in 1977.
L3500
1. Church and state in Italy—Congresses. 2. Catholic Church in Italy—Congresses. I. Baget Bozzo, Gianni, 1925-
BX1545.2.Q47 78-384982
*78 MARC

—HISTORY

L'Autunno del Concordato : Chiesa cattolica e Stato in Italia, il dibattito politico (1929-1977) / a cura di Mario Cordero ; introduzione di Francesco Traniello e Mario Cordero. — 2. ed. aggiornata. — Torino : Claudiana, 1977.
319 p. ; 21 cm. — (Collana Nostro tempo ; 22) It79-Feb
Includes bibliographies.
L4200
1. Concordat of 1929 (Italy) 2. Church and state in Italy—History. 3. Catholic Church in Italy—History. I. Cordero, Mario.
BX1545.A9 1977 79-351091
*79 MARC

Caravale, Mario.
Lo Stato pontificio da Martino V a Pio IX / [di] Mario Caravale, Alberto Caracciolo. — Torino : UTET, [1978]
xvi, 793 p., [36] leaves of plates : ill. ; 24 cm. — (Storia d'Italia ; v. 14) It79-Mar
Includes bibliographies and index.
ISBN 8802025193 : L30000
1. Papal States—History. 2. Papacy—History. 3. Church and state in Italy—History. 4. Catholic Church in Italy—History. I. Caracciolo, Alberto, joint author. II. Title. III. Series: Storia d'Italia (Torino) ; v. 14.
DG796.C34 945'.6 79-362829
*79 MARC

Isastia, Anna Maria.
Roma nel 1859 / Anna Maria Isastia. — Roma : Istituto per la storia del Risorgimento italiano, 1978.
325 p., [2] leaves of plates : facsims. ; 25 cm. — (Biblioteca scientifica - Istituto per la storia del Risorgimento italiano : Serie 2, Memorie ; v. 32) It***
Includes bibliographical references and index.
1. Rome (City)—History—1798-1870. 2. Italy—History—War of 1859. 3. Church and state in Italy—History. I. Title. II. Series: Istituto per la storia del Risorgimento italiano. Biblioteca scientifica : Serie 2, Memorie ; v. 32.
DG812.9.I7 79-372602
79 MARC

CHURCH AND STATE IN ITALY (Continued)

—HISTORY—SOURCES

Castro, Scipione di, 16th cent.
La politica come retorica / [di] Scipio Di Castro. — Roma : Istituto della Enciclopedia italiana, 1978.
171 p. ; 18 cm. — (Biblioteca biographica ; 19) It78-July
"A cura di Roberto Zapperi."
Bibliography: p. 43-44.
L2800
1. Castro, Scipione di, 16th cent. 2. Italy—History—1559-1789—Sources. 3. Church and state in Italy—History—Sources. 4. Statesmen—Italy—Correspondence. 5. Spies—Italy—Correspondence. I. Zapperi, Roberto. II. Title.
DG540.8.C32A36 78-381159
*79 MARC

CHURCH AND STATE IN JAPAN

—ADDRESSES, ESSAYS, LECTURES

Fujiwara, Fujio.
(Nihon seishin to Kirisutokyō).
日本精神と基督教 藤原藤男著 [堺] ともしび社 [昭和14 i. e. 1939]
74 p. 19 cm.
今年の紀元節の日に話した講演の筆記を少しく訂正増補したもの
1. Church and state in Japan—Addresses, essays, lectures. 2. Nationalism and religion—Japan—Addresses, essays, lectures. 3. Kokutai—Addresses, essays, lectures. 4. Christianity and other religions—Japanese—Addresses, essays, lectures. I. Title.
BR1305.5.F83 79-809091

CHURCH AND STATE IN KITZINGEN

—HISTORY

Stadt und Kirche in Kitzingen : Darst. u. Quellen zu Spätmittelalter u. Reformation / von Dieter Demandt u. Hans-Christoph Rublack ; [diese Arbeit ist im Sonderforschungsbereich 8, Tübingen entstanden]. — 1. Aufl. — Stuttgart : Klett-Cotta, 1978.
338 p. ; 24 cm. — (Spätmittelalter und frühe Neuzeit ; Bd. 10) GFR78-A
Bibliography: p. 324-330.
Includes indexes.
ISBN 3-12-911590-0 : DM83.00
1. Church and state in Kitzingen—History. 2. Kitzingen—Church history. 3. Kitzingen—History. I. Demandt, Dieter. II. Rublack, Hans-Christoph. III. Series.
BR858.K52S7 79-350391
79 MARC

CHURCH AND STATE IN LATIN AMERICA

Comblin, Joseph, 1923-
The church and the national security state / José Comblin. — Maryknoll, N.Y. : Orbis Books, c1979.
xiii, 236 p. ; 24 cm.
Bibliography: p. 235-236.
ISBN 0-88344-082-2
1. Liberation theology. 2. Catholic Church in Latin America. 3. Church and state in Latin America. I. Title.
BT83.57.C65 261.7'098 79-10881
79 MARC

CHURCH AND STATE IN MALTA

Great Britain. Malta Royal Commission, 1931.
L'inchiesta della R. [i.e. Regia] Commissione britannica sulla situazione politica di Malta : relazione stenografica della deposizione resa dall'avv. Enrico Mizzi davanti alla Regia Commissione di inchiesta nelle quattro sedute tenutesi dall'8 al 13 maggio 1931. — Malta : Tip. del quotidiano politico Malta, 1931.
91 p. ; 25 cm.
The translation from English of the Commission's interrogation with Mazzi's responses as originally given in Italian.
1. Malta—Politics and government. 2. Church and state in Malta. 3. Catholic Church in Malta. 4. Malta—Languages—Political aspects. 5. Strickland, Gerald Strickland, baron, 1861-1940. I. Mizzi, Enrico, 1885- II. Title.
DG994.G73 1931 320.9'458'5 79-383009
79 MARC

CHURCH AND STATE IN MATO GROSSO, BRAZIL (STATE)

Casaldáliga, Pedro.
La muerte que da sentido a mi credo : diario 1975-1977 / Pedro Casaldáliga. — [Bilbao] : Desclée de Brouwer, c1977.
86 p. ; 21 cm. — (Tercer mundo/tercera iglesia ; 1) Sp***
ISBN 8433003860 : 100ptas
1. Casaldáliga, Pedro. 2. Catholic Church—Bishops—Biography. 3. Bishops—Brazil—Mato Grosso (State)—Biography. 4. Mato Grosso, Brazil (State)—Biography. 5. Church and state in Mato Grosso, Brazil (State) 6. Mato Grosso, Brazil (State)—Politics and government. I. Title.
BX4705.C33547A35 78-377399
78 MARC

CHURCH AND STATE IN MEISENHEIM

—HISTORY

Adams, Carl Peter.
Kirche und Stadt Meisenheim / Carl Peter Adams. — Köln : Rheinland-Verlag ; Bonn : In Kommission bei R. Habelt, c1978.
423 p. : facsims. ; 23 cm. — (Schriftenreihe des Vereins für Rheinische Kirchengeschichte ; Bd. 55) GFR***
Includes bibliographical references.
CONTENTS: A. Kirche und Stadt Meisenheim im Spiegel der reformierten Presbyterial-Protokolle 1639-1868.—B. Hundert Jahre Rheinische Synode Meisenheim.
ISBN 3-7927-0387-4
1. Meisenheim—Church history. 2. Church and state in Meisenheim—History. I. Title. II. Series: Verein für Rheinische Kirchengeschichte. Schriftenreihe ; Bd. 55.
BR857.R5V4 Bd. 55 79-356998
[BR858.M38] 79 MARC

CHURCH AND STATE IN MEXICO

Stanchina, Peer Christopher.
Das Verhältnis von Staat und Kirche in Mexiko : seit d. Revolution von 1910/1917 bis heute / von Peer Christopher Stanchina. — 1. Aufl. — München : Beck, 1978.
xv, 179 p. ; 23 cm. — (Münchener Universitätsschriften) (Reihe des Juristischen Fachbereichs ; Bd. 38) GFR78-A
Bibliography: p. [171]-179.
ISBN 3-406-05438-2
1. Ecclesiastical law—Mexico. 2. Church and state in Mexico. I. Title. II. Series: Munich. Universität. Juristischer Fachbereich. Reihe des Juristischen Fachbereichs ; Bd. 38.
262.9'0972 79-364305
79 MARC

Villaseñor, Guillermo.
Estado e Iglesia : el caso de la educación / Guillermo Villaseñor. — 1. ed. — México : Editorial Edicol, 1978.
198 p. ; 20 cm. — (Sociología investigaciones)
Bibliography: p. 195-198.
ISBN 9684080050
1. Church and state in Mexico. 2. Catholic Church—Education. 3. Education and state—Mexico. 4. Mexico—Church history. I. Title. II. Series.
BR610.V54 78-112396
79 MARC

CHURCH AND STATE IN MODENA

—HISTORY

Peyronel Rambaldi, Susanna, 1942-
Speranze e crisi nel Cinquecento modenese : tensioni religiose e vita cittadina ai tempi di Giovanni Morone / Susanna Peyronel Rambaldi. — Milano : F. Angeli, c1979.
291 p. : map ; 22 cm. — (Studi e ricerche storiche ; 3) It***
Includes bibliographical references and index.
ISBN 8820414821 : L8000
1. Catholic Church in Modena—History. 2. Church and state in Modena—History. 3. Modena—Church history. 4. Morone, Giovanni, Cardinal, 1509-1580. I. Title. II. Series.
BX1548.M58P48 79-367963
79 MARC

CHURCH AND STATE IN MOSELLE, FRANCE (DEPT.)

—HISTORY—ADDRESSES, ESSAYS, LECTURES

Églises et État en Alsace et en Moselle : changement ou fixité? / Études publiées sous la direction de Jean Schlick. — Strasbourg : Cerdic-publications, 1979.
354 p. ; 22 cm. — (Hommes et Église ; 9) F***
Includes bibliographical references.
ISBN 2-85097-009-3 : 85.00F
1. Church and state in Alsace—History—Addresses, essays, lectures. 2. Church and state in Moselle, France (Dept.)—History—Addresses, essays, lectures. 3. Alsace—Politics and government—Addresses, essays, lectures. 4. Moselle, France (Dept.)—Politics and government—Addresses, essays, lectures. I. Schlick, Jean. II. Series.
BR847.A6E34 261.7'0944'383 79-376035
79 MARC

CHURCH AND STATE IN OLDENBURG

—HISTORY

Pohlschneider, Johannes, Bp., 1899-
Der nationalsozialistische Kirchenkampf in Oldenburg : Erinnerungen u. Dokumente / Johannes Pohlschneider. — Kevelaer : Butzon und Bercker, 1978.
147 p. : ill. ; 21 cm. GFR78-A36
Includes bibliographical references.
ISBN 3-7666-9006-X : DM16.80
1. Church and state in Oldenburg—History. 2. Catholic Church in Oldenburg—History. 3. Oldenburg—Church history. I. Title.
BX1537.O42P64 78-389978
79 MARC

CHURCH AND STATE IN OLSZTYN, POLAND (DIOCESE)

Reifferscheid, Gerhard.
Das Bistum Ermland und das Dritte Reich / von Gerhard Reifferscheid. — Köln ; Wien : Böhlau, 1975.
xxxi, 351 p. : ill., maps ; 24 cm. — (Bonner Beiträge zur Kirchengeschichte ; Bd. 7) GFR75-A
Bibliography: p. [xv]-xxviii.
Includes index.
ISBN 3-412-10874-X : DM68.00
1. Olsztyn, Poland (Diocese) 2. Church and state in Olsztyn, Poland (Diocese) I. Title. II. Series.
BX1568.O47R44 282'.438'3 75-521700
76[r79]rev2 MARC

CHURCH AND STATE IN POLAND

Godlewski, Jerzy, 1921-
Kościół rzymskokatolicki w Polsce wobec sekularyzacji życia publicznego : (1944-1974 / Jerzy F. Godlewski. — Wyd. 1. — Warszawa : Państ. Wydaw. Naukowe, 1978 (Łódź : ŁZG 1)
356, [3] p. ; 20 cm. P78-2282
Summary in English and Russian.
Bibliography: p. 323-[351]
zł35.00
1. Church and state in Poland. I. Title.
BR956.3.G62 78-375025
78 MARC

Pietrzak, Michał.
Prawo wyznaniowe : [skrypt dla studentów wydziałów prawa i administracji] / Michał Pietrzak. — Wyd. 1. — Warszawa : Państ. Wydaw. Naukowe, 1978 (Łódź : ZGWN)
133, [2] p. ; 24 cm. P78-10719
Bibliography: p. 130-[132]
zł22.00
1. Ecclesiastical law. 2. Church and state. 3. Church and state in Poland. I. Title.
K3280.P54 79-365647
79 MARC

—HISTORY

Engelmann, Henri.
Printemps de l'Église en Pologne / Henri Engelmann ; préf. de Gabriel Matagrin. — Paris : Éditions du Cerf, 1979.
127 p. : ill. ; 20 cm. — (Rencontres ; nouv. sér., 12) F***
Includes bibliographical references.
ISBN 2-204-01348-X : 34.00F
1. Catholic Church in Poland—History. 2. Church and state in Poland—History. 3. Poland—History. I. Title.
BX1566.2.E53 282'.438 79-366206
79 MARC

Kamiński, Franciszek.
Religione e Chiesa in Polonia, 1945-1975 : saggio storico-istituzionale / Franciszek Kamiński. — [Padova] : CESEO : Liviana, [1976?]
157 p. ; 23 cm. — (Collana di studi sull'Europa orientale ; 14a) It78-July
Bibliography: p. 137-157.
L6000
1. Church and state in Poland—History. 2. Catholic Church in Poland—History. I. Title. II. Series.
BX1566.2.K35 78-399564
*79 MARC

CHURCH AND STATE IN PORTUGAL

—HISTORY

Freire, José Geraldes.
Resistência católica ao salazarismo-marcelismo / José Geraldes Freire ; com carta-prefácio pelo bispo do Porto, António Ferreira Gomes. — Porto : Telos, 1976.
258 p. ; 22 cm. — (Colecção Teologia e sociedade ; 5)
Bibliography: p. [247]-252.
Includes index.
1. Catholic Church in Portugal—History. 2. Church and state in Portugal—History. 3. Portugal—Church history. 4. Portugal—History—1910-1974. I. Title. II. Series.
BX1555.F73 77-155576
79 MARC

CHURCH AND STATE IN PORTUGAL
—HISTORY (Continued)

Miller, Samuel J 1919-
Portugal and Rome c. 1748-1830 : an aspect of the Catholic enlightenment / Samuel J. Miller. — Roma : Università Gregoriana, 1978.
xii, 412 p. ; 26 cm. — (Miscellanea historiae pontificiae ; v. 44) It***
Bibliography: p. ⌈397⌉-404.
Includes index.
1. Catholic Church—Relations (diplomatic) with Portugal. 2. Portugal—Foreign relations—Catholic Church. 3. Catholic Church in Portugal—History. 4. Church and state in Portugal—History. I. Title. II. Series: Pontificia Università gregoriana. Facoltà di storia ecclesiastica. Miscellanea historiae pontificiae ; v. 44.
BX1555.M54 322'.1'09469 79-350360
79 MARC

CHURCH AND STATE IN QUÉBEC (PROVINCE)

Eid, Nadia F
Le clergé et le pouvoir politique au Québec : une analyse de l'idéologie ultramontaine au milieu du XIXe siècle / Nadia F. Eid. — Montréal : Hurtubise HMH, c1978.
viii, 318 p. : ill., facsims., ports. ; 21 cm. — (Collection Histoire) (Cahiers du Québec ; 41) C79-1162-X
Bibliography: p. ⌈289⌉-311.
Includes index.
ISBN 0-7758-0157-7 : $10.95
1. Church and state in Québec (Province) 2. Catholic Church in Québec (Province) 3. Ultramontanism. I. Title.
BX1422.Q4E46 261.7 79-359726
79 MARC

CHURCH AND STATE IN ROME

—HISTORY

Ramsay, William Mitchell, Sir, 1851-1939.
The church in the Roman Empire before A.D. 170 / by W. M. Ramsay. — Boston : Longwood Press, 1978.
xxiii, 510 p., ⌈3⌉ leaves of plates : ill. ; 22 cm.
Reprint of the 8th (1904) ed. published by Hodder and Stoughton, London. Original ed. issued as the Mansfield College lectures, 1892.
Includes bibliographical references and index.
ISBN 0-89341-216-3 : $50.00
1. Church history—Primitive and early church, ca. 30-600. 2. Church and state in Rome—History. I. Title. II. Series: Mansfield College lectures, 1892.
BR170.R3 1978 270.1 77-6997
78 MARC

CHURCH AND STATE IN RUSSIA

Codevilla, Giovanni.
Le comunità religiose nell'URSS : la nuova legislazione sovietica / Giovanni Codevilla. — Milano : Cooperativa editoriale La case di Matriona, 1978.
141 p. ; 18 cm. — (Ricerche ; 4) It79-Apr
Includes bibliographical references.
L2500
1. Ecclesiastical law—Russia. 2. Church and state in Russia. I. Title.
262.9'0947 79-362121
*79 MARC

Filaret, Metropolitan of Moscow, 1782-1867.
(Gosudarstvennoe uchenie Filareta, mitropolita Moskovskago)
Государственное ученіе Филарета, митрополита Московскаго / ⌈составитель⌉ В. Н. — Изд. 3., вновь доп. — Москва : Тип. Е. Г. Потапова, 1888.
x, 199 p., ⌈1⌉ leaf of plates : port. ; 18 cm.
Cover title: Gosudarstvennoe uchenie mitropolita Filareta.
Includes bibliographical references.
1. State, The. 2. Monarchy, Russian. 3. Russia—Politics and government—19th century. 4. Church and state in Russia. I. V. N. II. N., V. III. Nazarevskiĭ, Vladimir Vladimirovich. IV. Title. V. Title: Gosudarstvennoe uchenie mitropolita Filareta.
JC11.F54 1888 Russian Imperial Coll. 79-408403

Hebly, J A
The Russians and the World Council of Churches : documentary survey of the accession of the Russian Orthodox Church to the World Council of Churches, with commentary / by J. A. Hebly. — 1st ed. — Belfast : Christian Journals, 1978.
181 p. ; 18 cm. GB***
Includes biographical references and index.
ISBN 0-904302-39-3 : £1.80
1. World Council of Churches—History. 2. Orthodox Eastern Church, Russian—History. 3. World Council of Churches—Controversial literature. 4. Orthodox Eastern Church, Russian—Controversial literature. 5. Communism and Christianity. 6. Church and state in Russia. I. Title.
BX6.W78H35 270.82 79-312685
79 MARC

—1917-

Krasnov-Levitin, A 1915-
(Likhie gody)
Лихие годы : 1925-1941 : воспоминания / А. Краснов-Левитин. — Paris : YMCA-Press, 1977.
456 p. ; 22 cm.
1. Krasnov-Levitin, A., 1915- 2. Russian Republic—Biography. 3. Theater—Russian Republic. 4. Church and state in Russia—1917- I. Title.
CT1218.K664A34 79-414202

Makey, Walter.
The Church of the Covenant, 1637-1651 : revolution and social change in Scotland / Walter Makey. — Edinburgh : John Donald Publishers, 1979.
vii, 216 p. ; 24 cm. GB***
Bibliography: p. ⌈203⌉-210.
Includes index.
ISBN 0-85976-035-9 : £12.00
1. Church of Scotland—History. 2. Church and state in Scotland—History. 3. Scotland—Church history—17th century. I. Title.
BX9071.M28 285'.2'411 79-314845
79 MARC

—1917- —ADDRESSES, ESSAYS, LECTURES

(Ateizm, religiia, sovremennost')
Атеизм, религия, современность : Сборник трудов / Гос. музей истории религии и атеизма ; ⌈Редколлегия Я. Я. Кожурин (отв. ред.) ... et al.⌉. — Ленинград : ⌈s. n.⌉, 1977.
166 p. ; 21 cm. USSR 77
At head of title: Ministerstvo kul'tury RSFSR.
Includes bibliographical references.
0.50rub
1. Atheism—Addresses, essays, lectures. 2. Church and state in Russia—1917- —Addresses, essays, lectures. I. Kozhurin, Iakov Iakovlevich. II. Leningrad. Muzeĭ istorii religii i ateizma.
BL2747.3.A795 79-411352

—HISTORY—SOURCES

Khristianskiĭ komitet zashchity prav veruiushchikh v SSSR.
Documents of the Christian Committee for the Defense of Believers' Rights in the USSR = Dokumenty Khristianskogo komiteta zashchity prav veruiushchikh v SSSR. — San Francisco : Washington Street Research Center, 1978-
v. : ill. ; 28 cm.
Prefatory matter, tables of contents, and summaries in English.
1. Dissenters, Religious—Russia—History—Sources. 2. Persecution—Russia—History—Sources. 3. Church and state in Russia—History—Sources. 4. Religious liberty—Russia—History—Sources. 5. Russia—Church history—1917- —Sources. 6. Russia—History—1917- —Sources. I. Title. II. Title: Dokumenty Khristianskogo komiteta zashchity prav veruiushchikh v SSSR.
BR1608.R85K47 1978 323.44'2'0947 79-349076
79 MARC

CHURCH AND STATE IN SALVADOR

Latin America Bureau.
Violence and fraud in El Salvador : a report on current political events in El Salvador / compiled ... by Latin America Bureau. — London : The Bureau, 1977.
37 p. : maps ; 30 cm. GB***
Includes bibliographical references.
£2.00
1. Civil rights—Salvador. 2. Land reform—Salvador. 3. Church and state in Salvador. 4. Elections—Salvador—Corrupt practices. I. Title.
JC599.S2L37 1977 320.9'7284'05 78-317571
78 MARC

CHURCH AND STATE IN SPAIN

Aradillas Agudo, Antonio.
Yo acuso / Antonio Aradillas. — ⌈1. ed.⌉. — ⌈Madrid⌉ : Sedmay Ediciones : ⌈Distribuidora Maydi, 1977⌉
178 p. ; 21 cm. Sp***
ISBN 847380239X
1. Catholic Church in Spain. 2. Church and state in Spain. 3. Catholic Church—Doctrinal and controversial works—Catholic authors. I. Title.
BX1585.2.A72 78-388755
79 MARC

Castro Albarrán, Aniceto de.
Muchas gracias ... pero el problema sigue en pie : manifiesto por los hermanos / ⌈por⌉ A. de Castro Albarrán. — ⌈Madrid : El autor⌉, 1977 printing.
125 p. ; 24 cm.
ISBN 8440026331 : 200ptas
1. Catholic Church in Spain. 2. Church and state in Spain. 3. Catholic Church—Clergy—Salaries, pensions, etc. I. Title. II. Title: Manifiesto por los hermanos.
BX1585.2.C375 79-365518
79 MARC

—ADDRESSES, ESSAYS, LECTURES

La Iglesia en España sin concordato : una hipótesis de trabajo : ponencias de las I jornadas de reflexión de Profesores ... (22-24 enero 1976). — Madrid : Eapsa, 1976.
290 p. ; 24 cm. — (Derecho canónico ; I, 2) (Publicaciones de la Universidad Pontificia Comillas, Madrid : Serie I, Estudios ; 11) Sp78-Sept
On cover: Corral ... ⌈et al.⌉
"Organizadas por la Universidad Pontificia Comillas, Facultad de Derecho Canónico, Madrid."
Includes bibliographical references.
ISBN 8485281128
1. Church and state in Spain—Addresses, essays, lectures. 2. Catholic Church in Spain—Addresses, essays, lectures. 3. Concordats—Addresses, essays, lectures. I. Corral Salvador, Carlos Manuel, 1929- . II. Universidad Pontificia de Comillas. Facultad de Derecho Canónico. III. Series.
BX1583.I35 79-113104
79 MARC

—CONGRESSES

Semana de Historia Eclesiástica de España Contemporánea, 2d, Madrid, 1977.
Aproximación a la historia social de la Iglesia española contemporánea / M. Andrés ... ⌈et al.⌉. — El Escorial, Madrid : Biblioteca La Ciudad de Dios, 1978.
xxv, 321 p., 1 leaf ; 21 cm. — (Biblioteca La Ciudad de Dios : I, Libros ; 27) Sp***
Includes bibliographies.
ISBN 844005159X
1. Catholic Church in Spain—Congresses. 2. Church and state in Spain—Congresses. 3. Spain—Church history—Congresses. I. Title.
BX1583.S46 1977 79-110833
79 MARC

—HISTORY

Benavides, Domingo.
Democracia y cristianismo en la España de la Restauración, 1875-1931 / Domingo Benavides Gómez. — Madrid : Editora Nacional, ⌈c1978⌉
391 p. ; 22 cm. — (España en 3 ⌈i.e. tres⌉ tiempos) Sp***
Bibliography: p. 375-385.
ISBN 8427604513 : 600ptas
1. Catholic Church in Spain—History. 2. Church and state in Spain—History. 3. Christianity and democracy—Spain—History. I. Title.
BX1585.B38 79-376042
79 MARC

Benito Ruano, Eloy.
Los orígenes del problema converso / Eloy Benito Ruano. — Barcelona : Ediciones El Albir, 1976.
164 p. ; 18 cm. — (El Albir universal ; 4) Sp***
Includes bibliographical references and indexes.
ISBN 8473700201
1. Persecution—Spain—Toledo. 2. Church and state in Spain—History. 3. Converts, Catholic—Spain—Toledo. 4. Toledo, Spain—Church history. 5. Sarmiento, Pero. I. Title.
⌈DNLM: PN6490.R6 1975⌉
BR1608.S8B46 78-384985
79 MARC

Garriga, Ramón.
El Cardenal Segura y el Nacional-Catolicismo / Ramón Garriga. — 1. ed. — Barcelona : Editorial Planeta, 1977.
357 p. : ill. ; 25 cm. — (Espejo de España ; 37) Sp***
Bibliography: 349-352.
Includes index.
ISBN 8432056375
1. Church and state in Spain—History. 2. Catholic Church in Spain—History. 3. Segura y Sáenz, Pedro, Cardinal, 1880-1957. 4. Cardinals—Spain—Biography. 5. Spain—Politics and government—20th century. I. Title. II. Series.
BX1585.G33 78-345798
78⌈r79⌉rev MARC

Molina Meliá, Antonio.
Iglesia y Estado en el siglo de oro español : el pensamiento de Francisco Suárez / ⌈por⌉ Antonio Molina Meliá. — Valencia : Universidad, Secretariado de Publicaciones, D.L. 1977.
253 p. ; 24 cm. Sp78-Jan
Bibliography: p. 17-27.
ISBN 8437000203
1. Church and state in Spain—History. 2. Catholic Church in Spain—History. 3. Suárez, Francisco, 1548-1617. I. Title.
BX1584.M58 78-386638
79 MARC

Santidrián, Pedro R
España ha dejado de ser católica : las razones de Azaña, las razones de hoy : aportaciones a la nueva Constitución Española / ⌈por⌉ Pedro R. Santidrián. — Madrid : Centro de Investigación y Publicaciones, 1978.
179 p. ; 18 cm. Sp***
ISBN 848549900X
1. Church and state in Spain—History. 2. Catholic Church in Spain—History. I. Title.
BX1583.S26 79-113155
79 MARC

CHURCH AND STATE IN SWEDEN

Gustafsson, Göran, 1936-
Mellan religion och politik : studier av kyrkofullmäktigeinstitutionen / Göran Gustafsson. — Stockholm : Verbum ; Lund : H. Ohlsson, 1977.
247 p. ; 21 cm. S78-1/2
Bibliography: p. 243-247.
kr79.00
1. Svenska krykan—Government. 2. Church and state in Sweden. 3. Sweden—Church history. 4. Parish councils. I. Title.
BX8039.G87 78-380110
78 MARC

—ADDRESSES, ESSAYS, LECTURES

En levande kyrka / Kjell Ove Nilsson, red. — Stockholm : Verbum/Studiebokförl., 1978.
182 p. ; 21 cm. — (En SKS-bok) S78-15
1. Svenska kyrkan—Addresses, essays, lectures. 2. Church and state in Sweden—Addresses, essays, lectures. I. Nilsson, Kjell Ove.
BX8039.L446 79-382515
79 MARC

CHURCH AND STATE IN SWITZERLAND

Bräm, Werner Kurt.
Religionsunterricht als Rechtsproblem im Rahmen der Ordnung von Kirche und Staat : unter besonderer Berücksichtigung der Schulgesetzgebung in den Kantonen und der Unterrichtsordnungen der evangelisch-reformierten Kirchen der Schweiz / Werner Kurt Bräm. — Zürich : Theologischer Verlag, cop. 1978.
xlvii, 456 p. ; 23 cm. — (Forschung TVZ) Sw78-8279
Originally presented as the author's thesis, Basel, 1977.
Bibliography: p. xiv-xxvii.
ISBN 3-290-11397-3 : 36.00F
1. Religious education—Law and legislation—Switzerland. 2. Ecclesiastical law—Switzerland. 3. Church and state in Switzerland. I. Title.
79-352554
79 MARC

—HISTORY—ADDRESSES, ESSAYS, LECTURES

Verein für Reformationsgeschichte.
Stadt und Kirche im 16. Jahrhundert / hrsg. von Bernd Moeller. — 1. Aufl. — Gütersloh : Gütersloher Verlagshaus Mohn, 1978.
191 p. ; 23 cm. — (Schriften des Vereins für Reformationsgeschichte ; Bd. 190) GFR78-A
Papers of the second symposium of the Verein für Reformationsgeschichte held in Reinhausen, March 24-26, 1977.
Includes bibliographical references and index.
ISBN 3-579-04334-X : DM48.00
1. Church and state in Germany—History—Addresses, essays, lectures. 2. Church and state in Switzerland—History—Addresses, essays, lectures. 3. Reformation—Germany—Addresses, essays, lectures. 4. Reformation—Switzerland—Addresses, essays, lectures. I. Moeller, Bernd, 1931- II. Series: Verein für Reformationsgeschichte. Schriften ; Bd. 190.
BR300.V5 Bd. 190 78-379478
[BR309] 78 MARC

CHURCH AND STATE IN THE BASQUE PROVINCES

Rodríguez de Coro, Francisco.
País Vasco, iglesia y revolución liberal / por Francisco Rodríguez de Coro. — Vitoria : Obra Cultural de la Caja de Ahorros Municipal, D.L. 1978.
422 p. ; 19 cm. — (Biblioteca alavesa Luis de Ajuria ; 22) Sp78-Dec
"Apéndice documental": p. 331-381.
Bibliography: p. 383-410.
ISBN 8450026393 : 350ptas
1. Basque Provinces—Church history. 2. Church and state in the Basque Provinces. 3. Basque Provinces—History. I. Title. II. Series.
BR1027.B3R62 79-365397
79 MARC

CHURCH AND STATE IN THE UNITED STATES

Annese, Lucius.
The purpose of authority? : A recent emphasis in American political thought in the United States and in American churches / Lucius Annese. — Andover, Mass. : Charisma Press, c1978.
iv, 120 p. ; 22 cm.
Bibliography: p. 104-116.
Includes index.
1. Authority. 2. Authority (Religion) 3. Church and state in the United States. 4. Political science—United States. I. Title.
JC571.A548 301.15'52 78-72295
79 MARC

King, James Marcus, 1839-1907.
Facing the twentieth century / James M. King. — New York : Arno Press, 1977 [c1899]
640 p., [35] leaves of plates : ill. ; 24 cm. — (Anti-movements in America)
Reprint of the ed. published by American Union League Society, New York.
ISBN 0-405-09958-4
1. United States—Civilization. 2. United States—Politics and government. 3. Church and state in the United States. 4. Catholic Church—Doctrinal and controversial works—Protestant authors. I. Title. II. Series.
E169.1.K53 1977 973 76-46085
76[r79]rev MARC

O'Neill, James Milton, 1881-
Catholicism and American freedom / by James M. O'Neill. — Westport, Conn. : Greenwood Press, 1979, c1952.
xii, 287 p. ; 24 cm.
Reprint of the ed. published by Harper, New York.
Bibliography: p. 273-281.
Includes index.
ISBN 0-313-21153-1
1. Catholic Church in the United States. 2. Church and state in the United States. 3. Church and state—Catholic Church. 4. Blanshard, Paul, 1892- —American freedom and Catholic power. 5. Blanshard, Paul, 1892- —Communism, democracy, and Catholic power. I. Title.
[BX1406.2.O53 1979] 282'.73 78-21495
78 MARC

Peterson, Walfred Hugo.
Thy liberty in law / Walfred H. Peterson. — Nashville : Broadman Press, c1978.
185 p. ; 20 cm.
Includes bibliographical references and index.
ISBN 0-8054-6928-1
1. Religious liberty—United States. 2. Church and state in the United States. I. Title.
KF4783.P45 342'.73'085 78-59278
79 MARC

Shuster, George Nauman, 1894-
The Catholic spirit in America / George N. Shuster. — New York : Arno Press, 1978 [c1927]
xi, 296 p. ; 21 cm. — (The American Catholic tradition)
Reprint of the 1928 ed. published by the Dial Press, New York.
ISBN 0-405-10857-5
1. Catholic Church in the United States. 2. Church and state in the United States. I. Title. II. Series.
BX1406.S58 1978 282'.73 77-11312
77 MARC

CHURCH AND STATE IN TUNISIA

—HISTORY

O'Donnell, Joseph Dean, 1943-
Lavigerie in Tunisia : the interplay of imperialist and missionary / J. Dean O'Donnell, Jr. — Athens : University of Georgia Press, c1979.
xvii, 300 p., [1] leaf of plates : ill. ; 24 cm.
Bibliography: p. [267]-285.
Includes index.
ISBN 0-8203-0456-5
1. Lavigerie, Charles Martial Allemand, Cardinal, 1825-1892. 2. Church and state in Tunisia—History. 3. Tunisia—Politics and government. I. Title.
BX4705.L4O36 266'.2'0924 78-9041
78 MARC

CHURCH AND STATE IN UMBRIA, ITALY

—CONGRESSES

Convegno su laicato cattolico e chiesa locale in Umbria dal fascismo alla Resistenza, Foligno, Italy, 1975.
Cattolici e fascisti in Umbria 1922-1945 / a cura di Alberto Monticone. — Bologna : Il mulino, 1978.
483 p. ; 21 cm. — (Temi e discussioni) It78-Aug
Includes bibliographical references and index.
L10000
1. Umbria, Italy—Politics and government—Congresses. 2. Fascism—Italy—Umbria—Congresses. 3. Church and state in Umbria, Italy—Congresses. 4. Catholics—Italy—Umbria—Political activity—Congresses. I. Monticone, Alberto. II. Title.
DG975.U5C7 1975 78-384067
*78 MARC

CHURCH AND STATE IN VENEZUELA

—HISTORY

Ocando Yamarte, Gustavo.
Historia político-eclesiástica de Venezuela (1830-1847) / Gustavo Ocando Yamarte. — Caracas : Academia Nacional de la Historia, 1975-
v. : ill. ; 23 cm. — (Fuentes para la historia republicana de Venezuela) (Biblioteca de la Academia Nacional de la Historia ; [nueva ser.])
Includes bibliographical references and indexes.
1. Venezuela—Politics and government—1830-1935. 2. Church and state in Venezuela—History. 3. Catholic Church in Venezuela—History. 4. Venezuela—Church history. I. Title. II. Series. III. Series: Biblioteca de la Academia Nacional de la Historia (Caracas) ; nueva ser.
F2325.O22 322'.1'0987 79-110528
00 MARC

CHURCH AND STATE IN WEST GERMANY

—HISTORY—CONGRESSES

Arbeitskreis Deutscher Katholizismus im 19. und 20. Jahrhundert.
Katholizismus im politischen System der Bundesrepublik : 1949-1963 / hrsg. von Albrecht Langner. — Paderborn ; München ; Wien ; Zürich : Schöningh, 1978.
142 p. ; 22 cm. — (Beiträge zur Katholizismusforschung : Reihe B, Abhandlungen) GFR78-A
Papers originally presented at a meeting of the Arbeitskreis Deutscher Katholizismus im 19. und 20. Jahrhundert, and held in Augsburg, Oct. 5-7, 1977.
Includes bibliographical references and index.
ISBN 3-506-70733-7 : DM19.80
1. Catholic Church in West Germany—History—Congresses. 2. Church and state in West Germany—History—Congresses. 3. Germany, West—Politics and government. I. Langner, Albrecht. II. Title.
BX1536.2.A73 1978 79-353835
79 MARC

CHURCH AND STATE IN YUGOSLAVIA

Alexander, Stella.
Church and state in Yugoslavia since 1945 / Stella Alexander. — Cambridge ; New York : Cambridge University Press, 1979.
xxi, 351 p. : maps ; 23 cm. — (Soviet and East European studies)
Bibliography: p. 316-322.
Includes index.
ISBN 0-521-21942-6
1. Church and state in Yugoslavia. 2. Yugoslavia—Church history. 3. Communism and Christianity—Yugoslavia. I. Title. II. Series.
BR966.3.A4 261.7'09497 77-88668
78 MARC

Kurtović, Todo.
Crkva i religija u socijalističkom samoupravnom društvu / Todo Kurtović ; [izbor Dragan Bartolović i Franc Cengle]. — Beograd : Rad, 1978.
viii, 386 p. ; 21 cm. Yu***
Includes bibliographical references.
1. Church and state in Yugoslavia. 2. Yugoslavia—Church history. I. Bartolović, Dragan. II. Cengle, Franc. III. Title.
BR966.3.K867 78-399446
79 MARC

CHURCH AND THE POOR

Cussiánovich, Alejandro.
Desde los pobres de la tierra : perspectivas de vida religiosa / Alejandro Cussiánovich. — Salamanca : Sígueme, 1977.
219 p. ; 18 cm. — (Pedal ; 66) Sp***
Includes bibliographical references.
ISBN 8430104771 : 250ptas
1. Monastic and religious life. 2. Catholic Church in Latin America. 3. Church and the poor. I. Title.
BX2435.C88 1977 79-365514
79 MARC

Cussiánovich, Alejandro.
Religious life and the poor : liberation theology perspectives / Alejandro Cussiánovich ; translated by John Drury. — Maryknoll, N.Y. : Orbis Books, c1979.
vii, 168 p. ; 24 cm.
Translation of Desde los pobres de la tierra.
Includes bibliographical references and index.
ISBN 0-88344-429-1
1. Monastic and religious life. 2. Church and the poor. 3. Liberation theology. I. Title.
BX2435.C8813 255 78-16740
78 MARC

Les Sans-pouvoir au pays des Appalaches : ce pays est pour moi une demeure ... : lettre pastorale des évêques catholiques des Appalaches sur les sans-pouvoir qui peuplent cette région. — Bruxelles, r. des Deux Églises, 17 : Commission Justice et paix francophone, [1976]
60 p. ; 21 cm. — (Témoins de la justice) Be78-1
Translation of This land is home to me.
60.00F
1. Catholic Church—Pastoral letters and charges. 2. Church and the poor. 3. Appalachian region—Social conditions. 4. Appalachian region—Economic conditions. I. Commission Justice et paix francophone. II. Series.
BX874.A2T4714 261.8'5 78-382489
*79 MARC

Santa Ana, Julio de.
El desafío de los pobres a la Iglesia / Julio de Santa Ana. — 1. ed. — Ciudad Universitaria Rodrigo Facio, Costa Rica : Editorial Universitaria Centroamericana, 1977.
142 p. ; 21 cm. — (Colección DEI)
Includes bibliographical references.
1. Church and the poor. I. Title. II. Series: Universidad Nacional (Costa Rica). Departamento Ecuménico de Investigaciones. Colección DEI.
BV639.P6S26 78-367747
78[r79]rev MARC

Toulat, Jean, 1915-
Les Forces de l'amour : de Jean Vanier à Mère Teresa / Jean Toulat. — Paris : Éditions S.O.S., 1976.
135 p., [8] p. of plates ; 21 cm. F77-2936
Includes bibliographical references.
ISBN 2-7185-0820-5 : 34.00F
1. Church work with the mentally handicapped. 2. Vanier, Jean, 1928- . 3. Church work with the sick. 4. Teresa, Mother, 1910- . 5. Church and the poor. 6. Catholic Church—Missions. 7. Catholic Church—Charities. I. Title.
BX2347.8.M4T68 361.7'5 78-399927
79 MARC

CHURCH AND THE POOR (Continued)

—ADDRESSES, ESSAYS, LECTURES

Ministry with the poor : a world consultation in Latin America. — Geneva : Theological Education Fund, World Council of Churches, c1977.
80 p. : ill. ; 25 cm. Sw***
"First published as the January 1977 number of the International review of mission."
Includes bibliographical references.
1. Liberation theology—Addresses, essays, lectures. 2. Pastoral theology—Addresses, essays, lectures. 3. Church and the poor—Addresses, essays, lectures. 4. Pentecostalism—Latin America—Addresses, essays, lectures. I. Theological Education Fund.
BT83.57.M56 261.8 79-311945
79 MARC

—CONGRESSES

Together through the wilderness : report of an Asian workshop on rural community motivation. — Geneva : World Council of Churches, 1977.
1 portfolio (35, [23] p. : 1 ill.) ; 33 cm. — (CCPD documents ; 12)
CONTENTS: Report of a workshop "Towards a theological understanding of community".—Case studies on Sri Lanka, India, the Philippines, CCA programme.—Reflections on the workshop: a letter to Pastor Smith.—List of participants.
1. Sociology, Christian—Congresses. 2. Church and the poor—Congresses. 3. Community development—Asia—Congresses. 4. Rural development—Asia—Congresses. I. Series: World Council of Churches. Commission on the Churches' Participation in Development. CCPD documents ; 12.
BT738.T62 261.8'34'41 79-309514
79 MARC

—HISTORY

Santa Ana, Julio de.
Good news to the poor : the challenge of the poor in the history of the church / Julio de Santa Ana ; [transl. from the Spanish by Helen Whittle]. — Geneva : World Council of Churches, Commission on the Churches' Participation in Development, c1977.
xi, 124 p. ; 22 cm. Sw78-1466
Translation on El desafio de los pobres a la iglesia.
Includes bibliographical references and index.
ISBN 2-8254-0540-X : 9.90F
1. Church and the poor—History. 2. Poor—Biblical teaching. I. Title.
BV639.P6S2613 1977 261.8'34'41 79-337549
79 MARC

Santa Ana, Julio de.
Good news to the poor : the challenge of the poor in the history of the church / Julio de Santa Ana ; [translated from the Spanish by Helen Whittle]. — Maryknoll, N.Y. : Orbis Books, 1979, c1977.
xi, 124 p. ; 21 cm.
Part of a study undertaken by the Commission on the Churches' Participation in Development, World Council of Churches.
Reprint of the ed. published by the CCPD, Geneva.
Includes index.
ISBN 0-88344-158-6 (pbk.)
1. Church and the poor—History. 2. Poor—Biblical teaching. I. World Council of Churches. Commission on the Churches' Participation in Development. II. Title.
BV639.P6S2613 1979 261.8'34'41 78-18763
78 MARC

—HISTORY—ADDRESSES ESSAYS, LECTURES

Separation without hope? : essays on the relation between the Church and the poor during the industrial revolution and the Western colonial expansion / [authors, Julio A. Barreiro ... et al.] ; [ed., Julio de Santa Ana]. — Geneva : World Council of Churches, Commission on the Churches' Participation in Development, c1978.
192 p. ; 22 cm. Sw78-10106
Includes bibliographical references.
ISBN 2-8254-0576-0 : 12.50F
1. Church and the poor—History—Addresses essays, lectures. 2. Church and social problems—History—Addresses essays, lectures. I. Barreiro, Julio. II. Santa Ana, Julio de.
BV639.P6S46 261.8 79-315194
79 MARC

CHURCH AND THE PRESS

see also Journalism, Religious

CHURCH AND THE WORLD

see also Church and underdeveloped areas

Gogarten, Friedrich, 1887-1967.
Despair and hope for our time. Translated by Thomas Wieser. Philadelphia, Pilgrim Press [1970]
v, 170 p. 22 cm. 7.95
Translation of Verhängnis und Hoffnung der Neuzeit.
Includes bibliographical references.
ISBN 0-8298-0159-6
1. Church and the world. 2. Secularism. I. Title.
BR115.W6G613 230 72-115305
70[r78]rev MARC

Krasicki, Ignacy.
Watykan na rozdrożu (1960-1970) ; tło i wyniki Soboru Watykańskiego II. [Wyd. 1. Warszawa] Książka i Wiedza, 1971.
445, [1] p. illus. 20 cm. zł50.00
Bibliography: p. 445-[446]
1. Vatican Council. 2d, 1962-1965. 2. Church and social problems—Catholic Church. 3. Church and the world. I. Title.
BX830 1962.K65 77-479384
78 MARC

Webber, George W
Today's church : a community of exiles and pilgrims / George W. Webber. — Nashville : Abingdon, c1979.
160 p. ; 21 cm.
ISBN 0-687-42320-1
1. Mission of the Church. 2. Church and the world. I. Title.
BV601.8.W396 262'.001 78-32084
79 MARC

—ADDRESSES, ESSAYS, LECTURES

Hong, Harold S., 1911-
(Chayŏn kwa ŭnch'ong ŭi kyoch'aro)
自然과恩寵의交叉路：敎會와社會 / 洪顯卨著；[編輯趙善出]. — 서울：大韓基督敎書會, 1973.
191 p. ; 18 cm. — (現代新書 ; 47)
On verso of t. p.: The cross-road of nature and grace.
Series romanized: Hyŏndae sinsŏ.
Bibliography: p. 189-191.
1. Church and the world—Addresses, essays, lectures. I. Cho, Sŏn-ch'ul. II. Title. III. Title: The cross-road of nature and grace.
BR115.W6H644 79-826273

König, Franz, Cardinal, 1905-
Kirche und Welt : Ansprachen, Referate, Aufsätze / Franz Kardinal König ; [red. Betreuung, Friedrich Müller]. — Wien ; München : Herold, c1978.
316 p. ; 21 cm. Au79-3-12
Includes index.
ISBN 3-7008-0153-X : S280.00
1. Church and the world—Addresses, essays, lectures. 2. Catholic Church—Doctrinal and controversial works—Catholic authors—Addresses, essays, lectures. 3. Family—Religious life—Addresses, essays, lectures. I. Müller, Friedrich, Mag. II. Title.
BR115.W6K59 79-347441
*79 MARC

CHURCH AND UNDERDEVELOPED AREAS

Arnulf, Ingrid.
Lev mitt håp / Ingrid Arnulf. — [Oslo] : Luther, 1978 (Drammen : Tangen-trykk)
210 p., [8] leaves of plates : ill. ; 21 cm. N78-Oct.
ISBN 8253141262 : kr78.00
1. Sri Lanka—Social conditions. 2. Thailand—Social conditions. 3. Africa—Social conditions—1945- 4. Church and underdeveloped areas. I. Title.
HN670.8.A8A73 79-346418
79 MARC

Krauss, Heinrich, 1922-
Mission und Entwicklung : d. Jesuitenstaat in Paraguay : fünfteiliger Kurs im Medienverbund : für Erwachsenenbildung, Schule u. Jugendarbeit / Heinrich Krauss, Anton Täubl ; [Kt., Peter Schimmel]. — München : Kösel, 1979.
191 p. : ill. ; 21 cm. GFR79-A
Includes bibliographical references.
ISBN 3-466-36051-X : DM19.80
1. Indians of South America—Paraguay—Missions. 2. Guarani Indians—Missions. 3. Jesuits—Paraguay. 4. Missions—Study and teaching. 5. Missions—Theory. 6. Church and underdeveloped areas. I. Täubl, Anton, joint author. II. Title.
F2684.K82 79-381449
79 MARC

Lefever, Ernest W
Amsterdam to Nairobi : the World Council of Churches and the Third World / Ernest W. Lefever ; foreword by George F. Will. — Washington : Ethics and Public Policy Center, of Georgetown University, c1979.
xii, 114 p. ; 23 cm.
Includes bibliographical references and index.
ISBN 0-89633-024-9 (pbk.) : $5.00
1. World Council of Churches. 2. Church and underdeveloped areas. I. Georgetown Univrsity, Washington, D.C. Ethics and Public Policy Center. II. Title.
BX6.W78L43 261.8 79-2607
79 MARC

Månson, Thorsten, 1924-
En utmaning. Det internationella ansvaret i blickpunktet. Teckningar av Staffan Lindén. Stockholm, Verbum, 1973.
62 p. illus. 19 cm. (Triangelserien) kr.8.25 S73-42
1. Christianity and international affairs. 2. Church and underdeveloped areas. I. Title.
BR115.I7M28 79-384578
79 MARC

Senft, Josef.
Entwicklungshilfe oder Entwicklungspolitik : e. interessenpolit. Spannungsfeld, dargest. am Kirchl. Hilfswerk Misereor / Josef Senft ; mit e. Vorw. von Wilhelm Dreier. — Münster : Regensberg, 1978.
227 p. ; 21 cm. — (Theorie und Praxis der Sozialethik ; Bd. 2) GFR78-A
Bibliography: p. 206-225.
ISBN 3-7923-0424-4 : DM24.00
1. Church and underdeveloped areas. 2. Misereor, Aktion gegen Hunger und Krankheit in der Welt. I. Title. II. Series.
BR115.U6S46 79-353933
79 MARC

—CONGRESSES

The Ministry of development in evangelical perspective : a symposium on the social and spiritual mandate / Carl F. H. Henry, convenor ; Robert Lincoln Hancock, editor. — Pasadena, Calif. : William Carey Library, c1979.
viii, 109 p. : ill. ; 22 cm.
Articles presented at the symposium sponsored by Development Assistance Services, inc., held in Nov. 1977, in Colorado Springs.
Includes bibliographical references.
ISBN 0-87808-164-X
1. Church and social problems—Congresses. 2. Missions, Foreign—Congresses. 3. Church and underdeveloped areas—Congresses. I. Henry, Carl Ferdinand Howard, 1913- II. Hancock, Robert Lincoln, 1939- III. Development Assistance Services.
HN31.M64 261.8'3 78-27821
78 MARC

Proclamation and human development : documentation from a Lutheran World Federation Consultation, Nairobi, Kenya, October 21-25, 1974. — [Geneva] : Lutheran World Federation, c1975.
160 p. : ill. ; 23 cm. Sw78-20
1. Church and underdeveloped areas—Congresses. 2. Evangelical Church Mekane Yesus—Congresses. I. Lutheran World Federation.
BR115.U6P76 261.8'5 79-304952
79 MARC

—AFRICA—CONGRESSES

"Learning to choose"; the report of a conference held at the Continuing Education Centre of the University of Nigeria, Nsukka, from 3rd-8th September, 1971, with a view to helping Christians learn how to work out for themselves the priorities for churches in rural development. Ibadan, Institute of Church and Society [1971]
68 p. 19 cm. (Church and society books, no. 6)
Sponsored by Christian Rural Fellowship of Nigeria and Institute of Church and Society, Christian Council of Nigeria.
1. Rural development—Africa—Congresses. 2. Church and underdeveloped areas—Africa—Congresses. I. Christian Rural Fellowship of Nigeria. II. Christian Council of Nigeria. Institute of Church and Society.
HN772.L4 309.2'63'096 77-366059
78 MARC

—PHILIPPINE ISLANDS

Asian Social Institute. Research Dept.
The church and development in the Philippines. — Manila : Asian Social Institute, Research Dept., 1975.
177 leaves ; 33 cm.
1. Church and underdeveloped areas—Philippine Islands. 2. Catholic Church in the Philippine Islands. 3. Philippine Islands—Religious life and customs. I. Title.
BX1658.2.A84 1975 282'.599 79-312862
79 MARC

CHURCH ANTIQUITIES see Christian antiquities

CHURCH ARCHITECTURE

see also Abbeys; Architecture, Cistercian; Architecture, Gothic; Architecture, Jesuit; Basilicas; Cathedrals; Chapels; Church decoration and ornament; Churches; Crossing (Architecture); Crypts; Half-timbered churches; Liturgy and architecture; Mosques; Orientation (Architecture); Stave churches; Temples; Towers

Conti, Flavio, 1943-
The closed faith / Flavio Conti ; translated by Patrick Creagh. — Boston : HBJ Press, c1979.
168 p. : ill. ; 31 cm. — (His)
Translation of I santuari del misticismo.
ISBN 0-15-003733-3 : $14.95
1. Church architecture. 2. Temples. I. Title. II. Series.
NA4600.C6413 726 78-19642
78 MARC

CHURCH ARCHITECTURE (Continued)

Conti, Flavio, 1943-
Tribute to religion / Flavio Conti ; translated by Patrick Creagh. — New York : HBJ Press, c1979.
168 p. : ill. (some col.) ; 31 cm. — (His The grand tour)
Translation of Il fasto della religione.
ISBN 0-15-003735-X : $14.95
1. Church architecture. 2. Temples. 3. Mosques. 4. Monasteries. I. Title. II. Series: Conti, Flavio, 1943- Grand tour.
NA4600.C6313 726'.09 78-61757
78 MARC

Smith, Peter Frederick.
Third millennium churches [by] Peter F. Smith. London, Galliard Ltd; New York, Galaxy Music Corporation, 1972.
xi, 98 p. illus. 23 cm. £3.00 GB73-13001
ISBN 0-85249-159-X
1. Church architecture. 2. Architecture, Modern—20th century. I. Title.
NA4825.S57 726'.5'0904 74-157398
74[r79]rev MARC

—DETAILS

Bentmann, Reinhard, 1939-
Kirchen des Mittelalters / Text von Reinhard Bentmann u. Heinrich Lickes. — Wiesbaden : Ebeling, 1978.
192 p. : 100 ill. (chiefly col.) ; 34 cm. GFR78-A
Bibliography: p. 191-192.
ISBN 3-921195-21-7 : DM48.00
1. Churches. 2. Architecture, Medieval. 3. Church architecture—Details. 4. Church decoration and ornament. I. Lickes, Heinrich, joint author. II. Title.
NA4800.B38 726'.5'094 78-372118
78 MARC

—PHILOSOPHY

Lambert, Élie, 1888-
L'architecture des templiers / Élie Lambert. — Paris : A. et J. Picard, 1955, 1978 printing.
104 p. : ill. ; 23 cm. F***
Includes bibliographical references.
ISBN 2-7084-0025-8 : 48.00F
1. Church architecture—Philosophy. 2. Christian art and symbolism—Medieval, 500-1500. 3. Templars. I. Title.
NA4828.L35 729'.58'2 79-379107
79 MARC

—AUSTRIA

Wonisch, Othmar, 1884-
Wallfahrtskirche Mariazell : Steiermark/Österreich, Bezirk Bruck u. d. Mur / [Othmar Wonisch ; Fotos, Ellen Traubenkraut ... et al.]. — 10., neu bearb. Aufl. — München ; Zürich : Schnell und Steiner, 1977.
13 p. : numerous ill. ; 17 cm. — (Kunstführer ; Nr. 478) GFR78-A
Caption title.
Includes bibliographical references.
1. Wallfahrtskirche, Mariazell, Austria. 2. Church architecture—Austria. 3. Architecture, Baroque—Austria. I. Title. II. Series: Kleine deutsche Kirchenführer ; Nr. 478.
NA5564.K53 Nr. 478 914.3'04'8708 s 79-356413
[NA5510.W34] 79 MARC

—AUSTRIA—KLAGENFURT—GUIDE-BOOKS

Langhold, Norbert.
Kirchen in Klagenfurt : ein kleiner Führer / N. u. M. Langhold ; [Fotos, N. u. M. Langhold]. — Klagenfurt : Heyn, 1977.
86 p. : ill. ; 20 cm. Au78-1/2-291
ISBN 3-85366-231-5 : S48.00
1. Church architecture—Austria—Klagenfurt—Guide-books. 2. Churches—Austria—Klagenfurt—Guide-books. 3. Klagenfurt, Austria—Churches. I. Langhold, Margerita, joint author. II. Title.
NA5510.K57L36 914.36'6 79-378242
*79 MARC

—ENGLAND

Jones, Lawrence Elmore.
The beauty of English churches / [by] Lawrence E. Jones. — London : Constable, 1978.
391 p. : ill., 2 coats of arms, maps (on lining papers), ports. ; 18 cm. GB78-15096
Includes indexes.
ISBN 0-09-461730-9 : £3.95
1. Church architecture—England. 2. Churches—England. I. Title.
NA5461.J588 1978 726'.5'0942 78-318138
78 MARC

Swarbrick, John, 1879-
England's tribute to the architects & craftsmen of France / John Swarbrick ; foreword by Albert Richardson. — Cheam, Surrey : Wykeham Press, [1957?]
48 p., [1] leaves of plates : ill. ; 22 cm.
1. Church architecture—England. 2. Architecture, Norman—England. I. Title.
NA5463.S9 726'.5'0942 79-312848
79 MARC

—ENGLAND—LONDON—GUIDE-BOOKS

Blatch, Mervyn.
A guide to London's churches / [by] Mervyn Blatch. — London : Constable, 1978.
xxx, 434 p. : ill., maps ; 18 cm. GB78-40605
Bibliography: p. 424.
Includes index.
ISBN 0-09-462220-5 : £4.50
1. London—Churches—Guide-books. 2. Church architecture—England—London—Guide-books. I. Title.
NA5470.A1B52 914.21'04'857 79-317351
79 MARC

—FINLAND—SATAKUNTA

Sastamalan seitsemän kirkkoa. (Tyrvää), Tryvään sanomat, 1968.
72 p. illus. 22 cm. Fi
1. Church architecture—Finland—Satakunta. 2. Churches—Finland—Satakunta.
NA5955.F52S3717 79-345410
79 MARC

—FRANCE

L'Art religieux : abbayes, églises et cathédrales. — Paris : Larousse, 1978.
480 p. : col. ill. ; 30 cm. — (Beautés de la France) (Découvrir) F78-20420
Includes index.
ISBN (invalid) 2-03-013275-9 : 197F
1. Christian art and symbolism—France. 2. Church architecture—France. 3. France—Civilization.
N7949.A1A77 914.4'04'83 79-361629
79 MARC

Richards, James Maude, Sir.
800 years of Finnish architecture / J. M. Richards. — Newton Abbot [Eng.] : David & Charles, c1978.
191 p. : ill. ; 30 cm. GB***
Based on the author's 1966 ed. of Guide to Finnish architecture.
Bibliography: p. 188.
Includes index.
ISBN 0-7153-7512-1 : £9.50
1. Architecture—Finland. 2. Church architecture—Finland. I. Richards, James Maude, Sir. Guide to Finnish architecture. II. Title.
NA1455.F5R5 720'.9471 78-319214
79 MARC

Schäfke, Werner, 1944-
Frankreichs gotische Kathedralen : e. Reise zu d. Höhepunkten mittelalterl. Architektur in Frankreich / Werner Schäfke. — Köln : DuMont, 1979.
294 p. : ill. (some col.) ; 21 cm. — (DuMont Dokumente : DuMont Kunstreiseführer) GFR79-A
Bibliography: p. 285-287.
Includes indexes.
ISBN 3-7701-0975-9
1. Church architecture—France. 2. Architecture, Gothic—France. I. Title.
NA5543.S3 79-363499
79 MARC

—FRANCE—AUVERGNE

Leriche-Andrieu, Françoise.
Itinéraires romans en Auvergne / Françoise Leriche-Andrieu. — [St Léger Vauban] : Zodiaque, 1978.
108 p. : ill. ; 26 cm. — (Les Travaux des mois ; 17) F***
Includes index.
30.00F
1. Architecture, Romanesque—France—Auvergne. 2. Church architecture—France—Auvergne. 3. Sculpture, Romanesque—France—Auvergne. 4. Sculpture, French—France—Auvergne. I. Title.
NA5549.A8L47 914.4'59'0483 79-351718
79 MARC

—FRANCE—PÉRIGORD

Secret, Jean.
Itinéraires romans en Périgord / Jean Secret. — La Pierre-qui-Vire : Zodiaque, c1977.
93 p. : ill., maps ; 26 cm. — (Les Travaux des mois ; 16) F***
Includes index.
1. Church architecture—France—Périgord. 2. Architecture, Romanesque—France—Périgord. I. Title.
NA5549.P4S39 914.4'72 78-369022
78 MARC

—FRANCE—RAUCOURT-ET-FLABA

Sartelet, Alain.
Les Églises anciennes des cantons de Sedan et de Raucourt / Alain Sartelet. — [Sedan] (2, rue Norbet, 08200) : A. Sartelet, 1977.
265 p. : ill. ; 30 cm. F77-13272
Cover title.
Bibliography: p. 263-265.
50.00F
1. Church architecture—France—Sedan. 2. Church architecture—France—Raucourt-et-Flaba. I. Title.
NA5549.S38S27 726'.5'094431 79-368403
79 MARC

—FRANCE—SEDAN

Sartelet, Alain.
Les Églises anciennes des cantons de Sedan et de Raucourt / Alain Sartelet. — [Sedan] (2, rue Norbet, 08200) : A. Sartelet, 1977.
265 p. : ill. ; 30 cm. F77-13272
Cover title.
Bibliography: p. 263-265.
50.00F
1. Church architecture—France—Sedan. 2. Church architecture—France—Raucourt-et-Flaba. I. Title.
NA5549.S38S27 726'.5'094431 79-368403
79 MARC

—GEORGIA (TRANSCAUCASIA)

Tumanishvili, Dmitriĭ Grigor'evich.
(Tema "svobodnogo kresta" v srednevekovoĭ gruzinskoĭ arkhitekture)
Тема "свободного креста" в средневековой грузинской архитектуре / Дм. Туманишвили ; II Междунар. симпозиум по груз. искусству. — Тбилиси : Мецниереба, 1977.
12 p. ; 20 cm. USSR 77
Cover title.
At head of title: Akademiia nauk Gruzinskoĭ SSR. Institut istorii gruzinskogo iskusstva im. G. N. Chubinashvili.
Bibliography: p. 10–12.
0.07rub
1. Church architecture—Georgia (Transcaucasia) 2. Architecture, Medieval—Georgia (Transcaucasia) 3. Crossing (Architecture) I. Title.
NA5992.9.T85 79–412373

—GEORGIA (TRANSCAUCASIA)—CONGRESSES

Beridze, Vakhtang Vukolovich.
L'architecture religieuse géorgienne des IVe-VIIe siècles : IIe Symposium international sur l'art géorgien / Vakhtang Beridze. — Tbilisi : Met͡sniereba, 1977.
39 p. ; 20 cm. USSR***
At head of title: Académie des sciences de la R.S.S. de Géorgie, Institut Tschubinaschvili d'histoire de l'art géorgien.
Cover title.
1. Church architecture—Georgia (Transcaucasia)—Congresses. 2. Architecture, Medieval—Georgia (Transcaucasia)—Congresses. I. Title.
NA5992.9.B46 726'.5'094795 79-379184
79 MARC

—GERMANY

Otto, Christian F
Space into light : the church architecture of Balthasar Neumann / Christian F. Otto. — New York : Architectural History Foundation, c1979.
296 p. : ill. (some col.) ; 26 cm.
Bibliography: p. 275-288.
Includes index.
ISBN 0-262-15019-0
1. Neumann, Balthasar, 1687-1754. 2. Church architecture—Germany. 3. Architecture, Modern—17th-18th centuries—Germany. 4. Light in architecture. 5. Space (Architecture) I. Title.
NA1088.N4O87 726'.5'0924 78-27782
78 MARC

—GERMANY, WEST

Altmann, Lothar.
Stephanus-Gemeindezentrum der Evangelischen Kirchengemeinde Hilgen-Neuenhaus, Rheinisch-Bergischer Kreis / [Lothar Altmann]. — 1. Aufl. — München : Schnell & Steiner, 1977.
14, [1] p. : ill. ; 17 cm. — (Kunstführer ; Nr. 1053) GFR***
Caption title.
Bibliography: p. [15]
1. Stephanus-Gemeindezentrum. 2. Architecture, Modern—20th century—Germany, West. 3. Church architecture—Germany, West. I. Title. II. Series: Kleine deutsche Kirchenführer ; Nr. 1053.
NA5564.K53 Nr. 1053 914.3'04'8708 s 78-375515
[NA5586.S684] 78 MARC

—GERMANY, WEST—BRUNSWICK (CITY)

Dorn, Reinhard.
Mittelalterliche Kirchen in Braunschweig / Reinhard Dorn ; mit e. Vorw. von Gerhard Heintze ; [Aufnahmen von Willi Birker ... et al.]. — Hameln : Niemeyer, 1978.
256 p. : chiefly ill. ; 28 cm. GFR78-A46
Leaf with English legends inserted.
Bibliography: p. 254.
Includes index.
ISBN 3-87585-043-2 : DM53.00
1. Brunswick (City)—Churches. 2. Church architecture—Germany, West—Brunswick (City) 3. Architecture, Medieval—Germany, West—Brunswick (City) I. Title.
NA5586.B75D67 79-363855
79 MARC

CHURCH ARCHITECTURE (Continued)

—GERMANY, WEST—PASSAU

Schäffer, Gottfried.
Wallfahrten im Passauer Land / Gottfried Schäffer ; Gregor Peda. — Freilassing : Pannonia, 1978.
48 p. : 25 ill. ; 17 cm. — (Kleine Pannonia-Reihe ; 69) GFR78-A
Cover title.
ISBN 3-7897-0069-X : DM3.80
1. Church architecture—Germany, West—Passau. 2. Christian art and symbolism—Germany, West—Passau. I. Peda, Gregor. II. Title.
NA5576.P35S26 914.3'35 78-388781
79 MARC

—GREAT BRITAIN

Harvey, John Hooper.
The perpendicular style, 1330-1485 / ₍by₎ John Harvey. — London : Batsford, 1978.
308 p. : ill., maps ; 26 cm. GB78-37060
Bibliography: p. 271-275.
Includes indexes.
ISBN 0-7134-1610-6 : £12.50
1. Architecture, Gothic—Great Britain. 2. Church architecture—Great Britain. I. Title.
NA5463.H3 720'.942 78-326626
79 MARC

—GREECE, MEDIEVAL

Panagopoulos, Beata Maria, 1925-
Cistercian and mendicant monasteries in medieval Greece / Beata Kitsiki Panagopoulos. — Chicago : University of Chicago Press, 1979.
xx, 194 p. : ill. ; 21 x 22 cm.
Bibliography: p. ₍179₎-188.
Includes index.
ISBN 0-226-64544-4
1. Monasteries—Greece, Medieval. 2. Architecture, Cistercian—Greece, Medieval. 3. Achitecture, Gothic—Greece, Medieval. 4. Church architecture—Greece, Medieval. I. Title.
NA5593.P36 726'.7'09495 78-10769
78 MARC

—ITALY

Argan, Giulio Carlo.
L'architettura protocristiana, preromanica e romanica / Giulio Carlo Argan. — Bari : Dedalo libri, 1978.
64 p., ₍26₎ leaves of plates : ill. ; 18 cm. — (Universale di archittettura ; 13/14) It78-Dec
L4000
1. Church architecture—Italy. 2. Architecture—Italy. 3. Architecture, Early Christian—Italy. 4. Architecture, Romanesque—Italy. I. Title. II. Series.
NA5613.A73 726'.5'0945 79-352856
•79 MARC

—ITALY—LIGURIA

Chierici, Sandro.
Il Piemonte, la Val d'Aosta, la Liguria / ₍testo di Sandro Chierici, Duilio Citi ; fotografie di Zodiaque₎. — 1. ed. — Milano : Jaca book, 1979.
485 p., ₍6₎ leaves of plates : ill. ; 23 cm. — (Italia romanica ; v. 2) (Già a non ancora : Arte ; 2) It79-July
Issued also in French and Spanish.
Includes bibliographies.
L24000
1. Church architecture—Italy—Valle d'Aosta (Region) 2. Architecture, Romanesque—Italy—Valle d'Aosta (Region) 3. Church architecture—Italy—Piemonte. 4. Architecture, Romanesque—Italy—Piemonte. 5. Church architecture—Italy—Liguria. 6. Architecture, Romanesque—Italy—Liguria. I. Citi, Duilio, joint author. II. Title. III. Series.
NA5619.A57C47 726'.5'09451 79-376174
•79 MARC

—ITALY—PIEMONTE

Chierici, Sandro.
Il Piemonte, la Val d'Aosta, la Liguria / ₍testo di Sandro Chierici, Duilio Citi ; fotografie di Zodiaque₎. — 1. ed. — Milano : Jaca book, 1979.
485 p., ₍6₎ leaves of plates : ill. ; 23 cm. — (Italia romanica ; v. 2) (Già a non ancora : Arte ; 2) It79-July
Issued also in French and Spanish.
Includes bibliographies.
L24000
1. Church architecture—Italy—Valle d'Aosta (Region) 2. Architecture, Romanesque—Italy—Valle d'Aosta (Region) 3. Church architecture—Italy—Piemonte. 4. Architecture, Romanesque—Italy—Piemonte. 5. Church architecture—Italy—Liguria. 6. Architecture, Romanesque—Italy—Liguria. I. Citi, Duilio, joint author. II. Title. III. Series.
NA5619.A57C47 726'.5'09451 79-376174
•79 MARC

—ITALY—VALLE D'AOSTA (REGION)

Chierici, Sandro.
Il Piemonte, la Val d'Aosta, la Liguria / ₍testo di Sandro Chierici, Duilio Citi ; fotografie di Zodiaque₎. — 1. ed. — Milano : Jaca book, 1979.
485 p., ₍6₎ leaves of plates : ill. ; 23 cm. — (Italia romanica ; v. 2) (Già a non ancora : Arte ; 2) It79-July
Issued also in French and Spanish.
Includes bibliographies.
L24000
1. Church architecture—Italy—Valle d'Aosta (Region) 2. Architecture, Romanesque—Italy—Valle d'Aosta (Region) 3. Church architectute—Italy—Piemonte. 4. Architecture, Romanesque—Italy—Piemonte. 5. Church architecture—Italy—Liguria. 6. Architecture, Romanesque—Italy—Liguria. I. Citi, Duilio, joint author. II. Title. III. Series.
NA5619.A57C47 726'.5'09451 79-376174
•79 MARC

—JAPAN—NAGASAKI (PREFECTURE)

Pacheco, Diego, 1922-
(Nagasaki no tenshudō)
長崎の天主堂 / 著・撮影 パチェコ・ディエゴ. — 福岡：西日本文化協会, 1976.
126 p. : col. ill. ; 19 cm. Ja 78-24052
¥980

1. Church architecture—Japan—Nagasaki (Prefecture) 2. Churches, Catholic—Japan—Nagasaki (Prefecture) I. Title.
NA6056.N34P32 79-803774

—MEXICO—CHIHUAHUA (STATE)

Roca, Paul M
Spanish Jesuit churches in Mexico's Tarahumara / Paul M. Roca. — Tucson : University of Arizona Press, c1979.
xxiv, 369 p. : ill. ; 24 cm.
Bibliography: p. 341-359.
Includes index.
ISBN 0-8165-0651-5. ISBN 0-8165-0572-1 pbk.
1. Architecture, Jesuit—Mexico—Chihuahua (State) 2. Church architecture—Mexico—Chihuahua (State) 3. Architecture, Colonial—Mexico—Chihuahua (State) 4. Tarahumare Indians—Architecture. 5. Indians of Mexico—Mexico—Chihuahua (State)—Architecture. I. Title.
NA5256.C45R6 972'.1 78-14467
78 MARC

—NETHERLANDS—GRONINGEN (PROVINCE)

Meulen, Tony van der.
Oude kerken in Groningen / Tony van der Meulen. — Baarn : Bosch & Keuning, ₍1979₎
63 p. : with ill. ; 24 cm. — (Publicatie van de Stichting Oude Groninger Kerken ; 21) Ne79-24
ISBN 902464321X : fl 8.75
1. Churches—Netherlands—Groningen (Province) 2. Church architecture—Netherlands—Groningen (Province) I. Title. II. Series: Stichting Oude Groninger Kerken. Publicatie van de Stichting Oude Groninger Kerken ; 21.
NA5649.G76M48 79-376856
•79 MARC

—NORWAY

see also Stave churches

—ONTARIO—CONSERVATION AND RESTORATION

Kalman, Harold D
The conservation of Ontario churches : a programme for funding religious properties of architectural and historical significance : prepared for the Ministry of Culture and Recreation, Province of Ontario / by Harold D. Kalman ; with contributions by Stewart Crysdale, Curtis Fahey, and William J. Hemmerick. — ₍s.l. : s.n.₎, 1977 (Ottawa : Tri-Graphic Print.)
vi, 130 p. : ill. ; 22 x 27 cm. C78-4098-8
Includes bibliographies.
1. Church architecture—Ontario—Conservation and restoration. I. Ontario. Ministry of Culture and Recreation. II. Title.
NA5246.O6K34 338.4'7'06953 78-318659
78 MARC

—PORTUGAL

Tavares Chicó, Mário.
A arquitectura gótica em Portugal / Mário Tavares Chicó ; ₍capa de Moura George and partners/England₎. — 2. ed. — Lisboa : Livros Horizonte, 1968.
239 p., ₍21₎ leaves of plates : ill. ; 24 cm.
Bibliography: p. 220-222.
Includes index.
1. Church architecture—Portugal. 2. Architecture, Gothic—Portugal. I. Title.
NA5823.T38 1968 726'.5'09469 79-113589
79 MARC

—QUÉBEC (PROVINCE)

Noppen, Luc.
Les églises du Québec, 1600-1850 / Luc Noppen, avec la collaboration de Achille Murphy ... ₍et al.₎. — ₍Québec₎ : Éditeur officiel du Québec, c1977.
ix, 298 p. : ill. ; 20 cm. — (Collection Loisirs et culture) (Connaissance du Québec) C***
Bibliography: p. 64.
Includes indexes.
ISBN 0-7754-2869-8
1. Churches—Québec (Province)—Guide-books. 2. Church architecture—Québec (Province) I. Title.
NA5246.Q4N66 726'.5'09714 79-345535
79 MARC

Tardif-Painchaud, Nicole.
Dom Bellot et l'architecture religieuse au Québec / Nicole Tardif-Painchaud. — Québec : Presses de l'Université Laval, 1978.
xxi, 262 p. : ill. ; 23 cm. C***
Bibliography: p. ₍xiii₎-xxi.
ISBN 0-7746-6828-8 : $17.75
1. Bellot, Paul, 1876-1944. 2. Bellot, Paul, 1876-1944—Influence. 3. Church architecture—Québec (Province) I. Title.
NA749.B44T37 726'.5'0924 79-345541
79 MARC

—RUSSIA—BIBLIOGRAPHY

Pili͡avskii, Vladimir Ivanovich.
(Russkai͡a arkhitektura odinnadt͡satogo—nachala dvadt͡satogo v.)
Русская архитектура XI—начала XX в. : (Указ. избр. лит. на рус. яз. за 1811–1975 гг.) / В. И. Пилявский, Н. Я. Горшкова. — Ленинград : Изд-во ЛГУ, 1978.
158 p. : ill. ; 21 cm. USSR 78
At head of title: Ministerstvo vysshego i srednego spet͡sial'nogo obrazovanii͡a RSFSR.
Includes index.
0.90rub
1. Architecture—Russia—Bibliography. 2. Church architecture—Russia—Bibliography. I. Gorshkova, N. I͡A., joint author II. Title.
Z5944.R9P53 79–412204
[NA1181]

—RUSSIAN, REPUBLIC

Nikolaev, Ivan Sergeevich, 1901–
(Tvorchestvo drevnerusskikh zodchikh)
Творчество древнерусских зодчих / И. С. Николаев ; Под ред. К. Н. Афанасьева. — Москва : Стройиздат, 1978.
105 p. : ill. ; 21 cm. USSR 78
Includes bibliographical references.
0.55rub
1. Architecture—Russian Republic. 2. Church architecture—Russian Republic. 3. Architecture, Medieval—Russian Republic. I. Afanas'ev, Kirill Nikolaevich. II. Title.
NA1183.N54 79–410360

—RUSSIAN REPUBLIC—PSKOV, RUSSIA (CITY)

Spegal'skii, I͡Urii Pavlovich.
(Pskov)
Псков / Ю. Спегальский. — Изд. 2., доп. — Ленинград : Искусство, 1978.
246 p. : ill. (part col.) ; 15 cm. — (Архитектурно-художественные памятники XII–XVII веков) USSR***
Added t. p. in French; list of illustrations also in French.
Series romanized: Arkhitekturno-khudozhestvennye pami͡atniki XII–XVII vekov.
Bibliography: p. 237–₍238₎
1.70rub
1. Architecture — Russian Republic — Pskov, Russia (City) 2. Church architecture—Russian Republic—Pskov, Russia (City)
NA1197.P78S6 1978 79–409270

—RUSSIAN REPUBLIC—ZVENIGOROD

Nikolaeva, Tat'i͡ana Vasil'evna.
(Drevnii Zvenigorod)
Древний Звенигород = Old Zvenigorod : Архитектура, искусство : ₍Альбом₎ / Т. В. Николаева. — Москва : Искусство, 1978.
207 p. : col. ill. ; 26 cm. USSR 78
Summary in English; list of illustrations also in English.
Bibliography: p. ₍201₎–203.
1. Church architecture—Russian Republic—Zvenigorod. 2. Architecture—Russian Republic—Zvenigorod. 3. Christian art and symbolism—Russian Republic—Zvenigorod. I. Title. II. Title: Old Zvenigorod.
NA5697.Z84N54 79–410972

—SCOTLAND

Hay, George, 1911-
The architecture of Scottish post-Reformation churches, 1560-1843. Oxford, Clarendon Press, 1957.
xvi, 299 p. illus., 47 plates, plans. 25 cm.
Bibliography: p. 283-286.
1. Church architecture—Scotland. 2. Churches, Reformed—Scotland. 3. Church decoration and ornament—Scotland. I. Title.
NA5475.H3 726'.5'09411 57-2871
₍r79₎rev MARC

CHURCH ARCHITECTURE (Continued)

—SPAIN—CANTABRIA

Sá, María Eálo de.
El románico de Cantabria en sus cinco colegiatas / ₍por₎ María Eálo de Sá. — Santander : Diputación Provincial, Institución Cultural de Cantabria, Centro de Estudios Montañeses, 1978.
324 p. : ill. ; 24 cm. Sp***
Summary of thesis—Madrid, 1977.
Bibliography: p. 321-324.
ISBN 8450029309
1. Architecture, Romanesque—Spain—Cantabria. 2. Church architecture—Spain—Cantabria. 3. Churches—Spain—Cantabria. I. Title.
NA5809.C28S2 79-112708
79 MARC

—SPAIN—CATALONIA

Göttgens, René.
Romaanse kerken in het Catalaans Pyreneeëngebied / René Göttgens. — Nijmegen : Historie, 1978.
176 p. : with ill. ; 24 cm. Ne79-3
Bibliography: p. 175-176.
ISBN 9063130139 : fl 34.75
1. Church architecture—Spain—Catalonia. 2. Architecture, Romanesque—Spain—Catalonia. I. Title.
NA5809.C3G63 79-359165
*79 MARC

—SPAIN—CORDOBA—CONSERVATION AND RESTORATION—CONGRESSES

Nieto Cumplido, Manuel.
La mezquita-catedral de Córdoba y el ICOMOS / ₍por₎ Manuel Nieto Cumplido. — Córdoba : Servicio de Publicaciones del Excmo. Ayuntamiento, 1976.
131 p., ₍4₎ leaves of plates : ill. ; 28 cm. Sp78-Jan
"Conferencias del ICOMOS en Córdoba" ₍1973₎: p. ₍54₎-128.
ISBN 8450014492
1. Córdoba. Catedral—Congresses. 2. Church architecture—Spain—Cordoba—Conservation and restoration—Congresses. 3. Architecture, Islamic—Spain—Conservation and restoration—Congresses. 4. Monuments—Preservation—Congresses. 5. International Council of Monuments and Sites—Congresses. I. International Council of Monuments and Sites. II. Title.
NA5811.C62N53 78-382819
78 MARC

—SWEDEN—LINKÖPING

Cnattingius, Bengt Jacob, 1899-
Studier i Linköpings domkyrkas byggnadshistoria / av Bengt Cnattingius. — Stockholm : Almqvist & Wiksell international, 1977.
112 p. : ill. ; 25 cm. — (Kungl. Vitterhets-, historie- och antikvitetsakademiens handlingar : Antikvariska serien ; 29 ISSN 0083-6761) S77-26/28
Summary in German.
Bibliography: p. 109-112.
ISBN 917402048X : kr66.50
1. Linköping, Sweden. Domkyrkan. 2. Church architecture—Sweden—Linköping. I. Title. II. Series: Vitterhets-, historie- och antikvitetsakademien, Stockholm. Handlingar : Antikvariska serien ; 29.
NA5791.L5C56 726'.5'09486 79-341366
79 MARC

—SWITZERLAND—BROYE VALLEY

Martin, Vio.
Vieilles églises de la Broye / Vio Martin. — Lausanne : Éditions des Terreaux, 1977.
79 p. : ill. ; 24 cm. Sw77-11827
ISBN 2-88033-002-5
1. Church architecture—Switzerland—Broye Valley. 2. Churches—Switzerland—Broye Valley. I. Title.
NA5849.B76M38 914.94'3 78-383242
*78 MARC

—SWITZERLAND—FRANENKAPPELEN

Bellwald, Ulrich.
Frauenkapplen, Kirche und Dorf / ₍Ulrich Bellwald₎. — Basel ₍Münzgässlein 16₎ : Gesellschaft für Schweizerische Kunstgeschichte, 1975.
19 p. : ill. ; 20 cm. — (Schweizerische Kunstführer) Sw76-A-870
Cover title.
Bibliography: p. 19.
1. Kirche, Frauenkappelen, Switzerland. 2. Church architecture—Switzerland—Franenkappelen. I. Title. II. Series.
NA5851.F73B44 914.94'5 78-382437
*78 MARC

—SWITZERLAND—MAGDENAU

Anderes, Bernhard.
Magdenau / Bernhard Anderes. — Basel, Münzgässlein 16 : Gesellschaft für schweizerische Kunstgeschichte, 1977.
31 p. : ill. ; 20 cm. — (Schweizerische Kunstführer) Sw77-10195
Cover title.
Bibliography: p. 31.
1. Church architecture—Switzerland—Magdenau. 2. Religious articles—Switzerland—Magdenau. I. Title. II. Series.
NA5851.M33A5 914.94'7 78-381827
*78 MARC

—TURKEY—ANI

Cuneo, Paolo.
L'architettura della scuola regionale di Ani nell'Armenia medievale : relazione svolta nella seduta del 1° marzo 1976 / Paolo Cuneo. — Roma : Accademia Nazionale dei Lincei, 1977.
80 p., ₍35₎ leaves of plates (3 fold.) : ill. ; 27 cm. — (Problemi attuali di scienza e di cultura ; quaderno n. 235) It***
Includes bibliographical references.
L9500
1. Church architecture—Turkey—Ani. 2. Architecture, Medieval—Turkey—Ani. 3. Christian antiquities—Turkey—Ani. I. Title. II. Series: Accademia nazionale dei Lincei, Rome. Problemi attuali di scienza e di cultura ; quaderno n. 235.
AS222.R53 n. 235 085 s 78-378233
₍NA5871.A54₎ 78 MARC

—TURKEY, EASTERN

Marzpanian, Hovhannēs.
(Baghdatakan aknark mē krōnakan chartarapetowt'ean vray ew hay chartarapetowt'ean teghě anor mēj)
Բաղդատական ակնարկ մը կրօնական ճարտարապետութեան վրայ եւ հայ ճարտարապետութեան տեղը անոր մէջ / Յովհաննէս Մարզպանեան. — Իսթանպուլ : Հրատարակչութիւն Մխիթարեան Սանուց Միութեան, 1970.
198 p. : ill. ; 20 cm.
1. Church architecture—Turkey, Eastern. 2. Architecture, Armenian—Turkey, Eastern. 3. Architecture, Medieval—Turkey, Eastern. I. Title.
NA5869.E18M37 78-970773

—UKRAINE

Carmazinu Cacovschi, V
(Ukraïns'ka narodna arkhitektura)
Українська народна архітектура; хати і дерев'яні церкви XVIII сторіччя з 40 ілюстраціями. Ukrainian folk architecture; with 40 illustrations. За ред. С. Гординського. Рим, 1972.
58 p. illus. (part col.) 24 cm. (Видання "Богословії," ч. 41 И.) It***
At head of title: В. Кармазин-Каковський.
"Extractum e 'Bohoslovia,' t. 34 (1970."
Summary in Romanian, English, and German.
Includes bibliographical references.
1. Architecture—Ukraine. 2. Architecture, Domestic—Ukraine. 3. Church architecture—Ukraine. 4. Building, Wooden—Ukraine. 5. Vernacular architec- ture—Ukraine. I. Hordyns'kyĭ, Svíatoslav, 1906- ed. II. Title. III. Title: Ukrainian folk architecture.
NA1195.U4C37 74-317765

—VIRGINIA

Perdue-Davis, Vernon.
Virginia's ante-bellum churches : an introduction with particular attention to their furnishings / by Vernon Perdue Davis and James Scott Rawlings. — Richmond, Va. : Dietz Press, 1978.
xv, 129 p. : ill. ; 25 cm.
Includes index.
1. Church architecture—Virginia. 2. Architecture, Modern—19th century—Virginia. 3. Church decoration and ornament—Virginia. I. Rawlings, James Scott, joint author. II. Title.
NA5230.V8P47 726'.5'09755 79-100708
79 MARC

CHURCH ARCHIVES

—NETHERLANDS—CATALOGS

Kerkelijke archieven : inventarissen / onder red. van J. P. van Dooren. — Den Haag : Archiefdienst Nederlandse Hervormde Kerk, 19
v. ; 30 cm. Ne***
Includes bibliographical references and indexes.
1. Church archives—Netherlands—Catalogs. 2. Church records and registers—Netherlands—Bibliography—Catalogs. 3. Netherlands—Church history—Sources—Bibliography—Catalogs. I. Dooren, J. P. van. II. Nederlandse Hervormde Kerk. Archiefdienst.
CD1709.N4K47 79-367313
79 MARC

CHURCH ATTENDANCE

see also Public worship

CHURCH BANNERS see Church pennants

CHURCH BELLS see Bells

CHURCH BIOGRAPHY see Christian biography

CHURCH BUILDINGS see Church architecture; Churches; Church controversies

CHURCH CALENDAR

see also Chronology, Ecclesiastical; Church year; Fasts and feasts; and subheading Liturgy and ritual —Calendar under particular church bodies, e.g. Catholic Church. Liturgy and ritual—Calendar

De Ordinarii van Kampen uit de 15de en 16de eeuw / uitg. door W. Jappe Alberts. — Groningen : J. B. Wolters, 1961.
x, 47 p. ; 19 cm. — (Fontes minores Medii Aevi ; 12)
Includes bibliographical references and index.
CONTENTS: De oudste Ordinarius van Kampen.—De 16de eeuwse Ordinarius van Kampen.
1. Ordinances, Municipal—Netherlands—Kampen. 2. Church calendar. 3. Kampen—Religious life and customs. I. Jappe Alberts, Wybe. II. Oudste ordinarius van Kampen. 1961. III. Kampen. Ordinances, local laws, etc. Ordinarius antiquus. 1961. IV. Series.
79-375924
79 MARC

(Troiťskiĭ kalendar')
Троицкій календарь, 1943–
Jordanville, N. Y., Изд. Св. Троицкаго монастыря.
v. ill. 24 cm. annual.
Continued by: Troiťskiĭ pravoslavnyĭ russkiĭ kalendar'.
1. Orthodox Eastern Church, Russian, in the United States—Yearbooks. 2. Church calendar. I. Jordanville, N. Y. Holy Trinity Monastery.
BX496.A5T7 78–647599
(MARC–S)

CHURCH CAMPS

Eshbach, Theresa C 1940-
Strength to lead, faith to follow : a series of resources for outdoor camping / by Theresa C. Eshbach and Nancy F. Knepper. — ₍Elgin, Ill.₎ : Brethren Press, c1978.
125, ₍1₎ p. : ill. ; 28 cm.
Bibliography: p. ₍126₎
ISBN 0-87178-817-9
1. Church camps. I. Knepper, Nancy F., 1948- joint author. II. Title.
BV1650.E83 259 78-21492
78 MARC

Genné, Elizabeth.
Church family camps and conferences : an administrative and program manual / Elizabeth and William Genné. — Rev. ed. — Valley Forge, PA. : Judson Press, c1979..
77 p. ; 22 cm.
ISBN 0-8170-0818-7 : $2.95
1. Church work with families. 2. Church camps. 3. Family—Religious life. I. Genné, William H., joint author. II. Title.
BV4438.G46 1979 259 78-24395
78 MARC

Introduction to Christian camping / edited by Werner C. Graendorf and Lloyd D. Mattson. — Chicago : Moody Press, c1979.
223 p. : ill. ; 22 cm.
Includes bibliographies.
ISBN 0-8024-4131-9 : $4.95
1. Church camps. 2. Camps—Administration. I. Graendorf, Werner C. II. Mattson, Lloyd D.
BV1650.I53 254'.6 78-10248
78 MARC

CHURCH CHARITIES

(Kirisutokyō shakai fukushi gaisetsu)
キリスト教社会福祉概説 / 日本基督教社会福祉学会，日本基督教社会事業同盟編．— 東京：日本基督教団出版局，1978.
335 p. ; 19 cm. Ja 79-4655
Bibliography: p. 317-330.
¥1800.
1. Church charities. 2. Church and social problems. 3. Social service—Japan. I. Nihon Kirisutokyō Shakai Fukushi Gakkai. II. Nihon Kirisutokyō Shakai Jigyō Dōmei.
HV530.K57 79-807703

CHURCH CHARITIES (Continued)

—JUVENILE LITERATURE

Pike, Morris D 1928-
Your need for bread is mine : resources for helping the hungry / Morris D. Pike ; illustrated by Karen Tureck. — New York : Friendship Press, c1977.
64 p. : ill. ; 25 cm.
SUMMARY: Poems, prayers, stories, and songs describe efforts to alleviate world food shortages.
ISBN 0-377-00067-1 : $2.45
1. Food relief—Juvenile literature. 2. Church charities—Juvenile literature. [1. Food relief. 2. Food supply] I. Tureck, Karen. II. Title.
HV696.F6P5 362.5 77-2352
77[r78]rev MARC

—GERMANY—BIOGRAPHY

Wer mir dienen will : 24 Lebensbilder von Männern u. Frauen in Dienst d. Liebe / hrsg. im Auftr. von Innere Mission u. Hilfswerk d. Evang. Kirchen in d. DDR von Gerhard Bosinski u. Paul Toaspern ; [Porträtzeichn. von Werner Juza]. — 1. Aufl. — Berlin : Evangelische Verlagsanstalt, 1978.
440 p. : ports. ; 22 cm. GDR78-A
Includes bibliographical references.
14.80M
1. Christian biography—Germany. 2. Inner missions—Biography. 3. Church charities—Germany—Biography. I. Bosinski, Gerhard. II. Toaspern, Paul.
BR859.W47 79-382239
79 MARC

—INDIA

Karokaran, Anto, 1941-
Evangelization and diakonia / Anto Karokaran. — Bangalore : Dharmaram Publications, 1978.
xvii, 285 p. : map ; 23 cm.
"A study in the Indian perspective."
Bibliography: p. [265]-280.
Includes index.
Rs19.00 ($5.00 U.S.)
1. Evangelistic work—India. 2. Missions—India. 3. Church charities—India. 4. Church and social problems—India. 5. Theology, Doctrinal—India—History. I. Title.
BV3777.I4K37 269'.2'0954 78-905126
79 MARC

CHURCH CHRONOLOGY see Chronology, Ecclesiastical

CHURCH COLLEGES

see also Church and college

—CANADA—DIRECTORIES

White, Alex Sandri.
Guide to religious education : a directory of seminaries, Bible colleges, and theological schools covering the USA and Canada : includes resident and correspondence courses for all major faiths and denominations / by A. Sandri White. — New ed. — Allenhurst, N.J. : Aurea Publications, c1979.
85 p. ; 28 cm.
$7.95
1. Theological seminaries—United States—Directories. 2. Theological seminaries—Canada—Directories. 3. Church colleges—United States—Directories. 4. Church colleges—Canada—Directories. 5. Bible colleges—United States—Directories. 6. Bible colleges—Canada—Directories. I. Title.
BV4030.W46 1979 207'.73 79-107811
79 MARC

—UNITED STATES

Church related higher education : perceptions and perspectives / Robert Rue Parsonage, editor ; contributors, Merrimon Cuninggin ... [et al.]. — Valley Forge, PA : Judson Press, c1978.
344 p. ; 22 cm.
Includes bibliographical references.
ISBN 0-8170-0831-4 : $12.95
1. Church colleges—United States. 2. Higher education and state—United States. I. Parsonage, Robert Rue.
LC427.C48 377 78-10823
78 MARC

Danforth Foundation, St. Louis. Commission on Church Colleges and Universities.
Eight hundred colleges face the future; a preliminary report of the Danforth Commission on Church Colleges and Universities. Manning M. Pattillo, Jr. [director] Donald M. Mackenzie [associate director] St. Louis, Danforth Foundation, 1965.
74 p. 23 cm.
1. Church colleges—United States. I. Title.
LC368.D27 1965 377'.8 66-5553
[r79]rev MARC

—UNITED STATES—DIRECTORIES

White, Alex Sandri.
Guide to religious education : a directory of seminaries, Bible colleges, and theological schools covering the USA and Canada : includes resident and correspondence courses for all major faiths and denominations / by A. Sandri White. — New ed. — Allenhurst, N.J. : Aurea Publications, c1979.
85 p. ; 28 cm.
$7.95
1. Theological seminaries—United States—Directories. 2. Theological seminaries—Canada—Directories. 3. Church colleges—United States—Directories. 4. Church colleges—Canada—Directories. 5. Bible colleges—United States—Directories. 6. Bible colleges—Canada—Directories. I. Title.
BV4030.W46 1979 207'.73 79-107811
79 MARC

CHURCH COMMITTEES

CHURCH CONFERENCE CENTER

see also Parish councils; Retreats

CHURCH CONTROVERSIES

Dittes, James E
When the people say no : conflict and the call to ministry / James E. Dittes. — 1st ed. — San Francisco : Harper & Row, c1979.
ix, 150 p. ; 21 cm. — (Harper's ministers paperback library)
ISBN 0-06-061923-6
1. Church controversies. 2. Clergy—Office. 3. Pastoral theology. I. Title.
BV652.9.D57 1979 253 77-15879
79 MARC

Faber, Heije, 1907-
Kiezen of delen. Gedachten over polarisatie. [Door] H. Faber. Meppel, Boom, [1973].
136 p. 23 cm. fl 10.90 Ne73-47
Bibliography: p. 135-136.
ISBN 9060091116
1. Church controversies. 2. Pastoral psychology. 3. Polarization (Social sciences) I. Title.
BV652.9.F32 77-572880
78 MARC

Miller, John Michael, 1939-
The contentious community : constructive conflict in the church / by John M. Miller. — 1st ed. — Philadelphia : Westminster Press, c1978.
107 p. ; 21 cm.
ISBN 0-664-24198-0
1. Church controversies. I. Title.
BV652.9.M54 273 78-3682
78 MARC

CHURCH COSTUME see Church vestments

CHURCH COUNCILS see Councils and synods

CHURCH COURTS see Ecclesiastical courts

CHURCH DECORATION AND ORNAMENT

see also Altarpieces; Altars; Choir-stalls; Christian art and symbolism; Church furniture; Flower arrangement in churches; Glass painting and staining; Liturgical objects; Mosaics; Mural painting and decoration

Bentmann, Reinhard, 1939-
Kirchen des Mittelalters / Text von Reinhard Bentmann u. Heinrich Lickes. — Wiesbaden : Ebeling, 1978.
192 p. : 100 ill. (chiefly col.) ; 34 cm. GFR78-A
Bibliography: p. 191-192.
ISBN 3-921195-21-7 : DM48.00
1. Churches. 2. Architecture, Medieval. 3. Church architecture—Details. 4. Church decoration and ornament. I. Lickes, Heinrich, joint author. II. Title.
NA4800.B38 726'.5'094 78-372118
78 MARC

King, Bucky.
Ecclesiastical crafts / Bucky King, Jude Martin. — New York : Van Nostrand Reinhold, 1978.
132 p. : ill. (some col.) ; 24 cm.
Bibliography: p. 128-129.
Includes index.
ISBN 0-442-22966-6
1. Liturgical objects. 2. Church vestments. 3. Christian art and symbolism. 4. Church decoration and ornament. I. Martin, Jude, joint author. II. Title.
NK1650.K47 745.5 77-25033
77 MARC

—GERMANY

Koch, Rudolf, 1876-1934.
Das Kirchengerät im evangelischen Gottesdienst / von Rudolf Koch. — Hamburg : Agentur des Rauhen Hauses, [1933]
30 p. : 19 cm.
1. Liturgical objects. 2. Church decoration and ornament—Germany. I. Title.
BV195.K6 78-381536
78 MARC

—GERMANY, WEST—MARIENSTATT—ADDRESSES, ESSAYS, LECTURES

750 [i.e. Siebenhundertfünfzig] Jahre Abteikirche Marienstatt : Festschrift zur Kirchweihe 1977 / hrsg. von d. Mönchen d. Abtei Marienstatt. — Marienstatt : Buch- u. Kunstverl. Abtei Marienstatt, 1977.
150 p. : ill. ; 21 cm. — (Marienstatter Aufsätze ; 5) GFR77-B21
Includes bibliographical references.
DM8.50
1. Marienstatt, Ger. (Cistercian abbey). Abteikirche—Addresses, essays, lectures. 2. Church decoration and ornament—Germany, West—Marienstatt—Addresses, essays, lectures. 3. Marienstatt, Ger.—History—Addresses, essays, lectures. I. Marienstatt, Ger. (Cistercian abbey)
NA5586.M2683S56 78-377801
78 MARC

—NEW ENGLAND—EXHIBITIONS

Benes, Peter.
New England meeting house and church, 1630-1850 : a loan exhibition at the Currier Gallery of Art, Manchester, New Hampshire / Peter Benes & Philip D. Zimmerman. — Boston : Published by Boston University for the Dublin Seminar for New England Folklife, c1979.
xi, 177 p. : ill. ; 22 cm.
Bibliography: p. 110.
ISBN 0-87270-050-X : $7.95
1. Christian art and symbolism—Modern, 1500- —New England—Exhibitions. 2. Church decoration and ornament—New England—Exhibitions. 3. Churches, Friend—New England—Exhibitions. 4. Churches, Protestant—New England—Exhibitions. I. Zimmerman, Philip D., joint author. II. Currier Gallery of Art, Manchester, N.H. III. Dublin Seminar for New England Folklife. IV. Title.
N7905.B46 79-53115
709.948'2'097407401428
00 MARC

—QUÉBEC (PROVINCE)—BERTHIERVILLE, QUÉBEC

Les Églises et le trésor de Berthierville / [recherches et rédaction, Gaétan Chouinard ; relevés et dessins architecturaux, Monique Cloutier, Raynald Lemieux ; photos., Jean-Paul Body]. — Québec : Ministère des affaires culturelles, 1977, c1978.
22 p. : ill. ; 22 cm. — (Collection Les Retrouvailles ; no 1) C***
Bibliography: p. 22.
1. Église Sainte-Geneviève-de-Berthier, Berthierville, Québec. 2. Church decoration and ornament—Québec (Province)—Berthierville, Québec. I. Chouinard, Gaétan. II. Series.
NA5247.B47E35 1978 726'.5'0971443 79-350493
79 MARC

—QUÉBEC (PROVINCE)—QUÉBEC

Les Églises et le trésor de Saint-Pierre de la Rivière-du-Sud / [recherches et rédaction, Gaétan Chouinard ; photos., Jean-Paul Body ; relevé architectural, J. C. Zérounian, Raynald Lemieux]. — Québec : Ministère des affaires culturelles, 1978.
30 p. : ill. ; 22 cm. — (Collection Les Retrouvailles ; no 4) C***
Bibliography: p. 29-30.
1. Église Saint-Pierre-du-Sud, Saint-Pierre-de-la-Rivière-du-Sud, Québec. 2. Church decoration and ornament—Québec (Province)—Saint-Pierre-de-la-Rivière-du-Sud, Québec. I. Chouinard, Gaétan. II. Series.
NA5247.S33E35 726'.5'09714735 79-350492
79 MARC

—RUSSIA—EXHIBITIONS

Redin, Egor Kuz'mich, 1863–1908.
(Katalog dvi͡enadt͡satago Arkheologicheskago s"i͡ezda v g. Khar'kovi͡e)
Каталогъ XII Археологическаго съѣзда въ г. Харьковѣ : Отдѣлъ церковныхъ древностей / составилъ Е. К. Рѣдинъ. — Харьковъ : Тип. Губ. Правленія, 1902.
iii, 162, 28 p. ; 24 cm.
Includes index.
Exhibition catalog.
1. Church decoration and ornament—Russia—Exhibitions. 2. Orthodox Eastern Church and art—Exhibitions. 3. Christian antiquities—Russia—Exhibitions. I. Arkheologicheskiĭ s"ezd, 12th, Kharkov, 1903. II. Kharkov. Universitet. Istoriko-filologicheskoe obshchestvo. Otdel t͡serkovnykh drevnosteĭ. III. Title.
NK1653.R9R42 79–414620

CHURCH DECORATION AND ORNAMENT (Continued)

—SCOTLAND

Hay, George, 1911-
The architecture of Scottish post-Reformation churches, 1560-1843. Oxford, Clarendon Press, 1957.
xvi, 299 p. illus., 47 plates, plans. 25 cm.
Bibliography: p. 283-286.
1. Church architecture—Scotland. 2. Churches, Reformed—Scotland. 3. Church decoration and ornament—Scotland. I. Title.
NA5475.H3 726'.5'09411 57-2871
[r79]rev MARC

—SWEDEN—NORRLAND—HISTORY—18TH CENTURY

Telhammer, Ingrid.
Predikstolsmakare och predikstolskonst före 1777 i norra Hälsingland, Medelpad och Ångermanland / Ingrid Telhammer. — Umeå : Skytteanska samf., 1978.
248 p. : ill. ; 24 cm. — (Kungl. Skytteanska samfundets handlingar ; no. 20 ISSN 1560-2416) S78-34
Thesis—Umeå.
Summaries in English and German.
Bibliography: p. 236-241.
Includes indexes.
kr60.00
1. Pulpits—Sweden—Norrland. 2. Wood-carving, Baroque—Sweden—Norrland. 3. Wood-carving—Sweden—Norrland—History—18th century. 4. Church decoration and ornament, Baroque—Sweden—Norrland. 5. Church decoration and ornament—Sweden—Norrland—History—18th century. 6. Christian art and symbolism—Modern period, 1500- —Sweden—Norrland. 7. Wood-carvers—Sweden—Biography. I. Title. II. Series: Skytteanska samfundet. Handlingar ; no. 20.
DL601.S54 no. 20 748.5 s 79-362532
[NK9761.A3] 79 MARC

—SWITZERLAND—BINN VALLEY

Ruppen, Walter.
Das Binntal / Walter Ruppen. — Basel (Münzgässlein 16) : Gesellschaft für schweizerische Kunstgeschichte, c1977.
15 p. : ill. ; 20 cm. — (Schweizerische Kunstführer) Sw78-3746
Cover title.
Bibliography: p. 15.
1. Churches—Switzerland—Binn Valley. 2. Church decoration and ornament—Switzerland—Binn Valley. 3. Architecture—Switzerland—Binn Valley. 4. Historic buildings—Switzerland—Binn Valley. I. Title. II. Series.
NA5849.B562R86 914.94'7 78-377937
78 MARC

—VIRGINIA

Perdue-Davis, Vernon.
Virginia's ante-bellum churches : an introduction with particular attention to their furnishings / by Vernon Perdue Davis and James Scott Rawlings. — Richmond, Va. : Dietz Press, 1978.
xv, 129 p. : ill. ; 25 cm.
Includes index.
1. Church architecture—Virginia. 2. Architecture, Modern—19th century—Virginia. 3. Church decoration and ornament—Virginia. I. Rawlings, James Scott, joint author. II. Title.
NA5230.V8P47 726'.5'09755 79-100708
79 MARC

CHURCH DECORATION AND ORNAMENT, BAROQUE

—SILESIA

Heyer, Karl Johannes.
Das barocke Chorgestühl in Schlesien : e. Darst. d. Chorgestühle u.e. Beitr. zur Geschichte von Kunst u. Kunsthandwerk im Barock / von Karl Johannes Heyer ; mit e. musikgeschichtl. Beitr. von Johannes Aengenvoort. — Frankfurt am Main : Weidlich, 1977.
xvi, 247, [106] p. : 286 ill. ; 27 cm. — (Bau- und Kunstdenkmäler des deutschen Ostens : Reihe C, Schlesien ; Bd. 6 ISSN 0522-490X) GFR78-A37
A revision of the author's thesis, Breslau, 1929.
Bibliography: p. 226-231.
Includes indexes.
ISBN 3-8035-8609-7 : DM98.00
1. Wood-carving, Baroque—Silesia. 2. Choir-stalls, Baroque—Silesia. 3. Church decoration and ornament, Baroque—Silesia. I. Aengenvoort, Johannes. II. Title. III. Series.
NK9771.P6H49 1977 79-357549
79 MARC

—SWEDEN—NORRLAND

Telhammer, Ingrid.
Predikstolsmakare och predikstolskonst före 1777 i norra Hälsingland, Medelpad och Ångermanland / Ingrid Telhammer. — Umeå : Skytteanska samf., 1978.
248 p. : ill. ; 24 cm. — (Kungl. Skytteanska samfundets handlingar ; no. 20 ISSN 1560-2416) S78-34
Thesis—Umeå.
Summaries in English and German.
Bibliography: p. 236-241.
Includes indexes.
kr60.00
1. Pulpits—Sweden—Norrland. 2. Wood-carving, Baroque—Sweden—Norrland. 3. Wood-carving—Sweden—Norrland—History—18th century. 4. Church decoration and ornament, Baroque—Sweden—Norrland. 5. Church decoration and ornament—Sweden—Norrland—History—18th century. 6. Christian art and symbolism—Modern period, 1500- —Sweden—Norrland. 7. Wood-carvers—Sweden—Biography. I. Title. II. Series: Skytteanska samfundet. Handlingar ; no. 20.
DL601.S54 no. 20 748.5 s 79-362532
[NK9761.A3] 79 MARC

CHURCH DEDICATION

see also Corner stones, Laying of; Christian patron saints; Corner stones, Laying of

Catholic Church. Liturgy and ritual.
[De benedictione et impositione primarii lapidis pro ecclesia aedificanda]
Laying of the corner-stone, and Blessing of a church performed by a bishop, according to the Roman pontifical; with preparations, instructions, decrees. By Aurelius Stehle; rev. by Emmeran A. Rettger. Rev. ed. Latrobe, Pa., Archabbey Press, 1957.
31 p. 23 cm. .
Liturgical texts in Latin; editorial matter in English.
"The blessing of a church performed by a bishop (juxta Rituale Romanum, mutatis mutandis)": p. 19-27. Also issued separately.
Includes bibliographical references.
1. Corner stones, Laying of. 2. Church dedication. I. Stehle, Aurelius, 1877-1930. II. The Blessing of a church performed by a bishop. 1957. III. Title.
BX2031.D2 1957 57-12378
[r79]rev MARC

CHURCH DEDICATION SERMONS

Chang, Sŏng-sik.
(Ch'uksa taejŏn)
祝辭大典 / 張晟栻著. — 서울 : 聖光文化社 : 총판정음서림, 1978.
298 p. ; 20 cm.
W2000
1. Ordination sermons. 2. Installation sermons. 3. Church dedication sermons. 4. Occational sermons. 5. Sermons, Korean. I. Title.
BV4285.C46 79-825821

Zepeda, Pedro de.
Sermon qvinto qve dixo el M.R.P.M. Fr. Pedro de Zepeda ... à veinte de Abril dia de Santa Ynes de Monte Policiano. [Puebla, Mexico, 1690]
131-150 p. 19 cm.
Caption title.
Bound with Espinosa, J. de. Trivmpho festivo. [Puebla, 1690]
Detached, as is Espinosa's Triumpho, from the commemorative volume Octava maravilla del Nvevo Mundo.
1. Church dedication sermons. 2. Catholic Church—Sermons. 3. Sermons, Spanish. I. Octava maravilla del Nuevo Mundo. II. Title: Sermon quinto.
BX4610.P8S354 1690 78-365486
[BV4528] 78 MARC

CHURCH DISCIPLINE

see also Celibacy; Confession; Ecclesiastical courts; Inquisition; Penance; Purgatory; Visitations, Ecclesiastical; and subdivision Discipline under names of religions, religious denominations, etc., e.g. Buddhism—Discipline; Catholic Church—Discipline

CHURCH DISPUTES see Church controversies

CHURCH EMBROIDERY see Ecclesiastical embroidery

CHURCH ENTERTAINMENTS

Barker, Kenneth S 1932-
Seven dramatic moments in the life of Christ : plays for church events / by Kenneth S. Barker. — Atlanta : John Knox Press, c1978.
ca 200 p. ; 28 cm.
ISBN 0-8042-1432-8
1. Jesus Christ—Drama. 2. Church entertainments. I. Title.
PN6120.J35B37 812'.5'4 78-52443
78 MARC

CHURCH EXTENSION see Missions, Home

CHURCH FACILITIES

—DESIGNS AND PLANS

Kommunale und kirchliche Zentren = Community and church centres / mit Beitr. von Dieter Brezger ... [et al.]. — Stuttgart : Krämer, 1977.
4, xvi, 116 p. : numerous ill. ; 30 cm. — (Architektur-Wettbewerbe ; 92 ISSN 0341-2784) GFR78-A
"Wettbewerbsübersicht" [4] p.) inserted in pocket.
ISBN 3-7828-0192-X : DM23.50
1. Community centers—Designs and plans. 2. Church facilities—Designs and plans. 3. Architecture, Modern—20th century—Designs and plans. I. Brezger, Dieter. II. Title: Community and church centres.
NA4510.C7K65 78-389850
79 MARC

—PLANNING

Norske kirke. Bispedømmerådenes fellesråd.
Utfordringer og behov : håndbok i kirkelig struktureringsarbeid / Bispedømmerådenes fellesråd, Kirkerådet. — Oslo : Kirkerådet, 1976.
103 p. : ill. ; 30 cm. N78-Sept.
Bibliography: p. 99-103.
1. Church facilities—Planning. I. Title.
BV604.N67 1976 79-375356
79 MARC

Smith, Roland A
Before you build your church / Roland A. Smith. — Nashville : Broadman Press, c1979.
64, [1] p. ; 19 cm.
Bibliography: p. [65]
ISBN 0-8054-3511-5
1. Church facilities—Planning. I. Title.
BV604.S64 254'.7 78-73134
79 MARC

CHURCH FATHERS see Fathers of the church

CHURCH FESTIVALS see Fasts and feasts

CHURCH FIGHTS see Church controversies

CHURCH FINANCE

see also Christian giving; Church charities; Church fund raising; Church lands; Stewardship, Christian; Taxation, Exemption from; and subdivision Finance under church denominations, and under special topics, e.g. Church of England—Finance; Missions—Finance

Cunningham, Richard B
Creative stewardship / Richard B. Cunningham. — Nashville : Abingdon, c1979.
128 p. ; 20 cm. — (Creative leadership series)
Includes bibliographical references.
ISBN 0-687-09844-0
1. Stewardship, Christian. 2. Christian giving. 3. Church finance. I. Title. II. Series.
BV772.C779 248'.6 79-973
79 MARC

Swenson, Vernon, 1923-
Basic bookkeeping for young churches : a programmed study guide for individual or classroom use / Vernon Swenson. — [Pasadena, Calif. : William Carey Library], c1979.
118, [40] p. : ill. ; 28 cm.
ISBN 0-87808-170-4
1. Church finance. I. Title.
[BV773.S95] 657'.832 79-15812
79 MARC

CHURCH FINANCE (Continued)

White, John, 1924 (Mar. 5)-
The golden cow : materialism in the twentieth-century church / John White. — Downers Grove, Ill. : Inter-Varsity Press, c1979.
175 p. ; 21 cm.
ISBN 0-87784-490-9 : $3.50
1. Christianity—20th century. 2. Church finance. I. Title.
BR121.2.W45 270.8'2 78-13884
79 MARC

CHURCH FLOWER ARRANGEMENT
see Flower arrangement in churches

CHURCH FUND RAISING

McGinty, John, 1911-
How to raise the level of giving in your church / by John McGinty. — St. Louis : Bethany Press, c1979.
111 p. ; 23 cm.
ISBN 0-8272-1418-9 : $5.50
1. Church fund raising. 2. Stewardship, Christian. I. Title.
BV772.5.M82 254.8 78-12994
78 MARC

Weenink, Allan J
The art of church canvass / Allan J. Weenink ; ⌜introd. by Manfred Holck, Jr.⌝. — New Canaan, Conn. : Keats Pub., 1978.
142 p. ; 21 cm.
Bibliography: p. 137-140.
Includes index.
ISBN 0-87983-172-3 : $4.95
1. Church fund raising. I. Title.
BV772.5.W4 254.8 78-69998
79 MARC

—HISTORY

Kraus, Henry, 1905-
Gold was the mortar : the economics of cathedral building / Henry Kraus. — London ; Boston : Routledge & Kegan Paul, 1979.
292 p. : ill. ; 26 cm. GB***
ISBN 0-7100-8728-4 : £12.50
1. Cathedrals—France. 2. Cathedrals—England—York. 3. Church fund raising—History. I. Title.
BX4629.A1K7 282'.44 78-40526
79 MARC

CHURCH FURNITURE

see also Altarpieces; Altars; Choir-stalls; Pulpits

—BELGIUM—ANTWERP (PROVINCE) —CATALOGS

Brussels. Institut royal du patrimoine artistique.
Fotorepertorium van het meubilair van de Belgische bedehuizen : Provincie Antwerpen : Kanton ... / Jaak Jansen. — Brussel : Koninklijk Instituut voor het Kunstpatrimonium, 1973-
v. ; 23 cm. Be***
Cover title.
Catalog of photographs on file in the Département des archives photographiques of the Institut.
Consists of one or more vols. for each canton in the province.
1. Church furniture—Belgium—Antwerp (Province)—Catalogs. I. Jansen, Jaak. II. Brussels. Institut royal du patrimoine artistique. Département des archives photographiques. III. Title.
NA5669.A5B78 1973 78-378589
78 MARC

—BELGIUM—FLANDERS, EAST —CATALOGS

Brussels. Institut royal du patrimoine artistique.
Fotorepertorium van het meubilair van de Belgische bedehuizen : Provincie Oost-Vlaanderen : Kanton ... / Christina Vandenbussche-Van den Kerkhove, Hugonne Verschraegen. — Brussel : Koninklijk Instituut voor het Kunstpatrimonium, 1975-
v. ; 23 cm. Be***
Cover title.
Catalog of photographs on file in the Département des archives photographiques of the Institut.
Consists of one or more vols. for each canton in the province.
1. Church furniture—Belgium—Flanders, East—Catalogs. I. Vandenbussche-Van den Kerkhove, Christina. II. Verschraegen, H. III. Brussels. Institut royal du patrimoine artistique. Département des archives photographiques. IV. Title.
NA5669.E2B78 1975 78-378592
78 MARC

—BELGIUM—FLANDERS, WEST —CATALOGS

Brussels. Institut royal du patrimoine artistique.
Fotorepertorium van het meubilair van de Belgische bedehuizen : Provincie West-Vlaanderen : Kanton ... / Agnes de Vos-Stockman, Bernadette Roose. — Brussel : Koninklijk Instituut voor het Kunstpatrimonium, 1973-
v. ; 23 cm. Be***
Cover title.
Catalog of photographs on file in the Département des archives photographiques of the Institut.
Consists of one or more vols. for each canton in the province.
1. Church furniture—Belgium—Flanders, West—Catalogs. I. Vos-Stockman, Agnes de. II. Roose, Bernadette. III. Brussels. Institut royal du patrimoine artistique. Département des archives photographiques. IV. Title.
NA5669.W4B78 1973 78-378590
78 MARC

CHURCH GOVERNMENT see Church polity

CHURCH GROUP WORK

Anstösse zum Umgang mit biblischen Texten in Gruppen / hrsg. vom Bayer. Mütterdienst d. Evang.-Luth. Kriche in Bayern, Stein bei Nürnberg ; ⌜Red. u. Ms., Wolfram Jokisch, unter Mitarb. von Herta Atzkern ... et al.⌝. — Als Ms. gedr. — Stein/Mfr. : Laetare-Verlag, 1977.
128 p. ; 21 cm. — (Steiner Arbeitshilfen ; 2) GFR78-A
Bibliography: p. 127-128.
ISBN 3-7839-3016-2 : DM10.00
1. Bible—Study. 2. Church group work. I. Jokisch, Wolfram. II. Atzkern, Herta. III. Evangelisch-Lutherische Kirche in Bayern. Bayerischer Mütterdienst. IV. Series.
BS600.2.A55 78-389331
79 MARC

Bonham, Betty, 1926-
Open the door to the Bible / Betty Bonham. — Minneapolis : Bethany Fellowship, c1979.
127 p. ; 21 cm.
Includes bibliographical references.
ISBN 0-87123-415-7 : $2.95
1. Bible—Study. 2. Church group work. I. Title.
BS600.2.B66 220'.07 79-16734
79 MARC

Geiger, Lura Jane.
Finding hidden treasure : TA groups in the church / Lura Jane Geiger ; ill. by Sara Travis. — Sacramento, Calif. : Jalmar Press, c1979.
xiii, 103 p. : ill. ; 26 cm.
Bibliography: p. 99-103.
ISBN 0-915190-16-8
1. Transactional analysis. 2. Church group work. I. Title.
RC489.T7G44 616.8'914 78-71607
79 MARC

Posterski, Don.
The Bible meets you where you are ... : personal and group studies for using the Bible in everyday life / Don Posterski. — Waco, Tex. : Word Books, c1979.
140 p. ; 24 cm.
ISBN 0-8499-2857-5 : $4.95
1. Bible—Study. 2. Church group work. 3. Bible—Meditations. I. Title.
BS600.2.P67 78-65803
00 MARC

Roberts, Clarence.
Sharing of scripture / by Clarence and Edith Roberts. — New York : Paulist Press, c1978.
220 p. ; 19 cm. — (A Deus book)
ISBN 0-8091-2141-7 : $2.95
1. Bible—Study—Catholic Church. 2. Church group work. I. Roberts, Edith, joint author. II. Title.
BS587.R6 220'.07 78-61724
79 MARC

Vigeveno, H S
The group / H. S. Vigeveno. — Santa Ana, Calif. : Vision House, c1978.
216 p. ; 21 cm.
ISBN 0-88449-073-4
1. Church group work. 2. Pastoral counseling. I. Title.
BV652.2.V5 252.5 77-86533
78 MARC

CHURCH GROWTH

see also Small churches

Bartel, Floyd G
A new look at church growth / by Floyd G. Bartel, with Richard Showalter. — Newton, Kan. : Faith and Life Press, c1979.
viii, 143 p. ; 21 cm.
Includes bibliographical references.
ISBN 0-87303-027-3
1. Church growth. I. Showalter, Richard, joint author. II. Title.
⌜BV652.25.B37⌝ 254'.5 79-53523
00 MARC

DuBose, Francis M
How churches grow in an urban world / Francis M. DuBose. — Nashville : Broadman Press, c1978.
181 p. ; 20 cm.
Bibliography: p. 172-175.
Includes index.
ISBN 0-8054-2531-4
1. Church growth. 2. City churches. I. Title.
BV652.25.D8 254'.5 78-66365
79 MARC

Gibbs, Eddie.
Urban church growth : clues from South America and Britain / by Eddie Gibbs. — Bramcote : Grove Books, 1977.
24 p. ; 22 cm. — (Grove booklet on ministry and worship ; no. 55 ISSN 0305-3067) GB78-08798
Includes bibliographical references.
ISBN 0-905422-24-4 : £0.35
1. Church growth. 2. City churches. I. Title.
BV652.25.G5 1977 254.2'2 78-314491
78 MARC

Hunter, George G
The contagious congregation : frontiers in evangelism and church growth / George G. Hunter, III. — Nashville : Abingdon, c1979.
160 p. ; 20 cm.
Includes bibliographical references.
ISBN 0-687-09490-9
1. Evangelistic work. 2. Church growth. I. Title.
BV3790.H89 254'.5 78-12322
78 MARC

Miller, Calvin.
A view from the fields / Calvin Miller. — Nashville : Broadman Press, c1978.
136 p. ; 20 cm.
ISBN 0-8054-5520-5
1. Evangelistic work. 2. Church growth. I. Title.
BV3793.M5 253.7 78-62399
79 MARC

Morikawa, Jitsuo.
Biblical dimensions of church growth / Jitsuo Morikawa. — Valley Forge, PA : Judson Press, c1979.
95 p. ; 22 cm.
Includes bibliographical references.
ISBN 0-8170-0839-X : $3.95
1. Church growth. I. Title.
BV652.25.M67 262'.7 79-4040
79 MARC

Redford, Jack.
Planting new churches / Jack Redford. — Nashville : Broadman Press, c1978.
152 p. : ill. ; 21 cm.
ISBN 0-8054-6314-3
1. Evangelistic work. 2. Church growth. I. Title.
BV3793.R38 254 78-55694
79 MARC

Thomas, Roy L
Planting and growing a fundamental church / by Roy L. Thomas. — Nashville : Randall House Publications, 1979.
270 p. ; 23 cm.
Bibliography: p. 259-270.
ISBN 0-89265-055-9 : $7.95
1. Church growth. 2. Evangelistic work. 3. Theology, Practical. I. Title.
BV652.25.T47 253 79-113960
00 MARC

Wagner, C Peter.
Your church can be healthy / C. Peter Wagner. — Nashville : Abingdon, c1979.
128 p. ; 20 cm. — (Creative leadership series)
Includes bibliographical references and index.
ISBN 0-687-46870-1
1. Church growth. I. Title. II. Series.
BV652.25.W33 253'.5 79-974
79 MARC

Watkins, Morris, 1923-
Literacy, Bible reading, and church growth through the ages / by Morris Watkins. — South Pasadena, Calif. : William Carey Library, c1978.
xiii, 224 p. : ill. ; 22 cm.
Bibliography: p. 189-210.
Includes index.
ISBN 0-87808-325-1
1. Missions—History. 2. Church growth. 3. Bible—Reading. 4. Illiteracy. I. Title.
BV2105.W33 270 78-15315
78 MARC

—ADDRESSES, ESSAYS, LECTURES

Understanding church growth an decline, 1950-1978 / edited by Dean R. Hoge and David A. Roozen. — New York : Pilgrim Press, c1979
398 p. : ill. ; 23 cm.
Bibliography: p. 374-391.
Includes index.
ISBN 0-8298-0358-0 (pbk.) : $8.95
1. Church membership—Addresses, essays, lectures. 2. Church growth—Addresses, essays, lectures. I. Hoge, Dean R., 1937- II. Roozen, David A.
BV820.U5 301.5'8 79-4166
79 MARC

CHURCH GROWTH (Continued)

—CASE STUDIES

Walrath, Douglas Alan, 1933-
Leading churches through change / Douglas Alan Walrath. — Nashville : Abingdon, c1979.
124 p. ; 21 cm. — (Creative leadership series) (An Original Abingdon paper-back)
ISBN 0-687-21270-7
1. Church growth—Case studies. 2. Pastoral theology—Case studies. I. Title.
BV652.25.W34 254'.5 79-4456
79 MARC

—CONGRESSES

Explaining the Gospel in today's world. — London : Scripture Union [for the] Lausanne Committee for World Evangelization, [1978]
48 p. ; 22 cm. — (Lausanne occasional papers ; 1 & 2) GB78-23978
Reports based on meetings held May 31-June 2, 1977, Pasadena, Calif. and in Jan. 1978 sponsored by the Lausanne Committee's Strategy Working Group and Theology and Education Group.
Bibliography: p. 10.
CONTENTS: Church planting.—Gospel & culture.
ISBN 0-85421-720-7 : £0.75
1. Missions—Theory—Congresses. 2. Church growth—Congresses. 3. Evangelistic work—Congresses. I. Lausanne Committee for World Evangelization. Strategy Working Group. II. Lausanne Committee for World Evangelization. Theology and Education Group. III. Series.
BV2063.E9 269'.2 79-312052
79 MARC

National Church Growth Convention, Kansas City, Mo., 1978.
Higher goals : National Church Growth Convention digest / editorial committee, Gwen Jones, chairman, Ron Rowden, Mel Surface. — Springfield, Mo. : Gospel Pub. House, c1978.
viii, 346 p. ; 22 cm.
"A condensation of plenary sessions and seminars conducted at the National Church Growth Convention in Kansas City, Missouri, August 15-17, 1978."
1. Church growth—Congresses. I. Jones, Gwen, 1911- II. Rowden, Ron. III. Surface, Mel. IV. Title.
BV652.25.N37 1978 254'.5 78-74529
00 MARC

—INDIA

McGavran, Donald Anderson, 1897-
Ethnic realities and the church : lessons from India / Donald A. McGavran. — South Pasadena, Calif. : William Carey Library, c1979.
262 p. ; 22 cm.
Bibliography: p. 260-262.
ISBN 0-87808-168-2 : $8.95
1. Church growth—India. 2. Missions—India. 3. Missions—Sociological aspects. 4. Ethnology—India. I. Title.
BV652.25.M27 301.5'8 78-11517
78 MARC

—INDIA—MADRAS

Sargunam, M Ezra.
Multiplying churches in modern India : an experiment in Madras / by M. Ezra Sargunam. — 1st ed. — [Madras?] : Federation of Evangelical Churches in India ; Madras : copies can be had from ECI, Literature Dept., 1974.
v, 183, [1] p. : ill., maps ; 23 cm.
Originally presented as the author's thesis (M.A.), Fuller Theological Seminary, 1973, under the title: Multiplying churches in urban India.
Bibliography: p. 179-[184]
1. Church growth—India—Madras. 2. Oriental Missionary Society—History. 3. Missions—India—Madras. 4. Madras—Church history. I. Title.
BR1175.M3S27 1974 275.4 78-905973
79 MARC

—UNITED STATES

Wagner, C Peter.
Our kind of people : the ethical dimensions of church growth in America / by C. Peter Wagner. — Atlanta : J. Knox Press, c1979.
184 p. : ill. ; 21 cm.
Includes bibliographical references and indexes.
ISBN 0-8042-0838-7
1. Church growth—United States. 2. Ethnicity—Moral and religious aspects. 3. McGavran, Donald Anderson, 1897- I. Title.
BR526.W25 262'.22 77-15754
78 MARC

Williamson, Wayne B 1918-
Growth & decline in the Episcopal Church / Wayne B. Williamson. — Pasadena, Calif. : William Carey Library, 1979.
xii, 179 p. : ill. ; 28 cm.
Bibliography: p. 174-176.
ISBN 0-87808-328-6
1. Protestant Episcopal Church in the U.S.A. 2. Church growth—United States. I. Title.
BX5933.W55 283'.73 79-12303
79 MARC

CHURCH HISTORIANS

—BIOGRAPHY

Mailhet, E André.
Jacques Basnage théologien, controversiste, diplomate et historien : sa vie et ses écrits / E.-André Mailhet. — Genève : Mégariotis, 1978.
86 p. ; 22 cm. Sw78-3783
Reprint of the 1880 ed. published by C. Schuchardt, Geneva.
Includes bibliographical references.
20.00F
1. Basnage, Jacques, sieur de Beauval, 1653-1723. 2. Church historians—Biography. 3. Huguenots—Biography. I. Title.
BR139.B36M34 1978 230'.4'50924 78-383156
78 MARC

—NETHERLANDS—BIOGRAPHY

Kamerling, J
De kerkhistoricus Willem Moll (1812-1879) : fragmenten uit de correspondentie / J. Kamerling. — Leeuwarden : Tille, [1979]
98 p. : with ill. ; 22 cm. Ne79-21
Includes bibliographical references.
ISBN 9070010763 : fl 19.50
1. Moll, Willem, 1812-1879. 2. Church historians—Netherlands—Biography. I. Moll, Willem, 1812-1879. II. Title.
BR139.M57K35 79-380227
*79 MARC

CHURCH HISTORY

see also Abbeys; Church and state; Convents and nunneries; Creeds; Dioceses; Ecclesiastical geography; Episcopacy; Fathers of the church; Inquisition; Monasticism and religious orders; Papacy; Protestant churches;

Protestantism; Revivals; and also headings beginning with the word Christian; names of denominations, sects, churches, councils, etc.; and subdivision Church history under names of countries, cities, etc.

Agnoletto, Attilio.
Storia del cristianesimo / Attilio Agnoletto. — Milano : Istituto propaganda libraria, 1978.
428 p. ; 22 cm. — (Il Sestante) It79-June
Bibliography: p. 391-413.
Includes index.
L10000
1. Church history. I. Title.
BR150.A36 79-371397
*79 MARC

Bednarczyk, Krzysztof.
Formuła mocy / Krzysztof Bednarczyk. — Wyd. 1. — Warszawa : Słowo Prawdy, 1977 (Pozn. : Pallottinum)
220, [4] p. : ill. ; 17 cm. P78-929
Bibliography: p. 220.
1. Christianity—20th century. 2. Church history. 3. Christian life—1960- . I. Title.
BR121.2.B36 79-341547
79 MARC

Bible. German. Einheitsübersetzung. 1975-76.
Die Bibel : Altes und Neues Testament in neuer Einheitsübersetzung / hrsg. v. Mirjam Prager OSB u. Günter Stemberger. — Salzburg : Andreas, 1975-
v. : ill. ; 30 cm. & lexicon (112 p.) Au76-3-26
Vols. 1-5 contain the text of the Bible with a commentary which was published in French under title: La bible.
Vols. -7 have title: Die Bibel und das Christentum.
ISBN 3-85012-037-6 : S3990.00. ISBN 3-85012-038-4 (de luxe)
1. Church history. 2. Religions. 3. Bible—Dictionaries—German. I. Prager, Mirjam, 1906- II. Stemberger, Günter, 1940- . III. Title.
BS237 1975b 220.5'3'1 76-485526
*76[r79]rev MARC

Booty, John E
The church in history : written by John E. Booty ; with the assistance of a group of editorial advisors under the direction of the Church's teaching series committee. — New York : Seabury Press, 1979.
xiii, 297 p. : ill. ; 21 cm. — (The Church'steaching series ; v. 3)
Includes bibliographical references and index.
ISBN 0-8164-0420-8 : $9.50
1. Church history.
BR145.2.B66 270 78-26014
78 MARC

González, Justo L
Y hasta lo último de la tierra : una historia ilustrada del cristianismo / Justo L. González. — Miami, Fla. : Editorial Caribe, c19
v. : ill. ; 22 cm.
CONTENTS:
—v. 3. La era de las tinieblas.
1. Church history. I. Title.
BR145.2.G66 79-114394
79 MARC

Ishihara, Ken, 1882-
(Kirisutokyō shi)
基督教史 / 石原謙. — 増訂版. — 東京：岩波書店, 昭和23 [1948]
[7], 302, 11 p. ; 18 cm. — (岩波全書 ; 32)
Colophon inserted.
Series romanized: Iwanami zensho.
Bibliography: p. 6–8 (4th group)
Includes index.
1. Church history. I. Title.
BR145.I78 78–817200

Ishihara, Ken, 1884–
(Kirisutokyō shi)
キリスト教史 / [著者 石原謙]. — 東京：岩波書店, 1979.
8, 404 p. ; 22 cm. — (石原謙著作集 ; 第7巻) Ja***
Series romanized: Ishihara Ken chosaku shū.
Includes bibliographies.
CONTENTS: 基督教史—キリスト教思想史
¥4500
1. Church history. 2. Theology, Doctrinal—History. I. Ishihara, Ken, 1884– Kirisutokyō shisō shi. 1979. II. Title.
BR83.I8 vol. 7 79–809527
[BR145.2]

Kee, Howard Clark.
Christianity / by Howard Clark Kee. — Niles, Ill. : Argus Communications, c1979.
ix, 117 p. : col. ill. ; 23 cm. — (Major World religions series)
"Study questions for Christianity": [4] p. inserted.
Bibliography: p. 117.
ISBN 0-89505-021-8 : $2.95
1. Church history.
BR150.K43 270 78-68769
79 MARC

Khonglah, J
Ki khanatang jong ka balang Kristan / [J. Khonglah]. — [Shillong : Don Bosco Press], 1971.
175 p. ; 20 cm.
In Khasi.
Rs2.30
1. Church history. 2. Christianity—Influence—India. I. Title.
BR145.2.K48 78-903869
79 MARC

Moeller, Bernd, 1931-
Geschichte des Christentums in Grundzügen / Bernd Moeller. — 2., völlig neubearb. Aufl. — Göttingen : Vandenhoeck und Ruprecht, 1979.
414 p. ; 19 cm. — (Uni-Taschenbücher ; 905 : Theologie) GFR79-A
Includes bibliographical references and index.
ISBN 3-525-03253-6 : DM26.80
1. Church history. I. Title.
BR145.2.M6 1979 79-373529
79 MARC

Pak, Yŏng-gwan, 1941–
(Yŏksa sinhak kangŭi)
歷史神學講義 / 朴英官著. — 서울：예수교문서선교회, 1977, 1978 printing.
349 p. ; 22 cm.
On verso of t. p.: Lectures on the historical theology.
Bibliography: p. 329–335.
Includes index.
₩2500
1. Theology, Doctrinal—History. 2. Church history. I. Title. II. Title: Lectures on the historical theology.
BT21.2.P32 78–826326

Pierrard, Pierre.
Histoire de l'Église catholique / Pierre Pierrard. — Nouvelle éd. .... — Paris : Desclée, 1978.
327 p. ; 22 cm. F78-8504
Bibliography: p. [303]-307.
Includes index.
ISBN 2-7189-0117-9 : 49.00F
1. Catholic Church—History. 2. Church history. I. Title.
BX945.2.P46 1978 282 78-386143
79 MARC

Richardson, Cyril Charles, 1909-1976.
The church through the centuries / by Cyril Charles Richardson. — New York : AMS Press, 1979, c1938.
x, 255 p. ; 23 cm.
Reprint of the ed. published by Scribner, New York.
Bibliography: p. 247-248.
Includes index.
ISBN 0-404-10645-5
1. Church history. 2. Religious thought—History. I. Title.
BR148.R56 1979 270 72-6726
77 MARC

Roberts, Brigham Henry, 1857-1933.
Outlines of ecclesiastical history : a text book / by B. H. Roberts. — Salt Lake City : Deseret Book Co., 1979.
458 p. ; 24 cm. — (Classics in Mormon literature)
Includes bibliographical references and index.
ISBN 0-87747-748-5 : $4.95
1. Church history. 2. Mormons and Mormonism. I. Title. II. Series.
BR148.R6 1979 289.3 79-9744
79 MARC

CHURCH HISTORY (Continued)

Vergilius, Polydorus, d. 1555.
De rerum inuentoribus libri octo, per authroem quarto iam, ac diligentius recogniti & locupletati. Eiusdem in Dominicam precem commentariolum. Basileae ⌊J. Bebell⌋ 1536.
590 p. 18 cm.
Bebel's device on t.p. and at end.
1. Civilization—History. 2. Church history. I. Title.
CB23.V47 78-413636
78 MARC

—PRIMITIVE AND EARLY CHURCH, CA. 30-600

see also Arianism; Catacombs; Christianity and culture—Early church, ca. 30-600; Church polity —Early church, ca. 30-600; Fathers of the church; Jewish Christians —Early church, ca. 30-600; Monasticism and religious orders—Early church, ca. 30-600; Neoplatonism; Persecution —Early church, ca. 30-600

Arnot, William, 1808-1875.
Studies in Acts : the church in the house / by William Arnot. — Grand Rapids, Mich. : Kregel Publications, c1978.
x, 464 p. ; 21 cm.
Reprint of the 1883 ed. published by R. Carter, New York, under title: The Church in the house; with new pref.
ISBN 0-8254-2120-9 : $10.95
1. Bible. N.T. Acts—Criticism, interpretation, etc. 2. Church history—Primitive and early church, ca. 30-600. I. Title. II. Title: The church in the house.
BS2625.A7 1978 226'.6'07 78-59141
78 MARC

Bowder, Diana.
The age of Constantine and Julian / Diana Bowder. — London : Elek, 1978.
xiii, 230 p., ⌊16⌋ leaves of plates : geneal. tab., map, ports. ; 25 cm. GB***
Bibliography: p. 194-212.
Includes index.
ISBN 0-236-30922-6 : £12.50
1. Rome—History—Empire, 284-476. 2. Church history—Primitive and early church, ca. 30-600. I. Title.
DG315.B68 937'.08 78-320163
79 MARC

Bowder, Diana.
The age of Constantine and Julian / Diana Bowder. — New York : Barnes & Noble Books, 1978.
xiii, 230 p., ⌊16⌋ leaves of plates : ill. ; 25 cm.
Includes index.
Bibliography: p. 194-212.
ISBN 0-06-490601-9
1. Rome—History—Empire, 284-476. 2. Church history—Primitive and early church, ca. 30-600. I. Title.
DG315.B68 1978 937'.08 79-112742
00 MARC

Bruijne, P J O de.
Vaste grond : de Club van Rome en het uitzicht van de bijbel / P. J. O. de Bruijne. — Franeker : Wever, ⌊1978⌋
154 p. ; 22 cm. Ne78-27
ISBN 9061352649 : fl 17.50
1. Faith. 2. Church history—Primitive and early church, ca. 30-600. I. Title.
BT771.2.B78 79-346041
*79 MARC

Chrēstou, Panagiōtēs, 1917-
(Hellēnikē patrologia)
Ἑλληνικὴ πατρολογία / Παναγ. Κ. Χρήστου. — Θεσσαλονίκη : Πατριαρχικὸν Ἵδρυμα Πατερικῶν Μελετῶν, 1976-
v. : ill. ; 25 cm. — (Χριστιανικὴ γραμματολογία ; 2)
Bibliography: v. 1, p. ⌊355⌋-392.
Includes index.
CONTENTS: τ. 1. Εἰσαγωγή.
1. Church history—Primitive and early church, ca. 30-600. 2. Orthodox Eastern Church—History. 3. Fathers of the church Greek. I. Title. II. Series: Christianikē grammatologia ; 2.
BR160.C46 1976 78-419990

Colloque d'études historiques et patristiques, 4th, Chantilly, 1976.
Lactance et son temps : recherches actuelles : actes du IVe Colloque d'études historiques et patristiques, Chantilly, 21-23 september 1976 / édités par J. Fontaine et M. Perrin. — Paris : Éditions Beauchesne, c1978.
312 p. ; 22 cm. — (Théologie historique ; 48) F***
Includes bibliographical references.
87.00F
1. Lactantius, Lucius Caecilius Firmianus—Congresses. 2. Church history—Early church, ca. 30-600. I. Fontaine, Jacques. II. Perrin, Michel. III. Title. IV. Series.
BR65.L26C65 1976 270.1'092'4 78-385252
79 MARC

Eusebius Pamphili, Bp. of Caesarea.
Historia ecclesiastica. Latin. Rome, Joannes Philippus de Lignamine, 15 May, 1476.
⌊220⌋ l. fo. 34 cm.
Translation by Rufinus Tyrannius.
Two issues exist of this ed., the first with a dedication to Pope Sixtus IV, the second with one to Cardinal D'Estouteville. This substitution necessitated a reprinting of the first signature which shows slight textual differences. The register corresponds to the first issue. L. C. copy is of the second issue. However, the first 2 leaves (a blank and the dedication to the cardinal) are wanting, replaced by a ms. dedication, in a contemporary hand, to the pope (3 leaves).
Brit. Mus. Cat. (XV cent.) IV, p. 34 (IB.17419); Gesamtkat. d. Wiegendr., 9436 variant; Goff. Third census, E-126.
Large initial and dependent border illuminated in gold and color on leaf ⌊3a⌋, other initials supplied in red or blue throughout (including the ms. part).
1. Church history—Primitive and early church, ca. 30-600. I. Rufinus Tyrannius, Aquileiensis, 345 (ca.)-410, tr.
Incun. 1476.E916 74-227802
⌊BR165⌋ 78⌊r78⌋rev MARC

Fabro, Nando.
Volto e ventura di un'amicizia : i cristiani nel tempo / Nando Fabro. — Roma : A.V.E., c1974.
215 p. ; 19 cm. — (Ave minima ; 85) It75-Sept
Includes bibliographical references.
L2200
1. Church history—Primitive and early church, ca. 30-600. 2. Friendship. 3. Catholic Church—History. I. Title.
BR165.F17 78-419162
*79 MARC

Geffcken, Johannes, 1861-1935.
The last days of Greco-Roman paganism / by Johannes Geffcken ; translated by Sabine MacCormack. — Amsterdam ; New York : North Holland Pub. Co. ; New York : Distributors for the U.S.A. and Canada, Elsevier North-Holland, 1978.
xii, 343 p. ; 23 cm. — (Europe in the Middle Ages ; v. 8)
Translation of Der Ausgang des griechisch-römischen Heidentums.
This translation of the 1929 rev. ed. of the original 1920 ed. has up-dated references and index added by S. MacCormack.
Includes bibliographical references and index.
ISBN 0-444-85005-8
1. Paganism. 2. Church history—Primitive and early church, ca. 30-600. I. Title. II. Series.
BR166.G4313 291 77-23029
77 MARC

Haehling, Raban von.
Die Religionszugehörigkeit der hohen Amtsträger des Römischen Reiches seit Constantins I. Alleinherrschaft bis zum Ende der Theodosianischen Dynastie : (324-450 bzw. 455 n. Chr.) / von Raban von Haehling. — Bonn : Habelt, 1978.
xii, 680 p. ; 28 cm. — (Antiquitas : Reihe 3 (Serie in 4to), Abhandlungen zur Vor- und Frühgeschichte, zur klassischen und provinzial-römischen Archäologie und zur Geschichte des Altertums ; Bd. 23) GFR78-A
A revision of the author's thesis, Bonn, 1975.
Bibliography: p. 664-680.
Includes index.
ISBN 3-7749-1414-1
1. Roman emperors—Religious life. 2. Rome—Officials and employees—Religious life. 3. Church history—Primitive and early church, ca. 30-600. 4. Rome—History—Empire, 284-476. I. Title. II. Series.
DG274.H33 1978 78-359263
78 MARC

Haendler, Gert.
Von Tertullian bis zu Ambrosius : d. Kirche im Abendland vom Ende d. 2. bis zum Ende d. 4. Jh. / Gert Haendler. — 1. Aufl. — Berlin : Evangelische Verlagsanstalt, 1978.
135 p. ; 25 cm. — (Kirchengeschichte in Einzeldarstellungen : 1, pt. 3) GFR79-A8
Bibliography: p. 7-17.
9.00M
1. Church history—Primitive and early church, ca. 30-600. I. Title. II. Series.
BR145.2.K54 vol. 1, pt. 3 79-367888
⌊BR165⌋ 79 MARC

Hahn, Georg, Dr.
Die Kirche der Märtyrer und Katakomben, in Werkstunden zusammengestellt von Dr. Georg Hahn ; mit 47 Werkbildern und entsprechenden Erklärungen als Beilagen. Freiburg im Breisgau, Herder & Co., g.m.b.h.b., 1939.
xi, 474 p. 47 pl. (in pocket) 23 cm. (His Werkbuch der Kirchengeschichte 1. t)
"Literaturverzeichnis: p. ⌊470⌋-474.
1. Church history—Primitive and early church, ca. 30-600. I. Title.
BR4165.H2 270.1 39-23774
⌊r79⌋rev MARC

Hinson, E Glenn.
Jesus Christ / E. Glenn Hinson. — ⌊Wilmington, N.C. : Consortium Books⌋, c1977.
187 p. ; 23 cm. — (Faith of our fathers ; v. 1)
ISBN 0-8434-0620-8 : $9.50. ISBN 0-8434-0633-X pbk.
1. Jesus Christ—Person and offices. 2. Church history—Primitive and early church, ca. 30-600. I. Title. II. Series.
BT202.H53 232 77-9548
77⌊r79⌋rev MARC

Ishihara, Ken, 1882-
(Kirisutokyō no genryū)
キリスト教の源流 / ⌊著者　石原謙⌋. — 東京：岩波書店, 1979.
33, 624 p. ; 22 cm. — (His ヨーロッパ・キリスト教史 ; 1巻) (石原謙著作集 ; 第8巻) Ja***
Series 2 romanized: Ishihara Ken chosaku shū.
Includes bibliographical references.
¥4500
1. Church history—Primitive and early church, ca. 30-600. I. Title. II. Series: Ishihara, Ken, 1882- Yōroppa Kirisutokyō shi ; 1-kan.
BR83.I8 vol. 8 79-811847
[BR162.2]

Kenyon, Frederic George, Sir, 1863-1952.
The Bible and modern scholarship / by Sir Frederic G. Kenyon. — 1st ed. — Westport, Conn. : Greenwood Press, 1979.
ix, 53 p. ; 23 cm.
Reprint of the 1948 ed. published by J. Murray, London.
ISBN 0-313-21009-8
1. Bible—Evidences, authority, etc. 2. Barnes, Ernest William, Bp. of Birmingham, 1874-1953. The rise of Christianity. 3. Church history—Primitive and early church, ca. 30-600. 4. Bible. N.T.—History of contemporary events, etc. I. Title.
BS480.K4 1979 220.1 78-9892
78 MARC

McBirnie, William Steuart, 1920-
The search for the early church : New Testament principles for today's church / William Steuart McBirnie. — Wheaton, Ill. : Tyndale House Publishers, c1978.
165 p. : ill. ; 21 cm.
ISBN 0-8423-5834-X : $3.95
1. Church history—Primitive and early church, ca. 30-600. 2. Church. 3. Christianity—20th century. I. Title.
BR165.M18 281'.2 78-55982
79 MARC

Pallas, D
Les monuments paléochrétiens de Grèce découverts de 1959 à 1973 / D. Pallas. — Città del Vaticano : Pontificio Istituto di archeologia cristiana, 1977.
327 p., ⌊2⌋ fold. leaves of plates : ill. ; 24 cm. — (Sussidi allo studio delle antichità cristiane ; 5) VC***
Date in title printed as 1958, corrected by label.
Includes bibliographical references and indexes.
Errata slip inserted.
1. Greece, Modern—Antiquities. 2. Christian antiquities—Greece, Modern. 3. Excavations (Archaeology)—Greece, Modern. 4. Church history—Primitive and early church, ca. 30-600. I. Title. II. Series.
DF129.P34 937 78-394663
79 MARC

Raevskiĭ, Nikolaĭ Fedorovich, d. 1857.
(Nfeskol'ko urokov iz istorii drevneĭ khristianskoĭ fserkvi)
Нѣсколько уроковъ изъ исторіи древней христіанской церкви / составлено по программѣ одобренной святѣйшимъ правительствующимъ Сѵнодомъ для военно-учебныхъ заведеній, сообразно съ наставленіемъ высочайше утвержденнымъ 24-го декабря 1848 года Николаемъ Раевскимъ. — Санктпетербургъ : Въ Тип. Военно-учебныхъ заведеній, 1849.
48 p. ; 27 cm.
1. Church history—Primitive and early church, ca. 30-600. 2. Persecution—Early church, ca. 30-600. I. Title. II. Title: Nfeskol'ko urokov iz istorii fserkvi.
BR162.R27 Russian Imperial Coll. 79-409350

Ramsay, William Mitchell, Sir, 1851-1939.
The church in the Roman Empire before A.D. 170 / by W. M. Ramsay. — Boston : Longwood Press, 1978.
xxiii, 510 p., ⌊3⌋ leaves of plates : ill. ; 22 cm.
Reprint of the 8th (1904) ed. published by Hodder and Stoughton, London. Original ed. issued as the Mansfield College lectures, 1892.
Includes bibliographical references and index.
ISBN 0-89341-216-3 : $50.00
1. Church history—Primitive and early church, ca. 30-600. 2. Church and state in Rome—History. I. Title. II. Series: Mansfield College lectures, 1892.
BR170.R3 1978 270.1 77-6997
78 MARC

Richardson, Peter, 1935-
Israel in the apostolic Church. London, Cambridge U.P., 1969.
xiii, 257 p. 23 cm. (Society for New Testament Studies. Monograph series, 10) 80/-; ($12.50) B70-02130
Originally presented as the author's thesis, Cambridge, 1965.
Includes bibliographical references.
ISBN 0-521-07592-0
1. Judaism (Christian theology) 2. Judaism—Relations—Christianity. 3. Church history—Primitive and early church I. Title. II. Series: Studiorum Novi Testamenti Societas. Monograph series, 10.
BM535.R48 1969 261.2 74-79055
70⌊r79⌋rev MARC

Stevenson, J
The catacombs : rediscovered monuments of early Christianity / ⌊by⌋ J. Stevenson. — London : Thames and Hudson, 1978.
179 p. : ill., maps, plans, ; 25 cm. — (Ancient peoples and places) GB78-38564
Bibliography: p. 169-170.
Includes index.
ISBN 0-500-02091-4 : £7.50 ($16.95 U.S.)
1. Catacombs. 2. Christian art and symbolism. 3. Church history—Primitive and early church, ca. 30-600. I. Title.
BR132.S74 270'.1 79-317142
79 MARC

—PRIMITIVE AND EARLY CHURCH, CA. 30-600—ADDRESSES, ESSAYS, LECTURES

Brown, Peter Robert Lamont.
The making of late antiquity / Peter Brown. — Cambridge, Mass. : Harvard University Press, 1978.
viii, 135 p. ; 22 cm. — (The Carl Newell Jackson lectures ; 1976)
Includes bibliographical references and index.
ISBN 0-674-54320-3
1. Rome—History—Empire, 284-476—Addresses, essays, lectures. 2. Church history—Primitive and early church, ca. 30-600—Addresses, essays, lectures. I. Title. II. Series.
DG312.B76 937'.06 78-6844
78 MARC

CHURCH HISTORY
—PRIMITIVE AND EARLY CHURCH, CA. 30-600—ADDRESSES, ESSAYS, LECTURES (Continued)

L'Empereur Julien : études / rassemblées par René Braun et Jean Richer. — Paris : Belles Lettres, 1978-
v. : ill. ; 24 cm. F***
At head of title: Groupe de recherches de Nice et associés de Paris, Grenoble et Montpellier.
Includes bibliographical references.
CONTENTS: ₍1₎ De l'histoire à la légende (331-1715)

140F (v. 1)
1. Julianus, Apostata, Emperor of Rome, 331-363—Addresses, essays, lectures. 2. Julianus, Apostata, Emperor of Rome, 331-363—Legends—Addresses, essays, lectures. 3. Rome—History—Julian, 361-363—Addresses, essays, lectures. 4. Church history—Primitive and early church, ca. 30-600—Addresses, essays, lectures. I. Braun, René. II. Richer, Jean, 1915-
DG317.E46 937'.08'0924 79-357370
79 MARC

Gottlieb, Gunther, 1935-
Ost und West in der christlichen Kirche des 4. und 5. Jahrhunderts / Gunther Gottlieb. — München : Vögel, 1978.
26 p. ; 24 cm. — (Schriften der Philosophischen Fachbereichs der Universität Augsburg ; Nr. 12) GFR79-A
Includes bibliographical references.
ISBN 3-920896-44-0 : DM6.50
1. Church history—Primitive and early church, ca. 30-600—Addresses, essays, lectures. I. Title. II. Series: Augsburg. Universität. Philosophischer Fachbereich I. Schriften der Philosophischen Fachbereichs der Universität Augsburg ; Nr. 12.
BR205.G65 79-359311
79 MARC

Die Kirche des früheren Mittelalters / hrsg. von Knut Schäferdiek. — 1. Aufl. — München : Kaiser, 1978-
v. ; 24 cm. — (Kirchengeschichte als Missionsgeschichte ; Bd. 2-) GFR78-A
Bibliography: v. 1, p. ₍507₎-542.
Includes indexes.
ISBN 3-459-01107-6 (v. 1)
1. Church history—Primitive and early church, ca. 30-600—Addresses, essays, lectures. 2. Church history—Middle Ages, 600-1500—Addresses, essays, lectures. 3. Missions—History—Addresses, essays, lectures. I. Schäferdiek, Knut. II. Series.
BR203.K57 78-389564
78 MARC

—PRIMITIVE AND EARLY CHURCH, CA. 30-600—BIBLIOGRAPHY

Altaner, Berthold, 1885-1964.
Patrologie : Leben, Schriften u. Lehre d. Kirchenväter / Berthold Altaner, Alfred Stuiber. — 8., durchges. u. erw. Aufl. — Freiburg im Breisgau ; Basel ; Wien : Herder, 1978.
xxii, 672 p. ; 22 cm. GFR79-A
Includes bibliographical references and index.
ISBN 3-451-13580-9 : DM78.00
1. Christian literature, Early—History and criticism. 2. Fathers of the church—Bio-bibliography. 3. Church history—Primitive and early church, ca. 30-600—Bibliography. I. Stuiber, Alfred, joint author. II. Title.
BR67.A37 1978 79-353776
79 MARC

—PRIMITIVE AND EARLY CHURCH, CA. 30-600—COLLECTED WORKS

Schneemelcher, Wilhelm, 1914-
Gesammelte Aufsätze zum Neuen Testament und zur Patristik / Wilhelm Schneemelcher ; hrsg. von W. Bienert und K. Schäferdiek. — Thessaloniki : Patriarchal Institute for Patristic Studies, 1974.
406 p. ; 25 cm. — (Analekta Vlatadōn ; 22)
Bibliography of the author's works: p. ₍390₎-396.
Includes indexes.
1. Bible. N.T.—Criticism, interpretation, etc.—Collected works. 2. Church history—Early church, ca 30-600—Collected works. 3. Fathers of the Church—Collected works. I. Bienert, Walther, 1909- II. Schäferdiek, Knut. III. Title. IV. Series.
BS2395.S278 78-383417
79 MARC

—PRIMITIVE AND EARLY CHURCH, CA. 30-600—FICTION

Sienkiewicz, Henryk, 1846–1916.
₍Quo vadis? Armenian₎
Յո՞ երթաս = Քուօ՞ վատիս = Quo vadis? / *Հ. Սէնքէվիչ ; թարգմանեց Օշին (Ալ. Վ. Միրունեան)*. — 2. *տպագրութիւն*. — *Կ. Պոլիս : Հրատարակութիւն "Ծող" հանդիսի*, 1912.
3 v. : ill. ; 25 cm.
1. Rome — History — Nero, 54–68 — Fiction. 2. Church history — Primitive and early church—Fiction. I. Title. II. Title: K'uo vatis? III. Title: Quo vadis?
Title romanized: Ho ert'as?
PG7158.S4Q42 1912 78–971078

—PRIMITIVE AND EARLY CHURCH, CA. 30-600—SOURCES

Giles, Edward, ed.
Documents illustrating papal authority, A.D. 96-454 / edited and introduced by E. Giles. — Westport, Conn. : Hyperion Press, 1979.
xxi, 344 p. ; 22 cm.
Reprint of the 1952 ed. published by S.P.C.K., London.
Includes bibliographical references and indexes.
ISBN 0-88355-696-0 : $25.00
1. Papacy—History—Sources. 2. Church history—Primitive and early church, ca. 30-600—Sources. 3. Christian literature, Early. I. Title.
₍BX953.G5 1979₎ 262'.13 78-59023
78 MARC

—MIDDLE AGES, 600-1500

see also Albigenses; Crusades; Inquisition; Monasticism and religious orders—Middle Ages, 600-1500; Papacy; Religious thought—Middle Ages, 600-1500

Adamus Bremensis, 11th cent.
Hamburgische Kirchengeschichte / Adam von Bremen. — 3. Aufl. / hrsg. von Bernhard Schmeidler. Unveränd. Nachdr. ₍d. Ausg.₎ 1917. — Hannover : Hahn, 1977.
lxvii, ₍1₎, 353 p. ; 23 cm. GFR78-A26
Cover title: Magistri Adam Bremensis Gesta hammaburgensis ecclesiae pontificum.
First ed. published in 1846 under title: Adami Gesta hammaburgensis ecclesiae pontificum.
Text in Latin, with introd. and notes in German.
Reprint of the ed. published in the series: Scriptores rerum Germanicarum in usum scholarum ex Monumentis Germaniae historicis separatim editi.
Bibliography: p. ₍lxviii₎
Includes indexes.
1. Hamburg—Church history. 2. Germany—Church history—Middle Ages, 843-1517. 3. Scandinavia—Church history. 4. Church history—Middle Ages, 600-1500. I. Schmeidler, Bernhard, 1879-1959. II. Title. III. Title: Gesta hammaburgensis ecclesiae pontificum. IV. Series: Scriptores rerum Germanicarum in usum scholarum ex Monumentis Germaniae historicis separatim editi.
BR858.H3A3 1977 79-352558
79 MARC

Bainton, Roland Herbert, 1894-
The medieval church / Roland H. Bainton. — Huntington, N.Y. : R. E. Krieger Pub. Co., 1979, c1962.
191 p. ; 19 cm. — (The Anvil series)
Reprint of the ed. published by Van Nostrand, Princeton, N.J., in series: An Anvil original.
Bibliography: p. 188-189.
Includes index.
ISBN 0-88275-786-5 : $3.95
1. Church history—Middle Ages, 600-1500. I. Title.
₍BR252.B26 1978₎ 270 78-11433
78 MARC

Chambers, James.
The Devil's horsemen / by James Chambers. — 1st American ed. — New York : Atheneum, 1979.
xii, 190 p., ₍4₎ leaves of plates : ill. ; 23 cm.
Bibliography p. ₍173₎-177.
Includes index.
ISBN 0-689-10942-3
1. Mongols—History. 2. Europe—History—476-1492. 3. Military art and science—History. 4. Military history, Medieval. 5. Church history—Middle Ages, 600-1500. I. Title.
DS19.C45 1979 909'.04'94202 78-22055
78 MARC

González, Justo L
La era de las tinieblas / Justo L. González. — Miami, Fla. : Editorial Caribe, c1978.
199 p. : ill. ; 22 cm. — (His Y hasta lo último de la tierra ; t. 3)
1. Church history—Middle Ages, 600-1500. I. Title.
BR145.2.G66 t. 3 78-111532
79 MARC

Hahn, Georg, Dr.
Die abendländische Kirche im Mittelalter, ihr äusserer Aufbau; in Werkstunden zusammengestellt von Dr. Georg Hahn. Mit 67 Werkbildern nebst Erklärungen als Beilagen. Freiburg im Breisgau, Herder & Co. G.m.b.h., 1942.
v. 23 cm. (His Werkbuch der Kirchengeschichte. 2. T., 1.-Hälfte)
"Bilder" (₍36₎ p. of illus.) and "Bilderklärungen" (26 p.) in pocket at end of v. 1.
"Literatur": v. 1, p. ₍412₎-415.
1. Church history—Middle Ages, 600-1500. I. Title.
BR162.H33 270 45-17391
₍r79₎rev MARC

Ishihara, Ken, 1882-
(Chūsei Kirisutokyō kenkyū)
中世キリスト教研究 / ₍著者 石原謙₎. — 東京：岩波書店, 1979.
9, 543 p. ; 22 cm. — (石原謙著作集 ; 第4巻) Ja***
Series romanized: Ishihara Ken chosaku shū.
Includes bibliographies.
¥4500
1. Church history—Middle Ages, 600–1500. I. Title.
BR83.I8 vol. 4 79–806605
[BR252]

Visions of the end : apocalyptic traditions in the Middle Ages / Bernard McGinn. — New York : Columbia University Press, 1979.
xvii, 377 p. ; 24 cm. — (Records of civilization, sources and studies ; no. 96)
Bibliography: p. 347-363.
Includes indexes.
ISBN 0-231-04594-8
1. End of the world—History of doctrines. 2. Millennialism—History of doctrines. 3. Apocalyptic literature. 4. Church history—Middle Ages, 500-1500. I. McGinn, Bernard, 1937- II. Series.
BT876.V58 236 79-4303
79 MARC

Werner, Ernst, 1920-
Konstantinopel und Canossa : lateinisches Selbstverständnis im 11. Jahrhundert / Ernst Werner. — Berlin : Akademie-Verlag, 1977.
35 p. ; 22 cm. — (Sitzungsberichte der Akademie der Wissenschaften der DDR ; Jahrg. 1977, Nr. 4 G : Gesellschaftswissenschaften) GDR***
Includes bibliographical references.
2.50M
1. Church history—Middle Ages, 600-1500. 2. Church and state—History. 3. Heinrich IV, Emperor of Germany, 1050-1106. 4. Gregorius VII, Saint, Pope, 1015 (ca.)-1085. I. Title. II. Series: Akademie der Wissenschaften der DDR. Sitzungsberichte der Akademie der Wissenschaften der DDR ; Jahrg. 1977, Nr. 4 G.
BR270.W49 79-349244
79 MARC

—MIDDLE AGES, 600-1500—ADDRESSES, ESSAYS, LECTURES

Foreville, Raymonde.
Gouvernement et vie de l'Église au Moyen-Age : recueil d'études / Raymonde Foreville. — London : Variorum Reprints, 1979.
₍418₎ p. in various pagings : ill. ; 24 cm. — (Collected studies series ; CS95) GB79-08508
Reprinted articles previously published 1963-1977.
French or English.
Includes bibliographical references and index.
ISBN 0-86078-040-6 : £20.00
1. Church history—Middle Ages, 600-1500—Addresses, essays, lectures. I. Title.
BR252.F75 282'.09'02 79-371579
79 MARC

Die Kirche des früheren Mittelalters / hrsg. von Knut Schäferdiek. — 1. Aufl. — München : Kaiser, 1978-
v. ; 24 cm. — (Kirchengeschichte als Missionsgeschichte ; Bd. 2-) GFR78-A
Bibliography: v. 1, p. ₍507₎-542.
Includes indexes.
ISBN 3-459-01107-6 (v. 1)
1. Church history—Primitive and early church, ca. 30-600—Addresses, essays, lectures. 2. Church history—Middle Ages, 600-1500—Addresses, essays, lectures. 3. Missions—History—Addresses, essays, lectures. I. Schäferdiek, Knut. II. Series.
BR203.K57 78-389564
78 MARC

Xenia medii aevi historiam illustrantia oblata Thomae Kaeppeli O.P. / ediderunt Raymundus Creytens, Pius Künzle. — Roma : Edizioni di storia e letteratura, 1978.
2 v. (xxiv, 906 p.) : ill., port. ; 25 cm. — (Storia e letteratura ; 141-142) It78-Dec.
English, French, German, Italian, Latin, or Spanish.
Bibliography of Thomas Käppeli's works: p. ₍xvii₎-xxiv.
Includes bibliographical references and indexes.
1. Church history—Middle Ages, 600-1500—Addresses, essays, lectures. 2. Theology—Middle Ages, 600-1500—Addresses, essays, lectures. 3. Europe—Church history—Addresses, essays, lectures. 4. Europe—History—476-1492—Addresses, essays, lectures. 5. Käppeli, Thomas, 1900- —Addresses, essays, lectures. I. Käppeli, Thomas, 1900- II. Creytens, Raymond. III. Künzle, Pius.
BR251.X46 282'.09'02 79-372675
*79 MARC

—MIDDLE AGES, 600-1500—SOURCES

Catholic Church. Pope, 1119-1124 (Calixtus II)
Bullaire du Pape Calixte II / Ulysse Robert. — Hildesheim ; New York : G. Olms, 1979.
c, 397, 533 p. : ill. ; 23 cm. GFR***
Text in Latin; introd. in French.
Reprint of the 1891 ed. published in Paris.
ISBN 3-487-06765-X
1. Catholic Church—History—Middle Ages, 600-1500—Sources. I. Robert, Ulysse Léonard Léon, 1845-1903. II. Title.
BX1195.C37 1979 262.9'1 79-371992
79 MARC

—MODERN PERIOD, 1500

see also Catholic Church—History—Modern period, 1500; Counter-Reformation; Missions—History

CHURCH HISTORY (Continued)

—REFORMATION, 1517-1648 see Reformation

—ADDRESSES, ESSAYS, LECTURES

Histoire de l'Église par elle-même / ₍textes choisis₎ sous la direction de Jacques Loew et Michel Meslin. — Paris : Fayard, 1978.
679 p. : maps ; 24 cm. F79-1836
Bibliography: p. ₍589₎-600.
Includes index.
ISBN (invalid) 2-213-00649-3 : 109F
1. Church history—Addresses, essays, lectures. 2. Catholic Church—History—Addresses, essays, lectures. I. Loew, Jacques, 1908- II. Meslin, Michel.
BR155.H48 282 79-113849
79 MARC

Religious motivation, biographical and sociological problems for the church historians : papers read at the sixteenth summer meeting and the seventeenth winter meeting of the Ecclesiastical History Society / edited by Derek Baker. — Oxford : Pub. for the Ecclesiastical History Society by B. Blackwell, 1978.
xvi, 516 p. ; 22 cm. — (Studies in church history ; 15) GB***
Includes bibliographical references.
ISBN 0-631-19250-6 : £12.50
1. Church history—Addresses, essays, lectures. 2. Psychology, Religious—Addresses, essays, lectures. 3. Motivation (Psychology)—Addresses, essays, lectures. I. Baker, Derek. II. Ecclesiastical History Society, London. III. Series.
BR141.S84 vol. 15 270'.08 s 79-306938
₍BR148₎ 79 MARC

Vier, Jacques.
Littérature à l'emporte-pièce : huitième série / Jacques Vier. — Paris : Éditions du Cèdre, ₍1976₎
220 p. ; 23 cm. F***
Includes bibliographical references.
39.00F
1. Church history—Addresses, essays, lectures. 2. Theology—Addresses, essays, lectures. 3. French literature—History and criticism—Addresses, essays, lectures. I. Title.
BR155.V53 274 78-387332
79 MARC

Wolf, Gunther.
Grenzüberschreitungen : Gedanken u. Anregungen zu kirchl., gesellschaft. u. histor. Problemen / Gunther G. Wolf. — Frankfurt am Main : Haag und Herchen, 1978.
117 p. ; 21 cm. GFR78-A
Includes bibliographical references.
ISBN 3-88129-170-9 : DM19.80
1. History, Modern—Historiography—Addresses, essays, lectures. 2. Church history—Addresses, essays, lectures. I. Title.
D206.W64 909.08 79-350480
79 MARC

—BIBLIOGRAPHY

Lo Menzo Rapisarda, Grazia.
Guida allo studio della storia del cristianesimo / ₍di₎ Grazia Lo Menzo. — Catania : Giannotta, 1976.
62 p. ; 25 cm. It78-Dec
Bibliography: p. ₍61₎-62.
L2500
1. Church history—Bibliography. I. Title.
Z7777.L6 79-373972
₍BR145.2₎ *79 MARC

—COLLECTED WORKS

The Christian centuries : a new history of the Catholic Church / editorial committee, Louis J. Rogier ... ₍et al.₎. — New York : Paulist Press, 1978-
v. : ill. ; 24 cm.
Issued also in French under title: Nouvelle histoire de l'Église, and in German under title: Geschichte der Kirche.
Bibliography: v. 5, p. ₍639₎-676.
Includes indexes.
CONTENTS:
—v. 5. Aubert, R. The church in a secularised society.

ISBN 0-8091-0244-7 (v. 5)
1. Catholic Church—History—Collected works. 2. Church history—Collected works. I. Rogier, L. J.
BX945.2.C45 1978 282 79-102152
79 MARC

Ishihara, Ken, 1882-
(Ishihara Ken chosakushū)
石原謙著作集 / ₍編集 山谷省吾 ..., et al.₎. — 東京 : 岩波書店, 1978-
v. : port. ; 22 cm. Ja 79-3259 (v. 1)
Includes bibliographies.
CONTENTS: 第1巻 初期の著作—第2巻 旧・新約聖書—第3巻 初期キリスト教研究—第4巻 中世キリスト教研究—第5-6巻 宗教改革
¥4500 per vol.
1. Theology, Doctrinal—Collected works—20th century. 2. Church history—Collected works. I. Yamaya, Seigo, 1889-
BR83.I8 79-806620

Kirchengeschichte in Einzeldarstellungen / hrsg. von Gert Haendler, Kurt Meier und Joachim Rogge. — Berlin : Evangelische Verlagsanstalt, 19
v. ; 25 cm. GFR79-A8
Bibliography: v. 1, pt. 3, p. 7-17.
9.00M (v. 1, pt. 3)
1. Church history—Collected works. I. Haendler, Gert. II. Meier, Kurt, 1927- . III. Rogge, Joachim.
BR145.2.K54 79-367526
79 MARC

—CHRONOLOGY see Chronology, Ecclesiastical

—HISTORIOGRAPHY

see also Church historians

—MISCELLANEA

Heymann, Dietrich von.
Text Theologie I. ₍i.e. eins₎ : Grundwissen in Frage u. Antwort / Dietrich von Heymann. — Frankfurt/Main ; Bern ; Las Vegas : Lang, 1978.
270 p. ; 21 cm. — (Europäische Hochschulschriften : Reihe 23, Theologie ; Bd. 104) GFR78-A
Includes bibliographical references.
ISBN 3-261-02484-4
1. Bible—Examinations, questions, etc. 2. Church history—Miscellanea. 3. Theology—20th century—Miscellanea. I. Title. II. Series.
BS612.H45 78-387740
78 MARC

—PERIODICALS

(Kirisutokyō shigaku)
キリスト教史学
₍横浜₎ キリスト教史学会
v. 21 cm.
Organ of the Society of Historical Studies of Christianity, Japan.
Other title : The journal of history of Christianity.
In Japanese, with summaries in English.

1. Church history—Periodicals. I. Kirisutokyō Shigakukai. II. Title: Journal of history of Christianity.
BR140.K57 79-641448
(MARC-S)

—PHILOSOPHY

Mead, Sidney Earl, 1904-
History and identity / Sidney E. Mead. — Missoula, Mont. : Scholars Press, c1979.
63 p. ; 24 cm. — (American Academy of Religion studies in religion ; no. 19 ISSN 0084-6287)
Includes bibliographical references.
ISBN 0-89130-274-3
1. Identification (Religion) 2. History (Theology) 3. Church history—Philosophy. I. Title. II. Series: American Academy of Religion. AAR studies in religion ; 19.
BV4509.5.M38 230 78-26543
78 MARC

—POPULAR WORKS

Snelder, C
Riskante pelgrimage : beknopte kerkgeschiedenis / C. Snelder. — Hilversum : Gooi en Sticht, ₍1978₎
120 p. ; 21 cm. Ne78-38
ISBN 9030401559 : fl 12.50
1. Church history—Popular works. I. Title.
BR150.S57 78-394714
*79 MARC

—SOURCES

see also Church records and registers

Latinitas christiana : e. lat. Lesebuch mit Texten aus d. Geschichte d. christl. Kirchen / unter Mitarb. von Peter Heidrich ... ₍et al.₎ bearb. von Adolf Laminski ; hrsg. von Konrad von Rabenau. — 1. Aufl. — Berlin : Evangelische Verlagsanstalt, 1978.
2 v. ; 22 cm. GDR78-A
Text in Latin.
Bibliography: v. 2, p. 262-265.
Vol. 2 includes index.
1. Church history—Sources. I. Laminski, Adolf. II. Rabenau, Konrad von.
BR141.L35 79-357135
79 MARC

CHURCH LANDS

see also Parsonages

—AUSTRIA—ST. PAUL (BENEDICTINE ABBEY)

Benediktinerstift Sankt Paul.
Die mittelalterlichen Urbare des Benediktinerstiftes St. Paul in Kärnten, 1289/90 und 1371/72. Hrsg. von Walther Fresacher. ₍Mit Faksimiles₎ Wien; Graz, Köln, Böhlau in Kommission, 1968.
cviii, 354, 2 fold. tables, 3 maps (one fold.) 25 cm. (Österreichische Urbare. Abt. 3: Urbare geistlicher Grundherrschaften. Bd. 3: Die Mittelalterlichen Stiftsurbare Kärntens, T. 2) 396.00 Au69-5-299
Text of land titles in Latin.
"Abkürzungen" (bibliographical): p. ₍xiii₎-xvi. Bibliographical footnotes.
1. Church lands—Austria—St. Paul (Benedictine abbey) I. Fresacher, Walther, ed. II. Title.
HD634.U7 Bd. 3, T.2 75-428593
77₍r78₎rev MARC

—AUSTRIA, LOWER—HISTORY

Seitenstetten, Austria (Benedictine abbey)
Die mittelalterlichen Urbare des Benediktinerstiftes Seitenstetten, 1292/98 und 1386/98 / hrsg. v. Gerhard Flossmann. — Wien : Verl. d. Österr. Akad. d. Wiss., 1977.
cix, 300 p., ₍5₎ leaves of plates : ill. maps ; 25 cm. — (Österreichische Urbare : Abt. 3, Urbare geistlicher Grundherrschaften, Bd. 1 : Die Mittelalterlichen Stiftsurbare Niederösterreichs ; T. 3) Au78-13-287
Text in Latin and Middle High German; prefatory material in German.
Bibliography: p. ₍xi₎-xiv.
Includes indexes.
ISBN (invalid) 3-7001-0231-1 : S540.00
1. Church lands—Austria, Lower—History. 2. Seitenstettin, Austria (Benedictine abbey)—History. 3. Farm tenancy—Economic aspects—Austria, Lower—History. I. Flossmann, Gerhard. II. Title.
HD634.U7 Bd. 1, T. 3 79-375372
*79 MARC

—ENGLAND

₍Nichols, John₎, 1745-1826, comp.
Some account of the alien priories, and of such lands as they are known to have possessed in England and Wales / collected from the mss. of John Warburton and Dr. Ducarel. — New ed. 1st AMS ed. — New York : AMS Press, 1978.
494 p. in various pagings, ₍12₎ leaves of plates : ill. ; 19 cm.
Reprint of the 1786 ed. printed by and for J. Nichols, London.
Includes indexes.
ISBN 0-404-04689-4
1. Priories—France. 2. Priories—England. 3. Church lands—England. I. Warburton, John, 1682-1759. II. Ducarel, Andrew Coltee, 1713-1785. III. Title: Some account of the alien priories ...
BX2613.N5 1978 333.3'22 72-173079
76 MARC

—GERMANY, WEST—PADERBORN—HISTORY

Balzer, Manfred.
Untersuchungen zur Geschichte des Grundbesitzes in der Paderborner Feldmark / Manfred Balzer. — München : Fink, 1977.
xvii, 856 p. : 3 fold. maps (in pocket) ; 28 cm. — (Münstersche Mittelalter-Schriften ; Bd. 29) GFR78-A37
A revision of the author's thesis, Münster, 1975.
Bibliography: p. ₍810₎-826.
Includes index.
ISBN 3-7705-1245-6 : DM160.00
1. Church lands—Germany, West—Paderborn—History. 2. Paderborn—History. I. Title. II. Series.
HD660.P3B34 1977 333 79-342417
79 MARC

—ITALY—CALABRIA—HISTORY

Placanica, Augusto.
Alle origini dell'egemonia borghese in Calabria : la privatizzazione delle terre ecclesiastiche (1784-1815) / Augusto Placanica. — 1. ed. — Salerno ; Catanzaro : Società editrice meridionale, 1979.
606 p., ₍18₎ leaves of plates (9 fold.) : ill. ; 24 cm. — (Collezione meridionale : Studi, testi, strumenti ; 3) It79-July
Includes bibliographical references and index.
L20000
1. Church lands—Italy—Calabria—History. 2. Secularization—Italy—Calabria—History. 3. Calabria, Italy—Economic conditions. I. Title.
HD679.C3P54 79-381514
*79 MARC

CHURCH LANDS (Continued)

—QUÉBEC (PROVINCE)—HISTORY

Miller, J R
Equal rights : the Jesuits' Estates Act controversy / J. R. Miller. — Montreal : McGill-Queen's University Press, c1979.
xi, 223 p. : ill. ; 24 cm. C***
Bibliography: p. [203]-213.
Includes index.
ISBN 0-7735-0302-1
1. Canada—English-French relations. 2. Canada—Politics and government—1867-1914. 3. Church lands—Québec (Province)—History. 4. Jesuits—Québec (Province)—History. I. Title.
F1027.M57 320.9'71'05 79-308004
79 MARC

—RUSSIAN REPUBLIC
—MOSCOW (PROVINCE)

Zimin, Aleksandr Aleksandrovich.
(Krupnaia feodal'naia votchina i sotsial'no-politicheskaia bor'ba v Rossii)
Крупная феодальная вотчина и социально-политическая борьба в России : конец XV–XVI в. / А. А. Зимин. — Москва : Наука, 1977.
355 p. ; 21 cm. USSR***
At head of title: Akademiia nauk SSSR. Institut istorii SSSR.
Includes bibliographical references and index.
1.80rub
1. Land tenure—Russian Republic—Moscow (Province) 2. Church lands—Russian Republic—Moscow (Province) 3. Iosifo-Volokolanskii monastyr' (Province of Moscow) I. Title.
HD719.M6Z56 78-422174

—SPAIN—LOGROÑO (PROVINCE)—HISTORY

Lázaro Torres, Rosa María.
Desamortización de Espartero en la provincia de Logroño : [1840-1843 / por] Rosa María Lázaro Torres. — Logroño : Servicio de Cultura de la Excma. Diputación Provincial, Gonzalo de Berceo, 1977.
141 p. ; 16 cm. — (Biblioteca de temas riojanos ; 15) Sp78-Feb
Bibliography: p. 135-139.
ISBN 8485242165 : 390ptas
1. Church lands—Spain—Logroño (Province)—History. 2. Land tenure—Spain—Logroño (Province)—History. 3. Secularization—Spain—Logroño (Province)—History. I. Title.
HD779.L63L39 78-380483
78 MARC

CHURCH LAW see Ecclesiastical law

CHURCH LEADERSHIP see Christian leadership

CHURCH LIBRARIES see Libraries, Church

CHURCH MAINTENANCE AND REPAIR

Total Environmental Action, inc.
The energy-efficient church : how to save energy (and money) in your church / by Total Environmental Action, inc. ; Douglas R. Hoffman, editor ; foreword by Gordon Bucy. — New York : Pilgrim Press, c1979.
ix, 86 p. : ill. ; 21 cm.
Bibliography: p. 85-86.
ISBN 0-8298-0362-9 : $4.95
1. Church maintenance and repair. 2. Churches—Energy conservation. I. Hoffman, Douglas R., 1949- II. Title.
BV652.7.T6 1979 254'.7 79-10432
79 MARC

CHURCH MANAGEMENT

see also Church consultation; Church controversies; Church finance; Church maintenance and repair; Group ministry; Parish councils

Anderson, James Desmond.
The management of ministry / James D. Anderson, Ezra Earl Jones. — 1st ed. — San Francisco : Harper & Row, c1978.
xi, 202 p. : ill. ; 21 cm.
Bibliography: p. [199]-200.
Includes index.
ISBN 0-06-060235-X
1. Church management. 2. Clergy—Office. 3. Pastoral theology. I. Jones, Ezra Earl, joint author. II. Title.
BV652.A67 1978 254 76-62942
78 MARC

Dayton, Edward R
Strategy for leadership / Edward R. Dayton, Ted. W. Engstrom. — Old Tappan, N.J. : F. H. Revell Co., c1979.
240 p. : ill. ; 24 cm.
Bibliography: p. 233-235.
Includes index.
ISBN 0-8007-0994-2 : $8.95
1. Church management. 2. Christian leadership. I. Engstrom, Theodore Wilhelm, 1916- joint author. II. Title.
BV652.D38 658'.91'25 79-11180
79 MARC

Hutcheson, Richard G 1921-
Wheel within the wheel : confronting the management crisis of the pluralistic church / Richard G. Hutcheson, Jr. — Atlanta : John Knox Press, 1979.
272 p. ; 18 cm.
Includes bibliographical references.
ISBN 0-8042-1886-2
1. Church management. 2. Holy Spirit. I. Title.
BV652.H83 658'.91'25 79-11481
79 MARC

Knudsen, Raymond B
New models for church administration : the practical application of business principles / Raymond B. Knudsen. — Chicago : Association Press, c1979.
165 p. ; 22 cm.
Includes index.
ISBN 0-695-81254-8 : $5.95
1. Church management. I. Title.
BV652.K58 254 79-11908
79 MARC

Leas, Speed, 1937-
Time management : a working guide for church leaders / Speed B. Leas. — Nashville : Abingdon Press, c1978.
125 p. : ill. ; 20 cm. — (Creative leadership series)
Bibliography: p. 122-123.
ISBN 0-687-42120-9
1. Church management. 2. Time allocation. I. Title. II. Series.
BV652.9.L42 254 78-8628
78 MARC

Lovell, George.
Churches and communities : an approach to development in the local church / [by] George Lovell and Catherine Widdicombe. — London : Search Press, 1978.
218 p. : ill., form ; 22 cm. GB78-24116
Includes bibliographical references.
ISBN 0-85532-387-6 : £2.95
1. Church management. 2. Community development. I. Widdicombe, Catherine, joint author. II. Title.
BV652.9.L68 658'.91'25 78-323874
79 MARC

Primer, Ben, 1949-
Protestants and American business methods / by Ben Primer. — [Ann Arbor, Mich.] : UMI Research Press, c1979.
v, 223 p. : ill. ; 24 cm. — (Studies in American history and culture ; no. 7)
Bibliography: p. [195]-218.
Includes index.
ISBN 0-8357-0982-5 : $24.95. ISBN 0-8357-0983-3 pbk. : $21.95
1. Protestant churches—United States—History. 2. United States—Church history. 3. Church management. 4. Bureaucracy. I. Title. II. Series.
BR517.P74 1978 280'.4'0973 78-20879
78 MARC

Saloheimo, Mauno.
Kirkon yleishallinto / Mauno Saloheimo. — 3. uudistettu painos. — Helsinki : Suomen kirkon seurakuntatyön keskusliitto, 1975.
29 p. ; 21 cm. — (Seurakuntahallinnon opas ; no 2) Fi***
1. Church management. I. Title. II. Series.
BV652.S23 1975 78-386649
79 MARC

Schaller, Lyle E
Effective church planning / Lyle E. Schaller. — Nashville : Abingdon, c1979.
176 p. ; 22 cm.
Includes bibliographical references.
ISBN 0-687-11530-2
1. Church management. I. Title.
BV652.5.S32 658'.91'25 78-26462
78 MARC

Slaght, Lawrence T
A single board for churches : organizing for action / Lawrence T. Slaght. — Valley Forge, PA : Judson Press, c1979.
172 p. ; 18 cm.
Bibliography: p. 171-172.
ISBN 0-8170-0838-1 : $4.95
1. Church management. I. Title.
BV652.S54 658'.91'25 79-1224
79 MARC

—ADDRESSES, ESSAYS, LECTURES

Seurakuntasuunnittelu / toimittaneet Jouko Sihvo ja Heikki Mäkeläinen. — [s.l.] : Kirjapaja, 1971.
150 p. : ill. ; 21 cm. — (Kirkon Tutkimuslaitoksen julkaisuja ; 2) Fi***
Bibliography: p. 146-150.
1. Church management—Addresses, essays, lectures. 2. Parishes—Finland—Addresses, essays, lectures. I. Sihvo, Jouko. II. Mäkeläinen, Heikki. III. Series: Kirkon Tutkimuslaitos. Julkaisuja ; 2.
BV652.S45 78-391129
79 MARC

—CASE STUDIES

Cases in denominational administration : a management casebook for decision-making / edited by Harold R. Phillips and Robert E. Firth. — Berrien Springs, Mich. : Andrews University Press, 1978.
vi, 314 p. ; 23 cm.
1. Church management—Case studies. I. Phillips, Harold R. II. Firth, Robert E.
BV652.C33 658.4'03 79-101908
79 MARC

CHURCH MEMBERSHIP

see also Baptism; Confirmation; Lord's Supper; Non church-affiliated people

Dresselhaus, Richard L
The joy of belonging : a study in church membership / Richard L. Dresselhaus. — Springfield, Miss. : Gospel Pub. House, c1978.
128 p. ; 18 cm. — (Radiant books)
ISBN 0-88243-526-4 : $1.50
1. Church membership. 2. Christian life—1960- I. Title.
BV820.D73 260 78-66868
00 MARC

Dudley, Carl S 1932-
Where have all our people gone? : New choices for old churches / Carl S. Dudley. — New York : Pilgrim Press, c1979.
xiv, 136 p. : ill. ; 21 cm.
Includes bibliographies.
ISBN 0-8298-0359-9 : $5.95
1. Church membership. 2. Church work. 3. Christianity—20th century. I. Title.
BV820.D82 254'.5 79-525
79 MARC

—ADDRESSES, ESSAYS, LECTURES

Understanding church growth an decline, 1950-1978 / edited by Dean R. Hoge and David A. Roozen. — New York : Pilgrim Press, c1979
398 p. : ill. ; 23 cm.
Bibliography: p. 374-391.
Includes index.
ISBN 0-8298-0358-0 (pbk.) : $8.95
1. Church membership—Addresses, essays, lectures. 2. Church growth—Addresses, essays, lectures. I. Hoge, Dean R., 1937- II. Roozen, David A.
BV820.U5 301.5'8 79-4166
79 MARC

CHURCH MISSIONARY SOCIETY

—HISTORY

Foster, William Derek.
The Church Missionary Society and modern medicine in Uganda : the life of Sir Albert Cook, K.C.M.G., 1870-1951 / by W. D. Foster. — [Prestbury] : The author, 1978.
224 p., [8] p. of plates : ill., facsim, port. ; 22 cm. GB79-15213
Spine title : Sir Albert Cook.
Bibliography: p. 234.
ISBN 0-9506432-0-3 : £7.50
1. Cook, Albert Ruskin, Sir, 1870-1951. 2. Missionaries, Medical—Uganda—Biography. 3. Church Missionary Society—History. 4. Missions, Medical—Uganda—History—20th century. I. Title.
R722.32.C66F67 266'.3'0924 79-315710
79 MARC

CHURCH MISSIONARY SOCIETY
—HISTORY (Continued)

Yates, Timoty, E
Venn and Victorian bishops abroad : the missionary policies of Henry Venn and their repercussions upon the Anglican episcopate of the colonial period 1841-1872 / T. E. Yates. — Uppsala : Swedish inst. of missionary research ₍Sv. inst. för missionsforskning₎ ; London : SPCK, 1978.
212 p. ; 23 cm. — (Studia missionalia Upsaliensia ; 33 ISSN 0585-5373) S78-27
Thesis—Uppsala.
Bibliography: p. 203-208.
Includes index.
ISBN 918542403X : kr50.00
1. Venn, Henry, 1796-1873. 2. Church of England—Bishops. 3. Church of England—Missions. 4. Church Missionary Society—History. I. Title. II. Series.
BX5199.V443Y37 283'.092'4 79-302155
79 MARC

CHURCH MUSIC

see also Carillons; Carols; Choirs (Music); Choral music; Christmas music; Conducting, Choral; Liturgies; Motet; Music in Christian education; Organ music; Religion and music; Sequences (Liturgy); Wedding music

Dakers, Lionel.
Making church music work / by Lionel Dakers. — London : Mowbrays, 1978.
157, ₍60₎ p. : music ; 18 cm. — (Addington Press series ; no. 1) GB***
"Check list of anthems": p. ₍1₎-₍60₎
ISBN 0-264-66470-1 : £4.50
1. Church music. 2. Choirs (Music) 3. Conducting, Choral. 4. Church music—Bibliography. I. Title. II. Series.
MT88.D27 783'.026 79-302784
79 MARC

Koch, Eduard Emil, 1809-1871.
Geschichte des Kirchenlieds und Kirchengesangs der christlichen, insbesondere der deutschen evangelischen Kirche / Eduard Emil Koch. — Nachdruck der 3. umgearb. durchaus vermehrten Aufl. — Hildesheim ; New York : G. Olms, 1973.
8 v. ; 20 cm. GFR***
Some text in Latin.
Reprint of the 1866-1876 ed., published by C. Belser, Stuttgart.
Vol. 7 by A. W. Koch, v. 8 by R. Lauxmann.
Includes bibliographical references and indexes.
CONTENTS: Bd. 1-7. Erster Haupttheil, Die Dichter und Sänger.—Bd. 8. Zweiter Haupttheil, Die Lieder und Weisen.
ISBN 3-487-04740-3 (v. 1)
1. Hymns—History and criticism. 2. Hymns, German—History and criticism. 3. Protestant churches—Germany—Hymns—History and criticism. 4. Church music. I. Koch, Adolf Wilhelm, 1843-1912. II. Lauxmann, Richard, 1834-1890. III. Title.
BV310.K7 1973 79-344788
79 MARC

Layer, Adolf.
Musikgeschichte der Fürstabtei Kempten / von Adolf Layer. — 1. Aufl. — Kempten (Allgäu) : Verlag für Heimatpflege, 1975.
84 p. : ill., music ; 25 cm. — (Allgäuer Heimatbücher ; Bd. 76) GFR77-A
Includes bibliographical references and index.
DM19.80
1. Kempten, Ger. (Benedictine monastery) 2. Church music—Catholic Church. 3. Church music—Germany—Kempten (City) I. Title. II. Series.
ML3029.8.K44L4 783'.026'24337 79-368542
79 MARC

—BIBLIOGRAPHY

Dakers, Lionel.
Making church music work / by Lionel Dakers. — London : Mowbrays, 1978.
157, ₍60₎ p. : music ; 18 cm. — (Addington Press series ; no. 1) GB***
"Check list of anthems": p. ₍1₎-₍60₎
ISBN 0-264-66470-1 : £4.50
1. Church music. 2. Choirs (Music) 3. Conducting, Choral. 4. Church music—Bibliography. I. Title. II. Series.
MT88.D27 783'.026 79-302784
79 MARC

—CATHOLIC CHURCH

see also Catholic Church. Liturgy and ritual; Chants (Plain, Gregorian, etc.); Motet

Gavoty, Bernard.
La Crise de la musique religieuse : communication faite à la séance du 8 décembre 1976, Institut de France, Académie des Beaux-Arts / par M. Bernard Gavoty. — Paris : Institut de France, 1976.
16 p. ; 28 cm. F78-16453
"Institut, 1976, no 25."
1. Church music—Catholic Church. I. Title.
ML3007.G36 783'.026'2 79-337902
79 MARC

Lange, Francisco Curt, 1903-
Freguesia de Nossa Senhora do Pilar do Ouro Preto / Francisco Curt Lange. — Belo Horizonte : Arquivo Público Mineiro, 1979.
458 p. ; 21 cm. — (His História da música nas irmandades de Vila Rica ; v. 1) (Publicações do Arquivo Público Mineiro ; no. 2)
1. Church music—Brazil—Ouro Prêto—History and criticism. 2. Church music—Catholic Church. I. Title.
ML232.8.O95L3 vol. 1 780'.981'5 s 79-107716
79 MARC

Liban, Jerzy, 1464-ca. 1546.
De accentuum ecclesiasticorum exquisita ratione / Jerzy Liban. — Kraków : Polskie Wydawnictwo Muzyczne, 1975 i.e. 1976.
₍116₎ p. ; 16 cm. — (Monumenta musicae in Polonia : Bibliotheca antiqua ; 6) P***
Reprint of the 1539 ed. published by M. Szarffenberg, Cracow.
zł33.00
1. Music—Theory—16th-17th centuries. 2. Church music—Catholic Church. I. Title. II. Series: Monumenta musicae in Polonia : Seria D, Bibliotheca antiqua ; 6.
ML171.L43 1976 783'.026'2 79-344414
79 MARC

Ludwig, Friedrich, 1872–1930.
Repertorium organorum recentioris et motetorum vetustissimi stili. Hrsg. von Luther A. Dittmer. New York, Institute of Mediaeval Music, 1964–1978.
2 v. in 3. 25 cm. (Wissenschaftliche Abhandlungen, Bd. 7, 17, 26)
"Reprografischer Nachdruck der Ausgabe Halle 1910"; v. 1: "2 erw. Aufl."
CONTENTS: Bd. 1. Catalogue raisonné der Quellen. Abt. 1. Handschriften in Quadrat-Notation. Abt. 2. Handschriften in Mensural -Notation.—Bd. 2. Vollständiges musikalisches Anfangs-Verzeichnis des systematisch nach den Tenores geordneten.
ISBN 3-912024-87-9
1. Music—History and criticism—Medieval, 400–1500. 2. Church music—Catholic Church. 3. Motet. 4. Organum. I. Dittmer, Luther, A., 1927- II. Title. III. Series: Institute of Mediaeval Music, Brooklyn. Wissenschaftliche Abhandlungen, Bd. 7, 17, 26.
ML128.M3L9 1964 65–2290

Zaccaria, Sante.
Musica sacra in Italia dal 1925 al 1975 / Sante Zaccaria. — Padova : G. Zanibon, 1975.
55 p. ; 21 cm. It76-Sept
Includes bibliographical references and index.
1. Church music—Italy. 2. Church music—Catholic Church. I. Title.
ML3033.5.Z3 783'.026'245 79-346231
*79 MARC

—CATHOLIC CHURCH—BIBLIOGRAPHY
—CATALOGS

Paoli, Marco.
I corali della Biblioteca statale di Lucca / Marco Paoli. — Firenze : L. S. Olschki, 1977.
143 p., ₍24₎ leaves of plates : ill. ; 26 cm. — (Biblioteca di bibliografia italiana ; 83) It78-May
Includes bibliographies and indexes.
L11000
1. Church music—Catholic Church—Bibliography—Catalogs. 2. Music—Bibliography—Manuscripts. 3. Lucca. Biblioteca statale. I. Title. II. Series.
ML136.L88B57 79-345006
*79 MARC

—CATHOLIC CHURCH—HISTORY AND CRITICISM

Fellerer, Karl Gustav, 1902-
The history of Catholic Church music / Karl Gustav Fellerer ; translated by Francis A. Brunner. — Westport, Conn. : Greenwood Press, 1979, c1961.
235 p. : music ; 23 cm.
Translation of Geschichte der katholischen Kirchenmusik.
Reprint of the ed. published by Helicon Press, Baltimore.
Bibliography: p. 221-224.
Includes indexes.
ISBN 0-313-21147-7
1. Church music—Catholic Church—History and criticism. I. Title.
₍ML3002.F32 1979₎ 783'.026'209 78-21637
78 MARC

Giesel, Helmut.
Studien zur Symbolik der Musikinstrumente im Schrifttum der alten und mittelalterlichen Kirche : (von den Anfängen bis zum 13. Jahrhundert) / von Helmut Giesel. — Regensburg : G. Bosse, 1978.
307 p. ; 21 cm. — (Kölner Beiträge zur Musikforschung ; Bd. 94) GFR***
"Sammlung ausgewählter Deutungen zu jedem Musikinstrument" (Greek or Latin): p. 208-307.
Bibliography: p. 198-206.
ISBN 3-7649-2173-0
1. Church music—Catholic Church—History and criticism. 2. Church music—Catholic Church (Byzantine rite)—History and criticism. 3. Music—History and criticism—To 400. 4. Music—History and criticism—Medieval, 400-1500. 5. Musical instruments. I. Title. II. Series.
ML178.G43 79-346970
79 MARC

Hayburn, Robert F
Papal legislation on sacred music, 95 A.D. to 1977 A.D. / Robert F. Hayburn. — Collegeville, Minn. : Liturgical Press, c1979.
xiv, 619 p., ₍5₎ leaves of plates : ill. ;
Bibliography: p. 579-603.
Includes index.
ISBN 0-8146-1012-9
1. Church music—Catholic Church—Papal documents. 2. Bulls, Papal. 3. Church music—Catholic Church—History and criticism. I. Title.
ML3002.H43 783'.026'82 79-63340
79 MARC

Nemmers, Erwin Esser, 1916-
Twenty centuries of Catholic church music / by Erwin Esser Nemmers. — Westport, Conn. : Greenwood Press, 1978, c1949.
xvii, 213 p., ₍2₎ leaves of plates : ill. ; 23 cm.
Reprint of the ed. published by Bruce Pub. Co., Milwaukee.
"Bibliography of bibliographies": p. 193.
Includes index.
ISBN 0-313-20542-6
1. Church music—Catholic Church—History and criticism. I. Title.
₍ML3002.N35 1978₎ 783'.026'209 78-17248
78 MARC

Weinmann, Karl, 1873-1929.
History of church music / by Karl Weinmann ; translated from the German. — Westport, Conn. : Greenwood Press, 1979.
v, 216 p. ; 23 cm.
Translation of Geschichte der Kirchenmusik.
Reprint of the 1910 ed. published by F. Pustet, Ratisbon, New York.
Bibliography: p. ₍205₎-210.
Includes index.
ISBN 0-8371-4363-2
1. Church music—History and criticism. 2. Church music—Catholic Church—History and criticism. I. Title.
ML3002.W34 1979 783'.026'09 78-31681
78 MARC

—CATHOLIC CHURCH—HISTORY AND CRITICISM—ADDRESSES, ESSAYS, LECTURES

Contributi e studi di liturgia e musica nella regione padana. — Bologna : A.M.I.S., 1971.
386 p. : ill., music ; 25 cm. — (Miscellanee saggi convegni ; 6)
"Un tributo all'opera didattica svolta per un quarto di secolo dal Prof. Giuseppe Vecchi."
Includes bibliographical references.
1. Church music—Italy—Po Plain—History and criticism—Addresses, essays, lectures. 2. Church music—Catholic Church—History and criticism—Addresses, essays, lectures. 3. Catholic Church. Liturgy and ritual—History—Addresses, essays, lectures. 4. Vecchi, Giuseppe. I. Vecchi, Giuseppe. II. Series.
ML3033.C63 783.2'0945'2 79-379488
79 MARC

—CATHOLIC CHURCH—PAPAL DOCUMENTS

Hayburn, Robert F
Papal legislation on sacred music, 95 A.D. to 1977 A.D. / Robert F. Hayburn. — Collegeville, Minn. : Liturgical Press, c1979.
xiv, 619 p., ₍5₎ leaves of plates : ill. ;
Bibliography: p. 579-603.
Includes index.
ISBN 0-8146-1012-9
1. Church music—Catholic Church—Papal documents. 2. Bulls, Papal. 3. Church music—Catholic Church—History and criticism. I. Title.
ML3002.H43 783'.026'82 79-63340
79 MARC

—CATHOLIC CHURCH (BYZANTINE RITE)—HISTORY AND CRITICISM

Giesel, Helmut.
Studien zur Symbolik der Musikinstrumente im Schrifttum der alten und mittelalterlichen Kirche : (von den Anfängen bis zum 13. Jahrhundert) / von Helmut Giesel. — Regensburg : G. Bosse, 1978.
307 p. ; 21 cm. — (Kölner Beiträge zur Musikforschung ; Bd. 94) GFR***
"Sammlung ausgewählter Deutungen zu jedem Musikinstrument" (Greek or Latin): p. 208-307.
Bibliography: p. 198-206.
ISBN 3-7649-2173-0
1. Church music—Catholic Church—History and criticism. 2. Church music—Catholic Church (Byzantine rite)—History and criticism. 3. Music—History and criticism—To 400. 4. Music—History and criticism—Medieval, 400-1500. 5. Musical instruments. I. Title. II. Series.
ML178.G43 79-346970
79 MARC

CHURCH MUSIC (Continued)

—CONGRESSES

Religiöse Musik in nicht-liturgischen Werken von Beethoven bis Reger / gemeinsam mit Günther Massenkeil und Klaus Wolfgang Niemöller, hrsg. von Walter Wiora. — Regensburg : G. Bosse, 1978.
285 p. ; 25 cm. — (Studien zur Musikgeschichte des 19. Jahrhunderts ; Bd. 51) GFR***
Papers based on a conference organized by the Arbeitskreis Musikwissenschaft of Forschungsunternehmen Neunzehntes Jahrhundert of Fritz Thyssen Stiftung, held at the Musikwissenschaftliches Institut of Cologne University, Feb. 28-Mar. 2, 1974.
Includes bibliographical references and index.
ISBN 3-7649-2135-8
1. Church music—Congresses. 2. Religion and music. I. Massenkeil, Günther, 1926- II. Niemöller, Klaus Wolfgang. III. Wiora, Walter. IV. Forschungsunternehmen "Neunzehntes Jahrhundert." Arbeitskreis Musikwissenschaft. V. Cologne. Universität. Musikwissenschaftliches Institut. VI. Series.
ML2900.R44 783'.026'09034 78-389927
79 MARC

—HISTORY AND CRITICISM

see also Music--History & criticism--Medieval, 400–1500

Routley, Erik.
Church music and the Christian faith / Erik Routley ; forward by Martin E. Marty. — Carol Stream, Ill. : Agape, c1978.
vi, 153 p. : music ; 22 cm.
Includes bibliographies.
ISBN 0-916642-10-0 : $3.95.
1. Church music—History and criticism. I. Title.
ML3001.R83 783'.026 78-110219
79 MARC

Weinmann, Karl, 1873-1929.
History of church music / by Karl Weinmann ; translated from the German. — Westport, Conn. : Greenwood Press, 1979.
v, 216 p. ; 23 cm.
Translation of Geschichte der Kirchenmusik.
Reprint of the 1910 ed. published by F. Pustet, Ratisbon, New York.
Bibliography: p. ⌊205⌋-210.
Includes index.
ISBN 0-8371-4363-2
1. Church music—History and criticism. 2. Church music—Catholic Church—History and criticism. I. Title.
ML3002.W34 1979 783'.026'09 78-31681
78 MARC

—LUTHERAN CHURCH

Giebel, Alfred.
Die Pflege der Kirchenmusik in der Schwalmstadt Treysa / Alfred Giebel. — Erw. Sonderdr. — ⌊s. l. : s. n.⌋, 1974.
22 p. : ill. ; 24 cm. GFR74-A
1. Church music—Germany, West—Treysa. 2. Church music—Lutheran Church. I. Title.
ML3129.8.T7G5 783'.026'414341 79-365114
79 MARC

—PERIODICALS

Koorbode.
Bergen-op-Zoom ⌊Netherlands, s. n., 1908⌋-1915.
v. 29 cm. monthly.
"Maandblad tot bevordering van den Kerkzang en het Kerkelijk Orgelspel."
1. Church music—Periodicals. 2. Chants (Plain, Gregorian, etc.)—Periodicals.
ML5.K67 79-646436
MARC-S

—PROTESTANT CHURCHES

Krummacher, Friedhelm.
Die Choralbearbeitung in der protestantischen Figuralmusik zwischen Praetorius und Bach / Friedhelm Krummacher. — Kassel ; Basel ; Tours ; London : Bärenreiter, 1978.
xvi, 546 p. : music ; 24 cm. — (Kieler Schriften zur Musikwissenschaft ; Bd. 22)
Includes bibliographical references.
ISBN 3-7618-0526-8
1. Chorale. 2. Church music—Protestant churches. I. Title. II. Series.
ML3184.K76 78-388503
00 MARC

Le Huray, Peter.
Music and the Reformation in England, 1549-1660 / Peter le Huray. — Reprinted with corrections. — Cambridge ; New York : Cambridge University Press, 1978.
xv, 454 p. : music ; 24 cm. — (Cambridge studies in music) (Reprint of the 1967 ed. published by Oxford University Press, New York, in series: Studies in church music)
Bibliography: p. 427-442.
Includes indexes.
ISBN 0-521-21958-2. ISBN 0-521-29418-5 pbk.
1. Church music—England—History and criticism. 2. Church music—Protestant churches. 3. Music—History and criticism—16th century. 4. Music—History and criticism—17th century. I. Title. II. Series. III. Series: Studies in church music.
ML3131.L44 1978 783'.0942 77-87383
78 MARC

Michel, Josef, 1928-
Der Weg der Kirchenmusik zwischen Tradition und Mode : 10 Jahre Diskussion um e. neues Leitbild für d. Kirchenmusikerberuf / Josef Michel. — Berlin : Merseburger, 1978.
62 p. ; 24 cm. — (Edition Merseburger ; 1147) GFR79-A
ISBN 3-87537-164-X : DM10.50
1. Church music—Protestant churches. I. Title.
ML3100.M48 783'.026 79-379736
79 MARC

—AFRICA—HISTORY AND CRITICISM

Jones, A M
African hymnody in Christian worship : a contribution to the history of its development / A. M. Jones. — Gwelo, Rhodesia : Mambo Press, 1976.
64 p. ; 21 cm. — (Mambo occasional papers, Missio-pastoral series ; no. 8)
Bibliography: p. 59-64.
1. Hymns, African—History and criticism. 2. Church music—Africa—History and criticism. I. Title. II. Series.
ML3186.J65 783.9'0967 78-322798
79 MARC

—BRAZIL—OURO PRÊTO—HISTORY AND CRITICISM

Lange, Francisco Curt, 1903-
Freguesia de Nossa Senhora do Pilar do Ouro Preto / Francisco Curt Lange. — Belo Horizonte : Arquivo Público Mineiro, 1979.
458 p. ; 21 cm. — (His História da música nas irmandades de Vila Rica ; v. 1) (Publicações do Arquivo Público Mineiro ; no. 2)
1. Church music—Brazil—Ouro Prêto—History and criticism. 2. Church music—Catholic Church. I. Title.
ML232.8.O95L3 vol. 1 780'.981'5 s 79-107716
79 MARC

—ENGLAND—HISTORY AND CRITICISM

Le Huray, Peter.
Music and the Reformation in England, 1549-1660 / Peter le Huray. — Reprinted with corrections. — Cambridge ; New York : Cambridge University Press, 1978.
xv, 454 p. : music ; 24 cm. — (Cambridge studies in music) (Reprint of the 1967 ed. published by Oxford University Press, New York, in series: Studies in church music)
Bibliography: p. 427-442.
Includes indexes.
ISBN 0-521-21958-2. ISBN 0-521-29418-5 pbk.
1. Church music—England—History and criticism. 2. Church music—Protestant churches. 3. Music—History and criticism—16th century. 4. Music—History and criticism—17th century. I. Title. II. Series. III. Series: Studies in church music.
ML3131.L44 1978 783'.0942 77-87383
78 MARC

—GERMANY—HISTORY AND CRITICISM

Orf, Wolfgang.
Die Musikhandschriften Thomaskirche Mss. 49/50 und 51 in der Universitätsbibliothek Leipzig / komm. u. bibliograph. erschlossen von Wolfgang Orf. — 1. Aufl. — Wilhelmshaven : Heinrichshofen, 1977.
175 p. : facsims., music ; 31 cm. — (Quellenkataloge zur Musikgeschichte ; 13) GFR77-A
Includes bibliographical references.
ISBN 3-7959-0243-6 : DM110.00
1. Leipzig. Universität. Bibliothek. MSS. (Th.-K. Ms. 49/50) 2. Leipzig. Universität. Bibliothek. MSS. (Th.-K. Ms. 51) 3. Church music—Germany—History and criticism. I. Title. II. Series.
ML136.L4U56 79-376154
79 MARC

—GERMANY—KEMPTEN (CITY)

Layer, Adolf.
Musikgeschichte der Fürstabtei Kempten / von Adolf Layer. — 1. Aufl. — Kempten (Allgäu) : Verlag für Heimatpflege, 1975.
84 p. : ill., music ; 25 cm. — (Allgäuer Heimatbücher ; Bd. 76) GFR77-A
Includes bibliographical references and index.
DM19.80
1. Kempten, Ger. (Benedictine monastery) 2. Church music—Catholic Church. 3. Church music—Germany—Kempten (City) I. Title. II. Series.
ML3029.8.K44L4 783'.026'24337 79-368542
79 MARC

—GERMANY, WEST—TREYSA

Giebel, Alfred.
Die Pflege der Kirchenmusik in der Schwalmstadt Treysa / Alfred Giebel. — Erw. Sonderdr. — ⌊s. l. : s. n.⌋, 1974.
22 p. : ill. ; 24 cm. GFR74-A
1. Church music—Germany, West—Treysa. 2. Church music—Lutheran Church. I. Title.
ML3129.8.T7G5 783'.026'414341 79-365114
79 MARC

—ITALY

Zaccaria, Sante.
Musica sacra in Italia dal 1925 al 1975 / Sante Zaccaria. — Padova : G. Zanibon, 1975.
55 p. ; 21 cm. It76-Sept
Includes bibliographical references and index.
1. Church music—Italy. 2. Church music—Catholic Church. I. Title.
ML3033.5.Z3 783'.026'245 79-346231
*79 MARC

—ITALY—PO PLAIN—HISTORY AND CRITICISM—ADDRESSES, ESSAYS, LECTURES

Contributi e studi di liturgia e musica nella regione padana. — Bologna : A.M.I.S., 1971.
386 p. : ill., music ; 25 cm. — (Miscellanee saggi convegni ; 6)
"Un tributo all'opera didattica svolta per un quarto di secolo dal Prof. Giuseppe Vecchi."
Includes bibliographical references.
1. Church music—Italy—Po Plain—History and criticism—Addresses, essays, lectures. 2. Church music—Catholic Church—History and criticism—Addresses, essays, lectures. 3. Catholic Church. Liturgy and ritual—History—Addresses, essays, lectures. 4. Vecchi, Giuseppe. I. Vecchi, Giuseppe. II. Series.
ML3033.C63 783.2'0945'2 79-379488
79 MARC

—NEW ENGLAND—HISTORY AND CRITICISM

Daniel, Ralph T
The anthem in New England before 1800 / Ralph T. Daniel. — New York : Da Capo Press, 1979, c1966.
xvi, 282 p. : ill. ; 24 cm. — (Da Capo Press music reprint series)
Reprint of the ed. published by Northwestern University Press, Evanston, Ill., in series: Pi Kappa Lambda studies in American music.
"Musical supplement": 175-272.
Bibliography: p. 156-171.
Includes index.
ISBN 0-306-79511-6 : $22.50
1. Church music—New England—History and criticism. I. Title.
⌊ML2911.D35 1979⌋ 783.4'0974 78-31121
78 MARC

—POLAND—HISTORY AND CRITICISM

Oloff, Ephraim, 1685-1735.
Ephraim Oloffs polnische Liedergeschichte : von poln. Kirchen-Gesängen u. dererselben Dichtern u. Übers., nebst einigen Anm. aus d. poln. Kirchen- u. Gelahrten-Geschichte. — Fotomechan. Neudr. d. Orig.-Ausg. Dantzig 1744. — Leipzig : Zentralantiquariat d. DDR, 1976.
586 p. : port. ; 19 cm. GDR77-A
Includes indexes.
120.00M
1. Church music—Poland—History and criticism. 2. Hymns, Polish—History and criticism. 3. Songs, Polish—History and criticism. I. Title. II. Title: Polnische Liedergeschichte.
ML3051.P64O4 1976 783'.026'09438 79-340070
79 MARC

—SWEDEN—PERIODICALS

Svensk kyrkomusik uppl. A/B. jan. 1970–
Vagnhärad, Kyrokomusikernas riksförbund.
v. ill. 27 cm. 10 no. a year.
Continues: KMT/KFS. ⌊Uppl.⌋ A/B.
Organ of Krykomusikernas riksförbund and, Jan. 1970–Jan. 1972 of Sveriges kyrkosångsförbund; Feb. 1972– of Riksförbundet Svensk kyrkomusik.
Vols. for Jan. 1970–Jan. 1972 issued with Sveriges kyrkosångsförbundet; Feb. 1972– with Riksförbundet Svensk kyrkomusik.
1. Church music—Sweden—Periodicals. I. Kyrkomusikernas riksförbund. II. Sveriges kyrkosångsförbund. III. Riksförbundet svensk kyrkomusik.
ML5.K12 783'.026'41485 78-647785
MARC-S

CHURCH MUSIC
—SWEDEN—PERIODICALS (Continued)

Svensk kyrkomusik uppl. B. 20 jan. 1970–
Vagnhärad, Kyrkomusikernas riksförbund.
v. ill. 27 cm. 10 no. a year.
Continues: KMT/KSF. ⌈Uppl.⌉ B.
Organ of Kyrkomusikernas riksförbund and, Jan. 20, 1970–Dec. 20, 1971 of Sveriges kyrkosångsförbund; Jan. 20, 1972– of Riksförbundet Svensk kyrkomusik.
Vols. for Jan. 20, 1970–Dec. 20, 1971 issued with Sveriges kyrkosångsförbundet; Jan. 20, 1972– with Riksförbundet Svensk kyrkomusik.
1. Church music—Sweden—Periodicals. I. Kyrkomusikernas riksförbund. II. Sveriges kyrkosångsförbund. III. Riksförbundet svensk kyrkomusik.
ML5.K13 783'.026'41485 78-647786
MARC-S

—UNITED STATES—ECONOMIC ASPECTS

Steinmetz, C Martin.
The work and compensation of the church musician / ⌈prepared for the Boston Chapter, A.G.O., by C. Martin Steinmetz⌉. — ⌈Wellesley, Mass.⌉ : Boston Chapter, American Guild of Organists, c1970.
15 p. : graphs ; 22 cm.
1. Organists—United States—Salaries, pensions, etc. 2. Church music—United States—Economic aspects. I. American Guild of Organists. Boston Chapter. II. Title.
ML3795.S833 331.2'81'7830973 79-111544
79 MARC

—WALES—HISTORY AND CRITICISM

Paul, Leslie Douglas.
Music at Bangor Cathedral Church : some historical notes / by Leslie D. Paul. — ⌈Bangor : Bangor Cathedral⌉, 1975.
24 p. ; 21 cm. — (Bangor Cathedral monographs ; no. 1) GB76-24542
Cover title.
First published in Welsh music, Spring 1971.
Includes bibliographical references.
ISBN 0-905229-00-2 : £0.25
1. Church music—Wales—History and criticism. 2. Bangor Cathedral. I. Title. II. Series: Bangor Cathedral. Bangor Cathedral monographs ; no. 1.
ML3131.P38 783'.026'0942925 79-314621
79 MARC

CHURCH OF CHRIST OF LATTER-DAY SAINTS see Church of Jesus Christ of Latter-Day Saints; Mormons and Mormonism

CHURCH OF CREATIVE BIO-DYNAMICS

Edwards, Floyd M
Manual of disciplines / prepared by Floyd M. Edwards and Jacquelyn J. Edwards. — Georgetown, CA : Church of Creative Bio-dynamics, c1979.
iii, 187 p. ; 19 cm.
1. Church of Creative Bio-dynamics. I. Edwards, Jacquelyn J., joint author. II. Church of Creative Bio-dynamics. III. Title.
BP605.C53E34 289.1 79-51168
79 MARC

CHURCH OF ENGLAND

see also Anglicans; Cathedrals; Church and state in Great Britain; Dissenters, Religious—England; Pentecostalism—Church of England; Puritans

—ARCHIVES—CATALOGS

Kitching, C J
The central records of the Church of England : a report and survey presented to the Pilgrim and Radcliffe Trustees / ⌈by C. J. Kitching⌉. — London : CIO Publishing, 1976.
iv, 100 p. ; 22 cm. GB***
ISBN 0-7151-3678-X : £3.00
1. Church of England—Archives—Catalogs. 2. Church records and registers—England. I. Church of England. II. Pilgrim Trust. III. Radcliffe Trust. IV. Title.
CD1069.L72K57 026.283'42 77-352399
77⌈r79⌉rev MARC

—BIOGRAPHY

see also Church of England—Clergy

—BISHOPS

Yates, Timoty, E
Venn and Victorian bishops abroad : the missionary policies of Henry Venn and their repercussions upon the Anglican episcopate of the colonial period 1841-1872 / T. E. Yates. — Uppsala : Swedish inst. of missionary research ⌈Sv. inst. för missionsforskning⌉ ; London : SPCK, 1978.
212 p. ; 23 cm. — (Studia missionalia Upsaliensia ; 33 ISSN 0585-5373)
S78-27
Thesis—Uppsala.
Bibliography: p. 203-208.
Includes index.
ISBN 918542403X : kr50.00
1. Venn, Henry, 1796-1873. 2. Church of England—Bishops. 3. Church of England—Missions. 4. Church Missionary Society—History. I. Title. II. Series.
BX5199.V443Y37 283'.092'4 79-302155
79 MARC

—BISHOPS—BIOGRAPHY

Cosin, John, 1595-1672.
The correspondence of John Cosin, D.D., lord bishop of Durham: together with other papers illustrative of his life and times ... Durham ⌈Eng.⌉ Pub. for the Society by Andrews & co.; ⌈etc., etc.⌉ 1869-72.
2 v. 23 cm. (The publications of the Surtees society ... vol. LII, LV)
Edited by George Ornsby.
Consists chiefly of letters to Bishop Cosin.
"Corrigenda": v. 1, p. xlv; v. 2, p. ⌈lvi⌉
1. Cosin, John, 1595-1672. 2. Church of England—Bishops—Biography. 3. Bishops—England—Biography. I. Ornsby, George, 1809-1886, ed.
DA20.S9 vol. 52, etc. 942.0062 s cs 35-13
⌈BX5199.C66⌉ ⌈r79⌉rev MARC

Huntley, Frank Livingstone, 1902-
Bishop Joseph Hall, 1574-1656 : a biographical and critical study / Frank Livingstone Huntley. — Cambridge ⌈Eng.⌉ : D. S. Brewer, 1979.
viii, 180 p. ; 22 cm. GB***
Includes bibliographical references and index.
ISBN 0-85991-035-0 : £7.00
1. Hall, Joseph, Bp. of Norwich, 1574-1656. 2. Church of England—Bishops—Biography. 3. Bishops—England—Biography. I. Title.
BX5199.H25H86 283'.092'4 79-309300
79 MARC

Ross Williamson, Hugh, 1901-
Jeremy Taylor / Hugh Ross Williamson. — Philadelphia : R. West, 1978.
179 p. : ill. ; 23 cm.
Reprint the 1952 ed. published by D. Dobson, London, in series: A Pegasus biography.
Bibliography: p. 176.
Includes index.
ISBN 0-8492-3027-6 : $25.00
1. Taylor, Jeremy, Bp. of Down and Connor, 1613-1667. 2. Church of England—Bishops—Biography. 3. Bishops—England—Biography.
⌈BX5199.T3R65 1978⌉ 283'.092'4 78-882
78 MARC

—CLERGY

see also Anglican orders

—CLERGY—BIOGRAPHY

Carroll, Peter N
The other Samuel Johnson : a psychohistory of early New England / Peter N. Carroll. — Rutherford ⌈N.J.⌉ : Fairleigh Dickinson University Press, c1978.
247 p. ; 22 cm.
Bibliography: p. 230-244.
Includes index.
ISBN 0-8386-2059-0 : $16.50
1. Johnson, Samuel, 1696-1772. 2. Church of England—Clergy—Biography. 3. Clergy—New England—Biography. 4. New England—Biography. I. Title.
BX5995.J59C37 283'.092'4 77-74413
78 MARC

Dicken, Hélène.
Women and the apostolic ministry / ⌈by⌉ Hélène Dicken. — London : Church Literature Association, 1978.
9, ⌈3⌉ p. ; 21 cm. — (The Ordination of women) GB78-27857
Cover title.
£0.28
1. Ordination of women—Church of England. 2. Church of England—Clergy. I. Title. II. Series.
BV676.D5 262'.14 79-300127
79 MARC

Distad, N Merrill.
Guessing at truth : the life of Julius Charles Hare (1795-1855) / by N. Merrill Distad. — Shepherdstown, W. Va. : Patmos Press, 1979.
xii, 258 p. ; 22 cm.
Bibliography: p. 239-247.
ISBN 0-915762-07-2 : $19.95
1. Hare, Julius Charles, 1795-1855. 2. Church of England—Clergy—Biography. 3. Clergy—England—Biography. I. Title.
BX5199.H35D57 283'.092'4 78-11625
78 MARC

Falconer, Ronald.
The kilt beneath my cassock / Ronnie Falconer. — Edinburgh : Handsel Press, 1978.
viii, 254 p., ⌈9⌉ leaves of plates : ill. ; 23 cm. GB***
Includes index.
ISBN 0-905312-07-4 : £4.75. ISBN 0-905312-02-3 pbk.
1. Falconer, Ronald. 2. Church of England—Clergy—Biography. 3. Clergy—Scotland—Biography. 4. Radio in religion—Scotland. 5. Television in religion—Scotland. I. Title.
BX5199.F26A33 283'.092'4 78-325343
79 MARC

Jordan, Hugh, 1906-
Born under a lucky star : the memoirs of Hugh Jordan. — Bramcote : Grove Books, 1977.
84 p. ; 22 cm. GB78-06648
ISBN 0-905422-20-1 : £1.75
1. Jordan, Hugh, 1906- 2. Church of England—Clergy—Biography. 3. Clergy—England—Biography. I. Title.
BX5199.J65A33 283'.092'4 79-315515
79 MARC

Le Quesne, A Laurence.
After Kilvert / A. L. Le Quesne. — Oxford ⌈Eng.⌉ ; New York : Oxford University Press, 1978.
ix, 233 p. : ill. ; 23 cm. GB78
Includes index.
ISBN 0-19-211748-3 : £4.95
1. Kilvert, Robert Francis, 1840-1879. Kilvert's diary. 2. Kilvert, Robert Francis, 1840-1879. 3. Church of England—Clergy—Biography. 4. Clergy—England—Biography. 5. Le Quesne, A. Laurence. 6. Clyro, Wales—Description. I. Title.
BX5199.K49A32 941.08 78-40173
78 MARC

Lock, John.
A man of sorrow : the life, letters, and times of the Rev. Patrick Brontë, 1777-1861 / by John Lock and W. T. Dixon ; foreword by the Archbishop of Canterbury (formerly the Archbishop of York). — London : I. Hodgkins ; Westport, Conn. : Meckler Books, 1979.
xiv, 566 p., ⌈7⌉ leaves of plates : ill. ; 24 cm.
Reprint of the ed. published by Nelson, London.
Bibliography: p. 541-549.
Includes index.
ISBN 0-930466-15-2 : $35.95
1. Brontë, Patrick, 1777-1861—Biography. 2. Brontë family. 3. Authors, English—19th century—Biography. 4. Clergy—England—Biography. 5. Church of England—Clergy—Biography. I. Dixon, William Thomas, joint author. II. Title.
PR4174.B2L6 1979 828'.7'09 78-27130
78 MARC

Witts, Francis Edward, 1783-1854.
The diary of a Cotswold parson / ⌈by Francis Edward Witts⌉ ; chosen, edited, and introduced by David Verey. — Dursley : Alan Sutton ; Wakefield : distributed by EP Publishing, 1978.
⌈2⌉, 189 p. : ill., facsim., ports. ; 23 cm. GB78-32628
Includes bibliographical references and index.
ISBN 0-904387-19-4 : £6.95
1. Witts, Francis Edward, 1783-1854. 2. Church of England—Clergy—Biography. 3. Clergy—England—Gloucestershire—Biography. 4. Gloucestershire, Eng.—Biography. I. Verey, David, 1913- II. Title.
BX5199.W733A33 283'.092'4 79-305811
79 MARC

Wollaston, John Ramsden, 1791-1856.
Journals and diaries, 1841-1856, of Revd. John Ramsden Wollaston, M.A. / collected by A. Burton. — ⌈Nedlands, Australia : University of Western Australia Press ; Portland, Or. : distributed by International Scholarly Book Services, 1975⌉
2 v. : ill. ; 22 cm. Aus***
Vol. 1 first published in 1975 by Paterson Brokensha; vol. 2 first published in 1948 by Pitman.
Bibliography: v. 1, p. xvi.
Includes indexes.
CONTENTS: v. 1. Wollaston's Picton journal, 1841-1844.—v. 2. Wollaston's Albany journals, 1841-1856.
ISBN 0-85564-103-7 (v. 1)
1. Wollaston, John Ramsden, 1791-1856. 2. Church of England—Clergy—Biography. 3. Clergy—Western Australia—Biography. 4. Western Australia—Biography. I. Burton, Alfred, 1865-1947. II. Title.
BX5720.W64A34 1975 283'.092'4 79-309465
79 MARC

Wollaston, John Ramsden, 1791-1856.
Wollaston's Albany journals, 1848-1856 / edited with notes by Percy U. Henn ; introd. by Malcolm J. L. Uren. — Nedlands, Australia : University of Western Australia Press ; Portland, Or. : ⌈distributed by⌉ International Scholarly Book Services, 1975.
252 p. : ill. ; 22 cm. — (His Journals and diaries, 1841-1856, of Revd. John Ramsden Wollaston, M.A. ; v. 2)
"First published in 1948 by C. H. Pitman & Son."
Includes index.
ISBN 0-85564-104-5
1. Wollaston, John Ramsden, 1791-1856. 2. Church of England—Clergy—Biography. 3. Clergy—Australia—Western Australia—Biography. 4. Western Australia—Biography. I. Henn, Percy U. II. Title. III. Title: Albany journals, 1848-1856.
BX5720.W64A34 1975 vol. 2 79-309463
283'.092'4 R
79 MARC

CHURCH OF ENGLAND
—CLERGY—BIOGRAPHY (Continued)

Wollaston, John Ramsden, 1791-1856.
Wollaston's Picton journal, 1841-1844 / edited with introd. and notes by Canon Burton and Percy U. Henn. — ⌜Nedlands, Australia : University of Western Australia Press ; Portland, Or. : distributed by International Scholarly Book Services, 1975⌝
xxvi, 289 p. : ill. ; 22 cm. — (His Journals and diaries, 1841-1856, of Revd. John Ramsden Wollaston, M.A. ; v. 1) Aus***
"First published by Paterson Brokensha Pty. Ltd. 1975."
Bibliography: p. xvi.
Includes indexes.
ISBN 0-85564-103-7
1. Wollaston, John Ramsden, 1791-1856. 2. Church of England—Clergy—Biography. 3. Clergy—Australia—Western Australia—Biography. 4. Western Australia—Biography. I. Burton, Alfred, 1865-1947. II. Henn, Percy U. III. Title. IV. Title: Picton journal, 1841-1844.
BX5720.W64A34 1975 vol. 1 283'.092'4 79-309464
79 MARC

—CLERGY—BIOGRAPHY—ADDRESSES, ESSAYS, LECTURES

The View from the pulpit : Victorian ministers and society / edited and with an introd. by P. T. Phillips. — Toronto : Macmillan of Canada, c1978.
326 p. : ports. ; 24 cm. C***
Bibliography: p. ⌜299⌝-317.
Includes index.
ISBN 0-7705-1590-8
1. Church of England—Clergy—Biography—Addresses, essays, lectures. 2. Dissenters, Religious—England—Biography—Addresses, essays, lectures. 3. Catholic Church—Clergy—Biography—Addresses, essays, lectures. 4. Clergy—Great Britain—Biography—Addresses, essays, lectures. I. Phillips, Paul T., 1942-
BX5619.V53 280'.092'2 78-321384
79 MARC

—COLLECTED WORKS

Cudworth, Ralph, 1617-1688.
Collected works of Ralph Cudworth : facsimile editions / prepared by Bernhard Fabian. — Hildesheim ; New York : G. Olms, 19
v. : ill. ; 24 cm. GFR***
Photoreprint ed.
CONTENTS:
—v. 2. A treatise concerning eternal and immutable morality (1731) A treatise of freewill (1838)

Includes bibliographical references.
ISBN 3-487-06010-8 (v. 2)
1. Theology—Collected works—17th century. 2. Church of England—Collected works. I. Fabian, Bernhard. II. Title.
BR75.C82 1979 230'.3 79-313924
00 MARC

Lewis, Clive Staples, 1898-1963.
The joyful Christian : 127 readings from C. S. Lewis. — Boston : G. K. Hall, 1978, c1977.
xvii, 421 p. ; 24 cm.
Large print ed.
Bibliography: p. 417-421.
ISBN 0-8161-6634-X
1. Theology—Collected works—20th century. 2. Church of England—Collected works. 3. Christian life—Anglican authors—Collected works. 4. Large type books. I. Title.
⌜BX5037.L4 1978⌝ 230'.092'4 78-10549
78 MARC

—DOCTRINAL AND CONTROVERSIAL WORKS—ANGLICAN AUTHORS

Edwards, David Lawrence.
Your faith / ⌜by⌝ David L. Edwards. — London ⌜etc.⌝ : Mowbrays, 1978.
155 p., ⌜8⌝ p. of plates : col. ill. ; 23 cm. GB78-31908
Includes index.
ISBN 0-264-66399-3 : £3.75
1. Church of England—Doctrinal and controversial works—Anglican authors. 2. Anglican Communion—Doctrinal and controversial works—Anglican authors. I. Title.
BX5131.2.E32 230'.3 79-301562
79 MARC

—DOCTRINAL AND CONTROVERSIAL WORKS—CATHOLIC AUTHORS

Fisher, John, 1579-1641.
The answere unto the nine points of controversy 1626 / John Percy ⌜i.e. J. Fisher⌝. — Ilkley, ⌜Eng.⌝ : Scolar Press, 1978.
⌜6⌝, 160. 400 p. ; 22 cm. — (English recusant literature, 1558-1640 ; v. 379) GB***
Reprint of the 1626 ed.
"Reproduced (original size) from a copy in the British Library ... references: Allison and Rogers 605; STC 10911."
ISBN 0-85967-468-1
1. Catholic Church—Doctrinal and controversial works—Catholic authors. 2. Church of England—Doctrinal and controversial works—Catholic authors. 3. White, Francis, bp. of Ely, 1564?-1638. 4. James I, King of Great Britain, 1566-1625. I. Title. II. Series: Rogers, David Morrison, comp. English recusant literature, 1558-1640 ; v. 379.
BX1750.A1E5 vol. 379 230'.2 s 78-322029
⌜BX1750⌝ 78 MARC

Osorio, Jeronymo, Bp. of Silves, 1506-1580.
Amplissimi atqve doctissimi viri D. Hieronymi Osorii, Episcopi Sylvensis, In Gvaltearvm Haddonvm Magistrvm Libellorvm Svpplicvm apud clarissimam Principe Helisabetham, Angliae, Franciae, & Hiberniae Reginam, libri tres. Olissipone, Excudebat F. Correa, Regius Typographus, 1567.
157 l. 21 cm.
A reply to W. Haddon's Pro reformatione Anglicana epistola apologetica ad Hier. Osorium.
1. Catholic Church—Doctrinal and controversial works—Catholic authors. 2. Church of England—Doctrinal and controversial works—Catholic authors. I. Haddon, Walter, 1516-1572. Pro reformatione Anglicana epistola apologetica. II. Title: In Gualterum Haddonum.
BX1750.O86 78-381539
78 MARC

—DOCTRINAL AND CONTROVERSIAL WORKS—CHURCH OF ENGLAND AUTHORS

Wansey, Christopher.
The clockwork church : reform in the Church of England at worm's eye level / by Christopher Wansey. — Oxford : Becket Publications, c1978.
138 p., ⌜8⌝ leaves of plates : ill. ; 22 cm. GB***
£1.95
1. Church of England—Doctrinal and controversial works—Church of England authors. 2. Church renewal—Church of England. 3. Wansley, Christopher. I. Title.
BX5131.2.W36 262'.001 79-309118
79 MARC

—DOCTRINAL AND CONTROVERSIAL WORKS—CHURCH OF ENGLAND AUTHORS—ADDRESSES, ESSAYS, LECTURES

Ryle, John Charles, Bp. of Liverpool, 1816-1900.
Charges and addresses / ⌜by⌝ J. C. Ryle. — Edinburgh ; Carlisle, Pa. : Banner of Truth Trust, 1978.
xiv, 384 p. : map (on lining papers) ; 23 cm. GB79-07019
Includes bibliographical references and index.
ISBN 0-85151-267-4 : £4.50
1. Church of England—Pastoral letters and charges. 2. Church of England—Doctrinal and controversial works—Church of England and authors—Addresses, essays, lectures. I. Title.
BX5034.R9 1978 252'.03 79-312726
79 MARC

—DOCTRINAL AND CONTROVERSIAL WORKS—COLLECTED WORKS

Clarke, Samuel, 1675-1729.
The works, 1738 / Samuel Clarke. — New York : Garland Pub., 1978.
4 v. : port. ; 29 cm. — (British philosophers and theologians of the 17th & 18th centuries)
Reprint of the 1738 ed. printed for J. and P. Knapton, London, under title: The works of Samuel Clarke.
CONTENTS: v. 1. Sermons on several subjects.—v. 2. Sermons on several subjects. Eighteen sermons on several occasions. Sixteen sermons on the being and attributes of God, the obligations of natural religion, and the truth and certainty of the Christian revelation.—v. 3. A paraphrase on the four Evangelists. Three practical essays on baptism, confirmation, and repentance. An exposition of the church catechism. A letter to Mr. Dodwell concerning the immortality of the soul. Reflections on Amyntor.—v. 4. The scripture doctrine of the Trinity. Several tracts relating to the subject of the Trinity. A collection of papers which passed between the late learned M. Leibnitz and Dr. Clarke. A letter to Benjamin Hoadly.
ISBN 0-8240-1762-5 : $25.00 per vol.
1. Church of England—Sermons. 2. Sermons, English. 3. Bible. N.T. Gospels—Paraphrases, English. 4. Church of England—Doctrinal and controversial works—Collected works. 5. Trinity—Early works to 1800—Collected works. 6. Clarke, Samuel, 1675-1729. I. Series.
BX5037.C5 1978 230'.3 75-11207
76 MARC

—DOCTRINAL AND CONTROVERSIAL WORKS—CONGRESSES

National Evangelical Anglican Congress, 2d, Nottingham, Eng., 1977.
The Nottingham statement : the official statement of the second National Evangelical Anglican Congress held in April 1977. — London : Church Pastoral Aid Society, 1977.
79 p. ; 18 cm. — (A Falcon booklet) GB78-07316
Cover title.
Addendum inserted.
Sponsored by the Church of England Evangelical Council, the Church Society, and the Church Pastoral Aid Society.
ISBN 0-85491-117-0 : £0.48
1. Evangelicalism—Church of England—Congresses. 2. Church of England—Doctrinal and controversial works—Congresses. I. Church of England. Evangelical Council. II. Church Society. III. Church Pastoral Aid Society. IV. Title.
BX5125.N33 1977 230'.3'41 78-314503
78 MARC

—FINANCE

Church of England. Joint Liaison Committee.
A resourceful Church? : a report on the finances of the Church of England, 1976-1979 / prepared on behalf of the Central Board of Finance and the Church Commissioners ⌜by the Joint Liaison Committee⌝. — London : Church Information Office, 1978.
⌜4⌝, 23 p. ; 21 cm. GB78-37458
"GS 381."
ISBN 0-7151-3685-2 : £0.65
1. Church of England—Finance. I. Church of England. Central Board of Finance. II. Church of England. Church Commissioners. III. Title.
BX5165.C455 1978 262 79-309858
79 MARC

—GOVERNMENT

Davies, Ebenezer Thomas.
Episcopacy and the royal supremacy in the Church of England in the XVI century / by E. T. Davies. — Westport, Conn. : Greenwood Press, 1978.
137 p. ; 23 cm.
Reprint of the 1950 ed. published by B. Blackwell, Oxford.
Includes bibliographical references.
ISBN 0-313-20626-0
1. Royal supremacy (Church of England) 2. Episcopacy. 3. Church of England—Government. I. Title.
⌜BX5157.D3 1978⌝ 262'.12 78-13202
78 MARC

—HISTORY

Wilkinson, Alan B
The Church of England and the First World War / Alan Wilkinson. — London : SPCK, 1978.
xiv, 370 p., ⌜4⌝ leaves of plates : ill. ; 23 cm. GB***
Bibliography: p. ⌜345⌝-357.
Includes index.
ISBN 0-281-03616-0 : £10.00
1. European War, 1914-1918—Church of England. 2. Church of England—History. I. Title.
D639.C54W54 283'.42 79-300124
79 MARC

—MISSIONS

Hilliard, David Lockhart, 1941-
God's gentlemen : a history of the Melanesian Mission, 1849-1942 / ⌜by⌝ David Hilliard. — St. Lucia, Q. : University of Queensland Press, 1978.
xv, 342 p. : ill. ; 23 cm. Aus
Distributed in the United Kingdom by Prentice-Hall International, Hemel Hempstead, Eng.
Bibliography: p. ⌜303⌝-328.
Includes index.
ISBN 0-7022-1066-8
1. Missions—Melanesia. 2. Church of England—Missions. 3. Church of the Province of New Zealand. Missionary dioceses. Melanesia. I. Title.
BV3675.H54 266'.3'93 78-319010
78 MARC

Yates, Timoty, E
Venn and Victorian bishops abroad : the missionary policies of Henry Venn and their repercussions upon the Anglican episcopate of the colonial period 1841-1872 / T. E. Yates. — Uppsala : Swedish inst. of missionary research ⌜Sv. inst. för missionsforskning⌝ ; London : SPCK, 1978.
212 p. ; 23 cm. — (Studia missionalia Upsaliensia ; 33 ISSN 0585-5373) S78-27
Thesis—Uppsala.
Bibliography: p. 203-208.
Includes index.
ISBN 918542403X : kr50.00
1. Venn, Henry, 1796-1873. 2. Church of England—Bishops. 3. Church of England—Missions. 4. Church Missionary Society—History. I. Title. II. Series.
BX5199.V443Y37 283'.092'4 79-302155
79 MARC

—PASTORAL LETTERS AND CHARGES

Ryle, John Charles, Bp. of Liverpool, 1816-1900.
Charges and addresses / ⌜by⌝ J. C. Ryle. — Edinburgh ; Carlisle, Pa. : Banner of Truth Trust, 1978.
xiv, 384 p. : map (on lining papers) ; 23 cm. GB79-07019
Includes bibliographical references and index.
ISBN 0-85151-267-4 : £4.50
1. Church of England—Pastoral letters and charges. 2. Church of England—Doctrinal and controversial works—Church of England and authors—Addresses, essays, lectures. I. Title.
BX5034.R9 1978 252'.03 79-312726
79 MARC

—PRAYER-BOOKS AND DEVOTIONS—ENGLISH

Cosin, John, 1595-1672.
A collection of private devotions; edited by P. G. Stanwood with the assistance of Daniel O'Connor. London, Oxford U.P., 1967.
lv, 371 p. front. (port.), 2 plates (facsims.), col. tables. 23 cm. 75/- B67-26310
Originally published, London, 1627.
1. Church of England—Prayer-books and devotions—English. I. Stanwood, P. G., ed. II. O'Connor, Daniel. III. Title.
BX5145.A55C64 1967 242'.1 68-76791
⌜r79⌝rev MARC

CHURCH OF ENGLAND (Continued)

—RELATIONS

Leonard, Graham.
The Church of England and the ten propositions / by the Bishop of Truro and Peter Boulton. — London : Church Literature Association for the Church Union, 1977.
[2], 10 p. ; 21 cm. GB78-08069
ISBN 0-85191-105-6 : £0.12
1. Christian union—England. 2. Church of England—Relations. I. Boulton, Peter, joint author. II. The Church Union. III. Title.
BR759.L37 274.1 78-315581
78 MARC

—RELATIONS—CATHOLIC CHURCH

Albion, Gordon, 1906-
Charles I and the Papacy / by Gordon Albion. — Ilford : Royal Stuart Society, 1974.
19 p. ; 21 cm. — (Royal Stuart papers ; 6) GB76-07982
Includes bibliographical references.
ISBN 0-9502443-6-8 : £0.50
1. Charles I, King of Great Britain, 1600-1649—Religion. 2. Catholic Church—Relations—Church of England. 3. Church of England—Relations—Catholic Church. I. Title.
DA396.A22A63 322'.1'0942 78-316444
78 MARC

—RENEWAL see Church renewal—Church of England

—SERMONS

Clarke, Samuel, 1675-1729.
The works, 1738 / Samuel Clarke. — New York : Garland Pub., 1978.
4 v. : port. ; 29 cm. — (British philosophers and theologians of the 17th & 18th centuries)
Reprint of the 1738 ed. printed for J. and P. Knapton, London, under title: The works of Samuel Clarke.
CONTENTS: v. 1. Sermons on several subjects.—v. 2. Sermons on several subjects. Eighteen sermons on several occasions. Sixteen sermons on the being and attributes of God, the obligations of natural religion, and the truth and certainty of the Christian revelation.—v. 3. A paraphrase on the four Evangelists. Three practical essays on baptism, confirmation, and repentance. An exposition of the church catechism. A letter to Mr. Dodwell concerning the immortality of the soul. Reflections on Amyntor.—v. 4. The scripture doctrine of the Trinity. Several tracts relating to the subject of the Trinity. A collection of papers which passed between the late learned M. Leibnitz and Dr. Clarke. A letter to Benjamin Hoadly.
ISBN 0-8240-1762-5 : $25.00 per vol.
1. Church of England—Sermons. 2. Sermons, English. 3. Bible. N.T. Gospels—Paraphrases, English. 4. Church of England—Doctrinal and controversial works—Collected works. 5. Trinity—Early works to 1800—Collected works. 6. Clarke, Samuel, 1675-1729. I. Series.
BX5037.C5 1978 230'.3 75-11207
76 MARC

Ford, Douglas William Cleverley.
New preaching from the New Testament / [by] D. W. Cleverley Ford. — London : Mowbrays, 1977.
x, 114 p. ; 22 cm. GB78-02980
ISBN 0-264-66401-9 : £2.50
1. Church of England—Sermons. 2. Sermons, English. I. Title.
BX5133.F73N48 252'.03 79-313608
79 MARC

Glanvill, Joseph, 1636-1680.
Some discourses, sermons, and remains : 1681 / Joseph Glanvill. — New York : Garland Pub., 1979.
422, 25 p. ; 23 cm. — (British philosophers and theologians of the 17th & 18th centuries)
Reprint of the 1681 ed. printed for H. Mortlock and J. Collins, London.
ISBN 0-8240-1775-7 : $29.50
1. Church of England—Sermons. 2. Sermons, English. I. Title. II. Series.
BX5133.G53S65 1979 252'.03 75-11221
78 MARC

CHURCH OF ENGLAND. BOOK OF COMMON PRAYER

Church of England. Book of common prayer.
[The booke of common prayer, and administration of the sacraments. London, B. Norton and J. Bill, 1621]
[91] p. 23 cm.
Signatures: [ ]⁴, A⁸, B¹⁰, C-E⁸.
The Psalter (sig. C1b-E6a) is followed by Godly prayers (E6b-E8a); the Ordinal is not included.
STC (2d ed.) 16357.7.
L.C. copy imperfect: p. [1-8] and [33-36] (the 4 unsigned leaves, incl. t.p., and signatures B5-6) wanting.
Bound with Speed, John. The genealogies. [London, 1624 or 5]; Bible. English. 1625. Authorized. The Holy Bible. London, 1625; Herrey, R. F. Two right profitable and frvitfvll concordances. London, 1622; and Sternhold, Thomas. The whole booke of psalmes. London, 1625.
I. Title.
BX5145.A4 1621 79-305176
79 MARC

Church of England. Book of Common Prayer.
The Book of common prayer and administration of the sacraments and other rites and ceremonies of the Church according to the use of the Church of England together with the Psalter or Pslams of David pointed as they are to be sung or said in churches. London, J. Sturt, 1717.
xxii, 166, [1] p. illus., ports. 21 cm.
Engraved throughout.
Imprint reads: London, Engraven and printed by the permission of Mr. John Baskett printer to the Kings most Excellent Majesty 1717. Sold by John Sturt engraver ..
L.C. copy imperfect: movable disk on p. v wanting (supplied in photocopy)
"Books and plates engraven and sold by John Sturt": p. [167]
Provenance: I. White, New Coll., Ox. (bookplate); Arthur Haddaway.
Contemporary gold-tooled black morocco.
I. Sturt, John, 1658-1730.
BX5145.A4 1717a Rosenwald Coll. 78-319857
78 MARC

Church of England. Book of common prayer.
The liturgy of the Church of England. [London, ca. 1717]
85 p. illus., port. 18 cm.
Engraved throughout by J. Sturt.
A shortened ed. of the Book of common prayer, omitting Epistle and Gospel texts. Sturt used many of the same engraved pages, surrounded by ornamental borders, in his complete ed. published in 1717 under title: The Book of common prayer ... In the shortened ed. the frontispiece port. of Queen Anne (instead of King George) and the collects are on totally different plates.
Label mounted on t. p.: Sold by Joseph Hazard, at the Bible against Stationers-Hall.
I. Sturt, John, 1658-1730. II. Title.
BX5145.A4 1717 78-319105
78 MARC

Hassel, Rudolph Chris, 1939-
Renaissance drama & the English church year / R. Chris Hassel, Jr. — Lincoln : University of Nebraska Press, c1979.
ix, 215 p. : ill. ; 22 cm.
Bibliography: p. 205-210.
Includes index.
ISBN 0-8032-2304-8
1. English drama—Early modern and Elizabethan, 1500-1600—History and criticism. 2. English drama—17th century—History and criticism. 3. Christianity in literature. 4. Church year. 5. Church of England. Book of common prayer. 6. Theater—England—History. I. Title.
PR658.R43H3 822'.3'09 78-24233
78 MARC

Morison, Stanley, 1889-1967.
English liturgical books / by Stanley Morison. — [Liverpool] : Radcliffe Liturgical Library, 1942.
xi, 123 p. ; 25 cm.
An unpublished preliminary version of the work issued in 1943 under title: English prayer books. Cf. J. Carter's note in Handover, P. M. Stanley Morison, a second handlist.
Imprint reads: Cambridge, At the University Press; printed by W. Lewis, M.A. for the Radcliffe Liturgical Library, 1942.
1. Liturgies. 2. Church of England. Book of common prayer. 3. Church of England. Liturgy and ritual. I. Title.
BV175.M6 1942 264'.00941 78-323064
79 MARC

CHURCH OF ENGLAND. BOOK OF COMMON PRAYER. PSALTER

see also Psalters

CHURCH OF ENGLAND. LITURGY AND RITUAL

see also Church dedication

Morison, Stanley, 1889-1967.
English liturgical books / by Stanley Morison. — [Liverpool] : Radcliffe Liturgical Library, 1942.
xi, 123 p. ; 25 cm.
An unpublished preliminary version of the work issued in 1943 under title: English prayer books. Cf. J. Carter's note in Handover, P. M. Stanley Morison, a second handlist.
Imprint reads: Cambridge, At the University Press; printed by W. Lewis, M.A. for the Radcliffe Liturgical Library, 1942.
1. Liturgies. 2. Church of England. Book of common prayer. 3. Church of England. Liturgy and ritual. I. Title.
BV175.M6 1942 264'.00941 78-323064
79 MARC

CHURCH OF ENGLAND IN AMERICA

—BISHOPS

Mills, Frederick V 1931-
Bishops by ballot : an eighteenth century ecclesiastical revolution / Frederick V. Mills, Sr. — New York : Oxford University Press, 1978.
xii, 367 p. ; 22 cm.
"The Frank S. and Elizabeth D. Brewer prize essay for 1975, the American Society of Church History."
Bibliography: p. 339-355.
Includes index.
ISBN 0-19-502411-7 : $14.95
1. Church of England in America—Bishops. 2. Bishops—Appointment, call, and election—History. I. Title.
BX5881.M55 262'.12 78-857
78 MARC

CHURCH OF ENGLAND IN AUSTRALIA

—BISHOPS—BIOGRAPHY

Shaw, George Peter, 1934-
Patriarch and patriot : William Grant Broughton 1788-1853 / [by] G. P. Shaw. — Carlton, Vic. : Melbourne University Press, 1978.
xii, 347 p. ; 25 cm. Aus
Bibliography: p. 318-325.
Includes index.
ISBN 0-522-84122-8 : $18.60
1. Broughton, William Grant, Bp. of Sydney, 1788-1853. 2. Church of England in Australia—Bishops—Biography. 3. Bishops—Australia—Biography. I. Title.
BX5720.B7S5 283'.092'4 78-320451
79 MARC

CHURCH OF ENGLAND IN AUSTRALIA IN GIPPSLAND, AUSTRALIA

—HISTORY

Maddern, Ian Theodore.
Light and life : a history of the Anglican church in Gippsland / compiled by I. T. Maddern. — Morwell, Vic. : I. T. Maddern, 1977.
119 p., 20 p. of plates : ill. ; 22 cm. Aus
ISBN 0-909625-11-5
1. Church of England in Australia in Gippsland, Australia—History. 2. Gippsland, Australia—Church history. I. Title.
BX5713.G56M32 283'.945 79-311099
79 MARC

CHURCH OF ENGLAND IN CANADA see Anglican Church of Canada

CHURCH OF ENGLAND IN CONNECTICUT

—HISTORY

Samuel Seabury, 1729-1796 : his election, consecration, and reception : the documentary history / re-edited with an index by Kenneth Walter Cameron. — Hartford : Transcendental Books, c1978.
80, 37 leaves, [10] leaves of plates : ill. ; 28 cm.
Leaves 1-75 are a reprint of Documentary history of the Protestant Episcopal church in the United States of America, edited by F. L. Hawks and W. S. Perry, published in 1863-1864 by J. Pott, New York. Leaves 10-25 (2d group) are a reprint of Kingston and the loyalists of the "Spring fleet" of A.D. 1783, by W. Bates, published in 1889 by Barnes, Saint John, N.B.
1. Seabury, Samuel, Bp., 1729-1796. 2. Church of England in Connecticut—History. 3. Protestant Episcopal Church in the U.S.A.—Connecticut—History. 4. Kingston, N.B.—History. 5. American loyalists—New Brunswick. 6. Connecticut—Church history. 7. New Brunswick—Church history. I. Cameron, Kenneth Walter, 1908- II. Hawks, Francis Lister, 1798-1866, ed. Documentary history of the Protestant Episcopal church in the United States of America. 1978. III. Bates, Walter, 1760-1842. Kingston and the loyalists of the "Spring fleet" of A.D. 1783. 1978.
BX5995.S3S25 283'.092'4 79-102266
79 MARC

CHURCH OF ENGLAND IN CORNWALL, ENG.

—EXHIBITIONS

Hull, P L
A thousand years : Kenstec to Benson / [by P. L. Hull]. — [Truro] : Truro Cathedral, 1977.
[7], 28 p. ; 30 cm. GB78-08764
Cover title.
Catalogue of an exhibition.
1. Cornwall, Eng.—Church history—Exhibitions. 2. Church of England in Cornwall, Eng.—Exhibitions. I. Title.
BR765.C6H84 274.23'78 78-325346
79 MARC

CHURCH OF ENGLAND IN INDIA

—BISHOPS—BIOGRAPHY

Graham, Carol.
Azariah of Dornakal / Carol Graham. — Rev. ed. — Delhi : Indian Society for Promoting Christian Knowledge ; Madras : Christian Literature Society, 1972.
vi, 118 p., [1] leaf of plates : ill. ; 20 cm.
1. Azariah, Vedanayakam Samuel, bp., 1874-1945. 2. Church of England in India—Bishops—Biography. 3. Bishops—India—Biography. I. Title.
BX5680.A9G7 1972 283'.092'4 78-912234
79 MARC

CHURCH OF ENGLAND IN IRELAND see Church of Ireland

CHURCH OF ENGLAND IN LITERATURE

Rataboul, Louis J
Le pasteur anglican dans le roman victorien : aspects sociaux et religieux / Louis J. Rataboul. — Paris : Didier Érudition, c1978.
495 p. ; 24 cm. — (Études anglaises ; 70) F***
Bibliography: p. [449]-474.
Includes indexes.
ISBN 2-208-03264-0 : 180F
1. English fiction—19th century—History and criticism. 2. Clergy in literature. 3. Church of England in literature. I. Title. II. Series.
PR1.E8 no. 70 820'.9 s 79-367666
[PR878.C56] 79 MARC

CHURCH OF GOD (ANDERSON, IND.)

—DOCTRINAL AND CONTROVERSIAL WORKS—ADDRESSES, ESSAYS, LECTURES

A Time to remember—projections / edited by Barry L. Callen. — Anderson, Ind. : Warner Press, c1978.
128 p. ; 21 cm. — (Church of God heritage series) (Pages also numbered 720-846 continuing the paging of other vols. in the series)
"Series cumulative index": p. 122-128.
Includes index.
ISBN 0-87162-209-2
1. Church of God (Anderson, Ind.)—Doctrinal and controversial works—Addresses, essays, lectures. I. Callen, Barry L. II. Series.
BX7026.T55 230'.9'9 79-105164
79 MARC

—HISTORY

Reardon, Robert H 1919-
The early morning light : a friendly reflection on some of the main events in the life of the Church of God reformation movement during the first fifty years / by Robert H. Reardon. — Anderson, Ind. : Warner Press, c1979.
96 p. : ill. ; 21 cm.
ISBN 0-87162-217-3
1. Church of God (Anderson, Ind.)—History. I. Title.
BX7025.R4 289.9 79-119968
79 MARC

CHURCH OF GOD (CLEVELAND, TENN.)

—HISTORY

Conn, Charles W
Like a mighty army : a history of the Church of God, 1886-1976 / Charles W. Conn. — Rev. ed. — Cleveland, Tenn. : Pathway Press, 1977.
xxxi, 477 p. : ill. ; 24 cm.
Bibliography: p. 457-460.
Includes index.
ISBN 0-87148-510-9
1. Church of God (Cleveland, Tenn.)—History. I. Title.
BX7032.C66 1977 289.9 77-82067
78[r79]rev MARC

CHURCH OF IRELAND

—HISTORY

Akenson, Donald H
The Church of Ireland : ecclesiastical reform and revolution, 1800-1885 / by Donald Harman Akenson. — New Haven : Yale University, 1971.
xiii, 413 p. : map ; 25 cm.
Bibliography: p. 379-408.
Includes index.
ISBN 0-300-01448-1
1. Church of Ireland—History. I. Title.
BX5500.A37 283'.415 76-151565
79 MARC

CHURCH OF IRELAND. LITURGY AND RITUAL

Pyle, Hilary.
You can say that again : common prayer in the Church of Ireland / by Hilary Pyle ; with line drawings by Alicia Boyle. — Dublin : A.P.C.K. [for] the Mothers' Union of the United Dioceses of Dublin and Glendalough, 1977.
64 p. : ill., facsims., plan, ports. ; 27 cm. GB78-10671
ISBN 0-900010-16-9 : £1.80
1. Church of Ireland. Liturgy and ritual. 2. Prayer-books—History. I. Title.
BX5535.P94 264'.03 78-320317
78 MARC

CHURCH OF IRELAND. REPRESENTATIVE CHURCH BODY. LIBRARY

—CATALOGS

Church of Ireland. Representative Church Body. Library.
The Watson collection : prayer books and related liturgical works given by Edward John Macartney Watson, M.D. : catalogue / compiled by Newport B. White. — Dublin: A.P.C.K., 1948.
52 p. ; 21 cm.
At head of title: Representative Church Body, Library.
Includes index.
1. Liturgics—Anglican Communion—Bibliography—Catalogs. 2. Church of Ireland. Representative Church Body. Library—Catalogs. I. White, Newport Benjamin. II. Title.
Z7813.C48 1948 016.264'03 79-310361
[BX5145] 79 MARC

CHURCH OF JESUS CHRIST OF LATTER-DAY SAINTS

Davies, Joseph Kenneth, 1925-
Deseret's sons of toil : a history of the worker movements of Territorial Utah, 1852-1896 / J. Kenneth Davies. — Salt Lake City : Olympus Pub. Co., c1977.
264 p. : ill. ; 24 cm.
Bibliography: p. 227-252.
Includes indexes.
ISBN 0-913420-64-6 : $12.50
1. Church and labor—Utah—History. 2. Church of Jesus Christ of Latter Day Saints. 3. Trade-unions—Utah—History. I. Title.
HD6338.2.U52U82 261.8'5 76-26459
76 MARC

—BIOGRAPHY

DeMille, Janice Force, 1946-
The St. George Temple, first 100 years / Janice Force DeMille. — Hurricane, Utah : Homestead Publishers, c1977.
427 p., [1] leaf of plates : ill. ; 24 cm.
Includes bibliographical references.
1. St. George Temple, St. George, Utah. 2. Church of Jesus Christ of Latter-Day Saints—Biography. I. Title.
BX8685.S34D45 288.3'792'48 78-113153
79 MARC

—DOCTRINAL AND CONTROVERSIAL WORKS

Gospel principles. — Salt Lake, Utah : Deseret Book Co., 1978.
vi, 362, 16 p. : ill. (some col.) ; 24 cm.
Manual prepared for use by new members of the Church of Jesus Christ of Latter-day Saints.
Bibliography: p. 361-362.
1. Church of Jesus Christ of Latter-day Saints—Doctrinal and controversial works. I. Church of Jesus Christ of Latter-day Saints.
BX8635.2.G67 230'.9'33 79-103909
79 MARC

Kaiser, Edgar P
How to respond to ... the Latter Day Saints / Edgar P. Kaiser. — St. Louis : Concordia Pub. House, c1977.
40 p. ; 23 cm. — (The Response series)
Bibliography: p. 40.
ISBN 0-570-07680-3 : $1.25
1. Church of Jesus Christ of Latter-Day Saints—Doctrinal and controversial works. I. Title. II. Title: Latter-Day Saints.
BX8645.K28 230'.9'3 77-10155
77 MARC

McElveen, Floyd.
The Mormon revelations of convenience / Floyd McElveen. — Minneapolis : Bethany Fellowship, c1978.
108 p. ; 18 cm.
Bibliography: p. 107-108.
ISBN 0-87123-385-1 : $1.75
1. Revelation (Mormonism)—Controversial literature. 2. Church of Jesus Christ of Latter-day Saints—Doctrinal and controversial works. I. Title.
BX8643.R4M32 230'.9'3 78-72945
79 MARC

—DOCTRINAL AND CONTROVERSIAL WORKS—ADDRESSES, ESSAYS, LECTURES

God the Father / compiled by Gordon Allred. — Salt Lake City : Deseret Book Co., 1979.
ix, 316 p. ; 24 cm.
Includes bibliographical references and index.
ISBN 0-87747-746-9 : $6.95
1. God—Addresses, essays, lectures. 2. Church of Jesus Christ of Latter-Day Saints—Doctrinal and controversial works—Addresses, essays, lectures. I. Allred, Gordon T.
BT102.A1G55 231 79-545
79 MARC

—HISTORY

Arrington, Leonard J
The Mormon experience : a history of the Latter-Day Saints / Leonard J. Arrington and Davis Bitton. — 1st ed. — New York : Knopf : distributed by Random House, 1979.
xiv, 404 p., [10] leaves of plates : ill. ; 25 cm.
Bibliography: p. [341]-344.
Includes index.
ISBN 0-394-46566-0 : $15.00
1. Church of Jesus Christ of Latter-day Saints—History. I. Bitton, Davis, 1930- joint author. II. Title.
BX8611.A78 289.3'09 78-20561
78 MARC

Taylor, Samuel Woolley, 1907-
Rocky Mountain Empire : the Latter-day Saints today / Samuel W. Taylor. — New York : Macmillan, 1978.
xi, 273 p. ; 24 cm.
Bibliography: p. 261-266.
Includes index.
ISBN 0-02-616610-0
1. Church of Jesus Christ of Latter-day Saints—History. 2. Mormons and Mormonism in Utah—History. 3. Utah—Church history. I. Title.
BX8611.T37 289.3'3 78-9655
78 MARC

—HISTORY—JUVENILE LITERATURE

Bickerstaff, George.
The church story / by George Bickerstaff ; ill. by Douglas Jordan. — Salt Lake City, Utah : Bookcraft, 1974-
v. : col. ill. ; 29 cm.
SUMMARY: Presents the history of the Church of Jesus Christ of Latter-Day Saints from Joseph Smith's first vision to the death of Brigham Young.
CONTENTS: book 1. From Sacred Grove to Land of Zion.

ISBN 0-88494-274-0 (v. 1)
1. Church of Jesus Christ of Latter-Day Saints—History—Juvenile literature. [1. Church of Jesus Christ of Latter-Day Saints—History. 2. Mormons and Mormonism] I. Jordan, Douglas, ill. II. Title.
BX8611.B4 289.3'3 74-21758
75[r79]rev MARC

—JUVENILE LITERATURE

Jacobson, Marsha T
Through the eyes of little saints / written by Marsha T. Jacobson ; ill. by Bill Kuhre. — Salt Lake City, Utah : Bookcraft, c1978.
[23] p. : col. ill. ; 26 cm.
SUMMARY: Labeled pictures and brief text explain activities at the Church of Jesus Christ of Latter-Day Saints' meetinghouses.
ISBN 0-88494-355-0
1. Church of Jesus Christ of Latter-Day Saints—Juvenile literature. [1. Church of Jesus Christ of Latter-Day Saints. 2. Mormons and Mormonism] I. Kuhre, William. II. Title.
BX8638.J3 264'.083 78-72277
79 MARC

—MISSIONS

Durrant, George D
Get ready! Get called! Go! / George D. Durrant. — Salt Lake City, Utah : Bookcraft, c1979.
xi, 72 p. ; 24 cm.
ISBN 0-88494-363-1 : $3.95
1. Church of Jesus Christ of Latter-Day Saints—Missions. I. Title.
BX8661.D87 266'.9'33 78-74966
00 MARC

Palmer, Spencer J
The expanding church / Spencer J. Palmer. — Salt Lake City : Deseret Book Co., 1978.
xi, 232 p. ; 24 cm.
Includes bibliographical references and index.
ISBN 0-87747-732-9
1. Church of Jesus Christ of Latter-day Saints—Missions. I. Title.
BX8661.P34 266'.9'33 78-26082
78 MARC

—AUSTRALIA

Australia Brisbane Mission.
Sketches of Australia and the Church of Jesus Christ of Latter-Day Saints, 1840-1975 / Australia Brisbane Mission. — Windsor, Q. : Australia Brisbane Mission, 1975.
vi, 54 p. : ill., map ; 17 cm.
Aus
ISBN 0-9597106-0-4
1. Australia. 2. Church of Jesus Christ of Latter-day Saints—Australia. I. Title.
DU110.A87 1975 994 78-326698
79 MARC

CHURCH OF JESUS CHRIST OF LATTER-DAY SAINTS. RELIEF SOCIETY

Hoole, Daryl Van Dam.
Our own society / Daryl V. Hoole ; photography by Libby Frech. — Salt Lake City : Bookcraft, c1979.
81 p. : ill. ; 24 cm.
ISBN 0-88494-366-6 : $4.50
1. Church of Jesus Christ of Latter-Day Saints. Relief Society. 2. Women—Religious life. I. Title.
BX8643.W4H66 267'.44'933 79-63403
79 MARC

CHURCH OF SCIENTOLOGY. MINISTRY OF PUBLIC RELATIONS

The Dossier disease : the truth about government agency false files. — [Hollywood, Calif.] : Church of Scientology, Ministry of Public Relations, c1979.
31 p. : ill. ; 28 cm.
ISBN 0-915598-27-2 : $3.50
1. Privacy, Rights of—United States. 2. Public records—United States. 3. Church of Scientology. Ministry of Public Relations.
JC599.U5D68 323.44 79-111954
00 MARC

CHURCH OF SCIENTOLOGY OF CALIFORNIA, LOS ANGELES

Council of Scientology Ministers.
A study of religious discrimination : a submission to the U.S. Commission on Civil Rights / presented by the Council of Scientology Ministers. — Los Angeles, CA : The Council, c1979.
48 p. : ill. ; 28 cm.
ISBN 0-915598-29-9
1. Church of Scientology of California, Los Angeles. 2. Persecution—United States—Case studies. I. United States. Commission on Civil Rights. II. Title.
BP605.S2C68 1979 322'.1 79-117006
79 MARC

CHURCH OF SCOTLAND

see also Presbyterian Church

—CLERGY

Blakey, Ronald S
The man in the manse / Ronald S. Blakey. — Edinburgh : Handsel Press, 1978.
x, 150 p. ; 23 cm. GB***
Bibliography: p. [137]-142.
Includes index.
ISBN 0-905312-05-8 : £4.50
1. Church of Scotland—Clergy. 2. Clergy—Scotland. I. Title.
BX9079.B58 253 79-308669
79 MARC

—HISTORY

Makey, Walter.
The Church of the Covenant, 1637-1651 : revolution and social change in Scotland / Walter Makey. — Edinburgh : John Donald Publishers, 1979.
vii, 216 p. ; 24 cm. GB***
Bibliography: p. [203]-210.
Includes index.
ISBN 0-85976-035-9 : £12.00
1. Church of Scotland—History. 2. Church and state in Scotland—History. 3. Scotland—Church history—17th century. I. Title.
BX9071.M28 285'.2'411 79-314845
79 MARC

—RELATIONS—CATHOLIC CHURCH

Church of Scotland. Panel on Doctrine.
Agreement and disagreement : the common ground and differences in fundamental belief between the Church of Scotland and the Roman Catholic Church / [Church of Scotland Panel on Doctrine]. — Edinburgh : St. Andrew Press, 1977.
[2], 16 p. ; 22 cm. GB78-02983
Cover title.
ISBN 0-7152-0369-X : £0.30
1. Church of Scotland—Relations—Catholic Church. 2. Catholic Church—Relations—Church of Scotland. I. Title.
BX1789.P73C48 1977 230'.2 79-306069
79 MARC

CHURCH OF THE BRETHREN

—MIDDLE ATLANTIC STATES—HISTORY—SOURCES

MacMaster, Richard Kerwin, 1935-
Conscience in crisis : Mennonites and other peace churches in America, 1739-1789 : interpretation and documents / Richard K. MacMaster, with Samuel L. Horst and Robert F. Ulle. — Scottdale, Pa. : Herald Press, 1979.
575 p. ; 23 cm. — (Studies in Anabaptist and Mennonite history ; no. 20)
Bibliography: p. [549]-560.
Includes index.
ISBN 0-8361-1213-X : $17.95
1. Mennonites in the Middle Atlantic States—History—Sources. 2. Friends, Society of—Middle Atlantic States—History—Sources. 3. Schwenkfelders in the Middle Atlantic States—History—Sources. 4. Church of the Brethren—Middle Atlantic States—History—Sources. 5. Moravian Church—Middle Atlantic States—History—Sources. 6. Nonviolence—Moral and religious aspects. 7. Pacifism—History—Sources. 8. Middle Atlantic States—Church history—Sources. 9. Middle Atlantic States—History—Sources. I. Horst, Samuel, 1919- joint author. II. Ulle, Robert F., 1948- joint author. III. Title. IV. Series.
BX8116.42.M33 261.8'73 78-27530
78 MARC

CHURCH OF THE NAZARENE

—BIOGRAPHY

Kiemel, Ann.
Yes / Ann Kiemel. — Wheaton, Ill. : Tyndale House Publishers, c1978.
127 p. : ill. ; 23 cm.
ISBN 0-8423-4653-8 : $5.95
1. Kiemel, Ann—United States. 2. Church of the Nazarene—Biography. I. Title.
BX8699.N38K534 248'.2 78-58743
79 MARC

—CLERGY—BIOGRAPHY

Moore, Paul, 1942-
Shepherd of Times Square / by Paul Moore & Joe Musser. — Nashville : T. Nelson, c1979.
249 p. ; 21 cm.
ISBN 0-8407-5166-4 : $9.95
1. Moore, Paul, 1942- . 2. Church of the Nazarene—Clergy—Biography. 3. Clergy—United States—Biography. I. Musser, Joe, joint author. II. Title.
BX8699.N38M665 289.9 79-16249
79 MARC

CHURCH OF THE PROVINCE OF NEW ZEALAND. MISSIONARY DIOCESES. MELANESIA

Hilliard, David Lockhart, 1941-
God's gentlemen : a history of the Melanesian Mission, 1849-1942 / [by] David Hilliard. — St. Lucia, Q. : University of Queensland Press, 1978.
xv, 342 p. : ill. ; 23 cm. Aus
Distributed in the United Kingdom by Prentice-Hall International, Hemel Hempstead, Eng.
Bibliography: p. [303]-328.
Includes index.
ISBN 0-7022-1066-8
1. Missions—Melanesia. 2. Church of England—Missions. 3. Church of the Province of New Zealand. Missionary dioceses. Melanesia. I. Title.
BV3675.H54 266'.3'93 78-319010
78 MARC

CHURCH OF THE REDEEMER, HOUSTON, TEX.

Harper, Michael, 1931-
A new way of living; how the Church of the Redeemer, Houston, found a new life-style. Plainfield, N.J., Logos International [1973]
144 p. 21 cm. $4.95
Includes bibliographical references.
ISBN 0-88270-066-9
1. Church of the Redeemer, Houston, Tex. I. Title.
BX5980.H66C53 283'.764'1411 73-180328
74[r79]rev MARC

CHURCH OF UGANDA, RWANDA, BURUNDI AND BOGA-ZAIRE

—ADDRESSES, ESSAYS, LECTURES

Taylor, John Vernon, 1914-
The secret people : an address by the Bishop of Winchester, the Rt Rev. John V. Taylor, DD, in celebration of the centenary of the Church of Uganda. — London : Church Missionary Society, 1977.
[12] p. ; 21 cm. GB78-03015
"Given at 157 Waterloo Road, London ... on Tuesday, June 21, 1977."
ISBN 0-900287-31-4 : £0.15
1. Church of Uganda, Rwanda, Burundi and Boga-Zaire—Addresses, essays, lectures. 2. Uganda—Church history—Addresses, essays, lectures. I. Title.
BX5700.8.T39 283'.676'1 79-312232
00 MARC

—BISHOPS—BIOGRAPHY

Ford, Margaret.
Janani, the making of a martyr / Margaret Ford. — London : Marshall, Morgan and Scott, 1978.
93 p. : map ; 18 cm. — (Lakeland ; 369) GB***
ISBN 0-551-00794-X : £0.95
1. Luwum, Janani, 1922-1977. 2. Church of Uganda, Rwanda, Burundi and Boga-Zaire—Bishops—Biography. 3. Bishops—Uganda—Biography. I. Title.
BX5700.8.L88F67 283'.092'4 79-300535
79 MARC

Kivengere, Festo.
I love Idi Amin : the story of triumph under fire in the midst of suffering and persecution in Uganda / [by] Festo Kivengere ; with Dorothy Smoker. — London : Marshall, Morgan and Scott, 1977.
63 p. : ports. ; 18 cm. — (New life ventures) GB78-12475
ISBN 0-551-05577-4 : £0.50
1. Kivengere, Festo. 2. Bishops—Uganda—Biography. 3. Church of Uganda, Rwanda, Burundi, and Boga—Zaire—Bishops—Biography. 4. Amin, Idi, 1925- I. Smoker, Dorothy, joint author. II. Title.
BX5700.8.Z8K58 1977b 283'.092'4 78-319506
78 MARC

CHURCH OFFICERS

see also Church secretaries; Church trustees; Deacons

CHURCH ORDERS, ANCIENT

see also Canon law; Church history—Primitive and early church, ca. 30-600; Church polity—Early church, ca. 30-600; Liturgies, Early Christian

CHURCH ORNAMENT see Church decoration and ornament

CHURCH PENNANTS

Lauckner, Edie.
Signs of celebration : how to design church banners / Edie Lauckner. — St. Louis : Concordia Pub. House, c1978.
32 p. : ill. ; 28 cm.
ISBN 0-570-03770-4 : $2.50
1. Wall hangings. 2. Church pennants. I. Title.
TT850.2.L38 746.3 77-13111
77 MARC

CHURCH PLATE

—EXHIBITIONS

Museo civico Giovanni Fattori.
I paramenti sacri della Chiesa greco-ortodossa della SS. Trinità di Livorno / a cura di Antonella Capitanio. — Pisa : Giardini, 1978.
36 p. : ill. ; 25 cm. — (Collezione monografica del Museo civico Giovanni Fattori di Livorno ; 2) It78-Oct
1. Ecclesiastical embroidery—History—18th century—Exhibitions. 2. Ecclesiastical embroidery—History—19th century—Exhibitions. 3. Church plate—Exhibitions. 4. Chiesa della Santissima Trinità, Livorno, Italy. I. Capitanio, Antonella. II. Title. III. Series: Museo civico Giovanni Fattori. Collezione monografica del Museo civico Giovanni Fattori di Livorno ; 2.
NK9310.M87 1978 746.9 78-397071
*78 MARC

—SCOTLAND

Burns, Thomas, fl. 1892.
Old Scottish communion plate, by the Rev. Thomas Burns ... with a preface by the Right Rev. James Macgregor .. and chronological tables of Scottish hall-marks prepared by Alexander J. S. Brook ... Edinburgh, R. & R. Clark, 1892.
xxx, 651, [1] p. front., illus., 84 pl. 30 cm.
"The edition of this book is limited to one hundred and seventy-five copies on large paper and five hundred copies on small paper".
"Impression—one hundred and seventy-five copies, this copy is no. 108.".
1. Church plate—Scotland. 2. Hall-marks. I. Macgregor, James, 1832-1910. II. Brook, Alexander J. S. III. Title.
NK7215.B9 739.2'3'8209411 10-20788
[r79]rev MARC

CHURCH POLITY

see also Cardinals; Church finance; Church membership; Clergy; Ecclesiastical law; Episcopacy; Indigenous Church administration; Investiture; Laity; Liberty of conscience; Puritans; and subdivision Government under church denominations, e.g. Church of England—Government

Harper, Michael, 1931-
Let my people grow! : ministry and leadership in the Church / by Michael Harper. — London : Hodder and Stoughton, 1977.
254 p. ; 22 cm. — (Ecclesia books) GB77-19209
Bibliography: p. [247]-248.
Includes index.
ISBN 0-340-21573-9 : £2.95
1. Church renewal—Church of England. 2. Church polity. 3. Christian leadership. I. Title.
BV600.2.H35 1977b 262'.001 77-379727
78[r79]rev MARC

CHURCH POLITY (Continued)

Harper, Michael, 1931-
Let my people grow : ministry and leadership in the church / Michael Harper. — Plainfield, N.J. : Logos International, c1977.
254 p. ; 21 cm.
Bibliography: p. [247]-248.
Includes index.
ISBN 0-88270-236-X : $3.50
1. Church renewal—Church of England. 2. Church polity. 3. Christian leadership.
BV600.2.H35 1977 262'.001 77-73840
77[r79]rev MARC

—EARLY CHURCH, CA. 30-600

Teaching of the Twelve Apostles. French & Greek.
La doctrine des Douze Apôtres = Didachè / introd., traduction, notes, appendice et index par Willy Rordorf et André Tuilier. — Paris : Éditions du Cerf, 1978.
228 p. ; 20 cm. — (Sources chrétiennes ; no 248) F***
Bibliography: p. [129]-135.
Includes indexes.
ISBN 2-204-01278-5 : 169F
1. Christian ethics—Early church, ca. 30-600. 2. Church polity—Early church, ca. 30-600. I. Rordorf, Willy. II. Tuilier, André. III. Title.
BS2940.T4F8 1978 230'.1'1 79-374963
79 MARC

—EARLY CHURCH, CA. 30-600—ADDRESSES, ESSAYS, LECTURES

Das Kirchliche Amt im Neuen Testament / hrsg. von Karl Kertelge. — Darmstadt : Wissenschaftliche Buchgesellschaft, [Abt. Verl.], 1977.
vii, 574 p. ; 20 cm. — (Wege der Forschung ; Bd. 439) GFR77-A41
Bibliography: p. [565]-574.
ISBN 3-534-06683-9 : DM89.00
1. Church polity—Early church, ca. 30-600—Addresses, essays, lectures. 2. Bible. N.T.—Criticism, interpretation, etc.—Addresses, essays, lectures. I. Kertelge, Karl.
BV648.K57 78-374112
78 MARC

CHURCH PROPERTY

see also Church lands; Convents and nunneries; Monasteries; Parsonages; Privileges and immunities, Ecclesiastical; Secularization

—TAXATION

see also Taxation, Exemption from

—ITALY—CONGRESSES

La Chiesa e i beni culturali : atti del Convegno di studio organizzato dalla Regione ecclesiastica toscana in collaborazione con la Commissione pontificia centrale per l'arte sacra in Italia : Lucca, 3-5 giugno 1977. — 1. ed. — Milano : Massimo, 1978.
99 p. ; 24 cm. It79-Apr
Includes bibliographical references.
1. Church property—Italy—Congresses. 2. Cultural property, Protection of—Law and legislation—Italy—Congresses. 3. Ecclesiastical law—Italy—Congresses. I. Catholic Church. Regione ecclesiastica toscana. II. Catholic Church. Pontificia commissione centrale per l'arte sacra in Italia.
79-369104
*79 MARC

—SPAIN—GRANADA

Casas, mezquitas y tiendas de los habices de las iglesias de Granada / edición, introd. e índices por Ma. del Carmen Villanueva Rico. — Madrid : Instituto Hispano-Arabe de Cultura, 1966.
170 p. ; 25 cm. — (Libros de habices del Reino de Granada ; 2)
Half title: Traslado del libro de la medida de las casas y mezquitas y tiendas de las yglesias de Granada y de sus habizes de ellas y su dotación por los Señores Reyes Católicos.
Introd. includes bibliographical references.
Includes indexes.
1. Churches—Spain—Granada. 2. Granada—Church history. 3. Church property—Spain—Granada. I. Villanueva Rico, María del Carmen. II. Instituto Hispano-Arabe de Cultura. III. Series.
BR1027.G7C37 79-108934
79 MARC

—SPAIN—SANTO DOMINGO DE LA CALZADA

Ubieto Arteta, Agustín.
Notas sobre el patrimonio calceatense en los siglos XII y XIII / [por] Agustín Ubieto Arteta. — Logroño : Gonzalo de Berceo, 1978.
93 p. : ill. ; 24 cm. — (Biblioteca de temas riojanos ; 19) Sp78-Jul
On t.p.: Servicio de Cultura de la Excma. Diputación Provincial.
"Conferencia pronunciada en el Instituto de Estudios Riojanos, el día 26 de febrero de 1976."
Includes bibliographical references.
ISBN 8485242211 : 300ptas
1. Santo Domingo de la Calzada, Spain—History. 2. Santo Domingo de la Calzada, Spain—Church history. 3. Church property—Spain—Santo Domingo de la Calzada. I. Title.
DP402.S26U24 946'.354 79-358275
79 MARC

CHURCH PUBLICITY

Sumrall, Velma, 1927-
Telling the story of the local church : the who, what, when, where, and why of communication / Velma Sumrall, Lucille Germany. — New York : Seabury Press, 1979.
vii, 117 p. : ill. ; 28 cm.
"A Crossroad book."
Bibliography: p. 116
ISBN 0-8164-2193-5
1. Church publicity. 2. Mass media in religion. I. Germany, Lucille, joint author. II. Title.
BV653.S92 254.4 79-3934
79 MARC

CHURCH RECORDS AND REGISTERS

see also Church archives; Church statistics

Schletterer, Joseph.
Ueber die Führung der Pfarrbücher, oder der Geburts-, Trauungs- und Sterbeprotokolle, und ihre Beweiskraft. Ein Versuch nach den k. k. oesterreichischen Gesetzen. Innsbruck, Gedruckt auf Kosten des Verfassers mit Wagner'schen Schriften, 1824.
xii, 271 p. forms. 21 cm.
Bibliographical footnotes.
1. Church records and registers. 2. Church records and registers—Austria. I. Title.
BV707.S34 79-343462
79 MARC

—AUSTRIA

Schletterer, Joseph.
Ueber die Führung der Pfarrbücher, oder der Geburts-, Trauungs- und Sterbeprotokolle, und ihre Beweiskraft. Ein Versuch nach den k. k. oesterreichischen Gesetzen. Innsbruck, Gedruckt auf Kosten des Verfassers mit Wagner'schen Schriften, 1824.
xii, 271 p. forms. 21 cm.
Bibliographical footnotes.
1. Church records and registers. 2. Church records and registers—Austria. I. Title.
BV707.S34 79-343462
79 MARC

—BELGIUM—FLANDERS, EAST—BIBLIOGRAPHY—CATALOGS

Belgium. Archives de l'État, Ghent.
Archieven van OOstvlaamse Kerkfabrieken, Oud regiem. [Door] L. van Buyten en J. Verhelst. Met voorwoord door J. Buntinx. Brussel, Algemeen Rijksarchief, [Ravensteingalerij 78], 1968-76.
3 v. map. 30 cm. 150F (v. 2) Be69-4687
Includes bibliographical references and indexes.
1. Church records and registers—Belgium—Flanders, East—Bibliography—Catalogs. 2. Belgium. Archives de l'État, Ghent—Catalogs. I. Buyten, Leon van. II. Verhelst, Juul. III. Title.
CD1689.F55B44 1968 68-124369
[r79]rev2 MARC

—BELGIUM—LIÈGE (PROVINCE)—BIBLIOGRAPHY—CATALOGS

Belgium. Archives de l'État, Liège.
Répertoire général des registres paroissiaux antérieurs au Concordat de 1801 conservés dans la province de Liège / par Georges Hansotte et Richard Forgeur, avec la collaboration de Jean Pieyns et Marie-Paule Lux-Mossay. — Bruxelles : Archives générales du Royaume, 1977.
ix, 453 p. ; 30 cm. Be***
At head of title: Ministère de l'éducation nationale et de la culture française et Ministère de l'éducation nationale et de la culture néerlandaise. Archives générales du Royaume et Archives de l'État dans les provinces. Archives de l'État à Liège.
Bibliography: p. viii-ix.
Includes indexes.
1. Church records and registers—Belgium—Liège (Province)—Bibliography—Catalogs. 2. Liège (Province)—Genealogy—Bibliography—Catalogs. 3. Liège (Province)—Church history—Sources—Bibliography—Catalogs. 4. Belgium. Archives de l'État, Liège—Catalogs. I. Hansotte, Georges. II. Forgeur, Richard. III. Title.
CD1689.L532B44 1977 016.929'3493'4 79-343606
79 MARC

—BELGIUM—NAMUR (PROVINCE)—BIBLIOGRAPHY—CATALOGS

Belgium. Archives de l'État, Namur.
Inventaire général sommaire des archives ecclésiastiques de la province de Namur, par J. Bovesse, conservateur. Bruxelles, 1962.
xxiv, 339 p. map. 25 cm.
Includes bibliographies.
1. Church records and registers—Belgium—Namur (Province)—Bibliography—Catalogs. 2. Namur (Province)—Church history—Sources—Bibliography—Catalogs. 3. Monasticism and religious orders—Belgium—Namur (Province)—History—Sources—Bibliography—Catalogs. 4. Belgium. Archives de l'État, Namur—Catalogs. I. Bovesse, Jean. II. Title.
CD1689.N35B44 1962 016.282'4934 64-32595
[r79]rev MARC

Belgium. Archives de l'État, Namur.
Inventaire général sommaire des archives ecclésiastiques de la province de Namur : supplément I, accroissements 1962-1975 / par J. Bovesse, chef de département. — Bruxelles : Archives générales du Royaume, 1977.
183 p. : map ; 25 cm. Be***
At head of title: Ministère de l'éducation nationale et de la culture française et Ministère de l'éducation nationale et de la culture néerlandaise, Archives générales du Royaume et Archives de l'État dans les provinces, Archives de l'État à Namur.
Bibliography: p. [6]-10.
Includes index.
1. Church records and registers—Belgium—Namur (Province)—Bibliography—Catalogs. 2. Namur (Province)—Church history—Sources—Bibliography—Catalogs. 3. Monasticism and religious orders—Belgium—Namur (Province)—History—Sources—Bibliography—Catalogs. 4. Belgium. Archives de l'État, Namur—Catalogs. I. Bovesse, Jean. II. Title.
CD1689.N35B44 1977 016.27493'4 79-368690
79 MARC

—DENMARK—FYN—DATA PROCESSING

Thestrup, Poul.
Projekt Langeland : den demografiske database ved Landsarkivet for Fyn : dokumentation / Poul Thestrup. — Odense [Landsarkivet for Fyn, Jernbanegade 36, 5000 C] : Arkivserien, 1978-
v. ; 29 cm. — (Arkivserien-Landsarkivet for Fyn ; 5) D79-8
Includes bibliographical references.
CONTENTS: 1. Magleby sogn 1682-1901.

kr36.10 (v. 1)
1. Church records and registers—Denmark—Fyn—Data processing. 2. Registers of births, etc.—Denmark—Fyn—Data processing. 3. Fyn, Denmark—Genealogy—Data processing. 4. Fyn, Denmark—History, Local—Sources—Data processing. I. Title. II. Series: Denmark. Landsarkivet for Fyn. Arkivserien-Landsarkivet for Fyn ; 5.
CD1787.F9T48 79-382685
79 MARC

—ENGLAND

Kitching, C J
The central records of the Church of England : a report and survey presented to the Pilgrim and Radcliffe Trustees / [by C. J. Kitching]. — London : CIO Publishing, 1976.
iv, 100 p. ; 22 cm. GB***
ISBN 0-7151-3678-X : £3.00
1. Church of England—Archives—Catalogs. 2. Church records and registers—England. I. Church of England. II. Pilgrim Trust. III. Radcliffe Trust. IV. Title.
CD1069.L72K57 026.283'42 77-352399
77[r79]rev MARC

—ENGLAND—LONDON (DIOCESE)

Greater London Record Office.
A survey of the parish registers of the Diocese of London, Inner London area / Great London Record Office. — [New ed.]. — London : Greater London Council, 1972.
80 p. ; 21 cm. GB72-12274
Cover title: Parish registers, Diocese of London, Inner London area.
Includes index.
ISBN 0-7168-0401-8 : £0.625
1. London (Diocese)—Genealogy—Bibliography. 2. Church records and registers—England—London (Diocese) I. Title. II. Title: Parish registers, Diocese of London, Inner London area.
Z5305.G7G73 1972 016.929'3421 79-305105
[CS436.L7] 79 MARC

CHURCH RECORDS AND REGISTERS
(Continued)

—FRANCE—CÔTES-DU-NORD (DEPT.)—BIBLIOGRAPHY —CATALOGS

Côtes-du-Nord, France (Dept.) Archives.
Répertoire numérique de la série V (cultes) : les Églises, les congrégations et l'État dans le département des Côtes-du-Nord de l'an VIII à 1940 / par Régis Le Saulnier de Saint-Jouan. — Saint-Brieuc : Archives des Côtes-du-Nord, 1977.
civ, 406 p. ; 25 cm. F***
At head of title: Archives départementales des Côtes-du-Nord.
Includes bibliographical references.
ISBN 2-86022-000-3 : 78.00F
1. Church records and registers—France—Côtes-du-Nord (Dept.)—Bibliography—Catalogs. 2. Côtes-du-Nord, France (Dept.) Archives—Catalogs. 3. Côtes-du-Nord, France (Dept.)—Church history—Sources—Bibliography—Catalogs. I. Le Saulnier de Saint-Jouan, Régis. II. Title.
CD1218.C68C67 1977 016.944'12 78-388279
79 MARC

—GERMANY, WEST—MAINZ (ARCHDIOCESE)—BIBLIOGRAPHY—CATALOGS

Günther, Barbara.
Verzeichnis der katholischen Kirchenbücher und Elenchen der Diözese Mainz sowie der Elenchen des ehemaligen Erzbistums Mainz / bearb. von Barbara Günther. — Koblenz : Selbstverlag der Landesarchivverwaltung Rheinland-Pfalz, 1977.
xii, 215 p. ; 24 cm. — (Veröffentlichungen der Landesarchivverwaltung Rheinland-Pfalz ; Bd. 28) GFR***
Includes index.
1. Church records and registers—Germany, West—Mainz (Diocese)—Bibliography—Catalogs. 2. Church records and registers—Germany, West—Mainz (Archdiocese)—Bibliography—Catalogs. 3. Mainz (Diocese)—Church history—Sources—Bibliography—Catalogs. 4. Mainz (Archdiocese)—Church history—Sources—Bibliography—Catalogs. 5. Rhineland-Palatinate. Landesarchiv—Catalogs. I. Title: Verzeichnis der katholischen Kirchenbücher ... II. Series: Rhineland-Palatinate. Landesarchivverwaltung. Veröffentlichungen ; Bd. 28.
CD1377.M34G83 78-398042
79r79rev MARC

—GERMANY, WEST—MAINZ (DIOCESE) —BIBLIOGRAPHY—CATALOGS

Günther, Barbara.
Verzeichnis der katholischen Kirchenbücher und Elenchen der Diözese Mainz sowie der Elenchen des ehemaligen Erzbistums Mainz / bearb. von Barbara Günther. — Koblenz : Selbstverlag der Landesarchivverwaltung Rheinland-Pfalz, 1977.
xii, 215 p. ; 24 cm. — (Veröffentlichungen der Landesarchivverwaltung Rheinland-Pfalz ; Bd. 28) GFR***
Includes index.
1. Church records and registers—Germany, West—Mainz (Diocese)—Bibliography—Catalogs. 2. Church records and registers—Germany, West—Mainz (Archdiocese)—Bibliography—Catalogs. 3. Mainz (Diocese)—Church history—Sources—Bibliography—Catalogs. 4. Mainz (Archdiocese)—Church history—Sources—Bibliography—Catalogs. 5. Rhineland-Palatinate. Landesarchiv—Catalogs. I. Title: Verzeichnis der katholischen Kirchenbücher ... II. Series: Rhineland-Palatinate. Landesarchivverwaltung. Veröffentlichungen ; Bd. 28.
CD1377.M34G83 78-398042
79r79rev. MARC

—GERMANY, WEST—RHINE VALLEY—BIBLIOGRAPHY—CATALOGS

Schüler, Heinz.
Verzeichnis der Kirchenbücher der Archivstelle Koblenz des Archivs der Evangelischen Kirche im Rheinland / aufgestellt von Heinz Schüler. — Düsseldorf ; [Koblenz : Archivstelle Koblenz d. Evang. Kirche im Rheinland] 1971.
94 p.; 24 cm. GDB72-B17
DM12.00
1. Evangelische Kirche im Rheinland. Archivstelle Koblenz—Catalogs. 2. Rhine Valley—Church history—Sources—Bibliography—Catalogs. 3. Church records and registers—Germany, West—Rhine Valley—Bibliography—Catalogs. 4. Rhine Valley—Genealogy—Bibliography—Catalogs. I. Evangelische Kirchen in Deutschland. Archivstelle Koblenz. II. Title.
CD1377.K623S38 79-356868
79 MARC

—GREAT BRITAIN—BIBLIOGRAPHY

Jackson, Ronald Vern.
Inventory of church records of the British Isles / editors, Ronald Vern Jackson, Gary Ronald Teeples, David Schaefermeyer. — Bountiful, Utah : Accelerated Indexing Systems, c1976.
109 p., [2] leaves of plates : maps ; 28 cm. — (Encyclopedia of local history and genealogy ; ser. 2, v. 1)
Schaefermeyer's name covered by label.
1. Church records and registers—Great Britain—Bibliography. 2. Great Britain—Genealogy—Bibliography. 3. Great Britain—Church history—Sources—Bibliography. I. Teeples, Gary Ronald, joint author. II. Title. III. Series.
CD1068.A2J3 929'.341 78-113522
79 MARC

—GREAT BRITAIN—INDEXES

Steel, Donald John.
National index of parish registers / by D. J. Steel. — 3d corr. impression. — London : Published for the Society of Genealogists [by] Phillimore, 1976-
v. ; 24 cm. GB***
Vol. 5, compiled by D. J. Steel, assisted by A. E. F. Steel and C. W. Field, has subtitle: A guide to Anglican Roman Catholic and Nonconformist registers before 1837, together with information on marriage licenses, bishop's transcripts and modern copies.
CONTENTS: v. 1. General sources of births, marriages and deaths before 1837. v. 5. South Midlands and Welsh Border Counties comprising the counties of Gloucestershire, Herefordshire, Oxfordshire, Shropshire, Warwickshire and Worcestershire.
ISBN 0-900592-41-9 (v. 1) : 6.00 (v. 1)
1. Church records and registers—Great Britain—Indexes. I. Title.
CD1068.A2S8 1976 929'.341 79-311064
79 MARC

—ITALY—ROME (CITY)—BIBLIOGRAPHY

Sbrana, Carla.
Gli stati delle anime a Roma dalle origini al secolo XVII : origini, consistenza, contenuti : con appendice sulle altre registrazioni parrocchiali / [di] Carla Sbrana, Rosa Traina, Eugenio Sonnino. — Roma : La goliardica, 1977.
651 p., [6] leaves of plates : facsims. ; 24 cm. It79-Jan
At head of title: Comitato italiano per lo studio dei problemi della popolazione (CISP). Fonti per lo studio della popolazione di Roma.
Includes bibliographical references.
L18000
1. Church records and registers—Italy—Rome (City)—History. 2. Church records and registers—Italy—Rome (City)—Bibliography. I. Traina, Rosa, joint author. II. Sonnino, Eugenio, joint author. III. Title. IV. Title: Fonti per lo studio della popolazione di Roma.
BX1548.R6S27 79-348466
*79 MARC

—ITALY—ROME (CITY)—HISTORY

Sbrana, Carla.
Gli stati delle anime a Roma dalle origini al secolo XVII : origini, consistenza, contenuti : con appendice sulle altre registrazioni parrocchiali / [di] Carla Sbrana, Rosa Traina, Eugenio Sonnino. — Roma : La goliardica, 1977.
651 p., [6] leaves of plates : facsims. ; 24 cm. It79-Jan
At head of title: Comitato italiano per lo studio dei problemi della popolazione (CISP). Fonti per lo studio della popolazione di Roma.
Includes bibliographical references.
L18000
1. Church records and registers—Italy—Rome (City)—History. 2. Church records and registers—Italy—Rome (City)—Bibliography. I. Traina, Rosa, joint author. II. Sonnino, Eugenio, joint author. III. Title. IV. Title: Fonti per lo studio della popolazione di Roma.
BX1548.R6S27 79-348466
*79 MARC

—NETHERLANDS—BIBLIOGRAPHY—CATALOGS

Kerkelijke archieven : inventarissen / onder red. van J. P. van Dooren. — Den Haag : Archiefdienst Nederlandse Hervormde Kerk, 19
v. ; 30 cm. Ne***
Includes bibliographical references and indexes.
1. Church archives—Netherlands—Catalogs. 2. Church records and registers—Netherlands—Bibliography—Catalogs. 3. Netherlands—Church history—Sources—Bibliography—Catalogs. I. Dooren, J. P. van. II. Nederlandse Hervormde Kerk. Archiefdienst.
CD1709.N4K47 79-367313
79 MARC

—NETHERLANDS—DE RIJP—BIBLIOGRAPHY—CATALOGS

Netherlands (Kingdom, 1815-). Rijksarchief in Noord-Holland, Haarlem.
Archieven van de Nederlandse Hervormde gemeenten en instellingen te Graft-De Rijp, 1593-1951 / door R. M. de Raat. — Haarlem : Rijksarchief in Noord-Holland, 1977.
126 p., [6] leaves of plates : ill. ; 29 cm. — (Rijksarchieven in Holland ; nr. 13) Ne***
Includes bibliographical references and index.
1. Church records and registers—Netherlands—Graft—Bibliography—Catalogs. 2. Church records and registers—Netherlands—De Rijp—Bibliography—Catalogs. 3. Graft, Netherlands—Church history—Sources—Bibliography—Catalogs. 4. De Rijp, Netherlands—Church history—Sources—Bibliography—Catalogs. 5. Netherlands (Kingdom, 1815-). Rijksarchief in Noord-Hooland, Haarlem—Catalogs. I. Raat, R. M. de. II. Title. III. Series.
CD1708.5.G7N47 1977 78-396984
79 MARC

—NETHERLANDS—GRAFT—BIBLIOGRAPHY—CATALOGS

Netherlands (Kingdom, 1815-). Rijksarchief in Noord-Holland, Haarlem.
Archieven van de Nederlandse Hervormde gemeenten en instellingen te Graft-De Rijp, 1593-1951 / door R. M. de Raat. — Haarlem : Rijksarchief in Noord-Holland, 1977.
126 p., [6] leaves of plates : ill. ; 29 cm. — (Rijksarchieven in Holland ; nr. 13) Ne***
Includes bibliographical references and index.
1. Church records and registers—Netherlands—Graft—Bibliography—Catalogs. 2. Church records and registers—Netherlands—De Rijp—Bibliography—Catalogs. 3. Graft, Netherlands—Church history—Sources—Bibliography—Catalogs. 4. De Rijp, Netherlands—Church history—Sources—Bibliography—Catalogs. 5. Netherlands (Kingdom, 1815-). Rijksarchief in Noord-Hooland, Haarlem—Catalogs. I. Raat, R. M. de. II. Title. III. Series.
CD1708.5.G7N47 1977 78-396984
79 MARC

—OHIO—WOOD CO.—BIBLIOGRAPHY

Angel, James L
The Wood County historical church records survey / James L. Angel. — Bowling Green, Ohio : Center for Archival Collections, Bowling Green State University, 1979.
xv, 56 p. ; 29 cm.
Includes index.
ISBN 0-932690-01-7
1. Church records and registers—Ohio—Wood Co.—Bibliography. 2. Wood Co., Ohio—Church history—Sources—Bibliography. 3. Wood Co., Ohio—History, Local—Sources—Bibliography. 4. Wood Co., Ohio—Genealogy—Bibliography. I. Title.
CD3449.W66A53 016.28'009771'6 79-50390
79 MARC

CHURCH REFORM see Church renewal

CHURCH REGISTERS see Registers of births, etc.

CHURCH-RELATED COLLEGES see Church colleges

CHURCH RENEWAL

see also Church consultation; Mission of the church

Commissaris, Jan.
Profetische planning : planning, vernieuwing en beleid in de kerkelijke gemeente / Jan Commissaris. — Bloemendaal : Nelissen, [1978]
91 p. ; 21 cm. — (Proces-materiaal) Ne78-27
Bibliography: p. 86-91.
ISBN 9024410398 : fl 24.90
1. Church renewal. I. Title.
BV600.2.C63 79-380759
*79 MARC

Drummond, Lewis A
The awakening that must come / Lewis A. Drummond. — Nashville : Broadman Press, c1978.
149 p. ; 19 cm.
Includes bibliographical references.
ISBN 0-8054-6535-9
1. Church renewal. 2. Revivals. I. Title.
BV600.2.D7 269'.2 78-59239
79 MARC

Gish, Arthur G
Living in Christian community / Arthur G. Gish. — Scottdale, Pa. : Herald Press, 1979.
379 p. ; 22 cm.
Bibliography: p. [371]-373.
Includes index.
ISBN 0-8361-1887-1 : $8.95
1. Christian communities. 2. Church renewal. 3. Christian life—Church of the Brethren authors. 4. Church. I. Title.
BV4405.G57 262 79-11848
79 MARC

Lissenburg, D
Samen thuis—samen uit : over de begeleiding van veranderingsprocessen in kerken / D. Lissenburg. — Assen : Van Gorcum, 1975.
xv, 441 p. : ill. ; 22 cm. Ne75-28
Errata slip inserted.
Bibliography: p. [407]-419.
ISBN 9023212908 : fl 58.00
1. Church renewal. I. Title.
BV600.2.L57 78-383984
*78 MARC

CHURCH RENEWAL (Continued)

Lovelace, Richard F
Dynamics of spiritual life : an evangelical theology of renewal / Richard F. Lovelace. — Downers Grove, Ill. : Inter-Varsity Press, c1979.
455 p. ; 21 cm.
Includes bibliographical references and index.
ISBN 0-87784-626-X : $8.95
1. Revivals—United States—History. 2. Church renewal. 3. Evangelicalism—United States. 4. Spiritual life—History of doctrines. I. Title.
BV3773.L63 269'.2 78-24757
79 MARC

Marshall, Michael, 1936-
Glory under your feet : the challenge of catholic renewal today / ₍by₎ Michael Marshall. — London : Darton, Longman, and Todd, 1978.
137 p. ; 22 cm. GB78-29210
Includes bibliographical references.
ISBN 0-232-51306-5 : £2.95
1. Church renewal. 2. Christianity—20th century. 3. Church. I. Title.
BV600.2.M345 1978 260 79-301452
79 MARC

Schaller, Claude.
L'Église en quête de dialogue / Claude Schaller. — Bern ; Las Vegas : P. Lang, c1977.
373 p. : ill. ; 21 cm. — (Études d'histoire interculturelle du christianisme ; 13) Sw***
Two hundred copies printed.
Bibliography: p. 355-373.
ISBN 3-261-02003-8
1. Church renewal. 2. Missionaries—Switzerland—Attitudes. I. Title. II. Series: Studien zur interkulturellen Geschichte des Christentums ; Bd. 13.
BV600.2.S29 266'.023 78-396918
79 MARC

Tucker, Michael R
The church, change or decay / Michael R. Tucker. — Wheaton, Ill. : Tyndale House Publishers, c1978.
192 p. ; 21 cm.
Includes bibliographical references.
ISBN 0-8423-0277-8 : $3.95
1. Church renewal. I. Title.
BV600.2.T82 254 77-93750
79 MARC

—ADDRESSES, ESSAYS, LECTURES

Barr, Browne.
The well church book / Browne Barr. — New York : Seabury Press, c1976.
116 p. : ill. ; 22 cm.
"A Crossroad book."
ISBN 0-8164-0304-X : $7.95
1. Church renewal—Addresses, essays, lectures. 2. Church—Addresses, essays, lectures. 3. Pastoral theology—Addresses, essays, lectures. I. Title.
BV600.2.B36 262'.001 76-18291
76 MARC

Frisch, Helga.
Kirche im Abseits : die bekannte Berliner Pastorin plädiert für Reform / Helga Frisch. — München : R. Piper, c1978.
219 p. ; 21 cm. GFR***
Includes bibliographical references.
ISBN 3-492-02402-5
1. Church renewal—Addresses, essays, lectures. I. Title.
BV600.2.F73 78-392625
79 MARC

Jewett, Dick.
Let's fan the flame : an eleventh-hour challenge for Seventh-Day Adventists to finish their task / Dick Jewett ; ₍cover ill. Jack Pardue₎. — Washington : Review and Herald Pub. Association, c1979.
142 p. ; 20 cm.
$4.95
1. Seventh-Day Adventists—Doctrinal and controversial works—Seventh-Day Adventist authors—Addresses, essays, lectures. 2. Church renewal—Addresses, essays, lectures. I. Title.
BX6151.J48 262'.001 79-12033
79 MARC

—ANGLICAN CHURCH OF CANADA

Stuchbery, Ian.
Growing in Christ : new patterns of initiation and education in the parish community / Ian Stuchbery. — Toronto : Anglican Book Centre, c1978.
109 p. ; 22 cm. C78-6567-0
Bibliography: p. ₍105₎-107.
ISBN 0-919030-23-8 : £3.25
1. Christian education. 2. Church renewal—Anglican Church of Canada. I. Title.
BV1471.2.S86 262'.001 79-312815
79 MARC

—ANGLICAN COMMUNION

Bishops' move / edited by Michael Harper. — London ₍etc.₎ : Hodder and Stoughton, 1978.
160 p. ; 22 cm. — (Ecclesia books) GB78-32716
"Six Anglican bishops share their experience of renewal."
ISBN 0-340-22798-2 : £2.50
1. Anglican Communion—Bishops—Biography. 2. Bishops—Biography. 3. Pentecostalism—Anglican Communion. 4. Church renewal—Anglican Communion. I. Harper, Michael Claude.
BX5009.B57 283'.092'2 79-309879
79 MARC

—CATHOLIC CHURCH

see also Catholic traditionalist movement

Schruers, Paul.
Tekenen van hoop : elementen voor kerkopbouw / Paul Schruers. — ₍3000₎ Leuven, ₍Blijde Inkomststraat 79₎ : Davidsfonds, ₍1978₎
251 p. ; 21 cm. — (Keurreeks ; nr. 138) Be78-4
Includes bibliographical references.
ISBN 9061521378 : 210F
1. Church renewal—Catholic Church. I. Title.
BX1746.S3 79-366008
*79 MARC

Schulz, Heinz-Manfred, 1933-
Ein Jahr in Gottes Werkstatt : e. Gemeinde macht neue Erfahrungen / Heinz-Manfred Schulz. — 1. Aufl. — Mainz : Matthias-Grünewald-Verlag, 1978.
145 p. : 10 ill. ; 21 cm. GFR78-A
ISBN 3-7867-0691-3 : DM15.80
1. Christkönigsgemeinde in Eschborn. 2. Church renewal—Catholic Church. I. Title.
BX4630.E83C487 79-340850
79 MARC

Serrano, Vicenç.
Tierra de exilio ₍por₎ Vicente Serrano. Madrid, Rialp ₍1972₎
167 p. 19 cm. (Libros de bolsillo Rialp, 60) Sp***
1. Church renewal—Catholic Church. 2. Jews—History—Babylonian captivity, 598-515 B.C. I. Title.
BX1746.S44 79-414184
79 MARC

—CATHOLIC CHURCH—CONGRESSES

The Catholic Church : the United States experience : a symposium jointly sponsored by Holy Names College, Oakland, California, and Prospective International, Brussels, Belgium : ₍papers₎ / edited by Irene Woodward. — New York : Paulist Press, c1979.
134 p. ; 21 cm. — (An Exploration book)
Includes bibliographical references.
ISBN 0-8091-2148-4 : $4.95
1. Catholic Church in the United States—Congresses. 2. Church renewal—Catholic Church—Congresses. I. Woodward, Irene. II. Spokane. Holy Names College. III. Prospective International.
BX1404.C26 282'.73 78-61730
79 MARC

—CHURCH OF ENGLAND

Gunstone, John Thomas Arthur.
A people for his praise : renewal and congregational life / John Gunstone. — London : Hodder and Stoughton, 1978.
190 p. ; 22 cm. — (Ecclesia books) GB***
Includes bibliographical references.
ISBN 0-340-22808-3 : £3.25
1. Pentecostalism—Church of England. 2. Church renewal—Church of England. I. Title.
BX5125.5.G86 262'.001 78-322901
79 MARC

Harper, Michael, 1931-
Let my people grow! : ministry and leadership in the Church / by Michael Harper. — London : Hodder and Stoughton, 1977.
254 p. ; 22 cm. — (Ecclesia books) GB77-19209
Bibliography: p. ₍247₎-248.
Includes index.
ISBN 0-340-21573-9 : £2.95
1. Church renewal—Church of England. 2. Church polity. 3. Christian leadership. I. Title.
BV600.2.H35 1977b 262'.001 77-379727
78₍r79₎rev MARC

Harper, Michael, 1931-
Let my people grow : ministry and leadership in the church / Michael Harper. — Plainfield, N.J. : Logos International, c1977.
254 p. ; 21 cm.
Bibliography: p. ₍247₎-248.
Includes index.
ISBN 0-88270-236-X : $3.50
1. Church renewal—Church of England. 2. Church polity. 3. Christian leadership.
BV600.2.H35 1977 262'.001 77-73840
77₍r79₎rev MARC

Wansey, Christopher.
The clockwork church : reform in the Church of England at worm's eye level / by Christopher Wansey. — Oxford : Becket Publications, c1978.
138 p., ₍8₎ leaves of plates : ill. ; 22 cm. GB***
£1.95
1. Church of England—Doctrinal and controversial works—Church of England authors. 2. Church renewal—Church of England. 3. Wansley, Christopher. I. Title.
BX5131.2.W36 262'.001 79-309118
79 MARC

—MISCELLANEA

Ravenhill, Leonard.
America is too young to die / Leonard Ravenhill. — Minneapolis : Bethany Fellowship, c1979.
123 p. : port. ; 21 cm.
ISBN 0-87123-013-5 : $2.95
1. Christianity—United States—Miscellanea. 2. Church renewal—Miscellanea. I. Title.
BR526.R38 262'.001 79-19229
79 MARC

—REFORMED CHURCH

Lissenburg, D
Veranderen in de kerk ... hoe doe je dat? / D. Lissenburg. — Assen : Van Gorcum, 1978.
156 p. ; 22 cm. Ne78-15
Bibliography: p. ₍140₎-148.
ISBN 9023215281 : fl 19.50
1. Church renewal—Reformed Church. I. Title.
BV600.2.L573 79-381624
*79 MARC

—EUROPE

Henley, Wallace.
Europe at the crossroads : a reporter looks at Europe's spiritual crisis / by Wallace Henley. — Westchester, Ill. : Good News Publishers, c1978.
118 p. ; 21 cm.
Includes bibliographical references.
ISBN 0-89107-156-3 : $3.50
1. Evangelicalism—Europe. 2. Church renewal—Europe. 3. Europe—Religion. I. Title.
BR1642.E85H46 269'.2'094 78-68409
79 MARC

CHURCH SALVATION UNIVERSALISM

see also Salvation outside the Catholic Church

CHURCH SCHOOLS

see also Church colleges

—LAW AND LEGISLATION—UNITED STATES

Tavel, David.
Church-state issues in education / by David Tavel. — Bloomington, Ind. : Phi Delta Kappa Educational Foundation, c1979.
45 p. ; 18 cm. — (Fastback - Phi Delta Kappa Educational Foundation ; 123)
ISBN 0-87367-123-6 : $0.75
1. Religious education—Law and legislation—United States. 2. Church schools—Law and legislation—United States. 3. State aid to private schools—United States. I. Title. II. Series: Phi Delta Kappa Educational Foundation. Fastback ; 123.
KF4124.Z9T38 344'.73'0796 79-83627
79 MARC

—LAW AND LEGISLATION—UNITED STATES—CASES

Education, religion, and the Supreme Court / edited by Richard C. McMillan. — Danville, VA : Association of Baptist Professors of Religion, c1979.
129 p. ; 22 cm. — (Perspectives in religious studies : Special studies series ; no. 6)
Page 129 added.
ISBN 0-932180-05-1 : $4.95
1. Religious education—Law and legislation—United States—Cases. 2. Church schools—Law and legislation—United States—Cases. I. McMillan, Richard C. II. Series.
KF4124.A59E38 344'.73'0796 78-74196
79 MARC

CHURCH SCHOOLS
—LAW AND LEGISLATION
—UNITED STATES—CASES (Continued)

Twomley, Dale E
Parochiaid and the courts / by Dale E. Twomley. — Berrien Springs, Mich. : Andrews University Press, 1979.
ix, 165 p. ; 24 cm. — (Studies in education) (Andrews University monographs)
Bibliography: p. 158-165.
1. Church schools—Law and legislation—United States—Cases. 2. Federal aid to private schools—United States—Cases. I. Title. II. Series: Studies in education (Berrien Springs, Mich.) III. Series: Andrews University. Monographs.
KF4124.A7T85 344'.73'076 79-115963
79 MARC

—BELGIUM

Billiet, Jaak.
Secularisering en verzuiling in het onderwijs : een sociologisch onderzoek naar de vrije schoolkeuze als legitiematieschema en als sociaal proces / door J. Billiet. — 3000 Leuven (Krakenstr. 3) : Universitaire Pers, 1977.
xiv, 244 p. ; 24 cm. — (Sociologische verkenningen ; v. 9) Be78-2
Bibliography: p. 215-224.
ISBN 9061860636 : 600F
1. School, Choice of. 2. Church schools—Belgium. 3. Public schools—Belgium. 4. Educational sociology. I. Title. II. Series.
LC53.B4B55 78-385663
•78 MARC

—PHILIPPINE ISLANDS—MANILA METROPOLITAN AREA—FINANCE

Villavicencio, Veronica Roco.
The prediction of total family income : a socioeconomic survey : final report / prepared for the Faura Research Center, inc., and the Philippine Social Science Council by the Institute of Philippine Culture on 15 September 1973 ; Veronica Roco Villavicencio. — Quezon City : Ateneo de Manila, Institute of Philippine Culture, 1973.
vii, 65 leaves ; 28 cm.
Bibliography: leaf 65.
1. Income—Philippine Islands—Manila metropolitan area. 2. Church schools—Philippine Islands—Manila metropolitan area—Finance. I. Ateneo de Manila University. Institute of Philippine Culture. II. Title.
HC460.I5V54 339.4'1'09599 79-303781
79 MARC

CHURCH SCHOOLS, WEEK-DAY see Week-day church schools

CHURCH SLAVIC LANGUAGE

Bagin, Anton.
Kresťanská archeológia. Úvod do staroslovienčiny, časť historická / Anton Bagin. — Bratislava : Spolok sv. Vojtecha v CN, 1977.
175 p. ; 29 cm. CzS77
Cover title.
At head of title: Rímskokatolícka Cyrilometodejská bohoslovecká fakulta v Bratislave.
Includes bibliographies.
Kčs58.00
1. Christian antiquities. 2. Church Slavic language. I. Bagin, Anton. Úvod do staroslovienčiny, časť historická. 1977. II. Title.
BR131.B26 79-379916
•79 MARC

Nikolić, Svetozar.
Staroslovenski jezik / sastavio Svetozar Nikolić. — 3. izd. — Beograd : Naučna knjiga, 19
v. ; 20 cm. — (Univerzitetski udžbenici) Yu***
CONTENTS:
—2. Primeri sa rečnikom.
1. Church Slavic language. I. Title.
PG613.N54 1976 79-376211
79 MARC

—ADDRESSES, ESSAYS, LECTURES

Jagić, Vatroslav, 1838–1923.
(Rassuzhdenīi͡a i͡uzhnoslavi͡anskoĭ i russkoĭ stariny o t͡serkovno-slavi͡anskom i͡azyke)
Рассуждения южнославянской и русской старины о церковно-славянском языке = Codex slovenicus rerum grammaticarum / собрал и объяснил И. В. Ягич. — München : W. Fink, 1968.
xxiii, 782 p. ; 24 cm. — (Slavische Propyläen ; Bd. 25)
Preface in Latin; text in Church Slavic or Russian.
"Nachdruck des Separatdrucks, Berlin 1896."
Includes bibliographical references and indexes.
1. Church Slavic language—Addresses, essays, lectures. I. Title. II. Title: Codex slovenicus rerum grammaticarum. III. Series.
PG613.J3 1968 79–410656

—GRAMMAR

Aitzetmüller, Rudolf.
Altbulgarische Grammatik als Einführung in die slavische Sprachwissenschaft / Rudolf Aitzetmüller. — Freiburg i. Br. : Weiher, 1978.
x, 253 p. ; 25 cm. — (Monumenta linguae Slavicae dialecti veteris ; T. 12) GFR79-A
Some text in Church Slavic.
Bibliography: p. 251-253.
ISBN 3-921940-01-X : DM48.00
1. Church Slavic language—Grammar. 2. Slavic languages—Grammar, Comparative—Indo-European. 3. Indo-European languages—Grammar, Comparative—Slavic. I. Title. II. Series.
PG619.A36 491.8'1701 79-369004
79 MARC

Schmalstieg, William R
An introduction to Old Church Slavic / William R. Schmalstieg. — Cambridge, Mass. : Slavica Publishers, 1976.
xi, 291 p. ; 23 cm.
Bibliography: p. 280-286.
1. Church Slavic language—Grammar. I. Title.
PG623.S3 491.8'1701 78-111640
79 MARC

Smoczyński, Paweł.
Gramatyka opisowa języka staro-cerkiewnosłowiańskiego : przewodnik metodyczny dla słuchaczy studiów zaocznych II roku filologii polskiej i rosyjskiej / Paweł Smoczyński. — Lublin : Uniwersytet Marii Curi-Skłodowskiej, Instytut Filologii Rosyjskiej i Słowiańskiej, 1975.
29 p. ; 21 cm. P***
1. Church Slavic language—Grammar. I. Title.
PG623.S57 78-422541
79 MARC

Smotryts'kyĭ, Melitiĭ, Bp., 1578-1633.
[Grammatīki ili Pismēnnit͡sa i͡azyka Slovēnskagō]
Hrammatiki ili pismennica jazyka sloven'skaho, Kremjaneć, 1638 : eine gekürzte Fassung der kirchenslavischen Grammatik / von Melitij Smotryćkyj ; hrsg. und eingeleitet von Olexa Horbatsch. — München : Kubon & Sagner, 1977.
120 p. ; 21 cm. — (Specimina philologiae Slavicae ; Bd. 11) GFR***
Introductory matter in German; text in Church Slavic.
Includes bibliographical references and indexes.
1. Church Slavic language—Grammar. I. Horbach, Oleksa. II. Title. III. Series.
PG623.S64 1977 78-400827
79 MARC

—NOUN

Trost, Klaus.
Untersuchungen zur Übersetzungstheorie und -praxis des späteren Kirchenslavischen : d. Abstrakta in d. Hexaemeronübers, d. Zagreber Zbornik von 1469 / Klaus Trost. — München : Fink, 1978.
381 p. ; 22 cm. — (Forum Slavicum ; Bd. 43) GFR78-A33
Habilitationsschrift—Würzburg, 1972.
"Wortregister" (Church Slavic-Greek, Greek-Church Slavic): p. 315-375.
Bibliography: p. 296-313.
ISBN 3-7705-1311-8 : DM98.00
1. Zagrebski sbornik ot 1469. 2. Vladislav Gramatik, fl. 1456-1480. 3. Church Slavic language—Translating. 4. Greek language, Medieval and late—Translating into Church Slavic. 5. Church Slavic language—Noun. I. Title.
PG705.Z343T7 491.8'1701 79-364139
79 MARC

—READERS

(Istorii͡a na bŭlgarskii͡a ezik)
История на българския език : анотирани текстове, анализи и речник : [уч. помагало за студентите от Великотърновския унив. Кирил и Методий] / Ангел Давидов ... [et al.]. — В. Търново : Великотърновски унив. Кирил и Методий, 1978.
143 p. : ill. ; 22 cm. Bu 79-510
At head of title: VTU Kiril i Metodiĭ. Filologicheski fakultet. Katedra po bŭlgarsko ezikoznanie.
0.60 lv
1. Bulgarian language—Readers. 2. Church Slavic language—Readers. I. Davidov, Angel Asenov. II. Velikotŭrnovski universitet Kiril i Metodiĭ. Katedra po bŭlgarsko ezikoznanie.
PG838.I 8 79–413277

—READERS—EXHIBITIONS

Pekars'ka, I I
(Chotyr'i͡okhsotrichchi͡a "Bukvari͡a" Ivana Fedorova)
400—річчя "Букваря" Івана Федорова : каталог виставки / [склала І. Й. Пекарська]. — Львів : Львівська наук. бібліотека, 1977.
59 p., [14] leaves of plates : ill. ; 20 cm. USSR***
At head of title: Akademii͡a nauk Ukraïns'koï RSR. L'vivs'ka naukova biblioteka im. V. Stefanyka.
Added t. p.: Chetyrekhsotletie "Bukvari͡a" Ivana Fedorova.
Includes indexes.
0.30rub
1. Ivan Fedorov, printer, d. 1583. Hramatyka Ivana Fedorova—Exhibitions. 2. Church Slavic language—Readers—Exhibitions. 3. Slavic languages, Eastern—Readers—Exhibitions. 4. L'vivs'ka derzhavna naukova biblioteka—Catalogs. I. L'vivs'ka derzhavna naukova biblioteka. II. Title. II. Title: Chetyrekhsotletie "Bukvari͡a" Ivana Fedorova.
PG623.I 93P4 79–411546

—RUSSIAN RECENSION

Plähn, Jürgen.
Der Gebrauch des modernen russischen Kirchenslavisch in der russischen Kirche : mit zahlr. Notenbeispielen u. kirchenslav. Texten / Jürgen Plähn. — Hamburg : Buske, 1978.
224 p. : ill. ; 24 cm. GFR78-A
Bibliography: p. 215-223.
ISBN 3-87118-338-5 : DM30.00
1. Liturgical language—Church Slavic. 2. Orthodox Eastern Church, Russian. 3. Church Slavic language—Russian recension. I. Title.
BX350.P52 79-351315
79 MARC

—SERBIAN RECENSION

Costantini, Lionello.
Slavo ecclesiastico e volgare nella Grammatika italianskaja di Vikentije Ljuština / Lionello Costantini. — Firenze : Licosa, 1976.
207 p., [1] leaf of plates : facsim. ; 24 cm. — (Studia historica et philologica : III, Sectio Slavoromanica ; 1) It78-May
Italian or Serbo-Croatian with summary in English.
Text of Grammatika italīanskaja in appendices (p. [57]-203)
Includes bibliographical references.
1. Ljuština, Vikentije, 1761-1805. Grammatika italīanskai͡a. 2. Church Slavic language—Serbian recension. 3. Serbo-Croatian language—18th century—Grammar. I. Ljuština, Vikentije, 1761-1805. Grammatika italīanskai͡a. 1976. II. Title.
PG705.L57G733 78-384018
•78 MARC

—SUFFIXES AND PREFIXES

Keipert, Helmut, 1941–
Die Adjektive auf-telьnъ : Studien zu e. kirchenslav. Wortbildungstyp / Helmut Keipert. — 1. Aufl. — Wiesbaden : Harrassowitz, 1977–
v. ; 24 cm. — (Veröffentlichungen der Abteilung für Slavische Sprachen und Literaturen des Osteuropa-Instituts (Slavisches Seminar) an der Freien Universität Berlin ; Bd. 45)
GFR 77–A (v. 1)
Habilitationsschrift—Bonn, 1974.
Bibliography : v. 1, p. [206]–223.
Includes indexes.
ISBN 3-447-01905-0(v. 1)
1. Church Slavic language—Word formation. 2. Church Slavic language—Suffixes and prefixes. 3. Slavic languages—Word formation. 4. Slavic languages—Suffixes and prefixes. I. Series: Berlin. Freie Universität. Osteuropa-Institut. Veröffentlichungen der Abteilung für Slavische Sprachen und Literaturen des Osteuropa-Instituts (Slavisches Seminar) ; Bd. 45.
78–360810
(MARC)

—TRANSLATING

Trost, Klaus.
Untersuchungen zur Übersetzungstheorie und -praxis des späteren Kirchenslavischen : d. Abstrakta in d. Hexaemeronübers, d. Zagreber Zbornik von 1469 / Klaus Trost. — München : Fink, 1978.
381 p. ; 22 cm. — (Forum Slavicum ; Bd. 43) GFR78-A33
Habilitationsschrift—Würzburg, 1972.
"Wortregister" (Church Slavic-Greek, Greek-Church Slavic): p. 315-375.
Bibliography: p. 296-313.
ISBN 3-7705-1311-8 : DM98.00
1. Zagrebski sbornik ot 1469. 2. Vladislav Gramatik, fl. 1456-1480. 3. Church Slavic language—Translating. 4. Greek language, Medieval and late—Translating into Church Slavic. 5. Church Slavic language—Noun. I. Title.
PG705.Z343T7 491.8'1701 79-364139
79 MARC

—WORD FORMATION

Keipert, Helmut, 1941–
Die Adjektive auf-telьnъ : Studien zu e. kirchenslav. Wortbildungstyp / Helmut Keipert. — 1. Aufl. — Wiesbaden : Harrassowitz, 1977–
v. ; 24 cm. — (Veröffentlichungen der Abteilung für Slavische Sprachen und Literaturen des Osteuropa-Instituts (Slavisches Seminar) an der Freien Universität Berlin ; Bd. 45)
GFR 77–A (v. 1)
Habilitationsschrift—Bonn, 1974.
Bibliography : v. 1, p. [206]–223.
Includes indexes.
ISBN 3-447-01905-0(v. 1)
1. Church Slavic language—Word formation. 2. Church Slavic language—Suffixes and prefixes. 3. Slavic languages—Word formation. 4. Slavic languages—Suffixes and prefixes. I. Series: Berlin. Freie Universität. Osteuropa-Institut. Veröffentlichungen der Abteilung für Slavische Sprachen und Literaturen des Osteuropa-Instituts (Slavisches Seminar) ; Bd. 45.
78–360810
(MARC)

CHURCH–SLAVIC LEGENDS see Legends, Church-Slavic

CHURCH SLAVIC LITERATURE

Mincheva, Angelina.
(Starobŭlgarski kirilski otkŭslet͡si)
Старобългарски кирилски откъслеци : [изследване] / Ангелина Минчева. — София : БАН, 1978.
259 p., [2] leaves of plates : ill. ; 21 cm. — (Български езикови паметници ; т. 1) Bu 78–1789
At head of title: Bŭlgarska akademii͡a na naukite. Institut za bŭlgarski ezik.
Errata slip inserted.
Includes bibliographical references and index.
8.50 lv
1. Church Slavic literature. 2. Manuscripts, Cyrillic. I. Title. II. Series: Bŭlgarski ezikovi pametnit͡si ; t. 1.
PG703.M54 79–408083

—TRANSLATIONS FROM FOREIGN LANGUAGES—HISTORY AND CRITICISM

Meshcherskiĭ, Nikita Aleksandrovich, 1906–
(Istochniki i sostav drevneĭ slavi͡ano-russkoĭ perevodnoĭ pis'mennosti 9.–15. vekov)
Источники и состав древней славяно-русской переводной письменности IX–XV веков : (Учеб. пособие) / Н. А. Мещерский. — Ленинград : Изд-во ЛГУ, 1978.
112 p. ; 21 cm. USSR 78
At head of title: Leningradskiĭ gosudarstvennyĭ universitet imeni A. A. Zhdanova.
Includes bibliographical references and index.
0.32rub
1. Church Slavic literature—Translations from foreign languages—History and criticism. 2. Russian literature—To 1700—Translations from foreign languages—History and criticism. 3. Byzantine literature—Translations into Church Slavic—History and criticism. I. Title.
PG701.45.M47 79–414037

—TRANSLATIONS FROM GREEK—HISTORY AND CRITICISM—ADDRESSES, ESSAYS, LECTURES

Izbornik Svi͡atoslava 1073 g. : Sbornik stateĭ / AN SSSR, Nauch. sovet po istorii mirovoĭ kul'tury ; [Otv. red. B. A. Rybakov]. — Moskva : Nauka, 1977.
342 p. : ill. ; 22 cm. USSR77
Includes bibliographical references.
2.04rub
1. Izbornik Svi͡atoslava 1073 g.—Addresses, essays, lectures. 2. Church Slavic literature—Russia—History and criticism—Addresses, essays, lectures. 3. Church Slavic literature—Bulgaria—History and criticism—Addresses, essays, lectures. 4. Manuscripts, Chruch Slavic—Addresses, essays, lectures. 5. Church Slavic literature—Translations from Greek—History and criticism—Addresses, essays, lectures. 6. Greek literature—Translations into Church Slavic—History and criticism—Addresses, essays, lectures. I. Rybakov, Boris Aleksandrovich, 1908- . II. Akademii͡a nauk SSSR. Nauchnyĭ sovet po istorii mirovoĭ kul'tury. III. Izbornik Svi͡atoslava 1073 g. Selections. 1977.
PG703.I973I9 79-379332
79 MARC

—TRANSLATIONS INTO BULGARIAN

(Iz starata bŭlgarska literatura)
Из старата българска литература : [антология] / под ред. [с предг.] на Петър Динеков. — [8. изд.] — София : Бълг. писател, 1977 (Перник : печ. В. Андреев)
288 p. ; 20 cm. — (Библиотека за ученика) Bu 78–543
Series romanized: Biblioteka za uchenika.
Includes bibliographical references.
0.40 lv
1. Bulgarian literature—To 1762. 2. Church Slavic literature—Translations into Bulgarian. 3. Bulgarian literature—Translations from Church Slavic. I. Dinekov, Petŭr Nikolov, 1910–
PG1020.5.I 9 1977 79–411544

—BULGARIA—HISTORY AND CRITICISM—ADDRESSES, ESSAYS, LECTURES

Izbornik Svi͡atoslava 1073 g. : Sbornik stateĭ / AN SSSR, Nauch. sovet po istorii mirovoĭ kul'tury ; [Otv. red. B. A. Rybakov]. — Moskva : Nauka, 1977.
342 p. : ill. ; 22 cm. USSR77
Includes bibliographical references.
2.04rub
1. Izbornik Svi͡atoslava 1073 g.—Addresses, essays, lectures. 2. Church Slavic literature—Russia—History and criticism—Addresses, essays, lectures. 3. Church Slavic literature—Bulgaria—History and criticism—Addresses, essays, lectures. 4. Manuscripts, Chruch Slavic—Addresses, essays, lectures. 5. Church Slavic literature—Translations from Greek—History and criticism—Addresses, essays, lectures. 6. Greek literature—Translations into Church Slavic—History and criticism—Addresses, essays, lectures. I. Rybakov, Boris Aleksandrovich, 1908- . II. Akademii͡a nauk SSSR. Nauchnyĭ sovet po istorii mirovoĭ kul'tury. III. Izbornik Svi͡atoslava 1073 g. Selections. 1977.
PG703.I973I9 79-379332
79 MARC

—ROMANIA

534 [i.e. Cinci sute treizeci și patru] documente istorice slavo-române din Țara Românească și Moldova : privitoare la legăturile cu Ardealul, 1346-1603 : din arhivele orașelor Brașov și Bistrița, în text original slav însoțit de traducere românească, tipărite la Viena în 1905-1906 / prin îngrijirea lui Gr. G. Tocilescu. — București : Libr. "Cartea Românească", 1931.
xxxvi, 551 p. ; 31 cm.
Old Slavonic text with Romanian translation.
Includes bibliographical references.
1. Romania—History—To 1711—Sources. 2. Church Slavic literature—Romania. I. Tocilescu, Grigore George, 1845-1909.
DR240.C56 79-365940
79 MARC

—RUSSIA—HISTORY AND CRITICISM—ADDRESSES, ESSAYS, LECTURES

Izbornik Svi͡atoslava 1073 g. : Sbornik stateĭ / AN SSSR, Nauch. sovet po istorii mirovoĭ kul'tury ; [Otv. red. B. A. Rybakov]. — Moskva : Nauka, 1977.
342 p. : ill. ; 22 cm. USSR77
Includes bibliographical references.
2.04rub
1. Izbornik Svi͡atoslava 1073 g.—Addresses, essays, lectures. 2. Church Slavic literature—Russia—History and criticism—Addresses, essays, lectures. 3. Church Slavic literature—Bulgaria—History and criticism—Addresses, essays, lectures. 4. Manuscripts, Chruch Slavic—Addresses, essays, lectures. 5. Church Slavic literature—Translations from Greek—History and criticism—Addresses, essays, lectures. 6. Greek literature—Translations into Church Slavic—History and criticism—Addresses, essays, lectures. I. Rybakov, Boris Aleksandrovich, 1908- . II. Akademii͡a nauk SSSR. Nauchnyĭ sovet po istorii mirovoĭ kul'tury. III. Izbornik Svi͡atoslava 1073 g. Selections. 1977.
PG703.I973I9 79-379332
79 MARC

CHURCH SOCIABLES see Church entertainments

CHURCH STATISTICS

see also Ecclesiastical geography

CHURCH TAX

—GERMANY, WEST—STATES

Giloy, Jörg.
Kirchensteuerrecht und Kirchensteuerpraxis in den Bundesländern / von Jörg Giloy, unter Mitarb. von Amtsrat König. — Stuttgart ; Wiesbaden : Forkel, 1978.
288 p. ; 21 cm. — (Forkel-Reihe Recht und Steuern) GFR79-A7
Includes legislation.
"Das vorliegende Werk ist zugleich als Sonder-Lieferung der 'Rechts- und Wirtschafts-Praxis' erschienen."
Bibliography: p. 13.
Includes index.
ISBN 3-7719-6204-8 : DM39.00
1. Church tax—Germany, West—States. I. König, Amtsrat, joint author. II. Title.
343'.43'052 79-365616
79 MARC

CHURCH TOWERS see Towers

CHURCH TRUSTEES

Tibbetts, Orlando L
The work of the church trustee / Orlando L. Tibbetts. — Valley Forge, PA : Judson Press, c1979.
125 p. ; 18 cm.
Includes bibliographical references.
ISBN 0-8170-0825-X : $3.95
1. Church trustees. I. Title.
BV705.T5 254 79-9882
79 MARC

CHURCH UNITY see Christian union

CHURCH VESTMENTS

see also Pallium

King, Bucky.
Ecclesiastical crafts / Bucky King, Jude Martin. — New York : Van Nostrand Reinhold, 1978.
132 p. : ill. (some col.) ; 24 cm.
Bibliography: p. 128-129.
Includes index.
ISBN 0-442-22966-6
1. Liturgical objects. 2. Church vestments. 3. Christian art and symbolism. 4. Church decoration and ornament. I. Martin, Jude, joint author. II. Title.
NK1650.K47 745.5 77-25033
77 MARC

CHURCH WORK

see also Chaplains, Industrial; Church and social problems; Church group work; Church management; City churches; City missions; Deaconesses; Evangelistic work; Fish movement (Christianity); Indigenous Church administration; Mass media in religion;

Missions; Pastoral counseling; Pastoral psychology; Pastoral theology; Radio in religion; Revivals; Rural churches; Small churches; Visitations (Church work); Women in church work

Brister, C W
Take care / C. W. Brister. — Nashville : Broadman Press, c1978.
164 p. ; 20 cm.
Continues People who care.
Includes bibliographical references.
ISBN 0-8054-5578-7
1. Pastoral theology. 2. Church work. I. Title.
BV4011.B685 253.5 76-51022
79 MARC

Dudley, Carl S 1932-
Where have all our people gone? : New choices for old churches / Carl S. Dudley. — New York : Pilgrim Press, c1979.
xiv, 136 p. : ill. ; 21 cm.
Includes bibliographies.
ISBN 0-8298-0359-9 : $5.95
1. Church membership. 2. Church work. 3. Christianity—20th century. I. Title.
BV820.D82 254'.5 79-525
79 MARC

—ADDRESSES, ESSAYS, LECTURES

Kirchliche Erwachsenenarbeit / Jürgen Lott (Hrsg.). — 1. Aufl. — Stuttgart ; Berlin ; Köln ; Mainz : Kohlhammer, 1977.
138 p. ; 19 cm. — (Urban-Taschenbücher ; Bd. 633 : T-Reihe) GFR77-A46
Bibliography: p. 131-137.
ISBN 3-17-004430-3 : DM12.00
1. Christian education of adults—Addresses, essays, lectures. 2. Church work—Addresses, essays, lectures. I. Lott, Jürgen, 1943- .
BV1488.K57 79-366276
79 MARC

—HANDBOOKS, MANUALS, ETC.

Mueller, Charles S
God's people at work in the parish / Charles S. Mueller. — St. Louis : Concordia Pub. House, c1979.
63 p. ; 23 cm.
ISBN 0-570-08025-8 : $2.50
1. Church work—Handbooks, manuals, etc. I. Title.
[BV4400.M8] 254 79-11541
79 MARC

—NEDERLANDSE HERVORMDE KERK

Scholten, R G 1937-
Vrijwilligersarbeid in hervormde gemeenten : verslag van een in februari 1977 in opdracht van de Raad voor de Herderlijke Zorg van de Nederlandse Hervormde Kerk gehouden onderzoek / R. G. Scholten. — 's-Gravenhage : Boekencentrum, [1979]
78 p. ; 23 cm. Ne79-8
ISBN 9023914074 : fl 11.90
1. Church work—Nederlandse Hervormde Kerk. I. Nederlandse Hervormde Kerk. Raad voor de Herderlijke Zorg. II. Title.
BV4404.N4S36 79-373877
*79 MARC

CHURCH WORK, RURAL see Rural churches

CHURCH WORK WITH ADOLESCENTS

see also Christian education of adolescents

Explore; resources for junior highs in the Church, edited by Janice M. Corbett. ⌈Valley Forge, Pa., Judson Press, 1974-
v. illus. 28 cm.
Vol. 2 edited by J. E. Grant; v. 3 by B. Middleton.
Includes bibliographies.
ISBN 0-8170-0646-X (v. 1)
1. Church work with adolescents. 2. Youth—Religious life. I. Corbett, Jancie M. II. Grant, James E. III. Middleton, Barbara.
BV1475.9.E88 268'.433 74-8574
74⌈r78⌉rev2 MARC

CHURCH WORK WITH ADULTS see Church work; Church work with women

CHURCH WORK WITH CHILDREN

see also Church camps

—ADDRESSES, ESSAYS, LECTURES

Children : of such is the kingdom / edited by Barbara Howard. — Independence, Mo. : Herald House, c1979.
175 p. : ill. ; 20 cm.
ISBN 0-8309-0243-0
1. International Year of the Child, 1979—Addresses, essays, lectures. 2. Church work with children—Addresses, essays, lectures. I. Howard, Barbara.
HQ767.9.C45 261.8'34'314 79-9102
79 MARC

—FINLAND—TAMPERE

Laitinen, Tuure, 1907-
Tampereen poikia / Tuure Laitinen. — ⌈Tampere⌉ : Tampere-seura, 1977 (⌈s.l.⌉ : Tamprint)
135 p., ⌈1⌉ p. : ill. ; 22 cm. — (Tampere-Seuran julkaisuja ; n:o 44) F
ISBN 9519080163 : Fmk25.00
1. Church work with children—Finland—Tampere. 2. Boys—Societies and clubs—Finland—Tampere. I. Title. II. Series: Tampere-Seura. Julkaisuja ; n:o 44.
DK465.T25T25 no. 44 78-379223
⌈BV639.B7⌉ 79 MARC

—NETHERLANDS

Roelfs, Jan.
Kinderbescherming op het gereformeerde erf; het "hoe en waarom" van (her)opvoeding en begeleiding in deze tijd. Samengesteld in opdracht van het Gereformeerd Centrum voor Kinderbescherming van de Stichting Raad voor Gereformeerd Sociale Arbeid. Utrecht, Schotanus & J. Utrecht, 1965.
58 p. illus. 24 cm.
1. Church work with children—Netherlands. 2. Church work with youth—Netherlands. 3. Church work with young adults—Netherlands. 4. Child welfare—Netherlands. 5. Unmarried mothers—Netherlands. 6. Gereformeerde kerken in Nederland—Charities. I. Title.
BV639.C4R63 79-413530
79 MARC

CHURCH WORK WITH CRIMINALS

see also Church work with prisoners

—GERMANY, WEST

Stubbe, Ellen.
Seelsorge im Strafvollzug : histor., psychoanalyt. u. theol. Ansätze zu e. Theoriebildung / von Ellen Stubbe. — Göttingen : Vandenhoeck und Ruprecht, 1978.
275 p. ; 24 cm. — (Arbeiten zur Pastoraltheologie ; Bd. 15) GFR78-A
A revision of the author's thesis, Kiel, 1977, presented under the title: Identifikation-Projektion.
Bibliography: p. 242-249.
Includes indexes.
ISBN 3-525-62166-3
1. Church work with criminals—Germany, West. 2. Church work with prisoners—Germany, West. I. Title. II. Series.
BV4464.7.S78 1978 79-338863
79 MARC

CHURCH WORK WITH FAMILIES

Genné, Elizabeth.
Church family camps and conferences : an administrative and program manual / Elizabeth and William Genné. — Rev. ed. — Valley Forge, PA. : Judson Press, c1979..
77 p. ; 22 cm.
ISBN 0-8170-0818-7 : $2.95
1. Church work with families. 2. Church camps. 3. Family—Religious life. I. Genné, William H., joint author. II. Title.
BV4438.G46 1979 259 78-24395
78 MARC

Pro Mundi Vita (Society)
The Western family and the future of the Church. — ⌈Brussels⌉ : Pro Mundi Vita, Centrum Informationis, 1974.
40 p. ; 27 cm. — (⌈Bulletin⌉ - Pro Mundi Vita ; 51) Be***
Bibliography: p. 40.
1. Family. 2. Marriage. 3. Church work with families. I. Title. II. Series: Pro Mundi Vita (Society). Bulletin ; 51.
HQ728.P73 1974 301.42 78-323196
79 MARC

CHURCH WORK WITH FOREIGNERS

—GERMANY, WEST

Gastarbeiter werden Bürger : Handbuch zur evang. Ausländerarbeit / hrsg. von Jürgen Micksch. — Frankfurt am Main : Lembeck, 1978.
189 p. ; 21 cm. — (EPD-Paperback ; 1) GFR78-A
Bibliography: p. 183-186.
Includes index.
ISBN 3-87476-103-7 : DM9.80
1. Church work with foreigners—Germany, West. 2. Church work with migrant labor—Germany, West. I. Micksch, Jürgen, 1941- . II. Series: Evangelischer Pressedienst. EPD-Paperback ; 1.
BV639.I4G37 78-400893
79 MARC

CHURCH WORK WITH GIRLS see Church work with children

CHURCH WORK WITH HOMOSEXUALS

—CASE STUDIES

French, Joel, 1938-
Straight is the way / Joel & Jane French. — Minneapolis : Bethany Fellowship, c1979.
159 p. ; 21 cm.
ISBN 0-87123-497-1 : $2.95
1. Homosexuals, Male—Case studies. 2. Church work with homosexuals—Case studies. 3. Homosexuality—Moral and religious aspects. I. French, Jane, 1948- joint author. II. Title.
HQ76.F75 301.41'57 79-169
79 MARC

CHURCH WORK WITH JUVENILE DELINQUENTS

—CALIFORNIA—LOS ANGELES—CASE STUDIES

McGuire, Hilary.
Hōpie and the Los Homes Gang : a gangland pimer / Hilary McGuire. — Canfield, Ohio : Alba Books, c1979.
xiv, 152 p. : ill. ; 18 cm.
ISBN 0-8189-1150-6 : $1.95
1. Gangs—California—Los Angeles—Case Studies. 2. Church work with juvenile delinquents—California—Los Angeles—Case studies. 3. Lopez, Hopie. 4. Mexican American youth—California—Los Angeles—Case studies. I. Title.
HV6439.U7L74 301.44'94'0979494 78-56771
79 MARC

CHURCH WORK WITH LABORING CLASSES

see also Chaplains, Industrial; Church work with migrant labor

—AUSTRIA—CONGRESSES

Österreichische Pastoraltagung, Vienna, 1978.
Arbeiterpastoral in der Pfarre : Österreichische Pastoraltagung 27.-29. Dez. 1978 / im Auftrag des Österr. Pastoralinst. hrsg. v. Josef Wiener u. Helmut Erharter. — Wein ; Freiburg ; Basel : Herder, 1979.
184 p. ; 23 cm. Au79-14-20
Includes bibliographical references.
ISBN 3-210-24581-9 : S158.00
1. Church work with laboring classes—Austria—Congresses. 2. Pastoral theology—Catholic Church—Congresses. I. Wiener, Josef. II. Erharter, Helmut. III. Österreichisches Pastoralinstitut.
BX2347.8.L3O87 1978 79-380125
*79 MARC

—GERMANY, WEST

Friedberger, Walter.
Seelsorge in der Arbeitswelt / Walter Friedberger. — Würzburg : Seelsorge Verlag Echter, 1978.
183 p. : ill. ; 21 cm. — (Pastorale Handreichungen ; Bd. 18) GFR79-A
Bibliography: p. 164-166.
ISBN 3-429-00557-4 : DM16.80
1. Church work with laboring classes—Germany, West. 2. Catholic Church—Pastoral theology. I. Title. II. Series.
BX1913.P33 Bd. 18 79-376698
⌈BV2695.W6⌉ 79 MARC

—ITALY—TURIN

Ugo, Pisano
L'evangelizzazione del mondo operaio a Torino nell'ambiente della fabbrica dal 1942 ad oggi : extractum / Pisano Ugo. — Roma : Pontificia università lateranense, 1977.
134 p. ; 24 cm. It***
Extract of thesis—Pontificia università lateranense.
Bibliography: p. 125-131.
1. Church work with laboring classes—Italy—Turin. 2. Chaplains, Industrial—Italy—Turin. I. Title.
BV2695.W6U35 79-410779
79 MARC

CHURCH WORK WITH MENTALLY HANDICAPPED CHILDREN

—GERMANY, WEST

Kobbe, Ursula.
Sie brauchen Liebe : religiöse Erfahrungen mit geistig behinderten Kindern im Licht d. Vaterunsers / Ursula Kobbe. — Freiburg im Breisgau ; Basel ; Wien : Herder, 1977.
188 p. ; 21 cm. — (Veröffentlichungen der Stiftung Oratio Dominica) GFR78-A
ISBN 3-451-17639-4 : DM24.00
1. Kobbe, Ursula. 2. Catholics—Germany, West—Biography. 3. Church work with mentally handicapped children—Germany, West. 4. Lord's Prayer. I. Title. II. Series: Stiftung Oratio Dominica. Veröffentlichungen der Stiftung Oratio Dominica.
BX4705.K616A37 79-367394
79 MARC

CHURCH WORK WITH MIGRANT LABOR

—GERMANY, WEST

Gastarbeiter werden Bürger : Handbuch zur evang. Ausländerarbeit / hrsg. von Jürgen Micksch. — Frankfurt am Main : Lembeck, 1978.
189 p. ; 21 cm. — (EPD-Paperback ; 1) GFR78-A
Bibliography: p. 183-186.
Includes index.
ISBN 3-87476-103-7 : DM9.80
1. Church work with foreigners—Germany, West. 2. Church work with migrant labor—Germany, West. I. Micksch, Jürgen, 1941- . II. Series: Evangelischer Pressedienst. EPD-Paperback ; 1.
BV639.I4G37 78-400893
79 MARC

CHURCH WORK WITH MILITARY PERSONNEL

see also Chaplains, Military

CHURCH WORK WITH NARCOTIC ADDICTS

—CHINA—HONGKONG

Møller, Arvid, 1939-
Agnar Espegren / Arvid Møller og Lasse Thorseth. — Oslo : Luther, 1978 (Stavanger : Gjøstein boktr.)
171 p., [2] leaves : ill. ; 21 cm. N78-Oct.
ISBN 8253132182 : kr69.00
1. Espegren, Agnar, 1911- . 2. Missionaries—Norway—Biography. 3. Missionaries—Hong Kong—Biography. 4. Church work with narcotic addicts—Hong Kong. 5. Missions—China. I. Thorseth, Lasse, joint author.
BV3427.E78M63 78-395008
79 MARC

CHURCH WORK WITH PRISONERS

see also Church work with criminals

Colbaugh, Lloyd N
The gospel behind bars / Lloyd N. Colbaugh. — Springfield, Mo. : Radiant Books, c1979.
94 p. ; 18 cm.
ISBN 0-88243-503-5 : $1.50
1. Church work with prisoners. I. Title.
[BV4465.C59] 259 79-53942
79 MARC

Pederson, Duane.
How to establish a jail and prison ministry / Duane Pederson ; foreword by Corrie ten Boom. — Nashville : T. Nelson, c1979.
126 p. ; 23 cm.
Includes bibliographical references.
ISBN 0-8407-5675-5 : $2.95
1. Church work with prisoners. I. Title.
BV4465.P4 253.7'5 78-27438
78 MARC

—ENGLAND—CONGRESSES

Prison Chaplains' Conference, Oxford, 1977.
Prison Chaplains' Conference report. — [s.l. : s.n.], 1977.
32 p. ; 30 cm. GB***
Cover title.
1. Church work with prisoners—England—Congresses.
BV4465.P74 1977 365'.66 79-300892
79 MARC

—GERMANY, WEST

Stubbe, Ellen.
Seelsorge im Strafvollzug : histor., psychoanalyt. u. theol. Ansätze zu e. Theoriebildung / von Ellen Stubbe. — Göttingen : Vandenhoeck und Ruprecht, 1978.
275 p. ; 24 cm. — (Arbeiten zur Pastoraltheologie ; Bd. 15) GFR78-A
A revision of the author's thesis, Kiel, 1977, presented under the title: Identifikation-Projektion.
Bibliography: p. 242-249.
Includes indexes.
ISBN 3-525-62166-3
1. Church work with criminals—Germany, West. 2. Church work with prisoners—Germany, West. I. Title. II. Series.
BV4464.7.S78 1978 79-338863
79 MARC

CHURCH WORK WITH REFUGEES

—ADDRESSES, ESSAYS, LECTURES

Frieden, Versöhnung und Menschenrechte. — Gütersloh : Gütersloher Verlagshaus Mohn, 1978-
v. ; 19 cm. — (Die Denkschriften der Evangelischen Kirche in Deutschland ; Bd. 1, T. 1-) (Gütersloher Taschenbücher ; 13- : Siebenstern) GFR78-A
Includes bibliographical references.
ISBN 3-579-04801-5 (v. 1, pt. 1) : DM12.80 (v. 1, pt. 1)
1. Sociology, Christian—Addresses, essays, lectures. 2. Church work with refugees—Addresses, essays, lectures. 3. Church and race relations—Addresses, essays, lectures. 4. Christianity and international affairs—Addresses, essays, lectures. 5. Peace (Theology)—Addresses, essays, lectures. I. Series.
BT738.F68 78-380301
79 MARC

—AFRICA

All Africa Conference of Churches. Refugee Dept.
Africa's one million refugees : from where they escape, where they find asylum. — [Nairobi : Refugee Dept., All Africa Conference of Churches, 1975?]
22 p. : ill. ; 16 x 24 cm.
Cover title: The African refugee story.
1. Africa—Politics and government—1960- . 2. Refugees—Africa. 3. Church and refugee problems—Africa. 4. Church and state in Africa. 5. Church work with refugees—Africa. I. Title. II. Title: The African refugee story.
DT30.A467 1975 361.5'3'096 79-314375
79 MARC

CHURCH WORK WITH SINGLE PEOPLE

Brown, Raymond Kay, 1936-
Reach out to singles : a challenge to ministry / by Raymond Kay Brown. — 1st ed. — Philadelphia : Westminster Press, c1979.
191 p. ; 21 cm.
Bibliography: p. 185-191.
ISBN 0-664-24270-7
1. Church work with single people. I. Title.
BV639.S5B76 261.8'34 79-15495
79 MARC

—HANDBOOKS, MANUALS, ETC.

Reed, Bobbie.
Single on Sunday : a manual for successful single adult ministries / by Bobbie Reed. — St. Louis : Concordia Pub. House, c1979.
103 p. ; 28 cm.
Bibliography: p. 102-103.
ISBN 0-570-03781-6 : $4.95
1. Church work with single people—Handbooks, manuals, etc. I. Title.
BV639.S5R43 253 78-8297
78 MARC

CHURCH WORK WITH STUDENTS

see also Chaplains, School

Bourdoncle, Pierre.
Aumôneries de l'enseignement public : une lente révolution / Pierre Bourdoncle et Pierre Moitel. — Paris : Secrétariat national de L'Aumônerie de l'enseignement public, c1978.
119 p. ; 18 cm. — (Dossiers libres Cerf) F***
Includes bibliographical references.
ISBN 2-204-01306-4 : 18.00F
1. Church work with students. 2. Catholic Church in France. I. Moitel, Pierre, 1931- joint author. II. Title.
BV1610.B584 259 79-375764
79 MARC

CHURCH WORK WITH THE AGED

Alderman, Rosalie Givens.
Take heart / Rosalie Givens Alderman. — Nashville, Tenn. : Broadman Press, c1978.
128 p. : ill. ; 20 cm.
ISBN 0-8054-5534-5
1. Alderman, Rosalie Givens. 2. Baptists—United States—Biography. 3. Church work with the aged. I. Title.
BX6495.A36A7 286'.1'0924 78-59782
78 MARC

Clements, William M 1943-
Care and counseling of the aging / William M. Clements. — Philadelphia : Fortress Press, c1979.
xiv, 82 p. : ill. ; 22 cm. — (Creative pastoral care and counseling series)
Bibliography: p. 81-82.
ISBN 0-8006-0561-6 : $2.95
1. Church work with the aged. I. Title.
BV4435.C57 253.5 78-54547
78 MARC

Corso-base animatori e responsabili della pastorale anziani, 1st, Rome, 1977.
Gli anziani nella comunità ecclesiale / [atti del 1. Corso-base animatori e responsabili della pastorale anziani tenutosi a Roma presso la Domus Mariae dal 21 al 25 novembre 1977]. — Roma : A.V.E., c1978.
149 p. ; 19 cm. — (Nuovo impegno ; 4) It***
"Documenti": p. [103]-145.
L1500
1. Aged—Religious life. 2. Church work with the aged. I. Title.
BV4580.C58 1977 79-360369
79 MARC

Handbuch kirchlicher Altenarbeit / Jochen Schmauch (Hg.). — Mainz : Matthias Grünewald-Verlag, 1978.
256 p. ; 21 cm. — (Grünewald Praxis) GFR78-A
Includes bibliographical references.
ISBN 3-7867-0690-5 : DM29.50
1. Church work with the aged. I. Schmauch, Jochen.
BX2347.8.A4H36 79-341694
79 MARC

Hildemann, Klaus D
Altenarbeit in der Kirchengemeinde / Klaus D. Hildemann. — Heidelberg : Quelle und Meyer, 1978.
155 p. ; 21 cm. GFR79-A
Bibliography: p. 114-152.
ISBN 3-494-00970-8
1. Church work with the aged. I. Title.
BV4435.H54 79-353769
79 MARC

Hungs, Franz-Josef, 1932-
Altenbildung, Altenpastoral : Erfahrungen in d. theol. Erwachsenenbildung mit älteren Menschen / Franz-Josef Hungs. — München : Kösel, 1978.
96 p. ; 22 cm. GFR78-A
Bibliogrpahy: p. 88-90.
ISBN 3-466-36032-3
1. Church work with the aged. 2. Aged. 3. Christian education of adults. I. Title.
BV4435.H86 79-377618
79 MARC

Kline, Harvey, 1921-
A future with hope : aging creatively in christian community / Harvey Kline, Warren Eshback ; [drawings by Cindy Staub]. — Elgin, Ill. : Brethren Press, [1978]
93 p. : ill. ; 21 cm.
Bibliography: p. [87]-88.
ISBN 0-87178-298-7
1. Old age. 2. Aged. 3. Church work with the aged. I. Eshbach, Warren, 1940- joint author. II. Title.
HV1451.K55 261.8'34'35 78-17720
78 MARC

CHURCH WORK WITH THE BEREAVED

see also Funeral service

CHURCH WORK WITH THE DEAF

Lawrence, Edgar D
Ministering to the silent minority / Edgar D. Lawrence. — Springfield, Mo. : Gospel Pub. House, c1978.
92 p. ; 18 cm. — (Radiant books)
Includes bibliographical references.
ISBN 0-88243-738-0 : $1.50
1. Church work with the deaf. I. Title.
BV4463.L38 253 78-55679
79 MARC

CHURCH WORK WITH THE HANDICAPPED

Andersen, Asger, 1921-
Kirken og de handicappede. — København : Unitas, 1978.
159 p. ; 20 cm. D78-50
Bibliography: p. 155-[158]
Includes index.
ISBN 8775171074 : kr69.75
1. Church work with the handicapped. I. Title.
BV4460.A52 79-384274
79 MARC

Hämer, Andreas.
Rehabilitation von unten : d. Platz d. Körperbehinderten im Aufgabenfeld d. Kirche / Andreas Hämer ; mit einem Geleitwort von Theodor Strohm. — 1. Aufl. — München : Kaiser ; Mainz : Matthias-Grünewald-Verlag, 1978.
215, [1] p. ; 21 cm. — (Gesellschaft und Theologie : Abteilung Praxis der Kirche ; Nr. 28) GFR78-A
A revision of the author's Magisterschrift, Kirchliche Hochschule Berlin, 1975.
Bibliography: p. 202-[216]
ISBN 3-459-01140-8
1. Church work with the handicapped. I. Title. II. Series.
BV4460.H26 79-370189
79 MARC

CHURCH WORK WITH THE INSANE see Church work with the mentally ill

CHURCH WORK WITH THE MENTALLY HANDICAPPED

Toulat, Jean, 1915-
Les Forces de l'amour : de Jean Vanier à Mère Teresa / Jean Toulat. — Paris : Éditions S.O.S., 1976.
135 p., [8] p. of plates ; 21 cm. F77-2936
Includes bibliographical references.
ISBN 2-7185-0820-5 : 34.00F
1. Church work with the mentally handicapped. 2. Vanier, Jean, 1928- . 3. Church work with the sick. 4. Teresa, Mother, 1910- . 5. Church and the poor. 6. Catholic Church—Missions. 7. Catholic Church—Charities. I. Title.
BX2347.8.M4T68 361.7'5 78-399927
79 MARC

CHURCH WORK WITH THE MENTALLY ILL.

Casera, Dominique.
Les psychotiques : à l'écoute de la désespérance / Dominique Casera. — Montréal : Éditions paulines, c1978.
158 p. ; 20 cm. — (Notre temps ; 18) C***
Bibliography: p. 152-156.
ISBN 0-88840-652-5 : $12.00
1. Psychoses. 2. Church work with the mentally ill. I. Title.
RC512.C37 616.8'9'00242 79-345764
79 MARC

Oates, Wayne Edward, 1917-
The religious care of the psychiatric patient / by Wayne E. Oates. — 1st ed. — Philadelphia : Westminster Press, c1978.
xiv, 237 p. ; 21 cm.
ISBN 0-664-21365-0
1. Church work with the mentally ill. 2. Pastoral theology. 3. Pastoral psychology. 4. Mentally ill—Care and treatment. I. Title.
BV4461.O18 1978 259 78-18454
78 MARC

Vanderpool, James A
People in pain : a guide to pastoral care / by James A. Vanderpool. — Springfield, Ill. : Thomas, c1979.
xii, 190 p. ; 24 cm.
Bibliography: p. 179-185.
Includes index.
ISBN 0-398-03846-5
1. Pastoral counseling. 2. Church work with the mentally ill. I. Title.
BV4012.2.V27 253.5 78-13045
78 MARC

CHURCH WORK WITH THE SICK

see also Pastoral medicine

Pfender, Marcel.
Les Malades parmi nous, le ministère de l'Église auprès des malades. [78 Saint-Germain-en-Laye,] les Bergers et les mages, [10, rue de Villars,] 1971.
248 p. 20 cm. 28.00F F72-9292
Bibliography: p. [243]-244.
1. Pastoral medicine. 2. Church work with the sick. 3. Sick—Psychology. I. Title.
BV4335.P46 79-413796
79 MARC

Toulat, Jean, 1915-
Les Forces de l'amour : de Jean Vanier à Mère Teresa / Jean Toulat. — Paris : Éditions S.O.S., 1976.
135 p., [8] p. of plates ; 21 cm. F77-2936
Includes bibliographical references.
ISBN 2-7185-0820-5 : 34.00F
1. Church work with the mentally handicapped. 2. Vanier, Jean, 1928- . 3. Church work with the sick. 4. Teresa, Mother, 1910- . 5. Church and the poor. 6. Catholic Church—Missions. 7. Catholic Church—Charities. I. Title.
BX2347.8.M4T68 361.7'5 78-399927
79 MARC

—INDIA—ADDRESSES, ESSAYS, LECTURES

Chandy, Jacob, 1910-
Rethinking the healing ministry of the Indian Church. J. C. Madras, Published for the Christian Institute for the Study of Religion and Society by Christian Literature Society [1970]
45 p. 19 cm. (Pamphlets on social concern (new series), no. 2) Rs1.25
Includes bibliographical references.
1. Church work with the sick—India—Addresses, essays, lectures. I. Title. II. Series.
BV4460.C48 361.7'5'0954 75-916817
79 MARC

CHURCH WORK WITH TOURISTS, TRAVELERS, ETC.

Appendino, Filippo Natale.
Turismo, lavoro e pastorale della comunità / Filippo Natale Appendino. — Torino Leumann : Elle Di Ci, 1977.
380 p. ; 21 cm. — (Collana Evangelizzazione e promozione umana)
It78-Mar
Includes bibliographies.
L5200
1. Church work with tourists, travelers, etc. I. Title.
BV4467.A66 78-366159
•78 MARC

—CONGRESSES

Simposio europeo sulla pastorale del turismo, Vatican City, 1975.
Educazione al turismo = Pastorale Vorbereitung auf den Tourismus : atti del Symposium europeo, Città del Vaticano, 22-25 settembre 1975 / a cura di mons. Edmondo De Panfilis, delegato nazionale CEI per la pastorale del turismo. — Leumann/[Rivoli] : Elle Di Ci, 1978.
255 p. ; 21 cm. — (Collana Evangelizzazione e promozione umana)
It78-May
At head of title: Ufficio cattolico italiano per il turismo, Roma.
ISBN 8801119526
1. Church work with tourists, travelers, etc.—Congresses. I. De Panfilis, Edmondo. II. Ufficio cattolico italiano per il turismo. III. Title. IV. Title: Pastorale Vorbereitung auf den Tourismus.
BX2347.8.T69S57 1975 78-380785
•79 MARC

CHURCH WORK WITH WOMEN

see also Women in church work

—GERMANY

Klens, Hermann, 1880-1972.
Hermann Klens : Anwalt d. Frauen : Leben u. Werk / [Ingeborg Rocholl-Gärtner hat d. Material d. Klens-Memoiren ausgew., bearb. sowie mit e. Biographie u. Interpretationen vers.]. — 1. Aufl. — Düsseldorf : Klens-Verlag, 1978.
199 p. ; 25 cm. GFR78-A39
Errata slip inserted.
ISBN 3-87309-035-X : DM24.60
1. Klens, Hermann, 1880-1972. 2. Catholic Church—Clergy—Biography. 3. Clergy—Germany—Biography. 4. Church work with women—Germany. I. Rocholl-Gärtner, Ingeborg. II. Title: Anwalt der Frauen.
BX4705.K536A34 78-396982
79 MARC

—GERMANY, WEST

Weigle, Maria, 1893-
Bibelarbeit mit Frauen : Methoden u. Auslegungen sowie e. Beitr. zum Thema "Machen Frauen eine andere Theologie als Männer"? / Maria Weigle ; hrsg. von Ilse Ueckert. — Stein/Mfr. : Laetare-Verlag ; Gelnhausen ; Berlin : Burckhardthaus-Verlag, 1979.
141 p. ; 23 cm. — (Lese-Zeichen) GFR79-A
"Ein Laetare Buch."
Bibliography: p. 141.
ISBN 3-7839-8003-8
1. Bible—Study. 2. Church work with women—Germany, West. I. Ueckert, Ilse. II. Title.
BS600.2.W4 79-360185
79 MARC

—KOREA

Yŏnse Taehakkyo, Seoul, Korea. In'gu mit Kajok Kyehoek Yŏn'guso.
(Kyohoe hwaltong ŭl t'onghan tosi yŏngsemin ŭi kajok kyehoek saŏp e kwanhan yŏn'gu)
교외 [i. e. 회] 활동을통한도시영세민의가족계획사업에관한연구 = A study on family planning programme in urban slum area through church activities / 양재모 ... [et al.]. — [서울] : 연세대학교인구및가족계획연구소, 1975.
88 p. ; 26 cm.
In Korean; summary in English.
1. Birth control—Korea. 2. Church work with women—Korea. I. Yang, Chae-mo. II. Title. III. Title: A study on family planning programme in urban slum area through church activities.
HQ766.5.K6Y66 1975 362.8'2 79-825695

CHURCH WORK WITH YOUNG ADULTS

Praxisfeld "Kirchliche Jugendarbeit" : soziales Umfeld, Arbeits- u. Lebensformen, Beitr. zur Konzeption / hrsg. von Martin Affolderbach. — Orig.-Ausg., 1. Aufl. — Gütersloh : Gütersloher Verlagshaus Mohn, 1978.
204 p. ; 19 cm. — (Gütersloher Taschenbücher : Siebenstern ; 280)
GFR78-A
Includes bibliographical references.
ISBN 3-579-03680-7 : DM9.80
1. Church work with youth. 2. Church work with young adults. I. Affolderbach, Martin, 1947- .
BV4447.P67 79-341745
79 MARC

—CASE STUDIES

MacCollam, Joel A 1946-
Carnival of souls : religious cults and young people / by Joel A. MacCollam. — New York : Seabury Press, 1979.
188 p. ; 21 cm.
"A Crossroad book."
Bibliography: p. 185-188.
ISBN 0-8164-2211-7
1. Church work with young adults—Case studies. 2. Church work with youth—Case studies. 3. Cults—Controversial literature. 4. Christian sects—Controversial literature. 5. Pastoral theology—Case studies. I. Title.
BV4446.M26 259 78-25976
78 MARC

—NETHERLANDS

Roelfs, Jan.
Kinderbescherming op het gereformeerde erf; het "hoe en waarom" van (her)opvoeding en begeleiding in deze tijd. Samengesteld in opdracht van het Gereformeerd Centrum voor Kinderbescherming van de Stichting Raad voor Gereformeerd Sociale Arbeid. Utrecht, Schotanus & J. Utrecht, 1965.
58 p. illus. 24 cm.
1. Church work with children—Netherlands. 2. Church work with youth—Netherlands. 3. Church work with young adults—Netherlands. 4. Child welfare—Netherlands. 5. Unmarried mothers—Netherlands. 6. Gereformeerde kerken in Nederland—Charities. I. Title.
BV639.C4R63 79-413530
79 MARC

CHURCH WORK WITH YOUNG MARRIED COUPLES see Church work with young adults

CHURCH WORK WITH YOUTH

see also Church camps

Eggum, Tom.
Feeling good / Tom Eggum. — Nashville : T. Nelson Publishers, c1979.
120 p. ; 21 cm.
ISBN 0-8407-5693-3 : $3.95
1. Eggum, Tom. 2. Evangelists—United States—Biography. 3. Church work with youth. I. Title.
BV3785.E43A33 269.2'092'4 79-16014
79 MARC

Fackre, Gabriel J
Youth ministry : the Gospel and the people / Gabriel Fackre, Jan Chartier. — Valley Forge, Pa. : Judson Press, c1979.
135 p. : ill. ; 22 cm.
Bibliography: p. 133-135.
ISBN 0-8170-0829-2 : $5.95
1. Church work with youth. 2. Youth—Religious life. I. Chartier, Jan, joint author. II. Title.
BV4447.F32 259 79-4605
79 MARC

Martinelli, Raffaello.
Rinnovamento psicopedagogico della pastorale giovanile : ricerche e orientamenti / Raffaello Martinelli. — Roma : Pontificia Universitas lateranensis, 1977.
190 p. ; 25 cm. It***
Bibliography: p. 17-36.
Abstract of thesis—Pontificia Universitas lateranensis.
1. Church work with youth. 2. Christian education of young people. I. Title.
BV4447.M33 79-356993
79 MARC

Praxisfeld "Kirchliche Jugendarbeit" : soziales Umfeld, Arbeits- u. Lebensformen, Beitr. zur Konzeption / hrsg. von Martin Affolderbach. — Orig.-Ausg., 1. Aufl. — Gütersloh : Gütersloher Verlagshaus Mohn, 1978.
204 p. ; 19 cm. — (Gütersloher Taschenbücher : Siebenstern ; 280) GFR78-A
Includes bibliographical references.
ISBN 3-579-03680-7 : DM9.80
1. Church work with youth. 2. Church work with young adults. I. Affolderbach, Martin, 1947-
BV4447.P67 79-341745
79 MARC

Rice, Wayne.
Junior high ministry : a guidebook for the leading and teaching of early adolescents / Wayne Rice. — Grand Rapids : Zondervan Pub. House, c1978.
201 p. : ill. ; 21 cm.
Bibliography: p. 201.
ISBN 0-310-34971-0
1. Church work with youth. I. Title.
BV4447.R45 259 78-23473
78 MARC

—CASE STUDIES

MacCollam, Joel A 1946-
Carnival of souls : religious cults and young people / by Joel A. MacCollam. — New York : Seabury Press, 1979.
188 p. ; 21 cm.
"A Crossroad book."
Bibliography: p. 185-188.
ISBN 0-8164-2211-7
1. Church work with young adults—Case studies. 2. Church work with youth—Case studies. 3. Cults—Controversial literature. 4. Christian sects—Controversial literature. 5. Pastoral theology—Case studies. I. Title.
BV4446.M26 259 78-25976
78 MARC

—CATHOLIC CHURCH

Bund der Deutschen Katholischen Jugend. Diözesanstelle Münster. Autorengruppe Bischöfliches Jugendamt.
Glaubensverkündigung in der kirchlichen Jugendarbeit : Beispiele, Impulse, Erfahrungen / ꞅVerf., Autorengruppe Bischöfl. Jugendamt, Diözesanstelle BDKJ Münster ; Hrsg., Bischöfl. Generalvikariat Münster, Hauptabt. Seelsorgeꞅ. — München : Kösel ; Düsseldorf : Verlag Haus Altenberg, 1978.
88 p. ; 22 cm. GFR78-A
ISBN 3-466-36028-5
1. Church work with youth—Catholic Church. I. Title.
BX2347.8.Y7B86 1978 268'.433 78-375565
78 MARC

Pro Mundi Vita (Society)
The social context of youth catechesis. — ꞅBrusselsꞅ : Pro Mundi Vita, 1976.
23 p. ; 27 cm. — (Pro Mundi Vita bulletin ; 63) Be***
Drafted by Pierre Delooz in consultation with other staff members.
1. Church work with youth—Catholic Church. 2. Evangelistic work. I. Delooz, Pierre. II. Title. III. Series: Pro Mundi Vita (Society). Bulletin ; 63.
BX2347.8.Y7P76 1976 268'.433 78-320323
78 MARC

Steinkamp, Hermann, 1938-
Jugendarbeit als soziales Lernen : Ziele u. Aufgaben kirchl. Jugendarbeit : zum Beschluss d. Gemeinsamen Synode d. Bistümer in d. Bundesrepublik Deutschland / Hermann Steinkamp ; mit e. Nachwort von Christof Bäumler. — 1. Aufl. — München : Kaiser ; Mainz : Grünewald, 1977.
125 p. ; 21 cm. — (Gesellschaft und Theologie : Abteilung Praxis der Kirche ; Nr. 27) GFR77-A
Bibliography: p. 115-117.
ISBN 3-459-01121-1
1. Church work with youth—Germany, West. 2. Church work with youth—Catholic Church. I. Title. II. Series.
BX2347.8.Y7S75 78-344533
78 MARC

—LUTHERAN CHURCH

Ludwig, Glenn E 1946-
Building an effective youth ministry / Glenn E. Ludwig. — Nashville : Abingdon, c1979.
125 p. ; 21 cm. — (Creative leadership series) (An Abingdon original paperback)
Bibliography: p. 120-123.
ISBN (invalid) 0-687-16300-5
1. Church work with youth—Lutheran Church. I. Title. II. Series.
BV4447.L825 268'.433 79-12282
79 MARC

—BELGIUM

Jeugd kijk over je muur : een animatieboek voor jongeren en jeugdleiders / samengesteld door de leefgroep van het Centrum voor Religieuze Cultuur en Animatie Eigentijdse Jeugd, Robert Kino ... ꞅet al.ꞅ. — Tielt : Lannoo, ꞅ1976ꞅ
155 p. : ill. ; 30 x 15 cm. Ne76-47
ISBN 9020906445 : fl 17.50
1. Youth—Belgium. 2. Church work with youth—Belgium. I. Kino, Robert. II. Eigentijdse Jeugd (Youth center)
HQ799.B4J48 79-359326
*79 MARC

—GERMANY, WEST

Steinkamp, Hermann, 1938-
Jugendarbeit als soziales Lernen : Ziele u. Aufgaben kirchl. Jugendarbeit : zum Beschluss d. Gemeinsamen Synode d. Bistümer in d. Bundesrepublik Deutschland / Hermann Steinkamp ; mit e. Nachwort von Christof Bäumler. — 1. Aufl. — München : Kaiser ; Mainz : Grünewald, 1977.
125 p. ; 21 cm. — (Gesellschaft und Theologie : Abteilung Praxis der Kirche ; Nr. 27) GFR77-A
Bibliography: p. 115-117.
ISBN 3-459-01121-1
1. Church work with youth--Germany, West. 2. Church work with youth—Catholic Church. I. Title. II. Series.
BX2347.8.Y7S75 78-344533
78 MARC

—NETHERLANDS

Drimmelen, Rob van.
Zakendoen en jeugdprojecten : de twee kanten van het verpakkingsconcern Van Leer / Rob van Drimmelen, Sami Faltas, Anje Valk. — Amsterdam : Oecumenisch Studie en Actiecentrum voor Investeringen, ꞅ1977?ꞅ
ii, 77 p. : ill. ; 30 cm. Ne***
Bibliography: p. 54-56.
1. Welfare work in industry—Netherlands. 2. Social work with youth—Netherlands. 3. Emballage Industrie Van Leer. 4. Church work with youth—Netherlands. 5. Gereformeerde kerken in Nederland. I. Faltas, Sami, joint author. II. Valk, Anje, joint author. III. Title.
HD8516.5.D74 78-352352
78ꞅr79ꞅrev MARC

Roelfs, Jan.
Kinderbescherming op het gereformeerde erf; het "hoe en waarom" van (her)opvoeding en begeleiding in deze tijd. Samengesteld in opdracht van het Gereformeerd Centrum voor Kinderbescherming van de Stichting Raad voor Gereformeerd Sociale Arbeid. Utrecht, Schotanus & J. Utrecht, 1965.
58 p. illus. 24 cm.
1. Church work with children—Netherlands. 2. Church work with youth—Netherlands. 3. Church work with young adults—Netherlands. 4. Child welfare—Netherlands. 5. Unmarried mothers—Netherlands. 6. Gereformeerde kerken in Nederland—Charities. I. Title.
BV639.C4R63 79-413530

CHURCH YEAR

see also Christmas; Eastertide; Holy Week; Lent; Pentecost festival

Bessière, Gérard.
Le livre du dimanche / G. Bessière, H. Vulliez. — ꞅParisꞅ : Chalet, ꞅc1978ꞅ
261 p. : ill. ; 24 cm. F***
Added t.p.: Almanach du chrétien 1979.
ISBN 2-7023-0307-2
1. Church year. I. Vulliez, Hyacinthe, joint author. II. Title. III. Title: Almanach du chrétien 1979.
BV30.B48 242'.3 79-350515
79 MARC

Catholic Church. Consilium ad Exsequendam Constitutionem de Sacra Liturgia.
ꞅNormae universales de anno liturgico. Italianꞅ
Norme generali sull'anno liturgico e sul calendario; commento a cura del Consilium. 2. ed. Milano, Edizioni O.R., 1969.
117 p. 19 cm. It***
Includes Italian translation of the Normae and of the new Calendarium Romanum generale.
Includes bibliographical references.
1. Catholic Church. Liturgy and ritual—Calendar. 2. Church year. I. Catholic Church. Liturgy and ritual. Calendarium Romanum generale. Italian. 1969. II. Title.
BX2045.C343N6716 1969 74-332533
79ꞅr79ꞅrev MARC

Hassel, Rudolph Chris, 1939-
Renaissance drama & the English church year / R. Chris Hassel, Jr. — Lincoln : University of Nebraska Press, c1979.
ix, 215 p. : ill. ; 22 cm.
Bibliography: p. 205-210.
Includes index.
ISBN 0-8032-2304-8
1. English drama—Early modern and Elizabethan, 1500-1600—History and criticism. 2. English drama—17th century—History and criticism. 3. Christianity in literature. 4. Church year. 5. Church of England. Book of common prayer. 6. Theater—England—History. I. Title.
PR658.R43H3 822'.3'09 78-24233
78 MARC

Irwin, Kevin W
A celebrant's guide to the new sacramentary : C cycle / Kevin W. Irwin. — New York : Pueblo Pub. Co., c1977.
244 p. ; 22 cm.
1. Catholic Church. Liturgy and ritual. Sacramentary (U.S.) 2. Bible—Liturgical lessons, English. 3. Church year.
BX2015.I78 264'.023 79-100573
79 MARC

Jaworski, Jan.
Tradycje siłą narodu / Jan Jaworski. — London : Veritas, ꞅpref. 1975ꞅ
208 p. : ill. ; 21 cm. GB***
Summary in English.
1. Church year. 2. Fasts and feasts—Catholic Church. 3. Poland—Church history. I. Title.
BV30.J38 79-370011
79 MARC

Ordóñez Márquez, Juan.
Teología y espiritualidad del año litúrgico / por Juan Ordóñez Márquez ; prólogo por Marcelo González Martín. — Madrid : EDICA, 1978.
xv, 415 p. ; 20 cm. — (Biblioteca de autores cristianos ; 403) Sp***
Includes bibliographical references.
ISBN 8422008866 : 675ptas
1. Church year. I. Title.
BV30.O72 79-379601
79 MARC

Russell, Joseph P
Sharing our Biblical story : a guide to using liturgical readings as the core of church and family education / Joseph P. Russell. — Minneapolis : Winston Press, c1979.
289 p. (p. 287-289 blank) : ill. ; 23 cm.
Bibliography: p. 283-286.
ISBN 0-03-043926-4
1. Bible stories. 2. Bible—Study. 3. Church year. 4. Bible—Liturgical use. 5. Story-telling (Christian theology) I. Title.
BS546.R87 268'.6 78-62823
00 MARC

—CARICATURES AND CARTOONS

Charlot, Jean, 1898-
Cartoons Catholic : mirth and meditation from the brush and brain of Jean Charlot / with commentary by F. J. Sheed. — Huntington, Ind. : Our Sunday Visitor, c1978.
ꞅ107ꞅ p. : ill. ; 22 cm.
ISBN 0-87973-826-X : $4.95
1. Church year—Meditations. 2. Church year—Caricatures and cartoons. 3. American wit and humor, Pictorial. I. Sheed, Francis Joseph, 1897- II. Title.
BX2170.C55C45 242'.3 78-62242

CHURCH YEAR (Continued)

—HISTORY

Deddens, K
Annus liturgicus? : Een onderzoek naar de betekenis van Cyrillus van Jeruzalem voor de ontwikkeling van het "kerkelijk jaar" / door Karel Deddens. — Goes : Oosterbaan & Le Cointre, 1975.
272 p. ; 24 cm. Ne***
At head of title: Theologische Hogeschool van de Gereformeerde Kerken in Nederland te Kampen.
Thesis—Theologische Hogeschool, Kampen.
"Stellingen" : [3] p. inserted.
Vita.
Summary in English, French, and German.
Bibliography: p. 237-260.
Includes index.
1. Liturgics—History. 2. Church year—History. 3. Cyrillus, Saint, Bp. of Jerusalem, 315 (ca.)-386. I. Title.
BV185.D37 78-380305
79 MARC

—MEDITATIONS

Charlot, Jean, 1898-
Cartoons Catholic : mirth and meditation from the brush and brain of Jean Charlot / with commentary by F. J. Sheed. — Huntington, Ind. ; Our Sunday Visitor, c1978.
[107] p. : ill. ; 22 cm.
ISBN 0-87973-826-X : $4.95
1. Church year—Meditations. 2. Church year—Caricatures and cartoons. 3. American wit and humor, Pictorial. I. Sheed, Francis Joseph, 1897- . II. Title.
BX2170.C55C45 242'.3 78-62242
79 MARC

Houselander, Frances Caryll.
Lift up your hearts / Caryll Houselander. — New York : Arena Lettres, 1979.
99 p. ; 20 cm.
ISBN 0-88479-020-7 : $2.25
1. Mary, Virgin—Meditations. 2. Church year—Meditations. I. Title.
BT602.H65 1979 242 79-50862
00 MARC

Thielicke, Helmet, 1908-
Being a Christian when the chips are down / Helmut Thielicke ; translated by H. George Anderson. — Philadelphia : Fortress Press, c1979.
125 p. ; 23 cm.
Translation of Der Christ im Ernstfall.
ISBN 0-8006-0541-1 : $5.95
1. Christian life—Lutheran authors. 2. Church year—Meditations. I. Title.
BV4503.T5213 1979 248'.48'41 78-54562
78 MARC

—POETRY

Trümpy, Hans, 1917-
Die Fasti des Baptista Mantuanus von 1516 als volkskundliche Quelle : Textauswahl, Uebersetzung und Kommentar / von Hans Trümpy. — Nieuwkoop : B. de Graaf, 1979.
111 p. : port. ; 25 cm. — (Bibliotheca humanistica et reformatorica ; v. 26) Ne***
Selections from B. Mantuanus' Fastorum libri duodecim in German and Latin.
Includes bibliographical references and indexes.
ISBN 9060043588
1. Baptista Mantuanus, 1448-1516. Fastorum libri duodecim. 2. Church year—Poetry. I. Baptista Mantuanus, 1448-1516. Fastorum libri duodecim. German & Latin. Selections. 1979. II. Title. III. Series.
PA8463.B7F338 871'.04 79-375490
79 MARC

—PRAYER-BOOKS AND DEVOTIONS

Catholic Church. Liturgy and ritual.
Celebrating the saints : readings and prayers in honor of the saints and the solemnities of our Lord and our Lady / adapted from the Liturgy of the hours. — New York : Pueblo Pub. Co., c1978.
384 p. ; 22 cm.
An unofficial collection combining selections from the Liturgy of the hours (U.S.) and the Sacramentary (U.S.)
Includes index.
ISBN 0-916134-30-X
1. Church year—Prayer-books and devotions. I. Catholic Church. Liturgy and ritual. Liturgy of the hours (U.S.) II. Catholic Church. Liturgy and ritual. Sacramentary. (U.S.) III. Title.
BX2170.C55C34 1978 264'.02 79-100546
79 MARC

CHURCH YEAR SERMONS

see also Festival-day sermons

Assoziationen : Gedanken zu bibl. Texten / hrsg. von Walter Jens. — Stuttgart : Radius-Verlag, 1978-
v. ; 21 cm. — (Radius Bücher) GFR78-A(v.
ISBN 3-87173-535-3(v. 1.) : DM24.00(v. 1)
1. Church year sermons. 2. Sermons, German. I. Jens, Walter, 1923-
BV30.A85 78-394215
79 MARC

Bishop, John, 1908-
A word in season / John Bishop. — Nashville : Abingdon, c1979.
109 p. ; 19 cm.
ISBN 0-687-46127-4
1. Church year sermons. 2. Sermons, American. 3. Methodist Church—Sermons. I. Title.
BV30.B53 252'.6 79-958
79 MARC

Coniaris, Anthony M
Treasures from Paul's letters Sunday Epistle lessons as read in the Easter Orthodox Church / by Anthony M. Coniaris. — Minneapolis : Light and Life Pub. Co., c1978-
v. ; 21 cm.
Includes bibliographical references.
CONTENTS: v. 1. September through February.
1. Orthodox Eastern Church—Sermons. 2. Sermons, American. 3. Church year sermons. I. Title.
BX330.C632 252'.6 78-56638
79 MARC

Csőgl, János.
Igehirdetésünk a szentmisében : (homilia vázlatok) / Csőgl János. — Budapest : Szent István Társulat, 1977.
487 p. ; 18 cm. H77
ISBN 9633600545 : 95.00Ft
1. Church year sermons. 2. Catholic Church—Sermons. 3. Sermons, Hungarian. I. Title.
BX1756.C77137 78-381365
*78 MARC

Donders, Joseph G
Jesus, the way : reflections on the Gospel of Luke / Joseph G. Donders. — Maryknoll, N.Y. : Orbis Books, c1979.
viii, 307 p. ; 21 cm.
Includes index.
ISBN 0-88344-240-X
1. Bible. N.T. Luke—Sermons. 2. Jesus Christ—Person and offices—Sermons. 3. Catholic Church—Sermons. 4. Sermons, American. 5. Church year sermons. I. Title.
BS2595.4.D66 252'.6 79-4167
79 MARC

Isaac of Stella, d. 1169.
Sermons on the Christian year / Isaac of Stella ; translated by Hugh McCaffrey ; introd. by Bernard McGinn. — Kalamazoo, Mich. : Cistercian Publications, 1979-
v. : ill. ; 22 cm. — (Cistercian Fathers series ; no. 11-)
Translation of the original Latin text as published in Sermons, Paris, in 1967 and 1974.
Includes bibliographical references and index.
ISBN 0-87907-611-9
1. Church year sermons. 2. Catholic Church—Sermons. 3. Sermons, English—Translations from Latin. 4. Sermons, Latin—Translations into English. I. McCaffrey, Hugh. II. Title.
BX1756.I76S4713 252'.6 78-868
78 MARC

Káldy, Zoltán.
Hanem hogy ő szolgáljon / Káldy Zoltán igehirdetései. — Budapest : Magyarországi Evangélikus Egyház Sajtóosztálya, 1977.
465 p. ; 20 cm. H77
ISBN 963702011X : 136.00Ft
1. Church year sermons. 2. Lutheran Church—Sermons. 3. Sermons, Hungarian. I. Title.
BX8066.K3H36 79-351785
*79 MARC

Seiss, Joseph Augustus, 1823-1904.
Lectures on the Epistles for Sundays and the chief festivals / by Joseph A. Seiss. — Philadelphia : Lutheran Bookstore, 1885-
v. ; 23 cm.
1. Church year sermons. 2. Sermons, English. 3. Lutheran Church—Sermons. I. Title.
BX8066.S4L4 252'.6 78-322347
78 MARC

Vincentius Ferrerius, Saint, 1350 (ca.)-1419.
Sermons ... A cura de Josep Sanchis Sivera, prev. Barcelona [La Renaixença] 1932. Barcelona, Edit. Barcino [1971]-
v. 26 cm. (Els Nostres clàssics. Col·lecció B, v. 3, 5,) 275ptas (v. 1-2) Sp72-Jan
At head of title: Sant Vicent Ferrer.
Vol. 4: A cura de Gret Schib.
ISBN 8472260593 (v. 4)
1. Church year sermons. 2. Catholic Church—Sermons. 3. Sermons, Spanish. I. Sanchis y Sivera, José, 1867-1937, ed. II. Schib, Gret. III. Series.
BX1756.V54S47 1971 74-301568
79[r79]rev MARC

Záň, Augustín J
Pokrm na cestu / napísal Augustín J. Záň. — Cleveland, Ohio : Prvá katolícka slovenská jednota, 1977.
224 p. ; 24 cm.
1. Church year sermons. 2. Catholic Church—Sermons. I. Title.
BX1756.Z33P64 79-373708
79 MARC

—OUTLINES

Voigt, Gottfried.
Der schmale Weg : homiletische Auslegung der Predigttexte, neue Folge, Reihe I / Gottfried Voigt. — Göttingen : Vandenhoeck und Ruprecht, 1978.
518 p. ; 25 cm. GFR78-A
Includes index.
ISBN 3-525-60244-8
1. Church year sermons—Outlines. I. Title.
BV4223.V594 78-400384
79 MARC

CHURCHES

see also Cathedrals; Chapels; Christian patron saints; Church decoration and ornament; Corner stones, Laying of; Half-timbered churches; Oratories

Bentmann, Reinhard, 1939-
Kirchen des Mittelalters / Text von Reinhard Bentmann u. Heinrich Lickes. — Wiesbaden : Ebeling, 1978.
192 p. : 100 ill. (chiefly col.) ; 34 cm. GFR78-A
Bibliography: p. 191-192.
ISBN 3-921195-21-7 : DM48.00
1. Churches. 2. Architecture, Medieval. 3. Church architecture—Details. 4. Church decoration and ornament. I. Lickes, Heinrich, joint author. II. Title.
NA4800.B38 726'.5'094 78-372118
78 MARC

Conti, Flavio, 1943-
Shrines of power / by Flavio Conti ; translated [from the Italian] by Patrick Creagh. — London : Cassell, 1978.
168 p. : ill., facsim., plans, ports. ; 31 cm. — (His The grand tour) GB78-26859
Translation of I templi del potere.
ISBN 0-304-30018-7 : £7.95
1. Castles. 2. Churches. 3. Temples. I. Title. II. Series.
NA7710.C6713 1978 720'.9 79-313926
00 MARC

—ADMINISTRATION see Church management

—ENERGY CONSERVATION

Total Environmental Action, inc.
The energy-efficient church : how to save energy (and money) in your church / by Total Environmental Action, inc. ; Douglas R. Hoffman, editor ; foreword by Gordon Bucy. — New York : Pilgrim Press, c1979.
ix, 86 p. : ill. ; 21 cm.
Bibliography: p. 85-86.
ISBN 0-8298-0362-9 : $4.95
1. Church maintenance and repair. 2. Churches—Energy conservation. I. Hoffman, Douglas R., 1949- . II. Title.
BV652.7.T6 1979 254'.7 79-10432
79 MARC

—FUND RAISING see Church fund raising

—LIBRARIES see Libraries, Church

—MANAGEMENT see Church management

CHURCHES (Continued)

—OUTLINES, SYLLABI, ETC.

Reid, Richard.
Places of worship / ⌜Richard Reid⌝. — ⌜London : Mills & Boon⌝ ; Levittown, N.Y. : ⌜distributed by⌝ Transatlantic Arts, ⌜c1977⌝
⌜33⌝ p. : ill. ; 24 cm. — (Picture panorama of world building) (Owlet books) GB***
Distributor from label on t.p.
ISBN 0-263-06306-2
1. Temples—Outlines, syllabi, etc. 2. Churches—Outlines, syllabi, etc. 3. Mosques—Outlines, syllabi, etc. I. Title. II. Series.
NA4600.R36 726'.09 78-325991
79 MARC

—ALASKA—DIRECTORIES

The Lutheran Church in Alaska : brief histories of the Lutheran churches in Alaska in chronological order of their founding / J. L. Maakestad, editor. — 2d ed. — ⌜s.l. : s.n.⌝, c1978 (⌜s.l.⌝ : K. Wray's Print Shop)
xiv, 57 p. : ill. ; 22 cm.
1. Lutheran Church in Alaska—History. 2. Churches—Alaska—Directories. I. Maakestad, J. L.
BX8042.A4L87 1978 284'.1798 79-101683
79 MARC

—AUSTRIA—KLAGENFURT—GUIDE-BOOKS

Langhold, Norbert.
Kirchen in Klagenfurt : ein kleiner Führer / N. u. M. Langhold ; ⌜Fotos, N. u. M. Langhold⌝. — Klagenfurt : Heyn, 1977.
86 p. : ill. ; 20 cm. Au78-1/2-291
ISBN 3-85366-231-5 : S48.00
1. Church architecture—Austria—Klagenfurt—Guide-books. 2. Churches—Austria—Klagenfurt—Guide-books. 3. Klagenfurt, Austria—Churches. I. Langhold, Margerita, joint author. II. Title.
NA5510.K57L36 914.36'6 79-378242
*79 MARC

—AUSTRIA—STYRIA

Dieman, Kurt, 1923-
Magna Mater Styriae : ⌜Marienheiligentümer d. Steiermark⌝ / Kurt Dieman ; mit Fotos v. Peter Philipp. — Graz ; Wien ; Köln : Verl. Styria, c1977.
118 p., ⌜30⌝ leaves of plates : ill. (some col.) ; 22 cm. Au77-15-277
Includes index.
ISBN 3-222-11007-7 : S360.00
1. Mary, Virgin—Cultus—Austria—Styria. 2. Churches—Austria—Styria. 3. Mary, Virgin—Art. I. Philipp, Peter, 1946- . II. Title.
BT652.A9D53 79-344893
*79 MARC

—BERLIN—DIRECTORIES—GERMANY

Kühne, Günther, 1917-
Evangelische Kirchen in Berlin / Günther Kühne, Elisabeth Stephani ; mit e. Einf. von Oskar Söhngen. — Berlin : C.Z.V.-Verlag, 1978.
500 p. : numerous ill., maps ; 20 cm. GFR79-A
Bibliography: p. 485.
Includes indexes.
ISBN 3-7674-0158-4 : DM24.00
1. Evangelische Kirche von Berlin-Brandenburg—Directories. 2. Berlin—Churches—Directories. 3. Churches—Berlin—Directories—Germany. I. Stephani, Elisabeth, joint author. II. Title.
BX8022.B7K83 79-371394
79 MARC

—CORFU

Stamatopoulos, Nondas.
Old Corfu : history and culture / by Nondas Stamatopoulos. — 2d ed., rev. and completed. — Corfu : Stamatopoulos, 1978.
294 p., ⌜5⌝ leaves of plates : ill. ; 21 cm.
Errata sheet inserted.
Bibliography: p. 265-271.
Includes index.
1. Corfu—History. 2. Corfu—Antiquities. 3. Churches—Corfu. 4. Greece, Modern—Antiquities. I. Title.
DF901.C7S83 1978 949.5'5 79-670097
79 MARC

—ENGLAND

Jones, Lawrence Elmore.
The beauty of English churches / ⌜by⌝ Lawrence E. Jones. — London : Constable, 1978.
391 p. : ill., 2 coats of arms, maps (on lining papers), ports. ; 18 cm. GB78-15096
Includes indexes.
ISBN 0-09-461730-9 : £3.95
1. Church architecture—England. 2. Churches—England. I. Title.
NA5461.J588 1978 726'.5'0942 78-318138
78 MARC

—ENGLAND—GUIDE-BOOKS

Simmons, Jack, 1915-
A selective guide to England / Jack Simmons. — Edinburgh : Bartholomew, 1979.
xiv, 521 p. : ill. ; 19 cm. GB***
Bibliography: p. 496-497.
Includes index.
ISBN 0-7028-1098-3 : £7.50
1. England—Description and travel—1971- —Guide-books. 2. Historic buildings—England—Guide-books. 3. Churches—England—Guide-books. I. Title.
DA650.S57 1979 914.2'04'857 79-310388
79 MARC

—ENGLAND—RECREATIONAL USE—CONGRESSES

The Recreational use of churches, 1978 / editor, J. G. Davies. — ⌜Birmingham⌝ : University of Birmingham, ⌜1978⌝
47, ⌜4⌝ p., 6 p. of plates, leaf of plate : ill., plans ; 30 cm. GB78-11675
Papers presented at a meeting organized by the Institute of Worship and Religious Architecture, University of Birmingham, and the West Midlands Sports Council, Birmingham, 1977.
Includes bibliographical references.
ISBN 0-7044-0288-2 : £0.80
1. Churches—England—Recreational use—Congresses. I. Davies, John Gordon, 1919- . II. Birmingham, Eng. University. Institute of Worship and Religious Architecture. III. West Midlands Sports Council.
BV604.R4 333.7'8 79-301132
79 MARC

—ENGLAND—YORKSHIRE—GUIDE-BOOKS

Carter, Robert Alistair.
A visitor's guide to Yorkshire churches / compiled by Robert A. Carter ; with an introduction by Derek Linstrum. — Bradford : Watmoughs Ltd. ⌜for⌝ the Yorkshire Arts Association, 1976.
viii, 124 p., fold. plate : ill., coat of arms, facsim., map, ports. ; 15 x 21 cm. GB77-05761
Bibliography: p. 121.
Includes index.
ISBN 0-903775-02-6 : £1.40
1. Churches—England—Yorkshire—Guide-books. 2. Yorkshire, Eng.—Description and travel—Guide-books. I. Title.
DA670.Y6C37 914.28'1'04857 78-321450
79 MARC

—FINLAND—SATAKUNTA

Sastamalan seitsemän kirkkoa. (Tyrvää), Tryvään sanomat, 1968.
72 p. illus. 22 cm. Fi
1. Church architecture—Finland—Satakunta. 2. Churches—Finland—Satakunta.
NA5955.F52S3717 79-345410
79 MARC

—FRANCE

Églises et basiliques / ⌜Jacques Renoux, Jean-Pierre Moguilevitch, Pierre Gallerey, Gabriel Sacrez, etc.⌝. — Paris : Larousse, 1978.
1 v. (unpaged) : col. ill. ; 30 cm. — (Beautés de la France) (Découvrir) F78-12263
Includes index.
ISBN 2-03-013914-9 : 62.00F
1. Churches—France. 2. Basilicas—France.
NA5541.E36 914.4'04'83 79-342112
79 MARC

—FRANCE—ISÈRE (DEPT.)—GUIDE-BOOKS

Hamon, Paul, 1918-
Églises de l'Isère / par Paul Hamon. — Paris : Nouvelles éditions latines, ⌜1977?⌝
30 p. : ill. ; 19 cm. — (Art et tourisme) F***
ISBN 2-7233-0014-5 : 7.50F
1. Churches—France—Isère (Dept.)—Guide-books. I. Title.
NA5549.I8H35 914.4'99 79-374404
79 MARC

—FRANCE—YVETOT REGION

Cochet, Jean Benoît Désiré, 1812-1875.
Les Églises de l'arrondissement d'Yvetot / par M. l'abbé Cochet, ... — Saint-Pierre-de-Salerne : G. Monfort, 1977.
2 v. : ill. ; 21 cm. F77-13242
Reprint of the 1852 ed. published by Didron, Paris.
Includes bibliographical references and index.
250F
1. Churches—France—Yvetot region. 2. Yvetot region, France—Church history. I. Title.
BX1539.Y85C6 1977 282'.44'25 78-381705
78 MARC

—FRANCE—HAVRE

Cochet, Jean Benoît Désiré, 1812-1875.
Les Églises de l'arrondissement du Havre / par M. l'abbé Cochet, ... — Saint-Pierre-de-Salerne : G. Monfort, 1977.
2 v. (viii, 279 p., ⌜7⌝ leaves of plates, 376 p., ⌜4⌝ leaves of plates) ; 21 cm. F78-6478
Reprint of the 1845-1846 ed. published by Gaffney frères, Ingouville.
Includes bibliographical references.
250F
1. Catholic Church in Havre—History. 2. Havre—Church history. 3. Churches—France—Havre. I. Title.
BX1533.H38C62 1977 282'.44'25 79-350540
79 MARC

—GERMANY, WEST—EAST FRIESLAND—GUIDE-BOOKS

Rödiger, Hans-Bernd.
Friesische Kirchen im Jeverland und Harlingerland / Hans-Bernd Rödiger, Klaus Wilkens. — Jever : Mettcker, 1978.
87 p. : chiefly ill. ; 27 cm. GFR79-A4
Includes bibliographical references and index.
DM26.80
1. Churches—Germany, West—East Friesland—Guide-books. 2. Christian art and symbolism—Germany, West—East Friesland. I. Wilkens, Klaus, joint author. II. Title.
NA5574.E23R63 79-361947
79 MARC

—GERMANY, WEST—SCHLESWIG-HOLSTEIN

Welchert, Hans Heinrich, 1905-
Wanderungen zu den Domen und Schlössern in Schleswig-Holstein / Hans-Heinrich Welchert. — Frankfurt am Main : Societäts-Verlag, 1978.
262 p. : ill., facsims., maps ; 21 cm. GFR78-A
ISBN 3-7973-0310-6 : DM28.00
1. Castles—Germany, West—Schleswig-Holstein. 2. Churches—Germany, West—Schleswig-Holstein. 3. Schleswig-Holstein—History, Local. I. Title.
DD491.S65W44 78-380278
79 MARC

—GREAT BRITAIN—CONSERVATION AND RESTORATION

Binney, Marcus.
Churches at risk / ⌜compiled⌝ by Marcus Binney. — London : David Pearce for Save Britain's Heritage, 1977.
16 p. : chiefly ill. ; 30 cm. GB78-31573
Cover title.
£0.60
1. Churches—Great Britain—Conservation and restoration. I. Save Britain's Heritage (Association) II. Title.
NA5461.B5 726'.5'0942 79-302128
79 MARC

CHURCHES (Contin ued)

—GREAT BRITAIN—MISCELLANEA

Alexander, Marc, 1929-
Haunted churches and abbeys of Britain / ₍by₎ Marc Alexander. — London : A. Barker, 1978.
178 p., ₍4₎ leaves of plates : ill. ; 23 cm. GB***
Includes index.
ISBN 0-213-16677-1 : £5.95
1. Ghosts—Great Britain. 2. Churches—Great Britain—Miscellanea. 3. Abbeys—Great Britain—Miscellanea. I. Title.
BF1472.G7A43 133.1'29'41 79-309198
79 MARC

—GREAT BRITAIN—REMODELING FOR OTHER USE

Great Britain. Dept. of the Environment.
New life for old churches / Department of the Environment, Scottish Development Department ₍and₎ Welsh Office. — London : H.M.S.O., 1977.
₍52₎ p. : ill., maps, plans ; 25 cm. — (Aspects of conservation ; 3)
GB77-23591
Bibliography: p. ₍7₎
ISBN 0-11-751065-3 : £2.00
1. Churches—Great Britain—Remodeling for other use. I. Scotland. Scottish Development Dept. II. Great Britain. Welsh Office. III. Title.
NA5461.G68 1977 726'.5'028 78-319174
79 MARC

—GREECE

Janin, Raymond.
Les églises et les monastères des grands centres byzantins : Bithynie, Hellespont, Latros, Galèsios, Trébizonde, Athènes, Thessalonique / par Raymond Janin. — Paris : Institut français d'études byzantines, 1975.
xvi, 492 p. : maps ; 28 cm. — (Géographie ecclésiastique de l'Empire byzantin) (Publications de l'Institut français d'études byzantines) F***
Includes bibliographical references and indexes.
1. Orthodox Eastern Church in Turkey. 2. Orthodox Eastern Church in Greece. 3. Churches—Turkey. 4. Monasteries—Turkey. 5. Churches—Greece. 6. Monasteries—Greece. 7. Turkey—Church history. 8. Greece—Church history. I. Title. II. Series. III. Series: Paris. Institut français d'études byzantines. Publications.
BX752.T9J36 281.9'561 79-350269
79 MARC

—INDIA

Hargreaves, Cecil.
25 Indian churches / drawings by Mathew, K. G. ; descriptive notes by Cecil Hargreaves. — Calcutta : Bishop College ; Delhi : distributors, I.S.P.C.K., 1975.
xiii, 61 p. : ill. ; 15 x 23 cm.
1. Churches—India. 2. Churches in art. I. Mathew, K. G. II. Title.
NA6001.H37 726'.5'0954 78-906440
79 MARC

—ITALY—FLORENCE (PROVINCE)—GUIDE-BOOKS

Fognani, Ugo.
Andar per chiese / Ugo Fognani ; con testo introduttivo di Piero Bargellini. — ₍Firenze₎ : La ginestra, 1976.
179 p. : col. ill. ; 31 cm. It78-May
L6000
1. Fognani, Ugo. 2. Churches in art. 3. Churches—Italy—Florence (Province)—Guide-books. I. Title.
ND623.F58A4 1976 759.5 79-344348
*79 MARC

—ITALY—MOLISE

Di Iorio, Eduardo.
I cappuccini nel Molise (1530-1975) : arte e ricordi storici nelle loro chiese e conventi / ₍di₎ p. Eduardo Di Iorio (Francesco Saverio Di Iorio). — Campobasso : Arti grafiche La regione, 1976.
462 p., ₍25₎ leaves of plates : ill. ; 24 cm. It78-Mar
Bibliography: p. 401-413.
Includes index.
L7850
1. Capuchins—Italy—Molise. 2. Churches—Italy—Molise. 3. Monasteries—Italy—Molise. 4. Molise, Italy—Church history. I. Title.
BX3138.M64D54 78-379069
*78 MARC

—ITALY—REGGIO CALABRIA (PROVINCE)—HISTORY—SOURCES

Minuto, Domenico.
Catalogo dei monasteri e dei luoghi di culto tra Reggio e Locri / Domenico Minuto. — Roma : Edizioni di storia e letteratura, 1977.
xxix, 508 p., ₍20₎ leaves of plates : ill., fold. map in pocket ; 25 cm. — (Chiese e monasteri nella Calabria greca ; 1) (Thesaurus ecclesiarum Italiae ; XVII, 1)
It79-June
Bibliography: p. ₍xv₎-xxix.
Includes indexes.
L3000
1. Monasteries—Italy—Reggio Calabria (Province)—History—Sources. 2. Churches—Italy—Reggio Calabria (Province)—History—Sources. 3. Reggio Calabria (Province)—Church history—Sources. I. Title. II. Series. III. Series: Thesaurus ecclesiarum Italiae ; XVII, 1.
BX1543.A2T5 vol. 17, pt. 1 79-377874
₍BX2623.R43₎ *79 MARC

—ITALY—SARDINIA—GUIDE-BOOKS

Botteri, Mauro.
Guida alle chiese medioevali di Sardegna / Mauro Botteri. — 2. ed. — Sassari : Libreria Dessì, 1978.
xxxiv, 194 p. : ill. ; 29 cm. It***
Bibliography: p. 189-190.
1. Churches—Italy—Sardinia—Guide-books. 2. Architecture, Medieval—Italy—Sardinia. I. Title.
NA5619.S3B67 1978 914.59'04'92 78-376823
*78 MARC

—ITALY—TUSCANY

Negri, Daniele.
Chiese romaniche in Toscana / Daniele Negri. — Pistoia : Libreria editrice Tellini, 1978.
415 p. : ill. ; 24 cm. It78-Dec
Bibliography: p. 411-412.
L15000
1. Churches—Italy—Tuscany. 2. Architecture, Romanesque—Italy—Tuscany. I. Title.
NA5619.T84N43 726'.5'09455 79-351719
*79 MARC

—NEBRASKA

Czech churches in Nebraska = České kostely v Nebrasce / ₍editor, Vladimír Kučera₎. — ₍s.l. : Nebraska Czechs Inc., 1976₎
176 p. : ill. ; 24 cm.
Cover title.
1. Churches—Nebraska. 2. Nebraska—Church history. 3. Czechs in Nebraska. I. Kučera, Vladimír, 1906- . II. Title: České kostely v Nebrasce.
BR555.N2C95 280'.09782 79-316342
79 MARC

—NETHERLANDS—BRABANT (PROVINCE)—GUIDE-BOOKS

Leeuwen, Wies van.
Langs de oude Brabantse kerken : westelijk Brabant / Wies van Leeuwen. — Baarn : Bosch & Keuning, c1976.
160 p. : ill. ; 21 cm. — (De Oude Nederlandse kerken) Ne***
Includes index.
ISBN 9024641454
1. Churches—Netherlands—Brabant (Province)—Guide-books. I. Title.
NA5649.B7L43 78-394681
79 MARC

—NETHERLANDS—GRONINGEN (PROVINCE)

Meulen, Tony van der.
Oude kerken in Groningen / Tony van der Meulen. — Baarn : Bosch & Keuning, ₍1979₎
63 p. : with ill. ; 24 cm. — (Publicatie van de Stichting Oude Groninger Kerken ; 21) Ne79-24
ISBN 902464321X : fl 8.75
1. Churches—Netherlands—Groningen (Province) 2. Church architecture—Netherlands—Groningen (Province) I. Title. II. Series: Stichting Oude Groninger Kerken. Publicatie van de Stichting Oude Groninger Kerken ; 21.
NA5649.G76M48 79-376856
*79 MARC

—NETHERLANDS—HOLLAND, NORTH (PROVINCE)—GUIDE-BOOKS

Hoogewoud, Guido.
Langs de oude Noordhollandse kerken : Kennemerland, De Zaan-Het Gooi, Amsterdam / Guido Hoogewoud. — Baarn : Bosch & Keuning, ₍1979₎
64 p., 96 p. of photos : ill. ; 21 cm. — (De Oude Nederlandse kerken)
Ne79-11
Includes bibliographical references and index.
ISBN 9024643015 : fl 12.90
1. Churches—Netherlands—Holland, North (Province)—Guide-books. I. Title.
NA5649.N6H66 79-373518
*79 MARC

—NETHERLANDS—LEEUWARDEN—CONSERVATION AND RESTORATION

Terpstra, Rienk.
De Jacobijnerkerk te Leeuwarden : vijf jaar restauratie / Rienk Terpstra. — Leeuwarden : Tille, 1978.
₍58₎ p. : ill. ; 30 cm. Ne***
Bibliography: p. ₍58₎.
ISBN 9070010704
1. Jacobijnerkerk te Leeuwarden. 2. Churches—Netherlands—Leeuwarden—Conservation and restoration.
NA5651.L43T37 78-391389
78 MARC

—NETHERLANDS—VELUWE—GUIDE-BOOKS

Kolks, Zeno
Langs de oude Gelderse kerken : de Veluwe / Zeno Kolks. — Baarn : Bosch & Keuning, ₍1978₎
160 p. : ill. ; 21 cm. — (De Oude Nederlandse kerken) Ne78-36
Includes indexes.
ISBN 9024641446
1. Churches—Netherlands—Veluwe—Guide-books. I. Title.
NA5649.V44K64 78-394337
*79 MARC

—NORWAY—KARASJOK

Blix, Erik Schytte.
Kirker og kirkeliv i Karasjok / Erik Schytte Blix. — Tromsø : Karasjok kommune, 1974.
93 p. : ill. ; 21 cm. N75-Apr
Bibliography: p. 93.
1. Karasjok, Norway—Religious life and customs. 2. Churches—Norway—Karasjok. I. Title.
BR1008.K3B56 75-531438
*79₍r79₎rev MARC

—POLAND

Bohdziewicz, Piotr.
Studia z dziejów sztuki polskiej w okresie baroku i rokoka : kościół i klasztor PP. Wizytek w Warszawie, kościół fundacyjny w Kobyłce pod Warszawą i jego dekoracja malarska, kościół i klasztor OO. Bernardynów w Dubnie / Piotr Bohdziewicz. — Wyd. 1. — Lublin : Wydawn. Towarzystwa Naukowego Katolickiego Uniwersytetu Lubelskiego, 1973.
159 p., ₍56₎ leaves of plates : ill. ; 24 cm. — (Towarzystwo Naukowe Katolickiego Uniwersytetu Lubelskiego. Rozprawy Wydziału Historyczno-Filologicznego ; 37)
Includes bibliographical references.
1. Art, Polish. 2. Art, Baroque—Poland. 3. Art, Rococo—Poland. 4. Christian art and symbolism—Poland. 5. Churches—Poland. 6. Monasteries—Poland. I. Title.
N7255.P6B63 79-413686
79 MARC

Kneifel, Eduard.
Die evangelisch-augsburgischen Gemeinden in Polen, 1555-1939 : eine Parochialgeschichte in Einzeldarstellungen / Eduard Kneifel. — Vierkirchen über München : Kneifel, ₍1971₎
358 p., ₍9₎ leaves of plates : ill. ; 23 cm. GFR***
Bibliography: p. 345-351.
Bibliography of the author's works: p. 353-357.
Includes index.
1. Lutheran Church in Poland—History. 2. Churches—Poland. 3. Poland—Church history. I. Title.
BX8026.5.K53 79-354434
79 MARC

CHURCHES (Continued)

—POLAND—GLATZ (GRAFSCHAFT)—PICTORIAL WORKS

Grafschafter Kirchen in heutiger Zeit : Aufnahmen 1976-1978 / Bildausw. u. Zs.-Stellung Jörg Marx. — Leimen/Heidelberg : Marx, 1978.
340 p. : chiefly ill. ; 29 cm. GFR79-A8
Includes bibliographical references and index.
DM69.00
1. Churches—Poland—Glatz (Grafschaft)—Pictorial works. 2. Christian art and symbolism—Poland—Glatz (Grafschaft)—Pictorial works. I. Marx, Jörg.
NA5955.P62G454 79-379398
79 MARC

—QUÉBEC (PROVINCE)—GUIDE-BOOKS

Noppen, Luc.
Les églises du Québec, 1600-1850 / Luc Noppen, avec la collaboration de Achille Murphy ... [et al.]. — [Québec] : Éditeur officiel du Québec, c1977.
ix, 298 p. : ill. ; 20 cm. — (Collection Loisirs et culture) (Connaissance du Québec) C***
Bibliography: p. 64.
Includes indexes.
ISBN 0-7754-2869-8
1. Churches—Québec (Province)—Guide-books. 2. Church architecture—Québec (Province) I. Title.
NA5246.Q4N66 726'.5'09714 79-345535
MARC

—RHODE ISLAND—FOSTER

Matthews, Margery I
Churches of Foster : a history of religious life in rural Rhode Island / Margery I. Matthews, Virginia I. Benson, Arthur E. Wilson ; [ill. by Eve Small ; photography by Janice Shehan ... et al.]. — Foster, R.I. : North Foster Baptist Church, c1978.
165, [3] p. : ill. ; 21 x 26 cm.
Bibliography: p. [167]-[168]
1. Foster, R.I.—Church history. 2. Churches—Rhode Island—Foster. I. Benson, Virginia I., joint author. II. Wilson, Arthur Edward, joint author. III. Title.
BR560.F74M37 277.45'1 79-109020
00 MARC

—ROMANIA—DIRECTORIES

Orthodox Eastern Church, Romanian. Eparhia Rîmnicului şi Argeşului.
Eparhia Rîmnicului şi Argeşului : monografie / elaborată de Dumitru Sandu ... [et al] ; apare cu arhiereasca binecuvîntare a Prea Sfinţitului Iosif, Episcopul Rîmnicului şi Argeşului. — Rîmnicu Vîlcea, : Eparhia Rîmnicului şi Argeşului, 1976.
2 v. (1140 p., [6] leaves of plates) : ill. (some col.) ; 25 cm. R79-236
Includes bibliographical references and index.
lei400
1. Orthodox Eastern Church, Romanian. Eparhia Rîmnicului şi Argeşului—Directories. 2. Churches—Romania—Directories. I. Sandu, Dumitru, 1916- . II. Title.
BX691.O77 1976 79-357975
*79 MARC

—ROMANIA—TRANSYLVANIA

Vilhelm, Károly.
Festett famennyezetek : alakos ábrázolások a XV-XVIII. századi erdélyi templomokban / Vilhelm Károly. — Bukarest : Kriterion, 1975.
39 p., [20] leaves of plates : ill. (some col.) ; 20 cm.
Bibliography: p. 39.
lei16.00
1. Mural painting and decoration—Romania—Transylvania. 2. Primitivism in art—Romania—Transylvania. 3. Churches—Romania—Transylvania. I. Title.
ND2807.T73V53 78-390947
*79 MARC

—SPAIN—CANTABRIA

Sá, María Eálo de.
El románico de Cantabria en sus cinco colegiatas / [por] María Eálo de Sá. — Santander : Diputación Provincial, Institución Cultural de Cantabria, Centro de Estudios Montañeses, 1978.
324 p. : ill. ; 24 cm. Sp***
Summary of thesis—Madrid, 1977.
Bibliography: p. 321-324.
ISBN 8450029309
1. Architecture, Romanesque—Spain—Cantabria. 2. Church architecture—Spain—Cantabria. 3. Churches—Spain—Cantabria. I. Title.
NA5809.C28S2 79-112708
79 MARC

—SPAIN—CATALONIA—GUIDE-BOOKS

Gavin i Barceló, Josep Maria.
Inventari d'esglésies. — [1. ed.] — [s.l.] : Arxiu Gavin, [1977-
v. : chiefly ill. ; 23 cm. Sp***
CONTENTS: v. 1. Montsià, Baix Ebre, Terra Alta, Ribera d'Ebre, Priorat, Matarranya.—2. Baixa Ribagorça, Alta Ribagorça, Vall d'Aran.—3. Capcir, Cerdanya, Conflent, Vallespir, Rossello.
ISBN 8485180070 (v. 1)
1. Churches—Spain—Catalonia—Guide-books. I. Title.
NA5809.C3G38 914.6'7'0483 78-363039
78[r79]rev MARC

—SPAIN—GRANADA

Casas, mezquitas y tiendas de los habices de las iglesias de Granada / edición, introd. e índices por Ma. del Carmen Villanueva Rico. — Madrid : Instituto Hispano-Arabe de Cultura, 1966.
170 p. ; 25 cm. — (Libros de habices del Reino de Granada ; 2)
Half title: Traslado del libro de la medida de las casas y mezquitas y tiendas de las yglesias de Granada y de sus habizes de ellas y su dotazión por los Señores Reyes Católicos.
Introd. includes bibliographical references.
Includes indexes.
1. Churches—Spain—Granada. 2. Granada—Church history. 3. Church property—Spain—Granada. I. Villanueva Rico, María del Carmen. II. Instituto Hispano-Arabe de Cultura. III. Series.
BR1027.G7C37 79-108934
79 MARC

—SWITZERLAND

Speich, Klaus.
Kirchen und Klöster in der Schweiz / Klaus Speich, Hans R. Schläpfer ; [hrsg. von Niklaus Flüeler ; an d. Texten arbeiteten mit, Henriette Bon ... [et al.] ; die Grundrisse u. Isometrien zeichnete Andreas Adam ; Kt. Heinz Engel]. — München : Beck, 1978.
344 p. : numerous ill. (some col.), maps ; 30 cm. GFR79-A
Bibliography: p. 340-341.
Includes index.
ISBN 3-406-04111-6 : DM88.00
1. Churches—Switzerland. 2. Monasteries—Switzerland. 3. Convents and nunneries—Switzerland. 4. Christian art and symbolism—Switzerland. I. Schläpfer, Hans Ruedi, 1929- . II. Flüeler, Niklaus, 1934- . III. Title.
NA5841.S63 726'.5'09494 79-358419
79 MARC

—SWITZERLAND—BINN VALLEY

Ruppen, Walter.
Das Binntal / Walter Ruppen. — Basel (Münzgässlein 16) : Gesellschaft für schweizerische Kunstgeschichte, c1977.
15 p. : ill. ; 20 cm. — (Schweizerische Kunstführer) Sw78-3746
Cover title.
Bibliography: p. 15.
1. Churches—Switzerland—Binn Valley. 2. Church decoration and ornament—Switzerland—Binn Valley. 3. Architecture—Switzerland—Binn Valley. 4. Historic buildings—Switzerland—Binn Valley. I. Title. II. Series.
NA5849.B562R86 914.94'7 78-377937
78 MARC

—SWITZERLAND—BROYE VALLEY

Martin, Vio.
Vieilles églises de la Broye / Vio Martin. — Lausanne : Éditions des Terreaux, 1977.
79 p. : ill. ; 24 cm. Sw77-11827
ISBN 2-88033-002-5
1. Church architecture—Switzerland—Broye Valley. 2. Churches—Switzerland—Broye Valley. I. Title.
NA5849.B76M38 914.94'3 78-383242
*78 MARC

—TURKEY

Janin, Raymond.
Les églises et les monastères des grands centres byzantins : Bithynie, Hellespont, Latros, Galèsios, Trébizonde, Athènes, Thessalonique / par Raymond Janin. — Paris : Institut français d'études byzantines, 1975.
xvi, 492 p. : maps ; 28 cm. — (Géographie écclesiastique de l'Empire byzantin) (Publications de l'Institut français d'études byzantines) F***
Includes bibliographical references and indexes.
1. Orthodox Eastern Church in Turkey. 2. Orthodox Eastern Church in Greece. 3. Churches—Turkey. 4. Monasteries—Turkey. 5. Churches—Greece. 6. Monasteries—Greece. 7. Turkey—Church history. 8. Greece—Church history. I. Title. II. Series. III. Series: Paris. Institut français d'études byzantines. Publications.
BX752.T9J36 281.9'561 79-350269
79 MARC

—UNITED STATES—DIRECTORIES

Beaver, Robert Pierce, 1906-
The native American Christian community : a directory of Indian, Aleut, and Eskimo churches / R. Pierce Beaver, editor. — Monrovia, Calif. : MARC, c1979.
395 p. ; 24 cm.
ISBN 0-912552-25-5
1. Indians of North America—Missions—Directories. 2. Eskimos—Alaska—Missions—Directories. 3. Aleuts—Missions—Directories. 4. Churches—United States—Directories. 5. Indians of North America—Religion—Directories. 6. Eskimos—Alaska—Religion—Directories. 7. Aleuts—Religion—Directories. I. Missions Advanced Research and Communication Center. II. Title.
E98.M6B32 277.3 78-70999
79 MARC

—UNITED STATES—GUIDE-BOOKS

Vuilleumier, Marion.
America's religious treasures : a spiritual heritage travel guide / text by Marion Rawson Vuilleumier ; ill. by Pierre DuPont Vuilleumier. — 1st ed. — New York : Harper & Row, c1976.
xviii, 286 p. : ill. ; 21 cm.
Includes index.
ISBN 0-06-068940-4 : $9.95. ISBN 0-06-068941-2 pbk. : $4.95
1. United States—Description and travel—1960- —Guide-books. 2. Churches—United States—Guide-books. I. Title.
E158.V84 1976 917.3'04'925 76-9342
76[r79]rev MARC

CHURCHES, CATHOLIC

Schubert, Guilherme.
Arte para a fé : igrejas e capelas depois do Concilio Vaticano II / Guilherme Schubert. — Petrópolis, Brasil : Vozes, 1979.
xvi, 194 p., [15] leaves of plates : ill. ; 23 cm. — (Pastoral ; 5) (Publicações CID)
Includes bibliographical references.
Cr$200.00
1. Churches, Catholic. 2. Liturgy and architecture. I. Title. II. Series.
NA4828.S38 79-111108
79 MARC

—BRAZIL—RIO DE JANEIRO

Maurício, Augusto.
Igrejas históricas do Rio de Janeiro / Augusto Maurício. — Nova ed., atualizada. — [Rio de Janeiro] : Livraria Kosmos Editora, [1978?]
286 p. : ill. ; 23 cm. — (Coleção Estado do Rio de Janeiro)
"Publicada com a colaboração do Governo Faria Lima, Secretaria Estadual de Educação e Cultura, Departamento de Cultura, Instituto Estadual de Livro."
1. Churches, Catholic—Brazil—Rio de Janeiro. 2. Rio de Janeiro—Churches. 3. Rio de Janeiro—Church history. I. Title. II. Series.
BX4622.R55A2 1978 79-108771
79 MARC

—JAPAN—NAGASAKI (PREFECTURE)

Kimura, Nobuo.
(Nagasaki no tenshudō)
長崎の天主堂：その信仰と美 / 木村信郎・写真：監修　村松貞次郎，片岡弥吉；解説　片岡弥吉 ... [et al]. — 東京：技報堂出版，1977.
317 p. : col. ill. ; 30 cm. Ja 78-8508
¥15000
1. Churches, Catholic—Japan—Nagasaki (Prefecture) 2. Catholic church—Japan—Nagasaki (Prefecture) I. Muramatsu, Teijirō, 1924- II. Kataoka, Yakichi, 1908- III. Title.
NA6056.N33K55 78-810611

Pacheco, Diego, 1922-
(Nagasaki no tenshudō)
長崎の天主堂 / 著・撮影　パチェコ・ディエゴ. — 福岡：西日本文化協会，1976.
126 p. : col. ill. ; 19 cm. Ja 78-24052
¥980
1. Church architecture—Japan—Nagasaki (Prefecture) 2. Churches, Catholic—Japan—Nagasaki (Prefecture) I. Title.
NA6056.N34P32 79-803774

CHURCHES, CATHOLIC (Continued)

—NORTH AMERICA

Török, István.
Katolikus magyarok Észak-Amerikában / az anyagot gyűjtötte, összeállította, Török István. — Youngstown, Ohio : A Katolikus Magyarok Vasárnapja, 1978.
489 p. : ill. ; 22 cm.
Bibliography: p. 483-489.
1. Catholic Church in North America—History. 2. Churches, Catholic—North America. 3. Hungarians in North America. I. Title.
BX1403.2.T67 79-366679
79 MARC

—QUÉBEC (PROVINCE)—RIMOUSKI CO.

Session Ross, 8th, Université du Québec à Rimouski, 1978.
De notre patrimoine : églises et croix d'églises du diocèse de Rimouski : 8ème Session Ross. — Rimouski : [Session Ross '78], 1978.
137 leaves : ill. ; 23 x 31 cm. C***
$10.75
1. Catholic Church. Archdiocese of Rimouski. 2. Churches, Catholic—Québec (Province)—Rimouski Co. 3. Crosses—Québec (Province)—Rimouski Co. 4. Rimouski Co., Que.—Church history. I. Title.
BX1423.R5S47 1978 282'.714'77 79-358986
79 MARC

—SASKATCHEWAN

Baran, Anna Mariia.
(Ukraïns'ki katolyts'ki tserkvy Saskachevanu)
Українські католицькі церкви Саскачевану / Анна Марія Баран ; англійський переклад Христина Т. Пастершанк. — Saskatoon : [The Ukrainian Catholic Council of Saskatchewan], 1977.
389 p., [10] leaves of plates : ill. (some col.) ; 30 cm. C***
Added t. p.: Ukrainian Catholic churches of Saskatchewan.
English and Ukrainian.
Bibliography: p. 382–383.
Includes indexes.
$40.00
1. Churches, Catholic—Saskatchewan. 2. Ukrainians in Saskatchewan. I. Title. II. Title: Ukrainian Catholic churches of Saskatchewan.
NA5246.S37B37 79–408175

CHURCHES, CITY see City churches

CHURCHES, COUNTRY see Rural churches

CHURCHES, FRIEND

—NEW ENGLAND—EXHIBITIONS

Benes, Peter.
New England meeting house and church, 1630-1850 : a loan exhibition at the Currier Gallery of Art, Manchester, New Hampshire / Peter Benes & Philip D. Zimmerman. — Boston : Published by Boston University for the Dublin Seminar for New England Folklife, c1979.
xi, 177 p. : ill. ; 22 cm.
Bibliography: p. 110.
ISBN 0-87270-050-X : $7.95
1. Christian art and symbolism—Modern, 1500- —New England—Exhibitions. 2. Church decoration and ornament—New England—Exhibitions. 3. Churches, Friend—New England—Exhibitions. 4. Churches, Protestant—New England—Exhibitions. I. Zimmerman, Philip D., joint author. II. Currier Gallery of Art, Manchester, N.H. III. Dublin Seminar for New England Folklife. IV. Title.
N7905.B46 79-53115
709.948'2'097407401428
00 MARC

CHURCHES, PROTESTANT

—NEW ENGLAND—EXHIBITIONS

Benes, Peter.
New England meeting house and church, 1630-1850 : a loan exhibition at the Currier Gallery of Art, Manchester, New Hampshire / Peter Benes & Philip D. Zimmerman. — Boston : Published by Boston University for the Dublin Seminar for New England Folklife, c1979.
xi, 177 p. : ill. ; 22 cm.
Bibliography: p. 110.
ISBN 0-87270-050-X : $7.95
1. Christian art and symbolism—Modern, 1500- —New England—Exhibitions. 2. Church decoration and ornament—New England—Exhibitions. 3. Churches, Friend—New England—Exhibitions. 4. Churches, Protestant—New England—Exhibitions. I. Zimmerman, Philip D., joint author. II. Currier Gallery of Art, Manchester, N.H. III. Dublin Seminar for New England Folklife. IV. Title.
N7905.B46 79-53115
709.948'2'097407401428
00 MARC

CHURCHES, REFORMED

—SCOTLAND

Hay, George, 1911-
The architecture of Scottish post-Reformation churches, 1560-1843. Oxford, Clarendon Press, 1957.
xvi, 299 p. illus., 47 plates, plans. 25 cm.
Bibliography: p. 283-286.
1. Church architecture—Scotland. 2. Churches, Reformed—Scotland. 3. Church decoration and ornament—Scotland. I. Title.
NA5475.H3 726'.5'09411 57-2871
[r79]rev MARC

CHURCHES, RURAL see Rural churches

CHURCHES, SMALL see Small churches

CHURCHES, TOWN see City churches

CHURCHES, URBAN see City churches

CHURCHES IN ART

Fognani, Ugo.
Andar per chiese / Ugo Fognani ; con testo introduttivo di Piero Bargellini. — [Firenze] : La ginestra, 1976.
179 p. : col. ill. ; 31 cm. It78-May
L6000
1. Fognani, Ugo. 2. Churches in art. 3. Churches—Italy—Florence (Province)—Guide-books. I. Title.
ND623.F58A4 1976 759.5 79-344348
*79 MARC

Hargreaves, Cecil.
25 Indian churches / drawings by Mathew, K. G. ; descriptive notes by Cecil Hargreaves. — Calcutta : Bishop College ; Delhi : distributors, I.S.P.C.K., 1975.
xiii, 61 p. : ill. ; 15 x 23 cm.
1. Churches—India. 2. Churches in art. I. Mathew, K. G. II. Title.
NA6001.H37 726'.5'0954 78-906440
79 MARC

Pazzi, Luciano.
Chiese parrocchiali del Comune di Brisighella linoleografie di Luciano Pazzi ; note storiche di Piero Malpezzi. — Brisighella : Tip. Valgimigli, 1976-[1977]
3 v. : ill. ; 35 cm. It77-Apr
Cover title.
Edition limited to 99 copies.
Includes bibliographies.
1. Pazzi, Luciano. 2. Brisighella, Italy—Churches—Pictorial works. 3. Churches in art. 4. Brisighella, Italy—Churches. I. Malpezzi, Piero. II. Title.
NE1336.P39A4 1976 769'.92'4 77-562344
*77[r78]rev MARC

Vallerini, Fernando.
Monte Sant'Angelo sul Gargano : la Basilica di S. Michele Arcangelo e gli altri monumenti, la città, Pulsano, Siponto / Fernando Vallerini ; con 60 disegni, 16 incisioni, 3 illustrazioni, una pianta e la mappa del Gargano. — Pisa : A cura dell'A., 1977.
124 p. : ill. ; 39 cm. It78-Feb
Bibliography: p. 117-118.
Includes indexes.
1. Vallerini, Fernando. 2. Monte Sant'Angelo, Italy, in art. 3. Basilicata di San Michele, Monte Sant'Angelo, Italy—Pictorial works. 4. Churches in art. I. Title.
NC257.V26A4 1977 741.9'45 78-376038
*78 MARC

CHURCHES' UNITY COMMISSION. VISIBLE UNITY

Catholic Church. Bishops' Conference of England and Wales.
Official response to the ten propositions of the Churches' Unity Commission / the Catholic Bishops' Conference of England and Wales. — Abbots Langley : Catholic Information Services (Bishops' Conference of England and Wales), 1977.
[1], iii, 9, [1] p. ; 21 cm. GB78-07314
£0.20
1. Christian union—England. 2. Churches' Unity Commission. Visible unity. 3. Christian union—Catholic Church. 4. Catholic Church in England. I. Title.
BR759.C37 1977 262'.001 78-317292
78 MARC

Catholic Church. Ecumenical Commission for England and Wales.
The Roman Catholic Church in England and the Ten Propositions of the Churches' Unity Commission : a paper / by the Ecumenical Commission of England and Wales. — Abbots Langley : Catholic Information Office, 1976.
[1], 8 p. ; 21 cm. GB77-223
Cover title.
A provisional response to the report of the Churches' Unity Commission entitled: Visible unity: ten propositions.
ISBN 0-905241-04-5 : £0.15
1. Christian union—England. 2. Churches' Unity Commission. Visible unity. 3. Christian union—Catholic Church. 4. Catholic Church in England. I. Title: The Roman Catholic Church in England and the Ten Propositions ...
BR759.C38 1976 262'.001 77-359592
77[r78]rev MARC

CHURCHILL, CHARLES, 1731-1764

—CRITICISM AND INTERPRETATION

Lockwood, Thomas.
Post-Augustan satire : Charles Churchill and satirical poetry, 1750-1800 / by Thomas Lockwood. — Seattle : University of Washington Press, c1979.
198 p. ; 24 cm.
Includes bibliographical references and index.
ISBN 0-295-95612-7 : $17.50
1. Churchill, Charles, 1731-1764—Criticism and interpretation. 2. Satire, English—History and criticism. 3. English poetry—18th century—History and criticism. I. Title.
PR3346.C8L6 821'.6'09 78-4366
78 MARC

CHURCHILL, WINSTON, 1871-1947

Barker, Elisabeth.
Churchill and Eden at war / [by] Elisabeth Barker. — London : Macmillan, 1978.
346 p. ; 23 cm. GB78-40637
Bibliography: p. 335-336.
Includes index.
ISBN 0-333-23467-7 : £7.95
1. World War, 1939-1945—Diplomatic history. 2. World War, 1939-1945—Great Britain. 3. Churchill, Winston, 1871-1947. 4. Eden, Anthony, Earl of Avon, 1897- . 5. Great Britain—Foreign relations—1936-1945. I. Title.
D750.B38 940.53'22'41 79-314303
79 MARC

CHURCHILL, WINSTON LEONARD SPENCER, SIR, 1874-1965

Barbirolli, John, Sir, 1899–1970.
Sir John Barbirolli speaks. [Sound recording] [Retford, Eng.] Barbirolli Society, p1973–1974. BS01–02.
2 discs. 33⅓ rpm. mono. 12 in.
Descriptive notes by M. Kennedy on containers.
CONTENTS: v. 1. In conversation with Michael Kennedy, 1963. Tribute to Sir Winston Churchill, Houston, 1965.—v. 2. Golden Jubilee speech, Manchester Town Hall, 12th December 1961. Bach. Unaccompanied 'cello suite no. 3 in C major (Fourth movement: Sarabande) (John Barbirolli, violoncello) Saint-Saëns. Le cygne (John Barbirolli, violoncello; Ruth Fermoy, piano)
1. Barbirolli, John, Sir, 1899–1970. 2. Churchill, Winston Leonard Spencer, Sir, 1874–1965. 3. Suites (Violoncello) 4. Violoncello and piano music, Arranged. I. Kennedy, Michael, 1926– II. Bach, Johann Sebastian, 1685–1750. Suite, violoncello, S. 1009, C major. Sarabande. [Sound recording] 1973. III. Saint-Saëns, Camille, 1835–1921. Carnaval des animaux. Le cygne; arr. [Sound recording] 1973. IV. Fermoy, Ruth. V. Title.
[ML422] 79–740373

CHURCHILL, WINSTON LEONARD SPENCER, SIR, 1874-1965 (Continued)

Barker, Elisabeth.
Churchill and Eden at war / Elisabeth Barker. — New York : St. Martins, 1978.
346 p. ; 23 cm.
Bibliography: p. 335-336.
Includes index.
ISBN 0-312-13484-3 : $15.95
1. World War, 1939-1945—Great Britain. 2. Great Britain—History—George VI, 1936-1952. 3. Churchill, Winston Leonard Spencer, Sir, 1874-1965. 4. Eden, Anthony, Earl of Avon, 1897- . I. Title.
D759.B36 1978 940.53'41'0922 78-62958
79 MARC

Churchill, Winston Leonard Spencer, Sir, 1874-1965.
Memoirs of the Second World War : an abridgement of the six volumes of The Second World War / Winston S. Churchill ; with an epilogue by the author on the postwar years written for this volume ; abridgement by Denis Kelly. — New York : Bonanza Books, 1978, c1959.
xiv, 1065 p. : maps ; 24 cm.
Includes index.
ISBN 0-517-27032-3
1. World War, 1939-1945—Sources. 2. World War, 1939-1945—Great Britain—Sources. 3. Churchill, Winston Leonard Spencer, Sir, 1874-1965. 4. Prime ministers—Great Britain—Biography. I. Kelly, Denis. II. Title.
D743.C484 1978 940.53 78-25660
78 MARC

Churchill, Winston Leonard Spencer, Sir, 1874-1965.
To Pretoria and back. New York, Longmans, Green, 1900.
2 v. (274 p.) 18 cm.
Cover title. Same title used as caption and running title.
A shorter version, deposited for copyright Mar. 14 and 27, 1900, of the work published under title: London to Ladysmith via Pretoria. Cf. F. Woods. A bibl. of the works of Sir Winston Churchill. ⌜2d rev. ed., 1975⌝ p. 30-31.
Cover title: London to Ladysmith via Pretoria. Deposited for copyright May 14, 1900.
——— Variant. Cover title: London to Ladysmith via Pretoria. Deposited for copyright May 14, 1900.
DT932.C46 1900b
——— Microfilm. Washington, Library of Congress, 1969. 1 reel. 35 mm.
Microfilm 21864 DT
1. South African War, 1899-1902—Personal narratives. 2. Churchill, Winston Leonard Spencer, Sir, 1874-1965. 3. South African War, 1899-1902—Prisoners and prisons, Boer. 4. Escapes. 5. Buller, Redvers Henry, Sir, 1839-1908. 6. South African War, 1899-1902—Journalists—Biography. I. Title. II. Title: London to Ladysmith via Pretoria.
DT932.C46 1900 42-51332
⌜r79⌝rev MARC

Hägglöf, Gunnar.
Tre män i Jalta / Gunnar Hägglöf. — Stockholm : Norstedt, 1977.
248 p., ⌜3⌝ leaves of plates : ill. ; 22 cm. S77-41
Bibliography: p. 240-244.
Includes index.
ISBN 9117702026 : kr85.00
1. Crimea Conference, Yalta, 1945. 2. Churchill, Winston Leonard Spencer, Sir, 1874-1965. 3. Roosevelt, Franklin Delano, Pres. U.S., 1882-1945. 4. Stalin, Iosif, 1879-1953. 5. Heads of state—Biography. 6. Biography—20th century. 7. Hägglöf, Gunnar. I. Title.
D734.C7 1945zc 78-393456
79 MARC

Moore, Robin James.
Churchill, Cripps, and India, 1939-1945 / R. J. Moore. — Oxford : Clarendon Press ; New York : Oxford University Press, 1979.
viii, 152 p., ⌜1⌝ leaf of plates : port. ; 23 cm. GB79
Sequel to: The crisis of Indian unity, 1917-1940.
Includes bibliographical references and index.
ISBN 0-19-822485-0 : £6.25
1. India—Politics and government—1919-1947. 2. Churchill, Winston Leonard Spencer, Sir, 1874-1965. 3. Cripps, Richard Stafford, Sir, 1889-1952. I. Title.
DS480.82.M63 320.9'54'035 79-40188
79 MARC

Pilpel, Robert H
Churchill in America, 1895-1961 : an affectionate portrait / Robert H. Pilpel. — 1st ed. — New York : Harcourt Brace Jovanovich, c1976.
xiii, 318 p. : ill. ; 22 cm.
Bibliography: p. 305-309.
Includes index.
ISBN 0-15-117880-1
1. Churchill, Winston Leonard Spencer, Sir, 1874-1965. 2. Prime ministers—Great Britain—Biography. I. Title.
DA566.9.C5P54 941.082'092'4 76-21239
76⌜r79⌝rev MARC

Roskill, Stephen Wentworth.
Churchill and the admirals / Stephen Roskill. — London : Collins, 1977.
351 p., ⌜6⌝ leaves of plates : ill. ; 24 cm. GB***
Includes bibliographical references and index.
ISBN 0-00-216127-3 : £8.50
1. World War, 1939-1945—Naval operations, British. 2. European War, 1914-1918—Naval operations, British. 3. Churchill, Winston Leonard Spencer, Sir, 1874-1965. 4. Prime ministers—Great Britain—Biography. 5. Great Britain. Admiralty. I. Title.
D771.R67 1977 940.54'59'41 78-325327
79 MARC

Trukhanovskiĭ, Vladimir Grigor'evich.
Winston Churchill / V. G. Trukhanovsky ; ⌜translated from the Russian by Kenneth Russell, Alex Miller, and Christopher English⌝. — Moscow : Progress Publishers, 1978.
390 p. ; 21 cm. USSR
Translation of the 1978 rev. ed. of Uinston Cherchill'.
Bibliography: p. ⌜385⌝-390.
1. Churchill, Winston Leonard Spencer, Sir, 1874-1965. 2. Prime ministers—Great Britain—Biography. 3. Great Britain—History—Victoria, 1837-1901. 4. Great Britain—History—20th century.
DA566.9.C5T7813 941.082'092'4 79-370372
79 MARC

Weidhorn, Manfred, 1931-
Sir Winston Churchill / by Manfred Weidhorn. — Boston : Twayne Publishers, 1979.
174 p. : port. ; 21 cm. — (Twayne's English authors series ; TEAS 264)
Bibliography: p. 170-172.
Includes index.
ISBN 0-8057-6760-6
1. Churchill, Winston Leonard Spencer, Sir, 1874-1965. 2. Prime ministers—Great Britain—Biography. I. Title.
DA566.9.C5W367 941.082'092'4 79-580
79 MARC

—LITERARY ART

Starcke, Grå.
Churchills taler / ⌜af Grå Starcke⌝. — ⌜København, S. Njalsgade 84-96 : Københavns Universitet, Engelsk Institut, 1976⌝.
ii, 107 leaves ; 29 cm. D76-36
"Analyse af W. Churchills radiotaler mellem 1939 og 1945."
Errata slip inserted.
Bibliography: leaves 105-107.
1. Churchill, Winston Leonard Spencer, Sir, 1874-1965—Literary art. I. Title.
DA566.9.C5S7 79-346808
79 MARC

CHURCHILL, MANITOBA

—CITY PLANNING

Manitoba. Municipal Planning Branch.
Churchill townsite redevelopment proposal. ⌜Winnipeg⌝ 1971.
167 p. illus. 22 x 36 cm. C***
1. Churchill, Manitoba—City planning. 2. City planning—Manitoba. 3. Urban renewal—Manitoba—Churchill. I. Title.
HT169.C32C55 711'.4'0971271 72-172083
73⌜r79⌝rev MARC

—SOCIAL CONDITIONS

Spearman, Leonard B
Comparison of social welfare needs with service network in two northern Manitoba communities / by Leonard B. Spearman. — Winnipeg : Center for Settlement Studies, University of Manitoba, 1975.
vii, 89 p. ; 28 cm. — (Series 2, research report - Center for Settlement Studies, University of Manitoba ; no. 24) C***
Bibliography: p. 83-84.
1. Social service—Manitoba—Churchill. 2. Social service—Manitoba—The Pas. 3. Churchill, Man.—Social conditions. 4. The Pas, Man.—Social conditions. I. Title. II. Series: University of Manitoba. Centre for Settlement Studies. Series 2, research reports ; no. 24.
HV110.C58S63 361'.97127'1 79-305466
79 MARC

CHURCHILL FALLS (LABRADOR) CORPORATION

Smith, Philip, 1925-
Brinco : the story of Churchill Falls / Philip Smith. — Toronto : McClelland and Stewart, c1975.
viii, 392 p., ⌜16⌝ leaves of plates : ill. (some col.) ; 25 cm. C***
Includes index.
ISBN 0-7710-8184-7 : $15.95
1. Churchill Falls (Labrador) Corporation. 2. Brinco Limited. I. Title.
HD1696.C32L327 77-372723
338.7'62'1312134097182
77⌜r79⌝rev MARC

CHURCHILL RIVER REGION, SACK. AND MAN

—ANTIQUITIES

Meyer, David.
Archaeology / by David Meyer and S. J. Smailes. — ⌜Saskatoon, Sask. : Churchill River Study ; Ottawa : available from Information Canada⌝, 1975.
xii, 121 p. : ill. ; 28 cm. — (Final report - Churchill River Study ; 19) C***
"Prepared for the Churchill River Study Board."
Bibliography: p. 117-121.
1. Churchill River region, Sask. and Man.—Antiquities. 2. Excavations (Archaeology)—Churchill River region, Sask. and Man. 3. Saskatchewan—Antiquities. 4. Manitoba—Antiquities. I. Smailes, S. J., joint author. II. Churchill River Study. Board. III. Title. IV. Series: Churchill River Study. Final report - Churchill River Study ; 19.
F1074.C58M49 971.27'1 78-326719
79 MARC

—ECONOMIC CONDITIONS

Socio-economic (Saskatchewan) / by J. C. Stabler ... ⌜et al.⌝. — Saskatoon, Sask. : Churchill River Study, 1975.
xxii, 465 p. : ill., maps ; 28 cm. — (Final report - Churchill River Study ; 25) C***
Bibliography: p. 263-284.
1. Churchill River region, Sask. and Man.—Economic conditions. 2. Churchill River region, Sask. and Man.—Social conditions. 3. Dams—Economic aspects—Churchill River region, Sask. and Man. I. Stabler, Jack C. II. Series: Churchill River Study. Final report - Churchill River Study ; 25.
HC117.S3S63 330.9'7124'1 78-320082
78 MARC

Stewart, W Don.
Transportation / by W. D. Stewart and Judy Foraie. — Saskatoon, Sask. : Churchill River Study, 1975.
xxiii, 256 p. : maps ; 28 cm. — (Final report - Churchill River Study ; 6) C***
Bibliography: p. 205-211.
1. Transportation—Churchill River region, Sask. and Man. 2. Churchill River region, Sask. and Man.—Economic conditions. I. Foraie, Judy, joint author. II. Title. III. Series: Churchill River Study. Final report - Churchill River Study ; 6.
HE215.Z7C587 380.5'097124'1 78-325609
79 MARC

—POWER UTILIZATION—COST EFFECTIVENESS

Cost-benefit analysis (Saskatchewan) / by S. ⌜i.e. J.⌝ C. Stabler ... ⌜et al.⌝. — ⌜Saskatoon, Sask. : Churchill River Study⌝, 1975.
xxii, 155 p. : graphs ; 28 cm. — (Final report - Churchill River Study ; 24) C***
"Prepared for the Churchill River Study Board."
Bibliography: p. 97-99.
1. Water-power electric plants—Churchill River, Sask. and Manitoba—Cost effectiveness. 2. Churchill River, Sask. and Manitoba—Power utilization—Cost effectiveness. 3. Saskatchewan Power Corporation. I. Stabler, Jack C. II. Churchill River Study. Board. III. Series: Churchill River Study. Final report - Churchill River Study ; 24.
TK1427.C46C67 363.6'2 78-376514
78 MARC

—RECREATIONAL USE

EPEC Consulting Western Ltd.
Outdoor recreation and tourism demand / by EPEC Consulting Western Limited. — Saskatoon, Sask. : Churchill River Study, 1975.
xxii, 197 p. : 5 maps ; 28 cm. — (Final report- Churchill River Study. ; 29) C***
"Prepared for the Churchill River Study Board."
Bibliography: p. 149-150.
1. Churchill River, Sask, and Manitoba—Recreational use. 2. Outdoor recreation—Churchill River, Sask. and Manitoba. 3. Recreational surveys—Churchill River, Sask. and Manitoba. I. Churchill River Study. Board. II. Title. III. Series: Churchill River Study. Final report - Churchill River Study ; 29.
GV191.46.C47E2 1975 333.7'8'097124 78-319732
78 MARC

—SOCIAL CONDITIONS

Socio-economic (Saskatchewan) / by J. C. Stabler ... ⌜et al.⌝. — Saskatoon, Sask. : Churchill River Study, 1975.
xxii, 465 p. : ill., maps ; 28 cm. — (Final report - Churchill River Study ; 25) C***
Bibliography: p. 263-284.
1. Churchill River region, Sask. and Man.—Economic conditions. 2. Churchill River region, Sask. and Man.—Social conditions. 3. Dams—Economic aspects—Churchill River region, Sask. and Man. I. Stabler, Jack C. II. Series: Churchill River Study. Final report - Churchill River Study ; 25.
HC117.S3S63 330.9'7124'1 78-320082
78 MARC

CHURCHYARDS see Cemeteries

CHURNET VALLEY, ENG.

—DESCRIPTION AND TRAVEL—GUIDE-BOOKS

Porter, Lindsey.
The Churnet Valley and Alton Towers / ₍by₎ Lindsey Porter & Laurie Landon. — Hartington : Moorland Publishing Co., 1977.
48 p. : ill., maps ; 21 cm. GB77-32958
£0.80
1. Churnet Valley, Eng.—Description and travel—Guide-books. 2. Alton Towers, Eng. I. Landon, Laurie, joint author. II. Title.
DA670.C628P67 914.24'61 79-318388
79 MARC

CHUSCHI, PERU

Isbell, Billie Jean.
To defend ourselves : ecology and ritual in an Andean village / by Billie Jean Isbell. — ₍Austin₎ : Institute of Latin American Studies, University of Texas at Austin : distributed by University of Texas Press, c1978.
xv, 289 p. : ill. ; 24 cm. — (Latin American monographs ; no. 47)
Bibliography: p. ₍259₎-271.
Includes index.
ISBN 0-292-78030-3
1. Quechua Indians. 2. Chuschi, Peru. I. Title. II. Series: Latin American monographs (Austin, Tex.) ; no. 47.
F2230.2.K4I8 301.45'19'80852 78-620051
79 MARC

CHŪSHŌ KIGYŌ SHINKŌ JIGYŌDAN

—HISTORY

Chūshō Kigyō Shinkō Jigyōdan.
(Jūnen no ayumi)
十年のあゆみ. — 東京：中小企業振興事業団, 1977.
255 p., ₍12₎ leaves of plates : ill. ; 27 cm. Ja 78-10231

1. Chūshō Kigyō Shinkō Jigyōdan—History.
HD2346.J3C367 1977 78-812772

CHŪTŌ KOKUGO KYŌIKU KENKYŪKAI

Yoshida, Kenkō, 1282?-1350.
₍Tsurezuregusa. Selections₎
抄本徒然草 / ₍吉田兼好₎ ; 中等國語教育研究會編. — 東京：鍵文社, 昭和21 ₍1946₎, 昭和23 ₍1948₎ printing.
2, 2, 3, 71 p. ; 21 cm.

1. Chūtō Kokugo Kyōiku Kenkyūkai. II. Title.
Title romanized: Shōhon Tsurezuregusa.
PL791.6.T72 1948 79-801586
J

CHUVASH ARTS see Arts, Chuvash

CHUVASHES

—RELIGION—ADDRESSES, ESSAYS, LECTURES

(Sovremennye sot͡sial'nye i ėtnicheskie prot͡sessy v Chuvashskoĭ ASSR)
Современные социальные и этнические процессы в Чувашской АССР : сборник статей / ₍редакторы А. В. Изоркин, А. П. Сидорова, П. П. Фокин₎. — Чебоксары : Научно-иссл. ин-т языка, лит-ры, истории и экономики при Совете министров Чувашской АССР, 1978.
155 p. ; 21 cm. USSR***
Includes bibliographical references.
0.63rub
1. Chuvashes—Religion—Addresses, essays, lectures. 2. Chuvashia—Social conditions—Addresses, essays, lectures. 3. Ethnology—Chuvashia—Addresses, essays, lectures. I. Izorkin, A. V. II. Sidorova, A. P. III. Fokin, P. P. IV. Nauchno-issledovatel'skiĭ institut ëazyka, literatury, istorii i ėkonomiki.
BL975.C46S67 79-408517

—SOCIAL LIFE AND CUSTOMS

(Baldran-bazar v Kozmodem'ëanskom uezdě)
Балдранъ-базаръ въ Козмодемьянскомъ уѣздѣ. ₍Переводчикъ чувашскаго языка Спиридонъ Михайловъ. Казань, 1852.₎
6 p. 21 cm.
Caption title.
"Изъ Каз. губ. вѣд. 1852 г., № 21."
1. Holidays — Russian Republic — Mariyskaya ASSR. 2. Chuvashes—Social life and customs. 3. Cheremisses—Social life and customs.
GT4856.A3M372 79-412239

CHUVASHIA

Urvanov, Rim Urvanovich.
S vysokoĭ meroĭ otvetstvennosti i trebovatel'nosti / R. U. Urvanov ; ₍Lit. zapis' V. S. Poduzova₎. — Cheboksary : Chuvash. kn. izd-vo, 1978.
56 p. ; 20 cm. — (Reshenii͡a XXV ₍i.e. Dvadt͡sat' piatogo₎ s"ezda KPSS vypolnim) (V pomoshch' sekretarëam pervichnykh partiĭnykh organizat͡siĭ)
USSR78
0.05rub
1. Kommunisticheskai͡a partii͡a Sovetskogo Soi͡uza—Party work. 2. Chuvashia. I. Poduzov, V. S. II. Title.
JN6598.K7U63 79-379934
79 MARC

—ANTIQUITIES

(Issledovanii͡a po arkheologii Chuvashii)
Исследования по археологии Чувашии : ₍Сб. статей₎. — Чебоксары : НИИ яз., лит. истории и экономики при Совете Министров ЧувашАССР, 1978.
172 p. : ill. ; 22 cm. — (Труды - Научно-исследовательский институт языка, литературы, истории и экономики при Совете Министров Чувашской АССР ; вып. 80) USSR 78
Includes bibliographical references.
0.72rub
1. Chuvashia—Antiquities. 2. Russian Republic—Antiquities. I. Series: Nauchno-issledovatel'skiĭ institut ëazyka, literatury, istorii i ėkonomiki. Trudy - Nauchno-issledovatel'skiĭ institut ëazyka, literatury, istorii i ėkonomiki ; vyp. 80.
DK511.C5C48 vyp. 80 79-413555

—COLLECTED WORKS

Nauchno-issledovatel'skiĭ institut ëazyka, literatury, istorii i ėkonomiki.
(Trudy - Nauchno-issledovatel'skiĭ institut ëazyka, literatury, istorii i ėkonomiki)
Труды - Научно-исследовательский институт языка, литературы, истории и экономики. вып. 72–
Чебоксары, Научно-иссл. ин-т языка, лит-ры, экономики, 1977–
no. 23 cm.
Continues: Cheboksary, Russia. Nauchno-issledovatel'skiĭ institut. Uchenye zapiski.
1. Chuvashia—Collected works.
DK511.C5C48 79-642420
(MARC-S)

—ECONOMIC CONDITIONS

(Itogi, perspektivy, zadachi)
Итоги, перспективы, задачи : хозяйственное и культурное строительство Чувашской АССР в девятой и десятой пятилетках / ₍редакционная коллегия С. М. Ислюков ... et al.₎. — Чебоксары : Чувашское книжное изд-во, 1977.
284 p. : graphs ; 21 cm. USSR***
At head of title: Nauchno-issledovatel'skiĭ institut pri Sovete Ministrov Chuvashskoĭ ASSR.
Includes bibliographical references.
0.54rub
1. Chuvashia—Economic conditions. 2. Chuvashia—Social conditions. I. Isli͡ukov, S. M. II. Cheboksary, Russia. Nauchno-issledovatel'skiĭ institut.
HC337.C45 I 83 78-422328

—HISTORY—ADDRESSES, ESSAYS, LECTURES

(Chetyresta dvadt͡satipi͡atiletie dobrovol'nogo vkhozhdenii͡a Chuvashii v sostav Rossii)
425-летие добровольного вхождения Чувашии в состав России / ₍редакторы В. Д. Димитриев, Н. Е. Егоров, В. А. Прохорова₎. — Чебоксары : Науч.-иссл. ин-т при Совете Министров Чувашской АССР, 1977.
157 p. ; 22 cm. — (Труды - Научно-исследовательский институт при Совете Министров Чувашской АССР ; вып. 71) USSR***
Includes bibliographical references.
1.12rub
1. Chuvashia—History—Addresses, essays, lectures. I. Dimitriev, Vasiliĭ Dimitrievich. II. Egorov, N. E. III. Prokhorova, V. A. IV. Series: Cheboksary, Russia. Nauchno-issledovatel'skiĭ institut. Uchenye zapiski ; vyp. 71.
DK511.C5C48 vyp. 71 78-423234

—POLITICS AND GOVERNMENT—ADDRESSES, ESSAYS, LECTURES

(Istorii͡a i ėtnografii͡a Chuvashskoĭ ASSR)
История и этнография Чувашской АССР : ₍Сборник статей₎. — Чебоксары : Чуваш. НИИ, 1976.
160 p. ; 22 cm. — (Труды - Научно-исследовательский институт при Совете Министров Чувашской АССР ; вып. 69) USSR 77
Includes bibliographical references.
0.85rub
1. Chuvashia — Politics and government — Addresses, essays, lectures. 2. Arts—Chuvashia—Addresses, essays, lectures. I. Series: Cheboksary, Russia. Nauchno-issledovatel'skiĭ institut. Uchenye zapiski ; vyp. 69.
DK511.C5C48 vyp. 69 78-421841

—SOCIAL CONDITIONS

(Itogi, perspektivy, zadachi)
Итоги, перспективы, задачи : хозяйственное и культурное строительство Чувашской АССР в девятой и десятой пятилетках / ₍редакционная коллегия С. М. Ислюков ... et al.₎. — Чебоксары : Чувашское книжное изд-во, 1977.
284 p. : graphs ; 21 cm. USSR***
At head of title: Nauchno-issledovatel'skiĭ institut pri Sovete Ministrov Chuvashskoĭ ASSR.
Includes bibliographical references.
0.54rub
1. Chuvashia—Economic conditions. 2. Chuvashia—Social conditions. I. Isli͡ukov, S. M. II. Cheboksary, Russia. Nauchno-issledovatel'skiĭ institut.
HC337.C45 I 83 78-422328

—SOCIAL CONDITIONS—ADDRESSES, ESSAYS, LECTURES

(Sovremennye sot͡sial'nye i ėtnicheskie prot͡sessy v Chuvashskoĭ ASSR)
Современные социальные и этнические процессы в Чувашской АССР : сборник статей / ₍редакторы А. В. Изоркин, А. П. Сидорова, П. П. Фокин₎. — Чебоксары : Научно-иссл. ин-т языка, лит-ры, истории и экономики при Совете министров Чувашской АССР, 1978.
155 p. ; 21 cm. USSR***
Includes bibliographical references.
0.63rub
1. Chuvashes—Religion—Addresses, essays, lectures. 2. Chuvashia—Social conditions—Addresses, essays, lectures. 3. Ethnology—Chuvashia—Addresses, essays, lectures. I. Izorkin, A. V. II. Sidorova, A. P. III. Fokin, P. P. IV. Nauchno-issledovatel'skiĭ institut ëazyka, literatury, istorii i ėkonomiki.
BL975.C46S67 79-408517

CHUVASHIAN FICTION

—TRANSLATIONS INTO RUSSIAN

Ivanov-Paĭmen, V Z
[Kĕper. Russian]
Мост : роман / В. Иванов-Паймен ; перевод с чувашского В. Михайловой. — Москва : Современник, 1976.
430 p. : ill. ; 21 cm. USSR***
0.98rub
I. Title.
Title romanized: Most.
PL384.I 94K417 79–409113

CHUVASHSKIĬ GOSUDARSTVENNYĬ AKADEMICHESKIĬ DRAMATICHESKIĬ TEATR

—ADDRESSES, ESSAYS, LECTURES

(V tvorcheskom gorenii—vysshaîa radost!)
В творческом горении—высшая радость! : Жизнь и творчество К. И. Иванова, режиссера Чуваш. гос. акад. театра им. К. В. Иванова : [Сборник статей / составитель Фаина Александровна Романова]. — Чебоксары : Чуваш. кн. изд-во, 1976.
127 p. : ill. ; 21 cm. USSR 77
Includes bibliographical references.
0.35rub
1. Ivanov, Konstantin Ivanovich, 1911–1962—Addresses, essays, lectures. 2. Chuvashskiĭ gosudarstvennyĭ akademicheskiĭ dramaticheskiĭ teatr—Addresses, essays, lectures. 3. Theatrical producers and directors—Russian Republic—Chuvashia—Biography—Addresses, essays, lectures. I. Romanova, Faina Aleksandrovna.
PN2728.I 827V2 79–407184

CHVA (PHONETICS) see Shwa (Phonetics)

CHWEE LANGUAGE see Twi language

CHWISTEK, LEON, 1884-1944

Jakimowicz, Irena.
Witkacy, Chwistek, Strzemiński : myśli i obrazy / Irena Jakimowicz. — Wyd. 1. — Warszawa : Arkady, 1978 (Warsz. : DSP)
218, [2] p., [12] leaves of plates : ill. (some col.) ; 25 cm. P78-8351
Includes bibliographies.
zł120.00
1. Art, Modern—20th century—Poland. 2. Witkiewicz, Stanisław Ignacy, 1885-1939. 3. Chwistek, Leon, 1884-1944. 4. Strzemiński, Władysław, 1893-1952.
N7255.P6J34 79-368823
79 MARC

CHYATTE, SAMUEL B.

Chyatte, Samuel B
On borrowed time : living with hemodialysis / Samuel B. Chyatte. — Oradell, NJ : Medical Economics Co., Book Division, c1979.
x, 133 p. ; 24 cm.
ISBN 0-87489-213-9
1. Renal insufficiency—Biography. 2. Hemodialysis. 3. Chyatte, Samuel B. I. Title.
RC918.R4C62 616.6'1 79-109938
79 MARC

CHYDORIDAE

—CLASSIFICATION

Smirnov, Nikolaĭ Nikolaevich.
Chydoridae = Chydoridae fauny mira / N. N. Smirnov ; translated from Russian [by A. Mercado ; edited by O. Theodor]. — Jerusalem : Israel Program for Scientific Translations ; Springfield, Va. : available from the U.S. Dept. of Commerce, National Technical Information Service, 1974.
iv, 644 p. : ill. ; 25 cm. — (Crustacea ; v. 1, no. 2) (Fauna of the U.S.S.R. ; new ser. no. 101)
Bibliography: p. 190-237.
Includes index.
ISBN 0-7065-1362-2
1. Chydoridae—Classification. 2. Crustacea—Classification. I. Theodor, Oskar. II. Title. III. Series. IV. Series: Fauna of the U.S.S.R. ; new ser. no. 101.
QL435.C78 vol. 1, no. 2 595.3'08 s 78-377039
[QL444.B83] 78 MARC

CHYMOSIN see Rennin

CHYMOTRYPSIN

Kasche, Volker.
Specific protein—protein interaction and its application in studies on radiation-induced protein modification. Uppsala, Universitetet, Stockholm, Almqvists & Wiksell (distr.), 1971.
xiv, 132 p. 21 cm. (Uppsala dissertations from the Faculty of science, 2) (Acta Universitatis Upsaliensis) Kr36.00 S71-9
Akademisk avhandling-Uppsala.
Includes bibliographical references.
CONTENTS: Interaction of chymotrypsin with soybean trypsin inhibitor.—Interaction of chymotrypsin with soybean trypsin inhibitor bound to a polysaccharide.—Radiation-induced modification of α chymotrypsin.
1. Proteins, Effect of radiation on. 2. Chymotrypsin. I. Title. II. Series. III. Series: Uppsala. Universitet. Acta Universitatis Upsaliensis.
QP551.K295 79-300154
79 MARC

CIABANI, ROBERTO, 1939-

Ciabani, Roberto, 1939-
Roberto Ciabani / a cura di Mario Bucci. — Firenze : Giorgi & Gambi stampatori, 1977.
135 p. : ill. ; 30 cm. It78-Sept
L10000
1. Ciabani, Roberto, 1939- . I. Bucci, Mario.
N6923.C523A4 1977 709'.2'4 79-347769
*79 MARC

CIAFFI, VINCENZO

Ciaffi, Vincenzo.
Scritti inediti rari / Vincenzo Ciaffi. — Torino : G. Giappichelli, 1978.
173 p. ; 25 cm. — (Pubblicazioni della Facoltà di magistero, Università di Torino ; 41) It78-Mar
Includes bibliographical references.
L4000
1. Literature—History and criticism—Addresses, essays, lectures. 2. Ciaffi, Vincenzo. I. Series: Turin. Università. Facoltà di magistero. Pubblicazioni ; 41.
PN515.C49 78-389729
*79 MARC

CIAIA, IGNAZIO, 1766-1799

—BIOGRAPHY

Semeraro Herrmann, Marialuisa.
Ignazio Ciaia : impegno civile e opera poetica / Marialuisa Semeraro Herrmann. — Fasano : Schena, 1976.
251 p., [35] leaves of plates : ill. ; 24 cm. It78-Feb
Errata slip inserted.
Bibliography: p. 230-236.
L9000
1. Ciaia, Ignazio, 1766-1799—Biography. 2. Jacobins—Italy—Naples (Kingdom) 3. Parthenopean Republic. 4. Poets, Italian—18th century—Biography. 5. Revolutionists—Italy—Naples (Kingdom)—Biography. I. Ciaia, Ignazio, 1766-1799. II. Title.
PQ4688.C247Z867 78-370786
*78 MARC

CIANCABILLA, GIUSEPPE, 1872-1904

Fedeli, Ugo.
Giuseppe Ciancabilla / Ugo Fedeli. — Imola : Galeati, 1965.
81 p., [1] leaf of plates : port. ; 20 cm.
Imprint on cover : Cesena, Edizioni Antistato.
Includes bibliographical references.
1. Ciancabilla, Giuseppe, 1872-1904. 2. Anarchism and anarchists—Italy—Biography.
HX902.C5F4 335'.83'0924 79-385669
79 MARC

CIANO, GALEAZZO, CONTE, 1903-1944

Guerri, Giordano Bruno, 1950-
Galeazzo Ciano : una vita, 1903-1944 / Giordano Bruno Guerri. — 1. ed. — Milano : Bompiani, 1979.
718 p., [8] leaves of plates : ill. ; 22 cm. — (Saggi Bompiani) It79-Apr
Bibliography: p. 695-703.
Includes index.
L12500
1. Ciano, Galeazzo, conte, 1903-1944. 2. Italy—History—1922-1945. 3. Fascism—Italy—History. 4. Statesmen—Italy—Biography. 5. Fascists—Italy—Biography.
DG575.C52G83 945.091'092'4 79-367632
*79 MARC

CIBA-GEIGY LIMITED

Erni, Paul, 1917-
Die Basler Heirat : Geschichte der Fusion Ciba-Geigy / Paul Erni ; [Zeichnungen im Textteil, Hans Erni]. — Zürich : Buchverlag der Neuen Zürcher Zeitung, cop. 1979.
425 p. : ill. (some col.) ; 23 cm. Sw79-2559
23.80F
1. Ciba-Geigy Limited. 2. Ciba-Geigy Limited—History—Sources. I. Title.
HD9656.S84C56 79-361687
79 MARC

—HISTORY—SOURCES

Erni, Paul, 1917-
Die Basler Heirat : Geschichte der Fusion Ciba-Geigy / Paul Erni ; [Zeichnungen im Textteil, Hans Erni]. — Zürich : Buchverlag der Neuen Zürcher Zeitung, cop. 1979.
425 p. : ill. (some col.) ; 23 cm. Sw79-2559
23.80F
1. Ciba-Geigy Limited. 2. Ciba-Geigy Limited—History—Sources. I. Title.
HD9656.S84C56 79-361687
79 MARC

CIBINONG, INDONESIA (BOGOR)

—CITY PLANNING

Indonesia. Direktorat Tata Kota dan Tata Daerah.
Rencana tata kota Cibinong. — [Jakarta] : Pemerintah Daerah Tingkat II Kabupaten Bogor bekerja sama dengan Direktorat Tata Kota dan Tata Daerah, Direktorat Jenderal Cipta Karya, Departemen P.U.T.L., [1976]
2 v. : maps ; 30 x 42 cm.
Cover title.
Prepared by Direktorat Tata Kota dan Tata Daerah.
CONTENTS: 1. Kompilasi data.—2. Analisa dan rencana.
1. Cibinong, Indonesia (Bogor)—City planning. 2. City planning—Indonesia. 3. Cibinong, Indonesia (Bogor)—Statistics. I. Bogor, Indonesia (Kabupaten) II. Title.
HT169.I54C524 78-941028
78 MARC

—STATISTICS

Indonesia. Direktorat Tata Kota dan Tata Daerah.
Rencana tata kota Cibinong. — [Jakarta] : Pemerintah Daerah Tingkat II Kabupaten Bogor bekerja sama dengan Direktorat Tata Kota dan Tata Daerah, Direktorat Jenderal Cipta Karya, Departemen P.U.T.L., [1976]
2 v. : maps ; 30 x 42 cm.
Cover title.
Prepared by Direktorat Tata Kota dan Tata Daerah.
CONTENTS: 1. Kompilasi data.—2. Analisa dan rencana.
1. Cibinong, Indonesia (Bogor)—City planning. 2. City planning—Indonesia. 3. Cibinong, Indonesia (Bogor)—Statistics. I. Bogor, Indonesia (Kabupaten) II. Title.
HT169.I54C524 78-941028
78 MARC

CIBITUNG, INDONESIA

—STATISTICAL SERVICES

Indonesia. Direktorat Jenderal Pembangunan Masyarakat Desa.
Proses pengolahan data Kecamatan Cibitung. [Jakarta, 1974]
17 p. 21 cm.
1. Cibitung, Indonesia—Statistical services. 2. Cibitung, Indonesia—Statistics—Data processing. I. Title.
HA37.I47C54 74-940424
79 MARC

CIBITUNG, INDONESIA (Continued)

—STATISTICS—DATA PROCESSING

Indonesia. Direktorat Jenderal Pembangunan Masyarakat Desa.
Proses pengolahan data Kecamatan Cibitung. ₍Jakarta, 1974₎
17 p. 21 cm.
1. Cibitung, Indonesia—Statistical services. 2. Cibitung, Indonesia—Statistics—Data processing. I. Title.
HA37.I47C54 74-940424
79 MARC

CIBONEY INDIANS

—ANTIQUITIES

Veloz Maggiolo, Marcio.
Medioambiente y adaptación humana en la prehistoria de Santo Domingo / Marcio Veloz Maggiolo. — ₍2. ed.₎. — Santo Domingo : Ediciones de Taller, 1976-1977.
2 v. : ill. ; 22 cm. — (Biblioteca Taller ; 63) (v. 2: Colección Historia y sociedad ; no. 30) (v. 2: Publicaciones de la Universidad Autónoma de Santo Domingo ; v. 200)
Vol. 2 lacks ed. statement; published by Editora de la Universidad Autónoma de Santo Domingo.
Includes bibliographical references.
1. Indians of the West Indies—Hispaniola—Antiquities. 2. Hispaniola—Antiquities. 3. Ciboney Indians—Antiquities. 4. Indians of the West Indies—Antiquities. I. Title.
F1909.V447 1976 76-484466
76₍r79₎rev MARC

CICERO, MARCUS TULLIUS. DE FINIBUS BONORUM ET MALORUM. BOOK 3

Moscarini, Flaviana.
Cicerone e l'etica stoica nel III libro del De finibus / Flaviana Moscarini. — Roma : Tip. del senato del dott. G. Bardi, 1930.
50 p. ; 25 cm. — (Pubblicazioni della Scuola di filosofia della R. Università di Roma ; 2)
Includes bibliographical references.
1. Cicero, Marcus Tullius. De finibus bonorum et malorum. Book 3. 2. Stoics. I. Title. II. Series: Rome (City). Università. Scuola di filosofia. Pubblicazioni ; 2.
PA6296.D2M67 79-355236
79 MARC

CICERO, MARCUS TULLIUS. EPISTOLAE

Allegri, Giuseppina.
Bruto usuraio nell'epistolario ciceroniano / ₍di₎ Giuseppina Allegri. L'indipendenza di Lucilio / ₍di₎ Bruno Zucchelli. — Parma : Università degli studi, Istituto di lingua e letteratura latina, 1977.
141 p. ; 21 cm. — (₍Pubblicazioni₎ - Università degli studi di Parma, Istituto di lingua e letteratura latina ; 3) It78-Feb
Includes bibliographies and indexes.
L4000
1. Latin literature—History and criticism—Addresses, essays, lectures. 2. Brutus, M. Junius, 79 (ca.)-42 B.C. 3. Cicero, Marcus Tullius. Epistolae. 4. Lucilius, C.—Criticism and interpretation. I. Zucchelli, Bruno. L'indipendenza di Lucilio. 1977. II. Title. III. Series: Parma. Università. Istituto di lingua e letteratura latina. Pubblicazioni ; 3.
PA2014.P37 no. 3 79-347143
₍PA6011₎ *79 MARC

CICERO, MARCUS TULLUS. PRO ARCHIA POETA

Gotoff, Harold C
Cicero's elegant style : an analysis of the Pro Archia / Harold C. Gotoff. — Urbana : University of Illinois Press, c1979.
xiii, 255 p. ; 24 cm.
ISBN 0-252-00694-1 : $12.50
1. Cicero, Marcus Tullius. Pro Archia poeta. 2. Cicero, Marcus Tullius—Technique. I. Cicero, Marcus Tullius. Pro Archia poeta. 1979. II. Title.
PA6279.A93G6 875'.01 79-10245
79 MARC

CICERO, MARCUS TULLIUS. SOMNIUM SCIPIONIS

Favonius Eulogius, of Carthage.
Favonii Eulogii Disputatio de Somnio Scipionis. — Edizione critica, traduzione e note / a cura di Luigi Scarpa. — Padova : ₍s.n.₎, 1974.
xx, 75 p. ; 24 cm. — (Collana accademica ; 5) It76-May
At head of title: Accademia patavina di scienze, lettere ed arti, Università di Padova, Istituto di filologia latina.
Text in Latin and Italian, with introd. and notes in Italian.
Includes bibliographical references and indexes.
L5000
1. Cicero, Marcus. Tullius. Somnium Scipionis. 2. Symbolism of numbers. 3. Pythagoras and Pythagorean school. I. Scarpa, Luigi. II. Title: Disputatio de Somnio Scipionis. III. Series.
PA6304.S7F38 1974 78-398323
*79 MARC

CICERO, MARCUS TULLIUS

Mitchell, Thomas N 1939-
Cicero, the ascending years / Thomas N. Mitchell. — New Haven : Yale University Press, 1979.
xii, 259 p. ; 22 cm.
Bibliography: p. 243-252.
Includes indexes.
ISBN 0-300-02277-8 : $15.00
1. Cicero, Marcus Tullius. 2. Statesmen—Rome—Biography. 3. Orators—Rome—Biography. I. Title.
DG260.C53M57 937'.05'0924 78-31188
78 MARC

—BIOGRAPHY

Giebel, Marion.
Marcus Tullius Cicero in Selbstzeugnissen und Bilddokumenten / dargest. von Marion Giebel ; ₍den Anh. besorgte d. Autorin₎. — Reinbek bei Hamburg : Rowohlt, 1977.
155 p. : numerous ill. ; 19 cm. — (Rowohlts Monographien ; 261)
GFR78-A
Bibliography: p. 144-₍151₎
Includes index.
ISBN 3-499-50261-5 : DM6.80
1. Cicero, Marcus Tullius—Biography. 2. Authors, Latin—Biography. I. Title.
PA6320.G5 937'.05'0924 79-346812
79 MARC

—CORRESPONDENCE

Cicero, Marcus Tullius.
Cicero's Letters to Atticus / translated with an introd. by D. R. Shackleton Bailey. — Harmondsworth, Eng. ; New York : Penguin Books, 1978.
735 p. : maps ; 18 cm. — (Penguin classics) GB***
Originally published in parallel Latin and English texts under title: Letters to Atticus.
Bibliography: p. 25-26.
ISBN 0-14-044309-6 : £2.25 ($5.95)
1. Cicero, Marcus Tullius—Correspondence. 2. Atticus, Titus Pomponius. 3. Authors, Latin—Correspondence. I. Atticus, Titus Pomponius. II. Bailey, David Roy Shackleton. III. Title. IV. Title: Letters to Atticus.
PA6308.E6B3 876'.01 79-304440
79 MARC

Cicero, Marcus Tullius.
Cicero's letters to his friends / Cicero ; translated by D. R. Shackleton Bailey. — Harmondsworth, Eng. ; New York : Penguin Books, 1978.
2 v. : maps ; 19 cm. — (The Penguin classics) GB***
Includes bibliographical references.
ISBN 0-14-044340-1 (v. 1) : £2.25 ($5.95) (v. 1)
1. Cicero, Marcus Tullius—Correspondence. 2. Authors, Latin—Correspondence. I. Bailey, David Roy Shackleton. II. Title.
PA6308.E5B34 876'.01 79-310701
79 MARC

Cicero, Marcus Tullius.
Epistulae ad familiares / Cicero ; edited by D. R. Shackleton Bailey. — Cambridge ; New York : Cambridge University Press, 1977.
2 v. ; 22 cm. — (Cambridge classical texts and commentaries ; 16-17)
GB***
Bibliography: v. 1, p. ix-xii.
Includes indexes.
CONTENTS: v. 1. 62-47 B.C.—v. 2. 47-43 B.C.
ISBN 0-521-21152-2 (v. 1)
1. Cicero, Marcus Tullius—Correspondence. 2. Authors, Latin—Correspondence. I. Bailey, David Roy Shackleton. II. Title.
PA6297.A3 1977 937'.05'0924 76-11079
78 MARC

—INFLUENCE—HOLBERG

Kragelund, Aage, 1894-
Holberg og Cicero / af A. Kragelund. — København : Gad, 1978.
184 p. ; 23 cm. D78-48
Bibliography: p. 161-₍162₎
Includes indexes.
ISBN 8712474622 : kr70.80
1. Holberg, Ludvig, baron, 1684-1754—Criticism and interpretation. 2. Circero, Marcus Tullius—Influence—Holberg. 3. Holberg, Ludvig, Baron, 1684-1754. I. Title.
PT8090.K67 79-349246
79 MARC

—RELIGION AND ETHICS

Rothe, Christian.
Humanitas, Fides und Verwandtes in der römischen Provinzialpolitik : Unters. zur polit. Funktion röm. Verhaltensnormen bei Cicero / Christian Rothe. — Berlin : Akademie-Verlag, 1978.
107 p. ; 24 cm. — (Schriften zur Geschichte und Kultur der Antike ; 12)
GFR78-A29
A revision of the author's thesis, Rostock, 1973.
Bibliography: p. 100-101.
Includes index.
18.00M
1. Cicero, Marcus Tullius—Religion and ethics. 2. Rome—Officials and employees—Conduct of life. 3. Rome—Moral conditions. 4. Rome—Provinces—Administration. I. Title. II. Series.
DG260.C53R67 1978 79-370312
79 MARC

—TECHNIQUE

Berger, Dorothea.
Cicero als Erzähler : forens. u. literar. Strategien in d. Gerichtsreden / Dorothea Berger. — Frankfurt am Main ; Bern ; Las Vegas : Lang, 1978.
234 p. ; 21 cm. — (Europäische Hochschulschriften : Reihe 15, Klassische Philologie und Literatur ; Bd. 12) GFR78-A
A revision of the author's thesis, Constance, 1975.
"Anhang" (28 p. of tables) inserted.
Bibliography: p. 223-229.
ISBN 3-261-02448-8
1. Cicero, Marcus Tullius. Orationes. 2. Cicero, Marcus Tullius—Technique. I. Title. II. Series.
PA6285.B4 1978 875'.01 79-368181
79 MARC

Gotoff, Harold C
Cicero's elegant style : an analysis of the Pro Archia / Harold C. Gotoff. — Urbana : University of Illinois Press, c1979.
xiii, 255 p. ; 24 cm.
ISBN 0-252-00694-1 : $12.50
1. Cicero, Marcus Tullius. Pro Archia poeta. 2. Cicero, Marcus Tullius—Technique. I. Cicero, Marcus Tullius. Pro Archia poeta. 1979. II. Title.
PA6279.A93G6 875'.01 79-10245
79 MARC

CICERO, MARCUS TULLUS. ORATIONES

Berger, Dorothea.
Cicero als Erzähler : forens. u. literar. Strategien in d. Gerichtsreden / Dorothea Berger. — Frankfurt am Main ; Bern ; Las Vegas : Lang, 1978.
234 p. ; 21 cm. — (Europäische Hochschulschriften : Reihe 15, Klassische Philologie und Literatur ; Bd. 12) GFR78-A
A revision of the author's thesis, Constance, 1975.
"Anhang" (28 p. of tables) inserted.
Bibliography: p. 223-229.
ISBN 3-261-02448-8
1. Cicero, Marcus Tullius. Orationes. 2. Cicero, Marcus Tullius—Technique. I. Title. II. Series.
PA6285.B4 1978 875'.01 79-368181
79 MARC

CICERO, MARCUS TULLIUS. ORATIONES IN VERREM

Bartošek, Milan.
Verrinae : význam Ciceronových řečí proti Verrovi pro základní problémy státu a práva / Milan Bartošek. — Praha : Univerzita Karlova, 1977.
234 p., ₍8₎ leaves of plates : ill., map (on lining paper) ; 25 cm. Cz***
Includes bibliographical references.
Kčs44.50
1. Cicero, Marcus Tullius. Orationes in Verrem. 2. Roman law—Addresses, essays, lectures. I. Title.
79-371197
79 MARC

CICERONES see Tour guides (Persons)

CICHLIDAE

see also Haplochromis; Lethrinops; Tilapia

CICOGNANI, LUIGI

Cicognani, Luigi.
Pietro Antonelli e l'ambiente scioano nel diario inedito / di Luigi Cicognani ; [a cura di] Carlo Zaghi. — Ferrara : "Nuovi problemi", 1935.
90 p. ; 25 cm.
Appendix (p. 74-90) consists of correspondence.
1. Cicognani, Luigi. 2. Africa, Northeast—Colonization—History—Sources. 3. Shewa, Eghiopia—History—Sources. 4. Antonelli, Pietro, 1853-1901. 5. Menelik II, negus of Ethiopia, 1844-1913. 6. Explorers—Italy—Biography. 7. Explorers—Africa, Northeast—Biography. I. Zaghi, Carlo. II. Title.
DT367.72.C5A33 78-420915
79 MARC

CICONIIDAE see Storks

CICONIIFORMES

—UNITED STATES

Chlendorf, Harry M
Environmental pollutants and eggshell thickness : anhingas and wading birds in the Eastern United States / by Harry M. Chlendorf, Erwin E. Klaas, T. Earl Kaiser. — Washington : U.S. Dept. of the Interior, Fish and Wildlife Service, 1979.
94 p. : ill. ; 26 cm. — (Special scientific report—wildlife ; no. 216)
Bibliography: p. 21-22.
Supt. of Docs. no.: I 49.15/3:216
1. Ciconiiformes—United States. 2. Anhinga anhinga. 3. Birds—Eggs and nests. 4. Pollution—Environmental aspects—United States. 5. Birds—United States. I. Klaas, Erwin E., joint author. II. Kaiser, T. Earl, joint author. III. Title. IV. Series: United States. Fish and Wildlife Service. Special scientific report—wildlife ; no. 216.
SK361.A256 no. 216 639'.97'90973 s 79-1517
[QL696.C5] 79 MARC

CIDER

see also Wine and wine making

Vechar, Alíaksandr Stsíapanavich, 1905–
(Sidry i íablochnye igristye vina)
Сидры и яблочные игристые вина : химия и технология / А. С. Вечер, Л. А. Юрченко. — Москва : Пищевая промышл., 1976.
134 p. : ill. ; 22 cm. USSR***
Bibliography : p. 124–134.
0.49rub
1. Cider. 2. Fruit wines. I. ÍUrchenko, Liliía Aleksandrovna, joint author. II. Title.
TP563.V4 76–528799

CIECHOCINEK, POLAND

Umiński, Janusz.
Ciechocinek i okolice / Janusz Umiński. — Wyd. 1. — Warszawa : Sport i Turystyka, 1978.
119 p. : ill., maps (1 fold. col.) ; 20 cm. P***
One map inserted.
Bibliography: p. 114-[115]
Includes index.
zł22.00
1. Ciechocinek, Poland. I. Title.
DK4800.C49U48 79-343675
79 MARC

CIECIERSKI, JAN, 1899-

Ciecierski, Jan, 1899-
Mistrzowie i koledzy / Jan Ciecierski. — Wyd. 1. — Warszawa : Wydaw-a Artystyczne i Filmowe, 1978 (Białyst. : BZG)
207, [1] p., [8] leaves of plates : ill. ; 20 cm. P79-1238
zł35.00
1. Actors—Poland—Biography. 2. Ciecierski, Jan, 1899- . I. Title.
PN2859.P65C5 79-384362
79 MARC

CIESZKOWSKI, AUGUST DOŁĘGA, HRABIA, 1814-1894

Hellwig, Jan.
Działalność pedagogiczna Augusta Cieszkowskiego / Jan Hellwig. — Poznań : Wydawn. Naukowe Uniwersytetu im. Adama Mickiewicza, 1978.
177 p., [3] leaves of plates : ill. ; 24 cm. — (Seria Psychologia i pedagogika ; Nr 41) P***
Summary in German.
Bibliography: p. 166-[172]
zł52.00
1. Cieszkowski, August Dołęga, hrabia, 1814-1894. 2. Education—Philosophy. I. Title. II. Series: Seria Psychologia-pedagogika ; nr. 41.
LB675.C53H44 78-382358
78 MARC

Klepacz, Michał, 1893-1967.
Idea Boga w historjozofji Augusta Cieszkowskiego na tle ówczesnych pradów umysłowych / Michał Klepacz. — Kielce : [s.n.], 1933.
429 p. ; 24 cm.
Bibliography: p. [410]-416.
Includes index.
1. Cieszkowski, August Dołęga, hrabia, 1814-1894. 2. God. 3. History—Philosophy. I. Title.
B4691.C54K55 79-376505
79 MARC

Liebich, André, 1948-
Between ideology and utopia : the politics and philosophy of August Cieszkowski / André Liebich. — Dordrecht ; Boston : D. Reidel Pub. Co., c1979.
vii, 390 p. ; 23 cm. — (Sovietica ; v. 39)
Bibliography: p. 301-313.
Includes indexes.
ISBN 9027709238
1. Cieszkowski, August Dołęga, hrabia, 1814-1894. I. Title. II. Series.
B4691.C54L53 199'.438 78-11297
78 MARC

CIGAR BOXES

—UNITED STATES

Hyman, Tony.
Handbook of American cigar boxes / by Tony Hyman ; drawings by the author. — Elmira, N.Y. : Arnot Art Museum, c1979.
ix, 166 p. : ill. ; 32 cm.
Title on spine: Cigar boxes.
Includes indexes.
ISBN 0-932780-00-8 : $25.00
1. Cigar boxes—United States. I. Title. II. Title: Cigar boxes.
NK4696.5.H95 621.7'57 78-67508
00 MARC

CIGAR MANUFACTURE AND TRADE

—BURMA

Hkun Nwe, Maung.
(Sēin myạ nyo ywet thạnat hpet)
စိမ်းမြညိုရွက်သနပ်ဖက် / ခွန်နွယ် ; အဖုံးပန်းချီ ဦးစံတိုး. — ရန်ကုန် : စာပေဗိမာန်, 1976.
82 p., [2] leaves of plates : ill. ; 17 cm. — (ပြည်သူ့လက်စွဲစာစဉ်)
In Burmese.
Series romanized: Pyi thụ let swē sa sin.
Includes bibliographical references.
K1.50
1. Cigar manufacture and trade—Burma. I. Title.
TS2235.B93H54 77–985666

CIGARETTE CARDS

—GREAT BRITAIN—CATALOGS

London Cigarette Card Company.
The London Cigarette Card Company's trade card and post-1945 cigarette card issues catalogue / compiled by Ian A. Laker. — 3rd ed. — Drayton : London Cigarette Card Company, 1978.
viii, 89 p. ; 22 cm. GB78-15742
Cover title: Catalogue of trade card and post-1945 cigarette card issues.
1. Advertising cards—Great Britain—Catalogs. 2. Cigarette cards—Great Britain—Catalogs. I. Laker, Ian Alex. II. Title.
NC1002.C4L66 1978a 769'.5 79-309113
79 MARC

London Cigarette Card Company.
The London Cigarette Card Company's trade card and post-1945 cigarette card issues catalogue / compiled by Ian A. Laker. — 3d ed. — London : The London Cigarette Card Co. Ltd, 1978.
viii, 89 p. ; 22 cm. GB***
Cover title: Catalogue of trade card and post-1945 cigarette card issues.
£1.81
1. Advertising cards—Great Britain—Catalogs. 2. Cigarette cards—Great Britain—Catalogs. I. Laker, Ian Alex. II. Title. III. Title: Trade card and post-1945 cigarette card issues catalogue. IV. Title: Catalogue of trade card and post-1945 cigarette card issues.
NC1002.C4L66 1978 769'.5 78-323266
79 MARC

CIGARETTE HABIT

see also Cigarette smokers

Burton, Dee.
The joy of quitting : how to help young people stop smoking / Dee Burton and Gary Wohl ; [ill. by John Maddalone]. — 1st Collier Books ed. — New York : Collier Books, 1979.
x, 100 p. : ill. ; 24 cm.
SUMMARY: Discusses the reasons why people smoke, the physical effects of smoking, and ways of giving up the habit.
ISBN 0-02-059870-X (pbk.) : $4.95
1. Smoking and youth. 2. Cigarette habit. [1. Smoking] I. Wohl, Gary, joint author. II. Maddalone, John. III. Title. IV. Title: How to help young people stop smoking.
HV5745.B8 1979b 613.8'5 79-4076
79 MARC

Burton, Dee.
The joy of quitting : how to help young people stop smoking / Dee Burton and Gary Wohl. — New York : Macmillan, c1979.
x, 100 p. : ill. ; 24 cm.
SUMMARY: Discusses the reasons why people smoke, the physical effects of smoking, and ways of giving up the habit.
ISBN 0-02-518600-0 : $8.95
1. Smoking and youth. 2. Cigarette habit. [1. Smoking] I. Wohl, Gary, joint author. II. Title. III. Title: How to help young people stop smoking.
HV5745.B8 1979 613.8'5 79-4075
79 MARC

Eckholm, Erik P
Cutting tobacco's toll / Erik Eckholm. — [Washington] : Worldwatch Institute, 1978.
40 p. ; 22 cm. — (Worldwatch paper ; 18)
Includes bibliographical references.
ISBN 0-916468-17-8 : $2.00
1. Tobacco—Toxicology. 2. Cigarette habit. 3. Smoking. 4. Tobacco manufacture and trade. I. Title.
RA1242.T6E25 362.1'9'6865 78-53446
78 MARC

Geisinger, David L
Kicking it : the new way to stop smoking permanently / David L. Geisinger. — 1st ed. — New York : Grove Press : distributed by Random House, 1979, c1978.
191 p. ; 24 cm.
Includes bibliographical references and index.
ISBN 0-394-50291-4 : $8.95
1. Cigarette habit. 2. Behavior therapy. I. Title.
RC567.G44 1979 362.2'9 78-56045
78 MARC

Li, Chia-ying.
(Ni tui hsiang yen chih to shao)
你對香煙知多少 / [著者李佳瑩]. — 台北：全民出版事業公司, 民國67 [1978]
280 p., [5] leaves of plates : ill. ; 20 cm.
"香煙的科學分析"
NT$45.00

1. Cigarette habit. 2. Tobacco—Physiological effect. I. Title.
HV5740.L48 78–843599

C

Lu, Ch'un-ling, fl. 1978–
(Chieh yen hsin li hsüeh)
戒煙心理學 / 盧春齡編著. — 台北：哲志出版社, 民國67 [1978]
3, 127 p. : ill. ; 19 cm. — (哲志叢書 ; 65)
"香煙的沿革及種種戒煙方法"
Series romanized Che chih ts'ung shu.
NT$35.00

1. Cigarette habit. 2. Smoking. I. Title.
HV5740.L8 79–835611

CIGARETTE HABIT (Continued)

O'Farrell, Allanah.
Socio-psychological aspects of cigarette smoking, August 1969-August 1972 : report to the Department of National Health and Welfare / Allanah O'Farrell, F. R. Wake. — Ottawa : Psychology Department, St. Patrick's College, Carleton University, ⌜1973?⌝
ii, 124 leaves ; 28 cm. C***
Bibliography: leaves ⌜114⌝-124.
1. Cigarette habit. 2. Cigarette habit—Bibliography. 3. Smoking—Psychological aspects. 4. Smoking and youth—Canada. 5. Cigarette smokers—Canada. I. Wake, F. R., joint author. II. Canada. Dept. of National Health and Welfare. III. Carleton University. Psychology Dept. IV. Title.
HV5740.O37 362.2'9 79-350028
79 MARC

Prause, Gerhard.
Wer raucht, liebt besser; oder, Das harte Los des Nichtrauchers ⌜von⌝ Gerd P. Ehestorf. Mit Zeichnungen von Dieter Lange. ⌜Hamburg⌝ C. Wegner ⌜c1970⌝
142, ⌜1⌝ p. illus. 20 cm. GDB***
Bibliography: p. ⌜143⌝
ISBN 3-8032-0138-1
1. Cigarette habit. 2. Tobacco—Physiological effect. I. Title. II. Title: Das harte Los des Nichtrauchers.
HV5751.P73 72-534913
79⌜r79⌝rev MARC

—BIBLIOGRAPHY

O'Farrell, Allanah.
Socio-psychological aspects of cigarette smoking, August 1969-August 1972 : report to the Department of National Health and Welfare / Allanah O'Farrell, F. R. Wake. — Ottawa : Psychology Department, St. Patrick's College, Carleton University, ⌜1973?⌝
ii, 124 leaves ; 28 cm. C***
Bibliography: leaves ⌜114⌝-124.
1. Cigarette habit. 2. Cigarette habit—Bibliography. 3. Smoking—Psychological aspects. 4. Smoking and youth—Canada. 5. Cigarette smokers—Canada. I. Wake, F. R., joint author. II. Canada. Dept. of National Health and Welfare. III. Carleton University. Psychology Dept. IV. Title.
HV5740.O37 362.2'9 79-350028
79 MARC

—CONGRESSES

Self-administration of abused substances : methods for study / editor Norman A. Krasnegor. — Rockville, Md. : Dept. of Health, Education, and Welfare, Public Health Service, Alcohol, Drug Abuse, and Mental Health Administration, National Institute on Drug Abuse, Division of Research, 1978.
ix, 246 p. : ill. ; 23 cm. — (NIDA research monograph series ; 20) (DHEW publication ; (ADM) 78-727)
"Based on papers presented at a technical review conducted by Plog Research, inc., Reseda, California, under NIDA contract no. 271-77-3413 ... on February 14 and 15, 1978, in Reston, Virginia."
Includes bibliographies.
1. Drug abuse—Congresses. 2. Obesity—Congresses. 3. Cigarette habit—Congresses. I. Krasnegor, Norman A. II. Plog Research, inc. III. National Institute on Drug Abuse. IV. Series: National Institute on Drug Abuse. NIDA research monograph series ; 20. V. Series: United States. Dept. of Health, Education, and Welfare. DHEW publication ; no. (ADM) 78-727.
RC564.S44 616.8'63 78-63094
79 MARC

—SOCIAL ASPECTS—HISTORY

Sobel, Robert, 1931 (Feb.19)-
They satisfy : the cigarette in American life / Robert Sobel. — 1st ed. — New York : Anchor Press/Doubleday, 1978.
255 p. ; 22 cm.
Bibliography: p. ⌜244⌝-246.
Includes index.
ISBN 0-385-12956-4 : $8.95
1. Cigarette habit—Social aspects—History. 2. United States—Social life and customs. I. Title.
HV5740.S63 394.1'4'0973 77-27618
78 MARC

—GREAT BRITAIN

Royal College of Physicians of London.
Smoking or health : the third report from the Royal College of Physicians of London. — Kent, Eng. : Pitman Medical, 1977, 1978 printing.
128 p. : ill. ; 22 cm. GB***
Includes bibliographical references.
ISBN 0-272-79412-0 : £3.00
1. Tobacco—Toxicology. 2. Smoking. 3. Tobacco habit—Great Britain. 4. Cigarette habit—Great Britain. I. Title.
RA1242.T6R68 1977 616.8'5'0941 78-318490
78 MARC

—SOUTH AFRICA

Van der Burgh, Chris.
Smoking and drinking patterns of Whites, 1975 / C. Van der Burgh. — Pretoria : Human Sciences Research Council, 1978.
v, 42 p. ; 21 cm. — (Report - South African Human Sciences Research Council, Institute for Sociological, Demographic and Criminological Research ; no. S-51) SA***
Summary in Afrikaans.
Bibliography: p. 41-42.
ISBN 0-86965-493-4 : R1.30
1. Cigarette habit—South Africa. 2. Tobacco habit—South Africa. 3. Liquor problem—South Africa. I. Title. II. Series: Human Sciences Research Council. Institute for Sociological, Demographic, and Criminological Research. Report - Institute for Sociological, Demographic, and Criminological Research ; no. S-51.
HV5770.S6V35 362.2'92 79-310650
79 MARC

—UNITED STATES

Rolfe, Burton Phillip.
How to stop smoking before smoking stops you! / by Burton Phillip Rolfe ; with ill. by Steve Menteer. — Chicago : Adams Press, c1978.
118 p. : ill. ; 23 cm.
Includes bibliographical references.
ISBN 0-932636-00-4 : $6.75
1. Cigarette habit—United States. I. Title.
⌜HV5763.R64⌝ 362.2'9 78-61906
79 MARC

United States. National Cancer Institute. Office of Cancer Communications.
The smoking digest : progress report on a nation kicking the habit / ⌜produced by the Office of Cancer Communications, National Cancer Institute⌝. — Bethesda, Md. : The Institute, ⌜1977⌝
vii, 127 p. : ill. ; 19 cm.
Bibliography: p. 109-112.
Includes index.
1. Cigarette smokers—United States. 2. Cigarette habit—United States. 3. Smoking. 4. Tobacco manufacture and trade—United States. I. Title.
HV5760.U55 1977 613.8'5 78-601078
79 MARC

CIGARETTE MANUFACTURE AND TRADE

—UNITED STATES

Doron, Gideon.
The smoking paradox : public regulation in the cigarette industry / Gideon Doron. — Cambridge, Mass. : Abt Books, c1979.
xv, 141 p. : ill. ; 24 cm.
Includes bibliographical references and index.
ISBN 0-89011-531-1
1. Cigarette manufacture and trade—United States. 2. Industry and state—United States. I. Title.
HD9149.C42U535 338.4'7'679730973 79-50400
00 MARC

United States. Congress. House. Committee on Ways and Means. Subcommittee on Miscellaneous Revenue Measures.
Miscellaneous measures to discourage cigarette smuggling : hearing before the Subcommittee on Miscellaneous Revenue Measures of the Committee on Ways and Means, House of Representatives, Ninety-fifth Congress, second session ... March 21, 1978. — Washington : U.S. Govt. Print. Off., 1978.
iii, 118 p. ; 24 cm.
Hearings on H.R. 9667, 9722, 9733, 9763, 9905, 10066, 10579, and 11152.
"Serial 95-80."
1. Cigarette manufacture and trade—United States. 2. Smuggling—United States. I. Title.
KF27.W3446 1978a 345'.73'0233 78-603135
79 MARC

CIGARETTE SMOKERS

—CANADA

O'Farrell, Allanah.
Socio-psychological aspects of cigarette smoking, August 1969-August 1972 : report to the Department of National Health and Welfare / Allanah O'Farrell, F. R. Wake. — Ottawa : Psychology Department, St. Patrick's College, Carleton University, ⌜1973?⌝
ii, 124 leaves ; 28 cm. C***
Bibliography: leaves ⌜114⌝-124.
1. Cigarette habit. 2. Cigarette habit—Bibliography. 3. Smoking—Psychological aspects. 4. Smoking and youth—Canada. 5. Cigarette smokers—Canada. I. Wake, F. R., joint author. II. Canada. Dept. of National Health and Welfare. III. Carleton University. Psychology Dept. IV. Title.
HV5740.O37 362.2'9 79-350028
79 MARC

—GERMANY, WEST—ATTITUDES

Röper, Burkhardt.
Rauchmotivationen Jugendlicher / Burkhardt Röper. — Göttingen : Vandenhoeck und Ruprecht, 1978.
xiii, 231 p. : graphs ; 24 cm. — (Wirtschaftspolitische Studien aus dem Institut für Europäische Wirtschaftspolitik der Universität Hamburg ; 50) GFR78-A
Includes bibliographical references.
ISBN 3-525-12250-0
1. Smoking and youth—Germany, West. 2. Cigarette smokers—Germany, West—Attitudes. 3. Advertising—Cigarettes. I. Title. II. Series: Wirtschaftspolitische Studien ; 50.
HV5745.R64 616.8'65'071 79-355838
79 MARC

—UNITED STATES

United States. National Cancer Institute. Office of Cancer Communications.
The smoking digest : progress report on a nation kicking the habit / ⌜produced by the Office of Cancer Communications, National Cancer Institute⌝. — Bethesda, Md. : The Institute, ⌜1977⌝
vii, 127 p. : ill. ; 19 cm.
Bibliography: p. 109-112.
Includes index.
1. Cigarette smokers—United States. 2. Cigarette habit—United States. 3. Smoking. 4. Tobacco manufacture and trade—United States. I. Title.
HV5760.U55 1977 613.8'5 78-601078
79 MARC

CIGARETTE TAX

—UNITED STATES

United States. Congress. House. Committee on the Judiciary. Subcommittee on Crime.
Cigarette bootlegging : hearings before the Subcommittee on Crime of the Committee on the Judiciary, House of Representatives, Ninety-fifth Congress, second session, on cigarette bootlegging, February 28, March 8, and April 19, 1978. — Washington : U.S. Govt. Print. Off., 1978.
iv, 236 p. ; 24 cm.
"Serial no. 57."
Supt. of Docs. no.: Y 4.J89/1:57
1. Cigarette tax—United States. 2. Smuggling—United States. I. Title.
KF27.J858 1978 364.1'33 79-601113
79 MARC

—VIRGINIA

Virginia. General Assembly. House of Delegates. Committee of Finance. Subcommittee on Tabacco and Cigarette Taxes.
Report of the House Finance Subcommittee on Tobacco and Cigarette Taxes to the Governor and the General Assembly of Virginia. — Richmond : Commonwealth of Virginia, Dept. of Purchases and Supply, 1978.
54 p. ; 25 cm. — (House document - Commonwealth of Virginia ; no. 20)
Includes bibliographical references.
1. Cigarette tax—Virginia. 2. Tobacco—Taxation—Virginia. I. Series: Virginia. General Assembly, 1978. House of Delegates. Document ; no. 20.
J87.V9 1978b, no. 20 328.755'07'208 s 78-622029
⌜HD9149.C42U6⌝ 79 MARC

CIGARETTES

see also Cigarette habit; Cigarette manufacture and trade; Cigarette tax

—ADVERTISING see Advertising —Cigarettes

CIGARS

see also Cigar boxes

Olrik, Paul C
Kunsten at ryge cigarer ⌜af⌝ Paul C. Olrik. ⌜København⌝ Spektrum ⌜1964⌝
125 p. illus. 22 cm.
1. Cigars. I. Title.
TS2260.O44 78-393843
79 MARC

ČIHÁNKOVÁ, JARMILA, 1925-

Čihánková, Jarmila, 1925-
Jarmila Čihánková-tapisérie / ₍text₎ Viera Luxová; ₍aut. reprod. Igor Bačík a arch. J. Cihánkovej₎. — 1. vyd. — Bratislava : Pallas, 1978.
111 p. : ill. (some col.) ; 23 cm. — (Profily ; zv. 43) CzS78
Summary in German and Russian.
Kčs45.00
1. Čihánková, Jarmila, 1925- I. Luxová, Viera. II. Title: Tapisérie.
NK3071.C92C532 1978 79-354479
*79 MARC

CII-10010 (COMPUTER)

—PROGRAMMING—ADDRESSES, ESSAYS, LECTURES

Programozási nyelvek. — 2. kiad. — Budapest : ₍INFELOR₎, 1977.
177 p. ; 20 cm. — (INFELOR közlemények ; 5) H77
Summaries in English and Russian.
CONTENTS: Dettrich, Á. A PL/I nyelv elemzése a fordítóprogram kidolgozásának szempontjából.—Bakos, T. et al. A CII 10010 számítógép alkalmazása, interakív üzemmódban, adatfeldolgozásban.
ISBN 963020682X
1. PL/I (Computer program language)—Addresses, essays, lectures. 2. Compiling (Electronic computers)—Addresses, essays, lectures. 3. CII-10010 (Computer)—Programming—Addresses, essays, lectures. 4. COBOl (Computer program language)—Addresses, essays, lectures. 5. Interactive computer systems—Addresses, essays, lectures. I. Dettrich, Árpád. A PL/I nyelv elemzése a fordítóprogram kidolgozásának szempontjából. 1977. II. Bakos, Tamás. III. A CII 10010 számítógép alkalmazása, interaktiv üzemmódban, adatfeldolgozásban. 1977. IV. Series: Rendszer Technikai Vállalat. INFELOR közlemények ; 5.
QA76.73.P25P76 1977 78-397341
*79 MARC

Č'IJAVAŽE, ALEK'SANDRE, 1917-1971

Finkel'zord, Aleksandra Iosifovna.
(Aleksandr Chidzhavadze)
Александр Чиджавадзе / А. Финкельзорд. — Тбилиси : Хеловнеба, 1976.
35 p., ₍12₎ leaves of plates : ill. ; 16 cm. USSR 77
"Концертный репертуар А. Чиджавадзе": p. 32–34.
Includes bibliographical references.
0.16rub
1. Č'ijavaže, Alek'sandre, 1917–1971. 2. Violinists, violoncellists, etc.—Georgia (Transcaucasia)—Biography.
ML418.C64F5 79–409306

CIKARANG, INDONESIA (BEKASI)

—CITY PLANNING

Indonesia. Direktorat Tata Kota dan Tata Daerah.
Rencana kota Cikarang, salah satu pusat pertumbuhan baru dalam wilayah Jabotabek. — ₍Jakarta₎ : Pemerintah Daerah Tingkat II Kabupaten Bekasi bekerjasama dengan Direktorat Tata Kota dan Tata Daerah, Direktorat Jenderal Cipta Karya, Departemen P.U.T.L., ₍1976₎
2 v. : maps ; 32 x 41 cm.
Cover title.
Prepared by Direktorat Tata Kota dan Tata Daerah.
CONTENTS: 1. Kompilasi data.—2 Analisa dan rencana.
1. Cikarang, Indonesia (Bekasi)—City planning. 2. City planning—Indonesia. I. Bekasi, Indonesia (Kabupaten) II. Title.
HT169.I54C544. 78-941026
79 MARC

CILENTO MOUNTAIN REGION

Rizzi, Filippo, 1769-1850.
Osservazioni statistiche sul Cilento / Filippo Rizzi. — Riedizione. — Casalvelino Scalo : Galzerano, 1978.
70 p. ; 21 cm. — (Edizioni di storia cilentana) (Passato e presente ; 1) It***
Reprint of the 1809 ed. published by Tip. Trani, Napoli.
Bibliography: p. 21-22.
Includes index.
L1500
1. Cilento Mountain region. I. Title. II. Series.
DG975.C4945R59 1978 945'.74 79-348114
79 MARC

—HISTORY

Molinaro, Anna.
Il Cilento tra le due guerre / Anna Molinaro ; nota introduttiva di Attanasio Mozzillo. — 1. ed. — Casalvelino Scalo/₍-Casalvelino₎ : Edizioni di storia cilentana ; Galzerano, 1977.
141 p., ₍4₎ leaves of plates : ill. ; 22 cm. It77-Dec
L2000
1. World War, 1939-1945—Italy—Cilento Mountain region. 2. Cilento Mountain region—History. I. Title.
D763.I82C545 940.53'45 79-350475
*79 MARC

CILIA AND CILIARY MOTIONS

see also Epithelium

CILIARY BODY

see also Pars plana

CILIATA

Corliss, John O
The ciliated protozoa : characterization, classification, and guide to the literature / John O. Corliss. — 2d ed. — Oxford, Eng. ; New York : Pergamon Press, 1979.
xiv, 455 p., ₍1₎ leaf of plates : ill. ; 26 cm. GB***
Bibliography: p. 329-430.
Includes index.
ISBN 0-08-018752-8
1. Ciliata. 2. Ciliata—Bibliography. I. Title.
QL368.A22C67 1979 593'.17 78-41075
79 MARC

—BIBLIOGRAPHY

Corliss, John O
The ciliated protozoa : characterization, classification, and guide to the literature / John O. Corliss. — 2d ed. — Oxford, Eng. ; New York : Pergamon Press, 1979.
xiv, 455 p., ₍1₎ leaf of plates : ill. ; 26 cm. GB***
Bibliography: p. 329-430.
Includes index.
ISBN 0-08-018752-8
1. Ciliata. 2. Ciliata—Bibliography. I. Title.
QL368.A22C67 1979 593'.17 78-41075
79 MARC

—POLAND—POZNAŃ REGION

Szczepanowski, Piotr.
Orzęski epizoiczne na Asellus aquaticus (L) Poznania i okolic / Piotr Szczepanowski. — Poznań : Państwowe Wydawn. Naukowe, 1978.
72 p. : ill. ; 24 cm. — (Prace Komisji Biologicznej, Poznańskie Towarzystwo Przyjaciół Nauk ; t. 46) P***
At head of title: Poznańskie Towarzystwo Przyjaciół Nauk. Wydział Matematyczno-Przyrodniczy.
Summary in English.
Bibliography: p. 65-67.
zł20.00
1. Ciliata—Poland—Poznań region. 2. Parasites—Crustacea. 3. Asellus aquaticus. 4. Isopoda—Poland—Poznań region. I. Title. II. Series: Poznańskie Towarzystwo Przyjaciół Nauk. Komisja Biologiczna. Prace ; t. 46.
QH7.P59 t. 46 574'.08 s 79-347844
₍QL444.M34₎ 79 MARC

CILICIA

—HISTORY—ARMENIAN KINGDOM, 1080-1375

The Cilician kingdom of Armenia / ₍contributors₎ T. S. R. Boase ... ₍et al.₎ ; edited by T. S. R. Boase. — Edinburgh : Scottish Academic Press, 1978.
xii, 206 p., ₍32₎ leaves of plates : ill. ; 22 cm. GB***
Bibliography: p. ₍187₎-196.
Includes index.
ISBN 0-7073-0145-9 : £7.50
1. Cilicia—History—Armenian Kingdom, 1080-1375. I. Boase, Thomas Sherrer Ross, 1898-1974.
DS186.C54 956'.4'01 79-308699
79 MARC

—HISTORY—ARMENIAN KINGDOM, 1080-1375—ADDRESSES, ESSAYS, LECTURES

The Cilician kingdom of Armenia / T. S. R. Boase ... ₍et al.₎ ; edited by T. S. R. Boase. — New York : St. Martin's Press, 1978.
xii, 206 p., ₍33₎ leaves of plates : ill. ; 22 cm.
Bibliography: p. ₍187₎-196.
Includes index.
ISBN 0-312-13895-4
1. Cilicia—History—Armenian kingdom, 1080-1375—Addresses, essays, lectures. I. Boase, Thomas Sherrer Ross, 1898-1974.
DS186.C54 1978b 956'.4 79-106495
79 MARC

CILIGA, ANTON, 1898-

Ciliga, Anton, 1898-
Sam kroz Europu u ratu, 1939-1945 / Ante Ciliga. — Rim : Na pragu sutrašnjice, 1978.
586 p. ; 23 cm. It79-Apr
1. World War, 1939-1945—Personal narratives, Croatian. 2. Ciliga, Anton, 1898- . 3. Croatia—Biography. 4. Yugoslavia—Politics and government. I. Title.
D811.5.C47 1978 79-364474
79 MARC

CILLERO, ANDRÉS, 1934-

Chávarri Porpeta, Raúl, 1929-
Cillero / ₍por₎ Raúl Chávarri. — ₍Madrid₎ : Dirección General del Patrimonio Artístico y Cultural, D.L. 1977.
104 p., 2 leaves : col. ill. ; 17 cm. — (Colección Artistas españoles contemporáneos ; 145 : Serie Pintores) Sp78-Mar
Bibliography: p. 97-104.
ISBN 8436901932 : 100ptas
1. Cillero, Andrés, 1934-
N7113.C55C45 709'.2'4 78-377035
78 MARC

CILUBA LANGUAGE see Luba language

CIMAHI, INDONESIA (BANDUNG)

—CITY PLANNING

Indonesia. Direktorat Tata Kota dan Tata Daerah.
Rencana tata kota Cimahi. — ₍Jakarta₎ : Pemerintah Daerah Tingkat II Kabupaten Bandung bekerjasama dengan Direktorat Tata kota dan Tata Daerah, Direktorat Jenderal Cipta Karya, Departemen P.U.T.L., ₍1977₎
2 v. : maps ; 31 x 42 cm.
Cover title.
1. Cimahi, Indonesia (Bandung)—City planning. 2. City planning—Indonesia. I. Bandung, Indonesia (Kabupaten) II. Title.
HT169.I54C554 78-941023
79 MARC

CIMARRON VALLEY

—BIOGRAPHY

Cimarron family legends. — Perkins, Okla. : Evans Publications, 1978.
432 p. : ill. ; 34 cm.
Contributing editors, P. Skaggs and others.
1. Cimarron Valley—Genealogy. 2. Cimarron Valley—Biography. I. Skaggs, Pat.
F702.C53C55 929'.2'0973 79-100803
79 MARC

—GENEALOGY

Cimarron family legends. — Perkins, Okla. : Evans Publications, 1978.
432 p. : ill. ; 34 cm.
Contributing editors, P. Skaggs and others.
1. Cimarron Valley—Genealogy. 2. Cimarron Valley—Biography. I. Skaggs, Pat.
F702.C53C55 929'.2'0973 79-100803
79 MARC

CIMATTI, VINCENZO

Cimatti, Vincenzo.
Lettere di un missionario / Vincenzo Cimatti ; a cura di Alfonso Crevacore. — Torino Leumann : Elle Di Ci, [1976?]
126 p. ; 21 cm. It77-Feb
L1300
1. Cimatti, Vincenzo. 2. Missionaries—Japan—Correspondence. 3. Missionaries—Italy—Correspondence. I. Crevacore, Alfonso. II. Title.
BV3457.C47A34 79-383673
*79 MARC

CIMETIDINE

—CONGRESSES

Cimetidine : The Westminster Hospital Symposium / edited by Christopher Wastell and Peter Lance. — Edinburgh ; New York : Churchill Livingstone ; New York : distributed by Longman, 1978.
xix, 302 p. : ill. ; 24 cm. GB78
"Symposium held in Westminster Medical School on May 4th and 5th, 1978."
Includes bibliographical references and index.
ISBN 0-443-01882-0 : £13.00
1. Duodenum—Ulcers—Chemotherapy—Congresses. 2. Cimetidine—Congresses. 3. Dyspepsia—Chemotherapy—Congresses. 4. Stomach—Ulcers—Chemotherapy—Congresses. I. Wastell, C. II. Lance, Peter. III. London. Westminster Hospital. IV. Title.
RC824.C55 615'.735 78-40801
78 MARC

International Symposium on Histamine H_2-Receptor Antagonists, Göttingen, Federal Republic of Germany, 1977.
Cimetidine : proceedings of an International Symposium on Histamine H_2-Receptor Antagonists, 10th-11th November, 1977, Göttingen, Federal Republic of Germany / editor, W. Creutzfeldt. — Amsterdam : Excerpta Medica ; New York : sole distributors for the USA and Canada, Elsevier North-Holland, 1978.
viii, 319 p. : ill. ; 24 cm. — (International congress series ; no. 443)
Includes bibliographical references.
ISBN 0-444-90048-9
1. Peptic ulcer—Chemotherapy—Congresses. 2. Cimetidine—Congresses. 3. Stomach—Secretions—Congresses. I. Creutzfeldt, Werner. II. Title. III. Series. IV. Series: International congress series.
RC821.I59 1977 616.3'43'061 78-19025
78 MARC

CIMINI HILLS, ITALY

—DESCRIPTION AND TRAVEL

Bertarelli, Franco.
I Colli Cimini / a cura di Franco Bertarelli e di Bruno Barbini. — Novara : Istituto geografico De Agostini, 1977.
64 p. : ill. ; 30 cm. — (I Documentari : Visioni d'Italia) It***
1. Cimini Hills, Italy—Description and travel. I. Barbini, Bruno, joint author. II. Title.
DG975.C49467B47 914.5'625'0492 78-387868
79 MARC

CIMIOTTI, EMIL, 1927-

—CATALOGS

Cimiotti, Emil, 1927-
Emil Cimiotti : Werkverz. d. Plastiken 1955-1977 / d. Werkverz. wurde erstellt u. bearb. von Dieter Blume ; Texte von Eduard Trier ... [et al.] ; hrsg. von Dieter Brusberg. — Hannover : Verlag der Galerie Brusberg, 1978.
177 p. : chiefly ill. (some col.) ; 34 cm. — (Brusberg Dokumente ; 10) GFR79-A
Bibliography: p. 174-176.
ISBN 3-87972-039-8 : DM60.00
1. Cimiotti, Emil, 1927- —Catalogs. I. Blume, Dieter, fl. 1978- . II. Brusberg, Dieter. III. Series: Galerie Dieter Brusberg, Hannover. Brusberg Dokumente ; 10.
NB588.C56A4 1978 730'.92'4 79-363516
79 MARC

CÎMPULUNG, ROMANIA, BATTLE OF, 1916

Atanasiu, Victor.
Bătălia din zona Bran-Cîmpulung / Victor Atanasiu. — [Bucureşti] : Editura Militară, 1976.
140 p., [13] leaves of plates (1 fold.) : ill., col. maps (1 fold.) ; 17 cm. R77-1262
Bibliography: p. 138-[139]
lei5.50
1. Bran, Romania, Battle of, 1916. 2. Cîmpulung, Romania, Battle of, 1916. I. Title.
D565.B7A85 79-340574
*79 MARC

CÎMPULUNG MOLDOVENESC, ROMANIA

—DESCRIPTION—GUIDE-BOOKS

Iosep, I
Cîmpulung Moldovenesc / I. Iosep, M. Iacobescu. — Bucureşti : Editura "Sport-Turism", 1978.
93 p., 13 leaves of plates (1 fold.) : ill., fold. col. map ; 17 cm. — (Mic îndreptar turistic) R78-4348
lei7.00
1. Cîmpulung Moldovenesc, Romania—Description—Guide-books. I. Iacobescu, M., joint author. II. Title.
DR296.C5I55 79-337914
*79 MARC

CINCARI see Arumanians

CINCINNATI

—DESCRIPTION

Langsam, Walter Consuelo, 1906-
Cincinnati in color / text and notes on the ill. by Walter Consuelo Langsam ; a collection of color photos. by Julianne Warren. — New York : Hastings House, 1978, 1979 printing.
94 p. : ill. ; 25 cm. — (Profiles of America)
ISBN 0-8038-1248-5 : $5.95
1. Cincinnati—Description. 2. Cincinnati—History. I. Warren, Julianne. II. Title.
F499.C54L35 1978 977.1'78 78-2543
78 MARC

—HISTORY

Langsam, Walter Consuelo, 1906-
Cincinnati in color / text and notes on the ill. by Walter Consuelo Langsam ; a collection of color photos. by Julianne Warren. — New York : Hastings House, 1978, 1979 printing.
94 p. : ill. ; 25 cm. — (Profiles of America)
ISBN 0-8038-1248-5 : $5.95
1. Cincinnati—Description. 2. Cincinnati—History. I. Warren, Julianne. II. Title.
F499.C54L35 1978 977.1'78 78-2543
78 MARC

CINCINNATI. ART MUSEUM

—CATALOGS

Burt, Stanley G
2,292 pieces of early Rookwood pottery in the Cincinnati Art Museum in 1916 : Mr. S. G. Burt's record book of ware at Art Museum. — Cincinnati : Cincinnati Historical Society, c1978.
193 p. (p. 193 blank) ; 22 cm.
1. Rookwood Pottery Company—Catalogs. 2. Pottery, American—Ohio—Cincinnati—Catalogs. 3. Cincinnati. Art Museum—Catalogs. I. Cincinnati. Art Museum. II. Title.
NK4210.R55A4 1978 738.3'7 78-75042
79 MARC

Cincinnati. Art Museum.
Spanish paintings in the Cincinnati Art Museum / Millard F. Rogers, Jr. — Cincinnati : The Museum, 1978.
79 p. : ill. (some col.) ; 26 cm.
1. Painting, Spanish—Catalogs. 2. Painting—Ohio—Cincinnati—Catalogs. 3. Cincinnati. Art Museum—Catalogs. I. Rogers, Millard F. II. Title.
ND801.C56 1978 759.6'074'017178 78-112887
79 MARC

CINCINNATI BENGALS (FOOTBALL CLUB)

Brown, Paul E 1908-
PB, the Paul Brown story / Paul Brown with Jack Clary. — 1st ed. — New York : Atheneum, 1979.
338 p. : ill. ; 24 cm.
ISBN 0-689-10985-7 : $12.95
1. Brown, Paul E., 1908- . 2. Football coaches—United States—Biography. 3. Cleveland. Football club (National League, Browns) 4. Cincinnati Bengals (Football club) I. Clary, Jack T., joint author. II. Title.
GV939.B77A35 1979 796.33'2'0924 79-7314
79 MARC

CINCINNATI FINANCIAL CORPORATION

—HISTORY—ADDRESSES, ESSAYS, LECTURES

Schiff, John J
Cincinnati Financial Corporation : keeping every promise / John J. Schiff, Thomas J. Smart, Robert B. Morgan. — New York : Newcomen Society in North America, 1978.
20 p. ; 23 cm. — (Newcomen publication ; no. 1083)
Addresses delivered at the 1978 Cincinnati dinner of the Newcomen Society in North America, April 26th, 1978.
1. Cincinnati Financial Corporation—History—Addresses, essays, lectures. I. Smart, Thomas J., joint author. II. Morgan, Robert B., joint author. III. Series.
HG8540.C5S34 368'.006'577178 78-61403
78 MARC

CINCINNATI SYMPHONY ORCHESTRA

—DISCOGRAPHY

Fellers, Frederick P
Discographies of commercial recordings of the Cleveland Orchestra (1924-1977) and the Cincinnati Symphony Orchestra (1917-1977) / compiled by Frederick P. Fellers and Betty Meyers. — Westport, Conn. : Greenwood Press, 1978.
xii, 211 p. ; 24 cm.
Includes bibliographies and indexes.
ISBN 0-313-20375-X : $22.50
1. Cleveland Orchestra—Discography. 2. Cincinnati Symphony Orchestra—Discography. I. Meyers, Betty, joint author. II. Title.
ML156.7.C48F4 016.7899'12 78-3122
78 MARC

CINECITTÀ, S.P.A

Cinecittà / [a cura di Maria Cecchi e Sandro Reanda]. — [Roma] : Italia industriale editrice, [1939]
98, [44] p. : ill. ; 22 cm.
1. Cinecittà, s.p.a. 2. Moving-pictures—Italy. I. Cecchi, Maria. II. Reanda, Sandro.
PN1999.C5C524 791.43'0945'63 79-366253
79 MARC

CINEMA see Moving-pictures

CINEMASCOPE see Wide-screen processing (Cinematography)

CINEMATOGRAPH FILM see Moving-picture film

CINEMATOGRAPHERS

—BRAZIL—BIOGRAPHY

Humberto Mauro : sua vida, sua arte, sua trajetória no cinema : depoimentos sobre a riqueza da filmografia maureana e sua importância na moderna cultura brasileira / pref. de Leandro Tocantins. — Rio de Janeiro : Editora Artenova, c1978.
360 p., [16] leaves of plates : ill. ; 21 cm. — (Cinebiblioteca Embrafilme ; v. 3)
"Em convênio com a Empresa Brasileira de Filmes S.A., EMBRAFILME."
1. Mauro, Humberto, 1897- . 2. Cinematographers—Brazil—Biography.
TR849.M38H85 778.5'3'0924 79-341368
79 MARC

—CALIFORNIA—BIOGRAPHY

Cortez, Stanley.
Stanley Cortez : an American Film Institute seminar on his work. — Glen Rock, N. J. : Microfilming Corp. of America, 1977.
1 sheet (76 p.) ; 11 x 15 cm. — (The American Film Institute seminars ; pt. 1, no. 35) (New York times oral history program)
Microfiche.
ISBN 0-667-00354-1
1. Cortez, Stanley. 2. Cinematographers—California—Biography. 3. The Magnificent Ambersons (Motion picture) I. Series.
Microfiche TR849.C67 778.5'3'0924 [B] 77-12749

Dunn, Linwood.
Linwood Dunn : an American Film Institute seminar on his work. — Glen Rock, N. J. : Microfilming Corp. of America, 1977.
1 sheet (86 p.) ; 11 x 15 cm. — (The American Film Institute seminars ; pt. 1, no. 47) (New York times oral history program)
Microfiche.
ISBN 0-667-00366-5
1. Dunn, Linwood. 2. Cinematography, Trick. 3. Cinematographers—California—Biography. I. Series.
Microfiche TR849.D85 778.5'3'0924 [B] 77-12750

Garmes, Lee.
Lee Garmes : an American Film Institute seminar on his work. — Glen Rock, N. J. : Microfilming Corp. of America, 1977.
2 sheets (105 p.) ; 11 x 15 cm. — (The American Film Institute seminars ; pt. 1, no. 64) (New York times oral history program)
Microfiche.
ISBN 0-667-00383-5
1. Garmes, Lee. 2. Cinematographers—California—Biography. I. Series.
Microfiche TR849.G34 778.5'3'0924 [B] 77-14413

Hoch, Winton C
Winton C. Hoch : an American Film Institute seminar on his work. — Glen Rock, N. J. : Microfilming Corp of America, 1977.
1 sheet (62 p.) ; 11 x 15 cm. — (The American Film Institute seminars ; pt. 1, no. 82) (New York times oral history program)
Microfiche.
ISBN 0-667-00401-7
1. Hoch, Winton C. 2. Cinematographers—California—Biography. I. Series.
Microfiche TR849.H6 778.5'3'0924 [B] 77-12752

Larner, Stevan.
Stevan Larner : an American Film Institute seminar on his work. — Glen Rock, N. J. : Microfilming Corp. of America, 1977.
1 sheet (37 p.) ; 11 x 15 cm. — (The American Film Institute seminars ; pt. 1, no. 105) (New York times oral history program)
Microfiche.
ISBN 0-667-00424-6
1. Larner, Stevan. 2. Cinematographers—California—Biography. I. Series.
Microfiche TR849.L37 778.5'3'0924 [B] 77-12754

Voigtlander, Ted.
Ted Voigtlander : an American Film Institute seminar on his work. — Glen Rock , N. J. : Microfilming Corp. of America, 1977.
1 sheet (44 p.) ; 11 x 15 cm. — (The American Film Institute seminars ; pt. 1, no. 182) (New York times oral history program)
Microfiche.
ISBN 0-667-00501-3
1. Voigtlander, Ted. 2. Cinematographers—California—Biography. I. Series.
Microfiche TR849.V64 77-12409
778.5'3'0924 [B]

Zsigmond, Vilmos.
Vilmos Zsigmond : an American Film Institute seminar on his work. — Glen Rock, N. J. : Microfilming Corp. of America, 1977.
2 sheets (75 p.) ; 11 x 15 cm. — (The American Film Institute seminars ; pt. 1, no. 198) (New York times oral history program)
Microfiche.
Sheet 2 is a reproduction of the Institute's Dialogue on film, v. 3, no. 7)
ISBN 0-667-00517-X
1. Zsigmond, Vilmos. 2. Cinematographers—California—Biography. I. Series.
Microfiche TR849.Z85 77-12236
778.5'3'0924 [B]

—CALIFORNIA—HOLLYWOOD —BIOGRAPHY

Clothier, William.
William Clothier : an American Film Institute seminar on his work. — Glen Rock, N. J. : Microfilming Corp. of America, 1977.
1 sheet (56 p.) ; 11 x 15 cm. — (The American Film Institute seminars ; pt. 1, no. 34) (New York times oral history program)
Microfiche.
ISBN 0-667-00353-3
1. Clothier, William. 2. Cinematographers — California — Hollywood—Biography. I. Series.
Microfiche TR849.C56 778.5'3'0924 [B] 77-12748

Guffey, Burnett.
Burnett Guffey: an American Film Institute seminar on his work. — Glen Rock, N. J. : Microfilming Corp. of America, 1977.
1 sheet (67 p.) ; 11 x 15 cm. — (The American Film Institute seminars ; pt. 1, no. 71) (New York times oral history program)
Microfiche.
ISBN 0-667-00390-8
1. Guffey, Burnett. 2. Cinematographers—California—Hollywood—Biography. I. Series.
Microfiche TR849.G84 77-12418
778.5'3'0924 [B]

Kline, Richard.
Richard Kline : an American Film Institute seminar on his work. — Glen Rock, N. J. : Microfilming Corp. of America, 1977.
1 sheet (71 p.) ; 11 x 15 cm. — (The American Film Institute seminars ; pt. 1, no. 98) (New York times oral history program)
Microfiche.
ISBN 0-667-00417-3
1. Kline, Richard. 2. Cinematographers—California—Hollywood—Biography. I. Series.
Microfiche TR849.K54 77-12415
778.5'3'0924 [B]

Kovacs, Laszlo, 1933–
Laszlo Kovacs : an American Film Institute seminar on his work. — Glen Rock, N. J. : Microfilming Corp. of America, 1977.
1 sheet (56 p.) ; 11 x 15 cm. — (The American Film Institute seminars ; pt. 1, no. 99) (New York times oral history program)
Microfiche.
ISBN 0-667-00418-1
1. Kovacs, Laszlo, 1933- 2. Cinematographers—California—Hollywood—Biography. I. Series.
Microfiche TR849.K68 77-12413
778.5'3'0924 [B]

Krasner, Milton.
Milton Krasner : an American Film Institute seminar on his work. — Glen Rock, N. J. : Microfilming Corp. of America, 1977.
1 sheet (51 p.) ; 11 x 15 cm. — (The American Film Institute seminars ; pt. 1, no. 102) (New York times oral history program)
Microfiche.
ISBN 0-667-00421-1
1. Krasner, Milton. 2. Cinematographers — California — Hollywood—Biography. I. Series.
Microfiche TR849.K7 778.5'3'0924 [B] 77-12094

Lang, Charles B
Charles B. Lang, Jr. : an American Film Institute seminar on his work. — Glen Rock, N. J. : Microfilming Corp. of America, 1977.
1 sheet (64 p.) ; 11 x 15 cm. — (The American Film Institute seminars ; pt. 1, no. 104) (New York times oral history program)
Microfiche.
ISBN 0-667-00423-8
1. Lang, Charles B. 2. Cinematographers—California—Hollywood—Biography. I. Series.
Microfiche TR849.L35 77-12411
778.5'3'0924 [B]

Mohr, Hal.
Hal Mohr : an American Film Institute seminar on his work. — Glen Rock, N. J. : Microfilming Corp. of America, 1977.
1 sheet (77 p.) ; 11 x 15 cm. — (The American Film Institute seminars ; pt. 1, no. 123) (New York times oral history program)
Microfiche.
ISBN 0-667-00442-4
1. Mohr, Hal. 2. Cinematographers — California — Hollywood — Biography. I. Series.
TR849.M64 778.5'3'0924 [B] 77-12301

Polito, Gene.
Gene Polito : an American Film Institute seminar on his work. — Glen Rock, N. J. : Microfilming Corp of America, 1977.
1 sheet (64 p.) ; 11 x 15 cm. — (The American Film Institute seminars ; pt. 1, no. 135) (New York times oral history program)
Microfiche.
ISBN 0-667-00454-8
1. Polito, Gene. 2. Cinematographers—California—Hollywood—Biography. I. Series.
Microfiche TR849.P64 77-12410
778.5'3'0924 [B]

Stanley, Frank, 1923?–
Frank Stanley : an American Film Institute seminar on his work. — Glen Rock, N. J. : Microfilming Corp. of America, 1977 [i. e. 1978]
1 sheet (79 p.) ; 11 x 15 cm. — (The American Film Institute seminars ; pt. 1, no. 158) (New York times oral history program)
Microfiche.
ISBN 0-667-00477-1
1. Stanley, Frank, 1923?- 2. Cinematographers—California—Hollywood—Biography. I. Series.
Microfiche TR849.S83 778.5'3'0924 [B] 78-14958

—CANADA—BIOGRAPHY

Reid, Alison.
Richard Leiterman / by Alison Reid, with an interview by P. M. Evanchuck. — Ottawa : Canadian Film Institute, c1978.
120 p. : ill. ; 21 cm. — (Canadian film series ; 3 ISSN 0705-548X) C***
Filmography: p. [93]-115.
Bibliography: p. [117]-119.
ISBN 0-919096-02-6 : $5.50
1. Leiterman, Richard. 2. Cinematographers—Canada—Biography. I. Evanchuck, P. M., joint author. II. Series.
TR849.L44R44 778.5'3'0924 79-309768
79 MARC

—JAPAN—BIOGRAPHY

Hani, Susumu, 1928–
(Hani Susumu no Afurika nikki)
羽仁進のアフリカ日記 / 羽仁進. — 東京：平凡社, 1978.
285 p., [7] leaves of plates : ill. ; 19 cm. Ja 78-4848
¥1000
1. Africa—Description and travel—1951- 2. Wildlife cinematography. 3. Hani, Susumu, 1928- 4. Cinematographers—Japan—Biography. I. Title.
DT12.2.H29 78-809283

—NETHERLANDS—BIOGRAPHY

Joris Ivens : cinquant'anni di cinema / a cura di Giacomo Gambetti. — 1. ed. — Venezia : Marsilio, 1979.
90 p. : ill. ; 23 cm. It79-Apr
At head of title: Comune di Modena, Assessorato ai servizi culturali; Festival dei popoli.
Catalog of the exhibition held in Modena in 1979.
Filmography: p. 73-79.
L5000
1. Ivens, Joris, 1898- —Exhibitions. 2. Cinematographers—Netherlands—Biography. I. Gambetti, Giacomo, 1932- . II. Assessorato ai servizi culturali, Modena.
TR849.I82J67 79-373191
*79 MARC

Laan, Dick.
Dick Laan over film : herinneringen en belevenissen van een oud-filmer. — Haarlem : Focus, [1964]
146 p., [1] fold. leaf of plates : ill. ; 20 cm.
Filmography: p. 141-142.
1. Laan, Dick. 2. Cinematographers—Netherlands—Biography. 3. Cinematography. I. Title.
TR849.L32A35 78-389516
79 MARC

—RUSSIA—BIOGRAPHY

(Desiât' operatorskikh biografiĭ)
Десять операторских биографий : [Сб. статей / Сост. М. Голдовская ; Предисл. В. Монахова]. — Москва : Искусство, 1978–
v. : ill. ; 17 cm. — (Мастера советского театра и кино)
USSR 78
Series romanized: Mastera sovetskogo teatra i kino.
1.20rub (v. 1)
1. Cinematographers — Russia — Biography. 2. Moving-pictures—Russia. I. Goldovskaiâ, Marina Evseevna.
TR849.A1D47 78-421551

CINEMATOGRAPHERS (Continued)

—UNITED STATES—BIOGRAPHY

Leacock, Richard.
Richard Leacock : an American Film Institute seminar on his work. — Glen Rock, N.J. : Microfilming Corp. of America, 1977 ⌜i.e. 1978⌝
1 sheet (37 p.) ; 11 x 15 cm. — (The American Film Institute seminars ; pt. 1, no. 107) (New York times oral history program)
ISBN 0-667-00426-2
1. Leacock, Richard. 2. Cinematographers—United States—Biography. I. Series.
TR849.L4 778.5'3'0924 78-14949
79 MARC

CINEMATOGRAPHIC FILM see Moving-picture film

CINEMATOGRAPHY

see also Amateur moving-pictures; Cinematography, Trick; Color cinematography; Moving-picture film; Wide-screen processing (cinematography)

Bacher, Lutz.
The mobile mise en scene : a critical analysis of the theory and practice of long-take camera movement in the narrative film / Lutz Bacher. — New York : Arno Press, 1978.
304 p. ; 24 cm. — (Dissertations of film series)
Originally presented as the author's thesis (M.A.), Wayne State University.
Bibliography: p. 265-292.
Includes indexes.
ISBN 0-405-10750-1
1. Cinematography. 2. Moving-pictures—Production and direction. I. Title. II. Series.
TR850.B3 1978 778.5'3 77-22904
77 MARC

Bomback, R H
Cine-film and how to expose it / R. H. Bomback. — London : Fountain Press, 1950.
p. 52-96 : ill. ; 19 cm. — (Cinefacts ; no. 2)
Includes bibliographical references.
1. Cinematography. I. Title. II. Series.
TR851.B59 78-323090
79 MARC

Cleave, Alan.
Your instant guide to movie making / ⌜by⌝ Alan Cleave. — Watford : Fountain Press, 1977.
63 p. : ill. ; 22 cm. GB78-15121
Bibliography: p. ⌜59⌝
Includes index.
ISBN 0-85242-499-X : £1.95
1. Cinematography. I. Title.
TR851.C52 778.5'3 79-301805
79 MARC

D'Agostino, Peter, 1945-
Alpha, Trans, Chung, a photographic model : semiotics, film, and interpretation / Peter D'Agostino. — Dayton, Ohio : University Art Galleries, Wright State University ; San Francisco : distributed by NFS Press, c1978.
64 p. : ill. ; 23 x 31 cm.
Bibliography: p. 62.
ISBN 0-917986-09-1 : $7.95
1. Photography, Artistic. 2. Cinematography. 3. Semiotics. I. Title.
TR654.D28 779'.092'4 78-73214
00 MARC

Durand, Philippe, 1932-
L'Acteur et la caméra / Philippe Durand ; schémas de Michel Dery. — Paris : Éditions techniques européennes, 1975.
257 p. : ill. ; 19 cm. — (Première syntaxe pour une direction des acteurs au cinéma et à la télévision ; 1) F75-4709
Includes index.
ISBN 2-85407-001-1 : 75.00F
1. Moving-pictures—Production and direction. 2. Cinematography. 3. Moving-picture acting. I. Title.
PN1995.9.P7D8 vol. 1 791.43 75-513734
75⌜r79⌝rev MARC

Durand, Philippe, 1932-
Première syntaxe pour une direction des acteurs au cinéma et à la télévision / Philippe Durand ; schémas de Michel Dery. — Paris : Éditions techniques européennes, 1975-
v. : ill. ; 19 cm. F75-4709
Includes index.
CONTENTS: 1. L'acteur et la caméra.—2. Mise en scène et imagination.

ISBN 2-85407-001-1 (v. 1) : 75.00F (v. 1)
1. Moving-pictures—Production and direction. 2. Cinematography. 3. Moving-picture acting. I. Title.
PN1995.9.P7D8 791.43 79-364449
79 MARC

Freytag, Heinrich.
Schmalfilmbuch der Büchergilde / Heinrich Freytag. — Frankfurt am Main : Büchergilde Gutenberg, c1970, 1975 printing.
227 p. : 80 ill. (some col.) ; 24 cm. GFR***
ISBN 3-7632-1485-2
1. Cinematography. I. Büchergilde Gutenberg, Frankfurt am Main. II. Title.
TR850.F74 1975 778.5'3 76-489901
77⌜r79⌝rev MARC

Hacker, Leonard.
Cinematic design / Leonard Hacker. — New York : Arno Press, 1978, ⌜1931⌝
193 p., ⌜13⌝ leaves of plates : ill. ; 21 cm. — (Aspects of film)
Reprint of the ed. published by American Photographic Pub. Co., Boston.
ISBN 0-405-11131-2
1. Cinematography. 2. Moving-picture plays. I. Title. II. Series.
TR850.H3 1978 778.5'3 77-11375
78 MARC

Harpole, Charles Henry.
Gradients of depth in the cinema image / Charles Henry Harpole. — New York : Arno Press, 1978.
vi, 293 p. ; 24 cm. — (Dissertations on film series) (The Arno Press cinema program)
Originally presented as the author's thesis, New York University, 1976.
Bibliography: p. 256-279.
Includes index.
ISBN 0-405-10753-6
1. Moving-pictures—Aesthetics. 2. Cinematography. 3. Depth perception. I. Title. II. Series. III. Series: The Arno Press cinema program.
PN1995.H367 1978 791.43'015 77-22907
77 MARC

Laan, Dick.
Dick Laan over film : herinneringen en belevenissen van een oud-filmer. — Haarlem : Focus, ⌜1964⌝
146 p., ⌜1⌝ fold. leaf of plates : ill. ; 20 cm.
Filmography: p. 141-142.
1. Laan, Dick. 2. Cinematographers—Netherlands—Biography. 3. Cinematography. I. Title.
TR849.L32A35 78-389516
79 MARC

Lafrance, André.
Initiation à l'équipe de tournage / André A. Lafrance. — Montréal : Presses de l'Université de Montréal, c1978.
139 p. : ill. ; 19 cm. C***
Bibliography: p. ⌜135⌝-136.
ISBN 0-8405-0398-9 : $7.75
1. Cinematography. I. Title.
TR851.L234 778.5'3 78-392155
79 MARC

Lipton, Leonard.
Lipton filmmaking / by Lenny Lipton ; edited by Chet Roaman. — New York : Simon and Schuster, c1979.
223 p. : ill. ; 24 cm. — (A Fireside book)
ISBN 0-671-24427-2
1. Cinematography. I. Roaman, Chet, 1939- . II. Title.
TR850.L622 778.5 79-215
79 MARC

Maltin, Leonard.
The art of the cinematographer : a survey and interviews with five masters / by Leonard Maltin. — Dover ed., corr. and enl. — New York : Dover Publications, c1978.
140 p. : ill. ; 28 cm.
First ed. published in 1971 under title: Behind the camera.
Includes index.
ISBN 0-486-23686-2 : $5.00
1. Cinematography. I. Title.
TR850.M277 1978 778.5'3'0922 78-56838
79 MARC

Natta, Enzo.
Il linguaggio dell'immagine : un programma di educazione alla lettura del film / Enzo Natta. — Torino : Edizioni paoline, ⌜1977⌝
175 p. : ill. ; 24 cm. — (Comunicazione/Sussidi ; 1) It***
L3500
1. Cinematography. I. Title.
TR850.N26 778.5'3 79-349826
79 MARC

Nespolo, Ugo, 1941-
La fugace vita dei fotogrammi / Ugo Nespolo ; a cura di Vittorio Fagone. — Padova : Mastrogiacomo-Images 70, 1978.
124 p. : ill. ; 22 cm. It78-June
English and Italian.
Filmography: p. 121
Bibliography: p. 113-115
L5000
1. Cinematography. 2. Experimental films. I. Fagone, Vittorio. II. Title.
TR899.N47 778.5'3'0924 78-377187
*78 MARC

(Pa mi li she ying chi ch'iao)
八米厘攝影技巧. — 台北：黃祈桊：總經銷上品圖書社, 民國67 ⌜1978⌝
152 p. : ill. ; 27 cm.
NT$200.00

1. Cinematography. 2. Amateur moving-pictures.
TR850.P3 79-837090

Pop, Rodica.
Filmul de amatori : elemente de tehnică și cultură cinematografică / Rodica Pop, Dumitru Codăus. — București : Editura tehnică, 1976.
2 v. : ill. (some col.) ; 20 cm. — (Colecția Foto-film ; 26-27) R76-2865
Bibliography: v. 2, p. 110-⌜111⌝
lei16.00
1. Cinematography. 2. Amateur moving-pictures. I. Codăuș, Dumitru, joint author. II. Title.
TR851.P67 78-371013
*78 MARC

Samuelson, David W
Motion picture camera techniques / David W. Samuelson. — London : Focal Press ; New York : Hastings House, 1978.
200 p. : ill. ; 22 cm. — (Media Manuals) GB***
Bibliography: p. 200.
ISBN 0-8038-4700-9 : £2.95
1. Cinematography. I. Title.
TR850.S315 778.5'3 79-304207
79 MARC

Schwartz, Howard, 1919–
Howard Schwartz : an American Film Institute seminar on his work. — Glen Rock, N. J. : Microfilming Corp. of America, 1977.
1 sheet (59 p.) ; 11 x 15 cm. — (The American Film Institute seminars ; pt. 1, no. 160) (New York times oral history program)
Microfiche.
ISBN 0-667-00479-3
1. Schwartz, Howard, 1919– 2. Cinematography. I. Series.
Microfiche TR849.S39 77–14249
778.5'3'0924 [B]

Tritapepe, Rodolfo.
Linguaggio e tecnica cinematografica / Rodolfo Tritapepe. — Alba : Edizioni paoline, 1978.
183 p. : ill. ; 21 cm. & supplementary material in pocket. — (Comunicazione : tecniche ; 2) It78-June
L2500
1. Cinematography. I. Title. II. Series.
TR850.T73 778.5'3 79-337033
*79 MARC

Watkins, Derek.
Cine photography made easy / Derek Watkins. — Newton Abbot ; North Pomfret, Vt. : David and Charles, 1977.
126 p. : ill. ; 22 cm. — (David & Charles leisure & travel series) GB77-24389
Bibliography: p. 124.
Includes index.
ISBN 0-7153-7437-0 : £1.95
1. Cinematography. I. Title.
TR851.W36 778.5'3 77-74354
79 MARC

—CONGRESSES

Union internationale des associations techniques cinématographiques.
11. ⌜i.e. Undicesimo⌝ congresso UNIATEC : ⌜atti del congresso⌝ : 9. Concorso tecnico del film, Salerno, Italia 1974. — Roma : ATIC, ⌜1975⌝
241 p. : ill. ; 20 cm. It***
English, French, German, Italian, or Russian.
Includes bibliographies.
1. Cinematography—Congresses. I. Concorso tecnico del film, 9th, Salerno, 1974. II. Title.
TR846.U59 1975 778.5'3 78-379943
78 MARC

CINEMATOGRAPHY (Continued)

—DICTIONARIES—JAPANESE

Yamagishi, Tatsuji, 1929–
(Eizō yōgo jiten)
映像用語事典：現場の知恵と戦略 / 山岸達児. — 東京：じゃこめてい出版, 1978.
227 p. : ill. ; 20 cm. Ja 78-9992
¥1500

1. Moving-pictures—Dictionaries—Japanese. 2. Cinematography—Dictionaries—Japanese. 3. Japanese language—Dictionaries—English. I. Title.
PN1993.45.Y35 78-812050

—DICTIONARIES—POLYGLOT

Pozsonyi, Gábor.
Filmgyártás és filmtechnika / Pozsonyi Gábor. — Budapest : Akadémiai Kiadó, 1975.
220 p. ; 21 cm. — (Műszaki értelmező szótár ; 36)
Hungarian, English, French, German and Russian.
Includes index.
ISBN 9630505924 : 50.00Ft
1. Cinematography—Dictionaries—Polyglot. 2. Hungarian language—Dictionaries—Polyglot. I. Title. II. Series.
TR847.P69 78-391538
79 MARC

—HISTORY

Ham, Piet van der.
Op het spoor van de cinema / Piet van der Ham ; voorwoord door Rico Bulthuis. — Den Haag : Leopold, [1978] Ne78-10
88 p. : ill. ; 21 cm.
ISBN 9025806015 : fl 16.90
1. Cinematography—History. 2. Moving-pictures—History. I. Title.
TR848.H35 78-368164
•78 MARC

Mikalauskas, Vytautas.
Trumpa, bet nuostabi istorija / V. Mikalauskas. — Vilnius : Mintis, 1977.
116 p., [24] leaves of plates : ill. ; 21 cm. USSR78-3762
Bibliography: p. 113-[114]
0.96rub
1. Cinematography—History. 2. Moving-pictures—History. I. Title.
TR848.M45 79-382682
79 MARC

Museo nazionale del cinema.
Il Museo nazionale del cinema, Torino / Maria Adriana Prolo, Luigi Carluccio. — [Torino] : Cassa di risparmio di Torino, [1978]. It***
xiv, 235 p. : ill. ; 31 cm.
Errata slip inserted.
1. Museo nazionale del cinema. 2. Cinematography—History. I. Prolo, Maria Adriana. II. Carluccio, Luigi.
TR6.I82T675 79-357647
79 MARC

—LIGHTING

Ritsko, Alan J
Lighting for location motion pictures / Alan J. Ritsko. — New York : Van Nostrand Reinhold, 1979.
224 p. : ill. ; 24 cm.
Includes index.
ISBN 0-442-26956-0. ISBN 0-442-26957-9 pbk.
1. Cinematography—Lighting. I. Title.
TR891.R57 778.5'34 77-28366
77 MARC

—PERIODICALS

Millimeter.
New York, Millimeter Magazine.
v. ill. 28 cm. monthly.
"The magazine of the motion picture and television production industries."
Began in 1973. Cf. New serial titles.
Key title: Millimeter, ISSN 0164-9655
1. Cinematography—Periodicals. 2. Moving-picture industry—Periodicals. 3. Television industry—Periodicals.
TR845.M54 778.5'05 78-648363
MARC-S

—PROCESSING

Institut national de l'audiovisuel. Bureau des productions pédagogiques.
Le laboratoire cinéma. — [Bry-sur-Marne] : Institut national de l'audiovisuel, Bureau des productions pédagogiques, [1978] F***
224 p. : ill. ; 21 x 30 cm. — (Collection Information/formation)
Issued in portfolio.
Bibliography: p. 191-197.
Includes index.
1. Cinematography—Processing. I. Title.
TR886.2.I53 1978 778.5'3 79-381544
79 MARC

Solow, Sid.
Sid Solow : an American Film Institute seminar on his work. — Glen Rock, N. J. : Microfilming Corp. of America, 1977.
1 sheet (68 p.) ; 11 x 15 cm. — (The American Film Institute seminars ; pt. 1, no. 168) (New York times oral history program)
Held November 22, 1971."
Microfiche.
ISBN 0-667-00487-4
1. Solow, Sid. 2. Cinematography—Processing. I. Series.
Microfiche TR849.S64 778.5'3'0924 [B] 77-12251

—SCIENTIFIC APPLICATIONS

see also Wildlife cinematography

Tosi, Virgilio.
Cinematography and scientific research / by Virgilio Tosi. — Paris : International Scientific Film Association, [1975?]
45 leaves ; 30 cm.
Bibliography: leaves 42-43.
1. Cinematography—Scientific applications. I. International Scientific Film Association. II. Title.
TR893.T67 778.5'38'5 78-322498
79 MARC

—SOCIAL ASPECTS—ITALY

Branca, Remo, 1897–
Tutto il cinema dappertutto? : responsabilità sociali nella organizzazione nazionale del passo ridotto / Remo Branca ; relazione approvata dal 1. Congresso nazionale del Passo ridotto tenutosi in Viareggio dal 12 al 16 febbraio 1947. — Roma : L'Italia che scrive editrice, [1947]
30 p. ; 23 cm. — (Quaderni del Passo ridotto ; 1)
1. Cinematography—Social aspects—Italy. I. Title. II. Series.
TR899.B69 79-368535
79 MARC

—SPECIAL EFFECTS

Film magic : the fantastic guide to special effects filmmaking / compiled and edited by Don Dohler. — Baltimore : Cinema Enterprises, c1979.
118 p. : ill. ; 28 cm.
Cover title.
1. Cinematography—Special effects. I. Dohler, Don.
TR858.F54 778.5'34 79-88915
79 MARC

—STUDY AND TEACHING

Huxley, Douglas R 1932–
Making films in super eight : a handbook for primary and secondary teachers / D. R. Huxley. — Cambridge ; New York : Cambridge University Press, 1978.
vii, 88 p. : ill. ; 21 x 22 cm.
Bibliography: p. 85.
ISBN 0-521-29300-6 : $4.50
1. Cinematography—Study and teaching. 2. Moving-pictures—Production and direction—Study and teaching. I. Title.
TR852.H89 778.5'3 77-84807
77 MARC

Lods, Jean, fl. 1951–
Professional training of film technicians. Paris, UNESCO, 1951.
155 p. 22 cm. (Press, film and radio in the world today)
Translation of La formation professionnelle des techniciens du film.
Bibliography: p. 141-155.
1. Cinematography—Study and teaching. I. Title. II. Series: United Nations Educational, Scientific and Cultural Organization. Press, film and radio in the world today.
TR852.L6 778.5'3'07 52-8284
[r79]rev MARC

—TERMINOLOGY

Raffaelli, Sergio.
Cinema, film, regia : saggi per una storia linguistica del cinema italiano / Sergio Raffaelli. — Roma : Bulzoni, [1978] It78-May
302 p. ; 21 cm. — (Studi cinematografici ; 5)
Bibliography: p. 9-11.
Includes indexes.
L7500
1. Cinematography—Terminology. 2. Italian language—Terms and phrases. I. Title. II. Series.
TR847.R33 778.5'3'014 79-358854
•79 MARC

CINEMATOGRAPHY, SUBMARINE

see also Photography, Submarine

CINEMATOGRAPHY, TRICK

Dunn, Linwood.
Linwood Dunn : an American Film Institute seminar on his work. — Glen Rock, N. J. : Microfilming Corp. of America, 1977.
1 sheet (86 p.) ; 11 x 15 cm. — (The American Film Institute seminars ; pt. 1, no. 47) (New York times oral history program)
Microfiche.
ISBN 0-667-00366-5
1. Dunn, Linwood. 2. Cinematography, Trick. 3. Cinematographers—California—Biography. I. Series.
Microfiche TR849.D85 778.5'3'0924 [B] 77-12750

Reff, Werner.
Filmtrick, Trickfilm / Werner Reff, Istvan Vasarhelyi. — 5., neubearb. Aufl., 26.-30. Tsd. — Leipzig : Fotokinoverlag, VEB, 1978.
340 p. : ill. ; 22 cm. GFR79-A8
First-3d ed. published under title: Der Filmtrick und der Trickfilm; 4th ed. published in 1972 under title: Filmtrick und Trickfilm.
Bibliography: p. 335-336.
Includes index.
18.00M
1. Cinematography, Trick. I. Vásárhelyi, István, 1910– joint author. II. Title.
TR858.R4 1978 778.5'34 79-377731
79 MARC

Wilkie, Bernard.
Creating special effects for TV and film / [by] Bernard Wilkie. — London ; New York : Focal Press, 1977.
158 p. : ill. ; 22 cm. — (Media manuals) GB78-16390
Bibliography: p. 153.
ISBN 0-240-50947-1 : £2.75
1. Cinematography, Trick. I. Title.
TR858.W54 1977 791.43'025 79-310295
79 MARC

CINEMATOGRAPHY, WIDE-SCREEN
see Wide-screen processing (Cinematography)

CINEMATOGRAPHY, WILDLIFE see Wildlife cinematography

CINERAMA see Wide-screen processing (Cinematography)

CINESIOLOGY see Kinesiology

CINGRIA, ALEXANDRE, 1879-1945

Ramuz, Charles Ferdinand, 1878-1947.
Lettres / C.-F. Ramuz, A. Cingria ; présentation, choix et notes de Gilbert Guisan et Doris Jakubec. — Lausanne [etc.] : La Bibliothèque des arts, cop. 1978.
2 v. : ill. ; 24 cm. Sw78-11334
Includes bibliographical references and index.
CONTENTS: 1. 1900-1904.—2. 1904-1914.
72.00F
1. Ramuz, Charles Ferdinand, 1878-1947—Correspondence. 2. Cingria, Alexandre, 1879-1945. 3. Novelists, Swiss—20th century—Correspondence. 4. Artists—Switzerland—Correspondence. I. Cingria, Alexandre, 1879-1945, joint author. II. Guisan, Gilbert. III. Jakubec, Doris.
PQ2635.A35Z543 1978 848'.9'1209 79-347579
79 MARC

CINNA, C. HELVIUS

—CRITICISM AND INTERPRETATION

Dahlmann, Hellfried, 1905-
Über Helvius Cinna / von Hellfried Dahlmann. — 1. Aufl. — Mainz : Akademie der Wiss. u. d. Literatur ; Wiesbaden : Steiner [in Komm.], 1977.
50 p. ; 24 cm. — (Abhandlungen der Geistes- und Sozialwissenschaftlichen Klasse, Akademie der Wissenschaften und der Literatur ; Jahrg. 1977, Nr. 8) GFR77-A15
Includes bibliographical references and indexes.
ISBN 3-515-02709-2 : DM13.00
1. Cinna, C. Helvius—Criticism and interpretation. I. Title. II. Series: Akademie der Wissenschaften und der Literatur, Mainz. Geistes- und Sozialwissenschaftliche Klasse. Abhandlungen ; Jahrg. 1977, Nr. 8.
AS182.M232 1977, Nr. 8 79-340545
[PA6371.C7] 79 MARC

CINNAMON

—INDONESIA—SUMATRA

Djohan, Rusli.
Inventarisasi produksi dan tata-niaga cassia vera di Prop. Sum. Barat, Kab. Kerintji (Prop. Djambi) dan Kab. Tapanuli Selatan (Prop. Sum. Utara) / oleh Rusli Djohan, Basjir Radja, Thamrin Nurdin. — Padang : Team Survey Cassia Vera, Fak. Pertanian, Univ. Andalas, 1969.
ii, 45 leaves ; 32 cm.
Summary in English.
1. Cinnamon—Indonesia—Sumatra. I. Radja, Basjir, joint author. II. Nurdin, Thamrin, joint author. III. Universitas Andalas. Team Survey Cassia Vera. IV. Title: Inventarisasi produksi dan tata-niaga cassia vera ...
SB307.C5D56 75-941392
79 MARC

CINNAMON INDUSTRY

—SRI LANKA

Moore, M P
The state and the cinnamon industry in Sri Lanka / by M. P. Moore. — Colombo : Agrarian Research & Training Institute, 1978.
51 p. : graphs ; 31 cm. — (Occasional paper - Agrarian Research & Training Institute ; no. 15)
Includes bibliographical references.
1. Cinnamon industry—Sri Lanka. 2. Agriculture and state—Sri Lanka. I. Title. II. Series: Agrarian Research and Training Institute. Occasional paper - Agrarian Research & Training Institute ; no. 15.
HD9211.C553S645 338.1'7'383 79-900839
79 MARC

CINO DA PISTOIA, 1270-1337

—CRITICISM AND INTERPRETATION—CONGRESSES

Colloquio Cino da Pistoia, Rome, 1975.
Colloquio Cino da Pistoia : Roma, 25 ottobre 1975. — Roma : Accademia nazionale dei Lincei, 1976.
153 p. ; 26 cm. — (Atti dei convegni lincei - Accademia nazionale dei Lincei ; 18) It78-Apr
Bibliography: p. 152.
L6000
1. Cino da Pistoia, 1270-1337—Criticism and interpretation—Congresses. 2. Law—History and criticism—Congresses. I. Series: Accademia nazionale dei Lincei, Rome. Atti dei convegni lincei ; 18.
PQ4299.C5Z6 1975 78-377230
*78 MARC

CINQUE PORTS

—DESCRIPTION—GUIDE-BOOKS

Whitney, C E
Discovering the Cinque Ports / [by] C. E. Whitney. — Princes Risborough : Shire Publications, 1978.
63 p. : ill., maps ; 18 cm. — (Discovering series ; no. 237) GB79-09883
Bibliography: p. 61.
Includes index.
ISBN 0-85263-416-1 : £0.60
1. Cinque Ports—Description—Guide-books. I. Title.
DA690.C6W48 914.22'352 79-315010
79 MARC

CINQUIN, EMMANUELLE

Cinquin, Emmanuelle.
Chiffonnière avec les chiffonniers / Emmanuelle Cinquin. — Paris : Éditions ouvrières, c1977.
189 p. ; 18 cm. — (Collection A pleine vie) F***
24.00F
1. Cinquin, Emmanuelle. 2. Nuns—Egypt—Cairo—Biography. 3. Rag-pickers—Egypt—Cairo—Biography. 4. Cairo—Biography. I. Title.
BX4705.C5435A33 271'.9 79-378965
79 MARC

CINZANO LIMITED

—ART COLLECTIONS—CATALOGS

The Cinzano glass collection / [texts by Peter Lazarus ; photos. by Derek Balmer]. — London : Cinzano, c1978.
[64] p. : chiefly ill. ; 24 cm. GB***
Cover title.
1. Glassware—Catalogs. 2. Cinzano Limited—Art collections—Catalogs. I. Lazarus, Peter. II. Cinzano Limited.
NK5103.C56C56 748.8 79-300221
79 MARC

CIONINI VISANI, MARIA

Per Maria Cionini Visani : scritti di amici. — Torino : Stabilimento tipolitografico G. Canale, 1977.
173 p., [25] leaves of plates : ill., 1 port. ; 22 cm. It78-June
Includes bibliographical references.
1. Art, Italian—Addresses, essays, lectures. 2. Art—Addresses, essays, lectures. 3. Cionini Visani, Maria. I. Cionini Visani, Maria.
N6911.P529 78-382379
*78 MARC

CIP PROGRAM see Cataloging in publication

CIPAMOKOAN, INDONESIA

—ECONOMIC CONDITIONS

Adiwilaga, R Anwas, 1911-
Laporan tentang Cipamokolan / oleh Adiwilaga. — Bandung : Lembaga Penelitian Kemasyarakatan, Universitas Pajajaran, 1975.
19 leaves ; 32 cm.
1. Cipamokoan, Indonesia—Economic conditions. I. Title.
HC448.C56A34 76-941740
79 MARC

CIPARIU, TIMOTEI, 1805-1887

Cipariu, Timotei, 1805-1887.
Jurnal. Ediţie îngrijită, prefaţă, note şi glosar de Maria Protase. Cluj, "Dacia", 1972.
192 p. 17 cm. (Restituiri) lei4.75 R72-3290
Includes bibliographical references.
1. Cipariu, Timotei, 1805-1887. 2. Philologists—Romania—Biography. I. Protase, Maria, ed.
PC617.C5A34 1972 79-364773
79 MARC

CIPEROVA, MARIE

Čiperová, Marie.
Teorie tvorby vyučovacích pomůcek / Marie Čiperová. — 1. vyd. — Praha : SNTL, 1977.
100, [1] p. ; 30 cm. — (Publikace T (Teoretické studie) Výzkumného ústavu odborného školství) Cz77
Bibliography: p. 97-100.
1. Ciperova, Marie. 2. Education—Philosophy. I. Title. II. Series: Výzkumný ústav odborného školství. Publikace T (Teoretické studie) Výzkumného ústavu odborného školství.
LB880.C48T45 78-378099
*78 MARC

CIPHER AND TELEGRAPH CODES

Novissimo cifrario Mengarini. — 5. ed. interamente riv. e aggiornata. — Roma : Mantegazza, 1924.
304 p. ; 21 cm.
1. Cipher and telegraph codes. I. Mengarini, Arnaldo.
HE7678.I8N68 1924 79-349274
79 MARC

—PATENTS

Derwent Publications, ltd., London.
Central patents index. Large company code, January 1971. London, 1971.
60 p. 27 cm. GB***
Cover title.
1. Cipher and telegraph codes—Patents. I. Title.
T257.C2D47 1971 602'.7 78-318821
78 MARC

Derwent Publications, ltd., London.
FARMDOC AGDOC company code, December 1970. London, 1971.
63 p. 27 cm. GB***
Cover title.
1. Cipher and telegraph codes—Patents. I. Title.
T257.C2D47 1971a 602'.7 78-318820
78 MARC

CIPHERS

see also Abbreviations; Cryptography; Telegraph; Writing

—JUVENILE LITERATURE

Albert, Burton.
More codes for kids / Burton Albert, Jr. ; illustrated by Jerry Warshaw. — Chicago : A. Whitman, c1979.
32 p. : col. ill. ; 24 cm.
SUMMARY: More codes for cryptanalysts to crack, many with jokes or riddles for answers. Includes, among others, a Code-a-Nut Tree and Back-Flip Triple Decker codes.
ISBN (invalid) 0-8075-5270-4 : $5.50
1. Ciphers—Juvenile literature. [1. Ciphers] I. Warshaw, Jerry. II. Title.
Z103.3.A535 001.54'36 79-245
79 MARC

CIPHERS (LETTERING) see Monograms

CIQUATERA see Poisonous fishes

CIRCADIAN RHYTHMS

Endocrine rhythms / editor, Dorothy T. Krieger. — New York : Raven Press, c1979.
xii, 332 p. : ill. ; 24 cm. — (Comprehensive endocrinology ; v. 1)
Includes bibliographies and index.
ISBN 0-89004-234-9 : $26.00
1. Endocrine glands. 2. Circadian rhythms. 3. Hormones. I. Krieger, Dorothy T. II. Series.
QP187.E53 591.1'42 77-75655
78 MARC

Schell, Hermann.
Biorhythmik und Epidermisproliferation / Hermann Schell. — Berlin : Grosse, 1978.
161 p. ; 21 cm. — (Grosse Scripta ; 4) GFR78-A
Bibliography: p. 117-139.
ISBN 3-88040-015-6 : DM19.00
1. Epidermis—Growth. 2. Circadian rhythms. 3. Hydrocortisone—Physiological effect. 4. Cell proliferation. I. Title.
QP88.5.S33 79-349376
79 MARC

Weston, Lee.
Body rhythm : the circadian rhythms within you / Lee Weston. — 1st ed. — New York : Harcourt Brace Jovanovich, c1979.
viii, 170 p. ; 22 cm.
Bibliography: p. 157-170.
ISBN 0-15-113338-7
1. Circadian rhythms. 2. Biological rhythms. I. Title.
QP84.6.W47 612 79-1854
79 MARC

Wever, Rütger A
The circadian system of man : results of experiments under temporal isolation / Rütger A. Wever. — New York : Springer-Verlag, c1979.
xi, 276 p. : graphs ; 27 cm. — (Topics in environmental physiology and medicine)
Bibliography: p. 256-265.
Includes indexes.
ISBN 0-387-90338-0
1. Circadian rhythms. 2. Human biology. I. Title.
QP84.6.W48 612 78-11543
78 MARC

—CONGRESSES

Colloque sur le travail de nuit et les horaires alternants, Université de Paris I, 1974.
Ergonomie du travail de nuit et des horaires alternants / Colloque sur le travail de nuit et les horaires alternants, 28-29 novembre 1974, Paris ; ₍publié par₎ P. Andlauer, J. Carpentier, P. Cazamian. — Paris : Éditions Cujas, 1977.
272 p. : graphs. ; 22 cm. — (Éducation permanente) F78-1508
Includes bibliographies.
1. Night work—Physiological aspects—Congresses. 2. Circadian rhythms—Congresses. 3. Shift system—Physiological aspects—Congresses. I. Andlauer, Pierre, 1911- . II. Carpentier, James. III. Cazamian, P. IV. Title.
QP301.C5873 1974 612 79-382362
79 MARC

CIRCLE

Pedoe, Daniel.
Circles, a mathematical view / D. Pedoe. — New York : Dover Publications, 1979.
viii, 87 p. : ill. ; 22 cm.
"Corrected and enlarged republication of the work ... published in 1957 by Pergamon Press."
Bibliography: p. viii.
ISBN 0-486-63698-4 : $2.75
1. Circle. I. Title.
QA484.P4 1979 516'.22 78-73522
79 MARC

CIRCLE (IN RELIGION, FOLK-LORE, ETC.)

Langham, Derald G
Circle gardening : producing food by Genesa principles / by Derald G. Langham. — Old Greenwich, Conn. : Devin-Adair Co., c1978.
xii, 93, ₍2₎ p. : ill. ; 23 cm.
Bibliography: p. ₍95₎
ISBN 0-8159-5217-1. ISBN 0-8159-5215-5 pbk. : $5.95
1. Vegetable gardening. 2. Circle gardening. 3. Circle (in religion, folk-lore, etc.) I. Title.
SB321.L29 635 77-92698
79 MARC

CIRCLE (SURVEYING INSTRUMENT)

—CALIBRATION

Weise, Hans, Dipl.-Ing.
Untersuchungen zur Rationalisierung und Genauigkeitssteigerung von Kreisteilungsprüfungen. ₍Potsdam₎ Geodätisches Institut Potsdam der Deutschen Akademie der Wissenschaften zu Berlin ₍1964₎
129 p. illus. 30 cm. (Arbeiten aus dem Geodätischen Institut Potsdam, Nr. 3)
Inaug.-Diss.—Technische Universität, Dresden.
Bibliography: p. 120-129.
1. Circle (Surveying instrument)—Calibration. I. Title. II. Series: Potsdam. Geodätisches Institut. Arbeiten, Nr. 3.
QB275.P6 Nr. 3 79-378165
₍TA575₎ 79 MARC

CIRCLE GARDENING

Langham, Derald G
Circle gardening : producing food by Genesa principles / by Derald G. Langham. — Old Greenwich, Conn. : Devin-Adair Co., c1978.
xii, 93, ₍2₎ p. : ill. ; 23 cm.
Bibliography: p. ₍95₎
ISBN 0-8159-5217-1. ISBN 0-8159-5215-5 pbk. : $5.95
1. Vegetable gardening. 2. Circle gardening. 3. Circle (in religion, folk-lore, etc.) I. Title.
SB321.L29 635 77-92698
79 MARC

CIRCLEVILLE AND WASHINGTON TURNPIKE COMPANY

Ohio. Laws, statutes, etc.
Charter of the Circleville and Washington Turnpike Company : ₍act₎ passed February 13, 1832, and revived and amended February 23, 1839. — Circleville ₍Ohio₎ : Printed by J. Case, 1839.
8 p. ; 23 cm.
1. Circleville and Washington Turnpike Company. 2. Highway law—Ohio. I. Title.
KFO444.9.C57A32 1839 343'.771'0942 78-319522
78 MARC

CIRCUIT COURTS

—UNITED STATES

Berkson, Larry Charles.
A study of the U.S. Circuit Judge Nominating Commission : executive summary : a project of the American Judicature Society / Larry Berkson, Susan Carbon, Alan Neff. — Chicago : American Judicature Society, 1979.
vii, 92 p. ; 23 cm.
Includes bibliographical references.
1. Judges—United States—Appointment, qualifications, tenure, etc. 2. United States. Circuit Judge Nominating Commission. 3. Circuit courts—United States. I. Carbon, Susan J., joint author. II. Neff, Alan, joint author. III. American Judicature Society. IV. Title.
KF8776.B4 347'.73'14 79-117292
79 MARC

CIRCUITRY, NEURAL see Neural circuitry

CIRCUITS, BRIDGE see Bridge circuits

CIRCUITS, ELECTRIC see Electric circuits

CIRCUITS, HYDRAULIC see Hydraulic circuits

CIRCUITS, INTEGRATED see Integrated circuits

CIRCUITS, LOGIC see Logic circuits

CIRCUITS, MICROWAVE see Microwave circuits

CIRCUITS, SWITCHING see Switching circuits

CIRCUITS, THICK-FILM see Thick-film circuits

CIRCUITS, THIN-FILM see Thin-film circuits

CIRCULAR DICHROISM

Charney, Elliot.
The molecular basis of optical activity : optical rotatory dispersion and circular dichroism / Elliot Charney. — New York : Wiley, c1979.
x, 364 p. : ill. ; 24 cm.
"A Wiley-Interscience publication."
Includes bibliographical references and index.
ISBN 0-471-14900-4
1. Optical rotatory dispersion. 2. Circular dichroism. I. Title.
QD473.C48 541'.22 79-9705
79 MARC

Optically active polymers / edited by Éric Sélégny. — Dordrecht, Holland ; Boston : D. Reidel, c1979.
xi, 417 p. : ill. ; 25 cm. — (Charged and reactive polymers ; v. 5)
Includes bibliographical references and index.
ISBN 9027709041
1. Polymers and polymerization—Optical properties. 2. Optical rotation. 3. Circular dichroism. I. Sélégny, Éric. II. Series.
QD381.8.O67 547'.84 79-16124
79 MARC

—CONGRESSES

Nato Advanced Study Institute, University of Sussex, 1978.
Optical activity and chiral discrimination : proceedings of the NATO Advanced Study Institute, held at the University of Sussex, Falmer, England, September 10-22, 1978 / edited by Stephen F. Mason. — Dordrecht, Holland ; Boston : D. Reidel Pub. Co., c1979.
xi, 372 p. : graphs ; 25 cm. — (NATO advanced study institutes series : Series C, Mathematical and physical sciences ; v. 48)
Includes bibliographical references and index.
ISBN 9027709823
1. Optical rotatory dispersion—Congresses. 2. Circular dichroism—Congresses. 3. Chirality—Congresses. I. Mason, Stephen Finney, 1923- . II. Title. III. Series.
QD473.N37 1978 541'.7 79-10532
79 MARC

CIRCULAR INSANITY see Manic-depressive psychoses

CIRCULAR VELOCITY OF MONEY

see also Credit; Inflation (Finance); Multiplier (Economics); Quantity theory of money

—HUNGARY

Tutzer, Lajos.
Aktuális kérdések a pénzforgalom köréből / irta, Tutzer Lajos. — Budapest : Magyar Nemzeti Bank, 1978.
34 p. ; 29 cm. — (MNB Közgazdasági Főosztály közleményei ; 82 ISSN 0324-6264) H***
ISBN 9637103252
1. Circular velocity of money—Hungary. 2. Money—Hungary. I. Title. II. Series: Magyar Nemzeti Bank. Közgazdasági Főosztly. MNB Közgazdasági Főosztály közleményei ; 82.
HG970.5.T85 79-381346

CIRCULAR VELOCITY OF MONEY (Continued)

—UNITED STATES

Garvy, George.
Deposit velocity and its significance / George Garvy. — Westport, Conn. : Greenwood Press, 1978.
88 p. : graphs ; 24 cm.
Reprint of the 1959 ed. published by the Federal Reserve Bank of New York.
Includes bibliographical references.
ISBN 0-313-21022-5
1. Circular velocity of money—United States. 2. Bank deposits—United States. I. Title.
[HG538.G34 1978] 332.1'752'0973 78-14435
78 MARC

CIRCULATION, OCEAN see Ocean circulation

CIRCULATION, REGIONAL see Regional blood flow

CIRCULATION OF THE BLOOD see Blood—Circulation

CIRCULATORY SYSTEM see Cardiovascular system

CIRCULO DE BELLAS ARTES, CARACAS

—EXHIBITIONS

Center for Inter-American Relations.
A century of Venezuelan landscape painting : la Escuela de Caracas. — New York : Center for Inter-American Relations, 1979.
[48] p. : ill. (some col.) ; 28 cm.
Exhibition held at the Center, Feb. 20-Apr. 15, 1979.
1. Circulo de Bellas Artes, Caracas—Exhibitions. 2. Landscape painting—20th century—Venezuela—Exhibitions. I. Title.
ND1352.V4C46 1979 79-109168
758'.1'098707401471
79 MARC

CÍRCULO DE BELLAS ARTES, MONTEVIDEO

Laroche, Walter E
Los setenta años del Círculo de Bellas Artes del Uruguay, 1905-1975 / W. E. Laroche. — Montevideo : Museo y Archivo Ernesto Laroche, 1976.
14 p. ; 23 cm. — (Publicación del Museo y Archivo Ernesto Laroche ; año 35, no. 31)
1. Círculo de Bellas Artes, Montevideo. I. Title. II. Series: Montevideo. Museo y Archivo Ernesto Laroche. Publicaciones ; año 35, no. 31.
N6720.M65a ano. 35, no. 31 78-397086
700'.9895 s
[N16] 78 MARC

CIRCUMCELLIONS

Frazão, Euza Rossi de Aguiar.
O donatismo e os circunceliões na obra de Santo Agostinho / Euza Rossi de Aquiar Frazão. — São Paulo : Faculdade de Filosofia, Letras e Ciências Humanas, Universidade de São Paulo, 1976.
253 p. ; 24 cm. — (Boletim - Curso de História Antiga ; no. 1) (Boletim - Departamento de História ; no. 4) (Boletim - Faculdade de Filosofia, Letras e Ciências Humanas, Universidade de São Paulo ; no. 5 (nova sér.))
Bibliography: p. 241-253.
1. Donatists. 2. Circumcellions. 3. Augustinus, Aurelius, Bp. of Hippo. I. Title. II. Series: São Paulo, Brazil (City). Universidade. Curso de História Antiga. Boletim - Curso de História Antiga ; no. 1. III. Series: São Paulo, Brazil (City). Universidade. Departamento de História. Boletim - Departamento de História ; no. 4. IV. Series: São Paulo, Brazil (City). Universidade. Faculdade de Filosofia, Letras e Ciências Humanas. Boletim - Faculdade de Filosofia, Letras e Ciências Humanas, Universidade de São Paulo ; no. 5 (nova ser.)
BT1370.F68 79-382049

CIRCUMCISION

Maertens, Thierry, 1921-
Le corps sexionné : essai d'anthropologie des inscriptions génitales / Jean-Thierry Maertens, avec la collaboration de Marguerite DeBilde. — Paris : Aubier Montaigne, c1978.
190 p. ; 22 cm. — (His Ritologiques ; 2) (Étranges étrangers) F***
Bibliography: p. [187]-190.
Includes index.
39.00F
1. Circumcision. 2. Initiations (in religion, folk-lore, etc.) 3. Mutilation. I. DeBilde, Marguerite, joint author. II. Title. III. Series.
GN484.M33 301.2'1 78-323076
79 MARC

CIRCUMNAVIGATION see Voyages around the world

CIRCUMSTANCE ETHICS see Situation ethics

CIRCUMVENTRICULAR ORGANS

see also Neurohypophysis

—CULTURES AND CULTURE MEDIA

Möller, Wilhelm, 1937-
Circumventriculäre Organe in der Gewebekultur = Circumventricular organs in cell culture / Wilhelm Möller. — Berlin ; New York : Springer-Verlag, 1978.
95 p. : ill. ; 25 cm. — (Ergebnisse der Anatomie und Entwicklungsgeschichte ; 54/1)
Summary in English.
Bibliography: p. 78-89.
Includes index.
ISBN 0-387-08578-5
1. Circumventricular organs—Cultures and culture media. 2. Tissue culture. I. Title. II. Title: Circumventricular organs in cell culture. III. Series: Advances in anatomy, embryology, and cell biology ; 54/1.
[DNLM: 1. Cerebral ventricles—Anatomy and histology. 2. Choroid plexus—Anatomy and histology. 3. Pineal body—Anatomy and histology. 4. Spinal cord—Anatomy and histology. 5. Glycogen—Metabolism. W1 AD433K v. 54. fasc. 1 / WL203 M693c]
QL801.E67 vol. 54/1 574.4'08 s 77-19030
[QP376] 78 MARC

CIRCUS

see also Acrobats and acrobatism; Aerialists; Amateur circus; Amusement parks; Animals, Training of; Circus owners; Circus performers

Ramirez, Francis.
Joies du cirque / [Francis Ramirez et Christian Rolot] ; [préface de Pierre Etaix]. — [Paris] : Hachette : "Réalités", 1977.
237 p. : ill. (some col.) ; 29 cm. — (Collection Joies et réalités)
F78-1951
ISBN 2-01-003400-7 : 130F
1. Circus. I. Rolot, Christian, joint author. II. Title.
GV1815.R35 791.3 78-365021
78 MARC

—COLLECTIBLES—TEXAS—SAN ANTONIO

San Antonio. Public Library.
A bibliography of 19 century clown songsters of the Harry Hertzberg Circus Collection of the San Antonio Public Library. — [s.l. : s.n., pref. 1962] (San Antonio : Carleton Print. Co.)
[4] p. ; 25 cm.
Pref. signed: Leonard V. Farley.
"No. 18."
Colored ill. inserted.
1. Songs, American— Bibliography—Catalogs. 2. Clowns—Songs and music—Bibliography—Catalogs. 3. San Antonio. Public Library—Catalogs. 4. Hertzberg, Harry—Circus collection. 5. Circus—Collectibles—Texas—San Antonio. I. Hertzberg, Harry. II. Farley, Leonard V., 1908- . III. Title.
ML136.S25H43 016.7846 78-321609

—HISTORY

Bose, Günter.
Circus : Geschichte u. Ästhetik e niederen Kunst / von Günter Bose u. Erich Brinkmann. — Berlin : Wagenbach, 1978.
202, [1] p. : ill. ; 18 cm. — (Wagenbachs Taschenbücherei ; 46)
GFR79-A
Bibliography: p. 199-[203]
ISBN 3-8031-2046-2 : DM9.50
1. Circus—History. 2. Circus—Germany—History. I. Brinkmann, Erich, joint author. II. Title.
GV1801.B66 791.3'09 79-358412
79 MARC

Hippisley Coxe, Antony.
A seat at the circus; with special illus. by John Skeaping. London, Evans Bros. [1951]
252 p. illus. 22 cm.
1. Circus—History. I. Title.
GV1801.H5 791.3 52-36313
[r79]rev MARC

Le Grand livre du cirque / [red. en chef, Monica J. Renevey]. — [Lausanne] : Bibliothèque des arts, cop. 1977.
2 v. (456, 449 p.) : ill. (some col.) ; 29 cm. Sw78-638
Includes indexes.
195F
1. Circus—History. 2. Circus performers—Biography. I. Renevey, Monica J.
GV1801.G78 791.3'09 78-401052
79 MARC

Thétard, Henry, 1884-
La merveilleuse histoire du cirque / Henry Thétard, suivie de Le cirque depuis la guerre / par L.-R. Dauven. — [Paris] : Julliard, c1978.
631, [8] p., [26] leaves of plates : ill. (some col.) ; 25 cm. F***
Bibliography: p. [634]
Includes index.
ISBN 2-260-00138-6 : 150F
1. Circus—History. I. Dauven, L. R. Le cirque depuis la guerre. 1978. II. Title.
GV1801.T5 1978 791.3 79-346351
79 MARC

Verney, Peter, 1930-
Here comes the circus / by Peter Verney. — New York : Paddington Press : distributed by Grosset & Dunlap, c1978.
287 p. : ill. ; 26 cm.
Includes index.
SUMMARY: Discusses the history of the circus from ancient times to today and such aspects as animals, clowns, aerialists, side shows, and the circus way of life.
ISBN 0-448-23115-8
1. Circus—History. [1. Circus] I. Title.
GV1801.V47 1978 791.3 77-13987
77[r79]rev MARC

—HISTORY—ADDRESSES, ESSAYS, LECTURES

Zirkus = Circus = Cirque : 28. Berliner Festwochen 1978, Ausstellung vom 9.9. bis 5.11.1978 : [Katalog] / hrsg. von Jörn Merkert. — Berlin : Nationalgalerie, c1978.
247 p. : ill. (1 fold. col. in pocket) ; 31 cm. GFR***
Includes bibliographies.
1. Circus—History—Addresses, essays, lectures. 2. Circus—Posters. 3. Berliner Festwochen, 1978—Addresses, essays, lectures. I. Merkert, Jörn. II. Berliner Festwochen, 1978. III. Title: Circus. IV. Title: Cirque.
GV1801.Z57 791.3'09 79-343528
79 MARC

—JUVENILE LITERATURE

Disney (Walt) Productions.
Disney's world of adventure presents The circus book. — New York : Random House, c1978.
59 p. : col. ill. ; 26 cm. — (Disney's world of adventure)
CONTENTS: Toby Tyler.—The greatest show on earth.—The high steppers of the horse world.—Be a clown!—The chimpanzee, our closest relative.—Circus big-shot.
ISBN 0-394-83597-2 : $3.95. ISBN 0-394-93597-7 lib. bdg. : $4.99
1. Circus—Juvenile literature. [1. Circus. 2. Circus stories] I. Title. II. Title: The circus book. III. Series.
GV1817.D57 1978 791.3 77-74462
78 MARC

Prelutsky, Jack.
Circus / by Jack Prelutsky ; pictures by Arnold Lobel. — New York : Collier Books, 1978, c1974.
[32] p. : col. ill. ; 23 x 25 cm.
SUMMARY: Poems and illustrations depict the sights and sounds of the circus.
ISBN 0-02-044760-4
1. Circus—Juvenile poetry. [1. Circus—Poetry] I. Lobel, Arnold. II. Title.
[PZ8.3.P9 Ci 1978] 811'.5'4 77-16496
77 MARC

CIRCUS
—JUVENILE LITERATURE (Continued)

Sedgwick, Paulita.
Circus ABC / by Paulita Sedgwick. — New York : Holt, Rinehart and Winston, c1978.
[32] p. : ill. (some col.) ; 21 cm.
SUMMARY: The Alphabet Man introduces a circus as well as the letters of the alphabet.
ISBN 0-03-042391-0 : $6.95 .
1. English language—Alphabet—Juvenile literature. 2. Circus—Juvenile literature. [1. Alphabet. 2. Circus] I. Title.
PE1155.S4 421'.1 78-4281
78 MARC

—LITERARY COLLECTIONS

Cirkusvärld : en antologi / av Kaj Beckman ; med bilder av Per Beckman. — Stockholm : Rabén & Sjögren, 1976.
157 p. : ill. ; 26 cm. S76-46
ISBN 9129460333 : kr85.00
1. Circus—Literary collections. I. Beckman, Kaj. 1913- . II. Beckman, Per, 1913-
PN6071.C63C57 78-360730
*78 MARC

—POETRY

Gammelgaard, Per, 1948-
Den store finale og andre digte om cirkus : digte / [af] Per Gammelgaard. — [Nykøbing Mors] : Attika, c1978.
48 p. ; 22 cm. D78-21
ISBN 8787597454 : kr48.00
1. Circus—Poetry. I. Title.
PT8176.17.A45S84 78-395423
79 MARC

—POSTERS

American circus posters in full color / edited by Charles Philip Fox. — New York : Dover Publications, 1978.
43 p. : chiefly col. ill. ; 37 cm.
ISBN 0-486-23693-5 : $6.95
1. Circus—Posters. I. Fox, Charles Philip, 1904-
GV1826.A46 1978 769'.5 78-54311
79 MARC

Menschen, Tiere, Sensationen : Zirkusplakate 1880-1930 : Ausstellung, 5. Mai bis 18. Juni 1978 : Katalog / hrsg. von Carl-Albrecht Haenlein und Wolfgang Till. — Hannover : Kestner-Gesellschaft, 1978.
187 p. : chiefly ill. (some col.) ; 21 cm. — (Katalog - Kestner-Gesellschaft Hannover ; 2/1978) GFR***
Bibliography: p. 178-179.
1. Circus—Posters. I. Haenlein, Carl Albrecht. II. Till, Wolfgang. III. Series: Kestner-Gesellschaft. Katalog ; 1978, 2.
N5070.H3K4 1978.2 708'.3'59 s 78-384259
[GV1826] 79 MARC

Zirkus = Circus = Cirque : 28. Berliner Festwochen 1978, Ausstellung vom 9.9. bis 5.11.1978 : [Katalog] / hrsg. von Jörn Merkert. — Berlin : Nationalgalerie, c1978.
247 p. : ill. (1 fold. col. in pocket) ; 31 cm. GFR***
Includes bibliographies.
1. Circus—History—Addresses, essays, lectures. 2. Circus—Posters. 3. Berliner Festwochen, 1978—Addresses, essays, lectures. I. Merkert, Jörn. II. Berliner Festwochen, 1978. III. Title: Circus. IV. Title: Cirque.
GV1801.Z57 791.3'09 79-343528
79 MARC

—BRAZIL

Militello, Dirce.
Picadeiro / Dirce (Tangará) Militello ; [capa e ilustrações, Humberto II]. — [1. ed.]. — [São Paulo] : Guarida, [1978]
144 p. : ill. ; 22 cm.
Cover title.
Cr$120.00
1. Circus—Brazil. I. Title.
GV1805.B6M54 79-114814
79 MARC

—GERMANY—HISTORY

Bose, Günter.
Circus : Geschichte u. Ästhetik e niederen Kunst / von Günter Bose u. Erich Brinkmann. — Berlin : Wagenbach, 1978.
202, [1] p. : ill. ; 18 cm. — (Wagenbachs Taschenbücherei ; 46) GFR79-A
Bibliography: p. 199-[203]
ISBN 3-8031-2046-2 : DM9.50
1. Circus—History. 2. Circus—Germany—History. I. Brinkmann, Erich, joint author. II. Title.
GV1801.B66 791.3'09 79-358412
79 MARC

—NETHERLANDS—HISTORY

Van binnen moet je wezen ... : tweehonderd jaar circus in Nederland / Marja Keyser ... [et al.]. — Nieuwkoop : Heureka, [1978]
112 p. : with ill., fold. cover ; 23 cm. Ne78-30
Bibliography: p. 101.
ISBN 9062621112 : fl 25.00
1. Circus—Netherlands—History. I. Keyser, Marja.
GV1805.N4V36 78-380332
*79 MARC

—RUSSIA

Bardian, Feodosiĭ Georgievich.
(Sovetskiĭ fsirk na piati kontinentakh)
Советский цирк на пяти континентах / Ф. Бардиан. — Москва : Искусство, 1977.
206 p. : ill. ; 22 cm. USSR 77
Includes bibliographical references.
2.00rub
1. Circus—Russia. I. Title.
GV1805.R9B37 78-422097

—RUSSIA—BIOGRAPHY

Abidov, Tulkun.
(Arena Akrama IUsupova)
Арена Акрама Юсупова : творческий портрет / Тулкун Абидов. — Ташкент : Изд-во литературы и искусства, 1976.
83 p., [15] leaves of plates : ill. ; 17 cm. USSR***
Includes bibliographical references.
0.64rub
1. IUsupov, Akram. 2. Circus—Russia—Biography. I. Title.
GV1811.I 97A24 79-409852

—UNITED STATES—HISTORY—JUVENILE LITERATURE

Burnside, Irving L 1924-
The first American circus ever / by Irving L. Burnside ; illustrated by Sam Viviano. — New York : C.P.I. ; Morristown, N.J. : distributor, Silver Burdett Co., c1978.
48 p. : col. ill. ; 24 cm.
SUMMARY: Follows the career of John Bill Ricketts after his arrival in this country where he established our first circus in 1793.
ISBN 0-89547-040-3 lib. bdg. : 5.58
1. Circus—United States—History—Juvenile literature. 2. Ricketts, John Bill—Juvenile literature. 3. Circus owners—Biography—Juvenile literature. [1. Ricketts, John Bill. 2. Circus owners. 3. Circus—History] I. Viviano, Sam. II. Title.
GV1803.B87 791.3'092'4 78-15552
78 MARC

CIRCUS ANIMAL TRAINERS see Animal trainers

CIRCUS ANIMALS

Leonardi, Ruggero, 1936-
Animali nel circo / Ruggero Leonardi ; fotografie di Silvano Bergamaschi. — Torino : Società editrice internazionale, 1977.
153 p., [8] leaves of plates : ill. ; 23 cm. It78-Nov
1. Circus animals. I. Title.
GV1829.L4 791.3'2 79-345322
*79 MARC

CIRCUS COLLECTING see Wild animal collecting

CIRCUS IN ART

Vesque, Marthe.
Le cirque en images / Marthe et Juliette Vesque ; dessins présentés par Edith Mauriange et Paul Bouissac. — Paris : G.-P. Maisonneuve et Larose, c1977.
207 p. : ill. ; 24 cm. — (Archives d'ethnologie française ; no 5) F***
Includes indexes.
ISBN 2-7068-0649-4 : 62.00F
1. Vesque, Marthe. 2. Circus in art. I. Vesque, Juliette, joint author. II. Title. III. Series.
NC248.V48A4 1977 741.9'44 79-341446
79 MARC

CIRCUS OWNERS

—BIOGRAPHY—JUVENILE LITERATURE

Burnside, Irving L 1924-
The first American circus ever / by Irving L. Burnside ; illustrated by Sam Viviano. — New York : C.P.I. ; Morristown, N.J. : distributor, Silver Burdett Co., c1978.
48 p. : col. ill. ; 24 cm.
SUMMARY: Follows the career of John Bill Ricketts after his arrival in this country where he established our first circus in 1793.
ISBN 0-89547-040-3 lib. bdg. : 5.58
1. Circus—United States—History—Juvenile literature. 2. Ricketts, John Bill—Juvenile literature. 3. Circus owners—Biography—Juvenile literature. [1. Ricketts, John Bill. 2. Circus owners. 3. Circus—History] I. Viviano, Sam. II. Title.
GV1803.B87 791.3'092'4 78-15552
78 MARC

CIRCUS PERFORMERS

see also Acrobats and acrobatism; Animal trainers; Clowns

—BIOGRAPHY

Le Grand livre du cirque / [red. en chef, Monica J. Renevey]. — [Lausanne] : Bibliothèque des arts, cop. 1977.
2 v. (456, 449 p.) : ill. (some col.) ; 29 cm. Sw78-638
Includes indexes.
195F
1. Circus—History. 2. Circus performers—Biography. I. Renevey, Monica J.
GV1801.G78 791.3'09 78-401052
79 MARC

—JUVENILE LITERATURE

Powledge, Fred.
Born on the circus / by Fred Powledge ; illustrated with photos. by the author. — 1st ed. — New York : Harcourt Brace Jovanovich, c1976.
94 p. : ill. ; 26 cm.
SUMMARY: Describes the hard work and excitement of circus life as seen through the eyes of an eleven-year-old who performs as juggler, trampoline artist, and horseback rider.
ISBN 0-15-209970-0
1. Cristiani, Armando—Juvenile literature. 2. Circus performers—Juvenile literature. [1. Cristiani, Armando. 2. Circus—Biography] I. Title.
GV1811.C74P68 791.3'092'4 76-2449
76[r79]rev MARC

—BIOGRAPHY

Møller Pedersen, P E
Elvira Madigan / [af] Poul-Erik Møller-Pedersen. — København : Hernov, 1978.
138 p. : ill. ; 25 cm. D78-22
ISBN 8772154047 : kr94.75. ISBN 8772154039 pbk.
1. Madigan, Elvira, 1867-1889. 2. Sparre, Sixten, 1854-1889. 3. Circus performers—Biography.
GV1811.M33M62 78-383704
78 MARC

CIRCUS PERFORMERS (Continued)

—UNITED STATES—BIOGRAPHY

Thumb, Tom, Mrs., 1841-1919.
The autobiography of Mrs. Tom Thumb : (some of my life experiences) / by Countess M. Lavinia Magri, formerly Mrs. General Tom Thumb, with the assistance of Sylvester Bleeker ; edited and introduced by A. H. Saxon. — Hamden, Conn. : Archon Books, 1979.
199 p., ⌈8⌉ leaves of plates : ill. ; 22 cm. — (An Archon book on popular entertainments)
Bibliography: p. 191-193.
Includes index.
ISBN 0-208-10760-7
1. Thumb, Tom, Mrs., 1841-1919. 2. Circus performers—United States—Biography. I. Bleeker, Sylvester, joint author. II. Saxon, A. H. III. Series: Archon book on popular entertainments.
GV1811.T55A33 791.3'092'4 78-26267
78 MARC

CIRCUS TRAINS

Parkinson, Tom.
The circus moves by rail / by Tom Parkinson and Charles Philip Fox. — 1st ed. — Boulder, Colo. : Pruett Pub. Co., c1978.
391 p. : ill. ; 22 x 27 cm.
Includes index.
ISBN 0-87108-515-1 : $35.00
1. Circus trains. I. Fox, Charles Philip, 1904- joint author. II. Title.
GV1822.P37 791.3'028 78-23457
78 MARC

CIRE-PERDUE PROCESS see Precision casting

CIRENCESTER ABBEY

—HISTORY—SOURCES

Cirencester Abbey.
The cartulary of Cirencester Abbey, Gloucestershire, edited by C. D. Ross. London, Oxford University Press, 1964-77.
3 v. facsims., port. 26 cm.
Volume 3 ed. by M. Devine.
"Based upon the two abbey registers now in the possession of the present Lord Vestey at Stowell Park, near Cirencester."
Bibliographical footnotes.
1. Cirencester Abbey—History—Sources. I. Ross, Charles Derek, ed. II. Devine, Mary, ed. III. Title.
DA690.C61C5 271'.08'042417 64-55757
⌈r79⌉rev MARC

CIRICI PELLICER, ALEJANDRO

Cirici Pellicer, Alejandro.
A cor batent / Alexandre Cirici. — 1. ed. — Barcelona : Edicions Destino, 1976.
231 p. ; 19 cm. — (Col·lecció El Dofí) Sp***
ISBN 8423309568
1. Cirici Pellicer, Alejandro. 2. Art historians—Spain—Biography. 3. Spain—History—Civil War, 1936-1939. I. Title.
N7483.C53A3 1976 709'.2'4 76-479437
76⌈r79⌉rev MARC

Cirici Pellicer, Alejandro.
Les hores clares / Alexandre Cirici. — 1. ed. — Barcelona : Destino, 1977.
314 p. ; 19 cm. — (Col·lecció El Dofí) Sp***
ISBN 8423306925
1. Cirici Pellicer, Alejandro. 2. Art historians—Spain—Biography. I. Title.
N7483.C53A34 709'.2'4 79-338882
79 MARC

Cirici Pellicer, Alejandro.
Nen, no t'enfilis ⌈per⌉ Alexandre Cirici. ⌈1. ed.⌉ Barcelona, Destino ⌈1972⌉
215 p. 20 cm. (Col·lecció El Dofí) Sp***
Sequel: El temps barrat.
1. Cirici Pellicer, Alejandro. 2. Art historians—Spain—Biography. I. Title.
N7483.C53A3 1972 709'.2'4 73-310526
79⌈r79⌉rev MARC

Cirici Pellicer, Alejandro.
El temps barrat ⌈per⌉ Alexandre Cirici. ⌈1. ed.⌉ Barcelona, Destino ⌈1973⌉
338 p. 20 cm. (Col·lecció El Dofí) Sp***
Sequel to Nen, no t'enfilis.
ISBN 8423307557
1. Cirici Pellicer, Alejandro. 2. Art historians—Spain—Biography. I. Title.
N7483.C53A3 1973 709'.2'4 73-332105
79⌈r79⌉rev MARC

CIRILLO, DOMENICO MARIA LEONE, 1739-1799

Lombardi, Felice.
La scienza e l'arte medica di Domenico Cirillo. Napoli, Tip. R. Licenziato, 1964.
54 p. port. 24 cm. (Scientia veterum, 72)
Bibliography: p. 54.
1. Physiology, Pathological. 2. Cirillo, Domenico Maria Leone, 1739-1799. I. Title. II. Series.
R131.A1S27 no. 72 78-383782
⌈RB113⌉ 78 MARC

CIRNE, ADOLPHO TAĆIO DA COSTA

Menezes, José Rafael de.
Paraibanos em distinções na Faculdade de Direito do Recife / José Rafael de Menezes. — João Pessoa : Editora Universitária/UFPb., 1977.
114 p. ; 21 cm. — (Coleção Sanhauá ; 2)
1. Pinheiro, Luís Ferreira Maciel, 1838 or 9-1889. 2. Cirne, Adolpho Tacio da Costa. 3. Lawyers—Brazil—Biography. 4. Recife, Brazil. Faculdade de Direito. I. Title. II. Series.
79-350353
79 MARC

CIRRHOSIS HEPATIS see Liver—Cirrhosis

CIRRHOSIS OF THE LIVER see Liver—Cirrhosis

CIRRIPEDIA

see also Acrothoracica; Barnacles

CIRY, MICHEL

Boran, Pierre.
Michel Ciry, cantate / Pierre Boran ; gravure de Michel Ciry. — ⌈Goudargues⌉ : Éditions Chambelland, c1975.
79 p. ; 21 cm. F***
1. Ciry, Michel. I. Title.
N6853.C49B67 759.4 79-377537
79 MARC

Ciry, Michel.
L'approche du soir : journal, 1975-1978 / Michel Ciry. — ⌈Paris⌉ : Plon, c1978.
455 p. ; 24 cm. F***
ISBN 2-259-00380-X : 55.00F
1. Ciry, Michel. 2. Artists—France—Biography. I. Title.
N6853.C49A2 1978 759.4 79-367265
79 MARC

Ciry, Michel.
Détruire la nuit : journal 1974-1975 / Michel Ciry. — ⌈Paris⌉ : Plon, c1977.
401 p. ; 25 cm. F***
ISBN 2-259-00275-7 : 50.00F
1. Ciry, Michel. 2. Artists—France—Biography. I. Title.
N6853.C49A2 1977 759.4 78-367784
78 MARC

Ciry, Michel.
Michel Ciry / introd. de Michel Droit ; suivi d'extraits du Journal de Michel Ciry. — Neuchâtel : Ides et Calendes ; Paris : Bibliothèque des arts, cop. 1977.
236 p. : ill. (some col.) ; 27 cm. Sw78-1110
Bibliography: 232-233.
Includes index.
120F
1. Ciry, Michel. 2. Painters—France—Biography. I. Droit, Michel, 1923-
ND553.C57A4 1977 759.4 78-382873
78 MARC

CISIC-CHARLES, JEANNE, 1874-1955

—ART COLLECTIONS—CATALOGS

Hodler, Ferdinand, 1853-1918.
Eine unbekannte Hodler-Sammlung aus Sarajewo / Jura Brüschweiler. — Bern : Benteli, cop. 1978.
119 p. : 120 ill. ; 25 cm. — (Hodler-Publikation ; 1) Sw78-4688
Bibliography: p. 119.
ISBN 3-7165-0263-4 : 38.00F
1. Hodler, Ferdinand, 1853-1918—Catalogs. 2. Cisic-Charles, Jeanne, 1874-1955—Art collections—Catalogs. 3. Art, Swiss—Yugoslavia—Sarajevo—Catalogs. I. Brüschweiler, Jura. II. Title. III. Series.
N7153.H6A4 1978 78-381247
79 MARC

CISKEIAN TERRITORY

—ECONOMIC POLICY

Davies, William J
Development planning for symbiotic communities / by W. J. Davies. — Port Elizabeth : Institute for Planning Research, University of Port Elizabeth, 1978.
39, ⌈2⌉ p. : ill. (some col.) ; 30 cm. — (Fact paper - University of Port Elizabeth, Institute for Planning Research ; no. 26) Aus***
"An abbreviated version of this paper was presented at the Ciskei Development Conference held at Zwelitsha on May 11th 1978."
Bibliography: p. ⌈40⌉-⌈41⌉
ISBN 0-86988-145-0
1. Ciskeian Territory—Economic policy. I. Title. II. Series: Port Elizabeth, South Africa. University. Institute for Planning Research. Fact paper ; no. 26.
HC517.S72C53 330.9'68'7 79-307268
79 MARC

CIST, JACOB, 1782-1825

Powell, Howard Benjamin, 1937-
Philadelphia's first fuel crisis : Jacob Cist and the developing market for Pennsylvania anthracite / H. Benjamin Powell. — University Park : Pennsylvania State University Press, c1978.
167 p. : ill. ; 24 cm.
Includes bibliographical references and index.
ISBN 0-271-00533-5
1. Anthracite coal—Pennsylvania—History. 2. Cist, Jacob, 1782-1825. 3. Businessmen—United States—Biography. I. Title.
HD9547.P4P63 338.2'7'2509748 77-88471
78 MARC

CISTERCIAN ARCHITECTURE see Architecture, Cistercian

CISTERCIANS

see also Benedictines

Abbaye de Timadeuc.
Vie cistercienne / Abbaye de Timadeuc. — ⌈Bréhan-Loudéac⌉ (56580 Rohan) : Abbaye de Timadeuc, 1977.
96 p. : ill. ; 20 cm. F78-2184
"La vie cistercienne, méditation sur le mystère de la vie monastique, par Thomas Merton": p. 45-93.
7.00F
1. Cistercians. I. Merton, Thomas, 1915-1968. La vie cistercienne. 1977. II. Title.
BX3402.2.A22 1977 271'.12 79-367772
79 MARC

CISTERCIANS (Continued)

L'Esprit de Cîteaux : Saint Bernard sur le Cantique des Cantiques / introd. de Raymond Oursel. — La Pierre-qui-Vire : Zodiaque, 1978.
196 p. : ill. (some col.) ; 26 cm. — (Les Points cardinaux ; 26) F***
Includes bibliographical references.
90.00F
1. Cistercians. 2. Monasteries—Europe. 3. Bible. O.T. Song of Solomon—Sermons. 4. Sermons, French. I. Bernard de Clairvaux, Saint, 1091?-1153. Sermons. Selections. French. 1978. II. Oursel, Raymond.
BX3402.2.E84 271'.12 79-362413
79 MARC

—BIBLIOGRAPHY

Rochais, Henri.
Bibliographie générale de l'Ordre cistercien : archivalia / par H. Rochais et E. Manning. — Rochefort : Documentation cistercienne, 1977.
64 p. ; 30 cm. — (Documentation cistercienne ; v. 21, fasc. 2) Be***
Cover title.
1. Cistercians—Bibliography. I. Manning, Eugène, joint author. II. Title. III. Series.
Z7840.C5R6 78-382150
[Z3402.2] 78 MARC

—COLLECTED WORKS

Zisterzienser-Studien. 1-
Berlin, Colloquium Verlag, 1975-
v. 24 cm. (Studien zur europäischen Geschichte)
1. Cistercians—Collected works. I. Series.
BX3401.Z57 271'.12'04 78-645980
MARC-S

—HISTORY—ADDRESSES, ESSAYS, LECTURES

Cistercian ideals and reality / edited by John R. Sommerfeldt. — Kalamazoo, Mich. : Cistercian Publications, 1978.
vii, 351 p. ; 22 cm. — (Cistercian studies series ; no. 60)
Includes bibliographical references.
ISBN 0-87907-860-X
1. Cistercians—History—Addresses, essays, lectures. I. Sommerfeldt, John R. II. Series.
BX3406.2.C57 271'.12'04 78-16615
78 MARC

CISTERCIANS IN CASTILE

—HISTORY

Alvarez Palenzuela, Vicente A
Monasterios cistercienses en Castilla : (siglos XII-XIII) / [por] Vicente-Angel Alvarez Palenzuela. — Valladolid : Universidad, Secretariado de Publicaciones, 1978.
292 p. ; 25 cm. Sp78-July
Thesis—Valladolid, 1974.
Bibliography: p. 259-270.
Includes indexes.
ISBN 8460010848
1. Cistercians in Castile—History. 2. Monasteries—Spain—Castile—History. 3. Castile—Church history. I. Title.
BX3444.C37A48 79-349390
79 MARC

CISTERCIANS IN DENMARK

—ADDRESSES, ESSAYS, LECTURES

Festskrift til Olga Bartholdy på 80-årsdagen 20. juli 1978 / [redaktion Niels Sterum, desuden Frede Gotthardsen og Jens Kr. Krarup]. — Løgumkloster : Refugiets kulturforening, 1978.
168 p. : ill. ; 21 cm. — (Løgumkloster-studier ; 1) D***
Summaries in English, French, and German.
Includes bibliographical references.
CONTENTS: Bork Hansen, A. Olga Bartholdy og Løgumkloster.—McGuire, B. P. Clairvaux og Nordens cistercienser i 1100-tallet.—Graae, R. Slottet i Løgumkloster.—Haaren, F. van. Klosterlivets kætteri.—Haase, N. Løgumskloster under tredveårskrigen.—Lindhardt, P. G. Præsteuddannelse.—Riising, A. Om Løgumkloster 1863-64.—Sterum, N. Om to kendte og een nyopdaget varmekælder i cistercienserklostret i Løgum.—Vessum, B. van. Lectio divina.—Wissing, J. A. Betragtninger vedrørende tekniske anlæg i Løgum kloster.
ISBN 8798022938
1. Løgum kloster—Addresses, essays, lectures. 2. Cistercians in Denmark—Addresses, essays, lectures. 3. Bartholdy, Olga—Addresses, essays, lectures. I. Bartholdy, Olga. II. Sterum, Niels. III. Gotthardsen, Niels. IV. Gotthardsen, Frede. V. Krarup, Jens Kr. VI. Løgumkloster refugium. Kulturforening.
BX2644.L63F47 79-381182
79 MARC

CISTERCIANS IN EUROPE

—HISTORY—PERIODICALS

Zisterzienser-Studien. 1-
Berlin, Colloqium Verlag, 1975-
v. 24 cm. (Studien zur europäischen Geschichte)
1. Cistercians in Europe—History—Periodicals. 2. Europe—Church history—Periodicals. 3. Europe—History—476-1492—Periodicals. I. Series.
BX3415.Z57 79-641572
MARC-S

CISTERCIANS IN FRANCE

—BIOGRAPHY

Luzsénsky, Guy.
Guy Luzsénszky, moine. — Paris (10, rue Mayet, 75006) : J.-P. Delarge, 1978.
211 p. ; 24 cm. F78-14110
ISBN 2-7113-0096-X : 48.00F
1. Luzsénszky, Guy. 2. Cistercians in France—Biography. I. Title.
BX4705.L953A33 271'.12'024 79-385549
79 MARC

CISTERCIANS IN ITALY

—CONGRESSES

I Cistercensi e il Lazio : atti delle giornate di studio dell'Istituto di storia dell'arte dell'Università di Roma, 17-21 maggio 1977 / Istituto nazionale di archeologia e storia dell'arte, Istituto di storia dell'arte dell'Università di Roma. — 1. ed. — Roma : Multigrafica, 1978.
x, 302 p. : ill. ; 28 cm. It78-Sept
Italian or German.
Includes bibliographical references and indexes.
1. Art, Cistercian—Italy—Lazio—Congresses. 2. Christian art and symbolism—Medieval, 500-1500—Italy—Lazio—Congresses. 3. Cistercians in Italy—Congresses. I. Rome (City). Università. Istituto di storia dell'arte. II. Istituto nazionale di archeologia e storia dell'arte.
N7952.A3L392 79-364090
*79 MARC

CISTERCIANS IN NORWAY

—HISTORY

Johnsen, Arne Odd, 1909-
De norske cistercienserklostre 1146-1264 sett i europeisk sammenheng / Arne Odd Johnsen. — Oslo : Universitetsforl., 1977 (Kristiansand : E. Høgfeldt boktr.)
96 p. ; 25 cm. — (Avhandlinger - Det Norske videnskaps-akademi. Hist.-filos. klasse ; ny ser., no. 15) N77-Dec.
Summary in English.
Includes bibliographical references and index.
ISBN 8200016536
1. Cistercians in Norway—History. 2. Norway—Church history. I. Title. II. Series: Norske videnskaps-akademi i Oslo. Historisk-filosofisk klasse. Avhandlinger ; ny ser., no. 15.
BX3443.N8J63 78-380526
78 MARC

CISTERNOGRAPHY, MYELOGRAPHIC see Myelography

CISTOPHORUS (COIN)

Kleiner, Fred S
The early cistophoric coinage / by Fred S. Kleiner and Sydney P. Noe. — New York : American Numismatic Society, 1977.
128 p., [20] leaves of plates : ill. ; 29 cm. — (Numismatic studies ; no. 14)
Bibliography: p. 3-9.
1. Cistophorus (Coin) I. Noe, Sydney Philip, joint author. II. Title. III. Series.
CJ359.K57 737.4 79-317946
79 MARC

CISTOTHORUS APOLINARI see Apolinar's marsh wren

CITATION INDEXES

Garfield, Eugene.
Citation indexing - its theory and application in science, technology, and humanities / Eugene Garfield. — New York : Wiley, c1979.
xxi, 274 p. : ill. ; 26 cm. — (Information sciences series)
"A Wiley-Interscience publication."
Bibliography: p. 235-239.
Includes indexes.
ISBN 0-471-02559-3
1. Science—Abstracting and indexing. 2. Technology—Abstracting and indexing. 3. Humanities—Abstracting and indexing. 4. Citation indexes. I. Title.
Z697.S5G37 029.5 78-9713
78 MARC

CITATION OF AUTHORITY IN THE NEWS see Attribution of news

CITATION OF LEGAL AUTHORITIES

—FLORIDA

Girtman, Ben E
Courts, reports, and digests : Girtman's compendium of Federal and Florida case law : where to find it and how to cite it / by Ben E. Girtman. — Tallahassee : Girtman, c1979.
x, 118 p. ; 22 cm.
Bibliography: p. 115-116.
$6.50
1. Legal research—Florida. 2. Citation of legal authorities—Florida. 3. Legal research—United States. 4. Citation of legal authorities—United States. I. Title. II. Title: Girtman's compendium of Federal and Florida case law.
KFF75.G57 347'.759'072 75-35468
79 MARC

—NEW YORK (STATE)

New York rules of citation. — 1st ed. — [Jamaica, N.Y.] : St. John's Law Review, [c1978]
32 p. : ill. ; 26 cm.
Cover title.
1. Citation of legal authorities—New York (State) I. St. John's Law Review.
KFN5074.5.N48 340'.07'20747 79-106585
79 MARC

—PHILIPPINE ISLANDS

Cortés, Irene R 1921-
Philippine manual of legal citations / by Irene R. Cortés, Merlin M. Magallona, Myrna S. Feliciano. — Rev. ed. — [Quezon City] : U.P. Law Center, 1977.
vi, 91 p. ; 20 cm.
1. Citation of legal authorities—Philippine Islands. I. Magallona, Merlin M., joint author. II. Feliciano, Myrna S., joint author. III. Title.
340'.01'48 79-102242
79 MARC

CITATION OF LEGAL AUTHORITIES
(Continued)

—TEXAS

Texas rules of form / Texas law review. — 3d ed. — Austin : Texas Law Review Publications, c1974.
18 p. ; 23 cm.
Cover title.
1. Citation of legal authorities—Texas. I. Texas law review.
KFT1275.T4 1974 347'.764'055 79-113027
79 MARC

—UNITED STATES

Bieber, Doris M
Dictionary of legal abbreviations used in American law books / compiled by Doris M. Bieber. — Buffalo : W. S. Hein, 1979.
337 p. ; 26 cm.
ISBN 0-930342-61-5
1. Law—United States—Abbreviations. 2. Citation of legal authorities—United States. I. Title.
KF246.B46 340.1'48 78-60173
00 MARC

Girtman, Ben E
Courts, reports, and digests : Girtman's compendium of Federal and Florida case law : where to find it and how to cite it / by Ben E. Girtman. — Tallahassee : Girtman, c1979.
x, 118 p. ; 22 cm.
Bibliography: p. 115-116.
$6.50
1. Legal research—Florida. 2. Citation of legal authorities—Florida. 3. Legal research—United States. 4. Citation of legal authorities—United States. I. Title. II. Title: Girtman's compendium of Federal and Florida case law.
KFF75.G57 347'.759'072 75-35468
79 MARC

CITATIONS, BIBLIOGRAPHICAL
see Bibliographical citations

CITATIONS (LAW) see Annotations and citations (Law)

CITHARA

Vogel, Martin, 1923-
Chiron der Kentaur mit der Kithara / Martin Vogel. — Bonn-Bad Godesberg : Verlag für Systematische Musikwissenschaft, 1978.
2 v. (764 p.) : ill. ; 25 cm. — (Orpheus, Schriftenreihe zu Grundfragen der Musik ; Bd. 25-26) GFR***
Bibliography: p. v. 2, p. 665-671.
Includes indexes.
1. Music, Greek and Roman. 2. Cithara. I. Title. II. Series.
ML169.V6 78-383153
79 MARC

CITIES, IMAGINARY see Geographical myths

CITIES AND STATE see Urban policy

CITIES AND TOWNS

see also Art, Municipal; Capitals (Cities); Central places; Community; Company towns; Cries; Education, Urban; Garden cities; Land use, Urban; Markets; Plazas; Sociology, Urban; Squatter settlements; Streets; Tenement—Houses; Urbanization; Villages; and headings beginning with the words City, Municipal and Urban

Bertrand, Michel Jean.
Pratique de la ville / par Michel-Jean Bertrand. — Paris ; New York : Masson, 1978.
210 p., [4] leaves of plates : ill. ; 24 cm. — (Collection Géographie) F***
Includes bibliographical references.
ISBN 2-225-48202-0 : 55.00F
1. Cities and towns. 2. City planning. I. Title.
HT151.B437 301.36 78-396400
79 MARC

Bookchin, Murray, 1921-
The limits of the city [by] Murray Bookchin. New York, Harper & Row [1974]
xi, 147 p. 19 cm. (Harper colophon books/CN 1013) $2.75
Includes bibliographical references.
ISBN 0-06-136134-8. ISBN 0-06-091013-5 (pbk.)
1. Cities and towns. 2. Cities and towns—Growth. I. Title.
HT151.B58 1974 301.36'3 73-17852
74[r79]rev2 MARC

Boyden, Stephen Vickers.
An integrative ecological approach to the study of human settlements / Stephen Boyden. — Paris : Unesco, 1979.
87 p. : ill. ; 27 cm. — (MAB technical notes ; 12) F***
Bibliography: p. 85-87.
ISBN 923101689X
1. Human ecology. 2. Cities and towns. 3. City planning. 4. Human ecology—Hongkong. I. Title. II. Series: International Co-ordinating Council of the Programme on Man and the Biosphere. MAB technical reports ; 12.
GF101.B63 301.34 79-317697
79 MARC

Castells, Manuel.
La Question urbaine / Manuel Castells. — Éd. revue / suivie d'une postface 1975. — Paris : F. Maspero, 1977.
529 p. ; 22 cm. — (Textes à l'appui : Sociologie) F78-3675
Includes bibliographical references.
ISBN 2-7071-0513-9 : 55.00F
1. Cities and towns. 2. Urbanization. I. Title.
HT151.C376 1977 301.36'1 78-396399
79 MARC

Le Città capitali nel XX secolo / Mario Manieri-Elia ; contributi di G. Ciucci ... [et al.]. — 1. ed. — Milano : Fratelli Fabbri, 1978.
128 p. : ill. ; 28 cm. — (L'Arte nella società) It78-Dec
Bibliography: p. 126-127.
L4500
1. Cities and towns. 2. City planning. 3. Capitals (Cities) I. Manieri-Elia, Mario, 1929- . II. Ciucci, Giorgio.
HT153.C55 79-346226
*79 MARC

Cucu, Vasile.
Urbanism : [curs] / Vasile Cucu, Cezar Popescu. — Bucureşti : Centrul de multiplicare al Universităţii din Bucureşti, 1975-
v. : ill. ; 24 cm. R75-4274
At head of title: Universitatea din Bucureşti. Facultatea de Geologie-Geografie. Catedra de Geografie Economică.
"Destinat studenţilor anului II curs de zi şi anului III curs fără frecvenţă de la Facultatea de drept a Universităţii din Bucureşti."
Bibliography: v. 1, p. 316-320.
lei15.00 (v. 1)
1. Cities and towns. 2. City planning. 3. Cities and towns—Romania. I. Popescu, Cezar, joint author. II. Title.
HT151.C82 78-393930
*79 MARC

Deligny, Jean Louis.
Mieux se déplacer dans votre ville : transports collectifs et individuels, circulation, sécurité, environnement, cadre de vie, stationnement, équipements, financement, gestion / Jean-Louis Deligny, Claude Gressier, avec la collaboration de Emmanuel Gambet ... [et al.]. — Paris : Éditions du Moniteur, 1978.
332 p. : ill. ; 24 cm. — (Collection Vie locale ISSN 0151-0185) F***
Bibliography: p. 323-332.
ISBN 2-86282-012-1 : 88.00F
1. Cities and towns. 2. Urbanization. I. Gressier, Claude, joint author. II. Title.
HT151.D425 301.36'1 79-348626
79 MARC

Ganassi Agger, Simona.
Autogestione urbana : l'urbanistica per una nuova società / Simona Ganassi Agger. — Bari : Dedalo libri, 1977.
275 p. ; 21 cm. — (Saggi ; 27) It78-Nov
Bibliography: p. 267-[273]
Includes indexes.
L4000
1. Cities and towns. 2. City planning. 3. Sociology, Urban. I. Title.
HT151.G318 301.36'1 79-351884
*79 MARC

Ganassi Agger, Simona.
Urban self-management : planning for a new society / Simona Ganassi Agger. — White Plains, N.Y. : M. E. Sharpe, c1979.
235 p. ; 24 cm.
Expanded and rev. version of the author's Autogestione urbana, published in 1977.
"Published simultaneously as vol. IX, no. 1-2 of International journal of sociology."
Includes bibliographical references.
ISBN 0-87332-125-1
1. Cities and towns. 2. City planning. 3. Sociology, Urban. I. Title.
HT151.G31813 309.2'62 78-73223
79 MARC

Garnier, Jean Pierre.
La comédie urbaine : ou, La cité sans classes / Jean-Pierre Garnier, Denis Goldschmidt. — Paris : F. Maspero, 1978.
334 p. ; 22 cm. — (Textes à l'appui) F***
Includes bibliographical references.
ISBN 2-7071-1037-X : 55.00F
1. Cities and towns. 2. Sociology, Urban. 3. Urbanization. I. Goldschmidt, Denis, joint author. II. Title.
HT151.G335 301.36'1 79-355812
79 MARC

Great cities of the world / [editor, Nicolas Wright ; written by Tony Aldous et al.]. — London ; New York : Marshall Cavendish, 1978.
viii, 5-580 p. : ill. (some col.), maps, ports. ; 28 cm. GB79-09875
Ill. on lining papers.
ISBN 0-85685-514-6 : £9.95
1. Cities and towns. I. Wright, Nicolas, fl. 1978- . II. Aldous, Tony.
G140.G69 909'.09'732 79-313973
79 MARC

Guest, Arthur.
Man and landscape : a practical handbook of case studies and techniques in geography with special reference to Europe / [by] Arthur Guest. — 2nd ed. — London : Heinemann Educational, 1978.
x, 116 p. : ill., maps (some col.) ; 31 cm. GB78-21725
Part of illustrative matter in pocket.
Bibliography: p. 111-112.
Includes indexes.
ISBN 0-435-35355-1 : £2.95
1. Anthropo-geography. 2. Geography, Economic. 3. Cities and towns. I. Title.
GF41.G83 1978 940.55 78-325967
79 MARC

Hajčíková, Marta.
Zoznam vžitých slovenských názvov miest sveta / [sprac. Marta Hajčíková]. — Bratislava : Slov. úrad geodézie a kartografie, 1974.
15 p. ; 24 cm. — (Kartografické informácie ; 4) CzS74
Bibliography: p. 15.
1. Names, Geographical—Slovak. 2. Cities and towns. I. Title.
G106.S57H326 78-380896
*78 MARC

Heineman, Hans-Erland, 1919-
Stad / [Hans-Erland] Heineman. — Stockholm : Esselte studium, 1976.
120 p. : ill. ; 23 cm. — (Scandinavian university books) S76-41
Bibliography: p. 119-120.
ISBN 9124261696 : kr69.00
1. Cities and towns. 2. Cities and towns—Sweden. I. Title.
HT151.H367 79-338833
*79[r79]rev MARC

Kakumoto, Ryōhei, 1920-
(Konnichi no toshi asu no toshi)
今日の都市明日の都市：都市固定化の時代 / [角本良平]. — 町田：玉川大学出版部, 1978.
240 p. : 19 cm. — (玉川選書 : 85) Ja 78-25871
Series romanized: Tamagawa sensho.
Bibliography: p. 233-237.
Includes index.
¥880
1. Cities and towns. 2. City planning. I. Title.
HT151.K3 79-802607

Kiuchi, Shinzō, 1910-
(Toshi chirigaku genri)
都市地理学原理 / 木内信蔵著. — 東京：古今書院, 1979.
ii, 8, 380 p., [3] leaves of plates : ill. ; 22 cm. Ja 79-5972
Title on verso of t. p.: Principles of urban geography.
Includes bibliographies and indexes.
¥4800
1. Cities and towns. 2. Land use, Urban. I. Title. II. Title: Principles of urban geography.
GF125.K49 79-808434

CITIES AND TOWNS (Continued)

Kupriiânov, Vsevolod Vladimirovich.
(Gidrologicheskie aspekty urbanizatsii)
Гидрологические аспекты урбанизации : (Гидрология городов и урбанизир. территорий) / В. В. Куприянов. — Ленинград : Гидрометеоиздат, 1977.
183 p. : ill. ; 19 cm. USSR 77
Table of contents and summary also in English.
Bibliography : p. 173–[181]
2.14rub
1. Cities and towns. 2. Human ecology. 3. Hydrology. I. Title.
GF125.K87 78–422079

Lavoie, Claude.
Initiation à l'urbanisme : gestion municipale / Claude Lavoie ; préf. de Guy Tardif. — Montréal : G. Le Pape, c1978.
150 p. : ill., maps ; 22 cm. C***
ISBN 0-88536-032-X : $15.00
1. Urbanization. 2. Cities and towns. I. Title.
HT151.L29 309.2'62 79-352181
79 MARC

Marinović-Uzelac, Ante.
Socijalni prostor grada / Ante Marinović-Uzelac ; [crteže izradio Zlatko Barović]. — Zagreb : SNL, 1978.
241 p. : ill. ; 24 cm. — (Udžbenici Sveučilišta u Zagrebu) Yu***
Includes bibliographical references and index.
1. Cities and towns. 2. Space in economics. 3. Human ecology. I. Title.
HT111.M29 78-399721
79 MARC

Mura, Pietro Mario, 1945-
Lo spazio geografico : introduzione alla geografia urbana e regionale / Pietro Mario Mura. — 1. ed. — Napoli : Liguori, 1978.
165 p. ; 22 cm. It78-Oct
Includes bibliographical references.
L4500
1. Cities and towns. 2. Space in economics. 3. Regional planning. I. Title.
HT151.M82 78-395893
*79 MARC

Muramatsu, Shigeki, 1905–
(Toshi no hattatsu)
都市の発達 / 村松繁樹著. — 東京 : 三省堂, 昭和25 [1950]
5, 225, 5 p., [1] leaf of plates : ill. ; 18 cm. — (社会科文庫)
Series romanized: Shakaika bunko.
Includes index.

1. Cities and towns. 2. Cities and towns—History. I. Title.
HT111.M86 79–810013

Northam, Ray M
Urban geography / Ray M. Northam. — 2d ed. — New York : Wiley, c1979.
x, 512 p. : ill. ; 26 cm.
Includes bibliographical references and index.
ISBN 0-471-03292-1
1. Cities and towns. 2. Cities and towns—United States. I. Title.
HT151.N64 1979 301.36'3'09 78-12335
78 MARC

Racionero, Luis.
Sistemas de ciudades y ordenación del territorio / Luis Racionero. — Madrid : Alianza, c1978.
168 p. : ill. ; 20 cm. — (Alianza universidad ; 210) Sp***
Bibliography: p. [165]-168.
ISBN 8420622109
1. Cities and towns. 2. Regional economics. 3. Cities and towns—Spain. I. Title.
HT151.R25 79-343937
79 MARC

Shakū'ī, Ḥusayn.
(Jughrāfiyā-yi shahrī)
جغرافيای شهری / تاليـــف حسين شكوئـــی. — تهران : كتابفروشی تهران ، [19 .13
v. : ill. ; 25 cm.
Added t. p.: Géographie urbaine.
In Persian.
Includes bibliographies and index.
1. Cities and towns. I. Title. II. Title: Géographie urbaine.
GF125.S52 79–971079

Tendências atuais na geografia urbano/regional : teorização e quantificação / Speridão Faissol ... [et al.]. — [Rio de Janeiro] : IBGE, 1978.
301 p. : ill. ; 27 cm.
At head of title: Secretaria de Planejamento da Presidência da República, Fundação Instituto Brasileiro de Geografia e Estatística, Diretoria Técnica—Superintendência de Estudos Geográficos e Sócio-Econômicos. Instituto Pan-Americano de Geografia e História, Comissão de Geografia.
Includes bibliographies.
1. Anthropo-geography—Methodology. 2. Cities and towns. 3. Central places. 4. Space and economics. I. Faissol, Speridão, 1923- . II. Fundação IBGE. Superintendência de Estudos Geográficos e Sócio-Econômicos. III. Pan American Institute of Geography and History. Commission on Geography.
GF125.T46 79-114030

Terranova, Antonino, 1942-
Città sognate / di Antonino Terranova. — 1. ed. — Firenze : La nuova Italia, 1977.
131 p. : ill. ; 18 cm. — (Italia nostra ; 6 : Educazione) It78-Feb
Bibliography: p. 131.
L1900
1. Cities and towns. 2. Sociology, Urban. I. Title.
HT151.T45 78-368820
*78 MARC

Urbani, Leonardo.
La città è sola : plurispazialità territoriale, obsolescenza del manufatto edilizio, mobilità della forma urbana / Leonardo Urbani. — Roma : Studium, 1978.
xxix, 273 p. ; 21 cm. — (La Cultura ; 14) It78-Nov
Includes bibliographical references.
L8500
1. Cities and towns. 2. City planning. 3. Architecture. I. Title.
HT153.U75 309.2'62 79-341774
*79 MARC

Watanuki, Isahiko, 1892–1943.
(Shūraku-chirigaku)
聚落地理學 / 綿貫勇彦著. — 増補版. — 東京 : 矢島書房, 昭和30 [1955]
2, 2, 4, 286 p., [2] leaves of plates : ill. ; 22 cm.
List of the author's works: p. 278–279.
Includes bibliographies and index.

1. Land settlement. 2. Cities and towns. I. Title.
GF101.W37 1955 79–803128

Yao, Jung-ling.
(Tu shih shih wu kai lun)
都市事務概論 / 姚榮齡著. — 臺北 : 永大書局, 民國67 [1978]
8, 434 p. ; 22 cm.
Includes bibliographical references.
NT$150.00

1. Cities and towns. I. Title.
HT151.Y33 79–835091

Yuge, Shichirō, 1891–
(Toshi seisaku)
都市政策 / 弓家七郎著. — 東京 : 廣文社, 昭和26 [1951]
4, 258 p. ; 22 cm.
Includes bibliographical references.

1. City planning. 2. Cities and towns. I. Title.
HT166.Y83 78–815799

Zimm, Alfred.
Zu einigen Tendenzen der Bevölkerungsverteilung und der Naherholung in agglomerierten Gebieten / Alfred Zimm. Zur Stellung der Städte der DDR im Siedlungssystem nach ihren In-, Umland- und Fernfunktionen / Frankdieter Grimm. — Berlin : Akademie-Verlag, 1977.
39 p. : graphs ; 22 cm. — (Sitzungsberichte der Akademie der Wissenschaften der DDR ; Jahrg. 1977, Nr. 19/N : Mathematik, Naturwissenschaften, Technik) GFR78-A28
Cover title: Tendenzen der Bevölkerungsverteilung und Stellung der Städte im Siedlungssystem.
Includes bibliographies.
4.00M
1. Cities and towns. 2. Recreation. 3. Cities and towns—Germany, East. I. Grimm, Frankdieter. Zur Stellung der Städte der DDR im Siedlungssystem nach ihren In-, Umland- und Fernfunktionen. 1977. II. Title. III. Title: Tendenzen der Bevölkerungsverteilung und Stellung der Städte im Siedlungssystem. IV. Series: Akademie der Wissenschaften der DDR. Sitzungsberichte der Akademie der Wissenschaften der DDR ; Jahrg. 1977, Nr. 19/N.
HT153.Z49 79-375429
79 MARC

—ADDRESSES, ESSAYS, LECTURES

Bailly, Antoine.
La perception de l'espace urbain : les concepts, les méthodes d'étude, leur utilisation dans la recherche urbanistique / Antoine S. Bailly. — Paris : Centre de recherche d'urbanisme, [1977]
264 p. : ill. ; 21 cm. F***
Bibliography: p. 215-249.
1. Cities and towns—Addresses, essays, lectures. I. Title.
HT151.B255 301.36 78-388242
79 MARC

Blumenfeld, Hans.
Metropolis and beyond : selected essays / by Hans Blumenfeld ; edited by Paul D. Speiregen. — New York : Wiley, c1978.
xiii, 420 p. : port. ; 26 cm.
"A Wiley-Interscience publication."
Includes bibliographical references and indexes.
ISBN 0-471-04281-1
1. Cities and towns—Addresses, essays, lectures. I. Spreiregen, Paul D. II. Title.
HT151.B56 301.36'2 78-17955
78 MARC

The Built environment / general editors, John Lenihan and William W. Fletcher. — New York : Academic Press, 1978.
xii, 171 p. : ill. ; 23 cm. — (Environment and man ; v. 8)
Includes bibliographies and index.
ISBN 0-12-443508-4
1. Human ecology—Addresses, essays, lectures. 2. Cities and towns—Addresses, essays, lectures. 3. Technology—Social aspects—Addresses, essays, lectures. 4. Pollution—Addresses, essays, lectures. I. Lenihan, J. M. A. II. Fletcher, William Whigham, 1918- . III. Series.
GF125.B84 301.36 79-108781
00 MARC

Cités géantes / sous la direction de Jean Planchais. — Paris : Fayard, c1978.
263 p. : maps ; 23 cm. F***
Articles published in Le Monde, whose name appears at head of title.
Includes bibliographical references.
ISBN 2-213-00656-3 : 49.00F
1. Cities and towns—Addresses, essays, lectures. I. Planchais, Jean, 1922- . II. Le Monde, Paris.
HT151.C567 301.36'3 79-371460
79 MARC

La Città e la crisi del capitalismo / [scritti di] Clara Cardia ... [et al.]. — Roma ; Bari : Laterza, 1978.
xiv, 206 p. ; 18 cm. — (Tempi nuovi ; 102) It78-June
Includes bibliographical references.
L3900
1. Sociology, Urban—Addresses, essays, lectures. 2. Cities and towns—Addresses, essays, lectures. 3. Economic history—1945- —Addresses, essays, lectures. I. Cardia, Clara.
HT108.C53 78-380869
*78 MARC

La Crisi del governo urbano : istituzioni, strutture economiche e processi politici nelle città del capitalismo maturo / a cura di Paolo Ceccarelli. — 1. ed. — Venezia : Marsilio, 1978.
299 p. ; 21 cm. — (Biblioteca Marsilio : Architettura e urbanistica ; 31) It78-Aug
Includes bibliographical references.
L7800
1. Cities and towns—Addresses, essays, lectures. 2. Municipal finance—Addresses, essays, lectures. 3. Municipal government—Addresses, essays, lectures. I. Ceccarelli, Paolo, 1934- .
HT153.C74 79-337964
*79 MARC

Davies, R J
Of cities and societies : a geographer's viewpoint / R. J. Davies. — [Cape Town] : University of Cape Town, 1976.
20 p. : ill. ; 22 cm. — (Inaugural lecture : New series - University of Cape Town ; no. 38) SA***
Cover title.
Delivered May 20, 1976.
Bibliography: p. 19-20.
ISBN 0-7992-0178-2
1. Anthropo-geography—Addresses, essays, lectures. 2. Cities and towns—Addresses, essays, lectures. 3. Cities and towns—South Africa—Addresses, essays, lectures. I. Title. II. Series: Cape Town. University of Cape Town. Inaugural lecture : New series ; no. 38.
GF125.D38 301.36 79-313616
79 MARC

L'Ecosistema urbano : scritti / di U. Cardarelli ... [et al.] : [a cura di] Manfredi Nicoletti. — Bari : Dedalo libri, 1978.
398 p. : ill. ; 22 x 24 cm. — (Il Politecnico ; 9) It79-Jan
Includes bibliographies.
L24000
1. Cities and towns—Addresses, essays, lectures. 2. City planning—Addresses, essays, lectures. 3. Human ecology—Addresses, essays, lecture. I. Nicoletti, Manfredi, 1930- . II. Cardarelli, Urbano.
HT151.E25 301.36'1 79-347094
*79 MARC

L'Homme et sa ville / [J. Chirac et al.]. — Paris : Tallandier, 1977.
132 p. ; 23 cm. — (La Nef ; 34. année, nouv. sér., cahier no 64) F***
18.00F
1. Cities and towns—Addresses, essays, lectures. 2. City and town life—Addresses, essays, lectures. I. Chirac, Jacques, 1932- . II. Series.
HT151.H642 909'.09'732 79-362311
79 MARC

(Rekishi-chiri kenkyū to toshi kenkyū)
歴史地理研究と都市研究 / 藤岡謙二郎先生退官記念事業会編. — 東京 : 大明堂, 1978.
2 v. : ill. ; 27 cm. Ja 79-4586
List of K. Fujioka's works: v. 2, p. [513]–516.
Includes bibliographical references.
¥18000
1. Japan—Historical geography—Addresses, essays, lectures. 2. Cities and towns—Addresses, essays, lectures. 3. Geography, Historical—Addresses, essays, lectures. 4. Fujioka, Kenjirō, 1914– I. Fujioka, Kenjirō, 1914– II. Fujioka Kenjirō Sensei Taikan Kinen Jigyōkai.
DS806.4.R44 79–809998

CITIES AND TOWNS
—ADDRESSES, ESSAYS, LECTURES
(Continued)

Die Stadt als Text. — Tübingen : Wasmuth, 1976.
123 p. : ill., map ; 22 cm. — (Konzept ; 3) GFR76-A
CONTENTS: Carlini, A. Die Stadt war ein Buch.—Barthes, R. Semiotik und Urbanismus.—Choay, F. Semiotik und Urbanismus.—Schneider, B. Die Stadt als Text.—Trabant, J. Die Stadt und die Sprache.—Hillier, B. und Leaman, A. Das Mensch-Umwelt-Paradigma und seine Paradoxien.—Schmidt, S. J. Ästhetiktheorie und Architekturästhetik.
ISBN 3-8030-0124-2 : DM24.00
1. Cities and towns—Addresses, essays, lectures. 2. City planning—Addresses, essays, lectures.
HT155.S664 78-380749
79 MARC

—BEAUTIFICATION see Urban beautification

—BIBLICAL TEACHING

Shriver, Donald W
Is there hope for the city? / Donald W. Shriver, Jr., and Karl A. Ostrom. — 1st ed. — Philadelphia : Westminster Press, c1977.
204 p. : ill. ; 20 cm. — (Biblical perspectives on current issues)
Includes bibliographical references and indexes.
ISBN 0-664-24147-6
1. Cities and towns—Biblical teaching. I. Ostrom, Karl A., 1939- joint author. II. Title. III. Series.
BS680.C5S5 220.8'301'363 77-22187
77[r79]rev MARC

—BIBLIOGRAPHY

Brzeski, Wladyslaw.
An annotated bibliography of the literature on filtering and related aspects of urban housing markets / Wladyslaw Brzeski. — Vancouver : Faculty of Commerce & Business Administration, University of British Columbia, 1977.
102 p. ; 22 cm. — (Bibliography series - Urban Land Economics ; 1) C***
1. Housing—Bibliography. 2. Cities and towns—Bibliography. I. Title. II. Series: University of British Columbia. Urban Land Economics Division. Bibliography series - Urban Land Economics ; 1.
Z7164.H8B79 016.3015'4 79-311779
[HD7287] 79 MARC

—CIVIC IMPROVEMENT

see also Community art projects; Community centers; Community organization; Municipal buildings; Urban beautification; Urban renewal

Isomura, Eiichi, 1903–
(Toshi kenshō)
都市憲章 / 磯村英一. — 東京：鹿島出版会, 1978.
223 p. : ill. ; 19 cm. — (SD選書 ; 136) Ja 78–14343
Series romanized: SD sensho.
¥980
1. City planning. 2. Cities and towns—Civic improvement. I. Title.
HT166.I85 78–812617

—CIVIC IMPROVEMENT—ADDRESSES, ESSAYS, LECTURES

Designing with community participation / edited by Henry Sanoff. — Stroudsburg, Pa. : Dowden, Hutchinson & Ross, c1978.
ix, 253 p. : ill. ; 23 x 29 cm. — (Community development series ; 40)
Bibliography: p. 249.
Includes index.
ISBN 0-87933-314-6 (McGraw-Hill)
1. City planning—Citizen participation—Addresses, essays, lectures. 2. Community development, Urban—Addresses, essays, lectures. 3. Cities and towns—Civic improvement—Addresses, essays, lectures. I. Sanoff, Henry.
HT166.D387 1978 309.2'62 78-3723
78 MARC

—COLLECTED WORKS

Geography and the urban environment. v. 1–
Chichester, New York, John Wiley, 1978–
v. 24 cm.
1. Cities and towns—Collected works. 2. Anthropo-geography—Collected works. 3. City planning—Collected works.
GF125.G46 301.36'05 78-647093
MARC-S

—CONGRESSES

Misure umane : un dibattito internazionale su borgo città quartiere comprensorio / [a cura di] Carlo Doglio, Luigina Fasoli, Paolo Guidicini. — Milano : F. Angeli, [1978]
380 p. ; 22 cm. — (Collana di sociologia urbana e rurale ; 10) It78-July
Proceedings of a meeting held in Bologna in 1977.
Includes bibliographical references.
L5000
1. Cities and towns—Congresses. 2. Sociology, Urban—Congresses. 3. City planning—Congresses. I. Doglio, Carlo, 1914- . II. Fasoli, Gina, 1905- . III. Guidicini, Paolo.
HT107.M57 78-384612
*78 MARC

Toshi Mondai Shimpojūmu, 1st, Tokyo, 1977.
(Tōkyō Rondon no kenkyū)
東京·ロンドンの研究：都市問題シンポジウムの記録 / 広岡治哉, 柴田徳衛編. — 東京：法政大学出版局, 1978.
x, 199 p. : ill. ; 20 cm. — (叢書·現代の社会科学) Ja 79–2363
Series romanized: Sōsho gendai no shakai kagaku.
Includes bibliographies.
¥1000
1. Cities and towns—Congresses. 2. Tokyo—Social conditions—Congresses. 3. London—Social conditions—Congresses. I. Hirooka, Haruya. II. Shibata, Tokue, 1924- III. Title.
HT107.T67 1977 79–805443

—DICTIONARIES—SPANISH

Landa, Horacio.
Terminología de urbanismo / Horacio Landa. — México : Centro de Investigación, Documentación e Información sobre la Vivienda, Instituto Nacional para el Desarrollo de la Comunidad Rural y de la Vivienda Popular, 1976.
iv, 82 p. ; 21 cm. — (Publicaciones del CIDIV)
Bibliography: p. 81-82.
1. Cities and towns—Dictionaries—Spanish. 2. Urbanization—Dictionaries—Spanish. I. Instituto Nacional para el Desarrollo de la Comunidad Rural y de la Vivienda Popular. Centro de Investigación, Documentación e Información sobre la Vivienda. II. Title. III. Series: Instituto Nacional para el Desarrollo de la Comunidad Rural y de la Vivienda Popular. Centro de Investigación, Documentación e Información sobre la Vivienda. Publicaciones del CIDIV.
HT108.5.L36 78-107748
79 MARC

—ECONOMIC ASPECTS see Urban economics

—ENVIRONMENTAL ASPECTS—EVALUATION

Urban environmental indicators. — Paris : Organisation for Economic Co-operation and Development ; [Washington, D.C. : sold by OECD Publications Center], c1978.
274 p. : ill. ; 24 cm. F***
Includes bibliographical references.
ISBN 9264117547 : $14.00 (U.S.)
1. Cities and towns—Environmental aspects—Evaluation. I. Organization for Economic Cooperation and Development.
HT155.U68 301.31 78-321251
79 MARC

—ENVIRONMENTAL ASPECTS—JAPAN—PUBLIC OPINION

Tokyo. Tomin Seikatsukyoku. Sanka Suishinbu. Kōchōka.
(Toshi kankyō ni kansuru seron chōsa)
都市環境に関する世論調査 / [編集 都民生活局参加推進部広聴課]. — [東京]：同課, 昭和53 [1978]
104 p. ; 25 cm. Ja***
Cover title.
1. Cities and towns—Environmental aspects—Japan—Public opinion. 2. Cost and standard of living—Japan—Public opinion. 3. City and town life—Japan—Public opinion. 4. Public opinion—Japan. I. Title.
HT147.J3T53 1978 79–809223

—GROWTH

see also Agricultural colonies; Annexation (Municipal government); Land subdivision; Metropolitan areas; Suburbs

Altezor, Carlos.
Historia urbanística y edilicia de la ciudad de Montevideo; [desde su fundación colonial hasta nuestros días, por] Carlos Altezor [y] Hugo Baracchini. Montevideo, Junta Departamental de Montevideo, Biblioteca José Artigas, 1971.
318 p. illus. 24 cm. LACAP72-3872
Bibliography: p. 309-317.
1. Montevideo—History. 2. Cities and towns—Growth. 3. Montevideo—City planning. 4. City planning—Uruguay. I. Baracchini, Hugo, joint author. II. Biblioteca José Artigas. III. Title.
HN360.M6A7 72-375032
79[r79]rev MARC

Bello-Morales Merino, Antonio.
Las nuevas ciudades residenciales ... / [por] Antonio Bello-Morales Merino ; prólogo de José Luis Zataraín Iglesias. — Madrid, [etc.] : Dossat, 1978.
xi, 367 p., [1] fold. leaves of plates : ill. ; 21 cm. Sp79-Apr
Bibliography: p. [365]-367.
ISBN 8423703762
1. New towns. 2. Cities and towns—Growth. 3. City planning. I. Title.
HT153.B47 79-110256
79 MARC

Blumenfeld, Hans.
Beyond the metropolis / Hans Blumenfeld. — [Toronto] : Dept. of Urban and Regional Planning, University of Toronto, 1977.
15 p. ; 28 cm. — (Papers on planning and design ; no. 12) C***
Cover title.
$1.50
1. Cities and towns—Growth. 2. Metropolitan areas. I. Title. II. Series.
HT371.B57 301.36'4 78-324006
79 MARC

Bookchin, Murray, 1921-
The limits of the city [by] Murray Bookchin. New York, Harper & Row [1974]
xi, 147 p. 19 cm. (Harper colophon books/CN 1013) $2.75
Includes bibliographical references.
ISBN 0-06-136134-8. ISBN 0-06-091013-5 (pbk.)
1. Cities and towns. 2. Cities and towns—Growth. I. Title.
HT151.B58 1974 301.36'3 73-17852
74[r79]rev2 MARC

Bourne, Larry S
Frameworks for locational decision-making in national urban strategies : a review / L. S. Bourne. — [Toronto] : Centre for Urban and Community Studies, University of Toronto, 1974.
41 p. : ill. ; 28 cm. — (Research paper - Centre for Urban and Community Studies, University of Toronto ; 66) C***
"Paper prepared for discussion at a Conference on National Urban Settlement Systems and Strategies, held at International Institute for Applied Systems Analysis, Schloss Laxenburg, Austria, December 16-19, 1974."
Bibliography: p. 38-41.
1. Cities and towns—growth. 2. Urban policy. I. Title. II. Series: University of Toronto. Centre for Union and Community Studies. Research paper ; 66.
HT166.B679 301.36'1 79-303989
79 MARC

Burchell, Robert W
The fiscal impact handbook : estimating local costs and revenues of land development / Robert W. Burchell, David Listokin, in collaboration with Ralph W. Lake ... [et al.], with the assistance of Rodney E. Engelen ... [et al.]. — New Brunswick, N.J. : Center for Urban Policy Research, c1978.
xxiv, 480 p. ; 26 cm.
Bibliography: p. 443-466.
Includes index.
ISBN 0-88285-045-8 : $20.00
1. Municipal services—Finance. 2. Tax revenue estimating. 3. Cities and towns—Growth. I. Listokin, David, joint author. II. Title.
HD4431.B85 352'.12 78-6216
78 MARC

(Gendai Nihon no toshi supurōru mondai)
現代日本の都市スプロール問題 / 山崎不二夫, 森滝健一郎 ... [et al.] 編. — 東京：大月書店, 1978.
2 v. : ill. ; 20 cm. — (現代資本主義叢書 ; 4–5) Ja 78–7247 (v. 1)
Includes bibliographical references.
¥1800 per vol.
1. Cities and towns—Growth. 2. Cities and towns—Japan. 3. Urbanization—Japan. 4. Land use—Japan. I. Yamazaki, Fujio, 1909- II. Moritaki, Ken'ichirō, 1932- III. Series: Gendai shihonshugi sōsho ; 4–5.
HT371.G38 79–800071

CITIES AND TOWNS
—GROWTH (Continued)

Loraut, Marie.
Urbanotropismes : essai d'interprétation de statistiques migratoires des villes / Marie Lorant. — Paris : Éditions Cujas : Centre d'étude des techniques économiques modernes, 1977.
205 p. : ill. ; 23 cm. — (Temespace ; no 12) F78-2674
Bibliography: p. 197-201.
45.00F
1. Cities and towns—Growth. 2. Urbanization. I. Title. II. Series.
HT371.L66 301.32'6 78-398466
79 MARC

Ortiz de Zevallos, Luis.
Lima, su evolución creadora / Luis Ortiz de Zevallos. — ₍Lima?₎ : Ortiz de Zevallos, ₍1978 or 1979₎
59 p. : ill. ; 29 cm.
Cover title.
Bibliography: p. 54, 57.
1. Lima—History. 2. Cities and towns—Growth. I. Title.
HN350.L5O77 309.1'85'2 79-110254
79 MARC

Pred, Allan Richard, 1936-
City systems in advanced economies : past growth, present processes, and future development options / Allan Pred. — New York : Wiley, c1977.
256 p. : ill. ; 23 cm.
"A Halsted Press book."
Bibliography: p. ₍231₎-247.
Includes index.
ISBN 0-470-99073-2
1. Urban economics. 2. Cities and towns—Growth. I. Title.
HT321.P73 1977 330.9'173'2 76-58376
77₍r79₎rev MARC

Rugg, Dean S
Spatial foundations of urbanism / Dean S. Rugg. — 2d ed. — Dubuque, Iowa : W. C. Brown Co., c1979.
viii, 375 p. : ill. ; 23 cm.
Bibliography: p. 357-366.
Includes index.
ISBN 0-697-07589-3
1. Cities and towns—Growth. 2. Urbanization. I. Title.
HT371.R83 1979 301.36'3 78-55767
79 MARC

Schneider, Kenneth R
On the nature of cities : toward enduring and creative human environments / Kenneth R. Schneider. — 1st ed. — San Francisco : Jossey-Bass Publishers, 1979.
xiv, 352 p. ; 24 cm. — (The Jossey-Bass social and behavioral science series)
Bibliography: p. 334-340.
Includes index.
ISBN 0-87589-391-0
1. Cities and towns—United States. 2. Urban policy—United States. 3. Cities and towns—Growth. 4. Sociology, Urban. I. Title.
HT123.S317 301.36'3'0973 78-62556
79 MARC

Telkamp, Gerard J
Urban history and European expansion : a review of recent literature concerning colonial cities and a preliminary bibliography / by Gerard J. Telkamp. — ₍Leiden₎ : Leiden Centre for the History of European Expansion, 1978.
76 p. : ill. ; 30 cm. — (Intercontinenta ; no. 1 ISSN 0165-2850) Ne***
Includes bibliographies.
1. Cities and towns—Growth. 2. Cities and towns—Growth—Bibliography. 3. Colonization. 4. Colonization—Bibliography. I. Title. II. Series.
HT371.T44 301.36'094 79-301978
79 MARC

—GROWTH —ADDRESSES, ESSAYS, LECTURES

International urban growth policies : new town contributions / Gideon Golany, editor. — New York : Wiley, c1978.
xx, 460 p., ₍1₎ fold. leaf of plates : ill. ; 29 cm.
"A Wiley-Interscience publication."
Includes bibliographies and index.
ISBN 0-471-03748-6
1. New towns—Addresses, essays, lectures. 2. Cities and towns—Growth—Addresses, essays, lectures. I. Golany, Gideon.
HT166.I65 309.2'62 77-28274
77 MARC

—GROWTH—BIBLIOGRAPHY

Telkamp, Gerard J
Urban history and European expansion : a review of recent literature concerning colonial cities and a preliminary bibliography / by Gerard J. Telkamp. — ₍Leiden₎ : Leiden Centre for the History of European Expansion, 1978.
76 p. : ill. ; 30 cm. — (Intercontinenta ; no. 1 ISSN 0165-2850) Ne***
Includes bibliographies.
1. Cities and towns—Growth. 2. Cities and towns—Growth—Bibliography. 3. Colonization. 4. Colonization—Bibliography. I. Title. II. Series.
HT371.T44 301.36'094 79-301978
79 MARC

—GROWTH—CASE STUDIES

Manchester and São Paulo : problems of rapid urban growth / edited by John D. Wirth and Robert L. Jones. — Stanford, Calif. : Stanford University Press, 1978.
234 p. ; 24 cm.
"The chapters of this book were initially papers given at a conference held at Stanford University in April 1977 under the sponsorship of the university's Center for Latin American Studies."
Includes bibliographical references and index.
ISBN 0-8047-1005-8
1. Cities and towns—Growth—Case studies. 2. Manchester, Eng.—Economic conditions. 3. São Paulo, Brazil (City)—Economic condition. I. Wirth, John D. II. Jones, Robert Lee, 1954- . III. Stanford University. Center for Latin American Studies.
HT371.M28 301.36'3'0942733 78-55321
78 MARC

Stadtentwicklungen in kapitalistischen und sozialistischen Ländern / Jürgen Friedrichs (Hg.) ; mit Beitr. von Michael Brenner ... ₍et al.₎. — Reinbek bei Hamburg : Rowohlt, 1978.
365 p. : ill., maps ; 19 cm. — (Rowohlts deutsche Enzyklopädie ; 378) GFR78-A
Bibliography: p. 345-₍364₎
Includes index.
ISBN 3-499-55378-3 : DM16.80
1. Cities and towns—Growth—Case studies. 2. City planning—Case studies. I. Friedrichs, Jürgen. II. Brenner, Michael, 1951-
HT151.S63 79-340080
79 MARC

—GROWTH—MATHEMATICAL MODELS

Urban residential location models / ₍edited by₎ Stephen H. Putman. — Boston : Martinus Nijhoff Pub., c1979.
viii, 169 p. : ill. ; 24 cm. — (Studies in applied regional science ; v. 13)
Includes bibliographical references.
ISBN 0-89838-011-1
1. Land use, Urban—Mathematical models. 2. Cities and towns—Growth—Mathematical models. 3. Residential mobility—Mathematical models. I. Putman, Stephen H. II. Series.
HD111.U72 333.7'7 78-12285
78 MARC

Webber, Michael John.
Diffusion of innovations and the growth of cities / M. J. Webber and A. E. Joseph. — Hamilton, Ont. : Dept. of Geography, McMaster University, 1976.
81 p. : ill. ; 28 cm. — (Discussion paper - Department of Geography, McMaster University ; no. 8) C***
Bibliography: p. 78-81.
1. Cities and towns—Growth—Mathematical models. 2. Cities and towns—Canada—Ontario—Growth—Case studies. I. Joseph, A. E. II. Title. III. Series: McMaster University. Dept. of Geography. Discussion paper - Department of Geography, McMaster University ; no. 8.
HT371.W35 301.36'1'0971 78-323136
79 MARC

—GROWTH—SIMULATION METHODS—CONGRESSES

Urban development models / edited by Richard Baxter, Marcial Echenique and Janet Owers. — Lancaster, Eng. : Construction Press, 1975.
335 p. : ill. ; 31 cm. — (LUBFS Conference proceedings ; no. 3) GB***
Proceedings of the 1974 conference.
Bibliography: p. 311-327.
ISBN 0-904406-11-3
1. Cities and towns—Simulation methods—Congresses. 2. Cities and towns—Growth—Simulation methods—Congresses. 3. Land use, Urban—Simulation methods—Congresses. I. Baxter, Richard Stephen. II. Echenique, Marcial. III. Owers, Janet. IV. Series: Cambridge. University. Centre for Land Use and Built Form Studies. LUBFS conference proceedings ; no. 3.
HT107.U715 301.36'1'0184 79-301020
79 MARC

—GROWTH—SENEGAL—DAKAR

Vernière, Marc.
Volontarisme d'État et spontanéisme populaire dans l'urbanisation du Tiers-monde : formation et évolution des banlieues dakaroises : le cas de Dagoudane Pikine / Marc Vernière. — Paris : Bibliothèque nationale, 1977.
278 p. : ill. ; 34 cm. — (Mémoires de la Section de géographie, Comité des travaux historiques et scientifiques ; 7) F***
On cover: Dakar et son double, Dagoudane Pikine.
Originally presented as the author's thesis, Paris, 1973.
Bibliography: p. ₍261₎-266.
ISBN 2-7177-1377-8
1. Dakar—Social conditions. 2. Dakar—Suburbs and environs. 3. Cities and towns—Growth—Senegal—Dakar. 4. Underdeveloped areas—Urbanization—Case studies. I. Title. II. Title: Dakar et son double, Dagoudane Pikine. III. Series: France. Comité des travaux historiques et scientifiques. Section de géographie. Mémoires ; 7.
H13.F7 t. 7 300'.8 s 79-359797
₍HN829.D34₎ 79 MARC

—HANDBOOKS, MANUALS, ETC.

Survival kit : processes and ideas for studying the urban environment / by Richard Harbeck ... ₍et al.₎. — Winnipeg, Man. : Printed by the Manitoba Teachers' Society, 1975.
154 p. : ill. ; 28 cm. C***
Part of illustrative matter in pocket.
Includes bibliographies.
1. Human ecology—Study and teaching. 2. Cities and towns—Study and teaching. 3. Human ecology—Handbooks, manuals, etc. 4. Cities and towns—Handbooks, manuals, etc. I. Harbeck, Richard. II. Manitoba Teachers' Society.
GF26.S89 301.31'07'1 79-313847
00 MARC

—HISTORY

Bortolotti, Lando.
Storia, città e territorio / Lando Bortolotti. — Milano : F. Angeli, ₍1979₎
199 p. : ill. ; 22 cm. — (Collana di Storia urbana ; 1) It79-Apr
Bibliography: p. ₍193₎-199.
ISBN 8820415313 : L6000
1. Cities and towns—History. 2. Land use, Urban—History. I. Title. II. Series.
HT161.B67 79-366512
79 MARC

Ciudad y utopía / introd., notas, y selección de textos, Alberto Sato. — Buenos Aires : Centro Editor de América Latina, c1977.
104 p. : ill. ; 19 cm. — (Los Fundamentos de las ciencias del hombre ; 3) (Biblioteca total ; 12)
Selections translated from French and English.
Bibliography: p. 22.
1. Cities and towns—History. 2. Utopias. I. Sato, Alberto. II. Series.
HT111.C57 78-113517
79 MARC

Muramatsu, Shigeki, 1905-
(Toshi no hattatsu)
都市の発達 / 村松繁樹著. — 東京 : 三省堂, 昭和25 ₍1950₎
5, 225, 5 p., ₍1₎ leaf of plates : ill. ; 18 cm. — (社会科文庫)
Series romanized: Shakaika bunko.
Includes index.

1. Cities and towns. 2. Cities and towns—History. I. Title.
HT111.M86 79-810013

Ţopa, Leon.
Istoria oraşului / Leon Ţopa. — Bucureşti : Editura Ştiinţifică, 1968.
99 p. : ill. ; 19 cm. — (Orizonturi)
Includes bibliographical references.
1. Cities and towns—History. I. Title.
HT111.T58 79-380322
79 MARC

—HISTORY—ADDRESSES, ESSAYS, LECTURES

Die Stadt : Gestalt u. Wandel bis zum industriellen Zeitalter / hrsg. von Heinz Stoob. — Köln ; Wien : Böhlau, 1979.
xxix, 273 p. : ill. ; 21 cm. — (Städtewesen ; 1) GFR78-A
Includes bibliographies.
ISBN 3-412-04477-6
1. Cities and towns—History—Addresses, essays, lectures. I. Stoob, Heinz. II. Series.
HT111.S68 301.36'09 79-357609
79 MARC

—HISTORY—PERIODICALS

Die Alte Stadt. 5.- Jahrg.
Stuttgart, W. Kohlhammer, 1978-
v. ill. 24 cm. quarterly.
Continues: Zeitschrift für Stadtgeschichte, Stadtsoziologie und Denkmalpflege.
1. Cities and towns—Periodicals. 2. Sociology, Urban—Periodicals. 3. Cities and towns—History—Periodicals. 4. Monuments—Preservation—Periodicals.
HT101.Z4 79-649612
MARC-S

CITIES AND TOWNS
—HISTORY—PERIODICALS
(Continued)

Urban History Group (U. S.)
Urban History Group newsletter. no. -39 -Apr. 1975.
[Milwaukee, University of Wisconsin-Milwaukee, Dept. of History, [etc.]
no. 28 cm. 2 no. a year.
Began with Nov. 1954 issue.
Superseded by: Urbanism past & present, ISSN 0160-2780.
Key title: Urban History Group newsletter, ISSN 0084-067X.
1. Cities and towns—History—Periodicals. 2. Sociology, Urban—Periodicals. 3. Urbanization—Periodicals. I. Title.
HT101.U674a 301.36'05 78-645137
MARC-S

—JUVENILE LITERATURE

Carter, Robert A 1930-
You, communities and change / Robert A. Carter, Edith McCall, Vernon Prinzing. — Westchester, Ill. : Benefic Press, c1978.
239 p. : ill. (some col.) ; 25 cm. — (People in a world of change)
SUMMARY: A social studies textbook discussing how communities are formed, how they reflect peoples needs, and how they relate to other communities.
ISBN 0-8175-5302-9
1. Community—Juvenile literature. 2. Cities and towns—Juvenile literature. 3. Social change—Juvenile literature. [1. Community life. 2. City and town life] I. McCall, Edith S., joint author. II. Prinzing, Vernon, joint author. III. Title. IV. Series.
HT111.C37 301.34 77-75872
79 MARC

Mathieu, Joseph.
Busy city / Joe Mathieu. — New York : Random House, c1978.
[36] p. : col. ill. ; 21 cm. — (The Best book club ever)
SUMMARY: A brief introduction to the people and places of a modern city.
ISBN 0-394-83700-2
1. Cities and towns—Juvenile literature. 2. City and town life—Juvenile literature. [1. City and town life. 2. Cities and towns] I. Title.
HT155.M28 301.36'3 77-86138
79 MARC

—LAW AND LEGISLATION see Municipal corporations

—LIGHTING see Street-lighting

—LITERARY COLLECTIONS

The Small town / edited by Flory Jones Schultheiss. — New York : Scribner, c1979.
vi, 296 p. : ill. ; 23 cm. — (Scribner student paperbacks ; SSP 44) (America in literature)
Includes index.
ISBN 0-684-16138-9
1. Cities and towns—Literary collections. 2. American literature—19th century. 3. American literature—20th century. I. Schultheiss, Flory Jones. II. Series: America in literature.
PS509.C57S6 810'.9'32 79-4209
79 MARC

—MAPS

Branch, Melville Campbell, 1913-
Comparative urban design : rare engravings, 1830-1843 / by Melville C. Branch. — New York : Arno Press, c1978.
[iv], 102 [i.e. 104] p. : ill., col. maps; 33 x 37 cm.
Reprint of maps published 1830-1844 by the Superintendence of the Society for the Diffusion of Useful Knowledge, London.
Includes bibliographical references.
ISBN 0-405-10524-X : $39.50
1. Cities and towns—Maps. 2. Cities and towns—Europe—Maps. 3. Cities and towns—United States—Maps. 4. Cities and towns—Asia—Maps. 5. Engravings—Europe—Maps. 6. Engravings—United States—Maps. 7. Engravings—Asia—Maps. I. Society for the Diffusion of Useful Knowledge, London. II. Title. III. Title: Rare engravings, 1830-1843.
G1028.B65 1978 711'.4 77-24511
77 MARC

—MAPS—INDEXES

Otness, Harold M
Index to early twentieth century city plans appearing in guidebooks, Baedeker, Muirhead-Blue guides, Murray, I. J. G. R., etc., plus selected other works to provide worldwide coverage of over 2,000 plans to over 1,200 communities, found in 74 guidebooks / Harold M. Otness. — Santa Cruz, CA : Western Association of Map Libraries, 1978.
xxx, 91 p. ; 23 cm. — (Occasional paper - Western Association of Map Libraries ; no. 4)
Bibliography: p. 87-91.
1. Cities and towns—Maps—Indexes. I. Title. II. Series: Western Association of Map Libraries. Occasional paper ; no. 4.
Z6026.C56O84 016.912 78-15094
[G1028] 78 MARC

—MATHEMATICAL MODELS

Lee, Russell.
Entropy models in spatial analysis / Russell Lee. — [Toronto] : University of Toronto, Dept. of Geography, 1974.
v, 64 p. ; 28 cm. — (Discussion paper - University of Toronto, Department of Geography ; no. 15) C***
Bibliography: p. 54-64.
$1.50
1. Cities and towns—Mathematical models. 2. Regional planning—Mathematical models. 3. Geography—Mathematics. 4. Spatial analysis (Statistics) I. Title. II. Series: University of Toronto. Dept. of Geography. Discussion paper - Department of Geography, University of Toronto ; no. 15.
HT153.L33 309.2'5'0182 75-315431
75[r79]rev MARC

Sayer, R Andrew.
A critique of urban modelling : from regional science to urban and regional political economy / R. Andrew Sayer. — [Oxford, Eng.] ; Elmsford, N.Y. : Pergamon Press, c1976]
187-254 p. : ill. ; 28 cm. — (Progress in planning ; v. 6, pt. 3 ISSN 0305-9006) GB***
Based upon part of the author's thesis, Sussex, which has title: Dynamic spatial models of urban and regional systems.
Bibliography: p. 251-254.
ISBN 0-08-021485-1
1. Cities and towns—Mathematical models. 2. Urban economics. 3. Regional economics. I. Title. II. Series.
HT153.S28 309.2'62'0151 78-318807
78 MARC

—PERIODICALS

Die Alte Stadt. 5.- Jahrg.
Stuttgart, W. Kohlhammer, 1978–
v. ill. 24 cm. quarterly.
Continues: Zeitschrift für Stadtgeschichte, Stadtsoziologie und Denkmalpflege.
1. Cities and towns—Periodicals. 2. Sociology, Urban—Periodicals. 3. Cities and towns—History—Periodicals. 4. Monuments—Preservation—Periodicals.
HT101.Z4 79-649612
MARC-S

—PLANNING see City planning

see also Cluster housing

—POETRY

Bergman, Bo, 1942-
Hela staden : prosadikter om staden, arkitekturen och politiken / Bo Bergman. — Göteborg : Korpen, 1978.
111 p. : ill. ; 16 cm. S78-19
ISBN 917374106X
1. Cities and towns—Poetry. I. Title.
PT9876.12.E6564H4 79-344070
79 MARC

—PSYCHOLOGICAL ASPECTS—
ADDRESSES, ESSAYS, LECTURES

The Urban environment / edited by Andrew Baum, Jerome E. Singer, Stuart Valins. — Hillsdale, N.J. : Lawrence Erlbaum Associates ; New York : distributed by the Halsted Press Division of Wiley, 1978.
xiii, 204 p. : ill. ; 24 cm. — (Advances in environmental psychology ; v. 1)
Includes bibliographies and indexes.
ISBN 0-470-26545-0
1. Cities and towns—Psychological aspects—Addresses, essays, lectures. 2. Sociology, Urban—Addresses, essays, lectures. I. Baum, Andrew. II. Singer, Jerome E. III. Valins, Stuart. IV. Series.
HT151.U66 301.36 78-13289
78 MARC

—SIMULATION METHODS

Ayeni, Bola, 1945-
Concepts and techniques in urban analysis / Bola Ayeni. — London : Croom Helm, c1979.
372 p. : ill. ; 23 cm. — (Croom Helm series in geography and environment) GB***
Includes bibliographies and index.
ISBN 0-85664-557-5 : £12.95
1. Cities and towns—Simulation methods. I. Title.
HT153.A9 1979 309.2'62'0184 79-317162
79 MARC

—SIMULATION METHODS—
CONGRESSES

Urban development models / edited by Richard Baxter, Marcial Echenique and Janet Owers. — Lancaster, Eng. : Construction Press, 1975.
335 p. : ill. ; 31 cm. — (LUBFS Conference proceedings ; no. 3) GB***
Proceedings of the 1974 conference.
Bibliography: p. 311-327.
ISBN 0-904406-11-3
1. Cities and towns—Simulation methods—Congresses. 2. Cities and towns—Growth—Simulation methods—Congresses. 3. Land use, Urban—Simulation methods—Congresses. I. Baxter, Richard Stephen. II. Echenique, Marcial. III. Owers, Janet. IV. Series: Cambridge. University. Centre for Land Use and Built Form Studies. LUBFS conference proceedings ; no. 3.
HT107.U715 301.36'1'0184 79-301020
79 MARC

—STATISTICS

Showers, Victor, 1910-
World facts and figures : a unique, authoritative collection of comparative information about cities, countries, and geographic features of the world / Victor Showers. — New York : Wiley, c1979.
xii, 757 p. : ill. ; 26 cm.
"A Wiley-Interscience publication."
Rev. and enl. ed. of the author's The world in figures, published 1973.
Bibliography: p. 679-688.
Includes index.
ISBN 0-471-04941-7
1. Geography—Tables, etc. 2. Cities and towns—Statistics. I. Title.
G109.S52 1979 910'.21'2 78-14041
78 MARC

—STUDY AND TEACHING

Survival kit : processes and ideas for studying the urban environment / by Richard Harbeck ... [et al.]. — Winnipeg, Man. : Printed by the Manitoba Teachers' Society, 1975.
154 p. : ill. ; 28 cm. C***
Part of illustrative matter in pocket.
Includes bibliographies.
1. Human ecology—Study and teaching. 2. Cities and towns—Study and teaching. 3. Human ecology—Handbooks, manuals, etc. 4. Cities and towns—Handbooks, manuals, etc. I. Harbeck, Richard. II. Manitoba Teachers' Society.
GF26.S89 301.31'07'1 79-313847
00 MARC

—SURVEYING see Surveying

—UNDERDEVELOPED AREAS see Underdeveloped areas—Cities and towns

—WATER-SUPPLY see Municipal water supply

—AFRICA—ADDRESSES, ESSAYS, LECTURES

Städte, Märkte, Zentren : Beiträge zur vergleichenden stadtgeographischen Forschung in Afrika. — Köln : Geographisches Institut der Universität zu Köln, 1973.
vi, 176 p. : ill., maps (3 fold.) ; 21 cm. — (Beiträge zur Länderkunde Afrikas ; Heft 5)
"Erscheint gleichzeitig als Veröffentlichung des Arbeitskreises für Afrikaforschung an der Universität Köln."
Summaries in English or French.
Three maps inserted.
Includes bibliographies.
1. Central places—Africa—Addresses, essays, lectures. 2. Cities and towns—Africa—Addresses, essays, lectures. I. Series.
HT148.A2S73 79-360493
79 MARC

CITIES AND TOWNS (Continued)

—AFRICA, NORTH

Roberts, M Hugh P
An urban profile of the Middle East / M. Hugh P. Roberts. — New York : St. Martin's Press, 1979.
239 p. : ill. ; 24 cm.
Includes bibliographies and index.
ISBN 0-312-83467-5
1. Cities and towns—Near East. 2. Cities and towns—Africa, North. 3. Urbanization—Near East. 4. Urbanization—Africa, North. 5. City planning—Near East. 6. City planning—Africa, North. I. Title.
HT147.N4R6 1979 309.2'62'0956 78-27185
78 MARC

—ALABAMA—HISTORY—BIBLIOGRAPHY

Brown, Lynda.
The history of Alabama's cities : a bibliography / Lynda Brown and Don Dodd. — Monticello, Ill. : Vance Bibliographies, 1979.
31 p. ; 28 cm. — (Public Administration series : Bibliography ; P-248)
Cover title.
$3.00
1. Cities and towns—Alabama—History—Bibliography. I. Dodd, Donald B., joint author. II. Title. III. Series.
Z7164.L8B93 301.36'09761 79-111927
[HT123.5.A2] 00 MARC

—ARAB COUNTRIES—CONGRESSES

L'Espace social de la ville arabe / par Dominique Chevallier ... [et al.]. — Paris : G.-P. Maisonneuve et Larose, 1979.
363 p., [12] leaves of plates : ill. ; 24 cm. — (Publications du Département d'islamologie de l'Université de Paris-Sorbonne (Paris IV) ; 7) F***
French or English.
On p. [4] of cover: Espaces socio-culturels et croissance urbaine dans le monde arabe : actes du colloque tenu les 24, 25 et 26 novembre 1977 : action thématique programmée du Centre national de la recherche scientifique.
Includes bibliographical references.
ISBN 2-7068-0766-0 : 68.00F
1. Cities and towns—Arab countries—Congresses. 2. Arab countries—Social conditions—Congresses. I. Chevallier, Dominique. II. France. Centre national de la recherche scientifique. III. Title: Espaces socio-culturels et croissance urbaine dans le monde arabe. IV. Series: Université de Paris IV: Paris-Sorbonne. Département d'islamologie. Publications du Département d'islamologie de l'Université de Paris-Sorbonne ; 7.
HT147.5.E78 79-375993
79 MARC

—ASIA—CONGRESSES

Pacific-Asian Congress of Municipalities.
Proceedings PACOM '74 : third general Conference, Taipei, Republic of China, May 27-30, 1974 / Pacific-Asian Congress of Municipalities. — [Taipei?] : The Congress, [1974?]
[19], 106 p., [3] leaves of plates : ill. ; 29 cm.
Cover title.
1. Cities and towns—Asia—Congresses. 2. Municipal government—Asia—Congresses. I. Title.
HT147.A2P3 1974 301.36'095 78-111815
79 MARC

—ASIA—MAPS

Branch, Melville Campbell, 1913-
Comparative urban design : rare engravings, 1830-1843 / by Melville C. Branch. — New York : Arno Press, c1978.
[iv], 102 [i.e. 104] p. : ill., col. maps ; 33 x 37 cm.
Reprint of maps published 1830-1844 by the Superintendence of the Society for the Diffusion of Useful Knowledge, London.
Includes bibliographical references.
ISBN 0-405-10524-X : $39.50
1. Cities and towns—Maps. 2. Cities and towns—Europe—Maps. 3. Cities and towns—United States—Maps. 4. Cities and towns—Asia—Maps. 5. Engravings—Europe—Maps. 6. Engravings—United States—Maps. 7. Engravings—Asia—Maps. I. Society for the Diffusion of Useful Knowledge, London. II. Title. III. Title: Rare engravings, 1830-1843.
G1028.B65 1978 711'.4 77-24511
77 MARC

—AUSTRALIA

Kilmartin, Leslie.
Cities unlimited : the sociology of urban development in Australia and New Zealand / Leslie Kilmartin, David C. Thorns. — Sydney ; Boston : Allen and Unwin, 1978.
195 p. : diagrs. ; 23 cm. — (Studies in society ; 1) Aus***
Bibliography: p. 168-185.
Includes indexes.
ISBN 0-86861-288-X : $18.25. ISBN 0-86861-296-0 pbk.
1. Cities and towns—Australia. 2. Cities and towns—New Zealand. 3. Sociology, Urban. I. Thorns, David C., joint author. II. Title. III. Series: Studies in society (Sydney) ; 1.
HT149.A8K54 301.36'09931 78-55230
79 MARC

—AUSTRALIA—ADDRESSES, ESSAYS, LECTURES

Seven cities of Australia. — Sydney : J. Ferguson, 1978.
182 p. : ill. ; 24 cm. Aus***
Based on a series appearing in the Bulletin March-April 1977.
ISBN 0-909134-10-3
1. Australia—Description and travel—1951- —Addresses, essays, lectures. 2. Cities and towns—Australia—Addresses, essays, lectures. I. The Bulletin (Sydney)
DU105.S47 994 79-306600
79 MARC

—AUSTRALIA—ILLAWARRA—GROWTH—ADDRESSES, ESSAYS, LECTURES

Urban Illawarra / edited by Ross Robinson. — Melbourne : Sorrett, 1977.
302 p. : ill., charts, graphs, maps ; 25 cm. Aus
Includes bibliographical references.
ISBN 0-909752-30-3
1. Urbanization—Australia—Illawarra—Addresses, essays, lectures. 2. Cities and towns—Australia—Illawarra—Growth—Addresses, essays, lectures. I. Robinson, Ross.
HN850.I44U7 301.36'1'09944 79-313534
79 MARC

—AUSTRALIA—SOUTH AUSTRALIA—ADDRESSES, ESSAYS, LECTURES

Urban South Australia : selected readings / edited by C. A. Forster and R. J. Stimson. — [Bedford Park, S.A.] : School of Social Sciences, Flinders University of South Australia, 1977.
iv, 201 p. : diagrs., maps, plans ; 29 cm. — (Centre for Applied Social and Survey Research monograph series ; no. 1) Aus
Bibliography: p. 195-201.
ISBN 0-7258-0085-2 : $6.50
1. Cities and towns—Australia—South Australia—Addresses, essays, lectures. 2. South Australia—Economic conditions—Addresses, essays, lectures. 3. South Australia—Social conditions—Addresses, essays, lectures. I. Forster, Clive A. II. Stimson, Robert John. III. Series: Flinders University of South Australia. Centre for Applied Social and Survey Research. Centre for Applied Social and Survey Research monograph series ; no. 1.
HT149.A8U694 309.2'62'099423 79-318273
79 MARC

—AUSTRIA

Österreichische Volkspartei. Ausschuss für Raumordnung.
Leben, aber wie? : [Analysen, Tendenzen, Perspektiven zur Entwicklung des ländlichen Raumes]. — Wien : Österr. Volkspartei, [1973]
55, [1] p. : ill. (some col.) ; 21 cm. Au74-23-98
Bibliography: p. 55-[56]
Not in trade.
1. Regional planning—Austria. 2. Cities and towns—Austria. 3. Austria—Economic conditions—1945- . 4. Austria—Social conditons. I. Title.
HT395.A94O378 1973 309.2'5'09436 75-585159
*79[r79]rev MARC

—AUSTRIA—CONGRESSES

Stadt und Dorf in der Region : Sammlung d. anlässl. des Symposiums Vorarlberg 1970 gehaltenen Referate. — Wien : Forschungsges. f. Wohnen, Bauen u. Planen, 1971.
88 p. : ill. ; 24 cm. — (Monographie - Forschungsgesellschaft für Wohnen, Bauen und Planen ; Nr. 16) Au76-22-94
Proceedings of a symposium sponsored by Gesprächskreis Die Erneuerung Unserer Städte, Märkte und Dörfer and Forschungsgesellschaft für Wohnen, Bauen und Planen, Sept. 30-Oct. 3, 1970 in Vorarlberg.
Includes bibliographical references.
S96.00
1. Cities and towns—Austria—Congresses. 2. Villages—Austria—Congresses. 3. Regional planning—Austria—Congresses. I. Gesprächskreis Die Erneuerung Unserer Städte, Märkte und Dörfer. II. Forschungsgesellschaft für Wohnen, Bauen und Planen. III. Series: Forschungsgesellschaft für Wohnen, Bauen und Planen. Monographie ; Nr. 16.
HT145.A8S73 79-362266

—AUSTRIA—GROWTH

Conditt, Georg.
Stadterneuerung und Stadterweiterung in den österreichischen Ballungsräumen : Expertengutachten f. den Unterausschuss d. Stellvertreterkommission d. Österr. Raumordnungskonferenz "Ballungsräume" / Georg Conditt. — Wien : Geschäftsstelle d. Österr. Raumordnungskonferenz, 1978.
ix, 357 p., [56] leaves of plates : ill., maps (some col.) ; 30 cm. — (Schriftenreihe - Österreichische Raumordnungskonferenz (ÖROK) ; Nr. 11)
Au78-17/18-505
Summaries and legends also in English, French, and Russian.
Includes bibliographical references.
S150.00
1. Urban renewal—Austria. 2. Urbanization—Austria. 3. Cities and towns—Austria—Growth. I. Österreichische Raumordnungskonferenz. Stellvertreterkommission. Unterausschuss Ballungsräume. II. Title. III. Series: Österreichische Raumordnungskonferenz. Schriftenreihe - Österreichische Raumordnungskonferenz ; Nr. 11.
HT178.A85C65 301.36'09436 79-359718
*79 MARC

—AUSTRIA—HISTORY

Bobek, Hans, 1903-
Das System der zentralen Orte Österreichs : eine empirische Untersuchung / Hans Bobek, Maria Fesl. — Wien ; Köln : Böhlau, 1978.
xix, 310 p. : graphs (some col.), 10 fold. col. maps (in pocket) ; 25 cm. — (Schriften der Kommission für Raumforschung der Österreichischen Akademie der Wissenschaften ; Bd. 3) Au79-10-277
Title on spine: Zentrale Orte.
Includes bibliographical references.
ISBN 3-205-08272-9 : S796.00
1. Cities and towns—Austria—History. 2. Central business districts—Austria—History. 3. Central business districts—Austria—Statistics. I. Fesl, Maria, joint author. II. Title. III. Title: Zentrale Orte. IV. Series: Akademie der Wissenschaften, Vienna. Kommission für Raumforschung. Schriften ; Bd. 3.
HT145.A8B6 79-371133
*79 MARC

Rausch, Wilhelm, Dr. Stadtarchivar.
Koordination der Forschung auf dem Gebiet der Stadtgeschichte mit besonderer Berücksichtigung der Sozialgeschichte / Wilhelm Rausch, Rautgundis Felser und Willibald Katzinger ; Ludwig-Boltzmann-Forschungsstelle für Stadtgeschichte. — Linz : Ludwig-Boltzmann-Forschungsstelle f. Stadtgeschichte, 1976.
76 leaves ; 30 cm. Au78-4-242
"Forschungsauftrag des Bundesministeriums für Wissenschaft und Forschung."
Includes bibliographical references.
Not in trade.
1. Cities and towns—Austria—History. I. Felser, Rautgundis, joint author. II. Katzinger, Willibald, joint author. III. Ludwig-Boltzmann-Forschungsstelle für Stadtgeschichte. IV. Title.
HT145.A8R38 301.36'3'09436 78-375717
*78 MARC

—BELGIUM—ADDRESSES, ESSAYS, LECTURES

Stadsgeografische studies. — [Leuven] : Geografisch Instituut, Katholieke Universiteit te Leuven, 1977.
145 p. : ill., maps (4 fold. col. in pocket) ; 28 cm. — (Acta geographica Lovaniensia ; v. 15) Be***
Dutch or English with summaries in Dutch, English, or French.
Includes bibliographies.
1. Cities and towns—Belgium—Addresses, essays, lectures. 2. Central places—Belgium—Addresses, essays, lectures. I. Series.
HT145.B4S72 78-378546
79 MARC

—BOLIVIA—RELIGIOUS LIFE

Valdivieso C , Roberto.
Creencias religiosas y escolaridad urbana / [descripción estadística y análisis sociológico por Roberto Valdivieso C., Giuseppe Calzavarini]. — La Paz, Bolivia : [CELAM] Departamento de Ecumenismo y de Catequesis, 1978.
51, [21] p. ; 30 cm. — (Estudios contemporáneos)
"Apéndice no. 2": p. [52]-[71]
1. Catholics—Bolivia. 2. Cities and towns—Bolivia—Religious life. 3. Bolivia—Religious life and customs. I. Calzavarini, Giuseppe, joint author. II. Title. III. Series.
BX1464.2.V34 79-113207
79 MARC

—BRAZIL

Andrade, Thompson Almeida.
Sistema urbano e cidades médias no Brasil / Thompson A. Andrade, Celsius A. Lodder. — Rio de Janeiro : IPEA/INPES, 1979.
146 p. ; 21 cm. — (Coleção Relatórios de pesquisa - Instituto de Planejamento Econômico e Social, Instituto de Pesquisas ; no. 43)
Bibliography: p. 143-146.
1. Cities and towns—Brazil. 2. Urbanization—Brazil. 3. Poor—Brazil. I. Lodder, Celsius A., joint author. II. Title. III. Series: Instituto de Planejamento Econômico e Social. Instituto de Pesquisas. Relatório de pesquisa ; no. 43.
HT129.B7A69 79-109768

CITIES AND TOWNS
—BRAZIL (Continued)

Paraná, Brazil (State). Universidade Federal
Percepção do espaço urbano / UNESCO, Organizaçaõ da O.N.U. para Educação, Ciência e Cultura, Universidade Federal do Paraná, Instituto de Pesquisa e Planejamento Urbano de Curitiba. — ₍Curitiba₎ : A Universidade, ₍1978₎
150 leaves : ill. ; 30 cm.
"O presente documento foi elaborado pelas equipes da UFPR e IPPUC/PMC, como relatório do projeto' Percepção do espaço urbano/centro,' para o V Simpósio Brasileiro de Psicologia Cognitiva."
Bibliography: leaf 148-150.
1. Geographical perception. 2. Cities and towns—Brazil. 3. Environmental psychology. 4. Child psychology. 5. Curitiba—Description. I. United Nations Educational, Scientific and Cultural Organization. II. Instituto de Pesquisa e Planejamento Urbano de Curitiba. III. Title.
GF125.P37 1978 79-102086
79 MARC

—BRAZIL—STATISTICS

Fundação IBGE.
Indicadores sociais para áreas urbanas. — Rio de Janeiro : Secretaria de Planejamento da Presidência da República, IBGE, 1977.
162 p. ; 27 cm.
1. Social indicators—Brazil. 2. Cities and towns—Brazil—Statistics. I. Title.
HN284.F86 1977 78-392168
79 MARC

—BRAZIL—RECIFE—GROWTH

Melo, Mário Lacerda de.
Metropolização e subdesenvolvimento : o caso do Recife / Mário Lacerda de Melo. — Recife : UFPE, Centro de Filosofia e Ciências Humanas, Departamento de Ciências Geográficas, 1978.
256 p., ₍4₎ leaves of plates : ill. ; 23 cm.
Bibliography: p. 253-256.
1. Recife, Brazil—Economic conditions. 2. Recife, Brazil—Social conditions. 3. Cities and towns—Brazil—Recife—Growth. I. Title.
HC189.R34M44 78-383129
79 MARC

—BRAZIL—RIO GRANDE DO SUL (STATE)—MAPS

Guia de Sylos do Rio Grande do Sul / editor e diretor, Himdemburg Ribeiro de Sylos. — ₍s.l. : s.n., 1975?₎ (Porto Alegre : Gráfica da Esc. Prof. Champagnat, 1975?)
₍136₎ p. : maps ; 16 x 24 cm.
Errata slip inserted.
1. Rio Grande do Sul (State), Brazil—Maps. 2. Cities and towns—Brazil—Rio Grande do Sul (State)—Maps. I. Sylos, Himdemburg Ribeiro de.
G1778.R5G8 1975 912'.81'6 78-417927
78 MARC

—BRAZIL, NORTHEAST

Davidovich, Fany.
Funções urbanas no Nordeste / elaborado por Fany Davidovich, Maria do Socorro Alves, Marília Carneiro Natal. — Fortaleza : Banco do Nordeste do Brasil, Departamento de Estudos Econômicos do Nordeste, 1977.
88 p., ₍6₎ leaves of plates : ill. ; 21 cm.
Bibliography: p. 87-88.
1. Cities and towns—Brazil, Northeast. 2. Urbanization—Brazil, Northeast. 3. Urban economics. I. Alves, Maria do Socorro, joint author. II. Natal, Marília Carneiro, joint author. III. Title.
HT129.B7D38 78-394500
79 MARC

—BULGARIA

Bulgaria. Komitet za otdikh i turizŭm.
Bulgaria museum towns / Committee of Recreation and Tourism. — ₍Sofia : Tourist Publicity Centre, 1976₎
35 p. ; 21 cm.
Cover title.
1. Cities and towns—Bulgaria. 2. Bulgaria—Description and travel—1970- I. Title.
DR98.A1B84 1976 949.77 78-321139
79 MARC

Devedzhiev, Marin.
(Selishtnite sistemi)
Селищните системи : глав. направление в селищното преустройство на Народна република България : ₍изследване₎ / Марин Деведжиев. — София : Наука и изкуство, 1978 (Варна : печ. Ст. Добрев-Странджата)
384 p. : ill. ; 23 cm. Bu 78-2952
Includes bibliographical references.
2.87 lv
1. Cities and towns—Bulgaria. 2. Bulgaria—Administrative and political divisions. 3. Bulgaria—Economic conditions—1944- I. Title.
HT145.B8D48 79-409925

—CALIFORNIA—CASE STUDIES

Hatch, Elvin.
Biography of a small town / Elvin Hatch. — New York : Columbia University Press, 1979.
293 p. : maps ; 24 cm.
Bibliography: p. ₍283₎-287.
Includes index.
ISBN 0-231-04694-4 : $20.00
1. Cities and towns—California—Case studies. 2. California—Rural conditions. 3. Community life—Case studies. I. Title.
HT123.5.C2H37 309.1'794 79-313
79 MARC

—CAMEROON—FAKO—DICTIONARIES

France. Office de la recherche scientifique et technique outre-mer. Centre de Yaounde. Section de géographie.
Dictionnaire des villages du Fako = Village dictionary of Fako Division / Office de la recherche scientifique et technique outre-mer, Centre O.R.S.T.O.M. de Yaounde, ₍Section de géographie₎. — ₍Yaoundé₎ : La Section, 1973.
71 p. : fold. map (in pocket) ; 30 cm.
1. Fako, Cameroon—Gazetteers. 2. Cities and towns—Cameroon—Fako—Dictionaries. I. Title. II. Title: Village dictionary of Fako Division.
DT581.F34F7 1973 916.7'11 79-355925
79 MARC

—CANADA

Nowlan, D M
Towards home rule for urban policy / David M. Nowlan. — ₍Toronto₎ : Centre for Urban and Community Studies and Dept. of Political Economy, University of Toronto, 1976.
27 p. ; 28 cm. — (Research paper - Centre for Urban and Community Studies, University of Toronto ; no. 83) C***
Bibliography: p. 26-27.
$3.00
1. Cities and towns—Canada. 2. Federal-city relations—Canada. I. Title. II. Series: University of Toronto. Centre for Urban and Community Studies. Research paper ; no. 83.
HT127.N68 352.071 78-322678
79 MARC

—CANADA—ADDRESSES, ESSAYS, LECTURES

Weaver, John C
Shaping the Canadian city : essays on urban politics and policy, 1890-1920 / John C. Weaver. — ₍Toronto₎ : Institute of Public Administration of Canada, ₍c1977₎
vii, 79 p. 26 cm. — (Monographs on Canadian urban government ; no. 1) C***
Bibliography: p. 77-79.
1. Cities and towns—Canada—Addresses, essays, lectures. 2. Municipal government—Canada—Addresses, essays, lectures. I. Title. II. Series.
HT127.W4 301.36'4'0971 78-322160
78 MARC

—CANADA—CASE STUDIES—JUVENILE LITERATURE

Knight, Lowry.
Cities are for people / by Lowry Knight and Leslie Richards. — Toronto : Oxford University Press, 1977.
185 p. : ill., col. maps ; 27 cm. C***
Includes index.
ISBN 0-19-540273-1
1. Cities and towns—Canada—Case studies—Juvenile literature. 2. Sociology, Urban—Juvenile literature. I. Richards, Les, 1924- joint author. II. Title.
HT127.K53 301.36'3'0971 79-303598
79 MARC

—CANADA—GROWTH

Careless, James Maurice Stockford, 1919-
The rise of cities in Canada before 1914 / J. M. S. Careless. — Ottawa : Canadian Historical Association, 1978.
26, ₍2₎ p. ; 21 cm. — (Canadian Historical Association booklets ; no. 32) C79-5046-3
Bibliography: p. ₍27₎
ISBN 0-88798-030-9 : $0.60
1. Cities and towns—Canada—History. 2. Cities and towns—Canada—Growth. I. Title. II. Series: Canadian Historical Association. Booklets ; no. 32.
F1001.C24 no. 32 971'.008 s 79-318034
₍HT127₎ 79 MARC

Population growth and urban problems / edited by Frank Kelly. — Ottawa : Science Council of Canada : available from Information Canada, 1975.
90 p. : ill. ; 28 cm. — (Perceptions ; 1) C***
At head of title: Science Council of Canada. Study on population and technology.
Includes bibliographical references.
$1.25 ($1.50 U.S.)
1. Cities and towns—Canada—Growth. 2. City planning—Canada. I. Kelly, Frank, 1935- . II. Series.
HT127.P66 301.36'1'0971 79-306794
79 MARC

Simmons, James William, 1936-
The growth of the Canadian urban system / James W. Simmons. — ₍Toronto₎ : Centre for Urban and Community Studies, University of Toronto, 1974.
28 leaves : ill. ; 28 cm. — (Research paper - Centre for Urban and Community Studies, University of Toronto ; no. 65) C***
Bibliography: leaf 28.
1. Cities and towns—Canada—Growth. 2. Urbanization—Canada. I. Title. II. Series: University of Toronto. Centre for Urban and Community Studies. Research paper ; no. 65.
HT127.S55 301.36'3'0971 78-320169
79 MARC

—CANADA—GROWTH—COLLECTED WORKS

Simmons, James William, 1936-
Migration and the Canadian urban system / J. W. Simmons. — Toronto : Centre for Urban and Community Studies, University of Toronto, 1977-
v. : ill. ; 28 cm. — (Research paper - Centre for Urban and Community Studies, University of Toronto ; no. 85) C***
Bibliography: v. 1, p. 44-45.
CONTENTS: pt. 1. Spatial patterns.
1. Cities and towns—Canada—Growth—Collected works. 2. Migration, Internal—Canada—Collected works. I. Title. II. Series: University of Toronto. Centre for Urban and Community Studies. Research paper - Centre for Urban and Community Studies, University of Toronto ; no. 85.
HT127.S555 301.36'1 79-318027
79 MARC

—CANADA—HISTORY

Careless, James Maurice Stockford, 1919-
The rise of cities in Canada before 1914 / J. M. S. Careless. — Ottawa : Canadian Historical Association, 1978.
26, ₍2₎ p. ; 21 cm. — (Canadian Historical Association booklets ; no. 32) C79-5046-3
Bibliography: p. ₍27₎
ISBN 0-88798-030-9 : $0.60
1. Cities and towns—Canada—History. 2. Cities and towns—Canada—Growth. I. Title. II. Series: Canadian Historical Association. Booklets ; no. 32.
F1001.C24 no. 32 971'.008 s 79-318034
₍HT127₎ 79 MARC

—CANADA—HISTORY—ADDRESSES, ESSAYS, LECTURES

Russwurm, Lorne H
Essays on Canadian urban process and form / Lorne H. Russwurm, Richard E. Preston, L. R. G. Martin. — Waterloo, Ont. : Dept. of Geography, Faculty of Environmental Studies, University of Waterloo, c1977.
ix, 377 p. : ill. ; 23 cm. — (Department of Geography publication series ; no. 10) C***
Includes bibliographies.
1. Cities and towns—Canada—History—Addresses, essays, lectures. 2. Urbanization—Canada—Addresses, essays, lectures. 3. Canada—Population—Addresses, essays, lectures. I. Preston, Richard E., joint author. II. Martin, Larry R. G., joint author. III. Title. IV. Series: University of Waterloo. Dept. of Geography. Department of Geography publication series ; no. 10.
HT127.R86 301.36'0971 78-322535
79 MARC

CITIES AND TOWNS (Continued)

—CANADA—ONTARIO—GROWTH
—CASE STUDIES

Webber, Michael John.
Diffusion of innovations and the growth of cities / M. J. Webber and A. E. Joseph. — Hamilton, Ont. : Dept. of Geography, McMaster University, 1976.
81 p. : ill. ; 28 cm. — (Discussion paper - Department of Geography, McMaster University ; no. 8) C***
Bibliography: p. 78-81.
1. Cities and towns—Growth—Mathematical models. 2. Cities and towns—Canada—Ontario—Growth—Case studies. I. Joseph, A. E. II. Title. III. Series: McMaster University. Dept. of Geography. Discussion paper - Department of Geography, McMaster University ; no. 8.
HT371.W35 301.36'1'0971 78-323136
79 MARC

—CENTRAL EUROPE

Born, Martin.
Geographie der ländlichen Siedlungen / von Martin Born. — 1. Aufl. — Stuttgart : Teubner, 1977-
v. : ill. ; 21 cm. — (Studienbücher der Geographie) (Teubner Studienbücher : Geographie) GFR77-A
Bibliography: v. 1, p. [204]-222.
Includes index.
CONTENTS: 1. Die Genese der Siedlungsformen in Mitteleuropa.

ISBN 3-519-03409-3 (v. 1) : DM26.80 (v. 1)
1. Anthropo-geography—Central Europe. 2. Cities and towns—Central Europe. 3. Land use—Central Europe. I. Title.
GF540.B67 79-346054
79 MARC

—CHILE—HISTORY

Guarda, Gabriel.
Historia urbana del reino de Chile / Gabriel Guarda. — Santiago : Editorial Andrés Bello, [1978]
509 p. : ill. ; 31 cm.
Includes bibliographical references.
1. Cities and towns—Chile—History. I. Title.
HT129.C45G8 79-107445
79 MARC

—CHINA

United States. Central Intelligence Agency.
China city brief[s]. — Washington : [Available from] Document Expediting Project, Exchange and Gift Division, Library of Congress, [197-]-
v. in : maps ; 27 cm.
Cover title.
1. Cities and towns—China. 2. China—Description and travel—1976- I. Title.
DS737.U54 1970z 951 79-600827
79 MARC

—COLOMBIA—ADDRESSES, ESSAYS, LECTURES

Urbanismo y vida urbana / compilación y prólogo, Carlos Castillo. — [Bogotá] : Instituto Colombiano de Cultura, Subdirección de Comunicaciones Culturales, 1977.
542 p. : ill. ; 21 cm. — (Biblioteca básica colombiana ; 30)
Cover title: Vida urbana y urbanismo.
Includes bibliographical references.
1. Cities and towns—Colombia—Addresses, essays, lectures. 2. Urbanization—Colombia—Addresses, essays, lectures. 3. Housing—Colombia—Addresses, essays, lectures. I. Castillo, Carlos. II. Title: Vida urbana y urbanismo. III. Series.
HT129.C7U68 301.36'09861 78-111472
79 MARC

—COLORADO

Crampon, L J
Business in the small cities of Colorado / by L. J. Crampon. — Boulder : Bureau of Business Research, University of Colorado, 1949.
vi, 29 p. ; 28 cm.
"Prepared for the Board of Industrial Development Research."
1. Colorado—Economic conditions. 2. Colorado—Industries. 3. Cities and towns—Colorado. I. Colorado. Board of Industrial Development Research. II. Title.
HC107.C7C7 79-311450
79 MARC

—CZECHOSLOVAK REPUBLIC—STATISTICS

Czechoslovakia. Federální statistický úřad.
Retrospektivní lexikon obcí Československé socialistické republiky 1850-1970 : počet obyvatelů a domů podle obcí a částí obcí podle správního členění k 1. lednu 1972 a abecední přehled obcí a částí obcí v letech 1850-1970 / [zprac. kol. ; předml. Jan Kazimour]. — Praha : SEVT, 1978.
2 v. (vii, 1184 p.) : 5 fold. col. maps ; 30 cm. Cz78
Maps inserted.
Summary in English, German, and Russian.
Includes bibliographies.
CONTENTS: sv. 1. Česká socialistická republika (1. část).—sv. 2. Česká socialistická republika (2. část). Slovenská socialistická republika.
Kčs198.00
1. Czechoslovak Republic—Census, 1972. 2. Cities and towns—Czechoslovak Republic—Statistics. I. Title.
HA1192.F42 1978 79-367886
*79 MARC

—DENMARK—HISTORY—BIBLIOGRAPHY

Jansen, Henrik M
A select bibliography of Danish works on the history of towns published 1960-1972 / Henrik M. Jansen & Thomas Riis, under medvirken af Knud Hannestad. — Odense : Institut for historie og samfundsvidenskab, Odense universitet, 1973.
46 leaves ; 30 cm. — (Skrifter fra Institut for historie og samfundsvidenskab, Odense universitet : Historie ; bind 3)
"Each Danish title is given with a translation of it into English."
1. Cities and towns—Denmark—History—Bibliography. I. Riis, Thomas, joint author. II. Hannestad, Knud. III. Title. IV. Series: Odense universitet. Institut for historie og samfundsvidenskab. Skrifter : Historie ; bind 3.
Z7165.D4J34 79-313810
[HT145.D4] 79 MARC

Jansen, Henrik M
A select bibliography of Danish works on the history of towns published 1960-1976 / af Henrik M. Jansen, under medvirken af Knud Hannestad & Thomas Riis. — Odense : [Odense Universitet, Institut for Historie og Samfundsvidenskab], 1977.
iii, 85 p. ; 21 cm. — (Byhistoriske hjælpemidler ; 2) (Skrifter fra Institut for historie og samfundsvidenskab, Odense universitet : Historie ; bd. 21) D78-7
Cover title.
Includes index.
Not in trade.
1. Cities and towns—Denmark—History—Bibliography. 2. Denmark—History—Bibliography. I. Hannestad, Knud, joint author. II. Riis, Thomas, joint author. III. Title. IV. Series. V. Series: Odense universitet. Institut for historie og samfundsvidenskab. Skrifter : Historie ; bd. 21.
Z7165.D4J35 016.9489 79-350669
[HT145.D4] 79 MARC

—DENMARK—FYN—GROWTH—MATHEMATICAL MODELS

Geleff, Palle.
Den analytiske løsning for Fyns Amtskommunes byvækstmodel / Palle Geleff. — [København] : Københavns universitets Økonomiske institut, Cykelafdelingen, 1977.
6 p. ; 30 cm. — (Memo - Københavns universitets Økonomiske institut, Cykelafdelingen ; nr. 41) D***
1. Cities and towns—Denmark—Fyn—Growth—Mathematical models. 2. Fyn, Denmark—City planning. 3. City planning—Denmark. I. Title. II. Series: Copenhagen. Universitet. Økonomiske institut. Cykelafdelingen. Memo - Københavns universitets Økonomiske institut, Cykelafdelingen ; nr. 41.
HT145.D4G44 78-399920
79 MARC

—EGYPT

Jouguet, Pierre, 1869-1949.
La vie municipale dans l'Égypte romaine, par Pierre Jouguet ... Paris, Fontemoing et cie, 1911.
xiii, 494 p. 24 cm. (Bibliothèque des Écoles françaises d'Athènes et de Rome, fasc. 104)
Series title also at head of t.-p.
Bibliography: p. [xv]-xxiii
"Les sources": p. [xxiv]-xlii.
1. Egypt—History—332-30 B.C. 2. Egypt—History—30 B.C.-640 A.D. 3. Municipal government—Egypt—History. 4. Cities and towns—Egypt. I. Series.
D5.B4 vol. 104 11-29753
[r79]rev MARC

Jouguet, Pierre, 1869-1941.
La Vie municipale dans l'Égypte romaine ... Paris, E. de Boccard, 1968.
xlii, 494 p. 25 cm. (Bibliothèque des Écoles françaises d'Athènes et de Rome, fasc. 104) 90F F***
"Réimpression conforme à la première édition parue en 1911."
Bibliographical footnotes.
1. Egypt—History—332-30 B.C. 2. Egypt—History—30 B.C.-640 A.D. 3. Municipal government—Egypt—History. 4. Cities and towns—Egypt. I. Title. II. Series.
D5.B4 fasc. 104 1968 932 68-118017
[r79]rev MARC

—ENGLAND—GROWTH—PICTORIAL WORKS—JUVENILE LITERATURE

Goodall, John S
The story of an English village / [by] John S. Goodall. — London : Macmillan, 1978.
[62] p. : of col. ill. ; 19 x 26 cm. GB78-30276
Page width varies.
ISBN 0-333-23835-4 : £2.95
1. Cities and towns—England—Growth—Pictorial works—Juvenile literature. 2. City and town life—England—History—Pictorial works—Juvenile literature. I. Title.
HT133.G66 1978 942'.009'734 79-301017
79 MARC

Goodall, John S
The story of an English village / John S. Goodall. — 1st American ed. — New York : Atheneum, 1979, c1978.
[61] p. : col ill. ; 19 x 26 cm.
"A Margaret K. McElderry book."
SUMMARY: Shows the growth of an English village from a medieval clearing to the urban congestion of the present day as seen from the same viewpoint approximately every hundred years.
ISBN 0-689-50125-0 : $7.95
1. Cities and towns—England—Growth—Pictorial works—Juvenile literature. 2. City and town life—England—History—Pictorial works—Juvenile literature. [1. Cities and towns—England—Growth—Pictorial works. 2. City and town life—England—History—Pictorial works. 3. England—Social life and customs—Pictorial works] I. Title.
HT133.G66 1979 942'.009'734 78-56242
00 MARC

—ENGLAND—SURREY

O'Connell, Martin.
Historic towns in Surrey / [by] Martin O'Connell. — Guildford : Surrey Archaeological Society, 1977.
vi, 50 p. : maps, plans (some col.) ; 30 cm. — (Research volume of the Surrey Archaeological Society ; no 5 ISSN 0308-342X) GB***
Bibliography: p. 49-50.
ISBN 0-9501345-2-X : £2.70
1. Surrey, Eng.—History, Local. 2. Cities and towns—England—Surrey. I. Title. II. Series: Surrey Archaeological Society. Research volume of the Surrey Archaeological Society ; no. 5.
DA670.S96O25 942.2'1 79-300986
79 MARC

—ESTONIA

Pullat, Raimo.
Linnad kodanlikus Eestis : ajaloolis-demograafiline käsitlus / Raimo Pullat. — Tallinn : Eesti Raamat, 1978.
228 [1] p., [4] leaves of plates : ill. ; 21 cm. USSR***
Bibliography: p. 214-[229]
0.70rub
1. Cities and towns—Estonia. I. Title.
HT145.E75P87 79-357626
79 MARC

—EUROPE

(Europe ichiryū burando no hon)
Europe 一流ブランドの本：ヨーロッパの一流店ショッピング全ガイド. — 東京：講談社，昭和52 [1977]
198 p. : col. ill. ; 29 cm. — (講談社 mook ; 第2巻, 第5号) Ja***
Cover title.
Series romanized: Kōdansha mook.
¥1500
1. Clothing trade—Europe—Catalogs. 2. Branded merchandise—Europe—Catalogs. 3. Cities and towns—Europe. I. Kōdansha. II. Title: Ichiryū burando no hon.
HD9940.E82E94 79-804515

—EUROPE—CONGRESSES

Cities and city regions in Europe : conference, 1973, Lanchester Polytechnic, Coventry / edited by John Holliday, David Liggins, William Ogden. — [Coventry : Lanchester Polytechnic, 1973?]
235 p. : ill. ; 30 cm.
Includes bibliographical references.
1. Cities and towns—Europe—Congresses. 2. City planning—Europe—Congresses. 3. Metropolitan areas—Europe—Congresses. I. Holliday, John C. II. Liggins, David. III. Ogden, William. IV. Title.
HT131.C496 309.2'62'094 78-317986
78 MARC

CITIES AND TOWNS
—EUROPE—CONGRESSES (Continued)

European Symposium of Historic Towns, 2d, Strasbourg, 1976.
Historic town centres in the development of present-day towns / 2d European Symposium of Historic Towns, Strasbourg, 30 September-2 October 1976. — Strasbourg : Council of Europe ; ‹New York : sold by Manhattan Pub. Co.›, 1977.
103 p. : maps ; 21 cm. F***
10.00F
1. Cities and towns—Europe—Congresses. 2. Historic sites—Europe—Congresses. I. Council of Europe. II. Title.
HT131.E94 1976 301.36'3'094 78-323263
79 MARC

—EUROPE—GROWTH—ADDRESSES, ESSAYS, LECTURES

Organization for Economic Cooperation and Development. Environment Directorate.
The management of urban growth : a report on policy instruments for influencing the form and structure of urban development and the location and distribution of urban growth / editor, Terence Bendixson. — Paris : Environment Directorate, Organisation for Economic Co-operation and Development, 1977.
ii, 171 p. : ill. ; 29 cm. F***
Includes bibliographical references.
1. Cities and towns—Europe—Growth—Addresses, essays, lectures. 2. Land use, Urban—Europe—Addresses, essays, lectures. I. Bendixson, Terence. II. Title.
HT131.O73 1977 301.36'1'094 78-316449
78 MARC

—EUROPE—GROWTH—CONGRESSES

Kuratorium für Vergleichende Städtegeschichte, 5th, Münster, 1974.
Probleme des Städtewesens im industriellen Zeitalter / hrsg. von Helmut Jäger. — Köln ; Wien : Böhlau, 1978.
xviii, 349 p. : ill., maps (6 fold. in pocket) ; 27 cm. — (Städteforschung : Reihe A, Darstellungen ; Bd. 5) GFR78-A
Meeting was held at the Institut für Vergleichende Städtegeschichte in Münster, April 8-11, 1974.
Includes bibliographies and index.
ISBN 3-412-03678-1
1. Cities and towns—Europe—History—Congresses. 2. Cities and towns—Europe—Growth—Congresses. 3. Central places—Europe—History—Congresses. I. Jäger, Helmut, 1923- . II. Title. III. Series.
HT131.K85 1974 301.36'1'094 79-347746
79 MARC

—EUROPE—GROWTH—HISTORY

Rottier, Honoré.
Stedelijke strukturen : een inleiding tot de ontwikkeling van de Europese stad / Honoré Rottier. — Muiderberg : Coutinho, 1978.
228 p. : with ill. ; 25 cm. Ne78-47
Bibliography: p. 224-225.
Includes index.
ISBN 906283518X : fl 29.50
1. Cities and towns—Europe—History. 2. Cities and towns—Europe—Growth—History. I. Title.
HT131.R67 79-346933
*79 MARC

—EUROPE—HISTORY

Ennen, Edith.
Die europäische Stadt des Mittelalters / Edith Ennen. — 3. überarb. u. erw. Aufl. — Göttingen : Vandenhoeck und Ruprecht, 1979.
331 p. : ill., maps ; 21 cm. — (Sammlung Vandenhoeck) GFR79-A
Bibliography: p. 271-321.
Includes index.
ISBN 3-525-01320-5
1. Cities and towns—Europe—History. 2. Cities and towns, Medieval—Europe. I. Title.
HT131.E56 1979 79-371138
79 MARC

Ennen, Edith.
The Medieval town / by Edith Ennen ; translated by Natalie Fryde. — Amsterdam ; New York : North-Holland Pub. Co. ; New York : distributors for the U.S.A. and Canada Elsevier North-Holland, 1979.
287 p., ‹1› leaf of plates : ill. 23 cm. — (Europe in the Middle Ages ; v. 15)
Translation of Die europäische Stadt des Mittelalters.
Bibliography: p. ‹215›-272.
Includes index.
ISBN 0-444-85133-X
1. Cities and towns—Europe—History. 2. Cities and towns, Medieval—Europe. I. Title. II. Series.
HT131.E5613 301.36'3'094 79-1362

Guidoni, Enrico, 1939-
La città europea : formazione e significato dal IV all'XI secolo / Enrico Guidoni. — Milano : Electa, ‹1978›
197 p. : ill. ; 24 cm. — (Saggi) It78-Nov
Bibliography: p. 177-187.
Includes index.
1. Cities and towns—Europe—History. I. Title.
HT131.G84 78-399600
*79 MARC

Martina, Enrichetto.
Genesi e lettura comparata dello sviluppo urbano di 12 città europee = (Vergleichsanalyse der Stadtentwicklung von 12 europäischen Städten) / Enrichetto Martina ; presentazione di Giacomo Donato. — Pisa : Giardini, 1976.
170 p., 41 leaves of plates (27 fold.) : ill., maps ; 21 cm. — (Architettura urbanistica, metodi di programmazione e progetti ; 4) It78-Sept
Bibliography: p. 157-166.
L6500
1. Cities and towns—Europe—History. I. Title. II. Title: Vergleichsanalyse der Stadtentwicklung von 12 europäischen Städten.
HT131.M28 301.36'094 79-341717
*79 MARC

Rottier, Honoré.
Stedelijke strukturen : een inleiding tot de ontwikkeling van de Europese stad / Honoré Rottier. — Muiderberg : Coutinho, 1978.
228 p. : with ill. ; 25 cm. Ne78-47
Bibliography: p. 224-225.
Includes index.
ISBN 906283518X : fl 29.50
1. Cities and towns—Europe—History. 2. Cities and towns—Europe—Growth—History. I. Title.
HT131.R67 79-346933
*79 MARC

—EUROPE—HISTORY—ADDRESSES, ESSAYS, LECTURES

Città, mercanti, dottrine nell'economia europea dal IV al XVIII secolo : saggi in memoria di Gino Luzzatto / Barbieri ... ‹et al.› ; raccolti e presentati da Amintore Fanfani. — Milano : Giuffrè, 1964.
346 p., ‹12› leaves of plates : ill. ; 23 cm. — (Biblioteca della rivista Economia e storia ; 11)
French, Italian, or Spanish.
Includes bibliographies and indexes.
CONTENTS: Fanfani, A. Gino Luzzatto nella vita, nelle lettere, nei saggi.—Vercauteren, F. La ville en Europe du IVe au XIe siècle.—Barbieri, G. Le dottrine economiche medioevali.—Cessi, R. Venezia, politica ed economia.—Heers, J. Gênes.—Melis, F. Firenze.—Bergier, J.-F. Genève.—Romani, M. Milano.—Mollat, M. Jacques Coeur.—Mira, G. La dottrina economica mercantilista.—Chaunu, P. Séville, pôle de croissance?.—Vázquez de Prada, V. Los Ruiz en la vida económica del siglo XVI.—Van Houtte, J.-A. Anvers.—Verlinden, Ch. Amsterdam.
1. Europe—Economic conditions—Addresses, essays, lectures. 2. Cities and towns—Europe—History—Addresses, essays, lectures. 3. Economics—Europe—History—Addresses, essays, lectures. I. Luzzatto, Gino. II. Barbieri, Gino. III. Fanfani, Amintore.
HC240.C497 79-345237
*79 MARC

—EUROPE—HISTORY—ADDRESSES, ESSAYS, LECTURES—COLLECTED WORKS

International Commission for the Histroy of Towns.
Guide international d'histoire urbaine / préparé par la Commission ; sous la direction de Philippe Wolff. — Paris : Éditions Klincksieck, 1977-
v. ; 24 cm. F***
At head of title: Commission internationale pour l'histoire des villes.
Includes bibliographies.
CONTENTS: 1. Europe.

ISBN 2-252-01836-4 : 196F (v. 1)
1. Cities and towns—Europe—History—Addresses, essays, lectures—Collected works. I. Wolff, Philippe, 1913- . II. Title.
HT131.I44 1977 909'.09'732 79-352352
79 MARC

—EUROPE—HISTORY—CONGRESSES

European towns : their archaeology and early history / edited by M. W. Barley. — London ; New York : Published for the Council for British Archaeology by Academic Press, 1977.
xxvii, 523 p. : ill. ; 27 cm. GB***
Papers presented at a conference on European towns held at Oxford, 1975.
Summaries in French and German.
Includes bibliographies and index.
ISBN 0-12-078850-0
1. Europe—Antiquities—Congresses. 2. Cities and towns, Ruined, extinct, etc.—Europe—History—Congresses. 3. Cities and towns—Europe—History—Congresses. I. Barley, Maurice Willmore, 1909- . II. Council for British Archaeology.
CC175.E94 301.36'1'0936 77-71807
78‹r79›rev MARC

Kuratorium für Vergleichende Städtegeschichte, 5th, Münster, 1974.
Probleme des Städtewesens im industriellen Zeitalter / hrsg. von Helmut Jäger. — Köln ; Wien : Böhlau, 1978.
xviii, 349 p. : ill., maps (6 fold. in pocket) ; 27 cm. — (Städteforschung : Reihe A, Darstellungen ; Bd. 5) GFR78-A
Meeting was held at the Institut für Vergleichende Städtegeschichte in Münster, April 8-11, 1974.
Includes bibliographies and index.
ISBN 3-412-03678-1
1. Cities and towns—Europe—History—Congresses. 2. Cities and towns—Europe—Growth—Congresses. 3. Central places—Europe—History—Congresses. I. Jäger, Helmut, 1923- . II. Title. III. Series.
HT131.K85 1974 301.36'1'094 79-347746
79 MARC

—EUROPE—MAPS

Branch, Melville Campbell, 1913-
Comparative urban design : rare engravings, 1830-1843 / by Melville C. Branch. — New York : Arno Press, c1978.
‹iv›, 102 ‹i.e. 104› p. : ill., col. maps; 33 x 37 cm.
Reprint of maps published 1830-1844 by the Superintendence of the Society for the Diffusion of Useful Knowledge, London.
Includes bibliographical references.
ISBN 0-405-10524-X : $39.50
1. Cities and towns—Maps. 2. Cities and towns—Europe—Maps. 3. Cities and towns—United States—Maps. 4. Cities and towns—Asia—Maps. 5. Engravings—Europe—Maps. 6. Engravings—United States—Maps. 7. Engravings—Asia—Maps. I. Society for the Diffusion of Useful Knowledge, London. II. Title. III. Title: Rare engravings, 1830-1843.
G1028.B65 1978 711'.4 77-24511
77 MARC

—EUROPE, EASTERN—ADDRESSES, ESSAYS, LECTURES

The Socialist city : spatial structure and urban policy / edited by R. A. French & F. E. Ian Hamilton. — Chichester ; New York : Wiley, c1979.
xviii, 541 p. : ill. ; 24 cm.
Includes bibliographies and index.
ISBN 0-471-99689-0
1. Cities and towns—Russia—Addresses, essays, lectures. 2. Cities and towns—Europe, Eastern—Addresses, essays, lectures. I. French, Richard Anthony. II. Hamilton, F. E. Ian.
HT145.R9S63 309.2'62'0947 78-16828
78 MARC

—FINLAND

Vuoristo, Kai Veikko.
Die Wirkung der Veränderung von Verkehrsverhältnissen auf die Entwicklung einiger zentralen Küstenorte in Finnland. Helsinki, 1966.
209 p. maps. 25 cm. (Fennia 94, n:o 1)
Bibliography: p. ‹205›-209.
1. Transportation—Finland. 2. Cities and towns—Finland. I. Title. II. Series.
‹G23.G4 vol. 94, no. 1› 947.1'008 s a 66-338
‹HE255.3› ‹r79›rev2 *MARC

—FINLAND—ADDRESSES, ESSAYS, LECTURES

Räsänen, Matti.
Kaupunkikansatieteen ongelmia : tutkimuskohde ja sen kokonaishahmo kulttuurin muutoksessa / Matti Räsänen, Juhani U. E. Lehtonen. — Jyväskylä : Jyväskylän yliopisto, 1976 (s.l. : Jyväskylä Kari).
4, 111 p. : ill. ; 21 cm. — (Tutkimuksia - Jyväskylän yliopisto, Etnologian laitos ; 7) Fi78-5
Bibliography: p. 103-111.
ISBN 9516779239
1. Cities and towns—Finland—Addresses, essays, lectures. 2. Sociology, Urban—Finland—Addresses, essays, lectures. I. Lehtonen, Juhani Urmas Eerikki, 1942- joint author. II. Title. III. Series: Jyväskylän yliopisto. Etnologian laitos. Tutkimuksia - Jyväskylän yliopisto, Etnologian laitos ; 7.
HT145.F5R33 79-346200
*79 MARC

—FINLAND—GROWTH

Rikkinen, Kalevi.
Suomen asutusmaantiede / Kalevi Rikkinen. — Helsinki : Otava, 1977.
163 p. : ill. ; 21 cm. — (Otavan korkeakoulukirjasto) Fi77
Bibliography: p. 199-163.
ISBN 9511042831 : Fmk54.00
1. Migration, Internal—Finland. 2. Cities and towns—Finland—Growth. 3. Rural-urban migration—Finland. 4. Land settlement—Finland. I. Title.
HB2068.3.A3R54 78-390248
79 MARC

CITIES AND TOWNS (Continued)

—FRANCE

La France des villes / sous la direction de Jacqueline Beaujeu-Garnier ; [publié par] la Documentation française. — Paris : la Documentation française, 1978-
v. : maps (some col.) ; 24 cm. F78-15028
Includes bibliographical references.
CONTENTS: 1. Le Bassin parisien, rédigé par Beaujeu-Garnier et al.

45.00F (v. 1)
1. Cities and towns—France. 2. Urbanization—France. I. Beaujeu-Garnier, Jacqueline.
HT135.F73 301.36'0944 79-359721
79 MARC

Garnier, Jean Pierre.
Le socialisme à visage urbain : essai sur la local-démocratie / Jean-Pierre Garnier, Denis Goldschmidt. — Paris : Éditions Rupture, c1978.
443 p. ; 21 cm. F***
Includes bibliographical references.
ISBN 2-86220-006-9 : 41.00F
1. Cities and towns—France. 2. Middle classes—France. 3. Civil rights—France. I. Goldschmidt, Denis, joint author. II. Title.
HT135.G37 301.36'1'0944 79-355809
79 MARC

Qui fait la ville? : Témoignages et suggestions / [présentés] par François Parfait, Association générale des hygiénistes et techniciens municipaux. — Paris : Technique & documentation, c1978.
124 p. : ill. ; 24 cm. F***
Includes bibliographical references.
ISBN 2-85206-027-2 : 55.00F
1. Municipal government—France. 2. Cities and towns—France. I. Parfait, F. II. Association générale des hygiénistes et techniciens municipaux.
JS4883.Q5 352'.008'0944 78-399891
79 MARC

—FRANCE—JUVENILE POETRY

Vingt mots pour ma ville : poèmes pour les enfants / Pierre Béarn, Luc Bérimont, Jacques Charpentreau, Andrée Chedid ... [etc.] ; écrits à l'initiative de Monique Bermond et Roger Boquié .... — Paris : Éditions ouvrières, 1978.
125 p. ; 21 cm. — (Collection Enfance heureuse) F78-10730
"A l'occasion du IXe Festival international du livre de Nice."
28.00F
1. Cities and towns—France—Juvenile poetry. I. Béarn, Pierre, 1902- . II. Bermond, Monique. III. Boquié, Roger. IV. Festival international du livre, 9th, Nice, 1977.
PZ24.3.V54 841'.9'1408 79-355297
79 MARC

—FRANCE—VIEWS

Rossel, André, comp.
Découverte de la Flandre [par] André Rossel [et] Jean Vidal. Avant-propos: Roger Desreumaux. [Paris] Éditions Hier et demain [1971]
1 portfolio ([8] p., 47 plates (part col.)) 58 cm. 195.00F F***
1. Flanders—Maps—To 1800. 2. Cities and towns—France—Views. 3. Maps, Early—Facsimiles. I. Vidal, Jean, fl. 1971- . II. Title.
G1843.N7R55 1971 912'.44'28 73-650045
79[r79]rev MARC

—FRANCE—LILLE (ARRONDISSEMENT)—MAPS

Ravet-Anceau (Firm)
Guide de la métropole : arrondissement de Lille et sa communauté urbaine. — Lille : Ravet-Anceau, [1977?]
208, 200 p. : ill. (some col.), maps (1 fold. col. inserted) ; 29 cm. F***
Cover title.
Includes indexes.
1. Lille, France (Arrondissement)—Description and travel. 2. Cities and towns—France—Lille (Arrondissement)—Maps. 3. Lille, France (Arrondissement)—Maps. 4. Lille metropolitan area, France—Maps. I. Title.
G1844.24.L5R3 1977 912'.44'28 78-381407
78 MARC

—GEORGIA—HISTORY—18TH CENTURY

Sears, Joan Niles.
The first one hundred years of town planning in Georgia / by Joan Niles Sears. — Atlanta : Cherokee Pub. Co., 1979.
xiii, 220 p. : ill. ; 24 cm.
Based on the author's thesis, Emory University, 1977.
Bibliography: p. 196-208.
Includes index.
ISBN 0-87797-046-7
1. Cities and towns—Georgia—History—18th century. 2. City planning—Georgia—History—18th century. I. Title.
HT123.5.G4S4 309.2'62'09758 78-74091
00 MARC

—GERMANY—HISTORY

Gööck, Roland.
Deutsche Städte vor 100 Jahren / Roland Gööck. — Gütersloh : Praesentverlag Peter, 1978.
253 p. : chiefly ill. ; 27 cm. GFR78-A
Includes index.
ISBN 3-87644-061-0 : DM38.00
1. Cities and towns—Germany—History. 2. Germany—Description and travel. I. Title.
DD91.G63 943 79-345557
79 MARC

Martin, Thomas Michael.
Die Städtepolitik Rudolfs von Habsburg / von Thomas Michael Martin. — Göttingen : Vandenhoeck und Ruprecht, 1976.
268 p. : 2 fold. maps (in pocket) ; 25 cm. — (Veröffentlichungen des Max-Planck-Instituts für Geschichte ; 44) GFR77-A6
A revision of the author's thesis, Giessen, 1971.
Bibliography: p. 211-235.
Includes indexes.
ISBN 3-525-35353-7 : DM62.00
1. Rudolf I, of Hapsburg, Emperor of Germany, 1218-1291. 2. Cities and towns—Germany—History. 3. Germany—Politics and government—1273-1517. I. Title. II. Series: Göttingen. Max-Planck-Institut für Geschichte. Veröffentlichungen ; 44.
DD159.M37 1976 943'.026'0924 79-342013
79 MARC

—GERMANY—HISTORY—ADDRESSES, ESSAYS, LECTURES

Die Deutsche Stadt im Industriezeitalter : Beitr. zur modernen dt. Stadtgeschichte / hrsg. von Jürgen Reulecke ; Beitr. von Wolfgang Hofmann ... [et al.]. — Wuppertal : Hammer, 1978.
151 p. : 3 ill., map ; 21 cm. GFR78-A23
Bibliography: p. 129-144.
Includes indexes.
ISBN 3-87294-124-0 : DM18.00
1. Cities and towns—Germany—History—Addresses, essays, lectures. 2. Germany—Industries—History—Addresses, essays, lectures. I. Reulecke, Jürgen.
HT137.D425 78-394898
79 MARC

—GERMANY—HISTORY—SOURCES

Die Urkunden und Akten der oberdeutschen Städtebünde vom 13. Jahrhundert bis 1549 / hrsg. durch d. Histor. Komm. bei d. Bayer. Akad. d. Wiss. — Göttingen : Vandenhoeck und Ruprecht, 1979-
v. ; 25 cm. GFR79-A(v.1)
Includes bibliographical references and indexes.
CONTENTS: Bd. 1. Ruser, K. Vom 13. Jahrhundert bis 1347.

ISBN 3-525-35913-6 (v. 1)
1. Germany—History—1273-1517—Sources. 2. Rhine Cities, League of, 1254—Sources. 3. Cities and towns—Germany—History—Sources. 4. Cities and towns—Switzerland—History—Sources. 5. Landfriede—History—Sources. I. Akademie der Wissenschaften, Munich. Historische Kommission.
DD156.A2U74 943 79-367818
79 MARC

—GERMANY, EAST

Zimm, Alfred.
Zu einigen Tendenzen der Bevölkerungsverteilung und der Naherholung in agglomerierten Gebieten / Alfred Zimm. Zur Stellung der Städte der DDR im Siedlungssystem nach ihren In-, Umland- und Fernfunktionen / Frankdieter Grimm. — Berlin : Akademie-Verlag, 1977.
39 p. : graphs ; 22 cm. — (Sitzungsberichte der Akademie der Wissenschaften der DDR ; Jahrg. 1977, Nr. 19/N : Mathematik, Naturwissenschaften, Technik) GFR78-A28
Cover title: Tendenzen der Bevölkerungsverteilung und Stellung der Städte im Siedlungssystem.
Includes bibliographies.
4.00M
1. Cities and towns. 2. Recreation. 3. Cities and towns—Germany, East. I. Grimm, Frankdieter. Zur Stellung der Städte der DDR im Siedlungssystem nach ihren In-, Umland- und Fernfunktionen. 1977. II. Title. III. Title: Tendenzen der Bevölkerungsverteilung und Stellung der Städte im Siedlungssystem. IV. Series: Akademie der Wissenschaften der DDR. Sitzungsberichte der Akademie der Wissenschaften der DDR ; Jahrg. 1977, Nr. 19/N.
HT153.Z49 79-375429
79 MARC

—GERMANY, WEST

Scholl, Robert.
Urban transformation : special survey / by Robert Scholl ; ed.-in-chief, Helmut Göbel ; responsible for contents, Rolf Gerlach. — Bonn-Bad Godesberg : Inter Nationes, [1976]
19 leaves ; 30 cm. — (Sonderdienst ; SO 1-76(e)) GFR***
Cover title.
1. Cities and towns—Germany, West. 2. City planning—Germany, West. 3. Urbanization. I. Goebel, Helmut, 1921- . II. Gerlach, Rolf. III. Title. IV. Series: Inter Nationes. Sonderdienst ; SO 1-76(e)
HT137.S353 309.2'62'0943 78-380044
78 MARC

—GERMANY, WEST—GROWTH—CASE STUDIES

Wolf, Helfried.
Grevenbroich, Würselen und Eschweiler, Entwicklungs- und Struktur-Vergleich dreier linksrheinischer Mittelstädte / von Helfreid Wolf. — Aachen : Geograph. Inst. d. RWTH, 1977.
vii, 383 p. : maps (18 in pocket) ; 21 cm. — (Aachener geographische Arbeiten ; Heft 11) GFR77-B20
Originally presented as the author's thesis, Technische Hochschule, Aachen, 1976.
Bibliography: p. 351-383.
1. Urbanization—Germany, West—Case studies. 2. Cities and towns—Germany, West—Growth—Case studies. 3. Grevenbroich, Ger.—History. 4. Würselen, Ger.—History. 5. Eschweiler, Ger. (Aachen)—History. I. Title. II. Series.
HT384.G3W64 1977 79-361688
79 MARC

—GERMANY, WEST—BADEN-WÜRTTEMBERG—STATISTICS

Baden-Württemberg. Statistisches Landesamt.
Regionalstruktur Baden-Württemberg 1977 [i.e. neunzehnhundertsiebenundsiebzig] : Gemeinden / hrsg. vom Statistischen Landesamt Baden-Württemberg. — Stuttgart : Das Landesamt, 1978.
219 p. : maps (some col.) ; 30 cm. — (Statistik von Baden-Württemberg ; Bd. 262) GFR***
Chiefly tables.
1. Baden-Württemberg—Statistics. 2. Cities and towns—Germany, West—Baden-Württemberg—Statistics. I. Title. II. Series: Baden-Württemberg. Statistisches Landesamt. Statistik von Baden-Württemberg ; Bd. 262.
HA1320.B2A32 Bd. 262 79-373985
79 MARC

—GERMANY, WEST—NORTH RHINE-WESTPHALIA—COLLECTED WORKS

Entscheidungshilfen für die Freiraumplanung : Verhaltungswissenschaftlich-empirischer Teil / Dietrich Garbrecht ... [et al.]. — Dortmund : Institut für Landes- und Stadtentwicklungsforschung des Landes Nordrhein-Westfalen (ILS) im Auftrage des Innenministers NW, 1978.
2 v. : ill. ; 30 cm. — (Schriftenreihe Landes- und Stadtentwicklungsforschung des Landes Nordrhein-Westfalen : Reihe 4, Materialien ; Bd. 10/1-2) GFR***
CONTENTS: Bd. 1. Text.—Bd. 2. Tabellen.
1. Cities and towns—Germany, West—North Rhine-Westphalia—Collected works. 2. Open spaces—Germany, West—North Rhine-Westphalia—Collected works. 3. Recreation areas—Germany, West—North Rhine-Westphalia—Collected works. I. Garbrecht, Dietrich. II. Institut für Landes- und Stadtentwicklungsforschung des Landes Nordrhein-Westfalen. III. Series.
HT137.E55 301.36'1'094355 79-349895
79 MARC

—GERMANY, WEST—OBERHARZ MOUNTAINS—HISTORY

Morich, Heinrich.
Kleine Chronik der Oberharzer Bergstädte und ihres Erzbergbaus / von H. Morich. — 4., erw. Aufl. der Chronik der Bergstadt Clausthal-Zellerfeld / überarb. u. erw. von Herbert Dennert. — Clausthal-Zellerfeld : Pieper, 1974.
xvi, 200 p. : numerous ill., maps ; 25 cm. GFR77-A
Bibliography: p. 199-200.
DM39.50
1. Clausthal-Zellerfeld, Ger.—History. 2. Cities and towns—Germany, West—Oberharz Mountains—History. 3. Mineral industries—Germany, West—Oberharz Mountains. I. Dennert, Herbert. II. Title.
DD901.C57M67 1974 79-337214

CITIES AND TOWNS (Continued)

—GERMANY, WEST—SWABIA —HISTORY

Laub, Joseph.
Geschichte der vormaligen fünf Donaustädte in Schwaben : mit übersichtl. Beschreibung d. polit. Verhältnisse Vorderoesterreichs, insbes. Schwäb.-Oesterreichs / bearb. von Joseph Laub. — Neuaufl. — [Mengen, Kr. Saulgau : Bürgermeisteramt, 1972] (Mengen : J. Goelz)
viii, 240 p. ; 22 cm. GFR75-B22
Reprint of the 1894 ed. published by K. Gruber, Mengen.
Bibliography: p. vi-viii.
Includes index.
DM8.00
1. Swabia—History. 2. Cities and towns—Germany, West—Swabia—History. 3. Württemberg—Foreign relations—Austria. 4. Austria—Foreign relations—Württemberg. 5. Austria, Upper—History. I. Title.
DD801.W92L38 1972 943'.37 79-364917
79 MARC

—GERMANY, WEST—WESTPHALIA

Stievermann, Dieter.
Städtewesen in Südwestfalen : d. Städte d. Märk. Sauerlandes im späten Mittelalter u. in d. frühen Neuzeit / von Dieter Stievermann. — 1. Aufl. — Stuttgart : Klett-Cotta, 1978.
263 p. ; 23 cm. — (Spätmittelalter und frühe Neuzeit ; Bd. 6) GFR78-A
Originally presented as the author's thesis, Münster, 1975 under title: Spätmittelalterlich-frühneuzeitliches Städtewesen in Südwestfalen.
Bibliography: p. 229-248.
Includes indexes.
ISBN 3-12-911580-3
1. Sauerland—Politics and government. 2. Cities and towns—Germany, West—Westphalia. 3. Westphalia—Politics and government. I. Title. II. Series.
DD491.W5S75 1978 320.9'43'56 78-389973
79 MARC

—GREAT BRITAIN

Civic Trust.
Urban wasteland : a report on land lying dormant in cities, towns and villages in Britain / Civic Trust ; [written and compiled by Timothy Cantell]. — London : The Trust, 1977.
56 p. : ill., map ; 30 cm. GB78-03092
ISBN 0-900849-87-8 : £2.00
1. Cities and towns—Great Britain. 2. Land use, Urban—Great Britain. 3. Land use—Planning—Great Britain. I. Cantell, Timothy. II. Title.
HT133.C52 1977 333.7'3 78-314120
78 MARC

Eyre, Reginald.
Hope for our towns & cities : the right approach to urban affairs / by Reginald Eyre. — London : Conservative Political Centre, 1977.
24 p. ; 21 cm. — (CPC [publications] ; no. 606) GB77-25409
ISBN 0-85070-600-9 : £0.40
1. Cities and towns—Great Britain. I. Title. II. Series: Conservative Political Centre, London. [Publications] ; no. 606.
HT133.E96 309.2'62'0941 78-322620
79 MARC

Millward, Roy.
Landscapes of Britain / Roy Millward and Adrian Robinson. — Newton Abbot ; North Pomfret, Vt. : David & Charles, 1977.
112 p. : ill. ; 25 cm. GB***
Includes bibliographical references.
ISBN 0-7153-7181-9 : $12.95
1. Landforms—Great Britain. 2. Cities and towns—Great Britain. 3. Industrial archaeology—Great Britain. I. Robinson, Adrian Henry Wardle, 1925- joint author. II. Title.
GB436.G7M54 911'.41 76-55884
79 MARC

Social problems and the city : geographical perspectives / edited by David T. Herbert and David M. Smith. — Oxford ; New York : Oxford University Press, 1979.
xi, 271 p. : ill. ; 24 cm. GB79
Includes bibliographies and indexes.
ISBN 0-19-874079-4 : £8.50
1. Cities and towns—Great Britain. 2. Social problems—Case studies. 3. Urban policy—Great Britain. I. Herbert, David T. II. Smith, David Marshall, 1936-
HT133.S65 362'.042'0941 78-41144
79 MARC

Urban deprivation and the inner city / edited by Colin Jones. — London : Croom Helm, c1979.
218 p. : ill. ; 23 cm. GB***
"Papers given to the SSRC Urban and Regional Economics Seminar Group."
Includes bibliographies and index.
ISBN 0-85664-713-6 : £9.95
1. Cities and towns—Great Britain. 2. Great Britain—Social conditions—1945- I. Jones, Colin.
HT133.U68 362'.042'0941 79-315585
79 MARC

—GREAT BRITAIN—BIBLIOGRAPHY

Filipovitch, Anthony J
Urban community : a guide to information sources / Anthony J. Filipovitch, Earl J. Reeves. — Detroit : Gale Research Co., c1978.
x, 286 p. ; 23 cm. — (Urban studies information guide series ; v. 4) (Gale information guide library)
Includes bibliographies and indexes.
ISBN 0-8103-1429-0
1. Community life—Bibliography. 2. Sociology, Urban—Bibliography. 3. Cities and towns—United States—Bibliography. 4. Cities and towns—Great Britain—Bibliography. I. Reeves, Earl J., joint author. II. Title. III. Series.
Z7164.S68F54 016.30136 78-13171
[HM131] 78 MARC

Lambert, Claire M
Inner cities : a select list of material / [Departments of the Environment and Transport] Library ; compiled by Claire M. Lambert. — London : The Library, 1977.
[1], iv, 69 p. ; 30 cm. — (Bibliography series ; no. 194) GB78-00662
Kept up to date by supplements.
1. Urban renewal—Great Britain—Bibliography. 2. Cities and towns—Great Britain—Bibliography. I. Great Britain. Departments of the Environment and Transport Library. II. Title. III. Series: Bibliography series (London) ; no. 194.
Z7164.U7L35 016.30136 79-304665
[HT178.G7] 79 MARC

—GREAT BRITAIN—CASE STUDIES

Champion, Anthony Gerard.
Facts about the new towns : a socio-economic digest / [by] A. G. Champion, K. Clegg and R. L. Davies. — Corbridge : Retailing and Planning Associates, 1977.
268 p. : ill., maps ; 21 cm. GB78-08107
Bibliography: p. 267-268.
ISBN 0-905269-03-9
1. New towns—Great Britain—Case studies. 2. Cities and towns—Great Britain—Case studies. I. Clegg, K., joint author. II. Davies, Rosser Llewelyn, joint author. III. Title.
HT169.G7C454 309.2'62'0941 78-322265
79 MARC

—GREAT BRITAIN—HISTORY

Treble, James H
Urban poverty in Britain, 1830-1914 / James H. Treble. — New York : St. Martin's Press, 1979.
216 p. ; 24 cm.
Includes bibliographical references and index.
ISBN 0-312-83463-2 : $20.00
1. Poor—Great Britain—History. 2. Cities and towns—Great Britain—History. I. Title.
HV4085.A4T73 1979 362.5'0941 78-23650
78 MARC

—GREAT BRITAIN—MAPS

Royal Automobile Club, London.
RAC United Kingdom town plans. — London : R.A.C., 1977-
v. : maps ; 26 cm. GB78-02828
Includes index.
CONTENTS: 1. Southern Ireland.

ISBN 0-902628-47-X (v. 1) : £1.75 (v. 1)
1. Cities and towns—Great Britain—Maps. I. Title. II. Title: United Kingdom town plans.
G1814.A1R6 1977 912'.41 79-318265
79 MARC

—GREAT BRITAIN—STATISTICS

Davies, Emily M
A classification of urban populations in England and Wales, 1951-1971 / [by] E. M. Davies. — [Birmingham] : University of Birmingham Environmental Modelling and Survey Unit, [1977]
[4], 298 p. (4 fold.) : ill. ; 30 cm. GB78-11631
ISBN 0-7044-0275-0 : £4.50
1. Great Britain—Statistics. 2. Great Britain—Population—Statistics. 3. Cities and towns—Great Britain—Statistics. I. Title.
HA1137.D38 312'.0942 79-307929
79 MARC

—GREECE—ADDRESSES, ESSAYS, LECTURES

Polis e economia nella Grecia antica : testi di Esiodo ... [et al.] / a cura di Mario Vegetti. — Bologna : Zanichelli, [1977], c1976.
109 p. ; 19 cm. — (Letture di filosofia e scienze umane ; 1) It78-Aug
Bibliography: p. [110]
L1400
1. Greece—Politics and government—Addresses, essays, lectures. 2. Greece—Economic conditions—Addresses, essays, lectures. 3. Cities and towns—Greece—Addresses, essays, lectures. I. Hesiodus. II. Vegetti, Mario.
DF82.P56 1977 320.9'38 79-371827
•79 MARC

—GREECE, MODERN—MEGARA REGION

Sakellariou, Michaēl B
(Megaris, Aigosthena, Ereneia)
Μεγαρίς, Αἰγόσθενα, Ἐρένεια [ἔγραψε] Μ. Σακελλαρίου [καὶ] Ν. Φαράκλας. [Ἀθῆναι, Ἀθηναϊκὸ Κέντρο Οἰκιστικῆς, 1972]
1 v. (various pagings) ill. 28 cm. (Ἀρχαῖες Ἑλληνικὲς πόλεις, 14)
A study conducted by the Athens Center of Ekistics.
Bibliography: p. 1–8 (8th group)
1. Megara region, Greece (Modern)—History. 2. Megara region, Greece (Modern)—Antiquities. 3. Cities and towns—Greece, Modern—Megara region. I. Faraklas, N., joint author. II. Athens Center of Ekistics. III. Title. IV. Series: Archaies Hellēnikes poleis, 14.
DF251.A7 no. 14 73–204525
[DF901.M7]

—HUNGARY—GROWTH

Urban development in the USA and Hungary / edited by György Enyedi. — Budapest : Akadémiai Kiadó, 1978.
311 p., [1] fold. leaf : ill., maps ; 25 cm. — (Studies in geography in Hungary ; 14) H***
Includes bibliographies.
ISBN 9630515148
1. Cities and towns—United States—Growth. 2. Cities and towns—Hungary—Growth. 3. Urbanization—United States. 4. Urbanization—Hungary. I. Enyedi, György. II. Series.
HT123.U744 301.36'1'09439 79-310628
79 MARC

—HUNGARY—HISTORY

Gerő, László, 1909-
Történelmi városmagok / Gerő László. — [Budapest] : Corvina, c1978.
135 p. : numerous ill. ; 24 cm. — (Építészeti hagyományok) H78
Bibliography: p. 133-134.
ISBN 9631302814 : 52.00Ft
1. Hungary—Description and travel—1945- . 2. Cities and towns—Hungary—History. I. Title.
DB917.G47 79-358962
•79 MARC

Granasztói, Pál.
Beautiful Hungarian towns through the eyes of an architect / Pál Granasztói ; photo. by Lajos Czeizing ; [translated by Lili Halapy]. — Budapest : Corvina Press, 1978.
263, [1] p. : ill. ; 24 cm. H***
Translation of Szép magyar városok egy építész szemével.
Bibliography: p. [264]
ISBN 9631302318
1. Hungary—Description and travel—1945- —Views. 2. Cities and towns—Hungary—History. I. Czeizing, Lajos. II. Title.
DB906.5.G713 943.9'00973'2 79-315093
79 MARC

Granasztói, Pál.
Szép magyar városok, egy építész szemével / Granasztói Pál ; fényképezte Czeizing Lajos. — [Budapest] : Corvina, 1978.
263 p. : numerous ill. ; 23 cm. H78
Bibliography: p. 11.
ISBN 9631301990 : 162.00Ft
1. Hungary—Description and travel—1945- —Views. 2. Cities and towns—Hungary—History. I. Czeizing, Lajos. II. Title.
DB906.5.G7 79-364672
•79 MARC

A Magyar városok címerei / [összeállította Castiglione Endre]. — Budapest : Közgazdasági és Jogi Könyvkiadó, 1975.
451 p. : col. ill. ; 24 cm.
Introduction and summary in English, French, German and Russian.
ISBN 9632202597
1. Heraldry—Hungary. 2. Cities and towns—Hungary—History. I. Castiglione, Endre.
CR539.5.M33 79-345408
79 MARC

CITIES AND TOWNS (Continued)

—HUNGARY—HAJDU-BIHAR

Hajdu-Bihar településeinek fejlődése : 1971-1975 / [főszerkesztő, Ónosi László ; bev., Angyal László]. — [Debrecen] : Hajdú-Bihar Megyei Tanács V. B. Építési, Közlekedési és Vízügyi Osztálya, [1977]
103 p. : ill. ; 19 x 20 cm. e-hu---
ISBN 9630305445
1. Hajdu-Bihar, Hungary—Economic conditions. 2. Cities and towns—Hungary—Hajdu-Bihar. I. Ónosi, László.
[DNLM: H77]
HC300.29.H34H34 79-338537
*79 MARC

—INDIA

Nilsson, Sten Åke, 1936-
Fem huvudstäder i Indien och Pakistan. En kritisk genomgång. Lund, Studentlitteratur, 1970.
70 p. illus. 23 cm. kr17.00 S71-05
Bibliography: p. 66-70.
1. Cities and towns—India. 2. Cities and towns—Pakistan. I. Title.
HT147.I5N54 78-399409
79 MARC

—INDIA—ADDRESSES, ESSAYS, LECTURES

Million cities of India / edited by R. P. Misra. — New Delhi : Vikas Pub. House, c1978.
ix, 405 p. : ill. ; 22 cm.
Bibliography: p. [381]-395.
Includes index.
ISBN 0-7069-0554-7 : Rs80.00
1. Cities and towns—India—Addresses, essays, lectures. 2. Urbanization—India—Addresses, essays, lectures. I. Misra, Rameshwar Prasad, 1930-
HT147.I5M55 301.36'0954 78-904602
79 MARC

—INDIA—CASE STUDIES

Prabhavalkar, B H
Trends of urbanisation in a new township : problems & prospects : a case study of Khopoli industrial complex / by B. H. Prabhavalkar, K. S. Bahuleyan. — Bombay : Regional Centre for Training and Research in Municipal Administration of All-India Institute of Local Self-government, [1977]
viii, 148, lix p. ; 29 cm.
Bibliography: p. lviii-lix.
Rs26.00
1. Cities and towns—India—Case studies. 2. Urbanization—India—Khopoli—Case studies. I. Bahuleyan, K. S. II. Title.
HT147.I5P68 301.36'1'0954792 78-904448
79 MARC

—INDIA—GROWTH

Jain, M K
Interstate variations in the trends of urbanization in India, 1951-71 / M. K. Jain. — Bombay : International Institute for Population Studies, 1977.
50, xix leaves ; 29 cm.
Includes bibliographical references.
1. India—Population. 2. Cities and towns—India—Growth. 3. Urbanization—India. I. International Institute for Population Studies. II. Title.
HB3639.J335 301.32'9'54 78-905505
79 MARC

—INDIA—GROWTH—CASE STUDIES

Dutta, Maya.
Jamshedpur : the growth of the city and its regions / Maya Dutta. — Calcutta : Asiatic Society, 1977.
xiv, 177 p., [22] leaves of plates : maps ; 22 cm.
Bibliography: p. [167]-177.
Rs40.00 ($7.00 U.S.)
1. Cities and towns—India—Growth—Case studies. 2. Jamshedpur, India—History.
HT147.I5D87 330.9'54'12 78-903508
79 MARC

—INDIA—HISTORY—BIBLIOGRAPHY

Van Willigen, John.
The Indian city : a bibliographic guide to the literature on urban India / John van Willigen. — New Haven : Human Relations Area Files, inc., 1979.
2 v. ; 23 cm. — (HRAFLex books ; AW1-001) (Bibliography series - Human Relations Area Files, inc)
1. City planning—India—Bibliography. 2. Urbanization—India—Bibliography. 3. Cities and towns—India—History—Bibliography. 4. Sociology, Urban—India—Bibliography. I. Human Relations Area Files, inc. II. Title. III. Series: Human Relations Area Files, inc. HRAFlex books ; AW1-001. IV. Series: Human Relations Area Files, inc. Bibliography series - Human Relations Area Files, inc.
Z5942.V36 016.30136'0954 79-117894
[HQ766] 79 MARC

—INDIA—PUNJAB (STATE)

Prabha, K 1946-
Towns, a structural analysis : a case study of Punjab / K. Prabha. — Delhi : Inter-India Publications, 1979.
214 p. : maps ; 22 cm.
Bibliography: p. [197]-209.
Includes index.
Rs60.00 ($12.00 U.S.)
1. Cities and towns—India—Punjab (State) 2. Urbanization—India—Punjab (State) I. Title.
HT147.I5P66 301.36'0954'552 79-902906
79 MARC

—INDONESIA

Boesriati B
Klassifikasi kota-kota di Indonesia / Boesriati B. — [s.l. : s.n., 1971]
iv, 25 leaves; 29 cm.
Thesis (sardjana)—Universitas Indonesia.
1. Cities and towns—Indonesia. 2. Indonesia—Administrative and political divisions. I. Title.
HT147.I54B63 79-950324
79 MARC

Djoko Sujarto.
Beberapa pengertian pokok mengenai kota/ oleh Djoko Soejarto. — [Yogyakarta: Pem. Da. Kotamadya Yogyakarta], 1974.
2 v. ; 29 cm.
Cover title.
1. Cities and towns—Indonesia. 2. Indonesia—Economic policy. I. Title.
HT147.I54D56 1974 301.36'1'09598 76-989579
79[r79]rev MARC

Sutami, 1928-
Masalah perumahan di-perkotaan (prasaran Symposium Urusan Perumahan). [Bandung] Lembaga Penjelidikan Masalah Bangunan [1970]
18 p. 29 cm. Rp280
Cover title.
1. Housing—Indonesia. 2. Cities and towns—Indonesia. I. Title.
HD7365.A3S97 301.5'4'09598 70-942885
79[r79]rev MARC

—INDONESIA—COLLECTED WORKS

Indonesia. Direktorat Tata Guna Tanah.
Kotamadya Balikpapan [dan kotamadya lainnya]. — [Jakarta] : Direktorat Tata Guna Tanah, Direktorat Jenderal Agraria, Departemen Dalam Negeri, 1978-
v. in : ill. ; 28 cm. — (Publikasi - Direktorat Tata Guna Tanah ; no. 103)
1. City planning—Indonesia—Collected works. 2. Cities and towns—Indonesia—Collected works. 3. Land use, Urban—Indonesia—Collected works. I. Title. II. Series: Indonesia. Direktorat Tata Guna Tanah. Publikasi - Direktorat Tata Guna Tanah ; no. 103 [etc.]
HT169.I53I54 1978 79-950351
79 MARC

—IRAN

Gaube, Heinz.
Iranian cities / Heinz Gaube. — New York : New York University Press, c1979, c1978.
xviii, 132 p. : ill. ; 28 cm. — (Hagop Kevorkian series on Near Eastern art and civilization)
Includes bibliographical references.
ISBN 0-8147-2971-1 : $22.50
1. Cities and towns—Iran. 2. Cities and towns, Islamic—Iran. 3. Herat, Afghanistan. 4. Iran—Social conditions.
HT147.I6G38 1979 301.36'3'0955 77-25748
78 MARC

—IRAN—HISTORY—COLLECTED WORKS

Bémont, Frédy.
Les Villes de l'Iran, des cités d'autrefois à l'urbanisme contemporain. Pairs, l'auteur, 1969-77
3 v. maps (some fold. col.), plates. 27 cm. 70.00F (v. 1) F***
Vol. 3 has title: Atlas de la civilisation iranienne.
Vol. 3 also issued separately.
Vol. 3 in French and English.
Bibliography: v. 1, p. 297-302.
1. Cities and towns—Iran—History—Collected works. 2. Iran—Economic conditions—Collected works. I. Title.
HT147.I6B45 79-399956
79[r79]rev2 MARC

—ISRAEL

Efrat, Elisha.
('Arim ye-'iyur be-Yiśrael)
ערים ועיור בישראל / אלישע אפרת. — תל־אביב : אחיאסף, [1976]
240 p. : ill. ; 24 cm.
On verso of t. p.: Towns and urbanization in Israel.
Bibliography: p. [231]-238.
Includes index.
1. Cities and towns—Israel. 2. Urbanization—Israel. I. Title.
HT147.I 7E37 79-951480

—ISRAEL—PICTORIAL WORKS

Meljon, Zeev.
... City for habitation / [editor, Zeev Meljon]. — [Tel-Aviv] : Union of Local Authorities, [1976]
355 p. : chiefly ill. ; 28 cm.
Added t.p.: 'Ir moshav ...
English and Hebrew.
1. Israel—Description and travel—Views. 2. Cities and towns—Israel—Pictorial works. I. Title. II. Title: 'Ir moshav.
DS108.5.M33 956.94'05'0222 78-951856
78 MARC

—ITALY

Fazio, Mario.
Il destino dei centri storici / di Mario Fazio. — 1. ed. — Firenze : La nuova Italia, 1977.
107 p. ; 18 cm. — (Italia nostra : Educazione ; 5) It78-Feb
Bibliography: p. 107.
L1900
1. Cities and towns—Italy. 2. City planning—Italy. I. Title.
HT164.I8F37 78-368823
*78 MARC

Guidicini, Paolo.
La partecipazione, l'uomo e la città / Paolo Guidicini. — Brescia : La scuola, c1978.
103 p. ; 20 cm. — (Partecipare oggi ; 1) It***
Bibliography: p. [93]-103.
L2500
1. Cities and towns—Italy. 2. Political participation—Italy. 3. Social participation—Italy. I. Title.
HT145.I8G82 79-371804
79 MARC

Le Splendide città d'Italia : guida ai centri più importanti del nostro paese. — [1. ed.]. — Milano : Selezione dal Reader's digest, 1979.
480 p. : ill. ; 27 cm. It79-Mar
L35000
1. Italy—Description and travel—1975- . 2. Cities and towns—Italy.
DG430.S62 945 79-362400
*79 MARC

—ITALY—ADDRESSES, ESSAYS, LECTURES

Le Città alpine : documenti e note / a cura di Bruno Parisi. — Milano : Vita e pensiero, c1975.
151 p. : ill., maps ; 24 cm. — (Corsi universitari) (Pubblicazioni dell'Università cattolica del Sacro Cuore, Milano)
Bibliography: p. 100-103.
L3700
1. Cities and towns—Italy—Addresses, essays, lectures. 2. Alpine regions—Addresses, essays, lectures. I. Parisi, Bruno.
HT145.I8C54 78-396571
79 MARC

CITIES AND TOWNS (Continued)

—ITALY—CONGRESSES

Interventi nel centro storico : confronto fra esperienze operative / promosso dall'assessorato per gli interventi nel centro storico del comune di Roma ... ; a cura di Mario Panizza — Bari : Dedalo libri, 1978.
244 p. : ill. ; 21 cm. — (Saggi ; 39) It78-Dec
Proceedings of a meeting held in Rome in 1977.
Bibliography: p. 15-16.
L5000
1. Urban renewal—Italy—Congresses. 2. City planning—Italy—Congresses. 3. Cities and towns—Italy—Congresses. I. Panizza, Mario.
HT178.I8I57 309.2'62'0945 79-352319
*79 MARC

—ITALY—HISTORY

Fubini, Alex.
Urbanistica in Italia / Alex Fubini. — 1. ristampa. — Milano : Clup, 1979.
431 p. : ill. ; 24 cm. It***
Includes bibliographical references.
ISBN 8870053784 : L7500
1. City planning—Italy—History. 2. Cities and towns—Italy—History. I. Title.
HT169.I83F8 79-362677
79 MARC

Martines, Lauro.
Power and imagination : city-states in Renaissance Italy / by Lauro Martines. — 1st ed. — New York : Knopf, 1979.
ix, 368 p., [8] leaves of plates : ill. ; 24 cm.
Bibliography: p. 345-351.
Includes index.
ISBN 0-394-50112-8 : $15.95
1. Italy—Politics and government—476-1268. 2. Italy—Politics and government—1268-1559. 3. Cities and towns—Italy—History. 4. Italy—Civilization—476-1268. 5. Italy—Civilization—1268-1559. I. Title.
DG494.M37 1979 945 78-11666
78 MARC

—JAPAN

Denis, Jacques, docteur.
Dynamique de l'urbanisation au Japon / par Jacques Denis. — Gand, Blandijnberg 2 : Société belge d'études géographiques, 1978.
p. 290-351 : ill. ; 25 cm. — (Mémoire - Société belge d'études géographiques ; no 31) Be78-5
Cover title.
"Extrait de 'Bulletin de la Société belge d'études géographiques', tome xlvi, 1977, no 2."
Summary in English, French, and Dutch.
Bibliography: p. 349-350.
130F
1. Urbanization—Japan. 2. Cities and towns—Japan. I. Title. II. Series: Société belge d'études géographiques. Mémoires ; no 31.
HT147.J3D4 301.36'10952 78-395506
*79 MARC

(Gendai Nihon no toshi supurōru mondai)
現代日本の都市スプロール問題 / 山崎不二夫, 森滝健一郎 ... [et al.] 編. — 東京 : 大月書店, 1978.
2 v. : ill. ; 20 cm. — (現代資本主義叢書 ; 4-5) Ja 78-7247 (v. 1)
Includes bibliographical references.
¥1800 per vol.
1. Cities and towns—Growth. 2. Cities and towns—Japan. 3. Urbanization—Japan. 4. Land use—Japan. I. Yamazaki, Fujio, 1909- II. Moritaki, Ken'ichirō, 1932- III. Series: Gendai shihonshugi sōsho ; 4-5.
''T371.G38 79-800071

Kitsugi, Torio.
(Watakushi no jōkamachi)
私の城下町 / 木継鳥夫編著. — 東京 : 金園社, 1978.
318 p., [4] leaves of plates : ill. ; 19 cm. Ja 78B-5815
"59の旅・ふるさとに歴史と伝統を訪ねて"
¥780
1. Japan—Description and travel—1945- 2. Castles—Japan. 3. Cities and towns—Japan. 4. Kitsugi, Torio. I. Title.
DS811.K516 79-802308

Matsuo, Toshirō, 1897-
(Tokai to inaka)
都会といなか / 松尾俊郎著. — 東京 : 三省堂, 昭和24 [1949]
5, 245, 6 p., [2] leaves of plates : ill. ; 19 cm. — (社會科文庫)
Colophon inserted.
Series romanized: Shakaika bunko.
Includes index.
1. Cities and towns—Japan. 2. Villages—Japan. I. Title.
HT147.J3M29 79-810011

(Nihon ni okeru toshika no chizu bunseki)
日本における都市化の地図分析 / 研究代表者 山本正三. — [東京 : トヨタ財団], 1977.
154 p. : ill. ; 30 cm. Ja 78-9390
Cover title.
"トヨタ財団研究助成"
1. Cities and towns—Japan. 2. City planning—Japan—Maps. 3. Urbanization—Japan. 4. Open spaces—Japan. I. Yamamoto, Shōzō, 1928-
HT147.J3N48 78-812126

(Nihon no toshi)
日本の都市 / 建設大臣官房広報室編. — 東京 : 建設広報協議会, 1965.
12, 271 p., [3] leaves of plates : ill. ; 21 cm. — (建設双書 ; 4)
1. City planning—Japan. 2. City planning and redevelopment law—Japan. 3. Cities and towns—Japan. I. Japan. Kensetsushō. Daijin Kambō. Kōhōshitsu. II. Series: Kensetsu sōsho ; 4.
HT169.J3N47 78-814950

Odauchi, Michitoshi, 1875-
(Inaka to tokai)
田舎と都会 / 小田内通敏著. — 東京 : 帝國書院, 昭和24 [1949]
4, 2, 164 p. ; 18 cm.
First ed. published in 1930 under title: Tokai to inaka.
1. Cities and towns—Japan. 2. Villages—Japan. I. Title.
HT147.J3O29 1949 79-808122

Sawada, Kiyoshi, 1920-
(Nihon no toshiken)
日本の都市圏 / 沢田清著. — 東京 : 古今書院, 1978.
8, 335 p. : ill. ; 22 cm. Ja 78-23382
Includes bibliographies and indexes.
¥3800
1. Cities and towns—Japan. 2. Metropolitan areas—Japan. I. Title.
HT147.J3S34 79-802340

(Shiro to jōkamachi)
城と城下町 / 藤本利治, 矢守一彦編 ; 浅野喜市写真 ; 監修 藤岡謙二郎. — 京都 : 淡交社, 1978.
201, [4] p. : ill. ; 31 cm. — (生きている近世 : 1) Ja 78-25356
Bibliography: p. [204]-[205]
¥6500
1. Castles—Japan. 2. Cities and towns—Japan. I. Fujimoto, Toshiharu, 1919- II. Yamori, Kazuhiko. III. Asano, Kiichi, 1914- IV. Series: Ikite iru kinsei ; 1.
DS806.3.S498 79-803021

Shōbidō, Tokyo. Henshūbu.
(Kaisei shinkyū taishō shichōson ichiran)
改正新舊對照市町村一覽 : 三府一廳四十三縣臺灣樺太朝鮮及關東州 / [編纂者鍾美堂編輯部]. — 訂正再版. — 東京 : 鍾美堂書店, 大正2 [1913]
44, 2, 2, 744 p. : maps ; 23 cm.
"諸官衙名稱 位置區劃附其里程圖"
1. Japan—Gazetteers. 2. Cities and towns—Japan. I. Title.
DS805.S53 1913 79-802718

Yamaga, Seiji, 1916-
(Hendōsuru toshi)
変動する都市 / 山鹿誠次. — 東京 : 東京書籍, 1979.
190 p. : ill. ; 19 cm. — (東書選書 ; 31) Ja 79-5734
Series romanized: Tōsho sensho.
Includes index.
¥820
1. Cities and towns—Japan. 2. City planning—Japan. 3. Land use, Urban—Japan. I. Title.
HT147.J3Y24 79-808605

Yokoyama, Sadao, 1918-
(Wareware no toshi to nōson)
われわれの都市と農村 / 横山定雄著. — 東京 : 高山書院, 昭和24 [1949]
2, 4, 211 p. : ill. ; 19 cm. — (若い人の文化叢書)
Series romanized: Wakai hito no bunka sōsho.
1. Cities and towns—Japan. 2. Villages—Japan. I. Title.
HT147.J3Y64 79-808815

Zaino, Hiroshi, 1939-
(Kaiwai)
かいわい : 日本の都心空間 / 材野博司. — 東京 : 鹿島出版会, 1978.
255 p. : ill. ; 19 cm. — (SD選書 ; 126) Ja 78-4129
Series romanized: SD sensho.
Includes bibliographical references.
¥980
1. City planning. 2. City planning—Japan. 3. Cities and towns—Japan. I. Title.
HT166.Z34 79-809683

—JAPAN—GROWTH

Glickman, Norman J
The growth and management of the Japanese urban system / Norman J. Glickman. — New York : Academic Press, 1979.
xiv, 370 p. : ill. ; 24 cm. — (Studies in urban economics)
Includes bibliographies and index.
ISBN 0-12-286950-8
1. Cities and towns—Japan—Growth. 2. Urbanization—Japan. 3. Urban economics—Case studies. I. Title. II. Series.
HT147.J3G55 301.36'0951 78-8831
78 MARC

Izeki, Jun, 1911-
(Chihō shōten han'ei no yōso)
地方商店繁栄の要素 / 井関純著. — 東京 : 誠文堂新光社, 昭和34 [1959]
2, 186 p. : ill. ; 19 cm. — (商店経営双書)
1. Store location—Japan. 2. Retail trade—Japan. 3. Rural economic development. 4. Cities and towns—Japan—Growth. I. Title. II. Series: Shōten keiei sōsho.
HF5429.6.J3I93 79-806275

Nakamura, Kōji, 1932-
(Jinkō kyūzō toshi)
人口急増都市 / 中村浩治. — [東村山] : 教育社 ; 東京 : 販売 教育社出版サービス, [1979]
205 p. : graphs ; 18 cm. — (時事問題解説 ; 201) (入門新書) Ja 79-5685
Cover title.
Series 2 romanized: Nyūmon shinsho.
Bibliography: p. 204-205.
¥400
1. Rural-urban migration—Japan. 2. Cities and towns—Japan—Growth. I. Title. II. Series: Jiji mondai kaisetsu ; 201.
HB2111.N36 79-809021

—JAPAN—HISTORY

(Daimyō ryōkoku to jōkamachi)
大名領国と城下町 / 魚澄惣五郎編. — 京都 : 柳原書店, 昭和32 [1957]
281 p. ; 22 cm.
Includes bibliographical references.
1. Japan—History—Period of civil wars, 1480-1603. 2. Cities and towns—Japan—History. 3. Castles—Japan. I. Uozumi, Sōgorō, 1889-1959.
DS868.2.D34 78-817187

CITIES AND TOWNS
—JAPAN—HISTORY (Continued)

Yamaga, Seiji, 1916-
(Toshi to mura)
都市と村 / [山鹿誠次著]. — 東京 : 文教書院, 昭和24 [1949]
2, 5, 138 p. : ill. ; 19 cm. — (樂しい社會科 ; 第2集)(中学教養文庫)
Colophon inserted.
Bibliography: p. 133-134.
Includes index.
1. Cities and towns—Japan—History. 2. Villages—Japan—History. I. Title. II. Series: Tanoshii shakaika ; dai 2-shū. III. Series: Chūgoku kyōyō bunko.
HT147.J3Y268 78-814018

—JAPAN—AOMORI (PREFECTURE)
—STATISTICS

Japan. Tōhoku Nōseikyoku. Aomori Tōkei Jōhō Jimusho.
(Shichōsonbetsu sōgō tōkei gaidobukku)
市町村別総合統計ガイドブック / 東北農政局青森統計情報事務所編集. — 青森 : 青森農林統計協会, 昭和52 [1977]
233 p. : ill. ; 25 cm. Ja 77-15856
Cover title.
Bibliography: p. 232-233.
Errata slip inserted.
1. Agriculture—Japan—Aomori (Prefecture)—Statistics. 2. Cities and towns—Japan—Aomori (Prefecture)—Statistics. 3. Aomori, Japan (Prefecture)—Statistics. I. Title.
HD2095.A56J36 1977 78-811983

—JAPAN—CHUGOKU REGION—GROWTH

Tomioka, Gihachi, 1924-
(Shiomichi to takasebune)
塩道と高瀬舟 陰陽交通路の発達と都市の構造変化 富岡儀八著 東京 古今書院 昭和48 (1973)
278 p. illus. 22 cm. Ja 73-17415
Includes bibliographical references.
1. Transportation—Japan—Chūgoku region—History. 2. Salt—Transportation—History. 3. Cities and towns—Japan—Chugoku region—Growth. I. Title.
HE277.T63 74-803882

—JAPAN—CHŪGOKU REGION
—GROWTH—MATHEMATICAL
MODELS

Nishimura, Mutsuo, 1915-
(Chūshinchi to seiryokuken)
中心地と勢力圏 / 西村睦男著. — 東京 : 大明堂, 昭和52 [1977]
iv, 194 p., [1] leaf of plates : ill. ; 22 cm. Ja 77-24418
Includes bibliographies.
Errata slip inserted.
¥2000
1. Cities and towns—Japan—Chūgoku region—Growth—Mathematical models. 2. Central business districts—Japan—Chūgoku region—Mathematical models. I. Title.
HT384.J3N54 79-801874

—JAPAN—NARA (PREFECTURE)
—HISTORY

Kita, Yoshiyuki, 1902-
(Yamato no kangō shūraku)
大和の環濠集落 / 喜多芳之著. — 大阪 : 日本古城友の会, 昭和51 [1976]
28 p. : ill. ; 26 cm. — (研究紀要 - 日本古城友の会 ; 2) Ja 76-19
Cover title.
1. Cities and towns—Japan—Nara (Prefecture)—History. 2. Moats—Social aspects—Japan—Nara (Prefecture)—History. 3. Anthropo-geography—Japan—Nara (Prefecture)—History. I. Title. II. Series: Nihon Kojō Tomo no Kai. Kenkyū kiyō - Nihon Kojō Tomo no Kai ; 2.
HT147.J3K48 78-812622

—JAPAN—TOKONAME—MAPS

(Tokoname-shi shi kinsei muraezu shū)
常滑市誌·近世村絵図集 / 編集 常滑市誌編集委員会 ; 監修 松井武敏, 安藤慶一郎. — 常滑 : 常滑市役所, 昭和52 [1977]
1 portfolio ([1] fold. leaf, 22 fold. leaves of plates : 22 fold. col. maps) ; 31 cm. Ja 77-22724
Title from colophon (inserted)
Title on label mounted on portfolio: Tokoname-shi muraezu shū.
"徳川林政史研究所所蔵の…現常滑市域の藩政村二二か村の絵図を原図そのままに復版"
Not in trade.
1. Tokoname, Japan—Maps. 2. Tokoname, Japan—Historical geography—Maps. 3. Villages—Japan—Tokoname—Maps. 4. Cities and towns—Japan—Tokoname—Maps. 5. Japan—History—1787-1868—Maps. 6. Maps, Early—Facsimiles. I. Tokoname Shishi Henshū Iinkai. II. Title: Tokoname-shi muraezu shū.
G2359.T65T6 1977 79-804940

—KOREA—YEARBOOKS

(Han'guk tosi yŏn'gam)
한국도시연감 Municipal yearbook of Korea.
[서울] 내무부
v. ill. 27 cm.
English and Korean.
1. Cities and towns—Korea—Yearbooks. I. Korea (Republic). Naemubu. II. Title: Municipal yearbook of Korea.
HT147.K6H36 78-647772 (MARC-S)

—LATIN AMERICA—ADDRESSES,
ESSAYS, LECTURES

Lecturas en problemas urbano-regionales / Miguel Morales A. [editor]. — Santiago de Chile : Instituto Latinoamericano de Investigaciones Sociales, 1973.
108 p. : ill. ; 21 cm. — (Estudios y documentos - Instituto Latinoamericano de Investigaciones Sociales ; 27)
Includes bibliographical references.
CONTENTS: Morse, R. Las ciudades latinoamericanas.—Morales, M., Parisi, L. Modo de producción, proceso de regionalización y relaciones ciudad-campo.—McDonald, J. La región: concepción, diseño y limitaciones.—Babarovic, I. Algunas notas sobre desarrollo regional y la planificación del espacio nacional.—Ginsburg, N. Las regiones de planificación y el concepto regional.
1. Regional planning—Addresses, essays, lectures. 2. Economic zoning—Addresses, essays, lectures. 3. Regional planning—Latin America—Addresses, essays, lectures. 4. Cities and towns—Latin America—Addresses, essays, lectures. I. Morales A. Miguel. II. Instituto Latinoamericano de Investigaciones Sociales. III. Series: Instituto Latinoamericano de Investigaciones Sociales. Estudios y documentos ; 27.
HT391.L43 79-362129
79 MARC

—LATIN AMERICA—HISTORY

Barrenechea, Anamaria.
The struggle for housing in Latin America / Anamaria Barrenechea, Miguel Lawner. — København (Peder Skrams gade 2d) : Kunstakademiets Arkitektskole, Institut for By- og Landskabsplanlægning, 1977-
v. : ill. ; 30 cm. — (FSC publication ; Series B ; no. 3) D77-42(v.
Bibliography: v. 1, 88-89.
1. Housing policy—Latin America—History. 2. Cities and towns—Latin America—History. I. Lawner, Miguel, joint author. II. Title. III. Series: Arkitektskolen i København. Institut for by- og landskapsplanlægning. FSC publication ; Series B ; no. 3.
HD7305.5.A3B37 301.5'4'098 79-313091
79 MARC

Martínez Lemoine, René.
El modelo clásico de ciudad colonial hispanoamericana : ensayo sobre los orígenes del urbanismo en América / René Martínez L. — Santiago de Chile : Departamento Planificación Urbano Regional F.A.U., Universidad de Chile, 1977.
102 p. : ill. ; 26 cm.
"500 ejemplares."
Includes bibliographical references.
1. Cities and towns—Latin America—History. 2. City planning—Latin America—History. 3. Urbanization—Latin America—History. I. Title.
HT127.5.M37 78-109673
79 MARC

—LATIN AMERICA—MAPS

Urbanismo español en América : [exposición itinerante por Hispanoamérica / organizada por el] Instituto de Cultura Hispánica ; [proyectada y dirigida por] Javier Aguilera Rojas, Joaquín Ibañez Montoya, Luis J. Moreno Rexach. — Madrid : Instituto de Cultura Hispánica, Comisaría de Exposiciones, [1976?]
[4], 43 p. : maps ; 22 cm. Sp***
Cover title.
1. Cities and towns—Latin America—Maps. I. Aguilera Rojas, Javier. II. Ibañez Montoya, Joaquín, 1947- . III. Moreno Rexach, Luis J., 1944- . IV. Spain. Instituto de Cultura Hispánica.
G1544.A1U73 1976 912'.8 79-383309
79 MARC

—MASSACHUSETTS—GROWTH

Perspectives on growth : excerpts from local growth policy statements : an interim report on the Massachusetts Growth policy development act, Chapter 807 of the Acts of 1975 / assembled by the Office of State Planning. — [Boston] : The Office, 1977.
74 p. : ill. ; 22 x 28 cm.
Includes index.
1. Urban policy—Massachusetts. 2. Cities and towns—Massachusetts—Growth. 3. Regional planning—Massachusetts. I. Massachusetts. Office of State Planning.
HT123.5.M4P47 309.2'5'09744 79-622313
79 MARC

—MAURITIUS—CONGRESSES

Séminaire d'étude sur l'action culturelle et sociale des municipalités mauriciennes, Port Louis, Mauritius, 1974.
Séminaire d'étude sur l'action culturelle et sociale des municipalités mauriciennes, 19 au 28 novembre 1974 / Association des Autorités Urbaines. — [Port Louis?] : L'Association, [1974]
2 v. ; 33 cm.
Cover title.
CONTENTS: ptie. 1. Questionnaire préparatoire.—ptie. 2. Compte-rendu des travaux du séminaire.
1. Cities and towns—Mauritius—Congresses. 2. Mauritius—Social conditions—Congresses. I. Association des autorités urbaines.
HT148.M37S45 1974 309.1'69'82 75-980191
78[r79]rev MARC

—MEDITERRANEAN REGION
—ADDRESSES, ESSAYS,
LECTURES

Megalopoli mediterranea / a cura di Calogero Muscarà ; [scritti di] Bellavitis ... [et al.]. — Milano : F. Angeli, [1978]
263 p. : ill. ; 22 cm. — (Geografia e società ; 7) It78-July
Includes bibliographies.
L8000
1. Metropolitan areas—Mediterranean region—Addresses, essays, lectures. 2. Cities and towns—Mediterranean region—Addresses, essays, lectures. I. Muscara, Calogero, 1929- . II. Bellavitis, Giorgio.
HT334.M35M43 78-384620
*78 MARC

—MINNESOTA—CIVIC IMPROVEMENT
—CONGRESSES

Heiberg, Martha Wallace.
Partners for a livable Minnesota : summary report of a statewide conference, June 16-17, 1978, Spring Hill Center, Wayzata, Minnesota / [prepared by Martha Wallace Heiberg ; editor and designer, John T. Ziegenhagen ; editorial assistant, Audrey Smith] ; co-sponsored by Spring Hill Center and Continuing Education in Public Policy at the University of Minnesota. — Wayzata : The Center, c1978.
22 p. ; 29 cm.
ISBN 0-932676-07-3
1. Cities and towns—Minnesota—Civic improvement—Congresses. 2. Environment policy—Minnesota—Congresses. 3. Landscape assessment—Minnesota—Congresses. I. Ziegenhagen, John. II. Smith, Audrey. III. Spring Hill Center. IV. Minnesota. University. Continuing Education in Public Policy. V. Title.
HT123.5.M56H43 301.31'09776 79-101688
79 MARC

—MOROCCO—BIBLIOGRAPHY

Cigar, Norman L
Modern Moroccan cities / Norman L. Cigar. — Monticello, Ill. : Council of Planning Librarians, 1978.
6 p. ; 28 cm. — (Exchange bibliography - Council of Planning Librarians ; 1464)
Cover title.
$1.50
1. Cities and towns—Morocco—Bibliography. 2. Urbanization—Morocco—Bibliography. I. Title. II. Series: Council of Planning Librarians. Exchange bibliography ; 1464.
Z5942.C68 no. 1464 016.3092 s 78-103384
[Z7164.A7] 79 MARC
[HT148.M8]

CITIES AND TOWNS (Continued)

—NEAR EAST

Roberts, M Hugh P
An urban profile of the Middle East / M. Hugh P. Roberts. — New York : St. Martin's Press, 1979.
239 p. : ill. ; 24 cm.
Includes bibliographies and index.
ISBN 0-312-83467-5
1. Cities and towns—Near East. 2. Cities and towns—Africa, North. 3. Urbanization—Near East. 4. Urbanization—Africa, North. 5. City planning—Near East. 6. City planning—Africa, North. I. Title.
HT147.N4R6 1979 309.2'62'0956 78-27185
78 MARC

—NEAR EAST—BIBLIOGRAPHY

Mahayni, Riad G
Urbanization in the Middle East / Riad G. Mahayni. — Monticello, Ill. : Vance Bibliographies, 1978.
18 p. ; 28 cm. — (Public administration series : Bibliography : P-23)
Cover title.
$2.00
1. Cities and towns—Near East—Bibliography. 2. Urbanization—Near East—Bibliography. I. Title. II. Series.
Z7165.N35M33 016.30136'1'0956 78-109748
ɼHT147.N4ɿ 79 MARC

—NETHERLANDS—GROWTH—CASE STUDIES

Hoekveld, G A
Baarn : schets van de ontwikkeling van een villadorp / G. A. Hoekveld. — Baarn : Bosch & Keuning, ɼ1964ɿ
233 p., ɼ4ɿ leaves of plates : ill. ; 25 cm.
Summary in English.
Bibliography: p. 206-210.
1. Baarn, Netherlands—City planning. 2. Baarn, Netherlands—History. 3. Urbanization—Netherlands—Amsterdam. 4. City planning—Netherlands. 5. Cities and towns—Netherlands—Growth—Case studies.
HT169.N42B323 79-340609
79 MARC

—NETHERLANDS—HISTORY—BIBLIOGRAPHY

Herwijnen, G van.
Bibliografie van de stedengeschiedenis van Nederland / samengesteld door G. van Herwijnen, met medewerking van W. G. van der Moer, M. Carasso-Kok, M. J. J. G. Chappin. — Leiden : Brill, 1978.
xxi, 355 p., 1 fold. leaf; 25 cm. — (Acta Collegii Historiae Urbanae Societatis Historicorum Internationalis) Ne78-39
Includes indexes.
ISBN 9004057005 : fl 80.00
1. Cities and towns—Netherlands—History—Bibliography. 2. Netherlands—History, Local—Bibliography. I. Title.
Z2423.H47 016.9492 79-340219
ɼDJ114ɿ *79 MARC

—NETHERLANDS—OLDAMBTEN (KWARTIER)

Groningen (Province). Provinciale Planologische Commissie.
Een beschouwing over de positie van de dorpen in het Oldambt. ɼGroningen, 1965?ɿ
27 p. illus. 25 cm.
1. Cities and towns—Netherlands—Oldambten (Kwartier) 2. Cities and towns—Netherlands—Oldambten (Kwartier)—Growth. I. Title.
HT145.N4G76 1965 78-396669
79 MARC

—NETHERLANDS—OLDAMBTEN (KWARTIER)—GROWTH

Groningen (Province). Provinciale Planologische Commissie.
Een beschouwing over de positie van de dorpen in het Oldambt. ɼGroningen, 1965?ɿ
27 p. illus. 25 cm.
1. Cities and towns—Netherlands—Oldambten (Kwartier) 2. Cities and towns—Netherlands—Oldambten (Kwartier)—Growth. I. Title.
HT145.N4G76 1965 78-396669
79 MARC

—NEW ZEALAND

Kilmartin, Leslie.
Cities unlimited : the sociology of urban development in Australia and New Zealand / Leslie Kilmartin, David C. Thorns. — Sydney ; Boston : Allen and Unwin, 1978.
195 p. : diagrs. ; 23 cm. — (Studies in society ; 1) Aus***
Bibliography: p. 168-185.
Includes indexes.
ISBN 0-86861-288-X : $18.25. ISBN 0-86861-296-0 pbk.
1. Cities and towns—Australia. 2. Cities and towns—New Zealand. 3. Sociology, Urban. I. Thorns, David C., joint author. II. Title. III. Series: Studies in society (Sydney) ; 1.
HT149.A8K54 301.36'09931 78-55230
79 MARC

—NIGERIA—OYO (STATE)—COLLECTED WORKS

Oyo, Nigeria (State). Ministry of Local Government and Information. Information Division.
Oyo State town series. — ɼIbadan : Information Division, Ministry of Local Government and Information, 1977-
no. in : ill. ; 23 cm.
Cover title; each no. has also special t.p.
CONTENTS: no. 1/14. Ẹdẹ.—no. 2/14. Ejigbo.—no. 3/14. Eruwa.—no. 4/14. Ibadan.—no. 5/14. Ikirun.—no. 6/14. Ila.—no. 7/14. Ile Ifẹ.—no. 8/14. Ilẹṣa.—No. 9/14. Iseyin.—no. 10/14. Iwo.—no. 11/14. Ogbomọṣọ.—no. 12/14. Oṣogbo.—no. 13/14. Ọyọ.
1. Cities and towns—Nigeria—Oyo (State)—Collected works. 2. Oyo, Nigeria (State)—Description and travel—Collected works. I. Title.
DT515.9.O9O95 1977 966.9'2 79-305070
79 MARC

—NORTH CAROLINA—MAPS

Champion Map Corporation.
Lake Norman 5***** ɼi.e. 5 starɿ atlas. Charlotte ɼ1963ɿ
19 p. (incl. cover) col. maps. 36 cm.
Cover title.
Covers Catawba, Lincoln, Mecklenburg, and Iredell Counties; includes detailed street maps of Charlotte, Lincolnton, Statesville, and other cities and towns.
1. Norman, Lake—Maps. 2. Cities and towns—North Carolina—Maps. I. Title.
G1302.N6C5 1963 map 65-690
ɼr79ɿrev MARC

—NORWAY—CASE STUDIES

Norges byer / redaktør, Brita Holtedahl. — Oslo : Leif Holtedahls forl., ɼ1977ɿ (Skien : Oluf Rasmussen)
108 p. : ill. ; 21 x 22 cm. N78-May
ISBN 8290157045 : kr55.00
1. Cities and towns—Norway—Case studies. I. Holtedahl, Britta.
HT145.N7N67 78-393777
79 MARC

—ONTARIO—ADDRESSES, ESSAYS, LECTURES

Bourne, Larry S
The form of cities in central Canada: selected papers; edited by L. S. Bourne, R. D. MacKinnon ɼandɿ J. W. Simmons. ɼTorontoɿ University of Toronto Press ɼ1973ɿ
ix, 246 p. illus. 23 cm. (University of Toronto. Dept. of Geography. Research publications, no. 11) $5.00 C***
Includes bibliographies.
ISBN 0-8020-3317-2
1. Cities and towns—Ontario—Addresses, essays, lectures. 2. Cities and towns—Quebec (Province)—Addresses, essays, lectures. I. MacKinnon, R. D., joint author. II. Simmons, James William, 1936- joint author. III. Title. IV. Series: University of Toronto. Dept. of Geography. Research publications, no. 11.
HT127.B67 301.36'3'09713 73-84353
74ɼr79ɿrev MARC

—PAKISTAN.

Nilsson, Sten Åke, 1936-
Fem huvudstäder i Indien och Pakistan. En kritisk genomgång. Lund, Studentlitteratur, 1970.
70 p. illus. 23 cm. kr17.00 S71-05
Bibliography: p. 66-70.
1. Cities and towns—India. 2. Cities and towns—Pakistan. I. Title.
HT147.I5N54 78-399409
79 MARC

—PALESTINE—HISTORY

Kempinski, Aharon.
The rise of an urban culture : the urbanization of Palestine in the Early Bronze Age, 3000-2150 B.C. / Aharon Kempinski ; ɼcover design, Rania Meljonɿ. — Jerusalem : Israel Ethnographic Society, 1978.
78 p. : ill. ; 25 cm. — (I.E.S. studies ; 4)
Added t.p.: Tsemiḥatah shel tarbut ʻironit.
Includes Hebrew summary.
Bibliography: p. 38-43.
1. Cities and towns—Palestine—History. 2. Palestine—Antiquities. I. Title. II. Title: Tsemiḥatah shel tarbut ʻironit. III. Series: haAgudah le-etnografyah be-Yisrael. I.E.S. studies ; 4.
HT147.P27K45 301.36'0933 79-111061
79 MARC

—PERU—HISTORY

Durán Montero, María Antonia.
Fundación de ciudades en el Perú durante el siglo XVI : estudio urbanístico / María Antonia Durán Montero. — 1. ed. — Sevilla : Escuela de Estudios Hispano-Americanos de Sevilla, 1978.
xxxviii, 210 p., ɼ19ɿ leaves of plates (3 fold.) : ill. ; 20 cm. — (Publicaciones de la Escuela de Estudios Hispano-Americanos de Sevilla ; 247) Sp***
Bibliography: p. ɼ167ɿ-191.
Includes index.
ISBN 8400043146
1. Cities and towns—Peru—History. 2. City planning—Peru—History. 3. City planning and redevelopment law—Peru—History. I. Title. II. Series: Seville. Universidad. Escuela de Estudios Hispano-Americanos. Publicaciones ; 247.
HT129.P5D87 79-112623
79 MARC

—PHILIPPINE ISLANDS—CASE STUDIES

Lopez-Nerney, Susan.
Socioeconomic profiles of six selected towns : final report / submitted to Camp, Dresser, and McKee International by the Institute of Philippine Culture on 9 March 1976 ; Susan Lopez-Nerney, Maria Elena Chiong-Javier, Cristina Montiel, with Pacifico Maghacot ... ɼet al.ɿ. — Quezon City : Ateneo de Manila University, Institute of Philippine Culture, 1976.
xiii, 125 leaves : ill. ; 27 cm.
Bibliography: leaves 121-125.
1. Social indicators—Philippine Islands. 2. Social surveys—Philippine Islands. 3. Cities and towns—Philippine Islands—Case studies. I. Chiong-Javier, Maria Elena, joint author. II. Montiel, Cristina, joint author. III. Camp, Dresser & McKee. IV. Ateneo de Manila University. Institute of Philippine Culture. V. Title.
HN714.L66 309.1'599'04 78-326250
79 MARC

—PHILIPPINE ISLANDS—MANILA

Reed, Robert Ronald, 1940-
Colonial Manila : the context of Hispanic urbanism and process of morphogenesis / by Robert R. Reed. — Berkeley : University of California Press, c1978.
xiv, 129 p. : 11 maps ; 26 cm. — (University of California publications in geography ; v. 22)
Bibliography: p. 107-119.
Includes index.
ISBN 0-520-09579-0
1. Urbanization—Philippine Islands—Manila. 2. Cities and town—Philippine Islands—Manila. 3. Manila—History. 4. Philippine Islands—History—1521-1898. I. Title. II. Series: California. University. University of California publications in geography ; v. 22.
G58.C3 vol. 22 301.36'1'095991 77-80476
ɼHN720.M33ɿ 79 MARC

—POLAND—CONGRESSES

Miejska polszczyzna mówiona : metodologia badań : materiały z konferencji naukowej w Sosnowcu w dniach 20-21 czerwca 1974 roku / ɼred. nauk. Władysław Lubaśɿ. — Katowice : UŚ, 1976 (Katow. : UŚ)
140 p. : ill. ; 25 cm. — (Prace naukowe Uniwersytetu Śląskiego w Katowicach ; nr 103) P76-3756
Includes bibliographical references.
zł23.00
1. Polish language—Dialectology—Congresses. 2. Polish language—Spoken Polish—Congresses. 3. Cities and towns—Poland—Congresses. I. Lubaś, Władysław. II. Series: Uniwersytet Śląski w Katowicach. Prace naukowe ; nr. 103.
PG6704.M5 78-422548
79 MARC

CITIES AND TOWNS (Continued)

—POLAND—HISTORY

Golińska, Zofia (Daszyńska) 1866-
Miasta i cechy w dawnej Polsce / Z. D. Golińska. — Warszawa : Skład Główny w Księgarni Naukowej, 1906.
100 p. : ill. ; 19 cm. — (Biblioteka spółczesna)
1. Cities and towns—Poland—History. 2. Poland—Economic conditions. 3. Gilds—Poland. I. Title.
HT145.P7G64 78-392257
79 MARC

Sporn, Thomas.
Die "Stadt zu polnischem Recht" und die deutschrechtliche Gründungsstadt / Thomas Sporn. — Frankfurt am Main ; Bern ; Las Vegas : Lang, 1978.
xxxiv, 222 p. ; 21 cm. — (Europäische Hochschulschriften : Reihe 2, Rechtswissenschaft ; Bd. 197) GFR78-A
Bibliography: p. xiii-xxxiv.
Includes index.
ISBN 3-261-02655-3
1. Poland—Constitutional history. 2. Cities and towns—Poland—History. 3. Law—Poland—Sources. 4. Law—Germany—History and criticism. I. Title. II. Series.
340'.09438 79-357236
79 MARC

—QUEBEC (PROVINCE)—ADDRESSES, ESSAYS, LECTURES

Bourne, Larry S
The form of cities in central Canada: selected papers; edited by L. S. Bourne, R. D. MacKinnon [and] J. W. Simmons. [Toronto] University of Toronto Press [1973]
ix, 246 p. illus. 23 cm. (University of Toronto. Dept. of Geography. Research publications, no. 11) $5.00 C***
Includes bibliographies.
ISBN 0-8020-3317-2
1. Cities and towns—Ontario—Addresses, essays, lectures. 2. Cities and towns—Quebec (Province)—Addresses, essays, lectures. I. MacKinnon, R. D., joint author. II. Simmons, James William, 1936- joint author. III. Title. IV. Series: University of Toronto. Dept. of Geography. Research publications, no. 11.
HT127.B67 301.36'3'09713 73-84353
74[r79]rev MARC

—QUÉBEC (PROVINCE)—BIBLIOGRAPHY

Angers, Majëlla.
Liste de publications reliées aux 63 [i.e. soixante-trois] principales agglomérations du Québec / par Majella Angers. — [Québec : Ministère des affaires municipales, 1975]
63 p. ; 25 cm. — (Collection du Ministère des affaires municipales ; v. 1, no 1) C***
1. Cities and towns—Québec (Province)—Bibliography. I. Title. II. Series: Québec (Province). Ministère des affaires municipales. Collection du Ministère des affaires municipales ; v. 1, no 1.
Z7165.C2A53 016.30136'09714 79-340873
[HT127] 79 MARC

—QUÉBEC (PROVINCE)—MONTRÉAL REGION

Martin, Fernand.
La satellisation économique dans la région de Montréal / Fernand Martin, avec la collab. de Jacques Nepveu. — Montréal : Centre de recherche en développement économique, 1977.
91 p. ; 28 cm. — (Cahier - Centre de recherche en développement économique ; no 15) C78-9480-8
Bibliography: p. 91.
$5.25 (U.S.)
1. Montréal metropolitan area, Québec—Economic conditions. 2. Cities and towns—Québec (Province)—Montréal region. I. Nepveu, Jacques, joint author. II. Title. III. Series: Université de Montréal. Centre de recherche en développement économique. Cahier - Centre de recherche en développement économique ; no 15.
HC118.M6M33 330.9'714'2 79-348620
79 MARC

Polèse, Mario, 1943-
Distance de Montréal et développement économique urbain / par Mario Polèse et Jean-Claude Thibodeau, en collab. avec Michel Boisvert ... [et al.]. — Montréal : Institut national de la recherche scientifique, INRS-Urbanisation, 1977.
140 p., [8] fold. leaves of plates : ill. ; 21 cm. — (Rapports de recherche - INRS-Urbanisation ; no 2) C78-5972-7
Bibliography: p. 135-140.
$2.50
1. Central places—Québec (Province)—Montréal region. 2. Cities and towns—Québec (Province)—Montréal region. 3. Industries, Location of—Québec (Province)—Montréal region. I. Thibodeau, Jean Claude, joint author. II. Title. III. Series: Institut national de la recherche scientifique (Québec). Urbanisation. Rapports de recherche - INRS-Urbanisation ; no 2.
HT127.P64 338.6'042 79-359731
79 MARC

—RHODESIA, SOUTHERN—HISTORY—DICTIONARIES

Smith, Robert G
Avondale to Zimbabwe : a collection of cameos of Rhodesian towns and villages / R. Cherer Smith ; ill. by M. M. Carlisle. — Borrowdale : Smith, [197-]
314 p. : ill., map (on lining paper) ; 21 cm.
ISBN 0-7974-0313-2
1. Rhodesia, Southern—History—Dictionaries. 2. Cities and towns—Rhodesia, Southern—History—Dictionaries. 3. Rhodesia, Southern—Gazetteers. I. Title.
DT962.5.S44 968.9'1'003 79-102946
79 MARC

—ROMANIA

Cucu, Vasile.
Urbanism : [curs] / Vasile Cucu, Cezar Popescu. — Bucureşti : Centrul de multiplicare al Universităţii din Bucureşti, 1975-
v. : ill. ; 24 cm. R75-4274
At head of title: Universitatea din Bucureşti. Facultatea de Geologie-Geografie. Catedra de Geografie Economică.
"Destinat studenţilor anului II curs de zi şi anului III curs fără frecvenţă de la Facultatea de drept a Universităţii din Bucureşti."
Bibliography: v. 1, p. 316-320.
lei15.00 (v. 1)
1. Cities and towns. 2. City planning. 3. Cities and towns—Romania. I. Popescu, Cezar, joint author. II. Title.
HT151.C82 78-393930
•79 MARC

—RUSSIA

Gustafson, Sture, 1938-
Leningrad, Tallinn, Riga, Vilnius / Sture Gustafson o. Ingelej Kindlund ; foto, Rein Välme. — Stockholm : Gen.-stab:s lit. anst., 1977.
[4], 278, [4] p. : ill., maps ; 20 cm. — (Semester) S78-20
Bibliography: p. 273-274.
Includes index.
ISBN 9170580367 : kr55.00
1. Russia—Description and travel—1970- —Guide-books. 2. Cities and towns—Russia. I. Kindlund, Ingelej, 1944- joint author. II. Title.
DK29.G87 79-337677
79 MARC

Khodzhaev, David Georgievich.
(Sistema rasselenii͡a v SSSR)
Система расселения в СССР : (Вопросы комплексного планирования) / Д. Г. Ходжаев, А. В. Кочетков, Ф. М. Листенгурт. — Москва : Экономика, 1977.
159 p. ; 20 cm. USSR 77
Bibliography: p. 152-[157]
1.30rub
1. Russia—Population. 2. Cities and towns—Russia. I. Kochetkov, Andreĭ Vasil'evich, joint author. II. Listengurt, Feliks Mikhaĭlovich, joint author. III. Title.
HB3607.K45 79-413147

Rabinovich, Mikhail Grigor'evich.
(Ocherki ėtnografii russkogo feodal'nogo goroda)
Очерки этнографии русского феодального города : Горожане, их обществ. и домаш. быт / М. Г. Рабинович. — Москва : Наука, 1978.
328 p., [1] leaf of plates : ill. (some col.) ; 22 cm. USSR 78
At head of title: Akademii͡a nauk SSSR. Institut ėtnografii im. N. N. Miklukho-Maklai͡a.
Includes bibliographical references and index.
1.70rub
1. Ethnology—Russia. 2. Russia—Social conditions. 3. Feudalism—Russia. 4. Cities and towns—Russia. I. Title.
DK33.R27 79-409267

—RUSSIA—ADDRESSES, ESSAYS, LECTURES

The Socialist city : spatial structure and urban policy / edited by R. A. French & F. E. Ian Hamilton. — Chichester ; New York : Wiley, c1979.
xviii, 541 p. : ill. ; 24 cm.
Includes bibliographies and index.
ISBN 0-471-99689-0
1. Cities and towns—Russia—Addresses, essays, lectures. 2. Cities and towns—Europe, Eastern—Addresses, essays, lectures. I. French, Richard Anthony. II. Hamilton, F. E. Ian.
HT145.R9S63 309.2'62'0947 78-16828
78 MARC

—RUSSIA—GROWTH

Trufanov, Ivan Pavlovich.
(Problemy byta gorodskogo naselenii͡a SSSR)
Проблемы быта городского населения СССР. Отв. ред. д-р ист. наук, проф. В. А. Ежов. Ленинград, Изд-во Ленингр. ун-та, 1973.
144 p. 21 cm. USSR 73
At head of title: Министерство высшего и среднего специального образования РСФСР. И. П. Труфанов.
Includes bibliographical references.
0.70rub
1. Cities and towns—Russia—Growth. 2. Sociology, Urban—Russia. 3. City and town life—Russia. I. Title.
HT145.R9T78 73-360015

—RUSSIA, NORTHERN—CONGRESSES

Sovetsko-finskiĭ simpozium po sot͡sial'no-ėkonomicheskoĭ istorii, 3d, Tallinn, 1974.
(Gorod v Severo-Vostochnoĭ Evrope v novoe i noveĭshee vremi͡a)
Город в Северо-Восточной Европе в новое и новейшее время : (XV–середина XX в.) : [Докл. симпоз., 3–5 сент. 1974 г.] / АН ЭССР, Ин-т истории ; [Отв. ред. Х. Арумяэ]. — Таллин : АН ЭССР, 1978.
93 p. ; 19 cm. USSR 78
Bibliography: p. [92]
0.65rub
1. Cities and towns—Russia, Northern—Congresses. 2. Cities and towns—Baltic Sea region—Congresses. I. Arumäe, Heino. II. Eesti NSV Teaduste Akadeemia. Ajaloo Instituut. III. Title.
HT169.R8S68 1974 79-411726

—RUSSIAN REPUBLIC—BALTIC SEA REGION—CONGRESSES

Sovetsko-finskiĭ simpozium po sot͡sial'no-ėkonomicheskoĭ istorii, 3d, Tallinn, 1974.
(Gorod v Severo-Vostochnoĭ Evrope v novoe i noveĭshee vremi͡a)
Город в Северо-Восточной Европе в новое и новейшее время : (XV–середина XX в.) : [Докл. симпоз., 3–5 сент. 1974 г.] / АН ЭССР, Ин-т истории ; [Отв. ред. Х. Арумяэ]. — Таллин : АН ЭССР, 1978.
93 p. ; 19 cm. USSR 78
Bibliography: p. [92]
0.65rub
1. Cities and towns—Russia, Northern—Congresses. 2. Cities and towns—Baltic Sea region—Congresses. I. Arumäe, Heino. II. Eesti NSV Teaduste Akadeemia. Ajaloo Instituut. III. Title.
HT169.R8S68 1974 79-411726

—RUSSIAN REPUBLIC—BURIAT-MONGOLIA

Mangataeva, Darima Dorzhievna.
(Gorodskie poselenii͡a Buri͡atskoĭ ASSR)
Городские поселения Бурятской АССР : (Развитие и размещение) / Д. Д. Мангатаева. — Улан-Удэ : Бурят. кн. изд-во, 1978.
111 p. : maps ; 22 cm. USSR 78
At head of title: Akademii͡a nauk SSSR. Sibirskoe otdelenie. Buri͡atskiĭ filial. Otdel ėkonomicheskikh issledovaniĭ.
Includes bibliographical references.
0.85rub
1. Buriat-Mongolia—Population. 2. Cities and towns—Russia Republic—Buriat-Mongolia. I. Title.
HB3608.B86M36 79-409715

—SAUDI ARABIA

'Aṣfūr, Maḥmūd 'Abd al-Laṭīf.
(al-Dirāsah al-maydānīyah fī jughrāfīyat al-'umrān)
الدراسة الميدانية فى جغرافية العمران : مع دراسة تطبيقية على المجمعة والدرعية والدمام فى المملكة العربية السعودية / تأليف محمود عبد اللطيف عصفور، السعيد ابراهيم البدوى. — القاهرة : مكتبة الأنجلو المصرية، 1976.
240, 7 p. : ill. ; 24 cm.
Bibliography: p. [3]–5 (2d group)
ISBN 977-266-094-6 : £E1.50
1. Urbanization—Saudi Arabia. 2. Cities and towns—Saudi Arabia. 3. ad Dammām, Saudi Arabia. 4. al Majma'ah, Saudi Arabia. 5. al Dir'īyah, Saudi Arabia. I. al-Badawī, al-Sa'īd Ibrāhīm, joint author. II. Title.
HT147.S33A83 77-961686

CITIES AND TOWNS (Continued)

—SAUDI ARABIA—KHAMIS MUSHAYT —GROWTH—CASE STUDIES

Tashkandi, Farhat Khourshid.
Urban dwelling environments in rapidly growing cities : case study, Khamis Mushait, Saudi Arabia / Farhat Khourshid Tashkandi. — Cambridge, Mass. : Urban Settlement Design in Developing Countries, School of Architecture and Planning, Massachusetts Institute of Technology, 1979.
56 p. : ill. ; 22 x 28 cm.
Bibliography: p. 56.
1. Cities and towns—Saudi Arabia—Khamis Mushayt—Growth—Case studies. 2. Land use, Urban—Saudi Arabia—Khamis Mushayt—Case studies. 3. Housing—Saudi Arabia—Khamis Mushayt—Case studies. I. Massachusetts Institute of Technology. Urban Settlement Design in Developing Countries. II. Title.
HT147.S33T37 333.7'7'09538 79-116298
79 MARC

—SCOTLAND—GROWTH—HISTORY

Adams, Ian H
The making of urban Scotland / Ian H. Adams. — London : Croom Helm, 1978.
303 p. : ill. ; 24 cm. GB78-05652
Includes bibliographical references and index.
ISBN 0-85664-518-4 : £20.00
1. Cities and towns—Scotland—Growth—History. 2. Urbanization—Scotland—History. I. Title.
HT133.A4 941.1'00973'2 78-317868
78 MARC

—SIBERIA

(Goroda Sibiri)
Города Сибири : эпоха феодализма и капитализма / ₍отв. редактор О. Н. Вилков₎. — Новосибирск : Наука, 1978.
334 p. : ill. ; 18 cm. USSR***
At head of title: Akademiiа nauk SSSR. Sibirskoe otdelenie. Institut istorii, filologii i filosofii.
Includes bibliographical references.
1.70rub
1. Siberia—Economic conditions. 2. Siberia—Social conditions. 3. Cities and towns—Siberia. I. Vilkov, Oleg Nikandrovich. II. Akademiiа nauk SSSR. Sibirskoe otdelenie. Institut istorii, filologii i filosofii.
HC337.S5G67 78-422580

—SILESIA, UPPER—HISTORY

Kwak, Jan.
Miasta księstwa opolsko-raciborskiego w XVI-XVIII wieku / Jan Kwak. — Wyd. 1. — Opole : Instytut Śląski, 1977 (Katow : ZG)
311 p., ₍6₎ leaves of plates : ill. ; 24 cm. P78-956
Bibliography: p. 280-₍307₎
Includes index.
zł60.00
1. Silesia, Upper—History. 2. Cities and towns—Silesia, Upper—History. I. Title.
DK4600.S467K87 78-394046
79 MARC

—SOUTH AFRICA—ADDRESSES, ESSAYS, LECTURES

Davies, R J
Of cities and societies : a geographer's viewpoint / R. J. Davies. — ₍Cape Town₎ : University of Cape Town, 1976.
20 p. : ill. ; 22 cm. — (Inaugural lecture : New series - University of Cape Town ; no. 38) SA***
Cover title.
Delivered May 20, 1976.
Bibliography: p. 19-20.
ISBN 0-7992-0178-2
1. Anthropo-geography—Addresses, essays, lectures. 2. Cities and towns—Addresses, essays, lectures. 3. Cities and towns—South Africa—Addresses, essays, lectures. I. Title. II. Series: Cape Town. University of Cape Town. Inaugural lecture : New series ; no. 38.
GF125.D38 301.36 79-313616
79 MARC

—SOUTH AFRICA—STATISTICS

Potgieter, J G
The houschold subsistence level in the major urban centres of the Republic of South Africa, October 1978 / by J. F. Potgieter. — Port Elizabeth ₍South Africa₎ : Institute for Planning Research, University of Port Elizabeth, 1978.
69 p. : ill. ; 30 cm. — (Fact paper - Institute for Planning Research, University of Port Elizabeth ; no. 28) SA***
Includes bibliographical references.
ISBN 0-86988-149-3
1. Cost and standard of living—South Africa—Statistics. 2. Households—South Africa—Statistics. 3. Cities and towns—South Africa—Statistics. I. Title. II. Series: Port Elizabeth, South Africa. University. Institute for Planning Research. Fact paper - Institute for Planning Research, University of Port Elizabeth ; no. 28.
HD7064.4.P686 339.4'2'0968 79-316897
79 MARC

—SOUTHERN STATES—GROWTH

Schwertz, Eddie L
The local growth management guidebook / by Eddie L. Schwertz, Jr. for the Southern Growth Policies Board. — ₍Research Triangle Park, N.C.₎ : The Board, 1979.
x, 191 p. : ill. ; 28 cm.
Includes bibliographical references.
1. Cities and towns—Southern States—Growth. 2. City planning—Southern States—Citizen participation. I. Southern Growth Policies Board. II. Title.
HT371.S33 309.2'62'0975 78-66185
00 MARC

—SOUTHERN STATES—GROWTH

Dusenbury, Patricia J
Southern urban trends, 1960-1977 / by Patricia J. Dusenbury and Thad L. Beyle. — ₍Research Traingle Park, N.C.₎ : Southern Growth Policies Board, 1978.
v, 37, ₍12₎ p. : ill. ; 28 cm. — (Southern Growth Policies Board research report)
Cover title.
Includes bibliographical references.
1. Cities and towns—Southern States—Growth. 2. Southern States—Population. I. Beyle, Thad L., 1934- joint author. II. Title. III. Series: Southern Growth Policies Board. Southern Growth Policies Board research report.
HT123.5.S6D87 301.36'0975 78-71004
79 MARC

—SPAIN

Martín Moreno, Jaime.
La estructura social de las ciudades españolas / Jaime Martín Moreno y Amando de Miguel, con la colaboración de M.a Victoria Fernández Allés, Anna Ubeda i Carulla, Josefina Núñez de Pedro. — Madrid : Centro de Investigaciones Sociológicas, 1978.
208 p. : maps ; 20 cm. — (Colección Monografías ; no. 6) Sp***
Includes bibliographical references.
ISBN 8474760038
1. Spain—Social conditions. 2. Cities and towns—Spain. I. Miguel, Amando de, joint author. II. Title.
HN583.5.M27 301.36'0946 79-363377
79 MARC

Racionero, Luis.
Sistemas de ciudades y ordenación del territorio / Luis Racionero. — Madrid : Alianza, c1978.
168 p. : ill. ; 20 cm. — (Alianza universidad ; 210) Sp***
Bibliography: p. ₍165₎-168.
ISBN 8420622109
1. Cities and towns. 2. Regional economics. 3. Cities and towns—Spain. I. Title.
HT151.R25 79-343937
79 MARC

—SPAIN—HISTORY

Torres Balbás, Leopoldo.
Ciudades hispanomusulmanas / por Leopoldo Torres Balbás ; advertencia preliminar, introd. y conclusión por Henri Terrasse. — ₍Madrid₎ : Ministerio de Asuntos Exteriores, Dirección General de Relaciones Culturales, Instituto Hispano-Arabe de Cultura, ₍1970?₎
2 v. (688 p.) : ill. ; 23 cm.
Bibliography: p. 661-688.
CONTENTS: t. 1. Historia e instituciones. Organización de las ciudades. Las calles.—t. 2. Las defensas urbanas.
1. Cities and towns—Spain—History. 2. Arabs in Spain. I. Terrasse, Henri, 1895-1971. II. Title.
HT145.S6T67 301.36'3'0946 78-375940
78 MARC

—SPAIN—BARCELONA—GROWTH

Ferras, Robert.
Barcelone : croissance d'une métropole / Robert Ferras. — Paris : Anthropos, ₍c1977₎
616 p., ₍8₎ leaves of plates : ill., maps (2 fold. in pocket) ; 22 cm. F***
Bibliography: p. 543-602.
ISBN 2-7157-0294-9 : 120F
1. Barcelona—Economic conditions. 2. Urbanization—Spain—Barcelona. 3. Cities and towns—Spain—Barcelona—Growth. I. Title.
HC388.B3F47 330.9'46'72083 78-376363
78 MARC

—SPAIN—GUIPÚZCOA—HISTORY

Arízaga Bolumburu, Beatriz.
El nacimiento de las villas guipuzcoanas en los siglos XIII y XIV ; morfología y funciones urbanas / Beatriz Arízaga Bolumburu. — San Sebastián : Sociedad Guipuzcoana de Ediciones y Publicaciones, 1978.
179 p., ₍1₎ fold. leaf of plates : ill. ; 24 cm. — (Monografías - Grupo Doctor Camino de Historia de San Sebastián ; 11) Sp78-July
Bibliography: p. ₍13₎-16.
ISBN 8471730553
1. Urbanization—Spain—Guipúzcoa—History. 2. Cities and towns—Spain—Guipúzcoa—History. 3. Guipúzcoa, Spain—Economic conditions. 4. Guipúzcoa, Spain—Social conditions. 5. Guipúzcoa, Spain—Politics and government. I. Title. II. Series: Sociedad Vascongada de Amigos del País. Grupo Dr. Camino de Historia Donostiarra. Monografías ; 11.
HT145.S6A73 301.36'3'094661 79-346879
79 MARC

—SWEDEN

Heineman, Hans-Erland, 1919-
Stad / ₍Hans-Erland₎ Heineman. — Stockholm : Esselte studium, 1976.
120 p. : ill. ; 23 cm. — (Scandinavian university books) S76-41
Bibliography: p. 119-120.
ISBN 9124261696 : kr69.00
1. Cities and towns. 2. Cities and towns—Sweden. I. Title.
HT151.H367 79-338833
*79₍r79₎rev MARC

—SWEDEN—GROWTH

Johannesson, Gösta.
Från köpstad till storkommun / Gösta Johannesson. — Stockholm : Natur o. kultur, 1978.
160 p. : tables ; 22 cm. — (Natur och kultur-serien ; 10) S78-13
Includes indexes.
ISBN 9127003426 : kr39.00
1. Cities and towns—Sweden—History. 2. Cities and towns—Sweden—Growth. I. Title.
HT145.S7J63 79-338054
79 MARC

—SWEDEN—HISTORY

Johannesson, Gösta.
Från köpstad till storkommun / Gösta Johannesson. — Stockholm : Natur o. kultur, 1978.
160 p. : tables ; 22 cm. — (Natur och kultur-serien ; 10) S78-13
Includes indexes.
ISBN 9127003426 : kr39.00
1. Cities and towns—Sweden—History. 2. Cities and towns—Sweden—Growth. I. Title.
HT145.S7J63 79-338054
79 MARC

—SWITZERLAND—HISTORY—SOURCES

Die Urkunden und Akten der oberdeutschen Städtebünde vom 13. Jahrhundert bis 1549 / hrsg. durch d. Histor. Komm. bei d. Bayer. Akad. d. Wiss. — Göttingen : Vandenhoeck und Ruprecht, 1979-
v. ; 25 cm. GFR79-A(v.1)
Includes bibliographical references and indexes.
CONTENTS: Bd. 1. Ruser, K. Vom 13. Jahrhundert bis 1347.

ISBN 3-525-35913-6 (v. 1)
1. Germany—History—1273-1517—Sources. 2. Rhine Cities, League of, 1254—Sources. 3. Cities and towns—Germany—History—Sources. 4. Cities and towns—Switzerland—History—Sources. 5. Landfriede—History—Sources. I. Akademie der Wissenschaften, Munich. Historische Kommission.
DD156.A2U74 943 79-367818
79 MARC

CITIES AND TOWNS (Continued)

—SWITZERLAND—MAPS

Kümmerly und Frey, Bern.
Schweiz : Strassenatlas mit Ortsverzeichnis und 35 Stadtplänen = Switzerland : road atlas with index and 35 town plans. — Bern : Kümmerly + Frey, 1978.
40, 88 p. : col. maps ; 26 cm. Sw78-6916
Explanatory text in French and German.
Includes Stadtplanatlas der Schweiz.
Scale of maps: 1:300,000.
Includes indexes.
15.80F
1. Switzerland—Road maps. 2. Cities and towns—Switzerland—Maps. I. Title. II. Title: Switzerland. III. Title: Stadtplanatlas der Schweiz.
G1896.P2K856 1978 912'.494 79-383476
79 MARC

—SWITZERLAND—BASEL-LAND (CANTON)

Meier, Eugen A
Rund um den Baselstab : 3 historische Bildbände über 200 Städte und Dörfer in der Regio Basiliensis / Eugen A. Meier ; mit einem Vorw. von Edmund Wyss. — 1. Aufl. — Basel : Birkhäuser, c1976-c1978.
3 v. : ill. (some col.) ; 27 cm. Sw77-380
Includes bibliographies.
CONTENTS: 1. Basels Landgemeinden, das Baselbiet.—2. Basels stadttore, Schwibbogen und Vorstadtlandschaften Schwarzbubenland, Dorneck, Leimental, Laufental, Fricktal.—3. Markgräflerland, Sundgau.
ISBN 3-7643-0791-9 (v. 1) : 92.00F (v. 1)
1. Basel-Land (Canton)—History. 2. Cities and towns—Switzerland—Basel-Land (Canton) I. Title.
DQ375.M44 77-477511
*77[r79]rev MARC

—TAIWAN

Liu, K'o-chih.
(Tu shih jen k'ou ting i chih yen chiu)
都市人口定義之研究 / 劉克智研究. — [台北] : 行政院經濟設計委員會都市規劃處, 民國63 [1974]
119 leaves : ill. ; 29 cm.
Caption title: T'ai-wan ti ch'ü tu shih, hsiang ts'un ting i chi fan wei hua ting chih yen chiu.
"行政院經濟設計委員會都市規劃處委託"
Reproduced from typescript.
Includes bibliographical references.

1. Cities and towns—Taiwan. 2. Taiwan—Population. I. Title. II. Title: T'ai-wan ti ch'ü tu shih, hsiang ts'un.
HT147.T3L58 1974 78-840904

—TAIWAN—ADDRESSES, ESSAYS, LECTURES

(Tu shih she hui fa chan chih yen chiu)
都市社會發展之研究 / 主編蔡勇美, 郭文雄. — 1版. — 臺北 : 巨流圖書公司, 民國67 [1978]
2, 304, 4 p. ; 22 cm.
Added t. p.: Modern urban society: a collection of essays.
Includes bibliographical references.
NT$120.00

1. Sociology, Urban—Addresses, essays, lectures. 2. Cities and towns—Taiwan—Addresses, essays, lectures. I. T'ai, Yung-mei, II. Kuo, Wen-hsiung. III. Title: Modern urban society.
HT151.T8 79-840627

—TANZANIA—BIBLIOGRAPHY

Tanganyika Library Service.
Bibliography of towns in Tanzania / Tanganyika Library Service [by M. Sumar, compiler]. — [Dar-es-Salaam] : The Service, 1972.
14 leaves ; 26 cm.
Cover title.
1. Tanzania—History, Local—Bibliography. 2. Tanzania—Administrative and political divisions—Bibliography. 3. Cities and towns—Tanzania—Bibliography. I. Sumar, M. II. Title.
Z3588.T36 1972 916.78 79-317927
[DT438] 79 MARC

—THE WEST—HISTORY

Reps, John William.
Cities of the American West : a history of frontier urban planning / John W. Reps. — Princeton, N.J. : Princeton University Press, c1979.
xii, 827 p., 32 [i.e. 16] leaves of plates : ill. ; 23 x 27 cm.
Bibliography: p. [747]-773.
Includes index.
ISBN 0-691-04648-4 : $75.00
1. Cities and towns—The West—History. 2. Urbanization—The West. I. Title.
HT123.5.A17R46 301.36'3'0978 78-51187
78 MARC

—TURKEY—HISTORY—19TH CENTURY

Höhfeld, Volker.
Anatolische Kleinstädte : Anlage, Verlegung u. Wachstumsrichtung seit d. 19. Jh. / Volker Höhfeld. — Erlangen : Fränkische Geographische Ges. : Palm und Enke [in Komm.], 1977.
220, [34] p. : ill., maps ; 24 cm. — (Erlanger geographische Arbeiten : Sonderband ; 6) GFR77-A45
Originally presented as the author's thesis, Erlangen-Nürnberg, 1976.
Summary also in English.
Bibliography: p. 195-208.
ISBN 3-920405-42-0 : DM30.00
1. Cities and towns—Turkey—History—19th century. 2. Urbanization—Turkey—History—19th century. I. Title. II. Series.
HT147.T9H63 1977 301.36'09561 79-374518
79 MARC

—UKRAINE

Gajecky, George.
The Cossack administration of the Hetmanate / George Gajecky. — Cambridge, Mass. : Harvard Ukrainian Research Institute, c1978.
2 v. (xv, 789 p.) : maps ; 22 cm. — (Sources and documents series)
Includes index.
Bibliography: p. [710]-730.
ISBN 0-916458-02-4
1. Cossacks. 2. Ukraine—Politics and government. 3. Cities and towns—Ukraine. 4. Ukraine—History, Local. I. Harvard Ukrainian Research Institute. II. Title. III. Series.
DK35.G25 947'.71'00491714 77-73708
00 MARC

—UKRAINE—L'VOVSKAYA OBLAST'

(L'vovskaíà oblast')
Львовская область / председатель редколлегии Д. А. Яремчук ... [et al.]. — Киев : Глав. редакция УСЭ, 1978.
794 p., [12] leaves of plates : ill. (some col.) ; 27 cm. — (История городов и сел Украинской ССР) USSR***
1. L'vovskaya oblast', Ukraine—History, Local. 2. Cities and towns—Ukraine—L'vovskaya oblast'. I. ÎAremchuk, D. A. II. Series: Istoriíà gorodov i sel Ukraínskoĭ SSR.
DK651.L915L86 79-410158

—UKRAINE—ODESSA (PROVINCE)

(Odesskaíà oblast')
Одесская область / Редкол. Чередниченко А. П. (пред.) ... [et al.]. — Киев : Ин-т истории АН УССР : Укр. сов. энциклопедия, 1978.
866 p., [12] leaves of plates : ill. ; 27 cm. — (История городов и сел Украинской ССР) USSR 78
"Переработанное и дополненное переиздание тома Одеська область, вышедшего в 1969 году."
Includes bibliographical references and indexes.
5.20rub
1. Odessa (Province)—History, Local. 2. Cities and towns—Ukraine—Odessa (Province) I. Cherednychenko, Anatoliĭ Pavlovych. II. Series: Istoriíà gorodov i sel Ukrainskoĭ SSR.
DK511.O32O35 79-412778

—UNITED STATES

Baldassare, Mark.
Residential crowding in urban America / Mark Baldassare. — Berkeley : University of California Press, c1979.
xiv, 250 p. ; 23 cm.
Bibliography: p. 225-239.
Includes index.
ISBN 0-520-03563-1 : $12.95
1. United States—Population density. 2. Personal space. 3. Crowding stress. 4. Cities and towns—United States. I. Title.
HT201.B34 301.32'9'73 77-83102
79 MARC

Burns, Jim, 1926-
Connections : ways to discover and realize community potentials / Jim Burns. — Stroudsburg, Pa. : Dowden, Hutchinson & Ross, c1979.
xi, 155 p. : ill. ; 24 cm. — (Community development series ; 43)
Bibliography: p. 147-150.
Includes index.
ISBN 0-07-009205-2
1. Cities and towns—United States. 2. City planning—United States. 3. Architecture—Human factors. I. Title.
HT123.B84 309.2'62'0973 78-14763
78 MARC

Caraley, Demetrios.
City governments and urban problems : a new introduction to urban politics / Demetrios Caraley. — Englewood Cliffs, N.J. : Prentice-Hall, c1977.
xv, 448 p. : ill. ; 24 cm.
Includes bibliographical references and index.
ISBN 0-13-134973-2
1. Municipal government—United States. 2. Cities and towns—United States. I. Title.
JS331.C33 352'.008'0973 76-28327
76[r79]rev MARC

Colker, David.
Running away from home : a guide to running trails in 32 cities across America / David Colker ; maps by Christina Haatainen. — New York : Jove Publications, 1979.
287 p. : maps ; 18 cm.
"A Jove/HBJ book."
$2.25
1. Running—United States—Guide-books. 2. United States—Description and travel—1960- —Guide-books. 3. Cities and towns—United States. I. Title.
GV1061.2.C64 917.3'04'4 78-20593
79 MARC

Congressional Quarterly, inc.
Urban America, policies and problems. — Washington : Congressional Quarterly, c1978.
105, 53 p. : ill. ; 28 cm.
Bibliography: p. 47-49 (2d group)
Includes index.
ISBN 0-87187-138-6 : $5.95
1. Cities and towns—United States. 2. United States—Social policy. 3. Urban Economics. 4. Community development, Urban—United States. I. Diamond, Robert A. II. Arieff, Irwin B. III. Title.
HT123.C633 1978 309.2'62'0973 78-13734
78 MARC

Curtis, Richard Farnsworth, 1931-
Inequality in American communities / Richard F. Curtis, Elton F. Jackson. — New York : Academic Press, 1977.
xii, 354 p. ; 24 cm. — (Quantitative studies in social relations)
Bibliography: p. 343-350.
Includes index.
ISBN 0-12-200250-4
1. Social classes—United States. 2. Social status. 3. Equality. 4. Cities and towns—United States. I. Jackson, Elton F., joint author. II. Title.
HN90.S6C87 301.44'0973 76-19485
76[r79]rev MARC

Northam, Ray M
Urban geography / Ray M. Northam. — 2d ed. — New York : Wiley, c1979.
x, 512 p. : ill. ; 26 cm.
Includes bibliographical references and index.
ISBN 0-471-03292-1
1. Cities and towns. 2. Cities and towns—United States. I. Title.
HT151.N64 1979 301.36'3'09 78-12335
78 MARC

Ostransky, Leroy.
Jazz city : the impact of our cities on the development of jazz / Leroy Ostransky. — Englewood Cliffs, N.J. : Prentice-Hall, c1978.
xvi, 274 p. ; 21 cm. — (A Spectrum book)
Bibliography: p. [249]-256.
Discography: p. [251]-264.
Includes index.
ISBN 0-13-509380-5. ISBN 0-13-509372-4 pbk.
1. Jazz music—United States. 2. Cities and towns—United States. I. Title.
ML3561.J3O84 785.4'2'0973 78-18492
78 MARC

Roeseler, W G
Alternative goals of urbanization for America / by W. G. Roeseler. — Bryan, Tex. : EMR Publications, c1977.
v, 6 p. ; 24 cm.
"Published in part in the Midwest review of public administration, March 1977."
ISBN 0-930308-02-6 : $0.95
1. Urbanization—United States. 2. Cities and towns—United States. I. Title.
HT123.R62 301.36'1'0973 78-16931
78 MARC

Schneider, Kenneth R
On the nature of cities : toward enduring and creative human environments / Kenneth R. Schneider. — 1st ed. — San Francisco : Jossey-Bass Publishers, 1979.
xiv, 352 p. ; 24 cm. — (The Jossey-Bass social and behavioral science series)
Bibliography: p. 334-340.
Includes index.
ISBN 0-87589-391-0
1. Cities and towns—United States. 2. Urban policy—United States. 3. Cities and towns—Growth. 4. Sociology, Urban. I. Title.
HT123.S317 301.36'3'0973 78-62556

CITIES AND TOWNS
—UNITED STATES (Continued)

Swanson, Bert E
Small towns and small towners : a framework for survival and growth / Bert E. Swanson, Richard A. Cohen, and Edith P. Swanson ; foreword by Harold S. Williams. — Beverly Hills, Calif. : Sage Publications, c1979.
280 p. : ill. ; 22 cm. — (Sage library of social research ; v. 79)
Includes bibliographical references and indexes.
ISBN 0-8039-1017-7. ISBN 0-8039-1018-5 pbk.
1. Cities and towns—United States. 2. Community. 3. Community development—United States. 4. Community leadership. 5. Regional planning—United States. I. Cohen, Richard A., joint author. II. Swanson, Edith, 1926- joint author. III. Title.
HT123.S843 301.35'0973 78-17697
78 MARC

United States. Congress. House. Committee on Banking, Finance, and Urban Affairs. Subcommitee on the City.
Small cities : how can the Federal and State governments respond to their diverse needs? : Hearings before the Subcommittee on the City of the Committee on Banking, Finance, and Urban Affairs, House of Representatives, Ninety-fifty Congress, second session, May 16, 19, and 26, 1978, Washington, D.C., Ada Okla., and Saratoga Springs, N.Y. — Washington : U.S. Govt. Print. Off., 1978.
iii, 584 p. : ill. ; 24 cm.
Includes bibliographical references.
1. Cities and towns—United States. 2. Cities and towns—United States—Growth. 3. Urban policy—United States. I. Title.
KF27.B52 1978 301.36'3'0973 78-603060
79 MARC

United States. Congress. House. Committee on Banking, Finance, and Urban Affairs. Subcommitee on the City.
Small cities : how can the Federal and State governments respond to their diverse needs? : Together with additional views : report / by the Subcommittee on the City of the Committee on Banking, Finance and Urban Affairs, House of Representatives, 95th Congress, second session, August 1978. — Washington : U.S. Govt. Print. Off., 1978.
vi, 37 p. ; 24 cm.
"Subcommittee print."
Includes bibliographical references.
1. Cities and towns—United States. 2. Federal-city relations—United States. 3. Local government—United States. 4. Urban policy—United States. I. Title.
HT123.U4 1978 301.36'0973 78-602806
79 MARC

Warren, Roland Leslie, 1915-
The community in America / Roland L. Warren. — 3d ed. — Chicago : Rand McNally College Pub. Co., c1978.
x, 448 p. ; 24 cm.
Includes bibliographical references and index.
ISBN 0-528-68096-X
1. Community life. 2. Cities and towns—United States. 3. United States—Social conditions. I. Title.
HT123.W25 1978 301.34'0973 77-77712
78 MARC

—UNITED STATES—ADDRESSES, ESSAYS, LECTURES

Current and future directions of real estate and urban development in the United States / edited by Frank G. Mittelbach. — Los Angeles : Housing, Real Estate, and Urban Land Studies Program, Urban Resources Study Center, University of California, 1978.
i, 100 p. : ill. ; 25 cm. — (Real Estate Chair lecture series)
"Under auspices of Housing, Real Estate, and Urban Land Studies Program, Urban Resources Study Center, University of California, Los Angeles."
Four public lectures delivered at UCLA during the 1977-1978 academic year.
Includes bibliographical references.
ISBN 0-911798-19-6
1. Cities and towns—United States—Addresses, essays, lectures. 2. Land use, Urban—United States—Addresses, essays, lectures. 3. Real estate development—United States—Addresses, essays, lectures. 4. Housing—United States—Addresses, essays, lectures. I. Mittelbach, Frank G. II. California. University, University at Los Angeles. Housing, Real Estate, and Urban Land Studies Program. III. Series.
HT123.C82 301.36'1'0973 79-622145
79 MARC

Current issues in urban economics / edited by Peter Mieszkowski and Mahlon Straszheim. — Baltimore : Johns Hopkins University Press, c1979.
xiv, 589 p. ; 24 cm.
Includes bibliographies and index.
ISBN 0-8018-2109-6
1. Cities and towns—United States—Addresses, essays, lectures. 2. Urban economics—Addresses, essays, lectures. I. Mieszkowski, Peter M. II. Straszheim, Mahlon H., 1939-
HT151.C87 330.9'173'2 78-14947
78 MARC

Feagin, Joe R comp.
The urban scene : myths and realities / edited by Joe R. Feagin ; [cover photo/design by Meryl Sussman Levavi]. — 2d ed. — New York : Random House, c1979.
305 p. ; 24 cm.
Includes bibliographical references.
ISBN 0-394-32225-8 : $5.95
1. Cities and towns—United States—Addresses, essays, lectures. I. Title.
HT123.F4 1979 301.36'3'0973 78-11742
78 MARC

Saving America's cities / edited by Evelyn Geller. — New York : H. W. Wilson, 1979.
208 p. ; 19 cm. — (The Reference shelf ; v. 51, no. 1)
Bibliography: p. 201-108.
ISBN 0-8242-0631-2
1. Cities and towns—United States—Addresses, essays, lectures. 2. Social problems—Addresses, essays, lectures. 3. Urban policy—United States—Addresses, essays, lectures. I. Geller, Evelyn. II. Series.
HT123.S28 301.36'0973 79-9719
79 MARC

The Urban predicament / edited by William Gorham, Nathan Glazer. — Washington : Urban Institute, c1976.
xix, 363 p. : graphs ; 24 cm.
Includes bibliographical references and index.
ISBN 0-87766-160-X : $4.95
1. Cities and towns—United States—Addresses, essays, lectures. 2. Housing—United States—Addresses, essays, lectures. 3. Urban policy—United States—Addresses, essays, lectures. 4. Housing—Washington, D.C.—Addresses, essays, lectures. 5. Washington, D.C.—Social conditions—Addresses, essays, lectures. I. Gorham, William, 1930- . II. Glazer, Nathan. III. Urban Institute.
HT123.U749 301.36'0973 76-16713
79 MARC

—UNITED STATES—BIBLIOGRAPHY

Filipovitch, Anthony J
Urban community : a guide to information sources / Anthony J. Filipovitch, Earl J. Reeves. — Detroit : Gale Research Co., c1978.
x, 286 p. ; 23 cm. — (Urban studies information guide series ; v. 4) (Gale information guide library)
Includes bibliographies and indexes.
ISBN 0-8103-1429-0
1. Community life—Bibliography. 2. Sociology, Urban—Bibliography. 3. Cities and towns—United States—Bibliography. 4. Cities and towns—Great Britain—Bibliography. I. Reeves, Earl J., joint author. II. Title. III. Series.
Z7164.S68F54 016.30136 78-13171
[HM131] 78 MARC

—UNITED STATES—CASE STUDIES

Halpern, Kenneth, 1944-
Downtown USA : urban design in nine American cities / by Kenneth Halpern. — New York : Whitney Library of Design, 1978.
256 p. : ill. ; 26 cm.
Bibliography: p. 239-244.
Includes index.
ISBN 0-8230-7154-5
1. Cities and towns—United States—Case studies. 2. City planning—United States—Case studies. I. Title.
HT123.H33 1978 309.2'62'0973 78-18339
78 MARC

Procter, Mary, 1942-
Gritty cities : a second look at Allentown, Bethlehem, Bridgeport, Hoboken, Lancaster, Norwich, Paterson, Reading, Trenton, Troy, Waterbury, Wilmington / Mary Procter and Bill Matuszeski. — Philadelphia : Temple University Press, 1978.
ix, 276 p. : ill. ; 26 cm.
Bibliography: p. [267]-276.
ISBN 0-87722-143-X : $17.50
1. Cities and towns—United States—Case studies. I. Matuszeski, William, 1941- joint author. II. Title.
HT123.P76 301.36'3'0973 78-15149
78 MARC

—UNITED STATES—COLLECTED WORKS

American communities tomorrow / [editor, Jack Kinton]. — Aurora, Ill. : Social Science Services & Resources, c1978-
v. : ill. ; 24 cm.
CONTENTS: v. 1. Neighborhood revivals & new towns.
ISBN 0-915574-08-X (v. 1)
1. Cities and towns—United States—Collected works. I. Kinton, Jack F.
HT123.A6624 301.36'0973 77-92885
79 MARC

—UNITED STATES—CONGRESSES

Central city economic development / edited by Benjamin Chinitz. — Cambridge, Mass. : Abt Books, c1979.
ix, 198 p. ; 24 cm.
Includes index.
ISBN 0-89011-524-9
1. Cities and towns—United States—Congresses. 2. Urban policy—United States—Congresses. 3. Urban economics—Congresses. I. Chinitz, Benjamin.
HT123.C42 301.36'1'0973 78-67847
79 MARC

Lethal aspects of urban violence / edited by Harold M. Rose. — Lexington, Mass. : Lexington Books, c1979.
viii, 100 p. : graphs ; 24 cm.
Includes bibliographical references.
ISBN 0-669-02117-2
1. Homicide—United States—Congresses. 2. Cities and towns—United States—Congresses. 3. Violence—United States—Congresses. I. Rose, Harold M.
HV6529.L47 364.1'52 77-18680
78 MARC

—UNITED STATES—ECONOMIC CONDITIONS—MAPS

United States. Bureau of the Census.
Maps of major concentrations of poverty : in standard metropolitan statistical areas of 250,000 or more population / prepared for the Office of Economic Opportunity by the U.S. Bureau of the Census in 1966 from 1960 census data. — [Washington : Govt. Print. Off., 1966]
3 v. : chiefly maps ; 27 cm.
CONTENTS: v. 1. Akron, Ohio, through Grand Rapids, Michigan.—v. 2. Harrisburg, Pa., through Providence, R.I.—v. 3. Reading, Pa., through Youngstown, Ohio.
1. United States—Economic conditions—Maps. 2. Untied States—Social conditions—Maps. 3. Cities and towns—United States—Social conditions—Maps. 4. Cities and towns—United States—Economic conditions—Maps. I. United States. Office of Economic Opportunity. II. Title.
G1201.E7U5 1966 912'.1'309173 78-324714
79 MARC

—UNITED STATES—GROWTH

Appelbaum, Richard P
Size, growth, and U.S. cities / Richard P. Appelbaum. — New York : Praeger, 1978.
xiv, 140 p. ; 25 cm.
Bibliography: p. 129-140.
ISBN 0-03-034621-5
1. Cities and towns—United States—Growth. I. Title.
HT123.A685 301.36'1'0973 78-61885
79 MARC

Bryce, Herrington J
Planning smaller cities / Herrington J. Bryce. — Lexington, Mass. : Lexington Books, c1979.
xxi, 214 p. ; 24 cm. — (Urban roundtable series)
Includes bibliographical references and index.
ISBN 0-669-02680-8
1. Cities and towns—United States—Growth. 2. City planning—United States. I. Title. II. Series.
HT123.B78 309.2'62'0973 78-14154
78 MARC

Goldfield, David R 1944-
Urban America : from downtown to no town / David R. Goldfield, Blaine A. Brownell. — Boston : Houghton Mifflin, c1979.
xi, 435 p. : ill. ; 24 cm.
Bibliography: p. 407-417.
Includes index.
ISBN 0-395-27397-8
1. Metropolitan areas—United States. 2. Cities and towns—United States—Growth. 3. Urbanization—United States. I. Brownell, Blaine A., joint author. II. Title.
HT123.G575 301.36'1'0973 78-69562
79 MARC

Sanders, Ralph A
American urban decline : updating evolutionary trends / Ralph A. Sanders. — Syracuse, N.Y. : Dept. of Geography, Syracuse University, 1978.
18 leaves : ill. ; 28 cm. — (Discussion paper series - Department of Geography, Syracuse University ; no. 48 ISSN 0363-6038)
Includes bibliographical references.
1. Cities and towns—United States—Growth. I. Title. II. Series: Syracuse University. Dept. of Geography. Discussion paper series - Department of Geography, Syracuse University ; no. 48.
HT384.U5S26 301.36'1'0973 78-109843
79 MARC

CITIES AND TOWNS
—UNITED STATES—GROWTH
(Continued)

Teaford, Jon C
City and suburb : the political fragmentation of metropolitan America, 1850-1970 / Jon C. Teaford. — Baltimore : Johns Hopkins University Press, c1979.
vii, 231 p. ; 24 cm. — (Johns Hopkins studies in urban affairs)
Bibliography: p. 209-224.
Includes index.
ISBN 0-8018-2202-5
1. Municipal government—United States. 2. Metropolitan government—United States. 3. Cities and towns—United States—Growth. I. Title. II. Series.
JS323.T4 352'.008'0973 78-20519
78 MARC

United States. Congress. House. Committee on Banking, Finance, and Urban Affairs. Subcommitee on the City.
Small cities : how can the Federal and State governments respond to their diverse needs? : Hearings before the Subcommittee on the City of the Committee on Banking, Finance, and Urban Affairs, House of Representatives, Ninety-fifty Congress, second session, May 16, 19, and 26, 1978, Washington, D.C., Ada Okla., and Saratoga Springs, N.Y. — Washington : U.S. Govt. Print. Off., 1978.
iii, 584 p. : ill. ; 24 cm.
Includes bibliographical references.
1. Cities and towns—United States. 2. Cities and towns—United States—Growth. 3. Urban policy—United States. I. Title.
KF27.B52 1978 301.36'3'0973 78-603060
79 MARC

Urban development in the USA and Hungary / edited by György Enyedi. — Budapest : Akadémiai Kiadó, 1978.
311 p., [1] fold. leaf : ill., maps ; 25 cm. — (Studies in geography in Hungary ; 14)
Includes bibliographies.
ISBN 9630515148
1. Cities and towns—United States—Growth. 2. Cities and towns—Hungary—Growth. 3. Urbanization—United States. 4. Urbanization—Hungary. I. Enyedi, György. II. Series.
HT123.U744 301.36'1'09439 79-310628
79 MARC

—UNITED STATES—GROWTH—
ADDRESSES, ESSAYS, LECTURES

Tolley, George S 1925-
Urban growth policy in a market economy / George S. Tolley, Philip E. Graves, John L. Gardner ; with contributions by Donald R. Haurin ... [et al.]. — New York : Academic Press, 1979.
xii, 220 p. : ill. ; 24 cm. (Studies in urban economics)
Includes bibliographies and index.
ISBN 0-12-692850-9
1. Cities and towns—United States—Growth—Addresses, essays, lectures. 2. Urban economics—Addresses, essays, lectures. I. Graves, Philip E., joint author. II. Gardner, John L., joint author. III. Title. IV. Series.
HT123.T63 301.36'3'0973 78-8838
78 MARC

—UNITED STATES—GROWTH
—CASE STUDIES

Norton, R D
City life-cycles and American urban policy / R. D. Norton. — New York : Academic Press, c1979.
xv, 182 p. ; 24 cm. — (Studies in urban economics)
Includes bibliographical references and indexes.
ISBN 0-12-521850-8
1. Cities and towns—United States—Growth—Case studies. 2. Industries, Location of—United States—Case studies. 3. Urban policy—United States—History. 4. Urban economics. 5. Federal aid to community development—United States. I. Title. II. Series.
HT123.N66 301.36'0973 79-51704
79 MARC

—UNITED STATES—GROWTH—
COLLECTED WORKS

Management & control of growth : issues, techniques, problems, trends / edited by Randall W. Scott, with the assistance of David J. Brower and Dallas D. Miner. — Washington : Urban Land Institute, 1975-
v. ; 25 cm.
Vol. 4 edited by Frank Schnidman, Jane A. Silverman, and Rufus C. Young, Jr. has subtitle: Techniques in application.
Includes bibliographies.
1. City planning—United States—Collected works. 2. Cities and towns—United States—Growth—Collected works. 3. Land use—United States—Collected works. I. Scott, Randall W. II. Brower, David J. III. Miner, Dallas D. IV. Urban Land Institute.
HT167.M36 309.2'62'0973 74-83560
75[r79]rev MARC

—UNITED STATES—GROWTH—
JUVENILE LITERATURE

Barker, Albert, 1900-
From settlement to city / by Albert Barker. — New York : J. Messner, c1978.
64 p. : ill. ; 22 cm.
Includes index.
SUMMARY: Uses a midwest prototype to portray how American cities have grown.
ISBN 0-671-32865-4
1. Cities and towns—United States—Growth—Juvenile literature. 2. Urbanization—United States—Juvenile literature. [1. Cities and towns—Growth. 2. Urbanization] I. Title.
HT123.B29 301.36'1'0973 78-12896
78 MARC

How cities can grow old gracefully / [prepared for the] Subcommittee on the City of the Committee on Banking, Finance, and Urban Affairs, House of Representatives, 95th Congress, first session. — Washington : U.S. Govt. Print. Off., 1977.
xiv, 222 p. : ill. ; 23 cm.
At head of title: Committee print.
Includes bibliographical references.
1. City planning—United States. 2. Cities and towns—United States—Growth. 3. Urban renewal—United States. I. United States. Congress. House. Committee on Banking, Finance, and Urban Affairs. Subcommittee on the City.
HT167.H68 301.36'3'0973 78-602443
78 MARC

—UNITED STATES—HISTORY

Gluck, Peter R
Cities in transition : social changes and institutional responses in urban development / Peter R. Gluck & Richard J. Meister. — New York : New Viewpoints, c1979.
xiii, 237 p. ; 21 cm.
Bibliography: p. 219-229.
Includes indexes.
ISBN 0-531-05409-8. ISBN 0-531-05623-6 pbk.
1. Cities and towns—United States—History. 2. Urbanization—United States—History. 3. Municipal government—United States—History. 4. Sociology, Urban. I. Meister, Richard J., 1938- joint author. II. Title.
HT123.G568 301.36'0973 79-12214
79 MARC

Hollingsworth, Joseph Rogers, 1932-
Dimensions in urban history : historical and social science perspectives on middle-size American cities / J. Rogers Hollingsworth and Ellen Jane Hollingsworth. — Madison : University of Wisconsin Press, 1979.
viii, 184 p. ; 24 cm.
Includes index.
ISBN 0-299-07820-5 : $19.50
1. Cities and towns—United States—History. 2. Cities and towns—Wisconsin—Case studies. 3. United States—Social conditions—1865-1918. 4. Urbanization—United States—History. 5. Sociology, Urban—United States. I. Hollingsworth, Ellen Jane, joint author. II. Title.
HT123.H65 301.36'0973 78-65011
79 MARC

—UNITED STATES—INDEXES

Parker, J Carlyle.
City, county, town, and township index to the 1850 Federal census schedules / J. Carlyle Parker. — Detroit : Gale Research Co., c1979.
xvii, 215 p. ; 22 cm. — (Gale genealogy and local history series ; v. 6)
ISBN 0-8103-1385-5 : $22.00
1. United States—Genealogy—Indexes. 2. Cities and towns—United States—Indexes. 3. United States—Census, 1850—Indexes. I. United States. Census Office. 7th census, 1850. II. Title.
CS65.P37 929'.373 79-11644
79 MARC

—UNITED STATES—GROWTH
—MANAGEMENT

Burrows, Lawrence B
Growth management : issues, techniques, and policy implications / Lawrence B. Burrows. — New Brunswick, N.J. : Center for Urban Policy Research, Rutgers University, c1977.
ix, 141 p. ; 23 cm.
Bibliography: p. [135]-141.
ISBN 0-88285-043-1
1. Cities and towns—United States—Growth—Management. 2. Suburbs—United States. 3. City planning—United States. 4. Zoning—United States. I. Title.
HT123.B85 309.2'6'0973 77-24034
77 MARC

—UNITED STATES—GUIDE-BOOKS

Scheerer, R Penelope.
The traveling runner's guide : where to run in 21 cities around the U.S. / R. Penelope Scheerer & John R. Schwanbeck. — 1st ed. — New York : Dutton, [1979] c1978.
xii, 259 p. : maps ; 21 cm.
ISBN 0-525-47530-3 : $5.95 ($7.50 Can)
1. Running—United States—Directories. 2. Cities and towns—United States—Guide-books. I. Schwanbeck, John R., joint author. II. Title.
GV1061.S33 1979 917.3'04'926 78-14824
78 MARC

—UNITED STATES—HISTORY

Crockett, Norman L
The Black towns / Norman L. Crockett. — Lawrence : Regents Press of Kansas, c1979.
xv, 244 p. : ill. ; 23 cm.
Bibliography: p. 221-231.
Includes index.
ISBN 0-7006-0185-6
1. Afro-Americans—History—1877-1964. 2. Black nationalism—United States—History. 3. Cities and towns—United States—History. 4. Afro-Americans—Segregation. I. Title.
E185.6.C93 973'.04'96073 78-15099
78 MARC

—UNITED STATES—MAPS

Branch, Melville Campbell, 1913-
Comparative urban design : rare engravings, 1830-1843 / by Melville C. Branch. — New York : Arno Press, c1978.
[iv], 102 [i.e. 104] p. : ill., col. maps; 33 x 37 cm.
Reprint of maps published 1830-1844 by the Superintendence of the Society for the Diffusion of Useful Knowledge, London.
Includes bibliographical references.
ISBN 0-405-10524-X : $39.50
1. Cities and towns—Maps. 2. Cities and towns—Europe—Maps. 3. Cities and towns—United States—Maps. 4. Cities and towns—Asia—Maps. 5. Engravings—Europe—Maps. 6. Engravings—United States—Maps. 7. Engravings—Asia—Maps. I. Society for the Diffusion of Useful Knowledge, London. II. Title. III. Title: Rare engravings, 1830-1843.
G1028.B65 1978 711'.4 77-24511
77 MARC

—UNITED STATES—PERIODICALS

The Urban interest. v. 1-
spring 1979-
[Lawrence, Dept. of Political Science, University of Kansas]
v. 26 cm.
"A journal of policy and administration."
Key title: The Urban interest, ISSN 0192-4974.
1. Municipal government—United States—Periodicals. 2. Cities and towns—United States—Periodicals. I. Kansas. University. Dept. of Political Science.
JS39.U75 352'.008'0973 79-643034
MARC-S

Urban world.
Teacher's ed.
[Columbus, Ohio, Xerox Education Center, etc.]
v. ill. 29 cm. biweekly (during the school year)
Key title: Urban world. Teacher's edition (Columbus),
ISSN 0160-3833
1. Sociology, Urban—United States—Periodicals. 2. Social problems—Periodicals. 3. Cities and towns—United States—Periodicals. I. Xerox Education Publications. Education Center. II. Key title.
HT101.U723 301.36'0973 78-640715
MARC-S

—UNITED STATES—SOCIAL
CONDITIONS—MAPS

United States. Bureau of the Census.
Maps of major concentrations of poverty : in standard metropolitan statistical areas of 250,000 or more population / prepared for the Office of Economic Opportunity by the U.S. Bureau of the Census in 1966 from 1960 census data. — [Washington : Govt. Print. Off., 1966]
3 v. : chiefly maps ; 27 cm.
CONTENTS: v. 1. Akron, Ohio, through Grand Rapids, Michigan.—v. 2. Harrisburg, Pa., through Providence, R.I.—v. 3. Reading, Pa., through Youngstown, Ohio.
1. United States—Economic conditions—Maps. 2. Untied States—Social conditions—Maps. 3. Cities and towns—United States—Social conditions—Maps. 4. Cities and towns—United States—Economic conditions—Maps. I. United States. Office of Economic Opportunity. II. Title.
G1201.E7U5 1966 912'.1'309173 78-324714
79 MARC

CITIES AND TOWNS (Continued)

—UNITED STATES—STATISTICS

Feinleib, Manning.
Mortality from cardiovascular and non-cardiovascular diseases for U.S. cities, 1949-1950, 1959-1961, 1969-1971, with selected environmental descriptors / compiled by Manning Feinleib, Richard Fabsitz, A. Richey Sharrett, Epidemiology Branch, Division of Heart and Vascular Diseases, National Heart, Lung, and Blood Institute. — Bethesda, Md. : U.S. Dept. of Health, Education, and Welfare, Public Health Service, National Institutes of Health ; Washington : for sale by the Supt. of Docs., U.S. Govt. Print. Off., ⌊1979⌋
vi, 480, 100 p. : map ; 36 cm. — (DHEW publication ; no. (NIH) 79-1453)
Tables.
1. Mortality—United States. 2. United States—Statistics, Medical. 3. United States—Statistics, Vital. 4. Cardiovascular system—Diseases—United States—Mortality. 5. Cancer—United States—Mortality. 6. Cities and towns—United States—Statistics. I. Fabsitz, Richard, joint author. II. Sharrett, A. Richey, joint author. III. National Heart, Lung, and Blood Institute. Epidemiology Branch. IV. Title. V. Series: United States. Dept. of Health, Education, and Welfare. DHEW publication ; no. (NIH) 79-1453.
RA407.3.F38 312'.26'0973 79-602032
79 MARC

—VENEZUELA—HISTORY

Morón, Guillermo.
Historia de la Provincia de Venezuela / Guillermo Morón. — 1. ed. — Caracas : Concejo Municipal del Distrito Federal, 1977.
390 p. : ill. ; 23 cm.
Includes bibliographical references.
1. Venezuela—History. 2. Cities and towns—Venezuela—History. I. Title.
F2322.M588 78-105389
79 MARC

—WHITE RUSSIA

Shtykhaŭ, H V
(Goroda Polotskoĭ zemli)
Города Полоцкой земли : (IX–XIII вв.) / Г. В. Штыхов ; Науч. ред. В. П. Даркевич. — Минск : Наука и техника, 1978.
159 p. : ill. ; 25 cm. USSR 78
At head of title: Akademiia nauk Belorusskoĭ SSR. Institut istorii.
Bibliography: p. 144–⌊157⌋
1.40rub
1. Polotsk region, White Russia—Antiquities. 2. White Russia—Antiquities. 3. Cities and towns—White Russia. I. Darkevich, Vladislav Petrovich. II. Title.
DK511.P67S55 79-412210

—WISCONSIN—CASE STUDIES

Hollingsworth, Joseph Rogers, 1932-
Dimensions in urban history : historical and social science perspectives on middle-size American cities / J. Rogers Hollingsworth and Ellen Jane Hollingsworth. — Madison : University of Wisconsin Press, 1979.
viii, 184 p. ; 24 cm.
Includes index.
ISBN 0-299-07820-5 : $19.50
1. Cities and towns—United States—History. 2. Cities and towns—Wisconsin—Case studies. 3. United States—Social conditions—1865-1918. 4. Urbanization—United States—History. 5. Sociology, Urban—United States. I. Hollingsworth, Ellen Jane, joint author. II. Title.
HT123.H65 301.36'0973 78-65011
79 MARC

—YUGOSLAVIA

Josip Broz Tito : 1892—1937—1977 / ⌊glavni i odgovorni urednik Branko Nedović⌋. — ⌊Beograd⌋ : Stalna konferencija gradova Jugoslavije, 1977.
288 p. : ill. ; 32 cm. — (Komuna ; g. 24, dvobroj 4/5) Yu***
Cover title: Tito, prvi i najdraži sugrađanin.
1. Tito, Josip Broz, Pres. Yugoslavia, 1892- . 2. Yugoslavia—Presidents—Biography. 3. Cities and towns—Yugoslavia. I. Nedović, Branko. II. Title: Tito, prvi i najdraži sugrađanin. III. Series.
HT101.K6 g. 24, dvobroj 4/5 78-391347
⌊DR359.T5⌋ 79 MARC

—YUGOSLAVIA—HISTORY

Grebenshchikov, Oleg Sergeevich.
(Zhemchuzhiny IUgoslavii)
Жемчужины Югославии : Этюды о городах / О. С. Гребенщиков. — Москва : Мысль, 1978.
155 p., ⌊4⌋ leaves of plates : ill. (some col.) ; 20 cm. USSR 78
Bibliography: p. 153–⌊154⌋
0.70rub
1. Yugoslavia—History. 2. Cities and towns—Yugoslavia—History. I. Title.
DR317.G69 79-411186

—YUGOSLAVIA—MACEDONIA (FEDERATED REPUBLIC, 1945-)

Čipan, Boris.
Makedonskite gradovi vo XIX vek i nivnata urbana perspektiva / Boris Čipan. — Skopje : Makedonska akademija na naukite i umetnostite, 1978.
133 p. : ill. ; 24 cm. Yu***
Summary in English.
Bibliography: p. 131-133.
1. Cities and towns—Yugoslavia—Macedonia (Federated Republic, 1945-) I. Title.
HT145.Y8C56 79-364518
79 MARC

—YUGOSLAVIA—SLOVENIA—STATISTICS

Slovenia (Federated Republic, 1945-). Zavod za statistiko.
Seznam sprememb naselij SR Slovenije od leta 1948 do 1972. (Preimenovanja, pristavki, združitve, razdružitve in razglasitve). Priredil Peter Cerovac. <Franta Komel: Uvod>. Ljubljana, Zavod SR Slovenije za statistiko, 1973.
99 p. 28 cm. (Its Statistično gradivo, št. 5/73) Yu73-3799
1. Cities and towns—Yugoslavia—Slovenia—Statistics. 2. Slovenia—Administrative and political divisions. I. Cerovac, Peter, ed. II. Title. III. Series: Slovenia (Federated Republic, 1945-). Zavod za statistiko. Statistično gradivo SR Slovenije, št. 5/73.
HT145.Y8S57 1973 73-979618
79 MARC

CITIES AND TOWNS, ANCIENT

—ADDRESSES, ESSAYS, LECTURES

Civilization : readings from Scientific American / with introductions by Brian M. Fagan. — San Francisco : W. H. Freeman, c1979.
158 p. : ill. (some col.) ; 30 cm.
CONTENTS: Adams. The origin of cities.—Kramer. The Sumerians.—Lamberg-Karlovsky and Lamberg-Karlovsky. An early city in Iran.—Millon. Teotihuacán.—Maddin, Muhly, and Wheeler. How the Iron Age began.—Piggott. The beginnings of wheeled transport.—Humphrey and Pedley. Roman Carthage.—Biddle. The archaeology of Winchester.—Hammond. The planning of a Maya ceremonial center.—Sjoberg. The origin and evolution of cities.—Brown. Human food production as a process in the biosphere.—Nef. An early energy crisis and its consequences.—Davis. The migrations of human populations.
Bibliography: p. ⌊151⌋-153.
Includes index.
ISBN 0-7167-1024-2. ISBN 0-7167-1023-4 pbk.
1. Civilization, Ancient—Addresses, essays, lectures. 2. Cities and towns, Ancient—Addresses, essays, lectures. I. Fagan, Brian M. II. Scientific American.
CB311.C68 909 78-15780
78 MARC

Gottes Stadt und unsere Städte : d. urbane Herausforderung d. Christentums : 4 Reden / Christof Bäumler ... ⌊et al.⌋. — 1. Aufl. — München : Kaiser, 1978.
60 p. ; 21 cm. — (Kaiser Traktate ; 34) GFR78-A
ISBN 3-549-01174-2
1. Cities and towns, Ancient—Addresses, essays, lectures. I. Bäumler, Christof, 1927-
HT114.G67 301.36'3'093 79-349284
79 MARC

—COLLECTED WORKS

(Uzumoreta kodai toshi)
埋もれた古代都市 / 森本哲郎編 ; ⌊編集 NHKサービスセンター⌋. — 東京 : 集英社, 1978-
v. : ill. ; 22 cm. — (NHK文化シリーズ : 歴史と文明)
Ja 78-23377 (v. 1)
"昭和五十二年十一月から十二月末まで、八回にわたって、NHK教育テレビで放映された「歴史と文明<埋もれた古代都市>」を、新しく編集したもの"
¥1300 (v. 1)
1. Cities and towns, Ancient—Collected works. 2. Cities and towns, Ruined, extinct, etc.—Collected works. 3. Civilization, Ancient—Collected works. I. Morimoto, Tetsurō, 1925- II. NHK Sābisu Sentā. III. Rekishi to bummei (Television program) IV. Series: NHK bunka shirizu : Rekishi to bummei.
D62.U97 79-801538

—CONGRESSES

Thèmes de recherches sur les villes antiques d'Occident : ⌊actes du colloque⌋, Strasbourg, 1er-4 octobre 1971 / publié sous la direction de P.-M. Duval et Ed. Frézouls. — Paris : Éditions du Centre national de la recherche scientifique, 1977.
429 p., ⌊1⌋ fold. leaf of plates : ill. ; 28 cm. — (Colloques internationaux du Centre national de la recherche scientifique ; no 542) F***
French, English, Italian, or German.
Held at the Université des sciences humaines de Strasbourg.
Includes bibliographical references and indexes.
ISBN 2-222-01595-2 : 195F
1. Cities and towns, Ancient—Congresses. I. Duval, Paul Marie. II. Frézouls, Edmond. III. Université des sciences humaines de Strasbourg. IV. Series: France. Centre national de la recherche scientifique. Colloques internationaux ; no 542.
HT114.T53 301.36'3'093 78-376530
78 MARC

—GREAT BRITAIN—BIBLIOGRAPHY

Stamas, Joan D
Roman towns in Britain : a bibliography / Joan Stamas. — Monticello, Ill. : Council of Planning Librarians, 1978.
21 p. ; 28 cm. — (Exchange bibliography - Council of Planning Librarians ; 1473)
Cover title.
$2.00
1. Great Britain—Antiquities, Roman—Bibliography. 2. Romans—Great Britain—Bibliography. 3. Great Britain—History—Roman period, 55 B.C.-449 A.D.—Bibliography. 4. Cities and towns, Ancient—Great Britain—Bibliography. I. Title. II. Series: Council of Planning Librarians. Exchange bibliography ; 1473.
Z5942.C68 no. 1473 016.3092 s 78-103355
⌊DA145⌋ 79 MARC

—JAPAN—NARA (PREFECTURE)—HISTORY

Nakamura, Haruhisa.
(Nikkan kodai toshi keikaku)
日韓古代都市計画 / 中村春壽. — 東京 : 六興出版, 1978.
213 p., ⌊2⌋ leaves of plates : ill. ; 19 cm. — (ロッコウブックス)
Ja 78-22365
Series romanized: Rokkō bukkusu.
¥980
1. Cities and towns, Ancient—Korea—History. 2. Cities and towns, Ancient—Japan—Nara (Prefecture)—History. 3. City planning—East Asia—History. I. Title.
HT147.K6N34 78-816004

—JORDAN

Segal, Arthur.
(Tikhnunan shel he-'arim le-orekh Via Traiana Nova ba-tekufah ha-romit)
תיכנונן של הערים לאורך Via Traiana Nova בתקופה הרו־מית / מאת ארתור סגל. — ⌊s. l. : s. n., 1975?⌋
2 v. : ill. ; 28 cm.
Added t. p.: The planning of the cities along the Via Traiana Nova in the Roman period.
Thesis—ha-Universiṭah ha-'Ivrit.
Summary in English.
Bibliography: v. 1, leaves 158—161.
1. City planning—Jordan. 2. Architecture, Ancient—Jordan. 3. Cities and towns, Ancient—Jordan. 4. Jordan—Antiquities. I. Title. II. Title: The planning of the cities along the Via Traiana Nova in the Roman period. III. Title: Via Traiana Nova.
NA9246.J6S43 711'.4'09394 79-951378

—KOREA—HISTORY

Nakamura, Haruhisa.
(Nikkan kodai toshi keikaku)
日韓古代都市計画 / 中村春壽. — 東京 : 六興出版, 1978.
213 p., ⌊2⌋ leaves of plates : ill. ; 19 cm. — (ロッコウブックス)
Ja 78-22365
Series romanized: Rokkō bukkusu.
¥980
1. Cities and towns, Ancient—Korea—History. 2. Cities and towns, Ancient—Japan—Nara (Prefecture)—History. 3. City planning—East Asia—History. I. Title.
HT147.K6N34 78-816004

CITIES AND TOWNS, ISLAMIC

—IRAN

Gaube, Heinz.
Iranian cities / Heinz Gaube. — New York : New York University Press, c1979, c1978.
xviii, 132 p. : ill. ; 28 cm. — (Hagop Kevorkian series on Near Eastern art and civilization)
Includes bibliographical references.
ISBN 0-8147-2971-1 : $22.50
1. Cities and towns—Iran. 2. Cities and towns, Islamic—Iran. 3. Herat, Afghanistan. 4. Iran—Social conditions.
HT147.I6G38 1979 301.36'3'0955 77-25748
78 MARC

CITIES AND TOWNS, MEDIEVAL

Bertelli, Sergio.
Il potere oligarchico nello Stato-città medievale / Sergio Bertelli. — 1. ed. — Firenze : La nuova Italia, 1978.
174 p. : ill. ; 20 cm. — (Storia) (Strumenti ; 88) It78-July
Bibliography: p. 169-171.
L3000
1. Cities and towns, Medieval. 2. Oligarchy. I. Title.
D134.B42 945'.04 78-395752
*79 MARC

Germany (Federal Republic, 1949-). Deutsche UNESCO-Kommission. Arbeitskreis Historische Stadtkerne.
The renewal of historic town centres in nine European countries : France, Great Britain, Italy, the Netherlands, Poland, short reports, Austria, CSSR, Hungary, Switzerland. — ⌊Bonn : Federal German Ministry for Regional Planning, Building, and Urban Development⌋, 1975.
325 p. : ill. ; 30 cm. — (Set of works "Town planning research" of the Federal German Ministry for Regional Planning, Building, and Urban Development ; 02.002) GFR***
"On behalf of the Federal Ministry for Regional Planning, Building, and Urban Development, Bonn-Bad Godesberg by the Working Party on 'Historic Town Centres' of the German Commission for UNESCO."
Includes bibliographies.
1. Urban renewal—Europe. 2. Cities and towns, Medieval. I. Title. II. Series: Schriftenreihe Stadtentwicklung ; 002.
HT178.E8G33 1975 309.2'62'094 79-303765
79 MARC

Mundy, John Hine, 1917-
The medieval town / John H. Mundy and Peter Riesenberg. — Huntington, N.Y. : Krieger, 1979, c1958.
190 p. ; 19 cm. — (The Anvil series ISSN 0570-1062)
Reprint of the ed. published by Van Nostrand, Princeton, N.J., which was issued as no. 30 of An Anvil original.
Bibliography: p. 185-187.
Includes index.
ISBN 0-88275-906-X : $4.95
1. Cities and towns, Medieval. I. Riesenberg, Peter N., 1925- joint author. II. Title. III. Series.
⌊HT115.M8 1979⌋ 301.36'3'094 79-9718
79 MARC

Viollet, Paul Marie, 1840-1914.
Les communes françaises au Moyen Age / Paul Viollet. — Genève : Slatkine-Megariotis Reprints, 1977.
158 p. ; 22 cm. Sw77-6461
Reprint of the 1900 ed., published by Klincksieck, Paris.
Pages also numbered 346-502, the pages of v. 36 of the Mémoires de l'Académie des inscriptions et belles-lettres from which the original ed. was reprinted.
Includes bibliographical references.
40.00F
1. Cities and towns, Medieval. 2. Municipal government—France—History. I. Title.
JS4831.C7V5 1977 352'.008'0944 78-390542
*79 MARC

—ADDRESSES, ESSAYS, LECTURES

Thrupp, Sylvia Lettice, 1903-
Society and history : essays / by Sylvia L. Thrupp ; edited by Raymond Grew and Nicholas H. Steneck. — Ann Arbor : University of Michigan Press, c1977.
363 p. ; 24 cm.
Chronological list of the author's works: p. 347-349.
Includes bibliographical references and index.
ISBN 0-472-08880-7
1. Cities and towns, Medieval—Addresses, essays, lectures. 2. Civilization, Medieval—Addresses, essays, lectures. 3. Social sciences and history—Addresses, essays, lectures. I. Title.
HT115.T48 1977 301.36'3'094 75-31056
79 MARC

—HISTORY

Hall, Thomas, 1939-
Mittelalterliche Stadtgrundrisse : Versuch einer Übersicht der Entwicklung in Deutschland und Frankreich / von Thomas Hall. — Stockholm : Vitterhets-, historie- o. antikvitetsakad. : Almqvist & Wiksell international, 1978.
⌊4⌋, 160 p. : ill., maps ; 25 cm. — (Antikvariskt arkiv ; 66 ISSN 0083-6737) S78-18
Bibliography: p. 146-154.
Includes index.
ISBN 9174020587
1. Cities and towns, Medieval—History. 2. Cities and towns, Medieval—Germany—History. 3. Cities and towns, Medieval—France—History. I. Title. II. Series: Vitterhets-, historie- och antikvitetsakademien, Stockholm. Antikvariskt arkiv ; 66.
DL601.V795 nr. 66 79-356168
⌊HT115⌋ 79 MARC

—MAPS

Campbell, James.
Historic towns, Norwich / James Campbell ; maps compiled by W. H. Johns. — London : Scolar Press, c1975.
25 p. : 7 col. maps (1 fold. in pocket) ; 41 cm. GB75-31413
Published by the Scolar Press in conjunction with the Historic Towns Trust.
ISBN 0-85967-184-4 : £5.75
1. Norwich, Eng.—Maps. 2. Norwich, Eng.—History. 3. Cities and towns, Medieval—Maps. I. Johns, William Henry. II. Historic Towns Trust. III. Title.
G1819.N5C3 1975 912'.426'15 78-318527
78 MARC

—ASIA, CENTRAL

Nagrodzka-Majchrzyk, Teresa.
Geneza miast u dawnych ludów tureckich, VII-XII w. / Teresa Nagrodzka-Majchrzyk. — Wrocław : Zakład Narodowy im. Ossolińskich, 1978.
168 p. : ill. ; 24 cm. P***
At head of title: Polska Akademia Nauk. Instytut Historii Kultury Materialnej.
Summary in French.
Bibliography: p. ⌊150⌋-156.
Includes indexes.
zł40.00
1. Turks—History. 2. Cities and towns, Medieval—Asia, Central. 3. Asia, Central—History. I. Title.
DS26.N33 78-395865
79 MARC

—EUROPE

Ennen, Edith.
Die europäische Stadt des Mittelalters / Edith Ennen. — 3. überarb. u. erw. Aufl. — Göttingen : Vandenhoeck und Ruprecht, 1979.
331 p. : ill., maps ; 21 cm. — (Sammlung Vandenhoeck) GFR79-A
Bibliography: p. 271-321.
Includes index.
ISBN 3-525-01320-5
1. Cities and towns—Europe—History. 2. Cities and towns, Medieval—Europe. I. Title.
HT131.E56 1979 79-371138
79 MARC

Ennen, Edith.
The Medieval town / by Edith Ennen ; translated by Natalie Fryde. — Amsterdam ; New York : North-Holland Pub. Co. ; New York : distributors for the U.S.A. and Canada Elsevier North-Holland, 1979.
287 p., ⌊1⌋ leaf of plates : ill. 23 cm. — (Europe in the Middle Ages ; v. 15)
Translation of Die europäische Stadt des Mittelalters.
Bibliography: p. ⌊215⌋-272.
Includes index.
ISBN 0-444-85133-X
1. Cities and towns—Europe—History. 2. Cities and towns, Medieval—Europe. I. Title. II. Series.
HT131.E5613 301.36'3'094 79-1362
79 MARC

—EUROPE—ADDRESSES, ESSAYS, LECTURES

Altständisches Bürgertum : 2 Bde. / hrsg. von Heinz Stoob. — Darmstadt : Wissenschaftliche Buchgesellschaft, ⌊Abt. Verl.⌋, 1978.
2 v. ; 20 cm. — (Wege der Forschung ; Bd. 352, 417) GFR78-A
Includes bibliographical references and indexes.
CONTENTS: Bd. 1. Herrschaft und Gemeinverfassung.—Bd. 2. Erwerbsleben und Sozialgefüge.
ISBN 3-534-05715-5 (v. 1) : DM74.00 (v. 1)
1. Cities and towns, Medieval—Europe—Addresses, essays, lectures. I. Stoob, Heinz.
HT115.A44 78-380753
79 MARC

—EUROPE—HISTORY

Schuerl, Wolfgang F
Medieval castles and cities / text by Wolfgang F. Schuerl ; translated by Francisca Garvie. — London : Cassell, 1978.
189 p. : ill. (chiefly col.), plans ; 34 cm. — (Monuments of civilization)
Translation of Burgen und Städte des Mittelalters.
ISBN 0-304-30192-2 : £9.95
1. Europe—History—476-1492. 2. Middle Ages—History. 3. Castles—Europe—History. 4. Cities and towns, Medieval—Europe—History. I. Title. II. Series.
D117.S3713 1978 940.1 79-300375
79 MARC

—EUROPE, EASTERN—CONGRESSES

Powszechny Zjazd Historyków Polskich, 11th, Toruń, Poland, 1974.
Miasta doby feudalnej w Europie środkowowschodniej : przemiany społeczne a układy przestrzenne : prace XI Powszechnego Zjazdu Historyków Polskich, Toruń, wrzesień, 1974 r. : ⌊materiały z kolokwium Komisji Historii Miast KNH i Komisji Mediewistycznej PTH⌋ / pod red. Aleksandra Gieysztora i Tadeusza Rosłanowskiego. — Warszawa : Państwowe Wydawn. Naukowe, 1976.
325 p. : ill. ; 21 cm. P77-1130
At head of title: Towarzystwo Naukowe w Toruniu.
Includes bibliographical references.
zł54.00
1. Cities and towns, Medieval—Europe, Eastern—Congresses. I. Gieysztor, Aleksander. II. Rosłanowski, Tadeusz. III. Towarzystwo Naukowe w Toruniu.
HT131.P68 1974 78-367294
78 MARC

—FRANCE

L'Atlas des villes et villages fortifiés en France (Moyen Age) / réalisé par Charles-Laurent Salch, avec la collaboration de Joëlle Burnouf ; préf. de Philippe Dollinger ; cartes de Maurice Frey ; plans par Claire Chantelouve et Nicolas du Saulcy ; photos de Dominique Martinez. — Strasbourg : Publitotal, 1978.
495 p. : ill. (some col.) ; 31 cm. F***
Errata slip inserted.
Bibliography: p. 489-491.
Includes indexes.
295F
1. Cities and towns, Medieval—France. 2. City walls—France. 3. France—Description and travel—1975- . I. Salch, Charles Laurent. II. Burnouf, Joëlle.
DC20.A78 944'.02 79-346280
79 MARC

—FRANCE—HISTORY

Hall, Thomas, 1939-
Mittelalterliche Stadtgrundrisse : Versuch einer Übersicht der Entwicklung in Deutschland und Frankreich / von Thomas Hall. — Stockholm : Vitterhets-, historie- o. antikvitetsakad. : Almqvist & Wiksell international, 1978.
⌊4⌋, 160 p. : ill., maps ; 25 cm. — (Antikvariskt arkiv ; 66 ISSN 0083-6737) S78-18
Bibliography: p. 146-154.
Includes index.
ISBN 9174020587
1. Cities and towns, Medieval—History. 2. Cities and towns, Medieval—Germany—History. 3. Cities and towns, Medieval—France—History. I. Title. II. Series: Vitterhets-, historie- och antikvitetsakademien, Stockholm. Antikvariskt arkiv ; 66.
DL601.V795 nr. 66 79-356168
⌊HT115⌋ 79 MARC

—GERMANY—HISTORY

Hall, Thomas, 1939-
Mittelalterliche Stadtgrundrisse : Versuch einer Übersicht der Entwicklung in Deutschland und Frankreich / von Thomas Hall. — Stockholm : Vitterhets-, historie- o. antikvitetsakad. : Almqvist & Wiksell international, 1978.
⌊4⌋, 160 p. : ill., maps ; 25 cm. — (Antikvariskt arkiv ; 66 ISSN 0083-6737) S78-18
Bibliography: p. 146-154.
Includes index.
ISBN 9174020587
1. Cities and towns, Medieval—History. 2. Cities and towns, Medieval—Germany—History. 3. Cities and towns, Medieval—France—History. I. Title. II. Series: Vitterhets-, historie- och antikvitetsakademien, Stockholm. Antikvariskt arkiv ; 66.
DL601.V795 nr. 66 79-356168
⌊HT115⌋ 79 MARC

CITIES AND TOWNS, MEDIEVAL (Continued)

—ITALY—HISTORY

Piccinato, Luigi.
Urbanistica medievale / Luigi Piccinato. — Bari : Dedalo libri, 1978.
91 p. : ill. ; 18 cm. — (Universale di architettura ; 6) It78-June
Includes bibliographical references.
L2500
1. Cities and towns, Medieval—Italy. 2. City planning—Italy—History. I. Title. II. Series.
HT115.P46 78-378485
•78 MARC

Waley, Daniel Philip.
The Italian city-republics / Daniel Waley. — 2d ed. — London ; New York : Longman, 1978.
xi, 163 p. : ill. ; 22 cm. — (A Longman paperback)
Bibliography: p. [157]-160.
Includes index.
ISBN 0-582-48698-X : £2.95
1. Italy—History—1268-1492. 2. Cities and towns, Medieval—Italy—History. I. Title.
DG530.W34 1978 945'.05 77-21757
77 MARC

—ITALY—SICILY—HISTORY

Peri, Illuminato.
Uomini, città e campagne in Sicilia dall'XI al XIII secolo / Illuminato Peri. — Roma ; Bari : Laterza, 1978.
358 p. ; 21 cm. — (Collezione storica) It79-Feb
Includes bibliographical references and indexes.
L12000
1. Sicily—History—1016-1194. 2. Sicily—History—1194-1282. 3. Cities and towns, Medieval—Italy—Sicily—History. 4. Social classes—Italy—Sicily—History. I. Title.
DG867.21.P47 945.8'03 79-373228
•79 MARC

CITIES AND TOWNS, MOVEMENT TO see Cities and towns; Rural-urban migration; Urbanization

CITIES AND TOWNS, MUSLIM see Cities and towns, Islamic

CITIES AND TOWNS, RENAISSANCE

—HISTORY

Arbore, Grigore.
Cetatea ideală în viziunea Renașterii : eseu asupra tipologiei formelor urbane / Grigore Arbore. — București : "Meridiane", 1978.
181 p., [20] leaves of plates : ill. ; 17 cm. — (Curente și sinteze ; 21) R79-1168
Includes bibliographical references.
lei11.00
1. Cities and towns, Renaissance—History. 2. City planning—History. I. Title.
NA9094.3.A72 79-378713
•79 MARC

CITIES AND TOWNS, RUINED, EXTINCT, ETC.

see also Excavations (Archaeology); and also names of individual ruined or extinct cities and towns

Jankuhn, Herbert, 1905-
Einführung in die Siedlungsarchäologie / Herbert Jankuhn. — Berlin ; New York : De Gruyter, 1977.
xiii, 202 p. : graphs, maps ; 21 cm. — (De Gruyter Studienbuch) GFR77-A23
Bibliography: p. [195]-202.
ISBN 3-11-004752-7 : DM38.00
1. Archaeology—Methodology. 2. Excavations (Archaeology) 3. Land settlement. 4. Cities and towns, Ruined, extinct, etc. I. Title.
CC75.J28 78-379075
78 MARC

—COLLECTED WORKS

(Uzumoreta kodai toshi)
埋もれた古代都市 / 森本哲郎編 ; [編集 NHKサービスセンター]. — 東京 : 集英社, 1978-
v. : ill. ; 22 cm. — (NHK文化シリーズ : 歴史と文明)
Ja 78-23377 (v. 1)
"昭和五十二年十一月から十二月末まで、八回にわたって、NHK教育テレビで放映された「歴史と文明〈埋もれた古代都市〉」を、新しく編集したもの"
¥1300 (v. 1)
1. Cities and towns, Ancient—Collected works. 2. Cities and towns, Ruined, extinct, etc.—Collected works. 3. Civilization, Ancient—Collected works. I. Morimoto, Tetsurō, 1925- II. NHK Sābisu Sentā. III. Rekishi to bummei (Television program) IV. Series: NHK bunka shirizu : Rekishi to bummei.
D62.U97 79-801538

—AISA—ADDRESSES, ESSAYS, LECTURES

(Ajia bummei no michi)
アジア文明の道 / 森本哲郎編 ; [編集 NHKサービスセンター]. — 東京 : 集英社, 1978.
247 p. : ill. ; 22 cm. — (埋もれた古代都市 ; 1) (NHK文化シリーズ : 歴史と文明) Ja 78-23377
"NHK教育テレビで放映された「歴史と文明〈埋もれた古代都市〉」... [の中の]「殷の巨大王墓」「幻の都・長岡京」に、新しく「楼蘭の悲史」を加えた。"
¥1300
1. Cities and towns, Ruined extinct, etc.—Asia—Addresses, essays, lectures. 2. Asia—Civilization—Addresses, essays, lectures. 3. Civilization, Ancient—Addresses, essays, lectures. I. Morimoto, Tetsurō, 1925- II. NHK Sābisu Sentā. III. Rekishi to bummei (Television program) IV. Series: Uzumoreta kodai toshi ; 1. V. Series: NHK bunka shirizu : Rekishi to bummei.
D62.U97 vol. 1 79-801487
[DS62.2]

—ALGERIA

Bouchenaki, Mounir.
Cités antiques d'Algérie / Mounir Bouchenaki. — Alger : Ministère de l'information et de la culture : diffusion, SNED, 1978.
109 p. : ill. ; 32 cm. — (Collection "Art et culture" ; 12)
Bibliography: p. 105.
ISBN 8439979045
1. Algeria—Antiquities. 2. Cities and towns, Ruined, extinct, etc.—Algeria. I. Title. II. Series.
DT281.B66 965 79-101609
79 MARC

—BLACK HILLS, S.D. AND WYO.

Miller, Donald C 1933-
Ghosts of the Black Hills / by Donald C. Miller. — Missoula, Mont. : Pictorial Histories Pub. Co., 1979.
70 p. : ill. ; 15 x 23 cm.
Bibliography: p. 67-68.
Includes index.
ISBN 0-933126-07-7 : $3.95
1. Black Hills, S.D. and Wyo.—History, Local. 2. Cities and towns, Ruined, extinct, etc.—Black Hills, S.D. and Wyo. 3. Black Hills, S.D. and Wyo.—Gold discoveries. I. Title.
F657.B6M55 978.3'9 79-87462
79 MARC

—BULGARIA

Sofia. Gradski naroden sŭvet. Otdel Kompleksno administrativno obsluzhvane.
Spravochnik na zalichenite i preimenuvani selishta v Narodna ripublika [i.e. republika] Bŭlgariia. — Sofiia : Sofiiski gradski naroden sŭvet, 1978.
121 p. ; 21 cm. Bu***
1. Cities and towns, Ruined, extinct, etc.—Bulgaria. 2. Names, Geographical—Bulgaria. 3. Bulgaria—History, Local. I. Title.
DR98.A1S63 1978 79-381705

—CALIFORNIA—HISTORY

Miller, Donald C 1933-
Ghost towns of California / by Donald C. Miller. — 1st ed. — Boulder, Colo. : Pruett Pub. Co., c1978.
202 p. : ill. ; 29 cm.
Includes bibliographical references and index.
ISBN 0-87108-517-8
1. Cities and towns, Ruined, extinct, etc.—California—History. 2. California—History, Local. I. Title.
F861.M56 979.4 78-7788
78 MARC

—CALIFORNIA—MAPS

Johnson, Robert Neil.
California-Nevada ghost town atlas / by Robert Neil Johnson. — 8th ed. — Susanville, Calif. : C. Johnson, 1974.
48 p. : ill., col. maps ; 23 cm.
"Copyright 1967 by H. Cyril Johnson."
$2.00
1. Cities and towns, Ruined, extinct, etc.—California—Maps. 2. Cities and towns, Ruined, extinct, etc.—Nevada—Maps. I. Johnson, Humphrey Cyril, 1905- II. Title.
G1526.A1J6 1974 912'.793 78-324699
79 MARC

—COLORADO—HISTORY

Shaffer, Ray, 1929-
A guide to places on the Colorado prairie, 1540-1975 / by Ray Shaffer. — 1st ed. — Boulder, Colo. : Pruett Pub. Co., c1978.
386 p. : ill. ; 23 cm.
Errata slip inserted.
Bibliography: p. 365-369.
Includes index.
ISBN 0-87108-513-5
1. Cities and towns, Ruined, extinct, etc.—Colorado—History. 2. Colorado—History, Local. I. Title.
F776.S5 917.88 78-1741
78 MARC

—EUROPE—HISTORY—CONGRESSES

European towns : their archaeology and early history / edited by M. W. Barley. — London ; New York : Published for the Council for British Archaeology by Academic Press, 1977.
xxvii, 523 p. : ill. ; 27 cm. GB***
Papers presented at a conference on European towns held at Oxford, 1975.
Summaries in French and German.
Includes bibliographies and index.
ISBN 0-12-078850-0
1. Europe—Antiquities—Congresses. 2. Cities and towns, Ruined, extinct, etc.—Europe—History—Congresses. 3. Cities and towns—Europe—History—Congresses. I. Barley, Maurice Willmore, 1909- II. Council for British Archaeology.
CC175.E94 301.36'1'0936 77-71807
78[r79]rev MARC

—FLORIDA

Warnke, James R
Ghost towns & side roads of Florida / [Warnke]. — Combined and updated ed. / by Roving Photographers & Associates, inc. — [Boynton Beach, FL. : Roving Photographers & Assoc., c1978]
114, 90 p. : ill. ; 21 cm.
Cover title.
A revision of the author's Ghost towns of Florida and Side roads of Florida which were originally published separately in 1971 and 1973.
$5.95
1. Florida—Description and travel—1951- —Guide-books. 2. Cities and towns, Ruined, extinct, etc.—Florida. 3. Automobiles—Road guides—Florida. I. Warnke, James R. Ghost towns of Florida. II. Warnke, James R. Side roads of Florida. III. Roving Photographers & Associates. IV. Title.
F309.3.W36 917.59'04'6 78-111273
79 MARC

—IDAHO

Miller, Donald C 1933-
Ghost towns of Idaho / by Donald C. Miller. — 1st ed. — Boulder, Colo. : Pruett Pub. Co., c1976.
ix, 102 p. : ill. ; 29 cm.
Includes bibliographical references and index.
ISBN 0-87108-205-5
1. Cities and towns, Ruined, extinct, etc.—Idaho. 2. Idaho—History, Local. 3. Mines and mining—Idaho—History. I. Title.
F746.M5 979.6 76-42297
76[r79]rev MARC

CITIES AND TOWNS, RUINED, EXTINCT, ETC. (Continued)

—JAPAN

Murai, Yasuhiko, 1930-
(Nihon no kyūto)
日本の宮都：古代都市の原像 / 村井康彦. — 東京：角川書店, 1978.
230 p. : ill. ; 21 cm. — (季刊論叢日本文化 ; 9) Ja 78-24464
¥1900

1. Japan—Civilization—To 794. 2. Japan—Civilization—794-1185. 3. Japan—Capital and capitol. 4. Cities and towns, Ruined, extinct, etc—Japan. I. Title. II. Series: Kikan sōron Nihon bunka ; 9.
DS855.M86 79-800970

—KANSAS

Fitzgerald, Daniel.
Ghost towns of Kansas / by Daniel Fitzgerald. — [s.l. : s.n.], c1976-c1979.
2 v. : ill. ; 29 cm.
Vol. 2. published by D. Fitzgerald, Topeka, Kan.
Includes bibliographical references and indexes.
1. Cities and towns, Ruined, extinct, etc.—Kansas. 2. Kansas—History, Local. I. Title.
F681.F57 978.1'03 77-150780
77[r79]rev MARC

—MONTANA

Miller, Donald C 1933-
Ghost towns of Montana / by Don C. Miller. — 1st ed. — Boulder, Colo. : Pruett Pub. Co., 1974.
xiii, 177 p. : ill. ; 29 cm.
Includes bibliographical references.
ISBN 0-87108-070-2
1. Cities and towns, Ruined, extinct, etc.—Montana. 2. Montana—History. I. Title.
F731.M62 978.6 72-95496
74[r79]rev MARC

—NEAR EAST

Tweedie, William King, 1803-1863.
Ruined cities of the East / by Rev. Dr. Tweedie. — London New York : T. Nelson, 1865.
193 p., [8] leaves of plates : ill. ; 17 cm.
Includes bibliographical references.
1. Cities and towns, Ruined, extinct, etc—Near East. 2. Near East—Antiquities. I. Title.
DS56.T86 1865 939 79-306077
79 MARC

—NEVADA—MAPS

Johnson, Robert Neil.
California-Nevada ghost town atlas / by Robert Neil Johnson. — 8th ed. — Susanville, Calif. : C. Johnson, 1974.
48 p. : ill., col. maps ; 23 cm.
"Copyright 1967 by H. Cyril Johnson."
$2.00
1. Cities and towns, Ruined, extinct, etc.—California—Maps. 2. Cities and towns, Ruined, extinct, etc.—Nevada—Maps. I. Johnson, Humphrey Cyril, 1905- . II. Title.
G1526.A1J6 1974 912'.793 78-324699
79 MARC

—NEW MEXICO

Woods, Dora Elizabeth Ahern, 1904-
Ghost towns and how to get to them / Betty Woods. — [Rev. ed.]. — Santa Fe, N.M. : Sunstone Press, 1978.
30 p. : map ; 15 cm.
ISBN 0-913270-30-X : $2.95
1. Cities and towns, Ruined, extinct, etc.—New Mexico. 2. New Mexico—History, Local. I. Title.
F801.W6 1978 978.9'04 77-78518
77 MARC

—ONTARIO

Brown, Ron, 1945-
Ghost towns of Ontario / Ron Brown. — Langley, B.C. : Stagecoach, c1978-
v. : ill. ; 24 cm. C***
Bibliography: v. 1, p. 194-197.
Includes index.
CONTENTS: v. 1. Southern Ontario.
ISBN 0-88983-081-5 (v. 1) : $11.95 (v. 1). ISBN 0-88983-020-7 pbk. (v. 1)
1. Cities and towns, Ruined, extinct, etc.—Ontario. 2. Ontario—History, Local. I. Title.
F1058.B76 971.3'03 78-326817
79 MARC

—RHODESIA, SOUTHERN

see also Zimbabwe, Southern Rhodesia

—THE WEST—GUIDE-BOOKS

Moore, Thomas W 1921-
The ghost town automobile atlas : being a glove compartment guide to Western ghost camps, sites, and shadow towns / by Thomas W. Moore. — 1st ed. — Camino, CA : Pony Press, c1979.
ii, 46 p. (p 45-46 blank for "Notes") : ill. ; 22 cm.
1. Cities and towns, Ruined, extinct, etc.—The West—Guide-books. 2. The West—Description and travel—1951- —Guide-books. I. Title.
F595.2.M58 917.9 79-88196
79 MARC

CITIES AND TOWNS (IN RELIGION, FOLKLORE, ETC.)

—UNITED STATES

Laba, Martin.
Urban folklore. [Sound recording] De Land, Fla., Everett/Edwards, p1979. 1654.
1 cassette. (American folklore)
Lecture, read by the author.
Duration: 24 min.
SUMMARY. Explains that the consequences of urban life on the modes and methods of expression and the very nature of that expression are the prime concerns in the study of urban folklore. Emphasizes that city life involves more than the survival of traditional items, and that urban folklore exists because of the city, not in spite of it.
1. Cities and towns (in religion, folklore, etc.)—United States. 2. Folk-lore—United States. 3. City and town life—United States. I. Title. II. Series: American folklore. [Sound recording]
[GR105] 79-740127

CITIES AND TOWNS IN ART

Childs, William A P 1942-
The city-reliefs of Lycia / William A. P. Childs. — Princeton, N.J. : Princeton University Press, c1978.
xx, 114 p., [36] leaves of plates : ill. ; 26 cm. — (Princeton monographs in art and archaeology ; 42)
Includes bibliographical references and index.
ISBN 0-691-03544-7 : $27.50
1. Relief (Sculpture)—Turkey—Lycia. 2. Cities and towns in art. 3. Sepulchral monuments—Turkey—Lycia. 4. Lycia. I. Title. II. Series.
NB130.L9C48 732'.9'3928 78-51159
78 MARC

Jöckle, Clemens.
Friedrich Jossé : Bilder aus d. Leben e. Kleinstadt / Clemens Jöckle. — Speyer : Bezirksgruppe Speyer d. Histor. Vereins d. Pfalz, 1978.
146 p. : numerous ill. (some col.) ; 21 cm. — (Beiträge zur Speyerer Stadtgeschichte ; Heft 5) GFR79-B1
Includes bibliographical references.
1. Jossé, Friedrich, 1897- . 2. Cities and towns in art. 3. Germany—Social life and customs—Caricatures and cartoons. I. Jossé, Friedrich, 1897- . II. Series.
N6888.J65J63 79-365717
79 MARC

Pieck, Anton, 1895-
Iedere stad een toren. — Den Haag : Omniboek, [1978]
44 p. : with col. ill. ; 22 x 27 cm. — (De Wereld van Anton Pieck) Ne78-12
ISBN 9062074286 : fl 15.50
1. Pieck, Anton, 1895- . 2. Cities in art. 3. Towers in art. I. Title.
NC983.5.P53A4 1978 78-384214
*78 MARC

Vaglieri, Tino.
Il conflitto dell'uomo urbano / Tino Vaglieri. — Milano : Magma, 1977.
93 p. : chiefly ill. ; 21 cm. — (Quaderni di Artecontro ; 2) It78-Feb
L2500
1. Vaglieri, Tino. 2. Cities and towns in art. I. Title. II. Series.
NC257.V254A4 1977 78-365511
*78 MARC

—EXHIBITIONS

Adamson, Jeremy.
Lawren S. Harris : urban scenes and wilderness landscapes, 1906-1930 : Art Gallery of Ontario, January 14-February 26, 1978 / Jeremy Adamson. — [Toronto] : Art Gallery of Ontario, c1978.
231 p. : ill. (some col.) ; 24 cm. C***
Bibliography: p. 216-218.
ISBN 0-919876-35-8
1. Harris, Lawren Stewart, 1885-1970—Exhibitions. 2. Cities and towns in art—Exhibitions. 3. Landscape in art—Exhibitions. I. Harris, Lawren Stewart, 1885-1970. II. Art Gallery of Ontario.
ND249.H28A4 1978 759.11 79-310279
79 MARC

Die Industrialisierung der Stadt / Katalog, Lucie Schauer. — Berlin : Neuer Berliner Kunstverein, [1978?]
[80] p. : chiefly ill. ; 24 cm. GFR***
The first of a series of three exhibitions of art with an urban theme held at the Neue Berliner Kunstverein.
On verso of t.p. : "Stadt 1: Industrialisierung."
1. Industry in art—Exhibitions. 2. Cities and towns in art—Exhibitions. 3. Art, modern—19th century—Exhibitions. 4. Art, modern—20th century—Exhibitions. I. Schauer, Lucie. II. Neuer Berliner Kunstverein. III. Title: Stadt 1: Industrialisierung.
N8218.I52 79-355124
79 MARC

CITIES AND TOWNS IN LITERATURE

see also Poetry of places; and names of individual cities and towns, e.g. London in literature

Festa-McCormick, Diana.
The city as catalyst : a study of ten novels / Diana Festa-McCormick. — Rutherford [N.J.] : Fairleigh Dickinson University Press, c1979.
213 p. ; 22 cm.
Bibliography: p. 197-208.
Includes index.
ISBN 0-8386-2156-2 : $14.50
1. Fiction—History and criticism. 2. Cities and towns in literature. I. Title.
PN3352.C5F4 809.3'3 77-85595
78 MARC

Schwarzbach, F S
Dickens and the city / F. S. Schwarzbach. — London : Athlone Press ; [Atlantic Highlands, N.J.] : distributed by Humanities Press, 1979.
xii, 258 p., [4] leaves of plates : ill. ; 23 cm. GB***
Bibliography: p. [225]-234.
Includes index.
ISBN 0-485-11174-8 : £10.50
1. Dickens, Charles, 1812-1870—Homes and haunts—England—London. 2. London in literature. 3. Cities and towns in literature. 4. Novelists, English—19th century—Biography. I. Title.
PR4584.S3 823'.8 79-306781
79 MARC

CITIZEN (SHIP)

Bill, Erastus, 1826-1905.
Citizen : an American boy's early manhood aboard a Sag Harbor whale-ship chasing delirium and death around the world, 1843-1849 : being the story of Erastus Bill who lived to tell it : with notes on Erastus Bill by Robert Wesley Bills. — Anchorage, Alaska : O. W. Frost, 1978.
136 p. : ill. ; 24 cm. — (World discovery books ; 2)
Includes index.
ISBN 0-930766-01-6 : $10.00. ISBN 0-930766-02-4 pbk.
1. Whaling. 2. Whalemen—New Yorks (State)—Sag Harbor—Biography. 3. Citizen (Ship) I. Title. II. Series.
SH383.2.B54 1978 910'.41 78-50525
79 MARC

CITIZEN ASSOCIATIONS see Citizen' associations

CITIZEN PARTICIPATION IN ADMINISTRATION OF CRIMINAL JUSTICE see Criminal justice, Administration of—Citizen participation

CITIZEN PARTICIPATION IN CRIME PREVENTION see Crime prevention —Citizen participation

CITIZEN SUITS (CIVIL PROCEDURE)

see also Class actions (Civil procedure); Popular actions

CITIZENS ADVISORY COMMITTEES IN EDUCATION

—OHIO

Citizen guide to quality education : a report of the Citizens' Council for Ohio Schools. — Cleveland : The Council, 1978.
iv, 76 p. ; 28 cm.
Includes bibliographies.
$3.00
1. Citizens' Council for Ohio Schools. 2. Education—Ohio. 3. Citizens advisory committees in education—Ohio. I. Citizens' Council for Ohio Schools.
LC221.2.O3C57 379 78-106804
78 MARC

CITIZENS' ASSOCIATION

see also Neighborhood government

—HANDBOOKS, MANUALS, ETC.

Biagi, Bob.
Working together : a manual for helping groups, work more effertively / by Bob Biagi ; cover drawing by Damian Biagi and Gia Biagi. — Amherst : Citizen Involvement Training Project, c1978.
124 p. : ill. ; 28 cm.
1. Citizens' association—Handbooks, manuals, etc. I. Title.
JS303.5.B52 301.18'5 79-624736
79 MARC

—GERMANY, WEST—FRANKFURT AM MAIN

Handbuch, Bürgerinitiativen in Frankfurt / hrsg. von Ingrid Damian-Hesser u. Michael Damian. — Frankfurt : Verlag im Leseladen, 1978.
190 p. : numerous ill. ; 24 cm. GFR79-A
On spine: Bürgerinitiativen in Frankfurt.
Bibliography: p. 189-190.
ISBN (invalid) 3-922080-01-6 : DM15.80
1. Community organization. 2. Citizens' association—Germany, West—Frankfurt am Main. I. Damian-Hesser, Ingrid. II. Damian, Michael. III. Title: Bügerinitiativen in Frankfurt.
HN458.F7H36 309.1'43'41 79-354065
79 MARC

—JAPAN

(Tachiagaru nanchōsha)
立ちあがる難聴者 / 全国難聴者連絡協議会編. — 東京 : たいまつ社, 1978.
212 p. : ill. ; 17 cm. — (たいまつ新書 ; 43) Ja 79-5350
Series romanized: Taimatsu shinsho.
¥680

1. Deaf—Japan. 2. Civil rights—Japan. 3. Citizens' associations—Japan. I. Zenkoku Nanchōsha Renraku Kyōgikai.
HV2883.T3 79-808641

—UNITED STATES

Lancourt, Joan E
Confront or concede, the Alinsky citizen-action organizations / Joan E. Lancourt. — Lexington, Mass. : Lexington Books, c1979.
ix, 197 p. ; 24 cm.
Bibliography: p. 177-190.
Includes index.
ISBN 0-669-02715-4
1. Radicalism—United States. 2. Social action. 3. Community organization. 4. Alinsky, Saul David, 1909-1972. 5. Citizen's associations—United States. I. Title.
HN90.R3L36 361.8 78-19676
79 MARC

CITIZENS BAND RADIO

see also Mobile radio stations; Models and modelmaking —Radio control systems

Sands, Leo G
Leo Sands' Complete guide to CB radio / Leo G. Sands. — West Nyack, N.Y. : Parker Pub. Co., c1979.
374 p. : ill. ; 26 cm.
Includes index.
ISBN 0-13-530295-1
1. Citizens band radio. I. Title. II. Title: Complete guide to CB radio.
TK6570.C5S32 621.3845'4 79-10693
79 MARC

—DICTIONARIES, JUVENILE

Downing, Joan.
Junior CB picture dictionary / compiled & edited by Joan Downing ; pictures by Slug Signorino. — Chicago : Childrens Press, c1978.
95 p. : ill. ; 21 cm.
Bibliography: p. 95.
SUMMARY: An illustrated dictionary of familiar CB slang.
ISBN 0-516-03517-7 (lib. bdg.)
1. Citizens band radio—Slang—Juvenile literature. 2. Citizens band radio—Dictionaries, Juvenile. 3. Picture dictionaries, English. [1. Citizens band radio—Slang. 2. Citizens band radio—Dictionaries] I. Signorino, Slug. II. Title.
TK6570.C5D67 427'.09 77-17105
77 MARC

—EQUIPMENT AND SUPPLIES

Safford, Edward L
CBer's handy manual of base stations / Edward L. Safford. — 1st ed. — Blue Ridge Summit, Pa. : Tab Books, c1978.
96 p. : ill. ; 21 cm.
Includes index.
ISBN 0-8306-7969-3 pbk. : $2.50
1. Citizens band radio—Equipment and supplies. I. Title.
TK6570.C5S23 621.3845'4 78-12548
78 MARC

—EQUIPMENT AND SUPPLIES—DESIGN AND CONSTRUCTION—AMATEURS' MANUALS

Benrey, Ronald.
CB accessories you can build / Ronald M. Benrey. — Rochelle Park, N.J. : Hayden Book Co., c1979.
110 p. : ill. ; 21 cm.
Includes index.
ISBN 0-8104-0854-6 : $4.95
1. Citizens band radio—Equipment and supplies—Design and construction—Amateurs' manuals. I. Title.
TK6570.C5B39 621.3845'4'028 78-20734
79 MARC

—EQUIPMENT AND SUPPLIES—MAINTENANCE AND REPAIR

Hobbs, Marvin.
Modern CB radio servicing / Marvin Hobbs. — Rochelle Park, N.J. : Hayden Book Co., c1979.
163 p. : ill. ; 25 cm.
Includes index.
ISBN 0-8104-0865-1
1. Citizens band radio—Equipment and supplies—Maintenance and repair. I. Title.
TK6570.C5H62 621.3845'4'028 79-52761
00 MARC

—EQUIPMENT AND SUPPLIES —PERIODICALS

CB radio marketing. –v. 2, no. 6; –June 1977.
Port Washington, N. Y., Cowan Pub. Corp.
v. ill. 29 cm. monthly.
"The trade magazine of the CB industry."
Merged with: Merchandising 2-way radio, ISSN 0362-2606, to form: Personal communications, ISSN 0147-7145.
1. Radio supplies industry—United States—Periodicals. 2. Citizens band radio—Equipment and supplies—Periodicals.
HD9999.R153U533 658.8'09'6213841 78-646633
MARC-S

—PERIODICALS

CB times. v. 1–
(no. 1–18) ; June 1976–
Derby, Conn., Charlton Publications.
v. ill. 28 cm. bimonthly.
"A magazine written for and by CB'ers."
Monthly, June 1976– ; bimonthly, May 1977–
Key title: CB times, ISSN 0162-0215
1. Citizens band radio—Periodicals.
TK6570.C5C225 384.5'3 78-646695
MARC-S

—SLANG—JUVENILE LITERATURE

Downing, Joan.
Junior CB picture dictionary / compiled & edited by Joan Downing ; pictures by Slug Signorino. — Chicago : Childrens Press, c1978.
95 p. : ill. ; 21 cm.
Bibliography: p. 95.
SUMMARY: An illustrated dictionary of familiar CB slang.
ISBN 0-516-03517-7 (lib. bdg.)
1. Citizens band radio—Slang—Juvenile literature. 2. Citizens band radio—Dictionaries, Juvenile. 3. Picture dictionaries, English. [1. Citizens band radio—Slang. 2. Citizens band radio—Dictionaries] I. Signorino, Slug. II. Title.
TK6570.C5D67 427'.09 77-17105
77 MARC

—UNITED STATES—DIRECTORIES

ABM Service Corporation.
The national directory of CB radio channels / prepared by the ABM Service Corporation. — Palm Springs, Calif. : ETC Publications, c1979.
ix, 360 p. ; 22 cm.
ISBN 0-88280-064-7 : $15.00. ISBN 0-88280-065-5 pbk. : $7.95
1. Citizens band radio—United States—Directories. I. Title.
TK6570.C5A17 1979 384.5'3'5 78-12796
78 MARC

CITIZENS' COMMITTEES IN EDUCATION see Citizens' advisory committees in education

CITIZENS' COUNCIL FOR OHIO SCHOOLS

Citizen guide to quality education : a report of the Citizens' Council for Ohio Schools. — Cleveland : The Council, 1978.
iv, 76 p. ; 28 cm.
Includes bibliographies.
$3.00
1. Citizens' Council for Ohio Schools. 2. Education—Ohio. 3. Citizens advisory committees in education—Ohio. I. Citizens' Council for Ohio Schools.
LC221.2.O3C57 379 78-106804
78 MARC

CITIZEN'S DEFENDER see Ombudsman

CITIZENS RADIO SERVICE see Citizens band radio

CITIZENS STATE BANK OF CARRIZO SPRINGS, TEX.

United States. Congress. House. Committee on Banking, Currency and Housing. Subcommittee on Financial Institutions Supervision, Regulation and Insurance.
The failure of Citizens State Bank of Carrizo Springs, Texas, and related financial problems : hearings before the Subcommittee on Financial Institutions Supervision, Regulation, and Insurance of the Committee on Banking, Currency, and Housing, House of Reprentatives, Ninety-fourth Congress, second session, November 30, December 1, 1976. — Washington : U.S. Govt. Print. Off., 1977.
2 v. : ill. ; 24 cm.
$8.50 (v. 1). $9.00 (v. 2)
1. Citizens State Bank of Carrizo Springs, Tex. 2. Banks and banking—United States—State supervision. 3. Bank examinination—United States. I. Title.
KF27.B396 1976c 332.1'224'09764455 79-601356
79 MARC

CITIZENSHIP

see also Aliens; Allegiance; Civics; Corporations—Nationality; Deportation; Domicile; Naturalization; Patriotism; Political rights; Self-determination, National; Statelessness; Suffrage

Salisbury, Robert Holt, 1930-
Key concepts of citizenship : perspectives and dilemmas / by Robert H. Salisbury. — ₍Washington₎ : U.S. Dept. of Health, Education, and Welfare, Office of Education, ₍1978₎
vii, 59 p. ; 24 cm. — (HEW publication ; no. (OE) 78-07005)
Includes bibliographical references.
1. Citizenship. 2. Political participation. I. Title. II. Series: United States. Dept. of Health, Education, and Welfare. DHEW publication ; no. (OE) 78-07005.
JF2071.S24 323.6 78-603562
79 MARC

—CONFLICT OF LAWS see Conflict of laws—Citizenship

—BRUNEI (STATE)

Hecker, Hellmuth.
Das Staatsangehörigkeitsrecht von Brunei, Indonesien, Malaysia, Singapur und den Philippinen / von Hellmuth Hecker. — Frankfurt am Main : Metzner, 1978.
297 p. ; 21 cm. — (Sammlung geltender Staatsangehörigkeitsgesetze ; Bd. 36) GFR78-A
Some text in English.
Includes bibliographies.
ISBN 3-7875-1351-5
1. Citizenship—Brunei (State) 2. Citizenship—Indonesia. 3. Citizenship—Malaysia. 4. Citizenship—Singapore. 5. Citizenship—Philippine Islands. I. Title. II. Series.
342'.59'083 79-342408
79 MARC

—CANADA

Canada. Dept. of the Secretary of State.
The Canadian citizen. — ₍Ottawa₎ : Minister of Supply and Services Canada, 1978.
16, 16 p. : ill. ; 28 cm. C***
Cover title.
Added t.p.: Le citoyen canadien.
English and French.
ISBN 0-662-00464-7
1. Civics, Canadien. 2. Citizenship—Canada. I. Le citoyen canadien. II. Title.
JL187.C245 1978 320.4'71 79-304129
79 MARC

Neice, David C
Ethnicity and Canadian citizenship : a metropolitan study / David C. Neice. — ₍Ottawa₎ : Secretary of State, 1978, c1979. C***
xv, 188 p. ; 28 cm.
Bibliography: p. 186-188.
ISBN 0-662-10462-5
1. Citizenship—Canada. 2. Naturalization—Canada. 3. Canada—Foreign population. I. Title.
JL187.N43 323.6'0971 79-314855
79 MARC

—CHINA—MANCHURIA

(Manshūkoku kokuseki narabini kaisha kokuseki oyobi shihon hōsaku)
滿洲國國籍竝會社國籍及資本方策. 一 ₍新京₎ : 南滿洲鐵道株式會社經濟調查會, 昭和10 ₍1935₎
1, 221 p. ; 23 cm. — (立案調查書類 - 南滿洲鐵道株式會社經濟調查會 ; 第26編第1卷)
"極祕"
1. Citizenship—China—Manchuria. 2. Corporations, Foreign—China—Manchuria. 3. Corporations—Nationality. 4. Investments, Foreign—China—Manchuria. I. Minami Manshū Tetsudō Kabushiki Kaisha. Keizai Chōsakai. II. Series: Minami Manshū Tetsudō Kabushiki Kaisha. Keizai Chōsakai. Ritsuan chōsa shorui ; dai 26-hen, dai 1-kan.
78-813941

—CZECHOSLOVAKIA

Schmid, Karin.
Staatsangehörigkeitsprobleme der Tschechoslowakei : e. Unters. sowie Dokumente zum Staatsangehörigkeit d. dt. Volkszugehörigen / Karin Schmid. — Berlin : Berlin-Verlag, 1979.
133 p. ; 22 cm. — (Politologische Studien ; Bd. 15) GFR79-A
Bibliography: p. 132-133.
ISBN 3-87061-141-3 : DM18.80
1. Citizenship—Czechoslovakia. 2. Citizenship, Loss of—Czechoslovakia. 3. Germans in Czechoslovakia.—Legal status, laws, etc. 4. Minorities—Legal status, laws, etc.—Czechoslovakia. 5. Citizenship—Germany, West. I. Title.
342'.437'083 79-359639
79 MARC

—EGYPT

Egypt. Laws, statutes, etc.
(Qānūn al-jinsīyah al-Miṣrīyah)
قانون الجنسيــــة المصرية : القانون رقم ٢٦ لسنــة ١٩٧٥ / جمهورية مصر العربيــة. ـــ ₍القاهرة₎ : الهيئــة العامة لشئون المطابع الأميرية ، 1976.
14 p. ; 18 cm.
1. Citizenship—Egypt. I. Title.
78-960343

Ṣādiq, Hishām ʻAlī.
(al-Jinsīyah wa-al-mawṭin wa-markaz al-ajānib)
الجنسية والموطن ومركز الأجانب : دراســة مقارنة فى المبادئ العامة والحلول المقررة فى التشريـــع الوضعـــى / هشـــام على صادق. ـــ الاسكندرية : منشاة المعارف ، 1977.
2 v. ; 24 cm. — (الكتب القانونية)
Vol. 2: al-ṭabʻah 1.
Series romanized: al-Kutub al-qānūnīyah.
Includes bibliographies.
CONTENTS: مجلد 1. فى الجنسيــة والموطن. ـــ مجلد 2. فى مركز الأجانــب
ISBN 977-7006-49-7 (v. 1) : £E6.00 (v. 1)
£E4.00 (v. 2)
1. Citizenship—Egypt. 2. Aliens—Egypt. 3. Conflict of laws—Citizenship—Egypt. I. Title.
78-961713

—GERMANY, WEST

Schmid, Karin.
Staatsangehörigkeitsprobleme der Tschechoslowakei : e. Unters. sowie Dokumente zum Staatsangehörigkeit d. dt. Volkszugehörigen / Karin Schmid. — Berlin : Berlin-Verlag, 1979.
133 p. ; 22 cm. — (Politologische Studien ; Bd. 15) GFR79-A
Bibliography: p. 132-133.
ISBN 3-87061-141-3 : DM18.80
1. Citizenship—Czechoslovakia. 2. Citizenship, Loss of—Czechoslovakia. 3. Germans in Czechoslovakia.—Legal status, laws, etc. 4. Minorities—Legal status, laws, etc.—Czechoslovakia. 5. Citizenship—Germany, West. I. Title.
342'.437'083 79-359639
79 MARC

—GREAT BRITAIN

Dummett, Ann, 1930-
Citizenship and nationality / by Ann Dummett ; with a foreword by J. E. S. Fawcett. — London : Runnymede Trust, 1976.
88, ₍1₎ p. : ill. ; 21 cm. — (A Runnymede Trust publication) GB77-26912
Bibliography: p. 85-88.
£1.50
1. Citizenship—Great Britain. I. Title.
KD4050.D85 342'.41'083 79-302099
79 MARC

Giese, Hannes.
Das Staatsangehörigkeitsrecht von Grossbritannien / bearb. von Hannes Giese. — 2., neubearb. Aufl. — Frankfurt am Main : Metzner, 1978.
156 p. ; 21 cm. — (Sammlung geltender Staatsangehörigkeitsgesetze ; Bd. 6) GFR78-A
"Eine Neubearbeitung der Arbeit Hampes von 1951, 'Das Staatsangehörigkeitsrecht von Grossbritannien.'"
Bibliography: p. 155-156.
ISBN 3-7875-1349-3
1. Citizenship—Great Britain. 2. Aliens—Great Britain. I. Hampe, Karl Alexander. Das Staatsangehörigkeitsrecht von Grossbritannien. II. Title. III. Series.
KD4053.G53 1978 342'.41'083 79-346136
79₍r79₎rev MARC

—INDIA

India (Republic). Laws, statutes, etc.
Consul's Law of foreigners, citizenship, and passports : containing exhaustive commentaries on Foreigners Act, Registration of Foreigners Act, Passports Act, Citizenship Act, with orders, rules, notifications, and allied laws. — 3d ed. / rev. by V. Parshad. — Allahabad : Law Book Co., 1977.
xviii, 477 p. ; 25 cm.
Caption title: Law of foreigners, citizenship & passports.
Includes bibliographical references and index.
Rs45.00
1. Aliens—India. 2. Citizenship—India. 3. Passports—India. I. Consul, S. C. II. Parshad, V. III. Title. IV. Title: Law of foreigners, citizenship, and passports.
342'.54'083 78-901044
79 MARC

—INDONESIA

Beberapa hal mengenai kewarganegaraan. — ₍Djakarta₎ : Inspektorat Djenderal Territorial dan Perlawanan Rakjat, ₍1959₎
158 p. ; 21 cm.
Includes bibliographical references.
1. Citizenship—Indonesia. 2. Dual nationality. I. Indonesia. Angkatan Darat. Inspektorat Djenderal Territorial dan Perlawanan Rakjat.
JQ777.A2B4 77-986956
79 MARC

Hecker, Hellmuth.
Das Staatsangehörigkeitsrecht von Brunei, Indonesien, Malaysia, Singapur und den Philippinen / von Hellmuth Hecker. — Frankfurt am Main : Metzner, 1978.
297 p. ; 21 cm. — (Sammlung geltender Staatsangehörigkeitsgesetze ; Bd. 36) GFR78-A
Some text in English.
Includes bibliographies.
ISBN 3-7875-1351-5
1. Citizenship—Brunei (State) 2. Citizenship—Indonesia. 3. Citizenship—Malaysia. 4. Citizenship—Singapore. 5. Citizenship—Philippine Islands. I. Title. II. Series.
342'.59'083 79-342408
79 MARC

—INDONESIA—CASE STUDIES

Ringeling, A B 1942-
Beliedsvrijheid van ambtenaren : het spijtoptantenprobleem als illustratie van de activiteiten van ambtenaren bij de uitvoering van beleid / A. B. Ringeling. — Alphen aan den Rijn : Samsom, 1978.
273 p. ; 22 cm. — (Serie maatschappijbeelden) Ne78-45
Originally presented as the author's thesis, Nijmegen.
Summary in English.
Bibliography: p. ₍259₎-267.
Includes index.
ISBN 9014027915 : fl 45.00
1. Citizenship—Netherlands—Case studies. 2. Citizenship—Indonesia—Case studies. 3. Administrative discretion—Netherlands—Case studies. 4. Public administration—Decision making—Case studies. I. Title.
JN5935.R56 1978 79-372319
*79 MARC

—ISRAEL

Gera-Grünbaum, Jehuda.
Das Staatsangehörigkeitsrecht von Israel / von Jehuda Gera-Grünbaum u. Aharon Zwergbaum. — 2., neubearb. Aufl. — Frankfurt (am Main) : Metzner, 1974.
72 p. ; 21 cm. — (Sammlung geltender Staatsangehörigkeitsgesetze ; Bd. 13) GFR75-A
First ed. published in 1954 under title: Das Staatsangehörigkeitsrecht des Staates Israel und des ehemaligen Mandatsgebietes Palästina and entered under Israel. Laws, statutes, etc.
Bibliography: p. 70-72.
Includes indexes.
ISBN 3-7875-1345-0 : DM22.00
1. Citizenship—Israel. 2. Citizenship—Palestine. I. Zwergbaum, Aharon, joint author. II. Israel. Laws, statutes, etc. III. Title. IV. Series.
342'.5694'083 78-379153

CITIZENSHIP (Continued)

—ITALY

Italy. Laws, statutes, etc.
Stato civile, rifugiati e apolidi, cittadinanza. — Milano : Pirola, 1977.
216 p. ; 22 cm. — (Collezione legale Pirola ; 1301) It78-Nov
Italian or French.
L4000
1. Registers of births, etc.—Italy. 2. Refugees—Legal status, laws, etc. 3. Citizenship, Loss of. 4. Citizenship—Italy. I. Title.
346'.45'012 79-356780
*79 MARC

Medail, Giorgio, 1945-
L'onorevole cittadino e il suo segreto potere / Giorgio Medail. — Milano : SugarCo, 1978.
222 p. ; 21 cm. — (Fatti e misfatti ; 55) It78-Feb
Includes index.
L3000
1. Citizenship—Italy. 2. Political rights—Italy. 3. Bertuzzi, Alberto, 1913- 4. Italy—Biography. I. Title.
JN5591.M42 323.6'0945 79-367487
*79 MARC

—JAPAN

Hiraga, Kenta, 1913-
(Kokusekihō)
国籍法 / 平賀健太著. — 東京 : 帝國判例法規出版社, 昭和25-26 [1950-1951]
2 v. (2, 2, 478, 50 p.) ; 21 cm.
Errata slip inserted in v. 1.
Bibliography: p. 48-50 (4th group)
1. Citizenship—Japan. I. Japan. Laws, statutes, etc. Kokusekihō. 1950. II. Title.
79-800471

Nihon Bei-Fu Kyōkai. Henshūbu.
(Dainisei to kokuseki mondai)
第二世と國籍問題 / [編輯者 日本米布協會編輯部]. — 東京 : 日本米布協會, 昭和13 [1938], 昭和15 [1940] printing.
[8], 46, 4 p. ; 22 cm. — (第二世叢書 ; 第三輯)
Cover title.
"海外同胞東京大會參加記念"
Bibliography: p. [8] (1st group)
1. Citizenship—Japan. 2. Japanese Americans—Legal status, laws, etc.—Japan. I. Title. II. Series: Dainisei sōsho ; dai 3-shū.
78-815039

—MALAYSIA

Hecker, Hellmuth.
Das Staatsangehörigkeitsrecht von Brunei, Indonesien, Malaysia, Singapur und den Philippinen / von Hellmuth Hecker. — Frankfurt am Main : Metzner, 1978.
297 p. ; 21 cm. — (Sammlung geltender Staatsangehörigkeitsgesetze ; Bd. 36) GFR78-A
Some text in English.
Includes bibliographies.
ISBN 3-7875-1351-5
1. Citizenship—Brunei (State) 2. Citizenship—Indonesia. 3. Citizenship—Malaysia. 4. Citizenship—Singapore. 5. Citizenship—Philippine Islands. I. Title. II. Series.
342'.59'083 79-342408
79 MARC

—NETHERLANDS—CASE STUDIES

Ringeling, A B 1942-
Beliedsvrijheid van ambtenaren : het spijtoptantenprobleem als illustratie van de activiteiten van ambtenaren bij de uitvoering van beleid / A. B. Ringeling. — Alphen aan den Rijn : Samsom, 1978.
273 p. ; 22 cm. — (Serie maatschappijbeelden) Ne78-45
Originally presented as the author's thesis, Nijmegen.
Summary in English.
Bibliography: p. [259]-267.
Includes index.
ISBN 9014027915 : fl 45.00
1. Citizenship—Netherlands—Case studies. 2. Citizenship—Indonesia—Case studies. 3. Administrative discretion—Netherlands—Case studies. 4. Public administration—Decision making—Case studies. I. Title.
JN5935.R56 1978 79-372319
*79 MARC

—PAKISTAN

Bashir, Muhammad, Advocate.
The Constitution of the Islamic Republic of Pakistan & Citizenship act : amended up-to-date with commentary / by Muhammad Bashir. — Lahore : Khyber Law Publishers, [1977]
x, 154, 52 p. ; 25 cm.
Rs16.00
1. Pakistan—Constitutional law. 2. Citizenship—Pakistan. I. Pakistan. Laws, statutes, etc. Pakistan citizenship act, 1951. 1977. II. Title.
342'.5491 77-930836
79 MARC

—PALESTINE

Gera-Grünbaum, Jehuda.
Das Staatsangehörigkeitsrecht von Israel / von Jehuda Gera-Grünbaum u. Aharon Zwergbaum. — 2., neubearb. Aufl. — Frankfurt (am Main) : Metzner, 1974.
72 p. ; 21 cm. — (Sammlung geltender Staatsangehörigkeitsgesetze ; Bd. 13) GFR75-A
First ed. published in 1954 under title: Das Staatsangehörigkeitsrecht des Staates Israel und des ehemaligen Mandatsgebietes Palästina and entered under Israel. Laws, statutes, etc.
Bibliography: p. 70-72.
Includes indexes.
ISBN 3-7875-1345-0 : DM22.00
1. Citizenship—Israel. 2. Citizenship—Palestine. I. Zwergbaum, Aharon, joint author. II. Israel. Laws, statutes, etc. III. Title. IV. Series.
342'.5694'083 78-379153
78 MARC

—PHILIPPINE ISLANDS

Hecker, Hellmuth.
Das Staatsangehörigkeitsrecht von Brunei, Indonesien, Malaysia, Singapur und den Philippinen / von Hellmuth Hecker. — Frankfurt am Main : Metzner, 1978.
297 p. ; 21 cm. — (Sammlung geltender Staatsangehörigkeitsgesetze ; Bd. 36) GFR78-A
Some text in English.
Includes bibliographies.
ISBN 3-7875-1351-5
1. Citizenship—Brunei (State) 2. Citizenship—Indonesia. 3. Citizenship—Malaysia. 4. Citizenship—Singapore. 5. Citizenship—Philippine Islands. I. Title. II. Series.
342'.59'083 79-342408
79 MARC

—PUERTO RICO

Cabranes, José A
Citizenship and the American Empire : notes on the legislative history of the United States citizenship of Puerto Ricans / José A. Cabranes. — New Haven : Yale University Press, 1979.
viii, 101 p. ; 24 cm.
"Originally published as an article in the University of Pennsylvania law review, 127 U.Pa. L. Rev. 391 (1978)."
ISBN 0-300-02325-1
1. Citizenship—United States. 2. Citizenship—Puerto Rico. 3. Puerto Rico—Politics and government—1898-1952. I. Title.
KF4720.P83C3 323.6'29 78-65479
79 MARC

—RUSSIA

Safronov, Vladimir Mikhaĭlovich.
(Pravovoe polozhenie grazhdan SSSR)
Правовое положение граждан СССР / В. М. Сафронов. — Москва : Знание, 1969.
61 p. ; 22 cm. — (Новое в жизни, науке, технике : Серия Государство и право ; 1969, 8)
Includes bibliographical references.
1. Citizenship—Russia. 2. Conflict of laws—Citizenship—Russia. I. Title. II. Series: Novoe v zhizni, nauke, tekhnike : Serii͡a Gosudarstvo i pravo ; 1969, 8.
79-411166

Safronov, Vladimir Mikhaĭlovich.
(Ty—grazhdanin Sovetskogo Soi͡uza)
Ты—гражданин Советского Союза / В. М. Сафронов. — Москва : Юрид. лит., 1978.
127 p. : ill. ; 16 cm. — (Советский закон и я) USSR 78
Includes bibliographical references.
0.15 lv
1. Citizenship—Russia. 2. Passports—Russia. 3. Conflict of laws—Citizenship—Russia. 4. Russia—Constitutional law. I. Title. II. Series: Sovetskiĭ zakon i i͡a.
79-408593

—SINGAPORE

Hecker, Hellmuth.
Das Staatsangehörigkeitsrecht von Brunei, Indonesien, Malaysia, Singapur und den Philippinen / von Hellmuth Hecker. — Frankfurt am Main : Metzner, 1978.
297 p. ; 21 cm. — (Sammlung geltender Staatsangehörigkeitsgesetze ; Bd. 36) GFR78-A
Some text in English.
Includes bibliographies.
ISBN 3-7875-1351-5
1. Citizenship—Brunei (State) 2. Citizenship—Indonesia. 3. Citizenship—Malaysia. 4. Citizenship—Singapore. 5. Citizenship—Philippine Islands. I. Title. II. Series.
342'.59'083 79-342408
79 MARC

—UNITED STATES

Cabranes, José A
Citizenship and the American Empire : notes on the legislative history of the United States citizenship of Puerto Ricans / José A. Cabranes. — New Haven : Yale University Press, 1979.
viii, 101 p. ; 24 cm.
"Originally published as an article in the University of Pennsylvania law review, 127 U.Pa. L. Rev. 391 (1978)."
ISBN 0-300-02325-1
1. Citizenship—United States. 2. Citizenship—Puerto Rico. 3. Puerto Rico—Politics and government—1898-1952. I. Title.
KF4720.P83C3 323.6'29 78-65479
79 MARC

Webster, Prentiss, 1851-
A treatise on the law of citizenship in the United States. Treated historically by Prentiss Webster ... Albany, N. Y., M. Bender, 1891.
xxiii, [1], 338 p. 24 cm.
——— Microfilm.
Microfilm LL-1485 KF
1. Citizenship—U. S.
KF4700.W42 12-39897

—UNITED STATES—HISTORY

Kettner, James H
The development of American citizenship, 1608-1870 / by James H. Kettner. — Chapel Hill : Published for the Institute of Early American History and Culture, Williamsburg, Va., by the University of North Carolina Press, c1978.
xi, 391 p. ; 24 cm.
Includes bibliographical references and index.
ISBN 0-8078-1326-5
1. Naturalization—United States—History. 2. Citizenship—United States—History. I. Institute of Early American History and Culture, Williamsburg, Va. II. Title.
JK1814.K47 323.6'0973 78-954
78 MARC

CITIZENSHIP, LOSS OF

see also Statelessness

Italy. Laws, statutes, etc.
Stato civile, rifugiati e apolidi, cittadinanza. — Milano : Pirola, 1977.
216 p. ; 22 cm. — (Collezione legale Pirola ; 1301) It78-Nov
Italian or French.
L4000
1. Registers of births, etc.—Italy. 2. Refugees—Legal status, laws, etc. 3. Citizenship, Loss of. 4. Citizenship—Italy. I. Title.
346'.45'012 79-356780
*79 MARC

—CZECHOSLOVAKIA

Schmid, Karin.
Staatsangehörigkeitsprobleme der Tschechoslowakei : e. Unters. sowie Dokumente zum Staatsangehörigkeit d. dt. Volkszugehörigen / Karin Schmid. — Berlin : Berlin-Verlag, 1979.
133 p. ; 22 cm. — (Politologische Studien ; Bd. 15) GFR79-A
Bibliography: p. 132-133.
ISBN 3-87061-141-3 : DM18.80
1. Citizenship—Czechoslovakia. 2. Citizenship, Loss of—Czechoslovakia. 3. Germans in Czechoslovakia—Legal status, laws, etc. 4. Minorities—Legal status, laws, etc.—Czechoslovakia. 5. Citizenship—Germany, West. I. Title.
342'.437'083 79-359639
79 MARC

CITIZENSHIP (INTERNATIONAL LAW)

see also Dual nationality

Ch'oe, Ch'ang-hwa, 1930-
(Kokuseki to jinken)
国籍と人権 / 崔昌華. — 東京 : 酒井書店, 1975.
2, 7, 357 p., [1] leaf of plates : ill. ; 22 cm. Ja 75-10555
"福岡大学大学院に修士論文として提出した「国際法からみた国籍に関する若干の考察」を基礎とした"
Bibliography: p. 341-345.
Includes legislation and indexes.
1. Citizenship (International law) 2. Koreans in Japan—Legal status, laws, etc. I. Title.
JX4211.C53 1975 75-808152
rev

—CONGRESSES

Staatsangehörigkeitsrecht und Völkerrecht in Deutschland : Referate e. Seminars / Univ. Hamburg ; hrsg. von Hellmuth Hecker. — Hamburg : Univ. Hamburg, 1977.
xv, 180 p. ; 30 cm. GFR78-B7
Bibliography: p. 172-180.
1. Conflict of laws—Citizenship—Germany, West—Congresses. 2. Citizenship (International law)—Congresses. 3. Statelessness—Congresses. I. Hecker, Hellmuth. II. Hamburg. Universität.
342'.43'042 79-341210
79 MARC

CITROËN AUTOMOBILE

Olyslager, Piet.
Vraagbaak voor uw Citroën : een complete handleiding voor de typen DS 19 en ID 19 / door Piet Olyslager. — Deventer : Æ. E. Kluwer, [1964]
62 p. : ill. ; 24 cm.
Cover title: Vraagbaak voor uw Citroën DS 19 en ID 19.
1. Citroën automobile. I. Title. II. Title: Vraagbaak voor uw Citroën DS 19 en ID 19.
TL215.C56O395 78-393859
79 MARC

CITRONELLA GRASS

—SRI LANKA—HAMBANTOTA

Perera, M P
The citronella industry in the Hambantota District / Mahinda Perera, Fredrick Abeyratne. — Colombo : Agrarian Research & Training Institute, 1978.
28 p., [2] leaves of plates : graphs ; 30 cm. — (Research study series - Agrarian Research & Training Institute ; no. 27)
Includes bibliographical references.
1. Citronella oil industry—Sri Lanka—Hambantota. 2. Citronella grass—Sri Lanka—Hambantota. I. Abeyratne, Fredrick, joint author. II. Title. III. Series: Agrarian Research and Training Institute. Research study series - Agrarian Research & Training Institute ; no. 27.
HD9211.C563S667 338.1'7'32 79-900872
79 MARC

CITRONELLA OIL INDUSTRY

—SRI LANKA—HAMBANTOTA

Perera, M P
The citronella industry in the Hambantota District / Mahinda Perera, Fredrick Abeyratne. — Colombo : Agrarian Research & Training Institute, 1978.
28 p., [2] leaves of plates : graphs ; 30 cm. — (Research study series - Agrarian Research & Training Institute ; no. 27)
Includes bibliographical references.
1. Citronella oil industry—Sri Lanka—Hambantota. 2. Citronella grass—Sri Lanka—Hambantota. I. Abeyratne, Fredrick, joint author. II. Title. III. Series: Agrarian Research and Training Institute. Research study series - Agrarian Research & Training Institute ; no. 27.
HD9211.C563S667 338.1'7'32 79-900872
79 MARC

CITRUS CODLING MOTH see Cryptophlebia leucotreta

CITRUS FRUIT INDUSTRY

—ESTIMATES—FLORIDA —PERIODICALS

Florida. University, Gainesville. Food and Resource Economics Dept.
Estimated costs of packing and selling fresh Florida citrus.
Gainesville, Food and Resource Economics Dept., Agricultural Experiment Stations, Institute of Food and Agricultural Sciences, University of Florida.
v. 28 cm. annual. (Economic information report)
1. Citrus fruit industry—Estimates—Florida—Periodicals.
I. Title. II. Series.
HD9259.C54U5865a 338.1'3 79-641526
MARC-S

Florida. University, Gainesville. Food and Resource Economics Dept.
Estimated costs of processing, warehousing and selling Florida citrus products.
Gainesville, Food and Resource Economics Dept., Agricultural Experiment Stations, Institute of Food and Agricultural Sciences, University of Florida.
v. 28 cm. (Economic information report)
1. Citrus fruit industry—Estimates—Florida—Periodicals.
I. Title. II. Series.
HD9259.C54U5865b 338.1'3 79-641503
MARC-S

—STATISTICS

United States. Foreign Agricultural Service (1953-). Fruit and Vegetable Division.
Information relating to world production and trade in fresh and processed citrus. — [Washington] : U.S. Dept. of Agriculture, Foreign Agricultural Service, Fruit and Vegetable Division, 1962.
73 p. ; 27 cm.
Cover title.
Tables.
1. Citrus fruit industry—Statistics. I. Title.
HD9259.C52U54 1962 79-310882
79 MARC

—ARGENTINE REPUBLIC

Instituto Nacional de Planificación Económica. Departamento Agropecuario.
La actividad citrícola argentina : bases para un programa nacional : documento de trabajo / Ministerio de Economía, Secretaria de Programación y Coordinación Económica, Instituto Nacional de Planificación Económica, Dirección Nacional de Planeamiento de los Sectores Productivos de Bienes, Departamento Agropecuario. — Buenos Aires : El Departamento, 1975.
v, 195, 45 leaves : ill. ; 29 cm.
1. Citrus fruit industry—Argentine Republic. I. Title.
HD9259.C53A74 79-106870
79 MARC

—EUROPE

Zampieri, Ugo.
Importazione europea della frutta e degli agrumi : itinerari e mercati per l'Italia : monografia destinata al Congresso nazionale dei dirigenti aziende commerciali / Ugo Zampieri. — Roma : Associazione nazionale fascista dirigenti aziende commerciali, 1934.
98, p., 11 leaves of plates : ill. ; 24 cm.
1. Fruit trade—Europe. 2. Citrus fruit industry—Europe. 3. Europe—Commerce—Italy. 4. Italy—Commerce—Europe. I. Title.
HD9255.E82Z35 382'.41'4094 79-347405
79 MARC

—ISRAEL

Hochman, Eitan.
The marketing of Israel's citrus fruit in major European markets / by Eitan Hochman, Meir G. Kohn, and Yakir Plessner. — Rehovot [Israel] : Center for Agricultural Economic Research, 1976.
x, 70 p. : ill. ; 24 cm.
Bibliography: p. 69-70.
1. Citrus fruit industry—Israel. 2. Israel—Commerce—Europe. 3. Europe—Commerce—Israel. I. Kohn, Meir G., joint author. II. Plessner, Yakir, 1935- joint author. III. ha-Merkaz le-meḥkar be-khalkalah ḥakla'it. IV. Title.
HD9259.C51584 382'.41'43094 78-109815
79 MARC

Lagin, Mishah.
(ha-Mediniyut shel neṭi'at hadarim le-ṭevaḥ arokh, mi-neḳudat re'uto shel ha-yatsran)
המדיניות של נטיעת הדרים לטווח ארוך, מנקודת ראותו של היצרן / מישה לגין. — [תל-אביב?] : משרד החקלאות, שירות ההדרכה והמקצוע, המחלקה לכלכלת הייצור החקלאי, 1976.
26 p. : diagrs. ; 24 cm.
Cover title.
1. Citrus fruit industry—Israel. I. Israel. ha-Maḥlaḳah le-khalkalat ha-yitsur ha-ḥakla'i. II. Title.
HD9259.C54P39 79-951078

—KOREA—CHEJU (PROVINCE) —PERIODICALS

(Kamgyul)
감귤
[서귀읍] 제주감귤협동조합
no. ill. 26 cm. bimonthly.

1. Citrus fruits—Korea—Cheju (Province)—Periodicals. 2. Citrus fruit industry—Korea—Cheju (Province)—Periodicals. I. Cheju Kamgyul Hyŏptong Chohap.
*SB369.5.K7K35 79-641907
(MARC-S)
K

—MEDITERRANEAN REGION —STATISTICS—PERIODICALS

Liaison Committee for Mediterranean Citrus Fruit Culture. Commission des études économiques.
Les exportations d'agrumes du bassin méditerranéen.
[s. l., s. n.] Paris, Impr. M. Faureau.
v. 27 cm.
"Statistiques, évaluations, répartitions."
1. Citrus fruit industry—Mediterranean region—Statistics—Periodicals. I. Title.
HD9259.C54M434 382'.41'43091822 79-645922
MARC-S

CITRUS FRUITS

see also Citrus fruit industry; Grapefruit; Orange

Citrus science and technology / edited by Steven Nagy, Philip E. Shaw, Matthew K. Veldhuis. — Westport, Conn. : Avi Pub. Co., c1977.
2 v. : ill. ; 24 cm.
Includes bibliographies and indexes.
CONTENTS: v. 1. Nutrition, anatomy, chemical composition, and bioregulation. —v. 2. Fruit production, processing practices, derived products, and personnel management.
ISBN 0-87055-221-X (v. 1) : $30.00 (v. 1)
1. Citrus fruits. 2. Citrus fruits—By-products. I. Nagy, Steven. II. Shaw, Philip E. III. Veldhuis, Matthew K.
SB369.C53 641.3'4'3 77-4438
77[r79]rev MARC

CITRUS FRUITS (Continued)

Nii, Naosuke, 1942-
(Kaju no hatsuiku to chūya no ondo jōken ni kansuru kenkyū)
果樹の発育と昼夜の温度条件に関する研究：ブドウと温州ミカンを中心として = Studies on the effect of different temperatures between day and night on the growth of fruit trees in particular grapes and Unshiu oranges / 新居直祐. — [静岡] : 静岡大学農学部園芸学科, 昭和46 [1971]
89 p. : ill. ; 26 cm. — (静岡大学農学部園芸研究報告 ; 第5号) Ja***
Cover title.
Summary in English.
Errata slip inserted.
Bibliography: p. 83-87.
1. Grapes. 2. Citrus fruits. 3. Plants, Effect of temperature on. I. Title. II. Title: Studies on the effect of different temperatures between day and night on the growth of fruit trees in particular grapes and Unshiu oranges. III. Series: Shizuoka Daigaku. Nōgakubu. Shizuoka Daigaku Nōgakubu Engei kenkyū hōkoku ; dai 5-gō.
SB398.J3N5 79-809076

—BY-PRODUCTS

see also Alcohol

Citrus science and technology / edited by Steven Nagy, Philip E. Shaw, Matthew K. Veldhuis. — Westport, Conn. : Avi Pub. Co., c1977.
2 v. : ill. ; 24 cm.
Includes bibliographies and indexes.
CONTENTS: v. 1. Nutrition, anatomy, chemical composition, and bioregulation. —v. 2. Fruit production, processing practices, derived products, and personnel management.
ISBN 0-87055-221-X (v. 1) : $30.00 (v. 1)
1. Citrus fruits. 2. Citrus fruits—By-products. I. Nagy, Steven. II. Shaw, Philip E. III. Veldhuis, Matthew K.
SB369.C53 641.3'4'3 77-4438
77[r79]rev MARC

—FLORIDA—STATISTICS

Brooke, Donald Lloyd, 1915-
Citrus production costs and returns in Florida, season 1969-70; with comparisons [by] D. L. Brooke. Gainesville, Dept. of Agricultural Economics, University of Florida, 1971.
19 l. 28 cm. (Ag[ricultural] econ[omics] report, 29)
Cover title.
1. Citrus fruits—Florida—Statistics. I. Title. II. Series.
SB369.2.F6B76 338.1'7'4309759 77-376372
78 MARC

—JAPAN

Kitagami, Mitsugu.
(Saishin kankitsu saibai sho)
最新柑橘栽培書 / 北神貢著. — 増訂改版. — 東京 : 裳華房, 明治43 [1910]
8, 456 p., [13] leaves of plates : ill. ; 23 cm.

1. Citrus fruits—Japan. I. Title.
SB369.5.J3K57 79-806037

—KOREA—CHEJU (PROVINCE)
—PERIODICALS

(Kamgyul)
감귤
[서귀읍] 제주감귤협동조합
no. ill. 26 cm. bimonthly.

1. Citrus fruits—Korea—Cheju (Province)—Periodicals. 2. Citrus fruit industry—Korea—Cheju (Province)—Periodicals. I. Cheju Kamgyul Hyŏptong Chohap.
SB369.5.K7K35 79-641907
(MARC-S)

—MEXICO—NUEVO LEÓN

Alanís Flores, Glafiro J
Estudio florístico-ecológico de las malezas en la región citrícola de Nuevo León, México / Glafiro J. Alanís Flores. — Monterrey, México : Universidad Autónoma de Nuevo León, Dirección General de la Investigación Científica y Facultad de Ciencias Biológicas, 1974.
p. 42-64 : ill. ; 23 cm. — (Publicaciones biológicas ; v. 1, no. 5)
Cover title.
Summary in Spanish and English.
Bibliography: p. 54.
1. Weeds—Mexico—Nuevo León. 2. Botany—Mexico—Nuevo León—Ecology. 3. Citrus fruits—Mexico—Nuevo León. I. Title. II. Series.
QH301.P79 vol. 1, no. 5 78-379790
[SB613.M4] 78 MARC

—NICARAGUA

Frutas, posibilidades de su desarrollo en Nicaragua. — [Managua] : Banco Central de Nicaragua, 1970.
63 p., [1] leaf of plates : map ; 23 cm.
CONTENTS: pt. 1. Malo, S. E. Frutas tropicales.—pt. 2. Krezdorn, A. H. Cítricos.
1. Fruit-culture—Nicaragua. 2. Citrus fruits—Nicaragua. 3. Tropical fruit—Nicaragua. I. Malo, Simón E. Frutas tropicales. 1970. II. Krezdorn, A. H. Cítricos. 1970.
SB354.6.N5F78 78-396118
79 MARC

—VENEZUELA—STATISTICS
—PERIODICALS

Venezuela. Ministerio de Agricultura y Cría. División de Estadística.
Estimación de cosechas: Cítricas.
Caracas, Ministerio de Agricultura y Cría, Dirección de Economía y Estadística Agropecuaria, División de Estadística.
v. 21 cm.
1. Citrus fruits—Venezuela—Statistics—Periodicals. I. Title.
SB369.5.V4V46a 79-645891
MARC-S

CITRUS INDUSTRY see Citrus fruit industry

CITRUS TRADE see Citrus fruit industry

CITTÀ DEL VATICANO see Vatican City

CITTÀ DI CASTELLO

—CHURCHES—GUIDE-BOOKS

Mancini, Giacomo.
Istruzione storico-pittorica per visitare le chiese e palazzi di Città di Castello : colle memorie di alcuni artefici del disegno che in detta città fiorirono / Giacomo Mancini. — [Bologna] : A. Forni, 1976.
326 p. ; 22 cm. It78-Mar
On spine: Chiese e palazzi di Città di Castello.
Edition limited to 300 copies.
Reprint of the 1832 ed. published by Tip. Baduel, Perugia.
Bibliography: p. 291-292.
Includes indexes.
1. Città di Castello—Description—Guide-books. 2. Città di Castello—Churches—Guide-books. 3. Città di Castello—Palaces—Guide-books. I. Title. II. Title: Chiese e palazzi di Città di Castello.
DG975.C5M34 1976 914.5'651 79-349164
*79 MARC

—DESCRIPTION—GUIDE-BOOKS

Mancini, Giacomo.
Istruzione storico-pittorica per visitare le chiese e palazzi di Città di Castello : colle memorie di alcuni artefici del disegno che in detta città fiorirono / Giacomo Mancini. — [Bologna] : A. Forni, 1976.
326 p. ; 22 cm. It78-Mar
On spine: Chiese e palazzi di Città di Castello.
Edition limited to 300 copies.
Reprint of the 1832 ed. published by Tip. Baduel, Perugia.
Bibliography: p. 291-292.
Includes indexes.
1. Città di Castello—Description—Guide-books. 2. Città di Castello—Churches—Guide-books. 3. Città di Castello—Palaces—Guide-books. I. Title. II. Title: Chiese e palazzi di Città di Castello.
DG975.C5M34 1976 914.5'651 79-349164

—PALACES—GUIDE-BOOKS

Mancini, Giacomo.
Istruzione storico-pittorica per visitare le chiese e palazzi di Città di Castello : colle memorie di alcuni artefici del disegno che in detta città fiorirono / Giacomo Mancini. — [Bologna] : A. Forni, 1976.
326 p. ; 22 cm. It78-Mar
On spine: Chiese e palazzi di Città di Castello.
Edition limited to 300 copies.
Reprint of the 1832 ed. published by Tip. Baduel, Perugia.
Bibliography: p. 291-292.
Includes indexes.
1. Città di Castello—Description—Guide-books. 2. Città di Castello—Churches—Guide-books. 3. Città di Castello—Palaces—Guide-books. I. Title. II. Title: Chiese e palazzi di Città di Castello.
DG975.C5M34 1976 914.5'651 79-349164
*79 MARC

CITTADELLA DI CAPO BONIFATI, ITALY

—HISTORY

Engels, Giovannino.
Fella benedettina : (memorie storiche) / Giovannino Engels. — Pieve S[anto] Stefano : Tip. Dalla Ragione, 1977.
140 p. : ill. ; 24 cm. It78-July
Italian or Latin.
Bibliography: p. [143]
L3500
1. Cittadella di Capo Bonifati, Italy—History. I. Title.
DG975.C513E53 945'.785 78-382310
*79 MARC

CITY, OPEN (INTERNATIONAL LAW) see Open city (International law)

CITY AND STATE, RELATION OF see Municipal home rule

CITY AND TOWN LIFE

Cousins, Albert N 1919-
Urban life : the sociology of cities and urban society / Albert N. Cousins, Hans Nagpaul. — New York : Wiley, c1979.
xvi, 608 p. : ill. ; 24 cm.
Includes bibliographies and indexes.
ISBN 0-471-03026-0
1. Sociology, Urban. 2. Urbanization. 3. Urbanization—United States. 4. City and town life. 5. City and town life—United States. I. Nagpaul, Hans, joint author. II. Title.
HT151.C65 301.36'3'0973 78-14427
78 MARC

Farallones Institute.
The integral urban house : self-reliant living in the city / by the Farallones Institute ; introd. by Sim Van der Ryn ; [photos. by Pat Gaidvis]. — San Francisco : Sierra Club Books, c1979.
x, 494 p. : ill. ; 28 cm.
Bibliography: p. 452-459.
Includes index.
ISBN 0-87156-213-8 : $12.95
1. Dwellings. 2. Environmental engineering (Buildings) 3. City and town life. I. Title.
TH4812.F37 1979 643 78-8592
78 MARC

—ADDRESSES, ESSAYS, LECTURES

L'Homme et sa ville / [J. Chirac et al.]. — Paris : Tallandier, 1977.
132 p. ; 23 cm. — (La Nef ; 34. année, nouv. sér., cahier no 64) F***
18.00F
1. Cities and towns—Addresses, essays, lectures. 2. City and town life—Addresses, essays, lectures. I. Chirac, Jacques, 1932- . II. Series.
HT151.H642 909'.09'732 79-362311
79 MARC

Sauberzweig, Dieter.
Bildungsreform und Stadtkultur / Dieter Sauberzweig. — 1. Aufl. — Braunschweig : Westermann, 1978.
215 p. ; 19 cm. — (Theorie und Praxis der Erwachsenenbildung) (Westermann-Taschenbuch ; 173) GFR79-A2
Includes bibliographical references.
ISBN 3-14-167173-7
1. Adult education—Germany, West—Addresses, essays, lectures. 2. City and town life—Addresses, essays, lectures. I. Title.
LC5256.G4S28 79-378309
79 MARC

CITY AND TOWN LIFE (Continued)

—ADDRESSES, ESSAYS, LECTURES —JUVENILE LITERATURE

Stories of the city / selected by Veronica Harvey. — London : Evans, 1977.
126 p. : ill. ; 23 cm. GB***
ISBN 0-237-44877-7 : £3.75
1. City and town life—Addresses, essays, lectures—Juvenile literature. I. Harvey, Veronica.
HT151.S694 301.36 79-314938
79 MARC

—BIBLIOGRAPHY

Diner, Hasia R
Women and urban society : a guide to information sources / Hasia R. Diner. — Detroit : Gale Research Co., c1979.
xii, 138 p. ; 23 cm. — (Urban studies information guide series ; v. 7) (Gale information guide library)
Bibliography: p. 113-116.
Includes indexes.
ISBN 0-8103-1425-8
1. Women—Bibliography. 2. Urbanization—Bibliography. 3. City and town life—Bibliography. 4. Sociology, Urban—Bibliography. I. Title. II. Series.
Z7961.D55 016.30141'2 78-13109
[HQ1154] 78 MARC

—COLLECTED WORKS

Guidicini, Paolo.
Gruppi e sub-unità spaziali nella città : quartiere, vicinato ed area naturale tra miti, utopie e valutazioni critiche / Paolo Guidicini. — Roma : Città nuova, 1978.
259 p. ; 18 cm. — (Idee ; 33) It78-June
"Testi antologici": p. [77]-243.
Bibliography: p. [63]-76.
Includes index.
L4700
1. Sociology, Urban—Collected works. 2. Community—Collected works. 3. City and town life—Collected works. I. Title.
HT153.G84 78-387850
*79 MARC

—CONGRESSES

International Federation for Housing and Planning.
Building the good city = La ville de nos aspirations = Die Stadt unserer Erwartungen. — The Hague : International Federation for Housing and Planning, [1968?]
292 p., [4] leaves of plates : ill. ; 23 cm.
At head of title: XXIXth World congress of the International Federation for Housing and Planning ... Philadelphia, 1968.
English, French, and German.
1. City planning—Congresses. 2. City and town life—Congresses. I. Title.
HT166.I615 1968 309.2'62 78-314230
78 MARC

International Federation for Housing and Planning.
Urban transportation and urban pattern = Circulation et structure urbaines = Stadtverkehr und Stadtstruktur ; Housing standards for low and lowest income groups ... = Standards de logement pour les groupes à revenus faibles et minima ... = Wohnungsstandards für Gruppen der niedrigen und niedrigsten Einkommenstufen .... — The Hague : International Federation for Housing and Planning, [1966?]
265 p., [8] leaves of plates : ill. ; 23 cm.
At head of title: XXVIIIth World congress of the International Federation for Housing and Planning ... Tokyo, 1966.
English, French, and German.
1. City planning—Congresses. 2. City and town life—Congresses. I. Title.
HT166.I615 1966 309.2'62 78-320487
78 MARC

The Itasca seminar : the value of urban life : cities as if people mattered, October 3-7, 1977. — [Minneapolis, : Minneapolis Foundation, c1978.]
81 p. : ports ; 23 cm.
Cover title.
Sponsored by the Minneapolis Foundation.
$4.50
1. City and town life—Congresses. 2. Urban economics—Congresses. 3. Local government—United States—Congresses. 4. Urban schools—Congresses. 5. Minneapolis Foundation. I. Title: The value of urban life.
HT123.I83 301.36'3 78-107009

—JUVENILE LITERATURE

Mathieu, Joseph.
Busy city / Joe Mathieu. — New York : Random House, c1978.
[36] p. : col. ill. ; 21 cm. — (The Best book club ever)
SUMMARY: A brief introduction to the people and places of a modern city.
ISBN 0-394-83700-2
1. Cities and towns—Juvenile literature. 2. City and town life—Juvenile literature. [1. City and town life. 2. Cities and towns] I. Title.
HT155.M28 301.36'3 77-86138
79 MARC

—JUVENILE POETRY

Crunk, Tony.
Two towns & the people who lived in them / Tony Crunk ; [Melissa Hamblin, ill.]. — Danville, Ky. : Vantage Point Press, 1978.
44 p. : ill. ; 19 cm.
"500 copies."
SUMMARY: Two towns, one inhabited by people who laughed all the time and the other by people who were always gloomy, discover each other and nothing is the same again.
1. City and town life—Juvenile poetry. [1. City and town life—Poetry. 2. American poetry] I. Hamblin, Melissa. II. Title.
PZ8.3.C88645 Tw 811'.5'4 78-57037
79 MARC

—LITERARY COLLECTIONS

The City / edited by Adele Stern. — New York : Scribner, c1979.
vi, 248 p. : ill. ; 23 cm. — (America in literature) (Scribner student paperbacks ; SSP 45)
Includes index.
ISBN 0-684-16139-7 : $6.50
1. American literature—20th century. 2. City and town life—Literary collections. I. Stern, Adele. II. Series.
PS509.C57C5 810'.8'032 79-13292
79 MARC

—ENGLAND—HISTORY—PICTORIAL WORKS—JUVENILE LITERATURE

Goodall, John S
The story of an English village / [by] John S. Goodall. — London : Macmillan, 1978.
[62] p. : of col. ill. ; 19 x 26 cm. GB78-30276
Page width varies.
ISBN 0-333-23835-4 : £2.95
1. Cities and towns—England—Growth—Pictorial works—Juvenile literature. 2. City and town life—England—History—Pictorial works—Juvenile literature. I. Title.
HT133.G66 1978 942'.009'734 79-301017
79 MARC

Goodall, John S
The story of an English village / John S. Goodall. — 1st American ed. — New York : Atheneum, 1979, c1978.
[61] p. : col ill. ; 19 x 26 cm.
"A Margaret K. McElderry book."
SUMMARY: Shows the growth of an English village from a medieval clearing to the urban congestion of the present day as seen from the same viewpoint approximately every hundred years.
ISBN 0-689-50125-0 : $7.95
1. Cities and towns—England—Growth—Pictorial works—Juvenile literature. 2. City and town life—England—History—Pictorial works—Juvenile literature. [1. Cities and towns—England—Growth—Pictorial works. 2. City and town life—England—History—Pictorial works. 3. England—Social life and customs—Pictorial works] I. Title.
HT133.G66 1979 942'.009'734 78-56242
00 MARC

—FINLAND—ANECDOTES, FACETIAE, SATIRE, ETC.

Onnimanni.
Hellästi hymyillen : pakinoita / Onnimanni ; kuv., Heikki Eklöf. — [Hämeenlinna] : Karisto, 1978.
143 [1] p. : ill. ; 21 cm. Fi78-10
ISBN 9512312654 : Fmk38.00
1. City and town life—Finland—Anecdotes, facetiae, satire, etc. 2. Finnish wit and humor. I. Title.
PH355.O53H4 79-383259
79 MARC

—FINLAND—CASE STUDIES

Ihminen ja pienyhteisö kulttuurin muutoksessa : metodiraportti / toimittanut Katriina Petrisalo. — Joensuu : Joensuun korkeakoulu, Karjalan tutkimuslaitos, 1977.
86 leaves : map ; 30 cm. — (Julkaisuja - Joensuun korkeakoulu Karjalan tutkimuslaitos ; n:o 24 ISSN 0355-6859) Fi***
"Tutkimusprojekti ... Kulttuurien vertaileva tutkimus."
Includes bibliographies.
ISBN 9516961304
1. City and town life—Finland—Case studies. 2. Community life—Case studies. I. Petrisalo, Katriina. II. Series: Joensuun korkeakoulu. Karjalan tutkimuslaitos. Julkaisuja - Joensuun korkeakoulu, Karjalan tutkimuslaitos ; n:o 24.
HT145.F5I46 78-394094
79 MARC

—FRANCE

Augoyard, Jean François, 1941-
Pas à pas : essai sur le cheminement quotidien en milieu urbain / Jean-François Augoyard. — Paris : Seuil, c1979.
185 p. : ill. ; 21 cm. — (Collection Espacements) F***
Bibliography: p. 181-[183]
Includes index.
ISBN 2-02-005170-2 : 47.00F
1. City and town life—France. 2. Urbanization—France. 3. City planning—France. I. Title.
HT135.A88 301.36'0944 79-375994
79 MARC

—GERMANY, WEST

Schulz, Wolfgang, 1940 (July 20)-
Sozialkontakte in der Grossstadt / Wolfgang Schulz. — Wien : Inst. f. Stadtforschung ; Wien ; München : Jugend & Volk in Komm., [1978]
173 p. ; 24 cm. — (Publikationen des Instituts für Stadtforschung ; 48) Au78-16-92
Bibliography: p. 161-166.
ISBN 3-7141-7827-9 : S280.00
1. Social interaction. 2. Sociology, Urban. 3. City and town life—Germany, West. I. Title. II. Series: Institut für Stadtforschung. Publikationen des Instituts für Stadtforschung ; 48.
HM131.S378 78-397303
*79 MARC

—GERMANY, WEST—BOCHUM—EXHIBITIONS

Über die moderne Art zu leben : oder, Rationalisierung des Lebens in der modernen Stadt / hrsg. vom Museum Bochum, Kunstsammlung ; Michael Fehr, Diethelm Koch ; mit Aufnahmen von Michael Wolf ... [et al.]. — 1. Aufl. — Lahn-Giessen : Anabas, 1977.
223 p. : ill. ; 26 cm. GFR77-A
Catalog of an exhibition held Sept. 3-Oct. 9, 1977, at the Museum Bochum.
Bibliography: p. 220.
ISBN 3-87038-051-9
1. Photography, Documentary—Exhibitions. 2. Bochum—Description—Views—Exhibitions. 3. City and town life—Germany, West—Bochum—Exhibitions. I. Fehr, Michael. II. Koch, Diethelm. III. Wolf, Michael, 1954- . IV. Bochum. Museum. V. Title: Rationalisierung des Lebens in der modernen Stadt.
TR820.5.U32 79-368533
79 MARC

—GREAT BRITAIN—HISTORY

Meller, Helen Elizabeth.
Leisure and the changing city, 1870-1914 / H. E. Meller. — London ; Boston : Routledge & Paul, 1976.
308 p., [4] leaves of plates : ill. ; 23 cm. GB***
Bibliography: p. 279-294.
Includes index.
ISBN 0-7100-8430-7 : £7.50
1. Urbanization—England—Bristol—History. 2. Bristol, Eng.—Amusements and recreational activities—History. 3. City and town life—Great Britain—History. I. Title.
HN398.B73M44 1976 942.3'93'081 76-383866
77[r79]rev MARC

—INDONESIA—YOGYAKARTA—STATISTICS

Indonesia. Direktorat Tata Kota dan Daerah.
Kompilasi data peningkatan kwalitas lingkungan pemukiman kotamadya Yogyakarta : tahap pendahuluan studi sosial. — [Jakarta] : Direktorat Tata Kota & Daerah, Dirjen. Cipta Karya, Dept. PUTL bekerjasama dengan UNICEF, [1975]
79 leaves in various foliations : maps ; 31 cm.
1. Yogyakarta, Indonesia—Social conditions—Statistics. 2. City and town life—Indonesia—Yogyakarta—Statistics. I. United Nations. Children's Fund. II. Title.
HN710.Y6I53 1975 76-940524

CITY AND TOWN LIFE (Continued)

—ITALY—HISTORY

Dupre Theseider, Eugenio.
Mondo cittadino e movimenti ereticali nel Medio Evo : (saggi) / ₍di₎ E. Dupré Theseider. — Bologna : Pàtron, 1978.
453 p. ; 24 cm. It79-Mar
Includes bibliographical references and index.
L13000
1. Heresies and heretics—Middle Ages, 600-1500. 2. City and town life—Italy—History. 3. Italy—Church history. 4. Italy—History—476-1492. I. Title.
BT1319.D86 79-358949
*79 MARC

—JAPAN—PUBLIC OPINION

Tokyo. Tomin Seikatsukyoku. Sanka Suishinbu. Kōchōka.
(Toshi kankyō ni kansuru seron chōsa)
都市環境に関する世論調査 / ₍編集 都民生活局参加推進部広聴課₎. — ₍東京₎ : 同課, 昭和53 ₍1978₎
104 p. ; 25 cm. Ja***
Cover title.
1. Cities and towns—Environmental aspects—Japan—Public opinion. 2. Cost and standard of living—Japan—Public opinion. 3. City and town life—Japan—Public opinion. 4. Public opinion—Japan. I. Title.
HT147.J3T53 1978 79-809223

—JAPAN—KINKI REGION—HISTORY

Sakudō, Yōtarō, 1924-
(Edo jidai no Kamigata chōnin)
江戸時代の上方町人 / 作道洋太郎. — ₍東村山₎ : 教育社 ; 東京 : 販売 教育社出版サービス, ₍1978₎
207 p., ₍1₎ leaf of plates : ill. ; 18 cm. — (歴史新書 ; 88)
Ja 78-25369
Bibliography: p. 203-207.
¥600
1. Merchants—Japan—Kinki region—History. 2. Kinki region, Japan—Commerce—History. 3. City and town life—Japan—Kinki region—History. I. Title. II. Series: Rekishi shinsho ; 88.
HF3829.K5S24 79-803971

—JAPAN—TOKYO
—PUBLIC OPINION

Tokyo. Tomin Seikatsukyoku. Sanka Suishinbu. Kōchōka.
(Tomin seikatsu ni kansuru seron chōsa)
都民生活に関する世論調査 / ₍編集 都民生活局参加推進部広聴課₎. — 東京 : 同課, 昭和53 ₍1978₎
166 p. : ill. ; 26 cm. Ja***
Cover title.
1. Cost and standard of living—Japan—Tokyo—Public opinion. 2. City and town life—Japan—Tokyo—Public opinion. 3. Public opinion—Japan—Tokyo. I. Title.
HD7057.T65 1978 78-813962

—KOREA

Sŏl, Pong-sik.
(Han'guk kagye ŭi sobi haengwi)
韓國家計의消費行爲 : 都市家計調査資料의分析 / 薛鳳植著. — 서울 : 創文閣, 1978.
103 p. ; 23 cm.
Bibliography: p. 67-77.
Appendices (p. ₍79₎-103): 1. 都市家計의消費支出變動.—2. 民間消費支出의變動.
₩1500
1. Consumers—Korea. 2. Cost and standard of living—Korea. 3. City and town life—Korea. I. Title.
HC470.C6S64 79-826267

—LITHUANIA

Daniliauskas, Antanas.
Lietuvos miesto gyventojų materialinė kultūra XX a. : etnografinė apybraiža / Antanas Daniliauskas. — Vilnius : Mokslas, 1978.
108, ₍4₎ p. : ill. ; 22 cm. USSR***
At head of title: Lietuvos TSR Mokslų akademija. Istorijos institutas.
Summary in Russian.
Bibliography: p. 103-₍109₎
0.75rub
1. Dwellings—Lithuania. 2. City and town life—Lithuania. 3. Lithuania—Social life and customs. I. Title.
GT342.L77D36 79-375121
79 MARC

—NETHERLANDS—BRABANT (DUCHY)
—CONGRESSES

Colloquium De Brabantse Stad, 5th, 's-Hertogenbosch, Netherlands, 1977.
Verslag Vijfde Colloquium De Brabantse Stad, 's-Hertogenbosch, 25-26 Nov. 1977. — 's-Hertogenbosch : Historische Sectie van het Provinciaal Genootschap van Kunsten en Wetenschappen in Noord-Brabant, 1978.
xxii, 402 p., 2 leaves of plates : ill. ; 22 cm. Ne***
Organized by the Belgisch-Nederlands Cultureel Verdrag.
Introd. previously published in Varia Historica Brabantica VI-VII.
Includes bibliographical references and index.
1. Brabant (Duchy)—Intellectual life—Congresses. 2. Brabant (Duchy)—Social life and customs—Congresses. 3. City and town life—Netherlands—Brabant (Duchy)—Congresses. I. Title.
DH801.B747C64 1977 79-361245
79 MARC

—RUSSIA—GROWTH

Trufanov, Ivan Pavlovich.
(Problemy byta gorodskogo naseleni͡a SSSR)
Проблемы быта городского населения СССР. Отв. ред. д-р ист. наук, проф. В. А. Ежов. Ленинград, Изд-во Ленингр. ун-та, 1973.
144 p. 21 cm. USSR 73
At head of title: Министерство высшего и среднего специального образования РСФСР. И. П. Труфанов.
Includes bibliographical references.
0.70rub
1. Cities and towns—Russia—Growth. 2. Sociology, Urban—Russia. 3. City and town life—Russia. I. Title.
HT145.R9T78 73-360015

—SOUTH AFRICA—ADDRESSES,
ESSAYS, LECTURES

South Africa's urban blacks : problems and challenges / G. Marais and R. van der Kooy, editors. — Pretoria : Centre for Management Studies, School of Business Leadership, University of South Africa, ₍1978?₎
370 p. : ill. ; 22 cm. SA***
Includes bibliographies and index.
CONTENTS: Swart, C. F. and Oosthuizen, A. J. G. Urbanisation.—Coetzee, J. H. A sociological view of a new society.—Du Preez, P. H. The psychology of the urban black.—Van der Reis, A. P. Urban black interests, attitudes, and consumer behaviour.—Möller, H. J. The dilemma of the church.—Gugushe, R. N. Education for free development.—Nel, P. The black consumer.—Kraft, R. L. and Motsepe, W. E. O. The urban black worker.—Mokoatle, B. N. The black entrepreneur in South Africa.—Marais, G. Economic constraints and black mobility.—Kotze, D. A. Political spectators.—Marais, G. Urban blacks and nationhood.—Tusenius, R. R. The problem, holistically, realistically, rationally, and objectively.—De Klerk, W. J. Fact, fiction, and philosophy.
ISBN 0-909063-13-3
1. Blacks—South Africa—Social conditions—Addresses, essays, lectures. 2. Blacks—South Africa—Economic conditions—Addresses, essays, lectures. 3. Urbanization—South Africa—Addresses, essays, lectures. 4. Blacks—South Africa—Psychology—Addresses, essays, lectures. 5. City and town life—South Africa—Addresses, essays, lectures. I. Marais, Georg, 1931- . II. Van der Kooy, R.
DT763.6.S69 301.45'19'6068 79-308532
79 MARC

—UNITED STATES

Cousins, Albert N 1919-
Urban life : the sociology of cities and urban society / Albert N. Cousins, Hans Nagpaul. — New York : Wiley, c1979.
xvi, 608 p. : ill. ; 24 cm.
Includes bibliographies and indexes.
ISBN 0-471-03026-0
1. Sociology, Urban. 2. Urbanization. 3. Urbanization—United States. 4. City and town life. 5. City and town life—United States. I. Nagpaul, Hans, joint author. II. Title.
HT151.C65 301.36'3'0973 78-14427
78 MARC

—UNITED STATES—ADDRESSES,
ESSAYS, LECTURES

Street, David.
Handbook of contemporary urban life / David Street and associates. — 1st ed. — San Francisco : Jossey-Bass, 1978.
xxii, 741 p. ; 24 cm. — (The Jossey-Bass social and behavioral science series)
Bibliography: p. 643-712.
Includes indexes.
ISBN 0-87589-372-4
1. City and town life—United States—Addresses, essays, lectures. 2. Urbanization—United States—Addresses, essays, lectures. I. Title.
HT123.S787 301.36'0973 78-1155
78 MARC

Urban life and the struggle to be human / ₍compiled by₎ Albert J. Mayer, Leonard Gordon. — Dubuque, Iowa : Kendall/Hunt Pub. Co., c1979.
vii, 233 p. : ill. ; 28 cm.
Includes bibliographical references and indexes.
ISBN 0-8403-2035-3
1. Urbanization—United States—Addresses, essays, lectures. 2. Sociology, Urban—Addresses, essays, lectures. 3. City and town life—United States—Addresses, essays, lectures. I. Mayer, Albert J., 1939- . II. Gordon, Leonard.
HT123.U746 301.36'0973 79-84620
00 MARC

—UNITED STATES—CASE STUDIES

Allon, Natalie.
Urban life styles / Natalie Allon. — Dubuque, Iowa : W. C. Brown Co. Publishers, c1979.
xvi, 215 p. ; 23 cm. — (Principal themes in sociology)
Bibliography: p. 197-204.
Includes indexes.
ISBN 0-697-07558-3
1. City and town life—United States—Case studies. 2. Reducing—Societies, etc.—Social aspects—United States. 3. Health resorts, watering-places, etc.—Social aspects—United States. 4. Hotels, taverns, etc.—Social aspects—United States. 5. Self-acceptance. 6. Interpersonal relations. 7. Middle classes—United States. I. Title.
HT123.A617 301.36'0973 78-74488
00 MARC

—UNITED STATES—PUBLIC OPINION

Louis Harris and Associates.
A survey of citizen views and concerns about urban life : final report / conducted for Department of Housing and Urban Development, contract #H-2835, conducted by Louis Harris and Associates, inc. — ₍Washington₎ : U.S. Dept. of Housing and Urban Development, Office of Policy Development and Research, 1978-
v. ; 26 cm.
"Study no. P2795"
"HUD-PDR-306"
Includes bibliographical references.
1. City and town life—United States—Public opinion. 2. Public opinion—United States. I. United States. Dept. of Housing and Urban Development. Office of Policy Development and Research. II. Title.
HT123.L67 1978 78-324617
301.15'43'301360973
79 MARC

CITY AND TOWN LIFE IN MOTION PICTURES

Belmans, Jacques.
La ville dans le cinéma : de Fritz Lang à Alain Resnais / Jacques Belmans. — Bruxelles : A. de Boeck, 1977.
285 p., ₍16₎ leaves of plates : ill. ; 19 cm. — (Collection Univers des sciences humaines ; 11) Be***
Bibliography: p. 284-285.
Includes indexes.
1. City and town life in motion pictures. I. Title.
PN1995.9.C513B44 791.43'0909'355 79-373634
79 MARC

CITY ATTORNEYS

—GEORGIA

Institute for City and County Attorneys, 24th, Athens, Ga., 1977.
Program materials for Twenty-fourth Institute for City and County Attorneys, September 9, 10, 1977, Georgia Center for Continuing Education, Athens, Georgia / ₍sponsored by the Institute of Continuing Legal Education in Georgia, in cooperation with City Attorneys' Section, Georgia Municipal Association, County Attorneys' Section, Association County Commissioners of Georgia, Local Government Section, State Bar of Georgia₎. — Athens : Institute of Continuing Legal Education in Georgia, c1977.
ca. 200 leaves in various foliations ; 29 cm.
1. Local government—Georgia. 2. City attorneys—Georgia. 3. County attorneys—Georgia. I. Institute of Continuing Legal Education in Georgia. II. Title.
KFG430.A75I5 1977 342'.758'09 78-108443

CITY ATTORNEYS' OPINIONS

—DISTRICT OF COLUMBIA

District of Columbia. Corporation Counsel.
Opinions of the Corporation Counsel. v. 1– 1976/77–
[Washington, Office of the Corporation Counsel]
v. 26 cm.
1. City attorneys' opinions—District of Columbia. I. Title.
KFD1640.A553 348'.753'05 78-640866 MARC-S

CITY BEAUTIFICATION see Urban beautification

CITY BLOCKS see Blocks (City planning)

CITY CHILDREN

Pollowy, Anne-Marie.
The urban nest / by Anne-Marie Pollowy. — Stroudsburg, Pa. : Dowden, Hutchinson & Ross, c1977.
xiv, 162 p. : ill. ; 24 cm. — (Community development series ; v. 26)
Includes bibliographies and index.
ISBN 0-87933-235-2
1. City children. 2. Child development. 3. Personal space. I. Title.
HT206.P64 301.36 76 76-22583 MARC

Ward, Colin.
The child in the city / Colin Ward ; with photos. by Ann Golzen and others. — London : Architectural Press, 1978, c1977.
ix, 221 p. : ill. ; 26 cm. GB****
Includes bibliographical references and index.
ISBN 0-85139-118-4 : £5.95
1. City children. I. Golzen, Ann. II. Title.
HT206.W37 1978b 301.43'14'091732 79 78-319288 MARC

—ADDRESSES, ESSAYS, LECTURES

La Ville & [i.e. et] l'enfant : [exposition, Centre national d'art et de culture Georges Pompidou, du 26 octobre 1977 au 13 février 1978 / organisée par le Centre de création industrielle]. — [Paris] : Centre Georges Pompidou, [c1977]
294 p. : ill. ; 24 cm. F***
Bibliography: p. 286-294.
ISBN 2-85850-043-6 : 50.00F
1. City children—Addresses, essays, lectures. I. Centre Georges Pompidou. II. Centre de création industrielle.
HT206.V54 301.43'1 79 78-392888 MARC

—PICTORIAL WORKS

New York street kids : 136 photographs / selected by the Children's Aid Society ; with captions by John von Hartz. — New York : Dover Publications, 1978.
106 p., [1] leaf of plates : chiefly ill. ; 28 cm.
ISBN 0-486-23692-7 : $14.50
1. City children—Pictorial works. 2. Children—New York (City)—Pictorial works. I. Von Hartz, John. II. Children's Aid Society, New York.
HT206.N48 1978 301.43'14'097471 79 78-57823 MARC

CITY CHURCHES

see also Church work with apartment dwellers; City missions

DuBose, Francis M
How churches grow in an urban world / Francis M. DuBose. — Nashville : Broadman Press, c1978.
181 p. ; 20 cm.
Bibliography: p. 172-175.
Includes index.
ISBN 0-8054-2531-4
1. Church growth. 2. City churches. I. Title.
BV652.25.D8 254'.5 79 78-66365 MARC

Gibbs, Eddie.
Urban church growth : clues from South America and Britain / by Eddie Gibbs. — Bramcote : Grove Books, 1977.
24 p. ; 22 cm. — (Grove booklet on ministry and worship ; no. 55 ISSN 0305-3067) GB78-08798
Includes bibliographical references.
ISBN 0-905422-24-4 : £0.35
1. Church growth. 2. City churches. I. Title.
BV652.25.G5 1977 254.2'2 78 78-314491 MARC

—ADDRESSES, ESSAYS, LECTURES

Discipling the city : theological reflections on urban mission / editor, Roger S. Greenway ; contributors, Harvie M. Conn ... [et al.]. — Grand Rapids : Baker Book House, c1979.
286 p. ; 22 cm.
Includes bibliographical references.
ISBN 0-8010-3727-1 : $5.95
1. City churches—Addresses, essays, lectures. 2. City missions—Addresses, essays, lectures. I. Greenway, Roger S. II. Conn, Harvie M.
BV637.D5 261.8 79 78-67165 MARC

—AFRICA

Monsma, Timothy, 1933-
An urban strategy for Africa / Timothy Monsma. — Pasadena, Calif. : W. Carey Library, c1979.
xv, 175 p. : ill. ; 22 cm.
Bibliography: p. 153-170.
Includes index.
ISBN 0-87808-430-4
1. City churches—Africa. 2. Missions—Africa. 3. Urbanization—Africa. 4. Africa—Social conditions. I. Title.
BV637.M59 254'.5'096 79 79-11273 MARC

—AFRICA—CONGRESSES

Stage de pastorale, Sarh, Chad, 1974.
La ville dans la société africaine d'aujourd'hui : questions posées à l'Église / Stage de pastorale, Sarh, 24 juin-13 juillet 1974. — [Sarh : s.n., between 1974 and 1978]
122 p. : ill. ; 27 cm.
1. City churches—Africa—Congresses. 2. Sociology, Christian—Africa—Congresses. I. Title.
BR1360.S7 1974 261.8 79 77-980948 MARC

—ILLINOIS—CHICAGO

Ziegenhals, Walter E 1928-
Urban churches in transition : reflections on selected problems and approaches to churches and communities in racial transition based on the Chicago experience / Walter E. Ziegenhals. — New York : Pilgrim Press, c1978.
201 p. ; 21 cm.
Bibliography: p. 195-201.
ISBN 0-8298-0355-6 : $5.95
1. City churches—Illinois—Chicago. 2. Church and race relations—Illinois—Chicago. 3. Chicago—Church history. 4. Chicago—Race relations. I. Title.
BV637.Z53 261.8'34'510420977311 78 78-16422 MARC

—SOUTHERN STATES

Fichter, Joseph Henry, 1908-
Dynamics of a city church / Joseph H. Fichter. — Reprint ed. — New York : Arno Press, 1978, c1951.
283 p., [1] leaf of plates : ill. ; 24 cm. — (The American Catholic tradition)
Reprint, with a new introd. by the author, of v. 1. of Southern parish, published by the University of Chicago Press.
Includes bibliographical references and index.
ISBN 0-405-10829-X
1. City churches—Southern States. 2. Parishes—Southern States. 3. Social surveys—Southern States. 4. Catholics—Southern States. I. Title. II. Series.
BX1407.S6F5 1978 282'.75 77 77-11286 MARC

—UNITED STATES—ADDRESSES, ESSAYS, LECTURES

Models of metropolitan ministry / B. Carlisle Driggers, compiler. — Nashville : Broadman Press, c1979.
196 p. ; 20 cm.
Includes bibliographical references.
ISBN 0-8054-2533-0
1. City churches—United States—Addresses, essays, lectures. 2. Pastoral theology—Addresses, essays, lectures. I. Driggers, B. Carlisle.
BV637.M56 254.2'2 00 78-67997 MARC

—UNITED STATES—CONGRESSES

Congress on the Urban Church, Chicago, 1978.
Metro-ministry / edited by David Frenchak and Sharrel Keyes. — Elgin, Ill. : D. C. Cook Pub. Co., c1979.
218 p. ; 22 cm.
A collection of addresses given in May of 1978 at the Congress on the Urban Church sponsored by the Seminary Consortium for Urban Pastoral Education.
Includes bibliographies.
ISBN 0-89191-101-4 : $6.95
1. City churches—United States—Congresses. I. Frenchak, David. II. Keyes, Sharrel. III. Seminary Consortium for Urban Pastoral Education. IV. Title.
BV637.C58 1978 253 79 78-72935 MARC

CITY CLERGY

see also City churches

CITY COUNCILMEN

—LEGAL STATUS, LAWS, ETC.—VIRGINIA

Wilkinson, Michaux H
Handbook for Virginia mayors and councilmembers / by Michaux H. Wilkinson and Mary Jo Fields. — x, 158 p. ; 28 cm.
x, 158 p. ; 28 cm. — (Joint report - Virginia Municipal League and Institute of Government, University of Virginia ; no. 37)
Edition for 1963 by J. D. Weeks published under title: Handbook for Virginia mayors and councilmen.
Bibliography: p. [117]-120.
Includes index.
1. Municipal corporations—Virginia. 2. City councilmen—Legal status, laws, etc.—Virginia. 3. Mayors—Legal status, laws, etc.—Virginia. I. Fields, Mary Jo, joint author. II. Weeks, J. Devereux. Handbook for Virginia mayors and councilmen. III. Title. IV. Series: Virginia Municipal League. Joint report - Virginia Municipal League and Institute of Government, University of Virginia ; no. 37.
KFV2831.Z9W4 1979 352'.008'09755 00 79-112050 MARC

—ENGLAND—LONDON—BIOGRAPHY

Stockdale, Edmund, Sir, bart., 1903-
Ptolemy Tortoise : pages from the life of an alderman / [by] Sir Edmund Stockdale ; with illustrations and some photographs by the author and others and a foreword by Lord Cullen of Ashbourne. — [London] : The author, 1978
xii, 235 p., [4] leaves of plates : ill., music, ports. (some col.) ; 25 cm. GB79-07854
Ill. on lining paper.
£4.50
1. Stockdale, Edmund, Sir, bart., 1903- . 2. London—Mayors—Biography. 3. City councilmen—England—London—Biography. 4. London—Politics and government. I. Title.
DA676.8.S75A36 352'.008'0924 79 79-315340 MARC

—GERMANY, WEST—BAVARIA

Handbuch für Stadt- und Gemeinderäte in Bayern / Hermann Büchner ... [et al.] ; Vorw., Paul Diethei ; Red., Manfred Wegmann. — München : Verlag für Verwaltungspraxis Rehm, 1978.
261 p. ; 22 cm. GFR78-A
Bibliography: p. 249-251.
Includes index.
ISBN 3-8073-0248-4 : DM28.80
1. City councilmen—Germany, West—Bavaria. 2. Municipal corporations—Germany, West—Bavaria. I. Büchner, Hermann, 1949- II. Wegmann, Manfred.
342'.43'09 78-384930

CITY COUNCILMEN (Continued)

—GERMANY, WEST—NORTHEIM
—DIRECTORIES

Bürgermeister und Ratsherren der Stadt Northeim : 1252 bis 1977 / von Adolf Hueg ... ⌊et al.⌋. — Göttingen : Reise, 1978.
191 p. : numerous ill., 2 maps ; 22 cm. — (Die Northeimer, Beiträge zur Northeimer Familien- und Bevölkerungsgeschichte) (Veröffentlichung der Familienkundlichen Kommission für Niedersachsen und Bremen)
GFR78-A6
1. Northeim, Ger.—Mayors—Directories. 2. City councilmen—Germany, West—Northeim—Directories. I. Hueg, Adolf, 1887-1950. II. Series. III. Series: Familienkundliche Kommission für Niedersachsen und Bremen sowie Angrenzende Ostfälische Gebiete. Veröffentlichung.
JS5437.B84 352'.005'2094359 79-378621
79 MARC

CITY CRIME see Crime and criminals

CITY ECOLOGY (BIOLOGY) see Urban ecology (Biology)

CITY-FEDERAL RELATIONS see Federal-city relations

CITY FLORA see Urban flora

CITY GEOLOGY see Urban geology

CITY GOVERNMENT see Municipal government

CITY LIBRARY, PORT LOUIS, MAURITIUS

Benoit, Gaetan M
The City Library of Port Louis, a brief history (1851-1968) / by Gaetan M. Benoit. — Port Louis, Mauritius : The Library, 1977.
18, ⌊1⌋ leaves ; 32 cm.
Bibliography: leaf ⌊19⌋
1. City Library, Port Louis, Mauritius. I. Title.
Z845.M38B45 027.4'69'82 78-110359
79 MARC

CITY LIFE see City and town life

CITY MANAGEMENT see Municipal government by city manager

CITY MANAGERS

see also Municipal government by city manager

—BIBLIOGRAPHY

Swanick, Eric L
The city manager : an introductory bibliography / Eric L. Swanick. — Monticello, Ill. : Vance Bibliographies, 1978.
4 p. ; 28 cm. — (Public administration series : Bibliography ; P-144)
Cover title.
$1.50
1. Municipal government by city manager—Bibliography. 2. Municipal government—United States—Bibliography. I. Title. II. Series.
Z7164.L8S95 016.352'0084'0973 79-105441
⌊JS344.C5⌋ 79 MARC

—MASSACHUSETTS

Massachusetts Federation of Taxpayers Associations.
The town manager plan in Massachusetts. — Boston : Massachusetts Federation of Taxpayers Associations, 1949.
24, iii leaves ; 28 cm.
Cover title.
Bibliography: leaves i-iii.
1. Municipal government by city manager—Massachusetts. I. Title.
JS344.C52M45 352'.0083'09744 79-313742
79 MARC

—MISSOURI—KANSAS CITY

Gilbert, Bill, 1931-
This city, this man : the Cookingham era in Kansas City / by Bill Gilbert ; with a supplement by Robert J. Saunders. — Washington : International City Management Association, c1978.
267 p. : ill. ; 21 cm.
Includes index.
ISBN 0-87326-021-X
1. Cookingham, L. Perry, 1896- . 2. City managers—Missouri—Kansas City—Biography. 3. Kansas City, Mo.—Politics and government. 4. Municipal government by city manager—Missouri—Kansas City. I. Title.
JS344.C53K364 352'.0084'0924 78-13401
78 MARC

—MISSOURI—KANSAS CITY—BIOGRAPHY

Gilbert, Bill, 1931-
This city, this man : the Cookingham era in Kansas City / by Bill Gilbert ; with a supplement by Robert J. Saunders. — Washington : International City Management Association, c1978.
267 p. : ill. ; 21 cm.
Includes index.
ISBN 0-87326-021-X
1. Cookingham, L. Perry, 1896- . 2. City managers—Missouri—Kansas City—Biography. 3. Kansas City, Mo.—Politics and government. 4. Municipal government by city manager—Missouri—Kansas City. I. Title.
JS344.C53K364 352'.0084'0924 78-13401
78 MARC

CITY MISSIONS

—ADDRESSES, ESSAYS, LECTURES

Discipling the city : theological reflections on urban mission / editor, Roger S. Greenway ; contributors, Harvie M. Conn ... ⌊et al.⌋. — Grand Rapids : Baker Book House, c1979.
286 p. ; 22 cm.
Includes bibliographical references.
ISBN 0-8010-3727-1 : $5.95
1. City churches—Addresses, essays, lectures. 2. City missions—Addresses, essays, lectures. I. Greenway, Roger S. II. Conn, Harvie M.
BV637.D5 261.8 78-67165
79 MARC

—ONTARIO—HAMILTON
—BIOGRAPHY

Golbeck, Kay.
"Lord, how will you get me out of this mess?" / Kay Golbeck, with Irene Burk Harrell. — Lincoln, Va. : Chosen Books ; Waco, Tex. : distributed by Word Books, c1978.
156 p. ; 23 cm.
ISBN 0-912376-34-1 : $6.95
1. Golbeck, Kay. 2. Baptists—Canada—Biography. 3. Arthritis—Biography. 4. Death, Apparent. 5. Faith-cure. 6. Missionaries—Canada—Biography. 7. City missions—Ontario—Hamilton—Biography. 8. Hamilton, Ont.—Biography. I. Harrell, Irene Burk, joint author. II. Title.
BX6495.G59A33 248'.2'0924 78-17395
78 MARC

CITY NATIONAL BANK & TRUST CO. OF ROCKFORD

Garson, Bill.
The knight on Broadway : the story of City National Bank & Trust Co. of Rockford and Sir Greenback—the bank's financial symbol / compiled, edited, and written by Bill Garson. — Rockford, Ill. : City National Bank & Trust Co. of Rockford, c1978.
100, ⌊7⌋ p. : ill. ; 28 cm.
ISBN 0-9602150-1-8
1. City National Bank & Trust Co. of Rockford. I. Title.
HG2613.R664C5713 332.1'223'0977331 78-64865
79 MARC

CITY NOISE

see also Noise control; Traffic noise

Sälzer, Elmar.
Städtebaulicher Schallschutz : Möglichkeiten u. Bewertung von städtebaul. Massnahmen zur Minderung d. Lärmbelästigung : Ergebnisse e. im Auftr. d. Niedersächs. Sozialministers, Hannover durchgeführten Untersuchung d. ITA Ingenieurges. für Techn. Akustik mbH / Elmar Sälzer. — 1. Aufl. — Wiesbaden ; Berlin : Bauverlag, 1975.
87 p. : 56 ill. ; 21 cm. GFR76-A
Bibliography: p. 83-87.
ISBN 3-7625-0642-6 : DM18.00
1. City noise. 2. City planning. I. Title.
TD892.S23 78-397850
79 MARC

—CONGRESSES

Lärm in der Stadt : Bericht über die Veranstaltung des Deutschen Instituts für Urbanistik in Zusammenarbeit mit dem Siedlungsverband Ruhrkohlenbezirk und dem Deutschen Städtetag am 29. und 30. November 1976 in Essen / ⌊redaktionelle Bearbeitung, Klaus Ernst⌋. — Berlin : Das Institut, 1977.
203, ⌊34⌋ p. : graphs ; 30 cm. GFR***
Includes bibliographical references.
1. City noise—Congresses. 2. Traffic noise—Congresses. I. Ernst, Klaus. II. Deutsches Institut für Urbanistik. III. Siedlungsverband Ruhrkohlenbezirk. IV. Deutscher Städtetag (Germany (Federal Republic)).
TD891.L28 363.6 78-369403
78 MARC

CITY OF LONDON POLYTECHNIC. LIBRARY AND LEARNING RESOURCES SERVICE

City of London Polytechnic. Library and Learning Resources Service.
A diary survey to establish time & cost data for library processes / City of London Polytechnic, Library & Learning Resources Service. — ⌊London⌋ : The Service, 1973.
81 leaves ; 30 cm.
Cover title.
1. Processing (Libraries) 2. City of London Polytechnic. Library and Learning Resources Service. I. Title. II. Title: Time and cost data for library processes.
Z792.C59C57 1973 025.2 78-101457
79 MARC

CITY OF THE VATICAN see Vatican City

CITY ORDINANCES see Ordinances, Municipal

CITY PLANNERS

—GERMANY, WEST—BIOGRAPHY

Fischer, Manfred Frithjof.
Fritz Schumacher, das Hamburger Stadtbild und die Denkmalpflege / Manfred F. Fischer. — Hamburg : H. Christians, 1977.
79 p. : ill., maps ; 30 cm. — (Arbeitshefte zur Denkmalpflege in Hamburg ; Nr. 4) GFR***
Bibliography: p. 69.
ISBN 3-7672-0521-1
1. Schumacher, Fritz, 1869-1947. 2. City planners—Germany, West—Biography. 3. City planning—Germany, West. 4. Hamburg—City planning. 5. Architecture—Germany, West. I. Title.
NA9085.S4F57 79-352429
79 MARC

CITY PLANNERS (Continued)

—GREAT BRITAIN—ATTITUDES

McLoughlin, Pete.
Regional policy and the inner areas : a study of planners' attitudes / ₍by₎ Pete McLoughlin. — Reading : Department of Geography, University of Reading, 1978.
₍2₎, iii, 31 p. : ill., map ; 21 cm. — (Geographical papers ; no. 64 ISSN 0305-5914) GB78-31964
Bibliography: p. 26.
ISBN 0-7049-0474-8 : £0.65
1. City planning—Great Britain. 2. Urban renewal—Great Britain. 3. City planners—Great Britain—Attitudes. I. Title. II. Series.
HT169.G7M27 309.2'62 79-314488
79 MARC

—UNITED STATES—BIOGRAPHY

Mumford, Lewis, 1895-
My works and days : a personal chronicle / Lewis Mumford. — 1st ed. — New York : Harcourt Brace Jovanovich, c1979.
545 p. ; 25 cm.
Includes index.
ISBN 0-15-164087-4 : $13.95
1. Mumford, Lewis, 1895- 2. Social reformers—United States—Biography. 3. City planners—United States—Biography. 4. Architects—United States—Biography. I. Title.
CT275.M734A35 818'.5'209 78-53893
78 MARC

CITY PLANNING

see also Aerial photography in city planning; Art, Municipal; Blocks (City planning); Central business districts; City planners; City planning; Community development, Urban; Flood damage prevention; Garden cities; Housing; Housing policy; Industries, Location of; Labor and laboring classes—Dwellings; Land subdivision; Megastructures; Municipal services; New towns; Parks; Playgrounds; Regional planning; Slums; Social surveys; Space (Architecture); Suburbs; Traffic surveys; Transportation planning; Urban beautification; Urban policy; Urban renewal; Urban transportation; Zoning

Bauer, Gérard, 1934-
La Morphologie architecturale à l'échelle du plan de masse / Ministère de la culture, Direction de l'architecture ... ₍Comité de la recherche et du développement en architecture₎ ; Atelier de recherche et d'études d'aménagement ; ₍rédigé par₎ Gérard Bauer ; ₍avec la collaboration de Robert Davreu et Sylvie Ragueneau₎. — Paris : AREA, 1977.
152 p. : ill. ; 30 cm. F78-5548
Bibliography: p. 149-152.
1. City planning. 2. Communication in architectural design. 3. Architecture—Research. I. France. Direction de l'architecture. II. Title.
NA9050.B328 711'.4 78-387021
79 MARC

Bello-Morales Merino, Antonio.
Las nuevas ciudades residenciales ... / ₍por₎ Antonio Bello-Morales Merino ; prólogo de José Luis Zatarain Iglesias. — Madrid, ₍etc.₎ : Dossat, 1978.
xi, 367 p., ₍13₎ fold. leaves of plates : ill. ; 21 cm. Sp79-Apr
Bibliography: p. ₍365₎-367.
ISBN 8423703762
1. New towns. 2. Cities and towns—Growth. 3. City planning. I. Title.
HT153.B47 79-110256
79 MARC

Bertrand, Michel Jean.
Pratique de la ville / par Michel-Jean Bertrand. — Paris ; New York : Masson, 1978.
210 p., ₍4₎ leaves of plates : ill. ; 24 cm. — (Collection Géographie) F***
Includes bibliographical references.
ISBN 2-225-48202-0 : 55.00F
1. Cities and towns. 2. City planning. I. Title.
HT151.B437 301.36 78-396400
79 MARC

(Blagoustroĭstvo sel)
Благоустройство сел : ₍Альбом / И. Л. Дабагян, Н. Н. Кассина, В. М. Свидерский, С. М. Соколов₎. — Киев : Будівельник, 1978.
168 p. : ill. ; 22 x 29 cm. USSR 78
At head of title: Gosudarstvennyĭ komitet Soveta Ministrov USSR po delam stroitel'stva. Ukrainskiĭ nauchno-issledovatel'skiĭ i proektnyĭ institut grazhdanskogo sel'skogo stroitel'stva.
3.40rub
1. Building. 2. City planning. I. Dabagi͡an, Ivan Lazarevich. II. Ukraine. Derzhavnyĭ komitet v spravakh budivnytstva. III. Ukrainskiĭ nauchno-issledovatel'skiĭ i proektnyĭ institut grazhdanskogo sel'skogo stroitel'stva.
TH153.B495 79-413077

Boyden, Stephen Vickers.
An integrative ecological approach to the study of human settlements / Stephen Boyden. — Paris : Unesco, 1979.
87 p. : ill. ; 27 cm. — (MAB technical notes ; 12) F***
Bibliography: p. 85-87.
ISBN 923101682X
1. Human ecology. 2. Cities and towns. 3. City planning. 4. Human ecology—Hongkong. I. Title. II. Series: International Co-ordinating Council of the Programme on Man and the Biosphere. MAB technical reports ; 12.
GF101.B63 301.34 79-317697
79 MARC

Brown, Howard.
Environmental design science primer / by Howard Brown, Robert Cook, Medard Gabel. — 2d ed. — New Haven, Conn. : Advocate Press : distributed in the United States by Earth Metabolic Design, 1978.
116 p. : ill. ; 22 x 28 cm.
Bibliography: p. 99-113.
1. Environmental engineering. 2. Man—Influence of environment. 3. Architectural design. 4. Cities and town—Planning. I. Cook, Robert, joint author. II. Gabel, Medard, joint author. III. Title.
TA170.B76 1978 624 78-56497
79 MARC

Camargo Júnior, Geraldo A
O arquiteto, o desenvolvimento e a qualidade de vida / ₍Geraldo A. Camargo Jr., Marcos A. Osello₎. — São Paulo : Instituto Roberto Simonsen, 1978.
47 p., ₍1₎ leaf of plates : ill. ; 21 cm.
Bibliography: p. 47.
1. Architecture. 2. City planning. I. Osello, Marcos A., joint author. II. Title.
NA2540.C35 79-110072
79 MARC

Chadwick, George F
A systems view of planning : towards a theory of the urban and regional planning process / by George Chadwick. — 2d ed. — Oxford ; New York : Pergamon Press, 1978.
xv, 432 p. (p. 431-432 publisher's list), ₍1₎ fold. leaf of plates : ill. ; 22 cm. — (Pergamon international library of science, technology, engineering, and social studies) (Urban and regional planning series)
Includes bibliographies and index.
ISBN 0-08-020626-3 : $15.00. ISBN 0-08-020625-5 flexicover : $9.00
1. City planning. 2. Regional planning. I. Title. II. Series: Urban and regional planning series.
HT166.C36 1978 309.2 77-30344
77 MARC

Chien chu kung ch'eng pu ch'eng shih she chi yüan. Tzu liao shih.
(Ch'eng shih tao lu kuei hua)
城市道路规划 / ₍建筑工程部城市设计院资料室编写₎. — 修订版 / ₍本社编辑部修订₎. — 北京：中国建筑工业出版社，1976.
5, 78 p. : ill. ; 19 cm. — (Its 城市规划知识小丛书 ; 6)
$0.19
1. Streets—Design and construction. 2. City planning. I. Title. II. Series: Chien chu kung ch'eng pu ch'eng shih she chi yüan. Tzu liao shih. Ch'eng shih kuei hua chih shih hsiao ts'ung shu ; 6.
TE175.C46 1976 78-841255

Chien chu kung ch'eng pu ch'eng shih she chi yüan. Tzu liao shih.
(Ch'eng shih yung ti hsüan tse chi fang an pi chiao)
城市用地选择及方案比较 / ₍建筑工程部城市设计院资料室编写₎. — 修订版 / ₍本社编辑部修订₎. — 北京：中国建筑工业出版社，1976.
6, 34 p. : ill. ; 19 cm. — (Its 城市规划知识小丛书 ; 5)
$0.11
1. Zoning. 2. City planning. I. Chung-kuo chien chu kung yeh ch'u pan she. Pien chi pu. II. Title. III. Series: Chien chu kung ch'eng pu ch'eng shih she chi yüan. Tzu liao shih. Ch'eng shih kuei hua chih shih hsiao ts'ung shu ; 5.
HT169.6.C48 1976 78-842249

Le Città capitali nel XX secolo / Mario Manieri-Elia ; contributi di G. Ciucci ... ₍et al.₎. — 1. ed. — Milano : Fratelli Fabbri, 1978.
128 p. : ill. ; 28 cm. — (L'Arte nella società) It78-Dec
Bibliography: p. 126-127.
L4500
1. Cities and towns. 2. City planning. 3. Capitals (Cities) I. Manieri-Elia, Mario, 1929- II. Ciucci, Giorgio.
HT153.C55 79-346226
*79 MARC

Crivelli, Daniele.
Comunicazione di massa e urbanistica : dall'informazione al territorio / Daniele Crivelli. — ₍Firenze₎ : Libreria editrice fiorentina, 1978.
238 p., ₍1₎ fold. leaf of plates : ill. ; 22 cm. It78-Nov
Bibliography: p. 229-238.
L4900
1. Mass media. 2. City planning. 3. Telecommunication—Social aspects. I. Title.
HM258.C74 301.16 79-339469
*79 MARC

Cucu, Vasile.
Urbanism : ₍curs₎ / Vasile Cucu, Cezar Popescu. — Bucureşti : Centrul de multiplicare al Universităţii din Bucureşti, 1975-
v. : ill. ; 24 cm. R75-4274
At head of title: Universitatea din Bucureşti. Facultatea de Geologie-Geografie. Catedra de Geografie Economică.
"Destinat studenţilor anului II curs de zi şi anului III curs fără frecvenţă de la Facultatea de drept a Universităţii din Bucureşti."
Bibliography: v. 1, p. 316-320.
lei15.00 (v. 1)
1. Cities and towns. 2. City planning. 3. Cities and towns—Romania. I. Popescu, Cezar, joint author. II. Title.
HT151.C82 78-393930
*79 MARC

Design Council.
Streets ahead. — 1st ed. — London : Design Council in association with the Royal Town Planning Institute, 1979.
111 p. : ill. ; 30 cm. GB***
Bibliography: p. 105.
Includes index.
ISBN 0-85072-081-8 : £6.50
1. Streets. 2. City planning. I. Title.
NA9053.S7D47 1979 713 79-312279
79 MARC

Dupuy, Gabriel.
Urbanisme et technique : chronique d'un mariage de raison / Gabriel Dupuy. — Paris : Centre de recherche d'urbanisme, ₍1978₎
420 p. : ill. ; 21 cm. F***
Bibliography: p. 401-416.
ISBN 2-85303-116-0 : 40.00F
1. City planning. 2. Technology. I. Title.
HT166.D85 309.2'62 78-398472
79 MARC

Ganassi Agger, Simona.
Autogestione urbana : l'urbanistica per una nuova società / Simona Ganassi Agger. — Bari : Dedalo libri, 1977.
275 p. ; 21 cm. — (Saggi ; 27) It78-Nov
Bibliography: p. 267-₍273₎
Includes indexes.
L4000
1. Cities and towns. 2. City planning. 3. Sociology, Urban. I. Title.
HT151.G318 301.36'1 79-351884
*79 MARC

Ganassi Agger, Simona.
Urban self-management : planning for a new society / Simona Ganassi Agger. — White Plains, N.Y. : M. E. Sharpe, c1979.
235 p. ; 24 cm.
Expanded and rev. version of the author's Autogestione urbana, published in 1977.
"Published simultaneously as vol. IX, no. 1-2 of International journal of sociology."
Includes bibliographical references.
ISBN 0-87332-125-1
1. Cities and towns. 2. City planning. 3. Sociology, Urban. I. Title.
HT151.G31813 309.2'62 78-73223
79 MARC

Garnier, Tony, 1869-1948.
Une cité industrielle, étude pour la construction des villes ₍par₎ Tony Garnier ... Paris, C. Massin & cie ₍1939₎
2 v. plates (part. col., part fold.) plans (part fold.) 34 x 43 cm.
On cover: Deuxième édition.
Issued in portfolios.
1. City planning. I. Title.
NA9030.G3 1939 711'.45 40-354
₍r79₎rev MARC

CITY PLANNING (Continued)

Gerosa, Pier Giorgio, 1945-
Le Corbusier - urbanisme et mobilité / Pier Giorgio Gerosa. — Basel ₍etc.₎ : Birkhäuser, 1978.
202 p. : ill. ; 25 cm. — (Eidgenössische Technische Hochschule, Zürich. Studien aus dem Institut für Geschichte und Theorie der Architektur ; Bd. 3) Sw78-3729
Originally presented as the author's thesis, Eidgenössische Technische Hochschule, Zürich.
"Bibliographie chronologique des écrits de Le Corbusier": p. 198-199.
Bibliography: p. 200-201.
ISBN 3-7643-0990-3 : 29.00F
1. Jeanneret-Gris, Charles Édouard, 1887-1965. 2. City planning. 3. Space (Architecture) 4. Social mobility. I. Title. II. Series: Zürich. Eidgenössische Technische Hochschule. Institut für Geschichte und Theorie der Architektur. Studien ; Bd. 3.
NA1053.J4G47 1978 720'.92'4 78-396214
79 MARC

Gertler, L O
Habitat and land / Len Gertler. — Vancouver : University of British Columbia Press : published in association with the Centre for Human Settlements at the University of British Columbia, 1978.
xi, 42 p., ₍4₎ leaves of plates : ill. ; 23 cm. — (Human settlement issues ; 1) C***
Includes bibliographical references.
ISBN 0-7748-0100-X
1. Land use, Urban—Planning. I. Title. II. Series.
HD111.G43 333.7 79-301021
79 MARC

Handbook of urban landscape / consultant editor Cliff Tandy ; edited by the Architects' journal, Technical Section. — London : Architectural Press, 1978.
275 p. : ill., maps, plans ; 30 cm. GB79-07438
Revised and updated version of what was originally a series in the Architects' journal.
Includes bibliographical references and index.
ISBN 0-85139-691-7 : £6.95
1. Landscape architecture. 2. City planning. 3. Urban beautification. 4. Landscape architecture—Great Britain. I. Tandy, Clifford R. V., ed. II. The Archtiects' journal. III. Title: Urban landscape.
SB472.7.H26 1978 711'.4 79-317256
79 MARC

Introduction to urban planning / edited by Anthony J. Catanese, James C. Snyder ; architectural drawings by Jeffrey Ollswang. — New York : McGraw-Hill, c1979.
xvii, 354 p. : ill. ; 24 cm.
Includes bibliographies and index.
ISBN 0-07-010228-7 : $22.50
1. City planning. 2. City planning—United States. I. Catanese, Anthony James. II. Synder, James C.
HT166.I67 309.2'62'0973 78-13275
78 MARC

Isomura, Eiichi, 1903-
(Toshi kenshō)
都市憲章 / 磯村英一. — 東京 : 鹿島出版会, 1978.
223 p. : ill. ; 19 cm. — (SD選書 ; 136) Ja 78-14343
Series romanized: SD sensho.
¥980

1. City planning. 2. Cities and towns—Civic improvement. I. Title.
HT166.I85 78-812617

Jägemann, Hans B
Zielbildungsprozesse in der Stadtplanung : Planungstechniken und -strategien aus planungstheoretischer Perspektive / Hans B. Jägemann. — Zürich ₍etc.₎ : H. Deutsch, 1977.
330, xlviii p. : ill. ; 21 cm. Sw78-3082
Originally presented as the author's thesis, Darmstadt, 1975.
Bibliography: p. i-xlvii.
ISBN 3-87144-339-5 : 33.60F
1. City planning. 2. Political planning. I. Title.
HT166.J27 1977 309.2'62 79-341775
79 MARC

Kakumoto, Ryōhei, 1920-
(Konnichi no toshi asu no toshi)
今日の都市明日の都市 : 都市固定化の時代 / ₍角本良平₎. — 町田 : 玉川大学出版部, 1978.
240 p. ; 19 cm. — (玉川選書 ; 85) Ja 78-25871
Series romanized: Tamagawa sensho.
Bibliography: p. 233-237.
Includes index.
¥880

1. Cities and towns. 2. City planning. I. Title.
HT151.K3 79-802607

Kaminsky, Jacob, 1933-
Environmental characteristics planning : an alternative approach to physical planning. — Baltimore, Md. : Regional Planning Council, 1972.
vii, 82 leaves : ill. ; 29 cm.
Cover title.
1. City planning. I. Regional Planning Council, Baltimore. II. Title.
TD159.3.D35 711'.13 79-302365
79 MARC

Karamanou, Zoé.
Au-delà de l'espace institutionnalisé : la notion de l'échelle à travers une analyse critique de l'aménagement architectonique-urbanistique / Z. Karamanou, N. Rodolakis. — Paris : Anthropos, ₍c1978₎
269 p. : ill. ; 22 cm. — (Collection Espaces et sociétés) F***
Bibliography: p. ₍259₎-266.
ISBN 2-7157-0296-5 : 60.00F
1. Space (Architecture) 2. City planning. 3. Architecture—Environmental aspects. I. Rodolakis, Nicolas, joint author. II. Title. III. Series.
NA9053.S6K37 78-384331
78 MARC

Kishi, Ichita, 1874-1937.
(Seibutsugakuteki toshi keiei ron)
生物學的都市經營論 / 岸一太著. — 東京 : 三正堂, 大正12 ₍1923₎
5, 6, 164 p. ; 23 cm.

1. City planning. 2. Urban ecology (Biology) I. Title.
HT166.K56 78-817007

Kugler-Gagnon, Marianne, 1948-
Computer processing of geological and geotechnical data for urban and land-use planning / Marianne Kugler-Gagnon. — Ottawa : University of Ottawa, Dept. of Geography and Regional Planning, 1975.
29 leaves : ill., diagrs., map ; 28 cm. — (Research notes - Geography and Regional Planning, University of Ottawa ; no. 6) C76-6384-2
Bibliography: leaves 28-29.
$1.50
1. Geology—Data processing. 2. Engineering geology—Data processing. 3. Cities and towns—Planning. 4. Land use—Planning. I. Title. II. Series: University of Ottawa. Dept. of Geography and Regional Planning. Research notes - Geography and Regional Planning, University of Ottawa ; no. 6.
QE48.8.K82 333.7'028'54 79-313631
79 MARC

Landzettel, Wilhelm.
Gestaltaspekte / Wilhelm Landzettel, Wolfram Goldapp, Hans Haas. — Wiesbaden : AVA, Arbeitsgemeinschaft z. Verbesserung d. Agrarstruktur in Hessen e.V., 1975.
92 p. : numerous ill., maps ; 30 cm. — (Beiträge zum ländlichen Bau- und Siedlungswesen ; Bericht 16) (Sonderheft - AVA, Arbeitsgemeinschaft zur Verbesserung der Agrarstruktur in Hessen e.V. ; 54) GFR75-B18
DM5.00
1. City planning. 2. Regional planning. 3. Architecture—Composition, proportion, etc. 4. City planning—Germany, West—Hesse. I. Goldapp, Wolfram, joint author. II. Haas, Hans, Bildhauer. III. Title. IV. Series. V. Series: Arbeitsgemeinschaft zur Verbesserung der Agrarstruktur in Hessen. Sonderheft ; 54.
HD1960.H5A7 Nr. 54 78-400000
₍NA9035₎ 79 MARC

Linstrum, Derek.
The conservation of historic towns and monuments : report on a Commonwealth Foundation lectureship in Africa, February-March 1976 / by Derek Linstrum. — London : Commonwealth Foundation, 1976.
39 p. ; 25 cm. — (Occasional paper - Commonwealth Foundation ; no. 38) GB***
Cover title.
1. Historic sites—Conservation and restoration. 2. Historic buildings—Conservation and restoration. 3. City planning. 4. Historic sites—Africa—Conservation and restoration. I. Title. II. Series: Commonwealth Foundation (British Commonwealth). Occasional paper ; no. 38.
CC135.L53 363 79-303644
79 MARC

Marescotti, Luca.
Urbanistica : saggio critico, testimonianze, documenti, bibliografia ragionata / Luca Marescotti. — Milano : Accademia, 1979.
467 p. : ill. ; 20 cm. — (Manuali Accademia) It79-Aug
"Testimonianze e documenti": p. 271-362.
Bibliography: p. 367-461.
L6000
1. City planning—Italy. 2. City planning. I. Title.
HT169.I83M36 79-383989
*79 MARC

Martins, Daniel Hugo, 1927-
El municipio contemporáneo : los gobiernos locales de Occidente y del Uruguay / Daniel Hugo Martins. — Montevideo : Fundación de Cultura Universitaria, ₍1978 or 1979₎
176 p. ; 25 cm. — (Colección mayor - FCU ; 3)
Includes bibliographical references.
1. Municipal government. 2. Municipal corporations. 3. Local government. 4. City planning. 5. Local government—Uruguay—History. I. Title.
JS141.M37 352'.008 79-112814
79 MARC

Miller, Brown, 1940-
Innovation in new communities ₍by₎ Brown Miller, Neil J. Pinney ₍and₎ William S. Saslow. Cambridge, Mass., M.I.T. Press ₍1972₎
x, 301, ₍79₎ p. illus. 23 x 29 cm. (M.I.T. report no. 23) $15.00
Last group of pages numbered A1-F3.
Includes bibliographical references.
ISBN 0-262-13082-3
1. City planning. I. Pinney, Neil J., joint author. II. Saslow, William S., joint author. III. Title. IV. Series: Massachusetts Institute of Technology. M.I.T. report no. 23.
HT166.M55 309.2'6'0973 74-175718
72₍r79₎rev MARC

Monestiroli, Antonio.
L'architettura della realtà / Antonio Monestiroli. — 1. ed. — Milano : Clup, 1979.
178 p., ₍6₎ leaves of plates : ill. ; 24 cm. It79-Aug
At head of title: Politecnico di Milano, Facoltà di architettura.
Includes bibliographies.
ISBN 8870054047 : L5000
1. Architecture—Environmental aspects. 2. City planning. I. Title.
NA2542.35.M66 79-381916
*79 MARC

Munson, Michael J
Planning : process and product / Michael Munson. — Princeton, N.J. : Research Center for Urban and Environmental Planning, School of Architecture and Urban Planning, Princeton University, 1974.
ii, 30 leaves ; 28 cm. — (Working paper - Research Center for Urban and Environmental Planning, Princeton University ; 5)
Bibliography: leaves 29-30.
1. City planning. 2. Planning. I. Title. II. Series: Princeton University. Research Center for Urban and Environmental Planning. Working paper - Research Center for Urban and Environmental Planning, Princeton University ; 5.
HT166.M833 309.2'12 79-307326
79 MARC

Nature in cities : the natural environment in the design and development of urban green space / edited by Ian C. Laurie. — Chichester ; New York : Wiley, c1979.
xix, 428 p. : ill. ; 25 cm.
Includes bibliographies and index.
ISBN 0-471-99605-X
1. Urban ecology (Biology) 2. City planning. 3. Urban beautification. I. Laurie, Ian C.
QH541.5.C6N37 309.2'62 77-20987
77 MARC

Prelog, Milan, 1919-
Prostor—vrijeme / Milan Prelog ; ₍naslovna strana Ante Kuduz₎. — Zagreb : Društvo historičara umjetnosti Hrvatske, 1973.
120 p., ₍19₎ leaves of plates : ill. ; 25 cm. — (₍Izdanja₎ - Društvo historičara umjetnosti Hrvatske ; knj. 21) Yu74-11120
Includes bibliographical references.
1. Space (Architecture) 2. City planning. 3. Architecture—Human factors. I. Title. II. Series: Društvo historičara umjetnosti Hrvatske. Izdanja ; knj. 21.
NA9053.S6P73 78-400341
79 MARC

Reid, Richard.
Designing for the community / ₍Richard Reid ; illustrated by Mike Tregenza₎. — ₍London : Mills & Boon₎ ; Levittown, N.Y. : sole distributor for the U.S.A. Transatlantic Arts, ₍c1977₎
₍32₎ p. : ill. ; 24 cm. — (Picture panorama of world building) (Owlet books) GB***
ISBN 0-263-06302-X
1. City planning. 2. Public buildings. I. Title. II. Series.
NA9035.R44 711'.4'09 78-326922
79 MARC

Riemer, Eberhard.
Verkehrserschliessung, Ver- und Entsorgung im Städtebau : Entwurfshilfen für d. Bauleitplanung / Eberhard Riemer. — Köln-Braunsfeld : Müller, 1978.
156 p. : ill. ; 30 cm. GFR78-A
Bibliography: p. 151-152.
Includes index.
ISBN 3-481-16891-8 : DM88.00
1. Streets—Design. 2. City planning. 3. Municipal engineering. I. Title.
TE175.R53 79-347593
79 MARC

Rossi, Aldo, 1931-
L'architettura della città / Aldo Rossi ; a cura di Daniele Vitale. — 1. ed. — Milano : Clup, 1978.
313 p., ₍12₎ leaves of plates : ill. ; 24 cm. It78-Dec
Bibliography: p. 305.
Includes indexes.
ISBN 8870053741 : L8600
1. City planning. I. Vitale, Daniele. II. Title.
HT166.R67 1978 79-345877
79 MARC

Rougerie, Jacques, 1945-
Thalassopolis : centres internationaux de recherche, de gestion et de contrôle du patrimoine marin / Jacques Rougerie, Édith Rougerie, Jacques Hirou ; ₍publié par l'₎ Unité pédagogique d'architecture no 7 ₍et l'₎Institut de l'environnement. — ₍Paris₎ : Institut de l'environnement, 1973.
₍61₎ p. : ill. ; 21 x 30 cm. F74-12195
10.00F
1. City planning. 2. Ocean engineering. I. Rougerie, Édith, joint author. II. Hirou, Jacques, joint author. III. Paris. École nationale supérieure des beaux-arts. Unité pédagogique d'architecture no 7. IV. Institut de l'environnement. V. Title.
HT166.R69 309.2'62 75-505544

CITY PLANNING (Continued)

Rowe, Colin.
Collage city / Colin Rowe & Fred Koetter. — Cambridge, Mass. : MIT Press, ₍1978₎
185 p. : ill. ; 31 cm.
Includes bibliographical references indexes.
ISBN 0-262-18086-3
1. City planning. 2. Functionalism (Architecture) 3. Eclecticism in architecture. 4. Architecture, Modern—20th century. I. Koetter, Fred, joint author. II. Title.
NA9050.R68 711'.4 78-60990
78 MARC

Ru‹enstein, Harvey M
Central city malls / Harvey M. Rubenstein. — New York : Wiley, c1978.
xii, 191 p. : ill. ; 29 cm.
"A Wiley-Interscience publication."
Bibliography: p. 187.
Includes index.
ISBN 0-471-03098-8
1. Shopping malls. 2. Central business districts. 3. Pedestrian facilities design. 4. City planning. 5. Landscape architecture. I. Title.
HF5430.R8 711'.552 78-7536
78 MARC

Sälzer, Elmar.
Städtebaulicher Schallschutz : Möglichkeiten u. Bewertung von städtebaul. Massnahmen zur Minderung d. Lärmbelästigung : Ergebnisse e. im Auftr. d. Niedersächs. Sozialministers, Hannover durchgeführten Untersuchung d. ITA Ingenieurges. für Techn. Akustik mbH / Elmar Sälzer. — 1. Aufl. — Wiesbaden ; Berlin : Bauverlag, 1975.
87 p. : 56 ill. ; 21 cm. GFR76-A
Bibliography: p. 83-87.
ISBN 3-7625-0642-6 : DM18.00
1. City noise. 2. City planning. I. Title.
TD892.S23 78-397850
79 MARC

Savoye, Philippe.
Terrain d'aventure : espace de vie / Philippe Savoye. — ₍Lyon₎ : Fédérop, ₍c1978₎
252 p. ; 19 cm. — (Collection Travailleurs sociaux ; 4) F***
40.00F
1. Recreation areas—Planning. 2. City planning. I. Title.
GV182.3.S28 711 79-354071
79 MARC

Scott, Allen John.
Urban planning in theory and practice : a re-appraisal / Allen J. Scott and Shoukry T. Roweis. — ₍Toronto₎ : Dept. of Urban and Regional Planning, University of Toronto, 1977.
52 p. : ill. ; 28 cm. — (Papers on planning and design ; no. 14) C***
Bibliography: p. 52.
$2.50
1. City planning. I. Roweis, Shoukry T., joint author. II. Title. III. Series.
HT166.S356 309.2'62 78-324055
79 MARC

Smith, Peter Frederick.
The dynamics of urbanism / Peter F. Smith. — London : Hutchinson, 1974.
256 p. : ill., plans ; 24 cm. GB75-04474
Includes bibliographical references and index.
ISBN 0-09-119780-5 : £4.45. ISBN 0-09-119781-3 pbk.
1. City planning. 2. Architecture—Psychological aspects. I. Title.
NA9050.S587 720'.1 75-314913
75₍r79₎rev MARC

Sokół, Adam.
Podstawy gospodarczo-społeczne programowania rozwoju miast / Adam Sokół, Adam Ginsbert-Gebert. — Wyd. 1. — Warszawa : Państ. Wydaw. Ekonomiczne, 1978 (Olszt. : OZGSP)
227, ₍1₎ p. : ill. ; 20 cm. P78-7029
Bibliography:p. 227-₍228₎
zł22.00
1. City planning. 2. Urbanization. I. Ginsbert-Gebert, Adam, joint author. II. Title.
HT166.S628 79-361433
79 MARC

Soulier, Louis.
Espaces verts et urbanisme / Louis Soulier. — 2. éd. rev. et complétée. — Paris : Centre de recherche d'urbanisme, ₍1977₎
329 p. : ill. ; 21 cm. F***
Bibliography: p. 313-315.
ISBN 2-85303-103-9
1. Landscape architecture. 2. City planning. 3. Urban beautification. I. Title.
SB472.7.S68 1977 712 78-390567
79 MARC

Stretton, Hugh.
Urban planning in rich and poor countries / Hugh Stretton. — Oxford ; New York : Oxford University Press, 1978.
220 p. ; 21 cm. GB78
Bibliography: p. ₍214₎-217.
Includes index.
ISBN 0-19-219134-9 : £3.95. ISBN 0-19-289121-9 pbk.
1. City planning. 2. Underdeveloped areas—City planning. I. Title.
HT166.S72 309.2'62 78-40576
78 MARC

Sugimura, Nobuji, 1929-
(Toshi no shōgyō)
都市の商業：その地誌的考察 / 杉村暢二著. — 東京：大明堂, 1978.
vii, 221 p. : ill. ; 22 cm. — (叢書都市 ; 5) Ja 78-8855
Series romanized: Sōsho toshi.
Includes bibliographies and index.
¥2300
1. Central business districts—Japan. 2. Central business districts. 3. Japan—Commerce. 4. City planning. I. Title.
HF5429.6.J3S93 79-800712
Library of Congress *79 J

Toshi Sekkei Kenkyūkai.
(Toshi sekkei no tame no atarashii sutorakuchā)
都市設計のための新しいストラクチャー / 久保貞監修；都市設計研究会編. — 東京：鹿島出版会, 1979.
234 p. : ill. ; 22 cm. Ja 79-5800
Includes bibliographies.
¥3300
1. City planning. I. Title. II. Title: Atarashii sutorakuchā.
HT166.T623 1979 79-808648
*79 J

University of Massachusetts at Amherst. Environment and Behavior Research Center.
Pedestrians and wind in the urban environment / by Howard Cohen ... ₍et al.₎ ; principal investigators, Stanley M. Moss and Ervin H. Zube ; Environment and Behavior Research Center, Institute for Man and Environment, University of Massachusetts/Amhe₍r₎st & Weather Dynamics, inc. — Amherst : The Center, 1977.
xii, 124 p. : ill. ; 28 cm. — (Publication - Institute for Man and Environment, University of Massachusetts, Amherst ; no. UMASS/IME/R-77/13)
Bibliography: p. 123-124.
1. Tall buildings—Aerodynamics. 2. Urban climatology. 3. City planning. I. Cohen, Howard. II. Moss, Stanley M. III. Zube, Ervin H. IV. Weather Dynamics, inc. V. Title. VI. Series: University of Massachusetts at Amherst. Institute for Man and Environment. Publication - Institute for Man and Environment, University of Massachusetts, Amherst ; no. UMASS/IME/R-77/13.
TA654.5.U56 1977 690 79-621238
00 MARC

Urbani, Leonardo.
La città è sola : plurispazialità territoriale, obsolescenza del manufatto edilizio, mobilità della forma urbana / Leonardo Urbani. — Roma : Studium, 1978.
xxix, 273 p. ; 21 cm. — (La Cultura ; 14) It78-Nov
Includes bibliographical references.
L8500
1. Cities and towns. 2. City planning. 3. Architecture. I. Title.
HT153.U75 309.2'62 79-341774
*79 MARC

La Ville et l'énergie : journées d'études, janvier 1977, Paris. — ₍Paris₎ : AGHTM, ₍1977₎
463 p. in various pagings : ill. ; 26 cm. F***
Cover title.
Includes bibliographies.
214F
1. Power resources. 2. City planning. I. Association générale des hygiénistes et techniciens municipaux.
TJ163.2.V55 333.7 78-380948
78 MARC

Von Eckardt, Wolf.
Back to the drawing board! : Planning livable cities / Wolf von Eckardt ; illustrated by Forrest Wilson. — Washington : New Republic Books ; New York : trade distribution by Simon and Schuster, 1978.
ix, 245 p. : ill. ; 22 cm.
Bibliography: p. ₍236₎-239.
Includes index.
ISBN 0-915220-45-8
1. City planning. 2. City planning—United States. 3. Architecture—Human factors. I. Title.
HT166.V66 309.2'62 78-12257
78 MARC

Wang, Chi-k'un.
(Tu shih chi hua kai lun)
都市計劃概論 / 王紀鯤著. — 臺北：東大圖書公司：總經銷三民書局, 1978.
2, 222 p. : ill. ; 22 cm. — (滄海叢刊 : 建築)
Series romanized: Ts'ang hai ts'ung k'an : Chien chu.
Bibliography: p. 221-222.
1. City planning. 2. Urban renewal. I. Title.
HT166.W24 79-836221

Watanabe, Shinzō, 1914-
(Toshi keikaku yōron)
都市計画要論 / 渡辺新三, 松井寛共著. — 東京：国民科学社：発売 オーム社書店, 1978.
2, 5, 247, 6 p. : ill. ; 22 cm. Ja 78-26851
Includes bibliographies and index.
¥2500
1. City planning. 2. City planning—Japan. I. Matsui, Hiroshi, 1941- joint author. II. Title.
HT166.W339 79-803735

Wegener, Michael.
Mensch-Maschine-Systeme für die Stadtplanung / Michael Wegener. — Basel ₍etc.₎ : Birkhäuser, 1978.
174 p. : 28 ill. ; 23 cm. — (Interdisziplinäre Systemforschung ; 61) (Raumplanung ; v. 5) Sw79-1238
Summary also in English.
Bibliography: p. 152-168.
Includes indexes.
ISBN 3-7643-1046-4 : 34.00F
1. City planning. 2. Man-machine systems. I. Title. II. Series: Interdisciplinary systems research ; 61. III. Series: Urban and regional planning ; v. 5.
HT166.W42 79-376958
79 MARC

Yoshimura, Motoo, 1937-
(Kūkan no seitaigaku)
空間の生態学 / 吉村元男. — 東京：小学館, 昭和51 ₍1976₎
219 p. ; 20 cm. — (100万人の創造選書 ; 18) Ja 76-15770
Series romanized: Hyakumannin no sōzō sensho.
¥850
1. Space (Architecture) 2. City planning. I. Title.
NA9053.S6Y67 79-805532

Yuge, Shichirō, 1891-
(Toshi seisaku)
都市政策 / 弓家七郎著. — 東京：廣文社, 昭和26 ₍1951₎
4, 258 p. ; 22 cm.
Includes bibliographical references.
1. City planning. 2. Cities and towns. I. Title.
HT166.Y83 78-815799

Zaino, Hiroshi, 1939-
(Kaiwai)
かいわい：日本の都心空間 / 材野博司. — 東京：鹿島出版会, 1978.
255 p. : ill. ; 19 cm. — (SD選書 ; 126) Ja 78-4129
Series romanized: SD sensho.
Includes bibliographical references.
¥980
1. City planning. 2. City planning—Japan. 3. Cities and towns—Japan. I. Title.
HT166.Z34 79-809683

—1945-

Bamberg, Michael.
Ein Bewertungsverfahren zur Bestimmung des Versorgungsgrades von Stadtteilen mit Gemeindebedarfseinrichtungen : e. Entscheidungshilfe für d. Stadtteilentwicklungsplanung am Beispiel d. Freien u. Hansestadt Hamburg / Michael Bamberg. — 1. Aufl. — Frankfurt am Main : R. G. Fischer, 1978.
vi, 186 p., ₍1₎ fold. leaf of plates : ill., fold. map ; 21 cm. GFR78-A
Bibliography: p. 171-183.
ISBN 3-88323-001-4 : DM38.00
1. Hamburg—City planning. 2. City planning—1945- 3. Infrastructure—Germany, West—Hamburg. I. Title.
HT169.G32H283 78-393362

CITY PLANNING
—1945- (Continued)

Magnan, René.
Architecte et urbaniste 1999 / René Magnan. — Paris : Centre de recherche d'urbanisme, [1978]
84 p. ; 21 cm. F***
ISBN 2-85303-123-3
1. City planning—France. 2. City planning—1945- I. Title.
NA9197.M33 711'.4'0944 79-374205
79 MARC

—1945- —ADDRESSES, ESSAYS, LECTURES

Farenholtz, Christian.
Neue Aufgaben für die City : Platz u. Bauwerk als Beitr. zur Erhaltung u. Gestaltung d. Innenstädte / Christian Farenholtz, Max Bächer, Alain Trapenard ; [hrsg. von d. GEWOS, Ges. für Wohnungs- u. Siedlungswesen mbH in Hamburg]. — Hamburg : Christians, 1978.
64 p. : ill. ; 24 cm. GFR78-A19
ISBN 3-7672-0545-9 : DM22.00
1. City planning—1945- —Addresses, essays, lectures. 2. Centre Georges Pompidou—Addresses, essays, lectures. 3. Adelaide—Plazas—Adelaide Festival Center—Addresses, essays, lectures. I. Bächer, Max, joint author. II. Trapenard, Alain, joint author. III. Title.
NA9040.F37 78-381833
78 MARC

I Temi dell'architettura della città / [scritti di] Rosaldo Bonicalzi ... [et al.]. — 1. ed. — Milano : Clup, 1976.
185 p. : ill. ; 24 cm. It78-Apr
At head of title: Politecnico di Milano, Facoltà di architettura.
Includes bibliographical references.
L2800
1. City planning—1945- —Addresses, essays, lectures. I. Bonicalzi, Rosaldo. II. Milan. Politecnico. Facoltà di architettura.
NA9040.T45 78-387348
*79 MARC

Utopia realizzabile attraverso l'integrazione delle arti / Ico Parisi ... [et al.]. — Milano : F. Angeli, c1978.
215 p. : ill. ; 22 cm. — (Serie di architettura ; 3) It***
L4500
1. Architecture, Modern—20th century—Addresses, essays, lectures. 2. Avant-garde (Aesthetics)—Addresses, essays, lectures. 3. City planning—1945- —Addresses, essays, lectures. I. Parisi, Ico, 1916- II. Series.
NA680.U86 78-391453
79 MARC

—1945- —CONGRESSES

Kunstkongress, 4th, Göttingen, 1974.
Die wachsende Hässlichkeit unserer Städte : Freiheiten u. Zwänge im Planungsprozess / 4. Kunstkongress Göttingen, 14.-16. Juni 1974 ; Hrsg., Stadt Göttingen ; [Red., Lucy Hillebrand ... et al.]. — 1. Aufl. — Göttingen : Stadt Göttingen, 1975.
247 p. : ill., maps ; 30 cm. — (Die Kunst, eine Stadt zu bauen ; 2) GFR77-B20
Includes bibliographical references.
1. City planning—1945- —Congresses. 2. Architecture—Conservation and restoration—Congresses. 3. Architecture—Human factors—Congresses. I. Hillebrand, Lucy. II. Göttingen. III. Title. IV. Series.
NA9040.K86 1974 78-394064
79 MARC

—ADDRESSES, ESSAYS, LECTURES

Aspects of human settlement planning / edited by Habitat Conference Secretariat. — New York : Pergamon Press, c1978.
vii, 288 p. : graphs ; 29 cm.
Includes bibliographical references and index.
ISBN 0-08-022011-8
1. City planning—Addresses, essays, lectures. 2. Land use—Planning—Addresses, essays, lectures. I. United Nations Conference on Human Settlements, Vancouver, B.C., 1976.
HT166.A84 1978 309.2'62 77-9877
77 MARC

Building the good city : building and planning for modern man / editor, German Federation for Housing, Town- and Regional Planning ; English and French translation, Helga-Ellen Dietrichs. — Köln-Mülheim : The Federation, [1968?]
135 p. : ill. ; 24 cm. — (Publications of the German Federation for Housing, Town- and Regional Planning ; no. 76)
Translation of Die Stadt unserer Erwartungen.
Includes bibliographical references.
1. City planning—Addresses, essays, lectures. I. Deutscher Verband für Wohnungswesen, Städtebau und Raumplannung. II. Series: Deutscher Verband für Wohnungswesen, Städtebau und Raumplanung. Schriften ; no. 76.
HD7339.A2D44 no. 76 309.2'12'0943 s 79-311984
[HT166] 79 MARC

Castells, Manuel.
City, class, and power / Manuel Castells ; translation supervised by Elizabeth Lebas. — London : Macmillan, 1978.
x, 198 p. ; 23 cm. — (Sociology, politics, and cities) GB***
Essays which originally appeared in French journals.
Includes bibliographical references and index.
ISBN 0-333-22554-6 : £8.95. ISBN 0-333-22555-4 pbk.
1. City planning—Addresses, essays, lectures. 2. Social classes—Addresses, essays, lectures. 3. Power (Social sciences)—Addresses, essays, lectures. I. Title. II. Series.
HT166.C33 1978 309.2'62 79-302072
79 MARC

Castells, Manuel.
City, class, and power / Manuel Castells ; translation supervised by Elizabeth Lebas. — New York : St. Martin's Press, c1978.
x, 198 p. ; 22 cm.
Essays which originally appeared in French journals.
Includes bibliographical references and index.
ISBN 0-312-13989-6 : $18.95
1. City planning—Addresses, essays, lectures. 2. Social classes—Addresses, essays, lectures. 3. Power (Social sciences)—Addresses, essays, lectures. I. Title.
HT166.C33 309.2'62 78-2978
78 MARC

L'Ecosistema urbano : scritti / di U. Cardarelli ... [et al.] : [a cura di] Manfredi Nicoletti. — Bari : Dedalo libri, 1978.
398 p. : ill. ; 22 x 24 cm. — (Il Politecnico ; 9) It79-Jan
Includes bibliographies.
L24000
1. Cities and towns—Addresses, essays, lectures. 2. City planning—Addresses, essays, lectures. 3. Human ecology—Addresses, essays, lecture. I. Nicoletti, Manfredi, 1930- . II. Cardarelli, Urbano.
HT151.E25 301.36'1 79-347094
*79 MARC

Festschrift für Heinrich Habekost / [Herausgeber, Mitarbeiter des Instituts für Stadtbauwesen der T. U. Braunschweig]. — Braunschweig : [Das Institut], 1976.
2 v. : ill. ; 21 cm. — (Veröffentlichungen des Instituts für Stadtbauwesen, Technische Universität Braunschweig ; Heft 18-19) GFR***
Includes bibliographies.
CONTENTS: T. 1. Stadt-, Regional- und Verkehrsplanung. Schütte, K. Gemeindereform und Städtebau. Heidemann. C. Skizzen zu einer pragmatischen Modelltheorie. Kirchhoff, P. Möglichkeiten der Aufstellung und Durchführung von Programmen anwendungsbezogener Forschung auf den Gebieten der Stadtplanung und der Verkehrsplanung. Kutter, E. Bedeutung der Zustandserfassung für die integrierte Verkehrsplanung. Doormann, J. Ermittlung von Parametern für Verkehrserzeugungs- und Verteilungsmodelle. Stöcker, K. Ermittlung verkehrstechnischer Kennwerte von Netzen des öffentlichen Personennahverkehrs mit Hilfe eines Simulationsprogrammes. Schwerdtfeger, W. Die Verkehrs-Konflikt-Technik. Neuwerth, K.-W. und Wirth, R. Gedanken zur Berufssituation von diplomierten Bauingenieuren und Folgerungen für ihre Ausbildung. Beckmann, K. und Schmeckthal, U. Ausbildung eines ingenieurwissenschaftlichen Planers. Budelmann, H. Die Rolle der elektronischen Datenverarbeitung im Tätigkeitsbereich des Planungsingenieurs. Weichbrodt, C. Modelle zur Stadtentwicklungsplanung.—T. 2. Siedlungswasserwirtschaft. Bahrs, D. Einige Aspekte zur Konditionierung von Abwasserschlämmen durch Gefrieren. Ehrig, H.-J. Untersuchungen an künstlich belüfteten, hochbelasteten Abwasserteichen. Kayser, R. Sanierung bestehender Abwasserverregnungen durch biologische Vorbehandlung des Abwassers. Knoch, J. Untersuchungen zur Bildung von Einwohnergleichwerten fester Abfallstoffe. Korn, K. Die elektrolytische Leitfähigkeit als analytisches Hilfsmittel. Stegmann, R. Auswertungen von Wassergütedaten des Okereinzugsgebietes. Walther, W. Der Stoffaustrag bei kleinen Einzugsgebieten mit ackerbaulicher Nutzung. Zingler, E. Planungskriterien für die Erweiterung eines bestehenden Klärwerkes
1. Transportation planning—Addresses, essays, lectures. 2. Sanitary engineering—Addresses, essays, lectures. 3. Cities and towns—Planning—Addresses, essays, lectures. 4. Habekost, Heinrich. I. Habekost, Heinrich. II. Technische Universität Carolo-Wilhelmina. Institut für Stadtbauwesen. III. Series: Technische Universität Carolo-Wilhelmina. Institut für Stadtbauwesen. Veröffentlichungen ; Heft 18-19.
HE193.F4 76-474855
79 MARC

Festschrift, Werner Jäger, Fritz Kastner : Probleme der Raumordnung / [Herausgeber, das Stadbauamt der Stadt Wien. — Wien : Jugend und Volk Verlagsges., 1975?]
303-346 p. : ill. ; 31 cm.
Summaries in English and French.
"Erschienen in 'Der Aufbau,' Heft 9/10-1974."
Includes bibliographical references.
CONTENTS: Breit, R. Vorwort.—Jolles, H. M. Zur Gestaltung des Mitspracherechtes in der Raumplanung.—Backe, B. Bevölkerungspolitische Aspekte in der Stadtentwicklungsplanung.—Helczmanovszki, H. Die Entwicklung der Fruchtbarkeit in Österreich nach Gemeindegrössenklassen.—Jeglitsch, H. Zur regionalen Volkseinkommensrechnung als Grundlage der Planung.—Stöhr, W. Raumplanung in unterentwickelten und in entwickelten Ländern.—Sorgo, K. Zur Entwicklung von Methoden für die Planung.—Breit, R. Wettbewerb im Planungsprozess.—Greif, F. Bauphysiognomie und Betriebsstättenstruktur im 15. Wiener Gemeindebezirk.—Kvarda, W. Gesundes Wohnen und Baurecht.
1. Regional planning—Addresses, essays, lectures. 2. City planning—Addresses, essays, lectures. 3. Jäger, Werner. 4. Kastner, Fritz, 1913- . I. Jäger, Werner. II. Kastner, Fritz, 1913- III. Vienna. Stadtbauamt. IV. Der Aufbau.
HT391.F45 79-374053
79 MARC

Hoshino, Mitsuo, 1919-
(Toshi jichi meicho bunshū)
都市自治名著文集 / 星野光男著. — 東京：広文社, 1978.
246 p. ; 19 cm. Ja 78-12059
¥1700
1. Local government—Addresses, essays, lectures. 2. Local government—Japan—Addresses, essays, lectures. 3. City planning—Addresses, essays, lectures. I. Title.
JS241.H67 78-815798

Isozaki, Arata.
(Kenchiku oyobi kenchiku gaiteki shikō)
建築および建築外的思考 / 磯崎新対談. — 東京：鹿島出版会, 昭和51 [1976]
298 p. : 21 cm. Ja 76-8155
¥2200
1. Architecture—Addresses, essays, lectures. 2. City planning—Addresses, essays, lectures. 3. Artists—Japan—Interviews. 4. Art critics—Japan—Interviews. 5. Critics—Japan—Interviews. I. Title.
NA2560.I8 79-806842

Kurokawa, Noriaki, 1934-
(Kūkan no shisō)
空間の思想 / 黒川紀章. — 東京：白馬出版, 昭和52 [1977]
275 p. ; 19 cm. Ja 77-5003
¥980
1. Architecture—Addresses, essays, lectures. 2. City planning—Addresses, essays, lectures. 3. Space (Architecture)—Addresses, essays, lectures. I. Title.
NA27.K87 79-805631

Launspach, J A
Een vogel in de boom : kommentaren over stedebouw en ruimtelijke ordening 1969-1975 / Jan Launspach. — Amsterdam : Ekologische Uitgeverij, 1976.
130 p. ; 22 cm. Ne76-45
Bibliography: p. 129-130.
ISBN 9062240216 : fl 5.50
1. City planning—Addresses, essays, lectures. 2. Regional planning—Addresses, essays, lectures. 3. City planning—Netherlands—Addresses, essays, lectures. I. Title.
HT166.L382 78-394825
79 MARC

Progettazione urbanistica e territoriale attraverso l'analisi della soglia : Ceep-Seminario di urbanistica / a cura di Francesco Forte ; [scritti] di Pasquale Ferraioli ... [et al.]. — 2. ed. — Milano : F. Angeli, 1978.
362 p. : ill. ; 22 cm. — (Collana CEEP : Sezione III, Territorio ; 4) It78-Apr
Includes bibliographical references.
L8000
1. City planning—Addresses, essays, lectures. 2. Regional planning—Addresses, essays, lectures. I. Forte, Francesco, 1939- . II. Ferraioli, Pasquale. III. Series: Centro studi di politica economica. Collana CEEP : Sezione III, Territorio ; 4.
HT166.P74 1978 78-394393
*79 MARC

Quality in urban planning and design / editor, Roy Cresswell. — London ; Boston : Newnes-Butterworths, 1979.
158 p. : ill. ; 26 cm. GB79
Based on the proceedings of a conference held in London in November 1977.
Includes bibliographical references and index.
ISBN 0-408-00363-4 : £10.00
1. City planning—Addresses, essays, lectures. 2. Urban renewal—Addresses, essays, lectures. I. Cresswell, Roy.
HT166.Q28 309.2'62 78-40209
78 MARC

Resources and planning / edited by Brian Goodall and Andrew Kirby. — 1st ed. — Oxford ; New York : Pergamon Press, 1979.
xviii, 373 p. : ill. ; 26 cm. — (Pergamon Oxford geographers) (Pergamon International library of science, technology, engineering, and social studies) GB79
Written by staff members of the Geography Dept. at the University of Reading.
Includes bibliographical references and indexes.
ISBN 0-08-023711-8 : £15.00 ($35.00)
1. City planning—Addresses, essays, lectures. 2. Regional planning—Addresses, essays, lectures. I. Goodall, Brian. II. Kirby, Andrew. III. Reading, Eng. University. Dept. of Geography.
HT166.R43 1979 309.2'12 78-40931
78 MARC

Rossi, Aldo, 1931-
Scritti scelti sull'architettura e la città, 1956-1972 / Aldo Rossi ; a cura di Rosaldo Bonicalzi. — 2. ed. — Milano : Clup, 1978.
xxxii, 539 p. : ill. ; 17 cm. It79-Feb
Includes bibliographical references and index.
L5400
1. Architecture—Addresses, essays, lectures. 2. City planning—Addresses, essays, lectures. I. Bonicalzi, Rosaldo. II. Title.
NA27.R67 1978 79-353874
*79 MARC

Servizi e gestione urbanistica : le esperienze di Firenze, Genova, Milano / introduzione e documentazione a cura di Mariella Zoppi ; prefazione di Edoardo Detti ; interventi Giuseppina C. Romby, Vittorio Borachia, Marino Bianco. — Firenze : Cooperativa editrice universitaria, 1978.
219 p. ; 19 cm. — (Materiali politici) It79-Mar
L4000
1. City planning—Addresses, essays, lectures. 2. Florence—City planning—Addresses, essays, lectures. 3. Genoa—City planning—Addresses, essays, lectures. 4. Milan—City planning—Addresses, essays, lectures. I. Zoppi, Mariella, 1946-
HT169.I83S42 79-372941
*79 MARC

CITY PLANNING
—ADDRESSES, ESSAYS, LECTURES
(Continued)

Social areas in cities : processes, patterns, and problems / edited by D. T. Herbert and R. J. Johnston. — Chichester ; New York: Wiley, c1978.
xi, 311 p. : ill. ; 23 cm.
Selected essays from the 2 v. ed. of 1976.
Includes bibliographies and index.
ISBN 0-471-99691-2
1. Sociology, Urban—Addresses, essays, lectures. 2. City planning—Addresses, essays, lectures. 3. Personal space—Addresses, essays, lectures. I. Herbert, David T. II. Johnston, Ronald John.
HT151.S62 1978 301.36'3 78-16940
78 MARC

Die Stadt als Text. — Tübingen : Wasmuth, 1976.
123 p. : ill., map ; 22 cm. — (Konzept ; 3) GFR76-A
CONTENTS: Carlini, A. Die Stadt war ein Buch.—Barthes, R. Semiotik und Urbanismus.—Choay, F. Semiotik und Urbanismus.—Schneider, B. Die Stadt als Text.—Trabant, J. Die Stadt und die Sprache.—Hillier, B. und Leaman, A. Das Mensch-Umwelt-Paradigma und seine Paradoxien.—Schmidt, S. J. Ästhetiktheorie und Architekturästhetik.
ISBN 3-8030-0124-2 : DM24.00
1. Cities and towns—Addresses, essays, lectures. 2. City planning—Addresses, essays, lectures.
HT155.S664 78-380749
79 MARC

Toshi Mondai Kenkyūkai (Founded 1975)
(Toshi Mondai Kenkyūkai kōen shū)
都市問題研究会講演集. — ［東京］: 都市計画協会, 1978.
154 p. ; 26 cm. Ja 78-18904
Cover title.
1. City planning—Addresses, essays, lectures. 2. City planning—Japan—Addresses, essays, lectures. 3. City planning—Social aspects—Addresses, essays, lectures. I. Title.
HT166.T6187 1978 79-802749

Urbanistica : dall'empirismo verso la teoria / a cura di Federico Gorio ; scritti di P. Albisinni ... [et al.]. — Milano : F. Angeli, [1979]
202 p. ; 22 cm. — (Urbanistica : Documenti ; 2) It79-Apr
ISBN 8820415127 : L6000
1. City planning—Addresses, essays, lectures. 2. Regional planning—Addresses, essays, lectures. I. Gorio, Federico. II. Albisinni, Piero.
HT166.U748 309.2'5 79-363417
79 MARC

—BIBLIOGRAPHY

Alexander, Ernest R
Urban planning : a guide to information sources / Ernest R. Alexander, Anthony James Catanese, David S. Sawicki. — Detroit : Gale Research Co., c1979.
xiv, 165 p. ; 23 cm. — (Urban studies information guide series ; v. 2) (Gale information guide library)
Includes indexes.
ISBN 0-8103-1399-5
1. City planning—Bibliography. I. Catanese, Anthony James, joint author. II. Sawicki, David S., joint author. III. Title. IV. Series.
Z5942.A45 016.3092'62 78-13462
[HT166] 78 MARC

Chan, Yupo.
Urban and regional planning models / Yupo Chan. — Monticello, Ill. : Vance Bibliographies, 1978.
27 p. : 28 cm. — (Public administration series : Bibliography ; P-7)
Cover title.
$2.50
1. City planning—Bibliography. 2. Regional planning—Bibliography. I. Title. II. Series.
Z5942.C48 016.3092 s 78-107979
[HT166] 78 MARC

Colokathis, Jane.
Comprehensive index to CPL exchange bibliographies / compiled by Jane Colokathis. — Chicago : CPL Bibliographies, 1979-
v. ; 29 cm.
Cover title: Comprehensive index to CPL exchange bibliographies no. 1-1565, 1958-July, 1978.
CONTENTS: 1. Subject index. —
3. Numerical index.
$25.00
1. City planning—Bibliography. 2. Regional planning—Bibliography. 3. Planning—Bibliography. 4. Bibliography—Bibliography—City planning. 5. Bibliography—Bibliography—Regional planning. 6. Bibliography—Bibliography—Planning. I. Title.
Z5942.C65 016.3092 79-110547
[HT166] 79 MARC

Frankena, Frederick.
Cities, regions, and the energy problem : a bibliography and index / Frederick Frankena. — Monticello, Ill. : Vance Bibliographies, 1979.
35 p. ; 23 cm. — (Public administration series : Bibliography ; P-157)
Cover title.
Includes indexes.
$3.50
1. Energy policy—Bibliography. 2. City planning—Bibliography. 3. Regional planning—Bibliography. I. Title. II. Series.
Z7164.E6F72 016.3337 79-105675
[HD9502.A2] 79 MARC

Indonesia. Direktorat Penyelidikan Masalah Bangunan. Seksi Dokumentasi dan Perpustakaan.
Bibliography on human settlements : a list of articles abstracted and indexed in the library of the Directorate of Building Research / compiled by Library and Documentation Section. — [Bandung] : Ministry of PUblic Works and Electric Power, Directorate General of Housing, Building, Planning, and Urban Development, Directorate of Building Research, [1975?-
v. ; 21 x 30 cm.
Cover title.
Vol. 2, has subtitle: A list of books and reports available in the library of the Directorate of Building Research.
1. Housing—Bibliography. 2. City planning—Bibliography. 3. Indonesia. Direktorat Penyelidikan Masalah Bangunan—Bibliography—Catalogs. I. Title.
Z7164.H8I5 1975 016.30134 78-104704
[HD7287] 79 MARC

Johnson, Jeremy J
Performance standards and industrial location : a selective bibliography / compiled by Jeremy J. Johnson. — Adelaide : South Australian Housing Trust, 1978.
[6] p., 7 leaves ; 30 cm. — (Library bibliography ; no. 24) Aus***
ISBN 0-909118-25-6 : $1.00
1. City planning—Bibliography. 2. Zoning—Bibliography. 3. Industries, Location of—Bibliography. I. Title. II. Series: Library bibliography (Adelaide) ; no. 24.
Z5942.J64 016.3386'042 79-309394
[HT166] 79 MARC

Mexico. Instituto del Fondo Nacional de la Vivienda para los Trabajadores.
Bibliografía / INFONAVIT. — México : Servicio de Información Documental, Subdirección Financiera, [1976]
2 v. ; 22 cm.
Includes indexes.
1. Housing—Bibliography. 2. City planning—Bibliography. 3. Urbanization—Bibliography.
Z7164.H8M49 1976 016.3015'4 77-571422
[HD7287] 78[r79]rev MARC

Pollak, Patricia Baron.
Urban planning and the social sciences : perspectives, processes, and approaches / Patricia Baron Pollak. — Monticello, Ill. : Council of Planning Librarians, 1978.
73 p. ; 29 cm. — (Exchange bibliography - Council of Planning Librarians ; 1524)
Cover title.
$7.50
1. City planning—Bibliography. 2. Sociology, Urban—Bibliography. I. Title. II. Series: Council of Planning Librarians. Exchange bibliography ; 1524.
Z5942.C68 no. 1524 016.3092 s 78-105371
[Z7164.C68] 79 MARC
[HT166]

Steiber, Steven R
The humanistic element in urban planning / Steven R. Steiber. — Monticello, Ill., : Vance Bibliographies, 1978.
15 p. ; 28 cm. — (Public administration series : Bibliography ; P-78)
Cover title.
$1.50
1. City planning—Bibliography. 2. Community development, Urban—Bibliography. I. Title. II. Series.
Z5942.S83 016.3092'62 78-110867
[HT166] 79 MARC

—BIBLIOGRAPHY—CATALOGS

Torre Villar, Ernesto de la.
Guía bibliográfica para la historia y desarrollo de la arquitectura y el urbanismo en México / Ernesto de la Torre Villar, con la colaboración de Jorge Guerra Ruiz ; reproducción fotográfica de los libros originales Armando Salas Portugal. — México : Universidad Nacional Autónoma de México, Coordinación de Humanidades e Instituto de Investigaciones Bibliográficas, 1978.
38 p., [61] leaves of plates : ill. ; 28 cm. — (Catálogos de exposiciones : no. 4)
Includes index.
1. Architecture—Bibliography—Catalogs. 2. City planning—Bibliography—Catalogs. 3. Mexico. Biblioteca Nacional, Mexico—Catalogs. I. Guerra Ruiz, Jorge, joint author. II. Salas Portugal, Armando. III. Mexico (City). Universidad Nacional. Coordinación de Humanidades. IV. Mexico (City). Universidad Nacional. Instituto de Investigaciones Bibliográficas. V. Title. VI. Series.
Z5945.T67 79-101090
[NA2515] 79 MARC

—BIO-BIBLIOGRAPHY

Hulchanski, John David.
Thomas Adams : a biographical and bibliographic guide / John David Hulchanski. — Toronto : Dept. of Urban and Regional Planning, University of Toronto, 1978.
42 p. ; 28 cm. — (Papers on planning and design ; no. 15) C***
$2.50
1. City planning—Bio-Bibliography. 2. Adams, Thomas, 1871-1940. 3. Regional planning—Bio-Bibliography. I. Series.
HT166.H84 016.3092'62 78-326836
79 MARC

—CASE STUDIES

Stadtentwicklungen in kapitalistischen und sozialistischen Ländern / Jürgen Friedrichs (Hg.) ; mit Beitr. von Michael Brenner ... [et al.]. — Reinbek bei Hamburg : Rowohlt, 1978.
365 p. : ill., maps ; 19 cm. — (Rowohlts deutsche Enzyklopädie ; 378) GFR78-A
Bibliography: p. 345-[364]
Includes index.
ISBN 3-499-55378-3 : DM16.80
1. Cities and towns—Growth—Case studies. 2. City planning—Case studies. I. Friedrichs, Jürgen. II. Brenner, Michael, 1951-
HT151.S63 79-340080
79 MARC

Strong, Ann Louise.
Land banking : European reality, American prospect / Ann L. Strong. — Baltimore : Johns Hopkins University Press, c1979.
xii, 312 p. : ill. ; 24 cm. — (Johns Hopkins studies in urban affairs)
Includes bibliographies and index.
ISBN 0-8018-2169-X
1. Land use, Urban—Case studies. 2. City planning—Case studies. 3. Public lands—Case studies. I. Title. II. Series.
HD111.S76 333.1'094 78-11804
78 MARC

—CITIZEN PARTICIPATION
—ADDRESSES, ESSAYS, LECTURES

Designing with community participation / edited by Henry Sanoff. — Stroudsburg, Pa. : Dowden, Hutchinson & Ross, c1978.
ix, 253 p. : ill. ; 23 x 29 cm. — (Community development series ; 40)
Bibliography: p. 249.
Includes index.
ISBN 0-87933-314-6 (McGraw-Hill)
1. City planning—Citizen participation—Addresses, essays, lectures. 2. Community development, Urban—Addresses, essays, lectures. 3. Cities and towns—Civic improvement—Addresses, essays, lectures. I. Sanoff, Henry.
HT166.D387 1978 309.2'62 78-3723
78 MARC

—CITIZEN PARTICIPATION—CASE STUDIES

Schultz, Karen Ingrid.
Krisen og beboerbevægelserne / [af] Karen Ingrid Schulz. — [København] : Aurora, c1977.
191 p. : ill. ; 20 cm. D78-10
Bibliography: p. [188]-[192]
ISBN 8787534037 : kr44.00
1. City planning—Citizen participation—Case studies. 2. City planning—Denmark—Citizen participation—Case studies. 3. Housing policy—Citizen participation—Case studies. 4. Housing policy—Denmark—Citizen participation—Case studies. I. Title.
HT166.S3168 78-394142
79 MARC

—COLLECTED WORKS

Geography and the urban environment. v. 1–
Chichester, New York, John Wiley, 1978–
v. 24 cm.
1. Cities and towns—Collected works. 2. Anthropo-geography—Collected works. 3. City planning—Collected works.
GF125.G46 301.36'05 78-647093
MARC-S

CITY PLANNING (Continued)

—COMPUTER PROGRAMS

see also BESI (Computer program); SIARSSY (Computer program)

—CONGRESSES

Colloque de recherche urbaine, Paris, 1978.
Aménagement urbain et mouvements sociaux / Colloque de recherche urbaine, Paris, du 4 au 7 avril 78. — Paris : Centre de recherche d'urbanisme, 1978.
365 p. ; 21 cm. F***
Sponsored by the Centre and the Mission de la recherche urbaine.
Includes bibliographical references.
ISBN 2-85303-118-7
1. City planning—Congresses. I. Centre de recherche d'urbanisme. II. Mission de la recherche urbaine. III. Title.
HT166.C548 1978 309.2'62 78-396423
79 MARC

Conference on Two World Cities: Paris and New York, Paris, 1978.
Survival strategies : Paris and New York : report on the Conference on Two World Cities: Paris and New York, Paris, May, 1978 / the French/American Foundation, the Council for International Urban Liaison ; edited by George G. Wynne. — New Brunswick, N.J. : Transaction Books, c1979.
iv, 200 p. : ill. ; 23 cm.
Includes bibliographical references.
ISBN 0-87855-749-0 : $5.95
1. City planning—Congresses. 2. New York (City)—City planning—Congresses. 3. Paris—City planning—Congresses. 4. Urban policy—Congresses. 5. City planning—France—Congresses. 6. City planning—New York (State)—Congresses. I. Wynne, George A. II. French-American Foundation. III. Council for International Urban Liaison. IV. Title.
HT166.C6217 1978 309.2'62'0944361 78-66205
79 MARC

Convegno internazionale di storia urbanistica, 2d, Lucca, 1977.
Le città di fondazione : atti del 2. Convegno internazionale di storia urbanistica : Lucca, 7-11 settembre 1977 / a cura di Roberta Martinelli e Lucia Nuti. — 1. ed. — Lucca : CISCU ; Venezia : Marsilio, 1978.
327 p. : ill. ; 21 x 23 cm. It78-Dec
Italian, English, or French.
Includes bibliographical references and index.
L15000
1. City planning—Congresses. I. Martinelli, Roberta. II. Nuti, Lucia. III. Title.
NA9010.C695 1977 711'.4 79-350133
*79 MARC

Environmental design research. Editor: Wolfgang F. E. Preiser. Stroudsburg, Pa., Dowden, Hutchinson & Ross ₍1973-74.
2 v. illus. 23 cm. (Community development series)
Erratum slip inserted in v. 1.
"Proceedings of the fourth international EDRA conference which was held on April 15-18, 1973. The College of Architecture at Virginia Polytechnic Institute and State University sponsored the event."
CONTENTS: v. 1. Selected papers.—v. 2. Symposia and workshops.
ISBN 0-87933-029-5 (v. 1)
1. Architecture—Environmental aspects—Congresses. 2. Architecture—Human factors—Congresses. 3. City planning—Congresses. I. Preiser, Wolfgang F. E., ed. II. Environmental Design Research Association.
NA2542.35.E69 720'.1 73-2010
73₍r79₎rev2 MARC

Ifac Workshop on Urban, Regional, and National Planning, 1st, Kyoto, 1977.
Urban, regional, and national planning (UNRENAP) : ₍environmental aspects₎ : proceedings of the IFAC Workshop, Kyoto, Japan, 5-6 August, 1977 / edited by T. Hasegawa and K. Inoue. — Oxford ; New York : Published for the International Federation of Automatic Control by Pergamon Press, 1978.
233 p. : ill. ; 31 cm. GB78
Includes bibliographical references and index.
Workshop sponsored by the International Federation of Automatic Control Systems Engineering Committee, its Environmental Systems Symposium Committee, and the International Institute for Applied Systems Analysis; organized by IFAC Workshop on UNRENAP Organizing Committee; prepared by IFAC SECOM Working Group on UNRENAP.
"IFAC 1977 conference proceedings."
Includes index.
ISBN 0-08-022013-4 : £14.00 ($28.00)
1. City planning—Congresses. 2. Regional planning—Congresses. 3. Environmental policy—Congresses. 4. System analysis—Congresses. I. Hasegawa, Toshiharu. II. Inoue, Koichi, 1940- . III. International Federation of Automatic Control. Systems Engineering Committee. IV. International Federation of Automatic Control. Environmental Systems Symposium Committee. V. International Institute for Applied Systems Analysis. VI. Ifac Workshop on Urban, Regional, and National Planning. Organizing Committee. VII. International Federation of Automatic Control. Systems Engineering Committee. Working Group on Unrenap. VIII. Title.
HT166.I37 1977 309.2'12 78-40573
78 MARC

₍International Congress for Housing and Town Planning, 18th, Hastings, 1946₎
Final report on the 18th congress of the International Federation for Housing & Town Planning. ₍London? 1946?₎
234 p. 24 cm.
1. City planning—Congresses. 2. Housing—Congresses. I. International Federation for Housing and Town Planning.
NA9185.I55 1946 55-34825
₍r79₎rev MARC

International Congress for Housing and Town Planning, 20th, Amsterdam, 1950.
Final report : City University, Amsterdam, 27th August-2nd September 1950. — ₍s.l. : s.n., 1950?₎ (₍Amsterdam : N. V. Van Munster's Drukkerijen₎)
116 p., ₍4₎ leaves of plates : ill. ; 24 cm.
"Final report of the XXth congress of the International Federation for Housing and Town Planning."
English or French.
1. City planning—Congresses. 2. Housing—Congresses. I. International Federation for Housing and Town Planning.
HT166.I612 1950a 78-320472
78 MARC

International Congress for Housing and Town Planning, 20th, Amsterdam, 1950.
The implementation of planning measures = La mise en application des mesures d'urbanisme. — ₍s.l. : s.n., 1950?₎ (₍Amsterdam₎ : N. V. Van Munster's Drukkerijen)
180 p. ; 24 cm.
English or French, with summaries in the other language.
1. City planning—Congresses. I. Title.
HT166.I612 1950 78-368714
78 MARC

International Congress for Housing and Town Planning, 21st, Lisbon, 1952.
Long term reconstruction = Reconstruction à long terme. — ₍s.l. : s.n., 1952?₎ (₍Amsterdam : N. V. Van Munster's Drukkerijen₎)
143 p., ₍4₎ leaves of plates : ill. ; 24 cm.
English or French, with summaries in the other language.
1. City planning—Congresses. 2. Urban renewal—Congresses. I. Title.
HT166.I612 1952 78-314233
78 MARC

International Congress for Housing and Town Planning, 21st, Lisbon, 1952.
The relation between dwelling type and plan and the layout of the residential quarter = La relation entre le type et le plan de l'habitation et l'aménagement du quartier résidentiel. — ₍s.l. : s.n., 1952?₎ (₍Amsterdam : N. V. Van Munster's Drukkerijen₎)
63 p., ₍4₎ leaves of plates : ill. ; 24 cm.
English or French, with summaries in the other language.
1. City planning—Congresses. 2. Dwellings—Congresses. 3. Architecture, Domestic—Congresses. 4. Housing—Congresses. I. Title.
HT166.I612 1952a 78-368716
78 MARC

International Federation for Housing and Planning.
Building the good city = La ville de nos aspirations = Die Stadt unserer Erwartungen. — The Hague : International Federation for Housing and Planning, ₍1968?₎
292 p., ₍4₎ leaves of plates : ill. ; 23 cm.
At head of title: XXIXth World congress of the International Federation for Housing and Planning ... Philadelphia, 1968.
English, French, and German.
1. City planning—Congresses. 2. City and town life—Congresses. I. Title.
HT166.I615 1968 309.2'62 78-314230
78 MARC

International Federation for Housing and Planning.
Research for implementation = La recherche et son application dans la pratique = Die Forschung und ihre Anwendung in der Praxis : proceedings report, 31st World Congress IFHP, Liverpool, 31.V-4.VI, 1972. — The Hague : The Federation, 1972.
2 v. : ill. ; 30 cm. Ne***
Discussion summaries also in French and German.
Errata slip inserted in v. 1.
Includes bibliographical references.
1. Regional planning—Congresses. 2. City planning—Congresses. 3. Housing policy—Congresses. I. Title. II. Title: La recherche et son application dans la pratique. III. Title: Die Forschung und ihre Anwendung in der Praxis.
HT391.I485 1972 309.2'5 78-365340
78 MARC

International Federation for Housing and Planning.
Urban transportation and urban pattern = Circulation et structure urbaines = Stadtverkehr und Stadtstruktur ; Housing standards for low and lowest income groups ... = Standards de logement pour les groupes à revenus faibles et minima ... = Wohnungsstandards für Gruppen der niedrigen und niedrigsten Einkommenstufen .... — The Hague : International Federation for Housing and Planning, ₍1966?₎
265 p., ₍8₎ leaves of plates : ill. ; 23 cm.
At head of title: XXVIIIth World congress of the International Federation for Housing and Planning ... Tokyo, 1966.
English, French, and German.
1. City planning—Congresses. 2. City and town life—Congresses. I. Title.
HT166.I615 1966 309.2'62 78-320487
78 MARC

International Federation for Housing and Town Planning.
Proceedings of the South East Asia regional conference, New Delhi, February, 1-7, 1954. ₍New Delhi, 1957₎
vi (i.e. viii), 489 p. illus., ports. 26 cm.
1. Housing—Congresses. 2. City planning—Congresses. 3. Community development, Urban—Congresses. I. Title.
HD7286.I515 1957 301.5'4 sa 66-4184
₍r79₎rev MARC

International Federation for Housing and Town Planning. 20th Congress, Amsterdam, 1950.
Reports and summaries received too late for distribution prior to the Congress / XX. International Congress for Housing and Town Planning. — Amsterdam : ₍The Federation₎, 1950.
69 p. ; 24 cm.
Cover title.
English or French.
1. Housing—Congresses. 2. City planning—Congresses. 3. Building—Congresses.
HD7286.I516 1950 301.5'4 79-313375
79 MARC

International Garden Cities Town-Planning Association.
Report of conference, London, 1922. London ₍Printed by Garden city press, Letchworth, Eng., 1922₎
27, ₍1₎ p. 24 cm.
1. City planning—Congresses.
NA9010.I623 1922 24-13970
₍r79₎rev2 MARC

International Housing and Town Planning Congress, 10th, Vienna, 1926.
International Housing and Town Planning Congress, Vienna, 1926 = Congrès international de l'habitation et de l'aménagement des villes, Vienne, 1926 Internationaler Wohnungs -und Städtebaukongress, Wien, 1926. — ₍s.l. : s.n., 1926?₎
3 v. : ill. ; 23 cm.
Congress of the International Federation for Housing and Town Planning.
English, French, or German.
CONTENTS: 1-2. Papers.—pt. 3. Report.
1. City planning—Congresses. I. International Federation for Housing and Town Planning.
NA9010.I625 1926 309.2'62 78-319098
78 MARC

Internaional Housing and Town Planning Congress, 11th, Paris, 1928.
International housing and town planning congress, Paris, 1928; Congrès international de l'habitation et de l'aménagement des villes, Paris, 1928; Internationaler wohnungs- und städtebaukongress, Paris, 1928 ... ₍Paris, Imprimerie Chaix, 1928- ₎
v. illus., diagrs. 24 cm.
At head of title: International federation for housing and town planning. Fédération internationale de l'habitation et de l'aménagement des villes. Internationaler verband für wohnungswesen und städtebau ...
English, French and German.
CONTENTS:
—pt. II. Programme and general reports.—pt. III. Report.
1. City planning—Congresses. 2. Housing—Congresses. I. International Federation for Housing and Town Planning.
NA9010.I625 1928 711.0621 agr 29-505
₍r79₎rev2 MARC

International Housing and Town Planning Congress, 13th, Berlin, 1931.
XIII International housing and town planning congress, Berlin, 1931. XIII Congrès international de l'habitation et de l'aménagement des villes, Berlin, 1931. XIII Internationaler Wohnungs- und städtebaukongress, Berlin, 1931 ... London, Bureau ₍of the federation, 193-₎
3 v. illus. (incl. port., maps, plans) diagrs. 24 cm.
At head of title: International federation for housing and town planning. Fédération internationale de l'habitation et de l'aménagement des villes. Internationaler verband für wohnungswesen und städtebau ...
Papers and reports are in English or German, with translations or summaries in the other two languages.
CONTENTS: I. Papers. Rapports. Vorberichte.—II. Programme and general reports. Programme et rapports généraux. Programm und generalberichte.—III. Report. Compte-rendu. Bericht.
1. City planning—Congresses. 2. Housing—Congresses. I. International Federation for Housing and Town Planning.
NA9010.I625 1931 711.0621 34-15918
₍r79₎rev MARC

International Housing and Town Planning Congress, 14th, London, 1935.
XIV International housing and town planning congress, London, 1935. XIV Congrès international de l'habitation et de l'aménagement des villes, Londres, 1935. XIV Internationaler wohnungs- und städtebaukongress, London, 1935 ... ₍London, 1935₎
2 v. illus. (incl. port., plans) diagrs. 24 cm.
At head of title: International federation for housing and town planning. Fédération internationale de l'habitation et de l'aménagement des villes. Internationaler verband für wohnungswesen und städtebau. Bureau: ... London ...
Papers and reports are in English, French or German, with translation or summaries in the other two languages.
CONTENTS: I. Papers and general reports.—II. Report.
1. City planning—Congresses. 2. Housing—Congresses. I. International Federation for Housing and Town Planning.
NA9010.I625 1935 711.0621 37-15642
₍r79₎rev MARC

CITY PLANNING
—CONGRESSES (Continued)

International Housing and Town Planning Congress, 16th, Mexico, 1938.
Informe final del XVIe Congreso Internacional de Planificación y de la Habitación. México, 1939.
103 p. 30 cm.
Congress of the Federación Internacional de Planificación y de la Habitación.
1. Housing—Congresses. 2. City planning—Congresses. I. International Federation for Housing and Town Planning.
HD7286.I55 1938 55-45902
[r79]rev MARC

International Town Planning Conference, Amsterdam, 1924.
Conférence internationale de l'aménagement des villes, Amsterdam, 1924 ...; Internationale städtebautagung, Amsterdam, 1924 ...; International town planning conference, Amsterdam, 1924 ...; Internationaal stedebouwcongres, Amsterdam, 1924 ... [Amsterdam, N. v. de erven M. Van Munster & zoon, 1924?]
2 v. illus., plans, diagrs. 24 cm.
Conference of the International Federation for Town and Country Planning and Garden Cities.
French, German, English, and Dutch.
CONTENTS: pt. I. Papers.—pt. II. Report.
1. City planning—Congresses. I. International Federation for Town and Country Planning and Garden Cities.
NA9010.I635 1924 a 30-567
[r79]rev2 MARC

International Town Planning Conference, New York, 1925.
International town planning conference, New York, 1925. Report. Cnoférence internationale de l'aménagement des villes, New York, 1925. Compte rendu. Internationale städtebautagung, New York, 1925. Bericht. [Baltimore? 1925]
1 p.l., v-xvi, 532 p. front. (port group) plates, fold. maps, plans. 24 cm.
Conference of the International Federation for Town and Country Planning and Garden Cities.
Contributions in English, French or German.
CONTENTS: Addresses of welcome.—Progress in planning, by G. B. Ford.—Discussion by representatives of various countries.—The traffic problem.—Decentralization within regions.—The New York regional plan.—A plan for the state of New York, by C. S. Stein.—How to lay out regions for planning, by Alfred Bettman.—The government of metropolitan areas, by T. H. Reed.—Typical city plans.—Selection of airplane fields by W. N. Hensley, jr.—Aerial photographic mapping, by G. H. Matthes.—Planning unbuilt areas.—Zoning in practice.—Acquisition of lands benefited by public improvements, by Dr. I. G. Gibbon.—The business value of city and regional planning, by Bancroft Gherardi.—Tour of American cities.
1. City planning—Congresses. I. International Federation for Town and Country Plannning and Garden Cities.
NA9010.I635 1925 a 34-31
[r79]rev MARC

Misure umane : un dibattito internazionale su borgo città quartiere comprensorio / [a cura di] Carlo Doglio, Luigina Fasoli, Paolo Guidicini. — Milano : F. Angeli, [1978]
380 p. ; 22 cm. — (Collana di sociologia urbana e rurale ; 10) It78-July
Proceedings of a meeting held in Bologna in 1977.
Includes bibliographical references.
L5000
1. Cities and towns—Congresses. 2. Sociology, Urban—Congresses. 3. City planning—Congresses. I. Doglio, Carlo, 1914- II. Fasoli, Gina. 1905- . III. Guidicini, Paolo.
HT107.M57 78-384612
*78 MARC

Solar energy, architecture, and planning : conference at the Polytechnic of North London, April 1975. — London : International Solar Energy Society, UK Section, [1975]
[3], 54 p. : ill., plans ; 30 cm. — ([Conference proceedings of the UK Section] ; C6) GB75-17036
Cover title.
Includes bibliographical references.
ISBN 0-904963-00-4 : £4.00
1. Solar energy—Congresses. 2. Solar heating—Congresses. 3. Architecture and solar radiation—Congresses. 4. City planning—Congresses. I. International Solar Energy Society. UK Section. II. Series: International Solar Energy Society. UK Section. Conference proceedings of the UK Section ; C6.
TJ810.S618 729'.2 76-353888
76[r79]rev MARC

Symposium of Studies on Urban and Regional Systems, University of Calabria, 1973.
Proceedings of the Symposium of Studies on Urban and Regional Systems : University of Calabria, April 24-27, 1973 / edited by S. De Julio, A. De Maio. — [s.l. : s.n., 1974] (Roma : Tip. artigiana Multistampa)
149 p., [3] leaves of plates : ill. ; 31 cm. It76-Jan
Cover title.
At head of title: Consiglio nazionale delle ricerche, Gruppo nazionale di automatica e sistematica.
Includes bibliographies.
1. Regional planning—Congresses. 2. City planning—Congresses. I. De Julio, Sergio. II. De Maio, A. III. Università degli studi della Calabria. IV. Gruppo nazionale di automatica e sistematica.
HT391.S93 1973 309.2'5 78-314648
78 MARC

United Nations Conference on Human Settlements, Vancouver, B.C., 1976.
Policies, planning, and institutions : item 10 of the provisional agenda / Habitat—United Nations Conference on Human Settlements, Vancouver, 31 May to 11 June 1976. — [New York? : United Nations], 1976.
71 p. ; 28 cm. — ([Document] United Nations ; A/CONF.70/A/2)
Includes bibliographical references.
1. Regional planning—Congresses. 2. City planning—Congresses. I. Title. II. Series: United Nations. Document ; A/CONF.70/A/2.
JX1977.A2 A/CONF.70/A/2 78-326680
301.34
——— Copy 2. HT391.U47 1976 MARC
79

World Congress for Housing and Planning, 26th, Paris, 1962.
Habitat et civilisation = Human environment and civilization = Die menschliche Behausung und die Zivilisation / XXVIme Congrès mondial, Paris, Sept. 1962. — [s.l. : s.n., 1962?] ([Colombes, France : Imprimerie Desseaux])
453 p. : ill. ; 25 cm.
At head of title: Fédération internationale pour l'habitation, l'urbanisme et l'aménagement des territoires. International Federation for Housing and Planning. Internationaler Verband für Wohnungswesen, Städtebau und Raumordnung.
French, English, and German.
1. Housing—Congresses. 2. City planning—Congresses. 3. Regional planning—Congresses. I. International Federation for Housing and Planning. II. Title. III. Title: Human environment and civilization.
HD7286.W58 1962 301.5'4 78-399997
79 MARC

—DATA PROCESSING

Morrison, William Ian.
Input-output methods in urban and regional planning : a practical guide / W. I. Morrison and P. Smith. — Oxford ; New York : Pergamon Press, 1977.
p. 59-151 ; 28 cm. — (Progress in planning ; v. 7, pt. 2 ISSN 0305-9006) GB***
Bibliography: p. 132-135.
ISBN 0-08-021858-X
1. City planning—Data processing. 2. Regional planning—Data processing. 3. Urban economics—Linear programming. 4. Regional economics—Linear programming. I. Smith, Peter, 1947- joint author. II. Title. III. Series.
HT166.M593 1977 309.2'5'02854 79-314911
79 MARC

—DICTIONARIES—POLISH

Siegel, Jaromír.
Polsko-český slovník odborných výrazů z oboru urbanismu a územního plánování / Jaromír Siegel. — 1. vyd. — Praha : Výzkum. ústav výstavby a architektury, 1978.
126, [1] p. ; 21 cm. Cz78
1. City planning—Dictionaries—Polish. 2. Polish language—Dictionaries—Czech. I. Title.
HT108.5.S53 79-363672
*79 MARC

—DICTIONARIES—JAPANESE

(Gendai toshi keikaku yōgoroku)
現代都市計画用語録 / 梶秀樹 ... [et al.] 著. — 東京 : 彰国社, 1978.
258 p. : ill. ; 19 cm. Ja 79-2517
Bibliography: p. 236-238.
Includes index.
¥1600
1. City planning—Dictionaries—Japanese. 2. City planning—Japan—Dictionaries—Japanese. I. Kaji, Hideki, 1942-
HT166.G46 79-804795

—DICTIONARIES—POLYGLOT

Instituto Nacional para el Desarrollo de la Comunidad Rural y de la Vivienda Popular. Centro de Investigación, Documentación e Información sobre la Vivienda.
Diccionario técnico de arquitectura y urbanismo / CIDIV, Centro de Investigación, Documentación e Información sobre la Vivienda. — México : Instituto Nacional para el Desarrollo de la Comunidad Rural y de la Vivienda Popular, 1976.
345 p. ; 22 cm. — (Publicaciones del CIDIV ; 3)
English, French, German, Italian, and Spanish.
$230.00
1. Architecture—Dictionaries—Polyglot. 2. City planning—Dictionaries—Polyglot. I. Title. II. Series: Instituto Nacional para el Desarrollo de la Comunidad Rural y de la Vivienda Popular. Centro de Investigación, Documentación e Información sobre la Vivienda. Publicaciones del CIDIV ; 3.
NA31.I57 1976 720'.3 78-107972
79 MARC

—ECONOMIC ASPECTS

Pałaszewski, Tadeusz.
Ekonomika projektowania architektonicznego i urbanistycznego / Tadeusz Pałaszewski. — Wyd. 3 popr. — Warszawa : Wydaw-a Politechniki Warszawskiej, 1978- (Warsz. : PW)
v. : ill. ; 25 cm. P78-1934
Previous editions published under title: Ekonomika projektowania w budownictwie (architektonicznego i urbanistycznego).
Bibliography: cz. 1, p. 219-220.
zł16.00 (v. 1)
1. Architectural design—Economic aspects. 2. City planning—Economic aspects. 3. Building—Estimates. I. Title.
NA2540.P32 1978 79-364404
79 MARC

—ENVIRONMENTAL ASPECTS

Nagayasu, Yukimasa, 1941-
(Gendai keizai bummei no seitaigaku)
現代経済文明の生態学 : 自然 経済 思想 / 永安幸正. — 東京 : 前野書店, 1978.
xiv, 280 p. : ill. ; 19 cm. Ja 78-14952
Includes bibliographical references.
¥1600
1. City planning—Environmental aspects. 2. Urban ecology (Biology) 3. Economic history—20th century. 4. Japan—Economic policy—1945- 5. Civilization, Modern—20th century. I. Title.
HT166.N3 78-812213

Willson, John S
Comprehensive planning and the environment : a manual for planners / John S. Willson, Philip Tabas, Marian Henneman. — Cambridge, Mass. : Abt Books, c1979.
ix, 283 p. : ill. ; 24 cm.
Bibliography: p. 259-276.
Includes index.
ISBN 0-89011-515-X
1. City planning—Environmental aspects. 2. Environmental policy. I. Tabas, Philip, joint author. II. Henneman, Marian, joint author. III. Title.
HT166.W535 301.31 78-66683
79 MARC

—ENVIRONMENTAL ASPECTS—
GERMANY, WEST—CONGRESSES

Ökologie und Planung im städtischen Siedlungsraum : 14 Aufsätze : [Kolloquium veranstaltet am 29.4.-30.4.1976 vom SVR und der Gesamthochschule Essen (GHS)] / Herausgeber, Siedlungsverband Ruhrkohlenbezirk. — Essen : Siedlungsverband Ruhrkohlenbezirk, 1978.
150 p. : ill., graphs, maps ; 30 cm. — (Schriftenreihe Siedlungsverband Ruhrkohlenbezirk ; 61) GFR***
Includes bibliographies.
1. City planning—Environmental aspects—Germany, West—Congresses. 2. Landscape assessment—Germany, West—Congresses. I. Siedlungsverband Ruhrkohlenbezirk. II. Gesamthochschule Essen. III. Series: Siedlungsverband Ruhrkohlenbezirk. Schriftenreihe ; 61.
HT395.G42R76 Nr. 61 79-359295
[HT169.G3] 79 MARC

—EVALUATION

Klein, Reinhard.
Nutzenbewertung in der Raumplanung : Überblick und praktische Anleitung / Reinhard Klein. — Basel [etc.] : Birkhäuser, 1978.
148 p. : ill. ; 23 cm. — (ISR, Interdisciplinary systems research ; 54) (Urban and regional planning ; v. 2) Sw78-8740
Revision of the author's thesis, Dortmund, 1976.
Summary also in English.
Bibliography: p. 137-145.
Includes index.
ISBN 3-7643-1008-1 : 26.00F
1. Regional planning—Evaluation. 2. City planning—Evaluation. I. Title. II. Series. III. Series: Urban and regional planning ; v. 2.
HT391.K58 1978 78-396767
78 MARC

—EXHIBITIONS

International Cities and Town Planning Exhibition.
English catalogue. Jubilee exhibition, Gothenburg, Sweden, 1923. [Göteborg, Wezäta, W. Zachrissons boktryckeri a.-b., 1923]
2 p. l., [7]-389, [1] p. incl. illus., plans. 24 cm.
Edited by Werner Hegemann.
1. City planning—Exhibitions. I. Hegemann, Werner, 1881-1936, ed. II. Title.
NA9016.S8G674 24-13964
[r79]rev2 MARC

CITY PLANNING
—EXHIBITIONS (Continued)

Neues Bauen in alter Umgebung : eine Ausstellung der Bayerischen Architektenkammer und der Neuen Sammlung München : [Katalog / Konzeption und Gestaltung, Michael Gaenssler, Rüdiger Möller]. — München : Die Neue Sammlung, c1978.
207 p. : chiefly ill. ; 25 cm. GFR***
"Herkunft der Zitate": p. 206.
Includes index.
1. City planning—Exhibitions. 2. Architecture—Conservation and Restoration—Exhibitions. 3. Architecture and history—Exhibitions. I. Gaenssler, Michael. II. Möller, Rüdiger. III. Bayerische Architektenkammer. IV. Munich. Neue Sammlung.
NA9016.G3M866 78-384974
78 MARC

—GRAPHIC METHODS

Cianchetta, Bruno.
Atlante di urbanistica : ricerca operativa / Bruno Cianchetta ; collaborazione di Rosella Pierotti. — Roma : Bulzoni, [1978]-
v. : ill. ; 24 cm. It78-July
CONTENTS: 1. I grafi e il territorio.—2. I modelli di assegnazione.—3. La composizione automatica.—4. I bacini idrografici.—5. L'equilibrio territoriale.

L3500 per vol.
1. City planning—Graphic methods. 2. Regional planning—Graphic methods. 3. Operations research. I. Pierotti, Rosella. II. Title.
HT166.C464 309.2'12'0184 79-339236
*79 MARC

—HISTORY

Arbore, Grigore.
Cetatea ideală în viziunea Renaşterii : eseu asupra tipologiei formelor urbane / Grigore Arbore. — Bucureşti : "Meridiane", 1978.
181 p., [20] leaves of plates : ill. ; 17 cm. — (Curente şi sinteze ; 21)
R79-1168
Includes bibliographical references.
lei11.00
1. Cities and towns, Renaissance—History. 2. City planning—History. I. Title.
NA9094.3.A72 79-378713
*79 MARC

Houghton-Evans, William.
Architecture and urban design / [by] W. Houghton Evans. — Lancaster ; New York [etc.] : Construction Press, 1978.
[6], 185 p. : ill., maps, plans ; 31 cm. GB78-33129
Includes index.
ISBN 0-904406-97-0 : £15.00
1. City planning—History. I. Title.
NA9090.H62 711'.4 79-304900
79 MARC

—HISTORY—BIBLIOGRAPHY

Sutcliffe, Anthony, 1942-
The history of modern town planning : a bibliographic guide / Anthony Sutcliffe. — Birmingham [Eng.] : University of Birmingham, Centre for Urban and Regional Studies, 1977.
iv leaves, 112 p. ; 30 cm. — (Research memorandum - Centre for Urban and Regional Studies, University of Birmingham ; no. 57 ISSN 0306-4034)
GB***
Includes indexes.
ISBN 0-7044-0255-6 : £1.75
1. City planning—History—Bibliography. I. Title. II. Series: Birmingham, Eng. University. Centre for Urban and Regional Studies. Research memorandum - Centre for Urban and Regional Studies, University of Birmingham ; no. 57.
Z5942.S9 016.3092'62'09 79-306580
[HT111] 79 MARC

—HYGIENIC ASPECTS—CONGRESSES

Fortbildungstagung Siedlungshygiene und Stadtplanung, Berlin, 1975.
Siedlungshygiene und Stadtplanung / K. Aurand und H. Ising, unter Mitarbeit von H. Becker ... [et al.]. — Stuttgart : G. Fischer, 1975.
357 p. : ill. ; 21 cm. — (Schriftenreihe des Vereins für Wasser-, Boden- und Lufthygiene ; 45) GFR***
Conference sponsored by the Institut für Wasser-, Boden- und Lufthygiene, Forschungsstätte für Allgemeine Hygiene und Gesundheitstechnik.
Includes bibliographical references.
ISBN 3-437-30224-8
1. Urban health—Congresses. 2. City planning—Hygienic aspects—Congresses. I. Aurand, K. II. Ising, H. III. Berlin. Institut für Wasser-, Boden- und Lufthygiene, Forschungsstätte für Allgemeine Hygiene und Gesundheitstechnik. IV. Title. V. Series: Verein für Wasser-, Boden- und Lufthygiene. Schriftenreihe ; 45.
RA566.7.F67 79-361129
79 MARC

—HYGIENIC ASPECTS
—RUSSIAN REPUBLIC
—KEMEROVO (PROVINCE)

(Gigienicheskie aspekty raĭonnoĭ planirovki i gradostroitel'stva v Kemerovskoĭ oblasti)
Гигиенические аспекты районной планировки и градостроительства в Кемеровской области / [А. А. Добринский, В. М. Пивкин, А. Ф. Бугреева ... et al. ; Редкол. Е. М. Горбачев (отв. ред.) ... et al.]. — Новосибирск : Наука, Сиб. отд-ние, 1978.
229 p. : ill. ; 22 cm. USSR 78
At head of title: Akademii͡a nauk SSSR. Sibirskoe otdelenie. Institut pochvovedenii͡a i agrokhimii. Ministerstvo zdravookhranenii͡a RSFSR. Novosibirskiĭ nauchno-issledovatel'skiĭ sanitarnyĭ institut.
Bibliography: p. 224-[227]
2.00rub
1. City planning—Hygienic aspects—Russian Republic—Kemerovo (Province) 2. Urban health — Russian Republic — Kemerovo (Province) I. Dobrinskiĭ, Anatoliĭ Abramovich. II. Gorbachev, Evgeniĭ Matveevich. III. Akademii͡a nauk SSSR. Sibirskoe otdelenie. Institut pochvovedenii͡a i agrokhimii. IV. Novosibirsk. Nauchno-issledovatel'skiĭ sanitarnyĭ institut.
RA566.5.R9G54 79-413416

—HYGIENIC ASPECTS—RUSSIAN REPUBLIC—NOVOSIBIRSK (PROVINCE)

(Sanitarno-gigienicheskie osnovy gradostroitel'stva v Novosibirskoĭ oblasti)
Санитарно-гигиенические основы градостроительства в Новосибирской области / [А. А. Добринский ... et al.] ; отв. ред. канд. мед. наук Е. М. Горбачев. — Новосибирск : Наука, Сиб. отд-ние, 1974.
116, [1] p. : maps ; 22 cm. USSR 74
At head of title: Novosibirskiĭ nauchno-issledovatel'skiĭ sanitarnyĭ institut Ministerstva zdravookhranenii͡a RSFSR. Obshchestvennyĭ nauchno-issledovatel'skiĭ institut problem okhrany prirody, zelenogo stroitel'stva i sadovo-parkovogo khozi͡aĭstva pri Novosibirskom gorispolkome.
Bibliography: p. 114-[117]
0.38rub
1. City planning—Hygienic aspects—Russian Republic—Novosibirsk (Province) 2. Urban health—Russian Republic—Novosibirsk (Province) I. Dobrinskiĭ, Anatoliĭ Abramovich. II. Gorbachev, E. M. III. Novosibirsk. Nauchno-issledovatel'skiĭ sanitarnyĭ institut. IV. Obshchestvennyĭ nauchno-issledovatel'skiĭ institut problem okhrany prirody, zelenogo stroitel'stva i sadovo-parkovogo khozi͡aĭstva.
RA566.5.R9S24 79-413327

—LAW AND LEGISLATION

see City planning and redevelopment law

—LAW AND LEGISLATION
—GERMANY, WEST

Deutsches Institut für Urbanistik.
Baufreiheit und Stadtentwicklung : Bericht über zwei Seminare des Deutschen Instituts für Urbanistik vom 26.-30. 9. 1977 und 24.-28. 4. 1978 in Berlin / [redaktionelle Bearbeitung, Rudolf Schäfer]. — Berlin : Deutsches Institut für Urbanistik, 1978.
262 p. ; 30 cm. GFR***
On cover: Tagungsbericht.
Includes bibliographical references.
ISBN 3-88118-050-8
1. City planning—Germany, West. 2. City planning—Law and legislation—Germany, West. I. Schäfer, Rudolf, Assessor Dipl.-Pol. II. Title.
HT169.G3D49 1978 79-362073
79 MARC

—LIBRARY RESOURCES—GEORGIA—DIRECTORIES

Thaxton, Lyn.
Resources for urban research in Atlanta and Athens libraries / Lyn Thaxton. — [Atlanta] : Thaxton, 1978, c1979.
[28] p. ; 23 cm.
1. City planning—Library resources—Georgia—Directories. 2. Library resources—Georgia—Directories. I. Title.
Z5942.T47 309.2'62'025758 79-109895
[HT167.5.G45] 79 MARC

—MATHEMATICAL MODELS

Bucher, Jürg.
Stadtentwicklung in kleinräumigen Regionen : ein Modellansatz und seine empirische Überprüfung am Beispiel der Region Bern / J. Bucher, W. Konanz, O. Messmer. — Bern [etc.] : P. Haupt, c1978.
88 p. : ill. ; 23 cm. — (Berner Beiträge zur Nationalökonomie ; Bd. 32)
Sw78-5291
Bibliography: p. 87.
ISBN 3-258-02674-2 : 24.00F
1. City planning—Switzerland—Bern region. 2. SIARSSY (Computer program) 3. Regional planning—Mathematical models. 4. City planning—Mathematical models. I. Konanz, Walter, joint author. II. Messmer, Otto, Dr. rer. pol., joint author. III. Title. IV. Series.
HT169.S92B473 78-383075
79 MARC

Cianchetta, Bruno.
La classificazione delle aree urbane / Bruno Cianchetta ; collaborazione di Rosella Pierotti. — Roma : Bulzoni, [1978]
[57] p. : ill. ; 24 cm. — (Atlante di urbanistica : Ricerca operativa ; 6)
It78-Sept
L3500
1. City planning—Mathematical models. I. Pierotti, Rosella, joint author. II. Series.
HT166.C465 301.36'1 78-398521
*79 MARC

Möllers, Heribert.
Infrastrukturausstattung und Entwicklung von Städten : Methoden der multivariaten Analyse / Heribert Möllers. — Münster : Institut für Siedlungs- und Wohnungswesen der Universität Münster, 1977.
260 p. : ill. ; 21 cm. — (Beiträge zum Siedlungs- und Wohnungswesen und zur Raumplanung ; Bd. 42) GFR***
Originally presented as the author's thesis, Münster.
Bibliography: p. 251-260.
1. City planning—Mathematical models. 2. Regional planning—Mathematical models. I. Title. II. Series.
HT166.M584 1977 78-368177
78 MARC

Snickars, Folke, 1944-
Some mathematical models for urban and regional planning / Folke Snickars. — Stockholm : Dept. of Mathematics, Royal Institute of Technology, [1978]
29 leaves : ill. ; 30 cm. S***
"TRITA-MAT-1978-4."
Summary of thesis—Royal Institute of Technology.
Bibliography: leaves 27-29.
1. City planning—Mathematical models. 2. Regional planning—Mathematical models. I. Title.
HT166.S59 309.2'5'0151 79-307269
79 MARC

—METHODOLOGY

Elementi di metodo della pianificazione territoriale e urbana / testi di Carlo Monti ... [et al.] ; presentazione di Fernando Clemente. — Bologna : Cooperativa libraria universitaria editrice, 1976.
380 p. : ill. ; 24 cm. It78-Aug
At head of title: Università di Bologna, Istituto di architettura ed urbanistica.
Bibliography: p. 379-380.
1. Regional planning—Methodology. 2. City planning—Methodology. I. Monti, Carlo, fl. 1975-
HT391.E53 309.2'12'018 79-342867
*79 MARC

Prentice, Richard.
The determination of worth : macrorationality in urban and regional planning / [by] Richard Prentice. — Reading : Department of Geography, University of Reading, 1977.
[2], ii, 30 p. : 2 ill. ; 21 cm. — (Geographical papers ; no. 61 ISSN 0305-5914)
GB78-21805
Bibliography: p. 24-28.
ISBN 0-7049-0471-3
1. City planning—Methodology. 2. Regional planning—Methodology. I. Title. II. Series.
HT166.P677 309.2'12 78-324679
79 MARC

United Nations. Dept. of Economic and Social Affairs.
Threshold analysis handbook / Department of Economic and Social Affairs. — New York : United Nations, 1977.
viii, 167, 48 p. : ill. ; 28 cm. — ([Document] - United Nations ; ST/ESA/64)
"United Nations publication. Sales no. E.78.IV.2."
Bibliography: p. 164-167.
$7.00
1. City planning—Methodology. I. Title. II. Series: United Nations. Document ; ST/ESA/64.
JX1977.A2 ST/ESA/64 300'.8 s 78-315830
—— —— Copy 2. HT166.U49 1977 MARC
78

CITY PLANNING (Continued)

—METHODOLOGY—CONGRESSES

Regional Science Association. British Section.
Theory and method in urban and regional analysis / edited by P. W. J. Batey. — London : Pion, 1978.
[7], 184 p. : ill., maps ; 23 cm. — (London papers in regional science ; 8) GB78-23217
"A selection of the papers presented at the ninth annual conference of the British Section of the Regional Science Association, held in London on September 2nd and 3rd, 1976."
Distributed by Academic Press, London, New York.
Includes bibliographies.
ISBN 0-85086-066-0 : £5.00
1. Regional planning—Methodology—Congresses. 2. City planning—Methodology—Congresses. 3. Central places—Mathematical models—Congresses. 4. Space in economics—Mathematical models—Congresses. I. Batey, P. W. J. II. Title.
HT391.R35 1978 309.2'5'0941 78-326371
79 MARC

—PERIODICALS

Built environment. v. 4-
June 1978-
[London] Kogan Page.
v. ill. 25 cm. 4 no. a year.
Vols. for June 1978- "published in association with the Regional Studies Association."
Continues: Built environment quarterly, ISSN 0308-1508.
1. Housing—Great Britain—Periodicals. 2. Regional planning—Periodicals. 3. City planning—Periodicals. 4. Urban renewal—Periodicals. I. Regional Studies Association.
HD7333.A3B797 301.5'4'0941 78-647819
MARC-S

Institut d'aménagement et d'urbanisme de la région d'Ile-de-France.
Cahiers de l'Institut d'aménagement et d'urbanisme de la région d'Ile-de-France.
Paris [Institut d'aménagement et d'aménagement et d'urbanisme de la région d'Ile-de-France]
v. ill. 30 cm. quarterly.
Continues: France. Institut d'aménagement et d'urbanisme de la région parisienne. Cahiers.
Issues for published in several parts in a binder.
1. City planning—Periodicals. 2. Urbanization—Periodicals.
HT166.F72a 301.36'1'05 78-644008
MARC-S

International review. v. 1-
Nov. 1978-
[Washington] U. S. Dept. of Housing and Urban Development, Office of International Affairs.
v. 21 cm.
Key title: International review (Washington), ISSN 0191-9679.
1. City planning—Periodicals. 2. Housing policy—Periodicals. 3. Urban policy—Periodicals. I. United States. Dept. of Housing and Urban Development. Office of International Affairs.
HT166.I 616 309.2'62'05 79-642902
MARC-S

—POLITICAL ASPECTS—BIBLIOGRAPHY

Simmie, J M
A bibliography on the political sociology of urban development / by J. M. Simmie. — [London] : [University College, London], 1978.
[1], 43 p. ; 30 cm. — (Town planning discussion paper ; no. 30) GB79-04768
£1.00
1. City planning—Political aspects—Bibliography. 2. Urban policy—Bibliography. I. Title. II. Series: London. University. University College. School of Environmental Studies. Discussion paper series ; no. 30.
Z5942.S53 016.3092'62 79-309396
[HT166] 79 MARC

—REMOTE SENSING

Remote sensing for planners / edited by Kristina Ford ; contributors, Leonard W. Bowden ... [et al.]. — New Brunswick, N.J. : Center for Urban Policy Research, c1979.
xv, 219 p., [4] leaves of plates : ill. ; 26 cm.
Includes bibliographical references and index.
ISBN 0-88285-058-X : $20.00
1. City planning—Remote sensing. 2. Regional planning—Remote sensing. I. Ford, Kristina, 1946-
HT166.R357 309.2'6 78-31594
78 MARC

—RESEARCH—ENGLAND

Faludi, Andreas.
The promotion and control of development in Leiden : interim report of the Oxford-Leiden comparative research project / [by] Andreas Faludi and Steve Hamnett, with the assistance of ... [others]. — [Oxford] : Oxford Polytechnic, Department of Town Planning, [1977]
[2], iv leaves, 145 [i.e. 146] p., [9] leaves of plates : maps, plans ; 30 cm. — (Working paper - Oxford Polytechnic, Department of Town Planning ; no. 27) GB78-12067
Bibliography: p. 144-145.
£3.02
1. Leyden—City planning—Case studies. 2. Urban renewal—Netherlands—Leyden—Case studies. 3. City planning—Research—Netherlands. 4. City planning—Research—England. I. Hamnett, Stephen, joint author. II. Title. III. Series: Oxford working papers in planning education and research ; no. 27.
HT169.N42L494 309.2'62'094923 78-325366
79 MARC

—RESEARCH—IRELAND

Foras Forbartha.
Urban and regional research projects in Ireland, 1976 : an annotated list / prepared by An Foras Forbartha. — [Dublin : An Foras Forbartha, 1977]
26 leaves : chiefly forms ; 30 cm. Ir***
Cover title.
1. Regional planning—Research—Ireland. 2. City planning—Research—Ireland. I. Title.
HT395.I7F67 1977 309.2'12'0720417 79-302204
79 MARC

—RESEARCH—NETHERLANDS

Faludi, Andreas.
The promotion and control of development in Leiden : interim report of the Oxford-Leiden comparative research project / [by] Andreas Faludi and Steve Hamnett, with the assistance of ... [others]. — [Oxford] : Oxford Polytechnic, Department of Town Planning, [1977]
[2], iv leaves, 145 [i.e. 146] p., [9] leaves of plates : maps, plans ; 30 cm. — (Working paper - Oxford Polytechnic, Department of Town Planning ; no. 27) GB78-12067
Bibliography: p. 144-145.
£3.02
1. Leyden—City planning—Case studies. 2. Urban renewal—Netherlands—Leyden—Case studies. 3. City planning—Research—Netherlands. 4. City planning—Research—England. I. Hamnett, Stephen, joint author. II. Title. III. Series: Oxford working papers in planning education and research ; no. 27.
HT169.N42L494 309.2'62'094923 78-325366
79 MARC

—RESEARCH—UNITED STATES

Recent contributions to the urban policy debate / Peter A. Morrison ... [et al.]. — Santa Monica, CA. : Rand Corp., 1979.
ix, 34 p. : ill. ; 28 cm. — ([Report] - Rand Corportation ; R-2394-RC)
"Contains the text of five briefings on research in Rand's urban policy analysis program that were presented to Rand's Board of Trustees at their annual meeting on April 13, 1978."
ISBN 0-8330-0072-1
1. Urban policy—Research—United States. 2. City planning—Research—United States. I. Morrison, Peter A. II. Rand Corporation. III. Series: Rand Corporation. Rand report ; R-2394-RC.
AS36.R3 R-2394 081 s 78-25950
[HT123] 78 MARC

—SIMULATION METHODS

Diekmann, Peter.
Planspiel und Planspiel-Simulation in der Raumplanung : Text, Spiele, Literatur / Peter Diekmann, Heribert Leppert. — Basel [etc.] : Birkhäuser, 1978.
xi, 168 p. ; 23 cm. — (ISR, Interdisciplinary systems research ; 60) (Urban and regional planning ; v. 4) Sw79-1096
Includes bibliographies.
ISBN 3-7643-1045-6 : 38.00F
1. City planning—Simulation methods. 2. Regional planning—Simulation methods. 3. City planning—Simulation methods—Bibliography. 4. Regional planning—Simulation methods—Bibliography. 5. Game theory. 6. Game theory—Bibliography. I. Leppert, Heribert, joint author. II. Title. III. Series. IV. Series: Urban and regional planning ; v. 4.
HT166.D547 79-368374
79 MARC

—SIMULATION METHODS—BIBLIOGRAPHY

Diekmann, Peter.
Planspiel und Planspiel-Simulation in der Raumplanung : Text, Spiele, Literatur / Peter Diekmann, Heribert Leppert. — Basel [etc.] : Birkhäuser, 1978.
xi, 168 p. ; 23 cm. — (ISR, Interdisciplinary systems research ; 60) (Urban and regional planning ; v. 4) Sw79-1096
Includes bibliographies.
ISBN 3-7643-1045-6 : 38.00F
1. City planning—Simulation methods. 2. Regional planning—Simulation methods. 3. City planning—Simulation methods—Bibliography. 4. Regional planning—Simulation methods—Bibliography. 5. Game theory. 6. Game theory—Bibliography. I. Leppert, Heribert, joint author. II. Title. III. Series. IV. Series: Urban and regional planning ; v. 4.
HT166.D547 79-368374
79 MARC

—SOCIAL ASPECTS

Pieper, Richard, 1945-
Soziologie im Städtebau : e. Einf. für Architekten, Stadtplaner u. Sozialwissenschaftler / Richard Pieper. — Stuttgart : Enke, 1979.
v, 170 p. : ill. ; 19 cm. — (Flexibles Taschenbuch : Soz) GFR78-A
Bibliography: p. 163-166.
Includes index.
ISBN 3-432-90191-7 : DM16.80
1. City planning—Social aspects. 2. Sociology, Urban. 3. Environmental policy—Social aspects. 4. Environmental policy—Psychological aspects. I. Title.
HT166.P5 301.36'1 79-352810
79 MARC

—SOCIAL ASPECTS —ADDRESSES, ESSAYS, LECTURES

Toshi Mondai Kenkyūkai (Founded 1975)
(Toshi Mondai Kenkyūkai kōen shū)
都市問題研究会講演集. — [東京] : 都市計画協会, 1978.
154 p. ; 26 cm. Ja 78-18904
Cover title.
1. City planning—Addresses, essays, lectures. 2. City planning—Japan—Addresses, essays, lectures. 3. City planning—Social aspects—Addresses, essays, lectures. I. Title.
HT166.T6187 1978 79-802749

—STANDARDS—ITALY

Falco, Luigi.
Gli standard urbanistici / Luigi Falco. — 2. ed. — Roma : Edizioni delle autonomie, 1978.
262 p. : ill. ; 21 cm. — (Amministrare l'urbanistica ; 3) It78-Aug
"Appendici" (legislative): p. [207]-262.
Includes bibliographical references.
L5500
1. City planning—Standards—Italy. I. Title.
HT169.I83F33 1978 78-398797
*79 MARC

—STUDY AND TEACHING

The Content of a planning theory course : papers from a seminar held at Oxford Polytechnic, June 1976 / edited at Oxford Polytechnic, Department of Town Planning. — [Oxford] : Oxford Polytechnic, 1976.
[1], vi leaves, 168 [i.e. 173], [1] p. : 2 ill. ; 30 cm. — (Working paper - Oxford Polytechnic, Department of Town Planning ; no. 25) GB78-08851
Includes bibliographical references.
1. City planning—Study and teaching. 2. Planning—Study and teaching. I. Oxford Polytechnic. Dept. of Town Planning. II. Series: Oxford working papers in planning education and research ; no. 25.
HT166.C6235 309.2'07'11 79-309639
79 MARC

International Federation for Housing and Town Planning.
Education in town planning, an international survey. The Hague, 1952.
139 p. 24 cm.
1. City planning—Study and teaching. I. Title.
NA9012.I55 1952 711.07 a 54-5012
[r79]rev2 MARC

CITY PLANNING (Continued)

—STUDY AND TEACHING—BRAZIL
—BAHIA (STATE)

II [i.e. Segundo] Curso de especialização em planejamento urbano —CEPU : anais : Salvador, 1975/76. — [Salvador : Universidade Federal da Bahia, 1976?]
183 p. : ill. (part fold.) ; 23 cm.
Cover title: Curso de especialização em planejamento urbano.
On t.p.: Convênio SUDENE/UFBa. Programa Nacional de Capacitação de Recursos Humanos.
Includes bibliographies.
1. City planning—Study and teaching—Brazil—Bahia (State) I. Bahia, Brazil (State). Universidade Federal. II. Title: Curso de especialização em planejamento urbano.
HT165.53.B7S43 79-372162
79 MARC

—STUDY AND TEACHING—NIGERIA

Vagale, L R
Proposals for an integrated programme of education and training in urban and regional planning in Nigeria : final report no. 2 [i.e. no. 1] / L. R. Vagale. — Ibadan, Nigeria : Vagale, 1972.
83 p. ; 33 cm.
At head of title: United Nations Office of Technical Cooperation.
Bibliography: p. 37-38.
1. City planning—Study and teaching—Nigeria. 2. Regional planning—Study and teaching—Nigeria. I. United Nations. Office of Technical Co-operation. II. Title.
HT109.V33 309.2'12'0710669 79-107203
79 MARC

—TERMINOLOGY

International Federation for Housing and Town Planning.
International glossary of technical terms used in housing and town planning. [London, 1934]
80 p. 24 cm. (Its Special report no. 2)
English, French, German, Italian.
1. Housing—Terminology. 2. City planning—Terminology. I. Title. II. Series.
HD7287.I5 1934 301.5'4'03 35-31074
[r79]rev2 MARC

International Federation for Housing and Town Planning.
International glossary of technical terms used in housing and town planning. 2d rev. and enl. ed. The editor: H. J. Spiwak. Amsterdam [1951]
144 p. 25 cm.
English, French, German, Italian and Spanish.
1. Housing—Terminology. 2. City planning—Terminology. I. Spiwak, H. J., ed. II. Title.
HD7287.I5 1951 309.2'62'03 53-39800
[r79]rev2 MARC

—UNDERDEVELOPED AREAS see
Underdeveloped areas—City planning

—ZONE SYSTEM see Zoning

—AEGEAN SEA REGION

(Ēgekai no mura to machi)
エーゲ海の村と街 = Aegean Sea / 企画・撮影 二川幸夫 ; 文 磯崎新 ; 解説・作図 鈴木恂 ; デザイン 細谷巖. — 東京 : A. D. A. EDITA Tokyo Co., 1973.
143 p. : chiefly ill. (some col.) ; 30 cm. — (世界の村と街 ; 1)
Ja 73-10303
¥3200
1. Architecture—Aegean Sea region. 2. City planning—Aegean Sea region. I. Futagawa, Yukio, 1932- II. Isozaki, Arata. III. Title: Aegean Sea. IV. Series: Sekai no mura to machi ; 1.
NA1099.A4E33 79-801495

—AFRICA, NORTH

Roberts, M Hugh P
An urban profile of the Middle East / M. Hugh P. Roberts. — New York : St. Martin's Press, 1979.
239 p. : ill. ; 24 cm.
Includes bibliographies and index.
ISBN 0-312-83467-5
1. Cities and towns—Near East. 2. Cities and towns—Africa, North. 3. Urbanization—Near East. 4. Urbanization—Africa, North. 5. City planning—Near East. 6. City planning—Africa, North. I. Title.
HT147.N4R6 1979 309.2'62'0956 78-27185
78 MARC

—AFRICA, WEST

Fassassi, Masudi Alabi.
L'architecture en Afrique noire : cosmoarchitecture / Masudi Alabi Fassassi. — Paris : F. Maspero, 1978.
189 p. : ill. ; 27 cm. — (Textes à l'appui) F***
Bibliography: p. 179-180.
ISBN 2-7071-0870-7
1. Architecture—Africa, West. 2. Architecture and society—Africa, West. 3. City planning—Africa, West. I. Title.
NA1598.F37 309.2'62'0966 78-377715
78 MARC

—ALASKA

CH2M Hill Alaska (Firm)
Waterfront plan for Juneau, Alaska : an element of the comprehensive plan for the city and borough of Juneau, Alaska ... — [s.l. : s.n.], 1973 ([Juneau] : Reproduction Dept. of Ch2M Hill Alaska)
161 p. : ill. (part col.) ; 22 x 28 cm.
"Prepared by the firm of CH2M Hill Alaska ... in collaboration with the Planning Department."
Bibliography: p. 155-156.
1. Juneau, Alaska—City planning. 2. City planning—Alaska. 3. Waterfronts—Alaska—Juneau. I. Juneau, Alaska. City & Borough Planning Dept. II. Title.
HT168.J86C13 1973 711'.4'0972982 78-100990
79 MARC

New capital city planning process : background report. — [Anchorage : Capital Site Planning Commission], 1978.
14 v. in 15 : ill. ; 22-28 x 22-29 cm.
Cover title.
Includes bibliographies.
CONTENTS: no. 1. Capital Site Planning Commission. Introduction and overview.—no. 2. L. Lane Associates et al. Analysis of State government.—no. 3. Rivkin Associates, CCC/HOK, Hitchins, D. R. M. Program components of the new city.—no. 4. R & M Consultants. Natural site conditions.—no. 5. R & M Consultants. Water source.—no. 6. M. Fryer & Associates et al. Energy & utilities.—no. 7. State of Alaska, Dept. of Transportation, Bull, Field, Volkman & Stockwell, Sedway-Cooke. Transportation.—no. 8. State of Alaska, Alaska Youth Conservation Corps. Outdoor recreation resources.—no. 9. American City Corp., D. Witter-Reynolds, inc., Hanscomb Associates. Financial plan and detailed economic projections. 2 v.—no. 10. Rivkin Associates. Socio-economic impact analysis for Juneau and the Matanuska-Susitna Borough.—no. 11. Dowell, T., Jr. Environmental guidelines & assessment, permits & procedures.—no. 12. American City Corp., W. King & Associates. Implementation plan.—no. 13. American City Corp., Ely, Guess & Rudd, Wohlforth & Flint. Legislative package.—no. 14. Design guidelines, R. Morehouse, ed.
1. City planning—Alaska. 2. Alaska—Capital—History. 3. Capitals (Cities) I. Alaska. Capital Site Planning Commission.
HT167.5.A4N48 309.2'62'09798 78-623989
79 MARC

—ALASKA—CASE STUDIES

American City Corporation.
Financing plan for new capital city of Alaska / prepared for New Capital Site Planning Commission by the American City Corporation. — Columbia, Md. : The Corporation, 1978.
ca. 200 leaves in various foliations ; 28 cm.
1. Willow, Alaska—City planning. 2. City planning—Alaska—Case studies. 3. New towns—Alaska—Willow. 4. Alaska—Capital and capitol. I. Alaska. Capital Site Planning Commission. II. Title.
HT168.W53A53 1978 309.2'62'097983 78-622545
79 MARC

—ALBERTA—CASE STUDIES

Alberta. Task Force on Urbanization and the Future.
Task Committee reports. — [Edmonton?] : Task Force on Urbanization and the Future, [1972].
92 p. : ill. ; 25 cm. C***
Cover title.
1. City planning—Alberta—Case studies. 2. Urbanization—Alberta—Case studies. 3. Regional planning—Alberta—Case studies.
HT169.C32A42 309.2'62'097123 78-314242
78 MARC

—ARAB COUNTRIES

Shiber, Ṡaba George, 1923–
(al-'Ilm wa-tanẓim al-mudun)
العلم وتنظيم المدن العربية ، تأليف سابا جورج شبر . [بيروت]
دار الكاتب العربي [التوطئة 1964]
302 p. 23 cm.
1. City planning—Arab countries. 2. City planning—Kuwait, Arabia (State) I. Title.
HT178.A65S54 N E 65-1484

—ARGENTINE REPUBLIC

Paolera, Carlos M della.
Buenos Aires y sus problemas urbanos / trabajos originales de Carlos María Della Paolera ; introd. y selección a cargo de Patricio H. Randle. — Buenos Aires : OIKOS Asociación para la Promoción de los Estudios Territoriales y Ambientales, 1977.
146 p. ; 24 cm.
Mostly articles previously published in La Nación and La Razón, Buenos Aires.
1. Buenos Aires—City planning. 2. Urbanization—Argentine Republic—Buenos Aires. 3. City planning—Argentine Republic. I. Randle, Patricio H. II. Title.
HT169.A75B865 78-109705
79 MARC

—ARGENTINE REPUBLIC
—BUENOS AIRES (PROVINCE)

Buenos Aires (Province). Secretaría de Planeamiento y Desarrollo.
Subdivisión parcelaria de la tierra urbana en la Provincia de Buenos Aires. — La Plata : Gobernación de la Provincia de Buenos Aires, Secretaría de Planeamiento y Desarrollo, 1978.
40 p. : ill. ; 26 cm.
"SPD-37-IV/78"
1. Land use, Urban—Argentine Republic—Buenos Aires (Province) 2. Land use, Urban—Planning—Argentine Republic—Buenos Aires (Province) I. Title.
HD479.B9B87 1978 79-109273
79 MARC

—ASIA—CONGRESSES

United Nations. Centre for Regional Development.
Urban development strategies in the context of regional development : report of the seminar held at Nagoya, Japan, from 28 October to 8 November 1974 / United Nations Centre for Regional Development. — 2d ed. — Nagoya, Japan : The Centre, 1975.
66 p. : ill. ; 29 cm. Ja***
"OTC/CRD/UIPD/3."
Includes bibliographical references.
1. City planning—Asia—Congresses. 2. Regional planning—Asia—Congresses. 3. Urbanization—Asia—Congresses. I. Title.
HT169.A77U54 1975 309.2'62'095 77-377909
78[r79]rev MARC

—ASIA, SOUTHEASTERN

Quality of life in human settlements : planning issues and space standards. — Mysore : Institute of Development Studies, University of Mysore, 1976.
63 p. : diagrs. ; 26 cm. — (Development studies ; no. 12)
Includes bibliographical references.
CONTENTS: Bhooshan, B. S. A fresh look on urban development.—Sundaram, K. V. Physical planning standards in selected countries in South-East Asia.
Rs10.00
1. City planning—India. 2. City planning—Asia, Southeastern. I. Bhooshan, B. S. A fresh look on urban development. 1976. II. Sundaram, K. V. Physical planning standards in selected countries in South-East Asia. 1976. III. Series: Development studies (Mysore) ; no. 12.
HT169.I5Q34 309.2'62'0959 78-900592
79 MARC

—AUSTRALIA

Burchett, Winston H
East Melbourne, 1837-1977 : people, places, problems / [by] Winston Burchett. — Hawthorn, Vic. : Craftsman Press, 1978.
xiii, 177 p. : ill. ; 23 x29 cm. Aus
Bibliography: p. 167-169.
Includes index.
ISBN 0-909108-09-9 : $14.95
1. Melbourne—City planning. 2. City planning—Australia.
HT169.A82M425 994.5 78-326505
79 MARC

CITY PLANNING
—AUSTRALIA (Continued)

Cocks & Carmichael.
Medium density cluster housing, Site G, Melrose Drive & Heckendorf Road, South West, Wodonga : town planning permit application : a submission on behalf of Albury Wodonga Development Corporation / prepared by Cocks & Carmichael. — ₍s.l. : s.n.₎, 1976.
14 leaves : ill. ; 30 cm. Aus***
1. Cluster housing—Australia—Wodonga. 2. Wodonga, Australia—City planning. 3. City planning—Australia. I. Albury Wodonga Development Corporation. II. Title.
NA9051.4.C62 1976 711'.4'09945 79-315068
79 MARC

MSJ Keys Young Planners.
Women and planning : women's attitudes / prepared for the Cities Commission by MSJ Keys Young Planners, November 1975. — Canberra : Dept. of Environment, Housing and Community Development, 1977.
₍10₎, 171, ₍34₎ leaves ; 30 cm. Aus
Companion to Women and planning: planners attitudes.
Bibliography: leaf 171.
ISBN 0-642-92320-5
1. City planning—Australia. 2. Women—Australia—Social conditions. 3. Community life. I. Australia. Cities Commission. II. Australia. Dept. of Environment, Housing and Community Development. III. Title.
HT169.A8M26 1977 301.36 78-324111
79 MARC

Paterson, John, PhD.
A mansion or no house : a report for UDIA on consequences of planning standards and their impact on land and housing / by John Paterson, David Yencken, Graeme Gunn. — Melbourne : Hawthorn Press, 1976.
iii, 135 p., 18 leaves of plates : ill., diagrs. ; 25 cm. Aus
Includes bibliographical references and index.
ISBN 0-7256-0176-0
1. Housing policy—Australia. 2. City planning—Australia. 3. Building laws—Australia. 4. Architecture, Domestic—Australia. I. Yencken, David, joint author. II. Gunn, Graem, joint author. III. Urban Development Institute of Australia (Victoria) IV. Title.
HD7379.A3P37 301.5'4'0994 79-306086
79 MARC

—AUSTRALIA—SOUTH AUSTRALIA—CONGRESSES

The Opportunity for choice : where shoud Australia locate her people? / edited by E. G. Hallsworth. — East Melbourne, Vic. : Commonwealth Scientific and Industrial Research Organization ; Forest Grove, Or. : distributed by ISBS, 1978.
71 p. : ill. ; 25 cm. Aus
"Collected papers of a symposium organized jointly by CSIRO and the Australian Institute of Agricultural Science, Adelaide, S.A., 1976."
Includes bibliographies.
1. City planning—Australia—South Australia—Congresses. I. Hallsworth, Ernest G. II. Australia. Commonwealth Scientific and Industrial Research Organization. III. Australian Institute of Agricultural Science.
HT169.A82S66 309.2'62'099423 79-670168
00 MARC

—AUSTRALIA—VICTORIA

Tract Consultants Pty. Ltd.
Cluster housing, Westerngate Estate, Deer Park : town planning permit application : a submission on behalf of the Victorian Urban Land Council / prepare₍d₎ by Tract Consultants Pty. Ltd., Cocks & Carmichael. — ₍s.l. : s.n.₎, 1976.
17 leaves, 10 fold. leaves of plates : ill. ; 30 cm. Aus***
Cover title: Westerngate Estate.
1. Real estate development—Australia—Victoria. 2. City planning—Australia—Victoria. I. Cocks & Carmichael. II. Victoria, Australia. Urban Land Council. III. Title. IV. Title: Westerngate Estate.
HD1039.V5T72 1976 333 79-312261
00 MARC

—AUSTRIA

Aymonino, CArlo.
Parigi e Vienna / Carlo Aymonino, Gianni Fabbri, Angelo Villa. — Roma : Officina, 1975.
315 p., ₍136₎ leaves of plates : ill. ; 25 cm. — (Le Città capitali del XIX secolo ; 1) (Collana di architettura ; 14) It***
Includes bibliographies.
L15000
1. Paris—City planning. 2. City planning—France. 3. Architecture, Modern—19th century—France—Paris. 4. Vienna—City planning. 5. City planning—Austria. I. Fabbri, Gianni, joint author. II. Fabbri, Gianni, joint author. III. Villa, Angelo, 1945- joint author. IV. Title. V. Series.
NA9053.C3C57 vol. 1 76-514625
₍NA9198.P2₎ 79₍r79₎rev MARC

Conditt, Georg.
Instrumente der örtlichen Raumplanung / Georg Conditt, Hugo Potyka. — Wien : Institut für Stadtforschung : Kommissionsverlag, Jugend und Volk, ₍19 ₎-
v. : ill. ; 21 x 30 cm. — (Publikationen des Instituts für Stadtforschung ; Nr. 59) Au***
Vol. 2, pt. ₍2₎ has subtitle: Gesetzetexte.
CONTENTS:
—T. 2. Bestimmungen über die Bodenordnung. 2 v.

ISBN 3-7141-7854-6 (v. 2, pt. 1)
1. City planning—Austria. 2. Land use, Urban—Austria. 3. Land use, Urban—Law and legislation—Austria. I. Potyka, Hugo, 1927- joint author. II. Title. III. Series: Institut für Stadtforschung. Publikationen des Instituts für Stadtforschung ; Nr. 59.
HT169.A9C66 79-382242
79 MARC

Heigl, Franz, Dipl.-Ing. Dr. tech.
Altstadt-Erhaltungskonzept Stadtgemeinde Hartberg / verf. v. Franz Heigl. — Hartberg : Stadtgemeinde Hartberg, ₍1975₎
171 p. : ill. (some col.) ; 20 x 21 cm. Au77-13-295
Bibliography: p. 170.
Not in trade.
1. Architecture—Austria—Hartberg—Conservation and restoration. 2. Hartberg, Austria—Buildings. 3. Hartberg, Austria—City planning. 4. City planning—Austria. I. Title.
NA9053.C6H44 78-381923
*78 MARC

—AUSTRIA—ADDRESSES, ESSAYS, LECTURES

Wien, Alternativen der Stadtentwicklung. — Wien : Sensen-Verl., 1978.
127 p. : maps ; 24 cm. — (Mitteilungen des Instituts für Gesellschaftspolitik, Wien ; Heft 23) (Papers presented during a lecture series sponsored by the Institut für Gesellschaftspolitik in the spring of 1977) Au78-19-115
Includes bibliographical references.
S80.00
1. Vienna—City planning—Addresses, essays, lectures. 2. Urban renewal—Austria—Vienna—Addresses, essays, lectures. 3. City planning—Austria—Addresses, essays, lectures. I. Institut für Gesellschaftspolitik. II. Series: Institut für Gesellschaftspolitik. Mitteilungen des Institut für Gesellschaftspolitik ; Heft 23.
HT169.A92V58 309.2'62'0943613 79-352849
*79 MARC

—AUSTRIA—BIBLIOGRAPHY—UNION LISTS

Literatur zur Raumforschung und Raumplanung in Österreich : ÖROK-Dokumentation / im Auftrag d. ÖROK bearb. vom Kommunalwissenschaftl. Dokumentationszentrum u. vom Österr. Inst. f. Raumplanung. — Wien : Geschäftsstelle d. Österr. Raumordnungskonferenz, 1978-
v. ; 30 cm. — (Schriftenreihe - Österreichische Raumordnungskonferenz ; Nr. 12) Au78-17/18-15
Introductory material in English, French, German, and Russian.
Includes indexes.
S90.00 (v. 1)
1. Regional planning—Austria—Bibliography—Union lists. 2. City planning—Austria—Bibliography—Union lists. 3. Austria—Social policy—Bibliography—Union lists. 4. Kommunalwissenschaftliches Dokumentationszentrum—Catalogs. 5. Österreichisches Institut für Raumplanung—Library—Catalogs. I. Kommunalwissenschaftliches Dokumentationszentrum. II. Österreichisches Institut für Raumplanung. III. Series: Österreichische Raumordnungskonferenz. Schriftenreihe - Österreichische Raumordnungskonferenz ; Nr. 12.
Z7164.R33L58 78-396268
₍HT395.A96₎ *79 MARC

—AUSTRIA—EXHIBITIONS

Mayer, Wolfgang, Dr.
Die städtebauliche Entwicklung Wiens bis 1945 : Ausstellung d. Geschäftsgruppe Stadtplanung u. des Wiener Stadt- u. Landesarchivs, Messepalasr Halle P, 6. Dez. 1978 bis 30. April 1979 / ₍Wissenschaftl. Bearb. u. Katalog, Wolfgang Mayer₎. — Wien : Verein f. Geschichte d. Stadt Wien, 1978.
162 p. : ill. (some col.) ; 23 cm. Au79-13-236
Bibliography: p. 137-138.
Not in trade.
1. Vienna—City planning—Exhibitions. 2. City planning—Austria—Exhibitions. I. Geschäftsgruppe Stadtplanung. II. Wiener Stadt- und Landesarchiv. III. Title.
NA9196.V5M39 79-379607
*79 MARC

—AUSTRIA—STYRIA—HANDBOOKS, MANUALS, ETC.

Styria. Landesbaudirektion. Örtliche Raumplanung.
Handbuch der örtlichen Raumplanung in der Steiermark / Verf. d. einzelnen Abschnitte, ₍Gunther₎ Hasewend ... ₍et al.₎, Mitarb., Ecker ... ₍et al.₎ ; Graph. Gestaltung, Bujatti. — Graz : Amt d. Steiermärk. Landesregierung, Fachabteilungsgruppe Landesbaudirektion, Fachabt. 1b, Örtliche Raumplanung, ₍1977-
1 v. ; 33 cm. Au77-7-97
Includes text of Steiermärkisches Raumordnungsgesetz 1974.
Loose-leaf for updating.
1. City planning—Austria—Styria—Handbooks, manuals, etc. 2. Regional planning—Austria—Styria—Handbooks, manuals, etc. I. Hasewend, Gunther. II. Styria. Laws, statutes, etc. Steiermärkisches Raumordnungsgesetz 1974. 1977. III. Title.
HT145.A8S8 1977 352'.96'094365 78-377430
78 MARC

—AUSTRIA, LOWER

Koepf, Hans.
Stadtbaukunst in Niederösterreich / von Hans Koepf ; hrsg. vom Amt d. Niederösterr. Landesregierung als Beitrag zur Aktion "Schöneres Bauen in Niederösterreich" u. vom Inst. f. Baukunst u. Bauaufnahmen d. Techn. Univ. Wien. — Wien : Amt d. Niederösterr. Landesregierung, 1977.
270 p. : chiefly ill. (some col.) ; 21 x 30 cm. Au78-24-465
Bibliography of the author's works: p. 270.
Not in trade.
1. Architecture—Austria, Lower—Conservation and restoration. 2. City planning—Austria, Lower. I. Austria, Lower. Landesamtsdirektion. II. Technische Universität Wien. Institut für Baukunst und Bauaufnahmen. III. Title.
NA1009.L68K63 79-374758
*79 MARC

—BELGIUM

Agglomération de Bruxelles. Service de l'urbanisme et de l'aménagement du territoire.
Plans généraux d'aménagement des 19 communes de l'agglomération de Bruxelles : guide synthétique de la deuxième phase. — ₍Bruxelles, r. de la Loi 15₎ : Agglomération de Bruxelles, Service de l'urbanisme et de l'aménagement du territoire, 1977.
21 p. : ill. ; 24 cm. Be77-3
Cover title.
1. Brussels—City planning. 2. City planning—Belgium. I. Title.
HT169.B42B792 309.2'62'094933 79-348591
*79 MARC

—BOLIVIA

Bolivia. Laws, statutes, etc.
₍Decreto ley no. 14665₎
Creación de nuevos centros urbanos = Creation of new urban centers. — La Paz : Instituto Nacional de Preinversión, ₍1977 or 1978₎
24 p. ; 23 cm.
Cover title.
At head title: República de Bolivia, Ministerio de Planeamiento y Coordinación de la Presidencia de la República.
Spanish, English, French, and German.
1. New towns—Bolivia. 2. City planning—Bolivia. I. Title. II. Title: Creation of new urban centers.
HT169.B6B64 1977 346'.84'045 78-111189
79 MARC

France. Bureau de recherches géologiques et minières.
Plan de desarrollo urbano, ciudad de La Paz / consultoras, Bureau de recherches géologiques et minières, Bureau central d'études pour les équipements d'outre-mer, Prudencio Claros y Asociados. — La Paz : Alcaldía Municipal de La Paz, ₍197 cover, 197
v. : ill. ; 30 cm.
CONTENTS:
—t. 3. Esquema urbano.— —t. ₍5₎
Resumen y gráficos de síntesis.
1. La Paz, Bolivia—City planning. 2. City planning—Bolivia. I. Bureau central d'études pour les équipements d'outre-mer. II. Prudencio Claros y Asociados. III. La Paz, Bolivia. Alcaldía. IV. Title.
HT169.B62L26 79-105450
79 MARC

—BRAZIL

Andrade, Manuel Correia de Oliveira.
Polarizações para João Pessoa / Manuel Correia de Andrade. — ₍Recife₎ : Universidade Federal de Pernambuco, 1975.
74 p. ; 22 cm. — (Série Estudos urbanológicos ; publicação no. 3)
"Trabalho escrito pelo autor como parte dos relatórios finais do Plano de Desenvolvimento Urbano de João Pessoa, elaborado sob assessoria do G.P.I. - Grupo de Planejamento Integrado."
Includes bibliographical references.
1. João Pessoa, Brazil—City planning. 2. City planning—Brazil. I. Title. II. Series.
HT169.B72J64 79-110089
79 MARC

Cavalcanti, Sandra.
Rio, viver ou morrer / Sandra Cavalcanti. — ₍1. ed.₎. — Rio de Janeiro : Editora Expressão e Cultura, ₍1978₎
141 p. : ill. ; 28 cm.
Cr$50.00
1. Rio de Janeiro—City planning. 2. City planning—Brazil. 3. Housing policy—Brazil—Rio de Janeiro. I. Title.
HT169.B72R473 79-345003
79 MARC

CITY PLANNING
—BRAZIL (Continued)

Empresa Metropolitana de Planejamento da Grande São Paulo.
O desafio metropolitano / [Secretaria de Estado dos Negócios Metropolitanos, EMPLASA, Empresa Metropolitana de Planejamento da Grande São Paulo]. — São Paulo : EMPLASA, 1976.
214 p., [16] fold. leaves of plates : ill. ; 23 cm. — (Série Documentos - EMPLASA ; 1)
At head of title: Governo do Estado de São Paulo, Sistema de Planejamento e de Administração Metropolitana.
Includes bibliographical references.
1. São Paulo, Brazil (City)—City planning. 2. City planning—Brazil. I. Title. II. Series: Empresa Metropolitana de Planejamento da Grande São Paulo. Série Documentos - EMPLASA ; 1.
HT169.B72S36 79-376639
79 MARC

Fundação João Pinheiro. Centro de Desenvolvimento Urbano.
Plano de desenvolvimento urbano, Vila Rondônia-Ro. — [Belo Horizonte] : Fundação João Pinheiro, Centro de Desenvolvimento Urbano, [1974 or 1975]
273 leaves : ill. ; 30 cm.
Cover title.
"O presente plano ... foi elaborado com base no convênio de cooperação técnica, celebrado em 22 de junho de 1973, entre a Superintendência do Desenvolvimento da Região Centro Oeste (SUDECO) e a Fundação Pinheiro."
1. Rondônia, Brazil—City planning. 2. City planning—Brazil. 3. Rondônia, Brazil—Economic conditions. 4. Rondônia, Brazil—Social conditions. I. Superintendência do Desenvolvimento da Região Centro-Oeste. II. Title.
HT169.B72R663 79-364766
79 MARC

Grupo Executivo da Região Metropolitana.
Plano de desenvolvimento metropolitano : documento. — [Porto Alegre : República Federativa do Brasil, Governo do Estado de Rio Grande do Sul, Conselho Metropolitano de Municípios, Grupo Executivo da Região Metropolitana, 19
v. : ill. ; 30 cm.
Cover title.
On cover: Região Metropolitano de Porto Alegre-CMM-GERM.
"Trabalho elaborado com ajuda técnica e financeira do Governo da República Federal da Alemanha."
Includes bibliographical references.
1. Regional planning—Brazil—Porto Alegra metropolitan area. 2. Porto Alegre metropolitan area—Economic conditions. 3. Porto Alegre metropolitan area—Social conditions. 4. Porto Alegre metropolitan area—City planning. 5. City planning—Brazil. I. Title.
HT395.B73P674 1973a 78-378749
78 MARC

Newplan-Consultoria, Planejamento Ltda.
Plano de desenvolvimento local integrado de Cuiabá : P.D.L.I. Cuiabá / NewPlan S/A. — [Cuiabá] : Newplan, [1973-
v. : maps ; 30 cm.
Cover title.
On cover: Prefeitura Municipal de Cuiabá.
CONTENTS: A. Plano de diretrizes para o desenvolvimento: v. 1. Plano a longo prazo. v. 2. Plano municipal de apoio ao desenvolvimento do setor secundário. v. 3. Plano orientador da expansão física da cidade.—B. Plano a médio prazo: v. 1. Educação, áreas verdes, recreação e esportes, cultura. v. 2. Saúde, assistência social. v. 3. Limpeza pública, iluminação pública, habitação. — —D. Programas e projetos específicos: v. 1. Programa de abastecimento de gêneros alimentícios. v. 4. Programa do setor institucional.
1. Cuiabá, Brazil—City planning. 2. City planning—Brazil. I. Title.
HT169.B72C835 79-108045
79 MARC

Olinda, Brazil.
Plano de desenvolvimento local integrado de Olinda. — [Olinda] : Prefeitura Municipal, Sociedade Civil de Planejamento, Serviço Federal de Habitação a Urbanismo, [1972 or 1973-
v. : ill. ; 30 cm.
Includes bibliographical references.
CONTENTS: v. 1. Análise da situação e tendências do desenvolvimento diagnóstico e prognóstico. 2 v.—v. 2. Diretrizes para o desenvolvimento programs setoriais.—v. 3. Plano de ação do governo municipal orçamento plurianual de investimentos.
1. Olinda, Brazil—City planning. 2. City planning—Brazil. 3. Olinda, Brazil—Economic conditions. 4. Olinda, Brazil—Social conditions. I. Title.
HT169.B72O44 1972 79-104950
00 MARC

Plano director de São Luís. — [São Luís : s.n.], 1975, 1977 printing.
99 p. : ill. ; 30 x 45 cm.
1. São Luís (Maranhão)—City planning. 2. City planning—Brazil.
HT169.B72S296 78-398280
79 MARC

—BRAZIL—ADDRESSES, ESSAYS, LECTURES

Dilemas do planejamento urbano e regional no Brasil : ensaios de homenagem póstuma a Paulo Camillo de Oliveira Penna / organizadores, Antônio Octávio Cintra e Paulo Roberto Haddad. — Rio de Janeiro : Zahar Editores, 1978.
253 p. ; 21 cm. — (Biblioteca de ciências sociais)
Includes bibliographical references.
Cr$120.00
1. Regional planning—Brazil—Addresses, essays, lectures. 2. City planning—Brazil—Addresses, essays, lectures. 3. Brazil—Economic policy—Addresses, essays, lectures. I. Pena, Paulo Camilo de Oliveira. II. Haddad, Paulo Roberto. III. Cintra, Antônio Octávio.
HT395.B7D54 78-380757

—BRAZIL—COLLECTED WORKS

Superintendência do Desenvolvimento da Região Sul.
Projeto do litoral sul de Sta. Catarina : plano de desenvolvimento integrado, Imbituba / Ministério do Interior, Superintêndencia do Desenvolvimento da Região Sul-SUDESUL, Governo do Estado de Santa Catarina, Prefeitura Municipal de Imbituba. — [Porto Alegre] : Secretaria Executiva do Acôrdo No. 03/72, [19 -
v. ca. 100 leaves ; 30 cm.
CONTENTS:
—pt. 5. Legislação básica.
1. Imbituba, Brazil—City planning—Collected works. 2. City planning—Brazil—Collected works. 3. City planning and redevelopment law—Brazil—Imbituba—Collected works. I. Imbituba, Brazil. II. Title.
HT169.B72I467 79-113538
79 MARC

—BULGARIA

(Tekhniko-ikonomicheski pokazateli za zhilishtnite teritorii)
Технико-икономически показатели за жилищните територии / [отговорен редактор Любен Тонев]. — София : Изд-во на БАН, 1978.
261 p., [8] leaves of plates : ill. ; 23 cm. Bu***
At head of title: Bŭlgarska akademii͡a na naukite. Institut za teorii͡a i istorii͡a na gradoustroĭstvoto i arkhitekturata (ITIGA) Ministerstvo na stroezhite i arkhitekturata. Kompleksen nauchno-izsledovatelski i proektantski institut po teritorialno ustroĭstvo, gradoustroĭstvo i arkhitektura (KNIPITUGA)
Added t. p.: Indices techniques et économiques pour les territoires résidentiels.
Summary in French and Russian.
Includes bibliographical references.
3.07 lv
1. Dwellings—Bulgaria. 2. City planning—Bulgaria. I. Tonev, L͡iuben. II. Bŭlgarska akademii͡a na naukite, Sofia. Institut po teorii͡a i istorii͡a na gradoustroĭstvoto i arkhitekturata. III. Kompleksen nauchnoizsledovatelski i proektantski institut po teritorialno ustroĭstvo, gradoustroĭstvo i arkhitektura. IV. Title: Indices techniques et économiques pour les territoires résidentiels.
TH4809.B9T44 79-413089

—CALIFORNIA

Jacobs, Allan B
Making city planning work / Allan B. Jacobs. — Chicago : American Society of Planning Officials, c1978.
xxv, 323 p. : ill. ; 23 cm.
Includes bibliographical references and index.
ISBN 0-918286-12-3
1. San Francisco—City planning. 2. City planning—California. I. Title.
HT168.S2J3 309.2'62'09794 78-72577
79 MARC

—CALIFORNIA—CITIZEN PARTICIPATION—CONGRESSES

Toward an urban design : linking citizen participation with civic purpose : a record of proceedings of three public discussion-conferences conducted in Berkeley, 1977-1978. — Berkeley, Calif. : Urban Care, c1978.
26 p. : ill. ; 28 cm. — (Concerns ; v. 8, no. 1)
Cover title.
CONTENTS: Public transportation and city cohesion, May 21, 1977.—The role of the neighborhood community, Nov. 10, 1977.—Emerging patterns for the city's design, Apr. 1, 1978.
$1.50
1. Berkeley, Calif.—City planning—Citizen participation—Congresses. 2. Urban transportation—California—Berkeley—Congresses. 3. City planning—California—Citizen participation—Congresses. I. Urban Care (Association) II. Series.
HT168.B47T68 309.2'62'0979467 78-113808
79 MARC

—CANADA

Coopersmith, Penina.
Heritage by design / Penina Coopersmith, with Richard C. Hall. — [Ottawa] : Published by Macmillan of Canada for the Ministry of State for Urban Affairs, 1976.
51 p. : ill. ; 23 cm. — (Urban prospects) C***
Bibliography: p. 51.
1. Historic sites—Canada—Conservation and restoration. 2. City planning—Canada. I. Hall, Richard C., joint author. II. Title.
F1010.C66 069'.53 78-321119
79 MARC

Edmonton, the emerging metropolitan pattern / edited by P. J. Smith. — Victoria, B.C. : Dept. of Geography University of Victoria, 1978.
xxvii, 291 p. : ill. ; 23 cm. — (Western geographical series ; vol. 15 ISSN 0315-2022) C***
Includes bibliographical references.
1. Edmonton, Alta.—Economic conditions. 2. Edmonton, Alta.—Social conditions. 3. Edmonton, Alta.—City planning. 4. City planning—Canada. I. Smith, Peter J., 1931- II. Series.
HC118.E35E35 309.1'7123'3 79-302803
79 MARC

Population growth and urban problems / edited by Frank Kelly. — Ottawa : Science Council of Canada : available from Information Canada, 1975.
90 p. : ill. ; 28 cm. — (Perceptions ; 1) C***
At head of title: Science Council of Canada. Study on population and technology.
Includes bibliographical references.
$1.25 ($1.50 U.S.)
1. Cities and towns—Canada—Growth. 2. City planning—Canada. I. Kelly, Frank, 1935- II. Series.
HT127.P66 301.36'1'0971 79-306794
79 MARC

—CANADA—ADDRESSES, ESSAYS, LECTURES

Canada, an urban agenda : a collection of papers / H. Peter Oberlander, editor, assisted by Hilda Symonds. — Ottawa : Community Planning Press, 1976.
ix, 234 p. : ill. ; 20 cm. C***
Includes bibliographies.
ISBN 0-88725-002-2 : $7.00
1. City planning—Canada—Addresses, essays, lectures. I. Oberlander, Heinz Peter, 1922- II. Symonds, Hilda.
HT169.C3C32 309.2'62'0971 78-319912
79 MARC

Urban and regional planning in a Federal State : the Canadian experience / edited by William T. Perks and Ira M. Robinson. — Stroudsburg, Pa. : Dowden, Hutchinson & Ross, c1979.
xiv, 363 p. : ill. ; 24 cm. — (Community development series ; v. 41)
Includes bibliographies and index.
ISBN 0-07-049310-3
1. City planning—Canada—Addresses, essays, lectures. 2. Regional planning—Canada—Addresses, essays, lectures. 3. Land use—Planning—Canada—Addresses, essays, lectures. I. Perks, William T. II. Robinson, Ira M.
HT169.C3U73 1979 309.2'62'0971 78-3648
78 MARC

—CANADA—CONGRESSES

ICSID Interdesign Seminar, 3d, Port Hope, Ont. and Cobourg, Ont., 1974.
Design for small communities : a report of Interdesign '74 Ontario / edited by David Fulton. — Toronto : Macmillan of Canada, c1975.
141 p. : ill. ; 22 x 28 cm. C***
"A project of the Association of Canadian Industrial Designers in co-operation with the government of Ontario."
"Sanctioned by the International Council of Societies of Industrial Design, Brussels, Belgium."
ISBN 0-7705-1311-5
1. Cobourg, Ont.—City planning—Congresses. 2. Port Hope, Ont.—City planning—Congresses. 3. City planning—Canada—Congresses. 4. Architecture—Human factors—Congresses. I. Fulton, David. II. Association of Canadian Industrial Designers. III. International Council of Societies of Industrial Design. IV. Title.
HT169.C32C634 1974 309.2'62'09713 77-372966
77[r79]rev MARC

—CHILE—MATHEMATICAL MODELS

Bastarreche Alfaro, Manuel.
Un modelo simple estático : aplicación a Santiago de Chile / [por] Manuel Bastarreche Alfaro. — Madrid : Fundación Juan March, D.L. 1977.
52 p., 4 leaves : 21 cm. — (Serie universitaria ; 23) Sp78-Apr
Bibliography: p. 51-52.
ISBN 8470750445
1. Santiago de Chile—City planning—Mathematical models. 2. City planning—Chile—Mathematical models. I. Title.
HT169.C52S353 78-385308
78 MARC

—CHILE—O'HIGGINS—CASE STUDIES

Rother, Klaus.
Gruppensiedlungen in Mittelchile : erl. am Beispiel d. Provinz O'Higgins / Klaus Rother. — Düsseldorf : Geograph. Inst. d. Univ., 1977.
80 p., [12] leaves of plates : 54 ill., maps ; 30 cm. — (Düsseldorfer geographische Schriften ; Heft 9) GFR78-B4
Bibliography: p. 78-79.
Summary also in Spanish.
DM15.00
1. City planning—Chile—O'Higgins—Case studies. I. Title. II. Series.
HT169.S52O367 79-384057

CITY PLANNING (Continued)

—CHINA

(Harupin toshi kensetsu hōsaku)
哈爾賓都市建設方策. — [新京]：南滿洲鐵道株式會社經濟調査會, 昭和10 [1935]
4, 450 p., [1] fold. leaf of plates : maps ; 23 cm. & portfolio (23 fold. leaves of plates : 23 maps). — (立案調査書類 - 南滿洲鐵道株式會社經濟調査會 ; 第20編, 第4卷)
"極祕"
1. Harbin—City planning. 2. Harbin—City planning—Maps. 3. City planning—China. I. Minami Manshū Tetsudō Kabushiki Kaisha. Keizai Chōsakai. II. Series: Minami Manshū Tetsudō Kabushiki Kaisha. Keizai Chōsakai. Ritsuan chōsa shorui ; dai 20-hen, dai 4-kan.
HT169.C62H374 78–812583

(Hokuanchin toshi kensetsu hōsaku)
北安鎮都市建設方策. — [新京]：南滿洲鐵道株式會社經濟調査會, 昭和11 [1936]
2, 28 p., [4] fold. leaves of plates : 4 col. maps ; 23 cm. — (立案調査書類 - 南滿洲鐵道株式會社經濟調査會 ; 第20編, 第6卷)
"極祕"
1. Pei-an chen, China—City planning. 2. City planning—China. I. Minami Manshū Tetsudō Kabushiki Kaisha. Keizai Chōsakai. II. Series: Minami Manshū Tetsudō Kabushiki Kaisha. Keizai Chōsakai. Ritsuan chōsa shorui ; dai 20-hen, dai 6-kan.
HT169.J32P454 78–812520

(Hōten toshi kensetsu hōsaku)
奉天都市建設方策. — [新京]：南滿洲鐵道株式會社經濟調査會, 昭和10 [1935]
4, 116 p., [1] leaf of plates : ill. ; 23 cm. — (立案調査書類 - 南滿洲鐵道株式會社經濟調査會 ; 第20編, 第3卷)
"極祕"
Accompanied by 13 fold. maps in a case.
1. Mukden—City planning. 2. Mukden—City planning—Maps. 3. City planning—China. I. Minami Manshū Tetsudō Kabushiki Kaisha. Keizai Chōsakai. II. Series: Minami Manshū Tetsudō Kabushiki Kaisha. Keizai Chōsakai. Ritsuan chōsa shorui ; dai 20-hen, dai 3-kan.
HT169.C62M944 78–813926

Manchuria. Ming cheng pu. T'u mu ssu.
(Hokuanchin toyū keikaku an)
北安鎮都邑計畫案 / 民政部土木司立案(寫). — [新京]：經濟調査會第三部, 康德1 [1934]
9 leaves ; 27 cm.
Cover title.
At head of title: Gokuhi (rubber stamped)
1. Pei-an chen, China—City planning. 2. City planning—China. I. Minami Manshū Tetsudō Kabushiki Kaisha. Keizai Chōsakai. Dai 3-bu. II. Title.
HT169.C62P445 79–801606
J

(Shinkyō toshi kensetsu hōsaku)
新京都市建設方策. — [新京]：南滿洲鐵道株式會社經濟調査會, 昭和10 [1935]
2 v. : ill., 7 fold. maps ; 23 cm. — (立案調査書類 - 南滿洲鐵道株式會社經濟調査會 ; 第20編, 第2卷)
"極祕"
1. Ch'ang-ch'ung, China—City planning. 2. Ch'ang-ch'ung, China—City planning—Maps. 3. City planning—China. I. Minami Manshū Tetsudō Kabushiki Kaisha. Keizai Chōsakai. II. Series: Minami Manshū Tetsudō Kabushiki Kaisha. Keizai Chōsakai. Ritsuan chōsa shorui ; dai 20-hen, dai 2-kan.
HT169.C62C577 78–812250
J

(Tomon toshi keikaku gaiyō an)
圖們都市計畫概要案. — [新京]：南滿洲鐵道株式會社經濟調査會, 昭和10 [1935]
3, 64 p. : 5 fold. maps (in pocket) ; 23 cm. — (立案調査書類 - 南滿洲鐵道株式會社經濟調査會 ; 第20編 第5卷)
Cover title: Tomon toshi kensetsu hōsaku.
"極祕"
1. T'u-men, China—City planning. 2. City planning—China. I. Minami Manshū Tetsudō Kabushiki Kaisha. Keizai Chōsakai. II. Title: Tomon toshi kensetsu hōsaku. III. Series: Minami Manshū Tetsudō Kabushiki Kaisha. Keizai Chōsakai. Ritsuan chōsa shorui ; dai 20-hen, dai 5-kan.
HT169.J32T938 78–812806

—CHINA—MANCHURIA

Koshizawa, Akira.
(Shokuminchi Manshū no toshi keikaku)
植民地満州の都市計画 / 越沢明著. — 東京：アジア経済研究所：発売所 アジア経済出版会, 1978.
x, 192 p. : ill. ; 25 cm. — (研究参考資料 - アジア経済研究所 ; 271)
Ja***
Includes bibliographical references.
1. City planning—China—Manchuria. I. Title. II. Series: Ajia Keizai Kenkyūjo, Tokyo. Kenkyū sankō shiryō ; 271.
HT169.C62M318 79–807445

(Manshū toshi kensetsu ippan hōsaku)
滿洲都市建設一般方策. — [新京]：南滿洲鐵道株式會社經濟調査會, 昭和10 [1935]
2, 45 p. : map ; 23 cm. — (立案調査書類 - 南滿洲鐵道株式會社經濟調査會 ; 第20編, 第1卷)
"極祕"
1. City planning—China—Manchuria. I. Minami Manshū Tetsudō Kabushiki Kaisha. Keizai Chōsakai. II. Series: Minami Manshū Tetsudō Kabushiki Kaisha. Keizai Chōsakai. Ritsuan chōsa shorui ; dai 20 -hen, dai 1-kan.
HT169.C62M32 78–812425

—COLOMBIA

Bogotá. Instituto de Desarrollo Urbano.
Estudio de talleres automotores y propuesto de renovación urbana en el sector Los Martires / Instituto de Desarrollo Urbano. — Bogotá : IDU, [1977]
198 leaves : ill. ; 25 x 35 cm.
1. Automobiles—Maintenance and repair. 2. Automobile supplies industry—Colombia—Bogotá. 3. Bogotá—City planning. 4. City planning—Colombia. I. Title.
HD9710.3.C33B633 79-107247
79 MARC

Contreras Niño, Víctor.
Cúcuta, bases socioeconómicas y administrativas para el planeamiento integral del municipio / por Víctor Contreras N. — Bogotá : Centro de Estudios sobre Desarrollo Económico, Universidad de los Andes, Facultad de Economía, 1970.
ix leaves, 222 p. : ill. ; 28 cm. — (Serie P - CEDE, Universidad de los Andes ; no. 5)
Includes bibliographical references.
1. Cúcuta, Colombia—City planning. 2. City planning—Colombia. 3. Cúcuta, Colombia—Economic conditions. 4. Cúcuta, Colombia—Social conditions. I. Title. II. Series: Bogotá. Universidad de los Andes. Centro de Estudios sobre Desarrollo Económico. Serie P - Universidad de los Andes, Centro de Estudios sobre Desarrollo Económico ; no. 5.
HT169.C72C823 79-370136
79 MARC

Contreras Niño, Víctor.
Girardot, alternativas socioeconómicas y administrativas para la planificación municipal / por Víctor Contreras N. — Bogotá : Centro de Estudios sobre Desarrollo Económico, Universidad de los Andes, Facultad de Economía, 1971.
xv leaves, 243 p. : ill. ; 28 cm. — (Serie P - CEDE, Universidad de los Andes ; no. 7)
Includes bibliographical references.
1. Girardot, Colombia—Economic conditions. 2. Girardot, Colombia—Social conditions. 3. Urbanization—Colombia—Girardot. 4. Girardot, Colombia—City planning. 5. City planning—Colombia. I. Title. II. Series: Bogotá. Universidad de los Andes. Centro de Estudios sobre Desarrollo Económico. Serie P - Universidad de los Andes, Centro de Estudios sobre Desarrollo Económico ; no. 7.
HC198.G57C66 79-111469
79 MARC

—COLOMBIA—SOCIETIES, ETC.—DIRECTORIES

Centro Nacional de Documentación sobre Desarrollo Urbana y Vivienda (Colombia)
Directorio institucional sobre desarrollo urbano y vivienda / Centro Nacional de Documentación sobre Desarrollo Urbano y Vivienda, CENDUVI. — Bogotá : CENAC, 19
v. ; 28 cm.
At head of title: Centro Nacional de Estudios de la Construcción. Fondo Colombiano de Investigaciones Científicas y Proyectos Especiales "Francisco José de Caldas." Sistema Nacional de Información.
"CEN-DOC-02-76" (v. 2). "CEN-DOC-03-76" (v. 3)
CONTENTS:
—[2] Medellín, Cali y Barranquilla.—[3] Bucaramanga, Cartagena, Cucuta, Manizales, Pereira.
1. City planning—Colombia—Societies, etc.—Directories. 2. Housing—Colombia—Societies, etc.—Directories. 3. Regional planning—Colombia—Societies, etc.—Directories. I. Centro Nacional de Estudios de la Construcción (Colombia) II. Fondo Colombiano de Investigaciones Científicas y Proyectos Especiales Francisco José de Caldas. III. Sistema Nacional de Información. IV. Title.
HT169.C7C46 1976 79-105457

—CONNECTICUT

Brown, Elizabeth Mills.
New Haven, a guide to architecture and urban design / Elizabeth Mills Brown ; [photos. by Gilbert Kenna, Ray Schmakel, and Lee Tabor]. — New Haven : Yale University Press, 1976.
ix, 228 p., [1] leaf of plates : ill. ; 23 cm.
Includes index.
ISBN 0-300-01842-8
1. New Haven—Buildings. 2. Architecture—Connecticut—New Haven—Guide-books. 3. New Haven—City planning. 4. City planning—Connecticut. I. Title.
NA735.N39B76 917.46'8'044 75-18166
79 MARC

New Haven.
New Haven development guide. [New Haven] 1961.
[40] p. illus. 28 cm.
Cover title.
1. New Haven—City planning. 2. City planning—Connecticut. I. Title.
HT168.N347N48 1961 78-311778
78 MARC

—COSTA RICA

Instituto Nacional de Vivienda y Urbanismo.
I.N.V.U. : organización, objetivos, el problema de vivienda, realizaciones, aspectos financieros, planes de vivienda, planeamiento físico / Instituto Nacional de Vivienda y Urbanismo. — [San José] : El Instituto, [between 1977 and 1979]
36 leaves, [1] leaf of plates : ill. ; 28 cm.
Cover title.
1. Instituto Nacional de Vivienda y Urbanismo. 2. Housing policy—Costa Rica. 3. Housing—Costa Rica—Finance. 4. City planning—Costa Rica. I. Title.
HD7308.A3I62 1977 79-111591
79 MARC

—CZECHOSLOVAKIA

Dušpekt, Zdeněk.
Zásady řešení průmyslového území ve městě a jeho regionu / Zdeněk Dušpekt. — 1. vyd. — Praha : Výzkum. ústav výstavby a architektury, 1978.
188, [1] p. ; 29 cm. Cz78
At head of title: Ministerstvo výstavby a techniky ČSR.
Bibliography: p. 183-186.
1. Regional planning—Czechoslovakia. 2. City planning—Czechoslovakia. 3. Industries, Location of—Czechoslovakia. I. Title.
HT395.C9D87 79-355671
*79 MARC

Šťastný, Jan, engineer.
Koncepce dopravy v obytných zónách. [Autoři:] Jan Šťastný a Jiří Soukup. 1. vyd. Praha, Výzkum. ústav výstavby a architektury, rozmn., 1968.
260, [1] p. illus., plans. 21 cm. Cz70-SKC
Summaries in Russian, English, French, and German.
Bibliography: p. 244-248.
1. Traffic engineering—Czechoslovak Republic. 2. Cities and towns—Planning—Czechoslovak Republic. I. Soukup, Jiří, joint author. II. Title.
HE363.C92S78 75-540077
79 MARC

—DENMARK

Allpass, John.
Helsingør bykerne 1970-1978 ; Holstebro bykerne 1960-1978 : europæisk konkurrence om boligmiljø : analyse af byfornyelsen / [af John Allpass]. — [København] [Holbergsgade 23, 1057 K] : Planstyrelsen, 1978.
59, 49 p. : ill., maps (2 fold. in pocket) D78-33
ISBN 8750325965 : kr25.00
1. Urban renewal—Denmark—Helsingør. 2. Urban renewal—Denmark—Holstebro. 3. Helsingør, Denmark—City planning. 4. Holstebro, Denmark—City planning. 5. City planning—Denmark. I. Allpass, John. Holstebro bykerne 1960-1978. II. Denmark. Planstyrelsen. III. Title.
HT178.D42H442 79-341678
79 MARC

Brandtoft, Arne.
Ressourcebesparende byudvikling : belyst ved en idéskitse for en sydlig vækstretning fra Århus : afgangsopgave / [af] Arne Brandtoft, Søren Klint Kistorp, Christian Rasmussen. — Århus : Husets Forlag, 1978.
229 p. : ill. ; 29 cm. D78-27/28
At head of title: Arkitektskolen i Aarhus. Afdeling for byplanlægning.
Includes bibliographies.
ISBN 8787582546 : kr80.00
1. Aarhus, Denmark—City planning. 2. City planning—Denmark. 3. Aarhus, Denmark—Suburbs and environs. I. Klint Kistorp, Soren, joint author. II. Rasmussen, Christian, 1948- joint author.
HT169.D42A3956 79-347367
79 MARC

CITY PLANNING
—DENMARK (Continued)

Copenhagen. Direktoratet for Generalplanlægning.
Christianshavn : dispositionsplan for Christianshavn / Magistratens 4. afd., Direktoratet for Københavns kommunes generalplanlægning. — [Nyt oplag]. — [København] [Østergade 26, 1100 K] : [Direktoratet for Københavns kommunes generalplanlægning], 1976.
115 leaves : ill. ; 30 cm. D77-15/16
kr30.00
1. Copenhagen—City planning. 2. Christianshavn, Copenhagen. 3. City planning—Denmark. I. Title.
HT169.D42C44 78-374779
78 MARC

Copenhagen. Magistraten. 4. afdeling.
Københavns inderhavn : redegørelse for de byplanmæssige forhold / Magistratens 4. afdeling. — [København] : Afdelingen, 1973.
70 leaves : ill., maps ; 30 cm.
Bibliography: leaves 62-63.
1. Urban renewal—Denmark—Copenhagen. 2. Copenhagen—City planning. 3. City planning—Denmark. 4. Copenhagen—Harbor. 5. Waterfronts—Denmark—Copenhagen. I. Title.
HT178.D42C6625 79-352474
79 MARC

Copenhagen. Magistratens 4. afdeling.
Træk af byens ansigt 1938-1978 / [redaktion og tilrættelegelse ved magistratens 4. afdeling]. — København : Magistratens 4. afdeling : [eksp. Haase], 1978.
96 p. : ill. ; 30 cm. D78-11
Based on sources and notes collected by V. Malling.
Includes bibliographical references.
ISBN 8785164062 : kr75.00
1. Copenhagen—City planning. 2. Copenhagen—History. 3. City planning—Denmark. I. Malling, Vilhelm Carl, 1901-1977. II. Title.
HT169.D42C49 78-375852
78 MARC

Dam, Ajs.
Træer i Hjørring / udarbejdet i sommeren 1976 for Hjørring kommune af Ajs Dam og Mogens Hjørne, i samarbejde med Børge Kahr Nielsen og byplanafdelingen ; [foto, Torben Pedersen]. — [Hjørring] : Hjørring Kommune : [eksp. Vendsyssel Historiske Museum] [Museumsgade 2, 9800], 1977.
192 p., [1] fold. leaf of plates : ill. ; 21 x 31 cm. D78-16
Cover title.
Includes indexes.
kr50.00
1. Trees in cities—Denmark—Hjørring (City). 2. Hjørring, Denmark (City)—City planning. 3. City planning—Denmark. I. Hjørne, Mogens, joint author. II. Title.
SB435.6.D42H533 78-370827
78 MARC

Geleff, Palle.
Den analytiske løsning for Fyns Amstkommunes byvækstmodel / Palle Geleff. — [København] : Københavns universitets Økonomiske institut, Cykelafdelingen, 1977.
6 p. ; 30 cm. — (Memo - Københavns universitets Økonomiske institut, Cykelafdelingen ; nr. 41) D***
1. Cities and towns—Denmark—Fyn—Growth—Mathematical models. 2. Fyn, Denmark—City planning. 3. City planning—Denmark. I. Title. II. Series: Copenhagen. Universitet. Økonomiske institut. Cykelafdelingen. Memo - Københavns universitets Økonomiske institut, Cykelafdelingen ; nr. 41.
HT145.D4G44 78-399920
79 MARC

—DENMARK—ADDRESSES, ESSAYS, LECTURES

Rødovre 1901-1976 [i.e. nittenhundrede og en-nittenhundrede seksoghalvfjerds] / af Michael Broch ... [et al.] ; redigeret af Sigurd Rambusch. — [Rødovre] : Rødovre Kommune : [eksp. Hovedbiblioteket, Rødovre Kommune] [Rødovre Parkvej 140, 2610], 1978.
347 p. : ill. ; 31 cm. D78-20
Includes bibliographical references and index.
ISBN 8787526085 : kr120.00
1. Rodovre, Denmark—City planning—History—Addresses, essays, lectures. 2. City planning—Denmark—Addresses, essays, lectures. I. Broch, Michael. II. Rambusch, Sigurd.
HT169.D42R486 78-397793
79 MARC

—DENMARK—CITIZEN PARTICIPATION

Beboerarbejde er blandt andet beboernes kamp imod overgreb : demokratiet i Århus sættes nu til debat / [udg. af Jeppe Gitz-Johansen] ; [i samarbejde med Ole Langballe]. — Århus : [Jeppe Gitz-Johansen], [1976]
[107] p. : ill. ; 30 cm. D77-17
Cover title.
"Trykt i 150 eks."
kr50.00
1. Aarhus, Denmark—City planning—Citizen participation. 2. City planning—Denmark—Citizen participation. I. Gitz-Johansen, Jeppe. II. Langballe, Ole.
HT169.D42A3954 78-397315

—DENMARK—CITIZEN PARTICIPATION—CASE STUDIES

Schultz, Karen Ingrid.
Krisen og beboerbevægelserne / [af] Karen Ingrid Schulz. — [København] : Aurora, c1977.
191 p. : ill. ; 20 cm. D78-10
Bibliography: p. [188]-[192]
ISBN 8787534037 : kr44.00
1. City planning—Citizen participation—Case studies. 2. City planning—Denmark—Citizen participation—Case studies. 3. Housing policy—Citizen participation—Case studies. 4. Housing policy—Denmark—Citizen participation—Case studies. I. Title.
HT166.S3168 78-394142
79 MARC

—DENMARK—EXHIBITIONS

Copenhagen. Stadsgartnerens kontor.
Det grønne København / [Stadsingeniørens direktorat, stadsgartnerens kontor]. — [København] : Stadsingeniørens Direktorat : [eksp. Stadsgartnerens Kontor] [Rådhuspladsen 45, 1550 V], 1976.
36 p. : ill. ; 20 x 21 cm. D77-47
Cover title.
1. Copenhagen—Parks—Exhibitions. 2. Trees in cities—Denmark—Copenhagen—Exhibitions. 3. Copenhagen—City planning—Exhibitions. 4. City planning—Denmark—Exhibitions. I. Title.
DL276.6.A2C66 1976 78-382810
78 MARC

—EAST ASIA—HISTORY

Nakamura, Haruhisa.
(Nikkan kodai toshi keikaku)
日韓古代都市計画 / 中村春壽. — 東京 : 六興出版, 1978.
213 p., [2] leaves of plates : ill. ; 19 cm. — (ロッコウブックス)
Ja 78-22365
Series romanized: Rokkō bukkusu.
¥980
1. Cities and towns, Ancient—Korea—History. 2. Cities and towns, Ancient—Japan—Nara (Prefecture)—History. 3. City planning—East Asia—History. I. Title.
HT147.K6N34 78-816004

—EGYPT

ʻAbd Allāh, Muḥammad Aḥmad.
(Takhṭīṭ al-manāṭiq al-ṣināʻīyah)
تخطيط المناطـــق الصناعيـــة / محمد أحمد عبد الله. —
[القاهرة] : مكتبة الأنجلو المصرية، 1975.
227, [1] p., [4] fold. leaves of plates : ill. ; 28 cm.
Bibliography: p. 225-[228]
1. Helwan, Egypt—City planning. 2. City planning—Egypt. 3. Helway, Egypt—Industries. I. Title.
HT169.E32H442 78-960296

—ENGLAND

Boaden, Noel.
Sample surveys and public participation / [by] Noel Boaden and Raymond Walker. — Sheffield : Linked Research Project into Public Participation in Structure Planning, 1976.
[1], ii, 30, [2] p. ; 30 cm. — (Interim research paper - Linked Research Project into Public Participation in Structure Planning ; 10 ISSN 0307-5699)
GB77-03142
Cover title.
Includes bibliographical references.
ISBN 0-904856-08-9 : £0.95
1. Social surveys—England—Teeside. 2. Social surveys—England—Merseyside. 3. City planning—England. I. Walker, Raymond, 1949- joint author. II. Title. III. Series: Linked Research Project into Public Participation in Structure Planning. Interim research paper - Linked Research Project into Public Participation in Structure Planning ; 10.
HN398.E5B6 711'.12 78-323038
79 MARC

Cambridge, Eng. Dept. of Architecture and Planning.
St. Matthew's draft local plan / [drawn up by the staff of the Cambridge City Architect and Planning Officer]. — [Cambridge] : The Department, 1976.
[7], 50, [9] p., [14] leaves of plates : ill., maps, plans ; 29 cm.
GB77-10866
Errata slip inserted.
Includes bibliographical references.
ISBN 0-902696-03-3
1. Cambridge, Eng.—City planning. 2. City planning—England. 3. Architecture—England—Cambridge—Conservation and restoration. I. Title.
HT169.G72C2932 711'.4'0942659 79-303835
79 MARC

Kelly, Frances, fl. 1971-
Classifications of the London boroughs. London, Greater London Council, 1971.
48 p. illus., map. 30 cm. (Greater London Council. Intelligence Unit. Research report no. 9) £0.40 B71-[15721
At head of title: Greater London Research.
Bibliography: p. [49]
ISBN 0-7168-0331-3
1. London—Population. 2. City planning—England. 3. London—City planning. I. Title. II. Series: Greater London Council. Intelligence Unit. Research report, no. 9.
HT169.G7G74 no. 9 309.2'62'09421 s 75-875959
[HB2046.L6] 72[r79]rev MARC

Reeve, F A
The Cambridge that never was / F. A. Reeve. — Cambridge : Oleander Press, c1976.
40 p. : ill. ; 21 cm. — (Cambridge town, gown, and county ; no. 8)
GB***
Bibliography: p. 40.
ISBN 0-902675-72-9
1. Cambridge, Eng.—City planning. 2. Cambridge, Eng.—History. 3. City planning—England. I. Title.
HT169.G72C2955 711'.4'0942659 79-301632
79 MARC

—ENGLAND—ADDRESSES, ESSAYS, LECTURES

Planning and the historic environment II : papers presented to a conference in Oxford, 1977 / edited by R. T. Rowley and M. Breakell. — Oxford : Oxford University Department for External Studies ; [Oxford] : [Oxford Polytechnic Department of Town Planning], 1977.
vi, 168 p. : ill., forms, maps, plan ; 30 cm. GB78-40630
Includes bibliographies.
ISBN 0-903736-05-5 : £2.50
1. Historic buildings—England—Conservation and restoration—Addresses, essays, lectures. 2. England—Antiquities—Addresses, essays, lectures. 3. Cities and towns—Planning—England—Addresses, essays, lectures. 4. Regional planning—England—Addresses, essays, lectures. 5. Monuments—England—Preservation—Addresses, essays, lectures. I. Rowley, Trevor, 1942- II. Breakell, Mike.
DA655.P59 363.6 79-308137
79 MARC

—ENGLAND—CASE STUDIES

Cambridge, Eng. Dept. of Architecture and Planning.
St. Matthew's district plan / City of Cambridge [Department of Architecture and Planning]. — [Cambridge] : [The Department], 1977.
[6], 87 [i.e. 98] p., [12] leaves of plates (8 fold.) : ill., maps, plans ; 30 cm.
GB77-27971
Errata slip inserted.
£2.50
1. Saint Matthew, Cambridge, Eng.—City planning. 2. City planning—England—Case studies. I. Title.
HT169.G72S353 711'.4'0942659 79-314580
79 MARC

Land use conflicts in the urban fringe : a case study of aggregate extraction in the London Borough of Havering / [project supervisor, Philip Lowe]. — London : University College, [1977?]
60 p. : maps ; 30 cm. — (Town planning discussion paper ; no. 26)
GB***
Caption title.
Includes bibliographical references.
£1.20
1. Land use—England—Havering. 2. Havering, Eng.—City planning. 3. City planning—England—Case studies. I. Lowe, Philip. II. Series: London. University. University College. School of Environment Studies. Discussion paper series ; no. 26.
HD610.H33L36 333.7'7'0942174 79-305233
79 MARC

—ENGLAND—STATISTICS —PERIODICALS

Chartered Institute of Public Finance and Accountancy. Statistical Information Service.
Planning and development statistics. Estimates. [London] CIPFA Statistical Information Service.
v. 30 cm. annual.
Each issue published in 2 or more vols.
1. City planning—England—Statistics—Periodicals. 2. Regional planning—England—Statistics—Periodicals. I. Title.
HT169.G72E588 309.2'12'0941 79-645065
MARC-S

—ENGLAND—LONDON METROPOLITAN AREA

Greater London Council.
Modified Greater London development plan : Greater London Council amendments and detailed comments. — London : The Council, [1976]
44 p. ; 30 cm. GB***
1. City planning—England—London metropolitan area. 2. Regional planning—England—London metropolitan area. I. Title.
HT169.G72L635 309.2'62'09421 79-300680
79 MARC

CITY PLANNING (Continued)

—ENGLAND—NOTTINGHAMSHIRE —CASE STUDIES

Smigielski, W [redacted] Konrad.
Self-supporting co-operative village at Stanford Hall, Nottinghamshire / by Konrad Smigielski. — ⌈Coalville⌉ : Building and Social Housing Foundation, ⌈1978⌉
71 p. : ill. (some col.), chart, maps (chiefly col.), plans (some col.) ; 28 x 34 cm. GB78-23218
1. Rural development—England—Nottinghamshire—Case studies. 2. Energy policy—England—Nottinghamshire—Case studies. 3. Cooperative societies—England—Nottinghamshire—Case studies. 4. City planning—England—Nottinghamshire—Case studies. 5. Autarchy—Case studies. I. Title.
HN398.N63S57 309.2'63'0942529 78-325445
79 MARC

—ENGLAND—SURREY

Surrey, Eng. County Council.
Surrey structure plan draft written statement / ⌈Surrey County Council⌉. — ⌈Kingston upon Thames⌉ : The Council, 1978.
⌈1⌉, vii, 184 p., ⌈2⌉ fold leaves of plates : ill., maps (2 col.) ; 30 cm. GB78-22142
Cover title.
£1.50
1. City planning—England—Surrey. I. Title.
HT169.G72S8448 309.2'5'094221 79-303665
79 MARC

—ESTONIA—ADDRESSES, ESSAYS, LECTURES

Tallinna arengu probleeme / ⌈toimetuskolleegium, I. Kallion ... et al.⌉ — Tallinn : Eesti Raamat, 1978.
267 p., ⌈4⌉ leaves of plates : ill. ; 21 cm. USSR***
Bibliography: p. 264-267.
0.70rub
1. Tallinn—Economic conditions—Addresses, essays, lectures. 2. Tallinn—City planning—Addresses, essays, lectures. 3. City planning—Estonia—Addresses, essays, lectures. I. Kallion, I.
HC337.E72T297 79-354323
79 MARC

—EUROPE—ADDRESSES, ESSAYS, LECTURES

Le Norme e gli strumenti di pianificazione urbanistica in Europa / a cura di Giuseppe Furitano. — Milano : A. Giuffrè, 1978.
196 p. : ill. ; 28 cm. — (Studi urbanistici ; v. 11) It78-Oct
At head of title: Fondazione Aldo Della Rocca.
Includes legislation in French.
Bibliography: p. 135-142.
CONTENTS: Sisi, E. Germania, Olanda, Danimarca, Belgio: una regione.—Pasqualini, C. M. I centri storici in Francia dalla carta di Atene alla legge Malraux.
L12000
1. City planning—Europe—Addresses, essays, lectures. 2. City planning and redevelopment law—Europe—Addresses, essays, lectures. I. Furitano, Giuseppe. II. Series.
NA9000.S78 vol. 11 79-341032
⌈HT169.E8⌉ *79 MARC

Rational architecture : the reconstruction of the European city = Architecture rationnelle : la reconstruction de la ville européenne. — Bruxelles : Éditions des Archives d'architecture moderne, 1978.
213 p. : ill. (some col.) ; 21 cm. Be***
French and English.
1. Functionalism (Architecture)—Europe—Addresses, essays, lectures. 2. City planning—Europe—Addresses, essays, lectures. I. Title: Architecture rationnelle. II. Title: The reconstruction of the European city. III. Title: La reconstruction de la ville européenne.
NA9053.F8R37 711'.4'094 79-372544
79 MARC

—EUROPE—CITIZEN PARTICIPATION—CASE STUDIES

Ulram, Peter Adolf, 1951-
Zwischen Bürokratie und Bürger : sozialistische Kommunalpolitik in Wien, Stockholm u. Bologna / Peter A⌈dolf⌉ Ulram. — Wien : Braumüller, 1978.
191 p. ; 23 cm. — (Studien zur österreichischen und internationalen Politik ; Bd. 1) Au79-1/2
Bibliography: p. 181-191.
ISBN 3-7003-0187-1 : S180.00
1. City planning—Europe—Citizen participation—Case studies. 2. Municipal government—Europe—Case studies. 3. Vienna—City planning. 4. Sweden—City planning. 5. Bologna—City planning. I. Title. II. Series.
HT169.8.E85U47 79-372859
*79 MARC

—EUROPE—CONGRESSES

Cities and city regions in Europe : conference, 1973, Lanchester Polytechnic, Coventry / edited by John Holliday, David Liggins, William Ogden. — ⌈Coventry : Lanchester Polytechnic, 1973?⌉
235 p. : ill. ; 30 cm.
Includes bibliographical references.
1. Cities and towns—Europe—Congresses. 2. City planning—Europe—Congresses. 3. Metropolitan areas—Europe—Congresses. I. Holliday, John C. II. Liggins, David. III. Ogden, William. IV. Title.
HT131.C496 309.2'62'094 78-317986
78 MARC

—FINLAND

Palonen, Tarmo.
Tavoitteellinen kuntasuunnittelu / Tarmo Palonen. — Helsinki : ⌈Suomen kunnallisliitto, 1976⌉
145 p. : ill. ; 21 cm. — (Kunnallistietoa) Fi76
Bibliography: p. 122-123.
ISBN 9517730349 : Fmk19.00
1. City planning—Finland. I. Title.
HT169.F5P34 79-367592
*79 MARC

—FINLAND—CONGRESSES

Growing space needs in the urbanized region / ⌈by Gunnar Lindman ... et al. ; translation from Swedish by Roger Waters⌉. — ⌈Stockholm : National Swedish Board of Building and Planning, 1965⌉
73 leaves : ill. ; 30 cm.
Cover title.
At head of title: International conference of the International Federation for Housing and Planning, Örebro, Sweden, 1965.
Bibliography: leaf 73.
1. City planning—Scandinavia—Congresses. 2. City planning—Finland—Congresses. 3. Land use, Urban—Scandinavia—Congresses. 4. Land use, Urban—Finland—Congresses. I. Lindman, Gunnar. II. International Federation for Housing and Planning.
HT169.S27G76 309.2'62'0948 78-320840
79 MARC

—FRANCE

Augoyard, Jean François, 1941-
Pas à pas : essai sur le cheminement quotidien en milieu urbain / Jean-François Augoyard. — Paris : Seuil, c1979.
185 p. : ill. ; 21 cm. — (Collection Espacements) F***
Bibliography: p. 181-⌈183⌉
Includes index.
ISBN 2-02-005170-2 : 47.00F
1. City and town life—France. 2. Urbanization—France. 3. City planning—France. I. Title.
HT135.A88 301.36'0944 79-375994
79 MARC

Aymonino, CArlo.
Parigi e Vienna / Carlo Aymonino, Gianni Fabbri, Angelo Villa. — Roma : Officina, 1975.
315 p., ⌈136⌉ leaves of plates : ill. ; 25 cm. — (Le Città capitali del XIX secolo ; 1) (Collana di architettura ; 14) It***
Includes bibliographies.
L15000
1. Paris—City planning. 2. City planning—France. 3. Architecture, Modern—19th century—France—Paris. 4. Vienna—City planning. 5. City planning—Austria. I. Fabbri, Gianni, joint author. II. Fabbri, Gianni, joint author. III. Villa, Angelo, 1945- joint author. IV. Title. V. Series.
NA9053.C3C57 vol. 1 76-514625
⌈NA9198.P2⌉ 79⌈r79⌉rev MARC

Bauer, Gérard, 1934-
Un Urbanisme pour les maisons / par Gérard Bauer ; ⌈avec la collaboration de Jean-Michel Roux et Vincent Renard⌉. — Paris : Union générale d'éditions, 1979.
255, ⌈1⌉ p. : ill. ; 18 cm. — (10/18 ⌈i.e. Dix/dix-huit⌉ ; 1293) F79-5827
Bibliography: p. 253-⌈256⌉
ISBN 2-264-00953-5 : 17.00F
1. Housing—France. 2. Housing policy—France. 3. City planning—France. I. Roux, Jean Michel, 1941- joint author. II. Renard, Vincent, joint author. III. Title.
HD7338.A3B35 301.5'4'0944 79-376026
79 MARC

Boudon, Philippe.
Richelieu, ville nouvelle : essai d'architecturologie / Philippe Boudon. — ⌈Paris⌉ : Dunod, c1978.
xiii, 186 p. : ill. ; 21 cm. — (Collection Aspects de l'urbanisme) F***
Bibliography: p. ⌈143⌉-152.
Includes index.
1. Richelieu, France—City planning. 2. City planning—France. I. Title.
NA9198.R5B68 711'.4'094454 78-379539
78 MARC

Canacos, Henry, 1928-
Sarcelles : ou, Le béton apprivoisé / Henry Canacos. — Paris : Éditions sociales, c1979.
267 p. ; 22 cm. — (Notre temps/société) F79-6454
ISBN 2-209-05320-X : 45.00F
1. Sarcelles, France—City planning. 2. Sarcelles, France—Social conditions. 3. City planning—France. I. Title.
HT169.F72S373 309.2'62'0944367 79-375949
79 MARC

Chevalier, Louis, 1911-
L'assassinat de Paris / Louis Chevalier. — Paris : Calmann-Lévy, c1977.
285 p. ; 23 cm. — (Archives des sciences sociales) F***
ISBN 2-7021-0176-3 : 55.00F
1. Historic buildings—France—Paris—Conservation and preservation. 2. Historic sites—France—Paris—Conservation and preservation. 3. Paris—City planning. 4. City planning—France. 5. Urban renewal—France—Paris. 6. Paris—Description—1975- I. Title.
DC771.C47 069'.53'0944361 78-394472
79 MARC

Colloque international de géographie appliquée, Liège, 1967.
Excursions. Liège, 8, 9 et 10 septembre 1967. 3me réunion de la Commission de géographie appliquée de l'Union géographique internationale. Liège (Société géographique de Liège et Séminaire de géographie de l'Université de Liège), (1968).
108, 84 p. illus. 25 cm. (Travaux géographiques de Liège, no. 153) Be***
At head of title: Université de Liège.
Proceedings of the 3d of a series of meetings; proceedings of the 2d are entered under International Geographical Union. Commission on Applied Geography; the 4th, under International Geographical Union. Commission on Applied Geography.
1. Liège—City planning. 2. City planning—France. 3. Regional planning—Belgium—Condroz. 4. Regional planning—Belgium—Hesbaye. I. International Geographical Union. Commission on Applied Geography. II. Series.
HT169.B42L53 1967 309.2'5'09493 70-438649
79⌈r79⌉rev MARC

Doubrère, Jean Claude.
Cours d'urbanisme appliqué / par Jean-Claude Doubrère ; préf. de Pierre Mayet. — 2. éd., rev. et corr. — Paris : Eyrolles, 1979, ⌈c1978⌉
260 p. : ill. ; 22 cm. F***
151F
1. City planning—France. I. Title.
HT169.F7D6 1979 309.2'62'0944 79-349591
79 MARC

Dujardin, Daniel.
Voyage au centre de Clermont-Fd / Daniel Dujardin, Françoise Lorgeoux. — Clermont-Ferrand : G. de Bussac, ⌈1977⌉
344 p. : ill. (some col.) ; 21 cm. F***
Bibliography: p. 281-288.
66.00F
1. Clermont-Ferrand, France—City planning. 2. City planning—France. 3. Urban renewal—France—Clermont-Ferrand. I. Lorgeoux, Françoise, joint author. II. Title.
HT169.F72C573 309.2'62'0944591 79-361727
79 MARC

Espaces extérieurs urbains : rencontre du Centre de recherche d'urbanisme / coordination, J. P. Muret, P. de Fouchier, M. C. Paoletti. — Paris : Centre de recherche d'urbanisme, ⌈1977⌉
196 p. : ill. (some col.) ; 21 x 30 cm.
Bibliography: p. 187-193.
Includes indexes.
ISBN 2-85303-107-1
1. Open spaces—France. 2. City planning—France. I. Muret, Jean Pierre. II. Fouchier, P. de. III. Paoletti, M. C. IV. Centre de recherche d'urbanisme.
HT169.F7E84 711'.4'0944 79-348613
79 MARC

France. Direction de l'aménagement foncier et de l'urbanisme.
Politiques d'action foncière : définition et mise en œuvre dans les extensions urbaines. — ⌈Paris⌉ : Ministère de l'équipement, Direction de l'aménagement foncier et de l'urbanisme, c1975.
vi, 151 p. ; 30 cm. F***
"Circulaire no 75-156 du 5 octobre 1975 de M. le ministre de l'Équipement et note technique sur les moyens de l'action foncière."
1. City planning—France. I. Title.
HT169.F7F7 1975b 309.2'62'0944 79-345196
79 MARC

Lacaze, Jean Paul.
Introduction à la planification urbaine : imprécis d'urbanisme à la française / Jean-Paul Lacaze. — Paris : Moniteur, 1979.
xiv, 302 p. : ill. ; 26 cm. F***
Includes bibliographical references.
ISBN 2-86282-047-4 : 98.00F
1. City planning—France. I. Title.
HT135.L33 309.2'62'0944 79-378701
79 MARC

Magnan, René.
Architecte et urbaniste 1999 / René Magnan. — Paris : Centre de recherche d'urbanisme, ⌈1978⌉
84 p. ; 21 cm. F***
ISBN 2-85303-123-3
1. City planning—France. 2. City planning—1945- I. Title.
NA9197.M33 711'.4'0944 79-374205
79 MARC

CITY PLANNING
—FRANCE (Continued)

Philipponneau, Michel.
Changer la vie, changer la ville : Rennes 1977 / Michel Philipponneau ; cartographie, Sylvain Philipponneau. — La Baule : Éditions Breiz, [c1976]
375 p. : ill. ; 23 cm. F***
Includes bibliographical references.
45.00F
1. Rennes—City planning. 2. City planning—France. 3. Rennes—Economic conditions. 4. Rennes—Politics and government. I. Title.
HT169.F72R446 309'.2'62'094415 78-396398
79 MARC

Procédures et institutions d'aménagement et d'urbanisme / [élaboré pour la Direction de l'aménagement foncier et de l'urbanisme par] Y. M. Danan ... [et al.]. — Paris : Direction de l'urbanisme et des paysages, Service technique de l'urbanisme, 1978-
v. : ill. ; 21 cm. — (Processor) F***
Includes bibliographies.
ISBN 2-11-081732-1 (v. 1) : 20.00F (v. 1)
1. City planning—France. 2. Regional planning—France. I. Danan, Yves Maxim. II. France. Direction de l'aménagement foncier et de l'urbanisme.
HT169.F7P76 309.2'5'0944 79-365667
79 MARC

Rothenbach, Jean Jacques.
Histoire et urbanisme : réflexions d'un Strasbourgeois / Jean-Jacques Rothenbach .... — [Strasbourg] (6, Av. du Général-de-Gaulle, 67000) : [Agence d'urbanisme pour l'agglomération strasbourgeoise], 1977.
37 p. : ill. (some col.) ; 21 x 27 cm. F78-11030
1. Strasbourg—City planning. 2. City planning—France. I. Title.
NA9198.S87R67 711'.4'09443835 79-378878
79 MARC

Truttmann, Philippe.
Fortification, architecture et urbanisme aux XVIIe et XVIIIe siècles : essai sur l'œuvre artistique et technique des ingénieurs militaires sous Louis XIV et Louis XV / Philippe Truttmann ; accompagné de figures, planches et cartes de la main de l'auteur ; préf. de Gabriel Stiller. — [Thionville] : Service culturel de la ville de Thionville, [1976]
77 p. : ill. ; 31 cm. — (Région de Thionville : Études historiques ; no 32) F***
Includes bibliographical references.
20.00F
1. Architecture, Military—France. 2. Architecture, Modern—17th-18th centuries—France. 3. City planning—France. I. Title. II. Series.
NA497.F7T78 720'.944 78-389057
79 MARC

Vilmorin, Catherine de.
La politique d'espaces verts / Catherine de Vilmorin. — Paris : Centre de recherche d'urbanisme, [1978?]
439 p. ; 21 cm. F***
Originally presented as the author's thesis, Paris II, 1976.
Bibliography: p. 423-435.
Includes index.
ISBN 2-85303-119-5 : 40.00F
1. Greenbelts—France. 2. City planning—France. 3. Urbanization—France. I. Title.
HT169.F7V54 1978 309.2'62 79-346892
79 MARC

—FRANCE—CONGRESSES

Conference on Two World Cities: Paris and New York, Paris, 1978.
Survival strategies : Paris and New York : report on the Conference on Two World Cities: Paris and New York, Paris, May, 1978 / the French/American Foundation, the Council for International Urban Liaison ; edited by George G. Wynne. — New Brunswick, N.J. : Transaction Books, c1979.
iv, 200 p. : ill. ; 23 cm.
Includes bibliographical references.
ISBN 0-87855-749-0 : $5.95
1. City planning—Congresses. 2. New York (City)—City planning—Congresses. 3. Paris—City planning—Congresses. 4. Urban policy—Congresses. 5. City planning—France—Congresses. 6. City planning—New York (State)—Congresses. I. Wynne, George A. II. French-American Foundation. III. Council for International Urban Liaison. IV. Title.
HT166.C6217 1978 309.2'62'0944361 78-66205
79 MARC

Qualité de l'habitat : comment améliorer la qualité de l'urbanisme quotidien : [colloque], 14-15 juin 1977. — Paris : Centre de recherche d'urbanisme, [1978]
207 p. : ill. ; 21 cm. — (Colloques de Marly ISSN 0150-0856) F***
Bibliography: p. 199-202.
ISBN 2-85303-121-7
1. Urbanization—France—Congresses. 2. City planning—France—Congresses. I. Centre de recherche d'urbanisme.
HT135.Q34 309.2'62'0944 79-340480
79 MARC

—FRANCE—STATISTICS

France. Direction de l'aménagement foncier et de l'urbanisme.
Recueil d'informations statistiques sur l'urbanisme / République française, Ministère de l'équipement et de l'aménagement du territoire, Direction de l'aménagement foncier et de l'urbanisme. — [3. éd.]. — [Paris] : La Direction, 1977.
173 p. ; 30 cm. F***
Tables.
Published in 1975 under title: Recueil d'informations statistiques utiles à l'urbanisme.
Includes bibliographical references.
25.00F
1. City planning—France—Statistics. I. Title.
HT169.F7F7 1977 309.2'62'0944 79-353102
79 MARC

—GEORGIA—ADDRESSES, ESSAYS, LECTURES

The Future of Atlanta's central city / [edited by] Edwin N. Gorsuch, Dudley S. Hinds. — Atlanta : Pub. Services Division, College of Business Administration, Georgia State University, 1977.
viii, 134 p. : map ; 28 cm. — (Research monograph - College of Business Administration, Georgia State University ; no. 73)
ISBN 0-88406-110-8
1. Atlanta—Economic conditions—Addresses, essays, lectures. 2. Atlanta—Social conditions—Addresses, essays, lectures. 3. Central business districts—Georgia—Atlanta—Addresses, essays, lectures. 4. Atlanta—City planning—Addresses, essays, lectures. 5. City planning—Georgia—Addresses, essays, lectures. I. Gorsuch, Edwin N., 1939- II. Hinds, Dudley S., 1926- III. Title. IV. Series: Georgia State University. College of Business Administration. Research monograph - College of Business Administration, Georgia State University ; no. 73.
HC108.A75F83 330.9'758'23104 77-9257
77 MARC

—GEORGIA—HISTORY—18TH CENTURY

Sears, Joan Niles.
The first one hundred years of town planning in Georgia / by Joan Niles Sears. — Atlanta : Cherokee Pub. Co., 1979.
xiii, 220 p. : ill. ; 24 cm.
Based on the author's thesis, Emory University, 1977.
Bibliography: p. 196-208.
Includes index.
ISBN 0-87797-046-7
1. Cities and towns—Georgia—History—18th century. 2. City planning—Georgia—History—18th century. I. Title.
HT123.5.G4S4 309.2'62'09758 78-74091
00 MARC

—GEORGIA (TRANSCAUCASIA)
—TIFLIS—ADDRESSES, ESSAYS, LECTURES

(Arkhitektura i gradostroitel'stvo)
Архитектура и градостроительство : [Сборник статей / Науч. ред. канд. архитектуры З. Кикнадзе]. — Тбилиси : ТбилЗНИИЭП, 1975.
120 p. : ill. ; 20 cm. — (Сборник научных трудов - Зональный научно-исследовательский и проектный институт типового и экспериментального проектирования жилых и общественных зданий в г. Тбилиси Госгражданстроя)
Includes bibliographies.
1. Architecture—Georgia (Transcaucasia)—Tiflis—Addresses, essays, lectures. 2. Tiflis—City planning—Addresses, essays, lectures. 3. City planning—Georgia (Transcaucasia)—Tiflis—Addresses, essays, lectures. I. Kiknadze, Z. II. Series: Tbilisskiĭ zonal'nyĭ nauchno-issledovatel'skiĭ i proektnyĭ institut tipovogo i ėksperimental'nogo proektirovanii͡a zhilykh i obshchestvennykh zdaniĭ. Sbornik nauchnykh trudov - Zonal'nyĭ nauchno-issledovatel'skiĭ i proektnyĭ institut tipovogo i eksperimental'nogo proektirovanii͡a zhilykh i obshchestvennykh zdaniĭ v g. Tbilisi Gosgrazhdanstroi͡a.
NA1492.9.A73 1975 79-413847

—GERMANY

Larsson, Lars-Olof, 1938-
Die Neugestaltung der Reichshauptstadt : Albert Speers Generalbebauungsplan für Berlin / von Lars Olof Larsson. — Stockholm : Almqvist & Wiksell International (distr.), 1977, 1978 printing.
196 p. : ill. ; 27 cm. — (Acta Universitatis Stockholmiensis : Stockholm studies in history of art ; 29) S78-17
Includes bibliographical references and index.
ISBN 912200131X
1. Speer, Albert, 1905- 2. Berlin—City planning. 3. City planning—Germany. 4. National socialism and architecture. I. Title. II. Series: Stockholm studies in history of art ; 29.
NA1088.S56L37 79-350512
79 MARC

—GERMANY—ADDRESSES, ESSAYS, LECTURES

Piccinato, Giorgio.
La costruzione dell'urbanistica : Germania 1871-1914 / Giorgio Piccinato. — 2. ed. — Roma : Officina, 1977.
193 p., [24] leaves of plates : ill. ; 25 cm. — (Collana di architettura ; 12) It78-Jan
Includes bibliographical references.
L6000
1. City planning—Germany—Addresses, essays, lectures. I. Title.
HT169.G3P53 1977 78-366203
*78 MARC

—GERMANY—HISTORY

Trieb, Michael.
Stadtgestaltungspolitik : Aufgaben, Instrumente, Strategien / Michael Trieb, Ursula Grammel, Alexander Schmidt, unter Mitarb. von Walter Kiess ... [et al.]. — Stuttgart : Deutsche Verlags-Anstalt, 1979.
223 p. : ill. ; 22 cm. GFR79-A
Includes bibliogrpahical references.
ISBN 3-421-02534-7 : DM48.00
1. City planning—Germany, West. 2. City planning—Germany, West—Case studies. 3. City planning—Germany—History. I. Grammel, Ursula, joint author. II. Schmidt, Alexander, Dipl.-Ing., joint author. III. Title.
HT169.G3T75 79-378061
79 MARC

—GERMANY, EAST

Institut für Denkmalpflege. Arbeitsstelle Schwerin.
Denkmale in Mecklenburg : ihre Erhaltung u. Pflege in d. Bezirken Rostock, Schwerin u. Neubrandenburg / erarb. im Inst. für Denkmalpflege, Arbeitsstelle Schwerin ; [Red., Hans Müller]. — 2., durchges. Aufl. — Weimar : Böhlau, 1977.
359 p. : ill. (some col.), maps ; 24 cm. — (Schriften zur Denkmalpflege in der Deutschen Demokratischen Republik) GFR78-A27
Includes index.
35.00M
1. Art—Conservation and restoration—Germany, East—Mecklenburg. 2. Monuments—Germany, East—Mecklenburg—Preservation. 3. City planning—Germany, East. I. Title. II. Series.
N6889.2.M42I57 1977 79-374681
79 MARC

—GERMANY, EAST—ADDRESSES, ESSAYS, LECTURES

La Residenza e la costruzione della città nella Repubblica democratica tedesca / [scritti di] Junghan[n]s, Schmidt, Zeuchner, Wessel. — 1. ed. — Milano : Clup, 1976.
vii, 7-254 p. : ill. ; 24 cm. It78-May
At head of title: Politecnico di Milano, Facoltà di architettura, Gruppo di ricerca Cislaghi, De Benedetti, Spagnoli, Spinelli.
Translated from articles in German originally published in Deutsche Architektur.
L4000
1. City planning—Germany, East—Addresses, essays, lectures. 2. Housing—Germany, East—Addresses, essays, lectures. 3. Architecture—Germany, East—Addresses, essays, lectures. I. Junghanns, Kurt. II. Milan. Politecnico. Facoltà di architettura.
HT169.G33R47 78-399148
*79 MARC

—GERMANY, EAST—CONGRESSES

Symposium Stadtverkehr, 3d, Berlin, 1977.
Generalverkehrsplanung und ihre Wechselbeziehung zum sozialistischen Städtebau : 3. Symposium Stadtverkehr des ZFIV und der KDT. — Berlin : Zentrales Forschungsinstitut des Verkehrswesens der DDR, 1977.
242, 9 p. : ill. ; 21 cm. — (Report - Zentrales Forschungsinstitut des Verkehrswesens der DDR ; 4. Jahrg., Heft 11) GDR***
1. Transportation planning—Germany, East—Congresses. 2. Urban transportation—Germany, East—Case studies. 3. City planning—Germany, East—Congresses. I. Zentrales Forschungsinstitut des Verkehrswesens der DDR. II. Kammer der Technik. III. Title. IV. Series: Zentrales Forschungsinstitut des Verkehrswesens der DDR. Report - Zentrales Forschungsinstitut des Verkehrswesens der DDR ; 4. Jahrg., Heft 11.
HE249.5.S95 1977 78-377131
78 MARC

CITY PLANNING (Continued)

—GERMANY, WEST

Bach, Heinz Willi, 1950-
Planungswertausgleich und Raumordnungspolitik : räuml. Wirkungen e. Ausgleichs entwicklungsbedingter Bodenwerterhöhungen (Planungswertausgleich) u. weiterer bodenpolit. Instrumente / Heinz Willi Bach. — München : Florentz, 1978.
233 p. : ill. ; 21 cm. — (Schriftenreihe Wirtschaftwissenschaftliche Forschung und Entwicklung ; Bd. 25) GFR79-A
Originally presented as the author's thesis, Marburg.
Bibliography: p. 200-223.
Includes index.
ISBN 3-88259-010-6 : DM31.80
1. Regional planning—Germany, West. 2. City planning—Germany, West. 3. Land use, Urban—Germany, West. 4. Land value taxation—Germany, West. 5. Infrastructure (Economics)—Germany, West. I. Title. II. Series.
HT395.G4B3 1978 309.2'12'0943 79-349684
79 MARC

Berlin (West Berlin). Senat.
Verkehrsentwicklungsplanung, Berlin : Bericht 1977 : Materialien. — Berlin : Der Senat von Berlin, 1978.
4 v. : ill. ; 30 cm. GFR***
Cover title.
Caption title: VEP-Verkehrsentwicklungsplanung, Berlin.
Includes bibliographical references.
1. Urban transporation—Berlin. 2. Transportation planning—Berlin. 3. Berlin—City planning. 4. City planning—Germany, West. I. Title. II. Title: VEP-Verkehrsentwicklungsplanung, Berlin.
HE311.G32B474 711'.73'09431554 79-343538
79 MARC

Deutsches Institut für Urbanistik.
Baufreiheit und Stadtentwicklung : Bericht über zwei Seminare des Deutschen Instituts für Urbanistik vom 26.-30. 9. 1977 und 24.-28. 4. 1978 in Berlin / ⌊redaktionelle Bearbeitung, Rudolf Schäfer⌋. — Berlin : Deutsches Institut für Urbanistik, 1978.
262 p. ; 30 cm. GFR***
On cover: Tagungsbericht.
Includes bibliographical references.
ISBN 3-88118-050-8
1. City planning—Germany, West. 2. City planning—Law and legislation—Germany, West. I. Schäfer, Rudolf, Assessor Dipl.-Pol. II. Title.
HT169.G3D49 1978 79-362073
79 MARC

(Doitsu no mura to machi)
ドイツの村と街 = Germany / 企画・撮影 二川幸夫 ; 文・解説 ルウドウイッグ・グレイサー ; 作図 ウェイン・藤井 ; デザイン 細谷巖. — 東京 : A. D. A. EDITA Tokyo Co., 1974.
133 p. : chiefly ill. (some col.) ; 30 cm. — (世界の村と街 ; 7) Ja 75-22720
Japanese.
Y 3500
1. Architecture—Germany, West. 2. City planning—Germany, West. I. Futagawa, Yukio, 1932- II. Glaeser, Ludwig. III. Title: Germany. IV. Series: Sekai no mura to machi ; 7.
NA1061.D64 79-801493

Fischer, Manfred Frithjof.
Fritz Schumacher, das Hamburger Stadtbild und die Denkmalpflege / Manfred F. Fischer. — Hamburg : H. Christians, 1977.
79 p. : ill., maps ; 30 cm. — (Arbeitshefte zur Denkmalpflege in Hamburg ; Nr. 4) GFR***
Bibliography: p. 69.
ISBN 3-7672-0521-1
1. Schumacher, Fritz, 1869-1947. 2. City planners—Germany, West—Biography. 3. City planning—Germany, West. 4. Hamburg—City planning. 5. Architecture—Germany, West. I. Title.
NA9085.S4F57 79-352429
79 MARC

Jobert, Bruno.
La planification urbaine : l'exemple de Stuttgart / Bruno Jobert. — ⌊Grenoble⌋ : Presses universitaires de Grenoble, ⌊1977?⌋
143 p. ; 21 cm. — (Collection État et société) F***
Includes bibliographical references.
35.00F
1. Urban renewal—Germany, West—Stuttgart. 2. Stuttgart—City planning. 3. City planning—Germany, West. I. Title.
HT178.G42S855 309.2'62'0943 78-391702
79 MARC

Kienzler, Herbert.
Der Alte Flecken Freudenberg : zsfassende. Darst. d. Unters. d. Kernes d. Stadt Freudenberg im Kreis Siegen nach d. Gesichtspunkten d. Denkmalpflege unter Betrachtung d. Einzelobjekte u.d. städtebaul. Leistung / Herbert Kienzler ; erarb. in Zusammenwirken mit d. Kultusminister von Nordrhein-Westfalen u.d. Landeskonservator von Westfalen-Lippe ; Hrsg., d. Kultusminister von Nordrhein-Westfalen in Verbindung mit d. Stadt Freudenberg. — 1. Aufl. — Hagen : v. d. Linnepe, 1978.
276 p. : numerous ill., maps (some col.) ; 31 cm. GFR78-A
Bibliography: p. 275-276.
ISBN 3-921297-21-4 : DM39.00
1. Architecture—Germany, West—Freudenberg (North Rhine-Westphalia) 2. Freudenberg, Ger. (North Rhine-Westphalia)—City planning. 3. City planning—Germany, West. I. Title.
NA9053.C6K53 79-356235

Krauss, Heinz Ulrich.
Stadtplanung und Politik : e. krit. Auseinandersetzung / Heinz Ulrich Krauss. — Stuttgart : Hatje, 1977.
246 p. ; 23 cm. GFR78-A11
Bibliography: p. 236-246.
ISBN 3-7757-0115-X
1. City planning—Germany, West. 2. Public administration—Decision making. I. Title.
HT169.G3K73 78-376080
78 MARC

Laage, Gerhart.
Planung, ein Märchen von einem, der auszog, das Fürchten zu lernen : 9 Gruselgeschichten, aus e. Arbeitstag d. Architekten Bähnelz : "nach d. Laben" aufgeschrieben, für diejenigen, d. gerne unheiml. Geschichten lesen...um sich eventuell nicht mehr fürchten zu müssen / von Gerhart Laage. — ⌊s.l. : s.n., 1975?⌋
146 p. : ill. ; 21 cm. GFR***
Erratum slip inserted.
1. City planning—Germany, West. I. Title.
HT169.G3L2 78-380457
79 MARC

Der Ludwigsplatz Karlsruhe : Forschung, Planung, Ausführung / Kunsthistor. Seminar an d. Staatl. Akad. d. Bildenden Künste Karlsruhe ; Leitung, Siegfried Wichmann in Verbindung mit d. Engelhorn-Stiftung ; Photogr., Thilo Mechau. — Karlsruhe : Kunsthistor. Seminar an d. Staatl. Akad. d. Bildenden Künste, ⌊1978?⌋
248 p. : chiefly ill. (some col.), maps ; 31 cm. GFR78-B15
Bibliography: p. 246-247.
Includes index.
1. Karlsruhe—Plazas—Ludwigsplatz. 2. Architecture—Germany, West—Karlsruhe—Conservation and restoration. 3. Karlsruhe—Buildings. 4. Karlsruhe—City planning. 5. City planning—Germany, West. I. Wichmann, Siegfried, 1921- . II. Staatliche Akademie der Bildenden Künste, Karlsruhe. Kunsthistorisches Seminar.
NA9072.K37L84 79-378117
79 MARC

Mischke, Manfred.
Faktorenökologische Untersuchung zur räumlichen Ausprägung der Sozialstruktur in Pforzheim / von Manfred Mischke. — Karlsruhe : Geograph. Inst., Univ. Karlsruhe, 1976.
x, 171 p., ⌊11⌋ leaves of fold. plates : graphs, maps (numerous fold.) ; 30 cm. — (Karlsruher Manuskripte zur mathematischen und theoretischen Wirtschafts- und Sozialgeographie ; Heft Nr. 15) GFR76-B19
Thesis—Universität Fridericiana Karlsruhe.
Vita (inserted leaf mounted on p. ⌊3⌋ of cover)
Bibliography: p. 165-171.
1. Pforzheim, Ger.—Social conditions. 2. Social surveys—Germany, West—Pforzheim. 3. City planning—Germany, West. 4. Anthropo-geography—Germany, West. I. Title. II. Series.
HN458.P46M57 78-384322
78 MARC

Modelle zur zukünftigen Siedlungs- und Regionalentwicklung : Endbericht / Cord Culemann ... ⌊et al.⌋, Mitverf., Volker Baehr, Ilse Lange. — 1. Aufl. — Baden-Baden : Nomos-Verlagsgesellschaft, 1977.
270 p. : ill. ; 23 cm. — (Planen ; Bd. 10) GFR77-A32
Includes bibliographical references.
ISBN 3-7890-0281-X : DM24.00
1. City planning—Germany, West. 2. Regional planning—Germany, West. 3. Housing—Germany, West. I. Culemann, Cord.
HT169.G3M6 78-396940
79 MARC

Naschold, Frieder.
Alternative Raumpolitik : e. Beitr. zur Verbesserung d. Arbeits- u. Lebensverhältnisse / Frieder Naschold. — 1. Aufl. — Kronberg : Athenäum-Verlag, 1978.
98 p. ; 22 cm. — (Sozialwissenschaft und Praxis ; Bd. 2) GFR78-A
Bibliography: p. 91-98.
ISBN 3-7610-8210-X : DM19.80. ISBN 3-7610-8217-7 pbk.
1. Regional planning—Germany, West. 2. City planning—Germany, West. 3. Germany, West—Social policy. I. Title. II. Series.
HT395.G4N37 309.2'5'0943 78-388912
79 MARC

Ohl, Herbert.
Stadtbausystem mit industriell gefertigten vierseitig geschlossenen Raumzellen-Segmenten, Beispiel "Saarlouis-Beaumarais" : Forschungsauftrag BMBau St II 4-704102-76 (1974) / im Auftr. d. Bundesmin. f. Raumordnung, Bauwesen u. Städtebau, Bonn-Bad Godesberg ; Forschungsbeauftragter, Herbert Ohl. — Bonn-Bad Godesberg : Bundesmin. f. Raumordnung, Bauwesen u. Städtebau, 1975.
293 p. : chiefly ill. ; 30 cm. — (Schriftenreihe Städtebauliche Forschung des Bundesministers für Raumordnung, Bauwesen und Städtebau ; Heft Nr. 34) GFR75-B20
1. Saarlouis—City planning. 2. City planning—Germany, West. 3. Modular coordination (Architecture) 4. Modular coordination industry. I. Title: Stadtbausystem mit industriell gefertigten vierseitig geschlossenen Raumzellen-Segmenten ... II. Series: Schriftenreihe Städtebauliche Forschung; Heft Nr. 34.
NA9200.S23O37 79-361760
79 MARC

Renovierung, Sanierung = Renovation and rehabilitation. — Stuttgart : K. Krämer, 1978.
iv, 104 p. : chiefly ill. ; 30 cm. — (Architektur Wettbewerbe ; 94 ISSN 0341-2784) GFR***
Cover title.
ISBN 3-7828-0194-6 : DM23.50
1. Architecture—Germany, West—Conservation and restoration. 2. City planning—Germany, West. 3. Architecture—Designs and plans. 4. Buildings—Germany, West—Remodeling for other use. I. Title: Renovation and rehabilitation.
NA9053.C6R46 720'.28 78-392161
79 MARC

Scholl, Robert.
Urban transformation : special survey / by Robert Scholl ; ed.-in-chief, Helmut Göbel ; responsible for contents, Rolf Gerlach. — Bonn-Bad Godesberg : Inter Nationes, ⌊1976⌋
19 leaves ; 30 cm. — (Sonderdienst ; SO 1-76(e)) GFR***
Cover title.
1. Cities and towns—Germany, West. 2. City planning—Germany, West. 3. Urbanization. I. Goebel, Helmut, 1921- . II. Gerlach, Rolf. III. Title. IV. Series: Inter Nationes. Sonderdienst ; SO 1-76(e)
HT137.S353 309.2'62'0943 78-380044
78 MARC

Stadt im Fluss : städtebaul. Entwicklung am Beispiel Hamburg / ⌊Hrsg. Baubehörde Hamburg in Zsarbeit mit d. Staatl. Pressestelle Hamburg ; Gesamtred., Klaus Müller-Ibold, Bernt Möller, Detmar Müller-Landré ; Fotos, Angel Medina⌋. — Hamburg : Christians, 1978.
91 p. : numerous ill. (some col.), maps ; 30 cm. GFR78-A48
Cover title.
ISBN 3-7672-0596-3 : DM24.00
1. Hamburg—City planning. 2. Urban renewal—Germany, West—Hamburg. 3. City planning—Germany, West. I. Müller-Ibold, Klaus. II. Möller, Bernt. III. Müller-Landré, Detmar. IV. Hamburg. Baubehörde. V. Hamburg. Staatliche Pressestelle.
HT169.G72H288 309.2'62'0943515 79-345751
79 MARC

Sulzer, Jürg, 1942-
Stadtentwicklung, Koordination von Raum- und Investitionsplanung : Analyse von 5 Beispielen in d. Bundesrepublik Deutschland / Jürg Sulzer. — Frankfurt am Main ; Bern ; Las Vegas : Lang, 1979.
x, 224 p. : ill., maps ; 21 cm. — (Beiträge zur kommunalen und regionalen Planung ; 2) GFR79-A
Bibliography: p. 214-224.
ISBN 3-261-02707-X
1. City planning—Germany, West. 2. City planning—Germany, West—Case studies. 3. Investments—Germany, West. I. Title. II. Series.
HT169.G3S9 79-365436
79 MARC

Trebbi, Giorgio, 1926-
La ricostruzione di una città : Berlino 1945-1975 / Giorgio Trebbi. — Milano : G. Mazzotta, 1978.
197 p. : ill. ; 24 cm. — (Planning & design ; 24) It78-Oct
Includes bibliographical references.
ISBN 8820202182 : L10000
1. Berlin—City planning. 2. City planning—Germany, West. I. Title.
HT169.G32B444 78-394902
*79 MARC

Trieb, Michael.
Stadtgestaltungspolitik : Aufgaben, Instrumente, Strategien / Michael Trieb, Ursula Grammel, Alexander Schmidt, unter Mitarb. von Walter Kiess ... ⌊et al.⌋. — Stuttgart : Deutsche Verlags-Anstalt, 1979.
223 p. : ill. ; 22 cm. GFR79-A
Includes bibliogrpahical references.
ISBN 3-421-02534-7 : DM48.00
1. City planning—Germany, West. 2. City planning—Germany, West—Case studies. 3. City planning—Germany—History. I. Grammel, Ursula, joint author. II. Schmidt, Alexander, Dipl.-Ing., joint author. III. Title.
HT169.G3T75 79-378061
79 MARC

—GERMANY, WEST—ADDRESSES, ESSAYS, LECTURES

Wehling, Hans-Werner, 1949-
Stadtsanierung : Verfahren, Probleme, Zielkonflikte / von Hans-Werner Wehling. — ⌊Frankfurt, Main⌋ : Hirschgraben-Verlag, 1977.
48 p. : ill., maps ; 23 cm. — (Geographische Zeitfragen in Einzelheften ; 20) GFR77-A
Cover title.
Bibliography: p. 48.
ISBN 3-454-11200-1 : DM4.00
1. Architecture—Germany, West—Conservation and restoration—Addresses, essays, lectures. 2. City planning—Germany, West—Addresses, essays, lectures. 3. Regional planning—Germany, West—Addresses, essays, lectures. I. Title.
NA9053.C6W4 78-378497
78 MARC

—GERMANY, WEST—BIBLIOGRAPHY

Ege, Reinhold.
Bibliographie zur Stadtplanung einführende Literatur / Bearbeiter, Reinhold Ege, Dieter Hezel. — Stuttgart : Städtebauliches Institut im Fachbereich Orts, Regional- und Landesplanung, 1978.
102 leaves ; 29 cm. — (Arbeitsbericht - Städtebauliches Institut, Universität Stuttgart ; 30) GFR***
Project organized by the Abteilung Dokumentation, Städtebauliches Institut, Universität Stuttgart.
1. City planning—Germany, West—Bibliography. I. Hezel, Dieter, joint author. II. Stuttgart. Universität. Städtebauliches Institut. Abteilung Dokumentation. III. Title. IV. Series: Stuttgart. Universität. Städtebauliches Institut. Arbeitsbericht - Städtebauliches Institut, Universität Stuttgart ; 30.
Z5942.E34 79-378424
⌊HT169.G3⌋ 79 MARC

CITY PLANNING (Continued)

—GERMANY, WEST—CASE STUDIES

Böhm, Hans Reiner.
Die Integration der Entwässerungsplanung in die Flächennutzungsplanung : dargest. an e. Flächengemeinde / Hans Reiner Böhm. — 1. Aufl. — Frankfurt/Main : Haag und Herchen, 1978.
235 p. : ill. ; 21 cm. GFR78-A
Originally presented as the author's thesis, Bonn.
Bibliography: p. 224-232.
ISBN 3-88129-136-9 : DM48.00
1. Drainage—Germany, West—Meerbusch. 2. Meerbusch, Ger.—City planning. 3. City planning—Germany, West—Case studies. I. Title.
HD1683.G3B63 1978 711'.4'094355 79-359653
79 MARC

Gross, Gerd.
Bürgernahe Stadtentwicklungsplanung gescheitert?! / Gerd Gross. — 1. Aufl. — Berlin : Verlag für Ausbildung u. Studium, (VAS), 1978.
207 p. : ill. ; 21 cm. GFR79-A
Bibliography: p. 203-206.
ISBN 3-88290-100-4 : DM14.80
1. Munich—City planning—Citizen participation—History. 2. City planning—Germany, West—Case studies. I. Title.
HT169.G32M94 309.2'62'094336 79-359687
79 MARC

Krause, Hans-Joachim, Dozent.
Kinder in der inneren Stadt : auf Strassen und Plätzen : zum Einfluss der sozialen und städtebaulichen Umwelt auf die Lebensverhältnisse und Sozialisationsbedingungen von Kindern in innerstädtischen Gebieten / Auftraggeber, Freie und Hansestadt Hamburg, Baubehörde Landesplanungsamt ; Gutachter, Hans-Joachim Krause, Timm Ohrt, Hille v. Seggern. — Hamburg : Freie und Hansestadt Hamburg, Baubehörde Landesplanungsamt, 1977.
2 v. : ill. (some col.) ; 30 cm. GFR***
Includes bibliographical references.
CONTENTS: T. 1. Interpretation der Daten aus pädagogischer Sicht unter Berücksichtigung städtebaulicher Kategorien.—T. 2. Interpretation der Daten aus städtebaulicher Sicht unter Berücksichtigung pädagogischer Kategorien.
1. Hamburg—City planning. 2. Children—Germany, West—Hamburg—Social conditions. (3) City planning—Germany, West—Case studies. I. Ohrt, Timm, joint author. II. Seggern, Hille von, joint author. III. Title. IV. Title: Auf Strassen und Plätzen.
HT169.G32H285 309.2'62'0943515 79-351961
79 MARC

Stuttgart. Universität. Städtebauliches Institut.
Vorbereitende Untersuchungen in der Stadterneuerungsplanung : Forschungsauftrag BMBau RS II 6 B - 704102-174 (1975) / im Auftr. d. Bundesministers für Raumordnung, Bauwesen u. Städtebau, Bonn-Bad Godesberg ; vorgelegt vom Städtebaul. Inst. d. Univ. Stuttgart ; Bearb., Gerd Baldauf ... [et al.]. — Bonn-Bad Godesberg : Bundesminister für Raumordnung, Bauwesen u. Städtebau, 1976.
125 p. : ill., graphs, maps (some col.) ; 30 cm. — (Schriftenreihe Stadtentwicklung ; 7) GFR77-B12
Bibliography: p. 123-124.
1. Stuttgart—City planning. 2. City planning—Germany, West—Case studies. 3. Urban renewal—Germany, West—Zuffenhausen. I. Baldauf, Gerd. II. Title. III. Series.
HT169.G32S87 79-377574
79 MARC

Sulzer, Jürg, 1942-
Stadtentwicklung, Koordination von Raum- und Investitionsplanung : Analyse von 5 Beispielen in d. Bundesrepublik Deutschland / Jürg Sulzer. — Frankfurt am Main ; Bern ; Las Vegas : Lang, 1979.
x, 224 p. : ill., maps ; 21 cm. — (Beiträge zur kommunalen und regionalen Planung ; 2) GFR79-A
Bibliography: p. 214-224.
ISBN 3-261-02707-X
1. City planning—Germany, West. 2. City planning—Germany, West—Case studies. 3. Investments—Germany, West. I. Title. II. Series.
HT169.G3S9 79-365436
79 MARC

Trieb, Michael.
Stadtgestaltungspolitik : Aufgaben, Instrumente, Strategien / Michael Trieb, Ursula Grammel, Alexander Schmidt, unter Mitarb. von Walter Kiess ... [et al.]. — Stuttgart : Deutsche Verlags-Anstalt, 1979.
223 p. : ill. ; 22 cm. GFR79-A
Includes bibiogrpahical references.
ISBN 3-421-02534-7 : DM48.00
1. City planning—Germany, West. 2. City planning—Germany, West—Case studies. 3. City planning—Germany—History. I. Grammel, Ursula, joint author. II. Schmidt, Alexander, Dipl.-Ing., joint author. III. Title.
HT169.G3T75 79-378061
79 MARC

Vogel, Helmut.
Das Einkaufszentrum als Ausdruck einer kulturlandschaftlichen Innovation, dargestellt am Beispiel des Böblinger Regionalzentrums / Helmut Vogel. — Trier : Zentralausschuss für deutsche Landeskunde, Selbstverlag, 1978.
214 p., [1] leaf of plates : ill. ; 24 cm. — (Forschungen zue deutschen Landeskunde ; Bd. 209) GFR***
Illustrative matter in pocket.
Summary in English.
Bibliography: p. 201-214.
ISBN 3-88143-002-4
1. Shopping centers—Germany, West—Böblingen. 2. Shopping centers—Germany, West—Stuttgart metropolitan area. 3. City planning—Germany, West—Case studies. I. Title. II. Series.
G58.F73 Bd. 209 79-367106
[HF5430.6.G4] 79 MARC

—GERMANY, WEST—CITIZEN PARTICIPATION

Borghorst, Hermann, 1947-
Bürgerbeteiligung in der Kommunal- und Regionalplanung : e. krit. Problem- u. Literaturanalyse / Hermann Borghorst. — Leverkusen : Heggen, 1976.
178 p. ; 18 cm. GFR77-A
Bibliography: p. 163-177.
ISBN 3-920430-70-0 : DM12.00
1. City planning—Germany, West—Citizen participation. 2. Regional planning—Germany, West—Citizen participation. I. Title.
HT169.G3B67 309.2'12'0943 79-339937
79 MARC

Lösche, Wolfgang.
Thema Stadtzerstörung : einiges über d. zeitgenöss. Zerstörung von Lebensbereichen mit d. Fallbeispiel Bielefeld / Wolfgang Lösche. — Bielefeld : AJZ-Druck und -Verlag, 1977.
172 p. : ill. ; 18 x 21 cm. GFR78-A19
Bibliography: p. 170.
ISBN 3-921680-07-7
1. City planning—Germany, West—Citizen participation. 2. Bielefeld—City planning—Case studies. 3. Environmental policy—Germany, West. I. Title.
HT169.G3L6 309.2'62'094355 79-346441
79 MARC

Wohnungsbaugenossenschaft Urbanes Wohnen Köln.
Bürgerinitiative bie der Planung von Wohnung, Wohnumwelt und Stadt : dargestellt am Modellfall Genossenschaft Urbanes Wohnen Köln : Teilprojekt, Gruppendynamische Phänomene bei der Mitarbeit von Bürgern an Planungsprozessen und ihre Auswirkungen auf die bauliche und städtebauliche Realisierung : Forschungsauftrag BMBau RS II 6 B-70 41 02-99 (1972) / im Auftrag des Bundesministers für Raumordnung, Bauwesen und Städtebau, Bonn-Bad Godesberg, Forschungsbeauftrage, Bürgerinitiative und Wohnungsbaugesnossenschaft Urbanes Wohnen Köln GmbH. — Bonn-Bad Godesberg : Bundesminister für Raumordnung, Bauwesen und Städtebau, 1975.
119 p. : ill., maps ; 30 cm. — (Schriftenreihe städtebauliche Forschung des Bundesministers für Raumordnung, Bauwesen und Städtebau ; 03.039) GFR***
Bibliography: p. 109-114.
Includes indexes.
1. Cologne—City planning—Citizen participation. 2. City planning—Germany, West—Citizen participation. I. Title. II. Series: Schriftenreihe städtebauliche Forschung ; 03.039.
NA9200.C6W63 1975 79-359198
79 MARC

—GERMANY, WEST—CONGRESSES

Deutscher Verband für Wohnungswesen, Städtebau und Raumplanung.
Stadtumbau, Aufgaben, Instrumente, Verfahren : Referate d. Jahrestagung d. Dt. Verbandes f. Wohnungswesen, Städtebau u. Raumplanung am 20. Juni 1975 in Düsseldorf / [Dt. Verband f. Wohnungswesen, Städtebau u. Raumplanung mit Inst. f. Wohnungs- u. Planungswesen, Dokumentationsstelle]. — Köln : Dt. Verband f. Wohnungswesen, Städtebau u. Raumplanung, 1975.
51 p. : ill., maps ; 29 cm. GFR75-B22
1. City planning—Germany, West—Congresses. I. Title.
HT169.G3D48 1975 78-377569
78 MARC

Kunstkongress, 5th, Göttingen, 1975.
Die Geschichtlichkeit des Menschen und der Stadt / 5. Kunstkongress Göttingen, 20.-22. Juni 1975 ; Veranst., Stadt Göttingen, Kulturverwaltung, in Verbindung mit d. Vereinigung d. Landesdenkmalpfleger in d. Bundesrepublik Deutschland ; Hrsg., Stadt Göttingen ; [Red., Konrad Schilling, Gerhard Steidl, Lucy Hillebrand]. — 1. Aufl. — Göttingen : Stadt Göttingen, 1975.
380 p. : ill., maps ; 30 cm. — (Die Kunst, eine Stadt zu bauen ; 3) GFR77-B20
"Vertrieb: Druckerei und Verlag Steidl, 34 Göttingen."
Includes bibliographical references.
1. Architecture—Germany, West—Conservation and restoration—Congresses. (2) City planning—Germany, West—Congresses. 3. Monuments—Germany, West—Preservation—Congresses. I. Schilling, Konrad. II. Steidl, Gerhard. III. Hillebrand, Lucy. IV. Göttingen. Kulturverwaltung. V. Vereinigung der Landesdenkmalpfleger in der Bundesrepublik Deutschland. VI. Göttingen. VII. Title. VIII. Series.
NA9053.C6K87 1975 78-394065
79[r79]rev MARC

—GERMANY, WEST—DATA PROCESSING

PROSAB : e. computergestütztes System zum Entwurf u. zur Bewertung in d. kommunalen Bauleit- u. Entwicklungsplanung : Grundversion / hrsg. vom Inst. für Landes- u. Stadtentwicklungsforschung d. Landes Nordrhein-Westfalen (ILS) ; im Auftr. d. Innenministers NW. Dieter Schindowski ... [et al.] Inst. für Raumplanung Univ. Dortmund (IRPUD). — Essen : Verlag für Wirtschaft u. Verwaltung Wingen [in Komm.], 1976.
2 v. : ill. ; 30 cm. — (Schriftenreihe Landes- und Stadtentwicklungsforschung des Landes Nordrhein-Westfalen : Reihe 2, Stadtentwicklung, Städtebau ; Bd. 11) GFR77-A33
Includes bibliographical references.
CONTENTS: Bd. 1. Wissenschaftliche Grundlagen und Systembeschreibung.—Bd. 2. Materialien.
1. City planning—Germany, West—Data processing. 2. PROSAB (Computer program) I. Schindowski, Dieter. II. Institut für Landes- und Stadtentwicklungsforschung des Landes Nordrhein-Westfalen. III. Universität Dortmund. Institut für Raumplanung. IV. Series.
HT169.G3P17 79-376926
79 MARC

Zentrum Berlin für Zukunftsforschung.
Berliner Simulationsmodell : (Besi) / Zentrum Berlin für Zukunftsforschung e.V. ; bearb. von Per Bock-Nussbaum ... [et al.]. — Berlin : Zentrum Berlin für Zukunftsforschung e.V., 1969-
v. : ill. ; 30 cm. — (ZBZ-Bericht ;) GFR76-B16
Bibliography: v. 2, p. 102-103.
DM40.00 (v. 2)
1. City planning—Germany, West—Simulation methods. 2. City planning—Germany, West—Data processing. 3. City planning—Germany, West—Mathematical models. 4. BESI (Computer program) I. Bock-Nussbaum, Per. II. Title. III. Series: Zentrum Berlin für Zukunftsforschung. ZBZ-Bericht ; 6.
HT169.G3Z46 1969 78-393239
79 MARC

—GERMANY, WEST—DIRECTORIES

Germany (Federal Republic, 1949-). Bundesministerium für Raumordnung, Bauwesen und Städtebau.
Städtebaupolitik und Denkmalschutz : Bericht über Programme u. Massnahmen d. Bundes zur Stadterhaltung / [d. Bundesminister für Raumordnung, Bauwesen u. Städtebau]. — Bonn-Bad Godesberg : Bundesminister für Raumordnung, Bauwesen u. Städtebau, 1976.
135 p. : numerous ill. (some col.) : 30 cm. — (Schriftenreihe Stadtentwicklung ; Heft Nr. 6) GFR77-B6
"BMBau-RS III 7-70 20 72-13/1-."
1. Architecture—Germany, West—Conservation and restoration—Directories. 2. City planning—Germany, West—Directories. I. Title. II. Series.
NA9053.C6G47 1976 78-382880
78 MARC

—GERMANY, WEST—MATHEMATICAL MODELS

Zentrum Berlin für Zukunftsforschung.
Berliner Simulationsmodell : (Besi) / Zentrum Berlin für Zukunftsforschung e.V. ; bearb. von Per Bock-Nussbaum ... [et al.]. — Berlin : Zentrum Berlin für Zukunftsforschung e.V., 1969-
v. : ill. ; 30 cm. — (ZBZ-Bericht ;) GFR76-B16
Bibliography: v. 2, p. 102-103.
DM40.00 (v. 2)
1. City planning—Germany, West—Simulation methods. 2. City planning—Germany, West—Data processing. 3. City planning—Germany, West—Mathematical models. 4. BESI (Computer program) I. Bock-Nussbaum, Per. II. Title. III. Series: Zentrum Berlin für Zukunftsforschung. ZBZ-Bericht ; 6.
HT169.G3Z46 1969 78-393239
79 MARC

—GERMANY, WEST—SIMULATION METHODS

Zentrum Berlin für Zukunftsforschung.
Berliner Simulationsmodell : (Besi) / Zentrum Berlin für Zukunftsforschung e.V. ; bearb. von Per Bock-Nussbaum ... [et al.]. — Berlin : Zentrum Berlin für Zukunftsforschung e.V., 1969-
v. : ill. ; 30 cm. — (ZBZ-Bericht ;) GFR76-B16
Bibliography: v. 2, p. 102-103.
DM40.00 (v. 2)
1. City planning—Germany, West—Simulation methods. 2. City planning—Germany, West—Data processing. 3. City planning—Germany, West—Mathematical models. 4. BESI (Computer program) I. Bock-Nussbaum, Per. II. Title. III. Series: Zentrum Berlin für Zukunftsforschung. ZBZ-Bericht ; 6.
HT169.G3Z46 1969 78-393239

CITY PLANNING (Continued)

—GERMANY, WEST—GELSENKIRCHEN

Mäcke, Paul Arthur.
Siedlungsschwerpunkt Gelsenkirchen-Buer : Untersuchung zur Verkehrserschliessung / Arbeitsgemeinschaft, P. A. Mäcke, D. Hölsken ; erstellt im Auftrage des Siedlungsverbandes Ruhrkohlenbezirk. — Essen : SVR, 1975.
63 p. : ill. ; 30 cm. GFR***
Includes bibliographical references.
1. Urban transportation—Germany, West—Gelsenkirchen. 2. City planning—Germany, West—Gelsenkirchen. 3. Transportation and state—Germany, West—Gelsenkirchen. I. Hölsken, Dieter, joint author. II. Title.
HE311.G32G45 78-373986
78 MARC

—GERMANY, WEST—HESSE

Landzettel, Wilhelm.
Gestaltaspekte / Wilhelm Landzettel, Wolfram Goldapp, Hans Haas. — Wiesbaden : AVA, Arbeitsgemeinschaft z. Verbesserung d. Agrarstruktur in Hessen e.V., 1975.
92 p. : numerous ill., maps ; 30 cm. — (Beiträge zum ländlichen Bau- und Siedlungswesen ; Bericht 16) (Sonderheft - AVA, Arbeitsgemeinschaft zur Verbesserung der Agrarstruktur in Hessen e.V. ; 54) GFR75-B18
DM5.00
1. City planning. 2. Regional planning. 3. Architecture—Composition, proportion, etc. 4. City planning—Germany, West—Hesse. I. Goldapp, Wolfram, joint author. II. Haas, Hans, Bildhauer. III. Title. IV. Series. V. Series: Arbeitsgemeinschaft zur Verbesserung der Agrarstruktur in Hessen. Sonderheft ; 54.
HD1960.H5A7 Nr. 54 78-400000
[NA9035] 79 MARC

—GERMANY, WEST—HESSE—CASE STUDIES

Holz, Helga.
Analyse der Institutionen im öffentlichen Bereich der inneren Bauverwaltung für das Gebiet der Stadtplanung unter der besonderen Fragestellung nach der Abgrenzung von Bau- und Stadtplanung / Helga Holz, Lilo Kagelmann, Gesine Ohrt. — Kassel : Arbeitsgruppe Modellversuch Entwicklung von Integrierten Studiengängen für Architekten, Planer beim Gründungspräsidenten d. Gesamthochschule Kassel, 1976.
182 p. : ill., maps ; 30 cm. — (Veröffentlichungen der Arbeitsgruppe des Modellversuchs ; Nr. 12) GFR77-B4
Bibliography: p. 171.
1. City planning—Germany, West—Hesse—Case studies. 2. Architecture and society—Germany, West—Hesse—Case studies. 3. Hesse—Administrative and political divisions. 4. Regional planning—Research—Germany, West—Hesse. I. Kagelmann, Lilo, joint author. II. Ohrt, Gesine, joint author. III. Title. IV. Series: Gesamthochschule Kassel. Arbeitsgruppe Modellversuch Entwicklung von Integrierten Studiengängen für Architekten, Planer. Veröffentlichungen der Arbeitsgruppe des Modellversuchs ; Nr. 12.
HT169.G32H465 78-379772
78 MARC

—GERMANY, WEST—NORTH RHINE-WESTPHALIA

Probleme der Stadtgestaltung : Veröff. e. Vortragsreihe d. Landesgruppe Nordrhein-Westfalen d. Dt. Akad. für Städtebau u. Landesplanung / Aloys Machtemes ... [et al.] ; hrsg. von Inst. für Landes- u. Stadtentwicklungsforschung d. Landes Nordrhein-Westfalen (Ils) im Auftr. d. Innenministers NW. — Essen : Verlag für Wirtschaft u. Verwaltung Wingen [in Komm.], 1977.
118 p. : ill., plans ; 30 cm. — (Schriftenreihe Landes- und Stadtentwicklungsforschung des Landes Nordrhein-Westfalen : Reihe 2, Stadtentwicklung, Städtebau ; Bd. 28) GFR78-A
Bibliography: p. 98.
1. City planning—Germany, West—North Rhine-Westphalia. 2. Urbanization—Germany, West—North Rhine—Westphalia. I. Machtemes, A. II. Deutsche Akademie für Städtebau und Landesplanung. Landesgruppe Nordrhein-Westfalen. III. Institut für Landes- und Stadtentwicklungsforschung des Landes Nordrhein-Westfalen. IV. Series.
HT137.P76 79-377887
79 MARC

—GERMANY, WEST —NORTH RHINE-WESTPHALIA —CITIZEN PARTICIPATION

Städtebauliche Planung : Mitwirkung des Bürgers / [Herausgeber, Der Innenminister des Landes Nordrhein-Westfalen]. — Düsseldorf : Der Innenminister, 1978.
67 p. : ill., maps (some col.) ; 21 cm. — (Zur Information ; 16) GFR***
Cover title.
1. City planning—North Rhine-Westphalia—Citizen participation. I. North Rhine-Westphalia. Innenministerium. II. Series.
HT169.G32N6727 79-360562
79 MARC

—GHANA

Keta study. — Kumasi : University of Science and Technology, Faculty of Architecture, [1971?]
613 p. : ill. ; 31 cm. — (Occasional report - University of Science and Technology, Faculty of Architecture ; no. 15)
Bibliography: p. 161-163.
1. City planning—Ghana. 2. Keta, Ghana—City planning. I. Series: University of Science and Technology, Kumasi. Faculty of Architecture. Occasional report ; no. 15.
HT169.G52K474 309.2'62'09667 78-318109
78 MARC

—GREAT BRITAIN

Chapman, Sydney.
Town and countryside : future planning policies for Britain / [by] Sydney Chapman. — London : Conservative Political Centre, 1978.
38 p. ; 21 cm. — (CPC [publications] ; no. 619) GB78-17583
Includes bibliographical references.
ISBN 0-85070-613-0 : £0.60
1. Regional planning—Great Britain. 2. City planning—Great Britain. 3. Environmental policy—Great Britain. 4. Land use—Planning—Great Britain. I. Title. II. Series: Conservative Political Centre, London. Publications ; no. 619.
HT395.G7C53. 309.2'12'0941 78-319219
79 MARC

Great Britain. Environmental Board.
Environmental Board : first progress report to the Secretary of the State for the Environment. — London : H.M. Stationery Off., 1977.
10 p. ; 22 cm. GB***
At head of title: Department of the Environment.
ISBN 0-11-751226-5 : £0.45
1. City planning—Great Britain. 2. Land use—Great Britain. 3. Environmental policy—Great Britain. I. Great Britain. Dept. of the Environment. II. Title.
HT169.G7G692 1977 301.31'0941 78-322115
79 MARC

McKay, David H
The politics of urban change / David H. McKay and Andrew W. Cox. — London : Croom Helm, c1979.
297 p. : maps ; 23 cm. GB***
Bibliography: p. 287-290.
Includes index.
ISBN 0-85664-436-6 : £11.50
1. City planning—Great Britain. 2. Urban policy—Great Britain. I. Cox, Andrew W., joint author. II. Title.
HT169.G7M247 309.2'62'0941 79-309024
79 MARC

McLoughlin, Pete.
Regional policy and the inner areas : a study of planners' attitudes / [by] Pete McLoughlin. — Reading : Department of Geography, University of Reading, 1978.
[2], iii, 31 p. : ill., map ; 21 cm. — (Geographical papers ; no. 64 ISSN 0305-5914) GB78-31964
Bibliography: p. 26.
ISBN 0-7049-0474-8 : £0.65
1. City planning—Great Britain. 2. Urban renewal—Great Britain. 3. City planners—Great Britain—Attitudes. I. Title. II. Series.
HT169.G7M27 309.2'62 79-314488
79 MARC

—GREAT BRITAIN—BIBLIOGRAPHY

Development control : the Dobry Report. — Reading : Centre for Advanced Land Use Studies, 1975.
[5], 39 p. ; 21 cm. — (Occasional papers in estate management ; no. 8) GB78-05350
"These papers were originally given at a CALUS conference held in March, 1975."
ISBN 0-902132-20-2 : £1.00
1. City planning—Great Britain. 2. Regional planning—Great Britain. 3. Dobry, George. Review of the development control system, final report. I. Centre for Advanced Land Use Studies. II. Series.
HT169.G7D48 309.2'12'0941 78-314135
78 MARC

Great Britain. Dept. of the Environment. Headquarters Library.
Town and country planning : a select list of materials / compiled by Yael Musleah : [for the Headquarters Library, Department of the Environment]. — Revised [ed.]. — [London] : [The Library], 1974.
iii, 34 p. ; 30 cm. — (Bibliography series ; no. 70) GB76-18670
ISBN 0-903197-84-7
1. City planning—Great Britain—Bibliography. 2. Regional planning—Great Britain—Bibliography. I. Musleah, Yael. II. Title. III. Series: Bibliography series (London) ; no. 70.
Z5942.G7 1974 016.3092'12'0941 78-324054
[HT169.G7] 79 MARC

—GREAT BRITAIN—CONGRESSES

International Congress for Housing and Town Planning, 18th, Hastings, 1946.
Implementing the plans. [Westminster? London, 1946?]
2 pamphlets (in portfolio) 23 cm. (Its Congress reports)
Title from portfolio.
English, with summaries in French.
CONTENTS: [1] Great Britain.—[2] Norway.
1. City planning—Great Britain—Congresses. 2. City planning—Norway—Congresses. I. Title.
NA9185.I55 1946a 301.5'4 51-38283
[r79]rev MARC

—GREAT BRITAIN—HISTORY

Calabi, Donatella.
Il male città, diagnosi e terapia : didattica e istituzioni nell'urbanistica inglese del primo '900 / Donatella Calabi. — Roma : Officina, 1979.
437 p., [44] leaves of plates : ill. ; 25 cm. — (Collana di architettura ; 19) It79-Aug
Includes bibliographical references and index.
L11500
1. City planning—Great Britain—History. I. Title.
HT169.G7C33 79-383994
*79 MARC

—GREAT BRITAIN—STATISTICS

Coppock, John Terence.
Land use / by J. T. Coppock and Town and country planning by L. F. Gebbett. — Oxford [Eng.] ; New York : Published for the Royal Statistical Society and the Social Science Research Council by Pergamon Press, 1978.
xii, 219 p. ; 26 cm. — (Reviews of United Kingdom statistical sources ; v. 8) GB78
Includes bibliographical references and index.
ISBN 0-08-022451-2 : £12.50 ($22.50 U.S.)
1. Land use—Great Britain—Statistics. 2. City planning—Great Britain—Statistics. I. Gebbett, L. F. Town and country planning. 1978. II. Title. III. Series.
HD596.C67 1978 333.7'0941 77-30349
77 MARC

—GUINEA-BISSAU

Guinea, Portuguese.
Ordenamento rural e urbano na Guiné Portuguesa / Província da Guiné. — Lisboa : Agência-Geral do Ultramar, 1973.
141 p., 61, [2] fold. leaves of plates : ill. ; 30 cm.
1. Zoning—Guinea-Bissau. 2. City planning—Guinea-Bissau. 3. Regional planning—Guinea-Bissau. 4. Economic zoning—Guinea-Bissau. I. Portugal. Agência Geral do Ultramar. II. Title.
HT169.8.G96G83 1973 79-371256
79 MARC

—ILLINOIS

Stampley, James O 1915-
Challenges with changes : a documentary of Englewood / James O. Stampley. — [Chicago] : Stampley, c1979.
ix, 278 p. : ill. ; 22 cm.
Bibliography: p. 277-278.
ISBN 0-9602618-9-3
1. Englewood, Chicago, Ill.—City planning—History. 2. Chicago—City planning—History. 3. Englewood, Chicago, Ill.—Description. 4. Chicago—Description—1951- 5. Community development, Urban—Illinois. 6. City planning—Illinois. I. Title.
HT168.E53S73 309.2'62'097731 79-117584
00 MARC

—INDIA

Bijlani, H U
Urban problems / H. U. Bijlani. — New Delhi : Centre for Urban Studies, Indian Institute of Public Administration, 1977.
xv, 171 p., [15] leaves of plates : ill., maps ; 25 cm.
Bibliography: p. [155]-163.
Includes index.
Rs70.00 ($20.00 U.S.)
1. Delhi—City planning. 2. City planning—India. 3. Gangtok, Sikkim—City planning. 4. City planning—Sikkim. 5. Underdeveloped areas—City planning. I. Title.
HT169.I52D415 309.2'62 78-900409
79 MARC

CITY PLANNING
—INDIA (Continued)

Quality of life in human settlements : planning issues and space standards. — Mysore : Institute of Development Studies, University of Mysore, 1976.
63 p. : diagrs. ; 26 cm. — (Development studies ; no. 12)
Includes bibliographical references.
CONTENTS: Bhooshan, B. S. A fresh look on urban development.—Sundaram, K. V. Physical planning standards in selected countries in South-East Asia.
Rs10.00
1. City planning—India. 2. City planning—Asia, Southeastern. I. Bhooshan, B. S. A fresh look on urban development. 1976. II. Sundaram, K. V. Physical planning standards in selected countries in South-East Asia. 1976. III. Series: Development studies (Mysore) ; no. 12.
HT169.I5Q34 309.2'62'0959 78-900592
79 MARC

—INDIA—BIBLIOGRAPHY

Van Willigen, John.
The Indian city : a bibliographic guide to the literature on urban India / John van Willigen. — New Haven : Human Relations Area Files, inc., 1979.
2 v. ; 23 cm. — (HRAFLex books ; AW1-001) (Bibliography series - Human Relations Area Files, inc)
1. City planning—India—Bibliography. 2. Urbanization—India—Bibliography. 3. Cities and towns—India—History—Bibliography. 4. Sociology, Urban—India—Bibliography. I. Human Relations Area Files, inc. II. Title. III. Series: Human Relations Area Files, inc. HRAFlex books ; AW1-001. IV. Series: Human Relations Area Files, inc. Bibliography series - Human Relations Area Files, inc.
Z5942.V36 016.30136'0954 79-117894
[HQ766] 79 MARC

—INDIA—CASE STUDIES

Singh, Kamaldeo Narain, 1943-
Urban development in India / Kamaldeo Narain Singh. — 1st ed. — New Delhi : Abhinav Publications, 1978.
xiii, 231 p. ; 23 cm.
Bibliography: p. [213]-227.
Includes index.
Rs55.00 ($11.00 U.S.)
1. City planning—India—Case studies. 2. Gayā, India—City planning—Case studies. 3. Gaya Improvement Trust. I. Title.
HT169.I5S56 354'.54'12008 78-907699
79 MARC

—INDIA—CONGRESSES

Growing cities and priorities : proceedings of the seminar on January 31, 1972, Bombay-1 / sponsored by the All-India Institute of Local Self-Government, Regional Centre for Research and Training in Municipal Administration. — Bombay : The Institute, 1972, [pref. 1973]
92 p. ; 22 cm.
1. City planning—India—Congresses. I. All India Institute of Local Self Government. Regional Centre for Research and Training in Municipal Administration.
HT169.I5G76 309.2'62'0954 78-907847
79 MARC

Seminar on City Development Plans and Their Implementation, New Delhi, India, 1975.
City development plans and their implementation : papers and proceedings of the seminar. — New Delhi : Centre for Urban Studies, Indian Institute of Public Administration, 1977.
xv, 282 p. ; 22 cm.
Organized by the Centre for Urban Studies, Indian Institute of Public Administration.
Includes bibliographical references.
Rs35.00 ($10.00 U.S.)
1. City planning—India—Congresses. I. Indian Institute of Public Administration. Centre for Urban Studies. II. Title.
HT169.I5S43 1975 309.2'62'0954 78-901540
79 MARC

Seminar on the Architect & the Community, New Delhi, India, 1965.
Seminar on the Architect & the Community, New Delhi, February 14 to 20, 1965. New Delhi, India International Centre [1967]
67 p. 23 cm. 12.50 PL480:I-E-9324
Cover title: The architect and the community.
Bibliography: p. 67.
1. Architects and community—Congresses. 2. City planning—India—Congresses. I. India International Centre. II. Title: The architect and the community.
NA9010.S37 1965 711'.4 sa 68-12357
[r79]rev MARC

Seminar on Urban Planning and Development Authorities, New Delhi, India, 1974.
Urban planning and development authorities : papers and proceedings of the seminar. — New Delhi : Centre for Urban Studies, Indian Institute of Public Administration, 1978.
xiv, 223 p. ; 22 cm.
Rs30.00 ($9.00 U.S.)
1. City planning—India—Congresses. I. Title.
HT169.I5S47 1974 309.2'62'0954 79-903573
79 MARC

—INDIA—HISTORY

Begde, Prabhakar V 1945-
Ancient and mediaeval town-planning in India / Prabhakar V. Begde. — 1st ed. — New Delhi : Sagar Publications, 1978.
xiii, 245 p. : ill., maps ; 29 cm.
Bibliography: p. 239-241.
Includes index.
Rs175.00
1. City planning—India—History. I. Title.
HT169.I5B43 309.2'62'0934 78-902598
78 MARC

—INDIA—KERALA (STATE)

Kerala, India (State). Town Planning Dept.
Draft detailed town planning scheme for Palayam area / prepared by the Department of Town Planning, Kerala State. — [Trivandrum] : Trivandrum Town Planning Trust, [1976?]
23, [90] p. ; 32 cm.
Chiefly tables.
Rs9.00
1. Regional planning—India—Kerala (State) 2. City planning—India—Kerala (State) I. Title.
HT395.I52K474 309.2'62'095483 77-907572
79 MARC

—INDONESIA

Bekasi, Indonesia (Kabupaten)
Rencana tata kota Pondokgede. — [Jakarta] : Pemerintah Kabupaten Daerah Tingkat II Bekasi bekerjasama dengan Direktorat Tata Kota dan Tata Daerah, Direktorat Jenderal Cipta Karya, Departemen P.U.T.L., [1978]
2 v. : maps ; 31 x 42 cm.
Cover title.
CONTENTS: 1. Kompilasi data.—2. Analisa dan rencana.
1. Pondokgede, Indonesia—City planning. 2. City planning—Indonesia. 3. Regional planning—Indonesia—Bekasi (Kabupaten) I. Indonesia. Direktorat Tata Kota dan Tata Daerah. II. Title.
HT169.I54P583 78-941955
79 MARC

Bogor, Indonesia.
Master plan Bogor : beberapa catatan dari hasil diskusi, tanggal 4 Desember 1972 s/d 8 Desember 1972 / [Pemerintah Daerah Kotamadya Bogor] — [Bogor : Pemerintah Daerah Kotamadya Bogor, 1973 or 1974]
[80] leaves in various foliations : ill. ; 22 x 34 cm.
1. Bogor, Indonesia—City planning. 2. City planning—Indonesia. I. Title.
HT169.I54B642 1973 74-941114
79 MARC

Bogor, Indonesia.
Rencana tata kota Bogor. — [Bogor] : Pemerintah Daerah Tingkat II Kotamadya Bogor bekerjasama dengan Direktorat Tata Kota dan Tata Daerah, Direktorat Jenderal Cipta Karya, Departemen P.U., [1978]
2 v. : maps ; 31 x 42 cm.
Cover title.
CONTENTS: 1. Kompilasi data.—2. Analisa dan rencana.
1. Bogor, Indonesia—City planning. 2. City planning—Indonesia. I. Indonesia. Direktorat Tata Kota dan Tata Daerah. II. Title.
HT169.I54B643 78-942692
79 MARC

Bogor, Indonesia. Badan Perancang Pembangunan Kota.
Analisa suatu lingkungan pemukiman padat di lingkungan Tanah Sareal, Kec. Kota Bogor Utara. — [Bogor] : Badan Perancang Pembangunan Kota, Kotamadya Bogor, 1976.
i, 39 leaves, [13] leaves of plates : ill., maps ; 29 cm.
"Suatu praktek sebagai kelanjutan Kursus Perencanaan Sosial Pembangunan Kota, kursus diadakan atas kerjasama DTKD, Dirjen. Cipta Karya dengan UNICEF."
1. Kota Bogor Utara, Indonesia—City planning. 2. City planning—Indonesia. I. Title.
HT169.I54K62 78-942216
79 MARC

Bogor, Indonesia (Kabupaten)
Rencana tata kota Parung. — [Jakarta] : Pemerintah Daerah Tingkat II Kabupaten Bogor bekerja sama dengan Direktorat Tata Kota dan Tata Daerah, Direktorat Jenderal Cipta Karya, Departemen P.U., [1977?]
2 v. : maps ; 29 x 42 cm.
Cover title.
CONTENTS: 1. Kompilasi data.—2. Analisa dan rencana.
1. Parung, Indonesia (Jawa Barat)—City planning. 2. City planning—Indonesia. I. Indonesia. Direktorat Tata Kota dan Tata Daerah. II. Title.
HT169.I54P4152 78-943682
79 MARC

Djoko Sujarto.
Pendekatan pembangunan perkotaan ditinjau dari segi perencanaan lokal / oleh Djoko Sujarto. — Yogyakarta : Balai Pembinaan Administrasi, Universitas Gadjah Mada, 1977.
32 p. : ill. ; 20 cm. — (Buletin - Balai Pembinaan Administrasi, Universitas Gadjah Mada ; no. 12/1977)
Bibliography: p. 31-32.
1. City planning—Indonesia. 2. Urban renewal—Indonesia. I. Title. II. Series: Universitas Gadjah Mada. Balai Pembinaan Administrasi. Bulletin ; no. 12/1977.
HD37.I52U53a 1977, no. 12 78-941620
[HT169.I53] 79 MARC

Indonesia. Departemen Pekerjaan Umum dan Tenaga Listrik.
Kumpulan bahan-bahan dalam rangka "meet the press," chusus city & regional planning / oleh Menteri Pekerdjaan Umum & Tenaga Listrik. — [Djakarta] : Departemen Pekerdjaan Umum & Tenaga Listrik, 1972.
6, [15] leaves : maps. ; 29 cm.
"Disusun sebagai pendjelasan lebih landjut atas kumpulan bahan' dalam rangka "meet the press' jang dikeluarkan oleh Departemen PUTL pada tanggal 25 Pebruari 1972."
1. City planning—Indonesia. 2. Regional planning—Indonesia. I. Title.
HT169.I53I515 1972a 72-942117
79 MARC

Indonesia. Departemen Sosial. Perwakilan Daerah Khusus Ibukota Jakarta.
Laporan kegiatan pelaksanaan urban survey Kelurahan Kota Bambu dan Kelurahan Mangga Dua Selatan, Daerah Khusus Ibu Kota Jakarta. — [Jakarta] : Perwakilan Departemen Sosial R.I., Daerah Khusus Ibu Kota Jakarta, 1972.
34 leaves in various foliations ; 34 cm.
1. Jakarta—City planning. 2. City planning—Indonesia. I. Title.
HT169.I54J374 73-940260
79 MARC

Indonesia. Direktorat Jenderal Cipta Karya.
Nota penjelasan rencana garis besar Depok. [Jakarta, 1972]
20, 15 l. maps. 29 x 43 cm.
Cover title.
English and Indonesian.
1. Depok, Indonesia (Bogor)—City planning. 2. City planning—Indonesia. I. Title.
HT169.I54D3735 309.2'62'095982 73-940845
79 MARC

Indonesia. Direktorat Jenderal Cipta Karya.
Peranan Departemen Pekerjaan Umum dan Tenaga Listrik dalam pengisian program pembagunan kota : disajikan dalam rangka MAKSI V, tgl. 21 Januari 1975 di Surabaya. — [Jakarta] : Direktorat Jenderal Cipta Karya, Departemen Pekerjaan Umum dan Tenaga Listrik, [1975]
13 leaves ; 29 cm.
Cover title.
1. City planning—Indonesia. 2. Indonesia. Departemen Rekerjaan Umum dan Tenaga Listrik. I. Title: Peranan Departemen Pekerjaan Umum dan Tenaga Listrik ...
HT169.I53I52 1975 76-940264
79 MARC

Indonesia. Direktorat Jenderal Cipta Karya.
Pokok-pokok pengembangan wilayah Jakarta, Bogor, Tangerang, dan Bekasi (JABOTABEK). — [Jakarta] : Direktorat Jenderal Cipta Karya, Departemen Pekerjaan Umum dan Tenaga Listrik, 1974.
4, [6], 24 leaves : maps ; 29 cm.
Cover title.
1. Jakarta metropolitan area, Indonesia—City planning. 2. City planning—Indonesia. 3. Central places—Indonesia—Jakarta metropolitan area. I. Title: Pokok-pokok pengembangan wilayah Jakarta, Bogor, Tangerang, dan Bekasi ...
HT169.I 54J3743 79–950354

Indonesia. Direktorat Jenderal Cipta Karya.
Rencana pengembangan wilayah inti Bandung Raya. — [Jakarta] : Direktorat Jenderal Cipta Karya, Departemen Pekerjaan Umum, [1978]
ii, 51 leaves, [28] leaves of plates : maps ; 31 x 42 cm.
Cover title.
1. Bandung, Indonesia—City planning. 2. City planning—Indonesia. I. Title.
HT169.I54B423 78-942691
79 MARC

CITY PLANNING
—INDONESIA (Continued)

Indonesia. Direktorat Jenderal Cipta Karya.
A short note on problems and policies for urban and regional development in Indonesia. — ₍Djakarta₎ : Republic of Indonesia, Dept. of Public Works & Electric Power, Directorate General of Housing, Building, Planning and Urban Development, ₍1971 or 1972₎
7 leaves ; 28 cm.
1. City planning—Indonesia. 2. Regional planning—Indonesia. I. Title.
HT169.I53I518 1971 309.2'12'09598 72-942116
79 MARC

Indonesia. Direktorat Tata Guna Tanah.
Masalah pembangunan kota di Indonesia. — ₍Jakarta₎ : Direktorat Tata Guna Tanah, Direktorat Jenderal Agraria, Departemen Dalam Negeri, 1976.
9 leaves : ill. ; 29 cm. — (Publikasi - Direktorat Tata Guna Tanah ; no. 53)
1. City planning—Indonesia. I. Title. II. Series: Indonesia. Direktorat Tata Guna Tanah. Publikasi - Direktorat Tata Guna Tanah ; 53.
HT169.I 53 I 54 1976 76-989599

Indonesia. Direktorat Tata Kota dan Daerah.
Banda Atjeh. — ₍Djakarta₎ : Direktorat Tata Kota & Daerah, Direktorat Djenderal Tjipta Karya, Departemen Pekerdjaan Umum & Tenaga Listrik, ₍1971₎
2 v. in 1 : ill. ; 21 x 32 cm.
CONTENTS: 1. Rentjana pendahuluan.—2. Kompilasi data.
1. Banda Aceh, Indonesia—City planning. 2. City planning—Indonesia.
HT169.I54B395 72-942302
79 MARC

Indonesia. Direktorat Tata Kota dan Daerah.
Kendari : rentjana garis besar. — ₍Djakarta₎ : Direktorat Tata Kota & Daerah, Direktorat Djenderal Tjipta Karya, Departemen Pekerdjaan Umum & Tenaga Listrik, ₍1972₎
ii, 17 leaves : maps ; 21 x 32 cm.
1. Kendari, Indonesia—City planning. 2. City planning—Indonesia.
HT169.I54K463 73-940054
79 MARC

Indonesia. Direktorat Tata Kota dan Daerah.
Kotamadya Balikpapan. — ₍Djakarta₎ : Direktorat Tata Kota & Daerah, Direktorat Djenderal Tjipta Karya, Departemen Pekerdjaan Umum & Tenaga Listrik, ₍1970₎
2 v. in 1 : maps ; 21 x 33 cm.
CONTENTS: 1. Analisa dan rentjana garis besar.—2. Kompilasi data.
1. Balikpapan, Indonesia—City planning. 2. City planning—Indonesia. I. Title.
HT169.I54B35 73-940476
79 MARC

Indonesia. Direktorat Tata Kota dan Daerah.
Master plan Palangkaraja. — ₍Djakarta₎ : Direktorat Tata Kota & Daerah, Direktorat Djenderal Tjipta Karya, Departemen Pekerdjaan Umum & Tenaga Listrik, ₍1972-
v. : ill. ; 21 x 32 cm.
CONTENTS: 1. Kompilasi data.—2. Analysa planologi.
1. Palangkaraya, Indonesia—City planning. 2. City planning—Indonesia. I. Title.
HT169.I54P384 72-942088
79 MARC

Indonesia. Direktorat Tata Kota dan Daerah.
Rentjana kota Palu. — ₍Djakarta₎ : Direktorat Tata Kota & Daerah, Direktorat Djenderal Tjipta Karya, Departemen Pekerdjaan Umum & Tenaga Listrik, ₍1970₎
ii, 148 leaves, ₍14₎ leaves of plates : ill. ; 21 x 33 cm.
1. Palu, Indonesia—City planning. 2. City planning—Indonesia. I. Title.
HT169.I54P38954 73-940668
79₍r79₎rev MARC

Indonesia. Direktorat Tata Kota dan Daerah.
Rentjana kota Pangkal Pinang. — ₍Djakarta₎ : Direktorat Tata Kota & Daerah, Direktorat Djenderal Tjipta Karya, Departemen Pekerdjaan Umum & Tenaga Listrik, ₍1972₎
3 v. : maps ; 21 x 32 cm.
Includes bibliographical references.
CONTENTS: ₍1₎ Rentjana.—₍2₎ Kompilasi data.—₍3₎ Tindjauan umum regional. Analisa kota.
1. Pangkalpinang, Indonesia—City planning. 2. City planning—Indonesia. I. Title.
HT169.I54P394 72-942089
79 MARC

Indonesia. Direktorat Tata Kota dan Daerah.
Rentjana Kotamadya Bandung : kompilasi data. — ₍Djakarta₎ : Direktorat Tata Kota & Daerah, Direktorat Djenderal Tjipta Karya, Departemen Pekerdjaan Umum & Tenaga Listrik, ₍1971?₎
viii, 298 leaves : maps ; 21 x 33 cm.
1. Bandung, Indonesia—City planning. 2. City planning—Indonesia. I. Title.
HT169.I54B424 74-940764
79 MARC

Indonesia. Direktorat Tata Kota dan Daerah.
Site and services = (Penyediaan tempat & pelayanan) / oleh Direktorat Tata Kota & Daerah. — ₍Jakarta₎ : Direktorat, 1973.
16 leaves ; 28 cm.
"Disusun dalam rangka Kursus Perencanaan Pembangunan kota."
In Indonesian.
Bibliography: leaf 16.
1. City planning—Indonesia. I. Title. II. Title: Penyediaan tempat dan pelayanan.
HT169.I53I55 1973 74-941128
79 MARC

Indonesia. Direktorat Tata Kota dan Tata Daerah.
The function and task of the Directorate of City and Regional Planning / by Directorate of City and Regional Planning. — ₍Jakarta₎ : The Directorate, 1976.
21 leaves : maps ; 29 cm.
1. Indonesia. Direktorat Tata Kota dan Tata Daerah. 2. City planning—Indonesia. 3. Regional planning—Indonesia. I. Title.
HT169.I53I55 1976 309.2'5'09598 77-940590
79 MARC

Indonesia. Direktorat Tata Kota dan Tata Daerah.
Rencana kota Cikarang, salah satu pusat pertumbuhan baru dalam wilayah Jabotabek. — ₍Jakarta₎ : Pemerintah Daerah Tingkat II Kabupaten Bekasi bekerjasama dengan Direktorat Tata Kota dan Tata Daerah, Direktorat Jenderal Cipta Karya, Departemen P.U.T.L., ₍1976₎
2 v. : maps ; 32 x 41 cm.
Cover title.
Prepared by Direktorat Tata Kota dan Tata Daerah.
CONTENTS: 1. Kompilasi data.—2 Analisa dan rencana.
1. Cikarang, Indonesia (Bekasi)—City planning. 2. City planning—Indonesia. I. Bekasi, Indonesia (Kabupaten) II. Title.
HT169.I54C544. 78-941026
79 MARC

Indonesia. Direktorat Tata Kota dan Tata Daerah.
Rencana kota Lhokseumawe. — ₍Jakarta₎ : Direktorat Tata Kota dan Tata Daerah, Direktorat Jenderal Cipta Karya, Departemen P.U.T.L., ₍1975-1976₎
2 v. : ill. ; 28 x 43 cm.
Cover title.
CONTENTS: 1. Kompilasi data dan analisa.—2. Rencana garis besar.
1. Lhokseumawe, Indonesia—City planning. 2. City planning—Indonesia. I. Title.
HT169.I54L484 77-986500
79 MARC

Indonesia. Direktorat Tata Kota dan Tata Daerah.
Rencana tata kota Baturaja. — ₍Jakarta₎ : Pemerintah Kabupaten Daerah Tingkat II Ogan Komering Ulu bekerjasama dengan Direktorat Tata Kota dan Tata Daerah, Direktorat Jenderal Cipta Karya, Departemen P.U.T.L., ₍1976?₎
2 v. : maps ; 31 x 41 cm.
Cover title.
1. Baturaja, Indonesia (Sumatera Selatan)—City planning. 2. City planning—Indonesia. I. Ogan Komering Ulu, Indonesia. II. Title.
HT169.I54B454 78-941010
79 MARC

Indonesia. Direktorat Tata Kota dan Tata Daerah.
Rencana tata kota Cibinong. — ₍Jakarta₎ : Pemerintah Daerah Tingkat II Kabupaten Bogor bekerja sama dengan Direktorat Tata Kota dan Tata Daerah, Direktorat Jenderal Cipta Karya, Departemen P.U.T.L., ₍1976₎
2 v. : maps ; 30 x 42 cm.
Cover title.
Prepared by Direktorat Tata Kota dan Tata Daerah.
CONTENTS: 1. Kompilasi data.—2. Analisa dan rencana.
1. Cibinong, Indonesia (Bogor)—City planning. 2. City planning—Indonesia. 3. Cibinong, Indonesia (Bogor)—Statistics. I. Bogor, Indonesia (Kabupaten) II. Title.
HT169.I54C524 78-941028
78 MARC

Indonesia. Direktorat Tata Kota dan Tata Daerah.
Rencana tata kota Cimahi. — ₍Jakarta₎ : Pemerintah Daerah Tingkat II Kabupaten Bandung bekerjasama dengan Direktorat Tata kota dan Tata Daerah, Direktorat Jenderal Cipta Karya, Departemen P.U.T.L., ₍1977₎
2 v. : maps ; 31 x 42 cm.
Cover title.
1. Cimahi, Indonesia (Bandung)—City planning. 2. City planning—Indonesia. I. Bandung, Indonesia (Kabupaten) II. Title.
HT169.I54C554 78-941023
79 MARC

Indonesia. Direktorat Tata Kota dan Tata Daerah.
Rencana tata kota Garut. — ₍Jakarta₎ : Pemerintah Daerah Tingkat II Kabupaten Garut bekerjasama dengan Direktorat Tata Kota dan Tata Daerah, Direktorat Jenderal Cipta Karya, Departemen P.U., ₍1978₎
2 v. : maps ; 30 x 42 cm.
Cover title.
Prepared by Direktorat Tata Kota dan Tata Daerah.
CONTENTS: 1. Kompilasi data.—2. Analisa dan rencana.
1. Garut, Indonesia—City planning. 2. City planning—Indonesia. I. Garut, Indonesia (Kabupaten) II. Title.
HT169.I54G374 78-943409
79 MARC

Indonesia. Direktorat Tata Kota dan Tata Daerah.
Rencana tata kota Kendal. — ₍Jakarta₎ : Pemerintah Kabupaten Daerah Tingkat II Kendal bekerjasama dengan Direktorat Tata Kota dan Tata Daerah, Direktorat Jenderal Cipta Karya, Departement P.U.T.L., ₍1977₎
2 v. : maps ; 31 x 42 cm.
Cover title.
1. Kendal, Indonesia (Kendal)—City planning. 2. City planning—Indonesia. I. Kendal, Indonesia (Kabupaten) II. Title.
HT169.I54K4574 78-940999
79 MARC

Indonesia. Direktorat Tata Kota dan Tata Daerah.
Rencana tata kota Lembang. — ₍Jakarta₎ : Pemerintah Kabupaten Daerah Tingkat II Bandung bekerjasama dengan Direktorat Tata Kota dan Tata Daerah, Direktorat Jenderal Cipta Karya, Departemen P.U.T.L., ₍1977₎
2 v. : maps ; 31 x 42 cm.
Cover title.
CONTENTS: 1. Kompilasi data.—2. Analisa dan rencana.
1. Lembang, Indonesia—City planning. 2. City planning—Indonesia. I. Bandung, Indonesia (Kabupaten) II. Title.
HT169.I54L464 78-940997
79 MARC

Indonesia. Direktorat Tata Kota dan Tata Daerah.
Rencana tata kota Serpong, salah satu pusat pertumbuhan baru dalam wilayah Jabotabek. — ₍Jakarta₎ : Pemerintah Daerah Tingkat II Kabupaten Tangerang bekerja sama dengan Direktorat Tata Kota dan Tata Daerah, Direktorat Jenderal Cipta Karya, Departemen P.U.T.L., ₍1977₎
2 v. : maps ; 30 x 42 cm.
Cover title.
Prepared by Direktorat Tata Kota dan Tata Daerah.
CONTENTS: 1. Kompilasi data.—2. Analisa dan rencana.
1. Serpong, Indonesia—City planning. 2. City planning—Indonesia. I. Tangerang, Indonesia (Kabupaten) II. Title.
HT169.I54S474 78-941024
78 MARC

Indonesia. Direktorat Tata Kota dan Tata Daerah.
Rencana tata kota Sumenep. — ₍Jakarta₎ : Pemerintah Kabupaten Daerah Tingkat II Sumenep bekerjasama dengan Direktorat Tata Kota dan Tata Daerah, Direktorat Jenderal Cipta Karya, Departemen P.U.T.L., ₍1977₎
2 v. : maps ; 30 x 42 cm.
Cover title.
CONTENTS: 1. Kompilasi data.—2. Analisa dan rencana.
1. Sumenep, Indonesia—City planning. 2. City planning—Indonesia. I. Sumenep, Indonesia (Kabupaten) II. Title.
HT169.I54S924 78-940998
79 MARC

Indonesia. Direktorat Tata Kota dan Tata Daerah.
Rencana tata kota Sungai Penuh. — ₍Jakarta₎ : Pemerintah Kabupaten Daerah Tingkat II Kerinci bekerjasama dengan Direktorat Tata Kota dan Tata Daerah, Direktorat Jenderal Cipta Karya, Departemen P.U.T.L., ₍1978?₎
2 v. : maps ; 31 x 42 cm.
Cover title.
CONTENTS: 1. Kompilasi data.—2. Analisa dan rencana.
1. Sungaipenuh, Indonesia—City planning. 2. City planning—Indonesia. I. Kerinci, Indonesia. II. Title.
HT169.I54S934 79-940312
79 MARC

Indonesia. Direktorat Tata Kota dan Tata Daerah.
Rencana tata kota Ungaran. — ₍Jakarta₎ : Pemerintah Kabupaten Daerah Tk. II Semarang bekerja sama dengan Direktorat Tata Kota dan Tata Daerah, Direktorat Jenderal Cipta Karya, Departemen P.U.T.L., ₍1977₎
2 v. : maps ; 31 x 42 cm.
Cover title.
CONTENTS: 1. Kompilasi data.—2. Analisa dan rencana.
1. Ungaran, Indonesia—City planning. 2. City planning—Indonesia. I. Semarang, Indonesia (Kabupaten) II. Title.
HT169.I54U534 78-941025
79 MARC

Indonesia. Direktorat Tata Kota dan Tata Daerah. Unit Perencanaan Daerah Istimewa Aceh.
Rencana tata kota Meulaboh. — ₍Jakarta₎ : Unit Perencanaan Daerah Istimewa Aceh bekerja sama dengan Direktorat Tata Kota dan Tata Daerah, Direktorat Jenderal Cipta Karya, Departemen P.U.T.L., ₍1976-1977₎
2 v. : maps ; 31 x 42 cm.
Cover title.
CONTENTS: 1. Kompilasi data.—2. Analisa dan rencana.
1. Meulaboh, Indonesia—City planning. 2. City planning—Indonesia. I. Indonesia. Direktorat Tata Kota dan Tata Daerah. II. Title.
HT169.I54M484 78-940995
79 MARC

Indonesia. Direktorat Tata Kota dan Tata Daerah. Unit Perencanaan Daerah Tingkat I Riau.
Rencana Kota Tanjung Pinang. — ₍s.l.₎ : Unit Perencanaan Daerah Tk. I. Propinsi Riau bekerjasama dengan Direktorat Tata Kota dan Tata Daerah, Ditjen. Cipta Karya, Dep. PUTL, ₍1977-
v. : ill. ; 31 cm.
Cover title.
CONTENTS: 1. Kompilasi data.
1. Tanjungpinang, Indonesia—City planning. 2. City planning—Indonesia. I. Indonesia. Direktorat Tata Kota dan Tata Daerah. II. Title.
HT169.I54T364 78-941001
79 MARC

CITY PLANNING
—INDONESIA (Continued)

Indonesia. Direktorat Tata Kota dan Tata Daerah. Unit Perencanaan Dati I Bali. Bagian Proyek Pembinaan Pusat Training.
Program perbaikan kampung di kota Denpasar. — ₍Denpasar₎ : Kerja sama Proyek Pembinaan Pusat Training, Unit Perencanaan Dati I Bali, Direktorat Tata Kota dan Tata Daerah, Direktorat Jendral Cipta Karya, Dep. P.U.T.L., Kabupaten Dati II Badung, Propinsi Daerah Tingkat I Bali, Departemen Dalam Negeri, 1977-
v. : fold. maps ; 22 x 34 cm.
"Daftar ralat & halaman-halaman tambahan, buku I, Kompilasi data" (26 leaves) inserted.
CONTENTS: buku 1. Kompilasi data.
1. Community development, Urban—Indonesia—Denpasar. 2. Denpasar, Indonesia—City planning. 3. City planning—Indonesia. I. Badung, Indonesia (Kabupaten) II. Title.
HN710.Z9C62223 1977 78-941848
79 MARC

Indonesia. Direktorat Tata Kota dan Tata Daerah. Unit Perencanaan Dati I Bali. Bagian Proyek Pembinaan Pusat Training.
Rencana kota Seririt. — ₍Sanur?₎ : Bagian Proyek Pembinaan Pusat Training, Unit Perencanaan Dati I Bali, Direktorat Tata Kota dan Tata Daerah, Dit. Jen. Cipta Karya, Dep. PUTL, ₍1977-
v. : maps ; 21 x 33 cm.
Cover title.
CONTENTS: 1. Kompilasi data.
1. Seririt, Indonesia—City planning. 2. City planning—Indonesia. I. Title.
HT169.I54S454 78-941022
78 MARC

Kalimantan Tengah.
Kompilasi data dalam rangka penyusunan rencana induk tata kota Palangka Raya. — ₍Palangka Raya₎ : Pemerintah Daerah Tingkat I Propinsi Kalimantan Tengah bekerjasama dengan Direktorat Tata Kota dan Tata Daerah, Direktorat Jenderal Cipta Karya, Departemen P.U., 1978-
v. : maps ; 31 x 42 cm.
Cover title.
1. Palangkaraya, Indonesia—City planning. 2. City planning—Indonesia. I. Indonesia. Direktorat Tata Kota dan Tata Daerah. II. Title.
HT169.I54P385 78-942690
79 MARC

Ketapang, Indonesia (Kabupaten)
Rencana tata kota Ketapang. — ₍Jakarta₎ : Permerintah Kabupaten Daerah Tingkat II Ketapang bekerjasama dengan Direktorat Tata Kota dan Tata Daerah, Direktorat Jenderal Cipta Karya, Departemen P.U., ₍1978₎
2 v. : ill. ; 31 x 42 cm.
Cover title.
CONTENTS: 1. Kompilasi data.—2. Analisa dan rencana.
1. Ketapang, Indonesia—City planning. 2. City planning—Indonesia. 3. Regional planning—Indonesia—Ketapang (Kabupaten) I. Indonesia. Direktorat Tata Kota dan Tata Daerah. II. Title.
HT169.I54K474 1978 79-940308
79 MARC

Lampung, Indonesia. Badan Perencanaan Pembangunan Daerah.
Rencana induk tata kota Tg. Karang-Tl. Betung. — ₍Jakarta₎ : Bappeda Tingkat I Lampung dan Pemerintah Kotamadya Dt. II Tg. Karang-Tl. Betung bekerjasama dengan Direktorat Tata Kota dan Tata Daerah, Direktorat Jenderal Cipta Karya, Departemen P.U.T.L., ₍1977₎
2 v. : maps ; 31 x 42 cm.
Cover title.
CONTENTS: 1. Kompilasi data.—2. Analisa dan rencana.
1. Tanjungkarang-Telukbetung, Indonesia—City planning. 2. City planning—Indonesia. I. Tanjungkarang-Telukbetung, Indonesia. II. Indonesia. Direktorat Tata Kota dan Tata Daerah. III. Title.
HT169.I54T368 78-941956
79 MARC

Nusa Tenggara Timur. Dinas Pekerjaan Umum.
Program perbaikan kampung di kota Kupang. — ₍s.l.₎ : Kerja sama Dinas Pekerjaan Umum, Propinsi Nusa Tenggara Timur, Departemen P.U.T.L., Proyek Pembinaan Pusat Training, Unit Perencanaan Daerah Tingkat I Bali, Dit. Tata Kota dan Tata Daerah, Dit. Jen. Cipta Karya, Departemen P.U.T.L., Kabupaten Dati II Kupang, Propinsi Dati I Nusa Tenggara Timur, Departemen Dalam Negeri, ₍1977-
v. : maps ; 21 x 33 cm.
1. Kupang, Indonesia—Suburbs and environs. 2. City planning—Indonesia. I. Indonesia. Direktorat Tata Kota dan Tata Daerah. Unit Perencanaan Dati I Bali. Bagian Proyek Pembinaan Pusat Training. II. Kupang, Indonesia (Kabupaten) III. Title.
HT169.I54K85 77-941713
79 MARC

Nusa Tenggara Timur. Dinas Pekerjaan Umum.
Rencana kota Maumere. — ₍s.l.₎ : Kerja sama Dinas Pekerjaan Umum, Propinsi Nusa Tenggara Timur, Departemen P.U.T.L., Proyek Pembinaan Pusat Training, Unit Perencanaan Daerah Tingkat I Bali, Dit. Tata Kota dan Tata Daerah, Dit. Jen. Cipta Karya, Departemen P.U.T.L., Badan Perencanaan Pembangunan Daerah, Propinsi Daerah Tingkat I Nusa Tenggara Timur, Departemen Dalam Negeri, ₍1977-
v. ; 21 x 33 cm.
"Lampiran peta[2] kota Maumere" (12 maps): issued separately in an accompanying case.
CONTENTS: 1. Kompilasi data.
1. Maumere, Indonesia—City planning. 2. City planning—Indonesia. I. Indonesia. Direktorat Tata Kota dan Tata Daerah. Unit Perencanaan Dati I Bali. Bagian Proyek Pembinaan Pusat Training. II. Nusa Tenggara Timur. Badan Perencanaan Pembangunan Daerah. III. Title.
HT169.I54M386 77-941712
79 MARC

Pasaribu, Bemfie.
Tanah dalam hubungannja dengan perentjanaan kota / oleh Bemfie Pasaribu. — ₍Djakarta : Direktorat Tata Kota & Daerah, 1972?₎
14 leaves ; 29 cm.
Cover title.
"Tjeramah didepan pengikut-pengikut Kursus Peningkatan Tata Agraria/-Landreform, angkatan ke II di Djakarta."
1. Land use—Planning—Indonesia. 2. City planning—Indonesia. I. Title.
HD896.P37 72-942350
79 MARC

Riau, Indonesia (Propinsi). Dinas Pekerjaan Umum.
Rencana tata kota Rengat. — ₍Pakanbaru₎ : Dinas Pekerjaan Umum, Daerah Tingkat I Riau bekerjasama dengan Direktorat Tata Kota dan Tata Daerah, Direktorat Jenderal Cipta Karya, Departemen P.U., ₍1978₎
2 v. : maps ; 30 x 42 cm.
Cover title.
CONTENTS: 1. Kompilasi data.—2. Analisa dan rencana.
1. Rengat, Indonesia—City planning. 2. City planning—Indonesia. I. Indonesia. Direktorat Tata Kota dan Tata Daerah. II. Title.
HT169.I54R467 78-94269:
79 MARC

Soegijoko, Sugijanto.
Beberapa pemikiran kearah kebijaksanaan nasional perkotaan / oleh Soegianto Soegijoko. — ₍Yogyakarta : Pem. Da. Kotamadya Yogyakarta₎, 1974.
30 leaves ; 29 cm.
Cover title.
1. City planning—Indonesia. I. Title.
HT169.I53S59 79-950330
79 MARC

Solok, Indonesia (Sumatera Barat)
Rencana tata kota Solok. — ₍Jakarta₎ : Pemerintah Daerah Tingkat II Kotamadya Solok bekerjasama dengan Direktorat Tata Kota dan Tata Daerah, Direktorat Jenderal Cipta Karya, Departemen P.U., ₍1978₎
2 v. : maps ; 30 x 42 cm.
Cover title.
CONTENTS: 1. Kompilasi data.—2. Analisa dan rencana.
1. Solok, Indonesia (Sumatera Barat)—City planning. 2. City planning—Indonesia. I. Indonesia. Direktorat Tata Kota dan Tata Daerah. II. Title.
HT169.I54S634 78-943410
79 MARC

Sulawesi Utara. Dinas Pekerjaan Umum.
Rencana kawasan pusat kota Manado. — ₍Manado₎ : Dinas Pekerjaan Umum Dt. I Propinsi Sulawesi Utara ; ₍Jakarta₎ : Direktorat Tata Kota dan Tata Daerah, Direktorat Jenderal Cipta Karya, Departemen P.U., ₍1978-
v. : maps ; 30 x 42 cm.
Cover title.
Includes bibliographical references.
CONTENTS: ₍1₎ Analisa dan rencana.
1. Manado, Indonesia—City planning. 2. City planning—Indonesia. I. Indonesia. Direktorat Tata Kota dan Tata Daerah. II. Title.
HT169.I54M357 1978 79-940313
79 MARC

Universitas Diponegoro. Fakultas Tehnik.
Kota Tegal : rencana garis besar. — ₍Semarang₎ : Fakultas Tehnik, Universitas Diponegoro, kerja sama dengan Departemen P.U.T.L., Dit. Jen. Cipta Karya, Dit. Tata Kota & Daerah, ₍1974 or 1975₎
vi, 92 leaves : ill. ; 32 cm.
1. Tegal, Indonesia—City planning. 2. City planning—Indonesia. I. Indonesia. Direktorat Tata Kota dan Daerah. II. Title.
HT169.I54T387 76-940140
79 MARC

Universitas Gadjah Mada. Bagian Arsitektur. Research Center.
Pengamatan pendahuluan program site & services, kasus sekitar kota Yogyakarta dan kota Samarinda. — ₍Yogyakarta₎ : Fakultas Teknik, Universitas Gajah Mada, Research Center, Bagian Arsitektur dalam kerja sama dengan Departemen Pekerjaan Umum dan Tenaga Listrik, Direktorat Jenderal Cipta Karya, Direktorat Tatakota & Daerah, ₍1975 or 1976₎
iii, 131 leaves ; 30 cm.
1. Yogyarkarta, Indonesia—City planning. 2. Samarinda, Indonesia—City planning. 3. City planning—Indonesia. I. Indonesia. Direktorat Tata Kota dan Daerah. II. Title: Pengamatan pendahuluan program site ...
HT169.I54Y638 1975 76-941726
79 MARC

Washington, D.C. Economic Development Institute.
Laporan peninjauan dari rombongan Economic Development Institute, Bank Dunia mengenai perbaikan kampung (Proyek Mohammad Husni Thamrin) di Jakarta. — ₍Jakarta₎ : BAPPEM Proyek Mohammad Husni Thamrin Daerah Khusus Ibukota Jakarta, ₍1976?₎
34 p. ; 22 cm.
In English.
"Terbatas."
1. Community development, Urban—Indonesia—Jakarta. 2. Jakarta—City planning. 3. City planning—Indonesia. I. Title.
HN710.Z9C687 309.2'62'095982 78-943362
79 MARC

—INDONESIA—COLLECTED WORKS

Indonesia. Direktorat Tata Guna Tanah.
Kotamadya Balikpapan ₍dan kotamadya lainnya₎. — ₍Jakarta₎ : Direktorat Tata Guna Tanah, Direktorat Jenderal Agraria, Departemen Dalam Negeri, 1978-
v. in : ill. ; 28 cm. — (Publikasi - Direktorat Tata Guna Tanah ; no. 103)
1. City planning—Indonesia—Collected works. 2. Cities and towns—Indonesia—Collected works. 3. Land use, Urban—Indonesia—Collected works. I. Title. II. Series: Indonesia. Direktorat Tata Guna Tanah. Publikasi - Direktorat Tata Guna Tanah ; no. 103 ₍etc.₎
HT169.I53I54 1978 79-950351
79 MARC

—INDONESIA—CONGRESSES

Kursus Aplikasi Tata-Kota dan Daerah, tingkat nasional / penjelenggara, Direktorat Tata-Kota dan Daerah, Lembaga Penjelidikan Masalah Bangunan, Direktorat Djenderal Tjipta Karya, Departemen Pekerdjaan Umum & Tenaga Listrik. — ₍Bandung : Lembaga Penjelidikan Masalah Bangunan, 1970₎
2 v. : ill. ; 30 cm.
Cover title.
1. City planning—Indonesia—Congresses. 2. Regional planning—Indonesia—Congresses. I. Kursus Aplikasi Tata-Kota dan Daerah, Bandung, Indonesia, 1970. II. Indonesia. Direktorat Tata Kota dan Daerah. III. Lembaga Penyelidikan Masalah Bangunan.
HT169.I53K87 76-941186
79 MARC

Seminar Pengembangan Perkotaan, 2d, Jakarta, Indonesia, 1977.
Seminar Pengembangan Perkotaan II : himpunan prasaran / penyelenggara Direktorat Pengembangan Perkotaan, Ditjen. PUOD Depdagri bekerja sama dengan LPP-ITB, Ditjen. Bina Marga-PUTL, dan BKS-AKSI, Jakarta, 28, 29 dan 30 Juni 1977. — ₍Jakarta : s.n., 1977?₎
ca. 300 leaves in various foliations : ill. ; 28 cm.
Includes a bibliography.
1. City planning—Indonesia—Congresses. I. Indonesia. Direktorat Pengembangan Perkotaan.
HT169.I53S45 1977 78-943222
79 MARC

—INDONESIA—JAWA BARAT

Indonesia. Departemen Pekerjaan Umum dan Tenaga Listrik.
Petunjuk pengarahan pengembangan JABOTABEK, Jakarta, Bogor, Tangerang, Bekasi. — ₍Jakarta₎ : Departemen Pekerjaan Umum & Tenaga Listrik, 1975.
8, 9 leaves : ill. ; 34 cm.
Cover title.
1. City planning—Indonesia—Jawa Barat. 2. Metropolitan areas—Indonesia—Jawa Barat. I. Title.
HT169.I54J424 75-940720
79 MARC

—INDONESIA—SRAGEN (KABUPATEN)

Indonesia. Direktorat Tata Kota dan Daerah.
Sragen : analisa dan rentjana garis besar. — ₍Djakarta₎ : Direktorat Tata Kota & Daerah, Direktorat Djenderal Tjipta Karya, Departemen Pekerdjaan Umum & Tenaga Listrik, ₍1971₎
98, 45, ₍18₎ leaves : maps ; 21 x 33 cm.
1. City planning—Indonesia—Sragen (Kabupaten)
HT169.I54S695 72-940036
79 MARC

—IRAQ

Muḥammad, Ṣabāḥ Maḥmūd.
(Madīnat al-Ḥillah al-kubrá, wa-ẓā'ifuhā wa-'alāqātuhā al-iqlīmīyah)
مدينة الحلة الكبرى، وظائفها وعلاقاتها الاقليمية : دراسة في الاسس الجغرافيــة للتخطيط الحضري والاقليمي / تاليف صباح محمود محمد. ـــ الطبعة 1. ـــ بغداد : مكتبة المنار، 1974.
255 p. : ill. ; 24 cm.
Added t. p.: Greater Hillah, its functions & regional relationships.
Originally presented as the author's thesis (risālat mājistīr), Jāmi'at Baghdād.
Summary in English.
Bibliography: p. 239–248.
1. al Ḥillah, Iraq—Economic conditions. 2. al Ḥillah, Iraq—City planning. 3. City planning—Iraq. I. Title. II. Title: Greater Hillah, its functions & regional relationships.
HC497.I 72H545 1974 309.2'62'095675 79–971005

CITY PLANNING (Continued)

—IRELAND

Taisce—the National Trust for Ireland.
Study of amenity planning issues in Dublin and Dun Laoghaire / An Taisce — Dublin : The Trust, 1967.
52 p., [4] leaves of plates : ill. ; 24 cm.
Cover title: Amenity study of Dublin and Dun Laoghaire.
1. City planning—Ireland. 2. Dublin—City planning. 3. Kingstown, Ire.—City planning. 4. Architecture—Conservation and restoration. I. Title. II. Title: Amenity study of Dublin and Dun Laoghaire.
HT169.I77T34 1967 309.2'62'0941835 78-324941
79 MARC

—ISRAEL

Berler, Alexander.
(Shalav ha-mar'eh be-hitpatḥutah shel 'ir be-Yiśrael)
שלב המראה בהתפתחותה של עיר בישראל : מחקר גישוש / אלכסנדר בדלר, בהשתתפות נחום בן אליא ... et al.] ; עריכה, נורית גורן]. — רחובות : המרכז לחקר התישבות כפרית ועירונית, 1977.
119 p. : ill. ; 24 cm. — (פרסומים בבעיות פיתוח אזורי ; 23)
At head of title: ha-Merkaz le-ḥeḳer hityashvut kafrit ṿe-'ironit.
Bibliography: p. 117–119.
1. Beersheba, Israel—City planning. 2. 'Afula, Israel—City planning. 3. Israel—Social policy. 4. City planning—Israel. I. Ben Ilya, Naḥum, joint author. II. Goren, Nurit. III. ha-Merkaz le-ḥeḳer hityashvut kafrit ṿe-'ironit. IV. Title. V. Series: Pirsumim bi-ve'ayot pituaḥ ezori ; 23.
HT169.I 82B323 78–952031

—ISRAEL—BIBLIOGRAPHY

Vidergar, John J
Israel : social and economic development / John J. Vidergar. — Monticello, Ill. : Vance Bibliographies, 1978.
4 p. ; 28 cm. — (Public administration series : Bibliography ; P-31)
Cover title.
$1.50
1. Israel—Politics and government—Bibliography. 2. Israel—Economic conditions—Bibliography. 3. Israel—Social conditions—Bibliography. 4. City planning—Israel—Bibliography. I. Title. II. Series.
Z7165.I7V5 016.3091'5694'05 78-109210
[JQ1825.P3] 79 MARC

—ITALY

Airaldi, Luigi.
Le ragioni dell'urbanistica / Luigi Airaldi. — 1. ed. — Venezia : Marsilio, 1978.
252 p. ; 21 cm. — (Architettura e urbanistica ; 33) (Le Scienze della nuova società) (Biblioteca Marsilio) It78-Dec
Includes bibliographical references.
L8500
1. City planning—Italy. I. Title. II. Series: Architettura e urbanistica (Venezia) ; 33.
HT169.I83A55 309.2'62'0945 79-346878
•79 MARC

Alisio, Giancarlo, 1930-
Lamont Young, utopia e realtà nell'urbanistica napoletana dell'Ottocento / Giancarlo Alisio. — Roma : Officina, 1978.
194 p., [77] leaves of plates : ill. ; 24 cm. — (Officina ; 17) It79-Jan
Includes bibliographical references and indexes.
L10000
1. Naples—City planning. 2. City planning—Italy. 3. Young, Lamont, 1851-1929. I. Title.
HT169.I84N32 309.2'62'094573 79-351890
•79 MARC

Amendola, Giandomenico, 1941-
Casa, quartiere, rinnovo urbano : le trasformazioni sociali dello spazio marginale / Giandomenico Amendola ; prefazione di Paolo Portoghesi. — Bari : Dedalo libri, 1977.
156 p., [8] leaves of plates : ill. ; 21 cm. — (Saggi ; 25) It78-Sept
Includes bibliographical references.
L4000
1. Urban renewal—Italy—Salerno. 2. Housing—Italy—Salerno. 3. Salerno—City planning. 4. City planning—Italy. I. Title.
HT178.I82S233 309.2'62'094574 79-339238
•79 MARC

Battaglini, Giuseppe M
Cosmopolis : Portoferraio medicea, storia urbana, 1548-1737 / Giuseppe M. Battaglini — 1. ed. — Roma : Multigrafica editrice, 1978.
294 p. : ill. ; 28 cm. It79-Feb
"Regesto dei documenti d'archivio" (p. 235-284) in Italian or Latin.
Bibliography: p. 285-286.
Includes index.
L40000
1. Portoferraio, Italy—City planning. 2. City planning—Italy. 3. Portoferraio, Italy—History. I. Title.
HT169.I84P673 309.2'62'094556 79-351876

Benevolo, Leonardo.
Brescia, S. [i.e. San] Polo : un quartiere di iniziativa pubblica / Leonardo Benevolo. — 2. ed. — Brescia : Morcelliana, [1978]
191 p. : ill. ; 23 cm. — (Collezione di studi urbanistici) It77-May
L10000
1. San Polo, Brescia—Buildings. 2. Brescia—Buildings. 3. Brescia—City planning. 4. City planning—Italy. I. Title. II. Series.
NA9053.N4B46 1978 79-383812
•79 MARC

Bergamo.
Piano regolatore generale comunale 1969 / Comune di Bergamo. — Bergamo : Comune di Bergamo, [1970?]
10 pieces in portfolio : maps ; 32 cm.
Title from portfolio.
CONTENTS: Relazione generale.—Norme per l'attuazione del P.R.G.—Tabelle delle operazioni di intervente.—Elaborati grafici.—Relazione economico-finanziaria per l'applicazione dell'articolo 18 della legge urbanistica.—Allegati vari.
1. Bergamo—City planning. 2. City planning—Italy.
HT169.I84B473 711'.4'094524 79-343341
79 MARC

Birindelli, Massimo.
Roma italiana : come fare una capitale e disfare una città / Massimo Birindelli ; per le ricerche fotografiche collaborazione di P. Leone e di P. Scaglione. — Roma : Savelli, 1978.
80 p. : ill. ; 30 cm. It79-Jan
Bibliography: p. 80.
L5000
1. Rome (City)—City planning. 2. City planning—Italy. 3. Italy—Capital. 4. Rome (City)—History. I. Title.
HT169.I84R624 79-345786
•79 MARC

Brusa, Carlo.
Geografia e percezione dell'ambiente : Varese vista dagli operatori dell'ente pubblico locale / Carlo Brusa. — Torino : G. Giappichelli, 1978.
174 p., [5] fold. leaves of plates : ill. ; 21 cm. It78-June
Includes bibliographical references and index.
L4400
1. Varese—City planning. 2. City planning—Italy. 3. Varese—Description. I. Title.
HT169.I84V353 309.2'62'094522 78-388774
•79 MARC

Caldaretti, Sergio, 1949-
Elementi di definizione di una struttura urbana : l'area di gravitazione di Cosenza-Rende / a cura di Sergio Caldaretti e Giuseppe Imbesi. — Reggio Calabria : Casa del libro, 1979.
120, [28] p. : ill. ; 24 cm. — (Collana di studi, documenti, ricerche) It79-May
"Sintesi, rielaborata, del rapporto ... Indagine sulla struttura urbana dell'area di gravitazione Cosenza-Rende, sviluppata per conto dell'Assessorato ai LL.-PP. della Regione Calabria nel 1974."
Includes bibliographical references.
L5000
1. Cosenza, Italy—City planning. 2. City planning—Italy. 3. Regional planning—Italy—Cosenza (Province) I. Imbesi, Giuseppe, joint author. II. Title.
HT169.I84C693 79-367030
•79 MARC

Castelnovi, Paolo.
Le proprietà culturali nel tessuto urbano : progetti d'uso : per la gestione dello sviluppo del proprio spazio storico : metodo di servizio alle scelte autonome dei gruppi d'utenza : il caso S. Frediano Oltrarno / [di] Paolo Castelnovi, Maurizio De Marco. — Firenze : Uniedit, 1978.
186 p. : ill. ; 22 cm. — (Biblioteca di architettura : Saggi e documenti ; 10) It78-Mar
L5000
1. City planning—Italy. 2. Architecture—Italy—Conservation and restoration. I. De Marco, Maurizio, joint author. II. Title.
HT169.I5C37 78-381131
•79 MARC

Charles-Picard, Gilbert.
Rome et les villes d'Italie, des Gracques à la mort d'Auguste / par Gilbert Charles-Picard. — Paris : Société d'édition d'enseignement supérieur, c1978.
221 p. : ill. ; 19 cm. — (Regards sur l'histoire ; 32 : Histoire générale ; 2) F***
Bibliography: p. 217-218.
ISBN 2-7181-1849-0 : 48.50F
1. Rome (City)—History—To 476. 2. Rome (City)—City planning. 3. City planning—Italy. 4. Rome—History—Republic, 265-30 B.C. 5. Rome—History—Augustus, 30 B.C.-14 A.D. I. Title.
DG63.C45 937'.6'02 79-373637
79 MARC

Cipollaro, Pietro, 1938-
Roma, territorio e qualità della vita / Pietro Cipollaro. — Roma : Fratelli Palombi, 1978.
109 p., [1] leaf of plates : 1 map ; 26 cm. It79-June
L6000
1. Rome (City)—City planning. 2. City planning—Italy. I. Title.
HT169.I84R628 79-374996
•79 MARC

Contributi per un'indagine sulla gestione del territorio / a cura di Augusto Clerici. — Milano : A. Giuffrè, 1976.
viii, 277 p. ; 25 cm. — (Studi del Laboratorio di economia politica S. Cognetti De Martiis, Università di Torino ; III ser., v. 19) It78-Mar
Bibliography: p. [247]-277.
L6500
1. Regional planning—Italy. 2. City planning—Italy. 3. Local government—Italy. I. Clerici, Augusto. II. Title. III. Series: Turin. Università. Laboratorio di economia politica S. Cognetti de Martiis. Studi ; ser. 3, v. 19.
HT395.I8C57 79-339620
•79 MARC

Costruzione, alterazione e recupero del centro storico di Cesena / [di] Romeo Bellardini, Tommaso Cantori, Oddone Pezzi, Pier Carlo Righetti, Corrado Tossani. — Rimini : B. Ghigi, 1977.
334 p., [5] fold. leaves of plates : ill. plans ; 30 cm. & portfolio (19 fold. leaves of plates : col. ill., plans) It78-Oct
Errata slip inserted.
Bibliography: p. 329-334.
1. Urban renewal—Italy—Cesena. 2. Cesena—City planning. 3. City planning—Italy. I. Bellardini, Romeo.
HT178.I82C4853 79-356587
•79 MARC

Crespi, Raffaella.
Elementi per un progetto : l'utenza e i servizi : indagine sulla struttura dell'utenza e sui servizi relativi all'area di intervento / [di] Raffaella Crespi, Giorgio Prevosti. — Milano : F. Angeli, [1978]
176, [19] p. : ill. ; 22 cm. — (Ricerche di tecnologia dell'architettura ; 8) It78-Oct
Includes bibliographical references.
L5000
1. Urban renewal—Italy. 2. City planning—Italy. 3. Municipal services—Italy. I. Prevosti, Giorgio, 1947- joint author. II. Title.
HT178.I8C73 78-393709
•79 MARC

Crocioni, Giovanni.
Il rapporto città-campagna nel dopoguerra : trasformazioni territoriali e ciclo economico fra il 1945 e il 1975 / Giovanni Crocioni. — Milano : F. Angeli, [1978]
109 p. ; 22 cm. — (Scienze sociali : Sezione 1, Economia ; 1) It78-June
Includes bibliographical references.
L3000
1. City planning—Italy. 2. Italy—Economic conditions—1945- I. Title.
HT169.I83C7 309.2'62'0945 78-376009
•78 MARC

Erba, Valeria, 1944-
L'attuazione dei piani urbanistici / Valeria Erba. — 3. ed. — Roma : Edizioni delle autonomie, 1978.
305 p. ; 21 cm. — (Amministrare l'urbanistica ; 1) It***
Appendices contain legislation.
Includes bibliographical references.
L5500
1. City planning—Italy. I. Title.
HT169.I83E7 1978 309.2'62'0945 78-392344
79 MARC

Fabbrichesi, Renato, 1887-
Urbanistica ed edilizia italiana / R. Fabbrichesi. — Padova : R. Zannoni, 1935-
v. : ill. ; 26 cm.
Bibliography: v. 1, p. 95.
CONTENTS: pt. 1. Urbanistica.
1. City planning—Italy. 2. Building—Italy. I. Title.
HT169.I83F3 309.2'62'0945 79-345884
79 MARC

Falco, Luigi.
Torino, un secola de sviluppo urbano : appunti per una lettura de una città del capitale / [di] Luigi Falco, Guido Morbelli. — Torino : CELID, 1976.
104 p., [8] leaves of plates : ill. ; 24 cm. It78-Jan
Errata slip inserted.
Includes bibliographical references.
L2500
1. Turin—City planning. 2. City plannint—Italy. 3. Turin—Maps. I. Morbelli, Guido, joint author. II. Title.
HT169.I84T83 79-343949
•79 MARC

Fazio, Mario.
Il destino dei centri storici / di Mario Fazio. — 1. ed. — Firenze : La nuova Italia, 1977.
107 p. ; 18 cm. — (Italia nostra : Educazione ; 5) It78-Feb
Bibliography: p. 107.
L1900
1. Cities and towns—Italy. 2. City planning—Italy. I. Title.
HT164.I8F37 78-368823
•78 MARC

Filippini, Luigi.
Indagine sul patrimonio edilizio del centro storico di Peruggia [sic] / di Luigi Filippini. — Perugia : [s.n.], 1975.
130 p. : ill. ; 24 cm. — (Quaderno - Camera di commercio, industria, artigianato e agricoltura, Perugia ; n. 31) It77-Nov
Bibliography: p. 125.
1. Perugia—City planning. 2. Perugia—Buildings—Conservation and restoration. 3. Housing—Italy—Perugia. 4. City planning—Italy. I. Title. II. Series: Camera di commercio, industria, artigianato e agricoltura, Perugia. Quaderno ; n. 31.
HT169.I84P473 79-374999
•79 MARC

Giacopetti, Massimo.
Una città nella metropoli : 1977, un anno di politica urbanistica in una circoscrizione di periferia : prime pietre di una Roma diversa / [di] M. Giacopetti. — Roma : Bulzoni, [1978]
193 p., 12 leaves of plates : ill. ; 24 cm. It78-Oct
At head of title: Istituto di urbanistica della Facoltà di architettura di Roma.
On cover: Verso le municipalità.
L7000
1. Rome (City)—City planning. 2. City planning—Italy. I. Title.
HT169.I84R6364 78-396537
•79 MARC

CITY PLANNING
—ITALY (Continued)

La Padula, Attilio.
Appunti metodologici : schema metodologico per un approccio sistemico alla pianificazione / Attilio La Padula ; contributi di Fabrizio Crisafulli, Nicola D'Errico, Carlo Lefebvre, Valerio Romani. — Roma : Bulzoni, 1978.
149 p. : ill. ; 24 cm. — (Quaderni dell'Istituto di urbanistica della Facoltà di architettura di Roma ; 5) It78-May
Cover title.
Includes bibliographies.
L3000
1. City planning—Italy. I. Title. II. Series: Rome (City). Università Istituto di urbanistica. Quaderni dell'Istituto di urbanistica della Facoltà di architettura di Roma ; 5.
NA9203.L36 78-380995
*78 MARC

Marescotti, Luca.
Urbanistica : saggio critico, testimonianze, documenti, bibliografia ragionata / Luca Marescotti. — Milano : Accademia, 1979.
467 p. : ill. ; 20 cm. — (Manuali Accademia) It79-Aug
"Testimonianze e documenti": p. 271-362.
Bibliography: p. 367-461.
L6000
1. City planning—Italy. 2. City planning. I. Title.
HT169.I83M36 79-383989
*79 MARC

Mastrorilli, Franco.
Centri antichi: una politica dei trasporti per il recupero dei valori ambientali. Palermo, Agatekne, 1973.
43 p. 24 cm. (Quaderni del Centro provinciale di studi urbanistici di Palermo, 5) L1200 It74-Apr
Bibliography: p. 41-43.
1. Urban transportation—Italy. 2. City traffic—Italy. 3. City planning—Italy. I. Title. II. Series: Centro provinciale di studi urbanistici di Palermo. Quaderni, 5.
HE311.I8M37 711'.7'0945 79-413085
79 MARC

Il Momento dei Sassi di Matera : centri storici, una proposta di recupero / ₍di₎ Leonardo Abbruzzese ... ₍et al.₎ ; introduzione di Luigi Piccinato. — Roma : Bulzoni, 1978.
107 p. : ill. ; 21 cm. It78-May
Bibliography: p. 105.
L7500
1. Matera, Italy—City planning. 2. City planning—Italy. 3. Urban renewal—Italy—Matera. I. Abbruzzese, Leonardo.
HT169.I84M345 78-396573
*79 MARC

Il Nodo di S₍an₎ Bernardo : una struttura urbana tra il centro antico e la Roma moderna / ₍di₎ Carlo Pietrangeli ... ₍et al.₎. — Milano : F. Angeli, ₍1977₎
317 p. : ill. ; 28 cm. & portfolio (25 fold. leaves of plates : 25 plans) It78-Sept
Includes bibliographies.
L18000
1. Rome (City)—City planning. 2. City planning—Italy. 3. City traffic—Italy—Rome (City) I. Pietrangeli, Carlo.
HT169.I84R6394 309.2'62'0945632 79-341638
*79 MARC

Operazione Arcevia : comunità esistenziale / ₍a cura di₎ Ico Parisi ; ₍scritti di₎ Italo Bartoletti ... ₍et al.₎. — Como : C. Nani, 1976.
483, ₍6₎ p. : ill. ; 11 x 17 cm. It77-Jan
Cover title.
Bibliography: p. ₍485₎-₍486₎
L3500
1. Arcevia, Italy—City planning. 2. Arcevia, Italy—Economic policy. 3. City planning—Italy. I. Parisi, Ico, 1916- II. Bartoletti, Italo, 1915-
HT169.I84A726 309.2'62'0945671 79-339915
*79 MARC

Peano, Attilia.
L'attuazione urbanistica : metodi e strumenti / Attilia Peano. — Torino : Celid, 1979.
181 p. : ; 21 cm. It79-May
At head of title: Istituto di programmazione territoriale e progettazioni, Politecnico di Torino.
Includes bibliographies.
L5000
1. City planning—Italy. I. Title.
HT169.I83P4 79-371966
*79 MARC

Selem, Hilda.
Il sistema dei forti di Roma nella logica dell'ecosistema urbano per il riuso del costruito / Hilda Selem. — Roma : Bulzoni, 1979.
148 p. : ill. ; 22 cm. It79-July
Added t.p., preface and summary in English.
Bibliography: p. 135-139.
L7500
1. Rome (City)—City planning. 2. City planning—Italy. 3. Architecture—Italy—Rome (City)—Conservation and restoration. 4. Rome (City)—Fortifications, military installations, etc. I. Title. II. Title: The system of Roman fortresses within the logic of the urban ecosystem for the reuse and rehabilitation of old buildings.
HT169.I84R645 79-383494
*79 MARC

Testa, Alfonso.
Intervista alla città : gestione urbana e partecipazione in un dialogo con i comitati di quartiere e con il sindaco di Roma G. C. Argan / Alfonso Testa ; presentazione di Giovanni Berlinguer. — Bari : De Donato, 1979.
271 p. ; 18 cm. — (Atti ; 49) It79-Mar
Bibliography: p. 269-₍270₎
L4500
1. Rome (City)—Politics and government. 2. Rome (City)—City planning. 3. City planning—Italy. I. Argan, Giulio Carlo. II. Title.
JS5873.A3T47 352'.96'0945632 79-373179
*79 MARC

Villani, Andrea, 1934-
Realtà e miti della progettazione : politiche alternative dell'habitat e forze sociali, diritto alla città e decisione dell'importanza dei problemi urbani, dall'architettura degli architetti nel Movimento moderno all'architettura della partecipazione, pianificazione urbanistica e pianificazione economico-sociale, modelli teorici ed esperienza italiana / Andrea Villani. — Milano : F. Angeli, ₍1978₎
201 p. ; 22 cm. — (Collana di sociologia urbana e rurale ; 11) It78-Oct
Bibliography: p. ₍191₎-201.
L6000
1. City planning—Italy. 2. Regional planning—Italy. I. Title.
HT169.I83V54 309.2'12'0945 79-343309
*79 MARC

Zampetti, Pietro.
Il problema di Venezia / Pietro Zampetti. — ₍Firenze₎ : Sansoni, 1976.
160 p. : ill. ; 19 cm. — (Scuola aperta ; 79 : Arte) It78-Mar
"Testimonianze": p. ₍68₎-114.
Bibliography: p. 65-67.
L2000
1. Venice—Threat of destruction. 2. Historic buildings—Italy—Venice—Conservation and restoration. 3. Venice—City planning. 4. City planning—Italy. 5. Venice, Lagoon of—Water level. I. Title.
DG672.5.Z34 309.2'62'094531 78-388909
*79 MARC

—ITALY—ADDRESSES, ESSAYS, LECTURES

Agricoltura e urbanistica : contributi per un dibattito / ₍di₎ Maspoli ... ₍et al.₎ ; a cura di M. Chiotasso, A. Fubini. — 2. ed. riv. e aggiornata. — Torino : Celid, 1978.
xi, 206 p. ; 24 cm. — (Quaderni della Sezione piemontese, Istituto nazionale di urbanistica ; 1) It78-Aug
Bibliography: p. 189-197.
L3000
1. City planning—Italy—Addresses, essays, lectures. 2. Agriculture—Italy—Addresses, essays, lectures. 3. Regional planning—Italy—Addresses, essays, lectures. I. Maspoli, Giuseppe. II. Chiotasso, M. III. Fubini, Alex. IV. Series: Istituto nazionale di urbanistica. Sezione piemontese. Quaderni della Sezione piemontese, Istituto nazionale di urbanistica ; 1.
HT169.I8A63 1978 78-391607
*79 MARC

Beguinot, Corrado.
Programmazione e sviluppo : venti saggi su questioni di urbanistica e di pianificazione (1952-1962) / Corrado Beguinot. — Napoli : F. Fiorentino, 1963.
236 p. ; 25 cm.
Includes bibliographical references.
1. City planning—Italy—Addresses, essays, lectures. 2. Regional planning—Italy—Addresses, essays, lectures. I. Title.
HT169.I83B423 79-345219
79 MARC

Benevolo, Leonardo.
Urbanistica e crisi economica / Leonardo Benevolo. — Roma ; Bari : Laterza, 1979.
viii, 206 p. ; 18 cm. — (Tempi Nuovi ; 106) It79-Mar
Articles previously published in Corriere della sera.
L4000
1. City planning—Italy—Addresses, essays, lectures. 2. Italy—Economic conditions—1945- —Addresses, essays, lectures. I. Title.
HT169.I83B46 79-360513
*79 MARC

I Centri storici : politica urbanistica e programma di intervento pubblico : Bergamo, Bologna, Brescia, Como, Gubbio, Pesaro, Vicenza / a cura di Sandro Ciardini e Paola Falini ; con i contributi di Sandro Bruschi ... ₍et al.₎. — Milano : G. Mazzotta, 1978.
287 p. : ill. ; 24 cm. — (Planning & design ; 25) It78-Dec
Includes bibliographical references.
ISBN 8820202328 : L12000
1. City planning—Italy—Addresses, essays, lectures. 2. Urban renewal—Italy—Addresses, essays, lectures. I. Ciardini, Francesco. II. Falini, Paola. III. Bruschi, Sandro.
HT169.I83C4 309.2'62'0945 79-346442
*79 MARC

Le Città, il fascismo / a cura di Mario Sanfilippo. — ₍Cosenza₎ : Lerici ; ₍s.l. : distribuzione Dielle, 1979?₎
168 p., ₍6₎ leaves of plates : ill. ; 22 cm. — (Architettura) (La Rivista ; 2-3) It***
Includes bibliographies.
L5000
1. City planning—Italy—Addresses, essays, lectures. 2. Fascism and architecture—Italy—Addresses, essays, lectures. I. Sanfilippo, Mario.
NA9203.C49 79-372513

Meccoli, Sandro, 1937-
La battaglia per Venezia / Sandro Meccoli. — Milano : SugarCo, 1977.
289 p. ; 21 cm. — (Fatti e misfatti ; 45) It77-June
"I testi ... sono già apparsi, come articoli, nel Corriere della sera."
L4000
1. Venice—Threat of destruction—Addresses, essays, lectures. 2. Pollution—Italy—Venice—Addresses, essays, lectures. 3. Venice—City planning—Addresses, essays, lectures. 4. City planning—Italy—Addresses, essays, lectures. 5. Venice, Lagoon of—Water level—Addresses, essays, lectures. I. Title.
DG672.5.M42 309.2'62'094531 78-388908
*79 MARC

Le Radici malate dell'urbanistica italiana / ₍scritti di₎ Giancarlo De Carlo ... ₍et al.₎. — Milano : Moizzi, 1976.
155 p., ₍4₎ leaves of plates : ill. ; 22 cm. — (Collana Spazio & ₍i.e. e₎ società ; 2) It78-Jan
Includes bibliographical references.
L3500
1. City planning—Italy—Addresses, essays, lectures. 2. Regional planning—Italy—Addresses, essays, lectures. I. Carlo, Giancarlo de.
HT169.I83R32 78-365803
*78 MARC

—ITALY—CONGRESSES

Centro storico : città, regione : idee ed esperienze di risanamento : confronto sui problemi di Torino : atti e documentazione del Convegno indetto dal Comune di Torino e dalla Sezione Piemonte-Valle d'Aosta dell'Ancsa e della Mostra documentaria promossa dagli assessorati alla casa e all'urbanistica di Torino, dall'Ancsa e dalla Camera di commercio di Torino, 27-28-29 maggio 1977 / a cura di Roberto Gambino e Giancarlo Massarella. — Milano : F. Angeli, ₍1978₎
343 p. : ill. ; 22 cm. — (Urbanistica ; 1) It78-Dec
Includes bibliographical references.
L6000
1. Urban renewal—Italy—Turin—Congresses. 2. Turin—City planning—Congresses. 3. City planning—Italy—Congresses. I. Gambino, Roberto. II. Massarella, Giancarlo.
HT178.I82T863 79-344888
*79 MARC

Interventi nel centro storico : confronto fra esperienze operative / promosso dall'assessorato per gli interventi nel centro storico del comune di Roma ... ; a cura di Mario Panizza — Bari : Dedalo libri, 1978.
244 p. : ill. ; 21 cm. — (Saggi ; 39) It78-Dec
Proceedings of a meeting held in Rome in 1977.
Bibliography: p. 15-16.
L5000
1. Urban renewal—Italy—Congresses. 2. City planning—Italy—Congresses. 3. Cities and towns—Italy—Congresses. I. Panizza, Mario.
HT178.I8I57 309.2'62'0945 79-352319
*79 MARC

La Riconversione urbanistica : atti del Convegno dell'Istituto nazionale di urbanistica, Milano, 20, 21, 22 febbraio 1976 / a cura di Marcello Fabbri e Augusto Cagnardi. — Bari : Dedalo libri, 1978.
292 p. : ill. ; 25 cm. — (Spazio, immagine, società ; 2) It78-July
At head of title: INU.
L8000
1. Urban renewal—Italy—Congresses. 2. City planning—Italy—Congresses. I. Fabbri, Marcello. II. Cagnardi, Augusto. III. Istituto nazionale di urbanistica.
HT178.I8R52 78-386170
*79 MARC

—ITALY—EXHIBITIONS

Progetti per l'area direzionale di Firenze : concorso nazionale per la progettazione planivolumetrica di un'area direzionale situata sul territorio fiorentino all'interno dell'area centrale metropolitana : Firenze, Orsanmichele, 15 aprile-31 maggio 1978 / catalogo a cura di Francesco Bandini. — Firenze : Centro Di, 1978.
₍247₎ p. : ill. ; 28 cm. It78-Oct
Half title: Comune di Firenze, Assessorato all'urbanistica.
1. Florence—City planning—Exhibitions. 2. City planning—Italy—Exhibitions. I. Bandini, Francesco.
NA9204.F6P76 79-368051
*79 MARC

—ITALY—HISTORY

Fubini, Alex.
Urbanistica in Italia / Alex Fubini. — 1. ristampa. — Milano : Clup, 1979.
431 p. : ill. ; 24 cm. It***
Includes bibliographical references.
ISBN 8870053784 : L7500
1. City planning—Italy—History. 2. Cities and towns—Italy—History. I. Title.
HT169.I83F8 79-362677
79 MARC

CITY PLANNING
—ITALY—HISTORY (Continued)

Piccinato, Luigi.
Urbanistica medievale / Luigi Piccinato. — Bari : Dedalo libri, 1978.
91 p. : ill. ; 18 cm. — (Universale di architettura ; 6) It78-June
Includes bibliographical references.
L2500
1. Cities and towns, Medieval—Italy. 2. City planning—Italy—History. I. Title. II. Series.
HT115.P46 78-378485
*78 MARC

—ITALY—PERIODICALS

Studi di urbanistica. v. 1–
[Bari, Italy] Dedalo libri, 1978–
v. ill. 25 cm.
1. City planning—Italy—Periodicals.
HT169.I 83S84 309.2′62′0945 78–647654
MARC-S

—ITALY—BOLOGNA

Gottarelli, Elena.
Urbanistica e architettura a Bologna agli esordi dell'unità d'Italia / Elena Gottarelli. — Bologna : Cappelli, 1978.
150 p. : ill. ; 27 cm. It79-Jan
Includes bibliographical references and index.
L9000
1. Bologna—City planning. 2. City planning—Italy—Bologna. 3. Architecture—Italy—Bologna. 4. Architecture, Modern—19th century—Italy—Bologna. I. Title.
NA9204.B6G67 79-356584
*79 MARC

—ITALY—CAMPANIA

Territorio e risorse in Campania / [di] Mariarosaria Abignente ... [et al.] — Napoli : Guida, 1978.
2 v. : ill. ; 18 cm. — (I Tascabili ; 49) It78-Dec
"Quest'opera è già apparsa in fascicoli su La Voce della Campania dall'ottobre 1977 al maggio 1978."
Bibliography: v. 2, p. 150-166.
CONTENTS: v. 1. Tra marginalità e crescita.—v. 2. Il potenziale urbano.

L6000
1. Campania—Economic conditions. 2. Regional planning—Italy—Campania. 3. City planning—Italy—Campania. I. Abignente, Mariarosaria, 1948-
HC307.C33T46 79-342130
*79 MARC

—ITALY—REGGIO EMILIA—HISTORY—EXHIBITIONS

Reggio Emilia, la città dall'età romana al XX secolo / testi di Walter Baricchi, Rolando Cavandoli, Attilio Marchesini. — Reggio Emilia : Tecnostampa, 1978.
63 p., 14 leaves of plates : ill. ; 21 x 21 cm. It78-Dec
At head of title: Comune di Reggio Emilia, Assessorato alle istituzioni culturali, Civici musei.
Published on the occasion of an exhibition held in Reggio Emilia in 1978.
Bibliography: p. 51-54.
1. City planning—Italy—Reggio Emilia—History—Exhibitions. 2. Reggio Emilia—History—Exhibitions. I. Baricchi, Walter. II. Cavandoli, Rolando. III. Marchesini, Attilio. IV. Reggio Emilia. Musei civici.
HT169.I84E57 309.2′5′094543 79-357320
*79 MARC

—ITALY, SOUTHERN—ADDRESSES, ESSAYS, LECTURES

Dalla regione al comprensorio : problemi di pianificazione urbanistica / scritti di Nicola Cacace ... [et al.] ; a cura di Francesco Forte. — Milano : F. Angeli, [1978]
415 p. : ill. ; 22 cm. — (Centro studi di politica economica. Collana CEEP : Sezione 3, Territorio ; 6) It78-Dec
Includes bibliographical references.
L10000
1. City planning—Italy, Southern—Addresses, essays, lectures. 2. Regional planning—Italy, Southern—Addresses, essays, lecture. I. Cacace, Nicola. II. Forte, Francesco, 1939- III. Series.
HT169.I83D34 309.2′12′09457 79-347196
*79 MARC

—JAPAN

Ichiura Toshi Kaihatsu Kenchiku Konsarutantsu.
(Narita nyū taun kihon keikaku hōkokusho)
成田ニュータウン基本計画報告書　市浦都市開発建築コンサルタンツ　宅地開発研究所　日本技術開発株式会社 [編　成田]　千葉県北総開発局 1969
250 p. maps. 30 cm. Ja 72-18548
1. Narita, Japan—City planning. 2. City planning—Japan. I. Takuchi Kaihatsu Kenkyūjo. II. Nihon Gijutsu Kaihatsu Kabushiki Kaisha. III. Title.
HT169.J32N44 79-806189

(Jōkamachi Chōfu)
城下町長府：下関市長府地区伝統的建造物群調査報告書 / [下関市教育委員会]. — [下関]：下関郷土の文化財を守る会, 1977.
40 p., [4] leaves of plates : ill. ; 26 cm. Ja 77-22855
Cover title.
Errata slip inserted.
1. Architecture—Japan—Shimonoseki—Conservation and restoration. 2. Shimonoseki, Japan—City planning. 3. City planning—Japan. 4. Shimonoseki, Japan—History. I. Shimonoseki, Japan. Kyōiku Iinkai.
NA1557.S46J64 79-803585

Kōbe Daigaku. Kōgakubu. Kenchiku Keikaku Kenkyūshitsu.
(Tamba Sasayama sono rekishi bunka kūkan no sosei)
『丹波篠山』その歴史・文化・空間の蘇生：兵庫県篠山町歴史的町並保全整備調査研究報告書 / [調査研究　神戸大学工学部建築史研究室・建築計画研究室；編集作成　神戸大学工学部建築計画研究室]. — 篠山町(兵庫県)：篠山町教育委員会, 1976.
76 p., [2] leaves of plates : ill. ; 31 cm. Ja 77-823
Cover title.
Commisioned by Sasayama-chō Kyōiku Iinkai.
Bibliography: p. 7.
1. Architecture, Japanese—Japan—Sasayama. 2. Architecture—Japan—Sasayama—Conservation and restoration. 3. Sasayama, Japan—City planning. 4. City planning—Japan. I. Kōbe Daigaku. Kōgakubu. Kenchikushi Kenkyūshitsu. II. Sasayama, Japan. Kyōiku Iinkai. III. Title.
NA1557.S27K6 1976 78-817014

(Kyō no machinami)
京の町並 / [編集　京都府文化財保護基金]. — 京都：同基金, 1977.
129 p., [4] leaves of plates : chiefly ill. (some col.) ; 20 x 22 cm. Ja 78-4124
"昭和47年8月に京都府教育庁文化財保護課が行った町並・集落調査をもとに作成"
1. Architecture, Japanese—Japan—Kyoto. 2. Architecture—Japan—Kyoto—Conservation and restoration. 3. City planning—Japan. 4. Kyoto—City planning. 5. Kyoto—Aerial photographs. I. Kyōto-fu Bunkazai Hogo Kikin. II. Kyoto (Prefecture). Kyōikuchō. Bunkazai Hogoka.
NA1557.K9K955 79-807257

Murataka, Mikihiro.
(Tōkyō-shi no kaizō)
東京市の改造 / 村高幹博著. — 東京：民友社, 大正11 [1922]
2, 14, 168, 54 p. : ill. ; 19 cm.
Cover title.

1. Tokyo—City planning. 2. City planning—Japan. I. Title.
HT169.J32T646 79-808210

Nara Kokuritsu Bunkazai Kenkyūjo.
(Gojō)
五條：町並調査の記録. — 奈良：奈良国立文化財研究所, 昭和52 [1977]
248 p. : ill. ; 30 cm. — (奈良国立文化財研究所学報；第30冊) Ja 77-26190
Summary in English.
Errata slip inserted.
1. Architecture, Japanese—Japan—Gojō. 2. Architecture—Designs and plans. 3. Architecture—Japan—Gojō—Conservation and restoration. 4. Gojō, Japan—City planning. 5. City planning—Japan. I. Series: Nara Kokuritsu Bunkazai Kenkyūjo. Nara Kokuritsu Bunkazai Kenkyūjo gakuhō ; dai 30-satsu.
NA1557.G64N37 1977 78-813759

(Nihon no toshi)
日本の都市 / 建設大臣官房広報室編. — 東京：建設広報協議会, 1965.
12, 271 p., [3] leaves of plates : ill. ; 21 cm. — (建設双書；4)

1. City planning—Japan. 2. City planning and redevelopment law—Japan. 3. Cities and towns—Japan. I. Japan. Kensetsushō. Daijin Kambō. Kōhōshitsu. II. Series: Kensetsu sōsho ; 4.
HT169.J3N47 78-814950

Library of Congress 78 J

Ōishi, Taizan, 1938–
(Kobetsu jūtakuchika ni yoru toshika to kinkō chiiki no hembō oyobi sore ni fuzuisuru ni san no mondai)
個別住宅地化による都市化と近郊地域の変貌およびそれに付随する2・3の問題 / 大石堪山. — 東京：東京都立大学都市研究組織委員会, 1974.
vii, 207 p. : ill. ; 21 cm. — (都市研究報告；43) Ja***
Cover title.
Errata slip inserted.
Includes bibliographical references.

1. Urbanization—Japan. 2. Land use, urban—Japan. 3. City planning—Japan. 4. Housing—Japan. I. Title. II. Series: Toshi kenkyū hōkoku ; 43.
HT151.O49 79-809245

J

Sendai, Japan (Miyagi Prefecture). Dobokubu. Toshi Keikakuka.
(Sendai-shi toshi keikaku)
仙台市都市計画 = Sendai city planning / 仙台市土木部都市計画課. — 仙台：仙台市役所, 昭和31 [1956]
2 v. : ill. ; 27 cm.
Vol. 2: Fuzu.
1. Sendai, Japan (Miyagi Prefecture)—City planning. 2. City planning—Japan. I. Title. II. Title: Sendai city planning.
HT169.J32S467 79-806806

Tokyo.
(Tōkyō-to ni okeru kodomo no asobiba no tekisei haichi ni kansuru chōsa hōkokusho)
東京都におけるこどもの遊び場の適正配置に関する調査報告書 / [東京都総務局青少年対策部計画課編]. — 東京：東京都総務局青少年対策部計画課, 昭和49 [1974]
182 p. ; 25 cm. Ja 74-17932
Cover title.
"この調査は東京都が昭和42年度から昭和45年度にわたって実施した"

1. Tokyo—City planning. 2. Tokyo—Playgrounds—Directories. 3. City planning—Japan. I. Tokyo. Sōmukyoku. Seishōnen Taisakubu. Keikakuka. II. Title.
HT169.J32T6558 79-806710

Tokyo. Sōmukyoku. Shōgai Kankōbu. Gaijika.
City planning of Tokyo / [edited by the Liaison and Protocol Section, Bureau of General Affairs, Tokyo Metropolitan Government]. — Tokyo : Tokyo Metropolitan Govt., 1978.
174 p. : ill. ; 20 cm. — (TMG municipal library ; no. 13) Ja***
1. Tokyo—City planning. 2. City planning—Japan. I. Title. II. Series.
HT169.J32T6618 309.2′62′0952135 79-305002
79 MARC

(Toshi to rinkaibu kaihatsu)
都市と臨海部開発 / 北見俊郎, 奥村武正編著. — 東京：成山堂書店, 昭和52 [1977]
5, 3, 216 p., [2] leaves of plates : ill. ; 22 cm. Ja 77-13461
Includes bibliographical references.
¥2500

1. Tokyo—City planning. 2. Harbors—Planning—Japan—Tokyo Bay. 3. Fills (Earthwork) 4. City planning—Japan. I. Kitami, Toshirō. II. Okumura, Takemasa, 1917–
HT169.J32T678 78-815646

CITY PLANNING
—JAPAN (Continued)

Watanabe, Shinzō, 1914-
(Toshi keikaku yōron)
都市計画要論 / 渡辺新三, 松井寛共著. — 東京 : 国民科学社 : 発売 オーム社書店, 1978.
2, 5, 247, 6 p. : ill. ; 22 cm. Ja 78-26851
Includes bibliographies and index.
¥2500

1. City planning. 2. City planning—Japan. I. Matsui, Hiroshi, 1941- joint author. II. Title.
HT166.W339 79-803735

Yamaga, Seiji, 1916-
(Hendōsuru toshi)
変動する都市 / 山鹿誠次. — 東京 : 東京書籍, 1979.
190 p. : ill. ; 19 cm. — (東書選書 ; 31) Ja 79-5734
Series romanized: Tōsho sensho.
Includes index.
¥820

1. Cities and towns—Japan. 2. City planning—Japan. 3. Land use, Urban—Japan. I. Title.
HT147.J3Y24 79-808605

Zaino, Hiroshi, 1939-
(Kaiwai)
かいわい : 日本の都心空間 / 材野博司. — 東京 : 鹿島出版会, 1978.
255 p. : ill. ; 19 cm. — (SD選書 ; 126) Ja 78-4129
Series romanized: SD sensho.
Includes bibliographical references.
¥980

1. City planning. 2. City planning—Japan. 3. Cities and towns—Japan. I. Title.
HT166.Z34 79-809683

—JAPAN—ADDRESSES, ESSAYS, LECTURES

Andō, Motoo, 1934-
(Kyojūten no shisō)
居住点の思想 : 住民·運動·自治 / 安藤元雄. — 東京 : 晶文社, 1978.
282 p. ; 20 cm. Ja 78-21506
Includes bibliographical references.
¥1600
1. Fujisawa, Japan (Kanagawa Prefecture)—Politics and government—Addresses, essays, lectures. 2. Political participation—Japan—Fujisawa (Kanagawa Prefecture)—Addresses, essays, lectures. 3. Fujisawa, Japan (Kanagawa Prefecture)—City planning—Addresses, essays, lectures. 4. City planning—Japan—Addresses, essays, lectures. I. Title.
JS7385.F823A2 1978 79-802940

(Asu no jūtaku to toshi)
明日の住宅と都市 / 建設省編. — 東京 : 彰國社, 昭和24 [1949]
363 p., [1] leaf of plates : ill. ; 19 cm.

1. Architecture, Domestic—Japan—Addresses, essays, lectures. 2. City planning—Japan—Addresses, essays, lectures. I. Japan. Kensetsushō.
NA7451.A84 79-808438

Toshi Mondai Kenkyūkai (Founded 1975)
(Toshi Mondai Kenkyūkai kōen shū)
都市問題研究会講演集. — [東京] : 都市計画協会, 1978.
154 p. ; 26 cm. Ja 78-18904
Cover title.

1. City planning—Addresses, essays, lectures. 2. City planning—Japan—Addresses, essays, lectures. 3. City planning—Social aspects—Addresses, essays, lectures. I. Title.
HT166.T6187 1978 79-802749

—JAPAN—CITIZEN PARTICIPATION

(Jūmin undō no kiroku)
住民運動の記録 / [編集 都市問題連絡会]. — 東京 : 都市問題連絡会事務局, 1977-
v. : 19 cm. Ja 78-17996 (v. 1)
CONTENTS: 1. まちはどのようにして変えられていくのか

1. Tokyo—City planning—Citizen participation. 2. City planning—Japan—Citizen participation. 3. Community development, Urban—Japan—Tokyo—Citizen participation. I. Toshi Mondai Renrakukai.
HT169.J32T644 78-815409

Kihara, Keikichi.
(Kankyō no shisō o motomete)
環境の思想を求めて : 対談集 / 木原啓吉編著 ; 対談者 横山桂次 ... [et al.]. — 東京 : 核心評論社, 1978.
450 p. ; 19 cm. — (争点シリーズ ; no. 1) Ja 79-7782
Originally published in Sōten, July 1974-Sept. 1977.
¥1600
1. City planning—Japan—Citizen participation. 2. Community development—Japan—Citizen participation. 3. Environmental protection—Japan—Citizen participation. I. Yokoyama, Keiji, 1920- II. Title. III. Series: Sōten shirizu ; no. 1.
HT169.J3K5 79-810142

J

—JAPAN—CITIZEN PARTICIPATION —ADDRESSES. ESSAYS, LECTURES

Hirano, Ken, 1907-1978.
(Kukaku seiri hō wa Kempō ihan)
区画整理法は憲法違反 / 平野謙. — 東京 : 潮出版社, 1978.
282 p. ; 20 cm. Ja 78-16123
¥1200

1. Tokyo—City planning—Citizen participation—Addresses, essays, lectures. 2. City planning—Japan—Citizen participation—Addresses, essays, lectures. 3. Zoning law—Japan—Tokyo—Addresses, essays, lectures. I. Title.
HT169.J32T642 78-812882

—JAPAN—DICTIONARIES—JAPANESE

(Gendai toshi keikaku yōgoroku)
現代都市計画用語録 / 梶秀樹 ... [et al.] 著. — 東京 : 彰国社, 1978.
258 p. : ill. ; 19 cm. Ja 79-2517
Bibliography: p. 236-238.
Includes index.
¥1600

1. City planning—Dictionaries—Japanese. 2. City planning—Japan—Dictionaries—Japanese. I. Kaji, Hideki, 1912-
HT166.G46 79-801795

J

—JAPAN—MAPS

(Nihon ni okeru toshika no chizu bunseki)
日本における都市化の地図分析 / 研究代表者 山本正三. — [東京 : トヨタ財団], 1977.
154 p. : ill. ; 30 cm. Ja 78-9390
Cover title.
"トヨタ財団研究助成"

1. Cities and towns—Japan. 2. City planning—Japan—Maps. 3. Urbanization—Japan. 4. Open spaces—Japan. I. Yamamoto, Shōzō, 1928-
HT147.J3N48 78-812126

—JAPAN—YEARBOOKS

Tokyo. Toshi Keikakukyoku.
(Shuto no seibi)
首都の整備 Planning of Tokyo.
[東京 東京都都市計画局]
v. col. ill., col. maps. 21 x 30 cm. annual.
Continues: Tokyo. Shuto Seibikyoku. Shuto no seibi.
In English and Japanese.

1. Tokyo—City planning—Yearbooks. 2. City planning—Japan—Yearbooks. I. Tokyo. Toshi Keikakukyoku. Planning of Tokyo. II. Title.
HT169.J32T659 78-642539
(MARC-S)

—JAPAN—GUMMA (PREFECTURE)

(Maebashi Takasaki Isezaki chiiki ni okeru toshi keisei ni kakawaru chōsa kenkyū hōkokusho)
前橋·高崎·伊勢崎地域における都市形成に係る調査研究報告書. — [前橋] : 群馬県土木部都市計画課, 日本システム開発研究所, 昭和51 [1976]
188 p. in various pagings : ill. ; 30 cm. Ja***
Cover title.
1. City planning—Japan—Gumma (Prefecture) 2. Rural development—Japan—Gumma (Prefecture) I. Gumma, Japan (Prefecture). Dobokubu. Toshi Keikakuka. II. Nihon Shisutemu Kaihatsu Kenkyūjo.
HT169.J32G85 79-802610

—JAPAN—TOKYO

(Shutoken wa kō kawaru)
首都圏はこう変わる : 首都圏整備委員会の長期整備計画の全貌. — [東京] : 企業組織研究会 : 取り扱い 日刊電気ジャーナル社, 昭和38 [1963]
213 p. : ill. ; 25 cm.
Cover title.

1. Tokyo—City planning. 2. City planning—Japan—Tokyo. 3. Tokyo metropolitan area. I. Japan. Shutoken Seibi Iinkai. II. Kigyo Soshiki Kenkyukai.
HT169.J32T656 79-803618

—JORDAN

Segal, Arthur.
(Tikhnunan shel he-'arim le-orekh Via Traiana Nova ba-tekufah ha-romit)
תיכנונן של הערים לאורך Via Traiana Nova בתקופה הרו־מית / מאת ארתור סגל. — [s. l. : s. n., 1975?]
2 v. : ill. ; 28 cm.
Added t. p.: The planning of the cities along the Via Traiana Nova in the Roman period.
Thesis—ha-Universitah ha-'ivrit.
Summary in English.
Bibliography: v. 1, leaves 158—161.
1. City planning—Jordan. 2. Architecture, Ancient—Jordan. 3. Cities and towns, Ancient—Jordan. 4. Jordan—Antiquities. I. Title. II. Title: The planning of the cities along the Via Traiana Nova in the Roman period. III. Title: Via Traiana Nova.
NA9246.J6S43 711'.4'09394 79-951373

HE

—KENTUCKY

Morgan, William, 1944-
Louisville : architecture and the urban environment / William Morgan ; introd. by Noel Perrin. — Dublin, N.H. : W. L. Bauhan, 1979.
96 p. : ill. ; 27 cm.
Based on author's "Urban environment" column in the Courier-journal, Louisville, Ky.
ISBN 0-87233-050-8 : $6.95
1. Louisville, Ky.—Buildings. 2. Architecture—Kentucky—Louisville. 3. Louisville, Ky.—City planning. 4. City planning—Kentucky. I. Title.
NA735.L6M67 720'.9769'44 79-631
79 MARC

CITY PLANNING (Continued)

—KENYA

Nairobi Urban Study Group.
Nairobi, metropolitan growth strategy / Nairobi Urban Study Group. — Nairobi : City Council of Nairobi, 1973.
2 v. : ill. ; 30 cm.
Includes a bibliography.
CONTENTS: v. 1. Main report.—v. 2. Technical appendices.
1. Nairobi—City planning. 2. City planning—Kenya. I. Title.
HT169.K42N346 711'.4'0967625 78-980063
79 MARC

Stren, Richard E
Urban development in Kenya and Tanzania : a comparative analysis / by Richard E. Stren. — Nairobi : Institute for Development Studies, University of Nairobi, 1976.
30 p. ; 30 cm. — (Discussion paper - University of Nairobi, Institute for Development Studies ; no. 232)
Bibliography: p. 28-30.
1. City planning—Kenya. 2. City planning—Tanzania. 3. Urbanization—Kenya. 4. Urbanization—Tanzania. I. Nairobi. University. Institute for Development Studies. II. Title. III. Series: Nairobi. University. Institute for Development Studies. Discussion paper ; no. 231.
HT169.K4S87 309.2'62'096762 77-980079
79 MARC

—KENYA—KAKAMEGA

Kenya. Dept. of Urban & Rural Physical Planning.
Kakamega municipality physical development plan report, to accompany the approved short term physical development plan map. — ⌊Nairobi⌋ : Dept. of Urban & Rural Physical Planning, Ministry of Lands & Settlement, 1973.
16 leaves, ⌊1⌋ leaf of plates : ill. ; 31 cm.
"Reference no. DURP.16/6/17."
1. Kakamega, Kenya—City planning. 2. City planning—Kenya—Kakamega. I. Title.
HT169.K42K344 309.2'62'0967628 78-102659
78 MARC

—KENYA—MARSABIT

Kenya. Town Planning Dept.
Marsabit township development plan drawing no. 276/68/2. — Nairobi : Town Planning Dept., Ministry of Lands and Settlement, 1969.
8, ⌊3⌋ leaves ; 34 cm.
1. Marsabit, Kenya—City planning. 2. City planning—Kenya—Marsabit. I. Title.
HT169.K42M364 309.2'62'0967624 78-102635
78 MARC

Kim, Hyŏng-man.
(Taedŏk yŏn'gu hagwŏn tosi chosa yŏn'gu mit kibon kyehoek e kwanhan yŏn'gu)
大德研究學園都市調査研究및基本計劃에關한研究：基本調査資料 / 研究責任者金烱萬. — ⌊서울⌋：科學技術處, ⌊1973⌋
213 p. : ill. ; 27 cm.
"R-73-17"
1. Regional planning—Korea—Taedŏk-gun. 2. City planning—Korea. 3. Taejŏn, Korea—City planning. 4. Industrial districts—Korea—Taedŏk-gun. I. Title.
HT395.K62T334 78-826811

Kim, Hyŏng-man.
(Taedŏk yŏn'gu hagwŏn tosi chosa yŏn'gu mit kibon kyehoek e kwanhan yŏn'gu)
大德研究學園都市調査研究및基本計劃에關한研究 = A study on the master plan of the science town at Dai-tok / 研究責任者金烱萬. — ⌊서울⌋：科學技術處, ⌊1973?⌋
vi, 649 p. : ill. ; 27 cm.
In Korean.
"R73-17"
1. Regional planning—Korea—Taedŏk-gun. 2. City planning—Korea. 3. Taejŏn, Korea—City planning. 4. Industrial districts—Korea—Taedŏk-gun. I. Title. II. Title: A study on the master plan of the science town at Dai-tok.
HT395.K62T333 78-826890

—KUWAIT, ARABIA (STATE)

Shiber, Saba George, 1923–
(al-'Ilm wa-tanẓīm al-mudun)
العلم وتنظيم المدن العربية ، تاليف سابا جورج شبر . ⌊بيروت⌋
دار الكاتب العربي ⌊التوطئة 1964⌋
302 p. 23 cm.
1. City planning—Arab countries. 2. City planning—Kuwait, Arabia (State) I. Title.
HT178.A65S54 N E 65–1484

—LATIN AMERICA—HISTORY

Martínez Lemoine, René.
El modelo clásico de ciudad colonial hispanoamericana : ensayo sobre los orígenes del urbanismo en América / René Martínez L. — Santiago de Chile : Departamento Planificación Urbano Regional F.A.U., Universidad de Chile, 1977.
102 p. : ill. ; 26 cm.
"500 ejemplares."
Includes bibliographical references.
1. Cities and towns—Latin America—History. 2. City planning—Latin America—History. 3. Urbanization—Latin America—History. I. Title.
HT127.5.M37 78-109673
79 MARC

—LATVIA—ADDRESSES, ESSAYS, LECTURES

Voprosy gradostroitel'stva / ⌊redakt͡sionnai͡a kollegii͡a V. I. Oleĭnik ... et al.⌋. — Riga : Zvaigzne, 1977.
95 p. : ill. ; 20 cm. — (Voprosy stroitel'stva ; 6)
At head of title: Latviĭskiĭ nauchno-issledovatel'skiĭ i ėksperimental'no-tekhnologicheskiĭ institut stroitel'stva Gosstroi͡a Latviĭskoĭ SSR.
Includes bibliographies.
0.45rub
1. City planning—Latvia—Addresses, essays, lectures. I. Oleĭnik, V. I. II. Latviĭskiĭ nauchno-issledovatel'skiĭ i ėksperimental'no-tekhnologicheskiĭ institut stroitel'stva. III. Series.
HT169.R82L378 79-380580
79 MARC

—LITHUANIA

Čekanauskas, Vytautas, 1930–
(Zhiloĭ raĭon Lazdinaĭ v Vil'ni͡use)
Жилой район Лаздинай в Вильнюсе / В. Чеканаускас, нар. архитектор СССР, лауреат Ленинской премии, В. Римкус, канд. искусствоведения. — Москва : Знание, 1977.
32 p., ⌊8⌋ leaves of plates : ill. ; 20 cm. — (Новое в жизни, науке, технике : Серия Строительство и архитектура ; № 9, 1977)
USSR 77
Bibliography : p. 32.
0.12rub
1. Lazdinay, Vilnius, Lithuanian S. S. R. 2. Vilnius, Lithuanian S. S. R.—City planning. 3. City planning—Lithuania. 4. Architecture, Modern—20th century—Lithuania—Vilnius, Lithuanian S. S. R. I. Rimkus, Vitenis, joint author. II. Title. III. Series: Novoe v zhizni, nauke, tekhnike : Serii͡a Stroitel'stvo i arkhitektura ; 1977, 9.
TH4.N64 1977, 9 79–408781
[NA9053.N4]

—MALAYSIA—CONGRESSES

Seminar on Public Participation in the Town and Country Planning Process, Pinang, Pulau Pinang, 1976.
Report of proceedings and technical papers, discussions and recommendations of the Seminar on Public Partic⌊i⌋pation in the Town and Country Planning Process, 5-7 February, 1976, Hotel Merlin, Penang. — ⌊Kuala Lumpur⌋ : Malaysian Institute of Planners, ⌊1976?⌋
70 p. ; 30 cm.
Cover title.
Includes bibliographical references.
1. Regional planning—Malaysia—Congresses. 2. City planning—Malaysia—Congresses. I. Malaysian Institute of Planners.
HT395.M39S43 1976 309.2'5'09595 77-941546
78 MARC

—MANITOBA

Damas and Smith Limited.
Cranberry Portage, local government district of Consol : community infrastructure analysis / prepared by Damas and Smith Limited, for the Manitoba Dept. of Municipal Affairs, Municipal Planning Branch. — ⌊Winnipeg⌋ : Damas and Smith, 1976.
xi, 55, 22 p., 8 leaves of plates (5 fold.) : ill. ; 29 cm. C***
1. Cranberry Portage, Man.—City planning. 2. City planning—Manitoba. I. Manitoba. Municipal Planning Branch. II. Title.
HT169.C32.C723 711'.4'0971272 79-302688

Manitoba. Municipal Planning Branch.
Churchill townsite redevelopment proposal. ⌊Winnipeg⌋ 1971.
167 p. illus. 22 x 36 cm. C***
1. Churchill, Manitoba—City planning. 2. City planning—Manitoba. 3. Urban renewal—Manitoba—Churchill. I. Title.
HT169.C32C55 711'.4'0971271 72-172083
73⌊r79⌋rev MARC

Metropolitan Corporation of Greater Winnipeg, Man. Planning Division.
The Metropolitan development plan. ⌊Winnipeg, 1968⌋
112, ⌊3⌋ p. col. illus., col. maps. 37 cm. $15.00 C69-1416
Cover title.
Includes bibliographical references.
1. Winnipeg—City planning. 2. Regional planning—Manitoba—Winnipeg metropolitan area. 3. City planning—Manitoba. I. Title.
HT169.C32W54 711'.4'0971274 74-573621
71⌊r79⌋rev MARC

—MANITOBA—CASE STUDIES

Metropolitan Corporation of Greater Winnipeg, Man. Planning Division.
Downtown Winnipeg. ⌊Winnipeg, 1969⌋
vi, 110 p. illus. (part fold., part col.) 31 cm. $6.00 C69-4403
1. Winnipeg—City planning. 2. City planning—Manitoba—Case studies. I. Title.
HT169.C32W53 711'.4'0971274 72-183037
72⌊r79⌋rev MARC

—MARYLAND—CASE STUDIES

Murphy/Williams Urban Planning and Housing Consultants.
Ellicott City : new life for an old town ... : summary planning report : prepared for Howard County Office of Planning and Zoning / prepared by Murphy/Williams Urban Planning and Housing Consultants. — Ellicott City, Md. : The Office, 1976.
iii, 72 p. : ill. ; 29 cm.
1. Ellicott City, Md.—City planning. 2. City planning—Maryland—Case studies. I. Howard Co., Md. Office of Planning and Zoning. II. Title.
HT168.E44M87 1976 711'.4'0975281 78-322667
79 MARC

—MASSACHUSETTS

Massachusetts. Office of Local Assistance.
Community needs survey analysis / Department of Community Affairs, Office of Local Assistance. — Boston : The Office, 1975.
151, 4 leaves : forms ; 28 cm.
1. Municipal services—Massachusetts. 2. City planning—Massachusetts. I. Title.
HD4606.M4M37 1975 363.5 77-620649
77⌊r79⌋rev MARC

—MASSACHUSETTS—HISTORY

Emmet, Alan.
Cambridge, Massachusetts, the changing of a landscape / by Alan Emmet, with Douglas C. Allen ... ⌊et al.⌋. — ⌊Cambridge, Mass. : Harvard University⌋, c1978.
v, 87 p. : ill. ; 22 x 28 cm.
Bibliography: p. 83-84.
Includes index.
1. Cambridge, Mass.—City planning—History. 2. City planning—Massachusetts—History. 3. Land use, Urban—Massachusetts—Cambridge—History. 4. Cambridge, Mass.—History. I. Title.
HT168.C28E47 309.2'62'097444 78-67793
79 MARC

—MEDITERRANEAN REGION

(Chichūkai no mura to machi)
地中海の村と街 = Mediterranean Sea / 企画・撮影 二川幸夫；文・解説 鈴木恂；作図 ウェイン・藤井；デザイン 細谷巖. — 東京：A.D.A. EDITA Tokyo Co., 1975.
135 p. : chiefly ill. (some col.) ; 31 cm. — (世界の村と街 : 3)
Ja 75–23230
¥3500
1. Architecture—Mediterranean region. 2. City planning—Mediterranean region. I. Futagawa, Yukio, 1932– II. Suzuki, Makoto, 1935– III. Title: Mediterranean Sea. IV. Series: Sekai no mura to machi ; 3.
NA202.C45 79–801489

CITY PLANNING (Continued)

—MEXICO

Instituto Nacional de la Vivienda, Mexico.
Aztlán, 2000 ⌈i.e. dos mil⌉ : espacio habitable social. — ⌈México⌉ : Instituto Nacional de la Vivienda, ⌈1970⌉
⌈47⌉ p. : ill. ; 31 cm.
1. Housing policy—Mexico—Mexico (City) metropolitan area. 2. Mexico (City) metropolitan area—City planning. 3. City planning—Mexico. I. Title.
HD7306.M4I52 1970 78-111510
79 MARC

Plan Nacional de Desarrollo Urbano : mayo de 1978. — ⌈México⌉ : Secretaría de Asentamientos Humanos y Obras Públicas, Comisión Nacional de Desarrollo Urbano, Secretaría de Programación y Presupuesto, ⌈1978⌉
7 v. : maps ; 28 cm. (v. 6-7, 28 x 41 cm.)
CONTENTS: 1. Nivel normativo.—2. Nivel estratégico.—3. Corresponsabilidad sectorial.—4. Instrumentos jurídicos.—⌈5⌉ Anexo documental.—⌈6⌉ Matrices de congruencia.—⌈7⌉ Anexo gráfico.
1. City planning—Mexico. I. Mexico. Secretaría de Asentamientos Humanos y Obras Públicas. II. Mexico. Comisión Nacional de Desarrollo Urbano. III. Mexico. Secretaría de Programación y Presupuesto.
HT169.M4P55 79-351636
79 MARC

—MEXICO—PERIODICALS

Instituto Nacional para el Desarrollo de la Comunidad Rural y de la Vivienda Popular.
INDECO edita.
México, Instituto Nacional para el Desarrollo de la Comunidad Rural y de la Vivienda Popular.
no. ill. 28 cm. monthly.
Numbers for Feb. 1976– also called v. 2–
Supplements accompany some issues.
1. Community development—Mexico—Periodicals. 2. Regional planning—Mexico—Periodicals. 3. City planning—Mexico—Periodicals. 4. Housing policy—Mexico—Periodicals. I. Title.
HN111.I 57a 309.2′5′0972 78-646868
MARC-S

—MINNESOTA—HISTORY

Vance, John E
Inside the Minnesota experiment : a personal recollection of experimental planning and development in the Twin Cities metropolitan area, including an introduction to the Vance papers / by John E. Vance. — Minneapolis : Center for Urban and Regional Affairs, 1977.
vii, 117 p. ; 28 cm.
"Publication no. CURA 77-7."
1. City planning—Minnesota—History. 2. Regional planning—Minnesota—History. 3. Metropolitan Council of the Twin Cities Area. 4. St. Paul—City planning—History. 5. Minneapolis—City planning—History. I. Minnesota. University. Center for Urban and Regional Affairs. II. Title.
HT167.5.M6V35 309.2′5′09776579 78-101602
79 MARC

—MISSOURI

Simonds, Robert C
Handbook for planning commissioners in Missouri / by Robert C. Simonds ; rev. by Henry Galetschky. — ⌈Columbia, Mo.⌉ : Governmental Affairs Program, Dept. of Political Science, College of Arts and Science, 1978.
iii, 44 p. : maps ; 22 x 28 cm. — (Missouri local government handbook series)
"UED 46. 12/77/1M."
Bibliography: p. 22.
1. Regional planning—Missouri. 2. City planning—Missouri. I. Galetschky, Henry. II. Title. III. Series.
HT393.M8S55 352′.96′09778 79-621271
79 MARC

—MOROCCO

ousing and town-planning in Morocco. — Rabat ⌈Morocco⌉ : Éditions Africaines Perceval, ⌈1952?⌉
⌈36⌉ p. : ill. ; 27 cm.
1. Housing—Morocco. 2. City planning—Morocco.
HD7372.3.A3H68 78-315235
78 MARC

(Morokko no mura to machi)
モロッコの村と街 = Morocco / 企画・撮影 二川幸夫 ; 文・作図 鈴木恂 ; デザイン 細谷巖. — 東京 : A. D. A. EDITA Tokyo Co., 1975.
135 p. : chiefly ill. (some col.) ; 30 cm. — (世界の村と街 ; 10)
Ja 75-22721
¥3500
1. Architecture—Morocco. 2. City planning—Morocco. I. Futagawa, Yukio, 1932- II. Suzuki, Makoto, 1935- III. Hosoya, Gan. IV. Title: Morocco. V. Series: Sekai no mura to machi ; 10.
NA1590.M67 79-801525

—NEAR EAST

Roberts, M Hugh P
An urban profile of the Middle East / M. Hugh P. Roberts. — New York : St. Martin's Press, 1979.
239 p. : ill. ; 24 cm.
Includes bibliographies and index.
ISBN 0-312-83467-5
1. Cities and towns—Near East. 2. Cities and towns—Africa, North. 3. Urbanization—Near East. 4. Urbanization—Africa, North. 5. City planning—Near East. 6. City planning—Africa, North. I. Title.
HT147.N4R6 1979 309.2′62′0956 78-27185
78 MARC

—NEAR EAST—CONGRESSES

International Seminar on Regional and Community Planning, Berlin, 1966.
Seminar über "Regional- und Städteplanung = Seminar on "Regional and Community Planning". — ⌈Berlin⌉ : Deutsche Stiftung für Entwicklungsländer, ⌈1966⌉
58 pieces (in portfolio) ; 32 cm.
Title from portfolio.
Held Feb. 26-Mar. 26, 1966, at Berlin-Tegel, Reiherwerder.
In English.
Consists of a report, list of delegates, programme, minutes of sessions, country reports, etc.
"S1/66."
1. Regional planning—Near East—Congresses. 2. City planning—Near East—Congresses. I. Deutsche Stiftung für Entwicklungsländer.
HT395.N3157 1966 309.2′5′0956 79-302459
79 MARC

—NETHERLANDS

Dien, Ab van.
Eerste palen slaan : rapportage van en kanttekening bij wenselijkheid en werkelijkheid van inspraak bij bouwen en wonen : ontwikkelingen in stedebouw en architectuur / Ab van Dien. — Deventer : Kluwer Technische Boeken, 1978.
188 p. : ill. ; 30 cm. Ne78-22
ISBN 9020110608 : fl 32.50
1. Functionalism (Architecture)—Netherlands. 2. City planning—Netherlands. 3. Architecture, Modern—20th century—Netherlands. I. Title.
NA9053.F8D53 79-370474
*79 MARC

Faludi, Andreas.
Flexibility in Dutch local planning / ⌈by⌉ Andreas Faludi and Stephen Hamnett. — ⌈Oxford⌉ : Oxford Polytechnic, Department of Town Planning, 1977.
⌈2⌉, iii leaves, 83 p. ; 30 cm. — (Working paper - Oxford Polytechnic, Department of Town Planning ; no. 28) GB78-13698
Bibliography: p. 80-83.
£1.88
1. City planning—Netherlands. 2. City planning—Netherlands—Decision making. 3. City planning and redevelopment law—Netherlands. I. Hamnett, Stephen, joint author. II. Title. III. Series: Oxford working papers in planning education and research ; no. 28.
HT169.N4F34 309.2′62′09492 78-320525
78 MARC

Hoekveld, G A
Baarn : schets van de ontwikkeling van een villadorp / G. A. Hoekveld. — Baarn : Bosch & Keuning, ⌈1964⌉
233 p., ⌈4⌉ leaves of plates : ill. ; 25 cm.
Summary in English.
Bibliography: p. 206-210.
1. Baarn, Netherlands—City planning. 2. Baarn, Netherlands—History. 3. Urbanization—Netherlands—Amsterdam. 4. City planning—Netherlands. 5. Cities and towns—Netherlands—Growth—Case studies.
HT169.N42B323 79-340609
79 MARC

Netherlands (Kingdom, 1815-). Rijksdienst voor de IJsselmeerpolders.
Structuurplan Lelystad 1975-2000. — ⌈Lelystad⌉ : Rijksdienst voor de IJsselmeerpolders, 1978.
3 v. : ill., 3 fold. maps (1 fold. in pocket) ; 30 cm. Ne***
Includes bibliographies.
CONTENTS: 1. Onderzoeksnota.—2. Structuurnota.—3. Bijlagen.
1. Lelystad, Netherlands—City planning. 2. City planning—Netherlands. I. Title.
HT169.N42L446 79-362127
79 MARC

Priemus, Hugo.
Stadsvernieuwing : problemen en perspectieven : analyse van de Rotterdamse aanpak / Hugo Priemus. — Alphen aan den Rijn : Samsom, 1978.
500 p. : with ill. ; 25 cm. Ne78-31
Bibliography: p. 488-498.
ISBN 9014027516 : fl 49.90
1. Urban renewal—Netherlands—Rotterdam. 2. Urban renewal—Netherlands—Rotterdam—Finance. 3. Rotterdam—City planning. 4. City planning—Netherlands. I. Title.
HT178.N42R676 78-384038
*79 MARC

Ruilverkaveling Brummen-Voorst : advies landschapsbouw / ⌈samenstelling, E. N. te Boekhorst-Van Maren, L. van Nieuwenhuijze, medewerking van J. A. J. Vervloet ; kartografie, R. Mooy⌉. — Utrecht : Afd. Landschapsarchitektuur van het Staatsbosbeheer, 1978.
76 p. : ill. ; 30 cm. — (Rapport - Staatsbosbeheer ; nr. 1978-6) Ne***
Cover title.
Includes bibliographical references.
1. Regional planning—Netherlands—Gelderland. 2. Brummen, Netherlands—City planning. 3. Voorst, Netherlands—City planning. 4. City planning—Netherlands. I. Boekhorst-Van Maren, E. N. te. II. Nieuwenhuijze, L. van. III. Vervloet, J. A. J. IV. Series: Netherlands (Kingdom, 1815-). Staatsbosbeheer. Rapport - Staatsbosbeheer ; 1978, 6.
HT395.N42G4565 79-356110
79 MARC

Stadsvernieuwing door socialisering : de gemeentelijke aankoopaktie in Rotterdam / door Johan Korteweg ... ⌈et al.⌉. — Delft : Delftse Universitaire Pers, 1978.
xiii, 224 p. ; 24 cm. Ne78-18/19
Pref. by J. G. van der Ploeg.
Includes bibliographical references.
ISBN 9062750079 : fl 29.90
1. Urban renewal—Netherlands—Rotterdam. 2. Urban renewal—Netherlands—Finance. 3. Rotterdam—City planning. 4. City planning—Netherlands. 5. Housing—Netherlands—Rotterdam. I. Korteweg, Johan.
HT178.N42R677 78-382235
*79 MARC

—NETHERLANDS—ADDRESSES, ESSAYS, LECTURES

Launspach, J A
Een vogel in de boom : kommentaren over stedebouw en ruimtelijke ordening 1969-1975 / Jan Launspach. — Amsterdam : Ekologische Uitgeverij, 1976.
130 p. ; 22 cm. Ne76-45
Bibliography: p. 129-130.
ISBN 9062240216 : fl 5.50
1. City planning—Addresses, essays, lectures. 2. Regional planning—Addresses, essays, lectures. 3. City planning—Netherlands—Addresses, essays, lectures. I. Title.
HT166.L382 78-394825
79 MARC

—NETHERLANDS—CASE STUDIES

Goedman, Jan, 1944-
Naar een maatschappelijke planologie en planning : perspectieven voor een ruimtelijke theorie en politiek van de arbeid / Goedman. — Amsterdam : Ekologische Uitgeverij, c1978.
591 p. : ill. ; 22 cm. Ne***
Originally presented as the author's thesis, Technische Hogeschool Eindhoven.
Vita.
Errata slip inserted.
Summary in English.
Bibliography: p. 577-589.
ISBN 9062240119
1. Moerdijk, Netherlands—City planning—Case studies. 2. City planning—Netherlands—Case studies. 3. Shell Oil Company—Case studies. 4. Industry—Social aspects—Netherlands—Moerdijk—Case studies. I. Title.
HT169.N42M633 1978 309.2′62′094924 78-399475
79 MARC

—NETHERLANDS—DECISION MAKING

Faludi, Andreas.
Flexibility in Dutch local planning / ⌈by⌉ Andreas Faludi and Stephen Hamnett. — ⌈Oxford⌉ : Oxford Polytechnic, Department of Town Planning, 1977.
⌈2⌉, iii leaves, 83 p. ; 30 cm. — (Working paper - Oxford Polytechnic, Department of Town Planning ; no. 28) GB78-13698
Bibliography: p. 80-83.
£1.88
1. City planning—Netherlands. 2. City planning—Netherlands—Decision making. 3. City planning and redevelopment law—Netherlands. I. Hamnett, Stephen, joint author. II. Title. III. Series: Oxford working papers in planning education and research ; no. 28.
HT169.N4F34 309.2′62′09492 78-320525
78 MARC

CITY PLANNING (Continued)

—NETHERLANDS—PERIODICALS

BOS, Berichten over stadsvernieuwing. 1–
jaarg., winter 1975/76–
Den Haag, Ministerie van Volkshuisvesting en Ruimtelijke Ordening, Centrale Afdeling Voorlichting.
v. ill. 29 cm.
1. Urban renewal—Netherlands—Periodicals. 2. City planning—Netherlands—Periodicals. 3. Housing—Netherlands—Periodicals. I. Netherlands (Kingdom, 1815–). Ministerie van Volkshuisvesting en Ruimtelijke Ordening. Centrale Afdeling Voorlichting. II. Title: Berichten over stadsvernieuwing.
HT178.N4B67 79–640562
MARC-S

—NEW YORK (CITY)

New York (City). City Planning Commission.
Managing land use in New York City / City of New York, Department of City Planning, City Planning Commission. — New York : The Commission, 1977.
v, 347 p., ⌊28⌋ leaves of plates : ill. ; 28 cm. — (Its Section 701, comprehensive planning assistance ; v. 1)
"NYC DCP 77-33."
Leaf "107A" inserted.
Includes bibliographical references.
1. City planning—New York (City) 2. Land use—New York (City) 3. New York (City)—City planning. I. Title. II. Series.
HT168.N5N45 1977 333.7'7'097471 78-107784
78 MARC

—NEW YORK (STATE)

Conference on Two World Cities: Paris and New York, Paris, 1978.
Survival strategies : Paris and New York : report on the Conference on Two World Cities: Paris and New York, Paris, May, 1978 / the French/American Foundation, the Council for International Urban Liaison ; edited by George G. Wynne. — New Brunswick, N.J. : Transaction Books, c1979.
iv, 200 p. : ill. ; 23 cm.
Includes bibliographical references.
ISBN 0-87855-749-0 : $5.95
1. City planning—Congresses. 2. New York (City)—City planning—Congresses. 3. Paris—City planning—Congresses. 4. Urban policy—Congresses. 5. City planning—France—Congresses. 6. City planning—New York (State)—Congresses. I. Wynne, George A. II. French-American Foundation. III. Council for International Urban Liaison. IV. Title.
HT166.C6217 1978 309.2'62'0944361 78-66205
79 MARC

—NEW YORK (STATE)—CONGRESSES

Urban Land Institute.
Ogdensburg, New York : an evaluation of the commercial, residential, and industrial development potential of the city of Ogdensburg, Ogdensburg, New York, October 30-November 3, 1978 : a panel advisory service report / by ULI-the Urban Land Institute. — Washington : ULI, c1979.
88 p. : ill. ; 28 cm.
1. Ogdensburg, N.Y.—City planning. 2. City planning—New York (State) 3. Real estate development—New York (State)—Ogdensburg. 4. Central business districts—New York (State)—Ogdensburg. 5. Housing—New York (State)—Ogdensburg. I. Title.
HT168.O4U7 1979 309.2'62'0974756 79-104002
79 MARC

—NEW ZEALAND

New Zealand. Town and Country Planning Act Review Committee.
Report to government / Town and Country Planning Act Review Committee. — ⌊Wellington : Town and Country Planning Act Review Committee⌋, 1973.
15, 10 leaves ; 29 cm. NZ74-9
1. Regional planning—New Zealand. 2. City planning—New Zealand. 3. Regional planning—New Zealand—Law and legislation. I. Title.
HT395.N45N5 1973 346'.931'045 78-323203
79 MARC

—NICARAGUA—BIBLIOGRAPHY

Benjamin, Aaron J
Earthquake reconstruction program for Managua, Nicaragua / Aaron J. Benjamin, with Michelle Swallow, co-author. — Monticello, Ill. : Vance Bibliographies, 1979.
48 p. ; 28 cm. — (Public administration series : Bibliography ; P-193)
Cover title.
$5.00
1. Disaster relief—Nicaragua—Managua—Bibliography. 2. Managua—City planning—Bibliography. 3. City planning—Nicaragua—Bibliography. 4. Managua—Economic policy—Bibliography. 5. Managua—Earthquake, 1972—Bibliography. I. Swallow, Michelle, joint author. II. Title. III. Series.
Z7165.N48B46 79-107113
016.3092'62'097285
⌊HV555.N46⌋ 79 MARC

—NORTH CAROLINA—PERIODICALS

Carolina planning. v. 1–
summer 1975–
⌊Chapel Hill, N. C., Dept. of City and Regional Planning, University of North Carolina⌋
v. ill. 28 cm. semiannual.
Key title: Carolina planning, ISSN 0164-0070
1. Regional planning—North Carolina—Periodicals. 2. City planning—North Carolina—Periodicals. I. University of North Carolina at Chapel Hill. Dept. of City and Regional Planning.
HT393.N8C37 309.2'12'09756 79-640045
MARC-S

—NORTHERN IRELAND—CASE STUDIES

Northern Ireland. Planning Appeals Commission.
Poleglass area public inquiry : reports / by Planning Appeals Commission. — Belfast: H.M. Stationery Off., ⌊1977⌋
ca. 250 p. in various pagings : maps ; 30 cm. GB***
Cover title.
1. Housing—Northern Ireland—Belfast. 2. City planning—Northern Ireland—Case studies. 3. Belfast—City planning. I. Title.
HD7335.5.B44N67 1977 711'.4'094167 78-325770
79 MARC

—NORWAY—CONGRESSES

International Congress for Housing and Town Planning, 18th, Hastings, 1946.
Implementing the plans. ⌊Westminster? London, 1946?⌋
2 pamphlets (in portfolio) 23 cm. (Its Congress reports)
Title from portfolio.
English, with summaries in French.
CONTENTS: ⌊1⌋ Great Britain.—⌊2⌋ Norway.
1. City planning—Great Britain—Congresses. 2. City planning—Norway—Congresses. I. Title.
NA9185.I55 1946a 301.5'4 51-38283
⌊r79⌋rev MARC

—NORWAY—HISTORY

Ei bok om Oslo : planlegging og byutvikling før 1950 : opptrykk av et semester ved Arkitekthøgskolen i Oslo / ⌊ved⌋ Bjørn Baltzersen ... ⌊et al.⌋. — Oslo : Arkitekthøgskolen, 1977 (Oslo : AHO-trykk).
399 p. : ill. ; 29 cm. N78-Apr.
Includes bibliographies and index.
ISBN 8254700419
1. Oslo—City planning—History. 2. City planning—Norway—History. 3. Oslo—Economic conditions. 4. Architecture—Norway—Oslo—History. I. Baltzersen, Bjørn. II. Arkitekthøyskolen i Oslo.
HT169.N72O82 78-367896
78 MARC

—ONTARIO

Ontario. Ministry of Housing.
North Pickering project : summary of recommended plan. — ⌊Toronto⌋ : Ontario Ministry of Housing, 1975.
36 p. : ill. ; 30 cm. C***
Cover title.
1. New towns—Ontario. 2. City planning—Ontario. 3. Ontario. North Pickering Project. I. Title.
HT169.C32O654 1975 711'.4'0971355 79-300686
79 MARC

Silberberg, Ted.
The public interest and the politics of shopping centre development in Ontario / Ted Silberberg. — Toronto : Project: Saving Small Business, c1977.
59 leaves ; 30 cm. C78-7826-8
Bibliography: leaves 56-59.
ISBN 0-920278-02-7 : $3.00
1. Shopping centers—Ontario—Case studies. 2. Store location—Ontario—Case studies. 3. City planning—Ontario.
HF5430.6.C2S54 381 79-315495
79 MARC

Toronto, Ont. Planning Board.
South of Carlton : preliminary neighbourhood plan / City of Toronto Planning Board. — Toronto : City of Toronto Planning Board, 1975.
145 p. : ill., maps ; 28 cm. C75-9865-1
Includes bibliographical references.
1. Toronto, Ont.—City planning. 2. City planning—Ontario. I. Title. II. Title: Preliminary neighbourhood plan.
HT169.C32T482 711'.4'09713541 79-307246
79 MARC

—ONTARIO—CASE STUDIES

Ontario. Central Ontario Lakeshore Urban Complex Task Force.
Report to the Advisory Committee on Urban and Regional Planning of the Central Ontario Lakeshore Urban Complex Task Force, December 1974. — ⌊Toronto : s.n.⌋ : available from the Ontario Govt. bookstore, ⌊1974?⌋
xi, 49 p., ⌊10⌋ fold. leaves of plates : col. maps (1 fold. col. in pocket) ; 30 cm. C***
On cover: Central Ontario Lakeshore Urban Complex.
$5.00
1. City planning—Ontario—Case studies. 2. Regional planning—Ontario—Case studies. 3. Metropolitan areas—Ontario—Case studies. I. Title: Report to the Advisory Committee on Urban and Regional Planning ... II. Title: Central Ontario Lakeshore Urban Complex.
HT169.C32O639 711'.3'097135 78-323208
79 MARC

—ONTARIO—CITIZEN PARTICIPATION —CONGRESSES

Symposium on Downtown Cores, Toronto, Ont., 1977.
New directions : proceedings of a Symposium on Downtown Cores, November 1 and 2, 1977 / edited by R. R. Nykor and G. A. Przybylowski. — ⌊Toronto⌋ : Ministry of Housing : ⌊obtainable from⌋ Ontario Govt. Bookstore, ⌊1978⌋
194 p. : ill. ; 30 cm. C***
$3.50
1. City planning—Ontario—Citizen participation—Congresses. 2. Central business districts—Ontario—Congresses. 3. Urban renewal—Ontario—Congresses. I. Nykor, R. R. II. Przybylowski, G. A. III. Title.
HT169.C32O69 1977 309.2'62'09713 79-305731
79 MARC

—ONTARIO—HISTORY

Smith, W Randy.
The early development of three upper Canadian towns : Barrie, Holland Landing, and Newmarket / by W. Randy Smith. — Toronto : Dept. of Geography, York University, 1977.
54 p. : ill. ; 28 cm. — (Discussion paper - Department of Geography, York University ; no. 16) C***
Bibliography: p. 52-54.
1. City planning—Ontario—History. 2. Barrie, Ont.—City planning—History. 3. Holland Landing, Ont.—City planning—History. 4. Newmarket, Ont.—City planning—History. I. Title. II. Series: York University, Toronto, Ont. Dept. of Geography. Discussion paper - York University, Department of Geography ; no. 16.
HT169.C32O67 309.2'62'0971317 79-311351
79 MARC

—ONTARIO—TORONTO METROPOLITAN AREA

Dakin, John.
Historical instrument for considering Toronto planning / John Dakin. — Toronto : Dept. of Urban and Regional Planning, University of Toronto, 1978.
65 p. ; 28 cm. — (Papers on planning & design ; no. 16) C***
Bibliography: p. 62-65.
$3.00
1. City planning—Ontario—Toronto metropolitan area. I. Title. II. Series.
HT169.C32T427 309.2'62'09713541 79-303993
79 MARC

CITY PLANNING (Continued)

—PAKISTAN—PUNJAB (PROVINCE)

Punjab, Pakistan (Province). Communications & Works Dept. Physical Planning Cell.
Low cost planning of resettlement projects : a preliminary report. — Lahore : Communications & Works Dept., Town Planning & Housing Wing, Physical Planning Cell, 1970.
4 leaves ; 27 cm.
"PPC-11."
1. Housing policy—Pakistan—Punjab (Province) 2. City planning—Pakistan—Punjab (Province) I. Title.
HD7361.5.P8P83 1970 301.5'4 77-930343
79 MARC

—PAKISTAN—PUNJAB (PROVINCE)—HANDBOOKS, MANUALS, ETC.

Punjab, Pakistan (Province). Directorate of Physical Planning Cell.
Town planning manual. — Lahore : Directorate of Physical Planning Cell, Housing & Physical Planning Dept., Govt. of the Punjab, 1974.
ii, 69 p. ; 28 cm.
"PPC-30."
1. City planning—Pakistan—Punjab (Province)—Handbooks, manuals, etc. I. Title.
HT169.P32P866 352'.96'0954914 76-930332
79 MARC

—PERU—HISTORY

Durán Montero, María Antonia.
Fundación de ciudades en el Perú durante el siglo XVI : estudio urbanístico / María Antonia Durán Montero. — 1. ed. — Sevilla : Escuela de Estudios Hispano-Americanos de Sevilla, 1978.
xxxviii, 210 p., ⌊19⌋ leaves of plates (3 fold.) : ill. ; 20 cm. — (Publicaciones de la Escuela de Estudios Hispano-Americanos de Sevilla ; 247) Sp***
Bibliography: p. ⌊167⌋-191.
Includes index.
ISBN 8400043146
1. Cities and towns—Peru—History. 2. City planning—Peru—History. 3. City planning and redevelopment law—Peru—History. I. Title. II. Series: Seville. Universidad. Escuela de Estudios Hispano-Americanos. Publicaciones ; 247.
HT129.P5D87 79-112623
79 MARC

—PHILIPPINE ISLANDS

Philippines (Republic). Dept. of Public Works, Transportation, and Communications. Planning & Project Development Office.
MMetroplan : Metro Manila transport, land use, and development planning project : final report, summary, and recommendations. — Quezon City : Planning and Project Development Office, Dept. of Public Works, Transportation, and Communications : Freeman Fox and Associates, 1977.
58 p. : ill. ; 28 cm.
1. Transportation planning—Philippine Islands—Manila metropolitan area. 2. Urban transportation—Philippine Islands—Manila metropolitan area. 3. Manila metropolitan area—City planning. 4. City planning—Philippine Islands. I. Freeman Fox and Associates. II. Title.
HE276.Z7M366 711'.7'095991 78-320028
78 MARC

—POLAND

Housing, building, and planning in Polish People's Republic / ⌊prepared in the Housing Institute, the Institute for Building Technique and the Institute for Town Planning, by E. Kuminek et al.⌋. — Warsaw : Ministry of Building and Building Materials Industry, ⌊1973⌋
97 p. : ill. ; 25 cm.
Cover title.
"The work ordered by the Head Centre for Building Information."
1. Housing policy—Poland. 2. Construction industry—Poland. 3. City planning—Poland. 4. Regional planning—Poland. I. Kuminek, E. II. Instytut Gospodarki Mieszkaniowej. III. Centralny Ośrodek Informacji Budownictwa.
HD7345.7.A3H68 301.5'4'09438 78-315238

—PORTUGAL

Montez, Paulino.
Peniche : plano geral da península, plano da zona cêntrica, plano do sítio Dos Remédios / Paulino Montez. — Lisboa : ⌊s.n.⌋, 1978.
47 p. : ill. ; 27 cm. — (Estudos de urbanismo en Portugal ; 19-21)
1. Peniche, Portugal—City planning. 2. City planning—Portugal. I. Title.
HT169.P82P465 79-113542
79 MARC

—PORTUGAL—CONGRESSES

International Congress for Housing and Town Planning, 21st, Lisbon, 1952.
Évolution de l'urbanisme au Portugal, exemples du passé, point actuel, perspectives de l'avenir. Lisbonne, 1952.
35 p. illus. 25 cm.
1. City planning—Portugal—Congresses. I. Title.
NA9225.I57 1952 55-32488
⌊r79⌋rev MARC

—PUERTO RICO

Kohr, Leopold, 1909-
The city of man : (the Duke of Buen Consejo) / Leopold Kohr ; foreword by Ivan Illich. — 1st ed. — ⌊Río Piedras⌋ : Editorial Universitaria, Universidad de Puerto Rico, 1976.
71 p. ; 23 cm. — (Planning series ; U-1)
Includes indexes.
ISBN 0-8477-2431-X pbk.
1. San Juan, P.R.—City planning. 2. Buen Consejo, San Juan, P.R.—Social conditions. 3. Slums—Puerto Rico—Case studies. 4. City planning—Puerto Rico. I. Title. II. Series.
HT169.P92S254 309.2'62'097295 76-10299
76⌊r79⌋rev MARC

—QUÉBEC (PROVINCE)

Québec (Province). Secrétariat à l'aménagement.
La décentralisation : une perspective communautaire nouvelle : ⌊document de travail / préparé par le⌋ Ministère du conseil exécutif, Secrétariat à l'aménagement et à la décentralisation. — ⌊Québec⌋ : Le Secrétariat, ⌊1979?⌋-
v. : maps ; 24 cm. C***
CONTENTS: fasc. 1. Une vue d'ensemble.—fasc. 2. Les expériences étrangères.—fasc. 3. L'aménagement et l'urbanisme.—fasc. 4. L'organisation politique des comtés renouvelés.
1. Decentralization in government—Québec (Province) 2. Decentralization in government. 3. Local government—Québec (Province) 4. City planning—Québec (Province) 5. Regional planning—Québec (Province) I. Title.
JS1721.Q4Q42 1979 354'.714'083 79-376644
79 MARC

—QUÉBEC (PROVINCE)—CONGRESSES

Community Planning Association of Canada. Quebec Division.
Une politique du sol, pourquoi? : conférence conjointe, mai, 1976, ⌊Montréal⌋ : L'Association canadienne d'urbanisme, Division du Québec, Conseil canadien de recherches urbaines et régionales. — ⌊Québec : L'Association, Division du Québec, 1976⌋
5 v. ; 28 cm. C***
Cover title.
Bibliography: v. 2, p. 107-110.
CONTENTS: ⌊1⌋ Divay, G. Présentation générale des différents thèmes.—⌊2⌋ Le gouvernement du Québec et l'utilisation du sol: état partiel du dossier "Macro-zonages."—⌊3⌋ Mesures d'intervention sur le sol qu Québec.—⌊4⌋ Hamel, P., et Léonard, J.-F. Dossier sur les luttes urbaines à Montréal: six mobilisations populaires sur le logement.—⌊5⌋ Dansereau, F., et Gaudreau, M. Commerce du sol et promoteurs à Montréal.
$7.50
1. City planning—Québec (Province)—Congresses. 2. Montréal, Québec—City planning—Congresses. I. Canadian Council on Urban and Regional Research. II. Title.
HT169.C32Q384 309.2'62'09714 79-360152
79 MARC

—ROMANIA

Defour, Dorel.
Sistematizarea localităţilor rurale / Dorel Defour, Ioan L. Baucher. — Bucureşti : "Ceres", 1977.
175 p. : ill. ; 21 cm. R78-1143
Bibliography: p. 171-⌊174⌋
lei9.50
1. City planning—Romania. 2. Land use, Rural—Romania—Planning. I. Baucher, Ioan L., joint author. II. Title.
HT169.R75D43 78-385627
*78 MARC

Urbanismul în România / lucrare elaborată de un colectiv ⌊format din dr. arh. Cezar Lăzărescu, dr. Ion Ciobotaru, arh. Doina Cristea, et al.⌋ sub conducerea prof. emerit dr. arh. Cezar Lăzărescu. — Bucureşti : Editura Tehnică, 1977.
223 p. : ill., plans ; 27 cm. R78-1455
Summary in English and French.
Includes bibliographical references.
lei33.00
1. City planning—Romania. I. Lăzărescu, Cezar.
HT169.R75U73 309.2'62'09498 78-389388
*79 MARC

—ROME

Castagnoli, Ferdinando.
Roma antica : profilo urbanistico / di F. Castagnoli. — Roma : Jouvence, 1978.
138 p., ⌊12⌋ leaves of plates inserted : ill. ; 19 cm. — (Guide allo studio della civiltà romana ; I, 3) (Guide ; 2) It79-Mar
Bibliography: p. 119-127.
Includes index.
L4000
1. Rome (City)—Description. 2. Rome (City)—History—To 476. 3. Rome (City)—City planning. 4. City planning—Rome. I. Title. II. Series.
DG63.C32 937'.6 79-353680
*79 MARC

—ROME—HISTORY

Heimberg, Ursula.
Römische Landvermessung : Limitatio / Ursula Heimberg. — Stuttgart : Gesellschaft für Vor- und Frühgeschichte in Württemberg und Hohenzollern, 1977.
84 p. : ill. ; 21 cm. — (Kleine Schriften zur Kenntnis der römischen Besetzungsgeschichte Südwestdeutschlands ; Nr. 17) GFR***
Bibliography: p. 83.
1. Surveying—Rome—History. 2. City planning—Rome—History. I. Title.
TA516.H44 79-351901
79 MARC

—RUSSIA

(Akademgorodok bliz Novosibirska—ėlement novoĭ sistemy rasseleniía)
Академгородок близ Новосибирска—элемент новой системы расселения / отв. редактор А. Г. Аганбегян. — ⌊Новосибирск⌋ : Наука, сибирское отд-ние, 1968.
33 p. : ill.
At head of title: Akademiía nauk SSSR. Sibirskoe otdelenie. Institut ėkonomiki i organizatsii promyshlennogo proizvodstva. Ministerstvo vysshego i srednego spetsial'nogo obrazovaniía RSFSR. Novosibirskiĭ gosudarstvennyĭ universitet.
LC copy on microfilm only.

—— —— Microfilm. On 1 reel with other items ; 35 mm.
Microfilm Slavic 12546 HN
1. Akademgorodok, Novosibirsk. 2. City planning—Russia. 3. Akademgorodok, Novosibirsk—City planning. I. Aganbegían, Abel Gezevich. II. Akademiía nauk SSSR. Sibirskoe otdelenie. Institut ėkonomiki i organizatsii promyshlennogo proizvodstva. III. Novosibirsk. Universitet.
[HN530.A38A38] 78-423640

Gutnov, Alekseĭ Ėl'brusovich.
(Budushchee goroda)
Будущее города / А. Э. Гутнов, И. Г. Лежава. — Москва : Стройиздат, 1977.
126 p. : ill. ; 20 cm. — (Творческая трибуна архитектора) USSR 77
Series romanized: Tvorcheskaía tribuna arkhitektora.
Bibliography: p. 124-⌊125⌋
0.75rub
1. City planning—Russia. 2. New towns—Russia. 3. Architecture—Environmental aspects—Russia. I. Lezhava, Il'ía Georgievich, joint author. II. Title.
NA9211.G87 78-421634

Nuove città industriali : esperienze e progettazione / ⌊di⌋ Karro ... ⌊et al.⌋. — Bologna : Pàtron, 1978.
156 p. : ill. ; 21 cm. — (Studi urbanistici ; 1) It79-Mar
Translation of Novye industrial'nye goroda.
Includes bibliographical references.
L4900
1. City planning—Russia. I. Karro, V. M. II. Series.
NA9000.S78 vol. 1 79-364656
⌊HT169.R8⌋ *79 MARC

(Perspektivnoe planirovanie ėkonomicheskogo i sotsial'nogo razvitiía goroda)
Перспективное планирование экономического и социального развития города : Метод. рекомендации / ⌊Руткевич М. Н., Шкаратан О. И., Аитов Н. А. ... et al.⌋. — Москва : Профиздат, 1977.
200 p. ; 21 cm. USSR 78
Includes bibliographical references.
0.75rub
1. City planning—Russia. I. Rutkevich, Mikhail Nikolaevich.
HT169.R8P44 78-421575

CITY PLANNING (Continued)

—RUSSIA—ADDRESSES, ESSAYS, LECTURES

(Malye naselennye mesta v sistemakh rasseleni͡ia)
Малые населенные места в системах расселения : Сборник науч. трудов / Центр. науч.-исслед. и проектный ин-т по градостроительству ; Под ред. канд. архитектуры Е. М. Маркова ; [Предисл. В. Н. Белоусова]. — Москва : ЦНИИП градостроительства, 1976, вып. дан. 1977.
103 p. : ill. ; 21 cm. USSR 77
Includes bibliographical references.
0.45rub
1. City planning—Russia—Addresses, essays, lectures. I. Markov, Evgeniĭ Manuĭlovich. II. Moscow. T͡Sentral'nyĭ nauchno-issledovatel'skiĭ i proektnyĭ institut po gradostroitel'stvu.
HT169.R8M355 79-408539

Pami͡atniki arkhitektury v strukture gorodov SSSR ; [A. V. Ikonnikov, N. F. Guli͡anit͡skiĭ, E. V. Mikhaĭlovskiĭ ... et al.] ; Pod red. A. V. Ikonnikova i N. F. Guli͡anit͡skogo. — Moskva : Stroĭizdat, 1978.
208 p. : ill. ; 27 cm. USSR78
At head of title: Gosudarstvennyĭ komitet po grazhdanskomu stroitel'stvu i arkhitekture pri Gosstroe SSSR. T͡Sentral'nyĭ nauchno-issledovatel'skiĭ institut teorii i istorii arkhitektury.
Bibliography: p. 200-[207]
2.90rub
1. City planning—Russia—Addresses, essays, lectures. 2. Architecture and history—Russia—Addresses, essays, lectures. I. Ikonnikov, Andreĭ Vladimirovich. II. Guli͡anit͡skiĭ, Nikolaĭ Feodos'evich. III. Russia (1923- U.S.S.R.). Gosudarstvennyĭ komitet po grazhdanskomu stroitel'stvu i arkhitekture. IV. T͡Sentral'nyĭ nauchno-issledovatel'skiĭ institut teorii i istorii arkhitektury.
NA9211.P33 79-372875
79 MARC

—RUSSIA—ADDRESSES, ESSAYS, LECTURES

Sot͡sial'nye problemy planirovanii͡a sot͡sialisticheskogo goroda : [Sb. stateĭ] / LGU im. A. A. Zhdanova, NII kompleks. sot͡sial'nykh issled. ; [Pod red. P. N. Lebedeva, V. A. Sukhina]. — Leningrad : Izd-vo LGU, 1977.
174 p. ; 21 cm. — (Chelovek i obshchestvo : vyp. 16) USSR78
"Uchenye zapiski."
Includes bibliographical references.
0.80rub
1. City planning—Russia—Addresses, essays, lectures. I. Lebedev, Pavel Nikolaevich. II. Sukhin, Vladimir Aleksandrovich. III. Leningrad. Universitet. Nauchno-issledovatel'skiĭ institut kompleksnykh sot͡sial'nykh issledovaniĭ. IV. Series.
H35.C396 vyp. 16 79-381461
[HT169.R8] 79 MARC

—RUSSIA—EXHIBITIONS

L'Espace urbain en URSS, 1917-1978 : [exposition], Centre national d'art et de culture Georges Pompidou, Centre de création industrielle, [du 14 juin au 18 septembre 1978 : catalogue / conçu sous le forme de dialogue entre Alexeï Gutnov et Jean-Louis Cohen]. — [Paris] : Centre national d'art et de culture Georges Pompidou, C.C.I., [c1978]
52 p. : ill. ; 24 cm. F***
Bibliography: p. 58.
ISBN 2-85850-070-3 : 12.00F
1. City planning—Russia—Exhibitions. 2. Urban renewal—Russia—Exhibitions. I. Gutnov, Alekseĭ Ėl'brusovich. II. Cohen, Jean Louis. III. Centre de création industrielle.
NA9211.E8 79-350494
711'.4'094707404361
79 MARC

—RUSSIA, NORTHERN

Vaĭtens, M E
Arkhitekturno-khudozhestvennyĭ oblik severnykh gorodov / [Sost. M. E. Vaĭtens]. — Leningrad : Stroĭizdat, Leningr. otd-nie, 1978.
56 p. : ill. ; 21 cm. USSR78
At head of title: Gosudarstvennyĭ komitet po grazhdanskomu stroitel'stvu i arkhitekture pri Gosstroe SSSR. Nauchno-issledovatel'skiĭ i proektnyĭ institut po razrabotke general'nykh planov i proektov zastroĭki gorodov.
0.25rub
1. Architecture—Russia, Northern. 2. Architecture and climate—Russia, Northern. 3. City planning—Russia, Northern. I. Russia (1923- U.S.S.R.). Gosudartsvennyĭ komitet po grazhdanskomu stroitel'stvu i arkhitekture. II. Leningradskiĭ nauchno-issledovatel'skiĭ i proektnyĭ institut po razrabotke general'nykh planov i proektov zastroĭki gorodov. III. Title.
NA1195.N57V34 79-370295
79 MARC

—RUSSIAN REPUBLIC

Kirichenko, Evgenii͡a Ivanovna.
(Moskva na rubezhe stoletiĭ)
Москва на рубеже столетий / Е. И. Кириченко. — Москва : Стройиздат, 1977.
183 p. : ill. ; 22 cm. USSR 78
Bibliography : p. 181-182.
1.75rub
1. Moscow—Buildings. 2. Architecture, Modern—19th century—Russian Republic—Moscow. 3. Architecture, Modern—20th century—Russian Republic—Moscow. 4. Moscow—City planning. 5. City planning—Russian Republic. I. Title.
NA1197.M6K57 79-407096

Promyslov, Vladimir Fedorovich.
(General'nyĭ plan Moskvy v deĭstvii)
Генеральный план Москвы в действии / В. Ф. Промыслов. — Москва : Знание, 1978.
47 p., [8] leaves of plates : ill. ; 21 cm. — (Новое в жизни, науке, технике : Серия Строительство и архитектура ; № 10, 1978 г.) USSR 78
0.12rub
1. Moscow — City planning. 2. City planning — Russian Republic. I. Title. II. Series: Novoe v zhizni, nauke, tekhnike : Serii͡a Stroitel'stvo i arkhitektura ; 1978, 10.
TH4.N64 1978, 10 79-410054
[NA9212.M6]

—RUSSIAN REPUBLIC—KOLA PENINSULA

Nerush, Igor' Aleksandrovich.
Goroda Kol'skogo Severa : Ocherk istorii str-va i formirovanii͡a gorodov na Kol. p-ove / I. A. Nerush. — Murmansk : Kn. izd-vo, 1978.
108 p. : ill. ; 21 cm. USSR79
0.40rub
1. City planning—Russian Republic—Kola Peninsula. 2. Architecture—Russian Republic—Kola Peninsula. I. Title.
NA9212.K64N47 79-380431
79 MARC

—SCANDINAVIA—CONGRESSES

Growing space needs in the urbanized region / [by Gunnar Lindman ... et al. ; translation from Swedish by Roger Waters]. — [Stockholm : National Swedish Board of Building and Planning, 1965]
73 leaves : ill. ; 30 cm.
Cover title.
At head of title: International conference of the International Federation for Housing and Planning, Örebro, Sweden, 1965.
Bibliography: leaf 73.
1. City planning—Scandinavia—Congresses. 2. City planning—Finland—Congresses. 3. Land use, Urban—Scandinavia—Congresses. 4. Land use, Urban—Finland—Congresses. I. Lindman, Gunnar. II. International Federation for Housing and Planning.
HT169.S27G76 309.2'62'0948 78-320840
79 MARC

—SCOTLAND—CITIZEN PARTICIPATION

Burton, A W
Public participation in planning : a review of experience in Scotland / [by] A. W. Burton and R. Johnson. — Glasgow : Planning Exchange, 1976.
74 p. : ill. ; 30 cm. GB78-30787
Includes index.
£1.50
1. City planning—Scotland—Citizen participation. 2. Regional planning—Scotland—Citizen participation. I. Johnson, Robina, joint author. II. Title.
HT169.G72S452 309.2'12'09411 78-325586
79 MARC

—SIKKIM

Bijlani, H U
Urban problems / H. U. Bijlani. — New Delhi : Centre for Urban Studies, Indian Institute of Public Administration, 1977.
xv, 171 p., [15] leaves of plates : ill., maps ; 25 cm.
Bibliography: p. [155]-163.
Includes index.
Rs70.00 ($20.00 U.S.)
1. Delhi—City planning. 2. City planning—India. 3. Gangtok, Sikkim—City planning. 4. City planning—Sikkim. 5. Underdeveloped areas—City planning. I. Title.
HT169.I52D415 309.2'62 78-900409
79 MARC

—SOUTH AFRICA—NATAL

Natal. Town and Regional Planning Commission.
Town & regional planning in Natal : a study of the Town and Regional Planning Commission in its 21st year. — Pietermaritzburg : The Commission, 1972.
54 p. : ill. ; 30 cm. — (Report - Town and Regional Planning Commission ; = Verslag - Stads-en Streeksbeplanningkommissie no. 21)
1. Regional planning—South Africa—Natal. 2. City planning—South Africa—Natal. I. Title. II. Series: Natal. Town and Regional Planning Commission. Natal town and regional planning reports ; no. 21.
HC517.N3A44 no. 21 309.2'5'09684 s 78-319009
[HT395.S6] 78 MARC

—SOUTHERN STATES—CITIZEN PARTICIPATION

Schwertz, Eddie L
The local growth management guidebook / by Eddie L. Schwertz, Jr. for the Southern Growth Policies Board. — [Research Triangle Park, N.C.] : The Board, 1979.
x, 191 p. : ill. ; 28 cm.
Includes bibliographical references.
1. Cities and towns—Southern States—Growth. 2. City planning—Southern States—Citizen participation. I. Southern Growth Policies Board. II. Title.
HT371.S33 309.2'62'0975 78-66185
00 MARC

—SPAIN

Bonet Correa, Antonio.
Morfología y ciudad : urbanismo y arquitectura durante el antiguo régimen de España / Antonio Bonet Correa. — Barcelona : Gustavo Gili, D.L. 1978.
221 p. : ill. ; 20 cm. — (Colección Arquitectura y crítica) Sp78-July
Includes bibliographical references.
ISBN 842520707X : 520ptas
1. City planning—Spain. 2. Plazas—Spain. I. Title.
NA9223.B66 78-394562
79 MARC

Fernández Gutiérrez, Fernando.
La planificación urbana en Granada / [por] Fernando Fernández. — Granada : Colegio Oficial de Aparejadores y Arquitectos Técnicos, 1978.
99 p., [12] fold. leaves of plates : ill. ; 22 cm. Sp78-Apr
Includes index.
ISBN 8440043732
1. Granada (City)—City planning. 2. City planning—Spain. 3. Urbanization—Spain—Granada (City) I. Title.
HT169.S652G734 78-392808
79 MARC

García-Pablos, Rodolfo.
Los problemas urbanísticos de Toledo ante su futuro desarrollo económico / Rodolfo García-Pablos. — 1. ed. — Madrid : Secretaría General Técnica, Ministerio de la Vivienda, 1964.
58 p., [3] leaves of plates : ill. ; 22 cm. — (Conferencias y discursos ; 15)
1. Toledo, Spain—City planning. 2. Toledo, Spain—Economic conditions. 3. City planning—Spain. I. Title.
HT169.S652T643 79-361787
79 MARC

—SPAIN—HISTORY

Terán, Fernando de.
Planeamiento urbano en la España contemporánea : historia de un proceso imposible / Fernando de Terán. — Barcelona : G. Gili, c1978.
662 p. : ill. ; 25 cm. — (Biblioteca de arquitectura) Sp***
Bibliography: p. [645]-653.
Includes index.
ISBN 8425207118
1. City planning—Spain—History. I. Title.
HT169.S65T46 78-377906
78 MARC

—SPAIN—ANDALUSIA

Bonet Correa, Antonio.
Andalucía barroca : arquitectura y urbanismo / Antonio Bonet Correa. — Barcelona : Ediciones Polígrafa, [1978?]
340 p. : ill. ; 28 cm. Sp***
Bibliography: p. 337-340.
ISBN 8434302748
1. Architecture, Baroque—Spain—Andalusia. 2. Architecture—Spain—Andalusia. 3. City planning—Spain—Andalusia. I. Title.
NA1309.A6B66 720'.946'8 79-345849
79 MARC

CITY PLANNING (Continued)

—SPAIN—BASQUE PROVINCES

Linazasoro, José Ignacio, 1947-
Permanencias y arquitectura urbana : las ciudades vascas de la época romana a la Ilustración / José Ignacio Linazasoro. — Barcelona : G. Gili, c1978.
235 p., ₍1₎ fold. leaf of plates : ill. ; 24 cm. — (Colección Arquitectura/perspectivas) Sp***
Includes bibliographical references.
ISBN 8425208955
1. City planning—Spain—Basque Provinces. I. Title.
NA9224.B35L56 79-385434
79 MARC

—SPAIN—VALENCIA (PROVINCE)

Valencia (Province). Diputación Provincial. Departamento de Urbanismo.
Plan comarcal de La Safor : informe de salida / A. Peñin ... ₍et al.₎. — ₍Valencia₎ : Diputación Provincial de Valencia, Departamento de Urbanismo, 1977.
38, 38 p. : maps ; 30 cm. Sp***
Catalan and Spanish.
Added t.p. in Catalan: Pla comarcal de La Safor.
On cover: Planes de ordenación comarcal; plan general de ordenación de La Safor.
1. City planning—Spain—Valencia (Province) I. Peñin, A. II. Title. III. Title: Pla comarcal de La Safor. IV. Title: Planes de ordenación comarcal.
HT169.S652V33 78-396080
79 MARC

—SWITZERLAND

Environnement suisse en période de récession : digest 1975/76 : mit deutscher Zusammenfassung : English summary included / réd. en chef J. Stvan ; collab. J.-P. Gillard, G. Rochat. — Genève : Institut Ecoplan, 1976.
236 p., ₍12₎ leaves of plates ; 30 cm. — (Publications - Institut Ecoplan ; 6) Sw77-481
Summaries in English and German.
Bibliography: p. 235-236.
30.00F
1. Environmental policy—Switzerland. 2. City planning—Switzerland. 3. Regional planning—Switzerland. I. Štván, Jaromír. II. Gillard, Jean Pierre. III. Rochat, Gilbert. IV. Institut Ecoplan.
HC400.E5E56 614.7 78-400617
*79 MARC

Jacobi, Henri.
Utilisation du sol : la prise en compte de l'incertitude à l'exemple de l'estimation de l'utilisation du sol dans la ville de Zurich / H. Jacobi, W. Tomasoni. — Zurich : Institut pour l'aménagement national, régional et local EPF, 1978.
vii, 295 p. : map, ; 30 cm. — (Documents d'étude pour l'aménagement national, régional et local ; no 35) Sw79-4008
Forward in German.
Includes bibliographies.
1. Zürich—City Planning. 2. City Planning—Switzerland. I. Tomasoni, Walter, joint author. II. Title. III. Series: Studienunterlagen zur Orts-, Regional- und Landesplanung ; no 35.
HT169.S92J85 79-367099
79 MARC

Ortsbild-Inventarisation, aber wie? : Methoden dargelegt am Beispiel von Beromünster / hrg. in Zusarb. mit dem Schweizer Heimatschutz und der Dienststelle Heimatschutz beim Eidgenössischen Oberforstinspektorat ; ₍Beitr. von₎ Mane Hering-Mitgau ... ₍et al.₎. — Zürich : Manesse-Verlag, c1976.
243 p. : ill., maps ; 22 x 31 cm. — (Veröffentlichungen des Instituts für Denkmalpflege an der Eidgenössischen Technischen Hochschule Zürich ; Bd. 2)
ISBN 3-7175-8030-2 : 32.00F
1. Architecture—Switzerland—Beromünster—Conservation and restoration. 2. City planning—Switzerland. 3. Beromünster, Switzerland—City planning. 4. Monuments—Switzerland—Beromünster—Preservation. I. Schweizer Heimatschutz. II. Switzerland. Eidgenössiches Oberforstinspektorat. Dienststelle Heimatschutz. III. Series: Zürich. Eidgenössische Technische Hochschule. Institut für Denkmalpflege. Veröffentlichungen des Instituts für Denkmalpflege an der Eidgenössischen Technischen Hochschule Zürich ; Bd. 2.
NA9053.C6O77 78-389491
*79 MARC

—SWITZERLAND—BERN REGION

Bucher, Jürg.
Stadtentwicklung in kleinräumigen Regionen : ein Modellansatz und seine empirische Überprüfung am Beispiel der Region Bern / J. Bucher, W. Konanz, O. Messmer. — Bern ₍etc.₎ : P. Haupt, c1978.
88 p. : ill. ; 23 cm. — (Berner Beiträge zur Nationalökonomie ; Bd. 32) Sw78-5291
Bibliography: p. 87.
ISBN 3-258-02674-2 : 24.00F
1. City planning—Switzerland—Bern region. 2. SIARSSY (Computer program) 3. Regional planning—Mathematical models. 4. City planning—Mathematical models. I. Konanz, Walter, joint author. II. Messmer, Otto, Dr. rer. pol., joint author. III. Title. IV. Series.
HT169.S92B473 78-383075

—TAIWAN

China (Republic of China). Ching chi she chi wei yüan hui. Tu shih kuei hua ch'u.
(Kai chin shih ti ch'ung hua pan fa chih yen chiu)
改進市地重劃辦法之研究. — ₍台北₎：行政院經濟設計委員會都市規劃處，民國65 ₍1976₎
65 leaves ; 28 cm.
1. City planning—Taiwan. 2. City planning and redevelopment law—Taiwan. 3. Land use—Taiwan. I. Title.
HT169.T28C47 1976 79-836653

China (Republic of China). Ching chi she chi wei yüan hui. Tu shih kuei hua ch'u.
(Tu shih chü chu mi tu chih yen chiu)
都市居住密度之研究. — ₍台北₎：行政院經濟設計委員會都市規劃處，₍1978?₎
5, 138 p. : ill. ; 27 cm.
At head of title: Chung-hua min kuo liu shih liu nien ssu yüeh.
1. Housing—Taiwan. 2. City planning—Taiwan. I. Title.
HD7370.A3C46 1978 79-835315

Kuo li chung hsing ta hsüeh. Tu shih chi hua yen chiu so.
(T'ai-pei shih Pei-t'ou feng ching t'e ting ch'ü kuei hua pao kao)
台北市北投風景特定區規劃報告 / 國立中興大學法商學院都市計劃研究所研究. — ₍台中₎：該所，民國67 ₍1978₎
90 p. : ill. ; 27 x 40 cm.
Cover title.
"台北市政府建設局委託"
Bibliography: p. 88-90.
1. T'ai-pei (City)—City planning. 2. City planning—Taiwan. 3. Tourist trade—Taiwan—T'ai-pei (City) I. Title.
HT169.T282T3347 79-838138

Schinz, Alfred.
Chinesischer Städtebau in der Mandschu-Dynastie 1644-1911, dargestellt am Beispiel der Ting-Stadt Hsinchu in Taiwan / vorgelegt von Alfred Schinz. — ₍s.l. : s.n., 1976?₎
21, ₍3₎ p. : ill. ; 21 cm. GFR***
An abridgement of the author's thesis, Technische Universität, Munich, 1975.
Bibliography: p. ₍22₎-₍24₎
1. Hsin-chu, Taiwan—City planning. 2. City planning—Taiwan. I. Title.
HT169.T282H7772 78-392943
79 MARC

T'ai-nan, Taiwan. Ti cheng k'o.
(T'ai-nan shih ti i, erh, san ch'i shih ti ch'ung hua tsung pao kao shu)
台南市第一·二·三期市地重劃總報告書 / ₍台南市政府地政科編印₎. — 台南：該科，民國65 ₍1976₎
₍40₎ p. : ill. ; 22 x 29 cm.
Cover title.
1. T'ai-nan, Taiwan—City planning. 2. City planning—Taiwan. 3. Land subdivision—Taiwan—T'ai-nan. I. Title.
HT169.T282T327 79-840440

Thelin, Mark C
Demographic characteristics, planned urbanization, and receptivity to change in the Taichung Harbor area along the west coast of central Taiwan / by Mark C. Thelin. — Taichung, Taiwan : Environmental Research Center, Tunghai University, 1976.
20 p. ; 27 cm.
Includes bibliographical references.
1. T'ai-chung, Taiwan—City planning. 2. City planning—Taiwan. 3. T'ai-chung, Taiwan—Population. I. Tung hai ta hsüeh. Huan ching k'o hsüeh yen chiu chung hsin. II. Title.
HT169.T282T277 301.32'9'5125 78-318010
78 MARC

Wu Chu, Li-fang.
(T'ai-chung kang t'e ting ch'ü chih i liao ch'ing k'uang yen chiu)
台中港特定區之醫療情況研究 = A medical care survey in the Taichung Harbor urban planning area / 吳朱麗芳. — ₍臺中₎：東海大學環境科學研究中心，民國65 ₍1976₎
xiii, 112 p. : ill. ; 27 cm. — (環境科學研究叢刊 ; 21)
In Chinese, with English summary.
1. Medical care—Taiwan—T'ai-chung. 2. T'ai-chung, Taiwan—City planning. 3. City planning—Taiwan. I. Title. II. Title: A medical care survey in the Taichung Harbor urban planning area. III. Series: Huan ching k'o hsüeh yen chiu ts'ung k'an ; 21.
RA412.5.T28W8 78-842419

—TAIWAN—ABSTRACTS

Kuo li chung hsing ta hsüeh. Tu shih chi hua yen chiu so.
(Ti liu wu hsüeh nien shuo shih lun wen che yao)
第65學年(第8屆)碩士論文摘要 / 國立中興大學法商學院都市計劃研究所；₍編輯該所圖書室₎. — ₍台中₎：該所，民國66 ₍1977₎
84 p. in various pagings : ill. ; 27 cm.
Cover title.
1. Kuo li chung hsing ta hsüeh. Tu shih chi hua yen chiu so—Dissertations—Abstracts. 2. Dissertations, Academic—Taiwan—Abstracts. 3. City planning—Taiwan—Abstracts.
HT169.T28K85 1977 79-837077

—TAIWAN—DATA PROCESSING

Ch'en, Ch'un-i.
(T'ai-wan ti ch'ü tu shih k'ung chien chieh kou chih yen chiu)
台灣地區都市空間結構之研究：因子分析和電腦繪圖之應用 / 陳春益. — ₍台北? : s. n.₎，民國66 ₍1977₎
190 p. (on double or fold. leaves) : ill. ; 27 cm.
Reproduced from typescript.
"國立中興大學都市計劃研究所第八屆碩士論文"
Bibliography: p. ₍183₎-190.
1. City planning—Taiwan—Data processing. 2. Factor analysis—Data processing. I. Title.
HT169.T28C42 1977 79-841623

—TAIWAN—T'AI-PEI (CITY)

Kuo li chung hsing ta hsüeh. Tu shih chi hua yen chiu so.
(T'ai-pei shih tu shih keng hsin shih shih pan fa yen chiu pao kao)
台北市都市更新實施辦法研究報告 / 研究單位國立中興大學法商學院都市計劃研究所；委託單位台北市政府工務局都市計畫處. — ₍台中₎：該所. 民國67 ₍1978₎
158 p. : ill. ; 26 cm.
"興都研(67) 03-711.59"
1. City planning—Taiwan—T'ai-pei (City) I. T'ai-pei (City). Kung wu chü. Tu shih chi hua ch'u. II. Title.
HT169.T282T335 79-835780

C

—TANZANIA

Mallander, Bo.
Mbeya master plan 1974-1994 / prepared in Urban Planning Division by Bo Mallander. — Dar es Salaam : The Division, 1975.
2 v. : ill., maps (1 fold in pocket) ; 30 cm.
Vol. 2 prepared by J. Berghäll and M. Tibwita.
Summary in Swahili.
CONTENTS: 1. Main report.—2. Appendix report: Mbeya traffic plan by 1994.
1. Mbeya, Tanzania—City planning. 2. City planning—Tanzania. I. Berghäll, Jouke, joint author. II. Tibwita, Matalisi, joint author. III. Tanzania. Urban Planning Division. IV. Title.
HT169.T332M385 711'.4'0967828 77-980261
79 MARC

CITY PLANNING
—TANZANIA (Continued)

Stren, Richard E
Urban development in Kenya and Tanzania : a comparative analysis / by Richard E. Stren. — Nairobi : Institute for Development Studies, University of Nairobi, 1976.
30 p. ; 30 cm. — (Discussion paper - University of Nairobi, Institute for Development Studies ; no. 232)
Bibliography: p. 28-30.
1. City planning—Kenya. 2. City planning—Tanzania. 3. Urbanization—Kenya. 4. Urbanization—Tanzania. I. Nairobi. University. Institute for Development Studies. II. Title. III. Series: Nairobi. University. Institute for Development Studies. Discussion paper ; no. 231.
HT169.K4S87 309.2'62'096762 77-980079
79 MARC

Tanzania. Town Planning Division.
Master plan for Morogoro, United Republic of Tanzania / prepared by Aaron Agbatekwe and Md. Mushtaq Hussain. — Dar es Salaam : Town Planning Division, Ministry of Lands, Housing & Urban Development, 1974.
ii, 39 leaves, [12] fold. leaves of plates : ill. ; 29 cm.
1. Morogoro, Tanzania—City planning. 2. City planning—Tanzania. I. Agbatekwe, Aaron. II. Hussain, Md. Mushtaq. III. Title.
HT169.T332M657 711'.4'0967825 77-980264
79 MARC

—TEXAS

Goals for Dallas (Association)
Achieving the goals for Dallas, 1978-83. — Dallas : Goals for Dallas, c1979.
404 p. ; 21 cm.
1. Dallas—Social policy. 2. Dallas—City planning. 3. City planning—Texas. 4. Education—Texas—Dallas. I. Title.
HN80.D25G59 1979 309.2'62'097642812 78-73504
00 MARC

—UNITED STATES

Bryce, Herrington J
Planning smaller cities / Herrington J. Bryce. — Lexington, Mass. : Lexington Books, c1979.
xxi, 214 p. ; 24 cm. — (Urban roundtable series)
Includes bibliographical references and index.
ISBN 0-669-02680-8
1. Cities and towns—United States—Growth. 2. City planning—United States. I. Title. II. Series.
HT123.B78 309.2'62'0973 78-14154
78 MARC

Burns, Jim, 1926-
Connections : ways to discover and realize community potentials / Jim Burns. — Stroudsburg, Pa. : Dowden, Hutchinson & Ross, c1979.
xi, 155 p. : ill. ; 24 cm. — (Community development series ; 43)
Bibliography: p. 147-150.
Includes index.
ISBN 0-07-009205-2
1. Cities and towns—United States. 2. City planning—United States. 3. Architecture—Human factors. I. Title.
HT123.B84 309.2'62'0973 78-14763
78 MARC

Burrows, Lawrence B
Growth management : issues, techniques, and policy implications / Lawrence B. Burrows. — New Brunswick, N.J. : Center for Urban Policy Research, Rutgers University, c1977.
ix, 141 p. ; 23 cm.
Bibliography: p. [135]-141.
ISBN 0-88285-043-1
1. Cities and towns—United States—Growth—Management. 2. Suburbs—United States. 3. City planning—United States. 4. Zoning—United States. I. Title.
HT123.B85 309.2'6'0973 77-24034
77 MARC

Chapin, Francis Stuart, 1916-
Urban land use planning / F. Stuart Chapin, Jr. and Edward J. Kaiser. — 3d ed. — Urbana : University of Illinois Press, c1979.
xvi, 656 p. : ill. ; 26 cm.
Includes index.
ISBN 0-252-00580-5
1. City planning—United States. 2. Regional planning—United States. 3. Land use, Urban—United States. I. Kaiser, Edward John, joint author. II. Title.
HT167.C45 1979 333.7'7'0973 78-31672
78 MARC

Gardiner, Richard A
Design for safe neighborhoods : the environmental security planning and design process / Richard A. Gardiner. — [Washington] : National Institute of Law Enforcement and Criminal Justice, Law Enforcement Assistance Administration, U.S. Dept. of Justice : for sale by the Supt. of Docs., U.S. Govt. Print. Off., 1978 [i.e. 1979]
83 p. : ill. ; 23 x 31 cm.
A project prepared with the assistance of the U.S. Dept. of Housing and Urban Development, Office of Policy Development and Research.
Bibliography: p. 83.
1. Crime prevention—United States. 2. Crime prevention—United States—City planning. 3. City planning—United States. 4. Neighborhood—Case studies. I. National Institute of Law Enforcement and Criminal Justice. II. United States. Dept. of Housing and Urban Development. Office of Policy Development and Research. III. Title.
HV7431.G35 364.4 79-601938
79 MARC

How cities can grow old gracefully / [prepared for the] Subcommittee on the City of the Committee on Banking, Finance, and Urban Affairs, House of Representatives, 95th Congress, first session. — Washington : U.S. Govt. Print. Off., 1977.
xiv, 222 p. : ill. ; 23 cm.
At head of title: Committee print.
Includes bibliographical references.
1. City planning—United States. 2. Cities and towns—United States—Growth. 3. Urban renewal—United States. I. United States. Congress. House. Committee on Banking, Finance, and Urban Affairs. Subcommittee on the City.
HT167.H68 301.36'3'0973 78-602443
78 MARC

Introduction to urban planning / edited by Anthony J. Catanese, James C. Snyder ; architectural drawings by Jeffrey Ollswang. — New York : McGraw-Hill, c1979.
xvii, 354 p. : ill. ; 24 cm.
Includes bibliographies and index.
ISBN 0-07-010228-7 : $22.50
1. City planning. 2. City planning—United States. I. Catanese, Anthony James. II. Synder, James C.
HT166.I67 309.2'62'0973 78-13275
78 MARC

Schiffman, Irving.
The politics of land use planning : a review essay and annotated bibliography / Irving Schiffman. — Davis : Institute of Governmental Affairs, University of California, 1977.
ii, 110 leaves ; 28 cm. — (Environmental quality series ; no. 28)
1. Land use—Planning—United States. 2. City planning—United States. 3. Local government—United States. 4. Land use—Planning—United States—Abstracts. 5. City Planning—United States—Abstracts. 6. Local government—United States—Abstracts. I. Title. II. Series.
HD205.1977.S34 333.7'0973 78-112696
79 MARC

Uhlig, Klaus Rolf.
Stadterhaltung in Beispielen / Klaus R. Uhlig ; Fotos, Monika Uhlig. — München : Frühmorgen, 1977-
v. : ill., maps ; 24 cm. GFR78-A
Appendix of v. 1 contains various legislation in English.
Bibliography: v. 1, p. 110-116.
CONTENTS: Bd. 1. Vereinigte Staaten von Amerika.

ISBN 3-921251-01-X (v. 1)
1. Architecture—United States—Conservation and restoration. 2. City planning—United States. 3. City planning and redevelopment law—United States. I. Title.
NA9053.C6U48 78-400315
79 MARC

Vasu, Michael Lee.
Politics and planning : a national study of American planners / by Michael Lee Vasu. — Chapel Hill : University of North Carolina Press, c1979.
xviii, 235 p. ; 24 cm. — (Institute for Research in Social Science monograph series)
Bibliography: p. 225-231.
Includes index.
ISBN 0-8078-1342-7
1. Land use, Urban—United States—Planning. 2. City planning—United States. 3. United States—Social policy. 4. Planning. I. Title. II. Series: University of North Carolina at Chapel Hill. Institute for Research in Social Science. Institute for Research in Social Science monograph series.
HD205 1979.V37 309.2'12'0973 78-10440
78 MARC

Von Eckardt, Wolf.
Back to the drawing board! : Planning livable cities / Wolf von Eckardt ; illustrated by Forrest Wilson. — Washington : New Republic Books ; New York : trade distribution by Simon and Schuster, 1978.
ix, 245 p. : ill. ; 22 cm.
Bibliography: p. [236]-239.
Includes index.
ISBN 0-915220-45-8
1. City planning. 2. City planning—United States. 3. Architecture—Human factors. I. Title.
HT166.V66 309.2'62 78-12257
78 MARC

Weinberg, Nathan.
Preservation in American towns and cities / Nathan Weinberg. — Boulder, Colo. : Westview Press, 1979.
xvii, 233 p. : ill. ; 24 cm.
Includes bibliographical references and index.
ISBN 0-89158-488-9 : $16.00
1. Historic buildings—United States—Conservation and restoration. 2. City planning—United States. I. Title.
E159.W4 069'.53'0973 79-83

Wright, Edward T
Your city and the future, by Edward T. Wright. St. Louis, Published by PSI Press for Practical Seminar Institute [1974]
137 p. illus. 26 cm. $2.95
1. City planning—United States. I. Title.
HT167.W74 309.2'62'0973 73-89561
79 MARC

—UNITED STATES—ABSTRACTS

Schiffman, Irving.
The politics of land use planning : a review essay and annotated bibliography / Irving Schiffman. — Davis : Institute of Governmental Affairs, University of California, 1977.
ii, 110 leaves ; 28 cm. — (Environmental quality series ; no. 28)
1. Land use—Planning—United States. 2. City planning—United States. 3. Local government—United States. 4. Land use—Planning—United States—Abstracts. 5. City Planning—United States—Abstracts. 6. Local government—United States—Abstracts. I. Title. II. Series.
HD205.1977.S34 333.7'0973 78-112696
79 MARC

—UNITED STATES—ADDRESSES, ESSAYS, LECTURES

Planning theory in the 1980s : a search for future directions / edited by Robert W. Burchell and George Sternlieb ; contributor, Robert A. Beauregard ... [et al.]. — New Brunswick, N.J. : Center for Urban Policy Research, Rutgers University, [1978]
liii, 365 p. ; 24 cm.
Bibliography: p. 349-365.
ISBN 0-88285-048-2 : $15.00
1. City planning—United States—Addresses, essays, lectures. 2. Regional planning—United States—Addresses, essays, lectures. 3. United States—Social policy—Addresses, essays, lectures. 4. United States—Economic policy—1961- —Addresses, essays, lectures. I. Burchell, Robert W. II. Sternlieb, George.
HT167.P56 309.2'12'0973 78-12929
78 MARC

—UNITED STATES—BIBLIOGRAPHY

Goehlert, Robert U
City and regional planning : a bibliography of journal literature, 1945-1975 / Robert Goehlert. — Monticello, Ill. : Vance Bibliographies, 1978.
53 p. ; 28 cm. — (Public administration series : Bibliography ; P-124)
Cover title.
$5.50
1. City planning—United States—Bibliography. 2. Regional planning—United States—Bibliography. I. Title. II. Series.
Z5942.G6 016.3092'12'0973 79-101606
[HT166] 79 MARC

Manning, Steve, 1954-
The neighborhood : a viable unit for planning? / Steve Manning. — Monticello, Ill. : Vance Bibliographies, 1978.
25 p. ; 28 cm. — (Public administration series : Bibliography ; P-73)
Cover title.
$2.50
1. City planning—United States—Bibliography. 2. Neighborhood—Bibliography. I. Title. II. Series.
Z5944.U5M36 016.3092'62 78-110945
[HT167] 79 MARC

—UNITED STATES—CASE STUDIES

Halpern, Kenneth, 1944-
Downtown USA : urban design in nine American cities / by Kenneth Halpern. — New York : Whitney Library of Design, 1978.
256 p. : ill. ; 26 cm.
Bibliography: p. 239-244.
Includes index.
ISBN 0-8230-7154-5
1. Cities and towns—United States—Case studies. 2. City planning—United States—Case studies. I. Title.
HT123.H33 1978 309.2'62'0973 78-18339
78 MARC

—UNITED STATES—COLLECTED WORKS

Management & control of growth : issues, techniques, problems, trends / edited by Randall W. Scott, with the assistance of David J. Brower and Dallas D. Miner. — Washington : Urban Land Institute, 1975-
v. ; 25 cm.
Vol. 4 edited by Frank Schnidman, Jane A. Silverman, and Rufus C. Young, Jr. has subtitle: Techniques in application.
Includes bibliographies.
1. City planning—United States—Collected works. 2. Cities and towns—United States—Growth—Collected works. 3. Land use—United States—Collected works. I. Scott, Randall W. II. Brower, David J. III. Miner, Dallas D. IV. Urban Land Institute.
HT167.M36 309.2'62'0973 74-83560
75[r79]rev MARC

CITY PLANNING (Continued)

—UNITED STATES—FILM CATALOGS —UNION LISTS

Tippman, Don.
Film, etc. : historic preservation and related subjects / compiled by Don Tippman for the Audiovisual Collections, Education Services Division, Office of Preservation Services. — Washington : Preservation Press, c1979.
126 p. : ill. ; 21 cm.
Includes index.
ISBN 0-89133-074-7
1. United States—Civilization—Film catalogs—Union lists. 2. Art—United States—Film catalogs—Union lists. 3. City planning—United States—Film catalogs—Union lists. 4. Art—Film catalogs—Union lists. 5. Catalogs, Union—United States. I. National Trust for Historic Preservation in the United States. Office of Preservation Services. Education Services Division. Audiovisual Collections. II. Title.
E169.1.T59 016.79143'0909'31 79-111814
00 MARC

—UNITED STATES—MICROFORM CATALOGS

Tilton, Doreen Buck.
Housing and urban affairs, 1965-1976 : a bibliographic guide to the microfilm collection / edited by Doreen Buck Tilton. — Glen Rock, N.J. : Microfilming Corp. of America, 1978.
xii, 342 p. ; 28 cm.
Includes indexes.
ISBN 0-667-00519-6
1. Housing—United States—Microform catalogs. 2. City planning—United States—Microform catalogs. I. Title.
Z7164.H8T54 016.3015'4'0973 78-108394
[HD7293] 78 MARC

—URUGUAY

Altezor, Carlos.
Historia urbanística y edilicia de la ciudad de Montevideo; [desde su fundación colonial hasta nuestros días, por] Carlos Altezor [y] Hugo Baracchini. Montevideo, Junta Departamental de Montevideo, Biblioteca José Artigas, 1971.
318 p. illus. 24 cm. LACAP72-3872
Bibliography: p. 309-317.
1. Montevideo—History. 2. Cities and towns—Growth. 3. Montevideo—City planning. 4. City planning—Uruguay. I. Baracchini, Hugo, joint author. II. Biblioteca José Artigas. III. Title.
HN360.M6A7 72-375032
79[r79]rev MARC

Montevideo. Comision Financiera de la Rambla Sur.
50 [i.e. Cincuenta] años de la Comision Financiera de la Rambla Sur : compilación de antecedentes relacionados con la actuación de la Comisión Financiera de la Rambla Sur, desde su creación, al cumplirse su cincuentenario, el día 8 de marzo de 1976. — [Montevideo] : Intendencia Muncipal de Montevideo, 1976.
98 p. : ill. ; 17 cm.
1. Montevideo—City planning. 2. Montevideo. Comision Financiera de la Rambla Sur. 3. City planning—Uruguay. I. Title.
HT169.U72M6 78-106597
79 MARC

Montevideo. Concejo Departamental. Comisión Especial Asesora.
Limitación territorial de la ciudad de Montevideo; [recopilación de documentos técnicos oficiales. Montevideo, 1957]
173 p. illus., maps. 23 x 33 cm. (Ediciones I.T.U)
At head of title: Instituto de Teoría de la Arquitectura y Urbanismo. Facultad de Arquitectura. Universidad de la República.
Cover title.
1. Montevideo—City planning. 2. City planning—Uruguay. I. Montevideo. Universidad. Instituto de Teoría de la Arquitectura y Urbanismo. II. Title.
HT169.U72M63 309.2'62'098951 66-86473
[r79]rev MARC

—VENEZUELA

Venezuela. Dirección de Planeamiento Urbano.
Isla de Margarita : ordenamiento urbano, Punta de Piedras : informe y ordenanza. — [Caracas] : Ministerio de Obras Públicas, Dirección General de Desarrollo Urbanístico, Dirección de Planeamiento Urbano, Concejo Municipal del Distrito Díaz, [1976]
85 p., [4] leaves of plates : ill., 6 fold. (inserted) ; 22 x 29 cm.
Cover title.
1. Punta de Piedras, Venezuela—City planning. 2. City planning—Venezuela. I. Title. II. Title: Ordenamiento urbano, Punta de Piedras.
HT169.V42P855 711'.4'09875 77-154644
79 MARC

—VIRGINIA—FAIRFAX CO.

Fairfax Co., Va. Office of Comprehensive Planning.
Annual plan review - Office of Comprehensive Planning. [Fairfax] Office of Comprehensive Planning.
v. ill. 28 cm.
1. City planning—Virginia—Fairfax Co.
HT167.5.V8F38a 711'.4'09755291 78-642238
MARC-S

—WYOMING

Oblinger-Smith Corporation.
A planning and management handbook for Wyoming local jurisdictions / prepared for Office of the chief of State Planning, Department of Economic Planning and Development, State of Wyoming, prepared by Oblinger-Smith Corporation. — Denver : The Corporation, 1978.
120 p. : ill. ; 30 cm. — (Planning information series ; v. 3)
Includes bibliographical references.
1. Regional planning—Wyoming. 2. City planning—Wyoming. I. Title. II. Series.
HT393.W8O24 1978 352'.96'09787 78-622766
78 MARC

—ZAIRE—CONGRESSES

L'Organisation de l'espace urbain au Zaïre : séminaire organisé à Kinshasa du 29 avril au 11 mai, 1974 : recueil des documents. — Kinshasa, République du Zaïre : Institut national d'études politiques, [1975?]
193 leaves in various foliations ; 34 cm.
Cover title.
1. City planning—Zaire—Congresses. I. Institut national d'études politiques.
HT169.Z3O73 309.2'62'096751 78-377233
78 MARC

CITY PLANNING AND REDEVELOPMENT LAW

see also Regional planning—Law and legislation

Les Instruments juridiques de la politique foncière des villes : études comparatives portant sur quatorze pays occidentaux / sous la direction de Michel Fromont. — Bruxelles : É. Bruylant, 1978.
754 p. ; 24 cm. Be***
At head of title: Institut de droit comparé de l'Université de Dijon.
Includes bibliographical references and index.
ISBN 2-8027-0201-7
1. Eminent domain. 2. City planning and redevelopment law. 3. Preemption. I. Fromont, Michel. II. Dijon. Université. Institut de droit comparé.
K3511.I57 346.043 79-367044
79 MARC

International Congress for Housing and Town Planning, 19th, Zürich, 1948.
Financial relation of planning to ownership rights = L'aspect financier de l'aménagement par rapport aux droits des propriétaires. — [s.l. : s.n., 1948?] ([Westminster, Eng. : Wightman])
10 pamphlets (in portfolio) ; 22 cm.
English or French, with summaries in the other language.
CONTENTS: [1] Denmark.—[2] England.—[3] Holland.—[4] Netherlands East Indies.—[5] Northern Ireland.—[6] Norway.—[7] Scotland.—[8] South Africa.—[9] Switzerland.—[10] United States.
1. Regional planning—Law and legislation. 2. City planning and redevelopment law. 3. Right of property. 4. Police power. I. Title.
K3531.I5 1948 346.045 78-370168
78 MARC

—DICTIONARIES—FRENCH

Gilli, Jean Paul.
Droit de l'urbanisme / Jean-Paul Gilli, Jacques de Lanversin. — 1. éd. — Paris : PUF, c1978.
143 p. ; 21 cm. — (Lexique) F***
ISBN 2-13-035595-1 : 29.00F
1. City planning and redevelopment law—France—Dictionaries. 2. Zoning law—France—Dictionaries. 3. City planning and redevelopment law—Dictionaries—French. 4. Zoning law—Dictionaries—French. I. Lanversin, Jacques de, joint author. II. Title.
346'.44'045 79-357183
79 MARC

—ARIZONA

Arizona. Laws, statutes, etc.
Arizona zoning and subdivision laws / editor, Porter B. Williamson. — 1979 ed. — Tucson, AZ : Management and Systems Consultants, [1979] c1974.
p. 510-610, 700-712 ; 19 cm.
Separately published parts of a larger work entitled Arizona real estate laws.
Includes index.
1. Zoning law—Arizona. 2. Land subdivision—Law and legislation—Arizona. 3. City planning and redevelopment law—Arizona. I. Williamson, Porter B. II. Title.
KFA2858.A3 1979 346'.791'045 79-114311
79 MARC

—BRAZIL—BRASILIA

Companhia Urbanizadora da Nova Capital do Brasil. Assessoria Jurídica.
Pareceres de 1968. Brasília, 1970.
xxvi, 418 p. 22 cm.
At head of title: NOVACAP.
On cover: Dario Délio Cardoso.
1. City planning and redevelopment law Brazil—Brasilia. I. Cardoso, Dario Délio. II. Title.
72-223207
79[r79]rev MARC

—BRAZIL—IMBITUBA—COLLECTED WORKS

Superintendência do Desenvolvimento da Região Sul.
Projeto do litoral sul de Sta. Catarina : plano de desenvolvimento integrado, Imbituba / Ministério do Interior, Superintêndencia do Desenvolvimento da Região Sul-SUDESUL, Governo do Estado de Santa Catarina, Prefeitura Municipal de Imbituba. — [Porto Alegre] : Secretaria Executiva do Acôrdo No. 03/72, [19
v. ca. 100 leaves ; 30 cm.
CONTENTS:
—pt. 5. Legislação básica.
1. Imbituba, Brazil—City planning—Collected works. 2. City planning—Brazil—Collected works. 3. City planning and redevelopment law—Brazil—Imbituba—Collected works. I. Imbituba, Brazil. II. Title.
HT169.B72I467 79-113538
79 MARC

—CONNECTICUT

Connecticut. Laws, statutes, etc.
Statutes governing municipal planning and zoning. 1976–
[Hartford, Connecticut Dept. of Community Affairs, Office of Local Government]
v. 28 cm. (Community development aid)
Supersedes: Connecticut. Laws, statutes, etc. A selected compilation of Connecticut laws concerning planning and zoning.
1. City planning and redevelopment law—Connecticut. 2. Regional planning—Law and legislation—Connecticut. 3. Zoning law—Connecticut. I. Connecticut. Office of Local Government. II. Title. III. Series: Connecticut. Dept. of Community Affairs. Community development aid.
KFC4058.A29S7 346'.746'045 78-646577
MARC-S

—DENMARK

Denmark. Boligministeriet.
Byfornyelsesrapporten. — København : Boligministeriet, 1978.
237 p. : ill. ; 24 cm. D***
Includes bibliographical references.
ISBN 8750325876
1. Urban renewal—Denmark. 2. City planning and redevelopment law—Denmark. 3. Housing—Denmark. I. Title.
HT178.D4D46 1978 78-395576
79 MARC

—DISTRICT OF COLUMBIA

United States. Congress. House. Committee on the District of Columbia. Subcommittee on Fiscal and Government Affairs.
Amend Redevelopment act of 1945 and transfer U.S. real property to RLA : hearings and markups before the Subcommittee on Fiscal and Government Affairs and the Committee on the District of Columbia, House of Representatives, Ninety-fifth Congress, second session, on H.R. 10311 ... and H.R. 12117 and 13224 ... April 18 and June 20 and 27, 1978. — Washington : U.S. Govt. Print. Off., 1978.
iv, 135 p. ; 24 cm.
"Serial no. 95-16."
1. City planning and redevelopment law—District of Columbia. 2. United States—Government property. I. United States. Congress. House. Committee on the District of Columbia. II. Title.
KF27.D5395 1978c 346'.753'045 79-601203
79 MARC

CITY PLANNING AND REDEVELOPMENT LAW
—DISTRICT OF COLUMBIA (Continued)

United States. Congress. Senate. Committee on Governmental Affairs. Subcommittee on Governmental Efficiency and the District of Columbia.
District of Columbia legislation : hearing before the Subcommittee on Governmental Efficiency and the District of Columbia of the Committee on Governmental Affairs, United States Senate, Ninety-fifth Congress, second session, on H.R. 10311, H.R. 12116, H.R. 13243, September 19, 1978. — Washington : U.S. Govt. Print. Off., 1978.
iv, 206 p. ; 24 cm.
1. City planning and redevelopment law—District of Columbia. 2. Home rule (District of Columbia) 3. Housing—District of Columbia—Finance. 4. Bonds—District of Columbia. I. Title.
KF26.G6735 1978c 348'.753'01 79-601283
79 MARC

—EUROPE—ADDRESSES, ESSAYS, LECTURES

Le Norme e gli strumenti di pianificazione urbanistica in Europa / a cura di Giuseppe Furitano. — Milano : A. Giuffrè, 1978.
196 p. : ill. ; 28 cm. — (Studi urbanistici ; v. 11) It78-Oct
At head of title: Fondazione Aldo Della Rocca.
Includes legislation in French.
Bibliography: p. 135-142.
CONTENTS: Sisi, E. Germania, Olanda, Danimarca, Belgio: una regione.—Pasqualini, C. M. I centri storici in Francia dalla carta di Atene alla legge Malraux.
L12000
1. City planning—Europe—Addresses, essays, lectures. 2. City planning and redevelopment law—Europe—Addresses, essays, lectures. I. Furitano, Giuseppe. II. Series.
NA9000.S78 vol. 11 79-341032
[HT169.E8] *79 MARC

—FRANCE

Blumann, Claude.
Droit de l'urbanisme / Claude Blumann. — 2. éd. — Paris : Dalloz, 1979.
169 p. ; 25 cm. — (Mémentos Dalloz ; 3498) (Droit public) F***
Includes index.
ISBN 2-247-00060-6 : 34.00F
1. City planning and redevelopment law—France. I. Title.
346'.44'045 79-377445
79 MARC

France. Laws, statutes, etc.
Code de l'urbanisme : partie législative. — Éd. mise à jour au 30 juin 1977. — Paris : Journal officiel de la République française, 1977.
iii, 121 p. ; 21 cm. F***
ISBN 2-11-070548-5 : 10.00F
1. City planning and redevelopment law—France. 2. Land use, Urban—France. I. Title.
346'.44'04502632 78-368491
78 MARC

Jacquignon, Louis.
Le droit de l'urbanisme / par Louis Jacquignon ; préf. de G. Liet-Veaux. — 6. éd. entièrement refondue et mise à jour / par Yves Maxime Danan. — Paris : Eyrolles, 1978.
xxvii, 410 p. ; 25 cm. — (Bibliothèque de l'Institut de topométrie du Conservatoire national des arts et métiers) F***
Includes bibliographical references and index.
170F
1. City planning and redevelopment law—France. 2. Zoning law—France. I. Danan, Yves Maxime. II. Title. III. Series.
346'.44'045 79-340699
79 MARC

—FRANCE—DICTIONARIES

Gilli, Jean Paul.
Droit de l'urbanisme / Jean-Paul Gilli, Jacques de Lanversin. — 1. éd. — Paris : PUF, c1978.
143 p. ; 21 cm. — (Lexique) F***
ISBN 2-13-035595-1 : 29.00F
1. City planning and redevelopment law—France—Dictionaries. 2. Zoning law—France—Dictionaries. 3. City planning and redevelopment law—Dictionaries—French. 4. Zoning law—Dictionaries—French. I. Lanversin, Jacques de, joint author. II. Title.
346'.44'045 79-357183
79 MARC

—FRANCE—PERIODICALS

France. Ministère de l'équipement et de l'aménagement du territoire.
Bulletin officiel du Ministère de l'équipement et de l'aménagement du territoire.
[Paris, Direction des journaux officiels]
v. 21 cm.
Continues: France. Ministère de l'équipement. Bulletin officiel du Ministère de l'équipement.
1. France—Public works—Periodicals. 2. Regional planning—Law and legislation—France—Periodicals. 3. City planning and redevelopment law—France—Periodicals. 4. Transportation—Law and legislation—France—Periodicals.
344'.44'06 78-645986
MARC-S

—GERMANY, WEST

Arbeitsblätter zum novellierten Bundesbaugesetz / Bearbeiter, Rudolf Schäfer ... et al.]. — Berlin : Deutsches Institut für Urbanistik, 1977.
292 leaves : ill. ; 30 cm. GFR***
ISBN 3-88118-031-1 : DM20.00
1. City planning and redevelopment law—Germany, West. I. Schäfer, Rudolf, Assessor Dipl.-Pol. II. Deutsches Institut für Urbanistik.
346'.43'045 78-368761
78 MARC

Beus, Hans-Bernd.
Rechtsprobleme bei der Ausgestaltung der Raumordnung und Landesplanung als Entwicklungsplanung / Hans-Bernd Beus. — Münster (Westf.) : Institut für Siedlungs- und Wohnungswesen der Universität Münster, 1978.
xxiii, 201, 11 p. ; 21 cm. — (Beiträge zum Siedlungs- und Wohnungswesen und zur Raumplanung ; Bd. 46) GFR***
Originally presented as the author's thesis, Münster.
Bibliography: p. vi-ixx.
1. City planning and redevelopment law—Germany, West. 2. Regional planning—Law and legislation—Germany, West. I. Title. II. Series.
346'.43'045 79-350732
79 MARC

Bröll, Helmut, 1936-
Rechts- und Steuerfragen bei Planung und Erstellung gewerblicher Bauten : gesetzl. u. vertragl. Grundlagen mit ausführl. praxisbezogenen Erl. für Architekten, gewerbl. Bauherren u. beratende Berufe / baurechtl. Teil von Helmut Bröll, vertragsrechtl. Teil von Günter Gerhard, steuerrechtl. Teil von Dieter Tewes. — 1. Aufl. — Kissing : WEKA-Verlag, Fachverlag für Verwaltung u. Industrie, 1978.
168 p. ; 21 cm. GFR78-A
Includes bibliographical references and index.
ISBN 3-8111-7230-1 : DM34.00
1. Building—Contracts and specifications—Germany, West. 2. City planning and redevelopment law—Germany, West. 3. Construction industry—Taxation—Germany, West. I. Gerhard, Günter, joint author. II. Tewes, Dieter, joint author. III. Title.
343'.43'078 78-397456
7? MARC

Brückmann, Heinz.
Der Überlassungsvertrag im Städtebauförderungsgesetz : Beteiligungsmodell für Bürger u. Gemeinden / von Heinz Brückmann. — Siegburg : Reckinger, 1975.
xii, 174 p. ; 21 cm. — (Schriften zum deutschen Kommunalrecht ; Bd. 9) GFR78-A
Bibliography: p. 161-174.
ISBN 3-7922-0039-2 : DM24.00
1. City planning and redevelopment law—Germany, West. 2. Transfer (Law)—Germany, West. 3. Municipal corporations—Germany, West. I. Title. II. Series.
79-339193
79 MARC

Dümichen, Paul.
Bau- und Planungsrecht / von Paul Dümichen. — 2., überarb. u. erw. Aufl. — Herford : Maximilian-Verlag, 1978, c1963.
216 p. : ill., forms ; 21 cm. — (Leitfaden für den öffentlichen Dienst ; 11) GFR79-A
First ed. published in 1963 under the title: Bau- und Siedlungswesen.
Includes index.
ISBN (invalid) 3-7860-0087-6 : DM24.80
1. Building laws—Germany, West. 2. City planning and redevelopment law—Germany, West. 3. Regional planning—Law and legislation—Germany, West. I. Title. II. Series.
346'.43'045 79-366830
79 MARC

Fickert, Hans Carl.
Planfeststellung für den Strassenbau : Kommentar : systemat. Erl. unter bes. Berücks. d. Umweltschutzes mit Rechts- u. Verwaltungsvorschriften zur Planfeststellung : insbes. Erl. d. Vorschriften über Planung, Planfeststellung u. Enteignung d. Bundesfernstrassengestzes (FStrG) u. d. Vorschriften über d. Planfeststellungsverfahren (Sections 72-78) d. Verwaltungsverfahrensgesetzes (VwVfG) / von Hans Carl Fickert. — Köln ; Stuttgart ; Berlin ; Hannover : Duetscher Gemeindeverl. : Kohlhammer, 1978.
xx, 668 p. ; 21 cm. GFR78-A
Bibliography: p. xvii-xx.
Includes index.
ISBN 3-555-00275-9 : DM86.00
1. Highway law—Germany, West. 2. Environmental law—Germany, West. 3. City planning and redevelopment law—Germany, West. I. Title.
346'.43'044 78-379143

Förster, Hans, 1914-
Baunutzungsverordnung : in d. Fassung vom 15. September 1977 / von Hans Förster. — 3., (neubearb.) Aufl. — Stuttgart ; Berlin ; Köln ; Mainz : Kohlhammer, [1978] c1964.
275 p. ; 21 cm. — (Kohlhammer Kommentare) GFR78-A
Text of the regulation: p. 9-49.
"Dieser Kommentar ist ein Sonderdruck aus den Anhang A des ebenfalls im Kohlhammer Verlag erschienenen grossen Loseblatt-Kommentars zum Bundesbaugesetz von Brügelmann, Förster, Grauvogel, Meyer, Pohl, Stahnke."
Includes index.
ISBN 3-17-004951-8
1. Zoning law—Germany, West. 2. City planning and redevelopment law—Germany, West. I. Germany (Federal Republic, 1949-). Bundesministerium für Wohnungswesen und Städtebau. Baunutzungsverordnung. 1978. II. Title.
346'.43'045 78-393510
79 MARC

Für Erhalt, gegen Abriss : für Betroffene in Werkssiedlungen / Autorenkollektiv, Ulrich Adler ... [et al.] ; fachliche Beratung und Vorwort, Wolfgang Kantzow, Wolfgang Richter, Helmut Weuthen ; Karikaturen, Felix Mareite, Bilder, Gerd Plückelmann. — Dortmund : Weltkreis-Verlags-GmbH, 1976.
143 p. : ill. ; 19 cm. GFR76-A50
"Aus einer Projektarbeit am Fachbereich Architektur der Fachhochschule Dortmund entstanden."
ISBN 3-88142-171-8 : DM12.80
1. City planning and redevelopment law—Germany, West. 2. Urban renewal—Germany, West. I. Adler, Ulrich.
346'.43'045 78-400375
79 MARC

Gaentzsch, Günter.
Öffentliches Baurecht : Grundriss für Verwaltung u. Wirtschaft / von Günter Gaentzsch. — Köln ; Stuttgart ; Berlin ; Hannover : Kohlhammer, 1978.
319 p. ; 21 cm. GFR78-A38
Bibliography: p. 277-293.
Includes index.
ISBN 3-555-00292-9 : DM29.80
1. Building laws—Germany, West. 2. City planning and redevelopment law—Germany, West. I. Title.
78-399009
79 MARC

Germany (Federal Republic, 1949-). Laws, statutes, etc.
Bundesbaugesetz : : mit Verordnung über Grundsätze für die Ermittlung des Verkehrswertes von Grundstücken, Baunutzungsverordnung ... : Textausg. mit Verweisungen u. Sachreg. — 14., neubearb. Aufl. — München : Beck, 1979.
798 p. ; 16 cm. — (Beck'sche Textausgaben) GFR79-A
Includes index.
ISBN 3-406-07136-8 : DM19.80
1. City planning and redevelopment law—Germany, West. 2. Zoning law—Germany, West. 3. Inflammable liquids—Law and legislation—Germany, West. I. Title.
79-348938
79 MARC

Grundfragen des Erschliessungsbeitragsrechts in der kommunalen Praxis / von Elisabeth Bauernfeind ... [et al.]. — 2., völlig überarb. u. erw. Aufl. — Siegburg : Reckinger, 1978.
xii, 231 p. ; 21 cm. — (Schriften zum deutschen Kommunalrecht ; Bd. 6) GFR78-A
Papers of two conferences sponsored by the Verwaltungs- und Wirtschaftsakademie Wuppertal, held May 15-16, 1973 and May 3-4, 1977.
Includes bibliographical references and index.
ISBN 3-7922-0035-X : DM39.60
1. City planning and redevelopment law—Germany, West. 2. Local finance—Law and legislation—Germany, West. 3. Building laws—Germany, West—Digests. I. Bauernfeind, Elisabeth. II. Verwaltungs- und Wirtschafts-Akademie Wuppertal. III. Series.
78-399451
79 MARC

Müller, Karl, 1941-
Die bauliche Fortentwicklung des bisher unbeplanten Innenbereichs / vorgelegt von Karl Müller. — [s.l. : s.n., 1978?] (Köln : R. J. Hundt)
xvii, 211 p. ; 21 cm. GFR***
Thesis—Cologne.
Vita.
Bibliography: p. viii-xvii.
1. City planning and redevelopment law—Germany, West. 2. Building laws—Germany, West. I. Title.
346'.43'045 79-365178
79 MARC

Prognos AG. Abteilung Stadtentwicklung und Regionalplanung.
Erfahrungen der Gemeindeverwaltungen mit dem Vollzug des Städtebauförderungsgesetzes bei der Vorbereitung und Durchführung städtebaulicher Sanierungsmassnahmen : Forschungsprojekt BMBau RS II 6-704102-204(1978) / im Auftrag des Bundesministers für Raumordnung, Bauwesen und Städtebau, Bonn-Bad Godesberg, durchgeführt von einer Arbeitsgruppe der PROGNOS AG. Basel, Abt. Stadtentwicklung und Regionalplanung ; Bearbeiter, Hartmut E. Arras, Anke Schillinger, Erik Schmidt ; rechtliche Beratung, Hans-Jörg Birk. — Bad Godesberg : Der Bundesminister, 1978.
151 p. : col. ill. ; 30 cm. — (Schriftenreihe Stadtentwicklung des Bundesministers für Raumordnung, Bauwesen und Städtebau ; 016) GFR***
Includes bibliographical references.
1. City planning and redevelopment law—Germany, West. I. Arras, Hartmut E. II. Schillinger, Anke. III. Schmidt, Erik, Dipl.-Volkswirt. IV. Title. V. Series: Schriftenreihe Stadtentwicklung ; 016.
346'.43'045 79-373040
79 MARC

CITY PLANNING AND REDEVELOPMENT LAW
—GERMANY, WEST (Continued)

Rosenbach, Hans Josef.
Einführung in das öffentliche Baurecht : jur. Kurzlehrbuch mit Fällen, Abbilungen und Übersichten / von Hans Josef Rosenbach. — Düsseldorf : Werner, 1978.
xx, 215 p. ; 21 cm. GFR78-A
Includes bibliographical references and index.
ISBN 3-8041-3030-5
1. Building laws—Germany, West. 2. City planning and redevelopment law—Germany, West. I. Title.
346'.43'045 78-392313
79 MARC

Schulze-Fielitz, Helmuth.
Sozialplanung im Städtebaurecht : am Beispiel d. Stadterneuerung / Helmuth Schulze-Fielitz. — Königstein/Ts. : Athenäum, 1978.
xxi, 458 p. ; 22 cm. — (Monographien zur rechtswissenschaftlichen Forschung : Öffentliches Recht) GFR79-A
Bibliography: p. 408-442.
Includes indexes.
ISBN 3-7610-6185-4 : DM130.00
1. City planning and redevelopment law—Germany, West. I. Title.
346'.43'045 79-353985
79 MARC

Waterhouse, Alan.
City planning for a social market economy : rationality and evolution in German planning legislation / Alan Waterhouse. — Toronto : Dept. of Urban and Regional Planning, University of Toronto, 1978.
29 p. ; 28 cm. — (Papers on planning and design ; paper no. 17) C***
Bibliography: p. 28-29.
$2.50
1. Urban policy—Germany, West. 2. City planning and redevelopment law—Germany, West. I. Title. II. Series: Papers on planning and design ; no. 17.
HT169.G7W39 309.2'62'0943 79-303178
79 MARC

Zoubek, Gerhard.
Das Raumordnungsverfahren : eine rechtsvergleichende Untersuchung des förmlichen landesplanerischen Abstimmungssingstrumentes / Gerhard Zoubek. — Münster : Selbstverlag des Instituts für Siedlungs- und Wohnungswesen der Universität Münster, 1978.
xxiv, 232, 11 p. ; 21 cm. — (Beiträge zum Siedlungs- und Wohnungswesen und zur Raumplanung ; Bd. 45) GFR***
Originally presented as the author's thesis, Münster.
Bibliography: p. v-xviii.
DM26.00
1. City planning and redevelopment law—Germany, West. I. Title. II. Series.
346'.43'045 79-352312
79 MARC

Zuck, Rüdiger.
Das Recht des Bebauungsplans : Verfahrensgang, Rechtsbindung, Plansicherung, Gerichtskontrolle / von Rüdiger Zuck, unter Mitarb. von Kalus-Peter Dolde. — Stuttgart ; München ; Hannover : Boorberg, 1978.
216 p. ; 21 cm. GFR78-A
Includes bibliographical references and index.
ISBN 3-415-00701-4 : DM26.50
1. Building laws—Germany, West. 2. City planning and redevelopment law—Germany, West. I. Dolde, Klaus-Peter, joint author. II. Title.
346'.43'045 79-348783
79 MARC

—GERMANY, WEST—CASES

Germany (Federal Republic, 1949-). Bundesverwaltungsgericht.
Die Rechtsprechung des Bundesverwaltungsgerichts zum Erschliessungs- und Erschliessungsbeitragsrecht : systemat. Darst. im Anschluss an d. Gesetzestext Hans-Joachim Driehaus. — Stand 1. Juni 1978. — Köln : Deutsches Volksheimstättenwerk, 1978.
viii, 205 p. ; 21 cm. GFR78-A45
Includes the text of the Vorschriften des 6. Teils des Bundesbaugesetzes.
Includes index.
ISBN 3-87941-802-0 : DM14.80
1. Building laws—Germany, West—Cases. 2. City planning and redevelopment law—Germany, West—Cases. I. Driehaus, Hans-Joachim, 1940- . II. Germany (Federal Republic, 1949-). Laws, statutes, etc. Bundesbaugesetz. 6. Erschliessung. 1978. III. Title.
346'.43'045 79-378263
79 MARC

—GREAT BRITAIN

Heap, Desmond, Sir.
Community Land Act, 1975 / [by] Sir Desmond Heap, Frank Knowles. — [London] : Law Society, 1976.
13 p. ; 22 cm. GB76-18058
Cover title.
ISBN 0-901064-65-3 : £0.75
1. Regional planning—Law and legislation—Great Britain. 2. City planning and redevelopment law—Great Britain. 3. Zoning law—Great Britain. 4. Land use—Law and legislation—Great Britain. I. Knowles, Frank Cecil, joint author. II. Great Britain. Laws, statutes, etc. Community land act 1975. III. Title.
KD1125.Z9H37 346'.41'045 77-371131
77r79rev MARC

Heap, Desmond, Sir.
An outline of planning law / by Desmond Heap. — 7th ed. — London : Sweet and Maxwell, 1978.
lix, 344 p. : diagr. ; 23 cm. GB***
First ed. published 1949 under title: An outline of the new planning law.
Includes bibliographical references and indexes.
ISBN 0-421-22790-7 : £8.00. ISBN 0-421-22800-8 pbk.
1. City planning and redevelopment law—Great Britain. 2. Regional planning—Law and legislation—Great Britain. 3. Land use—Law and legislation—Great Britain. I. Title.
KD1125.H4 1978 346'.42'045 78-324745
79 MARC

Helb, John Vincent.
The British development rights experience / John Vincent Helb. — Monticello, Ill. : Council of Planning Librarians, 1978.
30 p. ; 28 cm. — (Exchange bibliography - Council of Planning Librarians ; # 1560)
Caption title.
$3.00
1. Regional planning—Law and legislation—Great Britain. 2. City planning and redevelopment law—Great Britain. 3. Development rights transfer—Law and legislation—Great Britain. I. Title. II. Series: Council of Planning Librarians. Exchange bibliography ; # 1560.
Z5942.C68 No. 1560 016.3092 s 79-300815
[KD1125] 79 MARC

—GREAT BRITAIN—CONGRESSES

Development control—thirty years on : papers from a conference held at Oxford, September 1978 / organised by the Bar Council, the Law Society, and the Royal Institution of Chartered Surveyors. — [London] : Sweet & Maxwell, 1979.
v, 94 p. ; 25 cm. — (Journal of planning and environmental law : Occasional papers) GB***
Includes bibliographical references.
ISBN 0-421-25890-X : £4.00
1. Regional planning—Law and legislation—Great Britain—Congresses. 2. City planning and redevelopment law—Great Britain—Congresses. I. General Council of the Bar of England and Wales. II. Law Society (United Kingdom) III. Royal Institution of Chartered Surveyors. IV. Series.
KD1125.A75D48 346'.41'045 79-318354
79 MARC

—GREAT BRITAIN—OUTLINES, SYLLABI, ETC.

Royal Institution of Chartered Surveyors. Planning and Development Division.
Enforcement of planning control surveying, valuation and planning investigation : a guide to practitioners / prepared by the Planning and Development Division of the Royal Institution of Chartered Surveyors. — London : The Institution, 1978.
[2], 18 p. ; 30 cm. — (Practice note - Planning and Development Division of the Royal Institution of Chartered Surveyors ; no. 4 ISSN 0141-1462) GB79-14555
Cover title.
Bibliography: p. 18.
£1.30
1. Regional planning—Law and legislation—Great Britain—Outlines, syllabi, etc. 2. City planning and redevelopment law—Great Britain—Outlines, syllabi, etc. I. Title. II. Series: Royal Institution of Chartered Surveyors. Planning and Development Division. Practice note - Planning and Development Division of the Royal Institution of Chartered Surveyors ; no. 4.
KD1125.Z9R69 346'.42'045 79-318358
79 MARC

—ILLINOIS

Illinois. Laws, statutes, etc.
Illinois laws relating to planning and development / prepared by Office of Research and Planning, State of Illinois, Department of Local Government Affairs. — [Springfield] : The Office, 1978.
89 p. ; 28 cm. — (Reference series - Department of Local Government Affairs ; no. 3)
"An update and revision of the publication, 'Reference series no. 3, Laws relating to planning and development,' June 1976."
1. Regional planning—Law and legislation—Illinois. 2. City planning and redevelopment law—Illinois. 3. Zoning law—Illinois. I. Illinois. Dept. of Local Government Affairs. Office of Research and Planning. II. Title. III. Series: Illinois. Dept. of Local Government Affairs. Reference series - Department of Local Government Affairs ; no. 3.
KFI1658.A3 1978 346'.773'045 79-624379
79 MARC

—INDIA—CONGRESSES

Seminar on Law and Urban Land, New Delhi, India, 1977.
Law and urban land / edited by H. U. Bijlani, M. K. Balachandran. — New Delhi : Centre for Urban Studies, Indian Institute of Public Administration, 1978.
xvi, 242 p. ; 22 cm.
Includes bibliographical references.
Rs35.00 ($10.00 U.S.)
1. Land use—Law and legislation—India—Congresses. 2. City planning and redevelopment law—India—Congresses. I. Bijlani, H. U. II. Balachandran, M. K. III. Title.
346'.54'045 78-904610

—INDONESIA

Dirdjosisworo, Sudjono.
Segi-segi hukum tentang tata bina kota di Indonesia / Soedjono D. — Cet. 1. — Bandung : Karya Nusantara, 1978.
xvii, 194 p. : ill. ; 21 cm.
Bibliography: p. 193-194.
1. City planning and redevelopment law—Indonesia. I. Title.
78-942616
79 MARC

—IRELAND

O'Sullivan, Philip.
Irish planning and acquisition law / Philip O'Sullivan. — Dublin : Institute of Public Administration, 1978.
vii, 77 p. ; 22 cm. GB***
Errata slip inserted.
Includes bibliographical references and index.
ISBN 0-902173-80-4 : £2.97
1. Eminent domain—Ireland. 2. Regional planning—Law and legislation—Ireland. 3. City planning and redevelopment law—Ireland. I. Title.
KDK290.O9 346'.415'045 78-324317
79 MARC

—ITALY

Agnoli, Francesco Mario, 1934-
La legge sulla edificabilità dei suoli : guida pratica / Francesco Mario Agnoli. — Bologna : Edagricole, 1978.
vii, 99 p. ; 21 cm. — (Collana teorico-pratica di legislazione tecnica e agraria) It78-July
"Legge 28 gennaio 1977, n. 10": p. 90-98.
Bibliography: p. 99.
L4000
1. Building laws—Italy. 2. Agricultural laws and legislation—Italy. 3. City planning and redevelopment law—Italy. I. Italy. Laws, statutes, etc. Legge 28 gennaio 1977, n. 10. II. Title. III. Series.
346'.45'045 78-376394
*78 MARC

Alibrandi, Tommaso.
I beni culturali e ambientali / [di] Tommaso Alibrandi, Piergiorgio Ferri. — Milano : Giuffrè, 1978.
iv, 682 p. ; 22 cm. — (Teoria e pratica del diritto : Sezione 4, Diritto amministrativo ; 9) It78-Oct
Includes bibliographical references and index.
L16000
1. Cultural property, Protection of—Law and legislation—Italy. 2. Natural monuments—Law and legislation—Italy. 3. City planning and redevelopment law—Italy. I. Ferri, Piergiorgio, joint author. II. Title.
78-397811
*79 MARC

Barucchi, Alfredo.
Jus aedificandi e proprietà dei suoli : contributo allo studio della disciplina dell'edificabilità nella prospettiva della riforma del regime dei suoli / Alfredo Barucchi. — Torino : G. Giappichelli, 1976.
194 p. ; 24 cm. It78-Apr
Includes legislation and proposed legislation.
Includes bibliographical references.
L4500
1. Building laws—Italy. 2. City planning and redevelopment law—Italy. 3. Real property—Italy. I. Title.
346'.45'045 78-376818
*78 MARC

Bottino, Felicia.
Il nuovo regime dei scoli / Felicia Bottino, Vito A. Brunetti ; introduzione di Michele Martuscelli. — 3. ed. — Roma : Edizioni delle autonomie, 1978.
376 p. ; 21 cm. — (Amministrare l'urbanistica ; 2) It***
Includes bibliographical references.
L5500
1. City planning and redevelopment law—Italy. 2. Building laws—Italy. 3. Eminent domain—Italy. I. Brunetti, Vito Antonio, 1923- II. Martuscelli, Michele, 1918- III. Title.
79-350507
79 MARC

Codice dell'urbanistica / [a cura di] Achille Cutrera, Vittorio Italia. — 2. ed. — Milano : A. Giuffrè, 1978-
v. ; 17 cm. It78-Nov
CONTENTS: 1. Norme statali.

L27000 (v. 1)
1. City planning and redevelopment law—Italy. I. Cutrera, Achille. II. Italia, Vittorio.
346'.45'045 79-337276
*79 MARC

CITY PLANNING AND REDEVELOPMENT LAW
—ITALY (Continued)

Convenzioni urbanistiche e tutela nei rapporti tra privati / ⌐scritti di¬ Mario Nigro ... ⌐et al.¬ ; a cura di Michele Costantino. — Milano : A. Giuffrè, 1978.
ix, 764 p. ; 23 cm. — (Territorio e casa ; v. 4) It78-Apr
Includes bibliographical references.
L17500
1. City planning and redevelopment law—Italy. 2. Building laws—Italy. I. Nigro, Mario. II. Costantino, Michele.
346'.45'045 78-365794
*78 MARC

De Roberto, Alberto.
Le sanzioni urbanistiche amministrative e penali / ⌐di¬ Alberto de Roberto, Giovanni de Roberto. — Milano : A. Giuffrè, 1978.
v, 108 p. ; 23 cm. — (Territorio e casa ; quaderno 2) It78-Aug
Includes bibliographical references.
L3000
1. City planning and redevelopment law—Italy. 2. Sanctions, Administrative—Italy. 3. Punishment—Italy. I. De Roberto, Giovanni, joint author. II. Title.
346'.45'045 78-387103
*79 MARC

Giuffrè, Vincenzo.
Il regime delle trasformazioni urbanistiche ed edilizie : commentario della Legge n. 10, 1977 : con documentazione, legislazione, giurisprudenza e bibliografia / Vincenzo Giuffrè. — Napoli : Jovene, 1979.
xi, 778 p. ; 24 cm. It79-June
Bibliography: p. ⌐233¬-246.
Includes indexes.
L20000
1. Building laws—Italy. 2. City planning and redevelopment law—Italy. I. Italy. Laws, statutes, etc. Legge 28 gennaio 1977, n. 10. II. Title.
346'.45'045 79-381316
*79 MARC

Lucarelli, Francesco.
Politica della casa e legislazione urbanistica / Francesco Lucarelli. — Napoli : Edizioni scientifiche italiane, c1979.
388 p. ; 21 cm. — (Problemi del territorio ; 2) It***
Bibliography: p. 387-388.
L11000
1. City planning and redevelopment law—Italy. 2. Housing—Law and legislation—Italy. 3. Eminent domain—Italy. I. Title. II. Series.
346'.45'045 79-383873
79 MARC

Pacelli, Eugenio.
Nuova disciplina dell'edificabilità dei suoli : legge 28-1-1977 n. 10 / Eugenio Pacelli, Antonio Saraceni. — Napoli : Editoriale scientifica, c1977.
83 p. ; 24 cm. — (Ricerche giuridiche ; 7) It***
Includes bibliographical references.
L2500
1. City planning and redevelopment law—Italy. 2. Building laws—Italy. 3. Eminent domain—Italy. I. Saraceni, Antonio, joint author. II. Italy. Laws, statutes, etc. Legge 28 gennaio 1977, n. 10. III. Title.
346'.45'045 78-387354
79 MARC

Predieri, Alberto.
La Legge 28 gennaio 1977, n. 10 sulla edificabilità dei suoli / Alberto Predieri. — Milano : Giuffrè, 1977.
631 p. ; 22 cm. — (Territorio e casa : Intervento ; 4) It78-July
Includes bibliographical references and index.
L12000
1. Building laws—Italy. 2. City planning and redevelopment law—Italy. 3. Housing—Italy—Law and legislation. I. Italy. Laws, statutes, etc. Legge 28 gennaio 1977, n. 10.
78-399865
*79 MARC

Sebasti, Rinaldo, 1934-
Controllo edilizio e lotta agli abusi / Rinaldo Sebasti ; introduzione di Franca D'Alessandro Prisco. — 1. ed. — Roma : Edizioni delle autonomie, 1979.
229 p. ; 21 cm. — (Amministrare l'urbanistica ; 6) It79-Mar
L4500
1. Building laws—Italy. 2. City planning and redevelopment law—Italy. 3. City planning and redevelopment law—Italy—Rome. I. Title.
79-373474
*79 MARC

Strumenti urbanistici / ⌐di¬ Paolo Bettini ... ⌐et al.¬. — Firenze : Libreria editrice fiorentina, 1977.
311 p. : ill. ; 24 cm. It78-Dec
At head of title: Università di Firenze, Facoltà di architettura, Istituto di ricerche territoriali e urbane.
Cover title.
Bibliography: p. 111-112.
1. Regional planning—Law and legislation—Italy. 2. City planning and redevelopment law—Italy. I. Bettini, Paolo.
346'.45'045 79-348338
*79 MARC

Il Territorio / ⌐di¬ G. Cocco ... ⌐et al.¬ ; prefazione di Alberto Predieri. — 1. ed. — Roma : Edizioni delle autonomie, 1978.
165 p. ; 21 cm. — (Istituzioni e programmazione ; 3) It78-Nov
Includes bibliographical references.
L5500
1. City planning and redevelopment law—Italy. 2. Transportaton, Automotive—Law and legislation—Italy. 3. Environmental law—Italy. I. Cocco, Giuliano.
78-399703

—ITALY—ADDRESSES, ESSAYS, LECTURES

Il Ruolo delle regioni nella disciplina e gestione del territorio / a cura di Ardea Ferrero. — Milano : A. Giuffrè, 1979.
258 p. ; 29 cm. — (Studi urbanistici ; v. 12) It79-June
At head of title: Fondazione Aldo Della Rocca.
CONTENTS: Lapadula, B. F., Lefebvre, C. e Sersale, G. Il ruolo delle regioni nella disciplina e gestione del territorio.—Ferrara, G. Parchi ed aree protette nella dimensione regionale.—Ralli, M. Tutela del paesaggio ed urbanistica regionale.—Sisi, E. Tutela e valorizzazione dei beni culturali ed ambientali nel quadro generale della legislazione urbanistica nazionale e regionale.
Bibliography: p. 173-177.
L16000
1. City planning and redevelopment law—Italy—Addresses, essays, lectures. 2. National parks and reserves—Italy—Addresses, essays, lectures. 3. Landscape protection—Law and legislation—Italy—Addresses, essays, lectures. 4. Cultural property, Protection of—Law and legislation—Addresses, essays, lectures. I. Ferrero, Ardea. II. Fondazione Aldo Della Rocca. III. Series.
NA9000.S78 vol. 12 79-379606
*79 MARC

—ITALY—CONGRESSES

Convenzioni e centri storici : il quadro giuridico e le possibilità attuative / ⌐di¬ R. Airoldi ... ⌐et al.¬. — 1. ed. — Milano : Clup, 1978.
140 p. ; 24 cm. It79-May
Bibliography: p. 134-135.
ISBN 8870053911 : L2800
1. City planning and redevelopment law—Italy—Congresses. 2. Historic sites—Law and legislation—Italy—Congresses. I. Airoldi, Renato.
79-366468
79 MARC

Convivium notarile, 37th, Rapallo, Italy, 1977.
Atti del XXXVII Convivium notarile (Rapallo, 17 aprile 1977) : la Legge 28 gennaio 1977, n. 10 sul regime dei suoli considerata sotto il profilo dell'attività di notaio : in appendice, a) Atti della Tavola rotonda organizzata dall'UNI a Roma (26 marzo 1977), Dalla licenza di costruzione allo jus aedificandi, aspetti negoziali, b) Raccolta delle leggi richiamata dalla L. 28 gennaio 1977, n. 10 sul regime dei suoli. — Milano : A. Giuffrè, 1978.
221 p. ; 25 cm. — (Collana di studi - Consiglio nazionale del notariato ; 3) It78-Oct
At head of title: Commissione di studio in materia edilizio-urbanistica.
Includes bibliographical references.
L6000
1. City planning and redevelopment law—Italy—Congresses. 2. Building laws—Italy—Congresses. 3. Notaries—Italy—Congresses. I. Commissione di studio in materia edilizio-urbanistica. II. Title: La Legge 28 ⌐i.e. ventotto¬ gennaio 1977. III. Series: Consiglio nazionale del notariato. Collana di studi - Consiglio nazionale del notariato ; 3.
79-408969
*79 MARC

Seminario di studio sull'attuazione della nuova disciplina giuridica dei suoli, Bologna, 1977.
Diritto e urbanistica nella nuova disciplina dei suoli : prospettive di applicazione della Legge 28 gennaio 1977, n. 10 / Seminario di studio sull'attuazione della nuova disciplina giuridica dei suoli, Bologna, 24-25 ottobre 1977. — Bologna : Il mulino, 1978.
266 p. ; 24 cm. — (Collana dell'Istituto di studi giuridici della regione Emilia-Romagna) It78-Oct
"Legge 28 gennaio 1977, n.10": p. 249-281.
1. City planning and redevelopment law—Italy—Congresses. 2. Building laws—Italy—Congresses. I. Italy. Laws, statutes, etc. Legge 28 gennaio 1977, n. 10. II. Title.
346'.45'045 78-398125
*79 MARC

—ITALY—DIGESTS

Lorizio, M Athena.
Urbanistica e giudice amministrativo : rassegna critica di giurisprudenza del Consiglio di Stato sulla Legge ponte / M. Athena Lorizio. — Milano : A. Giuffrè, 1976.
447 p. ; 23 cm. — (Territorio e casa ; v. 2) It78-June
"Appendice di riferimento alla legge 28 gennaio 1977 n. 10 recante norme per la edificabilità dei suoli" (20 p.) inserted.
1. City planning and redevelopment law—Italy—Digests. 2. Italy. Consiglio di Stato. I. Title.
346'.45'045 78-382544
*78 MARC

—ITALY—LOMBARDY

Lombardy. Laws, statutes, etc.
Normativa urbanistica della Lombardia : (la legge urbanistica regionale e le norme regionali di attuazione della Legge n. 10 sull'edificabilità dei suoli, coordinate, commentate ed annotate) / ⌐a cura di¬ Vittorio Martinelli. — Brescia : F. Apollonio ; ⌐Cremona¬ : I.C.A., 1978.
206 p. ; 24 cm. — (Collana legislativa e amministrativa ; n. 241) It78-Nov
Includes index.
1. City planning and redevelopment law—Italy—Lombardy. I. Martinelli, Vittorio. II. Title.
346'.452'045 79-346339

—ITALY—ROME

Sebasti, Rinaldo, 1934-
Controllo edilizio e lotta agli abusi / Rinaldo Sebasti ; introduzione di Franca D'Alessandro Prisco. — 1. ed. — Roma : Edizioni delle autonomie, 1979.
229 p. ; 21 cm. — (Amministrare l'urbanistica ; 6) It79-Mar
L4500
1. Building laws—Italy. 2. City planning and redevelopment law—Italy. 3. City planning and redevelopment law—Italy—Rome. I. Title.
79-373474
*79 MARC

—JAPAN

(Nihon no toshi)
日本の都市 / 建設大臣官房広報室編. — 東京：建設広報協議会, 1965.
12, 271 p., ⌐3¬ leaves of plates : ill. ; 21 cm. — (建設双書 ; 4)
1. City planning—Japan. 2. City planning and redevelopment law—Japan. 3. Cities and towns—Japan. I. Japan. Kensetsushō. Daijin Kambō. Kōhōshitsu. II. Series: Kensetsu sōsho ; 4.
HT169.J3N47 78-814950

(Toshi kōen hō kaisetsu)
都市公園法解説 / 建設省都市局公園緑地課監修 ; ⌐編集・日本公園緑地協会¬. — 東京：同協会, 1978.
6, 458 p. ; 22 cm. Ja 78-22086
¥3000
1. Parks—Law and legislation—Japan. 2. City planning and redevelopment law—Japan. I. Japan. Kensetsushō. Toshikyoku. Kōen Ryokuchika. II. Nihon Kōen Ryokuchi Kyōkai. III. Japan. Laws, statutes, etc. Toshi kōen hō. 1978.
78-816738

—MALAYSIA—SABAH

Sabah. Laws, statutes, etc.
Town and country planning ordinance, (Sabah Cap. 141) : incorporating all amendments and modifications up to 31st May, 1969. — ⌐s.l. : s.n¬, 1969 (Jesselton? : Government Printer, Sabah)
ii, 35 p. ; 27 cm.
$1.77
1. Regional planning—Law and legislation—Malaysia—Sabah. 2. City planning and redevelopment law—Malaysia—Sabah. I. Title.
346'.5953'045 77-940986
78 MARC

—NETHERLANDS

Faludi, Andreas.
Flexibility in Dutch local planning / ⌐by¬ Andreas Faludi and Stephen Hamnett. — ⌐Oxford¬ : Oxford Polytechnic, Department of Town Planning, 1977.
⌐2¬, iii leaves, 83 p. ; 30 cm. — (Working paper - Oxford Polytechnic, Department of Town Planning ; no. 28) GB78-13698
Bibliography: p. 80-83.
£1.88
1. City planning—Netherlands. 2. City planning—Netherlands—Decision making. 3. City planning and redevelopment law—Netherlands. I. Hamnett, Stephen, joint author. II. Title. III. Series: Oxford working papers in planning education and research ; no. 28.
HT169.N4F34 309.2'62'09492 78-320525
78 MARC

—NETHERLANDS—HOLLAND, SOUTH (PROVINCE)

Holland, South (Province). Gedeputeerde Staten.
Nota stadsvernieuwing. — ⌐'s-Gravenhage¬ : Gedeputeerde Staten van Zuid-Holland, 1977.
⌐170¬ p. ; ill. ; 30 cm.
Ne77
Cover title.
1. Urban renewal—Netherlands—Holland, South (Province) 2. City planning and redevelopment law—Netherlands—Holland, South (Province) I. Title.
HT178.N42H643 79-373626
79 MARC

CITY PLANNING AND REDEVELOPMENT LAW (Continued)

—NIGERIA

Ola, C S
Town and country planning law in Nigeria / by C. S. Ola. — Ibadan, Nigeria : Oxford University Press, 1977.
xv, 134 p. ; 23 cm.
"Grew out of a small part of ... thesis ⌊University College, London⌋"
Bibliography: p. ⌊126⌋-129.
ISBN 0-19-575400-X
1. Regional planning—Law and legislation—Nigeria. 2. City planning and redevelopment law—Nigeria. I. Title.
346'.669'045 79-104886
79 MARC

—ONTARIO

Current problems in development, planning, and zoning : edited text from the programme held in October, 1976. — Toronto : Dept. of Continuing Education, Law Society of Upper Canada, c1977.
208 p. : ill. ; 28 cm. C***
Includes the Ontario Building code act 1974 and selections from other laws.
ISBN 0-88759-024-1
1. City planning and redevelopment law—Ontario. 2. Zoning law—Ontario. 3. Building laws—Ontario. I. Law Society of Upper Canada. Dept. of Continuing Education. II. Ontario. Laws, statutes, etc. Building code act, 1974. 1977.
KEO920.C87 346'.713'045 78-324385
79 MARC

Ontario. Ministry of Housing. Local Planning Policy Branch. Programs Section.
A planner's reference to legislation in Ontario. — ⌊2d ed.⌋. — Toronto : Ontario Govt. Bookstore, 1977.
⌊88⌋ p. ; 30 cm. C***
Includes index.
$2.00
1. Regional planning—Law and legislation—Ontario. 2. City planning and redevelopment law—Ontario. I. Title.
KEO920.A375H68 1977 346'.713'045 79-300719
79 MARC

—PERU—HISTORY

Durán Montero, María Antonia.
Fundación de ciudades en el Perú durante el siglo XVI : estudio urbanístico / María Antonia Durán Montero. — 1. ed. — Sevilla : Escuela de Estudios Hispano-Americanos de Sevilla, 1978.
xxxviii, 210 p., ⌊19⌋ leaves of plates (3 fold.) : ill. ; 20 cm. — (Publicaciones de la Escuela de Estudios Hispano-Americanos de Sevilla ; 247) Sp***
Bibliography: p. ⌊167⌋-191.
Includes index.
ISBN 8400043146
1. Cities and towns—Peru—History. 2. City planning—Peru—History. 3. City planning and redevelopment law—Peru—History. I. Title. II. Series: Seville. Universidad. Escuela de Estudios Hispano-Americanos. Publicaciones ; 247.
HT129.P5D87 79-112623
79 MARC

—SCOTLAND

Young, Eric S
The law of planning in Scotland / by Eric Young. — Glasgow : W. Hodge, 1978.
xxxii, 383 p. ; 22 cm. GB***
Bibliography: p. 366-369.
Includes index.
ISBN 0-85279-082-1 : £12.50
1. Regional planning—Law and legislation—Scotland. 2. City planning and redevelopment law—Scotland. I. Title.
KDC446.Y68 346'.411'045 78-319439
79 MARC

—SPAIN

Carceller Fernández, Antonio.
Instituciones de derecho urbanístico / por Antonio Carceller Fernández. — Madrid : Montecorvo, 1979.
367 p., ⌊1⌋ leaf ; 24 cm. Sp***
Includes bibliographies.
ISBN 847111108X
1. City planning and redevelopment law—Spain. I. Title.
346'.46'045 79-109887

González Pérez, Jesús.
Comentarios a la Ley del suelo / ⌊por⌋ Jesús González Pérez. — 3. ed. — Madrid : Civitas, 1977.
1332 p. ; 25 cm. Sp77-Nov
Includes bibliographical references and index.
ISBN 8473980352
1. City planning and redevelopment law—Spain. I. Spain. Laws, statutes, etc. Ley 19/1975, de 2 de mayo. 1977. II. Title.
346'.46'045 78-379410
78 MARC

Parejo Alfonso, Luciano.
La reforma de la ley del suelo y su derecho transitorio / ⌊por⌋ Luciano Parejo Alfonso. — Madrid : Consejo Superior de los Colegios de Arquitectos de España, 1977.
141 p., 1 leaf ; 21 cm. Sp77-Sept
Includes bibliographical references.
ISBN 8450020018
1. City planning and redevelopment law—Spain. I. Title.
346'.46'045 78-393188
79 MARC

Santos Díez, Ricardo.
El aprovechamiento medio, como instrumento de gestión urbanística / por Ricardo Santos Díez ... — 1. ed. — Madrid : Ministerio de Obras Públicas y Urbanismo, Secretaría General Técnica, Servicio de Publicaciones, 1978.
267 p. ; 25 cm. Sp78-Sept
At head of title: Tratado de derecho urbanístico.
Bibliography: p. ⌊235⌋-243.
ISBN 8474330335
1. Land use, Urban—Spain—Mathematical models. 2. City planning and redevelopment law—Spain. I. Title. II. Title: Tratado de derecho urbanístico.
HD778.S35 333.7'7'0946 79-344742
79 MARC

Spain. Laws, statutes, etc.
⌊Ley 19/1975, de 2 de mayo⌋
Régimen del suelo y ordenación urbana : texto refundido aprobado por Real Decreto 1346/1976, de 9 de abril. — Madrid : B.O.E., 1976.
90 p. ; 21 cm. — (Separatas del Boletín oficial del estado, Gaceta de Madrid) Sp***
Cover title.
90ptas
1. City planning and redevelopment law—Spain. I. Title. II. Series: Spain. Boletín oficial del estado : Separatas.
346'.46'045 76-486045
76⌊r79⌋rev MARC

—SPAIN—CONGRESSES

Ciclo de conferencias y mesa redonda sobre la reforma de la Ley del suelo. — ⌊Madrid⌋ : Colegio Oficial de Arquitectos de Madrid, ⌊1977?⌋
115 p. ; 30 cm. Sp***
Bibliography: p. 53.
1. City planning and redevelopment law—Spain—Congresses. 2. Building laws—Spain—Congresses. I. Colegio Oficial de Arquitectos de Madrid.
346'.46'045 79-340468
79 MARC

—SWITZERLAND

Bochsler, Werner, 1946-
Die Ortsplanungserlasse gemäss dem bündnerischen Raumplanungsgesetz vom 20. Mai 1973 / eingereicht von Werner Bochsler. — Basel : ⌊s.n.⌋, 1977 (Clausthal : Bönecke-Druck)
xxvi, 202 p. ; 21 cm. Sw77-10479
Thesis—Basel.
Bibliography: p. xiii-xix.
Vita.
1. Regional planning—Law and legislation—Switzerland. 2. City planning and redevelopment law—Switzerland. 3. Building laws—Switzerland. I. Title.
346'.494'045 79-346580
*79 MARC

—UNITED STATES

ALI-ABA Course of Study: Land Planning and Regulation of Development.
ALI-ABA course of study. Land planning and regulation of development: materials.
⌊Philadelphia⌋ American Law Institute-American Bar Association Committee on Continuing Professional Education.
v. ill. 29 cm.
Vols. for cosponsored by Japan Center for Human Environmental Problems, the Conservation Foundation, and Hawaii Institute of Continuing Legal Education.
Vols. for issued in 2 or more vols.
Title varies slightly.
Key title: ALI-ABA course of study. Land planning and regulation of development: materials, ISSN 0191-8125.
1. Land use—Law and legislation—United States. 2. Regional planning—Law and legislation—United States. 3. City planning and redevelopment law—United States. I. American Law Institute-American Bar Association Committee on Continuing Professional Education. II. Japan Center for Human Environmental Problems. III. Conservation Foundation. IV. Hawaii Institute of Continuing Legal Education. V. ALI-ABA Course of Study: Land Planning and Regulation of Development. Land Planning and Regulation of Development: materials. VI. Title.
KF5698.Z9A13 346'.73'045 79-641038
MARC-S

Uhlig, Klaus Rolf.
Stadterhaltung in Beispielen / Klaus R. Uhlig ; Fotos, Monika Uhlig. — München : Frühmorgen, 1977-
v. : ill., maps ; 24 cm. GFR78-A
Appendix of v. 1 contains various legislation in English.
Bibliography: v. 1, p. 110-116.
CONTENTS: Bd. 1. Vereinigte Staaten von Amerika.

ISBN 3-921251-01-X (v. 1)
1. Architecture—United States—Conservation and restoration. 2. City planning—United States. 3. City planning and redevelopment law—United States. I. Title.
NA9053.C6U48 78-400315
79 MARC

United States. Congress. House. Committee on Government Operations. Intergovernmental Relations and Human Resources Subcommittee.
Federal role in dealing with urban decline and the Intergovernmental coordination act of 1977 : hearings before a subcommittee of the Committee on Government Operations, House of Representatives, Ninety-fifth Congress, second session on H.R. 4406 ... February 7 and 9, 1978. — Washington : U.S. Govt. Print. Off., 1978.
iii, 104 p. ; 24 cm.
1. City planning and redevelopment law—United States. 2. Regional planning—Law and legislation—United States. I. Title.
KF27.G664 1978b 346'.73'045 79-600944
79 MARC

United States. Congress. Senate. Committee on Banking, Housing, and Urban Affairs. — Urban policy initiatives : hearings before the Committee on Banking, Housing, and Urban Affairs, United States Senate, Ninety-fifth Congress, second session, on S. 3209 ... S. 3210 ... S. 3211 ... July 11 and 12, 1978. — Washington : U.S. Govt. Print. Off., 1978.
iv, 273 p. : ill. ; 14 cm.
1. City planning and redevelopment law—United States. 2. Urban policy—United States. I. Title.
KF26.B39 1978aa 346'.73'045 79-600992
79 MARC

—UNITED STATES—CASES

American land planning law : cases and materials / ⌊edited by⌋ Norman Williams, Jr. — New Brunswick, N.J. : Center for Urban Policy Research, Rutgers University, c1978.
2 v. (1792) ; 24 cm.
ISBN 0-88285-041-5 : $25.00
1. Regional planning—Law and legislation—United States—Cases. 2. City planning and redevelopment law—United States—Cases. 3. Zoning law—United States—Cases. I. Williams, Norman.
KF5692.A4A47 346'.73'045 77-17009
77 MARC

Basic property law / by Olin L. Browder, Jr. ... ⌊et al.⌋. — St. Paul, Minn. : West Pub. Co., 1979.
xxxv, 1445 p. ; 26 cm. — (American casebook series)
First and 2d. editions are entered under: Browder, Olin L.
ISBN 0-8299-2037-4 : $23.95
1. Property—United States—Cases. 2. City planning and redevelopment law—United States—Cases. I. Browder, Olin L. Basic property law. II. Series.
KF560.B7 1979 346'.73'04 79-11898
79 MARC

Casenote legal briefs—and use : adaptable to courses utilizing Beuscher, Wright, and Gitelman's casebook on land use / Dana L. Blatt, senior editor, Norman S. Goldenberg, senior editor, Peter Tenen, managing editor ; staff writers, Terry Molloy, Sally Molloy, James Rosenthal. — Beverly Hills, CA : Casenotes Pub. Co. ; Gardena, Calif. : distributed by Law Distributors, c1978.
130 p. ; 28 cm.
$9.95
1. Zoning law—United States—Cases. 2. City planning and redevelopment law—United States—Cases. 3. Land use—Law and legislation—United States—Cases. I. Blatt, Dana L. II. Goldenberg, Norman S. III. Beuscher, Jacob Henry, 1907-1967. Cases and materials on land use. IV. Title: Land use.
KF5698.A4C37 346'.73'045 79-116885
79 MARC

Casenote legal briefs—land use : adaptable to courses utilizing Haar's casebook on land-use planning / Norman S. Goldenberg, senior editor, Peter Tenen, managing editor, Robert J. Switzer, associate editor ; staff writers, James Rosenthal, Jeff Kahn. — Beverly Hills, CA : Casenotes Pub. Co. ; Gardena, Calif. : distributed by Law Distributors, c1979.
89 p. ; 28 cm.
$8.95
1. City planning and redevelopment law—United States—Cases. 2. Zoning law—United States—Cases. 3. Eminent domain—United States—Cases. 4. Land use—Law and legislation—United States—Cases. I. Goldenberg, Norman S. II. Switzer, Robert J. III. Haar, Charles Monroe, 1921- Land-use planning. IV. Title: Land use.
KF5692.A4C37 346'.73'045 79-116874
79 MARC

CITY PLANNING AND REDEVELOPMENT LAW —UNITED STATES—CASES (Continued)

Casenote legal briefs property : adaptable to courses utilizing Browder, Cunningham and Julin's casebook on basic property law / Dana L. Blatt, senior editor, Norman S. Goldenberg, senior editor, Peter Tenen, managing editor ; staff writers, Terry Molloy ... [et al.]. — Beverly Hills, CA : Casenotes Pub. Co. ; Gardena, Calif. : distributed by Law Distributors, c1977.
114 p. ; 28 cm.
$9.50
1. Property—United States—Cases. 2. City planning and redevelopment law—United States—Cases. I. Blatt, Dana L. II. Goldenberg, Norman S. III. Browder, Olin L. Basic property law. IV. Title: Property.
KF560.C267 346'.73'04 79-116886
79 MARC

Rose, Jerome G
Legal foundations of land use planning : text/casebook and materials on planning law / written and edited by Jerome G. Rose. — New Brunswick, N.J. : Center for Urban Policy Research, Rutgers University—The State University of New Jersey, c1979.
xx, p. ; 25 cm.
Includes bibliographical references and index.
ISBN 0-88285-057-1
1. City planning and redevelopment law—United States—Cases. 2. Regional planning—Law and legislation—United States—Cases. I. Title.
KF5692.A4R67 346'.73'045 78-11527
78 MARC

—UNITED STATES—OUTLINES, SYLLABI, ETC.

Wright, Robert R
Land use in a nutshell / by Robert R. Wright and Susan Webber. — St. Paul, Minn. : West Pub. Co., 1978.
xxxv, 316 p. ; 19 cm. — (Nutshell series)
Includes index.
ISBN 0-8299-2014-5 : $5.95
1. Regional planning—Law and legislation—United States—Outlines, syllabi, etc. 2. City planning and redevelopment law—United States—Outlines, syllabi, etc. 3. Land use—United States—Outlines, syllabi, etc. I. Webber, Susan, 1948- joint author. II. Title.
KF5698.Z9W74 346'.73'045 78-13132
78 MARC

—VENEZUELA

Venezuela. Laws, statutes, etc.
Recopilación de disposiciones legales sobre vivienda y desarrollo urbano-regional. — [Caracas] : Instituto Nacional de la Vivienda, [between 1976 and 1978]-
2 v. ; 30 cm.
Loose leaf for updating.
CONTENTS: t. 1. Política habitacional. INAVI. Vivienda. Estímulos. Pequeños centros poblados.—t. 2. Desarrollo urbano regional. Tierras. Disposiciones varias. Financiamento.
1. Housing—Venezuela—Law and legislation. 2. City planning and redevelopment law—Venezuela. 3. Regional planning—Law and legislation—Venezuela. I. Instituto Nacional de la Vivienda (Venezuela) II. Title.
346'.87'045 78-384486
79 MARC

—WYOMING

Schaefer, Jay Dee.
A legal primer : comprehensive planning and land use regulation in Wyoming / by Jay Dee Schaefer and Margo Lamb. — [Cheyenne] : Wyoming Dept. of Economic Planning and Development, Office of the Chief of State Planning, 1977.
2 v. ; 30 cm. — (Planning information series ; v. 1-2)
Bibliography: v. 1, p. [120]-142.
CONTENTS: v. 1. Planning and regulation resource handbook, state level. —v. 2. Planning and regulation resource handbook, local level.
1. Regional planning—Law and legislation—Wyoming. 2. City planning and redevelopment law—Wyoming. 3. Land use—Law and legislation—Wyoming. I. Lamb, Margo, joint author. II. Wyoming. State Dept. of Economic Planning and Development. Office of the Chief of State Planning. III. Title. IV. Series: Planning information series (Cheyenne) ; v. 1-2.
KFW4658.S3 346'.787'045 78-622764
79 MARC

—YUGOSLAVIA

Bosanac, Milivoj.
Gradevinska regulativa '78 / Milivoj Bosanac, Vasja Simić. — Zagreb : Centar za informacije i publicitet, 1978.
681 p. : ill. ; 24 cm. — (Biblioteka stručnih izdanja) Yu***
Includes legislation.
1. Building laws—Yugoslavia. 2. City planning and redevelopment law—Yugoslavia. I. Simić, Vasja, joint author. II. Title.
78-394422

CITY SCHOOLS see Urban schools

CITY SOLICITORS see City attorneys

CITY SURVEYING see Surveying

CITY TRAFFIC

see also Pedestrians; Traffic noise

Polytechnic Institute of New York.
Traffic control in oversaturated street networks / William R. McShane ... [et al.] ; Polytechnic Institute of New York. — Washington : Transportation Research Board, National Research Council, 1978.
152 p. : ill. ; 28 cm. — (National Cooperative Highway Research Program report ; 194 ISSN 0077-5614)
Includes bibliographical references.
ISBN 0-309-02854-X : $9.60
1. Traffic flow. 2. City traffic. 3. Traffic engineering. I. McShane, William R. II. Title. III. Series: National Cooperative Highway Research Program. Report ; 194.
TE7.N25 no. 194 625.7'08 s 78-65950
[HE336.T7] 79 MARC

Šeštokas, Vaclovas.
Eismo organizavimas / V. Šeštokas. — Vilnius : Mokslas, 1978.
150 p. : ill. ; 23 cm. USSR***
Bibliography: p. 148-149.
0.80rub
1. Traffic engineering. 2. City traffic. 3. Traffic safety. I. Title.
HE333.S45 79-381872
79 MARC

—ECONOMIC ASPECTS

Dewees, Donald N
Congestion costs in urban motoring : some Toronto estimates / Donald N. Dewees. — Toronto : Centre for Urban and Community Studies, University of Toronto, 1976.
38 p. : ill. ; 28 cm. — (Research paper - Centre for Urban and Community Studies, University of Toronto ; no. 71) C***
Bibliography: p. 38.
1. Transportation, Automotive. 2. City traffic—Economic aspects. I. Title. II. Series: Toronto. University. Centre for Urban and Community Studies. Research paper ; no. 71.
HE5613.D45 388.4'131'4 78-323708
79 MARC

—ECONOMIC ASPECTS—GREAT BRITAIN—BIBLIOGRAPHY

Gomersall, Alan.
Traffic restraint techniques and their effect on transport planning and industry / compiled by A. Gomersall. — 2d ed. — London : Greater London Council : [sold by] GLC Bookshop, 1977.
31 p. ; 30 cm. — (Research bibliography - Greater London Council, Research Library ; no. 63) GB***
ISBN 0-7168-0918-4 : £1.75
1. Traffic engineering—Bibliography. 2. Transportation planning—Bibliography. 3. City traffic—Economic aspects—Great Britain—Bibliography. I. Title. II. Series: Greater London Council. Research Library. Research bibliography ; no. 63.
Z7164.T8G66 1977 016.3884'131 78-317961
[HE333] 78 MARC

—ENVIRONMENTAL ASPECTS

Neumaier, Herbert.
Die Steuerung des städtischen Strassenverkehrs unter dem Gesichtspunkt des Umweltschutzes : Luftreinhaltung und Lärmschutz an Hauptverkehrsadern durch ein anwendungsreifes, quantitatives Verfahren / Herbert Neumaier. — München : Minerva-Publikation, 1978.
vi, 198 p. : graphs ; 21 cm. — (Minerva-Fachserie Wirtschafts- und Sozialwissenschaften) GFR78-A
Originally presented as the author's thesis, Munich.
Bibliography: p. 193-198.
ISBN 3-597-10010-4 : DM32.00
1. City traffic—Environmental aspects. 2. Traffic engineering. I. Title.
TD195.T7N48 1978 363.6 78-371133
78 MARC

—FLORIDA—MIAMI

Miami, Fla. Dept. of Engineering.
Report on the Miami River study / prepared by the Department of Engineering, city of Miami ; Arthur E. Darlow, director. — [Miami] : The Dept., 1955.
13 leaves : ill. ; 28 cm.
Cover title.
1. City traffic—Florida—Miami. 2. Miami, Fla.—Bridges. I. Title. II. Title: Miami River study.
HE356.5.M45M5 1955 79-313260
388.4'131'409759381
79 MARC

—FRANCE

Centre d'études des transports urbains.
Catalogue de diffuseurs urbains / réalisé par le CETUR pour la Direction des routes et de la circulation routière. — [Bagneux] : Ministère de l'équipement et de l'aménagement du territoire, Centre d'études des transports urbains, 1978.
ca. 150 p. : ill., maps (17 in pockets) ; 31 cm. F***
1. Traffic surveys—France. 2. City traffic—France. 3. Roads—France—Interchanges and intersections. I. France. Direction des routes et de la circulation routière. II. Title.
HE373.F6C46 1978 388.4'131'40944 79-344288
79 MARC

—GREAT BRITAIN

Great Britain. Central Office of Information. Reference Division.
Town traffic in Britain / [prepared by Reference Division, Central Office of Information]. — Rev. — London : Central Office of Information, 1977.
23, [1] p. ; 24 cm. GB***
"R5860/77."
Bibliography: p. 23-[24]
1. City traffic—Great Britain. I. Title.
HE363.G7G68 1977 388.4'0941 79-300622
79 MARC

Great Britain. Central Office of Information. Reference Division.
Town traffic in Britain / [prepared by Reference Division, Central Office of Information]. — 2nd ed. — London : H.M.S.O., 1978.
[3], 24 p., [4] p. of plates : ill. ; 24 cm. — (Central Office of Information reference pamphlet ; 130) GB78-18502
Bibliography: p. 23-[24]
ISBN 0-11-700967-9 : £0.75
1. City traffic—Great Britain. I. Title. II. Series: Great Britain. Central Office of Information. Reference pamphlet ; 130.
HE363.G7G68 1978 388.4'0941 79-305565
79 MARC

Great Britain. Traffic Advisory Unit.
Some effects of flexible working hours on traffic conditions at a large office complex / [Traffic Advisory Unit]. — London : The Unit, 1977.
[2], 9 leaves : ill. ; 30 cm. GB78-10847
1. City traffic—Great Britain. 2. Automobile parking—Great Britain. 3. Hours of labor, Flexible—Great Britain. I. Title.
HE335.G68 1977 388.4'131'4 79-305559
79 MARC

—ITALY

Mastrorilli, Franco.
Centri antichi: una politica dei trasporti per il recupero dei valori ambientali. Palermo, Agatekne, 1973.
43 p. 24 cm. (Quaderni del Centro provinciale di studi urbanistici di Palermo, 5) L1200 It74-Apr
Bibliography: p. 41-43.
1. Urban transportation—Italy. 2. City traffic—Italy. 3. City planning—Italy. I. Title. II. Series: Centro provinciale di studi urbanistici di Palermo. Quaderni, 5.
HE311.I8M37 711'.7'0945 79-413085
79 MARC

—ITALY—ROME (CITY)

Il Nodo di S[an] Bernardo : una struttura urbana tra il centro antico e la Roma moderna / [di] Carlo Pietrangeli ... [et al.]. — Milano : F. Angeli, [1977]
317 p. : ill. ; 28 cm. & portfolio (25 fold. leaves of plates : 25 plans) It78-Sept
Includes bibliographies.
L18000
1. Rome (City)—City planning. 2. City planning—Italy. 3. City traffic—Italy—Rome (City) I. Pietrangeli, Carlo.
HT169.I84R6394 309.2'62'0945632 79-341638

CITY TRAFFIC (Continued)

—JAPAN—KAWASAKI (KANAGAWA PREFECTURE)

Kawasaki-shi Kyōiku Kenkyūjo.
(Kawasaki-shi no kōtsū)
川崎市の交通. — 川崎：川崎市教育研究所, 1954.
83 p. : ill. ; 22 cm. — (地域社会の学習資料シリーズ ; no. 9)
Cover title.

1. City traffic—Japan—Kawasaki (Kanagawa Prefecture) I. Title. II. Series: Chiiki shakai no gakushū shiryō shirizu ; no. 9.
HE311.J32K384 79-805687

—SASKATCHEWAN

Hack, C F
Urban traffic safety : [a review of the problems and suggested improvements / prepared by C. F. Hack for Saskatchewan Highways and Transportation, Traffic Safety Engineering Branch for use by urban municipalities in consideration of urban assistance for traffic safety improvements]. — [Regina] : Saskatchewan Highways and Transportation, [1977?]
26 p. : ill. ; 28 cm. C***
Cover title.
1. Traffic safety—Saskatchewan. 2. City traffic—Saskatchewan. I. Saskatchewan. Traffic Safety Engineering Branch. II. Title.
HE5614.5.S37H3 614.8'62'097124 79-305560
79 MARC

—UNITED STATES

Taebel, Del.
The political economy of urban transportation / Delbert A. Taebel, James V. Cornehls. — Port Washington, N.Y. : Kennikat Press, 1977.
ix, 218 p. ; 23 cm. — (Interdisciplinary urban series) (National university publications)
Bibliography: p. 207-214.
Includes index.
ISBN 0-8046-9178-9. ISBN 0-8046-9200-9 pbk.
1. Urban transportation policy—United States. 2. City traffic—United States. 3. Transportation—United States. I. Cornehls, James V., joint author. II. Title.
HE308.T33 388.4'0973 77-23150
77[r79]rev MARC

CITY TRANSIT see Local transit

CITY TRANSPORTATION see Urban transportation

CITY WALLS

—DESIGN AND CONSTRUCTION—EARLY WORKS TO 1800

Fiore, Francesco Paolo.
Città e macchine del '400 nei disegni di Francesco di Giorgio Martini / [di] F. Paolo Fiore. — Firenze : L. S. Olschki, 1978.
165 p., [48] leaves of plates : ill. ; 25 cm. — (Studi - Accademia toscana di scienze e lettere La Colombaria ; 49) It78-Sept
Includes bibliographical references and index.
1. Mechanical engineering—Early works to 1800. 2. City walls—Design and construction—Early works to 1800. 3. Martini, Francesco di Giorgio, 1439-1502. 4. Mechanical engineers—Italy—Biography. I. Martini, Francesco di Giorgio, 1439-1502. II. Title. III. Series: Accademia toscana di scienze e lettere La Colombaria. Studi ; 49.
TJ144.F56 78-391921
*79 MARC

—FRANCE

L'Atlas des villes et villages fortifiés en France (Moyen Age) / réalisé par Charles-Laurent Salch, avec la collaboration de Joëlle Burnouf ; préf. de Philippe Dollinger ; cartes de Maurice Frey ; plans par Claire Chantelouve et Nicolas du Saulcy ; photos de Dominique Martinez. — Strasbourg : Publitotal, 1978.
495 p. : ill. (some col.) ; 31 cm. F***
Errata slip inserted.
Bibliography: p. 489-491.
Includes indexes.
295F
1. Cities and towns, Medieval—France. 2. City walls—France. 3. France—Description and travel—1975- I. Salch, Charles Laurent. II. Burnouf, Joëlle.
DC20.A78 944'.02 79-346280
79 MARC

CIUCURENCU, ALEXANDRU, 1902-

Ciucurencu, Alexandru, 1902-
Ciucurencu : [album / Mircea Deac]. — [Bucureşti] : "Meridiane", 1978.
51 p., [62] leaves of plates : ill. (some col.) ; 33 cm.
Summary in English and French.
Includes bibliographical references.
lei100
1. Ciucurencu, Alexandru, 1902- I. Deac, Mircea.
ND933.C5A4 1978 78-382057
*78 MARC

CIUDAD BOLÍVAR, VENEZUELA

—HARBOR

Venezuela. Dirección de Navigación.
Guía de navegantes capitanía de Puerto de Ciudad Bolívar / República de Venezuela, Ministerio de Comunicaciones, Dirección General de Transporte y Tránsito Marítimos, Dirección de Navegación. — Ed. oficial. — Caracas : La Dirección, 1975.
207 p. : ill. ; 28 cm.
"Publicación no. T-62 G."
1. Ciudad Bolívar, Venezuela—Harbor. 2. Orinoco River—Navigation. I. Title.
HE556.C55V46 1975 79-386248
79 MARC

CIUDAD REAL, SPAIN (CITY)

—ECONOMIC CONDITIONS

Phillips, Carla Rahn, 1943-
Ciudad Real, 1500-1750 : growth, crisis, and readjustment in the Spanish economy / Carla Rahn Phillips. — Cambridge, Mass. : Harvard University Press, 1979.
viii, 190 p. : ill. ; 24 cm.
Bibliography: p. [171]-181.
Includes index.
ISBN 0-674-13285-8
1. Ciudad Real, Spain (City)—Economic conditions. 2. Ciudad Real region, Spain—Economic conditions. I. Title.
HC388.C58P5 330.9'46'45 78-9293
78 MARC

CIUDAD REAL, SPAIN (PROVINCE)

—DESCRIPTION AND TRAVEL

Quadrado, José María, 1819-1896.
Toledo y Ciudad Real / José M.a Quadrado y Vicente de la Fuente. — Barcelona : Ediciones El Albir, 1978.
508 p. : ill. ; 25 cm. — (His Castilla la Nueva ; 3) (Reprint of v. 3 of the 1885 ed. of the author's Castilla la Nueva, published by D. Cortezo, in series: España, sus monumentos y artes, su naturaleza e historia) Sp***
Limited ed. of 500 copies. No. 48.
ISBN 8473700325
1. Toledo, Spain (Province)—Description and travel. 2. Ciudad Real, Spain (Province)—Description and travel. I. Fuente, Vicente de la, 1817-1889, joint author. II. Title. III. Series. IV. Series: España, sus monumentos y artes, su naturaleza e historia.
DP302.T51Q33 1978 946'.43 79-350649
79 MARC

CIUDAD REAL REGION, SPAIN

—ECONOMIC CONDITIONS

Phillips, Carla Rahn, 1943-
Ciudad Real, 1500-1750 : growth, crisis, and readjustment in the Spanish economy / Carla Rahn Phillips. — Cambridge, Mass. : Harvard University Press, 1979.
viii, 190 p. : ill. ; 24 cm.
Bibliography: p. [171]-181.
Includes index.
ISBN 0-674-13285-8
1. Ciudad Real, Spain (City)—Economic conditions. 2. Ciudad Real region, Spain—Economic conditions. I. Title.
HC388.C58P5 330.9'46'45 78-9293
78 MARC

ČIULADA, JONAS, 1904-1971

Jonas Čiulada. — Vilnius : Mintis, 1976.
83 p., [1] leaf of plates : port. ; 21 cm. USSR***
0.36rub
1. Čiulada, Jonas, 1904-1971. 2. Lithuania—Biography.
DK511.L28C584 79-337955
79 MARC

CIUMBRUDEAN, DUMITRU

Ciumbrudean, Dumitru.
Jurnal de front / al caporalului Dumitru Ciumbrudean, 1914-1918. — Bucureşti : Editura Politică, 1969.
367 p. : ill. ; 20 cm.
1. European War, 1914-1918—Personal narratives, Romanian. 2. Ciumbrudean, Dumitru. 3. European War, 1914-1918—Campaigns—Eastern. 4. Soldiers—Romania—Biography. 5. Romania. Armata—Biography. I. Title.
D640.C566 79-358637
79 MARC

CIUPE, AUREL, 1900-

—CATALOGS

Ciupe, Aurel, 1900-
Aurel Ciupe : [album] / Mircea Deac ; [coperta şi grafica de Vasile Pop-Silaghi ; traducere de Andrei Bantaş]. — Cluj-Napoca : "Dacia", 1978.
19 p., [38] leaves of plates : ill. (some col.) ; 30 cm. R78-2792
Romanian and English.
Includes bibliographical references.
lei40.00
1. Ciupe, Aurel, 1900- —Catalogs. I. Deac, Mircea.
ND933.C53A4 1978 79-377530
*79 MARC

ČIURLIONIS, MIKALOJUS KONSTANTINAS, 1875-1911

—ANNIVERSARIES, ETC.

Čiurlioniui 100 [i.e. šimtas / sudarė Jonas Bruveris ; redakcinė komisija E. Balsys ... et al.]. — Vilnius : Vaga, 1977.
348 p., [25] leaves of plates : ill. ; 23 cm. USSR***
Errata slip inserted.
Includes bibliographical references and index.
2.40rub
1. Čiurlionis, Mikalojus Konstantinas, 1875-1911—Anniversaries, etc. I. Bruveris, Jonas. II. Balsys, Eduardas, 1919-
ML410.C6C58 780'.92'4 79-345119
79 MARC

ČIURLIONIS, MIKALOUJUS KONSTANTINAS, 1875-1911 (Continued)

—BIBLIOGRAPHY—CATALOGS

Būdienė, Teofilė.
M. K. Čiurlionis : Katalog der Literaturausstellung zum 100. Geburtstag des litauischen Malers und Komponisten, Berlin, 12 September - 19. Oktober 1975 / [Verfasser, Teofilė Būdienė, Laimė Lukošiūonienė, Eglė Marčenienė]. — [Vilna] : Lietuvos TSR Valst. respublikinė biblioteka, [1975?]
49 p. : ill. ; 21 cm. GFR***
Exhibit of the Staatsbibliothek der Litauischen SSR, and the Deutsche Staatsbibliothek.
Cover title: Katalog der Literaturausstellung M. K. Čiurlionis.
Includes bibliographical references.
1. Čiurlionis, Mikalojus Konstantinas, 1875-1911—Bibliography—Catalogs. I. Lukošiūnienė, Laimė, joint author. II. Marčenienė, Eglė, joint author. III. Vilna. Lietuvos TSR Valstybinė respublikinė biblioteka. IV. Berlin. Deutsche Staatsbibliothek. V. Title: Katalog der Literaturausstellung M. K. Čiurlionis.
Z8171.B8 79-368054
[NX556.Z9] 79 MARC

—CATALOGS

Čiurlionis, Mikalojus Konstantinas, 1875-1911.
Mikalojus Konstantinas Čiurlionas / leidinį parengė Valerija Čiurlionytė-Karužienė, Judita Grigienė. — Vilnius : Vaga, 1977.
308 p. : chiefly col. ill. ; 33 cm. USSR***
Lithuanian, English, French, German, and Russian.
Text by Lionginas Šepetys.
21.30rub
1. Čiurlionis, Mikalojus Konstantinas, 1875-1911—Catalogs. I. Šepetys, Lionginas. II. Čiurlionytė-Karužienė, Valerija. III. Grigienė, Judita.
ND699.C6A4 1977 78-379670
78 MARC

—MUSEUMS, RELICS, ETC.

Nedzelskis, Adelbertas.
M. K. Čiurlionio memorialinis muziejus Druskininkuose / A. Nedzelskis. — Vilnius : Mintis, 1976.
46 p. : chiefly ill. ; 16 cm. USSR***
Lithuanian and Russian.
1. Čiurlionis, Mikalojus Konstantinas, 1875-1911—Museums, relics, etc. I. Čiurlionio (M. K.) memorialinis muziejus Druskininkuose. II. Title.
ML410.C6N4 780'.92'4 79-343142
79 MARC

CIVAKALAI, INDIA

—RELIGIOUS LIFE AND CUSTOMS

(Civakaḷai aruḷmiku Cōranāta Vināyakar Tirukkōyil kuṭamuḷukku viḷā niṉaivu malar, 4-4-1976)
சிவகளை அருள்மிகு சோரநாத விநாயகர் திருக்கோயில் குடமுழுக்கு விழா நினைவு மலர், 4-4-1976. — சிவகளை : மலர் வெளியீட்டுக் குழு, [1976]
64, [34] p., [4] leaves of plates : ill. ; 26 cm.
In Tamil.
Includes bibliographical references.
Rs5.00
1. Gaṇeśa (Hindu deity)—Cult. 2. Civakaḷai, India—Religious life and customs. I. Cōranāta Vināyakar Kōyil.
BL1225.G34C6 77-902568

CIVAMANI, K. , 1932-

Subrahmanya Bharati, C 1882-1921.
(Kuyiṟ pāṭṭu)
குயிற் பாட்டு. [கவி சுப்பிரமணிய பாரதி. உரையாசிரியர்] கு. சிவமணி. [1. பதிப்பு] தஞ்சாவூர், சிவா நிலையம் [1968]
94 p. 19 cm. Rs1.80
In Tamil.
I. Civamaṇi, K., 1932- II. Title.
PL4758.9.S8437K8 1968 76-906204

CIVET, GÉRARD

Civet, Gérard.
Les trésors de la Colombie / Gérard Civet ; [photos. de Gérard Civet, André Chastel et Alain Coutois]. — Paris : Presses de la Cité, c1979.
121 p. : ill. (some col.) ; 27 cm. — (Connaissance du monde ISSN 0182-3671) F***
Bibliography: p. 121.
ISBN 2-258-00493-4 : 50.00F
1. Colombia—Description and travel—1951- 2. Civet, Gérard. I. Title.
F2264.C56 986.1'063 79-374301
79 MARC

CIVIC ART see Art, Municipal

CIVIC CENTERS

see also Municipal buildings; Public buildings

CIVIC CEREMONIAL see Municipal ceremonial

CIVIC RIGHTS see Political rights

CIVIC UNIVERSITIES see Public universities and colleges

CIVICA BIBLIOTECA NUMISMATICA DI MILANO

—CATALOGS

Civica biblioteca numismatica di Milano.
Le cinquecentine della Biblioteca archeologica e numismatica di Milano : (catalogo e cenni storici) / Rina La Guardia. — [Milano] : Comune di Milano, Ripartizione cultura e spettacolo, [1976?].
42 p., 70 leaves of plates : ill. ; 24 cm. It***
Bibliography: p. 17-20.
Includes index.
1. Numismatics, Ancient—Bibliography—Catalogs. 2. Civica biblioteca numismatica di Milano—Catalogs. I. La Guardia, Rina. II. Title.
Z6868.A5C58 1976 016.737 79-357993
[CJ233] 79 MARC

CIVICI MUSEI DI STORIA ED ARTE DI TRIESTE. BIBLIOTECA

Coen Foschiatti, Gabriella.
La Biblioteca dei Civici musei di storia ed arte. Trieste, 1968.
32 p. 25 cm. (Atti dei Civici musei di storia ed arte. Quaderno, 1) It69-Mar
Includes bibliographical references.
1. Civici musei di storia ed arte di Trieste. Biblioteca. I. Title. II. Series: Civici musei di storia ed arte di Trieste. Atti dei Civici musei di storia ed arte, 1.
AM55.T74C59a no. 1 79-355331
[Z810.C57] 79 MARC

CIVICO, GERMAN

—STUDY AND TEACHING (ELEMENTARY)—CONGRESSES

Minssen, Friedrich, 1909-
Political education in the schools of the Federal Republic of Germany / Friedrich Minssen ; [original translation by M. Winck] ; edited by Louis F. Helbig. — Bloomington : Institute of German Studies, Indiana University, 1976.
29 p. ; 28 cm. — (German studies notes)
Delivered at a conference held at Indiana University on Sept. 15-19, 1975.
1. Civico, German—Study and teaching (Elementary)—Congresses. 2. Social sciences—Study and teaching (Elementary)—Germany, West—Congresses. I. Helbig, Louis Ferdinand, 1935- . II. Title. III. Series.
LB1584.5.G4M56 372.8'3'0440943 78-622886
79

CIVICO MUSEO BIBLIOGRAFICO MUSICALE

Schnoebelen, Anne.
Padre Martini's collection of letters in the Civico museo bibliografico musicale in Bologna : an annotated index / Anne Schnoebelen. — New York : Pendragon Press, c1979.
xix, 721 p. : port. ; 29 cm. — (Annotated reference tools in music ; no. 2)
Bibliography: p. xiii-xvii.
Includes indexes.
ISBN 0-918728-11-8
1. Martini, Giovanni Battista, 1706-1784—Dictionaries, indexes, etc. 2. Civico museo bibliografico musicale. I. Civico museo bibliografico musicale. II. Title. III. Series.
ML134.M438S3 780'.92'4 79-10540
79 MARC

CIVICO MUSEO NAVALE, GENOA

Secchi, Laura.
Pegli, Museo navale di Villa Doria / a cura di Laura Secchi. — Genova : Sagep, 1977.
32 p. : ill. ; 21 cm. — (Guide di Genova ; 50) It79-Feb
Bibliography: p. 31.
L1000
1. Civico museo navale, Genoa. I. Title. II. Series.
DG644.5.C58S42 914.5'182 79-360134
*79 MARC

CIVICS

see also Americanization; Citizenship; Civil rights; Patriotism; Political ethics; Political rights

Kownslar, Allan O
Civics, citizens, and society / Allan Kownslar, Terry Smart ; consultants, Kenneth J. Barnabe ... [et al.]. — New York : Webster Division, McGraw-Hill, c1980.
448 p. : ill. ; 26 cm.
Includes index.
ISBN 0-07-035412-X : $13.64
1. Civics. 2. United States—Politics and government. I. Smart, Terry L., joint author. II. Title.
JK274.K693 320.4'73 79-534
79 MARC

—JUVENILE LITERATURE

Branson, Margaret Stimmann.
Civics for today / Margaret Stimmann Branson, Fred S. Coombs. — Boston : Houghton Mifflin Co., c1978.
576 p. : ill. ; 24 cm.
Includes index.
SUMMARY: A textbook exploring the structure, function, and duties of local, state, and federal government.
ISBN 0-395-26201-1 (student's ed.)
1. Civics—Juvenile literature. [1. Civics] I. Coombs, Fred S., joint author. II. Title.
JK1759.B8 320.4'73 78-78231
79 MARC

Hartley, William Harrison.
Americn civics / William H. Hartley and William S. Vincent. — 3d ed. — New York : Harcourt Brace Jovanovich, 1979.
xiv, 544 p. : ill. (some col.) ; 25 cm.
Includes index.
SUMMARY: A textbook describing the construction and duties of local, state, and federal government, the roles and responsibilities of the American economy.
ISBN 0-15-371460-3
1. Civics—Juvenile literature. [1. Civics] I. Vincent, William Shafer, 1907- joint author. II. Title.
H95.H29 1979 309.1'73'0926 79-102102
79 MARC

—STUDY AND TEACHING—CONGRESSES

Citizen education behavior variables : final report / submitted to National Institute of Education by Research for Better Schools, Inc. — Philadelphia : Research for Better Schools, 1978-
v. ; 29 cm.
"The colloquium took place on May 11 and 12, 1978, at the Sugar Loaf Conference Center in Philadelphia."
Includes bibliographical references.
CONTENTS: pt. A. Summary.
1. Civics—Study and teaching—Congresses. 2. Political socialization—United States—Congresses. I. National Institute of Education. II. Research for Better Schools, inc.
JA88.U6C52 320.4'07 79-111302

CIVICS
—STUDY AND TEACHING —CONGRESSES (Continued)

Shirley H. Engle Conference, Bloomington, Ind., 1977.
Decision-making, the heart of social studies instruction revisited : papers presented at the Shirley H. Engle Conference held in Bloomington, Indiana, in April, 1977. — Bloomington : Social Studies Development Center, Indiana University, c1978.
iv, 41 p. ; 23 cm. — (Occasional paper - Social Studies Development Center, Indiana University, Bloomington ; no. 1)
Includes bibliographical references.
1. Engle, Shirley H. Decision making, the heart of social studies instruction. 2. Civics—Study and teaching—Congresses. I. Title. II. Series: Indiana University, Bloomington. Social Studies Development Center. Occasional paper ; no. 1.
H62.5.U5E547 1977 300'.7 78-107549
78 MARC

—STUDY AND TEACHING —PERIODICALS

International journal of political education. v. 1– Sept. 1977–
Amsterdam, Elsevier Scientific Pub. Co.
v. 24 cm. quarterly.
1. Political science—Study and teaching—Periodicals. 2. Civics—Study and teaching—Periodicals.
JA1.A1 I 588 320'.07 78-647638
MARC-S

—STUDY AND TEACHING—GERMANY, WEST

Oppenheim, Abraham Naftali, 1924-
Civic education and participation in democracy : the German case / A. N. Oppenheim. — London ; Beverly Hills : Sage, 1977.
51 p. ; 22 cm. — (Sage professional papers in contemporary political sociology ; no. 06-023 ISSN 0305-3482) GB***
Bibliography: p. 50-51.
ISBN 0-8039-9874-0
1. Political socialization—Case studies. 2. Civics—Study and teaching—Germany, West. I. Title.
JA76.O66 301.5'92 77-79263
78 MARC

—STUDY AND TEACHING (ELEMENTARY)—HANDBOOKS, MANUALS, ETC.

Arangis, Louise M
The red, white, and blue art ideas handbook : patriotic themes for the elementary grades / Louise M. Arangis ; illustrated by James F. Madden. — West Nyack, N.Y. : Parker Pub. Co., c1979.
254 p. : ill. ; 26 cm.
Includes index.
ISBN 0-13-769828-3
1. Civics—Study and teaching (Elementary)—Handbooks, manuals, etc. 2. Patriotism—Study and teaching (Elementary)—United States—Handbooks, manuals, etc. I. Madden, James F. II. Title.
LB1584.A7 372.8'32 78-21657
78 MARC

—INDONESIA—ADDRESSES, ESSAYS, LECTURES

Gultom, R M S
Tanggung jawab orang Kristen dalam pengamalan Pancasila dan pembangunan / oleh R. M. S. Gultom. — [Salatiga? : s.n., 1976?]
36 p. ; 21 cm.
"Brosur."
Includes bibliographical references.
1. Church and state in Indonesia—Addresses, essays, lectures. 2. Civics—Indonesia—Addresses, essays, lectures. I. Title.
BR1220.G84 78-941055
79 MARC

CIVICS, AMERICAN see Civics

CIVICS, BRAZILIAN

Joffily, José.
Distorções e revisões / José Joffily. — [s.l. : s.n.], 1977 (Londrina [Brasil] : Gráfica Lider)
193 p., [20] leaves of plates : ill. ; 23 cm.
Bibliography: p. 123.
Includes index.
Cr$100.00
1. Civics, Brazilian. I. Title.
JL2483.J63 79-346609
79 MARC

CIVICS, BRITISH

—STUDY AND TEACHING

Stradling, Robert.
The political awareness of the school leaver / Robert Stradling. — London : Hansard Society, c1977.
iii, 81 p. ; 23 cm. GB***
Cover title.
At head of title: A Programme for political education.
Imprint from label on verso of t.p.
Includes bibliographical references.
£1.50
1. Civics, British—Study and teaching. 2. Political science—Study and teaching—Great Britain. 3. Political socialization—Great Britain. I. Title. II. Title: Programme for political education.
JA88.G7S83 320.4'41 79-302504
79 MARC

CIVICS, CANADIAN

Canada. Dept. of the Secretary of State.
The Canadian citizen. — [Ottawa] : Minister of Supply and Services Canada, 1978.
16, 16 p. : ill. ; 28 cm. C***
Cover title.
Added t.p.: Le citoyen canadien.
English and French.
ISBN 0-662-00464-7
1. Civics, Canadien. 2. Citizenship—Canada. I. Le citoyen canadien. II. Title.
JL187.C245 1978 320.4'71 79-304129
79 MARC

CIVICS, CHINESE

(Kung min chiao yü hsiao ts'ung shu)
公民教育小叢書 / [編輯者青年協會公民教育委員會]. — 彙訂本. — [香港?: s.n., 1977?]
310 p. in various pagings ; 17 cm.
Cover title.
Reprint of the 1928 ed. published by Ch'ing nien hsieh hui shu chü, Shanghai.

1. Civics, Chinese. I. Ch'ing nien hsieh hui. Kung min chiao yü wei yüan hui.
JQ1517.A2K86 1977 78–843135

CIVICS, FRENCH

France.
Rapport sur les mesures prises par le Gouvernement pour assurer le développement de l'instruction civique et la formation aux responsabilités du citoyen / [élaboré par un groupe de travail administratif à la demande du] Premier ministre. — [Paris] : Premier ministre, [1977?]
75 p. ; 30 cm. F***
1. Civics, French. I. France. Premier ministre. II. Title.
JN2919.F73 1977 320.4'44 78-368432
78 MARC

—HANDBOOKS, MANUALS, ETC.

Guide de la vie pratique / sous la direction de Denis Perier Daville. — Paris : Larousse, 1978.
xxxi, 1423 p. : ill. ; 24 cm. F78-19825
Includes bibliographical references and index.
ISBN 2-03-029612-0 : 140F
1. Social service—France—Handbooks, manuals, etc. 2. Civics, French—Handbooks, manuals, etc. 3. France—Social policy—Handbooks, manuals, etc. I. Perier Daville, Denis.
HV265.G79 361'.944 79-364954
79 MARC

CIVICS, GERMAN

Akademie für Staats- und Rechtswissenschaft der DDR.
DDR : Gesellschaft, Staat, Bürger / [Herausgeber, Akademie für Staats- und Rechtswissenschaft der DDR Potsdam-Babelsberg ; Autorenkollektiv unter Leitung von Gerhard Schulze]. — 3., bearb. Aufl. — Berlin : Staatsverlag der Deutschen Demokratischen Republik, 1979.
246 p. : ill. (some col.), col. graphs, map ; 24 cm. GDR***
Includes bibliographical references and index.
10.00M
1. Civics, German. 2. Germany, East—Politics and government. I. Schulze, Gerhard, 1926- II. Title.
JN3971.5.A92A4 1979 320.4'43 79-378620
79 MARC

—STUDY AND TEACHING

100 questions, 100 answers : GDR / [authors, Günter Albrecht ... et al. ; translated by Intertext Berlin]. — Berlin : Panorama DDR, 1978.
180 p., [16] leaves of plates : ill. ; 19 cm. GDR***
Translation of 100 Fragen, 100 Antworten.
1. Civics, German—Study and teaching. 2. Germany, East—Politics and government. I. Albrecht, Günter, 1930-
JN3971.5.A92H8613 1978 78-325103
320.4'43
79 MARC

CIVICS, HUNGARIAN

Tanulmányok az állampolgári nevelés módszereihez / [szerk. Závodszky Géza]. — Budapest : Tankönyvkiadó, 1977.
164 p. ; 20 cm. H77
Includes bibliographical references.
ISBN 963172655X : 13.50Ft
1. Civics, Hungarian. I. Závodszky, Géza.
JN2165.T35 78-386093
*79 MARC

CIVICS, INDONESIAN

Djahiri, A Kosasih.
Kearah memahami beberapa sistim & dasar kenegaraan Republik Indonesia dan Pancasila, oleh A. Kosasih Djahiri. Bandung, Institut Keguruan dan Ilmu Pendidikan, 1973.
2, 82 p. 28 cm.
1. Indonesia—Politics and government. 2. Civics, Indonesian. I. Title.
JQ768.D59 74-940020
79 MARC

Hatta, Mohammad, 1902-
Pantjasila djalan lurus / Mohammad Hatta. — Bandung : Angkasa, 1966.
16 p. ; 16 cm. PL480:Indo-4353
1. Civics, Indonesian. I. Title.
JQ777.A2H319 75-949789
79 MARC

Husin, Ahmad.
Pantjasila dan Komunisme / Ahmad Husin. — Djakarta : Pusaka, [1957]
79 p. ; 18 cm.
Includes bibliographical references.
1. Civics, Indonesian. 2. Communism—Indonesia. I. Title.
JQ777.A2H87 79-949519
79 MARC

Nasution, Abdul Haris, 1918-
Pantjasila democracy, today and tomorrow / by Abdul Haris Nasution. — Djakarta : Seruling Masa, 1971.
60 p. ; 21 cm.
1. Civics, Indonesian. I. Title.
JQ777.A2N34 320.9'598'03 77-941063
78 MARC

Notosoetardjo, Achmad, Hadji.
Bung Karno dan Pantja sila / oleh H. A. Notosoetardjo. — Djakarta : Endang, 1966.
63 p. : ill. ; 16 cm.
1. Sukarno, Pres. Indonesia, 1901-1970. 2. Civics, Indonesian. I. Title.
DS644.1.S8N58 77-950247
79 MARC

Partai Nasional Indonesia. Panitya Peringatan Lahirnja Pantjasila.
Pantja-sila sepandjang masa. — [Djakarta] : Panitya Peringatan Lahirnja Pantja-Sila, PNI-Front Marhaenis, [1967]
32 p. ; 21 cm.
Includes bibliographical references.
1. Civics, Indonesian. I. Title.
JQ777.A2P32 1967 77-950212
79 MARC

Prawiranegara, Sjafruddin.
Sejarah sebagai pedoman untuk membangun masa depan : peranan mu'jizat dalam perjoangan kemerdekaan bangsa Indonesia : ceramah pada tanggal 30 Juli 1975 di Gedung Kebangkitan Nasional Jakarta. — Jakarta : Yayasan Idayu, 1975.
52 p. ; 21 cm.
Includes bibliographical references.
1. Indonesia—Politics and government. 2. Religion and state—Indonesia. 3. Civics, Indonesian. I. Title.
JQ768.P69 77-941420
79 MARC

CIVICS, INDONESIAN (Continued)

Sakimi, Soemantri.
Pokok² pelaksanaan dari demokrasi Pantjasila / oleh Soemantri Sakimi. — ‹Medan› : SKARDA-B/SUMUT, ‹1972›
61 p. ; 21 cm.
At head of title: Departemen Pertahanan-Keamanan, Dewan Kekaryaan Daerah-B/Sumatera Utara, Staf Kekaryaan Daerah-B.
1. Civics, Indonesian. 2. Indonesia—Politics and government—1966– I. Title.
JQ777.A2S24 72–941498

Sentral Organisasi Buruh Pantjasila.
Tuntunan bagi kader buruh Pantjasila. — ‹Djakarta : Sentral Organisasi Buruh Pantjasila, 1960?-
v. ; 22 cm.
Bibliography: v. 1, p. 84-85
1. Civics, Indonesian. I. Title.
JQ777.A2S36 1960 78-950701
78 MARC

Sjahrir, Soetan, 1909-1966.
Pedoman kehidupan bangsa. — Djakarta : Sekretariat Dewan Partai, Partai Sosialis Indonesia, ‹1957?›
32 p. ; 18 cm. — (Seri didikan politik)
1. Civics, Indonesian. 2. Indonesian—Politics and government. I. Title.
JQ777.A2S54 76-989547
79 MARC

—ADDRESSES, ESSAYS, LECTURES

Indonesia. Penguasa Perang Daerah Swatantra I Djawa-Barat.
Instruksi Penguasa Perang no. Instr./Peperpu/077/1959 tentang Operasi kesadaran nasional : salinan. — ‹Bandung› : Staf Penguasa Perang Daerah Swatantra Tingkat I Djawa Barat, ‹1959›
47 p. ; 23 cm.
1. Civics, Indonesian—Addresses, essays, lectures. 2. Flags—Indonesia—Addresses, essays, lectures. 3. National songs, Indonesian—Addresses, essays, lectures. I. Title.
JQ777.A2I49 1959 77-986868
79 MARC

P.N.I. penegak Pantjasila! — Djakarta : Dep. PENPROP DPP. PNI, 1965.
25 p. ; 15 cm.
Cover title.
CONTENTS: Deklarasi Marhaenis.—Pidato Ketua Umum PNI pada Rapat umum "Dwikora" di Bandung.—Presiden Sukarno bergembira sekali dengan ditjetuskannja Deklarasi Marhaenis.—Sukarno. Amanat kepada Konperensi Kerdja dan Musjawarah Kerdja Keputrian G.M.N.I. Seluruh Indonesia di Pontianak.—Sifat politik luar negeri.—Situasi dunia pada masa sekarang.
1. Partai Nasional Indonesia—Addresses, essays, lectures. 2. Civics, Indonesian—Addresses, essays, lectures. I. Partai Nasional Indonesia.
JQ779.A56P26 77-986815
79 MARC

—CONGRESSES

Konperensi Gereja dan Masyarakat, 3d, Klender, Indonesia, 1976.
Melihat tanda-tanda jaman : pengamalan Pancasila dalam membangun masa depan : laporan Konperensi Gereja dan Masyarakat, Klender, 15-21 Maret 1976. — ‹Jakarta› : Diterbitkan oleh BPK Gunung Mulia untuk Dewan Gereja-Gereja di Indonesia, ‹1976›
295 p. ; 21 cm.
1. Church and social problems—Indonesia—Congresses. 2. Civics, Indonesian—Congresses. I. Dewan Gereja-Gereja di Indonesia. II. Title.
HN39.I53K63 1976 77-941827
79 MARC

—STUDY AND TEACHING—INDONESIA

Widjaja, A W
Pancasila dan humanitas (ilmu pengetahuan tentang kemanusiaan) / oleh A. W. Widjaja. — Palembang : Biro Hubungan Masyarakat, Universitas Sriwidjaya, 1976.
18 p. ; 21 cm.
Bibliography: p. 17-18.
1. Education and state—Indonesia. 2. Civics, Indonesian—Study and teaching—Indonesia. I. Title.
LC94.I5W53 77-940365
79 MARC

—STUDY AND TEACHING—INDONESIA —ADDRESSES, ESSAYS, LECTURES

Dekker, I Nyoman.
Beberapa masalah tentang ilmu pengetahuan sosial (IPS) : kumpulan karangan / oleh Nyoman Dekker. — Malang : Lembaga Penerbit Almamater, YPTP IKIP Malang, 1976.
iii, 44 p. ; 20 cm. — (Seri pengetahuan masyarakat ; no. 16)
Includes bibliographical references.
1. Civics, Indonesian—Study and teaching—Indonesia—Addresses, essays, lectures. I. Title. II. Series.
JQ777.A2D44 76-941363
79 MARC

—STUDY AND TEACHING (HIGHER)— INDONESIA—CONGRESSES

Rapat Dinas Pentavipan Kurikulum Perguruan Tinggi, Djakarta, 1965.
Rapat Dinas Pentavipan Kurikulum Perguruan Tinggi : bahan² dan kesimpulan / diselenggarakan dibawah koordinasi Biro Pendidikan, Pengadjaran, Penelitian dan Pengabdian Masjarakat (Biro P. 4) Departemen P.T.I.P. pada tanggal 25 dan 26 Juni 1965 di Wisma Hasta Senajan Djakarta. — Djakarta : Biro P. 3 Departemen Perguruan Tinggi dan Ilmu Pengetahuan, ‹1965›
iii, 143 leaves, ‹1› leaf of plates : ill. ; 33 cm.
Cover title.
"Seri B no. 002."
1. Civics, Indonesian—Study and teaching (Higher)—Indonesia—Congresses. I. Indonesia. Biro Pendidikan, Pengadjaran, Penelitian dan Pengabdian Masjarakat.
JQ777.A2R36 1965 77-950205
79 MARC

CIVICS, IRISH

—STUDY AND TEACHING—IRELAND— ADDRESSES, ESSAYS, LECTURES

Association of Teachers of Civics & Allied Subjects in Ireland.
A source text in civics / Association of Teachers of Civics & Allied Subjects in Ireland. — ‹Dublin› : The Association, ‹1975› Ir***
62 p. ; 21 cm.
Bibliography: p. 59-62.
£0.75
1. Civics, Irish—Study and teaching—Ireland—Addresses, essays, lectures. I. Title.
H62.5.I7A77 1975 78-319828
78 MARC

CIVICS, ISRAELI

—STUDY AND TEACHING

Lewy, Arieh, 1923–
(Hebeṭim ṿe-heśegim be-ma'arekhet ha-ḥinukh be-Yiśrael)
חיבטים והישגים במערכת החינוך בישראל : מחקר השוואתי בין־לאומי / אריה לוי, חנן רפפורט, מרדכי רימור ; בוצע במסגרת מכון הנרייטה סאלד. — ‹תל־אביב› : מערכות־חינוך רמות, אוניברסיטת תל־אביב, ~~1967~~ 1978 .738
204 p. : graphs ; 22 cm.
On verso of half title: Educational achievements in the Israeli school system.
Bibliography: p. 199–202.
1. Educational tests and measurements—Israel. 2. Reading (Elementary)—Israel. 3. English language—Study and teaching—Israel. 4. Civics, Israeli—Study and teaching. I. Rapaport, Chanan, joint author. II. Rimor, Mordechai, joint author. III. Mekhon Henri'eṭah Sold. IV. Title.
LB3058.I 78L48 79–951267

CIVICS, ITALIAN

Dragoni, Ugo.
Educazione civica dei lavoratori / Ugo Dragoni. — Roma : Edizioni ACLI, ‹1960?›
197 p. : ill. ; 21 cm.
1. Civics, Italian. 2. Labor and laboring classes—Education—Italy. I. Title.
JN5591.D68 320.4'45 79-357048
79 MARC

Insieme : elementi per la partecipazione / ‹di Gianni Galli ... et al.› ; a cura di Gianni Galli. — Firenze : Cultura, 1976.
323 p. : ill. ; 24 cm. It78-Mar
"Appendice" (legislative): p. ‹224›-323.
L3300
1. Civics, Italian. 2. Italy—Constitutional history. 3. Italy—Economic policy. 4. Agriculture and state—Italy. 5. Trade-unions—Italy. I. Galli, Gianni.
JN5591.I57 78-381341
*78 MARC

—STUDY AND TEACHING

Perucci, Carlo.
L'educazione politica nel quadro dell'educazione permanente / Carlo Perucci ; con una presentazione di Aldo Agazzi. — Firenze : F. Le Monnier, 1976.
ix, 317 p. ; 19 cm. It78-May
Bibliography: p. 303-309.
Includes index.
L5000
1. Civics, Italian—Study and teaching. 2. Education—Italy. I. Title.
JN5591.P43 320.4'45 78-393059
*79 MARC

CIVICS, JAPANESE

Kōmin Kyōiku Kenkyūkai.
(Kōmin kyōiku no kaisetsu to reiwa)
公民教育の解說と例話 / 公民教育研究會編. — 東京：厚生閣書店, 昭和7 ‹1932›
2, 10, 415 p. ; 19 cm.
1. Civics, Japanese. I. Title.
JQ1683.K65 1932 79–800408

—STUDY AND TEACHING—JAPAN

Shimane-ken Shihan Gakkō. Fuzoku Shōgakkō.
(Nihon seishin o kichō to seru shōgakkō no kōmin kyōiku)
日本精神を基調とせる小學校の公民教育 / 島根縣師範學校附屬小學校, 社會教育會共編. — 東京：社會教育會, 昭和10 ‹1935›
371 p. : ill. ; 23 cm.
"附錄 修身公民科に關する諸法令抜萃": p. 303–371.
1. Civics, Japanese—Study and teaching (Elementary)—Japan. 2. Civics, Japanese—Study and teaching—Japan. I. Shakai Kyōikukai. II. Title. III. Title: Shōgakkō no kōmin kyōiku.
LB1584.5.J3S547 1935 79–804056

—STUDY AND TEACHING (ELEMENTARY) —JAPAN

Miura, Tōsaku, 1887–
(Kodomo no tame no kōmin no hanashi)
子供のための公民の話 / 三浦藤作著. — 東京：青葉書房, 昭和22 ‹1947›
186 p. ; 18 cm.
1. Civics, Japanese—Study and teaching (Elementary)—Japan. 2. Fifth grade (Education) 3. Sixth grade (Education) I. Title. II. Title: Kōmin no hanashi.
LB1584.5.J3M48 79–805149

Shimane-ken Shihan Gakkō. Fuzoku Shōgakkō.
(Nihon seishin o kichō to seru shōgakkō no kōmin kyōiku)
日本精神を基調とせる小學校の公民教育 / 島根縣師範學校附屬小學校, 社會教育會共編. — 東京：社會教育會, 昭和10 ‹1935›
371 p. : ill. ; 23 cm.
"附錄 修身公民科に關する諸法令抜萃": p. 303–371.
1. Civics, Japanese—Study and teaching (Elementary)—Japan. 2. Civics, Japanese—Study and teaching—Japan. I. Shakai Kyōikukai. II. Title. III. Title: Shōgakkō no kōmin kyōiku.
LB1584.5.J3S547 1935 79–804056

CIVICS, MALAYSIAN

Buku panduan muhibbah. — Kuala Lumpur : Jabatan Chetak Kerajaan, 1971.
ix, 29 p. ; 15 cm.
1. Civics, Malaysian.
JQ719.A2B84 77-942868
79 MARC

—STUDY AND TEACHING (SECONDARY)—MALAYSIA

Awang Had Salleh, 1934-
Siviks menengah dewan / Awang Had bin Salleh, Ali bin Esa, Isahak bin Haron ; disunting oleh Mohd. Kasim bin Talib. — Cet. 1. — Kuala Lumpur : Dewan Bahasa dan Pustaka, Kementerian Pelajaran Malaysia, 1976-
v. : ill. ; 22 cm. — (Siri pelajaran menengah DBP ; bil. 136)
$2.50
1. Civics, Malaysian—Study and teaching (Secondary)—Malaysia. I. Ali bin Esa, joint author. II. Isahak bin Haron, joint author. III. Title. IV. Series: Dewan Bahasa dan Pustaka. Siri pelajaran menengah ; bil. 136.
JQ719.A2A96 76-989797
79 MARC

CIVIDALE, ITALY

—ANTIQUITIES—PERIODICALS

Museo archeologico nazionale di Cividale del Friuli.
Forum Iulii. 1- 1977-
[Cividale, Italy] Museo archeologico nazionale di Cividale del Friuli.
v. ill. 21 cm. annual.
"Annuario del Museo archeologico nazionale di Cividale del Friuli."
In Italian.
1. Museo archeologico nazionale di Cividale del Friuli. 2. Cividale, Italy—Antiquities—Periodicals. 3. Friuli—Antiquities—Periodicals. 4. Italy—Antiquities—Periodicals. 5. Cividale, Italy—History—Periodicals. I. Title.
AM101.C6827A28 79-645439
MARC-S

—HISTORY

Monumento alla Resistenza : 20 aprile 1975, Inaugurazione : note storiche : trentennale della liberazione / [a cura di Giuseppe Jacolutti]. — Udine : Tip. Filacorda, 1975.
89 p. : ill. ; 25 cm. It77-May
At head of title: Città di Cividale del Friuli, Comitato promotore.
L2000
1. World War, 1939-1945—Underground movements—Italy—Cividale. 2. Cividale, Italy—History. I. Jacolutti, Giuseppe.
D802.I82C585 940.53'45'391 79-344750
*79 MARC

—HISTORY—PERIODICALS

Museo archeologico nazionale di Cividale del Friuli.
Forum Iulii. 1- 1977-
[Cividale, Italy] Museo archeologico nazionale di Cividale del Friuli.
v. ill. 21 cm. annual.
"Annuario del Museo archeologico nazionale di Cividale del Friuli."
In Italian.
1. Museo archeologico nazionale di Cividale del Friuli. 2. Cividale, Italy—Antiquities—Periodicals. 3. Friuli—Antiquities—Periodicals. 4. Italy—Antiquities—Periodicals. 5. Cividale, Italy—History—Periodicals. I. Title.
AM101.C6827A28 79-645439
MARC-S

CIVIDALE, ITALY. SANTA MARIA IN VALLE (URSULINE CONVENT). TEMPIETTO

L'Orange, Hans Peter, 1903-
Il Tempietto longobardo di Cividale / Hans Peter L'Orange e Hjalmar Torp. — Roma : G. Bretschneider, 1977-
v. : ill., plans ; 31 cm. — (Acta ad archaelogiam et artium historiam pertinentia ; v. 7) It***
Includes bibliographical references and index.
CONTENTS: pt. 1. Tavole e rilievi.—pt. 2. L'architettura del Tempietto.
1. Cividale, Italy. Santa Maria in Valle (Ursuline convent). Tempietto. I. Torp, Hjalmar, 1898- joint author. II. Title. III. Series.
N1.A1A25 vol. 7 78-399487
[NA5621.C54] 79 MARC

CIVIL AERONAUTICS see Aeronautics, Commercial; Private flying

CIVIL AVIATION see Aeronautics, Commercial; Private flying

CIVIL CODE

—MEXICO—BAJA CALIFORNIA (STATE)

Baja California (State). Laws, statutes, etc.
[Código civil]
Código civil para el E.L. y S. de Baja California, con sus reformas. — 1. ed. — Puebla, Méx. : Editorial Cajica, c1978.
573 p. ; 14 p. — (Serie Leyes del Estado de Baja California) (Colección de leyes mexicanas)
"Esta edición estuvo al cuidado del Lic. José M. Cajica."
ISBN 9686012176
1. Civil code—Mexico—Baja California (State) I. Cajica, José María. II. Title.
346'.722'002632 79-116411
79 MARC

CIVIL COMMITMENT OF THE INSANE
see Insane—Commitment and detention

CIVIL DEATH

see also Infamy (Law)

CIVIL DEFENSE

see also Air defenses, Civil; Air raid shelters; Atomic bomb shelters; Auxiliary police; Disaster relief; Emergency housing; Evacuation of civilians; Rescue work; World War, 1939-1945—Evacuation of civilians

Chęciński, Józef.
Ochrona ludności we współczesnej wojnie / Józef Chęciński. — Wyd. 3 popr. i rozsz. — Warszawa : Wydaw. Min. Obrony Narodowej, 1978 (Łódź : WojsD)
387, [1] p. : ill. ; 20 cm. P78-6415
Includes bibliographical references.
zł30.00
1. Civil defense. I. Title.
UA926.C465 1978 78-399512
79 MARC

United States. Congress. House. Committee on Armed Services. Delegation to the Middle East and Europe.
Report of the Delegation to the Middle East and Europe of the Committee on Armed Services, House of Representatives, Ninety-sixth Congress, first session, March 8, 1979. — Washington : U. S. Govt. Print. Off., 1979.
iii, 21 p. ; 24 cm.
At head of title: 96th Congress, 1st session. Committee print no. 5.
1. Israel—Defenses. 2. Egypt—Defenses. 3. Europe—Defenses. 4. Civil defense. 5. Military assistance, American. 6. Strategic Arms Limitation Talks.
UA853.I8U54 1979 355.03'3056 79-602668
79 MARC

Violence, non-violence. — [Bruxelles, r. des Deux Églises 17] : Commission Justice et paix francophone, Pax Christi, [1976]
102 p. ; 25 cm. Be77-8
Includes bibliographical references.
75.00F
1. Nonviolence. 2. Civil defense. I. Commission Justice et paix francophone.
HM278.V57 78-348992
*78 MARC

Zimmermann, Dieter.
Grundwissen für die Einsatzkräfte / [Autoren, Dieter Zimmermann, Werner Schneider]. — 1. Aufl. — Berlin : Militärverlag d. Dt. Demokrat. Republic, 1977.
162 p. : ill. ; 19 cm. — (Ausbildungsanleitung der Zivilverteidigung der DDR ; Heft 1) GFR78-A17
Bibliography: p. 162.
1.75M
1. Civil defense. 2. Atomic warfare. 3. Chemical warfare. 4. Biological warfare. I. Schneider, Werner, Major Ingenieur, joint author. II. Title. III. Series.
UA926.Z54 79-371310
79 MARC

—EQUIPMENT AND SUPPLIES

see also Disaster hospitals

—HANDBOOKS, MANUALS, ETC.

United States. Defense Civil Preparedness Agency.
In time of emergency : a citizen's handbook on nuclear attack, natural disasters. — Rev. — [Baltimore] : Defense Civil Preparedness Agency, Dept. of Defense, 1977.
viii, 98 p. : ill. ; 22 cm.
First ed. (1968) by U.S. Office of Civil Defense.
"H-14."
1. Civil defense—Handbooks, manuals, etc. 2. Disasters—Handbooks, manuals, etc. I. United States. Office of Civil Defense. In time of emergency. II. Title.
UA927.U576 1977b 363.3'5 78-601316
79 MARC

—PERIODICALS

Journal of civil defense. v. 9-
Jan./Feb. 1976-
[Starke, Fla.] American Civil Defense Association [etc.]
v. ill. 28 cm. bimonthly.
Continues: Survive.
Issues for 1976- issued by the Civil Defense Forum and other similar bodies.
1. Civil defense—Periodicals. 2. United States—Civil defense—Periodicals. I. Civil Defense Forum.
UA926.A1J68 363.3'5'05 78-640267
MARC-S

Survive. v. 1-8; May/June 1968-Nov./Dec. 1975. [Starke, Fla., Association for Community-Wide Protection from Nuclear Attack, etc.]
8 v. ill. 28 cm. bimonthly.
"The American journal of civil defense" (varies slightly).
Vols. for 1968- issued by the Civil Defense Forum and other similiar bodies.
Continued by: Journal of civil defense.
1. United States—Civil defense—Periodicals. 2. Civil defense—Periodicals. I. Civil Defense Forum.
UA927.S93 363.3'5'0973 78-640266
MARC-S

—STUDY AND TEACHING
—RUSSIA

Bystrit͡skiĭ, S P
(Kompleksnye ob''ektovye uchenii͡a grazhdanskoĭ oborony v kolkhoze (sovkhoze)
Комплексные объектовые учения гражданской обороны в колхозе (совхозе) : [Учеб. пособие] / С. П. Быстрицкий. — Москва : Воениздат, 1977.
111 p. ; 20 cm. — (Гражданская оборона СССР) USSR 77
Series romanized: Grazhdanskai͡a oborona SSSR.
0.25rub
1. Civil defense—Study and teaching—Russia. 2. Russia—Civil defense. 3. Agriculture—Russia—Defense measures. I. Title.
UA929.R9B95 78-420668

—FRANCE

Broc, André Pierre.
La Protection civile : la sécurité civile / André-Pierre Broc, ... — [Paris] : PUF, 1977.
123, [5] p. ; 18 cm. — (Que sais-je? ; 1682) F77-12098
Bibliography: p. [127]
9.60F
1. Safety regulations—France. 2. Civil defense—France. I. Title.
344'.44'047 78-395246
79 MARC

—KOREA—HANDBOOKS, MANUALS, ETC.

Korea (Republic). Naemubu.
(Min Pangwidae tongwŏn p'yŏllam)
民防衛隊動員便覽. — [서울] : 内務部, 1976.
48 p. : ill. ; 21 cm.

1. Civil defense—Korea—Handbooks, manuals, etc. 2. Korea (Republic). Min Pangwidae. I. Title.
UA929.K6K67 1976 78-826243

CIVIL DISOBEDIENCE see Government, Resistance to

CIVIL DISORDERS see Riots

CIVIL ENGINEERING

see also Arches; Bridges; Coffer-dams; Construction equipment; Curves in engineering; Drainage; Dredging; Embankments; Excavation; Foundations; Guy anchors; Harbors; Highway engineering; Hydraulic engineering; Marine engineering; Masonry; Mechanical engineering; Military engineering; Mining engineering;

Piling (Civil engineering); Public works; Railroad engineering; Railroads —Construction; Reclamation of land; Roads; Sanitary engineering; Steel, Structural; Streets; Strength of materials; Structural dynamics; Structural engineering; Surveying; Walls

The Encyclopedia of how it's built / edited by Donald Clarke. — New York : A & W Publishers, 1979.
184 p. : ill. (some col.) ; 30 cm.
Spine title: How it's built.
ISBN 0-89479-047-1 : $16.95
1. Building. 2. Civil engineering. 3. Mechanical engineering. I. Clarke, Donald. II. Title: How it's built.
TH145.E53 1979 620 79-51558
79 MARC

Jaeger, Charles, 1901-
Rock mechanics and engineering / Charles Jaeger. — 2d ed. — Cambridge [Eng.] ; New York : Cambridge University Press, 1979.
xii, 523 p. : ill. ; 24 cm.
Bibliography: p. 463-493.
Includes indexes.
ISBN 0-521-21898-5
1. Rock mechanics. 2. Civil engineering. I. Title.
TA706.J3 1979 624'.1513 77-85700
77 MARC

Matsumoto, Yoshiji, 1929-
(Doboku kaiseki hō)
土木解析法　松本嘉司·伯野元彦著　東京　技報堂　昭和46-50 (1971-75)
2 v. 22 cm. Ja 71-19126 (v. 1)
Subtitle in colophon: Kōgi to enshū.
Includes bibliographies.
1. Engineering mathematics. 2. Civil engineering. I. Hakuno, Motohiko, joint author. II. Title.
TA330.M35 75-803177
Library of Congress 75 [r78]rev J

—BIBLIOGRAPHY

Mexico. Consejo Nacional de Ciencia y Tecnología. Departamento de Sistemas de Información.
Biblioteca básica a nivel profesional, ingeniería civil / elaborada por el Departamento de Sistemas de Información, CONACYT. — 1. ed. — México : Consejo Nacional de Ciencia y Tecnología, 1973, t.p. 1976, c1976.
viii, 88 leaves ; 28 cm.
Originally published under title: Acervo básico para bibliotecas de escuela de ingeniería civil, lista preliminar.
Includes indexes.
1. Civil engineering—Bibliography. 2. Bibliography—Best books—Civil engineering. I. Title.
Z5851.M39 1976 016.624 78-337953
[TA145] 78[r79]rev MARC

—COLD WEATHER CONDITIONS—CONGRESSES

Cold Regions Specialty Conference, Anchorage, Alaska, 1978.
Proceedings of the Conference on Applied Techniques for Cold Environments / Cold Regions Specialty Conference, Anchorage, Alaska, May 17-19, 1978 ; sponsored by Alaska Section, ASCE, ASCE Task Committee on Cold Regions, cosponsored by the Canadian Society for Civil Engineers. — New York : American Society of Civil Engineers, c1978-c1979.
2 v. (vii, 1167 p.) : ill. ; 22 cm.
On spine: Applied techniques for cold environments.
Includes bibliographical references and indexes.
1. Civil engineering—Cold weather conditions—Congresses. 2. Civil engineering—Alaska—Cold weather conditions—Congresses. 3. Cold regions—Congresses. I. American Society of Civil Engineers. Alaska Section. II. American Society of Civil Engineers. Task Committee on Cold Regions. III. Canadian Society for Civil Engineers. IV. Title: Applied techniques for cold environments.
TA713.C64 1978 620'.411 79-116292
79 MARC

—COLLECTED WORKS

Kungnip Kŏnsŏl Yŏn'guso.
(Yŏn'gu pogo - Kungnip Kŏnsŏl Yŏn'guso)
研究報告 - 國立建設研究所　第1-　輯
[서울]　建設部國立建設研究所 1967-
no. ill. 26 cm.
English or Korean.
1. Civil engineering—Collected works.
TA4.K79a 666'.8 78-648065
(MARC-S)

—COMPUTER PROGRAMS

Medearis, Kenneth.
Report on an investigation of the feasibility of establishing a national civil engineering software center to the American Society of Civil Engineers for the Research Council on Computer Practices. Sponsored by the National Science Foundation, Washington, D.C. Conducted by Kenneth Medearis in cooperation with Research Services Dept., American Society of Civil Engineers. [New York, American Society of Civil Engineers] 1973.
viii, 116 p. 28 cm. $4.00
1. Civil engineering—United States—Data processing. 2. Data processing service centers—United States. 3. Civil engineering—Computer programs. I. American Society of Civil Engineers. Research Services Dept. II. American Society of Civil Engineers. Research Council on Computer Practices. III. Title.
TA345.M42 624'.028'55 79-302366
79 MARC

—CONGRESSES

Conference on Civil Engineering in the Oceans, 4th, San Francisco, 1979.
Civil engineering in the oceans IV : proceedings of the specialty conference, Hyatt Regency Hotel, San Francisco, California, September 10-12, 1979 / sponsored by the Waterway, Port, Coastal, and Ocean Division of the American Society of Civil Engineers, co-sponsored by U.S. Naval Facilities Engineering Command, SNAME ; host, San Francisco Section, ASCE. — New York : ASCE, c1979.
2 v. (viii, 1086 p.) : ill. ; 22 cm.
Includes bibliographies and indexes.
$30.00
1. Ocean engineering—Congresses. 2. Civil engineering—Congresses. I. American Society of Civil Engineers. Waterway, Port, Coastal and Ocean Division. II. United States. Naval Facilities Engineering Command. III. Society of Naval Architects and Marine Engineers, New York. IV. Title.
TC1505.C6 1979 627'.09162 79-121221
79 MARC

Engineering Foundation Conference : Planning, Engineering, and Constructing the Superprojects, Pacific Grove, Calif., 1978.
Planning, engineering, and constructing the superprojects : proceedings of the Engineering Foundation research conference, Asilomar conference grounds, Pacific Grove, California, April 30-May 5, 1978. New York : American Society of Civil Engineers, c1979.
ix, 527 p. : ill. ; 22 cm.
Includes indexes.
$24.00
1. Civil engineering—Congresses. 2. Civil engineering—United States—Congresses. I. Engineering Foundation, New York. II. Title.
TA5.E53 1978 658'.92'4 79-115579
79 MARC

South African Institution of Civil Engineers.
Diamantjubileum-konvensie, 1963 : verhandelinge = Diamond jubilee convention, 1963 : proceedings / P. J. W Klintworth, honorary editor. — [Johannesburg] : Suid-Afrikaanse Instituut van Siviele Ingenieurs, [196-]
288 p. : ill. ; 28 cm.
Includes bibliographical references.
1. South African Institution of Civil Engineers. 2. Civil engineering—Congresses. 3. Civil engineering—South Africa—Congresses. I. Klintworth, P. J. W.
TA1.S67473 78-320910
79 MARC

Specialty Conference on Engineering for the Space Environment, Houston, Tex., 1970.
Engineering for the space environment; selected papers and abstracts. New York, American Society of Civil Engineers [1971?]
185 p. illus. 22 cm. $6.00
"Co-sponsored by Aero-Space Transport Division, American Society of Civil Engineers, and American Institute of Aeronautics and Astronautics."
Includes bibliographical references.
1. Astronautics—Congresses. 2. Civil engineering—Congresses. I. American Society of Civil Engineers. Aero-Space Transport Division. II. American Institute of Aeronautics and Astronautics. III. Title.
TL787.S78 1970 629.44 78-325926
79 MARC

—CONTRACTS AND SPECIFICATIONS—ENGLAND

Seeley, Ivor H
Civil engineering quantities, by Ivor H. Seeley. London, Macmillan [1965]
ix, 341 p. illus. 25 cm.
Includes bibliographies.
1. Engineering—Estimates—England. 2. Civil engineering—Contracts and specifications—England. I. Title.
TA183.S45 1965 624 73-239152
71[r79]rev MARC

—CONTRACTS AND SPECIFICATIONS—GREAT BRITAIN

Seeley, Ivor H
Civil engineering specification / Ivor H. Seeley. — 2nd ed. — London : Macmillan, 1976, 1978 printing. GB***
viii, 247 p. ; 24 cm.
Includes index.
ISBN 0-333-04227-1. ISBN 0-333-19870-0 (printed-case) : £6.95
1. Civil engineering—Contracts and specifications—Great Britain. I. Title.
TA180.S4 1976 624'.021'2 79-302878
79 MARC

—DATA PROCESSING—CONGRESSES

Conference on Computing in Civil Engineering, 1st, Atlanta, 1978.
Conference on Computing in Civil Engineering, at the Sheraton Atlanta Hotel, Atlanta, Georgia, June 26-29, 1978 / sponsored by American Society of Civil Engineers, Technical Council on Computer Practices (ASCE TCCP), Society for Computer Applications in Engineering, Planning, and Architecture (CEPA), ICES Users Group, Georgia Institute of Technology ; co-chairmen, Morton B. Lipetz, Leroy Z. Emkin. — New York, N.Y. : American Society of Civil Engineers, c1978.
viii, 855 p. : ill. ; 22 cm.
Includes bibliographies and indexes.
$40.00
1. Civil engineering—Mathematical models—Congresses. 2. Civil engineering—Data processing—Congresses. I. American Society of Civil Engineers. Technical Council on Computer Practices. II. Society for Computer Applications in Engineering, Planning, and Architecture. III. Ices Users Group. IV. Lipetz, Morton B. V. Emkin, Leroy Z.
TA329.C65 1978 624'.01'84 78-108567
78 MARC

—DICTIONARIES

Chao, Kuo-hua.
(Ying Han t'u mu kung ch'eng ming tz'u hui pien)
英漢土木工程名詞彙編 = English-Chinese dictionary of civil engineering：名詞33000 / 趙國華編著. — 香港：香港工程技術出版社, [1978?]
539 p. in various pagings ; 19 cm.
Cover title.
$15.00
1. Civil engineering—Dictionaries. 2. English language—Dictionaries—Chinese. I. Title. II. Title: T'u mu kung ch'eng ming tz'u hui pien. III. Title: English—Chinese dictionary of civil engineering.
TA9.C5 78-841099

Kuo li pien i kuan, T'ai-pei.
(T'u mu kung ch'eng ming tz'u)
土木工程名詞：續編本 / 國立編譯館編訂. — 臺初版. — 臺北：正中書局；九龍：海外總經銷集成圖書公司, 民國66 [1977]
1, 1, 2, 242 p. ; 27 cm.
Cover title.
"中華民國六十六年五月教育部公布"
English and Chinese.
1. Civil engineering—Dictionaries. 2. English language—Dictionaries—Chinese. I. Title.
TA9.K85 1977 79-841144

—ENVIRONMENTAL ASPECTS

Environmental impacts of international civil engineering projects and practices : proceedings of a session / sponsored by the Research Council on Environmental Impact Analysis of the ASCE Technical Council on Research at the ASCE national convention in San Francisco, California, October 17-21, 1977 ; Charles G. Gunnerson and John M. Kalbermatten, editors. — New York : American Society of Civil Engineers, c1978.
ix, 254 p. : ill. ; 22 cm.
Includes bibliographies.
$16.00
1. Civil engineering—Environmental aspects. 2. Underdeveloped areas—Public works—Environmental aspects. I. Gunnerson, Charles G. II. Kalbermatten, John M. III. ASCE Technical Council on Research. Environmental Impact Analysis Research Council.
TD195.C54E58 333.7 78-109929
78 MARC

CIVIL ENGINEERING (Continued)

—HANDBOOKS, MANUALS, ETC.

Chung-kuo t'u mu kung ch'eng hsüeh hui (Founded 1936)
(Chung-kuo kung ch'eng shih shou ts'e)
中國工程師手册：土木類 / 中國土木工程學會編. — 增訂本. — 臺北：中國土木水利工程學會：總經銷科技圖書股份有限公司, 民國66 [1977]
3 v. : ill. ; 20 cm.
NT$440.00
1. Civil engineering—Handbooks, manuals, etc. I. Title.
TA151.C558 1977 78-841586

Den Nye anleggsteknikk : anleggsmannens tekniske håndbok / [redaktør, Gunnar Aas]. — Oslo : Teknologisk forl., c1977 (Oslo : Aktietr.)
626 p. : ill. ; 22 cm. — (Industriens og håndverkets tekniske bibliotek)) N78-Jan/Feb.
Includes index.
ISBN 8251201438 : kr285
1. Civil engineering—Handbooks, manuals, etc. I. Aas, Gunnar.
TA151.N88 78-395863
79 MARC

Manuale del costruttore civile e del geometra. — 4. ed. riveduta. — Roma : Cremonese, 1977.
xxx, 1590 p. : ill. ; 18 cm. — (Manuali Cremonese) It78-July
Includes index.
L15500
1. Civil engineering—Handbooks, manuals, etc.
TA151.M174 1977 78-398139
*79 MARC

—MANAGEMENT see Engineering —Management

—MATHEMATICAL MODELS —CONGRESSES

Conference on Computing in Civil Engineering, 1st, Atlanta, 1978.
Conference on Computing in Civil Engineering, at the Sheraton Atlanta Hotel, Atlanta, Georgia, June 26-29, 1978 / sponsored by American Society of Civil Engineers, Technical Council on Computer Practices (ASCE TCCP), Society for Computer Applications in Engineering, Planning, and Architecture (CEPA), ICES Users Group, Georgia Institute of Technology ; co-chairmen, Morton B. Lipetz, Leroy Z. Emkin. — New York, N.Y. : American Society of Civil Engineers, c1978.
viii, 855 p. : ill. ; 22 cm.
Includes bibliographies and indexes.
$40.00
1. Civil engineering—Mathematical models—Congresses. 2. Civil engineering—Data processing—Congresses. I. American Society of Civil Engineers. Technical Council on Computer Practices. II. Society for Computer Applications in Engineering, Planning, and Architecture. III. Ices Users Group. IV. Lipetz, Morton B. V. Emkin, Leroy Z.
TA329.C65 1978 624'.01'84 78-108567
78 MARC

—PERIODICALS

American Society of Civil Engineers.
Journal of the technical councils of ASCE. v. 103– Dec. 1977–
[New York] American Society of Civil Engineers.
v. ill. 22 cm.
"Proceedings of the American Society of Civil Engineers."
"Codes and standards, computer practices, lifeline earthquake engineering, research."
Key title: Journal of the technical councils of ASCE,
ISSN 0148-9909
1. Civil engineering—Periodicals. 2. Civil engineering—United States—Periodicals. I. Title.
TA1.A52317 624'.05 78–645626
MARC–S

Doboku Kenkyujo, Tokyo.
(Jutaku kenkyūhi hōkoku)
受託研究費報告
[東京] 土木研究所
v. illus. 26 cm. (19 -74: 年報 - 建設省土木研究所(TA4.D624a))
1. Civil engineering—Periodicals. 2. Engineering—Research grants—Japan—Periodicals. I. Title. II. Series: Doboku Kenkyūjo, Tokyo. Nempō.
TA4.D624C 73–817699
(MARC–S)

—STUDY AND TEACHING—BRAZIL —CATALOGS

Brazil. Coordenação do Aperfeiçoamento de Pessoal de Nível Superior.
Pós-graduação catálogo de cursos: engenharia mecânica, civil.
[Brasília, Coordenação do Aperfeiçoamento de Pessoal de Nível Superior.
v. 23 cm.
1. Mechanical engineering—Study and teaching—Brazil—Catalogs. 2. Civil engineering—Study and teaching—Brazil—Catalogs. I. Title.
TJ158.B73a 79–645320
MARC–S

—VOCATIONAL GUIDANCE

Netherlands (Kingdom, 1815-). Directoraat-Generaal voor de Arbeidsvoorziening.
Civiel ingenieur / Directoraat-Generaal voor de Arbeidsvoorziening. — 2. geheel herz. druk. — 's-Gravenhage : Staatsuitgeverij, 1976.
vii, 52 p. ; 20 cm. — (Beroepenmonografieën) Ne***
ISBN 9012012740
1. Civil engineering—Vocational guidance. I. Title.
TA157.N375 1976 78-368155
78 MARC

—ALASKA—COLD WEATHER CONDITIONS—CONGRESSES

Cold Regions Specialty Conference, Anchorage, Alaska, 1978.
Proceedings of the Conference on Applied Techniques for Cold Environments / Cold Regions Specialty Conference, Anchorage, Alaska, May 17-19, 1978 ; sponsored by Alaska Section, ASCE, ASCE Task Committee on Cold Regions, cosponsored by the Canadian Society for Civil Engineers. — New York : American Society of Civil Engineers, c1978-c1979.
2 v. (vii, 1167 p.) : ill. ; 22 cm.
On spine: Applied techniques for cold environments.
Includes bibliographical references and indexes.
1. Civil engineering—Cold weather conditions—Congresses. 2. Civil engineering—Alaska—Cold weather conditions—Congresses. 3. Cold regions—Congresses. I. American Society of Civil Engineers. Alaska Section. II. American Society of Civil Engineers. Task Committee on Cold Regions. III. Canadian Society for Civil Engineers. IV. Title: Applied techniques for cold environments.
TA713.C64 1978 620'.411 79-116292
79 MARC

—DENMARK—HORSENS

Stadsingeniøren i Horsens 1875-1975 / [red. af H. Arnum ; udg. af Horsens Kommune]. — [Horsens : Horsens Kommune : eksp., Thems Boghandel, Søndergade 35, 1975]
32, [79] : ill. ; 24 cm. D
"Udarbejdet i anledning af stadsingeniørkontorets 100 års jubilæum."
ISBN 8798041401
1. Civil engineering—Denmark—Horsens. 2. Horsens, Denmark. Stadsingeniørens kontor.
TD70.H67S73 78-385356
78 MARC

—EGYPT—ADDRESSES, ESSAYS, LECTURES

(al-Mukhtārāt min Majallat al-muhandisīn)
المختارات من مجلة المهندسين ، السنــــة الاولى ١٩٤٥ . [القاهرة
1946?]
256 p. illus. 17 cm.
Cover title.
1. Engineering—Addresses, essays, lectures. 2. Civil engineering—Egypt—Addresses, essays, lectures. I. Majallat al-muhandisīn.
TA155.M84 78–970872

—FRANCE—NOMENCLATURE

Syndicat national du béton armé & des techniques industrialisées.
Nomenclature codifiée des ouvrages ou tâches élémentaires et temps moyens. Paris (16e), Société de diffusion des techniques du bâtiment et des travaux publics, 9, rue La Pérouse, 1970.
vi, 163 p. 27 cm. F71-4261
1. Civil engineering—France—Nomenclature. 2. Building—France—Nomenclature. 3. Civil engineering—France—Production standards. 4. Building—France—Production standards. I. Title.
TA153.S97 1970 624'.183 79-343420
79 MARC

—FRANCE—PRODUCTION STANDARDS

Syndicat national du béton armé & des techniques industrialisées.
Nomenclature codifiée des ouvrages ou tâches élémentaires et temps moyens. Paris (16e), Société de diffusion des techniques du bâtiment et des travaux publics, 9, rue La Pérouse, 1970.
vi, 163 p. 27 cm. F71-4261
1. Civil engineering—France—Nomenclature. 2. Building—France—Nomenclature. 3. Civil engineering—France—Production standards. 4. Building—France—Production standards. I. Title.
TA153.S97 1970 624'.183 79-343420
79 MARC

—GERMANY—HANOVER (PROVINCE, ETC.)—HISTORY

Scholl, Lars Ulrich.
Ingenieure in der Frühindustrialisierung : staatl. u. private Techniker im Königreich Hannover u. an d. Ruhr (1815-1873) / Lars Ulrich Scholl. — Göttingen : Vandenhoeck und Ruprecht, 1978.
482 p. ; 24 cm. — (Studien zu Naturwissenschaft, Technik und Wirtschaft im neunzehnten Jahrhundert ; Bd. 10) GFR78-A
A revision of the author's thesis, Technische Universität Hanover, 1977.
Bibliography: p. [432]-468.
Includes index.
ISBN 3-525-42209-1
1. Civil engineering—Germany—Hanover (Province, etc.)—History. 2. Civil engineers—Germany—Hanover (Province, etc.) 3. Hanover (Province, etc.)—Industries—History. 4. Civil engineering—Germany—Ruhr Valley region—History. 5. Civil engineers—Germany—Ruhr Valley region. 6. Ruhr Valley region, Ger.—Industries—History. I. Title. II. Series.
TA74.H3S36 1978 624'.0943'59 79-344143
79 MARC

—GERMANY—RUHR VALLEY REGION —HISTORY

Scholl, Lars Ulrich.
Ingenieure in der Frühindustrialisierung : staatl. u. private Techniker im Königreich Hannover u. an d. Ruhr (1815-1873) / Lars Ulrich Scholl. — Göttingen : Vandenhoeck und Ruprecht, 1978.
482 p. ; 24 cm. — (Studien zu Naturwissenschaft, Technik und Wirtschaft im neunzehnten Jahrhundert ; Bd. 10) GFR78-A
A revision of the author's thesis, Technische Universität Hanover, 1977.
Bibliography: p. [432]-468.
Includes index.
ISBN 3-525-42209-1
1. Civil engineering—Germany—Hanover (Province, etc.)—History. 2. Civil engineers—Germany—Hanover (Province, etc.) 3. Hanover (Province, etc.)—Industries—History. 4. Civil engineering—Germany—Ruhr Valley region—History. 5. Civil engineers—Germany—Ruhr Valley region. 6. Ruhr Valley region, Ger.—Industries—History. I. Title. II. Series.
TA74.H3S36 1978 624'.0943'59 79-344143
79 MARC

—GREAT BRITAIN—BIBLIOGRAPHY— CATALOGS

Institution of Civil Engineers, London. Library.
Early printed reports and maps (1665-1850) in the Library of the Institution of Civil Engineers / A. W. Skempton. — London : Institution of Civil Engineers : distributed by T. Telford, 1977.
xv, 84 p. : ill. ; 23 cm. GB***
Includes bibliographical references and index.
ISBN 0-7277-0047-2 : £6.00
1. Civil engineering—Great Britain—Bibliography—Catalogs. 2. Civil engineering—Great Britain—History—Sources—Bibliography—Catalogs. 3. Institution of Civil Engineers, London. Library. I. Skempton, A. W. II. Title.
Z5855.G7I57 1977 016.624'0941 79-300954
[TA57] 79 MARC

—GREAT BRITAIN—HISTORY— SOURCES—BIBLIOGRAPHY— CATALOGS

Institution of Civil Engineers, London. Library.
Early printed reports and maps (1665-1850) in the Library of the Institution of Civil Engineers / A. W. Skempton. — London : Institution of Civil Engineers : distributed by T. Telford, 1977.
xv, 84 p. : ill. ; 23 cm. GB***
Includes bibliographical references and index.
ISBN 0-7277-0047-2 : £6.00
1. Civil engineering—Great Britain—Bibliography—Catalogs. 2. Civil engineering—Great Britain—History—Sources—Bibliography—Catalogs. 3. Institution of Civil Engineers, London. Library. I. Skempton, A. W. II. Title.
Z5855.G7I57 1977 016.624'0941 79-300954
[TA57] 79 MARC

CIVIL ENGINEERING (Continued)

—JAPAN

Akita, Giichi, fl. 1835–1837.
(Sampō jikata taisei)
算法地方大成：[5巻] / 村上直, 荒川秀俊校訂；[長谷川寛閲；秋田義一編]. — 東京：近藤出版社, 昭和51 [1976]
21, 184, 12 p. : ill. ; 20 cm. — (日本史料選書；12) Ja 76–21891
Series romanized: Nihon shiryō sensho.
Includes index.
¥2800
1. Land value taxation—Japan. 2. Civil engineering—Japan. 3. Agriculture and state—Japan. I. Murakami, Tadashi, 1925- II. Arakawa, Hidetoshi, 1907- III. Hasegawa, Hiroshi, 1782–1838. IV. Title.
HD1295.J3A37 1976 78–813340

—JAPAN—HISTORY

Takahashi, Yutaka, 1927-
(Nihon doboku gijutsu no rekishi)
日本土木技術の歴史 / 高橋裕, 酒匂敏次共著. — 東京：地人書館, 昭和35 [1960]
208, 15 p., [1] leaf of plates : ill. ; 18 cm. — (日本技術史選書)
Series romanized: Nihon gijutsu shi sensho.
"文献解題": p. 204–205.
Includes index.
1. Construction industry—Japan—History. 2. Civil engineering—Japan—History. I. Sakō, Toshitsugu, 1934- joint author. II. Title.
HD9715.J22T34 79–801298

—KOREA—YEARBOOKS

(Kŏnsŏl ch'ongnam)
建設總覧 1976-
[서울] 建設技術研究會
v. ill. 27 cm. annual.
1. Construction industry—Korea—Yearbooks. 2. Civil engineering—Korea—Yearbooks. 3. Korea—Public works—Yearbooks. 4. Construction industry—Law and legislation—Korea. I. Kŏnsŏl Kisul Yŏn'guhoe.
HD9715.K8K65 78–647742 (MARC-S)

—NETHERLANDS—HISTORY

Metzelaar, W
Nederland, deltaland : inleiding tot de weg- en waterbouwkunde / door W. Metzelaar. — [Haarlem : Stam, 1962]
146 p. : ill. ; 25 cm. — (Weg- en waterbouwkunde ; deel 1)
Bibliography: p. 146.
1. Civil engineering—Netherlands—History. I. Title. II. Series.
TA77.M47 78-395151
79 MARC

—NEW ENGLAND—HISTORY

Parkman, Aubrey, 1918-
Army engineers in New England : the military and civil work of the Corps of Engineers in New England, 1775-1975 / by Aubrey Parkman. — Waltham, Mass. : U.S. Army Corps of Engineers, New England Division, 1978.
xi, 319 p., [4] leaves of plates : ill. ; 25 cm.
Bibliography: p. 299-309.
Includes index.
1. United States. Army. Corps of Engineers—History. 2. Military engineering—New England—History. 3. Civil engineering—New England—History. I. Title.
UG23.P37 358'.22'0974 78-110862
78 MARC

—RUSSIA

Pompeev, IUriĭ Aleksandrovich.
(Osnovy stroitel'noĭ spetsial'nosti)
Основы строительной специальности / Ю. А. Помпеев. — Ленинград : Изд-во Ленинградского университета, 1976.
87, [1] p. : ill. ; 22 cm. USSR***
At head of title: Ministerstvo vysshego i srednego spetsial'nogo obrazovaniia RSFSR.
Bibliography: p. [88]
1. Building—Russia. 2. Civil engineering—Russia. I. Title.
TH85.P64 79–407726

—SOUTH AFRICA—CONGRESSES

South African Institution of Civil Engineers.
Diamantjubileum-konvensie, 1963 : verhandelinge = Diamond jubilee convention, 1963 : proceedings / P. J. W Klintworth, honorary editor. — [Johannesburg] : Suid-Afrikaanse Instituut van Siviele Ingenieurs, [196-]
288 p. : ill. ; 28 cm.
Includes bibliographical references.
1. South African Institution of Civil Engineers. 2. Civil engineering—Congresses. 3. Civil engineering—South Africa—Congresses. I. Klintworth, P. J. W.
TA1.S67473 78-320910
79 MARC

—UNITED STATES—CONGRESSES

Engineering Foundation Conference : Planning, Engineering, and Constructing the Superprojects, Pacific Grove, Calif., 1978.
Planning, engineering, and constructing the superprojects : proceedings of the Engineering Foundation research conference, Asilomar conference grounds, Pacific Grove, California, April 30-May 5, 1978. New York : American Society of Civil Engineers, c1979.
ix, 527 p. : ill. ; 22 cm.
Includes indexes.
$24.00
1. Civil engineering—Congresses. 2. Civil engineering—United States—Congresses. I. Engineering Foundation, New York. II. Title.
TA5.E53 1978 658'.92'4 79-115579
79 MARC

—UNITED STATES—DATA PROCESSING

Medearis, Kenneth.
Report on an investigation of the feasibility of establishing a national civil engineering software center to the American Society of Civil Engineers for the Research Council on Computer Practices. Sponsored by the National Science Foundation, Washington, D.C. Conducted by Kenneth Medearis in cooperation with Research Services Dept., American Society of Civil Engineers. [New York, American Society of Civil Engineers] 1973.
viii, 116 p. 28 cm. $4.00
1. Civil engineering—United States—Data processing. 2. Data processing service centers—United States. 3. Civil engineering—Computer programs. I. American Society of Civil Engineers. Research Services Dept. II. American Society of Civil Engineers. Research Council on Computer Practices. III. Title.
TA345.M42 624'.028'55 79-302366
79 MARC

—UNITED STATES—JUVENILE LITERATURE

Olney, Ross Robert, 1929-
They said it couldn't be done / Ross R. Olney. — 1st ed. — New York : Dutton, c1979.
134 p. : ill. ; 19 x 24 cm.
Bibliography: p. 130.
Includes index.
SUMMARY: Describes 10 feats of engineering that were believed to be impossible including the construction of the Brooklyn Bridge, the Empire State Building, the Hoover Dam, and the landing on the Moon.
ISBN 0-525-41060-0
1. Civil engineering—United States—Juvenile literature. 2. Space flight to the moon—Juvenile literature. [1. Civil engineering. 2. Space flight to the moon] I. Title.
TA149.O46 1979 624'.0973 78-12405
78 MARC

—UNITED STATES—PERIODICALS

American Society of Civil Engineers.
Journal of the technical councils of ASCE. v. 103–Dec. 1977–
[New York] American Society of Civil Engineers.
v. ill. 22 cm.
"Proceedings of the American Society of Civil Engineers."
"Codes and standards, computer practices, lifeline earthquake engineering, research."
Key title: Journal of the technical councils of ASCE,
ISSN 0148–9909
1. Civil engineering—Periodicals. 2. Civil engineering—United States—Periodicals. I. Title.
TA1.A52317 624'.05 78–645626
MARC-S

CIVIL ENGINEERS

see also Highway engineers

—DIRECTORIES

International directory of consulting environmental and civil engineers.
Blue Bell, Penn., International Research Service.
v. 23 cm.
Key title: International directory of consulting environmental and civil engineers, ISSN 0191–9636.
1. Environmental engineers—Directories. 2. Civil engineers—Directories. 3. Consulting engineers—Directories. I. International Research Service (U. S.). II. Title: Directory of consulting environmental and civil engineers.
TD169.6.I 57 625'.025 79–642903
MARC-S

—EMPLOYMENT—BRAZIL—SANTA CATARINA

Ensslin, Leonardo.
Análise da tendência da oferta e demanda de engenheiros civis e arquitetos em Santa Catarina : para Engo. Carlos Calliari, presidente do CREA 10a Região / coordenador Leonardo Ensslin ; [trabalho realizado pelos seguintes professores do Departamento de Engenharia Industrial da UFSC, Leonardo Ensslin, David Borile, José Luiz Fonseca da Silva Filho]. — [Florianópolis?] : FEESC, 1976.
43 p. : graphs ; 21 cm.
On cover: Ministério do Trabalho, Conselho Regional de Engenharia, Arquitetura e Agronomia da Décima Região/FEESC.
1. Civil engineers—Employment—Brazil—Santa Catarina. 2. Architects—Employment—Brazil—Santa Catarina. I. Calliari, Carlos. II. Borile, David, joint author. III. Silva Filho, José Luiz Fonseca da, joint author. IV. Title.
TA157.E687 78-387431
79 MARC

—AUSTRALIA—WESTERN AUSTRALIA—BIOGRAPHY

Tauman, Merab.
The chief, C. Y. O'Connor / Merab Tauman. — Nedlands [Australia] : University of Western Australia Press ; Forest Grove, Or. : [sold by] International Scholarly Book Services, 1978.
xiv, 290 p., [7] leaves of plates : ill. ; 25 cm. Aus***
Includes bibliographical references and index.
ISBN 0-85564-123-1
1. O'Connor, C. Y., 1843-1902. 2. Civil engineers—Australia—Western Australia—Biography. I. Title.
TA140.O36T38 1978 624'.092'4 79-308921
79 MARC

—FRANCE—BIOGRAPHY

Fernández Ordóñez, José A 1933-
Eugène Freyssinet / José A. Fernández Ordóñez. — Barcelona : 2c Ediciones, [1978?]
417 p., [1] fold. leaf of plates : ill. ; 22 cm. Sp***
ISBN 8485434005
1. Freyssinet, Eugène, 1879-1962. 2. Civil engineers—France—Biography.
TA140.F85F47 1978 78-388280
79 MARC

Kérisel, Jean, 1908-
Albert Caquot, 1881-1976 : créateur et précurseur / par Jean Kérisel. — Paris : Eyrolles, 1978.
164 p. : ill. ; 24 cm. F***
Includes bibliographical references.
60.00F
1. Caquot, Albert Irénée, 1881-1976. 2. Civil engineers—France—Biography.
TA140.C365K47 624'.092'4 78-368843
78 MARC

Tagliaventi, Ivo.
Eiffel e la sua torre. [Bologna] Pàtron [1962]
50 p. illus. 25 cm.
Bibliography: p. 48-49.
1. Eiffel, Gustave, 1832-1923. 2. Eiffel Tower, Paris. 3. Civil engineers—France—Biography. I. Title.
TA140.E4T33 78-396675
79 MARC

—GERMANY—HANOVER (PROVINCE, ETC.)

Scholl, Lars Ulrich.
Ingenieure in der Frühindustrialisierung : staatl. u. private Techniker im Königreich Hannover u. an d. Ruhr (1815-1873) / Lars Ulrich Scholl. — Göttingen : Vandenhoeck und Ruprecht, 1978.
482 p. ; 24 cm. — (Studien zu Naturwissenschaft, Technik und Wirtschaft im neunzehnten Jahrhundert ; Bd. 10) GFR78-A
A revision of the author's thesis, Technische Universität Hanover, 1977.
Bibliography: p. [432]-468.
Includes index.
ISBN 3-525-42209-1
1. Civil engineering—Germany—Hanover (Province, etc.)—History. 2. Civil engineers—Germany—Hanover (Province, etc.) 3. Hanover (Province, etc.)—Industries—History. 4. Civil engineering—Germany—Ruhr Valley region—History. 5. Civil engineers—Germany—Ruhr Valley region. 6. Ruhr Valley region, Ger.—Industries—History. I. Title. II. Series.
TA74.H3S36 1978 624'.0943'59 79-344143
79 MARC

CIVIL ENGINEERS (Continued)

—GERMANY—RUHR VALLEY REGION

Scholl, Lars Ulrich.
Ingenieure in der Frühindustrialisierung : staatl. u. private Techniker im Königreich Hannover u. an d. Ruhr (1815-1873) / Lars Ulrich Scholl. — Göttingen : Vandenhoeck und Ruprecht, 1978.
482 p. ; 24 cm. — (Studien zu Naturwissenschaft, Technik und Wirtschaft im neunzehnten Jahrhundert ; Bd. 10) GFR78-A
A revision of the author's thesis, Technische Universität Hanover, 1977.
Bibliography: p. [432]-468.
Includes index.
ISBN 3-525-42209-1
1. Civil engineering—Germany—Hanover (Province, etc.)—History. 2. Civil engineers—Germany—Hanover (Province, etc.) 3. Hanover (Province, etc.)—Industries—History. 4. Civil engineering—Germany—Ruhr Valley region—History. 5. Civil engineers—Germany—Ruhr Valley region. 6. Ruhr Valley region, Ger.—Industries—History. I. Title. II. Series.
TA74.H3S36 1978 624'.0943'59 79-344143
79 MARC

—GREAT BRITAIN—BIOGRAPHY

Pearce, Rhoda Mary.
Thomas Telford : an illustrated life of Thomas Telford, 1757-1834 / [by] Rhoda M. Pearce. — [2nd ed.]. — [Aylesbury] : Shire Publications, [1978].
48 p. : ill., maps, plans, ports. ; 21 cm. — (Lifelines ; 10) GB78-1682
Bibliography: p. 47.
Includes index.
ISBN 0-85263-410-2 : £0.75
1. Telford, Thomas, 1757-1834. 2. Civil engineers—Great Britain—Biography.
TA140.T3P42 1978 624'.092'4 78-324489
79 MARC

—LITHUANIA—BIOGRAPHY

Gimbutas, Jurgis.
Steponas Kolupaila / Jurgis Gimbutas ir Juozas V. Danys. — Čikaga : Akademinės skautijos leidykla, 1974.
464 p. : ill., facsims., ports. ; 24 cm.
Summary in English.
Includes bibliographies.
1. Kolupaila, Steponas. 2. Civil engineers—Lithuania—Biography. I. Danys, Juozas V., joint author.
TA140.K59G55 79-370953
79 MARC

—POLAND—BIOGRAPHY

Wasiutyński, Zbigniew.
Pisma / Zbigniew Wasiutyński. — Wyd. 1. — Warszawa : Państ. Wydaw. Naukowe, 1977-
v. : ill., ports. ; 25 cm. P78-5453
At head of title, v. 1- : Polska Akademia Nauk.
Includes bibliographical references.
CONTENTS: t. 1. Życiorys, charakterystyka działalności naukowej, prace wybrane, bibliografia prac.—t. 2. O zagadnieniach optymalizacji konstrukcyj i o rozwijaniu tych zagadnień.

zł95.00 (v. 1) varies
1. Wasiutyński, Zbigniew. 2. Structural engineering—Addresses, essays, lectures. 3. Civil engineers—Poland—Biography.
TA140.W27A36 79-369641
79 MARC

—SPAIN—BIOGRAPHY

Martín Gaite, Carmen.
El conde de Guadalhorce, su época y su labor / [por] Carmen Martín Gaite. — Madrid : Colegio de Ingenieros de Caminos, Canales y Puertos, 1977.
285 p., [7] leaves of plates : ill. ; 22 cm. — (Colección Ciencias, humanidades e ingeniería ; 4)
Bibliography: p. 279-280.
ISBN 8438000231 : 500ptas
1. Benjumea y Burín, Rafael, conde de Guadalhorce. 2. Spain—Public works—History. 3. Civil engineers—Spain—Biography. 4. Cabinet officers—Spain—Biography. I. Title. II. Series.
HD4243.B46M37 79-346870
79 MARC

—TAJIKISTAN—BIOGRAPHY

Vishnevskiĭ, Aleksandr Íàkovlevich.
(Roħsoz)
Роҳсоз : дар бораи ҳаёт ва фаъолияти В. Ҷ. Карамов / А. Я. Вишневский. — Душанбе : Ирфон, 1966.
62 p. : port. ; 17 cm.
Includes bibliographical references.
1. Karamov, Vagarshak Javodovich, 1901-1956. 2. Civil engineers—Tajikistan—Biography. I. Title.
TA140.K37V57 79-971063

—UKRAINE—BIOGRAPHY

(Fedor Pavlovich Beli͡ankin)
Федор Павлович Белянкин / [вступ. статья и указатель печатных работ А. И. Стрельбицкой]. — Киев : Наук. думка, 1978.
84 p. ; 17 cm. — (Біобібліографія вчених Української РСР) USSR***
At head of title: Akademii͡a nauk USSR.
Series romanized: Biobibliohrafii͡a vchenykh Ukraïns'koï RSR.
0.07rub
1. Bi͡eli͡ankin, Fedor Pavlovych. 2. Civil engineers—Ukraine—Biography. I. Strel'bit͡skai͡a, Aleksandra Ivanovna.
TA140.B48F42 78-422940

CIVIL GOVERNMENT see Political science

CIVIL IMPROVEMENT see Cities and towns —Civil improvement

CIVIL LAW

see also Conditions (Law); Duress (Law); Infamy (Law); Liability (Law); Nullity; Obligations (Law); Reconciliation (Law); Roman laws; Unjust enrichment; Usufruct

'Abd Allāh, Sāmī.
(Naẓarīyat al-ṣūrīyah fī al-qānūn al-madanī)
نظرية الصورية في القانون المدني : دراســـة مقارنة / سامي عبد الله . ـــ الطبعة 1. ـــ بيروت : 1977, [s. n.]
568 p. ; 24 cm.
Bibliography: p. 539-558.
1. Civil law. I. Title.
K623.A2 78-971006

Baudry-Lacantinerie, Gabriel, 1837-1913.
Précis de droit civil, contenant dans une première partie l'exposé des principes et dans und deuxième les questions de détail et les controverses, par G. Baudry-Lacantinerie ... 14. éd., conforme au programme des facultés de droit refondue et mise au courant de la législation ... Par Paul Guyot ... Paris, Société anonyme du Recueil Sirey, 1926-
v. 23 cm.
CONTENTS: t. 1. Contenant les matières exigées pour l'examen de première année de la licence en droit.
1. Civil law. I. Guyot, Paul, ed. II. Title.
27-374
[r79]rev MARC

Donisi, Carmine.
Ricerche di diritto civile / Carmine Donisi. — [s.l.] : Edizioni scientifiche italiane, [1979]
280 p. ; 24 cm. It***
Includes bibliographical references.
L11000
1. Civil law—Italy. 2. Civil law. I. Title.
346'.45 79-373550
79 MARC

Ferid, Murad.
Gutachten zum internationalen und ausländischen Privatrecht, 1976 / veröffentlicht im Auftrage des Deutschen Rates für Internationales Privatrecht von Murad Ferid, Gerhard Kegel, Konrad Zweigert ; Redaktion Hilmar Krüger. — Tübingen : J. C. B. Mohr, 1977.
x, 630 p. ; 24 cm. — (Materialien zum ausländischen und internationalen Privatrecht ; 28 ISSN 0543-0194) GFR***
Includes bibliographical references and indexes.
ISBN 3-16-640392-2
1. Conflict of laws. 2. Civil law. I. Kegel, Gerhard, joint author. II. Zweigert, Konrad, 1911- joint author. III. Title. IV. Series.
K7040.F452 340.9 79-356284
79 MARC

Karakatsanēs, Iōannēs K
(Zētēmata astikou dikaiou)
Ζητήματα ἀστικοῦ δικαίου : ἀστικόν δίκαιον καὶ μεθηδολογία τοῦ δικαίου / Ἰωάννου Κ. Καρακατσάνη. — Ἀθῆναι : Ἀ. Ν. Σάκκουλα, 1978.
302 p. ; 25 cm.
Bibliography: p. [287]-294.
Includes index.
1. Civil law—Greece, Modern. 2. Civil law. I. Title.
79-411616

Khristoforov, Veselin G
Grazhdansko i tŭrgovsko pravo na kapitalisticheskite dŭrzhavi / Veselin Khristoforov. — Sofii͡a : Nauka i izkustvo, 1979.
447 p. ; 22 cm. Bu***
Bibliography: p. 436-437.
1.32lv
1. Civil law. 2. Commercial law. I. Title.
K623.K57 79-376000
79 MARC

(Minpŏp yŏnsŭp)
民法演習 / 執筆代表金曾漢, 李英俊 ; 執筆者賈在桓 ... [et al.]. — 서울 : 一潮閣, 1976-
v. ; 22 cm. — (民法演習시리이즈 ; 1-
Colophon inserted.
Includes bibliographical references.
CONTENTS: v. 1. 總則編.
₩2000 (v. 1)

1. Civil law—Korea. 2. Civil law. I. Kim, Chŭng-han, 1920- II. Yi, Yŏng-jun, 1938- III. Ka, Chae-hwan.
78-826282

Montes Panades, Vicente.
Apuntes de derecho civil sobre contratos en particular / [por] Vicente Montes Panades. — [Oviedo : Universidad, Servicio de Publicaciones], D.L. 1976.
121 p. ; 29 cm. Sp77-Sept
160ptas
1. Contracts—Spain. 2. Civil law—Spain. 3. Contracts. 4. Civil law. I. Title.
346'.46'02 79-340472
79 MARC

Preuss, Ulrich Klaus, 1939-
Die Internalisierung des Subjekts : zur Kritik d. Funktionsweise d. subjektiven Rechts / Ulrich K. Preuss. — 1. Aufl. — Frankfurt am Main : Suhrkamp, 1979.
355 p. ; 20 cm. — (Theorie) GFR79-A
Bibliography: p. 333-[351]
Includes index.
ISBN 3-518-06032-5
1. Sociological jurisprudence. 2. Law—Philosophy. 3. Civil law. 4. Public law. I. Title.
K372.P73 340.1 79-379035
79 MARC

Sancho Rebullida, Francisco de Asís.
Estudios de derecho civil / Francisco de Asís Sancho Rebullida. — Pamploma : Ediciones Universidad de Navarra, 1978.
2 v. ; 22 cm. Sp***
Includes bibliographical references.
ISBN 8431305363
1. Civil law—Spain. 2. Civil law. I. Title.
346'.46 79-354282
79 MARC

Sofwan, Sri Soedewi Masjchun.
Asas[2] hukum perdata : tahun 1958/1959 Fakultas Ekonomi / Machun Sofwan. — Jogja : Jajasan Badan Penerbit Gadjah Mada, [1959?]
100 p. ; 20 cm.
"Tidak diperdagangkan."
1. Civil law—Indonesia. 2. Civil law. I. Title.
77-986626
78[r79]rev MARC

Stewering, Hans Jürgen, 1947-
Die Anwendung öffentlichen Rechts durch Private / vorgelegt von Hans Jürgen Stewering. — [s.l. : s.n.], 1977, cover 1978.
vii, 202 p. ; 21 cm. GFR***
Thesis—Münster.
Vita.
Bibliography: p. 175-202.
1. Public law—Germany, West. 2. Civil law—Germany, West. 3. Public law. 4. Civil law. 5. Jurisprudence. I. Title.
340'.0943 79-377561
79 MARC

Wagner, Herbert, 1920-
Studien zur allgemeinen Rechtslehre des Gaius (ius gentium und ius naturale in ihrem Verhältnis zum ius civile) / Herbert Wagner. — Zutphen : Terra Pub. Co., 1978.
xxiv, 290 p. ; 29 cm. — (Studia Amstelodamensia ad epigraphicam, ius antiquum et papyrologicam pertinentia ; 15) Ne***
A revision and enlargement of the author's thesis, Humboldt Universität, Berlin, 1960, presented under the title: Ius gentium und ius naturale in ihrem Verhältnis zum ius civile bei Gaius.
Includes bibliographical references and indexes.
ISBN 9062552080
1. Gaius. 2. Jus gentium (Roman law) 3. International law. 4. Natural law. 5. Civil law. I. Title. II. Series.
340 79-367547
79 MARC

—ADDRESSES, ESSAYS, LECTURES

Modern polgári jogelméleti tanulmányok / szerk. Varga Csaba ; az előszót írta, Peschka Vilmos. — Budapest : Magyar Tudományos Akadémia Állam- és Jogtudományi Intézete, 1977.
145 p. ; 23 cm. H***
Translation of articles originally published in English and German.
Includes indexes.
1. Civil law—Addresses, essays, lectures. 2. Law—Philosophy—Addresses, essays, lectures. I. Varga, Csaba.
K623.Z9M6 79-350305
79 MARC

CIVIL LAW (Continued)

—BIBLIOGRAPHY

Soyke, Peter.
Bibliographie des deutschen Schrifttums zum internationalen und ausländischen Privatrecht : 1945-1970 / bearb. von Peter Soyke u. Eckart Wehser. — Berlin : Duncker und Humblot, 1978.
xliii, 616 p. ; 24 cm. — (Studien aus dem Institut für Internationales Recht an der Universität Kiel ; 5) GFR79-A
Includes index.
ISBN 3-428-04111-9 : DM56.00
1. Conflict of laws—Bibliography. 2. Civil law—Bibliography. I. Wehser, Eckart, joint author. II. Title. III. Series: Kiel. Universität. Institut für Internationales Recht. Studien aus dem Institut für Internationales Recht an der Universität Kiel ; 5.
K7000.S69 016.3409 79-347718
79 MARC

—COLLECTED WORKS

Katsumoto, Masaakira, 1895-
(Mimpō kenkyū)
民法研究 / 勝本正晃著. — 東京：巖松堂書店, 昭和7- [1932-
v. ; 22 cm.
Includes bibliographical references and indexes.

1. Civil law—Japan—Collected works. 2. Civil law—Collected works. I. Title.
79-805745

—HISTORY

Schlosser, Hans, 1934-
Grundzüge der neueren Privatrechtsgeschichte : e. Studienbuch / Hans Schlosser. — 3., völlig überarb. u. erw. Aufl. des von Erich Molitors begr. Werkes. — Heidelberg ; Karlsruhe : Müller, Juristischer Verl., 1979.
xiv, 177 p. ; 19 cm. — (Uni-Taschenbücher ; 882 : Rechtswissenschaft) GFR79-A
Includes bibliographies and indexes.
ISBN 3-8114-7178-3 : DM16.80
1. Civil law—History. I. Molitor, Erich, 1886-1963. Grundzüge der neueren Privatrechtsgeschichte. II. Title.
K623.S33 1979 79-369938
79 MARC

—INTERNATIONAL UNIFICATION —CONGRESSES

International Institute for the Unification of Private Law.
UNIDROIT proceedings and papers, 1828 [i. e. 1928]–1966. — [Dobbs Ferry, N. Y. : Oceana Publications, 1978?]
30 reels ; 35 mm.
Title from case.
1. Civil law—International unification—Congresses. I. Title. II. Title: Unidroit proceedings and papers.
Microfilm LL-2145K 79-335024
[K614]

—OUTLINES, SYLLABI, ETC.

Wu, I-hsin.
(Min fa)
民法 吳一心編 上海 中華書局 民國37 [1948]
19 p. 18 cm. (民衆教育：第一集)(中華文庫)
Colophon title.
Series 2 romanized: Chung-hua wen k'u.

1. Civil law—Outlines, syllabi, etc. I. Title. II. Series: Min chung chiao yü (Shang-hai) : Ti i chi.
K623.W8 79-840925
rev

—POPULAR WORKS

Mibuchi, Tadahiko, 1880-1950.
(Nichijō seikatsu to Mimpō)
日常生活と民法 / 三淵忠彥著；關根小鄉, 和田嘉子補修. — 東京：法曹会, 昭和25 [1950]
2, 2, 3, 242 p. ; 19 cm.

1. Civil law—Japan—Popular works. 2. Civil law—Popular works. I. Sekine, Kosato, 1905- II. Wada, Yoshiko. III. Title.
79-805158

—STUDY AND TEACHING—GERMANY, WEST

Diederichsen, Uwe.
Die BGB-Klausur / von Uwe Diederichsen. — 4., neubearb. u. erw. Aufl. — München : Beck, 1978.
xvi, 212 p. ; 23 cm. — (Schriftenreihe der Juristischen Schulung ; Heft 1) GFR78-A
Bibliography: p. [xv]-xvi.
Includes indexes.
ISBN 3-406-07007-8 : DM14.80
1. Law examinations—Germany, West. 2. Civil law—Study and teaching—Germany, West. 3. Civil law—Germany, West—Examinations, questions, etc. I. Title. II. Series: Juristische Schulung : Schriftenreihe ; Heft 1.
78-392358
79 MARC

—ARGENTINE REPUBLIC

Argentin Republic. Laws, statutes, etc.
[Código civil]
Código civil y leyes complementarias : comentado, anotado y concordado / director, Augusto C. Belluscio ; coordinador, Eduardo A. Zannoni ; autores, Guillermo E. Garbino ... [et al.]. — Buenos Aires : Editorial Astrea de A. R. Depalma, 1978.
v. ; 24 cm.
Bibliography: v. 1, p. [807]-811.
CONTENTS: t. 1. Artículos lo. a 239, Arts. 159 a 239 (Ley 2393), Leyes 14.394 y 18.248.
1. Civil law—Argentine Republic. I. Belluscio, Augusto César. II. Zannoni, Eduardo A. III. Garbino, Guillermo E. IV. Title.
346'.82 78-111277
79 MARC

—AUSTRIA

Austria. Laws, statutes, etc.
Allgemeines bürgerliches Gesetzbuch : ABGB : mit den wichtigsten einschlägigen Nebengesetzen und Verordnungen / mit einer Einführung und ausführlichem Register hrsg. von Herbert Spehar. — 5., unverändert Aufl. — München : Goldmann, 1978.
559 p. ; 18 cm. — (Goldmann Gesetze ; 8203) (Ein Goldmann-Taschenbuch) GFR***
At head of title: Österreichisches Recht.
Includes bibliographical references.
ISBN 3-442-08203-X : DM12.80
1. Civil law—Austria. I. Spehar, Herbert, ed. II. Title.
346'.43 79-377909
79 MARC

Jabornegg, Peter.
Privatrecht und Umweltschutz / Beitr. v. Peter Jabornegg, Peter Rummel, Rudolf Strasser ; hrsg. von Rudolf Strasser. — Linz : Inst. f. Kommunalwissenschaften u. Umweltschutz ; Wien ; München : Jugend & Volk in Komm., c1976.
191 p. ; 24 cm. — (Kommunale Forschung in Österreich ; 26) Au78-15
Includes bibliographical references.
ISBN 3-7141-7830-9 : S220.00
1. Environmental law—Austria. 2. Civil law—Austria. 3. Environmental law—Germany, West. 4. Civil law—Germany, West. I. Rummel, Peter, 1940- joint author. II. Strasser, Rudolf, joint author. III. Title. IV. Series.
344'.43'046 78-379823
*78 MARC

Koziol, Helmut.
Grundriss des bürgerlichen Rechts / von Helmut Koziol u. Rudolf Welser. — 4., durchges. u. erg. Aufl. — Wien : Manz, 1976-1978.
2 v. ; 23 cm. — (Manzsche Kurzlehrbuch-Reihe ; 1) Au77-6-35
Bibliography: v. 1, p. [xxiii]-xxiv; v. 2, p. [xxiii]-xxiv.
Includes indexes.
CONTENTS: Bd. 1. Allgemeiner Teil und Schuldrecht.—Bd. 2. Sachenrecht, Familienrecht, Erbrecht.
ISBN 3-214-24686-0 (v. 1) : S280.00 (v. 1)
1. Civil law—Austria. I. Welser, Rudolf, joint author. II. Title.
346'.43 77-464655
*77[r79]rev MARC

Krejci, Heinz.
Konsumentenschutzgesetz, ABGB und Verfahrensrecht / von Heinz Krejci, Bernd Schilcher, Viktor Steininger. — Wien : Manz, 1978.
104 p. ; 23 cm. Au79-4-59
Includes bibliographical references and index.
ISBN 3-214-06815-6 : S146.00
1. Consumer protection—Law and legislation—Austria. 2. Civil law—Austria. I. Schilcher, Bernd, 1940- joint author. II. Steininger, Viktor, joint author. III. Austria. Laws, statutes, etc. Konsumentenschutzgesetz. 1978. IV. Title.
343'.436'07 79-367500
*79 MARC

—AUSTRIA—CONGRESSES

Österreichischer Juristentag, 6th, Innsbruck, 1976.
Der Einfluss der Strafrechtsreform auf das Zivilrecht. — Wien : Manz, [1977]
87 p. ; 23 cm. — (Verhandlungen des sechsten Österreichischen Juristentages Innsbruck 1976 ; Bd. 2, T. 4) Au78-8-50
At head of title: Privatrechtliche Arbeitsgemeinschaft.
Includes bibliographical references.
CONTENTS: Mayer-Maly, T., Rummel, P., Steininger, H. Referate.—Bydlinski, F. et al. Diskussionsbeiträge.—Mayer-Maly, T. et al. Schlussworte.
S120.00
1. Criminal law—Austria—Congresses. 2. Civil law—Austria—Congresses. I. Mayer-Maly, Theo. II. Privatrechtliche Arbeitsgemeinschaft. III. Title. IV. Series: Österreichischer Juristentag, 6th, Innsbruck, 1976. Verhandlungen ; Bd. 2, T. 4.
347'.436 s 79-365605
*79 MARC

—AUSTRIA—HISTORY

Kocher, Gernot.
Österreichische und deutsche Privatrechtsgeschichte : Entwicklung d. Rechtswissenschaft u. ausgew. Kapitel zum Personen- u. Sachenrecht / Gernot Kocher. — Graz : Böhlau, 1978.
iv p., 89, 3 leaves of plates : ill. ; 24 cm. — (Pro studio) Au78-24-51
Bibliography: leaves 1-5.
S126.00
1. Civil law—Austria—History. 2. Civil law—Germany—History. I. Title.
346'.43 79-345725
*79 MARC

—BELGIUM

Baude, Jules.
Cours élémentaire de droit civil / Jules Baude, P. A. Belvaux. — 6. éd. rev. et mise à jour. — Bruxelles, r. P. de Champagne 7 : Ced-Samson, [1977]
314 p. ; 24 cm. Be77-12
Bibliography: p. [299]
400F
1. Civil law—Belgium. I. Belvaux, Paul Alzire, joint author. II. Title.
346'.493 78-396851
*79 MARC

Belgium. Law, statutes, etc.
[Code civil. Dutch]
Burgerlijk wetboek / bewerkt door S. de Borman. — Brussel : CED-Samsom, 1977.
497 p. ; 19 x 8 cm. — (Samsoms wettenserie) Be***
"Bijgewerkt tot 1 augustus 1977."
1. Civil law—Belgium. I. Borman, S. de. II. Title.
78-386376
78 MARC

Belgium. Laws, statutes, etc.
[Code civil. Dutch & French]
Belgisch burgerlijk wetboek = Code civil belge. — 3de druk — [2500] Lier, (Grote Markt 39) : J. Van In, 19
v. ; 20 cm. Be77-8(v.
Dutch and French.
CONTENTS:
—boek 2. Goederen en verschillende beperkingen van de eigendom.—boek 3. deel 1. Erfenissen, schenkingen, testamenten.

ISBN 9030602023 (v. 3) : 117F (v. 3)
1. Civil law—Belgium. I. Title. II. Title: Code civil belge.
78-394213
79 MARC

Belgium. Laws, statutes, etc.
[Code civil. Dutch & French]
Burgerlijk wetboek met bijzondere wetten en besluiten / Willy Delva, met medewerking van G. Baert ... [et al.]. — 2de druk. — [9000] Gent, [P. van Duyseplein 8] : E. Story-Scientia, 1977.
2 v. ; 26 cm. Be77-8
Dutch and French.
Includes indexes.
5,960F
1. Civil law—Belgium. 2. Civil law—Belgium—Interpretation and construction. I. Delva, Willy. II. Title.
340'.09493 78-382703
*79 MARC

Heerman, Jos.
Burgerlijk recht : inleidende begrippen / Jos Heerman. — 3000 Leuven, (Tiensestr. 134-136) : Acco, [1977]
ix, 107 leaves ; 27 cm. Be77-12
170F
1. Civil law—Belgium. I. Title.
78-392444
*79 MARC

CIVIL LAW (Continued)

—BELGIUM—INTERPRETATION AND CONSTRUCTION

Belgium. Laws, statutes, etc.
[Code civil. Dutch & French]
Burgerlijk wetboek met bijzondere wetten en besluiten / Willy Delva, met medewerking van G. Baert ... [et al.]. — 2de druk. — [9000] Gent, [P. van Duyseplein 8] : E. Story-Scientia, 1977.
2 v. ; 26 cm. Be77-8
Dutch and French.
Includes indexes.
5,960F
1. Civil law—Belgium. 2. Civil law—Belgium—Interpretation and construction. I. Delva, Willy. II. Title.
340'.09493 78-382703
*79 MARC

—BOLIVIA—ADDRESSES, ESSAYS, LECTURES

Libro homenaje a la Corte Suprema de Justicia de Bolivia : temas relativos al acto de su instalación y períodos históricos de la vida judicial en la República : comentarios referentes a la nueva legislación civil boliviana. — Sucre : Corte Suprema de Justicia de la Nación, 1977.
248 p. ; 22 cm. — (Biblioteca jurídica ; v. 2)
Includes bibliographical references.
CONTENTS: Abecia Baldivieso, V. La obra legislativa del mariscal Santa Cruz.—Condarco Morales, R. Administración y Estado antes de 1825.—Durán Padilla, M. Teoría y realidad del poder judicial.—Frerking Salas, O. El depósito, recensión de sus reformas en el nuevo Código civil boliviano.—Gantier, J. Manuel María Urcullu.—Mendieta, J. Breve historia de la Excma. Corte Suprema de Justicia.—Murillo Vacareza, J. Apuntes para la historia de la fundación de la Corte Superior de Justicia de Oruro.—Romero Linares, R. Reglas que gobiernan el título "De la posesión": arts. 87 al 93 del nuevo Código civil boliviano.—Tovar Gutzlaff, C. Dos rasgos en el nuevo Código de procedimiento civil.—Valencia Vega, A. La constitución bolivariana: su incomprensión y su destrucción.
1. Justice, Administration of—Bolivia—Addresses, essays, lectures. 2. Civil law—Bolivia—Addresses, essays, lectures. 3. Bolivia. Corte Suprema. I. Bolivia. Corte Suprema. II. Series: Biblioteca jurídica (Sucre) ; v. 2.
347'.84 78-382704
78 MARC

—BRAZIL

Brazil. Laws, statutes, etc.
[Código civil]
Código civil brasileiro : [texto completo, rigorosamente atualizado, com todas as alterações]. — Rio de Janeiro : Gráfica Auriverde, 1978.
414 p. ; 18 cm.
Cr$60.00
1. Civil law—Brazil. I. Title.
346'.81'002632 78-384808
78 MARC

Brazil. Laws, statutes, etc.
[Código civil]
Código civil brasileiro : Lei 3.071, de 1º de janeiro de 1916 : legislação complementar / Atualização com glosas e notas por Nirval Garcia da Silva. — 1. ed. — Rio de Janeiro : Forense, 1978.
xii, 680 p. ; 22 cm.
Includes bibliographical references and indexes.
Cr$300.00
1. Civil law—Brazil. I. Silva, Nirval Garcia da. II. Title.
340'.0981 78-379421
78 MARC

Dower, Nélson Godoy Bassil.
Curso moderno de direito civil / Nélson Godoy Bassil Dower ; pref. do Prof. Jorge Pereira de Andrade. — São Paulo : Editora Nelpa, [19
v. ; 24 cm.
Bibliography: v. 2, p. [321]-323.
CONTENTS:
—v. 2. Parte geral das obrigações.

Cr$300.00 (v. 2)
1. Civil law—Brazil. I. Title.
346'.81 78-384810
78 MARC

Levenhagen, Antônio José de Souza.
Código civil : parte geral : comentários didáticos / Antônio José de Souza Levenhagen. — 1. ed. — São Paulo : Editora Atlas, 1978.
268 p. ; 22 cm.
Includes text of the Código civil, parte geral.
Cr$180.00
1. Civil law—Brazil. I. Brazil. Laws, statutes, etc. Código civil. Parte geral. 1978. II. Title.
346'.81 79-340014
79 MARC

Levenhagen, Antônio José de Souza.
Código civil à luz da nova lei processual / Antônio José de Souza Levenhagen. — 5. ed., rigorosamente atualizada com a Lei no. 6.458, de 1.o/11/1977 (Duplicatas), Lei no. 6.015, dos Registros Públicos, e Lei no. 6.515, de 26/12/1977 (Regulamentadora do divórcio). — São Paulo : Editora Atlas, 1978.
384 p. ; 21 cm.
Includes index.
Cr$200.00
1. Civil law—Brazil. 2. Civil procedure—Brazil. I. Title.
347'.81'05 79-342579
79 MARC

—BRAZIL—CODIFICATION

Freitas, Augusto Teixeira de, 1816-1883.
Codificação do direito civil : carta de 20 de setembro de 1867 ao Ministro da Justiça : publicação comemorativa do Sesquicentenário dos Cursos Jurídicos, 1827-1977. — Rio de Janeiro : Ministério da Justiça, Arquivo Nacional, [1977]
12 p. ; 21 cm. — (Publicações históricas ; 1a. sér., no. 81)
1. Civil law—Brazil—Codification. I. Title. II. Series.
346'.81 79-347721
79 MARC

—BRAZIL—PERIODICALS

Revista de direito civil. ano 1–
julho/set. 1977–
[São Paulo, Editora Revista dos Tribunais]
v. 23 cm. quarterly.
"Imobiliário, agrário e empresarial."
1. Civil law—Brazil—Periodicals.
K19.D578 346'.81 78-648745
MARC-S

—CHILE

Alessandri Rodríguez, Arturo, 1895-
Curso de derecho civil : explicaciones basadas en las clases de los profesores de la Universidad de Chile, Arturo Alessandri Rodríguez y Manuel Somarriva Undurraga / redactadas, ampliadas y puestas al día por Antonio Vodanovic H. — 4. ed. — Santiago : Editorial Nascimento, 1971, spine 1972-
v. in ; 26 cm.
Includes bibliographies.
CONTENTS: [1] Parte general y los sujetos de derecho. 2 v.
1. Civil law—Chile. I. Somarriva Undurraga, Manuel, 1905- joint author. II. Vodanovic H., Antonio. III. Title.
78-101621
79 MARC

Alessandri Rodríguez, Arturo, 1895-
Parte general y los sujetos de derecho : explicaciones basadas en las clases de los profesores de la Universidad de Chile, Arturo Alessandri Rodríguez y Manuel Somarriva Undurraga / redactadas, ampliadas y puestas al día por Antonio Vodanovic H. — 4. ed. — Santiago : Editorial Nascimento, 1971, spine 1972.
2 v. ; 26 cm. — (His Curso de derecho civil ; [1])
Includes bibliographies.
1. Civil law—Chile. I. Somarriva Undurraga, Manuel, 1905- joint author. II. Vodanovic H., Antonio. III. Title.
78-389914
79 MARC

Chile. Laws, statutes, etc.
Código civil : edición oficial, al 31 de agosto de 1976, aprobada por Decreto no. 1.937, de 29 de noviembre de 1976 del Ministerio de Justicia. — 7. ed. — [Santiago de Chile] : Editorial Jurídica de Chile, [1977]
1075 p. ; 19 cm.
"Decreto ley no. 1.773" ([2] p.) inserted.
"Apéndice del Código civil" (p. [581]-1052) contains related legislation.
Includes bibliographical references.
1. Civil law—Chile.
346'.83 78-396850
79 MARC

—CHINA

Shao, Hsi.
(Min lü shih i)
民律釋義 / 編輯者邵羲；校勘者徐學詢；閲訂者徐元誥. — 上海：中華書局，民國6 [1917]
8, 662, 424 p. ; 22 cm.
Colophon title.

1. Civil law—China. I. Hsü, Hsüeh-hsün. II. Title.
78-842418

—COLOMBIA

Colombia. Laws, statutes, etc.
Código civil : con notas, concordancias, jurisprudencia de la Corte Suprema y normas legales complementarias / por Jorge Ortega Torres. — 12. ed., actualizada. — Bogotá : Editorial Temis, 1977.
1684 p. ; 21 cm.
Errata slip inserted.
Includes index.
1. Civil law—Colombia. I. Ortega Torres, Jorge. II. Colombia. Corte Suprema de Justicia.
346'.861 78-107361
79 MARC

—COLOMBIA—CASES

Colombia. Corte Suprema de Justicia.
Jurisprudencia civil, 1977-1978, de la Corte y Tribunales / [recopilación de] Jairo López Morales. — Bogotá : Ediciones Lex, [1979-
v. ; 22 cm.
Includes index.
1. Civil law—Colombia—Cases. I. López Morales, Jairo. II. Colombia. Tribunales Superiores. III. Title.
346'.861 79-109068
79 MARC

—DENMARK

Munch-Petersen, Hans Vilhelm, 1869-1934.
Borgerlig ret / Munch Petersen. — 23. reviderede udgave / ved W. E. v. Eyben, A. Vinding Kruse. — København : Gad, 1978.
2 v. ; 23 cm. D78-43
Previous editions published under title: Den borgerlige ret i hovedtræk.
Includes bibliographical references and indexes.
CONTENTS: bd. 1. Indledning til formueretten.—bd. 2. Person- familie- og arveret.
ISBN 8712588881 (v. 1) : kr135.70 (v. 1)
1. Civil law—Denmark. 2. Commercial law—Denmark. I. Eyben, William Edler von, 1912- II. Kruse, Anders Vinding. III. Title.
79-337518
79 MARC

—DENMARK—OUTLINES, SYLLABI, ETC.

Ziirsen, Marianne.
Hverdagens jura. — 4. udg., 1. oplag / af Marianne Ziirsen og Knud-Erik Ziirsen ; i samarbejde med Ella Blousgaard og Karen Zimsen. — [København] : A.O.F. : Fremad, 1977.
202 p. ; 20 cm. D77-36
Includes index.
ISBN 8774031090 : kr42.00
1. Civil law—Denmark—Outlines, syllabi, etc. I. Ziirsen, Knud-Erik, joint author. II. Title.
340'.09489 78-382770
79 MARC

—ECUADOR

Parraguez Ruíz, Luis.
Manual de derecho civil ecuatoriano / Luis Parraguez Ruíz. — [s.l. : s.n., 1977?]-
v. ; 21 cm.
Bibliography: v. 1, p. 315-319.
CONTENTS: v. 1. Personas y familia.
1. Civil law—Ecuador. I. Title.
346'.866 78-108678
79 MARC

—EUROPE

Elmelund, Nils.
Civilretlig forældelse i Vesteuropa / [af] Nils Elmelund. — [København] : Schul[t]z, 1978.
xxvii, 477 p. ; 24 cm. — (Schultz erhver[v]sbøger)) D79-8
Bibliography: p. 415-421.
Includes bibliographies.
ISBN 8756904738 : kr395.00
1. Civil law—Europe. 2. Limitation of actions—Europe. 3. Limitation of actions—Europe—Cases. I. Title. II. Series: Schultz erhvervsbøger.
79-375965
79 MARC

Gotō, Kiyoshi, 1902-
(Mimpōgaku josetsu)
民法學序說 / 後藤清著. — 東京：弘文堂書房，昭和11 [1936]
4, 6, 243 p. ; 23 cm.

1. Civil law—Europe. 2. Civil law—Japan. I. Title.
79-803273

CIVIL LAW
—EUROPE (Continued)

Hecke, Georges A van, 1915-
De invloed van het Westerse recht in de wereld / door G. van Hecke. — Brussel : Paleis der Academiën, 1978.
11 p. ; 27 cm. — (Mededelingen van de Koninklijke Academie voor Wetenschappen, Letteren en Schone Kunsten van België, Klasse der Letteren ; jaarg. 40, nr. 4) Be***
Summary in English and French.
Includes bibliographical references.
1. Civil law—Europe. 2. Common law. 3. Customary law. 4. Law. I. Title. II. Series: Vlaamse Academie voor Wetenschappen, Letteren en Schone Kunsten van België. Klasse der Letteren. Mededelingen ; jaarg. 40, nr. 4.
AS242.V6A25 jaarg. 40, nr. 4 78-395290
79 MARC

Mitsuma, Shinzō, 1879-1937.
(Mimpō sōsoku teiyo)
民法總則提要 / 三潴信三著. — 全訂. — 東京 : 有斐閣, 昭和6 [1931]
2 v. (5, 19, 549 p.) ; 23 cm.
Includes bibliographical references.
1. Civil law—Japan. 2. Civil law—Europe. I. Title.
79-801759

Wolf, A L de.
Veranderde omstandigheden / door A. L. de Wolf. — Zwolle : W. E. J. Tjeenk Willink, 1979.
vi, 118 p. ; 22 cm. — (Studiepockets privaatrecht ; nr. 18) Ne79-8
Includes bibliographical references.
ISBN 9027115109 : fl 17.00
1. Civil law—Europe. 2. Contracts—Europe. 3. Rebus sic stantibus clause—Europe. I. Title.
79-374874
*79 MARC

—EUROPE—ADDRESSES, ESSAYS, LECTURES

Kawashima, Takeyoshi, 1909-
(Mimpō kaishakugaku no shomondai)
民法解釋學の諸問題 / 川島武宜著. — 東京 : 弘文堂, 昭和24 [1949]
15, 275 p. ; 22 cm.
Colophon inserted.
Includes bibliographical references.
1. Civil law—Europe—Addresses, essays, lectures. 2. Civil law—Japan—Addresses, essays, lectures. I. Title.
79-804651
J

—EUROPE—HISTORY

Harada, Keikichi, 1903-1950.
(Nihon mimpōten no shiteki sobyō)
日本民法典の史的素描 / 原田慶吉著 ; 石井良助編. — 東京 : 創文社, 昭和29 [1954]
9, 3, 396, 11 p. ; 22 cm.
Includes bibliographical references and index.
1. Civil law—Japan—History. 2. Civil law—Europe—History. I. Ishii, Ryōsuke, 1907- II. Title.
79-803281
J

—EUROPE, EASTERN

Világhy, Miklós.
Gazdaságpolitika és polgári jog / Világhy Miklós. — Budapest : Akadémiai Kiadó, 1978.
120 p. ; 25 cm. H***
ISBN 9630515040 : 37.00Ft
1. Civil law—Europe, Eastern. 2. Europe, Eastern—Economic policy. 3. Sovet ėkonomicheskoĭ vzaimopomoshchi. I. Title.
79-380492
79 MARC

—EUROPEAN ECONOMIC COMMUNITY COUNTRIES

Semaine sociale wallonne, 59th, 1977.
Les travailleurs, la justice et le droit : [59ème Semaine sociale wallonne] / Battistoni, L. ... [et al.]. — Bruxelles, av. Van Volxem 305 : Vie ouvrière, 1977.
380 p. ; 18 cm. — (Collection Semaines sociales wallonnes) Be78-6
Errata slip inserted.
Includes bibliographical references.
ISBN 2-87003-119-X : 431F
1. Social legislation—European Economic Community countries. 2. Civil law—European Economic Community countries. 3. Labor laws and legislation—European Economic Community countries. I. Battistoni, L. II. Title.
344'.4'01 78-396039
*79 MARC

—FINLAND

Individen, företaget och samhället : inledning till den finska privaträtten / red. av Erkki Aurejärvi. — Helsinki : Juridica, 1978 (Helsinki : Monistuspalvelu)
2, x, 755 p. : ill. ; 22 cm. Fi78-5
Translation of Yksilö, yritys ja yhteiskunta.
Bibliography: p. 750-755.
ISBN 9519464018
1. Civil law—Finland. I. Aurejärvi, Erkki.
79-338826
79 MARC

—FRANCE

Carbonnier, Jean.
Droit civil / Jean Carbonnier. — Paris : Presses universitaires de France, 1969- c1955-1956.
v. ; 18 cm. — (Thémis : Droit) F***
Includes earlier editions of some volumes.
Includes bibliographical references and indexes.
CONTENTS: 1. Introduction. Les personnes. éd. 19 —2. La famille. Les incapacités. éd. 19 —3. Les biens. éd. 19 —4. Les obligations. éd. 19
1. Civil law—France. I. Title.
346'.44 70-401171
78[r79]rev2 MARC

Chevallier, Jean.
Droit civil : 1re année / Jean Chevallier, Louis Bach. — 7. éd. — [Paris] : Sirey, 1978-
v. ; 23 cm. — (Cours élémentaire-droit-économie) F***
CONTENTS: t. 1. Introduction à l'étude du droit, les personnes physiques, la famille, les biens, les obligations, les sûretés.
ISBN 2-248-00228-9 (v. 1) : 62.00F (v. 1)
1. Civil law—France. I. Bach, Louis, joint author. II. Title.
346'.44 78-390009
79 MARC

Guiho, Pierre.
Cours de droit civil / Pierre Guiho, ... — Lyon : Éditions l'Hermes, 1978-
v. ; 20 cm. — (Droit, économie, gestion : Bibliothèque ISSN 0181-5873) F78-1864
Bibliography: p. v. 1, p. [245]-248.
Includes indexes.
CONTENTS: 1. Introduction générale.—2. Les personnes.
ISBN 2-85934-053-X (v. 1) : 30.00F (v. 1)
1. Civil law—France. I. Title. II. Series.
346'.44 79-358119
79 MARC

Lambert, Guy, professeur.
Cours de droit civil / Guy Lambert, ... Gérard Barberoux, ... — Aix-en-Provence (12 A, rue de Nazareth, 13100) : Librairie de l'Université, 1977-
v. ; 30 cm. F78-64
Includes bibliographical references.
CONTENTS: Livre 1. Introduction générale au droit. Droit des personnes.
95.00F (v. 1)
1. Civil law—France. I. Barberoux, Gérard, joint author. II. Title.
340'.0944 78-375592
78 MARC

Lemeunier, Francis.
Principes et pratiques du droit civil [par] Francis Lemeunier, avec la collaboration du Centre international du droit des affaires. [4. éd. Paris] J. Delmas [1973]
1 v. (various pagings) 27 cm. (Ce qu'il vous faut savoir. Collection française) 30.00F F***
"Planning pour la préparation des examens" ([4] p.) and "Supplément: Tableaux synoptiques à l'usage des étudiants" ([32] p.) inserted in pocket.
Includes bibliographies.
ISBN 2-7034-0005-5
1. Civil law—France. I. Centre international du droit des affaires. II. Title.
346'.44 74-162086
74[r79]rev MARC

Lemeunier, Francis.
Principes et pratiques du droit civil / Francis Lemeunier, ... — 6e éd. à jour au 15 octobre 1977. — Paris : J. Delmas, 1977.
1 v. (various pagings) ; 27 cm. — (Ce qu'il vous faut savoir : Collection française) F78-4996
Includes bibliographies and index.
ISBN 2-7034-0108-6 : 69.00F
1. Civil law—France. I. Title. II. Title: Droit civil.
346'.44 79-352260
79 MARC

Mazeaud, Henri, 1898-
Leçons de droit civil [par] Henri et Léon Mazeaud [et] Jean Mazeaud. 2. éd. Paris, Montchrestien [1959-
v. in 25 cm.
Tome 4, v. 2 by M. de Juglart.
Kept up to date by supplements.
CONTENTS: t. 1. Nouvelle licence, première année. Introduction à l'étude du droit privé: droits, preuves, personnes, famille, incapacités. t. 3. Licence en droit, troisième année, Sûretés. Publicité foncière. Principaux contrats.—t. 4. Licence en droit, quatrième année. 1. t. Régimes matrimoniaux. 2. t. Successions. Libéralités.
1. Civil law—France. I. Mazeaud, Léon, 1900-1970, joint author. II. Mazeaud, Jean, joint author. III. Juglart, Michel de, ed. IV. Title.
63-58128
[r77]rev MARC

Peyrin, Annick.
Droit civil / Annick Peyrin. — [Paris] : Ministère de l'économie et des finances, Direction du personnel & des services généraux, Centre de formation professionnelle et de perfectionnement, 1976, 1977 printing.
2 v. ; 30 cm. F***
1. Civil law—France. I. France. Ministère de l'économie et des finances. Centre de formation professionnelle et de perfectionnement. II. Title.
346'.44 79-342819
79 MARC

Weill, Alex.
Droit civil / Alex Weill et François Terré. — 4. éd. — Paris : Dalloz, 197
v. ; 18 cm. — (Précis Dalloz : Droit) F***
Includes index.
CONTENTS:
—[2] Les personnes, la famille, les incapacités.
ISBN 2-247-01127-6 (v. 2) : 52.00F (v. 2)
1. Civil law—France. I. Terré, François, joint author. II. Title.
346'.44 78-394920
79 MARC

—FRANCE—DICTIONARIES

Répertoire de droit civil / Dalloz ; publié sous la direction de Pierre Raynaud, ... ; secrétaire de la rédaction, Marguerite Vanel, ... — 2e éd. — Paris : Dalloz, 19
v. ; 29 cm. — (Encyclopédie juridique) F75-2684
Loose-leaf for updating.
Kept up to date by replacement pages issued irregularly and by an annual "mise à jour."
Includes bibliographies and indexes.
CONTENTS:
t. 5. Médecine-Privilèges du Trésor.—t. 6. Promesse de vente-Sports.
ISBN 2-247-01512-3 (v. 5) : 400F per vol.
1. Civil law—France—Dictionaries. I. Raynaud, Pierre, 1910- . II. Vanel, Marguerite. III. Dalloz, firm, publishers, Paris. IV. Series.
346'.44'003 79-353986
79 MARC

—GERMANY

Schmidt, Folke Fredrik, 1909-
The German abstract approach to law; comments on the system of the Bürgerliches Gesetzbuch, by Folke Schmidt. [Stockholm, Almqvist & Wiksell, 1965]
134-158 p. 23 cm. (Studia juridica Stockholmiensia, 23)
Reprint from Scandinavian studies in law, 1965.
Includes bibliographical references.
1. Civil law—Germany. I. Title. II. Title: Comments on the system of the Bürgerliches Gesetzbuch. III. Series.
340'.0943 78-324380
79 MARC

Vangerow, Karl Adolph von, 1808-1870.
Lehrbuch der Pandekten / von Karl Adolph von Vangerow. — Marburg : R. G. Elwert, 1851-
3 v. ; 21 cm.
Previously published under title: Leitfaden für Pandekten-Vorlesungen.
Includes earlier ed. of some volumes.
Vol. 1 Aufl. ; v. 2 Aufl.
Includes bibliographical references.
1. Roman law. 2. Civil law—Germany. I. Title.
340.5'4 30-12893
MARC

—GERMANY—CODIFICATION

Fehrenbach, Elisabeth.
Traditionale Gesellschaft und revolutionäres Recht ; d. Einf. d. Code Napoléon in d. Rheinbundstaaten / von Elisabeth Fehrenbach. — 2. Aufl. — Göttingen : Vandenhoeck und Ruprecht, 1978.
226 p. ; 24 cm. — (Kritische Studien zur Geschichtswissenschaft ; Bd. 13) GFR78-A12
Habilitationsschrift—Giessen, 1973.
Bibliography: p. 214-223.
Includes index.
DM48.00
1. Civil law—Germany—Codification. 2. Rhine, Confederation of the, 1806-1813. I. Title. II. Series.
79-337211

CIVIL LAW (Continued)

—GERMANY—HISTORY

Kocher, Gernot.
Österreichische und deutsche Privatrechtsgeschichte : Entwicklung d. Rechtswissenschaft u. ausgew. Kapitel zum Personen- u. Sachenrecht / Gernot Kocher. — Graz : Böhlau, 1978.
iv p., 89, 3 leaves of plates : ill. ; 24 cm. — (Pro studio) Au78-24-51
Bibliography: leaves 1-5.
S126.00
1. Civil law—Austria—History. 2. Civil law—Germany—History. I. Title.
346'.43 79-345725
*79 MARC

Mitteis, Heinrich, 1889-1952.
Deutsches Privatrecht : e. Studienbuch / von Heinrich Mitteis ; neubearb. von Heinz Lieberich. — 8., durchges. u. erg. Aufl. — München : Beck, 1978.
ix, 190 p. ; 23 cm. — (Juristische Kurz-Lehrbücher) GFR78-A
Includes bibliographical references and index.
ISBN 3-406-06555-4
1. Civil law—Germany—History. I. Lieberich, Heinz, 1905- II. Title.
346'.43 79-338387
79 MARC

—GERMANY—HISTORY—ADDRESSES, HISTORY—ADDRESSES, ESSAYS,

Kanazawa, Masayasu, 1903–1948.
(Hōseishi ronshū)
法制史論集 / 金澤理康著. — 東京：成文堂, 1978.
5, 347 p., ⌊1⌋ leaf of plates : port. ; 22 cm. Ja 78-20420
Includes bibliographical references.
CONTENTS: 日本法制史編：我利息法に対する支那法の影響　我国中世の損害賠償制度　大名の封建負担　明治初期に於ける消費貸借法の変遷　氏名と姓名—西洋法制史論：独逸中世の損害賠償制度　独逸古法ザクセン・シュピーゲルに現はれたる封建制度　独逸新世襲農園法論
¥5000
1. Civil law—Japan—History—Addresses, essays, lectures. 2. Civil law—Germany—History—Addresses, essays, lectures. 3. Feudal law—History—Addresses, essays, lectures. I. Title.
78-816041

—GERMANY, EAST

Germany (Democratic Republic, 1949-). Laws, statutes, etc.
⌊Zivilgesetzbuch⌋
Zivilgesetzbuch der Deutschen Demokratischen Republik sowie angrenzende Gesetze und Bestimmungen : Textausgabe mit Anmerkungen und Sachregister / hrsg. vom Ministerium der Justiz ; ⌊zusammengestellt und bearb. von Carmen Rietz und Ruth Wüstneck⌋. — 1. Aufl. — Berlin : Staatsverlag der Deutschen Demokratischen Republik, 1978.
218 p. ; 24 cm. GDR***
Includes index.
6.50M
1. Civil law—Germany, East. I. Rietz, Carmen. II. Wüstneck, Ruth. III. Title.
346'.431 79-343718
79 MARC

Das Neue Zivilrecht der DDR nach dem Zivilgesetzbuch von 1975 / Klaus Westen (Hrsg.). — Baden-Baden : Nomos-Verlagsgesellschaft, in Komm., 1977.
318 p. ; 23 cm. — (Rechtswissenschaftliche Veröffentlichungen - Osteuropa-Institut an der Freien Universität Berlin ; Bd. 7) GFR78-A9
Includes index.
ISBN 3-7890-0329-8
1. Civil law—Germany, East. 2. Civil procedure—Germany, East. I. Westen, Klaus. II. Series: Berlin. Freie Universität. Osteuropa-Institut. Rechtswissenschaftliche Veröffentlichungen - Osteuropa-Institut an der Freien Universität Berlin ; Bd. 7.
346'.431 78-390606
79 MARC

—GERMANY, WEST

Brox, Hans.
Allgemeiner Teil des Bürgerlichen Gesetzbuchs / Hans Brox. — 3., verb. Aufl. — Köln ; Berlin ; Bonn ; München : Heymann, 1979.
xv, 330 p. ; 21 cm. — (Academia iuris, Lehrbücher der Rechtswissenschaft) GFR79-A
Includes bibliographical references and index.
ISBN 3-452-18540-0 : DM22.00
1. Civil law—Germany, West. I. Title. II. Series.
346'.43 79-364134
79 MARC

Einführungsgesetz zum Bürgerlichen Gesetzbuch. — 10./11., neubearb. Aufl. — Berlin : Schweitzer, 1978- c1977-
v. ; 25 cm. — (J. v. Staudingers Kommentar zum Bürgerlichen Gesetzbuch mit Einführungsgesetz und Nebengesetzen) GFR78-A13
Includes bibliographical references and index.
CONTENTS:
—T. 2b. Firsching, K. Internationales Schuldrecht I.

ISBN 3-8059-0467-3 (v. 2b) : DM46.00 (v. 2b)
1. Civil law—Germany, West. I. Series: Staudinger, Julius von, 1836-1902. Kommentar zum Bürgerlichen Gesetzbuch, mit Einführungsgesetz und Nebengesetzen.
346'.43 79-345391
79 MARC

Gerlach, Johann Wilhelm.
Radikalenfrage und Privatrecht / Johann W. Gerlach. — Tübingen : Mohr, 1978.
57 p. ; 23 cm. — (Recht und Staat in Geschichte und Gegenwart ; 482/483 ISSN 0340-7012) GFR78-A
Includes bibliographical references.
ISBN 3-16-640621-2 : DM8.10
1. Political rights—Germany, West. 2. Radicalism—Germany, West. 3. Civil law—Germany, West. 4. Political questions and judicial power—Germany, West. I. Title.
342'.43'085 78-389871
79 MARC

Germany (Federal Republic, 1949-). Laws, statutes, etc.
⌊Bürgerliches Gesetzbuch⌋
BGB mit Leitsätzen : Bürgerl. Gesetzbuch mit Leitsätzen aus d. höchstrichterl. Rechtsprechung / hrsg. von Hein Kötz, Wolfgang Eith, Diert Müller-Gindullis. — 2., überarb. u. erg. Aufl. — München : Beck, 1978.
xiv, 721 p. ; 20 cm. GFR78-A
Includes index.
ISBN 3-406-07111-2 : DM28.00
1. Civil law—Germany, West. I. Kötz, Hein. II. Eith, Wolfgang, 1943- III. Müller-Gindullis, Dierk. IV. Title.
346'.43'002632 79-347779
79 MARC

Germany (Federal Republic, 1949-). Laws, statutes, etc.
Bürgerliches Gesetzbuch : mit Einführungsgesetz, Beurkundungsgesetz, Gesetz zur Regelung des Rechts der allgemeinen Geschäftsbedingungen ... / Palandt ; bearb. von Peter Bassenge ... ⌊et al.⌋. — 38., neubearb. Aufl. — München : Beck, 1979.
xlv, 2477 p. ; 25 cm. — (Beck'sche Kurz-Kommentare ; Bd. 7) GFR78-A
"Benutzungshinweise" (1 leaf) inserted.
ISBN 3-406-07255-0 : DM138.00
1. Civil law—Germany, West. I. Palandt, Otto, 1877-1951. II. Bassenge, Peter. III. Title.
346'.43 79-342724
79 MARC

Germany (Federal Republic, 1949-). Laws, statutes, etc.
Bürgerliches Gesetzbuch und zugehörige Gesetze : Textausg. mit Verw. u. Sachverz. — 96., völlig neubearb. Aufl., nach d. Stand vom 1. April 1978. — München : Beck, 1978.
xxx, 1135 p. ; 17 cm. — (Beck'sche Textausgaben) GFR78-A
Includes index.
ISBN 3-406-07129-5 : DM14.80
1. Civil law—Germany, West. I. Title.
346'.43 78-390010
79 MARC

Jabornegg, Peter.
Privatrecht und Umweltschutz / Beitr. v. Peter Jabornegg, Peter Rummel, Rudolf Strasser ; hrsg. von Rudolf Strasser. — Linz : Inst. f. Kommunalwissenschaften u. Umweltschutz ; Wien ; München : Jugend & Volk in Komm., c1976.
191 p. ; 24 cm. — (Kommunale Forschung in Österreich ; 26) Au78-15
Includes bibliographical references.
ISBN 3-7141-7830-9 : S220.00
1. Environmental law—Austria. 2. Civil law—Austria. 3. Environmental law—Germany, West. 4. Civil law—Germany, West. I. Rummel, Peter, 1940- joint author. II. Strasser, Rudolf, joint author. III. Title. IV. Series.
344'.43'046 78-379823
*78 MARC

Jürgens, Günter.
Einführung in das bürgerliche Recht / von Günter Jürgens. — Regensburg : Walhalla-und-Praetoria-Verlag, 1978.
181 p. : ill. ; 21 cm. GFR79-A
ISBN 3-8029-7370-4 : DM19.00
1. Civil law—Germany, West. I. Title.
346'.43 79-361514
79 MARC

Klein, Hans-Joachim, 1927-
Wirtschaftsprivatrecht / Hans-Joachim Klein. — Wiesbaden : Gabler, 1978.
251 p. ; 23 cm. — (Moderne Wirtschaftsbücher : 3, Volkswirtschaft und Recht ; 09) GFR78-A
Includes bibliographies and index.
ISBN 3-409-72121-5 : DM32.00
1. Civil law—Germany, West. 2. Commercial law—Germany, West. 3. Civil procedure—Germany, West. I. Title.
78-377705
78 MARC

Merle, Werner.
Das Wohnungseigentum im System des bürgerlichen Rechts / von Werner Merle. — Berlin : Duncker und Humblot, 1979.
226 p. ; 24 cm. — (Schriften zum bürgerlichen Recht ; Bd. 52) GFR79-A
Habilitationsschrift—Mainz, 1977/78.
Bibliography: p. ⌊217⌋-226.
ISBN 3-428-04350-2 : DM76.00
1. Condominium (Housing)—Germany, West. 2. Civil law—Germany, West. I. Title. II. Series.
79-374868
79 MARC

Münchener Kommentar zum Bürgerlichen Gesetzbuch : in 7 Bd. / hrsg. von Kurt Rebmann, Franz-Jürgen Säcker. — 1. Aufl. — München : Beck, 1978-
v. ; 25 cm. GFR78-A
"Verzeichnis der Abkürzungen und der abgekürzt zitierten Literatur" (19 p.) inserted in v. 5.
CONTENTS: Bd. 1. Allgemeiner Teil (Paragraphen 1-240). AGB-Gesetz.

ISBN 3-406-06631-3 (v. 1) : DM248.00 (v. 1)
1. Civil law—Germany, West. I. Rebmann, Kurt, 1924- . II. Säcker, Franz-Jürgen, 1941- . III. Germany (Federal Republic, 1949-). Laws, statutes, etc. Bürgerliches Gesetzbuch. 1978.
346'.43 79-342723
79 MARC

Rebe, Bernd.
Privatrecht und Wirtschaftsordnung : zur vertragsrechtl. Relevanz d. Ordnungsfunktionen dezentraler Interessenkoordination in e. Wettbewerbswirtschaft / von Bernd Rebe. — Bielefeld : Gieseking, 1978.
xvi, 260 p. ; 23 cm. — (Industrie-Gesellschaft und Recht ; Bd. 13) GFR78-A
Habilitationsschrift—Bielefeld, 1977.
Bibliography: p. 231-257.
Includes index.
ISBN 3-7694-0713-X : DM76.00. ISBN 3-7694-0413-0 pbk.
1. Trade regulation—Germany, West. 2. Restraint of trade—Germany, West. 3. Civil law—Germany, West. I. Title. II. Series.
343'.43'08 79-348983
79 MARC

Schwab, Dieter.
Einführung in das Zivilrecht / Dieter Schwab. — 3. Aufl., mit e. Nachtr. 1977. — Kronberg/Ts. : Athenäum-Verlag, 1977.
xvi, 430 p. ; 19 cm. — (Athenäum Taschenbücher : Rechtswissenschaft) GFR78-A7
Includes bibliographical references and indexes.
ISBN 3-7610-6016-5 : DM25.80
1. Civil law—Germany, West. I. Title.
78-385403
79 MARC

Stewering, Hans Jürgen, 1947-
Die Anwendung öffentlichen Rechts durch Private / vorgelegt von Hans Jürgen Stewering. — ⌊s.l. : s.n.⌋, 1977, cover 1978.
vii, 202 p. ; 21 cm. GFR***
Thesis—Münster.
Vita.
Bibliography: p. 175-202.
1. Public law—Germany, West. 2. Civil law—Germany, West. 3. Public law. 4. Civil law. 5. Jurisprudence. I. Title.
340'.0943 79-377561
79 MARC

—GERMANY, WEST—ADDRESSES, ESSAYS, LECTURES

Festschrift für Fritz Hauss zum 70. Geburtstag / hrsg. von Ernst von Caemmerer ... ⌊et al.⌋. — Karlsruhe : VVW, Verlag Versicherungswirtschaft, 1978.
x, 410 p. : port. ; 24 cm. GFR79-A
Includes bibliographical references.
DM68.00
1. Civil law—Germany, West—Addresses, essays, lectures. 2. Law—Addresses, essays, lectures. 3. Hauss, Fritz. I. Hauss, Fritz. II. Caemmerer, Ernst von.
346'.43 79-353450
79 MARC

Gesetzgebungstheorie, juristische Logik, Zivil- und Prozessrecht : Gedächtnisschrift für Jürgen Rödig / hrsg. von U. Klug ... ⌊et al.⌋ ; mit Beiträgen von E. Baden ... ⌊et al.⌋. — Berlin ; New York : Springer, 1978.
viii, 396 p. : port. ; 25 cm. GFR78
ISBN 0-387-08642-0
1. Legislation—Germany, West—Addresses, essays, lectures. 2. Law—Methodology—Addresses, essays, lectures. 3. Civil law—Germany, West—Addresses, essays, lectures. 4. Law—Addresses, essays, lectures. 5. Rödig, Jürgen. I. Rödig, Jürgen. II. Klug, Ulrich.
340'.0943 78-4629
78 MARC

—GERMANY, WEST—CONGRESSES

International Congress of Comparative Law, 10th, Budapest, 1978.
Deutsche zivil-, kollisions- und wirtschaftsrechtliche Beiträge zum X. Internationalen Kongress für Rechtsvergleichung in Budapest : 1978 / hrsg. vom Max-Planck-Inst. für Ausländ. u. Internat. Privatrecht, Hamburg. — Tübingen : Mohr, 1978.
ix, 431 p. ; 24 cm. — (Materialien zum ausländischen und internationalen Privatrecht ; 29 ISSN 0543-0194) GFR78-A
Two contributions in English; summaries in English and French.
Includes bibliographical references.
ISBN 3-16-640892-4
1. Civil law—Germany, West—Congresses. 2. Conflict of laws—Germany, West—Congresses. 3. Commercial law—Germany, West—Congresses. I. Hamburg. Max-Planck-Institut für Ausländisches und Internationales Privatrecht. II. Title. III. Series.
346'.43 79-339199
79 MARC

CIVIL LAW (Continued)

—GERMANY, WEST—EXAMINATIONS, QUESTIONS, ETC.

Ballhorn, Detlef, 1941-
Rechtsfälle des täglichen Lebens / von Detlef Ballhorn u. Peter Engel. — Orig.-Ausg., 1. Aufl. — München : Goldmann, 1978.
240 p. ; 18 cm. — (Goldmann-Ratgeber) GFR78-A22
Includes index.
ISBN 3-442-10701-6 : DM7.80
1. Civil law—Germany, West—Examinations, questions, etc. I. Engel, Peter, 1949- joint author. II. Title.
78-391843
79 MARC

Diederichsen, Uwe.
Die BGB-Klausur / von Uwe Diederichsen. — 4., neubearb. u. erw. Aufl. — München : Beck, 1978.
xvi, 212 p. ; 23 cm. — (Schriftenreihe der Juristischen Schulung ; Heft 1) GFR78-A
Bibliography: p. ₍xv₎-xvi.
Includes indexes.
ISBN 3-406-07007-8 : DM14.80
1. Law examinations—Germany, West. 2. Civil law—Study and teaching—Germany, West. 3. Civil law—Germany, West—Examinations, questions, etc. I. Title. II. Series: Juristische Schulung : Schriftenreihe ; Heft 1.
78-392358
79 MARC

Diederichsen, Uwe.
Fälle, und Lösungen nach höchstrichterlichen Entscheidungen : BGB, Allgemeiner Teil / von Uwe Diederichsen. — 4., neubearb. Aufl. — Heidelberg ; Karlsruhe : Müller, Juristischer Verl., 1978.
ix, 176 p. ; 23 cm. GFR78-A
Bibliography: p. ₍viii₎-ix.
ISBN 3-8114-5878-7 : DM19.80
1. Civil law—Germany, West—Examinations, questions, etc. I. Title.
346'.43 79-365179
79 MARC

Lindacher, Walter F 1937-
Fälle zum allgemeinen Teil des BGB / von Walter F. Lindacher. — 1. Aufl. — München : Beck, 1978.
xv, 89 p. ; 23 cm. — (Schriftenreihe der Juristischen Schulung ; Heft 65) GFR78-A
Bibliography: p. xiii-xv.
Includes indexes.
ISBN 3-406-03745-3 : DM18.50
1. Civil law—Germany, West—Examinations, questions, etc. I. Title. II. Series: Juristische Schulung : Schriftenreihe ; Heft 65.
346'.43'0076 79-348942
79 MARC

Medicus, Dieter.
Bürgerliches Recht : e. nach Anspruchsgrundlagen geordnete Darstellung zur Examensvorbereitung / Dieter Medicus. — 8., neubearb. u. erw. Aufl. — Köln ; Berlin ; Bonn ; München : Heymann, 1978.
xiv, 466 p. ; 21 cm. — (Academia iuris, Lehrbücher der Rechtswissenschaft) GFR78-A
Includes bibliographical references and indexes.
ISBN 3-452-18415-3 : DM26.80
1. Civil law—Germany, West—Examinations, questions, etc. I. Title. II. Series.
78-392365
79 MARC

Müller, Klaus, 1933-
Bürgerliches Recht : (Rechtskunde) / von Klaus Müller. — Köln ; Berlin ; Bonn ; München : Heymann, 1977.
164 p. : ill. ; 21 cm. — (Der Verwaltungssekretär ; Heft 7) GFR77-A17
Includes index.
ISBN 3-452-18242-8 : DM15.00
1. Civil law—Germany, West—Examinations, questions, etc. I. Title. II. Series.
340'.0943 78-377721
78 MARC

Weimar, Wilhelm, 1903-
1000 ₍i.e. Tausend₎ Fragen und Antworten zum BGB und HGB / von Wilhelm Weimar. — 2. Aufl. — Sankt Augustin : Asgard-Verlag Hippe, 1977.
166 p. ; 21 cm. GFR78-A19
ISBN 3-537-67002-8
1. Civil law—Germany, West—Examinations, questions, etc. 2. Commercial law—Germany, West—Examinations, questions, etc. I. Title.
346'.43'07 79-353311
79 MARC

Werner, Olaf, 1939-
Fälle mit Lösungen für Anfänger im bürgerlichen Recht / von Olaf Werner. — Frankfurt am Main : Metzner, 1978.
131 p. ; 24 cm. GFR78-A
Bibliography: p. 130-131.
ISBN 3-7875-5270-7 : DM13.00
1. Civil law—Germany, West—Examinations, questions, etc. I. Title.
78-396322
79 MARC

—GERMANY, WEST—FORMS

Beck'sches Formularbuch zum bürgerlichen Handels- und Wirtschaftsrecht / hrsg. von Michael Hoffmann-Becking, Helmut Schippel, in Zusammenarbeit mit Klaus Anschütz ... ₍et al.₎. — 1. Aufl. — München : Beck, 1978.
xxix, 957 p. ; 25 cm. GFR78-A
Includes index.
ISBN 3-406-05052-2
1. Civil law—Germany, West—Forms. 2. Commercial law—Germany, West—Forms. 3. Industrial laws and legislation—Germany, West—Forms. I. Hoffmann-Becking, Michael. II. Schippel, Helmut. III. Anschütz, Klaus.
346'.43'00269 78-398331
79 MARC

—GERMANY, WEST—OUTLINES, SYLLABI, ETC.

Burkholz, Richard.
Lehrbuch der Rechtskunde / von Richard Burkholz. — 4., erneuerte Aufl. — Herne ; Berlin : Verlag Neue Wirtschafts-Briefe, 1978-
v. ; 21 cm. GFR78-A
Includes index.
CONTENTS: Bd. 1. Bürgerliches Recht.
ISBN 3-482-75214-1 (v. 1) : DM29.00 (v. 1)
1. Civil law—Germany, West—Outlines, syllabi, etc. 2. Commercial law—Germany, West—Outlines, syllabi, etc. I. Title.
79-339946
79 MARC

Diederichsen, Uwe.
Der allgemeine Teil des Bürgerlichen Gesetzbuches für Studienanfänger / von Uwe Diederichsen. — 3., überarb. Aufl. — Heidelberg ; Karlsruhe : C .F. Müller, 1978.
xiv, 191 p. ; 23 cm. GFR78-A
Bibliography: p. xi.
Includes index.
ISBN 3-8114-5778-0 : DM19.80
1. Civil law—Germany, West—Outlines, syllabi, etc. I. Title.
79-337050
79 MARC

Lüschen, Rolf.
Leitfaden zum bürgerlichen Recht : Grundriss für d. Ausbildung u. Fortbildung / von Rolf Lüschen. — Köln ; Stuttgart ; Hannover ; Kiel : Deutscher Gemeindeverlag, 1979.
viii, 112 p. ; 21 cm. — (Dgv-Skript) GFR79-A
Includes index.
ISBN 3-555-20108-5 : DM13.80
1. Civil law—Germany, West—Outlines, syllabi, etc. I. Title.
346'.43 79-368983
79 MARC

—GREECE, MODERN

Babouskos, Kōnstantinos A
(Encheiridion Astikou dikaiou)
'Εγχειρίδιον 'Αστικοῦ δικαίου / Κωνστ. 'Αναστ. Βαβούσκου. — Ἔκδ. 2. — Θεσσαλονίκη : 'Εκδοτ. Οἶκος Σάκκουλα, 1977.
448 p. ; 25 cm.
1. Civil law—Greece, Modern. I. Title.
79-408003

Geōrgiadēs, Apostolos S 1935-
(Astikos kōdix)
'Αστικὸς κῶδιξ : ἑρμηνεία κατ'ἄρθρον-νομολογία-βιβλιογραφεία / 'Αποστόλου Σ. Γεωργιάδη, Μιχαὴλ Π. Σταθοπούλου ; μὲ τὴν συνεργασίαν τῶν Σπύρου Βρέλλη ... ₍et al.₎. — 'Αθῆναι : 'Εκδοτ. Οἶκος 'Αφοὶ Σάκκουλα, 1978-
v. ; 25 cm.
Includes bibliographies.
CONTENTS: τ. 1. Γενικαὶ ἀρχαί.
1. Civil law — Greece, Modern. 2. Civil law — Greece, Modern — Bibliography. I. Stathopoulos, Michael P., joint author. II. Greece. Laws, statutes, etc. Astikos kōdix. 1978. III. Title.
78-421823

Karakatsanēs, Iōannēs K
(Zētēmata astikou dikaiou)
Ζητήματα ἀστικοῦ δικαίου : ἀστικόν δίκαιον καὶ μεθηδολογία τοῦ δικαίου / 'Ιωάννου Κ. Καρακατσάνη. — 'Αθῆναι : 'Α. Ν. Σάκκουλα, 1978.
302 p. ; 25 cm.
Bibliography: p. ₍287₎-294.
Includes index.
1. Civil law—Greece, Modern. 2. Civil law. I. Title.
79-411616

—GREECE, MODERN —BIBLIOGRAPHY

Geōrgiadēs, Apostolos S 1935-
(Astikos kōdix)
'Αστικὸς κῶδιξ : ἑρμηνεία κατ'ἄρθρον-νομολογία-βιβλιογραφεία / 'Αποστόλου Σ. Γεωργιάδη, Μιχαὴλ Π. Σταθοπούλου ; μὲ τὴν συνεργασίαν τῶν Σπύρου Βρέλλη ... ₍et al.₎. — 'Αθῆναι : 'Εκδοτ. Οἶκος 'Αφοὶ Σάκκουλα, 1978-
v. ; 25 cm.
Includes bibliographies.
CONTENTS: τ. 1. Γενικαὶ ἀρχαί.
1. Civil law — Greece, Modern. 2. Civil law — Greece, Modern — Bibliography. I. Stathopoulos, Michael P., joint author. II. Greece. Laws, statutes, etc. Astikos kōdix. 1978. III. Title.
78-421823

—HAITI—HISTORY

Delatour, Ferdinand.
Les 150 ₍i.e. cent cinquante₎ ans du régime du code civil dans le contexte social haïtien, 1826-1976 / Ferdinand Delatour. — Port-au-Prince : Les Éditions Fardin, ₍between 1976 and 1978₎
xiii, 301 p. ; 21 cm.
Bibliography: p. 271-281.
1. Civil law—Haiti—History. 2. Sociological jurisprudence—History. 3. Haiti—Social conditions. I. Title.
309.1'7294 79-110216
79 MARC

—INDONESIA

Mertokusumo, Sudikno.
Hukum acara perdata Indonesia / oleh Sudikno Mertokusumo. — Cet. 1. — Yogyakarta : Liberty, 1977.
xvi, 208 p. ; 22 cm.
Bibliography: p. 179-183.
Includes index.
1. Civil law—Indonesia. 2. Civil law—Indonesia—Digests.
79-102465
79 MARC

Sofwan, Sri Soedewi Masjchun.
Asas² hukum perdata : tahun 1958/1959 Fakultas Ekonomi / Machun Sofwan. — Jogja : Jajasan Badan Penerbit Gadjah Mada, ₍1959?₎
100 p. ; 20 cm.
"Tidak diperdagangkan."
1. Civil law—Indonesia. 2. Civil law. I. Title.
77-986626
78₍r79₎rev MARC

—INDONESIA—DIGESTS

Mertokusumo, Sudikno.
Hukum acara perdata Indonesia / oleh Sudikno Mertokusumo. — Cet. 1. — Yogyakarta : Liberty, 1977.
xvi, 208 p. ; 22 cm.
Bibliography: p. 179-183.
Includes index.
1. Civil law—Indonesia. 2. Civil law—Indonesia—Digests.
79-102465
79 MARC

Yurisprudensi tentang perbuatan melanggar hukum (onrechtmatige daad) / oleh Chidir Ali. — Cet. 1. — ₍Bandung₎ : Binacipta, 1978.
xix, 267 p. ; 22 cm.
Dutch or Indonesian.
Bibliography: p. xiii.
1. Civil law—Indonesia—Digests. I. Ali, Chidir.
78-942525
79 MARC

—IRAN

'Adl, Muṣṭafá.
(Ḥuqūq-i madanī)
حقوق مدنی / تألیف مصطفی عدل «منصور السلطنه». — چاپ 8. — ₍تهران₎ : مؤسسهٔ انتشارات امیر کبیر، ₍1975 or 1976₎ 1354.
627, 8 p. ; 22 cm.
In Persian.
1. Civil law—Iran. I. Title.
79-971500

Amīrī Qā'im'maqāmī, 'Abd al-Majīd.
(Ḥuqūq-i ta'ahhudāt)
حقوق تعهدات : نظریهٔ کلی تعهدات قانون مدنی : با بررسی امکان رسوخ راه حل‌های جدید از حقوق تعهدات آلمان و سویس و فرانسه در حقوق ایران / تألیف عبد المجید امیری قائم مقامی. — چاپ 2. — تهران : مؤسسهٔ انتشارات و چاپ دانشگاه تهران، ‎-1976₎ ‎-2535.
v. ; 24 cm. — (انتشارات دانشگاه تهران ؛ شمارهٔ 1475/1)
Title on p. ₍4₎ of cover: Partie générale du droit des obligations en droit civil iranien.
In Persian.
Includes bibliographical references.
CONTENTS: جلد 1. کلیات حقوق تعهدات. — وقایع حقوقی ؛ ضمایم : اقرار. — اجاره اشیاء.
1. Obligations (Law)—Iran. 2. Civil law—Iran. I. Title. II. Title: Partie générale du droit des obligations en droit civil iranien. III. Series: Teheran. Dānishgāh. Intishārāt ; shumārah-'i 1475/1
79-970826

Iran. Laws, statutes, etc.
(Majmū'ah-'i kāmil-i qavānīn va muqarrarāt-i ḥuqūqī)
مجموعهٔ کامل قوانین و مقررات حقوقی / تدوین غلامرضا حجتی اشرفی ؛ زیر نظر غلامعلی امیری. — ₍تهران₎ : کتابفروشی گنج دانش، ₍1976 or 1977₎ 2535.
88, 778 p. ; 18 cm.
In Persian.
400.00 IR
1. Civil law — Iran. 2. Civil procedure — Iran. I. Ḥujjatī Ashrafī, Ghulām Riẓā. II. Title.
79-970575

CIVIL LAW
—IRAN (Continued)

Iran. Laws, statutes, etc.
(Qavānīn-i ḥuqūqī)
قوانین حقوقی : با آخرین اصلاحات و بعضی آئین‌نامه‌ها : اصل تشکیلات دادگستری، سازمان قضائی، آئین دادرسی مدنی، قوانین ثبت و دفاتر رسمی جدید، شورای داوری و خانه انصاف، قوانین اوقاف و ملی شدن آب، امور حسبی، و قوانین‌های دیگر / مؤلف احمد کمانگر. — [تهران] : انتشارات جاویدان، [1977] 2536.
596 p. ; 18 cm.
In Persian.
«مجموعهٔ شماره ۱.»
300.00 IR
1. Civil law—Iran. 2. Courts—Iran. 3. Justice, Administration of—Iran. I. Kamāngar, Aḥmad. II. Title.
79-970831

—ITALY

Bertolucci, Alberto.
Diritto della navigazione ed elementi di diritto pubblico e privato / Alberto Bertolucci. — Milano : L. Trevisini, 1977.
205 p. ; 24 cm. It78-Nov
L3800
1. Maritime law—Italy. 2. Public law—Italy. 3. Civil law—Italy. I. Title.
343'.45'096 79-341372
*79 MARC

Bianca, C Massimo.
Diritto civile / C. Massimo Bianca. — Milano : A. Giuffrè, 1978-
v. ; 25 cm. It78-Dec
Includes bibliographical references.
CONTENTS: 1. La norma giuridica, I soggetti.
L11000 (v. 1)
1. Civil law—Italy. I. Title.
346'.45 79-340700
*79 MARC

Cupis, Adriano de.
Istituzioni di diritto privato / Adriano De Cupis. — Milano : A. Giuffrè, 1975-1978.
3 v. ; 25 cm. It76-Jan
CONTENTS: 1. Nozione generale del diritto privato, Famiglia.—2. Diritti reali e diritti su beni immateriali, diritti d'obbligazione.—3. Fonti delle obbligazioni, Impresa, Successione e causa di morte.
L3200 (v. 2)
1. Civil law—Italy. I. Title.
346'.45 75-411067
*77[r79]rev MARC

Cupis, Adriano de.
Nozioni elementari di diritto privato : ad uso degli istituti tecnici / Adriano De Cupis. — Milano : A. Giuffrè, 1978.
408 p. ; 25 cm. It***
L7000
1. Civil law—Italy. I. Title.
346'.45 79-354670
79 MARC

Donisi, Carmine.
Ricerche di diritto civile / Carmine Donisi. — [s.l.] : Edizioni scientifiche italiane, [1979]
280 p. ; 24 cm. It***
Includes bibliographical references.
L11000
1. Civil law—Italy. 2. Civil law. I. Title.
346'.45 79-373550
79 MARC

Grosso, Giuseppe, 1906-1973.
Le successioni : parte generale / [di] Giuseppe Grosso e Alberto Burdese. — Torino : Unione tipografico-editrice torinese, [1977]
xix, 574 p. ; 25 cm. — (Trattato di diritto civile italiano ; v. 12, t. 1) It78-Dec
Bibliography: p. [xi]-xix.
Includes indexes.
L18000
1. Civil law—Italy. 2. Inheritance and succession—Italy. I. Burdese, Alberto, joint author. II. Title. III. Series.
346'.45'052 79-356907
*79 MARC

Italy. Laws, statutes, etc.
Codice civile : annotato con la giurisprudenza della Corte costituzionale e della Corte di cassazione / G. Pescatore, C. Ruperto. — 7. ed. interamente rifatta. — Milano : A. Giuffrè, 1978.
3120 p. ; 19 cm. It***
Errata slip inserted.
Includes index.
L50000
1. Civil law—Italy. I. Pescatore, Gabriele. II. Ruperto, Cesare. III. Title.
346'.45 78-394386
79 MARC

Patti, Salvatore.
Profili della tolleranza nel diritto privato / Salvatore Patti. — Napoli : E. Jovene, 1978.
xxiv, 211 p. ; 24 cm. — (Biblioteca di diritto privato ; 6) It79-Ap
Includes bibliographical references and index.
L5500
1. Civil law—Italy. 2. Silence (Law)—Italy. 3. Good faith (Law)—Italy. I. Title. II. Series.
79-365348
*79 MARC

Trimarchi, Pietro.
Istituzioni di diritto privato / Pietro Trimarchi. — 3. ed. — Milano : Giuffrè, 1977.
xxxiii, 943 p. ; 23 cm. It79-Feb
Includes index.
L18000
1. Civil law—Italy. I. Title.
346'.45 79-384136
*79 MARC

—JAPAN

Aoyama, Michio, 1902-
(Mimpō nyūmon)
民法入門 / 青山道夫著. — 東京：日本學燈社, 昭和24 [1949]
282 p. ; 19 cm.
Colophon inserted.
Includes bibliographical references.
1. Civil law—Japan. I. Title.
79-804375

Aoyama, Michio, 1902-
(Mimpō taii)
民法大意 / 青山道夫著. — 東京：青林書院, 昭和30- [1955-
v. ; 19 cm.
Colophon inserted.
1. Civil law—Japan. I. Title.
79-803549

Ariizumi, Tōru, 1906-
(Mimpō)
民法 / 有泉亨. — 東京：法政大學通信教育部, 昭和24- [1949-
v. ; 21 cm.
CONTENTS: [1] 總則 I
1. Civil law—Japan. I. Title.
79-803394

Ariizumi, Tōru, 1906-
(Mimpō)
民法. — 改訂版. — 東京：高文社, 1974-
v. : ill. ; 21 cm. — (新法律学ハンドブック)
Includes bibliographies.
CONTENTS:
3. 有泉亨 西井龍夫 親族・相続
1. Civil law—Japan. I. Nishii, Tatsuo, 1920- joint author. II. Title. III. Series: Shin hōritsugaku handobukku.
79-803872

Azuma, Mitsutoshi, 1903-1973.
(Mimpō sōsoku)
民法総則 / 吾妻光俊著. — 東京：青林書院, 昭和29 [1954]
223 p. ; 22 cm. — (His 民法講義 ; 1)
Colophon inserted.
1. Civil law—Japan. I. Title. II. Series: Azuma, Mitsutoshi, 1903-1973. Mimpō kōgi ; 1.
79-803875

Azuma, Mitsutoshi, 1903-1973.
(Reikai Mimpō seigi)
例解民法精義 / 吾妻光俊著. — 東京：白桃書房, 昭和29-30 [1954-1955]
2 v. ; 22 cm.
Includes indexes.
CONTENTS: [1] 総則・物権法・担保物権法—[2] 債権総論・債権各論
1. Civil law—Japan. I. Title. II. Title: Mimpō seigi.
79-803399

Gotō, Kiyoshi, 1902-
(Mimpōgaku josetsu)
民法學序説 / 後藤清著. — 東京：弘文堂書房, 昭和11 [1936]
4, 6, 243 p. ; 23 cm.
1. Civil law—Europe. 2. Civil law—Japan. I. Title.
79-803273

Gotō, Kiyoshi, 1902-
(Zaisanhō nyūmon)
財産法入門 / 後藤清著. — 京都：法律文化社, 1957.
12, 256 p. ; 19 cm. — (新文化選書 ; 32)
Colophon inserted.
Series romanized: Shin bunka sensho.
1. Civil law—Japan. I. Title.
79-803301

Hayashi, Nobuo, 1905-
(Nihon Mimpō)
日本民法 / 林信雄, 石川利夫共著. — 東京：評論社, 昭和27- [1952-
v. ; 22 cm.
CONTENTS: [1] 財産編
1. Civil law—Japan. 2. Property—Japan. I. Ishikawa, Toshio, 1923- joint author. II. Title.
79-803053

Hayashi, Nobuo, 1905-
(Shin Mimpō gairon, zaisanhen)
新民法概論 財産編 / 林信雄著. — 東京：青也書店, 昭和23 [1948]
2 v. ; 22 cm.
First ed. published in 1938 under title: Zaisanhō gairon.
1. Civil law—Japan. 2. Property—Japan. I. Title.
79-803292

Ikeda, Eitarō, 1931-
(Keiei hōgaku no kiso)
経営法学の基礎 / 池田栄太郎. — 京都：晃洋書房, 1977-
v. ; 21 cm. Ja 78-1439 (v. 1)
¥700 (v. 1)
1. Civil law—Japan. 2. Business law—Japan. I. Title.
78-815985

Ishida, Bunjirō, 1892-
(Genkō Mimpō sōron)
現行民法總論 / 石田文次郎著. — 東京：弘文堂書房, 昭和9 [1934], 昭和15 [1940] printing.
4, 6, 15, 529, 12 p. ; 23 cm.
Bibliography: p. 1-5 (2d group)
Includes index.
1. Civil law—Japan. I. Title. II. Title: Mimpō sōron.
79-804029

Isogai, Kōjirō, 1865-
(Mimpō tsūron)
民法通論 / 磯谷幸次郎著. — 東京：巖松堂書店, 昭和 [193 昭和 [193 printing.
v. ; 23 cm.
CONTENTS:
2. 債権編
1. Civil law—Japan. I. Title.
79-805374

CIVIL LAW
—JAPAN (Continued)

Kainō, Michitaka, 1908–1975.
(Mimpō nyūmon)
民法入門：市民法としての民法 / 戒能通孝著. — 東京：巖松堂書店, 昭和22 [1947]
6, 186 p. ; 19 cm.
Colophon inserted.

1. Civil law—Japan. I. Title.
79-804821

Kainō, Michitaka, 1908–1975.
(Mimpōgaku gairon)
民法學概論 / 戒能通孝著. — 東京：日本評論新社, 昭和31 [1956]
2, 8, 332 p. ; 19 cm.
Bibliography: p. 324-332.

1. Civil law—Japan. I. Title.
79-804819

Kanayama, Masanobu, 1914–
(Mimpō sōsoku)
民法總則 / 金山正信著. — 京都：ミネルヴァ書房, 昭和31 [1956]
285 p. ; 22 cm.
Errata slip inserted.
Bibliography: p. [269]-285.

1. Civil law—Japan. I. Title.
79-804870

Kanayama, Masanobu, 1914–
(Zaisanhō yōsetsu)
財産法要説 / 金山正信著. — 京都：ミネルヴァ書房, 昭和29 [1954]
2, 2, 260 p. ; 21 cm.

1. Civil law—Japan. 2. Property—Japan. I. Title.
79-804894

Katsumoto, Masaakira, 1895–
(Shin mimpō sōsoku)
新民法總則 / 勝本正晃. — 1948年版. — 東京：弘文堂, 昭和23 [1948]
24, 354, 20 p. ; 21 cm.
Includes indexes.

1. Civil law—Japan. I. Title.
79-805167

Kawana, Kenshirō, 1875–1914.
(Nihon Mimpō sōron)
日本民法総論 / 川名兼四郎著. — 東京：金刺芳流堂, 明治45 [1912]
1, 8, 321, 11, 27 p. ; 22 cm.
Includes indexes.

1. Civil law—Japan. I. Title.
79-804762

Kawashima, Takeyoshi, 1909–
(Mimpō)
民法 / 川島武宜. — 改訂増補. — 東京：有斐閣, 昭和 [19
v. ; 19 cm. — (有斐閣全書)
Series romanized: Yūhikaku zensho.
Includes bibliographical references and index.
CONTENTS:
3. 人・親族・相続・法人

1. Civil law—Japan. I. Title.
79-809202

Kawazoe, Seikichi, 1898–
(Mimpō kōgi)
民法講義 / 川添清吉著. — 東京：巖松堂書店 昭和9- [1934-
v. ; 23 cm.
CONTENTS: 第1編 總則 訂正再版 1935.

1. Civil law—Japan. I. Title.
79-805155

Kawazoe, Seikichi, 1898–
(Mimpō sōron teiyō)
民法總論提要 / 川添清吉著. — 東京：明治大學出版部, 昭和27 [1952], 昭和33 [1958] printing.
5, 150 p. ; 22 cm.

1. Civil law—Japan. I. Title.
79-804869

Koike, Ryūichi, 1898–
(Mimpō)
民法 / 小池隆一著. — 東京：慶應出版社, 昭和16 [1941]
160 p. ; 21 cm. — (慶應義塾大學講座經濟學 ; 34)

1. Civil law—Japan. I. Title. II. Series: Keiō Gijuku Daigaku, Tokyo. Keio Gijuku Daigaku kōza keizaigaku ; 34.
79-804758

Koike, Ryūichi, 1898–
(Nihon Mimpō sōron)
日本民法總論 / 小池隆一著. — 改訂増補. — 東京：清水書店, 昭和16 [1941], 昭和19 [1944] printing.
2, 12, 439, 13 p. ; 22 cm.
Includes bibliographical references and index.

1. Civil law—Japan. I. Title.
79-804054

Mibuchi, Tadahiko, 1880–1950.
(Mimpō gaisetsu)
民法概説 / 三淵忠彦著. — 東京：慶應義塾出版局：賣捌所 東京堂書店, 大正13 [1924], 大正15 [1926] printing.
4, 10, 492, 12 p. ; 22 cm.
Includes index.

1. Civil law—Japan. I. Title.
79-804866

(Mimpō)
民法. — 東京：三省堂, 昭和52-53 [1977-1978]
5 v. ; 22 cm. — (判例コンメンタール ; 3-7) Ja 77-23108 (v. 1)
Includes indexes.
CONTENTS: 1. 篠塚昭次編 総則・物権—2. 篠塚昭次編 担保物件・債権總則・契約総則—3. 篠塚昭次編 契約各則・事務管理・不当利得・不当行為—4. 島津一郎編 親族—5. 島津一郎編 相続・渉外家族法

1. Civil law—Japan. I. Shinozuka, Shōji, 1928– II. Shimazu, Ichirō, 1921– III. Series: Hanrei kommentāru (Tokyo, Sanseidō) ; 3-7.
78-801285

(Mimpō kōgi)
民法講義. — 東京：有斐閣, 昭和51-53 [1976-1978]
8 v. ; 22 cm. — (有斐閣大学双書)
Series romanized: Yūhikaku daigaku sōsho.
Includes bibliographies and indexes.
CONTENTS: 1. 五十嵐清 et al. 総則—2. 原島重義 et al. 物権—3. 高木多喜男 et al. 担保物権—4. 森泉章 et al. 債権総論—5. 稲本洋之助 et al. 契約—6. 高木多喜男 et al. 不法行為等—7. 久貴忠彦 et al 親族—8. 泉久雄 et al. 相続
¥15000

1. Civil law—Japan. I. Igarashi, Kiyoshi, 1925–
77-801983

(Mimpō kōgi)
民法(総則・物権)講議 / 水本浩, 甲斐道太郎編 ; 執筆者・家崎宏 ... [et al.]. — 改訂. — 東京：青林書院新社, 1979.
13, 355, 11 p. ; 22 cm. — (青林講義シリーズ) Ja 79-9615
Series romanized: Seirin kōgi shirizu.
Includes index.
¥2400
1. Civil law—Japan. 2. Property—Japan. I. Mizumoto, Hiroshi, 1920– II. Kai, Michitarō. III. Iezaki, Hiroshi, 1930– IV. Title: Mimpō sōsoku bukken kōgi.
79-811853

(Mimpō sōsoku)
民法総則 / 中川善之助編. — 東京：青林書院, 昭和32 [1957]
282, 10 p. ; 19 cm. — (法律学ハンドブック)
Colophon inserted.
Series romanized: Hōritsugaku handobukku.
Includes index.

1. Civil law—Japan. I. Nakagawa, Zennosuke, 1897–1975.
79-805367

(Mimpō tokuhon)
民法読本 / 雨宮真也 ... [et al.]. — 東京；高文堂出版社, 1978.
3 v. ; 19 cm. — (高文堂読本シリーズ) Ja 78-21522 (v. 2)
Series romanized: Kōbundō tokuhon shirizu.
Includes bibliographies.
CONTENTS: 1. 雨宮真也 ... et al. 総則・物権・担保物権—2. 森田三男. 松嶋由紀子 債権—3. 石田雅男 家族法
¥4300

1. Civil law—Japan. I. Amemiya, Masaya, 1935–
78-816000

(Mimpō tokuhon)
民法読本 / 好美清光, 米倉明編. — 東京：有斐閣, 1978-1979.
3 v. : ill. ; 19 cm. — (有斐閣選書) Ja 78-7650 (v. 1)
Vol. 3: Yoshimi Seikō, Kuki Tadahiko, Yonekura Akira hen.
Series romanized: Yūhikaku sensho.
Includes bibliographies and indexes.
CONTENTS: 1. 総則・物権法—2. 債権法—3. 親族法・相続法
¥1200 per vol.
1. Civil law—Japan. I. Yoshimi, Seikō, 1929– II. Yonekura, Akira, 1934– III. Kuki, Tadahiko, 1931–
78-809092

rev

Mitsuma, Shinzō, 1879–1937.
(Mimpō sōsoku teiyō)
民法總則提要 / 三潴信三著. — 全訂. — 東京：有斐閣, 昭和6 [1931]
2 v. (5, 19, 549 p.) ; 23 cm.
Includes bibliographical references.

1. Civil law—Japan. 2. Civil law—Europe. I. Title.
79-801759

Miyakawa, Kiyoshi, 1917–
(Mimpō kōgi)
民法講義 宮川澄著 東京 青木書店 1956-59.
4 v. 22 cm.

Includes bibliographies and indexes.
CONTENTS: [1] 総論—[2] 物権・担保物語—[3] 総則—[4] 債権

1. Civil law—Japan. I. Title.
65-2459

rev

Nagashima, Hatasu, 1880–
(Mimpō sōron)
民法總論 / 長島毅著. — 訂正3版. — 東京：巖松堂書店, 大正11 [1922]
3, 30, 687, 9 p. ; 23 cm.
Includes index.

1. Civil law—Japan. I. Title.
79-805366

CIVIL LAW
—JAPAN (Continued)

Nagashima, Hatasu, 1880-
(Mimpō sōsoku kaigi)
民法總則解疑 / 長島毅著. — 東京：清水書店, 大正11 [1922]
4, 22, 602 p. ; 19 cm.

1. Civil law—Japan. I. Title.

79-805746

Nagata, Kikushirō, 1895-
(Kaisei Mimpō kōwa)
改正民法講話 / 永田菊四郎著. — 東京：櫻門出版社, 昭和23 [1948], 昭和25 [1950] printing.
190 p. ; 22 cm.
Colophon inserted.

1. Civil law—Japan. I. Title.

79-805351

Nakagawa, Zennosuke, 1897-1975.
(Mimpō taiyō)
民法大要 / 中川善之助著. — 全訂. — 東京：勁草書房, 昭和 [19
v. ; 19 cm.
Includes indexes.
CONTENTS: [2] 親族法 相續法

1. Civil law—Japan. I. Title.

79-805372

Nakagawa, Zennosuke, 1897-1975.
(Mimpō taiyō)
民法大要 / 中川善之助著. — 東京：勁草書房, 昭和 [19
v. ; 19 cm.
Includes indexes.
CONTENTS: [1] 總則 物權法 [3] 親族法 相續法

1. Civil law—Japan. I. Title.

79-805370

Nakajima, Hiromichi, 1885-1960.
(Mimpō sōsoku)
民法總則 / 中島弘道著. — 東京：駿河台出版社, 昭和31 [1956], 昭和32 [1957] printing.
7, 174 p. ; 21 cm.

1. Civil law—Japan. I. Title.

79-805369

Nakajima, Tamakichi, 1875-1961.
(Mimpō shakugi)
民法釋義 / 中島玉吉著. — 東京：金刺芳流堂, 大正
v. ; 22 cm.
Vol. 2, pt. 1: Teiseiban.
Includes index.
CONTENTS:
2. I-II. 物權篇 上下
1. Civil law—Japan. I. Japan. Laws, statutes, etc. Mimpō. 1915. II. Title.

79-805365

Nakamura, Muneo, 1894-1975.
(Mimpō)
民法 / 中村宗雄著. — 訂正6版. — 東京：敬文堂, 昭和29 [1954]
2, 12, 298, 12 p. ; 22 cm.
Includes bibliographical references and index.

1. Civil law—Japan. I. Title.

79-805744

Nakamura, Muneo, 1894-1975.
(Mimpō sōsoku)
民法總則 / 中村宗雄著. — 東京：早稲田大学出版部, 昭和31 [1956]
2, 11, 288, 13 p. ; 22 cm.
Includes index.

1. Civil law—Japan. I. Title.

79-805368

Namiki, Toshimori, 1925-
(Mimpō tokuhon)
民法読本 / 並木俊守著. — 東京：塙書房, 昭和31 [1956]
2, 10, 347, 13 p. ; 19 cm.
Includes index.

1. Civil law—Japan. I. Title.

79-805373

Nozu, Tsutomu, 1894-
(Mimpō nyūmon)
民法入門 / 野津務著. — 改訂. — 東京：有斐閣, 昭和23- [1948-
v. ; 19 cm.
CONTENTS: [1] 總則—[2] 物權法

1. Civil law—Japan. I. Title.

79-805696

Numa, Yoshio, 1883-1966.
(Mimpō yōron)
民法要論 / 沼義雄著. — 東京：巖松堂書店, 大正 [19
v. ; 23 cm.
CONTENTS: 第2編 物權

1. Civil law—Japan. I. Title.

79-805970

Numa, Yoshio, 1883-1966.
(Sōgō mimpō yōron)
綜合民法要論 / 沼義雄著. — 東京：巖松堂書店, 昭和13 [1938]
23, 879 p. ; 23 cm.
Includes bibliographical references.

1. Civil law—Japan. I. Title.

79-805767

Oho, Fujio, 1908-
(Mimpō sōsoku kōgi)
民法総則講義 / 於保不二雄著. — 東京：有信堂, 昭和26 [1951]
16, 316 p. ; 22 cm.
Colophon inserted.

1. Civil law—Japan. I. Title.

79-805747

Sōmiya, Shinji, 1894-
(Mimpō sōron)
民法總論 / 宗宮信次著. — 新訂. — 東京：有斐閣, 昭和30 [1955]
2, 1, 10, 268, 10 p. ; 22 cm.
Includes index.

1. Civil law—Japan. I. Title.

79-805637

Suehiro, Izutarō, 1888-1951.
(Mimpō kōwa)
民法講話 / 末弘嚴太郎著. — 東京：岩波書店, 昭和 [192
v. ; 20 cm.
Includes index in v. 2.

1. Civil law—Japan. I. Title.

79-805361

Suehiro, Izutarō, 1888-1951.
(Mimpō kōwa)
民法講話 / 末弘嚴太郎著. — [再版] / 戒能通孝改訂. — 東京：岩波書店, 昭和29 [1954], v. 1, 昭和30 [1955] printing.
3 v. (10, 4, 849 p.) ; 19 cm.
"参考書について": v. 3, p. 838-849.

1. Civil law—Japan. I. Title.

79-805362

Yamanaka, Yasuo, 1908-
(Mimpō)
民法：総論・総則・家族・相続 / 山中康雄著. — 京都：法律文化社, 1978.
6, 23, 444 p. ; 22 cm. Ja 78-14839
¥3300

1. Civil law—Japan. 2. Domestic relations—Japan. I. Title.

78-812279

—JAPAN—ADDRESSES, ESSAYS, LECTURES

Ishizaka, Otoshirō, 1877-1917.
(Mimpō kenkyū)
民法研究 / 石坂音四郎著. — 改纂. — 東京：有斐閣書房：賣捌所 有終閣, 大正13-15 [1924-1926]
2 v. ; 23 cm.

1. Civil law—Japan—Addresses, essays, lectures. I. Title.

79-804867

Katsumoto, Masaakira, 1895-
(Shūsen to saiken kankei)
終戦と債權關係 / 勝本正晃著. — 東京：有斐閣, 昭和23 [1948]
7, 440 p. ; 22 cm. — (**His** 民法研究 ; 6)
Colophon inserted.

1. Obligations (Law)—Japan—Addresses, essays, lectures. 2. Civil law—Japan—Addresses, essays, lectures. I. Title. II. Series: Katsumoto, Masaakira, 1895- Mimpō kenkyū ; 6.

79-805267

Kawashima, Takeyoshi, 1909-
(Mimpō kaishakugaku no shomondai)
民法解釋學の諸問題 / 川島武宜著. — 東京：弘文堂, 昭和24 [1949]
15, 275 p. ; 22 cm.
Colophon inserted.
Includes bibliographical references.

1. Civil law—Europe—Addresses, essays, lectures. 2. Civil law—Japan—Addresses, essays, lectures. I. Title.

79-804651

Koike, Ryūichi, 1898-
(Zaisanhō gaisetsu)
財産法概説 / 小池隆一著. — 東京：東洋書館, 昭和24 [1949]
282 p. ; 22 cm.
Colophon inserted.

1. Civil law—Japan—Addresses, essays, lectures. 2. Property—Japan—Addresses, essays, lectures. I. Title.

79-804893

CIVIL LAW
—JAPAN—ADDRESSES, ESSAYS, LECTURES (Continued)

Suehiro, Izutarō, 1888-1951.
(Mimpō zakkō)
民法雑考 / 末弘嚴太郎著. — 8版. — 東京：日本評論社, 昭和16 [1941]
267 p. ; 19 cm.
Includes bibliographical references.

1. Civil law—Japan—Addresses, essays, lectures. I. Title.
79-805375

—JAPAN—COLLECTED WORKS

Katsumoto, Masaakira, 1895-
(Mimpō kenkyū)
民法研究 / 勝本正晃著. — 東京：巖松堂書店, 昭和7- [1932-
v. ; 22 cm.
Includes bibliographical references and indexes.

1. Civil law—Japan—Collected works. 2. Civil law—Collected works. I. Title.
79-805745

—JAPAN—DIGESTS

Kai, Michitarō.
(Mimpō hanrei)
民法判例 甲斐道太郎 乾昭三 椿寿夫編 東京 有斐閣 昭和45-52 (1970-77)
3 v. 19 cm. (有斐閣双書) Ja 70-12761 (v. 1)
Series romanized: Yūhikaku sōsho.
CONTENTS: 1. 総則・物権—2. 債権—3. 親族・相続

1. Civil law—Japan—Digests. I. Inui, Shōzō, 1928- joint author. II. Tsubaki, Toshio, joint author. III. Title.
76-799665

rev

Kai, Michitarō.
(Mimpō hanrei)
民法判例 / 甲斐道太郎, 乾昭三, 椿寿夫編. — 新版. — 東京：有斐閣, 1978-
v. ; 19 cm. — (有斐閣双書) Ja 78-5991 (v. 1)
Series romanized: Yūhikaku sōsho.
Includes indexes.
CONTENTS: 1. 総則物権—2. 債権
¥1400 (v. 1)
1. Civil law—Japan—Digests. I. Inui, Shōzō, 1928- joint author. II. Tsubaki, Toshio, joint author. III. Title.
79-800192

Suekawa, Hiroshi, 1892-1977.
(Hanrei Mimpō no rironteki kenkyū)
判例民法の理論的研究 / 末川博著. — 東京：日本評論社, 昭和17- [1942-
v. ; 22 cm.

1. Civil law—Japan—Digests. I. Title.
79-805349

Takashima, Heizō, 1922-
(Hanrei kyōshitsu Mimpō)
判例教室 民法 / 高島平蔵著. — 東京：法学書院, 1978.
6, 325, 2 p. ; 19 cm. Ja 79-581
Includes bibliographies and index.
¥1200

1. Civil law—Japan—Digests. I. Title.
79-803533

—JAPAN—EXAMINATIONS, QUESTIONS, ETC.

Taniguchi, Tomohei, 1906-
(Shimpan Mimpō enshū)
新版・民法演習 / 谷口知平, 加藤一郎編. — 東京：有斐閣, 1978-
v. ; 22 cm. — (有斐閣ブックス) Ja 78-12683 (v. 1)
Second ed. published in 1967 under title: Shin Mimpō enshū.
Series romanized: Yūhikaku bukkusu.
Includes indexes.
CONTENTS: 1. 総則
¥1200 (v. 1)

1. Civil law—Japan—Examinations, questions, etc. I. Katō, Ichirō, 1922- joint author. II. Title.
78-815636

*78 J

—JAPAN—HISTORY

Harada, Keikichi, 1903-1950.
(Nihon mimpōten no shiteki sobyō)
日本民法典の史的素描 / 原田慶吉著 ; 石井良助編. — 東京：創文社, 昭和29 [1954]
9, 3, 396, 11 p. ; 22 cm.
Includes bibliographical references and index.

1. Civil law—Japan—History. 2. Civil law—Europe—History. I. Ishii, Ryōsuke, 1907- II. Title.
79-803281

79 J

—JAPAN—HISTORY—ADDRESSES, ESSAYS, LECTURES

Kanazawa, Masayasu, 1903-1948.
(Hōseishi ronshū)
法制史論集 / 金澤理康著. — 東京：成文堂, 1978.
5, 347 p., [1] leaf of plates : port. ; 22 cm. Ja 78-20420
Includes bibliographical references.
CONTENTS: 日本法制史編：我利息法に対する支那法の影響 我国中世の損害賠償制度 大名の封建負担 明治初期に於ける消費貸借法の変遷 氏名と姓名—西洋法制史論：独逸中世の損害賠償制度 独逸古法ザクセン・シュピーゲルに現はれたる封建制度 独逸新世襲農園法論

¥5000
1. Civil law—Japan—History—Addresses, essays, lectures. 2. Civil law—Germany—History—Addresses, essays, lectures. 3. Feudal law—History—Addresses, essays, lectures. I. Title.
78-816041

Makino, Eiichi, 1878-1970.
(Mimpō no gojūnen to Keihō no yonjūnen)
民法の五十年と刑法の四十年 / 牧野英一. — 東京：中央公論社, 昭和24 [1949]
165 p. ; 19 cm. — (國民學術選書 ; 1)
Series romanized: Kokumin gakujutsu sensho.

1. Civil law—Japan—History—Addresses, essays, lectures. 2. Criminal law—Japan—History—Addresses, essays, lectures. I. Title.
79-804868

79

—JAPAN—INTERPRETATION AND CONSTRUCTION

Wagatsuma, Sakae, 1897-1973.
(Rengōbu hanketsu junreki)
聯合部判決巡歴 / 我妻栄著. — 東京：有斐閣, 昭和33- [1958-
v. ; 22 cm. — (ジュリスト選書)
Series romanized: Jurisuto sensho.
Includes indexes.
CONTENTS: 1. 総則・物権
1. Civil law—Japan—Interpretation and construction. 2. Property—Japan—Interpretation and construction. 3. Japan. Taishin'in. I. Title.
79-809848

—JAPAN—POPULAR WORKS

(Anata no hōritsu sōdan)
あなたの法律相談 / 樋口幸子, 坂本福子, 平山知子編. — 東京：新日本出版社, 1978.
253 p. : ill. ; 19 cm. Ja 78B-7233
¥950

1. Civil law—Japan—Popular works. 2. Domestic relations—Japan—Popular works. I. Higuchi, Sachiko. II. Sakamoto, Fukuko, 1932- III. Hirayama, Tomoko, 1939-
79-801872

Chūō Hōritsukai.
(Katei no hōritsu komon)
家庭の法律顧問 / 栗林敏夫鑑修 ; 中央法律倉編. — 東京：鷺ノ宮書房, 昭和30 [1955]
2, 426 p. : forms ; 19 cm.
Colophon inserted.
Previous ed. published under title: Sugu ni yakudatsu hōritsu komon.

1. Domestic relations—Japan—Popular works. 2. Civil law—Japan—Popular works. I. Kuribayashi, Toshio. II. Title.
79-804640

Chūō Hōritsukai.
(Katei no hōritsu komon)
家庭の法律顧問 / 栗林敏雄監修 ; 中央法律會編. — 東京：鷺の宮書房, 昭和32 [1957]
434 p. : forms ; 19 cm.
Colophon inserted.

1. Domestic relations—Japan—Popular works. 2. Civil law—Japan—Popular works. I. Kuribayashi, Toshio. II. Title.
79-804641

(Hōritsu sōdan)
法律相談 / 川島武宜 ... [et al.] 編. — 東京：岩波書店, 昭和32 [1957]
xi, 206 p. ; 19 cm. — (岩波新書 ; 291)
Series romanized: Iwanami shinsho.
Bibliography: p. 204-205.

1. Domestic relations—Japan—Popular works. 2. Civil law—Japan—Popular works. I. Kawashima, Takeyoshi, 1909-
79-804634

Kamba, Wataru.
(Ningen isshō no hōritsu to tetsuzuki shū)
人間一生の法律と手続集：イザというとき役に立つ / 樺且純著. — 東京：海南書房, 1978.
270 p. : ill. ; 19 cm. Ja 78B-6443
Colophon inserted.
¥1200

1. Domestic relations—Japan—Popular works. 2. Civil law—Japan—Popular works. I. Title.
79-802125

Kumakura, Takeshi, 1910-
(Shimin no hōritsu)
市民の法律 / 熊倉武, 宮川澄. — 東京：青木書店, 昭和32 [1957]
262 p. ; 18 cm. — (青木講座法律 ; 2)
Bibliography: p. 260-262.

1. Law—Japan—Popular works. 2. Civil law—Japan—Popular works. I. Miyakawa, Kiyoshi, 1917- joint author. II. Title. III. Series: Aoki kōza hōritsu ; 2.
79-804615

Mibuchi, Tadahiko, 1880-1950.
(Nichijō seikatsu to Mimpō)
日常生活と民法 / 三淵忠彦著 ; 關根小郷, 和田嘉子補修. — 東京：法曹会, 昭和25 [1950]
2, 2, 3, 242 p. ; 19 cm.

1. Civil law—Japan—Popular works. 2. Civil law—Popular works. I. Sekine, Kosato, 1905- II. Wada, Yoshiko. III. Title.
79-805158

CIVIL LAW
—JAPAN—POPULAR WORKS
(Continued)

Ogawa, Kagehito, 1923-
(Shūen no nou hau)
終焉のノウハウ：もしもの時のための知的対応 / 小川景士, 東藤生. — 東京：日本工業新聞社, 昭和52 [1977]
282 p. ; 19 cm. — (Ohtemachi books) Ja 77-19416
¥1000

1. Civil law—Japan—Popular works. 2. Social security—Law and legislation—Japan—Popular works. I. Azuma, Fujio, 1947- joint author. II. Title.

78-814599

Shimomitsu, Gunji, 1914-
(Minji hōritsu sōdan)
民事法律相談：わかりやすい解説付の書式と手続 / [著者 下光軍二] ; 朝日新聞法律相談所編. — 東京：元々社, 昭和30 [1955]
4, 8, 227, 12 p. ; 19 cm.
Includes index.

1. Civil law—Japan—Popular works. I. Asahi Shimbun Hōritsu Sōdanjo. II. Title.

79-803388

Tomiyama, Yoshiaki, 1934-
(Mimpō ni tsuyokunaru hon)
民法に強くなる本：難解な条文も簡単、カンタン / 富山義昭. — 東京：エール出版社, [1978]
203 p. : ill. ; 18 cm. — (Yell books) Ja 78-21523
¥780

1. Civil law—Japan—Popular works. I. Title.

79-803873

J

—KOREA

Kwŏn, Tae-yŏl.
(Minpŏp)
民法 / 權大烈, 劉哲均, 車漢成共編著. — 서울：進明文化社, 1978.
480 p. : 23 cm. — (司法·行政考試準備總整理書 : 2)
Series romanized: Sabŏp, haengjŏng kosi chunbi ch'ong chŏngnisŏ.
₩3000

1. Civil law—Korea. I. Yu, Ch'ŏl-gyun, joint author. II. Ch'a, Han-sŏng, joint author. III. Title.

78-826699

(Minpŏp yŏnsŭp)
民法演習 / 執筆代表金曾漢, 李英俊 ; 執筆者賈在桓 ... [et al.]. — 서울：一潮閣, 1976-
v. ; 22 cm. — (民法演習시리이즈 ; 1-
Colophon inserted.
Includes bibliographical references.
CONTENTS: v. 1. 總則編.
₩2000 (v. 1)

1. Civil law—Korea. 2. Civil law. I. Kim, Chŭng-han, 1920- II. Yi, Yŏng-jun, 1938- III. Ka, Chae-hwan.

78-826282

K

Yi, Chae-ch'ŏl, 1923-
(Minpŏp ch'ongnon)
民法總論 / 李在澈, 金起東共著. — 서울：博英社, 1974.
291 p. ; 23 cm. — (大學全書 : 法律學講義)
₩1600

1. Civil law—Korea. I. Kim, Ki-dong, joint author. II. Title.

75-825000

—KOREA—ADDRESSES, ESSAYS, LECTURES

(Minsa chaep'an ŭi chemunje)
民事裁判의諸問題 / 民事實務研究會編. — 서울：韓國司法行政學會, 1977-
v. ; 22 cm.
"限定版"
Includes legislation and bibliographical references.
₩4500 (v. 1)

1. Civil law—Korea—Addresses, essays, lectures. I. Minsa Silmu Yŏn'guhoe.

78-826603

—KOREA—EXAMINATIONS, QUESTIONS, ETC.

Yi, Chong-ch'an.
(Minsapŏp)
民事法：五·四級公務員採用·昇進試驗對策 / 李鍾燦, 南勝吉共編. — 서울：華學社：供給處韓國出版協同組合, 1977-
v. : 22 cm.
At head of cover title: Kaejŏng sihŏmnyŏng, sae ch'ulche kyŏnghyang e ttarŭn ch'oe sinp'an.
Colophon inserted.
"民法": v. 1, p. 393-452.
CONTENTS: 上. 主·客觀式民法.
₩1800 (v. 1)

1. Civil law—Korea—Examinations, questions, etc. 2. Civil procedure—Korea—Examinations, questions, etc. I. Nam, Sŭng-gil, joint author. II. Korea (Republic). Laws, statutes, etc. Minpŏp. 1977. III. Title.

79-825274

—LEBANON—DIGESTS

Bāz, Jamīl.
(Khulāṣat al-aḥkām al-ṣādirah 'an Maḥkamat al-Tamyīz al-Madanīyah khilāla sanat 1972)
خلاصة الاحكام الصادرة عن محكمة التمييز المدنيـة خلال سنة ١٩٧٢ / لخصها وجمعها جميـل باز. — [s. l. : s. n., 1973?]
(بيروت : مطبعة جوزف سليم صيقلي)
368 p. ; 28 cm.
«المجموعة العشرون.»
1. Civil law—Lebanon—Digests. 2. Lebanon. Maḥkamat al-Tamyīz. I. Lebanon. Maḥkamat al-Tamyīz. II. Title.
79-970046

—LUXEMBURG

Luxemburg. Laws, statutes, etc.
Code civil en vigueur dans le Grande-Duché de Luxembourg : annoté d'après la jurisprudence luxembourgeoise : mise à jour, législation, jusqu'au 31 décembre 1975, jurisprudence, pasicrisie lux. T. 22 inclus. — [Luxembourg] : Ministère de la justice, 1975.
505 p. ; 22 cm. Lu***
Includes index.
1. Civil law—Luxemburg. I. Luxemburg. Ministère de la justice. II. Title.
346'.4934 79-114815
79 MARC

—MEXICO

Peniche López, Edgardo.
Introducción al derecho y lecciones de derecho civil : adaptada al programa oficial aprobado por el Consejo Técnico y la Junta de Profesores de la Escuela Nacional de Comercio y Administración de la Universidad Nacional Autónoma de México, y puesto en vigor a partir del año lectivo de 1960 / Edgardo Peniche López. — 11. ed. — México : Editorial Porrúa, 1977.
320 p. : facsim. ; 22 cm.
1. Civil law—Mexico. I. Title.
340'.0972 79-370091
79 MARC

—MEXICO—CHIHUAHUA (STATE)

Chihuahua, Mexico (State). Laws, statutes, etc.
[Código civil (1973)]
Código civil del Estado de Chihuahua. — 1. ed. — México : Editorial Porrúa, 1978.
531 p. ; 17 cm. — (Leyes y códigos de México) (Colección Porrúa)
1. Civil law—Mexico—Chihuahua (State) I. Title.
346'.72 79-107159
79 MARC

—MEXICO—MEXICO (FEDERAL DISTRICT)

Mexico (Federal District). Laws, statutes, etc.
[Código civil (1928)]
Código civil para el Distrito Federal. — 44. ed. — México : Editorial Porrúa, 1978.
638 p. ; 17 cm. — (Leyes y códigos de México) (Colección Porrúa)
Includes index.
$30.00
1. Civil law—Mexico—Mexico (Federal District) 2. Civil law—Mexico—Quintana Roo (Ter.) I. Title.
346'.725 79-110301
79 MARC

Mexico (Federal District). Laws, statutes, etc.
[Código civil (1928)]
Código civil para el Distrito Federal. — 45. ed. — México : Editorial Porrúa, 1978.
638 p. ; 17 cm. — (Leyes y códigos de México) (Colección Porrúa)
Includes index.
$35.00
1. Civil law—Mexico—Mexico (Federal District) 2. Civil law—Mexico—Quintana Roo (Ter.)
346'.725 79-108211
79 MARC

—MEXICO—MEXICO (STATE)

Mexico (State). Laws, statutes, etc.
[Código civil (1956)]
Código civil del Estado de México. — 1. ed. — México : Editorial Porrúa, 1978.
525 p. ; 17 cm. — (Leyes y códigos de México) (Colección Porrúa)
1. Civil law—Mexico—Mexico (State) I. Title.
346'.725 79-107664
79 MARC

—MEXICO—NUEVO LEÓN

Nuevo León, Mexico. Laws, statutes, etc.
[Código civil (1935)]
Código civil para el Estado de Nuevo León. — 1. ed. — México : Editorial Porrúa, 1978.
509 p. ; 17 cm. — (Leyes y códigos de México) (Colección Porrúa)
1. Civil law—Mexico—Nuevo León. I. Title.
346'.721 79-107156
79 MARC

—MEXICO—QUINTANA ROO (TER.)

Mexico (Federal District). Laws, statutes, etc.
[Código civil (1928)]
Código civil para el Distrito Federal. — 44. ed. — México : Editorial Porrúa, 1978.
638 p. ; 17 cm. — (Leyes y códigos de México) (Colección Porrúa)
Includes index.
$30.00
1. Civil law—Mexico—Mexico (Federal District) 2. Civil law—Mexico—Quintana Roo (Ter.) I. Title.
346'.725 79-110301
79 MARC

Mexico (Federal District). Laws, statutes, etc.
[Código civil (1928)]
Código civil para el Distrito Federal. — 45. ed. — México : Editorial Porrúa, 1978.
638 p. ; 17 cm. — (Leyes y códigos de México) (Colección Porrúa)
Includes index.
$35.00
1. Civil law—Mexico—Mexico (Federal District) 2. Civil law—Mexico—Quintana Roo (Ter.)
346'.725 79-108211
79 MARC

—NEPAL

Nepal. Laws, statutes, etc.
(Mulukī aina)
मुलुकी ऐन : सातौं संशोधनसमेत मिलाइएको रूपमा. — [काठमाडौं] : श्री ५ को सरकार, कानून तथा न्याय मन्त्रालय, कानून किताब व्यवस्था समिति, 2034 [1977]
4, 214 p. ; 23 cm.
In Nepali.
Rs10.00

1. Civil law—Nepal. I. Title.

78-903597

CIVIL LAW (Continued)

—NETHERLANDS

Feenstra, Robert, 1920-
Inleidende hoofdstukken betreffende de Romeinsrechtelijke grondslagen van het Nederlands privaatrecht / R. Feenstra. — 2e uitgebreide en herz. druk. — Den Haag : Universitaire Pers Leiden, 1978.
xxiv, 410 p. ; 23 cm. Ne78-48
Bibliography: p. xxiii.
Includes indexes.
ISBN 9060214374 : fl 42.31
1. Law—Netherlands—Roman influence. 2. Civil law—Netherlands. I. Title.
79-358319
*79 MARC

Fluitman, J P M
Sprekend recht / J. P. M. Fluitman en J. F. Scheers, met medewerking van E. Boelen, F. H. L. Schut, F. Th. N. C. Twaalhoven. — 10e gew. dr. — Culemborg : Tjeenk Willink-Noorduijn, ₍1977₎
279 p. ; 23 cm. & suppl. (15 p.) Ne77
Includes index.
ISBN 9011822005 : fl 17.90
1. Civil law—Netherlands. I. Scheers, J. F., joint author. II. Title.
78-397466
*79 MARC

Netherlands (Kingdom, 1815-). Laws, statutes, etc.
Burgerlijk wetboek : internationale regelingen / bew. door E. Lukács. — Zwolle : W. E. J. Tjeenk Willink, 1978.
ix, 406 p. ; 19 x 8 cm. — (Nederlandse wetboeken) Ne78-45
Dutch, French, and/or English.
"Behoort bij de 29e druk van het Burgerlijk wetboek."
"Bijgewerkt tot 1 januari 1978."
ISBN 9027114994 : fl 24.50
1. Civil law—Netherlands. I. Lukács, E. II. Title. III. Series.
346'.492 79-337705
79 MARC

Ploeg, T J van der.
Het bergerlijk recht en de vrijwillege organisaties : beschouwingen n. a. v. de regeling van de vereniging en de stichting in Boek 2 B. W. / door T. J. van der Ploeg. — Deventer : Kluwer, 1978.
x, 226 p. ; 25 cm. — (Vennootschaps- en rechtspersonenrecht ; deel 17) Ne78-49
Summary in German.
Bibliography: p. 209-211.
Includes indexes.
ISBN 9026810407 : fl 39.00
1. Associations, institutions, etc.—Law and legislation—Netherlands—History. 2. Corporations, Nonprofit—Netherlands. 3. Civil law—Netherlands. I. Title.
79-354224
*79 MARC

Tak, Antonius Quirinus Cornelis, 1942-
Overheidsbestuur en privaatrecht / A. Q. C. Tak. — Alphen aan den Rijn : Samsom, 1978.
424 p. ; 26 cm. — (Hand- en leerboek der bestuurswetenschappen ; deel 23) Ne***
Includes bibliographical references and indexes.
ISBN 9014027621
1. Administrative law—Netherlands. 2. Civil law—Netherlands. 3. Judicial review of administrative acts—Netherlands. I. Title. II. Series.
79-337802
79 MARC

—NETHERLANDS—DIGESTS

Netherlands (Kingdom, 1815-). Hooge Raad.
Arresten over burgerlijk recht en handelsrecht voor beknopt privaatrecht / H. J. N. Boskamp, J. L. P. Cahen. — Alphen aan den Rijn : H. D. Tijeenk Willink, 1978.
376 p. ; 20 cm. Ne78-23
Includes indexes.
ISBN 9060920236 : fl 22.50
1. Civil law—Netherlands—Digests. 2. Commercial law—Netherlands—Digests. I. Boskamp, H. J. N. II. Cahen, J. L. P. III. Title.
78-374967
*78 MARC

—NETHERLANDS—HANDBOOKS, MANUALS, ETC.

Croes, A L
Privaatrecht voor de makelaar o.g. / A. L. Croes. — Alphen aan den Rijn : Samsom, 1978.
283 p. ; 25 cm. Ne78-39
Issued with the collaboration of the Instituut voor Sociale Wetenschappen te Den Haag.
ISBN 9014027737 : fl 42.50
1. Civil law—Netherlands—Handbooks, manuals, etc. I. Instituut voor Sociale Wetenschappen te Den Haag. II. Title.
78-394328
*79 MARC

—PAKISTAN—DIGESTS

Farani, Merajuddin, 1935-
Twenty years' index of cases overruled, reversed, dissented from, not approved, disapproved, not followed, etc. and, Annual digest, 1966, civil, criminal, and revenue / by M. Farani. — Lahore : Shah Nawaz Bros., 1967.
xiv, 115 p. ; 25 cm.
Cover title.
1. Civil law—Pakistan—Digests. 2. Criminal law—Pakistan—Digests. 3. Taxation—Pakistan—Digests. I. Title. II. Title: Annual digest, 1966, civil, criminal, and revenue.
348'.5491'046 78-984057
78 MARC

Masood, M
Thirty one years civil refrencer, 1947-77 / M. Masood, Muhammad Bashir. — Lahore : Eastern Law House, ₍197-₎
270 p. ; 23 cm.
Rs30.00
1. Civil law—Pakistan—Digests. I. Bashir, Muhammad, Advocate. II. Title.
348'.5491'026 78-930430
79 MARC

—PANAMA—DIGESTS

Jurisprudencia civil : recensiones de fallos dictados por la Corte Suprema de Justicia y Tribunales Superiores en materia civil tomados del Registro judicial (1940-1957) y del Repertorio jurídico (1961-1964). — Panamá : Universidad de Panamá, Facultad de Derecho y Ciencias Políticas, Sección de Investigación Jurídica, 1968.
180 p. ; 23 cm.
Cover title.
Includes index.
1. Civil law—Panama—Digests. I. Panama. Corte Suprema de Justicia. II. Panama (City). Universidad. Sección de Investigación Jurídica.
346'.7287 79-101412
79 MARC

—PERU

Peru. Laws, statutes, etc.
₍Código civil (1936)₎
Código civil : actualizado con las últimas disposiciones, legislación modificatoria y conexa : D.L. 19983 (13/4/1973) no precisa testigos, inscripción de nacimientos ni defunción, D.L. 19987 (13/4/1973) inscripción de nacimientos de oficio / ₍compilado₎ por F. Bonilla. — Lima : Editorial Mercurio : ₍Importadora y Distribuidora Rivera, 1978₎
453 p. ; 20 cm. — (Legislación peruana)
S/400.00
1. Civil law—Peru. I. Bonilla, F.
346'.85 78-111017
79 MARC

Peru. Laws, statutes, etc.
₍Código civil (1936)₎
Código civil : concordancias y jurisprudencia de la Corte Suprema al día / ₍editor₎ Jorge Eugenio Castañeda. — 6. ed. — Lima : ₍s.n.₎, 1978-
2 v. ; 25 cm.
Includes indexes.
1. Civil law—Peru. I. Castañeda, Jorge Eugenio. II. Peru. Corte Suprema. III. Title.
346'.85 79-102569
79 MARC

—PHILIPPINE ISLANDS

Philippines (Republic). Laws, statutes, etc.
Civil code of the Philippines (Republic Act 386 as amended) : with related special laws and Presidential decrees / prepared by NBS editorial staff. — 1977 ed. — ₍Manila₎ : National Book Store, 1977.
xi, 528 p. ; 18 cm.
Appendices: p. 429-528 : A. Child and youth welfare code.—B. Presidential decree no. 604.—C. Civil registry law.—D. Condominium act.—E. Uniform currency act.—F. Maceda law.
1. Civil law—Philippine Islands. I. National Book Store. II. Title.
346'.599'002632 79-102161
79 MARC

—POLAND

Kopff, Andrzej.
Konstrukcje cywilistyczne w prawie wynalazczości / Andrzej Kopff. — Wyd. 1. — Warszawa : Państ. Wydaw. Naukowe, 1978 (Łódź : ŁZG 3)
282, ₍2₎ p. ; 20 cm. P79-56
Summary in French.
Includes bibliographical references.
zł42.00
1. Patent laws and legislation—Poland. 2. Inventions—Poland. 3. Civil law—Poland. I. Title.
79-381246
79 MARC

Poland. Laws, statutes, etc.
Kodeks cywilny ; Kodeks rodzinny i opiekuńczy ; Kodeks postępowania cywilniego; Prawo prywatne międzynarodowe : stan prawny na dzień 1 września 1977 r. / ₍redaktor Sabina Chorążkowa₎. — Wyd. 3. — Warszawa : Wydaw. Prawnicze, 1978 (Katow.: ZG 3)
519, ₍1₎ p., ; 22 cm. P78-1890
zł90.00
1. Civil law—Poland. 2. Domestic relations—Poland. 3. Conflict of laws—Poland. I. Chorążkowa, Sabina. II. Title.
78-384094
78 MARC

Polish civil law. Edited by D. Lasok. Leiden, Sijthoff, 1973-75.
4 v. 24 cm. (Law in Eastern Europe, no. 184 (1-4)) fl52.00 (v. 1) Ne73-29
Bibliography: v. 1, p. 203-212.
Vol. 2 by W. J. Wagner, has title: Obligations in Polish law.
ISBN 9028603026 (v. 1)
1. Civil law—Poland. I. Lasok, Dominik, ed. II. Wagner, Wenceslas J. III. Series.
340'.09438 72-89167
74₍r79₎rev MARC

Sośniak, Mieczysław.
Podstawy prawa cywilnego i rodzinnego : zarys dla Studium Administracyjnego / Mieczysław Sośniak. — Katowice : Uniwersytet Śląski, 1977 (Katow. : UŚ)
₍2₎, 503 p. ; 25 cm. P78-2303
"155"
Bibliography: p. ₍491₎-494.
zł61.00
1. Civil law—Poland. 2. Domestic relations—Poland. I. Title.
79-344501
79 MARC

—PORTUGAL

Machado, João Baptista, fl. 1968-
Sobre a aplicação no tempo do novo Código Civil; casos de aplicação imediata, critérios fundamentais ₍por₎ J. Baptista Machado. Coimbra, Livraria Almedina, 1968.
374 p. 24 cm.
Bibliographical footnotes.
1. Civil law—Portugal. 2. Time (Law)—Portugal. I. Title.
75-401605
79₍r79₎rev MARC

Portugal. Laws, statutes, etc.
₍Código civil (1966)₎
Código civil anotado : doutrina, jurisprudência, vocabulário jurídico / Abílio Neto, Herlânder A. Martins. — Lisboa : Livraria Petrony, 1978.
1250 p. ; 25 cm.
Includes index.
1. Civil law—Portugal. I. Neto, Abílio. II. Martins, Herlânder A.
346'.469 79-112741
79 MARC

Portugal. Laws, statutes, etc.
₍Código civil (1966)₎
Código civil : aprovado pelo Decreto lei no. 47344, de 25 de novembro de 1966 : inclui legislação complementar / textos revisto, notas e índice alfabético remissivo por Helder Martins Leitão. — Porto : Portugal Jurídico-Económico, 1978.
1063 p. ; 21 cm. — (Colecção Quid juris?)
Includes index.
1. Civil law—Portugal. I. Leitão, Helder Martins. II. Series.
346'.469 78-109855
79 MARC

Portugal. Laws, statutes, etc.
₍Código civil (1966)₎
Código civil e legislação complementar / actualizados e anotados com notas explicativas, jurisprudência e doutrina, Rabindranath Capelo de Sousa, José António de França Pitão. — Coimbra : Livraria Almedina, 1978-
v. ; 24 cm.
Includes indexes.
CONTENTS: v. 1. Das leis, relações jurídicas, obrigações em geral e contratos em especial, art.os 1o. a 1250.
1. Civil law—Portugal. I. Sousa, Rabindranath Capelo de. II. Pitão, José António de França. III. Title.
346'.469 78-109845
79 MARC

Portugal. Laws, statutes, etc.
₍Código civil (1966)₎
Código civil português : anotado e actualizado / Jacinto Fernandes Rodrigues Bastos. — 4. ed. — Coimbra : Livraria Almedina, 1978.
876 p. ; 23 cm.
1. Civil law—Portugal. I. Bastos, Jacinto Fernandes Rodrigues.
346'.469 78-109840
79 MARC

CIVIL LAW

—PORTUGAL (Continued)

Portugal. Laws, statutes, etc.
⌊Código civil (1966)⌋
Código civil português : com as novas alterações introduzidas pelo Decreto-lei no. 496/77 de 25 de novembro / anotações de Carlos Adrião Rodrigues. — 1. ed. — Lisboa : Moraes, 1977.
782 p. ; 23 cm. — (Colecção Códigos)
1. Civil law—Portugal. I. Rodrigues, Carlos Adrião.
346'.469 78-109857
79 MARC

Portugal. Laws, statutes, etc.
Código civil e legislação complementar : direito de associação, arrendamento urbano e rural ... / dirigida e rev. por Artur Anselmo de Castro (Filho), José António Pereira da Silva, Rui Afonso. — Lisboa : ⌊Leiabril, Distribuidora Editorial⌋, 1978.
623 p. ; 24 cm.
1. Civil law—Portugal. I. Castro, Artur Anselmo de. II. Silva, José António Pereira da. III. Afonso, Rui.
346'.469 78-109842
79 MARC

—PRUSSIA

Koch, Christian Friedrich, 1798-1872.
Lehrbuch des preussischen gemeinen Privatrechts / von C. F. Koch. — Berlin : T. Trautwein, 1845-1846.
2 v. ; 22 cm.
Includes bibliographical references and index.
1. Civil law—Prussia. I. Title.
340'.0943 79-362539
79 MARC

Koch, Christian Friedrich, 1798-1872.
Lehrbuch des preussischen gemeinen Privatrechts / von C. F. Koch. — 2. verb. und verm. Aufl. — Berlin : T. Trautwein, 1851-1852.
2 v. ; 22 cm.
Includes bibliographical references and index.
1. Civil law—Prussia. I. Title.
340'.0943 79-362538
79 MARC

—ROME see Roman law

—RUSSIA—ADDRESSES, ESSAYS, LECTURES

(Pravovye voprosy grazhdanskogo prava i prot͡sessa)
Правовые вопросы гражданского права и процесса : Темат. сборник / Дальневост. гос. ун-т ; Под ред. проф. Н. И. Овчинникова. — Владивосток : ДВГУ, 1976.
111 p. ; 20 cm. USSR 77
At head of title: Ministerstvo vysshego i srednego spet͡sial'nogo obrazovanii͡a RSFSR.
Includes bibliographical references.
0.65rub
1. Civil law—Russia—Addresses, essays, lectures. 2. Civil procedure—Russia—Addresses, essays, lectures. I. Ovchinnikov, Nikolaĭ Ivanovich. II. Vladivostok. Dal'nevostochnyĭ universitet.
78-422895

(Voprosy grazhdanskogo prava i prot͡sessa)
Вопросы гражданского права и процесса : ⌊Сборник статей⌋ / АН СССР, Ин-т государства и права ; ⌊Ред. коллегия д-р юрид. наук А. Ю. Кабалкин ... et al.⌋. — Москва : Ин-т государства и права, 1977.
106 p. ; 20 cm. USSR 77
Includes bibliographical references.
0.60rub
1. Civil law—Russia—Addresses, essays, lectures. 2. Civil procedure—Russia—Addresses, essays, lectures. I. Kabalkin, Aleksandr I͡Ur'evich. II. Akademii͡a nauk SSSR. Institut gosudarstva i prava.
78-421564

—RUSSIA—OUTLINES, SYLLABI, ETC.

Gribanov, Veniamin Petrovich.
(Chto nado znat' kazhdomu o grazhdanskom zakonodatel'stve)
Что надо знать каждому о гражданском законодательстве : (Пособие для слушателей) / В. П. Грибанов. — Москва : Знание, 1978.
127 p. ; 21 cm. — (Народный университет : Факультет правовых знаний) USSR 78
Series romanized: Narodnyĭ universitet : Kakul'tet pravovykh znaniĭ.
Bibliography: p. ⌊126⌋
0.25rub
1. Civil law—Russia—Outlines, syllabi, etc. I. Title.
78-421852

—SPAIN

Albaladejo García, Manuel.
Curso de derecho civil español común y foral / Manuel Albaladejo. — Barcelona : Bosch, 1977.
2 v. ; 24 cm. Sp***
Includes bibliographies.
CONTENTS: 1. Introducción y parte general.—2. Derecho de obligaciones.
ISBN 8472940756 (v. 1). ISBN 8472940691 (v. 2)
1. Civil law—Spain. I. Title.
346'.46 78-349310
78⌊r79⌋rev MARC

Albaladejo García, Manuel.
Derecho civil / ⌊por⌋ Manuel Albaladejo. — 5. ed. — Barcelona : Bosch, 19
v. ; 24 cm. Sp78-Jan
Includes bibliographical references.
CONTENTS: t. 1. Introducción y parte general, v. 1. v. 2. La relación, las cosas y los hechos jurídicos.

ISBN 8472941035 : 1000ptas (t. 1, v. 2)
1. Civil law—Spain. I. Title.
340'.0946 78-396849
79 MARC

Albaladejo García, Manuel.
Derecho civil / ⌊por⌋ Manuel Albaladejo. — 6. ed. — Barcelona : Bosch, 1978-
v. ; 24 cm. Sp78-Jan
Bibliography: t. 1, v. 1, p. 15.
CONTENTS: t. 1. Introducción y parte general, v. 1. Introducción y derecho de la persona.

ISBN 8472941043 : 1000ptas (t. 1, v. 1)
1. Civil law—Spain. I. Title.
78-394271
79 MARC

Badenes Gasset, Ramón.
Conceptos fundamentales del derecho : las relaciones jurídicas patrimoniales / Ramón Badenes Gasset. — 3. ed. — Barcelona : Marcombo, D.L. 1977.
325 p., 1 leaf ; 22 cm. Sp78-June
ISBN 8426702163
1. Civil law—Spain. I. Title.
78-391852
79 MARC

Castán Tobeñas, José.
Derecho civil español, común y foral / ⌊por⌋ José Castán Tobeñas ... — 12. ed. rev. y puesta al día / por José Luis de los Mozos ... — Madrid : Reus, 1978-
v. ; 24 cm. Sp78-Oct
Includes bibliographical references.
CONTENTS: t. 1. Introducción y parte general. v.

ISBN 842901019X : 2200ptas (v. 1, pt. 2)
1. Civil law—Spain. I. Mozos, José Luis de los. II. Title.
340'.0946 79-370241
79 MARC

Comentarios al Código civil y compilaciones forales / dirigidos por Manuel Albaladejo. — Jaén : Editorial Revista de Derecho Privado, c1978-
v. ; 22 cm. Sp***
Includes bibliographical references.
ISBN 8471302160
1. Civil law—Spain. I. Albaladejo García, Manuel. II. Spain. Laws, statutes, etc. Código civil. 1978.
78-394005
79 MARC

Díez-Picazo y Ponce de León, Luis.
Sistema de derecho civil / ⌊por⌋ Luis Díez-Picazo y Antonio Gullón. — 2. ed. — Madrid : Tecnos, 1977-
v. ; 22 cm. — (Biblioteca universitaria de Editorial Tecnos) Sp78-Jan
Includes bibliographies.
CONTENTS: v. 1. Introducción. Derecho de la persona. Negocio jurídico.

ISBN 8430907270
1. Civil law—Spain. I. Gullón Ballesteros, Antonio, joint author. II. Title.
346'.46 78-389656
79 MARC

Díez-Picazo y Ponce de León, Luis.
Sistema de derecho civil / Luis Díez-Picazo y Antonio Gullón. — Madrid : Editorial Tecnos, 1975-1978.
4 v. ; 22 cm. — (Biblioteca universitaria de Editorial Tecnos) Sp*
Includes bibliographies.
CONTENTS: v. 1. Introducción. Derecho de la persona. Negocio jurídico. —v. 2. Teoría general del contrato. La relación obligatoria en general. Las relaciones obligatorias en particular—v. 3. Derecho de cosas.—v. 4. Derecho de familia. Derecho de sucesiones.
ISBN 843090588X
1. Civil law—Spain. I. Gullón Ballesteros, Antonio, joint author. II. Title.
346'.46 75-522421
76⌊r79⌋rev MARC

López Sanz, Salvador.
Elementos de derecho : especialmente civil y mercantil / Salvador López Sanz. — 6 ed. — Valencia : Editorial Bello, 1978.
494 p. ; 24 cm. Sp***
ISBN 8421200461
1. Civil law—Spain. 2. Commercial law—Spain. 3. Law—Spain. I. Title.
340'.0946 79-374137
79 MARC

Puig Peña, Federico.
Compendio de derecho civil español / ⌊por⌋ Federico Puig Peña. — 3. ed., rev. y puesta al día. — Madrid : Pirámide, D.L. 1976.
6 v. ; 24 cm. Sp76-Dec
Bibliography: v. 6, p. ⌊747⌋-788.
Includes index.
CONTENTS: 1. Parte general.—2. Derechos reales.—3. Obligaciones y contratos.—4. Contratos.—5. Familia y sucesiones.—6. Sucesiones.
ISBN 8436800435
1. Civil law—Spain. I. Title.
346'.46 79-371578
79 MARC

Sancho Rebullida, Francisco de Asís.
Estudios de derecho civil / Francisco de Asís Sancho Rebullida. — Pamploma : Ediciones Universidad de Navarra, 1978.
2 v. ; 22 cm. Sp***
Includes bibliographical references.
ISBN 8431305363
1. Civil law—Spain. 2. Civil law. I. Title.
346'.46 79-354282
79 MARC

—SPAIN—ADDRESSES, ESSAYS, LECTURES

Estudios de derecho civil en honor del Prof. Batlle Vázquez. — Madrid : Editoriales de Derecho Reunidas, D.L. 1978.
854 p. : port. ; 22 cm. Sp78-Apr
Errata slip inserted.
Includes bibliographical references.
ISBN 847130211X
1. Civil law—Spain—Addresses, essays, lectures. 2. Batlle y Vázquez, Manuel. I. Batlle y Vázquez, Manuel.
340'.0946 78-381849
78 MARC

Estudios sobre el título preliminar del código civil. — Madrid : Editoriales de Derecho Reunidas, 1977 printing.
2 v. ; 22 cm. — (Estudios - Academia Matritense del Notariado ; t. 1) Sp78-Jan
Includes bibliographical references.
CONTENTS: v. 1. Parte general.—v. 2. Derecho internacional privado.
ISBN 8471301954 : 1850ptas
1. Civil law—Spain—Addresses, essays, lectures. I. Series: Academia Matritense del Notariado. Estudios - Academia Matritense del Notariado ; t. 1.
346'.81 78-384809
79 MARC

—ST. LUCIA—HISTORY

Liverpool, N J O
The history and development of the St. Lucia civil code / N. J. O. Liverpool. — Cave Hill, Barbados : Institute of Social and Economic Research (Eastern Caribbean), University of the West Indies, c1977.
iv, 41 p. ; 21 cm. — (Occasional paper - Institute of Social and Economic Research (Eastern Caribbean), University of the West Indies ; no. 5)
Includes bibliographical references.
1. Civil law—St. Lucia—History. I. Title. II. Series: University of the West Indies, Cave Hill, Barbados. Institute of Social and Economic Research. Occasional paper - Institute of Social and Economic Research (Eastern Caribbean), University of the West Indies, Cave Hill, Barbados ; no. 5.
346'.729843'09 78-107601
79 MARC

—SWEDEN

Civilrätt : grundläggande lärobok / utarb. av Bertil Brinck ... ⌊et al.⌋. — 5., omarb. uppl. — Umeå : Brinck, 1978.
382 p. ; 25 cm. S78
Includes index.
ISBN 9197031801
1. Civil law—Sweden. I. Brinck, Bertil, 1916-
346'.485 79-354964
79 MARC

—SWITZERLAND

Schnyder, Bernhard.
Supplement Kindesrecht / von Bernhard Schnyder. — Zürich : Schulthess Polygraphischer Verlag, c1977.
xiii, 101 p. ; 23 cm. Sw***
Supplement to the 9th ed. of Tuor, Peter, 1876-1957. Das Schweizerische Zivilgesetzbuch, published in 1975.
Includes bibliographical references and index.
ISBN 3-7255-1857-2
1. Civil law—Switzerland. 2. Children—Law—Switzerland. I. Tuor, Peter, 1876-1957. Das Schweizerische Zivilgesetzbuch. II. Title. III. Title: Kindesrecht.
79-340248
79 MARC

CIVIL LAW
—SWITZERLAND (Continued)

Switzerland. Laws, statutes, etc.
⌐Zivilgesetzbuch¬
Schweizerisches Zivilgesetzbuch : (vom 10. Dezember 1907). — (Stand am 1. Okt. 1975). — Bern : Schweizerische Bundeskanzlei : Auslfg., Eidgen. Drucksachen- und Materialzentrale, 1975.
254 p. ; 21 cm. Sw76-1245
7.00F
1. Civil law—Switzerland. I. Title.
340′.09494 76-460658
*76⌐r79¬rev MARC

Switzerland. Laws, statutes, etc.
⌐Zivilgesetzbuch. French¬
Code civil suisse et Code des obligations : annotés / Georges Scyboz, ... Pierre-Robert Gilliéron, ... ; introd. historique de Jean Gauthier, ... — 2e éd. mise à jour / publ. sous la direction de Pierre Cavin. — Lausanne : Payot, 1977.
xx, 700 (p. 688-700 blank), 558 p. ; 19 cm. Sw78-58
Cover title: CC & CO annotés.
Includes indexes.
ISBN 2-601-00381-2 : 98.00F
1. Civil law—Switzerland. 2. Obligations (Law)—Switzerland. I. Scyboz, Georges. II. Gilliéron, Pierre Robert. III. Cavin, Pierre. IV. Switzerland. Laws, statutes, etc. Obligationenrecht (1911). French. 1977. V. Title.
346′.494 79-358313
79 MARC

Switzerland. Laws, statutes, etc.
⌐Zivilgesetzbuch¬
Schweizerisches Zivilgesetzbuch : Textausgabe mit Einleitung, Anmerkungen, Ausführungserlassen und Sachregister / W. Schönenberger. — 33. Aufl. — Zürich : Schulthess, 1979.
xxiii, 535, 44 p. ; 19 cm. Sw79-5090
Includes bibliographical references and index.
ISBN 3-7255-1965-X : 38.00F
1. Civil law—Switzerland. I. Schönenberger, Wilhelm, 1898- . II. Title.
346′.494 79-381211
79 MARC

Switzerland. Laws, statutes, etc.
⌐Zivilgesetzbuch¬
Schweizerisches Zivilgesetzbuch vom 10. Dezember 1907 mit einschlägigen Nebengesetzen und Verordnungen / mit Anmerkungen und Sachregister, hrsg. von W. Stauffacher und Heinz Aeppli. — 20., überarb. Aufl. — Zürich : Orell Füssli, cop. 1979.
xxxiv, 510 p. ; 16 cm. — (Taschenausgaben der Bundesgesetze)
Sw79-3105
Cover title: ZGB/OR, Schweizerisches Zivilgesetzbuch und Obligationenrecht.
Issued with Switzerland. Laws, statutes, etc. Obligationenrecht (1911). Zürich, c1979.
Includes index.
ISBN 3-280-01005-5 : 38.00F
1. Civil law—Switzerland. I. Stauffacher, Werner. II. Aeppli, Heinz, 1925- . III. Title. IV. Title: ZGB/OR. V. Title: Schweizerisches Zivilgesetzbuch und Obligationenrecht.
79-376165
79 MARC

Williams, Ivy.
The sources of law in the Swiss civil code / by Ivy Williams. — Zurich : ReMak Verlag, c1976.
199 p. ; 22 cm. Sw***
Reprint of the 1923 ed. published by Oxford University Press, London.
Label mounted on t.p.: Distributed in U.S.A. by Fred B. Rothman, Littleton, Conn.
Bibliography: p. ⌐190¬-191.
Includes index.
ISBN 3-85856-004-9
1. Civil law—Switzerland. I. Title.
346′.494 79-318359
79 MARC

—SWITZERLAND—ADDRESSES, ESSAYS, LECTURES

Oftinger, Karl, 1909-
Ausgewählte Schriften / Karl Oftinger ; hrsg. von den Mitgliedern der Juristischen Abteilung der Universität Zürich. — Zürich : Schulthess, 1978.
xv, 407 p., ⌐1¬ leaf of plates : port. ; 25 cm. Sw78-11247
Bibliography of K. Oftinger's works: p. 403-407.
Includes bibliographical references.
ISBN 3-7255-1912-9 : 65.00F
1. Civil law—Switzerland—Addresses, essays, lectures. 2. Oftinger, Karl, 1909- I. Zürich. Universität. Juristische Abteilung.
346′.494 79-345862
79 MARC

—SYRIA

Sawwār, Muḥammad Waḥīd al-Dīn.
(Sharḥ al-qānūn al-madanī)
شرح القانون المدني : الحقوق العينية / محمد وحيد الدين سوار. — الطبعة 1. — ⌐دمشق؟ : مديرية الكتب الجامعية¬، 1968/1969-.
v. ; 25 cm.
Bibliography: v. 1, p. 5-9 (1st group)
CONTENTS: 1. الحقوق العينية الأصلية. —
1. Civil law—Syria. 2. Property—Syria. I. Title. II. Title: al-Ḥuqūq al-'aynīyah.
79-970236

—TAIWAN

Chiang, I-chu.
(Kou wu chih nan)
購屋指南 / 編著者江儀助. — 台北：建宏出版社：建宏書局, 民國67 ⌐1978¬
2, 8, 216 p. ; 21 cm.
Colophon title.
NT$100.00
1. Vendors and purchasers—Taiwan. 2. Civil law—Taiwan. I. Title.
78-841282

Fan, Heng-sheng.
(Min fa kai yao)
民法概要 / 范衡生著. — 臺北：五南圖書出版公司, 民國67 ⌐1978¬
5, 513, 8 p. ; 21 cm.
"大學用書"
NT$140.00
1. Civil law—Taiwan. I. Title.
79-836747

P'an, Wei-ho.
(Chung-kuo min fa ching i)
中國民法精義 / 潘維和著. — 台北：漢林出版社, 民國66 ⌐1977¬
2, 22, 566 p. ; 22 cm. — (東亞法律叢書)
At head of title: Ta hsüeh yung shu.
Series romanized: Tung Ya fa lü ts'ung shu.
Bibliography: p. 563-566.
1. Civil law—Taiwan. I. Title.
78-843316

Sun, Ting-fang.
(Min fa yao i)
民法要義 / 孫定方編著. — 台北：孫定方：總經銷三民書局, 民國67 ⌐1978¬
2, 20, 360 p. ; 22 cm.
Bibliography: p. 359-360.
NT$100.00
1. Civil law—Taiwan. I. Title.
79-837807

(Tsui hsin fa lü shih wu wen t'i hui pien)
最新法律實務問題彙編 / 夏正台編著. — 再版. — ⌐台北¬：五南圖書出版公司, 民國66 ⌐1977¬
737 p. ; 22 cm.
"本書所蒐集之資料係從司法專刊自民國五十七年七月起至民國六十三年六月止所載各法院舉行司法座談會所討論之法律問題而成"
NT$200.00
1. Civil law—Taiwan. 2. Criminal law—Taiwan. I. Hsia, Cheng-t'ai.
78-842890

—TAIWAN—DIGESTS

China (Republic of China). Tsui kao fa yüan.
(Tsui kao fa yüan min shih ts'ai p'an yao chih hui pien)
最高法院民事裁判要旨彙編 / 司法行政部民事司編輯. — 台北：司法通訊社, 民國66 ⌐1977¬
656 p. ; 21 cm.
Includes index.
"最高法院六十一年至六十六年十月民庭庭推總會議紀錄": p. 535-656.
NT$100.00
1. Civil law—Taiwan—Digests. I. China (Republic of China). Ssu fa hsing cheng pu. Min shih ssu. II. Title.
78-843943

(Chung-hua min kuo min hsing t'e pieh fa p'an chieh shih i ch'üan shu)
中華民國民刑特別法判解釋義全書 / 編輯者新陸書局編輯部. — 台北：新陸書局, 民國65 ⌐1976¬
12, 1466 p. ; 20 cm.
NT$230.00
1. Civil law—Taiwan—Digests. 2. Criminal law—Taiwan—Digests. I. Hsin lu shu chü, T'ai-pei. Pien chi pu. II. China (Republic of China). Ssu fa yüan. III. China (Republic of China). Tsui kao fa yüan. IV. China (Republic of China). Laws, statutes, etc. Min fa. 1976. V. China (Republic of China). Laws, statutes, etc. Hsing fa. 1976.
78-842493
78 C

—TAIWAN—EXAMINATIONS, QUESTIONS, ETC.

Li, Mo.
(Min fa wen t'i yen chiu)
民法問題研究 / 李模著. — 再版. — ⌐台北?¬：李模, 民國67 ⌐1978¬
2, 12, 256 p. ; 22 cm.
NT$100.00
1. Civil law—Taiwan—Examinations, questions, etc. I. Title.
78-843524
78 C

—TURKEY

Akıntürk, Turgut.
Bankacılar için medeni hukuk bilgisi / Turgut Akıntürk. — 3. bası. — Cebeci-Ankara : Banka ve Ticaret Hukuku Araştırma Enstitüsü, Türkiye İş Bankası Tesisi-Hukuk Fakültesi, 1973.
xxxvi, 309 p. ; 17 cm. — (Bankacılar serisi ; no. 14) (Yayın - Banka ve Ticaret Hukuku Araştırma Enstitüsü ; no. 96)
Bibliography: p. xxxiii-xxxiv.
1. Civil law—Turkey. I. Title. II. Title: Medeni hukuk bilgisi. III. Series: Ankara. Üniversite. Banka ve Ticaret Hukuku Araştırma Enstitüsü. Bankacılar serisi ; no. 14. IV. Series: Ankara. Üniversite. Banka ve Ticaret Hukuku Araştırma Enstitüsü. Banka ve Ticaret Hukuku Araştırma Enstitüsü yayınları ; no. 96.
79-356738
79 MARC

Turkey. Laws, statutes, etc.
⌐Türk kanunu medenîsi¬
Sözlüklü-içtihatlı Türk kanunu medenîsi ve Borçlar kanunu / Rıdvan Bülent Erciyes. — Ankara : Doruk Kitap ve Yayınevi, 1975.
xvi, 463 p. ; 17 cm. — (Yargıtay içtihadı birleştirme kararlariyle notlu kanunlar ; dizi 2) (Kanun dizisi ; 2) (Doruk yayınları ; 2)
Spine title: Türk kanunu medenîsi; borçlar kanunu.
1. Civil law—Turkey. 2. Debtor and creditor—Turkey. I. Erciyeş, Rıdvan Bülent. II. Turkey. Yargıtay. III. Turkey. Laws, statutes, etc. Borçlar kanunu. 1975. IV. Title. V. Title: Türk kanunu medenîsi. VI. Series: Yargıtay içtihat birleştirme kararlariyle notlu kanunlar serisi ; dizi 2.
78-970727
79 MARC

—TURKEY—ADDRESSES, ESSAYS, LECTURES

Medenî kanun'un 50. ⌐i.e. ellinci¬ yılı : bilimsel hafta, 15-17 Nisan 1976. — Ankara : ⌐Ankara Üniversitesi, Hukuk Fakültesi¬, 1977.
335 p. ; 24 cm. — (Ankara Üniversitesi, Hukuk Fakültesi yayınları ; no. 408)
Includes bibliographical references.
70.00TL
1. Civil law—Turkey—Addresses, essays, lectures. I. Ankara. Üniversite. Hukuk Fakültesi.
79-105723
79 MARC

—TURKEY—DIGESTS

Özkan, Hasan.
İzahlı-içtihatlı sulh hukuk dâvaları ve tatbikatı / Hasan Özkan. — İlâveli 3. baski. — Ankara : Seçkin Kitabevi, 1975.
xi, 731 p. ; 25 cm.
Second ed. published in 1970 under title: İzahlı-içtihatlı asliye hukuk davaları ve tatbikatı.
Bibliography: p. 728-730.
1. Civil law—Turkey—Digests. I. Title. II. Title: Sulh hukuk dâvaları ve tatbikatı.
79-359481
79 MARC

CIVIL LAW (Continued)

—TURKEY—EXAMINATIONS, QUESTIONS, ETC.

Oğuzman, M Kemal.
Medenî hukuk pratik çalışmaları [yazanlar] M. Kemal Oğuzman [ve] Aydın Aybay. Genişletilmiş 2. bası. İstanbul, İstanbul Üniversitesi, Hukuk Fakültesi, 1970–
v. ; 25 cm. (Hukuk Fakültesi [yayınları] no. 343, 375) (İstanbul Üniversitesi yayınlarından, no. 1575, 1775)
Includes bibliographical references.
CONTENTS: [1] Umumî esaslar, şahsın hukuku, aile hukuku.—2. Oğuzman, M. K. Borçlar hukuku, miras hukuku.
1. Civil law—Turkey—Examinations, questions, etc. I. Aybay, Aydın, joint author. II. Title. III. Series: Istanbul. Üniversite. Hukuk Fakültesi. Hukuk Fakültesi yayınları, no. 343 [etc.]
74–211422

—URUGUAY

Código civil : textos, antecedentes y discusión — Montevideo : República Oriental del Uruguay, Poder Legislativo, Consejo de Estado, 1978.
403 p. ; 27 cm.
"Modificación de los artículos 148, 164, 167, 187, 189, 285, 311, 396 y 1756 y derogación del artículo 182, Ley no. 14.766, de 18 de abril de 1978."
Includes text of Ley no. 14.759.
1. Civil law—Uruguay. I. Uruguay. Consejo de Estado. II. Uruguay. Laws, statutes, etc. Ley no. 14.766. 1978. III. Uruguay. Laws, statutes, etc. Ley no. 14.759. 1978.
346'.895 79-103999
79 MARC

—UZBEKISTAN

Uzbek S. S. R. Laws, statutes, etc.
[Grazhdanskiĭ kodeks Uzbekskoĭ Sovetskoĭ Sofšialisticheskoĭ Respubliki. Uzbek]
Ўзбекистон Совет Социалистик Республикасининг Гражданлик кодекси. — Тошкент : Ўзбекистон ССР давлат нашриёти, 1963.
245 p. ; 21 cm.
Uzbek.
1. Civil law—Uzbekistan. I. Title.
Title romanized: Ŭzbekiston Sovet Sofšialistik Respublikasining Grazhdanlik kodeksi.
78–970797

—YUGOSLAVIA

Privredno pravo sa osnovama građanskog prava / Karlo Kovač ... [et al.]. — 7., izm. i dop. izd. — Beograd : Savremena administracija, 1978.
xiv, 425 p. ; 24 cm. Yu***
Includes bibliographical references.
1. Commercial law—Yugoslavia. 2. Civil law—Yugoslavia. I. Kovač, Karlo.
78-391109
79 MARC

CIVIL LAW (ISLAMIC LAW) see Islamic law

CIVIL LAW (JEWISH LAW) see Jewish law

CIVIL LAW (ROMAN LAW) see Roman law

CIVIL LIBERTY see Liberty

CIVIL-MILITARY RELATIONS see Militarism; Sociology, Military

CIVIL OBEDIENCE see Government, Resistance to

CIVIL PROCEDURE

see also Actions and defenses; Agricultural courts and procedure; Appellate procedure; Arbitration and award; Attachment and garnishment; Class actions (Civil procedure); Code pleading; Complaints (Civil procedure); Compromise (Law); Contumacy; Costs (Law); Defense (Civil procedure); Dismissal and nonsuit; Divorce suits; Equity pleading and procedure; Executions (Law); Extraordinary remedies; Injunctions; Instructions to juries; Intervention (Civil procedure); Judgments; Judicial assistance; Jury; Legal advertising; Matrininial actions; Motions (Law); New trials; Non-contentious jurisdiction; Parties to actions; Pleading; Pre-trial procedure; Prejudicial actions; Process; Provisional remedies; Remedies (Law); Set-off and counterclaim; Small claims courts; Summary proceedings; Time (Law); Trial practice

Couture, Eduardo J 1906-1956.
Estudios de derecho procesal civil. Buenos Aires, EDIAR [1948-50]
3 v. 24 cm.
Vol. 2 has also special title: Pruebas en materia civil.
1. Civil procedure—Argentine Republic. 2. Civil procedure. I. Title.
347'.82'05 48-25860
[r79]rev MARC

Couture, Eduardo J 1906-1956.
Estudios de derecho procesal civil / Eduardo J. Couture. — 2. ed. — Buenos Aires : Ediciones Depalma, 1978.
3 v. ; 23 cm.
Includes bibliographical references and indexes.
CONTENTS: t. 1. La Constitución y el proceso civil.—t. 2. Pruebas en materia civil.—t. 3. El juez, las partes y el proceso.
1. Civil procedure—Argentine Republic. 2. Civil procedure. I. Title.
347'.82'05 79-100255
79 MARC

Hosono, Nagayoshi, 1883–
(Minji soshō hō yōgi)
民事訴訟法要義 / 細野長良著. — 東京：巖松堂書店, 昭和5-12 [1930-1937], 昭和12-16 [1937-1941] printing.
5 v. ; 23 cm.
Includes bibliographical references and indexes.
1. Civil procedure—Japan. 2. Civil procedure. I. Title.
79–80239

Kaissis, Anthanassios, 1947-
Die Verwertbarkeit materiell-rechtswidrig erlangter Beweismittel im Zivilprozess / vorgelegt von Athanassios Kaissis. — Frankfurt am Main : P. Lang, 1978.
265 p. ; 21 cm. GFR***
"Erschienen als Band 203 der Reihe II, Rechtswissenschaft, der 'Europäischen Hoshschulschriften' im Verlag Peter Lang GmbH in Frankfurt am Main."
Thesis—Heidelberg.
Vita.
Bibliography: p. 231-265.
1. Evidence (Law) 2. Illegality. 3. Civil procedure. I. Title.
K2261.K34 347.06 79-376842
79 MARC

Kaneko, Hajime, 1906–1973.
(Jittaihō to soshōhō)
實體法と訴訟法：民事訴訟の基礎理論 / 兼子一著. — 東京：有斐閣, 昭和32 [1957]
5, 3, 166, 5 p. ; 22 cm.
Includes bibliographical references and index.
1. Civil procedure. I. Title.
K2205.4.K36 79–803209

Nakajima, Hiromichi, 1885–1960.
(Saiban no sōzōsei genri)
裁判の創造性原理：殊に民事訴訟と權利の關係に就て / 中島弘道著. — 東京：岩波書店, 昭和16 [1941]
4, 13, 541 p. ; 22 cm.
Includes bibliographical references.
1. Civil procedure. 2. Civil procedure—Japan. 3. Procedure. I. Title.
K2205.4.N33 79–802247

Nakamura, Muneo, 1894–1975.
(Soshō hōgaku)
訴訟法學 / 中村宗雄著. — 東京：評論社, 昭和25 [1950]
267 p. : 19 cm. — (法律學全書 ; 10)
1. Civil procedure. 2. Civil procedure—Japan. I. Title. II. Series: Hōritsugaku zensho (Tokyo) ; 10.
K2205.4.N34 79–802355

Ramos, Geōrgios Th 1902–
(Encheiridion astikou dikonomikou dikaiou)
Ἐγχειρίδιον ἀστικοῦ δικονομικοῦ δικαίου. / Γεωργίου Θ. Ράμμου. — Ἀθῆναι : Ἐκδοτ. Οἶκος Α. Ν. Σάκκουλα, 1978–
v. ; 24 cm.
Includes bibliographies and indexes.
1. Civil procedure—Greece, Modern. 2. Civil procedure. I. Title.
79–411623

Theodoro Júnior, Humberto.
Processo de conhecimento / Humberto Theodoro Júnior. — 1. ed. — Rio de Janeiro : Forense, 1978.
2 v. (xii, 879 p.) : diagrs. ; 22 cm.
Bibliography: p. [812]-825.
Includes indexes.
Cr$580.00
1. Civil procedure—Brazil. 2. Civil procedure. I. Title.
347'.81'05 79-346091
79 MARC

—ADDRESSES, ESSAYS, LECTURES

(Jittaihō to tetsuzukihō no kōsaku)
実体法と手続法の交錯：山木戸克己教授還暦記念 / [編集代表 山木戸克己教授還暦記念論文集刊行発起人会]. — 東京：有斐閣, 昭和49-53 [1974-1978]
2 v. : port. : 22 cm. Ja 74-19983 (v. 1)
Chronological list of Kokuki Yamakido's works: v. 1, p. 3-10 (3d group)
Includes bibliographical references.
¥4500 (v. 1) ¥6000 (v. 2)
1. Civil procedure—Japan—Addresses, essays, lectures. 2. Civil procedure—Addresses, essays, lectures. 3. Yamakido, Kokuki, 1910– I. Yamakido, Kokuki, 1910– II. Yamakido Kokuki Kyōju Kanreki Kinen Rombunshū Kankō Hokkininkai.
75–801436

Kijimoto, Rōzō, 1876–1922.
(Minji soshō hō no shomondai)
民事訴訟法の諸問題 / 雉本朗造著. — 東京：有斐閣, 昭和30 [1955]
2, 2, 785, 18 p. ; 22 cm. — (學術選書 ; 5)
Series romanized: Gakujutsu sensho.
"舊民事訴訟法": p. 687–785.
Includes bibliographical references and indexes.
1. Civil procedure—Japan—Addresses, essays, lectures. 2. Civil procedure—Addresses, essays, lectures. I. Japan. Laws, statutes, etc. Minji soshō hō. 1955. II. Title.
79–801842

Nakamura, Muneo, 1894–1975.
(Jittaihōgaku to soshōhōgaku)
實體法學と訴訟法學 / 中村宗雄著. — 東京：朝倉書店, 昭和23 [1948]
293 p. ; 19 cm.
1. Civil procedure—Addresses, essays, lectures. 2. Procedure (Law)—Addresses, essays, lectures. 3. Justice, Administration of—Addresses, essays, lectures. I. Title.
K2205.6.N33 79–803208

Nakamura, Muneo, 1894–1975.
(Minji soshō hōgaku no kiso riron)
民事訴訟法学の基礎理論 / 中村宗雄著. — 東京：敬文堂書店, 昭和32 [1957]
2, 2, 371, [1], 18 p. ; 22 cm.
Chronological list of the author's works: p. 371–[372]
Includes bibliographical references and index.
CONTENTS: 中村宗雄 実体法学と訴訟法学との間隙に在るもの 民事訴訟における基本的な諸原理 系譜的にみた「訴訟」の構造分析 訴訟遂行権の系譜的考察 「一事不再理」と「一事不再議」 ドイツ徴税制度に関する調査報告 家事審判の訴訟的性格と行政的性格—中村英郎 "訴訟の目的"概念の生成過程 訴因について—内田武吉 公訴権理論の一考察
1. Civil procedure—Addresses, essays, lectures. I. Nakamura, Hideo, 1926– II. Uchida, Takeyoshi, 1926– III. Title.
K2205.6.N34 79–802181

CIVIL PROCEDURE
—ADDRESSES, ESSAYS, LECTURES (Continued)

Nakata, Jun'ichi, 1908–1970.
(Soshō oyobi chūsai no hōri)
訴訟及び仲裁の法理 / 中田淳一著. — 東京 : 有信堂, 昭和28 [1953]
2, 7, 434 p. ; 22 cm. — (民事訴訟法研究 ; 1)
Colophon inserted.
Includes bibliographical references.
1. Civil procedure—Addresses, essays, lectures. 2. Arbitration and award—Addresses, essays, lectures. I. Title. II. Series: Minji soshō hō kenkyū ; 1.
K2205.6.N36 79–803349

Yanase, Yoshimoto, 1905–
(Gyōseihō ni okeru kōhō to shihō)
行政法に於ける公法と私法 / 柳瀬良幹著. — 東京 : 有斐閣, 昭和18 [1943]
4, 2, 158 p. ; 22 cm. — (公法叢書 ; 2)
Series romanized: Kōhō sōsho.
"「行政法に於ける公法と私法」及び「民事事件と行政事件」と題して昭和十五年及び十六年に雑誌「法學」に掲載したもの"
Includes bibliographical references.
1. Administrative law—Addresses, essays, lectures. 2. Administrative procedure—Addresses, essays, lectures. 3. Civil procedure—Addresses, essays, lectures. I. Title. .
79–801826

—CONGRESSES

International Congress on the Law of Civil Procedure, 1st, State University of Ghent, 1977.
Towards a justice with a human face : the first International Congress on the Law of Civil Procedure, Faculty of Law, State University of Ghent, 27 August-4, September, 1977 / edited by Marcel Storme and Hélène Casman. — Antwerpen : Kluwer, 1978.
x, 590 p. ; 24 cm. Ne***
Dutch, English, French, German, and/or Spanish; summaries in English.
Includes bibliographical references.
ISBN 9026809743
1. Civil procedure—Congresses. I. Storme, Marcel. II. Casman, Hélène. III. Ghent. Rijksuniversiteit. Fakulteit der Rechten. IV. Title.
K2204.6.I57 1977a· 347.05 78-324796
79 MARC

—HISTORY

Fiala, Josef.
Historický vývoj některých procesních principů, zásad a institutů civilního procesu / Josef Fiala. — 1. vyd. — Praha : Univ. Karlova, 1975.
102, [1] p. ; 24 cm. — (Acta Universitatis Carolinae : Iuridica : Monographia ; 21/1974) Cz75
Summary in German and Russian.
Table of contents also in German and Russian.
Includes bibliographical references.
1. Civil procedure—Czechoslovakia—History. 2. Civil procedure—History. I. Title. II. Series: Prague. Universita Karlova. Acta : Iuridica : Monographia ; 1974, 21.
79-372652
*79 MARC

—ARAB COUNTRIES

Qusṭū, Jalīl.
(Qawā'id al-'āmmah lil-murāfa'āt al-madanīyah fī al-qānūn al-'Arabī al-muqāran)
القواعد العامة للمرافعات المدنية في القانون العربي المقارن : دراسة مقارنة / جليل قسطو. — الطبعة 1. — بيروت : دار الثقافة ، 1975.
172 p. ; 25 cm.
1. Civil procedure—Arab countries. I. Title.
79–970761

—ARAB COUNTRIES—TERMS AND PHRASES

Niqābat al-Muḥāmīn fī al-'Irāq.
(Tawḥīd al-muṣṭalaḥāt fī qānūn al-murāfa'āt al-madanīyah fī al-aqṭār al-'Arabīyah)
توحيد المصطلحات في قانون المرافعات المدنية في الاقطار العربية : مشروع مقدم من نقابة المحامين فى العراق الى المؤتمر الثالث عشر لاتحاد المحامين العرب المنعقد في تونس ابتداء من ١٥ — ١١ — ١٩٧٦ . — 1976 ,[s. l. : s. n.] (بغداد : مطبعة الشعب)
20 p. ; 24 cm.
£E0.30
1. Civil procedure—Arab countries—Terms and phrases. I. Mu'tamar al-Huḥāmīn al-'Arab, 13th, Tunis, 1976. II. Title.
77–961038

—ARGENTINE REPUBLIC

Argentine Republic. Laws, statutes, etc.
[Código procesal civil y comercial]
Código procesal civil y comercial de la Nación (Ley no. 17.-454) : en apéndice, leyes complementarias, acordadas y resoluciones / con la exposición de motivos, versión directa del Boletín oficial e índice alfabético bajo la supervisión de Miguel Danielián. — 8. ed. — Buenos Aires : Abeledo-Perrot, [1978]
593 p. ; 24 cm.
Includes bibliographical references and indexes.
1. Civil procedure—Argentine Republic. I. Danielián, Miguel. II. Title.
347'.82'0502632 78-111425
79 MARC

Argentine Republic. Laws, statutes, etc.
[Código procesal civil y comercial]
Código procesal civil y comercial de la Nación : reglamento para la justicia nacional, jurisdicción y competencia, tasas judiciales, mandamientos y notificaciones, etc. : Aranceles de abogados y procuradores, ley 21.839 / textos actualizados y ordenados por Antonio José Giangrasso. — Buenos Aires : Ediciones Depalma, 1978.
xiv, 254 p. ; 23 cm.
Half title: Código procesal civil y comercial de la Nación y normas complementarias.
"Apéndice" (p. [137]-254) contains related legislation.
Includes bibliographical references.
1. Civil procedure—Argentine Republic. I. Giangrasso, A. J. II. Argentine Republic. Laws, statutes, etc. Aranceles y honorarios de abogados y procuradores (1978). 1978. III. Title.
347'.82'05 79-112106
79 MARC

Couture, Eduardo J 1906-1956.
Estudios de derecho procesal civil. Buenos Aires, EDIAR [1948-50]
3 v. 24 cm.
Vol. 2 has also special title: Pruebas en materia civil.
1. Civil procedure—Argentine Republic. 2. Civil procedure. I. Title.
347'.82'05 48-25860
[r79]rev MARC

Couture, Eduardo J 1906-1956.
Estudios de derecho procesal civil / Eduardo J. Couture. — 2. ed. — Buenos Aires : Ediciones Depalma, 1978.
3 v. ; 23 cm.
Includes bibliographical references and indexes.
CONTENTS: t. 1. La Constitución y el proceso civil.—t. 2. Pruebas en materia civil.—t. 3. El juez, las partes y el proceso.
1. Civil procedure—Argentine Republic. 2. Civil procedure. I. Title.
347'.82'05 79-100255
79 MARC

Creus, Carlos.
Influencias del proceso penal sobre el proceso civil / Carlos Creus. — Santa Fe, [Argentina] : Rubinzal y Culzoni, [1977]
169 p. ; 23 cm.
Bibliography: p. [5]-[8]
1. Civil procedure—Argentine Republic. 2. Criminal procedure—Argentine Republic. I. Title.
347'.82'05 79-372775
79 MARC

Fassi, Santiago Carlos, 1902-
Código procesal civil y comercial de la Nación, y demás normas procesales vigentes : comentado, anotado y concordado / Santiago C. Fassi. — 2. ed., actualizada y ampliada. — Buenos Aires : Editorial Astrea de A. y R. Depalma, 19
v. ; 24 cm.
1. Civil procedure—Argentine Republic. I. Title.
347'.82'05 79-110057
79 MARC

—ARKANSAS

Arkansas. Courts.
Arkansas rules of civil procedure, with rules of appellate procedure, and rules for inferior courts / drafted by Arkansas Supreme Court Committee on Civil Procedure Revision. — Indianapolis : Michie/Bobbs-Merrill Co., [1978]
287 p. ; 23 cm.
Includes index.
1. Court rules—Arkansas. 2. Civil procedure—Arkansas. 3. Appellate procedure—Arkansas. I. Arkansas. Supreme Court. Committee on Civil Procedure Revision. II. Title.
KFA4129.A2 1978 347'.767'05 79-113633
79 MARC

—AUSTRIA

Gustermann, Anton Wilhelm, 1760 (ca.)-1823.
Oesterreichische Privatrechts-Praxis, enthaltend die Theorie des bürgerlichen Processes, die Geschäfte des adelichen Richteramtes, und den Justiz-Geschäftstil. 2. verm., und verb. Aufl. Wien, J. G. Ritter von Mösle, 1805.
2 v. 19 cm.
1. Civil procedure—Austria. 2. Non-contentious jurisdiction—Austria. I. Title.
347'.436'05 76-283961
79[r79]rev MARC

—AZERBAIJAN

Azerbaijan. Laws, statutes, etc.
(Grazhdanskiĭ prot̄sessual'nyĭ kodeks Azerbaĭdzhanskoĭ SSR)
Гражданский процессуальный кодекс Азербайджанской ССР : (С изм. и доп. на 1-е янв. 1976 г.). — Баку : Азернешр, 1977.
190 p. ; 20 cm. USSR 78-121
0.63rub
1. Civil procedure—Azerbaijan. I. Title.
79–408886

—BELGIUM

Belgium. Laws, statutes, etc.
[Code judiciaire. Dutch and French]
Belgisch gerechtelijk wetboek = Code judiciaire belge. — 2e druk. — Lier : Van In, 1977.
707 p. ; 20 cm. Be77-6
Cover title.
Errata slip inserted.
ISBN 9030601728 : 619F
1. Courts—Belgium. 2. Judicial power—Belgium. 3. Civil procedure—Belgium. I. Title.
347'.493'0102632 78-394918
*79 MARC

—BOLIVIA

Bolivia. Laws, statutes, etc.
[Código de procedimiento civil]
Código de procedimiento civil, concordado y anotado y Ley de organización judicial, concordada con arreglo a las ediciones oficiales / Carlos Morales Guillén. — 1. ed. — La Paz : Gisbert, 1978.
xxiv, 804 p. ; 19 cm.
Bibliography: p. [791]-793.
1. Civil procedure—Bolivia. 2. Courts—Bolivia. I. Morales Guillén, Carlos. II. Bolivia. Laws, statutes, etc. Decreto ley no. 10267. III. Title.
347'.84'0502633 79-112346
79 MARC

—BRAZIL

Brazil. Laws, statutes, etc.
[Código de processo civil (1973)]
Anotações jurisprudenciais ao Código de processo civil / Sérgio Luiz Monteiro Salles. — São Paulo : Edição Saraiva, 1978.
534 p. ; 24 cm.
Includes index.
Cr$350.00
1. Civil procedure—Brazil. 2. Annotations and citations (Law)—Brazil. I. Salles, Sérgio Luiz Monteiro. II. Title.
347'.81'05 79-341370
79 MARC

Brazil. Laws, statutes, etc.
[Código de processo civil (1973)]
Código de processo civil : Lei n. 5.869, de 11 de janeiro de 1973 / organizada por Rubens B. Minguzzi. — São Paulo : Editora Resenha Universitária, 1976.
xiv, 450 p. ; 22 cm. — (Informática legislativa ; 13)
"Exposição do motivos do ministro Alfredo Buzaid" (p. [1]-32) includes bibliographical references.
Includes index.
Cr$150.00
1. Civil procedure—Brazil. I. Minguzzi, Rubens B. II. Buzaid, Alfredo. III. Title.
347'.81'0502632 79-341563
79 MARC

Brazil. Laws, statutes, etc.
[Código de processo civil (1973)]
Código de processo civil : Lei no. 5.869, de 11-1-1973, com as retificações da Lei no. 5.925, de 1-10-1973 : acompanhado das disposições em vigor do Decreto-lei no. 1.608, de 18-9-1939, de apêndice contendo leis processuais especiais com as adaptações introduzidas pelas Leis ns. 6.014, de 27-12-1973, e 6.071, de 3-7-1974 ... / organização dos textos, notas remissivas e índice alfabético e remissivo pelo desembargador Manoel Augusto Vieira Neto. — 8. ed. — [São Paulo] : Edição Saraiva, 1978.
xxix, 725 p. ; 17 cm. — (Legislação brasileira)
"Exposição de motivos" (p. [1]-31) by A. Buzaid.
Includes bibliographical references and index.
Cr$200.00
1. Civil procedure—Brazil. I. Vieira Neto, Manoel Augusto. II. Buzaid, Alfredo.
347'.81'0502632 78-378931
78 MARC

CIVIL PROCEDURE
—BRAZIL (Continued)

Brazil. Laws, statutes, etc.
[Código de processo civil (1973)]
Código de processo civil e legislação processual em vigor / organização, seleção e notas, Theotonio Negrão. — 7. ed., atualizada até 10.3.78. — São Paulo : Editora Revista dos Tribunais, 1978.
711 p. ; 22 cm.
Includes indexes.
Cr$350.00
1. Civil procedure—Brazil. 2. Procedure (Law)—Brazil. I. Negrão, Theotonio.
347'.81'0502632 78-394279
79 MARC

Brazil. Laws, statutes, etc.
[Código de processo civil (1973)]
Código do processo civil : contendo as modificações introduzidas pelas leis n.°s. 5.925, de 1°. de outubro de 1973, 6.014, de 27 de dezembro de 1973, 6.071, de 3 de julho de 1974, Emenda constitucional n.°9, de 28 de junho de 1977, e Lei n°. 6.515, de 26 de dezembro de 1977, que regulamenta o divórcio / [ed. rev. e atualizada por Vicente Sabino Júnior]. — São Paulo : Brasilivros, Editora e Distribuidora, 1979.
314 p. ; 22 cm.
Includes bibliographical references and indexes.
Cr$200.00
1. Civil procedure—Brazil. 2. Procedure (Law)—Brazil. I. Sabino Júnior, Vicente.
347'.81'0502633 79-109271
79 MARC

Campos, Ronaldo Cunha.
Comentários ao Código de processo civil / Ronaldo Cunha Campos. — 1. ed. — Rio de Janeiro : Forense, 1979-
v. ; 24 cm.
Bibliography: v. 1, p. 313-323.
Includes index.
CONTENTS: v. 1. Teoria geral do processo civil. t. 1. A norma processual.
1. Civil procedure—Brazil. I. Title.
347'.81'05 79-109195
79 MARC

Castro, José Antônio de.
Execução no C.P.C. : doutrina, prática, jurisprudência / José Antônio de Castro. — São Joaquim da Barra [Brasil] : O. Dip Editores, 1978.
x, 376 p. ; 24 cm.
"110 ementas inseridas nos textos e analisadas; 292 ementas distribuídas no final dos textos para melhor compreensão de cada matéria, dezenas de votos completos."
Includes bibliographical references.
Cr$450.00
1. Executions (Law)—Brazil. 2. Civil procedure—Brazil. I. Title.
347'.81'05 79-341562
79 MARC

Comentários ao Código de processo civil. — 1. ed. — Rio de Janeiro : Forense, 19
v. ; 24 cm.
Includes bibliographies and indexes.
CONTENTS:
—v. 5. Theodoro Júnior, H. Arts. 796 a 889.—v. 6. Santos, E. F. dos. Arts. 890 a 1.102.—v. 7. Prata, E. Arts. 1.103 a 1.220.

Cr$400.00 (v. 5) varies
1. Civil procedure—Brazil. I. Theodoro Júnior, Humberto. II. Santos, Ernane Fidelis dos. III. Prata, Edson G. IV. Brazil. Laws, statutes, etc. Código de processo civil (1973). 1978.
347'.81'05 79-105501
79 MARC

Levenhagen, Antônio José de Souza.
Código civil à luz da nova lei processual / Antônio José de Souza Levenhagen. — 5. ed., rigorosamente atualizada com a Lei no. 6.458, de 1.o/11/1977 (Duplicatas), Lei no. 6.015, dos Registros Públicos, e Lei no. 6.515, de 26/12/1977 (Regulamentadora do divórcio). — São Paulo : Editora Atlas, 1978.
384 p. ; 21 cm.
Includes index.
Cr$200.00
1. Civil law—Brazil. 2. Civil procedure—Brazil. I. Title.
347'.81'05 79-342579
79 MARC

Levenhagen, Antônio José de Souza.
Novo Código de processo civil : comentários didáticos e práticos, com amplo formulário exemplificativo de petições e outros atos e termos do processo civil / Antônio José de Souza Levenhagen. — 8. ed., rev. e atualizada segundo a lei regulamentadora do divórcio, a nova Lei dos registros públicos e a recente Lei 6.458, de 1/11/1977, que alterou a Lei das duplicatas. — São Paulo : Editora Atlas, 1978.
485 p. ; 21 cm.
Includes index.
Cr$180.00
1. Civil procedure—Brazil. I. Title.
347'.81'05 78-396262
79 MARC

Malta, Christovão Piragibe Tostes.
Você conhece processo civil? / C. P. Tostes Malta, João Carlos B. Balaguer. — 2. ed. — Rio de Janeiro : Editora Rio, 1978-
v. ; 21 cm. — (Coleção Você conhece ; 7)
"Em convênio com as Faculdades Integradas Estácio de Sá."
Includes bibliographical references and index.
1. Civil procedure—Brazil. I. Balaguer, João Carlos B., 1949- joint author. II. Title.
347'.81'05 79-355657
79 MARC

Miranda, Francisco Cavalcanti Pontes de, 1892-
Comentários ao Código de processo civil / Pontes de Miranda. — 2. ed., rev. e aumentada. — Rio de Janeiro : Forense, 1979-
v. ; 24 cm.
Includes indexes.
CONTENTS: t. 1. Arts 1o.-45.

Cr$390.00 (v. 1)
1. Civil procedure—Brazil. I. Brazil. Laws, statutes, etc. Código de processo civil (1973) 1979. II. Title.
347'.81'05 79-109251
79 MARC

Moreira, José Carlos Barbosa.
O novo processo civil brasileiro : exposição sistemática do procedimento / José Carlos Barbosa Moreira. — Rio de Janeiro : Forense, 1978.
2 v. ; 22 cm.
Includes index.
CONTENTS: 1. Processo de conhecimento. 3. ed., rev. e atualizada.—2. Processo de execução. Processo cautelar. 2. ed., rev.
Cr$320.00
1. Civil procedure—Brazil. I. Title.
347'.81'05 78-394919
79 MARC

Sevá, José.
Manual de prática forense civil : noções gerais, formulário e notas, de acordo com o novo C. P. civil / José Sevá. — 3. ed., rev. e ampliada. — [São Paulo] : Brasilivros Editora e Distribuidora, 1977.
356 p. ; 21 cm.
Includes index.
Cr$220.00
1. Civil procedure—Brazil. I. Title.
347'.81'05 79-365032
79 MARC

Theodoro Júnior, Humberto.
Processo de conhecimento / Humberto Theodoro Júnior. — 1. ed. — Rio de Janeiro : Forense, 1978.
2 v. (xii, 879 p.) : diagrs. ; 22 cm.
Bibliography: p. [812]-825.
Includes indexes.
Cr$580.00
1. Civil procedure—Brazil. 2. Civil procedure. I. Title.
347'.81'05 79-346091
79 MARC

—BRAZIL—DICTIONARIES

Montenegro, César.
Dicionário de prática processual civil / César Montenegro. — 8. ed. — São Paulo : Sugestões Literárias, 1978.
2 v. (790 p.) ; 24 cm.
Cr$580.00
1. Civil procedure—Brazil—Dictionaries. I. Title.
347'.81'0503 79-341367
79 MARC

—BRAZIL—DICTIONARIES—PORTUGUESE

Montenegro, César.
Dicionário de prática processual civil / César Montenegro. — 6. ed. — São Paulo : Sugestões Literárias, 1974.
2 v. (714 p.) ; 24 cm.
1. Civil procedure—Brazil—Dictionaries—Portuguese. I. Title.
347'.81'0503 79-104962
79 MARC

—BRAZIL—DIGESTS

Bussada, Wilson.
Código de processo civil interpretado pelos Tribunais : lei no. 5.869, de 11. 1. 1973 / Wilson Bussada. — [Rio de Janeiro] : Liber Juris, 1976-1978.
2 v. ; 24 cm.
Includes indexes.
Cr$180.00 (v. 1) Cr$400.00 (v. 2)
1. Civil procedure—Brazil—Digests. I. Title.
347'.81'05 77-568340
78[r79]rev MARC

Direito processual civil : jurisprudência pesquisada / coligida pelo doutor Sérgio Luiz Monteiro Salles. — São Paulo : Hemeron Editora, 1975-
v. ; 21 cm. — (Jurispesquisa)
Cr$40.00
1. Civil procedure—Brazil—Digests. I. Salles, Sérgio Luiz Monteiro.
347'.81'05 76-465282
76[r79]rev MARC

—BRAZIL—FORMS

Caldas, Gilberto.
Advocacia na prática (cível) / Gilberto Caldas. — S[ão] Paulo : Livraria e Editora Universitária de Direito, 1978.
210 p. ; 22 cm.
Cr$150.00
1. Civil procedure—Brazil—Forms. I. Title.
347'.81'055 79-341604
79 MARC

Silva, Ivan de Hugo.
Prática do processo civil / Ivan de Hugo Silva. — 1. ed. — Rio de Janeiro : Aide Editora e Comércio de Livros, 1978.
3 v. (1559 p.) ; 21 cm.
Includes indexes.
Cr$1000.00
1. Civil procedure—Brazil. 2. Civil procedure—Brazil—Forms. I. Title.
347'.81'05 79-338910
79 MARC

—BRAZIL—INDEXES

Peres, Alcides Conejeiro.
Vocabulário do Código de processo civil / Alcides Conejeiro Peres. — 1. ed. — Rio de Janeiro : Forense, 1978.
viii, 708 p. ; 22 cm.
Cr$400.00
1. Civil procedure—Brazil—Indexes. I. Title.
347'.81'05 78-382273
79 MARC

—CALIFORNIA

California. Legislature. Joint Committee on Tort Liability.
Transcript of hearing on procedural reform, Sacramento, California, November 15, 1977 / Joint Committee on Tort Liability. — [Sacramento] : The Committee, [1978?].
130 p. ; 28 cm.
$4.20
1. Torts—California. 2. Civil procedure—California. I. Title.
KFC10.T67 1977d 346'.794'03 78-622386
79 MARC

California family law : practice and procedure / [contributing authors, Kenneth J. Levy ... et al.] editorial consultant, Christian E. Markey, Jr. — New York : M. Bender, 1978-
v. ; 26 cm.
Loose-leaf for updating.
CONTENTS: 1. Marriage.—2. Dissolution of marriage: substantive law.—3. Dissolution of marriage: procedure.
1. Domestic relations—California. 2. Domestic relations courts—California. 3. Civil procedure—California. I. Levy, Kenneth J. II. Markey, Christian E.
KFC115.C34 346'.794'0150269 78-70343
79 MARC

Faber, Stuart J
Handbook of civil procedure, with forms / Stuart J. Faber. — 2d ed. — Los Angeles : Lega-Books, c1977.
353 p. : forms ; 22 cm. — (The Roadmap to the courthouse series)
Includes index.
ISBN 0-89074-050-X : $18.50
1. Civil procedure—California. I. Title. II. Series.
KFC995.F3 1977 347'.794'05 78-108599
78 MARC

—CALIFORNIA—FORMS

California State Sheriffs' Association.
Civil procedural manual. — [Updated and rev.]. — [Los Angeles : California State Sheriffs' Association], c1978.
1 v. : forms ; 28 cm.
Loose-leaf for updating.
1. Sheriffs—California. 2. Civil procedure—California—Forms. I. Title.
KFC985.S4C3 1978 347'.794'05 78-108768
78 MARC

—CANADA

Law Society of Upper Canada.
Civil procedure II : bar admission course, 1977-78. — [Toronto] : Law Society of Upper Canada, c1978.
107 p. ; 28 cm. — (Special lectures - Law Society of Upper Canada ; 1978)
C***
ISBN 0-88759-036-5
1. Civil procedure—Canada. I. Title. II. Title: Bar admission course, 1977-78. III. Series: Law Society of Upper Canada. Special lectures ; 1978.
347'.0971 s 78-319181
[KE8349] 79 MARC

CIVIL PROCEDURE
—CANADA (Continued)

Law Society of Upper Canada.
Civil procedure II : bar admission course, 1978-79. — [Toronto] : Law Society of Upper Canada, c1979.
93 p. ; 28 cm. — (Special lectures - Law Society of Upper Canada ; 1979)
C***
"This book contains the lectures presented to the candidates of the 1978-79 Bar Admission Course."
Includes bibliographical references.
ISBN 0-88759-054-3
1. Civil procedure—Canada. I. Title. II. Title: Bar admission course, 1978-79. III. Series: Law Society of Upper Canada. Special lectures ; 1979.
KE8349.2.L38 347'.71'05 79-318349
79 MARC

—CANADA—CASES

Watson, Garry D
Canadian civil procedure : cases and materials by Garry D. Watson, Neil J. Williams. — 2d ed. — Toronto : Butterworth & Co. (Canada), c1977.
840 p. in various pagings ; 25 cm. — (Canadian legal casebook series)
C78-5119-X
First ed., by G. D. Watson, S. Borins and N. J. Williams.
Includes bibliographical references and index.
ISBN 0-409-87568-6 : $55.00. ISBN 0-409-87569-4 pbk.
1. Civil procedure—Canada—Cases. I. Williams, Neil J., joint author. II. Title.
KE8348.5.W37 1977 347'.71'05 78-324315
79 MARC

—CANADA—DIGESTS

Power, W Kent.
Power's Western practice digest : an index-digest of western practice cases in civil actions. — 3d ed. to the end of 1976 / G. F. Butterwick. — Calgary, Alta. : Burroughs, 1977.
2 v. (x, 1151 p.) ; 25 cm.
C***
First-2d ed. published under title: An index of western practice cases in civil actions.
CONTENTS: 1. Action to joinder.—2. Judgements to words and phrases.
ISBN 0-459-31960-4
1. Civil procedure—Canada—Digests. I. Butterwick, G. F. II. Title. III. Title: Western practice digest.
KE8346.22.P6 1977 347'.71'0502638 78-320395
79 MARC

—CHINA

Hsieh, Sen.
(Min shih su sung ch'iang chih chih hsing shih wu)
民事訴訟強制執行實務 / [謝森編輯]. — 上海：上海法學編譯社, [民國24 i. e. 1935序]
2, 2, 26, 318 p. ; 27 cm.

1. Executions (Law)—China. 2. Civil procedure—China. I. Title.
79-836698

Shao, Hsi.
(Min shih su sung lü shih i)
民事訴訟律釋義 / 編輯者邵羲；校勘者徐學詢；閱訂者徐元誥. — 上海：中華書局, 民國6 [1917]
4, 546 p. ; 22 cm.
Colophon title.

1. Civil procedure—China. I. China. Laws, statutes, etc. Min shih su sung lü. 1917. II. Title.
78-842416

—CHINA—FORMS

Fu, Yu-hsin.
(Min hsing chuang shih ta ch'üan)
民刑狀式大全：訴訟南針 / [編輯者傅優新]. — 上海：大通書局, 民國14 [1925]
2 v. ; 19 cm.
Cover title.

1. Civil procedure—China—Forms. 2. Criminal procedure—China—Forms. I. Title.
78-842412

Wu, Nuan-sen.
(Wu Nuan-sen Wu Wei-huan lü shih su chuang hui pien)
伍暖森吳偉寰律師訴狀彙編. — [廣州?]：伍暖森吳偉寰律師事務所：總代售處光東書局, 民國19 [1930]
7, 10, 196, 4 p. ; 22 cm.
Cover title.

1. Civil procedure—China—Forms. 2. Criminal procedure—China—Forms. I. Wu, Wei-huan. II. Title.
78-843470

—COLOMBIA

Guzmán Díaz, Carlos A
Procedimiento penal aplicado / Carlos A. Guzmán Díaz. — 5. ed., corr. y aumentada. — Bogotá : Editorial Temis, 1978.
xxiii, 549 p. ; 24 cm.
ISBN 8482720503
1. Civil procedure—Colombia. I. Title.
347'.861'05 79-356830
79 MARC

—CONNECTICUT

Connecticut practice book : containing annotated rules of practice, judicial and professional codes, forms. — Revision of 1978. — [Hartford] : Commission on Official Legal Publications, c1979-
v. ; 26 cm.
Includes index.
CONTENTS: v. 1. Rules and codes.
1. Civil procedure—Connecticut. 2. Criminal procedure—Connecticut. 3. Court rules—Connecticut. 4. Forms (Law)—Connecticut. 5. Legal ethics—Connecticut. I. Connecticut. Courts.
KFC4130.C6 1979 347'.746'05 79-113305
79 MARC

—CZECHOSLOVAKIA—HISTORY

Fiala, Josef.
Historický vývoj některých procesních principů, zásad a institutů civilního procesu / Josef Fiala. — 1. vyd. — Praha : Univ. Karlova, 1975.
102, [1] p. ; 24 cm. — (Acta Universitatis Carolinae : Iuridica : Monographia ; 21/1974)
Cz75
Summary in German and Russian.
Table of contents also in German and Russian.
Includes bibliographical references.
1. Civil procedure—Czechoslovakia—History. 2. Civil procedure—History. I. Title. II. Series: Prague. Universita Karlova. Acta : Iuridica : Monographia ; 1974, 21.
79-372652
*79 MARC

—ECUADOR—INDEXES

Puig Vilazar, Carlos.
Indice de procedimiento civil ecuatoriano : a través del articulado de los códigos y de las ejecutorias del Tribunal Supremo / Carlos Puig Vilazar. — [Guayaquil] : Edit. Casa de la Cultura Ecuatoriana, Núcleo del Guayas, 19
v. ; 22 cm.
1. Civil procedure—Ecuador—Indexes. I. Title.
347'.866'0502638 78-389354
79 MARC

—EGYPT

Abū al-Wafā, Aḥmad.
(al-Murāfa'āt al madanīyah wa-al-tijārīyah bi-muqtaḍā Qānūn al-murāfa'āt al-jadīd raqm 18 li-sanat 1968, wa-Qānūn al-ithbāt raqm 25 li-sanat 1968)
المرافعات المدنية والتجارية بمقتضى قانون المرافعات الجديد رقم ١٣ لسنة ١٩٦٨ ، وقانون الاثبات رقم ٢٥ لسنة ١٩٦٨ / احمد ابو الوفا. — الطبعة 12. — الاسكندرية : منشاة المعارف ، [1978?]
868 p. ; 24 cm. — (الكتب القانونية)
Series romanized: al-Kutub al-qānūnīyah.
Bibliography: p. [4]–7.
1. Civil procedure—Egypt. I. Title.
77-962223

Abū al-Wafā, Aḥmad.
(Naẓarīyat al-dufū' fī qānūn al-murāfa'āt)
نظرية الدفوع فى قانون المرافعات / أحمد أبو الوفا. — الطبعة 5. — الاسكندرية : منشاة المعارف ، 1977.
867 p. ; 25 cm. — (الكتب القانونية)
Series romanized: al-Kutub al-qānūnīyah.
«الجزاء فى قانون المرافعات والشروط الاساسية لاعماله ، وصاحب المصلحة والصفة فى التمسك به ، ومدى سريانه على الماضى. الدفع بعدم الاختصاص وبتجاوز السلطة وبالاحالة وبالبطلان وبعدم قبول الدعوى ، وحالات وقف الخصومة وانقطاعها ، وسقوطها وتركها واعتبارها كأن لم تكن. التمسك بالجزاء فى قانون ايجار الاماكن ، وفى الطعن بالنقض.»
Bibliography: p. [6]–10.
£E5.00
1. Civil procedure—Egypt. 2. Exceptions (Law)—Egypt. I. Title.
77-962216

al-Sharqāwī, 'Abd al-Mun'im Aḥmad.
(al-Murāfa'āt al-madanīyah wa-al-tijārīyah)
المرافعات المدنية والتجارية : شرح القانون رقم ١٣ سنة ١٩٦٨ والتشريعات المكملة له / تاليف عبد المنعم الشرقاوى ، فتحى والى. — [القاهرة] : دار النهضة العربية ، 1976/1977.
320, 287 p. ; 24 cm.
Bibliography: p. 309–310.
£E4.00
1. Civil procedure—Egypt. 2. Commercial law—Egypt. I. Wālī, Fatḥī, joint author. II. Title.
77-960980

United Arab Republic. Laws, statutes, etc.
(Taqnīn al-murāfa'āt fī ḍaw' al-qaḍā' wa-al-fiqh)
تقنين المرافعات فى ضوء القضاء والفقه / محمد كمال عبد العزيز. — الطبعة 2. — القاهرة : الشركة المصرية ، يطلب من مكتبة وهبه ، 1978.
839 p. ; 24 cm.
«نصوص تقنين المرافعات ومايقابلها من نصوص القانون الملغى او ما يتصل بها من نصوص القوانين الخاصة والمعاهدات الدولية ، مذيلة بالمذكرات الايضاحية والاعمال التحضيرية ومعلقا عليها بأحكام محكمة النقض وآراء الفقه.»
£E5.50
1. Civil procedure—Egypt. I. 'Abd al-'Azīz, Muḥammad Kamāl. II. Title.
78-961316

—EUROPE, EASTERN

Wengerek, Edmund.
Postępowanie cywilne procesowe przed sądami pierwszej instancji w państwach socjalistycznych / Edmund Wengerek. — Warszawa : Wydaw. Prawnicze, 1978 (Warsz. : D1)
167, [1] p. ; 21 cm. — (Studia z dziedziny prawa cywilnego) P78-9634
On cover: Instytut Badania Prawa Sądowego.
Table of contents also in German and Russian.
Errata slip inserted.
Bibliography: p. 149-157.
zł40.00
1. Civil procedure—Europe, Eastern. I. Title. II. Series.
79-350228
79 MARC

—FLORIDA

Coleman, Ted P
Lawyer's edition of Florida summary claims handbook / by Ted P. Coleman. — Clearwater, Fla. : D & S. Publishers, c1978.
1 v. ; 24 cm. — (Practice manual series)
Loose-leaf for updating.
Published in 1975 under title: Florida summary claims handbook.
1. Civil procedure—Florida. 2. Summary proceedings—Florida. 3. Small claims courts—Florida. 4. County courts—Florida. I. Title. II. Series.
KFF530.C6 1978 347'.759'05 79-117706
79 MARC

Massey, M Minnette, 1929-
Florida civil discovery manual / by M. Minnette Massey. — Clearwater, Fla. : D & S Publishers, c1979.
348 p. ; 23 cm. — (Practice manual series)
1. Discovery (Law)—Florida. 2. Civil procedure—Florida. I. Title. II. Series.
KFF537.M37 347'.759'072 79-115981
79 MARC

Trawick, Henry P
Florida practice and procedure. 19 -77. Norcross, Ga., [etc.] Harrison Co.
v. 25 cm. annual.
Continued by: Trawick's Florida practice and procedure, ISSN 0191-7684.
Cover title, 19 -77: Trawick's Florida practice and procedure.
Supplements accompany some issues.
Key title: Florida practice and procedure, ISSN 0191-7676.
1. Civil procedure—Florida. 2. Court rules—Florida. I. Trawick, Henry P. Trawick's Florida practice and procedure. II. Title.
KFF530.T73 347'.759'05 79-641255
MARC-S

Trawick, Henry P
Trawick's Florida practice and procedures. 1978–
Norcross, Ga., Harrison Co.
v. 24 cm.
Continues: Florida practice and procedure, ISSN 0191-7676.
Running title, 1978– : Florida practice and procedure.
Key title: Trawick's Florida practice and procedure,
ISSN 0191-7684.
1. Civil procedure—Florida. 2. Court rules—Florida. I. Title. Florida practice and procedure. II. Key title.
KFF530.T73 347'.759'05 79-641355
MARC-S

CIVIL PROCEDURE (Continued)

—FLORIDA—FORMS

LaCoe, Norman, 1932-
LaCoe's Forms for pleading under the Florida rules of civil procedure / Norm LaCoe. — 1978 ed. — Norcross, Ga. : Harrison Co., 1978.
xl, 918 p. : forms ; 24 cm.
"This publication is a soft bound version of the 1978 revision of LaCoe's volume 11 of Sapp Florida pleading, practice, and legal forms."
Includes index.
1. Civil procedure—Florida—Forms. I. Sapp, Alfred E. Harrison's Sapp Florida pleading, practice, and legal forms. II. Title. III. Title: Forms for pleading under the Florida rules of civil procedure.
KFF530.A65L3 1978 347'.759'055 78-109204
78 MARC

—FRANCE

Japan. Hōmushō. Daijin Kambō. Shihō Hōsei Chōsa-bu.
(Chūshaku Furansu shin minji soshō hōten)
注釈フランス新民事訴訟法典 / 法務大臣官房司法法制調査部編. — 東京：法曹会, 1978.
24, 585 p. ; 21 cm. Ja 78-26787
Bibliography: p. 560-568.
Includes index.
¥2800
1. Civil procedure—France. I. France. Laws, statutes, etc. Code de procédure civile. Japanese. 1978. II. Title. III. Title: Furansu shin minji soshō hōten.
79-809779

Motulsky, Henri.
Écrits. Préf. de. Gérard Cornu [et al.] [Paris] Dalloz, 1973-78.
3 v. 24 cm. 50.00F (v. 1); 74.00F (v. 2) F***
Includes bibliographical references and indexes.
CONTENTS: [1] Études et notes de procédure civile.—2. Études et notes sur l'arbitrage.—3. Études et notes de droit international privé.
ISBN 2-247-00458-X (v. 1)
1. Civil procedure—France. 2. Arbitration and award—France. 3. Conflict of laws—France. 4. Conflict of laws.
73-327105
78[r78]rev2 MARC

Vincent, Jean, docteur en droit.
Procédure civile / Jean Vincent. — 19. éd. — Paris : Dalloz, 1978.
viii, 1140 p. ; 18 cm. — (Précis Dalloz : Droit) F***
First-12th ed. by P. Cuche published under title: Précis de procédure civile et commerciale; 13th ed. by P. Cuche and J. Vincent published under title: Précis de procédure civile; 14th-16th ed. by J. Vincent published under title: Précis de procédure civile.
Includes bibliographical references and index.
ISBN 2-247-03095-5 : 80.00F
1. Civil procedure—France. I. Cuche, Paul, 1868- Précis de procédure civile. II. Title.
347'.44'05 78-382275
78 MARC

—FRANCE—ADDRESSES, ESSAYS, LECTURES

Lassaque, Lucien.
Défense et illustration du nouveau code de procédure civile : discours de rentrée / prononcé par Lucien Lassaque. Allocutions de Paul Nougaret, Jean-Jacques Lajaunie. — Montpellier : Cour d'appel de Montpellier, 1977.
28 p. ; 25 cm. — (Audience solennelle de rentrée - Cour d'appel de Montpellier ; du 3 janvier 1977) F***
1. Civil procedure—France—Addresses, essays, lectures. 2. France. Cour d'appel, Montpellier. I. Nougaret, Paul. II. Lajaunie, Jean Jacques. III. Title. IV. Series: France. Cour d'appel, Montpellier. Audience solennelle ; du 3 janvier 1977.
347'.44'02 s 79-357001
79 MARC

—FRANCE—OUTLINES, SYLLABI, ETC.

Larguier, Jean.
Procédure civile : droit judiciaire privé / par Jean Larguier. — 7. éd. — Paris : Dalloz, 1978.
131 p. ; 24 cm. — (Mémentos Dalloz ; 3168) (Droit privé) F***
Includes index.
ISBN 2-247-03168-4 : 27.00F
1. Civil procedure—France—Outlines, syllabi, etc. I. Title.
347'.44'05 78-368660
78[r79]rev MARC

—FRANCE—ALSACE-LORRAINE

Lehmann, Jacques.
Organisation judiciaire et procédure civile en Alsace-Lorraine / Jacques Lehmann. — Paris : Librairies techniques, [1978] F***
108 p. ; 24 cm. — (Litec droit)
Bibliography: p. [97]
Includes index.
ISBN 2-7111-0109-6 : 30.00F
1. Courts—France—Alsace-Lorraine. 2. Civil procedure—France—Alsace-Lorraine. I. Title.
347'.44'01 79-347720
79 MARC

—GERMANY

Saitō, Hideo, 1909-
(Sensō to Minji soshō hō)
戰爭と民事訴訟法 / 齋藤秀夫著. — 東京：日本評論社, 昭和19 [1944]
3, 6, 453 p. ; 19 cm. — (法學叢書 ; 15)
Series romanized: Hōgaku sōsho.
Includes bibliographical references.
1. Civil procedure—Germany. 2. Civil procedure—Japan. 3. Trade regulation. 4. War and emergency legislation. I. Title.
79-803401

—GERMANY, EAST

Germany (Democratic Republic, 1949-). Laws, statutes, etc.
[Zivilprozessordnung]
Zivilprozessordnung und andere verfahrensrechtliche Bestimmungen : Textausg. mit Anmerkungen und Sachregister / hrsg. vom Ministerium der Justiz ; [zusammengestellt und bearb. von Gerhard Krüger, Peter Wallis]. — 1. Aufl. — Berlin : Staatsverlag der Deutschen Demokratischen Republik, 1978.
160 p. ; 24 cm. GDR***
Includes index.
3.80M
1. Civil procedure—Germany, East. I. Krüger, Gerhard, Dipl.-Jur. II. Wallis, Peter. III. Germany (Democratic Republic, 1949-). Ministerium der Justiz. IV. Title.
347'.431'05 79-343706
79 MARC

Das Neue Zivilrecht der DDR nach dem Zivilgesetzbuch von 1975 / Klaus Westen (Hrsg.). — Baden-Baden : Nomos-Verlagsgesellschaft, in Komm., 1977.
318 p. ; 23 cm. — (Rechtswissenschaftliche Veröffentlichungen - Osteuropa-Institut an der Freien Universität Berlin ; Bd. 7) GFR78-A9
Includes index.
ISBN 3-7890-0329-8
1. Civil law—Germany, East. 2. Civil procedure—Germany, East. I. Westen, Klaus. II. Series: Berlin. Freie Universität. Osteuropa-Institut. Rechtswissenschaftliche Veröffentlichungen - Osteuropa-Institut an der Freien Universität Berlin ; Bd. 7.
346'.431 78-390606
79 MARC

—GERMANY, WEST

Arens, Peter.
Zivilprozessrecht : Erkenntnisverfahren, Zwangsvollstreckung / von Peter Arens. — München : Beck, 1978.
xxiii, 428 p. ; 20 cm. — (Grundrisse des Rechts) GFR78-A
Bibliography: p. [xxiv]
Includes indexes.
ISBN 3-406-00903-4 : DM24.50
1. Civil procedure—Germany, West. I. Title.
347'.43'05 79-343703
79 MARC

Bruns, Rudolf.
Zivilprozessrecht : e. systemat. Darst. / von Rudolf Bruns. — 2., neubearb. Aufl. — München : Vahlen, 1979.
xv, 538 p. ; 23 cm. GFR79-A
Includes bibliographical references and indexes.
ISBN 3-8006-0752-2 : DM58.00
1. Civil procedure—Germany, West. I. Title.
347'.43'05 79-357559
79 MARC

Fricke, Gerhard.
Zivilprozess I : Gerichtsverfassung und 1. bis 2. Buch der ZPO / G. Fricke und J. Wiefels. — Neubearb. Aufl., 105-.107. Tsd. — Heidelberg : Verlagsgesellschaft Recht u. Wirtschaft, 1977.
163 p. ; 23 cm. — (Schaeffers Grundriss des Rechts und der Wirtschaft : Abt. 1, Privat- und Prozessrecht ; Bd. 6, T. 1) GFR77-A
Earlier editions by C. Schaeffer and J. Wiefels.
Includes index.
1. Civil procedure—Germany, West. I. Wiefels, Josef, 1893- joint author. II. Title.
347'.43'05 79-356294
79 MARC

Geigel, Reinhart, 1884-1963.
Der Haftpflichtprozess : mit Einschluss d. materiellen Haftpflichtrechts / Geigel ; [Verzeichnis der Bearbeiter, Robert Geigel ... et al.]. — 17., neubearb. Aufl. — München : Beck, 1979.
xxiii, 1669 p. ; 24 cm. GFR79-A
Includes texts of selected legislation.
Includes bibliographies and index.
ISBN 3-406-07219-4 : DM148.00
1. Torts—Germany, West. 2. Liability (Law)—Germany, West. 3. Civil procedure—Germany, West. I. Geigel, Robert. II. Title.
346'.43'03 79-355654
79 MARC

Germany (Federal Republic, 1949-). Kommission für das Zivilprozessrecht.
Bericht der Kommission für das Zivilprozessrecht / hrsg. vom Bundesministerium d. Justiz. — Bonn : Deutscher Bundes-Verlag, 1977.
383 p. ; 24 cm. GFR78-A9
Includes index.
1. Civil procedure—Germany, West. I. Germany (Federal Republic, 1949-). Bundesministerium der Justiz. II. Title.
347'.43'05 78-396264
79 MARC

Germany (Federal Republic, 1949-). Laws, statutes, etc.
Zivilprozessordnung : mit Gerichtsverfassungsgesetz u. d. Einführungsgesetzen / erl. von Heinz Thomas u. Hans Putzo. — 10., neubearb. Aufl. — München : Beck, 1978.
xxi, 1617 p. ; 17 cm. GFR78-A
Includes index.
ISBN 3-406-07114-7 : DM56.00
1. Civil procedure—Germany, West. I. Thomas, Heinz, 1920- . II. Putzo, Hans, 1926- . III. Title.
347'.43'0502632 78-395255
79 MARC

Germany (Federal Republic, 1949-). Laws, statutes, etc.
Zivilprozessordnung : mit Einführungsgesetz, Gerichtsverfassungsgesetz (Ausz.) ... : Textausg. mit ausführl. Sachreg. u. e. Einf. von Karl Heinz Schwab. — Sonderausg. 12., neubearb. Aufl., Stand 1. Juni 1978. — München : Deutscher Taschenbuch-Verlag, 1978.
456 p. ; 18 cm. — (Beck-Texte) GFR78-A38
Includes index.
ISBN 3-423-05005-5 : DM9.80
1. Civil procedure—Germany, West. I. Schwab, Karl Heinz. II. Title.
347'.43'0502632 79-373466
79 MARC

Germany (Federal Republic, 1949-). Laws, statutes, etc.
Zivilprozessordnung : mit Gerichtsverfassungsgesetz u. anderen Nebengesetzen / begr. von Adolf Baumbach ; fortgef. von Wolfgang Lauterbach ; nunmehr bearb. von Jan Albers, Peter Hartmann. — 37., neubearb. Aufl. — München : Beck, 1979.
56, 2423 p. ; 21 cm. — (Beck'sche Kurz-Kommentare ; Bd. 1) GFR78-A
Bibliography: p. 6-11 (2d group)
Includes index.
ISBN 3-406-07238-0 : DM118.00
1. Civil procedure—Germany, West. I. Baumbach, Adolf, 1874-1945. II. Albers, Jan. III. Hartmann, Peter, Dr. jur. IV. Title.
347'.43'0502632 79-356295
79 MARC

Germany (Federal Republic, 1949-). Laws, statutes, etc.
Zivilprozessordnung : mit Gerichtsverfassungsgesetz u. d. Einführungsgesetzen, mit internat. Zivilprozessrecht, Mustern für Anträge, Klagen u. Entscheidungen, Kostenanm. / begr. von Richard Zöller ; bearb. von Reinhold Geimer ... [et al.]. — 12., völlig neubearb. u. wesentl. erw. Aufl. — Köln : O. Schmidt, 1979.
xxiv, 2409 p. ; 24 cm. GFR79-A
Bibliography: p. xxiii-xxiv.
Includes index.
ISBN 3-504-47000-3 : DM158.00
1. Civil procedure—Germany, West. I. Zöller, Richard, 1905-1961. II. Geimer, Reinhold. III. Title.
347'.43'0502632 79-375609
79 MARC

Heller, Franz-Wilhelm, 1944-
Die gerichtliche Schadensermittlung nach 287 ZPO / vorgelegt von Franz-Wilhelm Heller. — [s.l. : s.n., 1977]
xi, 257 p. ; 21 cm. GFR***
Thesis—Erlangen-Nürnberg, Erlangen.
Vita.
Bibliography: p. [v2]-xi.
1. Damages—Germany, West. 2. Civil procedure—Germany, West. I. Title.
346'.43'03 78-377730
78 MARC

Klein, Hans-Joachim, 1927-
Wirtschaftsprivatrecht / Hans-Joachim Klein. — Wiesbaden : Gabler, 1978.
251 p. ; 23 cm. — (Moderne Wirtschaftsbücher : 3, Volkswirtschaft und Recht ; 09) GFR78-A
Includes bibliographies and index.
ISBN 3-409-72121-5 : DM32.00
1. Civil law—Germany, West. 2. Commercial law—Germany, West. 3. Civil procedure—Germany, West. I. Title.
78-377705
78 MARC

CIVIL PROCEDURE
—GERMANY, WEST (Continued)

Peters, Egbert, 1928-
Zivilprozessrecht einschliesslich Zwangsvollstreckung und Konkurs / von Egbert Peters. — 2., überarb. Aufl. — Frankfurt am Main : Metzner, 1978.
181 p. ; 26 cm. — (Wiederholungs- und Vertiefungskurs in den Kerngebieten des Rechts ; Bd. 6) GFR78-A
Includes index.
ISBN 3-7875-3315-X : DM20.00
1. Civil procedure—Germany, West. 2. Executions (Law)—Germany, West. 3. Bankruptcy—Germany, West. I. Title. II. Series.
78-376636
78 MARC

Schellhammer, Kurt, 1935-
Die Arbeitsmethode des Zivilrichters : prakt. Hilfen für Sachverhalt, Gutachten u. Urteil anhand von Fällen u.e. bearb. Musterakte / von Kurt Schellhammer. — 3., überarb. Aufl. — Heidelberg ; Karlsruhe : Müller Juristischer Verl., 1978.
xxv, 451 p. : ill. ; 24 cm. — (Jura Modellkurs) GFR78-A
Bibliography: p. xxv.
Includes index.
ISBN 3-8114-4978-8 : DM34.00
1. Judgments—Germany, West. 2. Civil procedure—Germany, West. 3. Judges—Germany, West. I. Title.
347'.43'077 79-357200
79 MARC

Schrader, Siegfried.
Zivilprozess / begr. von Siegfried Schrader ; fortgef. von Karl-Friedrich Steinert. — 6., neubearb. u. erw. Aufl. — München : Beck, 1979.
xvii, 397 p. ; 24 cm. — (Handbuch der Rechtspraxis ; Bd. 1a) GFR78-
First ed. published in 1953 under title: Zivilprozess und Mobilienzwangsvollstreckung; 5th ed. published in 1974 under title: Zivilprozess und Zwangsvollstreckung in das bewegliche Vermögen.
Bibliography: p. ⌊xviii⌋
Includes index.
ISBN 3-406-01524-7
1. Civil procedure—Germany, West. I. Steinert, Karl-Friedrich. II. Title.
347'.43 79-368241
79 MARC

Simon, Richard, 1935-
Die Behandlung einander widersprechender rechtskräftiger Zivilurteile : ein Beitrag zum Urteils- und Rechtskraftkonflikt im Zivilprozess / vorgelegt von Richard Simon. — ⌊s.l. : s.n., 1978?⌋ (Marburg : E. Mauersberger)
xii, 165 p. ; 21 cm. GFR***
Thesis—Cologne.
Vita.
Bibliography: p. iii-viii.
1. Judgments—Germany, West. 2. Civil procedure—Germany, West. 3. Conflict of judicial decisions—Germany, West. I. Title.
347'.43'077 79-363129
79 MARC

Wolf, Manfred, 1939-
Gerichtliches Verfahrensrecht : Darstellung d. Zivilprozesses mit vergleichender Betrachtung von Strafprozessen u. Verwaltungsgerichtsprozess / Manfred Wolf. — Reinbek bei Hamburg : Rowohlt, 1978.
381 p. : 1 ill. ; 19 cm. — (Rororo Studium ; 119 : Rechtswissenschaften)
Bibliography: p. 366-367.
Includes indexes.
ISBN 3-499-21119-X : DM16.80
1. Judicial process—Germany, West. 2. Civil procedure—Germany, West. 3. Criminal procedure—Germany, West. 4. Administrative procedure—Germany, West. I. Title.
347'.43'01 79-350328
79 MARC

Zeiss, Walter, 1933-
Zivilprozessrecht / Walter Zeiss. — 3., neubearb. Aufl. — Tübingen : Mohr, 1978.
viii, 420 p. ; 24 cm. GFR78-A
Bibliography: p. ⌊409⌋
Includes index.
ISBN 3-16-640501-1
1. Civil procedure—Germany, West. I. Title.
347'.43'05 78-387541
78 MARC

—GERMANY, WEST—CASES

Lüke, Gerhard.
Fälle zum Zivilverfahrensrecht / von Gerhard Lüke. — München : Beck, 1979-
v. ; cm. — (Schriftenreihe der Juristischen Schulung ; Heft 68)
GFR79-A(v.
CONTENTS: Bd. 1. Erkenntnis- und Vollstreckungsverfahren der ZPO.

Bibliography: v. 1, p. ⌊xi⌋
Includes indexes.
ISBN 3-406-05068-9 (v. 1)
1. Civil procedure—Germany, West—Cases. I. Title. II. Series: Juristische Schulung : Schriftenreihe ; Heft 68.
347'.43'05 79-356829
79 MARC

—GERMANY, WEST—EXAMINATIONS, QUESTIONS, ETC.

Weimar, Wilhelm, 1903-
Zivilprozess und Einzelzwangsvollstreckung : 500 Fragen mit Antworten / von Wilhelm Weimar. — Sankt Augustin : Asgard-Verlag Hippe, 1978.
120 p. (p. 119-120 advertisements) ; 21 cm. GFR78-A
ISBN 3-537-66401-X : DM18.00
1. Civil procedure—Germany, West—Examinations, questions, etc. 2. Executions (Law)—Germany, West—Examinations, questions, etc. I. Title.
347'.43'05076 79-364847
79 MARC

—GERMANY, WEST—FORMS

Schüller, Ulrich.
Formularbuch der Prozesspraxis : für d. Rechtsanwalt u. sein Büro / von Ulrich Schüller, Herbert Hempel. — 4. Aufl. — Essen : Juristischer Fachbuchverlag, 1978.
296 p. ; 22 cm. GFR78-A50
Includes index.
DM48.00
1. Procedure (Law)—Germany, West—Forms. 2. Civil procedure—Germany, West—Forms. I. Hempel, Herbert, librarian, joint author. II. Title.
347'.43'055 79-382158
79 MARC

Wagner, Hermann, doctor of law.
Das Prozessformularbuch : prakt. Erl. u. Muster für d. Zivilprozess, d. Zwangsvollstreckung einschl. Lohnpfändung, d. Vergleichs- u. Konkursverfahren sowie d. finanzgerichtl. Verfahren / von Hermann Wagner, Hugo Zartmann. — 4., aktualisierte Aufl. — Köln : O. Schmidt, 1978.
xxvii, 800 p. ; 22 cm. GFR78-A
Includes index.
ISBN 3-504-07012-9
1. Civil procedure—Germany, West—Forms. 2. Bankruptcy—Germany, West—Forms. 3. Composition (Law)—Germany, West—Forms. I. Zartmann, Hugo, joint author. II. Title.
347'.43'055 79-347703
79 MARC

—GREAT BRITAIN

Collins, Hugh C
Notes on county court practice and procedure / Hugh C. Collins. — 5th ed. — London : Oyez Pub., 1978.
xi, 219 p. ; 22 cm. GB***
Includes index.
ISBN 0-85120-373-6 : £6.50
1. County courts—Great Britain. 2. Civil procedure—Great Britain. I. Title. II. Title: County court practice and procedure.
KD7239.C64 1978 347'.41'025 79-312452
79 MARC

—GREECE, MODERN

Koutourisēs, Kōnstantinos I
(Prosōpikē kratēsis kata ton kōdika politikēs dikonomias)
Ἡ προσωπικὴ κράτησις κατὰ τὸν κώδικα πολιτικῆς δικονομίας / Κωνσταντίνου Ἰ. Κουτουρίση. — Ἀθῆναι : Ἐκδοτ. Οἶκος Ἀφοὶ Π. Σάκκουλα, 1978.
209 p. ; 24 cm.
Includes bibliographical references and index.
1. Executions (Law)—Greece, Modern. 2. Debt, Imprisonment for—Greece, Modern. 3. Civil procedure—Greece, Modern. I. Title.
79-413742

Ramos, Geōrgios Th 1902-
(Encheiridion astikou dikonomikou dikaiou)
Ἐγχειρίδιον ἀστικοῦ δικονομικοῦ δικαίου. / Γεωργίου Θ. Ράμμου. — Ἀθῆναι : Ἐκδοτ. Οἶκος Α. Ν. Σάκκουλα, 1978-
v. ; 24 cm.
Includes bibliographies and indexes.
1. Civil procedure—Greece, Modern. 2. Civil procedure. I. Title.
79-411623

—HAITI

Haiti (Republic). Laws, statutes, etc.
Code de procédure civile, 1963 : voté le 17 septembre, 1963 par l'Assemblee legislative, promulgué le 17 janvier, 1964, exécutoire a partir du premier avril, 1964. — ⌊Port-au-Prince⌋ : Barreau de Port-au-Prince, 1965.
263 p. ; 21 cm.
1. Civil procedure—Haiti.
347'.7294'0502632 78-107531
79 MARC

—INDIA

Chakraverti, Suranjan.
Suranjan Chakraverti & B. Nath's Cases and materials on Code of civil procedure, with state amendments. — 2d ed. / Bholeshwar Nath. — Lucknow : Eastern Book Co., 1978.
1, 1369 p. ; 25 cm.
First ed. published in 1971 under title: Cases and materials on Code of civil procedure.
Cover title: Cases and materials on Code of civil procedure.
Running title: Code of civil procedure, 1908.
Includes index.
Rs85.00
1. Civil procedure—India. I. Bholeshwar Nath, 1928- joint author. II. India (Republic). Laws, statutes, etc. Code of civil procedure, 1908. 1978. III. Title: Cases and materials on Code of civil procedure. IV. Title: Code of civil procedure, 1908.
347'.54'05 78-904192
79 MARC

Farani, Merajuddin, 1935-
Code of civil procedure / by M. Farani. — ⌊Rev. ed.⌋. — Lahore : Lahore Law Times Publications, 1978.
xxi, 966 p. ; 25 cm.
Includes index.
Rs85.00
1. Civil procedure—Pakistan. 2. Civil procedure—India. I. India (Republic). Laws, statutes, etc. Code of civil procedure, 1908. 1978. II. Title.
345'.54'05 78-931170
79 MARC

India (Republic). Laws, statutes, etc.
The Code of civil procedure, 1908 (5 of 1908), as modified up to the 1st May 1977. — ⌊New Delhi⌋ : Govt. of India, Ministry of Law, Justice, and Company Affairs, ⌊1977⌋
vii, 347 p. ; 30 cm.
Rs9.50 ($3.42 U.S.)
1. Civil procedure—India. 2. Court rules—India. I. Title.
347'.54'0502632 78-903319
79 MARC

—INDONESIA

Mertokusumo, Sudikno.
Hukum atjara perdata (kuliah kerdja). Jogjakarta, Jajasan Badan Penerbit Gadjah Mada ⌊1964?⌋
68 p. 21 cm.
Cover title.
1. Jurisdiction—Indonesia. 2. Civil procedure—Indonesia. I. Title.
76-949875
79⌊r79⌋rev MARC

—IRAN

Iran. Laws, statutes, etc.
(Majmū'ah-'i kāmil-i qavānīn va muqarrarāt-i ḥuqūqī)
مجموعهٔ کامل قوانین و مقررات حقوقی / تدوین غلامرضا حجتی اشرفی ؛ زیر نظر غلامعلی امیری. — ⌊تهران⌋ : کتابفروشی گنج دانش ، ⌊1976 or 1977⌋ 2535.
88, 778 p. ; 18 cm.
In Persian.
400.00 IR
1. Civil law—Iran. 2. Civil procedure—Iran. I. Ḥujjatī Ashrafī, Ghulām Riżā. II. Title.
79-970575

—IRELAND

Woods, James V
A district court guide in civil licensing, and family law / by James V. Woods. — ⌊s.l. : s.n., 1977?⌋
xxvi, 289 p. ; 23 cm. Ir***
On spine: Vol. II.
Companion vol. to the author's A district court guide in offence cases.
Includes index.
£9.35
1. Civil procedure—Ireland. 2. District courts—Ireland. 3. Summary proceedings—Ireland. 4. Licenses—Ireland. 5. Domestic relations—Ireland. I. Title.
KDK1642.W66 347'.417'05 79-312647
79 MARC

—ITALY

Bongiorno, Girolamo.
I provvedimenti del tribunale fallimentare / Girolamo Bongiorno. — Milano : A. Giuffrè, 1979.
viii, 393 p. ; 25 cm. It79-June
Bibliography: p. ⌊363⌋-382.
Includes index.
L10000
1. Bankruptcy—Italy. 2. Courts—Italy. 3. Civil procedure—Italy. I. Title.
79-377505
*79 MARC

CIVIL PROCEDURE
—ITALY (Continued)

Dini, Mario.
La domanda riconvenzionale nel diritto processuale civile (art. 36 cod. proc. civ.) / Mario Dini. — 3. ed. riveduta ed aggiornata. — Milano : Giuffrè, 1978.
555 p. ; 22 cm. — (Teoria e pratica del diritto : Sezione 1, Diritto e procedura civile ; 13) It78-Oct
"Massime di giurisprudenza (dal 21 aprile 1942)": p. [403]-502.
Bibliography: p. [505]-515.
Includes indexes.
L12000
1. Set-off and counterclaim—Italy. 2. Civil procedure—Italy. I. Title.
347'.45'05 79-342143
•79 MARC

Padiglione, Giovanni.
Le notificazioni civili e penali : pratica esposizione dell'istituto con ampi riferimenti a dottrina ed oltre 600 massime di giurisprudenza, formulario e indici / Giovanni Padiglione. — 2. ed. ampliata e aggiornata. — Latina : Bucalo, 1978.
236 p. ; 24 cm. — (Collana Servizi giudiziari ; 4) It78-Dec
Bibliography: p. 231-232.
Includes index.
L6000
1. Notice (Law)—Italy. 2. Civil procedure—Italy. 3. Criminal procedure—Italy. I. Title.
347'.45'05 79-372362
•79 MARC

Verde, Giovanni.
Profili del processo civile : parte generale / Giovanni Verde. — Napoli : Jovene, 1978.
281 p. ; 24 cm. It***
Bibliography: p. [269]-278.
L7500
1. Civil procedure—Italy. I. Title.
347'.45'05 79-377599
79 MARC

—ITALY—FORMS

Berri, Mario.
Formulario della procedura civile / a cura di M. Berri, ... [et al.] ; prefazione del prof. Enrico Redenti. — Nuova ed. aggiornata al 31 marzo 1978. — Milano : A. Giuffrè, 1978.
xix, 699 p. ; 17 cm. It78-Nov
Includes indexes.
L13000
1. Civil procedure—Italy—Forms. I. Title.
347'.45'055 79-348395
•79 MARC

—JAPAN

Fuse, Tatsuji, 1880-1953.
(Kosaku sōgi hōtei senjutsu kyōkasho)
小作爭議法廷戰術教科書 / 布施辰治著. — 東京：希望閣, 昭和5 [1930]
2, 2, 2, 350 p. ; 19 cm.
Colophon title: Kosaku sōgi to hōtei senjutsu.
1. Farm tenancy—Japan. 2. Landlord and tenant—Japan. 3. Civil procedure—Japan. I. Title.
79-805359

Hōsōkai.
(Minji soshō hō)
民事訴訟法　例題解説　法曹会編　東京　法曹会　昭和46-52 (1971-77)
2 v. 18 cm. (法曹新書 17, 30) Ja 72-3881 (v. 1)
Originally published in Hōsō, dai 187-go-dai 209-gō, dai 218-go-dai 246-gō under title: Hōsō kōza minji soshō hō.
Includes indexes.
¥400 (v. 1) ¥500 (v. 2)
1. Civil procedure—Japan. I. Title. II. Series: Hōsō shinsho, 17 [etc.]
73-800442

Hosono, Nagayoshi, 1883-
(Minji soshō hō yōgi)
民事訴訟法要義 / 細野長良著. — 東京：巖松堂書店, 昭和5-12 [1930-1937], 昭和12-16 [1937-1941] printing.
5 v. ; 23 cm.
Includes bibliographical references and indexes.
1. Civil procedure—Japan. 2. Civil procedure. I. Title.
79-802393

Imamura, Nobuyuki, 1840-1909.
(Minji soshō hō chūkai)
民事訴訟法註解 / 今村信行増著. — 増訂再版. — 東京：明治大學出版部講法會, 明治36-37 [1903-1904]
4 v. ; 21 cm.
LC copy incomplete: v. 1 wanting.
1. Civil procedure—Japan. I. Japan. Laws, statutes, etc. Minji soshō hō. 1903. II. Title.
79-802236

Itakura, Matsutarō, b. 1868.
(Minji soshō hō kōyō)
民事訴訟法綱要 / 板倉松太郎著. — 新訂. — 東京：巖松堂書店, 大正2 [1913], 昭和2 [1927] printing.
657, [29] p. ; 23 cm.
1. Civil procedure—Japan. I. Title.
79-803392

Itakura, Matsutarō, b. 1868.
(Minji soshō hō kōyō)
民事訴訟法綱要 / 板倉松太郎著. — 新訂11版. — 東京：巖松堂書店, 昭和2 [1927], 昭和4 [1929] printing.
657, [33] p. ; 23 cm.
1. Civil procedure—Japan. I. Title.
79-803391

Itō, Makoto, 1945-
(Minji soshō no tōjisha)
民事訴訟の当事者 / 伊藤真. — 東京：弘文堂, 1978.
xi, 222 p. : ill. ; 19 cm. — (弘文堂法学選書 ; 6) Ja 78-23771
Series romanized: Kōbundō hōgaku sensho.
Includes bibliographical references.
¥1200
1. Parties to actions—Japan. 2. Civil procedure—Japan. I. Title.
79-800751

Iwamoto, Yūjirō, 1880-
(Shin Minji soshō hō yōron)
新民事訴訟法要論 / 岩本勇次郎, 三ケ尻好人共著. — 東京：巖松堂書店, 昭和3 [1928], 昭和4 [1929] printing.
2 v. (21, 1647 p.) ; 23 cm.
1. Civil procedure—Japan. I. Mikajiri, Yoshito, joint author II. Title.
79-802579

Japan. Laws, statutes, etc.
[Minji soshō hō. German]
Die japanische ZPO in deutscher Sprache / übers. von Hideo Nakamura u. Barbara Huber ; mit Einf. in d. japan. Zivilprozessrecht von Hideo Nakamura. — Köln ; Berlin ; Bonn ; München : Heymann, 1978.
x, 135 p. ; 21 cm. — (Japanisches Recht ; Bd. 4) GFR78-A35
Includes bibliographical references.
ISBN 3-452-18429-3 : DM24.00
1. Civil procedure—Japan. I. Nakamura, Hideo, 1926- . II. Title. III. Series.
347'.52'0502633 79-377559
79 MARC

Japan. Laws, statutes, etc.
(Saishin tōki roppō)
最新登記六法 / 戸田修三監修 ; 津島一雄編集代表. — 昭和53年版. — 東京：暁出版, 1978.
1368 p. ; 19 cm. Ja 78-26832
Errata slip inserted.
¥2800
1. Recording and registration—Japan. 2. Civil procedure—Japan. I. Tsushima, Kazuo. II. Title.
79-803339

Japan. Saikō Saibansho. Jimusōkyoku. Minjikyoku.
(Minji sosho tokuhon)
民事訴訟讀本 / 最高裁判所事務總局民事局編. — 東京：法曹會, 昭和24 [1949]
9, 185 p. ; 22 cm.
1. Civil procedure—Japan. I. Title.
79-803279

Japan. Saikō Saibansho. Jimusōkyoku.
(Minji soshō kisoku no kaisetsu)
民事訴訟規則の解説 / 最高裁判所事務総局編. — 東京：法曹会, 昭和31 [1956]
1, 1, 191 p. ; 21 cm.
Cover title.
Includes bibliographical references.
1. Civil procedure—Japan. 2. Court rules—Japan. I. Title.
79-804818

(Jinji soshō gakusetsu jitsurei tetsuzuki sōran)
人事訴訟學說實例手續綜覧 / 法律研究會編. — 東京：酒井書店, 大正15 [1926], 昭和3 [1928] printing.
2, 1, 32, 446, 66 p. ; 23 cm.
Includes legislation.
1. Domestic relations courts—Japan. 2. Civil procedure—Japan. I. Hōritsu Kenkyūkai.
79-802543

Kaneko, Hajime, 1906-1973.
(Jōkai Minji soshō hō)
條解民事訴訟法 / 兼子一著. — 東京：弘文堂, 昭和26-27 [1951-1952]
3 v. ; 19 cm.
1. Civil procedure—Japan. I. Japan. Laws, statutes, etc. Minji soshō hō. 1951. II. Title.
79-802768

Kaneko, Hajime, 1906-1973.
(Minji soshō hō)
民事訴訟法 / 兼子一. — 東京：弘文堂, 昭和27- [1952-
v. ; 21 cm. — (法律學講座 ; 2, 4-5, 8-)
Series romanized: Hōritsugaku kōza.
Vol. 3 has subtitle: Hanketsu tetsuzuki ; v. 4: Kyōsei shikkō.
1. Civil procedure—Japan. I. Title.
79-802563

Kaneko, Hajime, 1906-1973.
(Minji soshō hō)
民事訴訟法 / 兼子一, 染野義信共著. — 東京：青林書院, 昭和33- [1958-
v. ; 22 cm. — (判例法学全集 ; 11-)
Colophon inserted.
Includes bibliographical references and indexes.
1. Civil procedure—Japan. I. Someno, Yoshinobu, 1918- joint author. II. Title. III. Series: Hanrei hōgaku zenshū ; 11 [etc.]
79-803827

Kaneko, Hajime, 1906-1973.
(Minji soshō hō taikei)
民事訴訟法体系 / 兼子一著. — 新修. — 東京：酒井書店, 昭和34 [1959]
497, 26 p. ; 22 cm.
Colophon inserted.
Includes bibliographical references and indexes.
1. Civil procedure—Japan. I. Title.
79-803225

CIVIL PROCEDURE —JAPAN (Continued)

Katayama, Michio.
(Hanrei o chushin to shitaru Minji sosho ho no shomondai)
判例を中心としたる民事訴訟法の諸問題　片山通夫著. — 大阪：大同書院：東京：大賣捌所巖松堂書店, 昭和7-8 [1932-1933]
2 v. : 19 cm.
Includes bibliographical references.
1. Civil procedure—Japan. I. Title. II. Title: Minji sosho ho no shomondai.
79-802535

Katō, Masaharu, 1871-1952.
(Minji soshō hō yōron)
民事訴訟法要論 / 加藤正治著. — 東京：有斐閣, 昭和21 [1946]
10, 553, 19, 10, [1] p. ; 22 cm.
Includes bibliographical references and indexes.
List of the author's works: p. [1] (4th group)
1. Civil procedure—Japan. I. Title.
79-802454

Kawamoto, Kiyoshi, 1904-
(Minji soshō hō teiyō)
民事訴訟法提要 / 河本喜與之著. — 東京：南郊社, 昭和10 [1935], 昭和17 [1942] printing.
1, 16, 605 p. ; 22 cm.
Bibliography: p. 599-605.
1. Civil procedure—Japan. I. Title.
79-802565

Kikkawa, Daijirō, 1901-
(Minji soshō hō)
民事訴訟法 / 吉川大二郎, 山木戸克己共著. — 東京：評論社, 昭和26 [1951]
288 p. ; 19 cm. — (法律學全書 ; 23)
1. Civil procedure—Japan. I. Yamakido, Kokuki, 1910- joint author. II. Title. III. Series: Hōritsugaku zensho (Tokyo)
79-802894

Kikui, Tsunahiro, 1899-
(Minji soshō hō)
民事訴訟法 / 菊井維大. — [s. l. : s. n., 192-?]
404 p. ; 22 cm.
Includes bibliographical references.
1. Civil procedure—Japan. I. Title.
79-802235

Kikui, Tsunahiro, 1899-
(Minji soshō hō)
民事訴訟法 / 菊井維大. 刑事訴訟法 / 小野清一郎. — [s. l. : s. n., 192-?]
404, 132 p. : 22 cm.
Cover title.
Includes bibliographies.
1. Civil procedure—Japan. 2. Criminal procedure—Japan. I. Ono, Seiichirō, 1891- Keiji soshō hō. 192-? II. Title.
79-802233

Kikui, Tsunahiro, 1899-
(Minji soshō hō)
民事訴訟法 / 菊井維大. — 東京：弘文堂, 昭和33- [1958-
v. ; 22 cm. — (法律学講座叢書)
Colophon inserted.
Series romanized: Hōritsugaku kōza sōsho.
Includes bibliographical references.
1. Civil procedure—Japan. I. Title.
79-802234

Kikui, Tsunahiro, 1899-
(Minji soshō hō kōgi)
民事訴訟法講義 / 菊井維大著. — 東京：弘文堂, 昭和30 [1955]
7, 517 p. ; 22 cm.
Colophon inserted.
1. Civil procedure—Japan. I. Title.
79-802237

Kikui, Tsunahiro, 1899-
(Minji soshō hō kyōzai)
民事訴訟法教材 / 菊井維大, 三ケ月章編. — 東京：有斐閣, 昭和29- [1954-
v. ; 22 cm. — (判例教材叢書)
Includes index.
1. Civil procedure—Japan. I. Mikazuki, Akira, 1921- joint author. II. Title. III. Series: Hanrei kyōzai sōsho.
79-802453

Kikui, Tsunahiro, 1899-
(Zentei minji soshō hō)
全訂民事訴訟法 / 菊井維大, 村松俊夫著 ; 鈴木重信 ... [et al. 解説]. — 東京：日本評論社, 1978-
v. : 19 cm. Ja 78-24957 (v. 1)
Bibliography: v. 1, p. 10-14 (1st group)
¥6200 (v. 1)
1. Civil procedure—Japan. I. Muramatsu, Toshio, 1902- joint author. II. Suzuki, Shigenobu, 1924- III. Japan. Laws, statutes, etc. Minji soshō hō. 1978. IV. Title. V. Title: Minji soshō hō.
79-803025

Maeda, Naonosuke, 1874-1946.
(Minji soshō hō kaigi)
民事訴訟法解疑 / 前田直之助, 阿部文二郎, 細野長良合輯. — 東京：清水書店, 大正11 [1922]
13, 268 p. : 20 cm.
Includes bibliographical references.
1. Civil procedure—Japan. I. Abe, Bunjirō, 1880- joint author. II. Hosono, Nagayoshi, 1883- joint author. III. Title.
79-802334

Maeda, Naonosuke, 1874-1946.
(Minji soshō hō kōgi)
民事訴訟法講義 / 前田直之助述. — 東京：巖松堂書店, 大正11-12 [1922-1923]
2 v. ; 23 cm.
Cover title.
1. Civil procedure—Japan. I. Title.
79-803390

Maeno, Jun'ichi, 1889-
(Minji soshō hō ron)
民事訴訟法論 / 前野順一著. — 東京：松華堂, 昭和12- [1937-], 昭和13 [1938] printing.
v. ; 22 cm.
Bibliography: v. 1, p. 1-2 (2d group)
1. Civil procedure—Japan. I. Title.
79-802997

(Minji shōji ni kansuru arayuru shurui no so to sono saiban)
民事商事に關するあらゆる種類の訴と其裁判 / 尾高武治著. — 改訂増補. — 東京：清水書店, 昭和 - [19
v. ; 23 cm.
1. Civil procedure—Japan. I. Odaka, Takeji, 1889- II. Title: So to sono saiban.
79-802561

(Minji shōji ni kansuru arayuru shurui no so to sono saiban)
民事商事に關するあらゆる種類の訴と其の裁判 / 尾高武治著. — 全訂改版. — 東京：清水書店, 昭和11 [1936]
2 v. (1560 p.) ; 23 cm.
1. Civil procedure—Japan. I. Odaka, Takeji, 1889- II. Title: So to sono saiban.
79-802560

(Minji soshō hō kōza)
民事訴訟法講座 / 民事訴訟法學会. — 東京：有斐閣, 昭和29-31 [1954-1956]
5 v. (1, 1551 p.) ; 22 cm.
Includes bibliographical references and indexes.
1. Civil procedure—Japan. I. Minji Soshō Hōgakkai.
79-802896

Mita, Takasaburō, 1899-
(Jinji soshō tetsuzuki hō yōron)
人事訴訟手續法要論 / 三田高三郎著. — 東京：帝國判例法規出版社：發賣所 帝國行政學會, 昭和15 [1940]
2, 25, 410 p. : 23 cm.
The 1952 ed. published under title: Jinji soshō tetsuzuki hō kaisetsu.
1. Domestic relations courts—Japan. 2. Civil procedure—Japan. I. Title.
79-802213

Miyake, Masao, 1909-
(Ryakkai Minji soshō hō, hanketsu tetsuzuki hen)
略解民事訴訟法·判決手続篇 / 三宅正雄著. — 東京：富山房, 1978.
11, 246, 8 p. ; 19 cm. Ja 78-12080
"付録　改正民事訴訟法の解説　奥野健一·三宅正雄": p. [179]-246.
¥1500
1. Civil procedure—Japan. I. Title. II. Title: Minji soshō hō, hanketsu tetsuzuki hen.
79-801683

Morita, Toyojirō, 1888-
(Minji soshō hō gaiyō)
民事訴訟法概要 / 森田豊次郎著. — 東京：巖松堂書店, 昭和11 [1936], 昭和18 [1943] printing.
1, 5, 407 p. ; 22 cm.
Includes bibliographical references.
1. Civil procedure—Japan. I. Title.
79-803275

Morita, Toyojirō, 1888-
(Minji soshō hō gaiyō)
民事訴訟法概要 / 森田豐次郎著. — 訂正版. — 東京：巖松堂, 昭和13- [1938-
v. ; 24 cm.
1. Civil procedure—Japan. I. Title.
79-804049

Muramatsu, Toshio, 1902-
(Minji saiban no kenkyū)
民事裁判の研究 / 村松俊夫著. — 東京：日光書院, 昭和15 [1940], 昭和24 [1949] printing.
8, 4, 347 p. ; 22 cm.
Colophon inserted.
Includes bibliographical references.
1. Civil procedure—Japan. I. Title.
79-802778

CIVIL PROCEDURE
—JAPAN (Continued)

Nakajima, Hiromichi, 1885-1960.
(Saiban no sōzōsei genri)
裁判の創造性原理：殊に民事訴訟と權利の關係に就て / 中島弘道著. — 東京：岩波書店, 昭和16 [1941]
4, 13, 541 p. ; 22 cm.
Includes bibliographical references.

1. Civil procedure. 2. Civil procedure—Japan. 3. Procedure. I. Title.
K2205.4.N33 79-802247

Nakamura, Muneo, 1894-
(Minji soshō yōron)
民事訴訟要論 / 中村宗雄著. — 東京：敬文堂, 昭和28- [1953-], 昭和30- [1955-] printing.
v. ; 23 cm.
Includes bibliographical references.

1. Civil procedure—Japan. I. Title.
79-802569

Nakamura, Muneo, 1894-1975.
(Minji soshō yōron)
民事訴訟要論 / 中村宗雄著. — 改訂增補. — 東京：敬文堂, 昭和32- [1957-
v. : ill. ; 22 cm.
Includes bibliography.

1. Civil procedure—Japan. I. Title.
79-802568

Nakamura, Muneo, 1894-1975.
(Soshō hōgaku)
訴訟法學 / 中村宗雄著. — 東京：評論社, 昭和25 [1950]
267 p. ; 19 cm. — (法律學全書 ; 10)

1. Civil procedure. 2. Civil procedure—Japan. I. Title. II. Series: Hōritsugaku zensho (Tokyo) ; 10.
K2205.4.N34 79-802355

Nakata, Jun'ichi, 1908-
(Minji soshō hō kōgi)
民事訴訟法講義 / 中田淳一著. — 東京：有信堂, 昭和29- [1954-
v. : ill. ; 22 cm.
Colophon inserted.
Includes bibliographical references.

1. Civil procedure—Japan. I. Title.
79-802452

Noma, Shigeru, 1902-
(Minji soshō hōgaku)
民事訴訟法學 / 野間繁著. — 東京：巖松堂書店, 昭和30 [1955]
3, 15, 618, 28 p. ; 22 cm.
Colophon inserted.
Includes indexes.

1. Civil procedure—Japan. I. Title.
79-802897

Noma, Shigeru, 1902-
(Minji soshō hōgaku gaisetsu)
民事訴訟法學概說 / 野間繁著. — 新訂. — 東京：巖松堂書店, 昭和 [19
v. in ; 21 cm.
Includes bibliographical references.

1. Civil procedure—Japan. I. Title.
79-805376

Noma, Shigeru, 1902-
(Minji soshō hōgaku gaisetsu)
民事訴訟法學概說 / 野間繁著. — 東京：巖松堂出版, 昭和25 [1950]
3, 15, 500 p. ; 22 cm.
Edition for 1941 published under title: Minji soshō hōgaku gairon.
Colophon inserted.
Includes bibliographical references.

1. Civil procedure—Japan. I. Title.
79-802566

Noma, Shigeru, 1902-
(Minji soshō hōgaku gaisetsu)
民事訴訟法學概說 / 野間繁著. — [新訂]. — 東京：巖松堂書店, 昭和26-28 [1951-1953]
2 v. ; 22 cm.
Includes bibliographical references.

1. Civil procedure—Japan. I. Title.
79-802238

Noma, Shigeru, 1902-
(Minji soshō hōgaku gaisetsu)
民事訴訟法學概說 / 野間繁著. — 東京：巖松堂書, 昭和30- [1955-
v. ; 21 cm.
Colophon inserted.
Includes bibliographical references.

1. Civil procedure—Japan. I. Title.
79-802513

Oda, Kyūzō.
(Jitsurei hon'i minji soshō no hōritsu chishiki)
実例本位民事訴訟の法律知識 / 小田久藏著. — 東京：東洋書館, 昭和29 [1954]
272 p. ; 19 cm. — (実例本位法律知識全書 ; 8)

1. Civil procedure—Japan. 2. Civil procedure—Japan—Forms. I. Title. II. Title: Minji soshō no hōritsu chishiki. III. Series: Jitsurei hon'i hōritsu chishiki zensho ; 8.
79-801889

Oda, Kyūzō.
(Minji soshō hō tokuhon)
民事訴訟法読本 / 小田久蔵, 雨宮真也著. — 改訂. — 東京：高文堂出版社, 1979.
246 p. ; 19 cm. — (高文堂入門選書) Ja 79-4453
Series romanized: Kōbundō nyūmon sensho.
¥1900

1. Civil procedure—Japan. I. Amemiya, Masaya, 1935- joint author. II. Title.
79-806631

Okuno, Ken'ichi, 1898-
(Kaisei Minji soshō hō no kaisetsu)
改正民事訴訟法の解說 / 奥野健一, 三宅正雄共著. — 東京：海口書店, 昭和23 [1948]
4, 163 p. ; 22 cm.
Colophon inserted.
Includes legislation.

1. Civil procedure—Japan. I. Miyake, Masao, 1909- joint author. II. Japan. Laws, statutes, etc. III. Title. IV. Title: Minji soshō hō no kaisetsu.
79-802549

Onogi, Jō, 1909-
(Minji soshō hō)
民事訴訟法 / 小野木常著. — 京都：有信堂高文社, 昭和24 [1949]
2, 5, 361 p. ; 19 cm.
Colophon inserted.

1. Civil procedure—Japan. I. Title.
79-802892

Onogi, Jō, 1909-
(Minji soshō hō kōgi)
民事訴訟法講義 / 小野木常著. — 東京：有斐閣, 昭和26 [1951]
1, 12, 219 p. ; 21 cm.

1. Civil procedure—Japan. I. Title.
79-803277

Onogi, Jō, 1909-
(Minji soshō hō kōgi)
民事訴訟法講義 / 小野木常, 中野貞一郎共著. — 增補版. — 東京：有斐閣, 昭和31 [1956]
2, 14, 398, 48 p. ; 22 cm.
"民事訴訟規則" (8 p.) inserted at end.
Includes bibliographical references and indexes.

1. Civil procedure—Japan. I. Nakano, Teiichirō, 1925- joint author. II. Japan. Laws, statutes, etc. Minji soshō hō. 1956. III. Title.
79-802895

Saitō, Hideo, 1909-
(Sensō to Minji soshō hō)
戰爭と民事訴訟法 / 齋藤秀夫著. — 東京：日本評論社, 昭和19 [1944]
3, 6, 453 p. ; 19 cm. — (法學叢書 ; 15)
Series romanized: Hōgaku sōsho.
Includes bibliographical references.

1. Civil procedure—Germany. 2. Civil procedure—Japan. 3. Trade regulation. 4. War and emergency legislation. I. Title.
79-803401

Saitō, Tsunesaburo, 1878-1951.
(Chushaku sosho kiroku)
註釋訴訟記錄 / 齋藤常三郎著. — 京都：弘文堂書房, 大正 [19]- 大正13 [1924] printing.
v. ; 22 cm.
Cover title.
CONTENTS: —[2] 破産手續

1. Civil procedure—Japan. 2. Bankruptcy—Japan—Forms. 3. Trial practice—Japan. I. Title. II. Title: Sosho kiroku.
79-804569

Saitō, Tsunesaburō, 1878-1951.
(Chūshaku soshō kiroku)
註釋訴訟記錄 / 雉本朗造先生序文 ; 齋藤常三郎著. — 京都：弘文堂書房, 大正9 [1920]- 大正11 [1922] printing.
2 v. in 1 ; 23 cm.
CONTENTS: [1] 第一審手續—[2] 第二審手續

1. Civil procedure—Japan. 2. Damages—Japan—Forms. 3. Appellate procedure—Japan—Forms. 4. Trial practice—Japan. I. Title. II. Title: Soshō kiroku.
79-804958

Shimoyama, Shirō, 1897-
(Minji soshō hō)
民事訴訟法 / 下山四郎著. — 東京：非凡閣, 昭和10 [1935]
16, 408 p. ; 22 cm. — (大衆法律講座 ; 第11巻)

1. Civil procedure—Japan. I. Title. II. Series: Taishū hōritsu kōza ; dai 11-kan.
79-803274

Someno, Yoshinobu, 1918-
(Minji soshō hō genron)
民事訴訟法原論 / 染野義信著. — 東京：勁草書房, 1957.
4, 5, 541, 11 p. ; 22 cm.
Includes index.

1. Civil procedure—Japan. I. Title.
79-803276

CIVIL PROCEDURE —JAPAN (Continued)

Uda, Takao, 1931-
(Minji soshō hō)
民事訴訟法 / 宇田尭雄著. — 京都：嵯峨野書院, 1978.
6, 515, 12 p. ; 22 cm. Ja 79-1182
Includes index.
¥3400
1. Civil procedure—Japan. I. Title.
79-806630

Yamada, Shōzō, 1882-
(Kaisei Minji soshō hō)
改正民事訴訟法 / 山田正三著. — 京都：弘文堂書房, 昭和2-6 ⌊1927-1931⌋, v. 1, 昭和4 ⌊1929⌋ printing.
4 v. ; 23 cm.
Errata slip inserted.
1. Civil procedure—Japan. I. Title.
79-803374

Yamada, Shōzō, 1882-
(Minji soshō hō)
民事訴訟法 / 山田正三著. — 京都：弘文堂書房, 大正10-11 ⌊1921-1922⌋
2 v. ; 22 cm.
1. Civil procedure—Japan. I. Title.
79-803389

Yamakido, Kokuki, 1910-
(Minji soshō hō kōgi)
民事訴訟法講義 / 山木戸克己著. — 京都：三和書房, 昭和29 ⌊1954⌋, 昭和30 ⌊1955⌋ printing.
3, 11, 306 p. ; 22 cm.
Colophon inserted.
Includes bibliographies.
1. Civil procedure—Japan. I. Title.
79-804098

—JAPAN—ADDRESSES, ESSAYS, LECTURES

(Jittaihō to tetsuzukihō no kōsaku)
実体法と手続法の交錯：山木戸克己教授還暦記念 / ⌊編集代表 山木戸克己教授還暦記念論文集刊行発起人会⌋. — 東京：有斐閣, 昭和49-53 ⌊1974-1978⌋
2 v. : port. ; 22 cm. Ja 74-19983 (v. 1)
Chronological list of Kokuki Yamakido's works: v. 1, p. 3-10 (3d group)
Includes bibliographical references.
¥4500 (v. 1) ¥6000 (v. 2)
1. Civil procedure—Japan—Addresses, essays, lectures. 2. Civil procedure—Addresses, essays, lectures. 3. Yamakido, Kokuki, 1910- I. Yamakido, Kokuki, 1910- II. Yamakido Kokuki Kyōju Kanreki Kinen Rombunshū Kankō Hokkininkai.
75-801436
rev

Kijimoto, Rōzō, 1876-1922.
(Minji soshō hō no shomondai)
民事訴訟法の諸問題 / 雉本朗造著. — 東京：有斐閣, 昭和30 ⌊1955⌋
2, 2, 785, 18 p. ; 22 cm. — (學術選書 ; 5)
Series romanized: Gakujutsu sensho.
"新民事訴訟法": p. 687-785.
Includes bibliographical references and indexes.
1. Civil procedure—Japan—Addresses, essays, lectures. 2. Civil procedure—Addresses, essays, lectures. I. Japan. Laws, statutes, etc. Minji soshō hō. 1955. II. Title.
79-801842

Muramatsu, Toshio, 1902-
(Minji saiban no shomondai)
民事裁判の諸問題 / 村松俊夫著. — 東京：有信堂, 昭和28 ⌊1953⌋
3, 4, 258 p. ; 22 cm.
Colophon inserted.
Includes bibliographical references.
1. Civil procedure—Japan—Addresses, essays, lectures. I. Title.
79-805751

Sawada, Naoya, 1912-
(Karishobun zuisō)
仮處分隨想 / 沢田直也. — 大阪：布井書房, 昭和27 ⌊1952⌋
3, 4, 219 p. ; 19 cm.
1. Provisional remedies—Japan—Addresses, essays, lectures. 2. Civil procedure—Japan—Addresses, essays, lectures. I. Title.
79-803383

Sumiyoshi, Hiroshi, 1935-
(Minji soshō hanrei hyōshaku)
民事訴訟判例評釈 / 住吉博著. — 東京：法学書院, 1978.
3, 288 p. ; 22 cm. Ja 79-1154
Includes bibliographical references.
¥3200
1. Civil procedure—Japan—Addresses, essays, lectures. I. Title.
79-803876

—JAPAN—DIGESTS

Kikui, Tsunahiro, 1899-
(Hanrei minji tetsuzuki hō)
判例民事手續法 / 菊井維大著. — 東京：弘文堂, 昭和26 ⌊1951⌋
1, 13, 23, 548, 7 p. ; 22 cm.
Colophon inserted.
Includes index.
1. Civil procedure—Japan—Digests. I. Title. II. Title: Minji tetsuzuki hō.
79-802373

Niida, Masutarō, 1868-1945.
(Shimpō junkyo Minji soshō hō hanrei shū)
新法準據民事訴訟法判例集 / 仁井田益太郎編. — 東京：巖松堂書店, 昭和4 ⌊1929⌋, 昭和5 ⌊1930⌋ printing.
2, 83, 710 p. ; 23 cm.
Includes bibliographical references.
1. Civil procedure—Japan—Digests. I. Japan. Laws, statutes, etc. Minji soshō hō. 1929. II. Title. III. Title: Minji soshō hō hanrei shū.
79-802399

Yamahata, Toshiki.
(Uttae ni kansuru gakusetsu hanrei jitsurei sōran)
訴に関する學説判例実例綜攬 / 山端年樹著. — 大阪：大同書院, 昭和27 ⌊1952⌋
2, 2, 31, 398 p. ; 22 cm.
Bibliography: p. 1-2 (2d group)
1. Civil procedure—Japan—Digests. 2. Complaints (Civil procedure)—Japan—Digests. 3. Pleading—Japan—Digests. I. Title.
79-803795

—JAPAN—FORMS

(Minji soshō kiroku)
民事訴訟記錄 / 民事訴訟法學會編, 責任編集 野間繁. — 東京：有斐閣, 昭和29 ⌊1954⌋
5, 182 p. : forms ; 19 cm.
Errata slip inserted.
1. Civil procedure—Japan—Forms. I. Noma, Shigeru, 1902- II. Minji Soshō Hōgakkai.
79-803278

Oda, Kyūzō.
(Jitsurei hon'i minji soshō no hōritsu chishiki)
実例本位民事訴訟の法律知識 / 小田久藏著. — 東京：東洋書館, 昭和29 ⌊1954⌋
272 p. ; 19 cm. — (実例本位法律知識全書 ; 8)
1. Civil procedure—Japan. 2. Civil procedure—Japan—Forms. I. Title. II. Title: Minji soshō no hōritsu chishiki. III. Series: Jitsurei hon'i hōritsu chishiki zensho ; 8.
79-801889

—JAPAN—HISTORY—ADDRESSES, ESSAYS, LECTURES

Kobayakawa, Kingo.
(Kinsei minji soshō seido no kenkyū)
近世民事訴訟制度の研究 / 小早川欣吾著. — 東京：有斐閣, 昭和32 ⌊1957⌋
2, 25, 6, 696, 24 p. ; 22 cm.
Includes bibliographical references and index.
1. Civil procedure—Japan—History—Addresses, essays, lectures. 2. Japan—Politics and government—1600-1868—Addresses, essays, lectures. I. Title.
79-803214

—KENTUCKY

Kentucky. Laws, statutes, etc.
The code of practice in civil and criminal cases for the State of Kentucky : with all amendments made prior to Jan'y 1, 1867, with notes of the decisions of the Court of Appeals of Kentucky, and of the courts of Ohio and New York / originally prepared by M. C. Johnson, James Harlan, and J. W. Stevenson, commissioners, and adopted by the Legislature ; edited by Harvey Myers. — Cincinnati : R. Clarke, 1867.
xii, 899 p. ; 24 cm.
Includes indexes.
1. Civil procedure—Kentucky. 2. Criminal procedure—Kentucky. I. Johnson, M. C. II. Harlan, James, 1820-1899. III. Stevenson, John White, 1812-1886. IV. Myers, Harvey, 1828-1874. V. Title.
KFK1728.A2 1867 347'.769'05 79-103860
79 MARC

—KOREA

Korea (Republic). Taebŏbwŏn.
Civil procedure of Korea. — ⌊Seoul⌋ : Supreme Court of Republic of Korea, 1972.
13 p., ⌊3⌋ leaves of plates : ill. ; 26 cm.
1. Civil procedure—Korea. I. Title.
347'.5195'0502632 77-374715
78⌊r79⌋rev MARC

Kukka Kosi Yŏn'guhoe.
(Kukka kosi minsa sosongpŏp)
國家考試民事訴訟法：客觀式 / 国家考試研究会編. — 서울：法志社, 1977.
At head of title: Pŏbwŏn sŏgibo, chusabo, samugwan ch'aeyong sihŏm chich'imsŏ.
Colophon inserted.
"民事訴訟法": p. 337-338.
₩2500
1. Civil procedure—Korea. 2. Civil procedure—Korea—Examinations, questions, etc. I. Korea (Republic). Laws, statutes, etc. Minsa sosongpŏp. II. Title. III. Title: Minsa sosongpŏp. IV. Title: Kaekkwansik kukka kosi minsa sosongpŏp.
78-826605

—KOREA—EXAMINATIONS, QUESTIONS, ETC.

Kukka Kosi Yŏn'guhoe.
(Kukka kosi minsa sosongpŏp)
國家考試民事訴訟法：客觀式 / 国家考試研究会編. — 서울：法志社, 1977.
At head of title: Pŏbwŏn sŏgibo, chusabo, samugwan ch'aeyong sihŏm chich'imsŏ.
Colophon inserted.
"民事訴訟法": p. 337-338.
₩2500
1. Civil procedure—Korea. 2. Civil procedure—Korea—Examinations, questions, etc. I. Korea (Republic). Laws, statutes, etc. Minsa sosongpŏp. II. Title. III. Title: Minsa sosongpŏp. IV. Title: Kaekkwansik kukka kosi minsa sosongpŏp.
78-826605

CIVIL PROCEDURE —KOREA—EXAMINATIONS, QUESTIONS, ETC. (Continued)

Yi, Chong-ch'an.
(Minsapŏp)
民事法：五·四級公務員採用·昇進試驗對策 / 李鍾燦, 南勝吉共編. — 서울：華學社：供給處韓國出版協同組合, 1977-
v. ; 22 cm.
At head of cover title: Kaejŏng sihŏmnyŏng, sae ch'ulche kyŏnghyang e ttarŭn ch'oe sinp'an.
Colophon inserted.
"民法": v. 1, p. 393-452.
CONTENTS: 上. 上·客觀式民法.
₩1800 (v. 1)
1. Civil law—Korea—Examinations, questions, etc. 2. Civil procedure—Korea—Examinations, questions, etc. I. Nam, Sŭng-gil, joint author. II. Korea (Republic). Laws, statutes, etc. Minpŏp. 1977. III. Title.
79-825274

—KOREA—CHALLA-NAMDO

Rim, Gwang Taek.
(Minsa chaep'an ŭl t'onghayŏ pon Chŏnnam chibang ŭi punjaeng ŭi yuhyŏng e kwanhan chosa punsŏk)
民事裁判을通하여본全南地方의紛爭의類型에關한調査分析 = An investigation about disputes Jeonnam region in point of civil suit / 林光澤, 曺秉甲, 李邦基. — [s. l. : s. n., 197-]
49 p. : ill. ; 25 cm.
In Korean; summary in English.
Includes bibliographical references.
1. Civil procedure—Korea—Chŏlla-namdo. I. Cho, Byong Gap joint author. II. Lee, Bang Ky, joint author. III. Title. IV. Title: An investigation about disputes Jeonnam region in point of civil suit.
79-825926

—KUWAIT (STATE)

Wālī, Fatḥī.
(Qānūn al-qaḍā' al-madanī al-Kuwaytī)
قانون القضاء المدنى الكويتى : دراسة لمجموعة المرافعات المدنية والتجارية واهم التشريعات المكملة لها / تاليف فتحى والى. — [الكويت] : جامعة الكويت، 1977.
477 p. ; 24 cm.
Bibliography: p. 12.
1. Civil procedure—Kuwait. I. Title.
78-971043

—MASSACHUSETTS

Federal court litigation in the first circuit / edited by S. Stephen Rosenfeld. — Boston : Massachusetts Continuing Legal Education-New England Law Institute, c1978.
ca. 500 p. in various pagings ; 23 cm.
On cover: Practice handbook.
Includes bibliographical references and index.
1. Civil procedure—United States. 2. Civil procedure—Massachusetts. 3. United States. Court of Appeals for the First Circuit. 4. United States. District Court. Massachusetts. I. Rosenfeld, Samuel Stephen. II. Massachusetts Continuing Legal Education-New England Law Institute, inc.
KF8840.F39 347'.73'225 78-61279
79 MARC

—MEXICO

Medina Ochoa, Valentín.
Nuestro enjuiciamiento civil / Valentín Medina Ochoa. — 1. ed. — México : Editorial Porrúa, 1974.
347 p. ; 22 cm.
1. Civil procedure—Mexico. I. Title.
347'.72'05 79-378393
79 MARC

Pina, Rafael de.
Instituciones de derecho procesal civil / Rafael de Pina, José Castillo Larrañaga. — 11. ed., rev., aumentada y actualizada / por Rafael de Pina Vara. — México : Editorial Porrúa, 1976.
669 p. ; 24 cm.
Includes bibliographies.
1. Civil procedure—Mexico. I. Castillo Larrañaga, José, joint author. II. Pina Vara, Rafael de. III. Title.
347'.72'05 79-351995
79 MARC

—MEXICO—NUEVO LEÓN

Nuevo León, Mexico. Laws, statutes, etc.
[Código de procedimientos civiles (1973)]
Código de procedimientos civiles del Estado de Nuevo León. — 1. ed. — México : Editorial Porrúa, 1978.
252 p. ; 17 cm. — (Leyes y códigos de México) (Colección Porrúa)
1. Civil procedure—Mexico—Nuevo León. I. Title.
347'.721'05 79-107787
79 MARC

—MINNESOTA

McFarland, Douglas D
Minnesota civil practice / Douglas D. McFarland, William J. Keppel. — Saint Paul, Minn. : Mason Pub. Co., c1979.
2 v. (xiii, 1476 p.) : ill. ; 26 cm.
Cover title: Mason's Minnesota civil practice.
ISBN 0-917126-08-4
1. Civil procedure—Minnesota. I. Keppel, William J., joint author. II. Title. III. Title: Mason's Minnesota civil practice.
KFM5930.M3 347'.776'05 79-88507
00 MARC

—NETHERLANDS

Netherlands (Kingdom, 1815-). Laws, statutes, etc.
Wetboek van burgerlijke rechtsvordering / bew. door M. A. P. van der Veere. — 16e dr. — Zwolle : W. E. J. Tjeenk Willink, 1978.
xxxv, 532 p. ; 19 x 8 cm. — (Nederlandse wetboeken) Ne78-49
Cover title.
"Bijgewerkt tot 15 juni 1978."
Includes indexes.
ISBN 9027115052 : fl 35.00
1. Civil procedure—Netherlands. I. Veere, M. A. P. van der. II. Title. III. Series.
347.492'0502632 79-343709
•79 MARC

—NETHERLANDS—DIGESTS

Netherlands (Kingdom, 1815-). Hooge Raad.
Arresten handelsrecht en burgerlijk procesrecht : met annotaties ten behoeve van het onderwijs / verzameld door C. J. van Zeben, T. A. W. Sterk. — 2e herz. en verm. dr. — Zwolle : W. E. J. Tjeenk Willink, 1978.
x, 1339 p. ; 22 cm. Ne78-44
Includes indexes.
ISBN 9027114579 : fl 42.50
1. Commercial law—Netherlands—Digests. 2. Civil procedure—Netherlands—Digests. I. Zeben, Christiaan Johannes van. II. Sterk, T. A. W. III. Title.
79-337786
•79 MARC

—NEW YORK (STATE)—CASES

Abbott, Benjamin Vaughan, 1830-1890.
Reports of practice cases determined in the courts of the State of New-York : with a digest of all points of practice embraced in the standard New-York reports issued during the period covered by this volume : new series / by Benjamin Vaughan Abbott and Austin Abbott. — New York : Diossy & Cockcroft, 1866-
v. ; 23 cm.
Subtitle of v. 3 varies slightly.
Includes indexes.
1. Law reports, digests, etc.—New York (State) 2. Civil procedure—New York (State)—Cases. I. Abbott, Austin, 1831-1896, joint author. II. Title.
KFN5994.A19 1865 347'.747'05 79-304401
79 MARC

—NORTH DAKOTA

Bucklin, Leonard H
Civil practice of North Dakota / Leonard H. Bucklin. — Rev. ed. — Bismarck, ND : Trial Lawyers Service, 1979.
1 v. ; 30 cm.
Loose-Leaf for updating.
Includes bibliographical references and indexes.
1. Civil procedure—North Dakota. I. Title.
KFN9130.B83 1979 347'.784'05 79-118599
79 MARC

—PAKISTAN

Farani, Merajuddin, 1935-
Code of civil procedure / by M. Farani. — [Rev. ed.]. — Lahore : Lahore Law Times Publications, 1978.
xxi, 966 p. ; 25 cm.
Includes index.
Rs85.00
1. Civil procedure—Pakistan. 2. Civil procedure—India. I. India (Republic). Laws, statutes, etc. Code of civil procedure, 1908. 1978. II. Title.
345'.54'05 78-931170
79 MARC

Sabri, Badur-ul-Hassan Khan.
The law of inherent powers of civil courts and appeal, revision, reference and review / [Badur-ul-Hassan]. — Lahore : Eastern Law House, [1974]
iv, 114, ii p. ; 22 cm.
Includes index.
Rs8.00
1. Courts—Pakistan. 2. Civil procedure—Pakistan. I. Title.
347'.5491'012 74-930352
75[r79]rev MARC

—PAKISTAN—DIGESTS

Abdul Hamid.
Thirty-one years C.P.C. digest : vol. of 1908, with supplement, 1978 / by Abdul Hamid. — Lahore : Nishat Book House, 1978.
vii, 178 p. ; 24 cm.
Running title: Civil procedure code.
Rs20.00
1. Civil procedure—Pakistan—Digests. I. Title. II. Title: Civil procedure code.
347'.5491'05 78-931092
79 MARC

—PAPAL STATES—INDEXES

Repertorio alfabetico : ossia, indice ragionato delle materie contenute nel regolamento legislativo e giudiziario per gli affari civili, emanato dalla santità di N.S. Gregorio papa XVI, con moto proprio del 10 novembre 1834 ; seguito da uno specchio delle tasse a forma della tariffa, contenuta nell'Editto di sua eminenza reverendissima il sig. cardinale segretario per gli affari di Stato interni del giorno 17 dicembre 1834. — Perugia : Tip. di V. Santucci, 1835.
127 p. ; 29 cm.
Bound with Papal States. Laws, statutes, etc. Regolamento legislativo e giudiziario. Roma, 1834. Papal States. Segreteria di Stato. Editto. Roma, 1834.
1. Law—Papal States—Indexes. 2. Courts—Papal States—Indexes. 3. Civil procedure—Papal States—Indexes. I. Papal States. Laws, statutes, etc. Regolamento legislativo e giudiziario per gli affari civili. II. Papal States. Segreteria di Stato. Editto.
79-338819
79 MARC

—PERU

Figueroa Estremadoyro, Hernán.
Código de procedimientos civiles / comentado y actualizado con los últimos decretos y leyes por Hernán Figueroa Estremadoyro ; concordado por Manuel Balarezo G. — 1. ed. — Lima : Editorial Inkari, 1977.
301 p. ; 22 cm.
S/400.00
1. Civil procedure—Peru. I. Balarezo Gamarra, Manuel. II. Peru. Laws, statutes, etc. Código de procedimientos civiles. 1977.
347'.85'05 78-107500
79 MARC

Peru. Laws, statutes, etc.
Código de procedimientos civiles : actualizado por Decreto ley 21773 (18/1/1977), 60 arts. modificados, 13 arts. derogados ... / [editor] F. Bonilla. — 10. ed. — Lima : Editorial Mercurio : [pedidos, Importadora y Distribuidora Rivera], 1978.
321 p. ; 19 cm. — (Legislación peruana)
"No es oficial."
S/300.00
1. Civil procedure—Peru. I. Bonilla, F. II. Title.
347'.85'05 79-104283
79 MARC

—PHILIPPINE ISLANDS

Laureta, Wenceslao Goloyugo.
Remedial law reviewer / by Wenceslao Goloyugo Laureta. — 4th ed. — Manila : Rex Book Store, 1978.
xxxvii, 656 p. ; 24 cm.
Third ed. published in 1975 under title: Reviewer in remedial law.
Includes bibliographical references and index.
1. Civil procedure—Philippine Islands. 2. Criminal procedure—Philippine Islands. 3. Evidence (Law)—Philippine Islands. I. Title.
347'.599 79-102156
79 MARC

—POLAND

Berutowicz, Włodzimierz.
Postępowanie cywilne w zarysie / Włodzimierz Berutowicz. — Wyd. 2., zmienione. — Warszawa : Państwowe Wydawn. Naukowe, 1978.
496 p. ; 24 cm. p***
Bibliography: p. 474-[489]
Includes index.
zł62.00
1. Civil procedure—Poland. I. Title.
79-337514
79 MARC

CIVIL PROCEDURE
—POLAND (Continued)

Krzemiński, Zdzisław.
Postępowanie odrębne w sprawach ze stosunków między rodzicami a dziećmi / Zdzisław Krzemiński. — Warszawa : Wydaw. Prawnicze, 1978 (Warsz.: D 1)
151, ⌊1⌋ p. ; 22 cm. P78-4401
Errata slip inserted.
Bibliography: p. 145-151.
zł40.00
1. Paternity—Poland. 2. Domestic relations—Poland. 3. Civil procedure—Poland. I. Title.
78-384106
78 MARC

Niejadlik, Jan.
Kontrola rewizyjna postanowień poprzedzających wydanie wyroku w procesie cywilnym / Jan Niejadlik. — Warszawa : Wydaw. Prawnicze, 1978 (⌊Warsz : D 1⌋)
135, ⌊1⌋ p. ; 21 cm. P78-2924
Summary in German.
Bibliography: p. 124-128.
zł35.00
1. Remedies (Law)—Poland. 2. Civil procedure—Poland. I. Title.
79-341814
79 MARC

Siedlecki, Władysław.
Postępowanie cywilne : zarys wykładu / Władysław Siedlecki. — Warszawa : Państ. Wydaw. Naukowe, 1978 (Ciesz : CDW)
514, ⌊2⌋ p. ; 25 cm. P78-2927
Includes bibliographical references.
zł80.00
1. Civil procedure—Poland. I. Title.
347'.438'05 78-382701
78 MARC

—PORTUGAL

Portugal. Laws, statutes, etc.
Código de processo civil / compilação organizada e actualizada por Joaquim F. Nogueira. — ⌊Coimbra : Coimbra Editora⌋, 1977 ⌊i.e. 1978⌋
665 p. ; 23 cm. — (Códigos Coimbra Editora ; 2)
Includes index.
1. Civil procedure—Portugal. I. Nogueira, Joaquim F.
347'.469'0502632 78-109836
79 MARC

Portugal. Laws, statutes, etc.
⌊Código de processo civil (1961)⌋
Código de processo civil : anotado e actualizado / Jacinto Fernandes Rodrigues Bastos. — Coimbra : Livraria Almedina, 1978.
763 p. ; 23 cm.
Includes index.
1. Civil procedure—Portugal. I. Bastos, Jacinto Fernandes Rodrigues.
347'.469'0502633 79-112740
79 MARC

Portugal. Laws, statutes, etc.
⌊Código de processo civil (1961)⌋
Código de processo civil anotado / Abílio Neto. — 2. ed. — Lisboa : Livraria Petrony, 1978.
1181 p. ; 24 cm.
Includes index.
1. Civil procedure—Portugal. I. Neto, Abílio. II. Title.
347'.469'0502632 78-109839
79 MARC

—QUEBEC (PROVINCE)—CASES

Manuel pratique de preuve civile / ⌊textes compilés par⌋ Gary-Q. Ouellet. — ⌊Montréal⌋ : Wilson & Lafleur, tirage de 1978.
270 p. ; 23 cm. C78-8474-8
Includes judicial material in English.
$7.00
1. Evidence (Law)—Quebec (Province)—Cases. 2. Civil procedure—Quebec (Province)—Cases. I. Ouellet, Gary Q., 1945-
KEQ1126.A47M35 347'.714'060264 79-379004
79 MARC

—ROME see Civil procedure (Roman law)

—RUSSIA

Hazard, John Newbold, 1909-
Settling disputes in Soviet society : the formative years of legal institutions / by John N. Hazard. — New York : Octagon Books, 1978, c1960.
xiv, 534 p. ; 24 cm.
Reprint of the edition published by Columbia University Press, New York, in series: Studies of the Russian Institute, Columbia University; with new pref.
Bibliography: p. ⌊492⌋-508.
Includes index.
ISBN 0-374-93758-3
1. Courts—Russia. 2. Lawyers—Russia. 3. Criminal procedure—Russia. 4. Civil procedure—Russia. 5. Justice, Administrative of—Russia. I. Title. II. Series: Columbia University. Russian Institute. Studies.
345'.47'01 78-62948
78 MARC

Oĭgenzikht, Viktor Arkad'evich.
Prezumpt͡sii v sovetskom grazhdanskom prave / V. A. Oĭgenzikht ; Tadzh. gos. un-t im. V. I. Lenina. — Dushanbe : Irfon, 1976.
190 p. ; 20 cm. USSR77
At head of title: Ministerstvo narodnogo obrazovanii͡a Tadzhikskoĭ SSR.
Includes bibliographical references.
1. Presumptions (Law)—Russia. 2. Civil procedure—Russia. I. Title.
79-384492
79 MARC

Russia (1923- U. S. S. R.). Laws, statutes, etc.
(Sbornik normativnykh aktov po grazhdanskomu prot͡sessu)
Сборник нормативных актов по гражданскому процессу / Составители Бурова С. Н. ... ⌊et al.⌋ ; Под ред. Н. Г. Юркевича. — Минск : Вышэйш. школа, 1978.
352 p. ; 21 cm. USSR 78
Includes bibliographies.
0.85rub
1. Civil procedure — White Russia. 2. Civil procedure — Russia. I. Burova, S. N. II. I͡Urkevich, Nikolaĭ Grigor'evich. III. Title.
78-421601

—RUSSIA—ADDRESSES, ESSAYS, LECTURES

(Pravovye voprosy grazhdanskogo prava i prot͡sessa)
Правовые вопросы гражданского права и процесса : Темат. сборник / Дальневост. гос. ун-т ; Под ред. проф. Н. И. Овчинникова. — Владивосток : ДВГУ, 1976.
111 p. ; 20 cm. USSR 77
At head of title: Ministerstvo vysshego i srednego spet͡sial'nogo obrazovanii͡a RSFSR.
Includes bibliographical references.
0.65rub
1. Civil law—Russia—Addresses, essays, lectures. 2. Civil procedure—Russia—Addresses, essays, lectures. I. Ovchinnikov, Nikolaĭ Ivanovich. II. Vladivostok. Dal'nevostochnyĭ universitet.
78-422895

(Voprosy grazhdanskogo prava i prot͡sessa)
Вопросы гражданского права и процесса : ⌊Сборник статей⌋ / АН СССР, Ин-т государства и права ; ⌊Ред. коллегия д-р юрид. наук А. Ю. Кабалкин ... et al.⌋. — Москва : Ин-т государства и права, 1977.
106 p. ; 20 cm. USSR 77
Includes bibliographical references.
0.60rub
1. Civil law—Russia—Addresses, essays, lectures. 2. Civil procedure—Russia—Addresses, essays, lectures. I. Kabalkin, Aleksandr I͡Ur'evich. II. Akademii͡a nauk SSSR. Institut gosudarstva i prava.
78-421564

—SPAIN

Spain. Laws, statutes, etc.
⌊Ley de enjuiciamiento civil (1881)⌋
Comentarios á la ley de enjuiciamiento civil reformada : conforme á las bases aprobadas por la Ley de 21 de junio de 1880 con los formularios correspondientes á todos los juicios / por José María Manresa y Navarro. — 3. ed., aumentada con la jurisprudencia posterior y arreglada á los últimos textos y disposiciones legales que derogan y modifican algunos de los preceptos de dicha ley / por Francisco de P. Rives y Martí. — Madrid : Impr. de la Revista de Legislación, 19
v. ; 23 cm.
1. Civil procedure—Spain. 2. Courts—Spain. 3. Civil procedure—Spain—Forms. I. Manresa y Navarro, José María. II. Rives y Martí, Francisco de Paula. III. Title.
347'.46'05 78-109132
79 MARC

Spain. Laws, statutes, etc.
Leyes procesales. — 2. ed. — Pamplona : Aranzadi, 1977.
2213 p. ; 20 cm. — (Legislación civil Aranzadi) Sp77-Nov
Includes index.
ISBN 8470161180 : 1975ptas
1. Civil procedure—Spain. I. Title.
347'.46'05 78-378935
78 MARC

—SPAIN—FORMS.

Prieto Castro, Leonardo.
Modelos para la práctica jurídica procesal civil : (proceso civil declarativo y ejecución) ... / compuesto por Leonardo Prieto-Castro y Ferrandiz. — 2. ed. — Madrid : Tecnos, 1977.
283 p., 2 leaves : port. ; 21 cm. — (Manuales universitarios españoles ; 3) Sp78-Feb
ISBN 8430905952 : 400ptas
1. Civil procedure—Spain—Forms. I. Title.
347'.46'055 78-390601
79 MARC

Spain. Laws, statutes, etc.
⌊Ley de enjuiciamiento civil (1881)⌋
Comentarios á la ley de enjuiciamiento civil reformada : conforme á las bases aprobadas por la Ley de 21 de junio de 1880 con los formularios correspondientes á todos los juicios / por José María Manresa y Navarro. — 3. ed., aumentada con la jurisprudencia posterior y arreglada á los últimos textos y disposiciones legales que derogan y modifican algunos de los preceptos de dicha ley / por Francisco de P. Rives y Martí. — Madrid : Impr. de la Revista de Legislación, 19
v. ; 23 cm.
1. Civil procedure—Spain. 2. Courts—Spain. 3. Civil procedure—Spain—Forms. I. Manresa y Navarro, José María. II. Rives y Martí, Francisco de Paula. III. Title.
347'.46'05 78-109132
79 MARC

—SWITZERLAND

Aschwanden, Paul, 1949-
Die Behandlung der Querulanten im Zivilprozess / von Paul Aschwanden. — Zürich : Juris-Verlag, 1978.
ix, 156 p. ; 23 cm. Sw78-11236
Thesis—Zürich.
Bibliography: p. iv-vii.
ISBN 3-260-04508-2 : 35.00F
1. Civil procedure—Switzerland. 2. Litigious paranoia. 3. Insanity—Jurisprudence—Switzerland. I. Title.
347'.494'05 79-363606
79 MARC

Guldener, Max, 1903-
Schweizerisches Zivilprozessrecht / von Max Guldener. — 3., verb. und vermehrte Aufl. — Zürich : Schulthess, 1979.
lxviii, 667 p. ; 25 cm. Sw79-6348
Bibliography: p. xli-lxvi.
Includes index.
ISBN 3-7255-1964-1 : 110.00F
1. Civil procedure—Switzerland. I. Title.
347'.494'05 79-375610
79 MARC

Kummer, Max.
Grundriss des Zivilprozessrechts : nach den Prozessordnungen des Kantons Bern und des Bundes / Max Kummer. — 3., auf den neusten Stand der Gesetzgebung gebrachte Aufl. — Bern : Stämpfli, 1978, c1970.
xi, 295 p. ; 24 cm. Sw78-11801
Bibliography: p. xi.
Includes index.
ISBN 3-7272-9383-7 : 52.00F
1. Civil procedure—Switzerland—Bern (Canton) 2. Civil procedure—Switzerland. I. Title.
347'.4945'05 79-356293
79 MARC

Nussbaum, Beat.
Die Wirtschaftlichkeit eines Zivilprozesses um eine Geldforderung nach bernischem Recht / Beat Nussbaum. — Bern ; Las Vegas : P. Lang, c1978.
126, ⌊18⌋ p. : ill. ; 21 cm. — (Europäische Hochschulschriften : Reihe 2, Rechtswissenschaft ; Bd. 167)
Bibliography: p. 11-12.
ISBN 3-261-02906-4
1. Costs (Law)—Switzerland—Bern (Canton) 2. Civil procedure—Switzerland. I. Title. II. Series.
347'.4945'077 79-346135
79 MARC

Thouvenin, André, 1946-
Die bundesrechtliche Nichtigkeitsbeschwerde in Zivilsachen / André Thouvenin. — Zürich : Schulthess, cop. 1978.
253 p. ; 23 cm. — (Zürcher Studien zum Verfahrensrecht ; Bd. 29) Sw78-6561
Originally presented as the author's thesis, Zürich.
Bibliography: p. 13-19.
Vita.
ISBN 3-7255-1893-9 : 40.00F
1. Nullity—Switzerland. 2. Civil procedure—Switzerland. 3. Appellate procedure—Switzerland. I. Title. II. Series.
347'.494'05 79-346088
79 MARC

—SWITZERLAND—BERN (CANTON)

Chatelain, Jean Pierre, 1916-
La prise à partie en procédure civile bernoise. Delémont, 1946.
179 p. 23 cm.
Thèse—Bern.
Bibliography: p. ⌊6⌋-8.
1. Civil procedure—Switzerland—Bern (Canton) 2. Remedies (Law)—Switzerland—Bern (Canton) 3. Administrative responsibility—Switzerland—Bern (Canton) I. Title.
65-76977
⌊r79⌋rev MARC

CIVIL PROCEDURE
—SWITZERLAND—BERN (CANTON)
(Continued)

Kummer, Max.
Grundriss des Zivilprozessrechts : nach den Prozessordnungen des Kantons Bern und des Bundes / Max Kummer. — 3., auf den neusten Stand der Gesetzgebung gebrachte Aufl. — Bern : Stämpfli, 1978, c1970.
xi, 295 p. ; 24 cm. Sw78-11801
Bibliography: p. xi.
Includes index.
ISBN 3-7272-9383-7 : 52.00F
1. Civil procedure—Switzerland—Bern (Canton) 2. Civil procedure—Switzerland. I. Title.
347'.4945'05 79-356293
79 MARC

—SWITZERLAND—SOLOTHURN (CANTON)—HISTORY

Schwaller, Konrad.
Die Stellung des Richters im ordentlichen Verfahren der solothurnischen Zivilprozessordnung vom 11. September 1966 / vorgelegt von Konrad Schwaller. — ⌜s.l. : s.n.⌝, 1976. (Solothurn : Reinhard Offset)
xxiii, 119 p. ; 21 cm. Sw***
Thesis—Bern.
Bibliography: p. xi-xvii.
1. Civil procedure—Switzerland—Solothurn (Canton)—History. 2. Judges—Switzerland—Solothurn (Canton)—History. I. Solothurn (Canton). Laws, statutes, etc. Zivilprozessordnung. II. Title.
347'.4943'05 78-378933
78 MARC

—SWITZERLAND—ZURICH (CANTON)

Sträuli, Hans, 1902-
Kommentar zur Zürcherischen Zivilprozessordnung : Gesetz über den Zivilprozess vom 13. Juni 1976 / Hans Sträuli, Georg Messmer. Schiedsgerichte und Schiedsgutachten / von Felix Wiget. — Zürich : Schulthess Polygraphischer Verlag, c1976.
xii, 644 p. ; 25 cm. Sw76-12036
Bibliography: p. ix-x.
ISBN 3-7255-1671-5 : 145.00F
1. Civil procedure—Switzerland—Zurich (Canton) 2. Arbitration and award—Switzerland—Zurich (Canton) I. Messmer, Georg, 1921- joint author. II. Wiget, Felix, 1913- joint author. III. Zürich (Canton). Laws, statutes, etc. Zivilprozessordnung 1976. 1976. IV. Title.
347'.4945'05 79-341211
*79 MARC

—TAIWAN

Lin, Jung-yao.
(Min shih su sung shih wu)
民事訴訟實務 / 林榮耀著. — 臺北 : 五南圖書出版公司, 民國67 ⌜1978⌝
5, 636 p. ; 22 cm.
Includes bibliographical references.
NT$170.00

1. Civil procedure—Taiwan. I. Title.
78-843232

Liu, Fa-yün.
(Min shih su sung fa kai yao)
民事訴訟法概要 / 劉發鋆著. — 臺北 : 大林出版社, 民國65 ⌜1976⌝
10, 362 p. ; 22 cm. — (大林社會學叢書 ; 5)
NT$60.00

1. Civil procedure—Taiwan. I. Title. II. Series: Ta lin she hui hsüeh ts'ung shu ; 5.
78-840808

Liu, Fa-yün.
(Min shih su sung shou hsü pien lan)
民事訴訟手續便覽 / 劉發鋆編著. — 臺北 : 水牛出版社, ⌜1978?⌝
189 p. ; 22 cm. — (現代青年叢書 ; 16)
Series romanized: Hsien tai ch'ing nien ts'ung shu.
Reprint ed.

1. Civil procedure—Taiwan. 2. Civil procedure—Taiwan—Forms. I. Title.
78-842481

—TAIWAN—ADDRESSES, ESSAYS, LECTURES

Liu, T'ieh-cheng.
(T'an min shih chiu fen ti chieh chüeh)
談民事糾紛的解決 / 劉鐵錚講 ; 教育部社會教育司編印. — ⌜台北⌝ : 該司, 民國66 ⌜1977⌝
2, 20 p. : port. ; 19 cm. — (文化講座專集 ; 110)

1. Civil procedure—Taiwan—Addresses, essays, lectures. I. Title. II. Series: Wen hua chiang tso chuan chi ; 110.
78-843098

—TAIWAN—FORMS

Liu, Fa-yün.
(Min shih su sung shou hsü pien lan)
民事訴訟手續便覽 / 劉發鋆編著. — 臺北 : 水牛出版社, ⌜1978?⌝
189 p. ; 22 cm. — (現代青年叢書 ; 16)
Series romanized: Hsien tai ch'ing nien ts'ung shu.
Reprint ed.

1. Civil procedure—Taiwan. 2. Civil procedure—Taiwan—Forms. I. Title.
78-842481

—TENNESSEE

Tennessee Judicial Conference. Committee on Pattern Jury Instructions (Civil)
Tennessee pattern jury instructions, civil : (T.P.I.—civil) / prepared and edited by the Committee on Pattern Jury Instructions (Civil) of the Tennessee Judicial Conference ; consultant, W. Harold Bigham. — 1st ed. — St. Paul : West Pub. Co., 1979.
xxvi, 426 p. ; 27 cm. — (Tennessee practice ; v. 8)
Includes indexes.
1. Instructions to juries—Tennessee—Forms. 2. Civil procedure—Tennessee. I. Title. II. Series.
KFT80.T4 vol. 8 347'.768'05 s 79-64681
⌜KFT542⌝ 79 MARC

—TEXAS

Texas litigation guide / William V. Dorsaneo III ... ⌜et al.⌝ editor⌜s⌝. — New York : M. Bender, 1977-
v. : forms ; 26 cm.
Loose-leaf for updating.
Accompanied by loose-leaf index volume.
Includes bibliographical references.
CONTENTS: v. 1-6. Unit 1, Civil trial procedures.—v. 7-8. Unit 2, Business entities.—v. 9-10. Unit 3, Pleading in commercial litigation.
1. Civil procedure—Texas. I. Dorsaneo, William V. II. Bender (Matthew) and Company, inc.
KFT1730.T4 347'.764'07 77-71200
77⌜r79⌝rev2 MARC

—TUNISIA

Tunisia. Laws, statutes, etc.
⌜Majallat al-murāfa'āt al-madanīyah wa-al-tijārīyah. French⌝
Code de procédure civile et commerciale : en annexe, procédure de l'injonction de payer, rapports entre propiétaires et locataires. — ⌜Tunis⌝ : Impr. officielle, 1977.
109 p. ; 21 cm.
At head of title: République tunisienne.
1.00AD
1. Civil procedure—Tunisia. I. Title.
347'.611'0502633 79-105515
79 MARC

—UNITED STATES

Cound, John J
1978 civil procedure supplement : containing selective Federal and State statutes, rules, and problems, forms, and recent decisions / by John J. Cound, Jack H. Friedenthal, Arthur R. Miller. — St. Paul : West Pub. Co., c1978.
iii, 419 p. ; 25 cm. — (American casebook series)
"For use with all pleading and procedure casebooks."
Includes index.
1. Civil procedure—United States. I. Friedenthal, Jack H., joint author. II. Miller, Arthur Raphael, 1934- joint author. III. Title. IV. Title: Civil procedure supplement. V. Series.
KF8839.C624 1978 suppl. 78-102610
347'.73'5
79 MARC

Current problems in Federal civil practice, 1979 / Barry H. Garfinkel, chairman. — New York : Practising Law Institute, c1978.
2 v. (1164 p., p. 1164 blank) ; 22 cm. — (Litigation and administrative practice series) (Litigation course handbook series ; no. 130)
"Prepared for distribution at the current problems in Federal civil practice, 1979 program, January-March, 1979."
Includes bibliographical references.
1. Civil procedure—United States. I. Garfinkel, Barry H., 1928- II. New York (City). Practising Law Institute.
KF8841.C87 347'.73'5 78-66156
79 MARC

Federal court litigation in the first circuit / edited by S. Stephen Rosenfeld. — Boston : Massachusetts Continuing Legal Education-New England Law Institute, c1978.
ca. 500 p. in various pagings ; 23 cm.
On cover: Practice handbook.
Includes bibliographical references and index.
1. Civil procedure—United States. 2. Civil procedure—Massachusetts. 3. United States. Court of Appeals for the First Circuit. 4. United States. District Court. Massachusetts. I. Rosenfeld, Samuel Stephen. II. Massachusetts Continuing Legal Education-New England Law Institute, inc.
KF8840.F39 347'.73'225 78-61279
79 MARC

Green, Milton Douglas, 1903-
Basic civil procedure / Milton D. Green. — 2d ed. — Mineola, N.Y. : Foundation Press, 1979.
xiv, 327 p. ; 27 cm. — (University textbook series)
Includes bibliographical references and indexes.
1. Civil procedure—United States. I. Title. II. Series.
KF8840.G7 1979 347'.73'5 79-51357
79 MARC

Kane, Mary Kay.
Civil procedure in a nutshell / by Mary Kay Kane. — St. Paul : West Pub. Co., 1979.
xxxix, 271 p. ; 19 cm. — (Nutshell series)
Includes index.
ISBN 0-8299-2040-4 : $5.95
1. Civil procedure—United States. I. Title.
KF8841.K36 347'.73'5 79-11278
79 MARC

Karlen, Delmar.
Civil litigation / by Delmar Karlen. — Indianapolis : Bobbs-Merrill Co., c1978.
xv, 310 p. ; 24 cm.
Includes bibliographical references and index.
ISBN 0-672-82898-7
1. Civil procedure—United States. I. Title.
KF8840.K35 347'.73'5 78-50661
79 MARC

Schurtman, William.
Der amerikanische Zivilprozess : e. Überblick für d. Praxis / von William Schurtman u. Otto L. Walter. — Frankfurt am Main : Metzner ; Deventer : Kluwer, 1978.
122 p. ; 24 cm. GFR78-A
Glossary in English and German.
ISBN 3-7875-5271-5
1. Civil procedure—United States. I. Walter, Otto L., joint author. II. Title.
KF8840.S35 347'.73'5 79-343711
79 MARC

Shepard's manual of Federal practice / editorial staff, editor in chief, Rudolph W. Fischer ... ⌜et al.⌝. — 2d ed. — Colorado Springs : Shepard's, c1979.
ix, 860 p. ; 26 cm.
Revised ed. of Manual of Federal practice, by R. A. Lavine and G. D. Horning.
Includes index.
ISBN 0-07-056706-9 : $45.00
1. Civil procedure—United States. I. Fischer, Rudolph W. II. Lavine, Richard Allen, 1917- Manual of Federal practice. III. Shepard's, inc. of Colorado Springs. IV. Title: Manual of Federal practice.
KF8840.L3 1979 347'.73'5 79-14133
79 MARC

United States. Laws, statutes, etc.
The Judicial code and rules of procedure in the Federal courts, with excerpts from the Criminal code / compiled by Henry M. Hart, Jr. and Herbert Wechsler. — Students' ed., 1978 revision. — Mineola, N.Y. : Foundation Press, 1978.
viii, 732 p. : forms ; 25 cm.
1. Courts—United States. 2. Civil procedure—United States. 3. Criminal procedure—United States. 4. Court rules—United States. I. Hart, Henry Melvin. II. Wechsler, Herbert, 1909- . III. United States. Supreme Court. IV. United States. Courts of Appeals. V. United States. District Courts. VI. Title.
KF8815.A2 1978 347'.73'205 78-109382
78 MARC

Volz, Marlin M 1917-
West's Federal practice manual, by Marlin M. Volz and a staff of Federal practice experts. 2d ed. St. Paul, West Pub. Co., 1970-
v. forms. 27 cm.
1. Procedure (Law)—United States. 2. Administrative procedure—United States. 3. Civil procedure—United States. 4. Criminal procedure—United States. I. West Publishing Company, St. Paul. II. Title. III. Title: Federal practice manual.
KF8840.V6 1970 347'.73'5 76-15882
70⌜r79⌝rev2 MARC

CIVIL PROCEDURE (Continued)

—UNITED STATES—CASES

Casenote legal briefs—civil procedure : adaptable to courses utilizing Carrington and Babcock's casebook on civil procedure / Dana L. Blatt, senior editor, Norman S. Goldenberg, senior editor ; Peter Tennen, managing editor ; staff writers, James Rosenthal ... [et al.]. — Beverly Hills, CA : Casenotes Pub. Co. ; Gardena, Calif. : distributed by Law Distributors, c1978.
120 p. ; 28 cm.
Includes indexes.
1. Civil procedure—United States—Cases. I. Blatt, Dana L. II. Goldenberg, Norman S. III. Carrington, Paul D., 1931- Civil procedure. IV. Title: Civil procedure.
KF8839.C32 347'.73'5 79-114935
79 MARC

Casenote legal briefs, civil procedure : adaptable to courses utilizing Field, Kaplan, and Clermont's casebook on civil procedure / Dana L. Blatt, senior editor, Norman S. Goldenberg, senior editor ; Peter Tenen, managing editor ; staff writers, Matt Hardy ... [et al.]. — Beverly Hills, CA : Casenotes Pub. Co. ; Gardena, Calif. : distributed by Law Distributors, c1978.
iii, 100 p. ; 28 cm.
1. Civil procedure—United States—Cases. I. Blatt, Dana L. II. Goldenberg, Norman S. III. Hardy, Matt. IV. Field, Richard H. Materials for a basic course in civil procedure.
KF8839.C343 347'.73'5 79-114879
79 MARC

Casenote legal briefs : civil procedure : adaptable to courses utilizing Karlen, Meisenholder, Stevens, and Vestal's Casebook on civil procedure / Dana L. Blatt, senior editor, Norman S. Goldenberg, senior editor ; Peter Tenen, managing editor ; staff writers, Matt Hardy ... [et al.]. — Beverly Hills, CA. : Casenotes Pub. Co. ; Gardena, Calif. : distributed by Law Distributors, c1976.
126 p. ; 28 cm.
Includes index.
$8.95
1. Civil procedure—United States—Cases. I. Blatt, Dana L. II. Goldenberg, Norman S. III. Tenen, Peter. IV. Title: Civil procedure.
KF8839.C347 347'.73'5 79-114884
79 MARC

Casenote legal briefs—civil procedure : adaptable to courses utilizing Louisell and Hazard's casebook on Pleading and procedure / Dana L. Blatt, senior editor, Norman S. Goldenberg, senior editor, Peter Tenen, managing editor ; staff writers, Jerry Smilowitz ... [et al.]. — Los Angeles : Casenotes Pub. Co. ; Gardena, Calif. : distributed by Law Distributors, c1976.
127 p. ; 28 cm.
1. Civil procedure—United States—Cases. I. Blatt, Dana L. II. Goldenberg, Norman S. III. Louisell, David W., ed. Cases and materials on pleading and procedure.
KF8839.C348 347'.73'5 79-114883
79 MARC

Casenote legal briefs—civil procedure : adaptable to courses utilizing Rosenberg, Weinstein, Smit and Korn's casebook on civil procedure / staff writers, Robert Switzer ... [et al.] ; Dana L. Blatt, senior editor, Norman S. Goldenberg, senior editor ; Peter Tenen, managing editor. — Beverly Hills, CA : Casenotes Pub. Co. ; Gardena, Calif. : distributed by Law Distributors, c1977.
153 p. ; 28 cm.
$9.50
1. Civil procedure—United States—Cases. I. Rosenberg, Maurice, 1919- . II. Switzer, Robert J. III. Blatt, Dana L. IV. Elements of civil procedure, cases and materials. V. Title: Civil procedure.
KF8839.C349 347'.73'5 79-114899
79 MARC

Casenote legal briefs—Federal courts : adaptable to courses utilizing McCormick, Chadbourn, and Wright's casebook on Federal courts / Dana L. Blatt, senior editor, Norman S. Goldenberg, senior editor, Peter Tenen, managing editor ; staff writers, Susan Ydelson ... [et al.]. — Beverly Hills, CA : Casenotes Pub. Co. ; Gardena, Calif. : distributed by Law Distributors, c1978.
129 p. ; 28 cm.
$9.95
1. Civil procedure—United States—Cases. 2. Jurisdiction—United States—Cases. I. Blatt, Dana L. II. Goldenberg, Norman S. III. McCormick, Charles Tilford, 1889-1963. Cases and materials on Federal courts. IV. Title: Federal courts.
KF8839.C35 347'.73'2 79-116899
79 MARC

Louisell, David W ed.
Cases and materials on pleading and procedure : state and Federal / by David W. Louisell and Geoffrey C. Hazard, Jr. — 4th ed. — Mineola, N.Y. : Foundation Press, 1979.
lxvii, 1192 p. ; 27 cm. — (University casebook series)
Cover title: Pleading and procedure, state and Federal.
On spine: Cases on pleading and procedure.
Includes bibliographical references and index.
1. Civil procedure—United States—Cases. I. Hazard, Geoffrey C. II. Title. III. Title: Pleading and procedure, state and Federal. IV. Title: Cases on pleading and procedure. V. Series.
KF8839.L6 1979 347'.73'5 79-52464
00 MARC

—UNITED STATES—OUTLINES, SYLLABI, ETC.

Civil procedure. — [Duluth, Ga. : Professional Impressions, inc., c1979]
xiii, 107 p. ; 23 cm. — (Law review series ; 3)
Date and place of publication from label mounted on recto of t.p.
ISBN 0-934098-03-4
1. Civil procedure—United States—Outlines, syllabi, etc. I. Professional Impressions, inc. II. Series: Law review series (Duluth, Ga.) ; 3.
KF8841.C54 347'.73 79-84911
79 MARC

Rutter, William A 1928–
Gilbert law summaries: civil procedure. 1st- ed. Gardena, CA [etc.] Harcourt Brace Jovanovich Law Group [etc.], 1968–
v. 28 cm. annual.
Key title: Gilbert law summaries. Civil procedure, ISSN 0161-830X
1. Civil procedure—United States—Outlines, syllabi, etc. I. Harcourt Brace Jovanovich Law Group. II. Title.
KF8841.R87 347'.73'5 78-641502
MARC-S

—WHITE RUSSIA

Russia (1923– U. S. S. R.). Laws, statutes, etc.
(Sbornik normativnykh aktov po grazhdanskomu protsessu)
Сборник нормативных актов по гражданскому процессу / Составители Бурова С. Н. ... [et al.] ; Под ред. Н. Г. Юркевича. — Минск : Вышэйш. школа, 1978.
352 p. ; 21 cm. USSR 78
Includes bibliographies.
0.85rub
1. Civil procedure — White Russia. 2. Civil procedure — Russia. I. Burova, S. N. II. Íùrkevich, Nikolaĭ Grigor'evich. III. Title.
78-421601

—YUGOSLAVIA

Triva, Siniša.
Građansko parnično procesno pravo / Siniša Triva. — 3., prer. i dop. izd. — Zagreb : Narodne novine, 1978.
xxiii, 701 p. ; 25 cm. — (Biblioteka Udžbenici ; 85) (Udžbenici Sveučilišta u Zagrebu) Yu***
First-2d eds. published under title: Građansko procesno pravo.
Includes bibliographies and index.
1. Civil procedure—Yugoslavia. I. Title.
79-342467
79 MARC

CIVIL PROCEDURE (CANON LAW)

see also Matrimonial actions (Canon law)

Chłopocka, Helena.
Procesy Polski z Zakonem Krzyżackim w XIV wieku; studium źródłoznawcze. Poznań [Państwowe Wydawn. Naukowe] 1967.
253 p. facsims. 25 cm. (Poznańskie Towarzystwo Przyjaciół Nauk. Wydział Historii i Nauk Społecznych. Prace Komisji Historycznej, t. 23, zesz. 1)
Summary in German.
Bibliography: p. [247]-249.
1. Poland—History—Jagellons, 1386-1572. 2. Teutonic Knights—History—Sources. 3. Civil procedure (Canon law) I. Title. II. Series: Poznańskie Towarzystwo Przyjaciół Nauk. Komisja Historyczna. Prace, t. 23, zesz. 1.
DK4010.P69a t. 23, zecz. 1 68-126963
[DK4250] [r79]rev MARC

CIVIL PROCEDURE (INTERNATIONAL LAW)

see also Judgments, Foreign

Council of Europe. European Committee on Legal Cooperation.
Explanatory report on the Additional protocol to the European convention on information on foreign law. — Strasbourg : Council of Europe, 1978.
13 p. ; 21 cm. F***
Prepared under the joint authority of the European Committee on Legal Co-operation and the European Committee on Crime Problems.
Includes text of the Additional protocol.
1. Civil procedure (International law) I. Council of Europe. European Committee on Crime Problems. II. European convention on information on foreign law. Protocols, amendments, etc., March 15, 1978. 1978. III. Title.
341.7'8 78-325003
79 MARC

Nagel, Heinrich, 1911-
Durchsetzung von Vertragsansprüchen im Auslandsgeschäft : Klage vor d. Schiedsoder Staatsgericht im Inland oder Ausland / Heinrich Nagel. — 1. Aufl. — Berlin : Berlin-Verlag, 1978.
viii, 389 p. ; 22 cm. — (Internationales Privatrecht ; Bd. 5) GFR78-A
Includes the texts of numerous treaties and agreements.
Bibliography: p. 375-389.
ISBN 3-87061-131-8 : DM48.00
1. Judgments, Foreign. 2. Civil procedure (International law) 3. Arbitration and award. 4. International business enterprises—Law and legislation. I. Title. II. Series.
K7680.N33 341.7'8 78-397483
79 MARC

CIVIL PROCEDURE (ROMAN LAW)

Bekker, Ernst Immanuel, 1827-1916.
Die prozessualische Konsumtion im klassischen Römischen Recht / Ernst Immanuel Bekker. — Neudr. d. Ausg. Berlin 1853. — Aalen : Scientia-Verlag, 1978.
339 p. ; 19 cm. GFR78-A
Originally published by W. Hertz.
Includes bibliographical references.
ISBN 3-511-00459-4 : DM62.00
1. Civil procedure (Roman law) I. Title.
340.5'4 78-382267
79 MARC

CIVIL RELIGION see Christianity and culture; Nationalism and religion; Religion and culture

CIVIL REMEDIES see Remedies (Law)

CIVIL RIGHTS

see also Assembly, Right of; Catholic Church and civil rights; Children's rights; Detention of persons; Due process of law; Equality before the law; Freedom of association; Freedom of information; Freedom of movement; Habeas corpus; Indians of North America—Civil rights; Jews—Legal status, laws, etc.; Liberty; Liberty of contract Petition, Right of; Political rights; Race discrimination—Law and legislation; Religious liberty; Right of property; Right of counsel; Right to education; Searches and seizures; Sex discrimination against women—Law and legislation; Speedy trial

al-Ghurayyib, Mīshāl.
(al-Ḥurrīyāt al-'āmmah fī Lubnān wa-al-'ālam)
الحريات العامة في لبنان والعالم / ميشال الغريب ؛ [تصميم وتنفيذ مؤسسة الخضري] . — الطبعة 1. — [بيروت ؟] : الاتحاد الوطني لطلاب الجامعة اللبنانية ، مجلس فرع كلية الحقوق والعلوم السياسية والاداريه ، امانة شؤون المطبوعات ، [ca. 1975]
270 p., [2] leaves of plates : ill. ; 24 cm.
Includes legislation.
Bibliography: p. 251-258.
1. Liberty. 2. Civil rights. 3. Civil rights—Lebanon. I. Title.
JC571.G46 79-970076

Allard, Jean Louis.
L'éducation à la liberté : ou, La philosophie de l'éducation de Jacques Maritain / par Jean-Louis Allard ; préf. de Georges Pascal. — [Ottawa] : Éditions de l'Université d'Ottawa, 1978.
152 p. ; 23 cm. — (Collection Philosophica ; 11) C***
Bibliography: p. [141]-149.
ISBN 0-7766-1021-X
1. Maritain, Jacques, 1882-1973. 2. Education—Philosophy. 3. Liberty. 4. Civil rights. I. Title.
LB880.M3332A44 370.1 79-369149
79 MARC

Alzamora Valdez, Mario.
Los derechos humanos y su protección / Mario Alzamora Valdez. — Lima : JUS, 1977.
267 p. ; 21 cm.
"Apéndice[s] I-VI" (p. 151-267) contain documents.
Includes bibliographical references.
1. Civil rights (International law) 2. Civil rights. I. Title.
K3240.4.A79 341.48'1 79-357445
79 MARC

Amnesty International. International Executive Committee.
Impartiality and the defence of human rights / Amnesty International, International Executive Committee. — London : Amnesty International, [1978].
Folder ([6] p.) ; 22 cm. GB78-21880
1. Amnesty International. 2. Civil rights. 3. Political prisoners. I. Title.
JC571.A44 1978 365'.45'0621 79-311946
79 MARC

CIVIL RIGHTS (Continued)

Behrendt, Ethel Leonore, 1938-
Recht auf Gehör : Grundrecht u. Grundwert / Ethel Leonore Behrendt. — 1. Aufl. — München : Minerva-Publikation, 1978.
xxxiv, 482 p. ; 21 cm. — (Minerva-Fachserie Rechts- und Staatswissenschaften) GFR78-A
Includes bibliographical references.
ISBN 3-597-10028-7 : DM60.00
1. Due process of law—Germany, West. 2. Due process of law. 3. Jesus Christ—Trial. 4. Civil rights. 5. Worth. I. Title.
347'.43 79-376843
79 MARC

Billings, Peggy.
Paradox and promise in human rights / Peggy Billings. — New York : Friendship Press, c1979.
126 p., [4] leaves of plates : ill. ; 21 cm.
Includes bibliographical references.
ISBN 0-377-00083-3 : $2.95
1. Civil rights. 2. Sociology, Christian. I. Title.
JC571.B544 323.4 79-10325
79 MARC

Bleckmann, Albert.
Allgemeine Grundrechtslehren / von Albert Bleckmann. — Köln ; Berlin ; Bonn; München : Heymann, 1979.
xv, 327 p. ; 21 cm. — (Academia iuris : Lehrbücher der Rechtswissenschaft) GFR79-A
Includes bibliographical references and index.
ISBN 3-452-18621-0 : DM29.50
1. Civil rights. I. Title. II. Series.
K3240.4.B55 342.08'5 79-383840
79 MARC

Castro, Juventino V
Lecciones de garantías y amparo / Juventino V. Castro. — 1. ed. — México : Editorial Porrúa, 1974.
595 p. ; 24 cm.
Bibliography: p. 585-587.
1. Civil rights—Mexico. 2. Due process of law—Mexico. 3. Amparo (Writ)—Mexico. 4. Civil rights. 5. Due process of law. 6. Amparo (Writ) I. Title.
342'.72'085 77-155617
79 MARC

Derechos humanos / [compilador], Fernando Levi Ruffinelli. — [s.l. : s.n.], 1977 (Asunción : Orbis)
viii, 142 p. ; 20 cm.
1. Civil rights. I. Levi Ruffinelli, Fernando.
JC571.D39 323.4 79-100227
79 MARC

Los Derechos humanos. — [Madrid : Ministerio de Asuntos Exteriores, Oficina de Información Diplomática], D.L. 1977.
104 p., 2 leaves ; 27 cm. — (Cuadernos de documentación - Oficina de Información Diplomática) Sp78-Mar
Bibliography: p. [35]-36.
1. Civil rights. I. Series: Spain. Oficina de Información Diplomática. Cuadernos de documentación - Oficina de Información Diplomática.
JC571.D393 79-356076
79 MARC

Derechos humanos en el Uruguay. — Montevideo : República Oriental del Uruguay, 1977.
183 p. ; 24 cm. — (Serie de temas nacionales ; 2)
"Selección, textos y compilación: Federico Fernández Prando."
Bibliography: p. 181-182.
1. Civil rights—Uruguay. 2. Civil rights. I. Series.
342'.895'085 78-103229
78 MARC

De Villiers, Casper Francois.
Human rights and homelands / Cas de Villiers. — [Pretoria : Foreign Affairs Association], 1978.
12 p. ; 21 cm. — (Focus ; no. 8) SA***
ISBN 0-908397-20-8
1. Homelands, South Africa. 2. Segregation. 3. Ethnicity. 4. Civil rights. I. Title. II. Series: Focus (Pretoria) ; no. 8.
DT763.6.D48 323.1 79-312811
79 MARC

Dispot, Laurent, 1950-
La machine à terreur / Laurent Dispot. — Paris : B. Grasset, c1978.
302 p. ; 21 cm. — (Figures) F***
Includes bibliographical references.
ISBN 2-246-00660-0 : 42.00F
1. Terrorism. 2. Civil rights. I. Title.
HV6431.D57 301.6'33 79-355036
79 MARC

Eide, Asbjørn.
Fattig, ufri og mishandlet : om det internasjonale menneskerettighetsvernet / Asbjørn Eide. — Oslo : Universitetsforl., c1978 (Oslo : Aktietr.i Trondhjem)
238 p., [1] leaf ; 18 cm. — (U-bok ; nr. 245) N78-Dec.
Bibliography: p. 187-[190]
ISBN 8200032558 : kr58.00
1. Civil rights (International law) 2. Civil rights—Latin America. 3. Civil rights. I. Title.
K3240.4.E37 79-369496
79 MARC

Ēliou, Ēlias Philippou, 1904–
(Hoi paraviaseis tōn anthrōpinōn dikaiōmatōn)
Οἱ παραβιάσεις τῶν ἀνθρωπίνων δικαιωμάτων / Ἠλία Ἠλιοῦ. — [Ἀθήνα] : Ἑρμείας, [1977?]
40 p. ; 21 cm.
1. Civil rights. I. Title.
JC585.E43 79–414393

Fletcher, Catherine.
Am I free? / Catherine Fletcher. — Niles, Ill. : Argus Communications, c1975.
70 p. : ill. ; 23 cm.
ISBN 0-913592-43-9 : $1.95
1. Freedom (Theology) 2. Civil rights. 3. Church and civil rights. I. Title.
BT810.2.F58 261.8 75-18972
00 MARC

Foundation for Conflict Analysis.
Points of view. — Västervik : Foundation for Conflict Analysis, [197-]
19 p. ; 21 cm. S***
1. Communism—1945- . 2. Civil rights. 3. Environmental policy. I. Title.
HX44.F66 1970z 309.1'47'085 79-305147
79 MARC

Frankel, Charles, 1917-
Human rights and foreign policy / by Charles Frankel. — New York : Foreign Policy Association, c1978.
63, [1] p. : port. ; 20 cm. — (Headline series ; 241)
Cover title.
Bibliography: p. 63-[64]
$1.40
1. Civil rights. 2. United States—Foreign relations—1977- . I. Title. II. Series.
E744.H43 no. 241 081 s 78-65548
[JC571] 79 MARC

Gastil, Raymond D
Freedom in the world : political rights and civil liberties, 1978 / Raymond D. Gastil ; with essays by Richard W. Cottam ... [et al.]. — New York : Freedom House, c1978.
xi, 335 p. : map ; 25 cm.
Includes bibliographical references and index.
ISBN 0-932088-00-7
1. Civil rights. I. Cottam, Richard W. II. Title.
JC571.G336 323.4 78-53867
78 MARC

Gastil, Raymond D
Freedom in the world : political rights and civil liberties, 1979 / Raymond D. Gastil ; with papers by Bohdan R. Bociurkiw ... [et al.]. — New York : Freedom House, c1979.
xii, 321 p. ; 24 cm.
Includes bibliographical references and index.
1. Civil rights. I. Bociurkiw, Bohdan R., joint author. II. Title.
JC571.G336 1979 323.4 79-87596
79 MARC

Giannas, Nikos.
(Koinōnikē politeiologia)
Κοινωνικὴ πολιτειολογία : ὁ κοινωνικὸς ἄνθρωπος καὶ ἡ κοινωνικὴ ἱστορία τῆς Ἑλλάδας καὶ τοῦ Δυτικοῦ Κόσμου : ἀνθρώπινες σχέσεις ... / Νίκου Γιάννα. — Ἀθήνα : [s. n.], 1977.
326 p. ; 24 cm. — (Κοινωνικὲς καὶ συνταγματικὲς μελέτες)
Bibliography: p. 318–321.
1. Civil rights. I. Title. II. Series: Koinōnikes kai syntagmatikes meletes.
JC571.G474 79–409140

Glaser, Kurt, 1914-
Victims of politics : the state of human rights / Kurt Glaser, Stefan T. Possony. — New York : Columbia University Press, 1979.
xxiv, 614 p. : ill. ; 26 cm.
Bibliograhy: p. [557]-587.
Includes indexes.
ISBN 0-231-04442-9
1. Civil rights. 2. Discrimination. 3. Oppression (Psychology) I. Possony, Stefan Thomas, 1913- joint author. II. Title.
JC571.G55 301.5'92 78-5591
78 MARC

Groupe de Paris.
Charte de la liberté : [prononcée au Musée social, Paris, 5 octobre 1977]. — Luzarches, France : Groupe de Paris, 1977.
72 p. ; 22 cm. F***
Text of the Charter in various languages.
1. Civil rights. I. Title.
JC571.G7835 1977 323.4 79-371221
79 MARC

Henkin, Louis.
The rights of man today / Louis Henkin. — Boulder, Colo. : Westview Press, 1978.
xiv, 173 p. ; 24 cm. — (The Gottesman lectures)
"Derives from lectures delivered at Yeshiva University as Benjamin Gottesman lectures."
Includes bibliographical references.
ISBN 0-89158-174-X : $15.00
1. Civil rights. I. Title. II. Series.
JC571.H387 323.4 78-6722
78 MARC

Hollenbach, David.
Claims in conflict : retrieving and renewing the Catholic human rights tradition / by David Hollenbach. — New York : Paulist Press, c1979.
v, 219 p. ; 21 cm. — (Woodstock studies ; 4)
Includes bibliographical references and index.
ISBN 0-8091-0287-0
1. Church and social problems—Catholic Church. 2. Civil rights. I. Title. II. Series.
HN37.C3H584 261.8'3 79-84239
79 MARC

Ingram, Tolbert Robert, 1913-
What's wrong with human rights / by T. Robert Ingram. — Houston, Tex. : St. Thomas Press, 1978.
77 p. ; 21 cm.
1. Civil rights. I. Title.
JC571.I49 261.7 78-68732
00 MARC

Jambet, Christian.
Le monde : réponse à la question, Qu'est-ce que les droits de l'homme? / Christian Jambet, Guy Lardreau. — Paris : B. Grasset, c1978.
281 p. ; 21 cm. — (Their Ontologie de la Révolution ; 2) (Figures) F***
48.00F
1. Civil rights. 2. Social ethics. I. Lardreau, Guy, 1947- joint author. II. Title. III. Series.
JC571.J285 323.4'01 79-359345
79 MARC

Joyce, James Avery.
The new politics of human rights / James Avery Joyce. — London : Macmillan, 1978.
xi, 305 p. ; 23 cm. GB***
Bibliography: p. 295-296.
Includes index.
ISBN 0-333-24291-2 : £12.00
1. Civil rights. I. Title.
JC571.J66 1978 323.4 79-301986
79 MARC

Joyce, James Avery.
The new politics of human rights / James Avery Joyce. — New York : St. Martin's Press, 1979, c1978.
xi, 305 p. ; 22 cm.
Bibliography: p. 295-296.
Includes index.
ISBN 0-312-56880-0
1. Civil rights. I. Title.
JC571.J66 1979 323.4 78-13333
78 MARC

Kainō, Michitaka, 1908–
(Shimin no jiyū)
市民の自由：基本的人権と公共の福祉 / 戒能通孝著. — 東京：日本評論社，昭和26 [1951]
3, 2, 268 p. ; 19 cm.
Colophon inserted.
Includes bibliographical references.

1. Civil rights. I. Title.
K3240.4.K25 1951 79–803242

Kinloch, Graham Charles.
The sociology of minority group relations / Graham C. Kinloch. — Englewood Cliffs, N.J. : Prentice-Hall, c1979.
xi, 243 p. ; 23 cm. — (Prentice-Hall series in sociology)
Includes bibliographies and indexes.
ISBN 0-13-821017-9 : $6.95
1. Minorities—Social aspects. 2. Social groups. 3. Social interaction. 4. Civil rights. 5. Minorities—United States. I. Title.
HM133.K52 301.45 78-12274
78 MARC

CIVIL RIGHTS (Continued)

Klenner, Hermann.
Freiheit, Gleichheit und so weiter : dreizehn Streiflichter über die Menschenrechte / Hermann Klenner, ⌈Mitarbeit, Annelies Schwarzer⌉. — 1. Aufl. — Berlin : Staatsverlag der Deutschen Demokratischen Republik, 1978.
143 p. : ill. ; 17 cm. — (Schriftenreihe Recht in unserer Zeit ; 10) GDR***
Includes bibliographical references.
2.25M
1. Civil rights. 2. Civil rights—Communist countries. 3. Civil rights (International law) I. Schwarzer, Annelies, joint author. II. Title.
K3240.4.K56 323.4'09171'7 78-395441
79 MARC

Klenner, Hermann.
Menschenrechte : Anspruch und Wirklichkeit / ⌈H. Klenner, B. Graefrath, V. A. Sorin⌉. — Berlin : DDR-Komitee für Menschenrechte, 1978.
56 p. ; 21 cm. — (Schriften und Informationen - DDR-Komitee für Menschenrechte ; Heft 4/78) GDR***
Cover title.
Includes bibliographical references.
1. Civil rights (International law) 2. Civil rights. 3. Civil rights—Communist countries. I. Graefrath, Bernhard, joint author. II. Zorin, Valerian Aleksandrovich, 1902- joint author. III. Title. IV. Series: DDR-Komitee für Menschenrechte. Schriften und Informationen - DDR-Komitee für Menschenrechte ; Heft 1978, 4.
K3240.4.K566 341.48'1 79-365012
79 MARC

Langner, Albrecht.
Menschenrechte, Staat, Gesellschaft / von Albrecht Langner. — Köln : Bachem, 1975.
48 p. ; 21 cm. — (Katholische Soziallehre in Text und Kommentar ; Heft 3) GFR76-A
Bibliography: p. ⌈48⌉
ISBN 3-7616-0290-1 : DM1.00
1. Civil rights—Germany, West. 2. Civil rights. I. Title. II. Series.
342'.43'085 79-349627
79 MARC

Leenen, Henricus Jacobus Josephus, 1929-
Rechten van mensen in de gezondheidszorg : een gezondheidsrechtelijke studie / H. J. J. Leenen. — Alphen aan den Rijn : Samsom, 1978.
275 p. ; 26 cm. N***
Bibliography: p. 254-268.
Includes index.
ISBN 9014027362
1. Medical laws and legislation. 2. Civil rights. I. Title.
K3601.L43 79-337707
79 MARC

Mower, Alfred Glenn.
The United States, the United Nations, and human rights : the Eleanor Roosevelt and Jimmy Carter eras / A. Glenn Mower, Jr. — Westport, Conn. : Greenwood Press, 1979.
xii, 215 p. ; 22 cm. — (Studies in human rights ; no. 4 ISSN 0146-3586)
Bibliography: p. ⌈209⌉-212.
Includes index.
ISBN 0-313-21090-X
1. Civil rights (International law) 2. Civil rights. 3. United States—Foreign relations—1945- I. Title. II. Series.
K3240.4.M68 341.48'1 78-22134
79 MARC

Nishijima, Yatarō, 1889-
(Jinken to keiei rōmu)
人權と經營勞務 / 西島彌太郎著. — 東京 : 正統社, 昭和24 ⌈1949⌉
2, 4, 230 p. ; 19 cm.
Colophon inserted.

1. Industrial management. 2. Industrial relations. 3. Civil rights. I. Title.
HD37.J3N54 79-801175

O'Mahony, Patrick J
The fantasy of human rights / by Patrick O'Mahony. — Great Wakering : Mayhew-McCrimmon, 1978.
192 p. : ill. ; 18 cm. GB78-25821
Bibliography: p. 186-188.
Includes index.
ISBN 0-85597-256-4 : £2.00
1. Church and social problems—Great Britain. 2. Church and social problems—Catholic Church. 3. Civil rights. I. Title.
HN39.G7O43 1978 261.7 79-312169
79 MARC

Owen, David, 1938-
Human rights / David Owen. — London : J. Cape, 1978.
154 p. ; 23 cm. GB***
"Written or delivered between 1977 and 1978."
Includes index.
CONTENTS: Human rights in Britain.—Human rights abroad.—Soviet Communism.—Communism in Western Europe.—Race relations in Britain.—Apartheid.—The United Nations.—World poverty.—World peace.
ISBN 0-224-01659-8 : £4.50. ISBN 0-224-01667-9 pbk.
1. Civil rights. I. Title.
JC571.O94 1978b 323.4 78-323163
79 MARC

Owen, David, 1938-
Human rights / Daid Owen. — 1st ed. — New York : Norton, c1978.
154 p. ; 22 cm.
Includes index.
ISBN 0-393-01186-0
1. Civil rights. I. Title.
JC571.O94 1978 323.4 78-13668
78 MARC

Partridge, Astley Cooper.
Human rights : their origin, validity and implementation / A. C. Partridge. — Johannesburg : South African Institute of International Affairs, 1978.
14 p. ; 30 cm. — (Occasional paper - The South African Institute of International Affairs) SA78
Cover title.
Bibliography: p. 13-14.
ISBN 0-909239-36-3 : RO.50
1. Civil rights. 2. Civil rights—South Africa. I. Title. II. Series: South African Institute of International Affairs. Occasional papers - South African Institute of International Affairs.
JC571.P285 323.4 79-309562
79 MARC

(Pravda o pravakh cheloveka)
Правда о правах человека : деятели сов. культуры о правах человека / Константин Федин ... ⌈et al. ; составитель Борис Иванов⌉. — Москва : Политиздат, 1977.
158 p. ; 20 cm. USSR***
0.25rub
1. Civil rights—Russia. 2. Civil rights (International law) 3. Civil rights. I. Fedin, Konstantin Aleksandrovich, 1892- II. Ivanov, Boris Vladimirovich, 1917-
78-422937

Rechten van de mens. — Maarssen : Ars Aequi Libri, ⌈1978-
v. ; 25 cm. Ne***
Issued by Stichting Ars Aequi.
1. Civil rights. I. Stichting Ars Aequi.
K3240.4.R4 341.48'1 78-386862
78 MARC

Schwabe, Jürgen.
Probleme der Grundrechtsdogmatik / Jürgen Schwabe. — Darmstadt : J. Schwabe ⌈Selbstverl.⌉, 1977.
vii, 482 p. ; 21 cm. GFR78-B1
Bibliography: p. 448-467.
Includes indexes.
DM15.00
1. Civil rights—Germany, West. 2. Civil rights. I. Title.
342'.43'085 78-387927
79 MARC

Soziale Grundrechte : Arbeit, Bildung, Umweltschutz etc. / zsgest. u. mit e. Einl. vers. von Karl Hernekamp. — Berlin ; New York : de Gruyter, 1979.
256 p. ; 19 cm. — (Aktuelle Dokumente) GFR79-A
Bibliography: p. ⌈235⌉-252.
Includes index.
ISBN 3-11-007756-6 : DM19.80
1. Civil rights. 2. Civil rights (International law) I. Hernekamp, Karl.
K3240.4.S6 342.08'5 79-382164
79 MARC

Szabó, Imre.
Az emberi jogok / Szabó Imre. — 2., bőv. kiad. — Budapest : Akadémiai Kiadó, 1978.
163 p. ; 19 cm. — (Korunk tudománya) H***
ISBN 9630516225 : 18.00Ft
1. Civil rights. I. Title.
JC585.S9 1978 78-395819
79 MARC

Textos internacionales de derechos humanos / Javier Hervada, José M. Zumaquero. — Pamplona : Ediciones Universidad de Navarra, 1978.
1012 p. ; 24 cm. — (Colección jurídica ; 76) Sp***
English, French, or Spanish.
Bibliography: p. 19-20.
Includes index.
ISBN 8431305444
1. Civil rights (International law) 2. Civil rights. I. Hervada Xiberta, Francisco Javier. II. Zumaquero, José M. III. Series: Colección jurídica (Pamplona) ; 76.
K3238.A1T49 341.48'1 79-346011
79 MARC

United States. Congress. House. Committee on International Relations. Subcommittee on International Operations.
Institute for Human Rights and Freedom : hearings and markup before the Subcommittees on International Operations and on International Organizations of the Committee on International Relations, House of Representatives, Ninety-fifth Congress, second session, on H.R. 11326, April 13, 18, and 25, 1978. — Washington : U.S. Govt. Print. Off., 1978.
iii, 174 p. ; 24 cm.
At the markup session May 2, 1978, the subcommittees accepted a draft substitute in which the name of the institute was changed to Institute for International Human Rights; this substitute was later incorporated as Title 7 of H.R. 12598, the fiscal 1979 Foreign relations authorization act.
1. Institute for International Human Rights. 2. Civil rights. I. United States. Congress. House. Committee on International Relations. Subcommittee on International Organizations. II. Title.
KF27.I5493 1978c 342'.73'085 78-603324
79 MARC

United States. Congress. Senate. Committee on Foreign Relations.
Institute for Human Rights and Freedom : hearing before the Committee on Foreign Relations, United States Senate, Ninety-fifth Congress, second session, on S. 2894 ... June 29, 1978. — Washington : U.S. Govt. Print. Off., 1978.
iii, 26 p. ; 24 cm.
1. Institute for Human Rights and Freedom. 2. Civil rights. I. Title.
KF26.F6 1978b 342'.73'085 78-603171
78 MARC

United States. Dept. of State.
Report on human rights practices in countries receiving U.S. aid : report / submitted to the Committee on Foreign Relations, U.S. Senate, and Committee on Foreign Affairs, U.S. House of Representatives, by the Department of State. — Washington : U.S. Govt. Print. Off., 1979.
ix, 706 p. ; 24 cm.
At head of title: 96th Congress, 1st session. Joint committee print.
1. Civil rights. I. United States. Congress. Senate. Committee on Foreign Relations. II. United States. Congress. House. Committee on Foreign Affairs. III. Title.
JC571.U48 1979 323.4 79-601351
79 MARC

United States. Library of Congress. Foreign Affairs and National Defense Division.
Human rights conditions in selected countries and the U.S. response / prepared for the Subcommittee on International Organizations of the Committee on International Relations, U.S. House of Representatives, by the Foreign Affairs and National Defense Division, Congressional Research Service, Library of Congress. — Washington : U.S. Govt. Print. Off., 1978.
ix, 372 p. ; 24 cm.
Prepared by Brenda Branaman, and others.
At head of title: 95th Congress, 2d session. Committee print.
Includes bibliographical references.
1. Civil rights. 2. United States—Relations (general) with foreign countries. I. Branaman, Brenda. II. United States. Congress. House. Committee on International Relations. Subcommittee on International Organizations. III. Title.
JC571.U5 1978 323.4 78-602495
78 MARC

Volpe, Galvano della.
Umanesimo positivo e emancipazione marxista / Galvano della Volpe. — Milano : Sugar, 1964.
272 p. ; 20 cm. — (Argomenti ; 13)
Includes bibliographical references and index.
1. Humanism. 2. Civil rights. 3. Marx, Karl, 1818-1883. I. Title.
B821.V66 79-367353
79 MARC

Westphalen, Friedrich, Graf von.
Diskussion um die Grundwerte : Krisensymptom oder Trendwende? / Friedrich Graf von Westphalen. — ⌈1.-2. Tsd.⌉. — Köln : Bachem, 1979.
62 p. ; 19 cm. — (Beiträge zur Gesellschaftspolitik ; 15) GFR79-A
On spine: Grundwertediskussion.
Bibliography: p. 53-58.
ISBN 3-7616-0505-6
1. Civil rights. 2. Political rights. 3. Constitutional law. I. Title. II. Title: Grundwertediskussion. III. Series.
K3240.4.W44 323.4 79-382166
79 MARC

Wirak, Anders.
Om indikasjon av menneskerettighetene / Anders Helge Wirak. — Oslo : Universitetet i Oslo, Professoratet i konflikt- og fredsforskning, 1977.
81 p. ; 30 cm. — (World indicators program ; 14) (Papers - Professoratet i konflikt- og fredsforskning ; 49) N77-Dec.
Includes bibliographical references.
1. Civil rights. 2. Social indicators. I. Title. II. Series. III. Series: Oslo. Universitet. Professoratet i konflikt- og fredsforskning. Papers - Chair in Conflict and Peace Research, University of Oslo ; 49.
JC571.W49 79-341270
79 MARC

Z zagadnień wyższości socjalizmu nad kapitalizmem : materiały pomocnicze do szkolenia partyjnego / ⌈aut. Stefan Opara ... ⌈et al.⌉ ; wybór, oprac. i wstęp Franciszek Ociepka⌉. — Warszawa : Książka i Wiedza, 1978 (Łódź : PrasZG)
121, ⌈3⌉ p. ; 21 cm. P79-1977
Articles previously published in Ideologia i polityka, 1977-1978.
Includes bibliographical references.
zł10.00
1. Communism—Addresses, essays, lectures. 2. Capitalism—Addresses, essays, lectures. 3. Civil rights. 4. Competition, International. I. Opara, Stefan. II. Ideologia i polityka.
HX44.Z23 79-382596
79 MARC

Zischka, Anton, 1904-
Kampf ums Überleben : d. Menschenrecht auf Energie / Anton Zischka. — 1. Aufl. — Düsseldorf ; Wien : Econ-Verlag, 1979.
367 p. ; 22 cm. GFR79-A
Bibliography: p. 351-355.
Includes indexes.
ISBN 3-430-19964-6 : DM29.80
1. Atomic power—Law and legislation. 2. Liability for nuclear damages. 3. Civil rights. I. Title.
K3670.4.Z57 343.09'2 79-378830
79 MARC

CIVIL RIGHTS (Continued)

—ADDRESSES, ESSAYS, LECTURES

Alcalay, Milos.
Solidaridad y derechos humanos / por Milos Alcalay. — Caracas : Centro de Información, Documentación y Análisis Latinoamericano, ⌊between 1976 and 1978⌋
35 p. ; 30 cm. — (Documentos - Centro de Información, Documentación y Análisis Latinoamericano ; entrega no. 79)
"La presente ... recoge unos artículos de prensa escritos en los dos últimos años."
CONTENTS: Las organizaciones solidarias y los derechos humanos.—Las internacionales políticas y los derechos humanos.—Las internacionales sindicales y los derechos humanos.—La iglesia y la lucha en pro de los derechos humanos.—Los refugiados y los derechos humanos.
1. Civil rights—Addresses, essays, lectures. I. Title. II. Series: Centro de Información, Documentación y Análisis Lationamericano. Documentos - Centro de Información, Documentación y Análisis Latinoamericano ; entrega no. 79.
JC585.A35 323.4 78-107709
79 MARC

Bioethics and human rights : a reader for health professionals / edited by Elsie L. Bandman, Bertram Bandman. — 1st ed. — Boston : Little, Brown, c1978.
xix, 386 p. ; 24 cm.
Includes bibliographical references and index.
ISBN 0-316-07998-7
1. Medical ethics—Addresses, essays, lectures. 2. Bioethics—Addresses, essays, lectures. 3. Civil rights—Addresses, essays, lectures. I. Bandman, Elsie L. II. Bandman, Bertram.
R724.B46 174'.2 78-57419
78 MARC

Der Bürgerliche Rechtsstaat / hrsg. von Mehdi Tohidipur. — Erstausgabe, 1. Aufl. — Frankfurt am Main : Suhrkamp, 1978.
2 v. (616 p.) ; 18 cm. — (Edition Suhrkamp ; 901) GFR***
Contains in part articles which first appeared in various publications, 1930-1976.
Includes bibliographical references.
ISBN 3-518-10901-4 : DM20.00
1. Rule of law—Addresses, essays, lectures. 2. Civil rights—Addresses, essays, lectures. 3. Democracy—Addresses, essays, lectures. I. Mehdi Tohidipur.
K3171.B87 340.1'1 79-347714
79 MARC

Enhancing global human rights / Jorge I. Domínguez ... ⌊et al.⌋ ; introd. by Richard H. Ullman. — New York : McGraw-Hill, c1979.
xi, 270 p. ; 23 cm. — (1980s project/Council on Foreign Relations)
Bibliography: p. 259-260.
Includes index.
ISBN 0-07-017397-4. ISBN 0-07-017398-2 pbk.
1. Civil rights—Addresses, essays, lectures. I. Domínguez, Jorge I. II. Series: Council on Foreign Relations. 1980s project/Council on Foreign Relations.
JC571.E683 341.48'1 78-13002
78 MARC

Gladwyn, Hubert Miles Gladwyn Jebb, Baron, 1900-
The four freedoms; the background of the broadcasts. ⌊London⌋, ATV Library ⌊i.e. Associated Television Limited, 1962?⌋
26 p. 23 cm.
Cover title.
"From 11th February 1962 a series of one-hour programmes under the chairmanship of Lord Gladwyn will be broadcast."
1. Civil rights—Addresses, essays, lectures. I. Title.
JC571.G54 323.4 79-313464
79 MARC

Gronemeyer, Reimer, 1939-
Frieden / Reimer Gronemeyer. — Baden-Baden : Signal-Verlag, 1978.
160 p. ; 22 cm. — (Reihe Grundwerte ; Bd. 3) GFR78-A
Bibliography: p. 154-160.
ISBN 3-7971-0179-1 : DM16.80
1. Peace—Addresses, essays, lectures. 2. Civil rights—Addresses, essays, lectures. I. Title. II. Series.
JX1963.G85 327'.172 79-374563
79 MARC

Human rights / edited by Eugene Kamenka and Alice Erh-Soon Tay. — New York : St. Martin's Press, 1978.
viii, 148 p. ; 23 cm. — (Ideas and ideologies)
Bibliography: p. ⌊133⌋-134.
Includes index.
ISBN 0-312-39960-X :
1. Civil rights—Addresses, essays, lectures. I. Kamenka, Eugene. II. Tay, Alice Erh-Soon. III. Series.
JC571.H768 1978 323.4 78-4469
78 MARC

Human rights : cultural and ideological perspectives / edited by Adamantia Pollis and Peter Schwab. — New York : Praeger, 1979.
xvi, 165 p. ; 25 cm.
Includes bibliographies and index.
ISBN 0-03-046631-8 : $18.95
1. Civil rights—Addresses, essays, lectures. 2. Comparative government—Addresses, essays, lectures. I. Pollis, Adamantia. II. Schwab, Peter, 1940-
JC571.H7694 1979 323.4 78-19771
78 MARC

Human rights / edited by Eugene Kamenka and Alice Erh-Soon Tay. — Port Melbourne, Australia : E. Arnold (Australia), 1978.
viii, 148 p. ; 22 cm. — (Ideas and ideologies) Aus***
Includes bibliographical references and index.
ISBN 0-7267-2036-4
1. Civil rights—Addresses, essays, lectures. I. Kamenka, Eugene. II. Tay, Alice Erh-Soon. III. Series.
JC571.H768 1978b 323.4 79-306083
79 MARC

Human rights and American foreign policy / edited by Donald P. Kommers and Gilburt D. Loescher. — Notre Dame, Ind. : University of Notre Dame Press, c1979.
xii, 333 p. ; 24 cm.
Includes bibliographical references and index.
ISBN 0-268-01071-4 : $14.95
1. Civil rights—Addresses, essays, lectures. 2. United States—Foreign relations—1974- —Addresses, essays, lectures. I. Kommers, Donald P. II. Loescher, Gil.
JC571.H7687 323.4 78-62966
78 MARC

Human rights and U.S. foreign policy / edited by Barry M. Rubin and Elizabeth P. Spiro ; with a foreward by Muriel S. Humphrey. — Boulder, Colo. : Westview Press, 1979.
xix, 283 p. ; 24 cm.
Includes bibliographical references.
ISBN 0-89158-476-5
1. Civil rights—Addresses, essays, lectures. 2. United States—Foreign relations—1945- —Addresses, essays, lectures. 3. Civil rights (International law)—Addresses, essays, lectures. I. Rubin, Barry M. II. Spiro, Elizabeth P.
K3240.6.H86 341.48'1 79-872
79 MARC

Human rights and U.S. foreign policy : principles and applications / edited by Peter G. Brown, Douglas MacLean. — Lexington, Mass. : Lexington Books, c1979.
xxvii, 310 p. ; 24 cm.
Includes bibliographical references and index.
ISBN 0-669-02807-X
1. Civil rights—Addresses, essays, lectures. 2. United States—Foreign relations—1977- —Addresses, essays, lectures. 3. Civil rights (International law)—Addresses, essays, lectures. I. Brown, Peter G. II. MacLean, Douglas, 1947-
K3240.6.H87 341.48'1 78-24647
79 MARC

Human rights and world order / edited by Abdul Aziz Said. — New Brunswick, N.J. : Transaction Books, c1978.
xii, 170 p. ; 23 cm.
Includes bibliographies.
ISBN 0-87855-718-0
1. Civil rights—Addresses, essays, lectures. I. Said, Abdul Aziz.
JC571.H769 1978b 323.4 78-62438
78 MARC

Kainō, Michitaka, 1908–1975.
(Saiban)
裁判 / 戒能通孝著. — 東京 : 岩波書店, 1951, 1975 printing.
3, 2, 236 p. ; 18 cm. — (岩波新書 ; 63) Ja***
Series romanized: Iwanami shinsho.
¥230
1. Courts—Addresses, essays, lectures. 2. Civil rights—Addresses, essays, lectures. 3. Justice, Administration of—Addresses, essays, lectures. I. Title.
78–814589

MacCormick, Neil.
Civil liberties and the law / ⌊by⌋ Neil MacCormick. — Edinburgh : Heriot-Watt University, 1977.
⌊4⌋, 20 p. ; 21 cm. — (A Heriot-Watt University lecture) GB79-00173
Includes bibliographical references.
£0.60
1. Civil rights—Great Britain—Addresses, essays, lectures. 2. Civil rights—Addresses, essays, lectures. I. Title. II. Series: Heriot-Watt University. A Heriot-Watt University lecture.
KD4080.Z9M3 342.08'5 79-318361
79 MARC

Morgenthau, Hans Joachim, 1904-
Human rights & foreign policy / Hans J. Morgenthau. — New York : Council on Religion & International Affairs, c1979.
vii, 43 p. ; 22 cm. — (Distinguished CRIA lecture on morality & foreign policy ; 1st)
"In symposium with Hans J. Morgenthau: William Bradley ... ⌊et al.⌋": p. 9-43.
$4.00
1. International relations—Moral and religious aspects—Addresses, essays, lectures. 2. Civil rights—Addresses, essays, lectures. I. Bradley, William Lee. II. Council on Religion and International Affairs. III. Title. IV. Series: Council on Religion and International Affairs. Distinguished CRIA lecture on morality & foreign policy ; 1st.
JX1255.M59 172'.4 79-53084
79 MARC

Of life and hope : toward effective witness in human rights / edited by Mia Adjali. — New York : Friendship Press, c1979.
90 p. ; 26 cm.
Includes bibliographical references.
ISBN 0-377-00084-1 : $2.95
1. Civil rights—Addresses, essays, lectures. I. Adjali, Mia, 1939-
JC571.O37 323.4 79-10324
79 MARC

Political prisoners : — New York : Facts on File, c1978.
285 p. ; 24 cm. — (Checkmark books)
Includes index.
ISBN 0-87106-295-0 : $13.95
1. Political prisoners—Addresses, essays, lectures. 2. Civil rights—Addresses, essays, lectures. I. Grant, Alexander, 1948- . II. Sobel, Lester A. III. Facts on File, inc., New York.
HV8706.P64 365'.45 78-24188
78 MARC

Prawa i obowiązki obywatelskie w Polsce i świecie / redaktor naukowy Marian Szczepaniak. — Wyd. 2., zmienione. — Warszawa : Państwowe Wydawn. Naukowe, 1978.
492 p. ; 25 cm. P***
Table of contents also in English and Russian.
Includes bibliographical references.
zł75.00
1. Civil rights—Poland—Addresses, essays, lectures. 2. Civil rights—Addresses, essays, lectures. I. Szczepaniak, Marian.
79-359123
79 MARC

Rights / ⌊edited by⌋ David Lyons. — Belmont, Calif. : Wadsworth Pub. Co., c1979.
iv, 188 p. ; 23 cm. — (The Wadsworth series in social philosophy)
Bibliography: p. 187-188.
CONTENTS: Hart, H. L. A. Are there any natural rights?—Rawls, J. Constitutional liberty and the concept of justice.—Wasserstrom, R. Rights, human rights, and racial discrimination.—Lyons, D. Rights, claimants, and beneficiaries.—Feinberg, J. The nature and value of rights.—Dworkin, R. Taking rights seriously.—Hill, T. E., Jr. Servility and self-respect.—Hart, H. L. A. Bentham on legal rights.—Nozick, R. The entitlement theory.—Lyons, D. Human rights and the general welfare.
ISBN 0-534-00600-0 : $6.95
1. Civil rights—Addresses, essays, lectures. I. Lyons, David, 1935-
JC571.R528 323.4 78-17497
78 MARC

Russell-Tribunal pro und contra : Dokumente zu e. gefährl. Kontroverse / hrsg. von Freimut Duve u. Wolf-Dieter Narr. — Orig.-Ausg. — Reinbek bei Hamburg : Rowohlt, 1978.
157 p. ; 19 cm. — (Rororo aktuell) GFR78-A23
ISBN 3-499-14278-3 : DM4.80
1. Civil rights—Germany, West—Addresses, essays, lectures. 2. Civil rights—Addresses, essays, lectures. 3. Internationales Russell-Tribunal, Frankfurt am Main, 1978—Addresses, essays, lectures. I. Duve, Freimut, 1936- . II. Narr, Wolf-Dieter, 1937- .
342'.43'085 79-378245
79 MARC

Suekawa, Hiroshi, 1892–1977.
(Fuhō kōi narabini kenri ran'yō no kenkyū)
不法行爲竝に權利濫用の研究 / 末川博著. — 東京 : 岩波書店, 昭和8 ⌊1933⌋
3, 7, 374 p. ; 23 cm.
Includes bibliographical references.
1. Torts—Addresses, essays, lectures. 2. Chicanery (Law)—Addresses, essays, lectures. 3. Civil rights—Addresses, essays, lectures. I. Title.
79–805839

—CASES

Jurisprudentie over constitutioneel recht / verzameld door J. de Meyer ... ⌊et al.⌋. — 2e, herz. uitg. — Leuven : Acco, 1976.
461 p. ; 24 cm. Be77-6
Dutch, English, or French.
Includes bibliographical references.
590F
1. Civil rights—Belgium—Cases. 2. Belgium—Constitutional law—Cases. 3. Civil rights—Cases. 4. Constitutional law—Cases. I. Meyer, Jan de.
78-341614
*78 MARC

—CONGRESSES

30 ⌊i. e. Trzydziesta⌋ rocznica zwycięstwa nad faszyzmem = 30th anniversary of the victory over fascism = Тридцатая годовщина победы над фашизмом : ⌊Międzynarodowe Forum, 20–23 października 1975 r. / redaktorzy wydawn. Elżbieta Kasper, Lidia Wiśniakowska, Antoni Szumski.⌋. — Warszawa : Państ. Wydaw. Naukowe, 1977 (⌊Białyst. : BZG)⌋
305, ⌊1⌋ p. ; 24 cm. P 78–1078
At head of title: Polski Komitet do Spraw UNESCO.
English, Polish, and Russian.
Errata slip inserted.
1. World politics—1945- —Congresses. 2. United Nations Educational, Scientific and Cultural Organization—Congresses. 3. Civil rights—Congresses. I. Kasper, Elżbieta. II. Wiśniakowska, Lidia. III. Szumski, Antoni. IV. Poland. Polski Komitet do Spraw UNESCO. V. Title: 30th anniversary of the victory over fascism. VI. Title: Tridtsataia godovshchina pobedy nad fashizmom.
D839.2.T78 78–375819
(MARC)

CIVIL RIGHTS
—CONGRESSES (Continued)

Association of Korean Christian Scholars in North America.
Human rights in minority perspectives / edited by Dong Soo Kim and Byong-suh Kim. — Montclair, N.J. : Association of Korean Christian Scholars in North America, c1979.
xiii, 292, xv-xviii p. ; 22 cm. — (Publication series ; The Association of Korean Christian Scholars in North America ; no. 5)
Includes seven articles presented at the 12th annual meeting of the Association of Korean Christian Scholars in North America, held May 25-27, 1978 in Toronto.
Bibliography: p. 269-286.
$8.00
1. Church and civil rights—Congresses. 2. Civil rights—Congresses. 3. Civil rights—Korea—Congresses. I. Kim, Dong Soo. II. Kim, Byong-suh. III. Title. IV. Series: Association of Korean Christian Scholars in North America. Association of Korean Christian Scholars in North America publication series ; no. 5.
BT738.15.A84 1979 261.8 79-88571
79 MARC

Essays on human rights : contemporary issues and Jewish perspectives / edited by David Sidorsky, in collaboration with Sidney Liskofsky and Jerome J. Shestack ; Salo W. Baron ... [et al.]. — 1st ed. — Philadelphia : Jewish Publication Society of America, c1979.
xlv, 359 p. ; 24 cm.
Originated out of the McGill International Colloquium on Judaism and Human Rights held April 1974 at McGill University.
Includes bibliographical references.
ISBN 0-8276-0107-7 : $12.00
1. Jews—Politics and government—Congresses. 2. Civil rights—Congresses. 3. Civil rights (Jewish law)—Congresses. 4. Jews—Legal status, laws, etc.—Congresses. I. Sidorsky, David. II. Liskofsky, Sidney. III. Shestack, Jerome J. IV. Baron, Salo Wittmayer, 1895- . V. McGill International Colloquium on Judaism and Human Rights, 1974.
DS143.E76 323.1'19'24 78-1170
00 MARC

Human rights in the administration of criminal justice : report on the United Nations course, Canberra, Australia, 29 November-17 December 1976. — Canberra : Australian Institute of Criminology, 1978.
51 p. ; 21 cm. Aus***
Bibliography: p. [46]-51.
ISBN 0-642-91587-3
1. Criminal justice, Administration of—Congresses. 2. Civil rights—Congresses. 3. Prisoners—Legal status, laws, etc.—Congresses. I. Australian Institute of Criminology.
K5001.A3H86 345.05 79-315931
79 MARC

Menschenrechte : Aspekte ihrer Begründung und Verwirklichung : Vorträge und Diskussionen des Tübinger interdisziplinären Kolloquiums vom 23. bis 25. September 1977 / Johannes Schwartländer (Hrsg.) ; mit Beiträgen von Ernst-Otto Czempiel ... [et al.]. — Tübingen : Attempto, 1978.
241 p. ; 21 cm. — (Tübinger Universitätsschriften ; Bd. 1) GFR78-A4
Includes bibliographical references and indexes.
ISBN 3-921552-04-4 : DM17.80
1. Civil rights—Congresses. 2. Civil rights—Germany, West—Congresses. I. Schwartländer, Johannes, 1922- II. Series.
K3239.6.M46 323.4 79-344488
79 MARC

Menschenrechte im Spannungsfeld : eine Herausforderung für die Schweiz / mit Beitr. von Pierre Aubert ... [et al.] ; Hrsg., Amnesty International, Schweizer Sektion. — Bern [etc.] : P. Haupt, cop. 1978.
146 p. ; 21 cm. Sw79-719
Chiefly in German with one contribution in French.
ISBN 3-258-02816-8 : 19.00F
1. Civil rights—Switzerland—Congresses. 2. Civil rights—Congresses. I. Aubert, Pierre, 1927- II. Amnesty International. Schweizer Sektion.
342'.43'085 79-359741
79 MARC

—HISTORY

Oestreich, Gerhard.
Geschichte der Menschenrechte und Grundfreiheiten im Umriss / von Gerhard Oestreich. — 2., durchges. u. erg. Aufl. — Berlin : Duncker und Humblot, 1978.
158 p. ; 24 cm. — (Historische Forschungen ; Bd. 1) GFR78-A
Bibliography: p. [130]-158.
ISBN 3-428-02092-8 : DM24.80
1. Civil rights—History. I. Title.
K3239.4.O38 1978 323.4'09 79-356125
79 MARC

Seki, Itaru, 1903–
(Jinken sengen konjaku monogatari)
人権宣言今昔物語 / 關之著. — 東京 : 聯合出版社, 昭和25 [1950]
3, 4, 201 p. ; 19 cm.

1. Civil rights—Japan—History. 2. Civil rights—History. I. Title.
JQ1681.S432 79-807249

Suzuki, Yasuzō, 1904–
(Kihonteki jinken)
基本的人権 / 鈴木安藏著. — 東京 : 実教出版, 昭和26 [1951]
2, 7, 390 p. ; 19 cm.
Includes bibliographical references.

1. Civil rights—History. 2. Civil rights—Japan—History. I. Title.
K3240.4.S89 79-801124

—JUVENILE LITERATURE

Loescher, Gil.
Human rights : a global crisis / by Gil Loescher, with Ann Loescher. — 1st ed. — New York : Dutton, c1978.
xiii, 130 p. : ill. ; 24 cm.
Bibliography: p. 119-123.
Includes index.
SUMMARY: Discusses what human rights are, their violation throughout the world today with specific examples in various countries, and the efforts of organizations and individuals to protect these rights.
ISBN 0-525-32485-2 : $8.95
1. Civil rights—Juvenile literature. [1. Civil rights] I. Loescher, Ann Dull, joint author. II. Title.
JC571.L587 1978 323.4'09 78-8710
78 MARC

Meltzer, Milton, 1915-
The human rights book / Milton Meltzer. — 1st ed. — New York : Farrar, Straus, Giroux, 1979.
264 p. ; 22 cm.
Bibliography: p. [157]-162.
Includes index.
SUMMARY: Discusses the evolution of a concept of human rights, how these rights are defined and interpreted throughout the world, to what degree they are observed or violated, and the international institutions and agencies responsible for gaining and maintaining these rights for individuals.
ISBN 0-374-33514-1 : $9.95
1. Civil rights—Juvenile literature. [1. Civil rights] I. Title.
JC571.M398 1979 323.4 79-13017
79 MARC

—PERIODICALS

Center for International Policy.
Human rights and the U. S. foreign assistance program. 1977/78–
Washington, Center for International Policy.
v. 22 x 28 cm. annual.
Report for 1977/78– issued in 2 vols.
Key title: Human rights and the U. S. foreign assistance program, ISSN 0161-6684
1. Civil rights—Periodicals. 2. Economic assistance, American—Periodicals. 3. Military assistance, American—Periodicals. I. Title.
JC585.C417a 323.4'05 78-643551
MARC-S

Human Rights Internet.
Human Rights Internet newsletter. v. 1–
July 1976–
Washington [etc.] Human Rights Internet.
v. 28 cm. 9 no. a year.
Title varies slightly.
Key title: Human Rights Internet newsletter, ISSN 0163-9048.
1. Civil rights—Periodicals. I. Title.
JC571.H7696a 323.4'05 79-643675
MARC-S

—SOCIETIES, ETC.

Garling, Marguerite.
The human rights handbook : a guide to British and American international human rights organisations / compiled by Marguerite Garling for the Writers and Scholars Educational Trust. — London : Macmillan Press, 1979.
xv, 299 p. ; 23 cm. GB***
Bibliography: p. 290-294.
Includes index.
ISBN 0-333-25817-7 : £12.00
1. Civil rights—Societies, etc. I. Writers and Scholars Educational Trust. II. Title.
JC571.G32 1979 323.4'06 79-310154
79 MARC

Garling, Marguerite.
The human rights handbook : a guide to British and American international human rights organisations / compiled by Marguerite Garling for the Writers and Scholars Educational Trust. — New York : Facts on File, 1979.
xv, 299 p. ; 23 cm.
Bibliography: p. 290-299.
Includes index.
ISBN 0-87196-403-1 : $25.00
1. Civil rights—Societies, etc. I. Title.
JC571.G32 323.4'06'2 78-26169
78 MARC

—STUDY AND TEACHING
—JAPAN—HOKKAIDO

(Minshūshi undō)
民衆史運動 : その歴史と理論 / 小池喜孝監修 ; オホーツク民衆史講座編. — 東京 : 現代史出版会 : 発売 徳間書店, 1978.
279 p., [2] leaves of plates : ill. ; 20 cm. Ja 78-19597
Bibliography: p. 257-258.
Y1500
1. Civil rights—Japan—Hokkaido—History. 2. Civil rights—Study and teaching—Japan—Hokkaido. 3. Labor and laboring classes—Japan—Hokkaido—History. I. Koike, Yoshitaka, 1916- II. Ohōtsuku Minshūshi Kōza.
JS7384.H643A75 78-815375

—UNDERDEVELOPED AREAS see Underdeveloped areas—Civil rights

—YEARBOOKS

International League for Human Rights.
Annual review - International League for Human Rights.
[New York] International League for Human Rights.
v. 28 cm.
Continues: Annual review - International League for the Rights of Man, ISSN 0363-9347.
1. International League for Human Rights. 2. Civil rights—Yearbooks. I. Title.
JC571.I 594 323.4'062'1 78-645590

—AFRICA

Okolie, Charles Chukwuma
International law perspectives of the developing countries : the relationship of law and economic development to basic human rights / Charles Chukwuma Okolie. — New York : NOK Publishers, c1978.
xii, 369 p. ; 24 cm. — (NOK legal studies series)
Includes bibliographical references and index.
ISBN 0-88357-011-4
1. Africa—Foreign economic relations. 2. Civil rights—Africa. 3. Economic assistance. 4. Underdeveloped areas—Foreign economic relations. 5. Underdeveloped areas—Civil rights. 6. International law. I. Title.
341.7'59 74-81841
00 MARC

—AFRICA—CONGRESSES

International Seminar on Human Rights, Their Protection, and the Rule of Law in a One-party State, Dar es Salaam, Tanzania, 1978.
Human rights in a one-party state / International Seminar on Human Rights, Their Protection, and the Rule of Law in a One-party State : convened by the International Commission of Jurists. — London : Search Press in conjunction with the International Commission of Jurists, 1978.
133 p. ; 22 cm. GB***
Seminar held Sept. 23-28, 1976 at Dar es Salaam.
Includes bibliographical references.
1. Civil rights—Africa—Congresses. I. International Commission of Jurists (Founded 1952). II. Title.
JC599.A36I57 1978 323.4'096 78-316219
78 MARC

—AFRICA, SOUTHWEST

Dugard, C J R
"Namibia and human rights" : a report on the Dakar conference and its implications for the South West Africa issue and détente / John Dugard. — [Braamfontein : South African Institute of International Affairs, 1976]
19 p. ; 30 cm. SA***
Cover title.
Report prepared for the South African Institute of International Affairs, including the Declaration of Dakar and Programme of Action of the Dakar Conference on Namibia and Human Rights, held 5-8 Jan., 1976.
1. Namibia—Foreign relations—South Africa. 2. South Africa—Foreign relations—Namibia. 3. Civil rights—Africa, Southwest. I. South African Institute of International Affairs. II. Dakar Conference on Namibia and Human Rights, 1976. III. Title.
JX1586.S64D85 327'.0968 76-383559
77[r79]rev MARC

—ALABAMA

see also Selma-Montgomery Rights March, 1965

CIVIL RIGHTS (Continued)

—AMERICA—PERIODICALS

Inter-American Commission on Human Rights.
Annual report of the Inter-American Commission on Human Rights to the General Assembly.
Washington, General Secretariat of the Organization of American States.
v. 29 cm. (₍Official records₎ - Organization of American States, OEA/ser.L/V/II)
1. Civil rights—America—Periodicals. I. Series: Organization of American States. Official records, OEA/ser.L/V/II.
F1405.5 1959.O7 subser. 341.24'5 s 78-647534
[JC599.A5] [341.48'1] MARC-S

Inter-American Commission on Human Rights.
Informe anual de la Comisión Interamericana de Derechos Humanos a la Asamblea General.
Washington, Secretaría General de la Organización de los Estados Americanos.
v. 28 cm. (₍Documentos oficiales₎ - Organización de los Estados Americanos, OEA/ser.L/V/II)
1. Civil rights—America—Periodicals. I. Series: Organization of American States. Documentos oficiales, OEA/ser.L/V/II.
F1402.A169 subser. 341.24'5 s 78-645503
[JC599.A5] [323.4'098] MARC-S

—ARGENTINE REPUBLIC

Amnesty International.
Informe de una misión de Amnistía Internacional a la República Argentina : 6-15 de noviembre de 1976. — 1. ed. — Londres : Amnesty International Publications, 1977.
124 p. ; 22 cm. Sp***
ISBN 8470310410 : 175ptas
1. Political prisoners—Argentine Republic. 2. Torture—Argentine Republic. 3. Refugees—Argentine Republic. 4. Civil rights—Argentine Republic. I. Title.
HV9583.A55 1977 78-394388
79 MARC

Argentina : genocidio y resistencia. — Bilbao : Zero ; Madrid : distribuidor exclusivo, ZYX, c1977.
63 p. : ill. ; 25 cm. — (América Latina ; 1) (Cuadernos Aesla) Sp**
ISBN 8431704098
1. Argentine Republic—Economic conditions—1945- . 2. Argentine Republic—Politics and government—1955- . 3. Political prisoners—Argentine Republic. 4. Civil rights—Argentine Republic. I. Series: América Latina (Bilbao) ; 1.
HC175.A862 79-346031
79 MARC

Argentine Information Service Center.
Argentina today : a dossier on repression and the violation of human rights / prepared by the Argentine Information Service Center. — 2d ed. — ₍s.l.₎ : The Center, 1977.
₍34₎ p. : ill. ; 28 cm.
Cover title.
1. Civil rights—Argentine Republic. 2. Political prisoners—Argentine Republic. 3. Persecution—Argentine Republic. I. Title.
JC599.A7A73 1977 323.4'0982 79-300795
79 MARC

Inter-Church Committee on Chile.
Le cone sud de l'Amérique latine : une prison gigantesque : mission d'observation au Chili, en Argentine et en Uruguay, 30 septembre au 10 octobre 1976 / présentée par the Inter-Church Committee on Chile, en collaboration avec Comité chrétien pour les droits de l'homme en Amérique latine. — ₍Montréal : Comité chrétien pour les droits de l'homme en Amérique latine, 1976₎
93 p. : ill. ; 28 cm. C***
Cover title.
Bibliography: p. 92.
1. Civil rights—Chile. 2. Civil rights—Argentine Republic. 3. Civil rights—Uruguay. I. Comité chrétien pour les droits de l'homme en Amérique latine. II. Title. III. Title: Une prison gigantesque.
JC599.C5I57 1976 323.4'098 79-369296
79 MARC

—AUSTRIA

Pernthaler, Péter.
Verfassungsrechtliche Probleme des Arbeitsverfassungsgesetzes : Rechtsgutachten über die Verfassungsmässigkeit des Ministerialentwurfes eines Arbeitsverfassungsgesetzes / von Peter Pernthaler. — Wien : Österr. Wirtschaftsverl., ₍1973₎
76 p. ; 24 cm. — (Schriftenreihe der Bundeskammer der Gewerblichen Wirtschaft ; 21) Au78-13-57
Includes bibliographical references.
S54.00
1. Works councils—Austria. 2. Employees' representation in management—Austria. 3. Austria—Constitutional law. 4. Civil rights—Austria. I. Title. II. Series: Austria. Bundeskammer der Gewerblichen Wirtschaft, Vienna. Schriftenreihe ; 21.
79-338818
*79 MARC

—BELGIUM

De Belgische grondwet, De rechten van de mens, De Belgische provinciewet, De Belgische gemeentewet = La constitution belge, Les droits de l'homme, La loi provinciale belge, La loi communale belge. — 2500 Lier, (Grote Markt 39) : J. van In, 1977.
253 p. ; 20 cm. Be77-5
Dutch and French.
ISBN 9030604735 : 225F
1. Belgium—Constitutional law. 2. Civil rights (International law) 3. Civil rights—Belgium. 4. Local government—Belgium. I. Title: La constitution belge.
79-351447
*79 MARC

—BELGIUM—CASES

Jurisprudentie over constitutioneel recht / verzameld door J. de Meyer ... ₍et al.₎. — 2e, herz. uitg. — Leuven : Acco, 1976.
461 p. ; 24 cm. Be77-6
Dutch, English, or French.
Includes bibliographical references.
590F
1. Civil rights—Belgium—Cases. 2. Belgium—Constitutional law—Cases. 3. Civil rights—Cases. 4. Constitutional law—Cases. I. Meyer, Jan de.
78-341614
*78 MARC

—BELGIUM—SOCIETIES, ETC.

Guide de la Belgique des luttes. — Bruxelles, av. Van Volxem 305 : Vie ouvrière, 1977-
v. : ill. ; 19 cm. Be77-8
ISBN 2-87003-113-0 (v. 1) : 495F (v. 1)
1. Socialism in Belgium—Societies, etc. 2. Civil rights—Belgium—Societies, etc.
HX301.G84 335'.006'2493 78-375953
78 MARC

—BOLIVIA—ADDRESSES, ESSAYS, LECTURES

Morales Dávila, Manuel.
Los derechos humanos en Bolivia 1971/1977 / Manuel Morales Dávila. — 2. ed. — ₍La Paz₎ : Confederación Nacional de Profesionales de Bolivia, 1978.
183 p. ; 21 cm.
Collection of articles and memoranda addressed to various presidents and organizations.
1. Civil rights—Bolivia—Addresses, essays, lectures. I. Title.
JC599.B6M67 1978 323.4'0984 78-110013
79 MARC

—BRAZIL

Barbosa, Ruy, 1849-1923.
República, teoria e prática : textos doutrinários sobre direitos humanos e políticos consagrados na primeira constituição republicana / Rui Barbosa ; seleção e coordenação de Hildon Rocha. — Petrópolis, Brasil : Editora Vozes, 1978.
343 p. ; 21 cm. — (Dimensões do Brasil ; 12)
Includes bibliographical references.
1. Civil rights—Brazil. 2. Brazil—Constitutional history. I. Rocha, Hildon. II. Title.
JC599.B7B37 1978 79-343940
79 MARC

Brazil. Laws, statutes, etc.
Direitos e garantias individuais, liberdade de manifestação do pensamento e de informação, lei de segurança nacional. Rio de Janeiro, Gráfica Editora Aurora, 1973.
111 p. 19 cm. (Coleção Lex, no. 14) Cr$10.00
Half title: Direitos do homem.
Cover title: Lei de segurança nacional, liberdade de manifestação do pensamento e informação, direitos e garantias individuais.
1. Civil rights—Brazil. 2. Internal security—Brazil. I. Title. II. Title: Direitos do homem. III. Title: Lei de segurança nacional.
77-483151
78 MARC

Russell Tribunal on Repression in Brazil, Chile, and Latin America, 2d, Rome and Brussels, 1974-1975.
Brasile, violazione dei diritti dell'uomo / Tribunale Russell II ; a cura di Linda Bimbi. — 1. ed. — Milano : Feltrinelli, 1975.
319 p. ; 18 cm. — (I Nuovi testi ; 84) It76-Sept
Includes bibliographical references.
L2500
1. Civil rights—Brazil. 2. Torture—Brazil. 3. Political crimes and offenses—Brazil. I. Bimbi, Linda. II. Title.
323.4'0981 77-473860
*77₍r79₎rev MARC

—BRITISH COLUMBIA

Rigg, Kit.
Civil liberties / ₍Kit Rigg₎. — 2d ed., rev. and expanded / by David Garling. — Vancouver : Vancouver People's Law School, 1978.
39 p. : ill. ; 27 cm. — (Vancouver People's Law School public legal education series) C***
Includes bibliographical references.
ISBN 0-88847-028-2
1. Civil rights—British Columbia. I. Garling, David. II. Title. III. Series: Public legal education series.
KEB458.Z82R5 1978 342'.711'085 79-318576
79 MARC

—CANADA

Ontario. Human Rights Commission.
Life together : a report on human rights in Ontario. — Toronto : Ontario Human Rights Commission, 1977.
139 p. ; 26 cm. C***
Issued also in French.
1. Civil rights—Canada. I. Title.
JC599.C2O65 1977 323.4'09713 78-316220
78 MARC

Pearson, Mary.
International decade for women, 1976-1985 : what it means to Canadian women / by Mary Pearson. — Ottawa : Advisory Council on the Status of Women, 1977.
40 p. ; 28 cm. C***
Errata slip, ₍1₎ leaf inserted.
1. Women's rights—Canada. 2. Civil rights—Canada. 3. Women—Legal status, laws, etc.—Canada. I. Canada. Advisory Council on the Status of Women. II. Title.
HQ1457.P4 323.4 79-305868
79 MARC

—CANADA—BIBLIOGRAPHY

Dworaczek, M
Human rights : government documents held in the Research Library / Marian Dworaczek. — 2d ed. — Toronto : Ontario Ministry of Labour, Research Library, 1975.
61 p. ; 30 cm. — (Bibliography series - Ministry of Labour, Research Branch ; no. 1) C***
1. Civil rights—Canada—Bibliography. 2. Civil rights—Great Britain—Bibliography. 3. Civil rights—United States—Bibliography. 4. Canada—Government publications—Bibliography. 5. Great Britain—Government publications—Bibliography. 6. United States—Government publications—Bibliography. I. Ontario. Ministry of Labour. Library. II. Title. III. Series: Ontario. Ministry of Labour. Research Branch. Bibliography series - Ministry of Labour, Research Branch ; no. 1.
K3236.D9 1975 016.3234 79-301630
79 MARC

—CHILE

Inter-Church Committee on Chile.
Le cone sud de l'Amérique latine : une prison gigantesque : mission d'observation au Chili, en Argentine et en Uruguay, 30 septembre au 10 octobre 1976 / présentée par the Inter-Church Committee on Chile, en collaboration avec Comité chrétien pour les droits de l'homme en Amérique latine. — ₍Montréal : Comité chrétien pour les droits de l'homme en Amérique latine, 1976₎
93 p. : ill. ; 28 cm. C***
Cover title.
Bibliography: p. 92.
1. Civil rights—Chile. 2. Civil rights—Argentine Republic. 3. Civil rights—Uruguay. I. Comité chrétien pour les droits de l'homme en Amérique latine. II. Title. III. Title: Une prison gigantesque.
JC599.C5I57 1976 323.4'098 79-369296
79 MARC

World Council of Peace. Information Centre.
Chile yes, Pinochet no. — Helsinki : Information Centre of the World Peace Council, ₍1978?₎
35 p. ; 21 cm.
1. Civil rights—Chile. 2. Torture—Chile. 3. Political prisoners—Chile. 4. Fascism—Chile. I. Title.
JC599.C5W67 1978 323.4'0983 78-322336
78 MARC

—CHILE—CONGRESSES

International Commission of Enquiry into the Crimes of the Military Junta in Chile.
First session of the International Commission of Enquiry into the Crimes of the Military Junta in Chile : Dipoli, Finland, March 21-24, 1974 : documents / edited by the International Commission of Enquiry into the Crimes of the Military Junta in Chile. — Helsinki : Secretariat, ₍1974?₎
47 p. ; 20 cm. Fi***
Statutes of the International Commission: p. 8-₍11₎
Errata slip inserted.
1. Chile—History—Coup d'état, 1973—Congresses. 2. Civil rights—Chile—Congresses. 3. Political prisoners—Chile—Congresses. 4. Atrocities—Chile—Congresses.
F3100.I56 1974 323.4'0983 79-306646
*79 MARC

CIVIL RIGHTS (Continued)

—CHILE—PERIODICALS

Chile. no. [1]– July 1976–
Prague [Czechoslovakia, International Trade Union Committee of Solidarity with the Workers and People of Chile]
no. ill. 30 cm. 3 or 4 no. a year.
"Bulletin of the International Trade Union Committee of Solidarity with the Workers and People of Chile."
1. Trade-union—Chile—Periodicals. 2. Civil rights—Chile—Periodicals. 3. Trade-unions—Political activity—Periodicals. I. International Trade Union Committee of Solidarity with the Workers and People of Chile.
HD6616.A14 331.88′0983 79–642192
MARC-S

—CHINA

United States. Library of Congress. Far Eastern Law Division.
Some recent developments related to human rights in the People's Republic of China / prepared for the Committee on Foreign Relations, United States Senate, by the Far Eastern Law Division, Library of Congress. — Washington : U.S. Govt. Print. Off., 1979.
iii, 20 p. ; 24 cm.
At head of title: 96th Congress, 1st session. Committee print.
Includes bibliographical references.
1. Civil rights—China. I. United States. Congress. Senate. Committee on Foreign Relations. II. Title.
323.4′0951 79-601588
79 MARC

World Anti-Communist League. China Chapter.
Documented evidence how the Chinese Communists impinge on human rights. — [Taipei] : World Anti-Communist League, China Chapter, Asian Peoples' Anti-Communist League, Republic of China, 1977.
85 p. : ill. ; 26 cm.
Cover title.
Includes bibliographical references.
$2.00 (U.S.)
1. Civil rights—China. I. Asian Peoples' Anti-Communist League. China. II. Title.
JC599.C6W67 1977 323.4′0951 79-313286
79 MARC

—COLOMBIA—ADDRESSES, ESSAYS, LECTURES

Así piensa la clase emergente / [Carlos Ayala Jiménez ... et al.]. — [Bogotá?] : Sociedad de Integración Liberal, [1978]
227 p. ; 20 cm.
"Conferencias organizadas por la Sociedad ... en la sede de la Asociación Democrática Colombiana, durante el último semestre del año pasado."
1. Latin America—Economic integration—Addresses, essays, lectures. 2. Civil rights—Colombia—Addresses, essays, lectures. 3. Territorial waters—Colombia—Addresses, essays, lectures. I. Ayala Jiménez, Carlos. II. Sociedad de Integración Liberal.
HC125.A737 338.91′8 78-108018
79 MARC

—COLORADO—PERIODICALS

Colorado. Civil Rights Commission.
Activities - Colorado Civil Rights Commission.
[Denver] Colorado Civil Rights Commission.
v. ill. 28 cm.
Continues: Colorado. Civil Rights Commission. Activities report.
1. Civil rights—Colorado—Periodicals. 2. Discrimination—Colorado—Periodicals.
KFC2211.A833 323.4′09788 78–645080
MARC-S

—COMMUNIST COUNTRIES

Klenner, Hermann.
Freiheit, Gleichheit und so weiter : dreizehn Streiflichter über die Menschenrechte / Hermann Klenner, [Mitarbeit, Annelies Schwarzer]. — 1. Aufl. — Berlin : Staatsverlag der Deutschen Demokratischen Republik, 1978.
143 p. : ill. ; 17 cm. — (Schriftenreihe Recht in unserer Zeit ; 10)
GDR***
Includes bibliographical references.
2.25M
1. Civil rights. 2. Civil rights—Communist countries. 3. Civil rights (International law) I. Schwarzer, Annelies, joint author. II. Title.
K3240.4.K56 323.4′09171′7 78-395441
79 MARC

Klenner, Hermann.
Menschenrechte : Anspruch und Wirklichkeit / [H. Klenner, B. Graefrath, V. A. Sorin]. — Berlin : DDR-Komitee für Menschenrechte, 1978.
56 p. ; 21 cm. — (Schriften und Informationen - DDR-Komitee für Menschenrechte ; Heft 4/78) GDR***
Cover title.
Includes bibliographical references.
1. Civil rights (International law) 2. Civil rights. 3. Civil rights—Communist countries. I. Graefrath, Bernhard, joint author. II. Zorin, Valerian Alexandrovich, 1902- joint author. III. Title. IV. Series: DDR-Komitee für Menschenrechte. Schriften und Informationen - DDR-Komitee für Menschenrechte ; Heft 1978, 4.
K3240.4.K566 341.48′1 79-365012
79 MARC

Klotsch, Helmut.
Freiheit, ein Vorzug des Sozialismus / Helmut Klotsch, Alfred Kohlsdorf, Heinrich Opitz. — Berlin : Dietz Verlag, 1978.
84 p. ; 22 cm. — (Sozialismus, Erfahrungen, Probleme und Perspektiven)
GDR***
Includes bibliographical references.
1.80M
1. Liberty. 2. Communist countries. 3. Civil rights—Communist countries. I. Kohlsdorf, Alfred, joint author. II. Opitz, Heinrich, joint author. III. Title.
HM271.K59 79-363869
79 MARC

—CZECHOSLOVAKIA

Rusinow, Dennison I
Some observations regarding "Charter 77" / by Dennison I. Rusinow. — [Hanover, N.H.] : American Universities Field Staff, [1977-
2 v. ; 28 cm. — (Southeast Europe series ; v. 22, no. 1-2) (Fieldstaff reports : Europe)
Cover title.
"DIR-1-'77."
"DIR-2-'77."
Includes bibliographical references.
CONTENTS: pt. 1. The manifesto: action, reaction, and counterreaction.—pt. 2. "Freedoms great and small."

$1.50 per vol.

1. Civil rights—Europe, Eastern. 2. Civil rights—Czechoslovakia. I. Title. II. Series. III. Series: American Universities Field Staff. Fieldstaff reports.
D901.A54 vol. 22, no. 1-2 79-116555
081 s
[JC599.E92] 79 MARC

Since the Prague spring : the continuing struggle for human rights in Czechoslovakia / edited by Hans-Peter Riese ; translated from the German by Eugen Loebl. — 1st American ed. — New York : Random House, c1979.
xiv, 208 p. ; 22 cm.
Translation of Bürgerinitiative für die Menschenrechte.
Includes index.
ISBN 0-394-50056-3 : $10.00
1. Civil rights—Czechoslovakia. I. Riese, Hans-Peter.
323.4′09437 78-57113
79 MARC

Since the Prague spring : the continuing struggle for human rights in Czechoslovakia / edited by Hans-Peter Riese ; translated from the German by Eugen Loebl. — 1st Vintage Books ed. — New York : Vintage Books, 1979.
xiv, 208 p. ; 20 cm.
Translation of Bürgerinitiative für die Menschenrechte.
Includes index.
ISBN 0-394-72556-5
1. Civil rights—Czechoslovakia. I. Riese, Hans-Peter.
342′.437′085 78-51410
78 MARC

—DENMARK

Døygaard, Heino.
Privatlivets treude fred / Heino Døygaard. — [København] : Gyldendal, 1976.
129 p. : ill. ; 22 cm. — (Faglig orientering) D76-40
Bibliography: p. 127-[130]
ISBN 8701234919 : kr40.00
1. Privacy, Rights of—Denmark. 2. Civil rights—Denmark. I. Title.
JC599.D4D63 79-355751
79 MARC

Smidt, Poul.
Undergravende virksomhed : beretning fra Kommissionsdomstolen af 13. juli 1977 / [udgivet af] Poul Smidt. — [København] : Demos, c1977.
70 p. ; 20 cm. D78-9
Includes official report prepared by F. Poulsen.
"Hvorledes borgerskabet inddæmmer revolutionen og bekæmper de revolutionære ; 2."
ISBN 8787337908 : kr26.00
1. Intelligence service—Denmark. 2. Governmental investigations—Denmark. 3. Political crimes and offenses—Denmark. 4. Civil rights—Denmark. I. Poulsen, Frank, 1927- . II. Denmark. Kommissionsdomstolen af 13. juli 1977. III. Title.
78-381954
78[r79]rev MARC

—ESTONIA

Petrov, V fl. 1961–
(Dva berega—dva obraza zhizni)
Два берега—два образа жизни / В. Петров. — Таллин : Ээсти раамат, 1977.
200 p. ; 20 cm. USSR 77
Includes bibliographical references.
0.29rub
1. Finland—Social conditions. 2. Estonia—Social conditions. 3. Civil rights—Finland. 4. Civil rights—Estonia. I. Title.
HN533.5.P48 78–421981

—ETHIOPIA

Amnesty International.
Human rights violations in Ethiopia. — London : Amnesty International, 1978.
17 p. : map ; 30 cm. GB***
Cover title.
1. Ethiopia—Politics and government—1974- . 2. Political crimes and offenses—Ethiopia. 3. Civil rights—Ethiopia. 4. Atrocities—Ethiopia. 5. Amnesty International. I. Title.
DT387.95.A46 1978 323.4′9′0963 79-314514
79 MARC

—EUROPE

Alkema, E A
Studies over Europese grondrechten : de invloed van de Europese Conventie op het Nederlands recht / door Evert Albert Alkema. — Deventer : Kluwer, 1978.
xxi, 312 p. ; 25 cm. Ne78-44
Thesis—Leyden.
Vita.
Summary in English.
Bibliography: p. 299-305.
Includes index.
ISBN 9026810415 : fl 40.00
1. Civil rights—Europe. 2. Convention for the protection of human rights and fundamental freedoms. I. Title.
79-357180
*79 MARC

Brändle, Dieter, 1950-
Vorbehalte und auslegende Erklärungen zur europäischen Menschenrechtskonvention / vorgelegt von Dieter Brändle. — Zürich : Juris-Verlag, 1978.
xxii, 188 p. ; 23 cm. Sw78-9745
Thesis—Zürich.
Vita.
Bibliography: p. x-xviii.
ISBN 3-260-04493-0 : 40.00F
1. Civil rights—Europe. 2. Treaties—Reservations. 3. Convention for the protection of human rights and fundamental freedoms. I. Title.
341.48′1 78-399577
79 MARC

Craenen, Godelieve.
Elementen van publiek recht / G. Craenen. — Herwerkte uitg. 1976. — 3000 Leuven, (Tiensestr. 134-136) : Acco, 1977.
ix, 242 leaves ; 27 cm. Be78-6
Bibliography: leaves i-ii.
340F
1. Public law—Belgium. 2. Public law. 3. Civil rights—Europe. 4. International agencies. I. Title.
342′.493 79-347438
*79 MARC

Evans, Vincent, Sir, 1915-
Report by Sir Vincent Evans on the practice of European countries where direct effect is given to the European convention on human rights in internal law. — [s.l. : s.n., 1977?]
129 leaves ; 30 cm. GB***
1. Civil rights—Europe. I. Title. II. Title: On the practice of European countries where direct effect is given to the European convention on human rights in internal law.
342′.4′085 79-302416
79 MARC

Geiser, Hanspeter.
Rechtsschutz im Verwaltungsvollstreckungsverfahren / Hanspeter Geiser. — St. Gallen : Schweizerisches Institut für Verwaltungskurse an der Hochschule St. Gallen, 1978.
xxviii, 145 p. ; 23 cm. — (St. Galler Beiträge zum öffentlichen Recht ; Bd. 7) Sw78-10676
Thesis—Hochschule für Wirtschafts- und Sozialwissenschaften, St. Gall, Switzerland.
Bibliography: p. xiii-xxii.
1. Executions (Administrative law)—Switzerland. 2. Remedies (Law)—Switzerland. 3. Civil rights—Europe. I. Title. II. Series: Sankt Galler Beiträge zum öffentlichem Recht ; Bd. 7.
342′.494 79-358865
79 MARC

Jinken Shisō Kenkyūkai.
(Sekai kakkoku jinken sengen no kenkyū)
世界各國人權宣言の研究 / 人權思想研究會編 ; [代表者 関之]. — 東京 : 巖松堂書店, 昭和25-27 [1950-1952]
3 v. ; 19 cm. — (人權思想研究叢書 ; 第1-3巻)
CONTENTS: [1] 英米の部—[2] フランス・スイスの部—[3] ドイツ・ソ連・イタリヤの部
1. Civil rights—Europe. 2. Civil rights—United States. I. Seki, Itaru, 1903- II. Title. III. Series: Jinken shisō kenkyū sōsho ; dai 1-3-kan.
79-801146

CIVIL RIGHTS
—EUROPE (Continued)

Nedjati, Zaim M
Human rights under the European Convention / by Zaim M. Nedjati. — Amsterdam ; New York : North-Holland Pub. Co. ; New York : sole distributors for the U.S.A. and Canada, Elsevier North-Holland Inc., 1978.
xviii, 298 p. ; 23 cm. — (European studies in law ; v. 8)
Includes bibliographical references and index.
ISBN 0-444-85218-2
1. Civil rights—Europe. I. Convention for the protection of human rights and fundamental freedoms. II. Title.
341.48'1 78-23469
78 MARC

Poncet, Dominique.
La protection de l'accusé par la Convention européenne des droits de l'homme : étude de droit comparé / Dominique Poncet ; pref. du professeur François Clerc et de Jean Raymond. — Genève : Georg, c1977.
xx, 269 p. ; 24 cm. — (Mémoires publiés par la Faculté de droit de Genève ; no 52) Sw***
Revised and enlarged from the author's paper presented at the XV Journée juridique, Nov. 8, 1975.
Bibliography: p. [255]-261.
1. Criminal procedure—Europe. 2. Civil rights—Europe. I. Title. II. Series: Geneva. Université. Faculté de droit. Mémoires ; no 52.
345'.4'056 79-350734
79 MARC

Rechten van de mens in mundiaal en europees perspectief. — Maarssen : Ars Aequi Libri, [1978]
x, 229 p. ; 25 cm. — (Rechten van de mens ; 1) Ne***
Dutch or English.
Includes bibliographical references.
ISBN 9070094231
1. Civil rights (International law) 2. Civil rights—Europe. I. Series.
K3240.4.R4 vol. 1 341.48'1 s 78-386861
78 MARC

—EUROPE—BIBLIOGRAPHY

Council of Europe.
Bibliographie concernant la Convention européenne des droits de l'homme - Bibliography relating to the European convention on human rights. — Strasbourg : Conseil de l'Europe, 1978.
ix, 173 p. ; 29 cm. F***
French and English.
Includes index.
1. Civil rights—Europe—Bibliography. I. Title. II. Title: Bibliography relating to the European convention on human rights.
341.48'1 79-383839
79 MARC

—EUROPE—CONGRESSES

Die Ausübung staatlicher Gewalt in Ost und West nach Inkrafttreten der UN-Konvention über zivile und politische Rechte : Juristenkonferenz vom 23.-25. September 1977 in Ludwigshafen am Rhein / Dt. Sekt. d. Internat. Juristen-Komm ; mit Vorträgen von Gottfried Zieger ... [et al.]. — Heidelberg ; Karlsruhe : Müller, Juristischer Verl., 1978.
103 p. ; 23 cm. — (Rechtsstaat in der Bewährung ; Bd. 6) GFR79-A
Conference organized by the Deutsche Sektion of the International Commission of Jurists.
"Entwurf einer Ergänzung zur UN-Menschenrechtskonvention": p. 101-103.
Includes bibliographical references.
ISBN 3-8114-6678-X : DM32.00
1. Civil rights (International law)—Congresses. 2. Civil rights—Germany, East—Congresses. 3. Civil rights—Europe—Congresses. 4. International covenant on civil and political rights—Congresses. I. Zieger, Gottfried. II. International Commission of Jurists (Founded 1952). Deutsche Sektion. III. Series.
K3239.6.A9 341.48'1 79-356493
79 MARC

Conference of European Churches.
Europe after Helsinki and the developing regions : spiritual, moral, and practical responsibilities of our churches : report of a consultation, 7th to 11th March 1977, Gallneukirchen near Linz. — Geneva : Conference of European Churches ; Edinburgh : distributor, Saint Andrew Press, 1977.
100 p. ; 24 cm. — (Occasional paper - Conference of European Churches ; no. 9) Sw***
Includes bibliographical references.
1. Church and social problems—Europe—Congresses. 2. Civil rights—Europe—Congresses. 3. Europe—Economic policy—Congresses. I. Title. II. Series: Conference of European Churches. Occasional papers - Conference of European Churches ; no. 9.
HN39.E9C66 1977 261.8'094 79-303997
79 MARC

—EUROPE—PERIODICALS

Council of Europe.
Activities of the Council of Europe in the field of human rights.
Strasbourg, Council of Europe.
v. 30 cm.
1. Civil rights—Europe—Periodicals. I. Title.
341.48'1 78-643990
MARC-S

—EUROPE, EASTERN

CDU/CSU-Fraktion des Deutschen Bundestages.
White paper on the human rights situation in Germany and of the Germans in Eastern Europe. — Bonn : CDU/CSU Group in the German Bundestag, 1977.
114 p. : ill. ; 30 cm. GFR***
1. Civil rights—Germany, East. 2. Civil rights—Europe, Eastern. 3. Germans in Eastern Europe. I. Title.
JC599.G32C13 1977 323.4'0943 78-323042
79 MARC

Michael-Titus, C
A testimony for Europe / [by] C. Michael-Titus. — Upminster : Panopticum Press, 1977.
15 p. ; 21 cm. GB77-33643
Cover title: Yalta, Helsinki, Belgrade, a testimony for Europe.
1. Conference on Security and Cooperation in Europe. 2. Civil rights—Europe, Eastern. 3. Civil rights—Russia. 4. Dissenters—Europe, Eastern. 5. Dissenters—Russia. 6. Authors, Romanian. I. Title. II. Title: Yalta, Helsinki, Belgrade.
D1050.5.M5 322.4'4'0947 78-319995
78 MARC

Rusinow, Dennison I
Some observations regarding "Charter 77" / by Dennison I. Rusinow. — [Hanover, N.H.] : American Universities Field Staff, [1977-
2 v. ; 28 cm. — (Southeast Europe series ; v. 22, no. 1-2) (Fieldstaff reports : Europe)
Cover title.
"DIR-1-'77."
"DIR-2-'77."
Includes bibliographical references.
CONTENTS: pt. 1. The manifesto: action, reaction, and counterreaction.—pt. 2. "Freedoms great and small."

$1.50 per vol.
1. Civil rights—Europe, Eastern. 2. Civil rights—Czechoslovakia. I. Title. II. Series. III. Series: American Universities Field Staff. Fieldstaff reports.
D901.A54 vol. 22, no. 1-2 79-116555
081 s
[JC599.E92] 79 MARC

United States. Congress. House. Committee on the Judiciary.
Human rights and U.S. consular activities in Eastern Europe : report of the Committee on the Judiciary, House of Representatives, Ninety-fifth Congress, first session, based on a factfinding mission to four Eastern European nations. — Washington : U.S. Govt. Print. Off., 1978.
vi, 74 p. ; 24 cm.
At head of title: Committee print 95th Congress, 1st session, no. 12.
Includes bibliographical references.
1. Europe, Eastern—Emigration and immigration. 2. Civil rights—Europe, Eastern. I. Title.
JV7590.Z8 1978 325.47 78-602771
78 MARC

—EUROPEAN ECONOMIC COMMUNITY COUNTRIES

Pernice, Ingolf.
Grundrechtsgehalte im europäischen Gemeinschaftsrecht : e. Beitr. zum gemeinschaftsimmanenten Grundrechtsschutz durch d. Europ. Gerichtshof / Ingolf Pernice. — 1. Aufl. — Baden-Baden : Nomos-Verlagsgesellschaft, 1979.
270 p. ; 23 cm. — (Schriftenreihe europäische Wirtschaft ; Bd. 96) GFR79-A
A revision of the author's thesis, Augsburg, 1977/78.
Bibliography: p. 245-258.
Includes indexes.
ISBN 3-7890-0432-4 : DM74.00
1. Court of Justice of the European Communities. 2. Civil rights—European Economic Community countries. I. Title.
341.5'5 79-375952
79 MARC

—EUROPEAN ECONOMIC COMMUNITY COUNTRIES—CONGRESSES

Die Grundrechte in der Europäischen Gemeinschaft / mit Beitr. von Jochen Abr. Frowein ... [et al.]. — 1. Aufl. — Baden-Baden : Nomos Verlagsgesellschaft, 1978.
68 p. ; 23 cm. — (Schriftenreihe des Arbeitskreises Europäische Integration e.V. ; Bd. 2) GFR78-A
Papers presented at a meeting sponsored by the Arbeitskreis Europäische Integration June 2-4, 1977 in Luxembourg.
Includes bibliographical references.
ISBN 3-7890-0380-8
1. Civil rights—European Economic Community countries—Congresses. I. Frowein, Jochen Abraham, 1934- . II. Arbeitskreis Europäische Integration. III. Series: Arbeitskreis Europäische Integration. Schriftenreihe des Arbeitskreises Europäische Integration e.V. ; Bd. 2.
342'.4'085 78-384927
78 MARC

—FINLAND

Petrov, V fl. 1961–
(Dva berega—dva obraza zhizni)
Два берега—два образа жизни / В. Петров. — Таллин : Ээсти раамат, 1977.
200 p. ; 20 cm. USSR 77
Includes bibliographical references.
0.29rub
1. Finland—Social conditions. 2. Estonia—Social conditions. 3. Civil rights—Finland. 4. Civil rights—Estonia. I. Title.
HN533.5.P48 78-421981

—FRANCE

Brousse, Jean.
Contrôle juridictionnel des atteintes à la liberté / Jean Brousse, Lucien Remplon. — [Bordeaux] : École nationale de la magistrature, [1977]
150 p. ; 30 cm. F***
Cover title.
Bibliography: p. 149.
15.00F
1. Police—France. 2. Arrest—France. 3. Criminal procedure—France. 4. Civil rights—France. I. Remplon, Lucien, joint author. II. Title.
344'.44'052 79-338492
79 MARC

Brunois, Albert.
La liberté judiciaire, honneur des hommes / Albert Brunois. — Versailles : APIL, [1978]
283 p. ; 19 cm. F***
ISBN 2-85365-019-7 : 55.00F
1. Justice, Administration of—France. 2. Civil rights—France. I. Title.
342'.44'085 79-358294
79 MARC

Garnier, Jean Pierre.
Le socialisme à visage urbain : essai sur la local-démocratie / Jean-Pierre Garnier, Denis Goldschmidt. — Paris : Éditions Rupture, c1978.
443 p. ; 21 cm. F***
Includes bibliographical references.
ISBN 2-86220-006-9 : 41.00F
1. Cities and towns—France. 2. Middle classes—France. 3. Civil rights—France. I. Goldschmidt, Denis, joint author. II. Title.
HT135.G37 301.36'1'0944 79-355809
79 MARC

Roche, Jean, 1924-
Libertés publiques / par Jean Roche. — 5. éd. — Paris : Dalloz, 1978.
121 p. ; 24 cm. — (Mémentos Dalloz ; 3520) F***
Includes index.
ISBN 2-247-02170-0 : 26.00F
1. Civil rights—France. I. Title.
342'.44'085 79-339586
79 MARC

Vincent, François, fl. 1946-
Libertés publiques / François Vincent. — [Paris] : Ministère de l'économie et des finances, Direction du personnel et des services généraux, Centre de formation professionnelle et de perfectionnement, 1977.
198 p. ; 30 cm. F***
1. Civil rights—France. I. Title.
342'.44'085 79-341456
79 MARC

—FRANCE—ADDRESSES, ESSAYS, LECTURES

L'Exclusion sociale / Jean-Marie Albertini, Elie Alfandari, Jacques Attali, Jean-Michel Belorgey ... [etc.] ; sous la direction de Jean-Michel Belorgey et Jean-Jacques Dupeyroux. — Paris : Librairie sociale et économique, 1974.
194 p. : graphs ; 30 cm. — (Droit sociale ; 1974, no 11) F75-4208
Cover title.
"Numéro spécial."
Includes bibliographical references.
1. Public welfare—Law and legislation—France—Addresses, essays, lectures. 2. Minorities—France—Addresses, essays, lectures. 3. Civil rights—France—Addresses, essays, lectures. I. Albertini, Jean Marie. II. Belorgey, Jean Michel. III. Dupeyroux, Jean Jacques. IV. Series.
344'.44 s 79-378283
79 MARC

CIVIL RIGHTS (Continued)

—FRANCE—PUBLIC OPINION

La Liberté a la parole / Jean Sanitas. — Paris : Éditions sociales, c1978.
294 p. ; 22 cm. F***
ISBN 2-209-05291-2 : 45.00F
1. Civil rights—France—Public opinion. 2. France—Politics and government—1969- —Public opinion. 3. Parti communiste français—Public opinion. 4. Public opinion—France. I. Sanitas, Jean.
JC599.F8L52 79-350708
301.15'43'32340944
79 MARC

—GERMANY

Köppl, Peter, 1947-
Meinungsfreiheit und Arbeitsverhält[t]nis : eine Studie zur Grundrechtsdrittwirkung in der Weimarer Republik / vorgelegt von Peter Köppl. — [s.l. : s.n.], 1978 (München : Dissertationsdruck Schön)
viii, 218 p. ; 21 cm. GFR***
Thesis—Frankfurt am Main.
Vita.
Bibliography: p. 205-217.
1. Civil rights—Germany. 2. Liberty of speech—Germany. 3. Employees, Dismissal of—Germany. I. Title.
342'.43'085 79-372442
79 MARC

Laufs, Adolf.
Recht und Gericht im Werk der Paulskirche : [Vortrag 27. April 1978] / von Adolf Laufs. — 1. Aufl. — Heidelberg ; Karlsruhe : Müller Juristischer Verl., 1978.
52 p. ; 20 cm. — (Schriftenreihe - Juristische Studiengesellschaft Karlsruhe ; Heft 139) GFR78-A
Includes bibliographical references.
ISBN 3-8114-5378-5 : DM14.00
1. Rule of law—Germany. 2. Justice, Administration of—Germany. 3. Civil rights—Germany. I. Title. II. Series: Juristische Studiengesellschaft, Karlsruhe. Schriftenreihe ; Heft 139.
340.1'1 79-348116
79 MARC

—GERMANY—HISTORY

Gaile, Jochen.
Menschenrecht und bürgerliche Freiheit : Ideologiebildung im Zeichen d. entstehenden bürgerl. Gesellschaft 1789-1848 / Jochen Gaile. — Marburg : Verlag Arbeiterbewegung und Gesellschaftswiss., 1978.
309 p. ; 21 cm. — (Schriftenreihe für Sozialgeschichte und Arbeiterbewegung ; Bd. 15) GFR78-A
Bibliography: p. 289-309.
ISBN 3-921630-14-2 : DM19.80
1. Civil rights—Germany—History. 2. Natural law. 3. Sociological jurisprudence. I. Title. II. Series.
340.1'1 79-347840
79 MARC

—GERMANY, EAST

CDU/CSU-Fraktion des Deutschen Bundestages.
White paper on the human rights situation in Germany and of the Germans in Eastern Europe. — Bonn : CDU/CSU Group in the German Bundestag, 1977.
114 p. : ill. ; 30 cm. GFR***
1. Civil rights—Germany, East. 2. Civil rights—Europe, Eastern. 3. Germans in Eastern Europe. I. Title.
JC599.G32C13 1977 323.4'0943 78-323042
79 MARC

Freedom and democracy : for whom and for what? : the GDR presents its views / [translation, Intertext Berlin]. — Berlin : Panorama DDR, 1978.
78 p. : ill. ; 19 cm. — (First-hand information) GDR***
1. Civil rights—Germany, East. 2. Germany, East—Politics and government.
JC599.G32F73 323.4'0943'1 79-311569
79 MARC

Gegen kalten Krieg, für Förderung der Menschenrechte / [verantwortlich für die Redaktion, Siegfried Forberger]. — Berlin : DDR-Komitee für Menschenrechte, 1978.
60 p. ; 21 cm. — (Schriften und Informationen - DDR-Komitee für Menschenrechte ; Heft 3/78) GDR***
Cover title.
Includes bibliographical references.
1. Civil rights (International law) 2. Peace. 3. Civil rights—Germany, East. I. Forberger, Siegfried. II. Series: DDR-Komitee für Menschenrechte. Schriften und Informationen - DDR-Komitee für Menschenrechte ; 1978, Heft 3.
K3240.4.G44 341.48'1 79-363378
79 MARC

—GERMANY, EAST—CONGRESSES

Die Ausübung staatlicher Gewalt in Ost und West nach Inkrafttreten der UN-Konvention über zivile und politische Rechte : Juristenkonferenz vom 23.-25. September 1977 in Ludwigshafen am Rhein / Dt. Sekt. d. Internat. Juristen-Komm ; mit Vorträgen von Gottfried Zieger ... [et al.]. — Heidelberg ; Karlsruhe : Müller, Juristischer Verl., 1978.
103 p. ; 23 cm. — (Rechtsstaat in der Bewährung ; Bd. 6) GFR79-A
Conference organized by the Deutsche Sektion of the International Commission of Jurists.
"Entwurf einer Ergänzung zur UN-Menschenrechtskonvention": p. 101-103.
Includes bibliographical references.
ISBN 3-8114-6678-X : DM32.00
1. Civil rights (International law)—Congresses. 2. Civil rights—Germany, East—Congresses. 3. Civil rights—Europe—Congresses. 4. International covenant on civil and political rights—Congresses. I. Zieger, Gottfried. II. International Commission of Jurists (Founded 1952). Deutsche Sektion. III. Series.
K3239.6.A9 341.48'1 79-356493
79 MARC

Deutsch-Israelisches Juristensymposion, 2d, Munich, Ger., 1976.
Menschenrechte in Israel und Deutschland : ein Symposion der Gesellschaft zur Förderung der wissenschaftlichen Zusammenarbeit mit der Universität Tel Aviv / Gustav Stein (Hrsg.). — Köln : Verlag Wissenschaft und Politik, c1978.
175 p. ; 23 cm. GFR***
"Referate und die Diskussion beim zweiten deutsch-israelischen Symposion am 18./19. November in München."
Two articles in English.
Includes bibliographical references.
ISBN 3-8046-8550-1
1. Civil rights—Israel—Congresses. 2. Civil rights—Germany, West—Congresses. 3. Civil rights—Germany, East—Congresses. I. Stein, Gustav. II. Gesellschaft zur Förderung der Wissenschaftlichen Zusammenarbeit mit der Universität Tel Aviv. III. Title.
K3239.6.D48 1976 342.08'5 78-397418
79 MARC

—GERMANY, WEST

Berufsverbote, ein Staat sieht rot / [von Hans-Ulrich Knies, Klaus-Peter Löhr u. Hans-Peter Ruess]. — 3., durchges. u. erg. Aufl. — Berlin : DVK-Verlag, 1977.
75 p. : ill. ; 21 cm. GFR78-A13
"Sonderdruck 2"
Cover title.
Bibliography: p. 75.
ISBN 3-88107-010-9 : DM1.50
1. Discrimination in employment—Law and legislation—Germany, West. 2. Civil rights—Germany, West. 3. Justice and politics—Germany, West. I. Knies, Hans-Ulrich. II. Löhr, Klaus-Peter. III. Ruess, Hans-Peter.
79-353359
79 MARC

Bethge, Herbert.
Staatshaftung für den staatsfreien Rundfunk? : verfassungsverwaltungsrechtliche Probleme namentlich des Schicksals der Verbindlichkeiten des Norddeutschen Rundfunks als Folge der Kündigung des Staatsvertrages über den Norddeutschen Rundfunk / von Herbert Bethge. — Frankfurt am Main : A. Metzner, 1978.
103 p. ; 24 cm. — (Beiträge zum Rundfunkrecht ; Heft 18) GFR***
ISBN 3-7875-0418-4
1. Norddeutscher Rundfunk. 2. Radio—Law and legislation—Germany, West. 3. Civil Rights—Germany, West. I. Title. II. Series.
79-356299
79 MARC

Bethge, Herbert.
Zur Problematik von Grundrechtskollisionen / Herbert Bethge. — 1. Aufl. — München : Vahlen, 1977.
l, 431 p. ; 23 cm. — (Studien zum öffentlichen Recht und zur Verwaltungslehre ; Bd. 17) GFR77-A
Habilitationsschrift—Cologne, 1976.
Bibliography: p. xvii-[l]
ISBN 3-8006-0650-X
1. Civil rights—Germany, West. 2. Germany, West—Constitutional law. I. Title.
342'.43'085 78-377351
78 MARC

Cramer, Dettmar, 1929-
Bürgerrechte 77 [i.e. siebenundsiebzig] / Dettmar Cramer. — Köln : Verlag Wissenschaft u. Politik, 1977.
215 p. ; 19 cm. — (Bibliothek Wissenschaft und Politik) GFR78-A
Includes bibliographical references.
ISBN 3-8046-8540-4
1. Civil rights (International law) 2. Civil rights—Russia. 3. Civil rights—Germany, West. I. Title.
K3240.4.C66 341.48'1 78-379577
78 MARC

Grundrecht nach Bedarf : zur Situation d. Kriegsdienstverweigerer in d. Bundesrepublik Deutschland / hrsg. von Manfred Gillig u. Hansjörg Schultz. — 1. Aufl. — Freiburg i. Br. : Dreisam-Verlag, 1978.
286 p. ; 20 cm. GFR78-A
ISBN 3-921472-15-6 : DM9.80
1. Conscientious objectors—Law and legislation—Germany, West. 2. Civil rights—Germany, West. 3. Judicial review—Germany, West. I. Gillig, Manfred. II. Schultz, Hansjörg.
343'.43'012 79-364084
79 MARC

Herde, Klaus-Dieter.
Probleme der Pfandverfolgung / Klaus-Dieter Herde. — 1. Aufl. — Göttingen : Schwartz, 1978.
x, 182 p. ; 23 cm. — (Göttinger rechtswissenschaftliche Studien ; Bd. 101) GFR78-A
A revision of the author's thesis, Göttingen, 1977.
Bibliography: p. [171]-182.
ISBN 3-509-01029-9 : DM36.00
1. Pawnbroking—Germany, West. 2. Executions (Law)—Germany, West. 3. Civil rights—Germany, West. I. Title. II. Series.
78-396046
79 MARC

Internationales Russell Tribunal, Frankfurt am Main, 1978.
Zur Situation der Menschenrechte in der Bundesrepublik Deutschland : Dokumente, Verhandlungen, Ergebnisse / 3. Internat. Russell-Tribunal ; hrsg. vom Dt. Beirat u. Sekretariat d. 3. Internat. Russell-Tribunals. — Berlin : Rotbuch-Verlag, 1978-
v. ; 20 cm. — (Rotbuch ; 185-) GFR78-A
ISBN 3-88022-185-5 (v. 1) : DM8.00 (v. 1)
1. Civil rights—Germany, West. 2. Discrimination in employment Law and legislation—Germany, West. 3. Justice and politics—Germany, West. I. Internationales Russell Tribunal. Deutscher Beirat.
342'.43'085 78-393545
79 MARC

Katz, Alfred, Dr.
Grundkurs im öffentlichen Recht / Alfred Katz. — 3., völlig neu bearb. u. erw. Aufl., 7.-10. Tsd. — Königstein/Ts. : Athenäum-Verlag, 1978.
xx, 351 p. ; 21 cm. GFR78-A
Includes bibliographical references and indexes.
ISBN 3-7610-6187-0 : DM32.00. ISBN 3-7610-6217-6 pbk.
1. Public law—Germany, West. 2. Germany, West—Constitutional law. 3. Civil rights—Germany, West. I. Title.
342'.43 79-342581
79 MARC

Klas, Helmut.
Die Zulässigkeit der Verfassungsbeschwerde gegen Tarifnormen / Helmut Klas. — Frankfurt/Main : R. G. Fischer, 1979.
xli, 208 p. ; 21 cm. GFR79-A
Originally presented as the author's thesis, Münster, 1978.
Bibliography: p. xii-xli.
ISBN 3-88323-043-X : DM36.00
1. Collective labor agreements—Germany, West. 2. Appellate procedure—Germany, West. 3. Civil rights—Germany, West. I. Title.
79-372216
79 MARC

Lameyer, Johannes.
Streitbare Demokratie : e. verfassungshermeneut. Unters. / von Johannes Lameyer. — Berlin : Duncker und Humblot, 1978.
226 p. ; 24 cm. — (Schriften zum öffentlichen Recht ; Bd. 336) GFR78-A
Originally presented as the author's thesis, Göttingen, 1977.
Bibliography: p. [210]-226.
ISBN 3-428-04062-7 : DM76.00
1. Civil rights—Germany, West. 2. Judicial review—Germany, West. 3. Germany, West—Constitutional law—Interpretation and construction. 4. Internal security—Germany, West. I. Title.
342'.43'02 78-379147
78 MARC

Langner, Albrecht.
Menschenrechte, Staat, Gesellschaft / von Albrecht Langner. — Köln : Bachem, 1975.
48 p. ; 21 cm. — (Katholische Soziallehre in Text und Kommentar ; Heft 3) GFR76-A
Bibliography: p. [48]
ISBN 3-7616-0290-1 : DM1.00
1. Civil rights—Germany, West. 2. Civil rights. I. Title. II. Series.
342'.43'085 79-349627
79 MARC

Mikinovic, Stephan, 1949-
Strafprozess und Herrschaft : e. empir. Unters. zur Korrektur richterl. Entscheidungen / von Stephan Mikinovic, Wolfgang Stangl. — Neuwied ; Darmstadt : Luchterhand, 1978.
190 p. ; 18 cm. — (Demokratie und Rechtsstaat ; Bd. 40) GFR78-A
Bibliography: p. 176-187.
ISBN 3-472-16011-X : DM16.80
1. Judgments, Criminal—Germany, West. 2. Judicial power—Germany, West. 3. Civil rights—Germany, West. I. Stangl, Wolfgang, 1949- joint author. II. Title.
345'.43'077 79-364080
79 MARC

Moltmann, Jürgen.
Menschenwürde, Recht und Freiheit / Jürgen Moltmann. — 1. Aufl., (1-8. Tsd.). — Stuttgart ; Berlin : Kreuz-Verlag, 1979.
95 p. ; 19 cm. GFR79-A
ISBN 3-7831-0574-9 : DM6.80
1. Civil rights—Germany, West. 2. Dignity. I. Title.
342'.43'085 79-359638
79 MARC

Rechtskundlicher Leitfaden für den Werkschutz / hrsg. von Albert Riester ; bearb. von Eugen Gerbis u. Horst Neukirchner. — 1. Aufl. — Heidelberg : Kriminalistik-Verlag, 1978.
120 p. ; 21 cm. — (Kriminalistik Fachbücherei : Wirtschaftsschutz) GFR78-A
Includes index.
ISBN 3-7832-0578-6 : DM12.80
1. Civil rights—Germany, West. 2. Industrial safety—Law and legislation—Germany, West. 3. Criminal law—Germany, West. I. Riester, Albert. II. Gerbis, Eugen. III. Neukirchner, Horst.
78-394600
79 MARC

CIVIL RIGHTS
—GERMANY, WEST (Continued)

Schatzschneider, Wolfgang, 1945-
Ermittlungstätigkeit der Ämter für Verfassungsschutz und Grundrechte / Wolfgang Schatzschneider. — Frankfurt a.M. ; Bern ; Las Vegas : Lang, 1979.
xiii, 305 p. ; 21 cm. — (Frankfurter öffentlich-rechtliche Studien ; Bd. 4) GFR79-A
Bibliography: p. 300-305.
ISBN 3-8204-6327-5
1. Governmental investigations—Germany, West. 2. Civil rights—Germany, West. 3. Privacy, Right of—Germany, West. 4. Germany, West—Constitutional law. I. Title. II. Series.
79-368966
79 MARC

Schmidt-Bleibtreu, Bruno.
Kommentar zum Grundgesetz für die Bundesrepublik Deutschland / Bruno Schmidt-Bleibtreu, Franz Klein. — 4. Aufl. — Neuwied ; Darmstadt : Luchterhand, 1977.
1252 p. ; 24 cm. GFR77-A45
Includes the text of the Grundgesetz.
Includes bibliographical references and index.
ISBN 3-472-31003-0 : DM128.00
1. Germany, West—Constitutional law. 2. Civil rights—Germany, West. I. Klein, Friedrich, 1908- joint author. II. Germany (Federal Republic, 1949-). Constitution. 1977. III. Title.
342'.43'085 78-370650
78 MARC

Schneider, Harald, Dr.
Die Güterabwägung des Bundesverfassungsgerichts bei Grundrechtskonflikten : empir. Studie zu Methode u. Kritik e. Konfliktlösungsmodelles / Harald Schneider. — 1. Aufl. — Baden-Baden : Nomos-Verlagsgesellschaft, 1979.
263 p. ; 23 cm. — (Studien und Materialien zur Verfassungsgerichtsbarkeit ; Bd. 11)
Bibliography: p. 255-263.
ISBN 3-7890-0409-X
1. Civil rights—Germany, West. 2. Judicial review—Germany, West. 3. Germany (Federal Republic, 1949-). Bundesverfassungsgericht. I. Title. II. Series.
342'.43'085 79-359640
79 MARC

Schwabe, Jürgen.
Probleme der Grundrechtsdogmatik / Jürgen Schwabe. — Darmstadt : J. Schwabe ⌈Selbstverl.⌉, 1977.
vii, 482 p. ; 21 cm. GFR78-B1
Bibliography: p. 448-467.
Includes indexes.
DM15.00
1. Civil rights—Germany, West. 2. Civil rights. I. Title.
342'.43'085 78-387927
79 MARC

Wittkowski, Wolfram, 1949-
Der Schutz der Arbeitskraft durch das Grundgesetz : zugl. e. Beitr. zum Sozialstaatsprinzip / Wolfram Wittkowski. — Frankfurt am Main ; Bern ; Las Vegas : Lang, 1979.
xlvii, 249 p. ; 21 cm. — (Europäische Hochschulschriften : Reihe 2, Rechtswissenschaft ; Bd. 214) GFR79-A
Bibliography: p. vi-xlvii.
ISBN 3-8204-6523-5
1. Civil rights—Germany, West. 2. Labor supply—Germany, West. 3. Germany, West—Constitutional law. I. Title. II. Series.
342'.43'085 79-372853
79 MARC

World Council of Peace. Human Rights Commission.
Professional ban in the Federal Republic of Germany : an enquiry by the World Peace Council's Human Rights Commission. — Helsinki : Information Centre of the World Peace Council, ⌈1978?⌉
62 p. : ill. ; 16 cm. Fi***
1. Germany, West—Officials and employees—Appointments, qualifications, tenure, etc. 2. Discrimination in employment—Germany, West. 3. Civil rights—Germany, West. I. Title.
JN3971.A69A69 331.2 79-315548
79 MARC

—GERMANY, WEST—ADDRESSES, ESSAYS, LECTURES

Berufsverbote : Stimmen der internationalen Presse : Berichte und Analysen = International press views : reports and analyses = Vues de la presse internationale : reports et analyses / ⌈hrsg. vom Aktionskomitee gegen Berufsverbote an der Freien Universität Berlin und vom Komitee zur Verteidigung Demokratischer Grundrechte⌉. — Berlin : Aktionskomitee gegen Berufsverbote an der Freien Universität, 1976.
80 p. : ill. ; 30 cm. GFR***
Cover title.
Text also in Danish, Dutch, English, French, Italian, Norwegian, or Swedish.
Bibliography: p. 80.
1. Germany, West—Officials and employees—Appointment, qualifications, tenure, etc.—Addresses, essays, lectures. 2. Civil rights—Germany, West—Addresses, essays, lectures. 3. Germany, West—Foreign opinion—Addresses, essays, lectures. I. Aktionskomitee gegen Berufsverbote an der Freien Universität Berlin. II. Komitee zur Verteidigung Demokratischer Grundrechte. III. Title: International press views. IV. Title: Vues de la presse internationale.
JN3971.A69E453 354'.43'0013 78-366240
78 MARC

Russell-Tribunal pro und contra : Dokumente zu e. gefährl. Kontroverse / hrsg. von Freimut Duve u. Wolf-Dieter Narr. — Orig.-Ausg. — Reinbek bei Hamburg : Rowohlt, 1978.
157 p. ; 19 cm. — (Rororo aktuell) GFR78-A23
ISBN 3-499-14278-3 : DM4.80
1. Civil rights—Germany, West—Addresses, essays, lectures. 2. Civil rights—Addresses, essays, lectures. 3. Internationales Russell-Tribunal, Frankfurt am Main, 1978—Addresses, essays, lectures. I. Duve, Freimut, 1936- . II. Narr, Wolf-Dieter, 1937- .
342'.43'085 79-378245
79 MARC

Wo Recht zu Unrecht wird, wird Widerstand zur Pflicht : Protokoll e. Podiumsdiskussion zur polit. Unterdrückung in Deutschland / Konrad Boehmer ... ⌈et al.⌉. — 1.-3. Tsd. — Berlin : Oberbaumverlag, 1978.
98 p. : 1 ill. ; 20 cm. GFR78-A
ISBN 3-87628-139-3 : DM8.00
1. Opposition (Political science)—Addresses, essays, lectures. 2. Government, Resistance to—Addresses, essays, lectures. 3. Civil rights—Germany, West—Addresses, essays, lectures. I. Boehmer, Konrad, 1941- .
JC328.3.W6 322.4 78-393040
79 MARC

Zur Verfassung unserer Demokratie : 4 republikan. Reden / Max Güde ... ⌈et al.⌉. — Orig.-Ausg. — Reinbek bei Hamburg : Rowohlt, 1978.
92 p. ; 19 cm. — (Rororo aktuell) GFR78-A
A collection of speeches; L. Raiser's speech was given Nov. 9, 1977, the others in Jan. 1978.
CONTENTS: Güde, M. Die Verwirrung unseres Staatsschutzrechts.—Raiser, L. Krise der Demokratie?—Simon, H. Die rechts- und sozialstaatliche Demokratie.—Weizsäcker, C. F. von. Die Verteidigung der Freiheit.
ISBN 3-449-14279-1 : DM3.80
1. Germany, West—Constitutional history—Addresses, essays, lectures. 2. Civil rights—Germany, West—Addresses, essays, lectures. 3. Terrorism—Addresses, essays, lectures. I. Güde, Max, 1902- Die Verwirrung unseres Staatsschutzrechts. 1978.
342'.43 79-347722
79 MARC

—GERMANY, WEST—CONGRESSES

Deutsch-Israelisches Juristensymposion, 2d, Munich, Ger., 1976.
Menschenrechte in Israel und Deutschland : ein Symposion der Gesellschaft zur Förderung der wissenschaftlichen Zusammenarbeit mit der Universität Tel Aviv / Gustav Stein (Hrsg.). — Köln : Verlag Wissenschaft und Politik, c1978.
175 p. ; 23 cm. GFR***
"Referate und die Diskussion beim zweiten deutsch-israelischen Symposion am 18./19. November in München."
Two articles in English.
Includes bibliographical references.
ISBN 3-8046-8550-1
1. Civil rights—Israel—Congresses. 2. Civil rights—Germany, West—Congresses. 3. Civil rights—Germany, East—Congresses. I. Stein, Gustav. II. Gesellschaft zur Förderung der Wissenschaftlichen Zusammenarbeit mit der Universität Tel Aviv. III. Title.
K3239.6.D48 1976 342.08'5 78-397418
79 MARC

Menschenrechte : Aspekte ihrer Begründung und Verwirklichung : Vorträge und Diskussionen des Tübinger interdisziplinären Kolloquiums vom 23. bis 25. September 1977 / Johannes Schwartländer (Hrsg.) ; mit Beiträgen von Ernst-Otto Czempiel ... ⌈et al.⌉. — Tübingen : Attempto, 1978.
241 p. ; 21 cm. — (Tübinger Universitätsschriften ; Bd. 1) GFR78-A47
Includes bibliographical references and indexes.
ISBN 3-921552-04-4 : DM17.80
1. Civil rights—Congresses. 2. Civil rights—Germany, West—Congresses. I. Schwartländer, Johannes, 1922- II. Series.
K3239.6.M46 323.4 79-344488
79 MARC

—GERMANY, WEST—SAARLAND

Saarland. Constitution.
Die Verfassung des Saarlandes, erläutert von Alfons Senz und Hans Treinen. Das Grundgesetz für die Bundesrepublik Deutschland, erläutert von Friedrich Gries und Rolf-Joachim Sattler. ⌈5. verb. und erg. Aufl. Düsseldorf, Lehrmittelverlag W. Hagemann ⌈1961, c1960⌉
128 p. illus. 21 cm.
"Der Aufbau der Bundesrepublic Deutschland" and "Das Saarland": ⌈1⌉ fold. sheet in pocket.
Includes bibliographical references.
1. Saarland—Constitutional law. 2. Civil rights—Germany, West—Saarland. I. Senz, Alfons, ed. II. Gries, Friedrich, ed. III. Germany (Federal Republic, 1949-). Constitution. ⌈1961, c1960⌉ IV. Title.
77-480583
78 MARC

—GREAT BRITAIN

Street, Harry, 1919-
The state of civil liberties in England today / by Harry Street. — Manchester : John Rylands University Library of Manchester, 1978.
p. 232-249 ; 25 cm. GB***
"Reprinted from the Bulletin of the John Rylands University Library of Manchester, vol. 61, no. 1, autumn, 1978."
£0.70
1. Civil rights—Great Britain. I. John Rylands Library, Manchester. Bulletin. II. Title.
JC599.G7S76 323.4'0941 79-314131
00 MARC

—GREAT BRITAIN—ADDRESSES, ESSAYS, LECTURES

The Coming confrontation : will the open society survive to 1989? / W. H. Chaloner ... ⌈et al.⌉. — London : Institute of Economic Affairs ; Levittown, N.Y. : Transatlantic Arts, sole distributor for the U.S.A., 1978.
xxi, 250 p. ; 19 cm. — (Hobart paperback ; no. 12 ISSN 0309-1783) GB***
Includes bibliographical references and index.
ISBN 0-255-36115-7
1. Civil rights—Great Britain—Addresses, essays, lectures. 2. Great Britain—Economic policy—1945- —Addresses, essays, lectures. 3. Great Britain—Social policy—Addresses, essays, lectures. I. Chaloner, William Henry.
JC599.G7C6 309.1'41'0857 79-670077
79 MARC

In defence of freedom : essays / by Winston S. Churchill ... ⌈et al.⌉ ; edited by K. W. Watkins. — London : Cassell, 1978.
180 p. ; 23 cm. GB***
Includes bibliographical references.
CONTENTS: Lord De L'Isle. Freedom and the constitution.—Gouriet, J. Freedom and enterprise.—Haseler, S. Freedom and the trade unions.—McWhirter, N. Freedom of choice.—Lady Morrison of Lambeth. Freedom and the family.—Lewis, R. Freedom of speech and publication.—Churchill, W. S. Freedom and national security.—Saroop, N. Freedom and race.—Watkins, K. W. The challenge to freedom.—Moss, R. The defence of freedom.—Flew, A. The philosophy of freedom.
1. Civil rights—Great Britain—Addresses, essays, lectures. 2. Liberty—Addresses, essays, lectures. I. Churchill, Winston Spencer, 1940- . II. Watkins, K. W.
JC599.G715 323.4'0941 78-325731
79 MARC

MacCormick, Neil.
Civil liberties and the law / ⌈by⌉ Neil MacCormick. — Edinburgh : Heriot-Watt University, 1977.
⌈4⌉, 20 p. ; 21 cm. — (A Heriot-Watt University lecture) GB79-00173
Includes bibliographical references.
£0.60
1. Civil rights—Great Britain—Addresses, essays, lectures. 2. Civil rights—Addresses, essays, lectures. I. Title. II. Series: Heriot-Watt University. A Heriot-Watt University lecture.
KD4080.Z9M3 342.08'5 79-318361
79 MARC

—GREAT BRITAIN—BIBLIOGRAPHY

Dworaczek, M
Human rights : government documents held in the Research Library / Marian Dworaczek. — 2d ed. — Toronto : Ontario Ministry of Labour, Research Library, 1975.
61 p. ; 30 cm. — (Bibliography series - Ministry of Labour, Research Branch ; no. 1) C***
1. Civil rights—Canada—Bibliography. 2. Civil rights—Great Britain—Bibliography. 3. Civil rights—United States—Bibliography. 4. Canada—Government publications—Bibliography. 5. Great Britain—Government publications—Bibliography. 6. United States—Government publications—Bibliography. I. Ontario. Ministry of Labour. Library. II. Title. III. Series: Ontario. Ministry of Labour. Research Branch. Bibliography series - Ministry of Labour, Research Branch ; no. 1.
K3236.D9 1975 016.3234 79-301630
79 MARC

—GUADELOUPE

Les Libertés en Guadeloupe : ⌈colloque de la Ligue des droits de l'homme et du citoyen, Section de la Guadeloupe, Abymes, 25 avril 1976⌉. — ⌈s.l.⌉ : Jeunes Antilles, ⌈1976⌉
136 p. ; 18 cm. — (Collection Batailles ; no 2) F***
1. Civil rights—Guadeloupe. I. Ligue des droits de l'homme et du citoyen. Section de la Guadeloupe. II. Series.
JC599.G78L5 323.4'09729'76 79-337084
79 MARC

—INDIA

Sharma, Akhileshwar.
The Supreme Court of India, as the guardian of fundamental rights / Akhileshwar Sharma. — 1st ed. — Muzaffarpur : Manisha, 1977.
ii, ii, 196 p. ; 23 cm.
A revision of the author's thesis, University of Bihar, 1968.
Bibliography: p. ⌈186⌉-196.
Rs50.00
1. Civil rights—India. 2. India. Supreme Court. I. Title.
342'.54'085 78-904860
79 MARC

Sunder Raman.
Fundamental rights and the 42nd Constitutional amendment / Sunder Raman. — Calcutta : Minerva Associates (Publications), 1977.
ix, 104 p. ; 23 cm.
Bibliography: p. ⌈99⌉-101.
Includes index.
Rs25.00
1. Judicial review—India. 2. India—Constitutional law—Amendments. 3. Civil rights—India. I. Title.
342'.54'085 77-907434
79 MARC

CIVIL RIGHTS (Continued)

—INDIA—ADDRESSES, ESSAYS, LECTURES

Khanna, Hans Raj, 1912-
Constitution and civil liberties / H. R. Khanna. — New Delhi : Radha Krishna on behalf of the Institute of Constitutional and Parliamentary Studies, 1978.
94 p. ; 23 cm. — (Dr. B. R. Ambedkar memorial lectures)
Includes bibliographical references.
Rs25.00
1. Civil rights—India—Addresses, essays, lectures. 2. India—Constitutional law—Addresses, essays, lectures. I. Title. II. Series.
342'.54'085 79-900122
79 MARC

—INDIA—GOA

Cunha, T B
Pelo nacionalismo e pelos direitos elementares / T. Bragança-Cunha. — [Calcutta]: Bragança-Cunha, [pref. 1929]
ii, 18 p. ; 23 cm.
Cover title.
1. Civil rights—India—Goa. 2. Liberty of the press—India—Goa. I. Title.
JC599.I4C85 323.4'0954'799 78-912288
79 MARC

—IRAN

Lafue-Veron, Madeleine.
Voyage au pays de la peur, Iran 1978 / Madeleine Lafue-Veron. Vingt-huit intellectuels de Téhéran signent une lettre présentée au secrétaire général de l'Organisation des Nations Unies. — Genève : Comité suisse de défense des prisonniers politiques iraniens, [1978]
[36] p. ; 21 cm. Sw***
Cover title.
8.00F
1. Iran—Politics and government—1945- 2. Civil rights—Iran. I. Title.
DS318.L33 320.9'55'05 79-369489
79 MARC

—ISRAEL

Kerber, Nathalie Marguerite.
Les droits de l'homme dans les territoires administrés par Israël / Nathalie Marguerite Kerber. — Paris : A. Pedone, c1978.
239 p. ; 24 cm. F***
Bibliography: p. [231]-235.
ISBN 2-233-00055-2 : 60.00F
1. Civil rights—Israel. 2. Israel-Arab War, 1967—Occupied territories. I. Title.
342'.5694'085 79-352668
79 MARC

—ISRAEL—CONGRESSES

Deutsch-Israelisches Juristensymposion, 2d, Munich, Ger., 1976.
Menschenrechte in Israel und Deutschland : ein Symposion der Gesellschaft zur Förderung der wissenschaftlichen Zusammenarbeit mit der Universität Tel Aviv / Gustav Stein (Hrsg.). — Köln : Verlag Wissenschaft und Politik, c1978.
175 p. ; 23 cm. GFR***
"Referate und die Diskussion beim zweiten deutsch-israelischen Symposion am 18./19. November in München."
Two articles in English.
Includes bibliographical references.
ISBN 3-8046-8550-1
1. Civil rights—Israel—Congresses. 2. Civil rights—Germany, West—Congresses. 3. Civil rights—Germany, East—Congresses. I. Stein, Gustav. II. Gesellschaft zur Förderung der Wissenschaftlichen Zusammenarbeit mit der Universität Tel Aviv. III. Title.
K3239.6.D48 1976 342.08'5 78-397418
79 MARC

—ITALY

Rodotà, Stefano.
Alla ricerca delle libertà / Stefano Rodotà. — Bologna : Il mulino, 1978.
299 p. ; 19 cm. — (Universale paperbacks Il mulino ; 83) It78-De
L5000
1. Civil rights—Italy. 2. Internal security—Italy. 3. Judges—Italy. I. Title.
JN5591.R58 323.4'0945 79-350688
*79 MARC

—ITALY—ADDRESSES, ESSAYS, LECTURES

Stame, Federico.
Società civile e critica delle istituzioni / Federico Stame. — 1. ed. — Milano : Feltrinelli, 1977.
87 p. ; 17 cm. — (Opuscoli marxisti ; 19) It79-Jan
L1000
1. State, The—Addresses, essays, lectures. 2. Political rights—Italy—Addresses, essays, lectures. 3. Civil rights—Italy—Addresses, essays, lectures. 4. Partito comunista italiano—Addresses, essays, lectures. I. Title.
JC266.S68 320.1 79-357533
*79 MARC

—JAPAN

(Hōgaku yōsetsu)
法学要説 / 手塚豊監修. — 東京：慶応通信, 1978.
2, 8, 252, xi p. ; 21 cm. Ja 78-11463
Bibliography: p. 249-252.
Includes index.
"大日本帝国憲法・日本国憲法": p. 229-248.
¥1500
1. Japan—Constitutional law. 2. Civil rights—Japan. 3. Jurisprudence. I. Tezuka, Yutaka, 1911- II. Japan. Constitution. 1978.
78-812215

Ienaga, Saburō, 1913-
(Rekishi no naka no kempō)
歴史のなかの憲法 / 家永三郎. — 東京：東京大学出版会, 1977.
2 v. (810 p.) ; 19 cm. Ja 77-23120 (v. 1)
Includes bibliographical references.
¥1500 per vol.
1. Japan—Constitutional law. 2. Japan—Constitutional history. 3. Civil rights—Japan. I. Title.
78-800732

(Kempō)
憲法 / 芦部信喜編. — 東京：有斐閣, 1978-
v. ; 22 cm. — (有斐閣大学双書) Ja 78-25197 (v. 2)
Series romanized: Yūhikaku daigaku sōsho.
Bibliography: v. 2, p. vii-x.
Includes index.
CONTENTS: —2. 人権
¥3800 (v. 2)
1. Japan—Constitutional law. 2. Civil rights—Japan. I. Ashibe, Nobuyoshi, 1923-
79-801667

(Kihonteki jinken no kenkyū)
基本的人権の研究 / 鈴木安藏 ... [et al.] 著. — 東京：勁草書房, 1954.
2, 218 p. ; 21 cm. — (政治學研究叢書 ; 3)
Bibliography: p. [209]-218.
1. Civil rights—Japan. 2. Japan—Constitutional law. I. Suzuki, Yasuzō, 1904- II. Series: Seijigaku kenkyū sōsho ; 3.
79-800961

Mikami, Makio.
(Kihonteki jinken gairon)
基本的人権概論 / 三上眞幾雄著. — 東京：石崎書店, 昭和26 [1951]
2, 2, 5, 192 p. ; 19 cm.
1. Civil rights—Japan. 2. Japan—Constitutional law. I. Title.
79-800466

Ōnishi, Norishige, 1928-
(Kempō to gendaihō)
憲法と現代法 / 大西典茂, 前田繁一著. — 京都：晃洋書房, 1978.
xii, 191, 18 p. ; 22 cm. Ja 78-15492
"日本国憲法": p. 1-18 (2d group)
Includes bibliographies.
¥1900
1. Japan—Constitutional law. 2. Civil rights—Japan. I. Maeda, Shigekazu, 1932- joint author. II. Japan. Constitution. III. Title.
78-812592

Seki, Itaru, 1903-
(Jinken yōgo iin hō shakugi)
人権擁護委員法釈義 / 関之著. — 東京：新警察社, 昭和24 [1949]
4, 138 p. ; 18 cm.
1. Civil rights—Japan. 2. Civil rights workers—Japan. I. Japan. Laws, statutes etc. Jinken yōgo iin hō. 1949. II. Title.
79-801776

(Tachiagaru nanchōsha)
立ちあがる難聴者 / 全国難聴者連絡協議会編. — 東京：たいまつ社, 1978.
212 p. : ill. ; 17 cm. — (たいまつ新書 ; 43) Ja 79-5350
Series romanized: Taimatsu shinsho.
¥680
1. Deaf—Japan. 2. Civil rights—Japan. 3. Citizens' associations—Japan. I. Zenkoku Nanchōsha Renraku Kyōgikai
HV2883.T3 79-808641

Unno, Shinkichi, 1885-1968.
(Jinken no hōritsu sōdan)
人権の法律相談 / 海野普吉, 森川金壽著. — 東京：日本評論新社, 昭和28 [1953]
6, 4, 305 p. ; 19 cm.
Colophon inserted.
Includes bibliographical references.
1. Civil rights—Japan. 2. Remedies (Law)—Japan. 3. Justice, Administration of—Japan. I. Morikawa, Kinju, 1913- joint author. II. Title.
79-802881

Wagatsuma, Sakae, 1897-1973.
(Shin Kempō to kihonteki jinken)
新憲法と基本的人權 / 我妻榮 ; [編者 憲法普及會]. — 東京：國立書院, 昭和23 [1948]
4, 224 p. ; 19 cm.
Includes bibliographical references.
1. Civil rights—Japan. 2. Japan—Constitutional law. I. Title.
79-801859

Yamamoto, Yoshimori.
(Anata o futō na hakugai kara mamoru jinken no chishiki)
あなたを不当な迫害から護る人権の知識 / 山本嘉盛著 ; 吉川大二郎序 ; [編集 社会生活研究会]. — 東京：経済法律新聞社, 昭和34 [1959], 昭和35 [1960] printing.
4, 6, 254 p., [1] leaf of plates : ill. ; 18 cm.
1. Civil rights—Japan. 2. Justice, Administration of—Japan. I. Title. II. Title: Jinken no chishiki.
78-816527

—JAPAN—ADDRESSES, ESSAYS, LECTURES

(Enzai kokuhatsu! kenryoku no ronri to kōzō)
冤罪告発！権力の論理と構造 / [編集 民権人権確立委員会]. — 東京：谷沢書房, 1977, 1978 printing.
88 p. : ill. ; 21 cm. — (民権人権双書 ; 1) Ja***
"[1977年] 七月二十四日豊島公会堂で開催された「冤罪告発！民権人権確立大集会」...の講演の一部"
1. Criminal justice, Administration of—Japan—Addresses, essays, lectures. 2. Civil rights—Japan—Addresses, essays, lectures. 3. Malicious prosecution—Japan—Addresses, essays, lectures. I. Minken Jinken Kakuritsu Iinkai. II. Series: Minken jinken sōsho ; 1.
HV9813.E59 79-804188

Nihon Bengoshi Rengōkai. Jinken Yōgo Iinkai.
(Jinken jiken keikoku yōbō reishū)
人権事件警告・要望例集 / 日本弁護士連合会 [人権擁護委員会編]. — 東京：同委員会, 昭和52 [1977]
15, 980 p. ; 21 cm. Ja 78-24
1. Civil rights—Japan—Addresses, essays, lectures. 2. Criminal justice, Administration of—Japan—Addresses, essays, lectures. I. Title.
78-812410

CIVIL RIGHTS
—JAPAN—ADDRESSES, ESSAYS, LECTURES (Continued)

Numata, Inejirō, 1914–
(Gendai minshushugi ron)
現代民主主義論：人権と平和の思想 / 沼田稲次郎著. — 東京：労働旬報社, 1978.
271, 4 p. ; 20 cm. Ja 79-4424
Includes bibliographical references and index.
¥2000

1. Civil rights—Japan—Addresses, essays, lectures. 2. Democracy—Addresses, essays, lectures. 3. Peace—Addresses, essays, lectures. I. Title.

79-806778

Taketani, Mitsuo, 1911–
(Tokken to jinken)
特権と人権：不確実性を超える論理 / 武谷三男編著. — 東京：勁草書房, 1979.
9, 240 p. ; 19 cm. Ja 79-4939
¥1300

1. Civil rights—Japan—Addresses, essays, lectures. 2. Mass society—Addresses, essays, lectures. 3. Socialism in Japan—Addresses, essays, lectures. I. Title.

JQ1681.T34 79-808324

—JAPAN—HISTORY

(Jiyūtō gekika jiken to Koike Isamu)
自由党激化事件と小池勇 / 村上貢編著. — 名古屋：風媒社, 1976.
216 p. : ill. ; 20 cm. Ja***
Bibliography: p. 214–216.
CONTENTS: 小池勇　自叙伝　幽居余情　明治十九年日誌　塵の海—小池龍雄　祖父の思い出—小池勇年譜—小池勇関係史料—静岡事件参画者・小池勇の半生—『経世新誌』と『学事新報』について—自由党激化諸事件と一斉蜂起論
¥1500

1. Koike, Isamu, 1854–1940. 2. Politicians—Japan—Biography. 3. Civil rights workers—Japan—Biography. 4. Japan—Politics and government—1868–1912. 5. Civil rights—Japan—History. 6. Jiyūtō (Founded 1881)—History. I. Koike, Isamu, 1854–1940. II. Murakami, Mitsugi, 1926–

DS884.K565J59 78-812219

Murakawa, Ichirō, 1939–
(Nihon hoshutō shōshi)
日本保守党小史：自由民権と政党政治 / 村川一郎. — [東村山]：教育社；東京：販売　教育社出版サービス, [1978]
266 p., [1] leaf of plates : ill. ; 18 cm. — (歴史新書 ; 121) Ja 78-25187
Bibliography: p. 265–266.
¥600
1. Political parties—Japan—History. 2. Conservatism—Japan—History. 3. Civil rights—Japan—History. I. Title. II. Series: Rekishi shinsho ; 121.
JQ1698.A1M87 79-801239

Seki, Itaru, 1903–
(Jinken sengen konjaku monogatari)
人権宣言今昔物語 / 關之著. — 東京：聯合出版社, 昭和25 [1950]
3, 4, 201 p. ; 19 cm.

1. Civil rights—Japan—History. 2. Civil rights—History. I. Title.
JQ1681.S432 79-807249

Shimizu, Kimpachi.
(Minken hattatsu shi)
民権發達史 / 清水金八著. — 大垣：西濃商業學校出版部：發賣所　昭文社, 昭和8 [1933]
427 p. ; 23 cm.

1. Civil rights—Japan—History. 2. Japan—Politics and government—1868– I. Title.
JQ1681.S454 79-802288

Shōji, Kichinosuke, 1905–
(Kindai chihō minshū undō shi)
近代地方民衆運動史 / 庄司吉之助著. — 東京：校倉書房, 1978.
2 v. : ill. ; 20 cm. — (校倉歴史選書) Ja 79-1400 (v. 1)
¥2000 (v. 1) ¥2500 (v. 2)

1. Japan—History—1868– 2. Social movements—Japan—History. 3. Civil rights—Japan—History. I. Title. II. Series: Azekura rekishi sensho.
DS881.95.S47 79-804534

Suzuki, Yasuzō, 1904–
(Jiyū minken)
自由民權　鈴木安藏著　東京　白揚社　昭和23 [1948]
4, 4, 494 p. 22 cm.

1. Civil rights—Japan—History. 2. Japan—Constitutional history. I. Title.
J 63-406

Suzuki, Yasuzō, 1904–
(Jiyū minken, Kempō happu)
自由民權・憲法發布 / 鈴木安藏著. — 東京：白揚社, 昭和14 [1939]
9, 321, 15 p., [1] leaf of plates : ill. ; 23 cm. — (近代日本歴史講座 ; 第3冊)
Series romanized: Kindai Nihon rekishi kōza.

1. Civil rights—Japan—History. 2. Japan—Constitutional history. 3. Political parties—Japan—History. I. Title.
79-802216

Suzuki, Yasuzō, 1904–
(Kihonteki jinken)
基本的人権 / 鈴木安藏著. — 東京：実教出版, 昭和26 [1951]
2, 7, 390 p. ; 19 cm.
Includes bibliographical references.

1. Civil rights—History. 2. Civil rights—Japan—History. I. Title.
K3240.4.S89 79-801124

Takahashi, Tetsuo, 1918–
(Kōno Hironaka)
河野広中 / 高橋哲夫. — 会津若松：歴史春秋社, 昭和52 [1977]
2 v. : ill. ; 20 cm. — (福島人物の歴史 ; 第11-12巻) Ja 77-16677 (v. 1)
¥2200 per vol.

1. Kōno, Hironaka, 1849-1923. 2. Statesmen—Japan—Biography. 3. Japan—Politics and government—1868-1912. 4. Civil rights—Japan—History. I. Title. II. Series: Fukushima jimbutsu no rekishi ; dai 11-12-kan.
DS884.K6T34 78-813716

—JAPAN—PUBLIC OPINION

Japan. Sōrifu. Naikaku Sōri Daijin Kambō.
(Jinken yōgo ni kansuru seron chōsa)
人権擁護に関する世論調査：昭和53年2月調査. — [東京]：内閣総理大臣官房広報室, 昭和53 [1978]
123 p. : ill. ; 25 cm. — (世論調査報告書 - 内閣総理大臣官房) Ja***
Cover title.
1. Civil rights—Japan—Public opinion. 2. Legal aid—Japan—Public opinion. 3. Justice, Administration of—Japan—Public opinion. 4. Public opinion—Japan. I. Title. II. Series: Japan. Sōrifu. Naikaku Sōri Daijin Kambō. Seron chōsa hōkokusho.
JQ1681.J37 1978 79-800629

—JAPAN—POPULAR WORKS

Togano, Taiji, 1926–
(Hankenryoku no hōritsu chishiki)
反権力の法律知識 / 栂野泰二編著. — 東京：批評社：発売　社会評論社, 1979.
270 p. ; 19 cm. Ja 79-5265
¥1000

1. Civil rights—Japan—Popular works. I. Title.
79-810135

—JAPAN—HOKKAIDO—HISTORY

(Minshūshi undō)
民衆史運動：その歴史と理論 / 小池喜孝監修；オホーツク民衆史講座編. — 東京：現代史出版会：発売　徳間書店, 1978.
279 p., [2] leaves of plates : ill. ; 20 cm. Ja 78-19597
Bibliography: p. 257–258.
¥1500
1. Civil rights—Japan—Hokkaido—History. 2. Civil rights—Study and teaching—Japan—Hokkaido. 3. Labor and laboring classes—Japan—Hokkaido—History. I. Koike, Yoshitaka, 1916– II. Ohōtsuku Minshūshi Kōza.
JS7384.H643A75 78-815375

*78 J

—JAPAN—INA REGION—HISTORY

Masaki, Keiji, 1905–
(Tōkai to Ina)
東海と伊那：商品流通と政治・文化交流 / 正木敬二著. — 名古屋：正木, 1978.
xiii, 486 p., [2] leaves of plates : ill. ; 22 cm. Ja 79-1942
Includes bibliographies.
Errata slip inserted.

1. Chūbu region, Japan—Economic conditions. 2. Ina region, Japan—Economic conditions. 3. Civil rights—Japan—Ina region—History. 4. Chūbu region, Japan—Politics and government. I. Title.
HC463.C47M37 79-805497

*79 J

—JAPAN—KANTŌ REGION—HISTORY

Tamura, Kōichirō, 1923–
(Kabasan jiken shimatsuki)
加波山事件始末記：歩いて書いた民権激派の記録 / 田村幸一郎著. — 東京：伝統と現代社：発売　現代ジャーナリズム出版会, 1978.
250 p. : ill. ; 20 cm. Ja 78-12861
Bibliography: p. 250.
¥1500
1. Civil rights—Japan—Kantō region—History. 2. Kabasan Uprising, 1884. I. Title.
JS7384.5.K3T35 78-811944

*78 J

—JAPAN—OKINAUA (PREFECTURE)

Tōkyō Bengoshikai. Okinawa Mondai Tokubetsu Iinkai.
(Nichi-Bei chii kyōtei to jinken)
日米地位協定と人権 / [編集　東京弁護士会沖縄問題特別委員会. — 東京：東京弁護士会, 昭和51 [1976]
74 p. ; 21 cm. Ja 76-16993
Cover title.
"日本国とアメリカ合衆国との間の相互協力及び安全保障条約第六条に
1. Civil rights—Japan—Okinawa (Prefecture) 2. Justice, Administration of—Japan—Okinawa (Prefecture) 3. Military bases, American—Japan—Okinawa (Prefecture) I. U. S. Treaties, etc., 1953–1961 (Eisenhower) Agreement under article IV of the Treaty of mutual cooperation and security: facilities and areas and the status of United States Armed Forces in Japan between the United States of America and Japan. Japanese. 1976. II. Title.
78-812238

—JAPAN—OKAYAMA (PREFECTURE)—HISTORY

Yasuhara, Man.
(Mushiake kaidō)
虫明街道：渋染一揆物語 / [著者　やすはらまん]. — 東京：福武書店, 昭和52 [1977]
158 p. ; 19 cm. Ja 77-12459
¥900

1. Buraku people—Japan—Okayama (Prefecture)—Political activity—History. 2. Civil rights—Japan—Okayama (Prefecture)—History. 3. Peasant uprisings—Japan—Okayama (Prefecture)—History. I. Title.
HT725.J3Y34 78-815129

CIVIL RIGHTS (Continued)

—KOREA

Ch'oe, Ch'ang-yun.
(K'at'ŏ Haengjŏngbu ŭi todŏk oegyo chŏngch'aek kwa kŭgŏt i Han'guk e mich'inŭn yŏnghyang)
카터行政府의道德外交政策과그것이韓國에미치는影響 / 研究執筆責任崔昌潤. — ［서울］：國土統一院政策企劃室, 1977.
93 p. ; 26 cm. — (周邊環境與件)
"對外秘1978. 12. 31까지"
"국통정77-10-1192"
"300부발간"
Includes bibliographical references.

1. United States—Foreign relations—1977- 2. United States—Foreign relations—Korea. 3. Korea—Foreign relations—United States. 4. Civil rights—Korea. I. Title. II. Series: Chubyŏn hwan'gyŏng yŏkŏn.

E872.C474 79-825458

Ide, Guju, 1937-
(Kankoku no chishikijin to Kin Shika)
韓国の知識人と金芝河 / 井出愚樹著. — 東京：青木書店, 1977.
iv, 254 p. ; 20 cm. Ja 77-8599
¥1200

1. Kim, Chi-ha, 1941- —Criticism and interpretation. 2. Korea—Intellectuals. 3. Korea—Politics and government—1960- 4. Civil rights—Korea. I. Title.

PL992.415.C46Z7 77-815237

Kim, Chi-ha, 1941-
(Kugyō)
苦行：獄中におけるわが闘い / 金芝河；金芝河刊行委員会訳・編. — 東京：中央公論社, 1978.
673 p., ［1］ leaf of plates : port. ; 20 cm. Ja 78-25113
"「獄中メモ」全訳と〔?〕記録…法廷における…闘いの詳細な再現を試みたもの"
¥1900

1. Kim, Chi-ha, 1941- 2. Poets, Korean—Biography. 3. Civil rights—Korea. I. Kin Shika Kankō Iinkai. II. Title.

PL992.415.C46Z467 79-803653

(Nihon to Chōsen o kangaeru)
日本と朝鮮を考える / 鈴木二郎, 隅谷三喜男, 宮崎繁樹編著. — 東京：二月社, 1978.
324 p. : ill. ; 19 cm. Ja 78-14203
¥1000

1. Korea—Politics and government—1960- 2. Civil rights—Korea. 3. Korean reunification question (1945-) I. Suzuki, Jirō, 1916- II. Sumiya, Mikio, 1916- III. Miyazaki, Shigeki, 1925-

DS922.35.N537 78-816668

—KOREA—ADDRESSES, ESSAYS, LECTURES

(Gokuchū kara)
獄中から / 金芝河他著；井出愚樹編訳. — 東京：大月書店, 1977.
199 p. ; 15 cm. — (国民文庫 ; 831) Ja 77-6465
Series romanized: Kokumin bunko.
¥300

1. Korea—Politics and government—1960- —Addresses, essays, lectures. 2. Civil rights—Korea—Addresses, essays, lectures. I. Kim, Chi-ha, 1941- II. Ide, Guju, 1937-

DS922.26.G64 78-813862

(Kankoku kara no tsūshin)
韓国からの通信 / T.K生［著］；「世界」編集部編. — 東京：岩波書店, 1974-
v. ; 18 cm. — (岩波新書) Ja 74-16881 (v. 1)
Series romanized: Iwanami shinsho.
Originally published in Sekai since May, 1973.
¥230 (v. 1)

1. Korea—Politics and government —1960- —Addresses, essays, lectures. 2. Civil rights—Korea—Addresses, essays, lectures. I. T. K. II. K., T. III. Sekai.

DS922.35.K36 76-808252

—KOREA—CONGRESSES

Association of Korean Christian Scholars in North America.
Human rights in minority perspectives / edited by Dong Soo Kim and Byong-suh Kim. — Montclair, N.J. : Association of Korean Christian Scholars in North America, c1979.
xiii, 292, xv-xviii p. ; 22 cm. — (Publication series ; The Association of Korean Christian Scholars in North America ; no. 5)
Includes seven articles presented at the 12th annual meeting of the Association of Korean Christian Scholars in North America, held May 25-27, 1978 in Toronto.
Bibliography: p. 269-286.
$8.00
1. Church and civil rights—Congresses. 2. Civil rights—Congresses. 3. Civil rights—Korea—Congresses. I. Kim, Dong Soo. II. Kim, Byong-suh. III. Title. IV. Series: Association of Korean Christian Scholars in North America. Association of Korean Christian Scholars in North America publication series ; no. 5.
BT738.15.A84 1979 261.8 79-88571
79 MARC

—LATIN AMERICA

Eide, Asbjørn.
Fattig, ufri og mishandlet : om det internasjonale menneskerettighetsvernet / Asbjørn Eide. — Oslo : Universitetsforl., c1978 (Oslo : Aktietr.i Trondhjem)
238 p., ［1］ leaf ; 18 cm. — (U-bok ; nr. 245) N78-Dec.
Bibliography: p. 187-［190］
ISBN 8200032558 : kr58.00
1. Civil rights (International law) 2. Civil rights—Latin America. 3. Civil rights. I. Title.
K3240.4.E37 79-369496
79 MARC

—LATIN AMERICA—CONGRESSES

Russell Tribunal on Repression in Brazil, Chile and Latin America, 2d, Rome and Brussels, 1974-1975.
Found guilty : the verdict of the Russell Tribunal session in Brussels. — Nottingham : Bertland Russell Peace Foundation for The Spokesman, ［1976］
［2］, 36 p. ; 22 cm. — (Spokesman pamphlet ; no. 51) GB77-00058
Second session.
£0.40
1. Civil rights—Latin America—Congresses. 2. Torture—Latin America—Congresses. 3. Latin America—Foreign economic relations—Congresses. 4. Latin America—Politics and government—1948- —Congresses. I. Title.
JC599.L3R87 1975 323.4'098 77-361804
77［r79］rev MARC

Russell Tribunal on Repression in Brazil, Chile, and Latin America, 2d, Rome and Brussels, 1974-1975.
Repression in Latin America : a report on the first session of the second Russell Tribunal, Rome, April 1974 / edited and translated by William Jerman. — Nottingham : Bertrand Russell Peace Foundation for Spokesman Books, 1975.
［3］ p., 163 p. ; 22 cm.
1. Latin America—Politics and government—1948- —Congresses. 2. Civil rights—Latin America—Congresses. 3. Latin America—Armed Forces—Political activity—Congresses. 4. Russell Tribunaal Twee over Onderdrukking nia Brazilië, Chilien Latijns Amerika. I. Jerman, William. II. Title.
F1414.2.R87 1975 323.4'098 76-377143
76［r79］rev MARC

—LEBANON

al-Ghurayyib, Mīshāl.
(al-Ḥurrīyāt al-'āmmah fī Lubnān wa-al-'ālam)
الحريات العامة في لبنان والعالم / ميشال الغريب ؛ ［تصميم وتنفيذ مؤسسة الخضري］. — الطبعة 1. — ［بيروت ؟］ : الاتحاد الوطني لطلاب الجامعة اللبنانية ، مجلس فرع كلية الحقوق والعلوم السياسية والاداريه ، أمانة شؤون المطبوعات ، ［ca. 1975］
270 p., ［2］ leaves of plates : ill. ; 24 cm.
Includes legislation.
Bibliography: p. 251-253.
1. Liberty. 2. Civil rights. 3. Civil rights—Lebanon. I. Title.
JC571.G46 79-970076

—MALAYSIA—ADDRESSES, ESSAYS, LECTURES

Malaysia : an anthology / Tan Sri Haji Khalid. — 1st ed. — New York : Vantage Press, c1978.
97 p., ［1］ leaf of plates : ill. ; 22 cm.
Includes bibliographical references.
ISBN 0-533-03382-9 : $6.95
1. Civil rights—Malaysia—Addresses, essays, lectures. 2. Malaysia—Politics and government—Addresses, essays, lectures. 3. Communism—Malaysia—Addresses, essays, lectures. I. Khalid, Tan Sri Haji, 1925-
JC599.M35M34 320.9'595'05 79-103555
79 MARC

—MEXICO

Castro, Juventino V
Lecciones de garantías y amparo / Juventino V. Castro. — 1. ed. — México : Editorial Porrúa, 1974.
595 p. ; 24 cm.
Bibliography: p. 585-587.
1. Civil rights—Mexico. 2. Due process of law—Mexico. 3. Amparo (Writ)—Mexico. 4. Civil rights. 5. Due process of law. 6. Amparo (Writ) I. Title.
342'.72'085 77-155617
79 MARC

—MICHIGAN—DETROIT
—PERIODICALS

Detroit. Human Rights Dept.
Annual report - City of Detroit, Human Rights Department.
Detroit, City of Detroit, Human Rights Dept.
v. 28 cm.
1. Detroit. Human Rights Dept. 2. Civil rights—Michigan—Detroit.—Periodicals.
JC599.U52M484 352'.98'0977434 78-648432
MARC-S

—MISSOURI—PERIODICALS

Missouri. Commission on Human Rights.
Progress newsletter - Missouri Commission on Human Rights. Dec. 1966-
［Jefferson City］ Missouri Commission on Human Rights.
v. ill. 29 cm.
Bimonthly, Dec. 1966–Nov./Dec. 1967; monthly, Jan. 1968–
Issues for Dec. 1966–Nov. 1969 called also v. 1, no. 1–v. 3, no. 11.
1. Civil rights—Missouri—Periodicals.
KFM8211.A15H85 323.4'09778 78-643496
MARC-S

—NETHERLANDS—ASSEN

Zwartboek Assen, september 1977 / samenstelling, Landelijk Comité Zuid-Molukken. — Rotterdam : Ordeman, 1977.
36 p. : ill. ; 30 cm. — (Dossier - Landelijk Comité Zuid-Molukken ; nr. 2)
Ne***
ISBN 906330031X : fl 5.00
1. Moluccans, South in Assen, Netherlands. 2. Hijacking of trains—Netherlands—Assen. 3. Civil rights—Netherlands—Assen. 4. Moluccans, South in Assen, Netherlands—Legal status, laws, etc. 5. Assen, Netherlands—Foreign population. I. Landelijk Comité Zuid-Molukken. II. Series: Landelijk Comité Zuid-Molukken. Dossier - Landelijk Comité Zuid-Molukken ; nr. 2.
DJ411.A85Z82 79-344458
79 MARC

—NEW MEXICO—PERIODICALS

New Mexico. Human Rights Commission.
Annual report - State of New Mexico, Human Rights Commission.
［Santa Fe］ State of New Mexico, Human Rights Commission.
v. 28 cm.
Report year ends June 30.
1. Civil rights—New Mexico—Periodicals. 2. New Mexico. Human Rights Commission—Periodicals.
KFN4011.A84 353.9'789'00996 78-648366
MARC-S

—NEW YORK (STATE)

Fischer, Paul, 1948-
The human rights crisis in New York State : a time for radical change / prepared by Paul A. Fischer, Nancy Taddiken. — Rochester, N.Y. : Urban League of Rochester, 1979.
iv, 37 leaves ; 28 cm.
Includes bibliographical references.
1. New York (State). State Division of Human Rights. 2. Civil rights—New York (State) I. Taddiken, Nancy, joint author. II. Urban League of Rochester, N.Y. III. Title.
JC599.U52N54 353.9'747 79-112042
00 MARC

—NORTHERN IRELAND

Brady, Brian J
Corruption of law : memorandum to the Gardiner Committee on the working of emergency legislation in Northern Ireland / from Brian J. Brady, Denis Faul, Raymond Murray. — ［Dungannon］ : ［Fr. Denis Faul］, 1974.
31 p. ; 21 cm. GB78-13382
Text on cover.
1. Criminal procedure—Northern Ireland. 2. Civil rights—Northern Ireland. I. Faul, Denis, joint author. II. Murray, Raymond, joint author. III. Title.
KDE550.B7 342'.416'085 79-305534
79 MARC

CIVIL RIGHTS
—NORTHERN IRELAND (Continued)

Northern Ireland. Standing Advisory Commission on Human Rights.
The protection of human rights by law in Northern Ireland / presented to Parliament by the Secretary of State for Northern Ireland by command of Her Majesty, November 1977. — London : H.M.S.O., ₍1977₎
₍7₎, 141 p. ; 25 cm. — (Cmnd. ; 7009) GB78-03131
Bibliography: p. 139-141.
ISBN 0-10-170090-3 : £2.35
1. Civil rights—Northern Ireland. I. Title. II. Series: Great Britain. Parliament. Papers by command, cmnd. ; 7009.
KDE420.A825 342'.416'085 78-314711
78 MARC

—NORWAY

Lindebrække, Sjur.
Veien videre. Et konservativt utsyn. Oslo, Acshehoug, 1973.
145 p. 21 cm. kr24.50 N73-12
ISBN 8203056792
1. Norway—Politics and government—1945- . 2. Conservatism—Norway. 3. Civil rights—Norway. I. Title.
JN7445 1973.L55 73-342138
79₍r79₎rev MARC

Norsk samfunnsvitenskapelig datatjeneste.
Personvern og samfunnsforskning : seminarrapport / Norsk samfunnsvitenskapelig datatjeneste. — ₍Bergen₎ : NSD, 1977.
49 leaves ; 30 cm. — (NSD rapport ; nr 17) N77-Nov.
ISBN 8271700243
1. Civil rights—Norway. 2. Electronic data processing departments—Security measures. 3. Social science research—Norway. I. Title. II. Series: Norsk samfunnsvitenskapelig datatjeneste. NSD rapporter ; nr. 17.
JC599.N7N65 1977 78-386046
79 MARC

Politi, presse, overvåking : hvitbok om "liste-saka" / redigert av Halvor Elvik. — Oslo : Pax, 1977. (Hamar : A. Sæther)
188 p. ; 19 cm. — (En Pax-bok ; 561) N78-Jan
ISBN 8253009054 : kr37.00
1. Police—Norway—Political activity. 2. Civil rights—Norway. 3. Police ethics. I. Elvik, Halvor, 1947-
HV8230.A2P64 79-339056
79 MARC

—NORWAY—ADDRESSES, ESSAYS, LECTURES

Mathiesen, Thomas.
Det skjulte disiplinering : artikler om politisk kontroll / Thomas Mathiesen. — Oslo : Pax, 1978 (Hamar : A. Sæter)
163 p., ₍1₎ p. ; 19 cm. — (En Pax-bok ; 604) N79-May
Bibliography: p. ₍165₎
ISBN 8253009704 : kr45.00
1. Civil rights—Norway—Addresses, essays, lectures. 2. Police power—Norway—Addresses, essays, lectures. 3. Control (Psychology)—Addresses, essays, lectures. I. Title.
JC599.N7M37 79-378047
79 MARC

Overvåking i Norge / Ole Jakob Bae og Torstein Eckhoff (red.). — Oslo : Pax, 1979 (Tonsberg : Ekspresstrykk)
150 p. ; 19 cm. — (Pax-₍bokene₎ ; 632) N***
ISBN 8253010184 : kr53.00
1. Norway. Overvåkingspolitiet—Addresses, essays, lectures. 2. Intelligence service—Norway—Addresses, essays, lectures. 3. Internal security—Norway—Addresses, essays, lectures. 4. Civil rights—Norway—Addresses, essays, lectures. I. Bae, Ole Jakob. II. Eckhoff, Torstein Einang, 1916-
HV8230.A2O93 79-371800
79 MARC

—PERU

Ortecho Villena, Víctor Julio.
Las libertades públicas y el habeas corpus ₍por₎ Víctor Julio Ortecho V. ₍Trujillo, Perú, 1966?₎
53 p. 25 cm.
"Separata del no. 5 de la 'Revista de derecho y ciencias políticas y sociales' de Trujillo, Perú."
1. Civil rights—Peru. 2. Habeas corpus—Peru. I. Title.
342'.85'085 67-78641
₍r79₎rev MARC

—PHILIPPINE ISLANDS

International Commission of Jurists (Founded 1952)
The decline of democracy in the Philippines : a report of missions / by William J. Butler, John P. Humphrey, G. E. Bisson. — Geneva : International Commission of Jurists, 1977.
viii, 97 p. ; 23 cm. Sw***
ISBN 0-89192-193-1
1. Martial law—Philippine Islands. 2. Civil rights—Philippine Islands. 3. Political rights—Philippine Islands. 4. Justice, Administration of—Philippine Islands. I. Butler, William Jack, 1919- . II. Humphrey, John Thomas Peters, 1905- . III. Bisson, G. E. IV. Title.
320.9'599'04 78-323878
79 MARC

Makibaka : join us in the struggle : ₍a documentation of five years of resistance to martial law in the Philippines₎. — ₍London : Friends of the Philippines, 1978₎
173 p. : ill. ; 21 cm. GB***
ISBN 0-9506102-0-8 : £2.00
1. Civil rights—Philippine Islands. 2. Martial law—Philippine Islands. 3. Philippine Islands—Politics and government—1946- . I. Friends of the Philippines.
JC599.P5M34 323.4'09599 79-315084
79 MARC

United States. Congress. House. Committee on International Relations. Subcommittee on International Organizations.
Human rights in the Philippines : recent developments : hearing before the Subcommittee on International Organizations of the Committee on International Relations, House of Representatives, Ninety-fifth Congress, second session, April 27, 1978. — Washington : U.S. Govt. Print. Off., 1978.
vii, 28 p. ; 24 cm.
1. Civil rights—Philippine Islands. 2. United States—Foreign relations—Philippine Islands. 3. Philippine Islands—Foreign relations—United States. I. Title.
KF27.I5494 1978a 323.4'09599 78-602679
78 MARC

—PHILIPPINE ISLANDS—ADDRESSES, ESSAYS, LECTURES

Castro, Fred Ruiz.
Human rights in the Philippines / by Fred Ruiz Castro. — ₍Quezon₎ : University of the Philippines Law Center, ₍1977₎
15 p. ; 23 cm.
"Address to the World Law Conference (on World Law Day, August 21, 1977) ... Metro Manila, Philippines."
Cover title: The Chief Justice on human rights in the Philippines.
1. Civil rights—Philippine Islands—Addresses, essays, lectures. I. Title. II. Title: The Chief Justice on human rights in the Philippines.
JC599.P5C37 323.4'09599 78-106132
79 MARC

—POLAND

Bednarski, Stanisław.
Prawa i wolności obywatelskie w PRL / Stanisław Bednarski. — Wyd. 1. — Warszawa : Krajowa Agencja Wydawnicza, 1978.
99 p. ; 21 cm. — (Nasza Ojczyzna—Polska Ludowa) P***
Includes bibliographical references.
zł15.00
1. Civil rights—Poland. I. Title.
78-397788
79 MARC

Demokratyzm socjalistycznego systemu politycznego PRL ₍Polskiej Rzeczypospolitej Ludowej₎ : materiały z konferencji naukowej 23 maja 1977 r. oraz z konferencji teoretyczno-szkoleniowej 6 czerwca 1977 r.₎ / pod red. A. Łopatki, J. Wawrzyniaka. — Wyd. 1. — Warszawa : Książka i Wiedza, 1978 (Warsz. : DSP)
398, ₍2₎ p. ; 20 cm. — (Ideologia a współczesność)
Includes bibliographical references.
zł50.00
1. Poland—Politics and government—1945- . 2. Communist state. 3. Civil rights—Poland. 4. Polska Zjednoczona Partia Robotnicza. I. Łopatka, Adam. II. Wawrzyniak, Jan. III. Series.
JN6753 1978.D45 79-357800
79 MARC

In defence of Poland's freedom : facts and documents. — London : Polish Government in Exile, 1977.
31 p. ; 24 cm. GB***
1. Civil rights—Poland. 2. Poland—Politics and government—1945-
JC599.P6I5 323.4'09438 79-308736
79 MARC

Resich, Zbigniew.
Polska w obronie praw człowieka / Zbigniew Resich. — Wyd. 1. — Warszawa : Książka i Wiedza, 1978.
160 p. ; 20 cm. P***
Includes bibliographical references.
zł20.00
1. Civil rights—Poland. I. Title.
79-359663
79 MARC

—POLAND—ADDRESSES, ESSAYS, LECTURES

Dissent in Poland : reports and documents in translation ₍from the Polish₎, December 1975-July 1977. — London : Association of Polish Students and Graduates in Exile, 1977.
4, 200 p. ; 22 cm. GB78-14792
ISBN 0-9506048-0-1
1. Civil rights—Poland—Addresses, essays, lectures. 2. Poland—Politics and government—1945- —Addresses, essays, lectures.
JC599.P6D57 322.4'4'09438 79-313381
00 MARC

Prawa i obowiązki obywatelskie w Polsce i świecie / redaktor naukowy Marian Szczepaniak. — Wyd. 2., zmienione. — Warszawa : Państwowe Wydawn. Naukowe, 1978.
492 p. ; 25 cm. P***
Table of contents also in English and Russian.
Includes bibliographical references.
zł75.00
1. Civil rights—Poland—Addresses, essays, lectures. 2. Civil rights—Addresses, essays, lectures. I. Szczepaniak, Marian.
79-359123
79 MARC

—POLAND—SOURCES

Ruch oporu .... — Le Mesnil-le-Roi : Instytut Literacki ; ₍Paris₎ : ₍diffusion Libella₎, 1977.
285 p. ; 22 cm. — (Biblioteka "Kultury" ; t. 276) (Dokumenty ; 44) F78-5836
50.00F
1. Poland—Politics and government—1945- —Sources. 2. Civil rights—Poland—Sources. I. Series. II. Series: Dokumenty ; 44.
D410.D6 vol. 44 78-393712
₍DK4440₎ 79 MARC

—PORTUGAL

Carvalho, Orlando de.
Os direitos do homem no direito civil português. Coimbra, 1973.
45 p. 18 cm. (Textos Vértice, 7)
"Este trabalho reproduz, substancialmente, uma conferência que o autor fez em França, em janeiro de 1972, a convite do Comité Franco-Português do Rotary Internacional, sob o título Les droits de l'homme dans le droit civil portugais."
1. Civil rights—Portugal. I. Title.
77-480529
78 MARC

—RUSSIA

Bourdeaux, Michael.
White book on restrictions of religion in the U.S.S.R. / by Michael Bourdeaux. — Brussels : International Committee for the Defence of Human Rights in the U.S.S.R., ₍1977?₎
66 p. ; 28 cm. Be***
Caption title.
Translation of Livre blanc sur les restrictions religieuses en U.R.S.S.
Includes bibliographical references.
1. Religion and state—Russia. 2. Persecution—Russia. 3. Civil rights—Russia. 4. Communism and religion. 5. Russia—Religion—1917- I. Title.
BL65.S8B6813 322'.1'0947 78-323751
79 MARC

Chkhikvadze, Viktor Mikhaĭlovich.
(Sofŝialisticheskiĭ gumanizm i prava cheloveka)
Социалистический гуманизм и права человека : Ленинские идеи и современность / В. М. Чхиквадзе. — Москва : Наука, 1978.
303 p. ; 22 cm. USSR 79
At head of title: Akademiiâ nauk SSSR. Institut gosudarstva i prava.
Includes bibliographical references.
1.50rub
1. Civil rights—Russia. 2. Law and socialism. 3. Communist ethics. I. Title.
79-412611

Chkhikvadze, Viktor Mikhaĭlovich, 1912–
(Sovetskoe gosudarstvo i lichnostʹ)
Советское государство и личность / В. М. Чхиквадзе. — Москва : Юрид. лит., 1978.
103 p. ; 20 cm. — (Конституция общенародного государства) USSR 78
Includes bibliographical references.
0.10rub
1. Personality (Law)—Russia. 2. Russia—Constitutional law. 3. Socialist property—Russia. 4. Civil rights—Russia. I. Title. II. Series: Konstitufsiiâ obshchenarodnogo gosudarstva.
78–421828

Cramer, Dettmar, 1929-
Bürgerrechte 77 ₍i.e. siebenundsiebzig₎ / Dettmar Cramer. — Köln : Verlag Wissenschaft u. Politik, 1977.
215 p. ; 19 cm. — (Bibliothek Wissenschaft und Politik) GFR78-A
Includes bibliographical references.
ISBN 3-8046-8540-4
1. Civil rights (International law) 2. Civil rights—Russia. 3. Civil rights—Germany, West. I. Title.
K3240.4.C66 341.48'1 78-379577
78 MARC

(Dokumenty po delu Leonida Pliûshcha)
Документы по делу Леонида Плюща. — Frankfurt/Main : Посев, 1974.
76 p. ; 17 cm. — (Вольное слово ; вып. 11) GFR***
Includes index.
1. Civil rights—Russia. 2. Political crimes and offenses—Russia. 3. Pliûshch, Leonyd Ivanovych, 1939- I. Pliûshch, Leonyd Ivanovych, 1939- II. Series: Volʹnoe slovo ; vyp. 11.
JC591.V64 vyp. 11 79–414135

CIVIL RIGHTS
—RUSSIA (Continued)

Grigorenko, Andreĭ, 1945–
(A kogda my vernemsi͡a ...)
А когда мы вернемся ... / Андрей Григоренко ; обложка работы Ф. Юртера. — Нью Йорк : Фонд Крым, 1977.
210 p., [10] leaves of plates : ill. ; 22 cm.
On verso of t. p.: And when we return ...
"No. 1."
Bibliography: p. 201–202.
Includes index.
1. Abduldzhemil', Mustafa, 1943– 2. Grigorenko, Petr Grigor'evich. 3. Political crimes and offenses—Russia. 4. Tatars in the Crimea. 5. Civil rights—Russia. I. Title. II. Title: And when we return ...
78–670015

Grigorenko, Petr Grigor'evich, 1907-
Nashi budni : ili rasskaz o tom, kak fabrikui͡ut͡si͡a ugolovnye dela na sovetskikh grazhdan, vystupai͡ushchikh v zashchitu prav cheloveka / Petr Grigorenko. — [Mi͡unkhen] : Suchasnist', 1978.
117 p. ; 20 cm. — (Biblioteka prologa i suchasnisti ; n. 127) GFR***
1. Russia—Politics and government—1953- 2. Dissenters—Russia. 3. Civil rights—Russia. I. Title.
DK274.G733 79-374129
79 MARC

Grigori͡an, Levon Armenakovich.
(Sovetskoe obshchenarodnoe gosudarstvo i lichnost')
Советское общенародное государство и личность / Л. А. Григорян, д-р юрид. наук. — Москва : Знание, 1977.
47 p. ; 20 cm. — (Библиотечка Основной закон нашей жизни) USSR 77
Includes bibliographical references.
0.10rub
1. Persons (Law)—Russia. 2. Civil rights—Russia. 3. Russia—Constitutional law. I. Title. II. Series: Bibliotechka Osnovnoĭ zakon nasheĭ zhizni.
78–421593

Karpinskiĭ, Vi͡acheslav Alekseevich, 1880–1965.
[Osnovnye prava i obi͡azannosti grazhdan SSSR. Chinese]
蘇聯公民的基本權利與義務 / В. Карпинский : 易今譯. — 上海 : 中華書局, 1950.
2, 105 p. ; 18 cm. — (新時代小叢書 ; 10)
Title also in Russian.
Series romanized: Hsin shih tai hsiao ts'ung shu.
1. Civil rights—Russia. I. Title.
Title romanized: Su-lien kung min ti chi pen ch'üan li yü i wu.
JN6581.K312 79–836667

(Khronika tekushchikh sobytiĭ)
Хроника текущих событий.
Нью-Йорк, Изд-во Хроника.
v. 17–22 cm. irregular. (–Oct. 1972: Вольное слово (JC591.V64))
"Выступления в защиту прав человека в Советском Союзе продолжаются" (varies slightly).
Began with Apr. 1968 issue. Cf. New serial titles.
Other title, Dec. 1972– : A Chronicle of current events.
Imprint varies: –Oct. 1972, Frankfurt/Main, Possev-Verlag.
1. Political crimes and offenses—Russia. 2. Civil rights—Russia. I. Title: A Chronicle of current events. II. Series: Vol'noe slovo.
JC591.K48 78–643992
(MARC–S)

Koldaev, Vladislav M
Soviet citizens, their rights and duties : 40 questions answered / V. Koldayev. — Moscow : Novosti Press Agency Pub. House, 1976.
54 p., [8] leaves of plates : ill. ; 17 cm. USSR***
0.19rub.
1. Civil rights—Russia. 2. Russia—Social policy. I. Title.
JN6581.K64 323.4'0947 79-306047
79 MARC

Kozharov, Asen.
(Pravata na choveka i ideologicheskata borba dnes)
Правата на човека и идеологическата борба днес : [полит. очерк] / Асен Кожаров. — 1. изд. — София : Партиздат, 1977.
69 p. : ill. ; 20 cm. Bu 77–2555
Includes bibliographical references.
0.23 lv
1. Civil rights (International law) 2. Dissenters—Legal status, laws, etc. 3. Civil rights—United States. 4. Civil rights—Russia. I. Title.
K3240.4.K69 78–421597

Lapenna, Ivo.
Human rights : Soviet theory and practice : Helsinki and international law / by Ivo Lapenna. — London : Institute for the Study of Conflict, 1977.
15 p. ; 30 cm. — (Conflict studies ; no. 83) GB***
Cover title.
"Summary": [2] p. inserted.
Includes bibliographical references.
£2.00 ($4.50 U.S.)
1. Civil rights—Russia. 2. Civil rights (International law) I. Title. II. Series.
D839.3.C6 no. 83 327'.08 s 78-315542
78 MARC

Makarenko, Mikhail I͡Anovich, 1931–
(Iz moeĭ zhizni)
Из моей жизни : ответ провокаторам : дополнение к кассационной жалобе Верховному Суду РСФСР / М. Я. Макаренко. — Frankfurt/Main : Посев, 1974.
127 p. ; 17 cm. — (Вольное слово ; вып. 12) GFR***
Includes index.
1. Civil rights—Russia. 2. Political crimes and offenses—Russia. 3. Makarenko, Mikhail I͡Anovich, 1931– I. Title. II. Series: Vol'noe slovo ; vyp. 12.
JC591.V64 vyp. 12 79–414136

Michael-Titus, C
A testimony for Europe / [by] C. Michael-Titus. — Upminster : Panopticum Press, 1977.
15 p. ; 21 cm. GB77-33643
Cover title: Yalta, Helsinki, Belgrade, a testimony for Europe.
1. Conference on Security and Cooperation in Europe. 2. Civil rights—Europe, Eastern. 3. Civil rights—Russia. 4. Dissenters—Europe, Eastern. 5. Dissenters—Russia. 6. Authors, Romanian. I. Title. II. Title: Yalta, Helsinki, Belgrade.
D1050.5.M5 322.4'4'0947 78-319995
78 MARC

(Moe poslednee slovo)
Мое последнее слово : речи подсудимых на судебных процессах, 1966–1974. — Frankfurt/Main : Посев, 1974.
173 p. ; 17 cm. — (Вольное слово ; вып. 14–15) GFR***
Includes index.
1. Trials (Political crimes and offenses)—Russia. 2. Civil rights—Russia. I. Series: Vol'noe slovo ; vyp. 14–15.
JC591.V64 vyp. 14–15 79–414133

Morand, Bernadette.
L'URSS des profondeurs / Bernadette Morand ; préf. de Louis Leprince-Ringuet ; introd. de Anthony de Meeûs. — Brussels : Éditions Arts et voyages : diffusion, Gamma, c1978.
265 p., [8] leaves of plates : ill. ; 22 cm. — (Collection Inédits) Be***
"Textes et documents des Cahiers du samizdat."
Bibliography: p. [263]-265.
Includes index.
ISBN 2-8016-0005-9
1. Civil rights—Russia. I. Title.
JC599.R9M67 323.4'0947 79-357373
79 MARC

Pimenov, Revol't Ivanovich, 1931–
(Odin politicheskiĭ protsess)
Один политический процесс / Р. И. Пименов. — Frankfurt/Main : Посев, 1973.
107 p. ; 17 cm. — (Вольное слово ; вып. 8) GFR***
Includes index.
1. Civil rights—Russia. 2. Political crimes and offenses—Russia. 3. Pimenov, Revol't Ivanovich, 1931– I. Title. II. Series: Vol'noe slovo ; vyp. 8.
JC591.V64 vyp. 8 79–414137

(Pravda o pravakh cheloveka)
Правда о правах человека : деятели сов. культуры о правах человека / Константин Федин ... [et al.] ; составитель Борис Иванов]. — Москва : Политиздат, 1977.
158 p. ; 20 cm. USSR***
0.25rub
1. Civil rights—Russia. 2. Civil rights (International law) 3. Civil rights. I. Fedin, Konstantin Aleksandrovich, 1892– II. Ivanov, Boris Vladimirovich, 1917–
78–422937

Sparre, Victor, 1919-
The flame in the darkness : the Russian human rights struggle, as I have seen it / [by] Victor Sparre ; translated from the Norwegian by Alwyn and Dermot McKay ; foreword by Vladimir Maximov. — London : Grosvenor Books, 1979.
[7], 134 p., [4] p. of plates : ports. ; 22 cm. GB79-17301
Includes index.
ISBN 0-901269-35-2 : £4.50
1. Russia—Politics and government—1953- 2. Dissidents—Russia. 3. Civil rights—Russia. 4. Sparre, Victor, 1919- I. Title.
DK274.S68813 323.4'0947 79-317494
79 MARC

(Svidetel' po sobstvennomu delu i drugie dokumenty)
Свидетель по собственному делу и другие документы. — Frankfurt/Main : Посев, 1974.
108 p. ; 17 cm. — (Вольное слово ; вып. 13) GFR***
Includes index.
1. Civil rights—Russia. 2. Political crimes and offenses—Russia. I. Series: Vol'noe slovo ; vyp. 13.
JC591.V64 vyp. 13 79–414134

—RUSSIA—ADDRESSES, ESSAYS, LECTURES

Sakharov, Andreĭ Dmitrievich, 1921-
Alarm and hope / Andrei D. Sakharov ; edited by Efrem Yankelevich and Alfred Friendly, Jr. — 1st English language ed. — New York : Alfred A. Knopf : distributed by Random House, c1978.
xix, 200 p. ; 22 cm.
Translation of Trevoga i nadezhda.
Includes index.
ISBN 0-394-50369-4 : $8.95
1. International relations—Addresses, essays, lectures. 2. Civil rights—Russia—Addresses, essays, lectures. 3. Civil rights (International law)—Addresses, essays, lectures. I. I͡Ankelevich, Efrem. II. Friendly, Alfred. III. Title.
JX1395.S26513 323.4'0947 78-13845
78 MARC

—RUSSIA—BIBLIOGRAPHY

Woll, Josephine.
Soviet unofficial literature—samizdat : an annotated bibliography of works published in the West / compiled by Josephine Woll and Vladimir Treml. — Durham, N.C. : Duke University Center for International Studies, 1978.
xviii, 150 p. ; 29 cm.
Includes index.
1. Russian literature—20th century—Bibliography. 2. Underground literature—Russia—Bibliography. 3. United States—Imprints. 4. Europe—Imprints. 5. Civil rights—Russia—Bibliography. I. Treml, Vladimir G., joint author. II. Title.
Z2511.U5W64 016.8917 78-68417
[PG3026.U5] 78 MARC

—RUSSIA—HISTORY

Aims for Freedom and Enterprise.
The age of the GULAG : 1917-1977, sixty years of communism / Aims for Freedom and Enterprise. — London : Aims for Freedom and Enterprise, [1977]
[2], 13 p. ; 21 cm. GB78-17136
1. Political prisoners—Russia—History. 2. Civil rights—Russia—History. I. Title.
HV9712.A53 1977 947.084 79-300874
79 MARC

—RUSSIA—POPULAR WORKS

("Kak vesti sebi͡a na obyske" i drugie dokumenty 1976 goda)
"Как вести себя на обыске" и другие документы 1976 года. — Frankfurt/Main : Посев, 1977.
148 p. ; 16 cm. — (Вольное слово ; вып. 27) GFR***
Includes bibliographical references and index.
1. Searches and seizures — Russia — Popular works. 2. Political crimes and offences—Russia—Popular works. 3. Civil rights—Russia—Popular works. I. Series: Vol'noe slovo ; vyp. 27.
JC591.V64 vyp. 27 79–414132

—SALVADOR

Latin America Bureau.
Violence and fraud in El Salvador : a report on current political events in El Salvador / compiled ... by Latin America Bureau. — London : The Bureau, 1977.
37 p. : maps ; 30 cm. GB***
Includes bibliographical references.
£2.00
1. Civil rights—Salvador. 2. Land reform—Salvador. 3. Church and state in Salvador. 4. Elections—Salvador—Corrupt practices. I. Title.
JC599.S2L37 1977 320.9'7284'05 78-317571
78 MARC

—SOUTH AFRICA

Partridge, Astley Cooper.
Human rights : their origin, validity and implementation / A. C. Partridge. — Johannesburg : South African Institute of International Affairs, 1978.
14 p. ; 30 cm. — (Occasional paper - The South African Institute of International Affairs) SA78
Cover title.
Bibliography: p. 13-14.
ISBN 0-909239-36-3 : RO.50
1. Civil rights. 2. Civil rights—South Africa. I. Title. II. Series: South African Institute of International Affairs. Occasional papers - South African Institute of International Affairs.
JC571.P285 323.4 79-309562
79 MARC

—SOUTH AFRICA—ADDRESSES, ESSAYS, LECTURES

Buthelezi, Gatsha.
Human rights and constitutional developments in South Africa : 26th October, 1976 / by Mangosuthu Gatsha Buthelezi. — Lagos : Nigerian Institute of International Affairs, [1976?]
28 p. ; 22 cm. — (Lecture series - Nigerian Institute of International Affairs ; no. 11)
Cover title.
1. Civil rights—South Africa—Addresses, essays, lectures. 2. South Africa—Race relations—Addresses, essays, lectures. I. Title. II. Series: Nigerian Institute of International Affairs. Lecture series - Nigerian Institute of International Affairs ; no. 11.
JQ1981.B87 323.4'0968 78-106970
79 MARC

CIVIL RIGHTS
—SOUTH AFRICA—ADDRESSES, ESSAYS, LECTURES (Continued)

Menschenrechte in Südafrika : Perspektiven von Widerstand und Unterdrückung / hrsg. von Winfried Bassmann. — München : R. Piper, c1978.
240 p. ; 19 cm. — (Serie Piper ; 179) GFR***
Includes bibliographical references.
ISBN 3-492-00479-2 : DM18.00
1. South Africa—Race relations—Addresses, essays, lectures. 2. Civil rights—South Africa—Addresses, essays, lectures. I. Bassmann, Winfried, 1950-
DT763.M44 301.45'1'0968 78-399574
79 MARC

—SOUTHERN STATES—PERIODICALS

Southern changes. v. 1-
Sept. 1978-
[Atlanta, Ga., Southern Regional Council]
v. ill. 28 cm. monthly.
"A chronicle of the ongoing struggle for equality."
Key title: Southern changes, ISSN 0198-2446.
1. Civil rights—Southern States—Periodicals. I. Southern Regional Council.
JC599.U5S598 323.4'0975 79-643363
MARC-S

—SPAIN—ARAGON

Delgado Echeverría, Jesús.
Joaquín Costa y el derecho aragonés : (libertad civil, costumbre y codificación) / [por] Jesús Delgado Echeverría. — Zaragoza : Facultad de Derecho, 1978.
42 p., 1 leaf ; 22 cm. Sp78-July
Includes bibliographical references.
ISBN 8460011046
1. Costa y Martínez, Joaquín, 1846-1911. 2. Law—Spain—Aragon. 3. Civil rights—Spain—Aragon. I. Title.
78-400261
79 MARC

—SPAIN—BARCELONA

Foix, Pedro.
Los archivos del terrorismo blanco : el fichero Lasarte 1910-1930 / Pere Foix. — 3. ed. — Madrid : Las Ediciones de la Piqueta, 1978.
124 p. : ill. ; 20 cm. Sp***
"Este librito se publicó en catalán en el diario L'Opinió, de Barcelona ... 1931."
ISBN 8474430127 : 160ptas
1. Barcelona—Police—Complaints against. 2. Civil rights—Spain—Barcelona. 3. Anarchism and anarchists—Spain—Barcelona. 4. Political prisoners—Spain—Barcelona. I. Title. II. Title: El fichero Lasarte.
HV8238.B37F64 1978 78-378533
78 MARC

—SWITZERLAND

Mastronardi, Philippe.
Der Verfassungsgrundsatz der Menschenwürde in der Schweiz : e. Beitr. zu Theorie u. Praxis d. Grundrechte / von Philippe Mastronardi. — Berlin : Duncker und Humblot, 1978.
331 p. : ill. ; 23 cm. — (Schriften zum öffentlichen Recht ; Bd. 349)
GFR79-A
Originally presented as the author's thesis, Bern, 1977.
Bibliography: p. [322]-327.
Include index.
ISBN 3-428-04257-3 : DM78.00
1. Civil rights—Switzerland. 2. Switzerland—Constitutional law. 3. Dignity. I. Title.
342'.494'085 79-345992
79 MARC

—SWITZERLAND—CONGRESSES

Menschenrechte im Spannungsfeld : eine Herausforderung für die Schweiz / mit Beitr. von Pierre Aubert ... [et al.] ; Hrsg., Amnesty International, Schweizer Sektion. — Bern [etc.] : P. Haupt, cop. 1978.
146 p. ; 21 cm. Sw79-719
Chiefly in German with one contribution in French.
ISBN 3-258-02816-8 : 19.00F
1. Civil rights—Switzerland—Congresses. 2. Civil rights—Congresses. I. Aubert, Pierre, 1927- . II. Amnesty International. Schweizer Sektion.
342'.43'085 79-359741
79 MARC

—TAIWAN

T'ao, Pai-ch'uan, 1903-
(Jen ch'üan hu ying)
人權呼應 / 陶百川. — 台北：遠景出版社，民國68 [1979].
20, 302 p. ; 20 cm. — (遠景叢刊 ; 123)
Series romanized: Yüan ching ts'ung k'an.
NT$70.00
1. Criminal justice, Administration of—Taiwan. 2. Civil rights—Taiwan. I. Title.
79-840614

—TENNESSEE—MEMPHIS

United States. Commission on Civil Rights.
Hearing before the United States Commission on Civil Rights : hearing held in Memphis, Tennessee, May 9, 1977. — [Washington] : The Commission, [1978?]
v, 138 p. ; 23 cm.
1. Memphis—Police. 2. Public relations—Police. 3. Civil rights—Tennessee—Memphis.
HV8148.M382U54 1978 363.2 78-603041
79 MARC

—THAILAND

European Co-ordinating Committee for Solidarity with the Thai People.
Political repression in Thailand / European Co-ordinating Committee for Solidarity with the Thai People. — [s.l.] : The Committee ; [London : obtainable from Ad Hoc Group for Democracy in Thailand, 1978?]
64 p. : ill. ; 21 cm. GB***
Includes bibliographies.
£0.40
1. Civil rights—Thailand. 2. Political prisoners—Thailand. 3. Thailand—Politics and government. I. Title.
JC599.T5E89 1978 323.4'09593 79-307297
79 MARC

—TURKEY—ADDRESSES, ESSAYS, LECTURES

Yücel, Hasan Âli, 1897-1961.
Hürriyet gene hürriyet / Hasan-Âli Yücel. — Ankara : [Türkiye İş Bankası], 1960-1966.
2 v. ; 25 cm. — (Türkiye İş Bankası, Kültür yayınları ; seri 1, no. 14)
1. Civil rights—Turkey—Addresses, essays, lectures. 2. Liberty—Addresses, essays, lectures. 3. Turkey—Politics and government—1918-1960—Addresses, essays, lectures. I. Title.
JC599.T87Y83 79-349755
79 MARC

—UGANDA

Amnesty International.
Human rights in Uganda : report / [Amnesty International]. — London : Amnesty International, 1978.
[1], 25 p. ; 30 cm. GB78-31228
Cover title.
Errata slip inserted.
Includes bibliographical references.
1. Civil rights—Uganda. I. Title.
JQ2951.A91A48 1978 323.4'09676'1 79-302572
79 MARC

United States. Congress. Senate. Committee on Foreign Relations. Subcommittee on Foreign Economic Policy.
Uganda, the human rights situation : hearings before the Subcommittee on Foreign Economic Policy of the Committee on Foreign Relations, United States Senate, Ninety-fifth Congress, second session ... June 15, 21, 26, 1978. — Washington : U.S. Govt. Print. Off., 1978.
iii, 129 p. ; 24 cm.
1. Civil rights—Uganda. 2. United States—Foreign relations—Uganda. 3. Uganda—Foreign relations—United States. I. Title.
KF26.F6364 1978a 323.4'09676'1 79-601171
79 MARC

—UNITED STATES

see also Afro-Americans—Civil rights

American Civil Liberties Union.
1977 supplement to the American Civil Liberties Union policy guide / The American Civil Liberties Union. — Lexington, Mass. : Lexington Books, c1978.
xv, 189 p. ; 23 cm.
Includes indexes.
ISBN 0-669-02624-7
1. Civil rights—United States. 2. American Civil Liberties Union. I. American Civil Liberties Union. 1976 policy guide of the American Civil Liberties Union. II. Title.
KF4749.A74 Suppl 342'.73'085 78-19573
78 MARC

Buffle, Jean Claude.
Le piège à rois / Jean-Claude Buffle. — Paris : Duculot, c1978.
192 p. ; 24 cm. — (Document Duculot ISSN 0180-7641) Be***
Includes bibliographical references.
ISBN 2-8011-0207-5 : 325F
1. Civil rights—United States. 2. Conservatism—United States. 3. Decentralization in government—United States. I. Title.
JC599.U5B83 323.4'0973 79-380243
79 MARC

Congressional Quarterly, inc.
Editorial research reports on the rights revolution : timely reports to keep journalists, scholars, and the public abreast of developing issues, events, and trends. — Washington : Congressional Quarterly, c1978.
217 p. ; 23 cm.
Includes bibliographies and index.
ISBN 0-87187-144-0
1. Civil rights—United States. I. Congressional Quarterly, inc. II. Title: The rights revolution.
JC599.U5C635 1978 323.4'0973 78-31931
79 MARC

Corwin, Edward Samuel, 1878-1963.
Liberty against government : the rise, flowering, and decline of a famous juridical concept / by Edward S. Corwin. — Westport, Conn. : Greenwood Press, 1978, c1948.
xiii, 210 p. ; 23 cm.
Reprint of the ed. published by Louisiana State University Press, Baton Rouge.
Includes bibliographical references and index.
ISBN 0-8371-9589-6
1. Civil rights—United States. 2. United States—Constitutional history. 3. Liberty. I. Title.
KF4749.C67 1977 342'.73'085 77-4090
77 MARC

Evans, Sara.
Personal politics : the roots of women's liberation in the civil rights movement and the new left / by Sara Evans. — 1st ed. — New York : Knopf : distributed by Random House, 1979.
xii, 274 p. : 22 cm.
Bibliography: p. 257-268.
Includes index.
ISBN 0-394-41911-1 : $10.00
1. Women—United States—Social conditions. 2. Radicalism—United States. 3. Feminism—United States. 4. Civil rights—United States. I. Title.
HQ1426.E9 1979 301.41'2'0973 78-54929
78 MARC

Federal civil rights litigation.
[New York] Practising Law Institute.
v. 22 cm. annual. (Litigation course handbook series) (: Litigation and administrative practice series)
Key title: Federal civil rights litigation, ISSN 0192-8635.
1. Civil rights—United States. 2. Actions and defenses—United States. I. New York (City). Practising Law Institute. II. Series. III. Series: Litigation and administrative practice series.
KF4749.F42 342'.73'0850269 79-641057
MARC-S

Fiss, Owen M
The civil rights injunction / Owen M. Fiss. — Bloomington : Indiana University Press, c1978.
vi, 117 p. ; 22 cm.
Includes bibliographical references and indexes.
ISBN 0-253-31356-2
1. Injunctions—United States. 2. Civil rights—United States. I. Title.
KF9014.F5 342'.73'0850269 78-2052
78 MARC

Friends, Society of. Program on Government Surveillance and Citizens' Rights.
The police threat to political liberty : discoveries and actions of the American Friends Service Committee Program on Government Surveillance and Citizens' Rights, 1979. — Philadelphia : AFSC, 1979.
iv, 153 p. ; 21 cm.
Errata slip inserted.
Bibliography: p. 115-118.
1. Police—United States. 2. Privacy, Right of—United States. 3. Intelligence service—United States. 4. Civil rights—United States. I. Title.
HV8138.F73 1979 363.2 79-64537
00 MARC

CIVIL RIGHTS
—UNITED STATES (Continued)

Jinken Shisō Kenkyūkai.
(Sekai kakkoku jinken sengen no kenkyū)
世界各國人權宣言の研究 / 人權思想研究會編 ; ⌊代表者 関之⌋. — 東京 : 巖松堂書店, 昭和25-27 ⌊1950-1952⌋
3 v. ; 19 cm. — (人權思想研究叢書 ; 第1-3卷)
CONTENTS: ⌊1⌋ 英米の部—⌊2⌋ フランス・スイスの部—⌊3⌋ ドイツ・ソ連・イタリヤの部
1. Civil rights—Europe. 2. Civil rights—United States. I. Seki, Itaru, 1903- II. Title. III. Series: Jinken shisō kenkyū sōsho ; dai 1-3-kan.
79-801146

Kemerer, Frank R
Constitutional rights and student life : value conflict in law and education, cases and materials / by Frank R. Kemerer and Kenneth L. Deutsch. ; introductory comment by David L. Bazelon. — St. Paul : West Pub. Co., 1979.
xxix, 735 p. ; 27 cm.
Bibliography: p. 720-728.
Includes index.
ISBN 0-8299-2051-X : $15.95
1. Students—Legal status, laws, etc.—United States. 2. Educational law and legislation—United States. 3. Civil rights—United States. I. Deutsch, Kenneth L., joint author. II. Title.
KF4150.K46 344'.73'0793 79-13935
79 MARC

Kozharov, Asen.
(Pravata na choveka i ideologicheskata borba dnes)
Правата на човека и идеологическата борба днес : ⌊полит. очерк⌋ / Асен Кожаров. — 1. изд. — София : Партиздат, 1977.
69 p. : ill. ; 20 cm. Bu 77-2555
Includes bibliographical references.
0.23 lv
1. Civil rights (International law) 2. Dissenters—Legal status, laws, etc. 3. Civil rights—United States. 4. Civil rights—Russia. I. Title.
K3240.4.K69 78-421597

National Association of Attorneys General. Committee on the Office of Attorney General.
Awards against state defendants under the Civil rights attorneys' fees awards act. — Raleigh, N.C. : Committee on the Office of Attorney General, National Association of Attorneys General, 1978.
vii, 46 p. ; 28 cm.
Includes bibliographical references.
$4.00
1. Judgments—United States. 2. Lawyers—United States—Fees. 3. Civil rights—United States. I. Title.
KF8995.Z9N37 1978 347'.73'77 78-109580
79 MARC

Nikolaev, Ivan Petrovich.
(Pod Zheleznoĭ pi͡atoĭ)
Под Железной пятой / И. Николаев. — Москва : Мол. гвардия, 1978.
175 p. : ill. ; 20 cm. USSR 78
0.85rub
Includes bibliographical references.
1. United States—Politics and government—1945- 2. Anti-communist movements—United States. 3. Civil rights—United States. 4. Intelligence service—United States. I. Title.
E839.8.N54 79-409532

Perry, Richard L ed.
Sources of our liberties : documentary origins of individual liberties in the United States Constitution and Bill of Rights / edited by Richard L. Perry, under the general supervision of John C. Cooper. — Rev. ed. — Chicago : American Bar Foundation, 1978.
xxi, 466 p. ; 26 cm.
Bibliography: p. 443-456.
Includes index.
ISBN 0-910058-90-3
1. Civil rights—United States. I. Title.
KF4744 1978 342'.73'085 78-67316
79 MARC

Rise of the new right. ⌊Sound recording⌋ Washington, National Education Association, p1979.
4 cassettes. (Critical issues in education, v. 4)
Speeches, presented at the 17th annual conference of the National Education Association on human and civil rights in education, held in Washington, D. C., in 1979.
Duration : 3 hr., 11 min.
In container (19 x 27 cm.)
Descriptive and biographical notes inserted.
CONTENTS : Houston, J. Beyond right and left; education and its transformation.—Dellums, R. V. Minorities and the new right. 2 cassettes.—Herndon, T. Beyond right and left; future directions.
1. Minorities—Education—United States. 2. Educational equalization—United States. 3. Civil rights—United States. I. National Education Association of the United States. II. Series: Critical issues in education. ⌊Sound recording⌋ v. 4.
[LC3731] 79-740788

Rohrer, Daniel M
Freedom of speech and human rights : an international perspective / Daniel Morgan Rohrer. — Dubuque, Iowa : Kendall-/Hunt Pub. Co., c1979.
viii, 352 p. ; 23 cm.
Includes bibliographical references.
ISBN 0-8403-1987-8
1. Liberty of speech. 2. Civil rights (International law) 3. Liberty of speech—United States. 4. Civil rights—United States. I. Title.
K3254.R63 342'.73'085 78-71472
79 MARC

Slawson, John, 1896-
Unequal Americans : practices and politics of intergroup relations / John Slawson, in collaboration with Marc Vosk. — Westport, Conn. : Greenwood Press, 1979.
xiii, 249 p. ; 25 cm. — (Contributions in political science ; no. 24 ISSN 0147-1066)
Includes bibliographical references and index.
ISBN 0-313-21118-3
1. Minorities—United States. 2. United States—Social conditions—1960- 3. Civil rights—United States. 4. Social interaction—United States. I. Vosk, Marc, joint author. II. Title. III. Series.
E184.A1S63 301.45'1'0973 78-26318
78 MARC

United States. Commission on Civil Rights.
Age discrimination in Federally-assisted programs : hearing before the United States Commission on Civil Rights : hearing held in San Francisco, California, June 27-28, 1977. — ⌊Washington⌋ : The Commission, ⌊1978⌋
2 v. ; 23 cm.
On spine: Hearing on age discrimination, San Francisco, 1977.
Includes index.
CONTENTS: v. 1. Testimony.—v. 2. Exhibits.
1. Age discrimination—United States. 2. Civil rights—United States. I. Title. II. Title: Hearing on age discrimination, San Francisco, 1977.
JC599.U5U5 1978 323.4 78-601137
79 MARC

United States. Congress. Senate. Committee on the Judiciary. Subcommittee on the Constitution.
Civil rights improvements act of 1977 : hearings before the Subcommittee on the Constitution of the Committee on the Judiciary, United States Senate, Ninety-fifth Congress, second session on S. 35 .... — Washington : U.S. Govt. Print. Off., 1978.
vii, 766 p. : ill. ; 24 cm.
Hearings held Feb. 8-May 3, 1978.
Includes bibliographical references.
Appendices (p. 528-766): pt. 1. Additional statements and correspondence.—pt. 2. Law review articles.—pt. 3. Newspaper articles.—pt. 4. Court cases.
1. Civil rights—United States. 2. Government liability—United States. I. Title.
KF26.J8359 1978d 342'.73'085 79-600570
79 MARC

Wolfe, Alan, 1942-
The seamy side of democracy : repression in America / Alan Wolfe. — 2d ed. — New York : Longman, c1978.
xxiii, 247 p. ; 23 cm.
Includes bibliographical references and index.
ISBN 0-582-28052-4
1. Civil rights—United States. 2. Violence—United States. I. Title.
JC599.U5W64 1978 301.5'92 78-9410
78 MARC

—UNITED STATES—ADDRESSES, ESSAYS, LECTURES

Dellums, Ronald V 1935-
Minorities and the new right. ⌊Sound recording⌋ Washington, National Education Association, ⌊p1979⌋
2 cassettes. (Rise of the new right) (Critical issues in education, v. 4)
Title from accompanying notes.
Speech, presented by the author at the 17th annual conference of the National Education Association on human and civil rights in education, held in Washington, D.C., in 1979.
Duration : 1 hr., 16 min.
In container (19 x 27 cm.)
Descriptive and biographical notes inserted.
SUMMARY : Congressman Ronald Dellums charges the political right wing with fostering polarization at a time when human survival depends on unity. Suggests a political alternative based on human values that transcend differences of race, sex, class, and age.
1. Civil rights—United States—Addresses, essays, lectures. 2. Conservatism—United States—Addresses, essays, lectures. 3. Social values—Addresses, essays, lectures. I. National Education Association of the United States. II. Title. III. Series: Rise of the new right. ⌊Sound recording⌋ IV. Series: Critical issues in education. ⌊Sound recording⌋ v. 4.
[JC599] 79-740588

Immigration and public policy : human rights for undocumented workers and their families / edited by Antonio José Ríos-Bustamante: — 1978 rev. ed. — Los Angeles : Chicano Studies Center-Publications, ⌊1978⌋ c1977.
vi, 401 p. : ill. ; 28 cm. — (Chicano Studies Center document ; no. 5)
Includes bibliographies.
1. Alien labor—United States—Addresses, essays, lectures. 2. Mexicans in the United States—Addresses, essays, lectures. 3. Civil rights—United States—Addresses, essays, lectures. I. Ríos-Bustamante, Antonio José. II. Series: California. University. University at Los Angeles. Chicano Studies Center. Chicano Studies Center document ; no. 5.
HD8081.A5I49 1978 331.6'2'72073 79-622163
79 MARC

The New egalitarianism : questions and challenges / edited by David Lewis Schaefer. — Fort Washington, N.Y. : Kennikat Press, 1979.
248 p. ; 23 cm. — (National university publications)
Bibliography: p. 244-245.
Includes index.
ISBN 0-8046-9220-3 : $15.00
1. Civil rights—United States—Addresses, essays, lectures. 2. Affirmative action programs—United States—Addresses, essays, lectures. 3. Political science—United States—Addresses, essays, lectures. 4. Equality—Addresses, essays, lectures. 5. Social justice—Addresses, essays, lectures. I. Schaefer, David Lewis, 1943- .
JC599.U5N46 323.4'0973 78-27063
78 MARC

The William O. Douglas inquiry into the state of individual freedom / edited by Harry S. Ashmore ; with an introd. by Vern Countryman ; contributors, William H. Alsup ... ⌊et al.⌋. — Boulder, Colo. : Westview Press, 1979.
xvii, 250 p. : port. ; 24 cm.
Includes bibliographical references.
ISBN 0-89158-371-8 : $15.00. ISBN 0-89158-493-5 pbk. : $7.95
1. Civil rights—United States—Addresses, essays, lectures. I. Douglas, William Orville, 1898- . II. Ashmore, Harry S. III. Alsup, William H.
KF4749.A2W54 342'.73'085 78-21438
78 MARC

—UNITED STATES—BIBLIOGRAPHY

Dworaczek, M
Human rights : government documents held in the Research Library / Marian Dworaczek. — 2d ed. — Toronto : Ontario Ministry of Labour, Research Library, 1975.
61 p. ; 30 cm. — (Bibliography series - Ministry of Labour, Research Branch ; no. 1) C***
1. Civil rights—Canada—Bibliography. 2. Civil rights—Great Britain—Bibliography. 3. Civil rights—United States—Bibliography. 4. Canada—Government publications—Bibliography. 5. Great Britain—Government publications—Bibliography. 6: United States—Government publications—Bibliography. I. Ontario. Ministry of Labour. Library. II. Title. III. Series: Ontario. Ministry of Labour. Research Branch. Bibliography series - Ministry of Labour, Research Branch ; no. 1.
K3236.D9 1975 016.3234 79-301630
79 MARC

—UNITED STATES—CASES

Cushman, Robert Fairchild, 1918-
Cases in civil liberties / Robert F. Cushman. — 3d ed. — Englewood Cliffs, N.J. : Prentice-Hall, c1979.
xx, 441 p. ; 28 cm.
Consists of pt. 2, Relations between the individual and his government, of Cases in constitutional law, 5th ed., 1979, by R. F. Cushman.
ISBN 0-13-118596-9 : $9.95
1. Civil rights—United States—Cases. I. Title.
KF4748.C8 1979 342'.73'0850264 79-890
79 MARC

Gunther, Gerald, 1927-
1978 supplement, Cases and materials on constitutional law, ninth edition, Cases and materials on individual rights in constitutional law, second edition / by Gerald Gunther. — Mineola, N.Y. : Foundation Press, 1978.
xx, 418 p. ; 26 cm. — (University casebook series)
Cover title: Constitutional law and individual rights in constitutional law: cases and materials, 1978 supplement.
1. United States—Constitutional law—Cases. 2. Civil rights—United States—Cases. I. Gunther, Gerald, 1927- Cases and materials on constitutional law. 9th ed. II. Gunther, Gerald, 1927- Cases and materials on individual rights in constitutional law. 2d ed. III. Title. IV. Title: Constitutional law and individual rights in constitutional law: cases and materials, 1978 supplement. V. Series.
KF4549.G85 1975 Suppl. 2 78-111431
342'.73'085
79 MARC

Oddo, Gilbert Lawrence.
Freedom and equality : civil liberties and the Supreme Court / Gilbert L. Oddo. — Santa Monica, Calif. : Goodyear Pub. Co., c1979.
xi, 396 p. ; 24 cm.
Includes bibliographical references and index.
ISBN 0-87620-330-6
1. Civil rights—United States—Cases. 2. United States. Supreme Court. I. Title.
KF4748.O3 342'.73'085 78-27592
78 MARC

—UNITED STATES—DIGESTS

Human rights docket, US 1979 / Ann Fagan Ginger, editor ; Carolyn Mohr, associate editor ; foreword by Frank Newman. — Berkeley, Calif. : Meiklejohn Civil Liberties Institute, c1979.
xxxiv, 279 p. ; 28 cm.
Includes index.
ISBN 0-913876-11-9 : $35.00
1. Civil rights—United States—Digests. 2. Civil rights (International law)—Digests. I. Ginger, Ann Fagan. II. Mohr, Carolyn Curtis, 1919- III. Meiklejohn Civil Liberties Institute.
KF4746.3.H85 342'.73'08502648 79-4334
79 MARC

CIVIL RIGHTS (Continued)

—UNITED STATES—DIRECTORIES

Bundy, Mary Lee, 1927-
The national civil rights directory : an organizations directory / edited by Mary Lee Bundy and Irvin Gilchrist. — 1st ed. — College Park, Md. : Urban Information Interpreters, c1979.
iii, 183 p. ; 29 cm.
Bibliography: p. 165-183.
1. Civil rights—United States—Directories. I. Gilchrist, Irvin, joint author. II. Urban Information Interpreters. III. Title.
JC599.U5B84 323.4'025'73 79-113386
00 MARC

—UNITED STATES—HANDBOOKS, MANUALS, ETC.

Secretariat for Hispanic Affairs.
The farmworker family : a people on the move : program booklet in celebration of Farm Worker Week, May 1-7, 1978 / Secretariat for Hispanic Affairs, NCCB/USCC. — Washington : The Secretariat, c1978.
22, 22 p. : ill. ; 22 cm.
Added t.p. in Spanish has title: La familia campesina.
English and Spanish.
Bibliography: p. 19 (both groups)
1. Migrant agricultural laborers—United States—Handbooks, manuals, etc. 2. Civil rights—United States—Handbooks, manuals, etc. I. Title. II. Title: La familia campesina.
HD1525.S38 1978 331.5'44'0973 78-106416
79 MARC

—UNITED STATES—HISTORY

Goldstein, Robert Justin.
Political repression in modern America from 1870 to the present / Robert Justin Goldstein. — Boston : G. K. Hall, c1978.
xxii, 682 p. ; 24 cm.
Bibliography: p. 653-661.
Includes index.
ISBN 0-8467-0301-7
1. Civil rights—United States—History. I. Title.
JC599.U5G58 323.4'0973 76-9681
77 MARC

—UNITED STATES—HISTORY—ADDRESSES, ESSAYS, LECTURES

Greene, Jack P
All men are created equal : some reflections on the character of the American revolution : an inaugural lecture / by Jack P. Greene. — Oxford : Clarendon Press, 1976.
41 p. ; 22 cm. GB***
"Delivered before the University of Oxford on 10 February 1976."
Includes bibliographical references.
ISBN 0-19-951509-3
1. United States—History—Revolution, 1775-1783—Addresses, essays, lectures. 2. Civil rights—United States—History—Addresses, essays, lectures. I. Title.
E297.G68 323.4 79-303117
79 MARC

Konvitz, Milton Ridvas, 1908-
Judaism and the American idea / Milton R. Konvitz. — Ithaca, N.Y. : Cornell University Press, 1978.
223 p. ; 22 cm.
Includes bibliographical references and index.
ISBN 0-8014-1181-5
1. Civil rights—United States—History—Addresses, essays, lectures. 2. Political science—United States—History—Addresses, essays, lectures. 3. Judaism—History—To 70 A.D.—Addresses, essays, lectures. 4. Judaism and democracy—Addresses, essays, lectures. 5. Bible—Political science—Addresses, essays, lectures. I. Title.
JC599.U5K63 1978 323.4'01 78-58028
78 MARC

Roche, John Pearson, 1923-
Development of civil liberties in democratic societies ₍by₎ John P. Roche. ₍Rome?₎ Banco di Roma ₍1965?₎
21 p. illus. 23 cm.
"Discorso pronunciato a Roma, il 29 Octobre 1965 ... sotto gli auspici del Centro italiano di studi per la conciliazione internazionale."
Introductory text in Italian.
1. Civil rights—United States—History—Addresses, essays, lectures. I. Centro italiano di studi e pubblicazioni per la riconciliazione internazionale, Rome. II. Title.
JC599.U5R57 323.4'0973 79-312846
79 MARC

—UNITED STATES—JUVENILE LITERATURE

Sgroi, Peter P
Blue jeans and black robes : teenagers and the Supreme Court / Peter Sgroi. — New York : J. Messner, c1979.
192 p. ; 22 cm.
Bibliography: p. 187.
Includes index.
SUMMARY: Presents and defends the argument that minors are one of the last groups to receive their rights as evidenced in various Supreme Court decisions.
ISBN 0-671-32884-0
1. Children—Legal status, laws, etc.—United States—Juvenile literature. 2. Civil rights—United States—Juvenile literature. ₍1. Children's rights. 2. Civil rights. 3. Law₎
KF479.Z9S48 346'.73'013 79-115
79 MARC

—UNITED STATES—PERIODICALS

Human rights. v. 1–
Aug. 1970–
₍Chicago, Section of Individual Rights and Responsibilities of the American Bar Association₎
23 cm.
Microfilm. ₍South Hackensack, N. J., Rothman Reprints, Inc. 1975?₎– reel. 35 mm. (Contemporary legal periodical series)
1. Civil rights—United States—Periodicals. I. American Bar Association. Section of Individual Rights and Responsibilities. II. Series: Contemporary legal periodical series.
Microfilm LL–0256 K 78-646396
MARC-S

—UNITED STATES—POPULAR WORKS

Newman, Edwin S
Civil liberty and civil rights / by Edwin S. Newman. — 6th ed. / rev. by the Oceana editorial staff. — Dobbs Ferry, N.Y. : Oceana Publications, 1979.
vi, 122 p. ; 19 cm. — (Legal almanac series ; no. 13)
Includes index.
ISBN 0-379-11110-1 : $4.95
1. Civil rights—United States—Popular works. I. Oceana Publications, inc. II. Title.
KF4750.N44 1979 342'.73'085 78-12034
78 MARC

—URUGUAY

Derechos humanos en el Uruguay. — Montevideo : República Oriental del Uruguay, 1977.
183 p. ; 24 cm. — (Serie de temas nacionales ; 2)
"Selección, textos y compilación: Federico Fernández Prando."
Bibliography: p. 181-182.
1. Civil rights—Uruguay. 2. Civil rights. I. Series.
342'.895'085 78-103229
78 MARC

Inter-Church Committee on Chile.
Le cone sud de l'Amérique latine : une prison gigantesque : mission d'observation au Chili, en Argentine et en Uruguay, 30 septembre au 10 octobre 1976 / présentée par the Inter-Church Committee on Chile, en collaboration avec Comité chrétien pour les droits de l'homme en Amérique latine. — ₍Montréal : Comité chrétien pour les droits de l'homme en Amérique latine, 1976₎
93 p. : ill. ; 28 cm. C***
Cover title.
Bibliography: p. 92.
1. Civil rights—Chile. 2. Civil rights—Argentine Republic. 3. Civil rights—Uruguay. I. Comité chrétien pour les droits de l'homme en Amérique latine. II. Title. III. Title: Une prison gigantesque.
JC599.C5I57 1976 323.4'098 79-369296
79 MARC

—WYOMING

United States. Commission on Civil Rights. Wyoming Advisory Committee.
The emergence of civil rights in Wyoming : a report / prepared by the Wyoming Advisory Committee to the U.S. Commission on Civil Rights. — ₍Washington₎ : The Commission, 1978.
ix, 48 p. ; 27 cm.
1. Civil rights—Wyoming. 2. Discrimination—Wyoming. I. Title.
KFW4611.A83 323.4'09787 79-602145
79 MARC

—YUGOSLAVIA

Hoxha, Hajredin.
Titov doprinos nacionalnoj ravnopravnosti / Hajredin Hodža. — Priština : Pokrajinski centar za marksističko obrazovanje, 1977.
46 p. ; 20 cm. — (Biblioteka Tito i samoupravni socijalizam) Yu***
Issued also under title: Kontributi i Titos barabarësisë nacionale.
Includes bibliographical references.
1. Nationalism and socialism. 2. Minorities—Yugoslavia. 3. Civil rights—Yugoslavia. 4. Tito, Josip Broz, Pres. Yugoslavia, 1892- I. Title.
HX550.N3H6917 79-346108
79 MARC

Hrnjaz, Radovan.
Društvena samozaštita u teoriji i praksi samoupravnog društva / Radovan Hrnjaz. — Banjaluka : Glas, 1977.
250 p. ; 20 cm. Yu***
Errata slip inserted.
Bibliography: p. 241-244.
1. Labor laws and legislation—Yugoslavia. 2. Trade-unions—Yugoslavia. 3. Works councils—Yugoslavia. 4. Civil rights—Yugoslavia. I. Title.
79-350618
79 MARC

—YUGOSLAVIA—CONGRESSES

Seminar on the Promotion and Protection of Human Rights of National, Ethnic and Other Minorities, Ohrid, Yugoslavia, 1974.
The Ohrid Seminar on Minorities : United Nations Seminar on the Promotion and Protection of Human Rights of National, Ethnic, and Other Minorities, Ohrid, Yugoslavia, 25 June-8 July 1974 / ₍editor Boris Vishinski₎. — Skopje : Macedonian Review Editions, 1977.
287 p. ; 23 cm. Yu***
Includes bibliographical references.
1. Nationalism—Yugoslavia—Congresses. 2. Minorities—Yugoslavia—Congresses. 3. Civil rights—Yugoslavia—Congresses. I. Višinski, Boris. II. Title.
DR314.S43 1974 323.4'09497 79-316780
79 MARC

CIVIL RIGHTS (INTERNATIONAL LAW)

Alzamora Valdez, Mario.
Los derechos humanos y su protección / Mario Alzamora Valdez. — Lima : JUS, 1977.
267 p. ; 21 cm.
"Apéndice₍s₎ I-VI" (p. 151-267) contain documents.
Includes bibliographical references.
1. Civil rights (International law) 2. Civil rights. I. Title.
K3240.4.A79 341.48'1 79-357445
79 MARC

De Belgische grondwet, De rechten van de mens, De Belgische provinciewet, De Belgische gemeentewet = La constitution belge, Les droits de l'homme, La loi provinciale belge, La loi communale belge. — 2500 Lier, (Grote Markt 39) : J. van In, 1977.
253 p. ; 20 cm. Be77-5
Dutch and French.
ISBN 9030604735 : 225F
1. Belgium—Constitutional law. 2. Civil rights (International law) 3. Civil rights—Belgium. 4. Local government—Belgium. I. Title: La constitution belge.
79-351447
*79 MARC

Broch, Hermann, 1886-1951.
Menschenrecht und Demokratie : polit. Schriften / Hermann Broch ; hrsg. u. eingel. von Paul Michael Lützeler. — 1. Aufl. — Frankfurt am Main : Suhrkamp, 1978.
285 p. ; 19 cm. — (Bibliothek Suhrkamp ; Bd. 588) GFR78-A
Bibliography: p. 281-282.
ISBN 3-518-01588-5 : DM14.80
1. Civil rights (International law) 2. Democracy. 3. Political ethics. I. Lützeler, Paul Michael. II. Title.
K3240.4.B75 341.48'1 78-393572
79 MARC

Cramer, Dettmar, 1929-
Bürgerrechte 77 ₍i.e. siebenundsiebzig₎ / Dettmar Cramer. — Köln : Verlag Wissenschaft u. Politik, 1977.
215 p. ; 19 cm. — (Bibliothek Wissenschaft und Politik) GFR78-A
Includes bibliographical references.
ISBN 3-8046-8540-4
1. Civil rights (International law) 2. Civil rights—Russia. 3. Civil rights—Germany, West. I. Title.
K3240.4.C66 341.48'1 78-379577
78 MARC

Dadiani, Lionel' I͡Akovlevich.
(Mezhdunarodnai͡a zashchita prav cheloveka)
Международная защита прав человека / Л. Я. Дадиани, И. И. Яковлев. — Москва : Знание, 1968.
30 p. ; 22 cm. — (Новое в жизни, науке, технике : Серия Государство и право ; 10, 1968)
Includes bibliographical references.
1. Civil rights (International law) I. I͡Akovlev, Igor' Ivanovich, joint author. II. Title. III. Series: Novoe v zhizni, nauke, tekhnike ; Serii͡a Gosudarstvo i pravo ; 1968, 10.
79-411444
[K3240.4]

Derechos humanos y derechos de los estados. — Tegucigalpa : Consejo Asesor de la Jefatura de Estado, 1977.
38 p. ; 24 cm.
CONTENTS: Declaración universal de derechos humanos.—Declaración de los derechos del niño.—Convención americana sobre derechos humanos.—Carta de deberes y derechos económicos de los estados.
1. Civil rights (International law) I. Honduras. Jefatura de Estado. Consejo Asesor.
K3238.A1D45 341.48'1 78-104680
79 MARC

CIVIL RIGHTS (INTERNATIONAL LAW)
(Continued)

Eide, Asbjørn.
Fattig, ufri og mishandlet : om det internasjonale menneskerettighetsvernet / Asbjørn Eide. — Oslo : Universitetsforl., c1978 (Oslo : Aktietr.i Trondhjem)
238 p., [1] leaf ; 18 cm. — (U-bok ; nr. 245) N78-Dec.
Bibliography: p. 187-[190]
ISBN 8200032558 : kr58.00
1. Civil rights (International law) 2. Civil rights—Latin America. 3. Civil rights. I. Title.
K3240.4.E37 79-369496
79 MARC

Farace, Alessandro.
Diritti umani e distensione : commento a due voci tenuto a Roma il 13 aprile 1978, nella sede del Banco di Roma / Alessandro Farace, Leo J. Wollemborg ; sotto gli auspici del Centro italiano di studi per la conciliazione internazionale. — Roma : Banco di Roma, [1978?]
37 p., [3] leaves of plates : 3 group ports. ; 23 cm. It***
1. Civil rights (International law) 2. Pacific settlement of international disputes. I. Wollemborg, Leo J., joint author. II. Title.
K3240.4.F37 341.48'1 79-355160
79 MARC

Floretta, Hans.
Die Menschenrechtspakte der Vereinten Nationen : ein Beitrag zum Stand d. Grundrechte in Österreich, insbes. zu den sozialen Grundrechten / von Hans Floretta, Theo Öhlinger. — Wien : Manz, 1978.
70 p. ; 24 cm. — (Rechtswissenschaft und Sozialpolitik ; 13) Au78-15
Includes bibliographical references.
ISBN 3-214-06713-3 : S124.00
1. Civil rights (International law) 2. United Nations. I. Öhlinger, Theo, joint author. II. Title.
K3240.4.F55 341.48'1 78-382700
*79 MARC

Franchini Netto, M 1913-
Os direitos humanos na ONU / Miguel Franchini-Netto. — [Rio de Janeiro] : Faculdade de Direito Teixeira de Freitas em organização e mantida pela Sociedade Educadora Pedro II : distribuição da Livraria Freitas Bastos, [1979]
200 p. ; 24 cm.
Bibliography: p. [163]-171.
Appendices (p. [173]-195) contain U.N. documents in English or Spanish.
Cr$200.00
1. Civil rights (International law) I. Title.
K3240.4.F73 341.48'1 79-114126
79 MARC

Garimberti, Paolo.
Il dissenso nei paesi dell'Est prima e dopo Helsinki / Paolo Garimberti. — Firenze : Vallecchi, 1977.
114 p. ; 18 cm. — (Documenti e interventi) It77-Oct
Includes bibliographical references.
L3000
1. Dissenters—Legal status, laws, etc.—Russia. 2. Dissenters—Legal status, laws, etc.—Europe, Eastern. 3. Civil rights (International law) I. Title.
78-384977
*78 MARC

Gegen kalten Krieg, für Förderung der Menschenrechte / [verantwortlich für die Redaktion, Siegfried Forberger]. — Berlin : DDR-Komitee für Menschenrechte, 1978.
60 p. ; 21 cm. — (Schriften und Informationen - DDR-Komitee für Menschenrechte ; Heft 3/78) GDR***
Cover title.
Includes bibliographical references.
1. Civil rights (International law) 2. Peace. 3. Civil rights—Germany, East. I. Forberger, Siegfried. II. Series: DDR-Komitee für Menschenrechte. Schriften und Informationen - DDR-Komitee für Menschenrechte ; 1978, Heft 3.
K3240.4.G44 341.48'1 79-363378
79 MARC

Göttinger Arbeitskreis.
Die KSZE und die Menschenrechte : polit. u. rechtl. Überlegungen zur 2. Etappe / mit Beitr. von Dieter Blumenwitz ... [et al.]. — Berlin : Duncker und Humblot, 1977.
143 p. ; 24 cm. — (Studien zur Deutschlandfrage ; Bd. 2) (Veröffentlichung - Göttinger Arbeitskreis ; Nr. 420) GFR78-A
Includes bibliographical references.
ISBN 3-428-04042-2 : DM28.00
1. European cooperation. 2. Security, International. 3. Conference on Security and Cooperation in Europe. 4. Civil rights (International law) I. Blumenwitz, Dieter. II. Title. III. Series.
JX1542.Z5 1977g 78-379815
78 MARC

Grünbuch zu den Folgewirkungen der KSZE / hrsg. von Jost Delbrück, Norbert Ropers, Gerda Zellentin. — Köln : Verlag Wissenschaft u. Politik, 1977.
541 p. : ill. ; 23 cm. — (DGFK-Veröffentlichungen ; Bd. 3) GFR78-A
Includes bibliographical references.
ISBN 3-8046-8537-4
1. Conference on Security and Cooperation in Europe. 2. Europe, Eastern—International economic relations—European Economic Community countries. 3. European Economic Community countries—International economic relations—Europe, Eastern. 4. Human rights (International law) 5. Detente. I. Delbrück, Jost. II. Ropers, Norbert. III. Zellentin, Gerda, 1934- IV. Series: Deutsche Gesellschaft für Friedens- und Konfliktforsung. Veröffentlichungen der Deutschen Gesellschaft für Friedens- und Konfliktforschung ; Bd. 3.
JX1393.C65G78 78-381184
79 MARC

International human rights law and practice : the roles of the United Nations, the private sector, the government, and their lawyers / James C. Tuttle, editor. — Rev. ed. — [Chicago] : American Bar Association, c1978.
iii, 255 p. ; 23 cm. — (Professional education publications)
Papers presented at a national institute of continuing legal education, held Apr. 25-26, 1976 in Washington, D.C., conducted by the American Bar Association's Section of International Law.
Bibliography: p. 36-37.
1. Civil rights (International law) I. Tuttle, James C. II. American Bar Association. Section of International Law. III. Series.
K3240.4.I57 1978 341.48'4 78-61659
79 MARC

Japan. Fōmufu. Jinken Yōgokyoku.
(Sekai jinken sengen seiritsu no keii)
世界人権宣言成立の経緯. — [東京] : 法務府人権擁護局, 昭和26 [1951]
121, 34 p. ; 21 cm. — (自由人権叢書 ; 第九集)
Cover title.
1. United Nations. General Assembly. Universal declaration of human rights. 2. Civil rights (International law) I. Title. II. Series: Jiyū jinken sōsho ; dai 9-shū.
K3240.4.J36 79-800206

Klenner, Hermann.
Freiheit, Gleichheit und so weiter : dreizehn Streiflichter über die Menschenrechte / Hermann Klenner, [Mitarbeit, Annelies Schwarzer]. — 1. Aufl. — Berlin : Staatsverlag der Deutschen Demokratischen Republik, 1978.
143 p. : ill. ; 17 cm. — (Schriftenreihe Recht in unserer Zeit ; 10) GDR***
Includes bibliographical references.
2.25M
1. Civil rights. 2. Civil rights—Communist countries. 3. Civil rights (International law) I. Schwarzer, Annelies, joint author. II. Title.
K3240.4.K56 323.4'09171'7 78-395441
79 MARC

Klenner, Hermann.
Menschenrechte : Anspruch und Wirklichkeit / [H. Klenner, B. Graefrath, V. A. Sorin]. — Berlin : DDR-Komitee für Menschenrechte, 1978.
56 p. ; 21 cm. — (Schriften und Informationen - DDR-Komitee für Menschenrechte ; Heft 4/78) GDR***
Cover title.
Includes bibliographical references.
1. Civil rights (International law) 2. Civil rights. 3. Civil rights—Communist countries. I. Graefrath, Bernhard, joint author. II. Zorin, Valerian Aleksandrovich, 1902- joint author. III. Title. IV. Series: DDR-Komitee für Menschenrechte. Schriften und Informationen - DDR-Komitee für Menschenrechte ; Heft 1978, 4.
K3240.4.K566 341.48'1 79-365012
79 MARC

Kozharov, Asen.
(Pravata na choveka i ideologicheskata borba dnes)
Правата на човека и идеологическата борба днес : [полит. очерк] / Асен Кожаров. — 1. изд. — София : Партиздат, 1977.
69 p. : ill. ; 20 cm. Bu 77-2555
Includes bibliographical references.
0.23 lv
1. Civil rights (International law) 2. Dissenters—Legal status, laws, etc. 3. Civil rights—United States. 4. Civil rights—Russia. I. Title.
K3240.4.K69 78-421597

Lapenna, Ivo.
Human rights : Soviet theory and practice : Helsinki and international law / by Ivo Lapenna. — London : Institute for the Study of Conflict, 1977.
15 p. ; 30 cm. — (Conflict studies ; no. 83) GB***
Cover title.
"Summary": [2] p. inserted.
Includes bibliographical references.
£2.00 ($4.50 U.S.)
1. Civil rights—Russia. 2. Civil rights (International law) I. Title. II. Series.
D839.3.C6 no. 83 327'.08 s 78-315542
78 MARC

Lauterpacht, Hersch, Sir, 1897-1960.
An international bill of the rights of man, by H. Lauterpacht ... New York, Columbia university press, 1945.
x, 230 p. 24 cm.
———Microfilm
Microfilm LL-2422 K
1. Civil rights (International law) I. Title.
K3240.4.L37 341.48'1 45-5865

Lauterpacht, Hersch, Sir, 1897-1960.
International law and human rights. London, Stevens, 1950.
xvi, 475 p. 25 cm.
1. Civil rights (International law) I. Title.
K3240.4.L38 1950 341.48'1 54-25357
[r79]rev MARC

Lauterpacht, Hersch, Sir, 1897-1960.
International law and human rights. [1st ed.] New York, F. A. Praeger, 1950.
xvi, 475 p. 25 cm.
1. Civil rights (International law) I. Title.
K3240.4.L38 1950b 341.48'1 a 51-4824
Illinois. Univ. Library [r79]rev2 MARC

Lauterpacht, Hersch, Sir, 1897-1960.
International law and human rights, by Hersch Lauterpacht. With a new introd. for the Garland ed. by Isidore Silver. New York, Garland Pub., 1973.
24, xvi, 475 p. 22 cm. (The Garland library of war and peace)
Reprint of the 1950 ed. published by F. A. Praeger, New York.
Includes bibliographical references.
ISBN 0-8240-0494-9
1. Civil rights (International law) I. Title. II. Series.
K3240.4.L38 1973 341.48'1 72-147756
73[r79]rev MARC

Mower, Alfred Glenn.
The United States, the United Nations, and human rights : the Eleanor Roosevelt and Jimmy Carter eras / A. Glenn Mower, Jr. — Westport, Conn. : Greenwood Press, 1979.
xii, 215 p. ; 22 cm. — (Studies in human rights ; no. 4 ISSN 0146-3586)
Bibliography: p. [209]-212.
Includes index.
ISBN 0-313-21090-X
1. Civil rights (International law) 2. Civil rights. 3. United States—Foreign relations—1945- I. Title. II. Series.
K3240.4.M68 341.48'1 78-22134
79 MARC

Peleš, Aleksandar.
Rasna diskriminacija i međunarodno pravo / Aleksandar Peleš. — Sarajevo : Svjetlost, 1977.
254 p. ; 24 cm. Yu***
Bibliography: p. 239-250.
1. Race discrimination—Law and legislation. 2. Civil rights (International law) I. Title.
K3240.4.P45 78-380646
78 MARC

(Pravda o pravakh cheloveka)
Правда о правах человека : деятели сов. культуры о правах человека / Константин Федин ... [et al. ; составитель Борис Иванов]. — Москва : Политиздат, 1977.
158 p. ; 20 cm. USSR***
0.25rub
1. Civil rights—Russia. 2. Civil rights (International law) 3. Civil rights. I. Fedin, Konstantin Aleksandrovich, 1892- II. Ivanov, Boris Vladimirovich, 1917-
78-422937

Rechten van de mens in mundiaal en europees perspectief. — Maarssen : Ars Aequi Libri, [1978]
x, 229 p. ; 25 cm. — (Rechten van de mens ; 1) Ne***
Dutch or English.
Includes bibliographical references.
ISBN 9070094231
1. Civil rights (International law) 2. Civil rights—Europe. I. Series.
K3240.4.R4 vol. 1 341.48'1 s 78-386861
78 MARC

Rieber-Mohn, Hallvard, 1922-
Fra politikk til forbrytelse : legmannstanker om menneskerettighetene : med tillegg av Verdenserklæringen om menneskerettighetene / Hallvard Rieber-Mohn. — [Oslo] : Cappelen, 1977 Oslo : (Reistad & sønn).
p. 1-52, [1] leaf ; 22 cm. N77-Dec.
ISBN 8202039452 : kr20.00. ISBN 8202039444 pbk.
1. Civil rights (International law) I. United Nations. General Assembly. Universal declaration of human rights. Norwegian. 1977. II. Title.
K3240.4.R53 78-354851
79 MARC

Rohrer, Daniel M
Freedom of speech and human rights : an international perspective / Daniel Morgan Rohrer. — Dubuque, Iowa : Kendall-/Hunt Pub. Co., c1979.
viii, 352 p. ; 23 cm.
Includes bibliographical references.
ISBN 0-8403-1987-8
1. Liberty of speech. 2. Civil rights (International law) 3. Liberty of speech—United States. 4. Civil rights—United States. I. Title.
K3254.R63 342'.73'085 78-71472
79 MARC

Soziale Grundrechte : Arbeit, Bildung, Umweltschutz etc. / zsgest. u. mit e. Einl. vers. von Karl Hernekamp. — Berlin ; New York : de Gruyter, 1979.
256 p. ; 19 cm. — (Aktuelle Dokumente) GFR79-A
Bibliography: p. [235]-252.
Includes index.
ISBN 3-11-007756-6 : DM19.80
1. Civil rights. 2. Civil rights (International law) I. Hernekamp, Karl.
K3240.4.S6 342.08'5 79-382164
79 MARC

CIVIL RIGHTS (INTERNATIONAL LAW) (Continued)

Stavropoulos, Konstantin, 1947-
Das Verbot der Folter und der unmenschlichen oder erniedrigenden Behandlung oder Strafe im gegenwärtigen Völkerrecht : insbesondere nach Art. 3 der Europäischen Menschenrechts-Konvention / vorgelegt von Konstantin Stavropoulos. — ⌈s.l. : s.n.⌉, 1976. (Bamberg : M. Schadel)
139 p. ; 21 cm. GFR***
Thesis—Heidelberg.
Vita.
Bibliography: p. 128-139.
1. Torture. 2. Punishment. 3. Civil rights (International law) 4. Convention for the protection on human rights and fundamental freedoms. I. Title.
K5410.T6S73 79-371154
79 MARC

Steinmann, Wolfgang, 1936-
Grenzen einer völkerrechtlichen Garantie der individuellen Meinungsfreiheit. Würzburg, 1965.
x, 112 p. 21 cm.
Inaug.-Diss.—Würzburg.
Vita.
Bibliography: p. iv-ix.
1. Civil rights (International law) I. Title.
K3240.4.S7 77-483175
78 MARC

Textos internacionales de derechos humanos / Javier Hervada, José M. Zumaquero. — Pamplona : Ediciones Universidad de Navarra, 1978.
1012 p. ; 24 cm. — (Colección jurídica ; 76) Sp***
English, French, or Spanish.
Bibliography: p. 19-20.
Includes index.
ISBN 8431305444
1. Civil rights (International law) 2. Civil rights. I. Hervada Xiberta, Francisco Javier. II. Zumaquero, José M. III. Series: Colección jurídica (Pamplona) ; 76.
K3238.A1T49 341.48'1 79-346011
79 MARC

Truyol Serra, Antonio, comp.
Los derechos humanos : declaraciones y convenios internacionales / con un estudio preliminar de Antonio Truyol. — 2. ed. nueva ed. / con un epílogo del autor, Los derechos humanos, diez años después. — Madrid : Tecnos, 1977.
187 p. ; 20 cm. — (Colección Ventana abierta) Sp77-Nov
"Declaración universal de derechos humanos": p. 63-68.
Includes bibliographies.
ISBN 8430907106
1. Civil rights (International law) I. United Nations. General Assembly. Universal declaration of human rights. Spanish. 1977. II. Title.
K3238.A1D44 1977 341.48'1 78-397481
79 MARC

United Arab Republic. Wizārat al-Dākhilīyah. Idārat al-Shu'ūn al-'Āmmah.
(al-I'lān al-'ālamī li-ḥuqūq al-insān)
الاعلان العالمى لحقوق الانسان ودور الشرطة فى تطبيق مبادئه.
⌈القاهرة، وزارة الداخلية، ادارة الشئون العامة، 1968⌉
61 p. illus. 16 cm.
1. Civil rights (International law) I. United Nations. General Assembly. Universal declaration of human rights. Arabic. 1968. II. Title.
K3239.15 1968 79-971296

United Nations. Office of Public Information.
The United Nations and human rights / Office of Public Information, United Nations. — New York : UN, 1978.
xi, 166 p. ; 22 cm. — (United Nations publication ; OPI/621)
"Sales no. E.78.I.18."
$5.00
1. Civil rights (International law) 2. United Nations. I. Title. II. Series: United Nations. Office of Public Information. United Nations publication ; OPI 621.
K3240.4.U47 1978 341.48'1 79-107794
79 MARC

Vukas, Budislav.
Etničke manjine i međunarodni odnosi / Budislav Vukas. — Zagreb : Školska knjiga, 1978.
251 p. ; 20 cm. — (Biblioteka Suvremena misao) Yu***
Bibliography: p. 233-237.
Includes index.
1. Minorities—Yugoslavia. 2. Minorities. 3. Civil rights (International law). I. Title.
79-350686
79 MARC

Wagner, Wolfgang, 1946-
Die Verwirklichung der Menschenrechte im Pakt der Vereinten Nationen über bürgerliche und politische Rechte vom 16. Dezember 1966, im Grundvertrag und den ihn begleitenden Nebeninstrumenten / Wolfgang Wagner. — Frankfurt/M. : P. Lang, 1977.
viii, 122 p. ; 21 cm. — (Augsburger Schriften zum Staats- und Völkerrecht ; Bd. 7) GFR***
Bibliography: p. 109-119.
ISBN 3-261-01722-8
1. Civil rights (International law) I. Title. II. Series.
K3240.4.W3 341.48'1 78-388181
78 MARC

—ADDRESSES, ESSAYS, LECTURES

Carter y la lógica del imperialismo / Hugo Assmann, ed., colaboran, Noam Chomsky ... ⌈et al.⌉. — 1. ed. — Ciudad Universitaria Rodrigo Facio, Costa Rica : Editorial Universitaria Centroamericana, 1978.
2 v. ; 21 cm. — (Colección Dei)
Includes bibliographical references.
1. United States—Foreign relations—1977- —Addresses, essays, lectures. 2. Carter, Jimmy, 1924- —Addresses, essays, lectures. 3. United States—Foreign relations—South America—Addresses, essays, lectures. 4. South America—Foreign relations—United States—Addresses, essays, lectures. 5. Trilateral Commission—Addresses, essays, lectures. 6. Civil rights (International law)—Addresses, essays, lectures. I. Assmann, Hugo, 1933- . II. Chomsky, Noam. III. Series.
E873.C38 78-104449
78 MARC

Human dignity : the internationalization of human rights : with selected international human rights documents : essays based on an Aspen Institute workshop / sponsored by the Program on Justice, Society, and the Individual and the Program in International Affairs ; edited by Alice H. Henkin. — New York : Aspen Institute for Humanistic Studies ; Queenstown, Md. : copies available from the Institute, 1978, c1979.
xii, 203 p. ; 23 cm.
ISBN 0-915436-93-0 : $5.95
1. Civil rights (International law)—Addresses, essays, lectures. I. Henkin, Alice H. II. Aspen Institute for Humanistic Studies. Program on Justice, Society and the Individual. III. Aspen Institute Program on International Affairs.
K3240.6.H85 341.48'1 78-11817
78 MARC

Human rights : thirty years after the Universal declaration : commemorative volume on the occasion of the thirtieth anniversary of the Universal declaration of human rights / edited by B. G. Ramcharan. — The Hague ; Boston : M. Nijhoff ; Hingham, MA : distribution, Kluwer Boston,
xv, 274 p. ; 24 cm.
"Published under the auspices of the International Forum on Human Rights."
"The Universal declaration of human rights": p. ⌈263⌉-270.
Includes bibliographical references and index.
ISBN 9024721458
1. Civil rights (International law)—Addresses, essays, lectures. 2. United Nations—Civil rights—Addresses, essays, lectures. I. Ramcharan, B. G. II. International Forum on Human Rights. III. United Nations. General Assembly. Universal declaration of human rights. 1978.
K3240.6.H88 341.48'1 79-9932
79 MARC

Human rights and U.S. foreign policy / edited by Barry M. Rubin and Elizabeth P. Spiro ; with a foreward by Muriel S. Humphrey. — Boulder, Colo. : Westview Press, 1979.
xix, 283 p. ; 24 cm.
Includes bibliographical references.
ISBN 0-89158-476-5
1. Civil rights—Addresses, essays, lectures. 2. United States—Foreign relations—1945- —Addresses, essays, lectures. 3. Civil rights (International law)—Addresses, essays, lectures. I. Rubin, Barry M. II. Spiro, Elizabeth P.
K3240.6.H86 341.48'1 79-872
79 MARC

Human rights and U.S. foreign policy : principles and applications / edited by Peter G. Brown, Douglas MacLean. — Lexington, Mass. : Lexington Books, c1979.
xxvii, 310 p. ; 24 cm.
Includes bibliographical references and index.
ISBN 0-669-02807-X
1. Civil rights—Addresses, essays, lectures. 2. United States—Foreign relations—1977- —Addresses, essays, lectures. 3. Civil rights (International law)—Addresses, essays, lectures. I. Brown, Peter G. II. MacLean, Douglas, 1947-
K3240.6.H87 341.48'1 78-24647
79 MARC

The international law and policy of human welfare / ed. by Ronald St. John Mcdonald, Douglas M. Johnston and Gerald L. Morris. — Alphen aan den Rijn : Sijthoff & Noordhoff ⌈International Publishers⌉, 1978.
xviii, 690 p. ; 25 cm. Ne79-3
English or French.
Includes bibliographical references.
ISBN 9028608087 : fl 192.40
1. International law—Addresses, essays, lectures. 2. Persons (International law)—Addresses, essays, lectures. 3. Civil rights (International law)—Addresses, essays, lectures. 4. International economic relations—Addresses, essays, lectures. I. Macdonald, Ronald St. J., 1928- . II. Johnston, Douglas M. III. Morris, Gerald L.
JX3091.I66 341.48 79-353803
*79 MARC

Pour un droit des peuples : essais sur la déclaration d'Alger / publiés sous la direction de Antonio Cassese et Edmond Jouve ; avant-propos de Lelio Basso. — ⌈Paris⌉ : Berger Levrault, c1978.
220 p. ; 20 cm. — (Collection Tiers Monde en bref ISSN 0397-4327) F***
Includes bibliographical references.
ISBN 2-7013-0159-9 : 50.00F
1. Civil rights (International law)—Addresses, essays, lectures. I. Cassese, Antonio. II. Jouve, Edmond.
K3240.6.P68 341.48'1 78-399889
79 MARC

Sakharov, Andreĭ Dmitrievich, 1921-
Alarm and hope / Andrei D. Sakharov ; edited by Efrem Yankelevich and Alfred Friendly, Jr. — 1st English language ed. — New York : Alfred A. Knopf : distributed by Random House, c1978.
xix, 200 p. ; 22 cm.
Translation of Trevoga i nadezhda.
Includes index.
ISBN 0-394-50369-4 : $8.95
1. International relations—Addresses, essays, lectures. 2. Civil rights—Russia—Addresses, essays, lectures. 3. Civil rights (International law)—Addresses, essays, lectures. I. Ĭankelevich, Efrem. II. Friendly, Alfred. III. Title.
JX1395.S26513 323.4'0947 78-13845
78 MARC

—CONGRESSES

Die Ausübung staatlicher Gewalt in Ost und West nach Inkrafttreten der UN-Konvention über zivile und politische Rechte : Juristenkonferenz vom 23.-25. September 1977 in Ludwigshafen am Rhein / Dt. Sekt. d. Internat. Juristen-Komm ; mit Vorträgen von Gottfried Zieger ... ⌈et al.⌉. — Heidelberg ; Karlsruhe : Müller, Juristischer Verl., 1978.
103 p. ; 23 cm. — (Rechtsstaat in der Bewährung ; Bd. 6) GFR79-A
Conference organized by the Deutsche Sektion of the International Commission of Jurists.
"Entwurf einer Ergänzung zur UN-Menschenrechtskonvention": p. 101-103.
Includes bibliographical references.
ISBN 3-8114-6678-X : DM32.00
1. Civil rights (International law)—Congresses. 2. Civil rights—Germany, East—Congresses. 3. Civil rights—Europe—Congresses. 4. International covenant on civil and political rights—Congresses. I. Zieger, Gottfried. II. International Commission of Jurists (Founded 1952). Deutsche Sektion. III. Series.
K3239.6.A9 341.48'1 79-356493
79 MARC

—DIGESTS

Human rights docket, US 1979 / Ann Fagan Ginger, editor ; Carolyn Mohr, associate editor ; foreword by Frank Newman. — Berkeley, Calif. : Meiklejohn Civil Liberties Institute, c1979.
xxxiv, 279 p. ; 28 cm.
Includes index.
ISBN 0-913876-11-9 : $35.00
1. Civil rights—United States—Digests. 2. Civil rights (International law)—Digests. I. Ginger, Ann Fagan. II. Mohr, Carolyn Curtis, 1919- . III. Meiklejohn Civil Liberties Institute.
KF4746.3.H85 342'.73'08502648 79-4334
79 MARC

—POETRY

Nací al mundo pa'cantar. — Santiago de Chile : Arzobispado de Santiago, Vicaría de la Solidaridad, 1978.
⌈42⌉ p. : ill. ; 19 x 22 cm.
1. United Nations. General Assembly. Universal declaration of human rights—Poetry. 2. Civil rights (International law)—Poetry. I. Santiago de Chile (Archdiocese). Vicaría de la Solidaridad.
PQ8098.N3 861 79-109579
79 MARC

CIVIL RIGHTS (ISLAMIC LAW)

Qurbānī Lāhījī, Zayn al-'Ābidīn.
(Islām va ḥuqūq-i bashar)
اسلام و حقوق بشر / نگارش زين العابدين قربانى. —
تهران : كتابخانه صدر، ⌈پيشگفتار 1969 i. e. 1348⌉
510 p. ; 24 cm.
Added t. p.: Islam and human rights.
In Persian.
Includes bibliographical references.
1. Civil rights (Islamic law) I. Title. II. Title: Islam and human rights.
79-971363

—CONGRESSES

Conferences of Riyad, Paris, Vatican City, Geneva, and Strasbourg on Moslem doctrine and human rights in Islam, between Saudi canonist and eminent European jurists and intellectuals. — Riyad : Ministry of Justice, ⌈1974?⌉
221 p. ; 22 cm.
ISBN 8439964307
1. Civil rights (Islamic law)—Congresses. 2. Man (Islam)—Congresses. I. Saudi Arabia. Ministry of Justice.
340.5'9 78-107207
79 MARC

CIVIL RIGHTS (JEWISH LAW)

—CONGRESSES

Essays on human rights : contemporary issues and Jewish perspectives / edited by David Sidorsky, in collaboration with Sidney Liskofsky and Jerome J. Shestack ; Salo W. Baron ... [et al.]. — 1st ed. — Philadelphia : Jewish Publication Society of America, c1979.
xlv, 359 p. ; 24 cm.
Originated out of the McGill International Colloquium on Judaism and Human Rights held April 1974 at McGill University.
Includes bibliographical references.
ISBN 0-8276-0107-7 : $12.00
1. Jews—Politics and government—Congresses. 2. Civil rights—Congresses. 3. Civil rights (Jewish law)—Congresses. 4. Jews—Legal status, laws, etc.—Congresses. I. Sidorsky, David. II. Liskofsky, Sidney. III. Shestack, Jerome J. IV. Baron, Salo Wittmayer, 1895- . V. McGill International Colloquium on Judaism and Human Rights, 1974.
DS143.E76 323.1'19'24 78-1170
00 MARC

CIVIL RIGHTS WORKERS

—JAPAN

Seki, Itaru, 1903-
(Jinken yōgo iin hō shakugi)
人権擁護委員法釈義 / 関之著. — 東京：新警察社, 昭和24 [1949]
4. 138 p. : 18 cm.

1. Civil rights—Japan. 2. Civil rights workers—Japan. I. Japan. Laws, statutes etc. Jinken yōgo iin hō. 1949. II. Title.
79-801776

—JAPAN—BIOGRAPHY

(Jiyūtō gekika jiken to Koike Isamu)
自由党激化事件と小池勇 / 村上貢編著. — 名古屋：風媒社, 1976.
216 p. : ill. ; 20 cm. Ja***
Bibliography: p. 214–216.
CONTENTS: 小池勇 自叙伝 幽居余情 明治十九年日誌 塵の海—小池龍雄 祖父の思い出—小池勇年譜—小池勇関係史料—静岡事件参画者・小池勇の半生—『経世新誌』と『学事新報』について—自由党激化諸事件と—斉蜂起論
¥1500
1. Koike, Isamu, 1854–1940. 2. Politicians—Japan—Biography. 3. Civil rights workers—Japan—Biography. 4. Japan—Politics and government—1868–1912. 5. Civil rights—Japan—History. 6. Jiyūtō (Founded 1881)—History. I. Koike, Isamu, 1854–1940. II. Murakami, Mitsugi, 1926–
DS884.K565J59 78–812219

Kawai, Teikichi, 1901-
(Kanda Nishikichō Matsumototei)
神田錦町松本亭：[梁山泊の女・松本フミ] / 川合貞吉. — 東京：学芸書林, 1977.
302 p., [2] leaves of plates : ill. ; 20 cm. Ja 77–15991
¥1300

1. Matsumoto, Fumi, 1882–1970. 2. Civil rights workers—Japan—Biography. 3. Japan—Politics and government—20th century. I. Title.
DS885.5.M277K38 78–814327

Kikuchi, Shigeru, 1879–1932.
(Yanaka-mura mondai to gakusei undō)
谷中村問題と学生運動 / [菊地茂] ; 斎藤英子編. — 東京：早稲田大学出版部, 昭和52 [1977]
6, 632, 34 p., [3] leaves of plates : ill. ; 19 cm. — (菊地茂著作集 ; 第1巻) Ja***
Reprint of the author's Yanaka-mura mondai and Gakusei no kōdoku kyūsai undō which appeared in S. Tanaka's Gijin zenshū, v. 5 published in 1927, and related articles and letters.
Chronological list of the author's works: p. 541–542.
Bibliography: p. 631–632.
Includes indexes.
¥4300

1. Environmental policy—Japan—Yanaka-mura. 2. Kikuchi, Shigeru, 1879–1932. 3. Tanaka, Shōzō, 1841–1913. 4. Civil rights workers—Japan—Biography. 5. Students—Japan—Political activity. I. Saitō, Hideko. II. Title. III. Series: Kikuchi, Shigeru, 1879–1932. Kikuchi Shigeru chosaku shū ; dai 1-kan.
HC463.Y35K52 78–812205

—MISSISSIPPI—JACKSON—BIOGRAPHY

Salter, John R
Jackson, Mississippi : an American chronicle of struggle and schism / John R. Salter, Jr. ; with a foreword by R. Edwin King, Jr. — 1st ed. — Hicksville, N.Y. : Exposition Press, c1979.
xxi, 248 p. ; 24 cm. — (An Exposition-banner book)
ISBN 0-682-49353-8 : $10.00
1. Afro-Americans—Civil rights—Mississippi—Jackson. 2. Jackson, Miss.—Race relations. 3. Evers, Medgar Wiley, 1925-1963. 4. Salter, John R. 5. Civil rights workers—Mississippi—Jackson—Biography.
F349.J13S25 78-75356
323.1'19'6073076251
79 MARC

—SOUTHERN STATES—INTERVIEWS

Beardslee, William R
The way out must lead in : life histories in the civil rights movement / William R. Beardslee. — Limited 1st ed. — Atlanta : Center for Research in Social Change, Emory University, c1977.
vii, 169 p. ; 25 cm.
Bibliography: p. 164-169.
1. Civil rights workers—Southern States—Interviews. 2. Afro-Americans—Southern States—Interviews. 3. Afro-Americans—Civil rights—Southern States. I. Title.
E185.98.A1B4 323.4'092'2 77-77095
77 MARC

—UNITED STATES—BIOGRAPHY

Hedgeman, Anna Arnold.
The gift of chaos : decades of American discontent / Anna Arnold Hedgeman. — New York : Oxford University Press, 1977.
ix, 249 p. ; 22 cm.
Includes bibliographical references and index.
ISBN 0-19-502196-7 : $9.95
1. Afro-Americans—Civil rights. 2. Hedgeman, Anna Arnold. 3. Civil rights workers—United States—Biography. I. Title.
E185.61.H46 323.1'19'6073 76-9242
79 MARC

CIVIL SERVICE

see also Administrative law; Bureaucracy; Civil list; Civil service pensioners; Collective bargaining—Government employees; Correctional personnel; Employee-management relations in government; Local officials and employees; Public service employment; Trade-unions—Government employees; Uniform, Civil; Women in the civil service; and subdivision Officials and employees —Appointment, qualifications, tenure, etc. under names of countries, cities, etc.

International Labor Office.
Career problems in the public service : third item on the agenda : report / prepared by the International Labour Office. — Geneva : The Office, 1970, 1974 printing.
ii, 63 p. ; 24 cm.
At head of title: Report III, International Labour Organization, Joint Committee on the Public Service, first session, Geneva, 1970.
Includes bibliographical references.
ISBN 9221000672
1. Civil service. I. International Labor Organization. Joint Committee on the Public Service. II. Title.
JF1351.I534 1970 350'.6 79-312330
79 MARC

Shaṭā, Hammād Muḥammad.
(al-Niẓām al-qānūnī lil-ujūr wa-al-murattabāt fī al-waẓīfah al-ʻāmmah)
النظام القانونى للأجور والمرتبات فى الوظيفة العامة / حماد محمد شطا. — القاهرة : دار الفكر العربى ، [1977]
419 p. ; 24 cm.
Bibliography : p. 398–411.
£E2.00
1. Civil service. I. Title.
K3440.S5 77–961185

Vereinigung der Deutschen Staatsrechtslehrer.
Verfassungstreue und Schutz der Verfassung / Erhard Denninger u. Hans Hugo Klein. Der öffentliche Dienst im Staat der Gegenwart / Walter Rudolf u. Frido Wagener. — Berlin ; New York : de Gruyter, 1979.
350 p. ; 22 cm. — (Veröffentlichungen der Vereinigung der Deutschen Staatsrechtslehrer ; Heft 37) GFR79-A
"Berichte und Diskussionen auf der Tagung der Vereinigung der Deutschen Staatsrechtslehrer in Bonn vom 4.-7. Oktober 1978."
Includes bibliographical references and index.
ISBN 3-11-008025-7
1. Civil service—Germany, West. 2. Civil service. 3. Germany, West—Constitutional law. 4. Constitutional law. I. Denninger, Erhard. Verfassungstreue und Schutz der Verfassung. 1979. II. Rudolf, Walter. Der öffentliche Dienst im Staat der Gegenwart. 1979. III. Title. IV. Series: Vereinigung der Deutschen Staatsrechtslehrer. Veröffentlichungen ; Heft 37.
342'.43'068 79-377482
79 MARC

—BIBLIOGRAPHY

Tangier African Training and Research Centre in Administration for Development.
Preliminary bibliography on human resources in public administration = Bibliographie préliminaire sur les resources humaines de l'administration publique / African Training and Research Centre in Administration for Development. — Tanger : C.A.F.R.A.D. Documentation Centre, 1972.
39 p. ; 28 cm.
Cover title.
1. Civil service—Bibliography. 2. Public administration—Bibliography. I. Title. II. Title: Bibliographie préliminaire sur les ressources humaines de l'administration publique.
Z7164.C6T36 1972 016.35 78-317818
[JF1351] 78 MARC

—ETHICS see Civil service ethics

—IN-SERVICE TRAINING—NEPAL

Bhatta, Bhim Dev, 1944-
An evaluation of the executive development training program in Nepal / Bhim Dev Bhatta. — Kirtipur : Centre for Economic Development and Administration, Tribhuvan University, 1978
137 p. ; 29 cm.
Mimeographed.
Bibliography: p. 135-137.
1. Civil service—In-service training—Nepal. 2. Civil service—Study and teaching—Nepal. 3. Public administration—Study and teaching—Nepal. 4. Centre for Economic Development and Administration. I. Title.
JQ1825.N44B47 354'.549'600153 79-902115
79 MARC

—PERIODICALS—INDEXES

Singh, Mohinder, 1935-
Cumulative index to Public administration, journal of the Royal Institute of Public Administration, volumes 1-55, 1923-1977 / Mohinder Singh, R. N. Sharma. — Delhi : Concept, 1979.
144 p. ; 25 cm.
Rs80.00 ($16.00 U.S.)
1. Public administration—Indexes. 2. Public administration—Periodicals—Indexes. 3. Civil service—Periodicals—Indexes. I. Sharma, Rajendra Narayan, 1932- . II. Public administration, v. 1-55. Index. III. Title.
JA8.P82S56 016.35 79-901153
79 MARC

—POSITIONS see Civil service positions

—RECRUITING see Civil service recruiting

—STUDY AND TEACHING—EUROPE

Imoto, Hiroshi.
(Kenshū o chūshin to shita shokuin no nōryoku kaihatsu)
研修を中心とした職員の能力開発 / 井本洋. — [大阪]：大阪府職員研修所, [はしがき昭和50 i. e. 1975]
54 p. : ill. ; 26 cm. — (海外派遣研修報告書 ; 49–1) Ja***

1. Civil service—Study and teaching—United States. 2. Civil service—Study and teaching—Europe. 3. Manpower policy. I. Title. II. Series: Kaigai haken kenshū hōkokusho ; 49–1.
JK718.I46 78–812118

—STUDY AND TEACHING—NEPAL

Bhatta, Bhim Dev, 1944-
An evaluation of the executive development training program in Nepal / Bhim Dev Bhatta. — Kirtipur : Centre for Economic Development and Administration, Tribhuvan University, 1978
137 p. ; 29 cm.
Mimeographed.
Bibliography: p. 135-137.
1. Civil service—In-service training—Nepal. 2. Civil service—Study and teaching—Nepal. 3. Public administration—Study and teaching—Nepal. 4. Centre for Economic Development and Administration. I. Title.
JQ1825.N44B47 354'.549'600153 79-902115
79 MARC

CIVIL SERVICE (Continued)

—STUDY AND TEACHING
—UNITED STATES

Imoto, Hiroshi.
(Kenshū o chūshin to shita shokuin no nōryoku kaihatsu)
研修を中心とした職員の能力開発 / 井本洋. — [大阪] : 大阪府職員研修所, [はしがき昭和50 i. e. 1975]
54 p. : ill. ; 26 cm. — (海外派遣研修報告書 ; 49-1) Ja***

1. Civil service—Study and teaching—United States. 2. Civil service—Study and teaching—Europe. 3. Manpower policy. I. Title. II. Series: Kaigai haken kenshū hōkokusho ; 49-1.

JK718.I46 78-812118

—STUDY AND TEACHING (INTERNSHIP)—UNITED STATES—BIBLIOGRAPHY

Koehler, Cortus T
Public administration and public sector internships : 1960 through 1977 / Cortus T. Koehler, with the assistance of Patricia L. Hutchcraft. — Monticello, Ill. : Vance Bibliographies, 1979.
36 p. ; 28 cm. — (Public administration series : Bibliography ; P-190)
Cover title.
$3.50
1. Civil service—Study and teaching (Internship)—United States—Bibliography. I. Hutchcraft, Patricia L., joint author. II. Title. III. Series.
Z7165.U5K63 016.353006'07 79-107048
[JK718] 79 MARC

—VETERANS' PREFERENCE—NEW JERSEY

New Jersey. Legislature. General Assembly. State Government, Federal and Interstate Relations, and Veterans' Affairs Committee.
Public hearing before Assembly State Government, Federal and Interstate Relations, and Veterans' Affairs Committee on A-1675 (Civil Service reform act—veterans preference) held December 21, 1978, Assembly Chamber, State House, Trenton, New Jersey. — [Trenton, N.J.] : The Committee, [1979]
21 p. ; 28 cm.
Cover title.
Includes index.
1. Civil service—Veterans' preference—New Jersey. I. Title.
KFN1811.4.S78 1978c 342'.749'068 79-621853
79 MARC

—VOCATIONAL GUIDANCE
see Civil service positions

—AFRICA

see also Africanization

—AUSTRALIA

Australia. Royal Commission on Australian Government Administration.
Appendixes to report / Royal Commission on Australian Government Administration. — Canberra : Australian Government Publishing Service, 1976-
v. ; 25 cm. Aus
Chairman: H. C. Coombs.
Includes bibliographical references.
ISBN 0-642-02189-9 (v. 1)
1. Australia—Politics and government. 2. Civil service—Australia. I. Coombs, Herbert Cole, 1906-
JQ4031.A87 1976 354'.94 78-321172
79 MARC

—AUSTRIA

Austria. Laws, statutes, etc.
[Beamten-Dienstrechtsgesetz]
Beamtendienstrecht : (Bundesgesetz vom 2. Juni 1977, BGBl. Nr 329) : mit einschläg. Gesetzen, Verordnungen u. Erlässen sowie mit Entscheidungen der Verfassungsgerichtshofes u. des Verwaltungsgerichtshofes / hrsg. v. Viktor Hackl. — Stand 1. Jän. 1978. — Wien : Österr. Staatsdruckerei, 1977.
xix, 690 p. ; 18 cm. — (Handausgabe österreichischer Gesetze und Verordnungen ; n.F., Gruppe 1, Bd. 7a) Au78-11-32
"Beiheft" (23 p.) inserted.
Includes index.
S625.00
1. Civil service—Austria. 2. Austria—Officials and employees. I. Hackl, Viktor. II. Title. III. Series.
342'.436'068 78-396818
*79 MARC

—AUSTRIA—HISTORY

Quarthal, Franz.
Die Behördenorganisation Vorderösterreichs von 1753 bis 1805 und die Beamten in Verwaltung, Justiz und Unterrichtswesen / von Franz Quarthal u. Georg Wieland, in Zusammenarbeit mit Birgit Dürr. — Bühl/Baden : Verlag Konkordia, 1977.
591, 24 leaves of plates : ill., maps (1 fold. col. in pocket) ; 24 cm. — (Veröffentlichung des Alemannischen Instituts Freiburg i. Br. ; Nr. 43) GFR78-A
Errata slip inserted.
Bibliography: p. 27-41.
Includes index.
ISBN 3-7826-0043-6 : DM48.00
1. Austria—Officials and employees—History. 2. Civil service—Austria—History. 3. Austria—Politics and government—1740-1848. I. Wieland, Georg, joint author. II. Dürr, Birgit, joint author. III. Title. IV. Series: Freiburg i.B. Alemannisches Institut. Veröffentlichung ; Nr. 43.
JN1717.Q37 354'.436'00609 79-368188
79 MARC

—BOTSWANA

Botswana. Ministry of Finance and Development Planning.
Report of the public service manpower survey 1970. — Gaborone, Republic of Botswana : Ministry of Finance and Development Planning, 1970.
2 v. in 1 (xiii, 190 p.) ; 31 cm.
Cover title.
Includes bibliographical references.
1. Civil service—Botswana. I. Title. II. Title: Public service manpower survey, 1970.
JQ2760.A66B65 1970 354'.68'1006 79-108598
79 MARC

—BRAZIL

Brazil. Laws, statutes, etc.
O DASP e a política de pessoal; documentação básica. [Brasília] Departamento Administrativo do Pessoal Civil, CENDOC [i.e. Centro de Documentação e Informática] 1971.
59 p. illus. 23 cm.
1. Brazil. Departamento Administrativo do Pessoal Civil. 2. Civil service—Brazil. I. Title.
77-483142
78 MARC

Brazil. Laws, statutes, etc.
[Lei no. 1.711, de 28 de outubro de 1952]
Estatuto dos funcionários públicos civis da União como se acha em vigor : atualizada até janeiro de 1978, com referências a toda a legislação regulamentadora. — 3. ed. / atualização, referências e índices, José Serson. — São Paulo : Edições LTr, 1978.
170 p. ; 23 cm.
Includes indexes.
Cr$220.00
1. Brazil—Officials and employees. 2. Civil service—Brazil. I. Serson, José. II. Title.
342'.81'068 78-394506
79 MARC

—BRAZIL—CASES

Brazil. Departamento Administrativo do Serviço Público.
Ementário de decisões administrativas.
[Brasília] Serviço de Documentação,
v. 22 cm.
Began in 1956.
1. Brazil—Officials and employees—Cases. 2. Civil service—Brazil—Cases.
342'.81'068 79-641975
MARC-S

—BURMA

Burma (Union). Pyithu Wun-htān Sonsān yēi Kawmashin.
[Asiyin hkan sa]
ပြည်သူ့ဝန်ထမ်း စုံစမ်းရေးကော်မရှင်၏အစီရင်ခံစာ. — [ရန်ကုန်] : ပြည်ထောင်စုမြန်မာနိုင်ငံတော်အစိုးရပြည်သူ့ဝန်ထမ်း စုံစမ်းရေးကော်မရှင်, 1961.
177 p. ; 25 cm.
In Burmese.
1. Civil service—Burma. I. Title: Asiyin hkan sa.
Title romanized: Pyithu Wun-htān Sonsān yēi Kawmashin ǀ Asiyin hkan sa.
JQ452.A56 1961 76-984723

—BURMA—DIRECTORIES

(Ta hnit hnit kyein htok Myanma Naing Ngan asōya ayashi myā sayīn)
တနှစ်နှစ်ကြိမ်ထုတ်မြန်မာနိုင်ငံအစိုးရအရာရှိများစာရင်း. ရန်ကုန်မြို့, ပြည်ထဲရေးဝန်ကြီးဌာန.
v. 25 cm.
In Burmese.
Continues: Ta hnit lēi kyein htok Myanma Naing Ngan asōya ayashi myā sayīn.
1. Burma—Officials and employees—Directories. 2. Civil service—Burma—Directories. I. Burma (Union). Pyi Htō Yēi Wun-kyī Htana.
JQ452.B65
77-647745
MARC-S

(Ta hnit lēi kyein htok Myanma Naing Ngan asōya ayashi myā sayīn)
တနှစ်လေးကြိမ်ထုတ်မြန်မာနိုင်ငံအစိုးရအရာရှိများစာရင်း. ရန်ကုန်မြို့, ပြည်ထဲရေးဝန်ကြီးဌာန.
v. 25 cm.
In Burmese.
Continues: Thōn la pat sin Bama Naing Ngan-taw asōya ayashi sayīn.
Continued by: Ta hnit hnit kyein htok Myanma Naing Ngan asōya ayashi myā sayīn.
1. Burma—Officials and employees—Directories. 2. Civil service—Burma—Directories. I. Burma (Union). Ministry of Home Affairs.
JQ447.T48
77-647917
MARC-S
S A

—CALIFORNIA—PERSONNEL MANAGEMENT—PERIODICALS

California. State Personnel Board.
Annual report - California State Personnel Board. 1971–
[Sacramento] California State Personnel Board.
v. ill. 29 cm.
Continues: Annual report on state salaries and personnel management, ISSN 0193–290X.
Continued by: California State Personnel Board report to the Governor and legislature, ISSN 0193–0966.
Key title: Annual report - California State Personnel Board, ISSN 0193–0958.
1. Civil service—California—Personnel management—Periodicals. I. Key title.
JK8755.C344b 353.9'794'001 79–643282
MARC-S

California. State Personnel Board.
Annual report on state salaries and personnel management. 1970. [Sacramento] California State Personnel Board.
30 p. 29 cm.
Supersedes: California. State Personnel Board. Report.
Continued by: California. State Personnel Board. Annual report - California State Personnel Board.
1. Civil service—California—Personnel management—Periodicals. I. Title.
JK8755.C344b 353.9'794'001 79–643283
MARC-S

California. State Personnel Board.
California State Personnel Board report to the Governor and legislature. 1975–
[Sacramento, California State Personnel Board]
v. 29 cm. annual.
Continues: Annual report - California State Personnel Board, ISSN 0193–0958.
Key title: California State Personnel Board report to the Governor and legislature, ISSN 0193–0966.
1. Civil service—California—Personnel management—Periodicals. I. California. State Personnel Board. Report to the Governor and legislature. II. Title.
JK8755.C344b 353.9'794'001 79–643259
MARC-S

—CAMBODIA

Ouellette, James L
The civil service system of the Royal Khmer Government of Cambodia / James L. Ouellette, Marvin H. Murphy. — Phnom Penh : Public Administration Division, United States Agency for International Development, 1962.
56 leaves ; 28 cm. — (Public administration report ; 2)
1. Civil service—Cambodia. I. Title. II. Series: United States. Agency for International Development. Public Administration Division. Public administration report ; 2.
JQ936.Z1O93 354'.596'006 79-313072
79 MARC

CIVIL SERVICE (Continued)

—CAMEROON

Cameroon. Laws, statutes, etc.
La fonction publique en République unie du Cameroun : statut général, principaux statuts particuliers et textes d'application = The public service in the United Republic of Cameroon : general rules and regulation, principal special rules and regulations, texts of application. — ⸢Yaoundé⸣ : Bibliothèque de l'École nationale d'administration et de magistrature, ⸢1976⸣
194, 7 p. ; 21 cm. — (Série 1, Études et document - Library of the National School of Administration and Magistracy ; v. 1)
Cover title.
Text in French.
1. Civil service—Cameroon. I. Title. II. Title: The public service in the United Republic of Cameroon. III. Series: National School of Administration and Magistracy (Cameroon). Série 1, Études et document - Library of the National School of Administration and Magistracy ; v. 1.
342'.6711'068 79-342802
79 MARC

—CANADA

Canada. Treasury Board. Performance Measurement Policy Division.
A manager's guide to performance measurement / Performance Measurement Policy Division, Planning Branch, Treasury Board. — ⸢Ottawa⸣ : Treasury Board, 1976.
ii, 23, 23, ii p. : ill. ; 23 cm. C***
English and French.
Added t.p. in French: La mesure de la performance—guide du gestionnaire.
ISBN 0-662-00304-7
1. Civil service—Canada. 2. Employees, Rating of—Canada. 3. Performance standards. I. Title. II. Title: La mesure de la performance—guide du gestionnaire.
JL111.R38C36 1976 354'.71'00142 79-309981
79 MARC

—CANADA—ADDRESSES, ESSAYS, LECTURES

Public employment and compensation in Canada : myths and realities / edited by David K. Foot. — Toronto : Butterworths & Co. (Canada) for the Institute for Research on Public Policy, 1978.
xii, 188 p. ; 23 cm. — (IRPP series on public sector employment in Canada ; v. 1) C79-94083
Summary in French.
Includes bibliographies.
ISBN 0-409-88600-9
1. Civil service—Canada—Addresses, essays, lectures. 2. Canada—Officials and employees—Salaries, allowances, etc.—Addresses, essays, lectures. I. Foot, David K. II. Series: Institute for Research on Public Policy. IRPP series on public sector employment in Canada ; v. 1.
JL108.P8 331.2'041'35471 79-310993
79 MARC

—CHINA

Hu, Chien-kuo.
(Ch'ing tai wen kuan k'ao k'o chih tu chih yen chiu)
清代文官考課制度之研究 / 胡健國撰. — ⸢台北? : s. n., 1978?⸣
2 v. (467 p. on double leaves) ; 27 cm.
Reproduced from typescript.
Originally presented as the author's thesis (M.A.), Kuo li cheng chih ta hsüeh, 1971.
Bibliography: p. 461–467.
1. Civil service—China. 2. Employees, Rating of—China. 3. China—Politics and government—1644–1911. I. Title.
JQ1512.Z13R33 1978 79–835055

K'an, Tao-ming.
(Chung-kuo li tai han lin hsüeh shih tsuan yao)
中國歷代翰林學士纂要 / 著作者闞道明. — 新竹 : 三人出版社, 民國61 ⸢1972⸣
410 p. ; 21 cm.
Colophon title.

1. China—Registers. 2. Civil service—China. 3. Scholars—China. I. Title.
JQ1507.K35 79–835713

(Shao-hsing shih pa nien t'ung nien hsiao lu)
紹興十八年同年小錄. 中州人物考 / ⸢孫奇逢撰⸣. — ⸢台北 : 臺灣⸣商務⸢印書館, 1978?⸣
2 v. ; 20 cm. — (四庫全書珍本八集 ; 82–83)
Cover title.
Photoreprint ed. of two works, both of the Wen Yüan Ko copy.

1. China—Registers. 2. Civil service—China. 3. China—History—Sung dynasty, 960–1279—Biography. 4. China—Biography. 5. China—History—Ming dynasty, 1368–1644—Biography. 6. Honan, China (Province)—Biography. I. Sun, Ch'i-feng, 1585–1675. Chung chou jen wu k'ao. 1978?. II. Title: Chung chou jen wu k'ao. III. Series: Ssu k'u ch'uan shu. Ssu k'u ch'uan shu chen pen pa chi ; 82–83.
AC149.S7 1978, vol. 82–83 78–841087
[JQ1507]

Tseng, I-min.
(T'ang tai k'ao k'o chih tu yen chiu)
唐代考課制度研究 / 曾一民著. — 臺北 : 臺灣商務印書館, 民國67 ⸢1978⸣
8, 2, 3, 172 p. ; 20 cm. — (岫廬文庫 ; 52)
Series romanized: Hsiu-lu wen k'u.
Bibliography: p. 169–172.
NT$24.00
1. Civil service—China. 2. Employees, Rating of—China. 3. China—History—T'ang dynasty, 618–907. I. Title.
JQ1512.Z13R38 79–840571

—CHINA—EXAMINATIONS

Huang, Kuang-liang.
(Ch'ing tai k'o chü chih tu chih yen chiu)
清代科舉制度之研究 / 黃光亮著. — Taipei : 嘉新水泥公司文化基金會, 民國65 ⸢1976⸣
4, 2, 434 p. ; 22 cm. — (嘉新水泥公司文化基金會研究論文 ; 第311種) (嘉新水泥公司文化基金會叢書)
Series 1 romanized: Chia hsin shui ni kung ssu wen hua chi chin hui yen chiu lun wen. Series 2 romanized: Chia hsin shui ni kung ssu wen hua chi chin hui ts'ung shu.
Bibliography: p. 426–434.
1. Civil service—China—Examinations. 2. China—Politics and government—1644–1911. I. Title.
JQ1512.Z1H84 78–842398

Library of Congress 78

Huang, Ts'ung-lan, 19th cent.
(Kung chü k'ao lüeh)
貢舉考畧 / ⸢黃崇蘭輯⸣. — ⸢涇縣⸣ : 涇縣學署藏版, 嘉慶9 ⸢1804⸣ 新鐫.
4 v. (on double leaves) in case ; 30 cm.
Block print.
CONTENTS: ⸢1⸣–⸢2⸣ 明貢舉考畧 2卷.—⸢3⸣–⸢4⸣ 國朝貢舉考畧 2卷.
1. Civil service—China—Examinations. I. Title II. Title: Kuo ch'ao kung chü k'ao lüeh. III. Title: Ming kung chü k'ao lüeh.
JQ1512.H83 79–837119

—CHINA—HISTORY—SOURCES

China. Li pu. K'ao kung ssu.
(Li pu k'ao kung ssu t'i kao)
吏部考功司題稿 / 吏部考功司撰. — 臺北 : 偉文圖書出版社, 民國66 ⸢1977⸣
3 v. (1643 p.) ; 22 cm. — (明季史料集珍 ; 第二輯)
"起嘉靖二十年十一月迄嘉靖廿二年五月"
Series statement in colophon: Ming tai shih liao chi chen.
"本書係中央圖書館所藏"
Reprint ed.
NT$1000.00 ($26.40 U.S.)
1. Civil service—China—History—Sources. 2. China—History—Ming dynasty, 1368–1644—Sources. 3. Employees, Rating of—China—History—Sources. I. Title. II. Series: Ming chi shih liao chi chen : Ti erh chi.
JQ1512.Z1C49 1977 79–837285
79

—COLOMBIA

Younes Moreno, Diego.
Derecho administrativo laboral : función pública / Diego Younes Moreno. — Bogotá : Editorial Temis, 1978.
xiv, 338 p. ; 24 cm.
Appendix (p. ⸢143⸣-323) contains legislation.
Bibliography: p. ⸢325⸣
Includes indexes.
ISBN 8482720532
1. Civil service—Colombia. I. Title.
342'.861'068 79-101620
79 MARC

—CONNECTICUT—MINORITY EMPLOYMENT—STATISTICS

Connecticut. Commission on Human Rights and Opportunities.
Status of affirmative action in Connecticut State government : a report to the Governor and the Connecticut General Assembly. — ⸢Hartford⸣ : State of Connecticut, Commission on Human Rights and Opportunities, 1978.
ii, 121 p. ; 28 cm.
Cover title.
Bibliography: p. 119-121.
1. Affirmative action programs—Connecticut. 2. Civil service—Connecticut—Minority employment—Statistics. I. Title.
JK3360.A33C63 1978 331.2 78-623136
79 MARC

—EGYPT

Egypt. Laws, statutes, etc.
⸢al-Qānūn raqm 58 li-sanat 1971⸣
احكام نظام العاملين المدنيين بالدولة : القانــون رقم ٥٨ لسنة ١٩٧١ ، تشريعا وقضــاء وفتوى / اعداد اسماعيل فوزى احمد فوده . — ⸢القاهرة⸣ : الهيئة المصرية العامة للكتاب ، 1976.
540 p. ; 25 cm.
ISBN 977-201-244-8
1. Egypt—Officials and employees. 2. Civil service—Egypt. I. Fawdah, Ismā'īl Fawzī Aḥmad. II. Title.
Title romanized: Aḥkām niẓām al-'āmilīn al-madaniyīn bi-al-dawlah.
77-961100

—EUROPE

(Kakkoku kanri seido no kenkyū)
各國官吏制度の研究 / 蠟山政道 ... ⸢et al.⸣. / 東京 : プレブス社, 昭和23 ⸢1948⸣
304 p. ; 21 cm.
Includes bibliographies.

1. Civil service—Europe. 2. Civil service—United States. I. Rōyama, Masamichi, 1895-
JN96.A67K34 79–807300

—EUROPE—ADDRESSES, ESSAYS, LECTURES

Odpowiedzialność pracownika administracji / ⸢red. i wstęp Janusz Łętowski, Jan Pruszyński⸣. — Wrocław ; Warszawa : Zakład Narodowy im. Ossolińskich, 1978 (Białyst : BZG)
316 p. ; 25 cm. P78-2515
At head of title: Polska Adademia Nauk. Instytut Państwa i Prawa. Instytut Organizacji Zarządzania i Doskonalenia Kadr.
Papers presented at a conference held June 1975 ; some have been translated from various languages.
Includes bibliographies.
zł65.00
1. Civil service—Europe—Addresses, essays, lectures. 2. Administrative responsibility—Europe—Addresses, essays, lectures. 3. Misconduct in office—Europe—Addresses, essays, lectures. I. Łętowski, Janusz. II. Pruszyński, Jan. III. Polska Akademia Nauk. Instytut Państwa i Prawa. IV. Instytut Organizacji Zarządzania i Doskonalenia Kadr.
78-384629
78 MARC

—FINLAND

Finland. Valtion henkilöstöpolitiikkatoimikunta.
Valtion henkilöstöpolitiikkatoimikunnan mietintö = Betänkande / av kommissionen för statens personalpolitik. — Helsinki : ⸢Valtiovarainministeriö⸣, 1977.
⸢16⸣, 57, ⸢8⸣ p. : ill. ; 30 cm. — (Komiteanmietintö 1977:16) Fi77-July
Summary in Swedish.
Bibliography: p. ⸢5⸣-⸢8⸣ (3d group)
ISBN 9514622308
1. Civil service—Finland. I. Series: Finland. Komiteanmietintö ; 1977:-16.
J398.R13 1977:16 79-379699
⸢JN6712⸣ 79 MARC

—FLORIDA

Florida. Legislature. Senate. Governmental Operations Committee.
A report on positions exempt from the state career service system, chapter 110, Florida statutes / prepared pursuant to request of the President of the Florida Senate by staff of the Senate Governmental Operations Committee, Senator Dempsey J. Barron, chairman. — ⸢Tallahassee⸣ : The Committee, 1977.
1, 17, ⸢13⸣ p. ; 28 cm.
1. Civil service—Florida. I. Barron, Dempsey J. II. Title.
KFF11.72.G67 1977 353.9'759'006 79-620502
79 MARC

CIVIL SERVICE (Continued)

—FRANCE

Istituto per lo sviluppo della formazione professionale dei lavoratori.
Alcune proposte formative nella pubblica amministrazione. — Roma : L'Istituto, 1976.
137 p. ; 25 cm. — (Its Quaderni di formazione ISFOL ; n. 34) It***
1. Public administration—Study and teaching—Italy. 2. Public administration—Study and teaching—France. 3. Civil service—Italy. 4. Civil service—France. I. Title. II. Series.
JF1338.A2I87 1976 79-354929
79 MARC

Plantey, Alain.
Réformes dans la fonction publique : complément et mise à jour du "Traité pratique de la fonction publique" / Alain Plantey .... — Paris : L.G.D.J., 1978.
254, viii p. ; 23 cm. F78-6811
ISBN 2-275-01227-3 : 70.00F
1. Civil service—France. 2. International officials and employees. I. Plantey, Alain. Traité pratique de la fonction publique. II. Title.
342'.44'068 79-344339
79 MARC

—GEORGIA—PERSONNEL MANAGEMENT

Georgia. General Assembly. Senate. Governmental Operations Committee.
Report on Georgia'a merit system of personnel administration / by Senate Governmental Operations Committee and House Special Merit System Study Committee. — Atlanta : General Assembly, 1978.
10 leaves ; 36 cm.
1. Civil service—Georgia—Personnel management. I. Georgia. General Assembly. House of Representatives. Special Merit System Study Committee. II. Title.
JK4355.G46 1978 353.9'758'001 79-624488
79 MARC

—GERMANY, WEST

Boure, Robert.
Les interdictions professionnelles en Allemagne fédérale / Robert Boure. — Paris : F. Maspero, 1978.
135 p. ; 22 cm. — (Cahiers libres ; 339) F***
Bibliography: p. 127-128.
ISBN 2-7071-0998-3 : 26.00F
1. Interdiction (Civil law)—Germany, West. 2. Civil service—Germany, West. I. Title.
342'.43'068 78-385035
78 MARC

Hilg, Günter, 1939-
Allgemeines Beamtenrecht / Günter Hilg. — 1. Aufl. — München : Kommunalschriften-Verlag Jehle, 1978.
312 p. : ill. ; 23 cm. — (Beamtenrecht in Bayern ; Halbbd. 1) (Studienschriften für die öffentliche Verwaltung ; Bd. 2) GFR78-A
Includes bibliographical references and index.
ISBN 3-7825-0016-4
1. Civil service—Germany, West. I. Title. II. Series: Hilg, Günter, 1939- Beamtenrecht in Bayern ; Halbbd. 1. III. Series: Studienschriften für die öffentliche Verwaltung ; Bd. 2.
342'.433'068 s 78-382277
79 MARC

Kind, Hero.
Dienstrechtsreform und Laufbahnreform : zur Konzeption von Reformen d. öffentl. Dienstes / von Hero Kind. — Speyer : Hochsch. für Verwaltungswiss., 1978.
46 leaves ; 30 cm. — (Speyerer Arbeitshefte ; 24) GFR78-B13
Includes bibliographical references.
1. Civil service—Germany, West. 2. Germany, West—Officials and employees. I. Title. II. Series.
342'.43'068 79-360883
79 MARC

Lohmar, Ulrich.
Staatsbürokratie : d. hoheitl. Gewerbe : dt. Aspekte e. neuen Klassenkampfes / Ulrich Lohmar. — Orig.-Ausg., 2. Aufl. — München : Goldmann, 1978.
189 p. ; 18 cm. — (Goldmann-Sachbuch) GFR78-A
Includes index.
ISBN 3-442-11186-2 : DM5.80
1. Bureaucracy. 2. Civil service—Germany, West. I. Title.
JN3971.A67L63 1978 354'.43'01 78-395635
79 MARC

Starck, Christian.
Öffentlicher Dienst. — Baden-Baden : Nomos-Verlagsgesellschaft, 1977.
73 p. ; 23 cm. — (His Verfassungsrecht in Fällen ; 18) GFR78-A5
Cover title.
Includes bibliographical references.
1. Civil service—Germany, West. 2. Germany, West—Officials and employees. 3. Judicial review—Germany, West. I. Title.
342'.43'068 79-357268
79 MARC

Vereinigung der Deutschen Staatsrechtslehrer.
Verfassungstreue und Schutz der Verfassung / Erhard Denninger u. Hans Hugo Klein. Der öffentliche Dienst im Staat der Gegenwart / Walter Rudolf u. Frido Wagener. — Berlin ; New York : de Gruyter, 1979.
350 p. ; 22 cm. — (Veröffentlichungen der Vereinigung der Deutschen Staatsrechtslehrer ; Heft 37) GFR79-A
"Berichte und Diskussionen auf der Tagung der Vereinigung der Deutschen Staatsrechtslehrer in Bonn vom 4.-7. Oktober 1978."
Includes bibliographical references and index.
ISBN 3-11-008025-7
1. Civil service—Germany, West. 2. Civil service. 3. Germany, West—Constitutional law. 4. Constitutional law. I. Denninger, Erhard. Verfassungstreue und Schutz der Verfassung. 1979. II. Rudolf, Walter. Der öffentliche Dienst im Staat der Gegenwart. 1979. III. Title. IV. Series: Vereinigung der Deutschen Staatsrechtslehrer. Veröffentlichungen ; Heft 37.
342'.43'068 79-377482
79 MARC

—GERMANY, WEST—PERSONNEL MANAGEMENT

Koch, Rainer, Dr.
Organisationsplanung für den Personalbereich : e. theoret.-empir. Beitr. zur Verwaltungsreformlehre / von Rainer Koch. — Berlin : Duncker und Humblot, 1978.
127 p. ; 24 cm. — (Schriftenreihe der Hochschule Speyer ; Bd. 73) GFR78-A
Bibliography: p. [119]-127.
ISBN 3-428-04251-4 : DM48.00
1. Civil service—Germany, West—Personnel management. 2. Organizational change. 3. Manpower planning—Germany, West. I. Title. II. Series: Speyer. Hochschule. Schriftenreihe ; Bd. 73.
JN3971.A67K62 354'.43'001 79-351720
79 MARC

—GERMANY, WEST—BAVARIA

Bavaria. Laws, statutes, etc.
[Beamtengesetz]
Bayerisches Beamtengesetz : (BayBG) : in d. Fassung vom 17. November 1978 (GVBl. S. 831) : Handkommentar mit ausführl. Erl., Verweisungen, 2 Übers. u. e. Sachreg. / bearb. von Hans Hartmann, Fritz Janssen, Ulrich Kühn. — 5., völlig überarb. Aufl. — München : Kommunalschriften-Verlag Jehle, 1978.
xi, 243 p. ; 21 cm. — (Kova-Handkommentar) GFR79-A
Includes index.
ISBN 3-7825-0049-0
1. Civil service—Germany, West—Bavaria. I. Hartmann, Hans, Ministerialrat. II. Janssen, Fritz. III. Kühn, Ulrich, Regierungsdirektor. IV. Title.
342'.433'068 79-372673
79 MARC

Hilg, Günter, 1939-
Beamtenrecht in Bayern / Hilg, Müller. — 1. Aufl. — München : Kommunalschriften-Verlag Jehle, 1978-
v. ; 23 cm. — (Studienschriften für die öffentliche Verwaltung ; Bd. 2-) GFR78-A
Vol. 1 includes bibliographical references and index.
ISBN 3-7825-0016-4 (v. 1)
1. Civil service—Germany, West—Bavaria. I. Müller, Helmut, Oberamtsrat. II. Title. III. Series.
342'.433'068 78-382278
79 MARC

—GREAT BRITAIN

Brown, Ronald Gordon Sclater.
The administrative process in Britain [by] R. G. S. Brown. [London] Methuen [1970]
xiv, 349 p. 23 cm. 75/- B***
"Distributed in the USA by Barnes & Noble."
Bibliography: p. 325-339.
ISBN 0-416-15200-7
1. Civil service—Great Britain. 2. Great Britain—Politics and government. I. Title.
JN425.B717 354'.41'006 76-495048
70[r79]rev MARC

Brown, Ronald Gordon Sclater.
The administrative process in Britain / R. G. S. Brown and D. R. Steel. — 2d ed. — London : Methuen, 1979.
352 p. ; 23 cm. GB***
Bibliography: p. 338-341.
Includes indexes.
ISBN 0-416-85890-2 : £8.50
1. Civil service—Great Britain. 2. Great Britain—Politics and government. I. Steel, D. R., joint author. II. Title.
JN425.B717 1979 354'.41'006 79-313561
79 MARC

Emmerson, Harold Corti, Sir, 1896-
Masters and servants : a career in the Civil Service / by H. C. Emmerson. — Berkhamsted : The author, 1978.
[6], 112 p. : facsim. ; 30 cm. GB78-34461
1. Emmerson, Harold Corti, Sir, 1896- 2. Great Britain—Officials and employees—Biography. 3. Government executives—Great Britain—Biography. 4. Civil service—Great Britain. 5. Great Britain—Politics and government—20th century. I. Title.
JN425.E53 354'.41'0010924 79-304592
79 MARC

Great Britain. Civil Service Dept.
Application of race relations policy in the Civil Service. — London : H.M. Stationery Off., 1978.
v, 275 p. : ill. ; 30 cm. GB***
Includes bibliographical references.
ISBN 0-11-630226-7 : £5.00
1. Civil service—Great Britain. 2. Discrimination in employment—Great Britain. 3. Minorities—Employment—Great Britain. I. Title.
JN428.G73 1978 354'.41'001 79-314536
79 MARC

O'Broin, Nóirín.
The working and living conditions of civil service typists / [by] Nóirín O'Broin and Gillian Farren. — Dublin : Economic and Social Research Institute, 1978.
164 [i.e. 166] p. (1 fold.) ; 25 cm. — (Paper - Economic and Social Research Institute ; no. 93) GB78-27896
Bibliography: p. 164.
ISBN 0-7070-0015-7 : £3.50
1. Civil service—Great Britain. 2. Typists. I. Farren, Gillian, joint author. II. Title. III. Series: Economic and Social Research Institute. Paper ; no. 93.
HC257.I6E3 no.93 330.9'415 s 79-302715
[JN425] 79 MARC

—HONDURAS—CONGRESSES

Seminario sobre Administración de Personal y Reforma Administrativa, 1st, Tegucigalpa, 1976.
Primer Seminario sobre Administración de Personal y Reforma Administrativa, del 14 al 16 de julio de 1976, Tegucigalpa, D.C., Honduras, C.A. : conclusiones y recomendaciones. — [Tegucigalpa] : Centro de Adiestramiento del Ministerio de Hacienda y Crédito Público, [between 1976 and 1978]
13, 3 leaves ; 27 cm.
At head of title: Consejo Superior de Planificación Económica, (CONSUPLANE). Dirección General de Servicio Civil. Centro de Adiestramiento del Ministerio de Hacienda y Crédito Público (CENADIH).
1. Civil service—Honduras—Congresses. 2. Personnel management—Honduras—Congresses. 3. Civil service reform—Congresses. I. Honduras. Consejo Superior de Planificación Económica. II. Honduras. Direccion General de Servicio Civil. III. Honduras. Secretaría de Hacienda y Crédito Público (1971-). Centro de Adiestramiento.
JL1532.Z1S45 1976 78-105093
MARC

—INDIA

Azarkh, Anatoliĭ Ėmmanuilovich.
Gosudarstvennye sluzhashchie Indii : ėvoli͡ut͡sii͡a gosudarstvennoĭ bi͡urokratii v uslovii͡akh nezavisimosti / A. Ė. Azarkh. — Moskva : Nauka, 1979.
134 p. ; 20 cm. USSR***
At head of title: Akademii͡a nauk SSSR. Institut mirovoĭ ėkonomiki i mezhdunarodnykh otnosheniĭ.
Includes bibliographical references.
1.10rub
1. Civil service—India. I. Title.
JQ247.A96 79-377836
79 MARC

Dey, Bata K 1932-
Bureaucracy, development, and public management in India / Bata K. Dey. — 1st ed. — New Delhi : Uppal Pub. House, 1978.
xiv, 334 p. : ill. ; 23 cm.
Bibliography: p. [318]-325.
Includes index.
Rs80.00
1. Civil service—India. 2. Public administration—India. I. Title.
JQ247.D49 354'.54 78-903388
79 MARC

Hazarika, Niru, 1947-
Public service commissions : a study / Niru Hazarika. — Delhi : Leeladevi Publications, 1979, c1978.
200 p. ; 22 cm.
Bibliography: p. [193]-196.
Includes index.
Rs40.00 ($8.00 U.S.)
1. Civil service—India. 2. Assam. Public Service Commission. I. Title.
JQ247.H39 1979 354'.54'006 79-903328
79 MARC

Pai Panandiker, V A 1935-
Bureaucracy and development administration / V. A. Pai Panandiker, S. S. Kshirsagar. — New Delhi : Centre for Policy Research, c1978.
viii, 208 p. : ill. ; 25 cm.
Includes bibliographical references and index.
Rs60.00
1. Bureaucracy. 2. India—Politics and government—1947- 3. Civil service—India. I. Kshirsagar, S. S., joint author. II. Title.
JQ224.P25 354'.54'01 78-907574
79 MARC

Ram, N V Raghu, 1931-
Games bureaucrats play / N. V. R. Ram ; illustrated by Abu. — New Delhi : Vikas, c1978.
viii, 77 p. : ill. ; 23 cm.
Includes bibliographical references.
ISBN 0-7069-0631-4 : Rs24.00
1. Bureaucracy. 2. Civil service—India. I. Title.
JQ247.R33 354'.54'01 78-904894
79 MARC

CIVIL SERVICE
—INDIA (Continued)

Ray, Syamal Kumar.
Indian bureaucracy at the crossroads / Syamal Kumar Ray. — New Delhi : Sterling, c1979.
ix, 407 p. ; 22 cm.
Bibliography: p. [372]-400.
Includes index.
Rs90.00
1. Civil service—India. I. Title.
JQ247.R39 354'.54'006 79-900570
79 MARC

Secretariat Training School.
Notes on the Central civil services (conduct) rules, 1964; with appendices containing the Secretary of State's services (conduct) rules, 1942; the Government servants' conduct rules; the All-India services (conduct) rules, 1954 (with notes); the Indian Foreign Service (conduct & discipline) rules, 1961; the Railway services (conduct) rules, 1966; the Central civil services (conduct) rules, 1955; and the All India services (conduct) rules, 1968. Revised. Delhi, Manager of Publications, 1969.
x, 228 p. 21 cm. Rs5.50 ($1.98 U.S.)
"S.T.S. 10/3,000."
1. Civil service—India.
342'.54'066 77-924410
72[r79]rev MARC

Singh, Prabhu Narayan, 1943-
Grass-root personnel for community development / P. N. Singh ; foreword, K. N. Singh ; pref., B. B. Chatterjee. — Varanasi : Prakashan Sansthan, 1977.
xiii, 26 [i.e. 126] p. : ill., maps ; 22 cm.
A revision of the author's thesis, Kashi Vidyapith, 1971.
Bibliography: p. [98]-107.
Includes index.
Rs25.00 ($3.00 U.S.)
1. Community development—India. 2. Rural development—India. 3. Civil service—India. I. Title.
HN690.Z9C682 1977 309.2'63'0954 78-902596
79 MARC

—INDIA—CASE STUDIES

Mook, Byron T
Value and action in Indian bureaucracy / by Byron T. Mook. — Brighton, Eng. : Institute of Development Studies, University of Sussex, 1974.
187 p. ; 30 cm. — (IDS discussion paper ; no. 65) GB***
Revision of the author's thesis.
Includes bibliographical references.
1. Bureaucracy—Case studies. 2. Civil service—India—Case studies. I. Title. II. Series: Institute of Development Studies. IDS discussion paper ; no. 65.
JQ247.M6 1974 354'.54'01 78-321173
79 MARC

—INDIA—HANDBOOKS, MANUALS, ETC.

Secretariat Training School.
Notes on office procedure. [2d ed. New Delhi] 1964.
ix, 384 p. forms. 25 cm.
"For official use only."
Errata slip inserted.
1. Civil service—India—Handbooks, manuals, etc. 2. India (Republic) Parliament—Rules and practice. I. Title.
JQ247.S36 1964 354'.54 sa 66-8129
[r79]rev MARC

Secretariat Training School.
Notes on office procedure. [3d ed. New Delhi, 1970]
vii, 240 p. 24 cm. Rs5.75; ($2.70 U.S.)
"STS.8/3000."
1. Civil service—India—Handbooks, manuals, etc. 2. India (Republic). Parliament—Rules and practice. I. Title.
JQ247.S36 1970 354'.54 74-922828
72[r79]rev MARC

—INDIA—HISTORY

India. Public Service Commission.
British attitude towards the employment of Indians in civil service : report of the Public Service Commission, 1886-1887, headed by Sir Charles U. Aitchison / critical introd. by Bradford Spangenberg. — Delhi : Concept Pub. Co., 1977.
xxxvii, 153 p. ; 35 cm.
Subtitle on cover: Aitchison's report, 1886-87.
Reprint of the 1888 ed., printed by the Supt. of Govt. Print., Calcutta, under title: Report of the Public Service Commission, 1886-87.
Includes bibliographical references.
Rs100.00
1. Civil service—India—History. 2. India—Officials and employees—Appointment, qualifications, tenure, etc.—History. 3. India—Politics and government—19th century. I. Aitchison, Charles Umpherston, Sir, 1832-1896. II. Title. III. Title: Aichison's report, 1886-87.
JQ245.P8 1977 354'.54'006 78-911475
79 MARC

—INDIA—SALARIES, ALLOWANCES, ETC.

Secretariat Training School.
Notes on financial rules, October 1961. Delhi, Manager of Publications, 1961.
286 p. forms. 25 cm. PL480:I-E-1005
At head of title: Govt. of India, Secretariat Training School, Ministry of Home Affairs.
1. Civil service—India—Salaries, allowances, etc. I. Title.
342'.54'068 sa 68-6799
[r79]rev MARC

Secretariat Training School.
Notes on financial rules. [3d ed. Delhi, Manager of Publications] 1966.
vi, 363 p. forms. 24 cm. Rs3.90; $1.41 PL480:I-E-9430
At head of title: Govt. of India. Secretariat Training School, Ministry of Home Affairs.
1. Civil service—India—Salaries, allowances, etc. I. Title.
JQ249.S2S42 1966 354'.54'00123 sa 68-1462
[r79]rev MARC

Secretariat Training School.
Notes on financial rules. [4th ed. New Delhi] 1970 [i.e. 1971]
viii, 562 p. 21 cm. Rs7.00 ($2.52 U.S.)
"STS. 14/8,000."
1. Civil service—India—Salaries, allowances, etc. I. Title.
JQ249.S2S42 1971 354'.54'00123 70-926192
72[r79]rev MARC

—INDIA—ANDHRA PRADESH

Andhra Pradesh, India.
Andhra Pradesh State and subordinate services rules : as corrected upto 30th April 1977. — [Hyderabad] : Govt. of Andhra Pradesh, 1977.
75 p. ; 26 cm.
Rs2.00
1. Civil service—India—Andhra Pradesh. I. Title.
342'.5484'068 78-904829
79 MARC

Andhra Pradesh, India.
The fundamental rules and the subsidiary rules of the Andhra Pradesh Government : corrected upto 30th June 1974. — [s.l. : s.n.], 1977 (Hyderabad : Printed by the Director of Print. and Stationery at the Govt. Secretariat Press)
ii, 535 p. ; 25 cm.
Includes index.
Rs13.75
1. Civil service—India—Andhra Pradesh. I. Title.
342'.5484'068 78-901851
79 MARC

—INDIA—MADHYA PRADESH

Madhya Pradesh, India.
Madhya Pradesh services manual : containing various service rules of Madhya Pradesh / by Krishan Lal Sethi. — 2d ed. — Indore : Lawyers Home, 1976.
2, 2, 890 p. ; 25 cm.
Includes bibliographical references.
Rs60.00
1. Civil service—India—Madhya Pradesh. I. Sethi, Krishan Lal. II. Title.
342'.543'068 78-900826
79 MARC

—INDIA—RAJASTHAN

Rajasthan, India.
The Rajasthan service rules, made under Article 309 of the Constitution of India. — 5th ed., corr. and amended up to 31st March 1972. — Jaipur : Govt. Central Press, [1972-
v. in ; 26 cm.
Cover title.
At head of title: Govt. of Rajasthan, Dept. of Personnel, O. & M. (Codification) Section.
Pref. in Hindi.
Loose-leaf for updating.
Rs6.05 (v. 1, pt. A) varies
1. Civil service—India—Rajasthan. I. Rajasthan, India. O. & M. Codification Section. II. Title.
342'.544'068 75-901963
79 MARC

—INDIA—TAMIL NADU

Tamil Nadu.
Tamil Nadu services manual : incorporates amendments issued upto 31st October 1975. — [Madras] : Govt. of Tamil Nadu, 19
v. ; 29 cm.
Rs10.00 (v. 5)
1. Civil service—India—Tamil Nadu. I. Title.
342'.5482'068 78-904864
79 MARC

—INDIA—TAMIL NADU—STATISTICS—COLLECTED WORKS

Tamil Nadu. Director of Employment and Training. State Employment Market Information Unit.
Report on the occupational pattern of employees in the public sector in Tamil Nadu.
Madras, State Employment Market Information Unit, Directorate of Employment and Training.
v. 32 cm.
1. Tamil Nadu — Occupations — Statistics — Collected works. 2. Civil service—India—Tamil Nadu—Statistics—Collected works. I. Title.
HB2730.T35T35a 77-912314
MARC-S

—INDIANA—PERSONNEL MANAGEMENT

Indiana. Dept. of Administration. Personnel Division.
Affirmative action plan, FY-1977. — [Indianapolis : Dept. of Administration, Personnel Division, 1976?]
35 p. ; 28 cm.
At head of title: State of Indiana.
1. Affirmative action programs—Indiana. 2. Civil service—Indiana—Personnel management. I. Title.
JK5660.A33I53 1976 331.2 78-621580
79 MARC

—INDONESIA

Bachtiar, Harsja W
Bureaucracy and nation formation in Indonesia / by Harsja W. Bachtiar. — Djakarta : LEKNAS, [1971]
25 leaves ; 29 cm.
1. Civil service—Indonesia. 2. Bureaucracy. 3. Indonesia—Politics and government—20th century. I. Title.
JQ772.Z1B32 354'.598'01 78-943360
79 MARC

Himpunan lengkap undang-undang & [i.e. dan] peraturan-peraturan kepegawaian negara / disusun oleh Istomo Gatot. — Bandung : Karya Nusantara Cabang Bandung, [1978-
v. ; 22 cm.
1. Civil service—Indonesia. I. Gatot, Istomo.
78-943040
79 MARC

Indonesia. Departemen Dalam Negeri. Badan Pendidikan dan Latihan.
Beberapa aspek tentang dinamika pemerintahan daerah : pembinaan wilayah, kebijaksanaan politik polisionil, fungsi fungsi dan tinjauan sejarah kepamongprajaan. — [Cet. 2.]. — Jakarta : Badan Pendidikan dan Latihan, Departemen Dalam Negeri, 1977.
109 p. ; 21 cm.
Bibliography: p. 109.
1. Local government—Indonesia. 2. Indonesia—Politics and government—1966- . 3. Civil service—Indonesia. I. Title.
JS7192.I525 1977 78-943263
79 MARC

Pilihan peraturan-peraturan tentang kepegawaian negeri sipil / dihimpun oleh Syaukat Yusuf. — Cet. 1. — Jakarta : Percetakan Offset Dalco, 1977-
v. ; 21 cm.
1. Civil service—Indonesia. I. Yusuf, Syaukat.
78-940798
79 MARC

Sosromidjojo, Madenan.
Peraturan peremadjaan pegawai negeri : Peraturan Pemerintah no. 68 tahun 1958 dengan pendjelasan dan keterangan² / oleh R. Madenan Sosromidjojo. — [Jogjakarta? : s.n., 1958?]
20 p. ; 19 cm.
1. Civil service—Indonesia. I. Indonesia. Laws, statutes, etc. Peraturan Pemerintah tentang peremadjaan alat-alat negara. 1958? II. Title.
78-949697
79 MARC

—INDONESIA—HISTORY

Koesoemahatmadja, Djenal Hoesen.
Perkembangan fungsi dan struktur pamong praja ditinjau dari segi sejarah / Djenal Hoesen Koesoemahatmadja. — Bandung : Alumni, 1978.
xx, 248 p. : ill. ; 21 cm.
Cover title: Fungsi & struktur pamongpraja.
Originally presented as the author's thesis, Universitas Padjadjaran.
Bibliography: p. 221-227.
Includes index.
1. Civil service—Indonesia—History. I. Title. II. Title: Fungsi & struktur pamongpraja.
JQ772.Z1K66 1978 79-940477
79 MARC

CIVIL SERVICE (Continued)

—INDONESIA—STATISTICS
—COLLECTED WORKS

Indonesia. Biro Pusat Statistik.
Statistik pegawai pemerintah Sulawesi.
Jakarta, Biro Pusat Statistik.
v. 21 x 34 cm.
1. Civil service—Indonesia—Statistics—Collected works.
I. Title.
JQ772.I 53a 73-941159
MARC-S

—INDONESIA—BANDUNG (KABUPATEN)
—STATISTICS—COLLECTED
WORKS

Indonesia. Biro Pusat Statistik.
Statistik pegawai pemerintah KBN Bandung.
Jakarta, Biro Pusat Statistik.
v. 21 x 34 cm.
1. Civil service—Indonesia—Bandung (Kabupaten)—Statistics—Collected works. I. Title.
JQ779.B426C584 73-940424
MARC-S

—INDONESIA—SUMATRA
—STATISTICS—COLLECTED
WORKS

Indonesia. Biro Pusat Statistik.
Statistik pegawai pemerintah Sumatera.
Jakarta, Biro Pusat Statistik.
v. 21 x 34 cm.
1. Civil service—Indonesia—Sumatra—Statistics—Collected works.
I. Title.
JQ779.S864 I 5a 76-646698
MARC-S

—IOWA—CASES

Iowa. Publica Employment Relations Board.
Iowa Public Employment Relations Board decisions. — [Des Moines] : The Board, [1976?]
1 v. ; 30 cm.
Caption title.
Loose-leaf for updating.
1. Civil service—Iowa—Cases.
KFI4635.A56I68 1976 79-101371
344'.777'01890413539
79 MARC

—ISRAEL

Nachmias, David.
Bureaucratic culture : citizens and administrators in Israel / David Nachmias and David H. Rosenbloom. — London : Croom Helm, c1978.
212 p. ; 23 cm. GB***
Bibliography: p. 202-209.
Includes index.
ISBN 0-85664-791-8 : £8.95
1. Civil service—Israel. 2. Israel—Politics and government. 3. Political participation—Israel. I. Rosenbloom, David H., joint author. II. Title.
JQ1825.P34N3 1978b 301.18'32'095694 79-301769
79 MARC

Nachmias, David.
Bureaucratic culture : citizens and administrators in Israel / David Nachmias and David H. Rosenbloom. — New York : St. Martin's Press, 1978.
212 p. ; 22 cm.
Bibliography: p. 202-209.
Includes index.
ISBN 0-312-10808-7 : $15.95
1. Civil service—Israel. 2. Israel—Politics and government. 3. Political participation—Israel. I. Rosenbloom, David H., joint author. II. Title.
JQ1825.P34N3 1978 301.18'32'095694 78-17638
78 MARC

—ISRAEL—PERIODICALS

Israel. Netsivut sherut ha-medinah.
(Hoda'ot - Netsivut sherut ha-medinah)
הודעות - נציבות שירות המדינה.
ירושלים, נציבות שירות המדינה.
v. 17 x 24 cm.
1. Civil service—Israel—Periodicals.
76-643931
(MARC-S)

—ITALY

Istituto per lo sviluppo della formazione professionale dei lavoratori.
Alcune proposte formative nella pubblica amministrazione. — Roma : L'Istituto, 1976.
137 p. ; 25 cm. — (Its Quaderni di formazione ISFOL ; n. 34) It**
1. Public administration—Study and teaching—Italy. 2. Public administration—Study and teaching—France. 3. Civil service—Italy. 4. Civil service—France. I. Title. II. Series.
JF1338.A2187 1976 79-354929
79 MARC

—ITALY—ADDRESSES, ESSAYS,
LECTURES

L'Organizzazione e il personale delle regioni / a cura di Francesco Sidoti. — Milano : Edizioni di Comunità, 1979.
346 p. ; 18 cm. — (Quaderni di studi regionali ; 10) (Cultura e realtà ; 93) (Serie pubblicata in collaborazione con la Fondazione Adriano Olivetti ; 43)
It79-June
Includes bibliographical references.
L6000
1. Local government—Italy—Addresses, essays, lectures. 2. Regionalism—Italy—Addresses, essays, lectures 3. Civil service—Italy—Addresses, essays, lectures. I. Sidoti, Francesco. II. Series.
JS5735.O74 352.045 79-385478
*79 MARC

—JAPAN

Japan. Laws, statutes, etc.
[Kokka kōmuin hō]
新旧対照改正国家公務員法 / [編集 石見榮吉]. — 東京：大學書房, 昭和24 [1949]
104 p. : 20 cm.
Cover title.
1. Civil service—Japan. 2. Japan—Official and employees. I. Ishimi, Eikichi. II. Title. III. Title: Kaisei Kokka kōmuin hō.
Title romanized: Shinkyū taishō kaisei Kokka kōmuin hō.
79-800594

Muroi, Tsutomu, 1930-
(Kōmuin no kenri to hō)
公務員の権利と法 / 室井力著. — 東京：勁草書房, 1978.
10, 277 p. ; 19 cm. — (現代法選書 ; 4) Ja 78-17839
¥1400
1. Civil service—Japan. 2. Japan—Officials and employees. 3. Labor laws and legislation—Japan. I. Title. II. Series: Gendaihō sensho ; 4.
78-813227

Sase, Minoru, 1932-
(Nippon kōmuin jijō)
ニッポン公務員事情 / 佐瀬稔著. — 東京：日本実業出版社, 1978.
222 p. ; 20 cm. Ja 78-14218
¥980
1. Japan—Officials and employees—Salaries, allowances, etc. 2. Civil service—Japan. I. Title.
JQ1649.S2S27 78-814714

(Sengo Nihon no kokumin o dame ni shita kōmuin)
戦後日本の国民をダメにした公務員 / 週刊新潮編集部編. — 東京：山手書房, 1978.
295 p. ; 19 cm. Ja 79-6156
Cover title.
Originally published in Shūkan shinchō, Feb. 25, 1976-June 22, 1978.
¥960
1. Civil service—Japan. 2. Japan—Officials and employees—Discipline. 3. Misconduct in office—Japan. I. Shinchōsha. Shūkan Shinchō Henshūbu. II. Shūkan shinchō. III. Title: Nihon no kokumin o dame ni shita kōmuin.
JQ1629.C6S46 79-809543

(Shin Chihō kōmuin hō jitsurei hanrei shū)
新地方公務員法実例判例集 / 自治省公務員部公務員第一課編集. — 東京：第一法規出版, 1979.
92, 1013, 23 p. : 18 cm. Ja 79-1472
Edition for 1969 published under title: Chihō kōmuin hō jitsurei hanrei shū.
Includes indexes.
¥3000
1. Local officials and employees—Japan. 2. Civil service—Japan. I. Japan. Jichishō. Gyōseikyoku. Kōmu... Kōmuin Dai 1-ka. II. Japan. Laws, statutes, ... Chihō kōmuin hō. 1979. III. Chihō kōmuin hō jitsurei hanrei shū.
79-807195

(Shin kokka kōmuin tokuhon)
新國家公務員読本 / 内閣編. — 東京：白友社出版部, 昭和24 [1949]
200 p. ; 19 cm.
English or Japanese.
CONTENTS: マックァーサー元師書簡全文—國家公務員法の制定から改正まで—政令と憲法違反問題—改正國家公務員法逐條解說—マックァーサー元師声名全文—タフト・ハートレー法との対比—官公廳給與ベースの推移と全官公労の動向—アメリカとソ連の公務員制度
1. Civil service—Japan. 2. Japan—Officials and employees. I. Japan. Naikaku Kambō. II. Japan. Laws, statutes, etc. Kokka kōmuin hō. 1949.
79-803294

Tanaka, Motosuke.
(Chihō kōmuin seido)
地方公務員制度 / 田中基介. — 東京：ぎょうせい, 1978.
20, 485, 4 p. ; 22 cm. — (現代地方自治全集 ; 7) Ja 79-3093
Bibliography: p. 1-4 (3d group)
¥2000
1. Local officials and employees—Japan. 2. Civil service—Japan. I. Title. II. Series: Gendai chihō jichi zenshū ; 7.
79-805342

Usui, Toshirō, 1911-1968.
(Kōmuinhō to shinkyūyohō)
公務員法と新給與法：國家公務員法同第二次改正法・教育公務員特例法・新給與法 / 臼井俊郎著. — 東京：太平社, 昭和24 [1949]
3, 8, 220, 51 p. ; 22 cm.
Colophon inserted.
1. Civil service—Japan. 2. Japan—Officials and employees—Salaries, allowances, etc. I. Japan. Laws, statutes, etc. Kokka kōmuin hō. 1949. II. Japan. Laws, statutes, etc. Shinkyūyohō. 1949. III. Japan. Laws, statutes, etc. Kyōiku kōmuin tokurei hō. 1949. IV. Title.
79-801836

—JAPAN—EXAMINATIONS,
QUESTIONS, ETC.

(Gyōsei mondai no tetteiteki kenkyū)
行政問題の徹底的研究 / 高文社編集部編. — 東京：高文社, 昭和34 [1959]
3, 139 p. ; 18 cm. — (公務員試験研究叢書)
1. Civil service—Japan—Examinations, questions, etc. 2. Public administration—Examinations, questions, etc. I. Kōbunsha, Tokyo (Morikawachō). Henshūbu. II. Series: Kōmuin shiken kenkyū sōsho.
JQ1646.G96 78-812107

—JAPAN—HISTORY

Japan. Jinjiin. Jimu Sōkyoku.
(Jinji gyōsei sanjūnen no ayumi)
人事行政三十年の歩み / [編集 人事院事務総局]. — 東京：同局, 昭和53 [1978]
vii, 563 p., [3] leaves of plates : ill. ; 27 cm. Ja***
1. Civil service—Japan—History. 2. Japan. Jinjiin—History. I. Title.
JQ1646.J34 1978 79-806944

—KENTUCKY—MINORITY
EMPLOYMENT—PERIODICALS

Kentucky. Commission on Human Rights.
Black employment in Kentucky State agencies; report. 4th- 1969-
Frankfort, Ky., Kentucky Commission on Human Rights, Commonwealth of Kentucky.
v. 28 cm.
Continues: Kentucky. Commission on Human Rights. Negro employment in Kentucky State agencies; report.
1. Civil service—Kentucky—Minority employment—Periodicals. 2. Afro-Americans—Employment—Kentucky—Periodicals. I. Title.
JK5360.5.A34K45a 331.2'041'3539769 78-648426
MARC-S

CIVIL SERVICE (Continued)

—KENYA

Kenya Colony and Protectorate. Civil Service Commission.
Report of the Civil Service Commissioner (L. C. Hill ...). — Nairobi : Govt. Printer, 1945.
49 p. ; 25 cm.
Cover title.
1. Civil service—Kenya. I. Hill, Levi, 1883- .
JQ2947.A65K46 1945 354'.676'2006 79-318436
79 MARC

—KOREA—EXAMINATIONS

(Kukcho pangmok)
國朝榜目 / ⌈編輯大韓民國國會圖書館⌉. — ⌈서울⌉ : 同圖書館, 1971.
534, 319, ⌈14⌉ p. ; 26 cm.
Photoreprint ed.
"國朝榜目捷覽索引(性別字劃順)": p. 1–319 (2d group)

1. Korea—Registers. 2. Civil service—Korea—Examinations. 3. Korea—History—Yi dynasty, 1392–1910. I. Kukhoe Tosŏgwan, Seoul, Korea.
JQ1724.K8 1971 78–826629

—KOREA—EXAMINATIONS —PERIODICALS

(Chinch'ul)
進出 第1- 卷(通卷第1- 號); 1976年7月-
서울 企劃出版社
v. ill. 26 cm. monthly.
"就職試驗綜合受驗誌"

1. Civil service—Korea—Examinations—Periodicals. 2. Civil service—Korea—Examinations, questions, etc.—Periodicals. 3. Employment tests—Korea—Periodicals.
JQ1726.Z13E873 78–647725 (MARC-S)

—KOREA—HISTORY

Yi, Hyŏn-jong.
(Ch'ŏngbaengni chŏngsin kwa yŏlchŏn)
清白吏精神과列傳 / 李鉉淙著. — 서울 : 亞細亞文化社, 1977.
10, 261 p. ; 22 cm.
Includes index.
₩2200

1. Civil service—Korea—History. 2. Civil service ethics. 3. Korea—Officials and employees—Biography. I. Title.
JQ1726.Z1Y45 78–826665

—KOREA—HISTORY—SOURCES

(Yangjŏn p'yŏn'go)
兩銓便攷 / ⌈編輯法制處⌉. — ⌈서울⌉ : 同處, 1978.
210, 72, 43, ⌈5⌉ p. ; 22 cm. — (法制資料 ; 第96輯)
Includes photoreprint of the 1870 ed.
Not in trade.

1. Civil service—Korea—History—Sources. 2. Korea—Officials and employees—History—Sources. I. Korea (Republic). Pŏpchech'ŏ. II. Series: Pŏpche charyo ; che 96-chip.
JQ1726.Z1Y33 79–825498

—LEBANON

Moallem, Chafik.
Le régime disciplinaire de la fonction publique au Liban / par Chafik Moallem ; préf. de Antoine Baroud. — Beyrouth : Université libanaise, 1974.
vi, 342 p. ; 25 cm. — (Publications de l'Université libanaise : Section des études juridiques, politiques et administratives ; 3)
Title of p. ⌈4⌉ of cover: al-Niẓām al-ta'dībī lil-waẓīfah al-'āmmah fī Lubnān.
Bibliography: p. ⌈325⌉-334.
1. Civil service—Lebanon. 2. Labor discipline—Lebanon. I. Title. II. Series: Beirut. al-Jāmi'ah al-Lubnānīyah. Manshūrāt al-Jāmi'ah al-Lubnānīyah : Qism al-dirāsāt al-qānūnīyah, al-siyāsīyah wa-al-idārīyah ; 3.
354'.61'2001 76-960426
78⌈r79⌉rev MARC

—MALAYSIA

Puthucheary, Mavis.
The politics of administration : the Malaysian experience / Mavis Puthucheary. — Kuala Lumpur ; New York : Oxford University Press, 1978.
xi, 170 p. ; 26 cm. — (East Asian social science monographs)
A revision of the author's thesis, University of Manchester, 1973.
Bibliography: p. ⌈144⌉-163.
Includes index.
ISBN 0-19-580386-8. ISBN 0-19-580387-6 pbk.
1. Civil service—Malaysia. 2. Malaysia—Politics and government. I. Title.
JQ716.Z1P87 1978 354'.595'006 78-942725
79 MARC

—MASSACHUSETTS

Massachusetts. Division of Personnel Administration.
Information on 1977 legislation affecting Civil Service : prepared by the Massachusetts Division of Personnel Administration. — ⌈Boston⌉ : The Division, ⌈1978⌉
6, ⌈9⌉ p. ; 28 cm.
Cover title.
1. Civil service—Massachusetts. I. Title.
KFM2835.A86 342'.744'068 79-622487
79 MARC

—MASSACHUSETTS—PERSONNEL MANAGEMENT

Massachusetts. Governor's Management Task Force.
Personnel paperwork study. — ⌈Boston⌉ : Governor's Management Task Force, Commonwealth of Massachusetts, 1977.
41 p., ⌈1⌉ leaf of plates : diagrs. ; 28 cm.
Photocopy of original.
1. Massachusetts. Division of Personnel Administration. 2. Civil service—Massachusetts—Personnel management. I. Title.
JK3155.M39 1977 353.9'744'001 79-621125
79 MARC

—MASSACHUSETTS—PERSONNEL MANAGEMENT—CONGRESSES

University of Massachusetts (System). Institute for Governmental Services.
State-municipal personnel operations : a redefinition of roles : report of conference proceedings, September 30, 1976, University of Massachusetts-Amherst / prepared by Institute for Governmental Services ; sponsored by Division of Personnel Administration. — ⌈Boston⌉ : The Institute, 1976.
ii, 54 p. ; 28 cm.
1. Civil service—Massachusetts—Personnel management—Congresses. 2. Municipal government—Massachusetts—Personnel management—Congresses. I. Massachusetts. Division of Personnel Administration. II. Title.
JK3155.U54 1976 353.9'744'001 79-621669
79 MARC

—MASSACHUSETTS—PERSONNEL MANAGEMENT—HANDBOOKS, MANUALS, ETC.

Massachusetts. Division of Personnel Administration.
Chapter 778, a manual for program sponsors. — ⌈Boston : Commonwealth of Massachusetts, Executive Office for Administration and Finance, 1978⌉
iii, 96 leaves : forms ; 28 cm. — (Informational circular - Commonwealth of Massachusetts, Executive Office for Administration and Finance)
Developed by the Massachusetts Division of Personnel Administration and the Dept. of Manpower Development.
"Publication 10,377-97-3500-4-78-CR."
1. Affirmative action programs—Massachusetts—Handbooks, manuals, etc. 2. Civil service—Massachusetts—Personnel management—Handbooks, manuals, etc. I. Massachusetts. Dept. of Manpower Development. II. Title. III. Series: Massachusetts. Executive Office for Administration and Finance. Informational circular - Commonwealth of Massachusetts, Executive Office for Administration and Finance.
JK3160.A33M32 1978 331.2 79-622013
79 MARC

—MEXICO

Mexico. Laws, statutes, etc.
Legislación federal del trabajo burocrático : comentarios y jurisprudencia, disposiciones complementarias, legislación de seguridad social para las Fuerzas Armadas / Alberto Trueba Urbina, Jorge Trueba Barrera ⌈editores⌉. — 11. ed. actualizada. — México : Editorial Porrúa, 1978.
xvi, 506 p. ; 17 cm. — (Biblioteca jurídica Porrúa)
1. Civil service—Mexico. I. Trueba Urbina, Alberto, 1906- . II. Trueba Barrera, Jorge. III. Title.
342'.72'068 78-107735
78 MARC

—MEXICO—STATISTICS

Mexico. Comisión de Recursos Humanos.
Censo de recursos humanos del sector público federal : administración descentralizada y de participación estatal mayoritaria (1975). — ⌈1. ed.⌉. — ⌈México⌉ : Comisión de Recursos Humanos del Gobierno Federal, ⌈1976⌉
421 p. : ill. ; 28 cm.
1. Mexico—Officials and employees—Statistics. 2. Civil service—Mexico—Statistics. I. Title.
JL1247.M43 78-112274
79 MARC

—NEPAL

Nepal. Loka Sevā Āyoga.
Regarding Nepal Public Service Commission. ⌈Kathmandu, 1963⌉
35 p. 23 cm. Re0.90 PL480:N-161
Cover title.
Includes legislation.
1. Civil service—Nepal. I. Nepal. Laws, statutes, etc. II. Title.
JQ1825.N44N46 1963 354'.549'6006 sa 68-1162
⌈r79⌉rev MARC

—NETHERLANDS

Netherlands (Kingdom, 1815-). Laws, statutes, etc.
Ambtenarenwet 1929 / bew. door H. Bekker en H. A. A. G. Vermeulen. — 10e dr. — Zwolle : W. E. J. Tjeenk Willink, 1979.
2 v. ; 19 x 8 cm. — (Nederlandse staatswetten ; 43) Ne79-19
Cover title.
"Bijgewerkt tot december 1978."
Includes bibliographical references.
Vol. 1 includes index.
ISBN 9027115575 : fl 41.50 (v. 1) fl 31. 00 (v. 2)
1. Civil service—Netherlands. I. Bekker, H. II. Vermeulen, H. A. A. G. III. Title. IV. Series.
79-373442
*79 MARC

Spanningsvelden rondom de ambtenaar : congrespublikatie 1977 / ⌈onder eindred. van L. E. M. Klinkers⌉. — Alphen aan den Rijn : Samsom, 1978.
144 p. ; 25 cm. — (Geschriften van de Vereniging voor Bestuurskunde ; nr. 1) Ne78-16
"Met bijdragen van A. M. J. C. Aarts ... ⌈and others⌉; ingeleid door A. Hoogerwerf."
Papers presented at a congress organized by Vereniging voor Bestuurskunde.
Bibliography: p. 135-143.
ISBN 9014027591 : fl 18.00
1. Civil service—Netherlands. I. Klinkers, L. E. M. II. Vereniging voor Bestuurskunde. III. Series: Vereniging voor Bestuurskunde. Geschriften van de Vereniging voor Bestuurskunde ; nr. 1.
JN5851.S67 78-387206
*79 MARC

—NEW JERSEY

New Jersey. Legislature. General Assembly. State Government, Federal and Interstate Relations, and Veteran's Affairs Committee.
Public hearings before Assembly State Government, Federal and Interstate Relations, and Veterans Affairs Committee on A-1675 (Civil service reform act) : held November 2, 1978, Assembly Chamber, State House, Trenton, New Jersey. — Trenton : The Committee, ⌈1979?⌉
226 p. in various pagings : graphs ; 29 cm.
1. Civil service—New Jersey. 2. Civil service reform. I. Title.
KFN1811.4.S78 1978c 342'.749'0680262 79-621854
79 MARC

—NEW JERSEY—ADDRESSES, ESSAYS, LECTURES

Messick, Charles P
Reflections of a public servant : end of a journey / by Charles P. Messick ; foreword by William W. Boyer. — Newark : University of Delaware, c1978.
xiii, 76 p. ; 25 cm.
1. Public administration—Addresses, essays, lectures. 2. Civil service—United States—Addresses, essays, lectures. 3. Civil service—New Jersey—Addresses, essays, lectures. I. Title.
JF1351.M395 350 79-621699
79 MARC

CIVIL SERVICE (Continued)

—NEW MEXICO—MINORITY EMPLOYMENT

United States. Commission on Civil Rights. New Mexico Advisory Committee.
Indian employment in New Mexico State government : a report to the New Mexico Advisory Committee to the United States Commission on Civil Rights prepared for the information and consideration of the Commission. — ⌈Washington⌉ : The Commission, 1975.
ix, 71 p. : ill. ; 27 cm.
Cover title.
Includes bibliographical references.
1. Indians of North America—New Mexico—Employment. 2. Civil service—New Mexico—Minority Employment. I. Title.
E78.N65U59 1975 331.6'9'97 75-603191
76⌈r78⌉rev MARC

—NEW YORK (CITY)—EXAMINATIONS

Turner, David Reuben, 1915-
Administrative manager : the complete study guide for scoring high / by David R. Turner. — 1st ed. — New York : Arco, c1979.
224 p. (p. 223-224 publisher's list) : ill. ; 26 cm. — (Arco civil service test tutor)
ISBN 0-668-04813-1 : $8.00
1. Civil service—New York (City)—Examinations. 2. Public administration—Examinations, questions, etc. I. Arco Publishing Company, New York. II. Title.
JS1234.A4 1979b 352'.005'3097471 79-13381
79 MARC

Turner, David Reuben, 1915-
Parking enforcement agent; entering and supervisory grades, open to men and women, by David R. Turner. ⌈3d ed.⌉ New York ⌈1968⌉
384, 16 p. illus., forms. 26 cm. (Arco civil service test tutor)
At head of title: The complete study guide for scoring high.
1. Traffic regulations—New York (City) 2. Parking meter collectors—New York (City) 3. Civil service—New York (City)—Examinations. I. Arco Publishing Company, New York. II. Title.
HE372.N65 1968a 388.4'74'097471 68-4558
⌈r79⌉rev MARC

—NEW YORK (CITY)—MINORITY EMPLOYMENT

Viteritti, Joseph P 1946-
Bureaucracy and social justice : allocation of jobs and services to minority groups / Joseph P. Viteritti. — Port Washington, NY : Kennikat Press, 1979.
xiv, 199 p. ; 22 cm. — (National university publications) (Interdisciplinary urban series)
Bibliography: p.
Includes index.
ISBN 0-8046-9244-0 : $15.00
1. Civil service—New York (City)—Minority employment. 2. Municipal services—New York (City) I. Title.
JS1234.A4 1979 352.0474'1 79-10707
79 MARC

—NEW YORK (STATE)

New York (State). Legislative Commission on Expenditure Review.
Vacation credit exchange : program audit, 3.1.78, June 16, 1978 / Legislative Commission on Expenditure Review. — Albany : The Commission, ⌈1978⌉
3, iii, 26 p. : graphs ; 28 cm.
Cover title.
1. Vacations, Employee—New York (State) 2. Civil service—New York (State) 3. Non-wage payments—New York (State) I. Title.
HD5262.N7N48 1978 353.9'747'001233 78-624308
79 MARC

United States. Commission on Civil Rights. New York Committee.
The Puerto Rican and public employment in New York State : report of the New York State Advisory Committee to the U.S. Commission on Civil Rights / prepared for the information and consideration of the Commission. — ⌈New York⌉ : The Committee, 1973.
ii, 33 p. ; 27 cm.
Cover title.
Includes bibliographical references.
1. Minorities—Employment—New York (State) 2. Civil service—New York (State) 3. Local officials and employees—New York (State) 4. Puerto Ricans—Employment—New York (State) I. Title.
JK3460.5.M5U54 1973 79-313283
331.6'3'6872950747
79 MARC

—NEW YORK (STATE)—LABOR PRODUCTIVITY

New York (State). Legislature. Senate. Task Force on Critical Problems.
Productivity : a report / by the Task Force on Critical Problems for John R. Dunne, Martin S. Auer, Lloyd Paterson, to the Temporary State Commission on Management and Productivity in the Public Sector ; John M. Flynn, policy analyst. — Albany : ⌈The Task Force⌉, 1975.
iii, 107 p. ; 28 cm.
Bibliography: p. 101-107.
1. Civil service—United States—Labor productivity. 2. Civil service—United States—States—Labor productivity. 3. Civil service—New York (State)—Labor productivity. I. New York (State). Temporary State Commission on Management and Productivity in the Public Sector. II. Title.
JK768.4.N48 1975 353.9'747'00147 76-623425
79 MARC

—NEW YORK (STATE)—MINORITY EMPLOYMENT—STATISTICS PERIODICALS

New York (State). Dept. of Civil Service. Career Opportunities Division.
Report on the annual ethnic survey of the work force of New York State agencies.
Albany, New York State Dept. of Civil Service.
v. 30 cm.
1. Civil service—New York (State)—Minority employment—Statistics—Periodicals. 2. New York (State)—Officials and employees—Statistics—Periodicals. I. Title.
JK3460.5.M5N48a 331.2'041'3539747 78-643635
MARC-S

—NEW YORK (STATE)—ALBANY

Zenner, Walter P
Jewish State employees in Albany : a research report / by Walter P. Zenner. — ⌈Albany⌉ : Dept. of Anthropology, State University of New York at Albany, c1978.
37 leaves ; 28 cm. — (Working papers - Department of Anthropology, State University of New York at Albany ; no. 1)
Cover title.
Bibliography: leaves 36-37.
1. Jews in Albany—Social conditions. 2. Civil service—New York (State)—Albany. 3. New York (State)—Officials and employees. 4. Albany—Social conditions. I. Title. II. Series: New York (State). State University, Albany. Dept. of Anthropology. Working papers - Department of Anthropology, State University of New York at Albany ; no. 1.
F129.A39J58 78-623988
301.45'19'24074743
79 MARC

—NIGERIA

Katagum, Sule.
The development of the public service commission system and the problems of recruitment into the public service in a plural society / by Alhaji Sule Katagum. — ⌈Lagos⌉ : Federal Republic of Nigeria, ⌈1974⌉
14 p. ; 24 cm.
Cover title.
1. Civil service—Nigeria. I. Title.
JQ3092.Z1K37 354'.669'001 79-301670
79 MARC

—NIGERIA—OYO (STATE)—PERIODICALS

Oyo, Nigeria (State). Public Service Commission.
Report of the Public Service Commission.
⌈Ibadan, Public Service Commission, Oyo State of Nigeria⌉
v. ill. 25 cm. annual.
Report year ends Mar. 31.
1. Oyo, Nigeria (State). Public Service Commission—Periodicals. 2. Civil service—Nigeria—Oyo (State)—Periodicals. I. Title.
JQ3099.O924O93a 354'.669'2 79-649977
MARC-S

—NORTHERN IRELAND

Oliver, John, 1913-
Working at Stormont : memoirs / by John Andrew Oliver. — Dublin : Institute of Public Administration, 1978.
251 p., ⌈4⌉ leaves of plates : ill. ; 24 cm. Ir***
Includes index.
ISBN 0-902173-83-9 : £8.47
1. Civil service—Northern Ireland. 2. Northern Ireland—Officials and employees—Biography. 3. Oliver, John, 1913- I. Title.
JN1572.A66O45 354'.416'0010924 79-305789
79 MARC

—NORTHWESTERN STATES—MINORITY EMPLOYMENT

State government affirmative action in mid-America : a report / prepared by the Iowa, Kansas, Missouri, and Nebraska Advisory Committees to the United States Commission on Civil Rights. — Washington : The Commission : for sale by the Supt. of Docs., U.S. Govt. Print. Off., 1978.
xiii, 107 p. ; 26 cm.
Includes bibliographical references.
1. Civil service—Northwestern States—Minority employment. 2. Women in the civil service—Northwestern States. 3. Affirmative action programs—Northwestern States. I. United States. Commission on Civil Rights. Iowa State Advisory Committee.
JK2480.M5S7 331.2 78-602338
78 MARC

—OREGON

Oregon. Legislative Assembly. Legislative Counsel Committee. Subcommittee on Affirmative Action in State and Local Government.
Final report of the Subcommittee on Affirmative Action in State and Local Government, Oregon Legislative Assembly, Legislative Counsel Committee. — Salem : The Subcommittee, 1978.
iv, 17, 3 leaves ; 28 cm.
Includes bibliographical references.
1. Affirmative action programs—Oregon. 2. Civil service—Oregon. 3. Local officials and employees—Oregon.
JK9060.A33O73 1978 331.2 79-623555
79 MARC

—PAKISTAN

Ali, Hamid.
Combined set of F.R. & S.R. applicable to the central government, servants : with upto-date commentary / by Hamid Ali. — Rev. ed., amended & upto-date. — Karachi : Ideal Publishers, 1977.
2 v. in 1 ; 25 cm.
Includes index.
Rs68.00
1. Civil service—Pakistan. I. Title.
342'.5491'068 78-930000
79 MARC

Pannu, Nisar Ahmad.
New civil service laws / by Nisar Ahmad Pannoun. — Rev. & enl. ed. — Lahore : Mansoor Book House, ⌈1978⌉
viii, 259, 54 p. ; 22 cm.
1976 ed. published under title: New service laws.
Rs25.00
1. Civil service—Pakistan. I. Pakistan. Laws, statutes, etc. II. Title.
342'.5491'068 78-930576
79 MARC

—PAKISTAN—ADDRESSES, ESSAYS, LECTURES

Peshawar (City). Academy for Rural Development.
Administrative reforms, Chinese communes, and local government : a report on the training of divisional, and tehsil level officers / editor, S. M. Haider. — Peshawar : Pakistan Academy for Rural Development, 1976.
209 p. in various pagings ; 34 cm. — (Documentation & reports - Pakistan Academy for Rural Development)
Includes bibliographical references.
Rs35.00 ($7.00 U.S.)
1. Pakistan—Politics and government—Addresses, essays, lectures. 2. Civil service—Pakistan—Addresses, essays, lectures. 3. Communes (China)—Addresses, essays, lectures. I. Haider, S. M. II. Title. III. Series: Peshawar (City). Academy for Rural Development. Documentation & reports - Pakistan Academy for Rural Development.
JQ548.P47 1976 354'.549'1 78-930164
79 MARC

—PENNSYLVANIA—CASES

Pennsylvania. State Civil Service Commission.
Digest of adjudications of the Pennsylvania State Civil Service Commission, 1941-August, 1975 / ⌈prepared by Barbara M. Bonan ... et al.⌉. — ⌈Harrisburg⌉ : The Commission, c1976.
ca. 600 p. ; 26 cm.
Includes index.
1. Civil service—Pennsylvania—Cases. I. Bonan, Barbara M. II. Title.
KFP435.A57C59 342'.748'068 78-623162
79 MARC

CIVIL SERVICE (Continued)

—PERU

Manual práctico : legislación empleados públicos : leyes, decretos supremos, resoluciones supremas, resoluciones directorales / V. Jorge Urquizo Pérez ⌊editor⌋. — Arequipa, Peru : ⌊s.n.⌋, 19
v. ; 21 cm.
Includes index.
CONTENTS:
—pt. 2. Manuales normativos INAP.

S/600.00 (v. 2)
1. Civil service—Peru. 2. Peru—Officials and employees. I. Urquizo Pérez, V. Jorge.
342'.85'068 79-371094
79 MARC

—PHILIPPINE ISLANDS
—PERIODICALS

Philippines (Republic). Civil Service Commission.
Civil Service reporter.
⌊Manila, etc., Civil Service Commission⌋
v. ill. 31 cm. monthly.
Began in 1956. Cf. New serial titles.
Key title: Civil Service reporter (Quezon City), ISSN 0300-3620.
1. Civil service—Philippine Islands—Periodicals. I. Title.
JQ1412.P47a 354'.599'006 79-645149
MARC-S

—PRUSSIA

⌊**Ėrshtrem, Ė A** ⌋
(Ocherki prusskago administrativnago stroi͡a)
Очерки прусскаго административнаго строя. С.-Петербургъ : Гос. тип., 1904.
57, 11 p. ; 25 cm.
Name of author lettered on cover of binding of L.C. copy.
1. Civil service—Prussia. I. Title.
JN4535.E77 Russian Imperial Coll. 79-408228

—RHODE ISLAND

United States. Commission on Civil Rights. Rhode Island Advisory Committee.
Minorities and women in government : practice versus promise : a review of equal employment opportunity in state government and selected cities in Rhode Island / a report of the Rhode Island Advisory Committee to the United States Commission on Civil Rights. — ⌊Washington⌋ : The Committee, 1975.
ix, 141 p. ; 27 cm.
1. Discrimination in employment—Rhode Island. 2. Civil service—Rhode Island. 3. Minorities—Employment—Rhode Island. 4. Women—Employment—Rhode Island. I. Title.
HD4903.5.U6R48 331.1'33 79-312594
79 MARC

—SAUDI ARABIA

al-Ṣabbāb, Aḥmad.
(Shu'ūn al-muwaẓẓafīn wa-nuẓum al-khidmah al-madanīyah fī al-Mamlakah al-'Arabīyah al-Sa'ūdīyah)
شئون الموظفين ونظم الخدمة المدنية فى المملكة العربية السعودية / أحمد الصباب ، محمد محمد محجوب . — جدة : دار المجمع العلمى ، 1978.
2 v. ; 24 cm.
Bibliography: v. 2, p. 369-375.
1. Civil service—Saudi Arabia. 2. Saudi Arabia—Officials and employees. I. Maḥjūb, Muḥammad Muḥammad, joint author. II. Title.
JQ1825.S34S2 78-960232

—SEYCHELLES

Waller, Harold.
Seychelles public service : structural review : composition of the service on 1 January 1972, growth 1900-1972, localisation, training, salary structure, manpower / H. Waller. — ⌊s.l. : s.n.⌋, 1971 (Mahé, Seychelles : Govt. Printer)
v, 172 p. : ill. ; 34 cm.
Cover title.
Includes bibliographical references.
1. Civil service—Seychelles. I. Title.
JQ3185.A67W34 354'.69'6006 79-313254
79 MARC

—SOMALIA—PERSONNEL MANAGEMENT

Somali Institute of Public Administration.
Personnel administration in Somalia / Somali Institute of Public Administration. — Mogadiscio : Somali Democratic Republic, Somali Institute of Development Administration and Management, 1974.
23 p. ; 33 cm. — (Occasional papers - Somali Institute of Development Administration and Management)
1. Civil service—Somalia—Personnel management. I. Title. II. Series: Machadka Maamul-Horumarinta iyo Maareynta Soomaaliya. Occasional papers - Somali Institute of Development Administration and Management.
JQ3585.A67S65 1974 354'.67'73001 78-107630
79 MARC

—SPAIN

Carlos Esteban, Alvaro de.
Nociones de derecho administrativo / por Alvaro de Carlos Esteban ... y Manuel Galván Dopazo .... — 5. ed. — Madrid : A. Caballero, ⌊1978?⌋
235 p. ; 23 cm. Sp***
On t.p.: C-64.
On cover: Nociones de derecho administrativo para ingreso en cuerpos administrativos del estado.
ISBN 8470230980 : 800ptas
1. Administrative law—Spain—Handbooks, manuals, etc. 2. Civil service—Spain. I. Galván Dopazo, Manuel, joint author. II. Title.
342'.46'06 79-370276
79 MARC

—SUDAN

Sudan. Wizārat al-Khidmah al-'Āmmah wa-al-Iṣlāḥ al-Idārī.
Personnel : administrative and related reforms : report of an International Commission / Democratic Republic of the Sudan, Ministry of Public Service and Administrative Reform. — Khartoum : The Ministry, 1973.
37, ⌊16⌋ p. ; 30 cm.
1. Civil service—Sudan. I. Title.
JQ3981.S84S83 1973 354'.624'001 77-980880
78 MARC

—SWEDEN

Sweden. Utredningen angående den statliga personalutbildningens fortsatta inriktning och organisation.
Statlig personalutbildning : ett principförslag : ⌊betänkande från⌋ / Utredningen angående den statliga personalutbildningens fortsatta inriktning och organisation. — Stockholm : ⌊LiberFörlag/Allmänna förl.⌋, 1976.
183 p. ; 25 cm. — (Statens offentliga utredningar ; 1976:61 ISSN 0375-250X)
S77-4/5
1. Sweden—Officials and employees. 2. Employees, Training of—Sweden. 3. Civil service—Sweden. I. Title. II. Series: Sweden. Statens offentliga utredningar ; 1976:61.
J406.R15 1976:61 78-368477
⌊JN7903⌋ 78 MARC

—SWITZERLAND

Hangartner, Yvo.
Ausgewählte Rechtsfragen der öffentlichen Dienstverhältnisse : Referat = ⌊Choix de questions juridiques concernant les rapports de services publics : exposé⌋ / von Yvo Hangartner. — Bern, Postfach 61, 3025 : Zentralverband des Staats- und Gemeindepersonals der Schweiz, 1977.
30 p. ; 21 cm. Sw78-5031
"Anlässlich der Präsidenten-Konferenz des Zentralverbandes des Staats- und Gemeindepersonals der Schweiz 1977 in Aarau."
German and French.
3.00F
1. Civil service—Switzerland. I. Title. II. Title: Choix de questions juridiques concernant les rapports de services publics.
342'.494'068 78-383112
79 MARC

—TAIWAN

Ch'en, Pao-ming, 1938-
(Kung wu jen yüan tsai chih hsün lien chih yen chiu)
公務人員在職訓練之研究 / 陳寶明著. — ⌊台北⌋ : 嘉新水泥公司文化基金會, 民國57 ⌊1968⌋
2, 5, 142 p. ; 22 cm. — (嘉新水泥公司文化基金會研究論文 ; 第120種) (嘉新水泥公司文化基金會叢書)
Series 1 romanized: Chia hsin shui ni kung ssu wen hua chi chin hui yen chiu lun wen. Series 2 romanized: Chia hsin shui ni kung ssu wen hua chi chin hui ts'ung shu.
Bibliography: p. 137-142.
1. Civil service—Taiwan. 2. Employees, Training of—Taiwan. I. Title.
JQ1532.Z13I62 78-840586

—TAIWAN—EXAMINATIONS

Ch'u, Ying-jui.
(T'ai-wan chih k'ao shih yü hsün lien)
臺灣之攷試與訓練 / 編著者褚應瑞, 吳新元, 薛建吾. — ⌊臺北⌋ : 臺灣省政府教育廳 : 發行所臺灣書店, 民國37 ⌊1948⌋
3, 170 p. ; 18 cm. — (臺灣省教育叢刊 ; 2)
Colophon title.
Series romanized: T'ai-wan sheng chiao yü ts'ung k'an.
1. Teachers, Training of Taiwan. 2. Teachers—Certification—Taiwan. 3. Civil service—Taiwan—Examinations. I. Wu, Hsin-yüan, joint author. II. Hsüeh, Chien-wu, joint author. III. Title.
LB1727.T28C54 78-841363

Hsü, Hsiang-ming.
(Tsui hsin chiu yeh k'ao shih chih nan)
最新就業考試指南 / ⌊編著者徐祥明⌋. — 台北 : 環球半月雜誌社, ⌊民國⌋67 ⌊1978⌋
336 p. ; 22 cm.
Cover title.
NT$120.00
1. Civil service—Taiwan—Examinations. 2. Employment tests—Taiwan. 3. Universities and colleges—Taiwan—Admission. 4. Universities and colleges—Taiwan—Entrance requirements. I. Title. II. Title: Chiu yeh k'ao shih chih nan.
JQ1532.H77 78-843954

Shih, Chung-cheng.
(Kuo fu i chiao shih t'i ching chieh)
國父遺教試題精解 : 歷屆高普檢特考試試題 / 施仲政著. — 台南 : 光田出版社, 民國67 ⌊1978⌋
i, 312 p. ; 22 cm.
Bibliography: p. 312.
NT$70.00

1. Sun, Yat-sen, 1866-1925. 2. Civil service—Taiwan—Examinations. I. Title
DS777.S55 79-837458

—UNITED STATES

Beyer, Janice M
Implementing change : alcoholism policies in work organizations / Janice M. Beyer, Harrison M. Trice. — New York : Free Press, c1978.
xix, 347 p. ; 25 cm.
Bibliography: p. 321-336.
Includes index.
ISBN 0-02-903460-4
1. Alcoholism and employment—United States. 2. Civil service—United States. 3. Alcoholism—Treatment—United States. I. Trice, Harrison Miller, 1920- joint author. II. Title.
JK850.A4B49 658.38'2 78-54127
78 MARC

Cahn, Frances, 1907-
Federal employees in war and peace : selection, placement, and removal / by Frances T. Cahn. — Westport, Conn. : Greenwood Press, 1978, c1949.
xiii, 253 p. ; 22 cm.
Reprint of the ed. published by Brookings Institution, Washington.
Includes bibliographical references and index.
ISBN 0-313-20602-3
1. Civil service—United States. I. Title.
JK691.C25 1978 353.001 78-16400
78 MARC

Dwoskin, Robert P
Rights of the public employee / Robert P. Dwoskin. — Chicago : American Library Association, 1978.
x, 269 p. ; 25 cm.
Bibliography: p. 261-262.
Includes index.
ISBN 0-8389-0257-X : $12.50
1. Civil service—United States. I. American Library Association. II. Title.
KF5337.D87 342'.73'068 78-1658
78⌊r79⌋rev MARC

Goodman, Carl F 1937-
Handbook on public personnel law / Carl F. Goodman. — New York : Law Journal Seminar Press, c1978.
xviii, 366 p. ; 24 cm.
Includes index.
$22.00
1. Employee-management relations in government—United States. 2. Civil service—United States. I. Title.
KF5365.G66 342'.73'068 78-15710
78 MARC

Hyneman, Charles Shang, 1900-
Bureaucracy in a democracy / by Charles S. Hyneman. — 1st AMS ed. — New York : AMS Press, 1978, c1950.
xv, 586 p. ; 23 cm.
Reprint of the ed. published by Harper, New York.
Includes bibliographies and index.
ISBN 0-404-11549-7
1. Civil service—United States. I. Title.
JK42.H8 1978 353 74-6023
77 MARC

CIVIL SERVICE
—UNITED STATES (Continued)

(Kakkoku kanri seido no kenkyū)
各國官吏制度の研究 / 蠟山政道 ... [et al.]. / 東京：プレブス社，昭和23 [1948]
304 p. ; 21 cm.
Includes bibliographies.

1. Civil service—Europe. 2. Civil service—United States. I. Rōyama, Masamichi, 1895-
JN96.A67K34 79-807300

Lee, Robert D
Public personnel systems / by Robert D. Lee, Jr. — Baltimore : University Park Press, c1979.
xv, 434 p. ; 24 cm.
Bibliography: p. 421-424.
Includes index.
ISBN 0-8391-1452-4
1. Civil service—United States. 2. Civil service—United States—States. 3. Local officials and employees—United States. 4. Personnel management—United States. I. Title.
JK765.L43 353.006 79-12468
79 MARC

McDonald, Joan C
An historical survey of performance rating systems in the Federal service / Joan C. McDonald. — Washington : U.S. Civil Service Commission, Personnel Research and Development Center, 1974.
viii, 27 p. ; 27 cm. — (Professional series ; PS-74-1)
Bibliography: p. 23-24.
1. Civil service—United States. 2. Employees, Rating of—United States. I. Title. II. Series.
JK765.P73 no. 74-1 353.001'008 s 76-600887
[JK765] 76[r79]rev MARC

National Eight-hour Delegation.
The Eight-hour law in the Senate of the United States, Forty-sixth Congress, third session : argument submitted in favor of the passage of joint resolution H.R. 239 ... / by the National Eight-hour Delegation ; compiled by John G. Mills and Dyer D. Lum. — Washington : R. H. Darby, 1880 [i.e. 1881]
32 p. ; 24 cm.
LC copy replaced by microfilm.
1. Hours of labor—United States. 2. Civil service—United States. I. Mills, John G. II. Lum, Dyer Daniel. III. Title.
[KF3528.P8N3] 342'.73'068 79-307062
79 MARC

Reedy, Michael J 1905-
Play roles on Uncle's payrolls : doing one's thing / by Michael J. Reedy. — [Washington] : Reedy, c1976.
iv, 111 p. : ill. ; 23 cm.
Available from M. J. Reedy, 6424 Luzon Ave., N.W., Washington, D.C., 20012.
Includes index.
$4.00
1. Civil service—United States. 2. United States—Officials and employees—Biography. 3. Reedy, Michael J., 1905- I. Title.
JK691.R43 353'.00092'4 76-27579
77[r79]rev MARC

Rohr, John Anthony, 1934-
Ethics for bureaucrats : an essay on law and values / John A. Rohr ; foreword by Herbert J. Storing. — New York : M. Dekker, c1978.
xii, 292 p. ; 24 cm. — (Political science ; v. 8)
Includes bibliographical references and indexes.
ISBN 0-8247-6756-X
1. Civil service ethics. 2. Civil service—United States. 3. Government executives—United States. I. Title. II. Series: Political science (New York) ; v. 8.
JK468.E7R56 172'.2 78-15327
78 MARC

United States. Congress. House. Committee on Education and Labor. Subcommittee on Labor Standards.
A bill to amend the Fair labor standards act : hearings before the Subcommittee on Labor Standards of the Committee on Education and Labor, House of Representatives, Ninety-fifth Congress, second session, on H.R. 6256 ... February 7 and March 14, 1978. — Washington : U.S. Govt. Print. Off., 1978.
iii, 53 p. ; 24 cm.
1. United States—Armed Forces—Commissariat. 2. Labor laws and legislation—United States. 3. Civil service—United States. I. Title.
KF27.E348 1978d 344'.73'01 78-603384
79 MARC

United States. Congress. House. Committee on Government Operations. Legislation and National Security Subcommittee.
Reorganization plan no. 2 of 1978 (Federal Personnel Management System) : hearings before a subcommittee of the Committee on Government Operations, House of Representatives, Ninety-fifth Congress, second session, June 6, 13, and 15, 1978. — Washington : U.S. Govt. Print. Off., 1978.
iv, 319 p. ; 24 cm.
1. Civil service—United States. 2. Personnel management—United States. I. Title.
KF27.G6676 1978b 353.001 78-603131
79 MARC

United States. Congress. House. Committee on Post Office and Civil Service.
Civil service reform : hearings before the Committee on Post Office and Civil Service, House of Representatives, Ninety-fifth Congress, second session, on H.R. 11280 .... — Washington : U.S. Govt. Print. Off., 1978.
vii, 1025 p. ; 24 cm.
Hearings held Mar. 14-May 23, 1978.
"Serial no. 95-65."
1. Civil service—United States. I. Title.
KF27.P6 1978a 353.006 78-603259
79 MARC

United States. Congress. House. Committee on Post Office and Civil Service. Sucommittee on Investigations.
Discrimination in the Federal Government : hearing before the Subcommittee on Investigations of the Committee on Post Office and Civil Service, House of Representatives, Ninety-fifth Congress, second session, March 30, 1978. — Washington : U.S. Govt. Print. Off., 1978.
iii, 98 p. ; 24 cm.
"Serial no. 95-56."
1. Discrimination in employment—United States. 2. Civil service—United States. I. Title.
KF27.P646 1978 353.001'3 78-602708
78 MARC

United States. Congress. Senate. Committee on Governmental Affairs.
Civil Service reform act of 1978 and Reorganization Plan no. 2 of 1978 : hearings before the Committee on Governmental Affiars, United States Senate, Ninety-fifth Congress, second session, on S. 2640, S. 2707, and S. 2830 to reform the Civil Service laws, June 1978 : appendix. — Washington : U.S. Govt. Print. Off., 1978.
iv, 905 p. ; 24 cm.
Includes bibliographical references.
1. Civil service—United States. 2. Employee-management relations in government—United States. I. Title.
KF26.G67 1978c 342'.73'068 78-603393
79 MARC

United States. Congress. Senate. Committee on Governmental Affairs. Subcommittee on Civil Service and General Services.
Federal firefighter workweek and White House personnel authorization : hearing before the Subcommittee on Civil Service and General Services of the Committee on Governmental Affairs, United States Senate, Ninety-fifth Congress, second session, on S. 1163, H.R. 3161, H.R. 11003, May 2, 1978. — Washington : U.S. Govt. Print. Off., 1978.
iv, 158 p. ; 24 cm.
1. Fire fighters—Legal status, laws, etc.—United States. 2. Hours of labor—United States. 3. Civil service—United States. 4. United States. Executive Office of the President—Officials and employees. I. Title.
KF26.G6724 1978e 342'.73'068 78-603106
79 MARC

United States. Congress. Senate. Committee on Human Resources. Subcommittee on Labor.
Federal employees flexible and compressed work schedules act, 1978 : hearing before the Subcommittee on Labor of the Committee on Human Resources, United States Senate, Ninety-fifth Congress, second session, on S. 517 ... August 21, 1978. — Washington : U.S. Govt. Print. Off., 1978.
111, 101 p. ; 24 cm.
1. Hours of labor, Flexible—Law and legislation—United States. 2. Civil service—United States. I. Title.
KF26.H855 1978 342'.73'068 79-600755
79 MARC

United States. General Accounting Office.
Federal employee performance rating systems need fundamental changes : report to the Congress / by the Comptroller General of the United States. — [Washington : U.S. General Accounting Office], 1978.
vi, 142 p. : ill. ; 27 cm.
Cover title.
"FPCD-77-80."
"B-150411."
Bibliography: p. 109-111.
$1.00
1. Civil service—United States. 2. Employees, Rating of—United States. I. Title.
JK765.U57 1978 353.001'42 78-603670
79 MARC

—UNITED STATES—ADDRESSES, ESSAYS, LECTURES

Messick, Charles P
Reflections of a public servant : end of a journey / by Charles P. Messick ; foreword by William W. Boyer. — Newark : University of Delaware, c1978.
xiii, 76 p. ; 25 cm.
1. Public administration—Addresses, essays, lectures. 2. Civil service—United States—Addresses, essays, lectures. 3. Civil service—New Jersey—Addresses, essays, lectures. I. Title.
JF1351.M395 350 79-621699
79 MARC

—UNITED STATES
—BIBLIOGRAPHY—PERIODICALS

United States. Office of Personnel Management. Library.
Personnel literature. v. 38, no. 2-
Feb. 1979-
Washington, Office of Personnel Management, Library ; [for sale by the Supt. of Docs., U. S. Govt. Print. Off.]
v. 26 cm. monthly.
Continues: Personnel literature, ISSN 0031-5753, issued by: United States. Civil Service Commission. Library.
Key title: Personnel literature, ISSN 0031-5753.
1. Personnel management — Bibliography — Periodicals. 2. Civil service—United States—Bibliography—Periodicals. I. Title.
Z7164.C81U45683 016.3501 79-644985
[HF5549.A2] MARC-S

—UNITED STATES—CONGRESSES

Regional Civil Rights Conference, 2d, Boston, 1974.
Making public employment a model of equal opportunity : a report of the proceedings of Regional Civil Rights Conference II, sponsored by the U.S. Commission on Civil Rights in Boston, Massachusetts, September 22-24, 1974. — [Washington] ; The Commission, [1975]
v, 58 p. ; 27 cm.
1. Civil Service—United States—Minority employment—Congresses. 2. Discrimination in employment—United States—Congresses. 3. Affirmative action programs—United States—Congresses. 4. Civil service—United States—Congresses. I. United States. Commission on Civil Rights. II. Title.
JK723.M54R43 1974 353.001'3 75-602589
76[r79]rev MARC

—UNITED STATES—EXAMINATIONS

Arco Publishing.
Investigator-inspector : claims examiner—civil and criminal investigation (all fields of inspection—Federal, State, municipal) : complete study guide to pass high on your civil service test / by the Arco Editorial Board. — 4th ed. — New York : Arco, c1979.
320 p. : ill. ; 26 cm. — (Arco civil service test tutor)
Previous ed. issued by the body under its earlier name: Arco Publishing Company.
ISBN 0-668-01711-2
1. Civil service—United States—Examinations. 2. Investigations. I. Arco Publishing Company, New York. Investigator-inspector. II. Title.
JK716.A84 1979 364.12'076 78-24190
78 MARC

United States. General Accounting Office.
Federal employment examinations, do they achieve equal opportunity and merit principle goals? : Report to the Congress / by the Comptroller General of the United States. — [Washington : U.S. General Accounting Office], 1979.
xi, 128 p. : graphs ; 27 cm.
Cover title.
"FPCD-79-46."
"B-167015."
Bibliography: p. 70.
$1.00
1. Civil service—United States—Examinations. 2. Civil service—United States—Minority employment. I. Title.
JK716.U46 1979 353.003 79-602337
79 MARC

—UNITED STATES—HISTORY

Prince, Carl E
The Federalists and the origins of the U.S. civil service / Carl E. Prince. — New York : New York University Press, 1977.
xiii, 381 p. : ill. ; 24 cm.
Bibliography: p. 359-361.
Includes index.
ISBN 0-8147-6570-X : $16.00
1. Federal Party. 2. Civil service—United States—History. 3. Patronage, Political—United States—History. I. Title.
JK2306.P75 329'.1'009033 76-53708
77[r79]rev MARC

CIVIL SERVICE (Continued)

—UNITED STATES—LABOR PRODUCTIVITY

New York (State). Legislature. Senate. Task Force on Critical Problems.
Productivity : a report / by the Task Force on Critical Problems for John R. Dunne, Martin S. Auer, Lloyd Paterson, to the Temporary State Commission on Management and Productivity in the Public Sector ; John M. Flynn, policy analyst. — Albany : ₍The Task Force₎, 1975.
iii, 107 p. ; 28 cm.
Bibliography: p. 101-107.
1. Civil service—United States—Labor productivity. 2. Civil service—United States—States—Labor productivity. 3. Civil service—New York (State)—Labor productivity. I. New York (State). Temporary State Commission on Management and Productivity in the Public Sector. II. Title.
JK768.4.N48 1975 353.9'747'00147 76-623425
79 MARC

United States. Congress. House. Committee on the Budget. Task Force on Tax Expenditures, Government Organization, and Regulation.
Productivity in the Federal sector : hearings before the Task Force on Tax Expenditures, Government Organization, and Regulation of the Committee on the Budget, House of Representatives, Ninety-fifth Congress, second session, February 14, 16, and March 2, 1978. — Washington : U.S. Govt. Print. Off., 1978.
iv, 194 p. ; 24 cm.
1. Civil service—United States—Labor productivity. I. Title.
KF27.B884 1978 353.001'47 78-602834
78 MARC

United States. General Accounting Office.
Does the Federal incentive awards program improve productivity? : Report to the Congress / by the Comptroller General of the United States. — ₍Washington₎ : U.S. General Accounting Office, 1979.
iv, 88 p. : ill. ; 27 cm.
Cover title.
"FGMSD-79-9."
"B-163762."
$1.00
1. Performance awards—United States. 2. Civil service—United States—labor productivity. I. Title.
JK768.3.U54 1979 353.001'47 79-601765
79 MARC

—UNITED STATES—MINORITY EMPLOYMENT

United States. General Accounting Office.
Federal employment examinations, do they achieve equal opportunity and merit principle goals? : Report to the Congress / by the Comptroller General of the United States. — ₍Washington : U.S. General Accounting Office₎, 1979.
xi, 128 p. : graphs ; 27 cm.
Cover title.
"FPCD-79-46."
"B-167015."
Bibliography: p. 70.
$1.00
1. Civil service—United States—Examinations. 2. Civil service—United States—Minority employment. I. Title.
JK716.U46 1979 353.003 79-602337
79 MARC

—UNITED STATES—MINORITY EMPLOYMENT—CONGRESSES

Regional Civil Rights Conference, 2d, Boston, 1974.
Making public employment a model of equal opportunity : a report of the proceedings of Regional Civil Rights Conference II, sponsored by the U.S. Commission on Civil Rights in Boston, Massachusetts, September 22-24, 1974. — ₍Washington₎ ; The Commission, ₍1975₎
v, 58 p. ; 27 cm.
1. Civil Service—United States—Minority employment—Congresses. 2. Discrimination in employment—United States—Congresses. 3. Affirmative action programs—United States—Congresses. 4. Civil service—United States—Congresses. I. United States. Commission on Civil Rights. II. Title.
JK723.M54R43 1974 353.001'3 75-602589
76₍r79₎rev MARC

—UNITED STATES—MINORITY EMPLOYEES—STATISTICS—PERIODICALS

United States. Civil Service Commission. Bureau of Personnel Management Information Systems.
Equal employment opportunity statistics. Nov. 1976–
₍Washington₎ U. S. Civil Service Commission, Bureau of Personnel Management Information Systems ₍for sale by the Supt. of Docs., U. S. Govt. Print. Off.₎
v. 26 cm. semiannual. (₍Pamphlet₎ - U. S. Civil Service Commission)
At head of title, Nov. 1976– : Federal civilian workforce statistics.
Supersedes: Study of employment of women in the Federal Government, ISSN 0097-7764, and: Minority group employment in the Federal Government, ISSN 0090-3531.
Key title: Equal employment opportunity statistics, ISSN 0161-245X.
1. Civil service—United States—Minority employment—Statistics—Periodicals. 2. United States—Officials and employees—Statistics—Periodicals. I. United States. Civil Service Commission. Bureau of Personnel Management Information Systems. Federal civilian workforce statistics. II. Title. III. Series: United States. Civil Service Commission. Pamphlet.
JK723.M54U53a 331.2'041'353 78-642441
MARC-S

—UNITED STATES—PERIODICALS

Professional series.
Washington, Research Section, Personnel Research and Development Center, U. S. Civil Service Commission ₍etc.₎
no. 27 cm.
Vols. for issued by Test Services Section, United States Civil Service Commission; by Research Section, Personnel Research and Development Center, U. S. Civil Service Commission.
1. Civil service — United States — Periodicals. 2. Employment tests—Periodicals. I. United States. Civil Service Commission. Test Services Section. II. United States. Civil Service Commission. Personnel Research and Development Center. Research Section.
JK765.P73 353.001 75-646114
MARC-S

United States. Civil Service Commission. Bureau of Training.
Employee training in the federal service.
₍Washington₎ U. S. Civil Service Commission, Bureau of Training, ₍for sale by the Supt. of Docs., U. S. Govt. Print. Off.₎
v. ill. 28 cm. annual. (Pamphlet T)
Key title: Employee training in the federal service, ISSN 0565-1441
1. Civil service—United States—Periodicals. 2. Employees, Training of—United States—Periodicals. I. Title. II. Series: United States. Civil Service Commission. Bureau of Training. Pamphlet T.
JK718.U53b 353.001'5 78-646739
MARC-S

United States. Civil Service Commission. Office of Incentive Systems.
Progress through achievements.
1976/77–
Washington, U. S. Civil Service Commission, Office of Incentive Systems.
v. ill. 27 cm. annual.
Report year ends Sept. 30.
Continues: Achievements—U. S. Civil Service Commission, ISSN 0362-9988.
Other title, 1976/77 : Federal incentive awards program.
Key title: Progress through achievements, ISSN 0193-3361.
1. Civil service—United States—Periodicals. 2. Performance awards—United States—Periodicals. 3. Suggestion systems—United States—Periodicals. I. United States. Civil Service Commission. Office of Incentive Systems. Federal incentive awards program. II. Title.
JK768.3.U52a 353.006'05 79-643386
MARC-S

—UNITED STATES—PERSONNEL MANAGEMENT

United States. Personnel Management Project.
President's reorganization project : a comprehensive review of Federal personnel management principles, policies, processes, and organization to determine what improvements are required and to recommend appropriate legislation, policies, rules, regulations, processes, and organizational solutions / Personnel Management Project. — ₍Washington₎ : The Project, 1977-
v. : ill. ; 27 cm.
CONTENTS:
—v. 2. Appendices to the final staff report.—v. 3. Supplementary materials.
1. Civil service—United States—Personnel management. I. Title.
JK643 1977.C75 353.001 78-601811
79 MARC

—UNITED STATES—PERSONNEL MANAGEMENT—BIBLIOGRAPHY

United States. Civil Service Commission. Library.
Personnel policies and practices. — ₍Washington : U.S. Civil Service Commission, Library₎ : for sale by the Supt. of Docs., U.S. Govt. Print. Off., ₍1977₎
45 p. ; 27 cm. — (Personnel bibliography series ; no. 89)
Caption title.
Issued with cover to Personnel bibliography no. 88, Equal opportunity in employment.
"Updates Personnel bibliography no. 71 ... (1976)."
1. Personnel management—United States—Bibliography. 2. Civil service—United States—Personnel management—Bibliography. 3. Discrimination in employment—United States—Bibliography. I. Title. II. Series: United States. Civil Service Commission. Library. Personnel bibliography series ; no. 89.
Z7164.C81U4568 no. 89 016.353001 s 78-602001
₍HF5549.2.U5₎ 79 MARC

—UNITED STATES—PERSONNEL MANAGEMENT—CONGRESSES

Conference on Public Personnel Management Reform, Washington, D.C., 1978.
Conference report on public personnel management reform / sponsored by the United States Civil Service Commission ... ₍et al.₎ January 23, 1978 at the Hall of the States, Washington, D.C. — Washington : U.S. Civil Service Commission, Bureau of Intergovernmental Personnel Programs, 1978.
79 p. : ill. ; 27 cm. — (BIPP ; 152-94)
1. Civil service—United States—Personnel management—Congresses. 2. State governments—Personnel management—Congresses. 3. Local government—United States—Personnel management—Congresses. I. United States. Civil Service Commission. II. Title. III. Series: United States. Civil Service Commission. Bureau of Intergovernmental Personnel Programs. BIPP ; 152-94.
JK677 1978.C66 1978 350'.1'0973 78-603663
79 MARC

—UNITED STATES—STATES

Committee for Economic Development.
Improving management of the public work force : the challenge to State and local government : a statement on national policy / by the Research and Policy Committee of the Committee for Economic Development. — New York : CED, 1978.
138 p. ; 26 cm.
ISBN 0-87186-067-8
1. Civil service—United States—States. 2. Local officials and employees—United States. 3. Personnel management—United States. 4. Employee-management relations in government—United States. I. Title.
JK2465.C63 1978 353.001 78-11075
78 MARC

Lee, Robert D
Public personnel systems / by Robert D. Lee, Jr. — Baltimore : University Park Press, c1979.
xv, 434 p. ; 24 cm.
Bibliography: p. 421-424.
Includes index.
ISBN 0-8391-1452-4
1. Civil service—United States. 2. Civil service—United States—States. 3. Local officials and employees—United States. 4. Personnel management—United States. I. Title.
JK765.L43 353.006 79-12468
79 MARC

—UNITED STATES—STATES—LABOR PRODUCTIVITY

Downey, Edward H
Employee innovation and government productivity : a study of suggestion systems in the public sector / Edward H. Downey, Walter L. Balk. — Chicago : International Personnel Management Association, c1976.
v, 90 p. : ill. ; 28 cm. — (Personnel report ; no. 763)
Bibliography: p. 52.
Includes index.
1. Civil service—United States—States—Labor productivity. 2. Suggestion systems—United States. I. Balk, Walter L., joint author. II. Title. III. Series.
JF1601.P85 no. 763 350'.1'008 s 79-318258
₍JK2480.L24₎ 79 MARC

New York (State). Legislature. Senate. Task Force on Critical Problems.
Productivity : a report / by the Task Force on Critical Problems for John R. Dunne, Martin S. Auer, Lloyd Paterson, to the Temporary State Commission on Management and Productivity in the Public Sector ; John M. Flynn, policy analyst. — Albany : ₍The Task Force₎, 1975.
iii, 107 p. ; 28 cm.
Bibliography: p. 101-107.
1. Civil service—United States—Labor productivity. 2. Civil service—United States—States—Labor productivity. 3. Civil service—New York (State)—Labor productivity. I. New York (State). Temporary State Commission on Management and Productivity in the Public Sector. II. Title.
JK768.4.N48 1975 353.9'747'00147 76-623425
79 MARC

—UNITED STATES—STATES—PERSONNEL MANAGEMENT

Jarrett, James E
Permanent part-time employment : a staffing option for state government / prepared by James E. Jarrett. — Lexington, Ky. : Council of State Governments, ₍1978₎
₍11₎ p. ; 28 cm. — (Innovations) (RM ; 648)
Caption title.
Bibliography: p. ₍9₎-₍10₎
1. Part-time employment—United States—States. 2. Civil service—United States—States—Personnel management. I. Title. II. Series: Innovations transfer. III. Series: Council of State Governments. RM publications ; 648.
JS308.C6 no. 648 353.9'08 s 79-620944
₍JK2480.P37₎ 79 MARC

CIVIL SERVICE (Continued)

—UNITED STATES—STATISTICS —PERIODICALS

United States. Civil Service Commission. Bureau of Manpower Information Systems.
Federal civilian workforce statistics; monthly release. Nov. 1976–June 1977.
[Washington] U. S. Civil Service Commission, Bureau of Manpower Information Systems.
1 v. 26 cm. ([Pamphlet] SM-13)
"Employment and trends."
Caption title: Monthly release, federal civilian workforce statistics.
Continues: Federal civilian manpower statistics, ISSN 0090-7227.
Continued by: Federal civilian workforce statistics, ISSN 0163-8270, issued by: United States. Civil Service Commission. Bureau of Personnel Management Information Systems.
Key title: Federal civilian workforce statistics, ISSN 0163-8270
1. Civil service—United States—Statistics—Periodicals. 2. United States—Officials and employees—Statistics—Periodicals. I. United States. Civil Service Commission Bureau of Manpower Information Systems. Monthly release, federal civilian workforce statistics. II. Title. III. Series: United States. Civil Service Commission. Pamphlet.
JK639.A42 subser. 353.006'08 s 78-648454
[JK730] [331.2'81'353] MARC-S

—URUGUAY—PERIODICALS

Uruguay. Oficina Nacional del Servicio Civil.
Revista informativa - Oficina Nacional del Servicio Civil. Montevideo, Oficina Nacional del Servicio Civil.
v. ill. 27 cm.
1. Civil service—Uruguay—Periodicals. 2. Uruguay. Oficina Nacional del Servicio Civil—Periodicals.
J251.R987a 354'.895'006 78-646397
MARC-S

—VIRGINIA—PERSONNEL MANAGEMENT

Virginia. Dept. of Personnel and Training.
The plan for personnel management decentralization and the biennial report on personnel management to the Governor and the General Assembly of Virginia. — Richmond : Commonwealth of Virginia, Division of Purchases and Supply, 1979.
i, 25 p. ; 28 cm. — (House document - Commonwealth of Virginia ; no. 11)
1. Civil service—Virginia—Personnel management. I. Title. II. Series: Virginia. General Assembly, 1979. House of Delegates. Document ; no. 11.
J87.V9 1979c, no. 11 328.755'07'208 s 79-622554
[JK3955] 79 MARC

—WALES

Great Britain. Welsh Office. Staff Training Unit.
Annual report; training year.
[s. l.] Welsh Office.
v. 30 cm.
Period covered by reports ends July 31.
Vols. for issued in combined form.
1. Civil service—Wales. 2. Employees, Training of—Wales.
JN1156.Z13 I 64 354'.429'0015 79-641784
MARC-S

—WISCONSIN—PERSONNEL MANAGEMENT

Wisconsin. Bureau of Personnel.
Affirmative action report # 2 / prepared by the Department of Administration, State Bureau of Personnel. — [Madison] : The Bureau, 1974
70 p. : graphs ; 28 cm.
1. Affirmative action programs—Wisconsin. 2. Civil service—Wisconsin—Personnel management. I. Title.
JK6060.A33W57 1974 331.2 77-621633
79 MARC

—ZAMBIA

Zambia. Salaries Commission.
Report of the Commission of Inquiry into the Salaries, Salary Structures, and Conditions of Service of the Zambia public and teaching services, the Zambia Police Force and Prisons Service, the defence forces, and the staffs of local authorities, including casual and daily-paid employees, and of personnel employed by statutory boards and corporations, and by companies in which the State has a majority or controlling interest. — [Lusaka] : Republic of Zambia, 1975-
v. ; 30 cm.
Cover title.
CONTENTS: v. 1. The public services and the parastatal sector.
1. Civil service—Zambia. 2. Zambia—Officials and employees—Salaries. I. Title.
JQ2845.S24 1975 354'.689'4001 77-980749
78 MARC

CIVIL SERVICE, COLONIAL

see also Colonial administrators

CIVIL SERVICE, INTERNATIONAL
see International officials and employees

CIVIL SERVICE, MUNICIPAL see Municipal officials and employees

CIVIL SERVICE ETHICS

Chang, Yang-hao, 1269–1329.
(San shih chung kao)
三事忠告 / [張養浩撰]. 七國考 / [董説撰]. — [臺北：臺灣]商務[印書舘, 1979]
4 v. ; 22 cm. — (四庫全書珍本九集 ; 165–168)
Cover title.
"三事忠告四卷...七國攷十四卷"
Photoreprint ed. of the Wen Yüan Ko copy.
1. Local government—China. 2. Civil service ethics. 3. China—Politics and government—To 1643. 4. China—History—Warring States, 403–221 B. C. I. Tung, Shuo, 1620–1686. Ch'i kuo k'ao. 1979. II. Title. III. Series: Ssu k'u ch'üan shu. Ssu k'u ch'üan shu chen pen chiu chi ; 165–168.
AC149.S7 1979, vol. 165–168 79–840282
[JS7353.A3]

Kanamori, Tokujirō, 1886–1959.
(Kōmuin no rinri ni tsuite)
公務員の倫理について / 金森徳次郎著. — 東京：学陽書房, 昭和31 [1956]
2, 106 p. : 19 cm.
1. Japan—Officials and employees—Conduct of life. 2. Civil service ethics. I. Title.
JQ1629.E8K36 79–802331

Rohr, John Anthony, 1934-
Ethics for bureaucrats : an essay on law and values / John A. Rohr ; foreword by Herbert J. Storing. — New York : M. Dekker, c1978.
xii, 292 p. ; 24 cm. — (Political science ; v. 8)
Includes bibliographical references and indexes.
ISBN 0-8247-6756-X
1. Civil service ethics. 2. Civil service—United States. 3. Government executives—United States. I. Title. II. Series: Political science (New York) ; v. 8.
JK468.E7R56 172'.2 78-15327
78 MARC

Yi, Hyŏn-jong.
(Ch'ŏngbaengni chŏngsin kwa yŏlchŏn)
清白吏精神과列傳 / 李鉉淙著. — 서울：亞細亞文化社, 1977.
10, 261 p. ; 22 cm.
Includes index.
₩2200
1. Civil service—Korea—History. 2. Civil service ethics. 3. Korea—Officials and employees—Biography. I. Title.
JQ1726.Z1Y45 78–826665

CIVIL SERVICE FORM

—JAPAN

(Minshuteki gyōsei kaikaku no riron)
民主的行政改革の理論 / 渡辺佐平編. — 東京：月書店, 1978.
256 p. : ill. ; 20 cm. Ja 79-
Includes bibliographical references.
¥1600
1. Japan—Politics and government—1945- 2. Civil service form—Japan. 3. Public administration—Japan. I. Watanabe, Sa 1903-
JQ1626 1945.M58 79–806

CIVIL SERVICE JOBS see Civil service positions

CIVIL SERVICE PENSIONERS

—NEW JERSEY

New Jersey. Legislature. Office of Fiscal Affairs. Division of Program Analysis.
Survey of retired state employees : a background paper on the PUblic Employee' Retirement System / New Jersey State Legislature, Office of Fiscal Affairs, Division of Program Analysis. — [Trenton] : The Division, 1977.
ii, 36 p. ; 28 cm.
"77-2."
Includes bibliographical references.
1. Civil service pensioners—New Jersey. 2. Civil service pensions—New Jersey. I. Title.
JK3560.P4N48 1977a 353.9'749'005 77-623460
79 MARC

CIVIL SERVICE PENSIONS

see also Survivors' benefits

—FINANCE

Winklevoss, Howard E
Public pension plans : standards of design, funding, and reporting / Howard E. Winklevoss, Dan M. McGill, in collaboration with Glenn D. Allison, Roy S. Neff. — Homewood, Ill. : Dow Jones-Irwin, c1979.
xvi, 364 p. : graphs ; 24 cm.
Includes bibliographies and index.
ISBN 0-87094-174-7
1. Civil service pensions—Finance. I. McGill, Dan Mays, joint author. II. Title.
JF1671.W56 350'.5 78-70960
00 MARC

—TAXATION—UNITED STATES

United States. Congress. Senate. Committee on Finance. Subcommittee on Private Pension Plans and Employee Fringe Benefits.
Reporting requirements for State and local government pension plans and tax treatment of deferred amounts under nonqualified deferred compensation plans : hearing before the Subcommittee on Private Pension Plans and Employee Fringe Benefits of the Committee on Finance, United States Senate, Ninety-fifth Congress, second session, on S. 1587 ... S. 2627 ... March 15, 1978. — Washington : U.S. Govt. Print. Off., 1978.
iv, 179 p. ; 24 cm.
1. Civil service pensions—Taxation—United States. 2. Deferred compensation—Taxation—United States. I. Title.
KF26.F565 1978 343'.73'052 78-603214
79 MARC

—ALASKA

Winkenwerder, Richard A
Alternatives to the Social Security system for employees of the State of Alaska / prepared by Richard A. Winkenwerder for the Study Group on Withdrawal from Social Security, Alaska State Legislature. — [Juneau] : State of Alaska, Legislative Affairs Agency, 1978.
35 p. ; 28 cm.
1. Civil service pensions—Alaska. 2. Social Security—Alaska. I. Alaska. Study Group on Withdrawal from Social Security. II. Title.
JK9560.P4W56 353.9'798'005 78-622750
78 MARC

—BRAZIL

Brazil. Laws, statutes, etc.
Aposentadoria da funcionária pública e Sistema Nacional de Previdência (SINPAS) : Leis 6.438, de 1-9-77, 6.435, de 15-7-77, 6.462, de 9-11-77, Decreto 80.303, de 8-9-77. — Rio de Janeiro : Edições Trabalhistas, [1977 or 1978]
58 p. ; 21 cm.
Cr$40.00
1. Civil service pensions—Brazil. I. Brazil. Laws, statutes, etc. Lei no. 6.438, de 1 de setembro de 1977. 1977. II. Title.
79-348766
79 MARC

CIVIL SERVICE PENSIONS (Continued)

—CALIFORNIA

California. Dept. of Finance. Program Evaluation Unit.
A review of the internal structure of the State retirement program / prepared by Department of Finance, Program Evaluation Unit. — ₍Sacramento₎ : The Unit, 1979.
xxx, 252 p. : ill. ; 28 cm. — (Report - Department of Finance, Program Evaluation unit ; no. D79-3)
1. Civil service pensions—California. 2. Civil service retirement—California. I. Title. II. Series: California. Dept. of Finance. Program Evaluation Unit. Report - State of California, Department of Finance, Program Evaluation Unit ; no. D79-3.
JK8760.P4C3 1979 353.9'794'005 79-623191
79 MARC

—CALIFORNIA—COST-OF-LIVING ADJUSTMENTS

California. Legislature. Assembly. Public Employees and Retirement Committee.
State teachers' retirement system cost-of-living : transcript of proceedings / Assembly Public Employees and Retirement Committee, San Diego, California, November 15, 1977. — ₍Sacramento : Copies may be purchased from the Assembly Publications Office, 1977?₎
89 p. ; 28 cm.
Cover title.
"No. 692."
$3.40
1. Teachers—Salaries, pensions, etc.—California. 2. Civil service pensions—California—Cost-of-living adjustments. I. Title.
KFC10.4.P78 1977c 331.2'52 78-623679
79 MARC

—DISTRICT OF COLUMBIA

United States. Congress. House. Committee on the District of Columbia.
District of Columbia retirement reform act : report / by the Committee on the District of Columbia, including cost estimate of the Congressional Budget Office, to accompany H.R. 3939. — Washington : U.S. Govt. Print. Off., 1979.
v, 87 p. ; 23 cm. — (Report - House of Representatives, 96th Congress, 1st session ; no. 96-155)
At head of title: Union calendar no. 84.
1. Civil service pensions—District of Columbia. I. Title. II. Series: United States. 96th Congress. 1st session, 1979. House. Report ; no. 96-155.
KF32.D5 1979 344'.753'01252 79-602655
00 MARC

—GEORGIA

Denero, Walter A
Public pension plans : state and national issues / by Walter A. Denero, Mary E. Stakes. — ₍Athens₎ : Institute of Government, University of Georgia, 1978.
vi, 59 p. ; 29 cm. — (Legislative research series)
Bibliography: p. 57-59.
1. Civil service pensions—Georgia. 2. Social security—Georgia. 3. Civil service pensions—United States. 4. Social security—United States. I. Stakes, Mary E., joint author. II. Title. III. Series: Legislative research series (Athens)
JK4360.P4D46 353.9'758'005 79-108067
79 MARC

—GERMANY, WEST—EXAMINATIONS, QUESTIONS, ETC.

Bauer, Erwin, Oberamtsrat.
Versorgungsrecht : e. systemat. Querschnitt nach d. früheren Abh. d. Rechtsgebiets in d. Studienheften : mit Abdr. d. Beamtenversorgungsgesetzes / neu bearb. von Erwin Bauer. — ₍Stand 1. 1. 1978₎. — München : Verlag für Verwaltungspraxis Rehm, 1978.
vii, 235, 48 p. ; 21 cm. GFR78-A
Includes index.
ISBN 3-8073-0250-6 : DM34.00
1. Germany, West—Officials and employees—Examinations, questions, etc. 2. Civil service pensions—Germany, West—Examinations, questions, etc. I. Germany (Federal Republic, 1949-). Laws, statutes, etc. Beamtenversorgungsgesetz. 1978. II. Title.
78-386908
78₍r79₎rev MARC

—GREAT BRITAIN

Great Britain. Civil Service Dept.
The principal Civil Service pension scheme 1974, rules / Civil Service Department. — London : H.M.S.O., 1974 ₍i.e. 1975₎
viii, 139 p. ; 21 cm. GB75-07606
At head of title: Civil Service pension scheme.
ISBN 0-11-630407-3 : £1.30
1. Civil service pensions—Great Britain. I. Title.
KD4529.A33 1975 344'.41'01252 78-313927
78 MARC

—ILLINOIS—FINANCE—STATISTICS—PERIODICALS

Illinois. State Employees' Retirement System.
Annual report - State Employees' Retirement System of Illinois.
₍Springfield₎ State Employees' Retirement System of Illinois.
v. 28 cm.
Report year ends June 30.
1. Civil service pensions—Illinois—Finance—Statistics—Periodicals. 2. Illinois—Officials and employees—Pensions—Finance—Statistics—Periodicals.
JK5760.P4 I 43a 353.9'773'005 79-642933
MARC-S

—MASSACHUSETTS—FINANCE

Massachusetts. Retirement Law Commission.
Actuarial valuation report of the contributory retirement systems of the Commonwealth of Massachusetts as of January 1, 1976. — ₍Boston₎ : Massachusetts Retirement Law Commission, ₍1977₎
32 p. ; 28 cm.
Cover title.
1. Civil service pensions—Massachusetts—Finance. 2. Old age pensions—Massachusetts—Finance. I. Title.
JK3160.P4M34 1977 353.9'744'005 79-622338
79 MARC

—MISSISSIPPI

Moon, Jay Charles.
An analysis of the participation of Medicaid providers in the Mississippi State Government employees deferred compensation plan / information compiled and prepared by Jay Charles Moon, D. B. Ross, Frank Spencer. — ₍Jackson, Miss. : Legislative Audit Committee, 1977?₎
23 p. ; 29 cm.
1. Civil service pensions—Mississippi. 2. Medicaid—Mississippi. 3. Physicians—Salaries, pension, etc.—Mississippi. I. Ross, D. B., joint author. II. Spencer, Frank, 1947- joint author. III. Title.
JK4660.P4M66 353.9'762'005 78-621525
79 MARC

—NEW JERSEY

New Jersey. Legislature. Office of Fiscal Affairs. Division of Program Analysis.
Survey of retired state employees : a background paper on the PUblic Employee' Retirement System / New Jersey State Legislature, Office of Fiscal Affairs, Division of Program Analysis. — ₍Trenton₎ : The Division, 1977.
ii, 36 p. ; 28 cm.
"77-2."
Includes bibliographical references.
1. Civil service pensioners—New Jersey. 2. Civil service pensions—New Jersey. I. Title.
JK3560.P4N48 1977a 353.9'749'005 77-623460
79 MARC

—NEW YORK (CITY)

United States. Congress. Joint Committee on Taxation.
Application of pension provisions to New York City's financial problems / prepared for the use of the Committee on Finance by the staff of the Joint Committee on Taxation. — Washington : U.S. Govt. Print. Off., 1978.
iii, 6 p. ; 24 cm.
At head of title: Committee print.
"JCS-21-78."
1. Debts, Public—New York (City) 2. Civil service pensions—New York (City) I. United States. Congress. Senate. Committee on Finance. II. Title.
KFX2089.5.A16 336.3'63'097471 78-602632
78 MARC

—NEW YORK (CITY)—FINANCE

Gujarati, Damodar.
Pensions and New York City's fiscal crisis / Damodar Gujarati. — Washington : American Enterprise Institute for Public Policy Research, c1978.
72 p. ; 23 cm. — (AEI studies ; 212) (Studies in social security and retirement policy)
Includes bibliographical references.
ISBN 0-8447-3314-8
1. Civil service pensions—New York (City)—Finance. I. Title. II. Series: American Enterprise Institute for Public Policy Research. AEI studies ; 212. III. Series: Studies in social security and retirement policy.
JS1234.A4 1978 352'.005'5097471 78-11513
78 MARC

—PAKISTAN

Sind, Pakistan. Law Dept.
The West Pakistan Government servants benevolent fund rules, 1960. — Karachi : Sind Govt. Press, 1975.
45 p. ; 24 cm.
Cover title.
Rs1.50
1. Civil service pensions—Pakistan. I. Title.
342'.5491'068 77-938904
79 MARC

—UNITED STATES

Denero, Walter A
Public pension plans : state and national issues / by Walter A. Denero, Mary E. Stakes. — ₍Athens₎ : Institute of Government, University of Georgia, 1978.
vi, 59 p. ; 29 cm. — (Legislative research series)
Bibliography: p. 57-59.
1. Civil service pensions—Georgia. 2. Social security—Georgia. 3. Civil service pensions—United States. 4. Social security—United States. I. Stakes, Mary E., joint author. II. Title. III. Series: Legislative research series (Athens)
JK4360.P4D46 353.9'758'005 79-108067
79 MARC

Goodman, Isidore.
Retirement plans for employees in foreign service / by Isidore Goodman. — Chicago : Commerce Clearing House, c1979.
24 p. ; 24 cm.
Includes index.
$2.00
1. Civil service pensions—United States. 2. United States—Diplomatic and consular service. I. Title.
KF5380.G66 353.005 79-108367
00 MARC

Institute for Local Self Government.
A public safety employees contractual system : an alternative to traditional pension programs and to tenure-inducing retirement / Institute for Local Self Government. — Berkeley, Calif. : ILSG, 1977.
xvii, 86 leaves ; 23 cm. — (Its Alternatives to traditional public safety delivery systems ; PSDS report no. 1)
Bibliography: p. 86.
$3.00
1. Police—Salaries, pensions, etc.—United States. 2. Fire fighters—Salaries, pensions, etc.—United States. 3. Civil service pensions—United States. I. Title. II. Series.
HV8143.I57 1977 352.2 79-103708
79 MARC

National Governor's Association. Center for Policy Research.
Public pension reform : an overview. — Washington : National Governor's Association, Center for Policy Research, 1978.
24 p. ; 23 cm. — (Research notes - National Governors' Association, Center for Policy Research ; 2)
Prepared by Charles W. Fosler, and others.
$2.00
1. Civil service pensions—United States. 2. Pension trusts—United States. I. Fosler, Charles W. II. Title. III. Series. IV. Series: National Governor's Association. Center for Policy Research. Research notes - National Governor's Association, Center for Policy Research ; 2.
JK2474.N37 1978 353.005 79-105589
79 MARC

United States. Congress. House. Committee on Education and Labor. Subcommittee on Labor Standards.
Public employee retirement income security act of 1978 : hearing before the Subcommittee on Labor Standards of the Committee on Education and Labor, House of Representatives, Ninety-fifth Congress, second session, on H.R. 14138 ... Wheaton, Ill., on November 15, 1978. — Washington : U.S. Govt. Print. Off., 1979.
iv, 277 p. : graphs ; 24 cm.
Includes bibliographical references.
1. Civil service pensions—United States. 2. Pension trusts—United States. I. Title.
KF27.E348 1978i 342'.73'068 79-602198
00₍r79₎rev MARC

United States. Congress. House. Committee on Post Office and Civil Service. Subcommittee on Compensation and Employee Benefits.
Oversight hearing on cost of merging civil service and security retirement systems : hearings before the Subcommittee on Post Office and Civil Service, House of Representatives, Ninety-sixth Congress, first session, April 3 and 24, 1979. — Washington : U.S. Govt. Print. Off., 1979.
iii, 76 p. ; 24 cm.
"Serial no. 96-11."
1. Civil service pensions—United States. 2. Social security—United States. I. Title.
KF27.P638 1979b 353.005 79-602863
79 MARC

CIVIL SERVICE PENSIONS
—UNITED STATES (Continued)

United States. Congress. House. Committee on Post Office and Civil Service. Subcommittee on Compensation and Employee Benefits.
Special retirement policies as related to mandatory retirement for law enforcement officers, firefighters, and air traffic controllers / Subcommittee on Compensation and Employee Benefits of the Committee on Post Office and Civil Service, House of Representatives, Ninety-fifth Congress, second session. — Washington : U.S. Govt. Print. Off., 1978.
ii, 33 p. ; 24 cm.
At head of title: 95th Congress, 2d session. Committee print no. 95-18.
1. Civil service pensions—United States. 2. United States. Federal Bureau of Investigation—Officials and employees—Pensions. 3. Fire fighters—Salaries, pensions, etc.—United States. 4. Air traffic controllers—Salaries, pensions, etc.—United States. I. Title.
JK791.U557 1978 353.005 79-601618
79 MARC

United States. Congress. Senate. Committee on Governmental Affairs. Subcommittee on Civil Service and General Services.
Retirement and withholding tax legislation : hearing before the Subcommittee on Civil Service and General Services of the Committee on Governmental Affairs, United States Senate, Ninety-fifth Congress, second session ... May 17, 1978. — Washington : U.S. Govt. Print. Off., 1978.
iii, 78 p. ; 24 cm.
Hearing on S. 224, H.R. 4319, H.R. 4320, H.R. 8342, H.R. 8771, and H.R. 9491.
1. Civil service pensions—United States. 2. Withholding tax—United States. I. Title.
KF26.G6724 1978e 343'.73'052 78-602715
78 MARC

—UNITED STATES—ACCOUNTING

United States. Congress. House. Committee on Government Operations. Legislation and National Security Subcommittee.
Auditing Federal pension plans : hearing before a subcommittee of the Committee on Government Operations, House of Representatives, Ninety-fifth Congress, second session on H.R. 9701 ... July 24, 1978. — Washington : U.S. Govt. Print. Off., 1978.
iii, 51 p. ; 24 cm.
1. Civil service pensions—United States—Accounting. I. Title.
KF27.G6676 1978n 342'.73'068 79-600879
79 MARC

—UNITED STATES—FINANCE—ADDRESSES, ESSAYS, LECTURES

Financing the Civil Service Retirement System, a threat to fiscal integrity / edited by Dan M. McGill. — Homewood, Ill. : published for the Pension Research Council, Wharton School, University of Pennsylvania, by R. D. Irwin, c1979.
xix, 170 p. ; 24 cm.
Includes bibliographical references.
ISBN 0-256-02250-X
1. Civil service pensions—United States—Finance—Addresses, essays, lectures. I. McGill, Dan Mays. II. Pennsylvania. University. Wharton School of Finance and Commerce. Pension Research Council.
JK791.F55 353.005 78-72059
79 MARC

—UNITED STATES—STATES

Edward H. Friend & Company.
Report on State pension commissions, 1978. — Washington : E. H. Friend, c1978.
ii, 19 p. ; 28 cm.
Cover title.
At head of title: Pension Commission Clearinghouse.
1. Civil service pensions—United States—States. I. Title. II. Title: State pension commissions.
JK2474.E38 1978 353.9'3'5 79-100717
79 MARC

United States. Congress. House. Committee on Education and Labor. Subcommittee on Labor Standards. Pension Task Force.
Pension Task Force report on public employee retirement systems / Committee on Education and Labor, House of Representatives, Ninety-fifth Congress, second session, March 15, 1978. — Washington : U.S. Govt. Print. Off., 1978.
v, 858 p. ; 24 cm.
At head of title: Committee print.
Bibliography: p. 773-858.
1. Civil service pensions—United States—States. I. Title.
JK2474.U55 1978 353.9'3'5 78-602656
78 MARC

—UNITED STATES—STATES—CONGRESSES

Public sector decision making : legal fringe benefit and union parameters : fifth and sixth annual conference / contributors, Enid Beaumont ... [et al.] ; editor, Paul A. Weinstein. — [College Park, Md.] : Public Sector Labor Relations Conference Board, c1978.
iv, 106 p. ; 23 cm. — (PSLRCB publication ; no. 3)
ISBN 0-913400-02-5
1. Civil service pensions—United States—States—Congresses. 2. Non-wage payments—United States—States—Congresses. 3. Collective bargaining—Government employees—United States—States—Congresses. I. Beaumont, Enid. II. Weinstein, Paul A. III. Public Sector Labor Relations Conference Board. IV. Series: Public Sector Labor Relations Conference Board. PSLRCB publication ; no. 3.
JK2474.P8 353.001'234 78-70342
79 MARC

—UTAH

Dyner, Marthe F
Public pensions in Utah : report to the 42nd Legislature / by Marthe F. Dyner, Bert D. Hunsaker, Robert E. Wilcox. — [Salt Lake City] : Office of Legislative Research, State of Utah, 1978.
v, 140 leaves ; 28 cm. — (Research report - Office of Legislative Research ; no. 20)
Bibliography: leaf 140.
1. Civil service pensions—Utah. I. Hunsaker, Bert D., joint author. II. Wilcox, Robert E., joint author. III. Title. IV. Series: Utah. Office of Legislative Research. Research report - Office of Legislative Research, State of Utah ; no. 20.
JK8460.P4D9 353.9'792'005 78-622522
78 MARC

—VIRGINIA

Virginia. Retirement Study Commission.
Report of the Virginia Retirement Study Commission to the Governor and the General Assembly of Virginia. — Richmond : Commonwealth of Virginia, Division of Purchases and Supply, 1979.
46 p. : graphs ; 28 cm. — (House document - Commonwealth of Virginia ; no. 21)
1. Civil service pensions—Virginia. 2. Judges—Salaries, pensions, etc.—Virginia. 3. Police—Salaries, pensions, etc.—Virginia. I. Series: Virginia. General Assembly, 1979. House of Delegates. Document ; no. 21.
J87.V9 1979c, no. 21 328.755'07'208 s 79-623065
[JK3960.P4] 79 MARC

—VIRGINIA—MANAGEMENT

Virginia. General Assembly. Joint Legislative Audit & Review Commission.
Management review—Virginia supplemental retirement system / Joint Legislative Audit and Review Commission. — [Richmond] : The Commission, 1978.
vi, 96 p. : ill. ; 28 cm.
Cover title: Virginia supplemental retirement system.
1. Civil service pensions—Virginia—Management. 2. Pension trusts—Virginia—Management. I. Title. II. Title: Virginia supplemental retirement system.
JK3960.P4V53 1978 353.9'755'05 79-622557
79 MARC

—WEST VIRGINIA—STATISTICS—PERIODICALS

West Virginia. Public Employees Retirement System.
Annual report - West Virginia Public Employees Retirement System.
[Charleston] West Virginia Public Employees Retirement System.
v. 28 cm.
Report year ends June 30.
Continues: West Virginia. Public Employees Retirement System. Financial report.
1. Civil service pensions—West Virginia—Statistics—Periodicals.
JK4060.P4A3 353.9'755'005 79-643771
MARC-S

CIVIL SERVICE POSITIONS

—BRAZIL

Brazil. Secretaria de Modernização e Reforma Administrativa.
Cadastro da administração federal / Presidência da República, Secretaria de Planejamento, Secretaria de Modernização e Reforma Administrativa—SEMOR. — 2. ed. atualizada. — Brasília : SEMOR, 1978-
12 v. ; 26 cm.
Loose leaf for updating.
Includes indexes.
CONTENTS: [1] Presidência da República. Secretaria de Planejamento da Presidência da República-SEPLAN. Departamento Administrativo do Serviço Público—DASP.—[2] Ministério da Indústria e do Comércio.—[3] Ministério das Minas e Energia.—[4] Ministério da Agricultura.—[5] Ministério da Saúde.—[6] Ministério da Previdência e Assistência Social. Ministério do Trabalho.—[7] Ministério do Interior.—[8] Ministério da Justiça. Ministério das Relações Exteriores.—[9] Ministério da Educacão e Cultura.—[10] Ministério da Aeronaútica. Ministério do Exército. Ministério da Marinha.—[11] Ministério das Comunicações. Ministério dos Transportes.—[12] Ministério da Fazenda.
1. Brazil—Executive departments. 2. Civil service positions—Brazil. I. Title.
JL2442.B7 1978 354'.81'04 78-111853
79 MARC

—CANADA—CLASSIFICATION

Canada. Treasury Board. Personnel Policy Branch.
Classification and selection standard : welfare programmes : administrative and foreign service category = Norme de classification et de sélection : programmes de bien-être social : catégorie administrative et du service extérieur. — Ottawa : Personnel Policy Branch, Treasury Board : available from Print. and Pub., Supply and Services Canada, 1977.
130, 130 p. in various pagings ; 28 cm. C***
"T.B. 505-1, P.S.C. 600-310, Supplement 310."
English and French.
Opposite pages numbered in duplicate.
ISBN 0-660-01599-4 : $12.00
1. Civil service positions—Canada—Classification. 2. Social workers—Canada. I. Title. II. Title: Norme de classification et de sélection.
JL108.C36 1977 354'.71'001 78-317554
78 MARC

—CHINA

Kracke, Edward A
Translations of Sung civil service titles, classification terms, and governmental organ names / by E. A. Kracke, Jr. — 2d rev. ed. — San Francisco : Chinese Materials Center, 1978.
xiv, 35, [21] p. ; 21 cm. — (Asian library series ; no. 11)
Designations also in Chinese.
ISBN 0-89644-526-7
1. Civil service positions—China. 2. China—History—Sung dynasty, 960-1279—Sources. I. Title. II. Series.
JQ1512.Z1K7 1978 354'.51'00103 79-100815
79 MARC

—CHINA—HISTORY

Hsü, Shih-chung.
(Li tai kuan chih ping chi k'o chü chih ch'ang shih)
歷代官制兵制科舉制常识 / 徐師中編. — 第1版. — 澳門：爾雅出版社, 1977, 1978 printing.
ii, 278 p. ; 19 cm.
$10.00 (H.K.)
1. Civil service positions—China—History. 2. China—Armed Forces—Organization—History. I. Title.
JQ1512.Z1H77 79-837502

—ENGLAND—LONDON

Greater London Council.
Make London's future your future// Greater London Council, Inner London Education Authority. — [London : Public Information Branch of the Greater London Council, 1976?].
22 p. ; 21 cm. GB***
Cover title.
1. Civil service positions—England—London. I. London. Inner London Education Authority. II. Title.
JS3668.G73 1976 352'.005'10309421 78-322924
79 MARC

—JAPAN—CLASSIFICATION

Japan. Naikaku Kirokukyoku.
(Meiji shokkan enkaku hyō)
明治職官沿革表：合本 / 内閣記録局編. — 東京：原書房, 1978-
v. ; 22 cm. — (明治百年史叢書) Ja 78-17834 (v. 3)
Series romanized: Meiji hyakunenshi sōsho.
Reprint ed.
Includes indexes.
CONTENTS:
—3. 明治19年-明治22/職官·官廨
1. Japan—Officials and employees. 2. Civil service positions—Japan—Classification. I. Title.
78-812597

CIVIL SERVICE POSITIONS
—JAPAN—CLASSIFICATION (Continued)

(Shokkaisei jitsumu shiryō)
職階制実務資料 / 公務員編集部編. — ﹝s. l. : s. n., 1950?- .
v. ; 26 cm.

1. Civil service positions—Japan—Classification. I. Kōmuin Henshūbu.

78-815176

—KOREA

Yi, Wŏn-yŏng, 1918-
(Han'guk ssijok kyebo haesŏl)
韓國氏族系譜解說 ﹝이원영편저 서울 계민사 1972﹞
284 p. 21 cm. ₩800
Colophon inserted.

1. Names, Personal—Korea. 2. Korea—Genealogy. 3. Civil service positions—Korea. I. Title.
CS3012.Y5 73-834244

—NORTH CAROLINA—CLASSIFICATION

North Carolina. Office of State Personnel.
Position evaluation with the LGPCS factor ranking/benchmark/guidechart technique for North Carolina local governments : an operational manual. — Raleigh : Office of State Personnel, c1979.
v, 113 leaves : ill. ; 28 cm.
Includes index.
1. Local officials and employees—North Carolina—Salaries, allowances, etc. 2. Civil service positions—North Carolina—Classification. 3. Job evaluation—North Carolina. I. Title.
JS451.N87N57 1979 352'.005'10309756 79-118038
79 MARC

—NORTH DAKOTA—CLASSIFICATION

North Dakota. Dept. of Accounts and Purchases. Central Personnel Division.
North Dakota Central Personnel classification & pay index. — Bismarck : Dept. of Accounts & Purchases, Central Personnel Division, ﹝1978﹞
39, ﹝6﹞, 19 p. ; 28 cm.
Cover title.
1. Civil service positions—North Dakota—Classification. 2. North Dakota—Officials and employees—Salaries, allowances, etc. I. Title.
JK6457.N64 1978 353.9'784'00103 79-620777
79 MARC

—UNITED STATES

Moore, Donna J
Take charge of your own career : written especially for those who presently work for the Federal Government, and for those who would like to / by Donna J. Moore ; ﹝cover design and art work by Victoria Heemstra﹞. — Bainbridge Island, WA : Moore, c1979.
vi, 189 p. : ill. ; 29 cm.
Bibliography: p. 188-189.
$6.95
1. Vocational guidance. 2. Civil service positions—United States. I. Title.
HF5381.M68 650'.14'0243527 79-88808
00 MARC

United States. Congress. House. Committee on Post Office and Civil Service. Subcommittee on Employee Ethics and Utilization.
Administration requests for executive level positions : hearings before the Subcommittee on Employee Ethics and Utilization of the Committee on Post Office and Civil Service, House of Representatives, Ninety-fifth Congress, second session, May 9 and 11, 1978. — Washington : U.S. Govt. Print. Off., 1978.
iii, 59 p. ; 24 cm.
"Serial no. 95-70."
1. Civil service positions—United States. I. Title.
KF27.P6445 1978 353.001'3 78-603178
78 MARC

Waelde, David E
How to get a Federal civil service job and advance your Federal career / by David E. Waelde. — 3d ed. — Washington : Fedhelp Publications, 1978.
iv, 122 p. : forms ; 28 cm.
ISBN 0-917548-00-0 : $9.95
1. Civil service positions—United States. I. Title.
JK716.W24 1978 650'.14 78-72180
79 MARC

—UNITED STATES—CLASSIFICATION

A Preliminary analysis of the increase in the average grade of general schedule Federal employees : a report : prepared for Office of the Assistant Secretary of Defense/Manpower, Reserve Affairs, and Logistics / James H. Hayes ... ﹝et al.﹞. — Santa Monica, CA : Rand, 1978.
xiii, 38 p. ; 28 cm. — (﹝Report﹞ - Rand Corporation ; R-2329-MRAL)
Includes bibliographical references.
ISBN 0-8330-0051-9
1. Civil service positions—United States—Classification. 2. United States—Officials and employees—Salaries, allowances, etc. I. Hayes, James H. II. United States. Office of the Assistant Secretary of Defense/Manpower, Reserve Affairs, and Logistics. III. Series: Rand Corporation. Rand report ; R-2329-MRAL.
AS36.R3 R-2329 018 s 78-15086
﹝JK765﹞ 78 MARC

—UNITED STATES—STATES—DIRECTORIES

Phillips, Robert Lindsay.
The directory of psychology-related State civil service positions / R. Lindsay Phillips. — 2d ed. — Tallahassee, Fla. : Phillips Diversified Research Services, c1978.
vii, 132 leaves ; 28 cm.
1. Civil service positions—United States—States—Directories. 2. Social scientists in government—United States—States—Directories. 3. Psychologists—Employment—United States—Directories. I. Title.
JK2465.P45 1978 353.9'3'103 78-112686
79 MARC

—WISCONSIN—CLASSIFICATION—PERIODICALS

Wisconsin. State Bureau of Personnel.
Classification and compensation plan for positions in the classified service and selected positions in the executive salary schedule. 1969/70-
Madison, Dept. of Administration, State Bureau of Personnel.
v. 22 x 28 cm. annual.
Subtitle varies.
Continues: Classification and compensation plan for positions in the classified service, ISSN 0147-9709.
Continued by: Wisconsin. Division of Personnel. Classification and compensation plan.
1. Wisconsin—Officials and employees—Salaries, allowances, etc.—Periodicals. 2. Civil service positions—Wisconsin—Classification—Periodicals. I. Title.
JK6057.W57a 353.9'775'00103 79-645000
MARC-S

CIVIL SERVICE RECRUITING

—GREAT BRITAIN

Great Britain. Working Party on the Selection of Specialists.
Report of the Working Party on the Selection of Specialists ﹝of the﹞ Civil Service Commission. — ﹝Basingstoke﹞ : ﹝The Commission﹞, 1977.
﹝1﹞, 19 p. : form ; 30 cm. GB78-11732
Cover title.
Chairman: D. G. Daymond.
1. Civil service recruiting—Great Britain. I. Daymond, D. G. II. Title.
JN411.G78 1977 354'.41'0013 78-324757
79 MARC

—INDIA

India (Republic). Committee on Recruitment Policy and Selection Methods.
Civil services examination : report of the Committee on Recruitment Policy and Selection Methods. — ﹝New Delhi﹞ : Union Public Service Commission, 1976.
ix, 235 p. ; 22 cm.
"PUPSC.60/10,000."
1. Civil service recruiting—India. I. Title.
JQ247.I53 1976 354'.54'0013 78-905275
79 MARC

—PAKISTAN

Khan, Mohammad Mohabbat, 1949-
The recruitment and training systems in the civil service of Pakistan and the United States Federal Civil Service : a historical-analytical study / Mohammad Mohabbat Khan, Habib Mohammad Zafarullah. — ﹝Dacca﹞ : Center for Administrative Studies, University of Dacca, 1978.
31 p. ; 29 cm. — (CENTAS occasional paper ; 1)
Includes bibliographical references.
Tk5.00
1. Civil service recruiting—Pakistan. 2. Civil service recruiting—United States. 3. Employees, Training of—Pakistan. 4. Employees, Training of—United States. I. Zafarullah, Habib Mohammad, joint author. II. Title. III. Series: University of Dacca. Centre for Administrative Studies. CENTAS occasional paper ; 1.
JQ552.K52 354'.549'1001 78-905542
79 MARC

—UNITED STATES

Khan, Mohammad Mohabbat, 1949-
The recruitment and training systems in the civil service of Pakistan and the United States Federal Civil Service : a historical-analytical study / Mohammad Mohabbat Khan, Habib Mohammad Zafarullah. — ﹝Dacca﹞ : Center for Administrative Studies, University of Dacca, 1978.
31 p. ; 29 cm. — (CENTAS occasional paper ; 1)
Includes bibliographical references.
Tk5.00
1. Civil service recruiting—Pakistan. 2. Civil service recruiting—United States. 3. Employees, Training of—Pakistan. 4. Employees, Training of—United States. I. Zafarullah, Habib Mohammad, joint author. II. Title. III. Series: University of Dacca. Centre for Administrative Studies. CENTAS occasional paper ; 1.
JQ552.K52 354'.549'1001 78-905542
79 MARC

—UNITED STATES—STATES

Jarrett, James E
Assessing management potential in state government / prepared by James E. Jarrett. — Lexington, Ky. : Council of State Governments, ﹝1978﹞
12 p. : ill. ; 28 cm. — (Innovations) (RM ﹝publications﹞ ; 634)
Caption title.
Bibliography: p. 10.
$2.00
1. Government executives—United States—States. 2. Civil service recruiting—United States—States. I. Title. II. Series. III. Series: Council of State Governments. RM publications ; 634.
JS308.C6 no. 634 353.9'08 s 79-620941
﹝JK2471﹞ 79 MARC

CIVIL SERVICE REFORM

see also Corruption (in politics); Patronage, Political

Korea (Republic). Chungang Haengjŏng Kaesŏn Chagŏptan.
(Sŏjŏng swaesin ch'okchin ŭl wihan haengjŏng kaesŏn chagŏp kyŏlgwa pogosŏ purok)
庶政刷新促進을為한行政改善作業結果報告書附錄. — ﹝서울﹞ : 總務處中央行政改善作業團, 1976.
1109 p. : ill. ; 26 cm.

1. Korea—Executive departments—Management. 2. Public administration—Korea. 3. Civil service reform. I. Title. II. Title: Haengjŏng kaesŏn chagŏp kyŏlgwa pogosŏ purok.
JQ1725.5.K67 1976 79-825427

New Jersey. Legislature. General Assembly. State Government, Federal and Interstate Relations, and Veteran's Affairs Committee.
Public hearings before Assembly State Government, Federal and Interstate Relations, and Veterans Affairs Committee on A-1675 (Civil service reform act) : held November 2, 1978, Assembly Chamber, State House, Trenton, New Jersey. — Trenton : The Committee, ﹝1979?﹞
226 p. in various pagings : graphs ; 29 cm.
1. Civil service—New Jersey. 2. Civil service reform. I. Title.
KFN1811.4.S78 1978c 342'.749'0680262 79-621854
79 MARC

CIVIL SERVICE REFORM (Continued)

—CONGRESSES

Seminario sobre Administración de Personal y Reforma Administrativa, 1st, Tegucigalpa, 1976.
Primer Seminario sobre Administración de Personal y Reforma Administrativa, del 14 al 16 de julio de 1976, Tegucigalpa, D.C., Honduras, C.A. : conclusiones y recomendaciones. — ₍Tegucigalpa₎ : Centro de Adiestramiento del Ministerio de Hacienda y Crédito Público, ₍between 1976 and 1978₎
13, 3 leaves ; 27 cm.
At head of title: Consejo Superior de Planificación Económica, (CONSUPLANE). Dirección General de Servicio Civil. Centro de Adiestramiento del Ministerio de Hacienda y Crédito Público (CENADIH).
1. Civil service—Honduras—Congresses. 2. Personnel management—Honduras—Congresses. 3. Civil service reform—Congresses. I. Honduras. Consejo Superior de Planificación Económica. II. Honduras. Direccion General de Servicio Civil. III. Honduras. Secretaría de Hacienda y Crédito Público (1971-). Centro de Adiestramiento.
JL1532.Z1S45 1976 78-105093
MARC

CIVIL SERVICE RETIREMENT

see also Civil service pensioners; Civil service pensions

—CALIFORNIA

California. Dept. of Finance. Program Evaluation Unit.
A review of the internal structure of the State retirement program / prepared by Department of Finance, Program Evaluation Unit. — ₍Sacramento₎ : The Unit, 1979.
xxx, 252 p. : ill. ; 28 cm. — (Report - Department of Finance, Program Evaluation unit ; no. D79-3)
1. Civil service pensions—California. 2. Civil service retirement—California. I. Title. II. Series: California. Dept. of Finance. Program Evaluation Unit. Report - State of California, Department of Finance, Program Evaluation Unit ; no. D79-3.
JK8760.P4C3 1979 353.9'794'005 79-623191
79 MARC

—JAPAN

Kishimoto, Susumu, 1915–
(Kokka kōmuin tō taishoku teate zantei sochi hō shōkai)
国家公務員等退職手当暫定措置法詳解 / 岸本晋著. — 改訂新版. — 東京：学陽書房, 昭和32 ₍1957₎
4, 234 p. ; 22 cm.

1. Civil service retirement—Japan. I. Japan. Laws, statutes, etc. Kokka kōmuin tō taishoku teate zantei sochi hō. 1957. II. Title.
79–800965

Kishimoto, Susumu, 1915–
(Kōmuin no shin taishoku kyūyo seido kaisetsu)
公務員の新退職給与制度解説 / 岸本晋, 穂刈誠一, 曽我悟共著. — 東京：学陽書房, 昭和33 ₍1958₎
3, 144 p. ; 21 cm.

1. Civil service retirement—Japan. I. Hokari, Seiichi, joint author. II. Soga, Satoru, 1926– joint author. III. Title.
79–805740

—UNITED STATES

United States. Congress. House. Committee on Post Office and Civil Service. Subcommittee on Compensation and Employee Benefits.
Adverse effect created by early retirement because of RIF and reorganization : hearing before the Subcommittee on Compensation and Employee Benefits of the Committee on Post Office and Civil Service, House of Representatives, Ninety-sixth Congress, first session, March 6, 1979. — Washington : U.S. Govt. Print. Off., 1979.
iii, 31 p. ; 24 cm.
"Serial no. 96-3."
1. Civil service retirement—United States. I. Title.
KF27.P638 1979a 353.001'82 79-602444
79 MARC

United States. Congress. House. Committee on Post Office and Civil Service. Subcommittee on Compensation and Employee Benefits.
Civilian air traffic controllers and flight service station specialists : hearing before the Subcommittee on Compensation and Employee Benefits of the Committee on Post Office and Civil Service , House of Representatives, Ninety-fifth Congress, second session, on H.R. 8503 and H.R. 10633, April 17, 1978. — Washington : U.S. Govt. Print. Off., 1978.
iii, 63 p. ; 24 cm.
1. United States. Dept. of Defense—Officials and employees. 2. Air traffic control—United States. 3. Civil service retirement—United States. I. Title.
KF27.P638 1978c 344'.73'01252 78-602688
78 MARC

United States. Congress. Senate. Committee on Governmental Affairs. Subcommittee on Civil Service and General Services.
Military service credit under Civil Service retirement program : hearing before the subcommittee on Civil Service and General Services of the Committee on Governmental Affairs, United States Senate, Ninety-fifth Congress, second session, on S. 245 ... March 17, 1978. — Washington : U.S. Govt. Print. Off., 1978.
iii, 111 p. ; 24 cm.
1. Civil Service retirement—United States. I. Title.
KF26.G6724 1978c 342'.73'068 78-603412
79 MARC

United States. Congress. Senate. Committee on Governmental Affairs. Subcommittee on Civil Service and General Services.
National Guard employment credit for Civil Service retirement purposes : hearing before the Subcommittee on Civil Service and General Services of the Committee on Governmental Affairs, United States Senate, Ninety-fifth Congress, second session, on S. 1390 ... October 4, 1978. — Washington : U.S. Govt. Print. Off., 1979.
iii, 25 p. ; 24 cm.
1. Civil Service retirement—United States. 2. United States—National Guard. I. Title.
KF26.G6724 1978g 342'.73'068 79-601255
79 MARC

CIVIL SUPREMACY OVER THE MILITARY

—COMMUNIST COUNTRIES
—ADDRESSES, ESSAYS, LECTURES

Civil-military relations in communist systems / Dale R. Herspring and Ivan Volgyes. — Boulder, Colo. : Westview Press, ₍1978₎
xiii, 273 p. ; 24 cm. — (Westview special studies on the Soviet Union and Eastern Europe)
Includes bibliographical references and index.
ISBN 0-89158-165-0 : $20.00
1. Civil supremacy over the military—Communist countries—Addresses, essays, lectures. 2. Communist countries—Armed forces—Political activity—Addresses, essays, lectures. 3. Communist countries—Politics and government—Addresses, essays, lectures. I. Herspring, Dale Roy. II. Völgyes, Iván, 1936- . III. Series.
JF195.C5C55 1978 322'.5'091717 78-180
78 MARC

—JAPAN—HISTORY

Fujita, Tsuguo, 1885–1967.
(Meiji Kempō ron)
明治憲法論：舊憲法から新憲法へ / 藤田嗣雄著. — 東京：朝倉書店, 昭和23 ₍1948₎
5, 264 p. ; 22 cm.
Includes bibliographical references.

1. Japan—Constitutional history. 2. Civil supremacy over the military—Japan—History. 3. Japan—Politics and government—1868– I. Title.
79–80047

—PERU

Fernández Salvatteci, José Antonio.
Yo acuso : la Revolución Peruana / José Antonio Fernández Salvatteci. — ₍s.l. : s.n., 1978₎ (Tacna, Perú : Editorial El Siglo)
266 p. ; 21 cm.
1. Peru—History—1968- . 2. Revolutions—Peru. 3. Counterrevolutions—Peru. 4. Civil supremacy over the military—Peru. I. Title.
F3448.2.F47 985'.063 79-116723
79 MARC

—RUSSIA

Colton, Timothy J 1947-
Commissars, commanders, and civilian authority : the structure of Soviet military politics / Timothy J. Colton. — Cambridge, Mass. : Harvard University Press, 1979.
viii, 365 p. ; 25 cm. — (Russian Research Center studies ; 79)
Includes bibliographical references and index.
ISBN 0-674-14535-6
1. Civil supremacy over the military—Russia. 2. Russia—Armed Forces—Political activity. 3. Kommunisticheskaīa partīia Sovetskogo Soīuza. I. Title. II. Series: Harvard University. Russian Research Center. Studies ; 79.
JN6520.C58C64 322'.5'0947 78-23342
78 MARC

—UNITED STATES

Kanter, Arnold.
Defense politics : a budgetary perspective / Arnold Kanter. — Chicago : University of Chicago Press, 1979.
viii, 152 p. : ill. ; 23 cm.
Bibliography: p. 140-146.
Includes index.
ISBN 0-226-42373-5 : $13.50
1. United States—National security. 2. Budget—United States. 3. United States. Dept. of Defense—Appropriations and expenditures. 4. Civil supremacy over the military—United States. I. Title.
UA23.K34 353.6 78-21848
78 MARC

CIVIL UNIFORMS see Uniforms, Civil

CIVIL WAR

see also Insurgency

Euangelou, Christina S
To megalo mou pisteuō / Christinas St. Euangelou. — Athēna : Euangelou, 1978.
54 p. ; 21 cm.
1. Political science—Miscellanea. 2. War and society. 3. Civil war. I. Title.
JC273.E9 79-117236
79 MARC

Orígenes de las guerras civiles en Colombia. — ₍s.l. : Ediciones Historia y Sociedad, 19—₎.
70 p. ; 17 cm.
On cover: Un análisis marxista.
1. Colombia—History—Colombia—19th century. 2. Civil war. 3. Colombia—Economic conditions.
F2273.074 986.1 78-110011
79₍r79₎rev MARC

—GREAT BRITAIN see Great Britain—History—Civil War, 1642-1649

—UNITED STATES see United States—History—Civil War, 1861-1865

CIVIL WRONGS see Torts

CIVILETTI, BENJAMIN R.

United States. Congress. Senate. Committee on the Judiciary.
Nomination of Benjamin R. Civiletti to be Deputy Attorney General : hearings before the Committee on the Judiciary, United States Senate, Ninety-fifth Congress, second session .... — Washington : U.S. Govt. Print. Off., 1978-
v. ; 24 cm.
Hearings held Feb. 21-
1. Civiletti, Benjamin R. 2. United States. Dept. of Justice—Officials and employees. I. Title.
KF26.J8 1978 353.5 78-601878
78₍r79₎rev MARC

CIVILIAN CONTROL OF THE MILITARY see Civil supremacy over the military

CIVILIAN DEFENSE see Civil defense

CIVILIANS EVACUATION OF see Evacuation of civilians

CIVILIS, JULIUS, FL. 69 A.D.

Vaenius, Otto, 1556-1629.
Der Bataveraufstand 69/70 n. Chr. : Zerstörung u. Schlacht von Vetera / ꜀auctore Othone Vaenio ; Radierungen von Antonio Tempesta꜀ ; hrsg. u. mit e. Vorw. vers. von Wilhelm Müllers ; ꜀hrsg. aus Anlass d. 100jährigen Jubiläums d. Niederrhein. Altertumsvereins Xanten e. V. in 1977꜀. — Köln : Rheinland-Verlag ; Bonn : Habelt ꜀in Komm.꜀, 1978.
꜀107꜀ p. : 36 ill. ; 22 x 30 cm. GFR78-A
Reprint of the 1612 ed. published in Antwerp under the title: Batavorum cum Romanis Bellum.
Original t.p. has also Dutch title: De Batavische oft ovde Hollandtsche oorloghe teghen de Romeynen.
The plates, with letterpress in Dutch and Latin below, are accompained by corresponding passages of Tacitus selected from his Historiae, Books 4 and 5.
ISBN (invalid) 3-7927-0371-3 : DM33.00
1. Rome—History—Revolt of Civilis, 69-70. 2. Civilis, Julius, fl. 69 A.D. I. Tempesta, Antonio, 1555-1630. II. Tacitus, Cornelius. Historiae. Book 4. Selections. 1978. III. Tacitus, Cornelius. Historiae. Book 5. Selections. 1978. IV. Müllers, Wilhelm. V. Title. VI. Title: De Batavische oft ovde Hollandtsche oorloghe teghen de Romeynen.
DG288.V33 1978 937'.07 79-364833
79 MARC

CIVILIZATION

see also Acculturation; Animals and civilization; Anthropology; Archaeology; Art; Astronautics and civilization; Biculturalism; Culture; Education; Ethics; Ethnology; Humanism; Inventions; Learning and scholarship; Mundas; Plants and civilization; Progress; Religion and culture; Religions; Social evolution; Social problems; Social sciences; Society, Primitive; Technology and civilization

Asimov, Isaac, 1920-
Extraterrestrial civilizations / Isaac Asimov. — New York : Crown Publishers, c1979.
282 p. ; 24 cm.
Includes index.
ISBN 0-517-53075-9
1. Life on other planets. 2. Civilization. I. Title.
QB54.A84 1979 574.999 79-360
79 MARC

Charter, S P R
Man on earth : a preliminary evaluation of the ecology of man / by S. P. R. Charter ; foreword by Aldous Huxley. — Los Angeles : Guild of Tutors Press, 1979.
xxxii, 264 p. ; 23 cm.
Includes index.
ISBN 0-89615-019-4 : $8.00
1. Civilization. I. Title.
HM101.C52 1979 301.3'1 78-75232
00 MARC

Chédel, André.
Sur la terre comme au ciel : vues sur l'univers / André Chédel. — Neuchâtel : H. Messeiller, achevé d'impr. 1978.
470 p. ; 22 cm. Sw78-11756
Includes bibliographical references.
27.00F
1. Civilization. 2. Religions. I. Title.
CB78.C47 909 79-353125
79 MARC

Das Gupta, Surajit.
In quest of world-culture : Suniti Kumar Chatterji / by Surajit Dasgupta. — 1st ed. — Calcutta : Shankar Prakashan, 1977.
104 p. ; 22 cm.
Rs22.00
1. India—Civilization. 2. Civilization. 3. Chatterji, Suniti Kumar, 1890-1977. I. Title.
DS423.D1835 954 78-907772
79 MARC

Debré, Robert, 1882-
Ce que je crois / Robert Debré. — Paris : B. Grässet, c1976.
219 p. ; 21 cm. — (Ce que je crois) F***
ISBN 2-246-00367-9 pbk. : 34.00F. ISBN 2-246-00368-7 de luxe
1. Civilization. 2. Progress. I. Title.
HM101.D42 301.24 78-377216
78 MARC

Freud, Sigmund, 1856-1939.
Das Unbehagen in der Kultur / von Sigm. Freud. — 2., durchgesehene Aufl. 13.-27. Tausend. — Wien : Internationaler Psychoanalytischer Verlag, 1931, c1930.
136 p. ; 21 cm.
Includes bibliographical references.
1. Psychoanalysis. 2. Social psychology. 3. Civilization. I. Title.
BF173.F68 1931 78-381541
78 MARC

Iwamura, Shinobu, 1905-
(Bummei no keizai kōzō)
文明の経済構造 / 岩村忍著. — 東京 : 中央公論社, 1978.
212 p. ; 20 cm. — (中公叢書) Ja 78-24304
Series romanized: Chūkō sōsho.
¥950
1. Economic history. 2. Social history. 3. Civilization. I. Title.
HC21.I92 79-809229

Jaulin, Robert.
Les chemins du vide / Robert Jaulin. — Paris : C. Bourgois, c1977.
268 p. ; 20 cm. F***
Includes bibliographical references.
ISBN 2-267-00077-6 : 45.00F
1. Civilization. 2. Economic history. I. Title.
HM101.J26 309.1 78-376119
78 MARC

Miège, Bernard.
L'Appareil d'action culturelle / Bernard Miège, Jacques Ion, Alain-Noël Roux. — Paris : Éditions universitaires, 1974.
298 p. ; 20 cm. — (Citoyens ; 5) F75-3604
Includes bibliographical references.
39.95F
1. Civilization. 2. Social action. I. Ion, Jacques, joint author. II. Roux, Alain Noël, joint author. III. Title.
HM101.M493 301.2 79-363228
79 MARC

Ortiz, Tulio Eduardo.
Los ciclos históricos argentinos / Tulio Eduardo Ortiz. — Buenos Aires : Plus Ultra, c1977.
237 p. ; 20 cm. — (Colección Política e historia)
Bibliography: p. 230-237.
1. Argentine Republic—History—Philosophy. 2. Toynbee, Arnold Joseph, 1889-1975. A study of history. 3. Civilization. 4. History—Philosophy. I. Title.
F2829.O74 982 79-351441
79 MARC

Smith, N A fl. 1976.
The new enlightenment : an essay in political and social realism / by N. A. Smith. — London : J. Calder, 1976.
256 p. ; 22 cm. — (A Platform book) GB***
ISBN 0-7145-3593-1 : £6.50
1. Liberalism. 2. Civilization. I. Title.
JC571.S646 1976 320.5'1 77-355761
77꜀r79꜀rev MARC

—1950-

Shakhnazarov, Georgiĭ Khosroevich.
(Sotsialisticheskaia sud'ba chelovechestva)
Социалистическая судьба человечества / Г. Х. Шахназаров. — Москва : Политиздат, 1978.
462 p. ; 17 cm. USSR 78
Includes bibliographical references and indexes.
1.50rub
1. Forecasting. 2. Civilization—1950- 3. Socialism. I. Title.
CB158.S453 79-410407

—ADDRESSES, ESSAYS, LECTURES

Jenny, Hans, 1912-
Abendländische Gedanken / Hans Jenny ; ꜀hrsg. von der Stiftung für Abendländische Besinnung꜀. — Stäfa : T. Gut, 1977.
183 p. ; 19 cm. Sw78-2627
"Sammlung von veröffentlichten und unveröffentlichten Referaten, Aufsätzen und Briefen Hans Jennys."
Bibliography: p. 178-182.
15.00F
1. Economic history—1945- —Addresses, essays, lectures. 2. Civilization—Addresses, essays, lectures. I. Stiftung für Abendländische Besinnung. II. Title.
HC59.J36 330.9'04 79-355801
79 MARC

Kuwabara, Takeo, 1904-
(Chūgoku to tsukiau hō)
中国とつきあう法 / 桑原武夫, 加藤周一. — 東京 : 潮出版社, 1978.
194 p. ; 20 cm. Ja 78-10317
¥880
1. China—Civilization—Addresses, essays, lectures. 2. Civilization—Addresses, essays, lectures. I. Katō, Shūichi, 1919- joint author. II. Title.
DS721.K93 78-812158

Tradiŝiî v istorii kul'tury : ꜀Sb. stateĭ : Posviashchen 85-letiiu A. F. Loseva꜀ / AN SSSR, Nauch. sovet po istorii mirovoĭ kul'tury : ꜀Otv. red. V. A. Karpushin꜀. — Moskva : Nauka, 1978.
279 p. ; 22 cm. USSR79
"Bibliografiia pechatnykh trudov A. F. Loseva, 1973-1978 gg.": p. 275-꜀277꜀
Includes bibliographical references.
1.60rub
1. Civilization—Addresses, essays, lectures. 2. Culture—Addresses, essays, lectures. 3. Losev, Alekseĭ Fedorovich—Addresses, essays, lectures. I. Losev, Alekseĭ Fedorovich. II. Karpushin, V. A. III. Akademiia nauk SSSR. Nauchnyĭ sovet po istorii Mirovoĭ kul'tury.
CB35.R9T7 79-374466
79 MARC

—COLLECTED WORKS

Ferrario, Giulio, 1767-1847.
Aggiunte all'opera Il costume antico e moderno di tutti i popoli, cogli analoghi disegni / del dottore Giulio Ferrario. — Firenze : V. Batelli, 1833-1837.
3 v. : col. ill., fold. maps ; 24 cm.
Half title: Il costume antico e moderno di tutti i popoli.
Enlarged and updated from the earlier edition published in Milan in 1831-1834 under title Aggiunte e rettificazioni all' opera Il costume antico e moderno ...
Supplements the 4 works (on Asia, Africa, America, and Europe) published by Batelli in 1823-1832 in 26 volumes, without consecutive overall numbering but with identical half title, as 2d ed. of the whole.
Includes excerpts from other authors.
L.C. set imperfect: v. 2, p. 43-52, and approximately one-fourth of the illustrative matter wanting.
1. Civilization—Collected works. I. Title. II. Title: Costume antico e moderno di tutti i popoli.
CB7.F48 1833 909 79-387934
79 MARC

Toynbee, Arnold Joseph, 1889-1975.
Arnold Toynbee, a selection from his works / edited with an introd. by E. W. F. Tomlin. — Oxford ꜀Eng.꜀ ; New York : Oxford University Press, 1978.
xxi, 327 p. ; 23 cm. GB78
Includes bibliographical references and index.
ISBN 0-19-215259-9 : £4.95 ($14.95)
1. Civilization—Collected works. 2. History—Philosophy—Collected works. I. Tomlin, Eric Walter Frederick, 1913- . II. Title.
CB7.T59 907'.2 78-40203
78 MARC

—HISTORY

see also Civilization, Medieval; Civilization, Modern

Ben-JocHannan, Yosef.
African origins of civilization. ꜀Sound recording꜀ ꜀Wilmette, Ill.꜀ Films Inc., p1977. 417-8003.
1 cassette. ⅛-track. mono. (Roots of Black Americans, no. 3)
Sharon Thomas, host.
Duration: 29 min., 20 sec.
In container (24 cm.)
With transcript (6 p.)
SUMMARY: Dr. ben-YocHannan presents a history of the beginnings of civilization originating in the Nile Valley and dispels many misconceptions about the origins of civilization.
1. Civilization—History. 2. Blacks—History. I. Thomas, Sharon N. II. Title. III. Series: Roots of Black Americans. ꜀Sound recording꜀ no. 3.
[CB151] 77-740689

Bose, Pramatha Nath.
Epochs of civilization / Pramatha Nath Bose. — New Delhi : Asian Educational Services, 1978.
x, 339 p. ; 23 cm.
Reprint of the 1913 ed. published by W. Newman, Calcutta.
Includes bibliographical references and index.
Rs120.00
1. Civilization—History. I. Title.
CB67.B65 1978 909 79-912128
79 MARC

Bruller, Jean, 1902-
Sens et non-sens de l'histoire / Vercors ꜀i.e. J. Bruller꜀. — Paris : Éditions Galilée, 1978.
198 p. ; 22 cm. — (Débats ISSN 0152-3678) F78-15171
Includes index.
ISBN 2-7186-0101-9 : 49.00F
1. Civilization—History. 2. Civilization—Philosophy. I. Title.
CB78.B78 909 79-350149
79 MARC

CIVILIZATION
—HISTORY (Continued)

(Bunka no kōryū)
文化の交流 / [著者　岩村忍 ... et al.]. — 東京：毎日新聞社, 1978.
264 p. : ill. ; 20 cm. — (古代文明の謎と発見 ; 4)　Ja 78-15017
"参考図書解題": p. [257]-263.
¥1300

1. East and West. 2. Civilization, Ancient. 3. Civilization—History. I. Iwamura, Shinobu, 1905- II. Series: Kodai bummei no nazo to hakken ; 4.
CB151.B96　78-813807

Condorcet, Marie Jean Antoine Nicolas Caritat, marquis de, 1743-1794.
Sketch for a historical picture of the progress of the human mind / Antoine-Nicolas de Condorcet ; translated by June Barraclough ; with an introd. by Stuart Hampshire. — Westport, Conn. : Greenwood Press, 1979, c1955.
xvi, 202 p. ; 23 cm.
Translation of Esquisse d'un tableau historique des progrès de l'esprit humain.
Reprint of the ed. published by Noonday Press, New York, in series: Library of ideas.
Bibliography: p. [xv]-xvi.
ISBN 0-8371-7370-1
1. Civilization—Philosophy. 2. Civilization—History. 3. Progress. I. Title. II. Series: Library of ideas.
[CB19.C5613 1979]　190'.9　74-62
78　MARC

Elias, Norbert.
Über den prozess der zivilisation; soziogenetische und psychogenetische untersuchungen ... Basel, Haus zum Falken, 1939.
2 v. 23 cm.
Each volume has also special t.-p.
Bibliography included in "Anmerkungen" (v. 1, p. 302-327; v. 2, p. 455-490)
CONTENTS: 1. bd. Wandlungen des verhaltens in den weltlichen oberschichten des Abendlandes.—2. bd. Wandlungen der gesellschaft; entwurf zu einer theorie der zivilisation.
1. Civilization—Philosophy. 2. Civilization—History. I. Title.
CB83.E4 1939　901　42-10084
[r79]rev2　MARC

Garaudy, Roger.
Comment l'homme devint humain / Roger Garaudy. — [Paris] : Éditions J.A., [c1978]
335 p. : ill. (some col.) ; 28 cm. — (L'Épopée humaine)　F***
ISBN 2-85258-105-1 : 150F
1. Civilization—History. I. Title.
CB78.G37　909　79-340713
79　MARC

Grousset, René, 1885-1952.
The sum of history / René Grousset ; English version by A. and H. Temple Patterson. — Westport, Conn. : Hyperion Press, 1979.
vii, 254 p. ; 22 cm.
Translation of Bilan de l'histoire.
Reprint of the 1951 ed. published by Tower Bridge Publications, Hadleigh, Essex.
Includes index.
ISBN 0-88355-697-9 : $19.00
1. Civilization—History. I. Title.
CB73.G713 1979　901　78-59024
78　MARC

A Handbook of civilization / George Merrill ... [et al.]. — 2d ed. — Dubuque, Iowa : Kendall/Hunt Pub. Co., c1978.
2 v. (xxxii, 580 p.) : ill. ; 23 cm.
Volume 2 by W. S. Holley and others.
Includes index.
CONTENTS: v. 1. Earliest times to the seventeenth century.—v. 2. Seventeenth century to the present.
ISBN 0-8403-1833-2 (v. 1)
1. Civilization—History. I. Merrill, George D. II. Holley, William S.
CB69.H35 1978　909　77-91495
78[r79]rev　MARC

Hasumi, Shigehiko, 1936-
(Fūkō, Dōrūzu, Derida)
フーコー・ドゥルーズ・デリダ / 蓮實重彥. — 東京：朝日出版社, 1978.
216 p. ; 22 cm. — (エピステーメー選書)　Ja 78-6904
¥1500
1. Foucault, Michel. Les mots et les choses. 2. Learning and scholarship. 3. Civilization—History. 4. Deleuze, Gilles. Différence et répétition. 5. Difference (Philosophy) 6. Repetition (Philosophy) 7. Derrida, Jacques. De la Grammatologie. 8. Languages—Philosophy. I. Title. II. Series: Episutēmē sensho.
AZ101.F692H37　79-800172

Histoire des idéologies / sous la direction de Francois Châtelet et Gérard Mairet .... — [Paris] : Hachette, 1978.
3 v. ; 23 cm. — (Hachette littérature & [i.e. et] sciences humaines)
F78-7377
Includes bibliographies and indexes.
CONTENTS: 1. Brisson, L. et al. Les mondes divins jusqu'au VIIIe siècle de notre ère.—2. Cabat, O. et al. De l'Église à l'État, du IXe au XVIIe siècle.—3. Savoir et pouvoir du XVIIIe au XXe siècle.
ISBN 2-01-004802-4 (v. 1) : 116F (v. 1 & 2)
1. Civilization—History. 2. Religion and culture. 3. Ideology—History. I. Châtelet, François, 1925- . II. Mairet, Gérard, 1943- .
CB78.H5　901　78-375140
78[r79]rev　MARC

History and culture / [editors, John Tusa, Roger Hearn] ; introduced by Christopher Hill. — London : Mitchell Beazley, c1977.
2 v. : ill. (some col.) ; 30 cm. — (The Mitchell Beazley joy of knowledge library)　GB***
Includes index.
ISBN 0-85533-109-7 (v.1) : £12.50 per vol.
1. Civilization—History. I. Tusa, John. II. Hearn, Roger. III. Series.
CB69.H57　909　78-320091
79　MARC

Kamo, Giichi, 1899-
(Sekai bunka shi)
世界文化史 / 加茂儀一著. — 東京：三笠書房, 昭和25- [1950-
v. : ill. ; 19 cm. — (學藝全書 ; 3-)
Colophon inserted.
Series romanized: Gakugei zensho.

1. Civilization—History. I. Title.
CB113.J3K36　78-809421

Min, Sŏk-hong.
(Munhwasa)
文化史 / 閔錫泓, 閔斗基, 梁秉祐. — [서울]：서울大學校附設韓國放送通信大學, 1976.
iii, 249 p. ; 22 cm.
On cover: Sŏul Taehakkyo Pusŏl Han'guk Pangsong T'ongsin Taehak kyojae.

1. Civilization—History. I. Min, Tu-gi. II. Yang, Pyŏng-u, 1923- III. Title.
CB113.K6M5　79-826328

(Mūjaz tārīkh al-ḥaḍārah)
موجز تاريخ الحضارة ، تاليف نور الدين حاطوم [وآخرين. دمشق] مطبعة المال ، -1965.
v. 25 cm.
Bibliography: v. 1, p. 683-686.
CONTENTS: جزء 1. حضارات العصور القديمة . —
1. Civilization—History. I. Ḥāṭūm, Nūr al-Dīn.
CB113.A7M78　N E 67-1376

Nakao, Sasuke, 1916-
(Gendai bummei futatsu no genryū)
現代文明ふたつの源流：照葉樹林文化・硬葉樹林文化 / 中尾佐助著. — 東京：朝日新聞社, 1978.
iv, 228 p., [2] leaves of plates : ill. ; 19 cm. — (朝日選書 ; 110)
Ja 78-12843
Series romanized: Asahi sensho.
¥820
1. Comparative civilization. 2. Civilization—History. 3. Natural history. I. Title.
CB151.N32　78-811943

Namāzī, Muḥammad.
(Dāstān-i tamaddun)
داستان تمدن . — [s. l. : s. n., 1974?]
251 p., [2] leaves of plates : ill. ; 29 cm. — (مجموعة آثار محمد نمازى ؛ كتاب 1.)
Introd. dated 1353 [1974]
Series romanized: Majmū'ah-'i āsār-i Muḥammad Namāzī.
Includes bibliographical references.
1. Civilization—History. I. Title.
AC127.N33 1974, vol. 1　79-970764
[CB113.P4]

Paget, Reginald Thomas, Baron Paget, 1908-
The human journey / [by] Reginald Paget. — London : Davis-Poynter, 1979.
365 p. ; 23 cm.　GB79-14226
Includes index.
ISBN 0-7067-0231-X : £10.00
1. Civilization—History. 2. Religion and culture. I. Title.
CB69.P33　909　79-313572
79　MARC

Singer, Charles Joseph, 1876-1960, ed.
A history of technology, edited by Charles Singer [and others] Oxford, Clarendon Press, 1954-78.
7 v. illus. (part col.) port., maps. 26 cm.
Includes bibliographies.
Volumes 6-7 ed. by T. I. Williams.
CONTENTS: v. 1. From early times to fall of ancient empires.—v. 2. The Mediterranean civilizations and the Middle Ages, c. 700 B.C. to c. 1500 A.D.—v. 3. From the Renaissance to the industrial revolution, c. 1500-c. 1750.—v. 4. The industrial revolution, c. 1750 to c. 11850.—v. 5. The late nineteenth century, c. 1850 to c. 1900.—v. 6-7. The twentieth century, c. 1900 to c. 1950.
1. Technology—History. 2. Civilization—History. I. Williams, Trevor Illtyd. II. Title.
T15.S53　609　a 55-8645
[r79]rev2　MARC

Strayer, Joseph Reese, 1904-
The mainstream of civilization / Joseph R. Strayer, Hans W. Gatzke. — 3d ed. — New York : Harcourt Brace Jovanovich, c1979.
xxxi, 840 p., xxiv, p., [8] leaves of plates : ill. ; 27 cm.
Includes bibliographies and index.
SUMMARY: Traces the history of the world's major civilizations discussing their special characteristics and contributions.
ISBN 0-15-551562-4
1. Civilization—History. [1. Civilization—History] I. Gatzke, Hans Wilhelm, 1915- joint author. II. Title.
CB69.S88 1979　909　78-70777
79　MARC

Vergilius, Polydorus, d. 1555.
De rerum inuentoribus libri octo, per authroem quarto iam, ac diligentius recogniti & locupletati. Eiusdem in Dominicam precem commentariolum. Basileae [J. Bebell] 1536.
590 p. 18 cm.
Bebel's device on t.p. and at end.
1. Civilization—History. 2. Church history. I. Title.
CB23.V47　78-413636
78　MARC

The Western experience / [by] Mortimer Chambers ... [et al.] ; art essays by H. W. Janson. — 2d ed. — New York : Knopf, c1979.
xvi, 1119, xlii p., [32] leaves of plates : ill. (some col.) ; 24 cm.
Includes bibliographies and index.
ISBN 0-394-32135-9 : $16.95
1. Civilization—History. 2. Civilization, Occidental—History. I. Chambers, Mortimer.
CB59.W38 1979　909　78-12117
78　MARC

The Western experience / Mortimer Chambers ... [et al.] ; art essays by H. W. Janson ; advisory editor, Eugene Rice ; Consultant, Alvin Bernstein. — 2d ed. — New York : Knopf, c1979.
3 v. (xiv, 1119, xxi p., [32] leaves of plates) : ill. ; 24 cm.
CONTENTS: v. 1. Antiquity to the Middle Ages.—v. 2. The early modern period.—v. 3. The modern era.
Includes bibliographies and index.
Includes index.
ISBN 0-394-32138-3 (v. 1)
1. Civilization—History. 2. Civilization, Occidental—History. I. Chambers, Mortimer.
CB59.W38 1979b　909　78-12132
78　MARC

—HISTORY—ADDRESSES, ESSAYS, LECTURES

Muratori, Saverio.
Metodologia del sistema realtà-autocoscienza : dalle ultime lezioni 1972-1973 / Saverio Muratori ; a cura di Guido Marinucci. — Roma : Centro studi di storia urbanistica, 1978.
549 p. : ill. ; 25 cm.　It79-Mar
Limited edition of 500 numbered copies. No. 54.
Includes index.
L25000
1. Philosophy—History—Addresses, essays, lectures. 2. Civilization—History—Addresses, essays, lectures. 3. Reality—Addresses, essays, lectures. I. Marinucci, Guido. II. Title.
B87.M87　79-359419
*79　MARC

—HISTORY—COLLECTED WORKS

Eunsa historia universal. — Pamplona : Ediciones Universidad de Navarra, c1979-
v. : ill. ; 25 cm.　Sp***
Includes bibliographies.
CONTENTS: 1. Suárez Fernández, L. Las primeras civilizaciones.

ISBN 8431305525
1. Civilization—History—Collected works. I. Title: Historia universal.
CB104.E95　909　79-110514
79　MARC

The Story of man / [prepared by National Geographic Book Service]. — [Washington : National Geographic Society, c1977-
v. : col. ill. ; 27 cm.
Title from spine.
1. Civilization—History—Collected works. I. National Geographic Society, Washington, D.C. Book Service.
CB69.S86　909　77-151126
77[r79]rev　MARC

CIVILIZATION (Continued)

—HISTORY—JUVENILE LITERATURE

Carter, Robert A 1930-
You, world, and cultures / Robert A. Carter, Edith McCall, Vernon Prinzing. — Westchester, Ill. : Benefic Press, c1978.
432 p. : ill. (some col.) ; 27 cm. — (People in a world of change)
Includes indexes.
SUMMARY: A social studies textbook discussing forces that have influenced the development of various world cultures and how these cultures continue to affect each other.
ISBN 0-8175-5305-3
1. Civilization—History—Juvenile literature. [1. Civilization—History] I. McCall, Edith S., joint author. II. Prinzing, Vernon, joint author. III. Title. IV. Series.
CB69.2.C37 909 77-75871
79 MARC

Histoire et géographie avec le transatlas : 6e / Pierre Milza, ... Serge Berstein, ... Annette Buzacoux, ... Richard Dubreuil, ... — Paris : F. Nathan, 1977.
223 p. : col. ill. ; 24 cm. & atlas (30 p.) F78-10023
Includes bibliographical references.
28.00F
1. Civilization—History—Juvenile literature. 2. Anthropo-geography—Juvenile literature. I. Milza, Pierre.
CB78.H52 909 79-353550
79 MARC

Linder, Bertram L
A world history / Bertram L. Linder, Edwin Selzer, Barry M. Berk ; Miriam Greenblatt, contributing editor. — Chicago : Science Research Associates, c1979.
xi, 740 p. : ill. ; 25 cm.
Includes bibliographies and index.
SUMMARY: A world history concentrating on the people throughout the ages and how they lived and thought. Includes study questions.
ISBN 0-574-42004-5
1. Civilization—History—Juvenile literature. [1. Civilization—History. 2. World history] I. Selzer, Edwin, joint author. II. Berk, Barry M., joint author. III. Greenblatt, Miriam. IV. Title.
CB69.2.L56 909 79-101937
79 MARC

—HISTORY—MAPS

Cultuurhistorische atlas : het document in kaart / L. Th. Maes ... [et al.]. — Namen : Wesmael-Charlier, [1976]
54, 10 p. : col. maps ; 32 cm. Be77-9
Includes index.
265F
1. Geography, Historical—Maps. 2. Civilization—History—Maps. I. Maes, Louis Théo.
G1046.E6C8 1976 911 79-368296
*00 MARC

—JEWISH INFLUENCES

Finkelstein, Louis, 1895- ed.
The Jews : their history, culture, and religion / edited by Louis Finkelstein. — Westport, Conn. : Greenwood Press, 1979, c1960.
2 v. (xxxvi, 1859 p.), [15] leaves of plates : ill. ; 24 cm.
Reprint of the 3d ed. published by Harper, New York.
Fourth ed. (1970-71) published as a 3 v. work. Vol. 1 has title: The Jews: their history; v. 2: The Jews: their religion and culture; v. 3: The Jews: their role in civilization.
Includes bibliographies and indexes.
ISBN 0-313-21242-2
1. Jews—History. 2. Judaism—History. 3. Jewish literature—History and criticism. 4. Civilization—Jewish influences. I. Title.
[DS102.4.F5 1979] 909'.04'924 79-4193
79 MARC

—MISCELLANEA

(Maboroshi no tairiku to shi no miyako)
幻の大陸と死の都 / [著者 八幡一郎 ... et al.]. — 東京：毎日新聞社, 昭和52 [1977]
288 p. : ill.; 20 cm. — (古代文明の謎と発見 ; 2) Ja***
Bibliography: p. [279]-287.
¥1300

1. Civilization, Ancient. 2. Civilization—Miscellanea. I. Yawata, Ichirō. II. Series: Kodai bummei no nazo to hakken ; 2.
CB311.M16 78-813385

Rādhe Ajñānī.
(Mānava ati nikṛshṭa hai)
मानव अति निकृष्ट है. [लेखक] राधे 'अज्ञानी.' [1. संस्करण] मुजफ्फरपुर, हिमेशिया प्रकाशन [1973–
v. 22 cm. Rs2.00 (v. 1)
In Hindi.

1. Civilization—Miscellanea. I. Title.
CB113.H5R3 78-929554

—PERIODICALS

(Wŏlgan Tongsŏ munhwa)
月刊東西文化 Monthly East and West.

[서울 東西文化研究院
v. ill. 26 cm.
English or Korean.

1. Civilization—Periodicals. 2. Arts—Periodicals. 3. East and West—Periodicals. I. Tongsŏ Munhwa Yŏn'guwŏn. II. Title: Tongsŏ munhwa. III. Title: Monthly East and West.
CB3.W64 79-649985
(MARC-S)

—PHILOSOPHY

see also Philosophical anthropology

Ágh, Attila.
Tudományos-technikai forradalom és művelődés : filozófiai tanulmányok / Ágh Attila. — Budapest : Magvető, 1977.
400 p. ; 19 cm. — (Elvek és utak ISSN 0324-4318) H***
Includes bibliographical references.
CONTENTS: TTF.—Az elsajátítás marxi elmélete.—Az elsajátítás a TTF perspektívájában.—A TTF világtörténelmi folyamata.—Az idő ökonomiája.—Totalitáselmélet és rendszerelmélet.—A fogyasztói társadalom természetrajza.—Az ipari társadalom határai.—Társadalmi tervezés és szociológia.—A TTF rendszere.—Kant kultúrfogalma és embereszménye.—A marxi művelődéselmélet kialakulása.—A művelődéselmélet filozófiai alapvetéséhez.—A kultúrtörténet filozófiai alapvetéséhez.—Töprengések a nevelésfilozófiáról.
ISBN 9632705513 : 39.00Ft
1. Communism and science. 2. Communism and culture. 3. Civilization—Philosophy. I. Title.
HX541.A38 78-395474
79 MARC

Bahm, Archie J
The philosopher's world model / Archie J. Bahm. — Westport, Conn. : Greenwood Press, 1979.
viii, 295 p. ; 22 cm. — (Contributions in philosophy ; no. 12 ISSN 0084-926X)
Includes bibliographical references and index.
ISBN 0-313-21198-1
1. Civilization—Philosophy. 2. Civilization, Modern—1950- . 3. Philosophy. I. Title.
CB19.B272 901 78-67569
78 MARC

Batra, Raveendra N
The downfall of capitalism and communism : a new study of history / Raveendra N. Batra. — London : Macmillan, 1978. GB***
xi, 283 p. ; 23 cm.
Includes bibliographical references and index.
ISBN 0-333-21645-8 : £12.00
1. Civilization—Philosophy. 2. Capitalism. 3. Communism and culture. I. Title.
CB19.B294 901 79-306906
79 MARC

Becsi, Kurt, 1920-
Der Kosmos auf der Bühne : das galaktische Weltbild des künftigen Theaters / Kurt Becsi. — Wien ; Freiburg ; Basel : Herder, c1977.
327 p. : ill. ; 24 cm. Au77-15-265
Includes bibliographical references.
ISBN 3-210-24531-2 : S328.00
1. Theater. 2. Civilization—Philosophy. I. Title.
PN2039.B44 792 79-375682
*79 MARC

Blaga, Lucian, 1895-1961.
Orizzonte e stile / Lucian Blaga ; a cura di Antonio Banfi ; [traduzione di Mircea Popescu e Eugeniu Coseriu]. — Milano : A. Minuziano, 1946.
240 p. ; 17 cm. — (Estetica ; 9)
Translation of Orizont și stil.
Includes bibliographical references.
1. Aesthetics. 2. Civilization—Philosophy. I. Banfi, Antonio, 1886-1957. II. Title.
B4825.B553O7416 79-385269
00 MARC

Blaga, Lucian, 1895-1961.
Trilogia culturii. Cuvînt înainte de Dumitru Ghișe. București, Editora pentru Literatură Universală, 1969.
x, 396 p. 22 cm.
CONTENTS: Orizont și stil.—Spațiul mioritic.—Geneza metaforei și sensul culturii.
1. Civilization—Philosophy. I. Title.
CB19.B535 1969 79-376784
79 MARC

Bogoli͡ubova, Elena Vladimirovna.
(Kul'tura i obshchestvo)
Культура и общество : вопросы истории и теории / Е. В. Боголюбова. — Москва : Изд-во Московского университета, 1978.
230, [2] p. ; 20 cm. USSR***
Bibliography: p. 229–[231]
0.70rub
1. Civilization—Philosophy. I. Title.
CB19.B567 79–412171

Bril, Jacques.
Symbolisme et civilisation : essai sur l'éfficacité anthropologique de l'imaginaire ... / Jacques Bril. — Lille : Atelier Reproduction des thèses Université de Lille III ; Paris : diffusion H. Champion, 1977.
500 p., [18] p. of plates : ill. ; 24 cm. F78-7404
Thesis—Université de Grenoble II, 1975.
Bibliography: p. 450-475.
Includes indexes.
ISBN 2-7295-0032-4 : 55.00F
1. Symbolism. 2. Civilization—Philosophy. I. Title.
CB475.B74 301.29 78-395685
79 MARC

Bruller, Jean, 1902-
Sens et non-sens de l'histoire / Vercors [i.e. J. Bruller]. — Paris : Éditions Galilée, 1978.
198 p. ; 22 cm. — (Débats ISSN 0152-3678) F78-15171
Includes index.
ISBN 2-7186-0101-9 : 49.00F
1. Civilization—History. 2. Civilization—Philosophy. I. Title.
CB78.B78 909 79-350149
79 MARC

(Bunka to bummei no tetsugaku)
文化と文明の哲学 / 平井正編著. — 東京：学文社, 昭和51 [1976]
4, 226 p. ; 22 cm. — (現代哲学選書 ; 10) Ja 76-9181
Bibliography: p. 220–223.
Includes index.
¥2000

1. Civilization—Philosophy. 2. Culture. I. Hirai, Tadashi, 1929- II. Series: Gendai tetsugaku sensho ; 10.
CB19.B743 79–801103

Condorcet, Marie Jean Antoine Nicolas Caritat, marquis de, 1743-1794.
Sketch for a historical picture of the progress of the human mind / Antoine-Nicolas de Condorcet ; translated by June Barraclough ; with an introd. by Stuart Hampshire. — Westport, Conn. : Greenwood Press, 1979, c1955.
xvi, 202 p. ; 23 cm.
Translation of Esquisse d'un tableau historique des progrès de l'esprit humain.
Reprint of the ed. published by Noonday Press, New York, in series: Library of ideas.
Bibliography: p. [xv]-xvi.
ISBN 0-8371-7370-1
1. Civilization—Philosophy. 2. Civilization—History. 3. Progress. I. Title. II. Series: Library of ideas.
[CB19.C5613 1979] 190'.9 74-62
78 MARC

Crespi, Franco.
Esistenza e simbolico : prospettive per una cultura alternativa / Franco Crespi. — 1. ed. — Milano : Feltrinelli, [1978]
149 p. ; 22 cm. — (Biblioteca di filosofia) (I Fatti e le idee ; 406) It78-Nov
Includes bibliographical references.
L4500
1. Existentialism. 2. Symbolism. 3. Civilization—Philosophy. I. Title.
B819.C67 79-337190
*79 MARC

Cruz Vélez, Danilo.
Aproximaciones a la filosofía / Danilo Cruz Vélez. — [Bogotá] : Instituto Colombiano de Cultura, Subdirección de Comunicaciones Culturales, División de Publicaciones, c1977.
252 p. ; 21 cm. — (Colección Autores nacionales ; 19) (Biblioteca colombiana de cultura)
Includes bibliographical references.
1. Philosophy. 2. Civilization—Philosophy. 3. Languages—Philosophy. 4. Nietzsche, Friedrich Wilhelm, 1844-1900. I. Title.
B53.C75 79-103736
79 MARC

CIVILIZATION
—PHILOSOPHY (Continued)

Duerr, Hans Peter.
Traumzeit : über d. Grenze zwischen Wildnis u. Zivilisation / Hans Peter Duerr. — Frankfurt am Main : Syndikat, 1978.
415 p. ; 22 cm. GFR78-A
Bibliography: p. 347-405.
Includes index.
ISBN 3-8108-0077-5
1. Civilization—Philosophy. 2. Folk-lore. 3. Ethnopsychology. I. Title.
CB19.D77 901 78-399277
79 MARC

Elias, Norbert.
Über den prozess der zivilisation; soziogenetische und psychogenetische untersuchungen ... Basel, Haus zum Falken, 1939.
2 v. 23 cm.
Each volume has also special t.-p.
Bibliography included in "Anmerkungen" (v. 1, p. 302-327; v. 2, p. 455-490)
CONTENTS: 1. bd. Wandlungen des verhaltens in den weltlichen oberschichten des Abendlandes.—2. bd. Wandlungen der gesellschaft; entwurf zu einer theorie der zivilisation
1. Civilization—Philosophy. 2. Civilization—History. I. Title.
CB83.E4 1939 901 42-10084
ᵣr79ᵣrev2 MARC

Feng, Hu-hsiang, 1948–
(Wen hua che hsüeh mien mien kuan)
文化哲學面面觀 / 馮滬祥著. — 台北縣新店鎮：先知出版社；台北：總經銷先知書卡禮物中心，1976.
337 p. ; 20 cm. — (先知叢書 ; 36)
Title on verso of t. p.: Essays on the cultural philosophy.
Series romanized: Hsien chih ts'ung shu.
Includes bibliographical references.
NT$85.00
1. Civilization—Philosophy. 2. China—Civilization. I. Title.
CB19.F38 78–840890

Fischer, Klaus P 1942-
History and prophecy : Oswald Spengler and The decline of the West / Klaus P. Fischer. — Durham, N.C. : Moore Pub. Co., c1977.
275 p. ᵣ2ᵣ leaves of plates : ill. ; 24 cm.
Bibliography: p. 259-263.
Includes index.
ISBN 0-87716-080-5 : $9.95
1. Civilization—Philosophy. 2. Spengler, Oswald, 1880-1936. Der Untergang des Abendlandes. 3. Spengler, Oswald, 1880-1936. 4. Historians—Germany—Biography. I. Title.
CB19.F59 901 76-57812
79 MARC

Frängsmyr, Tore, 1938-
Svärmaren i vetenskapens hus : idéhistoriska essäer / Tore Frängsmyr. — ᵣStockholmᵣ : Rabén & Sjögren, 1977.
187 p. : ill. ; 20 cm. S77-14/16
Includes index.
ISBN 9129480833 : kr47.00
1. Science and civilization. 2. Civilization—Philosophy. 3. History—Philosophy. 4. Intellectual life. I. Title.
CB478.F68 78-394727
79 MARC

Freud, Sigmund, 1856-1939.
Reflections / by Sigmund Freud ; authorized English translation by A. A. Brill and Alfred B. Kuttner. — New York : Moffat, Yard, 1922.
71 p. ; 19 cm.
Translation of Zeitgemässes über Krieg und Tod.
First ed. published in 1918 under title: Reflections on war and death.
1. Psychoanalysis. 2. War—Psychological aspects. 3. Death—Psychological aspects. 4. Civilization—Philosophy. I. Title.
BF173.F7813 1922 79-342657
79 MARC

Friedrich, Heinz, 1922-
Kulturkatastrophe : Nachrufe auf d. Abendland / Heinz Friedrich. — 1. Aufl. — Hamburg : Hoffmann und Campe, 1979.
312, ᵣ1ᵣ p. ; 20 cm. GFR79-A
Bibliography: p. 307-ᵣ313ᵣ
ISBN 3-455-02255-3 : DM24.00
1. Civilization—Philosophy. 2. Civilization, Modern—20th century. I. Title.
CB19.F74 909.08 79-371692
79 MARC

Koneczny, Feliks, 1862-1949.
O ład w historii : z dodatkami o twórczości i wpływie Konecznego. — Londyn : Nakł. Komitetu Wydawniczego, 1977.
202 p. ; 23 cm. — (Wydawnictwa Towarzystwa im. Romana Dmowskiego ; Nr 13) GB***
Includes bibliographies by F. Koneczny and index.
1. Civilization—Philosophy. 2. Koneczny, Feliks, 1862-1949. I. Title.
CB19.K62 1977 79-353368
79 MARC

Landmann, Michael, 1913-
Alienatory reason / Michael Landmann ; translated by David J. Parent ; with a foreword by Zoltan Tar. — Normal, Ill. : Applied Literature Press, 1978.
xiii, 209 p. ; 25 cm. — (Illinois language and culture series ; v. 5) (Monograph publishing : Imprint series)
Translation of Entfremdende Vernunft.
Includes bibliographical references and index.
ISBN 0-8357-0333-9 : $15.75
1. Irrationalism (Philosophy) 2. Reason. 3. Civilization—Philosophy. 4. Alienation (Philosophy) I. Title. II. Series.
B824.2.L3613 149'.7 78-24106
78 MARC

Nishi, Shin'ichirō, 1873–1943.
(Bunkyōron)
文教論 / 西晋一郎著. — 新京：建國大學，康德8 ᵣ1941ᵣ
2, 194 p. ; 22 cm.
"非賣品"

1. Ethics, Japanese. 2. Humanities. 3. Civilization—Philosophy. I. Title.
BJ970.N566 79–806880

Ó Croidheáin, P
Uncharted reef : the Irish struggle viewed as a conflict between civilizations. — ᵣGaillimh, Ire. : Ó Croidheáin, 1977ᵣ
15 p. ; 21 cm. Ir***
Includes bibliographical references.
1. Civilization—Philosophy. 2. Irish question. I. Title.
CB19.O25 901 78-319483
78 MARC

Rivero San José, Jorge María.
Una alternativa al materialismo / Jorge Ma. Rivero San José. — Madrid : Ediciones de Cámara, 1978.
147 p. ; 20 cm. — (Colección Trazos de animación cultural) Sp***
ISBN 8485373022 : 260ptas
1. Idealism. 2. Materialism. 3. Civilization—Philosophy. 4. Ethics. I. Title.
B823.R53 79-342137
79 MARC

Schweitzer, Albert, 1875-1965.
Everyone needs a philosophy of life : Albert Schweitzer's philosophy, reverence for life / edited and arranged by Mary Louise R. O'Hara. — 1st ed. — ᵣGreat Barrington, Mass.ᵣ : Albert Schweitzer Friendship House, c1978.
74 p. ; 22 cm.
ISBN 0-89626-023-2
1. Schweitzer, Albert, 1875-1965—Philosophy. 2. Life. 3. Civilization—Philosophy. I. O'Hara, Mary Louise R. II. Title.
B2430.S374A32 193 78-73783
79 MARC

Schwientek, Alessandro, 1890-
Libertà e civiltà : studio filosofico sulla natura della civiltà nella vita umana / Alessandro Schwientek ; prefazione di Giorgio del Vecchio. — Roma : Editrice Studium, 1946.
148 p. ; 22 cm.
Includes bibliographical references and indexes.
1. Life. 2. Liberty. 3. Natural law. 4. Civilization—Philosophy. I. Title.
BD431.S288 79-375796
79 MARC

Sobol', Ol'ga Nikolaevna.
(Kritika filosofskikh kontšepfšiĭ amerikanskoĭ "kul'turnoĭ antropologii")
Критика философских концепций американской "культурной антропологии" / О. Н. Соболь. — Киев : Наук. думка, 1978.
99 p. ; 20 cm. USSR 78
At head of title: Akademiiā nauk Ukrainskoĭ SSR. Institut filosofiī.
Bibliography: p. 94–ᵣ98ᵣ
0.65rub
1. Philosophical anthropology. 2. Civilization—Philosophy. 3. Philosophy, American—20th century. I. Title.
BD450.S56 78–423409

Ṣubḥī, Aḥmad Maḥmūd.
(Fī falsafat al-ḥaḍārah)
فى فلسفة الحضارة / احمد محمود صبحى. — الاسكندرية : مؤسسة الثقافة الجامعية، ᵣ1976–
v. ; 24 cm.
Bibliography: v. 1, p. ᵣ243ᵣ–244.
CONTENTS: 1. الحضارة الافريقية. —
£E2.50 (v. 1)
1. Civilization—Philosophy. I. Title.
CB19.S78 77–961049

(Toimbī no shūkyōkan)
トインビーの宗教観 / 山本新編. — 東京：第三文明社，1974.
204 p. ; 18 cm. — (レグルス文庫 ; 39) Ja 74-17379
Includes bibliographical references.
¥480

1. Toynbee, Arnold Joseph, 1889–1975. 2. Civilization—Philosophy. 3. Religion—Historiography. I. Yamamoto, Shin, 1913–
CB19.T55 79–805662

(Toimbī no rekishikan)
トインビーの歴史観 / 山本新編. — 東京：第三文明社，1976.
190 p. ; 18 cm. — (レグルス文庫 ; 68) Ja 77–1679
Series romanized: Regurusu bunko.
Bibliography: p. ᵣ179ᵣ–190.
¥480

1. Toynbee, Arnold Joseph, 1889–1975. 2. Civilization—Philosophy. 3. History—Philosophy. I. Yamamoto, Shin, 1913–
CB19.T54 79–805274

Tschumi, Raymond.
Théorie de la culture / Raymond Tschumi. — Lausanne : Éditons l'Age d'homme, c1975.
261 p. : ill. ; 23 cm. — (Collection Dialectica) Sw76-1176
Includes bibliographical references.
39.50F
1. Civilization—Philosophy. 2. Language—Philosophy. I. Title.
CB19.T72 901 79-353551
*79 MARC

Ueyama, Shumpei, 1921-
(Shinsō bunkaron josetsu)
深層文化論序説 / 上山春平. — 東京：講談社，昭和51 ᵣ1976ᵣ
147 p. : ill. ; 15 cm. — (講談社学術文庫) Ja 77-2597
Series romanized: Kōdansha gakujutsu bunko.
CONTENTS: 深層文化論序説—人間学の位置—縄文の石笛
¥240

1. Civilization—Philosophy. 2. Man. 3. Jōmon culture. I. Title.
CB19.U34 79–800209

—PHILOSOPHY—ADDRESSES, ESSAYS, LECTURES

Cassirer, Ernst, 1874-1945.
Symbol, myth, and culture : essays and lectures of Ernst Cassirer, 1935-1945 / edited by Donald Phillip Verene. — New Haven : Yale University Press, 1979.
xii, 304 p. ; 25 cm.
Includes bibliographical references and index.
CONTENTS: The concept of philosophy as a philosophical problem.—Critical idealism as a philosophy of culture.—Descartes, Leibniz, and Vico.—Hegel's theory of the State.—The philosophy of history.—Language and art I.—Language and art II.—The educational value of art.—Philosophy and politics.—Judaism and the modern political myths.—The technique of our modern political myths.—Reflections on the concept of group and the theory of perception.
ISBN 0-300-02306-5 : $25.00
1. Philosophy—Addresses, essays, lectures. 2. Civilization—Philosophy—Addresses, essays, lectures. 3. Aesthetics—Addresses, essays, lectures. 4. Political science—Addresses, essays, lectures. I. Verene, Donald Phillip, 1937- . II. Title.
B3216.C31V47 1979 78-9887
78 MARC

(Gendai bunka tetsugaku no shomondai)
現代文化哲學の諸問題 / 谷川徹三 ... ᵣet al.ᵣ ; ᵣ中央大學公開講座委員會編ᵣ. — 東京：日本圖書，昭和23 ᵣ1948ᵣ
2, 1, 209 p. ; 19 cm.

1. Civilization—Philosophy—Addresses, essays, lectures. 2. Philosophy, Modern—Addresses, essays, lectures. I. Tanikawa, Tetsuzō, 1895– II. Chūō Daigaku, Tokyo. Kōkai Kōza Iinkai.
CB19.G365 78–815546

Hussain, Javed, 1946–
(Falsafah, saqāfat, aur tīsrī dunyā)
فلسفہ، ثقافت، اور تیسری دنیا / قاضی جاوید. — لاہور : مکتبہ میری لائبریری، 1976.
144 p. ; 18 cm. — (میری لائبریری ; 253)
In Urdu.
Series romanized: Merī Lā'ibrerī.
Rs8.00

1. Philosophy—Addresses, essays, lectures. 2. Civilization—Philosophy—Addresses, essays, lectures. 3. Underdeveloped areas—Addresses, essays, lectures. I. Title.
B53.H87 78–930217

Materialien zu Norbert Elias' Zivilisationstheorie / hrsg. von Peter Gleichmann, Johan Goudsblom und Hermann Korte. — 1. Aufl. — Frankfurt am Main : Suhrkamp, 1979.
436 p. ; 18 cm. — (Suhrkamp Taschenbuch Wissenschaft ; 233) GFR***
Bibliography of N. Elias' works: p. 432-435.
Includes bibliographical references.
ISBN 3-518-07833-X : DM16.00
1. Civilization—Philosophy—Addresses, essays, lectures. 2. Europe—Civilization—Addresses, essays, lectures. 3. Sociology—Addresses, essays, lectures. 4. Elias, Norbert—Addresses, essays, lectures. I. Gleichmann, Peter. II. Goudsblom, Johan. III. Korte, Hermann, 1937-
CB19.M36 901 79-362120
79 MARC

CIVILIZATION
—PHILOSOPHY—ADDRESSES, ESSAYS, LECTURES (Continued)

(Ningen to bunka)
人間と文化 / 島田豊編集. — 東京：青木書店, 1977.
viii, 358 p. ; 20 cm. — (講座史的唯物論と現代 ; 1) Ja***
Includes bibliographical references.

1. Civilization—Philosophy—Addresses, essays, lectures. I. Shimada, Yutaka, 1929- II. Series: Kōza shiteki yuibutsuron to gendai ; 1.
CB19.N56 78-811925

Ōki, Hideo, 1928-
(Gendaijin no Yudayajinka)
現代人のユダヤ人化：現代文明論集 / 大木英夫. — 東京：白水社, 1976.
326 p. ; 19 cm. — (白水叢書 ; 5) Ja 77-103
Series romanized: Hakusui sōsho.
Includes bibliographical references.
¥1200
1. Civilization, Modern—20th century—Addresses, essays, lectures. 2. Civilization—Philosophy—Addresses, essays, lectures. I. Title.
CB428.O35 78-813264

Oksala, Pellervo, 1939-
Ihminen, kulttuuri ja taide / Pellervo Oksala. — Jyväskylä : Gummerus, 1978.
96 p. ; 19 cm. — (Filosofisia yleisopintoja ; 4) Fi78-10
Bibliography: p. 95-96.
ISBN 9512015013 : Fmk18.00
1. Aesthetics—Addresses, essays, lectures. 2. Civilization—Philosophy—Addresses, essays, lectures. I. Title. II. Series.
BH39.O36 79-383682
79 MARC

París, Carlos.
El rapto de la cultura / Carlos París. — Madrid : Mañana, [1978?]
239 p. ; 20 cm. — (Colección La Nueva filosofía ; v. 2) Sp***
Includes bibliographical references.
ISBN 8474210437
1. Philosophy, Modern—20th century—Addresses, essays, lectures. 2. Civilization—Philosophy—Addresses, essays, lectures. I. Title. II. Series.
B804.P253 79-351198
79 MARC

Sawada, Tetsuo, 1931-
(Eien no jinrui de aru tame ni)
永遠の人類であるために：人類永遠化の論理と第三の生き方 / 沢田哲夫. — 東京：日新報道, 昭和50 [1975]
viii, 198 p. ; 19 cm. Ja 76-2709
¥980
1. Civilization—Philosophy—Addresses, essays, lectures. 2. Progress—Addresses, essays, lectures. I. Title.
CB19.S334 79-808958

Trottier, Pierre.
Un pays baroque / Pierre Trottier. — [Montréal] : La Presse, [c1979]
138 p. ; 21 cm. C***
ISBN 2-89043-004-9 : $7.25
1. Canada—Civilization—Addresses, essays, lectures. 2. Civilization—Philosophy—Addresses, essays, lectures. I. Title.
F1021.T76 971 79-373248
79 MARC

CIVILIZATION, AEGEAN

see also Civilization, Mycenaean; Minoans

Kōzu, Harushige, 1908-
(Girisha minzoku to bunka no seiritsu)
ギリシア民族と文化の成立 / 高津春繁著. — 東京：岩波書店, 昭和25 [1950]
4, 223, 15 p. ; 22 cm.
Includes bibliographical references and indexes.

1. Civilization, Aegean. 2. Civilization, Greek. I. Title.
DF220.K69 78-811463

CIVILIZATION, AMERICAN see Latin America—Civilization; United States—Civilization

CIVILIZATION, ANCIENT

see also Protohistory; Science, Ancient

Ancient civilizations / [editor, Frances M. Clapham, assistant editor, Abigail Frost ; contributors, Christopher Fagg ... et al. ; illustrators, Nigel Chamberlain ... et al.]. — New York : Warwick Press, [1978]
140 p. : some col. ill. ; 29 cm.
Bibliography: p. 139.
Includes index.
SUMMARY: Discusses civilizations of the ancient world, including Mesopotamia, Egypt, Greece, Persia, Phoenicia, Rome, India, and China.
ISBN 0-531-09091-4. ISBN 0-531-09073-6 lib. bdg.
1. Civilization, Ancient. [1. Civilization, Ancient] I. Clapham, Frances M. II. Fagg, Christopher. III. Chamberlain, Nigel.
CB311.A52 1978 930'.1 78-50676
79 MARC

Archeologia : culture e civiltà del passato nel mondo europeo ed extraeuropeo / [coordinamento scientifico di Leone Fasani]. — 1. ed. — Milano : A. Mondadori, 1978.
648 p. [14] leaves of plates : ill., tables ; 29 cm. — (Libri illustrati Mondadori) It78-Nov
Includes index.
L35000
1. Archaeology. 2. Civilization, Ancient. 3. Man, Prehistoric. I. Fasani, Leone.
CC165.A64 930'.1 79-346192
*79 MARC

Benesch, Kurt.
Rätsel der Vergangenheit : d. Abenteuer Archäologie heute / Kurt Benesch. — Gütersloh ; Berlin : Bertelsmann-Lexikon-Verlag ; Wien : Kremayr und Scheriau, 1977.
383 p. : col. ill., maps ; 23 cm. GFR77-A14
Includes bibliographies and index.
ISBN 3-570-02134-3 : DM34.00
1. Archaeology. 2. Civilization, Ancient. I. Title.
CC175.B46 930'.1 79-353446
79 MARC

(Bunka no kōryū)
文化の交流 / [著者 岩村忍 ... et al.]. — 東京：毎日新聞社, 1978.
264 p. : ill. ; 20 cm. — (古代文明の謎と発見 ; 4) Ja 78-15017
"参考図書解題": p. [257]-263.
¥1300

1. East and West. 2. Civilization, Ancient. 3. Civilization—History. I. Iwamura, Shinobu, 1905- II. Series: Kodai bummei no nazo to hakken ; 4.
CB151.B96 78-813807

Cazeau, Charles J
Exploring the unknown : great mysteries reexamined / Charles J. Cazeau and Stuart D. Scott, Jr. — New York : Plenum Press, c1979.
xi, 283 p. : ill. ; 24 cm.
Originally published under title: Great mysteries of the Earth.
Bibliography: p. 273-276.
Includes index.
ISBN 0-306-40210-6
1. Civilization, Ancient. 2. Curiosities and wonders. 3. Flying saucers. I. Scott, Stuart D., joint author. II. Title.
CB311.C37 001.9'42 78-27413
78 MARC

Cornell, James.
Lost lands and forgotten people / by James Cornell. — New York : Sterling Pub. Co., c1978.
224 p. : ill. ; 21 cm.
Includes index.
ISBN 0-8069-3926-5. ISBN 0-8069-3927-3 lib. bdg.
1. Civilization, Ancient. 2. Man, Prehistoric. I. Title.
CB311.C8 930 78-57795
79 MARC

[Goguet, Antoine Yves], 1716-1758.
The origin of laws, arts, and sciences, and their progress among the most ancient nations. — New York : AMS Press, c1976.
3 v. : ill. ; 23 cm. — (Language, man, and society)
By A. Y. Goguet, aided by A. C. Fugère. Cf. Pref.
Translation of De l'origine des loix, des arts, et des sciences, et leurs progrès chez les anciens peuples.
Reprint of the 1761 ed. printed by A. Donaldson and J. Reid, Edinburgh; with a new pref. by A. Cohn and H. D. Rosenbaum.
Bibliography: v. 3, p. 326-341.
ISBN 0-404-08230-0
1. Civilization, Ancient. 2. Law—History and criticism. 3. Art, Ancient. 4. Science, Ancient. I. Fugère, Alexandre Conrad, joint author. II. Title.
CB311.G613 1976 930 72-1630
75[r79]rev MARC

(Iseki no tabi)
遺跡の旅. — 東京：学習研究社, 昭和51- [1976-
v. : ill. (some col.) ; 23 cm. Ja 76-10078 (v. 1)
"NHK総合テレビで放映された「遺跡の旅—未来への遺産アルバムから」にもとづいて...編集"
CONTENTS: 1. エジプト・メソポタミア—2. シルクロード—3. マヤ・アステカ・インカ
¥980 per vol.

1. Civilization, Ancient. I. Gakushū Kenkyūsha, Tokyo.
CB311.I82 79-802269

James, Edward.
Our ancient heritage / [by] Edward James. — Dublin : Educational Company of Ireland, 1977.
viii, 280 p. : ill., facsims., maps, plan, ports. ; 21 x 22 cm. GB78-07561
Includes index.
£2.75
1. Civilization, Ancient. 2. Ireland—Civilization. I. Title.
CB311.J37 930 78-318882
78 MARC

Kondratov, Aleksandr Mikhaĭlovich.
Andres-Lemuriia? / Aleksandr Kondratov. — Leningrad : Gidrometeoizdat, 1978.
136 p., [4] leaves of plates : ill. ; 20 cm. USSR78
Bibliography: p. 134-[135]
0.50rub
1. Lemuria. 2. Civilization, Ancient. 3. Lost continents. 4. Indian Ocean region. I. Title.
GN751.K734 79-379296
79 MARC

Die Kulturen des Orients. — Frankfurt am Main,: Akademische Verlagsgesellschaft Athenaion, c1966.
103, 94, 76 p., 4 leaves of plates : ill. ; 30 cm. — (Handbuch der Kulturgeschichte (Frankfurt am Main) : 2. Abt., Kulturen der Völker)
Includes bibliographies and indexes.
CONTENTS: Schaeder, H. H. Die Kultur des vorderen Orients.— Wolf, W. Die Kultur Ägyptens. 2., neubearb. Aufl. — Kirfel, W. Die Kultur der Inder. 2., erw. Aufl.
1. Civilization, Oriental. 2. Civilization, Ancient. I. Schaeder, Hans Heinrich, 1896-1957. II. Wolf, Walther, 1900- III. Kirfel, Willibald, 1885-1964. IV. Series: Handbuch der Kulturgeschichte (Frankfurt am Main) : 2. Abt., Kulturen der Völker.
CB253.K8 930 79-341267
79 MARC

(Maboroshi no tairiku to shi no miyako)
幻の大陸と死の都 / [著者 八幡一郎 ... et al.]. — 東京：毎日新聞社, 昭和52 [1977]
288 p. : ill. ; 20 cm. — (古代文明の謎と発見 ; 2) Ja***
Bibliography: p. [279]-287.
¥1300

1. Civilization, Ancient. 2. Civilization—Miscellanea. I. Yawata, Ichirō. II. Series: Kodai bummei no nazo to hakken ; 2.
CB311.M16 78-813385

Matei, Horia.
Enigmele Terrei / Horia Matei ; [coperta de Gh. Marinescu]. — [București] : "Albatros", 1978.
367 p. : ill. ; 25 cm. R78-5163
Bibliography: p. [343]-346.
lei15.00
1. Civilization, Ancient. 2. Lost continents. 3. Antiquities. 4. Curiosities and wonders. I. Title.
CB311.M325 79-355894
*79 MARC

Mattson, Vernon W
The Dead Sea scrolls and other important discoveries / by Vernon W. Mattson, Jr. — Brandon, Fla. : Buried Records, c1978.
114 p. : ill. ; 23 cm.
Bibliography: p. 112-113.
$4.95
1. Civilization, Ancient. 2. Antiquities. I. Title.
CB311.M328 930 79-114600
00 MARC

(Miira wa kataru)
ミイラは語る / [著者 石田一良 ... et al.]. — 東京：毎日新聞社, 1978.
288 p. : ill. ; 20 cm. — (古代文明の謎と発見 ; 5) Ja 78-3982
Bibliography: p. 285-288.
¥1300

1. Mummies. 2. Civilization, Ancient. I. Ishida, Ichirō, 1913- II. Series: Kodai bummei no nazo to hakken ; 5.
GT3340.M53 78-811691

Moscati, Sabatino
Segreti del passato : alla scoperta del mondo sepolto / Sabatino Moscati. — 1. ed. — Milano : A. Mondadori, 1978.
311 p. : ill. ; 20 cm. — (Saggi ; 99) It**
1. Antiquities. 2. Civilization, Ancient. 3. Man, Prehistoric. I. Title.
CC165.M68 930 79-354749
79 MARC

CIVILIZATION, ANCIENT (Continued)

Mysteries of the ancient world / prepared by the Special Publications Division, National Geographic Society. — Washington : The Society, c1979.
223 p. : ill. ; 26 cm.
Bibliography: p. 220.
Includes index.
ISBN 0-87044-254-6 : $5.75
1. Civilization, Ancient. 2. Antiquities. I. National Geographic Society, Washington, D.C. Special Publications Division.
CB311.M92 001.9'4 77-93402
78 MARC

Nagle, D Brendan, 1936-
The ancient world : a social and cultural history / D. Brendan Nagle. — Englewood Cliffs, N.J. : Prentice-Hall, c1979.
xiv, 443 p. : ill. ; 24 cm.
Bibliography: p. 423-428.
Includes index.
ISBN 0-13-036400-2 : $12.95
1. Civilization, Ancient. I. Title.
CB311.N25 1979 930 77-11960
77 MARC

Nakajima, Ken'ichi, 1913-
(Kasen bummei no seitai shikan)
河川文明の生態史観 / 中島健一著. — 東京：校倉書房, 1977.
242 p. : ill. ; 22 cm. Ja 77-13649
Includes bibliographical references and indexes.
¥2800
1. Irrigation farming—History. 2. Reclamation of land—History. 3. Flood control—History. 4. Civilization, Ancient. I. Title.
SB112.N25 79-802108

Nippon Hōsō Kyōkai. Shuzaihan.
(Kizamareta jōnen)
刻まれた情念 / NHK取材斑. — 東京：日本放送出版協会, 昭和50 [1975]
352, vii, v p. : ill. ; 19 cm. — (NHK 未来への遺産取材記 ; 2) Ja***
Errata slip inserted.
Bibliography: p. i-vii (2d group)
¥950
1. Art, Prehistoric. 2. Art, Ancient. 3. Man, Prehistoric. 4. Civilization, Ancient. I. Title. II. Series: NHK mirai e no isan shuzaiki : 2.
N5310.N49 1975 79-800185

Noorbergen, Rene.
Secrets of the lost races : new discoveries of advanced technology in ancient civilizations / by Rene Noorbergen ; researched by Joey R. Jochmans. — Indianapolis : Bobbs-Merrill Co., c1977.
x, 228 p. : ill. ; 22 cm.
Bibliography: p. 211-215.
Includes index.
ISBN 0-672-52289-6
1. Man, Prehistoric. 2. Civilization, Ancient. 3. Interplanetary voyages. I. Title.
GN741.N66 001.9 77-76883
77 MARC

Ōta, Hidemichi, 1918-
(Higashi Chichūkai sekai)
東地中海世界：古代におけるオリエントとギリシア / 太田秀通著. — 東京：岩波書店, 1977.
x, 449, 16 p. ; 19 cm. — (世界歴史叢書) Ja 78-4783
Bibliography: p. 441-449.
Includes indexes.
¥1900
1. Civilization, Greek. 2. Civilization, Ancient. I. Title. II. Series: Sekai rekishi sōsho (Tokyo, Iwanami Shoten)
DF77.O68 78-812873

Ottolenghi, Aldo.
Orígenes de la civilización occidental / Aldo Ottolenghi. — [Buenos Aires] : Hachette, [1977?]
318 p. : ill. ; 20 cm.
1. Civilization, Occidental. 2. Civilization, Ancient. I. Title.
CB245.O83 78-109500
79 MARC

Pauwels, Louis, 1920 (Aug. 2)-
L'homme éternel / Louis Pauwels et Jacques Bergier. — Paris : Gallimard, 1974.
372 p. ; 18 cm. — (Their Embellissement de la vie ; 1) (Folio ; 356) F73-13760
Bibliography: p. [355]-367.
6.30F
1. Civilization, Ancient. 2. Man, Prehistoric. I. Bergier, Jacques, 1912- joint author. II. Title.
CB301.P18 930 74-186185
74[r79]rev MARC

[Poinsinet de Sivry, Louis], 1733-1804.
Origine des premières sociétés, des peuples, des sciences, des arts et des idiomes anciens et modernes. — Paris (76, rue Quincampoix, 75003) : Jobert, 1977.
612 p. ; 21 cm. — (Bibliotheca esoterica) F78-3599
By L. Poinsinet de Sivry. Cf. Bib. nat.; A. Cioranescu. Bibl. de la litt. fr. du 18e s.
Reprint of the 1769 ed. published by Lacombe, Paris.
180F
1. Protohistory. 2. Civilization, Ancient. I. Title. II. Series.
CB305.P64 1977 930 78-400730
79 MARC

(Seifuku to ensei)
征服と遠征 / [著者 橡川一朗 ... et al.]. — 東京：毎日新聞社, 1978.
272 p. : ill. ; 20 cm. — (古代文明の謎と発見 ; 10) Ja 78-13341
Bibliography: p. 265-272.
¥1300
1. Military history, Ancient. 2. Military history, Medieval. 3. Civilization, Ancient. I. Tochikawa, Ichirō, 1920- II. Series: Kodai bummei no nazo to hakken ; 10.
D25.S43 78-811755

Talbot, Laurence.
Le saut du puits / Laurence Talbot. — Paris : Centre des livres L.T., [c1978]
245 p. ; 19 cm. — (Collection Le Point sur l'i) F***
Includes bibliographical refrences.
35.00F
1. Mythology. 2. Civilization, Ancient. I. Title.
BL313.T32 291.1'3 78-399707
79 MARC

—ADDRESSES, ESSAYS, LECTURES

(Ajia bummei no michi)
アジア文明の道 / 森本哲郎編 ; [編集 NHKサービスセンター]. — 東京：集英社, 1978.
247 p. : ill. ; 22 cm. — (埋もれた古代都市 ; 1) (NHK文化シリーズ：歴史と文明) Ja 78-23377
"NHK教育テレビで放映された「歴史と文明 〈埋もれた古代都市〉」... [の中の]「殷の巨大王墓」「幻の都・長岡京」に、新しく「楼蘭の悲史」を加えた。"
¥1300
1. Cities and towns, Ruined extinct, etc.—Asia—Addresses, essays, lectures. 2. Asia—Civilization—Addresses, essays, lectures. 3. Civilization, Ancient—Addresses, essays, lectures. I. Morimoto, Tetsurō, 1925- II. NHK Sābisu Sentā. III. Rekishi to bummei (Television program) IV. Series: Uzumoreta kodai toshi ; 1. V. Series: NHK bunka shirizu : Rekishi to bummei.
D62.U97 vol. 1 79-801487
[DS62.2]

Chroniques des civilisations disparues : [anthologie d'articles extraits de la revue] Kadath / Patrick Ferryn et Ivan Verheyden. — Paris : R. Laffont, c1976.
298 p. : ill. ; 26 cm. — (Album Énigmes de l'univers) F***
Includes bibliographical references.
55.00F
1. Archaeology—Addresses, essays, lectures. 2. Civilization, Ancient—Addresses, essays, lectures. 3. Man, Prehistoric—Addresses, essays, lectures. I. Ferryn, Patrick. II. Verheyden, Ivan.
CC173.C48 930'.1 79-352868
79 MARC

Civilization : readings from Scientific American / with introductions by Brian M. Fagan. — San Francisco : W. H. Freeman, c1979.
158 p. : ill. (some col.) ; 30 cm.
CONTENTS: Adams. The origin of cities.—Kramer. The Sumerians.—Lamberg-Karlovsky and Lamberg-Karlovsky. An early city in Iran.—Millon. Teotihuacán.—Maddin, Muhly, and Wheeler. How the Iron Age began.—Piggott. The beginnings of wheeled transport.—Humphrey and Pedley. Roman Carthage.—Biddle. The archaeology of Winchester.—Hammond. The planning of a Maya ceremonial center.—Sjoberg. The origin and evolution of cities.—Brown. Human food production as a process in the biosphere.—Nef. An early energy crisis and its consequences.—Davis. The migrations of human populations.
Bibliography: p. [151]-153.
Includes index.
ISBN 0-7167-1024-2. ISBN 0-7167-1023-4 pbk.
1. Civilization, Ancient—Addresses, essays, lectures. 2. Cities and towns, Ancient—Addresses, essays, lectures. I. Fagan, Brian M. II. Scientific American.
CB311.C68 909 78-15780
78 MARC

Hamburg. Universität. Fachbereich Kulturgeschichte und Kulturkunde.
Kulturgeschichte und Kulturkunde. — Hamburg : Pressestelle der Universität Hamburg, 1976.
218 p. : ill. ; 30 cm. — (Uni hh Forschung ; 8) GFR***
Papers describing the activities of the various departments or seminars within the Fachbereich.
1. Civilization, Ancient—Addresses, essays, lectures. 2. Man, Prehistoric—Addresses, essays, lectures. 3. Antiquities—Addresses, essays, lectures. 4. Ethnology—Addresses, essays, lectures. 5. Hamburg. Universität. Fachbereich Kulturgeschichte und Kulturkunde—Addresses, essays, lectures. I. Title. II. Series.
CB311.H32 1976 930 79-337178
79 MARC

Historia de la antigüedad : selección de lecturas / Departamento de Historia General, Escuela de Historia, Facultad de Humanidades, Universidad de La Habana. — 2. ed. — La Habana : Instituto Cubano del Libro, 197
v. : ill. ; 26 cm.
1. Civilization, Ancient—Addresses, essays, lectures. 2. Mediterranean region—Civilization—Addresses, essays, lectures. I. Havana. Universidad. Departamento de Historia General.
CB311.H53 1975 930 78-110272
79 MARC

Marrou, Henri Irénée.
Christiana tempora : mélanges d'histoire, d'archéologie, d'épigraphie et de patristique / Henri-Irénée Marrou. — Roma : École française de Rome, 1978.
vi, 427 p., [1] fold. leaf of plates : ill. ; 25 cm. — (Collection de l'École française de Rome ; 35) It78-Oct
In French.
Reprint of previously published articles.
Includes bibliographical references.
1. Civilization, Ancient—Addresses, essays, lectures. 2. Inscriptions, Latin—Addresses, essays, lectures. 3. Inscriptions, Greek—Addresses, essays, lectures. 4. Christianity and culture—Early church, ca. 30-600—Addresses, essays, lectures. 5. Marrou, Henri Irénée—Addresses, essays, lectures. I. Title. II. Series: École française de Rome. Collection ; 35.
CB311.M3 78-393675
*79 MARC

El Mundo antiguo / selección, introd. y notas de José Luis Martínez. — 1. ed. — México : Secretaría de Educación Pública, 1976-
v. : ill. ; 22 cm. — (Panorama cultural) (Sep/documentos)
"Guía bibliográfica": v. 6, p. 415-448.
CONTENTS: 1. Mesopotamia. Egipto. India.—2. Grecia.—3. Hebreos y cristianos. Roma.—4. China. Japón.—6. America antigua.
1. Civilization, Ancient—History—Sources. 2. Civilization, Ancient—Addresses, essays, lectures. I. Martínez, José Luis. II. Series: Mexico. Secretaría de Educación Pública. Sep/Documentos.
CB311.M83 930 78-106778
79 MARC

Otárola, Alfredo J
Los orígenes y la filología : Tartessos-Ofir-Ophiussa. Las lenguas americanas y el vascuence : el sánscrito y el griego / Alfredo J. Otárola. — Buenos Aires : Casa Pardo, 1976.
281 p. ; 22 cm.
Includes bibliographical references.
CONTENTS: Tartessos.—Tartessos y otras antiguas civilizaciones.—África, cuna de la humanidad.—Bela Bella Bera.—Vero vascuence.—Nuevas reflecciones filológicas.—Breve resumen de sugerencias e hipotésis.—Historia y prehistoria.—Las lenguas americanas y el vascuence.—Algo más sobre algunos orígenes.—Juan de Garay y el enigma de su nacimiento.
1. Language and languages—Origin—Addresses, essays, lectures. 2. Civilization, Ancient—Addresses, essays, lectures. 3. Philology—Addresses, essays, lectures. I. Title.
P116.O86 409 79-366577
79 MARC

—COLLECTED WORKS

(Uzumoreta kodai toshi)
埋もれた古代都市 / 森本哲郎編 ; [編集 NHKサービスセンター]. — 東京：集英社, 1978-
v. : ill. ; 22 cm. — (NHK文化シリーズ：歴史と文明)
Ja 78-23377 (v. 1)
"昭和五十二年十一月から十二月末まで、八回にわたって、NHK教育テレビで放映された「歴史と文明〈埋もれた古代都市〉」を、新しく編集したもの"
¥1300 (v. 1)
1. Cities and towns, Ancient—Collected works. 2. Cities and towns, Ruined, extinct, etc.—Collected works. 3. Civilization, Ancient—Collected works. I. Morimoto, Tetsurō, 1925- II. NHK Sābisu Sentā. III. Rekishi to bummei (Television program) IV. Series: NHK bunka shirizu : Rekishi to bummei.
D62.U97 79-801538

—EXTRATERRESTRIAL INFLUENCES

Alexander, Marius.
Todos somos extraterrestres / [por] Marius Alexander. — Barcelona : Martínez Roca, D.L. 1978.
175 p. ; 18 cm. — (La Imposible verdad ; 4) Sp78-July
ISBN 8427004427 : 125ptas
1. Bible—Miscellanea. 2. Civilization, Ancient—Extraterrestial influences. I. Title.
BS534.A39 79-348465
79 MARC

Bernhardt, Karl-Heinz.
Sind wir Astronautenkinder? : die falschen Sensationen des Erich von Däniken im Lichte der biblischen Archäologie / Karl-Heinz Bernhardt. — 1. Aufl. — Berlin : Union Verlag, c1978.
127 p. ; 20 cm. — (Fakten, Argumente) GDR***
5.20M
1. Civilization, Ancient—Extraterrestrial influences. 2. Däniken, Erich von, 1935- 3. Bible—Antiquities. I. Title.
CB156.B48 79-370401
79 MARC

CIVILIZATION, ANCIENT
—EXTRATERRESTRIAL INFLUENCES (Continued)

Gladden, Lee.
Heirs of the gods : a space age interpretation of the Bible / by Lee Gladden and Vivianne Cervantes Gladden. — 1st ed. — New York : Rawson, Wade Publishers, c1978.
xi, 324 p. ; 22 cm.
Includes bibliographical references and index.
ISBN 0-89256-071-1 : $10.95
1. Bible—Miscellanea. 2. Civilization, Ancient—Extraterrestrial influences. 3. Life on other planets. 4. Religion and astronautics. I. Gladden, Vivianne Cervantes, joint author. II. Title.
BS534.G56 1978 236 78-53852
78 MARC

Lugo, Francisco Aniceto.
Desconocidos senderos de la vida : la dimensión total del universo / Francisco Aniceto Lugo. — 2. ed., 1. de Editorial Diana. — México : Editorial Diana, 1978.
163 p. ; 23 cm. — (His Las civilizaciones ignoradas ; t. 1)
1. Civilization, Ancient—Extraterrestrial influences. 2. Life on other planets. I. Title. II. Series.
CB156.L83 1978 001.9'4 79-109826
79 MARC

Lugo, Francisco Aniceto.
Emisarios de otros mundos : siempre hemos estado acompañados / Francisco Aniceto Lugo. — 2. ed., 1. de Editorial Diana. — México : Editorial Diana, 1978.
176 p. ; 23 cm. — (His Las civilizaciones ignoradas ; t. 3)
1. Civilization, Ancient—Extraterrestrial influences. I. Title. II. Series.
CB156.L84 1978 001.9'4 79-109832
79 MARC

Mitra, Birendra.
Dānikenatattva o Mahābhāratera svargadebatā / Bīrendra Mitra. — Kalikātā : Beṅgala Pābaliśārsa, 1978.
155 p. ; 22 cm.
In Bengali.
Rs10.00
1. Gods, Hindu. 2. Mahābhārata. Bhagavadgītā. 3. Däniken, Erich von, 1935- . 4. Civilization, Ancient—Extraterrestrial influences. I. Title.
BL1216.M57 79-901664
79 MARC

Pérez, Jorge Omar, 1943-
La historia antes de la historia / Jorge Omar Pérez. — Barcelona : A.T.E., 1977.
244 p., [4] leaves of plates : ill. ; 19 cm. Sp***
Bibliography: p. 233-244.
ISBN 8474420172
1. Civilization, Ancient—Extraterrestrial influences. 2. Lost continents. 3. Man, Prehistoric. I. Title.
CB156.P47 78-379662
78 MARC

Salló, Ervin.
Zöld emberkék, tollas kígyók, tüzes szekerek? : kritikai megjegyzések a paleoasztronautikai értelmezések elveiről és módszereiről / Salló Ervin, Mandics György. — Bukarest : Albatrosz, 1977.
171 p. ; 20 cm. R77-4973
Bibliography: p. 147-[148]
lei6.25
1. Civilization, Ancient—Extraterrestrial influences. I. Mandics, György, joint author. II. Title.
CB156.S24 78-370447
•78 MARC

Sasaki, Keigo, 1939-
(Uchū-kōkogaku nyūmon)
宇宙考古学入門 / 佐々木啓悟. — 東京：大陸書房, 1978.
220 p. : ill. ; 19 cm. Ja 79-1898
Bibliography: p. 215-217.
¥1300
1. Civilization, Ancient—Extraterrestial influences. 2. Astronautics and civilization. I. Title.
CB156.S27 79-808814

Schmitz, Emil Heinz.
Beweisnot : Glanz und Elend der Astronautengötter : das Ende einer Legende / Emil-Heinz Schmitz. — Genf : Ariston, c1978.
349 p. : ill. ; 23 cm. Sw78-7477
Includes bibliographical references.
ISBN 3-7205-1171-5 : 29.00F
1. Civilization, Ancient—Extraterrestrial influences. 2. Life on other planets. 3. Flying saucers. I. Title.
CB156.S35 78-384408
78 MARC

Siddhantashastree, Rabindra Kumar, 1918-
Debatā, pṛthibī, o mahābiśva / Rabīndra Kumāra Siddhāntaśāstrī. — Kalikātā : Saṃskṛta Pustaka Bhāṇḍāra, 1385 [1978]
125 p. ; 22 cm.
In Bengali.
Rs10.00
1. Gods, Hindu—Miscellanea. 2. Däniken, Erich von, 1935- . 3. Civilization, Ancient—Extraterrestial influences. 4. Hinduism—Sacred books—Miscellanea. I. Title.
BL1216.S52 79-901658
79 MARC

Thor, Antonio Jorge, 1950-
Amazônia : símbolos, enígmas e astronautas / Antonio Jorge Thor e Ararê. — Belém : Escola Salesiana do Trabalho, Oficinas de Artes Gráficas-Offset, 1977.
162 p., [45] leaves of plates : ill. ; 22 cm.
Includes bibliographical references.
Cr$90.00
1. Indians of South America—Amazon Valley—Antiquities. 2. Civilization, Ancient—Extraterrestrial influences. 3. Amazon Valley—Antiquities. 4. Brazil—Antiquities. I. Ararê, joint author. II. Title.
F2519.1.A6T47 981'.1 78-395799
79 MARC

Wegner, Willy.
Däniken i søgelyset / Willy Wegner. — København : FUFOS : i samarbejde med Dansk Ufo Center, 1977.
64 p. : ill ; 24 cm. D77-34
Bibliography: p. 60-64.
ISBN 8798003925 : kr22.50
1. Civilization, Ancient—Extraterrestrial influences. 2. Däniken, Erich von, 1935- . I. Title.
CB156.W43 78-379062
78 MARC

Wilson, Clifford A
War of the chariots / Clifford Wilson. — San Diego, Calif. : Master Books, c1978.
192 p. : ill. ; 21 cm.
Includes bibliographical references and index.
ISBN 0-89051-050-4 : $2.95
1. Civilization, Ancient—Extraterrestrial influences. 2. Life on other planets. 3. Däniken, Erich von, 1935- . I. Title.
CB156.W55 001.9'42 78-55211
79 MARC

Wilson, Edgar D
MEVS*, creator of the pyramids / by Edgar D. Wilson. — West Covina, Ca. : Astro-Research Pub., c1979.
144 p. : ill. ; 24 cm.
"*A visitor from outer space" at foot of t.p.
Bibliography: p. 143-144.
1. Pyramids—Egypt—Giza—Miscellanea. 2. Giza—Tombs—Miscellanea. 3. Egypt—Antiquities—Miscellanea. 4. Civilization, Ancient—Extraterrestrial influences. I. Title.
DT63.5.W74 001.9'4 79-50808
00 MARC

—EXTRATERRESTRIAL INFLUENCES—ADDRESSES, ESSAYS, LECTURES

Le Livre des anciens astronautes / [textes choisis et présentés par] Jacques Bergier et Georges H. Gallet. — Paris : A. Michel, 1977.
268 p. : ill. ; 21 cm. — (Les Chemins de l'impossible) F78-7867
Includes bibliographical references.
ISBN 2-226-00549-8 : 39.00F
1. Civilization, Ancient—Extraterrestrial influences—Addresses, essays, lectures. I. Bergier, Jacques, 1912- II. Gallet, Georges H.
CB156.L57 001.9'42 78-386025
79 MARC

—HISTORY—SOURCES

El Mundo antiguo / selección, introd. y notas de José Luis Martínez. — 1. ed. — México : Secretaría de Educación Pública, 1976-
v. : ill. ; 22 cm. — (Panorama cultural) (Sep/documentos)
"Guía bibliográfica": v. 6, p. 415-448.
CONTENTS: 1. Mesopotamia. Egipto. India.—2. Grecia.—3. Hebreos y cristianos. Roma.—4. China. Japón.—6. America antigua.
1. Civilization, Ancient—History—Sources. 2. Civilization, Ancient—Addresses, essays, lectures. I. Martínez, José Luis. II. Series: Mexico. Secretaría de Educación Pública. Sep/Documentos.
CB311.M83 930 78-106778
79 MARC

—PERIODICALS

Cahiers des études anciennes. 1-
1972-
Montréal, Presses de l'Université du Québec.
v. illus. 28 cm.
1. Civilization, Ancient—Periodicals.
CB311.C24 913'.38'0305 79-641175
MARC-S

Oikumene. 1-
Budapest, Akadémiai Kiadó, 1976-
v. 25 cm. annual.
English, French, German, or Russian.
"Studia ad historiam antiquam classicam et orientalem spectantia."
1. Civilization, Greek—Periodicals. 2. Rome—Civilization—Periodicals. 3. Civilization, Ancient—Periodicals.
DE1.O37 77-647530
MARC-S

—PERIODICALS—BIBLIOGRAPHY—UNION LISTS

Swift, Catherine Gordon, 1935-
Check-list of classical periodicals in New Zealand libraries / compiled by Catherine G. Swift. — [4th ed.]. — Dunedin : University of Otago, 1976.
31 p. ; 30 cm. NZ76-3
1. Civilization, Ancient—Periodicals—Bibliography—Union lists. 2. Catalogs, Union—New Zealand. I. Title.
Z5579.S94 1976 016.93'005 78-321386
[CB311] 79 MARC

CIVILIZATION, ANGLO-SAXON

Martin-Clarke, Daisy Elisabeth, 1891-
Culture in early Anglo-Saxon England / D. Elizabeth Martin-Clarke. — New York : Arno Press, 1979, c1947.
xi, 100 p., [10] leaves of plates : ill. ; 23 cm. — (Johns Hopkins University Press reprints)
Reprint of the ed. published by Johns Hopkins Press, Baltimore.
ISBN 0-405-10615-7 : $14.00
1. England—Antiquities. 2. Civilizaton, Anglo-Saxon. I. Title.
DA155.M3 1979 936.2 78-19259
78 MARC

—ADDRESSES, ESSAYS, LECTURES.

Bede and Anglo-Saxon England : papers in honour of the 1300th anniversary of the birth of Bede, given at Cornell University in 1973 and 1974 / edited by Robert T. Farrell. — London : British Archaeological Reports, 1978.
172 p. : ill. ; 30 cm. — (British archaeological reports ; 46) GB***
Includes bibliographical references.
ISBN 0-86054-005-7 : £3.50
1. Beda Venerabilis, 673-735—Addresses, essays, lectures. 2. Beda Venerabilis, 673-735—Contemporary England—Addresses, essays, lectures. 3. Civilization, Anglo-Saxon—Addresses, essays, lectures.. 4. Authors, Anglo-Saxon—Biography—Addresses, essays, lectures. I. Farrell, Robert T. II. Title. III. Series.
PR1578.B4 942.01'5 78-319454
79 MARC

—STUDY AND TEACHING—ADDRESSES, ESSAYS, LECTURES

Whitelock, Dorothy.
Changing currents in Anglo-Saxon studies : an inaugural lecture / by Dorothy Whitelock. — Norwood, Pa. : Norwood Editions, 1978.
31 p. ; 18 cm.
Reprint of the 1958 ed. published by Cambridge University Press, Cambridge, Eng.
Includes bibliographical references.
ISBN 0-8482-6959-4 lib. bdg. : $7.50
1. Civilization, Anglo-Saxon—Study and teaching—Addresses, essays, lectures. I. Title.
[CB216.W5 1978] 901 78-8293
78 MARC

CIVILIZATION, ARAB

see also Folk-songs, Brazilian--Arab influences

'Āmil, Mahdī.
(Azmat al-ḥaḍārah al-'Arabīyah am azmat al-burjuwāzīyāt al-'Arabīyah?)
ازمة الحضارة العربية ام ازمة البرجوازيات العربية؟ / مهدي عامل. — بيروت : دار الفارابي، 1974.
239 p. ; 19 cm.
1. Civilization, Arab. I. Title.
DS36.88.A44 78-970916

CIVILIZATION, ARAB (Continued)

Bāzāmah, Muḥammad Muṣṭafá.
(al-Thawrah al-thaqāfīyah al-'Arabīyah)
الثورة الثقافية العربيــة / تاليف محمد مصطفــى بازامه. —
الطبعة 2. — بيروت : مؤسسة ناصر للثقافة، 1975.
300 p. ; 18 cm. — (خزانة الفكر العربي)
Includes some speeches by M. al-Qadhdhāfī (p. 129-298)
1. Civilization, Arab. 2. Libya—Politics and government. I. al-Qadhdhāfī, Mu'ammar. II. Title. III. Series: Khizānat al-fikr al-'Arabī.
DS36.88.B39 1975 79-970092

Fāyid, 'Abd al-Ḥamīd.
(al-Mar'ah wa-atharuhā fī al-ḥayāh al-'Arabīyah)
المراة واثرها في الحياة العربية : تحية الى المراة العربية في عام
المراة العالمي، ١٩٧٥ / بقلم عبد الحميـد فايـد. — ₍بيروت₎ :
جامعة بيروت العربية، 1977.
122 p. ; 24 cm.
Summary in French.
Bibliography : p. 119-120.
1. Women—Arab countries. 2. Civilization, Arab. I. Title.
HQ1784.F38 78-970907

'Uzayzāt, Yūsuf Salīm al-Shuwayḥāt.
(Dawr al-'Arab fī thaqāfat al-'ālam wa-ḥaḍāratih)
دور العرب في ثقافــة العالم وحضارته / ₍يوســف سليم
الشويحات ـ العزيزات₎. — ₍s. l. : s. n., 197-₎
208 p. : ill. ; 24 cm.
Includes bibliographical references.
1. Near East—Civilization—To 622. 2. Civilization, Arab. I. Title.
DS57.U95 79-970943

—20TH CENTURY

Zurayq, Qusṭanṭīn, 1909-
(Naḥnu wa-al-mustaqbal)
نحن والمستقبل / قسطنطين زريق. — الطبعة 1. — بيروت :
دار العلم للملايين، 1977.
451 p. ; 24 cm.
Includes bibliographical references.
1. Civilization, Arab—20th century. 2. Forecasting. I. Title.
DS36.88.Z87 79-970848

—ADDRESSES, ESSAYS, LECTURES

Khālid, Manṣūr, 1931-
The socio-cultural determinants of Arab diplomacy / by Mansour Khalid. — Washington, D.C. : Embassy of the Democratic Republic of the Sudan, Office of the Cultural Counsellor, ₍1974?₎
13 p. ; 28 cm. — (Selections from Sudanese literature ; 2)
Cover title.
Includes bibliographical references.
1. Arab countries—Foreign relations—Addresses, essays, lectures. 2. Civilization, Arab—Addresses, essays, lectures. I. Title. II. Series.
DS62.85 1974.K46 327'.2'09174927 78-321228
79 MARC

—CONGRESSES

Nadwat Azmat al-Taṭawwur al-Ḥaḍārī fī al-Waṭan al-'Arabī, Kuwait, 1974.
(Azmat al-taṭawwur al-ḥaḍārī fī al-waṭan al-'Arabī)
ازمة التطور الحضاري في الوطن العربي : وقائع ندوة الكويت
ما بين ٧ـ١٢ نيســان، ابريل ١٩٧٤ / اعداد واشراف شاكر
مصطفى ؛ لوحة الغلاف لعبد الهادى الجزار. — الطبعـة 1. —
الكويت : جمعية الخريجين، جامعة الكويت، 1975.
598 p. ; 29 cm.
Includes bibliographical references.
1. Civilization, Arab—Congresses. I. Muṣṭafá, Shākir. II. Jāmi'at al-Kuwayt. Jam'īyat al-Khirrījīn. III. Title.
DS36.88.N32 1974 79-970269

—MISCELLANIA

Gabrieli, Francesco, 1904-
Testimonianze arabe ed europee / Francesco Gabrieli. — Bari : Dedalo libri, 1976.
155 p. ; 21 cm. — (Saggi ; 23) It78-Jan
1. Gabrieli, Francesco, 1904- . 2. Civilization, Arab—Miscellania. 3. Orientalists—Biography. I. Title.
DS37.5.G3A33 79-340799
•79 MARC

—PERIODICALS

(Iḥyā' al-turāth al-'Arabī al-Islāmī)
احياء التراث العربى الاسلامى.
بغداد ₍جمعية احياء التراث العربى الاسلامى₎
v. ill. 24 cm. quarterly.
Spine title, : Majallat iḥyā' al-turāth al-'Arabī al-Islāmī.
1. Civilization, Arab—Periodicals. 2. Civilization, Islamic—Periodicals. I. Jam'īyat Iḥyā' al-Turāth al-'Arabī al-Islāmī, Bagdad. II. Title: Majallat iḥyā' al-turāth al-'Arabī al-Islāmī.
DS36.8.I 44 78-961053
(MARC-S)

CIVILIZATION, ASSYRO-BABYLONIAN

see also Cuneiform inscriptions; Cuneiform writing

CIVILIZATION, BAROQUE

Assunto, Rosario.
Infinita contemplazione : gusto e filosofia dell'Europa barocca / Rosario Assunto. — Napoli : Società editrice napoletana, ₍1979₎
194 p., ₍21₎ leaves of plates : ill. ; 24 cm. — (Studi e testi di storia e critica dell'arte ; 7) It79-June
Includes bibliographical references and index.
L12000
1. Philosophy, Modern—17th century. 2. Philosophy, Modern—18th century. 3. Aesthetics, Modern—17th century. 4. Aesthetics, Modern—18th century. 5. Civilization, Baroque. I. Title.
B801.A83 79-377858
•79 MARC

—COLLECTED WORKS

Fagiolo dell'Arco, Maurizio, 1939-
L'effimero barocco : strutture della festa nella Roma del '600 / di Maurizio Fagiolo Dell'Arco, Silvia Carandini. — Roma : Bulzoni, 1977-
v. : ill. ; 21 cm. — (Biblioteca di storia dell'arte ; 10) It77-Nov
Bibliography: v. 1, p. xxix-xxxi.
CONTENTS: 1. Catalogo.

L20000 (v. 1)
1. Festivals—Italy—Rome (City)—History—Collected works. 2. Rome (City)—Carnival—History—Collected works. 3. Civilization, Baroque—Collected works. 4. Rome (City)—Social life and customs—Collected works. 5. Decoration and ornament, Baroque—Collected works. I. Carandini, Silvia, joint author. II. Title.
GT4852.R6F28 78-392383
•79 MARC

CIVILIZATION, BUDDHIST

(Bukkyō to bunka)
佛教と文化 / ₍編集者　浄土真宗本願寺派学校連合会₎. — 京都：百華苑, 昭和36 ₍1961₎
2, 115 p. : ill. ; 21 cm.

1. Buddhism—Japan—History. 2. Civilization, Buddhist. I. Jōdo Shinshū Honganjiha Gakkō Rengōkai.
BQ676.B84 78-813520

—CONGRESSES

International Seminar on Buddhism's Contribution to World Culture and Civilization, New Delhi, India, 1977.
Buddhism's contribution to the world culture and civilization / edited by Ananda W. P. Guruge, D. C. Ahir. — New Delhi : Maha Bodhi Society of India, ₍1977?₎
219 p., ₍8₎ leaves of plates : ill. ; 25 cm.
"Organised by the Maha Bodhi Society of India, with the co-operation of the Government of India, on the occasion of the 2600th birth anniversary of Lord Buddha."
Includes bibliographical references.
1. Buddhism—Congresses. 2. Buddhism—Influence—Congresses. 3. Civilization, Buddhist—Congresses. I. Guruge, Ananda W. P., 1928- II. Ahir, D. C. III. Maha Bodhi Society of India. IV. Title.
BQ100.I57 1977 294.3 78-905646
79 MARC

CIVILIZATION, CARLOVINGIAN
see France—Civilization—700-1000

CIVILIZATION, CELTIC

Filip, Jan.
Celtic civilization and its heritage / Jan Filip. — 2d rev. ed. — Praha : Academia, 1976.
231 p., ₍20₎ leaves of plates : ill. ; 21 cm. — (New horizons) Cz***
Translation of Keltská civilisace a její dědictví.
Bibliography: p. ₍200₎-203.
Includes indexes.
Kčs45.00
1. Civilization, Celtic. 2. Celts—Antiquities. I. Title.
CB206.F513 1976 936.4 79-348652
79 MARC

Markale, Jean.
Celtic civilization / J. Markale. — London ; New York : Gordon & Cremonesi, c1978.
320 p., ₍8₎ leaves of plates : ill. ; 26 cm. GB***
Translation of Les Celtes et la civilisation celtique.
Bibliography: p. ₍303₎-305.
Includes index.
ISBN 0-86033-045-1 : £12.50 ($24.95)
1. Celts. 2. Civilization, Celtic. I. Title.
D70.M3713 940'.04'916 78-40395
79 MARC

CIVILIZATION, CHRISTIAN

see also Chrstianity and culture

Senior, John, 1923-
The death of Christian culture / John Senior. — New Rochelle, N.Y. : Arlington House, c1978.
185 p. ; 24 cm.
Includes bibliographical references and index.
ISBN 0-87000-416-6 : $10.00
1. Civilization, Christian. 2. Christianity—20th century. I. Title.
BR115.C5S46 261.5 78-16154
78 MARC

CIVILIZATION, COMPARATIVE
see Comparative civilization

CIVILIZATION, DRAVIDIAN

see also Dravidians

CIVILIZATION, ETRUSCAN see Etruscans

CIVILIZATION, GERMANIC

Schlette, Friedrich.
Germanen zwischen Thorsberg und Ravenna; Kulturgeschichte der Germanen bis zum Ausgang der Völkerwanderung. ₍1. Aufl.₎ Leipzig, Urania-Verlag ₍1972₎
263 p. illus. (part col.) 25 cm. 18.00M GDNB***
Bibliography: p. 258-259.
1. Civilization, Germanic. 2. Art, Germanic. I. Title.
CB213.S26 1972 72-368108
78₍r78₎rev2 MARC

CIVILIZATION, GRECO-ROMAN

—CONGRESSES

Humanismus und Menschenbild im Orient und in der Antike / hrsg. von der Sektion Orient- und Altertumswissenschaften der Martin-Luther-Universität Halle-Wittenberg. — Halle (Saale) : Martin-Luther-Universität Halle-Wittenberg, Abt. Wissenschaftspublizistik, 1977.
2 v. : ill. ; 21-30 cm. — (Wissenschaftliche Beiträge - Martin-Luther-Universität Halle-Wittenberg ; 1977, 28 (I 2)) (Kongress- und Tagungsberichte der Martin-Luther-Universität Halle-Wittenberg) GDR***
Conference was held Nov. 1973.
Includes bibliographical references.
CONTENTS: Bd. ₍1₎ Konferenzvorträge.—Bd. ₍2₎ Bildtafeln.
₍B105.H8₎
1. Humanism—History—Congresses. 2. Civilization, Oriental—Congresses. 3. Civilization, Greco-Roman—Congresses. I. Halle. Universität. Sektion Orient- und Altertumswissenschaften. II. Series: Halle. Universität. Wissenschaftliche Beiträge ; 1977, 28. III. Series: Halle. Universität. Kongress- und Tagungsberichte der Martin-Luther-Universität Halle-Wittenberg.
AS182.H125 1977/28 78-384075
₍B105.H8₎ 79₍r79₎rev MARC

CIVILIZATION, GRECO-ROMAN (Continued)

—LITERARY COLLECTIONS

Jens, Walter, 1923-
Zur Antike / Walter Jens. — München : Kindler, 1978. GFR78-A
483 p. ; 21 cm.
Bibliography of the author's works: p. [481]-483.
CONTENTS: Die griechische Literatur.—Strukturgesetze der frühen griechischen Tragödie.—Euripides.—Verkleidete Götter.—Lessing und die Antike.—Der Gott der Diebe und sein Dichter.—Die Götter sind sterblich.—Das Testament des Odysseus.—Die Verschwörung.—Der tödliche Schlag.—Sophokles und Brecht.—Der Ajas des Sophokles.
ISBN 3-463-00730-4
1. Civilization, Greco-Roman—Literary collections. I. Title.
PT2619.E46Z455 880 79-370443
79 MARC

CIVILIZATION, GREEK

see also Civilization, Greco-Roman; Civilization, Homeric; Egypt—Civilization—Greek influences; Great Britain—Civilization—Greek influences; Hellenism

Agard, Walter Raymond, 1894-
The Greek mind / Walter R. Agard. — Huntington, N.Y. : R. E. Krieger Pub. Co., 1979, c1957.
190 p. ; 19 cm. — (The Anvil series)
Reprint of the ed. published by D. Van Nostrand, Princeton, N.J., as an Anvil original, no. 17.
Bibliography: p. 185-186.
Includes index.
ISBN 0-88275-811-X
1. Civilization, Greek. I. Title.
[DF77.A52 1979] 938 78-25755
78 MARC

Aspetti dell'opinione pubblica nel mondo antico / autori vari ; a cura di Marta Sordi. — Milano : Vita e pensiero, 1978.
vi, 216 p. : ill. ; 22 cm. — (Scienze storiche ; 17) (Contributi dell'Istituto di storia antica ; v. 5) It***
Includes bibliographical references.
L13000
1. Public opinion—History. 2. Civilization, Greek. 3. Rome—Civilization. I. Sordi, Marta. II. Series. III. Series: Milan. Università cattolica del Sacro Cuore. Istituto di storia antica. Contributi dell'Istituto di storia antica ; v. 5.
DE3.M48a vol. 5 79-352111
[HM261] 79 MARC

Barev, Ivan.
(Drevniíat svíat i klasicheskata demokratsiía)
Древният свят и класическата демокрация / Иван Барев. — Paris : L'Avenir, [1977].
189 p. ; 20 cm. F***
25.00F ($6.00U.S.)
1. Civilization, Greek. 2. Greece—History. I. Title.
DF77.B26 79-411466

Dickinson, Goldsworthy Lowes, 1862-1932.
The Greek view of life / by G. Lowes Dickinson ; pref. by E. M. Forster. — Westport, Conn. : Greenwood Press, 1979.
261 p. ; 22 cm.
Reprint of the 1958 ed. published by University of Michigan Press, Ann Arbor, which was issued as AA22 of Ann Arbor paperbacks.
Includes index.
ISBN 0-313-21195-7
1. Civilization, Greek. I. Title.
[DF77.D54 1979] 938 78-12661
78 MARC

Dreizehnter, Alois.
Die rhetorische Zahl : quellenkrit. Unters. anhand d. Zahlen 70 u. 700 / von Alois Dreizehnter. — 1. Aufl. — München : Beck, 1978.
113 p. ; 24 cm. — (Zetemata, Monographien zur klassischen Altertumswissenschaft ; Heft 73) GFR78-A
Includes bibliographical references and index.
ISBN 3-406-05163-4
1. Seventy (The number) 2. Seven hundred (The number) 3. Civilization, Greek. 4. Rome—Civilization. I. Title. II. Series.
BF1623.P9D73 79-358129
79 MARC

Elliger, Winfred.
Paulus in Griechenland : Philippi, Thessaloniki, Athen, Korinth / Winfried Elliger. — Stuttgart : Verlag Katholisches Bibelwerk, c1978.
276 p. ; 21 cm. — (Stuttgarter Bibelstudien ; 92/93) GFR***
"Bildteil" ([32] p.) inserted in pocket.
Bibliography: p. 252-259.
Includes indexes.
ISBN 3-460-03921-3
1. Civilization, Greek. 2. Paul, Saint, Apostle—Journeys. I. Title. II. Series.
DF78.E38 938'.09 79-358987
79 MARC

Frasnedi, Fabrizio.
Polis : etica e prassi politica nel pensiero antico / Fabrizio Frasnedi. — Milano : A.P.E. Mursia, c1978.
263 p. ; 23 cm. — (Prospettive filosofiche ; 2) It***
Bibliography: p. 48-51.
L4000
1. Civilization, Greek. 2. Greece—Politics and government. I. Title.
DF77.F69 938 79-382874
79 MARC

Gasser-Coze, Françoise.
La Grecia del Partenón / [por] Françoise Gasser-Coze. — Madrid : Círculo de Amigos de la Historia, D.L. 1977.
269 p., 1 leaf [8] leaves of plates : ill. ; 18 cm. — (Grandes civilizaciones desaparecidas ; 15) Sp77-Dec
ISBN 8422501481
1. Civilization, Greek. I. Title. II. Series.
DF77.G37 78-384326
78 MARC

Gentile, Panfilo.
Il genio della Grecia / Panfilo Gentile. — Roma : Faro, 1947.
186 p. ; 22 cm. — (Collezione storica ; 4)
1. Civilization, Greek. I. Title.
DF77.G45 1947 938 79-374063
79 MARC

Hara, Zuien, 1894–
(Girisha shi kenkyū yoteki)
ギリシア史研究余滴 / 原随園著. — 京都：同朋舎, 1976.
444 p. ; 22 cm. Ja 77–16603
¥5800

1. Civilization, Greek. 2. Historiography. 3. Political science—Greece—History. I. Title.
DF78.H37 78–809498

Hooper, Finley Allison, 1922-
Greek realities : life and thought in ancient Greece / Finley Hooper. — Detroit : Wayne State University Press, 1978.
xiv, 464 p. : ill. ; 23 cm.
Reprint of the 1967 ed. published by Scribner, New York.
Bibliography: p. [443]-452.
Includes index.
ISBN 0-8143-1596-8. ISBN 0-8143-1597-6.
1. Civilization, Greek. 2. Greece—History. I. Title.
[DF77.H67 1978] 938 78-19547
78 MARC

Knell, Heiner.
Perikleische Baukunst / Heiner Knell. — Darmstadt : Wissenschaftliche Buchgesellschaft, [Abt. Verl.], 1979. GFR79-A
v, 113 p., [6] leaves of plates : ill. ; 20 cm.
Includes bibliographical references.
ISBN 3-534-08019-X : DM27.50
1. Architecture—Greece. 2. Pericles, 499-429 B.C. 3. Greece—Civilization. I. Title.
NA270.K57 79-388436
79 MARC

Kōzu, Harushige, 1908–
(Girisha minzoku to bunka no seiritsu)
ギリシア民族と文化の成立 / 高津春繁著. — 東京：岩波書店, 昭和25 [1950]
4, 223, 15 p. ; 22 cm.
Includes bibliographical references and indexes.

1. Civilization, Aegean. 2. Civilization, Greek. I. Title.
DF220.K69 78–811463

Origini e sviluppo della città / di Giovanni Pugliese Carratelli ... [et al.]. — Milano : Bompiani, 1978.
2 v. : ill. ; 23 cm. — (Storia e civiltà dei greci ; 1-2) It79-Apr
Includes bibliographical references and index.
CONTENTS: [1] Il medioevo greco.—[2] L'arcaismo.
L1200 per vol.
1. Civilization, Greek. 2. Civilization, Homeric. 3. Civilization, Mycenaean. I. Pugliese Carratelli, Giovanni, 1911- . II. Series.
DF77.S72 vol. 1-2 938'.008 s 79-380718
[DF220] *79 MARC

Ōta, Hidemichi, 1918–
(Higashi Chichūkai sekai)
東地中海世界：古代におけるオリエントとギリシア / 太田秀通著. — 東京：岩波書店, 1977.
x, 449, 16 p. ; 19 cm. — (世界歴史叢書) Ja 78–4783
Bibliography: p. 441–449.
Includes indexes.
¥1900
1. Civilization, Greek. 2. Civilization, Ancient. I. Title. II. Series: Sekai rekishi sōsho (Tokyo, Iwanami Shoten)
DF77.O68 78–812873

Parandowski, Jan.
Z antycznego świata / Jan Parandowski. — Wyd. 3. — Warszawa : Iskry, 1978 (Warsz : DSP)
155, [3] p. ; 20 cm. P78-4540
zł25.00
1. Civilization, Greek. 2. Rome—Civilization. I. Title.
DE59.P37 1978 79-350716
79 MARC

Pavan, Massimiliano.
L'antichità classica / Massimiliano Pavan. — Roma : Studium, 1977.
xiii, 542 p. : maps ; 21 cm. — (La Cultura ; 12) It78-July
Bibliography: p. [425]-495.
Includes index.
L16000
1. Civilization, Greek. 2. Rome—Civilization. I. Title.
DE59.P38 938 78-387893
*79 MARC

Préaux, Claire.
Le monde hellénistique / Claire Préaux. — 1. éd. — Paris : Presses universitaires de France, c1978-
v. ; 19 cm. — (Nouvelle Clio ; 6) F***
Bibliography: v. 1, p. 14-76.
CONTENTS: t. 1. La Grèce et l'Orient de la mort d'Alexandre à la conquête romaine de la Grèce, 323-146 av. J.-C.

ISBN 2-13-035263-4 (v. 1) : 85.00F (v. 1)
1. Hellenism. 2. Civilization, Greek. I. Title.
DF78.P73 938'.08 79-341343
79 MARC

Stumpfe, Ortrud.
Die Heroen Griechenlands : Einübung d. Denkens von Theseus bis Odysseus / Ortrud Stumpfe. — Münster : Aschendorff, 1978.
282 p. : ill., map ; 24 cm. GFR78-A
Bibliography: p. 281-282.
ISBN 3-402-05803-0 : DM48.00
1. Mythology, Greek. 2. Heroes. 3. Civilization, Greek. I. Title.
BL785.S79 78-376773
78 MARC

Thomson, George Derwent.
Studies in ancient Greek society / by George Thomson. — 3d ed. with additions and corrections. — London : Lawrence and Wishart, 1978-
v. : ill., geneal. tables, maps, music ; 22 cm. GB78-15929
Bibliography: v. 1, p. [595]-609.
Includes index.
CONTENTS: [1] The prehistoric Aegean.

ISBN 0-85315-415-5 (v. 1) : £4.50 (v. 1)
1. Civilization, Greek. 2. Greece—Social conditions. 3. Bronze age—Greece. I. Title.
DF220.T54 1978 938 78-317365
78 MARC

Wilcoxon, George Dent, 1913-
Athens ascendant / George Dent Wilcoxon. — 1st ed. — Ames : Iowa State University Press, 1979.
ix, 309 p. ; 24 cm.
Bibliography: p. 293-302.
Includes index.
ISBN 0-8138-1935-0
1. Athens—History. 2. Greece—History—To 168 B.C. 3. Civilization, Greek. I. Title.
DF285.W55 938.5 78-10527
78 MARC

—ADDRESSES, ESSAYS, LECTURES

Baker, Howard, 1905-
Persephone's cave : cultural accumulations of the early Greeks / Howard Baker. — Athens : University of Georgia Press, c1979.
xv, 335 p. : ill. ; 24 cm.
Bibliography: p. [287]-325.
Includes index.
ISBN 0-8203-0438-7
1. Civilization, Greek—Addresses, essays, lectures. I. Title.
DF77.B25 938 77-11162
77 MARC

Bonner Festgabe : Johannes Straub zum 65. Geburtstag am 18. Oktober 1977 / dargebracht von Kollegen und Schülern ; [hrsg. von Adolf Lippold, Nikolaus Himmelmann]. — Bonn : Rheinland-Verlag ; In Kommission R. Habelt, 1977.
420 p. : ill. ; 27 cm. — (Beihefte der Bonner Jahrbücher ; Bd. 39) GFR***
Bibliography of J. A. Straub's works: p. [2]-6.
Includes bibliographical references.
ISBN 3-7927-0251-7
1. Civilization, Greek—Addresses, essays, lectures. 2. Rome—Civilization—Addresses, essays, lectures. 3. Classical philology—Addresses, essays, lectures. 4. Straub, Johannes, A., 1912- —Addresses, essays, lectures. I. Straub, Johannes A., 1912- . II. Lippold, Adolf, 1926- . III. Himmelmann-Wildschütz, Nikolaus. IV. Series: Bonner Jahrbücher. Beihefte ; Bd. 39.
DE3.B66 938 78-391835
79 MARC

CIVILIZATION, GREEK
—ADDRESSES, ESSAYS, LECTURES (Continued)

Dion, Roger, 1896-
Aspects politiques de la géographie antique / par Roger Dion. — Paris : Les Belles Lettres, 1977.
xxiii, 305 p. : ill. ; 25 cm. — (Collection d'études anciennes) F***
Bibliography: p. ⌈xiii⌉-xxiii.
Includes index.
105F
1. Classical geography—Addresses, essays, lectures. 2. Civilization, Greek—Addresses, essays, lectures. 3. Rome—Civilization—Addresses, essays, lectures. I. Title.
DE24.D56 938 78-388304
79 MARC

Farrington, Benjamin, 1891-
Head and hand in ancient Greece : four studies in the social relations of thought / by Benjamin Farrington. — Norwood, Pa. : Norwood Editions, 1978.
xi, 119 p. ; 26 cm.
Reprint of the 1947 ed. published by Watts, London, which was issued as no. 121 of The Thinker's library.
Includes bibliographical references.
ISBN 0-8482-0828-5 (lib. bdg.) : $17.50
1. Civilization, Greek—Addresses, essays, lectures. 2. Rome—Civilization—Addresses, essays, lectures. I. Title. II. Series: The Thinker's library ; no. 121.
⌈DF77.F27 1978⌉ 938 78-17678
78 MARC

Groningen, Bernhard Abraham van, 1894-
Bedwongen hartstocht : of, Hoe bouwde de Griek zijn beschaving op? / door B. A. van Groningen. — Leiden : E. J. Brill, 1964.
98 p., ⌈1⌉ leaf of plates : port. ; 23 cm.
1. Civilization, Greek—Addresses, essays, lectures. I. Title.
DF77.G867 79-363255
79 MARC

Janitor.
Disegno di una storia dell'età dell'oro / Janitor. — Spezia : Cooperativa tipografica, 1905.
76 p. ; 18 cm. — (Biblioteca del Libertario ; pubblicaz. n. 7)
1. Civilization, Greek—Addresses, essays, lectures. 2. Rome—Civilization—Addresses, essays, lectures. 3. Civilization, Occidental—Addresses, essays, lectures. 4. Utopias—Addresses, essays, lectures. 5. Socialism—Addresses, essays, lectures. I. Title.
DE60.J35 79-370765
79 MARC

Kreuzfahrten im östlichen Mittelmeer : Vorträge unserer Mentoren, gehalten auf Karawane-Mittelmeerkreuzfahrten / hrsg. im Karawane-Verlag Ludwigsburg mit Unterstützung der Karawane-Studienreisen des Büros für Länder- und Völkerkunde Ludwigsburg. — Ludwigsburg : Verlag Die Karawane, 1971.
48 p. : ill. ; 21 cm. — (Die Karawane ; Jahrg. 12, 1971, Heft 3) GFR**
Includes bibliographical references.
1. Civilization, Greek—Addresses, essays, lectures. 2. Classical antiquities—Addresses, essays, lectures. 3. Mediterranean region—Description and travel—Addresses, essays, lectures. I. Series.
DF77.K74 949.5 78-421481
79 MARC

Studi di storia antica offerti dagli allievi a Eugenio Manni. — Roma : G. Bretschneider, 1976.
304 p., 12 leaves of plates : ill. ; 25 cm. It78-Mar
Includes bibliographical references.
1. Civilization, Greek—Addresses, essays, lectures. 2. Rome—Civilization—Addresses, essays, lectures. 3. Manni, Eugenio. I. Manni, Eugenio.
DE3.S77 938 78-366655
*78 MARC

—AFRICAN INFLUENCES

Beardsley, Grace Maynard Hadley, 1896-
The Negro in Greek and Roman civilization / Grace Hadley Beardsley. — New York : Arno Press, 1979, c1929.
xii, 145 p., ⌈15⌉ leaves of plates : ill. ; 23 cm. — (Johns Hopkins University Press reprints)
Originally presented as the author's thesis, Johns Hopkins University, 1922.
Reprint of the ed. published by Johns Hopkins Press, Baltimore.
Bibliography: p. 135-143.
Includes index.
ISBN 0-405-11432-X : $14.00
1. Blacks—Greece. 2. Blacks—Rome. 3. Civilization, Greek—African influences. 4. Rome—Civilization—African influences. I. Title.
⌈DE73.2.B55B4 1979⌉ 938 78-19296
78 MARC

—BIBLIOGRAPHY

Hansen, Peter Allan, 1944-
A bibliography of Danish contributions to classical scholarship from the sixteenth century to 1970 / by Peter Allan Hansen. — Copenhagen : The Royal Library, Dept. of National Bibliography : distributed by Rosenkilde and Bagger, 1977.
xviii, 335 p. ; 25 cm. — (Danish humanist texts and studies ; v. 1) D77-43
Includes indexes.
ISBN 8770232326 : kr224.20
1. Civilization, Greek—Bibliography. 2. Rome—Civilization—Bibliography. 3. Classical antiquities—Bibliography. 4. Denmark—Imprints. I. Copenhagen. Kongelige Bibliotek. Nationalbibliografisk afdeling. II. Title. III. Series.
Z6207.G7H36 016.938 78-389889
⌈DE59⌉ 78 MARC

—BIBLIOGRAPHY—CATALOGS

Harvard University. Library.
Classical studies : classification schedules, classified listing by call number, chronological listing, author and title listing / Harvard University Library. — Cambridge, Mass. : The Library : distributed by the Harvard University Press, 1979.
215 p. ; 29 cm. — (Its Widener Library shelflist ; 57)
ISBN 0-674-13461-3 : $25.00
1. Civilization, Greek—Bibliography—Catalogs. 2. Rome—Civilization—Bibliography—Catalogs. 3. Classical literature—Bibliography—Catalogs. 4. Classification—Books—Greece. 5. Classification—Books—Rome. 6. Classification—Books—Classical literature. 7. Harvard University. Library—Catalogs. I. Title. II. Series.
Z6207.G7H37 1979 025.4'3 79-948
⌈DE3⌉ 79 MARC

—CONGRESSES

Schiavitù, manomissione e classi dipendenti nel mondo antico. — Roma : L'Erma di Bretschneider, 1979.
201 p., ⌈6⌉ leaves of plates : ill. ; 25 cm. — (Pubblicazioni dell'Istituto di storia antica, Università degli studi di Padova ; v. 13) It79-May
In English, French, German, or Italian.
"Atti del Colloquio internazionale tenuto a Bressanone/Brixen dal 25 al 27 novembre 1976 nell'ambito delle attività del Groupe international de recherches sur l'esclavage ancien (GIREA) e dei Corsi estivi dell'Università di Padova."
Includes bibliographical references.
1. Slavery—History—Congresses. 2. Slavery—Emancipation—History—Congresses. 3. Civilization, Greek—Congresses. 4. Rome—Civilization—Congresses. I. Groupe international de recherches sur l'esclavage ancien. II. Series: Padua. Università. Istituto di storia antica. Pubblicazioni ; v. 13.
HT855.S34 79-367029
*79 MARC

—INFLUENCE

Messina, Calogero, 1945-
Voltaire e il mondo classico / Calogero Messina. — Palermo : Herbita, 1976.
245 p. ; 22 cm. It78-Jan
"La mort de César": p. 173-226.
Includes bibliographical references and index.
L7000
1. Voltaire, François Marie Arouet de, 1694-1778—Knowledge and learning. 2. Classicism. 3. Civilization, Greek—Influence. 4. Rome—Civilization—Influence. I. Voltaire, François Marie Arouet de, 1694-1778. La mort de César. 1976. II. Title.
PQ2122.M4 78-381519
*78 MARC

—JUVENILE LITERATURE

Crosher, Judith.
The Greek / Judith Crosher ; ⌈illustrators, Peter Connolly ... et al.⌉. — London : Macdonald Educational ; Morristown, N.J. : Silver Burdett Co., c1974, 1977 printing.
61 p. : ill. (some col) ; 27 cm. — (Peoples of the past)
Includes index.
SUMMARY: Describes aspects of life in ancient Greece including the theater, government, religion, and domestic life. Also presents a portrait of life in Athens.
ISBN 0-382-06119-5
1. Civilization, Greek—Juvenile literature. ⌈1. Civlization, Greek⌉ I. Connolly, Peter. II. Title.
DF77.C94 1977 938 77-86190
79 MARC

Fagg, Christopher.
Ancient Greece / ⌈Christopher Fagg ; editor, Frances M. Clapham ; illustrators, Constance Dear ... et al.⌉. — New York : Warwick Press, 1979, c1978.
44 p. : col. ill. ; 28 cm. — (Modern knowledge library)
Includes index.
SUMMARY: Discusses the civilization of the ancient Greeks who were the first to develop a democratic way of life.
ISBN 0-531-09124-4
1. Civilization, Greek—Juvenile literature. ⌈1. Civilization, Greek⌉ I. Dear, Constance. II. Title.
DF77.F23 1978 938 78-68532
79 MARC

Powell, Anton.
Ancient Greeks / Anton Powell and Patricia Vanags ; W. A. Johnson, consultant ; illustrated by Ivan Lapper. — New York : Gloucester Press, 1978.
30 p. : col. ill. ; 27 cm. — (The Civilization Library) (A First look at)
Includes index.
SUMMARY: A history of Greece from 2000 B.C. to 300 B.C. including the customs, philosophy, science, military, trade, and government.
ISBN 0-531-01446-0
1. Civilization, Greek—Juvenile literature. ⌈1. Civilization, Greek⌉ I. Vanags, Patricia, joint author. II. Lapper, Ivan. III. Title.
DF77.P68 1978 938 78-2646
78 MARC

—PERIODICALS

The Ancient world. v. 1-
Mar. 1978-
⌈Chicago, Ares Publishers⌉
v. ill. 28 cm. quarterly.
Key title: The Ancient world, ISSN 0160-9645
1. Civilization, Greek—Periodicals. 2. Rome—Civilization—Periodicals. 3. Mediterranean region—Antiquities—Periodicals.
DE1.A375 930'.005 78-645202
MARC-S

Oikumene. 1-
Budapest, Akadémiai Kiadó, 1976-
v. 25 cm. annual.
English, French, German, or Russian.
"Studia ad historiam antiquam classicam et orientalem spectantia."
1. Civilization, Greek—Periodicals. 2. Rome—Civilization—Periodicals. 3. Civilization, Ancient—Periodicals.
DE1.O37 77-647530
MARC-S

CIVILIZATION, HINDU

see also Great Britain—Civilization—Hindu influences

Awasthi, Vishwambhar Dayal, 1937-
(Vaidika saṃskṛti aura darśana)
वैदिक संस्कृति और दर्शन / लेखक विश्वम्भर दयाल अवस्थी. — 1. संस्करण. — इलाहाबाद : सरस्वती प्रकाशन मन्दिर, 1978.
2, 3, 288 p. ; 22 cm.
In Hindi.
Bibliography: p. 284-288.
Rs50.00
1. Civilization, Hindu. 2. India—Civilization. 3. Philosophy, Hindu. I. Title.
DS425.A97 78-906541

Bhagavat Siṃhajī, Maharaja of Gondal, Sir, 1865-1944.
A short history of Aryan medical science / by Bhagvat Sinh Jee. — Delhi : New Asian Publishers, 1978.
206 p., 5 leaves of plates : ill. ; 23 cm.
Reprint of the 1896 ed. published by Macmillan, London.
Rs50.00
1. Medicine—India—History. 2. Medicine, Hindu—History. 3. Civilization, Hindu. I. Title.
R605.B57 1978 610'.954 78-912214
79 MARC

Bhārgava, Nirmalā.
(Vaidika sāhitya evaṃ saṃskṛti)
वैदिक साहित्य एवं संस्कृति, भृगु ऋषियों के योगदान के परिपार्श्व में. ⌈लेखिका⌉ निर्मला भार्गव. जयपुर, देवनागर प्रकाशन ⌈1972⌉
16, 277, 34 p. 23 cm. Rs30.00
In Hindi.
"आगरा विश्व-विद्यालय की पी. एच. डी. उपाधि के लिये स्वीकृत शोध-प्रबन्ध."
Bibliography: p. ⌈23⌉-34 (last group)
1. Hindu saints. 2. Vedic literature—History and criticism. 3. Civilization, Hindu. I. Title.
BL1171.B474 72-906696

Bowes, Pratima.
Hindu intellectual tradition / Pratima Bowes. — New Delhi : Allied Publishers, c1977.
vii, 218 p. ; 22 cm.
Imprint date covered by label: 1978.
Includes bibliographical references and index.
Rs48.00 ($9.60)
1. Civilization, Hindu. 2. India—Intellectual life. 3. Hinduism. I. Title.
DS423.B719 909'.09'7645 79-900374
79 MARC

Deleury, G A
Le modèle indou : essai sur les structures de la civilisation de l'Inde d'hier et d'aujourd'hui / Guy Deleury ; préf. de Roger Garaudy. — ⌈Paris⌉ : Hachette, c1978.
365 p. : ill. ; 23 cm. — (Le Temps & ⌈i.e. et⌉ les hommes) F***
Includes index.
ISBN 2-01-002958-5 : 72.00F
1. Civilization, Hindu. 2. India—Civilization. I. Title.
DS423.D35 954 79-362710
79 MARC

Iḷaṅkōvaṉ, K
(Urimaikkural)
உரிமைக்குரல் / எழுதியது க. இளங்கோவன். — 1. பதிப்பு. — வாழப்பாடி அஞ்சல், சேலம் மாவட்டம் : தமிழ்மணி பதிப்பகம், 1963.
82 p. ; 19 cm.
In Tamil.
1. Tamil Nadu—Politics and government. 2. Hinduism. 3. Civilization, Hindu. I. Title.
DS485.M28I4 77-905909

CIVILIZATION, HINDU (Continued)

Narayanan, M G S 1932-
Aspects of Aryanisation in Kerala / M. G. S. Narayanan. — 1st ed. — Trivandrum : Kerala Historical Society, 1973.
53 p. ; 21 cm.
Includes bibliographical references.
1. Kerala—Civilization. 2. Civilization, Hindu. I. Title.
DS485.K4N318 954'.83 78-904561
79 MARC

Ojha, V D
Spirit of ancient Hindu culture : a comparative study of the cultures and literatures of the East and West / V. D. Ojha. — New Delhi : Classical Publications, 1977.
vi, 112 p. ; 22 cm.
Rs40.00 ($8.00 U.S.)
1. Civilization, Hindu. 2. Philosophy, Hindu. 3. East and West. I. Title.
DS425.O38 934 78-902734
78 MARC

Sāhu, Keśaba Candra, 1917-
Dharma, saṃskṛti, darśana, o bijñāna / Keśaba Candra Sāhu. — Kaṭaka : Es. Pradīpa, 1977-
v. ; 22 cm.
In Oriya.
CONTENTS: 1. Utkaḷa darśana, alibhā atīta.
Rs15.00 (v. 1)
1. Civilization, Hindu. 2. India—Civilization. I. Title.
DS423.S2 78-907831
79 MARC

Tradition and politics in South Asia / edited by R. J. Moore. — New Delhi : Vikas, c1979.
xxii, 266 p. ; 22 cm.
Includes bibliographical references and index.
Rs75.00
1. India—Politics and government. 2. South Asia—Politics and government. 3. Civilizations, Hindu. I. Moore, Robin James.
DS445.T73 320.9'54 79-901842
79 MARC

—ADDRESSES, ESSAYS, LECTURES

Upādhyāya, Vishṇudeva, 1948–
(Dharma aura darśana)
धर्म और दर्शन : [महत्त्वपूर्ण धार्मिक और दार्शनिक शोधपरक विचार] / विष्णुदेव उपाध्याय. — 1. संस्करण. — नयी दिल्ली : नेशनल पब्लिशिंग हाउस, 1978.
194 p. ; 23 cm.
In Hindi.
Includes bibliographical references and index.
Rs25.00
1. Hinduism—Addresses, essays, lectures. 2. Civilization, Hindu—Addresses, essays, lectures. 3. Democracy—Addresses, essays, lectures. I. Title.
BL1210.U65 78–901563
SA

—COLLECTED WORKS

Athavale, Pandurang Vaijnath, 1920-
Thoughts on glorious heritage : from the discourses of ... Pandurang V. Athavale. — 2d ed. — Bombay : Sat Vichar Darshan, 1975.
xii, 264, ii p., [2] leaves of plates : ill. ; 22 cm.
Includes quotations in Sanskrit.
Bibliography: p. i-ii (3d group)
Rs12.00
1. Civilization, Hindu—Collected works. 2. India—Civilization—Collected works. I. Title.
DS425.A84 1975 954.05 76-903958
78 MARC

CIVILIZATION, HISPANIC

see also Latin America—Civilization; Spain—Civilization

Gil Munilla, Ladislao.
Geografía e historia de España y los países hispánicos / Ladislao Gil Munilla. — Zaragoza : Librería General, c1977.
557 p. : ill. ; 24 cm. Sp***
ISBN 8470780352
1. Spain. 2. Civilization, Hispanic. I. Title.
DP26.G54 946 78-390701
79 MARC

—ADDRESSES, ESSAYS, LECTURES

Les Cultures ibériques en devenir : essais publiés en hommage à la mémoire de Marcel Bataillon, 1895-1977. — Paris : Fondation Singer-Polignac, 1979.
783 p., 14 leaves of plates : ill. ; 25 cm. F***
French, English, or Spanish.
"Bibliographie de Marcel Bataillon": p. [21]-56.
Includes bibliographies.
ISBN 2-900927-01-3
1. Civilization, Hispanic—Addresses, essays, lectures. 2. Bataillon, Marcel—Addresses, essays, lectures. I. Bataillon, Marcel. II. Fondation Singer-Polignac.
CB226.C8 946 79-382090
79 MARC

Mélanges à la mémoire d'André Joucla-Ruau. — [Aix-en-Provence] : Éditions de l'Université de Provence, [1978]
2 v. (1260 p.) : port. ; 24 cm. — (Études littéraires ; 2) F***
French or Spanish.
Includes bibliographical references.
ISBN 2-85399-027-3 : 220F
1. Civilization, Hispanic—Addresses, essays, lectures. 2. Joucla-Ruau, André—Addresses, essays, lectures. I. Joucla-Ruau, André. II. Series: Études littéraires (Aix-en-Provence) ; 2.
CB226.M44 946 79-346076
79 MARC

—BIBLIOGRAPHY

Alberich, José.
Bibliografía anglo-hispánica 1801-1850 : ensayo bibliográfico de libros y folletos relativos a España e Hispanoamérica impresos en Inglaterra en la primera mitad del siglo diecinueve / José Alberich. — Oxford : Dolphin Book ; Barcelona : Gustavo Gili, 1978.
xl, 197 p. ; 22 cm. Sp78-June
Includes indexes.
ISBN 0-85215-063-6
1. Spain—Civilization—19th century—Bibliography. 2. Civilization, Hispanic—Bibliography. 3. Spain—Relations (general) with Great Britain. 4. Great Britain—Relations (general) with Spain. I. Title.
Z2707.C55A4 016.946 79-342111
[DP203.5] *79 MARC

—POETRY

Blanco, Andrés Eloy, 1898-1955.
Canto a España : marzo 1923 / Andrés Eloy Blanco. — [Caracas : Congreso de la República, 1973]
19 p. ; 26 cm.
Cover title.
"Homenaje del Congreso de la República y de la Asociación de Escritores Venezolanos, marzo 1973."
1. Civilization, Hispanic—Poetry. I. Title.
PQ8549.B47C3 1973 861 78-401109
79 MARC

CIVILIZATION, HOMERIC

Finley, Moses I
The world of Odysseus / by M. I. Finley. — 2nd ed. (revised and reset). — London : Chatto and Windus, 1977.
192 p. : map ; 23 cm. GB77-19904
Bibliography: p. 178-185.
Includes indexes.
ISBN 0-7011-2244-7 : £4.95
1. Civilization, Homeric. I. Title.
PA4037.F48 1977 883'.01 79-302985
79 MARC

Origini e sviluppo della città / di Giovanni Pugliese Carratelli ... [et al.]. — Milano : Bompiani, 1978.
2 v. : ill. ; 23 cm. — (Storia e civiltà dei greci ; 1-2) It79-Apr
Includes bibliographical references and index.
CONTENTS: [1] Il medioevo greco.—[2] L'arcaismo.
L1200 per vol.
1. Civilization, Greek. 2. Civilization, Homeric. 3. Civilization, Mycenaean. I. Pugliese Carratelli, Giovanni, 1911- . II. Series.
DF77.S72 vol. 1-2 938'.008 s 79-380718
[DF220] *79 MARC

—ADDRESSES, ESSAYS, LECTURES

Stagakis, George.
Studies in the Homeric society / by George Stagakis. — Wiesbaden : Steiner, 1975.
122 p. ; 25 cm. — (Historia : Einzelschriften ; Heft 26) GFR75-A
Includes bibliographical references and index.
ISBN 3-515-01988-X : DM38.00
1. Family—Greece—Addresses, essays, lectures. 2. Kinship—Greece—Addresses, essays, lectures. 3. Civilization, Homeric—Addresses, essays, lectures. 4. Homerus. Ilias. I. Title. II. Series: Historia (Wiesbaden) : Einzelschriften ; Heft 26.
DF93.S7 301.42'0938 79-313647
79 MARC

CIVILIZATION, ISLAMIC

see also Civilization, Occidental—Islamic influences

Alegre Peyrón, José María.
Influencias árabes en el pensamiento europeo y en las lenguas y literaturas romances / José Ma. Alegre Peyrón. — København : Universitet, Romansk Institut, 1976.
2 v. ; 24 cm. — (Romansk instituts duplikerede småskrifter ; nr. 39) D76-12
Cover title.
Includes bibliographical references.
kr3.50
1. Civilization, Medieval. 2. Europe—Civilization—Arab influences. 3. Civilization, Islamic. I. Title. II. Series: Copenhagen. Universitet. Romansk Institut. Romansk instituts duplikerede småskrifter ; nr. 39.
CB353.A45 940.1 79-363248
79 MARC

Bartol'd, Vasiliĭ Vladimirovich, 1869-1930.
Kul'tura musul'manstva : [obshchiĭ ocherk] / V. V. Bartol'd. — Petrograd : Ogni, 1918.
111 p., [1] fold. leaf : ill. ; 23 cm. — (Krug znaniĭa)
Bibliography: p. 108-109.
1. Civilization, Islamic. 2. Iran—Civilization. I. Title.
DS36.85.B37 79-363585
79 MARC

Djaït, Hichem.
L'Europe et l'Islam / Hichem Djaït. — Paris : Seuil, c1978.
186 p. ; 21 cm. — (Collection Esprit) F***
Includes bibliographical references.
ISBN 2-02-004816-7 : 39.00F
1. Near East—Relations (general) with Europe. 2. Europe—Relations (general) with The Near East. 3. Civilization, Islamic. 4. Europe—Civilization. I. Title.
DS63.2.E8D52 301.29'4'01767 78-376573
78 MARC

Geanakoplos, Deno John.
Medieval Western civilization and the Byzantine and Islamic worlds : interaction of three cultures / Deno J. Geanakoplos. — Lexington, Mass. : D. C. Heath, c1979.
xiii, 513 p. : ill. ; 25 cm.
Includes bibliographies and index.
ISBN 0-669-00868-0
1. Civilization, Medieval. 2. Civilization, Islamic. 3. Byzantine Empire—Civilization. I. Title.
CB351.G37 909.07 78-52827
79 MARC

Miquel, André.
L'Islam et sa civilisation : VII-XXe siècle / André Miquel. .... — 2e éd. revue et mise à jour. — Paris : A. Colin, 1977.
599 p., [32] p. of plates, [8] leaves of col. plates : ill. ; 24 cm. — (Destins du monde) F78-8915
Bibliography: p. 417-448.
Includes indexes.
1. Civilization, Islamic. 2. Islamic countries—History. I. Title.
DS36.85.M56 909'.097'671 79-362925
79 MARC

Nithi 'Īeosīwong.
('Itsalām samai rāēk)
อิสลามสมัยแรก / นิธิ เอียวศรีวงศ์. — พระนคร : สมาคมสังคมศาสตร์แห่งประเทศไทย, 2511 [1968]
21, 351 p. ; 23 cm.
In Thai.
Bibliography: p. 351.
1. Civilization, Islamic. 2. Islamic Empire—History. I. Title.
DS36.855.N57 77–985835

Patzelt, Erna, 1894-
Die fränkische Kultur und der Islam : mit besonderer Berücksichtigung der nordischen Entwicklung : eine universalhistorische Studie / Erna Patzelt. — 2., neubearb. Aufl. — Aalen : Scientia Verlag, 1978.
298 p. ; 22 cm. — (Veröffentlichungen des Seminars für Wirtschafts- und Kulturgeschichte an der Universität Wien ; 4) GFR***
"Les relations entre les Suédois et l'Orient": p. 273-298.
Includes bibliographical references and index.
ISBN 3-511-06934-3
1. Civilization, Medieval. 2. Franks. 3. Civilization, Islamic. I. Title. II. Series: Vienna. Universität. Seminar für Wirtschafts- und Kulturgeschichte. Veröffentlichungen ; 4.
CB353.P34 1978 909.07 78-399593
79 MARC

Shaukat Ali, Dr.
Intellectual foundations of Muslim civilization / by Shaukat Ali. -- Lahore : Publishers United, 1977.
xii, 191 p. ; 21 cm.
Includes bibliographical references.
Rs40.00 ($4.50 U.S.)
1. Civilization, Islamic. 2. Islamic learning and scholarship—History. 3. Ibn Khaldūn, 1332-1406. 4. Historians, Arab—Islamic Empire—Biography. I. Title.
DS36.855.S54 909'.09'7671 77-930815
79 MARC

CIVILIZATION, ISLAMIC (Continued)

—ADDRESSES, ESSAYS, LECTURES

Papers in oriental culture. — ₍Cairo₎ : al-Azhar University Bookshop, ₍1977₎
viii, 167 p. ; 24 cm.
"For undergraduate students, University College for Women, University of al-Azhar, 1977."
ISBN 977721703X : £E1.50
1. English language—Text-books for foreigners—Arabic. 2. College readers. 3. Civilization, Islamic—Addresses, essays, lectures.
PE1130.A8P3 428'.6'4927 77-960526
78 MARC

Schacht, Joseph, 1902–1969.
₍The legacy of Islam. Arabic₎
تراث الاسلام / تصنيف شاخت وبوزورث ؛ ترجمة محمد زهير السمهوري ؛ تعليق شاكر مصطفى ؛ مراجعه فؤاد زكريا. — الكويت : ₍المجلس الوطني للثقافة والفنون والآداب₎ ، 1978.
3 v. : ill. ; 22 cm. — (عالم المعرفة ؛ 8, 11–12)
Vols. 2–3 translated by Ḥ. Mu'nis and I. Ṣ. al-'Amad.
Includes bibliographies.
0.25KD per vol.
1. Civilization, Islamic — Addresses, essays, lectures. 2. Islam — Addresses, essays, lectures. I. Bosworth, Clifford Edmund, joint author. II. Shākir, Muṣṭafá. III. Title. IV. Series: 'Ālam al-ma'rifah ; 8 ₍etc.₎
Title romanized: Turāth al-Islām.
DS36.85.S312 79–970383

—OCCIDENTAL INFLUENCES

al-'Aql, Nāṣir ibn 'Abd al-Karīm.
(al-Taqlīd wa-al-taba'īyah wa-atharuhumā fī kiyān al-ummah al-Islāmīyah)
التقليد والتبعية واثرهما فى كيان الأمة الاسلامية : دراسة تاريخية تحليلية للأسباب تبعية الأمة الاسلامية وتقليدها لغيرها ، وبيان نتائج ذلك ، ومواقف الناس منه / تقديم ناصر بن عبد الكريم العقل. — الرياض : المملكة العربية السعودية ، جامعة الامام محمد بن سعود الاسلامية ، كلية الشريعة ، ₍1976?₎
239 p. ; 24 cm.
Thesis (al-shahādah al-'āliyah), Jāmi'at al-Imām Muḥammad ibn Sa'ūd al-Islāmīyah, 1974–1975.
Bibliography: p. 229–231.
1. Islam—20th century. 2. Civilization, Islamic—Occidental influences. I. Title.
BP163.A674 79–971262
NE

—PERIODICALS

(Iḥyā' al-turāth al-'Arabī al-Islāmī)
احياء التراث العربى الاسلامى.
بغداد ₍جمعية احياء التراث العربى الاسلامى₎
v. ill. 24 cm. quarterly.
Spine title, : Majallat iḥyā' al-turāth al-'Arabī al-Islāmī.
1. Civilization, Arab—Periodicals. 2. Civilization, Islamic—Periodicals. I. Jam'īyat Iḥyā' al-Turāth al-'Arabī al-Islāmī, Bagdad. II. Title: Majallat iḥyā' al-turāth al-'Arabī al-Islāmī.
DS36.8.I 44 78–961053
(MARC-S)

—SOURCES

Hitti, Philip Khuri, 1886-
Islam and the West : a historical cultural survey / Philip K. Hitti. — Huntington, N.Y. : R. E. Krieger Pub. Co., 1979, c1962.
191 p. ; 19 cm. — (The Anvil series ISSN 0570-1062)
Reprint of the ed. published by Van Nostrand, Princeton, N.J., in series: An Anvil original.
Bibliography: p. 188-189.
Includes index.
ISBN 0-88275-787-3 : $3.95
1. Near East—Relations (general) with Europe. 2. Europe—Relations (general) with the Near East. 3. Civilization, Islamic—Sources. 4. East and West. I. Title.
DS63.2.E8H57 1979 301.29'56'04 78-10793
78 MARC

CIVILIZATION, JEWISH see Jews
--Civilization

CIVILIZATION, MEDIEVAL

see also Chivalry; Education, Medieval; Feudalism; Middle Ages; Monasticism and religious orders; Renaissance; Science, Medieval

Alegre Peyrón, José María.
Influencias árabes en el pensamiento europeo y en las lenguas y literaturas romances / José Ma. Alegre Peyrón. — København : Universitet, Romansk Institut, 1976.
2 v. ; 24 cm. — (Romansk instituts duplikerede småskrifter ; nr. 39)
D76-12
Cover title.
Includes bibliographical references.
kr3.50
1. Civilization, Medieval. 2. Europe—Civilization—Arab influences. 3. Civilization, Islamic. I. Title. II. Series: Copenhagen. Universitet. Romansk Institut. Romansk instituts duplikerede småskrifter ; nr. 39.
CB353.A45 940.1 79-363248
79 MARC

Association des historiens médiévistes de l'enseignement supérieur public.
La mort au Moyen âge / Colloque de l'Association des historiens médiévistes français, réunis à Strasbourg en juin 1975 .... — Strasbourg : Librairie Istra, 1977.
2 v. (167 p., ₍3₎ leaves of plates) ; 24 cm. — (Publications de la Société savante d'Alsace et des régions de l'Est : Collection Recherches et documents ; t. 25)
F77-17645
Includes bibliographical references.
1. Death. 2. Civilization, Medieval. I. Title. II. Series: Société savante d'Alsace et des régions de l'Est. Publications : Collection Recherches et documents ; t. 25.
GT3180.A87 1977 78-392411
79 MARC

Daniel, Norman.
The Arabs and mediaeval Europe / Norman Daniel. — 2d ed. — London ; New York : Longman, 1979.
xiv, 385 p., ₍2₎ leaves of plates : ill. ; 23 cm. — (Arab background series)
GB78
Includes bibliographical references and index.
ISBN 0-582-78088-8 : £9.50
1. Europe—Relations (general) with the Islamic Empire. 2. Islamic Empire—Relations (general) with Europe. 3. Civilization, Medieval. I. Title.
CB353.D26 1979 301.29'4'0174927 78-40767
78 MARC

Dufourcq, Charles Emmanuel, 1914-
La vie quotidienne dans l'Europe médiévale sous domination arabe / Charles-Emmanuel Dufourcq. — ₍Paris₎ : Hachette, c1978.
287 p. : map ; 21 cm. — (₍La₎ V₍ie₎ q₍uotidienne₎) (Littérature & ₍i.e. et₎ sciences humaines)
F***
Bibliography: p. ₍264₎-265.
ISBN 2-01-003047-8 : 42.00F
1. Civilization, Medieval. 2. Arabs in Europe—History. I. Title.
CB353.D83 940.1 79-362540
79 MARC

Geanakoplos, Deno John.
Medieval Western civilization and the Byzantine and Islamic worlds : interaction of three cultures / Deno J. Geanakoplos. — Lexington, Mass. : D. C. Heath, c1979.
xiii, 513 p. : ill. ; 25 cm.
Includes bibliographies and index.
ISBN 0-669-00868-0
1. Civilization, Medieval. 2. Civilization, Islamic. 3. Byzantine Empire—Civilization. I. Title.
CB351.G37 909.07 78-52827
79 MARC

Harrison, Frederick, 1884-
Medieval man and his notions / by Frederick Harrison. — Philadelphia : R. West, 1978.
viii, 276 p., ₍13₎ leaves of plates : ill. ; 26 cm.
Reprint of the 1947 ed. published by J. Murray, London.
Includes index.
ISBN 0-8492-1070-4 lib. bdg. : $30.00
1. Great Britain—Civilization—Medieval period, 1066-1485. 2. Civilization, Medieval. I. Title.
₍DA185.H3 1978₎ 942 78-2883
78 MARC

Heer, Friedrich, 1916-
Mittelalter : vom Jahr 1000 bis 1350 / Friedrich Heer. — Erw. Ausg. — München : Kindler, 1977, c1961.
2 v. (816 p.) : ill. (some col.) ; 24 cm. — (Kindlers Kulturgeschichte des Abendlandes ; Bd. 9/10)
GFR77-A40
Bibliography: p. ₍768₎-816.
ISBN 3-463-13709-7 (v. 1) : DM78.00 (v. 1)
1. Civilization, Medieval. I. Title. II. Series.
CB351.H45 1977 78-371559
78 MARC

Little, Lester K
Religious poverty and the profit economy in medieval Europe / Lester K. Little. — Ithaca, NY : Cornell University Press, 1978.
xi, 267 p. ; 25 cm.
Includes bibliographical references and index.
ISBN 0-8014-1213-7
1. Poverty (Virtue)—History of doctrines. 2. Monastic and religious life—History of doctrines. 3. Civilization, Medieval. 4. Economic history—Medieval, 500-1500. I. Title.
BV4647.P6L57 1978 241'.4 78-58630
78 MARC

Kabayama, Kōichi.
(Goshikku sekai no shisōzō)
ゴシック世界の思想像 / 樺山紘一著. — 東京 : 岩波書店, 1976.
xi, 415, 22 p. ; 22 cm. Ja 76-19097
Bibliography: p. 11-22 (3d group)
Includes indexes.
¥3200
1. Philosophy, Medieval. 2. Civilization, Medieval. I. Title.
B721.K26 79-801502

Linskens, Roger, 1922-
Wat 'n leven! / Roger Linskens. — Antwerpen : Nederlandsche Boekhandel, 1974-
v. : ill. ; 22 cm. Ne74-43
Includes bibliographies and indexes.
CONTENTS: 1. Mannen en vrouwen in de middeleeuwen.—2. Straten en huizen, eten en drinken in de middeleeuwen.
ISBN 9022999841 : fl 19.90 (v. 1)
1. Civilization, Medieval. I. Title.
CB351.L454 74-357342
*79₍r79₎rev MARC

Manselli, Raoul.
Vita medioevale : la festa / Raoul Manselli. — Torino : G. Giappichelli, 1978.
110 p. ; 24 cm. — (Corsi universitari) It78-Aug
L3500
1. Festivals—Italy—History. 2. Festivals—History. 3. Civilization, Medieval. I. Title.
GT4852.A2M3 78-392382
*79 MARC

The Middle Ages / ₍prepared by National Geographic Book Service₎. — ₍Washington₎ : National Geographic Society, c1977.
378 p. : col. ill. ; 27 cm. — (The Story of man ; 3)
Published in 1969 under title: The Age of chivalry.
Bibliography: p. ₍377₎
Includes index.
ISBN 0-87044-075-6
1. Civilization, Medieval. I. National Geographic Society, Washington, D.C. Book Service. II. The Age of chivalry. III. Series.
CB69.S86 vol. 3 909 s 77-151034
₍CB353₎ 77₍r79₎rev MARC

Murray, Alexander, 1934-
Reason and society in the Middle Ages / by Alexander Murray. — Oxford : Clarendon Press ; New York : Oxford University Press, 1978.
xiv, 507 p., ₍9₎ leaves of plates : ill. ; 23 cm. GB***
Bibliography: p. ₍416₎-426.
Includes index.
ISBN 0-19-822540-7 : £15.00
1. Europe—Intellectual life. 2. Civilization, Medieval. I. Title.
CB203.M85 940.1'7 77-30294
79 MARC

Patzelt, Erna, 1894-
Die fränkische Kultur und der Islam : mit besonderer Berücksichtigung der nordischen Entwicklung : eine universalhistorische Studie / Erna Patzelt. — 2., neubearb. Aufl. — Aalen : Scientia Verlag, 1978.
298 p. ; 22 cm. — (Veröffentlichungen des Seminars für Wirtschafts- und Kulturgeschichte an der Universität Wien ; 4) GFR***
"Les relations entre les Suédois et l'Orient": p. 273-298.
Includes bibliographical references and index.
ISBN 3-511-06934-3
1. Civilization, Medieval. 2. Franks. 3. Civilization, Islamic. I. Title. II. Series: Vienna. Universität. Seminar für Wirtschafts- und Kulturgeschichte. Veröffentlichungen ; 4.
CB353.P34 1978 909.07 78-399593
79 MARC

Pernoud, Régine, 1909-
Pour en finir avec le Moyen Age / Régine Pernoud. — Paris : Éditions du Seuil, c1977.
159 p. ; 21 cm. F***
Includes bibliographical references and index.
ISBN 2-02-004593-1 : 32.00F
1. Civilization, Medieval. I. Title.
CB351.P414 940.1 79-353085
79 MARC

Szymański, Józef.
Pismo łacińskie i jego rola w kulturze / Józef Szymański. — Wrocław : Zakład Narodowy im. Ossolińskich, 1975.
158 p. : ill. ; 25 cm. — (Kultura Europy wczesnośredniowiecznej ; zesz. 12)
P***
At head of title: Polska Akademia Nauk. Instytut Historii Kultury Materialnej.
Includes bibliographical references and indexes.
zł45.00
1. Alphabet. 2. Civilization, Medieval. I. Title. II. Series.
P211.S97 78-396645
79 MARC

Toffanin, Giuseppe, 1891-
Il secolo senza Roma : (il rinascimento del secolo XIII) / Giuseppe Toffanin. — Bologna : N. Zanichelli, 1943.
179 p. ; 21 cm.
Bibliography: p. ₍173₎-174.
1. Civilization, Medieval. I. Title.
CB355.T63 79-385490
79 MARC

CIVILIZATION, MEDIEVAL (Continued)

Volpi, Domenico.
La vita e i costumi nel Medioevo / Domenico Volpi. Illustrato in nero e a colori con dipinti, stampe e documenti dell'epoca. — Milano : Mursia, [1978]
157 p. : ill. ; 29 cm. — (Collana Vita e costumi) It79-Jan
L9500
1. Civilization, Medieval. I. Title.
CB351.V6 940.1 79-348287
*79 MARC

Zimmer, Heinrich, 1851-1910.
The Irish element in mediaeval culture / by H. Zimmer ; translated by Jane Loring Edmands. — Folcroft, Pa. : Folcroft Library Editions, 1978 [c1891]
vii, 139 p. ; 22 cm.
Translation of an essay published in Preussische jahrbücher, Jan. 1887.
Reprint of the ed. published by Putnam, New York.
Includes index.
ISBN 0-8414-9809-1 lib. bdg. : $25.00
1. Irish in France—History. 2. Irish in Germany—History. 3. Civilization, Medieval. I. Title.
[DC34.5.I7Z54 1978] 943'.0004'9162 78-1573
78 MARC

—ADDRESSES, ESSAYS, LECTURES

Le Goff, Jacques.
Pour un autre Moyen Age : temps, travail et culture en Occident : 18 essais / Jacques Le Goff. — [Paris] : Gallimard, c1977.
424 p. ; 23 cm. — (Bibliothèque des histoires) F***
Includes bibliographical references.
1. Civilization, Medieval—Addresses, essays, lectures. I. Title.
CB351.L42 909.07 79-350473
79 MARC

Temi medievali e umanistici : cultura e teologia. — Pistoia : Memorie domenicane, [1978]
iv, 500 p., [5] leaves of plates : ill. ; 24 cm. — (Memorie domenicane ; nuova ser., n. 8-9) It***
English, Italian or Latin.
Includes bibliographical references and index.
L15000
1. Civilization, Medieval—Addresses, essays, lectures. 2. Humanism—Addresses, essays, lectures. 3. Ranzano, Pietro, bp. of Lucera, 1428 (ca.)-1492—Addresses, essays, lectures. 4. Catholic Church—Bishops—Biography—Addresses, essays, lectures. 5. Bishops—Italy—Lucera—Biography—Addresses, essays, lectures. 6. Lucera, Italy—Biography—Addresses, essays, lectures. I. Series.
BX3501.M45 n.s., n. 8-9 271'.2'008 s 79-350294
[CB353] 79 MARC

Thrupp, Sylvia Lettice, 1903-
Society and history : essays / by Sylvia L. Thrupp ; edited by Raymond Grew and Nicholas H. Steneck. — Ann Arbor : University of Michigan Press, c1977.
363 p. ; 24 cm.
Chronological list of the author's works: p. 347-349.
Includes bibliographical references and index.
ISBN 0-472-08880-7
1. Cities and towns, Medieval—Addresses, essays, lectures. 2. Civilization, Medieval—Addresses, essays, lectures. 3. Social sciences and history—Addresses, essays, lectures. I. Title.
HT115.T48 1977 301.36'3'094 75-31056
79 MARC

Transformation and continuity / edited by Paul Maurice Clogan. — Cambridge ; New York : Cambridge University Press, 1977.
xv, 261 p. ; 24 cm. — (Medievalia et humanistica ; new ser., no. 8 ISSN 0076-6127) GB***
Includes bibliographical references.
ISBN 0-521-21783-0 : £12.50
1. Renaissance—Addresses, essays, lectures. 2. Civilization, Medieval—Addresses, essays, lectures. I. Clogan, Paul Maurice. II. Series.
D111.M5 n.s., no. 8 940.1'05 s 79-302313
[CB365] 79 MARC

—BIBLIOGRAPHY

Simms, Norman Toby.
Ritual and rhetoric : intellectual and ceremonial backgrounds to Middle English literature : a critical survey of relevant scholarship / by Norman Toby Simms. — Philadelphia : R. West, 1977.
358 p. ; 24 cm.
Reprint of the 1973 ed. published by Norwood Editions, Norwood, Pa.
ISBN 0-8492-2480-2 : $35.00
1. English literature—Middle English, 1100-1500—Bibliography. 2. Civilization, Medieval—Bibliography. I. Title.
[Z2012.S55 1977] 016.941 77-28701
[PR255] 78 MARC

—CONGRESSES

Exclus et systèmes d'exclusion dans la littérature et la civilisation médiévales : [actes du colloque organisé par le C.U.E.R.M.A. a Aix-en-Provence, les 4-5-6 mars 1977]. — Aix-en-Provence : Édition CUERMA ; Paris : diffusion H. Champion, 1978.
421 p. ; 21 cm. — (Sénéfiance ; no 5) F***
Includes bibliographical references.
ISBN 2-901104-04-5 : 100F
1. Outsiders in literature—Congresses. 2. Literature, Medieval—History and criticism—Congresses. 3. Civilization, Medieval—Congresses. 4. Social isolation—Congresses. I. Centre universitaire d'études et de recherches médiévales d'Aix-en-Provence. II. Series.
PN56.5.O95E9 1977 809'.02 79-339289
79 MARC

Famille et parenté dans l'Occident médiéval : actes du Colloque de Paris (6-8 juin 1974) : organisé per l'École pratique des hautes études (VIe section) en collaboration avec le Collège de France et l'École française de Rome : communications et débats / présentés par Georges Duby et Jacques Le Goff ; ouvrage publié avec le concours de l'Istituto di storia medioevale e moderna della Facoltà di lettere dell'Università di Pisa. — Roma : École française de Rome, 1977.
447 p., [3] leaves of plates (1 fold.) : ill., geneal. tables ; 25 cm. — (Collection de l'École française de Rome ; 30) It78-Oct
Includes bibliographical references.
1. Family—Europe—History—Congresses. 2. Europe—Genealogy—Congresses. 3. Civilization, Medieval—Congresses. I. Duby, Georges. II. Le Goff, Jacques. III. Paris. École pratique des hautes études. IV. Paris. Collège de France. V. École française de Rome. VI. Series: École française de Rome. Collection ; 30.
HQ513.F35 301.42'094 78-398947
*79 MARC

Fifteenth century studies. v. 1-
1978-
[Ann Arbor, Mich.] University Microfilms International.
v. ill. 23 cm. (Monograph publishing on demand : Sponsor series)
Vols. for 1978- published for the Medieval Institute, Western Michigan University.
Includes a selection of papers presented at the Fifteenth-Century Symposia, 1977-
Key title: Fifteenth century studies, ISSN 0164-0933
1. Renaissance — Congresses. 2. Literature, Modern — 15th and 16th centuries—History and criticism—Congresses. 3. Civilization, Medieval—Congresses. 4. Literature, Medieval—History and criticism—Congresses. I. Fifteenth-Century Symposium. II. Michigan. Western Michigan University, Kalamazoo. Medieval Institute.
CB367.F53 909'.4 79-640105
MARC-S

Il Matrimonio nella società altomedievale : 22-28 aprile 1976. — Spoleto : Presso la sede del Centro, 1977.
2 v. (998 p., [26] leaves of plates) : ill. ; 22 cm. — (Settimane di studio del Centro italiano di studi sull'alto Medioevo ; 24) It78-July
French, German or Italian.
Includes bibliographical references.
L40000
1. Marriage—History—Congresses. 2. Civilization, Medieval—Congresses. I. Centro italiano di studi sull'alto Medioevo. II. Series: Centro italiano di studi sull'alto Medioevo. Settimane di studio ; 24.
HQ513.M3 301.42 78-384493
*79 MARC

—CONGRESSES—PERIODICALS

Reading, Eng. University. Graduate Centre for Medieval Studies.
Annual proceedings of the Graduate Centre for Medieval Studies in the University of Reading. v. 1-
1975-
[Reading, Graduate Centre for Medieval Studies in the University of Reading; Oxford, distributed by Parker and Son Ltd.]
v. 21 cm.
Cover title, 1975- : Reading medieval studies.
1. Civilization, Medieval—Congresses—Periodicals. I. Title. II. Title: Reading medieval studies.
CB351.R4a 940.1'05 77-641217
MARC-S

—PERIODICALS

Early medieval studies. 1-
[Stockholm, Almqvist & Wiksell, 1970-
v. illus. 25 cm. (Antikvariskt arkiv)
Vols. for 1970- issued by Vitterhets-, historie- och antikvitetsakademien.
1. Civilization, Medieval—Periodicals. I. Vitterhets-, historie- och antikvitetsakademien, Stockholm. II. Series: Vitterhets-, historie- och antikvitetsakademien, Stockholm. Antikvariskt arkiv.
DL601.V795 subser. 948.5'008 s 75-615896
[CB351] [940.1'05]

CIVILIZATION, MINOAH see Minoans

CIVILIZATION, MODERN

see also Civilization, Baroque; Detribalization; History, Modern; Renaissance

Ahlberg, Alf, 1892-
Människan och den moderna tekniken / Alf Ahlberg. — Stockholm : Natur o. kultur, 1978.
117, [1] p. ; 19 cm. S78-25
Includes bibliographical references.
ISBN 9127006069
1. Technology and civilization. 2. Civilization, Modern. I. Title.
HM221.A43 79-344377
79 MARC

Crespo, José Manuel.
Basuras del tiempo muerto / José Manuel Crespo. — [Bogotá] : Fundación para la Promoción de la Solidaridad Humana, [between 1976 and 1979]
31 p. : ill. ; 25 cm.
1. Catholic Church—Doctrinal and controversial works—Catholic authors. 2. Civilization, Modern. I. Title.
BX1751.2.C73 79-108302
79 MARC

Doridou, Mary France.
Histoire des faits et des idées / Mary-France Doridou. — [Paris] : Ministère de l'économie et des finances, Direction du personnel & des services généraux, Centre de formation professionnelle et de perfectionnement, 1976.
246 p. : ill. ; 30 cm. F***
1. History, Modern. 2. Civilization, Modern. I. Title.
D209.D67 909.8 79-350485
79 MARC

Little, Lester K
Religious poverty and the profit economy in medieval Europe / Lester K. Little. — London : P. Elek, 1978.
xi, 267 p., [1] leaf of plates : ill. ; 25 cm. GB***
Includes bibliographical references and index.
ISBN 0-236-40109-2 : £13.50
1. Poverty (Virtue)—History of doctrines. 2. Monastic and religious life—History of doctrines. 3. Civilization, Modern. 4. Economic history—Medieval, 500-1500. I. Title.
BV4647.P6L57 1978b 241'.4 79-306240
79 MARC

Mandrou, Robert.
From humanism to science, 1480 to 1700 / Robert Mandrou ; translated by Brian Pearce. — Atlantic Highlands, N.J. : Humanities Press, 1979.
329 p. : maps ; 23 cm.
Translation of Des humanistes aux hommes de science (XVIe et XVIIe siècles).
Bibliography: p. [303]-306.
Includes index.
ISBN 0-391-00541-3
1. Civilization, Modern. 2. Renaissance. 3. Humanism. I. Title.
CB401.M3613 1979 909'.08 78-10510
78 MARC

Plekhanov, Georgiĭ Valentinovich, 1856-1918.
(Ob ateizme i religii v istorii obshchestva i kul'tury)
Об атеизме и религии в истории общества и культуры : (Избр. произведения и извлеч. из трудов) / Г. В. Плеханов ; [Авт. примеч. Г. С. Лялина, Ю. Б. Пищик]. — Москва : Мысль, 1977, c1978.
355 p. : port. ; 21 cm. — (Научно-атеистическая библиотека) USSR 78
At head of title: Akademii͡a obshchestvennykh nauk pri T͡SK KPSS. Institut nauchnogo ateizma.
Series romanized: Nauchno-ateisticheskai͡a biblioteka.
Includes bibliographical references and index.
1.00rub
1. Atheism. 2. Religion. 3. Civilization, Modern. 4. Social sciences—Russia—History. I. Title.
BL2747.3.P54 1978 78-422888

Zink, Jörg.
Licht über den Wassern : Geschichten gegen d. Angst / Jörg Zink. — 1. Aufl., (1.-30. Tsd.). — Stuttgart ; Berlin : Kreuz-Verlag, 1978.
223 p. : ill. (some col), col. maps (on lining papers) ; 23 cm. GFR78-A47
Bibliography: p. 221-223.
ISBN 3-7831-0551-X : DM36.00
1. Civilization, Modern. 2. Bible. O.T.—Criticism, interpretation, etc. 3. Religion and culture. I. Title.
CB358.Z52 79-345747
79 MARC

—17TH CENTURY

Skrine, Peter N
The baroque : literature and culture in seventeenth-century Europe / Peter N. Skrine. — New York : Holmes & Meier, 1978.
ix, 176 p. ; 22 cm.
Bibliography: p. [165]-168.
Includes index.
ISBN 0-8419-0427-8. ISBN 0-8419-0443-X pbk.
1. Civilization, Modern—17th century. 2. Europe—Civilization. I. Title.
CB401.S56 1978 809'.032 78-15872
78 MARC

CIVILIZATION, MODERN (Continued)

—17TH CENTURY—ADDRESSES, ESSAYS, LECTURES

Recherches sur le XVIIème ⌐i.e. dix-septième¬ siècle / ⌐sous la direction de André Robinet¬. — Paris : Éditions du Centre national de la recherche scientifique, 1976-
v. ; 24 cm. — (Cahiers de l'Équipe de recherche 75, Centre d'histoire des sciences et des doctrines) F***
Includes bibliographical references.
ISBN 2-222-01963-X (v. 1) : 48.00F (v. 1)
1. Civilization, Modern—17th century—Addresses, essays, lectures. I. Robinet, André. II. Series: Centre d'histoire des sciences et des doctrines. Équipe de recherche 75. Cahiers de l'Équipe de recherche 75.
CB401.R42 940.2'52 79-365159
79 MARC

—17TH CENTURY—CONGRESSES

L'Europa cristiana nel rapporto con le altre culture nel secolo XVII : atti del Convegno di studio di Santa Margherita Ligure (19-21 maggio 1977) / ⌐relazioni di¬ Dino Pastine ... ⌐et al.¬. — 1. ed. — Firenze : La nuova Italia, 1978.
xvi, 172 p. ; 24 cm. — (Pubblicazioni del Centro di studi del pensiero filosofico del Cinquecento e del Seicento in relazione ai problemi della scienza del Consiglio nazionale delle ricerche : Serie I, Studi ; 12) It79-May
Includes bibliographical references and index.
L5000
1. Civilization, Modern—17th century—Congresses. 2. Europe—Civilization—Congresses. I. Pastine, Dino. II. Series: Centro di studi del pensiero filosofico del Cinquecento e del Seicento in relazione ai problemi della scienza. Pubblicazioni : Serie I, Studi ; 12.
CB401.E9 909'.6 79-382311
79 MARC

—18TH CENTURY

Anchor, Robert.
The modern Western experience / Robert Anchor. — Englewood Cliffs, N.J. : Prentice-Hall, c1978.
x, 339 p. : ill. ; 23 cm.
Includes bibliographies and index.
ISBN 0-13-599357-1
1. Civilization, Modern—18th century. 2. Civilization, Modern—19th century. 3. Civilization, Modern—20th century. 4. Civilization, Occidental. 5. Europe—Civilization. I. Title.
CB357.A55 940.2 77-16712
77⌐r79¬rev MARC

Napoli-Signorelli, Pietro, 1731-1815.
Vicende della coltura nelle Due Sicilie o sia, storia ragionata della loro legislazione e polizia, delle lettere, del commercio, delle arti e degli spettacoli dalle colonie straniere insino a noi, divisa in quattro parti. Napoli, Presso V. Flauto, 1784-86.
6 v. in 3. 19 cm.
On spine: Coltura delle Due Sicilie.
Includes bibliographical references.
1. Naples (Kingdom)—Civilization. 2. Civilization, Modern—18th century. I. Title. II. Title: Coltura delle Due Sicilie.
DG845.6.N34 1784 78-395972
79 MARC

Snyder, Louis Leo, 1907-
The age of reason / Louis L. Snyder. — Huntington, N.Y. : R. E. Krieger Pub. Co., 1979, c1955.
185 p. ; 18 cm. — (The Anvil series 0570-1062)
Reprint of the ed. published by Van Nostrand, New York, which was issued as no. 6 in the series, An Anvil original.
Bibliography: p. 181.
Includes index.
ISBN 0-88275-907-8 : $4.95
1. Civilization, Modern—18th century. 2. Enlightenment. I. Title.
⌐CB411.S58 . 1979¬ 909.7 79-4584
79 MARC

—18TH CENTURY—ADDRESSES, ESSAYS, LECTURES

American Society for Eighteenth-Century Studies.
Irrationalism in the eighteenth century. Edited by Harold E. Pagliaro. Cleveland, Press of Case Western Reserve University, 1972.
xxii, 393 p. 23 cm. (Studies in eighteenth-century culture, v. 2)
Papers presented at the second annual meeting of the American Society for Eighteenth-Century Studies.
Includes bibliographies.
ISBN 0-8295-0239-4
1. Civilization, Modern—18th century—Addresses, essays, lectures. 2. Eighteenth century—Addresses, essays, lectures. 3. Enlightenment—Addresses, essays, lectures. I. Pagliaro, Harold E., ed. II. Title. III. Series.
CB411.A58 1972 901.93'3 73-240
73⌐r79¬rev2 MARC

Verhaltenswandel in der industriellen Revolution : Beitr. z. Sozialgeschichte / hrsg. von August Nitschke. — Stuttgart ; Berlin ; Köln ; Mainz : Kohlhammer, 1975.
144 p. ; 21 cm. GFR75-A
Includes bibliographical references.
ISBN 3-17-002135-4 : DM20.00
1. Social history—19th century—Addresses, essays, lectures. 2. Social history—18th century—Addresses, essays, lectures. 3. Economic history—1750-1918—Addresses, essays, lectures. 4. Civilization, Modern—18th Century—Addresses, essays, lectures. 5. Civilization, Modern—19th century—Addresses, essays, lectures. I. Nitschke, August, 1926-
HN15.V37 79-356185
79 MARC

—18TH CENTURY—BIBLIOGRAPHY—PERIODICALS

The Eighteenth century. n. s. 1-
1975-
Philadelphia, American Society for Eighteenth-Century Studies.
v. 24 cm. annual.
"A current bibliography."
Previously issued in: Philological quarterly, ISSN 0031-7977.
Key title: The Eighteenth century, ISSN 0161-0996.
1. Civilization, Modern — 18th century — Bibliography — Periodicals. I. American Society for Eighteenth-Century Studies.
Z5579.6.E36 016.909 79-644920
[CB411] MARC-S

—18TH CENTURY—EUROPE

Corvisier, André.
Arts et sociétés dans l'Europe du XVIIIe siècle / André Corvisier. — 1. éd. — ⌐Paris¬ : Presses universitaires de France, c1978.
244 p. ; 21 cm. — (L'Historien ; 34) F***
Bibliography: p. ⌐231¬-236.
Includes index.
ISBN 2-13-035530-7 : 68.00F
1. Art and society—Europe. 2. Civilization, Modern—18th century—Europe. 3. Art, Modern—17th-18th century—Europe. I. Title.
N72.S6C595 700'.1 78-394399
79 MARC

—19TH CENTURY

Anchor, Robert.
The modern Western experience / Robert Anchor. — Englewood Cliffs, N.J. : Prentice-Hall, c1978.
x, 339 p. : ill. ; 23 cm.
Includes bibliographies and index.
ISBN 0-13-599357-1
1. Civilization, Modern—18th century. 2. Civilization, Modern—19th century. 3. Civilization, Modern—20th century. 4. Civilization, Occidental. 5. Europe—Civilization. I. Title.
CB357.A55 940.2 77-16712
77⌐r79¬rev MARC

Bjøl, Erling.
Vor tids kulturhistorie / ⌐af¬ Erling Bjøl. — ⌐København¬ : Politiken, 1978-
v. : ill. ; 25 cm. D78-17
CONTENTS: 1. Den tabte tid.

ISBN 8756730837 : kr130.00 (v. 1)
1. Civilization, Modern—19th century. 2. Civilization, Modern—20th century. I. Title.
CB415.B56 78-379268
78 MARC

(Kiki no jidai)
危機の時代 / ⌐編集者　金子武蔵，大塚久雄¬. — 東京：弘文堂，昭和34 ⌐1959¬
2, 334 p., ⌐2¬ leaves of plates : ports. : 22 cm. — (講座近代思想史 ; 6)
Colophon inserted.
Includes bibliographical references.
1. Philosophy, Modern—19th century. 2. Philosophy, Modern—20th century. 3. Civilization, Modern—19th century. 4. Civilization, Modern—20th century. I. Kaneko, Takezō, 1905- II. Ōtsuka, Hisao, 1907- III. Series: Kōza kindai shisō shi ; 6.
CB113.J3K68 vol. 6 79-805625
[B803]

—19TH CENTURY—ADDRESSES, ESSAYS, LECTURES

Il Decadentismo / a cura di Enrico Ghidetti. — 1. ed. — Roma : Editori riuniti, 1976.
277 p. ; 19 cm. — (Strumenti per la ricerca interdisciplinare ; 16) It76-Apr
Includes bibliographies.
L1800
1. Italian literature—19th century—History and criticism—Addresses, essays, lectures. 2. Literature, Modern—19th century—History and criticism—Addresses, essays, lectures. 3. Decadence (Literary movement)—Addresses, essays, lectures. 4. Italy—Civilization—1789-1900—Addresses, essays, lectures. 5. Civilization, Modern—19th century—Addresses, essays, lectures. I. Ghidetti, Enrico.
PQ4086.D38 809 79-351300
*79 MARC

Ernst, Fritz, 1889-1958.
Grösse des neunzehnten Jahrhunderts; ein komparatistischer Versuch. ⌐Aus dem Nachlass hrsg. von Elli Lehmann und Daniel Bodmer. Cologny-Genève¬ Bibliotheca Bodmeriana, 1962.
133 p. facsims., music. 25 cm.
Bibliography: p. 72-84.
1. Civilization, Modern—19th century—Addresses, essays, lectures. 2. Civilization, Modern—19th century—Bibliography. I. Lehmann, Elli, ed. II. Bodmer, Daniel, 1928- ed. III. Title.
CB417.E76 1962 79-376081
79 MARC

Verhaltenswandel in der industriellen Revolution : Beitr. z. Sozialgeschichte / hrsg. von August Nitschke. — Stuttgart ; Berlin ; Köln ; Mainz : Kohlhammer, 1975.
144 p. ; 21 cm. GFR75-A
Includes bibliographical references.
ISBN 3-17-002135-4 : DM20.00
1. Social history—19th century—Addresses, essays, lectures. 2. Social history—18th century—Addresses, essays, lectures. 3. Economic history—1750-1918—Addresses, essays, lectures. 4. Civilization, Modern—18th Century—Addresses, essays, lectures. 5. Civilization, Modern—19th century—Addresses, essays, lectures. I. Nitschke, August, 1926-
HN15.V37 79-356185
79 MARC

—19TH CENTURY—BIBLIOGRAPHY

Ernst, Fritz, 1889-1958.
Grösse des neunzehnten Jahrhunderts; ein komparatistischer Versuch. ⌐Aus dem Nachlass hrsg. von Elli Lehmann und Daniel Bodmer. Cologny-Genève¬ Bibliotheca Bodmeriana, 1962.
133 p. facsims., music. 25 cm.
Bibliography: p. 72-84.
1. Civilization, Modern—19th century—Addresses, essays, lectures. 2. Civilization, Modern—19th century—Bibliography. I. Lehmann, Elli, ed. II. Bodmer, Daniel, 1928- ed. III. Title.
CB417.E76 1962 79-376081
79 MARC

—20TH CENTURY

Anchor, Robert.
The modern Western experience / Robert Anchor. — Englewood Cliffs, N.J. : Prentice-Hall, c1978.
x, 339 p. : ill. ; 23 cm.
Includes bibliographies and index.
ISBN 0-13-599357-1
1. Civilization, Modern—18th century. 2. Civilization, Modern—19th century. 3. Civilization, Modern—20th century. 4. Civilization, Occidental. 5. Europe—Civilization. I. Title.
CB357.A55 940.2 77-16712
77⌐r79¬rev MARC

Bjøl, Erling.
Vor tids kulturhistorie / ⌐af¬ Erling Bjøl. — ⌐København¬ : Politiken, 1978-
v. : ill. ; 25 cm. D78-17
CONTENTS: 1. Den tabte tid.

ISBN 8756730837 : kr130.00 (v. 1)
1. Civilization, Modern—19th century. 2. Civilization, Modern—20th century. I. Title.
CB415.B56 78-379268
78 MARC

Boenders, Frans, 1942-
Denken in tweespraak / Frans Boenders. — Amsterdam : Bezige Bij, 1978.
290 p. : ill. ; 20 cm. — (Leven & letteren) Ne78-40
Bibliography: p. 283-290.
ISBN 902341523X : fl 32.50
1. Civilization, Modern—20th century. 2. Philosophy, Modern—20th century. I. Title.
CB425.B637 78-394613
*79 MARC

Braem, Renaat.
Naar een nieuwe visuele cultuur : (maar niet alleen visueel) / door R. Braem. — Brussel : Paleis der Academiën, 1978.
23 p. : ill. ; 26 cm. — (Medelingen van de Koninklijke Academie voor Wetenschappen, Letteren en Schone Kunsten van België, Klassse der Schone Kunsten ; jaarg. 40, nr. 5) Be***
Summary in French.
1. Art and society. 2. Visual perception. 3. Civilization, Modern—20th century. I. Title. II. Series: Vlaamse Academie voor Wetenschappen, Letteren en Schone Kunsten van België. Klasse der Schone Kunsten. Mededelingen ; jaarg. 40, nr. 5.
N13.V5 jaarg. 40, nr. 5 79-383855
⌐N72.S6¬ 79 MARC

Curtis, James M 1940-
Culture as polyphony : an essay on the nature of paradigms / James M. Curtis. — Columbia : University of Missouri Press, 1978.
xii, 194 p. ; 22 cm.
Bibliography: p. 179-185.
Includes index.
ISBN 0-8262-0251-9
1. Analogy. 2. McLuhan, Herbert Marshall. Understanding media. 3. Communication. 4. Communication and traffic. 5. Technology and civilization. 6. Civilization, Modern—20th century. I. Title.
BD190.C8 169 77-25242
77 MARC

CIVILIZATION, MODERN —20TH CENTURY (Continued)

Friedrich, Heinz, 1922-
Kulturkatastrophe : Nachrufe auf d. Abendland / Heinz Friedrich. — 1. Aufl. — Hamburg : Hoffmann und Campe, 1979.
312, [1] p. ; 20 cm. GFR79-A
Bibliography: p. 307-[313]
ISBN 3-455-02255-3 : DM24.00
1. Civilization—Philosophy. 2. Civilization, Modern—20th century. I. Title.
CB19.F74 909.08 79-371692
79 MARC

Horia, Vintilă.
Consideraciones sobre un mundo peor / Vintilă Horia. — 1. ed. — Esplugas de Llobregat : Plaza & Janés, 1978.
269 p. ; 22 cm. Sp***
Includes bibliographical references
ISBN 8401331374 : 575ptas
1. Social history—1945- . 2. Moral conditions. 3. Civilization, Modern—20th century. I. Title.
HN17.5.H66 79-340079
79 MARC

Hunter, Robert, 1941-
The enemies of anarchy; a gestalt approach to change. New York, Viking Press [1973, c1970]
x, 254 p. 22 cm. $7.95
Bibliography: p. 237-241.
ISBN 0-670-29497-7
1. Technology and civilization. 2. Civilization, Modern—20th century. I. Title.
HM221.H85 301.24'3 72-9739
73[r79]rev MARC

Jáuregui, José Antonio.
Las reglas del juego : las tribus / [por] José Antonio Jáuregui ; prólogo de José Luis Pinillos. — Madrid : Espasa-Calpe, 1978.
343 p. : ill. ; 17 cm. — (Selecciones Austral ; 40) Sp78-Sept
ISBN 8423920402 : 125ptas.
1. Civilization, Modern—20th century. 2. Culture. I. Title.
CB425.J356 1978 909.82 78-398969
79 MARC

(Kiki no jidai)
危機の時代 / [編集者　金子武蔵，大塚久雄]．— 東京：弘文堂，昭和34 [1959]
2, 334 p., [2] leaves of plates : ports. : 22 cm. — (講座近代思想史 ; 6)
Colophon inserted.
Includes bibliographical references.
1. Philosophy, Modern—19th century. 2. Philosophy, Modern—20th century. 3. Civilization, Modern—19th century. 4. Civilization, Modern—20th century. I. Kaneko, Takezō, 1905- II. Ōtsuka, Hisao. 1907- III. Series: Kōza kindai shisō shi ; 6.
CB113.J3K68 vol. 6 79-805625
[B803]

Kukarkin, Aleksandr Viktorovich.
(Po tu storonu rastsveta)
По ту сторону расцвета : Буржуазное общество : культура и идеология / А. В. Кукаркин. — 2-е изд. — Москва : Политиздат, 1977.
400 p. : ill. ; 26 cm. USSR 77
Bibliography : p. 395-[399]
1.31rub
1. Civilization, Modern—20th century. 2. Intellectual life. 3. Popular culture. 4. Arts and society. I. Title.
CB425.K82 1977 79-409840

Mabille, Pierre.
Égrégores : ou, La Vie des civilisations ... / Pierre Mabille ; [préface] par Radovan Ivsic. — Paris : le Sagittaire, 1977.
234 p. ; 21 cm. — (Contre-type ; 5 ISSN 0338-1161) F78-1778
ISBN 2-7275-0035-1 : 48.00F
1. Europe—Civilization—20th century. 2. Civilization, Modern—20th century. 3. Mabille, Pierre. I. Title.
CB203.M32 940.55 78-400732
79 MARC

Machwe, Prabhakar Balvant, 1917-
Modernity and contemporary Indian literature / Prabhakar Machwe. — 1st ed. — New Delhi : Chetana Publications, 1978, c1977.
viii, 186 p. ; 23 cm.
Includes index.
Rs50.00
1. Indic literature—20th century—History and criticism. 2. Civilization, Modern—20th century. I. Title.
PK5416.M3 1978 809'.8954 78-904093
78 MARC

Müller-Ramelsloh, Otthinrich, 1904-
Der neue Mensch : Weltanschauungen, sozilregulative, programmierte Selbstgesetzlichkeit, Darüberhinauserfahrung / Otthinrich Müller-Ramelsloh. — Darmstadt : Bläschke, [1977?]
185 p. ; 21 cm. GFR77-A
Bibliography: p. 180-181.
Includes index.
ISBN 3-87561-532-8 : DM19.80
1. Man. 2. Civilization, Modern—20th century. 3. Social psychology. I. Title.
BD450.M733 78-357632
78 MARC

Murphey, Dwight D
Understanding the modern predicament / by Dwight D. Murphey. — Wichita, Kan. : New Liberal Library, c1978.
221 p. ; 28 cm.
1. Civilization, Modern—20th century. 2. Alienation (Social psychology) I. Title.
CB425.M85 909.82 78-61119
78 MARC

Nagayasu, Yukimasa, 1941-
(Gendai keizai bummei no seitaigaku)
現代経済文明の生態学：自然 経済 思想 / 永安幸正. — 東京：前野書店，1978.
xiv, 280 p. : ill. ; 19 cm. Ja 78-14952
Includes bibliographical references.
¥1600
1. City planning—Environmental aspects. 2. Urban ecology (Biology) 3. Economic history—20th century. 4. Japan—Economic policy—1945- 5. Civilization, Modern—20th century. I. Title.
HT166.N3 78-812213

Nemésio, Vitorino, 1901-
Era do átomo, crise do homem / Vitorino Nemésio. — Amadora : Livraria Bertrand, c1976.
145 p. ; 19 cm. — (Colecção Ciências sociais e humanas)
1. Civilization, Modern—20th century. I. Title.
CB425.N38 909.82 79-376310
79 MARC

Nobile-Ventura, Attilio.
La maschera e il volto della nostra società. [2. ed.] Milano, C. Marzorati [1964]
146 p. 21 cm.
Bibliographical footnotes.
First ed. published under title: Il volto della nostra società.
1. Civilization, Modern—20th century. I. Title.
CB425.N58 1964 79-343458
79 MARC

Oraison, Marc.
L'apprenti sorcier / Marc Oraison. — Paris : Éditions du Seuil, c1976.
156, [4] p. ; 21 cm. F***
Bibliography: p. [158]
ISBN 2-02-004485-4 : 27.00F
1. Medicine—Philosophy. 2. Medical care. 3. Civilization, Modern—20th century. I. Title.
R723.O72 174'.2 78-376491
78 MARC

Picquart, Jean Gaston.
Les Terriens / J. G. Picquart. — Paris : la Pensée universelle, 1976.
171 p. ; 18 cm. F76-16697
27.90F
1. Civilization, Modern—20th century. I. Title.
CB425.P5174 909.82 78-399632
79 MARC

Posada, Paulino.
La esfinge sin secreto / Paulino Posada. — [1. ed.]. — [Madrid] : Sedmay Ediciones, [1976]
213 p. ; 21 cm. Sp***
ISBN 8473801938
1. Civilization, Modern—20th century. 2. Philosophy, Modern—20th century. 3. Philosophy, Modern—19th century. I. Title.
CB425.P59 909.82 79-352940
79 MARC

Rosselli, Renato.
I filantropi : ovvero, Resoconto al giudice ... / Renato Rosselli. — Firenze : R. Sandron, 1976.
189 p. ; 22 cm. It77-Jan
Three hundred numb. copies. L.C. copy no. 147.
Includes bibliographical references.
L2500
1. Civilization, Modern—20th century. I. Title. II. Title: Resoconto al giudice.
CB425.R78 309.1'09'04 78-389843
*79 MARC

Saint Marc, Philippe.
Progrès ou déclin de l'homme? / Philippe Saint Marc. — Paris : Stock, c1978.
414 p. ; 22 cm. — (Monde ouvert) F***
Bibliography: p. 411-[412]
ISBN 2-234-00857-3 : 49.00F
1. Progress. 2. Civilization, Modern—20th century. 3. Human ecology. I. Title.
HM101.S18 301.31 79-346185
79 MARC

Schumacher, Ernst Friedrich.
Good work / E. F. Schumacher. — 1st ed. — New York : Harper & Row, c1979.
xi, 223 p. : ill. ; 22 cm.
Includes index.
ISBN 0-06-013857-2 : $9.95
1. Technology and civilization. 2. Civilization, Modern—20th century. 3. Social action. 4. Technology—Social aspects. I. Title.
HM221.S365 1979 301.24'3 76-5528
78 MARC

Seymour, John, 1914-
Bring me my bow / [by] John Seymour. — London : Turnstone Books, 1977.
159 p. ; 22 cm. GB78-14782
ISBN 0-85500-078-3 : £3.95. ISBN 0-85500-079-1 pbk.
1. Agriculture—Economic aspects. 2. Civilization, Modern—20th century. 3. Social values. I. Title.
HD1415.S45 301.5'5 79-302259
79 MARC

Sprigg, Christopher St. John, 1907-1937.
The concept of freedom / by Christopher Caudwell [i.e. C. St. J. Sprigg]. — London : Lawrence and Wishart, 1977.
260 p. ; 22 cm. GB77-28270
Includes bibliographical references and index.
ISBN 0-85315-393-0 : £2.00
1. Civilization, Modern—20th century. 2. Physics—Philosophy. I. Title.
CB425.S623 1977 909'.09'812082 78-320316
79 MARC

Stuijvenberg, W van.
Tweede wereldoorlog en heden / W. van Stuijvenberg ; [red. medew. door Snežana Pejaković, Johan Jongma, Kick Sprangers]. — [Amsterdam] : Amsterdam Boek, [1977]
155 p. : ill. ; 28 cm. — (His 5000 [i.e. Vijfduizend] jaar dagelijks leven) Ne77-50
ISBN 9031898430 : fl 29.50
1. Civilization, Modern—20th century. 2. Manners and customs. I. Kićović-Pejaković, Snežana. II. Jongma, Johan. III. Sprangers, Kick. IV. Title. V. Series.
GT150.S8 78-394028
*79 MARC

Toraldo di Francia, Giuliano, 1916-
Il rifiuto : considerazioni semiserie di un fisico sul mondo di oggi e di domani / [di] G. Toraldo di Francia. — Torino : Einaudi, 1978.
129 p. ; 18 cm. — (Nuovo politecnico ; 99) It78-May
L2400
1. Social history—20th century. 2. Civilization, Modern—20th century. I. Title.
HN16.T647 309.1'04 78-390176
*79 MARC

Trotignon, Yves.
Le monde du XXe [i.e. vingtième] siècle : un outil d'investigation et de compréhension de notre époque / Yves Trotignon. — Paris : Fayard, c1978.
534 p. : ill. ; 25 cm. F***
Bibliography: p. [503]-[504]
Includes index.
ISBN 2-213-00646-6 : 89.00F
1. Civilization, Modern—20th century. I. Title.
CB425.T69 909.82 79-367927
79 MARC

Tsuchida, Kyōson.
(Bummei wa doko e yuku)
文明は何處へ行く / 土田杏村著. — 東京：千倉書房，1930.
6, 4, 338 p. ; 19 cm.
1. Civilization, Modern—20th century. 2. Twentieth century—Forecasts. I. Title.
CB425.T74 79-801763

Villella, Argo.
Una via sociale / Argo Villella. — Milano : Il falco, 1978.
269 p. ; 21 cm. — (Collana Il Ponte ; n. 3) It***
Includes bibliographical references and index.
L8500
1. Social history—20th century. 2. Civilization, Modern—20th century. I. Title.
HN16.V54 309.1'04 79-362560
79 MARC

Yabuki, Keiki, 1879-1939.
(Gendaijin to Bukkyō)
現代人と佛教　矢吹慶輝著　[東京]　三省堂 [昭和16 i. e. 1941]
2, 2, 7, 263 p. 19 cm. (青年佛教叢書　第9編)
1. Buddhism—Japan. 2. Civilization, Modern—20th century. I. Title.
BQ674.Y32 1941 79-809454

CIVILIZATION MODERN (Continued)

—20TH CENTURY—ADDRESSES, ESSAYS, LECTURES

Lin, Ch'ing-chiang.
(Hsin li chien she yü wen hua fa chan)
心理建設與文化發展 / 林清江講；教育部社會教育司編印. — ⌜台北⌝：該司, 民國67 ⌜1978⌝
14 p. ; 20 cm. — (文化講座專集 ; 128)

1. Civilization, Modern—20th century—Addresses, essays, lectures. I. China (Republic of China). Chiao yü pu. She hui chiao yü ssu. II. Title. III. Series: Wen hua chiang tso chuan chi ; 128.
CB428.L5 79-839518

(Minshushugi to tetsugaku)
民主主義と哲學 / 植田清次編. — 東京：統正社, 昭和23 ⌜1948⌝
2, 349 p. ; 19 cm.

1. Social sciences—Addresses, essays, lectures. 2. Democracy—Addresses, essays, lectures. 3. Civilization, Modern—20th century—Addresses, essays, lectures. I. Ueda, Seiji, 1902-
H35.M56 79-808074

Ōki, Hideo, 1928-
(Gendaijin no Yudayajinka)
現代人のユダヤ人化：現代文明論集 / 大木英夫. — 東京：白水社, 1976.
326 p. ; 19 cm. — (白水叢書 ; 5) Ja 77-103
Series romanized: Hakusui sōsho.
Includes bibliographical references.
¥1200

1. Civilization, Modern—20th century—Addresses, essays, lectures. 2. Civilization—Philosophy—Addresses, essays, lectures. I. Title.
CB428.O35 78-813264

Tendencias y movimientos actuales / J. Ma. González Ruiz ... ⌜et al.⌝ ; dirección y presentación José Antonio Merino. — Madrid : Cisneros, 1978.
172 p. ; 22 cm. Sp78-Mar
"Artículos aparecidos en la revista Verdad y vida, correspondientes a los núms. 139 y 140 de 1977."
Includes bibliographical references.
ISBN 8470470183
1. Christianity—20th century—Addresses, essays, lectures. 2. Civilization, Modern—20th century—Addresses, essays, lectures. I. González Ruiz, José María, 1916- . II. Merino, José Antonio.
BR481.T45 78-388705
79 MARC

Umesao, Tadao, 1920-
(Chikyū jidai no jinruigaku)
地球時代の人類学 / 梅棹忠夫. — 東京：中央公論社, 1978.
505 p. ; 20 cm. Ja 79-3751
"対談集"
Originally published in Hōsō Asahi, Jan.-Dec. 1975.
Includes bibliographies.
¥1500
1. Civilization, Modern—20th century—Addresses, essays, lectures. 2. Intellectuals—Japan—Interviews. I. Title.
CB430.U43 79-809062

Yi, Wŏn-sŏl, 1930-
(Sajo ŭi kyŏngnang sok esŏ)
思潮의激浪속에서 / 李元卨著. — 서울：聖光文化社, 1978.
515 p. ; 20 cm.
Includes bibliographical references.
₩2500

1. Civilization, Modern—20th century—Addresses, essays, lectures. 2. Korea—Civilization—20th century—Addresses, essays, lectures. I. Title.
CB425.Y5 79-825939

—20TH CENTURY—CONGRESSES

Människan i tekniksamhället : föredrag och diskussioner vid Vitterhetsakademiens konferens 25-27 januari 1977 / ⌜red. av Manfred Ribbing⌝. — Stockholm : Kungl. Vitterhets-, historie- o. antikvitetsakad. : Almqvist & Wiksell international (distr.), 1977.
⌜5⌝, 231, ⌜1⌝ p. : ill., tables ; 22 cm. — (Konferenser - Kungl. Vitterhets, historie och antikvitets akademien ; 1) S78-12
Includes bibliographical references.
ISBN 9174020315
1. Technology and civilization—Congresses. 2. Technology—Social aspects—Congresses. 3. Social history—20th century—Congresses. 4. Civilization, Modern—20th century—Congresses. I. Ribbing, Manfred. II. Vitterhets-, historie-, och antikvitetsakademien, Stockholm. III. Series: Vitterhets-, historie-, och antikvitetsakademien, Stockholm. Konferenser - Kungl. Vitterhets, historie och antikvitets akademien ; 1.
HM221.M283 78-392974
79 MARC

(Rekishi to bummei no tankyū)
歴史と文明の探求：文明問題懇談会全記録 / 桑原武夫, 中根千枝, 加藤秀俊編. — 東京：中央公論社, 昭和51- ⌜1976-
v. ; 20 cm. Ja 76-14210 (v. 1)
¥950 (v. 1)

1. Civilization, Modern—20th century—Congresses. I. Kuwabara, Takeo, 1904- II. Nakane, Chie, 1926- III. Katō, Hidetoshi, 1930- IV. Bummei Mondai Kondankai.
CB428.R45 78-812191

—20TH CENTURY—MISCELLANEA

Holman, Sona.
How to lie about your age / by Sona Holman & Lillian Friedman. — New York : Collier Books, c1979.
xii, 243 p. ; 21 cm.
ISBN 0-02-000550-4
1. Civilization, Modern—20th century—Miscellanea. I. Friedman, Lillian, joint author. II. Title.
CB427.H64 1979 909.82 78-32038
79 MARC

Mclan, Tomás.
Mentiras del mundo moderno : memorias del muy mirífico mundo moderno de Moloch y de Mamona / Tomás Mclan. — Buenos Aires : Cruz y Fierro, c1976.
214 p. ; 21 cm.
Includes bibliographical references and index.
1. Civilization, Modern—20th century—Miscellanea. I. Title.
CB427.M268 909.82 78-101835
78 MARC

—20TH CENTURY—PICTORIAL WORKS

Unser 20. ⌜i.e. zwanzigstes⌝ Jahrhundert / hrsg. von Herbert Reinoss ; die Texte schrieb Ludwig Knoll. — Gütersloh : Bertelsmann-Lexikon-Verlag, 1978.
234 p. : chiefly ill. (some col.) ; 30 cm. GFR78-A
Includes index.
ISBN 3-570-08293-8 : DM34.00
1. Civilization, Modern—20th century—Pictorial works. I. Reinoss, Herbert. II. Knoll, Ludwig.
CB427.U57 78-371563
78 MARC

—1945-

Le Compte, Herman.
Wir Kinder des Methusalem : Alters-Chancen d. jungen Gesellschaft / H. Le Compte, C. Northcote Parkinson ; ⌜Übers. d. engl. Ms. von Ursel Reinecke⌝. — Düsseldorf ; Wien : Econ-Verlag, 1978.
221 p. : ill. ; 21 cm. GFR78-A
Includes index.
ISBN 3-430-15908-3 : DM26.00
1. Social history. 2. Civilization, Modern—1945- . 3. Longevity. I. Parkinson, Cyril Northcote, 1909- joint author. II. Title.
HN8.L4 309.1'04 79-346184
79 MARC

—1950-

Arendt, Hannah.
Between past and future : eight exercises in political thought / Hannah Arendt. — Enlarged ed. — New York : Penguin Books, 1978, c1968.
306 p. ; 20 cm.
Reprint of the ed. published by Viking Press, New York.
Includes bibliographical references and index.
ISBN 0-14-004662-3
1. Civilization, Modern—1950- . 2. History—Philosophy. I. Title.
⌜CB427.A7 1978⌝ 909.82 77-7662
77 MARC

Backer, Rika de, 1923-
De groene jaren zeventig / Rika de Backer-van Ocken. — Antwerpen : De Nederlandsche Boekhandel, ⌜1979⌝
100 p. : with ill. ; 21 cm. Ne79-6
ISBN 9028903844 : 195F
1. Civilization, Modern—1950- . I. Title.
CB428.B28 79-362700
*79 MARC

Bahm, Archie J
The philosopher's world model / Archie J. Bahm. — Westport, Conn. : Greenwood Press, 1979.
viii, 295 p. ; 22 cm. — (Contributions in philosophy ; no. 12 ISSN 0084-926X)
Includes bibliographical references and index.
ISBN 0-313-21198-1
1. Civilization—Philosophy. 2. Civilization, Modern—1950- . 3. Philosophy. I. Title.
CB19.B272 901 78-67569
78 MARC

Baltinienė, Elvyra.
Buržuazinė "masinė kultūra" ir jaunimas / E. Baltinienė. — Vilnius : Mintis, 1976.
120 p. ; 16 cm. USSR76-27447
Includes bibliographical references.
0.16rub
1. Culture. 2. Popular culture. 3. Youth. 4. Civilization, Modern—1950- I. Title.
HM101.B255 78-399923
79 MARC

Bauer, François Charles, 1919-
Une année pas comme les autres / François Chalais ⌜i.e. F. C. Bauer⌝. — ⌜Paris⌝ : Stock, c1978.
396 p. ; 23 cm. F***
ISBN 2-234-00821-2 : 45.00F
1. Bauer, François Charles, 1919- —Biography. 2. Authors, French—20th century—Biography. 3. Civilization, Modern—1950- . I. Title.
PQ2603.A834Z463 791.43'09 79-382330
79 MARC

Champoiseau, Jacques.
Sur le chemin de l'esclavage : appel à la jeunesse / Jacques Champoiseau. — Versailles (9, Brd des Jeux Olympiques, 78000) : J. Champoiseau, 1977.
250 p. ; 22 cm. F77-13643
40.00F
1. Civilization, Modern—1950- . 2. World politics—1945- . I. Title.
CB428.C46 320'.904 79-358403
79 MARC

Chaunu, Pierre.
La mémoire et le sacré / Pierre Chaunu. — ⌜Paris⌝ : Calmann-Lévy, c1978.
284 p. : graph ; 23 cm. — (Archives des sciences sociales) F***
Includes bibliographical references.
ISBN 2-7021-0236-0 : 55.00F
1. Civilization, Occidental. 2. Civilization, Modern—1950- . I. Title.
CB245.C47 79-389200
79 MARC

Chaunu, Pierre.
La Peste blanche : comment éviter le suicide de l'Occident / Pierre Chaunu, Georges Suffert. — ⌜Paris⌝ : Gallimard, 1976.
263 p. ; 22 cm. — (L'Air du temps) F76-18025
37.00F
1. Civilization, Occidental. 2. Civilization, Modern,—1950- . I. Suffert, Georges, joint author. II. Title.
CB245.C513 909'.09'821082 79-356946
79 MARC

Chaunu, Pierre.
Le sursis : l'ardeur et la modération / Pierre Chaunu. — Paris : R. Laffont, c1979
335 p. ; 22 cm. — (Libertés 2000 ⌜i.e. deux mille⌝) F***
Includes bibliographical references.
ISBN 2-221-00179-6 : 54.00F
1. Civilization, Occidental. 2. Civilization, Modern—1950- . I. Title.
CB245.C514 909.82 79-355190
79 MARC

Davidov, Nesho.
("Moderni" utopii ili kŭde svŭrshva kulturata?)
"Модерни" утопии или къде свършва културата? : ⌜изследване⌝ / Нешо Давидов. — 1. изд. — София : Партиздат, 1977.
95 p. ; 20 cm. Bu 77-2216
Includes bibliographical references.
0.31 lv
1. Civilization, Modern—1950- I. Title.
CB428.D38 78-422048

Davis, William Jackson, 1942-
The seventh year : industrial civilization in transition / by W. Jackson Davis. — 1st ed. — New York : Norton, c1979.
xx, 296 p. : ill. ; 24 cm.
Includes bibliographical references and index.
ISBN 0-393-05693-7
1. Technology and civilization. 2. Civilization, Modern—1950- . 3. Natural resources. I. Title.
CB478.D275 1979 909.82 79-1324
79 MARC

CIVILIZATION, MODERN
—1950- (Continued)

Davy, Charles.
Towards a third culture / Charles Davy. — Rev. ed. — Edinburgh : Floris Books, 1978. GB***
178 p. ; 18 cm.
Includes bibliographical references and index.
ISBN 0-903540-14-2 : £2.25
1. Science and civilization. 2. Science and the humanities. 3. Civilization, Modern—1950- I. Title.
CB478.D28 1978 909 78-322467
79 MARC

Degregori, Pedro Mario.
El siglo que viene y el hombre que va / Pedro Mario Degregori. — [Buenos Aires] : Plus Ultra, 1977.
215 p. ; 20 cm.
Bibliography: p. 213-215.
1. Twentieth century—Forecasts. 2. Civilization, Modern—1950- I. Title.
CB161.D43 78-101747
79 MARC

Delaval, Albert.
L'homme en quête d'un destin / Albert Delaval. — Paris : Courrier du livre, c1975. F***
128 p. ; 22 cm.
ISBN 2-7029-0024-0
1. Civilization, Modern—1950- I. Title.
CB428.D43 909.8 79-340052
79 MARC

Deshusses, Jérôme.
Délivrez Prométhée / Jérôme Deshusses. — [Paris] : Flammarion, c1978. F***
392 p. ; 24 cm.
Includes bibliographical references.
ISBN 2-08-064083-6 : 68.00F
1. Civilization, Modern—1950- I. Title.
CB428.D46 909.82 79-352535
79 MARC

Dialoguri contemporane : [cu Remus Rădulet, Mircea Malița, Sabin Bălașa, et al.] / Romulus Balaban. — București : Editura Eminescu, 1976. R76-2707
197 p. ; 20 cm.
lei6.00
1. Civilization, Modern—1950- 2. Interviews. I. Balaban, Romulus.
CB428.D495 78-400432
*79 MARC

Drăgănescu, Mihai.
Sistem și civilizație : eseuri / Mihai Drăgănescu. — București : Editura Politică, 1976. R76-3962
502 p. : ill. ; 19 cm. — (Idei contemporane)
Summary in English.
Table of contents in Romanian, English, French, German, and Russian.
Includes bibliographical references.
lei14.00
1. Science and civilization. 2. Technology and civilization. 3. Civilization, Modern—1950- 4. Communism and society. I. Title.
CB478.D73 78-378253
*78 MARC

Edwards, David Lawrence.
A reason to hope / [by] David L. Edwards. — London ; New York : Collins, 1978. GB78-29007
252 p. ; 22 cm.
Includes index.
ISBN 0-529-05620-8 : £4.95 ($8.95)
1. Civilization, Modern, 1950- 2. Christianity—20th century. 3. Christianity and culture. I. Title.
CB430.E34 1978 909 78-56996
78 MARC

Elchinger, Léon Arthur, Bp., 1908-
Je plaide pour l'homme / Léon Arthur Elchinger. — Paris : Fayard, c1976. F***
301 p. ; 22 cm.
Includes bibliographical references.
ISBN 2-213-00357-2 : 37.00F
1. Civilization, Modern—1950- 2. Christianity—20th century. I. Title.
CB430.E42 909.82 79-341070
79 MARC

Gorney, Roderic, 1924-
The human agenda : how to be at home in the universe-without magic / Roderic Gorney. — [3d ed.]. — Los Angeles : Guild of Tutors Press, c1979.
698 p. ; 24 cm.
Bibliography: p. 605-627.
Includes index.
ISBN 0-89615-014-3 : $8.50
1. Values. 2. Civilization, Modern—1950- I. Title.
BF778.G64 1979 171'.7 78-75233
00 MARC

Gouraige, Ghislain.
Les semences de l'esprit : propos sur l'esclavage moderne / Ghislain Gouraige. — Sherbrooke, Québec : Éditions Naaman, [1978?]
58 p. : port. ; 20 cm. — (Collection Pour tous ; 5) C79-1164-6
$5.00
1. Civilization, Occidental. 2. Civilization, Modern—1950- I. Title.
CB245.G63 909'.09'821 79-357721
79 MARC

Grand'Maison, Jacques.
Quel homme? / Jacques Grand'Maison. — [Montréal] : Leméac, [c1978] C***
147 p. ; 20 cm. — (Collection Quelle?)
Includes bibliographical references.
ISBN 0-7761-0200-1 : $7.75
1. Civilization, Modern—1950- 2. Civilization, Occidental. I. Title.
CB428.G7 909.82 79-362111
79 MARC

Hunter, Robert, 1941-
The storming of the mind. Garden City, N.Y., Doubleday, 1971.
233 p. 22 cm. $6.95
Includes bibliographical references.
1. Social history—1945- 2. Civilization, Modern—1950- 3. Conflict of generations. I. Title.
HN18.H785 1971b 301.29'73 79-172519
72[r79]rev2 MARC

Kamiński, Bartłomiej, 1944-
Świat obecny, świat przyszły / Bartłomiej Kamiński, Marek Okólski. — Wyd. 1. — Warszawa : Wiedza Powszechna, 1978 (Byt : ZGK 5)
276 [4] p. : ill. ; 19 cm. — (Biblioteka wiedzy powszechnej Omega) P78-2711
Bibliography: p. 273-[277]
zł30.00
1. Forecasting. 2. Civilization, Modern—1950- I. Okólski, Marek, joint author. II. Title.
CB158.K35 78-391175
79 MARC

Kirimura, Yasuji, 1938-
(Asu no bummei o motomete)
明日の文明を求めて：西欧知識人との対話 / 桐村泰次. — 東京：第三文明社，昭和51 [1976]
238 p. ; 20 cm. Ja 76-17767
Originally published in Daibyakurenge during 1969 under title: Seiō to no taiwa.
¥960
1. Civilization, Modern—1950- 2. Intellectuals—Interviews. I. Title.
CB430.K53 1976 79-800390

Kraus, Wolfgang, 1924-
Die verratene Anbetung : Verlust und Wiederkehr der Ideale / Wolfgang Kraus. — München : R. Piper, c1978. GFR***
191 p. ; 22 cm.
Bibliography: p. [182]-191.
ISBN 3-492-02378-9
1. Civilization, Modern—1950- I. Title.
CB428.K73 78-393291
79 MARC

Krutch, Joseph Wood, 1893-1970.
Human nature and the human condition / by Joseph Wood Krutch. — Westport, Conn. : Greenwood Press, 1979, c1959.
211 p. ; 23 cm.
Reprint of the ed. published by Random House, New York.
ISBN 0-313-21010-1
1. Civilization, Modern—1950- I. Title.
[CB428.K78 1979] 909.8 78-12131
78 MARC

Laszlo, Ervin, 1932-
The inner limits of mankind : heretical reflections on today's values, culture, and politics / by Ervin Laszlo ; with a commentary by Aurelio Peccei. — Oxford [Eng.] ; New York : Pergamon Press, 1978.
xiii, 79 p. ; 22 cm. — (Pergamon international library of science, technology, engineering and social studies) (Systems science and world order library) GB78
Includes bibliographical references.
ISBN 0-08-023013-X : £6.50 ($12.00 U.S.). ISBN 0-08-023012-1 pbk. : £3.25 ($6.00 U.S.)
1. Civilization, Modern—1950- I. Title.
CB428.L27 1978 909.82 77-30732
78 MARC

Lewis, Peter, 1928-
The fifties / by Peter Lewis. — 1st American ed. — Philadelphia : Lippincott, c1978.
256 p. : ill. ; 26 cm.
Includes index.
ISBN 0-397-01355-8
1. Civilization, Modern—1950- I. Title.
CB425.L38 1978 909.82'5 78-32176
79 MARC

Loeser, Franz, 1924-
Durchbruch des neuen Geschlechts : Schöpfertum und Moral der Zukunft / Franz Loeser. — Berlin : Verlag Neues Leben, c1976.
174 p. : ill. ; 20 cm. — (NL konkret ; 30) GDR***
Includes bibliographical references.
4.00M
1. Civilization, Modern—1950- 2. Ethics, Modern—20th century. I. Title.
CB428.L62 909.82 79-352863
79 MARC

Mackay, Malcolm G 1919-
More than coincidence / Malcolm G. Mackay. — Edinburgh : Saint Andrew Press, 1979. GB***
242 p. ; 23 cm.
Includes bibliographical references.
ISBN 0-7152-0416-5 : £6.95
1. Christian life—1960- 2. Civilization, Modern—1950- I. Title.
BV4501.2.M24145 261 79-315306
79 MARC

Mumford, Lewis, 1895-
Lewis Mumford, an interview with the modern visionary. [Sound recording] N. Hollywood, Calif., Center for Cassette Studies [1973?]
1 cassette.
Duration: 26 min.
Notes and bibliography in container.
SUMMARY: Humanist Lewis Mumford discusses democracy and its inability to function in mass society. He points out that complaisance and tolerance of corruption can be corrected only through each person's rigorous examination of himself or herself. He notes the dangers of technology, concluding that explorations in the future will expand man's knowledge and awareness of the world and the inner self.
1. Civilization, Modern—1950- 2. Democracy. 3. Technology and civilization.
[CB428] 79-740600

Murray, Bruce C
Navigating the future / Bruce C. Murray. — 1st ed. — New York : Harper & Row, [1975]
xii, 175 p. ; 22 cm.
Includes bibliographical references.
ISBN 0-06-013122-5 : $7.95
1. Forecasting. 2. Civilization, Modern—1950- I. Title.
CB161.M87 1975 909.82 74-1839
75[r79]rev MARC

Powell, Robert, 1918-
The free mind; the inward path to liberation. New York, Julian Press [1972]
xi, 175 p. 22 cm. $6.00
1. Liberty. 2. Intellect. 3. Civilization, Modern—1950- I. Title.
B824.4.P66 128'.2 79-184317
72[r79]rev MARC

Pozzo, Gianni M
Riflessioni critiche sul nostro tempo / Gianni P. Pozzo. — Padova : CEDAM, 1977. It77-July
137 p. ; 22 cm.
L4200
1. Civilization, Modern—1950- 2. Science and civilization. I. Title.
CB428.P69 909.82'7 79-363407
*79 MARC

Ruyer, Raymond, 1902-
Le sceptique résolu : devant les discours intimidants / Raymond Ruyer. — Paris : R. Laffont, c1979. F***
315 p. ; 22 cm. — (Libertés 2000 [i.e. deux mille])
Includes bibliographical references.
ISBN 2-221-00256-3 : 54.00F
1. Civilization, Modern—1950- 2. Social reformers. 3. Impostors and imposture. I. Title.
CB428.R88 909.82 79-377710
79 MARC

Schwarz, Arturo, 1924-
Conversazione con Herbert Marcuse / [a cura di] Arturo Schwarz. — Milano : Multhipla, 1978. It78-May
47 p. ; 21 cm.
Bibliography: p. 38-45.
L1600
1. Social history—1945- 2. Civilization, Modern—1950- 3. Marcuse, Herbert, 1898- I. Marcuse, Herbert, 1898- II. Title.
HN17.5.S37 78-381442
*78 MARC

Sinai, I Robert.
The decadence of the modern world / I. Robert Sinai. — Cambridge, Mass. : Schenkman Pub. Co., 1977.
230 p. ; 23 cm.
Includes bibliographical references and index.
ISBN 0-87073-568-3
1. Civilization, Modern—1950- 2. Degeneration. I. Title.
CB425.S52 909.82 76-30409
76 MARC

Vadeboncoeur, Pierre, 1920-
L'autorité du peuple / Pierre Vadeboncoeur ; préf. de François Ricard. — Montréal : Éditions HMH, c1977. C***
132 p. ; 21 cm. — (Collection Constantes ; 34)
ISBN 0-7758-0110-0
1. Civilization, Modern—1950- 2. Capitalism. I. Title.
CB428.V32 909.82 79-353080
79 MARC

CIVILIZATION, MODERN —1950- (Continued)

Vajk, J Peter, 1942-
Doomsday has been cancelled / J. Peter Vajk. — Culver City, Calif. : Peace Press, 1978.
xvii, 238 p. : ill. ; 23 cm.
Bibliography: p. 229-234.
Includes index.
ISBN 0-915238-24-1 : $7.95
1. Twenty-first century—Forecasts. 2. Technology and civilization. 3. Civilization, Modern—1950- I. Title.
CB161.V35 909.83 78-62300
78 MARC

Wu, Te-yao.
Man and the global crisis / by Wu Teh-yao. — ₍Singapore₎ : Dept. of Political Science, University of Singapore, 1975.
26 p. ; 28 cm. — (Occasional paper series - Department of Political Science, University of Singapore ; no. 14)
1. World politics—1945- . 2. Civilization, Modern—1950- I. Title. II. Series: Singapore (City). University. Dept. of Political Science. Occasional paper series - Department of Political Science, University of Singapore ; no. 14.
D843.W8 309.1'09'047 76-940112
79 MARC

—1950- —ADDRESSES, ESSAYS, LECTURES

(Adhunātana pariveśa aura sṛjana kī samasyāyeṃ)
अधुनातन परिवेश और सृजन की समस्यायें : मधुमती परिचर्चा अंक / सम्पादक नवलकिशोर ... ₍et al.₎. — उदयपुर : राजस्थान साहित्य अकादमी प्रकाशन (संगम), 1975.
159 p. ; 26 cm.
"₍राजस्थान₎ साहित्य अकादमी की मुख पत्रिका 'मधुमती' ने ... अपने विशेषांक 'अधुनातन परिवेश और सृजन की समस्याएं' के रूप में एक व्यवस्थित मंच प्रस्तुत किया."
In Hindi.
Includes bibliographical references.
Rs5.00

1. Hindi literature—1947- —History and criticism—Addresses, essays, lectures. 2. Civilization, Modern—1950- —Addresses, essays, lectures. 3. India—Civilization—Addresses, essays, lectures. 4. Creation (Literary, artistic, etc.)—Addresses, essays, lectures. I. Nawal Kishore, 1932-

PK2039.A3 75-904044

Approaches to popular culture / edited by C. W. E. Bigsby. — Bowling Green, Ohio : Bowling Green University Popular Press, c1976.
viii, 280 p. : ill. ; 23 cm.
Includes bibliographies and index.
ISBN 0-87972-083-2
1. Civilization, Modern—1950- —Addresses, essays, lectures. 2. Popular culture—Addresses, essays, lectures. 3. Communication—Addresses, essays, lectures. I. Bigsby, C. W. E.
CB428.A7 1976 301.2 79-119755
79 MARC

Goals in a global community : the original background papers for Goals for mankind, a report to the Club of Rome / edited by Ervin Laszlo and Judah Bierman. — New York : Pergamon Press, c1977.
2 v. : ill. ; 27 cm.
Vol. 2 has subtitle: The original background studies for the goals for mankind.
Includes bibliographies and index.
CONTENTS: v. 1. Studies on the conceptual foundations.—v. 2. The international values and goals studies.
ISBN 0-08-022221-8 (v. 1). ISBN 0-08-022973-5 (v. 2)
1. Civilization, Modern—1950- —Addresses, essays, lectures. 2. International cooperation—Addresses, essays, lectures. I. Laszlo, Ervin, 1932- . II. Bierman, Judah. III. Club of Rome. IV. Goals for mankind.
CB430.G63 1977 909.82 77-79971
78₍r78₎rev MARC

Hosokawa, Takamoto, 1900-
(Jikyoku tembō)
時局展望 / 細川隆元. — 東京 : 尾崎行雄記念財団, 昭和51 ₍1976₎
27 p. ; 19 cm. — (討論集会シリーズ - 尾崎行雄記念財団 : no. 28)
Ja***
Cover title.
1. Japan—Civilization—1945- —Addresses, essays, lectures. 2. Civilization, Modern—1950- —Addresses, essays, lectures. I. Title. II. Series: Ozaki Yukio Kinen Zaidan. Tōron shūkai shirizu ; no. 28.
DS822.5.H66 79-800567

Où va la culture? = Hat Kultur Zukunft? / Par August Flammer ... ₍et al.₎. — Fribourg : Éditions universitaires, 1977.
76 p. ; 21 cm. — (Défis et dialogues ; 2) Sw78-379
French or German.
Includes bibliographical references.
ISBN 3-7278-0178-6 : 15.00F
1. Culture—Addresses, essays, lectures. 2. Civilization, Modern—1950- —Addresses, essays, lectures. I. Flammer, August. II. Title: Hat Kultur Zukunft? III. Series.
CB430.O9 78-384327
78 MARC

Rice, Moyle Q
An otherwise report / M. Q. Rice. — Logan : Utah State University Faculty Association, 1975.
29 p. ; 23 cm. — (Faculty honor lecture ; 52d)
1. Education—Addresses, essays, lectures. 2. Education, Higher—Addresses, essays, lectures. 3. Civilization, Modern—1950- —Addresses, essays, lectures. 4. Ethics—Addresses, essays, lectures. I. Title. II. Series: Utah. State University of Agriculture and Applied Science, Logan. Faculty Association. Annual faculty honor lecture ; 52d.
LB41.R535 370 76-374282
76₍r79₎rev MARC

Scalia, Gianni.
De anarchia : attorno al '68 poesia, follia, rivoluzione / Gianni Scalia. — Roma : Savelli, 1978.
218 p. ; 21 cm. — (Saggistica ; 84) It78-July
Bibliography: p. 218.
L4500
1. Civilization, Modern—1950- —Addresses, essays, lectures. I. Title.
CB428.S28 78-377975
*78 MARC

Solzhenitsyn, Aleksandr Isaevich, 1918-
A world split apart : commencement address delivered at Harvard University, June 8, 1978 / Aleksandr I. Solzhenitsyn ; ₍translated from the Russian by Irina Ilovayskaya Alberti₎. — 1st ed. — New York : Harper & Row, c1978.
61 p. ; 20 cm.
Translation of Raskolotyĭ mir.
English and Russian.
ISBN 0-06-014007-0 : $6.95. ISBN 0-06-090690-1 pbk. : $2.95
1. Civilization, Occidental—Addresses, essays, lectures. 2. Civilization, Modern—1950- —Addresses, essays, lectures. 3. Europe, Eastern—Civilization—Addresses, essays, lectures. 4. Communism—Addresses, essays, lectures. I. Title.
CB245.S56913 1978 909.82 78-19593
78 MARC

(Tokuhain no kokusai taiken)
特派員の国際体験 / 日本放送出版協会編. — 東京 : 同協会, 昭和52 ₍1977₎
259 p. : ill. ; 19 cm. Ja 77-25343
¥800

1. World politics—1965-1975—Addresses, essays, lectures. 2. Civilization, Modern—1950—Addresses, essays, lectures. I. Nippon Hōsō Shuppan Kyōkai.
D1058.T58 79-800162

Volpini, Valerio, 1923-
Cioro al ciero : imprudenze di un cattolico del consenso / Valerio Volpini. — 1. ed. — Milano : Rusconi, 1978.
245 p. ; 22 cm. — (Pamphlet) It79-Jan
Articles previously published in Il Tempo, 1977.
L5000
1. Italy—Civilization—1945- —Addresses, essays, lectures. 2. Civilization, Modern—1950- —Addresses, essays, lectures.
DG451.V64 945.092 79-353088
*79 MARC

Weizsäcker, Carl Friedrich, Freiherr von, 1912-
Deutlichkeit : Beiträge zu politischen und religiösen Gegenwartsfragen / Carl Friedrich von Weizsäcker. — München : Hanser, c1978.
182 p. ; 21 cm. GFR***
ISBN 3-446-12623-6
1. Civilization, Modern—1950- —Addresses, essays, lectures. I. Title.
CB430.W48 909.82 78-399904
79 MARC

Zukunftsperspektiven : Praxisberichte u. theoret. Ansätze zu e. integrierten Verständnis d. Lebenswelt / ₍hrsg. von₎ A. M. Klaus Müller ; mit Beitr. aus Wirtschaft u. Technik, Wiss. u. Theologie von Günter Altner ... ₍et al.₎ ; u.e. Einf. d. Hrsg. — Stuttgart : Steinkopf, 1976.
258 p. : ill. ; 21 cm. GFR77-A
Includes bibliographical references.
ISBN 3-7984-0324-4 : DM24.00
1. Twentieth century—Forecasts—Addresses, essays, lectures. 2. Civilization, Modern—1950- —Addresses, essays, lectures. I. Müller, A. M. Klaus.
CB161.Z82 78-401060
79 MARC

—1950- —ANECDOTES, FACETIAE, SATIRE, ETC.

Macaulay, David.
Motel of the mysteries / David Macaulay. — Boston : Houghton Mifflin, 1979.
95 p. : ill. ; 29 cm.
ISBN 0-395-28424-4 : $9.95
1. Civilization, Modern—1950- —Anecdotes, facetiae, satire, etc. 2. North America—Antiquities—Anecdotes, facetiae, satire, etc. I. Title.
PN6231.C46M3 818'.5'407 79-14860
79 MARC

—1950- —CONGRESSES

North American Congress of the Laity, Los Angeles, 1978.
At the edge of hope : Christian laity in paradox / ₍edited₎ by Howard Butt, with Elliott Wright with contributions by Peter L. Berger ... ₍et al.₎. — New York : Seabury Press, c1978.
viii, 211 p. ; 22 cm.
"A Crossroad book."
ISBN 0-8164-0414-3
1. Laity—Congresses. 2. Civilization, Modern—1950- —Congresses. 3. Social ethics—Congresses. I. Butt, Howard. II. Wright, H. Elliott, 1937- . III. Title.
BV687.N67 1978 262'.15 78-11729
78 MARC

—1950- —PERIODICALS

Nueva frontera. Documentos. 1-
feb. 1976-
₍Bogotá, Nueva Frontera₎
no. ill. 34 cm. monthly.
Other title, Febr. 1976- : Documentos Nueva frontera.
Issued as supplement to: Nueva frontera.
1. Colombia—Economic conditions—1970- —Periodicals. 2. Colombia—Civilization—Periodicals. 3. Civilizaton, Modern—1950- —Perodicals. I. Nueva frontera. II. Title: Documentos Nueva frontera.
HC196.N84 330.9'861 78-648733
MARC-S

—1950- —PHILOSOPHY

Jaquillard, Claude.
Requiem pour une révolution perdue : (l'errance helvétique) / Claude Jaquillard. — Lausanne (Pyramides 14) : L'Aire (M. Moret), cop. 1978.
73 p. ; 17 cm. Sw78-6112
10.50
1. Civilization, Modern—1950- —Philosophy. 2. Revolutions. I. Title.
CB430.J36 909.82 78-388073
78 MARC

—ADDRESSES, ESSAYS, LECTURES

En marge : l'Occident et ses "autres" / Alberto Asor Rosa ... ₍et al.₎. — Paris : Aubier Montaigne, c1978.
291 p. ; 22 cm. — (Présence et pensée) F***
Includes bibliographical references.
ISBN (invalid) 2-7007-0118-8 : 69.00F
1. Social history—20th century—Addresses, essays, lectures. 2. Civilization, Modern—Addresses, essays, lectures. I. Asor Rosa, Alberto.
HN16.E53 309.1'04 79-354357
79 MARC

Hertz, Paweł.
Miary i wagi / Paweł Hertz. — Wyd. 1. — Warszawa : Państ. Instytut Wydawniczy, 1978 (Wroc. : WZG)
220 p. ; 19 cm. P78-5711
Articles previously published in Tygodnik powszechny in 1974-1975.
Includes index.
zł20.00
1. Literature, Modern—History and criticism—Addresses, essays, lectures. 2. Civilization, Modern—Addresses, essays, lectures. I. Tygodnik powszechny. II. Title.
PN710.H44 78-398074
79 MARC

Igov, Svetlozar.
(Khumanizŭm i tvorchestvo)
Хуманизъм и творчество : проблеми на съвр. проза и критика / Светлозар Игов. — София : Бълг. писател, 1978 (В. Търново : печ. Д. Найденов)
313 p. ; 21 cm. Bu 79-519
Includes bibliographical references.
1.70 lv
1. Bulgarian literature—20th century—History and criticism—Addresses, essays, lectures. 2. Criticism—Bulgaria—Addresses, essays, lectures. 3. Humanism—Addresses, essays, lectures. 4. Civilization, Modern—Addresses, essays, lectures. I. Title.
PG1008.I 3 79-413288

Nelson, Benjamin N 1911-
Der Ursprung der Moderne : vergleichende Studien zum Zivilisationsprozess / Benjamin Nelson ; übers. von Michael Bischoff. — 1. Aufl. — Frankfurt am Main : Suhrkamp, 1977.
xv, 212 p. ; 20 cm. — (Theorie) GFR78-A5
Bibliography: p. 184-₍209₎
Includes index.
ISBN 3-518-06406-1 : DM30.00
1. Civilization, Modern—Addresses, essays, lectures. I. Title.
CB358.N4515 909.82 79-346362
79 MARC

CIVILIZATION, MODERN
—ADDRESSES, ESSAYS, LECTURES
(Continued)

Pavan, Massimiliano.
Antichità classica e pensiero moderno / Massimiliano Pavan. — 1. ed. — Firenze : La nuova Italia, 1977.
xiii, 380 p. ; 21 cm. — (Biblioteca di storia ; 22) It78-Feb
Includes bibliographical references and index.
L7500
1. Rome—Historiography—Addresses, essays, lectures. 2. Greece—Historiography—Addresses, essays, lectures. 3. Neoclassicism (Art)—Addresses, essays, lectures. 4. Neoclassicism (Literature)—Addresses, essays, lectures. 5. Civilization, Modern—Addresses, essays, lectures. I. Title.
DE8.P38 938'.07'2 78-383318
*79 MARC

Yi, Ŏ-nyŏng, 1934–
(Hyŏndaein i irhŏbŏrin kŏttŭl)
現代人이 읽어버린것들 (現代文明과人間) 李御寧著 [서울] 瑞文堂 [1971]
358 p. 20 cm. 700

1. Civilization, Modern—Addresses, essays, lectures. I. Title. II. Title: Hyŏndae munmyŏng kwa in'gan.
CB427.Y5 72–825552
rev
K

—FRENCH INFLUENCES—ADDRESSES, ESSAYS, LECTURES

Guide culturel, civilisations et littératures d'expression française / sous la direction de André Reboullet et Michel Tétu. — [Paris] : Hachette, 1977.
380 p. ; 21 cm. — (Collection F : Référence) F77-14935
Includes bibliographies.
ISBN 2-01-003770-7 : 35.00F
1. Civilization, Modern—French influences—Addresses, essays, lectures. 2. French language in foreign countries—Addresses, essays, lectures. I. Reboullet, André. II. Tétu, Michel. III. Title.
DC33.9.G84 944 79-362867
79 MARC

—HISTORY

(Shakai shisō to gendai bunka)
社會思想と現代文化. — 東京：春秋社，昭和26 [1951]
185, 14 p. ; 21 cm. — (現代社會思想講座；第5巻)
Cover title.
Colophon inserted.
Includes bibliographical references.

1. Social sciences—History. 2. Humanities—History. 3. Civilization, Modern—History. I. Series: Gendai shakai shisō kōza ; dai 5 -kan.
H51.S45 78–813010

—JUVENILE LITERATURE

Caleca, Giovanni.
Civiltà in cammino : corso di storia e di educazione civica per gli istituti professionali / Giovanni Caleca, Nico Ferrarone. — Milano : Santi editori La Prora, 1963.
3 v. : ill. ; 21 cm.
1. Civilization, Modern—Juvenile literature. 2. Italy—Civilization—Juvenile literature. I. Ferrarone, Nico. II. Title.
CB358.C34 909.08 79-374430
79 MARC

—MORAL AND RELIGIOUS ASPECTS

Deloria, Vine.
The metaphysics of modern existence / Vine Deloria, Jr. — 1st ed. — San Francisco : Harper & Row, c1979.
xiii, 233 p. ; 24 cm.
Bibliography: p. 218-221.
ISBN 0-06-450250-3
1. Religions (Proposed, universal, etc.) 2. Religion and culture. 3. Indians of North America—Religion and mythology. 4. Civilization, Modern—Moral and religious aspects. I. Title.
BL390.D43 1979 291.2 76-8708
78 MARC

—PHILOSOPHY—CONGRESSES

La Notion de moderne, modernité, modernisme / Colloque du 8 mars 1975, Université de Paris-Sorbonne, Lettres et civilisations, Institut de recherche sur les civilisations de l'Occident moderne .... — Paris : Université de Paris-Sorbonne, 1975.
iii, 143 leaves ; 30 cm. F76-17407
Cover title.
60.00F
1. Civilization, Occidental—Congresses. 2. Civilization, Modern—Philosophy—Congresses. 3. History—Periodization—Congresses. I. Université de Paris IV: Paris-Sorbonne. Institut de recherches sur les civilisations de l'Occident moderne.
CB245.N67 909'.09'821082 78-366129
78[r79]rev MARC

CIVILIZATION, MODERN, IN LITERATURE

Gupta, Ashok Kumar, 1951–
(Ādhunika yuga ke vātāyana se)
आधुनिक युग के वातायन से / अशोक कुमार गुप्त. — 1. संस्करण. — दिल्ली : पुस्तक प्रचार, 1976.
88 p. ; 22 cm.
In Hindi.
Includes bibliographical references.
Rs12.00

1. Hindi literature—20th century—History and criticism. 2. Civilization, Modern, in literature. I. Title.
PK2038.G75 76–903225

CIVILIZATION, MUSLIM see Civilization, Islamic

CIVILIZATION, MYCENAEAN

Bundgaard, Jens Andreas, 1897-
Parthenon and the Mycenaen City on the Heights / J. A. Bundgaard. — Copenhagen : The National Museum of Denmark : [eksp., Gyldendal], 1976.
194 p., [1] fold. leaves of plates : ill. ; 30 cm. — (Publications of the National Museum : Archaeological historical series ; v. 17) D76-40
Includes bibliographical references.
ISBN 8748067016 : kr218.50
1. Athens. Parthenon. 2. Athens. Acropolis. 3. Civilization, Mycenean. 4. Excavations (Archaeology)—Greece, Modern—Athens. 5. Greece, Modern—Antiquities. I. Title. II. Series: Copenhagen. Nationalmuseet. Nationalmuseets skrifter : Arkæologisk-historisk række ; v. 17.
DF287.P3B86 913.8'5 78-322151
79 MARC

Edwards, Ruth B
Kadmos the Phoenician : a study in Greek legends and the Mycenaean age / Ruth B. Edwards. — Amsterdam : Adolf M. Hakkert, 1979.
xiii, 265 p. : maps ; 23 cm. Ne***
A revision of the author's thesis, Cambridge, 1968 published under title: Greek legends and the Mycenaean age.
Bibliography: p. 209-244.
Includes index.
ISBN 9025607969
1. Cadmus (Greek mythology) 2. Mythology, Greek. 3. Civilization, Mycenaean. I. Title.
BL820.C13E38 1979 938'.01 79-311313
79 MARC

Kelly, Thomas, 1929-
A history of Argos to 500 B.C. / Thomas Kelly. — Minneapolis : University of Minnesota Press, c1976.
viii, 214 p., [1] leaf of plates : maps ; 23 cm. — (Minnesota monographs in the humanities ; v. 9)
Bibliography: p. 201-208.
Includes index.
ISBN 0-8166-0790-7
1. Argos, Greece—History. 2. Civilization, Mycenean. I. Title. II. Series.
DF221.A8K44 1976 938'.8 76-11500
77[r79]rev MARC

Origini e sviluppo della città / di Giovanni Pugliese Carratelli ... [et al.]. — Milano : Bompiani, 1978.
2 v. : ill. ; 23 cm. — (Storia e civiltà dei greci ; 1-2) It79-Apr
Includes bibliographical references and index.
CONTENTS: [1] Il medioevo greco.—[2] L'arcaismo.
L1200 per vol.
1. Civilization, Greek. 2. Civilization, Homeric. 3. Civilization, Mycenaean. I. Pugliese Carratelli, Giovanni, 1911- II. Series.
DF77.S72 vol. 1-2 938'.008 s 79-380718
[DF220] *79 MARC

—EGYPTIAN INFLUENCES

Vercoutter, Jean, 1911-
Essai sur les relations entre Égyptiens et Préhellènes. Paris, A. Maisonneuve [1954]
183 p. illus., maps. 19 cm. (L'Orient ancien illustré, 6)
Bibliography: p. [179]-183.
1. Egypt—Civilization—Mycenaean influences. 2. Civilization, Mycenaean—Egyptian influences. 3. Egypt—Antiquities. 4. Crete—Antiquities. I. Title. II. Series.
DT61.V4 301.29'32'0391 a 55-3361
[r79]rev MARC

—JUVENILE LITERATURE

Baumann, Hans, 1914-
Löwentor und Labyrinth : wie Troja, Mykenä u. Knossos entdeckt u. ausgegraben wurden / Hans Baumann ; [Ill. von Hans Peter Renner]. — München : Bertelsmann, 1978.
191, [1] p. : numerous ill. (some col.), maps ; 22 cm. GFR78-A
Bibliography: p. 191-[192]
ISBN 3-570-02153-X : DM19.80
1. Civilization, Mycenaean—Juvenile literature. 2. Crete—Antiquities—Juvenile literature. 3. Greece, Modern—Antiquities—Juvenile literature. I. Title.
DF220.B3 1978 913.3'8'031 78-375600
78 MARC

CIVILIZATION, OCCIDENTAL

see also The arts, Japanese —Occidental influences

Anchor, Robert.
The modern Western experience / Robert Anchor. — Englewood Cliffs, N.J. : Prentice-Hall, c1978.
x, 339 p. : ill. ; 23 cm.
Includes bibliographies and index.
ISBN 0-13-599357-1
1. Civilization, Modern—18th century. 2. Civilization, Modern—19th century. 3. Civilization, Modern—20th century. 4. Civilization, Occidental. 5. Europe—Civilization. I. Title.
CB357.A55 940.2 77-16712
77[r79]rev MARC

Chaunu, Pierre.
La mémoire et le sacré / Pierre Chaunu. — [Paris] : Calmann-Lévy, c1978.
284 p. : graph ; 23 cm. — (Archives des sciences sociales) F***
Includes bibliographical references.
ISBN 2-7021-0236-0 : 55.00F
1. Civilization, Occidental. 2. Civilization, Modern—1950- I. Title.
CB245.C47 78-389200
79 MARC

Chaunu, Pierre.
La Peste blanche : comment éviter le suicide de l'Occident / Pierre Chaunu, Georges Suffert. — [Paris] : Gallimard, 1976.
263 p. ; 22 cm. — (L'Air du temps) F76-18025
37.00F
1. Civilization, Occidental. 2. Civilization, Modern,—1950- I. Suffert, Georges, joint author. II. Title.
CB245.C513 909'.09'821082 79-356946
79 MARC

Chaunu, Pierre.
Le sursis : l'ardeur et la modération / Pierre Chaunu. — Paris : R. Laffont, c1979
335 p. ; 22 cm. — (Libertés 2000 [i.e. deux mille]) F***
Includes bibliographical references.
ISBN 2-221-00179-6 : 54.00F
1. Civilization, Occidental. 2. Civilization, Modern—1950- I. Title.
CB245.C514 909.82 79-355190
79 MARC

Garaudy, Roger.
Pour un dialogue des civilisations : l'Occident est un accident / Roger Garaudy. — Paris : Denoël, c1977.
233 p. ; 23 cm. — (Collection Coudées franches) F***
1. Civilization, Occidental. 2. East and West. I. Title.
CB245.G36 909.08 79-340051
79 MARC

Gouraige, Ghislain.
Les semences de l'esprit : propos sur l'esclavage moderne / Ghislain Gouraige. — Sherbrooke, Québec : Éditions Naaman, [1978?]
58 p. : port. ; 20 cm. — (Collection Pour tous ; 5) C79-1164-6
$5.00
1. Civilization, Occidental. 2. Civilization, Modern—1950- I. Title.
CB245.G63 909'.09'821 79-357721
79 MARC

CIVILIZATION, OCCIDENTAL (Continued)

Grand'Maison, Jacques.
Quel homme? / Jacques Grand'Maison. — ⌊Montréal⌋ : Leméac, ⌊c1978⌋
147 p. ; 20 cm. — (Collection Quelle?) C***
Includes bibliographical references.
ISBN 0-7761-0200-1 : $7.75
1. Civilization, Modern—1950- . 2. Civilization, Occidental. I. Title.
CB428.G7 909.82 79-362111
79 MARC

Grassi, Ernesto.
Die Macht der Phantasie : zur Geschichte abendländ. Denkens / Ernesto Grassi. — Königstein/Ts. : Anthenäum, 1979.
xix, 267 p. ; 22 cm. GFR79-A
Includes index.
ISBN 3-7610-8050-6 : DM38.00
1. Philosophy—History. 2. Civilization, Occidental. I. Title.
B82.G65 79-380976
79 MARC

James, Leonard Frank.
Western man and the modern world, by Leonard F. James. Content editor: Jean P. KixMiller. Project co ordinator: John R. L. Dent. Maps: David Cox and Jill Thompson. Picture research: Sandi Hughes-Jones and John Dent. Typography: Carl F. Zahn. New York, Pergamon, 1973-
5 v. illus. (part col.) 23 cm.
Volume 4: ed.
Includes earlier editions of some volumes.
CONTENTS: 1. Origins of Western civilization.—2. Rivalry, reason, and revolution.—3. Industrialism, imperialism, and war.—4. The Western world today.—5. Africa, Latin America, and the East.
ISBN 0-08-017198-2 (v. 1). ISBN 0-08-017199-0 (pbk.) (v. 1)
1. Civilization, Occidental. ⌊1. Civilization, Occidental. 2. World history⌋ I. Title.
CB245.J34 909'.09'821 72-10957
72⌊r79⌋rev2 MARC

Kagan, Donald.
The Western heritage / Donald Kagan, Steven E. Ozment, Frank M. Turner. — New York : Macmillan, c1979.
xxx, 987, xxx p. ⌊16⌋ leaves of plates : ill. ; 27 cm.
Includes bibliographies and index.
ISBN 0-02-361840-X. ISBN 0-02-361930-9 pbk.
1. Civilization, Occidental. I. Ozment, Steven E., joint author. II. Turner, Frank Miller, joint author. III. Title.
CB245.K28 909'.09'821 78-18450
78 MARC

Kagan, Donald.
The Western heritage, to 1715 / Donald Kagan, Steven Ozment, Frank M. Turner. — New York : Macmillan, c1979.
589 p. in various pagings : ill. (some col.) ; 25 cm.
"This book is a portion of The Western heritage, copyright 1979 by Macmillan Publishing Co."
Includes bibliographies and index.
ISBN 0-02-361850-7
1. Civilization, Occidental. I. Ozment, Steven E., joint author. II. Turner, Frank Miller, joint author. III. Title.
CB245K283 909'.09'821 78-16636
78 MARC

Ottolenghi, Aldo.
Orígenes de la civilización occidental / Aldo Ottolenghi. — ⌊Buenos Aires⌋ : Hachette, ⌊1977?⌋
318 p. : ill. ; 20 cm.
1. Civilization, Occidental. 2. Civilization, Ancient. I. Title.
CB245.O83 78-109500
79 MARC

Rudeng, Erik.
Patterns of Western history : unity in diversity / by Erik Rudeng. — Oslo : Universitetet i Oslo, Professoratet i konfliktog fredsforskning, 1975.
82 p. ; 30 cm. — (Trends in Western civilization program ; 8) (Papers - Chair in Conflict and Peace Research, University of Oslo ; no. 28) N77-Mar.
Includes bibliographical references.
1. Civilization, Occidental. I. Title. II. Series. III. Series: Oslo. Universitet. Professoratet i konflikt- og fredsforskning. Papers - Chair in Conflict and Reace Research, University of Oslo ; no. 28.
CB245.R8 78-379676
*78 MARC

Stassinopoulos, Arianna, 1950-
The other revolution / ⌊by⌋ Arianna Stassinopoulos. — London : Joseph, 1978.
x, 240 p. ; 24 cm. GB78-38253
Includes bibliographical references and index.
ISBN 0-7181-1720-4 : £6.95
1. Civilization, Occidental. 2. Europe—Intellectual life. 3. Conduct of life. I. Title.
CB245.S72 1978 323.44'09181'2 79-307096
79 MARC

—ADDRESSES, ESSAYS, LECTURES

Chodorow, Stanley, comp.
The other side of Western civilization : readings in everyday life / edited by Stanley Chodorow. — 2d ed. — New York : Harcourt Brace Jovanovich, c1979.
2 v. : ill. ; 24 cm.
Includes bibliographies.
CONTENTS: v. 1. The ancient world to the Reformation, edited by S. Chodorow.—v. 2. The sixteenth century to the present, edited by P. N. Stearns.
ISBN 0-15-567648-2 (v.1). ISBN 0-15-567649-0 (v.2)
1. Civilization, Occidental—Addresses, essays, lectures. 2. Europe—Civilization—Addresses, essays, lectures. I. Stearns, Peter N. II. Title.
CB245.C54 1979 940 78-60215
79 MARC

Downs, Robert Bingham, 1903-
Books that changed the world / Robert B. Downs. — 2d ed. — Chicago : American Library Association, 1978.
ix, 400 p. ; 24 cm.
Bibliography: p. 393-400.
ISBN 0-8389-0270-7 : $9.00
1. Civilization, Occidental—Addresses, essays, lectures. 2. Books and reading—Addresses, essays, lectures. 3. Literature—History and criticism—Addresses, essays, lectures. I. Title.
CB245.D67 1978 001.3 78-13371
78 MARC

Janitor.
Disegno di una storia dell'età dell'oro / Janitor. — Spezia : Cooperativa tipografica, 1905.
76 p. ; 18 cm. — (Biblioteca del Libertario ; pubblicaz. n. 7)
1. Civilization, Greek—Addresses, essays, lectures. 2. Rome—Civilization—Addresses, essays, lectures. 3. Civilization, Occidental—Addresses, essays, lectures. 4. Utopias—Addresses, essays, lectures. 5. Socialism—Addresses, essays, lectures. I. Title.
DE60.J35 79-370765
79 MARC

Solzhenitsyn, Aleksandr Isaevich, 1918-
A world split apart : commencement address delivered at Harvard University, June 8, 1978 / Aleksandr I. Solzhenitsyn ; ⌊translated from the Russian by Irina Ilovayskaya Alberti⌋. — 1st ed. — New York : Harper & Row, c1978.
61 p. ; 20 cm.
Translation of Raskolotyĭ mīr.
English and Russian.
ISBN 0-06-014007-0 : $6.95. ISBN 0-06-090690-1 pbk. : $2.95
1. Civilization, Occidental—Addresses, essays, lectures. 2. Civilization, Modern—1950- —Addresses, essays, lectures. 3. Europe, Eastern—Civilization—Addresses, essays, lectures. 4. Communism—Addresses, essays, lectures. I. Title.
CB245.S56913 1978 909.82 78-19593
78 MARC

Sallenave, Danièle, 1940-
Paysage de ruines avec personnages / Danièle Sallenave. — Paris : Aubier-Flammarion, c1975.
165 p. ; 22 cm. — (Digraphe) F***
ISBN 2-08-212000-7 : 65.00F
1. Europe—Antiquities—Addresses, essays, lectures. 2. Civilization, Occidental—Addresses, essays, lectures. 3. Classical antiquities—Addresses, essays, lectures. I. Title.
CC173.S24 930 79-354035
79 MARC

Zwischen den Kulturen : Festgabe für Georg Thürer zum 70. Geburtstag / dargebracht von der Kulturwissenschaftlichen Abteilung der Hochschule St. Gallen ; Felix Philipp Ingold (Hrsg.) ; mit einem Geleitw. von Hans Siegwart. — Bern ⌊etc.⌋ : P. Haupt, cop. 1978.
166 p. ; 23 cm. — (Veröffentlichungen der Hochschule St. Gallen für Wirtschafts- und Sozialwissenschaften, Schriftenreihe Kulturwissenschaft ; Bd. 4)
Sw78-9259
English, French, German, or Italian.
Includes bibliographical references.
CONTENTS: Beausire, P. Petits souvenirs quasi militaires.—Bridel, Y. Julien Gracq et la légende du graal.—Doerig, J. A. Iberoamerikanisches und angloamerikanisches Kulturverhalten Versuch einer Gegenüberstellung.—Fontana, P. L'episodio ariostesco di Olimpia.—Ingold, F. P. Rilke, Russland und die "russischen Dinge".—Leuenberger, T. Plädoyer für die Provinz.—Piguet, J.-C. Culture et cultures.—Roedel, R. Dante allora e poi.—Siebenmann, G. Poeta mediator, Vincente Aleixandre.—Tschumi, R. Poetry in four languages.—Wildermuth, A. und Kratz, P. Kultur in der Fachhochschule.
ISBN 3-258-02745-5
1. Civilization, Occidental—Addresses, essays, lectures. 2. Literature, Comparative—Addresses, essays, lectures. 3. Thürer, Georg, 1908- —Addresses, essays, lectures. I. Thürer, Georg, 1908- . II. Ingold, Felix Philipp, 1942- . III. St. Gall, Switzerland. Hochschule für Wirtschafts- und Sozialwissenschaften. Kulturwissenschaftliche Abteilung. IV. Series: St. Gall, Switzerland. Hochschule für Wirtschafts- und Sozialwissenschaften. Veröffentlichungen der Hochschule St. Gallen für Wirtschafts- und Sozialwissenschaften : Schriftenreihe Kulturwissenschaft ; Bd. 4.
CB245.Z86 909'.09'821 79-366056
79 MARC

—CONGRESSES

International Conference on Classical Influences, 3d, King's College, Cambridge, Eng., 1977.
Classical influences on Western thought A. D. 1650-1870 : proceedings of an international conference held at King's College, Cambridge, March 1977 / edited by R. R. Bolgar. — Cambridge ; New York : Cambridge University Press, 1979.
xiii, 394 p. ; 24 cm.
English, French or German.
Includes bibliographical references and index.
ISBN 0-521-21964-7
1. Civilization, Occidental—Congresses. 2. Europe—Intellectual life—Congresses. 3. Classical antiquities—Congresses. I. Bolgar, R. R. II. Title.
CB245.I57 1977 809'.02 77-91078
78 MARC

La Notion de moderne, modernité, modernisme / Colloque du 8 mars 1975, Université de Paris-Sorbonne, Lettres et civilisations, Institut de recherche sur les civilisations de l'Occident moderne .... — Paris : Université de Paris-Sorbonne, 1975.
iii, 143 leaves ; 30 cm. F76-17407
Cover title.
60.00F
1. Civilization, Occidental—Congresses. 2. Civilization, Modern—Philosophy—Congresses. 3. History—Periodization—Congresses. I. Université de Paris IV: Paris-Sorbonne. Institut de recherches sur les civilisations de l'Occident moderne.
CB245.N67 909'.09'821082 78-366129
78⌊r79⌋rev MARC

—HISTORY

Ch'en, Chi, fl. 1965-
(Hsi fang wen ming shih)
西方文明史：自先史人至西方文藝之復興 / 陳驥. — 台北：九思出版社, 民國67 ⌊1978⌋
4, 13, 475 p. : ill. ; 22 cm. — (新史學叢書 ; 2) (九思叢書 ; 27)
Series 2 romanized: Chiu-ssu ts'ung shu.
NT$160.00
1. Civilization, Occidental—History. I. Title. II. Series: Hsin shih hsüeh ts'ung shu ; 2.
CB245.C518 78-841195

McKay, John P
A history of Western society / John P. McKay, Bennett D. Hill, John Buckler. — Boston : Houghton Mifflin, c1979.
xv, 1027 p., ⌊8⌋ leaves of plates : ill. ; 26 cm.
Includes bibliographical references and index.
ISBN 0-395-27276-9
1. Civilization, Occidental—History. 2. Europe—Civilization. I. Hill, Bennett D., joint author. II. Buckler, John, joint author. III. Title.
CB245.M265 909'.09'821 78-69592
79 MARC

Satō, Kenji, 1889-
(Seiyō bunka shōshi)
西洋文化小史 / 佐藤堅司著. — 東京：講談社, 昭和24 ⌊1949⌋
303, 32, 34, ⌊10⌋ p., ⌊4⌋ leaves of plates : ill. ; 21 cm.
Includes index.
1. Civilization, Occidental—History. I. Title.
CB245.S25 78-811708

The Western experience / ⌊by⌋ Mortimer Chambers ... ⌊et al.⌋ ; art essays by H. W. Janson. — 2d ed. — New York : Knopf, c1979.
xvi, 1119, xlii p., ⌊32⌋ leaves of plates : ill. (some col.) ; 24 cm.
Includes bibliographies and index.
ISBN 0-394-32135-9 : $16.95
1. Civilization—History. 2. Civilization, Occidental—History. I. Chambers, Mortimer.
CB59.W38 1979 909 78-12117
78 MARC

The Western experience / Mortimer Chambers ... ⌊et al.⌋ ; art essays by H. W. Janson ; advisory editor, Eugene Rice : Consultant, Alvin Bernstein. — 2d ed. — New York : Knopf, c1979.
3 v. (xiv, 1119, xxi p., ⌊32⌋ leaves of plates) : ill. ; 24 cm.
CONTENTS: v. 1. Antiquity to the Middle Ages.—v. 2. The early modern period.—v. 3. The modern era.
Includes bibliographies and index.
Includes index.
ISBN 0-394-32138-3 (v. 1)
1. Civilization—History. 2. Civilization, Occidental—History. I. Chambers, Mortimer.
CB59.W38 1979b 909 78-12132
78 MARC

—HISTORY—PICTORIAL WORKS

Bettmann, Otto.
The Bettmann Archive picture history of the world : the story of Western civilization retold in 4460 pictures / by Otto L. Bettmann ; Manley Stolzman, editor. — New York : Random House, c1978.
223 p. : ill. ; 32 cm.
Includes index.
ISBN 0-394-41201-X
1. Civilization, Occidental—History—Pictorial works. I. Stolzman, Manley. II. Title.
CB245.B47 909'.09'821 77-5995
78 MARC

—ISLAMIC INFLUENCES

Ghānim, Ḥāmid Zayyān.
(Tārīkh al-ḥaḍārah al-Islāmīyah fī Ṣiqillīyah wa-atharuhā 'alá Ūrubbā)
تاريخ الحضارة الاسلامية فى صقلية واثرهاعلى اوربا / حامد زيان غانم . ـــ القاهرة : دار الثقافة ، 1977.
169 p. ; 24 cm.
Bibliography : p. 137-146.
Includes indexes.
ISBN 977-7257-19-8 : £E0.70
1. Sicily—Civilization. 2. Civilization, Occidental—Islamic influences. I. Title.
DG865.6.G45 77-961304

CIVILIZATION, OCCIDENTAL
—ISLAMIC INFLUENCES (Continued)

Vernet Ginés, Juan.
La cultura hispanoárabe an Oriente y Occidente / Juan Vernet. — Barcelona : Ariel, c1978.
395 p. ; 22 cm. — (Ariel historia ; 14) Sp***
Includes bibliographical references and indexes.
ISBN 8434478072
1. Spain—Intellectual life—711-1516. 2. Learning and scholarship—Spain—History. 3. Translation and interpreting—Spain—History. 4. Civilization, occidental—Islamic influences. 5. Arabs in Spain. I. Title.
DP103.V47 946'.02 78-387839
78 MARC

—ORIENTAL INFLUENCES
see East and West

CIVILIZATION, ORIENTAL

see also Asia—Civilization; Civilization, Buddhist

Die Kulturen des Orients. — Frankfurt am Main,: Akademische Verlagsgesellschaft Athenaion, c1966.
103, 94, 76 p., 4 leaves of plates : ill. ; 30 cm. — (Handbuch der Kulturgeschichte (Frankfurt am Main) : 2. Abt., Kulturen der Völker)
Includes bibliographies and indexes.
CONTENTS: Schaeder, H. H. Die Kultur des vorderen Orients. — Wolf, W. Die Kultur Ägyptens. 2., neubearb. Aufl. — Kirfel, W. Die Kultur der Inder. 2., erw. Aufl.
1. Civilization, Oriental. 2. Civilization, Ancient. I. Schaeder, Hans Heinrich, 1896-1957. II. Wolf, Walther, 1900- . III. Kirfel, Willibald, 1885-1964. IV. Series: Handbuch der Kulturgeschichte (Frankfurt am Main) : 2. Abt., Kulturen der Völker.
CB253.K8 930 79-341267
79 MARC

—ADDRESSES, ESSAYS, LECTURES

Hegel, Georg Wilhelm Friedrich, 1770-1831.
Philosophy and the oriental mind / Friedrich Wilhelm Georg Hegel. — Albuquerque, N.M. : American Classical College Press, [1979]
35 leaves, [6] leaves of plates : ill. ; 28 cm.
Translation of Die orientalische Welt.
ISBN 0-89266-212-3
1. Philosophy, Oriental—Addresses, essays, lectures. 2. Philosophy, Chinese—Addresses, essays, lectures. 3. Civilization, Oriental—Addresses, essays, lectures. I. Title.
[B2944.O742E5 1979] 181 79-20814
79 MARC

—CONGRESSES

Asian Scholars' Convention, Asian Cultural Center, APU, 2d, T'ai-pei, 1976.
(Ya-chou kuo hui i yüan lien ho hui, Ya-chou wen hua chung hsin, ti erh chieh Ya-chou)
亞洲國會議員聯合會, 亞洲文化中心第二屆亞洲學者會議彙刊 / 亞洲文化中心編印. — [台北] : 該中心, 民國66 [1977]
2, 40, 5, 342 p. : ill. ; 26 cm.
1. Civilization, Oriental—Congresses. 2. Asia—Civilization—Congresses. I. Asian Cultural Center. II. Title. II. Title: Ti erh chieh Ya-chou hsüeh che hui i hui k'an.
CB253.A84 1976 79-837280

Humanismus und Menschenbild im Orient und in der Antike / hrsg. von der Sektion Orient- und Altertumswissenschaften der Martin-Luther-Universität Halle-Wittenberg. — Halle (Saale) : Martin-Luther-Universität Halle-Wittenberg, Abt. Wissenschaftspublizistik, 1977.
2 v. : ill. ; 21-30 cm. — (Wissenschaftliche Beiträge - Martin-Luther-Universität Halle-Wittenberg ; 1977, 28 (I 2)) (Kongress- und Tagungsberichte der Martin-Luther-Universität Halle-Wittenberg) GDR***
Conference was held Nov. 1973.
Includes bibliographical references.
CONTENTS: Bd. [1] Konferenzvorträge.—Bd. [2] Bildtafeln.
[B105.H8]
1. Humanism—History—Congresses. 2. Civilization, Oriental—Congresses. 3. Civilization, Greco-Roman—Congresses. I. Halle. Universität. Sektion Orient- und Altertumswissenschaften. II. Series: Halle. Universität. Wissenschaftliche Beiträge ; 1977, 28. III. Series: Halle. Universität. Kongress- und Tagungsberichte der Martin-Luther-Universität Halle-Wittenberg.
AS182.H125 1977/28 78-384075
[B105.H8] 79[r79]rev MARC

—OCCIDENTAL INFLUENCES
see East and West

—PHILOSOPHY

Dijk, Klaas van.
Vanwaar en waarheen? : Geloofsbelijdenis van een liberaal / K. van Dijk. — Assen : Gorcum, 1979.
148 p. ; 20 cm. Ne79-7
Bibliography: p. 143-146.
Includes index.
ISBN 9023216636 : fl 23.95
1. Civilization, Occidental—Philosophy. I. Title.
CB245.D54 79-361602
*79 MARC

CIVILIZATION, PHENICIAN see Phenicians

CIVILIZATION, SCANDINAVIAN

Weinhold, Karl, 1823-1901.
Altnordisches Leben / Karl Weinhold ; Einl., Helmut Stumfohl. — Graz : Akadem. Druck- u. Verlagsanst., 1977.
xiii, 512 p. ; 20 cm. Au77-21-410
Reprint of the 1856 ed. published by Weidmann, Berlin; with new introd.
Bibliography: p. xi-xiii.
Includes index.
ISBN 3-201-01024-5 : $350.00
1. Civilization, Scandinavian. I. Stumfohl, Helmut. II. Title.
DL31.W42 1977 948'.02 79-364520
*79 MARC

CIVILIZATION, SEMITIC

see also Civilization, Arab; Jews—Civilization

CIVILIZATION, SLAVIC

see also Slavophilism

—CONGRESSES

International Conference on Slav Cultures and the Balkans, Varna, Bulgaria, 1975.
(Slavianskie kul'tury i Balkany)
Славянские культуры и Балканы : [докл. и науч. съобщения от междунар. науч. конф. Славянские культуры и Балканы, Варна, 15–20 септ. 1975 г.] / [редколлегия, Н. Тодоров ... et al.]. — София : БАН, 1978–
v. : ill. (some col.) ; 25 cm. Bu 78–2434
At head of title, v. 1– : Institut balkanistiki Bolgarskoĭ akademii nauk. Mezhdunarodnaia assotsiatsiia po izucheniiu i rasprostraneniiu slavianskikh kul'tur.
Added t. p., v. 1– : Les cultures slaves et les Balkans.
Contributions in Bulgarian, French, German, or Russian; table of contents also in French.
Includes bibliographical references.
CONTENTS: 1. IX–XVIII вв.
6.00 lv (v. 1)
1. Slavs in the Balkan Peninsula — Congresses. 2. Civilization, Slavic—Congresses. 3. Balkan Peninsula—Civilization—Congresses. I. Todorov, Nikolaĭ. II. Bŭlgarska akademiia na naukite, Sofia. Institut za balkanistika. III. Association internationale pour l'étude et la diffusion des cultures slaves. IV. Title. V. Title: Les cultures slaves et les Balkans.
DR25.I 57 1975 79–410971

CIVILIZATION, SPANISH see Civilization, Hispanic

CIVILIZATION, SUMERIAN see Sumerians

CIVILIZATION, WESTERN see Civilization, Occidental

CIVILIZATION AND ASTRONAUTICS see Astronautics and civilization

CIVILIZATION AND COMPUTERS see Computers and civilization

CIVILIZATION AND MACHINERY see Technology and civilization

CIVILIZATION AND PERSONALITY see Personality and culture

CIVILIZATION AND SCIENCE see Science and civilization

CIVILIZATION AND TECHNOLOGY see Technology and civilization

CIVILIZATION SPANISH see Spain—Civilization

CLAASSEN, JOHANN, 1820-1876

Klassen, Elizabeth Suderman, 1933-
Trailblazer for the Brethren : the story of Johann Classen, a leader in the early Mennonite Brethren Church / Elizabeth Suderman. — Scottdale, Pa. : Herald Press, 1978.
309 p. : ill. ; 19 cm.
Bibliography: p. 307-309.
ISBN 0-8361-1214-8 : $6.95
1. Claassen, Johann, 1820-1876. 2. Mennonite Brethren Church—Biography. I. Title.
BX8143.C55K57 289.7'3 78-11994
78 MARC

CLACHAN, SCOT.

—HOTELS, MOTELS, ETC.

Armstrong, Sybil.
A hotel by Clachan / Sybil Armstrong ; illustrated by the author. — London : Hutchinson, 1978.
144 p. : ill. ; 23 cm. GB***
ISBN 0-09-133210-9 : £3.95
1. Clachan, Scot.—Hotels, motels, etc. 2. Clachan, Scot.—Social life and customs. I. Title.
TX910.S3A75 78-324883
338.7'61'647940924
79 MARC

—SOCIAL LIFE AND CUSTOMS

Armstrong, Sybil.
A hotel by Clachan / Sybil Armstrong ; illustrated by the author. — London : Hutchinson, 1978.
144 p. : ill. ; 23 cm. GB***
ISBN 0-09-133210-9 : £3.95
1. Clachan, Scot.—Hotels, motels, etc. 2. Clachan, Scot.—Social life and customs. I. Title.
TX910.S3A75 78-324883
338.7'61'647940924
79 MARC

CLACK, MARY HAMPTON, 1859-

—COLLECTED WORKS

Pioneer days ... two views. — Abilene, Tex. : Zachry Associates, c1979.
234 p. : ill. ; 23 cm.
Early days in West Texas (by M. H. Clack) is a reprint of the 1932 ed.
CONTENTS: Duff, K. and Seibt, K. Introducing Miss Tommie.—Clack, T. H. Introducing Aunt Mollie.—Clack, M. H. Early days in West Texas.—Miss Tommie's recollections.—Clack, T. H. Epilogue.
1. Clack, Mary Hampton, 1859- —Collected works. 2. Clack, Tommie Hazlewood, 1882- —Collected works. 3. Abilene region, Tex.—Biography—Collected works. 4. Taylor Co., Tex.—Biography—Collected works. 5. Frontier and pioneer life—Texas—Abilene region—Collected works. 6. Frontier and pioneer life—Texas—Taylor Co.—Collected works. I. Clack, Mary Hampton, 1859- Early days in West Texas. 1979. II. Clack, Tommie Hazlewood, 1882- Miss Tommie's recollections. 1979.
F394.A15C55 1979 976.4'727 79-65793
79 MARC

CLACK, TOMMIE HAZLEWOOD, 1882-

—COLLECTED WORKS

Pioneer days ... two views. — Abilene, Tex. : Zachry Associates, c1979.
234 p. : ill. ; 23 cm.
Early days in West Texas (by M. H. Clack) is a reprint of the 1932 ed.
CONTENTS: Duff, K. and Seibt, K. Introducing Miss Tommie.—Clack, T. H. Introducing Aunt Mollie.—Clack, M. H. Early days in West Texas.—Miss Tommie's recollections.—Clack, T. H. Epilogue.
1. Clack, Mary Hampton, 1859- —Collected works. 2. Clack, Tommie Hazlewood, 1882- —Collected works. 3. Abilene region, Tex.—Biography—Collected works. 4. Taylor Co., Tex.—Biography—Collected works. 5. Frontier and pioneer life—Texas—Abilene region—Collected works. 6. Frontier and pioneer life—Texas—Taylor Co.—Collected works. I. Clack, Mary Hampton, 1859- Early days in West Texas. 1979. II. Clack, Tommie Hazlewood, 1882- Miss Tommie's recollections. 1979.
F394.A15C55 1979 976.4'727 79-65793
79 MARC

CLACKAMUS CO., OR.

—MAPS

Oregon Blue Print Company.
[Clackamas County quarter section plats. Portland, 1951-
v. maps 43 x 55 cm.
Title from copyright application.
Blue line print.
Scale of maps ca. 1: 2,400.
1. Clackamus Co., Or.—Maps. 2. Real property—Clackamas Co., Or.—Maps.
G1493.C4O7 1951 map 52-92
[r79]rev2 MARC

CLADDINGS, NUCLEAR FUEL see Nuclear fuel claddings

CLADOCERA

—KANSAS—LYON CO.

Fleming, Joyce F
Population dynamics of Physa anatina Lea in a natural spring community / by Joyce F. Fleming. Benthic species diversity and related physicochemical features of John Redmond Reservoir, 1971-72 / by Francis L. Funk and John D. Ransom. Winter fauna of Cladocera and Copepoda in ponds and ditches, Lyon County, Kansas / by David J. Palavanchuk. — Emporia, Kan. : School of Graduate and Professional Studies, Emporia State University, 1977.
50 p. : ill. ; 23 cm. — (Emporia State research studies ; v. 26, no. 1)
Includes bibliographies.
Each study originally presented as a thesis (M.S.), Emporia State University.
1. Physa anatina. 2. Mollusks—Kansas. 3. Lake ecology—Kansas—John Redmond Reservoir. 4. Cladocera—Kansas—Lyon Co. 5. Copepoda—Kansas—Lyon Co. 6. Crustacea—Kansas—Lyon Co. I. Funk, Francis L. Benthic species diversity and related physicochemical features of John Redmond Reservoir, 1971-72. 1977. II. Palavanchuk, David J. Winter fauna of Cladocera and Copepoda in ponds and ditches, Lyon County, Kansas. 1977. III. Title. IV. Series.
QL430.5.P37F57 1977 592'.09'29 78-620579
78 MARC

CLAERHOUT, MAY, 1939-

Claerhout, May, 1939-
De kunst van May Claerhout / poëtisch ingelijst door Jan Veulemans. — 2000 Antwerpen, (Eikenstr. 9) : Mercatorfonds & Arcade, 1977.
[92] p. : ill. ; 27 cm. Be78-9
ISBN 2-8005-0119-7 : 980F
1. Claerhout, May, 1939- I. Veulemans, Jan, 1928- . II. Title.
NB653.C56A4 1977 79-347961
*79 MARC

CLAESSENS, BOB

Fontaine, Colette.
Bob Claessens : le temps d'une vie ... / Colette Fontaine ; préf. de José Grotovitch ; avec des contributions d'Étienne Balibar et de Chantal De Smet. — Bruxelles : Éditions du Cercle d'éducation populaire, 1977.
292 p. : ill. ; 24 cm. Be77-12
Bibliography: p. 285-290.
490F
1. Claessens, Bob. 2. Belgium—Biography.
CT1178.C55F66 949.3'04'0924 78-384678
78 MARC

CLAIMING RACES

McKnight, Bob, 1906-
How to beat the claimers / Bob McKnight. — 1st ed. — Secaucus, N.J. : Citadel Press, c1979.
175 p. ; 22 cm.
ISBN 0-8065-0614-8 : $10.00
1. Horse race betting. 2. Claiming races. 3. Horse race betting—United States. 4. Claiming races—United States. I. Title.
SF333.4.M32 1979 798'.401 79-17489
79 MARC

—UNITED STATES

McKnight, Bob, 1906-
How to beat the claimers / Bob McKnight. — 1st ed. — Secaucus, N.J. : Citadel Press, c1979.
175 p. ; 22 cm.
ISBN 0-8065-0614-8 : $10.00
1. Horse race betting. 2. Claiming races. 3. Horse race betting—United States. 4. Claiming races—United States. I. Title.
SF333.4.M32 1979 798'.401 79-17489
79 MARC

CLAIMS

see also Alabama claims; Government liability (International law)

Bieniek, Gerard.
Roszczenia regresowe zakładu pracy wobec pracownika / Gerard Bieniek. — Warszawa : Wydaw. Prawnicze, 1978 (Olszt. : OZGSP)
124 p. ; 21 cm. P78-7260
Summary in Russian.
Bibliography: p. 114-[120]
zł35.00
1. Master and servant—Poland. 2. Liability (Law)—Poland. 3. Damages—Poland. 4. Third parties (Law)—Poland. 5. Claims. I. Title.
79-349771
79 MARC

CLAIRMONT, CLARA MARY JANE, 1798-1879

Shelley, Percy Bysshe, 1792-1822.
Letters from Percy Bysshe Shelley to Jane Clairmont. — Norwood, Pa. : Norwood Editions, 1978.
viii, 104 p. ; 22 cm.
Reprint of the 1889 ed., privately printed, London.
ISBN 0-8482-6261-1 : $20.00
1. Shelley, Percy Bysshe, 1792-1822—Correspondence. 2. Clairmont, Clara Mary Jane, 1798-1879. 3. Poets, English—19th century—Correspondence. I. Clairmont, Clara Mary Jane, 1798-1879. II. Title.
PR5431.A359 1978 821'.7 78-2792
78 MARC

CLAIRTONE SOUND CORPORATION

—HISTORY

Hopkins, Garth, 1933-
Clairtone : the rise and fall of a business empire / Garth Hopkins. — Toronto : McClelland and Stewart, c1978.
219 p. ; 24 cm. C***
ISBN (invalid) 0-7710-4120-8 : $15.95
1. Clairtone Sound Corporation—History. I. Title.
HD9696.A924C564 79-305727
338.7'62'1389330971
79 MARC

CLAIRVOYANCE

see also Clairvoyants; Divination; Extrasensory perception; Fortune-telling; Hypnotism; Mind-reading; Second sight

Reponen, Oskar.
Suomalaisia selvänäkijöitä / Oskar Reponen ; [valokuvat, Aarre Ekholm, Mauri Vuorinen]. — Helsingissä : Otava, 1978 (Keuruu : Otavan painolaitokset)
256 p., 8 leaves of plates : ill. ; 21 cm. Fi79-1
Bibliography: p. [255]-256.
ISBN 9511050060 : Fmk54.00
1. Clairvoyance. 2. Clairvoyants—Finland—Interviews. I. Title.
BF1325.R46 79-373359
79 MARC

CLAIRVOYANTS

Tervagne, Simone Dupont de, 1906-
Les hommes politiques et leurs voyantes / Simone de Tervagne. — [Paris] : J. Grancher, c1978.
245 p., [8] leaves of plates : ill. ; 23 cm. F***
Includes bibliographical references and index.
39.00F
1. Clairvoyants. 2. Astrologers. 3. Politicans—Miscellanea. I. Title.
BF1325.T47 133.3'092'2 79-348203
79 MARC

—AUSTRIA—BIOGRAPHY

Cziffra, Géza von.
Hanussen, Hellseher des Teufels : d. Wahrheit über d. Reichstagsbrand / Géza von Cziffra. — München ; Berlin : Herbig, 1978.
254 p. : 29 ill. ; 23 cm. GFR78-A
Includes index.
ISBN 3-7766-0879-X : DM28.00
1. Steinschneider, Hermann, 1889-1933. 2. Clairvoyants—Austria—Biography. 3. Berlin. Reichstagsgebäude—Fire, 1933—Miscellanea. I. Title.
BF1283.S83C9 78-394120
79 MARC

—FINLAND—INTERVIEWS

Reponen, Oskar.
Suomalaisia selvänäkijöitä / Oskar Reponen ; [valokuvat, Aarre Ekholm, Mauri Vuorinen]. — Helsingissä : Otava, 1978 (Keuruu : Otavan painolaitokset)
256 p., 8 leaves of plates : ill. ; 21 cm. Fi79-1
Bibliography: p. [255]-256.
ISBN 9511050060 : Fmk54.00
1. Clairvoyance. 2. Clairvoyants—Finland—Interviews. I. Title.
BF1325.R46 79-373359
79 MARC

—FRANCE—BIOGRAPHY

Edel, Marie Jeanne.
La croix, le bûcher, pour rien? / Marie-Jeanne Edel. — Marseille : Éditions Rijois, c1979.
192 p. ; 21 cm. — (Collection Carrefour de l'étrange ; 2) F***
ISBN 2-86269-022-8 : 38.00F
1. Edel, Marie Jeanne. 2. Clairvoyants—France—Biography. 3. Prophecies (Occult sciences) I. Title. II. Series.
BF1283.E33A33 133.8'092'4 79-376394
79 MARC

CLAM FISHERIES

—WASHINGTON (STATE)—PORT ANGELES

Washington (State). Dept. of Ecology.
Port Angeles Harbor biological studies, spring 1975. — [Olympia] : State of Washington, Dept. of Ecology, 1976.
i, 43 p. : ill. ; 28 cm.
"D.O.E. 76-4."
Includes bibliographies.
CONTENTS: Pine, R. E. Introduction and summary of studies.—Moore, A. W. Port Angeles Harbor field toxicity tests.—Kittle, L. Intertidal clam survey of Port Angeles Harbor.
1. Seashore ecology—Washington (State)—Port Angeles. 2. Clam fisheries—Washington (State)—Port Angeles. I. Pine, Roland E. II. Moore, Allen W. Port Angeles Harbor field toxicity tests. 1976. III. Kittle, Lew. Intertidal clam survey of Port Angeles Harbor. 1976. IV. Title.
QH541.5.S35W37 1976 363.6 76-622770
78 MARC

CLAMCOËT INDIANS see Karankawa Indians

CLAMS

see also Clam fisheries

CLAN GODS (SHINTO) see Tutelaries (Shinto)

CLANS AND CLAN SYSTEM

see also Kinship; Society, Primitive; and names of families, e.g. Cameron family; McGregor family

—IRELAND

Durning, William P
If you're a wee bit Irish : a chart of old Irish families collected from folk tradition / [by Ulliam O'Duirnin, i.e. W. P. Durning]. — La Mesa, Calif. : Irish Family Names Society, 1978.
xxvi, 166 p. : geneal. tables ; 22 cm.
Bibliography: p. 163-165.
Includes index.
ISBN 0-9601868-0-8 : $8.45
1. Ireland—Genealogy. 2. Folk-lore—Ireland. 3. Clans and clan system—Ireland. I. Title.
CS484.D87 929'.2'09415 78-61121
79 MARC

—JAPAN

Matsumoto, Seichō, 1909-
(Tennō to gōzoku)
天皇と豪族 / 松本清張. — 東京 : 講談社, 1978.
293 p. : ill. ; 20 cm. — (清張通史 ; 4) Ja 78-23383
Includes index.
¥850
1. Jinshin Revolt, 672. 2. Clans and clan system—Japan. 3. Japan—History—Taika period, 645-649. I. Title. II. Series: Matsumoto, Seichō, 1909- Seichō tsūshi ; 4.
DS855.6.M37 78-816018

—JAPAN—ADDRESSES, ESSAYS, LECTURES

Watanabe, Mitsuo, 1908-
(Seishi hyakuwa)
姓氏百話 / 渡辺三男. — 東京 : 新人物往来社 昭和51 [1976]
227 p. ; 20 cm. Ja 76-1063
Originally published in 54 issues of Rekishi tokuhon beginning with Jan. 1971.
¥1200
1. Japan—History—To 1868—Addresses, essays, lectures. 2. Clan and clan system—Japan—Addresses, essays, lectures. 3. Names, Personal—Japan—Addresses, essays, lectures. I. Title.
DS835.W377 79-80492

—MALAWI

Nurse, George T
Clanship in Central Malaŵi / G[eorge] T. Nurse. — Wien ; Föhrenau : Elisabeth Stiglmayr, 1978.
113 p. : ill., fig. ; 21 cm. — (Acta ethnologica et linguistica ; Nr. 41 : Series Africana ; 12) Au78-16-338
Bibliography: p. [104]-113.
S120.00
1. Clans and clan system—Malawi. 2. Malawi—Population. 3. Ethnology—Malawi. I. Title. II. Series: Acta ethnologica et linguistica ; Nr. 41.
GN657.M15N87 301.44'09689'7 78-326573
*79 MARC

—PERU

see also Ayllus

—ROME—HISTORY

Franciosi, Gennaro.
Clan gentilizio e strutture monogamiche : contributo alla storia della famiglia romana / Gennaro Franciosi. — 2. ed. — Napoli : Jovene, 1978-
v. ; 21 cm. It79-Apr
Includes bibliographical references.
L6000
1. Family—Rome—History. 2. Clans and clan system—Rome—History. 3. Marriage (Roman law) I. Title.
HQ511.F7 1978 79-378062
*79 MARC

—SCOTLAND

Martine, Roderick.
A guide to the clans and major families of Scotland / compiled by Roderick Martine ; with a foreword by Iain Moncreiffe. — [Perth : Holmes McDougall, [1977]
32 p. : ill. ; 30 cm. GB***
"Scottish field souvenir supplement."
1. Clans and clan system—Scotland. 2. Scotland—Genealogy. I. Scottish field. II. Title.
DA880.H6M4 929'.1'09411 78-320896
79 MARC

CLANDESTINE LITERATURE see Underground literature

CLAPP FAMILY

Clapp, Elmo F 1907-
The German Clapps in America / Elmo F. Clapp, Sr., author & compiler. — Evansville, Ind. : Unigraphic, 1978.
viii, 206 p., [1] leaf of plates : ill. ; 29 cm.
Includes indexes.
1. Clapp family. I. Title.
CS71.C59 1978 929'.2'0973 78-111438
79 MARC

CLARA, OF ASSISI, SAINT, D. 1253

Brogliato, Bortolo, 1926-
Assisi, incontro vivo con frate Francesco e sorella Chiara / Bortolo Brogliato. — Vicenza : L.I.E.F., 1976.
246 p., [14] leaves of plates : ill. ; 20 cm. It77-Oct
Bibliography: p. 236-240.
L3900
1. Francesco d'Assisi, Saint, 1182-1226. 2. Clara, of Assisi, Saint, d. 1253. 3. Christian saints—Italy—Assisi—Biography. 4. Assisi—Biography. 5. Art—Italy—Assisi. I. Title.
BX4700.F6B74 78-388702
*79 MARC

CLARE, JOHN, 1793-1864

—BIBLIOGRAPHY

Dendurent, H O
John Clare, a reference guide / H. O. Dendurent. — Boston : G. K. Hall, c1978.
135 p. ; 24 cm. — (A Reference publications in literature)
Includes index.
ISBN 0-8161-8071-7
1. Clare, John, 1793-1864—Bibliography. I. Title. II. Series.
PR4453.C6Z69 016.821'7 78-18010
78 MARC

CLARE CO., MICH

—HISTORY

Meek, Forest B
Michigan's heartland, 1900-1918 / by Forrest B. Meek. — Clare, MI : Edgewood Press, 1979.
xiii, 449 p. : ill. ; 27 cm.
Includes index.
ISBN 0-9602472-0-3
1. Clare Co., Mich.—History. I. Title.
F572.C65M43 977.4'71'04 79-63276
79 MARC

CLARENDON, EDWARD HYDE, 1ST EARL OF, 1609-1674

Clarendon, Edward Hyde, 1st earl of, 1609-1674.
Selections from the History of the rebellion and the Life by himself / Clarendon ; edited by G. Huehns ; with a new introd. by Hugh Trevor-Roper. — Oxford ; New York : Oxford University Press, 1978.
xxx, 492 p. ; 21 cm.
ISBN (invalid) 0-19-250544-0 : £4.95.
1. Great Britain—History—Civil War, 1642-1649. 2. Clarendon, Edward Hyde, 1st earl of, 1609-1674. 3. Statesmen—England—Biography. I. Clarendon, Edward Hyde, 1st earl of, 1609-1674. The history of the rebellion. Selections. 1978. II. Huehns, Gertrude. III. Title.
[DNLM: GB78]
DA415.C48 1978 941.06'2 78-40717
78 MARC

CLARENCE RIVER REGION, AUSTRALIA

—HANDBOOKS, MANUALS, ETC.

Hall, Glen, 1903-
The road to the river (1839-1939) : or, Draylines to deep water / by Glen Hall. — 2nd rev. ed. — Lismore, N.S.W. : Bording's Publications, 1977.
viii, 218 p. : ill. ; 22 cm. Aus
Includes bibliographical references and index.
ISBN 0-9598257-3-8
1. Transportation—Australia—Clarence River region—History. 2. Clarence River region, Australia—History. I. Title. II. Title: Draylines to deep water.
HE289.Z7C564 1977 994.4 78-321628
79 MARC

CLARETIAN MISSIONARIES

—HISTORY

Izquierdo, Mariano.
La comunidad claretiana de Medellín, 1925-1975 / por Mariano Izquierdo y Carlos E. Mesa. — Medellín : Editorial Granamérica, 1976.
272 p. ; 23 cm.
Includes bibliographical references.
1. Missions—Colombia—Medellín. 2. Claretian Missionaries—History. 3. Medellín, Colombia—Church history. I. Mesa, Carlos E., joint author. II. Title.
BV2853.C7I98 77-155335
79 MARC

CLARI, GIOVANNI CARLO MARIA, 1677-1754

Baggiani, Franco.
Giovan Carlo Maria Clari, musicista pisano del '700 / Franco Baggiani. — Pisa : Pacini, 1977.
87 p., [8] leaves of plates : ill. ; 22 cm. It78-Dec
List of works: p. 61-81.
Bibliography: p. 83-84.
Includes indexes.
L3500
1. Clari, Giovanni Carlo Maria, 1677-1754. 2. Composers—Italy—Biography. I. Title.
ML410.C606B2 784'.092'4 79-347063
*79 MARC

CLARINET

Heine, Alois.
Akustische Phänomene : Unters. u. Experimente mit d. Klarinette u. Einf. in d. Ansatztechniken zur Erzeugung von Überblasetönen, Untertönen u. Akkordklängen / Alois Heine. — München ; Salzburg : Katzbichler, 1978.
87 p. : ill. (chiefly music) ; 24 cm. — (Schriften der Hochschule Mozarteum Salzburg ; Heft 5) GFR79-A
"Publikation des Instituts für Musikalische Grundlagenforschung der Hochschule für Musik und Darstellende Kunst "ozarteum' in Salzburg."
Errata slip inserted.
Includes bibliographical references.
ISBN 3-87397-474-6 : DM24.00
1. Clarinet. 2. Music—Acoustics and physics. I. Title. II. Series: Hochschule für Musik und Darstellende Kunst Mozarteum in Salzburg. Schriften der Hochschule Mozarteum Salzburg ; Heft 5.
ML945.H44 788'.62 79-382365
79 MARC

Rutland, Jonathan.
The clarinet / by M. P. Rutland / illustrated by Bob Warburton. — London ; New York : F. Watts, 1978.
48 p. : ill., facsim., music, ports. ; 18 x 23 cm. GB78-12989
Bibliography: p. 46.
Includes index.
ISBN 0-85166-641-8 : £1.95
1. Clarinet. I. Title.
ML945.R87 788'.62 79-302687
79 MARC

CLARINETISTS

—ENGLAND—BIOGRAPHY

Brymer, Jack.
From where I sit / Jack Brymer. — London : Cassell, 1979.
vii, 183 p., [4] leaves of plates : ports. ; 23 cm. GB***
Includes index.
ISBN 0-304-30313-5 : £6.95
1. Brymer, Jack. 2. Clarinetists—England—Biography. I. Title.
ML419.B8A3 788'.62'0924 79-312668
79 MARC

CLARK, ANDREW HILL, 1911-1975

—ADDRESSES, ESSAYS, LECTURES

European settlement and development in North America : in honour and memory of Andrew Hill Clark / edited by James R. Gibson. — Folkestone, Eng. : Dawson, 1978.
viii, 230 p. : ill. ; 24 cm. GB***
Bibliography: p. 224-228.
CONTENTS: Meinig, D. W. Prologue.—Harris, R. C. The extension of France into rural Canada.—Gibson, J. R. Old Russia in the New World.—Mitchell, R. D. The formation of early American cultural regions.—Hilliard, S. B. Antebellum tidewater rice culture in South Carolina and Georgia.—Ray, A. J. The Hudson's Bay Company fur trade in the eighteenth century.—McQuillan, D. A. Territory and ethnic identity.—Ward, D. The early Victorian city in England and America.—Lemon, J. T. The weakness of place and community in early Pennsylvania.
ISBN 0-7129-0896-X : £7.50
1. United States—Historical geography—Addresses, essays, lectures. 2. Canada—Historical geography—Addresses, essays, lectures. 3. United States—Colonization—Addresses, essays, lectures. 4. Canada—Colonization—Addresses, essays, lectures. 5. Clark, Andrew Hill, 1911-1975—Addresses, essays, lectures. I. Clark, Andrew Hill, 1911-1975. II. Gibson, James. R.
E179.5.E94 1978b 970 79-305456
79 MARC

CLARK, ANDREW INGLIS, 1848-

—ARCHIVES—CATALOGS

The Clark papers. — Hobart : University of Tasmania, Dept. of History, 1968.
37 p., ₍1₎ leaf of plates : port. ; 26 cm. — (Report on the historical manuscripts of Tasmania ; no. 7)
Bibliography: p. 37.
1. Clark family—Archives—Catalogs. 2. Tasmania. University, Hobart. Library—Catalogs. 3. Clark, Andrew Inglis, 1848- —Archives—Catalogs. I. Tasmania. University, Hobart. Dept. of History. II. Series: Tasmania. University, Hobart. Dept. of History. Report on the historical manuscripts of Tasmania ; no. 7.
DU470.T28 no. 7 994.6 s 79-300749
₍CD2525.T37₎ 79 MARC

CLARK, CHARLES JOSEPH, 1939-

Nolan, Michael.
Joe Clark, the emerging leader / by Michael Nolan ; with photos. by Ted Grant. — Toronto : Fitzhenry & Whiteside, c1978.
139 p. : ill. ; 24 cm. C***
ISBN 0-88902-436-7
1. Clark, Charles Joseph, 1939- . 2. Legislators—Canada—Biography. 3. Canada. Parliament. House of Commons—Biography. I. Grant, Ted. II. Title.
F1034.3.C55N64 328.71'092'4 79-304720
79 MARC

CLARK, CHARLES MANNING HOPE

—ADDRESSES, ESSAYS, LECTURES

Clark, Charles Manning Hope.
A discovery of Australia / Manning Clark. — Sydney : Australian Broadcasting Commission, 1976.
56 p. ; 21 cm. — (Boyer lectures ; 1976) Aus***
Bibliography: p. 53.
ISBN 0-642-97394-6 : $1.75
1. Australia—Historiography—Addresses, essays, lectures. 2. Clark, Charles Manning Hope—Addresses, essays, lectures. I. Title. II. Series.
DU108.C56 994'.007'2 78-319765
79 MARC

CLARK, EUGENIE

—EXHIBITIONS

McGovern, Ann.
Shark lady : true adventures of Eugenie Clark / by Ann McGovern ; illustrated by Ruth Chew. — New York : Four Winds Press, c1978.
83 p. : ill. ; 22 cm.
Bibliography: p. 83.
SUMMARY: A biography of the ichthyologist whose interest in fish began at the age of nine during weekly trips to the Aquarium in New York City.
ISBN 0-590-07604-3
1. Clark, Eugenie—Juvenile literature. 2. Ichthyologists—United States—Juvenile literature. 3. Sharks—Research—Juvenile literature. ₍1. Clark, Eugenie. 2. Ichthyologists. 3. Zoologists₎ I. Chew, Ruth. II. Title.
QL31.C56M32 597'.009'24 78-22126
78 MARC

CLARK, GEORGIE WHITE

Clark, Georgie White.
Georgie Clark : thirty years of river of running / by Georgie White Clark and Duane Newcomb. — San Francisco : Chronicle Books, ₍1977₎
165 p. : ill. ; 27 cm.
ISBN 0-87701-105-2
1. Clark, Georgie White. 2. The West—Biography. 3. Rafting (Sports)—The West. I. Newcomb, Duane G., joint author.
CT275.C6265A34 917.8'04'30924 77-25448
77 MARC

CLARK, HUGH UPHAM

—LIBRARY—CATALOGS

Friedlaender, Marc, 1905-
Guide and index to the microfilm edition of the papers relating to the Quincy, Wendell, Holmes, and Upham families at the Massachusetts Historical Society, together with the Quincy, Wendell, Holmes, and Upham family papers in the collecion of Hugh Upham Clark, of Arlington, Virginia / Marc Friedlaender and Robert V. Sparks, editors. — Boston : Massachusetts Historical Society, 1977.
xvi, 123 p. : 2 geneal. tables (fold. in pocket) ; 23 cm.
1. Quincy family—Archives—Indexes. 2. Wendell family—Archives—Indexes. 3. Holmes family—Archives—Indexes. 4. Upham family—Archives—Indexes. 5. Clark, Hugh Upham—Library—Catalogs. 6. Massachusetts Historical Society, Boston. Library—Catalogs. 7. Documents on microfilm—Catalogs. I. Sparks, Robert V., joint author. II. The Quincy, Wendell, Holmes, and Upham family papers, 1633-1910. III. Massachusetts Historical Society, Boston. IV. Title.
CD3029.5.Q56F74 016.929'2'0973 78-100276
79 MARC

CLARK, MICHAEL, 1946-

—EXHIBITIONS

Michael Clark, picture windows : 24 May-29 June 1979, the Dimock Gallery, George Washington University, Washington, D.C., Lunn Gallery/Graphics International Ltd., Washington, D.C. — ₍s.l. : s.n. , 1979.₎
₍16₎ p. : ill. ; 28 cm.
1. Clark, Michael, 1946- —Exhibitions. 2. Windows in art—Exhibitions. I. Clark, Michael, 1946- . II. Dimock Gallery. III. Lunn Gallery.
ND237.C56A4 1979 759.13 79-88322
00 MARC

CLARK, WILLIAM, 1770-1838

Jackson, Donald Dean, 1919- ed.
Letters of the Lewis and Clark Expedition, with related documents, 1783-1854 / edited by Donald Jackson. — 2d ed., with additional documents and notes. — Urbana : University of Illinois Press, c1978.
2 v. (xxxi, 806 p., ₍12₎ leaves of plates : ill. ; 25 cm.
Bibliography: v. 2, p. 750-769.
Includes index.
ISBN 0-252-00697-6 : $30.00
1. Lewis and Clark Expedition—Sources. 2. The West—History—To 1848—Sources. 3. Lewis, Meriwether, 1774-1809. 4. Clark, William, 1770-1838. 5. Explorers—The West—Correspondence.
F592.7.J14 1978 978'.02'0924 78-15288
78 MARC

Muench, David.
Lewis & Clark country / photography by David Muench ; text by Archie Satterfield. — Portland, Or. : Beautiful America Pub. Co., c1978.
144 p. : col. ill., map ; 35 cm.
Bibliography: p. 141.
ISBN 0-915796-12-0 : $25.00
1. The West—Description and travel—1951- 2. Lewis and Clark Expedition. 3. Lewis, Meriwether, 1774-1809. 4. Clark, William, 1770-1838. I. Satterfield, Archie. II. Title.
F595.2.M8 917.8'04'2 78-8324
78 MARC

CLARK CO., OHIO

—BIOGRAPHY—PERIODICALS

Yester year in Clark County, Ohio. v. 1-
1947-
₍Springfield₎ Clark County Historical Society, 1978-
v. ill. 24 cm. annual.
Reprint of a periodical issued by the Clark County Historical Society.
Vols. for 1947-48 issued together.
1. Clark Co., Ohio—History—Periodicals. 2. Clark Co., Ohio—Biography—Periodicals. I. Clark County (Ohio) Historical Society.
F497.C5Y472 977.1'49'005 78-649078
MARC-S

—HISTORY

The History of Clark County, Ohio, containing a history of the county; its cities, towns, etc.; general and local statistics; portraits of early settlers and prominent men; history of the Northwest Territory; history of Ohio; map of Clark County; Constitution of the United States, miscellaneous matters, etc., etc. — Evansville, Ind. : Unigraphic, 1979.
1085, 92 p. : ill. ; 25 cm.
Reprint of the 1881 ed. published by W. H. Beers, Chicago.
Includes index.
1. Clark Co., Ohio—History.
F497.C5H6 1979 977.1'49 79-117895
79 MARC

—HISTORY—PERIODICALS

Yester year in Clark County, Ohio. v. 1-
1947-
₍Springfield₎ Clark County Historical Society, 1978-
v. ill. 24 cm. annual.
Reprint of a periodical issued by the Clark County Historical Society.
Vols. for 1947-48 issued together.
1. Clark Co., Ohio—History—Periodicals. 2. Clark Co., Ohio—Biography—Periodicals. I. Clark County (Ohio) Historical Society.
F497.C5Y472 977.1'49'005 78-649078
MARC-S

CLARK EQUIPMENT COMPANY

—HISTORY—ADDRESSES, ESSAYS, LECTURES

Phillips, Bert E 1919-
Plus faith unlimited : the story of Clark Equipment Company / B. E. Phillips. — New York : Newcomen Society in North America, 1979.
32 p. : ill. ; 23 cm. — (Newcomen publication ; no. 1087)
1. Clark Equipment Company—History—Addresses, essays, lectures. I. Title. II. Series.
HD9705.U64C536 338.7'68 78-71513
79 MARC

CLARK FAMILY

Carmichael, Dorothy.
The Clarke, Long & Gowan family saga / compiled by Dorothy Carmichael. — ₍s.l. : s.7n., pref. 1976₎ ; (Sydney : Seymour Office/Print Services
114 p. : ill. ; 25 cm.
1. Clark family. 2. Long family. 3. Gowen family. 4. Australia—Genealogy. I. Title.
CS2009.C56 1976 929.2'0994 78-310168
78 MARC

Crawford, William Lusk, 1907-
Ancestors & friends : a history & genealogy / William Lusk Crawford, Sr. — Dallas : Farmer Genealogy Co., c1978.
vii, 439 p. : ill. geneal. table (on lining papers), ; 24 cm.
Bibliography: p. 381-386.
Includes index.
1. Clark family. 2. Crawford family. 3. Lusk family. I. Title.
CS71.C6 1978 929'.2'0973 78-67813
00 MARC

—ARCHIVES—CATALOGS

The Clark papers. — Hobart : University of Tasmania, Dept. of History, 1968.
37 p., ₍1₎ leaf of plates : port. ; 26 cm. — (Report on the historical manuscripts of Tasmania ; no. 7)
Bibliography: p. 37.
1. Clark family—Archives—Catalogs. 2. Tasmania. University, Hobart. Library—Catalogs. 3. Clark, Andrew Inglis, 1848- —Archives—Catalogs. I. Tasmania. University, Hobart. Dept. of History. II. Series: Tasmania. University, Hobart. Dept. of History. Report on the historical manuscripts of Tasmania ; no. 7.
DU470.T28 no. 7 994.6 s 79-300749
₍CD2525.T37₎ 79 MARC

CLARK COLLEGE, ATLANTA

—HISTORY

Brawley, James P
The Clark College legacy : an interpretive history of relevant education, 1869-1975 / by James P. Brawley. — Atlanta : Clark College, c1977.
xiv, 347 p. : ill. ; 25 cm.
Bibliography: p. 339-342.
Includes index.
1. Clark College, Atlanta—History. I. Title.
LC2851.C7B72 378.758'231 78-108692
79 MARC

CLARKE, ARTHUR CHARLES, 1917-

Rabkin, Eric S
Arthur C. Clarke / Eric S. Rabkin. — West Linn, Or. : Starmont House, c1979.
80 p. : port. ; 21 cm. — (Starmont reader's guides to contemporary science fiction and fantasy authors ; 1)
Bibliography: p. 61-75
Includes index.
ISBN 0-916732-03-7 : $3.95
1. Clarke, Arthur Charles, 1917- . 2. Authors, English—20th century—Biography. I. Title. II. Series.
PR6005.L36Z86 823'.9'14 79-84709
00 MARC

—BIOGRAPHY—ADDRESSES, ESSAYS, LECTURES

Clarke, Arthur Charles, 1917-
The view from Serendip / Arthur C. Clarke. — London : Gollancz, 1978, c1977.
x, 273 p. ; 23 cm. GB***
Includes bibliographical references.
ISBN 0-575-02303-1 : £5.50
1. Clarke, Arthur Charles, 1917- —Biography—Addresses, essays, lectures. 2. Authors, English—20th century—Biography—Addresses, essays, lectures. 3. Technology and civilization—Addresses, essays, lectures. 4. Sri Lanka—Description and travel—Addresses, essays, lectures. I. Title.
PR6005.L36Z52 1978 823'.9'14 78-322958
79 MARC

—CRITICISM AND INTERPRETATION

Slusser, George Edgar.
The space odysseys of Arthur C. Clarke / George Edgar Slusser. — 1st ed. — San Bernardino, Calif. : Borgo Press, 1978.
64 p. ; 21 cm. — (Popular writers of today ; v. 8) (The Milford series)
ISBN 0-89370-212-9 : $1.95
1. Clarke, Arthur Charles, 1917- —Criticism and interpretation. I. Title.
PR6005.L36Z88 823'.9'14 77-24438
77 MARC

CLARKE, SAMUEL, 1675-1729

Clarke, Samuel, 1675-1729.
The works, 1738 / Samuel Clarke. — New York : Garland Pub., 1978.
4 v. : port. ; 29 cm. — (British philosophers and theologians of the 17th & 18th centuries)
Reprint of the 1738 ed. printed for J. and P. Knapton, London, under title: The works of Samuel Clarke.
CONTENTS: v. 1. Sermons on several subjects.—v. 2. Sermons on several subjects. Eighteen sermons on several occasions. Sixteen sermons on the being and attributes of God, the obligations of natural religion, and the truth and certainty of the Christian revelation.—v. 3. A paraphrase on the four Evangelists. Three practical essays on baptism, confirmation, and repentance. An exposition of the church catechism. A letter to Mr. Dodwell concerning the immortality of the soul. Reflections on Amyntor.—v. 4. The scripture doctrine of the Trinity. Several tracts relating to the subject of the Trinity. A collection of papers which passed between the late learned M. Leibnitz and Dr. Clarke. A letter to Benjamin Hoadly.
ISBN 0-8240-1762-5 : $25.00 per vol.
1. Church of England—Sermons. 2. Sermons, English. 3. Bible. N.T. Gospels—Paraphrases, English. 4. Church of England—Doctrinal and controversial works—Collected works. 5. Trinity—Early works to 1800—Collected works. 6. Clarke, Samuel, 1675-1729. I. Series.
BX5037.C5 1978 230'.3 75-11207
76 MARC

CLARKE, SHIRLEY

Clarke, Shirley.
Shirley Clarke : an American Film Institute seminar on her work. — Glen Rock, N. J. : Microfilming Corp. of America, 1977.
1 sheet (54 p.) ; 11 x 15 cm. — (The American Film Institute seminars ; pt. 1, no. 33) (New York times oral history program)
Microfiche.
ISBN 0-667-00352-5
1. Clarke, Shirley. 2. The Cool world. ⌈Motion picture⌉ 3. Harlem, New York (City) I. Series.
Microfiche PN1998.A3 791.43'0233'0924 77-12435

CLARTE, ABEL

—BIOGRAPHY

Clarté, Abel.
Je m'en souviendrai de cette planète : souvenirs / Abel Clarté. — Annonay (07103) : Éditions du Vivarais ; Juvisy-sur-Orge (B.P. 21, 91260) : H. Barrelle, 1978-
v. ; 21 cm. F79-414
Includes index.
CONTENTS: 1. 1904-1939.

ISBN 2-902592-21-3 (v. 1) : 50.00F (v. 1)
1. Clarté, Abel—Biography. 2. Authors, French—20th century—Biography. I. Title.
PQ2663.L29Z464 79-352988
79 MARC

CLARY-ALDRINGEN, ALFONS, 1887-

Clary-Aldringen, Alfons, 1887-
A European past : memoirs / by Prince Clary ; with a preface by Golo Mann ; translated by Ewald Osers. — London : Weidenfeld and Nicolson, c1978.
246 p., ⌈4⌉ leaves of plates : ill. ; 23 cm. GB***
Translation of Geschichten eines alten Österreichers.
ISBN 0-297-77547-2 : £8.95
1. Clary-Aldringen, Alfons, 1887- 2. Austria—Princes and princesses—Biography. I. Title.
DB90.C6A3313 1978 943.7'03'0924 79-308322
79 MARC

Clary-Aldringen, Alfons, 1887-
A European past : memoirs / by Prince Clary ⌈i.e. A. Clary-Aldringen⌉ ; with a pref. by Golo Mann ; translated by Ewald Osers. — New York : St. Martin's Press, 1978.
246 p., ⌈5⌉ leaves of plates : ill. ; 23 cm.
Translation of Geschichten eines alten Österreichers.
ISBN 0-312-27077-1 : $10.00
1. Clary-Aldringen, Alfons, 1887- 2. Austria—Princes and princesses—Biography. I. Title.
DB90.C6A3313 1978b 943.7'03'0924 78-63593
79 MARC

CLASPS

see also Buckles; Fibula (Archaeology)

CLASS, HELMUT

—ADDRESSES, ESSAYS, LECTURES

Zukunft aus dem Wort : Helmut Class zum 65. Geburtstag / hrsg. von Günther Metzger. — 1. Aufl. — Stuttgart : Calwer Verlag, 1978.
253 p. ; 22 cm. GFR78-A
Includes bibliographical references.
ISBN 3-7668-0587-8
1. Theology—Addresses, essays, lectures. 2. Class, Helmut—Addresses, essays, lectures. I. Class, Helmut. II. Metzger, Günther.
BR50.Z84 78-389260
79 MARC

CLASS ACTIONS (CIVIL PROCEDURE)

see also Parties to actions

—CALIFORNIA

ALI-ABA Course of Study: Class and Derivative Actions and Other Multiparty Complex Litigation, San Francisco, 1978.
ALI-ABA Course of Study—Class and Derivative Actions and Other Multiparty Complex Litigation : practical problems and new approaches : materials / cosponsored by California Continuing Education of the Bar. — ⌈Philadelphia⌉ : American Law Institute-American Bar Association Committee on Continuing Professional Education, c1978.
xii, 513 p. ; 29 cm.
"3301."
Includes bibliographical references.
1. Class actions (Civil procedure)—United States. 2. Actions and defenses—United States. 3. Class actions (Civil procedure)—California. 4. Actions and defenses—California. I. American Law Institute-American Bar Association Committee on Continuing Professional Education. II. California Continuing Education of the Bar. III. Title: Class and derivative actions and other multiparty complex litigation.
KF8896.Z9A16 1978 347'.73'5 78-110634
79 MARC

—UNITED STATES

ALI-ABA Course of Study: Class and Derivative Actions and Other Multiparty Complex Litigation.
ALI-ABA course of study. Class and derivative actions and other multiparty complex litigation: materials.
⌈Philadelphia, American Law Institute-American Bar Association Committee on Continuing Professional Education.
v. 29 cm.
Vols. for cosponsored by California Continuing Education of the Bar.
Key title: ALI-ABA course of study. Class and derivative actions and other multiparty complex litigations: materials, ISSN 0191-2011.
1. Class actions (Civil procedure)—United States. 2. Stockholders' derivative actions—United States. 3. Actions and defenses—United States. I. American Law Institute-American Bar Association Committee on Continuing Professional Education. II. California Continuing Education of the Bar. III. ALI-ABA Course of Study: Class and Derivative Actions and Other Multiparty Complex Litigation. Class and derivative actions and other multiparty complex litigation: materials. IV. Title.
KF8896.Z9A15 347'.73'5 79-642595
MARC-S

ALI-ABA Course of Study: Class and Derivative Actions and Other Multiparty Complex Litigation, San Francisco, 1978.
ALI-ABA Course of Study—Class and Derivative Actions and Other Multiparty Complex Litigation : practical problems and new approaches : materials / cosponsored by California Continuing Education of the Bar. — ⌈Philadelphia⌉ : American Law Institute-American Bar Association Committee on Continuing Professional Education, c1978.
xii, 513 p. ; 29 cm.
"3301."
Includes bibliographical references.
1. Class actions (Civil procedure)—United States. 2. Actions and defenses—United States. 3. Class actions (Civil procedure)—California. 4. Actions and defenses—California. I. American Law Institute-American Bar Association Committee on Continuing Professional Education. II. California Continuing Education of the Bar. III. Title: Class and derivative actions and other multiparty complex litigation.
KF8896.Z9A16 1978 347'.73'5 78-110634
79 MARC

ALI-ABA Course of Study: Litigation under the Federal Securities Laws.
ALI-ABA course of study. Litigation under the federal securities laws : materials.
⌈Philadelphia⌉ American Law Institute-American Bar Association Committee on Continuing Professional Education ⌈etc.⌉
v. 29 cm.
Vols. for cosponsored by Emory University School of Law.
Vols. for issued in 2 or more vols.
Key title: ALI-ABA course of study. Litigation under the Federal securities laws: Materials, ISSN 0191-2046.
1. Securities — United States. 2. Actions and defenses — United States. 3. Class actions (Civil procedure)—United States. I. American Law Institute-American Bar Association Committee on Continuing Professional Education. II. Emory University, Atlanta. School of Law. III. ALI-ABA Course of Study: Litigation under the Federal Securities Laws. Litigation under the federal securities laws: materials. IV. Title.
KF1440.A178 346'.73'0920269 79-640964
MARC-S

ALI-ABA Course of Study : the Private Antitrust Action : Recent Developments in Standing to Sue and Class Actions, Miami, Fla., 1978.
ALI-ABA Course of Study—the Private Antitrust Action—Recent Developments in Standing to Sue and Class Actions : materials. — ⌈Philadelphia⌉ : American Law Institute-American Bar Association Committee on Continuing Professional Education, c1978.
ix, 61, 50 p. ; 29 cm.
"3341."
1. Antitrust law—United States. 2. Class actions (Civil procedure)—United States. I. American Law Institute-American Bar Association Committe on Continuing Professional Education. II. Title: The Private Antitrust Action.
KF1657.T7A17 343'.73'0720269 78-110627
79 MARC

A Class-action suit that worked : the consumer refund at the antibiotic antitrust litigation / Thomas C. Bartsh ... ⌈et al.⌉. — Lexington, Mass. : Lexington Books, c1978.
xvii, 180 p. : ill. ; 24 cm.
Includes bibliographical references and index.
ISBN 0-669-02409-0
1. Class actions (Civil procedure)—United States. 2. Drugs—Laws and legislation—United States. 3. Antitrust law—United States. I. Bartsh, Thomas C.
KF8896.C53 347'.73'5 78-58814
78 MARC

Miller, Arthur Raphael, 1934-
An overview of Federal class actions : past, present, and future / by Arthur R. Miller. — Washington : Federal Judicial Center, 1977.
iii, 68 p. ; 27 cm. — (Education and training series)
"FJC-ETS-77-8."
Errata sheet inserted.
Includes bibliographical references.
1. Class actions (Civil procedure)—United States. I. Title. II. Series: Education and training series (Washington).
KF8896.M54 347'.73'52 78-111083
79 MARC

Schlurmann, Christa, 1946-
Die class action im Recht der USA : ein prozessuales Konzept nach Interessenstrukturen / vorgelegt von Christa Schlurmann. — ⌈s.l. : s.n., 1978?⌉ (Köln : Dissertationsdruck Hansen)
xxxvii, 175 p. ; 21 cm. GFR***
Thesis—Cologne.
Includes texts in English.
Vita.
Bibliography: p. iv-xix.
1. Class actions (Civil procedure)—United States. I. Title.
KF8896.S34 347'.73'5 79-378570
79 MARC

United States. Congress. Senate. Committee on the Judiciary. Subcommittee on Citizens and Shareholders Rights and Remedies.
Citizens' access to the courts act of 1978 : hearing before the Subcommittee on Citizens and Shareholders Rights and Remedies of the Committee on the Judiciary, United States Senate, Ninety-fifth Congress, second session, on S. 2390. — Washington : U.S. Govt. Print. Off., 1978-
v. ; 24 cm.
Includes bibliographical references.
1. Class actions (Civil procedure)—United States. 2. Justice, Administration of—United States. I. Title.
KF26.J8357 1978 347'.73'52 78-602931
78 MARC

CLASS ACTIONS (CIVIL PROCEDURE) —UNITED STATES (Continued)

United States. Congress. Senate. Committee on the Judiciary. Subcommittee on Improvements in Judicial Machinery.
Reform of class action litigation procedures : hearings before the Subcommittee on Judicial Machinery of the Committee on the Judiciary, United States Senate, Ninety-fifth Congress, second session ... November 29 and 30, 1978. — Washington : U.S. Govt. Print. Off., 1979.
iv, 516 p. ; 23 cm.
Includes bibliographical references.
1. Class actions (Civil procedure)—United States. I. Title.
KF26.J855 1978f 347'.73'52 79-601870
79 MARC

—UNITED STATES—CASES

Casenote legal briefs—consumer protection : adaptable to courses utilizing McCall's casebook on consumer protection / Norman S. Goldenberg, senior editor, Peter Tenen, managing editor, Robert J. Switzer, associate editor ; staff writers, Terry Molloy, Sally Molloy. — Beverly Hills, CA : Casenotes Pub. Co. ; Gardena, Calif. : distributed by Law Distributors, c1979.
iii leaves, 81 p. ; 28 cm.
Includes index.
$8.50
1. Consumer protection—Law and legislation—United States—Cases. 2. Class actions (Civil procedure)—United States—Cases. I. Goldenberg, Norman S. II. Switzer, Robert J. III. McCall, James R. Consumer protection : cases, notes, and materials.
KF1608.C36 343'.73'07 79-117468
79 MARC

—UNITED STATES—FORMS

American Bar Association. Committee for Class Action Notice Manual.
Manual of class action notice forms : a project of the Civil Practice and Procedure Committee, Antitrust Section, American Bar Association / Committee for Class Action Notice Manual ; Robert B. Gosline, chairman ... ⌈et al.⌉. — ⌈Chicago⌉ : ABA, c1979.
xi, 320 p. ; 23 cm.
Bibliography: p. 305.
Includes index.
1. Class actions (Civil procedure)—United States—Forms. I. Gosline, Robert B. II. Title.
KF8896.A65A34 347'.73'3 79-50160
00 MARC

CLASS CONFLICT see Social conflict

CLASS CRAFT

see also Glass blowing and working

CLASS DISTINCTION see Social classes

CLASS FIELD THEORY

Cohn, Harvey.
A classical invitation to algebraic numbers and class fields / Harvey Cohn. — New York : Springer-Verlag, c1978.
xiii, 328 p. : ill. ; 24 cm.
"With two appendices by Olga Taussky: 'Artin's 1932 Göttingen lectures on class field theory' and 'Connections between algebraic number theory and integral matrices.'"
Includes bibliographies and index.
ISBN 0-387-90345-3
1. Algebraic number theory. 2. Class field theory. I. Title.
QA247.C63 512'.74 78-13785
78 MARC

CLASS ORIENTED RING ASSOCIATED LANGUAGE see CORAL (Computer program language)

CLASS LIFE ADR SYSTEM

Commerce Clearing House.
1978 depreciation guide : including ADR system, system for pre-1971 assets. — Chicago : Commerce Clearing House, c1978.
263 p. ; 23 cm. — (CCH tax analysis series) (CCH editorial staff publication) "5312."
Includes index.
$7.00
1. Depreciation allowances—United States. 2. Class Life ADR System. I. Title. II. Title: Depreciation guide.
KF6386.Z9C64 343'.73'052 79-101092
79 MARC

CLASS SCHEDULES see Schedules, School

CLASS SIZE

Porwoll, Paul J
Negotiating the class size issue / ⌈study conducted and reported by Paul J. Porwoll⌉. — Arlington, Va. : Educational Research Service, c1978.
v, 93 p. ; 28 cm. — (ERS negotiation aid for school management)
Bibliography: p. 83-88.
$10.00
1. Class size. I. Educational Research Service, inc., Arlington, Va. II. Title. III. Series: Educational Research Service, inc., Arlington, Va. ERS negotiation aid for school management.
LB3013.P63 371.2'51 79-105846
79 MARC

Wright, Edgar Norman, 1932-
Effects of class size in the junior grades : a study / Edgar N. Wright, principal investigator ... ⌈et al.⌉. — Toronto : Ministry of Education : obtainable from Ontario Institute for Studies in Education, 1977.
ix, 130 p. : diagrs. ; 29 cm. C***
Bibliography: p. 127-130.
1. Class size. 2. Public schools—Ontario—Toronto. I. Ontario. Ministry of Education. II. Title.
LB3012.5.W74 371.2'51 78-317803
78 MARC

CLASS STRUGGLE see Social conflict

CLASSED CATALOGS see Catalogs, Classified

CLASSES (MATHEMATICS) see Set theory

CLASSICAL ANTIQUITIES

see also Archaeology; Architecture, Classical; Art, Classical; Art, Roman; Art objects, Classical; Fortification, Roman; Gems, Classical; Lamps, Classical; Mythology, Classical; Pottery, Greek; Pottery, Roman; Rings, Classical; Sculpture, Classical; Vases

Ginouvès, René.
La constitution des données en archéologie classique / par René Ginouvès et Anne-Marie Guimier-Sorbets. — Paris : Éditions du CNRS, c1978.
161 p. : ill. ; 30 cm. F***
Bibliography: p. 153-158.
ISBN 2-222-02321-1 : 75.00F
1. Archaeology—Methodology. 2. Classical antiquities. 3. Archaeology—Data processing. I. Guimier-Sorbets, Anne Marie, joint author. II. Title.
CC80.G5 930'.01'8 79-374037
79 MARC

Hafner, German.
Sternstunden der Archäologen : Wissenschaftler auf d. Spuren alter Kulturen / German Hafner. — 1. Aufl. — Düsseldorf ; Wien : Econ-Verlag, 1978.
400 p. : 164 ill. ; 22 cm. GFR78-A
Includes bibliographical references and indexes.
ISBN 3-430-13740-3 : DM38.00
1. Archaeology—Methodology. 2. Classical antiquities. I. Title.
CC75.H26 78-378112
78 MARC

Orlandini, Piero.
Introduzione all'archeologia e alla storia dell'arte greca e romana / Pietro Orlandini. — Milano : Cisalpino—Goliardica, 1975.
84 p. ; 24 cm. It77-Apr
At head of title: Università degli studi di Milano.
Bibliography: p. 82-83.
L1800
1. Archaeology. 2. Classical antiquities. 3. Art, Classical. I. Title.
CC175.O74 78-389202
*79 MARC

—ADDRESSES, ESSAYS, LECTURES

Kreuzfahrten im östlichen Mittelmeer : Vorträge unserer Mentoren, gehalten auf Karawane-Mittelmeerkreuzfahrten / hrsg. im Karawane-Verlag Ludwigsburg mit Unterstützung der Karawane-Studienreisen des Büros für Länder- und Völkerkunde Ludwigsburg. — Ludwigsburg : Verlag Die Karawane, 1971.
48 p. : ill. ; 21 cm. — (Die Karawane ; Jahrg. 12, 1971, Heft 3) GFR***
Includes bibliographical references.
1. Civilization, Greek—Addresses, essays, lectures. 2. Classical antiquities—Addresses, essays, lectures. 3. Mediterranean region—Description and travel—Addresses, essays, lectures. I. Series.
DF77.K74 949.5 78-421481
79 MARC

Sallenave, Danièle, 1940-
Paysage de ruines avec personnages / Danièle Sallenave. — Paris : Aubier-Flammarion, c1975.
165 p. ; 22 cm. — (Digraphe) F***
ISBN 2-08-212000-7 : 65.00F
1. Europe—Antiquities—Addresses, essays, lectures. 2. Civilization, Occidental—Addresses, essays, lectures. 3. Classical antiquities—Addresses, essays, lectures. I. Title.
CC173.S24 930 79-354035
79 MARC

Studies in classical art and archaeology : a tribute to Peter Heinrich von Blanckenhagen / edited by Günter Kopcke and Mary B. Moore. — Locust Valley, N.J. : J. J. Augustin, 1979.
xiv, 344 p., ⌈45⌉ leaves of plates : ill. ; 29 cm.
English, German, or Italian.
"Bibliography of Peter H. von Blanckenhagen": p. ⌈339⌉-340.
Includes bibliographical references.
1. Art, Classical—Addresses, essays, lectures. 2. Classical antiquities—Addresses, essays, lectures. 3. Blanckenhagen, Peter Heinrich von. I. Blanckenhagen, Peter Heinrich von. II. Moore, Mary B.
N5613.S88 709'.38 79-50203
00 MARC

—BIBLIOGRAPHY

Hansen, Peter Allan, 1944-
A bibliography of Danish contributions to classical scholarship from the sixteenth century to 1970 / by Peter Allan Hansen. — Copenhagen : The Royal Library, Dept. of National Bibliography : distributed by Rosenkilde and Bagger, 1977.
xviii, 335 p. ; 25 cm. — (Danish humanist texts and studies ; v. 1) D77-43
Includes indexes.
ISBN 8770232326 : kr224.20
1. Civilization, Greek—Bibliography. 2. Rome—Civilization—Bibliography. 3. Classical antiquities—Bibliography. 4. Denmark—Imprints. I. Copenhagen. Kongelige Bibliotek. Nationalbibliografisk afdeling. II. Title. .II. Series.
Z6207.G7H36 016.938 78-389889
⌈DE59⌉ 78 MARC

—CONGRESSES

Colloque sur les organisations internationales d'études classiques, Bibliothèque royale Albert Ier, 1976.
Aspects des études classiques : actes du Colloque ⌈sur les organisations internationales d'études classiques⌉, associé à la XVIe assemblée générale de la Fédération internationale des associations d'études classiques / édités par Jean Bingen et Guy Cambier. — Bruxelles : Éditions de l'Université de Bruxelles, 1977.
100 p. ; 24 cm. — (⌈Travaux⌉ - Université libre de Bruxelles, Faculté de philosophie et lettres ; 66) Be***
English or French.
Includes bibliographical references.
ISBN 2-8004-0646-1 : 250F
1. Classical antiquities—Societies, etc.—Congresses. 2. Classical antiquities—Congresses. I. Bingen, Jean. II. Cambier, Guy. III. International Federation of the Societies of Classical Studies. IV. Title. V. Series: Brussels. Université libre. Faculté de philosophie et lettres. Travaux ; 66.
DE2.C64 1976 938'.006'21 78-385450
79 MARC

International Conference on Classical Influences, 3d, King's College, Cambridge, Eng., 1977.
Classical influences on Western thought A. D. 1650-1870 : proceedings of an international conference held at King's College, Cambridge, March 1977 / edited by R. R. Bolgar. — Cambridge ; New York : Cambridge University Press, 1979.
xiii, 394 p. ; 24 cm.
English, French or German.
Includes bibliographical references and index.
ISBN 0-521-21964-7
1. Civilization, Occidental—Congresses. 2. Europe—Intellectual life—Congresses. 3. Classical antiquities—Congresses. I. Bolgar, R. R. II. Title.
CB245.I57 1977 809'.02 77-91078
78 MARC

CLASSICAL ANTIQUITIES (Continued)

—DICTIONARIES see Classical dictionaries

—PERIODICALS

Bern. Universität. Archäologisches Seminar.
Hefte des Archäologischen Seminars der Universität Bern. 1- 1975-
Bern, [Archäologisches Seminar der Universität Bern]
v. ill. 30 cm. annual.
French or German.
1. Classical antiquities—Periodicals.
DE1.B47a 938'.005 79-649785
MARC-S

—SOCIETIES, ETC.—CONGRESSES

Colloque sur les organisations internationales d'études classiques, Bibliothèque royale Albert Ier, 1976.
Aspects des études classiques : actes du Colloque [sur les organisations internationales d'études classiques], associé à la XVIe assemblée générale de la Fédération internationale des associations d'études classiques / édités par Jean Bingen et Guy Cambier. — Bruxelles : Éditions de l'Université de Bruxelles, 1977.
100 p. ; 24 cm. — ([Travaux] - Université libre de Bruxelles, Faculté de philosophie et lettres ; 66) Be***
English or French.
Includes bibliographical references.
ISBN 2-8004-0646-1 : 250F
1. Classical antiquities—Societies, etc.—Congresses. 2. Classical antiquities—Congresses. I. Bingen, Jean. II. Cambier, Guy. III. International Federation of the Societies of Classical Studies. IV. Title. V. Series: Brussels. Université libre. Faculté de philosophie et lettres. Travaux ; 66.
DE2.C64 1976 938'.006'21 78-385450
79 MARC

CLASSICAL ANTIQUITIES IN ART

Henares Cuéllar, Ignacio.
La teoría de las artes plásticas en España en la segunda mitad del siglo XVIII / Ignacio Henares Cuéllar. — [Granada] : Universidad de Granada, Departamento de Historia del Arte, 1977.
275 p. ; 21 cm. — (Colección monográfica - Universidad de Granada ; 58 : Arte) Sp***
Bibliography: p. [265]-275.
ISBN 8433800671
1. Art, Spanish. 2. Art, Modern—17th-18th centuries—Spain. 3. Neoclassicism (Art)—Spain. 4. Classical antiquities in art. I. Title. II. Series: Granada (City). Universidad. Colección monográfica ; 58.
N7106.H47 78-387306
79 MARC

CLASSICAL ARCHAEOLOGY see Classical antiquities

CLASSICAL ARCHITECTURE see Architecture, Classical

CLASSICAL ART see Art, Classical

CLASSICAL ATLASES see Classical geography

CLASSICAL BIOGRAPHY

see also Greece—Biography

Nepos, Cornelius.
Cornelii Nepotis Vitae cum fragmentis / edidit Peter K. Marshall. — 1. Aufl. — Leipzig : B. G. Teubner, 1977.
xvii, 122 p. ; 21 cm. — (Bibliotheca scriptorum Graecorum et Romanorum Teubneriana : [Scriptores Romani]) GDR***
Bibliography: p. xvi.
Includes index.
29.00M
1. Generals—Greece—Biography. 2. Generals—Rome—Biography. 3. Classical biography. I. Marshall, Peter K. II. Title. III. Series.
PA6515.A2 1977 78-393329
79 MARC

CLASSICAL CONCRETE POETRY see Concrete poetry, Classical

CLASSICAL DANCING see Modern dance

CLASSICAL DICTIONARIES

Lexikon der Antike / [hrsg. von Johannes Irmscher, in Zusammenarbeit mit Renate Johne]. — 2., neubearb. u. erw. Aufl. — Leipzig : Bibliographisches Institut, 1977.
638, 54 p. : ill. (some col.) ; 24 cm. GDR77-A
Bibliography: p. [634]-638.
32.00M
1. Classical dictionaries. I. Irmscher, Johannes. II. Johne, Renate.
DE5.L58 1977 938'.003 78-368519
78 MARC

Warrington, John.
Everyman's classical dictionary, 800 B.C.-A.D. 337 / by John Warrington. — 3d ed., rev. — London : J. M. Dent, c1970, 1978 printing.
xxxvii, 537 p. ; 20 cm. — (Everyman's reference library)
Bibliography: p. xvii-xx.
ISBN 0-460-03004-3 : £3.50
1. Classical dictionaries. I. Title.
DE5.W33 1978 938'.003 78-110947
79 MARC

CLASSICAL DRAMA

—COMEDY—HISTORY AND CRITICISM

Lefèvre, Eckard.
Der Phormio des Terenz und der Epidikazomenos des Apollodor von Karystos / Eckard Lefèver. — 1. Aufl. — München : Beck, 1978.
viii, 130 p. ; 24 cm. — (Zetemata, Monographien zur klassischen Altertumswissenschaft ; Heft 74) GFR78-A
Includes excerpts of the Latin text.
Bibliography: p. [122]-124.
Includes indexes.
ISBN 3-406-05164-2
1. Classical drama—Comedy—History and criticism. 2. Terentius Afer, Publius. Phormio. 3. Apollodorus, of Carystus. Epidikazomenos. I. Title. II. Series.
PA3163.L4 882 79-364148
79 MARC

—HISTORY AND CRITICISM

Butler, James Harmon, 1908-
The theatre and drama of Greece and Rome [by] James H. Butler. San Francisco, Chandler Pub. Co. [1972]
xv, 164 p. illus. 23 cm. (Chandler publications in history of theatre)
Bibliography: p. 145-146.
ISBN 0-8102-0439-8
1. Theater—Greece. 2. Theater—Rome. 3. Classical drama—History and criticism. I. Title.
PA3201.B8 792'.0938 74-164495
72[r79]rev MARC

—ILLUSTRATIONS—CATALOGS

Copenhagen. Nationalmuseet. Antiksamlingen.
Katalog over teatervidenskabelige kilder i Nationalmuseets antiksamling : med en analyse af den såkaldte Phlyaker-vase / Ernst Bræmme. — [København] : Det teatervidenskabelige Institut, Københavns universitet, 1976.
60, [3] leaves : ill. ; 25 cm. — (Opuscula - Det teatervidenskabelige institut, Københavns universitet ; 1) D***
Bibliography: leaf [61]
1. Art objects, Classical—Catalogs. 2. Classical drama—Illustrations—Catalogs. 3. Art objects—Denmark—Copenhagen—Catalogs. 4. Copenhagen. Nationalmuseet. Antiksamlingen—Catalogs. I. Bræmme, Ernst. II. Title. III. Series: Copenhagen. Universitet. Teatervidenskabeligt institut. Opuscula - Det teatervidenskabelige institut, Københavns universitet ; 1.
NK665.C66 1976 78-377327
78 MARC

CLASSICAL EDUCATION

see also Education, Humanistic; Humanism; Humanities; Universities and colleges

Yu, Sang-gŭn.
(Hyŏndae kyoyuk kwa kyoyang kyoyuk)
現代教育과教養教育 / 兪尚根著. — 第7版·再補正版. — [서울] : 明知大學出版部, 1977.
401 p. ; 17 cm. — (明知大學文庫 ; 1)
"附錄 東西洋古典目錄": p. [277]-400.
₩600
1. Classical education. 2. Education—1945- 3. Education—Korea. I. Title. II. Series: Myŏngje Taehak. Myŏngji Taehak mun'go ; 1.
LC1016.Y8 1977 79-825553

CLASSICAL FICTION

see also Latin fiction

CLASSICAL FIELD THEORY see Field theory (Physics)

CLASSICAL GEMS see Gems, Classical

CLASSICAL GEOGRAPHY

see also Geography, Ancient

—ADDRESSES, ESSAYS, LECTURES

Dion, Roger, 1896-
Aspects politiques de la géographie antique / par Roger Dion. — Paris : Les Belles Lettres, 1977.
xxiii, 305 p. : ill. ; 25 cm. — (Collection d'études anciennes) F***
Bibliography: p. [xiii]-xxiii.
Includes index.
105F
1. Classical geography—Addresses, essays, lectures. 2. Civilization, Greek—Addresses, essays, lectures. 3. Rome—Civilization—Addresses, essays, lectures. I. Title.
DE24.D56 938 78-388304
79 MARC

CLASSICAL JEWELRY see Jewelry, Classical

CLASSICAL LAMPS see Lamps, Classical

CLASSICAL LANGUAGES

see also Latin language

—ADDRESSES, ESSAYS, LECTURES

Étrennes de septantaine : travaux de linguistique et de grammaire comparée offerts à Michel Lejeune par un groupe de ses élèves. — Paris : Klincksieck, 1978.
238 p., [1] leaf of plates : port. ; 24 cm. — (Études et commentaires ; 91) F***
French or Italian.
"Bibliographie de Michel Lejeune": p. [9]-24.
Includes bibliographical references.
ISBN 2-252-01955-7 : 180F
1. Classical languages—Addresses, essays, lectures. 2. Lejeune, Michel, 1907- I. Lejeune, Michel, 1907-
PA26.L39 480 79-345814
79 MARC

(Leksikologicheskie i grammaticheskie issledovanii͡a)
Лексикологические и грамматические исследования : Романо-герман. и клас. филология : Межвуз. сб. / Кишин. гос. ун-т им. В. И. Ленина ; [Редкол. В. Я. Плоткин (отв. ред.) ... et al.]. — Кишинев : Штиинца, 1978.
158 p. : ill. ; 21 cm. USSR 78
At head of title: Ministerstvo narodnogo obrazovanii͡a Moldavskoĭ SSR.
Includes bibliographies.
1.50rub
1. Romance languages—Addresses, essays, lectures. 2. English language—Addresses, essays, lectures. 3. Classical languages—Addresses, essays, lectures. I. Plotkin, Vul'f I͡Akovlevich. II. Universitatii͡a de Stat V. I. Lenin.
PC57.L4 78-422860

CLASSICAL LANGUAGES (Continued)

—METRICS AND RHYTHMICS

Bione, Cesare.
Le forme poetiche dei greci e dei romani : compendio di metrica classica con riferimento alla versificazione italiana per le scuole e per gli studiosi / Cesare Bione. — Firenze : La nuova Italia, [1946]
viii, 191 p. ; 20 cm.
1. Classical languages—Metrics and rhythmics. I. Title.
PA186.B56 881'.009 79-355238
79 MARC

Hardie, William Ross, 1862-1916.
Res metrica / William Ross Hardie. — New York : Garland Pub., 1979.
xii, 275 p. ; 19 cm. — (The Garland library of Latin poetry)
Reprint of the 1920 ed. published at the Clarendon Press, Oxford.
Includes bibliographical references.
ISBN 0-8240-2970-4 : $25.00
1. Classical languages—Metrics and rhythmics. I. Title. II. Series.
PA186.H3 1979 486 77-70821
79 MARC

—METRICS AND RHYTHMICS—CONGRESSES

Giornate filologiche genovesi, 4th, Genoa, 1976.
Problemi di metrica classica : miscellanea filologica. — [Genova] : Istituto di filologia classica e medievale, 1978.
222 p. ; 22 cm. — (Pubblicazione dell'Istituto di filologia classica e medievale dell'Università di Genova ; 47 ISSN 0025-0852) It***
Includes bibliographical references.
1. Classical languages—Metrics and rhythmics—Congresses. I. Title. II. Series: Genoa. Università. Istituto di filologia classica e medioevale. Pubblicazioni ; 47.
PA185.G5 1976 480 79-367471
00 MARC

—PRONUNCIATION

Erasmus, Desiderius, d. 1536.
Desiderii Erasmi Roterodami de recta Latini Graecique sermonis pronuntiatione dialogus = Dialog über die richtige Aussprache der lateinischen und griechischen Sprache / als Lesetext hrsg., übers. u. kommentiert von Johannes Kramer. — Meisenheim am Glan : Hain, 1978.
xiii, 236 p. ; 23 cm. — (Beiträge zur klassischen Philologie ; Heft 98) GFR78-A
Latin and German.
Includes bibliographical references and indexes.
ISBN 3-445-01850-2 : DM35.00
1. Classical languages—Pronunciation. I. Kramer, Johannes, Dr. II. Title. III. Title: Dialog über die richtige Aussprache der lateinischen und griechischen Sprache. IV. Series.
PA131.E715 480 79-366138
79 MARC

—WORD ORDER

Weil, Henri, 1818-1909.
The order of words in the ancient languages compared with that of the modern languages / Henri Weil. — New ed. / with an introd. by Aldo Scaglione. — Amsterdam : J. Benjamins, 1978.
xxxix, 114 p. : port. ; 23 cm. — (Amsterdam studies in the theory and history of linguistic science : Series 1, Amsterdam classics in linguistics, 1800-1925 ; v. 14) Ne***
Translation of De l'ordre des mots dans les langues anciennes comparées aux langues modernes.
English or French.
Reprint of the 1887 ed. published by Ginn, Boston.
Bibliography of the author's works : p. xxv-xxix.
Bibliography: p. xv-xvii.
ISBN 9027208719
1. Languages, Modern—Word order. 2. Classical languages—Word order. I. Scaglione, Aldo D. II. Title. III. Series.
PB213.W413 1978 415 78-325278
78 MARC

CLASSICAL LITERATURE

see also Latin literature; Literature, Comparative—Modern and classical

Rerum Romanarum fontes ab anno CCXCII ad annum CCLXV a. Ch. n. / collegit atque notis illustravit Marina R. Torelli ; prefazione di Emilio Gabba. — Pisa : Giardini, 1978.
274 p. ; 25 cm. — (Biblioteca degli studi classici e orientali ; 14) It79-Jan.
Greek, Italian, or Latin.
Bibliography: p. [13]-23.
Includes index.
L18500
1. Rome—History—Republic, 510-265 B.C.—Sources. 2. Classical literature. I. Torelli, Marina R. II. Series.
DG235.R47 937'.03 79-366781
•79 MARC

—BIBLIOGRAPHY—CATALOGS

Harvard University. Library.
Classical studies : classification schedules, classified listing by call number, chronological listing, author and title listing / Harvard University Library. — Cambridge, Mass. : The Library : distributed by the Harvard University Press, 1979.
215 p. ; 29 cm. — (Its Widener Library shelflist ; 57)
ISBN 0-674-13461-3 : $25.00
1. Civilization, Greek—Bibliography—Catalogs. 2. Rome—Civilization—Bibliography—Catalogs. 3. Classical literature—Bibliography—Catalogs. 4. Classification—Books—Greece. 5. Classification—Books—Rome. 6. Classification—Books—Classical literature. 7. Harvard University. Library—Catalogs. I. Title. II. Series.
Z6207.G7H37 1979 025.4'3 79-948
[DE3] 79 MARC

—BIO-BIBLIOGRAPHY

Scriitori greci și latini / [autori, I. Banu, N. I. Barbu, N. Carandino, et al. ; coordonatori, N. I. Barbu, Adelina Piatkowski]. — București : Editura Științifică și Enciclopedică, 1978.
445 p. ; 22 cm. R79-737
In dictionary form.
Bibliography: p. [403]-[408]
lei29.00
1. Classical literature—Bio-bibliography. I. Barbu, Nicolae I., 1908- II. Piatkowski, Adelina.
PA31.S35 79-363154
•79 MARC

—COLLECTED WORKS

Verhoeven, Corn
De schaduw van één haar : lessen over antieke literatuur / Cornelis Verhoeven. — Baarn : Ambo, [1979] Ne79-12
208 p. ; 21 cm. — (Amboboeken)
Includes bibliographical references.
ISBN 9026304498 : fl 24.50
1. Classical literature—Collected works. I. Title.
PA3003.V4 79-367926
•79 MARC

—CRITICISM, TEXTUAL

Poliziano, Angelo, 1454-1494.
Miscellaneorum centuria secunda / [di] Angelo Poliziano ; per cura di Vittore Branca e Manlio Pastore Stocchi. — Editio minor. — Firenze : L. S. Olschki, 1978.
124 p. ; 25 cm. It78-Apr
Reprint of the 1972 ed. published by Alinari, Istituto di edizioni artistiche, Florence.
In Latin, with introduction and notes in Italian.
Bibliography: 4th prelim. page.
Includes indexes.
1. Classical literature—Criticism, Textual. I. Branca, Vittore. II. Pastore Stocchi, Manlio. III. Title.
PA3520.P57 1978 870'.9 79-351244
•79 MARC

—HISTORY AND CRITICISM

Ogilvie, Robert Maxwell.
The library of Lactantius / by R. M. Ogilvie. — Oxford [Eng.] : Clarendon Press, 1978.
viii, 121 p. ; 23 cm. GB***
Includes bibliographical references and index.
ISBN 0-19-826645-6 : £7.50
1. Lactantius, Lucius Caecilius Firmianus. 2. Classical literature—History and criticism. 3. Christian literature, Early—History and criticism. I. Title.
BR65.L26O35 270.1'092'4 78-40087
78 MARC

Rivers, Isabel.
Classical and Christian ideas in English Renaissance poetry : a students' guide / Isabel Rivers. — London ; Boston : G. Allen & Unwin, 1979.
231 p. ; 22 cm. GB***
Bibliography: p. [194]-227.
Includes index.
ISBN 0-04-807002-5 : £8.50 ($25.00)
1. English poetry—Early modern, 1500-1700—Sources. 2. Classical literature—History and criticism. 3. Christianity and literature. 4. Philosophy, Ancient. 5. Bible in literature. I. Title.
PR532.R5 821'.3'093 78-408053
79 MARC

Wilkins, Eliza Gregory, 1877-
"Know thyself" in Greek and Latin literature / Eliza Gregory Wilkins. — New York : Garland Pub., 1979.
104, [1] p. ; 23 cm. — (Ancient philosophy)
Originally presented as the author's thesis, University of Chicago, 1916.
Reprint of the 1917 ed. published by G. Banta Pub. Co., Menasha, Wis.
Bibliography: p. [105]
ISBN 0-8240-9572-3 : $13.00
1. Self-knowledge in literature. 2. Philosophy, Ancient. 3. Classical literature—History and criticism. I. Title. II. Series.
PA3015.S4W5 1979 880'.09'38 78-66584
79 MARC

—HISTORY AND CRITICISM—ADDRESSES, ESSAYS, LECTURES

Costa, Aída.
Temas clássicos / Aída Costa. — São Paulo : Editora Cultrix, 1978.
149 p. ; 20 cm.
"Esta obra é publicada em co-edição com a Secretaria da Cultura, Ciência e Tecnologia do Estado de São Paulo."
Includes bibliographical references.
1. Classical literature—History and criticism—Addresses, essays, lectures. I. Title.
PA3003.C63 880'.09 79-354869
79 MARC

—TRANSLATIONS INTO ENGLISH

Turnbull, Grace Hill, 1880-
Tongues of fire : a bible of sacred scriptures of the pagan world / Grace H. Turnbull. — New York : Arno Press, 1979, c1941.
xxvi, 416 p. ; 21 cm. — (Johns Hopkins University Press reprints)
Reprint of the 1929 ed. published by Macmillan, New York.
Includes bibliographical references.
ISBN 0-405-10634-3 : $28.00
1. Sacred books. 2. Oriental literature—Translations into English. 3. English literature—Translations from Oriental languages. 4. Classical literature—Translations into English. 5. English literature—Translations from Oriental languages. 6. Philosophy, Ancient. I. Title.
BL70.T7 1979 291.8'2 78-19278
78 MARC

CLASSICAL MYTHOLOGY see Mythology, Classical

CLASSICAL ORATIONS

see also Greek orations; Latin orations

CLASSICAL PHILOLOGY

see also Classical antiquities; Classicists; Greek literature; Hellenism; Humanism; Latin language; Latin literature; Latin philology

—ADDRESSES, ESSAYS, LECTURES

Bonner Festgabe : Johannes Straub zum 65. Geburtstag am 18. Oktober 1977 / dargebracht von Kollegen und Schülern ; [hrsg. von Adolf Lippold, Nikolaus Himmelmann]. — Bonn : Rheinland-Verlag ; In Kommission R. Habelt, 1977.
420 p. : ill. ; 27 cm. — (Beihefte der Bonner Jahrbücher ; Bd. 39) GFR***
Bibliography of J. A. Straub's works: p. [2]-6.
Includes bibliographical references.
ISBN 3-7927-0251-7
1. Civilization, Greek—Addresses, essays, lectures. 2. Rome—Civilization—Addresses, essays, lectures. 3. Classical philology—Addresses, essays, lectures. 4. Straub, Johannes, A., 1912- —Addresses, essays, lectures. I. Straub, Johannes A., 1912- . II. Lippold, Adolf, 1926- . III. Himmelmann-Wildschütz, Nikolaus. IV. Series: Bonner Jahrbücher. Beihefte ; Bd. 39.
DE3.B66 938 78-391835
79 MARC

Commentationes philologicae : volumen-homenaje al prof. Julio Campos Ruis. — Salamanca : Universidad Pontificia, 1977.
563 p., 1 leaf, [2] leaves of plates : ill. ; 24 cm. — (Bibliotheca salmanticensis : Estudios ; 19) Sp78-June
In English, French, German, Italian, or Spanish.
"Bibliografía del P. Julio Campos": p. [11]-16.
ISBN 8472990567 : 1000ptas
1. Classical philology—Addresses, essays, lectures. 2. Campos, Julio, 1906- . I. Campos, Julio, 1906- II. Series: Bibliotheca Salmanticensis : Estudios ; 19.
PA26.C28C65 79-383261
79 MARC

D'Eschyle à nos jours : leçons d'archéologie, de littérature, de philologie classiques / éditées par Guy Cambier. — Bruxelles, av. F. D. Roosevelt 50 : ULB-Faculté de philosophie et lettres, 1978.
147 p. ; 21 cm. Be78-9
French, Latin, or Greek.
Includes bibliographical references.
100F
1. Classical philology—Addresses, essays, lectures. I. Cambier, Guy.
PA25.D4 480 79-362035
79 MARC

Fra den klassisk filologiske hverdag : H. Friis Johansen quinquagenario / udg. af Vox populis red., P. S. Sørensen og G. Torresin. — Århus : [Vox populi], 1977.
56 leaves ; 30 cm. D77-34
Includes bibliographical references.
Not in trade.
1. Classical philology—Addresses, essays, lectures. I. Friis Johansen, H. II. Sørensen, Preben Steen, 1949- III. Torresin, Giuseppe.
PA26.F74 78-386299
79 MARC

CLASSICAL PHILOLOGY
—ADDRESSES, ESSAYS, LECTURES
(Continued)

Studi in onore di Anthos Ardizzoni / a cura di Enrico Livrea e G. Aurelio Privitera. — Roma : Edizioni dell'Ateneo & Bizzarri, 1978.
2 v. (xxiii, 1131 p.) ; 22 cm. — (Filologia e critica ; 25) It79-Mar
Italian, English, French, German, Greek, or Latin.
Bibliography: p. ⌐xix⌐-xxiii.
L44000
1. Classical philology—Addresses, essays, lectures. 2. Ardizzoni, Anthos. I. Ardizzoni, Anthos. II. Livrea, Enrico. III. Privitera, G. Aurelio. IV. Series.
PA26.A73S7 880'.9 79-354910
*79 MARC

—BIBLIOGRAPHY—CATALOGS

Newcastle-upon-Tyne. University. Dept. of Classics.
Pamphlets on classical subjects : a list of a collection made principally by J. Wight Duff, Professor of Literature, 1893-1898, Professor of Classical and Ancient History, 1898-1933 and housed in the Department of Classics of the University of Newcastle upon Tyne / compiled by Dorothy Monk ; with a prefatory note by G. B. A. Fletcher. — ⌐Newcastle-upon-Tyne⌐ : ⌐The Department⌐, ⌐1977⌐
⌐6⌐, 175 p. ; 30 cm. GB78-14030
Includes index.
1. Classical philology—Bibliography—Catalogs. 2. Pamphlets—Bibliography—Catalogs. 3. Newcastle-upon-Tyne. University. Dept. of Classics—Catalogs. I. Duff, John Wight, 1866-1944. II. Monk, Dorothy. III. Title.
Z7016.N48 1977 78-319841
⌐PA91⌐ 78 MARC

—BIOGRAPHY see Philologists

—COLLECTED WORKS

Lavagnini, Bruno, 1898-
Atakta : scritti minori di filologia classica, bizantina e neogreca / Bruno Lavagnini. — Palermo : Palumbo, 1978.
lxi, 796 p. : ill. ; 24 cm. It79-Apr
French, Italian, or Greek.
Reprint of the original editions.
"Bibliografia di Bruno Lavagnini": p. ⌐xxvi⌐-lviii.
Includes bibliographical references.
L40000
1. Classical philology—Collected works. 2. Greek philology, Medieval and late—Collected works. 3. Greek philology, Modern—Collected works. I. Title.
PA27.L34 1978 480'.9 79-367401
*79 MARC

—PERIODICALS

Debrecen, Hungary. Tudományegyetem.
Acta classica universitatis scientiarum Debreceniensis. t. 1- 1965-
Debrecini ⌐Kossuth Lajos Tudományegyetem Kiadói Bizottsága⌐
v. 24 cm. annual.
English, French, German, Hungarian, Italian or Latin.
1. Classical philology—Periodicals. 2. History, Ancient—Periodicals.
PA1.A1D4a 480'.05 78-649272
MARC-S

CLASSICAL POETRY

see also Concrete poetry, Classical

—HISTORY AND CRITICISM—
ADDRESSES, ESSAYS, LECTURES

Mills, Barriss
The unheroic muse : essays on five classical poets / Barriss Mills. — ⌐New Rochelle, N.Y.⌐ : Elizabeth Press ; Berkeley, CA : distributed by Serendipity Books, c1978.
49 p. ; 24 cm.
"500 copies printed."
1. Classical poetry—History and criticism—Addresses, essays, lectures. I. Title.
PA3020.M5 881'.009 79-109086
79 MARC

—HISTORY AND CRITICISM—
CONGRESSES

Liverpool Latin Seminar, 1976.
Papers of the Liverpool Latin Seminar, 1976 : classical Latin poetry, medieval Latin poetry, Greek poetry / edited by Francis Cairns. — Liverpool : Cairns, 1977.
305 p. ; 21 cm. — (Arca classical and medieval texts papers and monographs ; 2) GB***
Includes bibliographical references.
$20.00
1. Classical poetry—History and criticism—Congresses. I. Cairns, Francis. II. Title. III. Series.
PA3019.L58 1976 871'.009 79-304664
79 MARC

—TRANSLATIONS INTO ITALIAN

Carmi figurati greci e latini / ⌐a cura di⌐ Laura Simonini, Flaminio Gualdoni. — Pollenza : La nuova Foglio, 1978.
72 p. : ill. ; 24 cm. — (Collana Altro) It78-Nov
Greek or Latin with Italian translation.
L4000
1. Concrete poetry, Classical. 2. Classical poetry—Translations into Italian. 3. Italian poetry—Translations from classical languages. I. Simonini, Laura. II. Gualdoni, Flaminio.
PA3662.A2C3 881'.008 79-367248
*79 MARC

CLASSICAL RINGS see Rings, Classical

CLASSICAL SCULPTURE see Sculpture, Classical

CLASSICISM

see also Hellenism; Neoclassicism (Literature)

Buck, August, 1911-
Studien zu den "Volgarizzamenti" römischer Autoren in der italienischen Literatur des 13. und 14. Jahrhunderts / von August Buck, Max Pfister. — München : W. Fink, 1978.
87 p. ; 24 cm. — (Abhandlungen der Marburger Gelehrten Gesellschaft ; Jahrg. 1977, Nr. 1) GFR***
Includes texts in Italian and Latin.
Bibliography: p. 87.
ISBN 3-7705-1587-0
1. Italian literature—To 1400—History and criticism. 2. Italian literature—Translations from Latin. 3. Latin literature—Translations into Italian. 4. Classicism. I. Pfister, Max, 1932- joint author. II. Title. III. Series: Marburger Gelehrten Gesellschaft. Abhandlungen ; Jahrg. 1977, Nr. 1.
PQ4066.B8 850'.9'001 78-396223
79 MARC

Constantinescu, I
Introducere în literatura clasică / I. Constantinescu. — Iaşi : "Junimea", 1978.
336 p. ; 19 cm. R78-4093
Summary in French and German.
Includes bibliographical references and index.
lei9.50
1. Classicism. I. Title.
PN56.C6C6 79-347239
*79 MARC

Messina, Calogero, 1945-
Voltaire e il mondo classico / Calogero Messina. — Palermo : Herbita, 1976.
245 p. ; 22 cm. It78-Jan
"La mort de César": p. 173-226.
Includes bibliographical references and index.
L7000
1. Voltaire, François Marie Arouet de, 1694-1778—Knowledge and learning. 2. Classicism. 3. Civilization, Greek—Influence. 4. Rome—Civilization—Influence. I. Voltaire, François Marie Arouet de, 1694-1778. La mort de César. 1976. II. Title.
PQ2122.M4 78-381519
*78 MARC

Moskvicheva, G V
(Russkiĭ klassitsizm)
Русский классицизм / Г. В. Москвичева. — Москва : Просвещение, 1978.
128 p. : ill. ; 21 cm. — (Пособие для учителей) USSR 78
Series romanized: Posobie dlia uchitelei.
Includes bibliographical references.
0.35rub
1. Russian literature—18th century—History and criticism. 2. Classicism. I. Title.
PG3010.5.C53M67 79-412528

Przybylski, Ryszard.
To jest klasycyzm / Ryszard Przybylski ; wstęp Maria Janion. — Wyd. 1. — Warszawa : Czytelnik, 1978 (Szczec : SzZG)
241, ⌐2⌐ p. ; 21 cm. P78-8721
Includes index.
zł53.00
1. Polish poetry—20th century—History and criticism. 2. Poetry, Modern—20th century—History and criticism. 3. Classicism. I. Title.
PG7083.C55P78 79-356741
79 MARC

The Reception of classical German literature in England, 1760-1860 : a documentary history from contemporary periodicals / edited by John Boening. — New York : Garland Pub., 1977.
10 v. ; 32 cm.
A facsimile collection of reviews of German literature that appeared in contemporary English periodicals.
On spine: Classical German literature in England, 1760-1860.
Includes bibliographical references and index.
ISBN 0-8240-0990-8 (v. 1) : $900.00 (set)
1. German literature—18th century—Book reviews. 2. German literature—19th century—Book reviews. 3. German literature—18th century—History and criticism. 4. English literature—18th century—History and criticism. 5. Classicism. I. Boening, John. II. Title: Classical German literature in England, 1760-1860.
PT312.R4 1977 830'.9'006 77-24149
77 MARC

CLASSICISM IN ARCHITECTURE

see also Neoclassicism (Architecture)

Mertens, Jürgen.
Der Burgplatz am Ende des 16. Jahrhunderts / Jürgen Mertens. — Braunschweig : ⌐s.n.⌐, 1978.
36 p., ⌐4⌐ leaves of plates : ill., 1 col. map ; 21 cm. — (Arbeitsberichte aus dem Städtischen Museum Braunschweig ; 28) GFR***
Bibliography: p. 32-35.
1. Plazas. 2. Classicism in architecture. I. Title. II. Series: Brunswick (City). Städtisches Museum. Arbeitsberichte ; 28.
N2260.8.A25 Nr. 28 79-368053
⌐NA9070⌐ 79 MARC

—BIBLIOGRAPHY—CATALOGS

Texas. University at Austin. Humanities Research Center.
The sources of classicism : five centuries of architectural books from the collections of the Humanities Research Center : the Michener Galleries, Harry Ransom Center, 26 March-23 April, 1978 / by Drury Blakeley Alexander. — ⌐Austin⌐ : Humanities Research Center, The University of Texas at Austin, c1978.
⌐44⌐ p. : ill. ; 23 cm.
Bibliography: p. ⌐18⌐-⌐19⌐
1. Classicism in architecture—Bibliography—Catalogs. 2. Texas. University at Austin. Humanities Research Center—Catalogs. I. Alexander, Drury Blakeley. II. Texas. University at Austin. Art Museum. III. Title.
Z5942.5.C55T49 1978 016.72 78-623037
⌐NA260⌐ 79 MARC

—ITALY

Berger, Ursel.
Palladios Frühwerk : Bauten u. Zeichn. / von Ursel Berger. — Köln ; Wien : Böhlau ⌐in Komm.⌐, 1978.
x, 300, 26 p. : ill. ; 21 cm. — (Dissertationen zur Kunstgeschichte ; 5) GFR78-A
Originally presented as the author's thesis, Munich.
Bibliography: p. ⌐289⌐-297.
Includes indexes.
ISBN 3-412-00157-8
1. Palladio, Andrea, 1508-1580. 2. Classicism in architecture—Italy. I. Title.
NA1123.P2B367 1978 720'.92'4 78-400274
79 MARC

—NETHERLANDS—AMSTERDAM

Danckerts, Cornelis.
Een Amsterdams grachtenboekje uit de zeventiende eeuw : het plaatwerk van Cornelis Danckerts / ingeleid door I. H. van Eeghen. — ⌐Amsterdam⌐ : Geenootschap Amstelodamum, 1963.
24 p., ⌐17⌐ leaves of plates : ill., map (fold. in pocket) ; 20 x 28 cm.
1. Façades—Netherlands—Amsterdam. 2. Classicism in architecture—Netherlands—Amsterdam. 3. Amsterdam—Buildings. I. Eeghen, Isabella Henriette van. II. Title.
NA2840.D3 79-368067
79 MARC

CLASSICISM IN ART

see also Neoclassicism (Architecture); Neoclassicism (Art)

CLASSICISM IN ART (Continued)

Becatti, Giovanni, 1912-
L'età classica. ⌜Firenze⌝ Sansoni ⌜c1965⌝
421 p. illus. (part mounted col.) 32 cm. (Le Grandi epoche dell'arte, 3)
Bibliography: p. 387-407.
1. Art—History. 2. Classicism in art. I. Title. II. Series.
N5610.B33 709'.4 66-87423
⌜r79⌝rev MARC

Jones, Henri.
De l'esthétique classique, tirée du portrait et du nu / Henri Jones (Maxhim) ; préf. de Marcel Brion. — Montréal : Centre éducatif et culturel, c1969.
239 p. : ill. ; 22 cm. — (Reflets ; 4)
Bibliography: p. 229-236.
1. Classicism in art. 2. Art, Classical. 3. Arts. 4. Portrait. 5. Nude in art. I. Title.
NX600.C5J66 700 79-104167
79 MARC

—HISTORY

Greenhalgh, Michael.
The classical tradition in art / ⌜by⌝ Michael Greenhalgh. — London : Duckworth ; Dallas : distributed by Southwest Book Services, 1978.
271 p. : ill. ; 23 cm. GB***
Bibliography: p. 235-262.
Includes index.
ISBN 0-7156-1277-8 : £12.50
1. Classicism in art—History. 2. Art—History. I. Title.
N7432.5.C6G73 709'.4 79-309765
79 MARC

CLASSICISTS

—SPAIN—CONGRESSES

Pérez Rioja, José Antonio.
Humanismo español en el siglo XIX : conferencias pronunciadas en la Fundación Universitaria Española los días 10, 12 y 17 de mayo de 1976 / ⌜por⌝ José A. Pérez Rioja, Manuel Fernández-Galiano, Andrés Amorós. — Madrid : Fundación Universitaria Española, 1977.
94 p., 1 leaf ; 21 cm. — (Publicaciones de la Fundación Universitaria Española : Conferencias ; 83) Sp77-Nov
ISBN 8473920252
1. Classicists—Spain—Congresses. 2. Ranz Romanillos, Antonio—Congresses. 3. Pérez de Ayala, Ramón, 1880-1962—Congresses. I. Fernández-Galiano, Manuel, joint author. II. Amorós, Andrés, joint author. III. Title. IV. Series: Fundación Universitaria Española. Conferencias ; 83.
PA83.P47 78-401197
79 MARC

CLASSICS ANONYMOUS (CATALOGING)
see Cataloging of anonymous classics

CLASSIFICATION

see also Automatic classification; Classification of sciences; Enzymes —Classification; Ships—Rules for classification and construction

Katō, Shūkō, 1895-
(Tosho no bunrui).
図書の分類 / 加藤宗厚著. — 改稿版. — 東京：理想社，昭和41 ⌜1966⌝
158 p. ; 22 cm. — (図書館実務叢書 ; 3)
Includes bibliographical references and index.

1. Classification. 2. Cataloging. I. Title. II. Series: Toshokan jitsumu sōsho ; 3.
Z693.K34 79-802637

Kovář, Blahoslav.
Problémy teorie procesu věcného pořádání informací a selekčních jazyků / Blahoslav Kovář. — 1. vyd. — Praha : Univ. Karlova, 1976.
165 p. ; 24 cm. — (Acta Universitatis Carolinae, Philosophica et historica, Monographia ; 65/1976) Cz78
Summary also in English, German, and Russian.
Bibliography: p. 159-165.
Kčs20.00
1. Classification. 2. Indexing. 3. Subject headings. I. Title. II. Series: Prague. Universita Karlova. Acta, Philosophica et historica, Monographia ; 1976, 65.
AS141.A522 1976, no. 65 79-360021
909 s
⌜Z696⌝ *79 MARC

—ARCHIVES

McGill University, Montreal. Archives.
Outline of scheme for arranging the records / McGill University Archives. — ⌜Montreal⌝ : s.n.⌝, 1967.
⌜43⌝ leaves ; 28 cm.
1. Classification—Archives. I. Title.
Z697.A76M32 1967 025.4'3 78-316504
78 MARC

--AUDIOVISUAL MATERIALS see Classification--Nonbook materials

—BOOKS

see also Catalogs, Classified; Classification, BBK; Classification, Colon; Classification, Decimal; Classification, Dewer decimal; Classification, Library of Congress; Classification, Universal decimal

Abū al-Nūr, ʻAbd al-Wahhāb.
(al-Taṣnīf li-aghrāḍ istirjāʻ al-maʻlūmāt)
التصنيف لاغراض استرجاع المعلومات / اعداد عبد الوهاب عبد السلام ابو النور. — القاهرة : المنظمة العربية للتربية والثقافة والعلوم، ادارة التوثيق والاعلام، 1977.
195 p. ; 24 cm. — (سلسلة دراسات عن المعلومات ؛ 3)
Bibliography: p. 184-191.
1. Classification—Books. I. Title. II. Series: Silsilat dirāsāt ʻan al-maʻlūmāt ; 3.
Z696.A4A28 78-962326

Boráková, Drahomíra.
Stavění knižního fondu ve volném výběru : metodický materiál pro okresní a střediskové knihovny Sm. kraje / ⌜zprac.⌝ Drahomíra Boráková. — Olomouc : Státní vědecká knihovna, 1977.
20 p. ; 29 cm. — (Publikace Státní vědecké knihovny v Olomouci ; čís. 24 /1977) Cz***
1. Open and closed shelves. 2. Classification—Books. 3. Libraries—Czechoslovakia—Severomoravský kraj. I. Title. II. Series: Státní vědecká knihovna v Olomouci. Publikace Státní vědecké knihovny v Olomouci ; 1977/čís. 24.
Z796.S77S73a čís 24/1977 78-380563
025.8'1
⌜Z711⌝ 78⌜r79⌝rev MARC

British Library. Working Party on Classification and Indexing.
Final report / British Library Working Party on Classification and Indexing. — Wetherby : British Library Lending Division, 1976.
3-25 p. ; 30 cm. — (British Library research & development reports ; no. 5233) GB76-06246
"Appendix to British Library Working Party on Classifaction and Indexing final report": 7 sheets of microfiche in pocket.
ISBN 0-85350-133-5 : £2.00 ($4.00 U.S.)
1. Indexing. 2. Classification—Books. 3. British Library. I. Series: British Library. Research and Development Dept. British Library research & development reports ; no. 5233.
Z695.9.B84 1976 025.4 77-368528
77⌜r79⌝rev MARC

Collège de La Pocatière. Bibliothèque.
Guide du classificateur, rédigé par Réal-Léonce Lévesque. Texte provisoire. La Pocatiere, Bibliothèque, Collège de Sainte-Anne-de-la-Pocatière, 1966.
479, ⌜332⌝ p. 28 cm. 12.00 C69-2036
Includes bibliographical references.
1. Subject headings. 2. Classification—Books. I. Lévesque, Réal Léonce, 1924- ed. II. Title.
Z695.C692 1966 72-445612
79⌜r79⌝rev MARC

Dahlström, Asta, 1913-
On UCD with some malice / Asta Dahlström. — Stockholm : Royal Institute of Technology Library, 1976.
15 leaves ; 30 cm. — (Stockholm papers in library and information science ; report TRITA-LIB-6005) S***
Includes bibliographical references.
1. Classification, Universal decimal. 2. Classification—Books. I. Title. II. Series.
Z696.U86132 025.4 79-301256
79 MARC

(Edinaia skhema klassifikatsii literatury dlia knigoizdaniia v SSSR)
Единая схема классификации литературы для книгоиздания в СССР. — Москва : Книга, 1977.
467 p. ; 22 cm. USSR 77
At head of title: Gosudarstvennyĭ komitet po delam izdatel'stv, poligrafii i knizhnoĭ torgovli SSSR.
Includes index.
2.20rub
1. Classification—Books. I. Russia (1923- U. S. S. R.). Gosudarstvennyĭ komitet po delam izdatel'stv, poligrafii i knizhnoĭ torgovli.
Z696.A4E43 78-420957

Herdman, Margaret May, 1888-
Classification : an introductory manual / by Margaret M. Herdman. — 3d ed. / rev. by Jeanne Osborn. — Chicago : American Library Association, 1978.
44 p. ; 23 cm.
Bibliography: p. 38-44.
ISBN 0-8389-0277-4 : $3.50
1. Classification—Books. 2. Classification—Books—Bibliography. I. Osborn, Jeanne. II. American Library Association. III. Title.
Z696.A4H47 1978 025.4 78-11079
78 MARC

Moscow. Publichnaia biblioteka. Otdel sistematicheskikh i predmetnykh katalogov.
(Organizatsiia tekhnologicheskikh protsessov klassifitsirovaniia proizvedeniĭ pechati v Gosudarstvennoĭ biblioteke SSSR imeni V. I. Lenina)
Организация технологических процессов классифицирования произведений печати в Государственной библиотеке СССР имени В. И. Ленина : инструкции и методические рекомендации / Государственная библиотека СССР имени В. И. Ленина. Отдел систематических и предметных каталогов ; ⌜составитель В. И. Одарченко⌝. — Москва : Гос. библиотека СССР, 1977.
171 p. ; 22 cm. USSR***
Includes bibliographical references and index.
0.46rub
1. Cataloging. 2. Classification—Books. 3. Moscow. Publichnaia biblioteka. Otdel sistematicheskikh i predmetnykh katalogov. I. Odarchenko, V. I. II. Title.
Z693.M67 1977 79-407057

—BOOKS—ADDRESSES, ESSAYS, LECTURES

Bestandserschliessung und Bibliotheksstruktur : Rolf Kluth zum 10.2.1979 / hrsg. von Rainer Alsheimer. — Wiesbaden : Harrassowitz, 1979.
244 p., ⌜2⌝ leaves of plates : ill. ; 25 cm. — (Beiträge zum Buch- und Bibliothekswesen ; Bd. 17) GFR79-A
One contribution in French.
Bibliography of R. Kluth's works: p. 6-10.
Includes bibliographical references.
ISBN 3-447-02008-3
1. Library science—Addresses, essays, lectures. 2. Classification—Books—Addresses, essays, lectures. 3. Cataloging—Addresses, essays, lectures. 4. Kluth, Rolf—Addresses, essays, lectures. I. Kluth, Rolf. II. Alsheimer, Rainer. III. Series.
Z674.B45 79-368898
79 MARC

—BOOKS—AGRICULTURE

Hernández de Caldas, Angela.
Bibliotecas y centros de documentación bio-agrícolas latinoamericanos. Bogotá ⌜Ministerio de Educación Nacional, Sección de Servicios Bibliotecarios⌝ 1968.
10 l. 29 cm.
"Contribución a la II Reunión Interamericana de Bibliotecarios y Documentalistas Agrícola, Bogotá, diciembre 2-6 de 1968."
"Una revisión de Latin American bio-agricultural libraries and documentation centers, trabajo de base del 3rd. World Congress of Agricultural Librarians and Documentalists, Washington, D.C., 1965."
Bibliography: leaf 10.
1. Agricultural libraries—Latin America. 2. Agriculture—Latin America—Information services. 3. Classification—Books—Agriculture. I. Title.
Z675.A8H46 026'.63'098 78-414921
⌜S473.9⌝ 78 MARC

Koekebakker, F AE.
AGRICODE for agricultural libraries developed from UDC (Universal Decimal Classification) / by F. AE. Koekebakker ; translated and revised from Landbouwcode, 2d ed., by J. Stransky. — Wageningen ⌜Netherlands⌝ : Centre for Agricultural Pub. and Documentation, 1976.
136 p. ; 21 cm. Ne***
Translated with revisions from the author's Landbouwcode voor het klasseren van landbouwliteratuur.
Stamp on t.p.: Exclusive distributor, ISBS Inc., Forest Grove, OR.
Includes index.
ISBN 9022005674
1. Classification—Books—Agriculture. I. Stransky, J. II. Title.
Z697.A4K6313 1976 025.4'6'63 78-360260
78⌜r79⌝rev MARC

—BOOKS—ANTHROPOLOGY

Harvard University. Library.
Geography and anthropology : classification schedules, classified listing by call number, chronological listing, author and title listing / Harvard University Library. — Cambridge, Mass. : The Library : distributed by the Harvard University Press, 1979.
270 p. ; 29 cm. — (Widener Library shelflist ; 60)
ISBN 0-674-34855-9 : $30.00
1. Geography—Bibliography—Catalogs. 2. Anthropology—Bibliography—Catalogs. 3. Classification—Books—Geography. 4. Classification—Books—Anthropology. 5. Harvard University. Library. I. Title. II. Series: Harvard University. Library. Widener Library shelflist ; 60.
Z6009.H37 1979 016.91 79-617
⌜G116⌝ 79 MARC

CLASSIFICATION (Continued)

—BOOKS—BIBLIOGRAPHY

Herdman, Margaret May, 1888-
Classification : an introductory manual / by Margaret M. Herdman. — 3d ed. / rev. by Jeanne Osborn. — Chicago : American Library Association, 1978.
44 p. ; 23 cm.
Bibliography: p. 38-44.
ISBN 0-8389-0277-4 : $3.50
1. Classification—Books. 2. Classification—Books—Bibliography. I. Osborn, Jeanne. II. American Library Association. III. Title.
Z696.A4H47 1978 025.4 78-11079
78 MARC

—BOOKS—CANADIAN LITERATURE

National Library of Canada.
Class PS8000 : a classification for Canadian literature. — 2d ed. — Ottawa : National Library of Canada, 1978.
x, 21 leaves ; 26 cm. C***
Issued also in French.
Devised and updated by T. R. McCloy.
Includes index.
ISBN 0-660-01659-1 : $3.00
1. Classification—Books—Canadian literature. 2. Classification—Books—French-Canadian literature. 3. Classification, Library of Congress. I. McCloy, T. R. II. Title.
Z696.U5P84 1978a 025.4'6'81 79-308036
79 MARC

National Library of Canada.
Classe PS8000 : cadre de classification de la littérature canadienne. — 2. éd. — Ottawa : Bibliothèque nationale du Canada, 1978.
22 leaves ; 26 cm. C***
Issued also in English under title Class PS8000.
Includes index.
ISBN 0-660-01660-5 : $3.00
1. Classification—Books—Canadian literature. 2. Classification—Books—French-Canadian literature. 3. Classification, Library of Congress. I. Title. II. Title: Cadre de classification de la littérature canadienne.
Z696.U5P84 1978 025.4'6'8 79-358703
79 MARC

—BOOKS—CATALOGS, COLLEGE

Jarrell, Howard R
A classification scheme for college catalogs / by Howard R. Jarrell. — 3d ed. — Westminster, Colo. : Library Reports & Research Service, 1979.
78 p. ; 19 cm.
1. Classification—Books—Catalogs, College. I. Title.
Z697.C68J3 1979 025.4'6'37873 79-111449
00 MARC

—BOOKS—CHINESE LITERATURE

Chang, Shu-san.
(Chung wen t'u shu fen lei chih yüan li chi shih wu)
中文圖書分類之原理及實務 / 張樹三撰. — 臺北：臺灣中華書局, 民國66 [1977]
1, 3, 104 p. ; 22 cm.
Bibliography: p. 103.

1. Classification—Books—Chinese literature. I. Title.
Z697.C52C39 78-842832

—BOOKS—CLASSICAL LITERATURE

Harvard University. Library.
Classical studies : classification schedules, classified listing by call number, chronological listing, author and title listing / Harvard University Library. — Cambridge, Mass. : The Library : distributed by the Harvard University Press, 1979.
215 p. ; 29 cm. — (Its Widener Library shelflist ; 57)
ISBN 0-674-13461-3 : $25.00
1. Civilization, Greek—Bibliography—Catalogs. 2. Rome—Civilization—Bibliography—Catalogs. 3. Classical literature—Bibliography—Catalogs. 4. Classification—Books—Greece. 5. Classification—Books—Rome. 6. Classification—Books—Classical literature. 7. Harvard University. Library—Catalogs. I. Title. II. Series.
Z6207.G7H37 1979 025.4'3 79-948
[DE3] 79 MARC

—BOOKS—COMMERCE

International Trade Centre.
Classification de l'information commerciale. — Genève : Centre du commerce international CNUCED/GATT, 1976.
xlv, 131 p. ; 30 cm. Sw77-1701
$30.00 (U.S.)
1. Classification—Books—Commerce. I. Title.
Z697.C69157 1976 025.4'6'3801 78-378341
*78 MARC

—BOOKS—CONGRESSES

American Documentation Institute. Special Interest Group on Classification Research.
Minutes of organizational meeting of the Special Interest Group on Classification Research of the American Documentation Institute, October 3, 1966. — [Washington]: The Institute, [1966]
12 leaves in various foliations ; 28 cm.
Caption title.
1. Classification—Books—Congresses. 2. American Documentation Institute. Special Interest Group on Classification Research.
Z696.A2A45 1966 025.4 78-326125
79 MARC

—BOOKS—FRENCH-CANADIAN LITERATURE

National Library of Canada.
Classe PS8000 : cadre de classification de la littérature canadienne. — 2. éd. — Ottawa : Bibliothèque nationale du Canada, 1978.
22 leaves ; 26 cm. C***
Issued also in English under title Class PS8000.
Includes index.
ISBN 0-660-01660-5 : $3.00
1. Classification—Books—Canadian literature. 2. Classification—Books—French-Canadian literature. 3. Classification, Library of Congress. I. Title. II. Title: Cadre de classification de la littérature canadienne.
Z696.U5P84 1978 025.4'6'8 79-358703
79 MARC

National Library of Canada.
Class PS8000 : a classification for Canadian literature. — 2d ed. — Ottawa : National Library of Canada, 1978.
x, 21 leaves ; 26 cm. C***
Issued also in French.
Devised and updated by T. R. McCloy.
Includes index.
ISBN 0-660-01659-1 : $3.00
1. Classification—Books—Canadian literature. 2. Classification—Books—French-Canadian literature. 3. Classification, Library of Congress. I. McCloy, T. R. II. Title.
Z696.U5P84 1978a 025.4'6'81 79-308036
79 MARC

—BOOKS—GEOGRAPHY

Harvard University. Library.
Geography and anthropology : classification schedules, classified listing by call number, chronological listing, author and title listing / Harvard University Library. — Cambridge, Mass. : The Library : distributed by the Harvard University Press, 1979.
270 p. ; 29 cm. — (Widener Library shelflist ; 60)
ISBN 0-674-34855-9 : $30.00
1. Geography—Bibliography—Catalogs. 2. Anthropology—Bibliography—Catalogs. 3. Classification—Books—Geography. 4. Classification—Books—Anthropology. 5. Harvard University. Library. I. Title. II. Series: Harvard University. Library. Widener Library shelflist ; 60.
Z6009.H37 1979 016.91 79-617
[G116] 79 MARC

—BOOKS—GREEK LITERATURE

Harvard University. Library.
Ancient Greek literature : classification schedules, classified listing by call number, chronological listing, author and title listing / Harvard University Library. — Cambridge, Mass. : The Library : distributed by Harvard University Press, 1979.
638 p. ; 29 cm. — (Widener Library shelflist ; 58)
ISBN 0-674-03310-8 : $55.00
1. Greek literature—Bibliography—Catalogs. 2. Classification—Books—Greek literature. 3. Harvard University. Library—Catalogs. I. Title. II. Series: Harvard University. Library. Widener Library shelflist ; 58.
Z7025.H37 1979 016.88 79-9989
[PA3051] 79 MARC

—BOOKS—HOME ECONOMICS

Ústředí vědeckých, technických a ekonomických informací.
Mezinárodní desetinné třídění : třída 64 : domácí hospodářství / [odpovědný pracovník Dušan Simandl]. — 1. čs. úplné vyd. 13 mezinárodní vyd. — Praha : ÚVTEI, 1978.
x, 89 p. ; 30 cm. — (FID ; 419) Cz***
1. Classification—Books—Home economics. 2. Classification, Universal decimal. I. Simandl, Dušan. II. International Federation for Documentation. Classification décimale. III. Title. IV. Series: International Federation for Documentation. Publication ; 419.
Z696.U875H658 79-371375
79 MARC

—BOOKS—ISLAM

Sardar, Ziauddin.
Islam, outline of a classification scheme / Ziauddin Sardar. — London : C. Bingley ; New York : K. G. Saur, 1979.
81 p. ; 23 cm. GB***
Bibliography: p. 23-24.
Includes index.
ISBN 0-85157-285-5 : £4.00
1. Classification—Books—Islam. I. Title.
Z697.I77S27 025.4'6'909097671 79-314175
00 MARC

—BOOKS—LATIN LITERATURE

Harvard University. Library.
Latin literature : classification schedules, classified listing by call number, chronological listing, author and title listing / Harvard University Library. — Cambridge, Mass. : The Library : distributed by Harvard University Press, 1979.
610 p. ; 29 cm. — (Widener Library shelflist ; 59)
ISBN 0-674-51295-2 : $55.00
1. Latin literature—Bibliography—Catalogs. 2. Classification—Books—Latin literature. 3. Harvard University. Library—Catalogs. I. Title. II. Series: Harvard University. Library. Widener Library shelflist ; 59.
Z7030.H37 1979 016.87 79-9985
[PA6003] 79 MARC

—BOOKS—LAW

Carvalho, Dóris de Queiroz, 1920-
Classificação decimal de direito / Doris de Queiroz Carvalho. — 3. ed., rev. e aumentada. — Rio de Janeiro : Ministério da Fazenda, Departamento de Administração, Divisão de Documentação, 1977.
188 p. ; 30 cm.
At head of title: República Federativa do Brasil.
1. Classification—Books—Law. 2. Law—Classification. 3. Classification, Decimal. I. Brazil. Ministério da Fazenda. Divisão de Documentação. II. Title.
Z697.L4C35 1977 79-383021
79 MARC

Conklin, Curt E
Library of Congress subject headings, KF cross-reference / by Curt E. Conklin. — Provo, Utah : Brigham Young University Law Library, c1979.
215 p. ; 29 cm. — (BYU legal research series ; no. 5)
Bibliography: p. 214-215.
1. Classification—Books—Law. 2. Law—United States—Classification. 3. Classification, Library of Congress. 4. Subject headings—Law. I. Brigham Young University, Provo, Utah. Law Library. II. Title. III. Series: Brigham Young University, Provo, Utah. BYU legal research series ; no. 5.
Z696.U5K6 025.3'334 79-106943
79 MARC

United States. Library of Congress. Subject Cataloging Division.
Classification. Class K, subclass KE, Law of Canada / Subject Cataloging Division, Processing Department, Library of Congress. — Washington : The Library, 1976.
viii p., 134 leaves, 135-181 p. ; 26 cm.
Kept up to date by the division's L.C. classification—additions and changes, and by an irregular cumulative publication issued by Gale Research Company, entitled Library of Congress classificaton schedules: a cumulation of additions and changes, class K, subclass KE.
Includes index.
ISBN 0-8444-0224-9
1. Classification—Books—Law. 2. Law—Canada—Classification. 3. Classification, Library of Congress. I. Title.
Z696.U5K4 1976 025.4'6'34 76-54307
—— —— Copy 3. Z663.78.C5K4 1976 MARC
76[r79]rev

CLASSIFICATION (Continued)

—BOOKS—MEDICINE

Chevallier, Jacques, physician.
C.A.N.DO. médical et pharmaceutique : classification alphanumérique de la documentation d'après le code de classification mis au point en 1955 par J. Chevallier et Cl. Madier / par J. Chevallier, ... ; ⌊avec la collaboration de J. M. Alache, R. Alfieri, J. Archimbaud, Ch. Bachelerie, etc.⌋. — 3e éd. — Paris : Maloine, 1978.
1069 p. ; 20 cm. F78-13006
Addendum (9 p.) inserted.
Bibliography: p. 1067.
ISBN 2-224-00406-0 : 195F
1. Classification—Books—Medicine. 2. Classification—Books—Pharmacy. I. Title.
Z697.M4C45 1978 025.4'6'61 79-340691
79 MARC

United States. National Library of Medicine.
National Library of Medicine classification : a scheme for the shelf arrangement of books in the field of medicine and its related sciences. — 4th ed. — Bethesda, Md. : The Library, 1978.
xliii, 390 p. ; 29 cm. — (DHEW publication ; no. (NIH) 78-1535)
Third rev. ed. published in 1969 under title: Classification.
Bibliography: p. xvii.
Includes index.
1. Classification—Books—Medicine. I. Title. II. Series: United States. Dept. of Health, Education, and Welfare. DHEW publication ; no. (NIH) 78-1535.
Z697.M4U5 1978 025.4'6'61 78-56040
78 MARC

—BOOKS—MUSIC

United States. Library of Congress. Subject Cataloging Division.
Classification : Class M : music and books on music / Subject Cataloging Division, Processing Department, Library of Congress. — 3d ed. — Washington : The Library : for sale by Cataloging Distribution Services, Library of Congress, 1978.
x, 228 leaves ; 26 cm.
"A revision of the second edition, originally published in 1917, and includes additions and changes made by the Library of Congress through June 1977."
Includes index.
1. Classification—Music. 2. Classification—Books—Music. 3. Classification, Library of Congress. I. Title. II. Title: Music and books on music.
Z696.U5M 1978 025.4'6'78 78-606179
——— Copy 3. Z663.78.C5M 1978 MARC
78

—BOOKS—PHARMACY

Chevallier, Jacques, physician.
C.A.N.DO. médical et pharmaceutique : classification alphanumérique de la documentation d'après le code de classification mis au point en 1955 par J. Chevallier et Cl. Madier / par J. Chevallier, ... ; ⌊avec la collaboration de J. M. Alache, R. Alfieri, J. Archimbaud, Ch. Bachelerie, etc.⌋. — 3e éd. — Paris : Maloine, 1978.
1069 p. ; 20 cm. F78-13006
Addendum (9 p.) inserted.
Bibliography: p. 1067.
ISBN 2-224-00406-0 : 195F
1. Classification—Books—Medicine. 2. Classification—Books—Pharmacy. I. Title.
Z697.M4C45 1978 025.4'6'61 79-340691
79 MARC

—BOOKS—PHILOSOPHY

United States. Library of Congress. Subject Cataloging Division.
Classification. Class B, Subclasses B-B J : Philosophy, Psychology / Subject Cataloging Division, Processing Services, Library of Congress. — 3d ed. — Washington : The Library, 1979.
viii, 250 p. ; 26 cm.
Includes index.
ISBN 0-8444-0280-X
1. Classification—Books—Philosophy. 2. Classification—Books—Psychology. 3. Classification, Library of Congress. I. Title.
Z696.U5B1 1979 025.4'6'1 79-526
——— Copy 3. Z663.78.C5B1 1979 MARC
79

—BOOKS—PHOTOGRAMMETRY

Corten, François L
I.T.C. international bibliography, photogrammetry, photogrammetric abstracts / F. L. Corten. — 2d ed. — Delft : I.T.C., 1965.
48 p. ; 24 cm. — (I.T.C. information ; no. 8)
First published in 1962 under title: I.T.C. international bibliography of photogrammetry, photogrammetric abstracts.
Errata slip inserted.
1. Photogrammetry—Documentation. 2. Classification—Books—Photogrammetry. 3. Classification, Universal decimal. I. Title. II. Title: International bibliography, photogrammetry, photogrammetric abstracts. III. Series.
TA593.A2I2 No. 8 1965 526.9'823'08 s 78-317657
⌊Z695.1.P477⌋ 78 MARC
⌊TA593⌋

—BOOKS—PSYCHOLOGY

United States. Library of Congress. Subject Cataloging Division.
Classification. Class B, Subclasses B-B J : Philosophy, Psychology / Subject Cataloging Division, Processing Services, Library of Congress. — 3d ed. — Washington : The Library, 1979.
viii, 250 p. ; 26 cm.
Includes index.
ISBN 0-8444-0280-X
1. Classification—Books—Philosophy. 2. Classification—Books—Psychology. 3. Classification, Library of Congress. I. Title.
Z696.U5B1 1979 025.4'6'1 79-526
——— Copy 3. Z663.78.C5B1 1979 MARC
79

—BOOKS—SOCIAL SCIENCES

Kokuritsu Kokkai Toshokan, Tokyo. Chōsa oyobi Rippō Kōsakyoku. Hōritsu Seiji Toshokan.
(Kokusai jisshin bunrui hyō)
国際十進分類表：社会科学の部. — ⌊東京⌋：国立国会図書館調査立法考査局法律政治図書館, 昭和33 ⌊1958⌋
70 p. : 30 cm. — (国図調立資料 ; B154)
Cover title.
Errata slip inserted.
"国際十進分類表」...に拠って編さんしたもので...相当に取捨を加えてある"
1. Classification—Books—Social sciences. 2. Classification, Universal decimal. I. International Federation for Documentation. Classification décimale. Japanese. II. Title. III. Series: Kokuritsu Kokkai Toshokan, Tokyo. Chōsa oyobi Rippō Kosakyoku. Kokuto chōritsu shiryō ; B154.
Z697.S6K64 1958 79-804975

—BOOKS—TECHNOLOGY

Budapest. Országos Műszaki Könyvtár és Dokumentácios Központ.
Figyelőkarton-szolgálati tájékoztató. 6. módositott kiad. Budapest, 1961.
77 p. 20 cm.
At head of title: Országos Műszaki Könyvtár. Dokumentácios Osztály.
1. Classification—Books—Technology. 2. Classification, Universal decimal. I. Title.
Z696.U875T425 79-384761
00 MARC

—BOOKS—WELDING

International Institute of Welding.
Classification table of the documentation (bilingual, English-French) : supplement to the Bibliographical bulletin for welding and allied processes ... = Table de classement de la documentation (bilingue, français-anglais) : supplément au Bulletin de documentation de la soudure et de techniques connexes ... / International Institute of Welding. — Paris : Publications de la soudure autogène, S.A.R.L., ⌊1950?⌋
31, 31 p. ; 24 cm.
Cover title.
Includes index.
1. Classification—Books—Welding. I. International Institute of Welding. Bibliographical bulletin for welding and allied processes. II. Title. III. Title: Table de classement de la documentation.
Z697.T4I65 1950 026.4'6'67152 78-318203
78 MARC

—BOOKS—GREECE

Harvard University. Library.
Classical studies : classification schedules, classified listing by call number, chronological listing, author and title listing / Harvard University Library. — Cambridge, Mass. : The Library : distributed by the Harvard University Press, 1979.
215 p. ; 29 cm. — (Its Widener Library shelflist ; 57)
ISBN 0-674-13461-3 : $25.00
1. Civilization, Greek—Bibliography—Catalogs. 2. Rome—Civilization—Bibliography—Catalogs. 3. Classical literature—Bibliography—Catalogs. 4. Classification—Books—Greece. 5. Classification—Books—Rome. 6. Classification—Books—Classical literature. 7. Harvard University. Library—Catalogs. I. Title. II. Series.
Z6207.G7H37 1979 025.4'3 79-948
⌊DE3⌋ 79 MARC

—BOOKS—ROME

Harvard University. Library.
Classical studies : classification schedules, classified listing by call number, chronological listing, author and title listing / Harvard University Library. — Cambridge, Mass. : The Library : distributed by the Harvard University Press, 1979.
215 p. ; 29 cm. — (Its Widener Library shelflist ; 57)
ISBN 0-674-13461-3 : $25.00
1. Civilization, Greek—Bibliography—Catalogs. 2. Rome—Civilization—Bibliography—Catalogs. 3. Classical literature—Bibliography—Catalogs. 4. Classification—Books—Greece. 5. Classification—Books—Rome. 6. Classification—Books—Classical literature. 7. Harvard University. Library—Catalogs. I. Title. II. Series.
Z6207.G7H37 1979 025.4'3 79-948
⌊DE3⌋ 79 MARC

—BOOKS—URUGUAY

Peluffo de Chapt, Estela.
Proyecto de extensión del esquema 989.9, historia del Uruguay : contribución al estudio de la clase 9 de la C.D.U. / Estela Peluffo de Chapt, Sonia G. Mato. — Montevideo : Universidad de la República, División Publicaciones y Ediciones, ⌊1973⌋
90 p. ; 27 cm.
On cover: Universidad de la República, Escuela Universitaria de Bibliotecología y Ciencias Afines "Ing. Federico E. Capurro."
1. Classification—Books—Uruguay. 2. Classification, Universal decimal. I. Mato, Sonia G., joint author. II. Title.
Z696.U875U786 79-100951
79 MARC

—BOTANY see Botany—Classification

—DISEASES see Nosology

—DRAWING

Nason, Roger P
Plans & drawings classification system / Roger P. Nason. — ⌊Fredericton⌋ : Provincial Archives of New Brunswick, 1975.
9, ⌊3⌋ leaves ; 28 cm. CP75-81724-0
Cover title.
1. Classification—Drawings. I. Provincial Archives of New Brunswick. II. Title.
Z697.D73N37 025.3'4'6 78-385669
78 MARC

—MUSIC

Redfern, Brian L
Organising music in libraries / Brian Redfern. — Rev. and rewritten ed. — London : C. Bingley ; Hamden, Conn. : Linnet Books, 1978-
v. ; 23 cm.
Bibliography: v. 1, p. 99-101.
Includes index.
CONTENTS: v. 1. Arrangement and classification.

ISBN 0-208-01544-2
1. Music libraries. 2. Classification—Music. 3. Cataloging of music. 4. Cataloging of phonorecords. I. Title.
ML111.R4 1978 025.3'4'8 78-819
78 MARC

United States. Library of Congress. Subject Cataloging Division.
Classification : Class M : music and books on music / Subject Cataloging Division, Processing Department, Library of Congress. — 3d ed. — Washington : The Library : for sale by Cataloging Distribution Services, Library of Congress, 1978.
x, 228 leaves ; 26 cm.
"A revision of the second edition, originally published in 1917, and includes additions and changes made by the Library of Congress through June 1977."
Includes index.
1. Classification—Music. 2. Classification—Books—Music. 3. Classification, Library of Congress. I. Title. II. Title: Music and books on music.
Z696.U5M 1978 025.4'6'78 78-606179
——— Copy 3. Z663.78.C5M 1978 MARC
78

—NONBOOK MATERIALS—BIBLIOGRAPHY

Bogar, Candace W
Annotated bibliography of published literature on the cataloging and classification of films, pictures, and slides in architecture, city planning, and art ⌊by⌋ Candace W. Bogar. ⌊Monticello, Ill., Council of Planning Librarians⌋, 1973.
17 p. 29 cm. (Council of Planning Librarians. Exchange bibliography 405) $1.50
Cover title.
1. Cataloging of non-book materials—Bibliography. 2. Classification—Nonbook materials—Bibliography. I. Title. II. Series.
Z5942.C68 no. 405 016.3092 s 79-309672
⌊Z695.66⌋ 79 MARC

CLASSIFICATION (Continued)

—PLANTS see Botany—Classification

—SOUND RECORDINGS

Marschall, Friedrich.
Discographische Probleme der Nachkriegszeit und ihre Ursachen / von Friedrich Marschall. — Frankfurt : German Blues Circle, 1977.
33 p. ; 21 cm. GFR***
1. Classification—Phonorecords. 2. Cataloging—Phonorecords. I. Title.
ML111.5.M42 025.4'6'789912 79-337905
79 MARC

CLASSIFICATION, BBK

—ADDRESSES, ESSAYS, LECTURES

Teslenko, Ol'ga Pankrat'evna.
(Problemy vnedrenii͡a bibliotechno-bibliograficheskoĭ klassifikat͡sii)
Проблемы внедрения библиотечно - библиографической классификации : Сб. науч. тр. / ⌈О. П. Тесленко⌉. — Москва : ГБЛ, 1978.
104 p. ; 21 cm. USSR 78
At head of title: Gosudarstvennai͡a biblioteka SSSR imeni V. I. Lenina. Nauchno-issledovatel'skiĭ otdel bibliotechno-bibliograficheskoĭ klassifikat͡sii.
Includes bibliographical references.
0.60rub
1. Classification, BBK—Addresses, essays, lectures. I. Moscow. Publichanai͡a biblioteka. Nauchno-issledovatel'skiĭ otdel bibliotechno-bibliograficheskoĭ klassifikat͡sii. II. Title.
Z696.B2T47 79-409476

Usovershenstvovanie i primenenie bibliotechnobibliograficheskoĭ klassifikat͡sii : sbornik nauch. trudov / ⌈otv. redaktor A. A. Panov⌉. — Leningrad : BAN, 1978.
150 p. ; 20 cm. USSR***
At head of title: Biblioteka Akademii nauk SSSR.
Includes bibliographical references.
0.54rub
1. Classification, BBK—Addresses, essays, lectures. I. Panov, A. A., fl. 1976- . II. Akadmii͡a nauk SSSR. Biblioteka.
Z696.B2U86 79-384860
79 MARC

CLASSIFICATION, BIBLIOTHECAL-BIBLIOGRAPHICAL see Classification, BBK

CLASSIFICATION, DECIMAL

see also Classification, Dewey decimal; Classification, Universal decimal

Carvalho, Dóris de Queiroz, 1920-
Classificação decimal de direito / Doris de Queiroz Carvalho. — 3. ed., rev. e aumentada. — Rio de Janeiro : Ministério da Fazenda, Departamento de Administração, Divisão de Documentação, 1977.
188 p. ; 30 cm.
At head of title: República Federativa do Brasil.
1. Classification—Books—Law. 2. Law—Classification. 3. Classification, Decimal. I. Brazil. Ministério da Fazenda. Divisão de Documentação. II. Title.
Z697.L4C35 1977 79-383021
79 MARC

Mori, Kiyoshi.
(Nihon jisshin bunrui hō)
日本十進分類法 / もり・きよし原編. — 新訂8版 / 日本図書館協会分類委員会改訂. — 東京 : 同協会, 1978.
635 p. ; 22 cm. Ja 78-16083
On verso of t. p.: Nippon decimal classification.
Includes index.
¥3800
1. Classification, Decimal. I. Nihon Toshokan Kyōkai. Bunrui Iinkai. II. Title. III. Title: Nippon decimal classification.
Z696.D7M65 1978 78-812658

Vatican. Biblioteca vaticana. Scuola di biblioteconomia.
Sistemi di classificazione degli stampati / Biblioteca apostolica vaticana, Scuola vaticana di biblioteconomia. — Ed. 2. — Città del Vaticano : Biblioteca apostolica vaticana, 1941.
35 p. ; 25 cm.
1. Classification, Decimal. 2. Classification, Library of Congress. 3. Shelf-listing (Library science) I. Title.
Z696.V34 1941 025.4'3 79-378186
79 MARC

CLASSIFICATION, DEWEY DECIMAL

Dewey, Melvil, 1851-1931.
Dewey decimal classification and relative index / devised by Melvil Dewey. — Ed. 19 / edited under the direction of Benjamin A. Custer. — Albany : Forest Press, 1979.
3 v. ; 25 cm.
First published anonymously in 1876 under title: A classification and subject index. 2d ed. published under title: Decimal classification and relativ index.
CONTENTS: v. 1. Introduction. Tables.—v 2. Schedules.—v. 3. Relative index.
ISBN (invalid) 0-910608-19-9 (v. 1)
1. Classification, Dewey decimal. I. Custer, Benjamin Allen, 1912- . II. Title.
Z696.D519 1979 025.4'3 77-27967
77 MARC

Winslade, B A J
Introduction to the Dewey decimal classification for British schools / compiled by B. A. J. Winslade. — 3d ed., rev. and enlarged. — Albany, N.Y. : Forest Press for the School Library Association, 1977.
176 p. : maps (on lining papers) ; 24 cm.
Based on Abridged Dewey decimal classification and relative index, ed. 10. Second ed., published in 1968, by M. Chambers.
Includes indexes.
ISBN 0-910608-18-0
1. Classification, Dewey decimal. I. Dewey, Melvil, 1851-1931. Decimal classification and relative index. II. Chambers, Marjorie. Introduction to Dewey decimal classification for British schools. III. School Library Association. IV. Title.
Z696.D6C48 1977 025.4'3 78-110332
78 MARC

—PERIODICALS

United States. Library of Congress. Decimal Classification Division.
Dewey decimal classification additions, notes, and decisions.
Washington, Library of Congress, Processing Dept., Decimal Classification Division.
v. 25 cm.
Continues: Decimal classification additions, notes, and decisions, ISSN 0083-1573.
Key title: Dewey decimal classification additions, notes, and decisions, ISSN 0191-3646.
1. Classification, Dewey decimal—Periodicals. I. Title.
Z696.D5U6 025.4'3 78-648292
MARC-S

—PROGRAMMED INSTRUCTION

Elrod, J McRee.
Classification / by J. McRee Elrod. — 2d ed. — Metuchen, N.J. : Scarecrow Press, 1978.
viii, 79 p. : ill. ; 22 cm. — (His Modern library practices series ; no. 3)
ISBN 0-8108-1094-8
1. Classification, Dewey decimal—Programmed instruction. 2. Classification, Library of Congress—Programmed instruction. I. Title.
Z674.E47 no. 3 025 s 77-17282
⌈Z696.D51955⌉ 77 MARC

Piedade, Maria Antonietta Requião.
Introdução programada às 17a. e 18a. edições da Classificação Decimal de Dewey / Maria Antonietta Requião Piedade. — São Paulo : Editora McGraw-Hill do Brasil, 1975.
ca. 150 p. ; 21 cm.
Cover title: Classificação Decimal de Dewey.
Includes index.
1. Classification, Dewey decimal—Programmed instruction. I. Title. II. Title: Classificação Decimal de Dewey.
Z696.D8P53 025.4'3'02854 79-366104
79 MARC

CLASSIFICATION, FACETED

see also Classification, Colon

CLASSIFICATION, FOREST SITE see Forest site quality

CLASSIFICATION, LIBRARY OF CONGRESS

Conklin, Curt E
Library of Congress subject headings, KF cross-reference / by Curt E. Conklin. — Provo, Utah : Brigham Young University Law Library, c1979.
215 p. ; 29 cm. — (BYU legal research series ; no. 5)
Bibliography: p. 214-215.
1. Classification—Books—Law. 2. Law—United States—Classification. 3. Classification, Library of Congress. 4. Subject headings—Law. I. Brigham Young University, Provo, Utah. Law Library. II. Title. III. Series: Brigham Young University, Provo, Utah. BYU legal research series ; no. 5.
Z696.U5K6 025.3'334 79-106943
79 MARC

Cumulative subject index to the MARC data base, 1968-1978 / editor, Nancy B. Olson ; computer composition by SDC Search Service, Santa Monica, California. — Arlington, Va. : Carrollton Press, c1978-
v. (xvii, 6551 p.) ; 28 cm.
ISBN 0-8408-0250-1 (v. 1)
1. Machine-readable bibliographic data—Indexes. 2. MARC System. 3. Subject headings. 4. Classification, Library of Congress. I. Olson, Nancy B. II. System Development Corporation. Search Service.
Z699.4.M2C95 016.0253'028'54 78-63056
00 MARC

Library of Congress classification number index to the MARC data base, 1968-1978 / editor, Nancy B. Olson ; computer composition by SDC Search Service, Santa Monica, California. — Arlington, Va. : Carrollton Press, c1978
v. ; 27 cm.
ISBN 0-8408-0276-5
1. Machine-readable bibliographic data—Indexes. 2. MARC System—Indexes. 3. Classification, Library of Congress. 4. Subject headings. 5. Catalogs, Classified. I. Olson, Nancy B. II. SDC Search Service.
Z699.4.M2L52 016.0253'028'54 78-65752
79 MARC

National Library of Canada.
Classe PS8000 : cadre de classification de la littérature canadienne. — 2. éd. — Ottawa : Bibliothèque nationale du Canada, 1978.
22 leaves ; 26 cm. C***
Issued also in English under title Class PS8000.
Includes index.
ISBN 0-660-01660-5 : $3.00
1. Classification—Books—Canadian literature. 2. Classification—Books—French-Canadian literature. 3. Classification, Library of Congress. I. Title. II. Title: Cadre de classification de la littérature canadienne.
Z696.U5P84 1978 025.4'6'8 79-358703
79 MARC

National Library of Canada.
Class PS8000 : a classification for Canadian literature. — 2d ed. — Ottawa : National Library of Canada, 1978.
x, 21 leaves ; 26 cm. C***
Issued also in French.
Devised and updated by T. R. McCloy.
Includes index.
ISBN 0-660-01659-1 : $3.00
1. Classification—Books—Canadian literature. 2. Classification—Books—French-Canadian literature. 3. Classification, Library of Congress. I. McCloy, T. R. II. Title.
Z696.U5P84 1978a 025.4'6'81 79-308036
79 MARC

United States. Library of Congress. Subject Cataloging Division.
Classification. Class B, Subclasses B-B J : Philosophy, Psychology / Subject Cataloging Division, Processing Services, Library of Congress. — 3d ed. — Washington : The Library, 1979.
viii, 250 p. ; 26 cm.
Includes index.
ISBN 0-8444-0280-X
1. Classification—Books—Philosophy. 2. Classification—Books—Psychology. 3. Classification, Library of Congress. I. Title.
Z696.U5B1 1979 025.4'6'1 79-526
—— —— Copy 3. Z663.78.C5B1 1979 MARC
79

United States. Library of Congress. Subject Cataloging Division.
Classification. Class K, subclass KE, Law of Canada / Subject Cataloging Division, Processing Department, Library of Congress. — Washington : The Library, 1976.
viii p., 134 leaves, 135-181 p. ; 26 cm.
Kept up to date by the division's L.C. classification—additions and changes, and by an irregular cumulative publication issued by Gale Research Company, entitled Library of Congress classificaton schedules: a cumulation of additions and changes, class K, subclass KE.
Includes index.
ISBN 0-8444-0224-9
1. Classification—Books—Law. 2. Law—Canada—Classification. 3. Classification, Library of Congress. I. Title.
Z696.U5K4 1976 025.4'6'34 76-54307
—— —— Copy 3. Z663.78.C5K4 1976 MARC
76⌈r79⌉rev

United States. Library of Congress. Subject Cataloging Division.
Classification : Class M : music and books on music / Subject Cataloging Division, Processing Department, Library of Congress. — 3d ed. — Washington : The Library : for sale by Cataloging Distribution Services, Library of Congress, 1978.
x, 228 leaves ; 26 cm.
"A revision of the second edition, originally published in 1917, and includes additions and changes made by the Library of Congress through June 1977."
Includes index.
1. Classification—Music. 2. Classification—Books—Music. 3. Classification, Library of Congress. I. Title. II. Title: Music and books on music.
Z696.U5M 1978 025.4'6'78 78-606179
—— —— Copy 3. Z663.78.C5M 1978 MARC
78

United States. Library of Congress. Subject Cataloging Division.
LC classification outline / Subject Cataloging Division, Processing Department. — 4th ed. — Washington : Library of Congress, 1978.
32 p. ; 26 cm.
Editions of 1903-1934 published by Classification Division, Library of Congress, under title: Classification. Outline scheme of classes; rev. and enl. ed. (1942) and 2d ed. (1970) published by Subject Cataloging Division under title: Outline of the Library of Congress classification.
ISBN 0-8444-0255-9
1. Classification, Library of Congress. I. United States. Library of Congress. Classification Division. Classification. Outline scheme of classes. II. Title.
Z696.U4O 1978 025.4'3 77-608191
77 MARC

CLASSIFICATION, LIBRARY OF CONGRESS (Continued)

Vatican. Biblioteca vaticana. Scuola di biblioteconomia.
Sistemi di classificazione degli stampati / Biblioteca apostolica vaticana, Scuola vaticana di biblioteconomia. — Ed. 2. — Città del Vaticano : Biblioteca apostolica vaticana, 1941.
35 p. ; 25 cm.
1. Classification, Decimal. 2. Classification, Library of Congress. 3. Shelflisting (Library science) I. Title.
Z696.V34 1941 025.4'3 79-378186
79 MARC

—PROGRAMMED INSTRUCTION

Elrod, J McRee.
Classification / by J. McRee Elrod. — 2d ed. — Metuchen, N.J. : Scarecrow Press, 1978.
viii, 79 p. : ill. ; 22 cm. — (His Modern library practices series ; no. 3)
ISBN 0-8108-1094-8
1. Classification, Dewey decimal—Programmed instruction. 2. Classification, Library of Congress—Programmed instruction. I. Title.
Z674.E47 no. 3 025 s 77-17282
[Z696.D51955] 77 MARC

CLASSIFICATION, PRIMITIVE

Needham, Rodney.
Symbolic classification / Rodney Needham. — Santa Monica, Calif. : Goodyear Pub. Co., c1977.
xii, 78 p. ; 23 cm. — (The Goodyear perspectives in anthropology series)
Bibliography: p.
Includes index.
ISBN 0-87620-872-3
1. Classification, Primitive. 2. Anthropology—Classification. I. Title. II. Series.
GN470.N44 301.2'01'2 78-11508
78 MARC

—ADDRESSES, ESSAYS, LECTURES

Australian aboriginal concepts / edited by L. R. Hiatt. — Canberra : Australian Institute of Aboriginal Studies ; [Atlantic Highlands] N.J. : Humanities Press, 1978.
viii, 198 p. : ill. ; 25 cm. Aus***
Includes bibliographies and indexes.
ISBN 0-391-00887-0 (Humanities)
1. Philosophy, Australian aboriginal—Addresses, essays, lectures. 2. Australian aborigines—Psychology—Addresses, essays, lectures. 3. Classification, Primitive—Addresses, essays, lectures. I. Hiatt, Lester Richard.
GN666.A78 301.45'19'91094 78-58302
79 MARC

Hicks, David, 1939-
Structural analysis in anthropology, case studies from Indonesia and Brazil / David Hicks. — S[ank]t Augustin bei Bonn : Anthropos-Inst., 1978.
133 p. : ill. ; 23 cm. — (Studia Instituti Anthropos ; v. 30) GFR78-
Bibliography: p. 127-133.
ISBN 3-921389-50-X
1. Classification, Primitive—Addresses, essays, lectures. 2. Kaingangue Indians—Addresses, essays, lectures. 3. Shokleng Indians—Addresses, essays, lectures. 4. Tetum (Indonesian people)—Addresses, essays, lectures. I. Title. II. Series: Anthropos Institute. Studia ; v. 30.
GN468.4.H5 301.2'01'8 79-313819
00 MARC

—GHANA

Biebuyck, Daniel P 1925-
Symbolism of the Lega stool / Daniel P. Biebuyck. An ethnoscientific approach to Akan arts and aesthetics / D. M. Warren & J. Kweku Andrews. — Philadelphia : Institute for the Study of Human Issues, c1977.
30, 42 p., [3] leaves of plates : ill. ; 28 cm. — (Working papers in the traditional arts ; 2 & 3)
Includes bibliographies.
ISBN 0-915980-81-9
1. Waregas—Rites and ceremonies. 2. Stools—Social aspects—Zaire. 3. Art, Akan. 4. Classification, Primitive—Ghana. 5. Akan language—Glossaries, vocabularies, etc. I. Warren, Dennis M. An ethnoscientific approach to Akan arts and aesthetics. 1977. II. Title. III. Series.
DT650.W37B53 394 77-14571
77 MARC

CLASSIFICATION, UNIVERSAL DECIMAL

Budapest. Országos Műszaki Könyvtár és Dokumentácios Központ.
Figyelőkarton-szolgálati tájékoztató. 6. módositott kiad. Budapest, 1961.
77 p. 20 cm.
At head of title: Országos Műszaki Könyvtár. Dokumentácios Osztály.
1. Classification—Books—Technology. 2. Classification, Universal decimal. I. Title.
Z696.U875T425 79-384761
00 MARC

Comissão Brasileira de Classificação Decimal Universal.
Levantamento dos usuários da CDU no Brasil / Comissão Brasileira da Classificação Decimal Universal, Instituto Brasileiro de Bibliografia e Documentação. — Rio de Janeiro : O Instituto, 1975.
34 leaves ; 33 cm. — (Publicação avulsa - IBBD/CDU ; no. 1)
Cover title.
Photocopy.
1. Libraries—Brazil—Directories. 2. Classification, Universal decimal. I. Title. II. Series: Comissão Brasileira de Classificação Decimal Universal. Publicação avulsa - IBBD/CDU ; no. 1.
Z769.A1C59 1975 78-383433
79 MARC

Corbett, Lindsay.
Report on the FID Seminar on UDC in a Mechanized Retrieval System, held at the Danish Technical University, Copenhagen, 2-6 September 1968 / by Lindsay Corbett. — [Aldermaston, Eng. : AWRE Library], 1968.
31 leaves ; 31 cm. — (AWRE Library information note ; no. 68/9)
1. Classification, Universal decimal. 2. Automatic classification. 3. Information storage and retrieval systems. I. Title. II. Series: AWRE Library. AWRE Library information note ; no. 68/9.
Z696.U861265 025.4'3 77-154912
79 MARC

Corten, François L
I.T.C. international bibliography, photogrammetry, photogrammetric abstracts / F. L. Corten. — 2d ed. — Delft : I.T.C., 1965.
48 p. ; 24 cm. — (I.T.C. information ; no. 8)
First published in 1962 under title: I.T.C. international bibliography of photogrammetry, photogrammetric abstracts.
Errata slip inserted.
1. Photogrammetry—Documentation. 2. Classification—Books—Photogrammetry. 3. Classification, Universal decimal. I. Title. II. Title: International bibliography, photogrammetry, photogrammetric abstracts. III. Series.
TA593.A2I2 No. 8 1965 526.9'823'08 s 78-317657
[Z695.1.P477] 78 MARC
[TA593]

Czarnecka, Jadwiga.
Katalog rzeczowy w średnich bibliotekach publicznych : schemat układu oparty na UKD, poprzedzony wstępem metodycznym / oprac. J. Czarnecka, T. Turowska. — Warszawa : Biblioteka Narodowa, Instytut Książki i Czytelnictwa, 1977 (Warsz. : BN)
139, [1] p. ; 30 cm. P78-2071
Errata slip inserted.
Includes index.
zł40.00
1. Classification, Universal decimal. I. Turowska, Teresa, joint author. II. Title.
Z696.U86127 78-399925
79 MARC

Dafinov, Zdravko Nedi͡alkov.
(Tablitsi na desetichnata klasifikatsi͡a)
Таблици на десетичната класификация. [Съставили: Здравко Дафинов, Снежина Тошева и Лора Даскалова]. София, Наука и изкуство, 1967.
173 p. 29 cm.
At head of title: Комитет по културата и изкуството. Народна библиотека "Кирил и Методий."
1. Classification, Universal decimal. I. Tosheva, Snezhina. II. Daskalova-Ribarska, Lora. III. Sofia. Narodna biblioteka. IV. Title.
Z696.U8613 70-231775

Dahlström, Asta, 1913-
On UCD with some malice / Asta Dahlström. — Stockholm : Royal Institute of Technology Library, 1976.
15 leaves ; 30 cm. — (Stockholm papers in library and information science ; report TRITA-LIB-6005) S***
Includes bibliographical references.
1. Classification, Universal decimal. 2. Classification—Books. I. Title. II. Series.
Z696.U86132 025.4 79-301256
79 MARC

Kokuritsu Kokkai Toshokan, Tokyo. Chōsa oyobi Rippō Kōsakyoku. Hōritsu Seiji Toshokan.
(Kokusai jisshin bunrui hyō)
国際十進分類表：社会科学の部. — [東京]：国立国会図書館調査立法考査局法律政治図書館. 昭和33 [1958]
70 p. : 30 cm. — (国図調立資料：B154)
Cover title.
Errata slip inserted.
"国際十進分類表…に拠って編さんしたもので…相当に取捨を加えてあ
1. Classification—Books—Social sciences. 2. Classification, Universal decimal. I. International Federation for Documentation. Classification décimale. Japanese. II. Title. III. Series: Kokuritsu Kokkai Toshokan, Tokyo. Chōsa oyobi Rippō Kosakyoku. Kokuto chōritsu shiryō : B154.
Z697.S6K64 1958 79-804975

Peluffo de Chapt, Estela.
Proyecto de extensión del esquema 989.9, historia del Uruguay : contribución al estudio de la clase 9 de la C.D.U. / Estela Peluffo de Chapt, Sonia G. Mato. — Montevideo : Universidad de la República, División Publicaciones y Ediciones, [1973]
90 p. ; 27 cm.
On cover: Universidad de la República, Escuela Universitaria de Bibliotecología y Ciencias Afines "Ing. Federico E. Capurro."
1. Classification—Books—Uruguay. 2. Classification, Universal decimal. I. Mato, Sonia G., joint author. II. Title.
Z696.U875U786 79-100951
79 MARC

(UDC no tebiki)
UDCの手引き / 今まど子編. — 東京：日本ドクメンテーション協会, 1978.
131 p. ; 21 cm. Ja 78-16717
Bibliography: p. 124-131.
¥1200
1. Classification, Universal decimal. I. Kon, Madoko.
Z696.U8682 78-813027

Ústředí vědeckých, technických a ekonomických informací.
Mezinárodní desetinné třídění : třída 64 : domácí hospodářství / [odpovědný pracovník Dušan Simandl]. — 1. čs. úplné vyd. 13 mezinárodní vyd. — Praha : ÚVTEI, 1978.
x, 89 p. ; 30 cm. — (FID ; 419) Cz***
1. Classification—Books—Home economics. 2. Classification, Universal decimal. I. Simandl, Dušan. II. International Federation for Documentation. Classification décimale. III. Title. IV. Series: International Federation for Documentation. Publication ; 419.
Z696.U875H658 79-371375
79 MARC

—CONGRESSES

Seminario di studi sulla CDU, Rome, 1975.
Atti del seminario di studi sulla CDU : (Roma 22 settembra 1975) / a cura di Maria Pia Carosella. — Roma : Consiglio nazionale delle ricerche, Istituto di studi sulla ricerca e documentazione scientifica, 1977.
128 p. : ill., 24 cm. — (Note di bibliografia e di documentazione scientifica ; 30. ISSN 0085-2309) It***
Includes bibliographies.
L4000
1. Classification, Universal decimal—Congresses. I. Carosella, Maria Pia.
Z696.U867626 1975. 78-375022
78 MARC

UDC and EDP : lectures from a seminar in Oslo November 1.-3.1976 / edited by Bjørn L. Hegseth and Knut Thalberg ; [arranged by the Norwegian committee for Universal decimal classification and the Norwegian School of Librarianship]. — Oslo : Riksbibliotektjenesten, 1977.
[2] leaves, 84 p. ; 30 cm. — (Skrifter fra Riksbibliotektjenesten ; 10) N78-Mar.
English or Norwegian.
Includes bibliographical references.
ISBN 8271950134
1. Classification, Universal decimal—Congresses. 2. Automatic classification—Congresses. 3. Library science—Data processing—Congresses. I. Hegseth, Bjørn L. II. Thalberg, Knut, 1917- . III. Norsk UDK-udvalg. IV. Statens bibliotekskole. V. Series: Riksbibliotektjenesten. Skrifter fra Riksbibliotektjenesten ; 10.
Z696.U86768 025.4'3 78-371465
78 MARC

CLASSIFICATION OF CRIMES

—UNITED STATES

Smith, Charles P 1931-
Inconsistent labeling : a national assessment of case disposition and classification in the juvenile justice system / Charles P. Smith, T. Edwin Black, Fred R. Campbell ; submitted to the U.S. Office of Juvenile Justice and Delinquency Prevention ; prepared by the National Juvenile Justice System Assessment Center of the American Justice Institute. — Sacramento, Calif. : The Institute, 1979.
3 v. : ill. ; 29 cm.
Includes bibliographies.
CONTENTS: v. 1. Process description and summary.—v. 2. Smith, C. P., Black, T. E., Weir, A. W. Results of a literature search.—v. 3. Results of a survey.
1. Juvenile justice, Administration of—United States. 2. Classification of crimes—United States. 3. Prosecution—United States—Decision making. I. Black, T. Edwin, joint author. II. Campbell, Fred R., joint author. III. United States. Law Enforcement Assistance Administration. Office of Juvenile Justice and Delinquency Prevention. IV. National Juvenile Justice System Assessment Center. V. Title.
KF9779.S55 345'.73'08 79-111543
00 MARC

CLASSIFICATION OF RIVERS

see also Rivers—Classification

CLASSIFICATION OF SCHOOL CHILDREN

see Ability grouping in education

CLASSIFICATION OF SCIENCES

see also Numerical taxonomy

Kedrov, Bonifatiĭ Mikhaĭlovich.
La classification des sciences / B. Kédrov ; ₍traduit du russe par Jacques Lemagnen₎. — Moscou : Éditions du Progrès, c1977-
v. ; 22 cm. USSR***
Translation of Klassifikat͡sii͡a nauk.
Includes bibliographical references and indexes.
CONTENTS: 1. Engels et ses prédécesseurs.
1. Classification of sciences. I. Title.
Q177.K4314 501'.2 79-359654
79 MARC

Maioli, Bruno.
Teoria dell'essere e dell'esistente e classificazione delle scienze in M. S. Boezio : una delucidazione / Bruno Maioli. — Roma : Bulzoni, 1978.
82 p. ; 24 cm. — (Quaderni dell'Istituto di scienze filosofiche, Università degli studi di Siena, Sede di Arezzo ; 4) It78-Dec
Includes bibliographical references.
L2300
1. Boethius, d. 524 Ontology. 2. Ontology. 3. Classification of sciences. I. Title. II. Series: Siena. Università. Facoltà di magistero. Sede di Arezzo. Istituto si scienze filosofiche. Quaderni ; 4.
B659.Z7M34 79-341849
*79 MARC

Satō, Nobue, 1905-
(Kagaku no hōhō to bunrui)
科學の方法と分類 / 佐藤信衛著. — 東京 : 日本評論社, 昭和25 ₍1950₎
3, 8, 328 p. ; 22 cm. — (His 考 ; 巻2)
Includes bibliographical references.

1. Science—Methodology. 2. Classification of sciences. I. Title. II. Series: Satō, Nobue, 1905- Kangae ; maki 2.
Q175.S28 78-816709

—HISTORY

Frankowska-Terlecka, Małgorzata.
Idea jedności nauki w XII i XIII wieku / Małgorzata Frankowska-Terlecka. — Wyd. 1. — Wrocław : Zakład Narodowy im. Ossolińskich, 1976.
144 p., ₍6₎ leaves of plates : ill. ; 24 cm. — (Monografie z dziejów nauki i techniki ; t. 108) P***
Summary in English and Russian.
Bibliography: p. 124-₍129₎
Includes index.
zł40.00
1. Classification of sciences—History. 2. Science—Philosophy—History. 3. Science, Medieval. I. Title.
Q177.F72 78-387434
79 MARC

CLASSIFICATION OF VEGETATION
see Vegetation classification

CLASSIFICATION THEORY (STATISTICS)
see Discriminant analysis

CLASSIFICATION YARDS see Railroads —Yards

CLASSIFIED CATALOGS see Catalogs, Classified

CLASSIFIED DEFENSE INFORMATION see Defenses information, Classified

CLASSIFIED DOCUMENTS see Security classification (Government documents)

CLASSIFIED INFORMATION see Security classification (Government documents)

CLASSIFIERS (LINGUISTCS)

see also Grammar, Comparative and general—Determiners

CLASSROOM MANAGEMENT

Abbott, Jean.
Classroom strategies to aid the disabled learner : with glossary / Jean Abbott. — Cambridge, Mass. : Educators Pub. Service, c1978.
x, 71 p. ; 23 cm.
1. Learning disabilities. 2. Classroom management. I. Title.
LC4704.A22 371.9'26 78-74637
79 MARC

Boland, James M
Foundations of education in a field-based setting / James M. Boland, Judith Stevens, Stanley I. Adelman. — Washington : University Press of America, c1976.
193 p. : ill. ; 23 cm.
Includes bibliographies.
1. Learning, Psychology of. 2. Classroom management. I. Stevens, Judith, joint author. II. Adelman, Stanley I., joint author. III. Title.
LB1051.B467 371.1'02 76-26722
78 MARC

Bossert, Steven T 1948-
Tasks and social relationships in classrooms : a study of instructional organization and its consequences / Steven T. Bossert. — Cambridge ₍Eng.₎ ; New York : Cambridge University Press, 1979.
xi, 119 p. ; 24 cm. — (The Arnold and Caroline Rose monograph series of the American Sociological Association)
Bibliography: p. 117-119.
ISBN 0-521-22445-4 : $13.95
1. Teaching. 2. Interaction analysis in education. 3. Classroom management. I. Title. II. Series: Arnold and Caroline Rose monograph series in sociology.
LB1025.2.B675 371.1'02 78-67260
78 MARC

Carew, Jean V 1936-
Beyond bias : perspectives on classrooms / Jean V. Carew and Sara Lawrence Lightfoot. — Cambridge, Mass. : Harvard University Press, c1979.
291 p. ; 24 cm.
Bibliography: p. ₍285₎-288.
Includes index.
ISBN 0-674-06882-3
1. Classroom management. 2. Teacher-student relationships. 3. Interaction analysis in education. I. Lightfoot, Sara Lawrence, joint author. II. Title.
LB3013.C34 372.1'1'02 78-20997
78 MARC

Cooper, Barry.
Bernstein's codes, a classroom study / ₍by₎ Barry Cooper. — ₍Brighton₎ : ₍University of Sussex, Education Area₎, ₍1976₎
₍4₎, 46 p. ; 21 cm. — (Occasional paper - University of Sussex, Education Area ; 6) GB78-05959
Bibliography: p. 44-45.
ISBN 0-905414-05-5
1. Classroom management. I. Title. II. Series: Brighton, Eng. University of Sussex. Education Area. Occasional paper - University of Sussex, Education Area ; 6.
LB3013.C56 370.15'2 79-310198
79 MARC

Epstein, Charlotte.
Classroom management and teaching : persistent problems and rational solutions / Charlotte Epstein. — Reston, Va. : Reston Pub. Co., c1979.
xviii, 302 p. ; 24 cm.
Includes bibliographical references and index.
ISBN 0-8359-0824-0
1. Classroom management. 2. Teacher-student relationship. I. Title.
LB3013.E67 372.1'1'02 78-27124
78 MARC

Götz, Wolfgang.
Einstellungs- und Verhaltensänderung durch Unterricht : e. Unters. über d. Einfluss alternativer Unterrichtsprogramme auf d. pädag. Verhalten von Erziehern / Wolfgang Götz. — Weinheim ; Basel : Beltz, 1979.
xvi, 244 p. ; 21 cm. — (Studien zur pädagogischen Psychologie ; Bd. 13) GFR79-A
Bibliography: p. 234-244.
ISBN 3-407-20213-X
1. Teacher—Student relationships. 2. Classroom management. 3. Behavior modification. I. Title. II. Series.
LB1033.G6 79-362368
79 MARC

Gudmundsen, Abel.
You can control your class / by Abel Gudmundsen, Ellen Williams, and Rex B. Lybbert. — Salt Lake City : Class Control Associates, c1978.
xi, 53 p. ; 22 cm.
$3.50
1. Classroom management. I. Williams, Ellen, 1944- joint author. II. Lybbert, Rex B., joint author. III. Title.
LB3013.G82 371.1'02 78-111212
79 MARC

Gustafsson, Christina.
Classroom interaction : a study of pedagogical roles in the teaching process / Christina Gustafsson. — Lund : Liber-Läromedel, 1977.
241 p. : ill. ; 21 cm. — (Studies in education and psychology ; 1 ISSN 0348-291X) (Studies in curriculum theory and cultural reproduction ; 1) (₍Forskningsrapporter₎ - MAP-gruppen ; 1977:5) S78-20
Thesis—Stockholms universitet.
Bibliography: p. 221-232.
ISBN 9140046079
1. Classroom management. 2. Interaction analysis in education. I. Title. II. Series: Studies in education and psychology (Stockholm) ; 1. III. Series: Studies in curriculum theory and cultural reproduction ; 1. IV. Series: MAP-gruppen. Forskningsrapporter från MAP-gruppen ; 1977:5.
LB3013.G84 371.1'02 78-325406
79 MARC

House, Ernest R
Survival in the classroom : negotiating with kids, colleagues, and bosses / Ernest R. House, Stephen D. Lapan. — Boston : Allyn and Bacon, c1978.
xiv, 283 p. ; 25 cm.
Bibliography: p. 275-277.
Includes index.
ISBN 0-205-06093-5
1. Classroom management. 2. Teacher participation in administration. 3. Teacher—Student relationships. I. Lapan, Stephen D., joint author. II. Title.
LB3013.H672 371.1'04 77-25247
77 MARC

Hülsebusch, Herwig.
Pädagogische Autorität : ihr notwendiger Beitrag zum begründeten Reden u. gemeinsamen Handeln / Herwig Hülsebusch. — 1. Aufl. — Bad Honnef : Bock und Herchen, 1978.
190 p. ; 21 cm. GFR78-A
Bibliography: p. 181-190.
ISBN 3-88347-001-5 : DM28.00
1. Classroom management. 2. Teaching. 3. Authority. I. Title.
LB3013.H78 78-389407
79 MARC

Kohut, Sylvester, 1942-
Classroom discipline : case studies and viewpoints / by Sylvester Kohut, Jr., Dale G. Range. — Washington : National Education Association, c1979.
112 p. ; 25 cm. — (NEA aspects of learning)
Bibliography: p. 107-112.
ISBN 0-8106-1491-X pbk.
1. Classroom management. 2. Range, Dale G., joint author. I. Title. II. Series: National Education Association of the United States. NEA aspects of learning.
LB3013.K64 371.5 78-23392
78 MARC

Komiyama, Eiichi, 1913-
(Kyōshitsu no shinrigaku)
教室の心理学 / 小見山栄一, 井坂行男, 沢田慶輔共著. — 東京 : 牧書店, 昭和27 ₍1952₎
2, 4, 191 p. : ill. ; 19 cm.
Originally broadcasted in the program "Kyōshi no jikan" by NHK, Apr. 1951-Mar. 1952.

1. Classroom management. 2. Child psychology. I. Isaka, Yukio, joint author. II. Sawada, Keisuke, 1909- joint author. III. Title.
LB3011.K65 78-817206

Kostiuk, Nick.
Improving behavior / Nick Kostiuk, Eli Barkan, John Rocco ; ₍ill. by Richard Stalzer₎. — Morristown, N.J. : Silver Burdett Co., c1979.
xxviii, 131 p. : ill. ; 23 cm. — (Silver Burdett professional publications)
Bibliography: p. 130-131.
Includes index.
ISBN 0-382-02609-8
1. Classroom management. 2. School discipline. 3. Behavior modification. I. Barkan, Eli, joint author. II. Rocco, John, joint author. III. Title.
LB3013.K67 372.1'1'02 78-56760
79 MARC

Lemlech, Johanna.
Classroom management / Johanna Kasin Lemlech. — New York : Harper & Row, c1979.
xv, 297 p. : ill. ; 24 cm.
Bibliography: p. 287-290.
Includes indexes.
ISBN 0-06-043945-9
1. Classroom management. I. Title.
LB3013.L45 371.1'02 78-14593
78 MARC

CLASSROOM MANAGEMENT (Continued)

Lovitt, Thomas C
Managing inappropriate behaviors in the classroom : what research and experience say to the teacher of exceptional children / Thomas C. Lovitt. — Reston, Va. : Council for Exceptional Children, 1978.
xi, 44 p. : ill. ; 23 cm.
Bibliography: p. 41-44.
1. Exceptional children—Education. 2. Classroom management. 3. Behavior modification. I. Title.
LC3969.L6 371.9'3 78-54144
78 MARC

Moraleda, Mariano.
Sociodiagnóstico del aula / Mariano Moraleda. — Madrid : Marova, 1978.
239 p. : graphs ; 21 cm. — (Biblioteca del educador ; 20) Sp***
Bibliography: p. 235-[236]
ISBN 8426903762
1. Educational sociology. 2. Classroom management. I. Title.
LC191.M597 79-346152
79 MARC

Neulinger, Klaus Ulrich, 1946-
Muss die Schule krank machen? : Möglichkeiten d. Konfliktbegegnung in Unterricht u. Unterrichtsplanung / Klaus U. Neulinger. — Orig.-Ausg. — Freiburg im Breisgau ; Basel ; Wien : Herder, 1978.
141 p. ; 18 cm. — (Herderbücherei ; Bd. 9060 : Pädagogik) GFR78-
ISBN 3-451-09060-0 : DM10.90
1. Classroom management. 2. Educational psychology. I. Title.
LB3013.N45 78-389406
79 MARC

Ovchinnikova, Irina Grigor'evna.
(Chto takoe khoroshaia shkola)
Что такое хорошая школа / И. Г. Овчинникова. — Москва : Знание, 1977.
94 p. : ill. ; 16 cm. — (Народный университет : Педагогический факультет ; 8, 1977) USSR 77
0.14rub
1. Teaching. 2. Classroom management. 3. Education—Russia. I. Title. II. Series: Narodnyĭ universitet : Pedagogicheskiĭ fakul'tet ; 1977, 8.
AS261.N24 1977, no. 8 78-422736
[LB1025.2]

Paley, Vivian Gussin, 1929-
White teacher / Vivian Gussin Paley. — Cambridge : Harvard University Press, 1979.
xvi, 140 p. ; 22 cm.
ISBN 0-674-95185-9
1. Afro-Americans—Education. 2. Kindergarten. 3. Classroom management. I. Title.
LC2771.P34 371.9'7'96073 78-9841
78 MARC

Strategies for instructional management / Lanore A. Netzer ... [et al.]. — Boston : Allyn and Bacon, c1979.
xvii, 288 p. ; 25 cm.
Includes bibliographies and index.
ISBN 0-205-06448-5 : $14.95
1. Classroom management. 2. Interaction analysis in education. 3. Teacher-student relationships. I. Netzer, Lanore A. II. Title.
LB3013.2.S75 371.1'02 78-24433
78 MARC

Structured teaching : a design for classroom management & instruction / Howard N. Sloane ... [et al.]. — Champaign, Ill. : Research Press, c1979.
viii, 346 p. : ill. ; 23 cm.
Bibliography: p. 321-334.
ISBN 0-87822-195-6
1. Classroom management. I. Sloane, Howard Norman, 1932-
LB3013.S83 371.1'02 79-65089
00 MARC

Tanaka, Jack, 1943-
Classroom management : a guide for the school consultant / by Jack Tanaka. — Springfield, Ill. : Thomas, c1979.
xii, 199 p. : ill. ; 24 cm.
Bibliography: p. 194.
Includes index.
ISBN 0-398-03858-9
1. Classroom management. 2. Behavior modification. 3. Elementary school teachers—In-service training. I. Title.
LB3013.T35 372.1'1'02 78-9946
78 MARC

Teacher decision-making in the classroom : a collection of papers / edited by John Eggleston. — London ; Boston : Routledge & K. Paul, 1979.
x, 234 p. : ill. ; 23 cm. — (Routledge education books) GB***
Bibliography: p. 214-227.
Includes index.
ISBN 0-7100-0171-1
1. Classroom management. 2. Teaching. I. Eggleston, John. II. Series.
LB3013.T38 372.1'1'02 78-41187
79 MARC

Thompson, Charles Lowell.
Guidance activities for counselors and teachers / Charles L. Thompson, William A. Poppen. — Monterey, Calif. : Brooks/-Cole Pub. Co., c1979.
xi, 210 p. : ill. ; 23 cm.
Bibliography: p. 179-180.
Includes indexes.
ISBN 0-8185-0329-7 : $6.95
1. Personnel service in education. 2. Classroom management. 3. Student counselors. I. Poppen, William A., joint author. II. Title.
LB1027.5.T453 371.4 78-31338
78 MARC

Walker, Hill M
The acting-out child : coping with classroom disruption / Hill M. Walker. — Boston : Allyn and Bacon, c1979.
xi, 312 p. : ill. ; 24 cm.
Includes bibliographies and indexes.
ISBN 0-205-06569-4
1. Classroom management. 2. Problem children—Education. 3. Acting out (Psychology) I. Title.
LB3013.W327 371.1'02 79-1054
79 MARC

—ADDRESSES, ESSAYS, LECTURES

Classroom management / by the yearbook committee and associated contributors ; edited by Daniel L. Duke, editor for the Society, Kenneth J. Rehage. — Chicago : NSSE : distributed by University of Chicago Press, 1979.
xxi, 447 p. : ill. ; 24 cm. — (Yearbook of the National Society for the Study of Education ; 78th, pt. 2 ISSN 0077-5762)
Prepared by the Society's Committee on Classroom Management.
Includes bibliographical references and index.
1. Classroom management—Addresses, essays, lectures. I. Duke, Daniel Linden. II. National Society for the Study of Education. Committee on Classroom Management. III. Series: National Society for the Study of Education. Yearbook ; new ser., 78th, pt. 2.
LB5.N25 78th, pt. 2 371.1'02 78-66032
[LB3013] 00 MARC

Die Schulklasse, ein pädagogisches Handlungsfeld? / hrsg. von Hans-Jürgen Apel. — Kastellaun : Henn, 1978.
138 p. ; 20 cm. — (Henn Studientexte ; Bd. 10) GFR78-A
Includes bibliographical references.
ISBN (invalid) 3-450-00902-0 : DM12.80
1. Classroom management—Addresses, essays, lectures. I. Apel, Hans-Jürgen.
LB3013.S37 79-356425
79 MARC

Special education and development : perspectives on young children with special needs / edited by Samuel J. Meisels. — Baltimore : University Park Press, c1979.
xi, 324 p. : ill. ; 23 cm.
Bibliography: p. 297-319.
Includes index.
ISBN 0-8391-1351-X
1. Handicapped children—Education—Addresses, essays, lectures. 2. Classroom management—Addresses, essays, lectures. 3. Open plan schools—Addresses, essays, lectures. 4. Mainstreaming in education—Addresses, essays, lectures. I. Meisels, Samuel J.
LC4015.S675 371.9 79-4343
79 MARC

(ha-Talmid ha-be'ayati)
התלמיד הבעייתי : גישה חדשה לטיפול בכיתה / בעריכת אביגדור בנש"ק. — [תל-אביב] : אוצר המורה, 1977.
128 p. : graphs ; 24 cm.
On verso of t. p.: The problematic student.
Bibliography: p. 127-128.
1. Problem children—Education—Addresses, essays, lectures. 2. Problem children—Education—Israel—Addresses, essays, lectures. 3. Classroom management—Addresses, essays, lectures. I. Bonchek, Avigdor.
LC4801.T34 79-951658

—HANDBOOKS, MANUALS, ETC.

Mante, Daisy Rivers.
How to develop a model school and model classrooms for young children : a guide for administrators and teachers / Daisy Rivers Mante, Bonnie Warner Mathisen. — Redwood City, Calif. : Project Jifunza Educational Publications, c1977.
194 p. : ill. ; 23 cm.
Bibliography: p. 191.
ISBN 0-931310-00-8
1. Activity programs in education—Handbooks, manuals, etc. 2. Classroom management—Handbooks, manuals, etc. 3. Creative activities and seat work—Handbooks, manuals, etc. I. Mathisen, Bonnie Warner. II. Title.
LB1027.M318 372.21 78-58369
79 MARC

(Seishin hakujaku kyōiku gakkyū keiei jiten)
精神薄弱教育学級経営事典 / 編集 山口薫 ... [et al.]. — 東京 : 第一法規出版, 昭和52 [1977]
xvi, 422 p. ; 22 cm. Ja 77-3920
¥3500
1. Mentally handicapped children—Education—Handbooks, manuals, etc. 2. Classroom management—Handbooks, manuals, etc. I. Yamaguchi, Kaoru, 1924-
LC4601.S448 78-809530

—RESEARCH

Understanding classroom life / edited by Ray McAleese and David Hamilton. — Windsor [Eng.] : NFER Pub. Co. ; Atlantic Highlands, N.J. : distributed by Humanities Press, 1978.
131 p. ; 22 cm. GB***
"The papers and responses in this volume derive from a British Educational Research Association conference The Conceptualization of Classroom Life held at the Froebel Institute, London, on 24-26 March 1977."
Includes bibliographies and index.
ISBN 0-85633-158-9
1. Classroom management—Research. 2. School environment—Research. 3. Interaction analysis in education. I. McAleese, Ray. II. Hamilton, David, 1943- . III. British Educational Research Association.
LB3013.U5 371.1'02'072 79-314306
79 MARC

CLASSROOM OBSERVATION see Observation (Educational method)

CLASSROOM UTILIZATION

John, V V 1910-
The great classroom hoax & other reflections on India's education / V. V. John. — New Delhi : Vikas, c1978.
xvi, 229 p. ; 23 cm.
Includes index.
ISBN 0-7069-0630-6 : Rs60.00
1. Classroom utilization. 2. School buildings—India. I. Title.
LB3224.J63 378.54 78-903768
79 MARC

CLAUDEL, PAUL, 1868–1955. L'ANNONCE FAITE À MARIE

Baudouin, Marthe.
Les cycles bibliques de Violaine / Marthe Baudouin. — Ottawa : Éditions de l'Université d'Ottawa, 1978.
421 p. ; 20 cm. — (Cahier canadien Claudel ; 10) C***
Bibliography: p. [413]-416.
ISBN 0-7766-4210-3 : $18.00
1. Claudel, Paul, 1868-1955. La jeune fille Violaine. 2. Claudel, Paul, 1868-1955. L'annonce faite à Marie. 3. Claudel, Paul, 1868-1955—Religion and ethics. 4. Bible in literature. I. Title. II. Series.
PQ2605.L2J4433 842'.9'12 79-367188
79 MARC

CLAUDEL, PAUL, 1868-1955. CINQ GRANDES ODES

Carrière, Anne Lise.
Cinq grandes odes de Paul Claudel : étude des manuscrits / Anne-Lise Carrière. — Paris : Belles-Lettres, 1978.
114 p. ; 24 cm. — (Annales littéraires de l'Université de Besançon ; 222) ([Publication] - Centre de recherches de littérature française (XIXe et XXe siècles) ; v. 25) F***
"Compte-lignes pour Claudel, Œuvre poétique, Bibliothèque de la Pléiade": overlay inserted.
45.00F
1. Claudel, Paul, 1868-1955. Cinq grandes odes. 2. Claudel, Paul, 1868-1955—Manuscripts. I. Title. II. Series: Besançon, France. Université. Annales littéraires ; 222. III. Series: Centre de recherches de littérature française (XIXe et XXe siècles). [Publication] ; v. 25.
AS161.B39 vol. 222 840'.9 s 79-361693
[PQ2605.L2] 79 MARC

Maurocordato, Alexandre.
L'ode de Paul Claudel : essai de phénoménologie littéraire / Alexandre Maurocordato. — Paris : Lettres modernes, 1974-78.
2 v. ; 18 cm. — (Archives des lettres modernes ; no 152, 181 : Archives Paul Claudel ; no 11-12) F***
Vol. 2 has title: Thèmes et structures des Cinq grandes odes.
Includes bibliographical references.
ISBN 2-256-90343-5 (v. 1) : 18.00F (v. 1) 40.00F (v. 2)
1. Claudel, Paul, 1868-1955. Cinq grandes odes. I. Title.
PQ2605.L2C547 841'.9'12 75-501703
76[r79]rev MARC

CLAUDEL, PAUL, 1868–1955. LA JEUNE FILLE VIOLAINE

Baudouin, Marthe.
Les cycles bibliques de Violaine / Marthe Baudouin. — Ottawa : Éditions de l'Université d'Ottawa, 1978.
421 p. ; 20 cm. — (Cahier canadien Claudel ; 10) C***
Bibliography: p. [413]-416.
ISBN 0-7766-4210-3 : $18.00
1. Claudel, Paul, 1868-1955. La jeune fille Violaine. 2. Claudel, Paul, 1868-1955. L'annonce faite à Marie. 3. Claudel, Paul, 1868-1955—Religion and ethics. 4. Bible in literature. I. Title. II. Series.
PQ2605.L2J4433 842'.9'12 79-367188
79 MARC

CLAUDEL, PAUL, 1868-1955. LE PAIN DUR

Le Pain dur : ou, Co-naître Paul Claudel / [réalisé par le Service Information de la Maison de la culture d'Amiens ; sous la direction d'Anne Laurent ; dessins originaux de Augusto R. Da Costa]. — Amiens : Maison de la culture d'Amiens, [1978]
122 p. : ill. ; 30 cm. F***
Cover title: Claudel.
On spine: Paul Claudel.
Published on the occasion of the production of Claudel's Le pain dur by the Maison de la culture d'Amiens, Jan. 18, 1978.
Errata slip inserted.
Bibliography: p. 120-122.
1. Claudel, Paul, 1868-1955—Addresses, essays, lectures. 2. Claudel, Paul, 1868-1955. Le pain dur. 3. Authors, French—20th century—Biography—Addresses, essays, lectures. I. Claudel, Paul, 1868-1955. Le pain dur. II. Laurent, Anne. III. Maison de la culture d'Amiens. Service Information.
PQ2605.L2Z8273 848'.9'1209 78-379918
78 MARC

CLAUDEL, PAUL, 1868-1955. PROTÉE

Autrand, Michel.
Protée de Paul Claudel : introduction, fragments inédits, variantes et notes / Michel Autrand. — Paris : Les Belles lettres, 1977.
143 p. ; 24 cm. — (Annales littéraires de l'Université de Besançon ; 205) ([Publication] - Centre de recherches de littérature française (XIXe et XXe siècles) ; v. 20) F***
Originally presented as the author's thesis, Sorbonne, 1970.
Includes bibliographical references.
70.00F
1. Claudel, Paul, 1868-1955. Protée. I. Title. II. Series: Besançon, France. Université. Annales littéraires ; 205. III. Series: Centre de recherches de littérature française (XIXe et XXe siècles). Publication ; v. 20.
AS161.B39 vol. 205 840'.9 s 78-392495
[PQ2605.L2P7] 79 MARC

CLAUDEL, PAUL, 1868-1955

—ADDRESSES, ESSAYS, LECTURES

Le Pain dur : ou, Co-naître Paul Claudel / [réalisé par le Service Information de la Maison de la culture d'Amiens ; sous la direction d'Anne Laurent ; dessins originaux de Augusto R. Da Costa]. — Amiens : Maison de la culture d'Amiens, [1978]
122 p. : ill. ; 30 cm. F***
Cover title: Claudel.
On spine: Paul Claudel.
Published on the occasion of the production of Claudel's Le pain dur by the Maison de la culture d'Amiens, Jan. 18, 1978.
Errata slip inserted.
Bibliography: p. 120-122.
1. Claudel, Paul, 1868-1955—Addresses, essays, lectures. 2. Claudel, Paul, 1868-1955. Le pain dur. 3. Authors, French—20th century—Biography—Addresses, essays, lectures. I. Claudel, Paul, 1868-1955. Le pain dur. II. Laurent, Anne. III. Maison de la culture d'Amiens. Service Information.
PQ2605.L2Z8273 848'.9'1209 78-379918
78 MARC

—ARCHIVES

Jung, Jacqueline.
Donation Edwin Marie Landau : catalogue complet / établi par Jacqueline Jung, Edwin Marie Landau, Sven Siegrist. — Paris : Belles Lettres, 1977.
345 p. ; 21 cm. — (Annales littéraires de l'Université de Besançon ; 201) ([Publication] - Centre de recherches de littérature française (XIXe et XXe siècles) ; v. 19) F***
At head of title: Centre international de recherches claudéliennes. Université de Zurich.
On cover: Fonds Paul Claudel, Zurich.
1. Claudel, Paul, 1868-1955—Archives. 2. Claudel, Paul, 1868-1955—Bibliography—Catalogs. 3. Claudel, Paul, 1868-1955—Manuscripts—Catalogs. 4. Landau, Edwin Marie, 1904- —Library. I. Landau, Edwin Marie, 1904- joint author. II. Siegrist, Sven E., joint author. III. Centre international de recherches claudéliennes. IV. Title. V. Title: Fonds Paul Claudel, Zurich. VI. Series: Besançon, France. Université. Annales littéraires ; 201. VII. Series: Centre de recherches de littérature française (XIXe et XXe siècles). Publication ; v. 19.
AS161.B39 vol. 201 79-345154
[PQ2605.L2] 79 MARC

—BIBLIOGRAPHY—CATALOGS

Jung, Jacqueline.
Donation Edwin Marie Landau : catalogue complet / établi par Jacqueline Jung, Edwin Marie Landau, Sven Siegrist. — Paris : Belles Lettres, 1977.
345 p. ; 21 cm. — (Annales littéraires de l'Université de Besançon ; 201) ([Publication] - Centre de recherches de littérature française (XIXe et XXe siècles) ; v. 19) F***
At head of title: Centre international de recherches claudéliennes. Université de Zurich.
On cover: Fonds Paul Claudel, Zurich.
1. Claudel, Paul, 1868-1955—Archives. 2. Claudel, Paul, 1868-1955—Bibliography—Catalogs. 3. Claudel, Paul, 1868-1955—Manuscripts—Catalogs. 4. Landau, Edwin Marie, 1904- —Library. I. Landau, Edwin Marie, 1904- joint author. II. Siegrist, Sven E., joint author. III. Centre international de recherches claudéliennes. IV. Title. V. Title: Fonds Paul Claudel, Zurich. VI. Series: Besançon, France. Université. Annales littéraires ; 201. VII. Series: Centre de recherches de littérature française (XIXe et XXe siècles). Publication ; v. 19.
AS161.B39 vol. 201 79-345154
[PQ2605.L2] 79 MARC

—CONGRESSES

Journées Claudel, Brangues, France, 1976.
Actes / des Journées Claudel à Brangues, 23-24 juillet 1976 ; [organisées par l'] Association des amis du château de Brangues [et par l'] Université des langues et lettres de Grenoble. — Grenoble : Université de Grenoble III, 1977.
89 p. ; 26 cm. F77-15153
Includes bibliographical references.
1. Claudel, Paul, 1868-1955—Congresses. 2. Claudel, Paul, 1868-1955—Homes and haunts—United States—Congresses. 3. United States—History—1865-1933—Congresses. I. Association des amis du château de Brangues. II. Université des langues et lettres de Grenoble.
PQ2605.L2Z688 1976 78-383605
78 MARC

—CORRESPONDENCE

Claudel, Paul, 1868-1955.
Chroniques du "Journal de Clichy" / Claudel et Mauriac. Correspondance Claudel-Fontaine / textes établis et annotés par François Morlot et Jean Touzot. — Paris : les Belles lettres, 1978.
158 p. ; 24 cm. — (Annales littéraires de l'Université de Besançon ; 206) ([Publication] - Centre de recherches de littérature française (XIXe et XXe siècles) ; v. 21) F78-12911
Includes bibliographical references.
80.00F
1. Claudel, Paul, 1868-1955—Correspondence. 2. Fontaine, Daniel, 1862-1920. 3. Authors, French—20th century—Political and social views. 4. Clichy, France—Social conditions. 5. Authors, French—20th century—Correspondence. 6. Clergy—France—Clichy—Correspondence. 7. Catholic Church—Clergy—Correspondence. I. Mauriac, François, 1885-1970. II. Fontaine, Daniel, 1862-1920. III. Title. IV. Series: Besançon, France. Université. Annales littéraires ; 206. V. Series: Centre de recherches de littérature française (XIXe et XXe siècles). Publication ; v. 21.
AS161.B39 no. 206 848'.9'1209 79-350086
[PQ2605.L2] 79 MARC

—CRITICISM AND INTERPRETATION

Barrère, Jean Bertrand.
Claudel : le destin et l'œuvre / Jean-Bertrand Barrère. — Paris : Société d'édition d'enseignement supérieur, c1979.
278 p. : ill. ; 24 cm. F***
Bibliography: p. [271]-275.
ISBN 2-7181-0603-4
1. Claudel, Paul, 1868-1955—Criticism and interpretation.
PQ2605.L2Z5595 848'.9'1209 79-371424
79 MARC

Humes, Joy.
Two against time : a study of the very present worlds of Paul Claudel and Charles Péguy / by Joy Nachod Humes. — Chapel Hill : U.N.C. Dept. of Romance languages : [distributed by University of North Carolina Press], 1978.
171 p. ; 24 cm. — (North Carolina studies in the Romance languages and literatures ; no. 200)
Bibliography: p. [169]-171.
ISBN 0-8076-9200-9
1. Claudel, Paul, 1868-1955—Criticism and interpretation. 2. Péguy, Charles Pierre, 1873-1914—Criticism and interpretation. 3. Time in literature. I. Title. II. Series.
PQ2605.L2Z683 840'.9'00912 78-3474
78 MARC

Malicet, Michel.
Lecture psychanalytique de l'œuvre de Claudel / Michel Malicet. — Paris : Belles lettres, 19
v. ; 24 cm. — (Annales littéraires de l'Université de Besançon ;) ([Publication] - Centre de recherches de littérature française (XIXe et XXe siècles) ; v.) F***
Includes bibliographical references.
CONTENTS:
—2. Le monde imaginaire.

100F (v. 2)
1. Claudel, Paul, 1868-1955—Criticism and interpretation. 2. Psychoanalysis and literature. I. Title. II. Series: Besançon, France. Université. Annales littéraires ; 221. III. Series: Centre de recherches de littérature française (XIXe et XXe siècles). Publication ; v. 24.
AS161.B39 vol. 221 840'.9 79-363773
79 MARC

Malicet, Michel.
Le monde imaginaire / Michel Malicet. — Paris : Belles lettres, 1978.
250 p. ; 24 cm. — (Annales littéraires de l'Université de Besançon ; 221) ([Publication] - Centre de recherches de littérature française (XIXe et XXe siècles) ; v. 24) (His Licture psychanalytique de l'œuvre de Claudel ; 2) F***
Includes bibliographical references.
100F
1. Claudel, Paul, 1868-1955—Criticism and interpretation. 2. Psychoanalysis and literature. I. Title. II. Series: Besançon, France. Université. Annales littéraires ; 221. III. Series: Centre de recherches de littérature française (XIXe et XXe siècles). Publication ; v. 24.
AS161.B39 vol. 221, pt. 2 79-363772
840'.9 s
[PQ2605.L2] 79 MARC

—CRITICISM AND INTERPRETATION —ADDRESSES ESSAYS, LECTURES

Roberto, Eugène.
Claudel insolite / Eugène Roberto. — Ottawa : Éditions de l'Université d'Ottawa, 1977.
121 p. : ill. ; 20 cm. — (Cahier canadien Claudel ; 9) C***
Includes bibliographies.
ISBN 0-7766-4209-X : $5.25
1. Claudel, Paul, 1868-1955—Criticism and interpretation—Addresses, essays, lectures. I. Title. II. Series.
PQ2605.L2Z852 848'.9'1209 78-384519
79 MARC

—HOMES AND HAUNTS —UNITED STATES—CONGRESSES

Journées Claudel, Brangues, France, 1976.
Actes / des Journées Claudel à Brangues, 23-24 juillet 1976 ; [organisées par l'] Association des amis du château de Brangues [et par l'] Université des langues et lettres de Grenoble. — Grenoble : Université de Grenoble III, 1977.
89 p. ; 26 cm. F77-15153
Includes bibliographical references.
1. Claudel, Paul, 1868-1955—Congresses. 2. Claudel, Paul, 1868-1955—Homes and haunts—United States—Congresses. 3. United States—History—1865-1933—Congresses. I. Association des amis du château de Brangues. II. Université des langues et lettres de Grenoble.
PQ2605.L2Z688 1976 78-383605
78 MARC

—KNOWLEDGE—ASIA

Hue, Bernard.
Littératures et arts de l'Orient dans l'œuvre de Claudel / Bernard Hue. — [Paris] : C. Klincksieck, 1978.
433 p. ; 24 cm. — (Publications de l'Université de Haute-Bretagne ; 8) F78-15536
Bibliography: p. [389]-406.
Includes indexes.
ISBN 2-252-02022-9 : 150F
1. Claudel, Paul, 1868-1955—Knowledge—Asia. 2. Oriental literature—History and criticism. 3. Art, Oriental, in literature. 4. Exoticism in literature. I. Title.
PQ2605.L2Z682 848'.9'1209 79-342930
79 MARC

—MANUSCRIPTS

Carrière, Anne Lise.
Cinq grandes odes de Paul Claudel : étude des manuscrits / Anne-Lise Carrière. — Paris : Belles-Lettres, 1978.
114 p. ; 24 cm. — (Annales littéraires de l'Université de Besançon ; 222) ([Publication] - Centre de recherches de littérature française (XIXe et XXe siècles) ; v. 25) F***
"Compte-lignes pour Claudel, Œuvre poétique, Bibliothèque de la Pléiade": overlay inserted.
45.00F
1. Claudel, Paul, 1868-1955. Cinq grandes odes. 2. Claudel, Paul, 1868-1955—Manuscripts. I. Title. II. Series: Besançon, France. Université. Annales littéraires ; 222. III. Series: Centre de recherches de littérature française (XIXe et XXe siècles). [Publication] ; v. 25.
AS161.B39 vol. 222 840'.9 s 79-361693
[PQ2605.L2] 79 MARC

—MANUSCRIPTS—CATALOGS

Jung, Jacqueline.
Donation Edwin Marie Landau : catalogue complet / établi par Jacqueline Jung, Edwin Marie Landau, Sven Siegrist. — Paris : Belles Lettres, 1977.
345 p. ; 21 cm. — (Annales littéraires de l'Université de Besançon ; 201) ([Publication] - Centre de recherches de littérature française (XIXe et XXe siècles) ; v. 19) F***
At head of title: Centre international de recherches claudéliennes. Université de Zurich.
On cover: Fonds Paul Claudel, Zurich.
1. Claudel, Paul, 1868-1955—Archives. 2. Claudel, Paul, 1868-1955—Bibliography—Catalogs. 3. Claudel, Paul, 1868-1955—Manuscripts—Catalogs. 4. Landau, Edwin Marie, 1904- —Library. I. Landau, Edwin Marie, 1904- joint author. II. Siegrist, Sven E., joint author. III. Centre international de recherches claudéliennes. IV. Title. V. Title: Fonds Paul Claudel, Zurich. VI. Series: Besançon, France. Université. Annales littéraires ; 201. VII. Series: Centre de recherches de littérature française (XIXe et XXe siècles). Publication ; v. 19.
AS161.B39 vol. 201 79-345154
[PQ2605.L2] 79 MARC

—RELIGION AND ETHICS

Baudouin, Marthe.
Les cycles bibliques de Violaine / Marthe Baudouin. — Ottawa : Éditions de l'Université d'Ottawa, 1978.
421 p. ; 20 cm. — (Cahier canadien Claudel ; 10) C***
Bibliography: p. [413]-416.
ISBN 0-7766-4210-3 : $18.00
1. Claudel, Paul, 1868-1955. La jeune fille Violaine. 2. Claudel, Paul, 1868-1955. L'annonce faite à Marie. 3. Claudel, Paul, 1868-1955—Religion and ethics. 4. Bible in literature. I. Title. II. Series.
PQ2605.L2J4433 842'.9'12 79-367188
79 MARC

—SOCIETIES, PERIODICALS, ETC.

Société Paul Claudel (France)
Bulletin de la Société Paul Claudel.
Paris [Société Paul Claudel]
no. ill. 21-26 cm. quarterly (irregular)
Began in 1958? Cf. New serial titles.
1. Claudel, Paul, 1868-1955—Societies, periodicals, etc.
PQ2605.L2Z878 848'.9'1209 79-645621
MARC-S

CLAUDIUS, EMPEROR OF ROME, 10 B.C.-54 A.D.

—FICTION

Graves, Robert, 1895-
I, Claudius : from the autobiography of Tiberius Claudius, Emperor of the Romans, born 10 B.C., murdered and deified A.D. 54 / Robert Graves. — Harmondsworth ; New York : Penguin Books, 1978, c1934.
395 p. : geneal. table. ; 19 cm. GB***
ISBN 0-14-000318-5 : £0.95
1. Claudius, Emperor of Rome, 10 B.C.-54 A.D.—Fiction. I. Title.
PZ3.G7876 Iac 1978 823'.9'12 79-303223
[PR6013.R35] 79 MARC

—ICONOGRAPHY

Trillmich, Walter.
Familienpropaganda der Kaiser Caligula und Claudius : Agrippina Maior u. Antonia Augusta auf Münzen / von Walter Trillmich. — Berlin : de Gruyter, 1978.
xiv, 186 p., [21] leaves of plates : ill. ; 30 cm. — (Antike Münzen und geschnittene Steine ; Bd. 8) GFR78-A
At head of title: Deutsches Archäologisches Institut.
Based on part of the author's thesis, Tübingen, 1970, presented under title: Bildnis-Typen der Agrippina Maior und der Antonia Minor.
Bibliography: p. [xi]-xiv.
ISBN 3-11-007259-9 : DM140.00
1. Agrippina, wife of Germanicus Caesar, d. A.D. 33—Iconography. 2. Antonia Augusta, 36 B.C. (ca.)-38 A.D.—Iconography. 3. Caligula, Emperor of Rome, 12-41—Iconography. 4. Claudius, Emperor of Rome, 10 B.C.-54 A.D.—Iconography. 5. Propaganda, Roman. 6. Coins, Roman. 7. Roman emperors—Iconography. 8. Rome—Nobility—Iconography. 9. Rome—History—The five Julii, 30 B.C.-68 A.D. I. Title. II. Series.
DG281.T74 737.4'9'37 79-352702
79 MARC

CLAUS, HUGO, 1929-

—CRITICISM AND INTERPRETATION

Govaart, Th
Het geclausuleerde Beest : index en opstellen over F. Bordewijk, meesterschap over het monster en Hugo Claus, de krakende makelaar / Th. Govaart. — Hilversum : P. Brand, 1962.
224 p. ; 21 cm. — (Tweelicht)
1. Bordewijk, Ferdinand, 1884-1965—Criticism and interpretation. 2. Claus, Hugo, 1929- —Criticism and interpretation. I. Title.
PT5816.B7Z67 79-370511
79 MARC

CLAUSE, SENTENCE AND DISCOURSE PATTERNS IN SELECTED LANGUAGES OF NEPAL

—INDEXES

Seaward, Larry L 1930-
Index to word lists and comparative vocabularies / Larry L. Seaward and Austin Hale. — Kirtipur : Summer Institute of Linguistics, Institute of Nepal and Asian Studies, Tribhuvan University, 1974.
vii, 25 p. ; 29 cm.
Bibliography: p. vi-vii.
Rs15.00
1. Clause, sentence and discourse patterns in selected languages of Nepal—Indexes. 2. Hari, Maria. Comparative vocabularies of languages of Napel—Indexes. 3. Nepal—Languages—Addresses, essays, lectures—Indexes. I. Hale, Austin, joint author. II. Title.
P381.N37C52 490 75-905076
79 MARC

CLAUSEWITZ, KARL VON, 1780-1831

Marwedel, Ulrich.
Carl von Clausewitz : Persönlichkeit u. Wirkungsgeschichte seines Werkes bis 1918 / Ulrich Marwedel. — Boppard am Rhein : Boldt, 1978.
viii, 296 p. ; 24 cm. — (Wehrwissenschaftliche Forschungen : Abteilung militärgeschichtliche Studien ; 25) GFR78-A
Originally presented as the author's thesis, Münster.
Bibliography: p. 257-288.
Includes index.
ISBN 3-7646-1707-1 : DM40.00
1. Military art and science—History. 2. War. 3. Clausewitz, Karl von, 1780-1831. I. Title. II. Series.
U39.M37 1978 79-370003
79 MARC

—ADDRESSES, ESSAYS, LECTURES

Ryszka, Franciszek.
Lektury i przemyślenia : pięć szkiców o książkach i ich autorach / Franciszek Ryszka. — Wyd. 1. — Poznań : Wydaw. Poznańskie, 1978 (Szczec : SzZG)
184, [4] p. ; 20 cm.
Includes bibliographical references.
zł35.00
1. French literature—History and criticism—Addresses, essays, lectures. 2. Hitler, Adolf, 1889-1945—Addresses, essays, lectures. 3. Clausewitz, Karl von, 1780-1831—Addresses, essays, lectures. I. Title.
PQ141.R9 78-399845
79 MARC

—BIOGRAPHY

Cho, T'ae-hwan.
(Chŏnjaeng haksŏl)
戰爭學說 / 趙泰煥著. — [서울] : 陸軍士官學校, 1961.
93 p. ; 18 cm.
Bibliography: p. 92-93.
CONTENTS: 東洋兵學論.—Jomini.—Clausewitz.
1. Sun-tzu, 6th cent. B.C.—Biography. 2. Jomini, Henri, baron, 1779-1869—Biography. 3. Clausewitz, Karl von, 1780-1831—Biography. 4. Military art and science. 5. Military biography. I. Title.
U51.S95C47 78-826814

CLAUSILIIDAE

—ANATOMY

Loosjes, Fredrik Elisa.
Some anatomical and systematic data on Asiatic Clausiliidae (Gastropoda, Pulmonata) / F. E. Loosjes and A. C. W. Loosjes-van Bemmel. — Warszawa : Państwowe Wydawn. Naukowe, 1973.
p. 288-317 : ill. ; 24 cm. — (Annales zoologici ; t. 30, nr. 8)
Caption title.
At head of title: Polska Akademia Nauk Instytut Zoologiczny.
Summaries in Polish and Russian.
Bibliography: p. 313-315.
1. Clausiliidae—Classification. 2. Clausiliidae—Anatomy. 3. Mollusks—Classification. 4. Mollusks—Anatomy. 5. Mollusks—Asia. I. Loosjes-Van Bemmel, A. C. W., joint author. II. Title. III. Series.
QL1.W3 t. 30, nr. 8 591'.08 s 78-324236
[QL430.5.C57] 79 MARC

—CLASSIFICATION

Loosjes, Fredrik Elisa.
Some anatomical and systematic data on Asiatic Clausiliidae (Gastropoda, Pulmonata) / F. E. Loosjes and A. C. W. Loosjes-van Bemmel. — Warszawa : Państwowe Wydawn. Naukowe, 1973.
p. 288-317 : ill. ; 24 cm. — (Annales zoologici ; t. 30, nr. 8)
Caption title.
At head of title: Polska Akademia Nauk Instytut Zoologiczny.
Summaries in Polish and Russian.
Bibliography: p. 313-315.
1. Clausiliidae—Classification. 2. Clausiliidae—Anatomy. 3. Mollusks—Classification. 4. Mollusks—Anatomy. 5. Mollusks—Asia. I. Loosjes-Van Bemmel, A. C. W., joint author. II. Title. III. Series.
QL1.W3 t. 30, nr. 8 591'.08 s 78-324236
[QL430.5.C57] 79 MARC

CLAUSTHAL-ZELLERFELD, GER.

—ANTIQUITIES—ADDRESSES, ESSAYS, LECTURES

Westlicher Harz : Clausthal-Zellerfeld, Osterode, Seesen / mit Beiträgen von, H. W. Böhme ... [et. al]. — Mainz am Rhein : Philipp von Zabern, c1978.
xii, 218 p. : ill. ; 21 cm. — (Führer zu vor- und frühgeschichtlichen Denkmälern ; Bd. 36) GFR***
ISBN 3-8053-0305-X
1. Harz Mountains—Antiquities—Addresses, essays, lectures. 2. Clausthal-Zellerfeld, Ger.—Antiquities—Addresses, essays, lectures. 3. Osterode am Harz, Ger.—Antiquities—Addresses, essays, lectures. 4. Seesen, Ger.—Antiquities—Addresses, essays, lectures. 5. Germany, West—Antiquities—Addresses, essays, lectures. I. Böhme, Horst Wolfgang. II. Series.
DD51.F83 Bd. 36 79-365409
[DD801.H345] 79 MARC

CLAUSTRE, FRANÇOISE

Thiry, Raymond.
Ma mission au Tibesti à bord du D.C. 4 : [l'affaire Claustre] / Raymond Thiry. — Paris : Table ronde, c1977.
254 p. : ill. ; 21 cm. F***
42.00F
1. Chad—Politics and government. 2. Thiry, Raymond. 3. Guerrillas—Chad. 4. Daza (African people) 5. Claustre, Françoise. 6. Claustre, Pierre. 7. Diplomats—France—Biography. I. Title.
DT546.48.T48 967'.4304'0924 78-391484
79 MARC

CLAUSTRE, PIERRE

Thiry, Raymond.
Ma mission au Tibesti à bord du D.C. 4 : [l'affaire Claustre] / Raymond Thiry. — Paris : Table ronde, c1977.
254 p. : ill. ; 21 cm. F***
42.00F
1. Chad—Politics and government. 2. Thiry, Raymond. 3. Guerrillas—Chad. 4. Daza (African people) 5. Claustre, Françoise. 6. Claustre, Pierre. 7. Diplomats—France—Biography. I. Title.
DT546.48.T48 967'.4304'0924 78-391484
79 MARC

CLAUSULA REBUS SIC STANTIBUS see Rebus sic stantibus clause

CLAUTHAL-ZELLERFELD, GER.

—HISTORY

Morich, Heinrich.
Kleine Chronik der Oberharzer Bergstädte und ihres Erzbergbaus / von H. Morich. — 4., erw. Aufl. der Chronik der Bergstadt Clausthal-Zellerfeld / überarb. u. erw. von Herbert Dennert. — Clausthal-Zellerfeld : Pieper, 1974.
xvi, 200 p. : numerous ill., maps ; 25 cm. GFR77-A
Bibliography: p. 199-200.
DM39.50
1. Clauthal-Zellerfeld, Ger.—History. 2. Cities and towns—Germany, West—Oberharz Mountains—History. 3. Mineral industries—Germany, West—Oberharz Mountains. I. Dennert, Herbert. II. Title.
DD901.C57M67 1974 79-337214
79 MARC

CLAVÉ, ANTONI, 1913-

—EXHIBITIONS

Clavé, Antoni, 1913-
Antoni Clavé : œuvre graphique : [exposition] : École des beaux-arts, Bibliothèque municipale, Mulhouse, 11 juin-2 juillet 1977 / [organisée par Robert Simon et Danielle Taesch]. — Mulhouse : La Bibliothèque, [1977]
36 p. : chiefly ill. ; 30 cm. F***
1. Clavé, Antoni, 1913- —Exhibitions. I. Simon, Robert. II. Tasch, Danielle. III. Mülhausen, Alsace. École des beaux-arts. IV. Mülhausen, Alsace. Bibliothèque municipale.
NE702.C5A4 1977a 769'.92'4 79-352461
79 MARC

CLAVECIN see Harpsichord

CLAVEL, MAURICE

Gondal, Lise.
La Traversée de l'indifférence : dialogues à propos de Dieu avec R. Garaudy, B. Besret, M. Clavel / Lise Gondal .... — Paris : Éditions du Cerf, 1978.
192 p. ; 20 cm. — (Essais) F78-17156
Includes bibliographical references.
ISBN 2-204-01239-4 : 34.00F
1. God—Addresses, essays, lectures. 2. Garandy, Roger—Addresses, essays, lectures. 3. Besret, Bernard—Addresses, essays, lectures. 4. Clavel, Maurice,. I. Title.
BT102.G652 231 79-353420
79 MARC

CLAVICEMBALO see Harpsichord

CLAVICEPS PURPUREA

see also Erogtism; Shoulder Girdle

CLAVIEN, GERMAIN

—BIOGRAPHY

Clavien, Germain.
Châtaignerouge : souvenirs et portraits / Germain Clavien. — ⸢Pont-de-la-Morge⸣ : Éditions La Douraine : ⸢Germain Clavien, 1977⸣
271 p. ; 21 cm. — (His Lettre à l'imaginaire ; 8) Sw77-10592
24.00F
1. Clavien, Germain—Biography. 2. Authors, French—20th century—Biography. I. Title.
PQ2663.L333Z464 841'.9'14 78-391546
*79 MARC

CLAWS see Nails (Anatomy)

CLAY, HENRY, 1777-1852

Frings, Marie-Luise.
Henry Clays American System und die sektionale Kontroverse in den Vereinigten Staaten von Amerika 1815-1829 / Marie-Luise Frings. — Frankfurt am Main ; Bern ; Las Vegas : P. Lang, 1979.
xi, 466 p. ; 21 cm. — (Europäische Hochschulschriften : Reihe 3, Geschichte und ihre Hilfswissenschaften ; Bd. 117) GFR79-A
Originally presented as the author's thesis, Cologne, 1976.
Bibliography: p. 448-466.
ISBN 3-8204-6505-7
1. United States—Politics and government—1817-1825. 2. United States—Politics and government—1825-1829. 3. Clay, Henry, 1777-1852. 4. United States—Economic policy—To 1933. 5. Sectionalism (United States) 6. Bank of the United States (1816-1836) I. Title. II. Series.
E371.F74 1979 320.9'73'054 79-372139
79 MARC

Van Deusen, Glyndon Garlock, 1897-
The life of Henry Clay / by Glyndon G. Van Deusen. — Westport, Conn. : Greenwood Press, 1979, c1937.
viii, 448 p., ⸢15⸣ leaves of plates : ill. ; 24 cm.
Reprint of the ed. published by Little Brown, Boston.
Bibliography: p. ⸢427⸣-437.
Includes index.
ISBN 0-313-20717-8
1. Clay, Henry, 1777-1852. 2. United States—Politics and government—1783-1865. 3. United States. Congress—Biography. 4. Legislators—United States—Biography. I. Title.
E340.C6V3 1979 973.5'092'4 78-23688
78 MARC

CLAY, LANGDON, 1949-

Clay, Langdon, 1949-
Langdon Clay, color atlas : including flat lands & related views, July 21-September 23, 1979, The Corcoran Gallery of Art, Washington, D.C. — Washington : The Gallery, c1979.
⸢15⸣ p. : col. ill. ; 21 cm.
"Catalogue of the exhibition": p. ⸢12⸣
Bibliography: p. ⸢13⸣
1. Photography, Architectural—Exhibitions. 2. Photography—Interiors—Exhibitions. 3. Clay, Langdon, 1949- I. Corcoran Gallery of Art, Washington, D.C. II. Title.
TR659.C52 779'.4'0924 79-89032
79 MARC

CLAY, LUCIUS DU BIGNON, 1897-

—CONGRESSES

U.S. occupation in Europe after World War II : papers and reminiscences from the April 23-24, 1976 conference held at the George C. Marshall Research Foundation, Lexington, Virginia / edited with an introd. by Hans A. Schmitt. — Lawrence : Regents Press of Kansas, c1978.
viii, 172 p. : maps ; 24 cm.
"A George C. Marshall Research Foundation publication."
Includes bibliographical references and index.
ISBN 0-7006-0178-3 : $12.00. ISBN 0-7006-0179-3 pbk. : $6.95
1. Germany—History—Allied occupation, 1945- —Congresses. 2. Germany, West—Politics and government—Congresses. 3. Austria—History—Allied occupation, 1945-1955—Congresses. 4. Clay, Lucius Du Bignon, 1897- —Congresses. I. Schmitt, Hans A. II. George C. Marshall Research Foundation.
DD257.U54 355.4'9 78-51611
78 MARC

CLAY, RICHARD, 1789-1877

Moran, James.
Clays of Bungay / by James Moran. — Bungay : Richard Clay & Co. Ltd., 1978.
160 p, ⸢2⸣ leaves of plates : ill., coat of arms, facsims., geneal. table, ports. (1 col.) ; 24 cm. GB79-02476
Includes bibliographical references.
ISBN 0-9506104-0-2 : £3.50
1. Richard Clay and Co.—History. 2. Richard Clay (The Chaucer Press)—History. 3. Clay, Richard, 1789-1877. I. Title.
Z232.R594M67 79-309290
338.7'6'68620942641
79 MARC

CLAY

see also Argillite; Bricks; Ceramics; Clay soils; Kaolin; Modeling; Particles; Swelling soils

Shang-hai shih ti i chien chu ts'ai liao kung yeh kung ssu.
(Yüan liao)
原料 / 上海市第一建筑材料工业公司, 重庆市建筑材料工业公司. — 第1版. — 北京：中国建筑工业出版社, 1977.
102 p. : ill. ; 19 cm. — (粘土砖瓦工艺)
$0.23
1. Clay. I. Ch'ung-ch'ing shih chien chu ts'ai liao kung yeh kung ssu. II. Title. III. Series: Nien t'u chuan wa kung i.
TP811.S56 1977 79-837092

—ADDRESSES, ESSAYS, LECTURES

Food and Agriculture Organization of the United Nations. European Commission on Agriculture. Working Party on Water Resources and Irrigation.
Drainage of heavy soils / European Commission on Agriculture Working Party on Water Resources and Irrigation, Tel Aviv, Israel, 1970. — Rome : Water Resources and Development Service, Land and Water Development Division, Food and Agriculture Organization of the United Nations, 1971.
v, 109 p. : ill. ; 28 cm. — (Irrigation and drainage paper ; 6)
Includes bibliographies.
1. Drainage—Addresses, essays, lectures. 2. Clay—Addresses, essays, lectures. I. Title. II. Series.
TC970.F66 1971 631.6'2 78-323834
79 MARC

—ANALYSIS

Sudō, Toshio, 1911-
(Nendo kōbutsu)
粘土鑛物 / 須藤俊男著. — 東京；岩波書店, 1953.
10, 240 p. : ill. ; 18 cm. — (岩波全書 ; 178)
Series romanized: Iwanami zensho.
Bibliography: p. ⸢209⸣-219.
Includes indexes.
1. Clay minerals. 2. Clay—Analysis. I. Title.
QE389.625.S92 79-80405(

—CONGRESSES

All India Seminar on Raw Materials for Glass, Ceramic, and Refractory Industries, Shillong, India, 1977.
All India Seminar on Raw Materials for Glass, Ceramic, and Refractory Industries, Shillong, October 7-9, 1977 : key papers & abstracts / organised by Geological Survey of India. — ⸢s.l. : s.n., 1977 or 1978⸣ (⸢Shillong : Vivekananda Press⸣)
5, 39, 32, xxviii p. ; 30 cm.
Includes bibliographies.
1. Glass sand—Congresses. 2. Clay—Congresses. 3. Ceramic materials—Congresses. I. India (Republic). Geological Survey.
TP810.5.A44 1977 553'.6 78-912100
79 MARC

International Clay Conference, Stockholm, 1963.
Proceedings of the conference held at Stockholm, Sweden, August 12-16. Arranged by CIPÉA, Comité international pour l'étude des argiles. Sponsored by IUGS, International Union of Geological Sciences. Edited by I. Th. Rosenqvist ⸢and⸣ P. Graff-Petersen. New York, Macmillan, 1963-
v. illus., diagrs. 24 cm. (International series of monographs on earth sciences, v. 14)
English, French or German.
"A Pergamon Press book."
Includes bibliographies.
1. Clay—Congresses. I. Rosenqvist, Ivan Thoralf, 1916- ed. II. Graff-Petersen, Poul, ed. III. International Committee for the Study of Clays. IV. International Union of Geological Sciences.
QE471.3.I57 1963 553.61 63-18137
⸢r78⸣rev2 MARC

International Clay Conference, Jerusalem, 1966.
Proceedings, edited by L. Heller ⸢and⸣ A. Weiss. Jerusalem, Israel Program for Scientific Translations, 1966-67.
2 v. illus. 25 cm. PL480:Is-3500
English, French, or German.
Vol. 2, edited by L. Heller.
Arranged by CIPEA, Comitè international pour l'étude des argiles. Sponsored by IUGS, International Union of Geological Sciences and others.
Includes bibliographies.
1. Clay—Congresses. I. Heller, Liselotte, ed. II. Weiss, Armin, ed. III. International Committee for the Study of Clays. IV. International Union of Geological Sciences.
QE471.3.I57 1966 553.61 he 66-1992
⸢r79⸣rev2 MARC

International Clay Conference, Madrid, 1972.
Preprints. Editor: J. M. Serratosa. ⸢Madrid⸣ Sociedad Española de Arcillas ⸢1972-
v. illus. 24 cm. Sp***
English or French.
Includes bibliographies.
1. Clay—Congresses. I. Serratosa, J. M., ed.
QE471.3.I57 1972b 553'.61 73-178747
73⸢r79⸣rev MARC

International Clay Conference, Mexico (City), 1975.
Proceedings of the International Clay Conference, 1975, Mexico City, Mexico, July 16-23, 1975 / organized by Instituto de Geología, Universidad Nacional Autonoma de Mexico ; under the auspices of Association internationale pour l'étude des argiles ; editor-in Chief, S. W. Bailey. — Wilmette, Ill. : Applied Pub., c1976.
xvi, 691 p. : ill. ; 26 cm.
English or French.
Includes bibliographical references and indexes.
ISBN 0-915834-02-2
1. Clay—Congresses. I. Bailey, Sturges W., 1919- . II. Mexico (City). Universidad Nacional. Instituto de Geología. III. International Association for Clay Research.
QE471.3.I57 1975 553'.61 75-32132
76⸢r79⸣rev MARC

International Clay Conference, Tokyo, 1969.
Proceedings. Arr. by Association internationale pour l'étude des argiles and the science Council of Japan. Editor-in-Chief: Lisa Heller. Jerusalem, Israel Universities Press, 1969-
v. illus. 25 cm.
English, French, German, or Russian.
Sponsored by the Clay Science Society of Japan, the Geological Survey of Japan, and the Geographical Society of Japan.
Includes bibliographies.
1. Clay—Congresses. I. Heller, Liselotte, ed. II. International Association for Clay Research. III. Japan. Nihon Gakujutsu Kaigi.
QE471.I57 1969 553'.61 73-171520
73⸢r79⸣rev MARC

Laboratoire central des ponts et chaussées.
Soil mechanics and foundation engineering : contributions of the Laboratoires des ponts et chaussées ⸢at the⸣ ninth International Conference on Soil Mechanics and Foundation Engineering, Tokyo, 11-15 July, 1977, International Symposium on Soft Clay, Bangkok, 5-6 July, 1977, fifth Southeast Asian Conference on Soil Engineering, Bangkok, 2-4 July, 1977. — Paris : Ministère de l'équipement et de l'aménagement du territoire, 1978.
218 p. : ill. ; 30 cm. — (Bulletin de liaison des laboratoires des ponts et chaussées : Numéro spécial ; 6 E ISSN 0458-5860) F***
Includes bibliographies.
1. Soil mechanics—Congresses. 2. Foundations—Congresses. 3. Clay—Congresses. I. International Conference on Soil Mechanics and Foundation Engineering (1936-), 9th, Tokyo, 1977. II. International Symposium on Soft Clay, Bangkok, Thailand, 1977. III. Southeast Asian Conference on Soil Engineering, 5th, Bangkok, Thailand, 1977. IV. Title. V. Series.
TA710.A1L32 1978 624'.15 79-314944
79 MARC

—TESTING

Sanks, Robert L
Survey—suitability of clay beds for storage of industrial solid wastes : report to Texas Water Quality Board / by Robert L. Sanks, J. Michael LaPlante, Earnest F. Gloyna. — Austin : Center for Research in Water Resources, Environmental Health Engineering, University of Texas at Austin, 1975.
201 p. in various pagings : ill. ; 28 cm. — (CRWR ; 128)
"Interagency cooperation contract number 74-75-0940."
"EHE-76-04."
Includes bibliographies.
1. Sanitary landfills. 2. Clay—Testing. 3. Factory and trade waste. I. LaPlante, J. Michael, joint author. II. Gloyna, Earnest F., joint author. III. Texas. Water Quality Board. IV. Title. V. Series: Texas. University at Austin. Center for Research in Water Resources. CRWR ; 128.
TD795.7.S23 628.5'4 76-622785
76⸢r79⸣rev MARC

—FRANCE

Trauth, Norbert.
Argiles évaporitiques dans la sédimentation carbonatée continentale et épicontinentale tertiaire : bassins de Paris, de Mormoiron et de Salinelles (France), Jbel Ghassoul (Maroc) / par Norbert Trauth, ... — Strasbourg : Université Louis-Pasteur, 1977.
195 p., iii leaves of plates : ill. ; 27 cm. — (Sciences géologiques : Mémoire ; no 49 ISSN 0302-2684) F78-13320
Originally presented as the author's thesis, Strasbourg, 1974.
Bibliography: p. 171-184.
1. Clay—France. 2. Clay—Morocco. 3. Evaporites—France. 4. Evaporites—Morocco. 5. Geology, Stratigraphic—Tertiary. 6. Geology—France. 7. Geology—Morocco. I. Title. II. Series.
QE269.S863 no. 49 550'.8 s 79-373987
⸢QE471.3⸣ 79 MARC

—INDIA

Central Road Research Institute.
Puzzolanic clays of India: their industrial exploitation and use in engineering works. Delhi ⸢1964⸣
viii, 290 p. illus., maps (part fold.) 25 cm. (Road Research special reports, no. 1)
Bibliography: p. 259-290.
1. Clay—India. 2. Pozzuolanas—India. I. Title. II. Series.
TN943.I6C45 1964 sa 66-1560
MARC

CLAY (Continued)

—MOROCCO

Trauth, Norbert.
Argiles évaporitiques dans la sédimentation carbonatée continentale et épicontinentale tertiaire : bassins de Paris, de Mormoiron et de Salinelles (France), Jbel Ghassoul (Maroc) / par Norbert Trauth, ... — Strasbourg : Université Louis-Pasteur, 1977.
195 p., iii leaves of plates : ill. ; 27 cm. — (Sciences géologiques : Mémoire ; no 49 ISSN 0302-2684) F78-13320
Originally presented as the author's thesis, Strasbourg, 1974.
Bibliography: p. 171-184.
1. Clay—France. 2. Clay—Morocco. 3. Evaporites—France. 4. Evaporites—Morocco. 5. Geology, Stratigraphic—Tertiary. 6. Geology—France. 7. Geology—Morocco. I. Title. II. Series.
QE269.S863 no. 49 550'.8 s 79-373987
[QE471.3] 79 MARC

—OREGON—BENTON CO.

Schlicker, Herbert G
Rock material resources of Benton County, Oregon / Herbert G. Schlicker, Jerry J. Gray, James L. Bela ; in cooperation with Benton County Board of County Commissioners, Corvallis, Oregon, and Oregon Land Conservation and Development Commission. — Portland : State of Oregon, Dept. of Geology and Mineral Industries, 1978.
v, 49 p. : ill., maps (2 fold. col. in pocket) ; 29 cm. — (Short paper - Department of Geology and Mineral Industries ; 27)
"Conducted in conformance with ORS 516.030."
Bibliography: p. 32-33.
1. Building stones—Oregon—Benton Co. 2. Clay—Oregon—Benton Co. 3. Gravel—Oregon—Benton Co. I. Gray, Jerry J., joint author. II. Bela, James L., joint author. III. Oregon. State Dept. of Geology and Mineral Industries. IV. Title. V. Series: Oregon. State Dept. of Geology and Mineral Industries. Short paper ; 27.
TN24.O7A423 no. 27 553'.09795 s 78-624329
[TN951.O7] 79 MARC

—SOUTH AFRICA—WITWATERSRAND

Bredell, J H
Prospecting for refractory clay on the east Rand / by J. H. Bredell. — Pretoria : Dept. of Mines, Geological Survey, 1978.
iv, 20 p., [5] fold. leaves of plates : ill. ; 30 cm. — (Bulletin - Geological Survey ; 62) SA***
Summary in Afrikans.
ISBN 0-621-04341-9 : R5.80
1. Prospecting—South Africa—Witwatersrand. 2. Clay—South Africa—Witwatersrand. I. Series: South Africa. Geological Survey. Bulletin ; 62.
QE325.A53 no. 62 79-316281
[TN271.C53] 79 MARC

—SPAIN—CATALONIA—ANALYSIS

Martínez Manent, Salvador.
Estudio mineralógico y ceramidad de las arcillas de Monells (Baix Empordà) / autores, Salvador Martínez Manent, Carlos de la Fuente Cullell, José María Bosch Figueroa. — Girona : Delegación de Girona del Colegio Oficial de Aparejadores y Arquitectos Técnicos de Catalunya, 1977.
121 p. : ill., graphs ; 20 cm. Sp78-May
"Cerámica I."
Includes bibliographies.
1. Clay—Spain—Catalonia—Analysis. I. Fuente Cullell, Carlos de la, joint author. II. Bosch Figueroa, José María, joint author. III. Title.
TP811.M36 78-395566
79 MARC

—THAILAND—BANGKOK REGION—PERMEABILITY

Brand, Edward W
The permeability of Bangkok clay, by Edward W. Brand and Sopon Duangkhae. [Bangkok. Asian Institute of Technology] 1971.
xi, 57 l. illus. 28 cm. (Engineering characteristics of undisturbed soils in the Bangkok area; phased report no. 8) (Asian Institute of Technology, Research report no. 13)
Sopon Duangkhae's thesis (Master of Engineering), Asian Institute of Technology.
Bibliography: leaves 56-57.
1. Soil mechanics—Thailand—Bangkok region. 2. Clay—Thailand—Bangkok region—Permeability. I. Sopon Duangkhae, joint author. II. Title. III. Series: Asian Institute of Technology Research report no. 13.
TA710.4.T5B36 624'.1513 78-326744
79 MARC

—ZAMBIA

Legg, C
Brick clay in the Mansa area, with special reference to the Loshi clay deposit / by C. A. Legg, L. Maczka, M. Cap. — Lusaka : Republic of Zambia, Ministry of Mines and Industry, 1975.
vi, 73 p. : ill., maps (4 fold in pocket) ; 30 cm. — (Economic report of the Geological Survey ; no. 48)
Includes bibliographical references.
1. Clay—Zambia. I. Maczka, L., joint author. II. Cap, M., joint author. III. Zambia. Geological Survey Dept. IV. Title. V. Series: Zambia. Geological Survey Dept. Economic report ; no. 48.
TN119.R38A3 no. 48 553'.09689'4 s 77-980746
[TN943.Z33] 79 MARC

Reichwalder, P
Brick-clay deposits at Kasondi Dambo, Ngwerere River, and other localities near Lusaka / by P. Reichwalder ; with appendix 2 by M. Cap. — Lusaka : Republic of Zambia, Ministry of Mines and Mining Development, Geological Survey Dept., 1973.
v, 21 p. : 3 fold. maps (in pocket) ; 25 cm. — (Economic report ; no. 42)
Bibliography: p. 7.
1. Clay—Zambia. I. Title. II. Series: Zambia. Geological Survey Dept. Economic report ; no. 42.
TN119.R38A3 no. 42 553'.09689'4 s 76-980480
[TN943.Z33] 79 MARC

CLAY BUILDING see Buliding, Clay

CLAY CO., IND

—BIOGRAPHY

Blanchard, Charles, fl. 1882-1900, ed.
Counties of Clay and Owen, Indiana. Historical and biographical ... Charles Blanchard, editor. Chicago, F. A. Battey & co., 1884.
viii, [1]-966 p. incl. 48 port. on 47 pl. 25 cm.
Articles by various writers.
1. Clay Co., Ind.—History. 2. Clay Co., Ind.—Biography. 3. Owen Co., Ind.—History. 4. Owen Co., Ind.—Biography. I. Title.
F532.C6B6 5-21836
[r79]rev MARC

—HISTORY

Blanchard, Charles, fl. 1882-1900, ed.
Counties of Clay and Owen, Indiana. Historical and biographical ... Charles Blanchard, editor. Chicago, F. A. Battey & co., 1884.
viii, [1]-966 p. incl. 48 port. on 47 pl. 25 cm.
Articles by various writers.
1. Clay Co., Ind.—History. 2. Clay Co., Ind.—Biography. 3. Owen Co., Ind.—History. 4. Owen Co., Ind.—Biography. I. Title.
F532.C6B6 5-21836
[r79]rev MARC

CLAY CO., IOWA

—MAPS

Rockford Map Publishers, Rockford, Ill.
Land atlas & plat book, Clay County, Iowa : 1978. — Rockford, Ill. : Rockford Map Publishers ; Spencer, Iowa : distributed by Clay County Soil Conservation District, [c1978]
40 p. : maps ; 28 cm.
Cover title.
Cadastral maps of townships, scale 1:50,688 or 1 1/4 inch equals one mile.
Includes index.
1. Clay Co., Iowa—Maps. 2. Real property—Iowa—Clay Co.—Maps. I. Title.
G1433.C65R6 1978 912'.777'153 79-318249
79 MARC

CLAY CO., W. VA.

—BIOGRAPHY

Hickory & lady slippers : life and legend of Clay County people / [compiled by the Art Department of Clay County High School]. — [Clay, W. Va.] : The School, [c1978-
v. : ill. ; 22 cm.
CONTENTS:
—v.3. The Clay County clan. book 2. C-D.
1. Clay Co., W. Va.—Genealogy. 2. Clay Co., W. Va.—History. 3. Clay Co., W. Va.—Biography. I. Clay County High School.
F247.C5H52 975.4'67'04 78-111952
79 MARC

—GENEALOGY

Hickory & lady slippers : life and legend of Clay County people / [compiled by the Art Department of Clay County High School]. — [Clay, W. Va.] : The School, [c1978-
v. : ill. ; 22 cm.
CONTENTS:
—v.3. The Clay County clan. book 2. C-D.
1. Clay Co., W. Va.—Genealogy. 2. Clay Co., W. Va.—History. 3. Clay Co., W. Va.—Biography. I. Clay County High School.
F247.C5H52 975.4'67'04 78-111952
79 MARC

—HISTORY

Hickory & lady slippers : life and legend of Clay County people / [compiled by the Art Department of Clay County High School]. — [Clay, W. Va.] : The School, [c1978-
v. : ill. ; 22 cm.
CONTENTS:
—v.3. The Clay County clan. book 2. C-D.
1. Clay Co., W. Va.—Genealogy. 2. Clay Co., W. Va.—History. 3. Clay Co., W. Va.—Biography. I. Clay County High School.
F247.C5H52 975.4'67'04 78-111952
79 MARC

CLAY FAMILY

Berwind, Elizabeth B
Gateways to the past / Elizabeth B. Berwind. — Philadelphia : Berwind, 1978.
xiv, 562 p. : ill. ; 24 cm.
Includes index.
1. Berwind, Elizabeth B. 2. Philadelphia—Biography. 3. Pennsylvania—Biography. 4. Berwyn family. 5. Clay family. 6. Roberts family. I. Title.
F158.5.B42 974.8'11'040924 78-68845
79 MARC

CLAY INDUSTRIES

see also Brick trade; Bricks; Building, Clay; Titles

Butt, I͡Uriĭ Mikhaĭlovich.
(Obshchai͡a tekhnologii͡a silikatov)
Общая технология силикатов / Ю. М. Бутт, Г. Н. Дудеров, М. А. Матвеев. — Изд. 3., перер. и доп. — Москва : Стройиздат, 1976.
599 p. : ill. ; 21 cm. USSR***
"Допущено ... в качестве учебника для техникумов промышленности строительных материалов."
1.13rub
1. Silicates. 2. Clay industries. I. Duderov, G. N., joint author. II. Matveev, Mikhail Aleksandrovich, joint author. III. Title.
TP807.B94 1976 79-407984

—BRAZIL—SERGIPE (STATE)

Sergipe, Brazil (State). Conselho do Desenvolvimento Econômico.
Ocorrências de argilas : levantamento preliminar / Govêrno de Sergipe, Conselho do Desenvolvimento de Sergipe-CONDESE. — Aracaju : CONDESE, 1976.
580 p. : maps ; 31 cm.
At head of title: Ministério do Interior, Superintendência do Desenvolvimento do Nordeste-SUDENE.
Includes bibliographical references.
1. Clay industries—Brazil—Sergipe (State) 2. Argillite—Brazil—Sergipe (State) I. Brazil. Superintendência do Desenvolvimento do Nordeste. II. Title.
HD9595.B73S477 338.4'7'66642 79-338797
79 MARC

—UNITED STATES

Kitamura, Yaichirō.
(Hokubei Gasshūkoku yōgyō chōsa hōkoku)
北米合衆國窯業調査報告 / [北村彌一郎]. — 東京：農商務省商工局，大正5 [1916]
2, 130 p., [1] fold. leaf of plates : ill. ; 26 cm. — (商工彙纂；第45號)
Cover title.
1. Clay industries—United States. I. Japan. Nōshōmushō. Shōkōkyoku. II. Title. III. Series: Shōkō isan ; dai 45-gō.
TP798.K57 78-810647

CLAY MINERALS

Badyoczek, Halina.
Studium selektywnej rozpuszczalności minerałów żelaza i tytanu z surowców kaolinowych / Halina Badyoczek. Reakcje minerałów ilastych z $Ca(OH)_2$ i ich znaczenie praktyczne / Edeltrauda Helios Rybicka. — Kraków : Zakład Narodowy im. Ossolińskich, 1977.
84 p., [2] leaves of plates : ill. ; 24 cm. — (Prace mineralogiczne ; 52) P***
At head of title: Polska Akademia Nauk. Oddział w Krakowie. Komisja Nauk Mineralogicznych.
Summaries in English and Russian; legends also in English and Russian.
Includes bibliographies.
zł22.00
1. Kaolin. 2. Solubility. 3. Clay minerals. 4. Lime. I. Helios Rybicka, Edeltrauda. Reakcje minerałów ilastych z $Ca(OH)_2$ i ich znaczenie praktyczne. 1977. II. Title. III. Series: Polska Akademia Nauk. Komisja Nauk Mineralogicznych. Prace mineralogiczne ; 52.
QE351.P6 vol. 52 78-400686
[QE389.625] 79 MARC

Hübner, Manfred.
Untersuchungen zur Reaktion von Tonmineralen mit organischen Substanzen / von Manfred Hübner. — 1. Aufl. — Leipzig : Deutscher Verlag für Grundstoffindustrie, 1978.
74 p. : 17 ill. ; 22 cm. — (Freiberger Forschungshefte ; C, 332) GDR78-A
Summary also in English.
Includes bibliographies.
23.00M
1. Clay minerals. 2. Organic geochemistry. I. Title. II. Series: Freiberger Forschungshefte ; Reihe C, 332.
QE1.F73 vol. 332 550'.8 s 78-376423
[QE389.625] 78 MARC

Sudō, Toshio, 1911-
(Nendo kōbutsu)
粘土鑛物 / 須藤俊男著. — 東京 ; 岩波書店, 1953.
10, 240 p. : ill. ; 18 cm. — (岩波全書 ; 178)
Series romanized: Iwanami zensho.
Bibliography: p. [209]-219.
Includes indexes.
1. Clay minerals. 2. Clay—Analysis. I. Title.
QE389.625.S92 79-80405

Theng, B K G
Formation and properties of clay-polymer complexes / B. K. G. Theng. — Amsterdam ; New York : Elsevier Scientific Pub. Co., 1979.
xi, 362 p. : ill. ; 25 cm. — (Developments in soil science ; v. 9)
Includes bibliographies and indexes.
ISBN 0-444-41706-0
1. Clay minerals. 2. Organic geochemistry. 3. Soil chemistry. I. Title. II. Series.
QE389.625.T48 631.4'17 78-13704
78 MARC

—CONGRESSES

Conference on Clay Mineralogy and Petrology, 7th, Karlovy Vary, Czechoslovakia, 1976.
Seventh Conference on Clay Mineralogy and Petrology in Karlovy Vary, September 28-October 3, 1976 / edited by J. Konta. — Praha : Univerzita Karlova, [c1977]
497 p. : ill. ; 25 cm. — (Geologica - Universitas Carolina Pragensis) Cz***
English, French, German, or Russian.
Includes bibliographies.
Kčs70.00
1. Clay minerals—Congresses. 2. Petrology—Congresses. I. Konta, Jiří. II. Series: Prague. Universitá Karlova. Geologica.
QE389.625.C66 1976 549'.6 78-395297
79 MARC

International Clay Conference, 6th, Oxford, 1978.
International Clay Conference, 1978 : proceedings of the VI International Clay Conference 1978 held in Oxford, 10-14 July, 1978 / organized by the Clay Minerals Group, Mineralogical Society, London, under the auspices of Association internationale pour l'étude des argiles ; edited by M. M. Mortland and V. C. Farmer. — Amsterdam ; New York : Elsevier Scientific Pub. Co., ; New York : distributors for the U.S. and Canada, Elsevier/North-Holland,
xiv, 662 p. : ill. ; 25 cm. — (Developments in sedimentology ; 27)
Includes bibliographies and indexes.
ISBN 0-444-41773-7
1. Clay minerals—Congresses. I. Mortland, Max Merle, 1923- II. Farmer, Victor Colin. III. Mineralogical Society of Great Britain and Ireland. Clay Minerals Group. IV. Series.
QE389.625.I57 1978 549'.6 78-20909
78 MARC

Symposium in Geochemistry, Los Angeles, 1978.
Symposium in Geochemistry : low temperature metamorphism of kerogen and clay minerals, October 5, 1978, Los Angeles, Ca / convened and edited by Donald F. Oltz. — Los Angeles : Pacific Section, Society of Economic Paleontologists and Mineralogists, c1978.
101 p. : ill. ; 28 cm.
"Proceedings of a symposium sponsored by the Pacific Section, S.E.P.M."
Includes bibliographies.
1. Kerogen—Congresses. 2. Clay minerals—Congresses. 3. Rocks, Metamorphic—Congresses. 4. Geochemistry—Congresses. I. Oltz, Donald F. II. Society of Economic Paleontologists and Mineralogists. Pacific Section.
QE516.5.S95 1978 553'.2 79-108849
00 MARC

—CALIFORNIA—INYO CO.

Starkey, Harry C
Clay mineralogy of Pleistocene Lake Tecopa, Inyo County, California : in a saline alke with pH above 8.5 authegenic sepiolite and detrital lithium-bearing saponite were deposited in mudstones which were interlayered with volcanic tufts / by Harry C. Starkey and Paul D. Blackmon. — Washington : U.S. Govt. Print. Off., 1979.
iv, 34 p. : ill. 29 cm. — (Geological Survey professional paper ; 1061)
Bibliography: p. 33-34.
Supt. of Docs. no.: I 19.16:1061
1. Clay minerals—California—Inyo Co. 2. Meerschaum—California—Inyo Co. 3. Saponite—California—Inyo Co. 4. Saline waters—California—Inyo Co. I. Blackmon, Paul D., joint author. II. Title. III. Series: United States. Geological Survey. Professional paper ; 1061.
QE389.625.S73 549'.6 77-608318
77 MARC

—CZECHOSLOVAK REPUBLIC—CONGRESSES

Conference on Clay Mineralogy and Petrography, 3d, Prague, Czechoslovakia, 1965.
Third Conference on Clay Mineralogy and Petrology in Prague, September 3-5, 1965. — Praha : Universita Karlova, 1965.
135 p., [8] leaves of plates : ill. ; 25 cm. — (Acta Universitatis Carolinae : Geologica : Supplementum ; 2)
English, Czech, German, and Russian.
"Associated editor: Jiří Konta."
Includes bibliographies.
1. Clay minerals—Czechoslovak Republic—Congresses. 2. Petrology—Czechoslovak Republic—Congresses. I. Konta, Jiří. II. Series: Prague. Universita Karlova. Acta. Geologica. Supplementum ; 2.
QE1.A1262 vol. 2 550 s 79-368286
[QE389.625] 79 MARC

—GREAT PLAINS

Schultz, Leonard Gene, 1926-
Mixed-layer clay in the Pierre Shale and equivalent rocks, northern Great Plains region : x-ray diffraction techniques for interpretation of mixed-layer illite-smectite clay, chemical composition of the clay, and developmental history of the clay / by Leonard G. Schultz. — Washington : U.S. Govt. Print. Off., 1978.
iv, 28 p. : ill. ; 29 cm. — (Geochemistry of the Pierre Shale and equivalent rocks, northern Great Plains) (Geological Survey professional paper ; 1064-A)
Bibliography: p. 27-28.
1. Clay minerals—Great Plains. 2. Geology, Stratigraphic—Cretaceous. I. Title. II. Series. III. Series: United States. Geological Survey. Professional paper ; 1064-A.
QE389.625.S3 552'.5 77-26718
77 MARC

CLAY MODELING see Modeling

CLAY PIPE see Pipe, Clay

CLAY POT COOKERY

Clay cookery / by the editors of Consumer guide. — New York : Simon and Schuster, c1979.
160 p. : ill. ; 28 cm. — (A Fireside book)
Includes index.
ISBN 0-671-24820-0
1. Clay pot cookery. I. Consumer guide.
TX825.5.C52 641.5'88 79-10256
79 MARC

CLAY SOILS

see also Terra rossa

—RUSSIAN REPUBLIC—KOLA PENINSULA

Koff, Grigoriĭ L'vovich.
(Morskie glinistye grunty Kol'skogo poluostrova i stroitel'stvo na nikh)
Морские глинистые грунты Кольского полуострова и строительство на них / Г. Л. Кофф. — Мурманск : Кн. изд-во, 1976.
125 p. : ill. ; 20 cm. USSR 77
Bibliography : p. 122-[123]
0.34rub
1. Engineering geology—Russian Republic—Kola Peninsula. 2. Clay soils—Russian Republic—Kola Peninsula. 3. Building—Russian Republic—Kola Peninsula. I. Title.
TA705.K58 78-421244

CLAY TABLETS

see also Cuneiform inscriptions

CLAY TARGET SHOOTING see Trapshooting

CLAY TOBACCO-PIPES

Walker, Iain C
Clay tobacco-pipes, with particular reference to the Bristol industry / Iain C. Walker. — Ottawa : Parks Canada : available from Print. and Publ., Supply and Services Canada, 1977.
4 v. (xxii, 1839 p.) : ill., maps ; 28 cm. — (History and archaeology ; 11 A-D) C***
Errata slip inserted.
Bibliography: vol. B., p. 827-976.
ISBN 0-660-00869-6 : £25.00
1. Tobacco pipe industry. 2. Clay tobacco-pipes. 3. Tobacco pipe industry—England—Bristol. I. Title.
HD9999.P662W34 338.4'7'6884 79-303105
79 MARC

—GREAT BRITAIN

The Archaeology of the clay tobacco pipe / edited by Peter Davey. — Oxford : B.A.R., 1979-
v. : ill. ; 30 cm. — (BAR British series ; 63-) GB***
Includes bibliographies.
CONTENTS: 1. Britain: the Midlands and Eastern England.
ISBN 0-86054-051-0 (v. 1) : £8.50 (v. 1)
1. Clay tobacco-pipes—Great Britain. I. Davey, Peter J. II. Series.
TS2235.G7A72 688'.4 79-314598
79 MARC

Fletcher, Edward, fl. 1970-
Clay pipes / by Edward Fletcher ; edited and photographed by Michael J. Coffin. — Southampton : Southern Collectors Publications, 1977.
[3], 25 p. : ill. ; 22 cm. GB78-10329
ISBN 0-905438-09-4 : £0.95
1. Clay tobacco-pipes—Great Britain. I. Coffin, Michael John. II. Title.
NK4695.T55F58 688'.4 78-318774
78 MARC

CLAY TOYS see Ceramic toys

CLAYBROOKE MAGNA, ENG.

—BOUNDARIES

Phythian-Adams, Charles.
Continuity, fields and fission : the making of a Midland parish / by Charles Phythian-Adams. — Leicester : Leicester University Press, 1978.
53 p. : maps ; 25 cm. — (Occasional papers - Department of English Local History, University of Leicester : Third series ; no. 4) GB78-16684
Includes bibliographical references.
ISBN 0-7185-2034-3 : £4.00
1. Claybrooke Magna, Eng.—History. 2. Claybrooke Parva, Eng.—History. 3. Claybrooke Magna, Eng.—Boundaries. 4. Claybrooke Parva, Eng.—Boundaries. I. Title. II. Series: Leicester, Eng. University. Dept. of English Local History. Occasional papers : Third series ; no. 4.
DA690.C6148P47 942.5'44 78-324512
79 MARC

CLAYBROOKE MAGNA, ENG. (Continued)

—HISTORY

Phythian-Adams, Charles.
Continuity, fields and fission : the making of a Midland parish / by Charles Phythian-Adams. — Leicester : Leicester University Press, 1978.
53 p. : maps ; 25 cm. — (Occasional papers - Department of English Local History, University of Leicester : Third series ; no. 4) GB78-16684
Includes bibliographical references.
ISBN 0-7185-2034-3 : £4.00
1. Claybrooke Magna, Eng.—History. 2. Claybrooke Parva, Eng.—History. 3. Claybrooke Magna, Eng.—Boundaries. 4. Claybrooke Parva, Eng.—Boundaries. I. Title. II. Series: Leicester, Eng. University. Dept. of English Local History. Occasional papers : Third series ; no. 4.
DA690.C6148P47 942.5'44 78-324512
79 MARC

CLAYBROOKE PARVA. ENG.

—BOUNDARIES

Phythian-Adams, Charles.
Continuity, fields and fission : the making of a Midland parish / by Charles Phythian-Adams. — Leicester : Leicester University Press, 1978.
53 p. : maps ; 25 cm. — (Occasional papers - Department of English Local History, University of Leicester : Third series ; no. 4) GB78-16684
Includes bibliographical references.
ISBN 0-7185-2034-3 : £4.00
1. Claybrooke Magna, Eng.—History. 2. Claybrooke Parva, Eng.—History. 3. Claybrooke Magna, Eng.—Boundaries. 4. Claybrooke Parva, Eng.—Boundaries. I. Title. II. Series: Leicester, Eng. University. Dept. of English Local History. Occasional papers : Third series ; no. 4.
DA690.C6148P47 942.5'44 78-324512
79 MARC

—HISTORY

Phythian-Adams, Charles.
Continuity, fields and fission : the making of a Midland parish / by Charles Phythian-Adams. — Leicester : Leicester University Press, 1978.
53 p. : maps ; 25 cm. — (Occasional papers - Department of English Local History, University of Leicester : Third series ; no. 4) GB78-16684
Includes bibliographical references.
ISBN 0-7185-2034-3 : £4.00
1. Claybrooke Magna, Eng.—History. 2. Claybrooke Parva, Eng.—History. 3. Claybrooke Magna, Eng.—Boundaries. 4. Claybrooke Parva, Eng.—Boundaries. I. Title. II. Series: Leicester, Eng. University. Dept. of English Local History. Occasional papers : Third series ; no. 4.
DA690.C6148P47 942.5'44 78-324512
79 MARC

CLAYEY SOILS see Clay soils

CLAYTON, CHARLES PRITCHARD

Clayton, Charles Pritchard.
The hungry one / by C. P. Clayton ; edited by Michael Clayton. — Llandysul : Gomer Press, 1978.
244 p., ₍10₎ leaves of plates : ill. ; 22 cm. GB***
ISBN 0-85088-850-6 : £5.95
1. European War, 1914-1918—Personal narratives, Welsh. 2. Clayton, Charles Pritchard. 3. European War, 1914-1918—Campaigns—Western. 4. Great Britain. Army. Welsh Regiment—Biography. 5. Soldiers—Great Britain—Biography. I. Clayton, Michael. II. Title.
D640.C572 940.4'81'429 79-312688
79 MARC

CLAYTON CO., IOWA

—MAPS

Rockford Map Publishers, Rockford, Ill.
Land atlas and plat book, 1977, Clayton County, Iowa. — Rockford, Ill. : Rockford Map Publishers, ₍c1977₎
50 p. : maps ; 28 cm.
Cover title.
Cadastral maps of townships, scale 1:50,688 or 1 1/4 inch equals one mile.
Includes index.
$7.50
1. Clayton Co., Iowa—Maps. 2. Real property—Iowa—Clayton Co.—Maps. I. Title.
G1433.C7R6 1977 912'.777'36 78-319144
78₍r79₎rev MARC

CLEAN ROOMS

—CONGRESSES

International Symposium on Contamination Control, 4th, Washington, D.C., 1978.
4th International Symposium on Contamination Control, Washington, DC, USA, September 10-13, 1978 : proceedings. — Mt. Prospect, Ill. : Institute of Environmental Sciences, c1978.
xi, 342 p. : ill. ; 28 cm.
Sponsored by International Committee on Contamination Control Societies.
Includes bibliographical references and index.
ISBN 0-915414-58-9 : $25.00
1. Clean rooms—Congresses. 2. Contamination (Technology)—Congresses. I. International Committee on Contamination Control Societies.
TH7694.I57 1978 697.9'316 78-111481
78 MARC

Internationales Symposium für Reinraumtechnik, Zürich, 1972.
Reinraumtechnik I : Berichte des Internationalen Symposiums für Reinraumtechnik gehalten in Zürich, Schweiz, 18. bis 20. Oktober 1972 / vorbereitet von der Schweizerischen Gesellschaft für Reinraumtechnik (SRRT) ; hrsg. von W. H. Weihe und H. U. Wanner. — Baden : Schweizerische Gesellschaft für Reinraumtechnik, 1973.
175 p. : ill. ; 24 cm. Sw74-A-4107
English, French, or German.
Includes bibliographical references and index.
28.00F
1. Clean rooms—Congresses. I. Weihe, Wolf H. II. Wanner, Hans Urs. III. Schweizerische Gesellschaft für Reinraumtechnik. IV. Title.
TH7694.I58 1972 79-376082
*79 MARC

CLEAN WORK STATIONS see Clean rooms

CLEANING

see also Bleaching; Cleaning compounds; Dry cleaning; Dyes and dyeing; Laundry; Street cleaning

Lenormand, Louis Sébastien, b. 1757.
Manual práctico del arte de sacamanchas, año 1827 : instrucción acerca de los medios fáciles de quitar, cada uno por sí mismo. toda especie de manchas / Seb. Normand .... — ₍s.l.₎ : Revitec, 1976.
112 p. ; 23 cm. Sp***
Translation of Manuel pratique de l'art du dégraisseur.
1. Cleaning. 2. Spotting (Cleaning) I. Title.
TP932.L4618 79-376299
79 MARC

Treves, Ralph.
The all-around, upstairs-downstairs, indoors-outdoors, garage-to-living room, spic-and-span, shiny-bright, car-to-couch, diamond-to-doorknob, super-hygienic, all-inclusive and universal book of cleaning, with other household aids / by Ralph Treves. — 1st ed. — New York : Dutton, c1979.
xiii, 206 p. ; 21 cm.
Bibliography: p. 190-192.
ISBN 0-87690-238-7
1. House cleaning. 2. Cleaning. I. Title. II. Title: Book of cleaning.
TX324.T73 1979 648'.5 78-17290
78 MARC

CLEANING AND DYEING INDUSTRY

see also Dry cleaning

CLEANING COMPOUNDS

see also Detergents, Synthetic

Frost & Sullivan.
Hard surface cleaning products market. — New York : Frost & Sullivan, 1979.
a-j, iii, 281 p. ; 29 cm.
"#529."
Includes bibliographical references.
1. Soap trade—United States. 2. Cleaning compounds. 3. Market surveys—United States. I. Title.
HD9999.S73U52 1979 381'.45'6681 79-105785
79 MARC

Hauptmann, Heinz.
Grundlagen der Wäschereichemie / von Heinz Hauptmann. — Frankfurt/M. : Spohr, 1977.
451 p. : ill. ; 17 cm. — (Fachteil aus "Handbuch für Textilingenieure und Textilpraktiker" ; T 64) GFR77-A40
"Alleinvertrieb: Lorch-Verlag ... Frankfurt/M."
"Verzeichnis der zitierten Zeitschriften": p. 448-451.
Includes bibliographical references and indexes.
ISBN 3-87496-021-8 : DM58.00
1. Cleaning compounds. 2. Detergents, Synthetic. I. Title. II. Series: Handbuch für Textilingenieure und Textilpraktiker : Teildruck ; T 64.
TP990.H34 668'.1 78-387533
78 MARC

Schwartz, Anthony Max, 1908-
Surface active agents, their chemistry and technology, by Anthony M. Schwartz and James W. Perry. Huntington, N.Y., R. E. Krieger Pub. Co., 1977-78 ₍v. 1, 1978₎
2 v. illus. 24 cm.
Reprint of the 1949-58 ed. published by Interscience Publishers, New York.
Vol. 2 by A. M. Schwartz, J. W. Perry, and J. Berch has title: Surface active agents and detergents.
Includes bibliographical references.
ISBN 0-88275-684-2 (v. 1)
1. Surface active agents. 2. Cleaning compounds. I. Perry, James Whitney, 1907- joint author. II. Berch, Julian. III. Title.
TP994.S36 1977 668'.1 74-11051
74₍r79₎rev MARC

CLEANING MACHINERY AND APPLIANCES

see also Dust--Removal

CLEANING PREPARATIONS see Cleaning compounds

CLEANING OF FISH see Fishes, Dressing of

CLEANLINESS see Baths; Hygiene; Sanitation

CLEANNESS (MIDDLE ENGLISH POEM)

Davenport, William Anthony, 1935-
The art of the Gawain-poet / ₍by₎ W. A. Davenport. — London : Athlone Press ; ₍Atlantic Highlands₎ N.J. : ₍distributed by₎ Humanities Press, 1978.
xiii, 233 p. ; 23 cm. GB***
Bibliography: p. ₍xi₎-xiii.
Includes index.
ISBN 0-485-11173-X : £8.95
1. Pearl (Middle English poem) 2. Cleanness (Middle English poem) 3. Patience (Middle English poem) 4. Gawain and the Grene Knight. I. Title.
PR1972.G353D3 821'.1 79-305403
79 MARC

CLEAR CREEK, TEX.

—HISTORY—ANECDOTES, FACETIAE, SATIRE, ETC.

Odell, Bill.
Return to Clear Creek / Bill Odell ; ₍illus. by Leslie Carol Nicols₎. — ₍s.l. : s.n.₎, c1978-
v. : ill. ; 22 cm.
Cover title.
1. Clear Creek, Tex.—History—Anecdotes, facetiae, satire, etc. I. Title.
F394.C63O33 978.8'61 78-65169
79 MARC

CLEAR-CUTTING

—OREGON

Hydrologic factors and environmental impacts of mass soil movements in the Oregon Coast Range / by Gary Ketcheson and Henry A. Froehlich. — Corvallis : Water Resources Research Institute, Oregon State University, 1978.
94 p. : ill. ; 28 cm. — (WRRI ; 56)
"Project completion report, agreement no. 14-34-0001-7078, July 1976-September 1977, submitted to Water Resources Research Institute, Oregon State University, Corvallis, Oregon."
Bibliography: p. 84-88.
1. Mass-wasting—Oregon. 2. Hydrology—Oregon. 3. Clear-cutting—Oregon. I. Froehlich, Henry A., joint author. II. Oregon. State University, Corvallis. Water Resources Research Institute. III. Title. IV. Series: Oregon. State University, Corvallis. Water Resources Research Institute. WRRI ; 56.
HD1694.O7A13 no. 56 333.9'1'009795 s 78-622790
₍QE598.2₎ 79 MARC

CLEARANCE SALES see Special sales

CLEARING, INTERNATIONAL see International clearing

CLEARING AGREEMENTS see International clearing

CLEARING OF LAND

see also Clear-cutting; Shifting cultivation

—ENVIORNMENTAL ASPECTS—UTAH

Gifford, Gerald F
Hydrologic impact of burning and grazing on a chained pinyon-juniper site in southeastern Utah : completion report, Center for Water Resources Research, Utah State University Project A-022-Utah / by Gerald F. Gifford, John C. Buckhouse, Frank E. Busby. — Logan : Utah Water Research Laboratory, Utah State University, 1976.
vii, 22 p. : ill. ; 28 cm.
Bibliography: p. 22.
"PRJNRO12-1."
1. Grazing—Environmental aspects—Utah. 2. Burning of land—Environmental aspects—Utah. 3. Hydrology—Utah. 4. Clearing of land—Enviornmental aspects—Utah. I. Buckhouse, John C., joint author. II. Busby, Frank E., joint author. III. Title: Hydrologic impact of burning and grazing ...
SF85.35.U8G53 634.9'751'5 77-621112
79 MARC

—INDONESIA—NUNUKAN ISLAND

Universitas Mulawarman. Lembaga Pengabdian Masyarakat.
Beberapa pemikiran dalam konversi hutan Pulau Nunukan. — Samarinda : LAPAN Universitas Mulawarman, 1973.
x, 144, ₍4₎ leaves : ill., maps ; 28 cm.
"Lapan no. 018."
Bibliography: leaves ₍145₎-₍148₎
1. Nunukan Island, Indonesia—Economic conditions. 2. Clearing of land—Indonesia—Nunukan Island. 3. Forests and forestry—Economic aspects—Indonesia—Nunukan Island. I. Title.
HC448.N76U53 1973 76-940221
79 MARC

CLEARINGHOUSE

see also Check collection systems

—GREAT BRITAIN

Committee of London Clearing Bankers.
The London clearing banks : evidence by the Committee of London Clearing Bankers to the Committee to Review the Functioning of Financial Institutions, November, 1977. — London : Committee of London Clearing Bankers : distributed by Longman, 1978.
278 p. : ill. (some col.) ; 30 cm. GB***
Includes index.
ISBN 0-582-03029-3
1. Banks and banking—Great Britain. 2. Clearinghouse—Great Britain. I. Great Britain. Committee to Review the Functioning of Financial Institutions. II. Title.
HG2988.C64 1978 332.1'0941 78-326043
79 MARC

—POLAND

Latopolski, Kazimierz.
Rozliczenia między przedsiębiorstwami / Kazimierz Latopolski. — Wyd. 1. — Warszawa : Państ. Wydaw. Ekonomiczne, 1977 (Pab. : LZG 3)
171, ₍1₎ p. : ill. ; 20 cm. P78-2299
Includes bibliographical references.
zł16.00
1. Clearinghouse—Poland. 2. Payment—Poland. I. Title.
HG2341.P7L38 79-374651
79 MARC

—RUSSIA

Kosoĭ, Arkadiĭ Mironovich.
(Upravlenie beznalichnym platezhnym oborotom)
Управление безналичным платежным оборотом / А. М. Косой. — Москва : Финансы, 1978.
192 p. ; 20 cm. USSR 78
Includes bibliographical references.
0.60rub
1. Clearinghouse—Russia. I. Title.
HG2341.R8K6 79-409305

CLEARWING MOTHS

Duckworth, W Donald.
The clearwing moths of California (Lepidoptera: Sesiidae) / W. Donald Duckworth, Thomas D. Eichlin. — ₍Sacramento₎ : State of California, Dept. of Food and Agriculture, Division of Plant Industry-Laboratory Services, 1978.
80 p., ₍4₎ leaves of plates : ill. ; 23 cm. — (Occasional papers in entomology ; no. 27)
Cover title.
Bibliography: p. 50-52.
Includes index.
1. Clearwing moths. 2. Insects—California. I. Eichlin, Thomas D., 1938- joint author. II. Title. III. Series.
QL461.C15 no. 27 595.7'001'2 s 79-620729
₍QL561.S47₎ 79 MARC

CLEARY, TIMOTHY F.

United States. Congress. Senate. Committee on Labor and Human Resources.
Nomination : hearing before the Committee on Labor and Human Resources, United States Senate, Ninety-sixth Congress, first session, on Timothy F. Cleary, of Maryland, to be a member of the Occupational Safety and Health Review Commission, April 11, 1979. — Washington : U.S. Govt. Print. Off., 1979.
ii, 21 p. ; 24 cm.
1. Cleary, Timothy F. 2. United States. Occupational Safety and Health Review Commission—Officials and employees.
KF26.L27 1979a 353.008'3 79-602232
79 MARC

CLEAVER, ELDRIDGE, 1935-

Cleaver, Eldridge, 1935-
Soul on fire / Eldridge Cleaver. — Waco, Tex. : Word Books, c1978.
240 p. : ill. ; 23 cm.
ISBN 0-8499-0046-8 : $8.95
1. Cleaver, Eldridge, 1935- 2. Converts—United States—Biography. I. Title.
BV4935.C54A34 248'.2 77-83335
78 MARC

CLEFT LIP see Harelip

CLEFT PALATE

Cleft palate and cleft lip : a team approach to clinical management and rehabilitation of the patient / edited by Herbert K. Cooper, Sr.... ₍et al.₎ ; Steven M. Spencer, editorial consultant. — Philadelphia : Saunders, 1979.
xvi, 568 p. : ill. ; 27 cm.
Includes bibliographies and index.
ISBN 0-7216-2687-4
1. Cleft palate. 2. Harelip. I. Cooper, Herbert K. II. Spencer, Steven M. ₍DNLM: 1. Cleft lip—Rehabilitation. 2. Cleft lip—Therapy. 3. Cleft palate—Rehabilitation. 4. Cleft-palate—Therapy. WV440 C622₎
RD525.C53 617'.522 76-41536
78 MARC

—COMPLICATIONS AND SEQUELAE

Fletcher, Samuel Glen.
Diagnosing speech disorders from cleft palate / Samuel G. Fletcher. — New York : Grune & Stratton, c1978.
x, 222 p. : ill. ; 24 cm.
Includes bibliographies and indexes.
ISBN 0-8089-1074-4
1. Cleft palate—Complications and sequelae. 2. Speech disorders in children—Diagnosis. I. Title.
RD525.F48 617'.522 78-9792
78 MARC

—CONGRESSES

Clefts of lip, alveolus and palate. Yugoslav symposium with international participation, Maribor, May 16th-17th 1968. Edited by Franc Čelešnik. Ljubljana, Organizing Committee of the Third Yugoslav Congress for Plastic and Maxillofacial Surgery, 1968.
410, ₍1₎ p. illus. 25 cm. 187.50Din Yu69-5295
Includes bibliographies.
1. Harelip—Congresses. 2. Cleft palate—Congresses. I. Čelešnik, franc, ed.
₍RD524.C55₎ 617'.522 76-973413
79 MARC

CLEGG, JAMES, 1679-1755

Clegg, James, 1679-1755.
The diary of James Clegg of Chapel en le Frith, 1708-1755 / edited by Vanessa S. Doe. — Matlock : Derbyshire Record Society, 1978-
v. : ill. ; 22 cm. — (₍Publications₎ - Derbyshire Record Society ; v. 2) GB79-05431
Includes index.
CONTENTS: pt. 1. 1708-36.

ISBN 0-9505940-2-4 (pt. 1) : £8.00 (pt. 1)
1. Clegg, James, 1679-1755. 2. Presbyterian Church—Clergy—Biography. 3. Clergy—England—Biography. I. Doe, V. S. II. Title. III. Series: Derbyshire Record Society. Publications - Derbyshire Record Society ; v. 2.
BX9225.C57A32 285'.2'0924 79-312483
79 MARC

CLEIS, UGO, 1903-1976

—EXHIBITIONS

Cleis, Ugo, 1903-1976.
Ugo Cleis : œuvre gravé / ₍réd. du catalogue, Michel Terrapon₎. — Fribourg : Musée d'art et d'histoire : diffusion Office du livre, 1978.
70 p. : ill. ; 22 cm. Sw78-4730
Catalog of an exhibition held at the Musée d'art et d'histoire, Fribourg, March 18-April 23, 1978.
French or German.
Bibliography: p. 19-25.
1. Cleis, Ugo, 1903-1976—Exhibitions. I. Terrapon, Michel. II. Fribourg. Musée d'art et d'histoire.
NE1164.5.C55A4 1978 769'.92'4 78-394683
79 MARC

CLEM FAMILY

Climo, Frederic Hedley Barker, 1910-
Mabinogi an Clem : the story of the children of Clem / by Frederic Hedley Barker Climo and Percy Lloyd Climo. — ₍s.l. : s.n., 1975₎ ₍s.l.₎ : J. Deyell)
xvi, 239 p. : ill., maps, ports. ; 29 cm. C76-3824-4
Includes indexes.
$20.00
1. Clem family. 2. Canada—Genealogy. 3. England—Genealogy. 4. Cornwall, Eng.—History. I. Climo, Percy Lloyd, 1906- joint author. II. Title.
CS90.C58 1975 929'.2'0971 79-300090
79 MARC

CLEMEN, PAUL, 1866-1947

Lützeler, Heinrich, 1902-
Persönlichkeiten : Konrad Adenauer, Paul Clemen, Kardinal Frings, Johannes XXIII., Erich Rothacker, Max Scheler : auf d. anderen Seite Funktionäre, Bosse, Bürokraten / Heinrich Lützeler. — Orig.-Ausg. — Freiburg im Breisgau ; Basel ; Wien : Herder, 1978.
157 p. ; 18 cm. — (Herderbücherei ; Bd. 668) GFR78-A
Includes bibliographical references.
ISBN 3-451-07668-3 : DM5.90
1. Rothacker, Erich, 1888-1965. 2. Scheler, Max Ferdinand, 1874-1928. 3. Clemen, Paul, 1866-1947. 4. Philosophers—Germany—Biography. 5. Personality. I. Title.
B3327.R74L83 79-371626
79 MARC

CLEMENCEAU, GEORGES EUGÈNE BENJAMIN, 1841-1929

Davids, S
Clemenceau / Sem Davids. — Den Haag : Kruseman, ₍1964₎
93 p. ; 20 cm. — (Kopstukken uit de twintigste eeuw ; 19)
Bibliography: p. 93.
1. Clemenceau, Georges Eugène Benjamin, 1841-1929. 2. Statesmen—France—Biography.
DC342.8.C6D38 79-363254
79 MARC

CLEMENS, SAMUEL LANGHORNE, 1835-1910. FOLLOWING THE EQUATOR

Mutalik, Keshav.
Mark Twain in India / by Keshav Mutalik. — 1st ed. — Bombay : Noble Pub. House, 1978.
151 p., ₍9₎ leaves of plates : ill., maps ; 23 cm.
Bibliography: p. 143-151.
Rs75.00
1. Clemens, Samuel Langhorne, 1835-1910. Following the equator. 2. Clemens, Samuel Langhorne, 1835-1910—Journeys—India. 3. India—Description and travel—1859-1900. 4. Voyages around the world. 5. Authors, American—19th century—Biography. I. Title.
PS1310.M8 818'.4'09 78-904981
79 MARC

CLEMENS, SAMUEL LANGHORNE, 1835-1910

Macnaughton, William R 1939-
Mark Twain's last years as a writer / William R. Macnaughton. — Columbia : University of Missouri Press, 1979.
x, 254 p. ; 23 cm.
Bibliography: p. 243-247.
Includes index.
ISBN 0-8262-0264-0
1. Clemens, Samuel Langhorne, 1835-1910. 2. Clemens, Samuel Langhorne, 1835-1910—Biography—Last years and death. 3. Authors, American—19th century—Biography. I. Title.
PS1332.M34 818'.4'09 78-19846
78 MARC

—APPRECIATION—ENGLAND

Welland, Dennis Sydney Reginald.
Mark Twain in England / Dennis Welland. — London : Chatto and Windus, 1978.
267 p., ₍1₎ leaf of plates : ports. ; 23 cm. GB***
Includes bibliographical references and index.
ISBN 0-7011-2233-1 : £8.00
1. Clemens, Samuel Langhorne, 1835-1910—Appreciation—England. 2. Authors and publishers—England. 3. Authors, American—19th century—Biography. I. Title.
PS1333.W4 1978b 818'.4'09 78-324609
79 MARC

—BIOGRAPHY

Henderson, Archibald, 1877-1963.
Mark Twain / by Archibald Henderson ; with photos. by Alvin Langdon Coburn. — Philadelphia : R. West, 1977.
xiii, 230 p., ₍8₎ leaves of plates : ill. ; 23 cm.
Reprint of the 1912 ed. published by F. A. Stokes Co., New York.
Bibliography: p. ₍213₎-230.
ISBN 0-8492-1135-2 : $25.00
1. Clemens, Samuel Langhorne, 1835-1910—Biography. 2. Authors, American—19th century—Biography.
₍PS1331.H4 1977₎ 818'.4'09 77-16577
77 MARC

Storm, Ole.
Mark Twain og Mr. Samuel L. Clemens / ₍af₎ Ole Storm. — København : Lademann, c1978.
342 p. : ill. ; 26 cm. D79-3
Bibliography: p. 342-₍343₎
ISBN 8715083152
1. Clemens, Samuel Langhorne, 1835-1910—Biography. 2. Authors, American—19th century—Biography. I. Title.
PS1331.S76 79-377137
79 MARC

—BIOGRAPHY—CHARACTER

Pellowe, William Charles Smithson, 1890-
Mark Twain, pilgrim from Hannibal / by William C. S. Pellowe. — Norwood, Pa. : Norwood Editions, 1976 ₍c1945₎
xvi, 301 p., ₍9₎ leaves of plates : ill. ; 26 cm.
Reprint of the ed. published by Hobson Book Press, New York.
Bibliography: p. 277-285.
Includes index.
ISBN 0-8482-2081-1 lib. bdg. : $30.00
1. Clemens, Samuel Langhorne, 1835-1910—Biography—Character. 2. Clemens, Samuel Langhorne, 1835-1910—Religion and ethics. I. Title.
PS1331.P5 1976 818'.4'09 76-20548
76₍r79₎rev MARC

Pellowe, William Charles Smithson, 1890-
Mark Twain, pilgrim from Hannibal / by William C. S. Pellowe. — Philadelphia : R. West, 1977, ₍c1945₎.
xvi, 301 p., ₍10₎ leaves of plates : ill. ; 26 cm.
Reprint of the ed. published by Hobson Book Press, New York.
Bibliography: p. 277-285.
Includes index.
ISBN 0-8492-2105-6 : $30.00
1. Clemens, Samuel Langhorne, 1835-1910—Biography—Character. 2. Clemens, Samuel Langhorne—Religion and ethics. I. Title.
PS1331.P5 1977 818'.4'09 77-12469
77₍r79₎rev MARC

—BIOGRAPHY—LAST YEARS AND DEATH

Macnaughton, William R 1939-
Mark Twain's last years as a writer / William R. Macnaughton. — Columbia : University of Missouri Press, 1979.
x, 254 p. ; 23 cm.
Bibliography: p. 243-247.
Includes index.
ISBN 0-8262-0264-0
1. Clemens, Samuel Langhorne, 1835-1910. 2. Clemens, Samuel Langhorne, 1835-1910—Biography—Last years and death. 3. Authors, American—19th century—Biography. I. Title.
PS1332.M34 818'.4'09 78-19846
78 MARC

—BIOGRAPHY—YOUTH

Wecter, Dixon, 1906-1950.
Sam Clemens of Hannibal / by Dixon Wecter. — New York : AMS Press, 1979.
ix, 335 p. : port. ; 22 cm.
Reprint of the 1952 ed. published by Riverside Press, Cambridge, and by Houghton Mifflin, Boston.
Bibliography: p. 317-322.
Includes index.
ISBN 0-404-15328-3
1. Clemens, Samuel Langhorne, 1835-1910—Biography—Youth. 2. Authors, American—19th century—Biography. I. Title.
₍PS1332.W4 1979₎ 818'.4'09 76-6595
78 MARC

—CORRESPONDENCE

Clemens, Samuel Langhorne, 1835-1910.
Mark Twain's letters in the Muscatine journal / edited with an introd. by Edgar M. Branch. — ₍Philadelphia₎ : R. West, 1977 ₍c1942₎
28 p. : ill. ; 23 cm.
Reprint of the ed. published by the Mark Twain Association of America, Chicago.
ISBN 0-8492-0194-2 lib. bdg. : $7.50
1. Clemens, Samuel Langhorne, 1835-1910—Correspondence. 2. Authors, American—19th century—Correspondence. I. Branch, Edgar Marquess, 1913- . II. Muscatine journal and news-tribune. III. Mark Twain Association of America. IV. Title.
₍PS1331.A4 1977₎ 818'.4'07 77-18766
77 MARC

Clemens, Samuel Langhorne, 1835-1910.
Mark Twain's Letter to William Bowen, Buffalo, February sixth, 1870 / prefatory note by Clara Clemens Gabrilowitsch ; foreword by Albert W. Gunnison. — New York : Haskell House Publishers, 1975.
11 p. : ill. ; 22 cm.
Reprint of the 1938 ed. published by the Book Club of California, San Francisco.
ISBN 0-8383-2089-9
1. Clemens, Samuel Langhorne, 1835-1910—Correspondence. 2. Bowen, William, 1836-1893. 3. Authors, American—19th century—Correspondence. I. Bowen, William, 1836-1893. II. Title. III. Title: Letter to William Bowen, Buffalo, February sixth, 1870.
PS1331.A43 1975 818'.3'09 75-22382
75₍r79₎rev MARC

—CRITICISM AND INTERPRETATION

Sloane, David E E 1943-
Mark Twain as a literary comedian / David E. E. Sloane. — Baton Rouge : Louisiana State University Press, c1979.
221 p. ; 24 cm. — (Southern literary series)
Bibliography: p. ₍201₎-216.
Includes index.
ISBN 0-8071-0460-4 : $12.95
1. Clemens, Samuel Langhorne, 1835-1910—Criticism and interpretation. 2. Comic, The. 3. American wit and humor—History and criticism. I. Title. II. Series.
PS1338.S55 818'.4'09 78-11125
78 MARC

Smith, U S
Up a tree with Mark Twain : an oracular opus exposing a literary hoax so horrendous may God have mercy on the 1st amendment: / U. S. Smith. — Initial Hannibal, Missouri, ed. — Quincy, Ill. : Shondo-Shando Press, c1978.
531 p. : ill. ; 22 cm.
ISBN 0-9601754-1-5 : $13.75
1. Clemens, Samuel Langhorne, 1835-1910—Criticism and interpretation. 2. Stevenson, Robert Louis, 1850-1894. The strange case of Dr. Jekyll and Mr. Hyde. I. Title.
PS1331.S59 818'.5'407 78-56748
79 MARC

—JOURNEYS—HAWAII

Clemens, Samuel Langhorne, 1835-1910.
Letters from the Sandwich Islands / Mark Twain ₍i.e. S. H. Clemens₎ ; edited by Joan Abramson. — Norfolk Island, Australia : Island Heritage, c1975.
234 p. : ill. ; 21 x 23 cm. — (Hawaiian bicentennial library ; v. 1)
Aus***
"An Island Heritage limited edition."
1. Hawaii—Description and travel—To 1950. 2. Clemens, Samuel Langhorne, 1835-1910—Journeys—Hawaii. 3. Hawaii—Social life and customs. I. Abramson, Joan. II. Title. III. Series.
DU623.C6 1975 996.9 75-21943
79 MARC

—JOURNEYS—INDIA

Mutalik, Keshav.
Mark Twain in India / by Keshav Mutalik. — 1st ed. — Bombay : Noble Pub. House, 1978.
151 p., ₍9₎ leaves of plates : ill., maps ; 23 cm.
Bibliography: p. 143-151.
Rs75.00
1. Clemens, Samuel Langhorne, 1835-1910. Following the equator. 2. Clemens, Samuel Langhorne, 1835-1910—Journeys—India. 3. India—Description and travel—1859-1900. 4. Voyages around the world. 5. Authors, American—19th century—Biography. I. Title.
PS1310.M8 818'.4'09 78-904981
79 MARC

—LANGUAGE

Emberson, Frances Guthrie, 1912-
Mark Twain's vocabulary : a general survey / by Frances Guthrie Emberson. — Philadelphia : R. West, 1978
53 p. ; 33 cm.
Reprint of the 1935 ed. published at the University of Missouri, Columbia, which was issued as v. 10, no. 3 of the University of Missouri studies.
Originally presented as the author's thesis, University of Missouri, 1935.
Bibliography: p. 35-53.
ISBN 0-8492-0746-0 : $10.00
1. Clemens, Samuel Langhorne, 1835-1910—Language. 2. Vocabulary. I. Title. II. Series: Missouri. University. The University of Missouri studies ; v. 10, no. 3.
PS1344.E6 1978 818'.4'09 78-1908
78 MARC

—MANUSCRIPTS—CATALOGS

Simpson, Alan, 1912-
Mark Twain goes back to Vassar : an introduction to the Jean Webster McKinney family papers / by Alan Simpson ; published for the dedication of the Francis Fitz Randolph Rare Book Room in the Helen D. Lockwood Library. — ₍Poughkeepsie, N.Y.₎ : Vassar College, 1977.
20 p. : ill. ; 26 cm.
1. Clemens, Samuel Langhorne, 1835-1910—Manuscripts—Catalogs. 2. Webster, Jean, 1876-1916—Manuscripts—Catalogs. 3. Vassar College. Library—Catalogs. I. Vassar College. Library. II. Title.
PS1336.S5 818'.4'09 78-317659
78 MARC

—RELIGION AND ETHICS

Pellowe, William Charles Smithson, 1890-
Mark Twain, pilgrim from Hannibal / by William C. S. Pellowe. — Philadelphia : R. West, 1977, ₍c1945₎.
xvi, 301 p., ₍10₎ leaves of plates : ill. ; 26 cm.
Reprint of the ed. published by Hobson Book Press, New York.
Bibliography: p. 277-285.
Includes index.
ISBN 0-8492-2105-6 : $30.00
1. Clemens, Samuel Langhorne, 1835-1910—Biography—Character. 2. Clemens, Samuel Langhorne—Religion and ethics. I. Title.
PS1331.P5 1977 818'.4'09 77-12469
77₍r79₎rev MARC

Pellowe, William Charles Smithson, 1890-
Mark Twain, pilgrim from Hannibal / by William C. S. Pellowe. — Norwood, Pa. : Norwood Editions, 1976 ₍c1945₎
xvi, 301 p., ₍9₎ leaves of plates : ill. ; 26 cm.
Reprint of the ed. published by Hobson Book Press, New York.
Bibliography: p. 277-285.
Includes index.
ISBN 0-8482-2081-1 lib. bdg. : $30.00
1. Clemens, Samuel Langhorne, 1835-1910—Biography—Character. 2. Clemens, Samuel Langhorne, 1835-1910—Religion and ethics. I. Title.
PS1331.P5 1976 818'.4'09 76-20548
76₍r79₎rev MARC

CLEMENS, TITUS FLAVIUS, ALEXANDRINUS

Berciano, Modesto.
(Kairos)
Καιρός : tiempo humano e histórico-salvífico en Clemente de Alejandria / Modesto Berciano. — Burgos : Ediciones Aldecoa, ₍1976?₎
330 p. ; 26 cm. — (Publicaciones de la Facultad Teológica del Norte de España, sede de Burgos ; 34) Sp***
On spine: Καιρός en Clemente de Alejandria.
Bibliography: p. ₍7₎-24.
Includes indexes.
ISBN 84-7009-184-0
1. History (Theology)—History of doctrines. 2. Salvation—History of doctrines. 3. Kairós (The Greek word) 4. Clemens, Titus Flavius, Alexandrinus. I. Title. II. Title: Kairos en Clemente de Alejandria.
BR115.H5B435 77-560872
(MARC)

CLEMENS, TITUS FLAVIUS, ALEXANDRINUS (Continued)

Clark, Elizabeth, 1938-
Clement's use of Aristotle : The Aristotelian contribution to Clement of Alexandria's refutation of gnosticism / by Elizabeth A. Clark. — New York : E. Mellen Press, c1977.
vii, 182 p. ; 23 cm. — (Texts and studies in religion)
Bibliography: p. [153]-167.
Includes indexes.
ISBN 0-88946-985-7
1. Clemens, Titus Flavius, Alexandrinus. 2. Aristoteles—Influence. 3. Gnosticism—History. I. Title. II. Series.
B666.Z7C45 230'.1'30924 77-93913
79 MARC

CLEMENS WENZESLAUS, ABP. AND ELECTOR OF TRIER, 1739-1812

Wachter, Dorothea, 1912-
Degen & [i.e. und] Krummstab : Clemens Wenzeslaus, Prinz von Sachsen, Königl. Prinz von Polen & Litauen, Kurfürst u. Erzbischof von Trier, Fürstbischof von Augsburg : 1739-1812 / Dorothea Wachter. — Kempten : Allgäuer Zeitungsverlag, c1978.
295 p. : ill. ; 20 cm. — (Die Beliebten Allgäu-Schwaben Bücher) GFR79-A
Bibliography: p. 293-295.
ISBN 3-88006-046-0 : DM26.80
1. Clemens Wenzeslaus, Abp. and Elector of Trier, 1739-1812. 2. Germany—Princes and princesses—Biography. I. Title.
DD205.C54W32 943'.05'0924 79-369256
79 MARC

CLÉMENT, CHARLES, 1889-

Clément, Charles, 1889-
Pages de journal, 1926-1967 / Charles Clément ; choix et présentation par Gilbert Guisan. — Lausanne : La Bibliothèque des arts, [1976]
166 p. : ill. (some col.) ; 25 cm. Sw76-3956
Includes index.
66.00F
1. Clément, Charles, 1889- 2. Painters—Switzerland—Biography. I. Guisan, Gilbert. II. Title.
ND853.C59A2 1976 759.9494 79-347739
*79 MARC

CLÉMENT, GINETTE

Clément, Ginette.
Aventure mexicaine / Ginette Clément. — [Avignon] : Presses universelles, [1976?]
137 p. ; 18 cm. F***
1. Mexico—Description and travel—1951- . 2. Clément, Ginette. I. Title.
F1216.C53 917.2'04'82 79-357329
79 MARC

CLEMMENSEN FAMILY

Nordisk slægtsforskning.
En slægt Clemmensen fra Torning sogn (Lysgaard herred) : om Andreas Clemmentin J. Clemmensen, født 1840, hans forfædre og efterkommere / [Nordisk Slægtsforskning Aps]. — Skringstrup, Skals : Nordisk Slægtsforskning, 1977.
[116] leaves, [10] plates : ill. ; 29 cm. D78-18
Fold. geneal. table inserted.
Not in trade.
1. Clemmensen family. 2. Denmark—Genealogy. I. Title.
CS909.C59 1977 78-388602
79 MARC

CLEMENTS FAMILY

Watson, Robert Clement, 1890-
Watson family notes / Robert Clement Watson. — [s.l.] : Watson, [1978?]
157 p., [2] leaves of plates : ill. ; 28 cm.
Cover title.
1. Watson family. 2. Clements family. I. Title.
CS71.W34 1978 929'.2'0973 78-109304
79 MARC

CLEMM, MARIA

Poe, Edgar Allan, 1809-1849.
Letters and documents in the Enoch Pratt Free Library / Edgar Allan Poe ; edited by Arthur H. Quinn and Richard H. Hart. — Philadelphia : R. West, 1978 [c1941]
ix, 84 p., [22] leaves of plates : ill. ; 29 cm.
Reprint of the ed. published by Scholars' Facsimiles & Reprints, New York.
Includes bibliographical references and index.
ISBN 0-8492-2207-9 lib. bdg. : $20.00
1. Poe, Edgar Allan, 1809-1849—Correspondence. 2. Clemm, Maria. 3. Authors, American—19th century—Correspondence. 4. Enoch Pratt Free Library, Baltimore. I. Quinn, Arthur Hobson, 1875-1960. II. Hart, Richard Harry, 1908- . III. Enoch Pratt Free Library, Baltimore. IV. Title.
[PS2631.A335 1978] 818'.3'09 78-1582
78 MARC

CLEMMYS MUHLENBERGII see Bog turtle

CLEMSON UNIVERSITY

—FOOTBALL

Capps, Kerry.
The Orange machine / by Kerry Capps ; photos. by John Rowntree ; with special biographic profiles of senior Tigers by Steve Ellis. — Columbia, SC : R. L. Bryan Co., c1979.
167 p. : ill. ; 28 cm.
1. Clemson University—Football. 2. Football players—United States—Biography. 3. Football coaches—United States—Biography. I. Rowntree, John. II. Ellis, Steve, 1955- III. Title.
GV958.C56C36 796.33'263'0975723 79-63622
79 MARC

CLEP see College-Level Examination Program

CLERGY

see also Bishops; Celibacy; Deacons; Monasticism and religious orders; Parishes; Pastoral theology; Priests; Privileges and immunities, Ecclesiastical; Rabbis; Theologians; Women clergy

Mosley, Ernest E
Priorities in ministry / Ernest E. Mosley. — Nashville, Tenn. : Convention Press, c1978.
132 p. : ill. ; 21 cm.
Includes bibliographical references.
1. Clergy. I. Title.
BV660.2.M63 253 78-111210
78 MARC

—ANECDOTES, FACETIAE, SATIRE, ETC.

Wulf, Hans.
Pfarrer X und die Gesetze : Rechtskonflikte im Alltag e. Pfarrers / Hans Wulf, Albert Stein. — Neukirchen-Vluyn : Neukirchener Verlag, 1977.
84 p. : ill. ; 21 cm. GFR77-A
ISBN 3-7887-0530-2
1. Clergy—Anecdotes, facetiae, satire, etc. I. Stein, Albert, joint author. II. Title.
PN6231.C5W84 340'.02'42 79-379269
79 MARC

—APPOINTMENT, CALL, AND ELECTION

see also Missionaries--Appointment, call, and election; Vocation (in religious orders, congregations, etc.)

—BIOGRAPHY

Macleod, Paul, 1929-
Why do you strike me? : the life and challenge of Nikodems Serzants / [by] Paul Macleod. — Belmont, Vic. : Neptune Press, 1978.
159 p. : ill. ; 21 cm. Aus
ISBN 0-909131-08-2
1. Serzants, Nikodems, 1923- . 2. Catholic Church—Clergy—Biography. 3. Clergy—Biography. 4. Latvians in Australia—Biography. I. Title.
BX4705.S554M32 282'.092'4 79-309586
79 MARC

—BIOGRAPHY—JUVENILE LITERATURE

Daughters of St. Paul.
No greater love : the life of Father Damien of Molokai / written and illustrated by the Daughters of St. Paul. — Boston : St. Paul Editions, c1979.
92 p. : ill. ; 22 cm.
SUMMARY: A biography of the Belgian priest who was sent to Molokai, Hawaii, cared for lepers in a colony, and subsequently contracted the disease himself.
1. Damien, Father, 1840-1889—Juvenile literature. 2. Catholic Church—Clergy—Biography—Juvenile literature. 3. Clergy—Biography—Juvenile literature. 4. Missionaries—Belgium—Biography—Juvenile literature. 5. Missionaries—Hawaii—Biography—Juvenile literature. [1. Damien, Father, 1840-1889. 2. Missionaries] I. Title.
BX4705.D25D37 1979 266'.2'0924 79-10017
79 MARC

—CATHOLIC CHURCH—RELIGIOUS LIFE

John Paul II, Pope, 1920-
Sign of contradiction / by Karol Wojtyla (Pope John Paul II). — New York : Seabury Press, 1979.
xiv, 206 p. ; 22 cm.
"A Crossroad book."
Translation of Segno di contraddizione.
ISBN 0-8164-0433-X
1. Retreats for clergy. 2. Clergy—Catholic Church—Religious life. I. Title.
BX1912.5.J6313 242'.6'9 79-4606
79 MARC

—COSTUME

see also Church vestments

—DIVORCE

Bouma, Mary LaGrand.
Divorce in the parsonage / Mary LaGrand Bouma. — Minneapolis : Bethany Fellowship, c1979.
156, [3] p. ; 21 cm.
Bibliography: p. [159]
ISBN 0-87123-109-3 : $3.95
1. Clergy—Divorce. I. Title.
BV4395.5.B68 248'.892 79-16157
79 MARC

—EDUCATION see Clergy, Training of

--FAMILIES see Clergymen's families

—FICTION

Paredes Candia, Antonio.
Otros cuentos curas : folklore secreto / Antonio Paredes Candia. — 1. ed. — La Paz, Bolivia : Ediciones ISLA, 1978.
77 p. : ill. ; 15 cm. — (Colección Folclore secreto ; 6)
1. Clergy—Fiction. 2. Tales, Bolivian. I. Title.
PQ7820.P3O8 861 79-112234
79 MARC

—LEGAL STATUS, LAWS, ETC. —JAPAN

(Kokken to ryōshin)
国権と良心：種谷牧師裁判の軌跡 / 種谷牧師裁判を支援する会編. — 東京：新教出版社, 1975.
x, 419 p., [2] leaves of plates : ill. ; 19 cm. — (今日のキリスト教双書 ; 20) Ja 75-18546
Series romanized: Konnichi no Kirisutokyō sōsho.
¥2000
1. Tanetani, Shun'ichi, 1927- 2. Misprison—Japan. 3. Religious liberty—Japan. 4. Clergy—Legal status, laws, etc.—Japan. I. Tanetani Bokushi Saiban o Shiensuru Kai.
78-812837

—LIBRARIES see Libraries, Private; Theological libraries

CLERGY (Continued)

—MAJOR ORDERS see Bishops; Clergy

—OFFICE

Anderson, James Desmond.
The management of ministry / James D. Anderson, Ezra Earl Jones. — 1st ed. — San Francisco : Harper & Row, c1978.
xi, 202 p. : ill. ; 21 cm.
Bibliography: p. [199]-200.
Includes index.
ISBN 0-06-060235-X
1. Church management. 2. Clergy—Office. 3. Pastoral theology. I. Jones, Ezra Earl, joint author. II. Title.
BV652.A67 1978 254 76-62942
78 MARC

Catholic Church. Pastoralsynode der Jurisdiktionsbezirke in der DDR.
Dienste und Ordnungen im Leben der Gemeinde / [Pastoralsynode d. Jurisdiktionsbezirke in d. DDR ; hrsg. im Auftr. d. Berliner Ordinarienkonferenz]. — 1. Aufl. — Leipzig : S[ank]t-Benno-Verlag, 1976.
34 p. ; 21 cm. GFR78-A2
At head of title: Beschluss der Pastoralsynode.
1.50M
1. Catholic Church—Clergy. 2. Clergy—Office. 3. Lay ministry—Catholic Church. 4. Lay ministry—Germany, East. I. Title. II. Title: Beschluss der Pastoralsynode.
BX1912.C336 1976 79-376334
79 MARC

Dittes, James E
When the people say no : conflict and the call to ministry / James E. Dittes. — 1st ed. — San Francisco : Harper & Row, c1979.
ix, 150 p. ; 21 cm. — (Harper's ministers paperback library)
ISBN 0-06-061923-6
1. Church controversies. 2. Clergy—Office. 3. Pastoral theology. I. Title.
BV652.9.D57 1979 253 77-15879
79 MARC

Doty, William Lodewick, 1919-
Meet your pastor! / William L. Doty. — Canfield, Ohio : Alba Books, c1978
163 p. : ill. ; 18 cm.
ISBN 0-8189-1153-0 : $1.95
1. Catholic Church—Clergy. 2. Clergy—Office. I. Title.
[BX1912.D59] 253 78-56776
00 MARC

Hobbs, Herschel H
Preacher talk / Herschel H. Hobbs. — Nashville : Broadman Press, c1979.
168 p. ; 20 cm.
ISBN 0-8054-2705-8
1. Clergy—Office. I. Title.
BV660.2.H55 253 78-73346
79 MARC

Johnson, Merle Allison.
How to be happy in the non-electric church / Merle Allison Johnson ; illustrated with line drawings by Charles Fox. — Nashville : Abingdon, c1979.
112 p. : ill. ; 23 cm.
ISBN 0-687-17706-5 : $6.95
1. Small churches—Miscellanea. 2. Clergy—Office. 3. Pastoral theology. I. Title.
BV637.8.J63 253 78-26463
78 MARC

Kemper, Robert G
The new shape of ministry : taking accountability seriously / Robert G. Kemper. — Nashville : Abingdon, c1979.
142 p. ; 19 cm.
ISBN 0-687-27874-0
1. Clergy—Office. I. Title.
BV660.2.K47 253 78-12256
78 MARC

Keyes, Paul T 1936-
Pastoral presence and the diocesan priest / Paul T. Keyes. — 1st ed. — Whitinsville, Mass. : Affirmation Books, c1978.
142 p. ; 21 cm.
Includes bibliographical references.
ISBN 0-89571-004-8 : $4.95
1. Pastoral theology—Catholic Church. 2. Clergy—Office. I. Title.
BX1913.K44 253 78-22009
78 MARC

Leenhardt, Franz J 1902-
L'Église : questions aux protestants et aux catholiques / Franz-J. Leenhardt. — Genève : Labor et fides ; Paris : diffusion pour la France Librairie protestante, cop. 1978.
239 p. ; 21 cm. — (Lieux théologiques ; no 1) Sw79-1874
Includes bibliographical references.
38.00F
1. Church. 2. Clergy—Office. 3. Christian union. I. Title.
BV600.2.L42 262'.001 79-368994
79 MARC

Neuhaus, Richard John.
Freedom for ministry / Richard John Neuhaus. — 1st ed. — San Francisco : Harper & Row, c1979.
xiii, 232 p. ; 22 cm.
Includes bibliographical references and index.
ISBN 0-06-066094-5
1. Clergy—Office. 2. Pastoral theology. I. Title.
BV660.2.N42 1979 253 78-3352
78 MARC

Sorgi, Claudio.
Faccia da prete / Claudio Sorgi. — Torino : Società editrice internazionale, 1978.
183 p. ; 22 cm. — (Dossiers SEI) It78-Apr
L4000
1. Catholic Church—Clergy. 2. Clergy—Office. I. Title.
BX1912.S673 79-375863
*79 MARC

Switzer, David K 1925-
Pastor, preacher, person : developing a pastoral ministry in depth / David K. Switzer. — Nashville : Abingdon, c1979.
144 p. ; 21 cm.
ISBN 0-687-30137-8
1. Clergy—Office. 2. Pastoral theology. I. Title.
BV660.2.S88 253 78-24018
78 MARC

Young, Robert Doran, 1928-
Religious imagination : God's gift to prophets and preachers / by Robert D. Young. — 1st ed. — Philadelphia : Westminster Press, c1979.
176 p. ; 19 cm.
Includes bibliographical references.
ISBN 0-664-24239-1
1. Clergy—Office. 2. Imagination. 3. Creation (literary, artistic, etc.) 4. Identification (Religion) I. Title.
BV660.2.Y68 253 78-26843
78 MARC

—OFFICE—ADDRESSES, ESSAYS, LECTURES

In God's community : essays on the Church and its ministry / edited by David J. Ellis and W. Ward Gasque. — Wheaton, Ill. : H. Shaw Publishers, c1978, 1979 printing.
xiii, 186 p. ; 21 cm.
Bibliography: p. 169-171.
Includes indexes.
ISBN 0-87788-392-0 : $3.95
1. Church—Addresses, essays, lectures. 2. Clery—Office—Addresses, essays, lectures. 3. Mission of the Church—Addresses, essays, lectures. I. Ellis, David J. II. Gasque, W. Ward.
BV600.2.I5 260 79-12847
79 MARC

Smemo, Johannes, 1898-1973.
Kravet og kraften : ordinasjonstaler og andre bidrag til emnet kall og tjeneste / av Johannes Smemo ; samlet og utgitt av Tore Kopperud. — Oslo : Luther, 1978. (Sarpsborg : Centraltr.)
181 p. ; 21 cm. — (Teologi i dag) N78-Sept.
Includes bibliographical references.
ISBN 8253172524 : kr58.00
1. Ordination sermons. 2. Sermons, Norwegian. 3. Lutheran Church—Sermons. 4. Vocation, Ecclesiastical—Addresses, essays, lectures. 5. Clergy—Office—Addresses, essays, lectures. I. Kopperud, Tore. II. Title. III. Series.
BV4285.S6 1978 78-391781
79 MARC

—OFFICE—EVALUATION

Brekke, Milo
Ten faces of ministry : perspectives on pastoral and congregational effectiveness based on a survey of 5000 Lutherans / Milo L. Brekke, Merton P. Strommen, Dorothy L. Williams. — Minneapolis : Augsburg Pub. House, c1979.
256 p. ; 23 cm.
Includes bibliographical references.
ISBN 0-8066-1692-X : $7.50
1. Lutheran Churchin the United States—Clergy—Evaluation. 2. Clergy—Office—Evaluation. I. Strommen, Merton P., joint author. II. Williams, Dorothy Lowe, 1926- joint author. III. Title.
BX8041.B7 253 78-66956
00 MARC

—POST-ORDINATION TRAINING

Andriessen, Hermanus Crispinus Ignatius, 1927-
Leren aan ervaring en supervisie / door H. C. I. Andriessen. — Nijmegen : Dekker & Van de Vegt, 1975.
xx, 551 p. : diagrs. ; 20 cm.
Bibliography: p. [503]-546.
ISBN 9025598234 : fl 69.50
1. Pastoral psychology—Study and teaching. 2. Clergy, Training of. 3. Clergy—Post-ordination training. I. Title.
BV4012.A635 75-541594
*79[r79]rev MARC

Andriessen, Hermanus Crispinus Ignatius, 1927-
Pastorale Supervision : Praxisberatung in d. Kirche / Herman Andriessen ; [aus d. Niederländ. von Kurt Moritz]. — 1. Aufl. — München : Kaiser ; Mainz : Grünewald, 1978.
243, [1] p. ; 21 cm. — (Gesellschaft und Theologie : Abteilung Praxis der Kirche ; Nr. 24) GFR78-A
Translation of Leren aan ervaring en supervisie.
Bibliography: p. 230-[244]
ISBN 3-459-01090-8
1. Pastoral psychology—Study and teaching. 2. Clergy, Training of. 3. Clergy—Post-ordination training. I. Title. II. Series.
BV4012.A63515 79-383056
79 MARC

—PSYCHOLOGY

De Klerk, Willem Johannes.
Pastorale sensitiwiteit : fragmente uit die pastorale-psigologie / W. J. de Klerk. — Johannesburg : Perskor, 1975.
122 p. ; 22 cm. SA
Bibliography: p. 107-113.
ISBN 0-628-00722-1 : R3.00
1. Pastoral psychology. 2. Clergy—Psychology. I. Title.
BV4012.D353 78-383819
78 MARC

—RELIGIOUS LIFE

see also Retreats for clergy

[La Tour, Bertrand de] 1701-1780.
Des armoiries sur les bréviaires et les choses saintes. [Montauban? 1777?]
128 p. 19 cm.
Caption title.
For authorship and probable place and date of publication, cf. L. Bertrand. Hist. litt. Comp. de Saint-Sulpice.
Bookplate: Bibliotheca Lindesiana.
1. Catholic Church—Clergy. 2. Clergy—Religious life. 3. Devices—Moral and religious aspects. I. Title.
BX1912.5.L23 Rosenwald Coll. 78-383286
78[r79]rev MARC

Suaud, Charles.
La Vocation : conversion et reconversion des prêtres ruraux / Charles Suaud. — Paris : Éditions de Minuit, 1978.
280 p. : maps ; 22 cm. — (Collection Le Sens commun) F78-19933
Includes bibliographical references and index.
ISBN 2-7073-0228-7 : 45.00F
1. Catholic Church in France—Clergy. 2. Vocation, Ecclesiastical. 3. Clergy—Religious life. I. Title.
BX1530.2.S9 248'.89422 79-368019
79 MARC

—TAXATION—UNITED STATES

Abingdon clergy income tax guide. 6th- 1978-
Nashville, Abingdon.
v. 28 cm. annual.
Continues: Clergy's federal income tax guide, ISSN 0000-9866.
Key title: Abingdon clergy income tax guide, ISSN 0163-1241
1. Income tax—United States—Popular works. 2. Clergy—Taxation—United States.
KF6369.8.C5C54 343'.73'052 78-646504
MARC-S

Hungerford, Kenneth G
Federal income tax handbook for clergy for 1978 returns / Kenneth G. Hungerford II and William C. Slack. — Grand Rapids, Mich. : Baker Book House, c1979.
60, [22] p. : forms ; 28 cm.
Includes index.
ISBN 0-8010-4203-8 : $2.95
1. Clergy—Taxation—United States. 2. Taxation—Law and legislation—United States. I. Slack, William C., joint author. II. Title.
KF6369.8.C5H86 1978 343'.73'0520242 79-104185
79 MARC

CLERGY (Continued)

—ARMENIA—BIOGRAPHY

Agonts', Step'annos Giwvēr.
(Patmut'iwn kenats' ew varuts' Tearn Mkhit'aray Sebastats'woy Rabunapeti ew Abbayi)
Պատմութիւն կենաց եւ վարուց Տեառն Մխիթարայ Սեբաստացւոյ Ռաբունապետի եւ Աբբայի / յօրինեալ Ստեփաննոսի Գիւվէր Ագոնց. — Ի Վենետիկ : Ի վանս Սրբոյն Ղազարու, 1259, 1810.
15, 336 p., [3] leaves of plates : ill. ; 24 cm.
Includes index.
1. Mkhit'ar Sebastats'i, 1676–1749. 2. Mekhitarists. 3. Mkhit'arian Miabanowt'iwn i Venetik. 4. Armenian Church—Clergy—Biography. 5. Clergy—Armenia—Biography. I. Title.
BX129.M58A64 78–970960

Fērhat'ian, Barsegh.
(Mkhit'ar mardě, hayě, surbě)
Մխիթար մարդը, հայը, սուրբը / Բարսեղ Ֆէրհաթեան. — Վիեննա : Մխիթարեան Տպարան, 1976.
167 p., [1] leaf of plates : port. ; 21 cm.
1. Mkhit'ar Sebastats'i, 1676–1749. 2. Armenian Church—Clergy—Biography. 3. Clergy—Armenia—Biography. 4. Mekhitarists. I. Title.
BX129.M58F467 79–971274

—AUSTRALIA—BIOGRAPHY

Living responsibility. — Glen Irish, Vic. : Vital Publications, 1976.
142 p. : ill. ; 18 cm. Aus
CONTENTS: Williams, E. Lyall. Reflections.—White, A. E. Biography.
ISBN 0-909116-04-0
1. Williams, E. Lyall, 1906- . 2. Christian Church (Disciples of Christ)—Clergy—Biography. 3. Clergy—Australia—Biography. 4. Christian life—Christian Church (Disciples of Christ) authors. I. Williams, E. Lyall, 1906- . II. White, A. E.
BX7343.W53L58 286'.63 79-315513
79 MARC

—AUSTRALIA—SOUTH AUSTRALIA—BIOGRAPHY

Hussey, H
More than half a century of colonial life and Christian experience : with notes of travel, lectures, publications, etc. / by H. Hussey ; illustrated by Ernest Gall, Alma Chambers. — Adelaide : Libraries Board of South Australia, 1978.
504 p. : ill., portrs. ; 19 cm. — (Australiana facsimile editions ; no. 203) Aus
Spine title: Colonial life and Christian experience.
Reprint of 1st ed., Adelaide, imprinted by Hussey & Gillingham, 1897.
ISBN 0-7243-0062-7
1. Hussey, H. 2. Disciples of Christ—Clergy—Biography. 3. Clergy—Australia—South Australia—Biography. 4. South Australia—Biography. I. Title. II. Title: Colonial life and Christian experience.
BX7343.H87A34 1978 286'.6'0924 78-322811
79 MARC

—AUSTRALIA—WESTERN AUSTRALIA—BIOGRAPHY

Wollaston, John Ramsden, 1791-1856.
Journals and diaries, 1841-1856, of Revd. John Ramsden Wollaston, M.A. / collected by A. Burton. — [Nedlands, Australia : University of Western Australia Press ; Portland, Or. : distributed by International Scholarly Book Services, 1975]
2 v. : ill. ; 22 cm. Aus***
Vol. 1 first published in 1975 by Paterson Brokensha; vol. 2 first published in 1948 by Pitman.
Bibliography: v. 1, p. xvi.
Includes indexes.
CONTENTS: v. 1. Wollaston's Picton journal, 1841-1844.—v. 2. Wollaston's Albany journals, 1841-1856.
ISBN 0-85564-103-7 (v. 1)
1. Wollaston, John Ramsden, 1791-1856. 2. Church of England—Clergy—Biography. 3. Clergy—Australia—Western Australia—Biography. 4. Western Australia—Biography. I. Burton, Alfred, 1865-1947. II. Title.
BX5720.W64A34 1975 283'.092'4 79-309465
79 MARC

Wollaston, John Ramsden, 1791-1856.
Wollaston's Albany journals, 1848-1856 / edited with notes by Percy U. Henn ; introd. by Malcolm J. L. Uren. — Nedlands, Australia : University of Western Australia Press ; Portland, Or. : [distributed by] International Scholarly Book Services, 1975.
252 p. : ill. ; 22 cm. — (His Journals and diaries, 1841-1856, of Revd. John Ramsden Wollaston, M.A. ; v. 2)
"First published in 1948 by C. H. Pitman & Son."
Includes index.
ISBN 0-85564-104-5
1. Wollaston, John Ramsden, 1791-1856. 2. Church of England—Clergy—Biography. 3. Clergy—Australia—Western Australia—Biography. 4. Western Australia—Biography. I. Henn, Percy U. II. Title. III. Title: Albany journals, 1848-1856.
BX5720.W64A34 1975 vol. 2 79-309463
283'.092'4 s
79 MARC

Wollaston, John Ramsden, 1791-1856.
Wollaston's Picton journal, 1841-1844 / edited with introd. and notes by Canon Burton and Percy U. Henn. — [Nedlands, Australia : University of Western Australia Press ; Portland, Or. : distributed by International Scholarly Book Services, 1975]
xxvi, 289 p. : ill. ; 22 cm. — (His Journals and diaries, 1841-1856, of Revd. John Ramsden Wollaston, M.A. ; v. 1) Aus***
"First published by Paterson Brokensha Pty. Ltd. 1975."
Bibliography: p. xvi.
Includes indexes.
ISBN 0-85564-103-7
1. Wollaston, John Ramsden, 1791-1856. 2. Church of England—Clergy—Biography. 3. Clergy—Australia—Western Australia—Biography. 4. Western Australia—Biography. I. Burton, Alfred, 1865-1947. II. Henn, Percy U. III. Title. IV. Title: Picton journal, 1841-1844
BX5720.W64A34 1975 vol. 1 79-309464
283'.092'4 s
79 MARC

—AUSTRIA—TYROL

Liebenwein, Wolfgang, Pfarrer.
Ev. [i.e. Evangelische] Prediger in Tirol im 16. Jahrhundert. (Insbruck, 1967.)
15 p. 30 cm. N.T. Au67-23-25
Bibliography: p. 13-[16]
1. Protestants—Austria—Tyrol. 2. Clergy—Austria—Tyrol. 3. Tyrol—Church history. I. Title.
BX4841.L5 70-490653
79[r79]rev MARC

—BELGIUM

Verstegen, Raf
Geestelijken naar Belgisch recht : oude en nieuwe vragen / door Raf. Verstegen. — 2600 Berchem [Generaal Capiaumontstr. 15] : M. Kluwer, 1977.
120 p. ; 20 cm. Be78-5
"De juridische en sociale positie van de geestelijken in België."
Includes bibliographical references.
495F
1. Clergy—Belgium. I. Title.
78-387777
*78 MARC

—BRAZIL—BIOGRAPHY

Santos, João.
Monsenhor Frederico Costa, 1. [i.e. primeiro] prelado de Santarém / João Santos. — Belém : Conselho Estadual de Cultura, 1978.
129 p. : ill. ; 22 cm. — (Série "Arthur Vianna") (Coleção "História do Pará")
Bibliography: p. 127-129.
1. Costa, Frederico, 1875-1948. 2. Catholic Church—Clergy—Biography. 3. Clergy—Brazil—Biography. I. Title.
BX4705.C7786S25 78-377091
78 MARC

Souza, José Coelho de.
O sangue pela justiça : Pe. João Bosco Penido Burnier, S.J. / José Coelho de Souza. — São Paulo : Edições Loyola, 1978.
251 p. : ill. ; 21 cm.
Cr$150.00
1. Burnier, João Bosco Penido, 1917-1976. 2. Catholic Church—Clergy—Biography. 3. Clergy—Brazil—Biography. I. Title.
BX4705.B934B68 79-108704
79 MARC

—BRAZIL—CORRESPONDENCE

Borges, José Alaor.
Crendo no amor, escrevi estas cartas, que antes de seren escritas, foram vividas / José Alaor Borges. — [s.l. : s.n., 1975 (São Paulo : Escolas Profissionais Salesianas)
106 p. ; 21 cm. — (Biblioteca Pioneira de Ciências Sociais : educação)
Cr$20.00
1. Borges, José Alaor. 2. Catholic Church—Clergy—Correspondence. 3. Clergy—Brazil—Correspondence. I. Title.
BX4705.B598A4 1975 78-37729
78 MAR(

—BRAZIL—MINAS GERAIS

Carvalho, José Geraldo Vidigal de.
Ideologia e raízes sociais do clero da conjuração, século XVIII, Minas Gerais / José Geraldo Vidigal de Carvalho. — Viçosa, [Brasil] : Impr. Universitária, Universidade Federal de Viçosa, 1978.
86 p. ; 23 cm.
Bibliography: p. [77]-86.
1. Clergy—Brazil—Minas Gerais. 2. Catholic Church—Clergy. 3. Minas Gerais, Brazil—History. I. Title.
BX1467.M5C37 79-104043
79 MARC

—CAMEROON—BIOGRAPHY

Fotso Ledoux, Ambroise.
Diacre : le 24 avril 1977 / Ambroise Fotso Ledoux. — [New Bell, Cameroun] : Paroisse de New-Bell Aviation, [1977]
32 p. ; 21 cm.
1. Fotso Ledoux, Ambroise. 2. Deacons—Cameroon—Biography. 3. Catholic Church—Clergy—Biography. 4. Clergy—Cameroon—Biography. I. Title.
BX4705.F648A34 282'.092'4 78-384907
78 MARC

—CANADA—BIOGRAPHY

Bernier, Cyrille, 1926-
Joseph Bernier : prêtre parmi les hommes / Cyril Bernier. — Laval : [s.n.], 1978.
292 p. : ports. ; 22 cm. C***
Bibliography: p. 285-288.
$9.50
1. Bernier, Joseph, 1921-1977. 2. Catholic Church—Clergy—Biography. 3. Clergy—Canada—Biography.
BX4705.B3833B47 282'.092'4 79-340996
79 MARC

Mondésir, Edouard de.
Mondésir, Edouard de Mondésir ; avec une introd. par Gilbert Chinard. — New York : Arno Press, 1979.
60 p. ; 24 cm. — (Johns Hopkins University Press reprints)
Reprint of the 1942 ed. published by The Johns Hopkins Press, Baltimore, in series: Historical documents, Institut français de Washington.
ISBN 0-405-10594-0 : $10.00
1. Mondésir, Edouard de. 2. Catholic Church—Clergy—Biography. 3. Clergy—Canada—Biography. 4. Clergy—United States—Biography. 5. Jesuits—Canada—Biography. 6. Jesuits—United States—Biography. I. Chinard, Gilbert, 1881-1972. II. Title. III. Series: Institut français de Washington, Washington, D.C. Historical documents.
BX4705.M64A3 1979 271'.53'024 78-19280
78 MARC

Nakayama, Goichi, 1902-
(Nihon ga sukuwareru tame ni)
日本が救われるために：全日本救国伝道の旅 / 八代斌助序；中山吾一著. — Coaldale：日系人聖公会文書伝道部；名古屋：取次所　みたま社，昭和29 [1954]
192 p., [1] leaf of plates : port. ; 18 cm.
1. Nakayama, Goichi, 1902- . 2. Anglican Church of Canada—Clergy—Biography. 3. Clergy—Canada—Biography. 4. Missions—Japan. I. Title.
BX5620.N26A36 79–809989

Swanson, Cecil, 1889-
The days of my sojourning : a reminiscence / Cecil Swanson. — Calgary : Glenbow-Alberta Institute, c1977.
x, 141 p. : ill., ports. ; 24 cm. C78-5704-X
Includes index.
$7.95
1. Swanson, Cecil, 1889- . 2. Anglican Church of Canada—Clergy—Biography. 3. Clergy—Canada—Biography. I. Glenbow-Alberta Institute. II. Title.
BX5620.S96A33 283'.092'4 78-325748
79 MARC

—COLOMBIA

Bronx, Humberto.
Golconda y origen de ideas marxistas en minorías del clero colombiano y Templos de Medellín / Humberto Bronx. — [s.l. : s.n., 1978] (Medellín : Editorial A. Salazar)
82 p. ; 17 cm.
Cover title.
Spine title: Golconda e ideas marxistas en el clero.
1. Catholic church in Colombia—History. 2. Communism and christianity—Catholic Church—Colombia. 3. Clergy—Colombia. 4. Medellín, Colombia—Churches. 5. Liberation theology. 6. Colombia—Church history. I. Bronx, Humberto. Templos de Medellín. 1978. II. Title. III. Tit'e: Golconda e ideas marxistas en el clero.
BX1470.2.B76 1978 79-100902
79 MARC

—DENMARK—BIOGRAPHY

Bukh, Peder.
Folket formede mig : erindringer og meninger / Peder Bukh. — [Århus] : Aros ; [København] : [eksp. DBK], 1977.
165 p. ; 22 cm. D77-50
Includes index.
ISBN 8770033129 : kr59.00
1. Bukh, Peder. 2. Lutheran Church—Clergy—Biography. 3. Clergy—Denmark—Biography. I. Title.
BX8080.B82A34 79-350921
79 MARC

CLERGY (Continued)

—DENMARK—SKAGEN—BIOGRAPHY

Houkjær, Christian.
Mellem to have / ⌐af¬ Christian Houkjær. — Fredericia : Frimodt, 1978.
108 p. : ill. ; 20 cm. D79-9
ISBN 8774460641 : kr39.50
1. Houkjaer, Christian. 2. Lutheran Church—Clergy—Biography. 3. Clergy—Denmark—Skagen—Biography. 4. Skagen, Denmark—Biography. I. Title.
BX8080.H69A35 79-373225
79 MARC

—ENGLAND—BIOGRAPHY

Ayling, Stanley Edward.
John Wesley / ⌐by¬ Stanley Ayling. — London : Collins, 1979.
350 p., ⌐8¬ p. of plates : ill., facsim., ports. ; 23 cm. GB79-15812
Bibliography: p. ⌐337¬-339.
Includes index.
ISBN 0-00-216656-9 : £7.50
1. Wesley, John, 1703-1791. 2. Methodist Church—Clergy—Biography. 3. Clergy—England—Biography.
BX8495.W5A94 287'.092'4 79-315195
79 MARC

Clegg, James, 1679-1755.
The diary of James Clegg of Chapel en le Frith, 1708-1755 / edited by Vanessa S. Doe. — Matlock : Derbyshire Record Society, 1978-
v. : ill. ; 22 cm. — (⌐Publications¬ - Derbyshire Record Society ; v. 2)
GB79-05431
Includes index.
CONTENTS: pt. 1. 1708-36.

ISBN 0-9505940-2-4 (pt. 1) : £8.00 (pt. 1)
1. Clegg, James, 1679-1755. 2. Presbyterian Church—Clergy—Biography. 3. Clergy—England—Biography. I. Doe, V. S. II. Title. III. Series: Derbyshire Record Society. Publications - Derbyshire Record Society ; v. 2.
BX9225.C57A32 285'.2'0924 79-312483
79 MARC

De Rosa, Peter.
A Father before Christmas / ⌐by¬ Neil Boyd ⌐i.e. P. De Rosa¬. — London : Joseph, 1978.
192 p. ; 23 cm. GB78-10668
ISBN 0-7181-1692-5 : £3.95
1. De Rosa, Peter. 2. Catholic Church—Clergy—Biography. 3. Clergy—England—Biography. I. Title.
BX4705.D429A33 253 78-314819
78⌐r79¬rev MARC

Distad, N Merrill.
Guessing at truth : the life of Julius Charles Hare (1795-1855) / by N. Merrill Distad. — Shepherdstown, W. Va. : Patmos Press, 1979.
xii, 258 p. ; 22 cm.
Bibliography: p. 239-247.
ISBN 0-915762-07-2 : $19.95
1. Hare, Julius Charles, 1795-1855. 2. Church of England—Clergy—Biography. 3. Clergy—England—Biography. I. Title.
BX5199.H35D57 283'.092'4 78-11625
78 MARC

Edwards, Michael S
S. E. Keeble : the rejected prophet / by Michael S. Edwards. — Chester : Wesley Historical Society, 1977.
⌐3¬, 71 p. ; 22 cm. GB78-09350
Includes bibliographical references.
£0.50
1. Keeble, Samuel Edward, 1853-1946. 2. Methodist Church—Clergy—Biography. 3. Clergy—England—Biography.
BX8495.K38E43 287'.1'0924 78-320318
78 MARC

Goodall, Norman.
Second fiddle : recollections and reflections / ⌐by¬ Norman Goodall. — London : S.P.C.K., 1979.
viii, 168 p. ; 22 cm. GB79-13329
Includes bibliographical references and index.
ISBN 0-281-03675-6 : £4.25
1. Goodall, Norman. 2. United Reformed Church—Clergy—Biography. 3. Clergy—England—Biography. I. Title.
BX9890.U258G663 285'.2'0924 79-314669
79 MARC

Hastings, Adrian.
In filial disobedience / by Adrian Hastings. — Great Wakering : Mayhew-McCrimmon, 1978.
180 p. : ill., ports. ; 18 cm. GB78-24883
ISBN 0-85597-249-1 : £1.90
1. Hastings, Adrian. 2. Catholic Church—Clergy—Biography. 3. Clergy—England—Biography. 4. Catholic Church—Doctrinal and controversial works—Catholic Authors—Addresses, essays, lectures. I. Title.
BX4705.H333A34 1978 282'.092'4 78-324337
79 MARC

Hollings, Michael.
Living priesthood / by Michael Hollings. — Great Wakering : Mayhew-McCrimmon, 1977.
262 p. ; 18 cm. GB78-23175
ISBN 0-85597-110-X : £1.85
1. Hollings, Michael. 2. Catholic Church—Clergy—Biography. 3. Clergy—England—Biography. I. Title.
BX4705.H727A34 1977b 79-306466
253
79 MARC

Institute of Family Research.
John Lathrop (1584-1653) : reformer, sufferer, pilgrim, man of God. — Rev. — Salt Lake City : Institute of Family Research, 1979.
39 p., ⌐1¬ fold. leaf of plates : ill. ; 29 cm.
"The result of ... research by Richard W. Price ... and Arlene Eakle".
Includes bibliographies.
1. Lothropp, John, 1584-1653. 2. Clergy—England—Biography. 3. Clergy—Massachusetts—Biography. I. Price, Richard W. II. Eakle, Arlene H.
BR1725.L66I57 1979 285'.8'0924 79-106935
79 MARC

Jordan, Hugh, 1906-
Born under a lucky star : the memoirs of Hugh Jordan. — Bramcote : Grove Books, 1977.
84 p. ; 22 cm. GB78-06648
ISBN 0-905422-20-1 : £1.75
1. Jordan, Hugh, 1906- . 2. Church of England—Clergy—Biography. 3. Clergy—England—Biography. I. Title.
BX5199.J65A33 283'.092'4 79-315515
79 MARC

Le Quesne, A Laurence.
After Kilvert / A. L. Le Quesne. — Oxford ⌐Eng.¬ ; New York : Oxford University Press, 1978.
ix, 233 p. : ill. ; 23 cm. GB78
Includes index.
ISBN 0-19-211748-3 : £4.95
1. Kilvert, Robert Francis, 1840-1879. Kilvert's diary. 2. Kilvert, Robert Francis, 1840-1879. 3. Church of England—Clergy—Biography. 4. Clergy—England—Biography. 5. Le Quesne, A. Laurence. 6. Clyro, Wales—Description. I. Title.
BX5199.K49A32 941.08 78-40173
78 MARC

Lock, John.
A man of sorrow : the life, letters, and times of the Rev. Patrick Brontë, 1777-1861 / by John Lock and W. T. Dixon ; foreword by the Archbishop of Canterbury (formerly the Archbishop of York). — London : I. Hodgkins ; Westport, Conn. : Meckler Books, 1979.
xiv, 566 p., ⌐7¬ leaves of plates : ill. ; 24 cm.
Reprint of the ed. published by Nelson, London.
Bibliography: p. 541-549.
Includes index.
ISBN 0-930466-15-2 : $35.95
1. Brontë, Patrick, 1777-1861—Biography. 2. Brontë family. 3. Authors, English—19th century—Biography. 4. Clergy—England—Biography. 5. Church of England—Clergy—Biography. I. Dixon, William Thomas, joint author. II. Title.
PR4174.B2L6 1979 828'.7'09 78-27130
78 MARC

Mahan, Asa, 1800-1889.
Autobiography : intellectual, moral, and spiritual / Asa Mahan. — 1st AMS ed. — New York : AMS Press, 1979.
x, 458 p. ; 19 cm. — (Philosophy in America)
Reprint of the 1882 ed. published by T. Woolmer, London.
ISBN 0-404-59257-0
1. Mahan, Asa, 1800-1889. 2. Congregational churches—Clergy—Biography. 3. Clergy—England—Biography.
BX7260.M3A3 1979 285'.8'0924 75-3269
79 MARC

Müller-Bohn, Jost.
Spurgeon, ein Mensch von Gott gesandt / Jost Müller-Bohn. — Lahr-Dinglingen : Verlag der St.-Johannis-Druckerei, 1978.
128 p. ; 21 cm. GFR78-A
Includes bibliographical references.
ISBN 3-501-02021-X : DM10.80
1. Spurgeon, Charles Haddon, 1834-1892. 2. Baptists—Clergy—Biography. 3. Clergy—England—Biography. I. Title.
BX6495.S7M77 78-379026
78 MARC

Pak, Chong-gu.
(Sŭp'ŏjyŏn moksa)
스퍼전목사 = The life of C. H. Spurgeon : 그의생애 / 박종구편. — Seoul : Shin Mang Ae Press : 총판 보문서점, 1977.
331 p. : ill. ; 20 cm.
₩1700
1. Spurgeon, Charles Haddon, 1834-1892. 2. Baptists—Clergy—Biography. 3. Clergy—England—Biography. I. Title. II. Title: The life of C. H. Spurgeon.
BX6495.S7P34 78-826386

Pudney, John, 1909-1977.
John Wesley and his world / John Pudney. — New York : Scribner, c1978.
128 p. : ill. ; 24 cm.
Bibliography: p. 119.
Includes index.
ISBN 0-684-15922-8 : $10.95
1. Wesley, John, 1703-1791. 2. Methodist Church—Clergy—Biography. 3. Clergy—England—Biography. I. Title.
BX8495.W5P77 1978 287'.092'4 78-59110
00 MARC

—ENGLAND—CORRESPONDENCE

Lonsdale, John, 1737-1807.
Lonsdale documents / edited by Elizabeth Playne and G. de Boer. — ⌐Durham, Eng. : Surtees Society¬, 1976.
158 p. : ill. ; 23 cm. — (Publications of the Surtees Society ; v. 188)
GB***
Includes indexes.
1. Sunk Island, Eng.—History—Sources. 2. Lonsdale, John, 1737-1807. 3. Clergy—England—Correspondence. I. Title. II. Series: Surtees Society, Durham, Eng. Publications ; v. 188.
DA670.S94L66 ~~1976~~ 942.8'15 79-306841
79 MARC

—ENGLAND—GLOUCESTERSHIRE—BIOGRAPHY

Witts, Francis Edward, 1783-1854.
The diary of a Cotswold parson / ⌐by Francis Edward Witts¬ ; chosen, edited, and introduced by David Verey. — Dursley : Alan Sutton ; Wakefield : distributed by EP Publishing, 1978.
⌐2¬, 189 p. : ill., facsim., ports. ; 23 cm. GB78-32628
Includes bibliographical references and index.
ISBN 0-904387-19-4 : £6.95
1. Witts, Francis Edward, 1783-1854. 2. Church of England—Clergy—Biography. 3. Clergy—England—Gloucestershire—Biography. 4. Gloucestershire, Eng.—Biography. I. Verey, David, 1913- II. Title.
BX5199.W733A33 283'.092'4 79-305811
79 MARC

—ENGLAND—LEICESTERSHIRE

Pruett, John H 1947-
The parish clergy under the later Stuarts : the Leicestershire experience / John H. Pruett. — Urbana : University of Illinois Press, c1978.
203 p. : ill. ; 24 cm.
Bibliography: p. ⌐181¬-189.
Includes index.
ISBN 0-252-00662-3 : $8.95
1. Clergy—England—Leicestershire. 2. Leicestershire, Eng.—History. I. Title.
BR763.L4P78 301.44'44 78-8174
78 MARC

—FINLAND

Väänänen, Kyösti.
Pappissivistys vanhassa Suomessa / Kyösti Väisänen. — Helsinki : Suomen kirkkohistoriallinen seura, 1975.
408 p. ; 21 cm. — (Suomen Kirkkohistoriallisen Seuran toimituksia; 96)
Fi
Thesis—Helsinki.
Added t.p. with thesis statement inserted.
Summary in German.
Bibliography: p. 365-389.
Includes index.
ISBN 9519021132
1. Theology—Study and teaching—Finland. 2. Clergy—Finland. I. Title. II. Series: Suomen Kirkkohistoriallinen Seura. Toimituksia ; 96.
BV4140.F5V33 76-523034
79⌐r79¬rev MARC

—FINLAND—BIOGRAPHY

Faðir minn—presturinn / Hersteinn Pálsson bjó til prentunar. — ⌐Hafnarfjörður¬ : Skuggsjá, 1977.
228 p., ⌐8¬ leaves of plates : ports. ; 22 cm. Ic***
1. Suomen Evankelisluterilainen Kirkko—Clergy—Biography. 2. Clergy—Finland—Biography. I. Hersteinn Pálsson, 1916-
BX8079.F33 78-378366
78 MARC

Haavio, Jaakko.
Liperit ja lyyra / Jaakko Haavio. — ⌐Helsinki¬ : Kirjapaja, 1975.
238 p., ⌐8¬ leaves of plates : ill. ; 21 cm. Fi
Autobiographical.
ISBN 9516211445 : Fmk34.00
1. Haavio, Jaakko—Biography. 2. Authors, Finnish—20th century—Biography. 3. Lutheran Church—Clergy—Biography. 4. Clergy—Finland—Biography. I. Title.
PH355.H2337Z522 79-370902
*79 MARC

CLERGY
—FINLAND—BIOGRAPHY (Continued)

Haavio, Jaakko.
Vierähtävät vuodet / Jaakko Haavio. — [Helsinki] : Kirjapaja, 1978-
v : ill. ; 22 cm. Fi***
Autobiographical.
CONTENTS: 1. 1945-1959.

ISBN 9516212042
1. Haavio, Jaakko—Biography. 2. Authors, Finnish—20th century—Biography. 3. Lutheran Church—Clergy—Biography. 4. Clergy—Finland—Biography. I. Title.
PH355.H2337Z528 78-396157
79 MARC

Nousiainen, Tapio.
Kurikan honka : Juho Koivumäki lähikuvassa / Tapio Nousiainen. — Hämeenlinna : Päivä, 1978 (Saarijärvi : Ari-p.)
156, [1] p. : ill. ; 18 cm. Fi78-6
ISBN 9516221912
1. Koivumäki, Juho, 1899- . 2. Clergy—Finland—Biography. I. Title.
BR1725.K64N68 79-349741
79 MARC

Saarisalo, Aapeli, 1896-
Elämäni mosaiikkia : Rymättylän räätälin poika kotona ja kaukana / Aapeli Saarisalo. — Porvoo : W. Söderström, 1977.
270 p., [16] leaves of plates : ill. ; 22 cm. Fi77
Continues the author's Rymättylän räätälin poika.
ISBN 9510079839 : Fmk46.00
1. Saarisalo, Aapeli, 1896- . 2. Lutheran Church—Clergy—Biography. 3. Clergy—Finland—Biography. 4. Old Testament scholars—Finland—Biography. I. Title.
BX8080.S116A32 79-361095
79 MARC

Saarisalo, Aapeli, 1896-
Pietari Kurvinen, kaukokatseinen karjalainen / Aapeli Saarisalo. — Porvoo ; Helsinki ; Juva : Söderström, 1978 (Porvoo : WSOY:n graafiset laitokset)
205, [3] p., [8] leaves of plates : ill., maps ; 21 cm. Fi79-1
ISBN 9510087955 : Fmk49.00
1. Kurvinen, Pietari, 1842-1925. 2. Lutheran Church—Clergy—Biography. 3. Clergy—Finland—Biography. I. Title.
BX8080.K84S2 79-371856
79 MARC

Söderlund, Bertil.
Ombud för evigheten : Martin Vinbergs liv och verksamhet / Bertil Söderlund. — [Helsingfors] : Fria missionförbundets förlag, 1972.
133 p. : ill. ; 22 cm. Fi
1. Vinberg, Martin, 1874-1951. 2. Clergy—Finland—Biography. I. Title.
BR1725.V54S65 78-385569
*79 MARC

Stenwall, Edvin, 1899-1976.
Frid på jorden : ur minnets gömmor / Edvin Stenwall. — Närpes : Närpesnejdens fredsförening, 1976 (Vasa : Fram)
228 p., [8] leaves of plates : ill. ; 21 cm. Fi76-10
1. Stenwall, Edvin, 1899-1976. 2. Lutheran Church—Clergy—Biography. 3. Clergy—Finland—Biography. I. Title.
BX8080.S7372A35 79-344686
*79 MARC

—FINLAND—LAPIN LÄÄNI—
BIOGRAPHY

Itkonen, Tuomo.
Nuori Jaakko Fellman ja hänen ystävänsä Carl Gustaf Mannerheim / Tuomo Itkonen. — [Rovaniemi : Väinö Pekkala : Suomen kulttuurirahaston Lapin rahasto], 1977.
299 p. : ill. ; 26 cm. Fi77
Bibliography: p. 299.
1. Fellman, Jakob Esaiaksenpoika, 1795-1875. 2. Mannerheim, Carl Gustav, greve, 1797-1854. 3. Lutheran Church—Clergy—Biography. 4. Clergy—Finland—Lapin lääni—Biography. 5. Lapin lääni, Finland—Biography. I. Title.
BX8080.F39I84 79-351753
79 MARC

—FLORIDA—ST. PETERSBURG—
BIOGRAPHY

Davis, Enoch Douglas.
On the Bethel trail / by Enoch Douglas Davis. — St. Petersburg, Fla. : Valkyrie Press, c1979.
139 p. : ill. ; 22 cm.
Includes bibliographical references and index.
1. Davis, Enoch Douglas. 2. Baptists—Clergy—Biography. 3. Clergy—Florida—St. Petersburg—Biography. 4. St. Petersburg, Fla.—Biography. 5. Afro-Americans—Civil rights—Florida—St. Petersburg—History. 6. St. Petersburg, Fla.—Race relations. I. Title.
BX6455.D28A34 286'.1'0924 79-63135
79 MARC

—FRANCE

Potel, Julien.
Demain d'autres prêtres? : leur place et leurs rôles / Julien Potel. — Paris : le Centurion, 1977.
124 p. ; 21 cm. F77-12061
Includes bibliographical references.
ISBN 2-227-31527-X : 29.00F
1. Catholic Church—Clergy. 2. Clergy—France. I. Title.
BX1913.P65 78-397133
79 MARC

—FRANCE—BIOGRAPHY

Ball, Joseph.
L'Abbé Flory : 1886-1949 : documents et témoignages / recueillis par Joseph Ball ; images de Jean Garneret. — Lantenne-Vertière (25170 Recologne) : J. Garneret, 1978.
337 p., [4] p. of plates : ill. ; 24 cm. F79-4749
Includes bibliographical references.
60.00F
1. Flory, Jean, 1886-1949. 2. Catholic Church—Clergy—Biography. 3. Clergy—France—Biography. I. Title.
BX4705.F613B34 282'.092'4 79-377607
79 MARC

Bruckberger, Raymond Léopold, 1907-
Tu finiras sur l'échafaud : mémoires / R.-L. Bruckberger. — Paris : Flammarion, c1978.
436 p. ; 24 cm. F***
ISBN 2-08-064124-7 : 60.00F
1. Bruckberger, Raymond Léopold, 1907- . 2. Catholic Church—Clergy—Biography. 3. Clergy—France—Biography. I. Title.
BX4705.B84655A37 271'.2'024 79-359972
79 MARC

Bruniquel, Paul.
"Comme un veilleur attend l'aurore" : anecdotes et souvenirs pour un cinquantenaire, 1927-1977 / Paul Bruniquel, ... — Montpellier (Chapelle du refuge, 48, rue Lakanal, 34000) : P. Bruniquel, 1977.
132 p. : ill. ; 20 cm. F78-2564
1. Bruniquel, Paul. 2. Catholic Church—Clergy—Biography. 3. Clergy—France—Biography. I. Title.
BX4705.B878A33 282'.092'4 79-350537
79 MARC

Cardonnel, Jean.
Je ne serai jamais sous la terre / Jean Cardonnel. — Paris : R. Laffont, c1976.
230 p. ; 22 cm. — (La Vie selon ..) F***
32.00F
1. Cardonnel, Jean. 2. Catholic Church—Clergy—Biography. 3. Clergy—France—Biography. I. Title.
BX4705.C305A33 271'.2'024 78-399705
79 MARC

Dutot, Louis.
Le pèlerinage Notre-Dame de Montligeon : son fondateur, son histoire / Louis Dutot. — [s.l. : s.n., 1977] (La Chapelle-Montligeon : Impr. de Montligeon)
161 p., [6] leaves of plates : ill. ; 23 cm. F***
Includes bibliographical references.
1. Buguet, Paul Joseph, 1843-1918. 2. Catholic Church—Clergy—Biography. 3. Clergy—France—Biography. 4. Basilique Notre-Dame, La Chapelle-Montligeon, France—History. 5. Christian pilgrims and pilgrimages—France—Montligeon. I. Title.
BX4705.B8947D87 282'.092'4 79-350514
79 MARC

Galland, Daniel.
L'espérance maintenue : chronique d'un homme du commun / Daniel Galland ; préf. du pasteur Albert Nicolas. — Paris : Le Centurion, c1979.
231 p. ; 22 cm. F***
Includes bibliographical references.
ISBN 2-227-34602-7 : 42.00F
1. Galland, Daniel. 2. Reformed Church—Clergy—Biography. 3. Clergy—France—Biography. I. Title.
BX9459.G28A33 284'.2'0924 79-375182
79 MARC

Gascoin, Pierre, 1919-
Prêtre dans la tourmente / Pierre Gascoin. — Paris : Nouvelles éditions latines, 1977.
94 p. ; 19 cm. F***
Includes bibliographical references.
ISBN 2-7233-0015-3 : 15.00F
1. Gascoin, Pierre, 1919- . 2. Catholic Church—Clergy—Biography. 3. Clergy—France—Biography. I. Title.
BX4705.G256A35 282'.092'4 79-377798
79 MARC

Laval, Jacques, 1911-
Un homme partagé / Jacques Laval. — Paris : Julliard, c1978.
309 p. ; 23 cm. F***
ISBN 2-260-00127-0 : 42.00F
1. Laval, Jacques, 1911- —Biography. 2. Authors, French—20th century—Biography. 3. Catholic Church—Clergy—Biography. 4. Clergy—France—Biography. I. Title.
PQ2672.A87Z47 79-361689
79 MARC

Lochet, Louis.
Espérance de vie / Louis Lochet. — Paris : Éditions ouvrières, c1977.
180 p. ; 18 cm. — (Collection A pleine vie) F***
25.00F
1. Lochet, Louis. 2. Catholic Church—Clergy—Biography. 3. Clergy—France—Biography. I. Title.
BX4705.L678A33 282'.092'4 79-375330
79 MARC

Matray, André.
Mes visages de Dieu / André Matray ; préf. par A.-M. Besnard. — Paris : Éditions du Cerf, 1977.
110 p. ; 20 cm. — (L'Évangile au XXe [i.e. vingtième] siècle) F***
ISBN 2-204-01108-8 : 20.00F
1. Matray, André. 2. Catholic Church—Clergy—Biography. 3. Clergy—France—Biography. I. Title.
BX4705.M4277A35 282'.092'4 79-375830
79 MARC

Oliveira, José Fernandes de.
Por causa de um certo reino : história de Leão João Dehon e sua incrível paz inquieta / José Fernandes de Oliveira. — São Paulo : Edições Paulinas, 1978.
189 p., [11] leaves of plates : ill. ; 18 cm. — (Coleção Perspectivas ; 5)
"Em convênio com os Padres do Sagrado Coração de Jesus, Provincia BM."
Cr$40.00
1. Dehon, Léon Gustave, 1843-1925. 2. Catholic Church—Clergy—Biography. 3. Clergy—France—Biography. 4. Congregation of the Priests of the Sacred Heart—History. I. Title.
BX4705.D2947O43 79-341853
79 MARC

Oziol, Lucien, 1920-
Les enfants de saint Nicolas / Lucien Oziol ; récit recueilli par Anne Rouanet. — Paris : R. Laffont, c1976.
324 p. ; 25 cm. — (Collection Vécu) F***
Autobiographical.
42.00F
1. Oziol, Lucien, 1920- . 2. Catholic Church—Clergy—Biography. 3. Clergy—France—Biography. I. Rouanet, Anne. II. Title.
BX4705.O84A34 282'.092'4 78-387117
79 MARC

Popot, Jean, 1905-
Pour une brebis perdue / Jean Popot, avec la collaboration de Magdeleine Verrier. — [Paris] : Plon, c1976.
256 p. ; 21 cm. F***
ISBN 2-259-00092-4 : 35.00F
1. Popot, Jean, 1905- . 2. Catholic Church—Clergy—Biography. 3. Clergy—France—Biography. I. Verrier, Magdeleine, joint author. II. Title.
BX4705.P6574A32 261.8'3 78-388220
79 MARC

Regue, Raymonde.
Père Marie-Eugène de l'Enfant-Jésus, maître spirituel pour notre temps / Raymonde Regue. — Venasque : Éditions du Carmel, c1978.
125 p. : ill. ; 24 cm. F***
35.00F
1. Marie Eugène de l'Enfant-Jésus, Father. 2. Catholic Church—Clergy—Biography. 3. Clergy—France—Biography. I. Title.
BX4705.M3855R43 271'.73'024 78-387116
79 MARC

Rollet, Henri.
Jean Viollet, homme de l'avenir / Henri Rollet ; préf. de Jean Lacroix. — Paris : Beauchesne, c1978.
144 p., [1] leaf of plates : port. ; 19 cm. — (Figures d'hier et d'aujourd'hui) F***
Bibliography: p. [141]-142.
36.00F
1. Viollet, Jean. 2. Catholic Church—Clergy—Biography. 3. Clergy—France—Biography. I. Title.
BX4705.V63R64 282'.092'4 78-384868
78 MARC

Schmidt, Élisabeth.
Quand Dieu appelle des femmes : le combat d'une femme pasteur / Élisabeth Schmidt ; préf. d'André Dumas. — Paris : Cerf, 1978.
181 p. ; 20 cm. — (L'Évangile au XXe [i.e. vingtieme] siècle) F***
Includes bibliographical references.
ISBN 2-204-01303-X : 38.00F
1. Schmidt, Élisabeth. 2. Reformed Church—Clergy—Biography. 3. Clergy—France—Biography. I. Title.
BX9459.S35A36 284'.2'0924 79-359970
79 MARC

Six, Jean François.
Le courage de l'espérance : les dix ans qui ont suivi le Concile, 1965-1975 / Jean-François Six. — Paris : Éditions du Seuil, c1978.
330 p. ; 21 cm. F***
Includes bibliographical references and index.
ISBN 2-02-004878-7 : 55.00F
1. Six, Jean François. 2. Catholic Church—Clergy—Biography. 3. Clergy—France—Biography. I. Title.
BX4705.S658A34 282'.09'047 78-388844
79 MARC

CLERGY

—FRANCE—BIOGRAPHY (Continued)

Tavard, Georges Henri, 1922-
L'expérience de Jean-Martin Moye : mystique et mission, 1730-1793 / Georges Tavard. — Paris : Beauchesne, c1978.
230 p. ; 18 cm. — (Bibliothèque de spiritualité ; 12) F***
Bibliography: p. 14-15.
66.00F
1. Moÿe, Jean Martin, 1730-1793. 2. Catholic Church—Clergy—Biography. 3. Clergy—France—Biography. 4. Missionaries—China—Biography. 5. Missionaries—France—Biography. I. Title.
BX4705.M74T38 282'.092'4 78-384783
78 MARC

Toussaint, Joseph, chanoine.
Le Chanoine de Gouvets et les martyrs de l'Avranchin / J. Toussaint. — Avranches (20, rue de la Constitution, 50300) : Éditions l'Avranchin, [1977]
176 p. : ill. ; 24 cm. F78-4904
Includes bibliographies.
45.00F
1. Gouvets, Louis Georges de, d. 1796. 2. Catholic Church—Clergy—Biography. 3. Clergy—France—Biography. 4. Christian martyrs—France—Avranches—Biography. 5. Avranches, France—Biography. I. Title.
BX4705.G61752T68 282'.092'2 78-381708
78 MARC

Toussaint, Joseph, chanoine.
Monseigneur Bravard : le sauveur du Mont-Saint-Michel / Joseph Toussaint. — Avranches (20, rue de la Constitution, 50300) : Éditions de l'Avranchin, 1978.
332 p., [9] leaves of col. plates : graphs ; 23 cm. F79-4800
Bibliography: p. [325]-327.
70.00F
1. Bravard, Jean Pierre, 1811-1876. 2. Catholic Church—Clergy—Biography. 3. Clergy—France—Biography. 4. Mont St. Michel, France—History. I. Title.
BX4705.B824T68 282'.092'4 79-376009
79 MARC

—FRANCE—BRITTANY—BIOGRAPHY

Erlannig.
Pierre Nourry : recteur de Bignan, curé de la Cathédrale de Vannes et écrivain de langue bretonne : 1743-1804 : thèse de littérature celtique / Erlannig. — Saint-Brieuc : Presses bretonnes, 1978-
v. : ill. ; 24 cm. F***
At head of title: Au diocèse de Vannes, un prêtre de Lauzach.
Bibliography: v. 1, p. [23]-50.
Includes index.
CONTENTS: t. 1. Prolégomènes et biographie générale de Pierre Nourry.
1. Nourry, Pierre, 1743-1804. 2. Authors, Breton—18th century—Biography. 3. Catholic Church—Clergy—Biography. 4. Clergy—France—Brittany—Biography. 5. Brittany—Biography. I. Title. II. Title: Au diocèse de Vannes, un prêtre de Lauzach.
PB2903.N68Z64 79-351132
79 MARC

—FRANCE—CLICHY—CORRESPONDENCE

Claudel, Paul, 1868-1955.
Chroniques du "Journal de Clichy" / Claudel et Mauriac. Correspondance Claudel-Fontaine / textes établis et annotés par François Morlot et Jean Touzot. — Paris : les Belles lettres, 1978.
158 p. ; 24 cm. — (Annales littéraires de l'Université de Besançon ; 206) ([Publication] - Centre de recherches de littérature française (XIXe et XXe siècles) ; v. 21) F78-12911
Includes bibliographical references.
80.00F
1. Claudel, Paul, 1868-1955—Correspondence. 2. Fontaine, Daniel, 1862-1920. 3. Authors, French—20th century—Political and social views. 4. Clichy, France—Social conditions. 5. Authors, French—20th century—Correspondence. 6. Clergy—France—Clichy—Correspondence. 7. Catholic Church—Clergy—Correspondence. I. Mauriac, François, 1885-1970. II. Fontaine, Daniel, 1862-1920. III. Title. IV. Series: Besançon, France. Université. Annales littéraires ; 206. V. Series: Centre de recherches de littérature française (XIXe et XXe siècles). Publication ; v. 21.
AS161.B39 no. 206 848'.9'1209 79-350086
[PQ2605.L2] 79 MARC

—FRANCE—PARIS—CORRESPONDENCE

La Religion est perdue à Paris : lettres d'un vicaire parisien à son archevêque en date de 1849, suivies d'un mémoire adressé au même / présentation par Yvan Daniel. — Paris : Éditions Cana, c1978.
159 p. : ill. ; 23 cm. — (Collection Foi et histoire) F***
Bibliography: p. 153-155.
ISBN 2-86335-003-X : 40.00F
1. Daniel, Yvan, 1909- . 2. Catholic Church—Clergy—Correspondence. 3. Clergy—France—Paris—Correspondence. 4. Paris—Biography. 5. Paris—Social conditions. I. Daniel, Yvan, 1909- II. Series.
BX4705.D254R44 282'.092'4 79-352625
79 MARC

—FRANCE—REIMS—BIOGRAPHY

Jacobsen, Peter Christian.
Flodoard von Reims : sein Leben und seine Dichtung 'De triumphis Christi' / von Peter Christian Jacobsen. — Leiden : Brill, 1978.
xii, 298 p. ; 25 cm. — (Mittellateinische Studien und Texte ; Bd. 10) Ne78-22
Bibliography: p. [282]-293.
Includes index.
ISBN 9004054073 : fl 96.00
1. Flodoard, of Reims, 894-966. 2. Catholic Church—Clergy—Biography. 3. Clergy—France—Reims—Biography. 4. Reims—Biography. 5. Flodoard, of Reims, 894-966. De triumphis Christi.
BX4705.F6096J32 78-375479
*78 MARC

—GERMANY—BIOGRAPHY

Fuchs, Herbert.
Nur einer bringt zum Ziel : weitere Tatsachenberichte / von Herbert Fuchs. — Metzingen : Brunnquell-Verlag der Bibel und Missionsstiftung, c1977, 1978 printing.
192 p. ; 20 cm. GFR***
Companion vol. to the author's Nur einer kann helfen.
Bibliography of the author's works: p. 191-192.
ISBN 3-7656-0203-5 (linen)
1. Fuchs, Herbert. 2. Lutheran Church—Clergy—Biography. 3. Clergy—Germany—Biography. I. Title.
BX8080.F787A36 79-362993
79 MARC

Klens, Hermann, 1880-1972.
Hermann Klens : Anwalt d. Frauen : Leben u. Werk / [Ingeborg Rocholl-Gärtner hat d. Material d. Klens-Memoiren ausgew., bearb. sowie mit e. Biographie u. Interpretationen vers.]. — 1. Aufl. — Düsseldorf : Klens-Verlag, 1978.
199 p. ; 25 cm. GFR78-A39
Errata slip inserted.
ISBN 3-87309-035-X : DM24.60
1. Klens, Hermann, 1880-1972. 2. Catholic Church—Clergy—Biography. 3. Clergy—Germany—Biography. 4. Church work with women—Germany. I. Rocholl-Gärtner, Ingeborg. II. Title: Anwalt der Frauen. .
BX4705.K536A34 78-396982
79 MARC

Künneth, Walter, 1901-
Lebensführungen : d. Wahrheit verpflichtet / Walter Künneth. — Wuppertal : Brockhaus, 1979.
298 p. : 1 ill. ; 22 cm. — (ABC team) GFR79-A
Bibliography of the author's works: p. 275-292.
Includes index.
ISBN 3-417-12620-7 : DM28.00
1. Künneth, Walter, 1901- . 2. Lutheran Church—Clergy—Biography. 3. Clergy—Germany—Biography. I. Title.
BX8080.K79A34 79-362668
79 MARC

Munz, Alfred.
Philipp Matthäus Hahn : Pfarrer, Erfinder u. Erbauer von Himmelsmaschinen, Waagen, Uhren u. Rechenmaschinen / Alfred Munz. — 1. Aufl. — Sigmaringen : Thorbecke, 1977.
144 p. : ill. (some col.) ; 24 cm. GFR78-A
Bibliography of P. M. Hahn's works: p. 104.
Includes bibliographical references.
ISBN 3-7995-4025-3
1. Hahn, Philipp Matthäus, 1739-1790. 2. Clock and watchmakers—Germany—Biography. 3. Clergy—Germany—Biography. 4. Astronomical clocks.
TS544.8.H33M36 79-359760
79 MARC

—GERMANY—BIOGRAPHY—JUVENILE LITERATURE

Faccenda, Luigi M
Ascolta, ti parla padre Kolbe / [di] Luigi Faccenda. — Borgonuovo di Pontecchio Marconi [San Marconi] : Edizioni dell'Immacolata, 1977.
101 p. : ill. ; 17 cm. It78-Oct
Includes bibliographical references.
L3000
1. Kolbe, Maximilian, 1894-1941—Juvenile literature. 2. Catholic Church—Clergy—Biography—Juvenile literature. 3. Clergy—Germany—Biography—Juvenile literature. I. Title.
BX4705.K625F3 79-359274
*79 MARC

—GERMANY, EAST—BIOGRAPHY

Deszyk, Gerhard, 1899-
Ludwig Kirsch / Gerhard Deszyk. — 1. Aufl. — Berlin : Union Verlag, c1977.
26 p. : ill., ports. ; 18 cm. — (Reihe Christ in der Welt ; Heft 42) GDR***
Includes bibliographical references.
2.20M
1. Kirsch, Ludwig, 1891-1950. 2. Catholic Church—Clergy—Biography. 3. Clergy—Germany, East—Biography.
BX4705.K519D47 78-350446
78 MARC

—GERMANY, WEST—BIOGRAPHY

Brandenburg, Hans, 1895-
Mit Christus unterwegs : 60 Jahre im Dienst für Christus / Hans Brandenburg. — Giessen ; Basel : Brunnen Verlag, 1977.
79 p. ; 21 cm. — (ABC Team ; 102 : B, Berichte, Erzählungen, Lebensbilder) GFR77-A24
ISBN 3-7655-0415-7 : DM7.80
1. Brandenburg, Hans, 1895- . 2. Moravian Church—Clergy—Biography. 3. Clergy—Germany, West—Biography. I. Title.
BX8593.B7A35 79-363291
79 MARC

Davidson, Clarissa Start.
God's man : the story of Pastor Niemoeller / by Clarissa Start Davidson. — Westport, Conn. : Greenwood Press, 1979, c1959.
x, 242 p., [1] leaf of plates : port. ; 22 cm.
Reprint of the ed. published by I. Washburn, New York.
Bibliography: p. 241-242.
ISBN 0-313-21065-9
1. Niemoller, Martin, 1892- . 2. Lutheran Church—Clergy—Biography. 3. Clergy—Germany, West—Biography. I. Title.
[BX8080.N48D3 1979] 284'.1'0924 78-10131
78 MARC

Feddersen, Christian, 1786-1874.
Materialien zur Geschichte der Familie Feddersen aus Wester-Schnatebüll in Nordfriesland : Christian Feddersen (1786-1874), Pastor und Schriftsteller. — [s.l. : s.n., 1978?] (Kleiseerkoog : B. H. Feddersen)
150 p. : port. ; 29 cm. GFR***
Includes text by F. Feddersen; compiled and edited by B. H. Feddersen and F. Feddersen.
Limited ed. of 120 copies.
Includes bibliographical references.
DM30.00
1. Feddersen, Christian, 1786-1874. 2. Lutheran Church—Clergy—Biography. 3. Clergy—Germany, West—Biography. I. Feddersen, Friedrich. II. Feddersen, Berend Harke. III. Title.
BX8080.F38A35 79-381731
79 MARC

—GERMANY, WEST—POLITICAL ACTIVITY

Kirchliches Amt und politisches Mandat : Dokumentation zum Bremer Verfassungsrechtsstreit : Entscheidungen u. Texte / zsgest. u. eingel. von Detlef Dahrmann. — Hannover : Verlag des Amtsblattes d. Evang. Kirche in Deutschland, 1977.
227 p. ; 21 cm. GFR77-A37
Contains documents pertaining to the court case of the Bremische Bürgerschaft versus the Bremische Evangelische Kirche, in the Staatsgerichtshof der Freien Hansestadt Bremen and appealed in the Bundesverfassungsgericht, 1973-1976.
"Die Veröffentlichung erfolgt auf Veranlassung der Kirchenkanzlei der Evangelischen Kirche in Deutschland."
Includes bibliographical references.
ISBN 3-87843-007-8 : DM9.50
1. Bremen. Bürgerschaft. 2. Bremische Evangelische Kirche. 3. Incomactivity. I. Dahrmann, Detlef, 1937- . II. Evangelische Kirche in Deutschland. Kirchenkanzlei.
78-401069
79 MARC

—GERMANY, WEST—BAVARIA—BIOGRAPHY

Roser, Hans, 1931-
Im Dorf daheim / Hans Roser ; [Ill. vom Verf.]. — 1. Aufl. — Rothenburg ob d. Tauber : Peter, 1978.
252 p. : ill. ; 20 cm. GFR79-A
DM22.00
1. Roser, Hans, 1931- . 2. Lutheran Church—Clergy—Biography. 3. Clergy—Germany, West—Bavaria—Biography. 4. Bavaria—Biography. I. Title.
BX8080.R7A34 79-353768
79 MARC

—GERMANY, WEST—FRANCONIA—BIOGRAPHY

Barthels, Karl Josef, 1907-
Der Schatz im Acker : aus dem Leben eines würzburgischen Landpfarrers / Karl Josef Barthels. — [s.l.] : Barthels, 1976 (Würzburg : Fränkische Gesellschaftsdruckerei)
406 p., [8] leaves of plates : ill. ; 23 cm. GFR***
1. Barthels, Karl Josef, 1907- . 2. Catholic Church—Clergy—Biography. 3. Clergy—Germany, West—Franconia—Biography. I. Title.
BX4705.B234A37 78-375468
78 MARC

CLERGY (Continued)

—GERMANY, WEST— LINGEN —BIOGRAPHY

Tenfelde, Walter, Ing. f. Verm.
Die Prediger der reformierten Gemeinde der Stadt Lingen <Ems>. Lingen <Ems> (Bürgersohne-Aufzug "Die Kivelinge" [Auslieferung: R. van Acken]) 1968.
164 p. with illus. 24 cm. (Die Lingener Heimat, Heft 11) 7.50
GDNB68-A27-54
Includes bibliographies.
1. Reformed Church in Lingen—Clergy—Biography. 2. Clergy—Germany, West—Lingen—Biography. 3. Lingen. Reformierte Gemeinde. 4. Lingen—Biography. 5. Lingen—Church history. I. Title. II. Series: Bürgersöhne-Aufzug "Die Kivelinge." Schriftenreihe "Die Lingener Heimat," Heft 11.
BX9469.A1T4 76-534492
79[r79]rev MARC

—GERMANY, WEST—MEMMINGEN—BIOGRAPHY

Erhard, Hermann.
Memminger Pfarrerbuch / Hermann Erhard. — Neustadt a. d. Aisch : In Kommission bei Verlag Degener, 1977.
85 p., [4] leaves of plates : ill., geneal. tables ; 25 cm. — (Einzelarbeiten aus der Kirchengeschichte Bayerns ; Bd. 55) GFR***
Bibliography: p. [9]-12.
Includes indexes.
ISBN 3-7686-4079-5
1. Lutheran Church—Clergy—Biography. 2. Clergy—Germany, West—Memmingen—Biography. 3. Memmingen, Ger.—Biography. I. Title. II. Series.
BX8079.E73 78-382243
79 MARC

—GERMANY, WEST—WÜRTTEMBERG—BIOGRAPHY

Hahn, Philipp Matthäus, 1739-1790.
Die Kornwestheimer Tagebücher / Philipp Matthäus Hahn ; hrsg. von Martin Brecht u. Rudolf F. Paulus. — Berlin ; New York : de Gruyter, 1979-
v. ; 25 cm. — (Texte zur Geschichte des Pietismus : Abt. 8, Einzelgestalten und Sondergruppen ; Bd. 1) GFR79-A
Bibliography: v. 1, p. 40-41.
Includes indexes.
CONTENTS: [Bd. 1]. 1772-1777.

ISBN 3-11-007115-0 (v. 1) : DM175.00 (v. 1)
1. Hahn, Philipp Matthäus, 1739-1790. 2. Lutheran Church—Clergy—Biography. 3. Clergy—Germany, West—Württemberg—Biography. 4. Württemberg—Biography. I. Brecht, Martin. II. Paulus, Rudolf F. III. Title. IV. Series.
BX8080.H19A34 79-353972
79 MARC

Jäckh, Werner.
Blumhardt : Vater u. Sohn u. ihre Welt / Werner Jäckh. — Stuttgart : Steinkopf, 1977.
207 p. : ill. ; 19 cm. — (Zeugnisse und Bilder) GFR78-A
Includes bibliographical references and index.
ISBN 3-7984-0323-6 : DM14.80
1. Blumhardt, Johann Christoph, 1805-1880. 2. Blumhardt, Christoph, 1842-1919. 3. Evangelische kirche von Württemberg—Clergy—Biography. 4. Clergy—Germany, West—Württemberg—Biography. 5. Württemberg—Biography. I. Title.
BX8080.B615J33 78-371813
78 MARC

—GREAT BRITAIN—BIOGRAPHY—ADDRESSES, ESSAYS, LECTURES

The View from the pulpit : Victorian ministers and society / edited and with an introd. by P. T. Phillips. — Toronto : Macmillan of Canada, c1978.
326 p. : ports. ; 24 cm. C***
Bibliography: p. [299]-317.
Includes index.
ISBN 0-7705-1590-8
1. Church of England—Clergy—Biography—Addresses, essays, lectures. 2. Dissenters, Religious—England—Biography—Addresses, essays, lectures. 3. Catholic Church—Clergy—Biography—Addresses, essays, lectures. 4. Clergy—Great Britain—Biography—Addresses, essays, lectures. I. Phillips, Paul T., 1942-
BX5619.V53 280'.092'2 78-321384
79 MARC

—GREECE—BIOGRAPHY

Post-Byzantine ecclesiastical personalities : studies of ... / edited by Nomikos Michael Vaporis. — Brookline, Mass. : Holy Cross Orthodox Press, 1978.
110 p. ; 22 cm. — (The Byzantine Fellowship lectures ; no. 3)
Includes bibliographical references.
CONTENTS: Vaporis, N. M. Introduction.—Bebis, G. S. St. Nikodemos the Hagiorite.—Cavarnos, C. St. Nikephoros of Chios.—Vaporis, N. M. Neophytos Vamvas and the translation of the Scriptures in Greece.—Patsavos, L. J. Konstantinos Oikonomos of the Oikonomoi.—Chronis, T. Eusebios Matthopoulos, a contemporary prophet.
ISBN 0-916586-30-8
1. Orthodox Eastern Church, Greek—Clergy—Biography. 2. Clergy—Greece—Biography. I. Vaporis, Nomikos Michael. II. Series. III. Series: Byzantine Fellowship. The Byzantine Fellowship lectures ; no. 3.
BX619.A1P67 281.9'3 78-11037
78 MARC

—GREECE, MODERN—BIOGRAPHY

Papadopoulos, Iōannēs G
(Hosios Eugenios Giannoulēs ho Aitōlos)
Ὅσιος Εὐγένιος Γιαννούλης ὁ Αἰτωλός, 1597-1682 / Ἰωάννου Γ. Παπαδοπούλου. — [Ἀθῆναι : Παπαδόπουλος], 1976.
88 p. ; 21 cm.
1. Giannoulēs, Eugenios, 1597-1682. 2. Orthodox Eastern Church—Clergy—Biography. 3. Clergy—Greece, Modern—Biography. I. Title.
BX395.G5P36 79-413153

—HAWAII—BIOGRAPHY

Franco, Evaristo.
O homem que viveu no inferno; [biografia do Padre Damião] Lisboa, União Gráfica, 1966.
161 p. 19 cm.
1. Damien, Father, 1840-1889. 2. Catholic Church—Clergy—Biography. 3. Clergy—Hawaii—Biography. I. Title.
BX4705.D25F7 71-219190
79[r79]rev MARC

—HAWAII—BIOGRAPHY—JUVENILE LITERATURE

Christopher, Kenneth.
Damien and the island of sickness : a story about Father Damien / written by Kenneth Christopher ; illustrated by Judith Leo. — Minneapolis : Winston Press, c1979.
[32] p. : ill. (some col.) ; 21 cm. — (Christian heroes)
SUMMARY: A simple version of the life of Father Damien whose work among the lepers of Molokai focused the attention of the rest of the world on the dread disease.
ISBN 0-03-049416-8 : $1.50
1. Damien, Father, 1840-1889—Juvenile literature. 2. Catholic Church—Clergy—Biography—Juvenile literature. 3. Clergy—Hawaii—Biography—Juvenile literature. [1. Damien, Father, 1840-1889. 2. Missionaries] I. Leo, Judith. II. Title. III. Series.
BX4705.D25C43 266'.2'0924 78-65510
00 MARC

—HUNGARY—BIOGRAPHY

Rédey, Pál.
Az elfelejtett ember / írta, Rédey Pál. — Budapest : Magyarországi Evangélikus Egyház Sajtóosztálya, 1978.
109 p. : ill. ; 20 cm. H78
1. Remete, László, 1910-1945. 2. Reformed Church in Hungary—Clergy—Biography. 3. Clergy—Hungary—Biography. I. Title.
BX9449.R45R4 79-350410
*79 MARC

—HUNGARY—BIOGRAPHY—FICTION

Moldova, György, 1934-
Negyven prédikátor : regény / Moldova György. — 3. kiad. — Budapest : Szépirodalmi Könyvkiadó, 1978.
262 p. ; 21 cm. H78
ISBN 9631510654 : 23.50Ft
1. Reformed Church—Clergy—Biography—Fiction. 2. Lutheran Church—Clergy—Biography—Fiction. 3. Clergy—Hungary—Biography—Fiction. I. Title.
PH3281.M64N4 1978 79-366816
*79 MARC

—INDIA—BIOGRAPHY

Ratnam, S
(Dēvara kiṅkararu)
ದೇವರ ಕಿಂಕರರು / ಸಂಗ್ರಹ ಕರ್ತ ಎಸ್. ರತ್ನಂ. — ಬೆಂಗಳೂರು : ಬೆಂಗಳೂರು ಕೈಪತ್ರಿಕೆ ಮತ್ತು ಪುಸ್ತಕಗಳ ಸಂಘ, 1965.
v, 125 p. : ill. ; 19 cm.
In Kannada.
Rs1.80

1. Clergy—India—Biography. I. Title.

BR1155.R37 75-928198

—ITALY—BIOGRAPHY

Beltrame, Guido.
Giovanni Chiericato, oratoriano padovano, ministro e biografo di S. Gregorio Barbarigo : preti e fedeli di S. Gregorio / Guido Beltrame. — Padova : Erredici, 1976.
214 p. : ill. ; 24 cm. It78-Feb
Includes documents in Latin and Italian.
Bibliography: p. 41-45.
Includes indexes.
1. Chiericato, Giovanni, 1633-1717. 2. Catholic Church—Clergy—Biography. 3. Clergy—Italy—Biography. 4. Barbarigo, Gregorio, Saint, Cardinal, 1625-1697. I. Title.
BX4705.C474B44 78-387065
*79 MARC

Borrego, Jesús.
Giovanni Battista Baccino : estudio y edición de su biografía y epistolario / Jesús Borrego. — Roma : LAS, c1977.
435 p., [3] leaves of plates : ill. ; 24 cm. — (Studi e ricerche - Centro studi di storia delle missioni salesiane ; 4) It***
Bibliography: p. [8]-15.
Includes indexes.
1. Baccino, Giovanni Battista, 1843-1875. 2. Catholic Church—Clergy—Biography. 3. Clergy—Italy—Biography. I. Series: Centro studi di storia delle missioni salesiane. Studi e ricerche - Centro studi di storia delle missioni salesiane ; 4.
BX4705.B10193B67 79-360868
79 MARC

Carini Alimandi, Lia.
Luigi Guanella : per le vie del quarto mondo / Lia Carini Alimandi. — [Roma] : Città nuova, 1978.
189 p., [10] leaves of plates : ill. ; 18 cm. It79-Mar
Half title: Opera Don Guanella, Centro studi.
L2500
1. Guanella, Luigi, 1842-1915. 2. Catholic Church—Clergy—Biography. 3. Clergy—Italy—Biography. I. Title.
BX4705.G626C37 79-366163
*79 MARC

Costa, Giovanni, S J
Ma chi è stato Giovanni M. Boccardo? / Giovanni Costa. — Pinerolo : Alzani, 1976.
137 p., [8] leaves of plates : ill. ; 21 cm. It77-Nov
L1500
1. Boccardo, Giovanni Maria, 1848-1913. 2. Catholic Church—Clergy—Biography. 3. Clergy—Italy—Biography. I. Title.
BX4705.B5423C67 79-384246
*79 MARC

D'Amando, Filippo.
Accanto a due papi : p. Luigi Besi passionista, 1862-1923 / Filippo D'Amando ; con prefazione del prof. Paul Droulers. — S[an] Gabriele [dell'Addolorata] ; [Isola del Gran Sasso d'Italia] : Eco, 1975.
525 p., [10] leaves of plates : ill. ; 20 cm. — (Documenti e testimonianze di spiritualità ; 3) It77-Dec
Italian or Latin.
Appendix (p. 367-507) consists of documents.
Includes bibliographical references.
1. Besi, Luigi, 1862-1923. 2. Catholic Church—Clergy—Biography. 3. Clergy—Italy—Biography. I. Title. II. Series.
BX4705.B415D35 78-380589
*78 MARC

Donatelli, Armando.
Giovanni Bernardo Tolomei, padre e maestro di monaci / Armando Donatelli. — Siena : Cantagalli, 1977.
183 p., [8] leaves of plates : ill. ; 19 cm. It78-July
"Traduzione italiana del testo latino delle lettere del B. Bernardo Tolomei": p. 93-181.
Bibliography: p. 183-[184]
1. Tolomei, Bernardo, 1272-1348. 2. Catholic Church—Clergy—Biography. 3. Clergy—Italy—Biography.
BX4705.T674D66 271'.13'024 78-399098
*79 MARC

Falconi, Walter.
Il canonico Angelo Bughetti / Walter Falconi. — Imola : Grafiche Galeati, 1977.
383 p., [10] leaves of plates : ill. ; 22 cm. It78-July
Bibliography: p. 9-10.
1. Bughetti, Angelo, 1877-1935. 2. Catholic Church—Clergy—Biography. 3. Clergy—Italy—Clergy. I. Title.
BX4705.B8943F34 79-364204
*79 MARC

CLERGY

—ITALY—BIOGRAPHY (Continued)

Fallaci, Neera, 1932-
Dalla parte dell'ultimo : vita del prete Lorenzo Milani / Neera Fallaci. — 4. ed. riveduta, corretta, aggiornata. — [Milano] : Milano libri, 1977.
iv, 606 p. : ill. ; 23 cm. It79-Jan
Letters of Lorenzo Milani: p. 539-593.
Includes index.
L5000
1. Milani, Lorenzo. 2. Catholic Church—Clergy—Biography. 3. Clergy—Italy—Biography. I. Milani, Lorenzo. II. Title.
BX4705.M5526F34 1977 79-376680
*79 MARC

Fangareggi, Salvatore, 1933-
Il partigiano Dossetti / Salvatore Fangareggi ; prefazione di Benigno Zaccagnini. — Firenze : Vallecchi, 1978.
x, 112 p. ; 20 cm. — (Documenti e interventi) It78-Oct
Includes bibliographical references.
L2800
1. World War, 1939-1945—Underground movements—Italy—Reggio Emilia (Province)—Biography. 2. Dossetti, Giuseppe. 3. Reggio Emilia (Province)—History. 4. Guerrillas—Italy—Reggio Emilia (Province)—Biography. 5. Politicians—Italy—Biography. 6. Catholic Church—Clergy—Biography. 7. Clergy—Italy—Biography. I. Title.
D802.I82R42 940.53'45'430924 78-400612
*79 MARC

Fappani, Antonio.
Per amore di tutti : profilo e memorie di don Giuseppe Tedeschi / a cura de Antonio Fappani. — Brescia : La scuola : CE.DO.C., 1975-1976.
2 v. : ill. ; 21 cm. — (Per una storia del Movimento cattolico bresciano : Documenti e note) It77-Oct
CONTENTS: quaderno 1. La giovinezza (1883-1918).—quaderno 2. Novant'anni.
L2000 (v. 1) L3400 (v. 2)
1. Tedeschi, Giuseppe, 1883-1973. 2. Catholic Church—Clergy—Biography. 3. Clergy—Italy—Biography. 4. Educators—Italy—Biography. I. Tedeschi, Giuseppe, 1883-1973. II. Title. III. Series.
BX4705.T364F36 78-353351
*78[r79]rev MARC

Gecchele, Gaetano.
Biografia spirituale del servo di Dio don Giovanni Calabria : tesi di laurea / candidato, Gaetano Gecchele ; relatore, André Combes. — Roma : Pontificia universita lateranense, Facoltà di S. Teologia, 1966.
239 p. ; 24 cm. It***
On spine: Don Giovanni Calabria, biografia spirituale.
Includes bibliographies.
1. Calabria, Giovanni, 1873-1954. 2. Catholic Church—Clergy—Biography. 3. Clergy—Italy—Biography. I. Combes, André, 1899- II. Title.
BX4705.C245G4 79-411069
79 MARC

Mazzolari, Primo.
Lettere a un amico / Primo Mazzolari ; [a cura di R. Colla]. — Vicenza : La locusta, 1976.
189 p. ; 18 cm. It78-June
L3000
1. Mazzolari, Primo. 2. Catholic Church—Clergy—Correspondence. 3. Clergy—Italy—Correspondence. 4. Colla, Rienzo. I. Colla, Rienzo. II. Title.
BX4705.M472A4 1976 79-386281
*79 MARC

Morero, Vittorio.
Un profeta tra noi : esperienza e missione di Francesco Pianzola / Vittorio Morero. — Roma : Città nuova, 1979.
308 p., [9] leaves of plates : ill. ; 18 cm. It79-Aug
L3000
1. Pianzola, Francesco, 1881-1943. 2. Catholic Church—Clergy—Biography. 3. Clergy—Italy—Biography. I. Title.
BX4705.P44985M67 79-381698
*79 MARC

Neri, Ermanno, 1940-
... e il prete si fece uomo / Ermanno Neri. — Rimini ; Firenze : Guaraldi, c1974.
219 p. ; 21 cm. — (FC ; 22) It76-Feb
L3800
1. Neri, Ermanno, 1940- . 2. Catholic Church—Clergy—Biography. 3. Clergy—Italy—Biography. 4. Catholic Church—Clergy, Training of. I. Title.
BX4705.N3965A33 79-413793
*79 MARC

Nistri, Silvano.
Vita di don Giulio Facibeni / Silvano Nistri. — Firenze : Libreria editrice fiorentina, 1979.
501 p., [25] leaves of plates : ill. ; 24 cm. It79-Jan
Includes bibliographical references and index.
L9000
1. Facibeni, Giulio, 1884-1958. 2. Catholic Church—Clergy—Biography. 3. Clergy—Italy—Biography. I. Title.
BX4705.F2145N57 79-362278
*79 MARC

Pronzato, Alessandro.
... Non hanno più pane : profilo biografico di padre Annibale Difrancia / Alessandro Pronzato. — Torino : Gribaudi, c1977.
275 p. ; 19 cm. It***
Bibliography: p. 5-6.
L3000
1. Difrancia, Annibale Maria, 1851-1927. 2. Catholic Church—Clergy—Biography. 3. Clergy—Italy—Biography. I. Title.
BX4705.D538P76 78-393670
79 MARC

—ITALY—CORRESPONDENCE

Genocchi, Giovanni, 1860-1926.
Carteggio / Giovanni Genocchi ; a cura di Francesco Turvasi. — Roma : Edizioni di storia e letteratura, 1978-
v. : ill. ; 25 cm. It79-Feb
Italian or French.
Vol. 1 includes index.
CONTENTS: 1. (1877-1900).

L25000 (v. 1)
1. Genocchi, Giovanni, 1860-1926. 2. Catholic Church—Clergy—Correspondence. 3. Clergy—Italy—Correspondence. I. Turvasi, Francesco.
BX4705.G41747A4 1978 79-355097
*79 MARC

—ITALY—BIOGRAPHY

Pagnossin, Giuseppe.
Il calvario di Padre Pio / Giuseppe Pagnossin. — Padova : Pagnossin, [1978]
2 v. : ill. ; 31 cm. It***
Includes indexes.
1. Pio da Pietrelcina, Father. 2. Catholic Church—Clergy—Biography. 3. Clergy—Italy—Biography. 4. Stigmatization. I. Title.
BX4705.P49P27 79-359275
79 MARC

Rosa, Gabriele de.
Luigi Sturzo / Gabriele De Rosa. Con 24 tavole fuori testo. — Torino : Union tipografico-editrice torinese, 1977.
xiv, 515 p., [24] leaves of plates : ill. ; 24 cm. — (La Vita sociale della nuova Italia ; 27) It79-Jan
Bibliography: p. [483]-505.
Includes index.
L18000
1. Sturzo, Luigi, 1871-1959. 2. Italy—Politics and government—20th century. 3. Italy, Southern—Politics and government. 4. Statesmen—Italy—Biography. 5. Catholic Church—Clergy—Biography. 6. Clergy—Italy—Biography.
DG575.S8R66 945.091'092'4 79-348113
*79 MARC

—ITALY—SALARIES, PENSIONS, ETC.

Moriconi, Cherubino.
La previdenza sociale a favore degli ecclesiastici e dei religiosi / Cherubino Moriconi. — [Napoli] : Jovene, 1976.
144 p. ; 24 cm. — (Pubblicazioni della Facoltà di giurisprudenza dell'Università di Camerino ; 11) It78-July
Includes bibliographical references.
L3000
1. Social security—Italy. 2. Clergy—Italy—Salaries, pensions, etc. 3. Social security—Law and legislation—Italy. I. Title. II. Series: Camerino. Università. Facoltà giuridica. Pubblicazioni della Facoltà di giurisprudenza dell'Università di Camerino ; 11.
HD7184.M66 78-396618
*79 MARC

—ITALY—ASSISI—BIOGRAPHY

Ramati, Alexander, 1921-
While the Pope kept silent : Assissi and the Nazi occupation / Alexander Ramati, as told by Padre Rufino Niccacci. — London ; Boston : Allen & Unwin, 1978.
181 p. : map (on lining papers) ; 23 cm. GB***
American ed. published under title: The Assissi underground.
ISBN 0-04-940054-1 : £4.95
1. World War, 1939-1945—Jews—Rescue—Italy—Assisi. 2. Niccacci, Rufino. 3. Catholic Church—Clergy—Biography. 4. Clergy—Italy—Assisi—Biography. 5. Assisi—Biography. I. Niccacci, Rufino. II. Title.
D810.J4R35 1978b 940.53'45'651 78-324084
79 MARC

—ITALY—BARBIANA—BIOGRAPHY

Don Lorenzo Milani : dibattito aperto / a cura di Mario Lancisi ; interventi di Gaetano Arfé ... [et al.]. — Roma : Borla, 1979.
352 p. ; 19 cm. — (Frecce) It79-Apr
Includes bibliographical references.
L6000
1. Milani, Lorenzo. Esperienze pastorali. 2. Catholic Church—Clergy—Biography. 3. Clergy—Italy—Barbiana—Biography. 4. Barbiana, Italy—Biography. I. Lancisi, Mario.
BX4705.M5526E764 79-362886
*79 MARC

—ITALY—MARCHE

Brocanelli, Gianfranco.
Seminari e clero nelle Marche nella seconda metà dell'Ottocento / Gianfranco Brocanelli. — Roma : Herder editrice e libreria, 1978.
xi, 114 p. ; 25 cm. — (Quaderni della Rivista di storia della Chiesa in Italia ; 6) It79-May
"Pubblicato nella Rivista di storia della Chiesa in Italia, Anno XXXI, 1977."
Includes bibliographical references and index.
L3000
1. Catholic Church in Marche, Italy—Clergy, Training of. 2. Theology—Study and teaching—Italy—Marche. 3. Marche, Italy—Church history. 4. Clergy—Italy—Marche. I. Title. II. Series: Rivista di storia della Chiesa in Italia : Quaderni ; 6.
BX1546.M33B76 79-379413
*79 MARC

—ITALY—TREVISO REGION

Liberali, Giuseppe.
Lo stato personale del clero diocesano nel secolo XVI / Giuseppe Liberali. — [Treviso] : A cura della Biblioteca del Seminario vescovile, [1975]
304 p., [2] leaves of plates : ill. ; 25 cm. — (Documentari sulla riforma cattolica pre e post-tridentina a Treviso (1527-1577) ; 9) It***
Includes bibliographical references.
1. Catholic Church. Diocese of Treviso, Italy. 2. Clergy—Italy—Treviso region. 3. Treviso region, Italy—Church history. I. Title. II. Series.
BX1547.T75D62 vol. 9 79-379870
79 MARC

—KOREA—BIOGRAPHY

Chi, Tong-sik, 1910–
(Sinhak ŭi osolkil)
神學의오솔길 / 池東植著. — 서울 : 大韓基督教書會, 1976.
223 p. ; 18 cm. — (現代新書 ; 70)
Series romanized: Hyŏndae sinsŏ.
₩500
1. Chi, Tong-sik, 1910– 2. Kidokkyo Taehan Pogŭm Kyohoe—Clergy—Biography. 3. Clergy—Korea—Biography. 4. Theologians—Korea—Biography. I. Title.
BX7990.K538C463 79–826288

Kil, Sŏn-ju, 1869–1935.
(Kil Sŏn-ju Moksa sŏlgyo)
吉善宙牧師說教. — 서울 : 惠文社, 1977.
2, 272, 8, 277 p. : ports. ; 21 cm.
Colophon inserted.
Includes indexes.
CONTENTS: 略傳集.—강대보감.—다니엘사경안.
₩2300
1. Kil, Sŏn-ju, 1869–1935. 2. Presbyterian Church—Clergy—Biography. 3. Clergy—Korea—Biography. 4. Presbyterian Church—Sermons. 5. Sermons, Korean. I. Title.
BX9225.K54A25 1977 78–826627

Kim, Ch'ang-gŭn, 1908–
(Pul punnŭn kasi tŏmbul)
불붙는가시덤불 : 雪峰金昌根牧師著作選集. — 서울 : 聖光文化社, 1977.
441 p. ; 22 cm.
₩2000
1. Holiness churches—Sermons. 2. Sermons, Korean. 3. Kim, Ch'ang-gŭn, 1908– 4. Holiness churches—Clergy—Biography. 5. Clergy—Korea—Biography. I. Title.
BX7990.H69K55 78–826407

Pak, Yong-gyu, 1935–
(Han'guk i nahŭn tae puhŭngsa Kim Ik-tu Moksa chŏn'gi)
한국이낳은대부흥사김익두목사전기 : 안악산골 / 박용규지음. — 서울 : 혜성문화사, 1968, 1977 printing.
234 p. ; 21 cm.
Cover title: Anak sankol sun'gyoja Kim Ik-tu Moksa chŏn'gi.
Colophon inserted.
₩1300
1. Kim, Ik-tu, 1874–1950. 2. Presbyterian Church—Clergy—Biography. 3. Clergy—Korea—Biography. I. Title. II. Title: Kim Ik-tu Moksa chŏn'gi. III. Title: Anak sankol sun'gyoja Kim Ik-tu Moksa chŏn'gi.
BX9225.K568P34 79–825946

CLERGY
—KOREA—BIOGRAPHY (Continued)

Yi, Wŏn-sun.
(Kim Tae-gŏn ŭi sŏhan)
金大建의書翰 / 李元淳, 許寅編著. — 서울 : 正音社, 1975.
290 p. : ill. ; 20 cm.

1. Kim, Tae-gŏn, 1821-1846. 2. Catholic Church—Clergy—Biography. 3. Clergy—Korea—Biography. 4. Martyrs—Korea—Biography. 5. Catholic Church in Korea—History. I. Hŏ, In, joint author. II. Title.
BX4705.K494Y5 79-826548

—LATVIA—BIOGRAPHY

Donnert, Erich.
Johann Georg Eisen : (1717-1779) : e. Vorkämpfer d. Bauernbefreiung in Russland / Erich Donnert. — 1. Aufl. — Leipzig : Koehler und Amelang, 1978.
199 p. : ill. ; 20 cm. GFR78-A
Bibliography: p. 191-[195]
Includes index.
11.80M
1. Eisen, Johann Georg, 1717-1779. 2. Lutheran Church—Clergy—Biography. 3. Clergy—Latvia—Biography. 4. Social reformers—Latvia—Biography. 5. Latvia—Biography.
BX8080.E43D66 78-392407
79 MARC

Küng, Andres, 1945-
Lång väg till frihet / Andres Küng. — Örebro : Libris, 1979.
192 p. : ill. ; 19 cm. S79
Bibliography: p. 192.
ISBN 9171941657
1. Grivans, Hugo Maksimilians, 1901- . 2. Lutheran Church—Clergy—Biography. 3. Clergy—Latvia—Biography. 4. Latvia—Biography. I. Title.
BX8080.G73K83 79-373234
79 MARC

—LOUISIANA—NEW ORLEANS—BIOGRAPHY

Rogers, Peter V
Tragedy is my parish : working for God in the streets of New Orleans / Peter V. Rogers. — New York : Macmillan, c1979.
xii, 159 p., [6] leaves of plates : ill. ; 22 cm.
Autobiography.
ISBN 0-02-604390-4 : $7.95
1. Rogers, Peter V. 2. Catholic Church—Clergy—Biography. 3. Clergy—Louisiana—New Orleans—Biography. 4. New Orleans—Biography. 5. Chaplains, Police—Louisiana—New Orleans—Biography. I. Title.
BX4705.R648A34 1979 282'.092'4 78-25896
78 MARC

—MASSACHUSETTS—BIOGRAPHY

Andrews, Stephen M
Daniel Hix and the First Christian Church of Dartmouth, Mass., 1780-1880 / by S. M. Andrews ; with index by Leonard H. Smith, Jr. — Facsim. ed., with index / by Leonard H. Smith, Jr. — Clearwater, Fla. : Smith, c1978.
vi, 231 p. ; 29 cm.
Photoreprint of the 1880 ed. printed by E. Anthony, New Bedford, Mass., under title: A sketch of Elder Daniel Hix.
ISBN 0-932022-12-X
1. Hix, Daniel, 1755-1838. 2. Congregationalists—Clergy—Biography. 3. Clergy—Massachusetts—Biography. 4. Massachusetts—Biography. 5. Dartmouth, Mass. First Christian Church—History. 6. Dartmouth, Mass.—Genealogy. 7. Registers of births, etc.—Massachusetts—Dartmouth. I. Title.
BX7260.H52A73 1978 285'.8 78-69710
79 MARC

Institute of Family Research.
John Lathrop (1584-1653) : reformer, sufferer, pilgrim, man of God. — Rev. — Salt Lake City : Institute of Family Research, 1979.
39 p., [1] fold. leaf of plates : ill. ; 29 cm.
"The result of ... research by Richard W. Price ... and Arlene Eakle".
Includes bibliographies.
1. Lothropp, John, 1584-1653. 2. Clergy—England—Biography. 3. Clergy—Massachusetts—Biography. I. Price, Richard W. II. Eakle, Arlene H.
BR1725.L66I57 1979 285'.8'0924 79-106935
79 MARC

—MASSACHUSETTS—BOSTON—BIOGRAPHY

Norton, John, 1606-1663.
Abel being dead, yet speaketh : a biography of John Cotton / by John Norton ; a facsim. reproduction with an introd. by Edward J. Gallagher. — Delmar, N.Y. : Scholars' Facsimiles & Reprints, 1978.
xxvii, 51 p. : port. ; 22 cm.
Reprint of the 1658 ed. printed by T. Newcomb for L. Lloyd, London.
Includes bibliographical references.
ISBN 0-8201-1310-7
1. Cotton, John, 1585-1652. 2. Congregational churches—Clergy—Biography. 3. Clergy—Massachusetts—Boston—Biography. 4. Boston—Biography. I. Title.
BX7260.C79N6 1978 285'.8'0924 78-8184
78 MARC

—MASSACHUSETTS—BESTBOROUGH—BIOGRAPHY

Stearns, Raymond Phineas, 1904–
New England church "relations" and continuity in early Congregational history, by Raymond Phineas Stearns and David Holmes Brawner.
(In American Antiquarian Society, Worcester, Mass. Proceedings. Worcester. 25 cm. v. 75, pt. 1 (1965) p. [13]-45)
Bibliographical footnotes.
1. Congregational Churches in New England. 2. New England—Church history. 3. Parkman, Ebenezer, 1703–1782. 4. Congregational churches—Clergy—Biography. 5. Clergy—Massachusetts—Westborough—Biography.
E172.A35 vol. 75, pt. 1 A 67–95

—MASSACHUSETTS—DALTON—BIOGRAPHY

Higgins, George B
A pastoral profile : or, Parsons in the glen : being reflections on the clergy of the First Congregational Church (United Church of Christ), Dalton, Massachusetts / Geo B. Higgins. — [Dalton? Mass.] : Higgins, c1978.
142 p. : ports. ; 23 cm.
Errata slip inserted.
1. First Congregational Church, Dalton, Mass.—Biography. 2. United Church of Christ—Clergy—Biography. 3. Clergy—Massachusetts—Dalton—Biography. 4. Dalton, Mass.—Biography. I. Title.
BX9886.Z7D3444 285'.8'0922 78-62693
79 MARC

—NETHERLANDS—BIOGRAPHY

Berekamp, E van de, 1821-1904.
Veertigjarige levenservaring / door E. van de Berekamp. — [Groningen, Oude Kijk in 't Jatstraat 59 : Doornbos], 1978.
174 p. : with ill. ; 21 cm. Ne78-28
Reprint of the 1879 ed. published by G. J. Reits, Groningen.
ISBN 9061190029 : fl 14.50
1. Berekamp, E. van de, 1821-1904. 2. Reformed Church in the Netherlands—Clergy—Biography. 3. Clergy—Netherlands—Biography. I. Title.
BX9479.B46A38 1978 79-356547
•79 MARC

Gribling, J P
Willem Hubert Nolens 1860-1931 : uit het leven van een priester-staatsman / door J. P. Gribling. — Assen : Van Gorcum, 1978.
xxiv, 377 p., 8 p. of photos : ill. ; 23 cm. — (Maaslandse monografieën ; 26) Ne78-50
Summary in French and German.
Includes bibliographical references and index.
ISBN (invalid) 9023216433 : fl 49.50 (pbk)
1. Nolens, Willem Hubert, 1860-1931. 2. Catholic Church—Clergy—Biography. 3. Clergy—Netherlands—Biography. 4. Statesmen—Netherlands—Biography.
BX4705.N827G74 79-368043
•79 MARC

Miedema, P P 1915-
Wrakseling om frijheit / P. P. Miedema. — Boalsert : Osinga, 1978-
v. ; 20 cm. Ne78-49
CONTENTS: diel 1. Thús en om utens.

ISBN 9060663403 : fl 17.50 (v. 1)
1. Miedema, P. P., 1915- . 2. Reformed Church—Clergy—Biography. 3. Clergy—Netherlands—Biography. I. Title.
BX9479.M5A38 79-380581
•79 MARC

Sipkes, Jelle.
Korte schets van mijn leeven / Jelle Sipkes ; utwurke en taljochte troch J. S. Postma ; mei in bydrage oer it doarp Tearns en de skiednis dêrfan, gearstald mei help fan O. Santema. — Ljouwert : De Tille, [1978]
64 p. : with ill. ; 24 cm. — ([Utjeften] - Fryske Akademy ; nr. 546) Ne78-43
Published on the occasion of the 40th anniversary of the Fryske Akademy.
Dutch or Frisian.
1. Sipkes, Jelle. 2. Mennonites in the Netherlands—Clergy—Biography. 3. Clergy—Netherlands—Biography. 4. Teerns, Netherlands—History. 5. Sipkes Family. I. Postma, Johan Sjouke, 1910- . II. Santema, Oepke, 1904- . III. Fryske Akademy. IV. Series: Fryske Akademy. Utjeften ; nr. 546.
BX8143.S56A34 79-373501
•79 MARC

Vrijer, Marianus Johannes Antonie de, 1881-
Ds [i.e. Dominee] Bernardus Smytegelt en zijn "Gekrookte riet," door M. J. A. Vrijer. Vianen, De Banier, 1968.
xvi, 246 p. illus. 23 cm.
"Herleiding ... tot veel kleiner omvang ... in de taal van heden."
Bibliography: p. 241-246.
1. Smytegelt, Bernardus, 1665-1739. 2. Reformed Church—Clergy—Biography. 3. Clergy—Netherlands—Biography. 4. Pietism—Netherlands—History. 5. Smytegelt, Bernardus, 1665-1739. Gekrookte riet. 6. Bible. N.T. Matthew XII, 20-21—Sermons. I. Smytegelt, Bernardus, 1665-1739. Gekrookte riet. 1968. II. Title.
BX9479.S6V74 79-384749
00 MARC

Wely, Joseph van, 1882-
Schaepman; levensverhaal. Bussum, P. Brand, 1952.
636 p. plates, ports., facsim, 25 cm.
Bibliographical footnotes.
1. Schaepman, Herman Johan Aloysius Maria, 1844-1903. 2. Politicans—Netherlands—Biography. 3. Catholic Church—Clergy—Biography. 4. Clergy—Netherlands—Biography.
DJ219.S27W43 53-33703
[r79]rev MARC

Wely, Joseph van, 1882-
Schaepman [door] Jos. van Wely. Den Haag, Kruseman [1964]
96 p. 20 cm. (Kopstukken uit de twintigste eeuw, 17)
1. Schaepman, Herman Johan Aloysius Maria, 1844-1903. 2. Politicians—Netherlands—Biography. 3. Catholic Church—Clergy—Biography. 4. Clergy—Netherlands—Biography.
DJ219.S27W4 65-57546
[r79]rev MARC

—NEW BRUNSWICK—BIOGRAPHY

DeGrâce, Éloi, 1947-
Monseigneur Stanislas-J. Doucet, p.d., 1847-1925 / Éloi DeGrâce. — Shippagan, N.-B. : [s.n.], 1977.
159, [1] p. : ill., facsims., map, music, ports. ; 22 cm. C78-5065-7
Bibliography: p. 153-[160]
1. Doucet, Stanislas Joseph, 1847-1925. 2. Catholic Church—Clergy—Biography. 3. Clergy—New Brunswick—Biography. 4. New Brunswick—Biography.
BX4705.D668D43 971.5'03'0924 78-387113
79 MARC

—NEW ENGLAND—BIOGRAPHY

Carroll, Peter N
The other Samuel Johnson : a psychohistory of early New England / Peter N. Carroll. — Rutherford [N.J.] : Fairleigh Dickinson University Press, c1978.
247 p. ; 22 cm.
Bibliography: p. 230-244.
Includes index.
ISBN 0-8386-2059-0 : $16.50
1. Johnson, Samuel, 1696-1772. 2. Church of England—Clergy—Biography. 3. Clergy—New England—Biography. 4. New England—Biography. I. Title.
BX5995.J59C37 283'.092'4 77-74413
78 MARC

Parkes, Henry Bamford, 1904-
Jonathan Edwards, the fiery Puritan / by Henry Bamford Parkes. — 1st AMS ed. — New York : AMS Press, 1979, c1930.
271 p. ; 19 cm. — (Philosophy in America)
Reprint of the ed. published by Minton, Balch, New York.
Bibliography: p. 259-266.
Includes index.
ISBN 0-404-59144-2
1. Edwards, Jonathan, 1703-1758. 2. Congregationalists—Clergy—Biography. 3. Clergy—New England—Biography. I. Title.
BX7260.E3P3 1979 285'.8'0924 75-3135
79 MARC

—NEW MEXICO—BIOGRAPHY

De Aragon, Ray John.
Padre Martínez and Bishop Lamy / by Ray John de Aragon. — Las Vegas, N.M. : Pan-American Pub. Co., c1978.
vi, 141 p. : ill. ; 22 cm.
Includes bibliographies.
ISBN 0-932906-00-1 : $6.95
1. Martínez, Antonio José, 1793-1867. 2. Catholic Church—Clergy—Biography. 3. Clergy—New Mexico—Biography. 4. Lamy, John Baptist, Abp., 1814-1888. I. Title.
BX4705.M412553D4 282'.092'3 78-70565
79 MARC

CLERGY
—NEW MEXICO—BIOGRAPHY
(Continued)

Sánchez, Pedro, b. 1831.
Recollections of the life of the priest Don Antonio José Martínez = Memorias sobre la vida del presbítero Don Antonio José Martínez / by Pedro Sánchez ; translated by Ray John de Aragón. — Santa Fe, N.M. : Lightning Tree, c1978.
85 p. ; 18 cm.
English and Spanish.
ISBN 0-89016-044-9 : $9.95
1. Martínez, Antonio José, 1793-1867. 2. Catholic Church—Clergy—Biography. 3. Clergy—New Mexico—Biography. 4. Statesmen—New Mexico—Biography. 5. New Mexico—Biography. I. Title.
BX4705.M412553S2713 282'.092'4 78-51362
00 MARC

—NEW YORK (CITY)

Murnion, Philip J
The Catholic priest and the changing structure of pastoral ministry / Philip J. Murnion. — New York : Arno Press, 1978.
xi, 487 p. : ill. ; 24 cm. — (The American Catholic tradition)
Originally presented as the author's thesis, Columbia University, 1972, under title: Towards theopolitan ministry.
Bibliography: p. 466-487.
ISBN 0-405-10845-1
1. Catholic Church—Clergy. 2. Pastoral theology—Catholic Church—History. 3. Clergy—New York (City) I. Title. II. Series: American Catholic tradition.
BX1913.M83 1978 253'.2 77-11302
77 MARC

—NEW YORK (STATE)—BIOGRAPHY

Burtsell, Richard Lalor, 1840-1912.
The diary of Richard L. Burtsell, priest of New York : the early years, 1865-1868 / edited by Nelson J. Callahan. — New York : Arno Press, 1978.
xxvii, 422 p. ; 24 cm. — (The American Catholic tradition)
ISBN 0-405-10813-3
1. Burtsell, Richard Lalor, 1840-1912. 2. Catholic Church—Clergy—Biography. 3. Clergy—New York (State)—Biography. 4. New York (State)—Biography. 5. Catholic Church in New York (State)—History—Sources. I. Callahan, Nelson J. II. Title. III. Series: American Catholic tradition.
BX4705.B946A33 1978 282'.092'4 77-89146
77 MARC

Malone, Sylvester L ed.
Dr. Edward McGlynn / Sylvester L. Malone. — New York : Arno Press, 1978.
viii, 135 p., [23] leaves of plates : ill. ; 24 cm. — (The American Catholic tradition)
Reprint of the 1918 ed. published by Dr. McGlynn Monument Association, New York.
ISBN 0-405-10841-9
1. McGlynn, Edward, 1837-1900. 2. Catholic Church—Clergy—Biography. 3. Clergy—New York (State)—Biography. 4. New York (State)—Biography. I. Title. II. Series.
BX4705.M2M3 1978 282'.092'4 77-11298
77 MARC

—NORWAY—BIOGRAPHY

Apenes, Sverre Inge.
Rapport om Petter Dass : presten som diktet makt til folket / Sverre Inge Apenes : [notetegning, Stein Johannes Kolnes]. — Oslo : Gyldendal, 1978 (Oslo : Norbok)
3-257 p. : ill. ; 23 cm. N79-Feb.
Bibliography: p. 256-[258]
ISBN 8205114064 : kr97.00
1. Dass, Petter, 1647-1708. 2. Poets, Norwegian—1550-1700—Biography. 3. Lutheran Church—Clergy—Biography. 4. Clergy—Norway—Biography. I. Title.
PT8754.Z5A75 79-381131
79 MARC

Birkeli, Fridtjov.
Mellom himmel og jord / Fridtjov Birkeli ; [illustrasjoner, Ulf Aas]. — Oslo : Luther, 1978 (Stavanger : Gjøstein boktr.)
257 p. : ill. ; 21 cm. N78-May
ISBN 8253172222 : kr77.00
1. Brekke, Mauritz Andreas, 1878-1951. 2. Eika, Sverre, 1899- . 3. Wisløff, Hans Edvard, 1902-1969. 4. Enger, Asle. 5. Lutheran Church—Clergy—Biography. 6. Clergy—Norway—Biography. I. Title.
BX8079.B48 78-395590
79 MARC

Håland, Karthon, 1924-
Med krage og pistol : presten Erling Fagerheims dramatiske krigsopplevelser / Karthon Håland. — Oslo : Luther, 1978 (Stavanger : Gjøstein boktr.)
175 p., [4] leaves of plates ; 21 cm. N78-Oct.
ISBN 8253132190 : kr78.00
1. World War, 1939-1945—Norway. 2. Fagerheim, Erling. 3. Norway—History—German occupation, 1940-1945. 4. Lutheran Church—Clergy—Biography. 5. Clergy—Norway—Biography. I. Title.
D763.N6H22 79-350929
79 MARC

—PALESTINE—BIOGRAPHY

Ḥaydar, Ḥaydar.
(Kabūshī)
كبوشي : وثيقة حياة ونضال المطران ايلاريون كبوشي من الطفولة الى المنفى / اعداد وتقديم حيدر حيدر ؛ الغلاف لحلمي التونسي. — [s. l.] : دار ابن رشد للطباعة والنشر، 1978.
173 p., [9] leaves of plates : ill. ; 22 cm.
1. Kabūshī, Īlāryūn, 1922- 2. Catholic Church — Clergy — Biography. 3. Clergy—Palestine—Biography. 4. Statesmen—Palestine—Biography.
BX4705.K2H39 79-970884

—PENNSYLVANIA—BIOGRAPHY

Hotchkin, Samuel Fitch, 1833-1912.
Country clergy of Pennsylvania / by S. F. Hotchkin. — Philadelphia : P. W. Ziegler, 1890.
305 p., [25] leaves of plates : ill. ; 20 cm.
1. Protestant Episcopal Church in the U.S.A.—Clergy—Biography. 2. Clergy—Pennsylvania—Biography. 3. Pennsylvania—Biography. 4. Pennsylvania—Church history. I. Title.
BX5990.H67 283'.092'2 79-311457
79 MARC

—PERU—BIOGRAPHY

Valencia Vega, Alipio.
José Antonino Medina, el sacerdote de la revolución liberal en el Alto Perú / Alipio Valencia Vega. — La Paz : Librería Editorial Juventud, 1978.
103 p. ; 19 cm. — (Colección Tradición, historia)
Includes bibliographical references.
1. Medina, José Antonio, ca. 1773-ca. 1830. 2. Catholic Church—Clergy—Biography. 3. Clergy—Peru—Biography. I. Title. II. Series.
BX4705.M4819V34 79-339986
79 MARC

—POLAND—BIOGRAPHY

Byli wśród nas : praca zbiorowa / pod red. Feliksa Lenorta. — Wyd. 1. — Poznań : Księgarnia św. Wojciecha, 1978 (Bydg. : PrasZG)
451, [1] p., [6] leaves of plates : ports. ; 25 cm. P78-7437
Biographical sketches.
Includes bibliographies.
zł220.00
1. Catholic Church in Poland—Biography. 2. Catholic Church—Clergy—Biography. 3. Clergy—Poland—Biography. 4. Wielkopolska—Biography. I. Lenort, Feliks.
BX4690.P6B94 79-348759
79 MARC

Jakubowski, Zbigniew.
Działalność ks. F. Fabera CRL w Krakowie, 1853-1868 / Zbigniew Jakubowski. — Kraków : [s.n.], 1974.
92 p. ; 21 cm.
Summary in French.
Bibliography: p. 75-81.
Includes index.
1. Faber, Franciszek, 1821-1868. 2. Catholic Church—Clergy—Biography. 3. Clergy—Poland—Biography. I. Title.
BX4705.F198J34 79-382477
79 MARC

Syska, Henryk.
Przychodzień z góry ducha : o księdzu Antonim Szechu / Henryk Syska. — Wyd. 1. — Warszawa : Ludowa spółdzielnia wydawnicza, 1955.
143 p. ; 21 cm.
1. Szech, Antoni, 1869- . 2. Catholic Church—Clergy—Biography. 3. Clergy—Poland—Biography. I. Title.
BX4705.S94S95 78-397612
79 MARC

—POLAND—KAROLEWO—BIOGRAPHY

Dembowski, Hermann.
Heinz Dembowski : 1884-1945 : eine Beschreibung seines Lebens / von Hermann Dembowski. — [s.l. : s.n., 1977]
84 p., [10] leaves of plates : ill. ; 21 cm. GFR***
1. Dembowski, Heinz, 1884-1945. 2. Lutheran Church—Clergy—Biography. 3. Clergy—Poland—Karolewo—Biography. 4. Karolewo, Poland—Biography.
BX8080.D44D45 79-347181
79 MARC

—PRINCE EDWARD ISLAND—BIOGRAPHY

MacLennan, Jean M
From Shore to shore : the life and times of the Rev. John MacLennan of Belfast, P.E.I. / by Jean M. MacLennan. — Edinburgh : Knox Press, 1977.
vii, 96 p. [8] p. of plates : ill., facsims., maps, ports. ; 22 cm.
GB78-32717
Map on back cover.
Bibliography: p. 83-86.
Includes index.
£0.85
1. MacLennan, John, 1797-1852. 2. Presbyterian Church—Clergy—Biography. 3. Clergy—Prince Edward Island—Biography. I. Title.
BX9225.M2558M32 285'.2'0924 79-302695
79 MARC

—QUÉBEC (PROVINCE)—BIOGRAPHY

Longpré, Anselme.
Un mouvement spirituel au Québec (1931-1962) : un retour à l'Évangile / Anselme Longpré. — Montréal : Fides, c1976.
86 p. : ill. ; 22 cm. — (L'Église du Québec ; 4) C***
Includes bibliographical references.
ISBN 0-7755-0573-0
1. Lacouture, Onésime, 1881-1951. 2. Catholic Church—Clergy—Biography. 3. Clergy—Québec (Province)—Biography. I. Title.
BX4705.L232L66 271'.53'024 79-377300
79 MARC

—QUÉBEC (PROVINCE)—SAINTE-ANNE-DE-LA-PÉRADE—BIOGRAPHY

Chevalier, Jean Claude, abbé.
Prêtres natifs de Ste-Anne / Jean-Claude Chevalier. — [Trois-Rivières, Québec] : Éditions du Bien public, 1978-
v. : ports. ; 21 cm. — (Collection Notre passé ; no 24-) C79-2567-1
Cover title: Prêtres natifs de Sainte-Anne-de-la-Pérade.
1. Catholic Church—Clergy—Biography. 2. Clergy—Québec (Province)—Sainte-Anne-de-la-Pérade—Biography. 3. Sainte-Anne-de-la-Pérade, Québec (Province)—Biography. I. Title.
BX4671.C47 282'.092'2 79-362807
79 MARC

—RHODESIA, SOUTHERN—BIOGRAPHY

Jones, Neville, 1880-1954.
My friend Kumalo / by Mhlagazanhlansi (Neville Jones). — Bulawayo : Books of Rhodesia, 1972.
126 p., [10] leaves of plates : ill. ; 19 cm.
"Reprint of the 1944 edition with new publishers' introduction and portrait of the author."
1. Matabele—Social life and customs. 2. Kumalo, Mtompe, ca. 1875-1949. 3. Matabele—Biography. 4. Presbyterian Church in Southern Rhodesia—Clergy—Biography. 5. Clergy—Rhodesia, Southern—Biography. I. Title.
DT955.J66 1972 968.9'1 78-326418
79 MARC

—ROMANIA—BIOGRAPHY

Fröhlich, Hans.
In der vierten Nachtwache : Erlebnisberichte aus d. Deportation / Hans Fröhlich ; [Zeichn. von Eva Seltmann]. — München : Hilfskomitee d. Siebenbürger Sachsen im Diakon. Werk d. Evang. Kirche in Deutschland, 1977.
62 p. : ill. ; 21 cm. GFR78-B6
1. World War, 1939-1945—Prisoners and prisons, Russian. 2. World War, 1939-1945—Personal narratives, Romanian. 3. Fröhlich, Hans. 4. Prisoners of war—Romania—Biography. 5. Prisoners of war—Ukraine—Donetsk—Biography. 6. Donetsk, Ukraine—Biography. 7. Germans in Transylvania—Biography. 8. Clergy—Romania—Biography. 9. Lutheran Church—Clergy—Biography. I. Title.
D805.U4F76 940.54'72'430924 79-355161
79 MARC

Ratiu, Alexander.
Stolen church, martyrdom in Communist Romania / Alexander Ratiu & William Virtue. — Huntington, Ind. : Our Sunday Visitor, c1979.
192 p. ; 21 cm.
ISBN 0-87973-730-1 : $4.95
1. Ratiu, Alexander. 2. Catholic Church—Clergy—Biography. 3. Clergy—Romania—Biography. 4. Prisoners—Romania—Biography. 5. Persecution—Romania. 6. Catholic Church in Romania—Clergy. I. Virtue, William, joint author. II. Title.
BX4711.495.R37A37 272'.9 79-87926
79 MARC

CLERGY (Continued)

—RUSSIA

Immekus, Erwin.
Die russisch-orthodoxe Landpfarrei zu Beginn des XX. Jahrhunderts nach dem Gutachten der Diözesanbischöfe / Erwin Immekus. — Würzburg : Augustinus-Verlag, 1978.
xvi, 284 p. ; 23 cm. — (Das Östliche Christentum ; n.F., Bd. 30) GFR79-A
Originally presented as the author's thesis, Istituto orientale, Rome, 1960.
Bibliography: p. xiv-xvi.
ISBN 3-7613-0117-0 : DM52.80
1. Parishes—Russia. 2. Orthodox Eastern Church, Russian—History. 3. Clergy—Russia. I. Title. II. Series.
BX485.I49 1978 79-361861
79 MARC

—RUSSIA—BIOGRAPHY

Behr-Sigel, Élisabeth.
Alexandre Boukharev : un théologien de l'Eglise orthodoxe russe en dialogue avec le monde moderne : introduction et lettres à Valérien et Alexandre Lavrski / Élisabeth Behr-Sigel ; préf. d'Olivier Clément. — Paris : Beauchesne, c1977.
163 p. ; 27 cm. — (Beauchesne religions) F***
"Œeuvres d'Alexandre M. Boukharev": p. 17-19.
Bibliography: p. 20-24.
87.00F
1. Feodor, Archimandrite, 1822-1871. 2. Orthodox Eastern Church, Russian—Clergy—Biography. 3. Clergy—Russia—Biography. 4. Orthodox Eastern Church, Russian—Clergy—Correspondence. 5. Clergy—Russia—Correspondence. I. Feodor, Archimandrite, 1822-1871.
BX597.F34B43 230'.1'90924 78-371118
78 MARC

Blessed Paisius Velichkovsky. — Platina, Calif. : Saint Herman of Alaska Brotherhood, 1976-
v. : ill. ; 22 cm.
Vol. 1 consists chiefly of a translation of Zhitīe i pisanīīa moldavskago starīsa Paysīīa Vilichkovskago, published in 1847, and based on manuscript by schema-monk Mitrofan.
Bibliography: v. 1, p. 283-287.
Vol. 1 includes index.
CONTENTS: v. 1. The life and ascetic labors of our father, elder Paisius, archimandrite of the Holy Moldavian Monasteries of Niamets and Sekoul, Optina version, by schema-monk Metrophanes.
1. Paisiĭ Velichkovskiĭ, archimandrite, 1722-1794. 2. Orthodox Eastern Church, Russian—Clergy—Biography. 3. Clergy—Russia—Biography. I. Mitrofan, schema-monk.
BX597.P26B57 281.9'3 76-12010
79 MARC

Sokolov, Leonid.
(Episkop Ignatiĭ Brīanchaninov)
Епископъ Игнатій Брянчаниновъ : его жизнь, личность и морально-аскетическія воззрѣнія / Леонидъ Соколовъ. — Кіевъ : Тип. Первой Кіевской артели печатнаго дѣла, 1915.
2 v. : ill. ; 27 cm.
Includes articles and letters written by Bp. Ignatiĭ.
Includes bibliographical references and index.
1. Ignatiĭ, Bp. of Caucasus and Chernomor'e, 1807-1867. 2. Clergy—Russia—Biography. 3. Orthodox Eastern Church, Russian—Clergy—Biography. I. Ignatiĭ, Bp. of Caucasus and Chernomor'e, 1807-1867.
BX597.I 3S68 Russian Imperial Coll. 79-411555

—RUSSIA—CORRESPONDENCE

Behr-Sigel, Élisabeth.
Alexandre Boukharev : un théologien de l'Eglise orthodoxe russe en dialogue avec le monde moderne : introduction et lettres à Valérien et Alexandre Lavrski / Élisabeth Behr-Sigel ; préf. d'Olivier Clément. — Paris : Beauchesne, c1977.
163 p. ; 27 cm. — (Beauchesne religions) F***
"Œeuvres d'Alexandre M. Boukharev": p. 17-19.
Bibliography: p. 20-24.
87.00F
1. Feodor, Archimandrite, 1822-1871. 2. Orthodox Eastern Church, Russian—Clergy—Biography. 3. Clergy—Russia—Biography. 4. Orthodox Eastern Church, Russian—Clergy—Correspondence. 5. Clergy—Russia—Correspondence. I. Feodor, Archimandrite, 1822-1871.
BX597.F34B43 230'.1'90924 78-371118
78 MARC

—SCOTLAND

Blakey, Ronald S
The man in the manse / Ronald S. Blakey. — Edinburgh : Handsel Press, 1978.
x, 150 p. ; 23 cm. GB***
Bibliography: p. [137]-142.
Includes index.
ISBN 0-905312-05-8 : £4.50
1. Church of Scotland—Clergy. 2. Clergy—Scotland. I. Title.
BX9079.B58 253 79-308669
79 MARC

—SCOTLAND—BIOGRAPHY

Falconer, Ronald.
The kilt beneath my cassock / Ronnie Falconer. — Edinburgh : Handsel Press, 1978.
viii, 254 p., [9] leaves of plates : ill. ; 23 cm. GB***
Includes index.
ISBN 0-905312-07-4 : £4.75. ISBN 0-905312-02-3 pbk.
1. Falconer, Ronald. 2. Church of England—Clergy—Biography. 3. Clergy—Scotland—Biography. 4. Radio in religion—Scotland. 5. Television in religion—Scotland. I. Title.
BX5199.F26A33 283'.092'4 78-325343
79 MARC

—SOUTH CAROLINA—WHITE ROCK—BIOGRAPHY

Boozer, Mary Elizabeth.
Bethel's pastors : biographical sketches of the pastors of Bethel Lutheran Church, White Rock, South Carolina / by Mary Elizabeth Boozer. — White Rock, S.C. : The Church, 1978.
viii, 67 p. : ports. ; 23 cm.
Bibliography: p. 65-66.
Includes index.
1. Bethel Lutheran Church, White Rock, S.C.—Biography. 2. Lutheran Church—Clergy—Biography. 3. Clergy—South Carolina—White Rock—Biography. 4. White Rock, S.C.—Biography. I. Bethel Lutheran Church, White Rock, S.C. II. Title.
BX8076.W45B472 284'.1757'71 79-101493
79 MARC

—SPAIN—BIOGRAPHY

Alvarez, Baltasar, b. 1533.
Escritos espirituales / Baltasar Alvarez ; introd. biográfica y ed. Camilo Ma. Abad y Faustino Boado. — Barcelona : J. Flors, 1961.
x, 739 p., [3] leaves of plates : ill. ; 18 cm. — (Espirituales españoles : Serie A, Textos ; t. 4)
Includes bibliographical references and indexes.
1. Spiritual life—Catholic authors—Collected works. 2. Alvarez, Baltasar, b. 1533. 3. Catholic Church—Clergy—Biography. 4. Clergy—Spain—Biography. I. Abad, Camilo María. II. Boado, Faustino. III. Title.
BX2349.A66 1961 79-391977
00 MARC

Caballero, Nicolás.
Mi meditación yoga : una experiencia que transforma mi vida / por Nicolás Caballero. — Valencia : Comercial Editora de Publicaciones, [1977?]
298 p. ; 21 cm. — (His El camino de la libertad ; v. 6) (Colección CP ; no. 94-95) Sp***
Includes bibliographical references.
ISBN 8470500546
1. Caballero, Nicolás. 2. Catholic Church—Clergy—Biography. 3. Clergy—Spain—Biography. 4. Spiritual life—Catholic authors. 5. Yoga. I. Title.
BR45.C32 vol. 6 79-377580
[BX4705.C16] 79 MARC

Carner, Antonio.
Verdaguer / Antoni Carner i Borràs. — [Barcelona : Edicions de Nou Art Thor, 1978]
32 p. : ill. ; 21 cm. — (Gent nostra ; 2) Sp***
Cover title.
Bibliography: p. 17-18.
ISBN 8473270134
1. Verdaguer, Jacinto, 1845-1902. 2. Catholic Church—Clergy—Biography. 3. Clergy—Spain—Biography. 4. Poets, Catalan—19th century—Biography.
BX4705.V44C37 78-391774
79 MARC

Díez Alegría, José María, 1911-
Io creo nella speranza / José María Díez-Alegría ; a cura di Marilena Grasso. — 1. ristampa. — [Milano] : A. Mondadori, 1973, 1974 printing.
215 ; 20 cm. — (Idoc. Documentinuovi ; 25) It74-Sept
Translation of Yo creo en la esperanza!
Includes bibliographical references.
1. Díez Alegría, José María, 1911- . 2. Catholic Church—Clergy—Biography. 3. Clergy—Spain—Biography. 4. Catholic Church—Doctrinal and controversial works—Catholic authors. I. Title. II. Series.
BX4705.D537A316 78-400296
•79 MARC

González Ruiz, José María, 1916-
Contraint par la Parole, prêtre en Espagne [par] J. M. González Ruiz. Paris, Éditions du Cerf, 1972.
165 p. 20 cm. (L'Évangile au vingtième siècle) 17.00F F73-150:
Includes bibliographical references.
1. González Ruiz, José María, 1916- . 2. Catholic Church—Clergy—Biography. 3. Clergy—Spain—Biography. I. Title.
BX4705.G596A33 78-399030
79 MARC

Martín Hernández, Francisco.
Mosen Sol : vida de Manuel Domingo y Sol, fundador de la Hermandad de Sacerdotes Operarios Diocesanos / [por] Francisco Martín Hernández, Lope Rubio Parrado. — Salamanca : Sígueme, 1978.
540 p., [6] leaves of plates : ill. ; 24 cm. Sp78-Dec
Bibliography: p. [531]-534.
ISBN 843010738X
1. Domingo y Sol, Manuel, 1836-1909. 2. Catholic Church—Clergy—Biography. 3. Clergy—Spain—Biography. 4. Hermandad de Sacerdotes Operarios Diocesanos—History. I. Rubio Parrado, Lope, joint author. II. Title.
BX4705.D58253M37 79-376289
79 MARC

Montijano Chica, Juan.
Muñoz Garnica, polígrafo ubetense : (estudio biográfico) / [por] Juan Montijano Chica, Manuel López Pérez. — Salamanca : Instituto de Estudios Gienenses, 1978
181 p., 11 leaves of plates : ill. ; 20 cm. — (Colección Libros de bolsillo del Instituto de Estudios Gienenses ; t. 4) Sp
Bibliography: p. [179]-181.
ISBN 8400043413
1. Muñoz Garnica, Manuel, 1821-1876. 2. Catholic Church—Clergy—Biography. 3. Clergy—Spain—Biography. I. López Pérez, Manuel, joint author. II. Title.
BX4705.M9715M64 79-114158
79 MARC

Munar, Gaspar.
D. Miguel Maura y Montaner : rector del seminario de Mallorca y fundador de las Hnas. Celadoras del Culto Eucarístico / [por] Gaspar Munar Oliver. — Mallorca : [el autor], 1977.
244 p. : ill. ; 21 cm. Sp78-Jan
Bibliography: p. 241.
1. Maura y Montaner, Miguel, 1843-1915. 2. Catholic Church—Clergy—Biography. 3. Clergy—Spain—Biography. 4. Celadoras del Culto Eucarístico.
BX4705.M4535M86 78-400301
79 MARC

—SPAIN—AVILA—BIOGRAPHY

Taberna Martín, Mariano, 1913-
Glorias de Avila / Mariano Taberna Martín. — Avila : Obra Social y Cultural de la Caja Central de Ahorros y Préstamos de Avila, 1974.
137 p. ; 22 cm. — (Publicaciones de la Caja Central de Ahorros y Préstamos de Avila) Sp75-Apr
ISBN 8450065968
1. Gracía Mazo, Santiago José, 1768-1849. 2. Robles Degano, Felipe, 1863-1939. 3. Catholic Chruch—Clergy—Biography. 4. Clergy—Spain—Avila—Biography. 5. Avila, Spain—Biography. I. Title.
BX4688.G35T3 78-420888
•79 MARC

—SPAIN—CATALONIA—BIOGRAPHY

Balbuena, Silvio F
Xirinacs / Silvio F. Balbuena. — Madrid : Antonio San Román Yribas, 1978.
332 p., 1 leaf : ill. ; 20 cm. Sp78-May
ISBN 8485445007 : 385ptas
1. Xirinacs, Lluís M. 2. Social reformers—Spain—Catalonia—Biography. 3. Clergy—Spain—Catalonia—Biography. 4. Political prisoners—Spain—Catalonia—Biography.
HV28.X57B34 79-363363
79 MARC

—SWEDEN—BIOGRAPHY

Boëthius, Jacob, 1647-1718.
Mora-prosten mäster Jacob Boëthius' syndabekännelse av år 1707 : med en kort levernesbeskrivning och kommentar : samt Om synd och nåd, medveten strävan och omedvetet motstånd ... / av Hjalmar Sundén. — Stockholm : Proprius, 1977.
198 p. ; 19 cm. S***
ISBN 9171182381
1. Boëthius, Jacob, 1647-1718. 2. Lutheran Church—Clergy—Biography. 3. Clergy—Sweden—Biography. I. Title.
BX8080.B6364A35 79-376906
79 MARC

Eeg-Olofsson, Leif.
Johan Dillner : präst, musiker och mystiker / av Leif Eeg-Olofsson. — Stockholm : Almqvist & Wiksell international, 1978.
320 p., [6] leaves of plates : ill. ; 23 cm. — (Bibliotheca historico-ecclesiastica Lundensis ; 8) (Acta Universitatis Lundensis : Sectio ISSN I,) S78-29
"Johan Dillners tryckta skrifter": p. 295-297.
Bibliography: p. 299-314.
Includes index.
ISBN 9122001786
1. Dillner, Johan, 1785-1862. 2. Lutheran Church—Clergy—Biography. 3. Clergy—Sweden—Biography. I. Series. II. Series: Lund. Universitet. Acta : Sectio I, theologica, juridica, humaniora ; 27.
BX8080.D54E34 79-366678
79 MARC

CLERGY
—SWEDEN—BIOGRAPHY (Continued)

Hedlund, John, 1912-
Silvia och silverkronan : polemiska kåserier i aktuella ämnen / av John Hedlund. — Herrljunga : InterSkrift, 1976.
159 p. ; 19 cm. S78-6
"Kåserier från Svensk veckotidning."
ISBN 9173360686 : kr20.00
1. Hedlund, John, 1912- . 2. Clergy—Sweden—Biography. I. Title.
BR1725.H43A33 79-385160
79 MARC

Laestadius, Lars Levi, 1800-1861.
Huutavan ääni korvessa. Valikoima kirjoituksia hengellisestä aikakauslehdestä En ropandes röst i öknen vv. 1852-54. ⌈Kirj.⌉ L. L. Laestadius. Suom. J. F. Hellman. Hämeenlinna, Karisto, 1970.
358 p. 21 cm. Fmk23.00 HUL
1. Laestadius, Lars Levi, 1800-1861. 2. Lutheran Church—Clergy—Biography. 3. Clergy—Sweden—Biography. I. Title.
BX8080.L2A34 1970 78-411927
78 MARC

—SWEDEN—VÄSTERGÖTLAND

Bergstrand, Carl-Martin, 1899-
Ur västgötaprästers bouppteckningar från tiden före 1761 / av Carl-Martin Bergstrand. — ⌈Göteborg : Förf.⌉ ; Skövde : Skövde antikvariat (distr.), 1977.
189 p. ; 22 cm. S77-53
Includes indexes.
kr30.50
1. Svenska kyrkan—Clergy. 2. Clergy—Sweden—Västergötland. 3. Inventories of decedents' estates—Sweden—Västergötland. I. Title.
BX8040.V33B47 78-378154
78 MARC

—SWITZERLAND—BIOGRAPHY

Duss, Niklaus.
Josef Ignaz von Ah : 1834-1896 : der Weltüberblicker / Niklaus Duss. — Sarnen : Verlag des Historisch-Antiquarischen Vereins Obwalden, 1975.
239 p. : port. ; 23 cm. — (Obwaldner Geschichtsblätter ; Heft 13)
Sw76-1549
Bibliography: p. 208-227.
Includes index.
25.00F
1. Ah, Josef Ignaz von, 1834-1896. 2. Catholic Church—Clergy—Biography. 3. Clergy—Switzerland—Biography. I. Series.
DQ663.O26 Heft 13 79-360652
⌈BX4705.A38⌉ *79 MARC

Kaspar, Philipp, 1938-
Alois Gügler, 1782-1827 : ein bedeutender Luzerner Theologe im Spannungsfeld von Aufklärung und Romantik / ⌈Philipp Kaspar⌉. — Schüpfheim : P. Kaspar, 1977.
387 p. : 1 ill. ; 21 cm. Sw78-12312
Originally presented as the author's thesis, Fribourg, 1976.
Bibliography: p. 5-27.
Includes index.
33.00F
1. Gügler, Alois, 1782-1827. 2. Theologians—Switzerland—Biography. 3. Catholic Church—Clergy—Biography. 4. Clergy—Switzerland—Biography. I. Title.
BX4705.G6269K37 1977 79-378804
79 MARC

—UNITED STATES—BIOGRAPHY

Bissonnette, Georges.
Moscow was my parish / Georges Bissonnette. — Westport, Conn. : Greenwood Press, 1978, c1956.
272 p. ; 23 cm.
Reprint of the ed. published by McGraw-Hill, New York.
ISBN 0-313-20594-9
1. Bissonnette, Georges. 2. Catholic Church—Clergy—Biography. 3. Clergy—United States—Biography. 4. Catholic Church in Russia. I. Title.
⌈BX4705.B5235A34 1978⌉ 78-16489
282'.092'4
78 MARC

Bozarth-Campbell, Alla, 1947-
Womanpriest : a personal odyssey / Alla Bozarth-Campbell. — New York : Paulist Press, c1978.
ix, 229 p. ; 22 cm.
ISBN 0-8091-0243-9 : $9.95
1. Bozarth-Campbell, Alla, 1947- . 2. Protestant Episcopal Church in the U.S.A.—Clergy—Biography. 3. Clergy—United States—Biography. 4. Ordination of women. I. Title.
BX5995.B665A38 283'.092'4 78-58957
78 MARC

Carlin, Paul W 1933-
From pulpit to prison : the Paul Carlin story / Paul W. Carlin, Sr. — Dallas : Acclaimed Books c1978.
ix, 108 p., ⌈4⌉ leaves of plates : ill. ; 18 cm.
ISBN 0-932294-02-2
1. Carlin, Paul W., 1933- . 2. Baptists—Clergy—Biography. 3. Clergy—United States—Biography. 4. Prisoners—United States—Biography. I. Title.
⌈BX6495.C32A33⌉ 269'.2'0924 78-10513
78 MARC

Carothers, Merlin R
Victory on Praise Mountain / Merlin Carothers. — Plainfield, N.J. : Logos International, c1979.
viii, 177 p. ; 21 cm.
ISBN 0-88270-378-1 : $2.95
1. Carothers, Merlin R. 2. Pentecostals—Clergy—Biography. 3. Clergy—United States—Biography. I. Title.
BX8762.Z8C373 289'.9 79-88266
79 MARC

Chambers, Catherine, 1920-
The measure of a man : the biography of Stanley Warren Chambers / by Catherine Chambers. — Hazelwood, Mo. : Word Aflame Press, c1978.
191 p., ⌈16⌉ leaves of plates : ill. ; 22 cm.
1. Chambers, Stanley Warren, 1915- . 2. Pentecostals—Clergy—Biography. 3. Clergy—United States—Biography. I. Title.
BX8780.Z8C463 289.9 78-111263
79 MARC

Chappell, Wallace D
Clovis Chappell : preacher of the word / Wallace D. Chappell. — Nashville, Tenn. : Broadman Press, c1978.
96 p. : ill. ; 20 cm.
ISBN 0-8054-7223-1
1. Chappell, Clovis Gilliam, 1882-1972. 2. Methodist Church—Clergy—Biography. 3. Clergy—United States—Biography. I. Title.
BX8495.C48C48 287'.6'0924 78-59305
79 MARC

Clark, Clifford E 1941-
Henry Ward Beecher : spokesman for a Middle-class America / Clifford E. Clark, Jr. — Urbana : University of Illinois Press, c1978.
288 p., ⌈2⌉ leaves of plates : ill. ; 24 cm.
Includes bibliographical references and index.
ISBN 0-252-00608-9 : $11.95
1. Beecher, Henry Ward, 1813-1887. 2. Congregational churches—Clergy—Biography. 3. Clergy—United States—Biography.
BX7260.B3C6 285'.8'0924 78-1721
78 MARC

Gardner, Carl, 1931-
Andrew Young : a biography / Carl Gardner. — New York : Drake, 1978.
232 p. : ill. ; 24 cm.
Bibliography: p. 222-223.
Includes index.
ISBN 0-8473-1700-5 : $9.95
1. Young, Andrew J., 1932- . 2. United Church of Christ—Clergy—Biography. 3. Clergy—United States—Biography. 4. United States. Congress. House—Biography. 5. Legislators—United States—Biography. 6. Ambassadors—United States—Biography.
E840.8.Y64G37 973.92'092'4 77-88945
78 MARC

Hambrick-Stowe, Elizabeth A
Expecting : a Christian exploration of pregnancy and childbirth ; drawings by Don Bender / Elizabeth A. Hambrick-Stowe. — Valley Forge, Pa. : Judson Press, c1979.
94 p. : ill. ; 22 cm.
Includes bibliographical references.
ISBN 0-8170-0837-3 : $3.95
1. Hambrick-Stowe, Elizabeth A. 2. United Church of Christ—Clergy—Biography. 3. Clergy—United States—Biography. 4. Pregnancy—Biography. 5. Pregnancy—Moral and religious apsects. I. Title.
BX9886.Z8H353 285'.8'0924 79-11207
79 MARC

Harris, Irving D
The breeze of the spirit : Sam Shoemaker and the story of faith at work / Irving Harris. — New York : Seabury Press, 1978.
xv, 190 p. ; 24 cm.
"A Crossroad book."
Includes index.
ISBN 0-8164-0399-6 : $8.95
1. Shoemaker, Samuel Moor, 1893-1963. 2. Protestant Episcopal Church in the U.S.A.—Clergy—Biography. 3. Clergy—United States—Biography. 4 Faith at work. I. Title.
BX5995.S347H37 283'.092'4 78-18237
78 MARC

Henrichsen, Margaret.
Seven steeples / Margaret Henrichsen ; illustrated by William Barss. — Thorndike, Me. : Thorndike Press, ⌈1978⌉ c1953.
238 p. : ill. ; 20 cm.
Reprint of the ed. published by Houghton Mifflin, Boston.
ISBN 0-89621-023-5
1. Henrichsen, Margaret. 2. Methodist Church—Clergy—Biography. 3. Clergy—United States—Biography. I. Title.
⌈BX8495.H397A3 1978⌉ 287'.6'0924 78-26203
78 MARC

Hewitt, Joe.
I was raised a Jehovah's Witness / Joe Hewitt. — Denver : Accent Books, c1979.
191 p. ; 21 cm.
ISBN 0-89636-018-0 : $3.95
1. Jehovah's Witnesses—Doctrinal and controversial works. 2. Hewitt, Joe. 3. Baptists—Clergy—Biography. 4. Clergy—United States—Biography. I. Title.
BX8526.H396 289.9 78-73255
00 MARC

Huels, John M
Father Keane, servant of Mary / by John M. Huels. — 1st ed. — Berwyn, Ill. : Servite Press, 1979.
xiv, 143 p. : ill. ; 24 cm.
Includes bibliographical references.
1. Keane, James Mary, 1901-1975. 2. Catholic Church—Clergy—Biography. 3. Clergy—United States—Biography. I. Title.
BX4705.K329H83 271'.4'024 79-84871
79 MARC

Huie, William Bradford, 1910-
It's me O Lord! / William Bradford Huie. — Nashville : T. Nelson, c1979.
189 p., ⌈4⌉ leaves of plates : ill. ; 21 cm.
ISBN 0-8047-5141-9 : $7.95
1. Ronsisvalle, Daniel. 2. Pentecostal churches—Clergy—Biography. 3. Clergy—United States—Biography. I. Title.
BX8762.Z8R653 289.9 79-403
79 MARC

Hutchinson, Warner.
The Oral Roberts scrapbook / by Warner A. Hutchinson. — New York : Grosset & Dunlap, c1978.
124 p., ⌈2⌉ leaves of plates : ill. ; 28 cm.
Bibliography: p. 124.
ISBN 0-448-16257-1 : $14.95 ($16.95 Can)
1. Roberts, Oral. 2. Methodist Church—Clergy—Biography. 3. Clergy—United States—Biography. I. Title.
BX8495.R528H87 269'.2'0924 78-58611
00 MARC

Kotre, John N
The best of times, the worst of times : Andrew Greeley and America Catholicism, 1950-1975 / John N. Kotre. — Chicago : Nelson-Hall Co., c1978.
xvii, 274 p. ; 23 cm.
"Books by Andrew M. Greeley": p. 263-265.
Includes bibliographical references and index.
ISBN 0-88229-380-X. ISBN 0-88229-597-7 pbk.
1. Greeley, Andrew M., 1928- . 2. Catholic Church—Clergy—Biography. 3. Clergy—United States—Biography. 4. Catholic Church in the United States—History. I. Title.
BX4705.G6185K67 282'.092'4 78-14224
78 MARC

Leavell, William Hayne, 1850-
William Hayne Leavell, 1850-1930 : an autobiography. — ⌈Seattle : A. Bayless, c1979⌉
ii, 171 p. ; 28 cm.
1. Leavell, William Hayne, 1850- . 2. Presbyterian Church—Clergy—Biography. 3. Clergy—United States—Biography.
BX9225.L38A35 285'.1'0924 79-118371
79 MARC

Leynse, James P
Out of the sea : a young Zeelander / by James P. Leynse. — Westchester, Ill. : Good News Publishers, c1976.
190 p. ; 20 cm.
ISBN 0-89107-148-2 : $3.50
1. Leynse, James P. 2. Presbyterian Church—Clergy—Biography. 3. Clergy—United States—Biography. I. Title.
BX9225.L473A33 949.2'07'0924 76-45866
78 MARC

Long, Nat G
My long life / by Nat G. Long. — Conyers, GA. : T.H.P., 1978.
xii, 399 p., ⌈2⌉ leaves of plates : ill. ; 23 cm.
1. Long, Nat G. 2. Methodist Church—Clergy—Biography. 3. Clergy—United States—Biography. I. Title.
BX8495.L58A33 287'.6'0924 78-59556
79 MARC

Mondésir, Édouard de.
Mondésir, Édouard de Mondésir ; avec une introd. par Gilbert Chinard. — New York : Arno Press, 1979.
60 p. ; 24 cm. — (Johns Hopkins University Press reprints)
Reprint of the 1942 ed. published by The Johns Hopkins Press, Baltimore, in series: Historical documents, Institut français de Washington.
ISBN 0-405-10594-0 : $10.00
1. Mondésir, Édouard de. 2. Catholic Church—Clergy—Biography. 3. Clergy—Canada—Biography. 4. Clergy—United States—Biography. 5. Jesuits—Canada—Biography. 6. Jesuits—United States—Biography. I. Chinard, Gilbert, 1881-1972. II. Title. III. Series: Institut français de Washington, Washington, D.C. Historical documents.
BX4705.M64A3 1979 271'.53'024 78-19280
78 MARC

Moore, Paul, 1942-
Shepherd of Times Square / by Paul Moore & Joe Musser. — Nashville : T. Nelson, c1979.
249 p. ; 21 cm.
ISBN 0-8407-5166-4 : $9.95
1. Moore, Paul, 1942- . 2. Church of the Nazarene—Clergy—Biography. 3. Clergy—United States—Biography. I. Musser, Joe, joint author. II. Title.
BX8699.N38M665 289.9 79-16249
79 MARC

CLERGY
—UNITED STATES—BIOGRAPHY (Continued)

Pepper, George Whitfield, 1833-1899.
Under three flags; or, The story of my life as preacher, captain in the Army, chaplain, consul, with speechs and interviews, by George W. Pepper. Cincinnati : Printed for the author by Curts & Jennings, 1899.
542 p., 1 leaf front., plates, ports. 24 cm.
Errata slip inserted.
1. Pepper, George Whitfield, 1833-1899. 2. United States—Army—Biography. 3. Methodist Church—Clergy—Biography. 4. Clergy—United States—Biography. 5. Diplomats—United States—Biography. I. Title.
CT275.P565A38 a 34-1532
[r79]rev MARC

Reid, Thomas F
The exploding church / by Thomas F. Reid, with Doug Brendel. — Plainfield, N.J. : Logos International, c1979.
156 p. ; 21 cm.
ISBN 0-88270-299-8 : $3.50
1. Reid, Thomas F. 2. Clergy—United States—Biography. 3. Full Gospel Tabernacle, Orchard Park, N.Y. I. Brendel, Doug, joint author. II. Title.
BR1725.R39A34 289.9 78-73575
00 MARC

Ross, James, 1801-1878.
Life and times of Elder Reuben Ross / by James Ross ; with an introd. and notes by J. M. Pendleton. — [s.l. : s.n.], 1977 (Nashville : McQuiddy Print. Co.)
433 p., [2] leaves of plates : ill. ; 22 cm.
Reprint of the 1882? ed. printed by Grant, Faires & Rodgers, Philadelphia.
Includes index.
1. Ross, Reuben, 1776-1860. 2. Baptists—Clergy—Biography. 3. Clergy—United States—Biography. I. Title.
BX6495.R677R67 1977 286'.1'0924 79-116294
79 MARC

Shuba, Michael, 1903-
I'm just reminiscing / Michael Shuba. — Cleveland : First Catholic Slovak Union, 1978.
416 p. : ill. ; 23 cm.
1. Shuba, Michael, 1903- . 2. Catholic Church—Clergy—Biography. 3. Clergy—United States—Biography. 4. Slovak Americans—Biography. 5. Catholics, Slovak, in the United States—Biography. I. Title.
BX4705.S6147A34 282'.092'4 79-104230
79 MARC

Slosser, Bob.
Miracle in Darien / Bob Slosser. — Plainfield, N.J. : Logos International, c1979.
268 p. ; 22 cm.
ISBN 0-88270-355-2 : $6.95
1. Fullman, Terry. 2. Protestant Episcopal Church in the U.S.A.—Clergy—Biography. 3. Clergy—United States—Biography. 4. St. Paul's Episcopal Church, Darien, Conn. I. Title.
BX5995.F78S58 283'.092'4 79-83791
00 MARC

Strober, Gerald S
Jerry Falwell : a flame for God / Jerry Strober, Ruth Tomczak. — Nashville, TN : T. Nelson Pub. Co., c1979.
ix, 188 p., [4] leaves of plates : ill. ; 21 cm.
Includes index.
ISBN 0-8407-5172-9
1. Falwell, Jerry. 2. Baptists—Clergy—Biography. 3. Clergy—United States—Biography. 4. Thomas Road Baptist Church. I. Tomczak, Ruth, joint author. II. Title.
BX6495.F3S8 286'.1'0924 79-16248
79 MARC

Thompson-Peters, Flossie E
The shepherd : a biography of Dr. Arthur A. Peters / by Flossie E. Thompson-Peters. — Los Angeles : Atlas Press, c1979.
251 p. : ill. ; 22 cm.
Bibliography: p. 250-251.
$10.00
1. Peters, Arthur A. 2. Baptists—Clergy—Biography. 3. Clergy—United States—Biography. I. Title.
BX6495.P465T48 286'.13 79-50838
79 MARC

Twitchell, Thomas Donn.
"That they may be one" : convincing a puzzled world that God sent His Son / by Thomas Donn Twitchell. — Plainfield, N.J. : Logos International, c1979.
xii, 216 p. ; 21 cm.
ISBN 0-88270-360-9 : $3.95
1. Twitchell, Thomas Donn. 2. Presbyterian Church—Clergy—Biography. 3. Clergy—United States—Biography. 4. Christian union. I. Title.
BX9225.T84A36 285'.1'0924 79-83792
79 MARC

Wayman, Alexander Walker, Bp., 1821-1895.
Cyclopaedia of African Methodism / by Alexander W. Wayman. — Baltimore : Methodist Episcopal Book Depository, 1882.
viii, 190 p. ; 20 cm.
1. African Methodist Episcopal Church—Clergy—Biography. 2. Clergy—United States—Biography. 3. African Methodist Episcopal Church—Bishops—Biography. 4. Bishops—United States—Biography. I. Title.
BX8442.W39 287'.83 78-326107
79 MARC

—UNITED STATES—BIOGRAPHY—ADDRESSES, ESSAYS, LECTURES

An Approach to Christian ethics : the life, contribution, and thought of T. B. Maston / William M. Pinson, Jr., compiler/contributor. — Nashville : Broadman Press, c1979.
204 p. ; 21 cm.
Includes bibliographical references.
ISBN 0-8054-6120-5
1. Maston, Thomas Bufford, 1897- —Addresses, essays, lectures. 2. Baptists—Clergy—Biography—Addresses, essays, lectures. 3. Clergy—United States—Biography—Addresses, essays, lectures. I. Pinson, William M.
BX6495.M362A66 286'.1'0924 78-71465
79 MARC

—UNITED STATES—BIOGRAPHY—JUVENILE LITERATURE

Haskins, James, 1941-
Andrew Young, man with a mission / James Haskins. — 1st ed. — New York : Lothrop, Lee & Shepard Co., c1979.
192 p. : ill. ; 24 cm.
Includes index.
SUMMARY: An account of the life of Andrew Young, including his activities as a clergyman, civil rights worker, legislator, and United States Ambassador to the United Nations.
ISBN 0-688-41896-1. ISBN 0-688-51896-6 lib. bdg.
1. Young, Andrew J., 1932- —Juvenile literature. 2. Ambassadors—United States—Biography—Juvenile literature. 3. Legislators—United States—Biography—Juvenile literature. 4. United States. Congress. House—Biography—Juvenile literature. 5. United Church of Christ—Clergy—Biography—Juvenile literature. 6. Clergy—United States—Biography—Juvenile literature. [1. Young, Andrew J., 1932- . 2. Statesmen. 3. Afro-Americans—Biography]
E840.8.Y64H37 973.92'092'4 79-1046
79 MARC

Simpson, Janice Claire.
Andrew Young : a matter of choice / by Jan Simpson. — St. Paul : EMC Corp., 1978.
40 p. : ill. ; 23 cm. — (Headlines I)
SUMMARY: A biography of the black congressman who was appointed United States ambassador to the United Nations in 1977.
ISBN 0-88436-472-0 lib. bdg. : $4.95. ISBN 0-88436-473-9 pbk. : $2.95
1. Young, Andrew J., 1932- . 2. United Church of Christ—Clergy—Biography—Juvenile literature. 3. Clergy—United States—Biography—Juvenile literature. 4. Ambassadors—United States—Biography—Juvenile literature. 5. Legislators—United States—Biography—Juvenile literature. 6. United States. Congress. House—Biography—Juvenile literature. [1. Young, Andrew J., 1932- . 2. Statesmen. 3. Afro-Americans—Biography] I. Title. II. Series.
E840.8.Y64S55 341.23'3'0924 77-29229
78 MARC

—UNITED STATES—CORRESPONDENCE

Cameron, Kenneth Walter, 1908-
The papers of loyalist Samuel Peters : a survey of the contents of his notebooks—correspondence during his flight to England, exile, and the last years of his life / by Kenneth Walther Cameron. — Hartford : Transcendental Books, [c1978]
251 leaves, [1] leaf of plates : ill. ; 29 cm.
Includes index.
1. Peters, Samuel, 1735-1826. 2. American loyalists—History—Sources. 3. Protestant Episcopal Church in the U.S.A.—Connecticut—History—Sources. 4. Protestant Episcopal Church in the U.S.A.—Vermont—History—Sources. 5. Connecticut—Church history—Sources. 6. Vermont—Church history—Sources. 7. American loyalists—Correspondence. 8. Protestant Episcopal Church in the U.S.A.—Clergy—Correspondence. 9. Clergy—United States—Correspondence. I. Title.
E278.P47C35 283'.092'4 79-100013
79 MARC

Carey, William, 1761-1834.
Serampore letters : being the unpublished correspondence of William Carey and others with John Williams, 1800-1816 / edited by Leighton and Mornay Williams ; with an introd. by Thomas Wright. — New York : G. P. Putnam's Sons, 1892.
v, 150 p., [10] leaves of plates : ill. ; 20 cm.
1. Carey, William, 1761-1834. 2. Williams, John, 1767-1825. 3. Missionaries—India—Correspondence. 4. Clergy—United States—Correspondence. I. Williams, John, 1767-1825. II. Williams, Leighton. III. Williams, Mornay, 1856- . IV. Title.
BV3269.C3A37 79-312026
79 MARC

—WALES—BIOGRAPHY

Hughes, J Williams.
Troeon yr yrfa / [gan] J. Williams Hughes. — Llandysul : Gwasg Gomer, 1978.
219 p. ; 22 cm. GB78-21786
Originally published in weekly installments in Seren Cymru.
Bibliography: p. 205-206.
ISBN 0-85088-399-7 : £1.25
1. Hughes, J. Williams. 2. Baptists—Clergy—Biography. 3. Clergy—Wales—Biography. I. Title.
BX6495.H75A34 79-341469
79 MARC

Namyn bugail / golygwyd gan T. J. Davies. — Llandysul : Gwasg Gomer, 1978.
158 p. ; 19 cm. GB78-34305
ISBN 0-85088-760-7 : £1.25
1. Clergy—Wales—Biography. 2. Wales—Biography. I. Davies, Thomas James.
BR779.N35 79-301699
79 MARC

—WASHINGTON (STATE)—WHATCOM CO.—BIOGRAPHY—ADDRESSES, ESSAYS, LECTURES

Johnson, Dorothy H
John A. Tennant, early pioneer & preacher : essays / by Dorothy H. Johnson, Percival R. Jeffcott & Michael Sullivan. — Bellingham, Wash. : Fourth Corner Registry, 1978.
xii, 72 p. : ill. ; 23 cm. — (Builders of Whatcom County)
CONTENTS: Johnson, D. H. John Tennant.—Jeffcott, P. R. John A. Tennant.—Jeffcott, P. R. A sketch of the religious life of John Tennant.—Sullivan, M. John Tennant's Lake.
1. Tennant, John A., 1829-1893—Addresses, essays, lectures. 2. Methodist Church—Clergy—Biography—Addresses, essays, lectures. 3. Clergy—Washington (State)—Whatcom Co.—Biography—Addresses, essays, lectures. 4. Whatcom Co., Wash.—Biography—Addressses, essays, lectures. I. Jeffcott, Percival Robert. II. Sullivan, Michael, 1950- . III. Title.
BX8495.T43J63 287'.6'0924 79-100991
79 MARC

—YUGOSLAVIA—SLOVENIA—BIOGRAPHY

Jagodic, Jože.
Mojega življenja tek : spomini / Jože Jagodic. — Celovec : Jagodic, 1974.
327 p. : ill. ; 21 cm. Au77-8-276
1. Jagodic, Jože. 2. Catholic Church—Clergy—Biography. 3. Clergy—Yugoslavia—Slovenia—Biography. 4. Slovenia—Biography. I. Title.
BX4705.J3166A35 79-372091
79 MARC

CLERGY, AFRO-AMERICAN see Afro-American clergy

CLERGY, TRAINING OF

see also subdivision Clergy, Training of under church denominations, e. g. Catholic Church—Training of

Andriessen, Hermanus Crispinus Ignatius, 1927-
Leren aan ervaring en supervisie / door H. C. I. Andriessen. — Nijmegen : Dekker & Van de Vegt, 1975.
xx, 551 p. : diagrs. ; 20 cm.
Bibliography: p. [503]-546.
ISBN 9025598234 : fl 69.50
1. Pastoral psychology—Study and teaching. 2. Clergy, Training of. 3. Clergy—Post-ordination training. I. Title.
BV4012.A635 75-541594
*79[r79]rev MARC

Andriessen, Hermanus Crispinus Ignatius, 1927-
Pastorale Supervision : Praxisberatung in d. Kirche / Herman Andriessen ; [aus d. Niederländ. von Kurt Moritz]. — 1. Aufl. — München : Kaiser ; Mainz : Grünewald, 1978.
243, [1] p. ; 21 cm. — (Gesellschaft und Theologie : Abteilung Praxis der Kirche ; Nr. 24) GFR78-A
Translation of Leren aan ervaring en supervisie.
Bibliography: p. 230-[244]
ISBN 3-459-01090-8
1. Pastoral psychology—Study and teaching. 2. Clergy, Training of. 3. Clergy—Post-ordination training. I. Title. II. Series.
BV4012.A63515 79-383056
79 MARC

CLERGY IN LITERATURE

Ortega, José, 1933-
La figura del sacerdote en la moderna narrativa española / José Ortega, Francisco Carenas. — 1. ed. — Caracas : Casuz Editores, 1975.
145 p. ; 20 cm. — (Colección Varia)
Includes bibliographical references.
1. Spanish fiction—19th century—History and criticism. 2. Spanish fiction—20th century—History and criticism. 3. Clergy in literature. I. Carenas, Francisco, 1938- II. Title.
PQ6140.R4O78 863'.5'09352 79-339118
79 MARC

CLERGY IN LITERATURE (Continued)

Rataboul, Louis J
Le pasteur anglican dans le roman victorien : aspects sociaux et religieux / Louis J. Rataboul. — Paris : Didier Érudition, c1978.
495 p. ; 24 cm. — (Études anglaises ; 70) F***
Bibliography: p. [449]-474.
Includes indexes.
ISBN 2-208-03264-0 : 180F
1. English fiction—19th century—History and criticism. 2. Clergy in literature. 3. Church of England in literature. I. Title. II. Series.
PR1.E8 no. 70 820'.9 s 79-367666
[PR878.C56] 79 MARC

CLERGYMEN AS AUTHORS

see also Clergymen's writings

CLERGYMEN'S FAMILIES

see also Clergy—Divorce; Clergymen's wives

Bailey, Robert W
Coping with stress in the minister's home / Robert W. Bailey and Mary Frances Bailey. — Nashville : Broadman Press, c1979.
136 p. ; 20 cm.
Includes bibliographical references and index.
ISBN 0-8054-5266-4
1. Clergymen's families. I. Bailey, Mary Frances, joint author. II. Title.
BV4396.B34 301.42'72'2 79-51134
00 MARC

CLERGYMEN'S WIVES

see also Clergymen's families

Denton, Wallace.
The minister's wife as a counselor. Philadelphia, Westminster Press [c1965]
172 p. 21 cm.
Includes bibliographical references.
1. Peer counseling in the church. 2. Clergymen's wives. I. Title.
BV4409.D46 253.5 66-10704
[r78]rev MARC

Hickman, Martha, Whitmore, 1925-
How to marry a minister. [1st ed.] Philadelphia, Lippincott [1968]
191 p. 21 cm. $3.95
1. Clergymen's wives. I. Title.
BV4395.H5 253'.2 68-29730
[r79]rev MARC

Senter, Ruth Hollinger, 1944-
So you're the pastor's wife / Ruth Senter. — Grand Rapids, Mich. : Zondervan Pub. House, c1979.
114 p. ; 21 cm.
ISBN 0-310-38820-1 : $5.95
1. Clergymen's wives. I. Title.
BV4395.S46 248'.843 79-9340
79 MARC

—RELIGIOUS LIFE

Benjamin, Carol.
The rib section : a handbook for pastors' wives / Carol Benjamin ; illustrated by Barbara Roderick. — Anchorage, Alaska : Abbott Loop Community Chapel, c1979.
ix, 168 p. : ill. ; 22 cm.
1. Clergymen's wives—Religious life. I. Title.
BV4395.B46 248'.843 79-108477
79 MARC

CLERGYMEN'S WRITINGS, ITALIAN

Sacerdoti-poeti del 900 italiano / [a cura di] Santino Spartà. — Roma : M. Spada, 1978.
219 p. : ill. ; 24 cm. It78-Oct
Includes bibliographies.
L6000
1. Italian poetry—20th century. 2. Clergymen's writings, Italian. I. Spartà, Santino, 1933-
PQ4214.S26 79-383270
*79 MARC

CLERICAL CELIBACY see Celibacy

CLERICAL EMPLOYEES see Clerks

CLERICAL MEDICINE see Pastoral medicine

CLERICAL OCCUPATIONS

Turner, David Reuben, 1915-
Office aide / by David R. Turner. — 1st ed. — New York : Arco, c1978.
224 p. : ill. ; 26 cm. — (The Complete Arco test-tutor) (Arco civil service test tutor)
ISBN 0-668-04704-6 : $8.00
1. Clerks—Examinations, questions, etc. 2. Office practice—Examinations, questions, etc. 3. Clerical occupations. I. Title.
HF5501.T92 651'.374'076 78-16053
78 MARC

CLERICAL PSYCHOLOGY see Pastoral psychology

"CLERKLY" STYLE WRITING see Chinese language—Writing, Li style

CLERKS

see also Bank employees; Clerical occupations; Office practice

Finkelhor, Dorothy C 1902-
Occupational adjustments of beginning office workers; 1942 research award of Delta pi epsilon, national graduate fraternity of business education, by Dorothy C. Finkelhor ... Stillwater, Okl., Oklahoma agricultural and mechanical college, 1943.
ix, 86 p. 20 cm.
Bibliography: p. 84-86.
1. Clerks. 2. Office practice. I. Oklahoma. State University of agriculture and applied science, Stillwater. II. Title.
HF5501.F53 1943 651'.3743 44-42532
[r79]rev MARC

—EXAMINATIONS, QUESTIONS, ETC.

Turner, David Reuben, 1915-
Office aide / by David R. Turner. — 1st ed. — New York : Arco, c1978.
224 p. : ill. ; 26 cm. — (The Complete Arco test-tutor) (Arco civil service test tutor)
ISBN 0-668-04704-6 : $8.00
1. Clerks—Examinations, questions, etc. 2. Office practice—Examinations, questions, etc. 3. Clerical occupations. I. Title.
HF5501.T92 651'.374'076 78-16053
78 MARC

—LEGAL STATUS, LAWS, ETC.—SPAIN

Spain. Ministerio de Trabajo.
Ordenanza laboral : empresas consignatarias de buques / edición con introducción, referencias legales, sentencias, resoluciones administrativas y comentarios a cargo de Alejandro Mulas y Ramón Ayerra. — [Madrid] : Servicio de Publicaciones, Ministerio de Trabajo, D.L. 1976.
128 p. ; 18 cm. — (Serie Ordenanzas laborales ; RET 76-768) (Biblioteca de textos legales) Sp78-Jun
On cover: Empresas consignatarias de buques.
Includes index.
ISBN 8450017211 : 150ptas
1. Clerks—Legal status, laws, etc.—Spain. 2. White collar workers—Legal status, laws, etc.—Spain. 3. Labor laws and legislation—Spain. 4. Shipping—Spain. I. Mulas García, Alejandro. II. Ayerra, Ramón. III. Title: Empresas consignatarias de buques.
79-356512
79 MARC

—VOCATIONAL GUIDANCE see Clerical occupations

—GERMANY, WEST—SALARIES, PENSIONS, ETC.

Bundesversicherungsanstalt für Angestellte.
Die Versichertenältesten. Verzeichnis der in den Stadt- und Landkreisen tätigen Versichertenältesten. Stand: 1. März 1972. Berlin, [1972?]
44 p. 30 cm. (Its Schriftenreihe 20) GDB***
Cover title.
1. Clerks—Germany, West—Salaries, pensions, etc. 2. Social security—Germany, West. I. Title. II. Series.
HD716.M42G4334 Nr. 20 1972 73-338689
368.008 s
[HD4966.M4G3] 79[r79]rev MARC

—INDIA—MAHARASHTRA (STATE)—JOB DESCRIPTIONS

Maharashtra, India (State)
A brochure indicating the duties of general duty clerks in Secretariat Departments. — [Bombay] : Govt. Central Press, [1961]
34 p. ; 25 cm.
1. Maharashtra, India (State)—Officials and employees. 2. Clerks—India—Maharashtra (State)—Job descriptions. I. Title.
JQ299.M334M33 1961 354'.54'79200103 79-306954
79 MARC

—IRELAND

National Economic and Social Council.
Service-type employment and regional development / National Economic and Social Council. — Dublin : Stationery Off. : [sold by] Govt. Publications Sale Off., [1977]
142 p. : ill. ; 20 cm. — ([Report] - National Economic and Social Council ; no. 28) Ir***
Bibliography: p. 137-138.
£0.55
1. Regional planning—Ireland. 2. Service industries—Ireland. 3. Service industries workers—Ireland. 4. Offices—Location—Ireland. 5. Clerks—Ireland. I. Title. II. Series: National Economic and Social Council. Report — National Economic and Social Council ; no. 28.
HT395.I7N38 1977 309.2'5'09415 78-322724
79 MARC

—ITALY

Gasparini, Giovanni, 1940-
Gli impiegati : una analisi della condizione impiegatizia nel lavoro, nel sindacato, nella sfera extra-lavorativa / Giovanni Gasparini. — Milano : F. Angeli, 1978.
257 p. ; 22 cm. — (Collana di sociologia industriale ; sez. II, 4) It78-Dec
Includes bibliographical references.
ISBN 8820414996 : L7000
1. Clerks—Italy. 2. Trade-unions—Clerks—Italy. 3. Industrial relations—Italy. I. Title.
HD8039.M4I85 79-346224
*79 MARC

—MASSACHUSETTS—BROCKTON METROPOLITAN AREA—PERIODICALS

Massachusetts. Division of Employment Security.
Labor area trends in supply and demand: Brockton, Massachusetts, standard metropolitan statistical area. Boston [Division of Employment Security]
v. 28 cm.
1. Labor supply—Massachusetts—Brockton metropolitan area—Periodicals. 2. Machinists—Massachusetts—Brockton metropolitan area — Periodicals. 3. Practical nurses — Massachusetts — Brockton metropolitan area—Periodicals. 4. Clerks—Massachusetts—Brockton metropolitan area—Periodicals. I. Title.
HD5726.B76M35a 331.1'2'0974482 79-643342
MARC-S

—NORWAY

Norway. Laws, statutes, etc.
Lovsamling for grunnkurs i studieretning for handels- og kontorfag / ved Ragnvald Thilesen. — 6. [rev.] utg. — Oslo : Gyldendal, c1977.
351 p. ; 21 cm. N***
ISBN 8205096317
1. Commercial law—Norway. 2. Clerks—Norway. I. Thilesen, Ragnvald. II. Title.
78-389964
79 MARC

CLERKS (Continued)

—PERU—LIMA

León, Federico R
Factores considerados por empleados de oficina y trabajadores manuales al "ofrecérseles" un puesto / Federico R. León & Gerardo Pejerrey P. — Lima : ESAN, Departamento de Investigación, 1978.
32 p. : graphs ; 30 cm. — (Serie Documento de trabajo - ESAN, Departamento de Investigación ; no. 17)
Bibliography: p. 29-32.
1. Clerks—Peru—Lima. 2. Labor and laboring classes—Peru—Lima. 3. Job satisfaction. I. Pejerrey P., Gerardo, joint author. II. Title. III. Series: Escuela de Administración de Negocios para Graduados. Departamento de Investigación. Serie Documentos de trabajo - ESAN, Departamento de Investigación ; no. 17.
HD8039.M4P474 78-107475
79 MARC

—UNITED STATES

Palmer, Phyllis M
The status of clerical workers : a summary analysis of research findings and trends, with bibliography / Phyllis M. Palmer, Sharon L. Grant. — [Washington] : Women's Studies Program, George Washington University : [order from Business and Professional Women's Foundation], c1979.
69 p. ; 22 cm.
Bibliography: p. 52-69.
1. Clerks—United States. 2. Women—Employment—United States. I. Grant, Sharon L., joint author. II. Title.
HD6073.M392U555 79-115034
331.2'81'6513740973
79 MARC

CLERKS (RETAIL TRADE)

see also Drugstore employees; Sales personnel; Wages—Clerks (Retail trade)

—HEALTH AND HYGIENE

Pei-ching t'i yü hsüeh yüan.
(Shou huo yüan ts'ao)
售货员操：生产操 / 北京体育学院编. — 第1版. — [北京]：人民体育出版社，1975.
folder (16 p.) : ill. ; 19 cm.
Colophon title.
In case.
$0.80
1. Exercise. 2. Clerks (Retail trade)—Health and hygiene. I. Title.
RA781.P43 1975c 78-842787

—PROGRAMMED INSTRUCTION

Hartzler, F E
The retail salesperson : a programmed text / F. E. Hartzler. — 2d ed. — New York : Gregg Division, McGraw-Hill Book Co., c1979.
viii, 120 p. : ill. ; 28 cm. — (A Job skill manual in marketing)
ISBN 0-07-026967-X
1. Selling—Programmed instruction. 2. Clerks (Retail trade)—Programmed instruction. I. Title.
[HF5438.2.H37 1979] 658.85'07'7 78-15138
78 MARC

—CHILE—ÑUÑOA

Programa Regional del Empleo para América Latina y el Caribe.
Comercio informal en una comuna de Santiago / Organización Internacional del Trabajo, OIT, Programa Mundial del Empleo, Programa Regional del Empleo para América Latina y el Caribe (P.R.E.A.L.C.). — [Santiago, Chile] : P.R.E.A.L.C., 1978.
vii, 96 p. ; 28 cm. — (Documento de trabajo - PREALC ; 122)
Edited by Alois Möller.
Bibliography: p. 95-96.
1. Clerks (Retail trade)—Chile—Ñuñoa. 2. Retail trade—Chile—Ñuñoa. 3. Small business—Chile—Ñuñoa. I. Möller, Alois. II. Title. III. Series: Programa Regional del Empleo para América Latina y el Caribe. Documento de trabajo - PREALC ; 122.
HD8039.M4C476 381 79-100832
79 MARC

—HUNGARY

Molnár, György, fl. 1961-
A kereskedelem és a vendéglátóipar munkaügyi kérdései. [Írták: Molnár György, Sándor Gyula és Vándor Sándor, szerk. Koszó Károly] Budapest, Közgazdasági és Jogi Könyvkiadó, 1962.
446 p. 20 cm.
Bibliographical footnotes.
1. Clerks (Retail trade)—Hungary. 2. Restaurants, lunch rooms, etc.—Employees—Hungary. I. Sándor, Gyula, joint author. II. Vándor, Sándor, joint author. III. Koszó, Károly, ed. IV. Title.
HD8039.M4H84 71-207037
78[r78]rev2 MARC

—JAPAN

Japan. Chūshō Kigyōchō.
[Shōten rōmu no shishin]
商店の労務 / 中小企業庁編. — 東京：中小企業診断協会，昭和33 [1958]
2, 240 p. ; 22 cm.
Issued earlier under title: Shōten rōmu no shishin.
1. Clerks (Retail trade)—Japan. 2. Labor laws and legislation—Japan. I. Title.
Title romanized: Shōten no rōmu.
HD8039.M4J334 1958 79-808206

—NORWAY—ATTITUDES

Norsk produktivitetsinstitutt.
Arbeidsmiljø i detaljhandelen. — Oslo : Norsk produktivitetsinstitutt, 1978.
iv, 92, [26] p. ; 30 cm. — (NPI-prosjekt ; 1072) N***
Includes bibliography.
1. Clerks (Retail trade)—Norway—Attitudes. 2. Work environment—Norway. I. Title.
HD8039.M4N76 79-370825
79 MARC

CLERKS OF COURT

—FRANCE—HANDBOOKS, MANUALS, ETC.

Martinetti, Yves.
Règles pratiques à l'usage des secrétaires-greffiers des cabinets d'instruction / Yves Martinetti, ... ; [publié par le Ministère de la justice]. — [Paris] : [Ministère de la justice], 1976.
211 p. ; 26 cm. F78-662
Cover title: Le secrétaire-greffier d'instruction, règles pratiques.
Includes index.
1. Procedure (Law)—France—Compends. 2. Clerks of court—France—Handbooks, manuals, etc. I. Title. II. Title: Le secrétaire-greffier d'instruction, règles pratiques.
347'.44'05 78-372177
78 MARC

—GERMANY, WEST

Arnold, Egon.
Rechtspflegergesetz : Kommentar / von Egon Arnold u. Klaus Meyer-Stolte. — 3., neubearb. Aufl., zugl. 4. Aufl. von Arnold, Stellung und Aufgaben des Rechtspflegers. — Bielefeld : Gieseking, 1978.
xvi, 432 p. ; 23 cm. GFR78-A
Includes the text of the Rechtspflegergesetz.
Previous ed. published in 1970 under title: Rechtspflegergesetz 1970.
"Literatur- und Abkürzungsverzeichnis": p. ix-xvi.
Includes index.
ISBN 3-7694-0635-4 : DM88.00
1. Judges—Germany, West. 2. Clerks of court—Germany, West. I. Meyer-Stolte, Klaus, joint author. II. Germany (Federal Republic, 1949-). Laws, statutes, etc. Rechtspflegergesetz. 1978. III. Title.
347'.43'014 79-356827
79 MARC

—JAPAN

Utsuki, Yoshiharu.
(Keiji kōsoshin ni okeru shokikan jimu no jisshōteki kenkyū)
刑事控訴審における書記官事務の実証的研究 / 宇津木義治. — [東京]：裁判所書記官研修所，[はしがき 昭和44 i. e. 1969]
3, 11, 325 p. ; 21 cm. — (書記官実務研究報告書 - 裁判所書記官研修所 ; 第8巻第2号)
"昭和四四年度書記官実務研究"
Bibliography: p. 2-3 (1st group)
1. Appellate procedure—Japan. 2. Appellate procedure—Japan—Forms. 3. Clerks of court—Japan. I. Title. II. Series: Saibansho Shokikan Kenshūjo. Shokikan jitsumu kenkyū hōkokusho - Saibansho Shokikan Kenshūjo ; dai 8-kan, dai 2-gō.
79-801666

—UNITED STATES

Mason, Anthony K
Improving productivity in the courts : a primer for clerks of court / Anthony K. Mason. — [Washington] : National Institute of Law Enforcement and Criminal Justice, Law Enforcement Assistance Administration, U.S. Dept. of Justice : for sale by the Supt. of Docs., U.S. Govt. Print. Off., 1978.
ix, 58 p. : ill. ; 27 cm.
On cover: Prescriptive package.
Bibliography: p. 55-56.
1. Court administration—United States. 2. Clerks of court—United States. I. Title. II. Title: A primer for clerks of court.
KF8732.M37 347'.73'13 79-601939
79 MARC

CLERMONT-FERRAND, FRANCE

—ANTIQUITIES—EXHIBITIONS

Musée Bargoin.
Archéologie : guide / Musée Bargoin ; [rédigé] par Jean-Claude Ruiz .... — Clermont-Ferrand : Musée Bargoin, [1975]
[26] p. : ill. ; 22 cm. F76-17378
5.00F
1. Clermont-Ferrand, France—Antiquities—Exhibitions. 2. France—Antiquities—Exhibitions. 3. Musée Bargoin—Exhibitions. I. Ruiz, Jean Claude. II. Title.
DC801.C58M87 1975 936.4 78-389430
79 MARC

—CITY PLANNING

Dujardin, Daniel.
Voyage au centre de Clermont-Fd / Daniel Dujardin, Françoise Lorgeoux. — Clermont-Ferrand : G. de Bussac, [1977]
344 p. : ill. (some col.) ; 21 cm. F***
Bibliography: p. 281-288.
66.00F
1. Clermont-Ferrand, France—City planning. 2. City planning—France. 3. Urban renewal—France—Clermont-Ferrand. I. Lorgeoux, Françoise, joint author. II. Title.
HT169.F72C573 309.2'62'0944591 79-361727
79 MARC

CLERMONT-GANNEAU, CHARLES SIMON, 1846-1923

—ADDRESSES, ESSAYS, LECTURES

Dupont-Sommer, André.
Un Dépisteur de fraudes archéologiques, Charles Clermont-Ganneau : 1846-1923 ... : lecture faite dans la séance publique annuelle du 6 décembre 1974, Institut de France, Académie des inscriptions et belles-lettres / par M. André Dupont-Sommer, ... — Paris : Institut de France, 1974.
20 p. ; 28 cm. F75-12475
1. Clermont-Ganneau, Charles Simon, 1846-1923—Addresses, essays, lectures. 2. Orientalists—France—Biography—Addresses, essays, lectures. I. Title.
DS61.7.C56D86 950'.07'2024 79-346409
79 MARC

CLERT, IRIS

Clert, Iris.
Iris-time : l'artventure / Iris Clert. — Paris : Denoël, 1978.
363 p., [16] p. of plates ; 23 cm. F79-2613
Text in French.
Includes index.
58.00F
1. Clert, Iris. 2. Art dealers—France—Biography. I. Title.
N8660.C55A35 706'.5 79-375028
79 MARC

CLEVELAND, JOHN, 1613-1658

—BIBLIOGRAPHY

Morris, Brian Robert.
John Cleveland (1613-1658): a bibliography of his poems, by Brian Morris. London, Bibliographical Society, 1967.
vii, 61 p. front. (port.) 26 cm. (Bibliographical Society publication 1965)
40/- B
1. Cleveland, John, 1613-1658—Bibliography. I. Series: Bibliographical Society, London. Publication, 1965.
Z8176.5.M6 016.8214 67-107421
[PR3348.C7] MARC
——— Copy 3. Rosenwald Coll.
[r79]rev2

CLEVELAND

—BIOGRAPHY

Wick, Warren Corning.
My recollections of old Cleveland : manners, mansions, mischief / Warren Corning Wick ; Joanne M. Lewis, editor. — Limited 1st ed. — [Cleveland] : Wick, c1979.
120 p. : ill. ; 20 x 23 cm.
1. Cleveland—Social life and customs. 2. Upper classes—Ohio—Cleveland. 3. Wick, Warren Corning. 4. Cleveland—Biography. I. Lewis, Joanne M. II. Title.
F499.C65W52 977.1'32 79-114338
00 MARC

—DESCRIPTION—VIEWS

Vincent, Sidney Z
Merging traditions : Jewish life in Cleveland : a contemporary narrative, 1945-1975, a pictorial record, 1839-1975 / by Sidney Z. Vincent and Judah Rubinstein. — 1st ed. — Cleveland : Western Reserve Historical Society, c1978.
283 p. : ill. ; 25 cm. — ([Western Reserve Historical Society publication ; no. 144])
Includes bibliographical references.
ISBN 0-911704-18-3 : $10.00
1. Jews in Cleveland—History—20th century. 2. Jews in Cleveland—Pictorial works. 3. Cleveland—History—20th century. 4. Cleveland—Description—Views. I. Rubinstein, Judah, joint author. II. Title. III. Series: Western Reserve Historical Society, Cleveland. Publication ; no. 144.
F499.C69J548 977.1'32'004924 78-10307
78 MARC

—FOREIGN POPULATION

Sári Gál, Imre.
Clevelandi magyar múzeum : riportok, versek, fényképek a clevelandi magyarság életéből / Sári Gál Imre. — Toronto : Amerikai Magyar Írók, 1978.
208 p. : ill. ; 22 cm. C***
Errata slip inserted.
1. Hungarian Americans—Ohio—Cleveland. 2. Cleveland—Foreign population. I. Title.
F499.C69H957 78-393153
78 MARC

—HISTORY

Gartner, Lloyd P 1927-
History of the Jews of Cleveland / Lloyd P. Gartner. — Cleveland : Western Reserve Historical Society, c1978.
xv, 385 p. : ill. ; 25 cm. — (Regional history series of the American Jewish History Center of the Jewish Theological Seminary of America ; v. 4) ([The Western Reserve Historical Society publication ; no. 143])
Includes bibliographical references and index.
ISBN 0-911704-17-5
1. Jews in Cleveland—History. 2. Cleveland—History. I. Title. II. Series: Jewish Theological Seminary of America. American Jewish History Center. Regional history series ; v. 4. III. Series: Western Reserve Historical Society, Cleveland. Publication ; no. 143.
F499.C69J545 977.1'32'004924 78-5868
78 MARC

—HISTORY—20TH CENTURY

Vincent, Sidney Z
Merging traditions : Jewish life in Cleveland : a contemporary narrative, 1945-1975, a pictorial record, 1839-1975 / by Sidney Z. Vincent and Judah Rubinstein. — 1st ed. — Cleveland : Western Reserve Historical Society, c1978.
283 p. : ill. ; 25 cm. — ([Western Reserve Historical Society publication ; no. 144])
Includes bibliographical references.
ISBN 0-911704-18-3 : $10.00
1. Jews in Cleveland—History—20th century. 2. Jews in Cleveland—Pictorial works. 3. Cleveland—History—20th century. 4. Cleveland—Description—Views. I. Rubinstein, Judah, joint author. II. Title. III. Series: Western Reserve Historical Society, Cleveland. Publication ; no. 144.
F499.C69J548 977.1'32'004924 78-10307
78 MARC

—LAW see Law—Cleveland

—MAPS

Commercial Survey Company, Cleveland.
Official street atlas of Cleveland and Cuyahoga county including Avon, Avon Lake, Eastlake, Timberlake, Lakeline, Wickliffe, Willowick, Willoughby. — 1977-1978 ed. — Cleveland : Commercial Survey Co., c1976.
[13], 39 p. : chiefly maps ; 30 cm. — (Map - Commercial Survey Company ; no. 7381-W)
Cover title.
Scale of maps ca. 1:36,000
Includes index.
$2.00
1. Cleveland—Maps. 2. Cuyahoga Co., Ohio—Maps. I. Title.
G1399.C5C6 1976 912'.771'31 79-306732
79 MARC

—NIGHTCLUBS, DANCE HALLS, ETC.

Cleveland. Dance hall inspector.
Report of Robb O. Bartholomew, dance hall inspector. Cleveland, O., 1912.
Cover-title, 16 p. 23 cm.
Dated August 24th, 1912. A report of one year's regulation of Cleveland's dance halls conducted by virtue of Ordinance no. 20456, which became effective May 15th, 1911.
1. Cleveland—Nightclubs, dance halls, etc. I. Bartholomew, Robert O.
GV1623.C4 15-3320
MARC

—OFFICIALS AND EMPLOYEES—
SALARIES, ALLOWANCES, ETC.

United States. Bureau of Labor Statistics. North Central Regional Office.
Municipal government wage survey, Cleveland, Ohio, June 1977. — Chicago : U.S. Dept. of Labor, Bureau of Labor Statistics, North Central Region V, 1978.
v, 52, 16 p. ; 28 cm. — (Regional report - Bureau of Labor Statistics, North Central Region V ; 78-3)
Chiefly tables.
1. Cleveland, Ohio—Officials and employees—Salaries, allowances, etc. I. Title. II. Series: United States. Bureau of Labor Statistics. North Central Regional Office. Regional report - Bureau of Labor Statistics, North Central Regional Office ; 78-3.
HD8083.A4N63 no. 78-3 78-602395
331'.0977 s
[JS774.A4] 79 MARC

—PUBLIC WORKS

Humphrey, Nancy.
The future of Cleveland's capital plant / Nancy Humphrey, George E. Peterson, Peter Wilson. — Washington : Urban Institute, c1979.
xvi, 77 p. ; 23 cm. — (America's urban Capital Stock ; v. 2)
Includes bibliographical references.
ISBN 0-87766-258-4 : $4.00
1. Cleveland—Public works. 2. Capital investments—Ohio—Cleveland. 3. Infrastructure (Economic)—Ohio—Cleveland. 4. Finance, Public—Ohio—Cleveland. I. Peterson, George E., joint author. II. Wilson, Peter M., joint author. III. Title. IV. Series.
HD4606.C6H85 336.771'32 79-66670
79 MARC

—SOCIAL CONDITIONS

Alissi, Albert S
Boys in Little Italy : a comparison of their individual value orientations, family patterns, and peer group associations / Albert S. Alissi. — San Francisco : R & E Research Associates, 1978.
ix, 119 p. ; 28 cm.
Bibliography: p. 112-119.
1. Italian American children—Ohio—Cleveland. 2. Family—Ohio—Cleveland. 3. Social classes—Ohio—Cleveland. 4. Boys. 5. Cleveland—Social conditions. I. Title.
F499.C69I826 301.43'14'0977132 77-90360
79 MARC

—SOCIAL LIFE AND CUSTOMS

Wick, Warren Corning.
My recollections of old Cleveland : manners, mansions, mischief / Warren Corning Wick ; Joanne M. Lewis, editor. — Limited 1st ed. — [Cleveland] : Wick, c1979.
120 p. : ill. ; 20 x 23 cm.
1. Cleveland—Social life and customs. 2. Upper classes—Ohio—Cleveland. 3. Wick, Warren Corning. 4. Cleveland—Biography. I. Lewis, Joanne M. II. Title.
F499.C65W52 977.1'32 79-114338
00 MARC

CLEVELAND. CHAMBER OF COMMERCE

Cleveland. Chamber of Commerce.
Annual refection of the Cleveland Chamber of Commerce, Tuesday, April the 16, 1895. — [Cleveland : The Chamber of Commerce, 1895]
[4] p. (on double leaves) : 1 ill. ; 22 cm.
Menu and after-dinner program for a banquet celebrating the 47th anniversary of the Cleveland Chamber of Commerce.
The illustration (allegory of Cleveland) is on t.p.
An original layout in pen-and-ink and pencil accompanies L.C. copy.
1. Cleveland. Chamber of Commerce. I. Title.
HF296.C656 1895 78-320127
78 MARC

CLEVELAND. FOOTBALL CLUB (NATIONAL LEAGUE, BROWNS)

Brown, Paul E 1908-
PB, the Paul Brown story / Paul Brown with Jack Clary. — 1st ed. — New York : Atheneum, 1979.
338 p. : ill. ; 24 cm.
ISBN 0-689-10985-7 : $12.95
1. Brown, Paul E., 1908- . 2. Football coaches—United States—Biography. 3. Cleveland. Football club (National League, Browns) 4. Cincinnati Bengals (Football club) I. Clary, Jack T., joint author. II. Title.
GV939.B77A35 1979 796.33'2'0924 79-7314
79 MARC

CLEVELAND. PUBLIC LIBRARY

—HISTORY

Braverman, Miriam.
Youth, society, and the public library / by Miriam Braverman. — Chicago : American Library Association, 1979.
xvi, 289 p. ; 25 cm.
Bibliography: p. 267-282.
Includes index.
ISBN 0-8389-0260-X
1. Libraries, Young people's—United States—History—20th century. 2. New York (City). Public Library—History. 3. Cleveland. Public Library—History. 4. Enoch Pratt Free Library, Baltimore—History. I. Title.
Z718.5.B72 027.62'5 78-17267
78 MARC

CLEVELAND, ENG.

—DESCRIPTION—GUIDE-BOOKS

Woodhouse, Robert.
Discovering Cleveland / [by] R. Woodhouse. — Aylesbury : Shire Publications, 1978.
63 p. : ill., map ; 18 cm. — (Discovering series ; 234) GB78-29041
Bibliography: p. 61.
Includes index.
ISBN 0-85263-412-9 : £0.60
1. Cleveland, Eng.—Description—Guide-books. I. Title.
DA690.C6218W66 914.28'5'04857 79-300662
79 MARC

CLEVELAND METROPOLITAN AREA

—MAPS

Champion Map Corporation.
Metropolitan Cleveland atlas. Charlotte [1969]
54 p. col. maps. 36 cm.
Cover title.
Includes all of Cuyahoga County and built-up areas in Lorain and Lake Counties.
1. Cleveland metropolitan area—Maps. I. Title.
G1399.C5C5 1969 912.771'31 70-653771
70[r79]rev MARC

CLEVELAND MUSEUM OF ART

Cleveland Museum of Art.
Handbook. — [Cleveland] : Cleveland Museum of Art, 1978.
xiv, 424 p. : ill. ; 23 cm.
Includes index.
ISBN 0-910386-31-5. ISBN 0-910386-30-7 pbk.
1. Art—Ohio—Cleveland. 2. Cleveland Museum of Art.
N552.A6 1977 708'.171'32 76-54618
77 MARC

CLEVELAND MUSEUM OF ART (Continued)

Milliken, William Mathewson, 1889-
Born under the sign of Libra : an autobiography / William Mathewson Milliken. — Cleveland : Western Reserve Historical Society ; distributed by Howard Allen, c1977.
299 p. : ill. ; 24 cm.
Includes index.
ISBN 0-911704-16-7 : $10.00
1. Milliken, William Mathewson, 1889- . 2. Museum directors—United States—Biography. 3. Cleveland Museum of Art. I. Title.
N552.M543 708.00924 77-15426
77 MARC

CLEVELAND ORCHESTRA

—DISCOGRAPHY

Fellers, Frederick P
Discographies of commercial recordings of the Cleveland Orchestra (1924-1977) and the Cincinnati Symphony Orchestra (1917-1977) / compiled by Frederick P. Fellers and Betty Meyers. — Westport, Conn. : Greenwood Press, 1978.
xii, 211 p. ; 24 cm.
Includes bibliographies and indexes.
ISBN 0-313-20375-X : $22.50
1. Cleveland Orchestra—Discography. 2. Cincinnati Symphony Orchestra—Discography. I. Meyers, Betty, joint author. II. Title.
ML156.7.C48F4 016.7899'12 78-3122
78 MARC

CLEVELAND WAY, ENG.

—DESCRIPTION AND TRAVEL—GUIDE-BOOKS

Boyes, Malcolm.
A guide to the Cleveland Way and Missing Link / [by] Malcolm Boyes. — London : Constable, 1977.
166 p. : ill., maps ; 18 cm. GB78-01259
Bibliography: p. 160-161.
Includes index.
ISBN 0-09-461390-7 : £2.95.
1. Hiking—England—Cleveland Way—Guide-books. 2. Hiking—England—Missing Link Route—Guide-books. 3. Cleveland Way, Eng.—Description and travel—Guide-books. 4. Missing Link Route, England—Description and travel—Guide-books. I. Title.
GV199.44.G72C533 1977 79-307664
914.28'4
79 MARC

CLICHY, FRANCE

—SOCIAL CONDITIONS

Claudel, Paul, 1868-1955.
Chroniques du "Journal de Clichy" / Claudel et Mauriac. Correspondance Claudel-Fontaine / textes établis et annotés par François Morlot et Jean Touzot. — Paris : les Belles lettres, 1978.
158 p. ; 24 cm. — (Annales littéraires de l'Université de Besançon ; 206) ([Publication] - Centre de recherches de littérature française (XIXe et XXe siècles) ; v. 21) F78-12911
Includes bibliographical references.
80.00F
1. Claudel, Paul, 1868-1955—Correspondence. 2. Fontaine, Daniel, 1862-1920. 3. Authors, French—20th century—Political and social views. 4. Clichy, France—Social conditions. 5. Authors, French—20th century—Correspondence. 6. Clergy—France—Clichy—Correspondence. 7. Catholic Church—Clergy—Correspondence. I. Mauriac, François, 1885-1970. II. Fontaine, Daniel, 1862-1920. III. Title. IV. Series: Besançon, France. Université. Annales littéraires ; 206. V. Series: Centre de recherches de littérature française (XIXe et XXe siècles). Publication ; v. 21.
AS161.B39 no. 206 848'.91209 79-350086
[PQ2605.L2] 79 MARC

CLIENT-CENTERED PSYCHOTHERAPY

Bommert, Hanko, 1941-
Grundlagen der Gesprächspsychotherapie : Theorie, Praxis, Forschung / Hanko Bommert. — 1. Aufl. — Stuttgart ; Berlin ; Köln ; Mainz : Kohlhammer, 1977.
178 p. ; 21 cm. GFR77-A
Bibliography: p. 159-172.
Includes indexes.
ISBN 3-17-001341-6
1. Client-centered psychotherapy. 2. Group psychotherapy. I. Title.
RC481.B65 79-383279
79 MARC

Cladder, J M
Ouderkursussen op M.O.B.'s / door J. M. Cladder. — Amsterdam : Swets & Zeitlinger, 1979.
vii, 207 p. ; 25 cm. — (Social-wetenschappelijke proefschriften relevant voor de hulpverlening ; nr. 2) Ne79-4
Thesis—Utrecht.
Summary in English.
Bibliography: p. 157-163.
ISBN (invalid) 0-902650-29-8 : fl 31.20
1. Family psychotherapy—Netherlands. 2. Client-centered psychotherapy. 3. Behavior therapy. 4. Parent and child. I. Title. II. Series.
RC488.5.C55 79-372946
*79 MARC

Kirschenbaum, Howard.
On becoming Carl Rogers / by Howard Kirschenbaum. — New York : Delacorte Press, c1979.
xvii, 444 p., [4] leaves of plates : ill. ; 24 cm.
Bibliography: p. 435.
Includes index.
ISBN 0-440-06707-3
1. Rogers, Carl Ransom, 1902- . 2. Psychologists—United States—Biography. 3. Humanistic psychology. 4. Client-centered psychotherapy. I. Title.
BF109.R63K57 150'.19'2 78-13308
78 MARC

Paul, Irving H
The form and technique of psychotherapy / I. H. Paul. — Chicago : University of Chicago Press, 1978.
xix, 299 p. ; 24 cm.
Includes bibliographical references and index.
ISBN 0-226-64999-7 : $20.00
1. Psychotherapy. 2. Psychoanalysis. 3. Client-centered psychotherapy. I. Title.
RC480.5.P34 616.8'914 78-3181
78 MARC

CLIFF-DWELLINGS

—COLORADO—MESA VERDE

Nordenskiöld, Gustaf Erik Adolf, 1968-1895.
The cliff dwellers of the Mesa Verde, southwestern Colorado : their pottery and implements / by G. Nordenskiöld ; translated by D. Lloyd Morgan. — Glorieta, N.M. : Rio Grande Press, c1979.
36, 174, xi p., [74] leaves of plates : ill. (some col.) ; 29 cm. — (A Rio Grande classic)
Reprint of the 1893 ed. which was published by P. A. Norstedt & Söner, Stockholm.
Translation of Ruiner af klippboningar i Mesa Verde's cañons.
Includes bibliographical references and index.
ISBN 0-87380-127-X : $27.50
1. Cliff-dwellings—Colorado—Mesa Verde. 2. Mesa Verde. 3. Indians of North America—Colorado—Antiquities. 4. Colorado—Antiquities. I. Title.
E78.C6N6713 1979 978.8'27'00497 78-32107
79 MARC

CLIFFORD, GEORGE, 1685-1760

Linné, Carl von, 1707-1778.
Hortus Cliffortianus, plantas exhibens quas in hortis tam vivis quam siccis, Hartecampi in Hollandia, coluit ... Georgius Clifford ... reductis varietatibus ad species, speciebus ad genera, generibus ad classes, adjectis locis plantarum natalibus differentiisque specierum. Cum tabulis aeneis. Auctore Carolo Linnaeo. Amstelaedami, 1737 [i.e. 1738]
[32], x, 231, 301-501, [17] p. front., 36 plates. 43 cm.
Not published until about Sept. 1738. Cf. Hunt botanical cat., 504.
Plates engraved by J. Wandelaar (20 after designs by G. D. Ehret)
1. Botany—Classification. 2. Botany—Netherlands—Catalogs and collections. 3. Botnical illustration. 4. Hartekamp, Netherlands. 5. Clifford, George, 1685-1760. I. Wandelaar, Jan, 1690-1759. II. Ehret, Georg Dionysius, 1708-1770. III. Title.
QK91.H67 Rosenwald Coll. 6-21502
[r79]rev MARC

CLIFFORD, RICHARD L.

Clifford, Richard L
A human touch / by Richard L. Clifford ; [edited by Carol Mulvehill ; ill. by Robert Handville]. — Maryknoll, N.Y. : Maryknoll Fathers, c1979.
93 p., [1] leaf of plates : ill. ; 26 cm.
1. Clifford, Richard L. 2. Missionaries—Peru—Biography. 3. Missionaries—United States—Biography. I. Mulvehill, Carol. II. Title.
BV2853.P7C543 266'.2'0924 79-50933
79 MARC

CLIFFORD FAMILY

Clifford, Robert Laning.
Family histories : Clifford family, January 1946, updated April 1977, Kikendall family, January 1946, Lambert family, January 1946, Laning family, December 1967, Short family, January 1946 / researched, collated, and drafted by Robert L. Clifford. — Williamsburg, Va. : Clifford, [1977]
[17] leaves ; 28 cm.
1. Clifford family. I. Title.
CS71.C637 1977 929'.2'0973 79-313492
79 MARC

CLIMACTERIC

see also Climacteric, Male; Menopause

CLIMACTERIC, FEMALE see Menopause

CLIMACTERIUM VIRILE see Climacteric, Male

CLIMATERIC, MALE

Reitz, Rosetta.
Menopause : a positive approach / Rosetta Reitz. — New York : Penguin Books, 1979.
xi, 276 p. ; 20 cm.
Reprint of the 1977 ed. published by Chilton Book Co., Radnor, Pa.
Bibliography: p. [257]-265.
Includes index.
ISBN 0-14-005120-1 (pbk.)
1. Menopause. 2. Climateric, Male. I. Title.
[RG186.R38 1979] 618.1'75 78-24317
78 MARC

CLIMATE see Climatology

CLIMATE, INFLUENCE OF see Architecture and climate; Man—Influence of climate

CLIMATE AND ARCHITECTURE see Architecture and climate

CLIMATE AND CROPS see Crops and climate

CLIMATE AND FORESTS see Forest microclimatology

CLIMATIC CHANGES

see also Paleoclimatology

Climate change to the year 2000 : a survey of expert opinion / conducted by the Research Directorate of the National Defense University jointly with the U.S. Department of Agriculture ... [et al.]. — Washington : The University : [for sale by Supt. of Docs., U.S. Govt. Print. Off.], 1978.
xxi, 109 p., [10] leaves of plates : ill. ; 27 cm.
"Contract number MDA903-77-C-0914."
1. Climatic changes. 2. Long-range weather forecasting. I. National Defense University. Research Directorate.
QC981.8.C5C52 551.6 78-600020
79 MARC

Gribbin, John R
Our changing climate / John Gribbin. — London : Faber & Faber, 1975.
95 p. : ill. ; 22 cm. GB***
Bibliography: p. 93-94.
Includes index.
ISBN 0-571-10696-X
1. Climatic changes. I. Title.
QC981.8.C5G743 551.6 78-670202
78 MARC

Gribbin, John R
What's wrong with the weather? : the climatic threat of the 21st century / John Gribbin. — New York : Scribner, c1979.
x, 172 p., [4] leaves of plates : ill. ; 24 cm.
Bibliography: p. 161-172.
Includes index.
ISBN 0-684-15807-8 : $9.95
1. Climatic changes. 2. Weather. I. Title.
QC981.8.C5G73 1979 551.6 78-10562
78 MARC

Halacy, Daniel Stephen, 1919-
Ice or fire? : Surviving climatic change / D. S. Halacy, Jr. — 1st ed. — New York : Harper & Row, c1978.
ix, 212 p. ; 22 cm.
ISBN 0-06-011778-8 : $9.95
1. Climatic changes. 2. Glacial epoch. 3. Climatic changes—United States. I. Title.
QC981.8.C5H25 1978 551.6 78-2139
78 MARC

CLIMATIC CHANGES (Continued)

Huntington, Ellsworth, 1876-1947.
Climatic changes : their nature and causes / by Ellsworth Huntington and Stephen Sargent Visher. — 1st AMS ed. — New York : AMS Press, 1978.
xiii, 329 p. : ill. ; 19 cm.
Reprint of the 1922 ed. published by Yale University Press, New Haven.
Includes bibliographical references and index.
ISBN 0-404-16211-8
1. Climatic changes. 2. Paleoclimatology. I. Vicher, Stephen Sargent, 1887-1967, joint author. II. Title.
QC981.8.C5H85 1979 551.6 77-10231
78 MARC

Pearson, Ronald George.
Climate and evolution / Ronald Pearson. — London ; New York : Academic Press, 1978.
xi, 274 p. : ill. ; 24 cm. GB***
Bibliography: p. [231]-265.
Includes index.
ISBN 0-12-548250-7
1. Climatic changes. 2. Paleoclimatology. 3. Evolution. 4. Geology, Stratigraphic. I. Title.
QC981.8.C5P42 551.6 77-93200
79 MARC

Wright, Peter Brian.
Portrait of climatic change : a survey of the field and a review of recent books / by Peter B. Wright. — Norwich : Dr. T. D. Davies (PCC), 1977.
[2], 46 p. : 1 ill. ; 21 cm.
Cover title.
Obtainable from P. B. Wright (PCC), Dept. of Oceanography, University of Hawaii, Honolulu, Hawaii.
Bibliography: p. 41-[45]
ISBN 0-9505614-0-1 : £3.50 ($7.00 U.S.)
1. Climatic changes. 2. Climatic changes—Bibliography. I. Title.
QC981.8.C5W74 551.6 78-316217
78 MARC

—ADDRESSES, ESSAYS, LECTURES

(Issledovanie izmeneniĭ klimata i vlagooborota)
Исследование изменений климата и влагооборота : [Сборник статей] / Под ред. чл.-кор. АН СССР М. И. Будыко. — Ленинград : Гидрометеоиздат, 1977.
128 p. : ill. ; 21 cm. — (Труды Государственного гидрологического института ; вып. 247) USSR 77
At head of title: Glavnoe upravlenie gidrometeorologicheskoĭ sluzhby pri Sovete Ministrov SSSR.
Includes bibliographies.
0.70rub
1. Climatic changes—Addresses, essays, lectures. 2. Crops and climate—Addresses, essays, lectures. I. Budyko, Mikhail Ivanovich. II. Series: Leningrad. Gosudarstvennyĭ gidrologicheskiĭ institut. Trudy ; vyp. 247.
GB651.L38 vyp. 247 79-408928
[QC981.8.C5]

—BIBLIOGRAPHY

Wright, Peter Brian.
Portrait of climatic change : a survey of the field and a review of recent books / by Peter B. Wright. — Norwich : Dr. T. D. Davies (PCC), 1977.
[2], 46 p. : 1 ill. ; 21 cm.
Cover title.
Obtainable from P. B. Wright (PCC), Dept. of Oceanography, University of Hawaii, Honolulu, Hawaii.
Bibliography: p. 41-[45]
ISBN 0-9505614-0-1 : £3.50 ($7.00 U.S.)
1. Climatic changes. 2. Climatic changes—Bibliography. I. Title.
QC981.8.C5W74 551.6 78-316217
78 MARC

—CONGRESSES

Carbon dioxide, climate, and society : proceedings of a IIASA workshop / cosponsored by WMO, UNEP, and SCOPE, February 21-24, 1978 ; editor, Jill Williams. — Oxford ; New York : Pergamon Press, 1978.
x, 332 p. : ill. ; 26 cm. — (IIASA proceedings series : Environment ; v. 1) GB78
Includes bibliographical references.
ISBN 0-08-023252-3 : £15.00 ($30.00)
1. Carbon dioxide—Environmental aspects—Congresses. 2. Climatic changes—Congresses. I. Williams, Jill. II. International Institute for Applied Systems Analysis. III. World Meteorological Organization. IV. United Nations. Environment Programme. V. Scientific Committee on Problems of the Environment. VI. Series: International Institute for Applied Systems Analysis. IIASA proceedings series : Environment ; v. 1.
TD885.5.C3C37 333.7 78-40655
78 MARC

International Symposium on Recent Climatic Change and Food Production, Tsukuba and Tokyo, 1976.
Climatic change and food production : International Symposium on Recent Climatic Change and Food Production held October 4-8, 1976, Tsukuba and Tokyo / edited by Kōichirō Takahashi, Masatoshi M. Yoshino. — Tokyo : University of Tokyo Press, c1978.
xi, 433 p. : ill. ; 27 cm. Ja***
Stamped on flyleaf: Exclusive distributor, ISBS, Inc., Forest Grove, Or.
Includes bibliographies.
ISBN 0-86008-205-9 : Y10,000
1. Crops and climate—Asia—Congresses. 2. Climatic changes—Asia—Congresses. 3. Food crops—Asia—Congresses. 4. Crops and climate—Congresses. 5. Climatic changes—Congresses. 6. Food crops—Congresses. I. Takahashi, Kōichirō, 1913- . II. Yoshino, Masatoshi, 1928- . III. Title.
S600.7.C54I57 1976 630'.25'1695 78-324096
79 MARC

Man's impact on climate : proceedings of an international conference held in Berlin, June 14-16, 1978 / edited by Wilfrid Bach, Jürgen Pankrath, William Kellogg. — Amsterdam ; New York : Elsevier Scientific Pub. ; New York : sole distributors for the U.S.A. and Canada, Elsevier/North Holland, 1979.
xxiii, 327 p. : ill. ; 25 cm. — (Developments in atmospheric science ; 10)
Includes bibliographies and index.
ISBN 0-444-41766-4 (v. 10)
1. Climatic changes—Congresses. 2. Man—Influence on nature—Congresses. I. Bach, Wilfrid. II. Pankrath, Jürgen. III. Kellogg, William. IV. Series.
QC981.8.C5M35 551.6 78-23763
78 MARC

Seminar on Food and Climatic Change, Winnipeg, Man., 1977.
Living with climatic change : [proceedings of a Seminar on Food and Climatic Change, Winnipeg, Canada 13 January 1977]. — [Ottawa] : Science Council of Canada, [1977].
90 p. : ill. ; 28 cm. C***
Cover title.
Sponsored by the Science Council of Canada and the Canadian Meteorological Society.
Includes bibliographical references.
1. Crops and climate—Congresses. 2. Climatic changes—Congresses. I. Science Council of Canada. II. Canadian Meteorological Society. III. Title.
S600.2.S45 1977 630'.25'16 78-318311
78 MARC

—JUVENILE LITERATURE

Gallant, Roy A
Earth's changing climate / Roy A. Gallant. — New York : Four Winds Press, c1979.
xiii, 226 p. : ill. ; 24 cm.
Bibliography: p. 207-210.
Includes index.
SUMMARY: Discusses climate and explores reasons for changes in climate, including sunspots, cosmic dust clouds, and man-made factors. Analyzes the economic, social, and political effects of these changes.
ISBN 0-590-07447-4 : $9.95
1. Climatic changes—Juvenile literature. [1. Climatic changes] I. Title.
QC981.8.C5G34 551.6 78-22124
78 MARC

Weiss, Malcolm E
What's happening to our climate? / By Malcolm E. Weiss ; illustrated by Paul Plumer. — New York : Messner, c1978.
93 p. : ill. ; 22 cm.
Includes index.
SUMMARY: Discusses the changes in Earth's climate over the years and explains both natural and man-made reasons for them.
ISBN 0-671-32846-8
1. Climatic changes—Juvenile literature. [1. Climatic changes. 2. Climatology] I. Plumer, Paul, 1936- . II. Title.
QC981.8.C5W44 551.6 78-15684
78 MARC

—MATHEMATICAL MODELS

Semtner, Albert J
A model for the thermodynamic growth of sea ice in numerical investigations of climate / by Albert J. Semtner, Jr. — Santa Monica, Calif. : Climate Dynamics Program, Rand Corp., 1975.
25 p., [4] leaves of plates : graphs ; 28 cm. — (The Rand paper series ; P-5432)
1. Sea ice—Mathematical models. 2. Climatic changes—Mathematical models. 3. Numerical analysis. I. Title. II. Series.
AS36.R28 no. 5432 081 s 79-304875
[GB2501.72.M25] 79 MARC

United States. Central Intelligence Agency. National Foreign Assessment Center.
Relating climate change to its effects : a research paper / National Foreign Assessent Center. — Washington : The Center, 1978.
iii, 223 p. : ill. ; 27 cm.
Cover title.
"GS78-10154."
1. Climatic changes—Mathematical models. I. Title.
QC981.8.C5U513 1978 551.6 79-600807
79 MARC

Zilitinkevich, Sergeĭ Sergeevich.
Global'noe vzaimodeĭstvie atmosfery i okeana / S. S. Zilitinkevich, A. S. Monin. — Leningrad : Gidrometeoizdat, 1977.
24 p. ; 21 cm. — (Novosti i problemy nauki) USSR77
Bibliography: p. 22-[23]
0.20rub
1. Climatic changes—Mathematical models. 2. Ocean-atmosphere interaction—Mathematical models. 3. Atmospheric circulation—Mathematical models. I. Monin, Andreĭ Sergeevich, joint author. II. Title.
QC981.8.C5Z54 79-351354
79 MARC

—RESEARCH—UNITED STATES

Gates, William Lawrence, 1928-
Climatic change / by W. Lawrence Gates and John Imbrie. — [Santa Monica : Rand Corp.], 1975.
iii, 41 p. ; 28 cm. — (The Rand paper series ; P-5437)
Bibliography: p. 22-41.
1. Climatic changes—Research—United States. I. Imbrie, John, joint author. II. Title. III. Series.
AS36.R28 no. 5437 081 s 79-300735
[QC981.8.C5] 79 MARC

—SOCIAL ASPECTS

Roberts, Walter Orr.
The climate mandate / Walter Orr Roberts, Henry Lansford. — San Francisco : W. H. Freeman, c1979.
viii, 197 p. : ill. ; 24 cm.
Bibliography: p. 190-195.
Includes index.
ISBN 0-7167-1054-4
1. Climatic changes—Social aspects. 2. Crops and climate. 3. Demography. I. Lansford, Henry, 1929- joint author. II. Title.
QC981.8.C5R6 301.31 78-25677
78 MARC

—TROPICS—ADDRESSES, ESSAYS, LECTURES

Fukui, Hayao.
Climate variability and agriculture : the humid tropics / by Hayao Fukui. — Kyoto : Center for Southeast Asian Studies, Kyoto University, 1978.
ii, 26 p. : ill. ; 28 cm. — (Discussion paper - Center for Southeast Asian Studies ; no. 100) Ja***
Bibliography: p. 25-26.
1. Crops and climate—Tropics—Addresses, essays, lectures. 2. Climatic changes—Tropics—Addresses, essays, lectures. 3. Rice—Climatic factors—Addresses, essays, lectures. I. Title. II. Title: The humid tropics. III. Series: Kyōto Daigaku. Tōnan Ajia Kenkyū Sentā. Discussion paper ; no. 100.
S600.64.T76F84 79-316053
79 MARC

—AFRICA

Hervieu, Jean Paul.
La variation du climat en Afrique / J. Hervieu. — [N'Djamena] : ORSTOM, [1977]
21 leaves, [3] leaves of plates : maps ; 30 cm. — (Notes techniques du Centre ORSTOM de N'Djamena ; no. 17)
Cover title.
Bibliography: leaf 21.
1. Climatic changes—Africa. I. Title. II. Series: France. Office de la recherche scientifique et technique outre-mer. Centre de N'Djamena. Notes techniques du Centre ORSTOM de N'Djamena ; no. 17.
QC981.8.C5H47 551.6'9'6 77-980917
79 MARC

—ASIA—CONGRESSES

International Symposium on Recent Climatic Change and Food Production, Tsukuba and Tokyo, 1976.
Climatic change and food production : International Symposium on Recent Climatic Change and Food Production held October 4-8, 1976, Tsukuba and Tokyo / edited by Kōichirō Takahashi, Masatoshi M. Yoshino. — Tokyo : University of Tokyo Press, c1978.
xi, 433 p. : ill. ; 27 cm. Ja***
Stamped on flyleaf: Exclusive distributor, ISBS, Inc., Forest Grove, Or.
Includes bibliographies.
ISBN 0-86008-205-9 : Y10,000
1. Crops and climate—Asia—Congresses. 2. Climatic changes—Asia—Congresses. 3. Food crops—Asia—Congresses. 4. Crops and climate—Congresses. 5. Climatic changes—Congresses. 6. Food crops—Congresses. I. Takahashi, Kōichirō, 1913- . II. Yoshino, Masatoshi, 1928- . III. Title.
S600.7.C54I57 1976 630'.25'1695 78-324096
79 MARC

CLIMATIC CHANGES (Continued)

—CENTRAL AMERICA

Huntington, Ellsworth, 1876-1947.
The climatic factor as illustrated in arid America / by Ellsworth Huntington, with contributions by Charles Schuchert, Andrew E. Douglass, and Charles J. Kullmer. — 1st AMS ed. — New York : AMS Press, 1979.
vi, 341 p., ₍13₎ leaves : ill. ; 29 cm.
Reprint of the 1914 ed. published by the Carnegie Institution of Washington, which was issued as its Publication no. 192.
Bibliography: p. 297-298.
Includes index.
ISBN 0-404-16210-X
1. Climatic changes—North America. 2. Climatic changes—Central America. I. Schuchert, Charles, 1858-1942. II. Douglass, Andrew Ellicott, 1867-1962. III. Kullmer, Charles Julius, 1879-1942. IV. Title. V. Series: Carnegie Institution of Washington. Publication ; no. 192.
QC981.8.C5H86 1979 551.6'9'7 77-10230
78 MARC

—EUROPE

Meyer zu Düttingdorf, Anne-Marie.
Klimaschwankungen im maritimen und kontinentalen Raum Europas seit 1871 / Anne-Marie Meyer zu Düttingdorf. — Paderborn : F. Schöningh, 1978.
ix, 127 p. : graphs ; 30 cm. -- (Bochumer geographische Arbeiten ; Heft 32) GFR***
Originally presented as the author's thesis, Bochum, 1977.
Summary also in English.
Bibliography: p. 76-80.
ISBN 3-506-71212-8
1. Climatic changes—Europe. I. Title. II. Series.
QC981.8.C5M49 1978 551.6'9'4 79-365733
79 MARC

—FRANCE—CORBIÈRES

Brochier, Jacques Léopold.
Les modifications de l'environnement du würmien récent au postglaciaire en Languedoc / Jacques Léopold Brochier. — Paris : Éditions du Centre national de la recherche scientifique, 1978.
203 p. : ill. ; 29 cm. — (Paléoécologie de l'homme fossile ; 2) F***
Summary in French, English, and German.
Bibliography: p. 193-202.
ISBN 2-222-02166-9 : 60.00F
1. Climatic changes—France—Corbières. 2. Climatic changes—France—Massif Central. 3. Geology, Stratigraphic—Pleistocene. 4. Geology, Stratigraphic—Recent. 5. Sedimentation analysis. I. Title. II. Series.
QC884.2.C5B76 551.6'9'448 78-380338
78 MARC

—FRANCE—MASSIF CENTRAL

Brochier, Jacques Léopold.
Les modifications de l'environnement du würmien récent au postglaciaire en Languedoc / Jacques Léopold Brochier. — Paris : Éditions du Centre national de la recherche scientifique, 1978.
203 p. : ill. ; 29 cm. — (Paléoécologie de l'homme fossile ; 2) F***
Summary in French, English, and German.
Bibliography: p. 193-202.
ISBN 2-222-02166-9 : 60.00F
1. Climatic changes—France—Corbières. 2. Climatic changes—France—Massif Central. 3. Geology, Stratigraphic—Pleistocene. 4. Geology, Stratigraphic—Recent. 5. Sedimentation analysis. I. Title. II. Series.
QC884.2.C5B76 551.6'9'448 78-380338
78 MARC

—NORTH AMERICA

Huntington, Ellsworth, 1876-1947.
The climatic factor as illustrated in arid America / by Ellsworth Huntington, with contributions by Charles Schuchert, Andrew E. Douglass, and Charles J. Kullmer. — 1st AMS ed. — New York : AMS Press, 1979.
vi, 341 p., ₍13₎ leaves : ill. ; 29 cm.
Reprint of the 1914 ed. published by the Carnegie Institution of Washington, which was issued as its Publication no. 192.
Bibliography: p. 297-298.
Includes index.
ISBN 0-404-16210-X
1. Climatic changes—North America. 2. Climatic changes—Central America. I. Schuchert, Charles, 1858-1942. II. Douglass, Andrew Ellicott, 1867-1962. III. Kullmer, Charles Julius, 1879-1942. IV. Title. V. Series: Carnegie Institution of Washington. Publication ; no. 192.
QC981.8.C5H86 1979 551.6'9'7 77-10230
78 MARC

—UNITED STATES

Halacy, Daniel Stephen, 1919-
Ice or fire? : Surviving climatic change / D. S. Halacy, Jr. — 1st ed. — New York : Harper & Row, c1978.
ix, 212 p. ; 22 cm.
ISBN 0-06-011778-8 : $9.95
1. Climatic changes. 2. Glacial epoch. 3. Climatic changes—United States. I. Title.
QC981.8.C5H25 1978 551.6 78-2139
78 MARC

—YUGOSLAVIA—NIŠ (CITY)

Marković-Marjanović, Jelena.
Pleistocene klimatske promene u Niškoj kotlini = Les changements du climat au pléistocène dans le Bassin de Niš / Jelena Marković-Marjanović ; urednik Dušan Gavrilović. — Beograd : Srpsko geografsko društvo, 1978.
68 p. : ill. ; 24 cm. — (Posebna izdanja - Srpsko geografsko društvo ; sv. 48) Yu***
Summary in French.
Bibliography: p. 61-63.
1. Climatic changes—Yugoslavia—Niš (City) 2. Geology, Stratigraphic—Pleistocene. 3. Paleoclimatology—Yugoslavia—Niš (City) I. Title. II. Title: Les changements du climat au pléistocène dans le Bassin de Niš. III. Series: Srpsko geografsko društvo. Posebna izdanja ; sv. 48.
QC884.2.C5M37 79-365776
79 MARC

CLIMATIC CLASSIFICATION

see also Climatic zones

CLIMATIC GEOMORPHOLOGY

—CONGRESSES

Gordon-Konferenz, Munich, Ger., 1975.
Beiträge zur Reliefgenese in verschiedenen Klimazonen : erweiterte Vorträge der Gordon-Konferenz der Bayerischen Akademie der Wissenschaften vom 20.-23.2.1975 / hrsg. von J. Büdel. — Würzburg : Im Selbstverlag des Instituts für Geographie der Universität Würzburg in Verbindung mit der Geographischen Gesellschaft, 1977.
198 p., ₍28₎ leaves of plates : ill. ; 21 cm. — (Würzburger geographische Arbeiten ; Heft 45) (Arbeiten aus der Kommission für Geomorphologie der Bayerischen Akademie der Wissenschaften ; Nr. 2) GFR***
Includes bibliographies.
DM16.00
1. Climatic geomorphology—Congresses. I. Büdel, Julius. II. Akademie der Wissenschaften, Munich. III. Title. IV. Series. V. Series: Akademie der Wissenschaften, Munich. Kommission für Geomorphologie. Arbeiten aus der Kommission für Geomorphologie der Bayerischen Akademie der Wissenschaften ; Nr. 2.
G58.W8 Heft 45 78-380282
₍GB447₎ 79 MARC

CLIMATIC ZONES

—INDONESIA—SUMATRA

Moeljadi, 1939-
Distribusi hudjan dan iklim menurut Köppen dipulau Sumatera / Moeljadi. — Djakarta : Dinas Geografi, Direktorat Topografi Angkatan Darat, ₍1968₎
28 leaves, ₍19₎ leaves of plates : ill. ; 31 cm. — (Publikasi - Dinas Geografi ; no. 17)
Bibliography: leaf 17.
1. Rain and rainfall—Indonesia—Sumatra. 2. Climatic zones—Indonesia—Sumatra. I. Title. II. Series: Indonesia. Dinas Geografi. Publikasi ; no. 17.
G58.I45 No. 17 79-950309
₍QC925.8.I52S8₎ 79 MARC

CLIMATOLOGY

see also Architecture and climate; Atmospheric circulation; Atmospheric pressure; Atmospheric temperature; Bioclimatology; Climatic changes; Climatic geomorphology; Evaporation (Meteorology); Desertification; Man—Influence of climate; Meteorology; Microclimatology; Numerical weather forecasting; Paleoclimatology;
Rain and rainfall; Seasons; Statistical weather forecasting; Synoptic climatology; Urban climatology; Vegetation and climate; Weather; and subdivision Climate under names of countries, cities, etc., e. g. United States—Climate; also subdivision Climatic factors under subjects, e. g. Air conditioning—Climatic factors

Aguilar, Antonio.
El viento zonda / Antonio Aguilar. — ₍San Juan, Argentina₎ : Editorial Sanjuanina, 1978.
155, ₍2₎ p. ; 21 cm.
Bibliography: p. ₍156-157₎
1. Climatology. 2. Argentine Republic—Climate. I. Title.
QC981.A34 79-100514
79 MARC

Bessemoulin, Jean, 1892-
Vents, nuages & ₍i.e. et₎ tempêtes : météo / Jean Bessemoulin, Roger Clausse. — ₍Nouv. éd.₎. — ₍Paris₎ : Éditions Maritimes & d'outre-mer, ₍c1978₎
253 p., ₍4₎ leaves of plates : ill. (some col.) : 25 cm. F***
Bibliography: p. 247-248.
89.00F
1. Climatology. 2. Weather. 3. France—Climate. I. Clausse, Roger, joint author. II. Title.
QC981.B45 1978 551.6 78-385071
78 MARC

Brooks, Charles Ernest Pelham.
The evolution of climate / by C. E. P. Brooks ; with a pref. by G. C. Simpson. — New York : AMS Press, 1978.
173 p. ; 23 cm.
Reprint of the 1922 ed. published by Benn Brothers, London.
Includes bibliographies and index.
ISBN 0-404-16201-0
1. Climatology. 2. Paleoclimatology. I. Title.
QC981.B82 1978 551.6 77-10221
78 MARC

Escourrou, Gisèle.
Climatologie pratique / par Gisèle Escourrou, ... — Paris ; New York ; Barcelone : Masson, 1978.
172 p., 18 p. of plates (some col.) : ill. ; 24 cm. — (Collection Géographie ISSN 0338-2664) F78-7876
Bibliography: p. ₍8₎
Includes index.
ISBN 2-225-48280-2 : 54.00F
1. Weather forecasting. 2. Climatology. 3. France—Climate. I. Title.
QC995.E82 551.6'3 78-385072
78 MARC

Gavrilova, Mariia Kuz'minichna.
(Klimat i mnogoletnee promerzanie gornykh porod)
Климат и многолетнее промерзание горных пород / М. К. Гаврилова ; Отв. ред. В. Т. Балобаев. — Новосибирск : Наука, Сиб. отд-ние, 1978.
213 p. : ill. ; 22 cm. USSR 78
At head of title: Akademiia nauk SSSR. Sibirskoe otdelenie. Institut merzlotovedeniia.
Bibliography: p. 186-₍212₎
2.40rub
1. Climatology. 2. Frozen ground. 3. Weathering. I. Balobaev, Veniamin Tikhonovich. II. Title.
QC981.G36 79-412306

Gentilli, Joseph.
A geography of climate / by J. Gentilli. — 1st AMS ed. — New York : AMS Press, 1978.
107 p. : ill. ; 22 cm.
Reprint of the 1952 ed. published by the University of Western Australia Text Books Board, Crawley.
Bibliography: p. 104-105.
ISBN 0-404-16206-1
1. Climatology. I. Title.
QC981.G4 1979 551.5 77-10225
78 MARC

Gribbin, John R
Weather force : climate and its impact on our world / John Gribbin ; ₍editor, John Man₎. — New York : Putnam, c1979.
188 p. : ill. (some col.) ; 30 cm.
Includes index.
ISBN 0-399-12400-4 : $20.00
1. Climatology. 2. Weather. I. Man, John. II. Title.
QC981.G64 551.6 79-51033
79 MARC

CLIMATOLOGY (Continued)

Hann, Julius von, 1839-1921.
Handbook of climatology : Part I, General climatology / by Julius Hann ; translated with the author's permission from the 2d rev. and enl. German ed., with additional references and notes by Robert De Courcy Ward. — 1st AMS ed. — New York : AMS Press, 1979.
xiv, 437 p. ; 23 cm.
Translation of Handbuch der Klimatologie, Bd. 1, Allgemeine Klimatologie.
Reprint of the 1903 ed. published by Macmillan, New York.
Includes bibliographical references and index.
ISBN 0-404-16207-X
1. Climatology. I. Title.
QC981.H27 1979 551.5 77-10226
78 MARC

Koeppe, Clarence Eugene, 1891-
Weather and climate / Clarence E. Koeppe, George C. De Long. — Huntington, N.Y. : R. E. Krieger Pub. Co., 1979, c1958.
viii, 341 p., ₍2₎ leaves of plates : ill. ; 26 cm.
Reprint of the ed. published by McGraw-Hill, New York.
Bibliography: p. 303-307.
Includes index.
ISBN 0-88275-696-6
1. Meteorology. 2. Climatology. I. De Long, George Cass, 1910- joint author. II. Title.
QC863.K714 1979 551.5 78-8981
78 MARC

Lachinov, Dmitriĭ Aleksandrovich, 1842–1902.
(Kurs meteorologīi i klimatologīi)
Курсъ метеорологіи и климатологіи / составилъ Д. А. Лачиновъ. — С.-Петербургъ : Изд. Ф. Павленкова, 1889.
iii, 332 p., ₍6₎ leaves of plates : ill., col. maps ; 24 cm.
1. Meteorology. 2. Climatology. I. Title.
QC863.L27 78–421812

Lockwood, John George.
Causes of climate / John G. Lockwood. — London : E. Arnold, 1979.
vii, 260 p. : ill. ; 24 cm. GB***
Includes bibliographies and indexes.
ISBN 0-7131-6139-6
1. Climatology. I. Title.
QC981.L62 1979b 551.6 79-308823
79 MARC

Lockwood, John George.
Causes of climate / John G. Lockwood. — New York : Wiley, 1979.
vii, 260 p. : ill. ; 24 cm.
"A Halsted Press book."
bibliography: p. 250.
Includes indexes.
ISBN 0-470-26657-0. ISBN 0-470-26658-9 (pbk.)
1. Climatology. I. Title.
QC981.L62 1978 551.6 78-31907
79 MARC

Pleym, H
Miljøklimatologi / Harald Pleym. — ₍Trondheim₎ : Tapir, c1978 (₍s.l.₎ : Tapir)
viii, 115 p. : ill. ; 24 cm. N78-July
Bibliography: p. 110-115.
ISBN 8251902568 : kr36.00
1. Climatology. 2. Air—Pollution. 3. Aerosols. I. Title.
QC981.P53 78-395108
79 MARC

Saibene, Cesare.
Climi e paesaggi climatici / Cesare Saibene. — Milano : Vita e pensiero, 1975.
207 p. : ill. ; 24 cm. — (Corsi universitari) (Pubblicazioni dell'Università cattolica del Sacro Cuore, Milano) It***
Half-title: Problemi di geografia generale.
L3400
1. Climatology. I. Title.
QC981.S24 551.6 78-379925
78 MARC

Scherhag, Richard, 1907-1970.
Klimatologie / Richard Scherhag, Joachim Blüthgen, Wilhelm Lauer. — 9., verb. Aufl. — Braunschweig : Westermann, 1977, c1960.
202 p. : ill. (some col.) and maps (6 fold., chiefly col. in pocket) ; 24 cm. — (Das Geographische Seminar) GFR78-A
First-3d ed. published under title: Einführung in die Klimatologie.
Bibliography: p. 174-177.
Includes index.
ISBN 3-14-160284-0 : DM14.80
1. Climatology. I. Blüthgen, Joachim, joint author. II. Lauer, Wilhelm, joint author. III. Title.
QC981.S383 1977 551.6 78-364991
78 MARC

Schreiber, Detlev.
Meteorologie, Klimatologie / Detlef Schreiber. — 2., verb. Aufl. — Bochum : Studienverlag Brockmeyer, 1978.
211 p. : ill. ; 21 cm. GFR78-A
Includes bibliographical references.
ISBN 3-88339-013-5 : DM14.80
1. Meteorology. 2. Climatology. I. Title.
QC861.2.S37 1978 551.5 78-376047
78 MARC

Visher, Stephen Sargent, 1887-1967.
Climatic laws : ninety generalizations with numerous corollaries as to the geographic distribution of temperature, wind, moisture, etc. : a summary of climate / by Stephen Sargent Visher. — 1st AMS ed. — New York : AMS Press, 1978.
96 p. : ill. ; 23 cm.
Reprint of the 1924 ed. published by Wiley, New York.
Includes bibliographical references and index.
ISBN 0-404-16219-3
1. Climatology. I. Title.
QC981.V55 1978 551.6 77-10241
78 MARC

Winkless, Nels.
Climate and the affairs of men / Nels Winkless III and Iben Browning. — 1st ed. — New York : Harper's Magazine Press, ₍1975₎
228 p. : ill. ; 22 cm.
Bibliography: p. 223-228.
ISBN 0-06-129550-7 : $7.95
1. Climatology. 2. Paleoclimatology. 3. Volcanoes. 4. Man—Influence of climate. I. Browning, Iben, joint author. II. Title.
QC981.2.W56 1975 551.6 74-27305
75₍r79₎rev MARC

—ADDRESSES, ESSAYS, LECTURES

(Obshchaia i sinopticheskaia klimatologiia)
Общая и синоптическая климатология : ₍Сборник статей₎ / Под ред. д-ра геогр. наук О. А. Дроздова, канд. геогр. наук Е. В. Воробьевой. — Ленинград : Гидрометеоиздат, 1977.
148 p. : ill. ; 21 cm. — (Труды Главной геофизической обсерватории им. А. И. Воейкова ; вып. 386) USSR 77
At head of title: Glavnoe upravlenie gidrometeorologicheskoĭ sluzhby pri Sovete Ministrov SSSR.
Includes bibliographies.
0.76rub
1. Climatology—Addresses, essays, lectures. 2. Synoptic climatology—Addresses, essays, lectures. I. Drozdov, Oleg Alekseevich. II. Vorob'eva, Evgeniia Viktorovna. III. Series: Leningrad. Glavnaia geofizicheskaia observatoriia. Trudy ; vyp. 386.
QC801.L46 vyp. 386 78–422271
[QC981.4]

—BIBLIOGRAPHY

University of Toronto. Map Library.
Climate : a bibliography and index to maps, atlases, and reference books in the University of Toronto Map Library / ₍compiled by Jacqueline Woods and Mary Armstrong₎. — ₍Toronto : University of Toronto₎, 1976.
15 p. : map ; 28 cm. C***
Cover title.
Includes index.
1. Climatology—Bibliography. 2. Climatology—Charts, diagrams, etc.—Bibliography. 3. University of Toronto. Map Library—Indexes. I. Woods, Jacqueline. II. Armstrong, Mary. III. Title.
Z6683.C5U54 1976 016.912'1'5516 79-302899
₍QC981₎ 79 MARC

—CHARTS, DIAGRAMS, ETC.

Hastenrath, S
Climatic atlas of the Indian Ocean / Stefan Hastenrath, Peter J. Lamb. — Madison : University of Wisconsin Press, 1979.
2 v. : chiefly ill. ; 23 x 34 cm.
Includes bibliographies.
CONTENTS: pt. 1. Surface climate and atmospheric circulation.—pt. 2. The oceanic heat budget.
ISBN 0-299-07814-0 : $70.00
1. Indian Ocean—Climate—Charts, diagrams, etc. 2. Climatology—Charts, diagrams, etc. I. Lamb, Peter J., joint author. II. Title.
QC994.5.H37 551.6'9'165 78-65016
78 MARC

—CHARTS, DIAGRAMS, ETC.—BIBLIOGRAPHY

University of Toronto. Map Library.
Climate : a bibliography and index to maps, atlases, and reference books in the University of Toronto Map Library / ₍compiled by Jacqueline Woods and Mary Armstrong₎. — ₍Toronto : University of Toronto₎, 1976.
15 p. : map ; 28 cm. C***
Cover title.
Includes index.
1. Climatology—Bibliography. 2. Climatology—Charts, diagrams, etc.—Bibliography. 3. University of Toronto. Map Library—Indexes. I. Woods, Jacqueline. II. Armstrong, Mary. III. Title.
Z6683.C5U54 1976 016.912'1'5516 79-302899
₍QC981₎ 79 MARC

—CONGRESSES

Climate Diagnostics Workshop.
Proceedings of the annual Climate Diagnostics Workshop. ₍Washington, etc.₎ U. S. Dept. of Commerce, National Oceanic and Atmospheric Administration.
v. ill. 28 cm.
Cover title, 19 -77: Proceedings of the annual NOAA Climate Diagnostics Workshop.
Key title: Proceedings of the annual Climate Diagnostics Workshop, ISSN 0192-8759.
1. Climatology—Congresses. I. United States. National Oceanic and Atmospheric Administration. II. Climate Diagnostics Workshop. Proceedings of the annual NOAA Climate Diagnostics Workshop. III. Key title.
QC980.C54a 551.5'05 79–640997
MARC-S

Symposium/Workshop on Solar-Terrestrial Influences on Weather and Climate, Ohio State University, Columbus, 1978.
Solar-terrestrial influences on weather and climate : proceedings of a symposium/workshop held at the Fawcett Center for Tomorrow, the Ohio State University, Columbus, Ohio, 24-28 July 1978 / edited by Billy M. McCormac, Thomas A. Seliga. — Dordrecht, Holland ; Boston : D. Reidel Pub. Co., c1979.
xiii, 346 p. : ill. ; 25 cm.
Includes bibliographical references and index.
ISBN 9027709785
1. Solar activity—Congresses. 2. Atmosphere, Upper—Congresses. 3. Weather—Congresses. 4. Climatology—Congresses. I. McCormac, Billy Murray. II. Seliga, T. A. III. Title.
QC883.2.S6S94 1978 551.6'9 79-11425
79 MARC

—DATA PROCESSING

Boyd, Billie F
Data processing requirements for effective utilization of climatology in Indonesia / by Billie F. Boyd. — Jakarta : United Nations Development Programme, Departemen Perhubungan Republik Indonesia, 1975.
47, ₍15₎ leaves ; 32 cm.
"Project of the Government of Indonesia, meteorological applications, INS/73/029."
Includes bibliographical references.
Errata slip inserted.
1. Indonesia—Climate. 2. Climatology—Data processing. I. United Nations. Development Programme. II. Indonesia. Departemen Perhubungan. III. Title.
QC993.I5B69 551.6'9'598 76-941511
79 MARC

—MAPS

Canada. Meteorological Branch.
Atlas of climatic maps : series 1 to 10 = atlas de cartes climatologiques : série 1 à 10. — ₍s.l.₎ : Dept. of Transport, Meteorological Branch, 1967-1970.
₍62₎ p. : chiefly maps ; 44 x 57 cm. C***
$8.00
1. Canada—Climate—Maps. 2. Climatology—Maps. I. Title.
G1116.C8C3 1967 912'.1'5516971 79-318304
79 MARC

—MATHEMATICAL MODELS

Gates, William Lawrence, 1928-
The January global climate simulated by the two-level Mintz-Arakawa model : a comparison with observation : a report prepared for Advanced Research Projects Agency / W. L. Gates. — Santa Monica, CA : Rand, 1972.
ix, 107 p. : ill. ; 28 cm. — (₍Report₎ - Rand Corporation ; R-1005-ARPA)
Bibliography: p. 105-107.
1. Atmospheric circulation—Mathematical models. 2. Climatology—Mathematical models. I. Title. II. Series: Rand Corporation. Rand report ; R-1005-ARPA.
AS36.R3 R-1005 081 s 79-304286
₍QC880.4.A8₎ 79 MARC

—STUDY AND TEACHING—UNITED STATES

Pan American World Airways, inc. Communications Dept.
Correspondence course in meteorology / Pan American Airways System, Communications Department. — ₍New York₎ : The Department, ₍1939?₎
ca. 200 leaves : ill. ; 30 cm.
Caption title.
1. Climatology—Study and teaching—United States. 2. Correspondence schools and courses—United States. I. Title.
QC981.5.P36 1939 551.5 78-324030
79 MARC

CLIMATOLOGY (Continued)

—STUDY AND TEACHING (HIGHER)
—BRAZIL—SÃO PAULO (CITY)
—ADDRESSES, ESSAYS, LECTURES

Monteiro, Carlos Augusto de Figueiredo.
A climatologia do Brasil ante a renovação atual da geografia : um depoimento / Carlos Augusto de Figueiredo Monteiro. — São Paulo : Universidade de São Paulo, Instituto de Geografia, 1973.
15 p. ; 24 cm. — (Métodos en questão - Universidade de São Paulo, Instituto de Geografia ; 6)
"Comunicação apresentada no Simpósio sobre a Renovaçao da Geografia ... dia 10 de julho, foi parte integrante da XXV Reunião da Sociedade Braleira para o Progresso da Ciência, realizado no Rio de Janeiro, na Cidade Universitária da U.F.R.J., entre 8 e 14 de julho de 1973."
Bibliography: p. 14-15.
1. Climatology—Study and teaching (Higher)—Brazil—São Paulo (City)—Addresses, essays, lectures. 2. São Paulo, Brazil (City). Universidade. Instituto de Geografia—Addresses, essays, lectures. I. Title. II. Series: São Paulo, Brazil (City). Universidade. Instituto de Geografia. Métodos em questão ; 6.
QC981.5.M66 551.6'07'1281 78-393142
79 MARC

—TABLES

France. Secrétariat d'État auprès du Ministre de l'équipement et de l'aménagement du territoire (Transports)
Annales météorologiques, France d'outre-mer : décennie 1961-1970 / Secrétariat d'État auprès du Ministre de l'équipement et de l'aménagement du territoire (Transports), Direction de la météorologie. — Paris : La Direction, 197
v. : maps ; 30 cm. F***
CONTENTS:
—fasc. 2. Départements et territoires français de l'océan Indien et de l'Antarctique.—fasc. 3. Territoires français de l'Océan pacifique.
Errata slips inserted.
Chiefly tables.
1. Climatology—Tables, etc. I. France. Direction de la météorologie nationale. II. Title.
QC982.F73 1977 551.6'02'12 79-339095
79 MARC

—TURKEY—CLIMATE

Erinç, Sırrı.
Klimatoloji ve metodları / Sırrı Erinç. — Genişletilmiş 2. baskı. — [İstanbul] : İstanbul Universitesi, Coğrafya Enstitüsü, 1969.
ix, 538 p. : ill. ; 25 cm. — (Coğrafya Enstitüsü yayınları ; no. 35) (İstanbul Üniversitesi yayınları ; no. 994)
Bibliography: p. [521]-528.
Includes index.
1. Climatology—Turkey—Climate. I. Title. II. Series: Istanbul. Üniversite. Coğrafya Enstitüsü. Coğrafya Enstitüsü yayını ; no. 35.
QC981.E73 1969 79-371045
79 MARC

CLIMATOLOGY, AGRICULTURAL
see Crops and climate

CLIMATOLOGY, MEDICAL

see also Altitude, Influence of; Man —Influence of climate; Medical geography; Tropical medicine

Ardelean, I
Elemente de biometeorologie medicală. București, Editura medicală, 1972.
232 p. with figs. 20 cm. lei10.50 R72-3035
At head of title: I. Ardelean, M. Barnea.
Bibliography: p. 222-[232]
1. Climatology, Medical. 2. Weather—Mental and physiological effects. I. Barnea, Matei, joint author. II. Title.
RA793.A72 78-393891
79 MARC

Engelmann, Volker.
Todesursachen in Abhängigkeit vom Wettergeschehen. (München) Verlag UNI-Druck (1968).
49 p. with illus. 21 cm. 10.00 GDNB69-A2-473
Issued also as thesis, Munich.
Bibliography: p. 37-39.
1. Climatology, Medical. 2. Death—Causes. 3. Munich—Climate. 4. Mortality—Germany—Munich. I. Title.
RA793.E38 1968 75-565508
79 MARC

Fast, Julius, 1918-
Weather language / by Julius Fast. — 1st ed. — New York : Wyden Books : trade distribution by Simon and Schuster, c1979.
xii, 242 p. : ill. ; 22 cm.
Bibliography: p. [237]-242.
ISBN 0-88326-157-X
1. Weather—Mental and physiological effects. 2. Climatology, Medical. I. Title.
QP82.2.C5F37 613.1'1 78-21291
78 MARC

Gualtierotti, Roberto.
Elementi di climatologia ed ecologia medica : per il corso di idrologia e climatologia medica / Roberto Gualtierotti. — Milano : Edizioni della Libreria dello studente, 1974.
164 p. : ill. ; 24 cm. It74-Nov
At head of title: Università degli studi di Milano. Cattedra di idrologia e climatologia medica.
On spine: Climatologia ed ecologia medica.
Bibliography: p. 162.
1. Climatology, Medical. 2. Environmental health. I. Title. II. Title: Climatologia ed ecologia medica.
[DNLM: 1. Climate. 2. Ecology. 3. Environmental health. WB700 G912e 1974]
[RA793.G85] 613.1'1 74-595335
Shared Cataloging with *79[r79]rev MARC
DNLM

Lamarche, Maurice, 1924-
Hydrologie et climatologie médicale / M. Lamarche. — [Paris] : Édition Marketing, c1977.
222 p. : ill. ; 25 cm. F***
ISBN 2-7298-5001-5 : 50.00F
1. Hydrotherapy. 2. Climatology, Medical. 3. Thalassotherapy. 4. Health resorts, watering-places, etc.—France. 5. Mineral waters—France. I. Title.
RM811.L36 615'.853'0944 78-378658
78 MARC

Thomson, William Archibald Robson.
A change of air : climate and health / William A. R. Thomson. — 1st American ed. — New York : C. Scribner's Sons, 1979.
xiv, 225 p. : ill. ; 21 cm.
Bibliography: p. 217.
Includes index.
ISBN 0-684-16259-8 : $12.50
1. Climatology, Medical. I. Title.
RA793.T48 1979 613.1'1 79-63477
79 MARC

—ADDRESSES, ESSAYS, LECTURES

Martínez Domínguez, Rafael.
El clima, factores coadyuvantes y su influencia en la salud y en la enfermedad, con especial mención al clima de Sevilla : discurso de recepción / del académico electo Iltmo. Sr. Dr. D. Rafael Martínez Domínguez ; y discurso de contestación del académico numerario Iltmo. Sr. Dr. D. Ildefonso Camacho Baños. — Sevilla : Real Academia de Medicina de Sevilla, 1974.
152 p., [2] leaves of plates : ill. ; 25 cm. Sp75-Feb
Bibliography: p. [139]-141.
1. Climatology, Medical—Spain—Seville (Province)—Addresses, essays, lectures. 2. Seville (Province)—Climate—Addresses, essays, lectures. 3. Climatology, Medical—Addresses, essays, lectures. I. Camacho Banos, Ildefonso, joint author. II. Title: El clima, factores coadyuvantes y su influencia en la salud y en la enfermedad ...
RA880.S48M37 615'.834 78-418349
*78 MARC

Wpływ czynników meteorologicznych na organizm ludzi i zwierząt / [redaktor wydawn. Maria Tłustochowska. — Wyd. 1. — Kraków : Zakład Narodowy im. Ossolińskich, 1977.
45 p. : ill. ; 24 cm. P***
At head of title: Polska Akademia Nauk. Oddział w Krakowie. Komisja Ochrony Zdrowia Społecznego.
Summaries in English and Russian.
Includes bibliographies.
zł12.00
1. Climatology, Medical—Addresses, essays, lectures. I. Tłustochowska, Maria. II. Polska Akademia Nauk. Komisja Ochrony Zdrowia Społecznego.
RA793.W76 78-375067
78 MARC

—CONGRESSES

Symposium on Climate, Environment and Health, Mysore, 1955.
Symposium on Climate, Environment & Health : abstracts of papers. — [s.l. : s.n., 1955?] (Lucknow : Pioneer Press)
46 p. ; 22 cm.
Cover title.
Organized by the Council of the National Institute of Sciences of India.
1. Climatology, Medical—Congresses. 2. Climatology, Medical—India—Congresses. 3. Man—Influence of climate—Congresses. I. National Institute of Sciences of India.
RA793.S95 1955 79-316270
79 MARC

—INDIA—CONGRESSES

Symposium on Climate, Environment and Health, Mysore, 1955.
Symposium on Climate, Environment & Health : abstracts of papers. — [s.l. : s.n., 1955?] (Lucknow : Pioneer Press)
46 p. ; 22 cm.
Cover title.
Organized by the Council of the National Institute of Sciences of India.
1. Climatology, Medical—Congresses. 2. Climatology, Medical—India—Congresses. 3. Man—Influence of climate—Congresses. I. National Institute of Sciences of India.
RA793.S95 1955 79-316270
79 MARC

—SPAIN—SEVILLE (PROVINCE)—
ADDRESSES, ESSAYS, LECTURES

Martínez Domínguez, Rafael.
El clima, factores coadyuvantes y su influencia en la salud y en la enfermedad, con especial mención al clima de Sevilla : discurso de recepción / del académico electo Iltmo. Sr. Dr. D. Rafael Martínez Domínguez ; y discurso de contestación del académico numerario Iltmo. Sr. Dr. D. Ildefonso Camacho Baños. — Sevilla : Real Academia de Medicina de Sevilla, 1974.
152 p., [2] leaves of plates : ill. ; 25 cm. Sp75-Feb
Bibliography: p. [139]-141.
1. Climatology, Medical—Spain—Seville (Province)—Addresses, essays, lectures. 2. Seville (Province)—Climate—Addresses, essays, lectures. 3. Climatology, Medical—Addresses, essays, lectures. I. Camacho Banos, Ildefonso, joint author. II. Title: El clima, factores coadyuvantes y su influencia en la salud y en la enfermedad ...
RA880.S48M37 615'.834 78-418349
*78 MARC

—UNITED STATES

Forry, Samuel, 1811-1844.
The climate of the United States and its endemic influences : based chiefly on the records of the Medical Department and Adjutant General's Office, United States Army / by Samuel Forry. — 1st AMS ed. — New York : AMS Press, 1978.
378 p. : map ; 23 cm.
Reprint of the 1842 ed. published by J. & H. G. Langley, New York.
ISBN 0-404-16205-3
1. Climatology, Medical—United States. 2. United States—Climate. I. Title.
RA804.F6 1978 613.1'1'0973 77-10224
78 MARC

CLIMATOLOGY, PHENOLOGICAL see Phenology

CLIMATOTHERAPY

see also Health resorts, watering-places, etc. ; Sun-baths; Thalassotherapy

—ROMANIA

Munteanu, L
Ghidul stațiunilor balneoclimatice din România / Laviniu Munteanu, Constantin Stoicescu, Ludovic Grigore ; prefață de prof. dr. Nicolae Teleki. — București : "Sport-Turism", 1978.
301 p., [17] leaves of plates (1 fold.) : ill., 1 fold. col. map ; 21 cm. R78-4234
Bibliography: p. 293-[297]
lei17.00
1. Health resorts, watering-places, etc.—Romania—Directories. 2. Mineral waters—Romania—Guide-books. 3. Mineral waters—Therapeutic use. 4. Climatotherapy—Romania. I. Stoicescu, Constantin, joint author. II. Grigore, L., joint author. III. Title.
RA887.R8M87 79-338042
*79 MARC

CLIMBING PLANTS

see also Ornamental climbing plants

—AUSTRALIA

Jones, David Lloyd, B.Ag.Sc.
Australian climbing plants, with notes on their cultivation / by D. L. Jones and B. Gray ; photos. by B. Gray and E. R. Rotherham. — Terrey Hills, Australia : Reed, 1977.
166 p. : ill. (some col.) ; 26 cm. Aus***
Includes index.
ISBN 0-589-07221-8 : $9.95
1. Climbing plants—Australia. I. Gray, Bruce, joint author. II. Title.
QK431.J66 635.9'74 78-317793
78 MARC

CLIMBING PLANTS (Continued)

—NEW ZEALAND—PICTORIAL WORKS

Eagle, Audrey Lily.
Eagle's 100 shrubs & climbers of New Zealand : companion volume to Eagle's 100 trees of New Zealand : botanical paintings & notes / by Audrey Eagle. — Auckland : Collins, 1978.
144 p. : chiefly col. ill., col. maps ; 26 cm. NZ78-12
Paintings and notes first published 1975 in Eagle's trees and shrubs of New Zealand in colour.
Bibliography: p. 142.
Includes index.
ISBN 0-00-216932-0 : $15.00
1. Shrubs—New Zealand—Pictorial works. 2. Climbing plants—New Zealand—Pictorial works. I. Eagle, Audrey Lily. Eagle's 100 trees of New Zealand. II. Title. III. Title: 100 shrubs & climbers of New Zealand.
QK492.5.E18 582'.17'09931 79-317024
79 MARC

—VERMONT—IDENTIFICATION

Jones, Lewis Ralph, 1864-1945.
The handbook of Vermont shrubs and woody vines / by L. R. Jones and F. V. Rand ; illustrated by Mary Robinson ; with an introd. to the new ed. by Raymond T. Foulds, Jr. — 1st Tuttle ed. — Rutland : C. E. Tuttle Co., 1979.
v, 147 p. : ill. ; 22 cm.
Reprint of the 1909 ed. published by Free Press Print. Co., Burlington, as Bulletin no. 145 of the Vermont Agricultural Experiment Station, under title: Vermont shrubs and woody vines.
Includes index.
ISBN 0-8048-1316-7 : $3.95
1. Shrubs—Vermont—Identification. 2. Climbing plants—Vermont—Identification. I. Rand, Frederick Vernon, 1883-1961, joint author. II. Title. III. Series: Vermont. Agricultural Experiment Station, Burlington. Bulletin ; no. 145.
QK484.V5J66 1979 582'.17'09743 79-84806
00 MARC

CLIMBING ROSES

Thomas, Graham Stuart.
Climbing roses old and new / written and illustrated by Graham Stuart Thomas ... with a chapter on the botany and derivation of climbing roses, by Gordon D. Rowley. — Revised ed. — London [etc.] : Dent [for] the Royal Horticultural Society, 1978.
203 p., [24] p. of plates : ill. ; 23 cm. GB78-28515
Bibliography: p. 187-189.
Includes index.
ISBN 0-460-04346-3 : £8.95
1. Climbing roses. I. Title.
SB411.T488 1978 635.9'33'372 79-300561
79 MARC

CLIMENT, ENRIQUE, 1897-

—CATALOGS

Climent, Enrique, 1897-
Enrique Climent. — 1. ed. — México : Editorial J. Mortiz, 1977.
143 p. : chiefly ill. (some col.) ; 29 cm.
Spanish and/or English.
1. Climent, Enrique, 1897- —Catalogs.
ND813.C56A4 1977 759.972 78-109484
79 MARC

CLIMENT I AVINENT, JOSEP, 1706-1781

Tort Mitjans, Francesc.
El obispo de Barcelona Josep Climent i Avinent (1706-1781) : contribución a la historia de la teología pastoral tarraconense en el siglo XVIII / por Francesc Tort Mitjans. — Barcelona : Editorial Balmes, 1978.
xxvii, 452 p., [1] leaf of plates : port. ; 25 cm. — (Biblioteca histórica de la biblioteca Balmes : Serie II ; v. 29) Sp***
Bibliography: p. [411]-426.
Includes indexes.
ISBN 8421003615
1. Climent i Avinent, Josep, 1706-1781. 2. Catholic Church—Bishops—Biography. 3. Bishops—Spain—Barcelona—Biography. 4. Barcelona—Biography. I. Title. II. Series: Biblioteca histórica de la Biblioteca Balmes ; ser. 2, v. 29.
BX4705.C637T67 79-347950
79 MARC

CLINIC MANAGERS

—RHODE ISLAND—PROVIDENCE—BIOGRAPHY

McDougal, Gwynn.
Women's Medical Clinic of Providence : a story of private enterprise in the abortion arena / by Gwynn McDougal. — Providence : Scorpio Press Publishers, c1979.
101 p. ; 23 cm.
1. Women's Medical Clinic of Providence. 2. Abortion services—Rhode Island—Providence. 3. McDougal, Gwynn. 4. Clinic managers—Rhode Island—Providence—Biography. I. Title.
RG734.M28 362.1'9'888 79-108062
79 MARC

CLINICAL ALGORITHMS see Medical protocols

CLINICAL ASSISTANTS see Physicians' assistants

CLINICAL CHEMISTRY see Chemistry, Clinical

CLINICAL CHILD PSYCHOLOGY

Santostefano, Sebastian, 1929-
A biodevelopmental approach to clinical child psychology : cognitive controls and cognitive control therapy / Sebastiano Santostefano. — New York : Wiley, c1978.
xvi, 833 p. : ill. ; 26 cm. — (Wiley series on personality processes)
"A Wiley-Interscience publication."
Bibliography: p. 796-813.
Includes indexes.
ISBN 0-471-75380-7
1. Clinical child psychology. 2. Cognition in children. 3. Psychobiology. 4. Psychological tests for children. 5. Child psychotherapy. I. Title.
[DNLM: 1. Psychiatry—Bibliography. 2. Mental health—Bibliography. ZWM100 G798h]
RJ503.3.S26 618.9'28'9075 78-18485
78 MARC

CLINICAL EDUCATION, LEGAL
see Law—Study and teaching (Clinical education)

CLINICAL ENDOCRINOLOGY

see also Endocrine glands—Diseases

Clinical neuroendocrinology : a pathophysiological approach / editors, George Tolis ... [et al.]. — New York : Raven Press, c1979.
xviii, 474 p. : ill. ; 26 cm.
Includes bibliographies and index.
ISBN 0-89004-355-8
1. Endocrine glands—Diseases. 2. Neuroendocrinology. 3. Hypothalamic hormones. 4. Clinical endocrinology. I. Tolis, George.
[DNLM: 1. Endocrinology—Congresses. 2. Neurology—Congresses. WK100.3 C641 1978]
RC648.C59 616.4 78-64844
79 MARC

Cryer, Philip E
Diagnostic endocrinology / Philip E. Cryer. — 2d ed. — New York : Oxford University Press, 1979.
viii, 205 p. : ill. ; 22 cm.
Includes bibliographical references and index.
ISBN 0-19-502525-3. ISBN 0-19-502526-1 pbk.
1. Endocrine glands—Diseases—Diagnosis. 2. Clinical endocrinology. I. Title.
[DNLM: 1. Diagnosis, laboratory—Methods. 2. Endocrine diseases—Diagnosis. WK100.3 C957d]
RC649.C79 1979 616.4'07'5 78-9938
78 MARC

Davis, James Collinson.
Clinical endocrine pathology / J. C. Davis, L. J. Hipkin. — Oxford : Blackwell Scientific Publications ; Philadelphia : distributed by J. B. Lippincott Co., 1977.
viii, 392 p. : ill. ; 24 cm. GB***
Includes bibliographies and index.
ISBN 0-632-00106-2
1. Endocrine glands—Diseases. 2. Clinical endocrinology. I. Hipkin, L. J., joint author. II. Title.
RC648.D35 616.4'07 79-303239
79 MARC

Fletcher, Ronald F
Lecture notes on endocrinology / Ronald F. Fletcher. — 2d ed. — Oxford : Blackwell Scientific ; Philadelphia : distributed by J. B. Lippincott, 1978.
iv, 219 p. : ill. ; 22 cm. GB***
Includes bibliographies and index.
ISBN 0-632-00429-0 : £4.50
1. Clinical endocrinology. I. Title.
RC648.F56 1978 616.4 79-302893
79 MARC

Jubiz, William.
Endocrinology : a logical approach for clinicians / William Jubiz. — New York : McGraw-Hill, c1979.
xi, 418 p. : ill. ; 24 cm.
Includes bibliographies and index.
ISBN 0-07-033065-4 : $20.00
1. Endocrine glands—Diseases. 2. Clinical endocrinology. I. Title.
RC648.J8 616.4'2 78-2754
78 MARC

Korenman, Stanley G
Endocrine disease / Stanley G. Korenman, Daryl K. Granner, Barry M. Sherman. — Boston : Houghton Mifflin, c1978.
ix, 276 p. : ill. ; 19 cm. — (Practical diagnosis)
Includes bibliographies and index.
ISBN 0-89289-201-3 : $18.00
1. Endocrine glands—Diseases—Diagnosis. 2. Clinical endocrinology. 3. Endocrine diseases—Diagnosis. I. Granner, Daryl K., joint author. II. Sherman, Barry M., joint author. III. Title. IV. Series.
[DNLM: WK100.3 K84e]
RC649.K65 616.4'07'5 78-21765
78 MARC

Lee, Julius.
Essential endocrinology / Julius Lee, John Laycock. — Oxford ; New York : Oxford University Press, 1978.
ix, 227 p. : ill. ; 22 cm. — (Oxford medical publications) GB78
Bibliography: p. [213].
Includes index.
ISBN 0-19-261123-2 : £4.50 ($8.95)
1. Clinical endocrinology. I. Laycock, John F., joint author. II. Title.
RC648.L38 616.4 78-40066
78 MARC

CLINICAL ENZYMOLOGY

Clinical enzymology / edited by John C. Griffiths. — New York : Masson Pub. USA, c1979.
viii, 212 p. : ill. ; 24 cm.
Includes bibliographical references and index.
ISBN 0-89352-030-6
1. Clinical enzymology. I. Griffiths, John C.
RB48.C54 616.07'56 79-84461
00 MARC

—CONGRESSES

International Society for Clinical Enzymology.
Enzymes in health and disease : inaugural scientific meeting of the International Society for Clinical Enzymology, London, September 19-21, 1977 / editors, David M. Goldberg and J. Henry Wilkinson. — Basel ; New York : S. Karger, 1978.
xii, 237 p. : ill. ; 25 cm. S***
Includes bibliographical references.
ISBN 3-8055-2853-1 : 98.00F ($58.75)
1. Clinical enzymology—Congresses. 2. Enzymes—Congresses. I. Goldberg, David M. II. Wilkinson, John Henry. III. Title.
RB48.I57 1978 616.3'99 79-309776
79 MARC

CLINICAL GENETICS see Medical genetics

CLINICAL LABORATORY TECHNICIANS
see Medical technologists

CLINICAL LEGAL EDUCATION
see Law—Study and teaching (Clinical education)

CLINICAL MEDICINE see Medicine, Clinical

CLINICAL MICROBIOLOGY see Medical microbiology

CLINICAL NURSE SPECIALISTS see Nurse practitioners

CLINICAL PARASITOLOGY see Medical parasitology

CLINICAL PATHOLOGY see Diagnosis, Laboratory

CLINICAL PHYSIOLOGY see Physiology, Pathological

CLINICAL PROTOCOLS see Medical protocols

CLINICAL PSYCHOLOGY

see also Clinical child psychology; Psychodiagnostics

Golden, Charles J 1949-
The Luria-Nebraska neuropsychological battery : a manual for clinical and experimental uses / Charles J. Golden, Arnold D. Purisch, Thomas A. Hammeke. — Lincoln : University of Nebraska Press, c1979.
178 p. : ill. ; 29 cm. & cassette and 6 flash cards.
"Kit for Luria-Nebraska neuropsychological battery" (cassette and 6 flash cards) inserted.
Bibliography: p. 175-178.
ISBN 0-8032-7001-1
1. Neuropsychology. 2. Clinical psychology. I. Purisch, Arnold D., joint author. II. Hammeke, Thomas A., joint author. III. Title.
QP360.G64 616.8'04'75 78-78202
00 MARC

Hollis, Joseph William.
Psychological report writing : theory and practice / Joseph W. Hollis, Patsy A. Donn ; [editors, Trudy Pelligrini, Cindy Lyons ; corporate illustrator, Mary Blizzard]. — 2d ed. — Muncie, Ind. : Accelerated Development, c1979.
xv, 296 p. : ports. ; 23 cm.
Bibliography: p. 277-281.
Includes index.
ISBN (invalid) 0-915292-21-2
1. Clinical psychology. 2. Report writing. I. Donn, Patsy A., joint author. II. Pelligrini, Trudy. III. Lyons, Cindy. IV. Title.
RC467.H64 1979 808'.066'1579021 79-64499
79 MARC

K'o, Yung-ho.
(Lin ch'uang hsin li hsüeh)
臨床心理學 / 柯永河著. — 基隆：大洋出版社；台北：總經銷文笙書局,
v. : ill. ; 22 cm.
Includes bibliographies and index. 第2冊
CONTENTS:
心理治療
NT$150.00 (v. 2)
1. Clinical psychology. I. Title.
RC467.K6 78–841524

Phares, E Jerry.
Clinical psychology : concepts, methods, and profession / E. Jerry Phares. — Homewood, Ill. : Dorsey Press, 1979.
xvii, 609 p. : ill. ; 24 cm. — (The Dorsey series in psychology)
Bibliography: p. 539-589.
Includes indexes.
ISBN 0-256-02190-2
1. Clinical psychology. 2. Personality assessment. 3. Psychotherapy. I. Title.
RC467.P48 157'.9 78-62632
79 MARC

(Rinshō-shinrigaku)
臨床心理学 / 水山進吾, 江見佳俊編. — 東京：福村出版, 1978.
183 p. ; 22 cm. — (シリーズ現代心理学 ; 5) Ja 78–12001
Includes bibliographies and indexes.
¥1500
1. Clinical psychology. I. Mizuyama, Shingo. II. Emi, Yoshitoshi. III. Series: Shirizu gendai shinrigaku ; 5.
RC467.R516 78–813836

Sanocki, Włodzisław.
Koncepcja normy w psychologii klinicznej : studium metodologiczne / Włodzisław Sanocki. — Wyd. 1. — Gdańsk : Wydawn. Uniwersytetu Gdańskiego, 1978.
210 p. ; 24 cm. — (Zeszyty naukowe, rozprawy i monografie - Uniwersytet Gdański ; 1) P***
Summary in English.
Bibliography: p. 195-207.
zł25.00
1. Clinical psychology. I. Title. II. Series: Uniwersytet Gdański. Zeszyty naukowe, rozprawy i monografie - Uniwersytet Gdański ; 1.
RC467.S26 78-381096
79 MARC

—PICTORIAL WORKS

Geréb, György.
Pszichológiai atlasz / Geréb György. — 6. kiad. — Budapest : Tankönyvkiadó, 1977.
327 p. : ill. (some col.) ; 24 cm. H77
Summary in German.
Bibliography: p. 312-[313]
ISBN 9631711684 : 64.00Ft
1. Psychology, Physiological—Atlases. 2. Clinical psychology—Pictorial works. I. Title.
QP360.G47 1977 79-347762
*79 MARC

—PRACTICE

Pressman, Robert M
Private practice : a handbook for independent health practitioner / by Robert M. Pressman. — New York : Gardner Press : distributed by Halsted Press, c1979.
xiv, 230 p. : ill. ; 25 cm.
Bibliography: p. 139-146.
Includes index.
ISBN 0-470-26388-1
1. Psychiatry—Practice. 2. Clinical psychology—Practice. 3. Psychiatric social work—Practice. I. Title.
[DNLM: 1. Practice management, Medical—Handbooks. 2. Psychiatry—Handbooks. 3. Psychology, Clinical—Handbooks. 4. Social work, Psychiatric—Handbooks. WM21 P935p]
RC440.7.P73 658'.91'61 78-8008
78 MARC

Shimberg, Edmund, 1925-
The handbook of private practice in psychology / by Edmund Shimburg. — New York : Brunner/Mazel, c1979.
xiii, 104 p. ; 22 cm.
Includes index.
ISBN 0-87630-197-9
1. Clinical psychology—Practice. I. Title.
RC467.S483 658'.91'1579 79-1518
79 MARC

—PRACTICE—HANDBOOKS, MANUALS, ETC.

Browning, Charles H
Private practice handbook : the tools, tactics & techniques for successful practice development / by Charles H. Browning. — Los Alamitos, Calif. : Browning Therapy Group, c1979.
x, 156 p. ; 22 cm.
Includes bibliographical references.
1. Clinical psychology—Practice—Handbooks, manuals, etc. I. Title.
RC467.B76 658'.91'61689 79-113302
79 MARC

—GREAT BRITAIN

Great Britain. Sub-committee on the Role of Psychologists in the Health Services.
The role of psychologists in the Health Services : report of the Sub-committee. — London : H. M. Stationery Off., 1977.
iv, 32 p. ; 25 cm. GB***
At head of title: Department of Health and Social Security.
Chairman: W. H. Trethowan.
ISBN 0-11-320275-X : £0.70
1. Mental health services—Great Britain. 2. Clinical psychology—Great Britain. I. Trethowan, W. H. II. Title.
RA790.7.G7G74 1977 79-316229
79 MARC

CLINICAL READIOLOGY see Radiology, Medical

CLINICAL RECORDS see Medical records

CLINICAL TRIALS OF DRUGS see Drugs--Testing

CLINICS

see also Abortion services; Birth control clinics; Chiropractic clinics; Group medical practice; Medical offices

—ADMINISTRATION

see also Clinic managers

—PLANNING

Beales, Gerald.
Sick health centres and how to make them better / Gerald Beales. — Turnbridge Wells, Eng. : Pitman Medical, 1978.
vii, 147 p. : diagrs. ; 22 cm. GB***
"This book is based on research which was undertaken by members of the Organizational Analysis Research Unit, University of Bradford, from 1971 to 1976, under the direction of Professor D. J. Hickson."
"Catalogue Number 21 0217 81."
ISBN 0-272-79431-7 : £5.00
1. Clinics—Great Britain. 2. Health care teams—Great Britain. 3. Clinics—Planning. I. Title.
RA986.B35 362.1'2'0941 78-326323
79 MARC

—STANDARDS—UNITED STATES

Accreditation Council for Ambulatory Health Care.
Accreditation manual for ambulatory health care. — 1978 ed. — Chicago : Joint Commission on Accreditation of Hospitals, Accreditation Council for Ambulatory Health Care, c1978.
52 p. ; 28 cm.
$7.50
1. Clinics—Standards—United States. 2. Ambulatory medical care—Standards—United States. 3. Clinics—United States—Accreditation. I. Title.
RA966.A25 1978 362.1'2 79-104186
79 MARC

—BELGIUM—STATISTICS

Beckers, Robert.
Sanitair dossier van de gemeenten 31-12-'75 = Dossier sanitaire des communes 31-12-'75 / R. Beckers. — 1000 Brussel (Hertogstr. 81) : Ministerie van Volksgezondheid en van het Gezin, [1976?]
[66] leaves ; 30 cm. Be77-8
Dutch or French.
1. Environmental health—Belgium—Statistics. 2. Social service—Belgium—Statistics. 3. Medical personnel—Belgium—Statistics. 4. Clinics—Belgium—Statistics. I. Title. II. Title: Dossier sanitaire des communes.
RA566.5.B4B4 78-398174
*79 MARC

—GREAT BRITAIN

Beales, Gerald.
Sick health centres and how to make them better / Gerald Beales. — Turnbridge Wells, Eng. : Pitman Medical, 1978.
vii, 147 p. : diagrs. ; 22 cm. GB***
"This book is based on research which was undertaken by members of the Organizational Analysis Research Unit, University of Bradford, from 1971 to 1976, under the direction of Professor D. J. Hickson."
"Catalogue Number 21 0217 81."
ISBN 0-272-79431-7 : £5.00
1. Clinics—Great Britain. 2. Health care teams—Great Britain. 3. Clinics—Planning. I. Title.
RA986.B35 362.1'2'0941 78-326323
79 MARC

—NEW JERSEY

New Jersey. Commission of Investigation.
Report of the New Jersey State Commission of Investigation on the practices and procedures of practitioner groups participating in the New Jersey medicaid program. — [Trenton?] : The Commission, 1976.
77 leaves ; 29 cm.
1. Clinics—New Jersey. 2. Medicaid fraud—New Jersey. 3. Medical fees—New Jersey. 4. Pharmaceutical services—New Jersey.
RA981.N5N47 1976 364.1'63 77-620711
79 MARC

CLINICS (Continued)

—OHIO—CUYAHOGA CO.
—DIRECTORIES

Chisolm, Laura B
The Cleveland medical directory : a consumer's guide to family doctors, clinics and HMOs in Cuyahoga County / prepared by Laura B. Chisolm, with the assistance of Jill Baskind and Jenelle Foote. — 1st ed. — Cleveland : Cleveland Consumer Action Foundation, 1978.
224 p. ; 22 cm.
1. Physicians (General practice)—Ohio—Cuyahoga Co.—Directories. 2. Internists—Ohio—Cuyahoga Co.—Directories. 3. Pediatricians—Ohio—Cuyahoga Co.—Directories. 4. Clinics—Ohio—Cuyahoga Co.—Directories. 5. Health maintenance organizations—Ohio—Cuyahoga Co.—Directories. I. Baskind, Jill. II. Foote, Jenelle. III. Title.
R712.A2O32 362.1'025'77131 78-108062
78 MARC

—UNITED STATES—ACCREDITATION

Accreditation Council for Ambulatory Health Care.
Accreditation manual for ambulatory health care. — 1978 ed. — Chicago : Joint Commission on Accreditation of Hospitals, Accreditation Council for Ambulatory Health Care, c1978.
52 p. ; 28 cm.
$7.50
1. Clinics—Standards—United States. 2. Ambulatory medical care—Standards—United States. 3. Clinics—United States—Accreditation. I. Title.
RA966.A25 1978 362.1'2 79-104186
79 MARC

—UNITED STATES—STATISTICS

Agro, Gerlandino.
Study of ambulatory care facilities as a resource for medical education : final report / Gerlandino Agro. — Washington : Division of Operational Studies, Association of American Medical Colleges, 1977.
2 v. ; 30 cm.
Based on information made available to AAMC by Medical schools.
1. Medical colleges—United States—Statistics. 2. Ambulatory medical care—Study and teaching—United States—Statistics. 3. Clinics—United States—Statistics. 4. Health facilities—United States—Affiliations—Statistics. I. Association of American Medical Colleges. II. Title.
R745.A55 610'.7'1173 77-151046
77[r79]rev MARC

CLINTON CO., IOWA

—BIOGRAPHY

History of Clinton County, Iowa, 1976 : a bicentennial 1976 project / Clinton County Historical Society. — [s.l.] : Clinton County American Revolution Bicentennial Commission, c1978.
576 p. : ill. ; 28 cm.
Bibliography: p. 566-569.
Includes index.
1. Clinton Co., Iowa—History. 2. Clinton Co., Iowa—Biography. I. Clinton County Historical Society.
F627.C6H62 977.7'67 79-102029
79 MARC

—HISTORY

History of Clinton County, Iowa, 1976 : a bicentennial 1976 project / Clinton County Historical Society. — [s.l.] : Clinton County American Revolution Bicentennial Commission, c1978.
576 p. : ill. ; 28 cm.
Bibliography: p. 566-569.
Includes index.
1. Clinton Co., Iowa—History. 2. Clinton Co., Iowa—Biography. I. Clinton County Historical Society.
F627.C6H62 977.7'67 79-102029
79 MARC

—MAPS

Rockford Map Publishers, Rockford, Ill.
Atlas & plat book, Clinton County, Michigan. — 1978-1979 ed. — Rockford, Ill. : Rockford Map Publishers ; St. Johns, Mich. : available at Clinton County Cooperative Extension Service, [1978, c1969]
50 p. : maps ; 28 cm.
Cover title.
Cadastral maps of townships, scale 1:50,688 or 1 1/4 inch equals one mile.
Includes index.
1. Clinton Co., Mich.—Maps. 2. Real property—Michigan—Clinton Co.—Maps. I. Title.
G1413.C8R6 1978 912'.774'24 79-318190
79 MARC

Rockford Map Publishers, Rockford, Ill.
Land atlas & plat book, 1976, Clinton County, Iowa. — Rev. 1976. — Rockford : Rockford Map Publishers ; Clinton, Iowa : distributed by Clinton County Recorder, [1976, c1971]
46 p. : maps ; 28 cm.
Cover title.
Cadastral maps of townships, scale 1:50,688 or 1 1/4 inch equals one mile.
Includes index.
1. Clinton Co., Iowa—Maps. 2. Real property—Iowa—Clinton Co.—Maps. I. Title.
G1433.C8R6 1976 912'.777'67 78-319151
78[r79]rev MARC

Rockford Map Publishers, Rockford, Ill.
Land atlas & plat book, Clinton County, Iowa : 1979. — Rev. 1979. — Rockford, Ill. : Rockford Map Publishers, [1979, c1976]
46 p. : maps ; 28 cm.
Cover title.
Cadastral maps of townships, scale 1:50,688 or 1 1/4 inch equals one mile.
Edition for 1976 published under title: Land atlas & plat book, 1976, Clinton County, Iowa.
Includes index.
$10.50
1. Clinton Co., Iowa—Maps. 2. Real property—Iowa—Clinton Co.—Maps. I. Title.
G1433.C8R6 1979 912'.777'67 79-318218
79 MARC

CLINTON CO., MICH

—HISTORY, LOCAL

Ceasar, Ford Stevens.
Forgotten communities of Central Michigan / [by Ford Stevens Ceasar]. — 3d ed., rev. and enl. — [s.l.] : Ceasar, 1978.
xii, 83 p. : maps ; 23 cm.
Bibliography: p. 73-75.
Includes index.
$5.00
1. Clinton Co., Mich.—History, Local. 2. Ingham Co., Mich.—History, Local. 3. Gratiot Co., Mich.—History, Local. I. Title.
F572.C7C4 1978 977.4'2 78-59343
78 MARC

CLIPPER-SHIPS

—HISTORY

MacGregor, David Roy.
Clipper ships / David R. MacGregor. — Watford, Eng. : Argus Books, 1979.
v, 154 p. : ill. ; 23 cm. GB***
Includes index.
ISBN 0-85242-618-6 : £5.95
1. Clipper-ships—History. 2. Navigation—History—19th century. I. Title.
VM19.M22 387.2'2 79-309107
79 MARC

CLIPPINGS (BOOKS, NEWSPAPERS, ETC.)

see also Newspaper office libraries

CLIPSTON GRAMMAR SCHOOL

—HISTORY

Isham, Gyles, Sir, bart., 1903-
The story of Clipston Grammar School, Northamptonshire / by Sir Gyles Isham, bart. — [s.l. : s.n., 1956] ([Market Harborough : Green])
54 p. : ill. ; 22 cm.
Cover title.
Includes bibliographical references and index.
1. Clipston Grammar school—History. I. Title.
LF795.C5843I83 373.425'5 78-321198
79 MARC

CLITELLATA see Leeches; Oligochaeta

CLITORIS

The Classic clitoris : historic contributions to scientific sexuality / Thomas Power Lowry, editor. — Chicago : Nelson-Hall, c1978.
xiii, 120 p. : ill. ; 23 cm.
Includes bibliographical references and index.
ISBN 0-88229-387-7
1. Clitoris. 2. Cloaca (Zoology) 3. Generative organs, Female. 4. Anatomy, Comparative. I. Lowry, Thomas P.
[DNLM: 1. Sex—Essays. 2. Sex behavior—Essays. HQ21 C614]
QL881.C5 618.1'6 78-18298
78 MARC

CLIVE, ROBERT CLIVE, BARON, 1725-1774

Chaudhuri, Nirad C 1897-
Clive of India : a political and psychological essay / Nirad C. Chaudhuri. — London : Barrie & Jenkins, 1975.
vii, 446 p. ; 24 cm. GB***
Bibliography: p. 440-441.
Includes index.
ISBN 0-214-20021-3 : £7.50
1. Clive, Robert Clive, Baron, 1725-1774. 2. India—History—18th century. 3. Bengal—Governors—Biography. I. Title.
DS471.C53 954.02'9'0924 75-330510
75[r79]rev MARC

CLIVEDEN, ENG.

National Trust for Places of Historic Interest or Natural Beauty.
Cliveden, Buckinghamshire / [National Trust]. — London : The Trust, 1978.
38 p. : ill., map, ports. ; 25 cm. GB78-26467
Bibliography: p. 38.
£0.45
1. Cliveden, Eng. 2. Buckingham, George Villiers, 2d Duke of, 1628-1687. I. Title.
DA664.C5N38 1978 914.25'98 79-307251
79 MARC

CLOACA (ZOOLOGY)

see also Rectum; Rectum—Cancer

The Classic clitoris : historic contributions to scientific sexuality / Thomas Power Lowry, editor. — Chicago : Nelson-Hall, c1978.
xiii, 120 p. : ill. ; 23 cm.
Includes bibliographical references and index.
ISBN 0-88229-387-7
1. Clitoris. 2. Cloaca (Zoology) 3. Generative organs, Female. 4. Anatomy, Comparative. I. Lowry, Thomas P.
[DNLM: 1. Sex—Essays. 2. Sex behavior—Essays. HQ21 C614]
QL881.C5 618.1'6 78-18298
78 MARC

CLOCHE GARDENING

Shewell-Cooper, Wilfred Edward, 1900-
Basic book of cloche & frame gardening / [by] W. E. Shewell-Cooper. — London : Barrie and Jenkins, 1977.
160 p. : ill. ; 22 cm. GB78-10295
Spine title: Dr. Shewell-Cooper's basic book of cloche and frame gardening.
Includes index.
ISBN 0-214-20301-8 : £4.25
1. Cloche gardening. 2. Cold-frames. I. Title. II. Title: Cloche & frame gardening. III. Title: Dr. Shewell-Cooper's basic book of cloche and frame gardening.
SB127.S49 635'.04'8 79-305936
79 MARC

CLOCK AND WATCH MAKERS

Abeler, Jürgen.
Ullstein-Uhrenbuch : e. Kulturgeschichte d. Zeitmessung / Jürgen Abeler ; [Zeichn., Dieter Messerschmidt]. — 2., neu bearb. Aufl. — Berlin ; Frankfurt/M. ; Wien : Ullstein, 1979.
324 p. : numerous ill. (some col.) ; 24 cm. GFR79-A
Bibliography: p. 309-311.
Includes index.
ISBN 3-550-17462-4 : DM58.00
1. Clocks and watches—History. 2. Clock and watch makers. I. Title.
TS542.A23 1979 79-384045
79 MARC

—ENGLAND—BIOGRAPHY

Shenton, R K
Christopher Pinchbeck and his family / by Rita Shenton. — Ashford, Kent : Brant Wright Associates, 1976.
71 p. : ill., facsims., geneal. tab., ports. ; 22 cm. GB77-07977
Bibliography: p. 71
ISBN 0-903512-06-8 : £6.75
1. Pinchbeck family. 2. Clock and watch makers—England—Biography. I. Title.
TS544.8.P55S53 681'.11'0922 78-317450
78 MARC

CLOCK AND WATCH MAKERS (Continued)

—ENGLAND—EXETER—BIOGRAPHY

Ponsford, C N
Time in Exeter : a history of 700 years of clocks and clock-making in an English provincial city, with details of more than 300 former makers / by Clive N. Ponsford. — Exeter ⌈Eng.⌉ : Headwell Vale Books, 1978.
200 p., ⌈16⌉ leaves of plates : ill., ports. ; 23 cm. GB***
Includes bibliographical references and index.
ISBN 0-9806361-0-X : £6.00
1. Clock and watch making—England—Exeter—History. 2. Clock and watch makers—England—Exeter—Biography. I. Title.
TS543.G7P68 681'.11'0942356 79-303011
79 MARC

—ENGLAND—TIVERTON

Ponsford, C N
Clocks and clockmakers of Tiverton / by C. N. Ponsford, J. G. M. Scott and W. P. Authers. — ⌈Exeter⌉ : The authors, 1977.
64 p., ⌈20⌉ p. of plates : ill., facsim., ports. ; 22 cm. GB78-03512
Bibliography: p. 60-61.
Includes index.
1. Clock and watch making—England—Tiverton—History. 2. Clock and watch makers—England—Tiverton. 3. Tiverton, Eng.—Industries. I. Scott, John Gilbert Mortimer, 1925- , joint author. II. Authers, William Percival, joint author. III. Title.
TS543.G7P66 681'.113'0942354 78-315219
78 MARC

—FINLAND

Elfström, Taimi.
Kellosepät Suomessa vuosina 1600-1875 / Taimi Elfström. — Helsinki : Suomen Kelloseppäliitto, ⌈1976⌉
77 p. ; 21 cm. — (Suomen Kelloseppäliiton matrikkelisarja ; osa 1) Fi***
1. Clock and watch makers—Finland. I. Title. II. Series: Suomen Kelloseppäliitto. Suomen Kelloseppäliiton matrikkelisarja ; osa 1.
TS543.F5E43 79-349529
79 MARC

—GERMANY—BIOGRAPHY

Munz, Alfred.
Philipp Matthäus Hahn : Pfarrer, Erfinder u. Erbauer von Himmelsmaschinen, Waagen, Uhren u. Rechenmaschinen / Alfred Munz. — 1. Aufl. — Sigmaringen : Thorbecke, 1977.
144 p. : ill. (some col.) ; 24 cm. GFR78-A
Bibliography of P. M. Hahn's works: p. 104.
Includes bibliographical references.
ISBN 3-7995-4025-3
1. Hahn, Philipp Matthäus, 1739-1790. 2. Clock and watchmakers—Germany—Biography. 3. Clergy—Germany—Biography. 4. Astronomical clocks.
TS544.8.H33M36 79-359760
79 MARC

—GERMANY, WEST
—BLACK FOREST

Tyler, Eric John.
Black Forest clocks / ⌈by⌉ E. John Tyler. — London : N.A.G. Press, 1977.
vii, 120 p., ⌈16⌉ p. of plates : ill. (some col.), facsim., map (on lining papers) ; 22 cm. GB78-10072
Bibliography: p. 118.
Includes index.
ISBN 0-7198-0100-1 : £6.50
1. Clock and watch making—Germany, West—Black Forest—History. 2. Clock and watch makers—Germany, West—Black Forest. 3. Black Forest—Industries. I. Title.
TS543.G3T94 78-314440
338.4'7'68111309434
78 MARC

—NEW HAMPSHIRE—BIOGRAPHY

Parsons, Charles Sumner, 1904-
New Hampshire clocks & clockmakers / by Charles S. Parsons. — Exeter, N.H. : Adams Brown Co., 1976.
356 p. : ill. ; 29 cm.
Includes bibliographical references and index.
ISBN 0-917900-00-6 : $22.00
1. Clock and watch making—New Hampshire—History. 2. Clock and watch makers—New Hampshire—Biography. 3. New Hampshire—Industries. I. Title.
TS543.U6P35 681'.113'09742 76-28189
76 MARC

—SOUTHERN STATES

Gibbs, James W
Dixie clockmakers / James W. Gibbs. — Gretna, La. : Pelican Pub. Co., 1979.
191 p. : ill. ; 29 cm.
Bibliography: p. 179-181.
Includes index.
ISBN 0-88289-059-X : $19.95
1. Clock and watch making—Southern States—History. 2. Clock and watch makers—Southern States. I. Title.
TS543.U6G52 681'.113'0922 78-2204
78 MARC

—SWITZERLAND

Kunz, Bernard.
L'Emploi dans la région horlogère : étude réalisée sur mandat de l'OFIAMT et de la Communauté de travail des cantons horlogers / Bernard Kunz ; direction de l'étude, D. Maillat avec la collab. de J.-L. Juvet. — Neuchâtel : Université, Groupe d'études économiques, 1978.
57, 44, 32 p., ⌈1⌉ leaf of map : ill. ; 21 cm. Sw78-6590
1. Clock and watch makers—Switzerland. I. Maillat, Denis. II. Juvet, Jean Louis. III. Title.
HD5718.W22S94 79-365450
331.7'68'111309493
79 MARC

CLOCK AND WATCH MAKING

see also Clocks and watches

Romanov, Andreĭ Dmitrievich.
(Proektirovanie priborov vremeni)
Проектирование приборов времени : допущено ... в качестве учеб. пособия для студентов приборостроительных специальностей высших учеб. заведений / А. Д. Романов. — Москва : Высшая школа, 1975.
223 p. : ill. ; 23 cm. USSR***
Bibliography : p. ⌈222⌉
0.73rub
1. Clock and watch making. I. Title.
TS545.R65 79-407154

—HISTORY

Pomella, Fulgido, 1901-
L'orologio da portare addosso : arte e tecnica nell'orologio tascabile dalle origini al 1820-30 : ad uso di amatori e collezionisti / Fulgido Pomella. — 1. ed. — Ivrea : Priuli & Verlucca, 1978.
289 p. : ill. ; 25 cm. — (Ricerche 2 ⌈i.e. due⌉) It79-Jan
Bibliography: p. 278-283.
Includes index.
1. Clock and watch making—History. 2. Clocks and watches—History. I. Title.
TS542.P65 79-346763
*79 MARC

—TOOLS AND IMPLEMENTS—
CATALOGS

Wyke, John.
A catalogue of tools for watch and clock makers / by John Wyke of Liverpool ; introd. and technical commentary by Alan Smith. — Charlottesville : Published for the Henry Francis du Pont Winterthur Museum by the University Press of Virginia, 1978.
153 p. : ill. ; 23 x 31 cm.
"A facsimile edition of the tool catalogue of John Wyke of Liverpool."
ISBN 0-8139-0751-9
1. Clock and watch making—Tools and implements—Catalogs. I. Smith, Alan, 1925 (Oct. 31)- II. Henry Francis du Pont Winterthur Museum. III. Title.
TS549.W94 1978 681'.113'028 77-12219
77 MARC

—VOCATIONAL GUIDANCE

Lipton, Benjamin H
Aim for a job in watchmaking / Benjamin H. Lipton. — Rev. ed. — New York : R. Rosen Press, 1978.
126 p. : ill. ; 22 cm. — (Aim high vocational series)
Includes bibliographical references.
ISBN 0-8239-0453-9
1. Clock and watch making—Vocational guidance. I. Title.
TS545.9.L5 1978 681'.114'023 78-7438
78 MARC

—ENGLAND—EXETER—HISTORY

Ponsford, C N
Time in Exeter : a history of 700 years of clocks and clock-making in an English provincial city, with details of more than 300 former makers / by Clive N. Ponsford. — Exeter ⌈Eng.⌉ : Headwell Vale Books, 1978.
200 p., ⌈16⌉ leaves of plates : ill., ports. ; 23 cm. GB***
Includes bibliographical references and index.
ISBN 0-9806361-0-X : £6.00
1. Clock and watch making—England—Exeter—History. 2. Clock and watch makers—England—Exeter—Biography. I. Title.
TS543.G7P68 681'.11'0942356 79-303011
79 MARC

—ENGLAND
—NEWCASTLE-UNDER-LYME
—EXHIBITIONS

Treherne, Alan.
The Massey family : clock, watch, chronometer and nautical instrument makers : catalogue of an exhibition at the Museum, Newcastle-under-Lyme / by Alan Treherne. — Newcastle-under-Lyme : The Borough Museum, ⌈1977⌉
⌈2⌉, 36, ⌈1⌉ p. : ill., 2 facsims., geneal. table ; 30 cm. GB78-02530
Bibliography: p. 3 of cover.
1. Clocks and watches—Exhibitions. 2. Clock and watchmaking—England—Newcastle-under-Lyme—Exhibitions. I. Newcastle-under-Lyme Borough Museum. II. Title.
TS541.G72N487 681'.11'0922 78-314468
78 MARC

—ENGLAND—TIVERTON—HISTORY

Ponsford, C N
Clocks and clockmakers of Tiverton / by C. N. Ponsford, J. G. M. Scott and W. P. Authers. — ⌈Exeter⌉ : The authors, 1977.
64 p., ⌈20⌉ p. of plates : ill., facsim., ports. ; 22 cm. GB78-03512
Bibliography: p. 60-61.
Includes index.
1. Clock and watch making—England—Tiverton—History. 2. Clock and watch makers—England—Tiverton. 3. Tiverton, Eng.—Industries. I. Scott, John Gilbert Mortimer, 1925- , joint author. II. Authers, William Percival, joint author. III. Title.
TS543.G7P66 681'.113'0942354 78-315219
78 MARC

—GERMANY, WEST—BLACK FOREST
—HISTORY

Tyler, Eric John.
Black Forest clocks / ⌈by⌉ E. John Tyler. — London : N.A.G. Press, 1977.
vii, 120 p., ⌈16⌉ p. of plates : ill. (some col.), facsim., map (on lining papers) ; 22 cm. GB78-10072
Bibliography: p. 118.
Includes index.
ISBN 0-7198-0100-1 : £6.50
1. Clock and watch making—Germany, West—Black Forest—History. 2. Clock and watch makers—Germany, West—Black Forest. 3. Black Forest—Industries. I. Title.
TS543.G3T94 78-314440
338.4'7'68111309434
78 MARC

—NETHERLANDS

Zeeman, J
De Nederlandse stoelklok / J. Zeeman. — 2e, bijgewerkte en herz. dr. — Assen : Gorcum, 1978.
264 p. : with ill. ; 26 cm. Ne78-35
"Met bijdragen van A. Doedens ..."
Includes bibliographies and index.
ISBN 9023214161 : fl 135.00
1. Clock and watch making—Netherlands. 2. Clocks and watches—Netherlands. I. Doedens, A. II. Title.
TS543.N4Z43 1978 78-386050
*79 MARC

—NEW HAMPSHIRE—HISTORY

Parsons, Charles Sumner, 1904-
New Hampshire clocks & clockmakers / by Charles S. Parsons. — Exeter, N.H. : Adams Brown Co., 1976.
356 p. : ill. ; 29 cm.
Includes bibliographical references and index.
ISBN 0-917900-00-6 : $22.00
1. Clock and watch making—New Hampshire—History. 2. Clock and watch makers—New Hampshire—Biography. 3. New Hampshire—Industries. I. Title.
TS543.U6P35 681'.113'09742 76-28189
76 MARC

CLOCK AND WATCH MAKING (Continued)

—SOUTHERN STATES—HISTORY

Gibbs, James W
Dixie clockmakers / James W. Gibbs. — Gretna, La. : Pelican Pub. Co., 1979.
191 p. : ill. ; 29 cm.
Bibliography: p. 179-181.
Includes index.
ISBN 0-88289-059-X : $19.95
1. Clock and watch making—Southern States—History. 2. Clock and watch makers—Southern States. I. Title.
TS543.U6G52 681'.113'0922 78-2204
78 MARC

—SWITZERLAND—DIRECTORIES

Quid horloger. 81.- éd; 1978-
[Genève] Chapalay & Mottier.
v. 21 cm. annual.
Continues: Annuaire de l'horlogerie suisse. Guide général de horlogerie, bijouterie, lunetterie et mécanique de précision de Suisse.
English, French, German or Spanish.
1. Clock and watch making—Switzerland—Directories.
TS540.A65 78-646060
338.4'7'681113025494 MARC-S

—UNITED STATES

Frost & Sullivan.
The United States watch market. — New York : Frost & Sullivan, c1978.
iii, 256 p. : ill. ; 29 cm.
1. Clock and watch making—United States. 2. Market surveys—United States. I. Title.
HD9999.C6U525 1978 79-112076
381'.45'6811140973
79 MARC

—UNITED STATES—HISTORY

Terry, Henry, 1801-1877.
American clock making; its early history. Also, Illustrated catalogue of clocks, manufactured by the Terry Clock Company, Pittsfield, Mass. [Bristol, Conn., American Clock and Watch Museum, 1974]
94 p. illus. 23 cm.
Reprint of the 1885 ed., with additional information.
Includes bibliographical references.
ISBN 0-9600264-7-9
1. Clocks and watches—Catalogs. 2. Clock and watch making—United States—History. I. Terry Clock Company. Illustrated catalogue of clocks. 1974. II. Title.
TS549.T47 1974 681'.113'0974 73-94174
74[r79]rev MARC

—VIRGIN ISLANDS OF THE UNITED STATES

The Use of "low-labor" components in the insular possessions watch industry : a compilation of written comments solicited by the Subcommittee on International Trade, Abraham Ribicoff, chairman of the Committee on Finance, United States Senate, Russell B. Long, chairman. — Washington : U.S. Govt. Print. Off., 1978.
iii, 104 p. ; 24 cm.
At head of title: 95th Congress, 2d session. Committee print.
Includes bibliographical references.
1. Clock and watch making—Virgin Islands of the United States. 2. Tariff on watches—United States. I. United States. Congress. Senate. Committee on Finance. Subcommittee on International Trade.
HD9999.C6V578 79-601629
382'.45'68111409729722
79 MARC

CLOCK MUSIC

see also Chiming clocks

CLOCK-TOWERS see Towers

CLOCKS, ASTRONOMICAL see Astronomical clocks

CLOCKS, CHIMING see Chiming clocks

CLOCKS AND WATCHES

see also Astronomical clocks; Banjo clocks; Chiming clocks; Clock and watch making; Pocket watches; Time measurements; Tower-clocks

Brepohl, Erhard.
Schmuck und Uhren / Erhard Brepohl, Rudi Koch. — 3., neubearb. Aufl. — Leipzig : Fachbuchverlag, VEB, 1977.
287 p., [1] leaf of plates : ill. (some col.) ; 24 cm. GFR77-A20
Includes index.
18.00M
1. Jewelry. 2. Clocks and watches. I. Koch, Rudi, joint author. II. Title.
TS740.B8 1977 739.27 78-366428
78 MARC

Britten, Frederick James, 1843-1913.
Britten's watch & clock maker's handbook : dictionary and guide / F. J. Britten ; illustrated by John W. Wood. — 16th ed., rev. / by Richard Good. — New York : Arco Pub. Co., c1978.
viii, 459 p. : ill. ; 24 cm.
Bibliography: p. 431-434.
Includes indexes.
ISBN 0-668-04638-4 : $69.95
1. Clocks and watches. I. Good, Richard. II. Title. III. Title: Watch & clock maker's handbook.
TS540.B68 1978 681'.113'03 78-3539
78 MARC

Erbrich, Klaus.
Präzisionspendeluhren : von Graham bis Riefler / Klaus Erbrich ; [Redaktion, Christian Pfeiffer-Belli] — München : Callwey, 1978.
254 p. : numerous ill. (1 col.) ; 29 cm. GFR78-A46
ISBN 3-7667-0429-X : DM98.00
1. Clocks and watches. 2. Pendulum. I. Pfeiffer-Belli, Christian. II. Title.
TS545.E73 681'.113 79-347645
79 MARC

Weinfeld, Stefan.
Czas / S. Weinfeld ; ilustrował Zdzisław Milach. — Wyd. 2., rozsz. — Warszawa : Nasza Księgarnia, 1977.
77 p. : ill. (some col.) ; 17 cm. P***
zł22.00
1. Clocks and watches. 2. Time. I. Milach, Zdzisław. II. Title.
TS545.W44 1977 78-386521
79 MARC

—ADDRESSES, ESSAYS, LECTURES

Coleman, Jesse E 1897-1975.
The best of J. E. Coleman : from his "Questions & answers" and "Clockwise & otherwise" columns over the past forty years, published in the American horologist & jeweler / compiled and edited by Orville R. Hagans. — Denver, CO : Hagans, c1979.
ix, 528 p. : ill. ; 24 cm.
1. Clocks and watches—Addresses, essays, lectures. 2. Clocks and watches—Repairing and adjusting—Addresses, essays, lectures. I. Hagans, Orville Roberts, 1897- . II. American horologist and jeweler. III. Title.
TS545.C64 1979 681'.113'08 79-88671
79 MARC

—CATALOGS

Ingraham (E.) & Co.
Illustrated catalogue and price list of clocks manufactured by E. Ingraham & Co., Bristol, Conn., 1860-1880. — Bristol, CT : American Clock & Watch Museum, c1979.
53 p. : ill. ; 15 x 23 cm.
Text on p. [3] of cover.
Includes the 1880 Ingraham trade catalog and various other materials.
Bibliography: p. 53.
ISBN 0-930476-03-4
1. Clocks and watches—Catalogs. 2. Ingraham (E.) & Co.—Catalogs. I. American Clock & Watch Museum. II. Title.
TS549.I54 1979 79-120615
381'.45'681113097462
79 MARC

Maddison, Francis Romeril.
Watches / F. R. Maddison and A. J. Turner. — Oxford : University of Oxford, Museum of the History of Science, 1973.
vi, 91 p. : ill. ; 24 cm. — (Museum of the History of Science catalogue ; 2)
GB74-10243
Bibliography: p. 81-84.
Includes indexes.
1. Oxford. University. Museum of the History of Science. 2. Clocks and watches—Catalogs. I. Turner, Anthony John, joint author. II. Title. III. Series: Oxford. University. Museum of the History of Science. Catalogue ; 2.
TS541.G72O934 681'.114'07402574 78-314422
78 MARC

New Haven Clock Company.
Illustrated catalogue of clocks manufactured by the New Haven Clock Company. — Bristol, CT : American Clock and Watch Museum, c1976.
80 p. : chiefly ill. ; 18 x 25 cm.
Includes bibliographical references.
ISBN 0-9600264-9-5
1. Clocks and watches—Catalogs. I. Title: Illustrated catalogue of clocks
TS549.N48 1976 681'.113 76-27111
79 MARC

Terry, Henry, 1801-1877.
American clock making; its early history. Also, Illustrated catalogue of clocks, manufactured by the Terry Clock Company, Pittsfield, Mass. [Bristol, Conn., American Clock and Watch Museum, 1974]
94 p. illus. 23 cm.
Reprint of the 1885 ed., with additional information.
Includes bibliographical references.
ISBN 0-9600264-7-9
1. Clocks and watches—Catalogs. 2. Clock and watch making—United States—History. I. Terry Clock Company. Illustrated catalogue of clocks. 1974. II. Title.
TS549.T47 1974 681'.113'0974 73-94174
74[r79]rev MARC

—COLLECTORS AND COLLECTING—CATALOGS

Ehrhardt, Roy.
Clock identification and price guide / author and compiler, Roy Ehrhardt ; text and research by Malvern "Red" Rabeneck. — Kansas City, Mo. : Heart of America Press, [1979] c1977.
198, 4, 14 p. : ill. ; 29 cm.
"Index: Clock identification and price guide, book 1 (pub. 1977)": p. 1-4 (2d group)
"New up-dated prices for 1979 for the Clock identification & price guide, book 1 (published in 1977)": p. 1-14 (3d group)
ISBN 0-913902-23-3 : $15.00
1. Clocks and watches—Collectors and collecting—Catalogs. I. Rabeneck, Malvern, joint author. II. Title.
NK7485.E362 1979 681'.113'075 79-116552
79 MARC

—DICTIONARIES

Britten, Frederick James, 1843-1913.
Britten's watch & clock maker's handbook : dictionary and guide / [by] F. J. Britten. — 16th ed. / revised by Richard Good ; illustrated by John W. Wood. — London : Eyre Methuen : Spon, 1978.
ix, 465 p. : ill. (some col.), music ; 25 cm. GB79-02897
Bibliography: p. 431-434.
Includes indexes.
ISBN 0-413-28280-5 : £40.00
1. Clocks and watches—Dictionaries. I. Good, Richard. II. Title. III. Title: Watch & clock maker's handbook.
TS540.7.B68 1978b 681'.11'03 79-309987
79 MARC

The Country life international dictionary of clocks / consultant editor Alan Smith. — New York : Putnam, 1979.
350 p. : ill. (some col.) ; 31 cm.
Bibliography: p. 339-342.
Includes index.
ISBN 0-399-12338-5 : $45.00
1. Clocks and watches—Dictionaries. I. Smith, Alan, 1925 (Oct. 31)- II. Country life.
TS540.7.C68 681'.113'03 78-65713
79 MARC

De Carle, Donald.
Watches and their value / Donald de Carle. — London : N.A.G. Press, 1978.
159 p. : ill. ; 22 cm. GB***
Bibliography: p. 159.
ISBN 0-7198-0130-3 : £5.95
1. Clocks and watches—Dictionaries. I. Title.
TS540.7.D42 681'.114'03 79-309199
79 MARC

CLOCKS AND WATCHES (Continued)

—DICTIONARIES—POLYGLOT

Ébauches, s.a., Neuchâtel.
Technological dictionary of watch parts. — 2d ed. rev. and enl. — Neuchatel : Ébauches s.a., 1953.
xli, 225 p. : ill. ; 23 cm.
Translation of Dictionnaire technologique des parties de la montre.
Technical terms in English, French, German, Italian, and Spanish.
Includes index.
1. Clocks and watches—Dictionaries—Polyglot. 2. Dictionaries, Polyglot. I. Title.
TS540.7.E2313 1953 681'.112'014 79-358258
79 • MARC

—EXHIBITIONS

Treherne, Alan.
The Massey family : clock, watch, chronometer and nautical instrument makers : catalogue of an exhibition at the Museum, Newcastle-under-Lyme / by Alan Treherne. — Newcastle-under-Lyme : The Borough Museum, [1977]
[2], 36, [1] p. : ill., 2 facsims., geneal. table ; 30 cm. GB78-02530
Bibliography: p. 3 of cover.
1. Clocks and watches—Exhibitions. 2. Clock and watchmaking—England—Newcastle-under-Lyme—Exhibitions. I. Newcastle-under-Lyme Borough Museum. II. Title.
TS541.G72N487 681'.11'0922 78-314468
78 MARC

—HISTORY

Abeler, Jürgen.
Ullstein-Uhrenbuch : e. Kulturgeschichte d. Zeitmessung / Jürgen Abeler ; [Zeichn., Dieter Messerschmidt]. — 2., neu bearb. Aufl. — Berlin ; Frankfurt/M. ; Wien : Ullstein, 1979.
324 p. : numerous ill. (some col.) ; 24 cm. GFR79-A
Bibliography: p. 309-311.
Includes index.
ISBN 3-550-17462-4 : DM58.00
1. Clocks and watches—History. 2. Clock and watch makers. I. Title.
TS542.A23 1979 79-384045
79 MARC

Good, Richard.
Watches in colour / [by] Richard Good. — Poole : Blandford Press, 1978.
218 p. : ill. (chiefly col.) ; 20 cm. GB78-39752
Bibliography: p. 207-208.
Includes indexes.
1. Clocks and watches—History. I. Title.
TS542.G66 1978 681'.114'09 79-307671
79 MARC

Jagger, Cedric, 1920-
Clocks / [by] Cedric Jagger. — Revised English ed. — London : Orbis Books, 1975.
64 p. : ill. (chiefly col.), port. ; 31 cm. — (Connoisseur's library) GB78-25313
Bibliography: p. 16.
ISBN 0-85613-197-0 : £2.50
1. Clocks and watches—History. I. Title.
TS542.J3 1975 681'.113 78-325069
79 MARC

Meis, Reinhard.
Die alte Uhr : Geschichte, Technik, Stil : ein Handbuch für Sammler und Liebhaber / Reinhard Meis. — Braunschweig : Klinkhardt & Biermann, c1978.
2 v. : numerous ill. (some col.) ; 25 cm. — (Bibliothek für Kunst- und Antiquitätenfreunde ; Bd. 53) GFR***
Bibliography: v. 1, p. 322-324; v. 2, p. 293-295.
Includes indexes.
1. Clocks and watches—History. I. Title.
NK7486.M4 681'.11'09 79-346256
79 MARC

Pomella, Fulgido, 1901-
L'orologio da portare addosso : arte e tecnica nell'orologio tascabile dalle origini al 1820-30 : ad uso di amatori e collezionisti / Fulgido Pomella. — 1. ed. — Ivrea : Priuli & Verlucca, 1978.
289 p. : ill. ; 25 cm. — (Ricerche 2 [i.e. due]) It79-Jan
Bibliography: p. 278-283.
Includes index.
1. Clock and watch making—History. 2. Clocks and watches—History. I. Title.
TS542.P65 79-346763
•79 MARC

—JUVENILE LITERATURE

Breiter, Herta S
Time and clocks / words by Herta S. Breiter. — Milwaukee : Raintree Childrens Books, c1978.
47 p. : ill. (some col.) ; 24 cm. — (Read about)
Bibliography: p. 47.
Includes index.
SUMMARY: Discusses several concepts of time and various timekeeping devices.
ISBN 0-8393-0088-3 lib. bdg. : $6.60
1. Time—Juvenile literature. 2. Clocks and watches—Juvenile literatue. [1. Time. 2. Clocks and watches] I. Title.
QB209.5.B73 529'.7 77-19007
77 MARC

Gibbons, Gail.
Clocks and how they go / by Gail Gibbons. — 1st ed. — New York : Crowell, c1979.
[32] p. : col. ill. ; 21 x 23 cm.
SUMMARY: Describes the weight clock and the spring clock and how they work.
ISBN 0-690-03973-5 : $6.95
1. Clocks and watches—Juvenile literature. [1. Clocks and watches] I. Title.
TS542.5.G52 1979 681'.113 78-22498
79 MARC

Trivett, Daphne Harwood.
Time for clocks / by Daphne and John Trivett ; pictures by Giulio Maestro. — 1st ed. — New York : Crowell, c1979.
33 p. : col. ill. ; 24 cm.
SUMMARY: Provides instructions for several easy-to-make clocks that demonstrate the concept of time.
ISBN 0-690-03896-8 lib. bdg. : $6.89
1. Time—Juvenile literature. 2. Clocks and watches—Juvenile literature. [1. Time. 2. Clocks and watches] I. Trivett, John, joint author. II. Maestro, Giulio. III. Title.
QB209.5.T74 1979 529 78-4782
78 MARC

—REPAIRING AND ADJUSTING

Ch'en, Chia-ming.
(Shou piao wei hsiu chih shih)
手表维修知识 / 陈嘉明编. — 第1版. — [广州] : 广东人民出版社, 1977.
145 p., [3] fold. leaves of plates : ill. ; 19 cm.
$0.35

1. Clocks and watches—Repairing and adusting. I. Title.
TS547.C46 78–843093

Huang, Yü-hsien.
(Piao chih kou tsao yü hsiu li)
錶之構造與修理 / 著作人黃郁賢. — 台北 : 天工書局, 民國66 [1977]
203 p. : ill. ; 22 cm.
Colophon title.
NT$150.00

1. Clocks and watches—Repairing and adjusting. I. Title.
TS547.H8 78–841922

Smith, Eric, 1940-
Clocks : repair and maintenance / Eric Smith. — New York : Arco Pub. Co., c1977.
224 p. : ill. ; 23 cm.
Bibliography: p. 216-218.
Includes index.
ISBN 0-668-04187-0 : $12.95
1. Clocks and watches—Repairing and adjusting. I. Title.
TS547.S58 1977 681'.113'028 76-53567
76[r79]rev MARC

—REPAIRING AND ADJUSTING—ADDRESSES, ESSAYS, LECTURES

Coleman, Jesse E 1897-1975.
The best of J. E. Coleman : from his "Questions & answers" and "Clockwise & otherwise" columns over the past forty years, published in the American horologist & jeweler / compiled and edited by Orville R. Hagans. — Denver, CO : Hagans, c1979.
ix, 528 p. : ill. ; 24 cm.
1. Clocks and watches—Addresses, essays, lectures. 2. Clocks and watches—Repairing and adjusting—Addresses, essays, lectures. I. Hagans, Orville Roberts, 1897- II. American horologist and jeweler. III. Title.
TS545.C64 1979 681'.113'08 79-88671
79 MARC

--TARIFF see Tariff on watches

—AUSTRIA—EXHIBITIONS

Hellich, Erika.
Alt-Wiener Uhren : die Sammlung Sobeck im Geymüller-Schlössl, 1750-1900 / Erika Hellich ; Herausgeber, Österreichisches Museum für angewandte Kunst, Wien. — München : Callwey, 1978.
131 p. : ill. ; 25 cm. GFR78-A
English and German.
Bibliography: p. 131.
ISBN 3-7667-0444-3
1. Clocks and watches—Austria—Exhibitions. I. Vienna. Österreichisches Museum für Angewandte Kunst. II. Title.
TS541.A92V573 79-357360
79 MARC

—AUSTRIA—VIENNA

Ortenburger, Rick.
The Vienna and German regulator = Die Wiener und Deutsche Pendeluhr / by Rick Ortenburger ; translated into German by Elfi Ortenburger and Brigitte Laing ; photos. by the author. — Agoura, Calif. : Pendel Uhr Publications, c1979.
70 p. : ill. ; 28 cm.
English and German.
Bibliography: p. [7]
1. Clocks and watches—Austria—Vienna. 2. Clocks and watches—Germany. I. Title. II. Title: Die Wiener und Deutsche Pendeluhr.
TS543.A9O77 681'.133'0943 79-62914
00 MARC

—ENGLAND—HISTORY

Rose, Ronald E
English dial clocks / Ronald E. Rose. — Woodbridge [Eng.] : Antique Collectors' Club, c1978.
198 p. : ill. (some col.) ; 29 cm. GB***
Includes index.
ISBN 0-902028-78-2 : £15.00
1. Clocks and watches—England—History. I. Title.
TS543.G7R67 681'.113'0942 78-322227
78 MARC

—EUROPE

Mühe, Richard.
Alte Uhren : e. Handbuch europ. Tischuhren, Wanduhren u. Bodenstanduhren / Richard Mühe, Horand M. Vogel. — 2., korrigierte u. erg. Aufl. — München : Callwey, 1978.
306 p. : chiefly ill. (some col.), map ; 29 cm. GFR78-A
Bibliography: p. 300-303.
Includes index.
DM98.00
1. Clocks and watches—Europe. I. Vogel, Horand M., joint author. II. Title.
NK7495.E85M83 1978 681'.113'094 78-376827
78 MARC

—EUROPE—HISTORY

Brusa, Giuseppe.
L'arte dell'orologeria in Europa : sette secoli di orologi meccanici / Giuseppe Brusa. — Busto Arsizio : Bramante, [1978]
491 p. : ill. ; 31 cm. — (Arte e tecnica) It79-Jan
Bibliography: p. 473-479.
Includes index.
1. Clocks and watches—Europe—History. I. Title.
NK7495.E85B78 79-360210
•79 MARC

—FRANCE—HISTORY

Bollen, Ton, 1935-
Franse lantaarnklokken / Ton Bollen. — Haarlem : De Haan, [1978]
42, 121 p. : chiefly ill. ; 25 cm. Ne***
Bibliography: p. 41 (1st group)
ISBN 9022842630
1. Clocks and watches—France—History. I. Title.
TS543.F8B65 79-342888
79 MARC

CLOCKS AND WATCHES (Continued)

—FRANCE—FRANCHE-COMTÉ

Schmitt, Gustav.
Die Comtoiser Uhr / Gustav Schmitt. — Villingen/Schwarzwald : Müller, c1977.
400 p. : ill. (some col.) ; 24 cm. GFR78-A
Bibliography: p. 396.
Includes indexes.
ISBN 3-920662-02-4 : DM78.00
1. Clocks and watches—France—Franche-Comté. I. Title.
TS543.F8S35 79-344294
79 MARC

—GERMANY

Ortenburger, Rick.
The Vienna and German regulator = Die Wiener und Deutsche Pendeluhr / by Rick Ortenburger ; translated into German by Elfi Ortenburger and Brigitte Laing ; photos. by the author. — Agoura, Calif. : Pendel Uhr Publications, c1979.
70 p. : ill. ; 28 cm.
English and German.
Bibliography: p. [7]
1. Clocks and watches—Austria—Vienna. 2. Clocks and watches—Germany. I. Title. II. Title: Die Wiener und Deutsche Pendeluhr.
TS543.A9O77 681'.133'0943 79-62914
00 MARC

—GERMANY, WEST—BLACK FOREST

Kochmann, Karl, 1921-
The Black Forest cuckoo clock : European clockmaking / [Karl Kochmann]. — Concord, Calif. : Antique Clock Pub., c1976.
240 p. : ill. ; 22 cm.
$9.50
1. Clocks and watches—Germany, West—Black Forest. I. Title.
TS543.G3K57 681'.113 76-25944
00 MARC

—GREAT BRITAIN

Loomes, Brian.
Complete British clocks / Brian Loomes. — Newton Abbot [Eng.] ; North Pomfret, Vt. : David & Charles, c1978.
256 p. : ill. ; 24 cm. GB***
Bibliography: p. 251-252.
Includes index.
ISBN 0-7153-7567-9 : £7.95
1. Clocks and watches—Great Britain. I. Title.
NK7495.G7L66 681'.113'0941 78-66804
78 MARC

—GREAT BRITAIN—PATENTS

Great Britain. Patent Office.
Patents for inventions : abridgements of specifications, class 139 : watches, clocks, and other timepieces. — Facsim. ed. — Seattle : Odmark, 1979.
2 v. : ill. ; 27 cm.
Reprint of 1905-1907 ed. published at the Patent Office, London.
CONTENTS: v. 1. Period—A.D. 1855-1866.—v. 2. Period—A.D. 1901-4.
1. Clocks and watches—Great Britain—Patents. I. Title.
TS543.G7G73 1979 681'.11'027 79-87655
00 MARC

—NETHERLANDS

Zeeman, J
De Nederlandse stoelklok / J. Zeeman. — 2e, bijgewerkte en herz. dr. — Assen : Gorcum, 1978.
264 p. : with ill. ; 26 cm. Ne78-35
"Met bijdragen van A. Doedens ..."
Includes bibliographies and index.
ISBN 9023214161 : fl 135.00
1. Clock and watch making—Netherlands. 2. Clocks and watches—Netherlands. I. Doedens, A. II. Title.
TS543.N4Z43 1978 78-386050
*79 MARC

CLOCKS AND WATCHES, ELECTRIC

see also Electronic watches, Digital

Li, Ch'ao-lun.
(Shu tzu tien tzu chung piao)
數字電子鐘表 / 李超倫編著. — 香港：香港萬里書店, 1978.
2, 116 p. : ill. ; 19 cm.
Title in English on cover: Digital clock and watch.
$6.00

1. Electronic watches, Digital. 2. Clocks and watches, Electric. I. Title. II. Title: Digital clock and watch.
TS544.L5 78-843787

—EXHIBITIONS

Aked, Charles Kenneth.
Electrifying time : [catalogue of] an exhibition at the Science Museum to commemorate the centenary of the death of Alexander Bain, 2nd January 1877 ; open 15th December 1976 to 11th April 1977 / compiled by Charles K. Aked. — Ticehurst : Antiquarian Horological Society, 1976.
106 p. : ill., plan, ports. ; 25 cm. — (Monograph - Antiquarian Horological Society ; no. 10) GB77-25068
Includes bibliographies.
ISBN 0-901180-14-9
1. Clocks and watches, Electric—Exhibitions. I. Bain, Alexander, 1810-1877. II. Title. III. Series: Antiquarian Horological Society. Monograph ; no. 10.
TS541.G72L662 681'.116'07402134 79-303090
79 MARC

CLOCKS AND WATCHES, SELF-WINDING

Horamatic : montres à remontage automatique de 1770 à 1978 / [textes, J.-C. Beuchat et al.]. — Le Locle : Édition du Château des Monts, Musée d'horlogerie, 1978.
100 p. : ill ; 21 cm. — (Histoire de l'horlogerie ; fasc. 4) Sw78-9635
1. Clocks and watches, Self-winding. I. Beuchat, J. C. II. Musée d'horlogerie du Locle.
TS548.H67 681'.114 79-364519
79 MARC

CLOES, NICOLAS, 1889-

Schmits, Georges.
Nicolas Cloes, peintre naïf / Georges Schmits. — [Verviers, r. Renkin 10] : Temps mêlés, [1976]
87 p. : ill. ; 20 cm. Be78-9
Bibliography: p. 83-87.
1. Cloes, Nicolas, 1889- . 2. Painters—Belgium—Biography. 3. Primitivism in art—Belgium. I. Title.
ND673.C558S35 759.9493 79-352462
*79 MARC

CLOG-DANCING

see also Tap dancing

Brooks, Floyd.
Carolina mountain clogging : clog and hoedown basics / by Floyd and Eloise Brooks ; cover and photos. by Scott Trees. — [1st ed.]. — [Marion, N.C. : Brooks' School of Dance, c1977]
57 p. : ill. ; 22 cm.
1. Clog-dancing. I. Brooks, Eloise, joint author. II. Title.
GV1793.B76 793.3 79-119623
79 MARC

CLOISONNÉ

Kai, Yukinori, 1937-
(Shippō)
七宝：七宝焼の魅力のすべて / 甲斐征典著. — 東京：櫂書房, 1978.
237 p. : ill. (some col.) ; 27 cm. — (ニューライフ選書 ; 2)
Ja 78B-6295
¥3800

1. Cloisonné. I. Title. II. Series: Nyū raifu sensho ; 2.
TT382.6.K34 79-802688

Sano, Toshiko.
(Zoku Shippōyaki)
続七宝焼 / 佐野登志子. — 東京：主婦と生活社, 昭和53 [1978]
147 p. : ill. (some col.) ; 24 cm. Ja***
¥1800

1. Cloisonné. I. Title. II. Title: Shippōyaki.
TT382.6.S27 79-807992

—GEORGIA (TRANSCAUCASIA)—CONGRESSES

Khuskivadze, Leila.
Émaux cloisonnés géorgiens : IIe symposium international sur l'art géorgien / L. Khouskivadze. — Tbilissi : Metsniereba, 1977.
10 p. ; 20 cm. USSR***
At head of title: Académie des sciences de la R.S.S. de Géorgie, Institut Tschubinaschvili d'histoire de l'art géorgien.
1. Enamel and enameling—Georgia (Transcaucasia)—Congresses. 2. Cloisonné—Georgia (Transcaucasia)—Congresses. I. Title.
NK5004.R9K8414 738.4 79-370386
79 MARC

—JAPAN

(Gyoku garasu)
玉・ガラス / 関忠夫 ... [et al. ; 写真 大橋治三 ... et al.]. — 京都：淡交社, 昭和53 [1978]
218 p. : col. ill. ; 22 cm. — (カラー日本の工芸 ; 9) Ja***
Colophon inserted.
¥2500

1. Glassware—Japan. 2. Glass painting and staining—Japan. 3. Gems—Japan. 4. Ink-stones—Japan. 5. Cloisonné—Japan. I. Seki, Tadao, 1914- II. Ōhashi, Haruzō, 1927- III. Series: Karā Nihon no kōgei ; 9.
NK5184.A1G95 78-813467

(Nihon no shippō)
日本の七宝 / 鈴木規夫, 榊原悟編・著. — 京都：マリア書房, 1979.
266 p. : chiefly ill. (some col.) ; 26 x 27 cm. Ja 79-5445
Bibliography: p. 266.
¥9500

1. Cloisonné—Japan. I. Suzuki, Norio. II. Sakakibara, Satoru, 1948-
NK5015.N54 79-808079

CLOISTER HOTEL

Martin, Harold H
This happy isle : the story of Sea Island and the Cloister / Harold H. Martin. — Sea Island, Ga. : Sea Island Co., c1978.
xiv, 236 p., [34] leaves of plates : ill. ; 22 cm.
Includes index.
1. Cloister Hotel. I. Title.
TX941.C48M37 647'.94758'7 78-14320
78 MARC

CLOISTERS see Convents and nunneries; Monasteries

CLONE CELLS

see also Molecular cloning

—CONGRESSES

Society for Developmental Biology.
The clonal basis of development / Stephen Subtelny, editor, Ian M. Sussex, co-editor. — New York : Academic Press, 1978.
viii, 251 p. : ill. ; 24 cm. — (Symposium of the Society for Developmental Biology ; 36th)
Proceedings of the symposium held in Raleigh, N.C., June 13-15, 1977.
Includes bibliographies and index.
ISBN 0-12-612982-7
1. Developmental biology—Congresses. 2. Clone cells—Congresses. I. Subtelny, Stephen Stanley, 1925- . II. Sussex, Ian M., 1927- . III. Title. IV. Series: Society for Developmental Biology. Symposium ; 36th.
[DNLM: 1. Clone cells—Congresses. QH585 S678c]
QH511.S6 no. 36 574.3'08 s 78-23508
[QH491] 78 MARC

CLONFERT, IRE.

—GENEALOGY

Index to Clonfert and Kilmacduagh wills / edited by Patrick Smythe-Wood. — ⌈Ballycastle⌉ : ⌈The editor⌉, 1977.
⌈3⌉, 19 p. ; 21 cm. GB77-33011
£1.20
1. The Irish ancestor—Indexes. 2. Clonfert, Ire.—Genealogy. 3. Kilmacduagh, Ire.—Genealogy. I. Smythe-Wood, Patrick.
CS480.I712I53 016.929'34174 79-307959
79 MARC

CLONING

see also Clone cells; Molecular cloning; Plant propagation

McKinnell, Robert Gilmore.
Cloning : a biologist reports / Robert Gilmore McKinnell. — Minneapolis : University of Minnesota Press, c1979.
viii, 130 p. : ill. ; 24 cm.
Bibliography: p. 119-121.
Includes index.
ISBN 0-8166-0883-0 : $8.95
1. Cloning. 2. Cell nuclei—Transplantation. 3. Embryology, Experimental. I. Title.
QH442.2.M32 596'.01'6 79-10569
79 MARC

Rorvik, David M
In his image : the cloning of a man / David M. Rorvik. — London : H. Hamilton, 1978.
207 p. ; 23 cm. GB***
Bibliography: p. 197-207.
ISBN 0-241-10010-0 : £4.95
1. Cloning. 2. Human reproduction. I. Title.
QH442.2.R67 1978b 612.6 78-317366
78 MARC

—SOCIAL ASPECTS

Ebon, Martin.
The cloning of man : a brave new hope—or horror? / Edited by Martin Ebon. — New York : New American Library, 1978.
201 p. ; 18 cm. — (A Signet book)
Bibliography: p. 199-201.
$1.95
1. Cloning—Social aspects. 2. Genetic engineering—Social aspects. I. Title.
QH442.2.E26 612.6 78-111487
79 MARC

CLONING, MOLECULAR see Molecular cloning

CLOSE CORPORATIONS

see also Family corporations; Stock purchase aggreements (Close corporations)

—TAXATION—GREAT BRITAIN

Association of Certified Accountants.
Corporation tax / the Association of Certified Accountants. — ⌈Revised ed.⌉. — ⌈London⌉ : Certified Accountants Educational Trust, ⌈1977⌉
3 v. ; 24 cm. GB79-01780
Cover title.
CONTENTS: v. 1. General provisions of the Income and corporation taxes act 1970, as amended by the Finance acts, 1970 to 1977.—v. 2. Close companies, as amended by the Finance acts, 1970 to 1977.—v. 3. Double tax relief, as amended by the Finance acts, 1970 to 1977.
£4.00
1. Corporations—Taxation—Great Britain. 2. Close corporations—Taxation—Great Britain. 3. Taxation, Double—Great Britain. I. Title.
KD5504.A98 1977 343'.41'067 79-315479
79 MARC

—TAXATION—UNITED STATES

Commerce Clearing House.
Life insurance as estate planning tool for close corporation stockholders. — Chicago : Commerce Clearing House, c1978.
32 p. ; 23 cm. — (Tax angles and tax savings series) (CCH editorial staff publication)
"5313."
Includes index.
$1.50
1. Insurance, Life—United States. 2. Estate planning—United States. 3. Close corporations—Taxation—United States. I. Title. II. Series.
KF1175.Z9C65 346'.73'086 79-101052
79 MARC

Estate and financial planning for the closely held corporation, 1979 / Leslie S. Klinger, chairman. — New York : Practising Law Institute, c1979.
568 p. (p. 564-568 blank) : forms ; 22 cm. — (Tax law and estate planning series) (Estate planning and administration course handbook series ; no. 97)
"D4-5115."
"Prepared for distribution at the estate and financial planning for the closely held corporation program, January-March 1979."
1. Close corporations—United States. 2. Close corporations—Taxation—United States. 3. Estate planning—United States. I. Klinger, Leslie S. II. New York (City). Practising Law Institute. III. Series. IV. Series: Estate planning and administration course handbook series ; no. 97.
KF1466.Z9E82 1979 346'.73'0666 78-78042
79 MARC

—TAXATION—UNITED STATES—CONGRESSES

Tax and estate planning under the Tax reform act of 1976 for the owners of closely-held corporations : proceedings of a three-day special seminar / edited by Irving Schreiber ; panelists, Stuart Becker ... ⌈et al.⌉. — Greenvale, N.Y. : Panel Publishers, c1977.
ix, 222 p. ; 26 cm.
Includes index.
ISBN 0-916592-19-7
1. Close corporations—Taxation—United States—Congresses. 2. Inheritance and transfer tax—United States—Congresses. I. Schreiber, Irving.
KF6491.A75T37 343'.73'053 77-85161
79 MARC

—CALIFORNIA

Dunn, William H 1931-
Sales of California businesses : covering the practical, legal, and tax aspects of the sale of smaller closely held businesses where California law applies / by William H. Dunn. — San Jose, Calif. : Business Sale Institute, c1978.
ca. 200 leaves in various foliations ; 29 cm.
1. Business enterprises, Sale of—California. 2. Close corporations—California. 3. Small business—Law and legislation—California. I. Title.
KFC337.D86 346'.794'065 79-105403
79 MARC

—MASSACHUSETTS

Massachusetts Continuing Legal Education-New England Law Institute, inc.
Closely held corporations : problems of management and control. — Boston : Massachusetts Continuing Legal Education-New England Law Institute, c1978.
xxiv, 206 p. ; 22 cm.
Includes bibliographical references.
1. Close corporations—Massachusetts. 2. Close corporations—New England. I. Title.
KFM2615.C55M37 346'.744'0668 78-70701
79 MARC

—NEW ENGLAND

Massachusetts Continuing Legal Education-New England Law Institute, inc.
Closely held corporations : problems of management and control. — Boston : Massachusetts Continuing Legal Education-New England Law Institute, c1978.
xxiv, 206 p. ; 22 cm.
Includes bibliographical references.
1. Close corporations—Massachusetts. 2. Close corporations—New England. I. Title.
KFM2615.C55M37 346'.744'0668 78-70701
79 MARC

—UNITED STATES

Westwater, John C 1942-
The use and drafting of stock redemption agreements / by John C. Westwater. — Washington : Bureau of National Affairs, c1978.
88 p. in various pagings ; 28 cm. — (Corporate practice series ; no. 8 ISSN 0162-5691)
Includes bibliography and index.
1. Stock transfer—United States. 2. Stock transfer—Taxation—United States. 3. Close corporations—United States. I. Title. II. Series.
KF1448.W47 346'.73'0666 78-10249
78 MARC

CLOSE RANGE PHOTOGRAPHY see Photography, Close-up

CLOSE-UP PHOTOGRAPHY see Photography, Close-up

CLOSED CATEGORIES (MATHEMATICS)

Nel, Louis Daniel, 1934-
Cartesian closed coreflective hulls / by L. D. Nel. — Ottawa : Dept. of Mathematics, Carleton University, ⌈197-⌉
12 leaves ; 28 cm. — (Mathematical series ; no. 143 ISSN 0069-0600)
C***
Bibliography: leaves 11-12.
1. Closed categories (Mathematics) 2. Function spaces. 3. Embedding theorems. I. Title. II. Series: Carleton mathematical series ; no. 143.
QA169.N44 512'.55 78-322180
78 MARC

CLOSED-CIRCUIT TELEVISION

see also Television in education; Television in medicine

CLOSED ECOLOGICAL SYSTEMS

—CONGRESSES

Environmental control and agri-technology / edited by Eugene B. Konecci. — Tarzana, Calif. : American Astronautical Society : distributed by Univelt, c1976.
vi, 340 p. : ill. ; 25 cm. — (Science and technology ; v. 39) (An American Astronautical Society publication)
CONTENTS: Proceedings of the Conference on Controlled Environment—Food Technology.—Haile, R. The economics of hydroponic vegetable production in an environmentally controlled system.
Includes bibliographical references.
ISBN 0-87703-075-8
1. Greenhouses—Environmental engineering—Congresses. 2. Closed ecological systems—Congresses. 3. Truck farming—Congresses. 4. Hydroponics—Congresses. 5. Agriculture—Automation—Congresses. I. Konecci, Eugene B. II. Haile, Ramon. The economics of hydroponic vegetable production in an environmentally controlled system. 1976. III. Conference on controlled Environments—Food Technology, Winrock Farms, 1969. IV. Series: AAS science and technology series ; v. 39.
SB415.E58 631.5'8 78-323848
78 MARC

CLOSED GRAPH THEOREMS

Wilde, M de.
Closed graph theorems and webbed spaces / M. De Wilde. — London ; San Francisco : Pitman, 1978.
158 p. ; 25 cm. — (Research notes in mathematics ; 19) GB***
Bibliography: p. 149-156.
Includes index.
ISBN 0-273-08403-8 : £6.50
1. Webbed spaces. 2. Closed graph theorems. I. Title. II. Series.
QA322.W47 515'.7 78-324345
79 MARC

CLOSED SEA (MARE CLAUSUM) see Freedom of the seas

CLOSED SHELVES IN LIBRARIES see Open and closed shelves

CLOSED STACKS IN LIBRARIES see Open and closed shelves

CLOSETS see Clothes closets

CLOSING COSTS see Settlement costs

CLOSS, AUGUST

A Miscellany of tributes and translations presented to Professor August Closs, by his former students and his colleagues in the Department of German. — ⌜Bristol⌝ : University of Bristol, 1964.
40 p. : port. ; 25 cm.
Cover title: Germania, presented to Professor August Closs.
English and German.
1. German literature—Translations into English. 2. English literature—Translations from German. 3. Closs, August. I. Bristol, Eng. University. Dept. of German.
PT1160.E5M57 830.8'009'1 79-312568
79 MARC

CLOSTER, N.J.

—DWELLINGS

Historic homes in Closter / Jeanne Gurnee, editor ; architectural text by Joan T. Bastable ; ill. by M. Jane Bailey. — Closter, N.J. : Environmental Commission, ⌜c1979⌝
44 p. : ill. ; 14 x 22 cm.
Bibliography: p. 44.
1. Closter, N.J.—Dwellings. 2. Architecture, Domestic—New Jersey—Closter. 3. Historic buildings—New Jersey—Closter. I. Gurnee, Jeanne. II. Bastable, Joan T. III. Bailey, M. Jane. IV. Closter, N.J. Environmental Commission.
NA735.C62H57 917.49'21 79-111708
79 MARC

CLOSTRIDIUM

Tucker, Winston Harris.
Studies on Clostridium putrificum and Clostridium putrefaciens. 1930.
81 l. illus. 29 cm.
Typescript (Carbon copy)
Thesis—University of Chicago.
Bibliography: l. 81.
1. Clostridium putrefaciens. 2. Ham—Microbiology. 3. Clostridium. 4. Proteolytic enzymes. I. Title.
QR201.C54T82 79-305669
79 MARC

CLOSTRIDIUM PERFRINGENS

Giddings, Crandall Bland, 1915-
Properties of a new lecithinase, the active principle of alpha toxin produced by Cl. welchii (type A) ⌜by⌝ C. Bland Giddings. ⌜n.p.⌝ 1943.
⌜174⌝ l. illus. 29 cm.
Thesis—University of Cincinnati.
Bibliography: leaves ⌜171⌝-⌜172⌝
1. Lecithinase. 2. Clostridium perfringens. I. Title.
QP609.L43G53 78-326132
79 MARC

Torres-Anjel, Manuel J
A fluorescent-antibody technique for the detection of enterotoxin-producing cells of Clostridium perfringens type A / Manuel J. Torres-Anjel, Hans P. Riemann, and Che C. Tsai. — Washington : Pan American Health Organization, Pan American Sanitary Bureau, Regional Office of the World Health Organization, 1975.
26 p. : ill. ; 26 cm. — (Scientific publication - Pan American Health Organization ; no. 305)
Bibliography: p. 24-26.
1. Clostridium perfringens. 2. Enterotoxin—Analysis. 3. Fluorescent antibody technique. I. Riemann, Hans P., joint author. II. Tsai, Che C., joint author. III. Title. IV. Series: Pan American Health Organization. Publicaciones científicas ; no. 305.
RA10.P252 no. 305 79-316212
⌜QR201.C54⌝ 79 MARC

CLOSTRIDIUM PUTREFACIENS

Tucker, Winston Harris.
Studies on Clostridium putrificum and Clostridium putrefaciens. 1930.
81 l. illus. 29 cm.
Typescript (Carbon copy)
Thesis—University of Chicago.
Bibliography: l. 81.
1. Clostridium putrefaciens. 2. Ham—Microbiology. 3. Clostridium. 4. Proteolytic enzymes. I. Title.
QR201.C54T82 79-305669
79 MARC

CLOSURE, INTEGRAL see Integral closure

CLOTH see Inflammable textiles; Textile fabrics

CLOTH OF GOLD

Doléances des maîtres-ouvriers, fabricants en étoffes d'or, d'argent et de soie de la ville de Lyon : adressées au roi et à la nation assemblée / textes réunis et présentés par Fernand Rude. — Lyon (11, rue Ferrachat, 69005) : Fédérop, 1976.
1 v. (various pagings) ; 19 cm. F77-17654
Includes bibliographical references.
CONTENTS: Mémoire des électeurs fabricants d'étoffes en soie de la ville de Lyon.—Doléances des maîtres-ouvriers fabricants en étoffes d'or, d'argent & de soie de la ville de Lyon, addressées au Roi et à la Nation assemblée.—Recueil de mémoires et tarif.—Extrait des registres des délibérations des maîtres-ouvriers fabricants en étoffes d'or, d'argent et soie de la ville de Lyon, prises dans l'église cathédrale de Saint-Jean le 5 mai 1790.
19.00F
1. Weavers—France—Lyons—History—Sources. 2. Silk manufacture and trade—France—Lyons—History—Sources. 3. Cloth of gold. 4. Cloth of silver. I. Rude, Fernand.
HD8039.T42F823 79-343182
331.7'67'7390944582
79 MARC

CLOTH OF SILVER

Doléances des maîtres-ouvriers, fabricants en étoffes d'or, d'argent et de soie de la ville de Lyon : adressées au roi et à la nation assemblée / textes réunis et présentés par Fernand Rude. — Lyon (11, rue Ferrachat, 69005) : Fédérop, 1976.
1 v. (various pagings) ; 19 cm. F77-17654
Includes bibliographical references.
CONTENTS: Mémoire des électeurs fabricants d'étoffes en soie de la ville de Lyon.—Doléances des maîtres-ouvriers fabricants en étoffes d'or, d'argent & de soie de la ville de Lyon, addressées au Roi et à la Nation assemblée.—Recueil de mémoires et tarif.—Extrait des registres des délibérations des maîtres-ouvriers fabricants en étoffes d'or, d'argent et soie de la ville de Lyon, prises dans l'église cathédrale de Saint-Jean le 5 mai 1790.
19.00F
1. Weavers—France—Lyons—History—Sources. 2. Silk manufacture and trade—France—Lyons—History—Sources. 3. Cloth of gold. 4. Cloth of silver. I. Rude, Fernand.
HD8039.T42F823 79-343182
331.7'67'7390944582
79 MARC

CLOTHES CLOSETS

Jones, Peter, 1934-
Plan and build more storage space / Peter Jones. — New York : Butterick Pub., c1979.
176 p. : ill. ; 26 cm.
Includes index.
ISBN 0-88421-036-7 : $4.95
1. Cabinet-work. 2. Clothes closets. 3. Shelving (furniture) 4. Storage in the home. I. Title.
TT197.J655 684.1'6 79-10386
79 MARC

CLOTHES DESIGNERS see Costume designers

CLOTHES DRYERS

Chao, Chung-p'ing.
(Hsi i chi, kan i chi, hsi tieh chi chieh kou ho wei hsiu)
洗衣機，乾衣機，洗碟機結構和維修 / 趙仲平編著. — 香港：工藝技術雜誌社，1977.
2, 2, 130 p. : ill. ; 19 cm. — (家庭電器叢書)
Cover title: Electric washer, dryer & dishwasher.
Series romanized: Chia t'ing tien ch'i ts'ung shu.
$5.40
1. Washing-machines. 2. Clothes dryers. 3. Dishwashing machines. I. Title. II. Title: Electric washer, dryer & dishwasher.
TT997.C45 78-842274

CLOTHIER, WILLIAM

Clothier, William.
William Clothier : an American Film Institute seminar on his work. — Glen Rock, N. J. : Microfilming Corp. of America, 1977.
1 sheet (56 p.) ; 11 x 15 cm. — (The American Film Institute seminars ; pt. 1, no. 34) (New York times oral history program)
Microfiche.
ISBN 0-667-00353-3
1. Clothier, William. 2. Cinematographers — California — Hollywood—Biography. I. Series.
Microfiche TR849.C56 778.5'3'0924 [B] 77-12748

CLOTHIERS see Clothing trade

CLOTHING see Clothing and dress

CLOTHING, COLD WEATHER

Keeping warm : a guide for wintertime / edited by Barbara Kelman ... ⌜et al.⌝. — New York : Quick Fox, c1978.
181 p. : ill. ; 28 cm.
Bibliography: p. 171.
Includes index.
ISBN 0-8256-3079-7 : $7.95
1. Dwellings—Heating and ventilation. 2. Winter. 3. Clothing, Cold weather. 4. Diet. I. Kelman, Barbara.
TH7226.K43 640 77-78539
79 MARC

CLOTHING, DOLL see Doll clothes

CLOTHING, LEATHER see Leather garments

CLOTHING, PROTECTIVE

see also Clothing, Cold weather; Safety hats

Kupchin, A P
(Sredstva individual'noĭ zashchity rabotai͡ushchikh na proizvodstve)
Средства индивидуальной защиты работающих на производстве / А. П. Купчин. — Москва : Профиздат, 1977.
⌜112 p. ; 20 cm. USSR 77
Bibliography : p. 109–⌜110⌝
0.23rub
1. Clothing, Protective. I. Title.
HD7395.C5K86 79–407829

Profsoi͡uz rabochikh lesnoĭ, bumazhnoĭ i derevoobrabatyvai͡ushcheĭ promyshlennosti. T͡Sentral'nyĭ komitet.
(Sbornik norm vydachi spet͡sodezhdy, spet͡sobuvi i predokhranitel'nykh prisposobleniĭ rabotai͡ushchim v lesnoĭ, bumazhnoĭ i derevoobrabatyvai͡ushcheĭ promyshlennosti i lesnom khozi͡aĭstve)
Сборник норм выдачи спецодежды, спецобуви и предохранительных приспособлений работающим в лесной, бумажной и деревообрабатывающей промышленности и лесном хозяйстве / Центральный комитет Профсоюза рабочих лесной, бумажной и деревообрабатывающей промышленности ; ⌜составители М. В. Кулешов ... et al.⌝. — Москва : Лесная пром-сть, 1975.
264 p. : ill. ; 23 cm. USSR 75
1.12rub
1. Clothing, Protective. I. Kuleshov, Mikhail Vasil'evich. II. Title: Sbornik norm vydachi spet͡sodezhdy, spet͡sobuvi i predokhranitel'nykh prisposobleniĭ ...
HD7395.C5P66 79–407469

—STANDARDS—ROMANIA

Echipament de protecție și de lucru : (colecție STAS). — București : Editura Tehnică, 1977.
245 p. : ill. ; 21 cm. — (Biblioteca standardizării : Seria tehnică : A85) R78-411
lei15.00
1. Clothing, Protective—Standards—Romania. I. Series.
T59.B5 Ser. A, no. 85 78-382185
⌜HD7395.C5⌝ *78 MARC

CLOTHING AND DRESS

see also Aged—Clothing; Blouses; Buttons; Children's clothing; Clothing, Protective; Costume; Costume design; Fashion; Fashion shows; Fur garments; Garment cutting; Hand bags; Handicapped —Clothing; Handkerchiefs; Hats; Hosiery; Jeans (Clothing); Leather garments; Lederhosen; Men's clothing; Models, Fashion; Raincoats; Sweaters; Tailoring; Trousers; Underwear

Cho, Emily.
Looking terrific : how to express yourself through the language of clothing / by Emily Cho and Linda Grover ; photos. by Ishimuro ; illustrations by Catherine Clayton Purnell. — New York : Putnam, c1978.
205 p. : ill. ; 24 cm.
ISBN 0-399-12039-4 : $12.95
1. Clothing and dress. I. Grover, Linda, joint author. II. Title.
TT507.C56 1978 391'.07'2 78-2910
78 MARC

CLOTHING AND DRESS (Continued)

Dietch, Joan K
The success look—for women only / by Joan K. Dietch ; ₍ill. by Barbara Fleischer₎. — New York : Grosset & Dunlap, c1979.
137 p. : ill. ; 26 cm.
ISBN 0-448-16423-X : $6.95
1. Clothing and dress. I. Title.
TT507.D53 646'.34 78-67811
00 MARC

Erwin, Mabel Deane.
Clothing for moderns / Mabel D. Erwin, Lila A. Kinchen, Kathleen A. Peters. — 6th ed. — New York : Macmillan, c1979.
xiii, 493 p. : ill. ; 24 cm.
Includes bibliographies and index.
ISBN 0-02-334220-X
1. Dressmaking. 2. Clothing and dress. I. Kinchen, Lila A., joint author. II. Peters, Kathleen A., joint author. III. Title.
TT518.E7 1979 646.4'3'04 78-11350
78 MARC

Estate and financial planning for the closely held corporation, 1979 / Leslie S. Klinger, chairman. — New York : Practising Law Institute, c1979.
568 p. (p. 564-568 blank) : forms ; 22 cm. — (Tax law and estate planning series) (Estate planning and administration course handbook series ; no. 97)
"D4-5115."
"Prepared for distribution at the estate and financial planning for the closely held corporation program, January-March 1979."
1. Close corporations—United States. 2. Close corporations—Taxation—United States. 3. Estate planning—United States. I. Klinger, Leslie S. II. New York (City). Practising Law Institute. III. Series. IV. Series: Estate planning and administration course handbook series ; no. 97.
KF1466.Z9E82 1979 346'.73'0666 78-78042
79 MARC

Faragó, Ilona.
Az öltözködés ábécéje / Faragó Ilona. — Budapest : Minerva, 1977.
62 p., ₍8₎ leaves of plates : ill. (chiefly col.) ; 24 cm. — (Minerva családi könyvek ISSN 0133-2236) H***
ISBN 963223104X : 20.00Ft
1. Clothing and dress. 2. Costume—Hungary. I. Title.
TT507.F28 78-395226
79 MARC

Field, Leslie.
Look like a million / ₍by₎ Leslie Field ; illustrations by Lucy Su. — London ; New York : Hart-Davis, MacGibbon, 1978.
208 p. : ill. ; 24 cm. GB78-35396
Includes index.
ISBN 0-246-11043-0 : £3.95
1. Clothing and dress. 2. Beauty, Personal. I. Title.
TT507.F53 646.7'2 79-307433
79 MARC

Lane, Maggie.
Maggie Lane's Oriental patchwork / Maggie Lane. — New York : Scribner, c1978.
104 p., ₍8₎ leaves of plates : ill. (some col.) ; 28 cm.
Bibliography: p. 104.
Includes index.
ISBN 0-684-15621-0 : $16.95
1. Patchwork. 2. Clothing and dress. I. Title. II. Title: Oriental patchwork.
TT835.L33 746.9'2 78-7957
78 MARC

Levitt, Mortimer, 1907-
The executive look and how to get it : professional guidelines for putting together executive wardrobes for men and women / Mortimer Levitt. — New York : AMACOM, c1979.
76 p. : ill. ; 23 cm. — (An AMA management briefing)
ISBN 0-8144-2237-3 : $7.50
1. Clothing and dress. 2. Executives. I. Title. II. Series: American Management Associations. AMA management briefing.
TT507.L46 646'.3 79-18848
79 MARC

Milinaire, Caterine.
Cheap chic update / by Caterine Milinaire and Carol Troy ; designed by Bea Feitler. — Rev. ed. / rev. by Carol Troy. — New York : Harmony Books, c1978.
224 p. : ill. ; 29 cm.
ISBN 0-517-53460-6. ISBN 0-517-53456-8 pbk.
1. Fashion. 2. Clothing and dress. I. Troy, Carol, joint author. II. Title.
TT515.M59 1978 646'.34 78-4148
78 MARC

Nagata, Hisanori, 1924-
(Ifuku eiseigaku)
衣服衛生学　永田久紀著　東京　南江堂　昭和45 (1970)
138 p. 22 cm. Ja 70-2999
Bibliography: p. ₍134₎-138.
1. Clothing and dress. 2. Hygiene. I. Title.
RA779.N27 78-820879

Thourlby, William.
You are what you wear : the key to business success / William Thourlby. — Kansas City, Kan. : Sheed Andrews and McMeel, c1978.
xiv, 132 p., ₍5₎ leaves of plates : ill. ; 24 cm.
ISBN 0-8362-2902-9
1. Clothing and dress. 2. Men's clothing. 3. Grooming for men. I. Title.
TT507.T47 646'.32 78-12502
78 MARC

Thypin, Marilyn.
Try it on : buying clothing / by Marilyn Thypin and Lynne Glasner ; photos. by Ira Berger. — St. Paul : EMC Corp., 1979.
60 p. : ill. ; 23 cm.
SUMMARY: Two sisters offer pointers as they shop for clothes on a limited budget.
ISBN 0-88436-508-5 : $3.50
1. Clothing and dress. 2. Consumer education. ₍1. Clothing and dress. 2. Consumer education₎ I. Glasner, Lynne, joint author. II. Berger, Ira. III. Title.
TX340.T48 646'.3 78-12440
78 MARC

—ADDRESSES, ESSAYS, LECTURES

Gurel, Lois M comp.
Dimensions of dress and adornment : a book of readings / Lois M. Gurel, Marianne S. Beeson ; ill. by Clayton E. Peterson. — 3d ed. — Dubuque, Iowa : Kendall/Hunt Pub. Co., c1979.
ix, 219 p. : ill. ; 28 cm.
Includes bibliographies and index.
ISBN 0-8403-2038-8
1. Costume—Addresses, essays, lectures. 2. Clothing and dress—Addresses, essays, lectures. I. Beeson, Marianne S., joint comp. II. Title.
GT520.G87 1979 391'.008 79-115071
79 MARC

—ALTERATION

Goldsworthy, Maureen.
New fashion from old / ₍by₎ Maureen Goldsworthy. — London : Batsford, 1978.
128 p. : ill. ; 25 cm. GB78-24366
Includes index.
ISBN 0-7134-0913-4 : £1.99
1. Clothing and dress—Alteration. 2. Clothing and dress—Remaking. I. Title.
TT550.G64 646.4'04 79-303025
79 MARC

Smith, Patricia Burkhart.
Making your clothes fit / Patricia Burkhart Smith. — New York : Dolphin Books, 1979.
viii, 179 p. : ill. ; 26 cm.
Includes index.
ISBN 0-385-14666-3 : $8.95
1. Clothing and dress—Alteration. I. Title.
TT550.S64 646.4'3'04 78-20099
79 MARC

—CARE—HISTORY—19TH CENTURY

Walkley, Christina.
Crinolines and crimping irons : Victorian clothes : how they were cleaned and cared for / ₍by₎ Christina Walkley and Vanda Foster. — London : Owen, 1978.
199 p., ₍8₎ p. of plates : ill., facsims. ; 23 cm. GB78-26834
Bibliography: p. 188.
Includes index.
ISBN 0-7206-0500-8 : £7.95
1. Laundry—History—19th century. 2. Clothing and dress—Care—History—19th century. 3. Costume—Great Britain—History—19th century. I. Foster, Vanda, joint author. II. Title.
TT985.W23 1978 646.6'0941 79-305839
79 MARC

—CLEANING *see* Dry cleaning; Laundry

—CONGRESSES

Association of College Professors of Textiles and Clothing.
Proceedings / Eastern, Central, and Western Regions, Association of College Professors of Textiles and Clothing, New York, New York, New Orleans, Louisiana, Reno, Nevada. — Athens : Georgia Center for Continuing Education, University of Georgia, 1976.
250 p. : ill. ; 28 cm.
Includes bibliographical references.
1. Textile fabrics—Congresses. 2. Clothing and dress—Congresses.
TS1300.A85 1976 677 78-622166
79 MARC

—DICTIONARIES—GERMAN

Schierbaum, Wilfried.
Bekleidungslexikon / hrsg. von Wilfried Schierbaum. — Berlin : Schiele und Schön, 1978.
338 p. : ill. ; 24 cm. GFR78-A
ISBN 3-7949-0305-6 : DM58.00
1. Clothing trade—Dictionaries—German. 2. Clothing and dress—Dictionaries—German. I. Title.
TT494.S34 78-381510
78 MARC

—JUVENILE LITERATURE

Wiseman, Ann.
Cuts of cloth : quick classics to sew & wear, shells for the body / Ann Sayre Wiseman. — 1st ed. — Boston : Little, Brown, c1978.
64 p. : ill. ; 22 x 28 cm.
SUMMARY: Describes the evolution of several classic clothing styles and presents patterns and instructions for robes, dresses, bathing suits, unisex pants, and vest.
ISBN 0-316-94853-5. ISBN 0-316-94852-7 pbk.
1. Dressmaking—Juvenile literature. 2. Tailoring—Juvenile literature. 3. Clothing and dress—Juvenile literature. ₍1. Dressmaking. 2. Sewing. 3. Clothing and dress₎ I. Title.
TT515.W46 646'.36 78-16375
78 MARC

—PERIODICALS

Han'guk Ŭiryu Hakhoe.
(Han'guk Ŭiryu Hakhoe chi)
韓國衣類學會誌 Journal of the Korean Society of Clothing & Textiles.
₍서울₎ 韓國衣類學會
v. ill. 26 cm.
In Korean, with summaries in English.
1. Clothing and dress—Periodicals. I. Han'guk Ŭiryu Hakhoe. Journal.
TX340.H28a 79-640525
(MARC-S)

—PSYCHOLOGY—STUDY AND TEACHING

Binder, Harald, fl. 1974-
Mode im Unterricht : Aspekte d. Kunstunterrichts u. d. visuellen Kommunikation / Harald Binder, Hermann Burkhardt, Andreas Wiemann. — 1. Aufl. — Stuttgart : Klett, 1974.
165 p. : ill. ; 22 cm. GFR74-A37
Bibliography: p. 157-165.
ISBN 3-12-921200-0 : DM14.80
1. Clothing and dress—Social aspects—Study and teaching. 2. Clothing and dress—Psychology—Study and teaching. 3. Fashion—Study and teaching. I. Burkhardt, Hermann, joint author. II. Wiemann, Andreas, joint author. III. Title.
GT525.B55 77-511344
79₍r79₎rev MARC

—REMAKING

Goldsworthy, Maureen.
New fashion from old / ₍by₎ Maureen Goldsworthy. — London : Batsford, 1978.
128 p. : ill. ; 25 cm. GB78-24366
Includes index.
ISBN 0-7134-0913-4 : £1.99
1. Clothing and dress—Alteration. 2. Clothing and dress—Remaking. I. Title.
TT550.G64 646.4'04 79-303025
79 MARC

Lawson, Donna.
Kids' clothes for under $5 : how to save money and have your kids look great by learning to recycle clothes / by Donna Lawson ; ill. ; by Karen Feldgus ; ₍interior photos, Barry Wetcher₎. — New York : Quick Fox, c1978.
95 p. : ill. ; 28 cm.
ISBN 0-8256-3051-7 : $4.95
1. Children's clothing. 2. Clothing and dress—Remaking. I. Title.
TT635.L38 646.4'06 77-93715
78 MARC

—SOCIAL ASPECTS

Polhemus, Ted.
Fashion & anti-fashion : anthropology of clothing and adornment / Ted Polhemus, Lynn Procter. — ₍London₎ : Thames and Hudson, c1978.
102 p. : ill. (some col.) ; 28 cm. GB***
ISBN 0-500-27118-6 : £2.95
1. Fashion—Social aspects. 2. Clothing and dress—Social aspects. I. Procter, Lynn, joint author. II. Title.
GT525.P64 391 79-300280
79 MARC

CLOTHING AND DRESS (Continued)

—SOCIAL ASPECTS—ADDRESSES, ESSAYS, LECTURES

Sociologia dei fenomeni di moda / scritti di F. Alberoni ... ⌈et al.⌉ ; a cura di Gerardo Ragone. — Milano : F. Angeli, ⌈1976⌉
299 p. ; 21 cm. — (Comunicazione e società ; 4) It77-Mar
Includes bibliographical references.
L7000
1. Fashion—Social aspects—Addresses, essays, lectures. 2. Clothing and dress—Social aspects—Addresses, essays, lectures. I. Alberoni, Francesco. II. Ragone, Gerardo.
GT525.S64 391 78-399863
*79 MARC

—SOCIAL ASPECTS—STUDY AND TEACHING

Binder, Harald, fl. 1974-
Mode im Unterricht : Aspekte d. Kunstunterrichts u. d. visuellen Kommunikation / Harald Binder, Hermann Burkhardt, Andreas Wiemann. — 1. Aufl. — Stuttgart : Klett, 1974.
165 p. : ill. ; 22 cm. GFR74-A37
Bibliography: p. 157-165.
ISBN 3-12-921200-0 : DM14.80
1. Clothing and dress—Social aspects—Study and teaching. 2. Clothing and dress—Psychology—Study and teaching. 3. Fashion—Study and teaching. I. Burkhardt, Hermann, joint author. II. Wiemann, Andreas, joint author. III. Title.
GT525.B55 77-511344
79⌈r79⌉rev MARC

CLOTHING AND DRESS IN ART see Costume in art

CLOTHING FACTORIES

—AUTOMATION

American Apparel Manufacturers Association. Technical Advisory Committee.
A new look at apparel mechanization : closing the productivity gap in sewing and finishing : 1978 report of the Technical Advisory Committee, American Apparel Manufacturers Association. — ⌈Arlington, VA. : The Association, c1978⌉
ii, 46 p. : ill. ; 28 cm.
Cover title.
Includes index.
1. Clothing factories—Automation. I. Title.
TT498.A45 1978 338.4'568'70973 78-110745
79 MARC

Dubrovnyĭ, Viktor Afanas'evich.
(Avtomatizatsiia protsessov shveĭnogo proizvodstva)
Автоматизация процессов швейного производства / В. А. Дубровный, И. В. Орлов. — Киев : ⌈УкрНИИНТИ⌉, 1968.
41, ⌈3⌉ p. : ill.
At head of title: Gosudarstvennyĭ planovyĭ komitet Soveta Ministrov Ukrainskoĭ SSR. Ukrainskiĭ nauchno-issledovatel'skiĭ institut nauchno-tekhnicheskoĭ informatsii i tekhniko-ėkonomicheskikh issledovaniĭ. Ministerstvo legkoĭ promyshlennosti USSR.
L.C. copy on microfilm only.
Bibliography: p. ⌈43⌉
—— —— Microfilm. On 1 reel with other items ; 35 mm. Microfilm Slavic 12456 TT
1. Clothing factories—Automation. I. Orlov, Igor' Vasil'evich, joint author. II. Title.
[TT498.D82] 79-409219

CLOTHING FOR THE AGED see Aged—Clothing

CLOTHING FOR THE HANDICAPPED see Handicapped—Clothing

CLOTHING INDUSTRY

—TEXAS

Development Resources, inc.
Texas' textile & apparel industry development program : final report / sponsored by the Joint Committee on Textile Industry Development of the Texas Legislature. — ⌈s.l.⌉ : Development Resources, 1977.
iii, 76 leaves ; 28 cm. & appendix (iii, 44 leaves : ill.)
Includes bibliographical references.
1. Textile industry—Texas. 2. Clothing industry—Texas. I. Texas. Legislature. Joint Committee on Textile Industry Development. II. Title.
HD9857.T4D48 1977 338.4'7'677009764 79-622514
79 MARC

CLOTHING TRADE

see also Children's clothing industry; Men's clothing industry; Shirts, Men's; Tailoring; Women's clothing industry

Fröbel, Folker.
Die neue internationale Arbeitsteilung : strukturelle Arbeitslosigkeit in d. Industrieländern u. d. Industrialisierung d. Entwicklungsländer / Folker Fröbel, Jürgen Heinrichs, Otto Kreye. — Orig.-Ausg. — Reinbek bei Hamburg : Rowohlt, 1977.
653 p. : ill., maps ; 19 cm. — (Rororo aktuell) GFR77-A
Includes bibliographical references.
ISBN 3-499-14185-X : DM10.80
1. Textile industry—Germany, West. 2. Clothing trade—Germany, West. 3. Textile industry. 4. Clothing trade. 5. Foreign trade and employment. I. Heinrichs, Jürgen, joint author. II. Kreye, Otto, joint author. III. Title.
HD9863.5.F75 78-387961
79 MARC

Guseĭnova, Tamara Salekhovna.
(Tovarovedenie shveĭnykh i trikotazhnykh tovarov)
Товароведение швейных и трикотажных товаров : рекомендовано в качестве учебника для товароведных факультетов торговых вузов / Т. С. Гусейнова, Г. В. Жильцова, М. Ф. Коврижных. — Москва : Экономика, 1971.
335 p. : ill. ; 23 cm. USSR***
Bibliography: p. ⌈325⌉-⌈326⌉
Includes index.
1.00rub
1. Clothing trade. 2. Knit-goods industry. I. Zhil'tsova, G. V. II. Kovrizhnykh, Mikhail Fedorovich. III. Title.
TT497.G87 79-413959

Mathisen, Marilyn Purol.
Apparel and accessories / Marilyn Purol Mathisen ; consulting editor, Richard L. Lynch. — New York : Gregg Division, McGraw-Hill, c1979.
v, 250 p. : ill. ; 28 cm. — (Career competencies in marketing series)
ISBN 0-07-040905-6
1. Clothing trade. I. Title. II. Series.
TT497.M37 658.89'687 78-12321
78 MARC

—DICTIONARIES—GERMAN

Schierbaum, Wilfried.
Bekleidungslexikon / hrsg. von Wilfried Schierbaum. — Berlin : Schiele und Schön, 1978.
338 p. : ill. ; 24 cm. GFR78-A
ISBN 3-7949-0305-6 : DM58.00
1. Clothing trade—Dictionaries—German. 2. Clothing and dress—Dictionaries—German. I. Title.
TT494.S34 78-381510
78 MARC

—HISTORY—20TH CENTURY

Glynn, Prudence.
In fashion : dress in the twentieth century / by Prudence Glynn ; illustrated by Madeleine Ginsburg. — London ; Boston : G. Allen & Unwin, 1978.
243 p., ⌈4⌉ leaves of plates : ill. (some col.) ; 26 cm. GB***
Bibliography: p. 234-235.
Includes index.
ISBN 0-04-391003-3 : £8.95
1. Fashion—History—20thcentury. 2. Clothing trade—History—20th century. I. Title.
TT504.G55 1978b 391'.009'04 79-301247
79 MARC

Glynn, Prudence.
In fashion : dress in the twentieth century / by Prudence Glynn ; illustrated by Madeleine Ginsburg. — New York : Oxford University Press, 1978.
243 p. : ill. ; 25 cm.
Includes index.
ISBN 0-19-520072-1 : $19.95
1. Fashion—History—20th century. 2. Clothing trade—History—20th century. I. Title.
TT504.G55 391'.009'04 78-56372
78 MARC

Robinson, Julian.
Fashion in the '30s / ⌈by⌉ Julian Robinson. — London : Oresko Books, c1978.
104 p. : ill. (some col.) ; 30 cm. GB***
ISBN 0-905368-44-4 : £8.95. ISBN 0-905368-45-2 pbk.
1. Fashion—History—20th century. 2. Clothing trade—History—20th century. I. Title.
TT504.R59 1978 391'.2'09043 79-302257
79 MARC

—PRICES—EUROPEAN ECONOMIC COMMUNITY COUNTRIES

Rothe, Wolfgang, 1927-
Internationaler Preis- und Kaufkraftvergleich für Bekleidung in Ländern des Gemeinsamen Marktes und der Freihandelszone. Köln, Westdeutscher Verlag, 1962.
226 p. col. illus., forms, map. 30 cm. (Forschungsberichte des Landes Nordrhein-Westfalen, Nr. 1069)
Bibliography: p. 219-226.
1. Clothing trade—Prices—European Economic Community countries. 2. Clothing trade—European Economic Community countries. 3. Purchasing power—European Economic Community countries. I. Title.
T7.F65 Nr. 1069 083'.1 s 66-55464
⌈HD9940.E8562⌉ ⌈r79⌉rev MARC

—PRICES—UNITED STATES

Colberg, Marshall Rudolph, 1913-
The consumer impact of repeal of 14(b) / Marshall R. Colberg ; with a foreword by Sam J. Ervin. — Washington : Heritage Foundation, c1978.
44 p. ; 23 cm.
Includes bibliographical references.
1. Cost and standard of living—United States. 2. Food prices—United States. 3. Clothing trade—Prices—United States. 4. Open and closed shop—Economic aspects—United States. I. Title.
HD6983.C53 339.4'1'0973 77-17061
77 MARC

—STUDY AND TEACHING—ENGLAND —MANCHESTER

Egerton, Vikki.
A history of Hollings College, 1901-1976 / by Vikki Egerton ; foreword by Sir William Downward. — ⌈Manchester⌉ : ⌈The author⌉, ⌈1978⌉
viii, 116 p. : ill., facsims., plan, ports. ; 21 cm. GB78-28531
Bibliography: p. 98.
£2.50
1. Hollings College—History. 2. Home economics—Study and teaching—England—Manchester. 3. Food service—Study and teaching—England—Manchester. 4. Beauty culture—Study and teaching—England—Manchester. 5. Clothing trade—Study and teaching—England—Manchester. I. Title.
TX286.H64E38 640'.7'1142733 78-326260
79 MARC

—VOCATIONAL GUIDANCE

Fashion Group.
Your future in the new world of American fashion / by 15 famous members of the Fashion Group, inc. ; Christine LeVathes, editor. — 1st ed. — New York : Richards Rosen Press, 1979.
xii, 115 p. ; 22 cm.
ISBN 0-8239-0458-X
1. Fashion—Vocational guidance. 2. Clothing trade—Vocational guidance. I. LeVathes, Christine. II. Title.
TT507.F35 1979 687'.023 78-11905
78 MARC

Lulow, Jo Ann.
Your career in the fashion industry / Jo Ann Lulow. — New York : ARCO, c1979.
v, 138 p. : ill. ; 20 cm. — (Arco's career guidance series)
Includes bibliographies and index.
ISBN 0-668-04613-9 lib. bdg. : $5.95. ISBN 0-668-04620-1 pbk. : $2.95
1. Clothing trade—Vocational guidance. 2. Fashion—Vocational guidance. I. Title.
TT507.L84 338.4'7'687023 78-12724
78 MARC

CLOTHING TRADE (Continued)

—CANADA—HISTORY

Steed, Guy P F
An historical geography of the Canadian clothing industries : 1800-1930's / Guy P. F. Steed. — ⌊Ottawa⌋ : Dept. of Geography and Regional Planning, University of Ottawa, 1976.
52 p. : maps ; 29 cm. — (Research notes - Geography and Regional Planning, University of Ottawa ; no. 11) C77-8038-3
Abstract in English and French.
Bibliography: p. 49-52.
1. Clothing trade—Canada—History. I. Title. II. Series: University of Ottawa. Dept. of Geography and Regional Planning. Research notes - Geography and Regional Planning, University of Ottawa ; no. 11.
HD9940.C2S74 338.4'7'6870971 78-323970
79 MARC

—CANADA—JUVENILE LITERATURE

Mommy works on dresses : and, Madeleine's song / Louise de Grosbois ... ⌊et al.⌋ ; translated by Carloine Bayard ; ⌊Madeleine's song adapted by Lois Pike and Naomi Wall⌋. — 1st English ed. — Toronto, Ont. : Women's Press, c1977.
⌊24⌋ p. : ill. ; 21 cm. C***
Translation of Maman travaille dans la robe, published in Histoires vraies de tous les jours.
ISBN 0-88961-042-8 : $2.95
1. Clothing trade—Canada—Juvenile literature. I. Grosbois, Louise de. II. Pike, Louis. III. Wall, Naomi.
TT496.C2M6513 687'.12 79-342363
79 MARC

—DENMARK—HISTORY

Madsen, Allan, 1942-
Bidrag til tekstil- og beklædningsproduktionens historie i Danmark 1600-1850 / Allan Madsen. — København : Akademisk Forlag : ⌊eksp. DBK⌋, 1978.
323 p. : ill. ; 22 cm. — (His Tekstilproduktionens historie under overgangen fra feudalisme til kapitalisme ; 1) (Småskrifter fra Københavns universitet, Institut for kultursociologi ; hæfte 7) D79-15/16
Bibliography: p. 319-323.
ISBN 8750018639 : kr54.15
1. Textile industry—Denmark—History. 2. Clothing trade—Denmark—History. I. Title. II. Series. III. Series: Copenhagen. Universitet. Institut for kultursociologi. Småskrifter ; hæfte 7.
TS1369.M32 vol. 1 79-374894
79 MARC

—ENGLAND—LONDON—DIRECTORIES

O'Connor, Kaori.
The 1978 fashion guide / ⌊written and edited by⌋ Kaori O'Connor. — Sevenoaks : Coronet, 1978.
360 p. : ill., maps ; 18 cm. GB78-15513
Promotions and special projects by F. Kahn
Includes indexes.
ISBN 0-340-22302-2 : £1.50
1. Clothing trade—England—London—Directories. 2. Jewelry trade—England—London—Directories. 3. Shopping—England—London—Directories. I. Kahn, Farrol. II. Title.
HD9940.G82L65 79-304031
380.1'45'687025421
79 MARC

—EUROPE

International Trade Centre.
Men's and boy's shirts in selected western European countries, United Kingdom, Federal Republic of Germany, the Netherlands, Switzerland. — Geneva : International Trade Centre UNCTAD/GATT, 1977.
iv, 19 p. : ill. ; 30 cm. — (Monographs on trade channels) Sw***
"ITC/DTS/SMB/60(8)"
$10.00 (U.S.)
1. Shirts, Men's. 2. Clothing trade—Europe. I. Title. II. Series.
HD9969.S7E924 380.1'45'68711 78-314245
78 MARC

—EUROPE—CATALOGS

(Europe ichiryū burando no hon)
Europe 一流ブランドの本：ヨーロッパの一流店ショッピング全ガイド. — 東京：講談社, 昭和52 ⌊1977⌋
198 p. : col. ill. ; 29 cm. — (講談社 mook ; 第2巻, 第5号) Ja***
Cover title.
Series romanized: Kōdansha mook.
¥1500
1. Clothing trade—Europe—Catalogs. 2. Branded merchandise—Europe—Catalogs. 3. Cities and towns—Europe. I. Kōdansha. II. Title: Ichiryū burando no hon.
HD9940.E82E94 79-804515

—EUROPEAN ECONOMIC COMMUNITY COUNTRIES

Rothe, Wolfgang, 1927-
Internationaler Preis- und Kaufkraftvergleich für Bekleidung in Ländern des Gemeinsamen Marktes und der Freihandelszone. Köln, Westdeutscher Verlag, 1962.
226 p. col. illus., forms, map. 30 cm. (Forschungsberichte des Landes Nordrhein-Westfalen, Nr. 1069)
Bibliography: p. 219-226.
1. Clothing trade—Prices—European Economic Community countries. 2. Clothing trade—European Economic Community countries. 3. Purchasing power—European Economic Community countries. I. Title.
T7.F65 Nr. 1069 083'.1 s 66-55464
⌊HD9940.E8562⌋ ⌊r79⌋rev MARC

—FINLAND

Seppä, Ilpo.
Tekstiili- ja vaatetustavaroiden kulutus Suomessa v. 1954-1977 ⌊kirj.⌋ Ilpo Seppä ⌊ja⌋ Seppo Wallenius. Helsinki ⌊Elinkeinoelämän Tutkimuslaitos⌋ 1972.
95 p. illus. 25 cm. (Elinkeinoelämän Tutkimuslaitos. ⌊Julkaisuja. Sarja⌋ B, 3) Fi***
Bibliography: p. 94-95.
1. Clothing trade—Finland. 2. Textile industry—Finland. 3. Consumption (Economics)—Finland. I. Wallenius, Seppo, joint author. II. Title. III. Series: Elinkeinoelämän Tutkimuslaitos. Julkaisuja. Sarja B, 3.
HD9940.F52S46 79-343488
79 MARC

—GERMANY, WEST

Fröbel, Folker.
Die neue internationale Arbeitsteilung : strukturelle Arbeitslosigkeit in d. Industrieländern u. d. Industrialisierung d. Entwicklungsländer / Folker Fröbel, Jürgen Heinrichs, Otto Kreye. — Orig.-Ausg. — Reinbek bei Hamburg : Rowohlt, 1977.
653 p. : ill., maps ; 19 cm. — (Rororo aktuell) GFR77-A
Includes bibliographical references.
ISBN 3-499-14185-X : DM10.80
1. Textile industry—Germany, West. 2. Clothing trade—Germany, West. 3. Textile industry. 4. Clothing trade. 5. Foreign trade and employment. I. Heinrichs, Jürgen, joint author. II. Kreye, Otto, joint author. III. Title.
HD9863.5.F75 78-387961
79 MARC

Peters, Joachim H
Handbuch der Bekleidungsindustrie : Bekleidungstechnik in DOB u. HAKA / Joachim H. Peters. — 1. Aufl. — Frankfurt am Main : Deutscher Fachverlag, 1978.
282 p. : ill. ; 21 cm. — (Schriftenreihe der Textil-Wirtschaft) GFR78-A
ISBN 3-87150-112-3
1. Clothing trade—Germany, West. I. Title.
TT496.G3P47 79-358992
79 MARC

Schwarting, Uwe.
Direktinvestitionen, passive Lohnveredelung und Zukäufe als Unternehmensstrategien der deutschen Bekleidungsindustrie : e. unternehmensgrössenspezif. Unters. / Uwe Schwarting. — Göttingen : Schwartz, 1978.
xix, 273 p. : ill. ; 21 cm. — (Beiträge zur Mittelstandsforschung ; Heft 42) GFR78-A
Bibliography: p. 258-273.
ISBN 3-509-01075-2 : DM41.00
1. Clothing trade—Germany, West. 2. Germany, West—Economic policy—1974- . 3. Underdeveloped areas. 4. International economic relations. 5. Investments, German. I. Title. II. Series.
HD9940.G32S38 338.4'7'6870943 79-339330
79 MARC

—GREAT BRITAIN

Great Britain. Central Office of Information. Reference Division.
Textiles and clothing / prepared by Reference Division, Central Office of Information. — London : prepared for British Information Services by the Central Office of Information, 1978.
45, ⌊1⌋ p., ⌊2⌋ leaves of plates : ill. ; 24 cm. — (British industry today) GB***
Bibliography: p. ⌊46⌋
1. Textile industry—Great Britain. 2. Clothing trade—Great Britain. I. Great Britain. British Information Services. II. Title.
HD9861.5.G73 1978 338.4'7'67700941 78-325062
79 MARC

—GREAT BRITAIN—HISTORY—TWENTIETH CENTURY

Guppy, Alice.
Children's clothes, 1939-1970 : the advent of fashion / ⌊by⌋ Alice Guppy. — Poole : Blandford Press for the Pasold Research Fund, 1978.
⌊10⌋, 346 p., ⌊8⌋ p. of plates : ill. (some col.) ; 23 cm. GB78-37637
Includes index.
ISBN 0-7137-0896-4 : £9.50
1. Children's clothing—History—Twentieth century. 2. Clothing trade—Great Britain—History—Twentieth century. I. Title.
TT635.G78 391'.07'30941 79-318146
79 MARC

—HAWAII

Kurt Salmon Associates.
Garment manufacturing in Hawaii / prepared for the State Department of Planning and Economic Development by Kurt Salmon Associates, inc. — ⌊Honolulu⌋ : The Dept., 1979.
245 p. : graphs ; 28 cm.
Includes bibliographical references.
1. Clothing trade—Hawaii. I. Hawaii. Dept. of Planning and Economic Development. II. Title.
HD9940.U5H34 338.4'7'68709969 79-622407
79 MARC

—INDIA

Economic and Scientific Research Foundation.
Survey of India's export potential of textiles & made-up garments. Prepared under the guidance of the Export Promotion Division, U.S. Agency for International Development, New Delhi, for the Ministry of Commerce, Govt. of India. New Delhi ⌊1969⌋
5 v. in 6. 28 cm. 125.00
"For limited distribution only."
Commodity export survey, textile and man-made garments, an appraisal by the Research Unit, Economic and political weekly: v. 1, p. ⌊A1⌋-A23.
Bibliography: v. 1, p. 227-244.
1. Textile industry—India. 2. Clothing trade—India. I. India (Republic). Ministry of Commerce. II. United States. AID Mission to India. Export Promotion Division. III. Economic and political weekly. Commodity export survey. 1969. IV. Title.
HD9866.I62E36 1969 338.4'7'67700954 74-910163
70⌊r79⌋rev MARC

—INDIA—STATISTICS

Cotton Textiles Export Promotion Council.
Exports of readymade garments (fibrewise) from India in 1976. — Bombay : Cotton Textiles Export Promotion Council, ⌊1977⌋
116 p. ; 24 cm.
Tables.
Re1.00
1. Clothing trade—India—Statistics. I. Title.
HD9940.I42C67 1977 382'.45'6870954 78-906592
79 MARC

—IRELAND

Ireland (Eire). Committee on Industrial Organization.
Report on survey of the shirt industry / Committee on Industrial Organisation. — Dublin : Stationery Office : available from the Govt. Publications Sale Office, ⌊1963?⌋
169 p. ; 25 cm.
"(Pr. 6801)"
Bibliography: p. 47.
1. Shirts, Men's. 2. Clothing trade—Ireland. I. Title.
HD9969.S7I725 1963 338.8 79-302044
79 MARC

—ITALY

Ciabattoni, Amos.
Il sistema moda / a cura di Amos Ciabattoni ; presentazione di Mario Deaglio. — 1. ed. — Torino : Valentino, 1976.
227 p. : ill. ; 21 cm. It78-Feb
Bibliography: p. 221-224.
L5000
1. Clothing trade—Italy. 2. Textile industry—Italy. I. Title.
HD9940.I82C5 78-383071
*78 MARC

Frey, Luigi.
Occupazione e ristrutturazioni industriali : il caso dei settori tessile/abbigliamento / L. Frey, R. Livraghi, P. Lucà Trombetta. — Milano : F. Angeli, c1979.
183 p. ; 22 cm. — (Quaderni di economia del lavoro ; 10) It***
Includes bibliographical references.
ISBN 8820415216 : L5000
1. Textile workers—Italy. 2. Textile industry—Italy. 3. Clothing trade—Italy. I. Livraghi, Renata, joint author. II. Lucà Trombetta, P., joint author. III. Title.
HD8039.T42I84 79-371395
79 MARC

—ITALY—DIRECTORIES

Annuario dell'abbigliamento e del tempo libero. anno 1-1977/78-
⌊Torino, SEAT⌋
v. ill. 29 cm.
1. Clothing trade—Italy—Directories. 2. Textile industry—Italy—Directories. 3. Recreation—Italy—Apparatus and equipment—Directories. 4. Italy—Directories—Telephone.
HD9940.I8A54 79-645126
MARC-S

CLOTHING TRADE (Continued)

—ITALY—VENETO

Tessili e abbigliamento : organizzazione del lavoro, tecnologie e professionalità nelle fabbriche del Veneto / ⌈del⌉ Gruppo di ricerca Ial-Cisl nazionale, Livio Bortoloso ... ⌈et al.⌉ ; relazioni, Pier Luigi Marzola, Claudio Storti ; a cura di Annalisa Vittore. — Milano : F. Angeli, ⌈1979⌉
552 p. ; 22 cm. — (Quaderni di economia del lavoro ; 12) It79-Mar
Bibliography: p. ⌈551⌉-552.
ISBN 8820415070 : L8000
1. Textile industry—Italy—Veneto. 2. Clothing trade—Italy—Veneto. I. Bortoloso, Livio. II. Vittore, Annalisa.
HD9865.I83V437 79-354928
*79 MARC

—NETHERLANDS—MANAGEMENT

Economisch Instituut voor het Midden- en Kleinbedrijf.
Bedrijfsbeheer in de praktijk: detailhandel in damesconfectie; onderzoek naar het bedrijfsbeheer in een zevental bedrijven. 's-Gravenhage, 1964.
175 p. 25 cm. (Bedrijfseconomische publikaties)
1. Clothing trade—Netherlands—Management. I. Title.
HD9940.N42E27 1964 78-413849
78 MARC

—NEW YORK (CITY)—DIRECTORIES

Louie, Elaine.
The Manhattan clothes shopping guide / Elaine Louie. — New York : Macmillan, c1978.
xiii, 363 p. : maps ; 25 cm.
Includes indexes.
ISBN 0-02-575290-1
1. Shopping—New York (City)—Directories. 2. New York (City)—Stores, shopping centers, etc.—Directories. 3. Clothing trade—New York (City)—Directories. I. Title.
TX335.L66 1978b 382'.45'687097471 78-14781
78 MARC

Louie, Elaine.
The Manhattan clothes shopping guide / Elaine Louie. — New York : Collier Books, c1978.
xiii, 363 p. : maps ; 24 cm.
Includes indexes.
ISBN 0-02-080720-1
1. Shopping—New York (City)—Directories. 2. New York (City)—Stores, shopping centers, etc.—Directories. 3. Clothing trade—New York (City)—Directories. I. Title.
TX335.L66 1978 382'.45'687097471 78-14782
78 MARC

—POLAND—WIELKOPOLSKA

Mączak, Antoni.
Sukiennictwo wielkopolskie, XIV-XVII wiek / Antoni Mączek. — Wyd. 1. — Warszawa : Państwowe Wydawn. Naukowe, 1955.
324 p. : ill. ; 24 cm. — (Badania z dziejów rzemiosła i handlu w epoce feudalizmu ; 3) (Prace Instytutu Historycznego Uniwersytetu Warszawskiego)
Cover title varies slightly.
Bibliography: p. 300-⌈309⌉
Includes indexes.
1. Clothing trade—Poland—Wielkopolska. I. Title. II. Series: Warsaw. Uniwersytet. Instytut Historyczny. Prace.
HD9940.P63W545 78-388225
79 MARC

—SPAIN

Barella Miró, Alberto.
Aplicación de la fiabilidad a la industria de la confección : estudio / realizado en colaboración entre el Instituto de Tecnología Química y Textil del Patronato Juan de la Cierva, la Asociación de Investigación de Empresas Confeccionistas y la Asociación de Investigación Textil Algodonera, bajo la dirección del Prof. Alberto Barella, y Carlos Polo y Ramón Ma. Saurí como principales investigadores, y la colaboración de José Ma. Arús, José Ma. Etayo y Luis Viertel. — Barcelona : A.I.T.A., 19-
v. : ill. ; 30 cm. — (Estudios y documentos - Asociación de Investigación Textil Algodonera ; 1) (Cuadernos AITA)
On cover of pt. 3: Publicación especial.
Bibliography: pt. 3, leaf 86.
CONTENTS:
—pt. 3. Experiencias en el sub-sector de corsetería. Análisis de la distribución estadística del comportamiento de las prendas. Consecuencias de tipo práctico.
1. Clothing trade—Spain. 2. Reliability (Engineering) I. Polo, Carlos, joint author. II. Saurí, Ramón María, joint author. III. Spain. Consejo Superior de Investigaciones Científicas. Instituto de Tecnología Química y Textil. IV. Asociación de Investigación de Empresas Confeccionistas. V. Asociación de Investigación Textil Algodonera. VI. Title. VII. Series: Asociación de Investigación Textil Algodonera. Estudios y documentos - Asociación de Investigación Textil Algodonera ; 1.
TT496.S7B37 79-352719
79 MARC

—UNITED STATES

American Apparel Manufacturers Association. Apparel Political Education Committee.
The MFA and the American apparel industry / produced by the Apparel Political Education Committee of the American Apparel Manufacturers Association, c1978. — Arlington, Va. : The Committee, c1978.
24 p. : ill. ; 28 cm.
1. Clothing trade—United States. 2. Textile industry—United States. 3. Import quotas—United Stats. I. Title.
HD9940.U4A6 1978 382'.45'687 79-101876
79 MARC

Council on Wage and Price Stability.
A Study of the textile and apparel industries : including prices, wages, employment, foreign trade capacity : staff report / prepared by the Council on Wage and Price Stability. — Washington : Executive Office of the President : for sale by U.S. Govt. Print. Off., 1978.
viii, 134 p. : graphs ; 27 cm.
"Written principally by Frances Esposito."
Includes bibliographical references.
1. Textile industry—United States. 2. Clothing trade—United States. 3. Textile workers—United States. I. Esposito, Frances. II. Title.
HD9854.C7 1978 338.4'7'67700973 78-602809
79 MARC

International Ladies' Garment Workers' Union. Research Dept.
Conditions in the women's garment industry / Research Department, International Ladies' Garment Worker's Union. — ⌈New York?⌉ : The Department, 1975.
12 p. ; 28 cm.
1. Clothing trade—United States. I. Title.
HD9940.U4I62 1975 338.4'7'6870973 79-308099
79 MARC

Morton Research Corporation.
The rainwear industry : an economic, marketing, and financial investigation / project leader, B Tenenbaum ; research assistant, E. Stempel. — Merrick N.Y. : Morton Research Corp., 1978.
82 leaves : graphs ; 29 cm. — (A Morton report)
$95.00
1. Clothing trade—United States. 2. Raincoats. 3. Market surveys—United States. I. Tenenbaum, B. II. Stempel, E. III. Title.
HD9940.U4M6 1978 381'.45'68714 78-112083
79 MARC

Priestland, Carl H
Focus : economic profile of the apparel industry, with extensive comparative data on the American textile industry / compiled and written by Carl Priestland. — Arlington, Va. : American Apparel Manufacturers Association, c1979.
54 p. : graphs ; 22 x 28 cm.
Includes bibliographical references.
1. Clothing trade—United States. I. American Apparel Manufacturers Association. II. Title.
HD9940.U4P66 1979 338.4'7'6870973 79-109885
00 MARC

—UNITED STATES—CONGRESSES

Market information and research in fashion management / edited by Elizabeth A. Richards and David Rachman. — Chicago : American Marketing Association, ⌈c1978⌉
vii, 110 p. ; 23 cm. — (Proceedings series ; American Marketing Association)
Proceedings of a workshop sponsored by the New York Chapter of the American Marketing Association.
ISBN 0-87757-107-4
1. Clothing trade—United States—Congresses. 2. Clothing trade—United States—Management—Congresses. I. Richards, Elizabeth A. II. Rachman, David J. III. American Marketing Association. New York Chapter. IV. American Marketing Association. Proceedings series - American Marketing Association.
HD9940.U4M35 658.8'38'6870973 78-6963
78 MARC

—UNITED STATES—HISTORY

Masson, Irmalotte, 1904-
Die Levi-Strauss-Saga : d. märchenhafte Geschichte d. Mannes, d. d. Jeans erfand / Irmalotte Masson, Ursula von Wiese. — München : Kindler, 1978.
247 p. : ill. ; 23 cm. GFR78-A
ISBN 3-463-00726-6 : DM29.80
1. Strauss, Levi, 1829-1902. 2. Levi Strauss and Company—History. 3. Businessmen—United States—Biography. 4. Clothing trade—United States—History. I. Wiese und Kaiserswaldau, Ursula Renate von, 1905- joint author. II. Title.
HD9940.U6L456 338.7'68'70924 79-348550
79 MARC

—UNITED STATES—MANAGEMENT—CONGRESSES

Market information and research in fashion management / edited by Elizabeth A. Richards and David Rachman. — Chicago : American Marketing Association, ⌈c1978⌉
vii, 110 p. ; 23 cm. — (Proceedings series ; American Marketing Association)
Proceedings of a workshop sponsored by the New York Chapter of the American Marketing Association.
ISBN 0-87757-107-4
1. Clothing trade—United States—Congresses. 2. Clothing trade—United States—Management—Congresses. I. Richards, Elizabeth A. II. Rachman, David J. III. American Marketing Association. New York Chapter. IV. American Marketing Association. Proceedings series - American Marketing Association.
HD9940.U4M35 658.8'38'6870973 78-6963
78 MARC

—UNITED STATES—STATISTICS —PERIODICALS

Apparel outlook.
⌈New York⌉ Credit Clearing House.
v. 28 cm.
"Annual statistical issue."
Key title: Apparel outlook, ISSN 0191-1892.
1. Clothing trade—United States—Statistics—Periodicals.
I. Credit Clearing House.
HD9940.U3A655 381'.45'6870973 79-642548
MARC-S

CLOTHING WORKERS

see also Sweating system

International Textile, Garment & Leather Workers' Federation.
Employment of children and young people, textile, garment and leather sectors : first world congress, Amsterdam, Netherlands, 16-20 October 1972 / International Textile, Garment & Leather Workers' Federation. — London : The Federation, ⌈1972?⌉
22 p. ; 30 cm. GB***
1. Youth—Employment. 2. Textile workers. 3. Clothing workers. 4. Leather workers. I. Title.
HD6270.I57 1972 331.3'4 78-312501
78 MARC

—CHILE

Silva Donoso, María de la Luz.
El empleo femenino : algunos factores que inciden en la participación femenina en mercados de trabajo diferenciales / María de la Luz Silva Donoso. — ⌈Santiago de Chile?⌉ : Organización Internacional del Trabajo, Programa Regional del Empleo para América Latina y el Caribe, 1977.
vi, 140 p. : forms ; 28 cm. — (Monografía - PREALC ; 3)
Includes bibliographical references.
1. Women—Employment—Chile. 2. Clothing workers—Chile. I. Title. II. Series: Programa Regional del Empleo para América Latina y el Caribe. Monografía - PREALC ; 3.
HD6126.S53 78-113774
79 MARC

—GHANA—CASE STUDIES

Date-Bah, Eugenia.
Absenteeism and turnover in a Ghanaian garment factory / by Eugenia Date-Bah. — Geneva : International Institute for Labour Studies, 1976.
22 p. ; 30 cm. — (International educational materials exchange ; IEME 4138)
Sw***
Bibliography: p. 20-22.
1. Clothing workers—Ghana—Case studies. 2. Absenteeism (Labor)—Ghana—Case studies. 3. Labor turnover—Ghana—Case studies. I. Title. II. Series.
HD8039.C62G583 331.7'68'709667 79-304177
79 MARC

—TENNESSEE—ATTITUDES

Bragg, Emma W 1911-
A profile of work motivation attitudes of apparel workers / by Emma W. Bragg. — Ann Arbor, Mich. : Published for Tennessee State University by University Microfilms International, 1979.
xix, 265 p. : ill. ; 23 cm. — (Monograph publishing : Sponsor series)
Bibliography: p. 95-98.
ISBN 0-8357-0460-2 : $15.00
1. Clothing workers—Tennessee—Attitudes. 2. Employee motivation—Tennessee. I. Title.
HD8039.C62U6333 331.7'68'709768 79-17776
79 MARC

CLOTHING WORKERS
—TENNESSEE—ATTITUDES
(Continued)

Bragg, Emma W 1911-
The relationship between work motivation attitudes and demographic characteristics among apparel workers : a summary / by Emma W. Bragg. — Ann Arbor, Mich. : Published for Tennessee State University by University Microfilms International, 1979.
108 p. ; 23 cm. — (Monograph publishing : Sponsor series)
Bibliography: p. 99-108.
ISBN 0-8357-0474-2
1. Clothing workers—Tennessee—Attitudes. 2. Employee motivation—Tennessee. 3. Psychology, Industrial. I. Tennessee State University. II. Title.
HD8039.C62U6335 331.7'68'709768 79-21446
79 MARC

CLOUD FORECASTING

see also Weather forecasting

Forsdyke, A G
Cloud forecasting by means of the tephigram with particular reference to turbulence cloud / By A. G. Forsdyke. — ⌈Nairobi⌉ : Meteorological Dept., 1961.
11 p., ⌈7⌉ leaves of plates : graphs ; 26 cm. — (East African Meteorological Department memoirs ; v. 2, no. 4)
Cover title.
1. Cloud forecasting. I. Title. II. Series: East Africa High Commission. East African Meteorological Dept. Memoirs. Photo-offset reprints ; v. 2, no. 4.
QC921.F63 551.6'4'76 79-305377
79 MARC

CLOUD OF UNKNOWING

Curran, Thomas M
An introduction to "The cloud of unknowing" / by Thomas Curran. — Dublin : Carmelite Centre of Spirituality ; Bury : Koinonia Limited, 1978.
36 p ; 21 cm. — (Living flame series ; v. 4) GB79-00926
Bibliography: p. 35.
ISBN 0-86088-014-1 : £0.40
1. Cloud of unknowing. 2. Mysticism—Middle Ages, 600-1500. I. Title.
BV5080.C6C87 248'.22 79-314259
79 MARC

CLOUD PHYSICS

see also Precipitation (Meteorology)

(Kuchevye oblaka i svi͡azannai͡a s nimi deformat͡sii͡a poleĭ meteoėlementov)
Кучевые облака и связанная с ними деформация полей метеоэлементов / под ред. И. П. Мазина и С. М. Шметера. — Москва : Московское отд-ние Гидрометеоиздата, 1977.
126 p. : ill. ; 22 cm. — (Труды Центральной аэрологической обсерватории ; вып. 134) USSR***
At head of title: Glavnoe upravlenie gidrometeorologicheskoĭ sluzhby pri Sovete Ministrov SSSR.
Table of contents also in English.
Bibliography: p. 123-⌈125⌉
0.59rub
1. Cumulus—Measurement. 2. Cloud physics. I. Mazin, I. P. II. Shmeter, Solomon Moiseevich. III. Series: T͡Sentral'nai͡a aėrologicheskai͡a observatorii͡a. Trudy vyp. 134.
QC921.43.C8K82 79-407717

Pruppacher, Hans R 1930-
Microphysics of clouds and precipitation / by Hans R. Pruppacher and James D. Klett. — Dordrecht, Holland ; Boston : D. Reidel Pub. Co., c1978.
xiv, 714 p. : ill. ; 25 cm.
Bibliography: p. 657-697.
Includes index.
ISBN 9027705151
1. Cloud physics. 2. Precipitation (Meteorology) I. Klett, James D., 1940- joint author. II. Title.
QC921.5.P78 551.5'76 78-9686
78 MARC

—ADDRESSES, ESSAYS, LECTURES

Fizika oblakov i aktivnye vozdeĭstvii͡a : ⌈Sb. stateĭ⌉ / Pod red. V. P. Lominadze. — Leningrad : Gidrometeoizdat, 1977.
117 p. : ill. ; 21 cm. — (Trudy Zakavkazskogo nauchno-issledovatel'skogo gidrometeorologicheskogo instituta ; vyp. 64 (70)) USSR78
At head of title: Glavnoe upravlenie gidrometeorologicheskoĭ sluzhby pri Sovete Ministrov SSSR.
Includes bibliographies.
0.60rub
1. Cloud physics—Addresses, essays, lectures. 2. Weather control—Addresses, essays, lectures. I. Lominadze, Vasiliĭ Pavlovich. II. Series: Tiflis. Zakavkazskiĭ nauchno-issledovatel'skiĭ gidrometeorologicheskiĭ institut. Trudy ; vyp. 64.
QC851.T52 vyp. 64 79-356401
⌈QC921.55⌉ 79 MARC

—CONGRESSES

Advances in cloud physics and precipitation stimulation in Israel. — ⌈Jerusalem⌉ : Prime Minister's Office, National Council for Research and Development, 1975.
97 p. : ill. ; 28 cm.
Also sponsored by the Dept. of Atmospheric Sciences, Hebrew University of Jerusalem.
Includes bibliographies.
1. Cloud physics—Congresses. 2. Precipitation (Meteorology)—Israel—Simulation methods—Congresses. 3. Ice crystals—Congresses. I. Israel. ha-Mo'atsah ha-le'umit le-meḥkar ule-fituah. II. Jerusalem. Hebrew University. Dept. of Atmospheric Sciences.
QC921.48.A38 551.5'77 78-320033
78 MARC

—MATHEMATICAL MODELS

Koenig, Lloyd Randall, 1929-
Theoretical experiments on cumulus dynamics and glaciation / L. Randall Koenig, Francis W. Murray. — Santa Monica, Ca. : Rand, 1979.
xv, 116 p. : ill. ; 28 cm. — (⌈Report⌉ - Rand Corporation ; R-2424-NSF)
"Prepared under a grant from the National Science Foundation."
Bibliography: p. 113-116.
ISBN 0-8330-0104-3
1. Ice crystals—Mathematical models. 2. Cumulus—Mathematical models. 3. Cloud physics—Mathematical models. I. Murray, Francis W., 1921- joint author. II. United States. National Science Foundation. III. Title. IV. Title: Glaciation. V. Series: Rand Corporation. Rand report ; R-2424-NSF.
AS36.R3 R-2424 081 s 79-4484
⌈QC921.6.C6⌉ 79 MARC

Murray, Francis W 1921-
Simulation of convective cloudiness, rainfall, and associated phenomena caused by industrial heat released directly to the atmosphere / Francis W. Murray, L. Randall Koenig ; prepared for the U.S. Department of Energy. — Santa Monica, Calif. : Rand, 1979.
xiii, 93 p. : ill. ; 28 cm. — (⌈Report⌉ - Rand Corporation ; R-2456-DOE)
Bibliography: p. 87-93.
ISBN 0-8330-0133-7
1. Cloud physics—Mathematical models. 2. Cooling towers—Environmental aspects—Mathematical models. I. Koenig, Lloyd Randall, 1929- joint author. II. United States. Dept. of Energy. III. Title. IV. Series: Rand Corporation. Rand report ; R-2456-DOE.
AS36.R3 R-2456 081 s 79-15477
⌈QC921.5⌉ 79 MARC

CLOUD SEEDING see Rain-making

CLOUDS

see also Cloud physics; Cumulus

—ATLASES

Italy. Istituto idrografico.
Atlante delle nubi : appendice al volume 2. del Manuale dell'Ufficiale di rotta. — Genova : Tipo-lit. dell'Istituto idrografico della Marina, 1961.
6, 34 p., ⌈41⌉ leaves of plates : ill. ; 18 x 25 cm.
"I.I. 3100."
1. Clouds—Observers' manuals. 2. Clouds—Atlases. 3. Aids to naivigation. 4. Navigation. I. Italy. Istituto idrografico. Manuale dell'ufficiale di rotta. II. Title.
QC922.I8 1961 78-396649
79 MARC

—OBSERVERS' MANUALS

Italy. Istituto idrografico.
Atlante delle nubi : appendice al volume 2. del Manuale dell'Ufficiale di rotta. — Genova : Tipo-lit. dell'Istituto idrografico della Marina, 1961.
6, 34 p., ⌈41⌉ leaves of plates : ill. ; 18 x 25 cm.
"I.I. 3100."
1. Clouds—Observers' manuals. 2. Clouds—Atlases. 3. Aids to naivigation. 4. Navigation. I. Italy. Istituto idrografico. Manuale dell'ufficiale di rotta. II. Title.
QC922.I8 1961 78-396649
79 MARC

—PHOTOGRAPHS FROM SPACE

Breuer, Anton, 1945-
Die Bewölkungsverhältnisse des südhemisphärischen Südamerika und ihre klimageographischen Aussagemöglichkeiten : eine Untersuchung auf der Grundlage von Wettersatellitenbildern / vorgelegt von Anton Breuer. — Bonn : ⌈s.n.⌉, 1974.
194 p. : ill. ; 21 cm. GFR***
Thesis—Bonn.
Vita.
Bibliography: p. 156-168
1. Clouds—South America. 2. Clouds—Photographs from space. 3. Meteorological satellites. I. Title.
QC924.S65B73 79-372504
79 MARC

Bunsch-Makarewicz, Zofia.
O możliwości wprowadzenia informacji z satelitarnych zdjęć TV do ilościowej prognozy opadu deszczu dla obszaru Polski południowej / Zofia Bunsch-Makarewicz, Maria Feleksy-Bielak. — Warszawa : Instytut Meteorologii i Gospodarki Wodnej, 1978.
22 p. ; 21 cm. — (Materiały badawcze : Seria Meteorologia) P***
Summary also in English and Russian.
Bibliography: p. 18-19.
1. Precipitation forecasting—Poland. 2. Astronautics in meteorology—Poland. 3. Clouds—Photographs from space. I. Feleksy-Bielak, Maria, joint author. II. Title. III. Series.
QC925.4.P6B86 78-399949
79 MARC

—EUROPE—CHARTS, DIAGRAMS, ETC.

Malberg, Horst.
Ein Beitrag zur Bewölkungsklimatologie Europas und des Atlantiks anhand von Satellitenaufnahmen / von Horst Malberg. — Berlin : Reimer, 1977.
11 p. : ill. ; 30 cm. — (Meteorologische Abhandlungen : Serie A, Monographien ; n.F., Bd. 1, Heft 1) GFR78-A38
Summary also in English.
Bibliography: p. 22-23.
DM45.00
1. Clouds—Europe—Charts, diagrams, etc. 2. Clouds—North Atlantic region—Charts, diagrams, etc. 3. Clouds—South Atlantic region—Charts, diagrams, etc. 4. Astronautics in meteorology. I. Title. II. Series.
QC924.E85M34 551.5'76 79-340786
79 MARC

—HAWAII—DIURNAL VARIATION

Schroeder, Thomas A
Diurnal variation in rainfall and cloudiness / by Thomas A. Schroeder, Bernard J. Kilonsky, Bernard N. Meisner. — Honolulu : Water Resources Research Center, University of Hawaii, 1977.
viii, 67 p. : ill. ; 28 cm. — (Technical report - Water Resources Research Center, University of Hawaii ; no. 112)
"UHMET 77-03."
"Partial project completion report for Studies of mesoscale rainfall regimes utilizing satellite imagery, phase 1."
Includes bibliographies.
1. Rain and rainfall—Hawaii—Diurnal variation. 2. Clouds—Hawaii—Diurnal variation. I. Kilonsky, Bernard J., joint author. II. Meisner, Bernard N., joint author. III. Title. IV. Series: Hawaii. University, Honolulu. Water Resources Research Center. Technical report ; no 112.
TC1.H36 no. 112 553'.7'09969 s 79-620855
⌈QC925.1.U8⌉ 79 MARC

—JAPAN—TOCHIGI (PREFECTURE
—PICTORIAL WORKS

Suzuki, Shōichirō, 1916-
(Tochigi no kumo)
栃木の雲 / 鈴木正一郎. ― 鹿沼：月刊さつき研究社：発売元　栃の葉書房, 1976.
192 p. : chiefly ill. ; 17 x 19 cm. ―（しもつけ文庫 ; 3） Ja 77-5683
Bibliography: p. 192.
Series romanized: Shimotsuke bunko.
¥1200
1. Clouds—Japan—Tochigi (Prefecture)—Pictorial works. I. Title.
QC921.3.S93 78-811153

—NORTH ATLANTIC REGION—
CHARTS, DIAGRAMS, ETC.

Malberg, Horst.
Ein Beitrag zur Bewölkungsklimatologie Europas und des Atlantiks anhand von Satellitenaufnahmen / von Horst Malberg. — Berlin : Reimer, 1977.
11 p. : ill. ; 30 cm. — (Meteorologische Abhandlungen : Serie A, Monographien ; n.F., Bd. 1, Heft 1) GFR78-A38
Summary also in English.
Bibliography: p. 22-23.
DM45.00
1. Clouds—Europe—Charts, diagrams, etc. 2. Clouds—North Atlantic region—Charts, diagrams, etc. 3. Clouds—South Atlantic region—Charts, diagrams, etc. 4. Astronautics in meteorology. I. Title. II. Series.
QC924.E85M34 551.5'76 79-340786
79 MARC

CLOUDS (Continued)

—SOUTH AMERICA

Breuer, Anton, 1945-
Die Bewölkungsverhältnisse des südhemisphärischen Südamerika und ihre klimageographischen Aussagemöglichkeiten : eine Untersuchung auf der Grundlage von Wettersatellitenbildern / vorgelegt von Anton Breuer. — Bonn : ₍s.n.₎, 1974.
194 p. : ill. ; 21 cm. GFR***
Thesis—Bonn.
Vita.
Bibliography: p. 156-168
1. Clouds—South America. 2. Clouds—Photographs from space. 3. Meteorological satellites. I. Title.
QC924.S65B73 79-372504
79 MARC

—SOUTH ATLANTIC REGION—CHARTS, DIAGRAMS, ETC.

Malberg, Horst.
Ein Beitrag zur Bewölkungsklimatologie Europas und des Atlantiks anhand von Satellitenaufnahmen / von Horst Malberg. — Berlin : Reimer, 1977.
11 p. : ill. ; 30 cm. — (Meteorologische Abhandlungen : Serie A, Monographien ; n.F., Bd. 1, Heft 1) GFR78-A38
Summary also in English.
Bibliography: p. 22-23.
DM45.00
1. Clouds—Europe—Charts, diagrams, etc. 2. Clouds—North Atlantic region—Charts, diagrams, etc. 3. Clouds—South Atlantic region—Charts, diagrams, etc. 4. Astronautics in meteorology. I. Title. II. Series.
QC924.E85M34 551.5'76 79-340786
79 MARC

CLOUDS HILL, ENG.

National Trust for Places of Historic Interest or Natural Beauty.
Clouds Hill, Dorset / ₍National Trust₎. — ₍New ed.₎. — ₍London₎ : The Trust, 1977.
22 p., ₍4₎ p. of plates : ill., port. ; 22 cm. GB77-33528
£0.35
1. Lawrence, Thomas Edward, 1888-1935. 2. Clouds Hill, Eng. 3. Great Britain. Royal Air Force—Biography. 4. Air pilots, Military—Great Britain—Biography. 5. Air pilots, Military—Great Britain—Biography. 6. European War, 1914-1918—Biography. I. Title.
D568.4.L45N37 1977 940.4'15'0924 79-313106
79 MARC

CLOUGH, ARTHUR HUGH, 1819-1861

—CORRESPONDENCE

Arnold, Matthew, 1882-1888.
The letters of Matthew Arnold to Arthur Hugh Clough / edited with an introductory study by Howard Foster Lowry. — Philadelphia : R. West, 1977.
xi, 191 p. ; 26 cm.
Reprint of the 1932 ed. published by Oxford University Press, London and New York.
Includes bibliographical references and index.
ISBN 0-8492-1509-9 : $30.00
1. Arnold, Matthew, 1822-1888—Correspondence. 2. Clough, Arthur Hugh, 1819-1861—Correspondence. 3. Authors, English—19th century—Correspondence. I. Clough, Arthur Hugh, 1819-1861. II. Lowry, Howard Foster, 1901-1967. III. Title.
₍PR4023.A47C6 1977₎ 821'.8 77-14265
77 MARC

—CRITICISM AND INTERPRETATION

Houghton, Walter Edwards, 1904-
The poetry of Clough : an essay in revaluation / by Walter E. Houghton. — New York : Octagon Books, 1979, c1963.
xii, 236 p. ; 24 cm.
Reprint of the ed. published by Yale University Press, New Haven.
Bibliography: p. ₍229₎-230.
Includes index.
ISBN 0-374-93982-9 : $14.00
1. Clough, Arthur Hugh, 1819-1961—Criticism and interpretation. I. Title.
₍PR4459.H6 1979₎ 821'.8 78-27119
78 MARC

CLOUGH, CARMAN G.

Blough, Carman G
Carman G. Blough, his professional career and accounting thought / edited by William G. Shenkir ; with a pref. — New York : Arno Press, 1978.
ca. 350 p. in various pagings : ill. ; 24 cm. — (The Development of contemporary accounting thought)
Reprint of papers published 1937-1974.
ISBN 0-405-10931-8 : $25.00
1. Accounting—United States—History. 2. Clough, Carman G. 3. Accountants—United States—Biography. I. Shenkir, William G. II. Title. III. Series.
HF5616.U5B48 1978 657 77-87319
78 MARC

CLOUGHLEY, MAURICE R.

Cloughley, Maurice R
A world to the west : a voyage around the world / Maurice R. Cloughley. — New York : D. McKay, c1979.
xv, 366 p. : ill. ; 24 cm.
ISBN 0-679-51453-8
1. Nanook (Ketch) 2. Cloughley, Maurice R. 3. Voyages around the world—1951- I. Title.
G420.N33C57 910'.41'0924 79-261
79 MARC

CLOUTIER, FRANÇOIS

Cloutier, François.
L'enjeu : mémoires politiques, 1970-1976 / François Cloutier. — ₍Montréal₎ : Stanké, c1978.
171 p. ; 21 cm. C***
ISBN 0-88566-117-6 : $7.75
1. Québec (Province)—Politics and government—1960- . 2. Federal government—Canada. 3. Bilingualism—Canada. 4. Cloutier, François. I. Title.
F1053.2.C57 971.4'04 78-390544
79 MARC

CLOVE TRADE

—TANZANIA—ZANZIBAR

Clove Growers Association.
Cloves from Zanzibar. — Zanzibar : Clove Growers' Association, ₍19—₎
₍20₎ p. : ill. ; 28 cm.
Cover title.
1. Clove trade—Tanzania—Zanzibar. I. Title.
HD9211.T34C57 1900z 79-100277
79 MARC

CLOVERLEAF (HIGHWAY ENGINEERING)
see Roads—Interchanges and intersections

CLOW CORPORATION

—HISTORY—ADDRESSES, ESSAYS, LECTURES

Rinehart, Raymond G
Clow Corporation : 100 years of service to the water and wastewater industries / Raymond G. Rinehart. — New York : Newcomen Society in North America, 1979.
24 p. : ill. ; 23 cm. — (Newcomen publication ; no. 1092)
Address delivered at the "1978 Chicago dinner" of the Newcomen Society in North America, Sept. 28, 1978.
1. Clow Corporation—History—Addresses, essays, lectures. I. Title. II. Series.
TD23.R56 338.7'68'176 79-84286
00 MARC

CLOWNS

see also Comedians; Pierrot

Sanders, Toby.
How to be a compleat clown / Toby Sanders ; illustrated by David Schure and the author. — New York : Stein and Day, 1978.
xii, 270 p. : ill. ; 24 cm.
Bibliography: p. 259-265.
Includes index.
ISBN 0-8128-2508-X
1. Clowns. I. Title.
GV1828.S36 791.3'3 78-7520
78 MARC

T͡Sirk, tol'ko t͡sirk! / ₍sostavitel' L. G. Kukso₎. — Moskva : Sov. Rossii͡a, 1978.
69 p. : ill. ; 21 cm. — (Bibliotechka V pomoshch' khudozhestvennoĭ samodei͡atel'nosti ; No. 13) (Ėstrada ; vyp. 2) USSR***
0.30rub
1. Clowns. I. Kukso, Leonid. II. Series: Bibliotechka V pomoshch' khudozhestvennoĭ samodei͡atel'nosti ; 1978, no. 13. III. Series: Ėstrada ; 1978, vyp. 2.
PG3242.B5 1978, no. 13 79-354500
₍PN1955₎ 79 MARC

—BIOGRAPHY

Coryn, Émile, 1914-
Mimile le clown / Alain Laville, Émile Coryn. — ₍Paris₎ : Hachette, 1978.
286 p. ; 23 cm. F78-9509
On cover: Pierre Bonte présente.
"C'est Émile Coryn qui parle dans ces pages ... Avec la collaboration d'Alain Laville, il raconte son existence vagabonde, de cirque en cirque et de ville en ville."
"ISSN 0152-1594."
ISBN 2-01-004586-6 : 38.00F
1. Coryn, Émile, 1914- 2. Clowns—Biography. I. Laville, Alain. II. Title.
GV1811.C64A35 78-395872
79 MARC

—SONGS AND MUSIC—BIBLIOGRAPHY—CATALOGS

San Antonio. Public Library.
A bibliography of 19 century clown songsters of the Harry Hertzberg Circus Collection of the San Antonio Public Library. — ₍s.l. : s.n., pref. 1962₎ (San Antonio : Carleton Print. Co.)
₍4₎ p. ; 25 cm.
Pref. signed: Leonard V. Farley.
"No. 18."
Colored ill. inserted.
1. Songs, American— Bibliography—Catalogs. 2. Clowns—Songs and music—Bibliography—Catalogs. 3. San Antonio. Public Library—Catalogs. 4. Hertzberg, Harry—Circus collection. 5. Circus—Collectibles—Texas—San Antonio. I. Hertzberg, Harry. II. Farley, Leonard V., 1908- III. Title.
ML136.S25H43 016.7846 78-321609
79 MARC

—GREAT BRITAIN—BIOGRAPHY

Findlater, Richard, 1921-
Joe Grimaldi, his life and theatre / Richard Findlater. — 2d ed. — Cambridge, ₍Eng.₎ ; New York : Cambridge University Press, 1978.
260 p., ₍6₎ leaves of plates : ill. ; 22 cm.
First published in 1955 under title: Grimaldi, king of clowns.
Includes bibliographical references and index.
ISBN 0-521-22221-4 : $20.00. ISBN 0-521-29407-X pbk. : $8.00
1. Grimaldi, Joseph, 1779-1837. 2. Clowns—Great Britain—Biography. I. Title.
GV1811.G7F56 1978 791.3'3'0924 78-7465
78 MARC

—RUSSIA

Zguta, Russell.
Russian minstrels : a history of the skomorokhi / Russell Zguta. — ₍Philadelphia₎ : University of Pennsylvania Press, 1978.
xv, 160 p. : ill. ; 24 cm.
Includes bibliographical references and index.
ISBN 0-8122-7753-8
1. Clowns—Russia. I. Title.
PN1955.Z47 791.1'2 78-53331
78 MARC

CLOYES, FRANCE (EURE-ET-LOIR)

Beillard, Joseph.
Cahiers du temps jadis : Cloyes-sur-le-Loir, le prieuré d'Yron, Montigny-le-Gannelon / Joseph Beillard, ... ; ₍publié par le Syndicat d'initiative de Cloyes-sur-le-Loir₎. — Cloyes-sur-le-Loir (28220) : Syndicat d'initiative, 1974.
90 p. : ill. ; 22 cm. F77-10753
Cover title.
Includes bibliographical references.
20.00F
1. Cloyes, France (Eure-et-Loir) 2. Yron (Priory) 3. Montigny-le-Gannelon, France. I. Title.
DC801.C6B44 944'.51 79-353937
79 MARC

CLOYNE, IRE. (PARISH)

—HISTORY

The Book of Cloyne = Leabhar Chluain Uamha / compiled and published by Cloyne Historical and Archaeological Society ; edited by Padraigh O Loingsigh. — ₍Cloyne, Ire.₎ : The Society, 1977.
167, ₍4₎ p. : ill. ; 25 cm. Ir***
Bibliography: p. ₍171₎
£3.50
1. Cloyne, Ire. (Parish)—History. I. Ó Loingsigh, Pádraig. II. Cloyne Historical and Archaeological Society. III. Title: Leabhar Chluain Uamha.
DA995.C59B66 941.95 79-307787
79 MARC

CLUB ALPINO ITALIANO. SPEDIZIONE ITALIANA AL KARAKORÙM, 1954

Compagnoni, Achille, 1914-
Uomini sul K2. Testo: Achille Compagnoni. Fotografie: Achille Compagnoni. ⌈1 ed. Milano⌉ Veronelli ⌈1958⌉
76 p. illus., ports. 17 x 25 cm. (Collano "Gli sport e gli sportivi," 1)
1. Mountaineering—Pakistan—K2 (Mountain) 2. K2 (Mountain, Pakistan—Description. 3. Club alpino italiano. Spedizione italiana al Karakorùm, 1954. I. Title.
DS485.K2C65 59-30425
⌈r79⌉rev MARC

Desio, Ardito, 1897-
Ascent of K2, second highest peak in the world; translated by David Moore. London, Elek Books ⌈1955⌉
239 p. illus. 23 cm.
Translation of La conquista del K2.
1. Mountaineering—Pakistan—K2 (Mountain) 2. K2 (Mountain), Pakistan—Description. 3. Club alpino italiano. Spedizione italiana al Karakorùm, 1954. I. Title.
DS485.K2D422 1955 915.42 56-27608
⌈r79⌉rev MARC

Desio, Ardito, 1897-
Libro bianco; in margine alla conquista del K2. ⌈1. ed. Milano⌉ Garzanti ⌈1956⌉
122 p. 22 cm.
1. Mountaineering—Pakistan—K2 (Mountain) 2. K2 (Mountain), Pakistan—Description. 3. Club alpino italiano. Spedizione italiana al Karakorùm, 1954. I. Title.
DS485.K2D427 56-36015
⌈r79⌉rev MARC

Desio, Ardito, 1897-
Victory over K2, second highest peak in the world; translated by David Moore. New York, McGraw-Hill ⌈1956⌉
273 p. illus. 22 cm.
1. Mountaineering—Pakistan—K2 (Mountain) 2. K2 (Mountain), Pakistan—Description. 3. Club alpino italiano. Spedizione italiana al Karakorùm, 1954. I. Title.
DS485.K2D422 1956 915.46 56-8171
⌈r79⌉rev MARC

CLUB ATHLÉTIQUE BRIVISTE

—HISTORY

Degoulange, Thierry.
Brive, un grand parmi les grands du rugby / Thierry Degoulange .... — ⌈Brive⌉ (7, rue Rodin, 19100) : ⌈T. Degoulange⌉, 1976.
133 p., ⌈32⌉ p. of plates ; 24 cm. F77-12222
42.00F
1. Club athlétique briviste—History. I. Title.
GV945.6.C57D44 796.33'3'06 78-367316
78 MARC

CLUB DE FÚTBOL BARCELONA

—HISTORY

Calvet i Mata, Rossend.
Historia del F.C. Barcelona : de 1899 a 1977 / por Rossend Calvet i Mata. — Barcelona : Editorial Hispano Europea, c1978.
335 p., ⌈8⌉ leaves of plates : ill. ; 20 cm. — (Serie TF, Técnicas de fútbol) (Colección Herakles) Sp***
On cover: Historia del Fútbol Club Barcelona.
ISBN 8425504961
1. Club de Fútbol Barcelona—History. I. Title. II. Title: Historia del Fútbol Club Barcelona.
GV943.6.B3C34 796.33'463'094672 79-349512
79 MARC

CLUB DER BEGEGNUNG

—ADDRESSES, ESSAYS, LECTURES

Club der Begegnung.
5 ⌈i.e. Fünf⌉ Jahre CDB : Verein zur Förderung Zeitgenöss. Kunst : ⌈Ausstellung 1974/75 / hrsg. vom Verein zur Förderung Zeitgenöss. Kunst, Club d. Begegnung, Linz ; Red., Lydia Rauch⌉. — Linz : Stadtmuseum Linz im Nordico, ⌈1974⌉
⌈82⌉ p. : ill., ports. ; 20 x 20 cm. — (Kataloge zu Ausstellungen - Stadtmuseum Linz ; 13) Au77-9-17
Not in trade.
1. Club der Begegnung—Addresses, essays, lectures. 2. Verein zur Förderung Zeitgenössischer Kunst—Addresses, essays, lectures. 3. Arts, Austrian—Austria—Exhibitions. 4. Arts, Modern—20th century—Austria—Exhibitions. 5. Artists—Austria—Biography. I. Rauch, Lydia. II. Verein zur Förderung Zeitgenössischer Kunst. III. Linz, Austria. Stadtmuseum. IV. Title. V. Series: Linz, Austria. Stadtmuseum. Kataloge zu Ausstellungen ; 13.
AM101.L58552A3 no. 13 79-356434
069'.09436'2 s
⌈NX548.A1⌉ *79⌈r79⌉rev MARC

CLUB DES CORDELIERS

Mathiez, Albert, 1874-1932.
Le club des Cordeliers pendant la crise de Varennes et le massacre du Champ de Mars : documents en grande partie inédits, publiés avec des éclaircissements et des notes / Albert Mathiez. — Genève : Slatkine-Mégariotis Reprints, 1975.
392 p. : ill. ; 22 cm. Sw***
Reprint of the 1910 ed. published by H. Champion, Paris.
100F
1. Club des Cordeliers. 2. France—Politics and government—1774-1793. I. Title.
HS2385.C58M37 1975 944.04'1 79-352084
79 MARC

CLUB-HOUSES

see also Fishing lodges

CLUB SIRIO

Domínguez Dibb, Humberto.
Presencia y vigencia arabes en el Paraguay / Humberto Domínguez Dibb. — Asunción : ⌈s.n.⌉, 1977.
127 p. : ill. ; 20 cm.
1. Arabs in Paraguay. 2. Club Sirio.
F2699.A72D65 989.2'004'927 78-106314
79 MARC

CLUBS

see also Athletic clubs; Book clubs; Boys—Societies and clubs; Girls—Societies and clubs; Riding clubs; Social group work; Societies; Women—Societies and clubs; Youth—Societies and clubs; and subdivision Clubs under names of cities, countries, regions, etc.; and names of individual clubs

—ACCOUNTING

Kuusiola, Arto, 1952-
Perusyhdistysten kirjanpidon ja taloudenhoidon ohjeet / Arto Kuusiola. — 2. tarkistettu painos. — Helsinki : Kansalaiskasvatuksen keskus, ⌈1978⌉
78 p. : forms (4 fold.) ; 21 cm. — (Kansalaiskasvatuksen Keskuksen julkaisuja ; 27 ISSN 0356-7273) Fi***
Four forms inserted.
ISBN (invalid) 9519093381
1. Clubs—Accounting. 2. Youth—Finland—Societies and clubs. I. Title. II. Series: Kansaliskasvatuksen Keskus. Julkaisuja ; 27.
HF5686.C45K86 1978 79-375447
79 MARC

—MANAGEMENT

White, Ted E 1912-
Club operations and management / Ted E. White. — Boston : CBI Pub. Co., c1979.
vii, 253 p. : ill. ; 24 cm.
Includes index.
ISBN 0-8436-0783-1 : $14.95
1. Clubs—Management. I. Title.
TX911.3.M27W45 658'.91'6479 79-11864
79 MARC

—MATHEMATICAL MODELS

Berglas, Eitan.
Clubs, local public goods & transportation models / Eitan Berglas and David Pines. — Kingston, Ont. : Institute for Economic Research, Queen's University, 1978.
44 leaves ; 28 cm. — (Discussion paper - Institute for Economic Research, Queen's University ; 297 ISSN 0316-5078) C***
Bibliography: leaf 43-44.
1. Transportation—Mathematical models. 2. Public goods—Mathematical models. 3. Clubs—Mathematical models. I. Pines, David, joint author. II. Title. III. Series: Queen's University, Kingston, Ont. Institute for Economic Research. Discussion paper ; 297.
HE193.B43 363.6 79-304943
79 MARC

CLUBUL UNIREA

Morărașu, Marcel Gh
Clubul C.F.R. "Unirea" (1896-1976) : (o pagină din viața ceferiștilor pășcăneni) / ⌈lucrare redactată de conf. univ. dr. Marcel Gh. Morărașu⌉. — Iași : Comitetul Județean Iași al C.P.R., Secția de Propagandă, 1976.
95 p. : ill. ; 20 cm. R78-277
Includes bibliographical references.
lei2.10
1. Clubul Unirea. 2. Pașcani, Romania (Iași)—Intellectual life. I. Title.
DR296.P37M67 78-379232
*78 MARC

CLUJ-NAPOCA, ROMANIA

—DESCRIPTION

Șorobetea, Aurel, 1946-
Privire de pe cetățuie : ⌈oameni și locuri din Cluj-Napoca⌉ / Aurel Șorobetea. — Cluj-Napoca : "Dacia", 1978.
221 p. ; 20 cm. R78-3666
Bibliography: p. 219-⌈220⌉
lei5.50
1. Cluj-Napoca, Romania—Description. 2. Șorobetea, Aurel, 1946- I. Title.
DR296.C55S67 78-397254
*79 MARC

CLUNIA, SPAIN

Palol Salellas, Pedro de.
Clunia Sulpicia, ciudad romana; su historia y su presente, por Pedro de Palol. Burgos, Dirección General de Bellas Artes, Servicio Nacional de Excavaciones, Diputación Provincial, 1959.
115 p. illus., maps (part fold.) plans. 21 cm.
Bibliography: p. 108-114.
1. Clunia, Spain. I. Title.
DP402.C6314P34 79-377975
79 MARC

CLUNIACS

see also Benedictines

CLUNY, FRANCE (BENEDICTINE ABBEY)

—HISTORY—ADDRESSES, ESSAYS, LECTURES

Zerbi, Piero.
Tra Milano e Cluny : momenti di vita e cultura ecclesiastica nel secolo XII / Piero Zerbi. — Roma : Herder, 1978.
xiv, 417 p. : ill. ; 23 cm. — (Italia sacra ; 28) It78-Oct
Includes bibliographical references and index.
L18500
1. Milan (Archdiocese)—History—Addresses, essays, lectures. 2. Monasticism and religious orders—Middle Ages, 600-1500—Addresses, essays, lectures. 3. Cluny, France (Benedictine abbey)—History—Addresses, essays, lectures. I. Title.
BX1547.M5Z47 78-392377
*79 MARC

CLUPEA HARENGUS PALLASII see Herring

CLUPEA PALLASII see Herring

CLUPEI FORMES

see also Anchovies

CLUPEIDAE

see also Opisthonema

CLUSEN, RUTH C., 1922-

United States. Congress. Senate. Committee on Energy and Natural Resources.
Ruth C. Clusen, and Roger A. Markle nominations : hearing before the Committee on Energy and Natural Resources, United States Senate, Ninety-fifth Congress, second session ... August 18, 1978. — Washington : U.S. Govt. Print. Off., 1978.
iii, 44 p. : forms ; 24 cm.
"Publication no. 95-143."
1. Clusen, Ruth C., 1922- . 2. Markle, Roger A., 1933- . 3. United States. Dept. of Energy—Officials and employees. 4. United States. Bureau of Mines—Officials and employees. I. Title.
KF26.E55 1978k 353.008'23 78-603768
79 MARC

CLUSTER ANALYSIS

Bellacicco, Antonio.
Metodologia e tecnica della classificazione matematica / Antonio Bellacicco. — Roma : La goliardica, 1976.
224 p. ; 24 cm. It78-Aug
Errata slip inserted.
Bibliography: p. [219]-222.
L3900
1. Cluster analysis. 2. Partitions (Mathematics) I. Title.
QA278.B44 79-377362
*00 MARC

Kowalski, Adam.
Charakterystyka niektórych metod taksonomicznych / Adam Kowalski. — Warszawa : Państwowe Wydawn. Naukowe, 1977.
22 p. ; 24 cm. P***
At head of title: Polska Akademia Nauk. Instytut Podstaw Informatyki.
Summaries in English and Russian.
Bibliography: p. 21-22.
zł20.00
1. Cluster analysis. I. Title.
QA278.K68 78-377451
78 MARC

Sherman, Charles Roger, 1944-
Research and primary care : two dimensions of preference in medical school admissions : final report / Charles R. Sherman, Michael G. McShane. — Washington : Division of Operational Studies. Association of American Medical Colleges, 1978 [c1977]
vi, 33 p. ; 28 cm.
Bibliography: p. 29-30.
1. Medical colleges—United States—Classification—Statistical methods. 2. Medical colleges—United States—Admission. 3. Medical research. 4. Medical care. 5. Multidimensional scaling. 6. Cluster analysis. I. McShane, Michael G., joint author. II. Association of Medical Colleges. Division of Operational Studies. III. Title.
R745.S4923 610'.7'1173 78-108675
78 MARC

CLUSTER DWELLINGS see Cluster housing

CLUSTER HOUSING

—AUSTRALIA—WODONGA

Cocks & Carmichael.
Medium density cluster housing, Site G, Melrose Drive & Heckendorf Road, South West, Wodonga : town planning permit application : a submission on behalf of Albury Wodonga Development Corporation / prepared by Cocks & Carmichael. — [s.l. : s.n.], 1976.
14 leaves : ill. ; 30 cm. Aus***
1. Cluster housing—Australia—Wodonga. 2. Wodonga, Australia—City planning. 3. City planning—Australia. I. Albury Wodonga Development Corporation. II. Title.
NA9051.4.C62 1976 711'.4'09945 79-315068
79 MARC

—FINLAND

Mattinen, Maire.
Suomalaisia asuntoalueita 1960-75 : asuntoalueiden ympäristön laatu verrattuna Asuinympäristön suunnitteluperiaate -kirjan tavoitteisiin / Maire Mattinen. — [Helsinki] : Sisäasiainministeriö, Kaavoitus- ja rakennusosasto, [1978]
135 p. : ill. ; 30 cm. — (Tutkimus - Sisäasiainministeriö, Kaavoitus- ja rakennusosasto ; 1977:58) Fi***
Bibliography: p. 135.
1. Cluster housing—Finland. 2. Blocks (City planning)—Finland. I. Finland. Sisäasiainministeriö. Kaavoitus- ja rakennusosasto. Asuinympäristön suunnitteluperiaatteet. II. Title. III. Series: Finland. Sisäasiainministeriö. Kaavoitus- ja rakennusosasto. Tutkimus - Sisäasiainministeriö, Kaavoitus- ja rakennusosasto ; 1977:58.
NA9051.4.M37 79-380471
79 MARC

CLUSTER THEORY (NUCLEAR PHYSICS)

—CONGRESSES

International Conference on Clustering Aspects of Nuclear Structure and Nuclear Reactions, 3d, University of Manitoba, 1978.
Clustering aspects of nuclear structure and nuclear reactions (Winnipeg, 1978) / editors, W. T. H. Van Oers ... [et al.]. — New York : American Institute of Physics, 1978.
xxii, 792 p. : ill. ; 25 cm. — (AIP conference proceedings ; no. 47)
Sponsored by the International Union of Pure and Applied Physics, and others.
Includes bibliographical references and index.
ISBN 0-88318-146-0
1. Nuclear structure—Congresses. 2. Nuclear reactions—Congresses. 3. Cluster theory (Nuclear physics)—Congresses. I. Van Oers, W. T. H. II. International Union of Pure and Applied Physics. III. Title. IV. Series: American Institute of Physics. AIP conference proceedings ; no. 47.
QC793.3.S8156 1978 539.7'5 78-64942
78 MARC

CLUSTERED HOUSING see Cluster housing

CLUTCHES (MACHINERY)

see also Couplings; Magnetic clutches; and subdivision Clutches under subjects, e.g. Automobiles—Clutches; Motor-trucks—Clutches

CLUTTON-BROCK, ARTHUR, 1868-1924. SHAKESPEARE'S HAMLET

Robertson, John Mackinnon, 1856-1933.
"Hamlet" once more / by J. M. Robertson. — Philadelphia : R. West, 1978, [c1923].
191 p. ; 24 cm.
Reprint of the ed. published by R. Cobden-Sanderson, London.
Includes bibliographical references and index.
ISBN 0-8492-2345-8 : $25.00
1. Clutton-Brock, Arthur, 1868-1924. Shakespeare's Hamlet. 2. Shakespeare, William, 1564-1616. Hamlet. I. Title.
[PR2807.R58 1978] 822.3'3 78-8348
78 MARC

CLWYD, WALES, IN ART

—CATALOGS

Pratt, Derrick.
A handlist of the topographical prints of Clwyd (Denbighshire, Flintshire and Edeyrnion) / compiled by Derrick Pratt and A. G. Veysey. — Hawarden : Clwyd Record Office, 1977.
[1], 102 p., 24 p. of plates : ill., ports. ; 30 cm. — (Handlist - Clwyd Record Office ; no. 2 ISSN 0141-416X) GB78-6897
Bibliography: p. 90-94.
Includes index.
ISBN 0-904444-14-7 : £1.85
1. Clwyd, Wales, in art—Catalogs. 2. Prints—Catalogs. I. Veysey, A. G., joint author. II. Title. III. Series: Clwyd Record Office. Handlist - Clwyd Record Office ; no. 2.
NE954.3.G7P65 769'.4'99429307 79-300998
79 MARC

CLWYD RECORD OFFICE

—CATALOGS

Denbigh, Wales.
Handlist of the Denbigh Borough Records. — [Hawarden] : Clwyd Record Office, 1975.
[1], 70 p. : 1 ill., facsims. ; 30 cm. — (Handlist - Clwyd Record Office ; no. 1) GB78-20773
"County archivist: A. G. Veysey."
Includes index.
1. Denbigh, Wales—History—Sources—Bibliography—Catalogs. 2. Clwyd Record Office—Catalogs. I. Veysey, Arthur Geoffrey. II. Title. III. Series: Clwyd Record Office. Handlist - Clwyd Record Office ; no. 1.
Z2083.D38D46 1975 016.9429'37 78-326930
[DA745.D4] 79 MARC

CLYDE RIVER

—NAVIGATION

Banks, Desmond, Baron Banks, 1918-
The Clyde steamers. Edinburgh, Albyn Press [1947]
64 p. illus. 12 x 19 cm.
1. River steamers—Scotland—Clyde River. 2. Steam-navigation—Scotland—Clyde River. 3. Clyde River—Navigation. I. Title.
VM61.B3 386'.35'094141 48-23121
[r79]rev MARC

CLYRO, WALES

—DESCRIPTION

Le Quesne, A Laurence.
After Kilvert / A. L. Le Quesne. — Oxford [Eng.] ; New York : Oxford University Press, 1978.
ix, 233 p. : ill. ; 23 cm. GB78
Includes index.
ISBN 0-19-211748-3 : £4.95
1. Kilvert, Robert Francis, 1840-1879. Kilvert's diary. 2. Kilvert, Robert Francis, 1840-1879. 3. Church of England—Clergy—Biography. 4. Clergy—England—Biography. 5. Le Quesne, A. Laurence. 6. Clyro, Wales—Description. I. Title.
BX5199.K49A32 941.08 78-40173
78 MARC

CNEMIDOPHORUS LACERTOIDES

—GENETICS

Cole, Charles J
Karyotype of a South American Teiid lizard, Cnemidophorus lacertoides / Charles J. Cole, Clarence J. McCoy and Federico Achaval. — New York : American Museum of Natural History, 1979.
5 p. : ill. ; 26 cm. — (American Museum novitates ; no. 2671 ISSN 0003-0082)
Caption title.
Bibliography: p. 5.
$0.65
1. Cnemidophorus lacertoides—Genetics. 2. Karyotypes. 3. Reptiles—Genetics. 4. Reptiles—South America—Genetics. I. McCoy, Clarence J., joint author. II. Achaval, Federico, joint author. III. Title. IV. Series.
QL1.A436 no. 2671 500.9'08 s 79-103479
[QL666.L285] 79 MARC

CNIDARIA see Coelenterata

CO-EDS see Women college students

CO-HEIRS

see also Accretion (Law); Distribution of decedents' estates; Partition of decedents' estates

CO-MINISTRY see Group ministry

CO-OPERATIVE BANK OF KENYA

—PERIODICALS

Co-operative Bank of Kenya.
Annual report and statement of accounts - The Co-operative Bank of Kenya Limited.
Nairobi, Co-operative Bank of Kenya.
v. 27 cm.
Report year ends June 30.
1. Co-operative Bank of Kenya—Periodicals.
HG2089.K46C66a 334'.2'096762 79-646481
MARC-S

CO-OPERATIVE COMMONWEALTH FEDERATION

Avakumovic, Ivan.
Socialism in Canada : a study of the CCF-NDP in federal and provincial politics / Ivan Avakumovic. — Toronto : McClelland and Stewart, c1978.
316 p. ; 21 cm. C***
Bibliography: p. 295-298.
Includes index.
ISBN 0-7710-0978-X : $6.95
1. Co-operative Commonwealth Federation. 2. New Democratic Party (Canada) 3. Socialism in Canada. I. Title.
JL197.C6A92 329.9'71 78-323354
79 MARC

CO-ORDINATING COMMITTEE FOR EXPORT TO THE COMMUNIST AREA

United States. Congress. House. Committee on International Relations. Subcommittee on International Economic Policy and Trade.
Export licensing, Cocom list review proposals of the United States : hearings before the Subcommittee on International Economic Policy and Trade of the Committee on International Relations, House of Representatives, Ninety-fifth Congress, second session, June 14 and 26, 1978. — Washington : U.S. Govt. Print. Off., 1978.
iii, 122 p. : ill. ; 24 cm.
1. Export controls—United States. 2. Co-ordinating Committee for Export to the Communist Area. I. Title.
KF27.I54924 1978d 353.008'27 79-600901
79 MARC

CO-OWNERSHIP see Joint tenancy

COACH HORSES

see also Coaching

COACHELLA VALLEY, CALIF.

—ANTIQUITIES

Wilke, Philip J
Late prehistoric human ecology at Lake Cahuilla, Coachella Valley, California / Philip J. Wilke. — Berkeley : University of California, Dept. of Anthropology, 1978.
xii, 168 p. : ill. ; 28 cm. — (Contributions of the University of California Archaeological Research Facility ; no. 38)
Bibliography: p. 131-153.
1. Cahuilla Indians—Antiquities. 2. Indians of North America—California—Antiquities. 3. Coachella Valley, Calif.—Antiquities. 4. Cahuilla, Lake, Calif. 5. California—Antiquities. I. Title. II. Series: California. University. Archaeological Research Facilities. Contributions ; no. 38.
E51.C2 no. 38 930'.1 s 78-623848
[E99.C155] 79 MARC

—BIOGRAPHY

Shumway, Nina Paul.
Your desert and mine / Nina Paul Shumway ; with an introduction by Harold O. Weight. — Palm Springs, Calif. : ETC Publications, c1979.
xx, 336 p., [7] leaves of plates : ill. ; 21 cm.
ISBN 0-88280-068-X : $10.00
1. Shumway, Nina Paul. 2. Coachella Valley, Calif.—Biography. 3. Paul family. 4. Coachella Valley, Calif.—History. 5. Coachella Valley, Calif.—History. 6. Date—California—Coachella Valley. I. Title.
F868.R6S48 1979 979.4'97 78-32023
79 MARC

—HISTORY

Coachella Valley County Water District.
Coachella Valley's golden years : the early history of the Coachella Valley County Water District and stories about the discovery and development of this section of the Colorado Desert / [compiled and written by Ole J. Nordland]. — Rev. ed. — [Coachella, Calif.] : The District, 1978.
120 p. : ill. ; 28 cm.
Cover title.
$2.50
1. Coachella Valley, CA.—History. I. Nordland, Ole J. II. Title.
F868.C67C6 1978 979.4'9 78-112995
79 MARC

Shumway, Nina Paul.
Your desert and mine / Nina Paul Shumway ; with an introduction by Harold O. Weight. — Palm Springs, Calif. : ETC Publications, c1979.
xx, 336 p., [7] leaves of plates : ill. ; 21 cm.
ISBN 0-88280-068-X : $10.00
1. Shumway, Nina Paul. 2. Coachella Valley, Calif.—Biography. 3. Paul family. 4. Coachella Valley, Calif.—History. 5. Coachella Valley, Calif.—History. 6. Date—California—Coachella Valley. I. Title.
F868.R6S48 1979 979.4'97 78-32023
79 MARC

COACHES see Carriages and carts; Coaching

COACHES, TRUCK see Campers and coaches, Truck

COACHES (ATHLETICS)

see also Basketball coaches; Coaching (Athletics); Football coaches; Soccer—Coaching

—CERTIFICATION

Jeppson, Gordon D
Inservice education programs for coaches / compiled by Gordon D. Jeppson. — Washington : ERIC Clearinghouse on Teacher Education : order from AAHPER Publications Sales, 1978.
v, 42 p. ; 23 cm. — (AAHPER publications)
"SP 012 142."
1. Coaches (Athletics)—In-service training—United States. 2. Coaches (Athletics)—Certification. 3. Intramural sports. I. Title. II. Series: American Alliance for Health, Physical Education, and Recreation. AAHPER publications.
GV711.J46 796'.077 78-111725
79 MARC

—IN-SERVICE TRAINING—UNITED STATES

Jeppson, Gordon D
Inservice education programs for coaches / compiled by Gordon D. Jeppson. — Washington : ERIC Clearinghouse on Teacher Education : order from AAHPER Publications Sales, 1978.
v, 42 p. ; 23 cm. — (AAHPER publications)
"SP 012 142."
1. Coaches (Athletics)—In-service training—United States. 2. Coaches (Athletics)—Certification. 3. Intramural sports. I. Title. II. Series: American Alliance for Health, Physical Education, and Recreation. AAHPER publications.
GV711.J46 796'.077 78-111725
79 MARC

—LEGAL STATUS, LAWS, ETC.

see also Coaches (Athletics)—Certification

COACHING

see also Roads

—CALIFORNIA—HISTORY

Johnston, Francis J
The Bradshaw trail : narrative and notes / by Francis J. Johnston ; edited by John R. Brumgardt and Tom Patterson. — Riverside, Calif. : Historical Commission Press, [1976?]
vi, 215 p., [2] leaves of plates : ill. ; 20 cm.
Bibliography: p. 195-201.
Includes index.
1. Bradshaw, William D., 1826-1864. 2. Bradshaw Trail. 3. California—History—1850-1950. 4. Southwest, New—History—1848- . 5. Pioneers—California—Biography. 6. Coaching—California—History. 7. Coaching—Southwest, New—History. I. Title.
F864.B72J63 979.4'04 78-111457
79 MARC

—ENGLAND—HISTORY

Austen, Brian.
English provincial posts, 1633-1840 : a study based on Kent examples / by Brian Austen. — London [etc.] : Phillimore, 1978.
viii, 192 p. : 2 ill., facsims., maps ; 23 cm. GB78-29459
Bibliography: p. 174-179.
Includes index.
ISBN 0-85033-266-4 : £4.95
1. Postal service—England—History. 2. Coaching—England—History. I. Title.
HE6945.A97 383'.09422'3 78-325401
79 MARC

—ENGLAND—TIME-TABLES

Bates, Alan, 1929 (Sept.)-
Directory of stage coach services, 1836. New York, A. M. Kelley, 1969.
160 p. folded illus. 23 cm.
1. Coaching—England—Time-tables. I. Title.
HE5664.A45B36 388.3'22'0942 69-11233
[r79]rev2 MARC

Bates, Alan, 1929 (Sept.)-
Directory of stage coach services 1836. Newton Abbot, David & Charles, 1969.
[8], 160 p. plate, map. 23 cm. 50/- B69-12296
ISBN 0-7153-4379-3
1. Coaching—England—Time-tables. I. Title.
HE5664.A45B36 1969b 388.3'2 75-424413
69[r79]rev MARC

—SOUTHWEST, NEW—HISTORY

Johnston, Francis J
The Bradshaw trail : narrative and notes / by Francis J. Johnston ; edited by John R. Brumgardt and Tom Patterson. — Riverside, Calif. : Historical Commission Press, [1976?]
vi, 215 p., [2] leaves of plates : ill. ; 20 cm.
Bibliography: p. 195-201.
Includes index.
1. Bradshaw, William D., 1826-1864. 2. Bradshaw Trail. 3. California—History—1850-1950. 4. Southwest, New—History—1848- . 5. Pioneers—California—Biography. 6. Coaching—California—History. 7. Coaching—Southwest, New—History. I. Title.
F864.B72J63 979.4'04 78-111457
79 MARC

—WALES—HISTORY

Williams, Herbert, 1932-
Stage coaches in Wales / by Herbert Williams. — Barry : S. Williams, 1977.
120 p. : ill., facsims. ; 26 cm. GB78-05979
Bibliography: p. 118.
Includes index.
ISBN 0-900807-27-X : £4.50
1. Coaching—Wales—History. I. Title.
HE5664.A6W54 388.3'22'09429 78-319931
79 MARC

COACHING (ATHLETICS)

see also Baseball coaching; Basketball coaching; Coaches (Athletics); Field hockey coaching; Football coaching; Hockey coaching; Rugby football coaching; Soccer coaching; Softball coaching; Track-athletics; Wrestling coaching

Sabock, Ralph J
The coach / Ralph J. Sabock. — 2d ed. — Philadelphia : Saunders, 1979.
x, 309 p. : ill. ; 26 cm.
Bibliography: p. 301-305.
Includes index.
ISBN 0-7216-7891-2
1. Coaching (Athletics) 2. Coaching (Athletics)—Vocational guidance. I. Title.
GV711.S22 1979 796'.077 79-1522
79 MARC

Youth sports guide for coaches and parents / edited by Jerry R. Thomas ; editorial consultant, Albert Warson. — [Washington] : Manufactures Life Insurance Co. : available from AAHPER Publications, c1977.
134 p. : ill. ; 22 cm.
Bibliography: p. 131-134.
ISBN 0-9690538-1-9 : $3.00
1. Coaching (Athletics) 2. Physical education for children. 3. Sports for children. I. Thomas, Jerry R.
GV711.Y68 790.19'22 77-83343
78 MARC

COACHING (ATHLETICS) (Continued)

—ADDRESSES, ESSAYS, LECTURES

Toward an understanding of human performance : readings in exercise physiology for the coach and athlete / edited by Edmund J. Burke ; artwork by Tom Stearns, Deena Wickstrom, and Cheryl Babcock. — Ithaca, N.Y. : Mouvement Publications, c1977, 1978 printing.
92 p. : ill. ; 28 cm.
Includes bibliographies.
1. Physical education and training—Addresses, essays, lectures. 2. Sports—Physiological aspects—Addresses, essays, lectures. 3. Coaching (Athletics)—Addresses, essays, lectures. 4. Physical fitness—Addresses, essays, lectures. I. Burke, Edmund J.
GV711.5.T68 613.7'07 78-107650
78 MARC

—VOCATIONAL GUIDANCE

Sabock, Ralph J
The coach / Ralph J. Sabock. — 2d ed. — Philadelphia : Saunders, 1979.
x, 309 p. : ill. ; 26 cm.
Bibliography: p. 301-305.
Includes index.
ISBN 0-7216-7891-2
1. Coaching (Athletics) 2. Coaching (Athletics)—Vocational guidance. I. Title.
GV711.S22 1979 796'.077 79-1522
79 MARC

COAGULASE

see also Blood—Coagulation

COAGULATION

see also Proteins—Denaturation

COAGULATION DISORDERS see Blood—Coagulation, Disorders of

COAGULATION OF BLOOD see Blood—Coagulation

COAHUILA, MEXICO

—HISTORY—COLLECTED WORKS

Alessio Robles, Vito, 1879-1957.
Acapulco, Saltillo y Monterrey en la historia y en la leyenda / Vito Alessio Robles ; presentación de Vito Alessio Robles Cuevas. — 1. ed. en la Biblioteca Porrúa. — México : Editorial Porrúa, 1978.
xx, 670 p., ⌊41⌋ leaves of plates : ill. ; 24 cm. — (Biblioteca Porrúa ; 66)
Includes bibliographies.
CONTENTS: Acapulco en la historia y en la leyenda.—Saltillo en la historia y en la leyenda.—Monterrey en la historia y en la leyenda.—Bosquejos históricos.—Unas páginas traspapeladas de la historia de Coahuila y Texas.—La primera imprenta en Coahuila.—Heráldica coahuilense.
1. Acapulco, Mexico—History—Collected works. 2. Saltillo, Mexico—History—Collected works. 3. Monterey, Mexico—History—Collected works. 4. Coahuila, Mexico—History—Collected works. I. Title.
F1391.A15A25 1978 972'.1 78-113067
79 MARC

COAHUILA INDIANS see Cahuilla Indians

COAHUILA LANGUAGE see Cahuilla language

COAL

see also Briquets (Fuel); Carbonization; Coal mines and mining; Coke; Lignite

Berkovitch, Israel.
Coal : energy and chemical storehouse / by I. Berkovitch. — ⌊Redhill, Eng.⌋ : Portcullis Press, 1978.
vii, 140 p. : ill. ; 24 cm. GB***
Includes bibliographical references and index.
ISBN 0-86108-023-8 : £6.75
1. Coal. I. Title.
TP325.B457 662'.62 79-314005
00 MARC

Berkowitz, Norbert, 1923-
An introduction to coal technology / N. Berkowitz. — New York : Academic Press, 1979.
xiv, 345 p. : ill. ; 24 cm. — (Energy science and engineering)
Includes bibliographical references and index.
ISBN 0-12-091950-8
1. Coal. I. Title. II. Series.
TP325.B46 662'.62 78-19663
79 MARC

Buch, Alfred, 1902-
Kohle : Grundstoff d. Energie / Alfred Buch. — München : Pfriemer, 1979.
226 p. : ill., maps ; 23 cm. GFR79-A
Bibliography: p. 219-224.
Includes index.
ISBN 3-7906-0081-4 : DM42.00
1. Coal. I. Title.
TN800.B79 79-368717
79 MARC

Dybová-Jachowicz, Sonia.
Kiedy węgiel był zielony. ⌊Wyd. 1.⌋ Warszawa, Wydawn. Geologiczne, 1964.
191 p. illus., maps. 21 cm.
At head of title: S. ⌊i⌋ A. Jachowiczowie.
Bibliographical references included in "Wstęp" (p. 7-⌊8⌋)
1. Coal. I. Jachowicz, Aleksander, joint author. II. Title.
TN800.D93 65-51334
⌊r79⌋rev MARC

Ezra, Derek, Sir.
Coal and energy : the need to exploit the world's most abundant fossil fuel / Derek Ezra. — New York : Wiley, 1978.
viii, 182 p., ⌊2⌋ leaves of plates : ill. ; 23 cm.
"A Halstead Press book."
Includes index.
ISBN 0-470-26339-3
1. Coal. 2. Power resources. I. Title.
TN800.E97 1978b 333.8'2 78-5785
78⌊r79⌋rev MARC

International Energy Agency.
Steam coal : prospects to 2000. — Paris : Organisation for Economic Co-operation and Development ; ⌊Washington, D.C. : sold by OECD Publications and Information Center⌋, 1978.
157 p. : ill. ; 24 cm. F***
ISBN 9264118675 : $12.00
1. Coal. I. Title.
TN800.I49 1978 333.8'2 79-309952
79 MARC

Karolczuk, Henryk.
Racjonalna gospodarka węglem energetycznym / Henryk Karolczuk. — Wyd. 1. — Warszawa : Wydaw-a Naukowo-Techniczne, 1978 (Wroc. : WZG)
307 p. : ill. ; 21 cm. P78-10121
Errata slip inserted.
Bibliography: p. ⌊296⌋-299.
Includes index.
zł55.00
1. Coal. I. Title.
TN800.K3 79-357575
79 MARC

Kowalczuk, Jerzy M
Metody geofizyczne i geochemiczne w poszukiwaniu i rozpoznawaniu surowców stałych ze szczególnym uwzględnieniem węgla : wybrane działy z sejsmicznych metod poszukiwawczych i sejsmologii inżynierskiej / Jerzy Kowalczuk. — Wyd. 1. — Kraków : Akademia Górniczo-Hutnicza im. S. Staszica w Krakowie, 1976.
23 p. : ill. ; 24 cm. — (Materiały - Akademia Górniczo-Hutnicza im. S. Staszica w Krakowie, Wydział Geologiczno-Poszukiwawczy ; Nr 33) (Studia dyplomowe) P***
Bibliography: p. 23.
zł12.00
1. Seismic prospecting. 2. Coal. I. Title. II. Series: Kraków. Akademia Górniczo-Hutnicza. Wydział Geologiczno-Poszukiwawczy. Materiały - Akademia Górniczo-Hutnicza im. S. Staszica w Krakowie, Wydział Geologiczno-Poszukiwawczy ; nr. 33. III. Series: Studia dyplomowe.
TN271.C58K68 79-407162
79 MARC

Lindbergh, Kristina.
Coal, a contemporary energy story / by Kristina Lindbergh & Barry Provorse ; edited by Robert Conte. — Seattle : Scribe Pub. Corp., c1977.
207 p. : ill. (some col.) ; 29 cm.
Bibliography: p. 202-203.
Includes index.
ISBN 0-915748-02-9 : $22.50
1. Coal mines and mining. 2. Coal. I. Provorse, Barry L., joint author. II. Conte, Robert. III. Title.
TN802.L53 553'.2 79-103717
79 MARC

Raĭkhel', Boris L'vovich.
Ėkonomicheskaia͡ ot͡senka ugol'nykh mestorozhdeniĭ / B. L. Raĭkhel'. — Moskva : Nedra, 1979.
223, ⌊1⌋ p. ; 20 cm. USSR***
Bibliography: p. 223-⌊224⌋
0.60rub
1. Coal. 2. Mine valuation. I. Title.
TN800.R29 79-377564
79 MARC

Rohstoff Kohle : Eigenschaften, Gewinnung, Veredelung / von e. Autorenteam, F. Benthaus ... ⌊et al.⌋. — 1. Aufl. — Weinheim ; New York : Verlag Chemie, 1978.
viii, 289 p. : ill, graphs ; 25 cm. GFR78-A
Includes bibliographical references and index.
ISBN 3-527-25791-8 : DM64.00
1. Coal. I. Benthaus, Friedrich.
TP325.R65 662'.62 78-390042
79 MARC

Simeons, Charles.
Coal, its role in tomorrow's technology : a sourcebook on global coal resources / by Charles Simeons. — Oxford ; New York : Pergamon Press, 1978.
xv, 313 p. : ill. ; 26 cm. GB78
Bibliography: p. 306-308.
Includes index.
ISBN 0-08-022712-0 : £15.00 ($30.00)
1. Coal. I. Title.
TN800.S49 1978 553.2 78-40827
78 MARC

Tambang Batu Bara, P.N. Badan Pimpinan Umum.
Anda sudah kenal batubara? — Djakarta : B.P.U. Perusahaan² Tambang Batubara Negara, 1965.
⌊10⌋ p. : ill. ; 21 cm.
1. Coal. I. Title.
TP325.T27 1965 77-986606
79 MARC

Valorificarea prin ardere a combustibililor inferiori (solizi) / Nicolae Pănoiu ... ⌊et al.⌋. — București : Editura Tehnică, 1977.
349 p., ⌊1⌋ fold. leaf of plates : ill. ; 21 cm. — (Ciclul Dezvoltarea energeticii, economia de combustibil și energie) R78-1814
Bibliography: p. 341-349.
lei19.00
1. Combustion engineering. 2. Coal. I. Pănoiu, Nicolae.
TJ254.5.V34 78-397801
*79 MARC

—ANALYSIS

Analytical methods for coal and coal products / edited by Clarence Karr, Jr. — New York : Academic Press, 1978-
v. : ill. ; 24 cm.
Includes bibliographies.
ISBN 0-12-399901-4
1. Coal—Analysis. I. Karr, Clarence.
TP325.A58 662'.622 78-4928
78 MARC

Blagov, Igor' Sil'vestrovich.
(Ėkonomicheskaia͡ ėffektivnost' povysheniia͡ kachestva ugleĭ)
Экономическая эффективность повышения качества углей : ⌊учебник⌋ / И. С. Благов, В. К. Турченко. — Москва : Недра, 1978.
158 p. ; 20 cm. USSR***
0.45rub
1. Coal mines and mining—Quality control. 2. Coal—Analysis. I. Turchenko, Vasiliĭ Kuz'mich, joint author. II. Title.
TN803.B56 79-410473

Gabzdyl, Wiesław.
Wybrane zagadnienia z petrografii węgla : węgiel jako skała i surowiec / Wiesław Gabzdyl. — Gliwice : Politcehnika Śląska, 1978.
145 p. ; 24 cm. — (Skrypty uczelniane - Politechnika Śląska im. W. Pstrowskiego ; Nr 803) P***
Bibliography: p. 143-144.
zł15.00
1. Coal—Analysis. I. Title.
TP325.G32 79-351187
79 MARC

COAL

—ANALYSIS (Continued)

Guyot, R E

The application of coal petrography to hydrogenation research / by R. E. Guyot and C. F. K. Diessel. — North Ryde, N.S.W. : Australian Coal Industry Research Laboratories Ltd., 1979.

48 p. : ill. ; 25 cm. — (Published reports - Australian Coal Industry Research Laboratories Ltd. ; no. 3) Aus***

Bibliography: p. 17.

ISBN 0-909130-55-8

1. Coal—Analysis. 2. Hydrogenation. I. Diessel, C. F. K., joint author. II. Title. III. Series: Australian Coal Industry Research Laboratories Ltd. Published reports - Australian Coal Industry Research Laboratories Ltd. ; no. 3.

TP325.G85 662'.622 79-314597
79 MARC

Lau, Carlos Alberto.

Estudo estatístico da análise imediata de carvão mineral para fins de contrôle e previsão de resultados. Pôrto Alegre, Instituto Tecnológico do Rio Grande do Sul, 1971.

35 p. illus. 23 cm. (Instituto Tecnológico do Rio Grande do Sul. Boletim no. 55)

"Trabalho elaborado para o 17o. Congresso Brasileiro de Química, realizado em Pôrto Alegre de 21 a 27 de novembro de 1971."

Summary in Portuguese and English.

Bibliography: p. 35.

1. Coal—Analysis. I. Title. II. Series: Pôrto Alegre, Brazil. Instituto Tecnológico do Rio Grande do Sul. Boletim no. 55.

T4.P796 no. 55 78-385723
[TP325] 79 MARC

—ANALYSIS—CONGRESSES

Round Table Discussion Chemical and Physical Valorization of Coal, Rome, 1973.

Round-Table-Gespräch Chemische und Physikalische Veredlung von Kohle, Rom, 5. und 6. September 1973 = Round Table Discussion Chemical and Physical Valorization of Coal = Table ronde Valorisation chimique et physique du charbon. — Luxembourg : Amt für Amtliche Veröffentlichungen der Europäischen Gemeinschaften, 1974.

242 p. : ill. ; 30 cm. — (Forschungshefte Kohle ; Heft Nr. 60) Lu**

At head of title on cover: Kommission der Europäischen Gemeinschaften.

Contributions in English, French, or German.

"EUR 5215 d/e/f."

Includes bibliographical references.

DM33.80 (W. Ger.)

1. Coal—Analysis—Congresses. I. Commission of the European Communities. II. Series.

TP325.R69 1973 662'.622 78-395771
79 MARC

Symposium on Coal Evaluation, Calgary, Alta., 1974.

Symposium on Coal Evaluation, October 31-November 1, 1974, the Calgary Inn, Calgary, Alberta / sponsored by Alberta Research Council and the Coal Association of Canada ; editorial committee, J. F. Fryer, J. D. Campbell, J. G. Speight. — Edmonton : Alberta Research Council, [1976?]

182 p. : ill. ; 28 cm. — (Information series - Alberta Research Council ; 76) C77-81582-0

Includes bibliographical references.

$7.00

1. Coal—Analysis—Congresses. I. Fryer, J. F. II. Campbell, John Duncan, 1923- . III. Speight, J. G. IV. Alberta Research Council. V. Coal Association of Canada. VI. Series: Alberta Research Council. Information series - Alberta Research Council ; 76.

TP325.S95 1974 662'.622 79-307217
79 MARC

—BIBLIOGRAPHY

West Virginia. University. Coal Research Bureau.

Uranium in coal and carbonaceous rocks in the United States and Canada : an annotated bibliography / by John K. Alderman ... [et al.] ; edited by John K. Alderman. — Morgantown, W.Va. : Coal Research Bureau, College of Mineral and Energy Resources, West Virginia University, [1978?]

21 p. ; 28 cm. — (Technical report - Coal Research Bureau, West Virginia University ; no. 160)

Includes index.

1. Uranium—Bibliography. 2. Coal—Bibliography. 3. Rocks, Carbonate—Bibliography. I. Alderman, John K. II. Title. III. Series: West Virginia. University. Coal Research Bureau. Technical report - Coal Research Bureau, West Virginia University ; no. 160.

Z6679.U7W47 1978 016.553'493 79-622573
[TN490.U7] 79 MARC

—COMBUSTION—CONGRESSES

International Symposium on Coal-Oil Mixture Combustion, 1st, St. Petersburg Beach, Fla., 1978.

Proceedings of the First International Symposium on Coal-Oil Mixture Combustion, May 7-9, 1978, St. Petersburg Beach, Florida / sponsored by the United States Department of Energy and the General Motors Corporation ; proceedings editors, J. Coleman Blake, Alberto J. Sabadell. — [McLean, Va.] : Mitre Corp., Metrek Division, [1978]

vi, 417 p. : ill. ; 29 cm.

"CONF-7805141; M78-97."

Bibliography: p. 412-417.

1. Combustion engineering—Congresses. 2. Coal—Combustion—Congresses. 3. Petroleum as fuel—Combustion—Congresses. I. Blake, J. Coleman. II. Sabadell, Alberto J. III. United States. Dept. of Energy. IV. General Motors Corporation.

TJ254.5.I56 1978 621.4'023 78-78325
79 MARC

—COMBUSTION—ENVIRONMENTAL ASPECTS

Trace contaminants from coal / edited by S. Torrey. — Park Ridge, N.J. : Noyes Data Corp., 1978.

x, 294 p. : ill. ; 25 cm. — (Pollution technology review ; no. 50)

Includes bibliographical references.

ISBN 0-8155-0724-0 : $39.00

1. Trace elements—Environmental aspects. 2. Coal—Combustion—Environmental aspects. I. Torrey, S. II. Noyes Data Corporation. III. Series.

TD196.T7T72 662'.62 78-61890
79 MARC

—CONGRESSES

Coal, coke and the blast furnace : proceedings of an international conference / organized by the Metals Society and held in Middlesborough, England, on 15-17 June 1977. — London : Metals Society, 1978.

[5], 152 p. : ill. ; 31 cm. — (Book - Metals Society ; 191) GB79-01991

Includes bibliographical references and indexes.

ISBN 0-904357-15-5 : £22.50

1. Blast-furnaces—Congresses. 2. Coal—Congresses. 3. Coke—Congresses. I. Metals Society. II. Series: Metals Society. Book - Metals Society ; 191.

TN713.C7 669'.1413 79-314482
79 MARC

Institute on Scientific Problems Relevant to Coal Utilization, West Virginia University, 1977.

Scientific problems of coal utilization : proceedings of a conference at West Virginia University, Morgantown, West Virginia, May 23-25, 1977 / edited by Bernard R. Cooper ; sponsored by West Virginia University and Division of Materials and Exploratory Research, Fossil Energy, Energy Research and Development Administration. — Oak Ridge, Tenn. : Technical Information Center, U.S. Dept. of Energy ; Springfield, Va. : available from National Technical Information Service, U.S. Dept. of Commerce,

xii, 409 p. : ill. ; 23 cm. — (DOE symposium series ; 46)

"CONF-770509"

Includes bibliographical references and index.

ISBN 0-87079-400-0

1. Coal—Congresses. I. Cooper, Bernard R. II. West Virginia. University. III. United States. Energy Research and Development Administration. Division of Materials and Exploratory Research. IV. Title. V. Series: United States. Dept. of Energy. DOE symposium series ; 46.

TP325.I678 1977 662'.62 78-9553
78 MARC

International Coal Exploration Symposium, 1st, London, 1976; 2d, Denver, 1978.

Coal exploration : proceedings of the first International Coal Exploration Symposium, London, England, May 18-21, 1976 / edited by William L. G. Muir. — San Francisco : M. Freeman Publications, c1976-c1979.

2 v. : ill. ; 24 cm. — (A World coal book)

Proceedings of the second International Coal Exploration Symposium, Denver, Colorado, USA, October 1978, edited by George O. Argall, Jr.

Includes bibliographical references and indexes.

ISBN 0-87930-061-2 (v. 1)

1. Prospecting—Congresses. 2. Coal—Congresses. I. Muir, William L. G. II. Argall, George Oates. III. Title. IV. Series: World coal book.

TN271.C58I57 1978 622'.18'2 76-41151
77[r79]rev MARC

International Conference on Coal Science, 6th, Münster, 1965.

[Papers]. — [s.l. : s.n., 1965?]

62 pieces in 2 v. : ill. ; 30 cm.

English, French, or German.

Includes bibliographical references.

1. Coal—Congresses.

TN799.9.I59 1965 553'.2 78-325903
79 MARC

Symposium on Field Description of Coal, Ottawa, Ont., 1976.

Field description of coal : a symposium / sponsored by ASTM Committee D-5 on Coal and Coke, American Society for Testing and Materials, Ottawa, Ontario, Canada, 22-23 Sept. 1976 ; R. R. Dutcher, editor. — Philadelphia : The Society, c1978.

71 p. : ill. ; 23 cm. — (ASTM special technical publication ; 661)

Includes bibliographical references and index.

$7.50

1. Coal—Congresses. I. Dutcher, Russell R. II. American Society for Testing and Materials. Committee D-5 on Coal and Coke. III. Title. IV. Series: American Society for Testing and Materials. Special technical publication ; 661.

TN799.9.S95 1976 611'.18'2 78-59996
79 MARC

—GASIFICATION see Coal gasification

—GEOLOGY

see also Coal balls

Chiang-su sheng mei t'ien ti chih wu ts'e tui.

(Mei t'ien ti chen k'an t'an)

煤田地震勘探 / 江苏省煤田地质物测队，湖南省煤田地质物测队合编. — 第1版. — 北京：煤炭工业出版社，1977.

iii, 211 p. : ill. ; 19 cm.

$0.56

1. Seismic prospecting. 2. Coal—Geology. I. Hu-nan sheng mei t'ien ti chih wu ts'e tui. II. Title.

TN269.C44 1977 78-843100

Timofeev, Petr Petrovich.

(Ûrskaâ uglenosnaâ formaŝiâ Tuvinskogo mezhgornogo progiba)

Юрская угленосная формация Тувинского межгорного прогиба. Москва, Наука, 1964.

260 p. illus., maps, plates. 27 cm. (Академия наук СССР. Геологический институт. Труды, вып. 94)

At head of title: П. П. Тимофеев.

Added t. p.: Jurassic coal-bearing formation of Tuvinsk intermontane depression.

Table of contents also in English.

Bibliography: p. 236-[242]

1. Geology, Stratigraphic—Jurassic. 2. Coal—Geology. 3. Geology—Tannu-Tuva. I. Title. II. Series: Akademiâ nauk SSSR. Geologicheskiĭ institut (1956-) Trudy, vyp. 94.

QE1.A342 vyp. 94 66-98375
[QE681]

—GEOLOGY—POLAND—LUBLIN REGION—BIBLIOGRAPHY

Morawski, Jan.

Bibliografia Lubelskiego Zagłębia Węglowego / Jan Morawski, Jerzy Nowak. — Lublin : Uniwersytet Marii Curie-Skłodowskiej, Instytut Nauk o Ziemi, Zakład Geologii, 1977 (Lub.: UMCS)

58, [4] p. : ill. ; 21 cm. P77-7558

Includes bibliographical references and index.

1. Geology—Poland—Lublin region—Bibliography. 2. Coal—Geology—Poland—Lublin region—Bibliography. I. Nowak, Jerzy, joint author. II. Title.

Z6034.P7M67 78-384021
[QE276.5] 78 MARC

—GEOLOGY—POLAND—LUBUSKA REGION

Szwed-Lorenz, Jarosława.

Budowa petrograficzna i charakterystyka geologiczna węgli brunatnych złoża Sieniawa (Ziemia Lubuska) / Jarosława Szwed-Lorenz. — Wrocław : Wydaw. PW, 1976 (Wroc. : PW)

89 p. : ill. ; 25 cm. — (Prace naukowe Instytutu Górnictwa Politechniki Wrocławskiej ; 20) (Monografie Instytutu Górnictwa Politechniki Wrocławskiej ; 8) (Prace naukowe Politechniki Wrocławskiej) P76-3305

On p. [2] of cover: Petrographic and geologic characteristic of brown coal in Sieniawa.

Summary in English and Russian; table of contents also in English.

Bibliography: p. 82-87.

zł20.00

1. Lignite—Poland—Lubuska region. 2. Coal—Geology—Poland—Lubuska region. I. Title. II. Title: Petrographic and geologic characteristic of brown coal in Sieniawa. III. Series: Politechnika Wrocławska. Instytut Górnictwa. Prace naukowe ; 20. IV. Series: Politechnika Wrocławska. Instytut Górnictwa. Monografie ; 8.

TN275.A1B7 nr. 20 79-412720
[TN808.P72L87] 79 MARC

—GEOLOGY—PRINCE EDWARD ISLAND

Howie, R D

Geological studies and evaluation of MacDougall core hole 1A, Western Prince Edward Island / R. D. Howie. — Ottawa : Energy, Mines and Resources Canada, 1977.

26 p. : ill. ; 28 cm. — (Paper - Geological Survey of Canada ; 77-20) C***

Abstract in French.

Bibliography: p. 26.

ISBN 0-660-00843-2 : $3.00

1. Geology, Stratigraphic—Paleozoic. 2. Geology—Prince Edward Island. 3. Coal—Geology—Prince Edward Island. 4. Borings—Prince Edward Island. I. Title. II. Series: Canada. Geological Survey. Paper ; 77-20.

QE185.A42 No. 77-20 557.1'08 s 78-323014
[QE654] 79 MARC

COAL (Continued)

—GEOLOGY—ROCKY MOUNTAIN REGION—CONGRESSES

Symposium on the Geology of Rocky Mountain Coal, 2d, Colorado School of Mines, 1977.
Proceedings of the Second Symposium on the Geology of Rocky Mountain Coal, 1977 / sponsored by Colorado Geological Survey and U.S. Geological Survey ; Helen E. Hodgson, editor. — Denver : Colorado Geological Survey, Dept. of Natural Resources, State of Colorado, 1978.
viii, 219 p. : ill. ; 28 cm. — (Resource series - Colorado Geological Survey ; 4)
Includes bibliographies.
$5.00
1. Coal—Geology—Rocky Mountain region—Congresses. I. Hodgson, Helen E. II. Colorado. Geological Survey. III. United States. Geological Survey. IV. Series: Colorado. Geological Survey. Resource series - Colorado Geological Survey ; 4.
TN805.A5S96 1977 553'.24'0978 78-623149
79 MARC

—INDEXES

Hernández de Caldas, Angela.
Bibliografía sobre energética-carbón : índice de autores y geográfico / por Angela Hernández de Caldas y María Teresa Ramírez Vargas. — ₍s.l. : s.n., 1977?₎ (₍Bogotá : Editorial El Catolicismo₎)
15 p. ; 24 cm.
1. Power resources—Indexes. 2. Coal—Indexes. I. Ramírez Vargas, María Teresa, joint author. II. Title.
TJ163.2.H46 79-373683
79 MARC

—JUVENILE LITERATURE

Harter, Walter L
Coal : the rock that burns / Walter Harter. — 1st ed. — Nashville : Elsevier/Nelson Books, c1979.
128 p. : ill. ; 21 cm.
Includes index.
SUMMARY: Describes coal, where it's found, how its's recovered, who mines it, and its many uses.
ISBN 0-525-76609-2 : $6.95
1. Coal—Juvenile literature. ₍1. Coal. 2. Coal mines and mining₎ I. Title.
TN801.H18 553'.2 78-27745
78 MARC

—LIQUEFACTION see Coal liquefaction

—METHANE CONTENT

Ėttinger, Iosif L'vovich.
(Raspredelenie metana v porakh iskopaemykh ugleĭ)
Распределение метана в порах ископаемых углей / И. Л. Эттингер, Н. В. Шульман ; АН СССР, Сектор физ.-техн. горных проблем Ин-та физики Земли им. О. Ю. Шмидта. — Москва : Наука, 1975.
112 p. : ill. ; 21 cm. USSR 75
Bibliography: p. 108-111.
0.75rub
1. Coal—Methane content. I. Shul'man, Nadezhda Vladimirovna, joint author. II. Title.
TP325.E79 79-414379

—PERMEABILITY

Kuznet͡sov, Sergeĭ Vasil'evich.
(Prirodnai͡a pronit͡saemost' ugol'nykh plastov i metody ee opredelenii͡a)
Природная проницаемость угольных пластов и методы ее определения / С. В. Кузнецов, Р. Н. Кригман. — Москва : Наука, 1978.
122 p. : ill. ; 21 cm. USSR 78
At head of title: Akademii͡a nauk SSSR. Institut problem kompleksnogo osvoenii͡a nedr.
Bibliography: p. 114-₍120₎
0.95rub
1. Coal—Permeability. I. Krigman, Raisa Naumovna, joint author. II. Title.
TN800.K89 79-407725

—PIPE LINES

Systems study and assessment of methacoal as a coal transportation alternative : final report / edited by John H. Vanston, Jr., D. G. Hannan ; ₍contributions by₎ J. Wesley Barnes ... ₍et al.₎. — ₍Austin₎ : Center for Energy Studies, University of Texas at Austin, 1978.
vi, 139, ₍74₎ p. : ill. ; 28 cm. — (Research report - Center for Energy Studies, University of Texas at Austin ; no. 11)
Bibliography: p. 133-139.
1. Coal—Pipe lines. 2. Slurry. 3. Hydraulic conveying. 4. Methanol. I. Vanston, John H. II. Hannan, D. G. III. Barnes, J. Wesley. IV. Series: Texas. University at Austin. Center for Energy Studies. Research report - Center for Energy Studies, University of Texas at Austin ; no. 11.
TJ1405.S97 388.5 78-622863
79 MARC

United States. Congress. Office of Technology Assessment.
A technology assessment of coal slurry pipelines. — Washington : Congress of the United States, Office of Technology Assessment : for sale by the Supt. of Docs., U.S. Govt. Print. Off., 1978.
xv, 155 p. : ill. ; 26 cm.
"OTA-E-60."
Includes bibliographical references.
1. Coal—Pipe lines. I. Title.
TJ1405.U54 1978 388.5 78-600023
78 MARC

—PIPE LINES—CONGRESSES

International Technical Conference on Slurry Transportation, 3d, Las Vegas, Nev., 1978.
Proceedings / Third International Technical Conference on Slurry Transportation, March 29-31, 1978, Las Vegas, Nevada ; ₍sponsored by Slurry Transport Association, Battelle Memorial Institute, Electric Power Research Institute₎. — ₍Washington₎ : The Association, c1978.
vi, 223 p. : ill. ; 28 cm.
Includes bibliographical references.
ISBN 0-932066-03-8
1. Hydraulic conveying—Congresses. 2. Slurry—Transportation—Congresses. 3. Coal—Pipe lines—Congresses. I. Slurry Transport Association. II. Battelle Memorial Institute, Columbus, Ohio. III. Electric Power Research Institute.
TJ898.I58 1978 662'.624 78-52717
78 MARC

International Technical Conference on Slurry Transportation, 4th, Las Vegas, Nev., 1979.
Proceedings / Fourth International Technical Conference on Slurry Transportation, March 28-30, 1979, Las Vegas, Nevada. — ₍Washington : Slurry Transport Association, 1979₎
iv, 248 p. : ill. ; 28 cm.
Includes bibliographies.
ISBN 0-932066-04-6
1. Hydraulic conveying—Congresses. 2. Slurry—Transportation—Congresses. 3. Coal—Pipe lines—Congresses. I. Slurry Transport Association.
TJ898.I58 1979 662'.624 79-63391
00 MARC

—PIPE LINES—LAW AND LEGISLATION—UNITED STATES

United States. Congress. House. Committee on Interstate and Foreign Commerce. Subcommittee on Transportation and Commerce
Report of the Subcommittee on Transportation and Commerce on coal slurry pipeline and H.R. 1609, the Coal pipeline act of 1978, for the consideration of the Committee on Interstate Foreign Commerce : prepared for the use of the Committee on Interstate and Foreign Commerce, U.S. House of Representatives. — Washington : U.S. Govt. Print. Off., 1978.
iii, 63 p. ; 24 cm.
At head of title: 95th Congress, 2d session. Committee print, 95-56.
Includes bibliographical references.
1. Coal—Pipe lines—Law and legislation—United States. 2. Eminent domain—United States. I. Title.
KF2398.A25 1978a 343'.73'093 78-603589
79 MARC

United States. Congress. House. Committee on Interstate and Foreign Commerce. Subcommittee on Transportation and Commerce.
Transportation by slurry pipeline : hearings before the Subcommittee on Transportation and Commerce of the Committee on Interstate and Foreign Commerce, House of Representatives, Ninety-fifth Congress, second session, on H.R. 6248, H.R. 6643, and H.R. 10663 ... February 7 and 16, 1978. — Washington : U.S. Govt. Print. Off., 1978.
iv, 224 p. ; 23 cm.
"Serial no. 95-104."
1. Coal—Pipe lines—Law and legislation—United States. I. Title.
KF27.I5589 1978b 343'.73'093 78-602844
78 MARC

United States. Congress. House. Committee on Public Works and Transportation. Subcommittee on Surface Transportation.
Coal pipeline act of 1978 : hearings before the Subcommittee on Surface Transportation of the Committee on Public Works and Transportation, House of Representatives, Ninety-fifth Congress, second session, on H.R. 1609, to amend the Mineral leasing act of 1920, and for other purposes, March 22 and April 11, 1978. — Washington : U.S. Govt. Print. Off., 1978.
vi, 283 p. ; 24 cm.
"(95-59)"
1. Coal—Pipe lines—Law and legislation—United States. I. Title.
KF27.P8966 1978 343'.73'093 79-600903
79 MARC

United States. Congress. Senate. Committee on Energy and Natural Resources. Subcommittee on Public Lands and Resources.
Coal pipeline act : hearings before the Subcommittee on Public Lands and Resources of the Committee on Energy and Natural Resources, United States Senate, Ninety-fifth Congress, second session, on S. 707 ... S. 3046 ... May 17, 25, and June 19, 1978. — Washington : U.S. Govt. Print. Off., 1978.
iv, 539 p. : ill. ; 24 cm.
"Publication no. 95-136."
Includes bibliographical references.
1. Coal—Pipe lines—Law and legislation—United States. I. Title.
KF26.E558 1978a 343'.73'093 78-603326
79 MARC

—PRICES—GREAT BRITAIN

Great Britain. Price Commission.
Prices and margins in the distribution of coal and solid smokeless fuels for domestic consumption / Price Commission. — London : H.M.S.O., 1976.
vi, 29 p. ; 25 cm. — (Report - Price Commission ; no. 12) GB76-08924
Cover title: Distribution of domestic coal and solid smokeless fuel.
ISBN 0-11-700297-6 : £0.65
1. Coal trade—Great Britain. 2. Fuel trade—Great Britain. 3. Coal—Prices—Great Britain. 4. Fuel—Prices—Great Britain. I. Title. II. Title: Distribution of domestic coal and solid smokeless fuel. III. Series: Great Britain. Price Commission. Report - Price Commission ; no. 12.
HD9551.5.G64 1976 338.4'3'6626 78-326865
79 MARC

—PRICES—UNITED STATES

United States. Council on Wage and Price Stability.
A study of coal prices / Executive Office of the President, Council on Wage and Price Stability. — ₍Washington₎ : The Council : for sale by the Supt. of Docs., U.S. Govt. Print. Off., 1976.
xi, 97 p. : ill. ; 26 cm. — (Staff report - Council on Wage and Price Stability)
Cover title.
Includes bibliographical references.
1. Coal—Prices—United States. 2. Coal trade—United States. I. Title. II. Series: United States. Council on Wage and Price Stability. Staff report - Council on Wage and Price Stability.
HD9544.U5 1976 338.2'3 76-601379
76₍r79₎rev MARC

—RESERVES

Fettweis, Günter Bernhard.
World coal resources : methods of assessment and resources / by Günter B. Fettweis. — Amsterdam ; New York : Elsevier Scientific Pub. Co. ; New York : distributor for the U.S. and Canada, Elsevier North-Holland, 1979.
415 p. : ill. ; 25 cm. — (Development in economic geology ; v. 10)
Translation of Weltkohlenvorräte.
Includes bibliographies and index.
ISBN 0-444-99779-2
1. Coal—Reserves. I. Title. II. Series.
TN800.F4613 553.2 79-12167
79 MARC

—STATISTICS—PERIODICALS

Keystone news bulletin.
New York, Mining Informational Services of the McGraw-Hill Mining Publications.
v. 28 cm. monthly.
Key title: Keystone news bulletin, ISSN 0149-5801.
1. Coal trade—United States—Statistics—Periodicals. 2. Coal—Statistics—Periodicals.
HD9541.K45 338.2'7'20973 79-643452
MARC-S

COAL (Continued)

—TAXATION—GREAT PLAINS

Stinson, Thomas F 1942-
Coal development in the northern Great Plains : the impact on revenues of State and local governments / Thomas F. Stinson, Stanley W. Voelker ; Economics, Statistics, and Cooperatives Service, U.S. Dept. of Agriculture, in cooperation with the U.S. Environmental Protection Agency, the University of Minnesota, and North Dakota State University. — Washington : The Department, 1978.
ii, 66 p. : 26 cm. — (Agricultural economic report ; no. 394)
Cover title.
Supt. of Docs. no.: A 1.107: 394.
1. Coal trade—Great Plains. 2. Coal—Taxation—Great Plains. 3. Local finance—Great Plains. I. Voelker, Stanley W., joint author. II. United States. Dept. of Agriculture. Economics, Statistics, and Cooperatives Service. III. Title. IV. Series.
HD1751.A91854 no. 394 338.1'08 s 78-601276
[HD9547.A14] 78 MARC

—TAXATION—MONTANA

Montana. Legislative Assembly. Coal Tax Oversight Committee.
Report of the Coal Tax Oversight Committee : report to the Forty-sixth Legislature. — Helena : Montana Legislative Council, 1978.
78 p. in various pagings ; 28 cm.
1. Coal—Taxation—Montana. I. Montana. Legislative Assembly.
KFM9011.62.C6 1978 353.9'786'00722 79-622222
00 MARC

—TESTING

Zimmerman, Raymond E
Evaluating and testing the coking properties of coal / Raymond E. Zimmerman. — San Francisco : Miller Freeman Publications, c1979.
144 p. : ill. ; 27 cm. — (A World coal book)
Includes bibliographical references and index.
ISBN 0-87930-111-2
1. Coal—Testing. 2. Coke—Testing. I. Title. II. Series: World coal book.
TP325.Z55 662'.6224 79-84399
00 MARC

—TRANSPORTATION

see also Coal—Pipe lines

—TRANSPORTATION—CONGRESSES

Critical issues in coal transportation systems : proceedings of symposium / Maritime Transportation Research Board, Commission on Sociotechnical Systems, National Research Council. — Washington : National Academy of Sciences, 1979.
x, 358 p. : ill. ; 28 cm.
Includes bibliographical references.
ISBN 0-309-02869-8
1. Coal—Transportation—Congresses. 2. Coal—United States—Congresses. I. National Research Council. Maritime Transportation Research Board.
HE595.C6C68 380.5 78-78294
79 MARC

—ALASKA

McGee, Don L
Mineral resources of Alaska and the impact of Federal land policies on their availability : coal / by Don L. McGee and Kristina M. O'Connor. — Anchorage : State of Alaska, Dept. of Natural Resources, Division of Geological and Geophysical Surveys, Energy Production Section, 1975.
26 p. : maps ; 22 x 36 cm. — (Alaska open file report ; 51)
Bibliography: p. 26.
1. Coal—Alaska. 2. Land use—United States. I. O'Connor, Kristina M., joint author. II. Title. III. Series.
TN805.A7M2 553'.2'09798 79-624564
79 MARC

—ALASKA—ALASKA PENINSULA

Conwell, Cleland N
Herendeen Bay—Chignik coals, southern Alaska peninsula / by C. N. Conwell and D. M. Triplehorn. — College, Alaska : Division of Geological & Geophysical Surveys, 1978.
iii, 15 p. : ill. ; 28 cm. — (Special report - Division of Geological & geophysical surveys ; 8)
Bibliography: p. 15.
1. Coal—Alaska—Alaska Peninsula. I. Triplehorn, D. M., joint author. II. Title. III. Series: Alaska. Division of Geological and Geophysical Surveys. Special report - Division of Geological & Geophysical Surveys ; 8.
TN805.A7C73 553'.24'097984 78-623014
78 MARC

—ALBERTA

Alberta. Dept. of Energy and Natural Resources.
A coal development policy for Alberta. — [Edmonton, Alta.] : Dept. of Energy and Natural Resources, Government of Alberta, 1976.
38 p., [1] fold. leaf of plates : ill. ; 28 cm. C***
1. Coal—Alberta. 2. Coal trade—Alberta. 3. Power resources—Alberta. I. Title.
HD9554.C3A532 333.8'2 78-314241
78 MARC

Campbell, John Duncan, 1923-
Coal resources, Tofield-Donalda area, Alberta / J. D. Campbell. — Edmonton : Alberta Research Council, 1975.
64 p. : ill., maps (3 fold. in pocket) ; 25 cm. — (Report - Alberta Research Council ; 75-8) C***
Errata slip inserted.
Bibliography: p. 8-9.
$3.00
1. Coal—Alberta. I. Title. II. Series: Research Council of Alberta. Report ; 75-8.
TN806.C3A422 553'.24'097123 79-303895
79 MARC

—ALBERTA—TRANSPORTATION—COSTS—MATHEMATICAL MODELS

Krefting, Doug, 1941-
Rail systems operation/cost model of coal unit trains : southeastern British Columbia and western Alberta coal / prepared by D. Krefting. — [Vancouver] : Ministry of Economic Development, 1978.
135, 48, 34 p. : ill. ; 28 cm. C***
Report by Swan Wooster Engineering Co. for Transportation Sub-committee on North East Coal Development.
Includes bibliographical references.
ISBN 0-7718-8076-6
1. Coal—British Columbia—Transportation—Costs—Mathematical models. 2. Coal—Alberta—Transportation—Costs—Mathematical models. 3. Railroads—British Columbia—Cost of operation. 4. Railroads—Alberta—Cost of operation. I. Swan Wooster Engineering Co. II. British Columbia. Transportation Sub-committee on North East Coal Development. III. Title.
HE2321.C6K73 385'.1'09712 79-311392
79 MARC

—AMERICA—ADDRESSES, ESSAYS, LECTURES

Coal resources of the Americas : selected papers / edited by Frank E. Kottlowski, Aureal T. Cross, A. A. Meyerhoff. — Boulder, Colo. : Geological Society of America, c1978.
vi, 90 p. : ill. ; 28 cm. — (Special paper - Geological Society of America ; 179)
Papers presented to a symposium planned by the Coal Geology Division of the Geological Society of America held in Mexico City in 1968.
Includes bibliographies.
ISBN 0-8137-2179-2
1. Coal—America—Addresses, essays, lectures. I. Kottlowski, Frank Edward. II. Cross, Aureal T. III. Meyerhoff, Arthur A., 1928- . IV. Geological Society of America. Coal Geology Division. V. Series: Geological Society of America. Special papers ; 179.
TN805.A1C62 553'.2'098 78-57145
78 MARC

—AUSTRALIA—NEW SOUTH WALES—ANALYSIS

Marshall, Charles Edward.
Vitrain, fusain : a study of the variations in constitution, gas- and coke-making characteristics of important seam constituents and types of certain permian coal seams of New South Wales / by C. E. Marshall and A. Draycott. — [Sydney] : University of Sydney, Dept. of Geology and Geophysics, 1954.
iv, 68 p. : ill. ; 26 cm. — (Memoir - University of Sydney, Department of Geology and Geophysics ; 1954/3)
Bibliography: p. 34.
Includes index.
1. Coal—Australia—New South Wales—Analysis. I. Draycott, A., joint author. II. Title. III. Series: Sydney. University. Dept. of Geology and Geophysics. Memoir ; 1954/3.
TP326.A8M37 79-316565
79 MARC

—BRAZIL—RIO GRANDE DO SUL

Nahuys, Joanna.
Carvão de Candiota, suas características químicas e petrográficas / trabalho apresentado no 17o Congresso Brasileiro de Química, realizado em Pôrto Alegre de 21 a 27 de novembro de 1971 por Joanna Nahuys e Maria Regina de S. Câmara. — Pôrto Alegre, Brasil : Estado do Rio Grande do Sul, Instituto Tecnológico do Rio Grande do Sul, 1972.
46 p. : ill. ; 23 cm. — (Boletim - Instituto Tecnológico do Rio Grande do Sul ; no. 59)
Summary in French.
Bibliography: p. 45-46.
1. Coal—Brazil—Rio Grande do Sul. I. Câmara, Maria Regina de S. II. Title. III. Series: Pôrto Alegre, Brazil. Instituto Tecnológico do Rio Grande do Sul. Boletim ; no. 59.
T4.P796 no. 59 78-371161
[TP326.B7] 78 MARC

—BRITISH COLUMBIA

British Columbia. Resource Sub-committee on Northeast Development.
Report on coal resource evaluation studies, January, 1977 / Resource Sub-committee on Northeast Development. — [Victoria, B.C.] : Ministry of Mines and Petroleum Resources, [1977]
v, 126 p. : ill. ; 28 cm. C***
At head of title: Northeast Coal Study.
Bibliography: p. [50]
1. Coal—British Columbia. I. Title.
TN806.C3B742 553'.24'097111 79-301892
79 MARC

—BRITISH COLUMBIA—TRANSPORTATION—COSTS—MATHEMATICAL MODELS

Krefting, Doug, 1941-
Rail systems operation/cost model of coal unit trains : southeastern British Columbia and western Alberta coal / prepared by D. Krefting. — [Vancouver] : Ministry of Economic Development, 1978.
135, 48, 34 p. : ill. ; 28 cm. C***
Report by Swan Wooster Engineering Co. for Transportation Sub-committee on North East Coal Development.
Includes bibliographical references.
ISBN 0-7718-8076-6
1. Coal—British Columbia—Transportation—Costs—Mathematical models. 2. Coal—Alberta—Transportation—Costs—Mathematical models. 3. Railroads—British Columbia—Cost of operation. 4. Railroads—Alberta—Cost of operation. I. Swan Wooster Engineering Co. II. British Columbia. Transportation Sub-committee on North East Coal Development. III. Title.
HE2321.C6K73 385'.1'09712 79-311392
79 MARC

Swan Wooster Engineering Co.
A study of options for railroad access to the Peace River coal fields : final report / prepared for Ministry of Economic Development, Government of British Columbia by Swan Wooster Engineering Co., Ltd. — Rev. ed. — [Vancouver] : Province of British Columbia, Ministry of Economic Development, 1977.
2 v. : ill. ; 28 cm. C***
CONTENTS: v. 1. Main report and system analysis.—v. 2. Description of rail system operations/cost model.
ISBN 0-7718-8016-2 (v. 1)
1. Railroads—British Columbia—Cost of operation—Mathematical models. 2. Railroads—British Columbia—Cost of construction—Mathematical models. 3. Coal—British Columbia—Transportation—Costs—Mathematical models. I. British Columbia. Ministry of Economic Development. II. Title.
HE2809.B7S96 1977 385'.1'097111 78-319899
79 MARC

—CANADA

Canada. Dept. of Energy, Mines and Resources.
1976 assessment of Canada's coal resources and reserves. — [Ottawa] : Energy, Mines and Resources Canada, c1977.
vi, 20 p. : ill. ; 25 cm. — (Report EP ; 77-5) C***
Bibliography: p. 19.
ISBN 0-662-01265-8
1. Coal—Canada. I. Title. II. Series: Canada. Dept. of Energy, Mines and Resources. Report EP ; 77-5.
TN806.C2C35 1977 553'.2'0971 78-318433
78 MARC

—CHINA

Ikonnikov, A B
The coal industry of China / [by] A. B. Ikonnikov. — Canberra : Dept. of Economics, Research School of Pacific Studies, Australian National University, 1977.
xi, 204 p. : maps ; 24 cm. Aus
Includes bibliographical references.
ISBN 0-909524-14-9 : $9.50
1. Coal—China. 2. Coal mines and mining—China. I. Title.
HD9556.C52I38 338.2'72'0951 78-317060
79 MARC

COAL (Continued)

—CHINA—TRANSPORTATION

Minami Manshū Tetsudō Kabushiki Kaisha. Hokushi Jimukyoku, Peking. Chōsashitsu.
(Daidō tan zōsan ni taisuru Kei-Sui Tetsudō zōkyō taisaku)
大同炭増産ニ對スル京綏鐵道(特ニ南ロパス)増强對策　北京　滿鐵北支事務局調査室　昭和13 ⌊1938⌋
135 p. 27 cm.
Cover title.
At head of title: 極秘 (rubber stamped)
Prepared by 伊藤清司 and others.
L. C. copy imperfect: 2 maps wanting.
1. Coal—China—Transportation. 2. P'ing Sui t'ieh lu. I. Itō, Seiji. II. Title.
HE2321.C6M56 1938 79-803416

—COLUMBIA—CAUCA VALLEY

Auteco.
Cauca Valley coal survey. — Brussels : Auteco, ⌊196-⌋
xxiii, 693 p. : ill. ; 26 cm.
At head of title: Republic of Colombia.
1. Coal—Columbia—Cauca Valley. 2. Coal trade—Columbia—Cauca Valley. I. Title.
TN806.C72C38 338.2'7'2098615 78-325905
79 MARC

—EUROPEAN ECONOMIC COMMUNITY COUNTRIES—STATISTICS —PERIODICALS

Kulstatistik. Kohlestatistik. Coal statistics. 1975– ⌊Luxembourg, De europæiske fællesskabers statistiske kontor⌋
v. 30 cm. annual.
Continues: Statistical Office of the European Communities. Kohlestatistik.
English, French, and German.
1. Coal trade—European Economic Community countries—Statistics—Periodicals. 2. Coal—European Economic Community countries—Statistics—Periodicals. I. Statistical Office of the European Communities. II. Title: Kohlestatistik. III. Title: Coal statistics.
HD9555.E82S75b 338.2'7'2094 78-646437
MARC-S

—GREAT BRITAIN—CARBONIZATION— ADDRESSES, ESSAYS, LECTURES

Lee, G W
Improvement of the economy of the coke-oven process / G. W. Lee, J. P. Graham, and W. J. Pater. — ⌊s. l. : s. n., 1966⌋
9 p. ; 30 cm.
Caption title.
At head of title: Congrès international de Charleroi. Le coke en sidérurgie. B5.
Bibliography: p. 8-9.
1. Coal—Great Britain—Carbonization—Addresses, essays, lectures. 2. Coke-ovens—Addresses, essays, lectures. 3. Coke industry—Great Britain—Addresses, essays, lectures. I. Graham, J. P., joint author. II. Pater, W. J., joint author. III. Title.
TP336.L43 79-316502
79 MARC

—ILLINOIS

Simon, Jack Aaron.
Coal, Illinois' major fuel resource : a report to the Technical Committee of the Illinois Energy Resources Commission / by Jack A. Simon. — Springfield : The Commission, 1977.
vi, 166 p. : ill. ; 29 cm.
Includes bibliographical references.
1. Coal—Illinois. I. Illinois. Energy Resources Commission. Technical Committee. II. Title.
TP326.U6S88 333.8'2 78-622114
79 MARC

Smith, Janet Marian, 1952-
Who's mining the farm : a report on the energy corporations' ownership of Illinois land and coal reserves and its implications for rural people and their communities / by Janet M. Smith, David Ostendorf, and Mike Schechtman. — ⌊Herrin, Ill.⌋ : Illinois South Project, ⌊c1978⌋
76 p. : ill. ; 28 cm.
Cover title.
Bibliography: p. 72-75.
$4.00
1. Coal—Illinois. 2. Land tenure—Illinois. 3. Land use, Rural—Illinois. I. Ostendorf, David, joint author. II. Schechtman, Mike, joint author. III. Illinois South Project. IV. Title.
HD242.3.S58 333.8'2 79-105174
79 MARC

Smith, William Henking, 1919-
Coal and water resources for coal conversion in Illinois / William H. Smith and John B. Stall. — Urbana, Ill. : State Water Survey, 1975.
79 p. : ill., maps (3 fold. col. in pocket) ; 28 cm. — (Cooperative resources report ; 4)
Includes bibliographies.
1. Coal—Illinois. 2. Water-supply—Illinois. 3. Coal gasification—Illinois. 4. Coal liquefaction—Illinois. I. Stall, John Byron, joint author. II. Title. III. Series.
TN805.I3S62 553'.24'09773 75-623963
79 MARC

Treworgy, Colin G
Reserves and resources of surface-minable coal in Illinois / Colin G. Treworgy, Lawrence E. Bengal, and Amy G. Dingwell. — Urbana : Illinois State Geological Survey, 1978.
iv, 44 p. : ill. ; 28 cm. — (Circular - Illinois State Geological Survey ; 504)
Bibliography: p. 18.
1. Coal—Illinois. 2. Strip mining—Illinois. I. Bengal, Lawrence E., joint author. II. Dingwell, Amy G., joint author. III. Title. IV. Series: Illinois. State Geological Survey. Circular ; 504.
QE105.A45 no. 504 557.73'08 s 78-624072
⌊TN805.I3⌋ 79 MARC

—ILLINOIS—STATISTICS

Weil, Keith.
The Illinois coal digest / by Keith Weil, Alvin K. Grandys. — ⌊Springfield⌋ : Illinois Dept. of Business and Economic Development, Division of Energy, 1976.
ii leaves, 127 p. : ill. ; 29 cm.
Cover title.
"IDOE-RS-76-06."
1. Coal trade—Illinois—Statistics. 2. Coal—Illinois—Statistics. I. Grandys, Alvin K., joint author. II. Illinois. Division of Energy. III. Title.
HD9547.I3W44 338.2'7'2409773 78-622247
79 MARC

—INDIA

India (Republic). Geological Survey.
A glimpse into the past and present activities of the coal survey of Geological Survey of India / by B. Laskar, Dy. Director General. — ⌊Calcutta : G.S.I. Press⌋, 1976.
14 p., ⌊2⌋ leaves of plates (1 fold.) : ill., map ; 25 cm. — (GSI pamphlet ; no. 16/76)
Cover title.
"125th anniversary celebrations, Geological Survey of India (estd. 1851)."
Bibliography: p. 12-14.
1. Coal—India. I. Laskar, B. II. Title. III. Series: India (Republic). Geological Survey. GSI pamphlet ; no. 16/76.
TN809.I4I53 1976 622'.18'20954 78-911817
79 MARC

—IOWA—TRANSPORTATION

Baumel, C Phillip.
Economics of alternative coal transportation and distribution systems in Iowa / by C. Phillip Baumel, Thomas P. Drinka, and John J. Miller. — Ames : Agriculture and Home Economics Experiment Station, Iowa State University of Science and Technology, 1978.
60 p. : ill. ; 29 cm. — (Special report - Agriculture and Home Economics Experiment Station, Iowa State University of Science and Technology ; no. 81 ISSN 0578-6258)
Bibliography: p. 38-39.
1. Coal—Iowa—Transportation. I. Drinka, Thomas P., joint author. II. Miller, John Joseph, 1950- joint author. III. Title. IV Series: Iowa. Agricultura and Home Economics Experiment Station, Ames. Special report ; no. 81.
HE199.5.C6B38 380.5'2 79-621028
79 MARC

—MEXICO

Mexico. Comisión de Energéticos.
El carbón mineral / Comisión de Energéticos. — ⌊México : La Comisión⌋, 1975.
ii, 111 p. : ill. ; 21 cm.
Bibliography: p. 95.
1. Coal—Mexico. I. Title.
TN806.M6M49 1975 79-347657
79 MARC

Salas, Guillermo P
Geología, reservas y petrografía de los carbones mexicanos para la siderurgía / por Guillermo P. Salas, Jesús Ojeda Rivera. — México : Consejo de Recursos Naturales No Renovables, 1969.
viii, 22 p., ⌊5⌋ leaves of plates : maps ; 23 cm. — (Boletín - Consejo de Recursos Naturales No Renovables ; 70)
1. Coal—Mexico. 2. Metallurgy—Mexico. I. Ojeda Rivera, Jesús, joint author. II. Title. III. Series: Mexico. Consejo de Recursos Naturales No Renovables. Boletín ; 70.
TN28.A336 no. 70 79-346479
⌊TN806.M6⌋ 79 MARC

—NORTHWESTERN STATES— TRANSPORTATION

United States. Congress. Senate. Committee on Energy and Natural Resources.
Impact reduction of energy resources act of 1979 : hearing before the Committee on Energy and Natural Resources, United States Senate, Ninety-sixth Congress, first session, on S. 967 ... April 30, 1979. — Washington : U.S. Govt. Print. Off., 1979.
iii, 49 p. ; 23 cm.
"Publication no. 96-19."
1. Railroad law—United States. 2. Railroads—United States—Abandonment. 3. Coal—United States—Transportation. 4. Railroads—Northwestern States. 5. Coal—Northwestern States—Transportation. I. Title.
KF26.E55 1979g 343'.73'0958 79-603125
79 MARC

—OHIO

Couchot, Michael L
Resources of the Pittsburgh (no. 8) coal in the Belmont field, Ohio / by Michael L. Couchot ; ⌊cartographers, R. Anne Berry and James A. Brown⌋. — Columbus : State of Ohio, Dept. of Natural Resources, Division of Geological Survey, 1978.
iii, 12 p. : ill. ; 28 cm. — (Report of investigations - State of Ohio, Division of Geological Survey ; no. 105)
Four maps inserted.
1. Coal—Ohio. I. Berry, R. Anne. II. Brown, James Allen, 1936- . III. Title. IV. Series: Ohio. Division of Geological Survey. Report of investigations ; no. 105.
QE151.A186 no. 105 557.71'08 s 78-624295
⌊TN805.O3⌋ 79 MARC

—OHIO—ANALYSIS

Knapp, Norman F
Trace elements in Ohio coals / by Norman F. Knapp. — Columbus : State of Ohio, Dept. of Natural Resources, Division of Geological Survey, 1977.
iii, 12 p. ; 28 cm. — (Report of investigations - State of Ohio, Division of Geological Survey ; no. 103)
Bibliography: p. 11-12.
1. Coal—Ohio—Analysis. 2. Trace elements. I. Title. II. Series: Ohio. Division of Geological Survey. Report of investigations ; no. 103.
QE151.A186 no. 103 557.71'08 s 78-624293
⌊TP326.U6⌋ 79 MARC

—OHIO—ANALYSIS—TABLES

Botoman, George.
Analyses of Ohio coals / by George Botoman and David A. Stith. — Columbus : State of Ohio, Dept. of Natural Resources, Division of Geological Survey, 1978.
v, 148 p. ; 28 cm. — (Information circular - State of Ohio, Division of Geological Survey ; no. 47)
Chiefly tables (⌊5⌋ fold. in pocket)
Bibliography: p. 1.
1. Coal—Ohio—Analysis—Tables. I. Stith, David A., joint author. II. Title. III. Series: Ohio. Division of Geological Survey. Information circular ; no. 47.
QE151.A183 no. 47 557.71'08 s 78-624292
⌊TP326.U6O3⌋ 79 MARC

—OKLAHOMA

Ghebremedhin, Tesfa.
An analysis of Oklahoma's coal industry / Tesfa Ghebremedhin and Michael S. Salkin. — ⌊Stillwater⌋ : Agriculture ⌊sic⌋ Experiment Station, Oklahoma State University, 1978.
41 p. : graphs ; 28 cm. — (Research report - Agriculture ⌊sic⌋ Experiment Station, Oklahoma State University ; P-780)
Bibliography: p. 25.
1. Coal trade—Oklahoma. 2. Coal—Oklahoma. I. Salkin, Michael S., joint author. II. Title. III. Series: Oklahoma. Agricultural Experiment Station, Stillwater. Research report ; P-780.
HD9547.O5G47 338.2'7'2409766 79-621191
79 MARC

COAL (Continued)

—POLAND—SILESIA, LOWER —ADDRESSES, ESSAYS, LECTURES

Jasieńko, Stefan.
Dolnośląskie Zagłębie Węglowe : geneza, właściwości i struktura węgli kamiennych / Stefan Jasieńko, Adolf Pilawa. — Wrocław : Wydaw. PW, 1977 (Wroc. : PW)
48 p. : ill. ; 25 cm. — (Prace naukowe Instytutu Chemii i Technologii Nafty i Węgla Politechniki Wrocławskiej ; 29) (Seria Monografie ; 8) P77-6512
On p. [2] of cover: Lower Silesian Coal Basin.
Table of contents also in English; summaries in English and Russian.
Bibliography: p. 44-47.
zł20.00
1. Coal—Poland—Silesia, Lower—Addresses, essays, lectures. I. Pilawa, Adolf, joint author. II. Title. III. Title: Lower Silesian Coal Basin. IV. Series: Politechnika Wrocławska. Instytut Chemii i Technologii Nafty i Węgla. Prace naukowe ; 29. V. Series: Politechnika Wrocławska. Instytut Chemii i Technologii Nafty i Węgla. Monografie ; 8.
TP690.A1B74 nr. 29 78-367937
[TN808.P72S5] 78 MARC

—SASKATCHEWAN

Irvine, J A
Coal resources of southern Saskatchewan : a model for evaluation methodology / prepared by J. A. Irvine, S. H. Whitaker, P. L. Broughton ; with contributions by A. R. Cameron & A. R. Sweet. — [s.l. : s.n.], 1978.
2 v. : ill. ; 29 cm. and atlas (56 leaves of plates ; col. ill. ; 39 x 56 cm. — (Geological Survey of Canada Ec Geol report ; 30) (Department of Mineral Resources report ; 209) (Saskatchewan Research Council report ; 20) C***
Abstract also in French and Russian.
"Appendix 2, resource estimates by coal zone area": microfishe (2 sheets) in pocket.
Bibliography: v. 1, p. 129-130.
1. Coal—Saskatchewan. I. Whitaker, S. H., joint author. II. Broughton, P. L., joint author. III. Title. IV. Series: Canada. Geological Survey. Economic geology report ; 30. V. Series: Saskatchewan. Dept. of Mineral Resources. Report ; 209. VI. Series: Saskatchewan Research Council. Report ; 20.
QE185.A4 no. 30 553'.0971 s 79-310060
[TN806.C3S27] 79 MARC

—SOUTH AFRICA—ANALYSIS

Savage, W H D
Analyses of product samples taken by the Fuel Research Institute during 1972 / by W. H. D. Savage. — [Pretoria] : Fuel Research Institute of South Africa, [1973?]
14 p. ; 25 cm. — (Bulletin - Fuel Research Institute of South Africa ; no. 17)
Cover title.
Chiefly tables.
ISBN 0-620-00459-2
1. Coal—South Africa—Analysis. I. Title. II. Series: Fuel Research Institute of South Africa. Bulletin ; no. 17.
TP315.F82 no. 17 662'.6'08 s 79-310451
[TP326.S7] 79 MARC

—SUDETES

Lipiarski, Ireneusz.
Litologia pokładów węgla i petrografia węgla złoża Słupiec w Sudeckim Zagłębiu węgla kamiennego / Ireneusz Lipiarski. — Kraków : Zakład Narodowy im. Ossolińskich, 1976.
64 p., [7] leaves of plates : ill. ; 24 cm. — (Prace geologiczne ; 102) P***
At head of title: Polska Akademia Nauk. Oddział w Krakowie.
On p. facing t.p.: Russian title, Litologiíà ugol'nykh plastov i petrografiíà ugliâ mestorozhdeniíà Slupets v Sudetskom Kamennougol'nom basseĭne; English title, The lithology of coal seams and the petrography of coal from the Słupiec deposit in the Sudetic Coal Basin.
Summary in English; legends also in English.
Bibliography: p. [59]
zł20.00
1. Coal—Sudetes. I. Title. II. Title: Litologiíà ugol'nykh plastov i petrografiíà ugliâ mestorozhdeniíà Slupets w Sudetskom Kamennougol'nom basseĭne. III. Title: The lithology of coal seams and the petrography of coal from the Słupiec deposit in the Sudetic Coal Basin. IV. Series: Polska Akademia Nauk. Komisja Nauk Geologicznych. Prace geologiczne ; 102.
QE276.5.P62 vol. 102 79-379249
[TP326.P7] 79 MARC

—TEXAS

White, David M
An analysis of transportation alternatives for meeting Texas' industrial demand for western coal through the year 2000 / by David M. White. — [Austin] : Center for Energy Studies, University of Texas at Austin, 1978.
x, 124 p. : ill. ; 28 cm. — (Public information report ; no. 4)
Bibliography: p. 119-124.
1. Coal—Texas—Transportation. 2. Power resources—Texas. 3. Energy policy—Texas. 4. Coal—Texas. I. Title. II. Series.
HE595.C6W48 380.5'2 78-622860
79 MARC

—TEXAS—TRANSPORTATION

White, David M
An analysis of transportation alternatives for meeting Texas' industrial demand for western coal through the year 2000 / by David M. White. — [Austin] : Center for Energy Studies, University of Texas at Austin, 1978.
x, 124 p. : ill. ; 28 cm. — (Public information report ; no. 4)
Bibliography: p. 119-124.
1. Coal—Texas—Transportation. 2. Power resources—Texas. 3. Energy policy—Texas. 4. Coal—Texas. I. Title. II. Series.
HE595.C6W48 380.5'2 78-622860
79 MARC

—UKRAINE—L'VOVSKO-VOLYNSKIY BASIN

Kamennye ugli L'vovsko-Volynskogo basseĭna / G. P. Vyrvich, Ė. P. Gigashvili, Z. G. Dubik ... [et al.] ; Pod obshch. red. V. Z. Ershova. — L'vov : Vyshcha shkola, 1978.
174 p., [7] leaves of plates (3 fold.) : ill. ; 22 cm. USSR78
Bibliography: p. 168-[172]
1.80rub
1. Coal—Ukraine—L'vovsko-Volynskiy Basin. I. Vyrvich, Galina Petrovna. II. Ershov, Vasiliĭ Zakharovich.
TN808.R92U327 79-381917
79 MARC

—UNITED STATES

Hulett, David T
Projections of energy supply and demand for the 1978-1979 winter heating season : analysis memorandum / prepared by David T. Hulett. — [Washington] : Assistant Administrator for Applied Analysis, Energy Information Administration ; Springfield, Va. : available from National Technical Information Service, U.S. Dept. of Commerce, 1978, i.e. 1979.
40 p. ; 27 cm.
"AM/ES/79-10."
"DOE/EIA/33, dist. category UC-92."
$4.50
1. Energy consumption—United States. 2. Petroleum—United States. 3. Gas, Natural—United States. 4. Coal—United States. 5. Winter—United States. I. Title.
HD9502.U52H835 333.7 79-602014
79 MARC

United States. Congress. House. Committee on Interior and Insular Affairs. Subcommittee on Mines and Mining.
United States coal production and utilization : oversight hearings before the Subcommittee on Mines and Mining of the Committee on Interior and Insular Affairs, House of Representatives, Ninety-fifth Congress, second session ... hearings held in Washington, D.C., May 16 and 18, and July 27, 1978. — Washington : U.S. Govt. Print. Off., 1979.
iv, 616 p. ; 24 cm.
"Serial no. 95-22."
1. Coal—United States. 2. Coal mines and mining—United States. I. Title.
KF27.I536 1978b 338.2'7'20973 79-601244
79 MARC

United States. Congress. House. Committee on Interstate and Foreign Commerce. Subcommittee on Energy and Power.
National coal policy project : hearing before the Subcommittee on Energy and Power of the Committee on Interstate and Foreign Commerce, House of Representatives, Ninety-fifth Congress, second session, on the need to balance the energy needs of the United States against the environmental consequences of a renewed emphasis on coal use, April 10, 1978. — Washington : U.S. Govt. Print. Off., 1978.
iii, 347 p. ; 24 cm.
"Serial no. 95-138."
1. Coal—United States. 2. Energy policy—United States. 3. Environmental policy—United States. I. Title.
KF27.I5542 1978c 338.2'7'20973 79-600572
79 MARC

United States. Congress. Office of Technology Assessment.
The direct use of coal : prospects and problems of production and combustion. — Washington : Congress of the United States, Office of Technology Assessment : for sale by the Supt. of Docs., U.S. Govt. Print. Off., [1979]
xv, 411 p. : ill. ; 26 cm.
Includes bibliographical references.
1. Coal—United States. I. Title.
TN800.U54 1979 662'.62'0973 79-600071
79 MARC

United States. Congressional Budget Office.
Replacing oil and natural gas with coal : prospects in the manufacturing industries / the Congress of the United States, Congressional Budget Office. — Washington : The Office : for sale by the Supt. of Docs., U.S. Govt. Print. Off., 1978.
xix, 83 p. : graphs ; 26 cm. — (Background paper - Congressional Budget Office)
Written by Craig Roach.
Includes bibliographical references.
1. Coal—United States. 2. Petroleum—United States. 3. Gas, Natural—United States. 4. United States—Industries—Energy consumption. I. Roach, Craig. II. Title. III. Series: United States. Congressional Budget Office. Background paper - Congressional Budget Office.
HD9546.U65 1978 338.2'7'20973 78-603634
79 MARC

—UNITED STATES—CONGRESSES

Critical issues in coal transportation systems : proceedings of symposium / Maritime Transportation Research Board, Commission on Sociotechnical Systems, National Research Council. — Washington : National Academy of Sciences, 1979.
x, 358 p. : ill. ; 28 cm.
Includes bibliographical references.
ISBN 0-309-02869-8
1. Coal—Transportation—Congresses. 2. Coal—United States—Congresses. I. National Research Council. Maritime Transportation Research Board.
HE595.C6C68 380.5 78-78294
79 MARC

National Coal Policy Project.
A report of a seminar at the Colorado School of Mines, September, 1978 / the National Coal Policy Project ; edited by J. Charles Curran. — Washington : Center for Strategic and International Studies, Georgetown University, c1979.
vi, 83 p. ; 23 cm. — (CSIS report)
1. Coal trade—United States—Congresses. 2. Coal—United States—Congresses. 3. Energy policy—United States—Congresses. I. Curran, J. Charles. II. Title. III. Series: Georgetown University, Washington, D.C. Center for Strategic and International Studies. CSIS report.
HD9546.N34 1979 338.2'7'20973 79-66508
00 MARC

—UNITED STATES—TRANSPORTATION

United States. Congress. Senate. Committee on Energy and Natural Resources.
Impact reduction of energy resources act of 1979 : hearing before the Committee on Energy and Natural Resources, United States Senate, Ninety-sixth Congress, first session, on S. 967 ... April 30, 1979. — Washington : U.S. Govt. Print. Off., 1979.
iii, 49 p. ; 23 cm.
"Publication no. 96-19."
1. Railroad law—United States. 2. Railroads—United States—Abandonment. 3. Coal—United States—Transportation. 4. Railroads—Northwestern States. 5. Coal—Northwestern States—Transportation. I. Title.
KF26.E55 1979g 343'.73'0958 79-603125
79 MARC

—UNITED STATES—TRANSPORTATION —CONGRESSES

Conference on Coal Production and Transportation, 3d, Denver, 1977.
Coal production & transportation : third annual conference—1977 : April 13 and 14, 1977, the Brown Palace, Denver, Colorado : transcript of a conference / sponsored by PLM, inc. — [San Francisco] : PLM, inc., [c1977]
236 p. : ill. ; 23 cm.
Cover title.
1. Coal trade—United States—Congresses. 2. Coal—United States—Transportation—Congresses. I. PLM, inc. II. Title.
HD9545.C59 1977 338.3'7'20973 78-109835
78 MARC

—WYOMING—ANALYSIS

Glass, Gary B
Coal analyses and lithologic descriptions of five core holes drilled in the Carbon Basin of southcentral Wyoming / by Gary B. Glass. — Laramie : Geological Survey of Wyoming, 1978.
vii, 97 p. : ill., graphs (4 fold. in pocket) ; 28 cm. — (Report of investigations - The Geological Survey of Wyoming ; no. 16)
Bibliography: p. 31-32.
1. Coal—Wyoming—Analysis. I. Title. II. Series: Wyoming. Geological Survey. Report of investigations ; no. 16.
QE181.A26 no. 16 557.87'08 s 79-621333
[TP326.U5] 79 MARC

—WYOMING—TRANSPORTATION— COSTS

Cooper, Hal B H
Energy and environmental analysis of a proposed coal rail haul rate increase / Hal B. H. Cooper, Jr., Richard W. Miksad, E. Gus Fruh. — [Austin] : Center for Energy Studies, University of Texas at Austin, 1978.
vi, 53 leaves : ill. ; 28 cm. — (Environmental study ; no. 5)
Includes bibliographical references.
1. Coal—Wyoming—Transportation—Costs. 2. Electric power-plants—Texas—Fuel consumption. 3. Railroads—Wyoming—Rates. I. Miksad, Richard W., joint author. II. Fruh, E. Gus, 1939- joint author. III. Title. IV. Series: Environmental study (Austin, Tex.) ; no. 5.
HE2321.C6C63 385'.1 79-623880
79 MARC

COAL, PULVERIZED

see also Briquets (Fuel); Fly ash

Kuang-tung sheng Fo-shan shih chu tsao ch'ang.
(Mei fen hua t'ieh)
煤粉化铁 / 广东省佛山市铸造厂编. — 第1版. — 北京：机械工业出版社，1977.
viii, 64 p. : ill. ; 19 cm. — (机械工业技术革新技术改造选编)
$0.18

1. Cupola-furnaces. 2. Coal, Pulverized. I. Title. II. Series: Chi hsieh kung yeh chi shu ko hsin chi shu kai tsao hsüan pien.
TS231.K93 1977 78-842208

Pulverized-coal combustion and gasification : theory and applications for continuous flow processes / edited by L. Douglas Smoot and David T. Pratt. — New York : Plenum Press, c1979.
xvii, 333 p. : ill. ; 24 cm.
Includes bibliographical references and index.
ISBN 0-306-40084-7
1. Coal, Pulverized. 2. Combustion engineering. 3. Coal gasification. I. Smoot, Leon Douglas. II. Pratt, David T.
TP328.P84 662'.62 78 78-12564 MARC

—ADDRESSES, ESSAYS, LECTURES

Własności pyłów. Wrocław [Państwowe Wydawn. Naukowe] 1970.
64 p. illus. 24 cm. (Prace naukowe Instytutu Inżynierii Chemicznej i Urządzeń Cieplnych Politechniki Wrocławskiej, nr. 2) (Studia i materiały, 1) (Prace naukowe Politechniki Wrocławskiej) zł7.00
Summaries in English and Russian.
Includes bibliographies.
1. Coal, Pulverized—Addresses, essays, lectures. 2. Wood flour—Addresses, essays, lectures. 3. Dust—Addresses, essays, lectures. I. Series: Breslau. Politechnika. Instytut Inżynierii Chemicznej i Urządzeń Cieplnych. Prace naukowe, nr. 2. II. Series: Breslau. Politechnika. Instytut Inżynierii Chemicznej i Urządzeń Cieplnych. Studia i materiały, 1.
TP328.W55 79 79-357052 MARC

COAL APPLE see Coal balls

COAL ASH

see also Fly ash

Coal ash utilization : fly ash, bottom ash, and slag / edited by S. Torrey. — Park Ridge, N.J. : Noyes Data Corp., 1978.
xiv, 370 p. : ill. ; 25 cm. — (Pollution technology review ; no. 48)
Bibliography: p. 370.
ISBN 0-8155-0722-4 : $39.00
1. Coal ash. 2. Aggregates (Building materials) 3. Ash disposal. I. Torrey, S. II. Series.
TA455.A63C63 1978 662'.62 79 78-62525 MARC

COAL BALLS

—ILLINOIS—ILLINOIS BASIN

Phillips, Tom Lee, 1931-
Fossil peat of the Illinois basin : a guide to the study of coal balls of Pennsylvanian age / Tom L. Phillips, Matthew J. Avcin, Dwain Berggren. — Urbana : Illinois State Geological Survey, 1976.
39 p. : ill. ; 23 cm. — (Educational series ; 11)
Bibliography: p. 33.
1. Paleobotany—Pennsylvanian. 2. Coal balls—Illinois—Illinois Basin. I. Avcin, Matthew J., joint author. II. Berggren, Dwain, joint author. III. Title. IV. Series: Illinois. State Geological Survey. Educational series ; 11)
QE105.A3 no. 11 553'.08 s 79 78-622875 MARC
[QE919]

COAL-BURNING POWER PLANTS see Coal-fired power plants

Natusch, David.
The particulate problem. [Sound recording] [Washington] American Chemical Society [1975. 731]
1 cassette. mono. (Man and molecules)
With: Stewart, R. D. CO—the invisible enemy.
Stuart Finley, narrator.
Duration: ca. 15 min.
SUMMARY: Dr. Natusch describes research on the fly ash from coal-burning power plants. He notes that the smallest particles (and those which are inhaled deep into the lungs) were found to have the highest concentrations of toxic substances such as lead, arsenic, etc.
1. Fly ash—Toxicology. 2. Air—Pollution—Toxicology. 3. Coal-fired power plants—Environmental aspects. I. Finley, Stuart. II. Title. III. Series: Man and molecules. [Sound recording]
[RA578] 77-740329

COAL-FIRED POWER PLANTS

—ENVIRONMENTAL ASPECTS

COAL-GAS see Gas

COAL GASIFICATION

Cusumano, James A
Catalytic in coal conversion / Jame A. Cusumano, Ralph A. Dalla Betta, Ricardo Levy. — New York : Academic Press, 1978.
xiii, 272 p. : ill. ; 24 cm.
Includes bibliographical references and index.
ISBN 0-12-199935-1
1. Catalysis. 2. Catalysts. 3. Coal liquefaction. 4. Coal gasification. I. Dalla Betta, Ralph A., joint author. II. Levy, Ricardo, joint author. III. Title.
TP156.C35C87 662'.6622 78 77-25620 MARC

Moore, H F
Synthetic oil from coal : the economic impact of five alternatives for making hydrogen from coal : prepared for the Institute for Mining and Minerals Research, Kentucky Center for Energy Research Laboratory. / by H. F. Moore, E. T. Kim, R. I. Kermode. — Lexington : IMMR, 1978.
69 p. (p. 67-69 publisher's list) : ill. ; 28 cm.
"IMMR33-PD20-78."
Includes bibliographical references.
$4.50. $1.50 (microfiche)
1. Coal liquefaction. 2. Coal gasification. I. Kim, E. T., joint author. II. Kermode, R. I., joint author. III. Kentucky. University. Institute for Mining and Minerals Research. IV. Title.
TP352.M66 338.4'7'662662 79 78-662489 MARC

Pulverized-coal combustion and gasification : theory and applications for continuous flow processes / edited by L. Douglas Smoot and David T. Pratt. — New York : Plenum Press, c1979.
xvii, 333 p. : ill. ; 24 cm.
Includes bibliographical references and index.
ISBN 0-306-40084-7
1. Coal, Pulverized. 2. Combustion engineering. 3. Coal gasification. I. Smoot, Leon Douglas. II. Pratt, David T.
TP328.P84 662'.62 78 78-12564 MARC

United States. Library of Congress. Congressional Research Service.
Synthetic fuels from coal : status and outlook of coal gasification and liquefaction / printed at the request of the Committee on Energy and Natural Resources, United States Senate. — Washington : U.S. Govt. Print. Off., 1979.
x, 196 p. ; 23 cm.
At head of title: 96th Congress, 1st session. Committee print. "Publication no. 96-17."
1. Synthetic fuels. 2. Coal gasification. 3. Coal liquefaction. I. United States. Congress. Senate. Committee on Energy and Natural Resources. II. Title.
TP360.U59 1979 662'.6622 79 79-602911 MARC

United States. Technical Advisory Task Force on Synthesized Gaseous Hydrocarbon Fuels.
National gas survey : report to the Federal Energy Regulatory Commission / by the Technical Advisory Task Force on Synthesized Gaseous Hydrocarbon Fuels. — Washington : U.S. Dept. of Energy, Federal Energy Regulatory Commission ; for sale by the Supt. of Docs., U.S. Govt. Print. Off., 1978.
xiv, 291 p. : ill. ; 27 cm.
"DOE-FERC-008. Uc-13."
Includes bibliographical references.
1. Synthesis gas. 2. Coal gasification. 3. Oil gasification. I. United States. Federal Energy Regulatory Commission. II. Title.
TP759.U54 1978 665'.7 00 79-601236 MARC

—ADDRESSES, ESSAYS, LECTURES

Probleme der Vergasung von Brennstoffen / von e. Autorenkollektiv. — 1. Aufl. — Leipzig : Deutscher Verlag für Grundstoffindustrie, VEB, 1978.
132 p. : ill. ; 22 cm. — (Freiberger Forschungshefte ; [Reihe] A, 591 : Grundstoff-Verfahrenstechnik, Brennstofftechnik) GFR78-A
Includes bibliographical references.
39.50M
1. Coal gasification—Addresses, essays, lectures. I. Series: Freiberger Forschungshefte ; Reihe A, 591.
TN7.F7 no. 591 79 78-399604 MARC
[TP759]

—BIBLIOGRAPHY

McBride, John Raymond, 1946-
Low BTU gasification : an annotated bibliography of the technology : a report to the Montana Energy Advisory Council / by John R. McBride. — Helena : The Council, 1977.
43 p. : ill. ; 29 cm.
Includes bibliographical references.
1. Coal gasification—Bibliography. I. Montana. Energy Advisory Council. II. Title.
Z7914.F8M27 016.665'772 78[r79]rev 77-623395 MARC
[TP759]

—CONGRESSES

Coal Processing and Conversion Symposium, Morgantown, W. Va., 1976.
Proceedings of the Coal Processing and Conversion Symposium, June 1-3, 1976, Morgantown, WV / compilers, Carl J. Smith, Duane G. Nichols ; sponsors, West Viriginia Geological and Economic Survey, West Virginia University, College of Engineering. — [Morgantown, W. Va.] : The Survey, 1979.
xii, 115 p. : ill. ; 28 cm. — (Coal geology bulletin ; CGB-6)
Cover title.
Includes bibliographical references and index.
1. Coal gasification—Congresses. 2. Coal liquefaction—Congresses. 3. Coal preparation—Congresses. I. Smith, Carl J. II. Nichols, Duane G. III. West Virginia. Geological Survey. IV. West Virginia. University. College of Engineering. V. Series.
TP759.C62 1976 662'.625 79 79-624429 MARC

Fluidization : application to coal conversion processes / C. Y. Wen, editor. — New York : American Institute of Chemical Engineers, 1978.
vi, 242 p. : ill. ; 28 cm. — (AIChE symposium series ; v. 74, no. 176)
Includes bibliographical references.
1. Fluidization—Congresses. 2. Coal gasification—Congresses. I. Wen, Chin-Yung, 1928- . II. American Institute of Chemical Engineers. III. Series: American Institute of Chemical Engineers. AIChE symposium series ; v. 74, no. 176.
TP156.F65F575 662'.6622 78 78-8963 MARC

Symposium on Gasification and Liquefaction of Coal, Düsseldorf, 1976.
Symposium on Gasification and Liquefaction of Coal, Düsseldorf, 1976 = Colloque sur la gazéification et la liquéfaction du charbon = Симпозиум по вопросам газификации и сжижения угля / [hrsg. von Harald B. Giesel u. Werner Peters]. — Essen : Verlag Glückauf, 1976.
[420] p. in various pagings : ill., maps ; 21 cm. — (Rohstoffwirtschaft international ; Bd. 5) GFR 76-A
Summaries in French and Russian.
Includes 2 speeches in German.
Sponsored by the United Nations Economic Commission for Europe, Coal Committee.

On spine: ECE-Symposium Düsseldorf 1976.
Includes bibliographical references.
ISBN 3-7730-0181-X : DM46.00
1. Coal gasification—Congresses. 2. Coal liquefaction—Congresses. I. Giesel, Harald B. II. Peters, Werner. III. United Nations. Economic Commission for Europe. IV. Title: Colloque sur la gazéification et la liquéfaction du charbon. V. Title: Simpozium po voprosam gazifikatsii i szhizheniia uglia. VI. Title: ECE-Symposium Düsseldorf 1976.
TP759.S95 1976 662'.625 76-381723 (MARC)

Workshop on Materials Problems and Research Opportunities in Coal Conversion, Columbus, Ohio, 1974.
Conclusions and recommendations from Workshop on Materials Problems and Research Opportunities in Coal Conversion, 16-18 April 1974, Columbus, Ohio / sponsored by The National Science Foundation and The Office of Coal Research ; organized by The Department of Metallurgical Engineering, The Ohio State University, Columbus, Ohio ; workshop director, R. W. Staehle. — Columbus : Available from Corrosion Center, Ohio State University, [1974?]
vi, 67 p. ; 28 cm.
"Volume 1."
1. Coal liquefaction—Congresses. 2. Coal gasification—Congresses. I. Staehle, R. W., 1934- . II. United States. National Science Foundation. III. United States. Office of Coal Research.
TP343.W67 1974 662'.625 79 78-620972 MARC

COAL GASIFICATION (Continued)

—ENVIRONMENTAL ASPECTS—UNITED STATES

Environmental, health, and control aspects of coal conversion : an information overview : prepared for the Energy Research and Development Administration, Assistant Administrator for Environment and Safety / edited by H. M. Braunstein, E. D. Copenhaver, and H. A. Pfuderer. — Oak Ridge, Tenn. : Information Center Complex, Information Division, Oak Ridge National Laboratory, 1977.
2 v. : ill. ; 29 cm.
"ORNL/EIS-94."
"Contract no. W-7405-eng-26."
Includes bibliographies and index.
1. Coal liquifaction—Environmental aspects—United States. 2. Coal gasification—Environmental aspects—United States. I. Braunstein, H. M. II. Copenhaver, Emily D. III. Pfuderer, H. A. IV. United States. Energy Research and Development Administration. V. United States. National Laboratory, Oak Ridge, Tenn.
TD195.C58E59 662'.6622 77-604030
79 MARC

—ILLINOIS

Smith, William Henking, 1919-
Coal and water resources for coal conversion in Illinois / William H. Smith and John B. Stall. — Urbana, Ill. : State Water Survey, 1975.
79 p. : ill., maps (3 fold. col. in pocket) ; 28 cm. — (Cooperative resources report ; 4)
Includes bibliographies.
1. Coal—Illinois. 2. Water-supply—Illinois. 3. Coal gasification—Illinois. 4. Coal liquefaction—Illinois. I. Stall, John Byron, joint author. II. Title. III. Series.
TN805.I3S62 553'.24'09773 75-623963
79 MARC

COAL GASIFICATION, UNDERGROUND

Great Britain. National Coal Board. Operational Research Executive.
Underground gasification of coal : a National Coal Board reappraisal, 1976 / prepared for the National Coal Board by the Operational Research Executive ... [written by] P. N. Thompson, J. R. Mann, F. Williams. — London : N.C.B., 1976.
70 p. : ill. ; 30 cm. GB78-09665
£5.00
1. Coal gasification, Underground. I. Thompson, P. N. II. Mann, J. R. III. Williams, F. IV. Title.
TP759.G72 1976 665'.772 78-317783
78 MARC

COAL GEOLOGY see Coal—Geology

COAL-HANDLING

—CONGRESSES

Commission of the European Communities.
Underground transport in coal mines : information symposium, Luxembourg, May 24 to 26, 1978 / Commission of the European Communities. — Redhill [Eng.] : Colliery Guardian, c1978.
2 v. : ill. ; 24 cm. GB***
"EUR 5998 d, e, f."
Symposium organized by the Commission of the European Communities, Directorate-General for Energy and Directorate-General for Scientific and Technical Information and Information Management.
Includes bibliographical references.
ISBN 0-86108-014-9
1. Coal mines and mining—Congresses. 2. Coal-handling—Congresses. I. Commission of the European Communities. Directorate-General for Energy. II. Commission of the European Communities. Directorate-General for Scientific and Technical Information and Information Management. III. Title.
TN799.9.C66 1978 622'.6 79-310106
79 MARC

COAL-HANDLING MACHINERY

see also Conveying machinery

Curl, S J
Underground transport in coal mines / S. J. Curl. — London : IEA Coal Research, 1978.
54 p. : ill. ; 30 cm. — (Report - IEA Coal Research ; no. ICTIS/TR 01 ISSN 0142-1816) GB78-39131
Bibliography: p. 46-54.
ISBN 9290290099
1. Coal mines and mining—Equipment and supplies. 2. Mine haulage. 3. Coal-handling machinery. I. Title. II. Series: IEA Coal Research. Report - IEA Coal Research ; no. ICTIS/TR 01.
TN813.C87 622'.6 79-311804
79 MARC

COAL LANDS see Coal

COAL LEASES

—THE WEST

Cannon, James Spencer, 1949-
Mine control : western coal leasing and development / James S. Cannon ; research associate, Bart Walker ; editor, Wendy C. Schwartz ; pref. by Huey D. Johnson. — New York : Council on Economic Priorities, c1978.
193 p. :; 22 cm. — (CEP studies ; S8-2)
Bibliography: p. 187-193.
ISBN 0-87871-009-4
1. Coal leases—The West. I. Schwartz, Wendy C. II. Title. III. Series: Council on Economic Priorities. CEP studies ; S8-2.
KF1830.Z95C36 343'.73'077 78-58006
79 MARC

—UNITED STATES

United States. Congress. House. Committee on Interior and Insular Affairs. Subcommittee on Oversight and Investigations.
Illegal mining of Federal coal : a report / prepared by the Subcommittee on Oversight and Investigations of the Committee on Interior and Insular Affairs of the U.S. House of Representatives, Ninety-sixth Congress, first session. — Washington : U.S. Govt. Print. Off., 1979.
v, 9 p. ; 24 cm.
"Committee print no. 3, 96th Congress, 1st session."
1. Coal trade—United States. 2. Coal mines and mining—Law and legislation—United States. 3. Coal leases—United States. I. Title.
HD9546.U64 1979 364.1'62 79-603230
79 MARC

United States. Congress. Senate. Committee on Energy and Natural Resources. Subcommittee on Energy Production and Supply.
Coal leasing amendments : hearing before the Subcommittee on Energy Production and Supply of the Committee on Energy and Natural Resources, United States Senate, Ninety-fifty Congress, second session, on S. 3189 ... June 27, 1978. — Washington : U.S. Govt. Print. Off., 1978.
43 p. ; 24 cm.
"Publication no. 95-128."
1. Coal leases—United States. I. Title.
KF26.E553 1978d 346'.73'04682 78-603080
79 MARC

COAL LIQUEFACTION

Cusumano, James A
Catalytic in coal conversion / Jame A. Cusumano, Ralph A. Dalla Betta, Ricardo Levy. — New York : Academic Press, 1978.
xiii, 272 p. : ill. ; 24 cm.
Includes bibliographical references and index.
ISBN 0-12-199935-1
1. Catalysis. 2. Catalysts. 3. Coal liquefaction. 4. Coal gasification. I. Dalla Betta, Ralph A., joint author. II. Levy, Ricardo, joint author. III. Title.
TP156.C35C87 662'.6622 77-25620
78 MARC

Moore, H F
Synthetic oil from coal : the economic impact of five alternatives for making hydrogen from coal : prepared for the Institute for Mining and Minerals Research, Kentucky Center for Energy Research Laboratory. / by H. F. Moore, E. T. Kim, R. I. Kermode. — Lexington : IMMR, 1978.
69 p. (p. 67-69 publisher's list) : ill. ; 28 cm.
"IMMR33-PD20-78."
Includes bibliographical references.
$4.50. $1.50 (microfiche)
1. Coal liquefaction. 2. Coal gasification. I. Kim, E. T., joint author. II. Kermode, R. I., joint author. III. Kentucky. University. Institute for Mining and Minerals Research. IV. Title.
TP352.M66 338.4'7'662662 78-662489
79 MARC

Nowacki, Perry.
Coal liquefaction processes / Perry Nowacki. — Park Ridge, N.J. : Noyes Data Corp., 1979.
xii, 339 p. : ill. ; 25 cm. — (Chemical technology review ; no. 131) (Energy technology review ; no. 45)
Bibliography: p. 336-339.
Includes index.
ISBN 0-8155-0756-9 : $48.00
1. Coal liquefaction. I. Title. II. Series. III. Series: Energy technology review ; no. 45.
TP352.N68 662'.6622 79-14384
79 MARC

Thode, Edward F
Synthetic liquid fuels from coal and wood / prepared for the Governor's Energy Task Force, State of New Mexico by Edward F. Thode. — [Las Cruces] : New Mexico State University, 1974.
35 leaves : ill. ; 28 cm. — (Engineering Experiment Station technical report ; no. 86) (Position paper - New Mexico State University ; no. 15)
"NMSU-EES-86-74."
Bibliography: leaf 14.
1. Synthetic fuels. 2. Coal liquefaction. 3. Wood waste. I. New Mexico. Governor's Energy Task Force. II. Title. III. Series: New Mexico State University. Engineering Experiment Station. Technical report ; no. 86. IV. Series: New Mexico State University. Position paper ; no. 15.
TP360.T48 662'.66 74-623555
78 MARC

United States. Library of Congress. Congressional Research Service.
Synthetic fuels from coal : status and outlook of coal gasification and liquefaction / printed at the request of the Committee on Energy and Natural Resources, United States Senate. — Washington : U.S. Govt. Print. Off., 1979.
x, 196 p. ; 23 cm.
At head of title: 96th Congress, 1st session. Committee print. "Publication no. 96-17."
1. Synthetic fuels. 2. Coal gasification. 3. Coal liquefaction. I. United States. Congress. Senate. Committee on Energy and Natural Resources. II. Title.
TP360.U59 1979 662'.6622 79-602911
79 MARC

—CONGRESSES

Coal Processing and Conversion Symposium, Morgantown, W. Va., 1976.
Proceedings of the Coal Processing and Conversion Symposium, June 1-3, 1976, Morgantown, WV / compilers, Carl J. Smith, Duane G. Nichols ; sponsors, West Viriginia Geological and Economic Survey, West Virginia University, College of Engineering. — [Morgantown, W. Va.] : The Survey, 1979.
xii, 115 p. : ill. ; 28 cm. — (Coal geology bulletin ; CGB-6)
Cover title.
Includes bibliographical references and index.
1. Coal gasification—Congresses. 2. Coal liquefaction—Congresses. 3. Coal preparation—Congresses. I. Smith, Carl J. II. Nichols, Duane G. III. West Virginia. Geological Survey. IV. West Virginia. University. College of Engineering. V. Series.
TP759.C62 1976 662'.625 79-624429
79 MARC

Symposium on Gasification and Liquefaction of Coal, Düsseldorf, 1976.
Symposium on Gasification and Liquefaction of Coal, Düsseldorf, 1976 = Colloque sur la gazéification et la liquéfaction du charbon = Симпозиум по вопросам газификации и сжижения угля / [hrsg. von Harald B. Giesel u. Werner Peters]. — Essen : Verlag Glückauf, 1976.
[420] p. in various pagings : ill., maps ; 21 cm. — (Rohstoffwirtschaft international ; Bd. 5) GFR 76-A
Summaries in French and Russian.
Includes 2 speeches in German.
Sponsored by the United Nations Economic Commission for Europe, Coal Committee.

On spine: ECE-Symposium Düsseldorf 1976.
Includes bibliographical references.
ISBN 3-7780-0181-X : DM46.00
1. Coal gasification—Congresses. 2. Coal liquefaction—Congresses. I. Giesel, Harald B. II. Peters, Werner. III. United Nations. Economic Commission for Europe. IV. Title: Colloque sur la gazéification et la liquéfaction du charbon. V. Title: Simpozium po voprosam gazifikatsii i szhizhenia uglia. VI. Title: ECE-Symposium Düsseldorf 1976.
TP759.S95 1976 662'.625 76-381723
(MARC)

Workshop on Materials Problems and Research Opportunities in Coal Conversion, Columbus, Ohio, 1974.
Conclusions and recommendations from Workshop on Materials Problems and Research Opportunities in Coal Conversion, 16-18 April 1974, Columbus, Ohio / sponsored by The National Science Foundation and The Office of Coal Research ; organized by The Department of Metallurgical Engineering, The Ohio State University, Columbus, Ohio ; workshop director, R. W. Staehle. — Columbus : Available from Corrosion Center, Ohio State University, [1974?]
vi, 67 p. ; 28 cm.
"Volume 1."
1. Coal liquefaction—Congresses. 2. Coal gasification—Congresses. I. Staehle, R. W., 1934- . II. United States. National Science Foundation. III. United States. Office of Coal Research.
TP343.W67 1974 662'.625 78-620972
79 MARC

—ENVIRONMENTAL ASPECTS—UNITED STATES

Environmental, health, and control aspects of coal conversion : an information overview : prepared for the Energy Research and Development Administration, Assistant Administrator for Environment and Safety / edited by H. M. Braunstein, E. D. Copenhaver, and H. A. Pfuderer. — Oak Ridge, Tenn. : Information Center Complex, Information Division, Oak Ridge National Laboratory, 1977.
2 v. : ill. ; 29 cm.
"ORNL/EIS-94."
"Contract no. W-7405-eng-26."
Includes bibliographies and index.
1. Coal liquifaction—Environmental aspects—United States. 2. Coal gasification—Environmental aspects—United States. I. Braunstein, H. M. II. Copenhaver, Emily D. III. Pfuderer, H. A. IV. United States. Energy Research and Development Administration. V. United States. National Laboratory, Oak Ridge, Tenn.
TD195.C58E59 662'.6622 77-604030
79 MARC

COAL LIQUEFACTION (Continued)

—PERIODICALS

United States. Dept. of Energy. Division of Coal Conversion.
Coal liquefaction. Jan./Mar. 1977–
Washington, U. S. Dept. of Energy, Division of Coal Conversion.
v. ill. 28 cm. quarterly.
Continues: Coal liquefaction, ISSN 0362-7984, issued by: United States. Office of Fossil Energy.
Key title: Coal liquefaction, ISSN 0362-7984.
1. Coal liquefaction—Periodicals. I. Title.
TP352.U53a 662'.6622 78-643630
MARC-S

United States. Office of Fossil Energy.
Coal liquefaction.
Washington, Office of Fossil Energy, Energy Research and Development Administration.
v. ill. 28 cm. quarterly.
Key title: Coal liquefaction, ISSN 0362-7984.
Continued by: Coal liquefaction, ISSN 0362-7984, issued by: United States. Dept. of Energy. Division of Coal Conversion.
1. Coal liquefaction—Periodicals. I. Title.
TP352.U53a 662'.6622 76-643562
MARC-'

—ILLINOIS

Smith, William Henking, 1919-
Coal and water resources for coal conversion in Illinois / William H. Smith and John B. Stall. — Urbana, Ill. : State Water Survey, 1975.
79 p. : ill., maps (3 fold. col. in pocket) ; 28 cm. — (Cooperative resources report ; 4)
Includes bibliographies.
1. Coal—Illinois. 2. Water-supply—Illinois. 3. Coal gasification—Illinois. 4. Coal liquefaction—Illinois. I. Stall, John Byron, joint author. II. Title. III. Series.
TN805.I3S62 553'.24'09773 75-623963
79 MARC

COAL MINE WASTE

Van Voast, Wayne A
Hydrologic characteristics of coal-mine spoils, southeastern Montana : completion report : project A-077 MONT / by Wayne A. Van Voast, Robert B. Hedges, John J. McDermott. — Bozeman : Montana University Joint Water Resources Research Center, 1978.
34 p. : ill. ; 28 cm. — (MUJWRRC report ; no. 94)
Bibliography: p. 34.
1. Hydrology—Montana. 2. Coal mine waste. I. Hedges, Robert B., joint author. II. McDermott, John J., joint author. III. Title. IV. Series: Montana University Joint Water Resources Research Center. Report ; no. 94.
GB705.M9V35 551.4'8 79-621251
79 MARC

—ENVIRONMENTAL ASPECTS
—ILLINOIS

Huff, Linda L
Economic impact of dissolved solids regulation upon the coal mining industry (R76-7) / by Linda L. Huff, Gregg A. Jarrell. — Chicago : State of Illinois, Institute for Environmental Quality, 1977.
viii, 134 p. : ill. ; 28 cm. — (IIEQ document ; no. 77/28)
Includes bibliographical references.
1. Coal mines and mining—Illinois—Costs. 2. Coal mine waste—Enivronmental aspects—Illinois. 3. Water—Pollution—Law and legislation—Illinois. I. Jarrell, Gregg A., joint author. II. Title. III. Series. IV. Series: Illinois Institute for Environmental Quality IIEQ document ; no.77/28.
HD9547.I3H83 338.2'7'2409773 78-622262
78 MARC

—FIRES AND FIRE PREVENTION

Saranchuk, Viktor Ivanovich.
Bor'ba s goreniem porodnykh otvalov / V. I. Saranchuk. — Kiev : Nauk. dumka, 1978.
167 p. : ill. ; 22 cm. USSR79
At head of title: Akademiia nauk Ukrainskoĭ SSR. Institut fiziko-organicheskoĭ khimii i uglekhimii.
Bibliography: p. [163]-165.
1.80rub
1. Coal mine waste—Fires and fire prevention. 2. Spoil banks—Fires and fire prevention. I. Title.
TN315.S27 79-376986
79 MARC

—NORTH DAKOTA

Schroer, Fred W
Characterization of coal overburden and strip-mine spoils in North Dakota / F. W. Schroer. — Fargo : Soils Dept., North Dakota State University, 1978.
17, 76 p. : ill. ; 28 cm. — (North Dakota research report ; no. 68)
Chiefly tables.
Errata slip inserted.
Bibliography: p. 17.
1. Coal mine waste—North Dakota. 2. Spoil banks—North Dakota. 3. Strip mining—Environmental aspects—North Dakota. I. Title. II. Series.
TD899.M5S34 622'.18'2 78-624324
00 MARC

COAL-MINERS

see also Trade-unions—Coal miners

—DISEASES AND HYGIENE

see also Nystagmus

—DISEASES AND HYGIENE—
CONGRESSES

Minette, André.
Apport épidémiologique à l'étiologie de la bronchite chronique des mineurs de charbon / par A. Minette. — Lanaken, [Koning Albertlaan 123] : A. Minette, 1976, i.e. 1977.
203 p. : graphs ; 30 cm. Be77-11
At head of title: Université catholique de Louvain, Faculté de médecine et Institut d'hygiène des mines à Hasselt.
Summary in Dutch, English, French, German, and Italian.
Bibliography: p. 185-197.
400F
1. Bronchitis—Etiology—Congresses. 2. Coal-miners—Diseases and hygiene—Congresses. 3. Tobacco—Physiological effect—Congresses. 4. Air—Pollution—Physiological effect—Congresses. 5. Bronchitis—Belgium—Statistics. 6. Coal-miners—Diseases and hygiene—Belgium—Statistics. I. Title.
RC778.M56 1977 616.2'3 78-394345
*79 MARC

—DISEASES AND HYGIENE—BELGIUM
—STATISTICS

Minette, André.
Apport épidémiologique à l'étiologie de la bronchite chronique des mineurs de charbon / par A. Minette. — Lanaken, [Koning Albertlaan 123] : A. Minette, 1976, i.e. 1977.
203 p. : graphs ; 30 cm. Be77-11
At head of title: Université catholique de Louvain, Faculté de médecine et Institut d'hygiène des mines à Hasselt.
Summary in Dutch, English, French, German, and Italian.
Bibliography: p. 185-197.
400F
1. Bronchitis—Etiology—Congresses. 2. Coal-miners—Diseases and hygiene—Congresses. 3. Tobacco—Physiological effect—Congresses. 4. Air—Pollution—Physiological effect—Congresses. 5. Bronchitis—Belgium—Statistics. 6. Coal-miners—Diseases and hygiene—Belgium—Statistics. I. Title.
RC778.M56 1977 616.2'3 78-394345
*79 MARC

—DISEASES AND HYGIENE—FRANCE
—NORD (REGION)

Voogd, Abraham de.
J'étais médecin à Ratintout / A. de Voogd ; [postfaces du Dr Jacques Jouvie et de Gilbert Le Mouël]. — Paris : Éditions du Cerf, 1975.
200 p. ; 21 cm. F75-10529
26.00F
1. Voogd, Abraham de. 2. Physicians—France—Biography. 3. Coal-miners—Diseases and hygiene—France—Nord (Region) 4. Slums—France—Nord (Region) I. Title.
R507.V66A36 362.1'092'4 78-378653
78 MARC

—LEGAL STATUS, LAWS, ETC.—
UNITED STATES

United States. Congress. Senate. Committee on Appropriations. Subcommittee on Departments of Labor, and Health, Education, and Welfare, and Related Agencies.
Black lung supplemental and oversight on HEW positions : hearing before a subcommittee of the Committee on Appropriations, United States Senate, Ninety-fifth Congress, second session : special hearing [held at] Department of Health, Education, and Welfare, Department of Labor. — Washington : U.S. Govt. Print. Off., 1978.
ii, 117 p. ; 24 cm.
1. Insurance, Disability—United States. 2. Coal-miners—Legal status, laws, etc.—United States. 3. Lungs—Dust diseases. I. Title.
KF26.A653 1978a 368.3'86'0973 78-603085
79 MARC

United States. Congress. House. Committee on Education and Labor. Subcommittee on Labor Standards.
To amend the Federal coal mine health and safety act : hearing before the Subcommittee on Labor Standards of the Committee on Education and Labor, House of Representatives, Ninety-fifth Congress, first session on H.R. 4287 ... held in Washington, D.C., on March 29, 1977. — Washington : U.S. Govt. Print. Off., 1977.
iii, 329 p. : ill. ; 24 cm.
Includes bibliographical references.
1. Coal mines and mining—Safety regulations—United States. 2. Coal-miners—Legal status, laws, etc.—United States. I. Title.
KF27.E348 1977a 344'.73'047 77-604918
78[r79]rev MARC

—ALABAMA—BIOGRAPHY—
COLLECTED WORKS

Alabama coal miners / as told to Carl Elliott, Sr. ; Susan Crittenden, editor. — Jasper : Northwest Alabama Pub. Co., c1978-
v. : ill., facsims., ports. ; 29 cm.
Includes indexes.
1. Coal-miners—Alabama—Biography—Collected works. 2. Alabama—Social conditions—Collected works. 3. Lungs—Dust diseases—Collected works. I. Elliott, Carl, 1913- . II. Crittenden, Susan.
HD8039.M62U6113 331.7'62'2330922 79-100141
79 MARC

—ALABAMA—JEFFERSON CO.—
BIOGRAPHY

Lester D. Williams, Jefferson County / as told by him to Carl Elliott, Sr. — Jasper : Northwest Alabama Pub. Co., c1978.
107 leaves : facsims. ; 29 cm. — (Alabama coal miners ; v. 4)
Includes index.
1. Williams, Lester D., 1905- . 2. Coal-miners—Alabama—Jefferson Co.—Biography. 3. Jefferson Co., Ala.—Social conditions. 4. Lungs—Dust diseases. I. Elliott, Carl, 1913- . II. Series.
HD8039.M62U6113 vol. 4 79-102523
331.7'62'23340922 s
79 MARC

Robert C. Bice, Jefferson County / as told by him to Carl Elliott, Sr. — Jasper : Northwest Alabama Pub. Co., c1978.
iii, 220 leaves : ill., facsims., port. ; 29 cm. — (Alabama coal miners ; v. 3)
Includes index.
1. Bice, Robert C., 1926- . 2. Coal-miners—Alabama—Jefferson Co.—Biography. 3. Jefferson Co., Ala.—Social conditions. 4. Lungs—Dust diseases. I. Elliott, Carl, 1913- . II. Series.
HD8039.M62U6113 vol. 3 79-102528
331.7'62'23340922 s
79 MARC

—ALABAMA—MARION CO.
—BIOGRAPHY

Herbert South (1913-1975), Marion County / as told to Carl Elliott, Sr. — Jasper : Northwest Alabama Pub. Co., c1978.
ii, 240 leaves : facsims. 29 cm. — (Alabama coal miners ; v. 2)
Includes index.
1. South, Herbert, 1913-1975. 2. Coal-miners—Alabama—Marion Co.—Biography. 3. Marion Co., Ala.—Social conditions. 4. Lungs—Dust diseases. I. Elliott, Carl, 1913- . II. Series.
HD8039.M62U6113 vol. 2 79-102521
331.7'62'23340922 s
79 MARC

—ALABAMA—ST. CLAIR CO.—
BIOGRAPHY

William M. Warren, St. Clair County / as told to Carl Elliott, Sr. — Jasper : Northwest Alabama Pub. Co., c1979.
397 leaves : ill., facsims., port. ; 29 cm. — (Alabama coal miners ; v. 5)
Includes index.
1. Warren, William Melton, 1921- . 2. Coal-miners—Alabama—St. Clair Co.—Biography. 3. St. Clair Co., Ala.—Social conditions. 4. Lungs—Dust diseases. I. Elliott, Carl, 1913- . II. Series.
HD8039.M62U6113 vol. 5 79-102588
331.7'62'23340922 s
79 MARC

—BRITISH COLUMBIA

British Columbia. Manpower. Sub-commitee on North East Coal Development.
Report of the B.C. Manpower Sub-committee on N.E. Coal Development. — [Victoria?] : Ministry of Economic Development, 1976.
ix, 270 p. : ill. ; 28 cm. C***
Includes bibliographical references.
1. Coal-miners—British Columbia. 2. Labor supply—British Columbia. 3. Manpower policy—British Columbia.
HD5718.M616C22 331.1'1 78-314118
78 MARC

COAL MINERS
—BRITISH COLUMBIA (Continued)

British Columbia Research Council. Management Services Division.
Coal developments in northeastern British Columbia : construction phase manpower study / prepared for the Federal-Provincial Manpower Sub-committee on North East Coal Development by the Management Services Division, B.C. Research and the Trade Union Research Bureau. — [Ottawa] : Government of Canada, Regional Economic Expansion, 1978.
166 p. in various pagings : ill. ; 28 cm. C***
"Project no. 270060-1."
Includes bibliographical references.
1. Construction workers—British Columbia. 2. Labor supply—British Columbia. 3. Coal miners and mining—British Columbia. I. Federal-Provincial Manpower Sub-committee on North East Coal Development. II. Trade Union Research Bureau, Vancouver, B.C. III. Title.
HD8039.B92C22 331.1'1 79-309088
79 MARC

—CHINA—FU-SHUN HSIEN
—BIOGRAPHY—ADDRESSES, ESSAYS, LECTURES

(Li chin k'u nan chien yang kuang)
历尽苦难見阳光：矿工家史 / 撫順矿区矿工史編委会編. — [香港 : s. n., 1978?]
2, 146 p. : ill. ; 21 cm.
Reprint of the 1964 ed. published by Ch'un feng wen i ch'u pan she, Mukden.
1. Coal miners—China—Fu-shun hsien—Biography—Addresses, essays, lectures. 2. Fu-shun hsien, China—Biography—Addresses, essays, lectures. I. Fu-shun k'uang ch'ü k'uang kung shih pien wei hui.
HD8039.M62C637 1978 79-836449

—CHINA—HOPEI (PROVINCE)

Minami Manshū Tetsudō Kabushiki Kaisha. Hokushi Keizai Chōsajo, Peking.
(Seikei, Seihō ryōtankō rōdō gaiyō chōsa hōkoku)
井陘·正豊兩炭礦勞働概要調査報告 / [執筆者關戸嘉明]. — [北京]：滿鐵·北支經濟調査所, 昭和16 [1941]
4, 85 leaves ; 26 cm.
Cover title.
"第一次調査報告書(第六班)"
"極秘" (rubber stamped)
1. Coal miners—China—Hopei (Province) 2. Ching-hsing mei k'uang ku fen yu hsien kung ssu. 3. Cheng-feng mei k'uang kung ssu. I. Sekido, Yoshiaki, 1917- II. Title.
HD8039.M62C639 79-806639

—DONETS BASIN

Severinov, Kuz'ma Antipovich.
(Slovo o rabocheĭ chesti)
Слово о рабочей чести : [Рассказ бригадира шахты № 5–6 им. Г. М. Димитрова] / К. А. Северинов. — Донецк : Донбас, 1975.
88 p. : ill. ; 16 cm. USSR 76
Text partly in Ukrainian.
0.15rub
1. Coal-miners—Donets Basin. 2. Socialist competition. I. Title.
HD8039.M615S48 79-413137

—ENGLAND—CHESHIRE—HISTORY

Howarth, Ken.
Dark days : memoirs and reminiscences of the Lancashire and Cheshire coalmining industry up to nationalisation / by Ken Howarth. — Manchester : [The author], 1978.
[4], ii, 127 p. : ill., map, ports. ; 30 cm. GB79-06308
£2.50
1. Coal-miners—England—Lancashire—History. 2. Coal-miners—England—Cheshire—History. 3. Coal mines and mining—England—Lancashire—History. 4. Coal mines and mining—England—Cheshire—History. I. Title.
HD8039.M62G7574 331.2'042'2330922 79-317515
79 MARC

—ENGLAND—LANCASHIRE—HISTORY

Howarth, Ken.
Dark days : memoirs and reminiscences of the Lancashire and Cheshire coalmining industry up to nationalisation / by Ken Howarth. — Manchester : [The author], 1978.
[4], ii, 127 p. : ill., map, ports. ; 30 cm. GB79-06308
£2.50
1. Coal-miners—England—Lancashire—History. 2. Coal-miners—England—Cheshire—History. 3. Coal mines and mining—England—Lancashire—History. 4. Coal mines and mining—England—Cheshire—History. I. Title.
HD8039.M62G7574 331.2'042'2330922 79-317515
79 MARC

—FRANCE—CARMAUX—LANGUAGE

Gonzalez, Daniel.
L'Occitan parlat jos tèrra : los carboniès de Carmaus / per Daniel Gonzalez. — Toulouse : Université de Toulouse II, Le Mirail, [1977?]
2 v. (252 p.) : ill. ; 22 cm. — (Documents et archives pour la recherche sociolinguistique méridionale : Série E, Document de travail ; no 3-4) F***
At head of title: Université de Toulouse II, Le Mirail. Centre de sociolinguistique et de dialectologie sociale.
CONTENTS: t. 1. Tèxtes sul trabalh del Cròs.—t. 2. Vocabulari del Cròs.
1. Provençal language, Modern—Dialects—France—Carmaux. 2. Carmaux, France—Industries. 3. Coal-miners—France—Carmaux—Language. 4. Coal mines and mining—France—Carmaux. 5. Coal mines and mining—Terminology. I. Title. II. Series.
PC3447.C28G6 78-375700
78 MARC

—GERMANY—BIOGRAPHY

Werner, Georg, 1877-
Ein Kumpel, Erzählung aus dem Leben der Bergarbeiter. Berlin-Steglitz, Verlag "Die Knappschaft," 1930.
191 p. plates, port., facsim. 20 cm.
Sequel: Zwei Kumpel.
Republished in 1948 as the first part of the author's Kumpel.
1. Werner, Georg, 1877- 2. Coal-miners—Germany—Biography. 3. Supervisors, Industrial—Germany—Biography. I. Title.
HD8453.W44A33 1930 622'.33'0924 50-44387
[r78]rev MARC

—GERMANY, WEST
—STATISTICS
—PERIODICALS

Knappschaftliche Krankenversicherung: Statistische Ergebnisse.
[Bochum] Arbeitsgemeinschaft der Knappschaften der Bundesrepublik Deutschland.
v. 22 x 30 cm. annual.
1. Insurance, Health—Germany, West—Statistics—Periodicals. 2. Coal-miners—Germany, West—Statistics—Periodicals. I. Arbeitsgemeinschaft der Knappschaften der Bundesrepublik Deutschland.
HD7116.M616G84 78-647546
MARC-S

—GERMANY WEST—RUHR VALLEY REGION—HISTORY

Klessmann, Christoph, 1938-
Polnische Bergarbeiter im Ruhrgebiet, 1870-1945 : soziale Integration u. nationale Subkultur e. Minderheit in d. dt. Industriegesellschaft / von Christoph Klessmann. — Göttingen : Vandenhoeck und Ruprecht, 1978.
306 p. ; 24 cm. — (Kritische Studien zur Geschichtswissenschaft ; Bd. 30) GFR78-A
A revision of the author's Habilitationsschrift, Bochum, 1976.
Bibliography: p. 290-301.
Includes indexes.
ISBN 3-525-35982-9
1. Poles in the Ruhr Valley region, Ger.—History. 2. Coal-miners—Germany West—Ruhr Valley region—History. 3. Ruhr Valley region, Ger.—Social conditions. I. Title. II. Series.
HD8458.P65K57 1978 78-379460
78 MARC

—GREAT BRITAIN

Day, David, fl. 1973-
The Bevin boy / [by] David Day. — Kineton : Roundwood Press, 1975.
v, 122 p. ; 23 cm. GB75-23607
ISBN 0-900093-47-1
1. Day, David, fl. 1973- 2. Coal-miners—Great Britain. I. Title.
HD8393.D39A32 331.7'62'23350924 79-311776
79 MARC

Degras, Henry Ernst, 1910-
Charity main : a coalfield chronicle / by Mark Benney [i.e. H. E. Degras]. — Wakefield, Eng. : EP Pub., 1978.
176 p. ; 19 cm. — (Studies in British labour history) GB***
ISBN 0-7158-1343-9 : £4.95
1. Coal-miners—Great Britain. 2. Trade-unions—Coal-miners—Great Britain. I. Title. II. Series.
HD8039.M62G744 1978 338.2'7'209428 79-312899
79 MARC

McCormick, Brian Joseph.
Industrial relations in the coal industry / B. J. McCormick. — Hamden, Conn. : Archon Books, c1979.
xi, 263 p. : ill. ; 22 cm.
Includes bibliographical references and index.
ISBN 0-208-01772-0
1. Coal-miners—Great Britain. 2. Industrial relations—Great Britain. I. Title.
HD6976.M6152G75 1979 78-11465
331'.042'2330941
78 MARC

McCormick, Brian Joseph.
Industrial relations in the coal industry / B. J. McCormick. — London : Macmillan, 1979.
263 p. : ill. ; 23 cm. GB***
Includes bibliographical references and index.
ISBN 0-333-24108-8 : £15.00
1. Coal-miners—Great Britain. 2. Industrial relations—Great Britain. I. Title.
HD6976.M6152G75 1978 79-310408
331'.042'2330941
79 MARC

—GREAT BRITAIN—HISTORY—
ADDRESSES, ESSAYS, LECTURES

Independent collier : the coal miner as archetypal proletarian reconsidered / edited by Royden Harrison. — Hassocks [Eng.] : Harvester Press, 1978.
276 p. : ill. ; 25 cm. GB***
Includes bibliographical references and index.
ISBN 0-85527-702-5 : £12.50
1. Coal-miners—Great Britain—History—Addresses, essays, lectures. I. Harrison, Royden John.
HD8039.M62G7575 1978b 79-304181
331.7'62'2330941
79 MARC

Independent collier : the coal miner as archetypal proletarian reconsidered / edited by Royden Harrison. — New York : St. Martin's Press, 1978.
276 p. : maps ; 24 cm.
Includes bibliographical references and index.
ISBN 0-312-41290-8 : $18.95
1. Coal-miners—Great Britain—History—Addresses, essays, lectures. I. Harrison, Royden John.
HD8039.M62G7575 1978 78-3119
331.7'62'2330941
78 MARC

—JAPAN—BIOGRAPHY

Hayashi, Eidai, 1933-
(Chiku-Hō kōfuzuka)
筑豊坑夫塚 / 林えいだい. — 東京：晩声社, 1978.
244 p., [5] leaves of plates : ill. ; 20 cm. — (ルポルタージュ叢書 ; 8) Ja 78-26874
Series romanized: Ruporutāju sōsho.
¥1300
1. Katsura, Takumi, 1907-1977. 2. Coal-miners—Japan—Biography. I. Title.
HD8039.M62J263 79-804679

—JAPAN—HOKKAIDO—HISTORY

Hokkaidō Kaitaku Kinenkan.
(Hokkaidō ni okeru tankō no hatten to rōdōsha)
北海道における炭鉱の発展と労働者. — 札幌：北海道開拓記念館, 1978.
112 p. ; 26 cm. — (北海道開拓記念館研究報告 ; 第4号) Ja 78-16990
Cover title.
Title on p. [4] of cover: On the history of coal mine and colliery society in Hokkaido.
Summary in English.
"北海道石炭産業史関係文献資料目録": p. 99-112.
1. Coal trade—Japan—Hokkaido—History. 2. Coal miners—Japan—Hokkaido—History. I. Title. II. Title: On the history of coal mine and colliery society in Hokkaido. III. Series: Hokkaidō Kaitaku Kinenkan. Hokkaidō Kaitaku Kinenkan kenkyū hōkoku ; dai 4-gō.
HD9556.J33H6 78-812639

—NETHERLANDS—LIMBURG (PROVINCE)

Mijnwerkers : verhalen om te onthouden / Wim Nijsten, Jos Bours en Marlies Hautvast. — Nijmegen : Link, [1979]
207 p. : with ill. ; 20 cm. — (Boek van de arbeid ; 1) (Link-paperback ; 15) Ne79-5
1. Coal-miners—Netherlands—Limburg (Province) 2. Coal mines and mining—Netherlands—Limburg (Province) I. Nijsten, Wim. II. Bours, Jos, 1946- III. Hautvast, Marlies, 1952- IV. Series.
HD8039.M62N4647 79-359867
*79 MARC

—PENNSYLVANIA

see also Anthracite Coal Strike, 1897

COAL-MINERS (Continued)

—PENNSYLVANIA—CASE STUDIES

Blumberg, Melvin.
Job switching in autonomous groups : a descriptive and exploratory study in an underground coal mine / Melvin Blumberg. — San Francisco : R & E Research Associates, 1978.
ix, 164 p. : ill. ; 28 cm.
Bibliography: p. 122-129.
ISBN 0-88247-531-2 : $9.00
1. Coal-miners—Pennsylvania—Case studies. 2. Work groups—Pennsylvania—Case studies. 3. Industrial organization—Pennsylvania—Case studies. 4. Industrial sociology—Case studies. 5. Job satisfaction—Pennsylvania—Case studies. I. Title.
HD8039.M62U6173 78-62237
331.7'62'23340974879
79 MARC

—PENNSYLVANIA—HISTORY

Patch/work voices : the culture and lore of a mining people / compiled and written by Dennis F. Brestensky, Evelyn A. Hovanec, Albert N. Skomra. — Pittsburgh : University Center for International Studies, University of Pittsburgh, c1978.
xi, 83 p. : ill. ; 23 cm.
Bibliography: p. 80-81.
ISBN 0-916002-36-5 : $3.95
1. Coal-miners—Pennsylvania—History. I. Brestensky, Dennis F. II. Hovanec, Evelyn A. III. Skomra, Albert N.
HD8039.M62U6494 338.7'62'233409748 78-23824
78 MARC

—POLAND

see also Coal-miners' Strike, Krolewska Huta, Poland, 1871

—ROMANIA—OLTENIA—INTERVIEWS

Pospai, Mircea, 1948-
Amintiri din valea luminii : ⌈reportaje : viaţa şi activitatea minerilor din Oltenia⌉ / Mircea Pospai. — ⌈Craiova⌉ : "Scrisul Românesc", 1978.
189 p. ; 20 cm. R79-1217
Bibliography: p. 187-188.
lei6.75
1. Coal-miners—Romania—Oltenia—Interviews. I. Title.
HD8039.M62R843 79-373923
*79 MARC

—SILESIA, UPPER (PROVINCE)—HISTORY

Jończyk, Jan.
Der Bergarbeiterstreik im Jahre 1871 in Königshütte auf dem Hintergrund der Lage der Arbeiterklasse in Oberschlesien : aus Przegląd zachodni, Jg. 8 (1952) Sonderh. / Jan Jończyk ; übers. von Anne-Marie Griese. — Bremen : Universitätsbibliothek, 1977.
68 leaves ; 30 cm. GFR77-B11
Translation of an article in Polish from the journal Przegląd zachodni.
Includes bibliographical references.
1. Coal-miners' Strike, Królewska Huta, Poland, 1871. 2. Coal-miners—Silesia, Upper (Province)—History. I. Title: Der Bergarbeiterstreik im Jahre 1871 ...
HD5397.7.M6152 1871.K764 79-381310
79 MARC

—UKRAINE—DONETS BASIN

(Shakhtery Donbassa)
Шахтеры Донбасса : фотокнига / фото Л. С. Азриеля ; ⌈составитель Леонид Борисович Карнаухов⌉. — Киев : Мистецтво, 1977.
189 p. : ill. (some col.) ; 22 cm. USSR***
2.50rub
1. Coal-miners—Ukraine—Donets Basin. I. Azriel', L. S. II. Karnaukhov, Leonid Borisovich.
HD8039.M615S5 78-422481

—UNITED STATES

see also Coal Strike, United States, 1977-1978

Foster, William Z 1881-1961.
The coal miners, their problems in war and peace, by William Z. Foster. New York, New century publishers ⌈1945⌉
24 p. illus. 19 cm.
1. Coal-miners—United States. I. Title.
HD8039.M62U6236 331.7'62'2330973 45-3221
⌈r79⌉rev MARC

United States. Bureau of Labor Statistics.
Technological change and its labor impact in five energy industries : coal mining, oil and gas extraction, petroleum refining, petroleum pipeline transportation, electric and gas utilities. — Washington : U.S. Dept. of Labor, Bureau of Labor Statistics : for sale by the Supt. of Docs., U.S. Govt. Print. Off., 1979.
vi, 64 p. : ill. ; 28 cm. — (Bulletin - U.S. Bureau of Labor Statistics ; 2005)
Supt. of Docs. no. : L 2.3:2005
Bibliography: p. 63.
1. Machinery in industry—United States. 2. Technological innovations—Social aspects—United States. 3. Energy industries—United States—Employees. 4. Coal-miners—United States. 5. Petroleum workers—United States. 6. Electric utilities—United States—Employees. I. Title. II. Series: United States. Bureau of Labor Statistics. Bulletin ; 2005.
HD6331.2.U5U54 1979 331.1'0973 79-11188
79 MARC

—UNITED STATES—BIOGRAPHY

Husband, Joseph, 1885-1938.
A year in a coal-mine / Joseph Husband. — New York : Arno Press, 1977, c1911.
171 p. ; 21 cm. — (Work, its rewards and discontents)
Reprint of the ed. published by Houghton Mifflin, Boston.
ISBN 0-405-10178-3
1. Coal mines and mining. 2. Husband, Joseph, 1885-1938. 3. Coal-miners—United States—Biography. I. Title. II. Series.
TN803.H8 1977 338.2'7'240924 77-70508
77 MARC

Witt, Matt.
In our blood : four coal mining families / by Matt Witt ; photos. by Earl Dotter. — 1st ed. — Washington : Highlander Research and Education Center, c1979.
90 p. : ill. ; 29 cm.
ISBN 0-9602226-1-8 : $6.95
1. Coal mines and mining—United States. 2. Coal-miners—United States—Biography. I. Dotter, Earl. II. Title.
TN805.A5W67 338.2'7'20973 78-71518
79 MARC

—UNITED STATES—CASE STUDIES

Goodman, Paul S
Assessing organizational change : the Rushton quality of work experiment / Paul S. Goodman, with major contributions by Edward Conlon, Dennis Epple, Eduard Fidler. — New York : Wiley, c1979.
xii, 391 p. : ill. ; 24 cm. — (Wiley series on organizational assessment and change)
"A Wiley-Interscience publication."
Bibliography: p. 384-385.
Includes index.
ISBN 0-471-04782-1
1. Coal-miners—United States—Case studies. 2. Organizational change—Case studies. 3. Industrial organization—United States—Case studies. 4. Coal mines and mining—United States—Management—Case studies. I. Title. II. Series.
HD8039.M62U6258 658.31'52 78-31857
79 MARC

—UNITED STATES—HISTORY

Dix, Keith.
Work relations in the coal industry : the hand-loading era, 1880-1930 / Keith Dix. — ⌈Morgantown⌉ : Institute for Labor Studies, Division of Social and Economic Development, Center for Extension and Continuing Education, West Virginia University, 1977.
xvii, 127 p. : ill. ; 23 cm. — (West Virginia University bulletin ; ser. 78, no. 7-2)
Includes bibliographical references.
1. Coal-miners—United States—History. 2. Industrial relations—United States—History.
HD8039.M62U6232 331.7'62'23340973 78-622355
79 MARC

—WALES—CARMARTHENSHIRE

Town, Stephen W
After the mines : changing employment opportunities in a South Wales valley / by Stephen W. Town. — Cardiff : University of Wales Press, 1978.
⌈4⌉, vi, 138 p. : map ; 25 cm. — (Social science monographs ; no. 4)
GB78-26602
Bibliography: p. 137-138.
ISBN 0-7083-0660-8 : £2.50
1. Labor supply—Wales—Carmarthenshire. 2. Occupational mobility—Wales—Carmarthenshire. 3. Coal-miners—Wales—Carmarthenshire. 4. Carmarthenshire—Occupations. I. Title. II. Series: Social science monographs (Cardiff) ; no. 4.
HD5766.C37T68 331.1'09429'68 78-324959
79 MARC

COAL-MINERS' STRIKE, KRÓLEWSKA HUTA, POLAND, 1871

Jończyk, Jan.
Der Bergarbeiterstreik im Jahre 1871 in Königshütte auf dem Hintergrund der Lage der Arbeiterklasse in Oberschlesien : aus Przegląd zachodni, Jg. 8 (1952) Sonderh. / Jan Jończyk ; übers. von Anne-Marie Griese. — Bremen : Universitätsbibliothek, 1977.
68 leaves ; 30 cm. GFR77-B11
Translation of an article in Polish from the journal Przegląd zachodni.
Includes bibliographical references.
1. Coal-miners' Strike, Królewska Huta, Poland, 1871. 2. Coal-miners—Silesia, Upper (Province)—History. I. Title: Der Bergarbeiterstreik im Jahre 1871 ...
HD5397.7.M6152 1871.K764 79-381310
79 MARC

COAL MINES AND MINING

see also Coal—Reserves; Coal gasification, Underground; Coal mine waste; Radio in mining; Strikes and lockouts—Coal mining; and headings beginning with the word Mine and with the word Mining

Ch'ung-ch'ing mei t'an k'o hsüeh yen chiu so.
(Hsiao mei yao k'ai ts'ai chih shih)
小煤窑开采知识 / 重庆煤炭科学研究所编. — 北京：煤炭工业出版社，1976.
126 p. : ill. ; 19 cm.
"根据原燃料化学工业出版社纸型重印"
$0.30

1. Coal mines and mining. I. Title.
TN802.C49 1976 78-840966

Efremov, Alekseĭ Vladimirovich, kandidat ėkonomicheskikh nauk.
(Potochnye metody dobychi ugli͡a)
Поточные методы добычи угля / А. В. Ефремов, Э. И. Андрющенко, В. В. Середа. — Киев : Техніка, 1977.
148 p. : ill. ; 19 cm. — (Библиотека передового опыта : Технология производства) USSR 78
Series romanized: Biblioteka peredovogo opyta : Tekhnologii͡a proizvodstva.
Bibliography: p. ⌈142⌉
0.51rub
1. Coal mines and mining. I. Andri͡ushchenko, Ėduard Ivanovich, joint author. II. Sereda, Vladimir Vasil'evich, joint author. III. Title.
TN803.E35 78-420616

Glushko, Vasiliĭ Trofimovich.
(Geofizicheskiĭ kontrol' v ugol'nykh shakhtakh)
Геофизический контроль в угольных шахтах / В. Т. Глушко, В. С. Ямщиков, А. А. Яланский. — Киев : Наук. думка, 1978.
224 p. : ill. ; 21 cm. USSR 78
At head of title: Akademii͡a nauk Ukrainskoĭ SSR. Institut geotekhnicheskoĭ mekhaniki.
Bibliography: p. 213-⌈222⌉
2.00rub
1. Coal mines and mining. 2. Rock pressure. I. I͡Amshchikov, Valeriĭ Sergeevich, joint author. II. I͡Alanskiĭ, Anatoliĭ Aleksandrovich, joint author. III. Title.
TN803.G57 78-422430

Husband, Joseph, 1885-1938.
A year in a coal-mine / Joseph Husband. — New York : Arno Press, 1977, c1911.
171 p. ; 21 cm. — (Work, its rewards and discontents)
Reprint of the ed. published by Houghton Mifflin, Boston.
ISBN 0-405-10178-3
1. Coal mines and mining. 2. Husband, Joseph, 1885-1938. 3. Coal-miners—United States—Biography. I. Title. II. Series.
TN803.H8 1977 338.2'7'240924 77-70508
77 MARC

COAL MINES AND MINING (Continued)

Kilíachkov, Anatoliĭ Petrovich.
(Vskrytie i sistemy razrabotki ugol'nykh mestorozhdeniĭ)
Вскрытие и системы разработки угольных месторождений / А. П. Килячков. — Изд. 4., перер. и доп. — Москва : Недра, 1976.
359 p. : ill. ; 23 cm. USSR***
"Допущено ... в качестве учебника для горных техникумов."
Bibliography: p. ⌐356⌐
0.94rub
1. Coal mines and mining. I. Title.
TN802.K5 1976 78-423186

Líakhov, G M
(Osnovy razrabotki ugol'nykh mestorozhdeniĭ podzemnym sposobom)
Основы разработки угольных месторождений подземным способом : ⌐для сред. проф.-техн. учеб. заведений⌐ / Г. М. Ляхов. — Москва : Недра, 1974.
196 p. : ill. ; 22 cm. USSR 74
0.30rub
1. Coal mines and mining. I. Title.
TN802.L514 78-423181

Lindbergh, Kristina.
Coal, a contemporary energy story / by Kristina Lindbergh & Barry Provorse ; edited by Robert Conte. — Seattle : Scribe Pub. Corp., c1977.
207 p. : ill. (some col.) ; 29 cm.
Bibliography: p. 202-203.
Includes index.
ISBN 0-915748-02-9 : $22.50
1. Coal mines and mining. 2. Coal. I. Provorse, Barry L., joint author. II. Conte, Robert. III. Title.
TN802.L53 553'.2 79-103717
79 MARC

Rio Tinto North American Services Limited.
World coking coal supply study / prepared by Rio Tinto North American Services Limited for the Economic, Financial and Market Analysis Sub-committee on North East Coal Development. — ⌐Victoria⌐ : Province of British Columbia, Ministry of Economic Development, 1976.
iv, 94 p. ; 28 cm. C***
Bibliography: p. 92-94.
ISBN 0-7718-8024-3
1. Coal trade. 2. Coal mines and mining. I. British Columbia. Economic, Financial, and Market Analysis Sub-committee on North East Coal Development. II. Title.
HD9540.5.R56 1976 338.2'7'2 79-309089
79 MARC

Shan-tung k'uang yeh hsüeh yüan. Ts'ai mei chiao yen shih.
(Mei k'uang ts'ai chüeh chi pen chih shih)
煤矿采掘基本知识 / 山东矿业学院采煤教研室编. — 第1版. — 北京：煤炭工业出版社，1976.
268 p. : ill. ; 19 cm.
$0.59
1. Coal mines and mining. I. Title.
TN802.S37 1976 78-841022

Sitko, Włodzimierz.
Ekonomika górnictwa / Włodzimierz Sitko. — Gliwice : PŚ, 1976 (Gliw. : PŚ)
234 p. : ill. ; 25 cm. — (Skrypty uczelniane - Politechnika Śląska im. W. Pstrowskiego ; Nr 596) P76-6762
Bibliography: p. ⌐232⌐-234.
zł27.00
1. Coal mines and mining. I. Title.
TN802.S5263 79-344608
79 MARC

Ssu-ch'uan k'uang yeh hsüeh yüan. Ts'ai mei fang fa chiao yen shih.
(K'ai ts'ai fang fa)
开采方法 / 四川矿业学院采煤方法教研室编. — 第1版. — 北京：煤炭工业出版社，1976.
262 p., ⌐1⌐ fold. leaf of plates : ill. ; 19 cm. — (煤矿干部技术知识丛书)
$0.60
1. Coal mines and mining. I. Title. II. Series: Mei k'uang kan pu chi shu chih shih ts'ung shu.
TN802.S75 1976 78-841017

—ACCIDENTS

see also Coal—Reserves; Coal gasification, Underground; Coal mine waste; Mine explosions; Mine fires; Radio in mining; Strikes and lockouts—Coal mining; and headings beginning with the word Mine and with the word Mining

P'ang, Pai-hsien.
(Ju ching an ch'üan ch'ang shih)
入井安全常识 / 庞百先编写. — 新1版. — 北京：煤炭工业出版社，1976.
44 p. : ill. ; 19 cm. — (煤矿安全生产知识丛书)
$0.13
1. Coals mines and mining—Safety measures. 2. Coal mines and mining—Accidents. I. Title. II. Series: Mei k'uang an ch'üan sheng ch'an chih shih ts'ung shu.
TN295.P28 79-835971

—COLLECTED WORKS

Zenkoku Tankō Gijutsukai.
(Zenkoku Tankō Gijutsukai kenkyūkai hōkokusho)
全国炭鉱技術会研究会報告書 第14- 回; 1970-
⌐東京?⌐ 全国炭鉱技術会
v. ill. 30 cm.
Continues: Zenkoku Tankō Gijutsukai. Zentangi kenkyūkai hōkokusho.
1. Coal mines and mining—Collected works. 2. Coal mines and mining—Japan—Collected works. I. Title.
TN799.9.Z45b 78-644260
(MARC-S)

Zenkoku Tankō Gijutsukai.
(Zentangi kenkyūkai hōkokusho)
全炭技研究会報告書 第 -13回 ; 19 -69.
⌐東京⌐ 全国炭鉱技術会
v. ill. 26-30 cm.
Continued by: Zenkoku Tankō Gijutsukai. Zenkoku Tankō Gijutsukai kenkyūkai hōkokusho.
1. Coal mines and mining—Collected works. 2. Coal mines and mining—Japan—Collected works. I. Title.
TN799.9.Z45b 78-644250
(MARC-S)

—CONGRESSES

Commission of the European Communities.
Underground transport in coal mines : information symposium, Luxembourg, May 24 to 26, 1978 / Commission of the European Communities. — Redhill ⌐Eng.⌐ : Colliery Guardian, c1978.
2 v. : ill. ; 24 cm. GB***
"EUR 5998 d, e, f."
Symposium organized by the Commission of the European Communities, Directorate-General for Energy and Directorate-General for Scientific and Technical Information and Information Management.
Includes bibliographical references.
ISBN 0-86108-014-9
1. Coal mines and mining—Congresses. 2. Coal-handling—Congresses. I. Commission of the European Communities. Directorate-General for Energy. II. Commission of the European Communities. Directorate-General for Scientific and Technical Information and Information Management. III. Title.
TN799.9.C66 1978 622'.6 79-310106
79 MARC

Hlavní směry výzkumu v kamenouhelném hornictví do roku 1970 : materiály vydány pouze pro potřebu účastníků konference o technickém rozvoji 7LP. — Ostrava-Radvanice : Vědeckovýzkumný uhelmý ústav, ⌐1962?⌐
2 v. (330 p.) : ill. ; 21 cm.
Contributions by various authors.
1. Coal mines and mining—Congresses. I. Vědeckovýzkumný uhelný ústav.
TN799.9.H55 79-411613
79 MARC

International Symposium on Stability in Coal Mining, 1st, Vancouver, B.C., 1978.
Stability in coal mining : proceedings of the First International Symposium on Stability in Coal Mining, Vancouver, British Columbia, Canada, 1978 / edited by C. O. Brawner and Ian P. F. Dorling. — San Francisco : Miller Freeman Publications, c1979.
496 p. : ill. ; 24 cm. — (A World coal book)
Includes bibliographies and index.
ISBN 0-87930-108-2
1. Coal mines and mining—Congresses. 2. Mine subsidences—Congresses. I. Brawner, C. O. II. Dorling, Ian P. F. III. Title. IV. Series.
TN799.9.I635 1978 622'.33 78-73842
79 MARC

...onal Symposium on Thick Seam Mining, Dhanbad, India, 1977.
International Symposium on Thick Seam Mining, Dhanbad, 1977 / organised by Department of Mining Engineering, Indian School of Mines. — Dhanbad : The Dept., ⌐1977?⌐
⌐208⌐ p. : ill. ; 31 cm.
Cover title: Proceedings of International Symposium on Thick Seam Mining, Dhanbad, 1977.
"Golden jubilee, Indian School of Mines, 1926-76."
Includes bibliographies.
Rs150.00
1. Coal mines and mining—Congresses. I. Title. II. Title: Proceedings of International Symposium on Thick Seam Mining, Dhanbad, 1977.
TN799.9.I64 1977 622'.33 78-912980
79 MARC

—ELECTRIC EQUIPMENT

K'ai Luan mei k'uang.
(Mei k'uang tien kung chi ch'u chih shih)
煤矿电工基础知识 / 开滦煤矿编. — 第1版. — 北京：煤炭工业出版社，1975, 1977 printing.
228 p. : ill. ; 19 cm. — (煤矿井下电工丛书)
$0.52
1. Electricity in mining. 2. Coal mines and mining—Electric equipment. I. Title. II. Series: Mei k'uang ching hsia tien kung ts'ung shu.
TN343.K314 1975 79-841400

—ELECTRIC EQUIPMENT—ADDRESSES, ESSAYS, LECTURES

(Voprosy ėlektrifikat͡sii ugol'nykh shakht i sovershenstvovanii͡a ėlektroprivoda gornykh mashin)
Вопросы электрификации угольных шахт и совершенствования электропривода горных машин : ⌐Сборник статей / Ред. коллегия ... проф., д-р техн. наук И. Ф. Гончаревич (отв. ред.) ... et al.⌐. — Москва : ИГД, 1976.
70 p. : ill. ; 19 cm. — (Научные сообщения - Институт горного дела им. А. А. Скочинского ; вып. 144) USSR 77
At head of title: Ministerstvo ugol'noĭ promyshlennosti SSSR. Akademii͡a nauk SSSR.
Includes bibliographies.
0.48rub
1. Coal mines and mining—Electric equipment—Addresses, essays, lectures. 2. Mining machinery—Electric driving—Addresses, essays, lectures. I. Goncharevich, Igor' Fomich. II. Series: Moscow. Institut gornogo dela imeni A. A. Skochinskogo. Nauchnye soobshchenii͡a ; vyp. 144.
TN4.M642 vyp. 144 79-410430
[TN843]

—ELECTRIC EQUIPMENT—CONGRESSES

WVU Conference on Coal Mine Electrotechnology, 2d, Morgantown, W.Va., 1974.
Proceedings of the Second WVU Conference on Coal Mine Eletrotechnology, June 21 ⌐i.e. 12⌐-14, 1974, West Virginia University, Evansdale Campus, Morgantown, West Virginia / sponsored by Department of Electrical Engineering, College of Engineering, West Virginia University, and Industrial Applications Society, IEEE. — ⌐s.l.⌐ : Mining Industry Committee, ⌐1974⌐
307 p. in various pagings : ill. ; 28 cm.
Includes bibliographical references.
1. Coal mines and mining—Electric equipment—Congresses. 2. Coal mines and mining—Electronic equipment—Congresses. I. West Virginia. University. Dept. of Electrical Engineering. II. IEEE Industry Applications Society.
TN343.W14 1974 622'.48 79-317459
79 MARC

—ELECTRIC EQUIPMENT —MAINTENANCE AND REPAIR

Ch'en, Yen-shih.
(Mei tien tsuan ti chien hsiu)
煤电钻的检修 / 陈彦士编. — 修订本新1版. — 北京：煤炭工业出版社，1976.
102 p., ⌐1⌐ fold. leaf of plates : ill. ; 19 cm. — (煤矿机电设备维护检修丛书)
Series romanized: Mei k'uang chi tien she pei wei hu chien hsiu ts'ung shu.
First ed. published in 1963 under title: Tien tsuan ti chien hsiu.
$0.27
1. Drilling and boring machinery—Electric equipment—Maintenance and repair. 2. Coal mines and mining—Electric equipment—Maintenance and repair. I. Title.
TJ1263.C52 1976 78-841238

COAL MINES AND MINING (Continued)

—ELECTRONIC EQUIPMENT —CONGRESSES

WVU Conference on Coal Mine Electrotechnology, 2d, Morgantown, W.Va., 1974.
Proceedings of the Second WVU Conference on Coal Mine Eletrotechnology, June 21 ⌜i.e. 12⌝-14, 1974, West Virginia University, Evansdale Campus, Morgantown, West Virginia / sponsored by Department of Electrical Engineering, College of Engineering, West Virginia University, and Industrial Applications Society, IEEE. — ⌜s.l.⌝ : Mining Industry Committee, ⌜1974⌝
307 p. in various pagings : ill. ; 28 cm.
Includes bibliographical references.
1. Coal mines and mining—Electric equipment—Congresses. 2. Coal mines and mining—Electronic equipment—Congresses. I. West Virginia. University. Dept. of Electrical Engineering. II. IEEE Industry Applications Society.
TN343.W14 1974 622'.48 79-317459
79 MARC

—ENVIRONMENTAL ASPECTS—ALBERTA

Techman Ltd.
Materials handling methods for plains strip mines / prepared for Land Conservation and Reclamation Division of the Alberta Department of Environment and in cooperation with the Coal Association of Canada ⌜by⌝ Techman Ltd. — ⌜s.l.⌝ : Techman Ltd., 1977.
2 v. : ill. (some col.) ; 29 cm. C***
Bibliography: v. 1, p. 117-120.
CONTENTS: v. 1. Text.—v. 2. Figures, tables, appendices.
1. Coal mines and mining—Environmental aspects—Alberta. 2. Reclamation of land—Alberta. 3. Strip mining—Environmental aspects—Alberta. 4. Materials handling. I. Alberta. Land Conservation and Reclamation Division. II. Coal Association of Canada. III. Title.
TD195.C58T42 1977 622'.33 78-326850
79 MARC

—ENVIRONMENTAL ASPECTS—BRITISH COLUMBIA

Trenholme, Neil S
Aquatic migratory bird resources of the Northeast Coal Study area : a technical supplement to the Northeast Coal Study Preliminary environmental report on proposed transportation links and townsites : prepared for the Environment and Land Use Sub-committee on Northeast Coal Development / by Neil S. Trenholme ; in collaboration with the Resource Analysis Branch, Ministry of the Environment, Province of British Columbia. — Victoria, B.C. : Ministry of the Environment, 1978.
x, 43, ⌜5⌝ leaves : ill., 2 fold maps ; 28 cm. C78-16074-6
"This report is the outcome of a joint project of the Canadian Wildlife Service (Environment Canada) and the Resource Analysis Branch (B.C. Ministry of the Environment)"
Bibliography: leaf 44.
ISBN 0-7719-8098-1
1. Waterfowl—British Columbia. 2. Coal mines and mining—Environmental aspects—British Columbia. I. Canada. Wildlife Service. II. British Columbia. Ministry of the Environment. Resource Analysis Branch. III. British Columbia. Environment and Land Use Sub-committee on Northeast Coal Development. Preliminary environmental report of proposed transportation links and townsites. IV. Title.
QL696.A5T73 333.9'5 79-305928
79 MARC

—ENVIRONMENTAL ASPECTS—ILLINOIS

Illinois. Environmental Protection Agency.
An Illinois environmental issue : water resources and coal mining. — Springfield : Illinois Environmental Protection Agency, ⌜1978⌝
ii, 16, ⌜1⌝ p. : graphs ; 28 cm. — (Bulletin - Illinois Environmental Protection Agency ; no. 1)
Cover title.
Bibliography: p. ⌜17⌝
1. Coal mines and mining—Environmental aspects—Illinois. 2. Water—Pollution—Law and legislation—Illinois. I. Title. II. Series: Illinois. Environmental Protection Agency. Bulletin - Illinois Environmental Protection Agency ; no. 1.
TD428.M56I44 1978 333.9'102'09773 79-621101
79 MARC

Illinois Institute for Environmental Quality.
Problem sites : lands affected by underground mining for coal in Illinois / by J. R. Nawrot ... ⌜et al.⌝ Cooperative Wildlife Research Laboratory, Southern Illinois University at Carbondale. — Chicago : Illinois Institute for Environmental Quality, 1977.
i, 556 p. : ill. ; 28 cm. — (IIEQ document ; no. 77/38)
Bibliography: p. 544-551.
1. Coal mines and mining—Environmental aspects—Illinois. 2. Reclamation of land—Illinois. I. Nawrot, Jack R. II. Illinois. Southern Illinois University, Carbondale. Cooperative Wildlife Research Laboratory. III. Title. IV. Series: Illinois Institute for Environmental Quality. IIEQ document ; no. 77/38.
TD195.C58I46 1977 614.7'76 78-623280
79 MARC

Our reclamation future : the missing bet on trees / by W. Clark Ashby ... ⌜et al.⌝. — Chicago : State of Illinois, Institute for Environmental Quality, 1978.
99 p. : ill. ; 28 cm. — (IIEQ document ; no. 78/04)
"Project no. 80.057."
Bibliography: p. ⌜87⌝-90.
1. Reclamation of land—Illinois. 2. Strip mining—Environmental aspects—Illinois. 3. Tree planting—Illinois. 4. Coal mines and mining—Environmental aspects—Illinois. I. Ashby, William Clark, 1922- . II. Illinois Institute for Environmental Quality. III. Series: Illinois Institute for Environmental Quality. IIEQ document ; no. 78/04.
S621.5.S8O94 631.6 78-623281
79 MARC

—ENVIRONMENTAL ASPECTS—MARYLAND

Maryland. Environmental Service.
Feasibility of returning fly ash to Western Maryland for mine reclamation / prepared by the Maryland Environmental Service, Department of Natural Resources .... — Rev. — ⌜Washington : Applachian Regional Commission⌝, 1975.
ca. 150 p. : maps ; 29 cm. — (Report - Appalachian Regional Commission ; ARC 74-139-MD-3202)
Prepared under study grant 74-139 from the Appalachian Regional Commission.
Includes a bibliography.
1. Coal mines and mining—Environmental aspects—Maryland. 2. Reclamation of land—Maryland. 3. Fly ash. I. Appalachian Regional Commission. II. Title. III. Series: Appalachian Regional Commission. Report ; ARC74-139-MD-3202.
HC107.A133P6372a ARC74-139-MD-3202 78-623551
330'.974'9
⌜TD195.C58⌝ 79 MARC

—ENVIRONMENTAL ASPECTS—MONTANA—COLSTRIP REGION

DePuit, Edward J
Plant response and forage quality for controlled grazing on coal mine spoils pastures : 1976 progress report / prepared by Edward J. DePuit, Waite H. Willmuth, Joe G. Coenenberg. — Bozeman : Reclamation Research Unit, Montana Agricultural Experiment Station, Montana State University, 1977.
viii, 73 p. : ill. ; 29 cm. — (Research report - Montana Agricultural Experiment Station ; 155)
Bibliography: p. 67-73.
1. Grazing—Environmental aspects—Montana—Colstrip region. 2. Pastures—Montana—Colstrip region. 3. Range management—Montana—Colstrip region. 4. Revegetation—Montana—Colstrip region. 5. Strip mining—Environmental aspects—Montana—Colstrip region. 6. Coal mines and mining—Environmental aspects—Montana—Colstrip region. 7. Colstrip region, Mont. I. Willmuth, Waite H., joint author. II. Coenenberg, Joe G., joint author. III. Title: Plant response and forage quality ... IV. Title: Coal mine spoils pastures. V. Series: Montana. Agricultural Experiment Station, Bozeman. Research report ; 155.
SF85.35.M9D46 633'.2 77-623919
79 MARC

—ENVIRONMENTAL ASPECTS—NARRAGANSETT BASIN

Frimpter, Michael H 1934-
Geohydrologic impacts of coal development in the Narragansett Basin, Massachusetts and Rhode Island / by Michael H. Frimpter and Anthony Maevsky ; prepared in cooperation with the Massachusetts Science and Technology Foundation. — Washington : U.S. Govt. Print. Off., 1979.
iv, 35 p. : ill. ; 24 cm. — (Geological Survey water-supply paper ; 2062)
Bibliography: p. 33-35.
Supt. of Docs. no.: I 19.13:2062
1. Water, Underground—Narragansett basin. 2. Water-supply—Narragansett basin. 3. Coal mines and mining—Environmental aspects—Narragansett basin. I. Maevsky, Anthony, joint author. II. Massachusetts Science and Technology Foundation. III. Title. IV. Series: United States. Geological Survey. Water-supply paper ; 2062.
GB1025.M4F73 553'.79'09744 79-607009
79 MARC

—ENVIRONMENTAL ASPECTS—NEW MEXICO

New Mexico. Governor's Energy Task Force. Committee on Coal Surface Mining.
Coal surface mining in New Mexico : a report of the Governor's Task Force Committee on Coal Surface Mining. — Santa Fe : Governor's Energy Task Force, 1975.
ii, 28 leaves : ill. ; 28 cm.
Bibliography; leaf 28.
1. Coal mines and mining—Environmental aspects—New Mexico. 2. Strip mining—Environmental aspects—New Mexico. I. Title.
TD195.C58N48 1975 333.7 78-624238
79 MARC

—ENVIRONMENTAL ASPECTS—NORTH DAKOTA

Moran, Stephen R
Geologic, hydrologic, and geochemical concepts and techniques in overburden characterization for mined-land reclamation / by Stephen R. Moran, Gerald H. Groenewold, John A. Cherry. — ⌜Grand Forks⌝ : North Dakota Geological Survey, 1978.
ix, 152 p. : ill., diagrs. (12 fold. in pocket) ; 28 cm. — (Report of investigation North Dakota Geological Survey ; no. 63)
Bibliography: p. 102-104.
1. Coal mines and mining—Environmental aspects—North Dakota. 2. Reclamation of land—North Dakota. I. Groenewold, Gerald H., joint author. II. Cherry, John A., joint author. III. Title. IV. Series: North Dakota. Geological Survey. Report of investigations ; no. 63.
TN24.N9A3 no. 63 333.8'2 78-622786
⌜TD195.C58⌝ 79 MARC

—ENVIRONMENTAL ASPECTS—YELLOWSTONE RIVER VALLEY

Harza Engineerig Company, Chicago.
Analysis of energy projections and implications for resource requirements : prepared for Missouri River Basin Commission, Yellowstone level B study. — Chicago : Harza Engineering Co., 1976.
ca. 120 leaves in various foliations : ill. ; 28 cm.
Includes bibliographical references.
1. Power resources—Yellowstone River Valley. 2. Coal mines and mining—Environmental aspects—Yellowstone River Valley. 3. Power resources—United States. I. Missouri. River Basin Commission. II. Title.
TJ163.25.U6H37 1976 333.7 78-622067
79 MARC

—ENVIRONMENTAL ASPECTS—YELLOWSTONE RIVER WATERSHED, WYO.-N.D.

United States. Missouri River Basin Commission.
Yellowstone River basin and adjacent coal area level B study : report and environmental assessment. — Omaha, Neb. : Missouri River Basin Commission, 1978.
8 v. : ill. ; 28 cm.
Vols. 2-8 have cover title: Report on the Yellowstone Basin and adjacent coal area level B study.
On spine: Yellowstone Basin and adjacent coal area level B study.
Subtitle varies.
Vol. 1 has no special title.
Includes bibliographical references.
PARTIAL CONTENTS:
—v. 2. Upper Yellowstone, Montana.—v. 3. Clarks Fork-Big Horn, Montana.—v. 4. Tongue-Powder, Montana.—v. 5. Lower Yellowstone, Montana.—v. 6. North Dakota tributaries.—v. 7. Wind-Bighorn-Clarks Fork Basin, Wyoming.—v. 8. Northeast Wyoming planning area.
1. Water resources development—Environmental aspects—Yellowstone River watershed, Wyo.-N.D. 2. Coal mines and mining—Environmental aspects—Yellowstone River watershed, Wyo.-N.D. I. Title. II. Title: Report on the Yellowstone Basin and adjacent coal area level B study. III. Title: Yellowstone Basin and adjacent coal area level B study.
TC425.Y4U54 1978 333.9'1'0097863 79-622642
79 MARC

—EQUIPMENT AND SUPPLIES

Curl, S J
Underground transport in coal mines / S. J. Curl. — London : IEA Coal Research, 1978.
54 p. : ill. ; 30 cm. — (Report - IEA Coal Research ; no. ICTIS/TR 01 ISSN 0142-1816) GB78-39131
Bibliography: p. 46-54.
ISBN 9290290099
1. Coal mines and mining—Equipment and supplies. 2. Mine haulage. 3. Coal-handling machinery. I. Title. II. Series: IEA Coal Research. Report - IEA Coal Research ; no. ICTIS/TR 01.
TN813.C87 622'.6 79-311804
79 MARC

Miskell, Jack T
Equipment and supplies for fuel exploration / Jack T. Miskell. — Stamford, Conn. : Business Communications Co., 1979.
vii, 128 leaves ; 29 cm. — (Business opportunity report ; E-010R) .
1. Oil fields—Equipment and supplies. 2. Gas industry—Equipment and supplies. 3. Coal mines and mining—Equipment and supplies. 4. Prospecting—Equipment and supplies. I. Title. II. Series.
TN871.5.M53 338.2'7'2 79-110959
00 MARC

T͡Setnarskiĭ, Igor' Aleksandrovich.
(Gornaíà mekhanika)
Горная механика : ⌜учеб. пособие для горных техникумов⌝ / И. А. Цетнарский, А. А. Кораблев, Л. Д. Борисенко. — Изд. 2-е, перераб. и доп. — Москва : Недра, 1975.
278 p. : ill. ; 28 cm. USSR 75
First ed. entered under title.
Bibliography : p. ⌜275⌝
0.87rub
1. Coal mines and mining—Equipment and supplies. I. Korablev, Anatoliĭ Aleksandrovich, joint author. II. Borisenko, L. D., joint author. III. Title.
TN813.G58 1975 79-411068

COAL MINES AND MINING (Continued)

—FIRES AND FIRE PREVENTION

(Kikai saitan to shizen hakka taisaku)
機械採炭と自然発火対策　自走支保・ドラムカッタ・ボーベル　[東京]　全国炭鉱技術会　1968.
98 p. illus. 26 cm. (全炭技研究会報告書　第12回) Ja***
Cover title.
Errata slips inserted.

1. Coal-mining machinery. 2. Coal mines and mining—Fires and fire prevention. I. Zenkoku Tankō Gijutsukai. II. Series: Zenkoku Tankō Gijutsukai. Zentangi kenkyūkai hōkokusho, dai 12-kai.
TN799.9.Z45b vol. 12 78–813994
[TN813]

Vsesoi͡uznyĭ nauchno-issledovatel'skiĭ institut gornospasatel'nogo dela.
Tushenie podzemnyhk pozharov na ugol'nykh shakhtakh / [G. G. Sobolev, V. P. Charkov, A. M. Kushnarev ... et al.] ; Vsesoi͡uz. nauch.-issled. in-t gornospasat. dela. — Moskva : Nedra, 1977.
248 p. : ill. ; 22 cm. USSR77
At head of title: Ministerstvo ugol'noĭ promyshlennosti SSSR.
1.14rub
1. Coal mines and mining—Fires and fire prevention. I. Sobolev, Georgiĭ Georgievich. II. Title.
TN315.V73 1977 79-368618
79 MARC

—HYGIENIC ASPECTS—CONGRESSES

Institute on Coal Mine Health and Safety, 4th, Colorado School of Mines, 1978.
Fourth Institute on Coal Mine Health and Safety, November 29-December 1, 1978 : proceedings / compiled by Robert T. Reeder ; sponsored by the Colorado School of Mines, in cooperation with Mine Safety and Health Administration and United States Bureau of Mines. — Golden, CO : The School, 1979.
vi, 285 p. : ill. ; 28 cm.
Includes bibliographical references.
1. Coal mines and mining—Hygienic aspects—Congresses. 2. Coal mines and mining—Safety measures—Congresses. I. Reeder, Robert T. II. Colorado. School of Mines, Golden. III. United States. Mine Safety and Health Administration. IV. United States. Bureau of Mines. V. Title.
RC965.M5157 1978 622'.8 79-624956
79 MARC

—LAW AND LEGISLATION—BRAZIL

Brazil. Laws, statutes, etc.
Legislação do carvão nacional. — Brasília : Conselho Nacional do Petróleo, Assessoria de Relações Públicas, 1977-
v. ; 23 cm. — (Coletânea Legislaçõ; 2-)
1. Coal mines and mining—Law and legislation—Brazil. I. Brazil. Conselho Nacional do Petróleo. Assessoria de Relações Públicas. II. Title. III. Series.
343'.81'077 78-381773
78 MARC

—LAW AND LEGISLATION —GREAT BRITAIN

Great Britain. National Coal Board. Industrial Training Branch.
Deputy's handbook / NCB Industrial Training Branch. — 3rd ed. — [London] : National Coal Board, 1977.
[4], 122, [1] p. ; 15 cm. GB78-04901
£0.75
1. Coal mines and mining—Law and legislation—Great Britain. 2. Mine sanitation—Law and legislation—Great Britain. I. Title.
KD3172.M53N37 1977 344'.41'01762233 78-313032
78 MARC

—LAW AND LEGISLATION—POLAND

Sadowski, Tadeusz.
Coal mining health standards in Poland / prepared by Tadeusz Sadowski. — Washington : Library of Congress, Law Library, 1978.
45 p. ; 27 cm. — (Law Library publications)
Includes bibliographical references.
1. Coal mines and mining—Law and legislation—Poland. 2. Mine sanitation—Law and legislation—Poland. I. United States. Library of Congress. Law Library. II. Title. III. Series: United States. Library of Congress. Law Library. Law Library publications.
344'.438'047 78-600080
——— Copy 2. Z663.5.C552 MARC
79

—LAW AND LEGISLATION—UNITED STATES

ALI-ABA Course of Study: Legal Issues in the Coal Industry.
ALI-ABA course of study. Legal issues in the coal industry: materials.
[Philadelphia] American Law Institute-American Bar Association Committee on Continuing Professional Education.
v. 29 cm.
Vols. for issued in 2 or more vols.
Key title: ALI-ABA course of study. Legal issues in the coal industry: materials, ISSN 0191-1589.
1. Coal mines and mining — Law and legislation — United States. I. American Law Institute-American Bar Association Committee on Continuing Professional Education. II. ALI-ABA Course of Study: Legal Issues in the Coal Industry. Legal issues in the coal industry: materials. III. Title.
KF1830.Z9A17 343'.73'077 79-642581
MARC-S

ALI-ABA Course of Study: Legal Issues in the Coal Industry, Washington, D.C., 1978.
ALI-ABA Course of Study—Legal Issues in the Coal Industry : materials. — Philadelphia : American Law Institute-American Bar Association Committee on Continuing Professional Education, c1978.
2 v. (xiv, 885 p.) ; 29 cm.
1. Coal mines and mining—Law and legislation—United States. I. American Law Institute-American Bar Association Committee on Continuing Professional Education. II. Title: Legal Issues in the Coal Industry.
KF1830.A72 1978 343'.73'077 79-101079
79 MARC

United States. Congress. House. Committee on Interior and Insular Affairs. Subcommittee on Oversight and Investigations.
Illegal mining of Federal coal : a report / prepared by the Subcommittee on Oversight and Investigations of the Committee on Interior and Insular Affairs of the U.S. House of Representatives, Ninety-sixth Congress, first session. — Washington : U.S. Govt. Print. Off., 1979.
v, 9 p. ; 24 cm.
"Committee print no. 3, 96th Congress, 1st session."
1. Coal trade—United States. 2. Coal mines and mining—Law and legislation—United States. 3. Coal leases—United States. I. Title.
HD9546.U64 1979 364.1'62 79-603230
79 MARC

—LAW AND LEGISLATION—WEST VIRGINIA

West Virginia. Laws, statutes, etc.
Underground mining laws of West Virginia : a study guide / by Mining Extension Service, College of Mineral and Energy Resources, West Virginia University. — Morgantown : The Service, 1975.
67 p. ; 28 cm. — (Study guide series ; no. 1)
Includes index.
1. Coal mines and mining—Safety regulations—West Virginia. 2. Coal mines and mining—Law and legislation—West Virginia. I. West Virginia University. Mining Extension Service. II. Title. III. Series: Study guide series (Morgantown, W.Va.) ; no. 1.
KFW1536.M5A32 1975 346'.754'04682 78-620713
79 MARC

—PRODUCTION CONTROL

Tkachenko, Ivan Ivanovich.
(Upravlenie proizvodstvom v ugol'nom ob"edinenii)
Управление производством в угольном объединении / И. И. Ткаченко, М. А. Сребный. — Москва : Недра, 1977.
230 p. : diagrs. ; 21 cm. USSR 77
Bibliography : p. 225–[226]
1.10rub
1. Coal mines and mining—Production control. I. Srebnyĭ, Mikhail Aleksandrovich, joint author. II. Title.
TN803.T56 78–422532

—QUALITY CONTROL

Blagov, Igor' Sil'vestrovich.
(Ėkonomicheskai͡a ėffektivnost' povyshenii͡a kachestva ugleĭ)
Экономическая эффективность повышения качества углей : [учебник] / И. С. Благов, В. К. Турченко. — Москва : Недра, 1978.
158 p. ; 20 cm. USSR***
0.45rub
1. Coal mines and mining — Quality control. 2. Coal — Analysis. I. Turchenko, Vasiliĭ Kuz'mich, joint author. II. Title.
TN803.B56 79–410473

—SAFETY MEASURES

(Bezopasnai͡a vyemka ugli͡a pod vodnymi ob"ektami)
Безопасная выемка угля под водными объектами / [Б. Я. Гвирцман, Н. Н. Кацнельсон, Е. В. Бошенятов ... et al.] ; Под общ. ред. Б. Я. Гвирцмана. — Москва : Недра, 1977.
175 p. : ill. ; 20 cm. USSR 77
Bibliography : p. 167–[173]
0.43rub
1. Coal mines and mining—Safety measures. 2. Mine water. I. Gvirt͡sman, Boris I͡Akovlevich.
TN295.B448 78–421144

Grit͡sko, Gennadiĭ Ignat'evich.
(Gornoe davlenie pri gruppovoĭ razrabotke plastov)
Горное давление при групповой разработке пластов / Г. И. Грицко, В. А. Шалауров. — Новосибирск : Наука, Сиб. отд-ние, 1978.
93 p. : ill. ; 21 cm. USSR 78
At head of title: Akademii͡a nauk SSSR. Sibirskoe otdelenie. Institut gornogo dela.
Bibliography : p. 91–[92]
0.90rub
1. Rock pressure. 2. Coal mines and mining—Safety measures. I. Shalaurov, Vissarion Aleksandrovich, joint author. II. Title.
TN292.G73 78–420611

Hebblewhite, B K
Evaluation of a selected roof bolting pattern / by B. K. Hebblewhite, A. J. Richmond. — [North Ryde] : Australian Coal Industry Research Laboratories, 1978.
28 p. : ill. ; 25 cm. — (Published reports - Australian Coal Industry Research Laboratories ; PR 78-14) Aus***
ISBN 0-909130-49-3
1. Mine roof bolting. 2. Coal mines and mining—Safety measures. I. Richmond, A. J., joint author. II. Title. III. Series: Australian Coal Industry Research Laboratories. Published reports - Australian Coal Industry Research Laboratories ; PR 78-14.
TN289.3.H4 622'.28 79-314593
79 MARC

Issledovanie prot͡sessov vozniknovenii͡a vnezapnykh vybrosov ugli͡a i gaza / [A. Ė. Petrosi͡an, M. F. I͡Anovskai͡a, B. M. Ivanov, V. G. Krupeni͡a ; Otv. red. N. I. Ustinov]. — Moskva : Nauka, 1978.
111 p. : ill. ; 21 cm. USSR78
At head of title: Akademii͡a nauk SSSR. Ministerstvo ugol'noĭ promyshlennosti SSSR. Institut gornogo dela im. A. A. Skochinskogo.
Bibliography: p. 105-[110]
0.75rub
1. Coal mines and mining—Safety measures. 2. Rock bursts. 3. Gas bursts. I. Petrosi͡an, A. Ė. II. Ustinov, N. I. III. Akademii͡a nauk SSSR. Institut gornogo dela.
TN295.I77 79-373819
79 MARC

Lazarov, Georgi Ĭordanov.
Okhrana na truda vŭv vŭglishtnite rudnit͡si / Georgi Lazarov. — Sofii͡a : Nauchnoizsledovatelski institut po profsŭi͡uzni problemi G. Dimitrov pri T͡SS na BPS, 1978.
71 p. ; 20 cm. Bu***
Summary in English and Russian.
Bibliography: p. 69-71.
1. Coal mines and mining—Safety measures. I. Title.
TN295.L39 79-381960
79 MARC

P'ang, Pai-hsien.
(Ju ching an ch'üan ch'ang shih)
入井安全常识 / 庞百先编写. — 新1版. — 北京：煤炭工业出版社, 1976.
44 p. : ill. ; 19 cm. — (煤矿安全生产知识丛书)
$0.13

1. Coals mines and mining—Safety measures. 2. Coal mines and mining—Accidents. I. Title. II. Series: Mei k'uang an ch'üan sheng ch'an chih shih ts'ung shu.
TN295.P28 79–835971

Peng, Syd S 1939-
Coal mine ground control / Syd S. Peng. — New York : Wiley, c1978.
xiii, 450 p. : ill. ; 24 cm.
"A Wiley-Interscience publication."
Includes bibliographical references and indexes.
ISBN 0-471-04121-1
1. Ground control (Mining) 2. Coal mines and mining—Safety measures. I. Title.
TN288.P46 622'.8 78-8965
78 MARC

COAL MINES AND MINING
—SAFETY MEASURES (Continued)

Shan-tung k'uang yeh hsüeh yüan. T'ung feng an ch'üan chiao yen shih.
(Mei k'uang t'ung feng an ch'üan chi pen chih shih)
煤矿通风安全基本知识 / 山东矿业学院通风安全教研室编. — 第1版. — 北京：煤炭工业出版社, 1976.
113 p. : ill. ; 19 cm.
"本书是以煤炭工业部组织十三个干部学校编写的煤矿基层干部培训教材《通风安全基本知识》为基础改编的"
$0.28
1. Mine ventilation. 2. Coal mines and mining—Safety measures. I. Title.
TN301.S384 1976 78-841018

Yang-ch'üan i k'uang t'ung feng ch'ü pien hsieh tsu.
(K'uang ching wa ssu)
矿井瓦斯 / 阳泉一矿通风区编写组. — 新1版. — 北京：煤炭工业出版社, 1976.
65 p. : ill. ; 19 cm. — (煤矿安全生产知识丛书)
$0.18
1. Fire-damp. 2. Coal mines and mining—Safety measures. I. Title. II. Series: Mei k'uang an ch'üan sheng ch'an chih shih ts'ung shu.
TN305.Y36 1976 78-842173

—SAFETY MEASURES—ADDRESSES, ESSAYS, LECTURES

(Rudnichnaia ventiliatsiia i tekhnika bezopasnosti)
Рудничная вентиляция и техника безопасности : респ. межвед. науч.-техн. сб. / М-во высш. и сред. спец. образования УССР ; [Редкол., ... Зборщик, М. П. (отв. ред.) ... et al.]. — Киев : Техніка, 1978.
112 p. : ill. ; 20 cm. — (Разработка месторождений полезных ископаемых ; вып. 49) USSR 78
Includes bibliographies.
1.40rub
1. Mine ventilation—Addresses, essays, lectures. 2. Coal mines and mining—Safety measures—Addresses, essays, lectures. I. Zborshchik, Mikhail Pavlovich. II. Series: Razrabotka mestorozhdenii poleznykh iskopaemykh ; vyp. 49.
TN275.A1R3 vyp. 49 78-422287
[TN301]

—SAFETY MEASURES—CONGRESSES

Institute on Coal Mine Health and Safety, 4th, Colorado School of Mines, 1978.
Fourth Institute on Coal Mine Health and Safety, November 29-December 1, 1978 : proceedings / compiled by Robert T. Reeder ; sponsored by the Colorado School of Mines, in cooperation with Mine Safety and Health Administration and United States Bureau of Mines. — Golden, CO : The School, 1979.
vi, 285 p. : ill. ; 28 cm.
Includes bibliographical references.
1. Coal mines and mining—Hygienic aspects—Congresses. 2. Coal mines and mining—Safety measures—Congresses. I. Reeder, Robert T. II. Colorado. School of Mines, Golden. III. United States. Mine Safety and Health Administration. IV. United States. Bureau of Mines. V. Title.
RC965.M5I57 1978 622'.8 79-624956
79 MARC

—SAFETY REGULATIONS—UNITED STATES

The Federal mine safety and health act / James M. Day, chairman. — [New York] : Practising Law Institute, c1979.
360 p. (p. 356-360 blanks) : ill. ; 22 cm. — (Commercial law and practice course handbook series ; no. 213)
"A4-3045."
"Prepared for distribution at the Federal mine safety and health act program, May-June 1979."
"Federal mine safety & health act of 1977, Public law 91-173, as amended by Public Law 95-164": p. 255-349.
1. Mine safety—Law and legislation—United States. 2. Mine sanitation—Law and legislation—United States. 3. Coal mines and mining—Safety regulations—United States. I. Day, James MacDonald, 1930- . II. New York (City). Practising Law Institute. III. United States. Laws, statutes, etc. Federal mine safety and health act of 1977. 1979. IV. Series.
KF3574.M5F4 344'.73'047 79-88396
00 MARC

United States. Congress. House. Committee on Education and Labor. Subcommittee on Labor Standards.
To amend the Federal coal mine health and safety act : hearing before the Subcommittee on Labor Standards of the Committee on Education and Labor, House of Representatives, Ninety-fifth Congress, first session on H.R. 4287 ... held in Washington, D.C., on March 29, 1977. — Washington : U.S. Govt. Print. Off., 1977.
iii, 329 p. : ill. ; 24 cm.
Includes bibliographical references.
1. Coal mines and mining—Safety regulations—United States. 2. Coal-miners—Legal status, laws, etc.—United States. I. Title.
KF27.E348 1977a 344'.73'047 77-604918
78[r79]rev MARC

United States. Federal Mine Safety and Health Review Commission.
Decisions - Federal Mine Safety and Health Review Commission. Mar. 1979–
[Washington] Federal Mine Safety and Health Review Commission, for sale by the Supt. of Docs., U. S. Govt. Print. Off.]
v. 27 cm. monthly.
Key title: Decisions - Federal Mine Safety and Health Review Commission, ISSN 0193-7987.
1. Mine safety—Law and legislation—United States—Cases. 2. Mine sanitation—Law and legislation—United States—Cases. 3. Coal mines and mining—Safety regulations—United States—Cases. I. Key title.
KF3574.M5A4914 344'.73'0465 79-643482
MARC-S

—SAFETY REGULATIONS—WEST VIRGINIA

West Virginia. Laws, statutes, etc.
Underground mining laws of West Virginia : a study guide / by Mining Extension Service, College of Mineral and Energy Resources, West Virginia University. — Morgantown : The Service, 1975.
67 p. ; 28 cm. — (Study guide series ; no. 1)
Includes index.
1. Coal mines and mining—Safety regulations—West Virginia. 2. Coal mines and mining—Law and legislation—West Virginia. I. West Virginia. University. Mining Extension Service. II. Title. III. Series: Study guide series (Morgantown, W.Va.) ; no. 1.
KFW1536.M5A32 1975 346'.754'04682 78-620713
79 MARC

—TERMINOLOGY

Gonzalez, Daniel.
L'Occitan parlat jos tèrra : los carbonièrs de Carmaus / per Daniel Gonzalez. — Toulouse : Université de Toulouse II, Le Mirail, [1977?]
2 v. (252 p.) : ill. ; 22 cm. — (Documents et archives pour la recherche sociolinguistique méridionale : Série E, Document de travail ; no 3-4) F***
At head of title: Université de Toulouse II, Le Mirail. Centre de sociolinguistique et de dialectologie sociale.
CONTENTS: t. 1. Tèxtes sul trabalh del Cròs.—t. 2. Vocabulari del Cròs.
1. Provençal language, Modern—Dialects—France—Carmaux. 2. Carmaux. France—Industries. 3. Coal-miners—France—Carmaux—Language. 4. Coal mines and mining—France—Carmaux. 5. Coal mines and mining—Terminology. I. Title. II. Series.
PC3447.C28G6 78-375700
78 MARC

—WASTE DISPOSAL

see also Coal mine waste

Li, Ming-i.
(Jui san k'uang yeh kung ssu hsi mei ch'ang wei k'uang fei shui wu jan ch'u li chih yen chiu)
瑞三鑛業公司洗煤場尾礦廢水污染處理之研究 / [李明誼, 翁林廷彬]. — [台北?]：工業技術研究院礦業研究所, 民國65 [1976]
16 p. : ill. ; 27 cm. — (MRSO report ; 156)
Cover title.
1. Coal mines and mining—Waste disposal. 2. Jui san k'uang yeh kung ssu. I. Weng Lin, T'ing-pin, joint author. II. Title. III. Series: Kung yeh chi shu yen chiu yüan. K'uang yeh yen chiu so. MRSO report ; 156.
TD899.M5L5 79-835695

—BELGIUM—SOUMAGNE—HISTORY

Monard, Dieudonné.
Le siège de José de la Société anonyme des charbonnages de Werister à Romsée : sa vie depuis ses origines : contribution à l'étude de l'histoire de l'origine des houillères dans le plateau de Herve / par Dieudonné Monard. — Fléron : Cercle archéo-historique des cantons de Fléron et Grivegnée, 1975.
49 p. ; 28 cm. Be***
70.00F
1. Coal trade—Belgium—Soumagne—History. 2. Coal mines and mining—Belgium—Soumagne—History. I. Cercle archéo-historique des cantons de Fléron et Grivegnée. II. Title.
HD9555.B43S685 78-391615
338.7'62'233094934
79 MARC

—CALIFORNIA—LIVERMORE—HISTORY

Mosier, Dan L
Harrisville and the Livermore coal mines / by Dan L. Mosier. — San Leandro, Calif. : Mines Road Books, c1978.
183 p. : ill. ; 22 cm.
Bibliography: p. 165-169.
Includes index.
1. Coal trade—California—Livermore—History. 2. Coal mines and mining—California—Livermore—History. 3. Livermore, Calif.—Industries—History. I. Title.
HD9548.L58M67 338.2'7'220979465 79-106164
79 MARC

—CANADA

Canada. Dept. of Energy, Mines, and Resources. Information EMR Branch.
Le charbon au Canada / [préparé pour le secteur de la politique de l'énergie par Information EMR]. — [Ottawa] : Énergie, mines et ressources Canada, c1976.
19 p. : ill. ; 28 cm. C***
Published in English under title: Coal in Canada.
ISBN 0-662-00636-4
1. Coal mines and mining—Canada. I. Canada. Energy Policy Sector. II. Title.
TN806.C2C35 1976 333.8'2 78-377794
78 MARC

Dames & Moore.
A report on the current status of the Canadian coal mining industry / by Dames & Moore, Toronto. — 4th ed. — Toronto : Dames & Moore, 1978.
xiii, 355 p. : ill. ; 26 cm. C***
1. Coal mines and mining—Canada. I. Title.
TN806.C2D35 1978 338.2'7'20971 79-309402
79 MARC

—CHINA

Ikonnikov, A B
The coal industry of China / [by] A. B. Ikonnikov. — Canberra : Dept. of Economics, Research School of Pacific Studies, Australian National University, 1977.
xi, 204 p. : maps ; 24 cm. Aus
Includes bibliographical references.
ISBN 0-909524-14-9 : $9.50
1. Coal—China. 2. Coal mines and mining—China. I. Title.
HD9556.C52I38 338.2'72'0951 78-317060
79 MARC

—CHINA—FUSHUN (SHIH)

Minami Manshū Tetsudō Kabushiki Kaisha. Shomubu. Chōsaka.
(Bujuntan no hanro)
撫順炭の販路 / [執筆者 阿部勇]. — [大連]：南満洲鐵道株式會社庶務部調査課, 大正14 [1925]
3, 77 p. ; 27 cm. — (調査報告書 - 南満洲鐵道株式會社庶務部調査課；第24卷)
Cover title.
1. Coal trade—China. 2. Coal trade—Asia. 3. Coal mines and mining—China—Fu-shun (Shih) I. Abe, Isamu. II. Title. III. Series: Minami Manshū Tetsudō Kabushiki Kaisha. Shomubu. Chōsaka. Chōsa hōkokusho ; dai 24-kan.
HD9556.C52M56 1925 79-800997

—CZECHOSLOVAKIA—ACCIDENTS

Košek, Zdeněk.
Nelsonská tragédie / Zdeněk Košek. — 1. vyd. — Ústí nad Labem : Severočes. nakl., 1977.
96, [1] p., [8] leaves of plates : ill. ; 21 cm. — (Edice Sever) Cz77
Bibliography: p. [104]-[105]
Kčs10.50
1. Mine explosions—Czechoslovakia. 2. Coal mines and mining—Czechoslovakia—Accidents. I. Title.
TN313.K67 78-380529
*78 MARC

—DONETS BASIN

Nekrasovskii, Iakov Ėl'konovich.
(Bezopasnaia i ėffektivnaia razrabotka krutykh plastov Donbassa)
Безопасная и эффективная разработка крутых пластов Донбасса / Я. Э. Некрасовский, В. Н. Кухарев, В. И. Салли. — Москва : Недра, 1977.
204 p. : ill. ; 20 cm. USSR 77
Bibliography: p. 201–[202]
0.56rub
1. Coal mines and mining—Donets Basin. I. Kukharev, Valentin Nikolaevich, joint author. II. Salli, Vladimir Il'ich, joint author. III. Title.
TN808.R92D62272 78-421220

COAL MINES AND MINING (Continued)

—ENGLAND—CHESHIRE—HISTORY

Howarth, Ken.
Dark days : memoirs and reminiscences of the Lancashire and Cheshire coalmining industry up to nationalisation / by Ken Howarth. — Manchester : [The author], 1978.
[4], ii, 127 p. : ill., map, ports. ; 30 cm. GB79-06308
£2.50
1. Coal-miners—England—Lancashire—History. 2. Coal-miners—England—Cheshire—History. 3. Coal mines and mining—England—Lancashire—History. 4. Coal mines and mining—England—Cheshire—History. I. Title.
HD8039.M62G7574 331.2'042'2330922 79-317515
79 MARC

—ENGLAND—HARTLEY—ACCIDENTS

McCutcheon, John Elliott.
The Hartley Colliery disaster, 1862. With a foreword by Lord Robens of Woldingham. Seaham [Eng.] E. McCutcheon, 1963.
176 p. facsim., plates, ports. 23 cm.
1. Coal mines and mining—England—Hartley—Accidents. I. Title.
TN311.M32 78-322334
79 MARC

—ENGLAND—LANCASHIRE—HISTORY

Howarth, Ken.
Dark days : memoirs and reminiscences of the Lancashire and Cheshire coalmining industry up to nationalisation / by Ken Howarth. — Manchester : [The author], 1978.
[4], ii, 127 p. : ill., map, ports. ; 30 cm. GB79-06308
£2.50
1. Coal-miners—England—Lancashire—History. 2. Coal-miners—England—Cheshire—History. 3. Coal mines and mining—England—Lancashire—History. 4. Coal mines and mining—England—Cheshire—History. I. Title.
HD8039.M62G7574 331.2'042'2330922 79-317515
79 MARC

—ENGLAND—YORKSHIRE—ACCIDENTS

Great Britain. Mines and Quarries Inspectorate.
Report on explosion at Houghton Main Colliery, Yorkshire : report on the causes of, and circumstances attending, the explosion which occurred at Houghton Main Colliery, South Yorkshire, on 12th June, 1975 / [by] J. Carver, H. M. Chief Inspector of Mines and Quarries ; [for the] Health and Safety Executive. — London : H.M.S.O., 1976.
27 p., 2 p. of plates : ill., plans (6 fold. in pocket) ; 25 cm. GB76-06743
ISBN 0-11-880328-X : £1.00
1. Coal mines and mining—England—Yorkshire—Accidents. 2. Mine explosions—England—Yorkshire. I. Carver, J. II. Great Britain. Health and Safety Executive. III. Title.
TN313.G75 1976 79-316047
79 MARC

—EUROPE, WESTERN—COLLECTED WORKS

Comité d'étude des producteurs de charbon d'Europe occidentale.
Bulletin d'information du Comité d'étude des producteurs de charbon d'Europe occidentale.
[Bruxelles, Comité d'étude des producteurs de charbon d'Europe occidentale]
no. in v. 30 cm. Irregular.
1. Coal mines and mining—Europe, Western—Collected works.
TN808.A1C66a 333.7 78-647003
MARC-S

—FRANCE—CARMAUX

Gonzalez, Daniel.
L'Occitan parlat jos tèrra : los carbonièrs de Carmaus / per Daniel Gonzalez. — Toulouse : Université de Toulouse II, Le Mirail, [1977?]
2 v. (252 p.) : ill. ; 22 cm. — (Documents et archives pour la recherche sociolinguistique méridionale : Série E, Document de travail ; no 3-4) F***
At head of title: Université de Toulouse II, Le Mirail. Centre de sociolinguistique et de dialectologie sociale.
CONTENTS: t. 1. Tèxtes sul trabalh del Cròs.—t. 2. Vocabulari del Cròs.
1. Provençal language, Modern—Dialects—France—Carmaux. 2. Carmaux, France—Industries. 3. Coal-miners—France—Carmaux—Language. 4. Coal mines and mining—France—Carmaux. 5. Coal mines and mining—Terminology. I. Title. II. Series.
PC3447.C28G6 78-375700
78 MARC

—GERMANY, WEST—COLLECTED WORKS

Gesamtverband des Deutschen Steinkohlenbergbaus.
Jahresbericht - Gesamtverband des Deutschen Steinkohlenbergbaus.
[Essen] Gesamtverband des Deutschen Steinkohlenbergbaus.
v. 26 cm.
Cover title, : Steinkohle.
1. Coal mines and mining—Germany, West—Collected works. I. Gesamtverband des Deutschen Steinkohlenbergbaus. Steinkohle.
TN808.G3G47a 78-646904
MARC-S

—GREAT BRITAIN—SAFETY MEASURES

National Committee on Steep Seam Working in British Coal Mines.
First report of the National Committee on Steep Seam Working in British coal mines. — London : H.M.S.O., 1975.
32 p. : ill ; 25 cm. GB75-29186
At head of title: Health and Safety Executive.
Chairman: J. R. Cowan.
ISBN 0-11-410248-1 : £0.60
1. Coal mines and mining—Great Britain—Safety measures. 2. National Committee on Steep Seam Working in British Coal Mines. I. Cowan, J. R.
TN295.N27 1975 622'.8 76-360236
76[r79]rev MARC

National Committee on Steep Seam Working in British Coal Mines.
Final report of the National Committee on Steep Seam Working in British Coal Mines. — London : H.M.S.O., 1978.
23 p. : ill. ; 25 cm. GB78-15654
At head of title: Health and Safety Executive.
Errata slip inserted.
ISBN 0-11-882010-9 : £0.60
1. Coal mines and mining—Great Britain—Safety measures.
TN295.N27 1978 622'.8 78-324839
79 MARC

—ILLINOIS—COSTS

Huff, Linda L
Economic impact of dissolved solids regulation upon the coal mining industry (R76-7) / by Linda L. Huff, Gregg A. Jarrell. — Chicago : State of Illinois, Institute for Environmental Quality, 1977.
viii, 134 p. : ill. ; 28 cm. — (IIEQ document ; no. 77/28)
Includes bibliographical references.
1. Coal mines and mining—Illinois—Costs. 2. Coal mine waste—Enivronmental aspects—Illinois. 3. Water—Pollution—Law and legislation—Illinois. I. Jarrell, Gregg A., joint author. II. Title. III. Series. IV. Series: Illinois Institute for Environmental Quality ILEQ document ; no.77/28.
HD9547.I3H83 338.2'7'2409773 78-622262
78 MARC

—ILLINOIS—STREATOR—HISTORY

Franklin, Paula Angle, 1928-
Biography in black; a history of Streator, Illinois, by Paula Angle. Introd. by Ted Weber. [Streator, Ill.] Weber Co. [1962]
ix, 158 p. illus., ports., map. 23 cm.
1. Streator, Ill.—History. 2. Coal mines and mining—Illinois—Streator—History. I. Title.
F549.S9F7 977.3'27 63-4775
[r79]rev MARC

—INDIA

Ghosh, Amiya Bhushan.
Coal industry in India : an historical and analytical account / A. B. Ghosh. — New Delhi : Sultan Chand, [pref. 1977-
v. ; 23 cm.
Bibliography: pt. 1, p. [305]-308.
Includes index.
CONTENTS: pt. 1. Pre-independence period.
Rs50.00 (v. 1)
1. Coal trade—India. 2. Coal mines and mining—India. I. Title.
HD9556.I42G48 338.2'7'20954 78-907451
79 MARC

India (Republic). Committee on Amalgamation of Collieries.
Report of the Committee on Amalgamation of Collieries, 1956. — Delhi : Manager of Publications, 1957-
v. : ill. ; 25 cm.
Cover title.
"MP.8/1000."
Chairman: B. G. Mehta.
1. Coal trade—India. 2. Coal mines and mining—India. I. Mehta, Balvantray Gopaljee. II. Title.
HD9556.I42I53 1957 338.2'7'20954 78-912402
79 MARC

—JAPAN—COLLECTED WORKS

Zenkoku Tankō Gijutsukai.
(Zenkoku Tankō Gijutsukai kenkyūkai hōkokusho)
全国炭鉱技術会研究会報告書 第14- 回; 1970-
[東京?] 全国炭鉱技術会
v. ill. 30 cm.
Continues: Zenkoku Tankō Gijutsukai. Zentangi kenkyūkai hōkokusho.
1. Coal mines and mining—Collected works. 2. Coal mines and mining—Japan—Collected works. I. Title.
TN799.9.Z45b 78-644260
(MARC-S)

Zenkoku Tankō Gijutsukai.
(Zentangi kenkyūkai hōkokusho)
全炭技研究会報告書 第 -13回 ; 19 -69.
[東京] 全国炭鉱技術会
v. ill. 26-30 cm.
Continued by: Zenkoku Tankō Gijutsukai. Zenkoku Tankō Gijutsukai kenkyūkai hōkokusho.
1. Coal mines and mining—Collected works. 2. Coal mines and mining—Japan—Collected works. I. Title.
TN799.9.Z45b 78-644250
(MARC-S)

—JAPAN—FINANCE—HISTORY

Sekitan Kōgyō Gōrika Jigyōdan.
(Danshi)
団史 [東京] 石炭鉱業合理化事業団 昭和 40-51 [1965-76]
2 v. illus. 27 cm.
Errata slip inserted.
CONTENTS: [1] 整備編—[2] 整備・近代化編
1. Sekitan Kogyo Gorika Jigyodan—History. 2. Coal mines and mining—Japan—Finance—History. 3. Energy Policy—Japan—History.
HD9556.J33S47 1965 71-817851

—KAZAKHSTAN—KARAGANDA BASIN—LABOR PRODUCTIVITY

Trudovaiā doblest' shakhterov Karagandy : Rasskaz ob opyte kollektiva ob"edineniiā Karagandaugol' v bor'be za naivysshuiū proizvoditel'nost' truda / [otv. redaktor A. F. Rumiāntsev]. — Moskva : Pravda, 1976.
32 p. : ill. ; 20 cm. — (Peredovoĭ opyt bor'by za ėffektivnost' i kachestvo) USSR77
0.05rub
1. Coal mines and mining—Kazakhstan—Karaganda Basin—Labor productivity. I. Rumiāntsev, Alekseĭ Fedorovich.
HD9556.R92T78 79-382418
79 MARC

—LUXEMBURG—CONGRESSES

Information Symposium on High-Output Coal Winning Districts, Luxembourg, 1976.
Information Symposium on High-Output Coal Winning Districts, Luxembourg, from 29th September to 1st October, 1976 / edited by the Directorate General "Scientific and technical information management". — Luxembourg : Commission of the European Communities ; [Washington, D.C. : sold by European Community Information Service], 1976-1977.
2 v. : ill. ; 24 cm. Lu***
On spine: High-output coal winning districts.
Includes bibliographical references.
132.00F
1. Coal mines and mining—Luxemburg—Congresses. I. Commission of the European Communities. Directorate General Scientific and Technical Information Management. II. Title.
TN808.L9I53 1976 622'.33'24094935 78-323132
79 MARC

—NETHERLANDS—LIMBURG (PROVINCE)

Mijnwerkers : verhalen om te onthouden / Wim Nijsten, Jos Bours en Marlies Hautvast. — Nijmegen : Link, [1979]
207 p. : with ill. ; 20 cm. — (Boek van de arbeid ; 1) (Link-paperback ; 15) Ne79-5
1. Coal-miners—Netherlands—Limburg (Province) 2. Coal mines and mining—Netherlands—Limburg (Province) I. Nijsten, Wim. II. Bours, Jos, 1946- . III. Hautvast, Marlies, 1952- . IV. Series.
HD8039.M62N4647 79-359867
*79 MARC

COAL MINES AND MINING (Continued)

—NEW ZEALAND—GREYMOUTH—ACCIDENTS

New Zealand. Commission to Inquire into Disaster at Strongman Mine.
Report of Commission to Inquire into Disaster at Strongman Mine. — Wellington, New Zealand : R. E. Owen, Govt. Printer, 1967.
83 p. : ill., diagrams (5 fold. in pocket) ; 24 cm.
1. Coal mines and mining—New Zealand—Greymouth—Accidents. 2. Mine explosions—New Zealand—Greymouth.
TN313.N48 1967 79-316396
79 MARC

—OKLAHOMA—HASKELL CO.

Lintz, Christopher.
An archaeological survey of the Kerr-Mcgee Choctaw coal mine facility, Haskell County, Oklahoma / by Christopher Lintz. — [Norman] : Archaeological Research and Management Center, University of Oklahoma, 1978.
vi, 191 p. : ill. ; 28 cm. — (Project report series ; no. 2 ISSN 0160-3078)
Bibliography: p. 187-191.
1. Haskell Co., Okla—Antiquities. 2. Choctaw Indians—Antiquities. 3. Excavations (Archaeology)—Oklahoma—Haskell Co. 4. Coal mines and mining—Oklahoma—Haskell Co. 5. Oklahoma—Antiquities. 6. Kerr-McGee Corporation. I. Oklahoma. University. Archaeological Research and Management Center. II. Title. III. Series.
F702.H4L56 976.6'77 79-620759
79 MARC

—POLAND

Dziunikowski, Kazimierz.
Mécanisation et automatisation dans l'industrie houillère polonaise. Warszawa, Państwowe Wydawn. Naukowe, 1971.
15 p. illus. 24 cm. (Académie polonaise des sciences. Centre scientifique à Paris. Conférences, fasc. 86)
"Conférence faite au Centre scientifique de l'Académie polonaise des sciences, à Paris ... le 21 janvier 1970."
1. Coal mines and mining—Poland. 2. Coal mining machinery. I. Title. II. Series: Polska Akademia Nauk. Centre scientifique, Paris. Conférences, fasc. 86.
AS162.P7 fasc. 86 79-378883
[TN808.P7] 79 MARC

Kulpiński, Jan.
Zadania przemysłu węglowego w świetle uchwał VII Zjazdu PZPR ; wykład na VII Sesję Naukową Wydziału Górniczego / Jan Kulpiński. — Gliwice : PŚ, 1976 (Gliw. : PŚ)
24, [3] p. ; 25 cm. — (Górnictwo ; z. 71a) (Zeszyty naukowe - Politechnika Śląska ; Nr 500) P77-3309
Summary also in English and Russian.
zł4.00
1. Coal mines and mining—Poland. I. Title. II. Series: Gliwice, Upper Silesia (City). Politechnika Śląska. Górnictwo ; zesz. 71a. III. Series: Gliwice, Upper Silesia (City). Politechnika Śląska. Zeszyty naukowe ; nr. 500.
TN4.G6 Zesz.71a 79-347201
[TN808.P7] 79 MARC

—RUSSIA

Report of the delegation to the U.S.S.R. on the invitation of the Central Committee of the Russian Mineworkers, Thursday, July 27th to Friday, August 17, 1956. — [Durham?] : National Union of Mineworkers, Durham Area Union, [1956?]
67 p. ; 25 cm.
Caption title.
1. Workers' travel programs. 2. Coal mines and mining—Russia. 3. Russia—Description and travel—1945- . I. National Union of Mineworkers. Durham Area Union.
HD7790.R46 79-313468
79 MARC

Skogorev, Viktor Alekseevich.
125 115 [i. e. Sto dvadt͡sat′ pi͡at′ tysi͡ach sto pi͡atnadt͡sat′] tonn antrat͡sita iz lavy za 31 rabochiĭ den′ / V. A. Skogorev, I. K. Kart͡sev, I. I͡A. Pavlenko. — Donet͡sk : Donbass, 1968.
23 p. : ill. — (Opyt peredovikov—kazhdomu rabochemu)
L.C. copy on microfilm only.
——— Microfilm. On 1 reel with other items ; 35 mm.
Microfilm Slavic 13019 TN
1. Coal mines and mining—Russia. I. Kart͡sev, Ivan Kliment′evich, joint author. II. Pavlenko, Ivan I͡Akovlevich, joint author. III. Title.
[TN808.R9S544] 79-357458

Skogorev, Viktor Alekseevich.
125 115 [i. e. Sto dvadt͡sat′ pi͡at′ tysi͡ach sto pi͡atnadt͡sat′] tonn antrat͡sita iz lavy za 31 rabochiĭ den′ / V. A. Skogorev, I. K. Kart͡sev, I. I͡A. Pavlenko. — Donet͡sk : Donbass, 1968.
23 p. : ill. — (Opyt peredovikov—kazhdomu rabochemu)
L.C. copy on microfilm only.
——— Microfilm. On 1 reel with other items ; 35 mm.
Microfilm Slavic 13019 TN
1. Coal mines and mining—Russia. I. Kart͡sev, Ivan Kliment′evich, joint author. II. Pavlenko, Ivan I͡Akovlevich, joint author. III. Title.
[TN808.R9S544] 79-357458

—RUSSIAN REPUBLIC—KUZNETSK BASIN—HISTORY

Rubezhi shakhterskoĭ slavy : ugol′nai͡a promyshlennost′ Kuzbassa v 1946-1976 gg. : sbornik dokumentov / [Redakt͡sionnai͡a kollegii͡a Z. G. Karpenko, L. A. Pashchenko, N. F. Tebenev]. — Kemerovo : Kemerovskoe knizhnoe izd-vo, 1977.
220 p. ; 21 cm. USSR***
Includes bibliographical references.
0.68rub
1. Coal mines and mining—Russian Republic—Kuznetsk Basin—History. I. Karpenko, Zinaida Georgievna. II. Pashchenko, L. A. III. Tebenev, N. F.
TN809.R92K897 79-377637
79 MARC

—RUSSIAN REPUBLIC—SOVIET FAR EAST

Popov, Vladimir Ivanovich, writer on coal mines and mining.
(Skoroprokhodchiki)
Скоропроходчики. Владивосток, Дальневосточное книжное изд-во, 1964.
59 p. illus., ports. 20 cm.
At head of title: В. Попов.
1. Coal mines and mining—Russian Republic—Soviet Far East. 2. Coal mining machinery. I. Title.
TN809.R92S76 79-299286

—SCOTLAND—HISTORY

The Coal industry : 17th-20th centuries / ... researched and edited by Lynda Wright. — Stirling : Central Regional Council Education Department, 1978.
[2], 65 p., 8 leaves of plates : ill., map ; 21 cm. — (Historical sources for central Scotland series ; 2) GB79-02465
Bibliography: p. 63-65.
£0.30
1. Coal trade—Scotland—History. 2. Coal mines and mining—Scotland—History. I. Wright, Lynda. II. Series.
HD9551.7.A3C55 338.2'7'209413 79-313189
79 MARC

—TEXAS

Texas. Legislature. House of Representatives. Study Group.
Strip mining in Texas. — Austin : House Study Group, Texas House of Representatives, 1978.
x, 53 p. : ill. ; 28 cm. — (Special legislative report - House Study Group ; no. 28)
Bibliography: p. 52-53.
1. Strip mining—Texas. 2. Coal mines and mining—Texas. 3. Uranium mines and mining—Texas. I. Title. II. Series: Texas. Legislature. House of Representatives. Study Group. Special legislatives report - House Study Group ; no. 28.
TN291.T49 1978 333.8'2 79-623124
79 MARC

—UKRAINE—L'VOVSKO-VOLYNSKIY BASIN

Kushniruk, Venedikt Alekseevich.
(Gazonosnost′ uglenosnoĭ tolshchi L′vovsko-Volynskogo ugol′nogo basseĭna)
Газоносность угленосной толщи Львовско-Волынского угольного бассейна / В. А. Кушнирук. — Киев : Наук. думка, 1978.
116, [4] p., [1] fold. leaf of plates : ill. ; 22 cm. USSR***
At head of title: Akademii͡a nauk Ukrainskoĭ SSR. Institut geologii i geokhimii gori͡uchikh iskopaemykh.
Bibliography: p. 112-[117]
1.20rub
1. Mine gases. 2. Coal mines and mining — Ukraine — L′vovsko-Volynskiy Basin. I. Title.
TN305.K87 79-411819

—UNITED STATES

United States. Congress. House. Committee on Interior and Insular Affairs. Subcommittee on Mines and Mining.
United States coal production and utilization : oversight hearings before the Subcommittee on Mines and Mining of the Committee on Interior and Insular Affairs, House of Representatives, Ninety-fifth Congress, second session ... hearings held in Washington, D.C., May 16 and 18, and July 27, 1978. — Washington : U.S. Govt. Print. Off., 1979.
iv, 616 p. ; 24 cm.
"Serial no. 95-22."
1. Coal—United States. 2. Coal mines and mining—United States. I. Title.
KF27.I536 1978b 338.2'7'20973 79-601244
79 MARC

Witt, Matt.
In our blood : four coal mining families / by Matt Witt ; photos. by Earl Dotter. — 1st ed. — Washington : Highlander Research and Education Center, c1979.
90 p. : ill. ; 29 cm.
ISBN 0-9602226-1-8 : $6.95
1. Coal mines and mining—United States. 2. Coal-miners—United States—Biography. I. Dotter, Earl. II. Title.
TN805.A5W67 338.2'7'20973 78-71518
79 MARC

—UNITED STATES—BIBLIOGRAPHY

Parkinson, George, 1943-
Guide to coal mining collections in the United States / by George Parkinson, with the assistance of Claire Collier and Joseph Preston. — [Morgantown] : West Virginia University Library, [c1978]
xxiv, 182 p., [1] leaf of plates : ill. ; 23 cm.
Includes index.
1. Coal mines and mining—United States—Bibliography. I. Collier, Claire. II. Preston, Joseph. III. Title.
Z6738.C6P37 016.622'33 79-102049
[TN802] 79 MARC

—UNITED STATES—FINANCE

Brown, Stuart E
Scrip : trade tokens issued by United States coal mining companies and company stores / compiled by Stuart E. Brown, Jr. — Berryville, VA. : Virginia Book Co., c1978.
vi, 370 p. : ill. ; 29 cm.
Bibliography: p. 350-355.
Includes index.
1. Tokens—United States. 2. Currency question—United States. 3. Company stores—United States. 4. Coal mines and mining—United States—Finance. I. Title.
CJ4904.B76 737'.3'0973 79-102015
79 MARC

Mann, Charles Edward, 1950-
Coal and profitability : an investor's guide / by Charles E. Mann and James N. Heller, and others of Energy and Environmental Analysis. — [Hightstown, N.J.] : Coal Week, [c1979]
vii, 126 p. : graphs ; 28 cm.
Bibliography: p. 119-126.
ISBN 0-07-606610-X : $63.50
1. Coal trade—United States—Finance. 2. Coal mines and mining—United States—Finance. 3. Investments—United States. I. Heller, James N., joint author. II. Energy and Environmental Analysis, inc. III. Title.
HD9544.M36 332.6'722 78-65618
79 MARC

—UNITED STATES—MANAGEMENT—CASE STUDIES

Goodman, Paul S
Assessing organizational change : the Rushton quality of work experiment / Paul S. Goodman, with major contributions by Edward Conlon, Dennis Epple, Eduard Fidler. — New York : Wiley, c1979.
xii, 391 p. : ill. ; 24 cm. — (Wiley series on organizational assessment and change)
"A Wiley-Interscience publication."
Bibliography: p. 384-385.
Includes index.
ISBN 0-471-04782-1
1. Coal-miners—United States—Case studies. 2. Organizational change—Case studies. 3. Industrial organization—United States—Case studies. 4. Coal mines and mining—United States—Management—Case studies. I. Title. II. Series.
HD8039.M62U6258 658.31'52 78-31857
79 MARC

—VICTORIA—HISTORY

Quilford, Arthur, 1909-
The state mine : a pictorial history of the Powlett coalfields / by Arthur Quilford. — Wonthaggi, Vic. : Gumnut Press, 1977.
144 p. : chiefly ill., diagrs., facs., portrs. ; 22 x 29 cm. Aus
ISBN 0-908058-00-4
1. Coal mines and mining—Victoria—History. I. Title.
TN811.A82V526 338.2'7'2409945 79-303534
79 MARC

COAL MINES AND MINING (Continued)

—VIRGINIA—HYGIENIC ASPECTS

Virginia. Coal Mine Health and Safety Commission.
Report of the Coal Mine Health and Safety Commission to the Governor and the General Assembly of Virginia. — Richmond : Commonwealth of Virginia, Dept. of Purchases and Supply, 1978.
61 p. ; 25 cm. — (Senate document ; no. 25)
1. Coal mines and mining—Virginia—Safety measures. 2. Coal mines and mining—Virginia—Hygienic aspects. I. Title. II. Series: Virginia. General Assembly, 1978. Senate. Document ; no. 25.
J87.V9 1978b, no. 25 328.755'07'108 s 78-622027
[TN295] 79 MARC

—VIRGINIA—SAFETY MEASURES

Virginia. Coal Mine Health and Safety Commission.
Report of the Coal Mine Health and Safety Commission to the Governor and the General Assembly of Virginia. — Richmond : Commonwealth of Virginia, Dept. of Purchases and Supply, 1978.
61 p. ; 25 cm. — (Senate document ; no. 25)
1. Coal mines and mining—Virginia—Safety measures. 2. Coal mines and mining—Virginia—Hygienic aspects. I. Title. II. Series: Virginia. General Assembly, 1978. Senate. Document ; no. 25.
J87.V9 1978b, no. 25 328.755'07'108 s 78-622027
[TN295] 79 MARC

—WALES—BLAENAVON

Haigh, Martin J 1950-
Evolution of slopes on artificial landforms—Blaenavon, U.K. / by Martin J. Haigh. — Chicago : University of Chicago, Dept. of Geography, 1978.
xiv, 293 p. : ill. ; 23 cm. — (Research paper - The University of Chicago, Department of Geography ; no. 183)
Summary in Russian.
Bibliography: p. 271-287.
Includes index.
ISBN 0-89065-090-X : $6.00
1. Slopes (Physical geography)—Wales—Blaenavon. 2. Coal mines and mining—Wales—Blaenavon. 3. Anthropo-geography—Wales—Blaenavon. I. Title. II. Series: Chicago. University. Dept. of Geography. Research paper ; no. 183.
H31.C514 no. 183 910 s 78-7267
[GB448] 78 MARC

—WALES—CYNON—DRAMA

Morgan, Robert, 1921-
Voices in the dark : a verse play / by Robert Morgan. — Todmorden : Arc Publications, 1976.
[1], 39 p. ; 23 cm. GB77-22818
Originally published in Anglo Welsh review, v. 17, no. 40.
Limited ed. of 400 copies of which 175 are cloth bound and 15 are numb. and signed by the author.
ISBN 0-902771-56-6
1. Coal mines and mining—Wales—Cynon—Drama. I. Title.
PR6063.O733V64 822'.9'14 79-309141
79 MARC

—WEST VIRGINIA

Lawall, Charles Elmer, 1891-
Mining methods in West Virginia / by C. E. Lawall, Ivan A. Given, H. G. Kennedy. — Morgantown : West Virginia University, 1929.
230 p. : ill. ; 27 cm. — (Research bulletin - Engineering Experiment Station ; no. 4) (Bulletin - West Virginia University, College of Engineering ; ser. 14, no. 6)
Errata slips inserted.
1. Coal mines and mining—West Virginia. I. Given, Ivan A., joint author. II. Kennedy, Harry G., joint author. III. Title. IV. Series: West Virginia. University. Engineering Experiment Station. Research bulletin ; no. 4.
TN805.W4L38 622'.33'409755 79-317457
79 MARC

COAL-MINING MACHINERY

Dziunikowski, Kazimierz.
Mécanisation et automatisation dans l'industrie houillère polonaise. Warszawa, Państwowe Wydawn. Naukowe, 1971.
15 p. illus. 24 cm. (Académie polonaise des sciences. Centre scientifique à Paris. Conférences, fasc. 86)
"Conférence faite au Centre scientifique de l'Académie polonaise des sciences, à Paris ... le 21 janvier 1970."
1. Coal mines and mining—Poland. 2. Coal mining machinery. I. Title. II. Series: Polska Akademia Nauk. Centre scientifique, Paris. Conférences, fasc. 86.
AS162.P7 fasc. 86 79-378883
[TN808.P7] 79 MARC

Handbuch der Mechanisierung der Kohlengewinnung / Gesamtbearb. Heinz Kundel, unter Mitw. von F. Dürrer ... [et al.] — 5., neubearb. u. erw. Aufl. — Essen : Verlag Glückauf, 1978.
245 p. : ill. ; 21 cm. — (Glückauf-Betriebsbücher ; Bd. 6) GFR78-A
Issued by the Fachausschuss Gewinnung und Versatz of the Steinkohlenbergbauverein.
Second ed. issued by the Steinkohlenbergbauverein's Aussvhuss für Mechanisierung der Kohlengewinnung.
ISBN 3-7739-0224-7 : DM46.80
1. Coal-mining machinery. I. Kundel, Heinz. II. Steinkohlenbergbauverein. Fachausschuss Gewinnung und Vorsatz.
TN813.S76 1978 622'.33'028 78-386551
79 MARC

(Kikai saitan to gasu tosshutsu bōshi)
機械採炭とガス突出防止 [東京] 全国炭鉱技術会 1969.
156 p. illus. 30 cm. (全炭技研究会報告書第13回) Ja***
Cover title.

1. Coal-mining machinery. 2. Coal mines and mining—Safety measures. 3. Mine gasses. I. Zenkoku Tankō Gijutsukai. II. Series: Zenkoku Tankō Gijutsukai. Zentangi kenkyūkai hōkokusho, dai 13-kai
TN799.9.Z45b vol. 13 79-800962

(Kikai saitan to shizen hakka taisaku)
機械採炭と自然発火対策 自走支保・ドラムカッタ・ボーベル [東京] 全国炭鉱技術会 1968.
98 p. illus. 26 cm. (全炭技研究会報告書 第12回) Ja***
Cover title.
Errata slips inserted.

1. Coal-mining machinery. 2. Coal mines and mining—Fires and fire prevention. I. Zenkoku Tankō Gijutsukai. II. Series: Zenkoku Tankō Gijutsukai. Zentangi kenkyūkai hōkokusho, dai 12-kai.
TN799.9.Z45b vol. 12 78-813994
[TN813]

Mining mechanically / by S. C. Walker ... [et al.]. — London : Mechanical Engineering Publications, 1978.
[3], 28 p. : ill. ; 30 cm. — (Isaac Newton lecture ; 12) GB79-05769
ISBN 0-85298-413-8 : £0.95
1. Coal-mining machinery. I. Walker, Sidney Charles. II. Institution of Mechanical Engineers, London. III. Series.
TN813.M4517 622'.33'0941 79-313524
79 MARC

Popov, Vladimir Ivanovich, writer on coal mines and mining.
(Skoroprokhodchiki)
Скоропроходчики. Владивосток, Дальневосточное книжное изд-во, 1964.
59 p. illus., ports. 20 cm.
At head of title: В. Попов.
1. Coal mines and mining—Russian Republic—Soviet Far East. 2. Coal mining machinery. I. Title.
TN809.R92S76 79-299286

—DESIGN AND CONSTRUCTION

Hawrylak, Henryk.
Teoretyczne podstawy obliczeń i doboru mechanizmów urabiania koparek wielonaczyniowych. Wrocław, 1964.
50 p. illus. 30 cm. (Dolnośląskie Biuro Projektów Górniczych. Branżowy Ośrodek Informacji Technicznej i Ekonomicznej Przemysłu Węgla Brunatnego. Zeszyty problemowe, nr. 36, 1964)
Rozprawa habilitacyjna—Politechnika Wrocławska.
"Do użytku służbowego."
Bibliography: p. 47-48.
1. Coal mining machinery—Design and construction. I. Title. II. Series.
TN813.H38 79-364378
79 MARC

COAL-OIL see Kerosene; Petroleum

COAL PREPARATION

Artiūshin, Stepan Petrovich.
(Obogashchenie ugleĭ)
Обогащение углей : допущено ... в качестве учебника для учащихся горных техникумов / С. П. Артюшин. — Москва : Недра, 1975.
880 p. : ill. ; 22 cm. USSR***
Bibliography: p. [871]-[878]
Includes index.
1.06rub
1. Coal preparation. I. Title.
TN816.A792 78-423262

—CONGRESSES

Coal Processing and Conversion Symposium, Morgantown, W. Va., 1976.
Proceedings of the Coal Processing and Conversion Symposium, June 1-3, 1976, Morgantown, WV / compilers, Carl J. Smith, Duane G. Nichols ; sponsors, West Virginia Geological and Economic Survey, West Virginia University, College of Engineering. — [Morgantown, W. Va.] : The Survey, 1979.
xii, 115 p. : ill. ; 28 cm. — (Coal geology bulletin ; CGB-6)
Cover title.
Includes bibliographical references and index.
1. Coal gasification—Congresses. 2. Coal liquefaction—Congresses. 3. Coal preparation—Congresses. I. Smith, Carl J. II. Nichols, Duane G. III. West Virginia. Geological Survey. IV. West Virginia. University. College of Engineering. V. Series.
TP759.C62 1976 662'.625 79-624429
79 MARC

—HANDBOOKS, MANUALS, ETC.

Coal age operating handbook of coal preparation / edited by Paul C. Merritt. — New York : Coal Age Mining Informational Services, c1978.
vi, 311 p. : ill. ; 29 cm. — (Coal age library of operating handbooks, ; v. 3)
Articles first published in Coal age and in Keystone coal industry manual.
Includes bibliographical references and index.
ISBN 0-07-099922-8 : $19.50
1. Coal preparation—Handbooks, manuals, etc. I. Merritt, Paul C. II. Coal age. III. Keystone coal industry manual. IV. Title: Operating handbook of coal preparation. V. Series.
TN816.C58 662'.623 78-21539
78 MARC

—MATHEMATICAL MODELS

Piecuch, Tadeusz.
Analityczno-empiryczny model procesu filtracji próżniowej zawiesin mułów węglowych / Tadeusz Piecuch. — Gliwice : Dział Wydawnictw Politechniki Śląskiej, 1975.
96 p. : graphs ; 24 cm. — (Górnictwo ; z. 67) (Zeszyty naukowe - Politechnika Śląska ; nr. 434)
Summary also in English and Russian.
Bibliography: p. 87-90.
zł10.00
1. Coal preparation—Mathematical models. 2. Filters and filtration—Mathematical models. I. Title. II. Series: Gliwice, Upper Silesia (City). Politechnika Śląska. Górnictwo ; zesz. 67. III. Series: Gliwice, Upper Silesia (City). Politechnika Śląska. Zeszyty naukowe ; nr. 434.
TN4.G6 z. 67 79-376777
[TN816] 79 MARC

COAL PREPARATION PLANTS

see also Fuel-briquet plants; Gravity concentrators

—DESIGN AND CONSTRUCTION

Kornienko, Mikhail Trofimovich.
(Osnovy sooruzheniīa ugleobogatitel'nykh i briketnykh fabrik)
Основы сооружения углеобогатительных и брикетных фабрик : [Учеб. пособие для горных техникумов] / М. Т. Корниенко. — 2-изд., перераб. и доп. — Москва : Недра, 1975.
160 p. : ill. ; 20 cm. USSR 75
Bibliography : p. [158]
0.28rub
1. Coal preparation plants—Design and construction. 2. Fuel-briquet plants—Design and construction. I. Title.
TH4561.K59 1975 79-409918

COAL RESEARCH

—AUSTRALIA

Australia. National Coal Research Advisory Committee.
Coal research in Australia, 1976 / prepared under the auspices of the National Coal Research Advisory Committee on behalf of the Australian Steering Committee for the Third International Conference on Coal Research held in Sydney, 6-8 October, 1976. — [Canberra] : [Dept. of National Resources], 1976.
52 p. ; 30 cm. Aus
ISBN 0-642-92258-6
1. Coal research—Australia. I. International Conference on Coal Research, 3d, Sydney, Australia, 1976. II. Title.
TP326.A8A94 1976 662'.62'072094 78-315220
78 MARC

COAL RESERVES see Coal—Reserves

COAL SLURRY PIPE LINES see Coal —Pipe lines

COAL STRIKE, UNITED STATES, 1977-1978

Moody, Kim.
Battle line : the coal strike of '78 / by Kim Moody and Jim Woodward. — Detroit : Sun Press, 1978.
126 p. : ill. ; 21 cm.
ISBN 0-932608-00-0 : $2.50
1. Coal Strike, United States, 1977-1978. I. Woodward, Jim, joint author. II. Title.
HD5325.M63 1977.U55 331.89'282'2330973 78-63265
79 MARC

United States. Congress. House. Committee on Interstate and Foreign Commerce. Subcommittee on Energy and Power.
Energy impacts of the coal strike : hearing before the Subcommittee on Energy and Power of the Committee on Interstate and Foreign Commerce, House of Representatives, Ninety-fifth Congress, second session ... February 16, 1978. — Washington : U.S. Govt. Print. Off., 1978.
iii, 52 p. ; 24 cm.
"Serial no. 95-132."
1. Coal Strike, United States, 1977-1978. 2. Power resources—United States. I. Title.
KF27.I5542 1978b 333.7 78-603723
79 MARC

United States. Congress. Senate. Committee on Energy and Natural Resources. Subcommittee on Energy Production and Supply.
Effects of coal strike : hearing before the Subcommitte on Energy Production and Supply of the Committee on Energy and Natural Resources, United States Senate, Ninety-fifth Congress, second session .. March 6, 1978. — Washington : U.S. Govt. Print. Off., 1978.
iii, 73 p. : ill. ; 24 cm.
"Publication no. 95-112."
1. Coal Strike, United States, 1977-1978. I. Title.
KF26.E553 1978a 330.9'73'0926 78-602867
78 MARC

United States. Congress. Senate. Committee on Energy and Natural Resources. Subcommittee on Energy Conservation and Regulation.
Emergency sales of electric power : hearing before the Subcommittee on Energy Conservation and Regulation of the Committee on Energy and Natural Resources, United States Senate, Ninety-fifth Congress, second session ... May 11, 1978. — Washington : U.S. Govt. Print. Off., 1978.
iii, 126 p. : ill. ; 24 cm.
"Publication no. 95-138."
Includes bibliographical references.
1. Electric utilities—United States. 2. Coal Strike, United States, 1977-1978. I. Title.
KF26.E5527 1978i 338.4'3 79-600554
79 MARC

COAL-TAR

—ANALYSIS

Spielmann, Percy Edwin, 1881-
The constitutents of coal tar, by Percy Edwin Spielmann ... with two folding charts. London, New York ₍etc.₎ Longmans, Green and co., 1924.
xii, 219, ₍1₎ p. 2 fold. diagr. 23 cm. (Monographs on industrial chemistry)
$4.25
Bibliography: p. 193-207.
1. Coal-tar—Analysis.
TP953.S64 662'.9 24-25436
₍r79₎rev MARC

—TESTING

Standardization of Tar Products Tests Committee, London.
Standard methods for testing tar and its products. — 5th ed. / edited by P. V. Watkins. — Gomersal, Leeds : Standardization of Tar Products Tests Committee, 1962.
vii, 608 p., ₍6₎ fold. leaves of plates : ill. ; 23 cm.
Includes index.
1. Coal-tar—Testing. 2. Coal-tar products—Testing. I. Watkins, P. V., ed. II. Title.
TP953.S7 1962 79-301300
79 MARC

COAL-TAR COLORS

see also Coloring matter; Dyes and dyeing

COAL-TAR PRODUCTS

see also Distillation, Destructive; Gas; Oils and fats

—TESTING

Standardization of Tar Products Tests Committee, London.
Standard methods for testing tar and its products. — 5th ed. / edited by P. V. Watkins. — Gomersal, Leeds : Standardization of Tar Products Tests Committee, 1962.
vii, 608 p., ₍6₎ fold. leaves of plates : ill. ; 23 cm.
Includes index.
1. Coal-tar—Testing. 2. Coal-tar products—Testing. I. Watkins, P. V., ed. II. Title.
TP953.S7 1962 79-301300
79 MARC

COAL TRADE

Rio Tinto North American Services Limited.
World coking coal supply study / prepared by Rio Tinto North American Services Limited for the Economic, Financial and Market Analysis Sub-committee on North East Coal Development. — ₍Victoria₎ : Province of British Columbia, Ministry of Economic Development, 1976.
iv, 94 p. ; 28 cm. C***
Bibliography: p. 92-94.
ISBN 0-7718-8024-3
1. Coal trade. 2. Coal mines and mining. I. British Columbia. Economic, Financial, and Market Analysis Sub-committee on North East Coal Development. II. Title.
HD9540.5.R56 1976 338.2'7'2 79-309089
79 MARC

—DIRECTORIES

World coal industry report and directory. v. 1- 1979/80-
₍San Francisco, Miller Freeman Publications₎
v. ill. 29 cm. biennial.
Key title: World coal industry report and directory, ISSN 0193-5453.
1. Coal trade—Directories. 2. Coal trade—Periodicals.
HD9540.3.W67 338.2'7'2025 79-643612
MARC-S

—ENVIRONMENTAL APSECTS—UNITED STATES

Rosenbaum, Walter A
Coal and crisis : the political dilemmas of energy management / Walter A. Rosenbaum. — New York : Praeger, 1978.
ix, 107 p. ; 25 cm.
Includes bibliographical references and index.
ISBN 0-03-042596-4 : $15.00
1. Coal trade—United States. 2. Coal trade—Environmental apsects—United States. 3. Energy policy—United States. I. Title.
HD9546.R67 1978 338.2'7'20973 78-8606
78 MARC

—LAW AND LEGISLATION—JAPAN

Uemura, Kōgorō, 1894-
(Sekitan Tōseikai no shinro)
石炭統制會の進路 / 植村甲午郎. — 東京 : 東邦社, 昭和18 ₍1943₎
2, 8, 320 p. ; 19 cm. — (統制會叢書 ; 第5輯)
Series romanized: Tōseikai sōsho.
"石炭配給統制法": p. 240-251.
1. Sekitan Tōseikai. 2. Coal trade—Japan. 3. Coal trade—Law and legislation—Japan. I. Japan. Laws, statutes, etc. Sekitan haikyū tōsei hō. 1943. II. Title.
HD9556.J33S538 79-80526

—PERIODICALS

World coal industry report and directory. v. 1- 1979/80-
₍San Francisco, Miller Freeman Publications₎
v. ill. 29 cm. biennial.
Key title: World coal industry report and directory, ISSN 0193-5453.
1. Coal trade—Directories. 2. Coal trade—Periodicals.
HD9540.3.W67 338.2'7'2025 79-643612
MARC-S

—ALBERTA

Alberta. Dept. of Energy and Natural Resources.
A coal development policy for Alberta. — ₍Edmonton, Alta.₎ : Dept. of Energy and Natural Resources, Government of Alberta, 1976.
38 p., ₍1₎ fold. leaf of plates : ill. ; 28 cm. C***
1. Coal—Alberta. 2. Coal trade—Alberta. 3. Power resources—Alberta. I. Title.
HD9554.C3A532 333.8'2 78-314241
78 MARC

—ASIA

Minami Manshū Tetsudō Kabushiki Kaisha. Shomubu. Chōsaka.
(Bujuntan no hanro)
撫順炭の販路 / ₍執筆者 阿部勇₎. — ₍大連₎ : 南滿洲鐵道株式會社庶務部調查課, 大正14 ₍1925₎
3, 77 p. ; 27 cm. — (調查報告書 - 南滿洲鐵道株式會社庶務部調查課 ; 第24卷)
Cover title.
1. Coal trade—China. 2. Coal trade—Asia. 3. Coal mines and mining—China—Fu-shun (Shih) I. Abe, Isamu. II. Title. III. Series: Minami Manshū Tetsudō Kabushiki Kaisha. Shomubu. Chōsaka. Chōsa hōkokusho ; dai 24-kan.
HD9556.C52M56 1925 79-800997

—AUSTRALIA

Wilson, J R
Back to black coal / by J. R. Wilson. — ₍Melbourne₎ : Committee for Economic Development of Australia, 1976.
iv, 34 p. ; 30 cm. — (Position paper - CEDA Energy Project ; no. 3) Aus***
Includes bibliographical references.
$8.00
1. Coal trade—Australia. 2. Power resources—Australia. I. Title. II. Series: Committee for Economic Development of Australia. Energy Project. Position paper - CEDA Energy Project ; no. 3.
HD9558.A82W54 338.2'7'240994 78-318073
78 MARC

—BELGIUM—SOUMAGNE—HISTORY

Monard, Dieudonné.
Le siège de José de la Société anonyme des charbonnages de Werister à Romsée : sa vie depuis ses origines : contribution à l'étude de l'histoire de l'origine des houillères dans le plateau de Herve / par Dieudonné Monard. — Fléron : Cercle archéo-historique des cantons de Fléron et Grivegnée, 1975.
49 p. ; 28 cm. Be***
70.00F
1. Coal trade—Belgium—Soumagne—History. 2. Coal mines and mining—Belgium—Soumagne—History. I. Cercle archéo-historique des cantons de Fléron et Grivegnée. II. Title.
HD9555.B43S685 78-391615
338.7'62'233094934
79 MARC

—CALIFORNIA—LIVERMORE—HISTORY

Mosier, Dan L
Harrisville and the Livermore coal mines / by Dan L. Mosier. — San Leandro, Calif. : Mines Road Books, c1978.
183 p. : ill. ; 22 cm.
Bibliography: p. 165-169.
Includes index.
1. Coal trade—California—Livermore—History. 2. Coal mines and mining—California—Livermore—History. 3. Livermore, Calif.—Industries—History. I. Title.
HD9548.L58M67 338.2'7'220979465 79-106164
79 MARC

COAL TRADE (Continued)

—CHINA

Minami Manshū Tetsudō Kabushiki Kaisha. Shomubu. Chōsaka.
(Bujuntan no hanro)
撫順炭の販路 / ⌈執筆者 阿部勇⌉. — ⌈大連⌉：南滿洲鐵道株式會社庶務部調査課，大正14 ⌈1925⌉
3, 77 p. ; 27 cm. — (調査報告書 - 南滿洲鐵道株式會社庶務部調査課；第24巻)
Cover title.
1. Coal trade—China. 2. Coal trade—Asia. 3. Coal mines and mining—China—Fu-shun (Shih) I. Abe, Isamu. II. Title. III. Series: Minami Manshū Tetsudō Kabushiki Kaisha. Shomubu. Chōsaka. Chōsa hōkokusho ; dai 24-kan.
HD9556.C52M56 1925 79-800997

—COLUMBIA—CAUCA VALLEY

Auteco.
Cauca Valley coal survey. — Brussels : Auteco, ⌈196-⌉
xxiii, 693 p. : ill. ; 26 cm.
At head of title: Republic of Colombia.
1. Coal—Columbia—Cauca Valley. 2. Coal trade—Columbia—Cauca Valley. I. Title.
TN806.C72C38 338.2'7'2098615 78-325905
79 MARC

—EUROPE—FINANCE

European Coal and Steel Community. Commission.
Investment in the Community coalmining and iron and steel industries : report on the 1975 survey, position as at 1 January 1975 in the nine countries of the enlarged Community / European Coal and Steel Community Commission. — Luxembourg : Office for Official Publications of the European Communities, 1975.
93 p. : graphs ; 30 cm. Lu***
400.00F
1. Steel industry and trade—Europe—Finance. 2. Coal trade—Europe—Finance. I. Title.
HD9525.A2E8883 1975 338.2'3 78-317295
78 MARC

—EUROPEAN ECONOMIC COMMUNITY COUNTRIES —PERIODICALS

Europe. no. 211- Jan./Feb. 1979-
⌈Washington, Delegation of the Commission of the European Communities⌉
no. ill. 28 cm. bimonthly.
"Magazine of the European Community."
Continues: European Community.
Key title: Europe (Washington), ISSN 0191-4545.
1. Steel industry and trade—European Economic Community countries—Periodicals. 2. Coal trade—European Economic Community countries—Periodicals. 3. European Economic Community—Periodicals. I. Commission of the European Communities. Delegation (U. S.). II. Key title.
HD9525.A2E52 382'.9142'05 79-642763
MARC-S

—EUROPEAN ECONOMIC COMMUNITY COUNTRIES —STATISTICS—PERIODICALS

Kulstatistik. Kohlestatistik. Coal statistics. 1975-
⌈Luxembourg, De europæiske fællesskabers statistiske kontor⌉
v. 30 cm. annual.
Continues: Statistical Office of the European Communities. Kohlestatistik.
English, French, and German.
1. Coal trade—European Economic Community countries—Statistics—Periodicals. 2. Coal—European Economic Community countries—Statistics—Periodicals. I. Statistical Office of the European Communities. II. Title: Kohlestatistik. III. Title: Coal statistics.
HD9555.E82S75b 338.2'7'2094 78-646437
MARC-S

—GERMANY

Weisbrod, Bernd.
Schwerindustrie in der Weimarer Republik : Interessenpolitik zwischen Stabilisierung und Krise / Bernd Weisbrod. — Wuppertal : P. Hammer, c1978.
552 p. ; 21 cm. GFR***
A revision of the author's thesis, Bochum, 1976.
Bibliography: p. 509-544.
Includes indexes.
ISBN 3-87294-123-2
1. Steel industry and trade—Germany. 2. Coal trade—Germany. 3. North Rhine-Westphalia—Industries. 4. Germany—Economic conditions—1918-1945. I. Title.
HD9523.5.W4 1978 78-398694
79 MARC

—GERMANY, WEST—RUHR VALLEY

Schaaf, Peter.
Ruhrbergbau und Sozialdemokratie : d. Energiepolitik d. Grossen Koalition 1966-1969 / Peter Schaaf. — Marburg/Lahn : Verlag Arbeiterbewegung u. Gesellschaftswiss., 1978.
474 p. ; 21 cm. — (Schriftenreihe für Sozialgeschichte und Arbeiterbewegung ; Bd. 11) GFR78-A
Originally presented as the author's thesis.
Bibliography: p. 464-474.
ISBN 3-921630-10-X : DM28.00
1. Coal trade—Germany, West—Ruhr Valley. 2. Coal trade—Germany, West—Ruhr Valley—Finance. 3. Ruhrkohle AG. 4. Industrial relations—Germany, West—Ruhr Valley. 5. Energy policy—Germany, West. I. Title. II. Series.
HD9553.7.R8S3 1978 79-373919
79 MARC

—GERMANY, WEST—RUHR VALLEY— FINANCE

Schaaf, Peter.
Ruhrbergbau und Sozialdemokratie : d. Energiepolitik d. Grossen Koalition 1966-1969 / Peter Schaaf. — Marburg/Lahn : Verlag Arbeiterbewegung u. Gesellschaftswiss., 1978.
474 p. ; 21 cm. — (Schriftenreihe für Sozialgeschichte und Arbeiterbewegung ; Bd. 11) GFR78-A
Originally presented as the author's thesis.
Bibliography: p. 464-474.
ISBN 3-921630-10-X : DM28.00
1. Coal trade—Germany, West—Ruhr Valley. 2. Coal trade—Germany, West—Ruhr Valley—Finance. 3. Ruhrkohle AG. 4. Industrial relations—Germany, West—Ruhr Valley. 5. Energy policy—Germany, West. I. Title. II. Series.
HD9553.7.R8S3 1978 79-373919
79 MARC

—GREAT BRITAIN

Great Britain. Price Commission.
Prices and margins in the distribution of coal and solid smokeless fuels for domestic consumption / Price Commission. — London : H.M.S.O., 1976.
vi, 29 p. ; 25 cm. — (Report - Price Commission ; no. 12) GB76-08924
Cover title: Distribution of domestic coal and solid smokeless fuel.
ISBN 0-11-700297-6 : £0.65
1. Coal trade—Great Britain. 2. Fuel trade—Great Britain. 3. Coal—Prices—Great Britain. 4. Fuel—Prices—Great Britain. I. Title. II. Title: Distribution of domestic coal and solid smokeless fuel. III. Series: Great Britain. Price Commission. Report - Price Commission ; no. 12.
HD9551.5.G64 1976 338.4'3'6626 78-326865
79 MARC

—GREAT BRITAIN—HISTORY

Buxton, N K
The economic development of the British coal industry from industrial revolution to the present day / by Neil K. Buxton. — London : Batsford Academic, 1978.
297 p. ; 24 cm. GB***
Includes bibliographical references and index.
ISBN 0-7134-1994-6 : £15.00
1. Coal trade—Great Britain—History. I. Title.
HD9551.5.B89 338.2'7'20941 79-313114
79 MARC

—GREAT PLAINS

Stinson, Thomas F 1942-
Coal development in the northern Great Plains : the impact on revenues of State and local governments / Thomas F. Stinson, Stanley W. Voelker ; Economics, Statistics, and Cooperatives Service, U.S. Dept. of Agriculture, in cooperation with the U.S. Environmental Protection Agency, the University of Minnesota, and North Dakota State University. — Washington : The Department, 1978.
ii, 66 p. : 26 cm. — (Agricultural economic report ; no. 394)
Cover title.
Supt. of Docs. no.: A 1.107: 394.
1. Coal trade—Great Plains. 2. Coal—Taxation—Great Plains. 3. Local finance—Great Plains. I. Voelker, Stanley W., joint author. II. United States. Dept. of Agriculture. Economics, Statistics, and Cooperatives Service. III. Title. IV. Series.
HD1751.A91854 no. 394 78-601276
338.1'08 s
⌈HD9547.A14⌉ 78 MARC

—ILLINOIS—STATISTICS

Weil, Keith.
The Illinois coal digest / by Keith Weil, Alvin K. Grandys. — ⌈Springfield⌉ : Illinois Dept. of Business and Economic Development, Division of Energy, 1976.
ii leaves, 127 p. : ill. ; 29 cm.
Cover title.
"IDOE-RS-76-06."
1. Coal trade—Illinois—Statistics. 2. Coal—Illinois—Statistics. I. Grandys, Alvin K., joint author. II. Illinois. Division of Energy. III. Title.
HD9547.I3W44 338.2'7'2409773 78-622247
79 MARC

—INDIA

Coal India Limited.
Coal plan, 1976-77 to 1985-86. — ⌈s.l.⌉ : Coal India, ⌈pref. 1977⌉
24 p. : ill. ; 19 cm.
Cover title.
1. Coal trade—India. I. Title.
HD9556.I42C62 1977 338.2'7'20954 78-905712
79 MARC

Ghosh, Amiya Bhushan.
Coal industry in India : an historical and analytical account / A. B. Ghosh. — New Delhi : Sultan Chand, ⌈pref. 1977-
v. ; 23 cm.
Bibliography: pt. 1, p. ⌈305⌉-308.
Includes index.
CONTENTS: pt. 1. Pre-independence period.

Rs50.00 (v. 1)
1. Coal trade—India. 2. Coal mines and mining—India. I. Title.
HD9556.I42G48 338.2'7'20954 78-907451
79 MARC

India (Republic). Committee on Amalgamation of Collieries.
Report of the Committee on Amalgamation of Collieries, 1956. — Delhi : Manager of Publications, 1957-
v. : ill. ; 25 cm.
Cover title.
"MP.8/1000."
Chairman: B. G. Mehta.
1. Coal trade—India. 2. Coal mines and mining—India. I. Mehta, Balvantray Gopaljee. II. Title.
HD9556.I42I53 1957 338.2'7'20954 78-912402
79 MARC

—INDIA—ASSAM—STATISTICS

Assam. Directorate of Economics and Statistics.
Index of coal production, Assam, 1970-1975. — Gauhati : Directorate of Economics & Statistics, Assam, 1976.
5 p., ⌈2⌉ leaves of plates : ill. ; 25 cm.
Cover title.
1. Coal trade—India—Assam—Statistics. I. Title.
HD9556.I43A882 338.2'7'20954162 78-901797
79 MARC

—JAPAN

Uemura, Kōgorō, 1894-
(Sekitan Tōseikai no shinro)
石炭統制會の進路 / 植村甲午郎. — 東京：東邦社，昭和18 ⌈1943⌉
2, 8, 320 p. ; 19 cm. — (統制會叢書；第5輯)
Series romanized: Tōseikai sōsho.
"石炭配給統制法": p. 240-251.
1. Sekitan Tōseikai. 2. Coal trade—Japan. 3. Coal trade—Law and legislation—Japan. I. Japan. Laws, statutes, etc. Sekitan haikyū tōsei hō. 1943. II. Title.
HD9556.J33S538 79-805262

COAL TRADE (Continued)

—JAPAN—STATISTICS—YEARBOOKS

(Energī tōkei nempō)
エネルギー統計年報 Yearbook of coal, petroleum and coke statistics. 1975–
[東京 通商産業調査会]
v. 26 cm.
"編集・通商産業[省]大臣官房調査統計部"
Formed by the union of: Sekitan kōkusu nempō, and: Sekiyu tōkei nempō.
In Japanese.
1. Coal trade—Japan—Statistics—Yearbooks. 2. Petroleum industry and trade—Japan—Statistics—Yearbooks. 3. Coke industry—Japan—Statistics—Yearbooks. 4. Power resources—Japan—Statistics—Yearbooks. I. Japan. Tsūshō Sangyōshō. Daijin Kambō. Chōsa Tōkeibu.
HD9556.J29E53 78-642754
(MARC-S

—JAPAN—FUKUOKA-HAN—HISTORY —SOURCES

(Oshitate sumiyama sadame)
御仕立炭山定：福岡藩山方史料：附 御山方御法令 / 福岡大学研究所編. — 福岡：同研究所, 1978.
306 p. ; 21 cm. — (福岡大学研究所資料叢書 ; 第2冊) Ja 78-22193
Not in trade.
1. Coal trade—Japan—Fukuoka-han—History—Sources. 2. Forestry law and legislation—Japan—Fukuoka-han—History Sources. 3. Fuel—Japan—Fukuoka-han—History—Sources. 4. Kuroda family—Archives. I. Fukuoka-han, Japan. II. Fukuoka Daigaku. Kenkyūjo. III. Series: Fukuoka Daigaku. Kenkyūjo. Fukuoka Daigaku Kenkyūjo shiryō sōsho ; dai 2-satsu.
HD9556.J32F856 79-804605

—JAPAN—HOKKAIDO—HISTORY

Hokkaidō Kaitaku Kinenkan.
(Hokkaidō ni okeru tankō no hatten to rōdōsha)
北海道における炭鉱の発展と労働者. — 札幌：北海道開拓記念館, 1978.
112 p. ; 26 cm. — (北海道開拓記念館研究報告 ; 第4号) Ja 78-16990
Cover title.
Title on p. [4] of cover: On the history of coal mine and colliery society in Hokkaido.
Summary in English.
"北海道石炭産業史関係文献資料目録": p. 99–112.
1. Coal trade—Japan—Hokkaido—History. 2. Coal miners—Japan—Hokkaido—History. I. Title. II. Title: On the history of coal mine and colliery society in Hokkaido. III. Series: Hokkaidō Kaitaku Kinenkan. Hokkaidō Kaitaku Kinenkan kenkyū hōkoku ; dai 4-gō.
HD9556.J33H6 78-812639

—OKLAHOMA

Ghebremedhin, Tesfa.
An analysis of Oklahoma's coal industry / Tesfa Ghebremedhin and Michael S. Salkin. — [Stillwater] : Agriculture [sic] Experiment Station, Oklahoma State University, 1978.
41 p. : graphs ; 28 cm. — (Research report - Agriculture [sic] Experiment Station, Oklahoma State University ; P-780)
Bibliography: p. 25.
1. Coal trade—Oklahoma. 2. Coal—Oklahoma. I. Salkin, Michael S., joint author. II. Title. III. Series: Oklahoma. Agricultural Experiment Station, Stillwater. Research report ; P-780.
HD9547.O5G47 338.2'7'2409766 79-621191
79 MARC

—SCOTLAND—HISTORY

The Coal industry : 17th-20th centuries / ... researched and edited by Lynda Wright. — Stirling : Central Regional Council Education Department, 1978.
[2], 65 p., 8 leaves of plates : ill., map ; 21 cm. — (Historical sources for central Scotland series ; 2) GB79-02465
Bibliography: p. 63-65.
£0.30
1. Coal trade—Scotland—History. 2. Coal mines and mining—Scotland—History. I. Wright, Lynda. II. Series.
HD9551.7.A3C55 338.2'7'209413 79-313189
79 MARC

—SWEDEN

Schlasberg, Johan, 1946-
Att förstå kol : några frågor kring en ökad svensk kolimport / Johan Schlasberg. — Stockholm : Sekretariatet för framtidsstudier : LiberFörlag (distr.), 1978.
[5], 64 p. : ill. ; 30 cm. — (Framtidsstudien Energi och samhälle ; 16) S78-17
Bibliography: p. 63-64.
ISBN 9901417562
1. Coal trade—Sweden. 2. Energy policy—Sweden. I. Title. II. Series.
HD9555.S85S34 79-356533
79 MARC

—TEXAS—FAYETTE CO.

Demographic and community service impacts of coal utilization and development in rural areas : an examination of the case of Fayette County, Texas / by Steve H. Murdock ... [et al.] with the assistance of Marilyn Hansen ... [et al.]. — College Station : Texas Agricultural Experiment Station, Texas A & M University System, 1979.
86 p. : ill. ; 28 cm. — (Departmental technical report ; no. 79-1)
Bibliography: p. 85-86.
1. Coal trade—Texas—Fayette Co. 2. Fayette Co., Texas—Economic conditions. 3. Fayette Co., Texas—Social conditions. 4. Social service—Texas—Fayette Co. I. Murdock, Steven H. II. Series.
HD9547.T4D45 79-623876
338.2'7'2209764251
79 MARC

—UNITED STATES

Greenberg, Gail.
The coal conversion business : realities, markets / Gail Greenberg. — Stamford, Conn. : Business Communications Co., 1978.
xi, 195 leaves ; 29 cm. — (Business opportunity report ; E-029)
1. Coal trade—United States. 2. Market surveys—United States. I. Business Communications Co. II. Title. III. Series.
HD9545.6.G73 338.2'7'20973 79-101918
79 MARC

Rosenbaum, Walter A
Coal and crisis : the political dilemmas of energy management / Walter A. Rosenbaum. — New York : Praeger, 1978.
ix, 107 p. ; 25 cm.
Includes bibliographical references and index.
ISBN 0-03-042596-4 : $15.00
1. Coal trade—United States. 2. Coal trade—Environmental apsects—United States. 3. Energy policy—United States. I. Title.
HD9546.R67 1978 338.2'7'20973 78-8606
78 MARC

United States. Congress. House. Committee on Interior and Insular Affairs. Subcommittee on Oversight and Investigations.
Illegal mining of Federal coal : a report / prepared by the Subcommittee on Oversight and Investigations of the Committee on Interior and Insular Affairs of the U.S. House of Representatives, Ninety-sixth Congress, first session. — Washington : U.S. Govt. Print. Off., 1979.
v, 9 p. ; 24 cm.
"Committee print no. 3, 96th Congress, 1st session."
1. Coal trade—United States. 2. Coal mines and mining—Law and legislation—United States. 3. Coal leases—United States. I. Title.
HD9546.U64 1979 364.1'62 79-603230
79 MARC

United States. Congress. Senate. Committee on the Judiciary. Subcommittee on Antitrust and Monopoly.
Horizontal divestiture : joint hearing before the Subcommittee on Antitrust and Monopoly of the Committee on the Judiciary, United States Senate, and the Subcommittee on Energy and the Environment of the Committee on the Interior [sic], House of Representatives, Ninety-fifth Congress, second session ... June 28, 1978. — Washington : U.S. Govt. Print. Off., 1978.
v, 162 p. ; 24 cm.
Includes bibliographical references.
1. Petroleum industry and trade—United States. 2. Coal trade—United States. 3. Corporate divestiture—United States. I. United States. Congress. House. Committee on Interior and Insular Affairs. Subcommittee on Energy and the Environment. II. Title.
KF26.J835 1978a 338.8'042 79-600706
79 MARC

United States. Council on Wage and Price Stability.
A study of coal prices / Executive Office of the President, Council on Wage and Price Stability. — [Washington] : The Council : for sale by the Supt. of Docs., U.S. Govt. Print. Off., 1976.
xi, 97 p. : ill. ; 26 cm. — (Staff report - Council on Wage and Price Stability)
Cover title.
Includes bibliographical references.
1. Coal—Prices—United States. 2. Coal trade—United States. I. Title. II. Series: United States. Council on Wage and Price Stability. Staff report - Council on Wage and Price Stability.
HD9544.U5 1976 338.2'3 76-601379
76[r79]rev MARC

—UNITED STATES—CONGRESSES

Conference on Coal Production and Transportation, 3d, Denver, 1977.
Coal production & transportation : third annual conference—1977 : April 13 and 14, 1977, the Brown Palace, Denver, Colorado : transcript of a conference / sponsored by PLM, inc. — [San Francisco] : PLM, inc., [c1977]
236 p. : ill. ; 23 cm.
Cover title.
1. Coal trade—United States—Congresses. 2. Coal—United States—Transportation—Congresses. I. PLM, inc. II. Title.
HD9545.C59 1977 338.3'7'20973 78-109835
78 MARC

National Coal Policy Project.
A report of a seminar at the Colorado School of Mines, September, 1978 / the National Coal Policy Project ; edited by J. Charles Curran. — Washington : Center for Strategic and International Studies, Georgetown University, c1979.
vi, 83 p. ; 23 cm. — (CSIS report)
1. Coal trade—United States—Congresses. 2. Coal—United States—Congresses. 3. Energy policy—United States—Congresses. I. Curran, J. Charles. II. Title. III. Series: Georgetown University, Washington, D.C. Center for Strategic and International Studies. CSIS report.
HD9546.N34 1979 338.2'7'20973 79-66508
00 MARC

—UNITED STATES—FINANCE

Coal, lead, and zinc industry sourcebook, May, 1975-April, 1976 : materials / selected by MSS Information Corporation from the journals, indexes, and profiles of Disclosure, inc. — New York : The Corporation ; Edison, N.J. : exclusive sales agent, Lasky-Lanouette, c1977.
209 p. ; 23 cm.
"A guide to corporate reports filed with the S.E.C. May 1975 to April 1976."
ISBN 0-8422-4022-5 : $50.00
1. Coal trade—United States—Finance. 2. Lead industry and trade—United States—Finance. 3. Zinc industry and trade—United States—Finance. I. MSS Information Corporation. II. Disclosure, inc.
HD9545.C58 338.2'0973 77-82782
77 MARC

Mann, Charles Edward, 1950-
Coal and profitability : an investor's guide / by Charles E. Mann and James N. Heller, and others of Energy and Environmental Analysis. — [Hightstown, N.J.] : Coal Week, [c1979]
vii, 126 p. : graphs ; 28 cm.
Bibliography: p. 119-126.
ISBN 0-07-606610-X : $63.50
1. Coal trade—United States—Finance. 2. Coal mines and mining—United States—Finance. 3. Investments—United States. I. Heller, James N., joint author. II. Energy and Environmental Analysis, inc. III. Title.
HD9544.M36 332.6'722 78-65618
79 MARC

—UNITED STATES—PERIODICALS

Data Resources, inc.
Coal review. v. 1–
Nov. 1978–
Lexington, Mass., Data Resources, inc.
v. 28 cm. semiannual.
Spine title, Nov. 1978– : The Data Resources coal review.
Key title: Coal review (Lexington), ISSN 0191-4103.
1. Coal trade—United States—Periodicals. 2. Market surveys—United States—Periodicals. I. Data Resources, inc. The Data Resources coal review. II. Title. III. Key title.
HD9541.D37a 338.2'7'20973 79-642758
MARC-S

United States. Dept. of Justice.
Competition in the coal industry; report of the United States Department of Justice pursuant to section 8 of the Federal coal leasing amendments act of 1975.
[Washington] Dept. of Justice; [for sale by the Supt. of Docs., U. S. Govt. Print. Off.]
v. 26 cm. annual.
Key title: Competition in the coal industry, ISSN 0162-5624
1. Coal trade—United States—Periodicals. 2. Trusts, Industrial—United States—Periodicals. 3. Antitrust law—United States—Periodicals. I. Title.
HD9541.U5a 338.8 78-645280
MARC-S

—UNITED STATES—STATISTICS —PERIODICALS

Keystone news bulletin.
New York, Mining Informational Services of the McGraw-Hill Mining Publications.
v. 28 cm. monthly.
Key title: Keystone news bulletin, ISSN 0149-5801.
1. Coal trade—United States—Statistics—Periodicals. 2. Coal—Statistics—Periodicals.
HD9541.K45 338.2'7'20973 79-643452
MARC-S

COAL TRADE (Continued)

—WALES—HISTORY

Walters, Rhodri Havard.
The economic and business history of the south Wales steam coal industry, 1840-1914 / R. H. Walters. — New York : Arno Press, 1977.
xvi, 375 p., [6] leaves of plates : ill. ; 24 cm. — (Dissertations in European economic history)
Originally presented as the author's thesis, Oxford, 1975.
Bibliography: p. 367-375.
ISBN 0-405-10801-X
1. Coal trade—Wales—History. I. Title. II. Series.
HD9551.7.A7W34 1977 338.2'7'209429 77-77188
77[r79]rev MARC

COAL WASHING

see also Coal preparation

COALA see Koalas

COALBROOKDALE COMPANY, LTD., WELLINGTON, ENG.

Ironbridge Gorge Museum Trust.
The Coalbrookdale ironworks : a short history / Ironbridge Gorge Museum Trust. — [Telford] : The Trust, 1975.
[2], 21 p. : ill., map ; 24 cm. GB75-22516
Cover title.
"SJ 667048."
Based on an original text by A. Raistrick.
ISBN 0-903971-07-0 : £0.30
1. Coalbrookdale Company, ltd., Wellington, Eng. 2. Wellington, Eng.—Industries. I. Raistrick, Arthur. II. Title.
TS330.W44I76 338.2'7'30942456 76-356130
76[r79]rev MARC

COALITION (SOCIAL SCIENCES)

see also Coalition governments

Schneider, Jerrold E
Ideological coalitions in Congress / Jerrold E. Schneider. — Westport, Conn. : Greenwood Press, 1979.
xvi, 270 p. ; 24 cm. — (Contributions in political science ; no. 16 ISSN 0147-1066)
Bibliography: p. [235]-258.
Includes indexes.
ISBN 0-313-20410-1 : $22.50
1. United States. Congress—Voting. 2. Coalition (Social sciences) 3. Ideology. I. Title. II. Series.
JK1051.S46 328.73'07'75 78-4019
78 MARC

—MATHEMATICAL MODELS

Coalition formation by sophisticated players / Amnon Rapoport ... [et al.]. — Berlin ; New York : Springer-Verlag, 1979.
vii, 170 p. ; 24 cm. — (Lecture notes in economics and mathematical systems ; 169) (Mathematical economics)
Bibliography: p. 76-81.
ISBN 0-387-09249-8
1. Game theory. 2. Coalition (Social sciences)—Mathematical models. 3. Negotiation—Mathematical models. I. Rapoport, Amnon. II. Series. III. Series: Mathematical economics.
HB144.C6 300'.1'5193 79-13521
79 MARC

COALITION GOVERNMENTS

see also Popular fronts

(Tōitsu sensen no shomondai)
統一戦線の諸問題 / 労働大学調査研究所編. — 東京：同大学, 昭和52- [1977-
v. ; 18 cm. — (労大新書 ; 60,) Ja 77-14101 (v. 1)
Series romanized: Rōdai shinsho.
Includes bibliographical references.
CONTENTS: 1. 歴史篇
¥450 (v. 1)
1. Popular fronts. 2. Socialist parties. 3. Coalition governments. I. Rōdō Daigaku. Chōsa Kenkyūjo.
HX40.T58 78-814016

—MATHEMATICAL MODELS

Pressacco, Flavio.
Teoria dei giochi e previsioni di coalizioni governative : un contributo alla teoria della distanza politica / di Flavio Pressacco. — 1. ed. — Trieste : CLUET, [1977?]
27 p. ; 21 cm. — (Contributi e ricerche scienze politiche Trieste ; 5) It***
Bibliography: p. 27.
1. Coalition governments—Mathematical models. 2. Game theory. I. Title. II. Series.
JF331.P72 329'.001'51 78-391486
79 MARC

—EUROPE

Fukuoka, Masayuki, 1945-
(Rengō seiken)
連合政権：保革伯仲下の政権構想 / 福岡政行. — [東村山]：教育社；東京：販売 教育社出版サービス, [1978]
170 p. : ill. ; 18 cm. — (時事問題解説 : 97) (入門新書) Ja 78-23739
Cover title.
Series 2 romanized: Nyūmon shinsho.
Bibliography: p. 169-170.
Y400
1. Japan—Politics and government—1945- 2. Coalition governments—Japan. 3. Coalition governments—Europe. I. Title. II. Series: Jiji mondai kaisetsu : 97.
JQ1626 1945.F84 79-806802

Pappalardo, Adriano, 1948-
Partiti e governi di coalizione in Europa / Adriano Pappalardo. — Milano : F. Angeli, [1978]
183 p. ; 22 cm. — (Collana di scienza politica e relazioni internazionali ; 5) It78-July
Includes bibliographical references.
L6000
1. Political parties—Europe. 2. Coalition governments—Europe. I. Title.
JN94.A979P36 329'.0094 78-378985
*78 MARC

—GERMANY, WEST

Zwoch, Gerhard.
Koalition : Analytisches u. Dokumentarisches / Gerhard Zwoch, Jürgen Lorenz. — Bonn : AZ-Studio, 1974.
13, [9] p. : ill. ; 21 cm. — (Bundestag von A-Z ; Nr. 9) GFR76-A49
Bibliography: p. [22]
1. Political parties—Germany, West. 2. Coalition governments—Germany, West. I. Lorenz, Jürgen, joint author. II. Title. III. Series.
JN3971.A979Z85 329'.00943 79-344791
79 MARC

—INDIA—PUNJAB (STATE)

Verma, Manju.
The coalition ministries in Punjab / Manju Verma. — Patiala : Sonu Sales Agency, 1978.
161 p. ; 23 cm.
Rs30.00
1. Punjab, India (State)—Politics and government. 2. Coalition governments—India—Punjab (State) I. Title.
JQ563 1978.V47 329'.00954'5 79-900825
79 MARC

—INDONESIA—ADDRESSES, ESSAYS, LECTURES

Diskusi Kita (Association)
Kumpulan kertas karja / Diskusi Kita. — Djakarta : Sekretariat Diskusi Kita, [1970]
44 p. ; 31 cm.
1. Indonesia—Politics and government—20th century—Addresses, essays, lectures. 2. Coalition governments—Indonesia—Addresses, essays, lectures.
JQ768.D57 1970 77-941398
79 MARC

—ITALY

Di Palma, Giuseppe.
Political syncretism in Italy : historical coalition strategies and the present crisis / Giuseppe Di Palma. — Berkeley : Institute of International Studies, University of California, c1978.
vi, 58 p. ; 24 cm. — (Policy papers in international affairs ; no. 7)
Includes bibliographical references.
ISBN 0-87725-507-5 : $2.00
1. Political parties—Italy. 2. Coalition government—Italy. 3. Italy—Politics and government—1945- I. California. University. Institute of International Studies. II. Title. III. Series.
JN5651.D47 329'.00945 78-620042
79 MARC

—JAPAN

Fukuoka, Masayuki, 1945-
(Rengō seiken)
連合政権：保革伯仲下の政権構想 / 福岡政行. — [東村山]：教育社；東京：販売 教育社出版サービス, [1978]
170 p. : ill. ; 18 cm. — (時事問題解説 : 97) (入門新書) Ja 78-23739
Cover title.
Series 2 romanized: Nyūmon shinsho.
Bibliography: p. 169-170.
Y400
1. Japan—Politics and government—1945- 2. Coalition governments—Japan. 3. Coalition governments—Europe. I. Title. II. Series: Jiji mondai kaisetsu : 97.
JQ1626 1945.F84 79-806802

—JAPAN—ADDRESSES, ESSAYS, LECTURES

Miyamoto, Kenji, 1908-
(Kakushin tōitsu sensen o)
革新統一戦線を / [著者 宮本顕治]. — 東京：新日本出版社, 1978.
344 p. ; 20 cm. — (宮本顕治現代政治論; 2) Ja***
Series romanized: Miyamoto Kenji gendai seiji ron.
¥1500
1. Nihon Kyōsantō—Addresses, essays, lectures. 2. Japan—Politics and government—1945- —Addresses, essays, lectures. 3. Coalition governments—Japan—Addresses, essays, lectures. I. Title.
HX411.M59 vol. 2 79-803380
[JQ1698.N5]

—THAILAND

Somporn Sangchai.
Coalition behaviour in modern Thai politics : a Thai perspective / by Somporn Sangchai. — Singapore : Institute of Southeast Asian Studies, 1976.
26 leaves : diagrs. ; 28 cm. — (Occasional paper - Institute of Southeast Asian Studies ; no. 41)
Includes bibliographical references.
$4.00
1. Coalition governments—Thailand. 2. Political parties—Thailand. 3. Thailand—Politics and government. I. Title. II. Series: Institute of Southeast Asian Studies. Occasional paper ; no. 41.
JQ1749.A795S66 329'.009593 76-941188
77[r79]rev MARC

COANDĂ, HENRI, 1886-1972

Garabedeanu, Antoaneta.
Henri Coandă inventatorul avionului cu reacție / [Antoaneta Garabedeanu]. — [Iași] : Muzeul Politehnic Iași, 1976, t.p. 1975.
15 p. : ill. ; 17 cm. R76-4467
At head of title: Complexul Muzeistic Iași.
lei1.50
1. Coandă, Henri, 1886-1972. 2. Aeronautical engineers—Romania—Biography. I. Title.
TL540.C588G37 78-397969
*79 MARC

—JUVENILE LITERATURE

Firoiu, V
Henri Coandă. Text: V. Firoiu. Ilustrații: Damian Petrescu. București, Editura Ion Creangă, 1973.
16 p. with col. illus. 17 cm. (ABC,10) lei0.75 R73-4399
1. Coandă, Henri, 1886-1972—Juvenile literature. 2. Engineers—Romania—Biography—Juvenile literature.
TA140.C57F57 78-390983
79 MARC

COAST CHANGES

see also Beach erosion; Erosion; Sedimentation and deposition; Shore-lines; Shore protection

Jones, Michael C E
Finland : daughter of the sea / Michael Jones. — Folkestone, Eng. : Dawson ; Hamden, Conn. : Archon Books, 1977.
247 p. : ill. ; 23 cm. — (Studies in historical geography ISSN 0308-6607) GB77
Includes bibliographical references and index.
ISBN 0-208-01623-6 (Archon Books)
1. Land tenure—Finland. 2. Coast changes. 3. Geology, Structural. 4. Man—Influence of environment—Finland. 5. Anthropo-geography—Finland. I. Title.
HD723.J66 1977 301.34'09471 76-30001
77[r79]rev MARC

COAST CHANGES (Continued)

Research techniques in coastal environments / H. J. Walker. — Baton Rouge : School of Geoscience, Louisiana State University, 1977.
320 p. : ill. ; 29 cm. — (Geoscience and man ; v. 18)
Includes bibliographies.
1. Coast changes. 2. Coast changes—Technique. 3. Coastal zone management. I. Walker, Harley Jessie, 1921- . II. Series.
GB451.2.R47 551.3'6 78-113301
79 MARC

Sokol'nikov, IUrii Mykolaiovych.
(Inzhenernaia morfodinamika beregov i ee prilozheniia)
Инженерная морфодинамика берегов и ее приложения / Ю. Н. Сокольников ; АН УССР, Ин-т гидромеханики. — Киев : Наук. думка, 1976.
227 p. : ill. ; 20 cm. USSR 76
Bibliography : p. 218-[225]
1.28rub
1. Shore protection. 2. Coast changes. I. Title.
TC330.S64 78-422908

—MAPS

Canada-Ontario Great Lakes Shore Damage Survey.
Coastal zone atlas / Canada-Ontario Great Lakes Shore Damage Survey. — [Ottawa] : Environment Canada ; [Toronto] : Ontario Ministry of Natural Resources, c1976.
circa 800 p. : col. ill., maps (some col.) ; 39 x 54 cm. CG76-73252-3
On all maps: Great Lakes Shore Damage Survey - 1973.
Scale of maps 1:20,000 or one inch to 1,6666 feet.
CONTENTS: Lake Huron.—St. Clair River, Lake St. Claire, Detroit River.—Lake Eire.—Lake Ontario.
$135.00
1. Great Lakes—Storm, 1972, 1973—Maps. 2. Coast changes—Maps. 3. Coastal zone management—Maps. 4. Great Lakes region—Maps. 5. Great Lakes region—Photo maps. 6. Great Lakes—Shore protection—Maps. I. Canada. Dept. of the Environment. II. Ontario. Ministry of Natural Resources. III. Title.
G1107.G7C3 1976 912'.1'3339170977 78-323616
79 MARC

—MATHEMATICAL MODELS

McClain, Charles R
Application of a radiation-type boundary condition to the wave-porous bed problem / by Charles R. McClain, Norden E. Huang, Leonard J. Pietrafesa. — [Raleigh] : University of North Carolina Sea Grant College, 1977.
vi, 95 p. : ill. ; 28 cm. — (Center for Marine and Coastal Studies report ; no. 77-8) (Sea grant publication ; UNC-SG-77-10)
Bibliography: p. [74]-77.
$2.50
1. Coast changes—Mathematical models. I. Huang, Norden Eh, 1937- joint author. II. Pietrafesa, Leonard J., joint author. III. Title. IV. Series: North Carolina. State University, Releigh. Center for Marine and Coastal Studies. Center for Marine and Coastal Studies publication ; no. 77-8. V. Series: Sea grant publication (Raleigh) ; UNC-SG-77-10.
GB451.2.M32 551.3'6 78-622677
79 MARC

—TECHNIQUE

Research techniques in coastal environments / H. J. Walker. — Baton Rouge : School of Geoscience, Louisiana State University, 1977.
320 p. : ill. ; 29 cm. — (Geoscience and man ; v. 18)
Includes bibliographies.
1. Coast changes. 2. Coast changes—Technique. 3. Coastal zone management. I. Walker, Harley Jessie, 1921- . II. Series.
GB451.2.R47 551.3'6 78-113301
79 MARC

—ATLANTIC COAST (NORTH AMERICA)—CONGRESSES

Coastal Research Symposium, Boston, 1978.
Barrier islands from the Gulf of St. Lawrence to the Gulf of Mexico / edited by Stephen P. Leatherman. — New York : Academic Press, 1979.
xiv, 325 p. : ill. ; 24 cm.
"The proceedings of a Coastal Research Symposium held on March 9, 1978 in Boston, Massachusetts for the Geological Society of America, Northeast Section and the Society of Economic Paleontologists and Mineralogists, Eastern Section."
Includes bibliographies and index.
ISBN 0-12-440260-7
1. Barrier islands—Atlantic coast (North America)—Congresses. 2. Coast changes—Atlantic coast (North America)—Congresses. I. Leatherman, Stephen P. II. Geological Society of America. Northeastern Section. III. Society of Economic Paleontologists and Mineralogists. Eastern Section. IV. Title.
GB473.C62 1978 551.4'2 2-Apr 79-15954
79 MARC

—LOUISIANA—GULF REGION

Shoreline erosion in coastal Louisiana : inventory and assessment : final report to Louisiana Department of Transportation and Development / by R. D. Adams ... [et al.]. — [Baton Rouge] : Louisiana State University Center for Wetland Resources, 1978.
vi, 139 p. : maps ; 28 cm.
Bibliography: p. 130-133.
1. Erosion—Louisiana—Gulf region. 2. Coast changes—Louisiana—Gulf region. 3. Coastal zone management—Louisiana—Gulf region. I. Adams, Rodney D., 1937- . II. Louisiana. Dept. of Transportation and Development. III. Louisiana State University, Baton Rouge. Center for Wetland Resources.
GB459.4.S53 333.9'17'09763 79-621598
79 MARC

—MICHIGAN, LAKE—GUIDE-BOOKS

Coastal geology, sedimentology, and management, Chicago and the Northshore : prepared for the 4th annual field conference, Great Lakes Section, Society of Economic Paleontologists and Mineralogists, September 21-22, 1974 / Charles Collinson ... [et al.]. — [Urbana : Illinois State Geological Survey, 1974]
55 p. : ill. ; 29 cm. — (Illinois State Geological Survey guidebook series ; 12)
Bibliography: p. 54-55.
1. Coast changes—Michigan, Lake—Guide-books. 2. Michigan, Lake—Water level—Guide-books. 3. Coastal engineering—Michigan, Lake—Guide-books. I. Collinson, Charles William. II. Society of Economic Paleontologists and Mineralogists. Great Lakes Section. III. Series: Illinois. State Geological Survey. Guidebook series ; 12.
QE105.I57a no. 12 557.73'08 s 78-107254
[GB459.5.M5] 79 MARC

—MISSISSIPPI

Waller, Thomas H
Temporal changes in the offshore islands of Mississippi / by Thomas H. Waller, Lynn P. Malbrough. — Mississippi State : Water Resources Research Institute, Mississippi State University, 1976.
viii, 109 p. : ill. ; 28 cm.
Bibliography: p. 91-94.
1. Barrier islands—Mississippi. 2. Coast changes—Mississippi. I. Malbrough, Lynn P., joint author. II. Mississippi. State University. Water Resources Research Institute. III. Title.
GB459.4.W35 551.3'6'07621 78-620599
78 MARC

—PACIFIC COAST (NORTH AMERICA)

Pierce, William Dwight, 1881-1967.
The sand dune weevils of the genus Trigonoscuta, with a correlation of their anatomy to the geological history of our coast lines / by William Dwight Pierce. — Los Angeles : Natural History Museum of Los Angeles County, 1975.
iii, 160 p., 62 leaves of plates : ill. ; 23 cm.
Bibliography: p. 76.
1. Trigonoscuta—Classification. 2. Trigonoscuta—Anatomy. 3. Insects—Classification. 4. Insects—Anatomy. 5. Insects—Pacific coast (North America) 6. Coast changes—Pacific coast (North America) 7. Indicators (Biology) I. Title: The sand dune weevils of the genus Trigonoscuta ...
QL596.C9P64 1975 595.7'68 78-324434
79 MARC

—POLAND

Bączyk, Józef.
Les mouvements des eaux et leurs influences sur les côtes polonaises / Józef Bączyk. — Warszawa : Panstwowe Wydawn. Naukowe, 1977.
28 p. : ill. ; 24 cm. — (Conférences - Académie polonaise des sciences, Centre scientifique à Paris ; fasc. 116) P***
Bibliography: p. 27-28.
1. Coast changes—Poland. 2. Ocean waves—Baltic Sea. I. Title. II. Series: Polska Akademia Nauk. Centre scientifique, Paris. Conférences ; fasc. 116.
AS162.P7 fasc. 116 084'.1 s 78-387855
[GB457.59] 79 MARC

—SURINAM

Augustinus, P G E F
The changing shoreline of Surinam (South America) / by P. G. E. F. Augustinus. — Utrecht : Natuurwetenschappelijke Studiekring voor Suriname en de Nederlandse Antillen, 1978.
x, 232 p., [9] leaves of plates : ill. ; 24 cm. — (Uitgaven "Natuurwetenschappelijke Studiekring voor Suriname en de Nederlandse Antillen" ; no. 95) Ne***
Summary in Dutch.
Thesis—University of Utrecht, 1978.
Bibliography: p. 223-232.
1. Shore-lines—Surinam. 2. Coast changes—Surinam. I. Title. II. Series: Natuurwetenschappelijke Studiekring voor Suriname en de Nederlandse Antillen. Uitgaven ; no. 95.
QH7.N2842 no. 95 574'.08 s 79-312666
[GB459.15] 79 MARC

—UNITED STATES

Kaufman, Wallace.
The beaches are moving : the drowning of America's shoreline / Wallace Kaufman and Orrin Pilkey. — 1st ed. — Garden City, N.Y. : Anchor Press, 1979.
326 p., [8] leaves of plates : ill. ; 22 cm.
Bibliography: p. [311]-316.
Includes index.
ISBN 0-385-14364-8 : $10.95
1. Beaches—United States. 2. Coast changes—United States. I. Pilkey, O. H., 1934- joint author. II. Title.
GB460.A2K38 333.9'17'0974 78-22422
79 MARC

—WASHINGTON (STATE)

Phipps, James B
Coastal accretion and erosion in southwest Washington / by James. B. Phipps and John M. Smith. — Olympia, Wash. : Dept. of Ecology, 1978.
75 p. : ill. ; 28 cm.
"78-12."
Bibliography: p. 48-50.
1. Coast changes—Washington (State) 2. Coastal zone management—Washington (State) I. Smith, John M., 1926- joint author. II. Title.
GB459.4.P45 551.3'6'097979 79-622614
79 MARC

COAST ECOLOGY see Coastal ecology; Seashore ecology

COAST GUARD (UNITED STATES) see United States. Coast Guard

COAST-PILOT GUIDES see Pilot guides

COAST PROTECTION see Shore protection

COAST PROTECTIVE WORKS see Shore protection

COAST PROVINCE, KENYA

—POLITICS AND GOVERNMENT—PERIODICALS

Coast Province, Kenya. Provincial Commissioner.
Coast Province annual report.
Mombasa, Provincial Commissioner.
v. 34 cm.
1. Coast Province, Kenya—Politics and government—Periodicals. I. Title.
J731.T3C66a 354'.676'23 77-980069
MARC-S

COAST RADIO STATIONS see Marine radio stations

COASTAL CUTTHROAT TROUT see Cutthroat trout, Sea-run

COASTAL ECOLOGY

see also Seashore ecology

—IRELAND—NORTH BULL ISLAND

North Bull Island, Dublin Bay : a modern coastal natural history / edited by D. W. Jeffrey, general editor ... [et al.] ; designer, Jan de Fouw. — [Dublin] : Royal Dublin Society, 1977.
158 p. : ill. ; 30 cm. Ir***
Bibliography: p. 125-129.
Includes index.
ISBN 0-86027-001-7. ISBN 0-86027-002-5 pbk.
1. Natural history—Ireland—North Bull Island. 2. Coastal ecology—Ireland—North Bull Island. 3. Island ecology—Ireland—North Bull Island. I. Jeffrey, D. W.
QH143.N67 500.9'418'3 78-317568
78 MARC

COASTAL ECOLOGY (Continued)

—JAPAN

Productivity of biocenoses in coastal regions of Japan / edited by K. Hogetsu ... [et al.]. — Tokyo : University of Tokyo Press, 1977.
viii, 394 p. : ill. ; 26 cm. — (JIBP synthesis ; v. 14) Ja***
Label mounted on t.p.: Exclusive distributor, ISBS, Forest Grove, Or.
Bibliography: p. 367-388.
Y8400
1. Marine ecology—Japan. 2. Coastal ecology—Japan. 3. Marine productivity—Japan. I. Hogetsu, Kinji, 1913- . II. Series: Japanese Committee for the International Biological Program. JIBP synthesis ; v. 14.
QH188.P75 574.5'2636'0952 79-301742
79 MARC

—LOUISIANA

Burk and Associates.
Unique ecological features of the Louisiana coast : prepared for the Louisiana State Planning Office, Coastal Resources Program / prepared by Burk and Associates, inc. — Rev. — Baton Rouge, La. : The Office, 1977.
39 p. : fold. map (in pocket) ; 29 cm.
Includes bibliographies.
1. Coastal ecology—Louisiana. I. Louisiana. State Planning Office. II. Title.
QH105.L8B87 1977 333.9'1'009763 78-620793
78 MARC

—MAINE

Miller, Dorcas S 1949-
The Maine coast, a nature lover's guide / Docas S. Miller ; with ill. by Cherie Hunter Day. — Charlotte, N.C. : East Woods Press, c1979.
188 p. : ill. ; 20 cm.
Includes bibliographies and index.
ISBN 0-914788-12-4 : $6.95
1. Natural history—Maine. 2. Coastal ecology—Maine. 3. Maine—Description and travel—Guide-books. I. Title.
QH105.M2M54 500.9741 79-10290
79 MARC

—NORTH CAROLINA—BIBLIOGRAPHY

Dubach, Harold W
The North Carolina coastal zone and its environment : a compilation of resource materials covering the coastal plain, estuaries, and offshore waters / compiled by Harold W. Dubach ; approved by T. V. Crawford. — Aiken, S.C. : E.I. du Pont de Nemours, Savannah River Laboratory ; [Springfield, Va. : available from National Technical Information Service], 1977.
2 v. (389 p.) ; 28 cm.
"DP-1423."
Includes indexes.
1. Coasts—North Carolina—Bibliography. 2. Coastal ecology—North Carolina—Bibliography. 3. Continental shelf—North Carolina—Bibliography. I. Savannah River Laboratory. II. Title.
Z6004.C6D84 016.5514'614'8 77-3695
[GB459.4] 77 MARC

—SOUTH CAROLINA

An Annotated checklist of the biota of the coastal zone of South Carolina / edited by Richard G. Zingmark ; editorial committee, Dale R. Calder ... [et al.]. — 1st ed. — Columbia: Published for the Belle W. Baruch Institute for Marine Biology and Coastal Researh by the University of South Carolina Press, c1978.
xii, 364 p. ; 29 cm.
Includes bibliographies and indexes.
ISBN 0-87249-373-3. ISBN 0-87249-374-1 pbk.
1. Coastal ecology—South Carolina. I. Zingmark, Richard G. II. Belle W. Baruch Institute for Marine Biology and Coastal Research.
QH105.S6A5 574.9'757 78-110686
78 MARC

—TEXAS

Lay, Daniel W
Wildlife of the Texas coastal zone / Daniel W. Lay, Kaye F. Culbertson. — [Austin : Texas Parks and Wildlife Dept., Wildlife Division], 1978.
72 p. (p. 72 blank) : ill. ; 28 cm.
Bibliography: p. 65-71.
1. Coastal ecology—Texas. 2. Animal populations—Texas. I. Culbertson, Kaye F., joint author. II. Title.
QH105.T4L38 333.9'5 79-623086
79 MARC

—TROPICS

Odum, William E
Ecological guidelines for tropical coastal development / William E. Odum. — Morges : International Union for Conservation of Nature and Natural Resources, 1976.
61 p. : ill. ; 24 cm. — (IUCN publications ; new ser., no. 42) Sw77-2130
1. Coastal ecology—Tropics. 2. Environmental protection—Tropics. 3. Regional planning—Tropics. I. Title. II. Series: International Union for the Conservation of Nature and Natural Resources. IUCN publications ; new ser., no. 26.
QH541.5.C65O38 333.9'17'0913 78-321847
*79 MARC

COASTAL ENGINEERING

see also Coastal zone management; Dikes (Engineering); Shore protection

—BIBLIOGRAPHY

Szuwalski, Andre.
Annotated bibliography of CERC publication / by Andre Szuwalski. — Fort Belvoir, Va. : U.S. Army, Corps of Engineers, Coastal Engineering Research Center ; Springfield, Va. : available from National Technical Information Service, Operations Division, 1978.
99 p. in various pagings ; 27 cm.
Cover title.
Includes index.
1. Coastal engineering—Bibliography. 2. Coastal engineering—United States—Bibliography. I. United States. Coastal Engineering Research Center. II. Title.
Z5853.C57S9 016.62'0414'6 78-602794
[TC205] 79 MARC

—CONGRESSES

Coastal structures 79 : a specialty conference on the design construction, maintenance, and performance of port and coastal structures, March 14-16, 1979, Alexandria, Virginia : a symposium / sponsored by the Waterways [sic], Port, Coastal, and Ocean Division of the American Society of Civil Engineers, co-sponsors, U.S. Army Coastal Engineering Research Center, U.S. Naval Facilities Engineering Command, Permanent International Association of Navigation Congresses. — New York : American Society of Civil Engineers, c1979.
2 v. (viii, 1211 p.) : ill. ; 22 cm.
Includes bibliographical references and indexes.
$30.00
1. Coastal engineering—Congresses. 2. Hydraulic structures—Congresses. I. American Society of Civil Engineers. Waterway, Port, Coastal and Ocean Division.
TC203.5.C63 627 79-106948
79 MARC

—AUSTRALIA

Institution of Engineers, Australia. National Committee on Coastal and Ocean Engineering.
Coastal and Ocean Engineering in Australia : a survey / by the National Committee on Coastal and Ocean Engineering of the Institution of Engineers, Australia. — Sydney : Institution of Engineers, Australia, 1973.
52 p. in various pagings ; 30 cm. Aus***
Includes bibliographies.
1. Ocean engineering—Australia. 2. Coastal engineering—Australia. I. Title.
TC1645.I57 1973 627'.58 78-324836
79 MARC

—MICHIGAN, LAKE—GUIDE-BOOKS

Coastal geology, sedimentology, and management, Chicago and the Northshore : prepared for the 4th annual field conference, Great Lakes Section, Society of Economic Paleontologists and Mineralogists, September 21-22, 1974 / Charles Collinson ... [et al.]. — [Urbana : Illinois State Geological Survey, 1974]
55 p. : ill. ; 29 cm. — (Illinois State Geological Survey guidebook series ; 12)
Bibliography: p. 54-55.
1. Coast changes—Michigan, Lake—Guide-books. 2. Michigan, Lake—Water level—Guide-books. 3. Coastal engineering—Michigan, Lake—Guide-books. I. Collinson, Charles William. II. Society of Economic Paleontologists and Mineralogists. Great Lakes Section. III. Series: Illinois. State Geological Survey. Guidebook series ; 12.
QE105.I57a no. 12 557.73'08 s 78-107254
[GB459.5.M5] 79 MARC

—UNITED STATES—BIBLIOGRAPHY

Szuwalski, Andre.
Annotated bibliography of CERC publication / by Andre Szuwalski. — Fort Belvoir, Va. : U.S. Army, Corps of Engineers, Coastal Engineering Research Center ; Springfield, Va. : available from National Technical Information Service, Operations Division, 1978.
99 p. in various pagings ; 27 cm.
Cover title.
Includes index.
1. Coastal engineering—Bibliography. 2. Coastal engineering—United States—Bibliography. I. United States. Coastal Engineering Research Center. II. Title.
Z5853.C57S9 016.62'0414'6 78-602794
[TC205] 79 MARC

COASTAL FLORA

see also Tidemarsh flora

—IDENTIFICATION

Gosner, Kenneth L 1925-
A field guide to the Atlantic seashore : invertebrates and seaweeds of the Atlantic Coast from the Bay of Fundy to Cape Hatteras ; text and illustrations / by Kenneth L. Gosner. — Boston : Houghton Mifflin, 1979, c1978.
xvi, 329, 64 leaves of plates : col. ill. ; 19 cm. — (The Peterson field guide series ; 24)
Bibliography: p. 287-293.
Includes index.
ISBN 0-395-24379-3 : $12.95
1. Seashore biology. 2. Marine invertebrates—Identification. 3. Coastal flora—Identification. I. Title.
QH95.7.G67 1979 574.909'4'6 78-14784
78 MARC

—FLORIDA

Carlton, Jedfrey M
A survey of selected coastal vegetation communities of Florida / Jedfrey M. Carlton. — St. Petersburg : Florida Dept. of Natural Resources, Marine Research Laboratory, 1977.
40 p. : ill. ; 28 cm. — (Florida marine research publications ; no. 30 ISSN -0095-0157) (Contribution - Florida Dept. of Natural Resources, Marine Research Laboratory ; no. 244)
Bibliography: p. 30-32.
1. Coastal flora—Florida. 2. Plant communities—Florida. I. Title. II. Series. III. Series: Marine Research Laboratory, St. Petersburg, Fla. Contribution ; no. 244.
QK154.C38 581.9'759 78-622120
79 MARC

—POLAND

Misiewicz, Józef.
Flora synantropijna i zbiorowiska ruderalne polskich portów morskich / Józef Misiewicz. — Słupsk : Wyższa Szkoła Pedagogiczna, 1976 (Słupsk : WSP)
321 p. : ill. ; 21 cm. P77-411
Bibliography: p. 301-321.
zł15.00
1. Coastal flora—Poland. 2. Harbors—Poland. I. Title.
QK322.M57 78-381721
78 MARC

COASTAL FLORA (Continued)

—SWEDEN—BOHUSLÄN

Utterström-Gustavsson, Ulla, 1921-
Algvegetationen i marina hällkar : en studie från mellersta Bohuslän / ava Ulla Utterströn-Gustafsson. — Göteborg : Avdelningen för marin botanik, Göteborgs universitet, 1976.
102 p. : ill. ; 24 cm. S76-45
Thesis—Gothenburg.
Bibliography: p. 96-102.
1. Marine algae—Sweden—Bohuslän. 2. Coastal flora—Sweden—Bohuslän. 3. Tide pool ecology—Sweden—Bohuslän. I. Title.
QK574.S8U85 79-376919
*79 MARC

COASTAL SHIPPING see Coastwise shipping

COASTAL SIGNALS see Signals and signaling

COASTAL VESSELS see Coasters (Ships)

COASTAL WATERWAYS see Intracoastal waterways

COASTAL ZONE MANAGEMENT

see also Coastal engineering; Marine pollution

Research techniques in coastal environments / H. J. Walker. — Baton Rouge : School of Geoscience, Louisiana State University, 1977.
320 p. : ill. ; 29 cm. — (Geoscience and man ; v. 18)
Includes bibliographies.
1. Coast changes. 2. Coast changes—Technique. 3. Coastal zone management. I. Walker, Harley Jessie, 1921- . II. Series.
GB451.2.R47 551.3'6 78-113301
79 MARC

—LAW AND LEGISLATION —CALIFORNIA

California. Legislature. Senate. Committee on Natural Resources and Wildlife.
Hearing on implementation of the California coastal act of 1976 : Friday, December 9, 1977, 9:00 o'clock, A.M., Monterey, California / Senate Committee on Natural Resources and Wildlife. — ͵Sacramento͵ : The Committee, ͵1977?͵
iii leaves, 394 p. ; 28 cm.
1. Coastal zone management—Law and legislation—California. I. Title.
KFC10.3.N37 1977e 346'.794'046917 78-623568
78 MARC

California. Legislature. Senate. Committee on Natural Resources and Wildlife.
Hearing on implementation of the California coastal act of 1976 : ... December 10, 1977 ... Monterey, California / Senate Committee on Natural Resources and Wildlife. — Sacramento, Calif. : Video/Audio Recording Systems, ͵1977?͵
101 p. ; 29 cm.
Includes index.
1. Coastal zone management—Law and legislation—California. I. Title.
KFC10.3.N37 1977c 333.9'17 78-623132
78 MARC

California. Legislature. Senate. Select Committee on Land Use Management Organization.
Hearing held in San Diego, December 19, 1977, on the implementation of the coastal of 1976 : submitted written testimony / California Legislature, Senate Select Committee on Land Use Management Organizations. — Sacramento : The Committee, ͵1978͵
529 p. : ill. ; 28 cm.
Cover title.
1. Coastal zone management—Law and legislation—California. I. Title: Implementation of the coastal of 1976.
KFC10.3.L36 1977b 333.9'17'09794 78-623510
78 MARC

California. Legislature. Senate. Select Committee on Land Use Management Organizations.
Implementation of the Coastal act of 1976, Monday, December 19, 1977 / Senate Select Committee on Land Use Management Organizations. — Sacramento : Video/Audio Recording Systems, ͵1978?͵
ii leaves, 274 p. ; 28 cm.
1. Coastal zone management—Law and legislation—California. I. Title.
KFC10.3.L36 1977c 333.9'17'09794 78-623517
78 MARC

—LAW AND LEGISLATION— CHESAPEAKE BAY REGION

Physical alterations of coastal shorelines : an analysis of Chesapeake Bay shore zone development and regulation / edited by William H. Queen. — Annapolis : Chesapeake Research Consortium, 1979.
101 p. : ill. ; 23 cm. — (CRC publication ; no. 64)
Includes bibliographical references.
1. Coastal zone management—Chesapeake Bay region. 2. Coastal zone management—Law and legislation—Chesapeake Bay region. I. Queen, William H. II. Chesapeake Research Consortium. III. Series: Chesapeake Research Consortium. CRC publication ; no. 64.
HT392.5.C45P5 333.9'17'0975518 78-20913
78 MARC

—LAW AND LEGISLATION—NEW YORK (STATE)

Allayaud, William Kent.
Integrated planning for water quality management: the Federal water pollution control act amendments of 1972 and coastal zone management / by William Kent Allayaud. — Ithaca, N.Y. : Environmental Health Planning Training Program, Dept. of City and Regional Planning, Cornell University : may be obtained from Program in Urban and Regional Studies, Cornell University, 1979.
x, 217 p. ; 28 cm. — (Discussions in environmental health planning)
Bibliography: p. 213-217.
$5.00
1. Water—Pollution—Law and legislation—United States. 2. Coastal zone management—Law and legislation—United States. 3. Water—Pollution—Law and legislation—New York (State) 4. Coastal zone management—Law and legislation—New York (State) I. Title. II. Series.
KF3790.A74 353.008'232 79-108073
79 MARC

—LAW AND LEGISLATION—TEXAS

Quinlan, Suzanne.
The 65th Texas Legislature, 1977, coastal and marine legislation / by Suzanne Quinlan. — Houston, Tex. : University of Houston, College of Law, ͵1977?͵
46 leaves ; 28 cm.
At head of title: Texas Law Institute of Coastal and Marine Resources.
Includes index.
1. Water—Pollution—Law and legislation—Texas. 2. Coastal zone management—Law and legislation—Texas. 3. Natural resources—Law and legislation—Texas. I. Texas Law Institute of Coastal and Marine Resources. II. Title.
KFT1556.Q56 344'.764'0463 79-622649
79 MARC

—LAW AND LEGISLATION—UNITED STATES

Allayaud, William Kent.
Integrated planning for water quality management: the Federal water pollution control act amendments of 1972 and coastal zone management / by William Kent Allayaud. — Ithaca, N.Y. : Environmental Health Planning Training Program, Dept. of City and Regional Planning, Cornell University : may be obtained from Program in Urban and Regional Studies, Cornell University, 1979.
x, 217 p. ; 28 cm. — (Discussions in environmental health planning)
Bibliography: p. 213-217.
$5.00
1. Water—Pollution—Law and legislation—United States. 2. Coastal zone management—Law and legislation—United States. 3. Water—Pollution—Law and legislation—New York (State) 4. Coastal zone management—Law and legislation—New York (State) I. Title. II. Series.
KF3790.A74 353.008'232 79-108073
79 MARC

Brower, David J
Access to the Nation's beaches : legal and planning perspectives / David J. Brower, principal investigator, with William Dreyfoos ... ͵et al.͵ ; drawings by Don Meserve. — ͵Raleigh͵ : UNC Sea Grant, North Carolina State University, 1978.
208 p. : ill. ; 21 cm. — (UNC sea grant publication ; UNC-SG-77-18)
Includes bibliographical references.
$3.00
1. Beaches—Law and legislation—United States. 2. Recreation areas—Law and legislation—United States. 3. Coastal zone management—Law and legislation—United States. 4. Right of way—United States. I. Dreyfoos, William, joint author. II. Meserve, Don. III. Title. IV. Series: Sea grant publication (Raleigh) ; UNC-SC-77-18.
KF5627.B76 346'.73'046917 78-623348
79 MARC

Ocean and coastal law teaching materials / principal investigator Thomas J. Schoenbaum ; contributing authors, Deborah A. Brian ... ͵et al.͵. — ͵Raleigh, N.C. : UNC Sea Grant͵, 1977.
2 v. ; 28 cm. — (Sea grant publication ; UNC-SG-77-09)
CONTENTS: v. 1. Ocean law.—v. 2. Coastal law.
$7.00 (v. 1); $5.50 (v. 2)
1. Maritime law. 2. Maritime law—United States. 3. Shore protection—Law and legislation—United States. 4. Coastal zone management—Law and legislation—United States. I. Schoenbaum, Thomas J. II. Brian, Deborah A. III. Series: Sea grant publication (Raleigh) ; UNC-SG-77-09.
JX4408.O33 341.44'8 78-622499
79 MARC

—MAPS

Canada-Ontario Great Lakes Shore Damage Survey.
Coastal zone atlas / Canada-Ontario Great Lakes Shore Damage Survey. — ͵Ottawa͵ : Environment Canada ; ͵Toronto͵ : Ontario Ministry of Natural Resources, c1976.
circa 800 p. : col. ill., maps (some col.) ; 39 x 54 cm. CG76-73252-3
On all maps: Great Lakes Shore Damage Survey - 1973.
Scale of maps 1:20,000 or one inch to 1,6666 feet.
CONTENTS: Lake Huron.—St. Clair River, Lake St. Claire, Detroit River.—Lake Eire.—Lake Ontario.
$135.00
1. Great Lakes—Storm, 1972, 1973 Maps. 2. Coast changes—Maps. 3. Coastal zone management—Maps. 4. Great Lakes region—Maps. 5. Great Lakes region—Photo maps. 6. Great Lakes—Shore protection—Maps. I. Canada. Dept. of the Environment. II. Ontario. Ministry of Natural Resources. III. Title.
G1107.G7C3 1976 912'.1'3339170977 78-323616
79 MARC

—ALASKA

Alaska. Division of Legislative Audit.
A review of Alaska coastal management program, grant no. 04-7-158-44011, Office of the Governor, Division of Policy Development and Planning : October 1, 1976 - December 31, 1977 / State of Alaska, Division of Legislative Audit. — Juneau : The Division, ͵1978͵
19f ͵i.e. 35͵ leaves ; 29 cm.
1. Coastal zone management—Alaska. I. Title.
HT393.A4A525 1978a 353.9'798'008232 79-620511
79 MARC

Alaska. Division of Legislative Audit.
A review of Alaska coastal management program, Office of the Governor, Division of Policy Development and Planning. — Juneau : State of Alaska, Division of Legislative Audit, ͵1978͵
26 leaves ; 29 cm.
1. Coastal zone management—Alaska. I. Title.
HT393.A4A525 1978 353.9'798'008232 78-623020
78 MARC

—ALASKA—KODIAK ISLAND

Woodward-Clyde Consultants.
Oil terminal and marine service base sites in the Kodiak Island borough : final report / prepared for Alaska Department of Community and Regional Affairs, Division of Community Planning. — Anchorage : Woodward-Clyde Consultants, 1977.
145 leaves in various foliations : ill. ; 29 cm.
1. Offshore oil industry—Alaska—Kodiak Island. 2. Petroleum—Alaska—Kodiak Island—Pipe lines. 3. Coastal zone management—Alaska—Kodiak Island. I. Alaska. Division of Community Planning. II. Title.
HD9567.A42K628 338'.09798'4 78-621862
79 MARC

—CALIFORNIA

California. Dept. of Water Resources. Southern District.
Central coastal land use study, 1977 : district report / State of California, the Resources Agency, Department of Water Resources, Southern District. — ͵Los Angeles͵ : The District, 1979.
vi, 25 p. (incl. cover) : ill. ; 29 cm.
Cover title.
Prepared by Robert E. Yoha, assisted by Paul DeLong, and others.
1. Coastal zone management—California. 2. Land use—California. I. Yoha, Robert E. II. DeLong, Paul. III. Title.
HT393.C3C28 1979 333.9'17'097949 79-623208
79 MARC

Protecting the Golden Shore : lessons from the California Coastal Commissions / Robert G. Healy, editor ... ͵et al.͵. — Washington : Conservation Foundation, c1978.
xxxii, 257 p. : ill. ; 23 cm.
Bibliography: p. ͵254͵-257.
ISBN 0-89164-052-5
1. California. Coastal Zone Conservation Commission. 2. Coastal zone management—California. I. Healy, Robert G.
HT393.C3P76 333.9'17'09794 78-65565
79 MARC

COASTAL ZONE MANAGEMENT (Continued)

—CHESAPEAKE BAY REGION

Physical alterations of coastal shorelines : an analysis of Chesapeake Bay shore zone development and regulation / edited by William H. Queen. — Annapolis : Chesapeake Research Consortium, 1979.
101 p. : ill. ; 23 cm. — (CRC publication ; no. 64)
Includes bibliographical references.
1. Coastal zone management—Chesapeake Bay region. 2. Coastal zone management—Law and legislation—Chesapeake Bay region. I. Queen, William H. II. Chesapeake Research Consortium. III. Series: Chesapeake Research Consortium. CRC publication ; no. 64.
HT392.5.C45P5 333.9'17'0975518 78-20913
78 MARC

—DENMARK—HELSINGØR

Helsingør, Denmark (Kommune). Havneudvalget.
Kyst og kystanlæg. Helsingør kommunes rapport om kystens og havnenes udnyttelse og udvikling. ₍Udarb. af₎ Havneudvalget. Helsingør, Helsingør Kommune, Eksp.: Stads- og Havneingeniørens Kontro, Rådhuset, ₍1972₎.
79 l. illus. 21 x 30 cm. N.T. D73-39
1. Coastal zone mangement—Denmark—Helsingør. 2. Coasts—Denmark—Helsingør. I. Title.
HT395.D42H443 73-364274
79₍r79₎rev MARC

—ENGLAND—BIBLIOGRAPHY

Kupfer, Romaine L
Comparative coastal zone management : a selective partially annotated bibliography on the biophysical, sociological, and legal aspects of development in four delta-estuary regions—the Mississippi (Louisiana, USA), Rhone (France), Rhine (The Netherlands), and Thames (England) / compiled by Romaine L. Kupfer and Kate Wallach, assisted by Sheryl Jolissaint. — Baton Rouge : Center for Wetland Resources, Louisiana State University, ₍1977?₎
x, 179 p. : ill. ; 28 cm. — (Sea Grant pub. ; no. LSU-L-77-001)
1. Coastal zone management—Louisiana—Bibliography. 2. Coastal zone management—France—Bibliography. 3. Coastal zone management—Netherlands—Bibliography. 4. Coastal zone management—England—Bibliography. I. Wallach, Kate, joint author. II. Jolissaint, Sheryl, joint author. III. Title. IV. Series: Sea grant publication (Baton Rouge) ; no. LSU-L-77-001.
Z7164.O7K86 016.3339'17 78-622283
₍HT393.L6₎ 79 MARC

—FLORIDA

Florida. Beach and Inlet Task Force.
Beach and Inlet Task Force report. — ₍Tallahassee₎ : Bureau of Comprehensive Planning, Division of State Planning, Dept. of Administration, 1978.
xi, 48 p. : ill. ; 28 cm.
"DSP-BCP-31-78."
Includes bibliographical references.
1. Coastal zone management—Florida. 2. Shore protection—Florida. I. Title.
HT393.F5F525 1978 333.9'17'09759 78-623749
79 MARC

Florida. Division of State Planning.
Powers and responsibilities of local governments, regional planning councils, and the Division of State Planning relative to coastal zone management. — ₍Tallahassee₎ : The Division, 1977.
78, ₍30₎ p. : ill. ; 28 cm.
1. Coastal zone management—Florida. 2. Regional planning districts—Florida. I. Title.
HT393.F5F57 1977 333.9'17'09759 78-620517
79 MARC

—FRANCE

Une Politique de la mer et du littoral : bilan 1974-1977 : perspectives / introd. de Valéry Giscard d'Estaing. — Paris, : Service d'information et de diffusion, 1977.
43 p. ; 27 cm. — (Actualités documents) F***
"Supplément du no 293 d'Actualités-Service."
1. Coastal zone management—France. 2. Marine resources and state—France. I. Actualités-Service. II. Series.
HT395.F7P56 309.2'5'0944 78-376532
78 MARC

—FRANCE—BIBLIOGRAPHY

Kupfer, Romaine L
Comparative coastal zone management : a selective partially annotated bibliography on the biophysical, sociological, and legal aspects of development in four delta-estuary regions—the Mississippi (Louisiana, USA), Rhone (France), Rhine (The Netherlands), and Thames (England) / compiled by Romaine L. Kupfer and Kate Wallach, assisted by Sheryl Jolissaint. — Baton Rouge : Center for Wetland Resources, Louisiana State University, ₍1977?₎
x, 179 p. : ill. ; 28 cm. — (Sea Grant pub. ; no. LSU-L-77-001)
1. Coastal zone management—Louisiana—Bibliography. 2. Coastal zone management—France—Bibliography. 3. Coastal zone management—Netherlands—Bibliography. 4. Coastal zone management—England—Bibliography. I. Wallach, Kate, joint author. II. Jolissaint, Sheryl, joint author. III. Title. IV. Series: Sea grant publication (Baton Rouge) ; no. LSU-L-77-001.
Z7164.O7K86 016.3339'17 78-622283
₍HT393.L6₎ 79 MARC

—HAWAII—CONGRESSES

Workshop on Government Permit Simplification, Coordination, and Streamlining, Honolulu, 1978.
Red tape vs. green light : a Workshop on Government Permit Simplification, Coordination, and Streamlining : proceedings, July 7, 1978, Honolulu. — ₍Honolulu₎ : Hawaii Coastal Zone Management Program, Dept. of Planning and Economic Management, 1978.
93 p. : ports ; 28 cm.
1. Coastal zone management—Hawaii—Congresses. 2. Government paperwork—Hawaii—Congresses. I. Hawaii. Coastal Zone Management Program. II. Title.
HT393.H3W67 1978 350'.71'09969 78-624137
79 MARC

—IRELAND

Quinn, Ann C M
Sand dunes : formation, erosion and management / ₍by₎ Ann C. M. Quinn. — Dublin : An Foras Forbartha, 1977.
₍7₎, 92 p. : ill., maps ; 29 cm. GB78-29538
Bibliography: p. 78-83.
ISBN 0-900115-91-2 : £2.00
1. Sand-dunes—Ireland. 2. Coastal zone management—Ireland. I. Title.
GB457.285.Q56 627'.58 79-305968
79 MARC

—ISRAEL—MAPS

Efrat, Elisha.
Outline scheme of Israel coastal strips = תכנית מיתאר ארצית לחופים / Elisha Efrat. — Jerusalem : Ministry of Interior, Planning Dept., 1974.
iii, 50 p. : col. maps ; 50 cm.
Scale of maps 1:20,000.
Legend and summaries in English and Hebrew.
1. Coastal zone management — Israel — Maps. 2. Regional planning—Israel—Maps. I. Israel. Misrad ha-penim. Agaf le-tikhnun. II. Title. III. Title: Tokhnit mit'ar artsit la-ḥofim.
G2237.C6G45E8 1974 77-465537
912'.1'333917095694 (MARC)

—LOUISIANA

Louisiana. State University in New Orleans. Urban Studies Institute.
OCS development in coastal Louisiana : a socio-economic impact assessment / by Anthony J. Mumphrey, Jr. ... ₍et al.₎. — ₍Baton Rouge₎ : Louisiana Dept. of Transportation and Development, Coastal Resources Program, 1977.
x, 272 p. : ill. ; 29 cm.
"Disseminated under the sponsorship of the Department of Transportation and Development."
"This study was completed under contract number SPO-77-21."
Includes bibliographies.
1. Offshore oil industry—Louisiana. 2. Coastal zone management—Louisiana. 3. Louisiana—Economic conditions. I. Mumphrey, Anthony J. II. Louisiana. Dept. of Transportation and Development. III. Title.
HD9567.L8L68 1977 330.9'763'06 78-623248
79 MARC

—LOUISIANA—BIBLIOGRAPHY

Kupfer, Romaine L
Comparative coastal zone management : a selective partially annotated bibliography on the biophysical, sociological, and legal aspects of development in four delta-estuary regions—the Mississippi (Louisiana, USA), Rhone (France), Rhine (The Netherlands), and Thames (England) / compiled by Romaine L. Kupfer and Kate Wallach, assisted by Sheryl Jolissaint. — Baton Rouge : Center for Wetland Resources, Louisiana State University, ₍1977?₎
x, 179 p. : ill. ; 28 cm. — (Sea Grant pub. ; no. LSU-L-77-001)
1. Coastal zone management—Louisiana—Bibliography. 2. Coastal zone management—France—Bibliography. 3. Coastal zone management—Netherlands—Bibliography. 4. Coastal zone management—England—Bibliography. I. Wallach, Kate, joint author. II. Jolissaint, Sheryl, joint author. III. Title. IV. Series: Sea grant publication (Baton Rouge) ; no. LSU-L-77-001.
Z7164.O7K86 016.3339'17 78-622283
₍HT393.L6₎ 79 MARC

—LOUISIANA—MAPS

Gagliano, Sherwood M
Environmental atlas and multiuse management plan for south-central Louisiana : prepared for Department of the Army, New Orleans District, Corps of Engineers, Contract No. DACW 29-71-C-0219 and Office of Sea Grant, National Oceanic and Atmospheric Administration of U.S. Department of Commerce, Institutional Support Grant No. 04-3-158-19 / by Sherwood M. Gagliano, project director and Penny Culley ... ₍et. al.₎. — Baton Rouge : Center for Wetland Resources, Louisiana State University, 1973.
2 v. : ill., maps ; 28-41 x 22-62 cm. — (Hydrologic and geologic studies of coastal Louisiana ; report no. 18)
Includes bibliographies.
CONTENTS: Approach.—Natural setting.—Activities.—Management plan.—Summary and conclusions.—Appendices.
1. Coastal zone management—Louisiana—Maps. 2. Land use—Planning—Louisiana—Maps. 3. Coasts—Louisiana—Maps. I. United States. Army. Corps of Engineers. II. United States. National Oceanic and Atmospheric Administration. Office of Sea Grant Programs. III. Louisiana State University, Baton Rouge. Center for Wetland Resources. IV. Title. V. Series.
G1361.G45G3 1973 912'.1'3337097634 75-624375
78 MARC

—LOUISIANA—GULF REGION

Cumulative impact studies in the Louisiana coastal zone : eutrophication, land loss : final report to Louisiana Department of Transportation and Development / edited N. J. Craig and J. W. Day, Jr. ; contributors, N. J. Craig ... ₍et al.₎. — Baton Rouge : Louisiana State University, Center for Wetland Resources, 1977.
ix, 157 p. : ill. ; 28 cm.
On cover: Coastal resources program.
Includes bibliographies.
1. Eutrophication—Louisiana—Gulf region. 2. Coastal zone management—Louisiana—Gulf region. 3. Wetland conservation—Louisiana—Gulf region. I. Craig, N. J. II. Day, John W. III. Louisiana. Dept. of Transportation and Development.
QH105.L8C85 363 78-623247
79 MARC

Shoreline erosion in coastal Louisiana : inventory and assessment : final report to Louisiana Department of Transportation and Development / by R. D. Adams ... ₍et al.₎. — ₍Baton Rouge₎ : Louisiana State University Center for Wetland Resources, 1978.
vi, 139 p. : maps ; 28 cm.
Bibliography: p. 130-133.
1. Erosion—Louisiana—Gulf region. 2. Coast changes—Louisiana—Gulf region. 3. Coastal zone management—Louisiana—Gulf region. I. Adams, Rodney D., 1937- . II. Louisiana. Dept. of Transportation and Development. III. Louisiana State University, Baton Rouge. Center for Wetland Resources.
GB459.4.S53 333.9'17'09763 79-621598
79 MARC

—MAINE—STATISTICS

Maine. Coastal Program.
The Maine coast, a statistical source. — ₍Augusta₎ : Maine Coastal Program, Natural Resource Planning Division, Maine State Planning Office, ₍1978₎
vi, 84 p. : ill. ; 28 cm.
Bibliography: p. 82.
Includes index.
1. Coastal zone management—Maine—Statistics. I. Title.
HT393.M3M27 1978 333.9'17'09741 79-621069
79 MARC

—MAINE—SEARS ISLAND—PUBLIC OPINION

Seekins, Milton D
An economic study of voter attitudes toward different proposals for the use of an undeveloped island, Sears Island, Maine / Milton D. Seekins, David A. Storey. — ₍Amherst₎ : Massachusetts Agricultural Experiment Station, University of Massachusetts at Amherst, College of Food and Natural Resources, 1977.
79 p. ; 23 cm. — (Research bulletin - Massachusetts Agricultural Experiment Station, University of Massachusetts at Amherst ; no. 651)
Bibliography: p. 50-51.
1. Land use—Maine—Sears Island—Public opinion. 2. Coastal zone management—Maine—Sears Island—Public opinion. 3. Public opinion—Maine. I. Storey, David A., joint author. II. Title. III. Series: Massachusetts. Agricultural Experiment Station, Amherst (1888-) Research bulletin - Massachusetts Agricultural Experiment Station ; no. 651.
HD211.M2S43 78-622501
301.15'43'3337097413
79 MARC

COASTAL ZONE MANAGEMENT (Continued)

—MARYLAND

United States. Office of Coastal Zone Management.
Final environmental impact statement : proposed coastal management program for the State of Maryland / prepared by Office of Coastal Zone Management, National Oceanic & Atmospheric Administration, Department of Commerce and Department of Natural Resources, Energy & Coastal Zone Administration. — Washington : The Office, 1978.
L, 464 p. ; 29 cm.
At head of title: United States Department of Commerce.
Cover title: State of Maryland Coastal Management program and final environmental impact statement.
1. Coastal zone management—Maryland. 2. Environmental protection—Maryland. I. Maryland. Energy & Coastal Zone Administration. II. Title. III. Title: State of Maryland coastal management program and final environmental impact statement.
HT393.M37U54 1978 333.9'17'09752 78-603473
79 MARC

—MARYLAND—ASSATEAGUE ISLAND NATIONAL SEASHORE, MD. AND VA.

Dolan, Robert.
Atlas of environmental dynamics, Assateague Island National Seashore / by Robert Dolan, Bruce Hayden, Jeffrey Heywood ; illustrations, J. G. Carswell, Richard Garber. — [Washington] : U.S. Dept. of the Interior, National Park Service, 1977.
40 p., [2] fold. leaves of plates : ill. (some col.), col. maps ; 28 x 36 cm. — (Natural resource report ; no. 11)
Bibliography: p. 39
Includes glossary.
PARTIAL CONTENTS: Geology.—Coastal processes.—Monitoring ocean dynamics.—Barrier island management.
1. Assateague Island National Seashore, Md. and Va.—Maps. 2. Barrier islands. 3. Coastal zone management—Maryland—Assateague Island National Seashore, Md. and Va. 4. Coastal zone management—Virginia—Assateaque Island National Seashore, Md. and Va. I. Haden, Bruce P. II. Heywood, Jeffrey. III. Carswell, J. G. IV. Garber, Richard. V. Title. VI. Series: Natural resources report ; no. 11.
G1272.A85D6 1977 79-311828
912'.1'333780975221
79 MARC

—MASSACHUSETTS

Massachusetts. Executive Office of Environmental Affairs.
Establishment of the Coastal zone management program / by the Executive Office of Environmental Affairs. — Boston : The Office, [letter 1978]
117 leaves ; 28 cm.
Caption title.
1. Coastal zone management—Massachusetts. I. Title.
HT393.M4M39 1978a 344'.744'046 79-622886
79 MARC

Massachusetts. Executive Office of Environmental Affairs.
Massachusetts Coastal zone management program and final environmental impact statement / Executive Office of Environmental Affairs, U.S. Department of Commerce, National Oceanic and Atmospheric Administration, Office of Coastal Zone Management. — [Boston] : Executive Office of Environmental Affairs, 1978.
v, 439 p. : maps ; 29 cm.
Cover title.
At head of title: CZM.
1. Coastal zone management—Massachusetts. I. United States. Office of Coastal Zone Management. II. Title. III. Title: CZM.
HT393.M4M39 1978 333.9'17'09744 79-621110
79 MARC

Massachusetts. Office of State Planning.
Offshore oil development : implications for Massachusetts communities / prepared by the Office of State Planning. — [Boston] : Commonwealth of Massachusetts, 1976.
68 p. : ill. ; 28 cm.
Bibliography: p. 68.
1. Offshore oil industry—Massachusetts. 2. Coastal zone management—Massachusetts. I. Title.
HD9567.M4M37 1976 338.2'7'28209744 78-622986
79 MARC

—MINNESOTA—MAPS

North Shore data atlas, Minnesota coastal zone. — [St. Paul, Minn.] : Minnesota State Planning Agency : available from the State Planning Agency or the Arrowhead Regional Development Commission, [1978?]
40 p. : col. ill., col. maps ; 22 x 28 cm.
Cover title.
"Describes the natural resources and cultural patterns of the North Shore."
1. Coasts—Minnesota—Maps. 2. Coastal zone management—Minnesota—Maps. 3. Minnesota—Maps. I. Minnesota. State Planning Agency. II. Arrowhead Regional Development Commission.
G1427.C6N6 1978 78-623095
912'.1'33391709776
79 MARC

—NETHERLANDS—BIBLIOGRAPHY

Kupfer, Romaine L
Comparative coastal zone management : a selective partially annotated bibliography on the biophysical, sociological, and legal aspects of development in four delta-estuary regions—the Mississippi (Louisiana, USA), Rhone (France), Rhine (The Netherlands), and Thames (England) / compiled by Romaine L. Kupfer and Kate Wallach, assisted by Sheryl Jolissaint. — Baton Rouge : Center for Wetland Resources, Louisiana State University, [1977?]
x, 179 p. : ill. ; 28 cm. — (Sea Grant pub. ; no. LSU-L-77-001)
1. Coastal zone management—Louisiana—Bibliography. 2. Coastal zone management—France—Bibliography. 3. Coastal zone management—Netherlands—Bibliography. 4. Coastal zone management—England—Bibliography. I. Wallach, Kate, joint author. II. Jolissaint, Sheryl, joint author. III. Title. IV. Series: Sea grant publication (Baton Rouge) ; no. LSU-L-77-001.
Z7164.O7K86 016.3339'17 78-622283
[HT393.L6] 79 MARC

—NEW JERSEY—PINE BARRENS—CONGRESSES

Research Conference on the New Jersey Pine Barrens, 1st, Atlantic City, 1978.
Natural and cultural resources of the New Jersey Pine Barrens : inputs and research needs for planning : proceedings and papers of the First Research Conference on the New Jersey Pine Barrens, Atlantic City, N.J., May 22-23, 1978 / John W. Sinton, editor. — Pomona, N.J. : Stockton State College, c1979.
xiii, 365 p., [1] leaf of plates : ill. ; 28 cm.
"Conference sponsored by the Center for Environmental Research, Stockton State College, co-sponsored by the Center for Coastal and Environmental Studies, Rutgers University, and New Jersey Department of Environmental Protection."
Includes bibliographies.
1. Natural resources—New Jersey—Pine Barrens—Congresses. 2. Ecology—New Jersey—Pine Barrens—Congresses. 3. Pine Barrens—Civilization—Congresses. 4. Land use—Planning—New Jersey—Pine Barrens—Congresses. 5. Coastal zone management—New Jersey—Pine Barrens—Congresses. I. Sinton, John W. II. Stockton State College. Center for Environmental Research. III. Rutgers University, New Brunswick, N.J. Center for Coastal and Environmental Studies. IV. New Jersey. Dept. of Environmental Protection. V. Title.
HC107.N52P567 1978 333 79-623956
79 MARC

—NEW YORK (STATE)

Harper, David B
Toward a planning and management program for the St. Lawrence-Eastern Ontario shoreline / David B. Harper and William E. Tyson ; editorial committee, David J. Robinson (editor) ... [et al.]. — Syracuse, N.Y. : Dept. of Geography, Syracuse University, 1976.
21 leaves : ill. ; 28 cm. — (Discussion paper series ; Department of Geography, Syracuse University ; no. 12)
1. Coastal zone management—New York (State) 2. Coastal zone management—Ontario. 3. St. Lawrence River. I. Tyson, William E., joint author. II. Title: Toward a planning and management program ... III. Series: Syracuse University. Dept of Geography. Discussion paper series - Department of Geography, Syracuse University ; no. 12.
HT393.N7H37 333.9'17'097137 79-300425
79 MARC

—NORTH CAROLINA—OUTER BANKS

Pilkey, O H 1934-
From Currituck to Calabash : living with North Carolina's barrier islands / by Orrin H. Pilkey, Jr., William J. Neal, Orrin H. Pilkey, Sr. — Research Triangle Park, N.C. : North Carolina Science and Technology Research Center, c1978.
x, 228 p. : ill. ; 23 cm.
Bibliography: p. 175-189.
Includes index.
$3.95
1. Shore protection—North Carolina—Outer Banks. 2. Coastal zone management—North Carolina—Outer Banks. 3. Beach erosion—North Carolina—Outer Banks. 4. Shore-lines—North Carolina—Outer Banks. 5. House construction. I. Neal, William J., joint author. II. Pilkey, Orrin H., 1905- joint author. III. Title.
TC224.N8P53 333.9'17'097561 78-110798
78 MARC

—ONTARIO

Harper, David B
Toward a planning and management program for the St. Lawrence-Eastern Ontario shoreline / David B. Harper and William E. Tyson ; editorial committee, David J. Robinson (editor) ... [et al.]. — Syracuse, N.Y. : Dept. of Geography, Syracuse University, 1976.
21 leaves : ill. ; 28 cm. — (Discussion paper series ; Department of Geography, Syracuse University ; no. 12)
1. Coastal zone management—New York (State) 2. Coastal zone management—Ontario. 3. St. Lawrence River. I. Tyson, William E., joint author. II. Title: Toward a planning and management program ... III. Series: Syracuse University. Dept of Geography. Discussion paper series - Department of Geography, Syracuse University ; no. 12.
HT393.N7H37 333.9'17'097137 79-300425
79 MARC

—OREGON

Oregon. State Soil and Water Conservation Commission.
Oregon coastal management program : shoreline erosion management policies and procedures / State Soil and Water Conservation Commission. — Salem, OR : The Commission, 1978-
v. ; 28 cm.
1. Coastal zone management—Oregon. I. Title.
HT393.O7O78 1978 333.9'17'09795 79-622791
79 MARC

—PUERTO RICO

United States. Office of Coastal Zone Management.
Coastal management program for the Commonwealth of Puerto Rico : United States Department of Commerce final environmental impact statement / prepared by Office of Coastal Zone Management, National Oceanic and Atmospheric Administration, Department of Commerce, and Commonwealth of Puerto Rico, Department of Natural Resources, Puerto Rico Planning Board. — Washington : The Office, [1978]
194, [101] p., [28] leaves of plates : ill. ; 28 cm.
On cover: Puerto Rico coastal management program and final environmental impact statement.
1. Coastal zone management—Puerto Rico. 2. Environmental impact analysis—Puerto Rico. I. Puerto Rico. Dept. of Natural Resources. II. Title. III. Title: Puerto Rico coastal management program ...
HT395.P9U54 1978 333.9'17'097295 78-603652
79 MARC

—UNITED STATES

Under new management : port growth and emerging coastal management programs / Marc Hershman ... [et al.]. — Seattle : Washington Sea Grant : distributed by University of Washington Press, c1978.
x, 212 p. : ill. ; 23 cm. — (A Washington sea grant publication)
Bibliography: p. [197]-207.
Includes index.
ISBN 0-295-95659-3 : $2.50
1. Harbors—United States. 2. Coastal zone management—United States. I. Hershman, Marc, 1942- . II. Series: A Washington sea grant publication (unnumb.
HE553.U52 387.1'0973 78-66066
78 MARC

United States. Congress. Senate. Committee on Commerce, Science, and Transportation.
Coastal zone management oversight : hearing before the Committee on Commerce, Science, and Transportation, United States Senate, Ninety-fifth Congress, second session ... April 3, 1978. — Washington : U.S. Govt. Print. Off., 1978.
iv, 231 p. : ill. ; 24 cm.
"Serial no. 95-94."
Includes bibliographical references.
1. Coastal zone management—United States. I. Title.
KF26.C69 1978u 353.008'232 78-603093
79 MARC

United States. National Oceanic and Atmospheric Administration.
Report to the Congress on ocean pollution, overfishing, and offshore development, July 1975 through September 1976, submitted in compliance with Section 202 (c), Title II, of the Marine protection, research, and sanctuaries act of 1972 (Public law 92-532). — [Washington?] : U.S. Dept. of Commerce, National Oceanic and Atmospheric Administration ; Washington : for sale by the Supt. of Docs., U.S. Govt. Print. Off., 1977.
62 p. in various pagings : ill. ; 27 cm.
Includes bibliography.
1. Marine pollution—United States. 2. Fishery management—United States. 3. Coastal zone management—United States. 4. Marine resources conservation—United States. I. Title.
GC1211.U54 1977 333.9'164'0973 78-601072
79 MARC

—UNITED STATES—BIBLIOGRAPHY

Sulek, Joan.
Transportation policy and coastal zone management : a selected bibliography / Joan Sulek and Roger Richman. — Monticello, Ill. : Vance Bibliographies, 1979.
14 p. ; 28 cm. — (Public administration series : Bibliography ; P-164)
Cover title.
$1.50
1. Transportation and state—United States—Bibliography. 2. Coastal zone management—United States—Bibliography. I. Richman, Roger, joint author. II. Title. III. Series.
Z7164.T8S87 016.3339'17 79-106192
[HE206.2] 79 MARC

COASTAL ZONE MANAGEMENT (Continued)

—UNITED STATES—CONGRESSES

National Conference on the Effects of OCS Oil and Gas Development on Coastal States, Galveston, 1977.
Offshore oil dialogues toward understanding : proceedings of a National Conference on the Effects of OCS Oil and Gas Development on Coastal States, June 22-24, 1977, Galveston, Texas / presented by the General Land Office of Texas ; co-sponsors, Environmental Policy Center, Inc., American Petroleum Institute, League of Women Voters of Texas. — ₍s.l.₎ : Research and Planning Consultants, ₍1978₎
vi, 188 p. ; 28 cm.
1. Offshore oil industry—United States—Congresses. 2. Coastal zone management—United States—Congresses. I. Texas. General Land Office. II. Title.
HD9567.T3N38 1977 333.9'17'0973 78-623261
MARC

Symposium on Coastal Recreation Resources in an Urbanizing Environment, Hyannis, Mass., 1976.
Coastal recreation resources in an urbanizing environment : a monograph. — Amherst : Cooperative Extension Service, University of Massachusetts, 1977.
v, 129 p. : ill. ; 28 cm.
Cover title.
Papers presented and discussed at a symposium held Apr. 12-14, 1976, and sponsored by the Cooperative Extension Service, University of Massachusetts, in cooperation with the Massachusetts Coastal Zone Program and other bodies.
Includes bibliographical references.
$3.00
1. Coasts—United States Recreational use—Management—Congresses. 2. Coasts—Environmental aspects—United States—Congresses. 3. Coastal zone management—United States—Congresses. I. University of Massachusetts at Amherst. Cooperative Extension Service. II. Massachusetts. Coastal Zone Management Office. III. Title.
GV53.S9 1976 658'.91'333780973 79-621751
79 MARC

—VIRGIN ISLANDS OF THE UNITED STATES

Heyman, Ira Michael.
Authorities and organization : legal analysis and recommendations for implementing a coastal zone management program / prepared by Ira M. Heyman, Donald O. Gralnek. — ₍Charlotte Amalie₎ : Virgin Islands Planning Office, 1977.
85 p. ; 28 cm. — (Technical supplement - Virgin Islands Planning Office, Coastal Zone Management Program ; no. 3)
Includes bibliographical references.
1. Coastal zone management—Virgin Islands of the United States. I. Gralnek, Donald O., joint author. II. Virgin Islands of the United States. Coastal Zone Management Program. III. Title. IV. Series: Virgin Islands of the United States. Coastal Zone Management Program. Technical supplement - Virgin Islands Planning Office, Coastal Zone Management Program ; no. 3.
HT395.V6H49 78-624390
354'.729'722008232
79 MARC

McElroy, Jerome L
Economic and social impacts of the Virgin Islands Coastal Zone Management Program / prepared by Jerome McElroy for Government of the Virgin Islands of the United States, Virgin Islands Planning Office, Coastal Zone Management Program. — ₍Charlotte Amalie, St. Thomas₎ : The Program, 1978.
47, ₍1₎ p. ; 28 cm. — (Technical supplement - Virgin Islands Planning Office, Coastal Zone Management Program ; no. 4)
Bibliography: p. 47-₍48₎
1. Coastal zone managment—Virgin Islands of the United States. I. Virgin Islands of the United States. Coastal Zone Management Program. II. Title. III. Series: Virgin Islands of the United States. Coastal Zone Management Program. Technical supplement - Virgin Islands Planning Office, Coastal Zone Management Program ; no. 4.
HT395.U6M23 333.9'17'09729722 79-621968
79 MARC

United States. Office of Coastal Zone Management.
Proposed coastal zone management program for the Virgin Islands : final environmental impact statement / prepared by Office of Coastal Zone Management, National Oceanic and Atmospheric Administration, U.S. Department of Commerce. — Washington : The Office, ₍1979₎
x, 185, ₍66₎ p. : ill. ; 28 cm.
Cover title: The Virgin Islands coastal management program and final environmental impact statement.
1. Coastal zone management—Virgin Islands of the United States. 2. Environmental impact statements—Virgin Islands of the United States. I. Title. II. Title: Virgin Islands coastal management program and final environmental impact statement.
HT393.V5U54 1979 333.9'17'09729722 79-602180
79 MARC

Virgin Islands of the United States. Planning Office.
Preliminary Virgin Islands coastal zone management program. — ₍Charlotte Amalie₎ : Government of the Virgin Islands of the United States, Virgin Islands Planning Office, 1977.
194, 56 p. : ill., maps (2 fold. in pocket) ; 28 cm.
Title on spine: Virgin Islands coastal zone management-preliminary program.
Bibliography: p. 52-56 (2d group)
1. Coastal zone management—Virgin Islands of the United States. I. Title. II. Title: Virgin Islands coastal zone management-preliminary program.
HT395.V6V66 1977 333.9'17'09729722 77-623728
79 MARC

—VIRGIN ISLANDS OF THE UNITED STATES—PUBLIC OPINION

Virgin Islands of the United States. Coastal Zone Management Program.
Public attitudes towards coastal development : technical supplement no. 2 : a report on two surveys of public perceptions and attitudes undertaken in developing the Virgin Islands Coastal Zone Management Program. — ₍Charlotte Amalie₎ : Government of the Virgin Islands of the United States, Virgin Islands Planning Office, 1977.
ix, 103 p. : ill. ; 28 cm.
Cover title: Public attitude survey, technical supplement no. 2.
1. Coastal zone management—Virgin Islands of the United States—Public opinion. 2. Public opinion—Virgin Islands of the United States. I. Title. II. Title: A report on two surveys of public perceptions and attitudes undertaken in developing the Virgin Islands Coastal Zone Management Program. III. Title: Public attitude survey, technical supplement no. 2.
HT395.V6V64 1977 78-624389
301.15'43'33391709729722
79 MARC

—VIRGINIA—ASSATEAQUE ISLAND NATIONAL SEASHORE, MD. AND VA.

Dolan, Robert.
Atlas of environmental dynamics, Assateague Island National Seashore / by Robert Dolan, Bruce Hayden, Jeffrey Heywood ; illustrations, J. G. Carswell, Richard Garber. — ₍Washington₎ : U.S. Dept. of the Interior, National Park Service, 1977.
40 p., ₍2₎ fold. leaves of plates : ill. (some col.), col. maps ; 28 x 36 cm. — (Natural resource report ; no. 11)
Bibliography: p. 39
Includes glossary.
PARTIAL CONTENTS: Geology.—Coastal processes.—Monitoring ocean dynamics.—Barrier island management.
1. Assateague Island National Seashore, Md. and Va.—Maps. 2. Barrier islands. 3. Coastal zone management—Maryland—Assateague Island National Seashore, Md. and Va. 4. Coastal zone management—Virginia—Assateaque Island National Seashore, Md. and Va. I. Haden, Bruce P. II. Heywood, Jeffrey. III. Carswell, J. G. IV. Garber, Richard. V. Title. VI. Series: Natural resources report ; no. 11.
G1272.A85D6 1977 79-311828
912'.1'333780975221
79 MARC

—WASHINGTON (STATE)

Phipps, James B
Coastal accretion and erosion in southwest Washington / by James. B. Phipps and John M. Smith. — Olympia, Wash. : Dept. of Ecology, 1978.
75 p. : ill. ; 28 cm.
"78-12."
Bibliography: p. 48-50.
1. Coast changes—Washington (State) 2. Coastal zone management—Washington (State) I. Smith, John M., 1926- joint author. II. Title.
GB459.4.P45 551.3'6'097979 79-622614
79 MARC

COASTERS (FOR DRINKS)

—GREAT BRITAIN—CATALOGS

Brittain, Dennis G
United Kingdom brewery mat catalogue : current breweries / compiled by Dennis G. Brittain. — ₍Northwich₎ : British Beermat Collectors' Society, 1976- ₍i.e. 1977-
v. ; 21 cm. GB77-12590
Caption title.
Based on the original beermat catalogue compiled by R. Bristow.
ISBN 0-9505645-0-8 (v. 1)>
1. Coasters (for drinks)—Great Britain—Catalogs. I. Bristow, Roger. II. Title.
NE965.3.G7B74 769'.5 78-318059
78 MARC

COASTERS (SHIPS)

Duckworth, Christian Leslie Dyce.
Clyde and other coastal steamers / by Christian Leslie Dyce Duckworth and Graham Easton Langmuir. — 2nd ed. — Prescot (Prescot, Merseyside) : T. Stephenson and Sons Ltd., 1977.
xvi, 218 p., leaf of plate, ₍18₎ p. of plates : ill., facsim. ; 22 cm.
GB78-19099
Bibliography: p. 202-203.
Includes index.
ISBN 0-901314-19-6 : £5.95
1. Coasters (Ships) 2. Coastwide navigation—Great Britain. 3. Steamboat lines—Great Britain. I. Langmuir, Graham Easton, joint author. II. Title.
VM57.D8 1977 387.2'4 78-324128
79 MARC

—FRANCE—HISTORY

Lacroix, Louis, 1877-1958.
Les Derniers voiliers caboteurs français / Louis Lacroix, ... ; préface de Jean Randier. — ₍Paris₎ : Éditions maritimes et d'outre-mer, 1978.
293 p., ₍32₎ p. of plates ; 21 cm. F79-4964
Previous ed. published in 1946 under title: Les écraseurs de crabes sur les derniers voiliers caboteurs.
65.00F
1. Sailing ships—France—History. 2. Coasters (Ships)—France—History. 3. Coastwise navigation—France—History. I. Title.
VK71.L32 1978 387.2'2'0944 79-380276
79 MARC

—UNITED STATES

Morris, Paul C
American sailing coasters of the North Atlantic / by Paul C. Morris ; with a foreword by Edouard A. Stackpole. — New York : Bonanza Books, 1979, c1973.
xv, 224 p., ₍1₎ fold. leaf of plates : ill. ; 28 cm.
Bibliography: p. 205-207.
Includes index.
ISBN 0-517-26190-1
1. Schooners—United States. 2. Coasters (Ships)—United States. 3. Coastwise shipping—United States. I. Title.
VM311.F7M67 1979 623.82'2 79-333
79 MARC

COASTS

see also Coast changes; Coastal ecology; Coastal zone management; Estuaries; Inlets; Shore protection; Storm surges

—ADDRESSES, ESSAYS, LECTURES

Coastal sedimentation / edited by Donald J. P. Swift, Harold D. Palmer. — Stroudsburg, Pa. : Dowden, Hutchinson & Ross ; ₍New York₎ : distributed worldwide by Academic Press, c1978.
xv, 339 p. : ill. ; 26 cm. — (Benchmark papers in geology ; 42)
Includes bibliographies and indexes.
ISBN 0-87933-330-8
1. Coasts—Addresses, essays, lectures. 2. Sedimentation and deposition—Addresses, essays, lectures. I. Swift, Donald J. P. II. Palmer, Harold Dean, 1934-
GB451.2.C63 551.3'6 78-18696
78 MARC

—CONGRESSES

Ocean Wave Climate Symposium, Herndon, Va., 1977.
Ocean wave climate / edited by Marshall D. Earle and Alexander Malahoff. — New York : Plenum Press, c1979.
xi, 368 p. : ill. ; 26 cm. — (Marine science ; v. 8)
Includes bibliographical references and index.
ISBN 0-306-40079-0
1. Ocean waves—Congresses. 2. Coasts—Congresses. I. Earle, Marshall Delph, 1913- . II. Malahoff, Alexander, 1939- . III. Title.
GC206.O25 1977 551.4'7022 78-24469
78 MARC

—DICTIONARIES—GERMAN

Lüders, Karl.
Kleines Küstenlexikon : Natur u. Technik an d. dt. Nordseeküste / von Karl Lüders, Günter Luck. — 3., neubearb. u. erw. Aufl. — Hildesheim : Lax, 1976.
xv, 240, ii p. : numerous ill., maps (some col.) ; 24 cm. GFR77-A
ISBN 3-7848-4030-2 : DM28.00
1. Coasts—Germany, Northern—Dictionaries—German. 2. Coasts—Dictionaries—German. I. Luck, Günter, joint author. II. Title.
GB457.2.L83 1976 77-566180
79 MARC

COASTS (Continued)

—ENVIRONMENTAL ASPECTS—UNITED STATES—CONGRESSES

Symposium on Coastal Recreation Resources in an Urbanizing Environment, Hyannis, Mass., 1976.
Coastal recreation resources in an urbanizing environment : a monograph. — Amherst : Cooperative Extension Service, University of Massachusetts, 1977.
v, 129 p. : ill. ; 28 cm.
Cover title.
Papers presented and discussed at a symposium held Apr. 12-14, 1976, and sponsored by the Cooperative Extension Service, University of Massachusetts, in cooperation with the Massachusetts Coastal Zone Program and other bodies.
Includes bibliographical references.
$3.00
1. Coasts—United States Recreational use—Management—Congresses. 2. Coasts—Environmental aspects—United States—Congresses. 3. Coastal zone management—United States—Congresses. I. University of Massachusetts at Amherst. Cooperative Extension Service. II. Massachusetts. Coastal Zone Management Office. III. Title.
GV53.S9 1976 658'.91'333780973 79-621751
79 MARC

—MAPS—CONGRESSES

Coastal Mapping Symposium, Rockville, Md., 1978.
The Potomac Region of the American Society of Photogrammetry presents a Coastal Mapping Symposium, August 14-16, 1978 / in cooperation with NOAA National Ocean Survey, U.S. Geological Survey. — Falls Church, Va. : American Society of Photogrammetry, c1978.
v, 190 p. : ill. ; 23 cm.
Cover title: Proceedings, Coastal Mapping Symposium.
Includes bibliographies and index.
1. Coasts—Maps—Congresses. 2. Geomorphological mapping—Congresses. I. American Society of Photogrammetry. Potomac Region. II. National Ocean Survey. III. United States. Geological Survey. IV. Title.
GB450.2.C62 1978 526'.0914'6 78-68426
78 MARC

—RECREATIONAL USE

see also Aquatic sports

—ALASKA—CHARTS, DIAGRAMS, ETC.

Brower, William A
Climatic atlas of the Outer Continental Shelf waters and coastal regions of Alaska / William A. Brower, Jr., Harold W. Searby, James L. Wise ₍with₎ Henry F. Diaz, Anton S. Prechtel. — Anchorage : Arctic Environmental Information and Data Center, University of Alaska, 1977.
3 v. : ill. ; 29 cm. — (AEIDC publication ; B-77)
Includes bibliographies.
CONTENTS: v. 1. Gulf of Alaska.—v. 2. Bering Sea.—v. 3. Chukchi-Beaufort Sea.
$5.00 per vol.
1. Continental shelf—Alaska—Climate—Charts, diagrams, etc. 2. Coasts—Alaska—Charts, diagrams, etc. 3. Alaska—Climate—Charts, diagrams, etc. I. Searby, Harold W., joint author. II. Wise, James L., joint author. III. Series: University of Alaska, Anchorage. Arctic Environmental Information and Data Center. AEIDC publication ; B-77.
QC984.A4B76 551.6'9'798 78-106788
78 MARC

—ALASKA—OBSERVATIONS

National Climatic Center.
Summary of synoptic meteorological observations (SSMO) : Alaska coastal marine area / National Climatic Center. — ₍Washington₎ : U.S. Dept. of Commerce, National Oceanic and Atmospheric Administration, Environmental Data Service ; ₍Springfield, Va. : obtainable from the National Technical Information Service₎, 1978.
viii, 158 p. : ill. ; 22 x 28 cm.
Spine title: SSMO: Valdez, Alaska coastal marine area.
Chiefly tables.
1. Meteorology—Alaska—Observations. 2. Coasts—Alaska—Observations. 3. Synoptic meteorology. I. Title.
QC984.A4N37 1978 551.6'9'798 78-602310
78 MARC

—ALASKA—CORDOVA REGION—RECREATIONAL USE

Alaska. Division of Parks.
Coastal recreation resources, Cordova, Alaska / ₍by Alaska Division of Parks, Department of Natural Resources, for Alaska Coastal Management Program ; Alan H. Meiners, with initial research by Alison C. Horton₎. — ₍Juneau₎ : Alaska Coastal Management Program, 1977.
iv, 192 p. : ill. ; 28 cm.
Bibliography: p. 189-192.
1. Coasts—Alaska—Cordova region—Recreational use. 2. Recreational surveys—Alaska—Cordova region. I. Meiners, Alan H. II. Alaska. Coastal Management Program. III. Title.
GV54.A42C673 790'.09798'3 77-622786
79 MARC

—AUSTRALIA

Semler, Bruce.
East coast holidays / ₍by₎ Bruce Semler ; photography by Publishing Art ; maps by Bob Fosbery, Alan Sutherland, and Paul Hamlyn Pty. Limited. — Sydney : Ure Smith, 1978.
80 p. : ill., maps ; 28 cm. — (Summit book) Aus
Includes index.
ISBN 0-7254-0416-7 : $1.95
1. Australia—Description and travel—1951- . 2. Coasts—Australia. I. Title.
DU105.S45 919.4 79-300577
79 MARC

Stodart, Eleanor Mary.
Seashores of Australia / ₍by₎ Eleanor Stodart ; photographs by Ederic Slater. — Adelaide : Rigby, 1977.
63 p., ₍2₎ leaves of plates : ill. (some col.) ; 25 cm. — (Pageant of Australia series)
Aus
ISBN 0-7270-0314-3
1. Seashore biology—Australia. 2. Coasts—Australia. I. Title. II. Series.
QH197.S83 574.994 78-314081
78 MARC

—BALTIC SEA

Meeresgrund- und Küstenfroschung im Bereich der Ost- und Nordsee / Akademie der Wissenschaften der DDR, Institut für Meereskunde. — Berlin : Akademie-Verlag, 19
v. : ill. ; 30 cm. — (Beiträge zur Meereskunde ; Heft 35) GDR***
Vol. 4 has added t.p. "Die submarinen Terrassen der südlichen Ost- und Nordsee und ihre Beziehung zum eustatischen Meeresanstieg, von Otto Kolp."
Summary also in English and Russian.
Bibliography: v. 4, p. 46-48.
CONTENTS:
—Stechrohrproben. Submarine Uferterrassen. Eustatik. Isostasie.

30.00M (v. 4)
1. Submarine topography—Baltic Sea. 2. Submarine topography—North Sea. 3. Coasts—Baltic Sea. 4. Coasts—North Sea. I. Kolp, Otto. II. Warnemünde. Institut für Meereskunde. III. Title: Die submarinen Terrassen der südlichen Ost- und Nordsee und ihre Beziehung zum eustatischen Meeresanstieg. IV. Series.
GC1.B35 Heft 35, etc. 79-342406
₍GC83.2.B3₎ 79 MARC

—BRITISH COLUMBIA—PICTORIAL WORKS

Steltzer, Ulli.
Coast of many face / Ulli Steltzer and Catherine Kerr ; ₍maps by Marta Farevaag₎. — Seattle : University of Washington Press, 1979.
xii, 212 p. : ill. ; 28 cm.
ISBN 0-295-95689-5 : $25.95
1. British Columbia—Description and travel—Views. 2. Coasts—British Columbia—Pictorial works. I. Kerr, Catherine, joint author. II. Title.
F1087.8.S73 1979 971.1 79-4916
79 MARC

—CALIFORNIA—LAND USE—MAPS

California. Dept. of Water Resources.
Land use within the California coastal zone. — ₍Sacramento₎ : State of California, Resources Agency, Dept. of Water Resources, 1978.
viii, 19, 161 ₍i.e. 322₎ p. : ill., 161 maps ; 28 cm. — (Bulletin - State of California, Department of Water Resources ; 207)
1. Coasts—California—Land use—Maps. 2. California—Land use—Maps. I. Title. II. Series: California. Dept. of Water Resources. Bulletin ; 207.
G1527.C6G4C3 1978 79-620687
912'.1'33891709794
79 MARC

—CALIFORNIA—PICTORIAL WORKS

Discovering the California coast / ₍edited by Jack McDowell ; maps and lighthouse drawings, Joe Seney ; ill. Greg Irons₎. — 2d ed. — Menlo Park, Calif. : Lane Pub. Co., 1978.
224 p. : ill. ; 28 cm. — (A Sunset pictorial)
Title on spine: The California coast.
Includes index.
ISBN 0-376-05182-5 : $6.95
1. California—Description and travel—1951- —Views. 2. Coasts—California—Pictorial works. I. McDowell, Jack. II. Sunset. III. Title: The California coast.
F862.D57 1978 979.4'05'0222 75-58504
78 MARC

Lewis, Paul M
Beautiful California coast / text by Paul M. Lewis. — Beaverton, Or. : Beautiful America Pub. Co., 1979.
71 p. : col. ill. ; 29 cm.
ISBN 0-915796-97-X : $12.95
1. California—Description and travel—1951- —Views. 2. Coasts—California—Pictorial works. I. Title.
F862.L54 917.94 79-13134
79 MARC

—CONNECTICUT

Hildebrandt, Barry.
Coastal Connecticut, eastern region / by Barry and Susan Hildebrandt. — Old Saybrook, Conn. : Peregrine Press, c1979.
114 p. : ill. ; 22 cm.
Includes index.
ISBN 0-933614-00-4 : $4.95
1. Connecticut—Description and travel—1951- —Guide-books. 2. Coasts—Connecticut. I. Hildebrandt, Susan, joint author. II. Title.
F92.3.H54 917.46 79-120422
79 MARC

—CONNECTICUT—RECREATIONAL USE

Schneider, Margaret N
Recreational demand, opportunities, and limitations in Conecticut's coastal area / prepared for the State of Connecticut Department of Environmental Protection, Coastal Area Management Program, by Margaret N. Schneider. — Hartford : The Program, 1978.
iv, 48, ₍154₎ p. : map ; 29 cm. — (Planning report - The Connecticut Coastal Area Management Program ; no. 25)
Cover title: Coastal recreation.
Bibliography: p. 47-48 (2d group)
1. Coasts—Connecticut—Recreational use. 2. Recreational surveys—Connecticut. I. Connecticut. Coastal Area Management Program. II. Title. III. Title: Coastal recreation. IV. Series: Connecticut. Coastal Area Management Program. Planning report - Connecticut Coastal Area Management Program ; no. 25.
GV191.42.C8S36 333.7'8'09746 78-622756
79 MARC

—CRIMEA

Serhieieva, H
(Alushta—Sudak)
Алушта — Судак. Путеводитель. Симферополь, Крым, 1970.
86 p. with illus. 12 cm. USSR 70
At head of title: Г. Сергеева.
0.12rub
1. Crimea—Description and travel. 2. Coasts—Crimea. I. Title.
DK511.C7S43 914.7'717 72-342399

—DENMARK—HELSINGØR

Helsingør, Denmark (Kommune). Havneudvalget.
Kyst og kystanlæg. Helsingør kommunes rapport om kystens og havnenes udnyttelse og udvikling. ₍Udarb. af₎ Havneudvalget. Helsingør, Helsingør Kommune, Eksp.: Stads- og Havneingeniørens Kontro, Rådhuset, ₍1972₎.
79 l. illus. 21 x 30 cm. N.T. D73-39
1. Coastal zone mangement—Denmark—Helsingør. 2. Coasts—Denmark—Helsingør. I. Title.
HT395.D42H443 73-364274
79₍r79₎rev MARC

—ECUADOR—PUERTO LÓPEZ

Cuenca Vargas, José Manuel.
Estudio de la morfología y sedimentos de la plataforma continental de Puerto López, Provincia de Manabí, Ecuador / por José Manuel Cuenca Vargas. — Guayaquil : Universidad de Guayaquil, Facultad de Ciencias Naturales, Escuela de Geología, 1976, cover 1977.
54 leaves : ill. ; 28 cm.
Bibliography: leaf 10.
S/30.00
1. Coasts—Ecuador—Puerto López. 2. Marine sediments—Ecuador—Puerto López. I. Title.
GB458.9.C83 78-109672
79 MARC

COASTS (Continued)

—ENGLAND—YORKSHIRE

Young, Stephen.
Geology of the Yorkshire coast : Whitby to Bridlington / by Stephen Young. — Clapham, N. Yorkshire : Dalesman, 1978.
104 p. : ill., maps ; 21 cm. — (A "Dalesman" paperback) GB78-26047
Bibliography: p. 102.
Includes index.
ISBN 0-85206-429-2 : £1.25
1. Geology—England—Yorkshire. 2. Coasts—England—Yorkshire. I. Title.
QE262.Y8Y63 554.28'47 78-325888
79 MARC

—EUROPE

see also Atlantic coast (Europe)

—FINLAND—PICTORIAL WORKS

Mäkinen, Vesa.
Suomen rannikko ja saaristomeri : 100 matkailukohdetta = Den finska kusten och arkipelagen : 100 turistmål = The Finnish coast and archipelago : 100 places for the tourist to see / toimitus ja teksti, Vesa Mäkinen. — Porvoo : W. Söderström, 1977.
112 p. : ill. ; 31 cm. Fi***
English, Finnish, German, and Swedish.
ISBN 9510079898 (large size)
1. Finland—Description and travel—1945- —Views. 2. Coasts—Finland—Pictorial works. 3. Archipelagoes—Finland—Pictorial works. I. Title. II. Title: Den Finska kusten och arkipelagen. III. Title: The Finnish coast and archipelago.
DK450.2.M32 914.71'04'3 79-348661
79 MARC

—FRANCE

see also Atlantic coast (France); Les Calanques, France

Rivages de France / ⌜Gérald Pechmèze, Ange Bastiani, Pierre Gallerey, Gilles Lambert, etc.⌝. — Paris : Larousse, 1977.
480, ⌜6⌝ p. : col. ill. ; 30 cm. — (Beautés de la France) (Découvrir) F78-7723
Includes index.
ISBN 2-03-013271-3 : 190F
1. Coasts—France. 2. France—Description and travel—1975- I. Pechmèze, Gérald.
DC29.3.R58 914.4'04'83 79-337981
79 MARC

—GERMANY, NORTHERN—DICTIONARIES—GERMAN

Lüders, Karl.
Kleines Küstenlexikon : Natur u. Technik an d. dt. Nordseeküste / von Karl Lüders, Günter Luck. — 3., neubearb. u. erw. Aufl. — Hildesheim : Lax, 1976.
xv, 240, ii p. : numerous ill., maps (some col.) ; 24 cm. GFR77-A
ISBN 3-7848-4030-2 : DM28.00
1. Coasts—Germany, Northern—Dictionaries—German. 2. Coasts—Dictionaries—German. I. Luck, Günter, joint author. II. Title.
GB457.2.L83 1976 77-566180
79 MARC

—GERMANY, WEST

Reineck, Hans-Erich.
Das Watt : Ablagerungs- u. Lebensraum / hrsg. von Hans-Erich Reineck ; mit wissenschaftl. Beitr. von Karl-Ernst Behre ... ⌜et al.⌝. — 2., neubearb. u. erw. Aufl. — Frankfurt am Main : Kramer, 1978.
185 p. : ill., maps ; 22 cm. — (Senckenberg-Buch ; 50) GFR79-A
Includes bibliographies and index.
ISBN 3-7829-1067-2 : DM19.80
1. Coasts—Germany, West. 2. Seashore biology—Germany, West. 3. Marine sediments—Germany, West. I. Behre, Karl-Ernst. II. Title.
GB457.5.R4 1978 79-376929
79 MARC

—GERMANY, WEST—SAXONY, LOWER

Sindowski, Karl-Heinz.
Zwischen Jadebusen und Unterelbe / von Karl-Heinz Sindowski. — Berlin ; Stuttgart : Borntraeger, 1979.
x, 145 p. : ill. ; 20 cm. — (Sammlung geologischer Führer ; Bd. 66) GFR79-A
Bibliography: p. ⌜125⌝-133.
Includes index.
ISBN 3-443-15025-X : DM35.00
1. Geology—Germany, West—Saxony, Lower. 2. Coasts—Germany, West—Saxony, Lower. I. Title. II. Series.
QE269.S57 79-360289
79 MARC

—GREAT BRITAIN

Steers, James Alfred, 1899-
The sea coast. With 10 colour photos. taken by the author. London, Collins, 1953.
xii, 276 p. illus. (part col.) maps. 23 cm. (The New naturalist ⌜25⌝)
Errata slip inserted.
Bibliography: p. 271-272.
1. Coasts—Great Britain. I. Title. II. Series: New naturalist, 25.
GB460.G7S73 551.4'5 53-32917
⌜r79⌝rev MARC

Steers, James Alfred, 1899-
The sea coast, by J. A. Steers. 4th ed. London, Collins, 1969.
xii, 292 p., 32 plates. illus. (some col.), maps. 23 cm. (The New naturalist ⌜25⌝) £1.80 B69-26308
Includes bibliographies.
1. Coasts—Great Britain. I. Title. II. Series: New naturalist, 25.
GB460.G7S73 1969 551.4'5 70-851386
71⌜r79⌝rev MARC

—LEBANON

Sanlaville, Paul.
Étude géomorphologique de la région littorale du Liban / par Paul Sanlaville. — Beyrouth : Université libanaise, 1977.
2 v. (vi, 859 p., ⌜22⌝ leaves of plates) : ill. ; 24 cm. — (Publications de l'Université libanaise : Section des études géographiques ; 1)
Bibliography: p. ⌜809⌝-830.
1. Geomorphology—Lebanon. 2. Physical geography—Lebanon. 3. Coasts—Lebanon. I. Title. II. Series: Beirut. al-Jāmi'ah al-Lubnānīyah. Manshūrāt al-Jāmi'ah al-Lubnānīyah : Qism al-dirāsāt al-jughrāfīyah ; 1.
GB438.L42S26 551.4'095692 78-108430
79 MARC

—LIBERIA

Behrendt, John C 1932-
Aeromagnetic and gravity investigations of the sedimentary basins on the continental shelf and coastal area of Liberia, West Africa / by John C. Behrendt and Cletus S. Wotorson. — Monrovia : Republic of Liberia, Ministry of Lands and Mines, Liberian Geological Survey, 1972.
v, 13 p. : ill., maps (4 fold. in pocket) ; 29 cm. — (Geological Survey special paper ; no. 2)
Bibliography: p. 12-13.
1. Magnetism, Terrestrial—Liberia. 2. Rocks, Sedimentary. 3. Continental shelf—Liberia. 4. Coasts—Liberia. I. Wotorson, Cletus S., joint author. II. Title: Aeromagnetic and gravity investigations ... III. Series: Liberia. Geological Survey. Special paper ; no. 2.
QC825.6.B44 538'.78666'2 78-322294
79 MARC

—LOUISIANA

Coastal Environments, inc.
Siting energy related facilities in Louisiana's coastal zone / prepared for Coastal Resources Program, Louisiana Department of Transportation and Development by Coastal Environments. — ⌜Baton Rouge⌝ : The Dept., 1978.
54 p. : ill. ; 22 x 36 cm.
Bibliography: p. 53.
1. Energy facilities—Louisiana—Location. 2. Coasts—Louisiana. I. Louisiana. Coastal Resources Program. II. Title.
TJ163.25.U6C6 1978 621 79-621599
79 MARC

—LOUISIANA—MAPS

Gagliano, Sherwood M
Environmental atlas and multiuse management plan for south-central Louisiana : prepared for Department of the Army, New Orleans District, Corps of Engineers, Contract No. DACW 29-71-C-0219 and Office of Sea Grant, National Oceanic and Atmospheric Administration of U.S. Department of Commerce, Institutional Support Grant No. 04-3-158-19 / by Sherwood M. Gagliano, project director and Penny Culley ... ⌜et. al.⌝. — Baton Rouge : Center for Wetland Resources, Louisiana State University, 1973.
2 v. : ill., maps ; 28-41 x 22-62 cm. — (Hydrologic and geologic studies of coastal Louisiana ; report no. 18)
Includes bibliographies.
CONTENTS: Approach.—Natural setting.—Activities.—Management plan.—Summary and conclusions.—Appendices.
1. Coastal zone management—Louisiana—Maps. 2. Land use—Planning—Loiusiana—Maps. 3. Coasts—Louisiana—Maps. I. United States. Army. Corps of Engineers. II. United States. National Oceanic and Atmospheric Administration. Office of Sea Grant Programs. III. Louisiana State University, Baton Rouge. Center for Wetland Resources. IV. Title. V. Series.
G1361.G45G3 1973 912'.1'3337097634 75-624375
78 MARC

—MAINE—RECREATIONAL USE—COLLECTED WORKS

St. Pierre, James A
The Maine coast : recreation and open space / prepared for the Committee on Coastal Development and Conservation by James A. St. Pierre. — ⌜Augusta⌝ : State of Maine, Dept. of Conservation, 1978.
2 v. : ill. ; 28 cm.
"Work for this project has been conducted as part of the Maine Coastal Program."
Includes bibliography.
CONTENTS: v. 1. Recommendations for consideration.—v. 2. Appendix.
1. Coasts—Maine—Recreational use—Collected works. 2. Outdoor recreation—Maine—Collected works. 3. Recreational surveys—Maine—Collected works. I. Maine. Committee on Coastal Development and Conservation. II. Maine. Coastal Program. III. Title.
GV191.42.M2S24 790'.09741 79-623379
79 MARC

—MINNESOTA—MAPS

North Shore data atlas, Minnesota coastal zone. — ⌜St. Paul, Minn.⌝ : Minnesota State Planning Agency : available from the State Planning Agency or the Arrowhead Regional Development Commission, ⌜1978?⌝
40 p. : col. ill., col. maps ; 22 x 28 cm.
Cover title.
"Describes the natural resources and cultural patterns of the North Shore."
1. Coasts—Minnesota—Maps. 2. Coastal zone management—Minnesota—Maps. 3. Minnesota—Maps. I. Minnesota. State Planning Agency. II. Arrowhead Regional Development Commission.
G1427.C6N6 1978 78-623095
912'.1'33391709776
79 MARC

—NEW ENGLAND

Clayton, Barbara, 1931-
Exploring coastal New England, Glouchester to Kennebunkport / Barbara Clayton and Kathleen Whitley ; illustrated with photos., maps by Marian Manfredi. — New York : Dodd, Mead, 1979.
xi, 276 p. : ill. ; 24 cm.
Bibliography: p. 267-272.
Includes index.
ISBN 0-396-07572-X
1. New England—Description and travel—1951- 2. Coasts—New England. 3. New England—History, Local. 4. Historic buildings—New England. 5. Architecture—New England. I. Whitley, Kathleen, joint author. II. Title.
F10.C45 917.4 79-624
79 MARC

Young, George, 1924-
The short triangle : a story of the sea and men who go down to it in ships, the place, the coasts of New England and Nova Scotia, the time, summer 1942 / by George Young. — 1st ed. — Lunenburg, N.S. : Lunenburg County Press, 1975.
79 p. : ill. ; 23 cm. C***
1. World War, 1939-1945—Naval operations, British. 2. World War, 1939-1945—Naval operations—Submarine. 3. Coasts—New England. 4. Coasts—Nova Scotia. I. Title.
D771.Y68 940.54'59'41 79-305458
79 MARC

—NEW ENGLAND—ADDRESSES, ESSAYS, LECTURES

Snow, Edward Rowe.
Adventures, blizzards, and coastal calamities / Edward Rowe Snow. — New York : Dodd, Mead, c1978.
x, 275 p., ⌜4⌝ leaves of plates : ill. ; 21 cm.
Includes index.
ISBN 0-396-07634-3 : $8.95
1. New York—History, Local—Addresses, essays, lectures. 2. Coasts—New England—Addresses, essays, lectures. I. Title.
F4.5.S66 974 78-23316
78 MARC

COASTS (Continued)

—NORTH CAROLINA—BIBLIOGRAPHY

Dubach, Harold W
The North Carolina coastal zone and its environment : a compilation of resource materials covering the coastal plain, estuaries, and offshore waters / compiled by Harold W. Dubach ; approved by T. V. Crawford. — Aiken, S.C. : E.I. du Pont de Nemours, Savannah River Laboratory ; ₍Springfield, Va. : available from National Technical Information Service₎, 1977.
2 v. (389 p.) ; 28 cm.
"DP-1423."
Includes indexes.
1. Coasts—North Carolina—Bibliography. 2. Coastal ecology—North Carolina—Bibliography. 3. Continental shelf—North Carolina—Bibliography. I. Savannah River Laboratory. II. Title.
Z6004.C6D84 016.5514'614'8 77-3695
₍GB459.4₎ 77 MARC

—NORTH SEA

Meeresgrund- und Küstenfroschung im Bereich der Ost- und Nordsee / Akademie der Wissenschaften der DDR, Institut für Meereskunde. — Berlin : Akademie-Verlag, 19
v. : ill. ; 30 cm. — (Beiträge zur Meereskunde ; Heft 35) GDR***
Vol. 4 has added t.p. "Die submarinen Terrassen der südlichen Ost- und Nordsee und ihre Beziehung zum eustatischen Meeresanstieg, von Otto Kolp."
Summary also in English and Russian.
Bibliography: v. 4, p. 46-48.
CONTENTS:
—Stechrohrproben. Submarine Uferterrassen. Eustatik. Isostasie.

30.00M (v. 4)
1. Submarine topography—Baltic Sea. 2. Submarine topography—North Sea. 3. Coasts—Baltic Sea. 4. Coasts—North Sea. I. Kolp, Otto. II. Warnemünde. Institut für Meereskunde. III. Title: Die submarinen Terrassen der südlichen Ost- und Nordsee und ihre Beziehung zum eustatischen Meeresanstieg. IV. Series.
GC1.B35 Heft 35, etc. 79-342406
₍GC83.2.B3₎ 79 MARC

—NOVA SCOTIA

Young, George, 1924-
The short triangle : a story of the sea and men who go down to it in ships, the place, the coasts of New England and Nova Scotia, the time, summer 1942 / by George Young. — 1st ed. — Lunenburg, N.S. : Lunenburg County Press, 1975.
79 p. : ill. ; 23 cm. C***
1. World War, 1939-1945—Naval operations, British. 2. World War, 1939-1945—Naval operations—Submarine. 3. Coasts—New England. 4. Coasts—Nova Scotia. I. Title.
D771.Y68 940.54'59'41 79-305458
79 MARC

—OREGON

Gibbs, James Atwood, 1922-
Oregon's salty coast / by Jim Gibbs. — 1st ed. — Seattle : Superior Pub. Co., c1978.
160 p. : ill. ; 28 cm.
ISBN 0-87564-223-3 : $14.95
1. Coasts—Oregon. 2. Oregon—History. I. Title.
GB458.8.G52 979.5 78-11899
78 MARC

—PACIFIC COAST (NORTH AMERICA)—RECREATIONAL USE

Dixon, Sarah.
West coast beaches : a complete guide, from Baja to Canada / Sarah and Peter Dixon. — 1st ed. — New York : Dutton, c1979.
132 p. : ill. ; 28 cm.
"A Sunrise book."
Includes index.
ISBN 0-87690-285-9 : $8.95
1. Coasts—Pacific coast (North America)—Recreational use. 2. Beaches—Pacific coast (North America) 3. Pacific coast (North America)—Description. I. Dixon, Peter L., joint author. II. Title.
GV191.4.D58 1979 917.8 78-8106
78 MARC

—SENEGAL—CASAMANCE RIVER REGION

Vieillefon, J
Les Sols des mangroves et des tannes de Basse-Casamance (Sénégal) / Jacques Vieillefon, ... — Paris : Office de la recherche scientifique et technique outre-mer, 1977.
xiv, 291 p. : ill., fold. col. map (in pocket) ; 27 cm. — (Mémoires ORSTOM ; no 83) F77-17768
Summary in English.
Bibliography: p. ₍265₎-278.
ISBN 2-7099-0446-2 : 140F
1. Soils—Senegal—Casamance River region. 2. Soil formation—Senegal—Casamance River region. 3. Mangrove—Senegal—Casamance River region—Ecology. 4. Soils—Sulphur content. 5. Coasts—Senegal—Casamance River region. 6. Casamance River region, Senegal. I. Title. II. Title: Tannes de Basse-Casamance (Sénégal) III. Series: France. Office de la recherche scientifique et technique outre-mer. Mémoires ; no 83.
S599.5.S382C378 631.4'966'3 78-389727
79 MARC

—TEXAS—RECREATIONAL USE

Texas. General Land Office.
Accessibility and protection of Texas shorefront areas / produced by the General Land Office. — ₍Austin₎ : The Office, 1978.
44 p. : ill., maps (7 fold. in pocket) ; 28 cm.
Prepared by H. Hees, M. H. McKann, and J. W. McAtee.
Bibliography: p. 41.
1. Coasts—Texas—Recreational use. I. Hees, Hal. II. McKann, Michael H. III. McAtee, Jerry W. IV. Title.
GV54.T59T49 1978 333.7'8'09764 79-622510
79 MARC

—UNITED STATES

see also Atlantic coast (United States)

—UNITED STATES—RECREATIONAL USE—ADDRESSES, ESSAYS, LECTURES

United States. Library of Congress. Environment and Natural Resources Policy Division.
Past Federal surveys of shoreline recreation potential / prepared by the Environment and Natural Resources Policy Division, Congressional Research Service, Library of Congress, at the request of Henry M. Jackson, chairman, Committee on Energy and Natural Resources, United States Senate. — Washington : U.S. Govt. Print. Off., 1978.
ix, 226 p. : ill. ; 23 cm.
At head of title: 95th Congress, 2d session. Committee print.
"Publication no. 95-121."
1. Coasts—United States—Recreational use—Addresses, essays, lectures. 2. Recreational surveys—United States—Addresses, essays, lectures. I. United States. Congress. Senate. Committee on Energy and Natural Resources. II. Title.
GV191.4.U57 1978 333.7'8'0973 78-602543
79 MARC

—UNITED STATES—RECREATIONAL USE—MANAGEMENT—CONGRESSES

Symposium on Coastal Recreation Resources in an Urbanizing Environment, Hyannis, Mass., 1976.
Coastal recreation resources in an urbanizing environment : a monograph. — Amherst : Cooperative Extension Service, University of Massachusetts, 1977.
v, 129 p. : ill. ; 28 cm.
Cover title.
Papers presented and discussed at a symposium held Apr. 12-14, 1976, and sponsored by the Cooperative Extension Service, University of Massachusetts, in cooperation with the Massachusetts Coastal Zone Program and other bodies.
Includes bibliographical references.
$3.00
1. Coasts—United States Recreational use—Management—Congresses. 2. Coasts—Environmental aspects—United States—Congresses. 3. Coastal zone management—United States—Congresses. I. University of Massachusetts at Amherst. Cooperative Extension Service. II. Massachusetts. Coastal Zone Management Office. III. Title.
GV53.S9 1976 658'.91'333780973 79-621751
79 MARC

COASTWISE NAVIGATION

see also Coastwise shipping; Intracoastal waterways

—LAW AND LEGISLATION—FRANCE

Baranger, Patrice.
Nouveau code maritime : les trois permis / Patrice Baranger ; dessins de Pierre Convard. — ₍Paris₎ : Arthaud, c1978.
155 p. : ill. (some col.) ; 20 cm. F***
ISBN 2-7003-0190-0 : 40.00F
1. Coastwise navigation—Law and legislation—France. 2. Boats and boating—Law and legislation—France. I. Title.
343'.44'0966 78-385122
78 MARC

—LAW AND LEGISLATION—UNITED STATES

United States. Congress. House. Committee on International Relations. Subcommittee on International Organizations.
Ports and waterways safety act amendments : hearing before the Subcommittees on International Organizations and on Inter-American Affairs of the Committee on International Relations, House of Representatives, Ninety-fifth Congress, second session, on H.R. 13311, July 27, 1978. — Washington : U.S. Govt. Print. Off., 1978.
iii, 75 p. ; 24 cm.
1. Harbors—United States—Regulations. 2. Coastwise navigation—Law and legislation—United States. 3. Oil pollution of rivers, harbors, etc.—Law and legislation—United States. 4. Ships—Safety regulations—United States. I. United States. Congress. House. Committee on International Relations. Subcommittee on Inter-American Affairs. II. Title.
KF27.I5494 1978b 344'.73'047 78-603173
79 MARC

—AUSTRALIA—NEW SOUTH WALES—HISTORY

Richards, Michael P
North coast run : men and ships of the N.S.W. north coast / ₍by₎ Mike Richards. — Killara, N.S.W. : Turton & Armstrong, 1977.
vii, 174 p. : ill., diagrs. on endpapers ; 15 x 24 cm. Aus
Bibliography: p. 174.
ISBN 0-908031-02-5
1. Coastwise navigation—Australia—New South Wales—History. I. Title.
HE916.R53 387.5'24'09944 79-300335
79 MARC

—FRANCE—HISTORY

Lacroix, Louis, 1877-1958.
Les Derniers voiliers caboteurs français / Louis Lacroix, ... ; préface de Jean Randier. — ₍Paris₎ : Éditions maritimes et d'outre-mer, 1978.
293 p., ₍32₎ p. of plates ; 21 cm. F79-4964
Previous ed. published in 1946 under title: Les écraseurs de crabes sur les derniers voiliers caboteurs.
65.00F
1. Sailing ships—France—History. 2. Coasters (Ships)—France—History. 3. Coastwise navigation—France—History. I. Title.
VK71.L32 1978 387.2'2'0944 79-380276
79 MARC

—GREAT BRITAIN

Duckworth, Christian Leslie Dyce.
Clyde and other coastal steamers / by Christian Leslie Dyce Duckworth and Graham Easton Langmuir. — 2nd ed. — Prescot (Prescot, Merseyside) : T. Stephenson and Sons Ltd., 1977.
xvi, 218 p., leaf of plate, ₍18₎ p. of plates : ill., facsim. ; 22 cm.
GB78-19099
Bibliography: p. 202-203.
Includes index.
ISBN 0-901314-19-6 : £5.95
1. Coasters (Ships) 2. Coastwide navigation—Great Britain. 3. Steamboat lines—Great Britain. I. Langmuir, Graham Easton, joint author. II. Title.
VM57.D8 1977 387.2'4 78-324128
79 MARC

COASTWISE SHIPPING

see also Coasters (Ships); Coastwise navigation; Intercoastal shipping

COASTWISE SHIPPING (Continued)

—LAW AND LEGISLATION—CANADA

Canada. Marine Transportation Administration.
The Coasting trade act : a position paper dealing with the policy implications of a proposed bill on the coasting trade of Canada ₍La Loi sur le cabotage : exposé de position sur les répercussions du bill proposé sur le cabotage au Canada₎ / Transport Canada, Marine. — ₍Ottawa₎ : Transport Canada, Marine, 1977.
8, ₍25₎, ₍27₎, 7 p. ; 28 cm. CG78-70939-XE
Cover title.
English and French.
"TP 938"
ISBN 0-662-01032-9
1. Coastwise shipping—Law and legislation—Canada. I. Title. II. Title: La Loi sur le cabotage.
KE2440.A72M37 354'.71'00877544 79-305866
79 MARC

—CHINA—RATES

China (People's Republic of China, 1949-). Chiao t'ung pu.
(Shui yün yün chia hui pien)
水运运价彙編. — 合訂本. — ₍香港? : s. n., 1978?₎
5 v. in 1 ; 21 cm.
At head of title: Chung-hua jen min kung ho kuo chiao t'ung pu.
Reprint of the 1956 ed. published by Jen min chiao t'ung ch'u pan she. Peking.
CONTENTS: 1編 水运货物运价計算規則. 水运货物运价分級表.—2編 海港費收規則. 長江港口裝卸費收办法.—3編 货物品名檢查表.—4編 运价里程表.—5編 沿海货运运价率表. 長江幹綫货运运价率表. 珠江水系货运运价率表. 黑龍江, 松花江, 嫩江, 烏苏里江幹支流货运运价率表.
1. Inland water transportation—China—Rates. 2. Coastwise shipping—China—Rates. I. Title.
HE698.C48 1978 79-835572

—ENGLAND—HISTORY

Merry, Ian D
The Westcotts and their times / by Ian D. Merry. — London : National Maritime Museum, c1977.
x, 147 p. : ill. ; 30 cm. — (Maritime monographs and reports ; no. 25, 1977 ISSN 0307-8590) GB***
ISBN 0-905555-02-3 : £4.00
1. Westcott family. 2. Coastwise shipping—England—History. 3. Plymouth, Eng.—Commerce—History. I. Title. II. Series.
HE826.M47 387'.09423'58 79-306052
79 MARC

—INDIA—STATISTICS—COLLECTED WORKS

India (Republic). Dept. of Commercial Intelligence and Statistics.
Statistics of the inland coasting trade consignments of India.
Calcutta, Government of India, Dept. of Commercial Intelligence & Statistics.
v. 30 cm.
1. India—Commerce—Collected works. 2. Coastwise shipping—India—Statistics—Collected works. I. Title.
HF3781.I 47a 380.1'0954 78-914375
MARC-S

—INDONESIA

Indonesia. Direktorat Djenderal Perhubungan Laut.
Instruksi Direktur Djenderal Perhubungan Laut no. th. 5/1/1 tanggal 11 April 1970, Surat keputusan Direktur Djenderal Perhubungan Laut no. KAB. 4/3/4 tanggal 11 April 1970 tentang struktur organisasi dan uraian tugas Direktorat Djenderal Perhubungan Laut. — ₍s.l. : s.n., 1970₎
91 leaves ; 33 cm.
Cover title.
1. Indonesia. Direktorat Djenderal Perhubungan Laut. 2. Merchant marine—Indonesia. 3. Coastwise shipping—Indonesia.
HE887.I47 1970 77-941162
79 MARC

Indonesia. Direktorat Jenderal Perhubungan Laut.
Proyek Armada Perintis / Direktorat Jenderal Perhubungan Laut. — ₍Jakarta₎ : Direktorat Jenderal, ₍1974 or 1975₎
32 leaves ; 33 cm.
1. Proyek Armada Perintis. 2. Coastwise shipping—Indonesia. 3. Merchant marine—Indonesia.
HE887.I47 1974b 76-940159
79 MARC

—INDONESIA—COLLECTED WORKS

Indonesia. Direktorat Jenderal Perhubungan Laut.
Hasil rapat kerja terbatas Direktorat Jenderal Perhubungan Laut. 19 -75.
₍Jakarta₎ Departemen Perhubungan, Direkorat Jenderal Perhubungan Laut.
v. 33 cm.
Continued by: Indonesia. Direktorat Jenderal Perhubungan Laut. Hasil raker operational position check Direktorat Jenderal Perhubungan Laut.
Supplements accompany some vols.
1. Indonesia. Direktorat Jenderal Perhubungan Laut—Collected works. 2. Coastwise shipping—Indonesia—Collected works. 3. Merchant marine—Indonesia—Collected works. I. Title.
HE887.I 47b 78-644885
MARC-S

—INDONESIA—CONGRESSES

Musjawarah Nasional Maritim, 1st, Djakarta, Indonesia, 1963.
Persepakatan Musjawarah Nasional Maritim I, 1963, 23 s/d 28 September 1963 di Djakarta. — ₍Djakarta₎ : Badan Musjawarah Bahari, 1964.
96 p. : ill. ; 22 cm.
1. Coastwise shipping—Indonesia—Congresses. 2. Merchant marine—Indonesia—Congresses.
HE887.M87 1963 78-950327
79 MARC

—INDONESIA—PERIODICALS

Indonesia. Direktorat Jenderal Perhubungan Laut. Bagian Umum.
Himpunan surat-surat keputusan Direktur Jenderal Perhubungan Laut.
₍Jakarta₎ Bagian Umum, Sekretariat Direktorat Jenderal Perhubungan Laut ₍etc.₎
v. 33 cm. quarterly.
Continues: Indonesia. Direktorat Jenderal Perhubungan Laut. Bagian Tata Usaha. Himpunan surat² keputusan dan sambutan² Direktur Jenderal Perhubungan Laut.
1. Merchant marine—Indonesia—Periodicals. 2. Coastwise shipping—Indonesia—Periodicals. I. Title.
HE887.I 47a 75-940772
MARC-S

—INDONESIA—YEARBOOKS

Indonesia. Direktorat Jenderal Perhubungan Laut.
Laporan tahunan Direktorat Jenderal Perhubungan Laut. 1971–
₍Jakarta, Direktorat Jenderal Perhubungan Laut₎
v.
Reports for issued in 2 or more vols.
Continues: Indonesia. Direktorat Djenderal Perhubungan Laut. Laporan tahunan.
1. Indonesia. Direktorat Jenderal Perhubungan Laut—Yearbooks. 2. Coastwise shipping—Indonesia—Yearbooks. 3. Merchant marine—Indonesia—Yearbooks.
HE887.I 47c 79-640794
MARC-S

Indonesian Shipowners Association. Unsur Nusantara.
Laporan tahunan Unsur Nusantara D. P. P.-I. N. S. A.
₍Jakarta, Unsur Nusantara D. P. P.-I. N. S. A.₎
v. 33 cm.
Reports for issued on 2 or more vols.
English or Indonesian.
1. Indonesian Shipowners Association. Unsur Nusantara—Yearbooks. 2. Coastwise shipping—Indonesia—Yearbooks. 3. Merchant marine—Indonesia—Yearbooks.
HE564.F3 I 55 387.5'09598 78-942327
MARC-S

—NEW ZEALAND—HISTORY

Furniss, Clifford A
Servants of the north : adventures on the coastal trade with the Northern Steam Ship Company / Cliff Furniss. — Wellington : Reed, 1977.
xv, 201 p. : ill., facsims., map, ports. ; 26 cm. NZ77-12
Includes index.
ISBN 0-589-01031-X : $12.95
1. Northern Steam Ship Company—History. 2. Coastwise shipping—New Zealand—History. I. Title.
NE945.N83F87 387.5'24'065931 79-300333
79 MARC

—NORTH AMERICA—HISTORY

Burgess, Robert H
Coasting schooner, the four-masted Albert F. Paul / Robert H. Burgess. — Charlottesville : Published for the Mariners Museum, Newport News, Va., by the University Press of Virginia, 1978.
xxii, 269 p. : ill. ; 24 cm. — (Museum publication ; no. 35)
Includes index.
ISBN 0-917376-31-5
1. Albert F. Paul (Schooner) 2. Coastwise shipping—North America—History. I. Title. II. Series: Mariners' Museum, Newport News, Va. Museum publication ; no. 35.
VM395.A38B87 387.2'2 77-10554
77 MARC

—SCOTLAND

Donaldson, Gordon.
Northwards by sea / Gordon Donaldson. — 2nd ed. — Edinburgh : P. Harris, 1978.
162 p. : ill. ; 23 cm. GB78-08925
Includes bibliographical references.
ISBN 0-904505-36-7 : £6.00
1. Coastwise shipping—Scotland. 2. Steamboat lines—Scotland. I. Title.
HE828.D6 1978 387.5'24'09411 78-323382
79 MARC

—SPAIN—STATISTICS—PERIODICALS

Tráfico de cabotaje.
₍Madrid, MOP, Secretaría General Técnica, Servicio de Publicaciones₎
v. 30 cm. annual.
"Origen y destino de mercancías."
Vols. for issued by Dirección General de Puertos y Señales Marítimas.
1. Coastwise shipping—Spain—Statistics—Periodicals. I. Spain. Dirección General de Puertos y Señales Marítimas.
HE859.T7 387.5'24'0946 79-641862
MARC-S

—TEXAS

Research and Planning Consultants.
A study to assess the feasibility of inland canals as an alternative to bay and river margin port and industrial development : Texas Coastal Management Program, General Land Office of Texas / prepared by Research and Planning Consultants. — ₍Austin₎ : The Program, ₍1977₎
441 p. in various pagings : ill. ; 28 cm.
Title on cover: Inland canals, an alternative for industry.
Includes bibliographies.
1. Intracoastal waterways—Texas. 2. Intracoastal waterways—Environmental aspects—Texas. 3. Coastwise shipping—Texas. I. Texas. Coastal Management Program. II. Title. III. Title: Inland canals, an alternative for industry.
TC624.T4R47 1977 386'.46 78-621245
79 MARC

—UNITED STATES

Morris, Paul C
American sailing coasters of the North Atlantic / by Paul C. Morris ; with a foreword by Edouard A. Stackpole. — New York : Bonanza Books, 1979, c1973.
xv, 224 p., ₍1₎ fold. leaf of plates : ill. ; 28 cm.
Bibliography: p. 205-207.
Includes index.
ISBN 0-517-26190-1
1. Schooners—United States. 2. Coasters (Ships)—United States. 3. Coastwise shipping—United States. I. Title.
VM311.F7M67 1979 623.82'2 79-333
79 MARC

—UNITED STATES—HISTORY—ADDRESSES, ESSAYS, LECTURES

Morgan, Charles S
Coastal shipping under sail, 1880-1920 / Charles S. Morgan. — Concord, Mass. : Morgan, c1979.
22 p. : ill. ; 23 cm.
"Thirty-seventh annual Newcomen lecture presented at the United States Coast Guard Academy, New London, Connecticut, November 3, 1978."
Fictitious imprint on t.p.: New York : Newcomen Society in North America.
Bibliography: p. 22.
1. Coastwise shipping—United States—History—Addresses, essays, lectures. 2. Sailing ships—United States—History—Addresses, essays, lectures. I. Newcomen Society in North America. II. Title.
HE751.M67 387.5'24'0973 79-101727
79 MARC

COASTWISE SHIPPING (Continued)

—WALES—HISTORY

The Unity of Barmouth / ₍edited and compiled by₎ Lewis Lloyd. — ₍Caernarfon₎ : Gwynedd Archives Service, 1977.
236 p. : ill., chart, facsims., maps ; 23 cm. GB78-10844
Includes the text of the Disbursement book of the Unity, compiled by Capt. R. Edwards.
Includes bibliographical references.
ISBN 0-901337-21-8 : £3.00
1. Coastwise shipping—Wales—History. 2. Unity (Ship)—History. 3. Barmouth, Wales—Commerce—History. I. Lloyd, Lewis William, 1939- . II. Edwards, Rees, 1757-1837. III. Unity (Ship)
HE826.U54 387.5'44 79-305268
79 MARC

COAT COLOR OF ANIMALS see Color of animals

COATED PAPER see Paper coatings

COATES, WELLS, 1895-1958

Cantacuzino, Sherban.
Wells Coates, a monograph / Sherban Cantacuzino. — London : G. Fraser, 1978.
119 p. : ill. ; 31 cm. GB***
Includes bibliographical references and index.
ISBN 0-900406-59-3 : £12.50
1. Coates, Wells, 1895-1958. I. Title.
NA997.C59C36 720'.91'4 79-307584
79 MARC

COATING

see also Metal coating

COATING, ELECTROPHORETIC see Electrophoretic deposition

COATING, METAL see Metal coating

COATING, PLASTIC see Plastic coating

COATING COMPOSITIONS see Protective coatings

COATING PROCESSES

see also Coatings; Enamel and enaveling; Laminated materials; Sealing (Technology); Surface preparation; Waterproofing

COATINGS

see also Enamel and enameling; Lacquer and lacquering; Paint; Plastic coatings; Protective coatings; Thin films; Varnish and Varnishing

Brasholz, Anton, 1920-
Handbuch der Anstrich- und Beschichtungstechnik : Untergründe, Applikationsmethoden, Beschichtungsaufbau u. Eigenschaften von Beschichtungsstoffen / Anton Brasholz. — 1. Aufl. — Wiesbaden ; Berlin : Bauverlag, 1978.
435 p. : ill. (some col.) ; 22 cm. GFR78-A
Includes bibliographical references and index.
ISBN 3-7625-0964-6
1. Paint. 2. Coatings. I. Title.
TP935.B83 667'.9 78-378348
78 MARC

Svarika, Anatoliĭ Anatol'evich.
(Pokrytiia liteĭnykh form)
Покрытия литейных форм / А. А. Сварика. — Москва : Машиностроение, 1977.
215 p. : ill. ; 22 cm. USSR***
Bibliography : p. 208–₍214₎
0.96rub
1. Molding (Founding) 2. Coatings. I. Title.
TS243.S78 78-421464

—DICTIONARIES

Federation of Societies for Coatings Technology. Definitions Committee.
Paint/coatings dictionary / compiled by Definitions Committee of the Federation of Societies for Coatings Technology. — Philadelphia : The Federation, c1978.
xviii, 613 p. ; 24 cm.
Bibliography: p. 569-610.
1. Paint—Dictionaries. 2. Coatings—Dictionaries. I. Title.
TP934.3.F42 1978 667'.9'03 78-72941
79 MARC

COATINGS, DIFFUSION see Diffusion coatings

COATINGS, PROTECTIVE see Protective coatings

COATS IN LITERATURE

Pinto, Annemarie.
Das Mantelmotiv in Kellers Kleider machen Leute und Gogols Der Mantel / Annemarie Pinto. — Bern ; Las Vegas : P. Lang, c1978.
72 p. ; 21 cm. — (Europäische Hochschulschriften : Reihe 18, Vergleichende Literaturwissenschaft ; Bd. 13) Sw***
Bibliography: p. 71-72.
ISBN 3-261-02987-0
1. Keller, Gottfried, 1819-1890. Kleider machen heute. 2. Gogol', Nikolaĭ Vasil'evich, 1809-1852. Shivel'. 3. Coats in literature. I. Title. II. Series.
PT2374.K33P5 809.3'83 78-382656
79 MARC

COATS OF ARMS see Devices; Heraldry

COATSWORTH, ELIZABETH JANE, 1893-

—BIOGRAPHY

Coatsworth, Elizabeth Jane, 1893-
Personal geography: almost an autobiography / by Elizabeth Coatsworth. — Boston : G. K. Hall, 1979, c1976.
xv, 287 p. ; 24 cm.
Large print ed.
ISBN 0-8161-6672-2
1. Coatsworth, Elizabeth Jane, 1893- —Biography. 2. Authors, American—20th century—Biography. 3. Large type books. I. Title.
₍PS3505.O136Z52 1979₎ 811'.5'2 78-27176
78 MARC

COBALT

—ISOTOPES—THERAPEUTIC USE

Lajnat al-Ṭāqah al-Dharrīyah.
(Mashrūʻ waḥdat al-kūbālt)
مشروع وحدة الكوبالت / الجمهورية العراقية ، وزارة الصناعة ، لجنة الطاقة الذرية . — ₍بغداد₎ : اللجنة ، ₍1961₎
22 p. : ill. ; 18 cm.
1. Cobalt—Isotopes—Therapeutic use. 2. Radiotherapy. I. Title.
RM858.L35 1961 78-971093

—STATISTICS

Bundesanstalt für Geowissenschaften und Rohstoffe.
Cobalt : a research report compiled in the German Federal Republic / by the Federal Institute for Earth Sciences and Raw Materials, Hanover, and the German Institute for Economic Research, Berlin ; authors and researchers, Manfred Kruszona ... ₍et al.₎. — English language ed. / edited and annotated by Robert J. M. Wyllie. — San Francisco : Miller Freeman Publications, c1979.
240, 8, 2 p., ₍6₎ fold. leaves of plates : ill. ; 30 cm. — (A World mining report)
An updated version of the German work Kobalt published in 1978 as vol. XI of Untersuchungen über Angebot und Nachfrage mineralischer Rohstoffe.
Summaries in French, German, Russian, and Spanish.
ISBN 0-87930-120-1
1. Cobalt industry. 2. Cobalt—Statistics. I. Kruszona, Manfred. II. Wyllie, Robert J. M. III. Berlin. Deutsches Institut für Wirtschaftsforschung. IV. Title. V. Series: World mining report.
HD9539.C462B83 1979 338.2'7'48 79-87840
79 MARC

COBALT, ONT.

—FIRE, 1977

Curry, Don, 1947-
Fire : Cobalt, May 23, 1977 / by Don Curry. — Cobalt, Ont. : Highway Book Shop, c1978.
102 p. : ill. ; 21 cm. C***
ISBN 0-88954-173-6
1. Cobalt, Ont.—Fire, 1977. I. Title.
TH9449.C62C87 971.3'144 79-304713
79 MARC

COBALT ALLOYS

see also Tungsten carbide-cobalt alloys

COBALT INDUSTRY

Bundesanstalt für Geowissenschaften und Rohstoffe.
Cobalt : a research report compiled in the German Federal Republic / by the Federal Institute for Earth Sciences and Raw Materials, Hanover, and the German Institute for Economic Research, Berlin ; authors and researchers, Manfred Kruszona ... ₍et al.₎. — English language ed. / edited and annotated by Robert J. M. Wyllie. — San Francisco : Miller Freeman Publications, c1979.
240, 8, 2 p., ₍6₎ fold. leaves of plates : ill. ; 30 cm. — (A World mining report)
An updated version of the German work Kobalt published in 1978 as vol. XI of Untersuchungen über Angebot und Nachfrage mineralischer Rohstoffe.
Summaries in French, German, Russian, and Spanish.
ISBN 0-87930-120-1
1. Cobalt industry. 2. Cobalt—Statistics. I. Kruszona, Manfred. II. Wyllie, Robert J. M. III. Berlin. Deutsches Institut für Wirtschaftsforschung. IV. Title. V. Series: World mining report.
HD9539.C462B83 1979 338.2'7'48 79-87840
79 MARC

COBALT OXIDES

Mrowec, Stanisław.
Kinetyka utleniania kobaltu i dyfuzja własna w tlenku kobaltawym = Kinetika okisleniia kobal'ta i samodiffuziia v okisi kobal'ta = Kinetics of oxidation of cobalt and self-diffusion in cobaltous oxide / Stanisław Mrowec, Kazimierz Przybylski. — Wrocław ; Kraków : Zakład Narodowy im. Ossolińskich, 1978 (Krak. : UJ)
76, ₍4₎ p. : ill. ; 24 cm. — (Polska Akademia Nauk, Oddział w Krakowie. Prace Komisji Ceramicznej : Ceramika ; 27) P78-9437
Summaries in English and Russian.
Bibliography: p. 74-76.
zł18.00
1. Cobalt oxides. 2. Oxidation. I. Przybylski, Kazimierz, joint author. II. Title. III. Title: Kinetika okisleniia kobal'ta i samodiffuziia v okisi kobal'ta. IV. Title: Kinetics of oxidation of cobalt and self-diffusion in cobaltous oxide. V. Series: Polska Akademia Nauk. Komisja Ceramiczna. Prace : Ceramika ; 27.
TP785.P62 no. 27 79-360908
₍QD181.C6₎ 79 MARC

COBALT-TUNGSTEN CARBIDE ALLOYS see Tungsten carbide-cobalt alloys

COBB, LORIE LYNN, 1974-

Cobb, Mary Ann.
Lorie : a story of hope / Mary Ann Cobb. — Nashville : T. Nelson, c1979.
181 p. ; 21 cm.
Bibliography: p. 179-181.
ISBN 0-8407-5677-1
1. Down's syndrome—Biography. 2. Christian life—1960- 3. Cobb, Lorie Lynn, 1974-
RJ506.D68C6 362.7'8'30926 78-20951
78 MARC

COBB FAMILY

A History of the Cobb-Carr family and letters of the Cobb family during Civil War days / compiled by Miriam Cobb Wallace & Blanche Haskell Matlock. — ₍Grand Junction, Colo.₎ : Matlock, c1978.
227 p. : ill. ; 29 cm.
On spine: Cobb-Carr history.
1. Cobb family. I. Wallace, Miriam Cobb. II. Matlock, Blanche Haskell, 1918- III. Title: Cobb-Carr history.
CS71.C653 1978 929'.2'0973 78-113737
79 MARC

COBBLESTONE BUILDINGS

—NEW YORK (STATE)

Shelgren, Olaf William.
Cobblestone landmarks of New York State / Olaf William Shelgren, Jr., Cary Lattin, Robert W. Frasch ; photos. by Gerda Peterich. — 1st ed. — Syracuse : Syracuse University Press, 1978.
x, 163 p. : ill. ; 21 cm. — (A York State Book)
ISBN 0-8156-2201-5. ISBN 0-8156-0149-2 pbk.
1. Cobblestone buildings—New York (State) 2. Vernacular architecture—New York (State) I. Lattin, Cary, joint author. II. Frasch, Robert W., joint author. III. Peterich, Gerda. IV. Title.
NA730.N4S53 721'.0441 78-13994
78 MARC

COBBLESTONES

see also Cobblestone buildings

COBBOLD, RICHARD, 1797-1877

The Biography of a Victorian village : Richard Cobbold's account of Wortham Suffolk, 1860 / edited and introduced by Ronald Fletcher. — London : Batsford, 1977.
168 p. : ill. ; 26 cm. GB***
Includes index.
ISBN 0-7134-0787-5 : £5.50
1. Wortham, Eng.—History. 2. Wortham, Eng.—Buildings. 3. Villages—England—History. 4. Cobbold, Richard, 1797-1877. I. Cobbold, Richard, 1797-1877. II. Fletcher, Ronald.
DA690.W926B56 942.6'45 78-321040
79 MARC

COBDEN-SANDERSON, THOMAS JAMES, 1840-1922

Cobden-Sanderson, Thomas James, 1840-1922.
Four lectures / by T. J. Cobden-Sanderson ; edited, with an introductory essay on Cobden-Sanderson's life and ideals, with details of his American pupils and his lectures in the United States in 1907, by John Dreyfus. — San Francisco : Book Club of California, 1974.
xi, 105 p. : ill., port. ; 29 cm. — (Publication - Book Club of California ; no. 147)
"Four hundred and fifty copies."
Includes bibliographical references.
CONTENTS: Bookbinding (1888)—The social problem is in the main an artistic problem (1888)—Address on taking the chair at the Bookbinders' Pension & Asylum Society (1889)—Introduction to a lecture by William Morris (1895)
1. Bookbinding—Addresses, essays, lectures. 2. Cobden-Sanderson, Thomas James, 1840-1922. I. Dreyfus, John. II. Series: Book Club of California, San Francisco. Publication no. 147.
Z266.C53 686.3 75-324000
——— Copy 3. Rosenwald Coll. MARC
75[r79]rev

COBELLI, GIANCARLO

Groppali, Enrico, 1948-
Il teatro di Trionfo, Missiroli, Cobelli : la disperazione travestita / Enrico Groppali. — 1. ed. — Venezia : Marsilio, 1977.
242 p., [8] leaves of plates : ill. ; 22 cm. — (Saggi ; n. 51) It77-Oc
Includes bibliographical references.
L6000
1. Theater—Italy—History. 2. Trionfo, Aldo. 3. Missiroli, Mario, 1934- 4. Cobelli, Giancarlo. I. Title.
PN2684.G76 792'.0945 79-359352
*79 MARC

COBHAM, ALAN JOHN, SIR, 1894-1973

Cobham, Alan John, Sir, 1894-1973.
A time to fly / by Alan J. Cobham ; edited by Christopher Derrick. ; pref. by the Duke of Edinburgh. — London : Shepheard-Walwyn, 1978.
x, 214 p., [13] leaves of plates : ill. ; 24 cm. GB***
ISBN 0-85683-037-2 : £6.50
1. Cobham, Alan John, Sir, 1894-1973. 2. Air pilots—Great Britain—Biography. I. Derrick, Christopher. II. Title.
TL540.C6A32 629.13'092'4 79-302936
79 MARC

COBHAM FAMILY

Scoufos, Alice-Lyle, 1921-
Shakespeare's typological satire : a study of the Falstaff-Oldcastle problem / Alice-Lyle Scoufos. — Athens : Ohio University Press, c1979.
xvii, 378 p. : ill. ; 24 cm.
Includes bibliographical references and index.
ISBN 0-8214-0390-7
1. Shakespeare, William, 1564-1616—Characters—Falstaff. 2. Shakespeare, William, 1564-1616—Histories. 3. Shakespeare, William, 1564-1616—Knowledge—History. 4. Shakespeare, William, 1564-1616—Humor, satire, etc. 5. Cobham family. 6. Typology (Theology) in literature. I. Title.
PR2993.F2S3 822.3'3 77-92256
78 MARC

COBITIS DELICATA

Niwa, Hisashi.
(Ajime)
あじめ：アジメドジョウの総合的研究 = Adime : Comprehensive studies on Niwaella delicata (Niwa), a Japanese species of Cobitidae / 丹羽弥著. — 岐阜：大衆書房, 昭和51 [1976]
226 p., 22 leaves of plates : ill. ; 22 cm. Ja 77-24302
Includes bibliographical references and index.
Appendices (p. 157-211): Niwa, Hisashi. 1937, a new species of
1. Cobitis delicata. I. Title. II. Title: Adime.
QL638.C647N58 78-813032

COBOL (COMPUTER PROGRAM LANGUAGE)

Bonnin, Christian.
Le COBOL A.N.S. : avec exercices et corrigés / par Christian Bonnin. — 5. éd. mise à jour suivant la norme COBOL A.N.S. 74. — Paris : Eyrolles, 1978.
xi, 186 p. : ill. ; 23 cm. — (Pratique de l'informatique) F***
Includes index.
81.00F
1. COBOL (Computer program language) I. Title.
QA76.73.C25B66 1978 001.6'424 78-400572
79 MARC

Bonnin, Christian.
Les extensions au COBOL A.N.S. : avec exercices et corrigés / par Christian Bonnin. — 3. éd. mise à jour et augm. — Paris : Eyrolles, 1979.
vii, 203 p. ; 22 cm. — (Pratique de l'informatique) F***
Includes index.
91.00F
1. COBOL (Computer program language) I. Title.
QA76.73.C25B67 1979 001.6'424 79-351949
79 MARC

Carver, D K
Introduction to business data processing : with Basic, Fortran, and Cobol programming / D. K. Carver. — 2d ed. — New York : Wiley, c1979.
x, 366 p. : ill. ; 24 cm.
Includes index.
ISBN 0-471-03091-0
1. Electronic data processing. 2. Business—Data processing. 3. Basic (Computer program language) 4. FORTRAN (Computer program language) 5. COBOL (Computer program language) I. Title.
QA76.C3619 1979 001.6'4 78-19131
78 MARC

Coddington, L
Quick COBOL / [by] L. Coddington. — 2nd ed. — London : Macdonald and Jane's, 1978.
xvii, 257 p. : ill., form ; 23 cm. — (Computer monographs ; 16) GB78-39995
Includes index.
ISBN 0-354-04266-1 : £6.50
1. COBOL (Computer program language) I. Title.
QA76.73.C25C63 1978 001.6'424 79-305547
79 MARC

Davis, William S 1943-
COBOL : introduction to structured logic and modular program design / William S. Davis, Richard H. Fisher. — Reading, Mass. : Addison-Wesley Pub. Co., c1979.
xix, 552 p. : ill. ; 24 cm.
Includes index.
ISBN 0-201-01431-9 : $11.59
1. Structured programming. 2. COBOL (Computer program language) I. Fisher, Richard H., joint author. II. Title.
QA76.6.D377 001.6'424 78-68773
79 MARC

Dock, V Thomas.
Structured COBOL, American National Standard / V. Thomas Dock. — St. Paul : West Pub. Co., c1979.
xv, 269 p. : ill. ; 25 cm. — (West series in data processing and information systems)
Includes index.
ISBN 0-8299-0181-7 : $9.95
1. COBOL (Computer program language) 2. Structured programming. I. Title. II. Series.
QA76.73.C25D623 001.6'424 78-13585
78 MARC

Gleason, Gary M
Structured COBOL / by Gary M. Gleason, Lister W. Horn. — San Francisco : Boyd & Fraser Pub. Co., c1979.
xv, 310 p. : ill. ; 24 cm. — (Computer science series)
Includes index.
ISBN 0-87835-079-9 : $9.95
1. COBOL (Computer program language) 2. Structured programming. I. Horn, Lister Wayne, joint author. II. Title.
QA76.73.C25G58 001.6'424 78-31566
78 MARC

Grauer, Robert T 1945-
The COBOL environment / Robert T. Grauer, Marshal A. Crawford. — Englewood Cliffs, N.J. : Prentice-Hall, c1979.
xv, 464 p. : ill. ; 25 cm. — (Prentice-Hall software series)
Includes index.
ISBN 0-13-139394-4 : $19.95
1. COBOL (Computer program language) I. Crawford, Marshal A., 1943- joint author. II. Title. III. Series.
QA76.73.C25G73 001.6'424 78-25631
78 MARC

ICIS-COBOL compiler operating instructions for ISIS-II users. — Santa Clara, Calif. : Intel Corp., c1979.
69 p. in various pagings ; 29 cm.
1. ISIS (Computer system) 2. COBOL (Computer program language) 3. Compiling (Electronic computers) I. Intel Corporation.
QA76.8.I3I34 001.6'425 79-109287
79 MARC

iCIS-COBOL language reference manual. — Santa Clara, Calif. : Intel Corp., c1979.
ca. 200 p. in various pagings ; 28 cm.
"Vendor supplied product."
1. INTEL 8080 (Computer)—Programming. 2. INTEL 8085 (Computer)—Programming. 3. COBOL (Computer program language) I. Intel Corporation.
QA76.8.I28I18 001.6'424 79-112523
79 MARC

Koutchouk, Michel.
Construction logique de programmes COBOL / par Michel Koutchouk. — Paris ; New York : Masson, 1979.
x, 154 p. ; 24 cm. F***
Includes index.
ISBN 2-225-62799-1 : 58.00F
1. COBOL (Computer program language) I. Title.
QA76.73.C25K67 001.6'424 79-375740
79 MARC

Learning to program in structured COBOL : parts 1 and 2 / Edward Yourdon ... [et al.]. — Englewood Cliffs, N.J. : Prentice-Hall, c1979.
xiv, 478 p. : ill. ; 24 cm. — (Prentice-Hall software series)
"A Yourdon book."
Includes index.
ISBN 0-13-527713-2 : $15.95
1. COBOL (Computer program language) 2. Structured programming. I. Yourdon, Edward. II. Series.
QA76.73.C25L4 1979 001.6'424 79-12427
79 MARC

Lister, Timothy R
Learning to program in structured COBOL / by Timothy R. Lister and Edward Yourdon. — New York : Yourdon Press, c19-
v. : ill. ; 23 cm.
Includes index.
ISBN 0-917072-03-0
1. COBOL (Computer program language) 2. Structured programming. I. Yourdon, Edward, joint author. II. Title.
QA76.73.C25L57 001.6'424 77-99232
79 MARC

McClure, Carma L
Reducing COBOL complexity through structured programming / Carma L. McClure. — New York : Van Nostrand Reinhold Co., c1978.
x, 192 p. : ill. ; 24 cm.
bibliography: p. 185-187.
Includes index.
ISBN 0-442-80446-0
1. COBOL (Computer program language) 2. Structured programming. I. Title.
QA76.73.C25M315 1978 78-2838
001.6'424
78 MARC

Mullish, Henry.
A business-like approach to COBOL / Henry Mullish. — New York : Crowell, c1979.
xviii, 346 p. : ill. ; 24 cm.
Includes index.
ISBN 0-7002-2506-4
1. COBOL (Computer program language) I. Title.
QA76.73.C25M83 001.6'424 78-16190
78 MARC

Parkin, Andrew.
COBOL for students / [by] Andrew Parkin. — London : Edward Arnold, 1975.
x, 211 p. : ill. ; 24 cm. GB75-30804
Stamped on t.p.: Exclusive distributor, ISBS, Forest Grove, Or.
Includes index.
ISBN 0-7131-2525-X
1. COBOL (Computer program language) I. Title.
QA76.73.C25P37 001.6'424 78-320577
78 MARC

COBOL (COMPUTER PROGRAM LANGUAGE) (Continued)

Programare structurată / Ilie Văduva ... ₍et al.₎. — Bucureşti : Editura Tehnică, 1978.
246 p. : ill. ; 21 cm. — (Seria Practică) (Automatică, informatică electronică, management) R78-5274
Bibliography: p. ₍194₎-196.
lei10.00
1. Structured programming. 2. COBOL (Computer program language) 3. FORTRAN (Computer program language) I. Vaduva, Ilie.
QA76.6.P75124 79-351343
*79 MARC

Sass, C Joseph.
COBOL programming and applications / C. Joseph Sass. — Boston : Allyn and Bacon, c1979.
xiv, 431 p. : ill. ; 24 cm.
Bibliography: p. ₍426₎-427.
Includes index.
ISBN 0-205-06550-3 : $10.95
1. COBOL (Computer program language) I. Title.
QA76.73.C25S28 001.6'424 78-11362
78 MARC

Shelly, Gary B
Advanced structured COBOL : program design and file processing / by Gary B. Shelly & Thomas J. Cashman. — Fullerton, CA : Anaheim Pub. Co., c1978.
ca. 550 p. : ill. ; 28 cm.
Tablet of pseudocode specification sheets and IPO charts inserted.
Includes index.
ISBN 0-88236-112-0
1. COBOL (Computer program language) I. Cashman, Thomas J., joint author. II. Title.
QA76.73.C25S423 001.6'424 78-62480
79 MARC

Structured ANS COBOL / Mike Murach, Paul Noll ; ₍writer/editor, Judy Taylor₎. — Fresno, Calif. : Murach, c1979-
v. : ill. ; 26 cm. — (Paul Noll COBOL series)
Includes index.
CONTENTS: pt. 1. A course for novices.
1. COBOL (Computer program language) I. Murach, Mike. II. Noll, Paul. III. Taylor, Judy, 1955- IV. Series: Noll, Paul. Paul Noll COBOL series.
QA76.73.C25S76 001.6'424 78-69920
79 MARC

Wohl, Gerald.
Structured COBOL : a direct approach / Gerald Wohl. — Chicago : Science Research Associates, c1979.
viii, 382 p. : ill. ; 21 x 26 cm.
Includes index.
ISBN 0-574-21230-2 : $12.95
1. COBOL (Computer program language) I. Title.
QA76.73.C25W63 001.6'424 79-17940
79 MARC

—ADDRESSES, ESSAYS, LECTURES

Programozási nyelvek. — 2. kiad. — Budapest : ₍INFELOR₎, 1977.
177 p. ; 20 cm. — (INFELOR közlemények ; 5) H77
Summaries in English and Russian.
CONTENTS: Dettrich, Á. A PL/I nyelv elemzése a fordítóprogram kidolgozásának szempontjából.—Bakos, T. et al. A CII 10010 számítógép alkalmazása, interakaív üzemmódban, adatfeldolgozásban.
ISBN 963020682X
1. PL/I (Computer program language)—Addresses, essays, lectures. 2. Compiling (Electronic computers)—Addresses, essays, lectures. 3. CII-10010 (Computer)—Programming—Addresses, essays, lectures. 4. COBOL (Computer program language)—Addresses, essays, lectures. 5. Interactive computer systems—Addresses, essays, lectures. I. Dettrich, Árpád. A PL/I nyelv elemzése a fordítóprogram kidolgozásának szempontjából. 1977. II. Bakos, Tamás. III. A CII 10010 számítógép alkalmazása, interaktív üzemmódban, adatfeldolgozásban. 1977. IV. Series: Rendszer Technikai Vállalat. INFELOR közlemények ; 5.
QA76.73.P25P76 1977 78-397341
*79 MARC

—PROGRAMMED INSTRUCTION

Ashley, Ruth.
ANS COBOL / Ruth Ashley, in consultation with Nancy B. Stern. — 2d. ed. — New York : Wiley, c1979.
ix, 265 p. : ill. ; 26 cm. — (A Self-teaching guide)
Includes index.
ISBN 0-471-05136-5 : $5.95
1. COBAL (Computer program language)—Programmed instruction. I. Stern, Nancy B. II. Title.
₍QA76.73.C25A83 1979₎ 001.6'424 78-27717
78 MARC

—STANDARDS—NEW ZEALAND

Monin, D John.
COBOL coding conventions : a proposal / by D. John Monin. — ₍Palmerston North, N.Z.₎ : Dept. of Business Studies, School of Business, Massey University, 1976.
22 p. ; 26 cm. — (Occasional paper - Department of Business Studies, Massey University ; no. 7) NZ***
Cover title.
Includes bibliographical references.
1. COBOL (Computer program language)—Standards—New Zealand. I. Title. II. Series: Massey University. Dept. of Business Studies. Occasional paper - Department of Business Studies, Massey University ; no. 7.
QA76.73.C25M66 001.6'424 79-301859
79 MARC

COBORDISM THEORY

Snaith, Victor Percy, 1944-
Algebraic cobordism and K-theory / Victor P. Snaith. — Providence, R.I. : American Mathematical Society, 1979.
vii, 152 p. ; 25 cm. — (Memoirs of the American Mathematical Society ; no. 221)
"Volume 21 ... second of 3 numbers."
Bibliography: p. 144-149.
ISBN 0-8218-2221-7
1. Cobordism theory. 2. K-theory. 3. Homotopy. I. Title. II. Series: American Mathematical Society. Memoirs ; no. 221.
QA3.A57 no. 221 510'.8 s 79-17981
₍QA612.3₎ 79 MARC

COBOURG, ONT.

—CITY PLANNING—CONGRESSES

ICSID Interdesign Seminar, 3d, Port Hope, Ont. and Cobourg, Ont., 1974.
Design for small communities : a report of Interdesign '74 Ontario / edited by David Fulton. — Toronto : Macmillan of Canada, c1975.
141 p. : ill. ; 22 x 28 cm. C***
"A project of the Association of Canadian Industrial Designers in co-operation with the government of Ontario."
"Sanctioned by the International Council of Societies of Industrial Design, Brussels, Belgium."
ISBN 0-7705-1311-5
1. Cobourg, Ont.—City planning—Congresses. 2. Port Hope, Ont.—City planning—Congresses. 3. City planning—Canada—Congresses. 4. Architecture—Human factors—Congresses. I. Fulton, David. II. Association of Canadian Industrial Designers. III. International Council of Societies of Industrial Design. IV. Title.
HT169.C32C634 1974 309.2'62'09713 77-372966
77₍r79₎rev MARC

COBRAM, AUSTRALIA

—HISTORY

Dunlop, Alan James.
Wide horizons : the story of Yarrawonga, Tungamah, and Cobram shires / by Alan J. Dunlop. — ₍Bentleigh, Vic.₎ : A. J. Dunlop, 1978.
203 p. : ill. ; 21 cm. Aus
Includes bibliographical references and index.
ISBN 0-9596096-0-1
1. Yarrawonga, Australia—History. 2. Tungamah, Australia—History. 3. Cobram, Australia—History. I. Title.
DU230.Y3D86 994.5 79-304794
79 MARC

COBURG (CITY)

—HISTORY—SOURCES

Coburg (City)
Das älteste Coburger Stadtbuch, 1388-1453 / bearb. von Klaus Freiherr von Andrian-Werburg. — Neustadt/Aisch : Degener ₍in Komm.₎, 1977.
xvii, 442 p. : ill. ; 24 cm. — (Veröffentlichungen der Gesellschaft für Fränkische Geschichte : 10. Reihe, Quellen zur Rechts- und Wirtschaftsgeschichte Frankens ; Bd. 9) GFR77-A23
Includes bibliographical references and index.
ISBN 3-7686-4030-2 : DM78.00
1. Coburg (City)—History—Sources. I. Andrian-Werburg, Klaus, Freiherr von. II. Title. III. Series: Gesellschaft für Fränkische Geschichte. Veröffentlichungen : 10. Reihe, Quellen zur Rechts- und Wirtschaftsgeschichte Frankens ; Bd. 9.
DD901.C65C62 1977 78-380293
79 MARC

COBURN, ALVIN LANGDON, 1882-1966

Coburn, Alvin Langdon, 1882-1966.
Alvin Langdon Coburn, photographer : an autobiography with over 70 reproductions of his works / edited by Helmut and Alison Gernsheim ; with a new introd. by Helmut Gernsheim. — New York : Dover Publications, 1978.
x, 145 p. : ill. ; 28 cm.
Bibliography: p. 143-145.
Includes index.
ISBN 0-486-23685-4 : $6.00
1. Coburn, Alvin Langdon, 1882-1966. 2. Photography, Artistic. I. Gernsheim, Helmut, 1913- II. Gernsheim, Alison. III. Title.
TR140.C56A27 1978 770'.92'4 78-55345
79 MARC

COCA

see also Cocaine

Mortimer, William Golden.
Peru : history of coca, "the divine plant" of the Incas : with an introductory account of the Incas, and of the Andean Indians of to-day / by W. Golden Mortimer. — 1st AMS ed. — New York : AMS Press, 1978.
xxxi, 576 p. : ill. ; 23 cm.
Reprint of the 1901 ed. published by J. H. Vail, New York.
Bibliography: p. 519-544.
Includes index.
ISBN 0-404-11980-8
1. Coca. 2. Incas. 3. Peru—Description and travel. I. Title. II. Title: History of coca.
RS165.C5M8 1978 394.1 74-15120
78 MARC

COCA-COLA COMPANY

Shartar, Martin.
The wonderful world of Coca-Cola / by Martin Shartar and Norman Shavin ; editorial supervisor, Bruce Galphin. — 1st ed. — Atlanta : Perry Communications, c1978.
64 p. : ill. ; 29 cm.
ISBN 0-933238-01-0 : $2.95
1. Coca-Cola Company. I. Shavin, Norman, joint author. II. Galphin, Bruce. III. Title.
TP630.S45 338.7'66'362 78-71521
79 MARC

—COLLECTIBLES

Murken-Altrogge, Christa, 1944-
Werbung, Mythos, Kunst am Beispiel Coca-Cola / Christa Murken-Altrogge. — Tübingen : Wasmuth, 1977.
63, 80 p. : chiefly ill. (some col.) ; 28 cm. GFR78-A
Bibliography: p. 59-60.
Includes index.
ISBN 3-8030-3023-4 : DM84.00
1. Coca-Cola Company—Collectibles. 2. Advertising specialities—Collectors and collecting. 3. Advertising—Carbonated beverages—Collectors and collecting. I. Title.
NK1125.M884 659.1'9'66362 78-376828
78 MARC

—HISTORY

Watters, Pat.
Coca-cola : an illustrated history / Pat Watters. — 1st ed. — Garden City, N.Y. : Doubleday, 1978.
xiv, 288 p. : ill. ; 27 cm.
ISBN 0-385-13499-1 : $12.95
1. Coca-Cola Company—History. 2. Atlanta—Industries. 3. Atlanta—History. I. Title.
HD9349.C62U58 338.7'66'3620973 77-16954
78 MARC

COCAINE

Antonil.
Mama Coca / Antonil. — London : Hassle Free Press, 1978.
xvii, 295 p. : ill. ; 23 cm. GB***
Bibliography: p. 285-295.
ISBN 0-86166-001-3
1. Paez Indians—Narcotics. 2. Indians of South America—Andes—Narcotics. 3. Cocaine. I. Title.
F2270.2.P3A57 1978 364.1'57 79-313931
00 MARC

United States. Congress. House. Select Committee on Narcotics Abuse and Control.
Cocaine and marihuana trafficking in southeastern United States : hearings before the Select Committee on Narcotics Abuse and Control, House of Representatives, Ninety-fifth Congress, second session, June 9 and 10, 1978. — Washington : U.S. Govt. Print. Off., 1978.
iii, 214 p. : graphs ; 24 cm.
"SCNAC-95-2-6."
1. Narcotics, Control of—Southern States. 2. Cocaine. 3. Marihuana. I. Title.
KF27.5.N3 1978 363.4'5'0975 79-600912
79 MARC

COCAINE HABIT IN LITERATURE

Tracy, Jack, 1945-
Subcutaneously, my dear Watson : Sherlock Holmes and the cocaine habit / by Jack Tracy, with Jim Berkey ; illustrated by Paul M. McCall, with Sidney Paget's original representations of Sherlock Holmes and other turn-of-the-century illustrations. — 1st ed. — Bloomington, Ind. : J. A. Rock, 1978.
91 p. : ill. ; 23 cm.
Two hundred and twenty-two hardback copies printed. No. 50.
Bibliography: p. 89-91.
ISBN 0-918736-02-1. ISBN 0-918736-03-X pbk.
1. Doyle, Arthur Conan, Sir, 1859-1930—Characters—Sherlock Holmes. 2. Cocaine habit in literature. I. Berkey, Jim, joint author. II. Title.
PR4624.T7 823'.8 78-112021
79 MARC

COCALICO VALLEY, PA.

—GENEALOGY—PERIODICALS

Historical Society of the Cocalico Valley.
Journal of the Historical Society of the Cocalico Valley. ₍s. l., Historical Society of the Cocalico Village₎
v. ill. 23 cm.
1. Cocalico Valley, Pa.—History—Periodicals. 2. Cocalico Valley, Pa.—Genealogy—Periodicals. I. Title.
F157.C67H57a 974.8'15 79-641511
MARC-S

—HISTORY—PERIODICALS

Historical Society of the Cocalico Valley.
Journal of the Historical Society of the Cocalico Valley. ₍s. l., Historical Society of the Cocalico Village₎
v. ill. 23 cm.
1. Cocalico Valley, Pa.—History—Periodicals. 2. Cocalico Valley, Pa.—Genealogy—Periodicals. I. Title.
F157.C67H57a 974.8'15 79-641511
MARC-S

COCAMA LANGUAGE

—TEXT-BOOKS FOR FOREIGNERS—SPANISH

Faust Wille, Norma.
Gramática cocama: lecciones para el aprendizaje del idioma cocama ₍por₎ Norma Faust W. · ₍Lima₎ Instituto Linguístico de Verano, 1972.
173 p. 22 cm. (Serie Linguística peruana. Publicación, no. 6)
1. Cocama language—Text-books for foreigners—Spanish. I. Title. II. Series.
PM5823.F38 79-341306
79 MARC

COCAO TRADE

—GHANA

Milburn, Josephine F
British business and Ghanaian independence / by Josephine F. Milburn. — Hanover, N.H. : Published for the University of Rhode Island by the University Press of New England, 1977.
ix, 156 p. ; 23 cm.
Includes bibliographical references and index.
ISBN 0-87451-138-0
1. Corporations, British—Ghana. 2. Cocao trade—Ghana. 3. Ghana—Politics and government. 4. Ghana—Economic policy. I. Title.
JQ3036.M55 1977b 338.8'8 76-50681
79 MARC

COCARCINOGENESIS

see also Carcinogenesis

COCCIDAE see Scale-insects

COCCIDIA

see also Eimeria

COCCIDIOSIS

Prastowo, Bambang Probo.
Pemeriksaan pendahuluan terhadap beberapa species tertentu coccidia pada ajam dari Bogor dan Bandung. A preliminary search of certain species of chicken coccidia in Bogor and Bandung, oleh Bambang Probo Prastowo, Gatut Ashadi dan Soeprapto Soekardono. Bogor, Pertjetakan IPB ₍1967₎
8 p. 22 cm. (Institut Pertanian Bogor. Buletin penelitian no. 8, 1967)
Cover title.
In Indonesian; with summary in English.
Bibliography: p. 8.
1. Coccidiosis. 2. Chickens—Disease. 3. Eimeria. I. Ashadi, Gatut, joint author. II. Soekardono, Soeprapto, joint author. III. Title. IV. Series: Institut Pertanian Bogor. Buletin penelitian, 1967, no. 8.
S405.I66 no. 8 79-949590
₍SF995.6.C6₎ 79 MARC

COCCIDIUM see Eimeria

COCCINELLIDAE see Ladybirds

COCCOLITHOPHORIDAE

see also Coccoliths, Fossil

COCCOLITHS, FOSSIL

Stradner, Herbert.
Tertiäre Discoasteriden aus Österreich und deren stratigraphische Bedeutung, mit Hinweisen auf Mexiko, Rumänien und Italien, von H. Stradner und A. Papp. Wien, 1961.
160 p. illus. 24 cm. (Jahrbuch der Geologischen Bundesanstalt. Sonderband 7)
Summary in German and English.
Bibliography: p. 155-159.
1. Coccoliths, Fossil. 2. Paleobotany—Tertiary. 3. Paleobotany—Austria. I. Papp, Adolf, 1915- joint author. II. Title.
QE955.S78 561'.93 gs 66-104
₍r79₎rev MARC

COCEA, NICOLAE D., 1880-1949

Cocea, Nicolae D 1880-1949.
Jurnal / N. D. Cocea. — București : Editura Politică, 1970.
220 p. : ill. ; 19 cm. — (Colecția Mica bibliotecă de istorie)
At head of title: Institutul de Studii Istorice și Social-Politice de pe lîngă C.C. al P.C.R.
Includes bibliographical references.
1. Cocea, Nicolae D., 1880-1949. 2. Authors, Romanian—20th century—Biography. 3. Journalists—Romania—Biography. I. Title.
PC839.C527Z467 1970 79-342196
79 MARC

COCHABAMBA, BOLIVIA

—POLITICS AND GOVERNMENT

Serrate Paz, José Mario.
Cochabamba, sede del Gobierno Nacional : imperativo de un traslado (estudio sobre la necesidad del traslado de la sede de Gobierno a los volles ₍i.e. valles₎ de Cochabamba) / José Mario Serrate Paz. — Cochabamba : Editorial Canelas, ₍1975₎
63 p. ; 19 cm.
1. Bolivia—Capital—History. 2. Cochabamba, Bolivia—Politics and government. I. Title.
F3321.9.S47 320.9'84'2 79-342545
79 MARC

—RURAL CONDITIONS

Whitehead, Christine.
Cochabamba landowners and the agrarian reform / by Christine Whitehead. — 1970.
113 p.
Thesis (Ph. B.)—Oxford.
Bibliography: p. 109-113.
Microfiche. Oxford, Eng. : Micromedia Ltd., 1977. — 2 sheets ; 11 x 15 cm. — (Latin American studies)
ISBN 0-904735-30-3
1. Land reform—Bolivia—Cochabamba (Dept.) 2. Cochabamba, Bolivia (Dept.)—Rural conditions. I. Title. II. Series: Latin American studies (Oxford, Eng.)
Microfiche HD489.C6 79-335004

COCHABAMBA, BOLIVIA (DEPT.)

—DESCRIPTION AND TRAVEL

Guzmán, Augusto.
Geografía de Cochabamba / Augusto Guzman. — La Paz : Librería Editorial "Juventud", 1978.
103, ₍6₎ p. : ill. ; 22 cm.
"Antología geográfica de Cochabamba, por Antonio Vázquez de Espinoza" ... ₍et al.₎: p. ₍65₎-₍104₎
1. Cochabamba, Bolivia (Dept.)—Description and travel. I. Vázquez de Espinosa, Antonio, d. 1630. II. Title.
F3341.C7G9 918.4'2 79-116104
79 MARC

COCHIM, AUGUSTIN, 1876-1916

Furet, François, 1927-
Penser la Révolution française / François Furet. — ₍Paris₎ : Gallimard, 1978.
259 p. ; 23 cm. — (Bibliothèque des histoires) F79-426
Includes bibliographical references.
59.00F
1. France—History—Revolution, 1789-1799—Causes and character. 2. France—History—Revolution, 1789-1799—Historiography. 3. Cochim, Augustin, 1876-1916. 4. Tocqueville, Alexis Charles Henri Maurice Clérel de, 1805-1859. I. Title.
DC138.F8 944.04 79-382348
79 MARC

COCHIMI LANGUAGE

—GLOSSARIES, VOCABULARIES, ETC.

Mixco, Mauricio J
Cochimí and proto-Yuman : lexical and syntactic evidence for a new language family in Lower California / by Mauricio J. Mixco ; foreword by Ernest J. Burrus. — Salt Lake City : University of Utah Press, 1978.
xiv, 125 p. : ill. ; 28 cm. — (University of Utah anthropological papers ; no. 101)
Based on Miguel del Barco's text, Historia natural y crónica de la antigua California.
"Reproduction of del Barco's Lower California texts, Fondo Gesuítico, Mss. 1413/1414": p. 106-118.
Bibliography: p. 119-125.
ISBN 0-87480-150-8 : $8.00
1. Cochimi language—Grammar. 2. Cochimi language—Texts. 3. Cochimi language—Glossaries, vocabularies, etc. 4. Proto-Yuman language—Glossaries, vocabularies, etc. I. Barco, Miguel del, 1706-1790. Historia natural y crónica de la antigua California. II. Utah. University. III. Title. IV. Series: Anthropological papers (Salt Lake City) ; no. 101.
PM3688.M5 497'.5 79-621635
79 MARC

—GRAMMAR

Mixco, Mauricio J
Cochimí and proto-Yuman : lexical and syntactic evidence for a new language family in Lower California / by Mauricio J. Mixco ; foreword by Ernest J. Burrus. — Salt Lake City : University of Utah Press, 1978.
xiv, 125 p. : ill. ; 28 cm. — (University of Utah anthropological papers ; no. 101)
Based on Miguel del Barco's text, Historia natural y crónica de la antigua California.
"Reproduction of del Barco's Lower California texts, Fondo Gesuítico, Mss. 1413/1414": p. 106-118.
Bibliography: p. 119-125.
ISBN 0-87480-150-8 : $8.00
1. Cochimi language—Grammar. 2. Cochimi language—Texts. 3. Cochimi language—Glossaries, vocabularies, etc. 4. Proto-Yuman language—Glossaries, vocabularies, etc. I. Barco, Miguel del, 1706-1790. Historia natural y crónica de la antigua California. II. Utah. University. III. Title. IV. Series: Anthropological papers (Salt Lake City) ; no. 101.
PM3688.M5 497'.5 79-621635
79 MARC

COCHIMI LANGUAGE (Continued)

—TEXTS

Mixco, Mauricio J
Cochimi and proto-Yuman : lexical and syntactic evidence for a new language family in Lower California / by Mauricio J. Mixco ; foreword by Ernest J. Burrus. — Salt Lake City : University of Utah Press, 1978.
xiv, 125 p. : ill. ; 28 cm. — (University of Utah anthropological papers ; no. 101)
Based on Miguel del Barco's text, Historia natural y crónica de la antigua California.
"Reproduction of del Barco's Lower California texts, Fondo Gesuitico, Mss. 1413/1414": p. 106-118.
Bibliography: p. 119-125.
ISBN 0-87480-150-8 : $8.00
1. Cochimi language—Grammar. 2. Cochimi language—Texts. 3. Cochimi language—Glossaries, vocabularies, etc. 4. Proto-Yuman language—Glossaries, vocabularies, etc. I. Barco, Miguel del, 1706-1790. Historia natural y crónica de la antigua California. II. Utah. University. III. Title. IV. Series: Anthropological papers (Salt Lake City) ; no. 101.
PM3688.M5 497'.5 79-621635
79 MARC

COCHIN, INDIA (CITY)

Venn, T W
Cochin-Malabar, palms and pageants / by T. W. Venn. — ⌈s.l. : s.n., 1948?⌉ (Calicut : Peirce, Leslie)
249 p., ⌈9⌉ leaves of plates : ill. ; 25 cm.
1. Cochin, India (City) I. Title.
DS486.C7V46 954'.83 78-912591
79 MARC

COCHINEAL

Donkin, R A
Spanish red : an ethnogeographical study of cochineal and the Opuntia cactus / R. A. Donkin. — Philadelphia : American Philosophical Society, 1977.
84 p. : ill. ; 29 cm. — (Transactions of the American Philosophical Society ; v. 67, pt. 5 ISSN 0065-9746)
Bibliography: p. 63-77.
Includes index.
ISBN 0-87169-675-4
1. Cochineal insect. 2. Cochineal. 3. Prickly-pear. 4. Ethnozoology—Latin America. 5. Indians—Industries. I. Title. II. Series: American Philosophical Society, Philadelphia. Transactions ; v. 67, pt. 5.
SF562.C6D66 338.1'7'83 77-76426
78⌈r79⌉rev MARC

COCHINEAL INSECT

Donkin, R A
Spanish red : an ethnogeographical study of cochineal and the Opuntia cactus / R. A. Donkin. — Philadelphia : American Philosophical Society, 1977.
84 p. : ill. ; 29 cm. — (Transactions of the American Philosophical Society ; v. 67, pt. 5 ISSN 0065-9746)
Bibliography: p. 63-77.
Includes index.
ISBN 0-87169-675-4
1. Cochineal insect. 2. Cochineal. 3. Prickly-pear. 4. Ethnozoology—Latin America. 5. Indians—Industries. I. Title. II. Series: American Philosophical Society, Philadelphia. Transactions ; v. 67, pt. 5.
SF562.C6D66 338.1'7'83 77-76426
78⌈r79⌉rev MARC

COCHISE CO., ARIZ.

—ROAD MAPS

Arizona. Highway Division. Photogrammetry and Mapping Services.
Atlas of Cochise County / prepared by the Arizona Department of Transportation, Highway Divison, Photogrammetry and Mapping Serv. in cooperation with the U.S. Department of Transportation, Federal Highway Administration. — ⌈Phoenix⌉ : Photogrammetry and Mapping Services : ⌈available for public use at nominal cost, Highways Division⌉, c1977.
v., 36 leaves : col. maps ; 47 x 63 cm.
Cover title.
Running title: General highway map Cochise County, Arizona.
Spine title: Cochise County.
Scale of maps ca. 1:130,000 and ca. 1:24,000.
Edition for 1966 published by the Photogrammetry and Mapping Division of the Arizona Highway Dept.
Includes indexes.
1. Cochise Co., Ariz.—Road maps. I. Arizona. Highway Dept. Photogrammetry and Mapping Division. Atlas of Cochise County. II. Title.
G1513.C5P2A7 1977 912'.791'53 79-318454
79 MARC

COCHLEA

—SURGERY

see also Cochlear implants

COCHLEAR AUDIOMETRY see Electrocochleography

COCHLEAR IMPLANTS

Ballantyne, John Chalmers.
Electrical auditory stimulation in the management of profound hearing loss / by John C. Ballantyne, Edward F. Evans and Andrew W. Morrison. — Ashford, kent : Headley, 1978.
⌈4⌉, 117 p. ; 24 cm. — (Journal of laryngology and otology : Supplement ; no. 1) GB79-04402
"Report to the Department of Health and Social Security on visits in October, 1977 to centres in the U.S.A. involved in cochlear implant prostheses."
Bibliography: p. 115-117.
£3.00
1. Cochlear implants. 2. Cochlear implants—Research—United States. I. Evans, Edward Frank, joint author. II. Morrison, Andrew W., joint author. III. Great Britain. Dept. of Health and Social Security. IV. Title. V. Series.
RF305.B34 617.8'8 79-317185
79 MARC

—RESEARCH—UNITED STATES

Ballantyne, John Chalmers.
Electrical auditory stimulation in the management of profound hearing loss / by John C. Ballantyne, Edward F. Evans and Andrew W. Morrison. — Ashford, kent : Headley, 1978.
⌈4⌉, 117 p. ; 24 cm. — (Journal of laryngology and otology : Supplement ; no. 1) GB79-04402
"Report to the Department of Health and Social Security on visits in October, 1977 to centres in the U.S.A. involved in cochlear implant prostheses."
Bibliography: p. 115-117.
£3.00
1. Cochlear implants. 2. Cochlear implants—Research—United States. I. Evans, Edward Frank, joint author. II. Morrison, Andrew W., joint author. III. Great Britain. Dept. of Health and Social Security. IV. Title. V. Series.
RF305.B34 617.8'8 79-317185
79 MARC

COCHLEAR STIMULATORS see Cochlear implants

COCHRANE, ALTA.

—BIOGRAPHY

Cochrane and Area Historical Society (Alta.)
Big hill country : Cochrane and area. — Cochrane, Alta. : Cochrane and Area Historical Society, 1977, 1978 printing.
804 p. : ill. ; 29 cm. C***
Includes index.
ISBN 0-919213-18-9
1. Cochrane, Alta.—History. 2. Cochrane, Alta.—Biography. I. Title.
F1079.5.C56C6 1977 971.3'142'03 79-304647
79 MARC

—HISTORY

Cochrane and Area Historical Society (Alta.)
Big hill country : Cochrane and area. — Cochrane, Alta. : Cochrane and Area Historical Society, 1977, 1978 printing.
804 p. : ill. ; 29 cm. C***
Includes index.
ISBN 0-919213-18-9
1. Cochrane, Alta.—History. 2. Cochrane, Alta.—Biography. I. Title.
F1079.5.C56C6 1977 971.3'142'03 79-304647
79 MARC

COCHRANE, ELIZABETH, 1867-1922

—BIOGRAPHY—JUVENILE LITERATURE

Lisker, Tom, 1928-
Nellie Bly : first woman of the news / by Tom Lisker illustrated by Jeffrey Lindberg. — New York : C.P.I., c1978.
48 p. : ill. (some col.) ; 24 cm.
SUMMARY: A biography of one of the first women reporters, whose trip around the world in less than 80 days made her an international celebrity.
ISBN 0-89547-057-8 : $5.58
1. Cochrane, Elizabeth, 1867-1922—Biography—Juvenile literature. 2. Journalists—United States—Biography—Juvenile literature. ⌈1. Cochrane, Elizabeth, 1867-1922. 2. Journalists⌉ I. Lindberg, Jeffrey. II. Title.
PN4874.C59L5 070'.92'4 78-14382
78 MARC

COCHRANE, ONT.

—HISTORY

Barnes, Michael, 1934-
Cochrane : the Polar bear town / Michael Barnes ; ill. by Muriel Newton White. — Cobalt, Ont. : Highway Book Shop, c1976.
59 p. : ill., maps ; 22 cm. C78-7217-0
$2.50
1. Cochrane, Ont.—History. I. Title.
F1059.5.C63.B37 971.3'142 78-326793
79 MARC

COCK-FIGHTING

see also Cockers (Persons)

COCKATEELS

Rogers, Cyril H
Cockatiels : their care and breeding / ⌈by⌉ Cyril H. Rogers. — Leicester : K and R Books Ltd., 1978.
3-75 p. : ill. (some col.), forms, plans ; 19 cm. — (A K & R handbook) GB78-32271
Bibliography: p. 74.
Includes index.
ISBN 0-903264-26-9 : £1.25
1. Cockateels. I. Title.
SF473.C6R63 636.6'865 79-309739
79 MARC

COCKE, JOHN HARTWELL, 1780-1866

"Dear Master" : letters of a slave family / edited by Randall M. Miller. — Ithaca, N.Y. : Cornell University Press, c1978.
281 p. : ill. ; 24 cm.
Bibliography p. ⌈265⌉-269.
Includes index.
ISBN 0-8014-1134-3
1. Skipwith family. 2. Cocke, John Hartwell, 1780-1866. 3. Afro-Americans—Colonization—Liberia—Sources. 4. Slavery in the United States—History—Sources. I. Miller, Randall M.
E444.D42 976.1'004'96073 77-90907
78 MARC

COCKER, KATHLEEN BENN MCALISTER

Cocker, Kathleen Benn McAlister.
Pioneer tapestry : buildings of early Tasmania / paintings by Kathleen Cocker ; written by G. Hawley Stancombe. — Western Junction ⌈Australia⌉ : G. H. Stancombe, Glendessary Press, 1977.
90 p. : ill. ; 26 cm. Aus***
Includes index.
ISBN 0-9599293-4-7 : $18.00
1. Cocker, Kathleen Benn McAlister. 2. Buildings in art. 3. Buildings—Australia—Tasmania—Pictorial works. I. Stancombe, George Hawley, 1914- II. Title.
ND2091.C6A4 1977 759.994 79-311385
79 MARC

COCKER, WELSH see Welsh springer spaniel

COCKER SPANIELS

Gordon, John Frazer.
All about the cocker spaniel / ⌈by⌉ John F. Gordon. — Revised ed. — London : Pelham, 1978.
144 p. : ill. ; 26 cm. GB78-29701
Bibliography: p. ⌈140⌉-141.
Includes index.
ISBN 0-7207-1088-X : £4.50
1. Cocker spaniels. I. Title.
SF429.C55G67 1978 636.7'52 79-303159
79 MARC

Kraeuchi, Ruth.
The new cocker spaniel / by Ruth M. Kraeuchi. — 1st ed. — New York : Howell Book House, 1979.
287, ⌈1⌉ p. : ill. ; 24 cm.
Bibliography: p. ⌈288⌉
ISBN 0-87605-104-2
1. Cocker spaniels. I. Title.
SF429.C55K73 636.7'52 78-12698
78 MARC

COCKERHAM FAMILY

Cockrum, Emmett E 1912-
A history and genealogy of the Cockrum family in America / by Emmett E. Cockrum. — ₍s.l.₎ : Cockrum, c1979.
p. a-e, 273, xxxvi p. : ill. ; 24 cm.
Includes index.
1. Cockerham family. I. Title.
CS71.C6643 1979 929'.2'0973 79-111960
79 MARC

COCKPITS, PRESSURIZED see Airplanes —Pressurization

COCKTAILS

Bérard, Léone, 1927-
Guide des alcools, des liqueurs et de leurs à-côtés / Léone Bérard. — ₍Paris₎ : Éditions de la Courtille, ₍1977₎
238 p. : ill. (some col.) ; 29 cm. — (G) F***
ISBN 2-7207-0044-4 : 125F
1. Liquors. 2. Liqueurs. 3. Cocktails. I. Title.
TP597.B47 641.2'5 78-398770
79 MARC

Boyd, Ina C
Cocktails & hors d'oeuvres : ₍a 2 in 1 cookbook₎ / by Ina C. Boyd ; illustrated by Mike Nelson, ₍Craig Torlucci₎. — Concord, Calif. : Nitty Gritty Productions, c1978.
88, 90 p. : ill. ; 14 x 21 cm.
Consists of 2 pts., Cocktails & hors d'oeuvres and Hors d'oeuvres & cocktails, inverted with respect to each other.
Includes indexes.
ISBN 0-911954-45-7
1. Cocktails. 2. Cookery (Appetizers) I. Boyd, Ina C. Hors d'oeuvres & cocktails. 1978. II. Title.
TX951.B772 641.8'74 78-108144
78 MARC

Brossard, Claude.
Cocktails en toutes occasions / Claude Brossard ; préf. de Raymond Oliver. — Paris : Livre de poche, c1977.
382 p. : ill. ; 17 cm. — (Le livre de poche pratique ; 7703)
Includes index.
10.00F
1. Cocktails. I. Title.
TX951.B779 641.8'74 78-384233
78 MARC

Charles.
The cocktail bar / by "Charles" ; edited by "Carlos". — London : Foulsham, 1977.
125 p. ; 22 cm. GB78-05246
On cover: A collection of 400 recipes.
ISBN 0-572-00000-6 : £1.95
1. Cocktails. 2. Alcoholic beverages. 3. Beverages. I. Title.
TX951.C46 641.8'74 78-325450
79 MARC

Ch'en, Han-ying.
(Hsin ch'ao ti t'iao chiu i shu)
新潮的調酒藝術 / 陳漢營編著. — 九龍：新光出版社，₍1979?₎
149 p., ₍8₎ leaves of plates : col. ill. ; 22 cm.
$16.00

1. Cocktails. 2. Alcoholic beverages. I. Title. II. Title: T'iao chiu i shu.
TX951.C49 79-839528

Dauven, L R
Le livre des cocktails / L.-R. Dauven, Jacques Morlaine. — ₍Paris₎ : Solar, c1976.
176 p. : ill. ; 20 cm. F***
Includes index.
ISBN 2-263-00074-7 : 28.00F
1. Cocktails. 2. Alcoholic beverages. I. Morlaine, Jacques, joint author. II. Title.
TX951.D38 641.8'74 78-391719
79 MARC

Doxat, John, 1914-
Booth's handbook of cocktails and mixed drinks / ₍by₎ John Doxat. — ₍New and revised ed.₎. — London : Pan Books, 1977.
191 p. ; 18 cm. GB77-32722
Bibliography: p. 180-182.
Includes index.
ISBN 0-330-10593-0 : £0.70
1. Cocktails. I. Booth's Distilleries Ltd. II. Title.
TX951.D69 1977 641.8'74 78-325934
79 MARC

Gööck, Roland.
Mixbrevier : über 1000 Rezepte f. Mischgetränke mit u. ohne Alkohol / zusammengestellt von Roland Gööck ₍i.e. R. Jeromin₎ ; ₍Zeichn., Jochen Bartsch₎. — Gütersloh : Praesentverlag Peter, 1973.
185 p. : ill. (some col.) ; 22 cm. GFR74-A
Includes index.
1. Cocktails. 2. Beverages. I. Title.
TX951.G62 641.8'74 78-418820
78 MARC

Martínez Llopis, M
Aguardientes y licores / Dr. M. Martínez Llopis. — Bilbao : Editorial Cantábrica, c1978.
266 p., ₍12₎ leaves of plates : ill. ; 23 cm. Sp***
Bibliography: p. 266.
ISBN 8422103737
1. Liquors. 2. Liqueurs. 3. Cocktails. I. Title.
TP590.M35 79-350098
79 MARC

Mixgetränke : 350 Rezepte zum Mixen mit und ohne Alkohol / ₍Rezepte, Burda-Kochstudio ; Red., Elisabeth Klapper₎. — Offenburg : Burda, 1974.
144 p. : numerous col. ill. ; 24 cm. — (Burda-Kochbuch ; Nr. 8) GFR72-A
ISBN 3-920158-10-5 : Dm12.00
1. Alcoholic beverages. 2. Cocktails. 3. Beverages.
TX951.M54 78-396129
79 MARC

Russell, Ralph.
Mix and be merry : how to be a good host / by Ralph Russell. — Vancouver : J. J. Douglas, 1974.
109 p. : ill. ; 22 cm. C***
Label mounted verso t.p.: Exclusive distributor: ISBS, Forest Grove, Or.
Includes index.
ISBN 0-88894-037-8 : $3.95 (U.S.)
1. Cocktails. 2. Beverages. I. Title.
TX951.R87 641.8'74 78-325932
79 MARC

Schur, Sylvia.
Seagram's complete party guide : how to succeed at party planning, drink mixing, the art of hospitality / by Sylvia Schur. — New York : Warner Books, c1979.
232 p. : ill. ; 18 cm.
On spine: Complete party guide.
Includes index.
ISBN 0-446-91181-X : $2.50
1. Entertaining. 2. Cookery. 3. Cocktails. 4. Beverages. I. Seagram-Distillers Corporation. II. Title. III. Title: Complete party guide.
TX731.S34 642'.41 79-111817
00 MARC

Toye, Nina.
Petits & grands verres : choix des meilleurs cocktails / recueillis par Nina Toye & A. H. Adair ; et mis en français par Ph.-Le Huby ; Les gravures sont de J.-E. Laboureur. — Paris : Au Sans pareil, 1927.
131 p., ₍10₎ leaves of plates : ill. ; 24 cm.
Translation of Drinks—long & short.
"De ce recueil, il a été fait une imposition dans le format in-4° couronne, avec dix gravures hors-texte de J.-E. Laboureur, achevée d'imprimer ... le quinze juin mil neuf cent vingt-sept. Ce tirage comprend: 20 exemplaires sur vergé d'Arches ... 25 exemplaires sur japon impérial ... 225 exemplaires sur vélin Montgolfier (nos 46 à 270); et quelques exemplaires hors commerce ... Exemplaire no 231."
Includes index.
1. Cocktails. 2. Beverages. I. Adair, Arthur Henry. II. Laboureur, Jean Émile, 1877-1943. III. Title.
TX951.T614 Rosenwald Coll. 79-348924
79 MARC

Vence, Céline.
Alcools et cocktails en 10 ₍i.e. dix₎ leçons : ₍l'art de recevoir pour l'apéritif₎ / Céline Vence. — ₍Paris₎ : Hachette, c1976.
204 p. : ill. ; 20 cm. — (Collection En 10 leçons) F***
Includes index.
ISBN 2-01-001869-9 : 29.00F
1. Cocktails. 2. Alcoholic beverages. I. Title.
TX951.V45 641.8'74 78-395181
79 MARC

Yang, Pen-li.
(Chiu tao)
酒道 / 楊本禮著. — 臺北：空中雜誌社，民國 66 ₍1977₎
2, 179 p., ₍8₎ leaves of plates : ill. ; 19 cm.
NT$60.00

1. Alchoholic beverages. 2. Cocktails. I. Title.
TX951.Y36 78-842017

Zingales, Franco.
Amicobar / Francesco Zingales. — 1. ed. — Milano : Fratelli Fabbri, 1974.
140, ₍4₎ p. : ill. ; 27 cm. It76-Sept
Bibliography: p. ₍144₎
Includes index.
L5000
1. Alcoholic beverages. 2. Cocktails. I. Title.
TP505.Z56 78-393953
*79 MARC

COCO, FRANCESCO, 1908-1976

Il Caso Coco : processo a Giuliano Naria .... — Milano : Collettivo editoriale Librirossi, 1978.
99 p. : ill. ; 24 cm. It78-July
On cover: 4.
L2500
1. Naria, Giuliano. 2. Coco, Francesco, 1908-1976. 3. Trials (Political crimes and offenses)—Italy. 4. Terrorism—Italy.
345'.45'0231 79-350278
*79 MARC

Ranedda, Domenica Bruna.
Francesco Coco : radiografia di un magistrato / Domenica Bruna Ranedda. — Cagliari : Editrice sarda Fossataro, 1976.
278 p. ; 21 cm. It78-Jan
1. Coco, Francesco. 2. Judges—Italy—Biography.
347'.45'014 78-375606
*78 MARC

COCOA

see also Chocolate

International Cocoa Organization.
Study of cocoa production and consumption capacity. — ₍London₎ : International Cocoa Organization, 1975.
105 p. in various pagings ; 30 cm. GB***
"ICC/5/8."
"STC/18."
International Cocoa Council, 5th session, London, 19-22 August, 1975, Item 14 of the provisional agenda.
Addendum, International Cocoa Council, 7th session, London, 27-30 July 1976, Item 14 of the draft agenda (51 p.) inserted.
Addendum, International Cocoa Council, 9th session, London, 26-29 July, 1977, Item 15 of the draft agenda (29 p.) inserted.
1. Cocoa trade. 2. Cocoa. I. Title.
HD9200.A2I57 1975 338.1'7'374 78-325214
79 MARC

—CONGRESSES

EMPA Seminar on Cocoa-Coconuts, Tawau, Sabah, 1976.
EMPA Seminar on Cocoa-Coconuts, 1976 : proceedings. — Sandakan : East Malaysia Planters Association, ₍1976?₎
xvi, 207 p. : ill. ; 30 cm.
Errata slip inserted.
Includes bibliographical references.
1. Cacao—Malaysia—Congresses. 2. Coconut palm—Malaysia—Congresses. 3. Cocoa—Malaysia—Congresses. 4. Cacao—Congresses. 5. Coconut palm—Congresses. 6. Cocoa—Congresses. I. East Malaysia Planters Association.
SB268.M4E18 1976 633'.74'095953 79-940202
79 MARC

International Cocoa Research Conference, 5th, Ibadan, Nigeria, 1975.
Proceedings / V International Cocoa Research Conference, Ibadan, Nigeria, 1-9 September, 1975. — Ibadan : Cocoa Research Institute of Nigeria, 1977.
vii, 642 p. : ill. ; 27 cm.
English, French or Spanish.
Includes bibliographies.
1. Cacao—Congresses. 2. Cocoa—Congresses.
SB267.I53 1975 633'.74 79-101040
79 MARC

—BRAZIL—STATISTICS —PERIODICALS

Brazil. Comissão Executiva do Plano da Lavoura Cacaueira. Assistência de Economia.
Boletim estatístico do cacau: comercialização da safra cacaueira do Brasil.
Brasília, CEPLAC, COGEP, ASSEC.
v. ill. 31 cm.
1. Cocoa—Brazil—Statistics—Periodicals. I. Title.
HD9200.B7B7a 338.1'7'3740981 79-640670
MARC-S

—MALAYSIA—CONGRESSES

EMPA Seminar on Cocoa-Coconuts, Tawau, Sabah, 1976.
EMPA Seminar on Cocoa-Coconuts, 1976 : proceedings. — Sandakan : East Malaysia Planters Association, ₍1976?₎
xvi, 207 p. : ill. ; 30 cm.
Errata slip inserted.
Includes bibliographical references.
1. Cacao—Malaysia—Congresses. 2. Coconut palm—Malaysia—Congresses. 3. Cocoa—Malaysia—Congresses. 4. Cacao—Congresses. 5. Coconut palm—Congresses. 6. Cocoa—Congresses. I. East Malaysia Planters Association.
SB268.M4E18 1976 633'.74'095953 79-940202
79 MARC

COCOA (Continued)

—MALAYSIA—HILIR PERAK, PERAK

Lembaga Pemasaran Pertanian Persekutuan.
Cocoa production survey, Hilir Perak and Sabak Bernam. — ₍Kuala Lumpur₎ : Federal Agricultural Marketing Authority, Ministry of Agriculture, Malaysia, 1978.
11 p. ; 26 cm. — (FAMA survey report ; no. 7)
1. Cacao—Malaysia—Hilir Perak, Perak. 2. Cacao—Malaysia—Sabak Bernam, Selangor. 3. Cocoa—Malaysia—Hilir Perak, Perak. 4. Cocoa—Malaysia—Sabak Bernam, Selangor. 5. Hilir Perak, Perak—Industries. 6. Sabak Bernam, Selangor—Industries. I. Title. II. Series: Lembaga Pemasaran Pertanian Persekutuan. FAMA survey report ; no. 7.
SB268.M4L45 1978 79-316413
79 MARC

—MALAYSIA—SABAK BERNAM, SELANGOR

Lembaga Pemasaran Pertanian Persekutuan.
Cocoa production survey, Hilir Perak and Sabak Bernam. — ₍Kuala Lumpur₎ : Federal Agricultural Marketing Authority, Ministry of Agriculture, Malaysia, 1978.
11 p. ; 26 cm. — (FAMA survey report ; no. 7)
1. Cacao—Malaysia—Hilir Perak, Perak. 2. Cacao—Malaysia—Sabak Bernam, Selangor. 3. Cocoa—Malaysia—Hilir Perak, Perak. 4. Cocoa—Malaysia—Sabak Bernam, Selangor. 5. Hilir Perak, Perak—Industries. 6. Sabak Bernam, Selangor—Industries. I. Title. II. Series: Lembaga Pemasaran Pertanian Persekutuan. FAMA survey report ; no. 7.
SB268.M4L45 1978 79-316413
79 MARC

COCOA PALM see Coconut palm

COCOA TRADE

Assoumou, Jean.
L'économie du cacao : agriculture d'exportation et bataille du développement en Afrique tropicale / Jean Assoumou ; préf. de J.-M. Albertini. — Paris : J.-P. Delarge, c1977.
351 p. : ill. ; 24 cm. F***
Bibliography: p. ₍345₎-348.
ISBN 2-7113-0063-3 : 60.00F
1. Cocoa trade—Cameroon. 2. Cocoa trade. I. Title.
HD9200.C182A84 78-360499
338.1'7'3730967113
78₍r79₎rev MARC

International Cocoa Organization.
Study of cocoa production and consumption capacity. — ₍London₎ : International Cocoa Organization, 1975.
105 p. in various pagings ; 30 cm. GB***
"ICC/5/8."
"STC/18."
International Cocoa Council, 5th session, London, 19-22 August, 1975, Item 14 of the provisional agenda.
Addendum, International Cocoa Council, 7th session, London, 27-30 July 1976, Item 14 of the draft agenda (51 p.) inserted.
Addendum, International Cocoa Council, 9th session, London, 26-29 July, 1977, Item 15 of the draft agenda (29 p.) inserted.
1. Cocoa trade. 2. Cocoa. I. Title.
HD9200.A2I57 1975 338.1'7'374 78-325214
79 MARC

—INFORMATION SERVICES

United Nations. Industrial Development Organization.
Information sources on the coffee, cocoa, tea, and spices industry / United Nations Industrial Development Organization. — New York : United Nations, 1977.
xii, 74 p. ; 21 cm. — (UNIDO guides to information sources ; no. 28) (₍Document₎ - United Nations ; ID/198 (UNIDO/LIB/SER.D/28))
"Bibliographical sources used in compiling the present guide": p. 73-74.
$4.00
1. Coffee trade—Information services. 2. Cocoa trade—Information services. 3. Tea trade—Information services. 4. Spice trade—Information services. I. Title. II. Series: United Nations. Industrial Development Organization. UNIDO guides to information sources ; no. 28. III. Series: United Nations. Document ; ID/198 (UNIDO/LIB/SER.D/28)
JX1977.A2 ID/198(UNIDO/LIB/SER.D/28) 79-306816
300'.8 s
——— Copy 2. HD9199.A2U45 1977 MARC
79₍r79₎rev

—AFRICA, WEST—HISTORY

Ehrler, Franz.
Handelskonflikte zwischen europäischen Firmen und einheimischen Produzenten in Britisch Westafrika : die "Cocoa-hold-ups" in der Zwischenkriegszeit / Franz Ehrler. — Zürich : Atlantis, c1977.
xix, 496 p. ; 24 cm. — (Beiträge zur Kolonial- und Überseegeschichte ; Bd. 16) Sw***
Originally presented as the author's thesis, Zürich, 1976.
Bibliography: p. 457-481.
Includes index.
1. Cocoa trade—Africa, West—History. 2. Cocoa trade—Ghana—History. 3. Competition, Unfair—Africa, West. 4. Competition, Unfair—Ghana. I. Title. II. Series.
HD9200.A42E37 1977 78-388959
79 MARC

—CAMEROON

Assoumou, Jean.
L'économie du cacao : agriculture d'exportation et bataille du développement en Afrique tropicale / Jean Assoumou ; préf. de J.-M. Albertini. — Paris : J.-P. Delarge, c1977.
351 p. : ill. ; 24 cm. F***
Bibliography: p. ₍345₎-348.
ISBN 2-7113-0063-3 : 60.00F
1. Cocoa trade—Cameroon. 2. Cocoa trade. I. Title.
HD9200.C182A84 78-360499
338.1'7'3730967113
78₍r79₎rev MARC

—GHANA

Ghana. State Cocoa Marketing Board.
At work / Ghana Cocoa Marketing Board. — 5th ed. — ₍Accra₎ : The Board, 1968.
34 p. : ill. ; 22 cm.
Cover title.
1. Ghana. State Cocoa Marketing Board. 2. Cocoa trade—Ghana. I. Title.
HD9200.G64G574 1968 354'.667'008233 79-311460
79 MARC

—GHANA—HISTORY

Ehrler, Franz.
Handelskonflikte zwischen europäischen Firmen und einheimischen Produzenten in Britisch Westafrika : die "Cocoa-hold-ups" in der Zwischenkriegszeit / Franz Ehrler. — Zürich : Atlantis, c1977.
xix, 496 p. ; 24 cm. — (Beiträge zur Kolonial- und Überseegeschichte ; Bd. 16) Sw***
Originally presented as the author's thesis, Zürich, 1976.
Bibliography: p. 457-481.
Includes index.
1. Cocoa trade—Africa, West—History. 2. Cocoa trade—Ghana—History. 3. Competition, Unfair—Africa, West. 4. Competition, Unfair—Ghana. I. Title. II. Series.
HD9200.A42E37 1977 78-388959
79 MARC

Gunnarsson, Christer, 1948-
The Gold Coast cocoa industry 1900-1939 : production, prices and structural change / by Christer Gunnarsson. — Lund : ₍s.n.₎, 1978.
₍7₎, 184 p. : tables ; 23 cm. — (Publications of the Economic History Association, Lund ; 23) S78-13
Thesis—Lund.
Extra t.p. with thesis statement inserted.
Bibliography: p. 179-184.
1. Cocoa trade—Ghana—History. I. Title. II. Series: Ekonomisk-historiska föreningen i Lund. Skrifter ; 23.
HD9200.G62G86 338.1'7'37409667 78-325319
79 MARC

—MALAYSIA

Sekhon, Isher Singh.
Markets for Felda produce / Isher Singh Sekhon. — ₍Kuala Lumpur?₎ : Felda, ₍1974?₎
25 p. : graphs ; 25 cm.
Caption title.
"Felda Officers Conference 1974, paper no. 8."
1. Palm-oil industry—Malaysia. 2. Rubber industry and trade—Malaysia. 3. Cocoa trade—Malaysia. 4. Lembaga Kemjuan Tanah Persekutuan. I. Title.
HD9490.M32S44 382'.41'509595 79-305817
79 MARC

—VENEZUELA

Consulpro.
Cacao, café, caña de azucar, tabaco : estudio integral del sector / elaborado por Consulpro. — Caracas : Banco Central de Venezuela, ₍1974?₎
425 p. : ill. ; 30 cm. — (Colección Agricultura en Venezuela ; 6)
Includes bibliographical references.
1. Cocoa trade—Venezuela. 2. Coffee trade—Venezuela. 3. Sugar trade—Venezuela. 4. Tobacco manufacture and trade—Venezuela. I. Title. II. Series.
HD9200.V42C65 1974 79-107520
79 MARC

—VENEZUELA—STATISTICS —PERIODICALS

Venezuela. Ministerio de Agricultura y Cría. División de Estadística.
Estimación de cosechas: Cacao.
Caracas, Ministerio de Agricultura y Cría, Dirección de Planificación Agropecuaria, División de Estadística.
v. ill. 29 cm.
1. Cocoa trade—Venezuela—Statistics—Periodicals. I. Title.
HD9200.V4V46a 79-642276
MARC-S

COCOIDEA see Scale-insects

COCONINO CO., ARIZ.

—HISTORY—INDEXES

Shook, Sally, 1947-
Name index to the Arizona champion / compiled by Sally Shook ; supervised by John Irwin. — ₍Flagstaff : Special Collections, Northern Arizona University Library₎, 1974-
v. ; 30 cm.
Vol. ₍8₎ has title: Name index to the Arizona champion-Coconino sun; v. ₍9₎-₍11₎ have title: Name index to the Coconino sun.
Vols. ₍5₎-₍6₎ compiled by S. Shook and S. L. R. Ash; v. ₍7₎-₍11₎ compiled by S. L. R. Ash.
1. Arizona—History—Indexes. 2. Flagstaff, Ariz.—History—Indexes. 3. Coconino Co., Ariz.—History—Indexes. 4. Arizona champion—Indexes. 5. Coconino sun—Indexes. I. Ash, Susan L. Rogers, joint author. II. Irwin, John, 1949- . III. Northern Arizona University. Library. Special Collections Division. IV. Title. V. Title: Subject index to the Coconino sun.
Z1257.S478 051'.09'791 78-111185
79 MARC

Shook, Sally, 1947-
Subject index to the Arizona champion / compiled by Sally Shook ; supervised by John Irwin. — ₍Flagstaff : Special Collections, Northern Arizona University Library₎, 1974-
v. ; 30 cm.
Vol. ₍8₎ has title: Subject index to the Arizona champion-Coconino sun; v. ₍9₎-₍11₎ have title: Subject index to the Coconino sun.
Vols. ₍4₎-₍6₎ compiled by S. Shook and S. L. R. Ash; v. ₍7₎-₍11₎ compiled by S. L. R. Ash.
1. Arizona—History—Indexes. 2. Flagstaff, Ariz.—History—Indexes. 3. Coconino Co., Ariz.—History—Indexes. 4. Arizona champion—Indexes. 5. Coconino sun—Indexes. I. Ash, Susan L. Rogers, joint author. II. Irwin, John, 1949- . III. Northern Arizona University. Library. Special Collections Division. IV. Title. V. Title: Subject index to the Coconino sun.
Z1257.S48 974.9 78-111182
₍F811₎ 79 MARC

COCONINO SUN

—INDEXES

Shook, Sally, 1947-
Name index to the Arizona champion / compiled by Sally Shook ; supervised by John Irwin. — ₍Flagstaff : Special Collections, Northern Arizona University Library₎, 1974-
v. ; 30 cm.
Vol. ₍8₎ has title: Name index to the Arizona champion-Coconino sun; v. ₍9₎-₍11₎ have title: Name index to the Coconino sun.
Vols. ₍5₎-₍6₎ compiled by S. Shook and S. L. R. Ash; v. ₍7₎-₍11₎ compiled by S. L. R. Ash.
1. Arizona—History—Indexes. 2. Flagstaff, Ariz.—History—Indexes. 3. Coconino Co., Ariz.—History—Indexes. 4. Arizona champion—Indexes. 5. Coconino sun—Indexes. I. Ash, Susan L. Rogers, joint author. II. Irwin, John, 1949- . III. Northern Arizona University. Library. Special Collections Division. IV. Title. V. Title: Subject index to the Coconino sun.
Z1257.S478 051'.09'791 78-111185
79 MARC

Shook, Sally, 1947-
Subject index to the Arizona champion / compiled by Sally Shook ; supervised by John Irwin. — ₍Flagstaff : Special Collections, Northern Arizona University Library₎, 1974-
v. ; 30 cm.
Vol. ₍8₎ has title: Subject index to the Arizona champion-Coconino sun; v. ₍9₎-₍11₎ have title: Subject index to the Coconino sun.
Vols. ₍4₎-₍6₎ compiled by S. Shook and S. L. R. Ash; v. ₍7₎-₍11₎ compiled by S. L. R. Ash.
1. Arizona—History—Indexes. 2. Flagstaff, Ariz.—History—Indexes. 3. Coconino Co., Ariz.—History—Indexes. 4. Arizona champion—Indexes. 5. Coconino sun—Indexes. I. Ash, Susan L. Rogers, joint author. II. Irwin, John, 1949- . III. Northern Arizona University. Library. Special Collections Division. IV. Title. V. Title: Subject index to the Coconino sun.
Z1257.S48 974.9 78-111182
₍F811₎ 79 MARC

COCONUT

see also Coconut industry

Nomura, Tokushichi, 1878-1945.
(Gomu to yashi)
護謨と椰子 / 野村徳七著. — 大阪 : 野村, 大正5 ₍1916₎
5, 6, 214 p., ₍39₎ leaves of plates (1 fold.) : ill. (1 fold.) : 20 cm.
1. Rubber—Brunei (State) 2. Coconut. I. Title.
SB290.5.B85N65 79-806887

COCONUT (Continued)

Woodroof, Jasper Guy, 1900-
Coconuts : production, processing, products / Jasper Guy Woodroof. — 2d ed. — Westport, Conn. : AVI Pub. Co., c1979.
xi, 307 p. : ill. ; 24 cm.
Includes bibliographies and index.
ISBN 0-87055-276-7 : $31.00
1. Coconut. 2. Coconut products. 3. Coconut-palm. I. Title.
SB401.C6W65 1979 634'.61 78-21068
78 MARC

—ASIA, SOUTHEASTERN

Tazawa, Shingo, b. 1878.
(Kainantō to yashi)
海南島と椰子 / 田澤震五著. — 臺北 : 田澤化學工業研究所, 昭和15 [1940]
8, 11, 163, 55 p., [4] leaves of plates : ill. ; 19 cm.
Cover title.
Bibliography: p. 5 (1st group)

1. Coconut—Taiwan. 2. Coconut—Asia, Southeastern. I. Title.
SB401.C6T34 79-806014

—TAIWAN

Tazawa, Shingo, b. 1878.
(Kainantō to yashi)
海南島と椰子 / 田澤震五著. — 臺北 : 田澤化學工業研究所, 昭和15 [1940]
8, 11, 163, 55 p., [4] leaves of plates : ill. ; 19 cm.
Cover title.
Bibliography: p. 5 (1st group)

1. Coconut—Taiwan. 2. Coconut—Asia, Southeastern. I. Title.
SB401.C6T34 79-806014

COCONUT INDUSTRY

—KENYA

Eijnatten, C L M van, 1930-

A study of coconuts in Kenya's coastal strip / C. L. M. van Eijnatten, A. M. Gurnah, K. H. Niederstucke. — Nairobi : Dept. of Crop Science, Faculty of Agriculture, University of Nairobi, 1977.
56 p. : ill. ; 31 cm. — (Technical communication - Department of Crop Science, University of Nairobi ; no. 19)
Bibliography: p. 53-56.
1. Coconut palm—Kenya. 2. Coconut industry—Kenya. I. Gurnah, Abdulla M., joint author. II. Niederstucke, K. H., joint author. III. Title. IV. Series: Nairobi. University. Dept. of Crop Science. Technical communication - Department of Crop Science, University of Nairobi ; no. 19.
SB401.C6E38 634'.61'096762 78-980611
79 MARC

—PHILIPPINE ISLANDS—HISTORY

Hicks, George L
The Philippine coconut industry : growth and change, 1900-1965 / by George L. Hicks. — [Washington] : Center for Development Planning, National Planning Association, 1967.
216, A46 p. ; 28 cm. — (Field work report - Center for Development Planning, National Planning Association ; no. 17)
Bibliography: p. A36-A46.
1. Coconut industry—Philippine Islands—History. I. Title. II. Series: National Planning Association. Center for Development Planning. Field work report - Center for Development Planning, National Planning Association ; no. 17.
HD9259.C6P54 338.1'7'46109599 79-302076
79 MARC

COCONUT OIL

—INDONESIA—SULAWESI SELATAN

Indonesia. Direktorat Jenderal Aneka Industri dan Kerajinan.
Regional study industri minyak kelapa/makan / kerjasama Direktorat Jendral Aneka Industri dan Kerajinan, Departemen Perindustrian dengan Departemen Teknologi Hasil Pertanian, FATEMETA, Institut Pertanian Bogor. — [s.l. : s.n.], 1976.
xix, 178 leaves : ill. ; 29 cm.
Summary in English.
Bibliography: leaves [49]-50.
1. Coconut oil—Indonesia—Sulawesi Selatan. 2. Coconut oil—Indonesia—Sulawesi Tengah. I. Institut Pertanian Bogor. Departemen Teknologi Hasil Pertanian. II. Title.
HD9259.C6I6437 338.4'7'6643 76-941493
79 MARC

—INDONESIA—SULAWESI TENGAH

Indonesia. Direktorat Jenderal Aneka Industri dan Kerajinan.
Regional study industri minyak kelapa/makan / kerjasama Direktorat Jendral Aneka Industri dan Kerajinan, Departemen Perindustrian dengan Departemen Teknologi Hasil Pertanian, FATEMETA, Institut Pertanian Bogor. — [s.l. : s.n.], 1976.
xix, 178 leaves : ill. ; 29 cm.
Summary in English.
Bibliography: leaves [49]-50.
1. Coconut oil—Indonesia—Sulawesi Selatan. 2. Coconut oil—Indonesia—Sulawesi Tengah. I. Institut Pertanian Bogor. Departemen Teknologi Hasil Pertanian. II. Title.
HD9259.C6I6437 338.4'7'6643 76-941493
79 MARC

COCONUT-PALM

Woodroof, Jasper Guy, 1900-
Coconuts : production, processing, products / Jasper Guy Woodroof. — 2d ed. — Westport, Conn. : AVI Pub. Co., c1979.
xi, 307 p. : ill. ; 24 cm.
Includes bibliographies and index.
ISBN 0-87055-276-7 : $31.00
1. Coconut. 2. Coconut products. 3. Coconut-palm. I. Title.
SB401.C6W65 1979 634'.61 78-21068
78 MARC

—CONGRESSES

EMPA Seminar on Cocoa-Coconuts, Tawau, Sabah, 1976.
EMPA Seminar on Cocoa-Coconuts, 1976 : proceedings. — Sandakan : East Malaysia Planters Association, [1976?]
xvi, 207 p. : ill. ; 30 cm.
Errata slip inserted.
Includes bibliographical references.
1. Cacao—Malaysia—Congresses. 2. Coconut palm—Malaysia—Congresses. 3. Cocoa—Malaysia—Congresses. 4. Cacao—Congresses. 5. Coconut palm—Congresses. 6. Cocoa—Congresses. I. East Malaysia Planters Association.
SB268.M4E18 1976 633'.74'095953 79-940202
79 MARC

—INDIA—KARNATAKA

Karnataka, India. Planning Dept. Evaluation Division.
Report on the evaluation study of Coconut Development Programme. — Bangalore : Evaluation Division, Planning Dept., Govt. of Karnataka, 1975 [i.e. 1976]
ii, iii, 124 p. ; 22 cm.
Cover title.
1. Coconut palm—India—Karnataka. 2. Agriculture and state—India—Karnataka. I. Title. II. Title: Coconut Development Programme.
SB401.C6K37 1976 634'.61'095487 76-904316
79 MARC

—INDIA—WEST BENGAL

West Bengal. Directorate of Agriculture.
The coconut in West Bengal. — [Calcutta] : Directorate of Agriculture, Govt. of West Bengal, 1976.
16, [1] p., [2] leaves of plates : ill. ; 26 cm.
Bibliography: p. [17]
1. Coconut palm—India—West Bengal. I. Title.
SB401.C6W47 1976 634'.61'095414 78-911688
79 MARC

—KENYA

Eijnatten, C L M van, 1930-

A study of coconuts in Kenya's coastal strip / C. L. M. van Eijnatten, A. M. Gurnah, K. H. Niederstucke. — Nairobi : Dept. of Crop Science, Faculty of Agriculture, University of Nairobi, 1977.
56 p. : ill. ; 31 cm. — (Technical communication - Department of Crop Science, University of Nairobi ; no. 19)
Bibliography: p. 53-56.
1. Coconut palm—Kenya. 2. Coconut industry—Kenya. I. Gurnah, Abdulla M., joint author. II. Niederstucke, K. H., joint author. III. Title. IV. Series: Nairobi. University. Dept. of Crop Science. Technical communication - Department of Crop Science, University of Nairobi ; no. 19.
SB401.C6E38 634'.61'096762 78-980611
79 MARC

—MALAYSIA—CONGRESSES

EMPA Seminar on Cocoa-Coconuts, Tawau, Sabah, 1976.
EMPA Seminar on Cocoa-Coconuts, 1976 : proceedings. — Sandakan : East Malaysia Planters Association, [1976?]
xvi, 207 p. : ill. ; 30 cm.
Errata slip inserted.
Includes bibliographical references.
1. Cacao—Malaysia—Congresses. 2. Coconut palm—Malaysia—Congresses. 3. Cocoa—Malaysia—Congresses. 4. Cacao—Congresses. 5. Coconut palm—Congresses. 6. Cocoa—Congresses. I. East Malaysia Planters Association.
SB268.M4E18 1976 633'.74'095953 79-940202
79 MARC

COCONUT PRODUCTS

Woodroof, Jasper Guy, 1900-
Coconuts : production, processing, products / Jasper Guy Woodroof. — 2d ed. — Westport, Conn. : AVI Pub. Co., c1979.
xi, 307 p. : ill. ; 24 cm.
Includes bibliographies and index.
ISBN 0-87055-276-7 : $31.00
1. Coconut. 2. Coconut products. 3. Coconut-palm. I. Title.
SB401.C6W65 1979 634'.61 78-21068
78 MARC

COCONUT TREE see Coconut palm

COCOPA INDIANS

see also Cocopa Indian Reservation

COCOS see Coconut palm

COCOS ISLAND

—DESCRIPTION AND TRAVEL

Vergnes, Robert.
La Dernière île au trésor / Robert Vergnes. — [Paris] : Balland, 1978.
328 p., [16] p. of plates ; 22 cm. F78-15052
ISBN 2-7158-0149-1 : 49.00F
1. Cocos Island—Description and travel. 2. Vergnes, Robert. I. Title.
F1549.C6V47 972.86 79-368880
79 MARC

COCOS NUCIFERA see Coconut palm

COCTEAU, JEAN, 1889-1963

—CRITICISM AND INTERPRETATION

Crowson, Lydia.
The esthetic of Jean Cocteau / Lydia Crowson. — Hanover : Published for the University of New Hampshire by the University Press of New England, 1978.
200 p. ; 24 cm.
Bibliography: p. [175]-178.
Includes index.
ISBN 0-87451-149-6
1. Cocteau, Jean, 1889-1963—Criticism and interpretation. I. Title.
PQ2605.O15Z665 848'.9'1209 77-95326
79 MARC

Filipowska, Irena.
Éléments tragiques dans le théâtre de Jean Cocteau / Irena Filipowska. — Poznań : UAM, 1976.
147 p. ; 24 cm. — (Seria Filologia romańska ; nr. 4) P***
Bibliography: p. 143-147.
zł71.00
1. Cocteau, Jean, 1889-1963—Criticism and interpretation. 2. Tragic, The. I. Title. II. Series.
PQ2605.O15Z6824 842'.9'12 79-348608
79 MARC

COCTEAU, JEAN, 1889-1963 (Continued)

—KNOWLEDGE—MUSIC

Avec les musiciens. — ⌈Paris⌉ : Gallimard, ⌈c1978⌉
172 p. ; 21 cm. — (Cahiers Jean Cocteau ; 7) F***
Cover title.
Includes texts by Jean Cocteau.
Bibliography: p. ⌈169⌉-170.
1. Cocteau, Jean, 1889-1963—Knowledge—Music. 2. Music and literature. I. Cocteau, Jean, 1889-1963. II. Series.
PQ2605.O15Z564 848'.9'1209 79-358650
79 MARC

—MANUSCRIPTS—FACSIMILES

Major, Jean Louis.
Radiguet, Cocteau, "Les joues en feu" : suivi de quatre gravures de Jean Hugo et de la reproduction de l'exemplaire des Joues en feu illustré et annoté par Jean Cocteau / Jean-Louis Major. — Ottawa : Éditions de l'Université d'Ottawa, 1977.
98 p. : ill. ; 22 cm. — (Cahiers d'inédits ; no 11) C***
Includes bibliographical references.
ISBN 0-7766-4241-3
1. Radiguet, Raymond, 1903-1923. Les joues en feu—Illustrations. 2. Cocteau, Jean, 1889-1963—Manuscripts—Facsimiles. I. Radiguet, Raymond, 1903-1923. Les joues en feu. 1977. II. Title.
PQ2635.A25J635 841'.9'12 78-392487
79 MARC

COD-FISHERIES

—ECONOMIC ASPECTS—HISTORY

Innis, Harold Adams, 1894-1952.
The cod fisheries : the history of an international economy / by Harold A. Innis. — Rev. ed. — Toronto ; Buffalo : University of Toronto Press, 1978.
xviii, 522 p. : maps ; 23 cm. — (Canadian university paperbooks ; 212) C***
First published in the series Canada and the United States.
Includes bibliographical references and index.
ISBN 0-8020-6344-6
1. Cod-fisheries—History. 2. Cod-fisheries—Economic aspects—History. 3. Fish trade—History. 4. International economic relations. I. Title.
SH351.C5I56 1978 338.3'72'758 78-325357
79 MARC

—HISTORY

Innis, Harold Adams, 1894-1952.
The cod fisheries : the history of an international economy / by Harold A. Innis. — Rev. ed. — Toronto ; Buffalo : University of Toronto Press, 1978.
xviii, 522 p. : maps ; 23 cm. — (Canadian university paperbooks ; 212) C***
First published in the series Canada and the United States.
Includes bibliographical references and index.
ISBN 0-8020-6344-6
1. Cod-fisheries—History. 2. Cod-fisheries—Economic aspects—History. 3. Fish trade—History. 4. International economic relations. I. Title.
SH351.C5I56 1978 338.3'72'758 78-325357
79 MARC

—NORWAY

Norway. Torskefiskeutvalget, 1957.
Utdrag av Brofoss-utvalgets instilling. ⌈Oslo⌉ Arbeidernes opplysningsforbund ⌈1959⌉
71 p. 21 cm.
1. Cod-fisheries—Norway. I. Arbeidernes opplysningsforbund, Oslo. II. Title.
SH351.C5N6 1959 77-469625
78 MARC

CODE NAMES

see also Acronyms; Ciphers

CODE OF ADVERTISING PRACTICE COMMITTEE

Institute of Practitioners in Advertising.
Code of advertising practice : the CAP Committee : how it works, and why / ⌈Institute of Practitioners in Advertising and the Code of Advertising Practice Committee⌉. — London : The Institute : The Committee, 1976.
⌈2⌉, 8, iv p. : ill. ; 21 cm. GB78-06949
Cover title.
Addendum: 5 p.
1. Advertising—Great Britain—Self-regulation. 2. Code of Advertising Practice Committee. I. Code of Advertising Practice Committee. II. Title.
HF5834.I57 1976 659.1'06'241 78-322735
79 MARC

CODE OF FEDERAL REGULATIONS

—INDEXES

Index to the Code of Federal regulations, 1938-1976. — Englewood, Colo. : Information Handling Services, c1978-
v. ; 28 cm.
ISBN (invalid) 0-910972-74-5
1. Code of Federal regulations—Indexes. 2. Delegated legislation—United States—Indexes. I. Code of Federal regulations. II. Information Handling Services.
KF70.A312 348'.73'25 78-379
78 MARC

CODE PLEADING

La Salle Extension University, Chicago.
Pleading and practice forms : common-law, code, and equity pleading with selected illustrative simplified forms. — Chicago : La Salle Extension University, c1944.
v, 82 p. ; 23 cm.
At head of title: American law and procedure; a systematic law course presented by means of special textbooks, instruction lectures, and legal problems. Lecture 50.
1. Pleading—Illinois—Forms. 2. Equity pleading and procedure—Illinois—Forms. 3. Code pleading—United States—Forms. I. Title. II. Title: American law and procedure.
KFI1735.A65L37 1944 347'.773'072 78-317330
78 MARC

—UNITED STATES

La Salle Extension University, Chicago.
Pleading and practice forms : common-law, code, and equity pleading with selected illustrative simplified forms. — Chicago : La Salle Extension University, c1944.
v, 82 p. ; 23 cm.
At head of title: American law and procedure; a systematic law course presented by means of special textbooks, instruction lectures, and legal problems. Lecture 50.
1. Pleading—Illinois—Forms. 2. Equity pleading and procedure—Illinois—Forms. 3. Code pleading—United States—Forms. I. Title. II. Title: American law and procedure.
KFI1735.A65L37 1944 347'.773'072 78-317330
78 MARC

CODE REMEDIES see Remedies (Law)

CODE SWITCHING (LINGUISTICS)

Valdés-Fallis, Guadalupe.
Code switching and the classroom teacher / Guadalupe Valdés-Fallis ; prepared by ERIC Clearinghouse on Languages and Linguistics. — Arlington, Va. : Center for Applied Linguistics, 1978.
26 p. ; 23 cm. — (Language in education ; 4)
Bibliography: p. 22-26.
ISBN 0-87281-080-1
1. Code switching (Linguistics) 2. Education, Bilingual. I. Eric Clearinghouse on Languages and Linguistics. II. Title. III. Series.
P115.3.V34 401 79-114290
79 MARC

CODES see Ciphers

CODES, ERROR-CORRECTING see Error-correcting codes (Information theory)

CODES FOR PUNCHED CARD SYSTEMS see Punched card systems

CODETERMINATION (INDUSTRIAL RELATIONS) see Employees' representation in management

CODEX MENDOZA

Vollemaere, Antoon Léon.
Codex Mendoza : topogliefen / Antoon L. Vollemaere. — 2800 Mechelen, Nieuw Beggaardenstraat 15 : Vlaams Instituut voor Amerikanistiek v.z.w., 1977.
⌈184⌉ leaves : ill. ; 30 cm. — (America antiqua) Be78-4
Bibliography: leaves 151-152.
375F
1. Codex Mendoza. 2. Aztec language—Glossaries, vocabularies, etc. 3. Aztecs—Names. 4. Indians of Mexico—Names. I. Title. II. Series.
F1219.V69 78-394545
*78 MARC

CODICES see Manuscripts

CODICILS see Wills

CODING, PRODUCT see Product coding

CODING THEORY

see also Data compression (Computer science); Input design, Computer; Programming (Electronic computers)

Billingsley, Patrick.
Ergodic theory and information / Patrick Billingsley. — Huntington, N.Y. : R. E. Krieger Pub. Co., 1978, c1965.
xiii, 193 p. ; 24 cm.
Reprint of the ed. published by Wiley, N.Y., in series: Wiley series on the science and technology of materials.
Bibliography: p. 181-185.
Includes indexes.
ISBN 0-88275-666-4
1. Ergodic theory. 2. Statistical communication theory. 3. Coding theory. I. Title.
⌈QA313.B54 1978⌉ 519.2 78-2442
78 MARC

Ekberg, Jan, 1940-
Koodausteoria / Jan Ekberg, Seppo J. Halme. — ⌈Espoo⌉ : Otakustantamo, 1977.
112 p. : ill. ; 21 cm. Fi78-2
Bibliography: p. 112.
ISBN 9516711537
1. Coding theory. I. Halme, Seppo J. II. Title.
QA268.E38 78-395105
79 MARC

Image transmission techniques / edited by William K. Pratt. — New York : Academic Press, 1979.
xv, 281 p. : ill. ; 24 cm. — (Advances in electronics and electron physics : Supplement ; 12)
Includes bibliographies and index.
ISBN 0-12-014572-3
1. Image transmission. 2. Coding theory. I. Pratt, William K. II. Series.
TK5105.2.I42 621.38'0414 79-118795
79 MARC

Perceptual coding / edited by Edward C. Carterette and Morton P. Friedman. — New York : Academic Press, 1978.
xviii, 417 p. : ill. ; 24 cm. — (Handbook of perception ; v. 8)
Includes bibliographies and indexes.
ISBN 0-12-161908-7
1. Perception. 2. Coding theory. 3. Human information processing. I. Carterette, Edward C. II. Friedman, Morton P.
BF311.P362 153.7 78-16753
78 MARC

Viterbi, Andrew J
Principles of digital communication and coding / Andrew J. Viterbi, Jim K. Omura. — New York : McGraw-Hill, c1979.
xiii, 560 p. : ill. ; 25 cm. — (Communications and information theory) (McGraw-Hill series in electrical engineering)
Bibliography: p. 547-552.
Includes index.
ISBN 0-07-067516-3
1. Digital communications. 2. Coding theory. I. Omura, Jim K., joint author. II. Title. III. Series. IV. Series: McGraw-Hill series in electrical engineering.
TK5103.7.V57 621.38 78-13951
78 MARC

CODY, WILLIAM FREDERICK, 1846-1917

Cody, William Frederick, 1846-1917.
The life of Hon. William F. Cody, known as Buffalo Bill, the famous hunter, scout, and guide : an autobiography / foreword by Don Russell. — Lincoln : University of Nebraska Press, [1978]
xviii, 365 p. : ill. ; 21 cm.
"A Bison book."
Reprint of the 1879 ed. published by F. E. Bliss, Hartford; with new introd.
ISBN 0-8032-1406-5. ISBN 0-8032-6303-1 pbk.
1. Cody, William Frederick, 1846-1917. 2. Pioneers—The West—Biography. 3. Scouts and scouting—The West—Biography. 4. Frontier and pioneer life—The West. I. Title.
F594.C6745 1978 978'.02'0924 78-18732
78 MARC

COE FAMILY

Coe, Paul E 1919-
Robert Elliot Coe and his ancestors, 1340-1978 / by Paul E. Coe, Jr. and Robert E. Coe. — New Orleans : Coe, 1978.
235 p., [2] leaves of plates (p. 234-235 blank) : ill. ; 29 cm.
1. Coe family. I. Coe, Robert E., 1963- joint author. II. Title.
CS71.C672 1978 929'.2'0973 78-58986
79 MARC

COEDUCATION

see also Education; Education of women

Galli, Norberto.
Pedagogia della coeducazione / Norberto Galli. — [Brescia] : La scuola, [1977]
512 p. ; 25 cm. — ([Educazione e cultura ; 19]) It78-May
Bibliography: p. 449-498.
Includes index.
L9600
1. Coeducation. I. Title.
LB1601.G34 78-390323
*79 MARC

COEFFICIENT OF ELASTICITY see Elasticity (Economics)

COEFFICIENT OF EXPANSION see Expansion (Heat)

COELENTERATA

see also Anthozoa; Medusae; Sponges

—CLASSIFICATION

Cornelius, Paul Frederick Sinel.
A revision of the species of Lafoeidae and Haleciidae (Coelenterata, Hydroida) recorded from Britain and nearby seas / by Paul Frederick Sinel. — London : British Museum (Natural History), 1975.
p. 376-426 : ill. ; 25 cm. — (Bulletin of the British Museum (Natural History) : Zoology ; v. 28, no. 8 ISSN 0007-1498) GB***
Bibliography: p. 418-423.
Includes index.
£3.60
1. Lafoeidae. 2. Haleciidae. 3. Coelenterata—Classification. 4. Coelenterata—Great Britain. I. Title: A revision of the species of Lafoeidae and Haleciidae (Coelenterata, Hydroida) ... II. Series: British Museum (Natural History) Bulletin : Zoology ; v. 28, no. 8.
QL1.B75 vol. 28, No. 8 591'.08 s 78-321398
[QL377.H9] 79 MARC

Gardiner, John Stanley, 1872-1946.
Madreporaria excluding Flabellidae and Turbinolidae, by J. Stanley Gardiner and Peggy Waugh. The ecology of solitary corals, by J. Stanley Gardiner. London, Printed by order of the Trustees of the British Museum, 1939.
226-250 p. illus. 32 cm. (The John Murray Expedition, 1933-34. Scientific reports, v. 6, no. 5)
At head of title: British Museum (Natural History)
Cover title.
Includes bibliographical references.
1. Oculinidae—Classification. 2. Madreporaria—Ecology. 3. Coelenterata—Classification. I. Waugh, Peggy, joint author. II. Gardiner, John Stanley, 1872-1946. The ecology of solitary corals. 1939. III. British Museum (Natural History) IV. Title. V. Series: John Murray Expedition, 1933-1934. Scientific reports, v. 6, no. 5.
Q115.J55 vol. 6, no. 5 508.3'165 s 79-304291
[QL377.C5] 79 MARC

Hickson, Sydney John, 1859-1940.
The Pennatulacea. London, Printed by Order of the Trustees of the British Museum, 1937.
110-130 p. illus. 32 cm. (The John Murray Expedition, 1933-34. Scientific reports, v. 4, no. 5)
At head of title: British Museum (Natural History)
Cover title.
Bibliography: p. 129-130.
1. Sea-pens—Classification. 2. Coelenterata—Classification. I. British Museum (Natural History) II. Series: John Murray Expedition, 1933-1934. Scientific reports, v. 4, no. 5.
Q115.J55 vol. 4, no. 5 508.3'165 s 79-304294
[QL337.C6] 79 MARC

—DEVELOPMENT

Widersten, Bernt.
Studies on development in Actiniaria (Anthozoa) and semaeostomeae (scyphozoa). Uppsala [Almqvist & Wiksell] 1972.
12 p. 24 cm. (Abstracts of Uppsala dissertations from the Faculty of Science, 217) (Acta universitatis Upsaliensis) S***
Akademisk avhandling—Uppsala.
Extra t.p. with thesis statement inserted.
"The thesis is based on ... [5] papers published by the author."
Bibliography: p. 12.
1. Sea-anemones—Development. 2. Semaeostomeae—Development. 3. Coelenterata—Development. I. Title. II. Series. III. Series: Uppsala. Universitet. Acta Universitatis Upsaliensis.
QL377.C5W53 593'.6 78-316015
78 MARC

—EVOLUTION

Schmidt, Hajo.
Die Spermien der Anthozoen und ihre phylogenetische Bedeutung / Hajo Schmidt und Dieter Zissler. — Stuttgart : E. Schweizerbart, 1979.
viii, 96 p. : ill. ; 30 cm. — (Zoologica, Originalabhandlungen aus dem Gesamtgebiet der Zoologie ; Bd. 44, Lfg. 3, Heft 129) GFR***
Summary also in English.
Bibliography: p. 40-46.
ISBN 3-510-55015-3
1. Anthozoa—Reproduction. 2. Anthozoa—Evolution. 3. Spermatozoa. 4. Coelenterata—Reproduction. 5. Coelenterata—Evolution. I. Zissler, Dieter, joint author. II. Title. III. Series: Zoologica, Original-Abhandlungen aus dem Gesamtgebiet der Zoologie ; Bd. 44, Lfg. 3, Heft 129.
QL377.C5S36 593'.6'04166 79-369126
79 MARC

—REPRODUCTION

Schmidt, Hajo.
Die Spermien der Anthozoen und ihre phylogenetische Bedeutung / Hajo Schmidt und Dieter Zissler. — Stuttgart : E. Schweizerbart, 1979.
viii, 96 p. : ill. ; 30 cm. — (Zoologica, Originalabhandlungen aus dem Gesamtgebiet der Zoologie ; Bd. 44, Lfg. 3, Heft 129) GFR***
Summary also in English.
Bibliography: p. 40-46.
ISBN 3-510-55015-3
1. Anthozoa—Reproduction. 2. Anthozoa—Evolution. 3. Spermatozoa. 4. Coelenterata—Reproduction. 5. Coelenterata—Evolution. I. Zissler, Dieter, joint author. II. Title. III. Series: Zoologica, Original-Abhandlungen aus dem Gesamtgebiet der Zoologie ; Bd. 44, Lfg. 3, Heft 129.
QL377.C5S36 593'.6'04166 79-369126
79 MARC

—ATLANTIC COAST (CANADA)

Shih, Chang-tai.
A guide to the jellyfish of Canadian Atlantic waters / C. T. Shih. — Ottawa : National Museum of Canada, c1977.
90 p. : ill. (some col.) ; 26 cm. — (Natural history series ; no. 5 ISSN 0317-5642) C***
Summaries in English and French.
Bibliography: p. 83-84.
Includes index.
ISBN 0-660-00017-2
1. Medusae—Identification. 2. Coelenterata—Atlantic coast (Canada) I. Title. II. Series: Natural history series (Ottawa) ; no. 5.
QL377.S4S54 593'.71 78-319021
78 MARC

—GREAT BRITAIN

Cornelius, Paul Frederick Sinel.
A revision of the species of Lafoeidae and Haleciidae (Coelenterata, Hydroida) recorded from Britain and nearby seas / by Paul Frederick Sinel. — London : British Museum (Natural History), 1975.
p. 376-426 : ill. ; 25 cm. — (Bulletin of the British Museum (Natural History) : Zoology ; v. 28, no. 8 ISSN 0007-1498) GB***
Bibliography: p. 418-423.
Includes index.
£3.60
1. Lafoeidae. 2. Haleciidae. 3. Coelenterata—Classification. 4. Coelenterata—Great Britain. I. Title: A revision of the species of Lafoeidae and Haleciidae (Coelenterata, Hydroida) ... II. Series: British Museum (Natural History) Bulletin : Zoology ; v. 28, no. 8.
QL1.B75 vol. 28, No. 8 591'.08 s 78-321398
[QL377.H9] 79 MARC

—UNITED STATES

Agassiz, Louis, 1807-1873.
Contributions to the natural history of the United States of America / Jean Louis Rodolphe Agassiz. — New York : Arno Press, 1978.
4 v. in 2 ; 24 cm. — (Biologists and their world)
Reprint of the 1857-1862 ed. published by Little, Brown, Boston.
ISBN 0-405-10675-0
1. Zoology—United States. 2. Turtles—United States. 3. Coelenterata—United States. 4. Zoology—Classification. I. Title. II. Series.
QL155.A53 1978 591 77-81094
78 MARC

COELMONT, JOANNES LEONARDUS, 1776-1865

Stevaux, K
J. L. Coelmont, "agent municipal" en "maire" te Brustem / K. Stevaux. — Hasselt : Provinciale Bibliotheek, 1976.
40 p. ; 24 cm. — (Mededelingen van het Centrum voor Studie van de Boerenkrijg ; nr. 89) Be***
Dutch or French.
Includes bibliographical references.
1. Coelmont, Joannes Leonardus, 1776-1865. 2. Brustem, Belgium—Biography. I. Title. II. Series: Centrum voor Studie van de Boerenkrijg. Mededelingen ; nr. 89.
DH631.C45 Nr. 89 79-356120
[DH811.B88] 79 MARC

COELOLEPIDA

Karatajūtė-Talimaa, Valentina, 1930-
(Telodonty silura i devona SSSR i Shpitsbergena)
Телодонты силура и девона СССР и Шпицбергена = Silurian and Devonian Thelodonts of the USSR and Spitsbergen / В. Н. Каратаюте-Талимаа. — Вильнюс : Мокслас, 1978.
384 p. : ill. ; 27 cm. USSR 78
At head of title: Upravlenie geologii pri SM Litovskoĭ SSR. Litovskiĭ nauchno-issledovatel'skiĭ geologorazvedochnyĭ institut.
Added t. p.: TSRS ir špicbergeno silūro ir devono telodontai.
Bibliography: p. 197-205.
2.10rub
1. Coelolepida. 2. Paleontology—Silurian. 3. Paleontology—Devonia. 4. Paleontology—Russia. 5. Paleontology—Svalbard. I. Title. II. Title: Silurian and Devonian Thelodonts of the USSR and Spitsbergen. III. Title: TSRS ir špicbergeno silūro ir devono telodontai.
QE852.C6K37 79-410422

COELOSTEMMA

—CLASSIFICATION

Thompson, Fred Gilbert, 1934-
Some Mexican land snails of the genera Coelostemma and Metastoma (Urocoptidae) [by] Fred G. Thompson. Gainesville, University of Florida, 1971.
267-302 p. illus. 23 cm. (University of Florida. Bulletin of the Florida State Museum. Biological sciences, v. 15, no. 5) $0.50
Cover title.
Bibliography: p. 302.
1. Coelostemma—Classification. 2. Metastoma—Classification. 3. Mollusks—Classification. 4. Mollusks—Mexico. I. Title. II. Series: Florida. University, Gainesville. State Museum. Bulletin. Biological sciences, v. 15, no. 5.
QH1.F6 vol. 15, no. 5 574'.08 s 78-321905
[QL430.5.U68] 79 MARC

COEN, CORNELIUS, JANSZOON

Coen, Cornelis Janszoon.
Voyage to Cathay, Tartary and the gold- and silver-rich islands east of Japan, 1643 : the journal of Cornelis Jansz. Coen relating to the voyage of Marten Gerritsz. Fries to the north and east of Japan / original Dutch text and English translation, edited, with an introd., other relevant material and documents, notes ... by Willem C. H. Robert. — Amsterdam : Philo Press, 1975.
xv, 294 p., [6] leaves of plates : ill., charts (1 fold. in pocket) ; 25 cm. — (Contributions to a bibliography of Australia and the Pacific : Supplement ; 4) Ne76-4
Bibliography: p. [285]-286.
Includes index.
ISBN 906022504X : fl 55.00
1. East Asia—Description and travel. 2. Fries, Maerten, Gerritszoon. 3. Coen, Cornelius, Janszoon. I. Robert, Willem Carel Hendrik. II. Title. III. Series: Contributions to a bibliography of Australia and the South Sea Islands : Supplement ; 4.
DS506.C58 915'.04'3 79-302078
*79 MARC

COEN CARDONE, LUCIANA

Coen Cardone, Luciana.
Il diritto di vivere / Luciana Coen Cardone. — Roma : Adda, 1976.
292 p. ; 21 cm. It77-Nov
Cover has subtitle: 44 anni di scandali dell'assistenza sociale.
L3500
1. Coen Cardone, Luciana. 2. Social workers—Italy—Biography. 3. Social service—Italy—History. I. Title.
HV28.C58A33 79-340360
*79 MARC

COENOBITA

Scermino, Daniele.
Fun with land hermit crabs : a guide to their well-being & general care / Daniele Scermino. — St. Petersburg, Fla. : Palmetto Pub. Co., [1978]
52 p. : ill. ; 22 cm. — (Mini pet reference series ; no. 3)
ISBN 0-915096-08-0 : $2.95
1. Hermit-crabs as pets. 2. Coenobita. I. Title.
SF459.H47S33 639'.54'2 78-16413
78 MARC

COENOBITIDAE

see also Coenobita

COERCION (LAW) see Duress (Law)

COESTER, OTTO, 1902-

—EXHIBITIONS

Coester, Otto, 1902-
Otto Coester : Graphik 1930-1960 : Von der Heydt-Museum, Wuppertal, 3.9.-22.10.1978. — Wuppertal : Von der Heydt-Museum, [1978]
[78] p. : chiefly ill. ; 26 cm. GFR***
Exhibition catalog.
Bibliography: p. [12]
1. Coester, Otto, 1902- —Exhibitions. I. Von der Heydt-Museum.
NE2050.5.C63A4 1978 79-360840
79 MARC

COEXISTENCE see Russia—Foreign relations—United States

COFÁN INDIANS

—RELIGION AND MYTHOLOGY

Lucena Salmoral, Manuel.
Las últimas creencias de los indios Kofan : mágia, selva y petróleo en al alto Putumayo / Manuel Lucena Salmoral. — Murcia : Universidad, Departamento de Historia de América, [1977]
119 p., [5] leaves of plates : ill. ; 22 cm. Sp78-Apr
Includes bibliographical references.
ISBN 8460008304
1. Cofán Indians—Religion and mythology. 2. Indians of South America—Religion and mythology. I. Title.
F2230.2.C64L8 79-343690
79 MARC

COFFEA ARABICA see Coffee

COFFEE

see also Cookery (Coffee)

Kolpas, Norman.
The coffee lovers' companion / by Norman Kolpas. — New York : Quick Fox, 1977.
157 p. : ill. ; 24 cm.
On cover: The history and romance, fact and legend, brewing tips and recipes about the world's favorite drink.
Originally published under title: The coffee connoisseur's companion.
ISBN 0-8256-3125-4 : $6.95
1. Coffee. I. Title.
TX415.K64 641.3'37'3 78-56218
78 MARC

Liao, Mao-k'un.
(K'a fei ju men)
咖啡入門 / [作者廖茂坤]. — [台]北 : 廖茂坤 ; [台北?] : 總經銷聯大圖書社, 民國67 [1978]
230 p., [3] leaves of plates : ill. ; 20 cm.
Title in English on p. [4] of cover: Coffee.
"紅茶與可可": p. 201-230.
NT$250.00

1. Coffee. I. Title. II. Title: Coffee.
TX415.L5 79-837155

Maronde, Curt.
Rund um den Kaffee : e. vergnügl. Weltbummel mit Kaffe nebst vielen Ratschlägen u. Rezepten für seine genussreichste Zubereitung / Curt Maronde. — Orig. -Ausg. — Frankfurt am Main : Fischer-Taschenbuch-Verlag, 1976.
125 p. : ill. ; 18 cm. GFR76-A
Includes index.
ISBN 3-436-02221-7 : DM3.80
1. Coffee. 2. Cookery (Coffee) I. Title.
GT2918.M37 641.3'37'3 79-377430
79 MARC

McDonald, John Holt, 1869-
Coffee growing: with special reference to East Africa, by J. H. McDonald ... London, East Africa, 1930.
xii, 205 p. front., illus., plates, diagr. 22 cm.
Plates printed on both sides.
Includes bibliographical references.
1. Coffee. 2. Coffee—Africa, East. I. Title.
SB269.M275 agr 30-587
[r79]rev MARC

Pulla, Armas Joosef, 1904-
Kahvi / Armas J. Pulla. — Helsinki : Otava, 1976.
127 p. : ill. ; 23 cm. Fi76
ISBN 9511024663 : Fmk46.00
1. Coffee. I. Title.
GT2918.P84 79-339651
*79 MARC

Ravohitrarivo,
Opération café : plan et étude sur l'économie rurale / Ravohitrarivo. — Tananarive : M.D.R.R.A., 1975.
[59] leaves ; 29 cm.
Cover title.
1. Coffee. 2. Coffee—Madagascar. I. Title.
SB269.R35 338.1'7'373 76-980464
78 MARC

—ANALYSIS—CONGRESSES

International Colloquium on the Chemistry of Coffee, 7th, Hamburg, 1975.
Septième Colloque international sur la chimie des cafés verts, torréfiés et leurs dérivés, Hambourg, 9-14 juin 1975. — Paris : Association international du café, [1978?]
576 p. : ill. ; 27 cm. F***
Articles in English, French, German, and Spanish, with summaries in the language of the text and in English.
Includes bibliographies.
1. Coffee—Analysis—Congresses. I. International Scientific Association of Coffee.
TX560.C63157 1975 663'.93'1 78-378972
78 MARC

—CONGRESSES

Congresso Brasileiro de Pesquisas Cafeeiras.
Resumos - Congresso Brasileiro de Pesquisas Cafeeiras. [Rio de Janeiro, Setor de Programação Visual e Gráfica, IBC, GERCA]
v. ill. 22 cm.
Conferences for sponsored by the Instituto Brasileiro do Café, Grupo Executivo de Racionalização da Cafeicultura.
1. Coffee—Brazil—Congresses. 2. Coffee—Research—Brazil—Congresses. 3. Coffee—Congresses. I. Instituto Brasileiro do Café. Grupo Executivo de Racionalização da Cafeicultura.
SB270.B7C65a 79-645218
MARC-S

Komisi Teknis Perkebunan.
Sidang Komisi Tehnis Perkebunan budidaja kopi-tjoklat. -3; 19 -71.
[Djember, Balai Penelitian Perkebunan Tjabang Djember]
v. in 28 cm. biennial.
Vols. for 19 -71 issued in 4 vols.
English or Indonesian.
Continued by: Komisi Teknis Perkebunan. Sidang Komisi Tehnis Perkebunan budidaya kopi-coklat.
1. Coffee—Indonesia—Congresses. 2. Coffee—Congresses. 3. Cacao—Indonesia—Congresses. 4. Cacao—Congresses. I. Title.
SB270.I 6K65a 78-643333
MARC-S

Komisi Teknis Perkebunan.
Sidang Komisi Tehnis Perkebunan budidaya kopi-coklat. 4.- 1973-
[s. l., s. n.]
v. in 30 cm. biennial.
Vols. for 1973- issued in 4 vols.
Continues: Komisi Teknis Perkebunan. Sidang Komisi Tehnis Perkebunan budidaja kopi-tjoklat.
English or Indonesian.
1. Coffee—Indonesia—Congresses. 2. Coffee—Congresses. 3. Cacao—Indonesia—Congresses. 4. Cacao—Congresses. I. Title.
SB270.I 6K65a 79-640097
MARC-S

—EARLY WORKS TO 1800

Lavedán, Antonio.
Tratado de los usos, abusos, propiedades y virtudes del tabaco, café, té y chocolate; extractado de los mejores autores que han tratado de esta materia, á fin de que su uso no perjudique á la salud, antes bien pueda servir de alivio y curación de muchos males. Madrid, En la Imprenta Real, 1796.
237 p. 20 cm.
On spine: Usos del tabaco.
1. Health—Early works to 1800. 2. Tobacco—Early works to 1800. 3. Coffee—Early works to 1800. 4. Tea—Early works to 1800. 5. Chocolate—Early works to 1800. I. Title. II. Title: Usos del tabaco.
RA775.L68 78-411600
78 MARC

—ECONOMIC ASPECTS—BRAZIL—MINAS GERAIS

Empresa de Pesquisa Agropecuária de Minas Gerais.
Diagnóstico da cafeicultura da zona sul de Minas Gerais / [os autores, Glória Zélia Teixeira Caixeta ... et al.]. — Belo Horizonte : Empresa de Pesquisa Agropecuária de Minas Gerais, 1975.
103 p. : ill. ; 30 cm. — (Série Programação ; no. 3)
"Acordo IBC-DAC/Governo do Estado de Minas Gerais."
Includes bibliographical references.
1. Coffee—Brazil—Minas Gerais. 2. Coffee—Economic aspects—Brazil—Minas Gerais. I. Caixeta, Glória Zélia Teixeira. II. Title. III. Series.
SB270.B7E46 1975 78-385731
78 MARC

—PROCESSING see Coffee processing

—RESEARCH—BRAZIL—CONGRESSES

Congresso Brasileiro de Pesquisas Cafeeiras.
Resumos - Congresso Brasileiro de Pesquisas Cafeeiras. [Rio de Janeiro, Setor de Programação Visual e Gráfica, IBC, GERCA]
v. ill. 22 cm.
Conferences for sponsored by the Instituto Brasileiro do Café, Grupo Executivo de Racionalização da Cafeicultura.
1. Coffee—Brazil—Congresses. 2. Coffee—Research—Brazil—Congresses. 3. Coffee—Congresses. I. Instituto Brasileiro do Café. Grupo Executivo de Racionalização da Cafeicultura.
SB270.B7C65a 79-645218
MARC-S

—TAXATION—ETHIOPIA

Teshome Mulat.
A short note on the revenue effectiveness of Ethiopian coffee export taxation / Teshome Mulat. — [Addis Ababa] : Dept. of Economics, Addis Ababa University, 1977.
10, [2] leaves ; 28 cm. — (Occasional paper - Department of Economics, Addis Ababa University ; no. 9)
1. Coffee—Taxation—Ethiopia. I. Addis Ababa. Haile Sellassie I University. Dept. of Economics. II. Title. III. Series: Occasional economic paper ; no. 9.
HD9199.E782T473 336.2'63 77-980770
78 MARC

COFFEE (Continued)

—TAXATION—KENYA

Westlake, M J
Taxation and control of the Kenyan coffee industry / by M. J. Westlake. — Nairobi : Institute for Development Studies, University of Nairobi, 1973.
16 p. ; 30 cm. — (Discussion paper - Institute for Development Studies, University of Nairobi ; no. 185)
Includes bibliographical references.
1. Coffee—Taxation—Kenya. 2. Coffee trade—Kenya. I. Title. II. Series: Nairobi. University. Institute for Development Studies. Discussion paper ; no. 185.
HD9199.K42W47 78-107567
336.2'78'63373096762
79 MARC

—AFRICA, EAST

McDonald, John Holt, 1869-
Coffee growing: with special reference to East Africa, by J. H. McDonald ... London, East Africa, 1930.
xii, 205 p. front., illus., plates, diagr. 22 cm.
Plates printed on both sides.
Includes bibliographical references.
1. Coffee. 2. Coffeee—Africa, East. I. Title.
SB269.M275 agr 30-587
[r79]rev MARC

—BRAZIL—BIBLIOGRAPHY

Brazil. Coordenação de Informação Rural.
Bibliografia do café : projeto PNUD, FAO, BRA, 72,020 : Sistema Nacional de Informação e Documentação Agrícola / Ministério da Agricultura, Secretaria Geral, Coordenação de Informação Rural. — Brasília : A Coordinação, 1975-
v. ; 29 cm.
"DOC/TEC/75/004" (v. 1) "DOC/TEC/77/038" (v. 2)
Vol. 2 issued by the Sistema Nacional de Informação Rural and the Instituto Brasileiro do Café.
Includes indexes.
1. Coffee trade—Brazil—Bibliography. 2. Coffee—Brazil—Bibliography. I. Instituto Brasileiro do Café. II. Title.
Z5601.B7 1975 76-471837
016.3381'7'3730981
[HD9199.B69] 76[r79]rev MARC

—BRAZIL—CONGRESSES

Congresso Brasileiro de Pesquisas Cafeeiras.
Resumos - Congresso Brasileiro de Pesquisas Cafeeiras. [Rio de Janeiro, Setor de Programação Visual e Gráfica, IBC, GERCA]
v. ill. 22 cm.
Conferences for sponsored by the Instituto Brasileiro do Café, Grupo Executivo de Racionalização da Cafeicultura.
1. Coffee—Brazil—Congresses. 2. Coffee—Research—Brazil—Congresses. 3. Coffee—Congresses. I. Instituto Brasileiro do Café. Grupo Executivo de Racionalização da Cafeicultura.
SB270.B7C65a 79-645218
MARC-S

—BRAZIL—HISTORY—ADDRESSES, ESSAYS, LECTURES

O café no Brasil / [capa de AGGS Indústrias Gráficas S/A]. — Rio de Janeiro : Ministério da Indústria e do Comércio, Instituto Brasileiro do Café, 1978.
180 p. ; 24 cm.
Includes bibliographies.
CONTENTS: Reis Júnior, J. C. G. dos. Prefácio.—Pessamilio, H. M. R. A dinâmica social do café.—Oliveira, F. de. O café e o norte do Paraná.—Boianovsky, M. A expansão cafeeira e o movimento imigratório.—Bittencourt, G. A. de M. O café na formação da infra-estrutura capixaba.
1. Coffee trade—Brazil—History—Addresses, essays, lectures. 2. Coffee—Brazil—History—Addresses, essays, lectures. 3. Brazil—Social conditions—Addresses, essays, lectures. I. Instituto Brasileiro do Café.
HD9199.B7C2 79-112000
79 MARC

—BRAZIL—MINAS GERAIS

Empresa de Pesquisa Agropecuária de Minas Gerais.
Diagnóstico da cafeicultura da zona sul de Minas Gerais / [os autores, Glória Zélia Teixeira Caixeta ... et al.]. — Belo Horizonte : Empresa de Pesquisa Agropecuária de Minas Gerais, 1975.
103 p. : ill. ; 30 cm. — (Série Programação ; no. 3)
"Acordo IBC-DAC/Governo do Estado de Minas Gerais."
Includes bibliographical references.
1. Coffee—Brazil—Minas Gerais. 2. Coffee—Economic aspects—Brazil—Minas Gerais. I. Caixeta, Glória Zélia Teixeira. II. Title. III. Series.
SB270.B7E46 1975 78-385731
78 MARC

—INDONESIA—CONGRESSES

Komisi Teknis Perkebunan.
Sidang Komisi Tehnis Perkebunan budidaja kopi-tjoklat.
-3; 19 -71.
[Djember, Balai Penelitian Perkebunan Tjabang Djember]
v. in 28 cm. biennial.
Vols. for 19 -71 issued in 4 vols.
English or Indonesian.
Continued by: Komisi Teknis Perkebunan. Sidang Komisi Tehnis Perkebunan budidaya kopi-coklat.
1. Coffee—Indonesia—Congresses. 2. Coffee—Congresses. 3. Cacao—Indonesia—Congresses. 4. Cacao—Congresses. I. Title.
SB270.I 6K65a 78-643333
MARC-S

Komisi Teknis Perkebunan.
Sidang Komisi Tehnis Perkebunan budidaya kopi-coklat.
4.- 1973-
[s. l., s. n.]
v. in 30 cm. biennial.
Vols. for 1973- issued in 4 vols.
Continues: Komisi Teknis Perkebunan. Sidang Komisi Tehnis Perkebunan budidaja kopi-tjoklat.
English or Indonesian.
1. Coffee—Indonesia—Congresses. 2. Coffee—Congresses. 3. Cacao—Indonesia—Congresses. 4. Cacao—Congresses. I. Title.
SB270.I 6K65a 79-640097
MARC-S

—INDONESIA—STATISTICS

Indonesia. Biro Pusat Statistik.
Sensus dan survey kopi Indonesia, tahun 1974. — [Jakarta] : Biro Pusat Statistik, 1975.
xiii, 34 p. ; 21 cm.
English and Indonesian.
1. Coffee—Indonesia—Statistics. I. Title.
SB270.I6I52 1975 77-940692
79 MARC

—KENYA—KISII (DISTRICT)

Barnes, Carolyn.
The introduction of coffee in Gusiiland, Kenya, 1933-1942 / by Carolyn Barnes. — Nairobi : Institute for Development Studies, University of Nairobi, 1974.
27 p. ; 30 cm. — (Discussion paper - Institute for Development Studies, University of Nairobi ; no. 205)
Includes bibliographical references.
1. Coffee—Kenya—Kisii (District) 2. Plant introduction—Kenya—Kisii (District) 3. Gusii (Bantu tribe) I. Title. II. Series: Nairobi. University. Institute for Development Studies. Discussion paper ; no. 205.
SB270.K4B37 78-106919
79 MARC

—LATIN AMERICA

Buzzanell, Peter J
Coffee production and trade in Latin America / by Peter J. Buzzanell. — [Washington] : Foreign Agricultural Service, U.S. Dept. of Agriculture, 1979.
82 p., : ill. ; 27 cm. — (FAS M ; 288)
Bibliography: p. 72.
1. Coffee trade—Latin America. 2. Coffee—Latin America. I. Title. II. Series: United States. Foreign Agricultural Service (1953-) FAS M ; 288.
HD9199.L382B89 338.1'7'373098 79-602711
79 MARC

—MADAGASCAR

Ravohitrarivo,
Opération café : plan et étude sur l'économie rurale / Ravohitrarivo. — Tananarive : M.D.R.R.A., 1975.
[59] leaves ; 29 cm.
Cover title.
1. Coffee. 2. Coffee—Madagascar. I. Title.
SB269.R35 338.1'7'373 76-980464
78 MARC

—SUDAN—AL MUDĪRĪYAH AL ISTIWĀ'ĪYAH AL GHABĪYAH

al-Munaẓẓamah al-'Arabīyah lil-Tanmiyah al-Zirā'īyah.
A feasibility study for development of coffee production in East and West Equatoria Provinces, Democratic Republic of the Sudan / League of Arab States, Arab Organization for Agricultural Development. — Khartoum : The Organization, 1977.
xii, 378 p. : maps ; 28 cm.
Includes bibliographical references.
1. Coffee—Sudan—al Mudīrīyah al Istiwā'īyah al Ghabīyah. 2. Coffee—Sudan—al Mudīrīyah al Istiwā'īyah ash Sharqīyah. 3. al Mudīrīyah al Istiwā'īyah al Gharbīyah, Sudan—Economic conditions. 4. al Mudīrīyah al Istiwā'īyah ash Sharqīyah, Sudan—Economic conditions. I. Title.
SB270.S73M86 338.1'7'373096295 78-980370
79 MARC

—SUDAN—AL MUDĪRĪYAH AL ISTIWĀ'ĪYAH ASH SHARQĪYAH

al-Munaẓẓamah al-'Arabīyah lil-Tanmiyah al-Zirā'īyah.
A feasibility study for development of coffee production in East and West Equatoria Provinces, Democratic Republic of the Sudan / League of Arab States, Arab Organization for Agricultural Development. — Khartoum : The Organization, 1977.
xii, 378 p. : maps ; 28 cm.
Includes bibliographical references.
1. Coffee—Sudan—al Mudīrīyah al Istiwā'īyah al Ghabīyah. 2. Coffee—Sudan—al Mudīrīyah al Istiwā'īyah ash Sharqīyah. 3. al Mudīrīyah al Istiwā'īyah al Gharbīyah, Sudan—Economic conditions. 4. al Mudīrīyah al Istiwā'īyah ash Sharqīyah, Sudan—Economic conditions. I. Title.
SB270.S73M86 338.1'7'373096295 78-980370
79 MARC

COFFEE BREAKS see Rest periods

COFFEE CO., GA.

—GENEALOGY

Ward, Warren P
Ward's History of Coffee County / by Warren P. Ward ; with new index by Margaret H. Cannon. — Spartanburg, S.C. : Reprint Co., 1978 [c1930]
xiv, 388 p. : ill. ; 22 cm.
Reprint of the ed. published by Foote & Davies Co., Atlanta.
Includes index.
ISBN 0-87152-290-X : $17.50
1. Coffee Co., Ga.—History. 2. Coffee Co., Ga.—Genealogy. I. Title. II. Title: History of Coffee County.
F292.C63W3 1978 975.8'823 78-13236
78 MARC

—HISTORY

Ward, Warren P
Ward's History of Coffee County / by Warren P. Ward ; with new index by Margaret H. Cannon. — Spartanburg, S.C. : Reprint Co., 1978 [c1930]
xiv, 388 p. : ill. ; 22 cm.
Reprint of the ed. published by Foote & Davies Co., Atlanta.
Includes index.
ISBN 0-87152-290-X : $17.50
1. Coffee Co., Ga.—History. 2. Coffee Co., Ga.—Genealogy. I. Title. II. Title: History of Coffee County.
F292.C63W3 1978 975.8'823 78-13236
78 MARC

—HISTORY—MISCELLANEA

Have a cup of coffee : courtesy of the Coffee Junior High School English students in A-1, '75-'76. — [Douglas? Ga.] : The School, [1976]
107 p. : ill. ; 22 cm.
1. Coffee Co., Ga.—History—Miscellanea. I. Coffee Junior High School.
F292.C63H38 975.8'823 78-104997
79 MARC

COFFEE-HOUSES

—AUSTRIA—VIENNA—HISTORY

Das Wiener Kaffeehaus / einleitender Essay v. Hans Weigel; Text-u. Bildausw. v. Christian Brandstätter u. Werner J. Schweiger. Mit 121 Abb. — Wien ; München ; Zürich : Molden Edition, c1978.
152 p. : ill. ; 30 cm. Au78-15-277
Bibliography: p. 150-151.
ISBN 3-217-00881-2 : S490.00
1. Coffee-houses—Austria—Vienna—History. 2. Vienna—Social life and customs. I. Weigel, Hans. II. Brandstätter, Christian. III. Schweiger, Werner J., 1949-
TX910.A8W53 647'.95436'13 79-338395
*79 MARC

—ITALY

Stecchi de Bellis, Maria Antonietta.
Tra cronache e immagini di vecchi caffè / Maria Antonietta Stecchi de Bellis. — Fasano : Schena, 1976.
197 p., [1] leaf of plates : ill. ; 25 cm. It78-Sept
Selections of works by various authors included in "Appendice" (p. [121]-197)
Includes indexes.
L6000
1. Italy—Intellectual life. 2. Coffee-houses—Italy. I. Title.
DG442.S68 945 79-364939
*79 MARC

COFFEE-HOUSES (Continued)

—JAPAN

(Kōhī no mise)
珈琲の店. — 東京 : 柴田書店, 昭和50 ₍1975₎
210 p. : chiefly ill. ; 27 cm. — (食堂建築秀作シリーズ ; 4) Ja***
¥2500

1. Coffee-houses—Japan. 2. Interior decoration—Japan. I. Shibata Shoten. II. Series: Shokudō kenchiku shūsaku shirīzu ; 4.
NA7855.S5 vol. 4 79-800520

COFFEE PROCESSING

Sivetz, Michael.
Coffee technology / Michael Sivetz, Norman W. Desrosier. — Westport, Conn. : AVI Pub. Co., c1979.
vii, 716 p. : ill. ; 24 cm.
Edition of 1963 published under title: Coffee processing technology.
Includes bibliographies and index.
ISBN 0-87055-269-4 : $31.00
1. Coffee processing. I. Desrosier, Norman W., joint author. II. Title.
TP645.S5 1979 663'.93 79-10538
79 MARC

COFFEE TRADE

—INFORMATION SERVICES

United Nations. Industrial Development Organization.
Information sources on the coffee, cocoa, tea, and spices industry / United Nations Industrial Development Organization. — New York : United Nations, 1977.
xii, 74 p. ; 21 cm. — (UNIDO guides to information sources ; no. 28) (₍Document₎ - United Nations ; ID/198 (UNIDO/LIB/SER.D/28))
"Bibliographical sources used in compiling the present guide": p. 73-74.
$4.00
1. Coffee trade—Information services. 2. Cocoa trade—Information services. 3. Tea trade—Information services. 4. Spice trade—Information services. I. Title. II. Series: United Nations. Industrial Development Organization. UNIDO guides to information sources ; no. 28. III. Series: United Nations. Document ; ID/198 (UNIDO/LIB/SER.D/28)
JX1977.A2 ID/198(UNIDO/LIB/SER.D/28) 79-306816
300'.8 s
——— Copy 2. HD9199.A2U45 1977 MARC
79₍r79₎rev

—BRAZIL

Escola Interamericana de Administração Pública.
Instrumentos da política cafeeira / Fundação Getúlio Vargas, Banco Interamericano de Desenvolvimento, Escola Interamericana de Administração Pública. — Rio de Janeiro : A Escola, 1967.
114 leaves : ill. ; 29 cm. — (Série monográfica : instrumentos administrativos de implementação econômica ; no. 9, v. 2)
1. Coffee trade—Brazil. 2. Brazil—Economic policy. I. Title.
HD9199.B7E62 1967 79-112957
79 MARC

International Coffee Organization (founded 1962)
Resumo de plano nacional de política cafeeira do Brasil. — ₍Londres?₎ : International Coffee Organization, 1970.
2 v. : ill. ; 28 cm.
Cover title.
On t.p.: Fundo de Diversificação-Directoria, 18-21 maio 1970, Londres, Inglaterra.
Half title: Plano nacional de política cafeeira.
CONTENTS: pt. 1. Resumo do Plano.—pt. 2. Relatório do diretor-executivo.
1. Coffee trade—Brazil. 2. Industry and state—Brazil. I. Title. II. Title: Plano nacional de política cafeeira.
HD9199.B7I62 1970 79-106322
79 MARC

—BRAZIL—BIBLIOGRAPHY

Brazil. Coordenação de Informação Rural.
Bibliografia do café : projeto PNUD, FAO, BRA, 72,020 : Sistema Nacional de Informação e Documentação Agrícola / Ministério da Agricultura, Secretaria Geral, Coordenação de Informação Rural. — Brasília : A Coordinação, 1975-
v. ; 29 cm.
"DOC/TEC/75/004" (v. 1) "DOC/TEC/77/038" (v. 2)
Vol. 2 issued by the Sistema Nacional de Informação Rural and the Instituto Brasileiro do Café.
Includes indexes.
1. Coffee trade—Brazil—Bibliography. 2. Coffee—Brazil—Bibliography. I. Instituto Brasileiro do Café. II. Title.
Z5601.B7 1975 76-471837
016.3381'7'3730981
₍HD9199.B69₎ 76₍r79₎rev MARC

—BRAZIL—HISTORY—ADDRESSES, ESSAYS, LECTURES

O café no Brasil / ₍capa de AGGS Indústrias Gráficas S/A₎. — Rio de Janeiro : Ministério da Indústria e do Comércio, Instituto Brasileiro do Café, 1978.
180 p. ; 24 cm.
Includes bibliographies.
CONTENTS: Reis Júnior, J. C. G. dos. Prefácio.—Pessamilio, H. M. R. A dinâmica social do café.—Oliveira, F. de. O café e o norte do Paraná.—Boianovsky, M. A expansão cafeeira e o movimento imigratório.—Bittencourt, G. A. de M. O café na formação da infra-estrutura capixaba.
1. Coffee trade—Brazil—History—Addresses, essays, lectures. 2. Coffee—Brazil—History—Addresses, essays, lectures. 3. Brazil—Social conditions—Addresses, essays, lectures. I. Instituto Brasileiro do Café.
HD9199.B7C2 79-112000
79 MARC

—BRAZIL—HISTORY—CONGRESSES

Congresso de História de São Paulo, 2d, Araraquara, Brazil, 1974.
O café : anais do II Congresso de História de São Paulo, Araraquara, 2 a 7 de setembro de 1974 / Associação Nacional dos Professores Universitários de História, Núcleo Regional de São Paulo ; organizado pelo Prof. Edgard Carone. — São Paulo : A Associação, 1975.
404 p. ; 24 cm. — (Coleção da Revista de História ; 59)
1. Coffee trade—Brazil—History—Congresses. I. Carone, Edgard. II. Associação Nacional dos Professores Universitários de História. Núcleo Regional de São Paulo. III. Title.
HD9199.B7C57 1974 338.1'7'3730981 79-352700
79 MARC

—BRAZIL—STATISTICS

Instituto Brasileiro do Café.
Movimento da comercialização de café segundo estados, municípios e micro-regiões homogêneas / Departamento Econômico, Divisão de Estatística (fil. à FIBGE). — ₍Rio de Janeiro₎ : Ministério da Indústria e do Comércio, Instituto Brasileiro do Café, ₍1976 or 1977₎
73 p. : maps ; 22 cm.
1. Coffee trade—Brazil—Statistics. I. Title.
HD9199.B7I58 1976 79-345605
79 MARC

—CANADA—STATISTICS —PERIODICALS

Instituto Brasileiro do Café. Escritório de New York.
O mercado de café. 1976/77–
New York, Instituto Brasileiro do Café, Escritório de New York.
v. ill. 28 cm.
1. Coffee trade—United States—Statistics—Periodicals. 2. Coffee trade—Canada—Statistics—Periodicals. 3. Market surveys—Statistics—Periodicals. I. Title.
HD9199.U47 I 57a 78-647862
MARC-S

—COLOMBIA

Pérez Zapata, Hernán.
Enjuiciamiento a la política agraria y cafetera / Hernán Pérez Zapata. — 1. ed. — Bogotá : Asociación Colombiana de Ingenieros Agrónomos, 1978.
185 ; 17 cm.
Includes bibliographical references.
1. Agriculture and state—Colombia. 2. Coffee trade—Colombia. 3. Land reform—Colombia. I. Title.
HD1883 1978 .P47 79-108384
79 MARC

—COLOMBIA—CONGRESSES

Convención Bancaria y de Instituciones Financieras, 15th, Cartagena, Colombia, 1977.
Banca 77 ₍i.e. setenta y siete₎ : memoria de la XV Convención Bancaria y de Instituciones Financieras, Cartagena, octubre 27, 28, y 29 de 1977. — ₍Bogotá?₎ : Redactores Asociados, Poligrupo Comunicación, ₍1977 or 1978₎
104 p. ; 24 cm.
Sponsored by the Asociación Bancaria de Colombia.
1. Banks and banking—Colombia—Congresses. 2. Colombia—Economic conditions—1970- —Congresses. 3. Coffee trade—Colombia—Congresses. 4. International economic relations—Congresses. I. Asociación Bancaria de Colombia. II. Title.
HG2904.C66 1977 79-115425
79 MARC

—COLOMBIA—HISTORY

Bergquist, Charles W
Coffee and conflict in Colombia, 1886-1910 / Charles W. Bergquist. — Durham, N.C. : Duke University Press, 1978.
x, 277 p. : ill. ; 25 cm.
Bibliography: p. ₍263₎-269.
Includes index.
ISBN 0-8223-0418-X : $16.75
1. Colombia—Politics and government—1886-1903. 2. Colombia—Politics and government—1903-1930. 3. Colombia—Economic conditions. 4. Coffee trade—Colombia—History. I. Title.
F2276.5.B47 320.9'861'061 78-59581
79 MARC

—HAITI—PERIODICALS

Institut haïtien de promotion du café et des denrées d'exportation.
Rapports statistiques sur l'exportation du café.
₍Port-au-Prince?₎ Institut haïtien de promotion du café et des denrées d'exportation.
v. 22 x 36 cm.
1. Coffee trade—Haiti—Periodicals. I. Title.
HD9199.H2 I 56a 382'.41'373097294 78-646131
MARC-S

—INDONESIA—JAWA TIMUR

Universitas Brawijaja. Fakultas Ekonomi.
Pola produksi dan prospeknja budidaja karet, kopi, tembakau di Djawa Timur. Malang, 1971.
1 v. (various pagings) 30 cm. Rp5150
Cover title.
1. Rubber industry and trade—Indonesia—Jawa Timur. 2. Coffee trade—Indonesia—Jawa Timur. 3. Tobacco—Djawa Timur, Indonesia. I. Title.
HD9161.I63J388 70-943725
78₍r78₎rev MARC

—KENYA

Westlake, M J
Taxation and control of the Kenyan coffee industry / by M. J. Westlake. — Nairobi : Institute for Development Studies, University of Nairobi, 1973.
16 p. ; 30 cm. — (Discussion paper - Institute for Development Studies, University of Nairobi ; no. 185)
Includes bibliographical references.
1. Coffee—Taxation—Kenya. 2. Coffee trade—Kenya. I. Title. II. Series: Nairobi. University. Institute for Development Studies. Discussion paper ; no. 185.
HD9199.K42W47 78-107567
336.2'78'63373096762
79 MARC

—LATIN AMERICA

Buzzanell, Peter J
Coffee production and trade in Latin America / by Peter J. Buzzanell. — ₍Washington₎ : Foreign Agricultural Service, U.S. Dept. of Agriculture, 1979.
82 p. : ill. ; 27 cm. — (FAS M ; 288)
Bibliography: p. 72.
1. Coffee trade—Latin America. 2. Coffee—Latin America. I. Title. II. Series: United States. Foreign Agricultural Service (1953-) FAS M ; 288.
HD9199.L382B89 338.1'7'373098 79-602711
79 MARC

—UNITED STATES—STATISTICS —PERIODICALS

Instituto Brasileiro do Café. Escritório de New York.
O mercado de café. 1976/77–
New York, Instituto Brasileiro do Café, Escritório de New York.
v. ill. 28 cm.
1. Coffee trade—United States—Statistics—Periodicals. 2. Coffee trade—Canada—Statistics—Periodicals. 3. Market surveys—Statistics—Periodicals. I. Title.
HD9199.U47 I 57a 78-647862
MARC-S

—VENEZUELA

Consulpro.
Cacao, café, caña de azucar, tabaco : estudio integral del sector / elaborado por Consulpro. — Caracas : Banco Central de Venezuela, ₍1974?₎
425 p. : ill. ; 30 cm. — (Colección Agricultura en Venezuela ; 6)
Includes bibliographical references.
1. Cocoa trade—Venezuela. 2. Coffee trade—Venezuela. 3. Sugar trade—Venezuela. 4. Tobacco manufacture and trade—Venezuela. I. Title. II. Series.
HD9200.V42C65 1974 79-107520
79 MARC

COFFER-DAMS

Nikolaev, IÛriĭ Grigor'evich.
(Propusk raskhodov reki pri gidrotekhnicheskom stroitel'stve)
Пропуск расходов реки при гидротехническом строительстве / Ю. Г. Николаев, А. Г. Якобсон. — Москва : Энергия, 1978.
319 p. : ill. ; 21 cm. USSR 78
Bibliography : p. 314-[317]
1.30rub
1. Coffer-dams. 2. Hydraulic engineering. I. IÂkobson, Andreĭ Genrikhovich, joint author. II. Title.
TC198.N495 79-408495

COFFERS see Chests

COFFIN, ROBERT PETER TRISTRAM, 1892-1955

—HOMES AND HAUNTS—MAINE

Coffin, Robert Peter Tristram, 1892-1955.
Maine doings / by Robert P. Tristram Coffin ; with decorations by the author. — Thorndike, Me. : Thorndike Press, [1979?] c1950.
266 p. : ill. ; 20 cm.
Reprint of the 1st ed. published by Bobbs-Merrill, Indianapolis, with a new introd.
ISBN 0-89621-025-1. ISBN 0-89621-024-3 pbk.
1. Maine—Social life and customs. 2. Coffin, Robert Peter Tristram, 1892-1955—Homes and haunts—Maine. 3. Authors, American—20th century—Biography. I. Title.
F25.C66 1979 974.1 78-26413
78 MARC

COFIELD, J. R.

—PHOTOGRAPH COLLECTIONS

Cofield, Jack.
William Faulkner : the Cofield collection / by Jack Cofield ; introd. by Carvel Collins ; editing by Lawrence Wells ; art direction by Barry Mitchell ; layout and design by William A. Smith ; published by Howard G. Duvall, Jr., and Lawrence Wells. — Oxford, Miss. : Yoknapatawpha Press, c1978.
xi, 196 p. : ill. ; 30 cm.
ISBN 0-916242-02-1 : $20.00
1. Faulkner, William, 1897-1962—Portraits, etc. 2. Cofield, J. R.—Photograph collections. 3. Novelists, American—20th century—Portraits.
PS3511.A86Z756 813'.5'2 78-54859
78 MARC

COFRADÍA DE SANTIAGO DE BURGOS

Cofradía de Santiago de Burgos.
El libro de la Cofradía de Santiago de Burgos : primera reproducción facsímil, en color, del más antiguo armorial ecuestre que se conoce, manuscrito y miniado durante los siglos XIV al XVII / edición preparada por Faustino Menéndez Pidal de Navascués ; prólogo de Dalmiro de la Valgoma y Díaz-Varela. — Bilbao : La Gran Enciclopedia Vasca, 1977.
243 p. : ill. (some col.) ; 28 cm. Sp***
Summaries in English, French, and German.
Includes bibliographical references and indexes.
ISBN 8424803116
1. Cofradía de Santiago de Burgos. 2. Heraldry—Spain. I. Menéndez-Pidal y de Navascués, Faustino. II. Title.
CR4529.S7C63 1977 929.6 78-397907
79 MARC

COFRESÍ Y RAMÍREZ DE ARELLANO, ROBERTO, 1791-1825

—DRAMA

Cuchí Coll, Isabel.
Un patriota y un pirata / Isabel Cuchí Coll. — 1. ed. — San Juan, P.R. : [s.n.], 1973.
21, 13 p. ; 21 cm.
Cover title.
CONTENTS: El patriota.—Cofresí.
1. Baldorioty de Castro, Román, 1822-1889—Drama. 2. Cofresí y Ramírez de Arellano, Roberto, 1791-1825—Drama. I. Cuchí Coll, Isabel. Cofresí. 1973. II. Title.
PQ7440.C8P3 862 78-386201
79 MARC

COG-WHEELS see Gearing

COGHLAN, WELSH & GUEST

Gale, William Daniel.
History of Coghlan, Welsh & Guest / [W. D. (Bill) Gale]. — [Salisbury, Rhodesia? : Coghlan, Welsh & Guest, 1974?]
92 p. : ill. ; 22 cm.
Cover title.
1. Coghlan, Welsh & Guest. I. Title.
78-315388
338.7'61'340096891
78 MARC

COGNAC see Brandy

COGNAC, FRANCE

—HISTORY

Marvaud, François.
Études historiques sur la ville de Cognac et l'arrondissement ... / François Marvaud. — Marseille : Laffitte, 1977.
2 v. in 1 (335, 458 p.) ; 23 cm. F77-13563
Reprint of the 1870 ed. published by L. Clouzot, Niort.
Five hundred copies printed.
French, Latin, or Old French.
Includes bibliographical references and index.
250F
1. Cognac, France—History. I. Title.
DC801.C65M33 1977 944'.65 78-387000
79 MARC

COGNAT, ANDRÉ, 1938-

Cognat, André, 1938-
Antecume : ou, Une autre vie : récit / André Cognat ; recueilli par Claude Massot. — Paris : R. Laffont, c1977.
337 p., [6] leaves of plates : ill. ; 24 cm. — (Collection Vécu) F***
48.00F
1. Oyana Indians—Social life and customs. 2. Cognat, André, 1938- 3. Indianists—Guyana—Biography. I. Massot, Claude. II. Title.
F2230.2.O8C59 980'.004'98 79-362714
79 MARC

COGNITION

see also Attribution (Social psychology); Awareness; Cognition and culture; Cognitive styles; Cognitive therapy; Knowledge, Theory of; Preception; Thought and thinking

Bourne, Lyle Eugene, 1932-
Cognitive processes / Lyle E. Bourne, Jr., Roger L. Dominowski, Elizabeth F. Loftus. — Englewood Cliffs, N.J. : Prentice-Hall, c1979.
viii, 408 p. : ill. ; 24 cm. — (Prentice-Hall series in experimental psychology)
Bibliography: p. 370-390.
Includes indexes.
ISBN 0-13-139634-X
1. Cognition. I. Dominowski, Roger L., 1939- joint author. II. Loftus, Elizabeth F., 1944- joint author. III. Title.
BF311.B64 153.4 78-16953
78 MARC

Bransford, John.
Human cognition : learning, understanding, and remembering / John D. Bransford ; [ill., Florence Fujimoto and Lori Gilbo]. — Belmont, Calif. : Wadsworth Pub. Co., c1979.
xii, 300 p. : ill. ; 24 cm.
Bibliography: p. 265-288.
Includes indexes.
ISBN 0-534-00699-X
1. Cognition. 2. Learning, Psychology of. 3. Comprehension. 4. Memory. I. Title.
BF311.B7155 153.4 78-31698
78 MARC

Caramelli, Nicoletta.
Epistemologia genetica e teoria della conoscenza in J. Piaget / Nicoletta Caramelli. — Milano : F. Angeli, [1979]
135 p. ; 22 cm. — (Collana di psicologia : Sezione V, Testi di ricerca ; 7) It79-Feb
Bibliography: p. [133]-135.
L4500
1. Cognition. 2. Knowledge, Theory of. 3. Piaget, Jean, 1896- I. Title.
BF311.C277 79-365713
*79 MARC

Das, Jagannath Prasad.
Simultaneous and successive cognitive processes / J. P. Das, John R. Kirby, Ronald F. Jarman, with a chapter by James P. Cummins. — New York : Academic Press, 1979.
xii, 247 p. : ill. ; 24 cm.
Bibliography: p. 229-244.
Includes index.
ISBN 0-12-203150-4
1. Cognition. 2. Cognition in children. I. Kirby, John R., joint author. II. Jarman, Ronald F., joint author. III. Title.
BF311.D34 153.4 78-20039
79 MARC

Dijk, T A van.
Taal en handelen : een interdisciplinaire inleiding in de pragmatiek / Teun A. van Dijk. — Muiderberg : Dick Coutinho, 1978.
174 p. ; 22 cm. — (Randgebieden ; nr. 2) Ne78-27
Bibliography: p. 163-170.
Includes index.
ISBN 9062835198 : fl 19.50
1. Pragmatics. 2. Speech acts (Linguistics) 3. Grammar, Comparative and general. 4. Cognition. 5. Social interaction. 6. Language and culture. I. Title.
P99.4.P72D5 78-377288
*78 MARC

I Due cervelli : neuropsicologia dei processi cognitivi / a cura di Franco Denes e Carlo Umiltà. — Bologna : Il mulino, 1978.
428 p. : ill. ; 21 cm. — (Serie di psicologia) (Problemi e prospettive) It78-Nov
Bibliography: p. [386]-428.
L10000
1. Cognition. 2. Neuropsychology. I. Denes, Franco. II. Umiltà, Carlo.
BF311.D82 79-343799
*79 MARC

Forgus, Ronald H
Personality : a cognitive view / Ronald Forgus, Bernard H. Shulman. — Englewood Cliffs, N.J. : Prentice-Hall, c1979.
xiii, 434 p. : ill. ; 24 cm.
Bibliography: p. 375-416.
Includes indexes.
ISBN 0-13-657882-9 : $15.95
1. Personality. 2. Cognition. I. Shulman, Bernard H., joint author. II. Title.
BF698.F64 155.2 78-23529
78 MARC

Forman, George E 1942-
Cognitive development : a life-span view / George E. Forman, Irving E. Sigel ; [cover ill., Jim Pinckney ; interior ill., Ann Geiger]. — Monterey, Calif. : Brooks/Cole Pub. Co., c1979.
x, 210 p. : ill. ; 23 cm. — (Life-span human development series)
Bibliography: p. 191-202.
Includes indexes.
ISBN 0-8185-0275-4 : $7.95
1. Cognition. 2. Developmental psychology. I. Sigel, Irving E., joint author. II. Title.
BF311.F64 153.4 78-31176
78 MARC

Glass, Arnold Lewis, 1951-
Cognition / Arnold Lewis Glass, Keith James Holyoak, John Lester Santa. — Reading, Mass. : Addison-Wesley Pub. Co., c1979.
xix, 521 p. : ill. ; 23 cm.
Bibliography: p. 449-491.
Includes index.
ISBN 0-201-02449-7
1. Cognition. I. Holyoak, Keith James, 1950- joint author. II. Santa, John Lester, joint author. III. Title.
BF311.G54 153.4 78-67941
79 MARC

Goldberg, Lazer.
Cognifective domain: ways of knowing. [Sound recording] New York, Children's Book Council, p1977.
on 1 side of 1 cassette. (Prelude, series 3)
With: Fritz, J. American history through fiction.
Duration: 29 min., 30 sec.
In container: 25 cm.
Notes and bibliography inserted in container.
SUMMARY: The author discusses the fusion of the cognitive and affective domains in human personality. He relates this concept to teaching, emphasizing the interrelationships between science and the humanities as one way of realizing this fusion for children.
1. Cognition. 2. Affect (Psychology) 3. Knowledge, Theory of. 4. Teaching. I. Title. II. Series: Prelude. [Sound recording] series 3.
[BF311] 78-740910

Guilford, Joy Paul, 1897-
Cognitive psychology with a frame of reference / J. P. Guilford. — San Diego, Calif. : EdITS Publishers, c1979.
xii, 173 p. : ill. ; 24 cm.
Bibliography: p. [158]-165.
Includes indexes.
ISBN 0-912736-22-4
1. Cognition. I. Title.
BF311.G79 153 78-74137
79 MARC

COGNITION (Continued)

Hayes-Roth, Barbara.
Cognitive processes in planning : a report / Barbara Hayes-Roth, Frederick Hayes-Roth ; prepared for the Office of Naval Research. — Santa Monica, Ca. : Rand, 1978.
ix, 58 p. : ill. ; 28 cm. — (₍Report₎ - Rand Corporation ; R-2366-ONR)
Bibliography: p. 53-58.
ISBN 0-8330-0087-X
1. Decision-making. 2. Cognition. I. Hayes-Roth, Frederick, joint author. II. United States. Office of Naval Research. III. Title. IV. Series: Rand Corporation. Rand report ; R-2366-ONR.
AS36.R3 R-2366 081 78-27571
₍BF441₎ 78 MARC

Kognitive Komplexität : Bedeutung, Weiterentwicklung, Anwendung / hrsg. von Heinz Mandl u. Günter L. Huber ; mit Beitr. von Margaret W. Cohen ... ₍et al.₎. — Göttingen ; Toronto ; Zürich : Verlag für Psychologie Hogrefe, 1978.
415 p. : ill. ; 24 cm. GFR78-A
A revision of papers presented at a workshop held at the Universität Augsburg, Apr. 28-May 1, 1977.
Includes bibliographies and indexes.
ISBN 3-8017-0144-1 : DM48.00
1. Cognition. 2. Human information processing. 3. System theory. I. Huber, Günter L. II. Mandl, Heinz.
BF311.K653 78-375753
78 MARC

Lachman, Roy.
Cognitive psychology and information processing : an introduction / Roy Lachman, Janet L. Lachman, Earl C. Butterfield. — Hillsdale, N.J. : Lawrence Erlbaum Associates ; New York : distributed by Halsted Press, 1979.
xiii, 573 p. : ill. ; 24 cm.
Bibliography: p. 533-556.
Includes indexes.
ISBN 0-470-26649-X
1. Cognition. 2. Human information processing. I. Lachman, Janet L., joint author. II. Butterfield, Earl, 1935- joint author. III. Title.
BF311.L2 153 78-24744
79 MARC

Linguistic functions in cognitive theory / edited by W. K. Estes. — Hillsdale, N.J. : Lawrence Erlbaum Associates ; New York : distributed by the Halsted Press Division of Wiley, 1978.
viii, 311 p. : ill. ; 24 cm. — (Handbook of learning and cognitive processes ; v. 6)
Includes bibliographies and indexes.
ISBN 0-470-26311-3
1. Cognition. 2. Psycholinguistics. 3. Memory. I. Estes, William Kaye.
BF311.H334 vol. 6 153 s 78-13661
78 MARC

Lorenz, Konrad.
Behind the mirror : a search for a natural history of human knowledge / Konrad Lorenz ; translated by Ronald Taylor. — London : Methuen, 1977.
261 p. : ill. ; 23 cm. GB***
Translation of Die Rückseite des Spiegels.
Bibliography: p. ₍251₎-257.
Includes index.
ISBN 0-416-94270-9 : £4.90. ISBN 0-416-94280-6 pbk.
1. Psychology, Comparative. 2. Cognition. I. Title.
BF673.L6313 1977b 156 79-301451
79 MARC

McCormack, James E
Early cognitive instruction for the moderately and severely handicapped : program guide / James E. McCormack, Amanda J. Chalmers. — Champaign, Ill. : Research Press Co., c1978.
vii, 165 p. : ill. ; 28 cm. &teaching sequences (iv, 334 p. : ill. ; 30 cm. Bibliography: p. 159-164 (1st group)
ISBN 0-87822-187-5
1. Mentally handicapped children—Education—Handbooks, manuals, etc. 2. Cognition. I. Chalmers, Amanda J., joint author. II. Title.
LC4601.M23 371.9'28 78-62903
00 MARC

McGuigan, Frank J
Psychophysiological measurement of covert behavior : a guide for the laboratory / F. J. McGuigan. — Hillsdale, N.J. : L. Erlbaum Associates ; New York : distributed by Halsted Press Division of Wiley, 1979.
x, 131 p. : ill. ; 24 cm.
Bibliography: p. 117-124.
Includes indexes.
ISBN 0-470-26814-X
1. Psychology, Physiological. 2. Cognition. 3. Physiological apparatus. 4. Psychological apparatus. I. Title.
₍DNLM: 1. Psychophysiology—Laboratory manuals. WL25 M145p₎
QP360.M33 152 79-18482
79 MARC

O'Keefe, John M
The hippocampus as a cognitive map / John O'Keefe and Lynn Nadel. — Oxford : Clarendon Press ; New York : Oxford University Press, c1978.
xiv, 570 p. : ill. ; 24 cm. GB***
Bibliography: p. ₍477₎-543.
Includes indexes.
ISBN 0-19-857206-9 : £25.00
1. Hippocampus (Brain) 2. Cognition. 3. Memory. I. Nadel, Lynn, joint author. II. Title.
QP383.2.O38 612'.825 77-30572
79 MARC

Piaget, Jean, 1896-
Success and understanding / Jean Piaget, collaborators M. Amann ... ₍et al.₎ ; translated by Arnold J. Pomerans. — Cambridge, Mass. : Harvard University Press, 1978.
vi, 236 p. ; 24 cm.
Translation of Réussir et comprendre.
Includes bibliographical references and index.
ISBN 0-674-85387-3
1. Cognition. I. Title.
BF311.P51913 153.4 78-16435
78 MARC

Rescher, Nicholas.
Cognitive systematization : a systems-theoretic approach to a coherentist theory of knowledge / Nicholas Rescher. — Totowa, N.J. : Rowman and Littlefield, 1979.
xii, 211 p. : ill. ; 23 cm.
Includes bibliographical references and indexes.
ISBN 0-8476-6094-X
1. Knowledge, Theory of. 2. Cognition. 3. System theory. I. Title.
BD161.R47 1979 121 78-18317
78 MARC

Russell, James, 1948-
The acquisition of knowledge / James Russell. — London : Macmillan, 1978.
x, 294 p. : ill. ; 22 cm. GB***
Bibliography: p. ₍277₎-290
Includes index.
ISBN 0-333-19813-1 : £7.95
1. Cognition. 2. Cognition in children. 3. Knowledge, Theory of. I. Title.
BF311.R87 1978b 155.4'13 79-309744
79 MARC

Russell, James, 1948-
The acquisition of knowledge / James Russell. — New York : St. Martin's Press, 1978.
x, 294 p. ; 22 cm.
Bibliography: p. ₍277₎-290.
Includes index.
ISBN 0-312-00273-4 : $20.00
1. Cognition. 2. Cognition in children. 3. Knowledge, Theory of. I. Title.
BF311.R87 1978 155.4'13 78-6881
78 MARC

Sex related differences in cognitive functioning : developmental issues / edited by Michele Andrisin Wittig, Anne C. Petersen. — New York : Academic Press, 1979.
xvi, 378 p. : 24 cm. — (Academic Press series in cognition and perception)
Includes bibliographies and index.
ISBN 0-12-761150-9
1. Cognition. 2. Sex differences. 3. Sex differences (Psychology) I. Wittig, Michele Andrisin. II. Petersen, Anne C.
₍DNLM: 1. Cognition. 2. Genetics, Behavioral. 3. Sex factors. BF311 S518₎
BF311.S455 153.4 79-9827
79 MARC

Seymour, Philip Herschel Kean, 1938-
Human visual cognition : a study in experimental cognitive psychology / by Philip H. K. Seymour. — New York : St. Martin's Press, 1979.
xi, 350 p. : ill. ; 23 cm.
Bibliography: p. ₍315₎-338.
Includes indexes.
ISBN 0-312-39966-9 : $25.00
1. Cognition. 2. Visual perception. 3. Symbolism (Psychology) 4. Lexicology. 5. Memory. I. Title.
₍DNLM: 1. Visual perception. WW105 S521h₎
BF311.S46 1979 153.7 78-27246
78 MARC

Solso, Robert L 1933-
Cognitive psychology / Robert L. Solso ; under the general editorship of Jerome Kagan. — New York : Harcourt, Brace, Jovanovich, c1979.
xi, 499 p. : ill. ; 25 cm.
Bibliography: p. ₍453₎-482.
Includes indexes.
ISBN 0-15-507870-4
1. Cognition. I. Kagan, Jerome. II. Title.
BF311.S653 153 79-65964
79 MARC

Strategies of information processing / edited by Geoffrey Underwood. — London ; New York : Academic Press, 1978.
x, 455 p. : ill. ; 24 cm. GB***
Includes bibliographies and index.
ISBN 0-12-708950-0 : £20.00
1. Human information processing. 2. Cognition. I. Underwood, Geoffrey.
BF455.S74 153.4 77-93215
79 MARC

Tripp, Günter Matthias, 1937-
Betr., Piaget, Philosophie oder Psychologie : Idee u. Grenzen d. genet. Epistemologie von Jean Piaget : e. erkenntnistheoret. Kritik / Günter Matthias Tripp. — Köln : Pahl-Rugenstein, 1978.
214 p. ; 19 cm. — (Kleine Bibliothek ; 104) GFR78-A19
Includes bibliographical references.
ISBN 3-7609-0318-5 : DM14.80
1. Piaget, Jean, 1896- 2. Cognition. 3. Knowledge, Theory of. 4. Genetic psychology. I. Title.
BF311.T74 78-376197
78 MARC

Wickelgren, Wayne A 1938-
Cognitive psychology / Wayne A. Wickelgren. — Englewood Cliffs, N.J. : Prentice-Hall, c1979.
xi, 436 p. : ill. ; 24 cm.
Bibliography: p. 391-424.
Includes indexes.
ISBN 0-13-139543-2 : $15.95
1. Cognition. I. Title.
₍DNLM: 1. Cognition. 2. Psychology. BF311 W636c₎
BF311.W573 153.4 78-11301
78 MARC

Wyer, Robert S
Social cognition, inference, and attribution / Robert S. Wyer, Jr., Donal E. Carlston. — Hillsdale, N.J. : Lawrence Erlbaum Associates ; New York : distributed by the Halsted Press, 1979.
xii, 388 p. : ill. ; 24 cm.
Bibliography: p. 366-378.
Includes indexes.
ISBN 0-470-26850-6
1. Social perception. 2. Attribution (Social psychology) 3. Cognition. I. Carlston, Donal E., joint author. II. Title.
HM132.W93 301.1 79-18402
79 MARC

—ADDRESSES, ESSAYS, LECTURES

Cognitive growth and development : essays in memory of Herbert G. Birch / edited by Morton Bortner. — New York : Brunner/-Mazel, c1979.
xxiv, 352 p. : ill. ; 24 cm.
"Publications, Herbert G. Birch": p. 323-333.
Includes bibliographies and indexes.
ISBN 0-87630-183-9
1. Cognition—Addresses, essays, lectures. 2. Cognition in children—Addresses, essays, lectures. 3. Neuropsychology—Addresses, essays, lectures. 4. Birch, Herbert George, 1918-1973. I. Birch, Herbert George, 1918-1973. II. Bortner, Morton.
₍DNLM: 1. Cognition—In infancy and childhood. 2. Perception—In infancy and childhood. 3. Child development. WS105.5.C7 C676₎
BF311.C5517 153.4 78-12036
78 MARC

Perception and cognition : issues in the foundations of psychology / edited by C. Wade Savage for the Minnesota Center for Philosophy of Science. — Minneapolis : University of Minnesota Press, c1978.
x, 502 p. : ill. ; 24 cm. — (Minnesota studies in the philosophy of science ; v. 9)
"Co-sponsored by the Minnesota Center for Research in Human Learning and the Minnesota Center for Philosophy of Science."
Includes bibliographies and index.
ISBN 0-8166-0841-5 : $25.00
1. Cognition—Addresses, essays, lectures. 2. Human information processing—Addresses, essays, lectures. 3. Psychology—Philosophy—Addresses, essays, lectures. I. Savage, C. Wade. II. Minnesota. University. Center for Research in Human Learning. III. Minnesota. University. Center for Philosophy of Science. IV. Series.
Q175.M64 vol. 9 501 s 78-12743
₍BF311₎ 78 MARC

I Processi cognitivi / di G. Airenti ... ₍et al.₎ ; a cura di R. Luccio e A. Senini. — Milano : F. Angeli, c1977.
232 p. : ill. ; 22 cm. — (Collana Ricerche di psicologia ; 2) It***
Summaries in English.
Includes bibliographies.
L5000
1. Cognition—Addresses, essays, lectures. I. Airenti, Gabriella. II. Luccio, Riccardo. III. Senini, Adriano. IV. Series.
BF311.P747 78-367933
78 MARC

COGNITION
—ADDRESSES, ESSAYS, LECTURES
(Continued)

Studier av tenkning i almenpsykologisk perspektiv / redigert av Tore Helstrup og Kjell Raaheim. — Bergen : Universitetsforl., c1978 (Bergen : Ed. B. Giertsen)
208 p. ; 22 cm. N78-May
Includes bibliographies.
CONTENTS: Raaheim, K. Kognisjon og tilpasning.—Bengtsson, G. Information, feedback och beteende.—Kaufmann, G. Kognitiv representasjon i problemløsing.—Helstrup, T. Eksperimentelle hukommelsesstudier i almenpsykologisk perspektiv.—Teigen, K. H. Generaliseringsproblemet i den generelle psykologi.—Kaufmann, G. Metodeproblemer i studiet av visuell forestillingsaktivitet.—Gjerde, P. og Sand, R. Om diagnoser som selvoppfyllende profetier.—Svendsen, D. Psykisk utviklingshemning sett i lys av kognitiv utviklingspsykologi.—Steinset, A. Språkproblem hos innflyttarar.—Helstrup, T. Noen synspunkter på utvikling av språk og tanke.
ISBN 8200017648 : kr74.00
1. Cognition—Addresses, essays, lectures. 2. Thought and thinking—Addresses, essays, lectures. I. Helstrup, Tore. II. Raaheim, Kjell.
BF311.S7315 79-382428
79 MARC

—JUVENILE LITERATURE

Haines, Gail Kay.
Brain power : understanding human intelligence / by Gail Kay Haines. — New York : F. Watts, 1979.
117 p. : ill. ; 24 cm. — (An Inpact book)
Bibliography: p. 109.
Includes index.
SUMMARY: Explores the human mind, focusing on our understanding of "intelligence" and discussing how learning occurs, how intelligence is measured, what creativity is, and current developments in brain research and its possible applications.
ISBN 0-531-02287-0
1. Intellect—Juvenile literature. 2. Cognition—Juvenile literature. 3. Brain —Juvenile literature. ⌈1. Intellect. 2. Brain⌋ I. Title.
BF431.H26 153 78-10805
78 MARC

—MATHEMATICAL MODELS—
CONGRESSES

Nato Advanced Study Institute on Structural/Process Theories of Complex Human Behavior, Banff, Alta., 1977.
Structural/process models of complex human behavior : ⌈proceedings of the NATO Advanced Study Institute on Structural/Process Theories of Complex Human Behavior, Banff, Alberta, Canada, June 18-26, 1977⌋ / edited by Joseph M. Scandura and Charles J. Brainerd. — Alphen aan den Rijn : Sijthoff & Noordhoff, 1978.
viii, 612 p. : with ill. ; 25 cm. — (NATO advanced study institutes series : Series E, Applied science ; no. 26) Ne78-49
Includes bibliographies.
ISBN 9028605789 : fl 101.90
1. Human behavior—Mathematical models—Congresses. 2. Cognition—Mathematical models—Congresses. I. Scandura, Joseph M. II. Brainerd, Charles J. III. Title. IV. Series: Nato advanced study institutes series : Series E, Applied sciences ; no. 26.
BF39.N35 1978 153 79-351177
79 MARC

—PERIODICALS

Cognitive theory. v. ⌈1⌋–
1975–
Hillsdale, N. J., L. Erlbaum Associates.
v. ill. 24 cm.
Key title: Cognitive theory, ISSN 0163-2035
1. Cognition—Periodicals.
BF309.C65 153.4'05 78-641590
MARC-S

COGNITION (CHILD PSYCHOLOGY)

Jones, John, 1941-
Cognitive studies with students in Papua New Guinea / by John Jones. — ⌈Port Moresby⌋ : University of Papua and New Guinea, 1973 ⌈1976 printing⌋
50 p. : ill. ; 26 cm. — (E.R.U. report ; 10)
Bibliography: p. 50.
1. Examinations—Papua New Guinea. 2. Cognition (Psychology) I. Title. II. Series: University of Papua New Guinea. Educational Research Unit. ERU research report ; 10.
LB3058.P5J66 371.2'71'0995 79-311997
79 MARC

Kitsos, Kōnstantinos I
(Hoi ennoies tōn phainomenōn tēs physeōs kai tēs zōēs stā paidia tēs nēpiakēs kai scholikēs ēlikias)
Οἱ ἔννοιες τῶν φαινομένων τῆς φύσεως καὶ τῆς ζωῆς στὰ παιδιὰ τῆς νηπιακῆς καὶ σχολικῆς ἡλικίας : συμβολὴ στὴν κατανόηση τῆς γενετικῆς ψυχολογίας τοῦ Jean Piaget / Κωνσταντίνου 'Ι. Κίτσου. — 'Αθήνα : Δίπτυχο, 1975.
457 p. ; 21 cm.
Bibliography : p. ⌈451⌋–452.
1. Cognition (Child psychology) I. Title.
BF723.C5K563 79–413800

Whitehead, Graham J
Enquiry learning in social studies / Graham Whitehead. — Hawthorn, Vic. : Australian Council for Educational Research, 1978.
xvi, 164 p. ; 22 cm. — (ACER research series ; no. 101) Aus
Bibliography: p. 155-164.
ISBN 0-85563-164-3
1. Social sciences—Study and teaching (Elementary)—Australia. 2. Cognition (Child psychology) 3. Problem solving (Child psychology) I. Title. II. Series: Australian Council for Educational Research. Research series ; no. 101.
LB5.A85 no. 101 370'.8 s 78-321432
⌈LB1584.5.A8⌋ 79 MARC

—ADDRESSES, ESSAYS, LECTURES

Children communicating : media and development of thought, speech, understanding / Ellen Wartella, editor. — Beverly Hills, Calif. : Sage Publications, c1979.
286 p. : ill. ; 22 cm. — (Sage annual reviews of communications research ; v. 7)
Includes bibliographies.
ISBN 0-8039-1171-8. ISBN 0-8039-1172-6 pbk.
1. Television and children—Addresses, essays, lectures. 2. Child development—Addresses, essays, lectures. 3. Cognition (Child psychology)—Addresses, essays, lectures. 4. Interpersonal communication—Addresses, essays, lectures. I. Wartella, Ellen.
HQ784.T4C5 155.4'18 78-25867
78 MARC

COGNITION AND CULTURE

—ADDRESSES, ESSAYS, LECTURES

Magie : d. sozialwissenschaftl. Kontroverse über d. Verstehen fremden Denkens / hrsg. von Hans G. Kippenberg u. Brigitte Luchesi. — 1. Aufl. — Frankfurt am Main : Suhrkamp, 1978.
299, ⌈1⌋ p. ; 20 cm. — (Theorie) GFR78-A
Chiefly translations from various English sources.
Bibliography: p. 297-⌈300⌋
ISBN 3-518-06107-0
1. Magic—Addresses, essays, lectures. 2. Cognition and culture—Addresses, essays, lectures. I. Kippenberg, Hans Gerhard. II. Luchesi, Brigitte, 1943-
BF1613.M32 79-351328
79 MARC

COGNITION DISORDERS

see also Communicative disorders; Contamination (Psychology); Learning disabilities; Memory, Disorders of; Perception, Disorders of

—CONGRESSES

Cogenital and acquired cognitive disorders / editor, Robert Katzman. — New York : Raven Press, c1979.
xii, 314 p. : ill. ; 24 cm. — (Research publications - Association for Research in Nervous and Mental Disease ; v. 57)
Based on the proceedings of the 57th annual meeting of the Association for Research in Nervous and Mental Disease.
Includes bibliographies and index.
ISBN 0-89004-255-1
1. Cognition disorders—Congresses. I. Katzman, Robert. II. Association for Research in Nervous and Mental Disease.
RC394.C64C63 616.8'9 77-87458
78 MARC

Kittay Scientific Foundation.
Cognitive defects in the development of mental illness / edited by George Serban. — New York : Brunner/Mazel, c1978.
xxviii, 404 p. : ill. ; 24 cm.
Proceedings of the Fifth International Symposium of the Kittay Scientific Foundation held Apr. 3-5, 1977, New York City.
Includes bibliographies and indexes.
ISBN 0-87630-167-7 : $25.00
1. Cognitive disorders—Congresses. 2. Cognition in children—Congresses. 3. Schizophrenics—Language—Congresses. I. Serban, George, 1926- II. Title.
RC553.C64K57 1978 616.8'5 77-94739
78 MARC

—TESTING

Psychiatric symptoms and cognitive loss in the elderly : evaluation and assessment techniques / edited by Allen Raskin and Lissy Jarvik. — Washington : Hemisphere Pub. Corp. ; New York : distributed solely by Halsted Press, c1979.
xii, 308 p. ; 24 cm.
Includes bibliographies and indexes.
ISBN 0-470-26579-5
1. Geriatric psychiatry. 2. Cognitive disorders—Testing. 3. Mental illness —Diagnosis. I. Raskin, Allen, 1926- II. Jarvik, Lissy.
⌈DNLM: 1. Cognition disorders—In old age—Congresses. 2. Psychiatric status rating scales—In old age—Congresses. 3. Evaluation studies—Congresses. 4. Depression—In old age—Congresses. 5. Psychopathology—In old age—Congresses. WT150 P967⌋
RC451.4.A5P76 618.9'76'89075 78-23705
78 MARC

—TESTING—CONGRESSES

Infants at risk : assessment of cognitive functioning / edited by Richard B. Kearsley, Irving E. Sigel. — New Jersey : Lawrence Erlbaum Associates ; New York : distributed by the Halsted Press, 1979.
xiii, 233 p. : ill. ; 24 cm.
"Outcome of a conference sponsored by the National Foundation-March of Dimes."
Includes bibliographies and indexes.
ISBN 0-470-26574-4
1. Mentally handicapped children—Testing—Congresses. 2. Cognitive disorders—Testing—Congresses. 3. Infant psychology—Testing—Congresses. I. Kearsley, Richard B. II. Sigel, Irving E.
RJ506.M4I52 618.9'28'588 78-23246
78 MARC

COGNITION IN ANIMALS

—CONGRESSES

Cognitive processes in animal behavior / edited by Stewart H. Hulse, Harry Fowler, Werner K. Honig. — Hillsdale, N.J. : L. Erlbaum Associates ; New York : distributed by Halsted Press, 1978.
xii, 465 p. : ill. ; 24 cm.
Based on papers from a conference held June 1976 at Dalhousie University, Halifax, Nova Scotia.
Includes bibliographies and indexes.
ISBN 0-470-26484-5
1. Cognition in animals—Congresses. I. Hulse, Stewart H. II. Fowler, Harry, 1934- III. Honig, Werner K.
QL785.C53 156'.3 78-14522
78 MARC

COGNITION IN CHILDREN

Bates, Elizabeth.
The emergence of symbols : cognition and communication in infancy / Elizabeth Bates, with the collaboration of Laure Benigni ... ⌈et al.⌋. — New York : Academic Press, 1979.
xvi, 387 p. ; 24 cm. — (Language, thought, and culture)
Bibliography: p. 371-381.
ISBN 0-12-081540-0
1. Infant psychology. 2. Cognition in children. 3. Children—Language. 4. Symbolism (Psychology) I. Title.
⌈DNLM: 1. Language developement. 2. Psycholinguistics. WS105.5.C8 E53⌋
BF723.I6B34 155.4'22 78-20040
79 MARC

Brainerd, Charles J
The origins of the number concept / Charles J. Brainerd. — New York : Praeger, 1979.
xii, 227 p. ; 25 cm.
Bibliography: p. 222-227.
ISBN 0-03-029306-5
1. Number concept. 2. Cognition in children. I. Title.
BF456.N7B7 512'.7'019 78-21223
78 MARC

Bringuier, Jean Claude.
Conversations libres avec Jean Piaget / Jean-Claude Bringuier. — Paris : R. Laffont, c1977.
221 p., ⌈2⌋ leaves of plates ; 21 cm. — (Collection Réponses) F***
35.00F
1. Cognition in children. 2. Intellect. 3. Knowledge, Theory of. 4. Piaget, Jean, 1896- I. Piaget, Jean, 1896- II. Title.
BF723.C5B65 155.4'13'0924 77-574109
78⌈r79⌋rev MARC

Children's thinking : what develops? / Edited by Robert S. Siegler. — Hillsdale, N.J. : L. Erlbaum Associates ; New York : distributed by Halsted Press Division of Wiley, 1978.
xi, 371 p. : ill. ; 24 cm.
An outgrowth of the 13th annual Carnegie Cognition Symposium, held in May 1977.
Includes bibliographies and index.
ISBN 0-470-26520-5
1. Cognition in children. I. Symposium on Cognition, 13th, Carnegie-Mellon University, 1977. II. Siegler, Robert S.
BF723.C5C5 155.4'13 78-13119
78 MARC

Damiano, Elio.
Jean Piaget / Elio Damiano. — Brescia : La scuola, ⌈1977?⌋
207 p. ; 20 cm. — (Pedagogisti ed educatori ; 29) It78-Aug
Bibliography: p. 205-206.
L3000
1. Piaget, Jean, 1896- 2. Psychologists—Switzerland—Biography. 3. Cognition in children.
BF109.P5D33 155.4'13'0924 79-357278
*79 MARC

COGNITION IN CHILDREN (Continued)

Das, Jagannath Prasad.
Simultaneous and successive cognitive processes / J. P. Das, John R. Kirby, Ronald F. Jarman, with a chapter by James P. Cummins. — New York : Academic Press, 1979.
xii, 247 p. : ill. ; 24 cm.
Bibliography: p. 229-244.
Includes index.
ISBN 0-12-203150-4
1. Cognition. 2. Cognition in children. I. Kirby, John R., joint author. II. Jarman, Ronald F., joint author. III. Title.
BF311.D34 153.4 78-20039
79 MARC

Elkind, David, 1931-
The child and society : essays in applied child development / David Elkind. — New York : Oxford University Press, 1979.
xii, 304 p. ; 22 cm.
Includes bibliographies and index.
ISBN 0-19-502371-4 : $10.00. ISBN 0-19-502372-2 pbk. : $3.95
1. Child psychology. 2. Socialization. 3. Cognition in children. I. Title.
BF721.E35 155.4'18 78-2758
78 MARC

Frankenstein, Carl.
They think again : restoring cognitive abilities through teaching / Carl Frankenstein. — New York : Van Nostrand Reinhold, c1979.
xiii, 295 p. ; 23 cm.
Bibliography: p. 289-291.
Includes index.
ISBN 0-442-21646-7
1. Socially handicapped children—Education. 2. Cognition in children. 3. Teachers of socially handicapped children, Training of. I. Title.
LC4069.F7 371.9'67 78-17676
78 MARC

The Functions of language and cognition / edited by Grover J. Whitehurst, Barry J. Zimmerman. — San Francisco : Academic Press, 1979.
xii, 313 p. : ill. ; 24 cm.
Includes bibliographies and indexes.
ISBN 0-12-747050-6
1. Psycholinguistics. 2. Children—Language. 3. Cognition in children. I. Whitehurst, Grover J. II. Zimmerman, Barry J.
BF455.F86 153 78-20004
79 MARC

Ginsburg, Herbert.
Piaget's theory of intellectual development / Herbert Ginsburg, Sylvia Opper. — 2d ed. — Englewood Cliffs, N.J. : Prentice-Hall, c1979.
xvi, 253 p. : ill. ; 24 cm.
Bibliography: p. 239-244.
Includes index.
ISBN 0-13-675140-7 : $10.95. ISBN 0-13-675132-6 pbk. : $6.95
1. Piaget, Jean, 1896- . 2. Cognition in children. 3. Intellect. 4. Education—Philosophy. I. Opper, Sylvia, joint author. II. Title.
BF723.C5P5354 1979 155.4'13 78-12020
78 MARC

Klausmeier, Herbert John, 1915-
Cognitive development of children and youth : a longitudinal study / Herbert J. Klausmeier, Patricia S. Allen. — New York : Academic Press, 1978.
xii, 327 p. : ill. ; 24 cm. — (Educational psychology)
Bibliography: p. 301-313.
Includes indexes.
ISBN 0-12-411355-9
1. Cognition in children. 2. Cognition in children—Research. I. Allen, Patricia S., joint author. II. Title.
BF723.C5K57 155.4'13 78-13306
78 MARC

Longeot, François.
Les stades opératoires de Piaget et les facteurs de l'intelligence / François Longeot ; avec la collaboration des psychologues scolaires, Bernard ... [et al] ; préf. de Maurice Reuchlin. — [Grenoble] : Presses universitaires de Grenoble, 1978.
401 p. : ill. ; 22 cm. F***
Bibliography: p. 391-395.
ISBN 2-7061-0134-2 : 50.00F
1. Cognition in children. 2. Thought and thinking. I. Title.
BF311.L66 155.4'13 79-359208
79 MARC

Morgan, Alice P
If you ask me ...! / By Alice P. Morgan ; ill. by Robert Duffy. — [Troy, Ala. : Troy State University Press, c1978.
[32] leaves : ill. ; 28 cm.
"The questions and answers in this book were taken from a thesis written by the author ...."
ISBN 0-916624-23-4
1. Cognition in children. 2. Children's writings.
BF723.C5M67 500.9 78-62901
79 MARC

Morrison, Delmont.
Sensory-motor dysfunction and therapy in infancy and early childhood / by Delmont Morrison, Patricia Pothier, Katy Horr. — Springfield, Ill. : Thomas, c1978.
viii, 261 p. : ill. ; 24 cm.
Includes bibliographies and indexes.
ISBN 0-398-03766-3
1. Psychomotor disorders in children. 2. Cognition in children. 3. Mentally ill children. I. Pothier, Patricia C., joint author. II. Horr, Katy, joint author. III. Title.
RJ506.P68M68 618.9'28 77-25853
77 MARC

Noda, Yoshikazu, 1918-
(Kyōikugakuteki ninshikiron)
教育学的認識論：現代教育改革への試論 / 野田義一著. — 東京：御茶の水書房, 1978.
7, 300 p. ; 22 cm. Ja 78-7788
Bibliography: p. 291-300.
¥3000
1. Cognition in children. 2. Educational psychology. I. Title.
BF723.C5N598 79-800295

O'Connor, Neil.
Seeing and hearing and space and time / N. O'Connor and B. Hermelin. — London ; New York : Academic Press, 1978.
viii, 157 p. : ill. ; 24 cm. GB***
Bibliography: p. [138]-152.
Includes index.
ISBN 0-12-524150-X
1. Perception in children. 2. Cognition in children. 3. Perceptually handicapped children—Testing. I. Hermelin, Beate, joint author. II. Title.
BF723.P36O26 155.4'13 77-94293
00 MARC

Penrose, William Orestus, 1914-
A primer on Piaget / by William O. Penrose. — Bloomington, Ind. : Phi Delta Kappa Educational Foundation, c1979.
26 p. ; 18 cm. — (Fastback - Phi Delta Kappa Educational Foundation ; 128)
Bibliography: p. 26.
ISBN 0-87367-128-7 : $0.75
1. Cognition in children. 2. Piaget, Jean, 1896- I. Title. II. Series: Phi Delta Kappa. Educational Foundation. Fastback ; 128.
BF723.C5P38 155.4'13'0924 78-61324
00 MARC

Piaget, Jean, 1896-
The development of thought : equilibration of cognitive structures / Jean Piaget ; translated by Arnold Rosin. — Oxford : B. Blackwell, 1978, c1977.
viii, 213 p. ; 22 cm. GB***
Translation of L'équilibration des structures cognitives.
Includes bibliographical references and index.
ISBN 0-631-18910-6 : £8.50
1. Cognition in children. I. Title.
BF723.C5P49313 1977b 155.4'13 79-310048
79 MARC

Piaget and education / edited by Jeanette McCarthy Gallagher and J. A. Easley, Jr. — New York : Plenum Press, c1978.
xix, 294 p. ; 24 cm. — (Knowledge and development ; v. 2)
Includes bibliographies and index.
1. Piaget, Jean, 1896- . 2. Cognition in children. 3. Education—Philosophy. I. Gallagher, Jeanette McCarthy, 1932- II. Easley, J. A. III. Series.
BF311.K6385 vol. 2 153.4 s 79-103886
[BF723.C5] 79 MARC

Russell, James, 1948-
The acquisition of knowledge / James Russell. — London : Macmillan, 1978.
x, 294 p. : ill. ; 22 cm. GB***
Bibliography: p. [277]-290
Includes index.
ISBN 0-333-19813-1 : £7.95
1. Cognition. 2. Cognition in children. 3. Knowledge, Theory of. I. Title.
BF311.R87 1978b 155.4'13 79-309744
79 MARC

Russell, James, 1948-
The acquisition of knowledge / James Russell. — New York : St. Martin's Press, 1978.
x, 294 p. ; 22 cm.
Bibliography: p. [277]-290.
Includes index.
ISBN 0-312-00273-4 : $20.00
1. Cognition. 2. Cognition in children. 3. Knowledge, Theory of. I. Title.
BF311.R87 1978 155.4'13 78-6881
78 MARC

Santostefano, Sebastian, 1929-
A biodevelopmental approach to clinical child psychology : cognitive controls and cognitive control therapy / Sebastiano Santostefano. — New York : Wiley, c1978.
xvi, 833 p. : ill. ; 26 cm. — (Wiley series on personality processes)
"A Wiley-Interscience publication."
Bibliography: p. 796-813.
Includes indexes.
ISBN 0-471-75380-7
1. Clinical child psychology. 2. Cognition in children. 3. Psychobiology. 4. Psychological tests for children. 5. Child psychotherapy. I. Title. [DNLM: 1. Psychiatry—Bibliography. 2. Mental health—Bibliography. ZWM100 G798h]
RJ503.3.S26 618.9'28'9075 78-18485
78 MARC

Singer, Dorothy G
A Piaget primer : how a child thinks / Dorothy G. Singer & Tracey A. Revenson. — New York : New American Library, 1978.
x, 148 p. : ill. ; 21 cm. — (A Plume book)
Bibliography: p. 135-143.
Includes index.
ISBN 0-452-25189-3 : $3.95
1. Cognition in children. 2. Child psychology. 3. Piaget, Jean, 1896- I. Revenson, Tracey A., joint author. II. Title.
BF723.C5S6 155.4'13 78-61227
79 MARC

Singer, Dorothy G
A Piaget primer : how a child thinks / Dorothy G. Singer & Tracey A. Revenson. — New York : International Universities Press, c1978.
x, 148 p. : ill. ; 21 cm.
Bibliography: p. 135-143.
Includes index.
ISBN 0-8236-4136-8
1. Piaget, Jean, 1896- . 2. Cognition in children. I. Revenson, Tracey A., joint author. II. Title.
BF723.C5S6 1978b 155.4'13 78-58483
79 MARC

Stage theories of cognitive and moral development : criticisms and applications. — Cambridge, Mass. : Harvard educational review, c1978.
xi, 204 p. ; 26 cm. — (Harvard educational review : Reprint ; no. 13)
Includes bibliographies.
CONTENTS: Ashton, P. T. Cross-cultural Piagetian research.—Gibbs, J. C. Kohlberg's stages of moral judgment.—Gilligan, C. In a different voice.—Elkind, D. Piagetian and psychometric conceptions of intelligence.—Duckworth, E. The having of wonderful ideas.—Kohlberg, L. & Mayer, R. Development as the aim of education.—Phillips, D. C. & Kelly, M. E. Hierarchical theories of development in education and psychology.—Selman, R. L. Education for cognitive development.
ISBN 0-916690-16-4 : $4.95
1. Cognition in children. 2. Moral development. 3. Piaget, Jean, 1896- . 4. Education—Philosophy. I. Series: The Harvard educational review : Reprint series ; no. 13.
BF723.C5S72 155.4'13'08 77-91693
79 MARC

Stenild, Mariane.
Spædbarnet og førskolebarnet : aspekter af den intellektuelle udvikling / [af] Mariane Stenild. — [Hellerup] [Onsgårdsvej 10, 2900] : Danak Psykologisk Forlag, 1978.
128 p. ; 22 cm. — (Legestueprojekt ; bog 1) D79-1
Bibliography: p. 127-128.
ISBN 8787580160 : kr40.00
1. Cognition in children. I. Title. II. Series.
BF723.C5S74 79-380517
79 MARC

Stephens, Sunny Courington.
ABC cookery / Sunny Courington Stephens. — 1st ed. — Niles, Ill. : Argus Communications, c1979.
123 p. : ill. ; 24 cm.
ISBN 0-89505-024-2 : $3.95
1. Cookery. 2. Perceptural motor learning. 3. Cognition (Child psychology) 4. Language acquisition. I. Title.
TX652.S715 641.5 79-63173
00 MARC

Takizawa, Takehisa, 1931-
(Kodomo no shikō to ninshiki)
子どもの思考と認識：新しい発達心理学の視点から / [著者 滝沢武久]. — 東京：童心社, 1977.
238 p. ; 19 cm. — (子どもの文化叢書) Ja 77-10729
Bibliography: p. 237-238.
¥800
1. Child psychology. 2. Thought and thinking. 3. Cognition in children. I. Title. II. Series: Kodomo no bunka sosho.
LB1117.T373 79-805247

Várkonyi, Zsuzsa F
Tájékozottság és kompetencia : óvodáskorúak személyiséglélektani vizsgálata / F. Várkonyi Zsuzsa. — Budapest : Akadémiai Kiadó, 1978.
127 p. ; 19 cm. — (Pszichológia a gyakorlatban ; 36. köt) H***
Bibliography: p. 114-118.
ISBN 9630514990 : 15.00Ft
1. Child psychology. 2. Cognition in children. 3. Learning, Psychology of. I. Title. II. Series.
LB1105.P7 Köt. 36 79-376437
[BF721] 79 MARC

Wadsworth, Barry J
Piaget's theory of cognitive development : an introduction for students of psychology and education / Barry J. Wadsworth ; with drawings by the author. — 2d ed. — New York : Longman, c1979.
xvi, 189 p. : ill. ; 21 cm.
Bibliography: p. 182-186.
Includes index.
1. Piaget, Jean, 1896- . 2. Cognition in children. I. Title. [DNLM: 1. Child psychology. 2. Cognition—In infancy and childhood. WS105.5.C7 W124p]
BF723.C5W33 1979 79-12347
79 MARC

COGNITION IN CHILDREN (Continued)

Wilcox, Mary M 1921-
Developmental journey : a guide to the development of logical and moral reasoning and social perspective / Mary M. Wilcox. — Nashville : Abingdon, c1979.
286 p. : ill. ; 22 cm.
Includes bibliographical references and index.
ISBN 0-687-10510-2
1. Cognition in children. 2. Reasoning (Psychology) 3. Moral development. 4. Social perception. I. Title.
BF723.C5W53 153.4 79-1340
79 MARC

—ADDRESSES, ESSAYS, LECTURES

Cognitive development in the school years / edited by Ann Floyd. — New York : Wiley, c1979.
383 p. : ill. ; 23 cm.
"A Halsted Press book."
Includes bibliographies and index.
ISBN 0-470-26429-2
1. Cognition in children—Addresses, essays, lectures. 2. Piaget, Jean, 1896- —Addresses, essays, lectures. 3. Educational psychology—Addresses, essays, lectures. I. Floyd, Ann.
BF723.C5C63 155.4'24 78-9155
78 MARC

Cognitive growth and development : essays in memory of Herbert G. Birch / edited by Morton Bortner. — New York : Brunner/-Mazel, c1979.
xxiv, 352 p. : ill. ; 24 cm.
"Publications, Herbert G. Birch": p. 323-333.
Includes bibliographies and indexes.
ISBN 0-87630-183-9
1. Cognition—Addresses, essays, lectures. 2. Cognition in children—Addresses, essays, lectures. 3. Neuropsychology—Addresses, essays, lectures. 4. Birch, Herbert George, 1918-1973. I. Birch, Herbert George, 1918-1973. II. Bortner, Morton.
[DNLM: 1. Cognition—In infancy and childhood. 2. Perception—In infancy and childhood. 3. Child development. WS105.5.C7 C676]
BF311.C5517 153.4 78-12036
78 MARC

Early cognitive development / edited by John Oates. — New York : Wiley, c1979.
396 p. : ill. ; 22 cm.
"A Halsted Press book."
Includes bibliographies and index.
ISBN 0-470-26431-4
1. Cognition in children—Addresses, essays, lectures. 2. Infant psychology—Addresses, essays, lectures. 3. Child psychology—Addresses, essays, lectures. I. Oates, John, 1946-
BF723.C5E2 155.4'22 78-17694
78 MARC

Werner, Heinz, 1890-1964.
Developmental processes : Heinz Werner's selected writings / edited by Sybil S. Barten and Margery B. Franklin. — New York : International Universities Press, c1978.
2 v. (xviii, 562 p.) : ill. ; 24 cm.
Bibliography: pp. 533-549.
Includes indexes.
CONTENTS: v. 1. General theory and perceptual experiences.—v. 2. Cognition, language, and symbolization.
ISBN 0-8236-1250-3 (v. 1) : $17.50 per vol.
1. Cognition in children—Addresses, essays, lectures. 2. Psycholinguistics—Addresses, essays, lectures. 3. Symbolism (Psychology)—Addresses, essays, lectures. 4. Mentally handicapped children—Addresses, essays, lectures. 5. Brain-damaged children—Addresses, essays, lectures. I. Barten, Sybil S., 1933- II. Franklin, Margery B. III. Title.
[DNLM: 1. Cognition—Collected works. 2. Symbolism—Collected works. BF311 W493d]
BF723.C5W47 1978 155.4'13 77-92187
78 MARC

—CASE STUDIES

Church, Joseph, ed.
Three babies : biographies of cognitive development / Joseph Church, editor. — Westport, Conn. : Greenwood Press, 1978, c1966.
ix, 323 p. ; 23 cm.
Reprint of the ed. published by Vintage Books, New York.
Includes bibliographies and index.
ISBN 0-313-21029-2
1. Infant psychology—Case studies. 2. Cognition in children—Case studies. I. Title.
[BF723.I6C5 1978] 155.4'22 78-14252
78 MARC

—CONGRESSES

The Impact of Piagetian theory : on education, philosophy, psychiatry, and psychology / edited by Frank B. Murray ; contributors, Millie Almy ... [et al.]. — Baltimore : University Park Press, c1979.
xiii, 232 p. ; 24 cm. — (A publication of the Jean Piaget Society)
"Based upon the authors' presentations at the plenary sessions of the fifth annual symposium of the Jean Piaget Society."
Includes bibliographies and indexes.
ISBN 0-8391-1293-9
1. Piaget, Jean, 1896- —Congresses. 2. Cognition in children—Congresses. 3. Knowledge, Theory of—Congresses. 4. Child mental health—Congresses. 5. Educational psychology—Congresses. I. Murray, Frank B. II. Almy, Millie Corinne, 1915- III. Series: Jean Piaget Society. A publication of the Jean Piaget Society.
[DNLM: 1. Child psychology—Congresses. 2. Philosophy—Congresses. 3. Psychiatry—Congresses. 4. Education—Congresses. WS105.3 I34]
BF723.C5I46 155.4'13'0124 78-13551
78 MARC

Kittay Scientific Foundation.
Cognitive defects in the development of mental illness / edited by George Serban. — New York : Brunner/Mazel, c1978.
xxviii, 404 p. : ill. ; 24 cm.
Proceedings of the Fifth International Symposium of the Kittay Scientific Foundation held Apr. 3-5, 1977, New York City.
Includes bibliographies and indexes.
ISBN 0-87630-167-7 : $25.00
1. Cognitive disorders—Congresses. 2. Cognition in children—Congresses. 3. Schizophrenics—Language—Congresses. I. Serban, George, 1926- II. Title.
RC553.C64K57 1978 616.8'5 77-94739
78 MARC

—EXPERIMENTS

Centre for Educational Research and Innovation.
Piagetian inventories : the experiments of Jean Piaget / Centre for Educational Research and Innovation. — Paris : Organisation for Economic Co-operation and Development ; [Washington, D.C. : sold by OECD Publications Center], 1977.
109 p. : ill. ; 21 cm. F***
Includes bibliographical references and index.
1. Piaget, Jean, 1896- 2. Cognition in children—Experiments. I. Title.
BF723.C5C48 1977 155.4'13 79-300657
79 MARC

—PERIODICALS

Piagetian theory and its implications for the helping professions.
[Los Angeles, s. n., etc.]
v. 22 cm. annual.
"Proceedings ... Interdisciplinary Seminar."
Cover title : Piagetian theory & the helping professions.
Vols. for co-sponsored by University Affiliated Program, Childrens Hospital of Los Angeles and the University of Southern California, School of Education.
Key title: Piagetian theory and its implications for the helping professions, ISSN 0192-5318.
1. Cognition in children—Periodicals. 2. Piaget, Jean, 1896- —Periodicals. 3. Educational psychology—Periodicals. I. Los Angeles. Children's Hospital. University Affiliated Program. II. Los Angeles. University of Southern California. School of Education. III. Title: Piagetian theory & the helping professions.
BF309.P5 155.4'13'0924 79-640991
MARC-S

—RESEARCH

Klausmeier, Herbert John, 1915-
Cognitive development of children and youth : a longitudinal study / Herbert J. Klausmeier, Patricia S. Allen. — New York : Academic Press, 1978.
xii, 327 p. : ill. ; 24 cm. — (Educational psychology)
Bibliography: p. 301-313.
Includes indexes.
ISBN 0-12-411355-9
1. Cognition in children. 2. Cognition in children—Research. I. Allen, Patricia S., joint author. II. Title.
BF723.C5K57 155.4'13 78-13306
78 MARC

—TESTING

Piaget, Jean, 1896-
Réussir et comprendre / par Jean Piaget, avec la collaboration de M. Amann ... [et al.]. — 1. éd. — Paris : Presses universitaires de France, 1974.
253 p. ; 22 cm. — (Psychologie d'aujourd'hui)
Includes bibliographical references.
40.00F
1. Cognition in children—Testing. I. Title.
BF723.C5P532 1974 155.4'13'028 74-189598
74[r79]rev MARC

Piaget, Jean, 1896-
Success and understanding / by Jean Piaget, in collaboration with M. Amann [et al.] ; translated from the French by Arnold J. Pomerans. — London : Routledge & K. Paul, 1978.
viii, 236 p. : ill. ; 22 cm.
Translation of Réussir et comprendre.
Includes bibliographical references and index.
ISBN 0-7100-8946-5 : £7.95
1. Cognition in children—Testing. I. Title.
BF723.C5P53213 155.4'13 79-355655
79 MARC

COGNITIVE DISSONANCE see Dissonance (Psychology)

COGNITIVE MAPS see Geographical perception

COGNITIVE STYLES

Witkin, Herman A
Cognitive style in personal and cultural adaptation / by Herman A. Witkin. — [Worcester, MA] : Clark University Press, 1978.
68 p. ; 21 cm. — (Heinz Werner lecture series ; 1977)
Bibliography: p. 67-68.
ISBN 0-914206-10-9
1. Cognitive styles. 2. Personality. 3. Personality and culture. I. Title. II. Series: Heinz Werner lectures ; 1977.
BF311.W5947 155.2'3 77-12426
77 MARC

—UNITED STATES

Smith, Alfred Goud, 1921-
Cognitive styles in law schools / by Alfred G. Smith ; Patrick A. Nester and Lynn H. Pulford, research associates. — Austin : University of Texas Press, c1979.
xii, 178 p. : ill. ; 23 cm.
Bibliography: p. [161]-167.
Includes index.
1. Law—Study and teaching—United States. 2. Cognitive styles—United States. I. Title.
KF272.S58 340'.07'1173 78-12521
78 MARC

COGNITIVE THERAPY

Cognitive behavioral interventions : theory, research, and procedures / edited by Philip C. Kendall, Steven D. Hollon. — New York : Academic Press, 1979.
xvii, 481 p. : ill. ; 24 cm. — (Personality and psychopathology)
Includes bibliographies and indexes.
ISBN 0-12-404480-8
1. Cognitive therapy. 2. Rational-emotive psychotherapy. 3. Behavior therapy. I. Kendall, Philip C. II. Hollon, D. Steven.
[DNLM: 1. Behavior theraphy. 2. Cognition. W1 PE861 / WM425 P679]
RC489.C63C64 616.8'914 79-6940
79 MARC

Cognitive behavior therapy : research and application / edited by John P. Foreyt and Diana P. Rathjen. — New York : Plenum Press, c1978.
xiv, 265 p. ; 24 cm.
Includes bibliographies and indexes.
ISBN 0-306-31145-3
1. Cognitive therapy. I. Foreyt, John Paul. II. Rathjen, Diana P.
[DNLM: 1. Behavior therapy—Collected works. 2. Cognition—Collected works. WM425 C676]
RC489.C63C63 616.8'914 78-15948
78 MARC

Horan, John J 1945-
Counseling for effective decision making : a cognitive-behavioral perspective / John J. Horan. — North Scituate, Mass. : Duxbury Press, c1979.
xiv, 266 p. : ill. ; 24 cm.
Bibliography: p. 235-256.
Includes index.
ISBN 0-87872-195-9
1. Counseling. 2. Behavior modification. 3. Cognitive therapy. 4. Decision making. I. Title.
BF637.C6H65 158 78-15071
78 MARC

COHABITATION see Unmarried couples

COHEN, ALBERT, 1895-

—DIARIES

Cohen, Albert, 1895-
Carnets 1978 / Albert Cohen. — [Paris] : Gallimard, c1979.
190 p. ; 21 cm. F***
35.00F
1. Cohen, Albert, 1895- —Diaries. 2. Authors, French—20th century—Biography.
PQ2605.O24Z464 848'.9'1203 79-382301
79 MARC

COHEN, HENRY, 1808-1880

Kritische Neuaufnahme der Fundmünzen der römischen Zeit in Deutschland (FMRD), Cohen-RIC-Konkordanz / erarb. von Heinrich Chantraine ... [et al.] ; hrsg. von Maria R.-Alföldi. — Bonn : Habelt, 1978-
v. ; 23 cm. — (Antiquitas : Reihe 1, Abhandlungen zur alten Geschichte ; Bd. 30-) GFR78-A
Cover title: Cohen-RIC-Konkordanz.
CONTENTS: T. 1. Von Augustus bis zur Follis-Reform Diocletians.

ISBN 3-7749-1265-3 (v. 1)
1. Fundmünzen der römischen Zeit in Deutschland—Indexes. 2. The Roman imperial coinage—Indexes. 3. Cohen, Henry, 1808-1880. 4. Numismatics, Roman—Indexes. 5. Coin hoards—Germany—Indexes. 6. Coinage—Rome—History—Indexes. I. Chantraine, Heinrich. II. Alföldi, Maria R. III. Title: Cohen-RIC-Konkordanz. IV. Series.
CJ1117.K73 79-351129
79 MARC

COHEN, HERMANN, 1842-1918

Lisser, Kurt.
Der Begriff des Rechts bei Kant : mit e. Anh. über Cohen u. Görland / von Kurt Lisser. — Unveränd. Neudr. d. Ausg. Berlin 1922. — Vaduz/Liechtenstein : Topos-Verlag, 1978.
v, 80 p. ; 25 cm. GFR79-A9
Reprint of the ed. published by Reuther & Reichard, which was issued as Nr. 58 of the Ergänzungshefte to the Kant-Studien.
Bibliography: p. 77-79.
Includes indexes.
ISBN 3-289-00191-1 : DM25.00
1. Law—Philosophy. 2. Kant, Immanuel, 1724-1804—Ethics. 3. Cohen, Hermann, 1842-1918. 4. Görland, Albert, 1869-1952. I. Title. II. Series: Kant-Studien : Ergänzungshefte ; Nr. 58.
K457.K32L57 1978 340.1 79-382144
79 MARC

COHEN, LEONARD, 1934-

—CRITICISM AND INTERPRETATION

Manzano Lizandro, Alberto.
Leonard Cohen / Alberto Manzano Lizandro. — 1. ed. — Barcelona : Unilibro, 1978.
160 p. : ill. ; 20 cm. — (Testimonio ; 1) Sp***
Bibliography: p. 153-154.
Discography: p. 157-160.
Filmography: p. 160.
ISBN 8485380053
1. Cohen, Leonard, 1934- —Criticism and interpretation. 2. Cohen, Leonard, 1934- —Interviews. 3. Poets, Canadian—20th century—Interviews.
PR9199.3.C57Z78 811'.5'4 79-351834
79 MARC

Scobie, Stephen.
Leonard Cohen / Stephen Scobie. — Vancouver : Douglas & McIntyre, c1978.
xii, 192 p. ; 22 cm. — (Studies in Canadian literature ; 12) C78-2033-2
Bibliography: p. 187-192.
ISBN 0-88894-194-3 : $3.95
1. Cohen, Leonard, 1934- —Criticism and interpretation.
PR9199.3.C57Z88 811'.5'4 79-317983
79 MARC

—INTERVIEWS

Manzano Lizandro, Alberto.
Leonard Cohen / Alberto Manzano Lizandro. — 1. ed. — Barcelona : Unilibro, 1978.
160 p. : ill. ; 20 cm. — (Testimonio ; 1) Sp***
Bibliography: p. 153-154.
Discography: p. 157-160.
Filmography: p. 160.
ISBN 8485380053
1. Cohen, Leonard, 1934- —Criticism and interpretation. 2. Cohen, Leonard, 1934- —Interviews. 3. Poets, Canadian—20th century—Interviews.
PR9199.3.C57Z78 811'.5'4 79-351834
79 MARC

COHEN, MARVIN S., 1931-

United States. Congress. Senate. Committee on Commerce, Science, and Transportation.
Nomination—Civil Aeronautics Board : hearing before the Committee on Commerce, Science, and Transportation, United States Senate, Ninety-fifth Congress, second session, on nomination of Marvin S. Cohen, to be a member, Civil Aeronautics Board, September 26, 1978. — Washington : U.S. Govt. Print. Off., 1978.
iii, 11 p. ; 24 cm.
"Serial no. 95-116."
1. Cohen, Marvin S., 1931- . 2. United States. Civil Aeronautics Board—Officials and employees. I. Title.
KF26.C69 1978ae 353.008'77'7 79-600577
79 MARC

COHEN (B.) AND SONS TRAFALGAR GALLERIES

—EXHIBITIONS

Cohen (B.) and Sons Trafalgar Galleries.
Trafalgar Galleries at the Royal Academy : [catalogue]. — London : Trafalgar Galleries, 1977.
86 p. : chiefly ill. ; 31 cm. GB***
Exhibition catalog.
Includes bibliographical references and index.
1. Painting—Exhibitions. 2. Cohen (B.) and Sons Trafalgar Galleries—Exhibitions. I. London. Royal Academy of Arts. II. Title.
N5054.C63 1977 750'.74'02132 78-324354
79 MARC

COHERENCE (OPTICS)

see also Holographic interferometry

Françon, Maurice.
Optical image formation and processing / M. Françon ; translated by Bernard M. Jaffe. — New York : Academic Press, 1979.
xviii, 213 p. : ill. ; 24 cm.
Translation of Optique, formation et traitement des images.
ISBN 0-12-264850-1
1. Optics, Physical. 2. Coherence (Optics) 3. Fourier transformations. I. Title.
QC395.2.F713 535'.2 78-19668
79 MARC

—CONGRESSES

Nato Advanced Study Institute on Coherence in Spectroscopy and Modern Physics, Versilia, Italy, 1977.
Coherence in spectroscopy and modern physics / edited by F. T. Arecchi, R. Bonifacio and M. O. Scully. — New York : Plenum Press, c1978.
viii, 402 p. : ill. ; 26 cm. — (NATO advanced study institutes series : Series B, Physics ; v. 37)
"Proceedings of the NATO Advanced Study Institute on Coherence in Spectroscopy and Modern Physics held at Villa LePianore, Versilia, Italy, July 17-30, 1977."
Includes bibliographical references and index.
ISBN 0-306-40050-2
1. Coherence (Optics)—Congresses. 2. Spectrum analysis—Congresses. 3. Physics—Congresses. I. Arecchi, F. T. II. Bonifacio, R. III. Scully, Marlan Orvil, 1939- . IV. Title. V. Series.
QC476.C6N37 1977 536 78-14474
MARC

Rochester Conference on Coherence and Quantum Optics, 4th, 1977.
Coherence and quantum optics IV : [proceedings of the fourth Rochester Conference on Coherence and Quantum Optics, held at the University of Rochester, June 8-10, 1977] / edited by Leonard Mandel and Emil Wolf. — New York : Plenum Press, c1978.
xiv, 1011 p. : ill. ; 26 cm.
Includes bibliographical references and indexes.
ISBN 0-306-40038-3
1. Coherence (Optics)—Congresses. 2. Quantum optics—Congresses. I. Mandel, Leonard. II. Wolf, Emil. III. Title.
QC476.C6R63 1977 535'.2 78-15470
78 MARC

COHERENT LIGHT COMMUNICATION SYSTEMS see Laser communication systems

COHESION

see also Adhesion; Plasticity

COHN, ALVIN W. CRIME AND JUSTICE ADMINISTRATION

Blessington, Mark.
Manual for the instructor to accompany Crime and justice administration by Alvin W. Cohn / prepared by Mark Blessington. — Philadelphia : Lippincott, c1976.
v, 93 p. ; 23 cm.
Includes bibliographies.
1. Cohn, Alvin W. Crime and justice administration. 2. Crime and criminals—Handbooks, manuals, etc. 3. Criminal justice, Administration of—Handbooks, manuals, etc. I. Cohn, Alvin W. Crime and justice administration. II. Title.
HV6025.C57 suppl. 364 78-108948
78 MARC

COHN, HELGE OSKAR, 1893-

Cohn, Helge Oskar, 1893-
Nu, nyligen och länge sen ... / Helge Cohn. — Stockholm : Norstedt, 1977.
266 p. [4] leaves of plates : ill. ; 22 cm. S77-48
Autobiographical.
Includes index.
ISBN 9117733529 : kr90.00
1. Cohn, Helge Oskar, 1893- 2. Sweden—Biography. I. Title.
CT1328.C63A35 79-349698
79 MARC

COHOES, N.Y.

—SOCIAL CONDITIONS

Walkowitz, Daniel J
Worker city, company town : iron and cotton-worker protest in Troy and Cohoes, New York, 1855-84 / Daniel J. Walkowitz. — Urbana : University of Illinois Press, c1978.
xv, 292 p. : ill. ; 24 cm. — (The Working class in American history)
Based on the author's thesis, University of Rochester, 1972.
Bibliography: p. 268-284.
Includes index.
ISBN 0-252-00667-4 : $12.95
1. Iron and steel workers—New York (State)—Troy—History. 2. Textile workers—New York (State)—Cohoes—History. 3. Cotton manufacture—New York (State)—Cohoes—History. 4. Troy, N.Y.—Social conditions. 5. Cohoes, N.Y.—Social conditions. I. Title. II. Series.
HD8039.I52U67 301.44'42'0974741 78-18305
78 MARC

COHOMOLOGY THEORY see Homology theory

COHORT ANALYSIS

see also Age groups

—ADDRESSES, ESSAYS, LECTURES

Cohort analysis : a collection of interdisciplinary readings / edited by Donald W. Hastings, Linda G. Berry. — Oxford, Ohio : Scripps Foundation for Research in Population Problems, c1979.
349 p. : ill. ; 27 cm.
Includes bibliographies.
1. Cohort analysis—Addresses, essays, lectures. I. Wilson, Donald W., 1940- . II. Berry, Linda G.
HB885.C56 300'.2'8 79-114199
79 MARC

COHUILLA INDIANS see Cahuilla Indians

COIFFURE see Hairdressing

COIGÜE see Nothofagus dombeyi

COILS, ELECTRIC see Electric coils

COIMBRA. UNIVERSIDADE

—BIBLIOGRAPHY—CATALOGS

Coimbra. Universidade. Biblioteca.
Catálogo da exposição bibliográfica, 1926-1966 : suplemento — ₍Coimbra₎ : Universidade de Coimbra, Biblioteca Geral da Universidade, 1968.
xxi, 184 p. ; 25 cm.
1. Coimbra. Universidade—Bibliography—Catalogs. I. Title.
Z5055.P8C675 suppl 79-378914
₍AS304.C72₎ 79 MARC

COIN DESIGN

Ehrend, Helfried.
Gegenstempel auf Münzen / Helfried Ehrend, Ulrich Schreier. — Speyer : Numismatische Gesellschaft, 1975.
40 p. : ill. ; 21 cm. GFR***
Bibliography: p. 38-39.
1. Coin design. I. Schreier, Ulrich, joint author. II. Title.
CJ125.E37 79-377310
79 MARC

COIN ERRORS see Coins—Errors

COIN HOARDS

—DANUBE VALLEY

Fitz, Jenő.
Der Geldumlauf der römischen Provinzen im Donaugebiet Mitte des 3. Jahrdunderts / von Jenő Fitz ; ₍übers. von Mátyás Esterházy₎. — Budapest : Akadémiai Kiadó, 1978.
2 v. (945 p.) : ill. ; 25 cm. H***
Includes bibliographical references and indexes.
ISBN 9630508583
1. Coins, Roman—Danube Valley. 2. Coin hoards—Danube Valley. 3. Danube Valley—Antiquities, Roman. 4. Romans—Danube Valley. I. Title.
CJ1106.F5715 79-352160
79 MARC

—GAUL

Fischer, Brigitte.
Les monnaies antiques d'Afrique du Nord trouvées en Gaule / Brigitte Fischer. — Paris : Centre national de la recherche scientifique, 1978.
171 p. : ill., maps ; 28 cm. — (Supplément à Gallia ; 36) F***
Bibliography: p. ₍19₎-24.
Includes indexes.
ISBN 2-222-02250-9
1. Coins, Punic—Gaul. 2. Coin hoards—Gaul. 3. France—Antiquities. I. Title. II. Series: Gallia, fouilles et monuments archéologiques en France métropolitaine : Supplément ; 36.
CJ1381.F5 737.4'9'364 79-362661
79 MARC

—GERMANY—INDEXES

Kritische Neuaufnahme der Fundmünzen der römischen Zeit in Deutschland (FMRD), Cohen-RIC-Konkordanz / erarb. von Heinrich Chantraine ... ₍et al.₎ ; hrsg. von Maria R.-Alföldi. — Bonn : Habelt, 1978-
v. ; 23 cm. — (Antiquitas : Reihe 1, Abhandlungen zur alten Geschichte ; Bd. 30-) GFR78-A
Cover title: Cohen-RIC-Konkordanz.
CONTENTS: T. 1. Von Augustus bis zur Follis-Reform Diocletians.

ISBN 3-7749-1265-3 (v. 1)
1. Fundmünzen der römischen Zeit in Deutschland—Indexes. 2. The Roman imperial coinage—Indexes. 3. Cohen, Henry, 1808-1880. 4. Numismatics, Roman—Indexes. 5. Coin hoards—Germany—Indexes. 6. Coinage—Rome—History—Indexes. I. Chantraine, Heinrich. II. Alföldi, Maria R. III. Title: Cohen-RIC-Konkordanz. IV. Series.
CJ1117.K73 79-351129
79 MARC

—GERMANY, EAST—EICHSFELD

Lauerwald, Paul.
Zur Münz- und Geldgeschichte des Eichsfeldes / Paul Lauerwald. — Als Sonderausg. d. "Eichsfelder Heimathefte" hrsg. vom Pädag. Kreiskabinett Worbis in Verbindung mit d. Kulturbund d. DDR Kreisleitung Worbis, Heiligenstadt. — Worbis : Pädag. Kreiskabinett, 1976.
87 p. : 21 ill. and maps ; 20 cm. GFR77-B2
Bibliography: p. 84-87.
1. Coin hoards—Germany, East—Eichsfeld. 2. Coins, German—Germany, East—Eichsfeld. I. Eichsfelder Heimathefte : Sonderausgabe. II. Title.
CJ393.E32L38 79-348233
79 MARC

—GERMANY, WEST—LINGENFELD

Ehrend, Helfried.
Der Münzschatz von Lingenfeld 1969 ₍i.e. neunzehnhundertneunundsechzig₎ / Helfried Ehrend ; mit Beiträgen von Günter Stein und Friedrich Wielandt. — Speyer : Numismatische Gesellschaft Speyer, 1975.
64 p. : ill. ; 21 cm. GFR***
Includes bibliographical references.
1. Coin hoards—Germany, West—Lingenfeld. 2. Lingenfeld, Ger.—Antiquities. 3. Germany, West—Antiquities. 4. Coins, German—Germany, West—Lingenfeld. I. Stein, Günter, joint author. II. Wielandt, Friedrich, 1906- joint author. III. Title.
CJ1647.L56E37 79-381992
79 MARC

—GERMANY, WEST—WELBERGEN

Berghaus, Peter.
Der Münzschatzfund von Welbergen Kr. Steinfurt 1965 / von Peter Berghaus. — Münster : Westfälisches Landesmuseum für Kunst und Kulturgeschichte, 1978.
20, ₍1₎ p. : ill. ; 30 cm. — (Numismatische Schriften des Westfälischen Landesmuseum für Kunst und Kulturgeschichte, Landschaftsverband Westfalen-Lippe ; Nr. 1) GFR***
Cover title.
Bibliography: p. 19-₍21₎
1. Coin hoards—Germany, West—Welbergen. 2. Coins, German—Germany, West—Welbergen. 3. Welbergen, Ger.—Antiquities. 4. Germany, West—Antiquities. I. Title. II. Series: Westfälisches Landesmuseum für Kunst und Kulturgeschichte Münster. Numismatische Schriften des Westfälischen Landesmuseums für Kunst und Kulturgeschichte, Landschaftsverband Westfalen-Lippe ; Nr. 1.
CJ2729.W44B47 78-399001
79 MARC

—GERMANY, WEST—XANTEN—EXHIBITIONS

Geld in Xanten : ₍Ausstellung, Regionalmuseum Xanten, 4. Juni-18. August₎ / Volker Zedelius. — Köln : Rheinland-Verlag ; Bonn : Habelt ₍in Komm., 1978₎
64 p. : ill., maps ; 20 x 21 cm. — (Führer des Regionalmuseums ; Nr. 11) GFR78-A
Bibliography: p. 63-64.
ISBN 3-7927-0395-5 : DM10.00
1. Coin hoards—Germany, West—Xanten—Exhibitions. 2. Xanten, Ger.—Antiquities—Exhibitions. 3. Germany, West—Antiquities—Exhibitions. I. Zedelius, Volker. II. Regionalmuseum Xanten. III. Series: Regionalmuseum Xanten. Führer des Regionalmuseums Xanten ; Nr. 11.
CJ153.X36G43 79-346151
79 MARC

—GREECE, MODERN—ATHENS

American School of Classical Studies at Athens.
Mediaeval and modern coins in the Athenian Agora. — Princeton, N.J. : American School of Classical Studies at Athens, 1978.
32 p. : ill. ; 22 cm. — (Excavations of the Athenian Agora : Picture book ; no. 18)
Prepared by Fred S. Kleiner.
ISBN (invalid) 0-87661-618-3 : $1.00
1. Coin hoards—Greece, Modern—Athens. 2. Athens. Agora. I. Kleiner, Fred S. II. Title. III. Series.
CJ1647.A85A44 1978 737.4'9495'12 79-103884
79 MARC

—INDIA

Gupta, Parmeshwari Lal, 1914-
A bibliography of the hoards of punch-marked coins of ancient India. Bombay, Numismatic Society of India, 1955.
iii, 23 p. 25 cm. (Numismatic notes and monographs, no. 2)
1. Coin hoards—India. I. Title. II. Series: Numismatic notes and monographs (Varanasi), no. 2.
CJ3530.N78 no. 2 737'.0934 s 71-13344
₍CJ1391₎ 70₍r78₎rev2 MARC

—ITALY—SANTA RESTITUA (L'AQUILA)

Squilla, Gaetano.
Un cippo terminale a Luco nei Marsi (L'Aquila), Una pioggia di monete a S. Restituta di Morrea (L'Aquila). ₍Veroli₎, Tip. di Casamari, 1973.
57, ₍21₎ p. illus. 25 cm. L1000 It73-Dec
Bibliography: p. ₍15₎ (2d group)
1. Boundary stones—Italy—Luco dei Marsi. 2. Coin hoards—Italy—Santa Restitua (L'Aquila) 3. Coins, Roman—Santa Restituta (L'Aquila) 4. Luco dei Marsi, Italy—Antiquities. 5. Santa Restituta, Italy (L'Aquila)—Antiquities. 6. Italy—Antiquities. I. Squilla, Gaetano. Una pioggia di monete a S. Restituta di Morrea. 1973. II. Title.
CC605.I8S68 937'.7 79-356857
79 MARC

—ITALY—VERONA—CATALOGS

Milani, Luigi Adriano, 1854-1914.
Il ripostiglio della Venèra : monete romane della seconda metà del terzo secolo / Luigi Adriano Milani. — ₍Bologna₎ : A. Forni, 1977.
213 p., ₍3₎ leaves of plates : ill. ; 31 cm. It79-Jan
Reprint of the 1880 ed. published by Salviucci, Rome.
"Il ripostiglio della Venèra, conservato quasi per intero nel Civico Museo di Verona."
Includes bibliographical references.
1. Coins, Roman—Italy—Verona—Catalogs. 2. Coin hoards—Italy—Verona—Catalogs. 3. Verona. Museo Civico—Catalogs. I. Verona. Museo civico. II. Title.
CJ815.I82V475 1977 79-380002
*79 MARC

—POLAND—ZALESIE (SŁUPCA)

Wczesnośredniowieczny skarb srebrny z Zalesia, powiat Słupca. / Anna Kmietowicz ... ₍et al.₎. — Wrocław : Zakład Narodowy im. Ossolińskich, 1969-1974.
2 v. : ill., map ; 26 cm.
At head of title v. 1: Instytut Historii Kultury Materialnej Polskiej Akademii Nauk ; v. 2. Polska Akademia Nauk. Instytut Historii Kultury Materialnej.
Summaries in French.
Includes bibliographical references.
CONTENTS: t. 1. Monety arabskie.—t. 2. Monety bizantyjskie, Monety Zachodnioeuropejskie, Ozodoby, Podsumowanie.
1. Coin hoards—Poland—Zalesie (Słupca) 2. Coins, Arab—Poland—Zalesie (Słupca) 3. Coins, Medieval—Poland—Zalesie (Słupca) 4. Jewelry, Medieval—Poland—Zalesie (Słupca) 5. Silverwork—Poland—Zalesie (Słupca) I. Kmietowicz, Anna.
CJ1647.Z34W37 74-256081
79₍r79₎rev2 MARC

—ROMANIA

Chirilă, Eugen.
Tezaure şi descoperiri monetare din colecţia Muzeului judeţean Maramureş. Baia Mare, 1971.
77 p., 11 l. of plates. 24 cm. lei6.00 R72-2730
At head of title: Muzeul Judeţean Maramureş. Eugen Chirilă, Aurel Socolan.
Summaries in German.
Includes bibliographical references.
1. Coin hoards—Romania. 2. Coins, Ancient. 3. Muzeul Judetean Maramres. I. Socolan, Aurel, joint author. II. Muzeul Judeţean Maramureş. III. Title.
CJ153.R6C48 78-418764
78 MARC

—ROMANIA—MĂGURA (BACĂU)

Mihăilescu-Bîrliba, Virgil.
Tezaurul de la Măgura / Virgil Mihăilescu-Bîrliba, Ioan Mitrea. — Bacău : Muzeul Judeţean de Istorie şi Artă, 1977.
79 p., ₍14₎ leaves of plates : ill. ; 24 cm. — (Bibliotheca Carpicae ; 1) R***
Summary in English and French.
Includes bibliographical references.
1. Coin hoards—Romania—Măgura (Bacău) 2. Coins, Roman—Romania—Măgura (Bacău) 3. Măgura, Romania (Bacău)—Antiquities. 4. Romania—Antiquities. I. Mitrea, Ioan, joint author. II. Title. III. Series.
CJ893.M33M53 78-400555
79 MARC

—ROMANIA—SATU-MARE (JUDEŢ)

Muzeul de Istorie Satu Mare.
Tezaure monetare din judeţul Satu-Mare = Münzhorte aus Kreis Sathmar. — ₍Satu Mare₎ : Muzeul de Istorie Satu-Mare, 1968.
135 p., ₍21₎ leaves of plates (1 fold.) : ill., maps (1 fold.) ; 25 cm.
Summaries in German.
Table of contents in Romanian and German.
Includes bibliographical references.
1. Numismatics—Romania—Satu-Mare (Judeţ) 2. Coin hoards—Romania—Satu-Mare (Judeţ) I. Title. II. Title: Münzhorte aus Kreis Sathmar.
CJ3349.S23M88 79-382479
79 MARC

—TURKEY—MARDIN

Lowick, N M
The Mardin hoard : Islamic countermarks on Byzantine folles / ₍by₎ N. M. Lowick, S. Bendall, P. D. Whitting. — ₍London₎ : A. H. Baldwin and Sons Ltd., 1977.
2-79 p. : ill., map ; 21 cm. GB79-02942
Includes bibliographical references.
£3.00
1. Follis (Coin)—Turkey—Mardin. 2. Coin hoards—Turkey—Mardin. 3. Coins, Islamic—Turkey—Mardin. 4. Counterstamp (Numismatics)—Turkey—Mardin. 5. Mardin, Turkey—Antiquities. 6. Turkey—Antiquities. I. Bendall, S., joint author. II. Whitting, Philip D., joint author. III. Title.
CJ1251.L68 737.4'9'5667 79-315534
79 MARC

COIN HOARDS (Continued)

—UKRAINE

Kotliar, Mykola Fedorovych.
Znaleziska monet z XIV-XVII ͏[i.e. czternastego do siedamnastego] w. na obszarze Ukraińskiej SRR : materiały / Mikołaj Kotlar. — Wrocław : Zakład Narodowy im. Ossolińskich, 1975.
320 p. ; 24 cm. p***
At head of title: Polskie Towarzystwo Archeologiczne i Numizmatyczne.
Summary in French.
Includes indexes.
zł60.00
1. Coin hoards—Ukraine. 2. Ukraine—Antiquities. I. Title.
CJ3028.U4K69 76-510740
79[r79]rev MARC

COIN OPERATED MACHINES

see also Pinball machines; Slot machines

—COLLECTORS AND COLLECTING

Rubin, Ken.
Drop coin here / Ken & Fran Rubin ; produced & designed by Howard Brookstein. — New York : Crown Publishers, c1979.
96 p. : ill. (some col.) ; 29 cm.
Includes index.
ISBN 0-517-53298-0
1. Coin operated —Collectors and collecting. I. Rubin, Fran, joint author. II. Title.
TJ1557.R8 1979 629.8'32 78-12753
78 MARC

COINAGE

see also Commenorative coins; Currency question; Gold; Mints; Quantity theory of money

—FRANCE—HISTORY

Bisson, Thomas N
Conservation of coinage : monetary exploitation and its restraint in France, Catalonia, and Aragon (c.A.D.1000-c.1225) / Thomas N. Bisson. — Oxford [Eng.] ; New York : Clarendon Press, 1979.
xxiii, 250 p., [7] leaves of plates : ill. ; 24 cm. GB79
Bibliography: p. [207]-229.
Includes indexes.
ISBN 0-19-828275-3 : £12.50
1. Coinage—France—History. 2. Coinage—Spain—History. 3. Monetary policy—France—History. 4. Monetary policy—Spain—History. I. Title.
HG976.B57 332.4'042'0944 77-30359
79 MARC

—JAPAN—HISTORY

Hara, Mitsumasa, 1916-
(Nihon kodai kahei shi no kenkyū)
日本古代貨幣史の研究 / 原三正著. — 東京 : ボナンザ, 1978.
5, 308, 21 p., [4] leaves of plates : ill. ; 22 cm. Ja 78-22699
Limited edition: 500 copies printed.
Bibliography: p. 1-7 (3d group)
Includes indexes.
¥6000
1. Coinage—Japan—History. 2. Coins, Ancient—History. I. Title.
HG1273.H37 78-816832

—ROME—HISTORY—INDEXES

Kritische Neuaufnahme der Fundmünzen der römischen Zeit in Deutschland (FMRD), Cohen-RIC-Konkordanz / erarb. von Heinrich Chantraine ... [et al.] ; hrsg. von Maria R.-Alföldi. — Bonn : Habelt, 1978-
v. ; 23 cm. — (Antiquitas : Reihe 1, Abhandlungen zur alten Geschichte ; Bd. 30-) GFR78-A
Cover title: Cohen-RIC-Konkordanz.
CONTENTS: T. 1. Von Augustus bis zur Follis-Reform Diocletians.

ISBN 3-7749-1265-3 (v. 1)
1. Fundmünzen der römischen Zeit in Deutschland—Indexes. 2. The Roman imperial coinage—Indexes. 3. Cohen, Henry, 1808-1880. 4. Numismatics, Roman—Indexes. 5. Coin hoards—Germany—Indexes. 6. Coinage—Rome—History—Indexes. I. Chantraine, Heinrich. II. Alföldi, Maria R. III. Title: Cohen-RIC-Konkordanz. IV. Series.
CJ1117.K73 79-351129
79 MARC

—SOUTH AFRICA

South Africa. Parliament. House of Assembly. Select Committee on Coinage.
Report of the Select Committee on Coinage : printed by order of the House of Assembly. — Pretoria : Printed for the Govt. Printer by Cape and Transvaal Printers, Cape Town, 1964.
xv, 254, xlv p. ; 25 cm.
At head of title: Republic of South Africa.
1. Coinage—South Africa.
HG1351.Z8S66 1964 79-312604
79 MARC

—SPAIN—HISTORY

Bisson, Thomas N
Conservation of coinage : monetary exploitation and its restraint in France, Catalonia, and Aragon (c.A.D.1000-c.1225) / Thomas N. Bisson. — Oxford [Eng.] ; New York : Clarendon Press, 1979.
xxiii, 250 p., [7] leaves of plates : ill. ; 24 cm. GB79
Bibliography: p. [207]-229.
Includes indexes.
ISBN 0-19-828275-3 : £12.50
1. Coinage—France—History. 2. Coinage—Spain—History. 3. Monetary policy—France—History. 4. Monetary policy—Spain—History. I. Title.
HG976.B57 332.4'042'0944 77-30359
79 MARC

—SWITZERLAND—BERN—HISTORY

Martin, Colin Henri.
Essai sur la politique monétaire de Berne, 1400-1798 / Colin Martin. — Lausanne : Bibliothèque historique vaudoise, 1978.
370 p., [8] p. of plates ; 24 cm. — (Bibliothèque historique vaudoise ; no 60) Sw78-11523
Includes bibliographical references.
50.00F
1. Money—Switzerland—Bern—History. 2. Monetary policy—Switzerland—Bern—History. 3. Coinage—Switzerland—Bern—History. I. Title. II. Title: La politique monétaire de Berne. III. Series.
DQ721.B5 no. 60 949.4'3 s 79-339921
[HG1160.B47] 79 MARC

—UNITED STATES

United States. Congress. House. Committee on Banking, Finance and Urban Affairs. Subcommittee on Historic Preservation and Coinage.
Proposed smaller one-dollar coin : hearings before the Subcommittee on Historic Preservation and Coinage of the Committee on Banking, Finance and Urban Affairs, House of Representatives, Ninety-fifth Congress, second session, on H.R. 12444 ... May 17 and 31, 1978. — Washington, : U.S. Govt. Print. Off., 1978.
iv, 248 p. : ill. ; 24 cm.
Bibliography: p. 211.
1. Coinage—United States. 2. Dollar, American (Coin) I. Title.
KF27.B545 1978a 343'.73'032 78-603130
79 MARC

United States. Congress. House. Committee on Banking, Finance and Urban Affairs. Subcommittee on Historic Preservation and Coinage.
To amend the One bank holding company act of 1970 (Carson City silver dollars) : hearing before the Subcommittee on Historic Preservation and Coinage of the Committee on Banking, Finance and Urban Afairs, House of Representatives, Ninety-fifth Congress, second session, April 3, 1978. — Washington : U.S. Govt. Print. Off., 1978.
iii, 67 p. : ill. ; 23 cm.
1. Coinage—United States. 2. Dollar, American (Coin) I. Title.
KF27.B545 1978 343'.73'032 78-602696
78 MARC

United States. Laws, statutes, etc.
Laws of the United States concerning money, banking, and loans, 1778-1909 / compiled by A. T. Huntington and Robert J. Mawhinney. — Westport, Conn. : Greenwood Press, 1978.
v, xxii, 812 p. ; 24 cm.
At head of title: National Monetary Commission.
Reprint of the 1910 ed. published by Govt. Print. Off., Washington, which was issued as Senate document no. 580 of the 61st Congress, 2d session.
Includes index.
ISBN 0-313-20681-3
1. Coinage—United States. 2. Paper money—United States. 3. Banking law—United States. 4. Debts, Public—United States. I. Huntington, Andrew Tyler, 1848-1915. II. Mawhinney, Robert J. III. United States. National Monetary Commission. IV. Title. V. Series: United States. 61st Congress, 1st session, 1909-1910. Senate. Document ; no. 580.
KF6202 1978 346'.73'082 78-10669
78 MARC

COINAGE OF WORDS see Words, New

COINCIDENCE IN PSYCHICAL RESEARCH

—CASE STUDIES

Vaughn, Alan.
Incredible coincidence : the baffling world of synchronicity / Alan Vaughan. — 1st ed. — Philadelphia : Lippincott, c1979.
256 p. ; 24 cm.
Bibliography: p. 249-250.
Includes index.
ISBN 0-397-01351-5
1. Coincidence in psychical research—case studies. I. Title.
BF1175.V38 001.9'4 79-10864
79 MARC

COINCIDENCE THEORY (MATHEMATICS)

see also Fixed point theory

COINING, ILLICIT see Counterfeits and counterfeiting

COINS

see also Bracteate coins; Coin design; Coin hoards; Commemorative coins; Gold coins; Obsidional coins; Silver coins; and names of coins, e. g. Cent, Dollar Guinea (Coin), Shilling, Sou marque

Andrews, Charles J
Fell's international coin book / by Charles J. Andrews. — 6th rev. ed. — New York : F. Fell Publishers, [1976]
310 p. : ill. ; 21 cm.
ISBN 0-8119-0051-7
1. Coins. I. Title. II. Title: International coin book.
CJ89.A5 1976 737.4 79-14730
79 MARC

Auctiones A.G.
Monnaies anciennes et modernes provenant de collections privées : exposition, le 25 novembre 1974 chez Monnaies et médailles S.A., Bâle. — Bâle : Auctiones S.A., [1974]
54 p., [21] leaves of plates : ill. ; 29 cm. — (Its Vente aux enchères publiques ; 4) Sw***
Added t.p.: Münzen der Antike und Neuzeit aus Privatbesitz.
French and German.
Catalog of the auction of Nov. 26-27, 1974.
1. Coins. 2. Auctiones A.G. I. Title. II. Title: Münzen der Antike und Neuzeit aus Privatbesitz. III. Series: Auctiones A.G. Auktion ; 4.
CJ47.A92 1974 737.4 78-382534
78 MARC

Clain-Stefanelli, Elvira Eliza.
Das grosse Buch der Münzen und Medaillen : mit Münzkatalog Europa von 1900 bis heute / von Elvira und Vladimir Clain-Stefanelli, Günter Schön. — 1. Aufl. — München : Battenberg, 1976.
308 p. : ill. (some col.) ; 26 cm. GFR76-A
"Aus dem Amerikanischen übersetzt von Alfred P. Zeller."
Bibliography: p. 162-164.
ISBN 3-87045-119-X : DM39.00
1. Coins. 2. Numismatics. 3. Coins, European. I. Clain-Stefanelli, Vladimir, joint author. II. Schön, Günter, joint author. III. Title.
CJ59.C5515 78-382176
78 MARC

Fujisawa, Masaru, 1923-
(Sekai no koin)
世界のコイン / 藤沢優著. — 大阪 : 保育社, 昭和40 [1965]
153 p. : ill. : 15 cm. — (カラーブックス : 92)
Series romanized: Karā bukkusu.

1. Coins. 2. Numismatics—Collectors' and collecting. I. Title.
CJ75.F83 79-804341

COINS (Continued)

Schön, Günter.
Simon and Schuster world coin catalogue, 1979-1980, twentieth century / by Günter Schön. — Rev. and expanded ed. — New York : Simon and Schuster, ₍1978?₎ c1972.
1290 p. : ill. ; 19 cm.
"Translated from the third edition of Weltmünzkatalog, XX. Jahrhundert."
Edition published in 1973 under title: World coin catalogue, twentieth century.
ISBN 0-671-24638-0 : $16.50. ISBN 0-671-24639-9 pbk. : $9.95
1. Coins. I. Title. II. Title: World coin catalogue, 1979-1980, twentieth century.
CJ1755.S3413 1978 737.4'09'04 78-13192
78 MARC

—CATALOGS

Auctiones A.G.
Monnaies anciennes et modernes, provenant de collections privées : exposition, 1er décembre 1975, chez Monnaies et médailles S.A. / Auctiones S.A. — Bâle : Auctiones, ₍1975?₎
60 p., ₍22₎ leaves of plates : ill. ; 29 cm. — (Auctiones S.A. Vente aux enchères publiques ; 5) Sw***
Added t.p. : Münzen der Antike und Neuzeit, aus Privatbesitz.
French or German.
1. Coins—Catalogs. 2. Auctiones A.G.—Catalogs. I. Münzen und Medaillen A.G. II. Title. III. Title: Münzen der Antike und Neuzeit. IV. Series: Auctiones A.G. Auktion ; 5.
CJ47.A92 1975 737.4 79-337181
79 MARC

Krause, Chester L.
Standard catalog of world coins.
₍Iola, Wis., Krause Publications₎
v. ill.
Vols. for issued jointly with Clifford Mishler.
Key title: Standard catalog of world coins, ISSN 0190-7689.
1. Coins—Catalogs. I. Mishler, Clifford. II. Title.
CJ1755.K72 737.4'021'6 79-640940
MARC-S

—COLLECTORS AND COLLECTING

Harrison, Peter, fl. 1979-
Coins & banknotes for profit / Peter & Mary Harrison. — London : Barrie & Jenkins, 1979.
112 p., ₍4₎ leaves of plates : ill. ; 19 cm. GB***
Bibliography: p. 112.
ISBN 0-214-20509-6 : £2.95
1. Coins—Collectors and collecting. 2. Bank-notes—Collectors and collecting. I. Harrison, Mary, fl. 1979- joint author. II. Title.
CJ81.H28 737.4'075 79-316769
79 MARC

—DICTIONARIES

Salzade, de.
Dictionnaire historique des monnoies tant anciennes que modernes / par M. de Salzade. — Clermont-Ferrand : Thimonier, ₍1978?₎
187 p. : ill. ; 21 cm. F***
Cover title.
Reprint of the 1767 ed. published by J. J. Boucherie, Bruxelles.
32.00F
1. Coins—Dictionaries. I. Title.
CJ69.S24 1978 737.4'03 79-359296
79 MARC

—ERRORS

Harsche, Bert.
How to detect altered & counterfeit coins and paper money / by Bert Hars₍c₎he. — 6th ed. — Orlando, Fla. : House of Collectibles, c1978.
60 p. : ill. ; 21 cm.
Second ed. published in 1963 under title: Detecting altered coins.
ISBN 0-87637-104-7 : $2.95
1. Coins, American. 2. Coins—Errors. 3. Paper money—United States—Evaluation. 4. Counterfeits and counterfeiting—United States. I. Title.
CJ1832.H38 1978 737.4'9'73 78-61860
00 MARC

Herbert, Alan.
The official price guide to mint errors and varieties / by Alan Herbert. — 2d ed. — Orlando, Fla. : House of Collectioles, c1978.
182 p. : ill. ; 21 cm.
First published in 1974 under title: The official guide to mint errors.
ISBN 0-87637-103-9 : $3.95
1. Coins—Errors. 2. Coins, American. I. Title.
CJ125.H47 1978 737.4'9'37 78-67027
79 MARC

Marvin, Paul.
The design cud / by Paul Marvin and Arnold Margolis. ₍Rev. ed.₎. — ₍Oceanside, N.Y.₎ : A. Margolis, c1979.
xi, 205 p. (p. 201-205 advertisements) : ill. ; 22 cm.
Previous ed. by Mort Goodman.
$10.95
1. Coins, American. 2. Coins—Errors. I. Margolis, Arnold, joint author. II. Goodman, Mort. Design cud. III. Title.
CJ1832.M28 1979 737.4'9'73 79-116906
79 MARC

—ERRORS—ADDRESSES, ESSAYS, LECTURES

Dollar varieties and errors scrapbook : a collection of VAM Club reports and other articles from the Numismatic Error Collectors of America monthly Errorscope magazine / compiled by Leroy Van Allen. — ₍New York?₎ : Numismatic Error Collectors of America, c1979.
ii, 217 p. : ill. ; 22 cm.
Articles which originally appeared 1973-1979.
Includes index.
1. Dollar, American (Coin)—Addresses, essays, lectures. 2. Coins—Errors—Addresses, essays, lectures. I. Van Allen, Leroy C. II. Numismatic Error Collectors of America. NECA errorscope.
CJ1832.D64 737.4'9'73 79-112251
00 MARC

—EXHIBITIONS

Culture through coins; Singapore numismatic exhibition 1972, organised by the Singapore Numismatic Association, with the assistance of the Ministry of Culture and the National Museum, sponsored by Shell. ₍Singapore, 1972₎
42 p. illus. 27 cm.
Cover title.
In English or Chinese with introd. in English, Chinese, Malay, and Tamil.
1. Numismatics—Singapore—Exhibitions. 2. Coins—Exhibitions. 3. Singapore—History—Exhibitions. I. Singapore Numismatic Association.
CJ3629.S58C84 72-940913
79 MARC

—HISTORY—EXHIBITIONS

Eberhardt, Helmut.
Münze, Gesellschaft, Geschichte : Aussteller- und Objekteverzeichnis der Zentralen Münzausstellung der DDR / Veranstalter, Deutscher Kulturbund, Zentrale Kommission Natur und Heimat des Präsidialrates, Fachausschuss Numismatik ; Zusammenstellung und Bearb., Helmut Eberhardt. — ₍Berlin₎ : Deutscher Kulturbund, Zentrale Kommission Natur und Heimat des Präsidialrates, Fachausschuss Numismatik, ₍1971?₎
15 p. ; 21 cm. GDR***
Cover title.
1. Coins—History—Exhibitions. 2. Numismatics—Exhibitions. 3. Deutscher Kulturbund. Fachausschuss Numismatik—Exhibitions. I. Deutscher Kulturbund. Fachausschuss Numismatik. II. Title.
CJ41.B47E33 78-418761
78 MARC

—MISCELLANEA

Allenbaugh, Carl.
Coins : questions and answers / by Carl Allenbaugh ; edited by Robert M. Poeschl. — 2d ed. — Iola, Wis. : Krause Publications, c1976.
128 p. : facsims. ; 21 cm.
Rev. ed. of Coins; questions and answers: what, when, why, published in 1964.
ISBN 0-87341-008-4 : $1.50
1. Numismatics—United States—Miscellanea. 2. Numismatics—Miscellanea. 3. Coins—United States—Miscellanea. 4. Coins—Miscellanea. I. Title.
CJ1832.A48 1976 737.4'075 76-1543
76₍r78₎rev MARC

—PRICES

Kim, Nak-ch'ŏn.
(Han'guk kŭndae hwap'ye kagyŏk mongnok)
韓國近代貨幣價格目錄：近代鑄貨類·紙幣類·私札券(於音)類 / 金洛天著. — 서울：高麗苑, 1978.
163 p. : ill. ; 22 cm.
₩2000
1. Coins, Korean. 2. Coins—Prices. 3. Numismatics—Korea—Collectors and collecting. I. Title.
CJ3747.K55 79-826292

—CZECHOSLOVAKIA—EXHIBITIONS

Šimek, Eduard.
Pražská mincovna : devět století české mince : katalog výstavy, Muzeum hlavního města Prahy, Praha 1977 / Eduard Šimek ; ₍fot. Miroslav Fokt₎. — Praha : Muzeum hl. m. Prahy, 1977.
38, ₍1₎ p., ₍24₎ leaves of plates : ill. ; 24 cm. — (Acta Musei Pragensis ; 77) Cz78
Bibliography: p. 34-36.
Kčs19.00
1. Coins—Czechoslovakia—Exhibitions. 2. Pražská mincovna—Exhibitions. I. Series: Muzeum hlavního města Prahy. Acta ; 77.
CJ2604.2.P7S55 79-337200
*79 MARC

—GREAT BRITAIN—DICTIONARIES

Bradley, Howard W
A handbook of coins of the British Isles / by Howard W. Bradley. — London : Hale, 1978.
vii, 210 p. : ill. ; 23 cm. GB78-23564
Bibliography: p. ₍197₎-203.
Includes index.
ISBN 0-7091-6747-4 : £5.50
1. Coins—Great Britain—Dictionaries. I. Title.
CJ2479.B7 1978 737.4'9'41 79-301460
79 MARC

—POLAND—CATALOGS

Kamiński, Czesław.
Ilustrowany katalog monet polskich : 1916-1977 / Czesław Kamiński. — Wyd. 5. — Warszawa : Krajowa Agencja Wydawnicza, 1977 (Pozn : PZGMK)
218, ₍6₎ p. : ill. ; 22 cm. P78-1873
Bibliography: p. ₍221₎
zł35.00
1. Coins—Poland—Catalogs. I. Title.
CJ3047.K29 1977 78-400162
79 MARC

—TAJIKISTAN

Davidovich, Elena Abramovna.
Klady drevnikh i srednevekovykh monet Tadzhikistana / E. A. Davidovich. — Moskva : Nauka, 1979.
462 p. : ill. ; 22 cm. USSR79
At head of title: Akademiia nauk SSSR. Institut vostokovedeniia. Akademiia nauk Tadzhikskoĭ SSR. Institut istorii.
Bibliography: p. 419-₍426₎
Includes index.
2.70rub
1. Coins—Tajikistan. 2. Tajikistan—Antiquities. I. Title.
CJ3028.T34D38 79-378445
79 MARC

—TURKEY—CILICIA

Bedoukian, Paul Z
Coinage of Cilician Armenia / by Paul Z. Bedoukian. — Rev. ed. — Danbury, Conn. : Bedoukiar, 1979.
xxxi, 494 p., ₍6₎ leaves of plates : ill. ; 24 cm.
Includes bibliographies and indexes.
1. Coins—Turkey—Cilicia. I. Title.
CJ3449.C54B42 1979 737.4'9'566 79-117885
79 MARC

—UNITED STATES—MISCELLANEA

Allenbaugh, Carl.
Coins : questions and answers / by Carl Allenbaugh ; edited by Robert M. Poeschl. — 2d ed. — Iola, Wis. : Krause Publications, c1976.
128 p. : facsims. ; 21 cm.
Rev. ed. of Coins; questions and answers: what, when, why, published in 1964.
ISBN 0-87341-008-4 : $1.50
1. Numismatics—United States—Miscellanea. 2. Numismatics—Miscellanea. 3. Coins—United States—Miscellanea. 4. Coins—Miscellanea. I. Title.
CJ1832.A48 1976 737.4'075 76-1543
76₍r78₎rev MARC

COINS, AMERICAN

see also Beaver (Coin); Dollar, American (Coin)

Harsche, Bert.
How to detect altered & counterfeit coins and paper money / by Bert Hars[c]he. — 6th ed. — Orlando, Fla. : House of Collectibles, c1978.
60 p. : ill. ; 21 cm.
Second ed. published in 1963 under title: Detecting altered coins.
ISBN 0-87637-104-7 : $2.95
1. Coins, American. 2. Coins—Errors. 3. Paper money—United States—Evaluation. 4. Counterfeits and counterfeiting—United States. I. Title.
CJ1832.H38 1978 737.4'9'73 78-61860
00 MARC

Herbert, Alan.
The official price guide to mint errors and varieties / by Alan Herbert. — 2d ed. — Orlando, Fla. : House of Collectibles, c1978.
182 p. : ill. ; 21 cm.
First published in 1974 under title: The official guide to mint errors.
ISBN 0-87637-103-9 : $3.95
1. Coins—Errors. 2. Coins, American. I. Title.
CJ125.H47 1978 737.4'9'37 78-67027
79 MARC

Marvin, Paul.
The design cud / by Paul Marvin and Arnold Margolis. [Rev. ed.]. — [Oceanside, N.Y.] : A. Margolis, c1979.
xi, 205 p. (p. 201-205 advertisements) : ill. ; 22 cm.
Previous ed. by Mort Goodman.
$10.95
1. Coins, American. 2. Coins—Errors. I. Margolis, Arnold, joint author. II. Goodman, Mort. Design cud. III. Title.
CJ1832.M28 1979 737.4'9'73 79-116906
79 MARC

—CATALOGS

Iacovo, James S
A comprehensive guide to United States commemorative coins / by James S. Iacovo. — Dallas : Ivy Press, c1979.
x, 130 p. : ill. ; 24 cm.
Includes bibliographical references.
ISBN 0-933372-00-0
1. Commemorative coins—United States—Catalogs. 2. Coins, American—Catalogs. I. Title.
CJ1839.I18 737.4'9'73 78-71956
00 MARC

Rome Reports, inc.
Rome's prices realized for U.S. coins, 1972-1978. — [Inaugural ed.]. — [New York] : Rome Reports, [c1979]
4 v. ; 22 cm.
Cover title.
CONTENTS: v. 1. Gold.—v. 2-3. Silver.—v. 4. Copper.
1. Coins, American—Catalogs. I. Title.
CJ1826.R65 1979 737.4'9'73 79-114877
79 MARC

—PRICES—CATALOGS

Wilhite, Robert.
Standard guide to U.S. coin and paper money valuations / by Robert Whilite and Clifford Mishler. — 4th ed. — Iola, WI : Krause Publications, c1977.
194 p. : ill. ; 18 cm.
Cover title.
Edition for 1975 by G. Wright.
ISBN 0-87341-015-7 : $1.50
1. Coins, American—Prices—Catalogs. 2. Paper money—United States—Catalogs. I. Mishler, Clifford, joint author. II. Wright, Glenn. Standard guide to U.S. coin and paper money valuations. III. Title.
CJ1826.W74 1977 737.4'9'73 77-93060
78[r79]rev MARC

Wilhite, Robert.
Standard guide to U.S. coin and paper money valuations / by Robert Wilhite and Clifford Mishler. — 5th ed. — [Iola, WI : Krause Publications, c1978]
194 p. (p. 194 advertisement) : ill. ; 18 cm.
Cover title.
ISBN 0-87341-021-1 : $1.50
1. Coins, American—Prices—Catalogs. 2. Paper money—United States—Catalogs. I. Mishler, Clifford, joint author. II. Title. III. Title: U.S. coin and paper money valuations.
CJ1826.W74 1978 737.4'9'73 78-60747
79 MARC

COINS, ANCIENT

Burnett, Andrew.
The coins of late antiquity AD 400-700 / Andrew Burnett. — [London : British Museum Publications, c1977]
16 p. : ill. ; 24 cm. — (British Museum keys to the past) GB***
Cover title.
Bibliography: p. 16.
ISBN 0-7141-0067-6 : £0.45
1. Coins, Ancient. I. Title. II. Series: British Museum. British Museum keys to the past.
CJ233.B9 737.4 78-323478
79 MARC

Chirilă, Eugen.
Tezaure şi descoperiri monetare din colecţia Muzeului judeţean Maramureş. Baia Mare, 1971.
77 p., 11 l. of plates. 24 cm. lei6.00 R72-2730
At head of title: Muzeul Judeţean Maramureş. Eugen Chirilă, Aurel Socolan.
Summaries in German.
Includes bibliographical references.
1. Coin hoards—Romania. 2. Coins, Ancient. 3. Muzeul Judetean Maramres. I. Socolan, Aurel, joint author. II. Muzeul Judeţean Maramureş. III. Title.
CJ153.R6C48 78-418764
78 MARC

Fava, Anna Serena.
Il medagliere delle raccolte numismatiche torinesi; esemplari scelti delle serie greca, romana, bizantina, sabauda, piemontese e di altre zecche italiane. A cura di Anna Serena Fava, Luigi Sachero [e] Vittorio Viale. Torino, Museo civico d'arte antica, novembre 1964. [Torino, 1964]
223 p. illus. (part col.) 24 cm.
"Catalogo a cura di Anna Serena Fava": p. [13]-223.
Bibliography: p. [14]
1. Coins, Ancient. 2. Coins, Italian—Italy—Turin. 3. Turin. Museo civico d'arte antica. I. Sachero, Luigi. II. Viale, Vittorio, 1891- III. Turin. Museo civico d'arte antica. IV. Title.
CJ215.F38 78-365944
78 MARC

Vaĭnberg, Bėlla Il'inichna.
(Monety drevnego Khorezma)
Монеты древнего Хорезма / Б. И. Вайнберг ; АН СССР, Ин-т этнографии им. Н. Н. Миклухо-Маклая. — Москва : Наука, 1977.
194 p., xxxl leaves of plates : ill. ; 26 cm. USSR 77
Catalog: p. 106-185.
Bibliography: p. 189-198.
Includes index.
2.85rub
1. Coins, Ancient. 2. Khorezm Oasis—History. I. Title.
CJ277.K47V33 78-422262

Zograf, Aleksandr Nikolaevich.
Ancient coinage / A. N. Zograph ; translated from the Russian by H. Bartlett Wells. — Oxford : British Archaeological Reports, 1977.
2 v. (421 p., [25] leaves of plates) : ill. ; 30 cm. — (BAR supplementary series ; 33) GB***
Translation of Antichnie moneti.
Bibliography: p. 341-383.
CONTENTS: pt. 1. The general problems of ancient numismatics.—pt. 2. The ancient coins of the northern Black Sea littoral.
ISBN 0-904531-87-2 : £9.00
1. Numismatics, Ancient. 2. Coins, Ancient. I. Title. II. Series.
CJ233.Z6313 737.4 78-320046
79 MARC

—HISTORY

Hara, Mitsumasa, 1916-
(Nihon kodai kahei shi no kenkyū)
日本古代貨幣史の研究 / 原三正著. — 東京 : ボナンザ, 1978.
5, 308, 21 p., [4] leaves of plates : ill. ; 22 cm. Ja 78-22699
Limited edition: 500 copies printed.
Bibliography: p. 1-7 (3d group)
Includes indexes.
¥6000
1. Coinage—Japan—History. 2. Coins, Ancient—History. I. Title.
HG1273.H37 78-816832

—BULGARIA

Gerasimov, Todor Dimitrov, 1903-
Antique coins, struck and in circulation in the Bulgarian lands / Todor Gerassimov ; [translated by Margarita Alexieva ; photo, Stoyan Madjarov]. — Sofia [Bulgaria] : Septemvri Pub. House, 1977.
93 p. : col. ill. ; 18 cm. — (Museum treasures series ; 3)
Translation of Antichni moneti, kovani i tsirkulirali po bŭlgarskite zemi.
2.50lv
1. Coins, Ancient—Bulgaria. 2. Coins, Bulgarian. I. Title. II. Series: Series Museum treasures ; 3.
CJ277.B9G4713 737.4'9'4977 79-337851
79 MARC

—EGYPT—QARARA SITE—CATALOGS

Kosack, Wolfgang.
Antike Kultur auf Münzen von Qarara : aus der Ägyptensammlung des Museums / Wolfgang Kosack. — Freiburg i. Br. : Städtische Museen, 1977.
49 p., [4] leaves of plates : ill. ; 30 cm. — (Veröffentlichungen des Museums für Völkerkunde ; 1977, 2) GFR***
At head of title: Museum für Völkerkunde der Stadt Freiburg im Breisgau.
Bibliography: p. 15.
1. Coins, Roman—Egypt—Qarara site—Catalogs. 2. Qarara site, Egypt—Catalogs. 3. Coins, Ancient—Egypt—Qarara site—Catalogs. 4. Freiburg I.B. Museum für Volkerkunde—Catalogs. I. Freiburg I. B. Museum für Völkerkunde. II. Title. III. Series: Freiburg i. B. Museum für Völkerkunde. Veröffentlichungen des Museums für Völkerkunde ; 1977, 2.
CJ1073.Q37K67 79-353380
79[r79]rev MARC

—TURKEY—LYCIA—CATALOGS

Mørkholm, Otto.
Die lykischen Münzlegenden / von Otto Mørkholm, Günter Neumann. — Göttingen : Vandehoeck und Ruprecht, 1978.
38 p., [2] leaves of plates : ill. ; 25 cm. — (Nachrichten der Akademie der Wissenschaften in Göttingen, Philologisch-Historische Klasse ; Jahrg. 1978, Nr. 1) GFR78-A26
Bibliography: p. 38.
Includes index.
ISBN 3-525-85116-2 : DM9.00
1. Coins, Ancient—Turkey—Lycia—Catalogs. 2. Lycia—Antiquities—Catalogs. 3. Turkey—Antiquities—Catalogs. I. Neumann, Günter, 1920- joint author. II. Title. III. Series: Akademie der Wissenschaften, Göttingen. Philologisch-Historische Klasse. Nachrichten ; Jahrg. 1978, Nr. 1.
AS182.G823 Jahrg. 1978, Nr. 1 79-377253
083'.1 s
[CJ277.L9] 79 MARC

COINS, ARABIC

—POLAND—ZALESIE (SŁUPCA)

Wczesnośredniowieczny skarb srebrny z Zalesia, powiat Słupca.
/ Anna Kmietowicz ... [et al.]. — Wrocław : Zakład Narodowy im. Ossolińskich, 1969-1974.
2 v. : ill., map ; 26 cm.
At head of title v. 1: Instytut Historii Kultury Materialnej Polskiej Akademii Nauk ; v. 2. Polska Akademia Nauk. Instytut Historii Kultury Materialnej.
Summaries in French.
Includes bibliographical references.
CONTENTS: t. 1. Monety arabskie.—t. 2. Monety bizantyjskie, Monety Zachodnioeuropejskie, Ozodoby, Podsumowanie.
1. Coin hoards—Poland—Zalesie (Słupca) 2. Coins, Arab—Poland—Zalesie (Słupca) 3. Coins, Medieval—Poland—Zalesie (Słupca) 4. Jewelry, Medieval—Poland—Zalesie (Słupca) 5. Silverwork—Poland—Zalesie (Słupca) I. Kmietowicz, Anna.
CJ1647.Z34W37 74-256081
79[r79]rev2 MARC

COINS, ARMENIAN

Bedoukian, Paul Z
(Artashēsian harstowt'ean dramnerĕ)
Արտաշէսեան հարստութեան դրամները, գրեց Զարեհ Պոտուկեան : Վիեննա, Մխիթարեան Տպարան, 1969 :
118 p., 8 plates. 20 cm. (Ազգային մատենադարան, հ. 210)
Added t. p.: Coinage of the Artaxiad dynasty of Armenia, by Paul Z. Bedoukian.
Summary in English.
Some chapters based on articles previously published in the American Numismatic Society's Museum notes.
Includes bibliographical references.
1. Coins, Armenian. 2. Armenia — History — Artaxiad (Artashesian) dynasty, 189 B. C.–1 A. D. I. Title. II. Title: Coinage of the Artaxiad dynasty of Armenia. III. Series: Azgayin matenadaran, h. 210.
CJ3481.B39 79–970723

COINS, AUSTRIAN

—CATALOGS

Jungwirth, Helmut.
Corpus nummorum Austriacorum (CNA) / neu gestaltet u. bearb. v. Helmut Jungwirth ; nach dem Werk, V. Müller zu Aichholz, A. Loehr, E. Holzmair, Österr. Münzprägungen 1519-1938. — Wien : Bundessammlung von Medaillen, Münzen und Geldzeichen, Kunsthist. Museum : [Auslft., Österr. Numismatische Gesellschaft], 1975-
v. : ill. ; 31 cm. Au77-13-261
Bibliography: v.5, p. v-viii.
CONTENTS: Bd. 5. Leopold I. -Karl V.

S380.00 (v. 5)
1. Coins, Austrian—Catalogs. I. Miller zu Aichholz, Viktor von, 1845-1910. Österreichische Münzprägungen 1519-1938. II. Title.
CJ2580.J86 1975 737.4'9'436 79-344730
*79 MARC

COINS, BELGIAN

Philippen, Jos 1912-
Benelux sinds 1815 : catalogus van de munten met hun actuele waarde = Catalogue des monnaies avec leur valeur actuelle = Katalog der Münzen mit dem heutigen Wert = Catalogue of the coins with their present value / Jos Philippen. — Diest : Europa, ₍1973₎
108 p. : ill. ; 19 cm. Be75-6
Dutch, English, French, and German.
ISBN 9061880017 : 250F
1. Coins, Belgian. 2. Coins, Dutch. 3. Coins, Luxemburg. I. Title.
CJ2934.P47 78-366149
78 MARC

COINS, BRITISH

—CATALOGS

Seaby's standard catalogue of British coins. 12th- ed.; 1973-
London, B. A. Seaby.
v. ill. 19 cm. annual.
Supersedes: Seaby's standard catalogue: British coins, England and the United Kingdom.
Continued by: Standard catalogue of British coins.
Vols. for 1973- issued in 4 pts.: pts. 1 and 2, Coins of England and the United Kingdom—pt. 3, Coins and tokens of Ireland.—pt. 4, Coins and tokens of Scotland.
1. Coins, British—Catalogs. I. Title: Coins of England and the United Kingdom. II. Title: Coins and tokens of Ireland. III. Title: Coins and tokens of Scotland.
CJ2476.S82 737.4'9'41 78-646891
MARC-S

Standard catalogue of British coins.
London, Seaby.
v. ill. 23 cm.
Continues: Seaby's standard catalogue of British coins.
Cover title, : Seaby standard catalogue of British coins.
Vols. for issued in 2 vols.: v. 1, Coins of England and the United Kingdom.—v. 2, Coins of Scotland, Ireland and the islands.
1. Coins, British—Catalogs. I. Title: Seaby standard catalogue of British coins. II. Title: Coins of England and the United Kingdom. III. Title: Coins of Scotland, Ireland and the islands.
CJ2476.S82 737.4'9'41 78-646896
MARC-S

—GREAT BRITAIN

Kent, John.
2000 years of British coins and medals / John Kent. — London : British Museum Publications, 1978.
78 p. : ill. ; 24 cm. GB77-07982
Bibliography: p. 74.
ISBN 0-7141-0842-1. ISBN 0-7141-0845-6 pbk.
1. Coins, British—Great Britain. 2. Medals—Great Britain. 3. Numismatics—Great Britain. I. Title.
CJ2480.K46 737.4'9'41 78-314224
78 MARC

—MAN, ISLE OF

Mackay, James Alexander.
The Pobjoy encyclopaedia of Isle of Man coins and tokens / by James A. Mackay. — 2nd ed. — Sutton : Pobjoy Mint Ltd., 1978.
₍3₎, 84 p. : ill. (some col.), col. port. ; 22 cm. GB79-03780
Spine title: Isle of Man coins and tokens.
Bibliography: p. 72-73.
£3.50
1. Coins, British—Man, Isle of. 2. Tokens—Man, Isle of. I. Pobjoy Mint Limited of England. II. Title. III. Title: Isle of Man coins and tokens.
CJ2498.M25M33 1978 737.4'9'4279 79-308854
79 MARC

COINS, BULGARIAN

Gerasimov, Todor Dimitrov, 1903-
Antique coins, struck and in circulation in the Bulgarian lands / Todor Gerassimov ; ₍translated by Margarita Alexieva ; photo, Stoyan Madjarov₎. — Sofia ₍Bulgaria₎ : Septemvri Pub. House, 1977.
93 p. : col. ill. ; 18 cm. — (Museum treasures series ; 3)
Translation of Antichni moneti, kovani i tsirkulirali po bŭlgarskite zemi.
2.50lv
1. Coins, Ancient—Bulgaria. 2. Coins, Bulgarian. I. Title. II. Series: Series Museum treasures ; 3.
CJ277.B9G4713 737.4'9'4977 79-337851
79 MARC

COINS, BURMESE

—BURMA—ARAKAN (DIVISION)

Phayre, Arthur Purves, Sir, 1812-1885.
Coins of Arakan, of Pegu, and of Burma / by Sir Arthur P. Phayre. — London : Trübner, 1882.
47 p., 5 leaves of plates : ill. ; 32 cm. — (The International numismata orientalia ; pt. 8)
Includes bibliographical references and index.
1. Coins, Burmese—Burma—Arakan (Division) 2. Coins, Burmese—Burma—Pegu (Division) 3. Numismatics—Burma—Arakan (Division) 4. Numismatics—Burma—Pegu (Division) I. Title. II. Series.
CJ3639.A72P47 737.4'9591 78-317925
78 MARC

—BURMA—PEGU (DIVISION)

Phayre, Arthur Purves, Sir, 1812-1885.
Coins of Arakan, of Pegu, and of Burma / by Sir Arthur P. Phayre. — London : Trübner, 1882.
47 p., 5 leaves of plates : ill. ; 32 cm. — (The International numismata orientalia ; pt. 8)
Includes bibliographical references and index.
1. Coins, Burmese—Burma—Arakan (Division) 2. Coins, Burmese—Burma—Pegu (Division) 3. Numismatics—Burma—Arakan (Division) 4. Numismatics—Burma—Pegu (Division) I. Title. II. Series.
CJ3639.A72P47 737.4'9591 78-317925
78 MARC

COINS, BYZANTINE

see also Hexagram (Coin)

—CATALOGS

Bendall, S
The later Palaeologan coinage, 1282-1453 / by S. Bendall and P. J. Donald. — ₍London₎ : A. H. Baldwin and Sons Ltd., 1979.
271 p. : ill. ; 21 cm. GB79-15539
Errata slip inserted.
Bibliography: p. 15-19.
Includes index.
£12.50
1. Coins, Byzantine—Catalogs. I. Donald, P. J., joint author. II. Title.
CJ1283.B46 737.4'9'495 79-315606
79 MARC

—GREECE—ATHENS

Thompson, Margaret, 1911-
Coins from the Roman through the Venetian period / by Margaret Thompson. — Meriden, Conn. : Meriden Gravure Co., 1961.
vii, 122 p., ₍2₎ leaves of plates : ill. ; 31 cm. — (The Athenian Agora ; v. 2)
(Reprint of the 1954 ed. published by the American School of Classical Studies at Athens, Princeton, N.J.)
Bibliography: p. ₍8₎
Includes indexes.
1. Coins, Roman—Greece—Athens. 2. Coins, Byzantine—Greece—Athens. 3. Athens. Agora. I. Title. II. Series: American School of Classical Studies at Athens. The Athenian Agora ; v. 2.
DF287.A23A5 vol. 2 938'.5 s 75-328019
₍CJ893.A8₎ 76₍r79₎rev MARC

COINS, CARLOVINGIAN

—BELGIUM

Frère, Hubert.
Le denier carolingien : spécialement en Belgique / par Hubert Frère. — Louvain-la-Neuve : Institut supérieur d'archéologie et d'histoire de l'art, Séminaire de numismatique Marcel Hoc, 1977.
x, 134 p., ₍2₎, 10 leaves of plates : ill., maps ; 25 cm. — (Numismatica Lovaniensia ; 1) (Publications d'histoire de l'art et d'archéologie de l'Université catholique de Louvain ; 8) Be***
Bibliography: p. ₍107₎-108.
Includes index.
750F
1. Coins, Carlovingian—Belgium. I. Title. II. Series. III. Series: Université catholique de Louvain, Louvain-la-Neuve, Belgium. Publications d'histoire de l'art et d'archéologie de l'Université catholique de Louvain ; 8.
CJ2964.F73 737.4'9'493 79-349997
79 MARC

COINS, CELTIC

Kolníková, Eva.
Keltské mince na Slovensku / Eva Kolníková ; ₍typogr. Ivan Kovačevič ; fot. Jozef Krátky₎. — 1. vyd. — Bratislava : Pallas, 1978.
106 p. : ill. (some col.) ; 23 cm. — (Dávnoveké umenie Slovenska ; zv. 2) CzS79
Summary in French, German, and Russian.
Bibliography: p. 84.
Kčs40.00
1. Coins, Celtic. 2. Celts in Slovakia. I. Title. II. Series.
CJ2681.K64 79-359648
*79 MARC

—CATALOGS

Zürich. Schweizerisches Landesmuseum.
Keltische Münzen : Katalog der Sammlung des Schweizerischen Landesmuseums Zürich / Karel Castelin. — Stäfa : T. Gut, ₍1978₎-
v. : ill. ; 29 cm. Sw78-6330
Bibliography: v.1, p. 18-21.
Includes index.
95.00F (v. 1)
1. Coins, Celtic—Catalogs. 2. Zürich. Schweizerisches Landesmuseum—Catalogs. I. Castelin, Karel O. II. Title.
CJ2681.Z83 1978 737.4'9'364 79-342393
79 MARC

COINS, CUFIC

Stockholm. Statens historiska museum. Myntkabinettet.
Numi cufici Regii numophylacii holmiensis, quos omnes in terra Sueciae repertos digessit et interpretatus est d:r Carolus Johannes Tornberg ... Upsaliae, excudebant Leffler et Sebell, 1848.
3 p.l., lxxxviii, 315, ₍1₎ p. xiv pl. 27 x 23 cm. (Kongl. svenska myntkabinett. ₍Skrifter₎ II. Kufiska mynt)
"Impensis Reg. Academiae Litterr. Humann. Historiae et Antiquitatum."
1. Coins, Cufic. I. Tornberg, Carl Johan, 1807-1877. II. Title. III. Series: Stockholm. Statens historiska museum. Myntkabinettet. Skrifter ; 2.
CJ3404.S78 1848 10-27943
₍r79₎rev2 MARC

COINS, CYPIOTE

Lampros, Paulos, 1819-1887.
(Anekdota nomismata tou mesaiōnikou vasileiou tēs Kyprou)
'Ανέκδοτα νομίσματα τοῦ μεσαιωνικοῦ βασιλείου τῆς Κύπρου / ὑπὸ Π. Λάμπρου. — 'Αθῆναι : Βιβλιοπωλεῖον Ν. Καραβία, 1970.
52, iii, 48 p., ₍9₎ leaves of plates : ill. ; 24 cm. — (Βιβλιοθήκη ἱστορικῶν μελετῶν ; 36) (Νομισματολογικὴ σειρά)
Cover title.
Added t. p.: Monnaies inedites du royaume de Chypre au moyen age.
French and Greek.
Reprint of the 1876 ed.
Series 1 romanized: Vivliothēkē historikōn meletōn. Series 2 romanized: Nomismatologikē seira.
1. Numismatics—Cyprus. 2. Coins, Cypriote. I. Title. II. Title: Monnaies inedites du royaume de Chypre au moyen age.
CJ3893.L35 1970 737.4'9'5937 79-413657

COINS, DANISH

—INDIA—TRANQUEBAR—CATALOGS

Barner Jensen, Uno.
Trankebar-mønter : ca. 1620-1845 = Danish India Tranquebar coins / ₍af₎ Uno Barner Jensen. — ₍Gentofte₎₍Plantagevej 8, 2820₎ : Flensborg, c1978.
47 p. : ill. ; 21 cm. D78-39
Pref. and introd. also in English.
Errata slip inserted.
Bibliography: p. 39-40
ISBN 8798013440 : kr58.00
1. Coins, Danish—India—Tranquebar—Catalogs. I. Title. II. Title: Danish India Tranquebar coins.
CJ3119.T72B37 737.4'9'489 79-370069
79 MARC

COINS, DUTCH

Gelder, Hendrik Enno van.
Klein geld, grof geld : munten van de Nederlandse steden / H. Enno van Gelder. — 's-Gravenhage : Stichting Gemeentelijk Cultuurfonds, 1974.
42 p. : ill. ; 30 cm. Ne***
1. Coins, Dutch. I. Stichting Gemeentelijk Cultuurfonds. II. Title.
CJ2985.G44 79-377978
79 MARC

COINS, DUTCH (Continued)

Gelder, Hendrik Enno van.
De Nederlandse munten / H. Enno van Gelder. — 6. bijgewerkte druk. — Utrecht : Het Spectrum, 1976.
272 p. : ill. ; 18 cm. — (Aula-boeken ; 213) Ne76-25
Bibliography: p. 254-255.
ISBN 9027449120 : fl 5.90
1. Coins, Dutch. I. Title.
CJ2976.G4 1976 79-346752
*79 MARC

Philippen, Jos 1912-
Benelux sinds 1815 : catalogus van de munten met hun actuele waarde = Catalogue des monnaies avec leur valeur actuelle = Katalog der Münzen mit dem heutigen Wert = Catalogue of the coins with their present value / Jos Philippen. — Diest : Europa, ₍1973₎
108 p. : ill. ; 19 cm. Be75-6
Dutch, English, French, and German.
ISBN 9061880017 : 250F
1. Coins, Belgian. 2. Coins, Dutch. 3. Coins, Luxemburg. I. Title.
CJ2934.P47 78-366149
78 MARC

COINS, ENGLISH

see also Coins, British

Kenworthy, Jane.
Coins of England, 600-1662 / ₍by₎ Jane Kenworthy. — London : Gibbons, 1977.
31 p. : ill. ; 20 cm. — (Stanley Gibbons guides) GB78-15721
ISBN 0-85259-905-6 : £0.75
1. Coins, English. 2. Numismatics—England. I. Title.
CJ2489.5.K46 737.4'9'42 78-319695
78 MARC

COINS, EUROPEAN

Clain-Stefanelli, Elvira Eliza.
Das grosse Buch der Münzen und Medaillen : mit Münzkatalog Europa von 1900 bis heute / von Elvira und Vladimir Clain-Stefanelli, Günter Schön. — 1. Aufl. — München : Battenberg, 1976.
308 p. : ill. (some col.) ; 26 cm. GFR76-A
"Aus dem Amerikanischen übersetzt von Alfred P. Zeller."
Bibliography: p. 162-164.
ISBN 3-87045-119-X : DM39.00
1. Coins. 2. Numismatics. 3. Coins, European. I. Clain-Stefanelli, Vladimir, joint author. II. Schön, Günter, joint author. III. Title.
CJ59.C5515 78-382176
78 MARC

Dietzel, Heinz.
Die Münzen Europas ab 1945 ₍neunzehnhundertfünfundvierzig₎ : Nettokatalog / Heinz Dietzel. — 8., überarb. u. erg. Aufl. — Berlin : Pröh, 1976.
146 p. : ill. ; 17 cm. GFR76-A
1. Coins, European. I. Title.
CJ2456.D53 1976 737.4'9'4 79-377312
79 MARC

COINS, FRENCH

Houyez, René.
Initiation aux monnaies de France / par René Houyez. — Paris : G.A.R.C.E.N., c1976.
106 p. : ill. ; 20 cm. F***
Cover title.
32.00F
1. Coins, French. 2. Numismatics—Handbooks, manuals, etc. I. Title.
CJ2670.H75 737.4'9'44 78-384313
78 MARC

—CATALOGS

Houyez, René.
Valeur des monnaies de France / René Houyez. — 2e éd. — Paris (96, rue Legendre, 75017) : Garcen, 1977.
3 v. (143 p.) : ill. ; 21 cm. — (Collection La Valeur numismatique) F78-9209
25.00F per vol.
CONTENTS: ₍1₎ Période révolutionnaire.—₍2₎ Période sémi-moderne.—₍3₎ Période moderne.
1. Coins, French—Catalogs. I. Title. II. Series.
CJ2666.H68 1977 737.4'9'44 79-337812
79 MARC

Houyez, René.
Valeur des monnaies royales de François Ier à Henri IV / René Houyez. — 1. éd. — Paris : Garcen, ₍1978₎
212, ₍1₎ p. : ill. ; 22 cm. — (Collection La Valeur numismatique) F***
Bibliography: p. ₍219₎
Includes index.
ISBN 2-902497-04-0 : 95.00F
1. Coins, French—Catalogs. I. Title. II. Series.
CJ2683.H68 737.4'9'44 79-360069
79 MARC

COINS, GERMAN

see also Coins, Vandal

Arnold, Paul, Dr. phil.
Catalogue of German coins; gold, silver and minor coins since 1800, with their valuations, by P. Arnold, D. Steinhilber and H. Kuthmann. New York, Sterling Pub. Co. ₍1972₎
320 p. illus. 26 cm.
Translation of Grosser deutscher Münzkatalog von 1800 bis heute.
ISBN 0-8069-6028-0. ISBN 0-8069-6029-9 (lib. bdg.)
1. Coins, German. I. Küthmann, Harald, joint author. II. Steinhilber, Dirk, joint author. III. Title.
CJ2725.A7513 1972 737.4'9'43 72-126855
72₍r79₎rev MARC

Arnold, Paul, Dr. phil.
Grosser deutscher Münzkatalog von 1800 bis heute / Paul Arnold, Harald Küthmann, Dirk Steinhilber. — 4., völlig überarb. u. erw. Aufl. — München : Battenberg, 1974.
v, 431 p. : ill., maps ; 25 cm. GFR74-A42
Bibliography: p. ₍433₎-₍435₎
ISBN 3-87045-087-8 : DM48.00
1. Coins, German. I. Küthmann, Harald, joint author. II. Steinhilber, Dirk, joint author. III. Title.
CJ2725.A75 1974 737.4'9'43 75-563361
79₍r79₎rev MARC

Arnold, Paul, Dr. Phil.
Grosser deutscher Münzkatalog : von 1800 bis heute / Paul Arnold, Harald Küthmann, Dirk Steinhilber. — 5., völlig überarb. u. erw. Aufl. — München : Battenberg, 1977.
v, 434, ₍4₎ p. : numerous ill., maps ; 25 cm. GFR77-A42
Bibliography: p. ₍436₎-₍438₎
ISBN 3-87045-128-9
1. Coins, German. I. Küthmann, Harald, joint author. II. Steinhilber, Dirk, joint author. III. Title.
CJ2725.A75 1977 737.4'9'43 78-355914
78₍r79₎rev MARC

Schön, Günter.
Altdeutschland : e. Katalog d. bekanntesten Münzen d. Röm.-Dt. Reiches von 768-1806 / Günter Schön. — 1.-10. Tsd. — München : Battenberg, 1976.
729 p. : ill. ; 19 cm. GFR76-A
Bibliography: p. 716-727.
ISBN 3-87045-113-0 : DM28.00
1. Coins, German. I. Title.
CJ2706.S36 737.4'9'43 79-377311
79 MARC

—CATALOGS

Arslan, Ermanno A
Le monete di Ostrogoti longobardi e vandali : catalogo delle civiche raccolte numismatiche di Milano / Ermanno A. Arslan. — Milano : Comune di Milano, Ripartizione cultura e spettacolo, 1978.
91 p., ₍22₎ leaves of plates : ill. ; 24 cm. It***
Cover title: I Longobardi e la Lombardia.
Includes bibliographical references.
1. Coins, German—Catalogs. 2. Coins, Lombard—Catalogs. 3. Coins, Gothic—Catalogs. 4. Coins, Vandal—Catalogs. I. Title. II. Title: I Longobardi e la Lombardia.
CJ1657.A77 737.4'9'43 79-377592
79 MARC

Divo, Jean-Paul.
Die deutschen Goldmünzen von 1800-1872 / Jean-Paul Divo, Hans-Joachim Schramm. — Frankfurt am Main : Numismatischer Verlag P. N. Schulten, 1979.
170 p. : ill. ; 21 cm. GFR***
Bibliography: p. 170.
ISBN 3-921302-22-6
1. Gold coins—Germany—Catalogs. 2. Coins, German—Catalogs. I. Schramm, Hans-Joachim, joint author. II. Title.
CJ2714.D58 737.4'9'43 79-372116
79 MARC

Jaeger, Kurt.
Die Zahlungsmittel in Deutschland seit 1948. Münzen und Banknoten mit Bewertungen. Basel, Münzen und Medaillen A. G., 1972.
224 p. illus. 11 x 16 cm. 15.00F Sw74-A-2468
Includes suppl. (8 p.) inserted at end.
1. Coins, German—Catalogs. 2. Bank-notes—Germany—Catalogs. I. Title.
CJ2725.J32 737.4'9'43 79-409940
79 MARC

—GERMANY, EAST—EICHSFELD

Lauerwald, Paul.
Zur Münz- und Geldgeschichte des Eichsfeldes / Paul Lauerwald. — Als Sonderausg. d. "Eichsfelder Heimathefte" hrsg. vom Pädag. Kreiskabinett Worbis in Verbindung mit d. Kulturbund d. DDR Kreisleitung Worbis, Heiligenstadt. — Worbis : Pädag. Kreiskabinett, 1976.
87 p. : 21 ill. and maps ; 20 cm. GFR77-B2
Bibliography: p. 84-87.
1. Coin hoards—Germany, East—Eichsfeld. 2. Coins, German—Germany, East—Eichsfeld. I. Eichsfelder Heimathefte : Sonderausgabe. II. Title.
CJ393.E32L38 79-348233
79 MARC

—GERMANY, EAST—MEISSEN (MARGRAVATE)—CATALOGS

Schwinkowski, Walter, b. 1884.
Münz- und Geldgeschichte der Mark Meissen und Münzen der weltlichen Herren nach meissnischer Art (Brakteaten) vor der Groschenprägung / W. Schwinkowski. — Fotomechan. Neudr. d. Orig. -Ausg. Frankfurt am Main 1931. T. 1, Abbildungstafeln. — Leipzig : Zentralantiquariat der Dt. Demokrat. Republik, 1976.
4 p., 63 leaves : all ill. ; 27 cm. GFR77-A51
On cover: Die meissnischen Brakteaten.
No more published?
Reprint of the ed. published by Verlag Adolph Hess, in series Aus den Schriften der Sächsischen Kommission für Geschichte.
182.00M
1. Coins, German—Germany, East—Meissen (Margravate)—Catalogs. 2. Bracteate coins—Germany, East—Meissen (Margravate)—Catalogs. I. Title. II. Title: Die meissnischen Brakteaten. III. Series: Saxony. Kommission für Geschichte. Schriften ; Bd. 36.
CJ2809.M44S38 737.4'9'431 78-394076
79 MARC

—GERMANY, EAST—SAXONY—CATALOGS

Dietzel, Heinz.
Die Münzen des Kurfürstentums und Königreichs Sachsen ab 1694 : Nettokatalog / Heinz Dietzel ; Anh., Die Münzen Sachsens nach Reichsgründung. — 1. Aufl. — Berlin : Pröh, 1976.
56 p. : chiefly ill. ; 17 cm. GFR76-A
DM10.00
1. Coins, German—Germany, East—Saxony—Catalogs. I. Title.
CJ2809.S38D53 737.4'9'43 79-352155
79 MARC

—GERMANY, WEST—HILDESHEIM

Mehl, Manfred.
Schöne Hildesheimer Münzen : Gepräge d. Stadt u. d. Bistums / Manfred Mehl. — Hildesheim : Bernward-Verlag, 1974.
48 p. : ill. ; 18 cm. — (Hildesheimer Miniaturen ; Bd. 1) GFR75-A
Bibliography: p. 47-48.
ISBN 3-87065-055-9 : DM5.00
1. Coins, German—Germany, West—Hildesheim. 2. Coins, German—Germany, West—Hildesheim (Diocese) 3. Numismatics—Germany, West—Hildesheim. 4. Numismatics—Germany, West—Hildesheim (Diocese) I. Title. II. Series.
CJ2729.H53M43 79-382500
79 MARC

—GERMANY, WEST—HILDESHEIM (DIOCESE)

Mehl, Manfred.
Schöne Hildesheimer Münzen : Gepräge d. Stadt u. d. Bistums / Manfred Mehl. — Hildesheim : Bernward-Verlag, 1974.
48 p. : ill. ; 18 cm. — (Hildesheimer Miniaturen ; Bd. 1) GFR75-A
Bibliography: p. 47-48.
ISBN 3-87065-055-9 : DM5.00
1. Coins, German—Germany, West—Hildesheim. 2. Coins, German—Germany, West—Hildesheim (Diocese) 3. Numismatics—Germany, West—Hildesheim. 4. Numismatics—Germany, West—Hildesheim (Diocese) I. Title. II. Series.
CJ2729.H53M43 79-382500
79 MARC

—GERMANY, WEST—LINGENFELD

Ehrend, Helfried.
Der Münzschatz von Lingenfeld 1969 ₍i.e. neunzehnhundertneunundsechzig₎ / Helfried Ehrend ; mit Beiträgen von Günter Stein und Friedrich Wielandt. — Speyer : Numismatische Gesellschaft Speyer, 1975.
64 p. : ill. ; 21 cm. GFR***
Includes bibliographical references.
1. Coin hoards—Germany, West—Lingenfeld. 2. Lingenfeld, Ger.—Antiquities. 3. Germany, West—Antiquities. 4. Coins, German—Germany, West—Lingenfeld. I. Stein, Günter, joint author. II. Wielandt, Friedrich, 1906- joint author. III. Title.
CJ1647.L56E37 79-381992
79 MARC

COINS, GERMAN (Continued)

—GERMANY, WEST—MÜNSTER (ECCLESIASTICAL PRINCIPALITY)

Schulze, Ingrid.
Die fürstbischöflich-münsterschen Münzen der Neuzeit / Ingrid u. Wolfgang-Georg Schulze ; [Fotos, Heinz Thormann]. — Münster (Westfalen) : Numismatischer Verlag Dombrowski, 1973.
60 p., xi leaves : ill. ; 27 cm. GFR75-A37
Bibliography: p. 3.
1. Coins, German—Germany, West—Münster (Ecclesiastical principality) 2. Numismatics—Germany, West—Münster (Ecclesiastical principality) I. Schulze, Wolfgang Georg, 1946- joint author. II. Title.
CJ2728.M83S33 78-379293
78 MARC

—GERMANY, WEST—RATISBON—CATALOGS

Beckenbauer, Egon.
Die Münzen der Reichsstadt Regensburg / Egon Beckenbauer. — Grünwald : H. Geiger, 1978.
443 p. : ill., map ; 25 cm. — (Bayerische Münzkataloge ; Bd. 5) GFR***
Bibliography: p. 92-94.
1. Coins, German—Germany, West—Ratisbon—Catalogs. I. Title. II. Series.
CJ2746.B35 Bd. 5 79-359270
[CJ2729.R38] 79 MARC

—GERMANY, WEST—WELBERGEN

Berghaus, Peter.
Der Münzschatzfund von Welbergen Kr. Steinfurt 1965 / von Peter Berghaus. — Münster : Westfälisches Landesmuseum für Kunst und Kulturgeschichte, 1978.
20, [1] p. : ill. ; 30 cm. — (Numismatische Schriften des Westfälischen Landesmuseum für Kunst und Kulturgeschichte, Landschaftsverband Westfalen-Lippe ; Nr. 1) GFR***
Cover title.
Bibliography: p. 19-[21]
1. Coin hoards—Germany, West—Welbergen. 2. Coins, German—Germany, West—Welbergen. 3. Welbergen, Ger.—Antiquities. 4. Germany, West—Antiquities. I. Title. II. Series: Westfälisches Landesmuseum für Kunst und Kulturgeschichte Münster. Numismatische Schriften des Westfälischen Landesmuseums für Kunst und Kulturgeschichte, Landschaftsverband Westfalen-Lippe ; Nr. 1.
CJ2729.W44B47 78-399001
79 MARC

—HOLY ROMAN EMPIRE—CATALOGS

Schön, Günter.
Katalog der Kupfermünzen des Römisch-Deutschen Reiches im 16., 17. und 18. Jahrhundert / hrsg. v. Günter Schön. — Graz : Verl. f. Sammler, c1978.
xi, 233 p. : ill. ; 19 cm. Au78-20-328
Bibliography: p. 229-233.
ISBN 3-85365-032-5 : S140.00
1. Copper coins—Holy Roman Empire—Catalogs. 2. Coins, German—Holy Roman Empire—Catalogs. I. Title.
CJ2716.S36 79-363984
*79 MARC

—PRUSSIA

Martin, Klaus.
Die preussischen Münzprägungen von 1701-1786 / Klaus Martin. — Berlin : Pröh, 1976.
308 p. : numerous ill. ; 22 cm. — (Schriftenreihe die Münze ; Bd. 60) GFR77-A21
"Bewertungsliste" (20 p.) in pocket.
DM67.00
1. Coins, German—Prussia. 2. Numismatics—Prussia. I. Title. II. Series.
CJ2728.P78M37 78-375559
78 MARC

—RUSSIAN REPUBLIC—KALINIGRAD

Meinhardt, Günther.
Gemünzt zu Königsberg : e. Beitr. zur Münz- u. Geldgeschichte Ostpreussens / Günther Meinhardt. — Leer : Rautenberg [in Komm.], 1977.
88 p., [3] leaves of plates : ill. ; 21 cm. — (Prussia-Schriftenreihe ; Werk 1) GFR78-A39
Bibliography: p. 79.
ISBN 3-7921-0197-1 : DM12.50
1. Coins, German—Russian Republic—Kalinigrad. I. Title. II. Series.
CJ3029.K34M44 79-365977
79 MARC

COINS, GOTHIC

Heiss, Aloïss, 1820-
Description générale des monnaies des rois wisigoths d'Espagne / par Aloïss Heiss. — Ristampa. — [Bologna] : A. Forni, 1976.
iii, 185 p., 13 leaves of plates : ill. ; 32 cm. It78-Apr
Reprint of 1872 ed., published by Impr. nationale, Paris.
Three hundred and fifty copies printed.
Includes indexes.
1. Coins, Gothic. 2. Coins, Spanish. 3. Spain—History—Gothic period, 414-711. I. Title.
CJ3194.H5 1976 737.4'9'46 79-347233
*79 MARC

—CATALOGS

Arslan, Ermanno A
Le monete di Ostrogoti longobardi e vandali : catalogo delle civiche raccolte numismatiche di Milano / Ermanno A. Arslan. — Milano : Comune di Milano, Ripartizione cultura e spettacolo, 1978.
91 p., [22] leaves of plates : ill. ; 24 cm. It***
Cover title: I Longobardi e la Lombardia.
Includes bibliographical references.
1. Coins, German—Catalogs. 2. Coins, Lombard—Catalogs. 3. Coins, Gothic—Catalogs. 4. Coins, Vandal—Catalogs. I. Title. II. Title: I Longobardi e la Lombardia.
CJ1657.A77 737.4'9'43 79-377592
79 MARC

COINS, GREEK

see also Cistophorus (Coin)

Burelli, Laura.
Appunti di numismatica greca / Laura Burelli. — Bologna : Pàtron, 1977.
58 p., [6] leaves of plates : ill. ; 22 cm. — (Contributi didattici di storia antica ; 2) It78-Nov
Includes bibliographies.
L3000
1. Coins, Greek. I. Title.
CJ335.B87 737.4'9'38 79-358204
*79 MARC

Carradice, Ian.
Ancient Greek portrait coins / Ian Carradice. — [London : British Museum Publications, c1978]
16 p. : ill. ; 24 cm. GB***
Cover title.
Bibliography: p. 16.
ISBN 0-7141-0849-9 : £0.60
1. Coins, Greek. 2. Kings and rulers (in numismatics)—Mediterranean region. I. Title.
CJ351.C34 737.4'9'38 79-308128
79 MARC

Erhart, Katherine Patricia, 1950-
The development of the facing head motif on Greek coins and its relation to classical art / Katherine Patricia Erhart. — New York : Garland, 1979.
xv, 523 p., [21] leaves of plates : ill. ; 21 cm. — (Outstanding dissertations in the fine arts)
Originally presented as the author's thesis, Harvard University, 1978.
Bibliography: p. 508-523.
ISBN 0-8240-3954-8 : $45.00
1. Facing head (in numismatics) 2. Art, Greek. 3. Coins, Greek. I. Title. II. Series.
CJ161.F3E73 1979 737.4'9'38 78-74366
79 MARC

Giovannini, Adalberto.
Rome et la circulation monétaire en Grèce au IIe siècle avant Jésus-Christ / par Adalberto Giovannini. — Basel : F. Reinhardt, 1978.
x, 129, [6] leaves of plates ; 24 cm. — (Schweizerische Beiträge zur Altertumswissenschaft ; Heft 15) Sw79-2825
Includes bibliographical references.
ISBN 3-7245-0412-8 : 38.00F
1. Coins, Greek. 2. Coins, Roman—Greece. 3. Greece—History—281-146 B.C. I. Title. II. Series.
CJ427.G54 737.4'9'38 79-383143
79 MARC

Pareja López, Enrique, comp.
Repertorio numismático. [Granada] Departamento de Prehistoria y Arqueología, Facultad de Filosofía y Letras, 1972.
[74] p. illus., maps. 25 cm. Sp***
"500 ejemplares."
ISBN 846005571X
1. Coins, Greek. 2. Coins, Roman. 3. Coins, Greek—Spain. 4. Coins, Roman—Spain. I. Title.
CJ335.P37 78-390100
79 MARC

—CATALOGS

Cambridge. University. Corpus Christi College. Lewis Collection.
The Lewis Collection in Corpus Christi College, Cambridge. London, Oxford University Press for the British Academy, 1972-
v. illus. 39 cm. (Sylloge nummorum graecorum, v. 6) £8.00 (v. 1) GB72-31254
Includes index.
CONTENTS: pt. 1. The Greek and Hellenistic coins (with Britain and Parthia)
ISBN 0-19-725924-3 (v. 1)
1. Coins, Greek—Catalogs. 2. Lewis, Samuel Savage, 1836-1891—Coin collections—Catalogs. I. Title. II. Series: British Academy, London (Founded 1901) Sylloge nummorum graecorum, v. 6.
CJ314.B7 vol. 6 737.4'9'38 s 74-174825
[CJ315.G672C34] 74[r79]rev MARC

Chéhab, Maurice H
Monnaies gréco-romaines et phéniciennes du Musée national, Beyrouth, Liban / Maurice Chéhab. — Paris : Librairie d'Amérique et d'Orient, A. Maisonneuve, 1977.
viii, 91 p., 83 leaves of plates : ill. (some col.), maps ; 26 cm. F***
Includes indexes.
1. Coins, Greek—Catalogs. 2. Coins, Roman—Catalogs. 3. Coins, Phenician—Catalogs. 4. Beirut. Musée national libanais—Catalogs. I. Title.
CJ315.L42B443 737.4'0938 78-394670
79 MARC

Dietzel, Heinz.
Die Münzen Griechenlands ab 1828 [i. e. achtzehnhundertachtundzwanzig] mit Kreta : Nettokatalog / Heinz Dietzel. — 1. Aufl. — Berlin : Pröh, 1974.
25 p. : ill. ; 17 cm. GFR75-A
DM4.00
1. Coins, Greek—Catalogs. 2. Bank-notes—Greece, Modern—Catalogs. I. Title.
CJ2887.D53 737.4'9'495 79-343792
79 MARC

Sear, David R
Greek coins and their values / by David R. Sear. — London : Seaby, 1978-
v. : ill. ; 23 cm. GB79-05847
Bibliography: v. 1, p. xxxvii-xxxviii.
Includes index.
CONTENTS: v. 1. Europe.
ISBN 0-900652-46-2 (v. 1) : £12.50 (v. 1) ($27.50 v. 1 U.S.)
1. Coins, Greek—Catalogs. I. Title.
CJ321.S42 737.4'9'38 79-309601
79 MARC

—COLLECTORS AND COLLECTING

Sear, David R
Collecting Greek coins / [by] David Sear. — London : Gibbons, 1977.
31 p. : ill. ; 20 cm. — (Stanley Gibbons guides) GB78-15720
Bibliography: p. 31.
ISBN 0-85259-910-2 : £0.75
1. Coins, Greek—Collectors and collecting. I. Title.
CJ351.S4 737.4'9'38 78-319694
78 MARC

—CYPRUS

Gardiakos, S
The coins of Cyprus / by Soterios Gardiakos. — 2d ed. — Chicago : Obol International, c1975.
36 p., [1] leaf of plates : ill. ; 22 cm.
Bibliography: p. 36.
1. Coins, Greek—Cyprus. 2. Numismatics—Cyprus. 3. Numismatics, Greek. I. Title.
CJ2884.G37 1975 737.4'9'38 79-311748
79 MARC

—GREECE—LAKONIA

Grunauer-von Hoerschelmann, Susanne.
Die Münzprägung der Lakedaimonier / von Susanne Grunauer-von Hoerschelmann. — Berlin : de Gruyter, 1978.
xxii, 207 p., 32 leaves : ill. ; 30 cm. — (Antike Münzen und geschnittene Steine ; Bd. 7) GFR78-A50
An enlargement of the author's thesis, Universität des Saarlandes, Saarbrücken, 1973.
Includes bibliographical references and index.
ISBN 3-11-007222-X : DM180.00
1. Coins, Greek—Greece—Lakonia. 2. Lakonia, Greece—Antiquities. 3. Greece—Antiquities. I. Title. II. Series.
CJ435.G78 1978 737.4'9'38 79-358937
79 MARC

COINS, GREEK (Continued)

—INDIA—CATALOGS

Uttar Pradesh, India. State Museum.
Catalogue of Śaka Pahlava coins of north India in the State Museum, Lucknow / A. K. Srivastava, numismatic officer. — Lucknow : The Museum, 1972.
103, ₍4₎ p., ₍15₎ leaves of plates : ill. ; 25 cm. — (State Museum catalogue series ; no. 4)
Bibliography: p. ₍105₎-₍107₎
1. Coins, Greek—India—Catalogs. 2. Saka—Catalogs. 3. Uttar Pradesh, India. State Museum—Catalogs. I. Srivastava, Arvind Kumar. II. Title. III. Series: Uttar Pradesh, India. State Museum. State Museum catalogue series ; no. 4.
CJ668.U87 1972 737.4'954 74-902211
79 MARC

—IRAQ—SELEUCIA BABYLONIA

Newell, Edward Theodore, 1886-1941.
The coinage of the eastern Seleucid mints, from Seleucus I to Antiochus III / by Edward T. Newell ; with a summary of recent scholarship additions and corrections by Otto Mørkholm. — New York : American Numismatic Society, 1978.
ix, 307 p., ₍28₎ leaves of plates : ill. ; 28 cm.
Reprint of the 1938 ed. issued as no. 1 of Numismatic studies.
Includes bibliographical references and indexes.
ISBN 0-89722-174-5
1. Coins, Greek—Iraq—Seleucia Babylonia. 2. Seleucids. 3. Seleucia Babylonia. I. Mørkholm, Otto. II. American Numismatic Society. III. Title. IV. Series: Numismatic studies ; no. 1.
CJ688.N48 1978 737.4'9'35 78-108001
78 MARC

—ITALY—MAGNA GRECIA

Holloway, R Ross, 1934-
Art and coinage in Magna Graecia / R. Ross Holloway. — Bellinzona ₍Switzerland₎ : Edizioni Arte e Moneta, ₍1978?₎
173 p. : chiefly ill. ; 27 cm. Sw***
Bibliography: p. 165.
Includes indexes.
1. Coins, Greek—Italy—Magna Grecia. 2. Art, Greek—Italy—Magna Grecia. 3. Magna Grecia. I. Title.
CJ517.H64 737.4'9'38 79-317977
79 MARC

—ITALY—SICILY—EXHIBITIONS

Arslan, Ermanno A
La moneta della Sicilia antica : catalogo delle civiche raccolte numismatiche di Milano / Ermanno A. Arslan. — ₍Milano₎ : Comune di Milano, Ripartizione cultura e spettacolo, 1976.
xxiii, 68 p., ₍52₎ leaves of plates : ill. ; 24 cm. It***
"Museo archeologico, ottobre-dicembre 1976."
Bibliography: p. xix-xxi.
1. Coins, Greek—Italy—Sicily—Exhibitions. 2. Milan. Medagliere Milanese—Catalogs. I. Museo archeologico, Milan. II. Milan. Medagliere milanese. III. Title.
CJ549.A77 79-363985
79 MARC

—MACEDONIA

Le Rider, Georges.
Le Monnayage d'argent et d'or de Philippe II ₍i.e. deux₎ frappé en Macédoine de 359 à 294 / Georges Le Rider. — Paris : Bourgey, 1977, achevé d'impr. 1978.
484 p., 95 p. of plates, ₍1₎ fold. col. leaf of plates ; 32 cm. F78-11214
Includes bibliographical references and indexes.
ISBN 2-902617-01-1 : 545F
1. Coins, Greek—Macedonia. 2. Philip II, King of Macedonia, 382-336 B.C.
CJ546.L47 737.4'9'381 79-339024
79 MARC

—SPAIN

Pareja López, Enrique, comp.
Repertorio numismático. ₍Granada₎ Departamento de Prehistoria y Arqueología, Facultad de Filosofía y Letras, 1972.
₍74₎ p. illus., maps. 25 cm. Sp***
"500 ejemplares."
ISBN 846005571X
1. Coins, Greek. 2. Coins, Roman. 3. Coins, Greek—Spain. 4. Coins, Roman—Spain. I. Title.
CJ335.P37 78-390100
79 MARC

—SPAIN—AMPURIAS

Maluquer de Motes y Nicolau, Juan.
Monedas ampurianas, massaliotas e ibéricas halladas en Ampurias / J. Maluquer de Motes. — Barcelona : Instituto de Arqueología y Prehistoria, Universidad de Barcelona, 1969.
67 p., ₍7₎ leaves of plates : ill. ; 25 cm. — (Publicaciones eventuales - Instituto de Arqueología y Prehistoria, Universidad de Barcelona ; no. 14) (His La colección arqueológica Víctor Catalá ; 1)
1. Coins, Greek—Spain—Ampurias. 2. Coins, Roman—Spain—Ampurias. 3. Ampurias. 4. Albert y Paradís, Catalina, 1869-1966—Archaeological collections. I. Title. II. Series: Barcelona. Universidad. Instituto de Arqueología y Prehistoria. Publicaciones eventuales ; no. 14.
GN700.B365 no. 14 78-399034
₍CJ555.A56₎ 79 MARC

—TURKEY—APHRODISIAS

MacDonald, David J
Greek and Roman coins from Aphrodisias / ₍by₎ David J. Macdonald. — Oxford : British Archaeological Reports, 1976.
₍10₎, 50 p., iii leaves of plates : ill., map ; 30 cm. — (BAR supplementary series ; 9) GB77-02474
Spine title: Coins from Aphrodisias.
Bibliography: p. ₍9₎-₍10₎
ISBN 0-904531-41-4 : £1.25 ($2.60 U.S.)
1. Coins, Greek—Turkey—Aphrodisias. 2. Coins, Roman—Turkey—Aphrodisias. 3. Aphrodisias, Turkey. I. Title. II. Title: Coins from Aphrodisias. III. Series.
CJ595.A63M3 737.4'9'38 79-317976
79 MARC

COINS, INDIC

Chattopadhyay, Bhaskar.
The age of the Kushāṇas, a numismatic study. With a foreword by Benoy Chandra Sen. ₍1st ed.₎ Calcutta, Punthi Pustak, 1967.
xxxiii, 289 p. illus. (part fold.) 23 cm. Rs48
"This work is substantially the thesis approved by the University of Calcutta for the Degree of Doctorate of Philosophy."
Bibliography: p. ₍254₎-265.
1. Numismatics—India. 2. Kushans. 3. Coins, Indic. I. Title.
CJ1391.C458 737.4'9'54 sa 68-21005
MARC

Gopal, Lallanji.
Early medieval coin-types of northern India. Varanasi, Numismatic Society of India, 1966.
v, 81 p. illus. 25 cm. (Numismatic notes and monographs no. 12)
Bibliographical footnotes.
1. Coins, Indic. I. Title. II. Series: Numismatic notes and monographs (Varanasi) no. 12.
CJ3530.N78 no. 12 737'.0934 s 79-929079
₍CJ3544₎ 79 MARC

Singh, Jai Prakash.
History and coinage of Skandagupta Kramāditya / by Jai Prakash Singh. — 1st ed. — Varanasi : Dept. of Ancient Indian History, Culture & Archaeology, Banaras Hindu University, 1976.
xv, 140 p. ; 25 cm. — (Monographs of the Department of Ancient Indian History, Culture & Archaeology ; no. 9)
Bibliography: p. ₍135₎
Includes index.
Rs20.00
1. Coins, Indic. 2. Skandagupta, Emperor of Northern India, fl. 455-467. I. Title. II. Series: Banaras Hindu University. Dept. of Ancient Indian History, Culture and Archaeology. Monographs ; no. 9.
CJ1391.S56 737.4'9'34 78-900836
79 MARC

Singh, Om Prakash, 1945-
Religion and iconography on early Indian coins / Om Prakash Singh. — 1st ed. — Varanasi : Bharati Prakashan, 1978.
xii, 112 p., ₍7₎ leaves of plates : ill. ; 23 cm.
Bibliography: p. ₍102₎-110.
Includes index.
Rs40.00
1. Coins, Indic. 2. India—Religion. 3. Icons—India. I. Title.
CJ1391.S57 737.4'9'34 78-907989
79 MARC

—CONGRESSES

Seminar on Coins as a Source of Economic History of Ancient India, Patna University, 1969.
Coins and early Indian economy : papers presented at a Seminar on Coins as a Source of Economic History of Ancient India, held at Patna University on 2nd October 1969 / edited by Ajay Mitra Shastri. — Varanasi : Numismatic Society of India, 1976.
2, ii, 179 p. ; 26 cm. — (Memoirs series - Numismatic Society of India ; no. 6)
"Held on the occasion of the Annual Conference of the Numisnatic Society of India under the auspices of the Patna University."
Includes bibliographical references.
Rs35.00
1. Coins, Indic—Congresses. 2. Numismatics—India—Congresses. 3. India—Economic conditions—To 1000—Congresses. I. Shastri, Ajay Mitra. II. Title. III. Series: Numisnatic Society of India. Memoirs series - Numismatic Society of India ; no. 6.
CJ1391.S44 1969 737.4'9'54 78-901172
78 MARC

—BENGAL

Bhattasali, Nalini Kanta.
Coins and chronology of the early independent sultans of Bengal / by Nalini Kanta Bhattasali. — New Delhi : Indological Book Corp., 1976.
vii, 184 p., ₍5₎ leaves of plates : ill. ; 22 cm.
Reprint of the 1922 edition.
Includes bibliographical references and index.
Rs40.00
1. Coins, Indic—Bengal. 2. Coins, Islamic—Bengal. 3. Bengal—History. I. Title.
CJ3549.B4B45 1976 737.4'9'5492 76-903783
79 MARC

—INDIA—GWALIOR (STATE)

Lingen, Jan.
Coins of the Sindhias / ₍by₎ Jan Lingen, Kenneth W. Wiggins. — Sanderstead : Hawkins Publications ; London : Distributed by Seaby, 1978.
108 p. : ill., coat of arms, geneal. table, map ; 31 cm. — (Monographs on Eastern coinage ; no. 1) GB79-09686
Spine title: The Sindhias.
Bibliography: p. 108.
ISBN 0-904173-17-8 : £12.50
1. Coins, Indic—India—Gwalior (State) 2. Sindhia dynasty, 1726-1947. I. Wiggins, Kenneth W., joint author. II. Title. III. Title: The Sindhias. IV. Series.
CJ3549.G87L56 737.4'9'543 79-313639
79 MARC

COINS, ISLAMIC

Mitchiner, Michael.
The world of Islam / by Michael Mitchiner. — London : Hawkins Publications ; distributed by B. A. Seaby, c1977.
511 p. : ill. ; 31 cm. — (Oriental coins and their values) GB***
Added t.p.: 'Ālam al-Islām.
Introductory matter and captions in English and Arabic, text in English.
Bibliography: p. 49-53.
ISBN 0-904173-15-1 : £20.00
1. Coins, Islamic. I. Title. II. Title: 'Ālam al-Islām. III. Series.
CJ3406.M57 737.4'9'17671 79-348130
79 MARC

—BENGAL

Bhattasali, Nalini Kanta.
Coins and chronology of the early independent sultans of Bengal / by Nalini Kanta Bhattasali. — New Delhi : Indological Book Corp., 1976.
vii, 184 p., ₍5₎ leaves of plates : ill. ; 22 cm.
Reprint of the 1922 edition.
Includes bibliographical references and index.
Rs40.00
1. Coins, Indic—Bengal. 2. Coins, Islamic—Bengal. 3. Bengal—History. I. Title.
CJ3549.B4B45 1976 737.4'9'5492 76-903783
79 MARC

—TURKEY—MARDIN

Lowick, N M
The Mardin hoard : Islamic countermarks on Byzantine folles / ₍by₎ N. M. Lowick, S. Bendall, P. D. Whitting. — ₍London₎ : A. H. Baldwin and Sons Ltd., 1977.
2-79 p. : ill., map ; 21 cm. GB79-02942
Includes bibliographical references.
£3.00
1. Follis (Coin)—Turkey—Mardin. 2. Coin hoards—Turkey—Mardin. 3. Coins, Islamic—Turkey—Mardin. 4. Counterstamp (Numismatics)—Turkey—Mardin. 5. Mardin, Turkey—Antiquities. 6. Turkey—Antiquities. I. Bendall, S., joint author. II. Whitting, Philip D., joint author. III. Title.
CJ1251.L68 737.4'9'5667 79-315534
79 MARC

COINS, ITALIAN

—ITALY—NOVELLARA

Fabbrici, Gabriele.
Ricerche sulla zecca di Novellara (1533-1728) / Gabriele Fabbrici. — Novellara : Tip. Ruozi, 1975.
90 p., ₍2₎ leaves of plates : ill. ; 25 cm. It77-Mar
"Documenti": p. 66-76.
Includes bibliographical references.
L2200
1. Coins, Italian—Italy—Novellara. 2. Mints—Italy—Novellara—History. I. Title.
CJ2929.N68F32 79-359294
*79 MARC

COINS, ITALIAN (Continued)

—ITALY—TURIN

Fava, Anna Serena.
Il medagliere delle raccolte numismatiche torinesi; esemplari scelti delle serie greca, romana, bizantina, sabauda, piemontese e di altre zecche italiane. A cura di Anna Serena Fava, Luigi Sachero [e] Vittorio Viale. Torino, Museo civico d'arte antica, novembre 1964. [Torino, 1964]
223 p. illus. (part col.) 24 cm.
"Catalogo a cura di Anna Serena Fava": p. [13]-223.
Bibliography: p. [14]
1. Coins, Ancient. 2. Coins, Italian—Italy—Turin. 3. Turin. Museo civico d'arte antica. I. Sachero, Luigi. II. Viale, Vittorio, 1891- III. Turin. Museo civico d'arte antica. IV. Title.
CJ215.F38 78-365944
78 MARC

—ITALY—VERCELLI (PROVINCE)

Ordano, Rosaldo.
La Zecca di Vercelli : storia, catalogo delle monete / Rosaldo Ordano. — [Vercelli] : Società storica vercellese, 1975, 1976 printing.
101 p. : ill. ; 25 cm. It77-Aug
Bibliography: p. 97-99.
L7000
1. Coins, Italian—Italy—Vercelli (Province) 2. Numismatics—Italy—Vercelli (Province) 3. Zecca di Vercelli. I. Title.
CJ2928.V53O7 737.4'9'4517 78-401105
79 MARC

—ITALY—VIGEVANO—EXHIBITIONS

Il Tesoretto di Vigevano : monete auree tra Gotico e Rinascimento : Museo archeologico, giugno-ottobre 1975. — Milano : Comune di Milano, Ripartizione cultura, [1975?]
9 p., 7 leaves of plates : ill. ; 24 cm. It***
1. Coins, Italian—Italy—Vigevano—Exhibitions. 2. Gold coins—Italy—Vigevano—Exhibitions. I. Museo archeologico, Milan.
CJ2929.V53T47 79-357395
79 MARC

COINS, JAPANESE

—CATALOGS

Cummings, Michael L
Modern Japanese coinage / by Michael L. Cummings. — 2d ed. — Tokyo : Far East Journal, 1978.
104 p. : ill. ; 23 cm. Ja***
Cover title: Nihon kindai kahei.
Errata slip inserted.
Bibliography: p. 103-104.
1. Coins, Japanese—Catalogs. I. Title.
CJ3717.C85 1978 737.4'9'52 79-311695
79 MARC

COINS, KOREAN

Kim, Nak-ch'ŏn.
(Han'guk kŭndae hwap'ye kagyŏk mongnok)
韓國近代貨幣價格目錄 : 近代鑄貨類·紙幣類·私札券(於音)類 / 金洛天著. — 서울 : 高麗苑, 1978.
163 p. : ill. ; 22 cm.
W2000
1. Coins, Korean. 2. Coins—Prices. 3. Numismatics—Korea—Collectors and collecting. I. Title.
CJ3747.K55 79-826292

COINS, LIECHTENSTEIN

Rittmann, Herbert.
Kleine Münz- und Geldgeschichte von Liechtenstein / Herbert Rittmann. — Hilterfingen : Helvetische Münzenzeitung, cop. 1977.
77 p. : ill. ; 21 cm. Sw78-264
Bibliography: p. 76-77.
1. Coins, Liechtenstein. 2. Heraldry—Liechtenstein. I. Title.
CJ2695.R57 79-351197
79 MARC

COINS, LOMBARD

—CATALOGS

Arslan, Ermanno A
Le monete di Ostrogoti longobardi e vandali : catalogo delle civiche raccolte numismatiche di Milano / Ermanno A. Arslan. — Milano : Comune di Milano, Ripartizione cultura e spettacolo, 1978.
91 p., [22] leaves of plates : ill. ; 24 cm. It***
Cover title: I Longobardi e la Lombardia.
Includes bibliographical references.
1. Coins, German—Catalogs. 2. Coins, Lombard—Catalogs. 3. Coins, Gothic—Catalogs. 4. Coins, Vandal—Catalogs. I. Title. II. Title: I Longobardi e la Lombardia.
CJ1657.A77 737.4'9'43 79-377592
79 MARC

COINS, LUXEMBURG

Philippen, Jos 1912-
Benelux sinds 1815 : catalogus van de munten met hun actuele waarde = Catalogue des monnaies avec leur valeur actuelle = Katalog der Münzen mit dem heutigen Wert = Catalogue of the coins with their present value / Jos Philippen. — Diest : Europa, [1973]
108 p. : ill. ; 19 cm. Be75-6
Dutch, English, French, and German.
ISBN 9061880017 : 250F
1. Coins, Belgian. 2. Coins, Dutch. 3. Coins, Luxemburg. I. Title.
CJ2934.P47 78-366149
78 MARC

—CATALOGS

Weiller, Raymond.
Les monnaies luxembourgeoises / par Raymond Weiller. — Louvain-la-Neuve : Institut supérieur d'archéologie et d'histoire de l'art, Séminaire de numismatique Marcel Hoc, 1977.
311 p., [21] leaves of plates : ill. ; 26 cm. — (Numismatica Lovaniensia ; 2) (Publications d'histoire de l'art et d'archéologie de l'Université catholique de Louvain ; 9) Be***
Bibliography: p. [1]-2.
1. Coins, Luxemburg—Catalogs. I. Title. II. Series. III. Series: Université catholique de Louvain, Louvain-la-Neuve, Belgium. Publications d'histoire de l'art et d'archéologie de l'Université catholique de Louvain ; 9.
CJ2996.W44 737.4'9493'4 79-337148
79 MARC

COINS, MĀLAVA

Dasgupta, Kalyan Kumar, 1933-
The Mālavas. Calcutta, Sanskrit College, 1966.
xiii, 36 p. 4 plates (incl. map) 25 cm. (Calcutta Sanskrit College Research series, no. 50. Studies, no. 29) 7.00 PL480:I-E-10504
Bibliography: p. [32]-34.
1. Mālavas. 2. Coins, Mālava. I. Title. II. Series: Calcutta. University. Sanskrit College. Calcutta Sanskrit College research series, no. 50. III. Series: Calcutta. University. Sanskrit College studies ; no. 29.
DS451.D38 934 sa 68-20776
[r79]rev MARC

COINS, MALAYSIAN

Ali, Mohd. Kassim Haji.
Panduan mengumpul duit lama / oleh Mohd. Kassim Haji Ali. — Cet. 1. — Kuala Lumpur : Muzium Negara, 1976.
viii, 60 p. : ill. ; 21 cm.
Bibliography: p. 58-60.
1. Numismatics—Malaysia. 2. Coins, Malaysian. I. Title.
CJ3626.A44 77-940694
79 MARC

—MALAYSIA—MELAKA (STATE)

Shaw, William Arthur, 1865-1943.
Malacca coins, by William Shaw and Mohd. Kassim Haji Ali. [1st ed.] Kuala Lumpur, Muzium Negara [1970]
21 p. illus., map. 22 cm. $1.00
Bibliography: p. 20.
1. Coins, Malaysian—Malaysia—Melaka (State) 2. Numismatics—Malaysia—Melaka (State) I. Ali, Mohd. Kassim Haji. II. Title.
CJ3629.M44S48 1970 737.4'9'5951 72-941536
79 MARC

COINS, MEDIEVAL

—GERMANY

Felder, Egon.
Germanische Personennamen auf merowingischen Münzen : Studien zum Vokalismus / Egon Felder. — Heidelberg : Winter, 1978.
97 p. ; 24 cm. — (Beiträge zur Namenforschung, n.F. : Beiheft ; 14) GFR78
Originally presented as the author's thesis, Munich, 1970.
Bibliography: p. [8]-10.
ISBN 3-533-02729-5
1. Names, Germanic. 2. Names, Personal—Germany. 3. Germanic languages—Etymology—Names. 4. Germanic languages—Vowels. 5. Coins, Medieval—Germany. I. Title. II. Series.
P769.B46 no. 14 79-384333
[PD576] 79 MARC

—POLAND—ZALESIE (SŁUPCA)

Wczesnośredniowieczny skarb srebrny z Zalesia, powiat Słupca. / Anna Kmietowicz ... [et al.]. — Wrocław : Zakład Narodowy im. Ossolińskich, 1969-1974.
2 v. : ill., map ; 26 cm.
At head of title v. 1: Instytut Historii Kultury Materialnej Polskiej Akademii Nauk ; v. 2. Polska Akademia Nauk. Instytut Historii Kultury Materialnej.
Summaries in French.
Includes bibliographical references.
CONTENTS: t. 1. Monety arabskie.—t. 2. Monety bizantyjskie, Monety Zachodnioeuropejskie, Ozodoby, Podsumowanie.
1. Coin hoards—Poland—Zalesie (Słupca) 2. Coins, Arab—Poland—Zalesie (Słupca) 3. Coins, Medieval—Poland—Zalesie (Słupca) 4. Jewelry, Medieval—Poland—Zalesie (Słupca) 5. Silverwork—Poland—Zalesie (Słupca) I. Kmietowicz, Anna.
CJ1647.Z34W37 74-256081
79[r79]rev2 MARC

COINS, MEXICAN

—MEXICO—SAN LUIS POTOSI (CITY)—WEIGHT

León D , José E
Acuñación en San Luis Potosí, por primera vez en México, de las monedas de plata fraccionaria con el sistema métrico decimal, 1863 / José E. León D. — San Luis Potosí [México] : Academia de Historia Potosina, 1973.
31 p., [2] leaves of plates : ill. ; 24 cm. — (Biblioteca de historia potosina : Serie cuadernos ; 26)
Includes bibliographical references.
1. Coins, Mexican—Mexico—San Luis Potosi (City)—Weight. I. Title.
CJ1909.S36L46 737.4'9'724 79-102959
79 MARC

COINS, NEPALI

Jośī, Harirāma.
(Uttaraprācīna kālīna mudrā)
उत्तरप्राचीन कालीन मुद्रा / हरिराम जोशी. — 1. संस्करण. — काठमाडौं : साझा प्रकाशन, वि. सं. 2033 [1976]
3, 4, 237 p., [12] leaves of plates : ill. ; 20 cm.
In Nepali.
Includes bibliographical references.
Rs12.15
1. Numismatics—Nepal. 2. Coins, Nepali. I. Title.
CJ1397.J67 76-904862

COINS, ORIENTAL

—CATALOGS

Mitchiner, Michael.
The ancient & classical world, 600 B.C.-A.D. 650 / by Michael Mitchiner. — London : Hawkins Publications ; distributed by B. A. Seaby, c1978.
760 p. : ill. ; 31 cm. — (Oriental coins and their values) GB***
Bibliography: p. 730-756.
Includes index.
ISBN 0-904173-16-X : £45.00
1. Coins, Oriental—Catalogs. I. Title. II. Series.
CJ1309.M48 737.4'095 79-310434
79 MARC

COINS, PERUVIAN

—CATALOGS

Fishman, César.
Monedas del Perú, de 1751 a 1978 / César Fishman. — 2. ed. — Lima : Numismática Libertad, 1979.
142 p. : ill. ; 21 cm.
Spanish and English.
1. Coins, Peruvian—Catalogs. 2. Coins, Spanish—Peru—Catalogs. I. Title.
CJ2386.F57 1979 737.4'9'85 79-113479
79 MARC

COINS, PHENICIAN

—CATALOGS

Chéhab, Maurice H
Monnaies gréco-romaines et phéniciennes du Musée national, Beyrouth, Liban / Maurice Chéhab. — Paris : Librairie d'Amérique et d'Orient, A. Maisonneuve, 1977.
viii, 91 p., 83 leaves of plates : ill. (some col.), maps ; 26 cm. F***
Includes indexes.
1. Coins, Greek—Catalogs. 2. Coins, Roman—Catalogs. 3. Coins, Phenician—Catalogs. 4. Beirut. Musée national libanais—Catalogs. I. Title.
CJ315.L42B443 737.4'0938 78-394670
79 MARC

COINS, POLISH

Mikołajczyk, Andrzej.
Naczynia datowane skarbami monet XIV-XVIII w. na ziemiach polskich / Andrzej Mikołajczyk. — Wrocław : Zakład Narodowy im. Ossolińskich, 1977.
183 p. : ill. ; 24 cm. — (Biblioteka archeologictna ; 24) P78-1876
At head of title: Polskie Towarzystwo Archeologiczne i Numizmatyczne.
Summary in English.
Includes bibliographical references.
zł45.00
1. Poland—Antiquities. 2. Pottery—Poland. 3. Coins, Polish. I. Title.
DK4092.M54 78-375006
78 MARC

COINS, PUNIC

—GAUL

Fischer, Brigitte.
Les monnaies antiques d'Afrique du Nord trouvées en Gaule / Brigitte Fischer. — Paris : Centre national de la recherche scientifique, 1978.
171 p. : ill., maps ; 28 cm. — (Supplément à Gallia ; 36) F***
Bibliography: p. ⌊19⌋-24.
Includes indexes.
ISBN 2-222-02250-9
1. Coins, Punic—Gaul. 2. Coin hoards—Gaul. 3. France—Antiquities. I. Title. II. Series: Gallia, fouilles et monuments archéologiques en France métropolitaine : Supplément ; 36.
CJ1381.F5 737.4'9'364 79-362661
79 MARC

COINS, ROMAN

see also Follis (Coin)

Kent, John Philip Cozens.
Roman coins / J. P. C. Kent ; photos. by Max and Albert Hirmer. — Rev. ed. — New York : H. N. Abrams, 1978.
368 p., ⌊100⌋ leaves of plates : ill. ; 32 cm.
Revised translation of Die römische Münze.
Bibliography: p. 360-361.
Includes index.
ISBN 0-8109-1584-7
1. Coins, Roman. I. Title.
CJ833.K4613 1978 737.4'9'37 77-77534
79 MARC

Pareja López, Enrique, comp.
Repertorio numismático. ⌊Granada⌋ Departamento de Prehistoria y Arqueología, Facultad de Filosofía y Letras, 1972.
⌊74⌋ p. illus., maps. 25 cm. Sp***
"500 ejemplares."
ISBN 846005571X
1. Coins, Greek. 2. Coins, Roman. 3. Coins, Greek—Spain. 4. Coins, Roman—Spain. I. Title.
CJ335.P37 78-390100
79 MARC

Schiavi, Luciana.
Evoluzione architettonica del Tempio di Giove Capitolino nelle monete del periodo repubblicano. Modena, Cooperativa tipografi, 1971.
15 p. illus. 25 cm. It72-Mar
Bibliography: p. 15.
1. Coins, Roman. 2. Architecture (in numismatics) 3. Rome (City). Temple of Jupiter. I. Title.
CJ915.S34 78-383806
78 MARC

Shotter, David Colin Arthur.
Roman coins / by D. C. A. Shotter. — Clapham, N. Yorkshire : Dalesman, 1978.
64 p. : ill., map ; 19 cm. — (A Dalesman mini-book) GB78-13703
Ill. on inside covers.
Bibliography: p. 64.
ISBN 0-85206-432-2 : £0.70
1. Coins, Roman—England. 2. Coins, Roman. I. Title.
CJ1102.S53 737.4'9'37 78-320615
78 MARC

Trillmich, Walter.
Familienpropaganda der Kaiser Caligula und Claudius : Agrippina Maior u. Antonia Augusta auf Münzen / von Walter Trillmich. — Berlin : de Gruyter, 1978.
xiv, 186 p., ⌊21⌋ leaves of plates : ill. ; 30 cm. — (Antike Münzen und geschnittene Steine ; Bd. 8) GFR78-A
At head of title: Deutsches Archäologisches Institut.
Based on part of the author's thesis, Tübingen, 1970, presented under title: Bildnis-Typen der Agrippina Maior und der Antonia Minor.
Bibliography: p. ⌊xi⌋-xiv.
ISBN 3-11-007259-9 : DM140.00
1. Agrippina, wife of Germanicus Caesar, d. A.D. 33—Iconography. 2. Antonia Augusta, 36 B.C. (ca.)-38 A.D.—Iconography. 3. Caligula, Emperor of Rome, 12-41—Iconography. 4. Claudius, Emperor of Rome, 10 B.C.-54 A.D.—Iconography. 5. Propaganda, Roman. 6. Coins, Roman. 7. Roman emperors—Iconography. 8. Rome—Nobility—Iconography. 9. Rome—History—The five Julii, 30 B.C.-68 A.D. I. Title. II. Series.
DG281.T74 737.4'9'37 79-352702
79 MARC

Wegner, Max, 1902-
Gordianus III ⌊i.e. der Dritte⌋ bis Carinus / von Max Wegner ; mit Beitr. von Jörgen Bracker u. Willi Real ; Dt. Archäolog. Inst. — Berlin : Mann, 1979.
172, 62 p. : ill. ; 28 cm. — (Das Römische Herrscherbild : Abt. 3 ; Bd. 3) GFR79-A
Includes bibliographical references and index.
ISBN 3-7861-2000-5
1. Rome—History—Period of military anarchy, 235-285—Historiography. 2. Roman emperors—Portraits. 3. Roman empresses—Portraits. 4. Portrait sculpture, Roman. 5. Coins, Roman. 6. Hairdressing—Rome. I. Bracker, Jörgen, joint author. II. Real, Willi, joint author. III. Deutsches Archäologisches Institut. IV. Title. V. Series: Römische Herrscherbild : Abt. 3 ; Bd. 3.
DG305.W43 937'.07 79-382354
79 MARC

—CATALOGS

Carson, Robert Andrew Glindinning.
The principal coins of the Romans / R. A. G. Carson. — London : Published for the Trustees of the British Museum by British Museum Publications, c1978-
v. : ill. ; 26 cm. GB***
Bibliography: p. v. 1, p. ⌊84⌋
Includes indexes.
CONTENTS: v. 1. The Republic, c. 290-31 B.C.

ISBN 0-7141-0839-8 (v. 1) : £15.00 (v. 1)
1. Coins, Roman—Catalogs. I. Title.
CJ833.C37 737.4'937 79-300078
79 MARC

Chéhab, Maurice H
Monnaies gréco-romaines et phéniciennes du Musée national, Beyrouth, Liban / Maurice Chéhab. — Paris : Librairie d'Amérique et d'Orient, A. Maisonneuve, 1977.
viii, 91 p., 83 leaves of plates : ill. (some col.), maps ; 26 cm. F***
Includes indexes.
1. Coins, Greek—Catalogs. 2. Coins, Roman—Catalogs. 3. Coins, Phenician—Catalogs. 4. Beirut. Musée national libanais—Catalogs. I. Title.
CJ315.L42B443 737.4'0938 78-394670
79 MARC

Seaby, Herbert Allen.
Roman silver coins / by H. A. Seaby ; rev. by David R. Sear & Robert Loosley. — 3d ed. — London : Seaby, 1978-
v. : ill. ; 23 cm. GB***
Includes index.
CONTENTS: v. 1. The Republic to Augustus.

ISBN 0-900652-44-6 (v. 1) : £7.50 ($16.00 U.S.) (v. 1)
1. Coins, Roman—Catalogs. 2. Silver coins—Catalogs. I. Sear, David R. II. Loosley, Robert. III. Title.
CJ851.S42 1978 737.4'9'37 79-301012
79 MARC

Varesi, Clelio.
Le monete d'oro e d'argento della Repubblica romana / ⌊di⌋ C. Varesi, A. Castellotti. — ⌊s.l. : s.n., 1978?⌋
330 p. : ill. ; 22 cm. It78-Oct
Bibliography: p. 5.
Includes index.
1. Coins, Roman—Catalogs. I. Castellotti, Alberto. II. Title.
CT909.V37 78-394039
*79 MARC

—DANUBE VALLEY

Fitz, Jenő.
Der Geldumlauf der römischen Provinzen im Donaugebiet Mitte des 3. Jahrdunderts / von Jenő Fitz ; ⌊übers. von Mátyás Esterházy⌋. — Budapest : Akadémiai Kiadó, 1978.
2 v. (945 p.) : ill. ; 25 cm. H***
Includes bibliographical references and indexes.
ISBN 9630508583
1. Coins, Roman—Danube Valley. 2. Coin hoards—Danube Valley. 3. Danube Valley—Antiquities, Roman. 4. Romans—Danube Valley. I. Title.
CJ1106.F5715 79-352160
79 MARC

—EGYPT—QARARA SITE—CATALOGS

Kosack, Wolfgang.
Antike Kultur auf Münzen von Qarara : aus der Ägyptensammlung des Museums / Wolfgang Kosack. — Freiburg i. Br. : Städtische Museen, 1977.
49 p., ⌊4⌋ leaves of plates : ill. ; 30 cm. — (Veröffentlichungen des Museums für Völkerkunde ; 1977, 2) GFR***
At head of title: Museum für Völkerkunde der Stadt Freiburg im Breisgau.
Bibliography: p. 15.
1. Coins, Roman—Egypt—Qarara site—Catalogs. 2. Qarara site, Egypt—Catalogs. 3. Coins, Ancient—Egypt—Qarara site—Catalogs. 4. Freiburg I.B. Museum für Volkerkunde—Catalogs. I. Freiburg I. B. Museum für Völkerkunde. II. Title. III. Series: Freiburg i. B. Museum für Völkerkunde. Veröffentlichungen des Museums für Völkerkunde ; 1977, 2.
CJ1073.Q37K67 79-353380
79⌊r79⌋rev MARC

—ENGLAND

Shotter, David Colin Arthur.
Roman coins / by D. C. A. Shotter. — Clapham, N. Yorkshire : Dalesman, 1978.
64 p. : ill., map ; 19 cm. — (A Dalesman mini-book) GB78-13703
Ill. on inside covers.
Bibliography: p. 64.
ISBN 0-85206-432-2 : £0.70
1. Coins, Roman—England. 2. Coins, Roman. I. Title.
CJ1102.S53 737.4'9'37 78-320615
78 MARC

—GAUL—EXHIBITIONS

Vigier, Jean René.
Le Trésor monétaire de la Société archéologique et historique de la Charente : période gallo-romaine à l'exposition numismatique d'Angoulême ; avec le concours des collections du Musée municipal, décembre 1972, ⌊catalogue⌋ / par Jean-René Vigier .... — ⌊Angoulême⌋ (14, rue de Montemoreau, 16000) : ⌊Société archéologique et historique de Charente⌋, 1972.
58 p. ; 25 cm. F74-1980
Includes bibliography.
1. Coins, Roman—Gaul—Exhibitions. 2. France—Antiquities, Roman—Exhibitions. 3. Romans—France—Exhibitions. 4. Société Archéologique et Historique de la Charente—Exhibitions. I. Société archéologique et historique de la Charente. II. Title.
CJ1113.V53 737.4'9'364 78-384603
78 MARC

—GREECE

Giovannini, Adalberto.
Rome et la circulation monétaire en Grèce au IIe siècle avant Jésus-Christ / par Adalberto Giovannini. — Basel : F. Reinhardt, 1978.
x, 129, ⌊6⌋ leaves of plates ; 24 cm. — (Schweizerische Beiträge zur Altertumswissenschaft ; Heft 15) Sw79-2825
Includes bibliographical references.
ISBN 3-7245-0412-8 : 38.00F
1. Coins, Greek. 2. Coins, Roman—Greece. 3. Greece—History—281-146 B.C. I. Title. II. Series.
CJ427.G54 737.4'9'38 79-383143
79 MARC

—GREECE—ATHENS

Thompson, Margaret, 1911-
Coins from the Roman through the Venetian period / by Margaret Thompson. — Meriden, Conn. : Meriden Gravure Co., 1961.
vii, 122 p., ⌊2⌋ leaves of plates : ill. ; 31 cm. — (The Athenian Agora ; v. 2) (Reprint of the 1954 ed. published by the American School of Classical Studies at Athens, Princeton, N.J)
Bibliography: p. ⌊8⌋
Includes indexes.
1. Coins, Roman—Greece—Athens. 2. Coins, Byzantine—Greece—Athens. 3. Athens. Agora. I. Title. II. Series: American School of Classical Studies at Athens. The Athenian Agora ; v. 2.
DF287.A23A5 vol. 2 938'.5 s 75-328019
⌊CJ893.A8⌋ 76⌊r79⌋rev MARC

COINS, ROMAN (Continued)

—ITALY—VERONA—CATALOGS

Milani, Luigi Adriano, 1854-1914.
Il ripostiglio della Venèra : monete romane della seconda metà del terzo secolo / Luigi Adriano Milani. — ₍Bologna₎ : A. Forni, 1977.
213 p., ₍3₎ leaves of plates : ill. ; 31 cm. It79-Jan
Reprint of the 1880 ed. published by Salviucci, Rome.
"Il ripostiglio della Venèra, conservato quasi per intero nel Civico Museo di Verona."
Includes bibliographical references.
1. Coins, Roman—Italy—Verona—Catalogs. 2. Coin hoards—Italy—Verona—Catalogs. 3. Verona. Museo Civico—Catalogs. I. Verona. Museo civico. II. Title.
CJ815.I82V475 1977 79-380002
•79 MARC

—ROMANIA—MĂGURA (BACĂU)

Mihăilescu-Bîrliba, Virgil.
Tezaurul de la Măgura / Virgil Mihăilescu-Bîrliba, Ioan Mitrea. — Bacău : Muzeul Județean de Istorie și Artă, 1977.
79 p., ₍14₎ leaves of plates : ill. ; 24 cm. — (Bibliotheca Carpicae ; 1) R***
Summary in English and French.
Includes bibliographical references.
1. Coin hoards—Romania—Măgura (Bacău) 2. Coins, Roman—Romania—Măgura (Bacău) 3. Măgura, Romania (Bacău)—Antiquities. 4. Romania—Antiquities. I. Mitrea, Ioan, joint author. II. Title. III. Series.
CJ893.M33M53 78-400555
79 MARC

—SANTA RESTITUTA (L'AQUILA)

Squilla, Gaetano.
Un cippo terminale a Luco nei Marsi (L'Aquila), Una pioggia di monete a S. Restituta di Morrea (L'Aquila). ₍Veroli₎, Tip. di Casamari, 1973.
57, ₍21₎ p. illus. 25 cm. L1000 It73-Dec
Bibliography: p. ₍15₎ (2d group)
1. Boundary stones—Italy—Luco dei Marsi. 2. Coin hoards—Italy—Santa Restituta (L'Aquila) 3. Coins, Roman—Santa Restituta (L'Aquila) 4. Luco dei Marsi, Italy—Antiquities. 5. Santa Restituta, Italy (L'Aquila)—Antiquities. 6. Italy—Antiquities. I. Squilla, Gaetano. Una pioggia di monete a S. Restituta di Morrea. 1973. II. Title.
CC605.I8S68 937'.7 79-356857
79 MARC

—SPAIN

Pareja López, Enrique, comp.
Repertorio numismático. ₍Granada₎ Departamento de Prehistoria y Arqueología, Facultad de Filosofía y Letras, 1972.
₍74₎ p. illus., maps. 25 cm. Sp***
"500 ejemplares."
ISBN 846005571X
1. Coins, Greek. 2. Coins, Roman. 3. Coins, Greek—Spain. 4. Coins, Roman—Spain. I. Title.
CJ335.P37 78-390100
79 MARC

—SPAIN—AMPURIAS

Maluquer de Motes y Nicolau, Juan.
Monedas ampurianas, massaliotas e ibéricas halladas en Ampurias / J. Maluquer de Motes. — Barcelona : Instituto de Arqueología y Prehistoria, Universidad de Barcelona, 1969.
67 p., ₍7₎ leaves of plates : ill. ; 25 cm. — (Publicaciones eventuales - Instituto de Arqueología y Prehistoria, Universidad de Barcelona ; no. 14) (His La colección arqueológica Víctor Català ; 1)
1. Coins, Greek—Spain—Ampurias. 2. Coins, Roman—Spain—Ampurias. 3. Ampurias. 4. Albert y Paradís, Catalina, 1869-1966—Archaeological collections. I. Title. II. Series: Barcelona. Universidad. Instituto de Arqueología y Prehistoria. Publicaciones eventuales ; no. 14.
GN700.B365 no. 14 78-399034
₍CJ555.A56₎ 79 MARC

—TURKEY—APHRODISIAS

MacDonald, David J
Greek and Roman coins from Aphrodisias / ₍by₎ David J. Macdonald. — Oxford : British Archaeological Reports, 1976.
₍10₎, 50 p., iii leaves of plates : ill., map ; 30 cm. — (BAR supplementary series ; 9) GB77-02474
Spine title: Coins from Aphrodisias.
Bibliography: p. ₍9₎-₍10₎
ISBN 0-904531-41-4 : £1.25 ($2.60 U.S.)
1. Coins, Greek—Turkey—Aphrodisias. 2. Coins, Roman—Turkey—Aphrodisias. 3. Aphrodisias, Turkey. I. Title. II. Title: Coins from Aphrodisias. III. Series.
CJ595.A63M3 737.4'9'38 79-317976
79 MARC

COINS, RUSSIAN

Zacher, Martin.
50 ₍i.e. Fünfzig₎ Jahre Münzen der Sowjetunion : eine Studie / von Martin Zacher ; hrsg. von der Kreisleitung Berlin-Friedrichshain des Kulturbundes der Deutschen Demokratischen Republik gemeinsam mit der Fachgruppe Numismatik. — Berlin-Friedrichshain : Kulturbund d. DDR, 1973.
40 p., ₍8₎ leaves of plates : ill. ; 21 cm. GDR73-B23
"Bewertungstabelle", ₍2₎ p. inserted.
Bibliography: p. 39.
10.00M
1. Coins, Russian. I. Title.
CJ3025.Z3 79-365808
•79 MARC

COINS, SAN MARINO

Molteni, Mario, 1935-
Nove monete per San Marino / Mario Molteni ; introduzione di Giuseppe Rossi. — Milano : All'insegna del pesce d'oro, 1976.
15 p., 10 leaves of plates : ill. ; 16 cm. — (Quaderni d'arte del pesce d'oro ; n. 19) It77-Jan
L3000
1. Coins, San Marino. 2. Money—San Marino (Republic) I. Title.
CJ2694.M64 78-386659
•79 MARC

COINS, SCOTTISH

—CONGRESSES

Oxford Symposium on Coinage and Monetary History, 2d, 1977.
Coinage in medieval Scotland (1100-1600) : the second Oxford Symposium on Coinage and Monetary History / edited by D. M. Metcalf. — Oxford : British Archaeological Reports, 1977.
viii, 198 p. : ill. ; 30 cm. — (British archaeological reports ; 45) GB***
Includes bibliographies and index.
ISBN 0-86054-002-2
1. Coins, Scottish—Congresses. I. Metcalf, David Michael. II. Title. III. Series.
CJ2534.O9 1977 737.4'9'41 78-326391
79 MARC

COINS, SPANISH

Cayón, Juan Ramón.
Las monedas de los Borbones : Felipe V-Isabel II : 1700-1868 / Juan R. Cayón, Carlos Castán. — Madrid : Carlos Castán, ₍etc.₎, 1977 printing.
332 p. : ill. ; 24 cm. Sp77-Sept
ISBN 8440034296 : 1000ptas
1. Coins, Spanish. 2. Bourbon, House of. I. Castán Ramírez, Carlos, joint author. II. Title.
CJ3195.C39 78-388711
79 MARC

Heiss, Aloïss, 1820-
Description générale des monnaies des rois wisigoths d'Espagne / par Aloïss Heiss. — Ristampa. — ₍Bologna₎ : A. Forni, 1976.
iii, 185 p., 13 leaves of plates : ill. ; 32 cm. It78-Apr
Reprint of 1872 ed., published by Impr. nationale, Paris.
Three hundred and fifty copies printed.
Includes indexes.
1. Coins, Gothic. 2. Coins, Spanish. 3. Spain—History—Gothic period, 414-711. I. Title.
CJ3194.H5 1976 737.4'9'46 79-347233
•79 MARC

—CATALOGS

Numinter.
Segunda subasta de 1978 : en la sala de Numinter de Madrid ... : 23 de mayo de 1978. — Madrid : Numinter, ₍1978?₎
50 p. : ill. ; 22 cm. Sp***
On cover: 2/78.
1. Coins, Spanish—Catalogs. I. Title.
CJ3184.N85 1978 78-390300
79 MARC

—PERU—CATALOGS

Fishman, César.
Monedas del Perú, de 1751 a 1978 / César Fishman. — 2. ed. — Lima : Numismática Libertad, 1979.
142 p. : ill. ; 21 cm.
Spanish and English.
1. Coins, Peruvian—Catalogs. 2. Coins, Spanish—Peru—Catalogs. I. Title.
CJ2386.F57 1979 737.4'9'85 79-113479
79 MARC

COINS, SWEDISH

100 ₍i.e. Hundra₎ svenska myntgrupper : på hundra uppslag och med ca. 700 åt- och frånsidesbilder av mynt och skådemynt under tusen år / sammanställda och kommenterade av Åke Ohlmarks. — Stockholm : Plus, 1978.
207 p. : ill. ; 22 cm. S78-31
ISBN 9174061127
1. Coins, Swedish. I. Ohlmarks, Åke, 1911-
CJ3166.H86 79-346794
79 MARC

COINS, SWISS

Cahn, Erich B
Schöne Münzen der Stadt Basel = Belles monnaies de la ville de Bâle = Nice coins of the city of Basle / Texte, Erich B. Cahn ; Bilder, Maurice Babey. — Basel : Stiftung für das Historische Museum, 1975.
₍61₎ p. : col. ill. ; 18 cm. — (Schriften des Historischen Museums Basel ; Bd. 3) Sw76-352
Selection of coins from the collection of the Historisches Museum, Basel.
English, French, and German.
Bibliography: p. ₍9₎
1. Coins, Swiss. 2. Basel. Universität. Historisches Museum. I. Babey, Maurice. II. Basel. Universität. Historisches Museum. III. Title. IV. Title: Belles monnaies de la ville de Bâle. V. Title: Nice coins of the city of Basle. VI. Series: Basel. Universität. Historisches Museum. Schriften ; Bd. 3.
GN37.B3212 Bd. 3 069'.09494'3 s 79-354195
₍CJ43.S82B3₎ •79 MARC

—SWITZERLAND—ZÜRICH—HISTORY

Humm, Richard.
Vom Turiaco-Triens zum Zürcher Rappen : zur Geschichte zürcherischer Münzprägungen / Richard Humm. — Zollikon : Museumskommission, 1977.
₍20₎ p. : ill. ; 21 cm. Sw78-6316
Includes bibliographical references.
1. Coins, Swiss—Switzerland—Zürich—History. I. Title.
CJ3269.Z8H82 737.4'9'494 79-370706
79 MARC

COINS, TUNISIAN

—CATALOGS

Schweikert, H
Les monnaies tunisiennes depuis 1859; de l'avènement de Mohamed Sadok Pacha Bey à la République tunisienne ₍par₎ H. Schweikert. Munich, E. Battenberg Verlag ₍1973₎
80 p. illus. 24 cm. GFR***
1. Coins, Tunisian—Catalogs. I. Title.
CJ4110.S38 737.4'9'611 79-357995
79 MARC

COINS, VANDAL

—CATALOGS

Arslan, Ermanno A
Le monete di Ostrogoti longobardi e vandali : catalogo delle civiche raccolte numismatiche di Milano / Ermanno A. Arslan. — Milano : Comune di Milano, Ripartizione cultura e spettacolo, 1978.
91 p., ₍22₎ leaves of plates : ill. ; 24 cm. It***
Cover title: I Longobardi e la Lombardia.
Includes bibliographical references.
1. Coins, German—Catalogs. 2. Coins, Lombard—Catalogs. 3. Coins, Gothic—Catalogs. 4. Coins, Vandal—Catalogs. I. Title. II. Title: I Longobardi e la Lombardia.
CJ1657.A77 737.4'9'43 79-377592
79 MARC

COIR INDUSTRY

—INDIA—PERIODICALS

India (Republic). Coir Board.
India's exports and internal consumption of coir and coir goods.
Ernakulam.
v. 26 x 33 cm. annual.
Period covered ends Mar. 31.
Continued by: India (Republic). Coir Board. India's production, exports and internal consumption of coir and coir goods.
1. Coir industry—India—Periodicals. I. Title.
HD9156.C63 I 525a 78-641521
380.1'45'677180954 MARC-S

COITUS see Sexual intercourse

EL COJO ILUSTRADO

—INDEXES

Perea Romero, Martín.
Catálogo de El Cojo ilustrado, 1892-1915 / Martín Perea Romero ; José Agustín Catalá, editor. — Caracas : Ediciones Centauro, 1975.
2 v. ; 24 cm.
1. El Cojo ilustrado—Indexes. I. Catalá, José Agustín, 1915- II. El Cojo ilustrado. III. Title.
AP63.C52P47 78-105840
79 MARC

COKE, EDWARD, SIR, 1552-1634

White, Stephen D 1945-
Sir Edward Coke and "the grievances of the Commonwealth," 1621-1628 / by Stephen D. White. — Chapel Hill : University of North Carolina Press, c1979.
xv, 327 p. ; 25 cm. — (Studies in legal history)
Bibliography: p. 293-311.
Includes index.
ISBN 0-8078-1335-4
1. Law—Great Britain—History and criticism. 2. Coke, Edward, Sir, 1552-1634. 3. Great Britain—Politics and government—1603-1649. I. Title. II. Series.
KD612.W47 328.41'092'4 78-16418
78 MARC

COKE

see also Carbonization

(Tekhnologiia koksokhimicheskogo proizvodstva)
Технология коксохимического производства : допущено в качестве учеб. пособия для коксохимических и металлургических специальностей техникумов / ₍Р. Е. Лейбович ... et al.₎. — Изд. 2., перер. и доп. — Москва : Металлургия, 1974.
420, ₍3₎ p. : ill. ; 23 cm. USSR***
First ed. (1966) by R. E. Leibovich, IA. M. Obukhovskiĭ, and S. IA. Satanovskiĭ.
1. Coke. 2. Coke industry—By-products. I. Leibovich, Ruvim Evseevich.
TP336.L44 1974 79-413616

—CONGRESSES

Coal, coke and the blast furnace : proceedings of an international conference / organized by the Metals Society and held in Middlesborough, England, on 15-17 June 1977. — London : Metals Society, 1978.
₍5₎, 152 p. : ill. ; 31 cm. — (Book - Metals Society ; 191) GB79-01991
Includes bibliographical references and indexes.
ISBN 0-904357-15-5 : £22.50
1. Blast-furnaces—Congresses. 2. Coal—Congresses. 3. Coke—Congresses. I. Metals Society. II. Series: Metals Society. Book - Metals Society ; 191.
TN713.C7 669'.1413 79-314482
79 MARC

—TESTING

Muchnik, Damir Abramovich.
(Nomogrammy dlia opredeleniia kachestva koksa)
Номограммы для определения качества кокса / Д. А. Мучник, А. И. Трикило. — Киев : Техніка, 1978.
52 p. : ill. ; 26 cm. USSR 78
Bibliography : p. ₍51₎
0.35rub
1. Coke—Testing. I. Trikilo, Alik Ivanovich, joint author. II. Title.
TP336.M8 79-412752

Zimmerman, Raymond E
Evaluating and testing the coking properties of coal / Raymond E. Zimmerman. — San Francisco : Miller Freeman Publications, c1979.
144 p. : ill. ; 27 cm. — (A World coal book)
Includes bibliographical references and index.
ISBN 0-87930-111-2
1. Coal—Testing. 2. Coke—Testing. I. Title. II. Series: World coal book.
TP325.Z55 662'.6224 79-84399
00 MARC

COKE CO., TEX.

—ANTIQUITIES

Tunnell, Curtis D
The Gibson lithic cache from west Texas / by Curtis Tunnell. — Austin : Office of the State Archeologist, Texas Historical Commission, c1978.
xi, 58 p., ₍19₎ leaves of plates : ill. ; 28 cm. — (Office of the State Archeologist report ; 30)
Bibliography: p. 55-58.
1. Indians of North America—Texas—Coke Co.—Implements. 2. Indians of North America—Texas—Coke Co.—Antiquities. 3. Coke Co., Tex.—Antiquities. 4. Texas—Antiquities. I. Title. II. Series: Texas. Office of the State archeologist. Reports ; 30.
E78.T4T86 976.4'723 79-624151
79 MARC

—HISTORY

Pritchett, Jewell G
From the top of old Hayrick : a narrative history of Coke County / compiled and written by Jewell G. Pritchett. — 1st ed. — ₍Abilene, Tex.₎ : Pritchett, c1979.
190 p. : ill. ; 22 cm.
$11.95
1. Coke Co., Tex.—History. I. Title.
F392.C54P73 976.4'723'06 79-84385
79 MARC

COKE INDUSTRY

see also Gas industry

Gorodetskiĭ, Nikolaĭ Il'ich.
(Organizatsiia i planirovanie na predpriiatiiakh koksokhimicheskoĭ promyshlennosti)
Организация и планирование на предприятиях коксохимической промышленности / Н. И. Городецкий, Г. П. Кулиш, Е. М. Рабинович ; под ред. Г. П. Кулиша. — Москва : Металлургия, 1975.
373, ₍3₎ p. : ill. ; 21 cm. USSR***
Bibliography : p. ₍375₎
0.85rub
1. Coke industry. I. Kulish, Grigoriĭ Prokof'evich, joint author. II. Rabinovich, Efim Markovich, joint author. III. Title.
TP336.G663 79-413505

—BY-PRODUCTS

(Tekhnologiia koksokhimicheskogo proizvodstva)
Технология коксохимического производства : допущено в качестве учеб. пособия для коксохимических и металлургических специальностей техникумов / ₍Р. Е. Лейбович ... et al.₎. — Изд. 2., перер. и доп. — Москва : Металлургия, 1974.
420, ₍3₎ p. : ill. ; 23 cm. USSR***
First ed. (1966) by R. E. Leibovich, IA. M. Obukhovskiĭ, and S. IA. Satanovskiĭ.
1. Coke. 2. Coke industry—By-products. I. Leibovich, Ruvim Evseevich.
TP336.L44 1974 79-413616

—GREAT BRITAIN—ADDRESSES, ESSAYS, LECTURES

Lee, G W
Improvement of the economy of the coke-oven process / G. W. Lee, J. P. Graham, and W. J. Pater. — ₍S. l. : s. n., 1966₎
9 p. ; 30 cm.
Caption title.
At head of title: Congrès international de Charleroi. Le coke en sidérurgie. B5.
Bibliography: p. 8-9.
1. Coal—Great Britain—Carbonization—Addresses, essays, lectures. 2. Coke-ovens—Addresses, essays, lectures. 3. Coke industry—Great Britain—Addresses, essays, lectures. I. Graham, J. P., joint author. II. Pater, W. J., joint author. III. Title.
TP336.L43 79-316502
79 MARC

—JAPAN—STATISTICS—YEARBOOK

(Enerugī tōkei nempō)
エネルギー統計年報 Yearbook of coal, petroleum and coke statistics. 1975–
₍東京 通商産業調査会₎
v. 26 cm.
"編集·通商産業₍省₎大臣官房調査統計部"
Formed by the union of: Sekitan kōkusu nempō, and: Sekiyu tōkei nempō.
In Japanese.
1. Coal trade—Japan—Statistics—Yearbooks. 2. Petroleum industry and trade—Japan—Statistics—Yearbooks. 3. Coke industry—Japan—Statistics—Yearbooks. 4. Power resources—Japan—Statistics—Yearbooks. I. Japan. Tsūshō Sangyōshō. Daijin Kambō. Chōsa Tōkeibu.
HD9556.J29E53 78-642754
(MARC-S)

—POLAND

Krause, Witold.
Współczesne problemy produkcji koksu metalurgicznego w krajowych warunkach surowcowych / Witold Krause. — Wyd. 1. — Kraków : Akademia Górniczo-Hutnicza, 1976.
78 p. : ill. ; 24 cm. — (Metalurgia i odlewnictwo ; zesz. 70) (Prace Instytutu Metalurgii ; 16) (Zeszyty naukowe Akademii Górniczo-Hutniczej im. Stanisława Staszica ; Nr 578) P***
Added t.p.: Present-day problems of metallurgical coke production and the raw material resources of Poland.
Summary also in English and Russian.
Bibliography: p. ₍75₎-78.
zł11.00
1. Coke industry—Poland. I. Title. II. Title: Present-day problems of metallurgical coke production and the raw material resources of Poland. III. Series. IV. Series: Krakow. Akademia Górniczo-Hutnicza. Instytut Metalurgii. Prace ; 16.
TN4.M36 zesz. 70 79-383023
₍TP336₎ 79 MARC

—UNITED STATES

British Coke Research Association.
The American coking industry, 1957 : report of a delegation appointed to investigate the American coking industry. — London : British Coke Research Association, 1958.
p. A-B, 64 p. ; 28 cm.
Cover title.
L. O'Connor, Chairman.
1. Coke industry—United States. I. O'Connor, Leslie. II. Title.
HD9559.C7U33 79-311125
79 MARC

—UNITED STATES—STATISTICS —PERIODICALS

Distribution of oven and beehive coke and breeze.
Washington, Dept. of Energy, Energy Information Administration.
v. ill. 27 cm. annual. (Energy data reports)
Continues: Coke distribution.
Other title, : Coke distribution.
Vols. for issued by the Office of Energy Data and Interpretation.
Key title: Distribution of oven and beehive coke and breeze, ISSN 0193-4782.
1. Coke industry—United States—Statistics—Periodicals. I. United States. Energy Information administration. Office of Energy Data and Interpretation. II. Series.
HD9559.C7U37 338.4'7'662720973 79-643547
MARC-S

COKE-OVEN GAS

—DESULPHURIZATION—CONGRESSES

Symposium on Treatment of Coke Oven Gas, McMaster University, 1977.
Treatment of coke oven gas : proceedings of Symposium on "Treatment of Coke Oven Gas" organized by Department of Metallurgy and Materials Science, McMaster University, and held in Hamilton, Ontario, Canada, on May 26th & 27th, 1977 / edited by W-K. Lu. — Hamilton, Ont. : McMaster University, ₍1977?₎
ca. 350 p. in various pagings : ill. ; 28 cm. — (McMaster Symposium on Iron and Steelmaking ; no. 5) C***
Includes bibliographical references.
1. Coke-oven gas—Desulphurization—Congresses. I. Lu, Wei-Kao, 1933- . II. McMaster University. Dept. of Metallurgy and Materials Science. III. Title. IV. Series: McMaster Symposium on Iron- and Steelmaking ; no. 5.
TP336.S955 1977 665'.772 79-313522
79 MARC

COKE-OVENS

see also Coke-oven gas

COKE-OVENS (Continued)

—ADDRESSES, ESSAYS, LECTURES

Lee, G W
Improvement of the economy of the coke-oven process / G. W. Lee, J. P. Graham, and W. J. Pater. — ₍s. l. : s. n., 1966₎
9 p. ; 30 cm.
Caption title.
At head of title: Congrès international de Charleroi. Le coke en sidérurgie. B5.
Bibliography: p. 8-9.
1. Coal—Great Britain—Carbonization—Addresses, essays, lectures. 2. Coke-ovens—Addresses, essays, lectures. 3. Coke industry—Great Britain—Addresses, essays, lectures. I. Graham, J. P., joint author. II. Pater, W. J., joint author. III. Title.
TP336.L43 79-316502
79 MARC

COL, GONTIER, 1354?-1418

Coville, Alfred, 1860-1942.
Gontier et Pierre Col et l'humanisme en France au temps de Charles VI / Alfred Coville. — Genève : Slatkine Reprints, 1977.
256 p. ; 23 cm. Sw77-5950
Reprint of the 1934 ed. published by Droz, Paris.
Includes bibliographical references and index.
60.00F
1. Col, Gontier, 1354?-1418. 2. Col, Pierre, fl. 1402. 3. Authors, French —To 1500—Biography. 4. Diplomats—France—Biography. 5. Humanism—History. 6. France—History—Charles VI, 1380-1422. 7. Roman de la Rose. I. Title.
PQ1559.C56C6 1977 78-394266
*79 MARC

COL, PIERRE, FL. 1402

Coville, Alfred, 1860-1942.
Gontier et Pierre Col et l'humanisme en France au temps de Charles VI / Alfred Coville. — Genève : Slatkine Reprints 1977.
256 p. ; 23 cm. Sw77-5950
Reprint of the 1934 ed. published by Droz, Paris.
Includes bibliographical references and index.
60.00F
1. Col, Gontier, 1354?-1418. 2. Col, Pierre, fl. 1402. 3. Authors, French —To 1500—Biography. 4. Diplomats—France—Biography. 5. Humanism—History. 6. France—History—Charles VI, 1380-1422. 7. Roman de la Rose. I. Title.
PQ1559.C56C6 1977 78-394266
*79 MARC

COLACICCHI, PAOLO

Colacicchi, Paolo.
L'ultimo fronte d'Africa : Tunisia, novembre 1942-maggio 1943 / Paolo Colacicchi. — Milano : Mursia, 1977.
204 p., ₍4₎ leaves of plates : ill. ; 21 cm. — (Testimonianze fra cronaca e storia ; 88) It78-July
Bibliography: p. 197.
Includes indexes.
L5000
1. World War, 1939-1945—Campaigns—Tunisia. 2. World War, 1939-1945 —Personal narratives, Italian. 3. Colacicchi, Paolo. 4. Soldiers—Italy—Biography. 5. Tunisia—History—French occupation, 1881-1956. I. Title.
D766.99.T8C64 79-340927
*79 MARC

COLAS see Cholas

COLBERT, EDOUARD CHARLES VICTURNIEN, COMTE DE MAULEVIER, 1758-1820

Colbert, Edouard Charles Victurnien, comte de Maulevrier, 1758-1820.
Voyage dans l'intérieur des États-Unis et au Canada / par le comte de Colbert Maulevrier ; avec une introd. et des notes par Gilbert Chinard. — New York : Arno Press, 1979.
xxiii, 87 p., ₍17₎ leaves of plates : ill. ; 22 cm. — (John Hopkins University Press reprints)
Reprint of the 1935 ed. published by The Johns Hopkins, Press, Baltimore, which was issued as Cahier 8 of Historical documents, Institut français de Washington.
Includes index.
ISBN 0-405-10596-7 : $12.00
1. United States—Description and travel—1783-1848. 2. Canada—Description and travel—1763-1867. 3. Susquehanna Valley—Description and travel. 4. Colbert, Edouard Charles Victurnien, Comte de Maulevier, 1758-1820. I. Chinard, Gilbert, 1881-1972. II. Title. III. Series.
E164.C74 1979 973 78-19291
78 MARC

COLBERT, JEAN-BAPTISTE, 1619-1683

Malettke, Klaus.
Jean-Baptiste Colbert : Aufstieg im Dienste d. Königs / Klaus Malettke. — Göttingen ; Zürich ; Frankfurt ₍Main₎ : Musterschmidt, 1977.
115 p., ₍4₎ leaves of plates : ill. ; 19 cm. — (Persönlichkeit und Geschichte ; Bd. 99/100) GFR77-A
Bibliography: p. 114-115.
ISBN 3-7881-0099-0
1. Colbert, Jean-Baptiste, 1619-1683. 2. Statesmen—France—Biography. I. Series.
DC130.C6M34 78-381786
79 MARC

COLCHESTER, ENG.

—HISTORY

Stephenson, David.
The book of Colchester : a portrait of the town / by David Stephenson. — Chesham : Barracuda Books, 1978.
144 p. : ill., facsims., maps, ports. ; 31 cm. GB78-25748
Map on lining paper.
Includes bibliography: p. 139-140.
Includes index.
ISBN 0-86023-056-2 : £9.50
1. Colchester, Eng.—History. I. Title.
DA690.C7S84 942.6'723 78-323927
79 MARC

COLCRETE see Colloidal concrete

COLD

see also Cold waves (Meteorology); Cryobiology; Heat; Low temperatures; Refrigeration and refrigerating machinery; Temperature; Winter

—JUVENILE LITERATURE

Chase, Catherine.
Hot & cold / by Catherine Chase ; illustrated by Gail Gibbons. — New York : Dandelion Press, 1979.
₍32₎ p. : col. ill. ; 23 cm. — (A Dandelion first reader)
SUMMARY: Identifies items that are hot, cold, and warm and observes that their temperatures can change.
ISBN 0-89799-110-9
1. Heat—Juvenile literature. 2. Cold—Juvenile literature. ₍1. Heat. 2. Cold₎ I. Gibbons, Gail. II. Title.
QC256.C42 536 78-72101
79 MARC

—PHYSIOLOGICAL EFFECT

see also Hypothermia; Temperature —Physiological effect

Antonenko, Vitaliĭ Tymofiĭovych, 1933–
₍Zhizn' v usloviĭakh nizkikh temperatur. Ukrainian₎
Життя в умовах низьких температур / В. Т. Антоненко ; Переклад з російської В. П. Плачинди. — Київ : Наук. думка, 1978.
169 p. : ill. ; 17 cm. — (Серія Горизонти науки і техніки) (Людина, суспільство, прогрес) USSR***
Series 1 romanized: Serii͡a Horyzonty nauky i tekhniky. Series 2 romanized: Li͡udyna, suspil'stvo, prohres.
Bibliography: p. 167–₍168₎
0.30rub
1. Cold — Physiological effect. 2. Hypothermia. 3. Cryobiology I. Title.
Title romanized: Zytti͡a v umovakh nyz'kykh temperatur.
QP82.2.C6A5719 79-412652

Calcott, P H
Freezing and thawing microbes / P. H. Calcott. — Shildon ₍Eng.₎ : Meadonfield Press, c1978.
68 p. : ill. ; 21 cm. — (Patterns of progress : Microbiology series ; PP/M/14) GB***
Bibliography: p. 64-68.
ISBN 0-904095-27-4 : £3.40
1. Freeze-drying. 2. Micro-organisms—Collection and preservation. 3. Micro-organisms—Drying. 4. Cold—Physiological effect. 5. Thawing. I. Title.
QR69.F7C34 576'.11 79-305881
79 MARC

Preusser, Wilhelm.
Fit durch den Winter. Bad Harzburg, Verlag für Wissenschaft, Wirtschaft und Technik, 1969.
60 p. illus. 21 cm. (Lebensführung und Leistungssteigerung, Bd. 2)
1. Influenza. 2. Cold—Physiological effect. I. Title.
RC150.P74 79-408423
79 MARC

—PHYSIOLOGICAL EFFECT—CONGRESSES

Cold Water Symposium, Toronto, Ont., 1976.
Proceedings of the Cold Water Symposium : held in Toronto, Ontario, May 8, 1976 / sponsored by the Royal Life Saving Society Canada ; in cooperation with Dept. of Physiology, University of Toronto ₍and₎ Sports Medicine Section, Ontario Medical Association. — Toronto : Royal Life Saving Society Canada, c1976.
45 p. : ill. ; 28 cm. C***
Cover title.
Bibliography: p. 44-45.
1. Cold—Physiological effect—Congresses. 2. Hypothermia—Congresses. 3. Life-saving—Congresses. 4. Drowning—Congresses. I. Royal Life Saving Society Canada. II. University of Toronto. Dept. of Physiology. III. Ontario Medical Association. Sports Medicine Section.
QP82.2.C6C63 1976 616.9'89 79-317556
79 MARC

Cryoimmunologie = Cryoimmunology : colloque, Dijon, 17-19 juin 1976 / organisé par INSERM ... ₍et al.₎ ; publié sous la responsabilité de Denise Simatos, Douglas M. Strong, Jean-Michel Turc. — ₍Paris₎ : Éditions de l'Institut national de la santé et de la recherche médicale, ₍c1977₎
423 p. : ill. ; 25 cm. — (Science et technique du froid) (Colloques et séminaires - Institut national de la santé et de la recherche médicale ; v. 62) F***
French or English.
Includes bibliographies.
ISBN 2-85598-149-2
1. Immunology—Congresses. 2. Cold—Physiological effect—Congresses. 3. Immunocompetent cells—Congresses. 4. Cells—Preservation—Congresses. 5. Preservation of organs, tissues, etc.—Congresses. I. Simatos, D. II. Strong, Douglas M. III. Turc, Jean Michel. IV. Institut national de la santé et de la recherche médicale. V. Title: Cryoimmunology. VI. Series: Institut national de la santé et de la recherche médicale. Colloques ; v. 62.
QR180.3.C79 79-356279
616.07'9'0157419165
79 MARC

Proteins at low temperatures : based on a symposium sponsored by the Division of Agricultural and Food Chemistry at the 175th Meeting of the American Chemical Society, Anaheim, California, March 12-17, 1978 / Owen Fennema, editor. — Washington : American Chemical Society, 1979.
vi, 233 p. : ill. ; 24 cm. — (Advances in chemistry series ; 180 ISSN 0065-2393)
Includes bibliographical references and index.
ISBN 0-8412-0484-5
1. Proteins—Congresses. 2. Cold—Physiological effect—Congresses. 3. Proteins—Denaturation—Congresses. I. Fennema, Owen R. II. American Chemical Society. Division of Agricultural and Food Chemistry. III. American Chemical Society.
QD1.A355 no. 180 540'.8 s 79-16561
₍QP551₎ 79 MARC

—THERAPEUTIC USE SURGERY

see also Cryosurgery

COLD (DISEASE)

see also Cough; Influenza

Bennett, Harold, 1936-
Cold comfort : colds and flu everybody's guide to the self treatment / Hal Zina Bennett. — New York : C. N. Potter ; distributing by Crown Publishers, c1979.
155 p. : ill. ; 23 cm.
"A Yolla Bolly Press book."
Bibliography: p. ₍147₎-150.
Includes indexes.
ISBN 0-517-53593-9. ISBN 0-517-53594-7 pbk.
1. Cold (Disease) 2. Influenza. I. Title.
RF361.B46 616.2'03 78-11240
78 MARC

COLD COOKERY see Cookery (Cold dishes)

COLD DISHES see Cookery (Cold dishes)

COLD FLOW see Creep of metals

COLD-FRAMES

Shewell-Cooper, Wilfred Edward, 1900-
Basic book of cloche & frame gardening / ₍by₎ W. E. Shewell-Cooper. — London : Barrie and Jenkins, 1977.
160 p. : ill. ; 22 cm. GB78-10295
Spine title: Dr. Shewell-Cooper's basic book of cloche and frame gardening.
Includes index.
ISBN 0-214-20301-8 : £4.25
1. Cloche gardening. 2. Cold-frames. I. Title. II. Title: Cloche & frame gardening. III. Title: Dr. Shewell-Cooper's basic book of cloche and frame gardening.
SB127.S49 635'.04'8 79-305936
79 MARC

COLD REGIONS

see also Frozen ground; Polar regions

—CONGRESSES

Cold Regions Specialty Conference, Anchorage, Alaska, 1978.
Proceedings of the Conference on Applied Techniques for Cold Environments / Cold Regions Specialty Conference, Anchorage, Alaska, May 17-19, 1978 ; sponsored by Alaska Section, ASCE, ASCE Task Committee on Cold Regions, cosponsored by the Canadian Society for Civil Engineers. — New York : American Society of Civil Engineers, c1978-c1979.
2 v. (vii, 1167 p.) : ill. ; 22 cm.
On spine: Applied techniques for cold environments.
Includes bibliographical references and indexes.
1. Civil engineering—Cold weather conditions—Congresses. 2. Civil engineering—Alaska—Cold weather conditions—Congresses. 3. Cold regions—Congresses. I. American Society of Civil Engineers. Alaska Section. II. American Society of Civil Engineers. Task Committee on Cold Regions. III. Canadian Society for Civil Engineers. IV. Title: Applied techniques for cold environments.
TA713.C64 1978 620'.411 79-116292
79 MARC

COLD STORAGE

see also Compressed air; Farm produce—Storage; Meat industry and trade; Refrigeration and refrigerating machinery; Refrigerator-cars

International Institute of Refrigeration.
Conditions recommandées pour le transport terrestre des denrées périssables = Recommended conditions for land transport of perishable foodstuffs / Institut international du froid. — 3. éd. — Paris : L'Institut, 1974.
41 p. ; 24 cm. — (Annexe au Bulletin de l'Institut international du froid ; 1974) F***
French and English.
1. Food—Transportation. 2. Cold storage. I. Title. II. Title: Recommended conditions for land transport of perishable foodstuffs. III. Series: International Institute of Refrigeration. Bulletin. Annexe ; 1974.
TP373.2.I57 1974a 79-378773
79 MARC

(Unshū mikan no teion chozō)
温州ミカンの低温貯蔵 / 農林水産技術会議事務局. — 東京 : 農林統計協会, 1975.
39 p., ₍1₎ leaf of plates : ill. ; 21 cm. — (実用化技術レポート : 18)
Ja 77-320
Errata slip inserted.
Bibliography: p. 36-38.

1. Tangerine—Storage. 2. Cold storage. 3. Tangerine—Japan. I. Japan. Nōrin Suisan Gijutsu Kaigi. Jimukyoku. II. Series: Jitsuyōka gijutsu repōto ; no. 18.
SB370.T3O57 79-805862

—CONGRESSES

International Institute of Refrigeration. Commission C1.
Congélation, conservation à l'état congelé et lyophilisation des substances biologiques et des aliments = Freezing, frozen storage and freeze-drying of biological materials and foodstuffs : ₍réunion des Commissions C1 et C2 de l'I.F.F., Ettlingen-Karlsruhe, 6-8 septembre 1977₎. — Paris : Institut international du froid, 1977.
494 p. : ill. ; 24 cm. — (Science et technique du froid) F***
English or French.
"BFRAAV 1-493 (1977)"
Includes bibliographical references.
1. Food, Frozen—Congresses. 2. Cold storage—Congresses. 3. Food, Freeze-dried—Congresses. I. International Institute of Refrigeration. Commission C2. II. Title. III. Title: Freezing, frozen storage and freeze-drying of biological materials and foodstuffs.
TP372.3.I57 1977 664'.0285 79-309520
79 MARC

Refrigeration applications to fish, fruit, and vegetables in South East Asia : an international short-term training course, Durgapur, India, January 14-24, 1974. — Rome : Food and Agriculture Organization of the United Nations, c1974.
249 p. : ill. ; 24 cm. It***
Sponsored by Food and Agriculture Organization of the United Nations and International Institute of Refrigeration.
Errata slip inserted.
Includes bibliographical references.
1. Cold storage—Congresses. 2. Food, Frozen—Congresses. I. Food and Agriculture Organization of the United Nations. II. International Institute of Refrigeration.
TP372.2.R43 664'.0285 79-303081
79 MARC

—INSULATION

Seiffert, Karl, 1903-
Der Wärmeschutz-Ingenieur; Grundlagen der Wärme- und Kälteschutztechnik für den Gebrauch in der Praxis. München, C. Hanser, 1954.
334 p. illus. 21 cm.
1. Insulation (Heat) 2. Cold storage—Insulation. 3. Heat—Conduction. I. Title.
TH1715.S44 621.4'024 a 54-4166
Illinois. Univ. Library ₍r79₎rev2 MARC

COLD-STORAGE LOCKERS

see also Home freezers

COLD STORAGE ON SHIPBOARD

see also Fishing boats—Refrigeration; Marine refrigeration

COLD WAR see World politics—1945-

COLD WAVES (METEOROLOGY)

—BAIKAL LAKE REGION

Chu, E Wing Ki, 1935-
A method for forecasting the arrival of cold surges in Hong Kong / by E. W. K. Chu. — Hong Kong : Royal Observatory, 1978.
v, 32 leaves : ill. ; 30 cm. — (Technical note - Royal Observatory ; no. 43)
Includes bibliographical references.
1. Cold waves (Meteorology)—Hongkong. 2. Fronts (Meteorology)—Hongkong. 3. Cold waves (Meteorology)—Baikal Lake region. 4. Fronts (Meteorology)—Baikal Lake region. I. Title. II. Title: Cold surges in Hong Kong. III. Series: Hongkong. Royal Observatory. Technical notes ; no. 43.
QC990.H63A35 no. 43 551.6'9'512508 s 78-109849
₍QC981.8.A5₎ 79 MARC

—CHINA—HONGKONG

Chu, E Wing Ki, 1935-
A method for forecasting the arrival of cold surges in Hong Kong / by E. W. K. Chu. — Hong Kong : Royal Observatory, 1978.
v, 32 leaves : ill. ; 30 cm. — (Technical note - Royal Observatory ; no. 43)
Includes bibliographical references.
1. Cold waves (Meteorology)—Hongkong. 2. Fronts (Meteorology)—Hongkong. 3. Cold waves (Meteorology)—Baikal Lake region. 4. Fronts (Meteorology)—Baikal Lake region. I. Title. II. Title: Cold surges in Hong Kong. III. Series: Hongkong. Royal Observatory. Technical notes ; no. 43.
QC990.H63A35 no. 43 551.6'9'512508 s 78-109849
₍QC981.8.A5₎ 79 MARC

COLD WEATHER CLOTHING see Clothing, Cold weather

COLD WORKING OF METALS see Metals—Cold working

COLDEVIN, ISACH, 1877-1968

Moe, Knut.
Drømmen om en grønn hage : ₍Isach Coldevin₎ / Knut Moe. — Oslo : Nordnorsk forfatterlag : Tiden, c1977 (Oslo : Kirstes boktr.)
159 p. ; 22 cm. N77-Nov.
ISBN 8210014862 : kr62.00. ISBN 8210014854 pbk.
1. Coldevin, Isach, 1877-1968. 2. Agriculturists—Norway—Biography. I. Title.
S417.C62M63 78-375210
78 MARC

COLDITZ, GER. CASTLE

Platt, Joseph Ellison, 1900-1973.
Badre in Colditz : the diary of J. Ellison Platt / edited by Margaret Duggan. — London : Hodder and Stoughton, 1978.
319 p. ; 23 cm. GB78-38103
ISBN 0-340-22807-5 : £5.50
1. Colditz, Ger. Castle. 2. World War, 1939-1945—Prisoners and prisons, German. 3. World War, 1939-1945—Personal narratives, English. 4. Platt, Joseph Ellison, 1900-1973. I. Duggan, Margaret. II. Title.
D805.G3P56 1978 940.54'72'430924 79-308496
79 MARC

COLE, CATHARINE

O'Connor, Patti.
One summer in Israel / Patti O'Connor, Catharine Cole. — Chicago : Academy Press, c1978.
227 p. : ill. ; 29 cm.
ISBN 0-915864-65-7 : $19.95
1. Israel—Description and travel. 2. O'Connor, Patti. 3. Cole, Catharine. I. Cole, Catharine, joint author. II. Title.
DS107.4.O27 956.94 78-16137
78 MARC

COLE, GEORGE DOUGLAS HOWARD, 1889-1959

Houseman, Gerald L
G. D. H. Cole / by Gerald L. Houseman. — Boston : Twayne Publishers, 1979.
152 p. : port. ; 21 cm. — (Twayne's English authors series ; TEAS 255)
Bibliography: p. 144-148.
Includes index.
ISBN 0-8057-6746-0
1. Cole, George Douglas Howard, 1889-1959. 2. Guild socialism.
HD8393.C57H68 1979 330'.092'4 78-21665
78 MARC

Wright, A W
G. D. H. Cole and socialist democracy / A. W. Wright. — Oxford : Clarendon Press ; New York : Oxford University Press, 1979.
301 p., ₍1₎ leaf of plates : port. ; 23 cm. GB79
Includes bibliographical references and index.
ISBN 0-19-827421-1 : £9.00
1. Cole, George Douglas Howard, 1889-1959. 2. Gild socialism. I. Title.
HD8393.C57W74 335'.1 78-40644
78 MARC

COLE, MARGARET RICE

Cole, Margaret Rice.
Never too old for God / Margaret Rice Cole, as told to Don Tanner and Bruce Anderson. — Old Tappan, N.J. : Revell, c1979.
127 p. ; 18 cm. — (Spire books)
ISBN 0-8007-8348-4
1. Cole, Margaret Rice. 2. Missionaries—United States—Biography. I. Tanner, Don, joint author. II. Anderson, Bruce, 1946- joint author. III. Title.
BV3705.C585A35 266'.0092'4 78-21953
78 MARC

COLE FAMILY

Pearce, Donald, 1920-
A history of the Coles family : Somerset to South Canterbury, 1874-1974 / ₍compiled and edited by Don Pearce in co-operation with Bill and Jean Traves₎. — Timaru : W. Traves, ₍1975?₎ (Timaru : Timaru Herald Company)
72 p., ₍24₎ p. of plates : ill., facsims., geneal. tables, map, ports. ; 22 cm.
NZ77-1532
1. Cole family. 2. New Zealand—Genealogy. I. Traves, William John. II. Traves, Jean Phyllis Coles, 1923- III. Title.
CS2179.C64 1975 929'.2'09931 79-308829
79 MARC

COLEG HARLECH

—HISTORY

Stead, Peter.
Coleg Harlech : the first fifty years / by Peter Stead. — Cardiff : University of Wales Press, 1977.
xv, 135 p., [16] p. of plates : ill., ports. ; 25 cm. GB77-33156
Includes index.
ISBN 0-7083-0657-8 (pbk.) : £4.00
1. Coleg Harlech—History. 2. Harlech, Wales—History. I. Title.
LF1225.S73 374.9'429'29 79-316303
79 MARC

COLÉGIO AGRÍCOLA VIDAL DE NEGREIROS

Vicente Filho, José.
Memórias do meu colégio / José Vicente Filho. — João Pessoa : [s.n., 1977?]
203 p. : ill. ; 22 cm.
1. Colégio Agrícola Vidal de Negreiros. 2. Vicente Filho, José. I. Title.
S539.B8C648 79-347016
79 MARC

COLEGIO DE FILIPINOS. BIBLIOTECA

—CATALOGS

Rodríguez, Isacio Rodríguez y
Updated checklist of Filipiniana at Valladolid / compiled by Isacio R. Rodríguez. — [Manila] : National Historical Institute, 1976.
2 v. (xiv, 836 p.) ; 26 cm.
Originally published as A catalog of Filipiniana at Valladolid, edited by Helen R. Tubangui
Includes index.
1. Philippine Islands—Bibliography—Catalogs. 2. Colegio de Filipinos. Biblioteca—Catalogs. I. Tubangui, Helen R. A catalog of Filipiniana at Valladolid. II. Colegio de Filipinos. Biblioteca. III. Title.
Z3299.R63 016.9599 79-105156
[DS655] 79 MARC

COLEGIO MAYOR DEL ARZOBISPO FONSECA

—HISTORY

Sendín Calabuig, Manuel, 1949-
El Colegio Mayor del Arzobispo Fonseca en Salamanca / [por] Manuel Sendín Calabuig. — Salamanca : Universidad, 1977.
367 p., 1 leaf, [56] leaves of plates (1 fold.) : ill. ; 25 cm. — (Acta Salmanticensia : Historia de la Universidad ; 28) Sp78-July
Bibliography: p. [333]-344.
ISBN 8460010309
1. Colegio Mayor del Arzobispo Fonseca—History. I. Title. II. Series.
AS301.A26 t. 28 79-362520
[LF4823.S29] 79 MARC

COLEGIO OFICIAL DE ARQUITECTOS DE MADRID

Colegio Oficial de Arquitectos de Madrid. Comisión de Cultura. Servicio Histórico.
El COAM ante la destrucción de la ciudad : [exposición, enero 1978 / Comisión de Cultura, Servicio Histórico]. — [Madrid] : COAM, D.L. 1978.
1 leaf, 85 p. ; 21 cm. Sp78-Feb
ISBN 8440043600
1. Madrid—Buildings—Conservation and restoration—Exhibitions. 2. Architecture—Spain—Madrid—Conservation and restoration—Exhibitions. 3. Colegio Oficial de Arquitectos de Madrid. I. Title.
NA109.S7C64 1978 79-360304
79 MARC

COLÉGIO SANTO ANTÔNIO, BELÉM, BRAZIL

Chaves, Maria Paula Ramos.
O Convento de Sto. [i.e. Santo] Antônio e as Dorotéias / Maria Paula Ramos Chaves. — [Belém] : AAD, 1977.
89 p. : ill. ; 24 cm.
"Comemoração do centenário da chegada das Irmãs de Sta. Dorotéia a Belém do Pará, 1877-1977."
Bibliography: p. 89.
1. Colégio Santo Antônio, Belém, Brazil. 2. Sisters of Saint Dorothy. I. Title.
BX4486.6.Z9C643 78-400832
79 MARC

COLEMAN, KIT, 1864-1915

Ferguson, Ted.
Kit Coleman, Queen of Hearts / Ted Ferguson. — 1st ed. — Toronto : Doubleday Canada ; Garden City, N.Y. : Doubleday, 1978.
x, 182 p., [10] leaves of plates : ill. ; 22 cm.
Bibliography: p. [181]-182.
ISBN 0-385-13447-9 : $8.95
1. Coleman, Kit, 1864-1915. 2. Journalists—Canada—Biography. I. Title.
PN4913.C6F4 070'.92'4 77-92211
78 MARC

COLEMAN, JOHN S., 1902-

Coleman, John S 1902-
Bataan and beyond : memories of an American P.O.W. / by John S. Coleman, Jr. ; foreword by Thomas Dooley. — 1st ed. — College Station : Texas A & M University Press, c1978.
xii, 210 p., [6] leaves of plates : ill. ; 24 cm. — (The Centennial series of the Association of Former Students ; no. 6)
Includes index.
ISBN 0-89096-055-0 : $11.50
1. World War, 1939-1945—Prisoners and prisons, Japanese. 2. World War, 1939-1945—Personal narratives, American. 3. Coleman, John S., 1902- . 4. World War, 1939-1945—Philippine Islands. 5. Philippine Islands—History—Japanese occupation, 1942-1945. 6. Prisoners of war—Philippine Islands—Biography. I. Title. II. Series: Texas. A & M University, College Station. Association of Former Students. The centennial series of the Association of Former Students, Texas A & M University. : no. 6.
D805.P6C64 940.54'83'73 78-6365
78 MARC

COLEMAN, TERRY

Coleman, Terry.
The scented brawl : selected articles and interviews / Terry Coleman. — London : Elm Tree Books, 1978.
xii, 191 p. ; 23 cm. GB***
ISBN 0-241-10042-9 : £6.95
1. Biography—20th century. 2. Coleman, Terry. I. Title.
CT120.C64 1978 082 79-302167
79 MARC

COLEMAN, THOMAS, 1945-

Nance, John, 1935-
The mud-pie dilemma : a master potter's struggle to make art and ends meet / John Nance. — Forest Grove, Or. : Timber Press, c1978.
vi, 116 p., [24] leaves of plates : ill. (some col.) ; 29 cm. — (Timber Press series in arts and crafts)
ISBN 0-917304-18-7
1. Coleman, Thomas, 1945- 2. Potters—United States—Biography. I. Title. II. Series.
NK4210.C63N37 738'.092'4 78-13888
79 MARC

COLEMAN, THOMAS W., 1849-1921

Coleman, Thomas W 1849-1921.
I buried Custer : the diary of Pvt. Thomas W. Coleman, 7th U.S. Cavalry / edited by Bruce R. Liddic ; introd. by John M. Carroll ; illustrated by James H. Earle. — 1st ed. — College Station, Tex. : Creative Pub. Co., c1979.
vi, 210 p. : ill. ; 24 cm. — (The Young West series)
Includes bibliographical references and index.
ISBN 0-932702-02-3
1. Coleman, Thomas W., 1849-1921. 2. Dakota Indians—Wars, 1876—Personal narratives. 3. Little Big Horn, Battle of the, 1876—Personal narratives. 4. Custer, George Armstrong, 1839-1876. 5. United States. Army. 7th Cavalry—Biography. I. Liddic, Bruce R. II. Title. III. Series: Young West series.
E83.876.C695 973.8'2'0924 78-67551
79 MARC

COLEMAN FAMILY

Griffin, Vera White.
Mountain climbers and bridge builders : the families of White, Coleman, and Nelson, 1788-1973. — [s.l. : s.n.], c1978.
382 p. : ill. ; 24 cm.
Includes bibliographical references.
1. White family. 2. Coleman family. 3. Nelson family. I. Title.
CS71.W585 1978a 929'.2'0973 79-109340
79 MARC

COLERAINE, IRE.

—HISTORY

Mullin, Thomas Hugh.
Coleraine in Georgian times / by T. H. Mullin. — Coleraine : The author, 1977.
[6], 207 p. : ill. ; 22 cm. GB78-25740
Includes bibliographical references and index.
£3.00
1. Coleraine, Ire.—History. I. Title.
DA995.C64M85 941.6'27 78-326404
79 MARC

COLERIDGE, HARTLEY, 1796-1849

—CORRESPONDENCE—CATALOGS

Stephens, Fran Carlock.
The Hartley Coleridge letters : a calendar and index / by Fran Carlock Stephens. — [Austin] : Humanities Research Center, University of Texas at Austin, c1978.
217 p., [1] leaf of plates : ill. ; 24 cm. — (Tower bibliographical series ; no. 15)
Based primarily on the holdings of the Humanities Research Center of the University of Texas at Austin.
ISBN 0-87959-078-5
1. Coleridge, Hartley, 1796-1849—Correspondence—Catalogs. 2. Authors, English—19th century—Correspondence—Catalogs. 3. Texas. University of Austin. Humanities Research Center. I. Texas. University at Austin. Humanities Research Center. II. Title. III. Series.
Z6616.C7S73 821'.7 76-620064
[PR4468.A4] 00 MARC

COLERIDGE, SAMUEL TAYLOR, 1772-1834. DEJECTION: AN ODE

Dekker, George.
Coleridge and the literature of sensibility / George Dekker. — New York : Barnes & Noble Books, 1978.
270 p. ; 23 cm. — (Barnes & Noble critical studies)
Includes bibliographical references and index.
ISBN 0-06-491655-3
1. Coleridge, Samuel Taylor, 1772-1834. Dejection: an ode. 2. English poetry—18th century—History and criticism. 3. Romanticism—England. I. Title.
PR4480.D43D4 1978b 821'.7 79-102065
79 MARC

COLERIDGE, SAMUEL TAYLOR, 1772-1834. THE RIME OF THE ANCIENT MARINER

Piper, Herbert Walter, 1915-
Nature and the supernatural in The ancient mariner / by H. W. Piper. — Norwood, Pa. : Norwood Editions, 1978.
30 p. ; 23 cm.
Reprint of an inaugural public lecture delivered in Armidale, New South Wales on Sept. 19, 1955.
Bibliography: p. 28-30.
ISBN 0-8482-2116-8 : $7.50
1. Coleridge, Samuel Taylor, 1772-1834. The rime of the ancient mariner. 2. Nature in literature—Addresses, essays, lectures. 3. Supernatural in literature—Addresses, essays, lectures. I. Title.
PR4479.P5 1978 821'.7 78-7851
78 MARC

COLERIDGE, SAMUEL TAYLOR, 1772-1834

—AESTHETICS—ADDRESSES, ESSAYS, LECTURES

Willey, Basil, 1897-
Coleridge on imagination and fancy / by Basil Willey. — Folcroft, Pa. : Folcroft Library Editions, 1978.
15 p. ; 26 cm.
Reprint of the 1946 Warton lecture on English poetry, published in the Proceedings of the British Academy, London, v. 32 (1951).
ISBN 0-8414-9267-7 lib. bdg. : $7.50
1. Coleridge, Samuel Taylor, 1772-1834—Aesthetics—Addresses, essays, lectures. 2. Imagination—Addresses, essays, lectures. I. Title. II. Series: British Academy, London (Founded 1901). Warton lecture on Enlish poetry ; 1946.
[PR4487.I4W5 1978] 801'.93 78-17774
78 MARC

COLERIDGE, SAMUEL TAYLOR, 1772-1834 (Continued)

—BIBLIOGRAPHY

Caskey, Jefferson D
Samuel Taylor Coleridge : a selective bibliography of criticism, 1935-1977 / compiled by Jefferson D. Caskey and Melinda M. Stapper. — Westport, Conn. : Greenwood Press, 1978.
xvii, 174 p. ; 24 cm.
Includes index.
ISBN 0-313-20564-7
1. Coleridge, Samuel Taylor, 1772-1834—Bibliography. I. Stapper, Melinda M., joint author.
Z8182.C37 016.821'7 78-57764
[PR4483] 78 MARC

—CRITICISM AND INTERPRETATION

Cooke, Katharine.
Coleridge / Katharine Cooke. — London ; Boston : Routledge & Kegan Paul, 1979.
266 p. ; 23 cm. GB***
Bibliography: p. 255-259.
Incudes index.
ISBN 0-7100-0141-X : £6.95
1. Coleridge, Samuel Taylor, 1772-1834—Criticism and interpretation.
PR4484.C65 821'.7 78-41237
79 MARC

Kessler, Edward, 1927-
Coleridge's metaphors of being / Edward Kessler. — Princeton, N.J. : Princeton University Press, c1979.
xi, 202 p. ; 23 cm. — (Princeton essays in literature)
Bibliography: p. 195-197.
Includes index.
ISBN 0-691-06394-X : $12.50
1. Coleridge, Samuel Taylor, 1772-1834—Criticism and interpretation. I. Title.
PR4484.K4 821'.7 78-70302
79 MARC

Snyder, Alice Dorothea, 1887-1943.
The critical principle of the reconciliation of opposites as employed by Coleridge / by Alice D. Snyder. — Philadelphia : R. West, 1977.
59 p. ; 26 cm.
Reprint of the 1918 ed. published in Ann Arbor, Mich., which was issued as no. 9 of Contributions to rhetorical theory.
Originally presented as the author's thesis, University of Michigan, 1915.
Bibliography: p. [57]-59.
ISBN 0-8492-2468-3 (lib. bdg.) : $10.00
1. Coleridge, Samuel Taylor, 1772-1884—Criticism and interpretation. 2. Coleridge, Samuel Taylor, 1772-1884—Philosophy. I. Title. II. Series: Contributions to rhetorical theory ; no. 9.
[PR4484.S6 1977] 821'.7 77-28874
78 MARC

—CRITICISM AND INTERPRETATION —ADDRESSES, ESSAYS, LECTURES

Richards, Ivor Armstrong, 1893-
Coleridge's minor poems : a lecture / by I. A. Richards. — Norwood, Pa. : Norwood Editions, 1978.
30 p. ; 22 cm.
"Delivered in honor of the fortieth anniversary of Professor Edmund L. Freeman at Montana State University on April 8, 1960."
Reprint of the 1960 ed.
ISBN 0-8482-2322-5 : $8.50
1. Coleridge, Samuel Taylor, 1772-1834—Criticism and interpretation—Addresses, essays, lectures. I. Freeman, Edmund L. II. Title.
[PR4484.R5 1978] 821'.7 78-3893
78 MARC

—HOMES AND HAUNTS —ENGLAND—KESWICK (CUMBERLAND)

Howe, H W
Greta Hall : home of Coleridge and Southey / [by] H. W. Howe. — [Revised ed.] / with revisions by Robert Woof. — [Sevenoaks] : [Mrs. R. M. Abbott], 1977.
x, 117 p., 8 p. of plates : ill., facsim., plans, ports. ; 22 cm. GB78-09151
ISBN 0-9505970-0-7 : £2.00
1. Coleridge, Samuel Taylor, 1772-1834—Homes and haunts—England—Keswick (Cumberland) 2. Southey, Robert, 1774-1843—Homes and haunts—England—Keswick (Cumberland). 3. Keswick, Eng. (Cumberland). Greta Hall. 4. Authors, English—19th century—Biography. I. Woof, Robert. II. Title.
PR4483.H63 1977 821'.7 78-318266
78 MARC

—LIBRARY

Coleridge, Samuel Taylor, 1772-1834.
Critical annotations : being marginal notes inscribed in volumes formerly in the possession of Coleridge / by S. T. Coleridge ; edited by William F. Taylor. — Philadelphia : R. West, 1977.
48 p. : port. ; 29 cm.
Reprint of the 1889 ed. published by W. F. Taylor, Harrow, Eng.
ISBN 0-8492-3816-1 : $10.00
1. Coleridge, Samuel Taylor, 1772-1834—Library. I. Title.
[PR4472.T3 1977] 821'.7 77-26219
77 MARC

—PHILOSOPHY

Snyder, Alice Dorothea, 1887-1943.
The critical principle of the reconciliation of opposites as employed by Coleridge / by Alice D. Snyder. — Philadelphia : R. West, 1977.
59 p. ; 26 cm.
Reprint of the 1918 ed. published in Ann Arbor, Mich., which was issued as no. 9 of Contributions to rhetorical theory.
Originally presented as the author's thesis, University of Michigan, 1915.
Bibliography: p. [57]-59.
ISBN 0-8492-2468-3 (lib. bdg.) : $10.00
1. Coleridge, Samuel Taylor, 1772-1884—Criticism and interpretation. 2. Coleridge, Samuel Taylor, 1772-1884—Philosophy. I. Title. II. Series: Contributions to rhetorical theory ; no. 9.
[PR4484.S6 1977] 821'.7 77-28874
78 MARC

—RELIGION AND ETHICS

Pym, David.
The religious thought of Samuel Taylor Coleridge / David Pym ; with a foreword by John Coulson. — Gerrards Cross [Eng.] : C. Smythe, 1978.
105 p. ; 23 cm. GB***
Includes bibliographical references and index.
ISBN 0-901072-65-6 : £4.75
1. Coleridge, Samuel Taylor, 1772-1834—Religion and ethics. 2. Christianity and literature. I. Title.
PR4487.R4P9 1978 230'.092'4 79-303215
79 MARC

Pym, David.
The religious thought of Samuel Taylor Coleridge / David Pym ; with a foreword by John Coulson. — New York : Barnes & Noble Books, 1979, c1978.
105 p. : port. ; 22 cm.
Includes bibliographical references and index.
ISBN 0-06-495750-0 : $14.50
1. Coleridge, Samuel Taylor, 1772-1834—Religion and ethics. 2. Christianity and literature. I. Title.
PR4487.R4P9 1978b 821'.7 78-14840
78 MARC

COLESCOTT, WARRINGTON, 1921-

—EXHIBITIONS

Colescott, Warrington, 1921-
A history of printmaking : a traveling exhibition of the Madison Art Center, June 15-July 29, 1979 / Warrington Colescott. — Madison, Wis. : The Center, c1979.
[32] p. : ill. (some col.) ; 22 x 29 cm.
1. Colescott, Warrington, 1921- —Exhibitions. I. Madison Art Center (Wis.) II. Title.
N6537.C595A4 1979 769'.92'4 79-65878
79 MARC

COLESVILLE, N. Y.

—HISTORY

Jacob, Ruth Leone Thompson, 1906-
Famines, fires & festivals : a history of Colesville, 1785-1978 / R. Leone Jacob. — [Colesville, N.Y.] : Town of Colesville, c1978.
180 p. : ill. ; 23 cm.
Bibliography: p. 180.
1. Colesville, N.Y.—History. I. Title.
F129.C683J32 974.7'75 78-66101
79 MARC

COLETTE, SIDONIE GABRIELLE, 1873-1954

—BIOGRAPHY

Sarde, Michèle.
Colette, libre et entravée / Michèle Sarde. — Paris : Stock, c1978.
483, [4] p., [8] leaves of plates : ill. ; 22 cm. — (Femmes dans leur temps) F***
Bibliography: p. 479-[484]
ISBN 2-234-00861-1 : 65.00F
1. Colette, Sidonie Gabrielle, 1873-1954—Biography. 2. Authors, French—20th century—Biography. I. Title.
PQ2605.O28Z83 848'.9'1209 78-381738
78 MARC

Viel, Marie Jeanne.
Colette au temps des Claudine : récit / Marie-Jeanne Viel. — Paris : Publications essen-tielles, c1978.
176 p. : ill. ; 22 cm. F***
Bibliography: p. 171-172.
ISBN 2-86315-001-4 : 33.00F
1. Colette, Sidonie Gabrielle, 1873-1954—Biography. 2. Gauthier-Villars, Henry, 1859-1931—Biography. 3. Authors, French—20th century—Biography. I. Title.
PQ2605.O28Z9 848'.9'120 78-388537
78 MARC

COLEY, ENG.

—HISTORY

Cusden, Phoebe E
Coley : portrait of an urban village / [by] Phoebe E. Cusden ; with a foreword by Lord Bullock. — Reading : Reading Branch of the Workers' Educational Association, 1977.
94 p. : ill., maps, ports. ; 24 cm. GB78-07985
Includes index.
ISBN 0-903810-01-8 : £4.00. ISBN 0-903810-02-6 pbk.
1. Coley, Eng.—History. I. Title.
DA690.C703C87 942.2'93 78-317233
78 MARC

COLFAX CO. , NEB.

—BIOGRAPHIES

Tri-Tabula, inc.
Atlas of Colfax & Stanton Counties, Nebraska / compiled and published by Tri-Tabula, Inc. — 1776, Bi-Centennial ed., 1976. — Lake Elmo, Minn. : Tri-Tabula, 1976.
ca. 250 p. : ill., maps, ports. ; 33 cm.
Cover title.
Cadastral maps of townships, scale ca. 1:45,000.
"Containing maps, plats of the townships, alphabetical rural directory, histories of churches, early settlers, pictures of families and general interest."
Includes index and directories.
$47.50
1. Colfax Co. Neb.—Maps. 2. Stanton Co., Neb.—Maps. 3. Colfax Co., Neb.—Biographies. 4. Stanton Co., Neb.—Biographies. 5. Colfax Co., Neb. —Directories. 6. Stanton Co., Neb.—Directories. 7. Colfax Co., Neb.—History. 8. Stanton Co., Neb.—History. 9. Real property—Nebraska—Colfax Co.—Maps. 10. Real property—Nebraska—Stanton Co.—Maps. I. Title. II. Title: Atlas of Stanton County, Nebraska.
G1453.C7T7 1976 912'.782'532 78-324601
78 MARC

—DIRECTORIES

Tri-Tabula, inc.
Atlas of Colfax & Stanton Counties, Nebraska / compiled and published by Tri-Tabula, Inc. — 1776, Bi-Centennial ed., 1976. — Lake Elmo, Minn. : Tri-Tabula, 1976.
ca. 250 p. : ill., maps, ports. ; 33 cm.
Cover title.
Cadastral maps of townships, scale ca. 1:45,000.
"Containing maps, plats of the townships, alphabetical rural directory, histories of churches, early settlers, pictures of families and general interest."
Includes index and directories.
$47.50
1. Colfax Co. Neb.—Maps. 2. Stanton Co., Neb.—Maps. 3. Colfax Co., Neb.—Biographies. 4. Stanton Co., Neb.—Biographies. 5. Colfax Co., Neb. —Directories. 6. Stanton Co., Neb.—Directories. 7. Colfax Co., Neb.—History. 8. Stanton Co., Neb.—History. 9. Real property—Nebraska—Colfax Co.—Maps. 10. Real property—Nebraska—Stanton Co.—Maps. I. Title. II. Title: Atlas of Stanton County, Nebraska.
G1453.C7T7 1976 912'.782'532 78-324601
78 MARC

COLFAX CO., NEB. (Continued)

—HISTORY

Tri-Tabula, inc.
Atlas of Colfax & Stanton Counties, Nebraska / compiled and published by Tri-Tabula, Inc. — 1776, Bi-Centennial ed., 1976. — Lake Elmo, Minn. : Tri-Tabula, 1976.
ca. 250 p. : ill., maps, ports. ; 33 cm.
Cover title.
Cadastral maps of townships. scale ca. 1:45,000.
"Containing maps, plats of the townships, alphabetical rural directory, histories of churches, early settlers, pictures of families and general interest."
Includes index and directories.
$47.50
1. Colfax Co. Neb.—Maps. 2. Stanton Co., Neb.—Maps. 3. Colfax Co., Neb.—Biographies. 4. Stanton Co., Neb.—Biographies. 5. Colfax Co., Neb. —Directories. 6. Stanton Co., Neb.—Directories. 7. Colfax Co., Neb.—History. 8. Stanton Co., Neb.—History. 9. Real property—Nebraska—Colfax Co.—Maps. 10. Real property—Nebraska—Stanton Co.—Maps. I. Title. II. Title: Atlas of Stanton County, Nebraska.
G1453.C7T7 1976 912'.782'532 78-324601
78 MARC

—MAPS

Tri-Tabula, inc.
Atlas of Colfax & Stanton Counties, Nebraska / compiled and published by Tri-Tabula, Inc. — 1776, Bi-Centennial ed., 1976. — Lake Elmo, Minn. : Tri-Tabula, 1976.
ca. 250 p. : ill., maps, ports. ; 33 cm.
Cover title.
Cadastral maps of townships. scale ca. 1:45,000.
"Containing maps, plats of the townships, alphabetical rural directory, histories of churches, early settlers, pictures of families and general interest."
Includes index and directories.
$47.50
1. Colfax Co. Neb.—Maps. 2. Stanton Co., Neb.—Maps. 3. Colfax Co., Neb.—Biographies. 4. Stanton Co., Neb.—Biographies. 5. Colfax Co., Neb. —Directories. 6. Stanton Co., Neb.—Directories. 7. Colfax Co., Neb.—History. 8. Stanton Co., Neb.—History. 9. Real property—Nebraska—Colfax Co.—Maps. 10. Real property—Nebraska—Stanton Co.—Maps. I. Title. II. Title: Atlas of Stanton County, Nebraska.
G1453.C7T7 1976 912'.782'532 78-324601
78 MARC

COLIAS

Ferris, Clifford D
Notes on certain species of Colias (Lepidoptera: Pieridae) found in Wyoming and associated regions ⌊by⌋ Clifford D. Ferris. Sarasota, Fla., Allyn Museum of Entomology, 1972.
23 p. illus. 23 cm. (Bulletin of the Allyn Museum, no. 5)
Caption title.
"Published with the approval of the director, Wyoming Agricultural Experiment Station, as Journal paper no. 467."
Bibliography: p. 11-12.
1. Colias. 2. Insects—Classification. 3. Insects—Wyoming. I. Title. II. Series: Allyn Museum of Entomology. Bulletin no. 5.
QL561.P5F47 595.7'89 78-306251
78 MARC

COLIBACILLOSIS see Escherichia coli infections

COLIC MYXONEUROSIS see Irritable colon

COLIC TENIA see Taenia coli

COLIMA, MEXICO (STATE)

—DESCRIPTION AND TRAVEL—GUIDE-BOOKS

Oseguera Velázquez, Juan, 1919-
Colima, a guide to fiesta country / Juan Oseguera Velázquez ; translated by Richard H. Hancock. — ⌊Norman⌋ : International Training Programs, University of Oklahoma, c1978.
vi, 91 p., ⌊1⌋ leaf of plates : ill. ; 26 cm.
1. Colima, Mexico (State)—Description and travel—Guide-books. I. Title.
F1271.O83 917.2'3 79-622080
79 MARC

COLINUS VIRGIANIANUS see Bobwhite

COLIOMORPHAE see Corvidae

COLITIS

see also Ulcerative colitis

COLLA, RIENZO

Mazzolari, Primo.
Lettere a un amico / Primo Mazzolari ; ⌊a cura di R. Colla⌋. — Vicenza : La locusta, 1976.
189 p. ; 18 cm. It78-June
L3000
1. Mazzolari, Primo. 2. Catholic Church—Clergy—Correspondence. 3. Clergy—Italy—Correspondence. 4. Colla, Rienzo. I. Colla, Rienzo. II. Title.
BX4705.M472A4 1976 79-386281
*79 MARC

COLLABORATION IN LITERATURE see Authorship—Collaboration

COLLABORATIONISTS see Treason; and subdivision Collaborationists under names of wars and military occupations, e. g. World War, 1939-1945—Collaborationists

COLLABORATIVE PERINATAL PROJECT

Friedman, Emanuel A
Pregnancy hypertension : a systematic evaluation of clinical diagnostic criteria / Emanuel A. Friedman, Raymond K. Neff. — Littleton, Mass. : PSG Pub. Co., 1977.
ix, 258 p. : ill. ; 26 cm.
Based on data collected by the Perinatal Research Branch of the National Institutes of Neurological and Communicative Disorders and Stroke.
Includes bibliographical references and index.
ISBN 0-88416-185-4 : $40.00
1. Hypertension in pregnancy—Diagnosis. 2. Toxemia of pregnancy—Diagnosis. 3. Collaborative Perinatal Project. ⌊1. Pregnancy toxemias. 2. Pregnancy complications, Cardiovascular. 3. Hypertension. 4. Child Development⌋ I. Neff, Raymond K., joint author. II. National Institute of Neurological and Communicative Disorders and Stroke. Perinatal Research Branch. III. Title.
⌊DNLM: WA215 F911p⌋
RG580.H9F74 618.3 76-45951
79 MARC

Hardy, Janet B
The first year of life / the Collaborative Perinatal Project of the National Institute of Neurological and Communicative Disorders and Stroke ; Janet B. Hardy, Joseph S. Drage, Esther C. Jackson. — Baltimore : Johns Hopkins University Press, 1979.
xx, 336 p. : ill. ; 26 cm.
Includes bibliographical references and index.
ISBN 0-8018-2167-3
1. Infants (Newborn)—United States—Longitudinal studies. 2. Infants (Newborn)—Diseases—United States—Longitudinal studies. 3. Infants—Diseases—United States—Longitudinal studies. 4. Pediatric neurology. 5. Collaborative Perinatal Project. I. Drage, Joseph S., joint author. II. Jackson, Esther C., joint author. III. National Institute of Neurological and Communicative Disorders and Stroke. IV. Title.
⌊DNLM: 1. Nervous system diseases—Etiology. 2. Nervous system diseases—In infancy and childhood. WS340.3 H269f⌋
RJ60.U5H37 618.9'28'0471 78-20528
78 MARC

Heinonen, Olli P
Birth defects and drugs in pregnancy / Olli P. Heinonen, Dennis Slone, Samuel Shapiro, with Leonard F. Gaetano ... ⌊et al.⌋ ; David W. Kaufman, editor. — Littleton, Mass. : Publishing Sciences Group, 1977.
xi, 516 p. ; 26 cm.
Includes bibliographical references and indexes.
ISBN 0-88416-034-3 : $75.00
1. Fetus, Effect of drugs on the. 2. Abnormalities, Human—Etiology. 3. Teratogenic agents. 4. Collaborative Perinatal Project. I. Slone, Dennis, joint author. II. Shapiro, Samuel, 1931- joint author. III. Title.
RG627.6.D79H44 618.3'2 75-16094
76 MARC

COLLAGE

see also Fabric pictures

Allen, Sadie.
Creative embroidery collage / Sadie Allen. — London : Bell, 1977.
117 p., ⌊2⌋ leaves of plates : ill., diags. ; 26 cm. GB***
Bibliography: p. 116.
Includes index.
ISBN 0-7135-0167-7 : £5.95
1. Embroidery. 2. Collage. 3. Fabric pictures. I. Title.
TT770.A44 746.4'4 78-319840
78 MARC

French, Brian.
Principles of collage / Brian French. — Buchanan, N.Y. : Emerson Books, 1978.
118 p., ⌊2⌋ leaves of plates : ill. ; 21 cm.
Bibliography: p. 117-118.
ISBN 0-87523-188-8
1. Collage. I. Title.
N7433.7.F743 1978 702'.8 78-67955
79 MARC

Harter's picture archive for collage and illustration : over 300 19th-century cuts / edited by Jim Harter. — New York : Dover Publications, 1978.
⌊4⌋, 90 leaves of plates : chiefly ill. ; 30 cm. — (Dover pictorial archive series)
ISBN 0-486-23659-5 : $4.50
1. Collage. 2. Pictures. I. Harter, Jim. II. Title: Picture archive for collage and illustration.
TT910.H36 769'.90'4 78-54868
78 MARC

Llinàs, Francesc, 1945-
El cadáver del tiempo : (el collage como transmisión narrativa/ideológica) / Francesc Llinàs, Javier Maqua ; introducción de Julio Pérez Perucha. — Valencia : F. Torres, c1976.
xxv, 142 p. : ill. ; 18 cm. — (Textos cinematográficos ; 15) Sp***
ISBN 8473660676
1. Moving-picture plays—History and criticism. 2. Moving-pictures—Plots, themes, etc. 3. Collage. I. Maqua, Javier, 1954- joint author. II. Collage como transmisión narrativa/ideológica. III. Title.
PN1995.L548 809'.2'3 79-364153
79 MARC

Satō, Kōseki, 1923-
(Washie no susume)
和紙絵のすすめ / 佐藤香石. — 東京 : 日貿出版社, 1978-
v. : ill. (some col.) ; 27 cm. Ja 78-23478 (v. 1)
CONTENTS: ⌊1⌋ 入門篇
¥3500 (v. 1)

1. Collage. 2. Satō, Kōseki, 1923- I. Title.
N7433.7.S27 79-800051

—EXHIBITIONS

Zeichnungen und Collagen des Kubismus : didaktischer Ausstellungsteil : Picasso, Braque, Gris : Kunsthalle Bielefeld, 11. März-29. April 1979 / ⌊Konzeption, Ulrich Weisner und Klaus Kowalski in Zusammenarbeit mit Erich Franz und Rüdiger Jörn⌋. — Bielefeld : Kunsthalle, c1979.
58 p. : ill. ; 29 cm. GFR***
1. Cubism—Exhibitions. 2. Drawing, Modern—20th century—Exhibitions. 3. Collages—Exhibitions. 4. Picasso, Pablo, 1881-1976—Exhibitions. 5. Braque, Georges, 1882-1963—Exhibitions. 6. Gris, Juan, 1887-1927—Exhibitions. I. Picasso, Pablo, 1881-1976. II. Braque, Georges, 1882-1963. III. Gris, Juan, 1887-1927. IV. Weisner, Ulrich. V. Kowalski, Klaus. VI. Richard Kaselowsky Haus, Kunsthalle der Stadt Bielefeld.
N6494.C8Z43 79-373196
79 MARC

Zeichnungen und Collagen des Kubismus : Picasso, Braque, Gris : ⌊Ausstellung⌋, Kunsthalle Bielefeld, 11. März bis 29. April 1979 : ⌊Katalog / Ulrich Weisner⌋. — Bielefeld : Kunsthalle, c1979.
378 p. : ill. ; 29 cm. GFR***
Bibliography: p. 373-378.
1. Cubism—Exhibitions. 2. Drawing, Modern—20th century—Exhibitions. 3. Collages—Exhibitions. 4. Picasso, Pablo, 1881-1976—Exhibitions. 5. Braque, Georges, 1882-1963—Exhibitions. 6. Gris, Juan, 1887-1927—Exhibitions. I. Picasso, Pablo, 1881-1976. II. Braque, Georges, 1882-1963. III. Gris, Juan, 1887-1927. IV. Weisner, Ulrich. V. Richard Kaselowsky Haus, Kunsthalle der Stadt Bielefeld.
N6494.C8Z44 709'.04 79-379617
79 MARC

—TECHNIQUE

Liebman, Oscar, 1918-
Collage fundamentals : two- and three-dimensional techniques for illustration and advertising / by Oscar Liebman. — New York : Stravon Educational Press, c1979.
108 p. : ill. (some col.) ; 28 cm.
Includes index.
ISBN 0-87396-079-3 : $10.95
1. Collage—Technique. I. Title.
N7433.7.L53 702'.8 79-262
79 MARC

COLLAGE, AMERICAN

—EXHIBITIONS

Montclair, N.J. Art Museum.
Collage, American masters : March 25-June 24, 1979, Montclair Art Museum, Montclair, New Jersey. — Montclair : The Museum, ⌊c1979⌋
⌊12 p.⌋ : ill. ; 18 x 26 cm.
1. Collage, American—Exhibitions. 2. Art, Modern—20th century—United States—Exhibitions. I. Title.
N6512.5.C55M66 1979 751.7'074'074931 79-87447
79 MARC

COLLAGEN

Banga, Ilona.
Structure and function of elastin and collagen / by Ilona Banga ; /[ms. rev. by Gy. Romhányi and M. Németh-Csóka ; translated by E. Farkas]. — Budapest : Akadémiai Kiadó, 1966.
xii, 224 p. : ill. (some col.) ; 25 cm.
Bibliography: p. 207-217.
Includes index.
1. Elastin. 2. Collagen. 3. Biochemorphology. I. Title.
QP552.E4B36 591.1'16 79-314381
00 MARC

COLLAGEN DISEASES

see also Infectious arthritis; Rheumatism; Rheumatoid arthritis

—IMMUNOLOGICAL ASPECTS

Koffler, David.
The immunology of rheumatoid diseases / David Koffler ; [illustrated by Frank H. Netter]. — [Summit, N.J. : CIBA Pharmaceutical Co.], 1979.
36 p. : ill. ; 25 cm.
Cover title.
Article which comprised v. 31, no. 4 of Clincial symposia.
Bibliography: p. 36.
1. Collagen diseases—Immunological aspects. I. Title. II. Series: Clinical symposia ; v. 31, no. 4.
RC924.K62 616.7'2 79-120003
79 MARC

COLLAGENOSIS see Collagen diseases

COLLAPSE OF BUILDINGS see Building failures

COLLART, JEAN

Varron, grammaire antique et stylistique latine : par-pour Jean Collart, ... — Paris : les Belles lettres, 1978.
xxvi, 415 p., [3] leaves of plates ; 24 cm. — (Publications de la Sorbonne : Série Études ; t. 14) F78-18422
French, Latin, English, or Italian.
"Bibliographie de Jean Collart": p. [xiii]-xxi.
Includes indexes.
ISBN 2-85944-008-9 : 210F
1. Varro, Marcus Terentius. De lingua Latina. 2. Latin language—Grammar—To 1500—Addresses, essays, lectures. 3. Collart, Jean. I. Collart, Jean. II. Series.
PA6792.A3V3 475 79-368182
79 MARC

COLLATERAL LOANS see Pawnbroking

COLLECTIBLES see Toilet paraphernalia; and subdivision Collectibles under names of events, e. g. United States—History—Civil War, 1861-1865—Collectibles; United States—Centennial celebrations, etc.—Collectibles; and headings for kinds of paraphernalia or specific collectibles, e. g. Coffee making paraphernalia; Furniture; and subdivision Museums, relics

etc. under names of persons or families, e. g. Lincoln, Abraham, Pres. United States, 1809-1865—Museum, relics, etc.

COLLECTIF D'ART SOCIOLOGIQUE

—ADDRESSES, ESSAYS, LECTURES

Art sociologique : dossier / [réuni et rédigé par] Fred Forest. — [Paris] : Union générale d'éditions, c1977.
441 p. ; 18 cm. — (10/18 [i.e. Dix-dix-huit] ; 1188 : Série Cause commune) F***
Bibliography: p. 435-[436]
ISBN 2-264-00811-3 : 17.00F
1. Collectif d'art sociologique—Addresses, essays, lectures. 2. Arts and society—Addresses, essays, lectures. 3. Arts, Modern—20th century—Addresses, essays, lectures. I. Forest, Fred, 1933-
NX180.S6A74 700'.1 79-355006
79 MARC

COLLECTING see Collectors and collecting

COLLECTING OF ACCOUNTS

Butterworth, John.
Debt collection letters in ten languages : with notes on terminology, practice and methods of payment in different countries / John Butterworth. — Farnborough, Eng. : Gower Press, c1978
viii, 21, [252] p. ; 23 x 33 cm. GB***
ISBN 0-566-02084-X : £15.00
1. Debts, External. 2. Collecting of accounts. I. Title.
HG3753.B8 651.7'5 79-303106
79 MARC

Credit Research Foundation.
Customer deductions : evaluation and resolution. — Lake Success, N.Y. : Credit Research Foundation, [c1977]
vii, 22 p. ; 28 cm.
1. Accounts receivable. 2. Collecting of accounts. 3. Credit management. I. Title.
HG3751.C73 1977 658.8'83 78-100240
78[r79]rev MARC

Malcomson, Fleury.
Debt collection by telephone / text by Fleury Malcomson ; cartoons by Kat. — [Barna, Ire. : Glenbrook Publications, 197-?]
[36] p. : ill. ; 22 cm. Ir***
Cover title.
£0.60
1. Collecting of accounts. I. Title.
HG3751.M34 658.8'8 79-304257
79 MARC

Marks, Robert A 1945-
An effective collection system for delinquent consumer accounts / by Robert A. Marks. — New York : Pilot Books, c1979.
39 p. ; 26 cm.
ISBN 0-87576-082-1 : $3.95
1. Collecting of accounts. I. Title.
[HG3751.M37] 658.8'83 79-13231
79 MARC

Muchmore, Pat.
Dental economics' Guide to collections / by Pat Muchmore. — Tulsa, Okla. : Dental economics/PPC Books, c1979.
vii, 216 p. : forms ; 23 cm.
Includes index.
ISBN 0-87814-113-8
1. Dental fees. 2. Collecting of accounts. I. Dental economics. II. Title. III. Title: Guide to collections.
RK58.M75 658.1'5244 79-88075
79 MARC

—DATA PROCESSING

Poole, Lon.
Accounts payable and accounts receivable / by Lon Poole, Mary Borchers. — Berkeley, Calif. : A. Osborne, c1978.
xi, 318 p. : diagrs. ; 28 cm.
ISBN 0-931988-13-6
1. Accounts receivable—Data processing. 2. Collecting of accounts—Data processing. 3. Credit management—Data processing. I. Borchers, Mary, joint author. II. Title.
HF5681.A3P66 657'.72 78-113750
79 MARC

—UNITED STATES

United States. Congress. Senate. Committee on Appropriations. Subcommittee on the Legislative Branch.
Improving Government debt collection operations : hearings before a Subcommittee of the Committee on Appropriations, United States Senate, Ninety-sixth Congress, first session : special hearing, Legislative Branch. — Washington : U.S. Govt. Print. Off., 1979.
ii, 62 p. ; 24 cm.
1. Collecting of accounts—United States. 2. Accounts receivable—United States. 3. Government lending—United States. I. Title.
KF26.A6 1979 353.007'2 79-603092
79 MARC

United States. Congress. Senate. Committee on Finance. Subcommittee on Taxation and Debt Management Generally.
Collection of debt owed Federal Government : hearing before the Subcommittee on Taxation and Debt Management Generally of the Committee on Finance, United States Senate, Ninety-fifth Congress, second session, December 18, 1978. — Washington : U.S. Govt. Print. Off., 1979.
iii, 59 p. ; 24 cm.
1. Collecting of accounts—United States. 2. Accounts receivable—United States. 3. Government lending—United States. I. Title.
KF26.F5695 1978l 336.3'63'0973 79-601287
79 MARC

United States. General Accounting Office.
The government needs to do a better job of collecting amounts owed by the public : report to the Congress / by the Comptroller General of the United States. — [Washington : General Accounting Office, 1978]
v, 53 p. ; 27 cm.
Cover title.
"FGMSD-78-61."
"B-159687."
$1.00
1. Collecting of accounts—United States. 2. Accounts receivable—United States. 3. Finance, Public—United States. I. Title.
HJ274.C64U54 1978 353.007'2 78-603489
79 MARC

COLLECTION BUILDING (LIBRARIES) see Collection development (Libraries)

COLLECTION DEVELOPMENT (LIBRARIES)

see also Acquisitions (Libraries); Book selection; Libraries—Special collections; and subdivision Collection development under types of libraries, e. g, Libraries, University and college—Collection development

American Library Association. Collection Development Committee.
Guidelines for collection development / Collection Development Committee, Resources and Technical Services Division, American Library Association ; David L. Perkins, editor. — Chicago : American Library Association, 1979.
vi, 78 p. ; 23 cm.
Bibliography: p. 65-78.
ISBN 0-8389-3231-2 : $3.75
1. Collection development (Libraries) I. Perkins, David, 1939- II. Title.
Z687.A518 1979 025.2 79-16971
79 MARC

Background readings in building library collections / edited by Phyllis Van Orden and Edith B. Phillips. — 2d ed. — Methuchen, N.J. : Scarecrow Press, 1979.
ix, 417 p. ; 23 cm.
First ed. compiled by M. V. Gaver.
Includes bibliographies and index.
ISBN 0-8108-1200-2
1. Collection development (Libraries) I. Gaver, Mary Virginia, 1906- comp. Background readings in building library collections. II. Van Orden, Phyllis. III. Phillips, Edith.
Z687.B33 1979 025.2'1 78-31263
78 MARC

Bonk, Wallace John, 1923-
Building library collections / by Wallace John Bonk and Rose Mary Magrill. — 5th ed. — Metuchen, N.J. : Scarecrow Press, 1979.
viii, 380 p. ; 22 cm.
A revision of the work by M. D. Carter and others.
Includes bibliographies and index.
ISBN 0-8108-1214-2
1. Collection development (Libraries) I. Magrill, Rose Mary, joint author. II. Carter, Mary Duncan, 1896- Building library collections. III. Title.
Z687.B66 1979 025.2'1 79-11151
79 MARC

Evans, G Edward, 1937-
Developing library collections / G. Edward Evans. — Littleton, Colo. : Libraries Unlimited, 1979.
340 p. ; 24 cm. — (Library science text series)
Includes bibliographies and index.
ISBN 0-87287-145-2 : $15.00
1. Collection development (Libraries) I. Title.
Z687.E92 025.2 78-27303
78 MARC

United States. National Agricultural Library.
Library collection development policy. — Beltsville, Md. : National Agricultural Library, 1977.
44 p. ; 26 cm.
Includes index.
1. United States. National Agricultural Library. 2. Collection development (Libraries) I. Title.
Z733.U65143 1977 027.5752'51 78-602380
78 MARC

—GREAT BRITAIN

Apted, Shelia M
Information materials selection : policy statement in libraries of England and Wales / [by] Shelia Apted. — [Birmingham] : City of Birmingham Polytechnic, Department of Librarianship, 1978.
[1], 12 leaves ; 30 cm. GB79-01674
Cover title.
Bibliography: leaf 12.
1. Book selection—Policy statements. 2. Collection development (Libraries)—Great Britain. I. Title.
Z689.A7 025.2'1'0942 79-318285
79 MARC

COLLECTION LAWS

—CALIFORNIA

Chuan, Victor L
California debt collection manual / edited by Victor L. Chuan. — Berkeley : California Continuing Education of the Bar, c1978.
xii, 858 p., [1] leaf of plates ; 25 cm.
Based on California debt collection practice, by E. N. Jackson.
Bibliography: p. 787-788.
Includes indexes.
1. Collection laws—California. I. Jackson, Edward N. California debt collection practice. II. Title.
KFC256.C6C48 346'.794'077 77-91934
79 MARC

—CANADA—POPULAR WORKS

Paulsen, Timothy R 1949-
Collecting debts successfully : from customers, clients, and friends / Timothy R. Paulsen. — Vancouver ; Seattle : International Self-Counsel Press, 1978.
viii, 83 p. : ill. ; 21 cm. — (Self-counsel series) C78-2150-9
ISBN 0-88908-045-3 : $3.50
1. Debtor and creditor—Canada—Popular works. 2. Collection laws—Canada—Popular works. I. Title.
KE1030.Z82P38 346'.71'077 79-305428
79 MARC

—GERMANY, WEST

Rutkowsky, Heinz.
So kommen ich zu meinem Geld : Mahnverfahren, Prozessführung, Zwangsvollstreckung / Heinz Rutkowsky, Max Repschläger. — Orig.-Ausg. — München : Zürich : Droemer-Knaur, 1978.
142 p. ; 18 cm. GFR78-A13
Includes index.
ISBN (invalid) 3-426-00511-6 : DM4.80
1. Executions (Law)—Germany, West. 2. Collection laws—Germany, West. 3. Procedure (Law)—Germany, West. I. Repschläger, Max, joint author. II. Title.
79-373667
79 MARC

—IRELAND

Osman, F C
The recovery of advances / by F. C. Osman. — [Dublin] : Institute of Bankers in Ireland, 1976.
78 p. ; 22 cm. — (Irish banking practice ; 5) GB78-31250
Includes index.
£0.35
1. Loans—Ireland. 2. Collection laws—Ireland. 3. Security (Law)—Ireland. I. Title. II. Series.
KDK412.O8 332.1'753 79-301450
79 MARC

—JAPAN

(Kashidaorekin no hōritsu chishiki)
貸倒金の法律知識：取立の手続と書式 / 河部宗一 ... [et al.] 共編. — 大阪：月刊経理編輯部, 昭和25 [1950]
9, 283 p. ; 19 cm.

1. Collection laws—Japan. 2. Collection laws—Japan—Forms. 3. Obligations (Law)—Japan. 4. Executions (Law)—Japan. I. Kawabe, Sōichi.
79–805847

—JAPAN—FORMS

(Kashidaorekin no hōritsu chishiki)
貸倒金の法律知識：取立の手続と書式 / 河部宗一 ... [et al.] 共編. — 大阪：月刊経理編輯部, 昭和25 [1950]
9, 283 p. ; 19 cm.

1. Collection laws—Japan. 2. Collection laws—Japan—Forms. 3. Obligations (Law)—Japan. 4. Executions (Law)—Japan. I. Kawabe, Sōichi.
79–805847

—NORTH CAROLINA

Schiro, Lilona S
Collection of accounts : the law in North Carolina / by Lilona S. Schiro. — Norcross, Ga. : Harrison Co., c1979.
ix, 109 p. : forms ; 24 cm.
Includes indexes.
1. Collection laws—North Carolina. I. Title.
KFN7567.C6S34 346'.756'077 79-115219
79 MARC

—WASHINGTON (STATE)

Armstrong, Seth.
Credit law and bankruptcy handbook for Washington / Seth Armstrong. — Vancouver, B.C. : Self-Counsel Press ; Seattle : distributed by Self-Counsel Press, 1977.
ix, 148 p. : forms ; 22 cm. — (Self-counsel series) C***
Includes bibliographical references.
ISBN 0-88908-703-2 : $3.95
1. Consumer credit—Law and legislation—Washington (State) 2. Collection laws—Washington (State) 3. Bankruptcy—United States. I. Title.
KFW171.A7 346'.797'077 78-314276
78 MARC

COLLECTION LETTERS see Collecting of accounts

COLLECTIONS BAUR

—CATALOGS

Collections Baur.
Chinese jades and other hardstones / by Pierre-F. Schneeberger ; (translated by Katherine Watson). — Genève : Collections Baur : Boston : distributed by Routledge & K. Paul, c1976.
52 p., [90] p., [38] leaves of plates : ill. (some col.) ; 30 cm. Sw***
Half title: The Baur Collection.
At head of title: Baur Collection, Geneva.
English and French.
Bibliography: p. [84]-[85]
1. Jade art objects—China—Catalogs. 2. Collections Baur—Catalogs. I. Schneeberger, Pierre Francis. II. Title.
NK5750.C64 1976 77-363000
736'.24'0951074094945
77[r79]rev MARC

COLLECTIVE AGREEMENTS see Collective labor agreements

COLLECTIVE BARGAINING

see also Arbitration, Industrial; Collective labor agreements; Employees' representation in management; Labor contract; Labor-management committees; Mediation and conciliation, Industrial; Productivity bargaining; Strikes and lockouts; Trade-unions; Trade-unions —Recognition

Carpenter, Walter Huil.
Small business and pattern bargaining / Walter H. Carpenter, Jr., Edward Handler. — New York : Arno Press, 1979.
xi, 243 p. : form ; 24 cm. — (Small business enterprise in America) (Reprint of the 1961 ed. published by Babson Institute Press, Babson Park, Mass., in series: Small business management research reports)
Bibliography: p. 221-224.
ISBN 0-405-11461-3 : $16.00
1. Collective bargaining. 2. Small business. I. Handler, Edward, joint author. II. Title. III. Series. IV. Series: Small business management research reports.
[HD6483.C29 1979] 331.89 78-18953
78 MARC

Cousineau, Jean Michel.
La détermination des salaires dans le monde des grandes conventions collectives : une analyse des secteurs privé et public / Jean-Michel Cousineau, Robert Lacroix, avec la collaboration de Paul Robillard. — [Ottawa] : Conseil économique du Canada, c1977.
xiv, 150 p. ; 25 cm. C***
Issued also under title: Wage determination in major collective agreements in the private and public sectors.
Bibliography: p. [145]-150.
"No de catalogue, EC22-55/1977F."
ISBN 0-660-01615-X : $4.00
1. Wages. 2. Collective bargaining. I. Lacroix, Robert, joint author. II. Robillard, Paul, joint author. III. Title.
HD4906.C6814 331.2'15 78-393083
79 MARC

Louche, Claude.
Pouvoir et délégation dans les négociations professionnelles / par Claude Louche. — Paris : Éditions du Centre national de la recherche scientifique, 1977.
107 p. : ill. ; 25 cm. — (Monographies françaises de psychologie ; no 40) F***
Bibliography: p. [105]-107.
ISBN 2-222-02149-9 : 38.00F
1. Collective bargaining. 2. Collective bargaining—France. 3. Negotiation. 4. Power (Social sciences) I. Title. II. Series.
HD6483.L68 301.6'34 79-353049
79 MARC

(Rōdō kumiai soshiki ron)
労働組合組織論：各国の労働組合組織と団体交渉 / 労働省労政局労働教育課編. — 東京：労働法令協会, 昭和33 [1958]
6, 259 p. ; 18 cm.
Includes bibliographical references.

1. Trade-unions. 2. Collective bargaining. I. Japan. Rōdōshō. Rōseikyoku. Rōdō Kyōikuka.
HD6483.R623 79–802792

Stevens, Carl M
Strategy and collective bargaining negotiation / Carl M. Stevens. — Westport, Conn. : Greenwood Press, 1978, c1963.
xiii, 192 p. ; 24 cm.
Reprint of the ed. published by McGraw Hill, New York, which was issued in the Publications of the Wertheim Committee.
Bibliography: p. [161]-165.
Includes index.
ISBN 0-313-20377-6
1. Collective bargaining. I. Title. II. Series: Wertheim Committee. Publications.
[HD6483.S68 1978] 331.89 78-5304
78 MARC

—ADDRESSES, ESSAYS, LECTURES

Collective bargaining : new dimensions in labor relations / edited by Franklin J. Havelick ; foreword by Theodore W. Kheel. — Boulder, Colo. : Westview Press, 1979.
xiv, 183 p. : graphs ; 24 cm. — (A Westview special study)
"A Giniger project."
Includes index.
ISBN 0-89158-386-6 : $16.00
1. Collective bargaining—United States—Addresses, essays, lectures. 2. Collective bargaining—Addresses, essays, lectures. I. Havelick, Franklin J.
HD6508.C657 331.89'0973 79-4234
79 MARC

—AGRICULTURE—CALIFORNIA

California. Legislature. Joint Committee to Oversee the Agricultural Labor Relations Board.
Initial report to the legislature on the administration of the Agricultural labor relations act of 1975 / prepared by the Joint Committee to Oversee the Agricultural Labor Relations Board. — Sacramento : The Committee, [letter 1978]
ii, 130 p. ; 28 cm.
$4.20
1. California. Agricultural Labor Relations Board. 2. Arbitration, Industrial—California. 3. Collective bargaining—Agriculture—California. I. Title.
KFC10.62.A35 1978 353.9'794'008233 78-622348
78 MARC

California. Legislature. Joint Committee to Oversee the Agricultural Labor Relations Board.
Review of the administration of the Agricultural labor relations act of 1975, State Capitol, Sacramento, March 4, 1977 / California Legislature, Joint Committee to Oversee the Agricultural Labor Relations Board. — Sacramento, Calif. : The Committee : [may be purchased from the Assembly Publications Office, 1977]
ii, 131 p. ; 28 cm.
$4.00
1. California. Agricultural Labor Relations Board. 2. Arbitration, Industrial—California. 3. Collective bargaining—Agriculture—California. I. Title: Review of the administration of the Agricultural labor relations act of 1975 ...
KFC10.A33 1977a 353.9'794'0083 77-624015
78[r79]rev MARC

—ALUMINUM INDUSTRY—NORWAY—ÅRDAL (SOGN OG FJORDANE)—CASE, STUDIES

Gulbrandsen, Ørnulf.
Samarbeidsideologi og interessekonflikt : analyse av forholdet mellom bedrift og fagorganisasjon / av Ørnulf Gulbrandsen ; [utg. av] Instituttet for sosiologi, Universitetet i Oslo og Norsk institutt for by- og regionforskning. — Oslo : Norsk institutt for by- og regionforskning, 1976.
iii, 311 p. : ill. ; 21 cm. — (Rapport - Årdalsprosjektet ; nr. 7) N76-Sept.
Bibliography: p. 307-311.
kr30.00
1. Industrial relations—Norway—Årdal (Sogn og Fjordane)—Case studies. 2. Collective bargaining—Aluminum industry—Norway—Årdal (Sogn og Fjordane)—Case, studies. I. Title. II. Series: Årdalsprosjektet. Rapport - Årdalsprosjektet ; nr. 7.
HN570.A27A27 1975 nr. 7 78-392371
[HD8569.A17] *79 MARC

COLLECTIVE BARGAINING (Continued)

—CIVIL SERVICE see Collective bargaining—Government employees

—COLLEGE TEACHERS—UNITED STATES

Carnegie Council on Policy Studies in Higher Education.
Faculty bargaining in public higher education : a report and two essays : a report of the Carnegie Council on Policy Studies in Higher Education / essays by Joseph W. Garbarino, David E. Feller & Matthew W. Finkin. — 1st ed. — San Fancisco : Jossey-Bass, 1977.
viii, 191 p. ; 24 cm. — (The Carnegie Council series)
"State experience in collective bargaining [by] Joseph W. Garbarino": p. [29]-72.
"Legislative issues in faculty collective bargaining [by] David E. Feller, Matthew W. Finkin": p. [73]-183.
Includes bibliographical references and index.
ISBN 0-87589-322-8
1. Collective bargaining—College teachers—United States. 2. State universities and colleges—United States. I. Garbarino, Joseph William, 1919- State experience in collective bargaining. 1977. II. Feller, David E. Legislative issues in faculty collective bargaining. 1977. III. Title.
LB2335.885.U6C37 1977 76-50728
331.89'041'378120973
00 MARC

Garfin, Molly.
Directory of faculty contracts and bargaining agents in institutions [sic] of higher education / [compiled by Molly Garfin]. — New York : National Center for the Study of Collective Bargaining in Higher Education, Baruch College, CUNY, 1979.
iii, 47 p. ; 28 cm.
$5.00
1. Collection bargaining—College teachers—United States. 2. College teachers' unions—United States—Directories. 3. Collective labor agreements—Education, Higher—United States—Directories.
LB2335.885.U6G37 79-113060
331.89'041'378120973
79 MARC

Hankin, Joseph N
Negotiating a better future : planning and organizing for collective negotiations in community and junior colleges / Joseph N. Hankin. — Washington : American Association of Community and Junior Colleges, c1977.
i, 34 p. ; 28 cm.
Bibliography: p. 34.
$5.00
1. Collective bargaining—College teachers—United States. 2. Community colleges—United States. 3. Junior colleges—United States. I. American Association of Community and Junior Colleges. II. Title.
LB2335.885.U6H36 77-154474
331.89'041'37815430973
00 MARC

—CONGRESSES

International Labor Office.
Collective bargaining in industrialised countries : recent trends and problems : summary of discussions of a symposium on collective bargaining in industrialised countries, Vienna, 2-9 November 1977. — Geneva : International Labor Office, 1978.
ii, 113 p. ; 24 cm. — (Labour-management relations series ; 56) Sw
Includes bibliographical references.
ISBN 9221019438
1. Collective bargaining—Congresses. I. Title. II. Series: International Labor Office. Labour-management relations series ; 56.
HD6961.A1I73 no. 56 331 s 79-311322
[HD6483] 79 MARC

—CONSTRUCTION INDUSTRY—BRITISH COLUMBIA

British Columbia. Special Commission of Inquiry into British Columbia Construction.
First report / Special Commission of Inquiry into British Columbia Construction. — [s.l. : s.n.], 1975.
87 p. : graphs ; 28 cm. C***
Cover title.
"James Kinnaird, Commissioner."
Includes bibliographical references.
1. Collective bargaining—Construction industry—British Columbia. 2. Employers' associations—British Columbia. I. Kinnaird, James.
HD6528.B9B74 1975 331'.042'409711 79-306956
79 MARC

—CONSTRUCTION INDUSTRY—ONTARIO

Ontario. Industrial Inquiry Commission into Bargaining Patterns in the Construction Industry.
Report of the Industrial Inquiry Commission into Bargaining Patterns in the Construction Industry. — [Toronto] : Ontario Ministry of Labour, 1976.
ii, 185 p. ; 30 cm. C76-83527-6
1. Collective bargaining—Construction industry—Ontario. I. Title.
HD6976.B92C36 331.89'042'409713 79-309082
79 MARC

—CORRECTIONAL PERSONNEL—NEW YORK (STATE)

Jacobs, James B
Guard unions and the future of the prisons / by James B. Jacobs, Norma Meacham Crotty. — Ithaca : Institute of Public Employment, New York State School of Industrial and Labor Relations, 1978.
xi, 48 p. ; 26 cm. — (IPE monograph ; no. 9)
Includes bibliographical references.
$3.50
1. Prisons—United States—Officials and employees. 2. Collective bargaining—Correctional personnel—New York (State) 3. Trade-unions—Correctional personnel—United States. I. Crotty, Norma Meacham, joint author. II. Title. III. Series: Cornell University. New York State School of Industrial and Labor Relations, Ithaca. Institute of Public Employment. IPE monograph ; no. 9.
HV9471.J3 331.88'11'365973 78-62007
79 MARC

—CORRECTIONS—UNITED STATES

Montilla, M Robert.
Prison employee unionism : management guide for correctional administrators / by M. Robert Montilla. — [Washington] : National Institute of Law Enforcement and Criminal Justice, Law Enforcement Assistance Administration, U.S. Dept. of Justice : for sale by the Supt. of Docs., U.S. Govt. Print. Off., 1978.
xii, 469, [26] p. : ill. ; 26 cm.
Includes bibliographical references.
1. Trade-unions—Correctional personnel—United States. 2. Collective bargaining—Corrections—United States. 3. Prison administration—United States. I. National Institute of Law Enforcement and Criminal Justice. II. Title.
HV9471.M66 331.88'11'365973 78-602534
79 MARC

Wynne, John M
Prison employee unionism : the impact on correctional administration and programs / by John M. Wynne, Jr. — [Washington] : National Institute of Law Enforcement and Criminal Justice, Law Enforcement Assistance Administration, U.S. Dept. of Justice : for sale by the Supt. of Docs., U.S. Govt. Print. Off., 1978.
x, 238 p. ; 26 cm.
Bibliography: p. 235-238.
1. Trade-unions—Correctional personnel—United States. 2. Corrections—United States. 3. Prison administration—United States. 4. Collective bargaining—Corrections—United States. I. National Institute of Law Enforcement and Criminal Justice. II. Title.
HV9471.W96 331.88'11'365973 78-601503
79 MARC

—EDUCATION—CANADA

Blount, Gail.
Collective bargaining in Canadian education : an annotated bibliography / compiled by Gail Blount. — Toronto : Ontario Institute for Studies in Education, c1975.
x, 38 p. ; 28 cm. — (OISE bibliography series ; no. 1) C75-10220-9
Includes index.
ISBN 0-7744-0114-1
1. Collective bargaining—Education—Canada. I. Title. II. Series: Ontario Institute for Studies in Education. OISE bibliography series ; no. 1.
Z5815.C3B56 79-313634
016.33189'041'371100971
[LB2844.57.C2] 79 MARC

—EDUCATION—UNITED STATES

Newby, Kenneth A
Collective bargaining : practices and attitudes of school management / Kenneth A. Newby. — [Washington] : National School Boards Association, [c1977]
38 p. ; 28 cm. — (Research report - National School Boards Association ; 1977-2)
Includes bibliographical references.
$3.50
1. Collective bargaining—Education—United States. I. Title. II. Series: National School Boards Association. Research report - National School Boards Association ; 1977-2.
LB2844.57.U6N48 77-152256
331.89'04137100973
00 MARC

—ELECTRIC UTILITIES—GREAT BRITAIN

Great Britain. Advisory Conciliation and Arbitration Service.
Electricity supply industry : report of an inquiry into the negotiating and representative arrangements for managerial and higher executive grades / [Advisory Conciliation and Arbitration Service]. — London : The Service, [1976]
iv, 26, [1] p. ; 21 cm. — (Report - Advisory Conciliation and Arbitration Service ; no. 4 ISSN 0309-717X) GB77-15732
ISBN 0-906073-03-0
1. Collective bargaining—Electric utilities—Great Britain. 2. Middle managers—Great Britain. 3. Executives—Great Britain. I. Title. II. Series: Great Britain. Advisory Conciliation and Arbitration Service. Report - Advisory Conciliation and Arbitration Service ; no. 4.
HD6976.E32G744 77-378287
331.89'042'1310941
78[r79]rev MARC

—GOVERNMENT EMPLOYEES

see also Collective bargaining—Municipal employees

—GOVERNMENT EMPLOYEES—CONGRESSES

Report of the Special Conference on Trade Union and Collective Bargaining Rights of Public Employees, held at Munich, October 1953. — London : International Federation of Unions of Employees in Public and Civil Services, 1953.
49 p. in various pagings ; 34 cm.
1. Trade-unions—Government employees—Congresses. 2. Collective bargaining—Government employees—Congresses. I. Special Conference on Trade Union and Collective Bargaining Rights of Public Employees, Munich, 1953. II. Public Services International.
HD8005.R46 79-311746
79 MARC

—GOVERNMENT EMPLOYEES—CALIFORNIA—BIBLIOGRAPHY

Shaughnessy, Marlene.
Information sources for California public sector collective bargaining / by Marlene Shaughnessy. — Los Angeles : Institute of Industrial Relations, University of California, Los Angeles, 1978.
140 p. ; 24 cm. — (Policy & practice publications)
Bibliography: p. 77-137.
Includes index.
1. Collective bargaining—Government employees—California—Information services. 2. Collective bargaining—Government employees—California—Bibliography. 3. Industrial relations—Dictionaries. I. California. University. University at Los Angeles. Institute of Industrial Relations. II. Title. III. Series.
HD8005.6.U53C27 79-311070
331.89'041'3500009794
79 MARC

—GOVERNMENT EMPLOYEES—CALIFORNIA—INFORMATION SERVICES

Shaughnessy, Marlene.
Information sources for California public sector collective bargaining / by Marlene Shaughnessy. — Los Angeles : Institute of Industrial Relations, University of California, Los Angeles, 1978.
140 p. ; 24 cm. — (Policy & practice publications)
Bibliography: p. 77-137.
Includes index.
1. Collective bargaining—Government employees—California—Information services. 2. Collective bargaining—Government employees—California—Bibliography. 3. Industrial relations—Dictionaries. I. California. University. University at Los Angeles. Institute of Industrial Relations. II. Title. III. Series.
HD8005.6.U53C27 79-311070
331.89'041'3500009794
79 MARC

—GOVERNMENT EMPLOYEES—CANADA

Public Service Alliance of Canada.
Brief on collective bargaining in the Federal public service / presented to Committee on Legislation Review ; submitted by the Public Service Alliance of Canada. — [Ottawa?] : The Alliance, 1971.
2 v. ; 22-28 x 22-36 cm.
Includes bibliographical references.
Cover title.
1. Collective labor agreements—Government employees—Canada. 2. Collective bargaining—Government employees—Canada. I. Canada. Committee on Legislation Review. II. Title.
KE3196.F4P8 79-312065
344'.71'0189041354
79 MARC

COLLECTIVE BARGAINING (Continued)

—GOVERNMENT EMPLOYEES—CANADA—ADDRESSES, ESSAYS, LECTURES

Public-sector bargaining / editorial board, Benjamin Aaron, Joseph R. Grodin, James L. Stern. — Washington : Bureau of National Affairs, c1979.
vii, 327 p. ; 24 cm. — (Industrial Relations Research Association series)
Includes index.
ISBN 0-87179-291-5
1. Collective bargaining—Government employees—United States—Addresses, essays, lectures. 2. Collective bargaining—Government employees—Canada—Addresses, essays, lectures. I. Aaron, Benjamin. II. Grodin, Joseph R. III. Stern, James L. IV. Bureau of National Affairs, Washington, D.C. V. Series: Industrial Relations Research Association. Industrial Relations Research Association series.
HD8005.6.U5P8 331.89'041'353 78-25655
78 MARC

—GOVERNMENT EMPLOYEES—CANADA—BIBLIOGRAPHY

Tice, Terrence N
Employee relations bibliography : public, non-profit, and professional employment : essay, annotated listing, indexes / submitted to Division of Public Employee Labor Relations, Office of Labor-Management Relations Services, United States Department of Labor, by Terrence N. Tice. — Washington : U.S. Dept. of Labor, Labor-Management Services Administration : for sale by the Supt. of Docs., U.S. Govt. Print. Off., 1978.
iii, 162 p. ; 28 cm.
1. Employee-management relations in government—United States—Bibliography. 2. Employee-management relations in government—Canada—Bibliography. 3. Collective bargaining—Government employees—United States—Bibliography. 4. Collective bargaining—Government employees—Canada—Bibliography. I. United States. Division of Public Employee Labor Relations. II. Title.
Z7164.C6T52 78-602761
016.331'041'350000973
[HD8008] 79 MARC

—GOVERNMENT EMPLOYEES—CONNECTICUT—PERIODICALS

Connecticut. Office of Labor Relations.
Annual report - Connecticut Office of Labor Relations. 1976/77-
[Hartford] Connecticut Office of Labor Relations.
v. 28 cm.
Report year ends June 30.
At head of title, 1976/77- : Connecticut collective bargaining.
1. Collective bargaining—Government employees—Connecticut—Periodicals. I. Connecticut. Office of Labor Relations. Connecticut collective bargaining.
HD8005.6.U53C83 353.9'746'0083 78-645635
MARC-S

—GOVERNMENT EMPLOYEES—GREAT BRITAIN

Thomson, A W J
Public sector bargaining : a study of relative gain / A. W. J. Thomson and P. B. Beaumont. — Farnborough, Eng. : Saxon House, 1978.
viii, 205 p. ; 23 cm. GB***
Includes bibliographical references and index.
ISBN 0-566-00167-5 : £7.50
1. Collective bargaining—Government employees—Great Britain. 2. Employee-management relations in government—Great Britain. 3. Trade-unions—Government employees—Great Britain. I. Beaumont, P. B., joint author. II. Title.
HD8005.6.G7T48 78-322788
331.89'041'350000941
79 MARC

—GOVERNMENT EMPLOYEES—ILLINOIS

Illinois. General Assembly. Senate. Industry and Labor Committee. Subcommittee on Collective Bargaining.
Report on collective bargaining for public employees / the Subcommittee on Collective Bargaining of the Senate Industry and Labor Committee, Seventy-eighth General Assembly, adopted June 26, 1974. — [Springfield] : The Subcommittee, [1974]
24 leaves in various foliations ; 28 cm.
Cover title.
1. Collective bargaining—Government employees—Illinois. I. Title.
HD8011.I4I4 1974 75-621237
344'.773'01890413539
79 MARC

—GOVERNMENT EMPLOYEES—ITALY

Istituto per lo sviluppo della formazione professionale dei lavoratori.
Pubblico impiego : dinamiche contrattuali, processi di cambiamento, bisogni formativi / indagine ISFOL-Fondazione Giulio Pastore su alcune linee di intervento formativo indotte dalla contrattazione collettiva nella pubblica amministrazione. — Roma : Tip. Detti Dario e Ugo, 1978.
105 p. ; 25 cm. — (Its Quaderni di formazione ISFOL ; n. 54) It79-Feb
1. Collective bargaining—Government employees—Italy. 2. Italy—Officials and employees. 3. Occupational training—Italy. I. Fondazione Giulio Pastore. II. Title. III. Series.
HD8005.6.I8I78 1978 331.89'041'35445 79-352265
*79 MARC

—GOVERNMENT EMPLOYEES—UNITED STATES

Bent, Alan Edward.
Collective bargaining in the public sector : labor-management relations and public policy / Alan Edward Bent, T. Zane Reeves. — Menlo Park, Calif. : Benjamin/Cummings Pub. Co., c1978.
xi, 355 p. ; 24 cm.
Includes bibliographical references and index.
ISBN 0-8053-0672-2
1. Trade-unions—Government employees—United States. 2. Collective bargaining—Government employees—United States. I. Reeves, T. Zane, joint author. II. Title.
HD8005.2.U5B45 78-70514
331.89'041'350000973
79 MARC

Levine, Marvin J 1930-
Public sector labor relations / Marvin J. Levine, Eugene C. Hagburg. — St. Paul : West Pub. Co., c1979.
xvi, 245 p. : ill. ; 26 cm.
Bibliography: p. 237-240.
Includes index.
ISBN 0-8299-0184-1 : $15.95
1. Collective bargaining—Government employees—United States. 2. Employee-management relations in government—United States. I. Hagburg, Eugene C., joint author. II. Title.
HD8005.6.U5L478 78-12331
331.89'041'350000973
78 MARC

Public Service Research Council.
Public sector bargaining and strikes / prepared by Public Service Research Council. — 3d ed. — Vienna, Va. : The Council, c1978.
61 p. : graphs ; 28 cm.
1. Strikes and lockouts—Civil service—United States. 2. Collective bargaining—Government employees—United States. I. Title.
HD8004.2.U5P8 1978 331.89'041'353 78-106291
79 MARC

—GOVERNMENT EMPLOYEES—UNITED STATES—ADDRESSES, ESSAYS, LECTURES

Public-sector bargaining / editorial board, Benjamin Aaron, Joseph R. Grodin, James L. Stern. — Washington : Bureau of National Affairs, c1979.
vii, 327 p. ; 24 cm. — (Industrial Relations Research Association series)
Includes index.
ISBN 0-87179-291-5
1. Collective bargaining—Government employees—United States—Addresses, essays, lectures. 2. Collective bargaining—Government employees—Canada—Addresses, essays, lectures. I. Aaron, Benjamin. II. Grodin, Joseph R. III. Stern, James L. IV. Bureau of National Affairs, Washington, D.C. V. Series: Industrial Relations Research Association. Industrial Relations Research Association series.
HD8005.6.U5P8 331.89'041'353 78-25655
78 MARC

—GOVERNMENT EMPLOYEES—UNITED STATES—BIBLIOGRAPHY

Tice, Terrence N
Employee relations bibliography : public, non-profit, and professional employment : essay, annotated listing, indexes / submitted to Division of Public Employee Labor Relations, Office of Labor-Management Relations Services, United States Department of Labor, by Terrence N. Tice. — Washington : U.S. Dept. of Labor, Labor-Management Services Administration : for sale by the Supt. of Docs., U.S. Govt. Print. Off., 1978.
iii, 162 p. ; 28 cm.
1. Employee-management relations in government—United States—Bibliography. 2. Employee-management relations in government—Canada—Bibliography. 3. Collective bargaining—Government employees—United States—Bibliography. 4. Collective bargaining—Government employees—Canada—Bibliography. I. United States. Division of Public Employee Labor Relations. II. Title.
Z7164.C6T52 78-602761
016.331'041'350000973
[HD8008] 79 MARC

—GOVERNMENT EMPLOYEES—UNITED STATES—STATES

Tanimoto, Helene S
Guide to statutory provisions in public sector collective bargaining-impasse resolution procedures / by Helene S. Tanimoto. — 2d issue. — [Honolulu] : Industrial Relations Center, College of Business Administration, University of Hawaii, 1977.
v, 170 p. ; 23 cm. — (Occasional publication - Industrial Relations Center, University of Hawaii ; no. 122)
$4.00
1. Collective bargaining—Government employees—United States—States. I. Title. II. Series: Hawaii. University, Honolulu. Industrial Relations Center. Occasional publication ; no. 122.
KF3409.P77Z9547 1977 78-622700
344'.73'0189041353
79 MARC

—GOVERNMENT EMPLOYEES—UNITED STATES—STATES—CONGRESSES

Public sector decision making : legal fringe benefit and union parameters : fifth and sixth annual conference / contributors, Enid Beaumont ... [et al.] ; editor, Paul A. Weinstein. — [College Park, Md.] : Public Sector Labor Relations Conference Board, c1978.
iv, 106 p. ; 23 cm. — (PSLRCB publication ; no. 3)
ISBN 0-913400-02-5
1. Civil service pensions—United States—States—Congresses. 2. Non-wage payments—United States—States—Congresses. 3. Collective bargaining—Government employees—United States—States—Congresses. I. Beaumont, Enid. II. Weinstein, Paul A. III. Public Sector Labor Relations Conference Board. IV. Series: Public Sector Labor Relations Conference Board. PSLRCB publication ; no. 3.
JK2474.P8 353.001'234 78-70342
79 MARC

—INTERNATIONAL BUSINESS ENTERPRISES—ABSTRACTS

Günter, Hans.
Multinational corporations and labour : a selected and annotated bibliography / by Hans Günter ; bibliography prepared in collaboration with Maryse Gaudier. — Geneva : International Institute for Labour Studies, [1973?]
xii, 37 p. ; 30 cm. — (International educational materials exchange ; IEME 7011) Sw***
"Prepared for the Research Meeting on Multinational Corporations and Labour, organised by the International Institute for Labour Studies, Geneva, 5-7 December 1973."
1. International business enterprises—Abstracts. 2. International business enterprises—Personnel management—Abstracts. 3. Collective bargaining—International business enterprises—Abstracts. I. Günter, Hans. II. Gaudier, Maryse. III. Title. IV. Series.
HD2755.5.G83 016.6583 77-351212
77[r79]rev MARC

—JOURNALISTS—UNITED STATES

Barwis, Gail Lund.
Contractual newsroom democracy / Gail Lund Barwis. — [Lexington, Ky.] : Association for Education in Journalism, 1978.
33 p. ; 23 cm. — (Journalism monographs ; no. 57)
Includes bibliographical references.
$2.50
1. Collective bargaining—Journalists—United States. 2. Collective labor agreements—Journalists—United States. I. Title. II. Series: Journalism monographs (Austin) ; no. 57.
PN4722.J6 no. 57 070'.08 s 78-112032
[KF3409.N3] 79 MARC

—LOCAL OFFICIALS AND EMPLOYEES

see also Collective bargaining—Municipal employees

—LOCAL OFFICIAL AND EMPLOYEES—UNITED STATES

O'Reilly, Robert C 1928-
Understanding collective bargaining in education : negotiations, contracts, and disputes between teachers and boards / by Robert C. O'Reilly. — Metuchen, N.J. : Scarecrow Press, 1978.
xii, 185 p. ; 22 cm.
Includes bibliographical references and indexes.
ISBN 0-8108-1167-7
1. Collective bargaining—Local official and employees—United States. 2. Collective bargaining—Teachers—United States. I. Title.
HD8005.6.U5O73 331.89'041'3711 78-64495
78 MARC

COLLECTIVE BARGAINING (Continued)

—LOCAL OFFICIALS AND EMPLOYEES —UNITED STATES—STATISTICS

United States. Bureau of the Census.
Labor-management relations in State and local governments, 1976. — ⌊Washington⌋ : U.S. Dept. of Commerce, Bureau of the Census : for sale by the Supt. of Docs., U.S. Govt. Print. Off., 1978.
166 p. ; 28 cm. — (State and local government special studies ; no. 88)
Chiefly table.
Supt. of Docs. no.: C 3.145:88
1. Employee-management relations in government—United States—States—Statistics. 2. Collective bargaining—State government employees—United States—Statistics. 3. Collective bargaining—Local officials and employees—United States—Statistics. 4. Strikes and lockouts—State governments—United States—Statistics. 5. Strikes and lockouts—Local officials and employees—United States—Statistics. I. Title. II. Series: United States. Bureau of the Census. State and local government special studies ; no. 88.
JK2403.A35 no. 88 353.9'08 s 78-601971
⌊HD8008.A5⌋ 79 MARC

—MATHEMATICAL MODELS

Güth, Werner.
Zur Theorie kollektiver Lohnverhandlungen / Werner Güth. — 1. Aufl. — Baden-Baden : Nomos-Verlagsgesellschaft, 1978.
178 p. ; 23 cm. GFR78-A
Habilitationsschrift—Münster.
Bibliography: p. 177-178.
ISBN 3-7890-0372-7
1. Collective bargaining—Mathematical models. 2. Wages—Mathematical models. I. Title.
HD6483.G8 78-380458
79 MARC

—MERCHANT MARINE—FRANCE

Segrestin, Denis.
La marine marchande française / Denis Segrestin. — ⌊Genève⌋ : Institut international d'études sociales, c1977.
iv, 86 p. ; 24 cm. — (Série de recherche - Institut international d'études sociales ; no 27) Sw***
At head of title: Facteurs stratégiques des systèmes de relations professionnelles.
Bibliography: p. 85-86.
ISBN 9290142294 : 12.00F
1. Collective bargaining—Merchant marine—France. 2. Industrial relations—France—Case studies. 3. Merchant marine—France. I. Title. II. Title: Facteurs stratégiques des systèmes de relations professionnelles. III. Series: Geneva. International Institute for Labour Studies. Research series - International Institute for Labour Studies ; no. 27.
HD6976.S42F87 79-362843
331.89'041'38750944
79 MARC

—MERCHANT MARINE—GREAT BRITAIN

McConville, J
The shipping industry in the United Kingdom / J. McConville. — Geneva : International Institute for Labour Studies, c1977.
87 p. ; 24 cm. — (Research series - International Institute for Labour Studies ; no. 26) Sw***
At head of title: Strategic factors in industrial relations systems.
Includes bibliographical references.
ISBN 9290141271 : 12.00F
1. Merchant seamen—Great Britain. 2. Merchant marine—Great Britain. 3. Collective bargaining—Merchant marine—Great Britain. I. Title. II. Title: Strategic factors in industrial relations systems. III. Series: Geneva. International Institute for Labour Studies. Research series - International Institute for Labour Studies ; no. 26.
HD6976.S42G745 79-308552
331.89'041'38750941
79 MARC

—MUNICIPAL EMPLOYEES—CONNECTICUT

Furbish, Lawrence.
Collective bargaining in Connecticut's municipalities : contract negotiations and impasse procedures / prepared by Lawrence K. Furbish, Dennis Meltzer. — ⌊Hartford⌋ : Connecticut General Assembly, 1974.
⌊47⌋ p. ; 28 cm. — (Interim reports - Office of Legislative Research, Connecticut General Assembly ; no. 5, 1974)
At head of title: Office of Legislative Research.
1. Collective bargaining—Municipal employees—Connecticut. I. Meltzer, Dennis, joint author. II. Connecticut. General Assembly. Office of Legislative Research. III. Title. IV. Series: Connecticut. General Assembly. Office of Legislative Research. Interim reports Office of Legislative Research, Connecticut General Assembly ; no. 5, 1974.
HD8005.6.U53C84 75-622873
331.89'041'3520746
75⌊r79⌋rev MARC

—NEWSPAPERS

see also Collective bargaining —Printing industry

—NEWSPAPERS—JAPAN—HISTORY

Tanaka, Tetsuya, 1930-
(Aru sengoshi)
或る戦後史：『朝日新聞』の軌跡 / 田中哲也. — 東京：汐文社, 1978.
216 p. ; 20 cm. — (同時代叢書) Ja 79-5356
Series romanized: Dōjidai sōsho.
¥1200
1. Asahi Shimbun Tōitsu Rōdō Kumiai—History. 2. Collective bargaining—Newspapers—Japan—History. 3. Strikes and lockouts—Newspapers—Japan—History. I. Title.
PN5407.S75T3 79-810127

—NEWSPAPERS—JAPAN—TOYAMA (TOYAMA PREFECTURE)—HISTORY

Kita Nihon Shimbun Rōdō Kumiai. Kita Nihon Rōkumi Sanjūnenshi Hensan Iinkai.
(Kita Nihon Rōkumi sanjūnenshi)
北日本労組30年史 / ⌊編集 北日本新聞労働組合『北日本労組30年史』編纂委員会⌋. — 富山：北日本新聞労働組合, 1978.
333 p., ⌊1⌋ leaf of plates : ill. ; 22 cm. Ja 79-4245
Not in trade.
1. Kita Nihon Shimbun Rōdō Kumiai—History. 2. Collective bargaining—Newspapers—Japan—Toyama (Toyama Prefecture)—History.
PN5407.S75K47 1978 79-810150

—POSTAL SERVICE

Breit, Ernst.
Collective bargaining rights / by Ernst Breit. — ⌊Geneva : Postal, Telegraph and Telephone International, 1975⌋
23 p. ; 30 cm. Sw***
At head of title: P.T.T.I. Postal, Telegraph and Telephone International, 22nd World Congress, Oslo (Norway), 30 June to 4 July 1975.
1. Collective bargaining—Telecommunication. 2. Collective bargaining—Postal service. 3. Postal, Telegraph and Telephone Workers' International. I. Postal, Telegraph and Telephone Workers' International. II. Title.
HD6976.T24B73 331.89'041'3803 78-317770
78 MARC

—PRINTING INDUSTRY—UNITED STATES

Graphic Arts Union Employers of America.
COLA, the CPI, and collective bargaining / prepared by the UEA staff. — 8th ed. — ⌊Arlington, Va.⌋ : Graphic Arts Union Employers of America, 1978.
85 p. ; 28 cm.
1. Wages—Printers—United States. 2. Wages—Cost of living adjustments—United States. 3. Consumer price index. 4. Collective bargaining—Printing industry—United States. I. Title.
Z243.U5G68 1978 79-112333
331.89'041'6862'0973
00 MARC

—STATE GOVERNMENT EMPLOYEES—UNITED STATES

Carlson, Richard J 1943-
State employee labor relations / Richard J. Carlson and Thomas Sedwick. — Lexington, Ky. : Council of State Governments, 1977.
v, 57 p. ; 23 cm. — (RM ; 624)
$3.50
1. Collective bargaining—State government employees—Wisconsin. 2. Collective bargaining—State government employees—United States. I. Sedwick, Thomas, joint author. II. Title. III. Series: Council of State Governments. RM publications ; 624.
JS308.C6 no. 624 353.9'08 s 78-621012
⌊HD8005.6.U53⌋ 79 MARC

—STATE GOVERNMENT EMPLOYEES—UNITED STATES—STATISTICS

United States. Bureau of the Census.
Labor-management relations in State and local governments, 1976. — ⌊Washington⌋ : U.S. Dept. of Commerce, Bureau of the Census : for sale by the Supt. of Docs., U.S. Govt. Print. Off., 1978.
166 p. ; 28 cm. — (State and local government special studies ; no. 88)
Chiefly table.
Supt. of Docs. no.: C 3.145:88
1. Employee-management relations in government—United States—States—Statistics. 2. Collective bargaining—State government employees—United States—Statistics. 3. Collective bargaining—Local officials and employees—United States—Statistics. 4. Strikes and lockouts—State governments—United States—Statistics. 5. Strikes and lockouts—Local officials and employees—United States—Statistics. I. Title. II. Series: United States. Bureau of the Census. State and local government special studies ; no. 88.
JK2403.A35 no. 88 353.9'08 s 78-601971
⌊HD8008.A5⌋ 79 MARC

—STATE GOVERNMENT EMPLOYEES—WISCONSIN

Carlson, Richard J 1943-
State employee labor relations / Richard J. Carlson and Thomas Sedwick. — Lexington, Ky. : Council of State Governments, 1977.
v, 57 p. ; 23 cm. — (RM ; 624)
$3.50
1. Collective bargaining—State government employees—Wisconsin. 2. Collective bargaining—State government employees—United States. I. Sedwick, Thomas, joint author. II. Title. III. Series: Council of State Governments. RM publications ; 624.
JS308.C6 no. 624 353.9'08 s 78-621012
⌊HD8005.6.U53⌋ 79 MARC

—STEVEDORES—JAPAN

(Kōwan rōdō hō tōsō shi)
港湾労働法闘争史：登録日雇港湾労働者10年の闘い / 全日本港湾労働組合編. — 東京：同組合, 1977.
279 p. ; 22 cm. Ja 78-3375
Not in trade.
1. Collective bargaining—Longshoremen—Japan. 2. Labor laws and legislation—Japan. I. Zen Nihon Kōwan Rōdō Kumiai.
HD6976.L82J36 78-809410

—TEACHERS—IOWA

Iowa Association of School Boards.
The current status of negotiability, 1977 / prepared by Iowa Association of School Boards. — Des Moines : The Association, 1977.
iv, 60 p. ; 28 cm.
Includes index.
1. Collective bargaining—Teachers—Iowa. I. Title.
LB2844.59.U62I85 77-155789
331.89'041'3711009777
00 MARC

—TEACHERS—NEW JERSEY

Otto, Cathleen.
An analysis of teachers' contracts in New Jersey, 1977-78 and 1976-77 / by Cathleen Otto. — ⌊Trenton⌋ : New Jersey School Boards Association, c1978.
85 p. : ill. ; 28 cm.
"A report prepared by the New Jersey School Boards Association."
1. Collective bargaining—Teachers—New Jersey. I. New Jersey School Boards Association. II. Title.
LB2842.2.O85 79-103289
331.89'041'3711009749
79 MARC

Wary, Curt.
An analysis of teachers' contracts in New Jersey, 1978-79 / by Curt Wary and Cathleen Otto. — ⌊Trenton⌋ : New Jersey School Boards Association, 1979.
101 p. : ill. ; 22 x 28 cm.
"A report prepared by the New Jersey School Boards Association."
A continuation and expansion of An analysis of teachers' contracts in New Jersey, 1977-78 and 1976-77 by C. Otto.
1. Collective bargaining—Teachers—New Jersey. I. Otto, Cathleen, joint author. II. Otto, Cathleen. An analysis of teachers' contracts in New Jersey, 1977-78 and 1976-77. III. New Jersey School Boards Association. IV. Title.
LB2844.57.U62N58 79-113671
331.89'041'3711009749
79 MARC

COLLECTIVE BARGAINING (Continued)

—TEACHERS—TURKEY

Aksoy, Muammer.
Devrimci öğretmenin kıyımı ve mücadelesi / Muammer Aksoy. — Ankara : [s.n.], 1975.
2 v. (vi, 1298 p.) ; 24 cm.
Includes bibliographical references and index.
1. Teachers—Turkey—Political activity. 2. Collective bargaining—Teachers—Turkey. 3. Strikes and lockouts—Teachers—Turkey. I. Title.
LA941.82.A38 79-102754
79 MARC

—TEACHERS—UNITED STATES

Flygare, Thomas J
Collective bargaining in the public schools / by Thomas J. Flygare. — Bloomington, Ind. : Phi Delta Kappa Educational Foundation, c1977.
49 p. ; 18 cm. — (Fastback - Phi Delta Kappa Educational Foundation ; 99)
ISBN 0-87367-099-X : $0.75
1. Collective bargaining—Teachers—United States. I. Title. II. Series: Phi Delta Kappa. Educational Foundation. Fastback ; 99.
LB2844.59.U6F56 77-84041
331.89'041'371100973
00 MARC

McDonnell, Lorraine, 1947-
Organized teachers in American schools / Lorraine McDonnell, Anthony Pascal ; prepared for the National Institute of Education. — Santa Monica, CA : Rand, 1979.
xviii, 103 p. ; 28 cm. — ([Report] - Rand Corporation ; R-2407 NIE)
Bibliography: p. 99-103.
ISBN 0-8330-0105-1
1. Collective bargaining—Teachers—United States. I. Pascal, Anthony H., joint author. II. National Institute of Education. III. Title. IV. Series: Rand Corporation. Rand report ; R-2407-NIE.
AS36.R3 R-2407 081 s 79-10929
[LB2842.2] 79 MARC

O'Reilly, Robert C 1928-
Understanding collective bargaining in education : negotiations, contracts, and disputes between teachers and boards / by Robert C. O'Reilly. — Metuchen, N.J. : Scarecrow Press, 1978.
xii, 185 p. ; 22 cm.
Includes bibliographical references and indexes.
ISBN 0-8108-1167-7
1. Collective bargaining—Local official and employees—United States. 2. Collective bargaining—Teachers—United States. I. Title.
HD8005.6.U5O73 331.89'041'3711 78-64495
78 MARC

—TEACHERS—UNITED STATES
—CONGRESSES

Public access : citizens and collective bargaining in the public schools / Robert E. Doherty, editor. — Ithaca : New York State School of Industrial and Labor Relations, Cornell University, 1979.
100 p. : ill. ; 23 cm.
"Papers from the symposium Citizen Participation in the Collective Bargaining Process in Public Education, New York State School of Industrial and Labor Relations, Cornell University, Ithaca, New York, October 2 and 3, 1978.
Includes bibliographical references.
ISBN 0-87546-073-9 : $7.95
1. Collective bargaining—Teachers—United States—Congresses. I. Doherty, Robert Emmett, 1923- II. Cornell University. New York State School of Industrial and Labor Relations.
LB2842.2.P82 79-13189
331.89'041'371100973
79 MARC

—TELECOMMUNICATION

Breit, Ernst.
Collective bargaining rights / by Ernst Breit. — [Geneva : Postal, Telegraph and Telephone International, 1975]
23 p. ; 30 cm. Sw***
At head of title: P.T.T.I. Postal, Telegraph and Telephone International, 22nd World Congress, Oslo (Norway), 30 June to 4 July 1975.
1. Collective bargaining—Telecommunication. 2. Collective bargaining—Postal service. 3. Postal, Telegraph and Telephone Workers' International. I. Postal, Telegraph and Telephone Workers' International. II. Title.
HD6976.T24B73 331.89'041'3803 78-317770
78 MARC

—AUSTRALIA

Niland, John R
Collective bargaining and compulsory arbitration in Australia / John Niland. — Kensington, Australia : New South Wales University Press, 1978, cover 1979.
vii, 174 p. ; 22 cm. — (A CEDA study ; M53) Aus***
Bibliography: p. 167-174.
ISBN 0-909465-88-6
1. Arbitration, Industrial—Australia. 2. Collective bargaining—Australia. 3. Industrial relations—Australia. I. Title. II. Series: Committee for Economic Development of Australia. Study ; M53.
HC601.C58 no. M53 330.9'94 s 78-60245
79 MARC

—CANADA

Forrest, Anne.
Unions and the collective bargaining process / research and writing, Anne Forrest ; Robert M. Laxer, general editor. — Toronto : Ontario Institute for Studies in Education, c1978.
x, 77 p. : ill. ; 23 cm. — (Curriculum series - Ontario Institute for Studies in Education ; 29) C***
Bibliography: p. 75-77.
ISBN 0-7744-0159-1
1. Collective bargaining—Canada. 2. Trade-unions—Canada. I. Laxer, Robert. II. Title. III. Series: Ontario Institute for Studies in Education. Curriculum series ; 29.
HD6524.F67 331.89'0971 79-302791
79 MARC

—CANADA—ADDRESSES, ESSAYS, LECTURES

The Zucker lectures, 1975-1977 / Russell L. Ackoff, Richard M. Cyert, H. D. Woods. — Hamilton, Ont. : McMaster University, c1977.
viii, 71 p. ; 22 cm. C***
Contains the annual lectures, sponsored by the McMaster University, Faculty of Business, presented in 1975-1977.
Includes bibliographical references.
CONTENTS: Cyert, R. M. Decision making under uncertainty and managerial leadership.—Woods, H. D. The survival of collective bargaining.—Ackoff, Russell L. On growth and development of social systems.
1. Production (Economic theory)—Addresses, essays, lectures. 2. Decision-making—Addresses, essays, lectures. 3. Collective bargaining—Canada—Addresses, essays, lectures. 4. Economic development—Addresses, essays, lectures. I. Ackoff, Russell Lincoln, 1919- On growth and development of social systems. 1977. II. Cyert, Richard Michael, 1921- Decision making under uncertainty and managerial leadership. 1977. III. Woods, H. D. The survival of collective bargaining. 1977. IV. McMaster University. Faculty of Business.
HD31.Z8 658.4'03 79-310985
79 MARC

—CANADA—CONGRESSES

Dalhousie Labour Institute for the Atlantic Provinces, Dalhousie University, 1977.
Dalhousie Labour Institute for the Atlantic Provinces, June 13-17, 1977. — Halifax : Institute of Public Affairs, Dalhousie University, 1978.
x, 77 p. ; 28 cm. C***
ISBN 0-88926-651-4 : $5.00
1. Trade-unions—Canada—Congresses. 2. Collective bargaining—Canada—Congresses. 3. Industrial relations—Canada—Congresses. I. Dalhousie University. Institute of Public Affairs.
HD6524.D34 1977 331'.0971 79-317761
79 MARC

Industrial relations in Canada, towards a better understanding : a round table discussion held in Toronto June 15 and 16, 1977 / edited by William Dodge, Leslie Ann Ferrari, and Ann E. Jepson. — Ottawa : Conference Board in Canada, c1977.
xxi, 138 p. ; 23 cm. — (Canadian studies ; no. 49 ISSN 0069-8342) C***
"A report from the Compensation Research Centre of the Conference Board in Canada, June 1978."
Includes bibliographical references.
ISBN 0-88763-010-3 : $15.00
1. Industrial relations—Canada—Congresses. 2. Collective bargaining—Canada—Congresses. 3. Trade-unions—Canada—Congresses. I. Dodge, William, 1911- II. Ferrari, Leslie Ann. III. Jepson, Ann E. IV. Conference Board in Canada. Compensation Research Centre. V. Series.
HD8102.I54 331'.0971 79-302768
79 MARC

—DENMARK—ADDRESSES, ESSAYS, LECTURES

Demokrati i organisationer / red. af Torben Agersnap og Egil Fivelsdal. — København : Nyt Nordisk Forlag, 1978.
281 p. : ill. ; 25 cm. — (Skrifter fra Institut for organisation og arbejdssociologi ; 8) D78-20
Bibliography: p. 271-[280]
1. Industrial relations—Denmark—Addresses, essays, lectures. 2. Employees' representation in management—Denmark—Addresses, essays, lectures. 3. Collective bargaining—Denmark—Addresses, essays, lectures. 4. Industrial organization—Denmark—Addresses, essays, lectures. I. Agersnap, Torben. II. Fivelsdal, Egil, 1929- III. Series: Copenhagen. Handelshøjskolen. Institut for organisation og arbejdssociologi. Skrifter ; 8.
HD8546.5.D45 78-385049
78 MARC

—EUROPE

Crispo, John H G
Industrial democracy in western Europe : a North American perspective / John Crispo. — Toronto ; New York : McGraw-Hill Ryerson, c1978.
181 p. ; 23 cm. C77-001703-7
Bibliography: p. 175-176.
Includes index.
ISBN 0-07-082700-1
1. Industrial relations—Europe. 2. Collective bargaining—Europe. 3. Employees' representation in management—Europe. 4. Industrial relations—North America. I. Title.
HD8376.5.C74 331.89'094 79-311389
79 MARC

—EUROPEAN ECONOMIC COMMUNITY COUNTRIES

Rojot, Jacques.
International collective bargaining : an analysis and case study for Europe / by Jacques Rojot. — Deventer : Kluwer, 1978.
183 p. ; 25 cm. Ne78-14
Bibliography: p. 171-183.
ISBN 9031200476 : fl 45.00
1. Collective bargaining—European Economic Community countries. 2. Industrial relations—European Economic Community countries.
HD8380.5.R64 658.31'54'094 78-321699
*78 MARC

—FINLAND

Suomen Ammattiliittojen Keskusjärjestö.
Katsaus työehtosopimustoimintaan ja työehtosopimuspoliittinen ohjelma. — Helsinki : Suomen ammattiliittojen keskusjärjestö, 1976.
103 p. : ill. ; 21 cm. Fi
"Hyväksytty SAK:n XI edustajakokouksessa 17.-20.6. 1976."
ISBN 9519521194
1. Collective bargaining—Finland. 2. Collective labor agreements—Finland. 3. Wages—Finland. 4. Suomen Ammattiliittojen Keskusjärjestö. I. Title.
HD6735.5.S77 1976 79-370836
*79 MARC

—FRANCE

Birien, Jean Louis.
Comment négocier avec les syndicats aujourd'hui et demain? / Jean-Louis Birien, Alain Buchet, Jean-Yves Granger. — Paris : Publi-union, 1978.
261 p. : graphs ; 26 cm. — (Techniques & [i.e. et] pratique des relations sociales) F78-13136
ISBN 2-85790-011-2 : 90.00F
1. Collective bargaining—France. 2. Trade-unions—France. I. Buchet, Alain, joint author. II. Granger, Jean Yves, joint author. III. Title. IV. Series.
HD6684.B55 658.31'54'0944 79-362189
79 MARC

Louche, Claude.
Pouvoir et délégation dans les négociations professionnelles / par Claude Louche. — Paris : Éditions du Centre national de la recherche scientifique, 1977.
107 p. : ill. ; 25 cm. — (Monographies françaises de psychologie ; no 40) F***
Bibliography: p. [105]-107.
ISBN 2-222-02149-9 : 38.00F
1. Collective bargaining. 2. Collective bargaining—France. 3. Negotiation. 4. Power (Social sciences) I. Title. II. Series.
HD6483.L68 301.6'34 79-353049
79 MARC

Reynaud, Jean Daniel.
Les Syndicats, les patrons et l'État : tendances de la négociation collective en France / J.-D. Reynaud. — Paris : Éditions Économie et humanisme, Éditions ouvrières, c1978.
192 p. ; 22 cm. — (Collection Relations sociales) F***
Includes bibliographical references.
ISBN 2-7082-1976-6 : 40.00F
1. Collective bargaining—France. 2. Trade-unions—France. I. Title.
HD6684.R483 331.89'0944 79-356385
79 MARC

—GERMANY, WEST

Külp, Bernhard, 1933-
Der Lohnfindungsprozess der Tarifpartner : Bestimmungsgründe, Auswirkungen, Reformvorschläge / Bernhard Külp. — 1. Aufl. — Darmstadt : Wissenschaftliche Buchgesellschaft, 1977.
v, 110 p. ; 20 cm. — (Erträge der Forschung ; Bd. 77) GFR78-A7
Bibliography: p. 102-106.
Includes indexes.
ISBN 3-534-07587-0 : DM19.50
1. Wages—Germany, West. 2. Collective bargaining—Germany, West. 3. Trade-unions—Germany, West. I. Title.
HD5029.K96 79-381557
79 MARC

COLLECTIVE BARGAINING
—GERMANY, WEST (Continued)

Pohl, Hans-Joachim.
Unternehmensführung und Mitbestimmung bei technologischem Wandel : Ergebnisse e. empir. Unters. / Hans-Joachim Pohl. — 1. Aufl. — München : Florentz, 1978.
187 p. ; 21 cm. — (Schriftenreihe wirtschaftswissenschaftliche Forschung und Entwicklung ; Bd. 17) GFR78-A
Bibliography: p. 171-183.
Includes index.
ISBN 3-921491-97-5 : DM27.80
1. Machinery in industry—Germany, West. 2. Technological innovations—Social aspects—Germany, West. 3. Collective bargaining—Germany, West. I. Title. II. Series.
HD6331.2.G28P63 78-391443
79 MARC

Projektgruppe Gewerkschaftsforschung.
Tarifpolitik 1977 [i.e. neunzehnhundertsiebenundsiebzig] : Darst. u. Analyse d. Tarifbewegung in d. metallverarbeitenden, der chem. u. d. Druckindustrie sowie im öffentl. Dienst / Projektgruppe Gewerkschaftsforschung. — 1. Aufl. — Frankfurt am Main ; New York : Campus-Verlag, 1978.
ix, 331 p. ; 21 cm. — (Forschungsberichte des Instituts für Sozialforschung Frankfurt am Main) GFR78-A
Errata slip inserted.
Includes bibliographical references.
ISBN 3-593-32345-1 : DM38.00
1. Collective bargaining—Germany, West. I. Title. II. Series: Frankfurt am Main. Institut für Sozialforschung. Forschungsberichte des Instituts für Sozialforschung Frankfurt am Main.
HD6694.P67 1978 79-364461
79 MARC

—GREAT BRITAIN

Page, G Terry.
An employer's guide to disclosure of information / G. Terry Page. — London : Kogan Page, 1978.
160 p. : ill. ; 23 cm. GB***
Includes bibliographical references and index.
ISBN 0-85038-084-7 : £8.00
1. Disclosure in accounting. 2. Corporation reports. 3. Collective bargaining—Great Britain. 4. Freedom of information—Great Britain. I. Title.
HF5657.P25 346'.41'066 79-315681
79 MARC

—GREAT BRITAIN—HISTORY

Currie, Robert.
Industrial politics / Robert Currie. — Oxford : Clarendon Press ; New York : Oxford University Press, 1979.
viii, 294 p. ; 23 cm. GB79
Includes bibliographical references and index.
ISBN 0-19-827419-X : £8.00
1. Collective bargaining—Great Britain—History. 2. Industrial relations—Great Britain—History. 3. Trade-unions—Great Britain—History. 4. Employees' representation in management—Great Britain—History. 5. Labour Party (Gt. Brit.)—History. I. Title.
HD6664.C83 331'.0941 78-40480
79 MARC

—INDONESIA—CONGRESSES

Lokakarya Perjanjian Perburuhan, Ciloto, Indonesia, 1975.
Hasil Lokakarya Perjanjian Perburuhan, tanggal 21 s/d 4 Nopember 1975 di Ciloto, Puncak. — [Jakarta] : Direktorat Jenderal Perlindungan dan Perawatan Tenaga Kerja, Departemen Tenaga Kerja, Transmigrasi dan Koperasi, RI, [1975?]
82 leaves ; 33 cm.
1. Collective bargaining—Indonesia—Congresses. I. Indonesia. Direktorat Jenderal Perlindungan dan Perawatan Tenaga Kerja.
HD6822.L64 1975 78-941884
79 MARC

—ITALY

Scardillo, Pietro.
I sindacati e la contrattazione collettiva / Pietro Scardillo. — 1. ed. — Milano : ETAS libri, 1977.
167 p. ; 19 cm. — (Enciclopedia del quadro intermedio ; 7) It78-Sep
Bibliography: p. 161-164.
L3000
1. Trade-unions—Italy. 2. Collective bargaining—Italy. I. Title. II. Series.
HD6709.S38 331.88'0945 78-389823
*79 MARC

—JAPAN—ADDRESSES, ESSAYS, LECTURES

Kōshiro, Kazuyoshi, 1932–
(Tenkanki no chingin kōshō)
転換期の賃金交渉 / 神代和欣著. — 東京 : 東洋経済新報社, 1978.
xi, 260 p. : ill. ; 19 cm. — (東経選書) Ja 78-13299
Series romanized: Tōkei sensho.
Includes bibliographical references.
¥1500
1. Wages—Japan—Addresses, essays, lectures. 2. Collective bargaining—Japan—Addresses, essays, lectures. I. Title.
HD5077.K65 78-813849

—UNITED STATES

Fossum, John A
Labor relations : development, structure, process / John A. Fossum. — Dallas : Business Publications, 1979.
xiii, 481 p. : ill. ; 24 cm.
Includes bibliographies and indexes.
ISBN 0-256-02088-4
1. Industrial relations—United States. 2. Collective bargaining—United States. 3. Trade-unions—United States. I. Title.
HD8072.F59 331'.0973 78-61198
79 MARC

Mills, Daniel Quinn.
Employment and unemployment statistics in collective bargaining / Daniel Quinn Mills. — Washington : National Commission on Employment and Unemployment Statistics : for sale by the Supt. of Docs., U.S. Govt. Print. Off., 1978 [i.e. 1979]
vi, 54 p. ; 27 cm. — (Background paper - National Commission, Employment and Unemployment Statistics ; no. 10)
Bibliography: p. 43.
1. Labor supply—United States—Statistical services. 2. Collective bargaining—United States. I. Title. II. Series: United States. National Commission on Employment and Unemployment Statistics. Background paper - National Commission, Employment and Unemployment Statistics ; no. 10.
HD5724.M555 331.89 79-602352
79 MARC

Mounce, Earl Winifield.
Collective bargaining, by E. W. Mounce and Robley D. Stevens. 6098-1. [Ed. 2] Scranton, International Correspondence Schools, c1960.
114 p. 19 cm.
1. Collective bargaining—United States. 2. Collective labor agreements—United States. I. Stevens, Robley D., joint author. II. International Correspondence Schools, Scranton, Pa.
KF3408.Z9M68 344'.73'0189 60-52006
[r79]rev MARC

United States. Congress. House. Committee on Education and Labor. Subcommittee on Labor-Management Relations.
Collective bargaining for retired employees : hearing before the Subcommittee on Labor-Management Relations of the Committee on Education and Labor, House of Representatives, Ninety-fifth Congress, second session, on H.R. 13535 ... hearing held in Washington, D.C., on September 19, 1978. — Washington : U.S. Govt. Print. Off., 1978.
iv, 64 p. ; 24 cm.
1. Collective bargaining—United States. 2. Old age pensions—United States. I. Title.
KF27.E347 1978b 344'.73'01252 79-601115
79 MARC

—UNITED STATES—ADDRESSES, ESSAYS, LECTURES

Collective bargaining : new dimensions in labor relations / edited by Franklin J. Havelick ; foreword by Theodore W. Kheel. — Boulder, Colo. : Westview Press, 1979.
xiv, 183 p. : graphs ; 24 cm. — (A Westview special study)
"A Giniger project."
Includes index.
ISBN 0-89158-386-6 : $16.00
1. Collective bargaining—United States—Addresses, essays, lectures. 2. Collective bargaining—Addresses, essays, lectures. I. Havelick, Franklin J.
HD6508.C657 331.89'0973 79-4234
79 MARC

Dunlap, John Thomas, 1914–
Speech at the National Press Club luncheon, Washington, Dec. 11, 1975. [Sound recording]
1 reel. 3¾ ips. mono. 7 in.
Reel 704.
Originally broadcast over National Public Radio.
SUMMARY: Secretary of Labor John Dunlap talks about the American economy and its relation to industrial relations and collective bargaining. A question and answer session follows.
1. United States—Economic conditions—1971– —Addresses, essays, lectures. 2. Industrial relations—United States—Addresses, essays, lectures. 3. Collective bargaining—United States—Addresses, essays, lectures. I. National Press Club of Washington, Washington, D. C.
[HC106.7] 78–741989

—UNITED STATES—CASES

Oberer, Walter E
Cases and materials on labor law : collective bargaining in a free society / by Walter E. Oberer, Kurt L. Hanslowe, Jerry R. Andersen. — 2d ed. — St. Paul : West Pub. Co., 1979.
xli, 1168 p. ; 26 cm. — (American casebook series)
Includes index.
ISBN 0-8299-2024-2 : $22.95
1. Collective bargaining—United States—Cases. 2. Labor laws and legislation—United States—Cases. I. Hanslowe, Kurt L., joint author. II. Andersen, Jerry R., joint author. III. Title. IV. Series.
KF3408.A4O2 1979 344'.73'0189 78-20988
78 MARC

—UNITED STATES—CONGRESSES

Current developments in labor disputes settlements : a symposium, October 8, 1976 / edited by Michael Dudra. — Loretto, Pa. : Graduate School of Industrial Relations, Saint Francis College, c1976.
39 p. ; 24 cm.
1. Collective bargaining—United States—Congresses. 2. Labor disputes—United States—Congresses. I. Dudra, Michael.
HD6508.C85 331.89'0973 79-308983
79 MARC

COLLECTIVE BARGAINING UNIT

see also Trade-unions--Recognition

COLLECTIVE BEHAVIOR

see also Crowds; Deviant behavior; Violence

Beckmann, Michael.
Theorie der sozialen Bewegung : Anwendung sozialpsycholog. Hypothesen zur Erklärung d. Entstehungsbedingungen sozialer Bewegungen / Michael Beckmann. — München : Minerva-Publikation, 1979.
370 p. : ill. ; 21 cm. — (Minerva-Fachserie Wirtschafts- und Sozialwissenschaften) GFR79-A
Originally presented as the author's thesis, Hamburg, 1977.
Bibliography: p. 352-370.
ISBN 3-597-10061-9 : DM48.00
1. Social movements. 2. Collective behavior. 3. Social psychology. I. Title.
HN8.B36 1979 301.1 79-373172
79 MARC

Roberts, Ron E
Social movements : between the balcony and the barricade / Ron E. Roberts, Robert Marsh Kloss. — 2d ed. — St. Louis : Mosby, 1979.
xiii, 245 p. : ill. ; 25 cm.
Bibliography: p. 227-239.
Includes index.
ISBN 0-8016-4135-7
1. Social movements. 2. Collective behavior. I. Kloss, Robert Marsh, 1938- joint author. II. Title.
HN8.R6 1979 301.24'2 78-31214
78 MARC

Rowan, John.
The structured crowd / John Rowan. — London : Davis-Poynter, 1978.
157 p. : ill. ; 23 cm. — (His Psychological aspects of society ; book 4) GB***
Includes bibliographical references and indexes.
ISBN 0-7067-0164-X : £5.50
1. Social psychology. 2. Collective behavior. 3. Race discrimination. I. Title. II. Series.
HM258.R69 1978 301.1 79-307675
79 MARC

Schelling, Thomas C 1921-
Micromotives and macrobehavior / Thomas Schelling. — 1st ed. — New York : Norton, c1978.
252 p. : ill. ; 22 cm. — (Fels lectures on public policy analysis)
Includes index.
ISBN 0-393-05701-1. ISBN 0-393-09009-4 pbk.
1. Policy sciences. 2. Choice (Psychology) 3. Social choice. 4. Collective behavior. 5. Social problems. I. Title. II. Series.
H61.S355 1978 301.11'3 78-17119
78 MARC

Wright, Sam.
Crowds and riots : a study in social organization / Sam Wright. — Beverly Hills, Calif. : Sage Publications, c1978.
207 p. : ill. ; 22 cm. — (Sociological observations ; v. 4)
Bibliography: p. 181-198.
Includes indexes.
ISBN 0-8039-0995-0. ISBN 0-8039-0996-9 pbk.
1. Crowds. 2. Riots. 3. Collective behavior. 4. Nonverbal communication. I. Title. II. Series.
HM281.W74 301.18'2 78-626
78 MARC

COLLECTIVE BEHAVIOR (Continued)

—ADDRESSES, ESSAYS, LECTURES

Collective behavior and social movements / Louis E. Genevie, editor. — Itasca, Ill. : F. E. Peacock Publishers, c1978.
xxii, 531 p. ; 24 cm.
Includes indexes.
Bibliography: p. 489-513.
ISBN 0-87581-228-7
1. Collective behavior—Addresses, essays, lectures. 2. Social movements—Addresses, essays, lectures. 3. Social movements—United States—Addresses, essays, lectures. I. Genevie, Louis E.
HM251.C653 301.1 77-83419
00 MARC

COLLECTIVE CHOICE see Social choice

COLLECTIVE EDUCATION

Broström, Stig.
Makarenkos kollektivopdragelse—dansk gruppeopdragelse / [af] Stig Broström ; [fotos, A. S. Makarenko og Stig Broström]. — 1.oplag. — København (Skt. Hans Torv 26, 2200 N.) : Børn & Unge, 1978.
111 p. : ill. ; 21 cm. D79-7
Bibliography: p. 110-111.
ISBN 8787433249 : kr38.00
1. Collective education. 2. Makarenko, Anton Semenovich, 1888-1939. 3. Education—Philosophy. I. Title.
LC1025.B76 79-381436
79 MARC

—ISRAEL

Alon, Moni.
(Ne'urim ba-ḳibuts)
נעורים בקיבוץ : בעיני מחנכים וחניכים / מוני אלון. — [תל־אביב] : ספרית פועלים, 1975.
320 p. ; 22 cm. — (מדור לחינוך ולהשכלה)
Series romanized: Mador le-ḥinukh ule-haśkalah.
Bibliography: p. [314]-320.
1. Collective education—Israel. I. Title.
LC5163.I 75A46 79-951838

—THAILAND

Thānin Kraiwichian.
(Khrū kap thāng yū rọ̄t khọ̄ng sangkhom Thai)
ครูกับทางอยู่รอดของสังคมไทย / ธานินทร์ กรัยวิเชียร. — พิมพ์ครั้งที่ 4. — [กรุงเทพมหานคร] : กรมวิชาการ กระทรวงศึกษาธิการ, 2520 [1977]
81 p. : ill. ; 26 cm.
In Thai.
9.00B
1. Communism—Thailand. 2. Collective education—Thailand. I. Title.
HX400.55.A6T48 1977 77-985834
[½] S A

COLLECTIVE FARMS

see also Collective settlements

—HANDBOOKS, MANUALS, ETC.

Comune agricola : un manuale d'uso per vivere in campagna. — Roma : Savelli, 1978.
128 p. : ill. ; 19 cm. — (Controcultura ; 24) It78-Sept
At head of title: Stampa alternativa.
Bibliography: p. [124]-128.
1. Collective settlements—Handbooks, manuals, etc. 2. Collective farms—Handbooks, manuals, etc. I. Stampa alternativa.
HX635.C64 335'.9 78-392783
*79 MARC

—LABOR PRODUCTIVITY see Agriculture—Labor productivity

—LAW AND LEGISLATION—HUNGARY

A Közös vagyon védelme a mezőgazdasági termelőszövetkezetekben / [szerk. Diczig István]. — Budapest : Mezőgazd. Kiadó, 1978.
181 p. ; 20 cm. H78
Includes bibliographical references.
ISBN 963230392X : 16.00Ft
1. Collective farms—Law and legislation—Hungary. 2. Offenses against socialist property—Hungary. I. Diczig, István.
79-344498
*79 MARC

—LAW AND LEGISLATION—RUSSIA

Kolkhozniku o pravakh i obi͡azannosti͡akh / [Shkorupilo G. M., Volkova N. A., Vovk I͡U. A. ... et al. ; Otv. red. Shkorupilo G. M.]. — Moskva : I͡Urid. lit., 1978.
263 p. ; 20 cm. USSR79
Includes bibliographical references.
0.65rub
1. Collective farms—Law and legislation—Russia. 2. Collective farms—Russia—Officials and employees. 3. Collective farms—Pensions—Russia. I. Shkorupilo, Grigoriĭ Matveevich.
79-372647
79 MARC

(Sovetskoe zemel'noe pravo)
Советское земельное право : [Учебник для юрид. фак. ун-тов / Доц. Балезин В. П., проф. Краснов Н. И., доц. Бережной В. А. ... et al. ; Отв. ред. Н. И. Краснов и В. П. Балезин]. — Москва : Юрид. лит., 1977.
447 p. ; 21 cm. USSR 78
Bibliography: p. 438-[440]
1.10rub
1. Land tenure—Russia—Law. 2. Agricultural laws and legislation—Russia. 3. Collective farms—Law and legislation—Russia. 4. Land, Nationalization of. I. Balezin, Vadim Pavlovich. II. Krasnov, Nikolaĭ Ivanovich.
78-423523

—PENSIONS—RUSSIA

Kolkhozniku o pravakh i obi͡azannosti͡akh / [Shkorupilo G. M., Volkova N. A., Vovk I͡U. A. ... et al. ; Otv. red. Shkorupilo G. M.]. — Moskva : I͡Urid. lit., 1978.
263 p. ; 20 cm. USSR79
Includes bibliographical references.
0.65rub
1. Collective farms—Law and legislation—Russia. 2. Collective farms—Russia—Officials and employees. 3. Collective farms—Pensions—Russia. I. Shkorupilo, Grigoriĭ Matveevich.
79-372647
79 MARC

—AZERBAIJAN

Makhmudov, Akhmed Akber ogly.
(Nekotorye ėkonomicheskie voprosy intensifikat͡sii zemledelii͡a)
Некоторые экономические вопросы интенсификации земледелия в колхозах Азербайджанской ССР. Баку, Изд-во Академии наук Азербайджанской ССР, 1965.
93 p. 20 cm.
At head of title: Академия наук Азербайджанской ССР. Институт экономики. А. А. Махмудов.
Bibliographical footnotes.
1. Collective farms—Azerbaijan. I. Title.
HD1492.R92A96 338.7'63'094791 65-82119

—BYELORUSSIAN S. S. R.—BRESTSKAYA OBLAST'

Romash, Mecheslav Vikent'evich.
(Khozraschetnye otnoshenii͡a kolkhozov s gosudarstvom)
Хозрасчетные отношения колхозов с государством : (По материалам колхозов Брест. обл.) / М. В. Ромаш ; Редактор А. С. Мышелова. — Минск : Наука и техника, 1977.
119 p. ; 20 cm. USSR 77
Includes bibliographical references.
0.60rub
1. Agriculture and state—Byelorussian S. S. R.—Brestskaya oblast'. 2. Collective farms—Byelorussian S. S. R.—Brestskaya oblast'. I. Myshelova, Antonina Stepanovna. II. Title.
HD1995.B93R65 79-414095

—CHINA

Économie rurale en Chine : de la réforme agraire aux communes populaires / texte traduit du chinois et édité par Éditions du Centenaire. — Paris : Éditions du Centenaire, 1977.
223 p. ; 19 cm. F78-14266
At head of title: Éditions du peuple, Pékin, janvier 1976.
Includes bibliographical references.
25.00F
1. Agriculture, Cooperative—China. 2. Collective farms—China. 3. Communes (China)
HD1491.C6E36 334'.683'0951 79-362761
79 MARC

—CZECHOSLOVAKIA

Choma, Dimitrij.
Sbližování dvou forem socialistického vlastnictví a problémy efektivnosti výroby v zemědělství : materiál pro lektory / Dimitrij Choma. — 1. vyd. — Praha : Horizont, 1978.
134, [2] p. ; 21 cm. — (Studijní materiály na pomoc lektorům) Cz78
Bibliography: p. 130-134.
Kčs8.00
1. Collective farms—Czechoslovakia. 2. Socialist property. I. Title.
HD1492.C95C448 78-385337
*78 MARC

—ESTONIA

Otsman, Elmina, 1924–
(Sporii͡u s sud'boĭ)
Спорю с судьбой : [Рассказ комбайнера колхоза "Тарвасту" Вильяндис. р-на] / Эльмина Отсман ; Лит. запись Л. Бутковской ; Вступ. статья Э. Г. Тынуриста]. — Москва : Профиздат, 1978.
206 p., [8] leaves of plates : ports. ; 17 cm. — (Повести о героях труда)
Series romanized: Povesti o geroi͡akh truda.
0.85rub
1. Otsman, Elmina, 1924– 2. Agricultural laborers—Estonia—Biography. 3. Collective farms—Estonia. I. Butkovskai͡a, L. II. Title.
HD1536.E77O87 79-410560

—ESTONIA—HISTORY—SOURCES

Eesti NSV põllumajanduse kollektiviseerimine : dokumentide ja materjalide kogumik / [koostaja, E. Laasi]. — Tallinn : Eesti Raamat, 1978.
664 p. ; 23 cm. USSR***
Includes bibliographical references and index.
1.90rub
1. Collective farms—Estonia—History—Sources. I. Laasi, E.
HD1492.R92E73 79-375021
79 MARC

—EUROPE, EASTERN—CONGRESSES

The Political economy of collectivized agriculture : a comparative study of communist and non-communist systems / edited by Ronald A. Francisco, Betty A. Laird, Roy D. Laird. — New York : Pergamon Press, c1979.
xii, 250 p. : graphs ; 24 cm. — (Pergamon policy studies ; no. 14)
Revisions of papers originally presented at the Sept. 1977 American Political Science Association meeting in Washington, D.C. and subsequently at a conference held at the University of Nebraska in April 1978.
Includes bibliographical references and index.
ISBN 0-08-023880-7
1. Collective farms—Europe, Eastern—Congresses. I. Francisco, Ronald A. II. Laird, Betty A. III. Laird, Roy D.
HD1492.E8P64 1979 334'.683'0947 78-26515
78 MARC

—HUNGARY

Berta, Béla.
Az adott szó kötelez / Berta Béla, Sulyán Pál. — Budapest : Termelőszövetkezetek Országos Tanácsa, 1978.
203 p. ; 20 cm. H78
ISBN 9630306581
1. Collective farms—Hungary. 2. Socialist competition. I. Sulyán, Pál, joint author. II. Title.
HD1492.H8B44 79-360690
*79 MARC

COLLECTIVE FARMS
—HUNGARY (Continued)

Házunk tája, aranybánya / szerk. Kozma György ; ⌊írta Bessenyei Zoltán ... et al.⌋. — 2. átdolg., kiad. — Budapest : Mezőgazdasági Könyvkiadó, 1977, c1972.
270 p. : numerous ill. ; 24 cm. H77
Bibliography: p. 264.
ISBN 963230263X : 34.00Ft
1. Family farms—Hungary. 2. Collective farms—Hungary. I. Kozma, György.
S562.H9H39 1977 78-396456
*79 MARC

Hungary. Termelőszövetkezetek Országos Tanácsa.
Agricultural co-operatives in Hungary. — ⌊Budapest⌋ : National Council of Agricultural Cooperatives, in care of Economical and Judicial Publisher, ⌊1974?⌋
ca. 150 p. : ill. ; 21 x 25 cm.
ISBN 9632202534
1. Collective farms—Hungary. I. Title.
HD1492.H8H87 1974 334'.683'09439 78-314549
78 MARC

Hungary. Termelőszövetkezetek Országos Tanácsa.
Érvényben levő TOT-ajánlások, -állásfoglalások és -irányelvek gyűjteménye, 1967-1973 / összeállította, Szaszkó László, Zsohár András. — Budapest : Termelőszövetkezetek Országos Tanácsa, 1973.
262 p. ; 21 cm.
Cover title: Tot dokumentumok.
"Lezárva: 1973. szeptember 30-án."
1. Collective farms—Hungary. I. Szaszkó, László. II. Zsohár, András. III. Title. IV. Title: Érvényben levő TOT-ajánlások, -állásfoglalások és irányelvek gyűjteménye ... V. Title: TOT documentumok.
HD1492.H8H87 1973 79-374700
79 MARC

Kovács, Mihály.
Tsz-tagok kézikönyve / ⌊összeállította, Kovács Mihály, Markó Lajos⌋. — ⌊Budapest⌋ : Kossuth Könyvkiadó, 1977.
522 p. ; 20 cm. H77
Previous ed. entered under : Almási, István.
"A fontosabb jogszabályok jegyzéke" : p. 497 -⌊506⌋
Includes index.
ISBN 9630909693 : 35.00Ft
1. Collective farms—Hungary. I. Markó, Lajos, joint author. II. Title.
HD1492.H8A6 1977 79-337624
*79 MARC

Szabó, Béla, Dr.
A termelőszövetkezetek üzemi tevékenysége és társulásai ⌊írta⌋ Szabó Béla ⌊és⌋ Zsuffa Ervin. ⌊Budapest⌋ Kossuth Könyvkiadó, 1968.
109 p. 15 cm. (TSZ-tagok kiskönyvtára) 2.50
At head of title: Termelőszövetkezetek Országos Tanácsa.
1. Collective farms—Hungary. I. Zsuffa, Ervin, joint author. II. Hungary. Termelőszövetkezetek Országos Tanácsa. III. Title.
HD1492.H8S923 75-233111
77⌊r79⌋rev MARC

—HUNGARY—ADDRESSES, ESSAYS, LECTURES

Erdei, Ferenc.
Erdei Ferenc válogatott írásai és beszédei / ⌊válogatta Fekete Ferenc⌋. — ⌊Budapest⌋ : Kossuth, 1973.
439 p. ; 20 cm.
1. Agriculture—Economic aspects—Hungary—Addresses, essays, lectures. 2. Cooperation—Hungary—Addresses, essays, lectures. 3. Collective farms—Hungary—Addresses, essays, lectures.
HD1940.5.E74 79-341184
79 MARC

—HUNGARY—CONGRESSES

A Mezőgazdasági szövetkezetek III. ⌊i. e. harmadik⌋ kongresszusa : ⌊beszámolók, felszólalások, határozatok, 1976. december 14-15 / szerk., Balogh András ; rend. a Termelőszövetkezetek Országos Tanácsa⌋. — ⌊Budapest⌋ : Kossuth, 1977.
236 p. ; 20 cm. H77
ISBN 9630909219 : 22.00Ft
1. Collective farms—Hungary—Congresses. I. Balogh, András. II. Hungary. Termelőszövetkezetek Országos Tanácsa.
HD1492.H8M4 78-399918
*79 MARC

—HUNGARY—FINANCE

Szakál, Pál.
A hitel szerepe a mezőgazdasági termelőszövetkezetek fejlődésében 1968-1974 között / Szakál Pál. — Budapest : ⌊Szövetkezeti Kutató Intézet⌋, 1976.
74 p. ; 29 cm. — (Közlemények - Szövetkezeti Kutató Intézet ; 109)
1. Collective farms—Hungary—Finance. 2. Agriculture credit—Hungary. I. Title.
HD1492.H8S9233 79-376534
79 MARC

—HUNGARY—HISTORY

Fazekas, Béla.
A mezőgazdasági termelőszövetkezeti mozgalom Magyarországon / Fazekas Béla. — ⌊Budapest⌋ : Kossuth, 1976.
314, ⌊1⌋ p. ; 20 cm.
Bibliography: p. 313-⌊315⌋
ISBN 9630906430 : 32.00Ft
1. Collective farms—Hungary—History. I. Title.
HD1492.H8F39 78-391713
79 MARC

—HUNGARY—INCOME DISTRIBUTION

Moak, Samuel K
Marketing system of the socialist farms, North Korea / Samuel K. Moak. — Richmond : University of Richmond, c1974.
v, 115 p. ; 23 cm.
Includes bibliographies and index.
1. Produce trade—Korea (Democratic People's Republic) 2. Agriculture—Economic aspects—Korea (Democratic People's Republic) 3. Collective farms—Korea (Democratic People's Republic) I. Title.
HD9016.K62M6 381'.41'095193 74-13265
79 MARC

—HUNGARY—MANAGEMENT

Fekete, Ferenc.
Célok és optimumok a termelőszövetkezeti gazdálkodásban / Fekete Ferenc, Earl O. Heady, Bob R. Holdren. — ⌊Budapest⌋ : Mezőgazdasági Kiadó, 1977.
134, ⌊1⌋ p. ; 24 cm. H77
Originally published in 1976 under the title: Economics of cooperative farming.
Bibliography: p. 133-⌊135⌋
ISBN 9632302265 : 23.00Ft
1. Agriculture, Cooperative—Mathematical models. 2. Agriculture, Cooperative—Hungary. 3. Collective farms—Hungary—Management. I. Heady, Earl Orel, 1916- joint author. II. Holdren, Bob R., joint author. III. Title.
HD1491.A3F4416 79-344809
*79 MARC

Horváth, Lajos, belügymin, számv, igazgató, ed.
A termelőszövetkezetek gazdálkodása ⌊írták : Csete László, et al.⌋ Budapest, Közgazdasági és Jogi Könyvkiadó ⌊1961⌋
374 p. illus., maps (part fold. col.) 24 cm.
Bibliography: p. 369-370.
1. Collective farms—Hungary—Management. I. Csete, László. II. Title.
S562.H9H6 67-33851
⌊r79⌋rev MARC

—HUNGARY—OFFICIALS AND EMPLOYEES

Bognár, László.
A termelőszövetkezeti vezetők társadalmi mobilitása / Bognár László, Simó Tibor. — Budapest : ⌊Szövetkezeti Kutató Intézet⌋, 1975.
249 p. : graphs ; 29 cm. — (Közlemények - Szövetkezeti Kutató Intézet ; 101)
Includes bibliographical references.
1. Collective farms—Hungary—Officials and employees. 2. Social mobility—Hungary. I. Simó, Tibor, joint author. II. Title.
HD1492.H8B64 79-375783
79 MARC

Cseri, Péter.
Munkaügyi tervezés és statisztika a mezőgazdasági termelőszövetkezetekben : a termelőszövetkezeti munkaügyi tanfolyam tananyaga : az 1973 -ban megjelent tankönyv átdolgozott és bővített kiadása / ⌊írta, illetve egyes részeit összeállította, Cseri Péter és Hulényi Miklós ; szerk., Székely József.⌋. — Budapest : MÉM Mérnök- és Vezetőtovábbképző Intézet, 1977.
120 p. ; 29 cm. H77
9.00Ft
1. Collective farms—Hungary—Officials and employees. I. Hulényi, Miklós, joint author. II. Székely, József, Dr. III. Title.
HD1492.H8C754 79-338563
*79 MARC

Simó, Tibor.
Termelőszövetkezeti tagok és alkalmazottak átrétegeződése, társadalmi mobilitása / Simó Tibor. — Budapest : ⌊s.n.⌋, 1977.
178 p. ; 29 cm. — (Közlemenyek - Szövetkezeti Kutató Intézet ; 126) H77
ISBN 9637003487
1. Collective farms—Hungary—Officials and employees. 2. Social mobility—Hungary. I. Title.
HD1492.H8S49 79-381349
*79 MARC

Tanulmányok a termelőszövetkezeti parasztságról / ⌊írták, Bakurecz Jenőné ... et al. ; szerk. Gyenis János⌋. — ⌊Budapest⌋ : Kossuth Könyvkiadó, 1978.
193 p. ; 18 cm. H78
Includes bibliographical references.
ISBN 9630910772 : 15.00Ft.
1. Collective farms—Hungary—Officials and employees. 2. Peasantry—Hungary. 3. Agricultural laborers—Hungary. I. Gyenis, János.
HD1492.H8T36 79-337585
*79 MARC

—HUNGARY—STATISTICS
—PERIODICALS

Hungary. Termelőszövetkezetek Országos Tanácsa.
Mezőgazdasági szövetkezetek gazdálkodása a számok tükrében.
Budapest, Termelőszövetkezetek Országos Tanácsa.
v. 20 cm.
1. Collective farms—Hungary—Statistics—Periodicals. 2. Agriculture—Hungary—Statistics—Periodicals. I. Title.
HD1492.H8H87a 78-649279
MARC-S

—LATVIA

Kauliņš, Edgars, 1903-
Kad migla krīt ... : Manas dzīves stāsts / Edgars Kauliņš ; Pētera Bauģa literārais atstāstījums. — 2. papildinātais izdevums. — Rīga : Liesma, 1977.
254 p., ⌊6⌋ leaves of plates : ill. ; 21 cm. USSR77-38387
0.94rub
1. Collective farms—Latvia. I. Baugis, Pēteris. II. Title.
HD1492.R92L365 1977 79-362050
79 MARC

—LITHUANIA—HISTORY

Efremenko, A P
Kolūkinės santvarkos pergalė Lietuvoje / A. Jefremenka. — Vilnius : Mokslas, 1977.
87 p. : ill. ; 17 cm. — (Lietuvos istorija) USSR***
Includes bibliographical references.
0.15rub
1. Collective farms—Lithuania—History. I. Title.
S562.R9E3415 78-389050
79 MARC

—MOLDAVIAN S. S. R.

Novakov, Saveliĭ Zakharovich.
(Ukrupnenie kolkhozov Moldavskoĭ SSR)
Укрупнение колхозов Молдавской ССР : 1950–1965 / С. З. Новаков. — Кишинев : Штиинца, 1978.
174 p. ; 20 cm. USSR***
At head of title: Akademiía nauk Moldavskoĭ SSR. Institut istorii.
Includes bibliographical references.
1.40rub
1. Collective farms—Moldavian S. S. R. 2. Farms, size of—Moldavian S. S. R. I. Title.
S562.R9N64 79-407069

—POLAND

Gorzelak, Eugeniusz.
Gospodarowanie indywidualne a zespołowe / Eugeniusz Gorzelak. — Wyd. 1. — Warszawa : Ludowa Spółdzielnia Wydawnicza, 1954.
103 p. ; 20 cm.
1. Farm management—Poland. 2. Collective farms—Poland. I. Title.
S562.P6G67 78-392707
79 MARC

—RUSSIA

Avramenko, Alekseĭ Ivanovich.
Teoreticheskoe obosnovanie diffrenfsiafsii zakupochnykh fsen / A. I. Avramenko ; redaktor S. V. San'ko. — Minsk : Nauka i tekhnika, 1979.
126 p. ; 17 cm. USSR***
At head of title: Akademiía nauk Belorusskoĭ SSR. Institut ėkonomiki.
Includes bibliographical references.
0.60rub
1. Agricultural prices—Russia. 2. Collective farms—Russia. I. San'ko, Stanislav Vikent' evich. II. Title.
HD9015.R92A96 79-369316
79 MARC

Dfukov, Viktor Vladimirovich.
(Sovershenstvovanie material'nykh uslovii vosproizvodstva rabocheĭ sily v kolkhozakh)
Совершенствование материальных условий воспроизводства рабочей силы в колхозах / В. В. Дюков. — Казань : Изд-во Казанского университета, 1977.
164 p. ; 20 cm. USSR***
Includes bibliographical references.
1.28rub
1. Agriculture—Russia—Labor productivity. 2. Collective farms—Russia. I. Title.
S564.D55 79-410605

COLLECTIVE FARMS
—RUSSIA (Continued)

Fedorov, Vi͡acheslav Prokop'evich.
(Vyravnivanie ėkonomicheskikh uslovii khozi͡aĭstvovanii͡a v kolkhozakh)
Выравнивание экономических условий хозяйствования в колхозах / В. П. Федоров. — Москва : Изд-во МГУ, 1978.
205 p. ; 20 cm. USSR 78
Includes bibliographical references.
1.50rub
1. Collective farms—Russia. I. Title.
HD1492.R9F43 79–409303

Hryniv, O I
Kolhospne seli͡anstvo—zrostanni͡a sotsial' noï aktyvnosti / O. I. Hryniv, V. M. Nikonenko. — Kyïv : Politvydav Ukraïny, 1978.
127 p. ; 20 cm. USSR***
Includes bibliographical references.
0.25rub
1. Agricultural laborers—Russia. 2. Collective farms—Russia. I. Nikonenko, V. M., joint author. II. Title.
HD1536.R9H79 79-369634
79 MARC

Medvedev, Vadim Andreevich, kandidat ėkonomicheskikh nauk.
(Puti razviti͡a i sblizhenii͡a kolkhoznoĭ i obshchenarodnoĭ form sobstvennosti)
Пути развития и сближения колхозной и общенародной форм собственности. Ленинград, 1959.
42 p. 21 cm.
At head of title: Общество по распространению политических и научных знаний РСФСР. Ленинградское отделение.
Includes bibliography.
1. Collective farms—Russia. I. Title.
HD1491.R9M37 61–27746

Mili͡utin, Vladimir Pavlovich, 1884–
(Kolkhozy)
Колхозы : основные задачи коллективных хозяйств в современный период / В. П. Милютин. — Москва : Гос. изд-во, 1929.
32 p. ; 17 cm.
1. Collective farms—Russia. I. Title.
HD1492.R9M54 79–410841

Osokina, Vera I͡Akovlevna.
(Sot͡sialisticheskoe stroitel'stvo v derevne i obshchina, 1920–1933)
Социалистическое строительство в деревне и община, 1920–1933 / В. Я. Осокина. — Москва : Мысль, 1978.
176 p. ; 20 cm. USSR 78
Includes bibliographical references.
0.85rub
1. Russia—Rural conditions. 2. Collective farms—Russia. I. Title.
HN523.O86 78–422773

Vylt͡san, Mikhail Avgustovich.
(Zavershai͡ushchiĭ ėtap sozdanii͡a kolkhoznogo stroi͡a)
Завершающий этап создания колхозного строя : 1935–1937 гг. / М. А. Вылцан. — Москва : Наука, 1978.
261 p. ; 22 cm. USSR***
At head of title: Akademii͡a nauk SSSR. Institut istorii SSSR.
Includes bibliographical references and indexes.
2.10rub
1. Collective farms—Russia. I. Title.
HD1492.R9V8 78–421744

—RUSSIA—ADDRESSES, ESSAYS, LECTURES

Karcz, Jerzy F
The economics of Communist agriculture : selected papers / Jerzy F. Karcz ; edited and with an introductory essay by Arthur W. Wright. — Bloomington, Ind. : International Development Institute, c1979.
ix, 494 p. : ill. ; 24 cm. — (Studies in East European and Soviet planning, development, and trade ; no. 25)
"The complete professional works of Jerzy F. Karcz:" p. [485]-488.
Bibliography: p. [481]-484.
Includes index.
ISBN 0-89249-022-5 : $22.50
1. Collective farms—Russia—Addresses, essays, lectures. 2. Agriculture and state—Russia—Addreesses, essays, lectures. I. Title. II. Series.
HD1492.R9K364 77-21186
77 MARC

—RUSSIA—FINANCE

Dvoret͡skova, Nadezhda Kuz'minichna.
(Osnovnye fondy kolkhozov)
Основные фонды колхозов / Н. К. Дворецкова. — Ростов н/Д : Изд-во Рост. ун-та, 1977.
83 p. ; 20 cm. USSR 77
Includes bibliographical references.
0.33rub
1. Collective farms—Russia—Finance. I. Title.
HD1492.R9D64 79–412182

—RUSSIA—INCOME DISTRIBUTION

Petrovich, Viktor Brankovich.
Problemy sovershenstvonanii͡a oplaty truda v kolkhoznokooperativnom sektore : Vopr. metodologii i praktiki / V. B. Petrovich. — Kiev ; Odessa : Vyshcha shkola, 1978.
255 p. ; 20 cm. USSR78
Includes bibliographical references.
2.10rub
1. Collective farms—Russia—Income distribution. I. Title.
HD1492.R9P47 79-379088
79 MARC

Shaĭbekov, Kimalkhan Asambekovich.
Teoreticheskie problemy pravovogo regulirovanii͡a oplaty truda v kolkhozakh / Shaĭbekov K. A. — Alma-Ata : Kazakhskiĭ gos. universitet im. S. M. Kirova, I͡Urid. fakul'tet, Kafedra sovetskogo grazhdanskogo prava, 1968.
469 p.
Bibliography: p. [446]-[467]
L.C. copy on microfilm only.
——— Microfilm. On 1 reel with other items ; 35 mm. Microfilm Slavic 12794
1. Collective farms—Russia—Income distribution. 2. Agricultural wages—Russia. I. Title.
79–374710

—RUSSIA—OFFICIALS AND EMPLOYEES

Kolkhozniku o pravakh i obi͡azannosti͡akh / [Shkorupilo G. M., Volkova N. A., Vovk I͡U. A. ... et al. ; Otv. red. Shkorupilo G. M.]. — Moskva : I͡Urid. lit., 1978.
263 p. ; 20 cm. USSR79
Includes bibliographical references.
0.65rub
1. Collective farms—Law and legislation—Russia. 2. Collective farms—Russia—Officials and employees. 3. Collective farms—Pensions—Russia. I. Shkorupilo, Grigoriĭ Matveevich.
79-372647
79 MARC

—RUSSIAN REPUBLIC
—ABKHASIA

Kuprava, Arvelod Ėrastovich.
(Abkhazskai͡a derevni͡a na puti sot͡sializma)
Абхазская деревня на пути социализма : (Канун коллективизации 1926–1929 гг.) / А. Э. Куправа ; [АН ГССР, Абхаз. ин-т яз., литературы и истории им. Д. Гулиа]. — Тбилиси : Мецниереба, 1977.
310 p. ; 22 cm. USSR 77
Added t. p. in Georgian.
Includes bibliographical references.
2.70rub
1. Land tenure—Russian Republic—Abkhasia. 2. Collective farms—Russian Republic—Abkhasia. I. Title.
HD719.A24K86 78–422014

—RUSSIAN REPUBLIC—BASHKIRIA

Galiev, Munavir Mukhametaminovich.
(Sila gektara)
Сила гектара : [Колхоз Уныш Дюртолин. р-на] / М. М. Галиев. — Уфа : Башк. кн. изд-во, 1977.
64 p. : ill. ; 20 cm. USSR 77
0.10rub
1. Collective farms—Russian Republic—Bashkiria. I. Title.
HD1492.R92B33 79–407762

—RUSSIAN REPUBLIC
—BASHKIRIA—INCOME DISTRIBUTION

(Valovoĭ dokhod kolkhozov i sovkhozov i ėffektivnost' ego ispol'zovanii͡a)
Валовой доход колхозов и совхозов и эффективность его использования : [Сборник статей] / АН СССР, Башк. филиал, Отд. экон. исследований ; [Ред. коллегия, М. И. Такумбетов (отв. ред.) ... et al.]. — Уфа : БФ АН СССР, 1977.
164 p. ; 21 cm. USSR 77
Includes bibliographical references.
1.11rub
1. Collective farms—Russian Republic—Bashkiria—Income distribution. 2. State farms—Russian Republic—Bashkiria—Income distribution. I. Takumbetov, Makhmut Iskhakovich. II. Akademii͡a nauk SSSR. Bashkirskiĭ filial. Otdel ėkonomicheskikh issledovaniĭ.
HD1492.R92B388 79–410648

—RUSSIAN REPUBLIC
—GORKI (PROVINCE)

Beli͡aev, Mikhail Feofilovich.
(Schast'e trudnoĭ dorogi)
Счастье трудной дороги : [О пред. колхоза им. Ленина П. М. Соколове] / Михаил Беляев. — Москва : Сов. Россия, 1978.
190 p., [8] leaves of plates : ill. ; 16 cm. — (Люди Советской России) USSR 78
Series romanized: Li͡udi Sovetskoĭ Rossii.
0.45rub
1. Collective farms—Russian Republic—Gorki (Province) 2. Sokolov, Petr Mikhaĭlovich, 1930– 3. Agriculturists—Russian Republic—Biography. I. Title.
HD1492.R92G52 79–412677

—RUSSIAN REPUBLIC
—MORDOVSKAYA A.S.S.R.

Gurin, D D
(Istoki nashego bogatstva)
Истоки нашего богатства / Д. Д. Гурин, пред. колхоза Россия Рузаев. р-на, В. М. Кряжева; [Лит. запись Т. Н. Кулыгиной]. — Саранск : Мордов. кн. изд-во, 1977.
38 p. ; 20 cm. USSR 77
0.05rub
1. Collective farms—Russian Republic—Mordovskaya A. S. S. R. I. Kri͡azheva, V. M., joint author. II. Kulygina, T. N. III. Title.
S562.R9G828 79–407077

—RUSSIAN REPUBLIC—SMOLENSK (PROVINCE)

Denisenkov, Ivan Antonovich.
Selo meni͡aet oblik : [Rasskaz. pred. kolkhoza im. Radishcheva Gagarin. r-na] / I. A. Denisenkov ; [Lit. zapis' I. P. Fomenko ; Predisl. I. E. Klimenko]. — Moskva : Kolos, 1978.
96 p. : ill. ; 20 cm. — (Peredovye kollektivy) USSR78
0.15rub
1. Collective farms—Russian Republic—Smolensk (Province) I. Fomenko, Ivan Pavlovich. II. Title.
HD1492.R92S63 79-363539
79 MARC

—SIBERIA

Lifshits, S
History of the Jewish kolkhoz in Siberia : 1926-34 / by Solomon Lifshitz. — Jerusalem : Hebrew University of Jerusalem, Soviet and East European Research Centre, 1975.
63 p. ; 27 cm. — (Soviet institutions series ; paper no. 4)
Cover title.
Text in Russian; introduction in English.
Caption title: Istorii͡a evreĭskogo kolkhoza v Sibiri.
1. Jews in Siberia—Economic conditions. 2. Farmers, Jewish—Siberia. 3. Collective farms—Siberia. 4. Siberia—Economic conditions. I. Title. II. Title: Istorii͡a evreĭskogo kolkhoza v Sibiri. III. Series.
DS135.R93S524 334'.683'0957 79-356558
79 MARC

—UKRAINE

Gordon, Elye, 1903–
אין די יידישע קאלווירטן. פאקטן און פארצייכענונגען. מאס־קווע, מעלוכע־פארלאג "דער עמעס." 1940.
79 p. 15 cm.
1. Collective farms — Ukraine. 2. Farmers, Jewish — Ukraine. 3. Jews in the Ukraine—Economic conditions. I. Title.
Title romanized: In di yidishe kolvirtn.
S469.R92U417 58–51301

—UKRAINE—VINNYCHCHYNA

Kashchuk, Natali͡a Omeli͡anivna.
(Zoloti zhnyva)
Золоті жнива : документальна повість / Наталя Кащук. — Київ : Молодь, 1977.
141 p., [4] leaves of plates : ill. ; 17 cm. — (Герої, роки, звершення) USSR***
Series romanized: Heroï, roky, zvershenni͡a.
0.26rub
1. Collective farms—Ukraine—Vinnychchyna. I. Title.
HD1492.R92V524 79–412114

COLLECTIVE FARMS (Continued)

—UKRAINE—ZAKARPATSKAYA OBLAST'

Blaha, I P
Rezervy zrostannia produktyvnosti pratsi v hirs'kykh i peredhirnykh kolhospakh / I. P. Blaha. — Uzhhorod : Karpaty, 1976.
78 p. ; 20 cm. USSR***
Includes bibliographical references.
1. Agriculture—Ukraine—Zakarpatskaya oblast'—Labor productivity. 2. Collective farms—Ukraine—Zakarpatskaya oblast'. I. Title.
S564.B52 79-377931
79 MARC

COLLECTIVE LABOR AGREEMENTS

see also Grievance arbitration

Ruprecht, Alfredo.
Derecho colectivo del trabajo / por Alfredo J. Ruprecht ; epílogo del Dr. Manuel Alonso Olea. — Madrid : Confederación Española de Cajas de Ahorros, D.L. 1978.
367 p. ; 21 cm. — (Publicaciones del Fondo para la Investigación Económica y Social ; 81) Sp78-May
Bibliography: p. ₍347₎-357.
ISBN 8472314049 : 600ptas
1. Collective labor agreements. I. Title.
K1703.R86 344.01'89 78-391882
79 MARC

—AERONAUTICS—UNITED STATES

Air and rail labor relations : a judicial history of the Railway labor act / compiled and edited by John W. Gohmann. — Dubuque, Iowa : Kendall/Hunt Pub. Co., c1979.
xiii, 373 p. ; 24 cm.
Earlier ed. (c1976) published under title: The Railway labor act.
"Text of the Railway labor act, as amended": p. 329-358.
Includes index.
ISBN 0-8403-1966-5
1. Railroads—United States—Employees. 2. Collective labor agreements—Railroads—United States. 3. Air lines—Employees—United States. 4. Collective labor agreements—Aeronautics—United States. I. Gohmann, John W. II. The Railway labor act. III. United States. Laws, statutes, etc. Railway labor act. 1979.
KF3580.R2A94 1979 344'.73'0189041385 78-71078
79 MARC

—AGRICULTURE—CALIFORNIA

California. Agricultural Labor Relations Board.
A handbook on the California agricultural labor relations law = Un folleto sobre la ley de relaciones del trabajo agrícola de California / prepared by the staff of the Agricultural Labor Relations Board. — Sacramento : The Board, 1978.
67 p. ; 18 cm.
English and Spanish.
1. Agricultural laborers—Legal status, laws, etc.—California. 2. Collective labor agreements—Agriculture—California. I. Title. II. Title: Un folleto sobre la ley de relaciones del trabajo agrícola de California.
KFC589.A4A82 344'.794'0189043 79-621311
79 MARC

—ARTISANS—GERMANY, WEST

Hermann, Heiner.
Kommentar zum Rahmentarifvertrag für das Maler- und Lackiererhandwerk vom 20. Dezember 1976 : in d. Fassung d. Änderungstarifvertrages vom 15. Juni 1978 / Heiner Hermann. — Stuttgart : Deutsche Verlags-Anstalt, 1978.
207 p. ; 21 cm. GFR78-A
Includes index.
ISBN 3-421-02526-6 : DM24.00
1. Collective labor agreements—Painters, Industrial—Germany, West. 2. Collective labor agreements—Artisans—Germany, West. I. Title.
79-346017
79 MARC

—BREWING INDUSTRY—SPAIN

Spain. Ministerio de Trabajo.
Industria cervecerca : ordenanza de trabajo / edición con introducción, referencias legales, sentencias resoluciones administrativas y comentarios a cargo de Alejandro Mulas y Ramón Ayerra. — Madrid : Ministerio de Trabajo, Servicio de Publicaciones, D.L. 1977.
191 p. ; 18 cm. — (Serie Ordenanzas laborales : RET. 77-780) (Biblioteca de textos legales)
On cover: Industrias cerveceras.
Includes index.
ISBN 8474340004 : 250ptas
1. Brewery workers—Legal status, laws, etc—Spain. 2. Collective labor agreements—Brewing industry—Spain. I. Mulas García, Alejandro. II. Ayerra, Ramón. III. Title. IV. Title: Industrias cerveceras.
79-340697
79 MARC

—CONSTRUCTION INDUSTRY—FRANCE

France. Laws, statutes, etc.
Conventions collectives nationales : bâtiment, travaux publics, carrières et matériaux et toutes professions concourant à l'acte de bâtir et d'aménager. — 4. éd. mise à jour. — ₍Paris : Éditions du Moniteur₎, 1978.
355 p. ; 30 cm. F***
Errata slip inserted.
"Moniteur des travaux publics et du bâtiment, numéro hors série."
ISBN 2-86282-040-7 : 116F
1. Collective labor agreements—Construction industry—France. I. Moniteur des travaux publics et du bâtiment. II. Title.
344'.44'.0189049 79-378279
79 MARC

—CONSTRUCTION INDUSTRY—GERMANY, WEST

Bundesrahmentarifvertrag : Anh., Berufsgruppen : TV Berufsbildung, Leistungslohntarifvertrag ... für d. gewerb. Arbeitnehmer d. Baugewerbes im Gebiet d. Bundesrepublik Deutschland / hrsg. von Udo Knitter. — Stand, 1. Juli 1978. — Düsseldorf : Werner, 1978.
212 p. ; 17 cm. — (Bau-Tarifverträge) GFR78-A
Includes index.
Includes text of the Bundesrahmentarifvertrag, an agreement between the Zentralverband des Deutschen Baugewerbes, the Hauptverband der Deutschen Bauindustrie and the Industriegewerkschaft Bau-Steine-Erden im Gebiet der Bundesrepublik Deutschland.
ISBN 3-8041-1163-7 : DM15.00
1. Collective labor agreements—Construction industry—Germany, West. I. Knitter, Udo, 1931- II. Zentralverband des Deutschen Baugewerbes. III. Hauptverband der Deutschen Bauindustrie. IV. Industriegewerkschaft Bau, Steine, Erden.
78-392174
79 MARC

—CONSTRUCTION INDUSTRY—ILLINOIS—CHICAGO—DIGESTS

Building Construction Employers' Association of Chicago.
Directory and guide - Building Construction Employers' Association of Chicago, Inc.
Chicago, Building Construction Employers' Association of Chicago.
v. 24 cm.
Spine title, : BCEA directory.
Key title: Directory and guide—Building Construction Employers' Association of Chicago, Inc., ISSN 0190-4787.
1. Collective labor agreements—Construction industry—Illinois—Chicago—Periodicals. 2. Collective labor agreements—Construction industry—Illinois—Chicago—Digests. 3. Construction industry—Illinois—Chicago—Directories. I. Building Construction Employers' Association of Chicago. BCEA directory. II. Title: BCEA directory. III. Key title.
KFX1238.1.C64B85 344'.77311'018819 79-640480
MARC-S

—CONSTRUCTION INDUSTRY—ILLINOIS—CHICAGO—PERIODICALS

Building Construction Employers' Association of Chicago.
Directory and guide - Building Construction Employers' Association of Chicago, Inc.
Chicago, Building Construction Employers' Association of Chicago.
v. 24 cm.
Spine title, : BCEA directory.
Key title: Directory and guide—Building Construction Employers' Association of Chicago, Inc., ISSN 0190-4787.
1. Collective labor agreements—Construction industry—Illinois—Chicago—Periodicals. 2. Collective labor agreements—Construction industry—Illinois—Chicago—Digests. 3. Construction industry—Illinois—Chicago—Directories. I. Building Construction Employers' Association of Chicago. BCEA directory. II. Title: BCEA directory. III. Key title.
KFX1238.1.C64B85 344'.77311'018819 79-640480
MARC-S

—CONSTRUCTION INDUSTRY—QUÉBEC (PROVINCE)

Hébert, Gérard.
Les relations du travail dans la construction au Québec / Gérard Hébert. — ₍Ottawa₎ : Conseil économique du Canada, c1977-
v. : ill. ; 25 cm. C***
Includes bibliographical references.
CONTENTS: 1. Régime des relations du travail.

"Préparé, dans sa première version, pour le Conseil économique du Canada on 1972 et 1973."
ISBN 0-660-00781-9 : 6.00F
1. Collective labor agreements—Construction industry—Québec (Province) 2. Labor laws and legislation—Québec (Province) 3. Construction industry—Law and legislation—Québec (Province) I. Economic Council of Canada. II. Title.
KEQ656.C65H4 344'.714'01769 79-357780
79 MARC

—CONSTRUCTION INDUSTRY—UNITED STATES

Contractors Mutual Association.
The Nuclear stabilization agreement : its potential impact / prepared by Contractors Mutual Association. — Washington : CMA, c1978.
4, 2, ₍38₎ leaves ; 28 cm.
Includes the text of the Nuclear power construction stabilization agreement.
1. Collective labor agreements—Construction industry—United States. 2. Atomic power-plants—Design and construction. 3. Public utilities—United States. I. Nuclear power construction stabilization agreement. 1978. II. Title.
KF3409.C65C66 343'.73'092 78-105101
79 MARC

United States. Congress. Senate. Committee on Human Resources. Subcommittee on Labor.
Equal treatment of craft and industrial workers, 1977 : hearings before theSubcommittee on Labor of the Committee on Human Resources, United States Senate, Ninety-fifth Congress, on S. 924 ... March 17, 18, and 22, 1977. — Washington : U.S. Govt. Print. Off., 1977.
viii, 830 p. : ill. ; 24 cm.
Includes bibliographical references.
1. Collective labor agreements—Construction industry—United States. 2. Construction workers—Legal status, laws, etc.—United States. 3. Picketing—United States. I. Title.
KF26.H855 1977f 344'.73'0189049 78-602090
00 MARC

—DAIRYING—SPAIN

Spain. Ministerio de Trabajo.
Ordenanza laboral : industrias lácteas y sus derivados / edición con introducción, referencias legales, sentencias, resoluciones administrativas y comentarios a cardo ₍sic₎ de Alejandro Mulas y Ramón Ayerra. — Madrid : Ministerio de Trabajo, Servicio de Publicaciones, D.L. 1977.
189 p. ; 18 cm. — (Serie Ordenanzas laborales ; RET. 77-788) (Biblioteca de textos legales) Sp77-Oct
ISBN 8474340233 : 225ptas
1. Dairy workers—Legal status, laws, etc.—Spain. 2. Collective labor agreements—Dairying—Spain. 3. Labor laws and legislation—Spain. I. Mulas García, Alejandro. II. Ayerra, Ramón. III. Title. IV. Title: Industrias lácteas y sus derivados.
78-387478
78 MARC

—EDUCATION—CALIFORNIA

California. Legislature. Assembly. Public Employees and Retirement Committee.
Implementation of SB 160 : transcript of proceedings : San Francisco, California, November 3, 1977 / Assembly Public Employees and Retirement Committee. — ₍Sacramento₎ : The Committee, ₍1977₎
69 p. ; 29 cm.
Cover title.
"No. 653."
Includes index.
$3.00
1. Collective labor agreements—Education—California. I. Title.
KFC10.4.P78 1977b 78-623515
344'.794'018904137
00 MARC

—EDUCATION, HIGHER—UNITED STATES—DIRECTORIES

Garfin, Molly.
Directory of faculty contracts and bargaining agents in institutions ₍sic₎ of higher education / ₍compiled by Molly Garfin₎. — New York : National Center for the Study of Collective Bargaining in Higher Education, Baruch College, CUNY, 1979.
iii, 47 p. ; 28 cm.
$5.00
1. Collection bargaining—College teachers—United States. 2. College teachers' unions—United States—Directories. 3. Collective labor agreements—Education, Higher—United States—Directories.
LB2335.885.U6G37 79-113060
331.89'041'378120973
79 MARC

—ENGINEERS—GERMANY, WEST

Schmidt, Friedrich W Ing.
Das Eingruppierungsrecht der technischen Angestellten im öffentlichen Dienst : Kommentar zur Neubewertung d. Ingenieur- u. Technikertätigkeiten nach d. Tarifvertrag für Angestellte in techn. Berufen im BAT-Bereich vom 15. Juni 1972 / bearb. von F. W. Schmidt. — 3., Aufl. / überarb. von K. H. Eger. — Regensburg : Walhalla-und-Praetoria-Verlag, 1978.
328 p. in various pagings ; 15 cm. GFR78-A14
ISBN 3-8029-5335-5
1. Collective labor agreements—Engineers—Germany, West. 2. Collective labor agreements—Government employees—Germany, West. I. Eger, Klaus H. II. Title.
79-345385
79 MARC

COLLECTIVE LABOR AGREEMENTS (Continued)

—ESCALATOR CLAUSES see Wages—Cost-of-living adjustments

—FOOD INDUSTRY—ARGENTINE REPUBLIC

Sindicato de Obreros Fideeros y Afines.
Trabajadores de la industria fideera; convención colectiva de trabajo no. 293/71. ⌈Partes intervinientes: Sindicato de Obreros Fideeros y Afines y Asociación de Cámaras de la Industria Fideera Argentina. Buenos Aires⌉ Ministerio de Trabajo, 1972.
⌈28⌉ l. 30 cm. LACAP72-3585
Cover title.
1. Collective labor agreements—Food industry—Argentine Republic. I. Asociación de Cámaras de la Industria Fideera Argentina. II. Argentine Republic. Ministerio de Trabajo. III. Title. IV. Title: Convención colectiva de trabajo no. 293/71.
77-570711
78 MARC

—GOVERNMENT EMPLOYEES

International Labor Office.
Freedom of association and procedures for determining conditions of employment in the public service : fifth item on the agenda. — Geneva : International Labour Office, 1977-1978.
2 v. ; 24 cm. Sw***
At head of title: International Labour Conference, 64th Session, 1978, Report V (1-2).
ISBN 9221017540 (v. 1) : 15.00F (v. 1) 12.50F (v. 2)
1. Trade-unions—Government employees. 2. Collective labor agreements—Government employees. 3. Employee-management relations in government. I. International Labour Conference, 64th, Geneva, 1978. II. Title.
K3450.I575 344.01'89041'351 78-315970
78 MARC

—GOVERNMENT EMPLOYEES—CANADA

Public Service Alliance of Canada.
Brief on collective bargaining in the Federal public service / presented to Committee on Legislation Review ; submitted by the Public Service Alliance of Canada. — ⌈Ottawa?⌉ : The Alliance, 1971.
2 v. ; 22-28 x 22-36 cm.
Includes bibliographical references.
Cover title.
1. Collective labor agreements—Government employees—Canada. 2. Collective bargaining—Government employees—Canada. I. Canada. Committee on Legislation Review. II. Title.
KE3196.F4P8 79-312065
344'.71'0189041354
79 MARC

—GOVERNMENT EMPLOYEES—GERMANY, WEST

Germany (Federal Republic, 1949-).
Vergütungsverordnung des BAT : Tätigkeitsmerkmale. — 1. Aufl., Stand, 1.1.1977. — Düsseldorf : Werner, 1977.
310 p. ; 17 cm. GFR77-A
ISBN 3-8041-3823-3 : DM17.00
1. Germany, West—Officials and employees—Salaries, allowances, etc. 2. Collective labor agreements—Government employees—Germany, West. I. Germany (Federal Republic, 1949-). Bundes-Angestelltentarifvertrag. II. Title.
78-355438
78 MARC

Germany (Federal Republic, 1949-). Laws, statutes, etc.
BAT : Bundes-Angestellten-Tarifvertrag und Vergütungstarifverträge : Bund, Länder, Gemeinden. — Stand, 1. März 1978, 18. Aufl. — Düsseldorf : Werner, 1978.
xiii, 351 p. ; 17 cm. GFR78-A
ISBN 3-8041-1164-5 : DM17.80
1. Collective labor agreements—Government employees—Germany, West. I. Title.
78-387641
344'.43'0189041354
79 MARC

Schmidt, Friedrich W Ing.
Das Eingruppierungsrecht der technischen Angestellten im öffentlichen Dienst : Kommentar zur Neubewertung d. Ingenieur- u. Technikertätigkeiten nach d. Tarifvertrag für Angestellte in techn. Berufen im BAT-Bereich vom 15. Juni 1972 / bearb. von F. W. Schmidt. — 3., Aufl. / überarb. von K. H. Eger. — Regensburg : Walhalla-und-Praetoria-Verlag, 1978.
328 p. in various pagings ; 15 cm. GFR78-A14
ISBN 3-8029-5335-5
1. Collective labor agreements—Engineers—Germany, West. 2. Collective labor agreements—Government employees—Germany, West. I. Eger, Klaus H. II. Title.
79-345385
79 MARC

—GOVERNMENT EMPLOYEES—GREAT BRITAIN

Glover, Audrey F L
Labor rights of employees of the British Government / prepared by Audrey F. Glover. — ⌈Washington⌉ : Library of Congress, Law Library, 1975.
9 p. ; 27 cm.
Bibliography: p. 9.
1. Employee-management relations in government—Great Britain. 2. Collective labor agreements—Government employees—Great Britain. I. United States. Library of Congress. Law Library. II. Title.
KD4524.Z9G6 344'.41'018904135 77-601669
—— Copy 3. Z663.5.L32 MARC
79

Glover, Audrey F L
Labor rights of employees of the British Government / prepared by Audrey F. Glover. — Rev. — Washington : Library of Congress, Law Library, 1977.
9 p. ; 27 cm.
Bibliography: p. 9.
1. Employee-management relations in government—Great Britain. 2. Collective labor agreements—Government employees—Great Britain. I. United States. Library of Congress. Law Library. II. Title.
KD4524.Z9G6 1977 78-600780
344'.41'0189041354
—— Copy 3. Z663.5L32 1977 MARC
79

—GOVERNMENT EMPLOYEES—ITALY

Severi, Fabio Severo.
La contrattazione collettiva nel pubblico impiego : saggi / Fabio Severo Severi. — Milano : A. Giuffrè, 1978.
vii, 144 p. ; 24 cm. — (⌈Pubblicazioni⌉ - Università di Trieste, Facoltà di scienze politiche ; 9) It78-June
Includes bibliographical references.
L4000
1. Collective labor agreements—Government employees—Italy. I. Title. II. Series: Trieste. Università. Facoltà di scienze politiche. Pubblicazioni - Università di Trieste, Facoltà di scienze politiche ; 9.
78-398131
*79 MARC

—GOVERNMENT EMPLOYEES—MASSACHUSETTS

Massachusetts. Governor's Task Force on Chapter 150E and Impasse Procedures.
Interim report of the Governor's Task Force on Chapter 150E and Impasse Procedures. — Boston : The Task Force, ⌈letter 1976⌉
11, 2 p. ; 28 cm.
"Publication no. 9357-16-150-12-76-CR."
1. Arbitration, Industrial—Massachusetts. 2. Collective labor agreements—Government employees—Massachusetts.
KFM2733.8.P8A83 79-622988
331.89'041'3539744
79 MARC

—GOVERNMENT EMPLOYEES—NEBRASKA

Nebraska. Court of Industrial Relations.
Rules of the Court of Industrial Relations : rules as amended and adopted by orders of the court. — ⌈Lincoln⌉ : The Court, ⌈1978⌉
i, 26 p. : forms ; 22 cm.
"Entered September 7, 1972 ... May 5, 1978."
1. Collective labor agreements—Government employees—Nebraska. 2. Employee-management relations in government—Nebraska. 3. Court rules—Nebraska. I. Title.
KFN332.8.P77A32 1978 79-623882
344'.782'01890413539
79 MARC

—GOVERNMENT EMPLOYEES—NEW JERSEY

Stark, Harry Faber.
Public sector bargaining in New Jersey : the statute and the negotiating process : a guide to the New Jersey public employer-employee relations act / Harry F. Stark. — ⌈New Brunswick⌉ : Institute of Management and Labor Relations, University Extension Division, Rutgers-The State University of New Jersey, ⌈1978⌉
20 p. ; 23 cm.
1. Collective labor agreements—Government employees—New Jersey. I. Title.
KFN2132.8.P77S7 79-620535
344'.749'0189041353
79 MARC

—GOVERNMENT EMPLOYEES—NEW YORK (STATE)

New York (State). Laws, statutes, etc.
⌈Public employees' fair employment law⌉
The Taylor law, 1978-79. — ⌈Albany⌉ : New York State, Public Employment Relations Board, ⌈1978⌉
21 p. ; 22 cm.
1. Collective labor agreements—Government employees—New York (State) I. Title.
KFN5562.P8A32 1978 79-621548
344'.747'01890413539
79 MARC

—GOVERNMENT EMPLOYEES—UNITED STATES

Public sector labor relations / Stuart Linnick, chairman. — ⌈New York⌉ : Practising Law Institute, c1978.
152 p. (p. 147-152 blank) ; 22 cm. — (Litigation and administrative practice series) (Litigation course handbook series ; no. 126)
"Prepared for distribution at the public sector labor relations program, October-December 1978."
"H4-3879."
1. Collective labor agreements—Government employees—United States. 2. Employee-management relations in government—United States. I. Linnick, Stuart. II. New York (City). Practising Law Institute. III. Series. IV. Series: Litigation course handbook series ; no. 126.
KF3409.P77P8 1978 344'.73'0189041353 78-70875
79 MARC

United States. Bureau of Labor Statistics.
Collective bargaining agreements for state and county government employees. — Washington : U.S. Dept. of Labor, Bureau of Labor Statistics : for sale by Supt. of Docs., U.S. Govt. Print. Off., 1976.
vi, 85 p. ; 28 cm. — (Bulletin - U.S. Bureau of Labor Statistics ; 1920)
Prepared by R. R. Nelson, assisted by R. J. Symkowiak and J. E. Mann.
Supt. of Docs. no.: L 2.3:1920
1. Collective labor agreements—Government Employees—United States. 2. State governments—Officials and employees. 3. County officials and employees—United States. I. Nelson, Richard R. II. Symkowiak, Ronald J. III. Mann, Jacquelyn E. IV. Title. V. Series: United States. Bureau of Labor Statistics. Bulletin ; 1920.
KF3409.P77A843 76-608200
344'.73'01890413539
76 MARC

—GOVERNMENT EMPLOYEES—UNITED STATES—ADDRESSES, ESSAYS, LECTURES

Government labor relations : trends and information for the future / edited by Hugh D. Jascourt. — ⌈Greenbelt, Md.⌉ : Public Employment Relations Research Institute, c1979-
v. ; 28 cm.
Includes bibliographical references.
CONTENTS: v. 1. 1975 to 1978.

$12.00 (v. 1)
1. Employee-management relations in government—United States—Addresses, essays, lectures. 2. Collective labor agreements—Government employees—United States—Addresses, essays, lectures. 3. Employee-management relations in government—United States—Cases. 4. Collective labor agreements—Government employees—United States—Cases. I. Jascourt, Hugh D.
KF3580.G6G68 344'.73'0189041353 79-10924
79 MARC

COLLECTIVE LABOR AGREEMENTS
(Continued)

—GOVERNMENT EMPLOYEES
—UNITED STATES—CASES

Government labor relations : trends and information for the future / edited by Hugh D. Jascourt. — ⌊Greenbelt, Md.⌋ : Public Employment Relations Research Institute, c1979-
v. ; 28 cm.
Includes bibliographical references.
CONTENTS: v. 1. 1975 to 1978.

$12.00 (v. 1)
1. Employee-management relations in government—United States—Addresses, essays, lectures. 2. Collective labor agreements—Government employees—United States—Addresses, essays, lectures. 3. Employee-management relations in government—United States—Cases. 4. Collective labor agreements—Government employees—United States—Cases. I. Jascourt, Hugh D.
KF3580.G6G68 344'.73'0189041353 79-10924
79 MARC

—GOVERNMENT EMPLOYEES
—UNITED STATES—CASES
—INDEXES

United States. Federal Labor Relations Council.
Subject matter index with cross-reference table.
Washington, Federal Labor Relations Council, Office of the Executive Director.
v. 27 cm. annual.
Index to: Decisions and interpretations of the Federal Labor Relations Council, ISSN 0145-1529.
Key title: Subject matter index with cross-reference table, ISSN 0164-1832.
1. United States. Federal Labor Relations Council. Decisions and interpretations of the Federal Labor Relations Council—Indexes. 2. Employee-management relations in government—United States—Cases—Indexes. 3. Collective labor agreements—Government employees—United States—Cases—Indexes. I. Title.
KF5365.A553 Suppl. 79-640142
344'.73'01890413530002648 MARC-S

—GOVERNMENT EMPLOYEES—
UNITED STATES—STATES

Baird, James, 1943-
Practical labor relations for public employers / James Baird, Donald Anderson ; contributing author, Stanislaw S. Damas. — Wheat Ridge : Colorado Municipal League, c1978.
iii, 226 p. : ill. ; 28 cm.
Includes bibliographical references and index.
$15.95
1. Collective labor agreements—Government employees—United States—States. 2. Trade-unions—Government employees—United States—States. I. Anderson, Donald, 1943- joint author. II. Damas, Stanislaw S. III. Colorado Municipal League. IV. Title.
KF3409.P77Z9517 79-108739
344'.73'0189041353
00 MARC

Najita, Joyce M
Guide to statutory provisions in public sector collective bargaining, scope of negotiations / by Joyce M. Najita. — 2d issue. — ⌊Honolulu⌋ : Industrial Relations Center, University of Hawaii, 1978.
v, 140 p. ; 23 cm. — (Occasional publications - Industrial Relation Center, University of Hawaii ; no. 126)
$4.00
1. Collective labor agreements—Government employees—United States—States. I. Title. II. Series: Hawaii. University, Honolulu. Industrial Relations Center. Occasional publications ; no. 126.
KF3409.P77Z954 1978 78-111170
344'.73'0189041353
79 MARC

—GOVERNMENT EMPLOYEES—
UNITED STATES—STATES—TERMS AND PHRASES

Ferguson, John Bodley.
Guide to statutory provisions in public sector collective bargaining : terms and definitions / by John B. Ferguson. — ⌊Honolulu⌋ : Industrial Relations Center, College of Business Administration, University of Hawaii, 1977.
v, 141 p. ; 23 cm. — (Occasional publication - Industrial Relations Center, University of Hawaii ; no. 120)
$3.50
1. Collective labor agreements—Government employees—United States—States—Terms and phrases. I. Title. II. Series: Hawaii. University, Honolulu. Industrial Relations Center. Occasional publications ; no. 120.
KF3409.P77Z9525 77-623603
344'.73'0189041353
78 MARC

—HANDBOOKS, MANUALS, ETC.

Simmons R , John G
La contratación colectiva de trabajo como una responsabilidad más de la empresa / autor, John G. Simmons R. — Lima : Centro Interamericano de Administración del Trabajo, 1974.
67 leaves : ill. ; 28 cm.
"Proyecto regional (OIT/PNUD) RLA/71/254."
"Reg. 97/E1-14."
Bibliography: leaf 67.
1. Collective labor agreements—Handbooks, manuals, etc. I. Centro Interamericano de Administración del Trabajo. II. Title.
HD6971.S476 79-101186
79 MARC

—HEALTH FACILITIES—UNITED STATES

Health care labor manual / by Skoler & Abbott ; editor-in-chief, Brian E. Hayes. — Germantown, Md. : Health Law Center, c1974-
v. : ill. ; 30 cm.
Loose-leaf for updating.
Includes bibliography.
ISBN 0-912862-11-4
1. Labor laws and legislation—United States. 2. Trade-unions—Health facilities—United States. 3. Collective labor agreements—Health facilities—United States. I. Hayes, Brian E. II. Skoler & Abbott. III. Aspen Systems Corporation. Health Law Center.
KF3580.H4H4 344'.73'01 74-24672
79 MARC

—HOSPITALS—ITALY

Chiti, Mario P
Problemi della contrattazione collettiva ospedaliera / Mario P. Chiti. — Milano : A. Giuffrè, 1978.
115 p. ; 23 cm. — (Medicina e salute ; 5) It78-Apr
Includes bibliographical references.
L3000
1. Collective labor agreements—Hospitals—Italy. I. Title.
78-382463
*78 MARC

—JANITORS—ITALY

Portieri : contratto nazionale 24 febbraio 1978 in vigore dal I gennaio 1978 al 30 novembre 1980. — Milano : Pirola, 1978.
42 p. ; 22 cm. — (Collezione contratti di lavoro ; 2009) It78-Aug
L1500
1. Collective labor agreements—Janitors—Italy.
78-390689
*78 MARC

—JOURNALISTS—UNITED STATES

Barwis, Gail Lund.
Contractual newsroom democracy / Gail Lund Barwis. — ⌊Lexington, Ky.⌋ : Association for Education in Journalism, 1978.
33 p. ; 23 cm. — (Journalism monographs ; no. 57)
Includes bibliographical references.
$2.50
1. Collective bargaining—Journalists—United States. 2. Collective labor agreements—Journalists—United States. I. Title. II. Series: Journalism monographs (Austin) ; no. 57.
PN4722.J6 no. 57 070'.08 s 78-112032
⌊KF3409.N3⌋ 79 MARC

—LEATHER INDUSTRY—SPAIN

Spain. Ministerio de Trabajo.
Industrias de la piel : ordenanza de trabajo / edición con introducción, referencias legales, sentencias, resoluciones administrativas y comentarios a cargo de Alejandro Mulas y Ramón Ayerra. — ⌊Madrid⌋ : Servicio de Publicaciones, Ministerio de Trabajo, D.L. 1977.
216 p. ; 18 cm. — (Serie Ordenanzas laborales ; RET. 77/828) (Biblioteca de textos legales) Sp78-Feb
ISBN 8474340462 : 300ptas
1. Leather workers—Legal status, laws, etc.—Spain. 2. Collective labor agreements—Leather industry—Spain. I. Mulas García, Alejandro. II. Ayerra, Ramón. III. Title.
79-353292
79 MARC

—MACHINE INDUSTRY—ITALY

Segatori, Roberto.
La professionalità difficile : nascita e sviluppo dell'inquadramento unico / ⌊di⌋ Roberto Segatori, Daniela Torresini ; prefazione di Giovanni B. Montironi. — Roma : CEDIS, ⌊1979⌋
118 p. ; 21 cm. — (Organizzazione industriale ⌊i.e. aziendale⌋ e relazioni industriali ; n. 3) It79-Mar
Includes bibliographical references.
ISBN 8885018106 : L4000
1. Collective labor agreements—Machine industry—Italy. 2. Employees, Rating of—Italy. I. Torresini, Daniela, joint author. II. Title.
HD6713.M2S43 79-371151
79 MARC

—MEDICAL PERSONNEL

see also Collective labor agreements
—Hospitals

—METAL INDUSTRY—ITALY

Tondi, Luigi.
Metalmeccanici : trattamento del personale sulla base dei contratti 1 maggio 1976 (industria) e 24 settembre 1976 (artigianato) / Luigi Tondi. — 9. ed. — Roma : Buffetti, 1978.
150 p. ; 21 cm. — (I Contratti collettivi commentati ; 1) It78-Aug
Includes index.
L2500
1. Collective labor agreements—Metal industry—Italy. I. Title.
78-384169
*78 MARC

—MOVING-PICTURE INDUSTRY
—UNITED STATES—PERIODICALS

Brooks, Stanley J
Brooks' standard rate book. 1956–
⌊Los Angeles, S. J. Brooks Co., etc.⌋
v. 23 cm. annual.
"Complete Hollywood theatrical television film and videotape salary schedules for studio and distant locations includes I. A. T. S. E. unions, A. F. L. unions, independent unions and guilds."
Issue for 1979 covers 6 months only: Feb. 1–July 31, 1979.
Vols. for 1956– include I. A. T. S. E. unions, A. F. L. unions and independent unions.
Key title: Brooks' standard rate book, ISSN 0193–2814.
1. Wages—Moving-picture industry—United States—Periodicals. 2. Collective labor agreements—Moving-picture industry—United States—Periodicals. I. Title: Salary schedules for studio and distant locations. II. Title.
HD4966.M79U53 79–643357
331.2'81'791450973 MARC–S

—MUNICIPAL EMPLOYEES—UNITED STATES

Rhyne, Charles S 1912-
The law of municipal labor relations / by Charles S. Rhyne and Robert H. Drummer. — Washington : National Institute of Municipal Law Officers, 1979.
ix, 249 p. ; 23 cm. — (NIMLO research report ; 163)
$25.00
1. Collective labor agreements—Municipal employees—United States. I. Drummer, Robert H., joint author. II. Title. III. Series: National Institute of Municipal Law Officers, Washington, D.C. NIMLO research report ; 163.
KF5305.N3 no. 163 342'.73'09 s 79-88737
⌊KF3409.P77⌋ 79 MARC

—NURSES—NEW BRUNSWICK

New Brunswick. Treasury Board.
Agreement between Treasury Board and the New Brunswick Association of Registered Nurses' Civil Service Nurses' Provincial Collective Bargaining Council. Group: Nursing. Expires: August 31, 1972. ⌊Fredericton, 1972?⌋
21 p. 23 cm.
Cover title. C***
1. Collective labor agreements—Nurses—New Brunswick. I. New Brunswick Association of Registered Nurses. Civil Service Nurses Provincial Collective Bargaining Council. II. Title.
KEN403.N85T78
78-314658
331.89'041'61073069209715
78 MARC

COLLECTIVE LABOR AGREEMENTS
(Continued)

—PAINTERS, INDUSTRIAL—GERMANY, WEST

Hermann, Heiner.
Kommentar zum Rahmentarifvertrag für das Maler- und Lackiererhandwerk vom 20. Dezember 1976 : in d. Fassung d. Änderungstarifvertrages vom 15. Juni 1978 / Heiner Hermann. — Stuttgart : Deutsche Verlags-Anstalt, 1978.
207 p. ; 21 cm. GFR78-A
Includes index.
ISBN 3-421-02526-6 : DM24.00
1. Collective labor agreements—Painters, Industrial—Germany, West. 2. Collective labor agreements—Artisans—Germany, West. I. Title.
79-346017
79 MARC

—PHYSICIANS

see also Collective labor agreements —Hospitals

—RAILROADS—UNITED STATES

Air and rail labor relations : a judicial history of the Railway labor act / compiled and edited by John W. Gohmann. — Dubuque, Iowa : Kendall/Hunt Pub. Co., c1979.
xiii, 373 p. ; 24 cm.
Earlier ed. (c1976) published under title: The Railway labor act.
"Text of the Railway labor act, as amended": p. 329-358.
Includes index.
ISBN 0-8403-1966-5
1. Railroads—United States—Employees. 2. Collective labor agreements—Railroads—United States. 3. Air lines—Employees—United States. 4. Collective labor agreements—Aeronautics—United States. I. Gohmann, John W. II. The Railway labor act. III. United States. Laws, statutes, etc. Railway labor act. 1979.
KF3580.R2A94 1979 344'.73'0189041385 78-71078
79 MARC

The Railway labor act : a judical history / compiled and edited by John W. Gohmann. — Chicago : Lake Shore Litho, c1976.
iv, 321 p. ; 30 cm.
Text of the Railway labor act, as amended: p. 273-301.
Includes index.
$13.95
1. Railroads—United States—Employees. 2. Collective labor agreements—Railroads—United States. I. Gohmann, John W. II. United States. Laws, statutes, etc. Railway labor act. 1976.
KF3580.R2R33 76-363735
344'.73'0189041385
76[r79]rev MARC

—RETAIL TRADE—UNITED STATES —PERIODICALS

Retail/services labor report. no. 1293–
May, 9, 1974–
Washington, Bureau of National Affairs.
no. 29 cm. weekly.
Continues: Retail labor report.
Key title: Retail - services labor report, ISSN 0148-7980.
1. Collective labor agreements—Retail trade—United States—Periodicals. 2. Collective labor agreements—Service industries—United States—Periodicals. I. Bureau of National Affairs, Washington, D. C.
KF3409.R37A137 344'.73'0189041 78-649098
MARC-S

—SERVICE INDUSTRIES—PERIODICALS

Retail/services labor report. no. 1293–
May, 9, 1974–
Washington, Bureau of National Affairs.
no. 29 cm. weekly.
Continues: Retail labor report.
Key title: Retail - services labor report, ISSN 0148-7980.
1. Collective labor agreements—Retail trade—United States—Periodicals. 2. Collective labor agreements—Service industries—United States—Periodicals. I. Bureau of National Affairs, Washington, D. C.
KF3409.R37A137 344'.73'0189041 78-649098
MARC-S

—STONE-MASONS see Collective labor agreements —Construction industry

—TRANSPORT WORKERS

see also Collective labor agreements —Aeronautics; Collective labor agreements—Railroads

—AUSTRIA

Weissenberg, Gerhard.
Arbeitsverfassungsgesetz / mit Erl. v. Gerhard Weissenberg, Josef Cerny. — 2. Aufl. — Wien : Verl. der Österr. Gewerkschaftsbundes, 1978.
655 p. ; 17 cm. — (Schriftenreihe des Österreichischen Gewerkschaftsbundes ; Nr. 116) Au78-12-51
Includes text of the Arbeitsverfassungsgesetz.
Includes index.
ISBN 4-7035-0210-X : S228.00
1. Works councils—Austria. 2. Employees' representation in management —Austria. 3. Collective labor agreements—Austria. I. Cerny, Josef, Dr., joint author. II. Austria. Laws, statutes, etc. Arbeitsverfassungsgesetz. 1978. III. Title. IV. Series: Österreichischer Gewerkschaftsbund. Schriftenreihe ; Nr. 116.
344'.436'0189 78-385380
*79 MARC

—BELGIUM

Swennen, R
De arbeidsovereenkomsten : wet van 3 juli 1978 / door R. Swennen. — Kortrijk : UGA, [1978]
176 p. ; 24 cm. Be***
Includes bibliographical references and index.
1. Collective labor agreements—Belgium. I. Title.
79-369299
79 MARC

—CANADA

Hayes, James K
Perspectives on management rights : the curious logic of the argument for reducing industrial discord by removing the mid-term strike bar from labour relations legislation / James K. Hayes. — Don Mills, Ont. : Ontario Federation of Labour, [1974?]
87 p. ; 28 cm. C75-9219-X
Cover title.
Includes bibliographical references.
1. Strikes and lockouts—Canada. 2. Collective labor agreements—Canada. 3. Labor laws and legislation—Canada. I. Ontario Federation of Labour. II. Title.
KE3213.H37 344'.71'01892 78-321809
79 MARC

Palmer, Earl E 1932-
Collective agreement arbitration in Canada / Earl Edward Palmer ; with the assistance of B. Hovius. — Toronto : Butterworth & Co. (Canada), c1978.
lxxxviii, 635 p. ; 24 cm. — (Canadian legal textbook series) C77-1738-X
Includes bibliographical references and indexes.
ISBN 0-409-85770-X : $65.00
1. Arbitration, Industrial—Canada. 2. Collective labor agreements—Canada. I. Hovius, B., joint author. II. Title.
KE3206.P34 344'.71'018914 78-321954
78 MARC

—CANADA—CASES

Adams, George W
Labour relations law : supplement, 1977-78 / G. W. Adams ; edited by John Page. — [Downsview, Ont.] : York University, Osgoode Hall Law School, [1978]
209 p. ; 37 cm. C***
"Materials are for the exclusive use of students of Osgoode Hall Law School."
1. Collective labor agreements—Canada—Cases. I. Osgoode Hall Law School. II. Title.
KE3193.A7A3 Suppl. 344'.71'0189 78-324622
79 MARC

—CANADA—STATISTICS

Canada. Labour Data Branch.
Provisions in major collective agreements covering employees in certain transportation, communications, trade, utilities, and service industries in Canada, 1975 = Dispositions de grandes conventions collectives concernant les employés dans certaines industries de transports, communications, commerce, services d'utilité et autres entreprises au Canada, 1975 / Labour Canada, Labour Data. — [Ottawa] : Labour Canada, Labour Data, 1975.
89 p. ; 22 x 26 cm. C77-71467-6E
English and French.
Chiefly tables.
1. Collective labor agreements—Canada—Statistics. I. Title. II. Title: Dispositions de grandes conventions collectives concernant les employés dans certaines industries de transports, communications, commerce, services d'utilité et autres entreprises au Canada, 1975.
HD6524.C36 1975 331.2'0971 79-311916
79 MARC

—COLOMBIA

Escobar Uribe, Ignacio.
Los conflictos colectivos de trabajo en Colombia / Ignacio Escobar Uribe. — Bogotá : Editorial Temis, 1975.
x, 290 p. ; 23 cm.
Includes indexes.
1. Labor disputes—Colombia. 2. Arbitration, Industrial—Colombia. 3. Collective labor agreements—Colombia. I. Title.
344'.861'0189 77-155586
79 MARC

—CONNECTICUT—STATISTICS

Samper, Maria-Luz.
The Connecticut labor contract analysis program : manufacturing sector / by Maria Luz Samper ; under direction of David Pinsky. — [Storrs] : Labor Education Center, University of Connecticut, 1978.
vi, 100, 7 p. ; 28 cm.
1. Collective labor agreements—Connecticut—Statistics. I. Pinsky, David. II. Connecticut. University. Labor Education Center. III. Title.
HD6517.C7S25 331.89'09746 78-622369
79 MARC

—FINLAND

Suomen Ammattiliittojen Keskusjärjestö.
Katsaus työehtosopimustoimintaan ja työehtosopimuspoliittinen ohjelma. — Helsinki : Suomen ammattiliittojen keskusjärjestö, 1976.
103 p. : ill. ; 21 cm. Fi
"Hyväksytty SAK:n XI edustajakokouksessa 17.-20.6. 1976."
ISBN 9519521194
1. Collective bargaining—Finland. 2. Collective labor agreements—Finland. 3. Wages—Finland. 4. Suomen Ammattiliittojen Keskusjärjestö. I. Title.
HD6735.5.S77 1976 79-370836
*79 MARC

—GERMANY—ADDRESSES, ESSAYS, LECTURES

Kubo, Keiji, 1920-
(Rōdō kyōyaku no hōriron)
労働協約の法理論 / 久保敬治著. — 東京 : 総合労働研究所, 1978.
vii, 190 p. ; 22 cm. Ja 78-25762
Includes index.
¥2000
1. Collective labor agreements—Germany—Addresses, essays, lectures. 2. Collective labor agreements—Germany, West—Addresses, essays, lectures. 3. Collective labor agreements—Japan—Addresses, essays, lectures. I. Title.
79-801911

—GERMANY, WEST

Achilles, Wolfgang, 1938-
Der tatsächliche Parteiwille bei der Auslegung von Tarifverträgen. [n.p., 1967]
xviii, 144 p. 21 cm.
Inaug.-Diss.—Cologne.
Vita.
Bibliography: p. v-xvi.
1. Collective labor agreements—Germany, West. I. Title.
77-483185
78 MARC

Klas, Helmut.
Die Zulässigkeit der Verfassungsbeschwerde gegen Tarifnormen / Helmut Klas. — Frankfurt/Main : R. G. Fischer, 1979.
xii, 208 p. ; 21 cm. GFR79-A
Originally presented as the author's thesis, Münster, 1978.
Bibliography: p. xii-xli.
ISBN 3-88323-043-X : DM36.00
1. Collective labor agreements—Germany, West. 2. Appellate procedure—Germany, West. 3. Civil rights—Germany, West. I. Title.
79-372216
79 MARC

Mummenhoff, Winfried, 1942-
Die Sperrwirkung des Tarifvertrags nach den §§ 56 und 59 Betriebsverfassungsgesetz. [Köln, Spezialdruckerei für Dissertationen: Gouder u. Hansen, 1968]
xlviii, 188 p. 21 cm.
Inaug.-Diss.—Cologne.
Vita.
Bibliography: p. xix-xxxvii.
1. Works councils—Germany, West. 2. Collective labor agreements—Germany, West. I. Title.
77-551517
(MARC)

COLLECTIVE LABOR AGREEMENTS
—GERMANY, WEST (Continued)

Muszack, Rolf.
Stellenbewertung nach dem BAT : Leitfaden über d. Durchführung von Stellenbewertungen nach dem BAT unter Berücks. d. 37. u. 38. Tarifvertrages zur Änderung u. Erg. d. BAT / von Rolf Muszack u. Christian Müsgen. — 5. überarb. Aufl. — Köln ; Stuttgart ; Hamburg ; Hannover : Deutscher Gemeindeverlag, 1978.
xi, 81 p. ; 21 cm. GFR78-A44
Includes excerpts from and amendments to the Bundes-Angestelltentarifvertrag.
ISBN 3-555-00329-1 : DM26.00
1. Collective labor agreements—Germany, West. 2. Job evaluation—Germany, West. I. Müsgen, Christian, joint author. II. Title.
79-337269
79 MARC

Weiss, Manfred, 1940-
Gewerkschaftliche Vertrauensleute : tarifvertragl. Verbesserungen ihrer Arbeit im Betrieb / Manfred Weiss. — Köln : Europäische Verlagsanstalt, 1978.
109 p. ; 21 cm. — (Schriftenreihe der Otto-Brenner-Stiftung ; 11)
GFR78-A28
Includes bibliographical references.
ISBN 3-434-10116-0
1. Collective labor agreements—Germany, West. 2. Business agents (Trade-union) I. Title. II. Series: Otto Brenner Stiftung. Schriftenreihe ; 11.
79-368961
79 MARC

Zöllner, Wolfgang, 1928-
Aussperrung und arbeitskampfrechtliche Parität / von Wolfgang Zöllner. — Düsseldorf : Handelsblatt-GmbH, 1974.
62 p. ; 24 cm. — (Schriftenreihe Der Betrieb) GFR74-A
Includes bibliographical references and index.
DM19.60
1. Strikes and lockouts—Germany, West. 2. Collective labor agreements—Germany, West. 3. Equality before the law—Germany, West. I. Title.
78-392363
79 MARC

—GERMANY, WEST—ADDRESSES, ESSAYS, LECTURES

Kubo, Keiji, 1920-
(Rōdō kyōyaku no hōriron)
労働協約の法理論 / 久保敬治著. — 東京 : 総合労働研究所, 1978.
vii, 190 p. ; 22 cm. Ja 78-25762
Includes index.
¥2000
1. Collective labor agreements—Germany—Addresses, essays, lectures. 2. Collective labor agreements—Germany, West—Addresses, essays, lectures. 3. Collective labor agreements—Japan—Addresses, essays, lectures. I. Title.
79-801911

—HUNGARY

Hágelmayer, Istvánné.
A kollektív szerződés alapkérdései / Hágelmayer Istvánné. — Budapest : Akadémiai Kiadó, 1979.
342 p. ; 20 cm. H***
Includes bibliographical references.
ISBN 9630516934 : 62.00Ft
1. Collective labor agreements—Hungary. I. Title.
79-369269
79 MARC

—ITALY

Pontarollo, Enzo, 1946-
Struttura dei costi del lavoro e contrattazione : dinamica temporale e ipotesi di riforma / Enzo Pontarollo. — Milano : Vita e pensiero, c1978.
222, ⌊10⌋ p. ; 20 cm. — (Problemi economici d'oggi ; 16) It***
"Appendice statistica" (tables): p. 123-⌊231⌋
Includes bibliographical references.
L5900
1. Labor costs—Italy. 2. Collective labor agreements—Italy. I. Title. II. Series.
HC10.P752 n. 16 79-357966
⌊HD5034⌋ 79 MARC

—JAPAN

Shin Mitsubishi Jūkōgyō Kabushiki Kaisha.
(Rōdō kyōyaku)
労働協約. — ⌊東京⌋ : 新三菱重工業株式会社・新三菱重工業労働組合, ⌊1953?⌋
49 p. ; 19 cm.
Cover title.
1. Collective labor agreements—Japan. 2. Shin Mitsubishi Jūkōgyō Kabushiki Kaisha. 3. Shin Mitsubishi Jūkōgyō Rōdō Kumiai. I. Shin Mitsubishi Jūkōgyō Rōdō Kumiai. II. Title.
79-809039

—JAPAN—ADDRESSES, ESSAYS, LECTURES

Kubo, Keiji, 1920-
(Rōdō kyōyaku no hōriron)
労働協約の法理論 / 久保敬治著. — 東京 : 総合労働研究所, 1978.
vii, 190 p. ; 22 cm. Ja 78-25762
Includes index.
¥2000
1. Collective labor agreements—Germany—Addresses, essays, lectures. 2. Collective labor agreements—Germany, West—Addresses, essays, lectures. 3. Collective labor agreements—Japan—Addresses, essays, lectures. I. Title.
79-801911

Nomura, Heiji, 1902-1979.
(Dantai kōshō to kyōyaku tōsō)
団体交渉と協約闘争 / ⌊著者 野村平爾⌋. — 東京 : 労働旬報社, 1978.
291 p. ; 22 cm. — (野村平爾著作集 ; 第3巻) Ja***
Includes bibliographical references.
¥4000
1. Collective labor agreements—Japan—Addresses, essays, lectures. 2. Collective bargaining—Japan—Addresses, essays, lectures. 3. Employee rules—Japan—Addresses, essays, lectures. I. Title. II. Series: Nomura, Heiji, 1902-1979. Nomura Heiji chosaku shū ; dai 3-kan.
79-806938

—MEXICO

Kaye, Dionisio J
Aplicación práctica de la Ley federal del trabajo en la administración de personal / Dionisio J. Kaye. — 1. ed. — México : Editorial IEE, 1977.
335, ⌊5⌋ p. : forms ; 23 cm.
Bibliography: p. ⌊337⌋-⌊339⌋
1. Labor laws and legislation—Mexico. 2. Collective labor agreements—Mexico. I. Title.
344'.72'01 77-155518
79 MARC

—MEXICO—ADDRESSES, ESSAYS, LECTURES

Control y luchas del movimiento obrero : dos ensayos / Martín Moro, Alfonso Bouzas, Isaac Palacios S. — 1. ed. en español. — México : Editorial Nuestro Tiempo, 1978.
215 p. ; 19 cm. — (Colección Los grandes problemas nacionales)
Includes bibliographical references.
CONTENTS: Moro, M. Política burguesa en el movimiento obrero.—Bouzas, A., Palacios S., I. El movimiento obrero mexicano.
ISBN 9684270135
1. Trade-unions—Mexico—Addresses, essays, lectures. 2. Labor and laboring classes—Mexico—Addresses, essays, lectures. 3. Collective labor agreements—Mexico—Addresses, essays, lectures. I. Moro, Martín. Política burguesa en el movimiento obrero. 1978. II. Bouzas, Alfonso. El movimiento obrero mexicano. 1978. III. Palacios S., Issac.
HD6532.C66 79-107386
79 MARC

—ONTARIO

Haywood, Lenard.
Cost-of-living provisions in Ontario collective bargaining agreements, October 1974 / by Lenard Haywood and Susan Markell. — Toronto : Research Branch, Ontario Ministry of Labour, 1975.
25 cm. ; 30 cm. — (Bargaining information series ; no. 3) C***
1. Wages—Cost of living adjustments—Ontario. 2. Collective labor agreements—Ontario. I. Markell, Susan, joint author. II. Title. III. Series.
HD4932.C7H38 331.2'15 78-312506
78 MARC

Ontario. Ministry of Labor. Research Branch.
Overtime compensation in Ontario collective agreements, August 1975 / by Research Branch, Ontario Ministry of Labour. — Toronto : Ministry of Labour, Research Branch, 1975.
17 p. ; 30 cm. — (Bargaining information series ; no. 9) C***
1. Overtime—Ontario. 2. Wages—Ontario. 3. Collective labor agreements—Ontario. I. Title. II. Series.
HD5111.C2O57 1975 331.2'162 79-309284
79 MARC

—PORTUGAL—PERIODICALS

Portugal. Ministério do Trabalho. Serviço de Informação Científica e Técnica.
Boletim do trabalho e emprego. 1a. ⌊i. e. Primeira⌋ série. Lisboa, Serviço de Informação Científica e Técnica, Ministério do Trabalho.
no. 31 cm.
1. Labor and laboring classes—Portugal—Periodicals. 2. Collective labor agreements — Portugal — Periodicals. 3. Trade-unions — Portugal—Periodicals. I. Title.
HD8591.A34a 78-647692
MARC-S

—RUSSIA

Arkhipova, Bella Arsent'evna.
(Kollektivnyĭ dogovor)
Коллективный договор / Б. Архипова. — Москва : Профиздат, 1978.
78 p. ; 20 cm. — (Пособие для профсоюзного активиста)
USSR 78
Series romanized: Posobie dlīā profsoīūznogo aktivista.
Includes bibliographical references.
0.15rub
1. Collective labor agreements—Russia. I. Title.
79-407566

—SASKATCHEWAN—STATISTICS

Saskatchewan. Dept. of Labour. Policy Planning and Research Division.
Study of Saskatchewan union agreements 1978. — ⌊Regina⌋ : Saskatchewan Labour, Policy Planning and Research Division, 1979.
xiii, 70 p. ; 28 cm. C***
"Gordon T. Snyder, Minister. Robert W. Mitchell, Deputy Minister."
Errata slip inserted.
1. Collective labor agreements—Saskatchewan—Statistics. I. Snyder, Gordon T. II. Mitchell, Robert W. III. Title.
HD6529.S2S24 1979 331.89'097124 79-315240
79 MARC

—SPAIN

García Fernández, Manuel, 1946-
Unidad de negociación del convenio colectivo, capacidad convencional y representación profesional en el derecho positivo español / Manuel García Fernández. — 1. ed. — ⌊Madrid⌋ : Servicio de Publicaciones, Ministerio de Trabajo, 1977.
248 p. ; 21 cm. — (Colección Tesis doctorales ; 6) Sp78-Apr
Thesis—Barcelona.
Includes bibliographical references.
ISBN 8474340098 : 400ptas
1. Collective labor agreements—Spain. I. Title.
344'.46'0189 78-393206
79 MARC

Spain. Laws, statutes, etc.
⌊Ley 38/1973, de 19 de diciembre⌋
Convenios colectivos sindicales de trabajo : Ley 38/1973 : disposiciones complementarias y normas sindicales / comentario por Angel Trujillo Cabrera. — ⌊Madrid⌋ : Ministerio de Trabajo, Servicio de Publicaciones, D.L. 1974.
101 p. ; 21 cm. — (Serie Relaciones laborales ; RET 74-661) (Biblioteca de textos legales) (Textos legales) Sp78-Feb
Includes index.
1. Collective labor agreements—Spain. I. Trujillo Cabrera, Angel. II. Title.
344'.46'018902632 79-348943
79 MARC

—SWEDEN

Schmidt, Folke Fredrik, 1909-
Die Reform des kollektiven Arbeitsrechts in Schweden / Folke Schmidt. — 1. Aufl. — Köln ; Frankfurt am Main : Europäische Verlagsanstalt, 1978.
262 p. ; 21 cm. — (Schriftenreihe der Otto Brenner Stiftung ; 10)
GFR78-A
Includes legislation.
Includes bibliographical references and index.
ISBN 3-434-10113-6
1. Collective labor agreements—Sweden. 2. Employees' reresentation in management—Sweden. 3. Trade-unions—Sweden. I. Title. II. Series.
344'.485'0189 79-337451
79 MARC

Uisk, Ahto.
Lagen om anställningsskydd / med komment. av Ahto Uisk. — Stockholm : Federativ, 1979.
96 p. : ill. ; 21 cm. S79
"Bygger på Lars Lunnings bok Anställningsskydd."
"Lag om anställningsskydd": p. 69-79.
Includes bibliographical references and index.
ISBN 9185016586
1. Labor laws and legislation—Sweden. 2. Collective labor agreements—Sweden. I. Lunning, Lars, 1934- Anställningsskydd. II. Sweden. Laws, statutes, etc. Lagen om anställningsskydd. 1979. III. Title.
79-385803
80 MARC

—TURKEY

Ekonomi, Münir, 1933-
İş hukuku / Münir Ekonomi. — İstanbul : Teknik Üniversite Matbaası, 1976-
v. ; 24 cm. — (İstanbul Teknik Üniversitesi kütüphanesi ; sayı 1061)
Bibliography: v. 1, p. ⌊301⌋-317.
Includes index.
CONTENTS: cilt 1. Ferdî iş hukuku.
1. Labor laws and legislation—Turkey. 2. Collective labor agreements—Turkey. I. Title. II. Series: İstanbul Teknik Üniversitesi. İstanbul Teknik Üniversitesi kütüphanesi ; sayı 1061.
TA7.I93 sayı 1061, etc. 79-370230
79 MARC

LABOR LAWS AND LEGISLATION (Continued)

—UNITED STATES

Cabot, Stephen J
Labor management relations act manual : a guide to effective labor relations / by Stephen J. Cabot. — Boston : Warren, Gorham & Lamont, c1978-
1 v. : forms ; 30 cm.
Loose-leaf for updating.
Includes text of Labor management relations act.
Includes index.
ISBN 0-88262-208-0
1. Labor laws and legislation—United States. 2. Collective labor agreements—United States. I. United States. Laws, statutes, etc. Labor management relations act, 1947. 1978. II. Title.
KF3369.C3 344'.73'0189 78-60446
79 MARC

Leslie, Douglas L 1942-
Labor law in a nutshell / by Douglas L. Leslie. — St. Paul : West Pub. Co., 1979.
xxxix, 403 p. ; 19 cm. — (Nutshell series)
Includes index.
ISBN 0-8299-2053-6
1. Labor laws and legislation—United States. 2. Trade-unions—United States. 3. Collective labor agreements—United States. I. Title.
KF3369.3.L39 344'.73'01 79-15396
79 MARC

Mounce, Earl Winifield.
Collective bargaining, by E. W. Mounce and Robley D. Stevens. 6098-1. ₍Ed. 2₎ Scranton, International Correspondence Schools, c1960.
114 p. 19 cm.
1. Collective bargaining—United States. 2. Collective labor agreements—United States. I. Stevens, Robley D., joint author. II. International Correspondence Schools, Scranton, Pa.
KF3408.Z9M68 344'.73'0189 60-52006
₍r79₎rev MARC

Taylor, Benjamin J
Labor relations law / Benjamin J. Taylor, Fred Witney. — 3d ed. — Englewood Cliffs, N.J. : Prentice-Hall, c1979.
xiv, 834 p. ; 25 cm. — (Prentice-Hall industrial relations and personnel series)
Bibliography: p. 793-801.
Includes indexes.
ISBN 0-13-329276-2
1. Collective labor agreements—United States. 2. Labor laws and legislation—United States. I. Witney, Fred, 1917- joint author. II. Title.
KF3408.T35 1979 344'.73'0189 78-23169
78 MARC

United States. Bureau of Labor Statistics.
Wage administration provisions / U.S. Department of Labor, Bureau of Labor Statistics. — Washington : U.S. Govt. Print. Off., 1978.
vii, 84 p. ; 28 cm. — (Major collective bargaining agreements) (Bulletin - U.S. Department of Labor, Bureau of Labor Statistics ; 1425-17)
Prepared by Homer R. Kemp, Jr., and others.
1. Wages—United States. 2. Collective labor agreements—United States. I. Kemp, Homer R. II. Title. III. Series: United States. Bureau of Labor Statistics. Major collective bargaining agreements. IV. Series: United States. Bureau of Labor Statistics. Bulletin ; 1425-17.
HD8051.A62 no. 1425-17 78-603566
331'.0973 s
₍HD4973₎ 79 MARC

United States. Congress. Senate. Committee on Human Resources.
Labor law reform act of 1978 : hearing before the Committee on Human Resources, United States Senate, Ninety-fifth Congress, second session on H.R. 8410 ... June 29, 1978. — Washington : U.S. Govt. Print. Off., 1978.
iii, 103 p. ; 24 cm.
1. Collective labor agreements—United States. 2. United States. National Labor Relations Board.
KF26.H8 1978g 344'.73'01 79-600508
79 MARC

—UNITED STATES—CASES

Leslie, Douglas L 1942-
Cases and materials on labor law : process and policy / Douglas L. Leslie. — Boston : Little, Brown, c1979.
xx, 776 p. ; 25 cm. — (Law school casebook series)
Includes bibliographical references and indexes.
1. Labor laws and legislation—United States—Cases. 2. Collective labor agreements—United States—Cases. I. Title.
KF3368.L47 344'.73'01 78-69772
79 MARC

—UNITED STATES—CONGRESSES

San Fernando Valley College of Law Symposium on Labor Law and Industrial Relations, Los Angeles, 1978.
San Fernando Valley College of Law Symposium on Labor Law and Industrial Relations / editor-in-chief, Libby L. Barrabbee, editors, Paul H. Libis ... ₍et al.₎ ; symposium chairman, Stanislaus Pulle. — Sepulveda, Calif. : San Fernando Valley Law Review, c1978.
230 p. ; 26 cm.
Includes bibliographical references and indexes.
$8.50
1. Labor laws and legislation—United States—Congresses. 2. Collective labor agreements—United States—Congresses. I. Barrabbee, Libby L. II. San Fernando Valley College of Law.
KF3369.A2S36 344'.73'01 79-107270
79 MARC

—UNITED STATES—FORMS

Stessin, Lawrence.
The encyclopedia of collective bargaining contract clauses / by Lawrence Stessin, Len Smedresman. — New York : Man & Manager, c1977.
vii, 334 p. ; 30 cm.
Includes index.
1. Collective labor agreements—United States—Forms. I. Smedresman, Len, joint author. II. Title.
KF3408.A3S7 1977 344'.73'0189 78-110770
79 MARC

COLLECTIVE PSYCHOTHERAPY
see Group psychotherapy

COLLECTIVE SECURITY see Security, International

COLLECTIVE SETTLEMENTS

see also Christian communities; Collective farms; Religious communities

Bonnard, Arthur de, 1805-1875.
La Marmite libératrice : ou, le Commerce transformé : simple entretien / Gallus ₍i.e. A. de Bonnard₎ ; préface de Henri Desroche. — Paris : Balland : Adel, 1978.
xxix, 246 p. ; 22 cm. — (Bibliothèque des utopies ISSN 0399-7529)
F78-12683
Includes bibliographical references.
ISBN 2-7158-0151-3 : 59.00F
1. Collective settlements. 2. Utopias. 3. Bonnard, Arthur de, 1805-1875. 4. Socialists—Biography. I. Title. II. Series.
HX632.B66 1978 335'.9 79-370854
79 MARC

Poulton, Robert.
Contemporary communes, a bibliographical and interpretative essay / Robert Poulton & Robin Cohen. — ₍Birmingham₎ : Faculty of Commerce & Social Science, University of Birmingham, 1977.
33 p. ; 30 cm. — (Discussion papers - Faculty of Commerce & Social Science, University of Birmingham : Series C ; no. 31) GB***
Bibliography: p. 30-33.
1. Collective settlements. 2. Collective settlements—Bibliography. 3. Subculture. I. Cohen, Robin, joint author. II. Title. III. Series: Birmingham, Eng. University. Faculty of Commerce and Social Science. Discussion papers : Series C, Social Administration : no. 31.
HX626.P68 301.2'2 79-301500
79 MARC

Woolley, Pat.
The art of living together / Pat Woolley. — Melbourne ; Norwalk, Conn. : Visa, 1978.
125 p. : ill. ; 23 cm. Aus***
ISBN 0-86932-050-5: $7.95
1. Collective settlements. 2. Home economics. I. Title.
HX635.W66 335'.9 79-300922
79 MARC

—ADDRESSES, ESSAYS, LECTURES

Gordon, Michael, 1940- comp.
The nuclear family in crisis: the search for an alternative. Edited and with an introd. by Michael Gordon. New York, Harper & Row ₍1972₎
224 p. 24 cm.
Includes bibliographical references.
CONTENTS: Plato on women and the family.—Family surrogates in colonial America: the Moravian experiment, by G. L. Gollin.—Experimental family organization: an historico-cultural report on the Oneida community, by W. M. Kephart.—Is the family universal? The Israeli case, by M. E. Spiro.—Children of the kibbutz, by L. and K. Rabkin.—Aging in Israel, a planned society, by Y. Talmon.—The changing Soviet family, by U. Bronfenbrenner.—Li Kuei-ying, woman pioneer, aged 32, by J. Myrdal.—The position of Finnish women: regional and cross-national comparisons, by E. Haavio-Mannila.—Communes, by R. M. Kanter.—The new arrangement, by G. Fonzi.—Natural child birth and cooperative child rearing in psychedelic communes, by D. E. Smith and J. L. Sternfield.—The group marriage, by L. L. and J. M. Constantine.
ISBN 0-06-042405-2
1. Family—Addresses, essays, lectures. 2. Collective settlements—Addresses, essays, lectures. I. Title.
HQ728.G62 301.42'12'08 72-184937
72₍r79₎rev MARC

—BIBLIOGRAPHY

Poulton, Robert.
Contemporary communes, a bibliographical and interpretative essay / Robert Poulton & Robin Cohen. — ₍Birmingham₎ : Faculty of Commerce & Social Science, University of Birmingham, 1977.
33 p. ; 30 cm. — (Discussion papers - Faculty of Commerce & Social Science, University of Birmingham : Series C ; no. 31) GB***
Bibliography: p. 30-33.
1. Collective settlements. 2. Collective settlements—Bibliography. 3. Subculture. I. Cohen, Robin, joint author. II. Title. III. Series: Birmingham, Eng. University. Faculty of Commerce and Social Science. Discussion papers : Series C, Social Administration : no. 31.
HX626.P68 301.2'2 79-301500
79 MARC

—EDUCATION see Collective education

—HANDBOOKS, MANUALS, ETC.

Comune agricola : un manuale d'uso per vivere in campagna. — Roma : Savelli, 1978.
128 p. : ill. ; 19 cm. — (Controcultura ; 24) It78-Sept
At head of title: Stampa alternativa.
Bibliography: p. ₍124₎-128.
1. Collective settlements—Handbooks, manuals, etc. 2. Collective farms—Handbooks, manuals, etc. I. Stampa alternativa.
HX635.C64 335'.9 78-392783
*79 MARC

—HISTORY

Arrington, Leonard J
Building the city of God : community & cooperation among the Mormons / Leonard J. Arrington, Feramorz Y. Fox, Dean L. May. — Salt Lake City : Deseret Book Co., 1976.
xiii, 497 p. : ill. ; 24 cm.
Includes bibliographical references and index.
ISBN 0-87747-590-3
1. Mormons and Mormonism—History. 2. Collective settlements—History. I. Fox, Feramorz Y., 1881-1957, joint author. II. May, Dean L., joint author. III. Title.
BX8611.A77 289.3'73 76-27141
76₍r79₎rev MARC

—ALBERTA

Alberta. Special Advisory Committee on Communal Property and Land Use.
Communal property in Alberta. — ₍Edmonton₎ : Dept. of Municipal Affairs, Special Advisory Office, ₍1973?₎
35 p. ; 28 cm. C***
Cover title.
Bibliography: p. 33-35.
1. Hutterite Brethren—Alberta. 2. Collective settlements—Alberta. I. Title.
BX8129.H8A69 1973 289.7'7123 78-314495
78 MARC

—CALIFORNIA

Dean, Barbara, 1946-
Wellspring, a story from the deep country / Barbara Dean ; ill. by Julie Miyasaki. — Covelo, Calif. : Island Press, c1979.
xi, 208 p. : ill. ; 20 cm.
ISBN 0-933280-01-7 : $6.00
1. Country life—California. 2. Collective settlements—California. I. Title.
HT421.D4 979.4 79-2606
79 MARC

—DENMARK—COPENHAGEN

Grau, Monika.
Christiania : Elendsviertel, soziales Experiment, oder Selbstorganisation Nicht-Angepasster? / Monika Grau & Heiner Gringmuth. — 1. Aufl. — Wetzlar : Neuland-Verlag, 1977.
181 p. : ill. ; 18 x 21 cm. GFR***
Includes bibliographical references.
ISBN 3-921764-00-9
1. Hippies—Denmark—Copenhagen. 2. Collective settlements—Denmark—Copenhagen. 3. Christianshavn, Copenhagen. I. Gringmuth, Heiner, joint author. II. Title.
HN547.G65 78-384312
79 MARC

COLLECTIVE SETTLEMENTS (Continued)

—ENGLAND

Dunn, Nell, 1936-
Living like I do / [compiled by] Nell Dunn ; photographs by Tina Tranter. — London : Futura Publications, 1977.
200 p., [16] p. of plates : ill., ports. ; 18 cm. GB78-05660
American ed. published under title: Different drummers.
ISBN 0-86007-511-7 : £0.85
1. Single-parent family—England. 2. Collective settlements—England. 3. Sex customs—England. I. Title.
HQ616.D86 1977b 301.42'0942 78-318067
78 MARC

—ENGLAND—HISTORY

Hardy, Dennis.
Alternative communities in nineteenth century England / Dennis Hardy. — London ; New York : Longman, 1979.
268 p. : ill. ; 23 cm.
Bibliography: p. [254]-259.
Includes index.
ISBN 0-582-50213-6 : £5.95 (UK). ISBN 0-85250-216-0 pbk.
1. Collective settlements—England—History. 2. Utopias—History. I. Title.
HX695.A3H37 335'.9'42 77-26279
77 MARC

—GERMANY—HISTORY—20TH CENTURY

Heineke, Gustav.
Frühe Kommunen in Deutschland : Versuche neuen Zusammenlebens Jugendbewegung & Novemberrevolution 1919-24 / Gustav Heineke. — 1. Aufl. — Herford : Zündhölzchen, 1978.
77 p. ; 21 cm. GFR***
DM4.90
1. Youth movement—Germany—History—20th century. 2. Collective settlements—Germany—History—20th century. I. Title.
HN19.H38 335'.9'43 79-354632
79 MARC

—GERMANY, WEST—CASE STUDIES

Wortmann, Raoul.
Heim, Knast, und dann? : Bericht aus e. Jugendwohnkollektiv / Raoul Wortmann. — 1. Aufl. — Frankfurt/Main ; New York : Campus-Verlag, 1978.
245 p. ; 21 cm. — (Campus soziale Probleme) GFR78-A
Bibliography: p. 245.
ISBN 3-593-32291-9 : DM24.00
1. Youth centers—Germany, West—Case studies. 2. Collective settlements—Germany, West—Case studies. 3. Socialization. 4. Juvenile delinquency—Germany, West. I. Title.
HQ799.G5W67 301.43'15'0943 79-340817
79 MARC

—GREAT BRITAIN—DIRECTORIES

Michele.
Directory of alternative communities [in the British Isles] / compiled by Michele & Kevin of The Teachers. — 2d ed., 1st issue. — Bangor, Wales : The Teachers, 1978.
113 p. : map ; 26 cm. GB***
On cover: A Teachers publication.
Bibliography: p. 110-112.
ISBN 0-86085-011-0
1. Collective settlement—Great Britain—Directories. 2. Cooperative societies—Great Britain—Directories. I. Kevin, joint author. II. Title.
HX695.A3M5 1978 335'.9'41 78-324297
79 MARC

—ISRAEL

'Atid, David.
(Ḳeliṭat gar'ine Naḥal ba-hityashvut)
קליטת גרעיני נח"ל בהתיישבות / דוד עתיד, אורה כהנוביץ, שונמית ליפשיץ ; [ערך, חנן רותם]. — רחובות : המרכז לחקר התישבות כפרית ועירונית, 1975.
196 p. : ill. ; 24 cm. — (פרסומים בבעיות פיתוח אזורי ; 16)
At head of title: ha-Merkaz le-ḥeḳer hityashvut kafrit ye-'ironit; Berit ha-tenu'ah ha-ḳibutsit.
Bibliography: p. 195–196.
1. Collective settlements—Israel. 2. Israel. Tseva haganah le-Yiśrael. Naḥal. 3. Motivation (Psychology) I. Kahanovits, Orah, joint author. II. Lifshits, Shunamit, joint author. III. Rotem, Hanan. IV. ha-Merkaz le-ḥeḳer hityashvut kafrit ye-'ironit. V. Berit ha-tenu'ah ha-ḳibutsit. VI. Title. VII. Series: Pirsumim bi-ve'ayot pituaḥ ezori ; 16.
HX742.2.A3A8 79–951153

Bedford, John.
Kibbutz volunteer / [by] John Bedford. — Oxford : Vacation Work, 1978.
[1], 112 p. : 2 maps, port. ; 22 cm. GB78-31242
Bibliography: p. 22.
ISBN 0-901205-52-4 : £2.95
1. Collective settlements—Israel. 2. Youth volunteers in social service—Israel. 3. Israel—Description and travel—Guide-books. I. Title.
HX765.P3B22 1978 331.1 79-305567
79 MARC

Bockenheimer, Philipp.
Sturktur und Entwicklung ausgewählter Kibbuzim in Israel / Philipp Bockenheimer. — Giessen : Geograph. Inst. d. Justus-Liebig-Univ., 1978.
x, 272 p. : ill., maps ; 30 cm. — (Giessener geographische Schriften ; Heft 43) GFR78-B23
Originally presented as the author's thesis, Giessen.
Summary in English and Hebrew.
Bibliography: p. 262-267.
DM32.00
1. Collective settlements—Israel. I. Title. II. Series.
HX742.2.A3B6 1978 334'.683'095694 79-368670
79 MARC

Breitbach, Irmgard.
Identitätsentwicklung im Kindergartenalter unter den Bedingungen kollektiver Erziehung : d. Analyse dreier Kindergärten in israel. Kibbutzim / Irmgard Breitbach. — München : Minerva-Publikation, 1979.
355 p. ; 21 cm. — (Minerva-Fachserie Pädagogik) GFR79-A
Originally presented as the author's thesis, Marburg, 1978.
Bibliography: p. 345-355.
ISBN 3-597-10071-6 : DM48.00
1. Kindergartens—Israel. 2. Collective settlements—Israel. 3. Child development. 4. Identity (Psychology) I. Title.
LB1438.I75B73 1979 79-371632
79 MARC

(Dor ha-sheni)
הדור השני : הקיבוץ בין המשך לתמורה / מ. רוזנר ... [et al.] ; עריכה, דקלה גולומב. — [תל־אביב] : ספרית פועלים, c1978.
620 p. ; 22 cm.
On verso of t. p.: The second generation, continuity and change in the Kibbutz.
Bibliography: p. [591]–608.
Includes index.
1. Collective settlements—Israel. 2. Conflict of generations. I. Rosner, Menahem. II. Golomb, Diḳlah.
HX742.2.A3D67 79–951866

Gerson, Menachem, 1908-
Family, women, and socialization in the kibbutz / Menachem Gerson. — Lexington, Mass. : Lexington Books, c1978.
xiii, 142 p. ; 23 cm.
Includes bibliographical references and index.
ISBN 0-669-02371-X
1. Collective settlements—Israel. 2. Family—Israel. 3. Women—Israel. I. Title.
HX742.2.A3G47 334'.683'095694 78-57188
78 MARC

Orchan, Eliette.
(Hashpa'at me'afyene ha-ḳibuts 'al tahalikh ha-ḳelitah shel "bodedim")
השפעת מאפייני הקיבוץ על תהליך הקליטה של "בודדים" / על ידי אליאט אורחן. — [תל־אביב] : אוניברסיטת תל־אביב, הפאקולטה למדעי החברה, החוג לסוציולוגיה ואנטרופולוגיה, 1977.
140 leaves : ill. ; 28 cm.
Added t. p.: Factors of kibbutz structure that influence the absorption of "non-movement joiners" into the kibbutz.
Summary in English.
Thesis (M. A.)—Universiṭat Tel-Aviv.
Errata slip inserted.
Bibliography: leaves 129–136.
1. Collective settlements — Israel. 2. Israel — Social conditions. I. Title. II. Title: Factors of kibbutz structure that influence the absorption of "non-movement joiners" into the kibbutz.
HX742.2.A3O73 79–951256

Shepher, Joseph.
(Mavo le-sotsyologyah shel ha-ḳibuts)
מבוא לסוציולוגיה של הקיבוץ / יוסף שפר. — [תל־אביב] : המדרשה החקלאית ע"ש רופין, 1977.
260 p. ; 24 cm.
On verso of t. p.: Introduction to the sociology of the kibbutz.
Includes bibliographies.
1. Collective settlements—Israel. 2. Israel—Social conditions. I. Title.
HX742.2.A3S53 79–951337

Spiro, Melford E
Gender and culture : kibbutz women revisited / Melford E. Spiro. — Durham, N.C. : Duke University Press, 1979.
xix, 116 p. ; 25 cm.
"A volume in a series based on the Howard Eikenberry Jensen lectures on sociology and social action."
Bibliography: p. [111-113.
Includes index.
ISBN 0-8223-0427-9 : $9.75
1. Collective settlements—Israel. 2. Women—Israel. I. Title.
HX742.2.A3S64 334'.683'095694 78-74846
00 MARC

Yalan, E
The design of agricultural settlements : technological aspects of rural community development / E. Yalan. — Jerusalem : Jewish Agency for Israel, Rural Settlement Dept., c1975.
xiii, 132 p., 133-321 leaves : ill. ; 27 cm.
Bibliography: p. 126-132.
1. Rural development—Israel. 2. Collective settlements—Israel. 3. Villages—Israel. I. Title.
HD2058.5.Y34 334'.683'095694 79-306868
79 MARC

—ISRAEL—BIBLIOGRAPHY

Erickson, Judith B
The Israeli kibbutz / Judith B. Erickson. — Monticello, Ill. : Council of Planning Librarians, 1978.
67 p. ; 29 cm. — (Exchange bibliography - Council of Planning Librarians ; 1454)
Cover title.
$6.50
1. Collective settlements—Israel—Bibliography. I. Series: Council of Planning Librarians. Exchange bibliography ; 1454.
Z5942.C68 no. 1454 016.3092 s 78-103311
[Z7165.I7] 79 MARC
[HX742.2]

—ISRAEL—PERIODICALS

Iḥud ha-ḳevutsot yeha-ḳibutsim.
(Igeret la-ḥaverim)
אגרת לחברים. מס. 1- ; 1.11.51-
תל־אביב.
no. illus., ports. 28 cm. weekly.
Continued by: Igeret.
1. Collective settlements—Israel—Periodicals. 2. Iḥud ha-ḳevutsot yeha-ḳibutsim—Periodicals. I. Title.
HX742.2.A3 I 34 334'.683'095694 58–51792
MARC-S

—SENEGAL—DAKAR

Osmont, Annick.
Une communauté en ville africaine : les castors de Dakar / Annick Osmont. — [Grenoble] : Presses universitaires de Grenoble, c1978.
193 p. : ill. ; 24 cm. — (Actualités-Recherche : Dossiers du développement) F***
Bibliography: p. 189-192.
ISBN 2-7061-0119-9 : 29.00F
1. Dakar—Social conditions. 2. Housing—Senegal—Dakar. 3. Collective settlements—Senegal—Dakar. I. Title. II. Title: Les castors de Dakar.
HN829.D34O85 309.1'66'3 78-397608
79 MARC

—TANZANIA

Karioki, James N
Tanzania's human revolution / James N. Karioki. — University Park : Pennsylvania State University Press, c1979.
229 p. ; 24 cm.
Includes bibliographical references and index.
ISBN 0-271-00206-9
1. Tanzania—Social policy. 2. Tanzania—Politics and government. 3. Collective settlements—Tanzania. 4. Socialism in Tanzania. I. Title.
HN797.A8K37 309.1'678 78-68158
78 MARC

Madjri, John.
Une approche organisée des réalités d'un pays pour le développement : l'ujamaa en Tanzanie / par John Madjri. — [Bobo-Dioulasso, Haute-Volta] : Centre d'études économiques & sociales d'Afrique de l'Ouest, 1978.
56 p. ; 30 cm.
On t.p.: C.E.S.A.O. Cycle: Conseillers F.J.A.
Bibliography: p. 55-56.
1. Collective settlements—Tanzania. I. Centre d'études économiques et sociales d'Afrique occidentale. II. Title. III. Title: L'ujamaa en Tanzanie.
HX771.3.A3M32 335'.9'678 79-101841
79 MARC

McHenry, Dean E 1939-
Tanzania's ujamaa villages : the implementation of a rural development strategy / Dean E. McHenry, Jr. — Berkeley : Institute of International Studies, University of California, c1979.
xi, 268 p. : ill. ; 24 cm. — (Research series - Institute of International Studies, University of California, Berkeley ; no. 39)
Includes bibliographical references and index.
ISBN 0-87725-139-8
1. Collective settlements—Tanzania. 2. Rural development—Tanzania. 3. Tanzania—Rural conditions. I. Title. II. Series: California. University. Institute of International Studies. Research series ; no. 39.
HX771.3.A3M3 334'.683'09678 79-84636
00 MARC

Tanzania. Office of the Prime Minister and Second Vice-President.
Kuandikishwa kwa vijiji, vijiji vya ujamaa na uongozi wake : hotuba ya Waziri Mkuu na Makamu wa Pili wa Rais katika Bunge kuhusu sheria ya kuandikisha vijiji, vijiji vya ujamaa na uongozi wa vijiji vilivyoandikishwa. — [Dodoma : Ofisi ya Waziri Mkuu na Makamu wa Pili wa Rais, 1975]
16 p. : ill. ; 21 cm.
Cover title.
1. Collective settlements—Tanzania. 2. Villages—Tanzania. I. Title.
HX771.3.A3T36 1975 78-980904
79 MARC

COLLECTIVE SETTLEMENTS
—TANZANIA (Continued)

Tanzania. Ujamaa and Co-operative Development Dept.
Maendeleo ya ujamaa na ushirika Tanzania / kimetayarishwa na kutolewa na Idara ya Maendeleo ya Ujamaa na Ushirika, Ofisi ya Waziri Mkuu na Makamu wa Pili wa Rais. — Dodoma : Idara, 1976.
vii, 43 p. ; 21 cm.
1. Cooperation—Tanzania. 2. Collective settlements—Tanzania. I. Title.
HD3561.3.A4T36 1976 77-980579
79 MARC

Ujamaa and Christian communities / edited by Christopher Mwoleka and Joseph Healey. — Eldoret, Kenya : Gaba Publications, 1976.
62 p. ; 24 cm. — (Spearhead ; no. 45)
Includes bibliographical references.
1. Collective settlements—Tanzania. 2. Christianity—Tanzania. I. Mwoleka, Christopher. II. Healey, Joseph. III. Series: Spearhead (Eldoret, Kenya) ; no. 45.
HX771.3.A3U37 335'.9'678 77-980736
79 MARC

Ujamaa villages : a collection of original manuscripts, 1969-70. — Dar es Salaam : [s.n.], 1970.
115 leaves in various foliations ; 37 cm.
1. Collective settlements—Tanzania.
HX771.3.A3U38 334'.683'09678 79-113296
79 MARC

—TANZANIA—ADDRESSES, ESSAYS, LECTURES

Nyerere, Julius Kambarage, Pres. Tanzania, 1922-
Wafanyakazi na ujamaa Tanzania. — [Dar es Salaam? : Makao Makuu ya NUTA, 1977]
84 p. ; 22 cm.
1. Socialism in Tanzania—Addresses, essays, lectures. 2. Labor and laboring classes—Tanzania—Addresses, essays, lectures. 3. Collective settlements—Tanzania—Addresses, essays, lectures. I. Title.
HX448.5.A6N9 78-980023
79 MARC

—TANZANIA—CONGRESSES

Agricultural extension in ujamaa village development : papers and proceedings of a workshop / organized by Faculty of Agriculture and Forestry with support of Prime Minister's Office, Ministry of Agriculture and Institute of Adult Education ; edited by H. Hänsel, J. de Vries & P. C. Ndedya. — [s.l. : s.n., 1975?] (Dar es Salaam : Ukulima Wa Kisasa)
292 p. ; 21 cm.
"Held at University of Dar es Salaam, Faculty of Agriculture and Forestry, Dept. of Rural Economy and Extension, Morogoro, 22-27 September 1975."
Includes bibliographical references.
1. Rural development—Tanzania—Congresses. 2. Agricultural extension work—Tanzania—Congresses. 3. Collective settlements—Tanzania—Congresses. I. Dar es Salaam. University. Faculty of Agriculture and Forestry. II. Hänsel, H. III. De Vries, J. IV. Ndedya, P. C.
HD2128.5.Z8 1975 630'.7'15 76-980438
79 MARC

—TANZANIA—HANDENI (DISTRICT)

Sumra, Suleman.
Problems of agricultural production in ujamaa villages in Handeni District / by Suleman Sumra. — [Dar es Salaam] : Economic Research Bureau, University of Dar es Salaam, 1975.
19 p. ; 30 cm. — (E.R.B. paper ; 75.3)
Includes bibliographical references.
1. Collective settlements—Tanzania—Handeni (District) 2. Agriculture—Economic aspects—Tanzania—Handeni (District) I. Title. II. Series: Dar es Salaam. University. Economic Research Bureau. E.R.B. paper ; 75.3.
HX771.3.A38H367 334'.683'0967822 76-980157
79 MARC

—TANZANIA—WEST LAKE REGION

Boesen, Jannik.
Ujamaa, socialism from above / Jannik Boesen, Birgit Storgaard Madsen, Tony Moody. — Uppsala : Scandinavian Institute of African Studies ; [Stockholm : distributed by Almquist & Wiksell International], 1977.
183 p. : ill. ; 22 cm. S***
Includes bibliographical references.
ISBN 9171061258. ISBN 917106124X pbk.
1. Collective settlements—Tanzania—West Lake region. 2. Socialism in Tanzania. I. Madsen, Birgit Storgaard, joint author. II. Moody, Tony, joint author. III. Title.
HX771.3.A38W455 78-368737
78 MARC

—UNITED STATES

Veysey, Laurence R
The communal experience : anarchist and mystical communities in twentieth-century America / Laurence Veysey. — Chicago : University of Chicago Press, 1978.
xiii, 495 p. ; 23 cm. — (A Phoenix book)
Includes bibliographical references and index.
ISBN 0-226-85458-2 : $7.95
1. Collective settlements—United States. 2. Radicalism—United States. I. Title.
HX654.V48 1978 335'.9'73 78-55045
79 MARC

—UNITED STATES—CASE STUDIES

Co-ops, communes & collectives : experiments in social change in the 1960s and 1970s / edited by John Case and Rosemary C. R. Taylor. — 1st ed. — New York : Pantheon Books, c1979.
326 p. ; 21 cm.
Includes bibliographical references and index.
ISBN 0-394-42007-1. ISBN 0-394-73621-4 pbk.
1. United States—Social conditions—1960- . 2. Social change—Case studies. 3. Collective settlements—United States—Case studies. I. Case, John, 1944- II. Taylor, Rosemary C. R., 1946-
HN65.C627 301.24 78-51798
78 MARC

COLLECTIVISM

see also Communism; Fascism; Government ownership; Individualism; Socialism; Totalitarianism

Benoist, Jean Marie.
Les Nouveaux primaires / Jean-Marie Benoist. — Paris (22, rue Huyghens, 75680, cedex 14) : Hallier, 1978.
219 p. ; 21 cm. F78-6268
Includes bibliographical references.
29.00F
1. Social history—20th century. 2. Collectivism. 3. Violence. I. Title.
HN16.B454 309.1'04 79-350779
79 MARC

Donati, Pierpaolo.
Pubblico e privato : fine di una alternativa? / Pierpaolo Donati. — Bologna : Cappelli, 1978.
315 p. ; 19 cm. — (La Ricerca sociale) (Nuova universale Cappelli ; 7)
It78-May
Bibliography: p. 305-314.
L4000
1. Individualism. 2. Collectivism. 3. Social change. I. Title. II. Series.
HM136.D655 301.24 78-375951
•78 MARC

Ferns, Henry Stanley, 1913-
The disease of government / H. S. Ferns. — New York : St. Martin's Press, 1978.
148 p. ; 23 cm.
Bibliography: p. [141]-144.
Includes index.
ISBN 0-312-21256-9 : $9.95
1. State, The. 2. Bureaucracy. 3. Collectivism. I. Title.
JC325.F38 1978 301.5'92 78-17637
78 MARC

—ADDRESSES, ESSAYS, LECTURES

Beloff, Max, 1913-
The tide of collectivism : can it be turned? / [by] Max Beloff. — London : Conservative Political Centre, 1978.
23 p. ; 21 cm. GB79-00106
Based on the Swinton Memorial Lecture given by the author on July 1, 1978 during the CPC Summer School at Christ Church, Oxford.
ISBN 0-85070-623-8 : £0.60
1. Great Britain—Economic policy—1945- —Addresses, essays, lectures. 2. Great Britain—Social policy—Addresses, essays, lectures. 3. Collectivism—Addresses, essays, lectures. 4. Welfare state—Addresses, essays, lectures. I. Title.
HC256.6.B46 320.5'3'0941 79-305904
79 MARC

Kunst er våben : socialistisk teaterarbejde : analyser, reportager, erindringer, interviews, erfaringer / red., Olav Harsløf. — [København] : Tiderne Skifter, c1976.
398 p. : ill. ; 20 cm. D77-9
Includes bibliographical references.
kr69.00
1. Theater—History—20th century—Addresses, essays, lectures. 2. Socialism and theater—Addresses, essays, lectures. 3. Experimental theater—Addresses, essays, lectures. 4. Radicalism—Addresses, essays, lectures. 5. Collectivism—Addresses, essays, lectures. I. Harsløf, Olav, 1945-
PN2051.K8 79-363160
79 MARC

COLLECTORS, SOLAR see Solar collectors

COLLECTORS AND COLLECTING

see also Antiques; Book collectors; Hobbies; Japan—History—Allied occupation, 1945-1952—Collectibles; Mineraloty—Collectors and collecting; Plant collectors; Planters Peanuts (Firm)—Collectibles; Rocks—Collectors and collecting; World War, 1939-1945—Collectibles; and names of objects collected, e. g. Cameras—Collectors and collecting

—DIRECTORIES

Wilson, José.
The collector's catalogue : thousands of the most sought-after collectibles you can order by mail ... / by José Wilson and Arthur Leaman. — 1st ed. — New York : Holt, Rinehart, and Winston, c1979.
224 p. : ill. ; 29 cm.
"Members of the Antiquarian Booksellers' Association of America, Inc." : 195-207.
Includes index.
ISBN 0-03-042756-8 : $12.95.
1. Collectors and collecting—Directories. I. Leaman, Arthur, joint author. II. Title.
AM213.W54 069'.5 78-10168
78 MARC

—CHINA

Wang, Su-ts'un.
(Hsien hua wan ku)
閒話玩古 / 王素存著. — 台北：視聽教育出版社：經銷處藝蘭書屋, 民國66 [1977]
12, 2, 355 leaves ; 22 cm.
Reproduced from Ms. copy.
NT$150.00

1. Collectors and collecting—China. 2. Collectors and collecting—Taiwan. I. Title.
AM383.W36 78-843061

—JAPAN

Ishiko, Junzō, 1928-
(Garakuta hyakka)
ガラクタ百科：身辺のことばとそのイメージ / 石子順造著. — 東京：平凡社, 1978.
143 p. : col. ill. ; 26 cm. Ja 79-1470
¥1600

1. Collectors and collecting—Japan. I. Title.
AM384.I8 79-804629

—POLAND

Sowiński, Janusz.
Wędrówki przedmiotów : kolekcje i kolekcjonerzy / Janusz Sowiński. — Wrocław : Zakład Narodowy im. Ossolińskich, 1977.
112 p., [8] leaves of plates : ill. ; 16 x 17 cm. P***
Bibliography: p. 108-110.
zł30.00
1. Collectors and collecting—Poland. I. Title.
AM355.5.S67 78-380897
78 MARC

—SPAIN—BIOGRAPHY

Marés Deulovol, Federico.
El mundo fascinante del coleccionismo y de las antigüedades : memorias de la vida de un coleccionista / Federico Marés Deulovol. — Barcelona : [s.n.], 1977.
xi, 593 p., [1] leaf of plates : ill. ; 27 cm. Sp***
ISBN 8440039034
1. Marés Deulovol, Federico. 2. Collectors and collecting—Spain—Biography. 3. Art—Collectors and collecting. I. Title.
N5277.2.M37A2 1977 704'.7 78-375635
78 MARC

COLLECTORS AND COLLECTING (Continued)

—TAIWAN

Wang, Su-ts'un.
(Hsien hua wan ku)
閒話玩古 / 王素存著. — 台北：視聽教育出版社：經銷處藝蘭書屋，民國66 [1977]
12, 2, 355 leaves ; 22 cm.
Reproduced from Ms. copy.
NT$150.00

1. Collectors and collecting—China. 2. Collectors and collecting—Taiwan. I. Title.
AM383.W36 78-843061

COLLEGE, CHOICE OF

Lapatine-Katcher, Barbara M
What every parent and student should know about college planning (but probably did not ask!) / By Barbara M. Lapatine-Katcher. — Queens, N.Y. : Career Planning Corp., c1976.
vii, 150 p. ; 23 cm.
$6.95 $10.95
1. College, Choice of. 2. Universities and colleges—United States—Directories. I. Title: What every parent and student should know about college planning ...
LB2350.5.L36 378.73 75-25332
79 MARC

(Tōnan Ajia ryūgakusei no Nihon sentaku)
東南アジア留学生の日本選択 / 川野重任編. — 東京：アジア経済研究所，1976.
202 p. ; 25 cm. — (研究参考資料 - アジア経済研究所 ; 250) Ja***
Includes bibliographical references.
¥1400

1. Students, Foreign—Japan. 2. College students—Asia, Southeastern. 3. College, Choice of. I. Kawano, Shigetō, 1911- II. Series: Ajia Keizai Kenkyūjo, Tokyo. Kenkyū sankō shiryō ; 250.
LB2376.6.J3T66 79-808325

COLLEGE ADMINISTRATORS

see also College presidents; Deans (in schools); Departmental chairmen (Universities)

Walker, Donald E 1921-
The effective administrator / Donald E. Walker. — 1st ed. — San Francisco : Jossey-Bass Publishers, c1979.
xviii, 208 p. ; 24 cm. — (The Jossey-Bass series in higher education)
Bibliography: p. 186-202.
Includes index.
ISBN 0-87589-399-6
1. College administrators. I. Title.
LB2341.W26 378.1'1 78-24749
00 MARC

—UNITED STATES—ADDRESSES, ESSAYS, LECTURES

Developing and evaluating administrative leadership / Charles F. Fisher, guest editor. — San Francisco : Jossey-Bass, 1978.
viii, 120 p. ; 24 cm. — (New directions for higher education ; no. 22)
Includes bibliographies and index.
1. College administrators—United States—Addresses, essays, lectures. 2. Leadership—Addresses, essays, lectures. I. Fisher, Charles Frederick, 1936- II. Series.
LB2341.D43 378.1'1'0973 77-56409
79 MARC

COLLEGE ADMISSIONS see Universities and colleges—Admission

COLLEGE AND CHURCH see Church and college

COLLEGE AND COMMUNITY see Community and college

COLLEGE AND SCHOOL DRAMA

see also Children's plays; Drama in education

Bácskay, Mihály.
Diákszínpad és diákszínjátszás / [írta Bácskay Mihály]. — Budapest : Tankönyvkiadó, 1974.
124 p. ; 17 cm. — (A Tanítás problémái)
Bibliography: p. 123-124.
ISBN 9631703274 : 6.50Ft
1. College and school drama. 2. Drama in education. I. Title.
PN3175.B22 79-365931
79 MARC

(Gakkōgeki kyakuhon shū)
学校劇脚本集：低学年用 / [落合聰三郎 ... et al. 共著]. — 東京：学習研究社，昭和25 [1950]
58 p. : ill. ; 18 cm. — (学研教育シリーズ ; 第5集)

1. Children's plays, Japanese. 2. College and school drama. I. Ochiai, Sōzaburō, 1910- II. Series: Gakken kyōiku shirizu ; dai 5-shū.
PL768.A35G3 79-809833

—CONGRESSES

Schulspiel : Berichte über die gesamtösterr. Arbeitstagung f. Lehrer an allgemeinbildenden höheren Schulen in Schloss Seggau bei Leibniz vom 28. Mai bis 1. Juni 1971 / hrsg. v. Erich Benedikt, Herbert Jungwirth u. Walter Klaus. — Wien : Österr. Bundesverl., 1977.
167 p. : ill. ; 21 cm. — (Beiträge zur Lehrerfortbildung ; Bd. 17) Au78-8
Conference sponsored by the Bundesministerium für Unterricht und Kunst and organized by the Österreichisches Bundesverband für Schulspiel, Jugendspiel und Amateurtheater.
Includes bibliographies.
ISBN 3-215-02260-5 : S110.00
1. Drama in education—Congresses. 2. College and school drama—Congresses. I. Benedikt, Erich. II. Jungwirth, Herbert. III. Klaus, Walter, fl. 1976- . IV. Austria. Bundesministerium für Unterricht und Kunst. V. Österreichischer Bundesverband für Schulspiel, Jugendspiel und Amateurtheater. VI. Gesamtösterreichische Arbeitstagung für Lehrer an Allgemeinbildenden Höheren Schulen. VII. Series.
PN3171.S28 78-366165
•78 MARC

—HANDBOOKS, MANUALS, ETC.

Stern, Lwarence, 1935-
School and community theater management : a handbook for survival / Lawrence Stern. — Boston : Allyn and Bacon, c1979.
xii, 340 p. : ill. ; 24 cm.
Includes index.
ISBN 0-205-06174-5 : $16.95
1. College and school drama—Handbooks, manuals, etc. 2. Community plays, etc.—Handbook, manuals, etc. 3. Theater management—Handbooks, manuals, etc. I. Title.
PN3178.T68S75 792'.02 79-1079
79 MARC

—JUVENILE LITERATURE

Murayama, Tomoyoshi, 1901-
(Watakushitachi no geki)
私たちの劇 / 村山知義著. — 東京：國民図書刊行会，昭和24 [1949]
121 p., [1] leaf of plates : ill. ; 19 cm. — (教室文庫)
Series romanized: Kyōshitsu bunko.
Includes index.

1. College and school drama—Juvenile literature. 2. Children's plays—Presentation, etc.—Juvenile literature. I. Title.
PN3175.M8 78-811105

COLLEGE AND SCHOOL DRAMA, JAPANESE

(Gakusei engeki gikyoku shū)
学生演劇戯曲集 / [悲劇喜劇編集部編]. — 東京：早川書房，昭和28- [1953-
v. ; 19 cm.

1. One-act plays, Japanese. 2. College and school drama, Japanese. 3. Japanese drama—20th century. I. Hayakawa Shobō. Higeki Kigeki Henshūbu.
PL764.5.G3 79-803804

COLLEGE AND SCHOOL DRAMA, MALAY

Chěrita lakunan / pěnyunting B. L. Milne dan H. R. Cheeseman ; di-Mělayukan oleh Pějabat Karang-Měngarang, Jabatan Pělajaran, Kuala Lumpur. — Chetak yang 1. — Singapore : Malaya Pub. House ; [Kuala Lumpur] : Published for the Dept. of Education, Federation of Malaya, 1949-
v. ; 19 cm. — (Malay school series ; 64)
1. College and school drama, Malay. I. Milne, B. Lumsden. II. Cheeseman, Harold Ambrose Robinson. III. Malaya (Federation). Dept. of Education. IV. Series.
LA1235.M3 no. 64 78-949712
[PL5137] 79 MARC

COLLEGE AND SCHOOL JOURNALISM

see also College and school periodicals

Gough, John.
Into print / John Gough. — London : Batsford, 1979.
120 p. : ill. ; 26 cm. GB***
Includes index.
ISBN 0-7134-0739-5 : £4.25
1. Printing, Practical. 2. College and school journalism. I. Title.
Z244.G64 686.2'2 79-313220
79 MARC

Nie znasz życia, bracie / wybór i oprac. Zdzisław Pietrasik. — Wyd. 1. — Warszawa : Iskry, 1978 (Gdań. : ZG)
185, [3] p. ; 20 cm. P78-7591
"Wybór wypowiedzi Czytelników na konkurs tygodnika 'ITD' pn. Moja praca wakacyjna."
zł18.00
1. College and school journalism. I. Pietrasik, Zdzisław.
LB3621.N615 79-343288
79 MARC

Presson, Hazel.
The student journalist and interviewing / by Hazel Presson. — Rev. ed. — New York : R. Rosen Press, 1979.
185 p. : ill. ; 22 cm. — (The Student journalist guide series)
Bibliography: p. 180-183.
SUMMARY: Offers advice on how to prepare for an conduct an interview and how to turn the raw material into a story.
ISBN 0-8239-0488-1
1. Interviewing (Journalism) 2. College and school journalism. [1. Interviewing (Journalism) 2. Journalism] I. Title. II. Title: Interviewing.
PN4784.I6P7 1979 070.4'3 78-21266
78 MARC

Reque, John.
The student journalist and staff management / by John Reque. — 1st ed. — New York : Richards Rosen Press, 1979.
159 p. : ill. ; 22 cm. — (The Student journalist guide series)
Bibliography: p. 157-159.
SUMMARY: A guide for organizing and managing the staff of a student newspaper or magazine.
ISBN 0-8239-0442-3 : $7.97
1. College and school journalism. [1. College and school journalism] I. Title. II. Title: Staff management.
LB3621.R47 371.8'97 78-27820
78 MARC

Tōkyō Shimbun Sha. Shakaibu.
(Shimbun to sono mikata tsukurikata)
新聞とその見方作り方：社会科教育 / 東京新聞社社会部編. — 東京：文治書院，昭和24 [1949]
2, 6, 170 p., [1] leaf of plates : ill. ; 19 cm.

1. Newspapers. 2. Newspapers—History. 3. College and school journalism. I. Title.
PN4731.T64 1949 79-802025

—LAW AND LEGISLATION—UNITED STATES

Hull, Steven D
You can print that (probably) : first amendment rights for public high school publications / [by Steven D. Hull]. — [Storrs] : Journalism Dept., University of Connecticut, 1978.
20 p. ; 19 cm.
Cover title.
Bibliography: p. 20.
1. College and school journalism—Law and legislation—United States. I. Connecticut. University. Dept. of Journalism. II. Title.
KF4165.Z9H84 344'.73'0793 79-622986
79 MARC

COLLEGE AND SCHOOL JOURNALISM (Continued)

—MORAL AND RELIGIOUS ASPECTS

Kopenhaver, Lillian Lodge.
Ethics and responsibilities of advising college student publications / Lillian Lodge Kopenhaver, J. William Click. — Athens, Ohio : National Council of College Publications Advisers, c1978.
vii, 52 p. ; 23 cm. — (NCCPA college student press series)
Includes index.
1. College and school journalism—Moral and religious aspects. I. Click, J. W., joint author. II. Title. III. Series: National Council of College Publications Advisers. NCCPA college student press series.
LB3621.K59 378.1'98'97 78-69924
79 MARC

—GERMANY—HISTORY

Bohrmann, Hans.
Strukturwandel der deutschen Studentenpresse : Studentenpolitik und Studentenzeitschriften 1848-1974 / Hans Bohrmann. — München : Verlag Dokumentation, 1975.
337 p. ; 22 cm. — (Kommunikation und Politik ; Bd. 4) GFR***
Originally presented as the author's thesis, Freie Universität, Berlin.
"Bibliographie der deutschen Studentenzeitschriften in den Westzonen und in der Bundesrepublik Deutschland seit 1945": p. 292-327. Bibliography: p. 190-291.
Includes index.
ISBN 3-7940-4020-1
1. College and school journalism—Germany—History. 2. College students—Germany—Political activity. I. Title. II. Series.
LA728.7.B6 378.1'98'970943 75-520231
75[r79]rev MARC

—KOREA

Oh, Jin Hwan.
(Han'guk taehak sinmun chugandŭl ŭi ŭisik kujo punsŏk yŏn'gu)
韓國大學新聞主幹들의意識構造分析研究：大學新聞의理論定立을위한接近 / 吳鎭煥. — [서울：漢陽大學校], 1977.
90 p. : ill. ; 26 cm. — (漢陽大學校論文集 ; 第11輯)
Summary in English with title: A typology study of chief editors of Korean college newspapers.
Bibliography: p. 36-37.

1. College and school journalism—Korea. I. Title. II. Hanyang Taehakkyo, Seoul, Korea. Hanyang Taehakkyo nonmunjip ; che 11-chip.
LB3621.O48 378.1'98'97 79-826375

COLLEGE AND SCHOOL PERIODICALS

see also Scholarly periodicals

—HANDBOOKS, MANUALS, ETC.

Morse, R C
An A, B, C of publishing literary magazines / by R. C. Morse ; [photography by Laurie Leifer]. — 1st ed. — East Orange, N.J. : Upsala College, 1978.
54 p. : ill. ; 21 cm.
Bibliography: p. 46-48.
ISBN 0-9601668-1-5 : $1.00
1. College and school periodicals—Handbooks, manuals, etc. 2. Magazine design—Handbooks, manuals, etc. 3. Periodicals, Publishing of—Handbooks, manuals, etc.
LB3621.M67 070.5'72 78-54855
79 MARC

—HUNGARY—BIBLIOGRAPHY

Viczián, János.
A magyar egyetemi és főiskolai diáklapok bibliográfiája, 1857-1972 / Viczián János. — Budapest : [Felsőoktatási Pedagógiai Kutatóközpont], 1978.
viii, 331 p. ; 24 cm. — (Felsőoktatástörténeti kiadványok ; 5) H78
Bibliography: p. 325-330.
Includes indexes.
ISBN 9634010520
1. College and school periodicals—Hungary—Bibliography. I. Title. II. Series.
Z6944.S8V52 79-366713
[LB3621] *79 MARC

—JAPAN

Ōki, Kaoru, 1927-
(Sugao no kōkōsei)
素顔の高校生：高校生新聞の中にみる実態 / 大木薫著. — 東京：青也書店, 1978.
239 p. : ill. ; 19 cm. Ja 78-13791
¥880

1. High school students—Japan. 2. College and school periodicals—Japan. I. Title.
LA1316.7.O38 78-813735

COLLEGE ATHLETICS see Athletics; College sports; Track-athletics

COLLEGE ATTENDANCE

see also College dropouts

—PLANNING—IOWA

Anderson, Duane D
Enrollment in post secondary education in Iowa : prospects and potential problems for the period 1977 to 1990 / by Duane D. Anderson, Robert S. Ellinger, and Robert E. Engel. — Iowa City : University of Iowa, College of Education, [1977?]
iii, 42 p. : graphs ; 23 cm.
1. College attendance—Iowa. 2. College attendance—Planning—Iowa. I. Ellinger, Robert S., joint author. II. Engel, Robert E., joint author. III. Title.
LC144.I8A53 378.1'05'9777 79-623165
00 MARC

—CANADA

Association of Universities and Colleges of Canada.
The role of the university with respect to enrolments and career opportunities, admission policies, continuing education and community colleges : submitted to the Board of Directors of the Association of Universities and Colleges of Canada, October 1977 / [Association of Universities and Colleges of Canada] ; prepared by the Task Force for the study, Maurice Barbeau ... [et al.]. — [Ottawa] : Association of Universities and Colleges of Canada, 1978.
ix, 110 p. ; 28 cm. — (AUCC policy studies ; no. 1 ISSN 0701-8909)
C77-172-6
Errata (6 leaves) inserted.
Includes bibliographical references.
ISBN 0-88876-052-3 : $3.00
1. College attendance—Canada. 2. Continuing education—Canada. 3. Community colleges—Canada. I. Barbeau, Maurice. II. Title. III. Series: Association of Universities and Colleges of Canada. AUCC policy studies ; no. 1.
LC148.A84 1978 378.71 79-307296
79 MARC

Gianelli, Anthony J
A comparison of provincial contributions to Canadian universities : a report submitted to the Council of Ontario Universities, March 1976 / by Anthony J. Giannelli. — [s.l. : s.n.], 1976.
16 p. ; 28 cm. C***
1. State aid to higher education—Canada. 2. College attendance—Canada. I. Title.
LB2342.2.C3G5 379'.1214'0971 79-300578
79 MARC

—EUROPE—RESEARCH—CONGRESSES

Consultation for the Preparation of a Study on Access to Higher Education in Europe, Bucharest, 1977.
Consultation for the Preparation of a Study on Access to Higher Education in Europe = Consultation pour la préparation d'une étude sur l'accès à l'enseignement supérieur en Europe = Konsul'tativnai͡a vstrecha v svi͡azi s podgotovkoĭ issledovanii͡a po voprosu dostupa k vysshemu obrazovanii͡u v Evrope / organized by the European Centre for Higher Education (Bucharest, 18-20 October 1977) ; [edited by H. Zeiler]. — Bucarest : CEPES, 1978.
191 p. : diagrs. ; 30 cm. — (Unesco [document] ; ED-78/WS/2)
"Studies and documents edited by the European Centre for Higher Education."
English, French, and Russian.
Bibliography: p. 165-191.
1. College attendance—Europe—Research—Congresses. I. Zeiler, H. II. Series: United Nations Educational, Scientific and Cultural Organization. Document ; ED-78/WS/2.
AS4.U8A15 ED-78/WS/2 79-354460
080 s
[LC148] 79 MARC

—FLORIDA

Carroll, Stephen J 1940-
The transition from high school to college : a study of freshman enrollments in Florida / Stephen J. Carroll, Daniel A. Relles. — Santa Monica, CA : Rand, 1975.
ix, 66 p. ; 29 cm. — ([Report] - Rand Corporation ; R-1735-NSF)
Bibliography: p. 65-66.
1. Articulation (Education)—Florida. 2. College attendance—Florida. I. Relles, D. A., joint author. II. Title. III. Series: Rand Corporation. Rand report ; R-1735-NSF.
AS36.R3 R-1735 378.1'05'9759 76-370181
[LB2350] 76[r79]rev MARC

—GERMANY, WEST—STATISTICS

Albach, Horst.
Hochschulplanung / Horst Albach, Günter Fandel, Wolfgang Schüler, unter Mitarbeit von Hans-Peter Feckler, Jörg Kempken, Hermann Pieper. — 1. Aufl. — Baden-Baden : Nomos Verlagsgesellschaft, 1978.
451 p. ; 23 cm. GFR79-A
Bibliography: p. 427-433.
ISBN 3-7890-0367-0
1. Education, Higher—Germany, West—Costs—Statistics. 2. College attendance—Germany, West—Statistics. I. Fandel, Günter, joint author. II. Schüler, Wolfgang, joint author. III. Title.
LB2342.2.G4A42 79-358396
79 MARC

—GERMANY, WEST—STATISTICS—PERIODICALS

Germany (Federal Republic, 1949-). Statistisches Bundesamt.
Bevölkerung und Kultur. Reihe 10: Bildungswesen. V. Hochschulen: Studenten an Hochschulen.
-Wintersemester 1974/75.
Stuttgart, W. Kohlhammer.
v. 30 cm. semiannual. (Fachserie - Statistisches Bundesamt, A)
Continued by: Germany (Federal Republic, 1949-). Statisisches Bundesamt. Bildung und Kultur. Reihe 4.1: Studenten an Hochschulen.
1. College attendance—Germany, West—Statistics—Periodicals. 2. College students—Germany, West—Statistics—Periodicals. I. Germany (Federal Republic, 1949-). Statistisches Bundesamt. Studenten an Hochschulen. II. Title. III. Series: Germany (Federal Republic, 1949-). Statistisches Bundesamt. Fachserie.
LC148.G48b 78-648556
MARC-S

—INDIA—STATISTICS

Sharma, G D 1934-
Enrolment in higher education : a trend analysis, 1961-75 / G. D. Sharma. — New Delhi : Research Cell in Economics of Education, Association of Indian Universities, [pref. 1977]
115 p. ; 25 cm. — (Occasional papers - Association of Indian Universities ; 1)
Bibliography: p. 114-115.
Rs20.00
1. College attendance—India—Statistics. I. Title. II. Series: Association of Indian Universities. Occasional papers - Association of Indian Universities ; 1.
LC148.S38 378.1'05'954 77-906343
78 MARC

—IOWA

Anderson, Duane D
Enrollment in post secondary education in Iowa : prospects and potential problems for the period 1977 to 1990 / by Duane D. Anderson, Robert S. Ellinger, and Robert E. Engel. — Iowa City : University of Iowa, College of Education, [1977?]
iii, 42 p. : graphs ; 23 cm.
1. College attendance—Iowa. 2. College attendance—Planning—Iowa. I. Ellinger, Robert S., joint author. II. Engel, Robert E., joint author. III. Title.
LC144.I8A53 378.1'05'9777 79-623165
00 MARC

COLLEGE ATTENDANCE (Continued)

—MONTANA—STATISTICS

Montana. Office of the Commissioner of Higher Education.
Enrollment in higher education institutions in the State of Montana, fall, 1977. — Helena, Mont. : Office of the Commissioner of Higher Education, 1977.
35 leaves : ill. ; 28 cm.
At head of title: Board of Regents of Higher Education, State of Montana.
"Prepared by Paul C. Dunham."
Chiefly tables.
1. College attendance—Montana—Statistics. I. Dunham, Paul C. II. Montana. Board of Regents of Higher Education. III. Title.
LC148.M66 1977 378.1'05'9786 78-621899
79 MARC

Montana. Office of the Commissioner of Higher Education.
Enrollment in higher education institutions in the State of Montana, fall 1978 / ₍prepared by Paul C. Dunham, director of planning and research₎. — Helena : Office of the Commissioner of Higher Education, 1978.
27 leaves ; 28 cm.
At head of title: Board of Regents of Higher Education, State of Montana.
Chiefly tables.
1. College attendance—Montana—Statistics. I. Dunham, Paul C. II. Title.
LC148.M66 1978 378.1'05'9786 79-622256
79 MARC

—NEBRASKA—STATISTICS

Nebraska. Coordinating Commission for Postsecondary Education.
Enrollment trends in postsecondary education for the State of Nebraska, 1967-1987. — Lincoln : Nebraska Coordinating Commission for Postsecondary Education, 1978.
xi, 115 p. (p. 115 blank) : ill. ; 28 cm.
1. College attendance—Nebraska—Statistics. I. Title.
LC148.N327 1978a 378.1'05'9782 78-624269
79 MARC

—PENNSYLVANIA—STATISTICS
—PERIODICALS

Upperdivision undergraduate and postbaccalaureate enrollment at Pennsylvania's institutions of higher education. 1975–
₍Harrisburg₎ Division of Education Statistics, Bureau of Information Systems, Pennsylvania Dept. of Education.
v. 28 cm. annual. (Our colleges and universities today)
Continues: Students enrolled for advanced degrees at Pennsylvania institutions of higher education.
Key title: Upperdivision undergraduate and postbaccalaureate enrollment at Pennsylvania's institutions of higher education, ISSN 0192-1428.
1. College attendance — Pennsylvania — Statistics — Periodicals. I. Pennsylvania. Division of Educational Statistics. II. Series.
LC148.S74 378.1'05'9748 79-642924
MARC-S

—TEXAS—STATISTICS

Texas. Coordinating Board, Texas College and University System.
Fall headcount enrollment forecasts : Texas postsecondary educational institutions, 1978-1987. — Austin : Coordinating Board, Texas College and University System, 1978.
v, 19 p. ; 28 cm. — (CB study paper ; 27)
1. College attendance—Texas—Statistics. I. Title. II. Series: Texas. Coordinating Board, Texas College and University System. CB study paper ; 27.
LA370.5.T43 no. 27 378.1'05'9764 79-621330
₍LC148₎ 79 MARC

—UNITED STATES

Parker, Gail, 1943-
College on your own : how you can get a college education at home / Gail Thain Parker and Gene R. Hawes. — New York : Bantam Books, c1978.
417 p. ; 23 cm.
"A Hudson group book."
Includes bibliographies.
ISBN 0-553-01092-1 : $6.95
1. Education, Higher—United States—Curricula. 2. College attendance—United States. I. Hawes, Gene R., joint author. II. Title.
LB2361.5.P37 378.1'99 78-109538
79 MARC

Shulman, Carol Herrnstadt.
Enrollment trends in higher education / Carol Herrnstadt Shulman ; prepared by the ERIC Clearinghouse on Higher Education, the George Washington University. — Washington : American Association for Higher Education, 1976.
45 p. ; 23 cm. — (ERIC/Higher Education research report ; no. 6, 1976)
Bibliography: p. 41-45.
$3.00
1. College attendance—United States. I. Title. II. Series: ERIC Clearinghouse on Higher Education. ERIC/Higher Education research report ; no. 6, 1976.
LC148.S43 378.1'05'973 79-300160
79 MARC

—UNITED STATES—ADDRESSES, ESSAYS, LECTURES

Reducing the dropout rate / Lee Noel, guest editor. — San Francisco : Jossey-Bass, 1978.
ix, 108 p. ; 24 cm. — (New directions for student services ; no. 3)
Bibliography: p. 99-105.
Includes index.
1. College attendance—United States—Addresses, essays, lectures. 2. College dropouts—United States—Addresses essays, lectures. I. Noel, Lee. II. Series.
LC148.R36 378.1'69'130973 78-56392
79 MARC

—UNITED STATES—HISTORY

Doermann, Humphrey.
Toward equal access / Humphrey Doermann. — New York : College Entrance Examination Board, 1978.
xiii, 143 p. : ill. ; 23 cm.
Includes bibliographical references.
$7.95
1. College attendance—United States—History. 2. Universities and colleges—United States—Admission—History. I. Title.
LC148.D63 378.1'05'973 78-59930
78 MARC

—UNITED STATES—PLANNING

Centra, John A
College enrollment in the 1980s : projections and possibilities / John A. Centra. — ₍New York₎ : College Entrance Examination Board ; Princeton, N.J. : can be ordered from College Board Publication Orders, c1978.
26 p. : graphs ; 28 cm.
Cover title.
Bibliography: p. 24-26.
1. College attendance—United States—Planning. I. College Entrance Examination Board. II. Title.
LC148.C43 378.1'05'973 79-101661
79 MARC

—UNITED STATES—STATISTICS

Bishop, John, 1941-
Income, ability, and the demand for higher education / John Bishop. — ₍Madison₎ : University of Wisconsin-Madison, 1975.
44 p. ; 29 cm. — (Discussion papers - Institute for Research on Poverty ; 323-75)
Bibliography: p. 43-44.
1. College attendance—United States—Statistics. I. Title. II. Series: Wisconsin. University—Madison. Institute for Research on Poverty. Discussion papers ; 323-75.
LC148.B47 1975b 378.1'05'973 78-624372
79 MARC

Epsilon Data Management.
SRS, student recruitment survey. — Burlington, Mass. : Epsilon Data Management, c1978.
30 p. ; 22 x 28 cm.
Cover title.
Chiefly tables.
1. Universities and colleges—United States—Admission—Statistics. 2. College attendance—United States—Statistics. I. Title.
LB2351.E67 1978 378.1'05'60973 79-100356
79 MARC

National college databank / Karen C. Hegener, editor. — Princeton, N.J. : Peterson's Guides, c1979.
876 p. ; 24 cm.
ISBN 0-87866-095-X : $7.95
1. College attendance—United States—Statistics. 2. Universities and colleges—United States—Curricula—Statistics. I. Hegener, Karen C. II. Peterson's Guides, inc.
LC148.N323 378.73 78-64410
79 MARC

United States. Dept. of Health, Education, and Welfare. Office for Civil Rights.
Racial, ethnic, and sex enrollment data from institutions of higher education, fall 1976 / Department of Health, Education, and Welfare, Office for Civil Rights. — ₍Washington₎ : OCR, 1978.
1356 p. : forms ; 28 cm.
Tables.
1. Minorities—Education (Higher)—United States—Statistics. 2. College attendance—United States—Statistics. 3. College students—United States—Statistics. I. Title.
LC3731.U53 1978 378.1'05'973 79-600842
79 MARC

—VIRGINIA—STATISTICS
—PERIODICALS

Enrollment report.
₍Richmond₎ State Council of Higher Education for Virginia.
v. 28 cm. annual. (Staff technical report)
1. College attendance—Virginia—Statistics—Periodicals. I. Virginia. State Council of Higher Education. II. Series.
LC148.E57 378.1'05'9755 79-643099
MARC-S

COLLEGE AUTONOMY see University autonomy

COLLEGE COMMUNITY CENTERS see Student unions

COLLEGE COOPERATION see University cooperation

COLLEGE COSTS

see also Student aid; Student loan funds

Weathersby, George.
Institutional goals and student costs / George B. Weathersby and Frederic Jacobs ; prepared by the ERIC Clearinghouse on Higher Education, George Washington University. — Washington : American Association for Higher Education, 1977.
48 p. : ill. ; 23 cm. — (ERIC/Higher education research report ; no. 2, 1977)
Bibliography: p. 42-48.
$3.50
1. College costs. I. Jacobs, Frederic, joint author. II. Title. III. Series: ERIC Clearinghouse on Higher Education. ERIC/Higher Education research report ; no. 2, 1977.
LB2342.W37 379'.1214 79-312969
79 MARC

—BIBLIOGRAPHY

Quay, Richard H
The cost and benefits (mainly economic) of a college degree : a selective bibliography of recent literature / Richard H. Quay. — Monticello, Ill. : Vance Bibliographies, 1979.
20 p. ; 28 cm. — (Public administration series : Bibliography ; P-263)
Cover title.
$2.00
1. College costs—Bibliography. 2. Degrees, Academic—Bibliography. I. Title. II. Series.
Z5814.U7Q38 016.3783 79-113406
₍LB2342₎ 79 MARC

—PERIODICALS

Australia. Dept. of Education and Science.
University and residential college fees.
Canberra, Australian Govt. Pub. Service.
v. 25 cm.
Continued by: Australia. Dept. of Education. Tertiary and residential college fees.
1. Universities and colleges—Australia—Directories. 2. College costs—Periodicals. I. Title.
L981.A83a 378.3'8'02594 72-186877
MARC-S

—MASSACHUSETTS

United States. Congress. Senate. Committee on Human Resources. Subcommittee on Education, Arts, and Humanities.
Tuition advance fund act, 1978 : hearing before the Subcommittee on Education, Arts, and Humanities of the Committee on Human Resources, United States Senate, Ninety-fifty Congress, second session, on S. 3044 ... October 25, 1978, Boston, Mass. — Washington : U.S. Govt. Print. Off., 1979.
iv, 112 p. ; 24 cm.
1. Student Law and legislation—United States. 2. College costs—Massachusetts. I. Title.
KF26.H838 1978 344'.73'079 79-601383
79 MARC

COLLEGE COSTS (Continued)

—UNITED STATES

Case, Joe Paul.
Student expenses at postsecondary institutions, 1979-1980 / Joe Paul Case, Edmund C. Jacobson. — New York : College Entrance Examination Board, 1979.
xi, 34 p. : map ; 28 cm.
$5.00
1. College costs—United States. I. Jacobson, Edmund C., joint author. II. Title.
LB2342.C33 378.3'0973 79-116209
00 MARC

—UNITED STATES—ADDRESSES, ESSAYS, LECTURES

Analyzing and constructing cost / Meredith A. Gonyea, issue editor. — San Francisco : Jossey-Bass, 1978.
x, 108 p. ; 24 cm. — (New directions for institutional research ; no. 17)
Bibliography: p. 91-99.
1. College costs—United States—Addresses, essays, lectures. I. Gonyea Meredith A. II. Series.
LB2342.A67 378.73 78-5641(
79 MAR(

—WASHINGTON (STATE)—STATISTICS

Washington (State). Council for Postsecondary Education.
Resident and nonresident undergraduate tuition and/or required fees : public universities and public colleges and state universities / project officers, Denis Curry, Jackie Johnson. — ₍Olympia, Wash.₎ : Council for Postsecondary Education, 1978.
16 p. : graphs ; 28 cm. — (Report - Council for Postsecondary Education ; no. 78-9)
1. College costs—Washington (State)—Statistics. I. Curry, Denis J. II. Johnson, Jacquelin. III. Title. IV. Series: Washington (State). Council for Postsecondary Education. Report - Council for Postsecondary Education, State of Washington ; no. 78-9.
LB2342.W318 1978 379'.1214'09797 78-623648
79 MARC

COLLÈGE DE LA POCATIÈRE

—HISTORY—PICTORIAL WORKS

Collège de La Pocatière.
Collège de Sainte-Anne-de-la-Pocatière : 150e anniversaire, 1827-1977 : album-souvenir. — ₍Sainte-Anne-de-la-Pocatière, Québec : s.n., 1977?₎
120 p. : ill., facsims., music, ports. ; 29 cm. C77-11775-9
$5.00
1. Collège de La Pocatière—History—Pictorial works.
LE3.S64A53 378.714'75 78-381334
78 MARC

—ADDRESSES, ESSAYS, LECTURES

Gagnon, François.
L'alma mater / François Gagnon. — ₍La Pocatière, Québec₎ : Collège de Sainte-Anne-de-la-Pocatière, 1976.
258 p. : ports. ; 23 cm. C77-5329-7
Includes indexes.
$5.00
1. Collège de Sainte-Anne-de-la-Pocatière—History. 2. Collège de Sainte-Anne-de-la Pocatière—Addresses, essays, lectures. I. Title.
LE3.L15G33 378.714'75 78-391604
79 MARC

COLLÈGE DE SAINTE-ANNE-DE-LA-POCATIÈRE

—HISTORY

Gagnon, François.
L'alma mater / François Gagnon. — ₍La Pocatière, Québec₎ : Collège de Sainte-Anne-de-la-Pocatière, 1976.
258 p. : ports. ; 23 cm. C77-5329-7
Includes indexes.
$5.00
1. Collège de Sainte-Anne-de-la-Pocatière—History. 2. Collège de Sainte-Anne-de-la Pocatière—Addresses, essays, lectures. I. Title.
LE3.L15G33 378.714'75 78-391604
79 MARC

COLLEGE DEGREES see Degrees, Academic

COLLEGE DISCIPLINE

—JAPAN

Jiyū Minshutō. Bunkyō Seido Chōsakai. Semmon Shōiinkai.
(Daigaku funsō ni kanrensuru genkōhōtō no setsumei)
大学紛争に関連する現行法等の説明：未定稿：文教制度調査会専門小委員会調査報告. — ₍東京₎：参議院自由民主党政策審議会, 昭和44 ₍1969₎
62 p. ; 25 cm. — (政審資料 ; no. 44)
Cover title.
1. College discipline—Japan. 2. School vandalism—Japan. 3. College facilities—Japan. 4. Universities and colleges—Law and legislation—Japan. I. Title. II. Series: Jiyū Minshutō. Sangiin Seisaku Shingikai. Seishin shiryō ; no. 44.
78-812400

—UNITED STATES—ADDRESSES, ESSAYS, LECTURES

Sims, Olyn Suthern, 1938-
New directions in campus law enforcement; a handbook for administrators. Edited by O. Suthern Sims, Jr. Athens, University of Georgia Center for Continuing Education ₍1971₎
ix, 79 p. 28 cm.
Includes bibliographies.
1. Campus police—Addresses, essays, lectures. 2. College students—Legal status, laws, etc.—United States—Addresses, essays, lectures. 3. College discipline—United States—Addresses, essays, lectures. I. Georgia. University. Georgia Center for Continuing Education. II. Title.
HV8290.S55 363.2 70-634693
74₍r79₎rev MARC

COLLEGE DRAMA see College and school drama

COLLEGE DROPOUTS

—QUEBEC (PROVINCE)—STATISTICS

Gaumond, Diane.
Relance 1973 ₍i.e. Dix-neuf cent soixante-treize₎ : abandon ou poursuite des études chez les étudiants sortants des niveaux secondaire et collégial pour l'année 1971-72 / Diane Gaumond, Daniel Monnier. — ₍Québec₎ : Gouvernement du Québec, Ministère de l'éducation, Direction générale de la planification, 19
v. ; 28 cm. — (Documents Éducation et emploi ; 4-16-)
Cover title.
1. High school dropouts—Quebec (Province)—Statistics. 2. College dropouts—Quebec (Province)—Statistics. I. Monnier, Daniel, joint author. II. Title. III. Series.
LC145.C2G38 378.1'69'1309714 79-109411
79 MARC

—UNITED STATES—ADDRESSES ESSAYS, LECTURES

Reducing the dropout rate / Lee Noel, guest editor. — San Francisco : Jossey-Bass, 1978.
ix, 108 p. ; 24 cm. — (New directions for student services ; no. 3)
Bibliography: p. 99-105.
Includes index.
1. College attendance—United States—Addresses, essays, lectures. 2. College dropouts—United States—Addresses essays, lectures. I. Noel, Lee. II. Series.
LC148.R36 378.1'69'130973 78-56392
79 MARC

COLLEGE EDUCATION COSTS see College costs

COLLEGE ENTRANCE EXAMINATION BOARD

Brouder, Kathleen.
Report on student involvement in the College Board / Kathleen Brouder. — New York : College Entrace Examination Board ; Princeton, N.J. : order from College Board Publication Orders, 1978.
vi, 25 p. ; 28 cm.
$2.25
1. College Entrance Examination Board. 2. Student participation in administration. I. Title.
LB2350.B74 378 78-105896
78 MARC

College Entrance Examination Board.
College placement and credit by examination : 1978 guide to institutional policies : Advanced placement program, College-level examination program, achievement tests, Test of standard written english. — New York : College Entrance Examination Board ; Princeton, N.J. : may be ordered from College Board Publication Orders, 1978.
xvi, 438 p. ; 28 cm.
"A follow-up edition to College placement and credit by examination, 1975."
Includes index.
$6.95
1. Universities and colleges—United States—Entrance examinations. 2. College Entrance Examination Board. I. Title.
LB2353.C84 1978 378.1'62'640973 78-67797
79 MARC

COLLEGE ENTRANCE EXAMINATION BOARD. ADMISSIONS TESTING PROGRAM

Educational Testing Service.
School guide to the ATP summary reports / ₍prepared and produced by Educational Testing Service₎. — ₍New York₎ : Admissions Testing Program of the College Board, c1979.
28 p. : ill. ; 28 cm.
Cover title.
1. Universities and colleges—United States—Entrance examinations—Statistics. 2. Scholastic aptitude test—Statistics. 3. College entrance examination Board. Admissions Testing Program. I. College Entrance Examination Board. Admissions Testing Program. II. Title.
LB2353.55.E32 1979 378.1'6'64 79-119999
79 MARC

COLLEGE ENTRANCE EXAMINATION BOARD. ADVANCED PLACEMENT PROGRAM

Kenelly, John W
Grading the advanced placement examination in mathematics, 1977 and 1978 : a report / by John W. Kenelly. — Princeton, N.J. : Advanced Placement Program of the College Board, c1978.
41 p. : ill. ; 28 cm.
$1.00
1. Mathematics—Examinations, questions, etc. 2. College Entrance Examination. Advanced Placement Program. 3. Grading and working (Students) I. College Entrance Examination Board. Advanced Placement Program. II. Title.
QA43.K42 510'.76 79-121344
79 MARC

COLLEGE ENTRANCE EXAMINATION BOARD ENGLISH COMPOSITION TEST see English composition test

COLLEGE ENVIRONMENT

Moos, Rudolf H 1934-
Evaluating educational environments / Rudolf H. Moos. — 1st ed. — San Francisco : Jossey-Bass Publishers, 1979.
xvi, 334 p. : ill. ; 24 cm. — (Jossey-Bass higher education) (Social and behavioral science series)
Bibliography: p. 229-324.
Includes index.
ISBN 0-87589-401-1
1. College environment. 2. College students—Socio-economic status. 3. Universities and colleges—Administration. I. Title.
LB2343.3.M66 378.1'98 79-83568
79 MARC

COLLEGE FACILITIES

—PLANNING

see also Campus planning

Dixon, J F Cory.
Building for & by degrees : a client's experiences on building for universities / by J. F. Cory Dixon. — ⌈Guildford⌉ : University of Surrey, ⌈1978?⌉
⌈6⌉, 207 p. : forms ; 30 cm. GB79-05212
Bibliography: p. 190.
£2.00
1. Universities and colleges—Buildings. 2. Campus planning. 3. College facilities—Planning. I. Title.
LB3223.D59 692 79-310199
79 MARC

—PLANNING—IOWA—STATISTICS

Iowa. Higher Education Facilities Commission.
Biennium report - Iowa Higher Education Facilities Commission.
Des Moines, Iowa Higher Education Facilities Commission.
v. 28 cm.
Period covered by reports ends June 30.
1. Student aid—Iowa—Statistics. 2. College facilities—Planning—Iowa—Statistics.
LB2337.5.I 8 I 69a 378.3'09777 78-645449
MARC-S

—PLANNING—JAPAN—PERIODICALS

Japan. Mombushō. Kanrikyoku. Kyōiku Shisetsu.
(Kokuritsu gakkō shisetsu jittai chōsa hōkoku)
国立学校施設実態調査報告
⌈東京⌉ 文部省管理局教育施設部
v. 26 cm.
"部内資料"
1. College facilities—Planning—Japan—Periodicals. 2. Universities and colleges—Japan—Statistics—Periodicals. I. Title.
LB3223.5.J3J36a 76-649434
(MARC-S)

—CANADA, NORTHERN

Koenig, Del M
Northern people and higher education : realities and possibilities : phase 2 of The university and the Canadian North / prepared for the Association of Universities and Colleges of Canada by Del M. Koenig. — Ottawa : Association of Universities and Colleges of Canada, 1975.
xix, 236 p. : graphs, maps ; 28 cm. C76-944-9
Issued also in French.
Phase 1 of The university and the Canadian North was prepared by Maryse Caillol.
Bibliography: p. 219-223.
ISBN 0-88876-036-1 : $5.00
1. Canada, Northern—Study and Teaching (Higher). 2. College facilities—Canada, Northern. I. Caillol, Maryse. The university and the Canadian North. II. Association of Universities and Colleges of Canada. III. Title. IV. Title: The university and the Canadian North.
F1090.5.K65 971.9'007'1171 78-306702
78 MARC

—JAPAN

Jiyū Minshutō. Bunkyō Seido Chōsakai. Semmon Shōiinkai.
(Daigaku funsō ni kanrensuru genkōhōtō no setsumei)
大学紛争に関連する現行法等の説明：未定稿：文教制度調査会専門小委員会調査報告．— ⌈東京⌉：参議院自由民主党政策審議会，昭和44 ⌈1969⌉
62 p. ; 25 cm. — (政審資料 ; no. 44)
Cover title.
1. College discipline—Japan. 2. School vandalism—Japan. 3. College facilities—Japan. 4. Universities and colleges—Law and legislation—Japan. I. Title. II. Series: Jiyū Minshutō. Sangiin Seisaku Shingikai. Seishin shiryō ; no. 44.
78-812400

COLLEGE FACULTY see College teachers

COLLEGE FRATERNITIES see Greek letter societies; Students' societies

COLLEGE FRESHMEN

—UNITED STATES—STATISTICS

American Council on Education.
Women and minorities in health fields : a trend analysis of college freshmen / prepared for the Manpower Analysis Branch, Bureau of Health Manpower, by the American Council on Education. — ⌈Rockville, Md.⌉ : U.S. Dept. of Health, Education, and Welfare, Public Health Service, Health Resources Administration, Bureau of Health Manpower ; Washington : for sale by the Supt. of Docs., U.S. Govt. Print. Off., 1977-
v. in : ill. ; 28 cm. — (DHEW publication ; no. (HRA) 77-43-)
"Health manpower references."
"Contract number 231-75-0020."
Includes bibliographies.
CONTENTS: v. 1. Freshmen interested in the health professions. 2 v.—v. 2. Freshmen interested in nursing and allied health professions. v.—v. 3. A comparison of minority aspirants to health careers.
1. Minorities in medicine—United States—Statistics. 2. Women in medicine—United States—Statistics. 3. College freshmen—United States—Statistics. 4. Vocational interests—United States—Statistics. 5. Medicine—Vocational guidance—United States. I. United States. Health Resources Administration. Bureau of Health Manpower. Manpower Analysis Branch. II. Title. III. Series: United States. Dept. of Health, Education, and Welfare. DHEW publication ; no. (HRA) 77-43-.
R693.A43 1977 331.7'02 77-603201
79 MARC

COLLEGE FRESHMEN ORIENTATION see College student orientation

COLLEGE GRADUATES

see also Business education graduates

—EMPLOYMENT

Chapman, Elwood N
From campus to career success / Edwood N. Chapman. — Chicago : Science Research Associates, c1978.
xi, 286 p. : ill. ; 23 cm.
ISBN 0-574-20580-2
1. College graduates—Employment. 2. Vocational guidance. 3. Success. I. Title. II. Title: Career success.
HD6277.C48 650'.14 78-9085
78 MARC

Thain, Richard J
The managers : career alternatives for the college educated / by Richard J. Thain. — Bethelhem, Pa. : College Placement Council, c1978.
xii, 114 p. : 23 cm.
ISBN 0-913936-11-1 : $4.95
1. Management—Vocational guidance. 2. College graduates—Employment. I. Title.
HD38.T43 658.4'0023 78-61296
78 MARC

—EMPLOYMENT—ARGENTINE REPUBLIC

Pontifícia Universidad Católica Argentina Santa María de los Buenos Aires. Departamento de Sociología.
Capacidades profesionales y su aprovechamiento en la Argentina : investigación / realizada por el Departamento de Sociología de la Universidad Católica Argentina ; bajo la dirección de José Enrique Miguens. — Buenos Aires : Fundación Bolsa de Comercio de Buenos Aires, 1964?-
v. ; 27 cm. — (Estudios sociales)
CONTENTS: 1. Problemas del aprovechamiento del estudiante secundario. —2. Abandono y frustración en las carreras universitarias.
1. College graduates—Employment—Argentine Republic. 2. Professions—Argentine Republic. 3. Argentine Republic—Occupations. I. Miguens, José Enrique. II. Title. III. Series: Estudios sociales (Buenos Aires, 1964-)
HD6278.A67P65 1964 78-102416
78 MARC

—EMPLOYMENT—BELGIUM

Belgium. Services de programmation de la politique scientifique. Service enseignement supérieur.
Insertion dans la vie professionnelle et emploi des diplômés universitaires : présentation et analyse des résultats d'une enquête auprès des diplômés de 1972 et 1973. — Bruxelles : Services de programmation de la politique scientifique, Service enseignement supérieur, ⌈1977 or 1978⌉
2 v. : graphs ; 30 cm. Be***
CONTENTS: t. 1. Présentation de l'étude et conclusions générales.—t. 2. Analyse par groupe de diplômés.
1. College graduates—Employment—Belgium. I. Title.
HD6278.B4B47 1978 331.1'142 79-375054
79 MARC

Bonte, André.
Werkloos na de universiteit? : De academische arbeidsmarkt vandaag en morgen / A. Bonte. — Brugge : Orion, ⌈1976⌉
85 p. ; 21 cm. Be76-5
ISBN 9026462514 : 155F
1. College graduates—Employment—Belgium. 2. Unemployed—Belgium. I. Title.
HD6278.B4B66 79-353385
79 MARC

Coetsier, Leo.
Toekomst van de Belgische gediplomeerden : vraag en aanbod 1961-1970 / door L. Coetsier. — ⌈Gent : Laboratorium voor Toegepaste Psychologie en de Dienst voor Studieadvies bij de Rijksuniversiteit te Gemt⌉, 1961.
45 p. : graphs ; 23 cm. — (Mededelingen van het Laboratorium voor Toegepaste Psychologie en de Dienst voor Studieadvies bij de Rijskuniversiteit te Gent ; nr. 7)
Includes bibliographical references.
1. College graduates—Employment—Belgium. 2. Labor supply—Belgium. I. Title. II. Series: Ghent. Rijksuniversiteit. Laboratorium en Seminarie voor Toegepaste Psychologie. Mededelingen ; nr. 7.
HD6276.52.B3C63 79-360077
79 MARC

—EMPLOYMENT—FINLAND

Vuorinen, Juha, 1946-
Yhteiskuntatieteilijöiden työelämään siirtyminen / Juha Vuorinen. — Helsinki : Korkeakoulu- ja tiedepoliittinen tutkimussäätiö, 1975.
137, 17 p. : ill. ; 25 cm. — (KTTS:n julkaisusarja ; 10) Fi
Summary in English.
Bibliography: p. 133-137.
ISBN 9519252045
1. College graduates—Employment—Finland. 2. Social scientists—Employment—Finland. I. Title. II. Series: Korkeakoulu- ja Tiedepoliittinen Tutkimussäätiö. KTTS:n julkaisusarja ; 10.
HD6278.F5V86 79-369021
*79 MARC

—EMPLOYMENT—GERMANY, EAST

Wolter, Werner.
Die langfristige Vorausbestimmung der Ausbildungsleistungen des Hoch- und Fachschulwesens in der DDR : theoret. u. methodolog. Fragen / Werner Wolter. — Als Ms. gedr., 1. Ausg., 1. Aufl. — Berlin : Inst. für Hochschulbildung, 1977.
151 p. : 11 graphs ; 21 cm. — (Studien zur Hochschulentwicklung ; Nr. 85)
GFR79-B1
Summary in English, French, and Russian.
Includes bibliographical references.
1. Higher education and state—Germany, East. 2. College graduates—Employment—Germany, East. 3. Labor supply—Germany, East. I. Title. II. Series.
LB2322.I53 nr. 85 378.43 s 79-370265
⌈LC178.G35⌉ 00 MARC

—EMPLOYMENT—GERMANY, WEST

Der Arbeitsmarkt für Akademiker in Baden-Württemberg bis 1990 : Modellberechnungen zur Entwicklung von Nachfrage und Angebot / gemeinsames Projekt des Kultusministeriums Baden-Württemberg und des Statistischen Landesamtes Baden-Württemberg unter Mitwirkung der Landesstelle für Erziehung und Unterricht, Stuttgart und des Instituts für angewandte Wirtschaftsforschung, Tübingen ; ⌈Gesamtredaktion, Hans-Dieter Teske, Gerhard Schaaf⌉. — Villingen-Schwenningen : Neckar-Verlag, 1978.
xv, 231 p. : graphs ; 24 cm. — (Bildung in neuer Sicht : Reihe A ; Nr. 35)
GFR***
Bibliography: p. 227-231.
ISBN 3-7883-0738-2
1. College graduates—Employment—Germany, West—Baden-Württemberg—Addresses, essays, lectures. 2. Labor supply—Germany, West—Baden-Württemberg—Addresses, essays, lectures. 3. Employment forecasting—Germany, West—Baden-Württemberg—Addresses, essays, lectures. I. Teske, Hans-Dieter. II. Schaaf, Gerhard. III. Baden-Württemberg. Kultusministerium. IV. Baden-Württemberg. Statistisches Landesamt. V. Series.
LB5.B53 Nr. 35 79-365577
⌈HD6278.G42B33⌉ 79 MARC

Germanisten ohne Zukunft? : Empfehlungen zur Erhöhung d. berufl. Flexibilität germanist. Studienabsolventen / hrsg. von Ulrich Gaier ; ⌈die nachfolgenden Darst. sind d. Ergebnisse e. Forschungsprojektes, d. unter d. Titel "Empfehlungen zur Erhöhung der beruflichen Flexibilität germanistischer Studienabsolventen" im Auftr. d. Bundesministers für Bildung u. Wiss. durchgef. worden ist⌉. — 1. Aufl. — Kronberg/Ts. : Scriptor-Verlag, 1978.
iv, 239 p. ; 21 cm. GFR78-A
Includes bibliographies.
ISBN 3-589-20677-2 : DM38.00
1. German philology—Study and teaching (Higher)—Germany, West. 2. College graduates—Employment—Germany, West. I. Gaier, Ulrich.
PF3068.G43G4 78-380881
79 MARC

COLLEGE GRADUATES
—EMPLOYMENT
—GERMANY, WEST (Continued)

Güther, Bernd.
Arbeitslosigkeit von Hoch- und Fachhochschulabsolventen in der BRD / Verf., Bernd Güther. — Frankfurt am Main : Inst. für Marxist. Studien u. Forschungen, 1977.
83, 21 p. ; 24 cm. — (Informationsbericht - Institut für Marxistische Studien und Forschungen, IMSF ; Nr. 28) GFR77-B14
Includes bibliographical references.
DM6.00
1. College graduates—Employment—Germany, West. 2. Unemployed—Germany, West. I. Title. II. Series: Institut für Marxistische Studien und Forschungen. Informationsbericht ; Nr. 28.
HB97.5.I53 Nr. 28 79-378834
[HD6278.G4] 79 MARC

—EMPLOYMENT—GERMANY, WEST—
CONGRESSES

Arbeitsmarkt für Hochschulabsolventen : Tendenzen und politische Lösungen : Symposium vom 27.-28. 10. 1977 / [zusammengestellt und bearb. von Renate Kunze ; im Auftrage des Präsidenten der Freien Universität Berlin hrsg. von der Presse- und Informationsstelle der FU Berlin]. — Berlin : Presse- und Informationsstelle der FU Berlin, 1978.
148 p. ; 21 cm. — (Dokumentationsreihe der Freien Universität Berlin ; 2) GFR***
Bibliography: p. 78-85.
Papers of a symposium held at the Freie Universität Berlin.
1. College graduates—Employment—Germany, West—Congresses. I. Kunze, Renate. II. Berlin. Freie Universität. Presse- und Informationsstelle. III. Series: Berlin. Freie Universität. Dokumentationsreihe der Freien Universität Berlin ; 2.
HD6278.G4A7 79-362666
79 MARC

—EMPLOYMENT—GERMANY, WEST—
BAVARIA

Bayerisches Staatsinstitut für Hochschulforschung und Hochschulplanung.
Bedarf an Absolventen der Fachhochschulen, Fachrichtung Sozialwesen in Bayern / Bayer. Staatsinst. für Hochshculforschung u. Hochschulplanung ; Bearb., Meinhard Loibl, Sibylle Gross, Prof. Finkenstaedt. — München : Bayer. Staatsinst. für Hochschulforschung u. Hochschulplanung, 1975.
vii, 155 p. : graphs ; 30 cm. — (Bayerische Hochschulforschung : Materialien ; Nr. 13) GFR76-B9
Bibliography: p. 100-102.
DM8.00
1. Social workers—Germany, West—Bavaria—Supply and demand. 2. College graduates—Employment—Germany, West—Bavaria. I. Loibl, Meinhard. II. Gross, Sibylle. III. Finkenstaedt, Thomas, 1930- IV. Title. V. Series.
HV279.B46B39 1975 78-397831
79 MARC

—EMPLOYMENT—GREAT BRITAIN

Great Britain. Central Statistical Office.
Qualified manpower in Great Britain : 1971 census of population / [prepared by Rosemary Butler]. — London : H.M.S.O., 1976.
iv, 32 p. : ill. ; 30 cm. — (Its Studies in official statistics ; no. 29) GB76-32976
ISBN 0-11-630150-3 : £1.10
1. Manpower—Great Britain. 2. Skilled labor—Great Britain. 3. College graduates—Employment—Great Britain. I. Butler, Rosemary. II. Title. III. Series: Great Britain. Central Statistical Office. Studies in official statistics ; no. 29.
HA37.G583 no. 29 314.1 s 79-311836
[HD5765.A6] 79 MARC

Industry, education and management : a discussion paper. — London : Department of Industry, 1977.
86 p. in various pagings : ill. ; 30 cm. GB78-08862
Issued under the auspices of the Department of Industry, the Department of Education and Science, the Scottish Office and the Welsh Office.
1. Executives—Great Britain. 2. Industry and education—Great Britain. 3. Industrial management—Great Britain. 4. College graduates—Employment—Great Britain. I. Great Britain. Dept. of Industry.
HF5500.3.G7I5 331.1'142 79-308603
79 MARC

—EMPLOYMENT—INDIA

India (Republic). Directorate General of Employment and Training.
Report on the employment pattern of graduates. — New Delhi : Issued by Directorate General of Employment & Training, Govt. of India, 1976 [i.e. 1977]
iii, 111 p. ; 22 cm.
Cover title.
"Series: GS/1/76."
1. College graduates—Employment—India. I. Title.
HD6278.I5I5 1976 331.1'1 76-904147
79 MARC

—EMPLOYMENT—INDIA—ANDHRA
PRADESH—STATISTICS

Andhra Pradesh, India. Finance and Planning (Planning Wing) Dept.
A study of the characteristics of the unemployed postgraduates in Andhra Pradesh : fifth five-year plan. — [Hyderabad] : Govt. of Andhra Pradesh, Finance and Planning (Planning Wing) Dept., [pref. 1977]
49 p. ; 26 cm. — (Manpower studies ; 15)
Rs4.10
1. College graduates—Employment—India—Andhra Pradesh—Statistics. 2. Unemployed—India—Andhra Pradesh—Statistics. I. Title. II. Series: Manpower studies (Hyderabad, India) ; 14.
HD6278.I52A53 331.1'37804 78-906689
79 MARC

—EMPLOYMENT—INDIA—PUNJAB
(STATE)—STATISTICS

Punjab, India (State). Planning Dept.
Fact book on scientific, technical, and professional manpower, 1976. — Chandigarh : Planning Dept., Govt. of Punjab, [pref. 1976]
iv, 44 p. ; 30 cm.
Rev. ed. of part 3 of the Fact book on manpower.
1. Labor supply—India—Punjab (State)—Statistics. 2. College graduates—Employment—India—Punjab (State)—Statistics. I. Title.
HD5820.P8P87 1976 331.1'10954552 78-902348
79 MARC

—EMPLOYMENT—INDIA—TAMIL
NADU

The Demand for graduates and post-graduates : annual man power requirements of graduates and post-graduates in arts, science, and commerce upto 2001 A.D. : a statistical study with reference to Tamil Nadu / by K. N. Venkataraman ... [et al.]. — [Madras] : University of Madras, 1976.
viii, 42 p. ; 24 cm. — (University of Madras monograph ; 4)
Includes bibliographical references.
Rs4.00
1. College graduates—Employment—India—Tamil Nadu. 2. Employment forecasting—India—Tamil Nadu. I. Venkataraman, K. N. II. Series: Madras. University. University of Madras monograph ; 4.
HD6278.I52T352 331.1'26'095482 77-904315
79 MARC

—EMPLOYMENT—ITALY

De Francesco, Corrado.
La laurea e il posto : istruzione superiore e mercato del lavoro in Italia e all'estero / [di] Corrado De Francesco, Paolo Trivellato. — Bologna : Il mulino, 1978.
220 p. : graphs ; 22 cm. — (Studi e ricerche ; 79) It78-Aug
Includes bibliographical references.
L6000
1. College graduates—Employment—Italy. 2. Labor supply—Italy. 3. Labor supply—Europe. I. Trivellato, Paolo, joint author. II. Title.
HD6278.I8D43 78-382758
*78 MARC

—EMPLOYMENT—ITALY
—CASE STUDIES

Asta, Leonardo.
Occupazione, disoccupazione intellettuale : il caso di una facoltà / [di] Leonardo Asta, Fedele Iacono. — 1. ed. — Padova : Cleup, 1977.
220 p. ; 23 cm. — (Materiali) It78-Jan
Includes bibliographical references.
L3400
1. College graduates—Employment—Italy—Case studies. 2. Padua. Università. Facoltà di scienze politiche. I. Iacono, Fedele, joint author. II. Title.
HD6278.I5A89 78-368182
*78 MARC

—EMPLOYMENT—ITALY—
CONGRESSES

Gli Sbocchi professionali dei laureati in scienze politiche : atti dell'Incontro di studio presso la Facoltà di scienze politiche C. Alfieri (Firenze, 5-6 aprile 1974) / a cura di Pasquale Scaramozzino. — Pavia : Istituto di scienze politiche dell'Università ; Milano : A. Giuffrè, 1976.
vi, 142 p. ; 24 cm. — (Quaderni della rivista Il Politico ; n. 14) It78-Sept
Includes bibliographical references.
1. College graduates—Employment—Italy—Congresses. 2. Political science students—Employment—Italy—Congresses. I. Scaramozzino, Pasquale. II. Florence. Università. Facoltà di scienze politiche Cesare Alfieri. III. Series: Il Politico. Quaderni ; no. 14.
HD6278.I8S23 79-367429
*79 MARC

—EMPLOYMENT—JAPAN

Kadowaki, Atsushi, 1940-
(Gendai no shussekan)
現代の出世観：高学歴化でどう変わったか / 門脇厚司. — 東京：日本経済新聞社, 1978.
188 p. : ill. ; 18 cm. — (日経新書 ; 284) Ja 78-5409
Series romanized: Nikkei shinsho.
¥550
1. Japan—Occupations. 2. Success. 3. College graduates—Employment—Japan. I. Title.
HT690.J3K24 78-813574

Nihon Keieisha Dantai Remmei. Kōgakurekika Mondai Shōiinkai.
(Kōgakurekika mondai ni tsuite)
高学歴化問題について：日経連高学歴化問題小委員会報告. — 東京：日本経営者団体連盟, 昭和52 [1977]
48 p. ; 26 cm. Ja 78-40
Cover title.
1. College graduates—Employment—Japan. 2. Industry and education—Japan. 3. Labor supply—Japan. I. Title.
HD6278.J3N53 1977 79-802934

(Shinshakaijin shokun)
新社会人諸君 / 日本実業出版社編. — 東京：同社, 昭和43 [1968]
220 p. : ill. ; 18 cm.

1. White collar workers—Japan—Conduct of life. 2. College graduates—Employment—Japan. I. Nihon Jitsugyō Shuppansha.
HD8039.M4J377 79-806869

Tokyo. Shōkō Kaigisho.
(Shōwa gojūsannen sangatsu shinki sotsugyōsha saiyō jōkyō to kakutei shoninkyū)
昭和53年3月新規卒業者採用状況と確定初任給. — 東京：東京商工会議所, 1978.
40 p. ; 25 cm. Ja 78-26937
Cover title.
Chiefly tables.
¥600
1. Wages—Japan. 2. High school graduates—Employment—Japan. 3. College graduates—Employment—Japan. 4. Recruiting of employees—Japan. I. Title.
HD5077.T59 1978a 79-807387

—EMPLOYMENT—NETHERLANDS

Netherlands (Kingdom, 1815-). Commissie van Advies voor de Arbeidsvoorziening van Academici en Hoger Leidinggevend Personeel.
Nota inzake de arbeidsmarktpositie van hoger opgeleiden : met het commentaar van de Raad voor de Arbeidsmarkt / uitgebracht door de Commissie van Advies voor de Arbeidsvoorziening van Academici en Hoger Leidinggevend Personeel aan de Raad voor de Arbeidsmarkt. — ['s-Gravenhage] ; Sociaal Economische Raad, 1977.
47 p. : graphs ; 29 cm. — (Uitgave van de Sociaal-Economische Raad ; 1977, no. 12) Ne***
Includes bibliographical references.
1. College graduates—Employment—Netherlands. I. Netherlands (Kingdom, 1815-). Raad voor de Arbeidsmarkt. II. Title. III. Series: Sociaal-Economische Raad. Publicaties ; 1977, no. 12.
HC321.S63 1977, no. 12 79-381460
[HD6278.N48] 79 MARC

—EMPLOYMENT—NIGERIA

Education & employment aspirations of Nigerian University students. — Lagos : University of Lagos, Human Resources Research Unit, 1973.
iii, 28 leaves ; 26 cm. — (Research bulletin - Human Resources Research Unit, University of Lagos ; no. 1/002)
Cover title.
"Research project no. 1: Employment aspirations and prospects of Nigerian University undergraduates & graduates."
1. College students—Employment—Nigeria. 2. College graduates—Employment—Nigeria. I. Series: Lagos (Federal Territory). University. Human Resources Research Unit. Research bulletin - Human Resources Research Unit, University of Lagos ; no. 1/002.
HD6276.52.N5E38 331.1'1 79-308979
79 MARC

COLLEGE GRADUATES
—EMPLOYMENT—NIGERIA
(Continued)

Nigeria. National Manpower Board.
University graduate employment : a sample survey of employment experience amongst Nigerian men and women graduating in 1972 / Federal Republic of Nigeria, National Manpower Board, Federal Ministry of Economic Development and Reconstruction. — ⌊Lagos⌋ : The Board, 1973-
v. ; 25 cm. — (Manpower studies ; no. 14)
Cover title.
1. College graduates—Employment—Nigeria. I. Title. II. Series: Manpower studies (Lagos) ; no. 14.
HD5841.N5A2 no. 14 331.1'1'09969 s 79-313001
⌊HD6278.N5⌋ 79 MARC

—EMPLOYMENT—PENNSYLVANIA—STATISTICS

Donny, William F
Postgraduation activities : all degree levels in Pennsylvania, 1976 / prepared by William F. Donny. — Harrisburg : Pennsylvania Dept. of Education, 1977.
viii, 81 p. ; 28 cm.
1. College graduates—Employment—Pennsylvania—Statistics. I. Pennsylvania. Dept. of Education. II. Title.
HD6278.U52P44 1977 331.1'1'09748 78-622002
79 MARC

—EMPLOYMENT—QUÉBEC (PROVINCE)
—STATISTICS

Québec (Province). Ministère de l'éducation. Service général des communications.
Relance / réalisé par le Service général des communications du Ministère de l'éducation ; ⌊étude réalisée par Jean-Guy Darveau⌋. — ⌊Québec⌋ : Gouvernement du Québec, Ministère de l'éducation, Politiques et plans⌋, 1978.
141 p. ; 23 cm. — (Documents Éducation et emploi) C***
"Étude de la situation des sortants de l'enseignement professionnel, aux niveaux secondaire et collégial, pour la période de 1973 à 1976."
Chiefly tables.
1. College graduates—Employment—Québec (Province)—Statistics. 2. High school graduates—Employment—Québec (Province)—Statistics. I. Darveau, Jean Guy. II. Title.
HD6278.C32Q45 331.1'09714 79-370782
79 MARC

—EMPLOYMENT—SOUTH AFRICA—
STATISTICS

Smit, Pieter Cornelis.
Occupational information / P. C. Smit. — Pretoria : Institute for Manpower Research, 1978.
v, 52 p. : ill. ; 30 cm. — (HSRC guidance series ; GS-6) (Report - Institute for Manpower Research ; MM-72) SA***
Bibliography: p. 51-52.
ISBN 0-86965-525-6 : R1.00
1. College graduates—Employment—South Africa—Statistics. 2. Skilled labor—South Africa—Statistics. 3. South Africa—Occupations—Statistics. 4. Professions—South Africa—Statistics. I. Title. II. Series: Human Sciences Research Council. HSRC guidance series ; GS-6. III. Series: Human Sciences Research Council. Institute for Manpower Research. Report ; MM-72.
HD6278.S68S54 331.1'1'0968 79-312706
79 MARC

—EMPLOYMENT—SPAIN

Spain. Dirección General de Empleo y Promoción Social.
Situación actual y perspectivas de empleo de doctores y licenciados en ciencias y letras / Ministerio de Trabajo, Dirección General de Empleo y Promoción Social. — ⌊Madrid⌋ : El Ministerio, Servicio de Publicaciones, D.L. 1977.
365 p. ; 24 cm. Sp77-Dec
1. College graduates—Employment—Spain. 2. Psychology—Vocational guidance—Spain. 3. Psychologists—Spain. I. Title.
HD6278.S7S63 1977 331.1'0946 79-337056
79 MARC

—EMPLOYMENT—SWEDEN
—STATISTICS

Sweden. Statistiska centralbyrån. Prognosinstitutet.
Yrkesval för examinerade från universitet och högskolor läsåren 1970/71 och 1971/72 = Occupational choice for graduates from universities and specialized colleges 1970/71 and 1971/72 / ⌊Statistiska centralbyrån, Prognosinstitutet⌋. — Stockholm : Statistiska centralbyrån ; Vällingby : Liber distr., 1977.
73 p. ; 30 cm. — (Information i prognosfrågor ; 1976:5) (Sveriges officiella statistik) S77-32/33
By L. Stigendal.
Summary in English.
1. College graduates—Employment—Sweden—Statistics. I. Stigendal, Lars. II. Title. III. Title: Occupational choice for graduates ... IV. Series: Sweden. Statistiska centralbyrån. Information i prognosfrågor ; 1976:5. V. Series: Sweden. Statistiska centralbyrån. Sveriges officiella statistik.
LB2833.S9 1976:5 78-370199
⌊HD6278.S9⌋ 78 MARC

—EMPLOYMENT—TAIWAN

P'an, Chen-ch'iu, 1918-
(Hsing cheng yüan ch'ing nien fu tao wei yüan hui kung tso pao kao)
行政院青年輔導委員會工作報告：當前專科以上學校畢業青年的就業輔導 / 報告人潘振球. — ⌊台北? : s. n.⌋, 民國66 ⌊1977⌋
2, 17 p. ; 26 cm.
Cover title.
1. College graduates—Employment—Taiwan. 2. Vocational guidance—Taiwan. I. China (Republic of China). Hsing cheng yüan. Ch'ing nien fu tao wei yüan hui. II. Title.
HD6278.T27P36 78-843385

Su, Te-li.
(T'ai-wan kao chi jen li kung hsü chih yen chiu)
台灣高級人力供需之研究 / 蘇德禮撰. — ⌊台北⌋ : 政治大學, 民國66 ⌊1977⌋
iv, 140 p. (on double leaves) ; 27 cm.
"國立政治大學經濟學研究所碩士論文"
Reproduced from the typescript.
Bibliography: p. 135-140.
1. College graduates—Employment—Taiwan. I. Title.
HD6278.T27S9 79-835481

—EMPLOYMENT—TAIWAN
—ADDRESSES, ESSAYS, LECTURES

(Kuo chia chien she yü ch'ing nien ch'ien t'u chiang yen chuan chi)
國家建設與青年前途講演專輯 / 行政院青年輔導委員會編印. — ⌊台北⌋ : 該會, 民國67 ⌊1978⌋
2, 123 p. ; 19 cm.
Cover title.
CONTENTS: 官蔚藍 新青年對於就業應有的認識與準備.—施俊文 經濟加速發展中青年所面對的挑戰.—范珍輝 十大經濟建設完成後之台灣社會發展.—馬鶴凌 當前大專學生對就業應有的認識與準備.—郭良生 談國家建設與青年就業.—廖季清 國家建設與青年前途.—潘皓 均富政策與經濟發展.—潘文海 大專畢業青年就業問題及其解決途徑.—劉人紀 工專青年之當前問題.—顧守之 國家考試與大專畢業生.
1. College graduates—Employment—Taiwan—Addresses, essays, lectures. 2. Taiwan—Economic conditions—Addresses, essays, lectures. I. China (Republic of China). Hsing cheng yüan. Ch'ing nien fu tao wei yüan hui.
HD6278.T27K86 79-840244

—EMPLOYMENT—UNITED STATES

Bisconti, Ann S
Who will succeed? : College graduates as business executives / by Ann Stouffer Bisconti. — Bethlehem, PA : CPC Foundation, c1978.
16 p. ; 28 cm. — (Special topic series - The CPC Foundation ; no. 3)
Bibliography: p. 16.
$2.00
1. Executives—United States. 2. College graduates—Employment—United States. 3. Success. I. Title. II. Series: CPC Foundation. Special topic series - CPC Foundation ; no. 3.
HF5500.3.U54B58 658.4 78-110998
79 MARC

Erdlen, John D
Job hunting for the college graduate / John D. Erdlen, Donald H. Sweet. — Lexington, Mass. : D. C. Heath, c1979.
xii, 210 p. ; 21 cm.
Bibliography: p. 203.
Includes index.
ISBN 0-669-02630-1. ISBN 0-669-01893-7 pbk. : $5.95
1. Applications for positions. 2. Employment interviewing. 3. College graduates—Employment—United States. I. Sweet, Donald H., joint author. II. Title.
HF5383.E7 650'.14 78-66878
79 MARC

Freeman, Richard Barry, 1943-
Forecasting the Ph.D. labor market : pitfalls for policy / Richard B. Freeman, David W. Breneman. — Washington : National Board on Graduate Education, 1974.
vii, 50 p. : graphs ; 23 cm. — (Technical report - National Board on Graduate Education ; no. 2)
Includes bibliographical references.
1. College graduates—Employment—United States. 2. Labor supply—United States. 3. Doctor of philosophy degree. I. Breneman, David W., joint author. II. Title. III. Series: National Board on Graduate Education. Technical report - National Board on Graduate Education ; no. 2.
HD6278.U5F69 331.1'2 79-311633
79 MARC

National Board on Graduate Education.
Doctorate manpower : forecasts and policy : a report with recommendations of the National Board on Graduate Education. — Washington : NBGE, 1973.
ix, 21 p. : graphs ; 23 cm.
"Number two."
Includes bibliographical references.
1. College graduates—Employment—United States. 2. Doctor of philosophy degree. 3. Employment forecasting—United States. 4. Manpower policy—United States. I. Title.
ND6278.U5N35 1973 331.1'1 79-308459
79 MARC

—EMPLOYMENT—UNITED STATES—
LONGITUDINAL STUDIES

Solmon, Lewis C
College education and employment, the recent graduates / by Lewis C. Solmon and Nancy L. Ochsner ; with the assistance of Jean G. Kessler. — Bethlehem, PA : CPC Foundation, c1979.
vii, 118 p. ; 23 cm.
Bibliography: p. 106-107.
$9.95
1. College graduates—Employment—United States—Longitudinal studies. I. Ochsner, Nancy L., joint author. II. Kessler, Jean G. III. Title.
HD6278.U5S65 331.1'142 79-115004
00 MARC

—EMPLOYMENT—UNITED STATES
—PERIODICALS

Michigan. State University, East Lansing. Placement Services.
Recruiting trends, a study of businesses, industries, governmental agencies, and educational institutions employing new college graduates.
East Lansing, Michigan State University, Placement Services.
v. 28 cm. annual.
Vols. for prepared by J. D. Shingleton and by L. P. Scheetz.
Key title: Recruiting trends (East Lansing), ISSN 0163-5611
1. College graduates—Employment—United States—Periodicals. 2. Recruiting of employees—United States—Periodicals. I. Shingleton, John D. II. Scheetz, L. Patrick. III. Title. IV. Key title.
HD6278.U5M52a 331.1'0973 78-641579
MARC-S

United States. Internal Revenue Service.
College graduates appointed by IRS.
⌊Washington⌋ Dept. of the Treasury, Internal Revenue Service.
v. 27 cm. (Publication - Internal Revenue Service)
1. United States. Internal Revenue Service—Appointments, promotions, salaries, etc.—Periodicals. 2. United States. Internal Revenue Service—Officials and employees—Periodicals. 3. College graduates—Employment—United States—Periodicals. I. Title. II. Series: United States. Internal Revenue Service. Publication.
HJ5018.U545a 353.007'24 79-643158
MARC-S

Yeomans, William N
Jobs. College edition.
⌊New York⌋ Berkley Pub. Corp.
v. 20 cm. (A Berkley windhover book)
Key title: Jobs, college edition, ISSN 0162-5470
1. Vocational guidance—United States—Periodicals. 2. College graduates—Employment—United States—Periodicals. I. Title.
HF5382.5.U5Y46 650'.14'05 78-641550
MARC-S

—ISRAEL

Rekhess, Eli.
A survey of Israeli Arab graduates from institutions of higher learning in Israel (1961-1971) / Eli Rekhess. — ⌊Tel Aviv⌋ : Tel-Aviv University, Shiloah Center for Middle Eastern and African Studies ; ⌊Jerusalem : distributed by Israel Office, Foreign Affairs Dept., American Jewish Committee⌋, 1974.
34 leaves ; 28 cm. — (Surveys - The Shiloah Center for Middle Eastern and African Studies)
1. Palestinian Arabs—Israel—Social conditions. 2. College graduates—Israel. 3. Israel—Social conditions. I. Title: A survey of Israeli Arab graduates ... II. Series: Mekhon Shiloh le-ḥeker ha-Mizraḥ ha-tikhon ve-Afrikah. Surveys - The Shiloah Center for Middle Eastern and African Studies.
DS113.7.R45 301.45'19'275694 79-302271
79 MARC

—LOUISIANA—STATISTICS

Blackmon, C Robert.
Doctoral graduates in education from Louisiana State University, 1975-78 : a follow-up study / by C. Robert Blackmon and Mary Louise Stearns. — Baton Rouge : Bureau of Educational Materials and Research, College of Education, Louisiana State University, 1979.
i, 22 p. ; 28 cm. — (Research report - Bureau of Educational Materials and Research, Louisiana State University ; v. 9, no. 1)
1. Louisiana State University, Baton Rouge—Degrees. 2. Doctor of philosophy degree. 3. College graduates—Louisiana—Statistics. I. Stearns, Mary Louise, joint author. II. Title. III. Series: Louisiana State University, Baton Rouge. Bureau of Educational Materials and Research. Research report ; v. 9, no. 1.
LB2386.B58 378'.24'09763452 79-623434
79 MARC

COLLEGE GRADUATES (Continued)

—MINNESOTA—STATISTICS

Minnesota. Office of State Demographer.
Educational attainment in Minnesota, 1977. — St. Paul : Minnesota State Planning Agency, Office of State Demographer, 1978.
20 p. : graphs ; 28 cm.
Cover title.
1. High school graduates—Minnesota—Statistics. 2. College graduates—Minnesota—Statistics. I. Title.
L164.H5O33 1978 378.1'69'1209776 79-621788
79 MARC

—NIGERIA

Nigeria. National Manpower Board.
Nigeria's professional manpower in selected occupations, 1966. — Lagos : National Manpower Board, Federal Ministry of Economic Development, 1967.
24 p. ; 23 cm. — (Manpower studies ; no. 5)
1. Nigeria—Professions. 2. College graduates—Nigeria. I. Title. II. Series: Manpower studies (Lagos) ; no. 5.
HD5841.N5A2 no. 5 331.1'1'09669 s 79-306720
[HB2756.7.A3] 79 MARC

—SOUTH AFRICA

Van Rensburg, F A J
Graduation trends for non-whites at South African universities 1960-1975, with projections to 1990 / F. A. J. van Rensburg. — Pretoria : South African Human Sciences Research Council, Institute for Statistical Research, 1977.
ix, 115 p. : ill. ; 30 cm. — (Report - Institute for Statistical Research ; no. WS-20) SA***
Summary in Afrikaans.
Bibliography: p. 113-115.
ISBN 0-86965-453-5 : R4.50
1. College graduates—South Africa. I. Title. II. Series: Human Sciences Research Council. Institute for Statistical Research. Report - Institute for Statistical Research ; no. WS-20.
LB2430.S6V36 378.1'6912'0968 78-324494
79 MARC

COLLEGE GRADUATES, BLACK

—EMPLOYMENT—SOUTH AFRICA

Van der Walt, Helena Elizabeth.
Die Beroepsmobiliteit van swart manlike gegradueerdes / H. E. van der Walt. — Johannesburg : Nasionale Instituut vir Personeelnavorsing, Wetenskaplike en Nywerheidnavorsingsraad, 1977.
ix leaves, 82 p. ; 30 cm. — (Spesiale verslag, PERS - Suid-Afrikaanse Wetenskaplike en Nywerheidnavorsingsraad ; 259) SA
Afrikaans and/or English, with summary in English.
Bibliography: p. 68
ISBN 0-7988-1024-6 : R3.00
1. College Graduates, Black—Employment—South Africa. 2. Occupational Mobility—South Africa. I. Title. II. Series: South African Council for Scientific and Industrial Research. CSIR special report : PERS ; 1259.
HD6278.S68V36 79-345944
79 MARC

COLLEGE HOUSING see Student housing

COLLEGE INVESTMENTS see University investments

COLLEGE JOURNALISM see College and school journalism

COLLEGE-LEVEL EXAMINATION PROGRAM

Hanson, Julie.
CLEP subject examinations, college composition and freshman English / by Julie Hanson. — 1st ed. — New York : Arco Pub. Co., c1979.
256 p. ; 26 cm.
"Arco catalog # 668-03798-9."
ISBN 0-668-03798-9 : $4.95
1. College-Level Examination Program. 2. Universities and colleges—United States—Examinations. I. College-Level Examination Program. II. Title. III. Title: Subject examinations, college composition and freshman English.
LB2367.H29 808'.042'076 78-11550
78 MARC

COLLEGE-LEVEL EXAMINATIONS

Hartman, Susan.
How to earn a college degree without attending college / by Susan and John Hartman. — Berkeley, Calif. : Unique Programs, c1978.
120 p. ; 28 cm.
Includes index.
1. Degrees, Academic—United States. 2. College level examinations. 3. Independent study. 4. University extension—United States. I. Hartman, John, joint author. II. Title.
LB2381.H37 378'.24'0973 78-68407
79 MARC

Hawes, Gene R
Getting college course credits by examination to save $$$ / by Gene R. Hawes. — 1st McGraw-Hill ed. — New York : McGraw-Hill, 1979.
xii, 180 p. : ill. ; 28 cm.
"A Hudson Group book."
Includes bibliographies and index.
ISBN 0-07-027280-8 : $6.95
1. College-level examinations. I. Title.
LB2367.H33 378.1'68 79-9703
79 MARC

—HANDBOOKS, MANUALS, ETC.

College Level Examination Program.
Introduction to humanities : a resource manual for the general examination. — New York : McGraw-Hill, c1979.
vii, 150 p. ; 28 cm.
ISBN 0-07-010307-0 : $5.95
1. College level examination—Handbooks, manuals, etc. 2. Study, Method of—Handbooks, manuals, etc. 3. Humanities—Examinations, questions, etc. I. Title.
LB2353.C86 1979 001.3'076 78-32182
79 MARC

Turner, David Reuben, 1915-
CLEP, the general examination in the humanities : the complete study guide for scoring high / by David R. Turner. — 1st ed. — New York : Arco, c1979.
224 p. : ill. ; 26 cm.
ISBN 0-668-04727-5 pbk. : $6.00
1. College level examinations—Handbooks, manuals, etc. 2. Study, Method of—Handbook, manuals, etc. 3. Humanities—Examinations, questions, etc. I. Title.
LB2353.T785 001.3'076 79-142
79 MARC

COLLEGE LIBRARIANS

—SALARIES, PENSIONS, ETC.—UNITED STATES

Cameron, Donald F
Librarians in higher education : their compensation structures for the academic year 1972-73 : a third survey for the Council on Library Resources, inc. / by Donald F. Cameron and Peggy Heim. — Washington : The Council, 1974.
24 p. ; 23 cm.
1. College librarians—Salaries, pensions, etc.—United States. I. Heim, Peggy, 1924- joint author. II. Council on Library Resources. III. Title.
Z682.3.C34 331.2'81'378120973 78-104442
79 MARC

—ALABAMA—BIOGRAPHY

Johnston, Mary Tabb.
Amelia Gayle Gorgas : a biography / by Mary Tabb Johnston, with Elizabeth Johnston Lipscomb. — University : University of Alabama Press, c1978.
xiv, 168 p., [8] leaves of plates : ill. ; 22 cm.
Bibliography: p. [150]-154.
Includes index.
ISBN 0-8173-5235-X : $11.50
1. Gorgas, Amelia Gayle, 1826-1913. 2. College librarians—Alabama—Biography. 3. Alabama. University. Library—Biography. I. Lipscomb, Elizabeth Johnston.
Z720.G67J64 027.7'092'4 77-18889
77 MARC

COLLEGE LIBRARIES see Libraries, University and college

COLLEGE LIFE see College students

COLLEGE NEWSPAPER see College and school journalism

COLLEGE OF LIBRARIANSHIP, WALES

—CATALOGS

College of Librarianship, Wales.
Casgliad arbennig o lyfrau o ddiddordeb llyfryddol Cymreig = Catalogue of a special collection of Welsh books of bibliographical interest / College of Librarianship, Wales. — Aberystwyth : The College, c1977.
12 p. ; 30 cm. — (Occasional paper = Papur achlysurol - College of Librarianship, Wales ; 9) GB***
Cover title.
English and Welsh.
Includes indexes.
ISBN 0-904020-06-1
1. Wales—Bibliography—Catalogs. 2. Welsh imprints—Bibliography—Catalogs. 3. College of Librarianship, Wales—Catalogs. I. Title. II. Title: Catalogue of a special collection of Welsh books of bibliographical interest. III. Series: Library Association. International and Comparative Librarianship Group. Occasional paper ; 9.
Z2071.C64 1977 016.9429 78-313784
[DA708] 78 MARC

COLLEGE OFFICIALS see College administrators

COLLEGE PERIODICALS see College and school periodicals

COLLEGE PLAYS see College and school drama

COLLEGE PRESIDENTS

—BIBLIOGRAPHY

Eells, Walter Crosby, 1886-1962.
The college presidency, 1900-1960 : an annotated bibliography / Walter Crosby Eells and Ernest V. Hollis. — Westport, Conn. : Greenwood Press, 1978.
v, 143 p. ; 24 cm.
Reprint of the 1961 ed. published by the U.S. Dept. of Health, Education, and Welfare, Office of Education, as its Bulletin, 1961, no. 9.
Includes index.
ISBN 0-313-20680-5
1. College presidents—Bibliography. I. Hollis, Ernest Victor, 1895-1965, joint author. II. Title. III. Series: United States. Office of Education. Bulletin ; 1961, no. 9.
Z5814.U7E297 1978 016.3781'12 78-12207
[LB2341] 78 MARC

—ISRAEL—BIOGRAPHY

Biale, David, 1949-
Judah L. Magnes, pioneer & prophet on two continents : a pictorial biography / text and picture selection, David Biale. — [Berkeley, CA] : Judah L. Magnes Memorial Museum, 1977.
56 p. : ill. ; 28 cm.
Bibliography: p. 56.
1. Magnes, Judah Leon, 1877-1948. 2. Jerusalem. Hebrew University—Presidents—Biography. 3. Rabbis—United States—Biography. 4. College presidents—Israel—Biography. I. Judah L. Magnes Memorial Museum. II. Title.
LG341.J45M354 296.6'1 79-103004
79 MARC

—JAPAN

Ikeda, Shin'ichi, 1936-
(Sengo Nihon no daigaku gakumon o dame ni shita kyōju hyaku-ichinin)
戦後日本の大学・学問をダメにした教授101人 / 池田信一. — 東京 : 山手書房, 1978.
334 p. ; 19 cm. Ja 79-6608
¥980
1. College teachers, Professional ethics for. 2. College teachers—Japan. 3. College presidents—Japan. I. Title.
LB1778.I35 79-808995

COLLEGE PRESIDENTS (Continued)

—JAPAN—BIOGRAPHY

Abe, Ken'ichi, 1890-
(Shimbun to daigaku no aida)
新聞と大学の間：[学究・記者・早大紛争] / 阿部賢一. — 東京：毎日新聞社, 昭和50 [1975]
244 p., [1] leaf of plates : port. ; 20 cm. Ja 76-2841
¥950

1. Abe, Ken'ichi, 1890- 2. Economists—Japan—Biography. 3. College presidents—Japan—Biography. I. Title.
HB126.J4A23 78-816784

—MICHIGAN—BIOGRAPHY

Van de Water, Peter E
Alexander Grant Ruthven of Michigan : biography of a university president / Peter E. Van de Water. — Grand Rapids, MI : Eerdmans Pub. Co., c1977.
251 p. : ill. ; 23 cm.
Includes bibliographical references and index.
ISBN 0-8028-3493-0
1. Ruthven, Alexander Grant, 1882- . 2. College presidents—Michigan—Biography. I. Title.
LD3275 1929.V36 378.1'12'0924 76-45094
76[r79]rev MARC

—SASKATCHEWAN—BIOGRAPHY

Gorman, Jack, 1929-
Père Murray and the hounds : the story of Saskatchewan's Notre Dame College / Jack Gorman. — Sidney, B.C. : Gray's Pub., 1977.
xi, 164 p., [4] leaves of plates : ill. ; 23 cm. C***
Includes index.
ISBN 0-88826-069-5
1. Murray, Père, 1892-1975. 2. College presidents—Saskatchewan—Biography. 3. Notre Dame College, Wilcox, Sask. I. Title.
LE3.N72M873 378.7124'4 79-312223
00 MARC

—UNITED STATES—BIOGRAPHY—JUVENILE LITERATURE

Patterson, Lillie.
Booker T. Washington: leader of his people. Illustrated by Anthony D'Adamo. Champaign, Ill., Garrard Press [1962]
80 p. illus. 23 cm. (A Discovery book)
SUMMARY: A biography of the black educator and organizer of Tuskegee Institute who spent his life working for equality for the people of his race.
1. Washington, Booker Taliaferro, 1859?-1915—Juvenile literature. 2. Afro-Americans—Biography—Juvenile literature. 3. College presidents—United States—Biography—Juvenile literature. [1. Washington, Booker Taliaferro 1859?-1915. 2. Teachers. 3. Afro-Americans—Biography] I. D'Adamo Anthony, illus. II. Title.
E185.97.W4P37 378.1'11'0924 62-7331
[r78]rev MARC

—VIRGINIA—BIOGRAPHY

Friddell, Guy.
Colgate Darden : conversations with Guy Friddell. — Charlottesville : University Press of Virginia, 1978.
xi, 256 p. : port. ; 24 cm.
Includes index.
ISBN 0-8139-0744-6
1. Darden, Colgate Whitehead, 1897- . 2. Virginia—Governors—Biography. 3. College presidents—Virginia—Biography. 4. Virginia. University—Presidents—Biography. I. Darden, Colgate Whitehead, 1897- joint author.
F231.D27F74 975.5'04'0924 78-7026
78 MARC

COLLEGE PRESSES see University presses

COLLEGE PROSE, AMERICAN

—NEW YORK (STATE)—SYRACUSE

Syracuse poems and stories. 1979-
Syracuse, N. Y., Dept. of English, Syracuse University.
v. 24 cm. annual.
Continues: Syracuse stories and poems, ISSN 0192-9704.
Vols. for 1979- published in limited editions.
Key title: Syracuse poems and stories (1979), ISSN 0194-1003.
1. College verse, American—New York (State)—Syracuse. 2. College prose, American—New York (State)—Syracuse. 3. Syracuse University. I. Syracuse University. Dept. of English. II. Key title.
PN6110.C7S84 810'.8'0054 79-643822
MARC-S

Syracuse stories & poems. 1978-
[Syracuse, N. Y., Dept. of English, Syracuse University]
v. ill. 28 cm. annual.
Continues: Syracuse poems and stories, ISSN 0161-0686.
1978- published in limited editions.
Key title: Syracuse stories & poems, ISSN 0192-9704.
1. College verse, American—New York (State)—Syracuse. 2. College prose, American — New York (State) — Syracuse. 3. Syracuse University. I. Syracuse University. Dept. of English.
PN6110.C7S84 810'.8'0054 79-643211
MARC-S

COLLEGE PROSE, CANADIAN

—ONTARIO—OTTAWA

University of Ottawa. Dept. of English. Writing Workshop.
Choice parts / Writing Workshop, University of Ottawa ; selected by Seymour Mayne. — [Ottawa : The workshop] ; Oakville, Ont. : distribution, Mosaic Press/Valley Ed., c1976.
69 p. ; 26 cm. C***
Selected writings.
1. College verse, Canadian—Ontario—Ottawa. 2. College prose, Canadian—Ontario—Ottawa. I. Mayne, Seymour, 1944- . II. Title.
PR9194.5.C6U5 1976 810'.8'0054 78-321695
78 MARC

University of Ottawa. Dept. of English. Writing Workshop.
Splices / Writing Workshop, University of Ottawa. — [Ottawa : The Workshop ; Oakville, Ont. : distribution, Valley Ed.-/Mosaic Press, c1975]
55 p. : ill. ; 22 cm. C***
Selected writings.
1. College verse, American—Ontario—Ottawa. 2. College prose, Canadian—Ontario—Ottawa. I. Title.
PR9194.5.C6U5 1975 78-322957
79 MARC

COLLEGE PROSE, CHINESE

—TAIWAN—CHUNG-KUO WEN HUA HSÜEH YUAN

(She shui chi)
涉水集 / 著作者中國文化學院新聞系 ; 編輯郭俊良 ... [et al.]. — [陽明山] : 該系, 民國64 [1975]
201 p. : ill. ; 20 cm.
Colophon title.
"中國文化學院新聞系第九屆畢業同學...合力寫成."

1. College prose, Chinese—Taiwan—Chung-kuo wen hua hsüeh yüan. 2. Chung-kuo wen hua hsüeh yüan. Hsin wen hsi—Directories. I. Kuo, Chün-liang. II. Chung-kuo wen hua hsüeh yüan. Hsin wen hsi.
PL3031.T32S488 78-841187

COLLEGE READERS

Adams, W Royce.
Think, read, react, plan, write, rewrite : TRRPWR / W. Royce Adams. — New York : Holt, Rinehart and Winston, c1975.
xv, 316 p. : ill. ; 24 cm.
ISBN 0-03-012991-5
1. English language—Rhetoric. 2. College readers. I. Title.
[PE1408.A318 1975] 808'.04275 74-22177
74 MARC

Adams, W Royce.
Think, read, react, plan, write, rewrite : TRRPWR / W. Royce Adams. — 2d ed. — New York : Holt, Rinehart aand Winston, c1979.
xiii, 354 p. : ill. ; 24 cm.
Includes index.
ISBN 0-03-040436-3
1. English language—Rhetoric. 2. College readers. I. Title.
[PE1408.A318 1979] 808'.04275 78-15792
78 MARC

Brock, Dee.
Writing for a reason / Dee Brock, C. Jeriel Howard. — New York : Wiley, c1978.
xviii, 241 p. : ill. ; 23 cm.
ISBN 0-471-03017-1
1. College readers. 2. English language—Rhetoric. 3. Exposition (Rhetoric) I. Howard, C. Jeriel, joint author. II. Title.
PE1417.B718 808'.0427 77-12617
77[r78]rev MARC

Casty, Alan.
Staircase to writing and reading : a rhetoric and anthology / Alan Casty, Donald J. Tighe. — 3d. ed. — Englewood Cliffs, N.J. : Prentice-Hall, c1979.
xi, 500 p. : ill. ; 24 cm.
Includes bibliographical references and index.
ISBN 0-13-840579-4
1. English language—Rhetoric. 2. College readers. I. Tighe, Donald J., joint author. II. Title.
[PE1408.C392 1979] 808'.042 78-25595
78 MARC

Clayes, Stanley A 1922- ed.
Contexts for composition / edited by Stanley A. Clayes, David G. Spencer, Martin S. Stanford. — 5th ed. — Englewood, N.J. : Prentice-Hall, c1979.
xv, 491 p. ; 23 cm.
Includes bibliographical references.
ISBN 0-13-171512-7
1. College readers. 2. English language—Rhetoric. I. Spencer, David Gelvin. II. Stanford, Martin S., 1923- . III. Title.
PE1417.C626 1979 808'.04275 78-23899
78 MARC

Contemporary English texts for various purposes : a workbook for students / Bernd Kahrmann, Heinz J. Vienken [Hrsg.] — Kronberg (Ts.) : Scriptor-Verlag, 1974.
148 p. ; 21 cm. — (Skripten Literaturwissenschaft ; 15) GFR75-A
Includes bibliographical references.
ISBN 3-589-20041-3
1. College readers. I. Kahrmann, Bernd. II. Vienken, Heinz J.
PE1122.C587 428'.6 79-309067
79 MARC

Cooley, Thomas, 1942-
The Norton sampler : short essays for composition / Thomas Cooley. — 1st ed. — New York : Norton, c1979.
xxii, 378 p. ; 21 cm.
Includes bibliographical references.
ISBN 0-393-09007-8
1. College readers. 2. Essays. 3. English language—Rhetoric. I. Title.
PE1417.C655 808'.0427 78-23328
78 MARC

Developing college reading / [selected by] Lee A. Jacobus. — 2d ed. — New York : Harcourt Brace Jovanovich, c1979.
viii, 343 p. ; 24 cm.
ISBN 0-15-517602-1
1. College readers. I. Jacobus, Lee A.
PE1122.D44 1979 428'.6 78-71194
79 MARC

Elements of the essay : a reader for college writers / [edited by] H. Wendell Smith. — Belmont, Calif. : Wadsworth Pub. Co., c1979.
x, 499 p. ; 24 cm.
Includes indexes.
ISBN 0-534-00701-5 : $7.95
1. College readers. I. Smith, H. Wendell, 1923- .
PE1417.E38 808'.0427 78-26189
78 MARC

Fencl, Shirley Crum.
The two R's : paragraph to essay / Shirley Crum Fencl, Susan G. Jager. — New York : Wiley, c1979.
xxii, 394 p. : ill. ; 26 cm.
Includes bibliographical references and index.
ISBN 0-471-01947-X
1. English language—Rhetoric. 2. English language—Grammar—1950- . 3. College readers. I. Jager, Susan G., joint author. II. Title.
PE1408.F44 808'.042 78-16026
78 MARC

Field, John P
Cases for composition / John P. Field, Robert H. Weiss. — Boston : Little, Brown, c1979.
xiv, 256 p. ; 21 cm.
1. English language—Rhetoric. 2. College readers. 3. Creative writing. I. Weiss, Robert H., joint author. II. Title.
PE1417.F46 808'.042 78-19246
79 MARC

Forms of the essay : the American experience / [edited by] Deanne K. Milan, Naomi Cooks Rattner. — New York : Harcourt Brace Jovanovich, c1979.
xiii, 344 p. ; 24 cm.
Includes index.
ISBN 0-15-527970-X
1. Essays. 2. College readers. I. Milan, Deanne K. II. Rattner, Naomi Cooks.
PE1471.F6 808.84 78-71605
79 MARC

COLLEGE READERS (Continued)

Hackett, Laura L 1916- ed.
Anatomy of reading / Laura Hackett, Richard Williamson. — 2d ed. — Huntington, N.Y. : R. E. Krieger Pub. Co., 1977, c1970.
255 p. : ill. ; 24 cm.
Reprint of the ed. published by McGraw-Hill, New York.
ISBN 0-88275-558-7
1. College readers. I. Williamson, Richard, 1930- . II. Title.
[PE1122.H3 1977] 428.6 77-8215
77 MARC

Hogins, James Burl, comp.
Reading, writing, and rhetoric / [compiled by] James Burl Hogins, Robert E. Yarber. — 4th ed. — Chicago : Science Research Associates, c1979.
xvi, 558 p. : ill. ; 24 cm.
Includes index.
ISBN 0-574-22045-3
1. College readers. I. Yarber, Robert E., joint comp. II. Title.
PE1122.H64 1979 808'.04275 78-13940
78 MARC

Human dimensions : in nonfiction, fiction, and poetry / edited by Leonard Woolf, George F. Edmonds, James C. Holland. — Englewood Cliffs, N.J. : Prentice-Hall, c1979.
xvii, 204 p. ; 24 cm.
ISBN 0-13-444935-5
1. College readers. I. Woolf, Leonard, 1916- . II. Edmonds, George F., 1936- . III. Holland, James C., 1940- .
PE1417.H78 810'.8'09283 78-2691
78 MARC

Kerrigan, William J
Reading for the point / William J. Kerrigan. — New York : Harcourt Brace Jovanovich, c1979.
xi, 141 p. ; 24 cm.
ISBN 0-15-575640-0
1. College readers. I. Title.
PE1122.K466 428'.6 78-71336
79 MARC

Kirby, Lee, 1940-
From writers to writing : a college reading and writing workbook / Lee Kirby, John Scarry. — New York : Wiley, c1978.
xix, 364 p. : ill. ; 24 cm.
Includes index.
ISBN 0-471-01766-3
1. English language—Rhetoric. 2. English language—Grammar—1950- . 3. College readers. I. Scarry, John, joint author. II. Title.
[PE1408K669 1978] 808'.04275 78-11939
78 MARC

Morgan, Fred.
Here and now; an approach to writing through perception. New York, Harcourt, Brace & World [1968]
x, 228 p. illus., 14 plates (part col.) 23 cm.
Includes bibliographical references.
1. College readers. 2. English language—Rhetoric. I. Title.
PE1417.M6198 808.04'275 68-20727
[r79]rev MARC

Morgan, Fred.
Here and now III : an approach to writing through perception / Fred Morgan. — New York : Harcourt Brace Jovanovich, c1979.
xi, 238 p. : ill. (some col.) ; 23 cm.
Includes bibliographical references and indexes.
ISBN 0-15-535624-0
1. College readers. 2. English language—Rhetoric. I. Title.
PE1417.M624 808'.0427 78-71067
79 MARC

Papers in oriental culture. — [Cairo] : al-Azhar University Bookshop, [1977]
viii, 167 p. ; 24 cm.
"For undergraduate students, University College for Women, University of al-Azhar, 1977."
ISBN 977721703X : £E1.50
1. English language—Text-books for foreigners—Arabic. 2. College readers. 3. Civilization, Islamic—Addresses, essays, lectures.
PE1130.A8P3 428'.6'4927 77-960526
78 MARC

Rorabacher, Louise Elizabeth, 1906-
Assignments in exposition / Louise E. Roabacher, Georgia Dunbar. — 6th ed. — New York : Harper & Row, c1979.
xxvi, 374 p. ; 24 cm.
Includes indexes.
ISBN 0-06-045575-6
1. Exposition (Rhetoric) 2. College readers. I. Dunbar, Georgia Dolfield Sherwood, 1919- joint author. II. Title.
PE1429.R5 1979 808'.042 78-11234
78 MARC

Schrank, Jeffrey.
Snap, crackle & write / Jeffrey Schrank. — Skokie, Ill. : National Textbook Co., c1979.
295 p. : ill. ; 24 cm.
Includes index.
1. English language—Rhetoric. 2. College readers. I. Title.
PE1417.S362 808'.0427 78-61081
79 MARC

Sherrill, Anne.
Four elements : a creative approach to the short story / Anne Sherrill, Paula Robertson. — New York : Holt, Rinehart and Winston, c1975.
viii, 320 p. : ill. ; 23 cm.
ISBN 0-03-011311-3.
1. College readers. 2. Short stories, American. I. Robertson-Rose, Paula, joint author. II. Title.
PE1122.S415 813'.01 74-23934
74 MARC

The Short prose reader / [compiled by] Gilbert H. Muller, Harvey S. Wiener. — New York : McGraw-Hill, c1979.
x, 389 p. ; 21 cm.
ISBN 0-07-043991-5
1. College readers. I. Muller, Gilbert H., 1941- . II. Wiener, Harvey S.
PE1417.S446 808'.042 78-13284
78 MARC

Shrodes, Caroline, comp.
Reading for rhetoric : applications / Caroline Shrodes, Clifford A. Josephson, James R. Wilson. — 4th ed. — New York : Macmillan, c1979.
xxi, 543 p. ; 21 cm.
ISBN 0-02-410240-7
1. College readers. 2. English language—Rhetoric. I. Josephson, Clifford Anthon, 1922- . II. Wilson, James R. III. Title.
PE1417.S45 1979 808'.04275 78-5641
78 MARC

Subject and strategy : a rhetoric reader / editors Paul Eschholz, Alfred Rosa. — New York : St. Martin's Press, c1978.
xvii, 410 p. : ill. ; 21 cm.
ISBN 0-312-77471-0
1. College readers. 2. English language—Rhetoric. I. Eschholz, Paul A. II. Rosa, Alfred F.
PE1417.S79 808'.0427 77-86017
78 MARC

Sullens, Idelle, comp.
The inquiring reader : with exercises / [compiled by] Idelle Sullens, Edith Karas, Raymond Fabrizio. — 3d ed. — Lexington, Mass. : Heath, c1979.
xii, 580 p. ; 24 cm.
ISBN 0-669-01923-2
1. College readers. I. Karas, Edith, joint comp. II. Fabrizio, Raymond, joint comp. III. Title.
PE1122.S86 1979 808'.04275 78-60356
79 MARC

Symes, Ken M
Composing the essay : prewriting, shaping, and revising / Ken M. Symes. — New York : Harper & Row, c1979.
xiii, 258 p. ; 21 cm.
Includes bibliographical references.
ISBN 0-06-046572-7
1. English language—Rhetoric. 2. College readers. I. Title.
PE1408.S87 808'.04275 78-11938
78 MARC

Troyka, Lynn Quitman, 1938-
Steps in composition / Lynn Quitman Troyka, Jerrold Nudelman. — Alternate 2d ed. — Englewood Cliffs, N.J. : Prentice-Hall, c1979.
xviii, 472 p. : ill. ; 23 cm.
Includes indexes.
ISBN 0-13-846550-9
1. English language—Rhetoric. 2. College readers. I. Nudelman, Jerrold, 1942- joint author. II. Title.
[PE1408.T697 1979] 808'.042 78-12219
78 MARC

Twice-told tales : an anthology of short fiction / [edited by] Gerard A. Barker. — Boston : Houghton Mifflin, c1979.
xiii, 689 p. ; 24 cm.
Includes index.
ISBN 0-395-26635-1
1. College readers. 2. Short stories. I. Barker, Gerard A.
PE1122.T87 808.3'1 78-69561
79 MARC

Voice, audience, content : a writer's reader / [compiled by] Susan Gibson. — New York : Longman, c1979.
xxiv, 294 p. : ill. ; 24 cm. — (Longman English and humanities series)
Includes bibliographical references.
ISBN 0-582-28108-3 : $7.95
1. English language—Rhetoric. 2. College readers. I. Gibson, Susan, 1946- II. Title.
PE1417.V6 808'.04275 79-10129
79 MARC

The Wiley reader : designs for writing / [compiled by] Caroline D. Eckhardt, David H. Stewart. — Brief ed. — New York : Wiley, c1979.
xviii, 334 p. : ill. ; 24 cm.
Includes bibliographical references.
ISBN 0-471-03499-1
1. College readers. 2. English language—Rhetoric. I. Eckhardt, Caroline D., 1942- . II. Stewart, David Hugh, 1926- .
PE1417.W48 1979 808'.04275 78-15326
78 MARC

Willis, Hulon.
Basic usage, vocabulary, and composition : lessons on usage, reading selections, writing assignments / Hulon Willis, Charles M. Cobb. — 3d ed. — New York : Holt, Rinehart and Winston, c1979.
ix, 383 p. ; 24 cm.
ISBN 0-03-044401-2
1. English language—Grammar—1950- . 2. College readers. 3. English language—Rhetoric. I. Cobb, Charles M., joint author. II. Title.
PE1112.W53 1979 428'.2 78-10363
78 MARC

Words in action : a rhetoric—reader / Martin Steinmann, Jr. — New York : Harcourt Brace Jovanovich, c1979.
xvii, 424 p. : ill. ; 24 cm.
Includes index.
ISBN 0-15-596700-2
1. English language—Rhetoric. 2. College readers. I. Steinmann, Martin, 1915-
PE1417.W64 808'.042 77-26538
79 MARC

A Writer's reader / [compiled by] Donald Hall, D. L. Emblen. — Boston : Little, Brown, c1976.
xii, 369 p. ; 23 cm.
1. College readers. 2. English language—Rhetoric. I. Hall, Donald, 1928- . II. Emblen, Donald Lewis, 1918- .
PE1417.W67 1976 808.4 75-36958
76[r79]rev MARC

A Writer's reader / [compiled by] Donald Hall, D. L. Emblen. — 2d ed. — Boston : Little, Brown, c1979.
xxii, 500 p. ; 23 cm.
1. College readers. 2. English language—Rhetoric. I. Hall, Donald, 1928- . II. Emblen, Donald Lewis, 1918- .
PE1417.W67 1979 808'.0427 78-73831
79 MARC

The Writing experience / Carol Schoen ... [et al.]. — Boston : Little, Brown, c1979.
xv, 240 p. : ill. ; 23 cm.
Includes index.
1. English language—Rhetoric. 2. College readers. I. Schoen, Carol.
PE1417.W69 808'.042 78-70851
79 MARC

COLLEGE SPORTS

see also Baseball; Coaching (Athletics); Football; Intramural sports; Rowing; Soccer; Sports; Track-athletics; and subdivision Athletics under names of colleges, universities, etc.

Kutassi, László.
A magyar egyetemi és főiskolai sportmozgalom a felszabadulás után / Kutassi László. — Budapest : FPK, 1975.
111 p. ; 21 cm.
Includes bibliographical references.
ISBN 9634010067
1. Sports—Hungary. 2. College sports. I. Title.
GV648.H8K87 79-370110
79 MARC

McCallum, John Dennis, 1924-
College basketball, U.S.A., since 1892 / John D. McCallum. — New York : Stein and Day, c1978.
viii, 308 p. : ill. ; 29 cm.
Includes index.
ISBN 0-8128-2503-9 : $18.95
1. Basketball—United States—History. 2. College sports. I. Title.
GV883.M3 796.32'363'0973 78-7466
78 MARC

—BIBLIOGRAPHY

Binnewies, Harald.
Bibliographie zum Hochschulsport / Red., Harald Binneweis, Joachim Neu. — 1. Aufl. — Ahrensburg bei Hamburg : Czwalina, 1978.
272 p. ; 21 cm. — (Dokumente zum Hochschulsport ; 4) GFR78-A
Includes indexes.
ISBN 3-88020-065-3 : DM24.00
1. College sports—Germany, West—Bibliography. 2. College sports—Bibliography. I. Neu, Joachim, joint author. II. Title. III. Series.
Z7515.G29B56 79-351116
[GV347] 79 MARC

COLLEGE SPORTS (Continued)

—PHILOSOPHY see Sports —Philosophy

—SOCIAL ASPECTS

Cady, Edwin Harrison.
The big game : college sports and American life / Edwin H. Cady. — Knoxville : University of Tennessee Press, c1978.
xi, 254 p. ; 24 cm.
Bibliography: p. 234-246.
Includes index.
ISBN 0-87049-254-3
1. College sports—Social aspects. I. Title.
GV347.C3 796'.0973 78-6794
78 MARC

—AUSTRALIA

Lynch, Robert L
Student participation in physical activities and goal attainment in an Australian university : a report of findings and recommendations for future planning / research consultants, Robert L. Lynch, Geoffrey G. Watson. — [Nedlands, W.A.] : Community Recreation Council of Western Australia, 1978.
vii, 176 p. ; 30 cm. Aus***
Presented to the Community Recreation Council of Western Australia.
Bibliography: p. 126-131.
1. Physical education and training—Australia. 2. College sports—Australia. 3. Academic achievement. 4. College students—Australia. I. Watson, Geoffrey Grainge, joint author. II. Community Recreation Council of Western Australia. III. Title.
GV315.L96 613.7'07'1194 79-311117
79 MARC

—GERMANY, WEST

Gutsche, Klaus-Jürgen.
Aktionsformen und Curriculum des Hochschulsports, ein Modell / Klaus-Jürgen Gutsche, Sandor Köris. — 1. Aufl. — Ahrensburg bei Hamburg : Czwalina, 1978.
274 p. ; 21 cm. — (Dokumente zum Hochschulsport : 2/3) GFR78-A
"Ausgewählte Literatur zur speziellen Trainingslehre": p. 241-252.
Bibliography: p. 258-274.
ISBN 3-88020-059-9 : DM24.00
1. College sports—Germany, West. 2. Physical education and training—Germany, West—Curricula. I. Köris, Sandor, joint author. II. Title. III. Series.
GV347.G87 78-367929
78 MARC

Thomas, Alexander, 1939-
Handlungspsychologische Analyse des Freizeitsports : e. empir. Unters. am Hochschulsport / Alexander Thomas. — 1. Aufl. — Ahrensburg bei Hamburg : Czwalina, 1978.
155 p. : ill. ; 21 cm. — (Dokumente zum Hochschulsport ; 5) GFR78-A
Bibliography: p. 150-153.
ISBN 3-88020-067-X : DM18.00
1. Sports—Germany, West—Psychological aspects. 2. College sports—Germany, West. 3. Leisure—Germany, West. 4. Recreational surveys—Germany, West. I. Title. II. Series.
GV611.T47 79-356699
79 MARC

—GERMANY, WEST—BIBLIOGRAPHY

Binnewies, Harald.
Bibliographie zum Hochschulsport / Red., Harald Binneweis, Joachim Neu. — 1. Aufl. — Ahrensburg bei Hamburg : Czwalina, 1978.
272 p. ; 21 cm. — (Dokumente zum Hochschulsport ; 4) GFR78-
Includes indexes.
ISBN 3-88020-065-3 : DM24.00
1. College sports—Germany, West—Bibliography. 2. College sports—Bibliography. I. Neu, Joachim, joint author. II. Title. III. Series.
Z7515.G29B56 79-351116
[GV347] 79 MARC

—UNITED STATES

Hull, Gordon.
The six insidious traps of college tennis and how to avoid them! / Gordon Hull. — Mineola, N.Y. : Goodworth Pub., c1979.
212 p. ; 25 cm.
$10.95
1. Tennis—Psychological aspects. 2. College sports—United States. 3. School, Choice of—United States. I. Title.
GV1002.9.P75H84 796.34'2'01 78-78177
79[7909] MARC

United States. Congress. House. Committee on Interstate and Foreign Commerce. Subcommittee on Oversight and Investigations.
NCAA enforcement program : hearings before the Subcommittee on Oversight and Investigations of the Committee on Interstate and Foreign Commerce, House of Representatives, Ninety-fifth Congress, second session .... — Washington : U.S. Govt. Print. Off., 1978 [i.e. 1979]
2 v. (vii, 1520 p.) ; 24 cm.
Hearings held Feb. 27-Oct. 4, 1978.
"Serial no. 95-159" and "Serial no. 95-160."
"Errata sheet" ([4] p.) inserted in v. 1.
Supt. of Docs. no.: Y 4.In8/4:95-159.
1. National Collegiate Athletic Association. 2. College sports—United States. I. Title.
KF27.I5547 1978p 796'.06'273 79-601251
79 MARC

—UNITED STATES—ADDRESSES, ESSAYS, LECTURES

The Future of collegiate athletics. — Princeton, N.J. : College Entrance Examination Board, c1978.
32 p. : ill. ; 27 cm. — (The College Board review ; no. 109)
Cover title.
Includes index.
1. College sports—United States—Addresses, essays, lectures. I. College Entrance Examination Board. II. Series.
GV347.F86 371.8'9 78-111339
79 MARC

—UNITED STATES—RECORDS —PERIODICALS

National Collegiate Athletic Association.
National collegiate championships. 1977/78– [Shawnee Mission, Kan., National Collegiate Athletic Association]
v. ill. 22 cm.
Continues: National collegiate championships records book. ISSN 0148–9798.
Key title: National collegiate championships (1978), ISSN 0190–4329.
1. College sports—United States—Records—Periodicals. 2. National Collegiate Athletic Association—Periodicals. I. Title. II. Key title.
GV741.N37a 796'.06'273 79–640882
MARC-S

COLLEGE STUDENT ORIENTATION

Cohen, Robert Douglas, 1938-
Freshman seminar : a new orientation / Robert D. Cohen and Ruth Jody. — Boulder, Colo. : Westview Press, 1978.
vii, 142 p. ; 24 cm. — (Westview special studies in higher education)
Bibliography: p. 137-142.
ISBN 0-89158-098-0 : $13.50. ISBN 0-89158-099-9 pbk. : $6.50
1. College student orientation. I. Jody, Ruth, joint author. II. Title.
LB2343.3.C63 378.1'98 78-1682
78 MARC

Duncan, Paul E
On campus : a Christian guide / by Paul E. Duncan. — [Cleveland, Tenn.] : Church of God General Dept. of Youth and Christian Education, c1977.
96 p. ; 18 cm.
Articles originally issued in Truthway, a supplement to The lighted pathway magazine.
ISBN 0-87148-655-5
1. College student orientation. 2. Study, Method of. I. Title.
LB2343.3.D86 378.1'98 77-75433
79 MARC

Schlösser, Kurt.
Studium der Naturwissenschaften : e. Einf. in d. wissenschaftl. Arbeiten / Kurt Schlösser. — Stuttgart ; New York : Fischer, 1978.
xii, 168 p. ; 19 cm. — (Uni-Taschenbücher ; 809) GFR78-A
Bibliography: p. 164-165.
Includes index.
ISBN 3-437-20192-1 : DM16.80
1. Studying, Method of. 2. College student orientation. I. Title.
LB2395.S356 79-346413
79 MARC

Voeks, Virginia.
On becoming an educated person; an orientation to college. Philadelphia, Saunders, 1957.
147 p. illus. 21 cm.
1. College student orientation. 2. Study, Method of. 3. Student adjustment. 4. Self-culture. I. Title.
LB2395.V6 371.3 57-7052
[r79]rev MARC

Voeks, Virginia.
On becoming an educated person : an orientation to college and life / Virginia Voeks. — 4th ed. — Philadelphia : Saunders, 1979.
xv, 249 p. : ill. ; 25 cm. — (Saunders survival series)
Includes bibliographies and index.
ISBN 0-7216-9069-6
1. College student orientation. 2. Study, Method of. 3. Student adjustment. 4. Self-culture. I. Title. II. Series.
LB2343.3.V63 1979 378.1'8 79-3931
00 MARC

Voeks, Virginia.
On becoming an educated person; the university and college. 2d ed. Philadelphia, Saunders, 1964.
xix, 206 p. illus. 18 cm.
Includes bibliographies.
1. College student orientation. 2. Study, Method of. 3. Student adjustment. 4. Self-culture. I. Title.
LB2395.V6 1964 371.3028 64-14266
[r79]rev MARC

Voeks, Virginia.
On becoming an educated person; the university and college. 3d ed. Philadelphia, Saunders, 1970.
xxi, 278 p. illus. 18 cm.
Includes bibliographies.
1. College student orientation. 2. Study, Method of. 3. Student adjustment. 4. Self-culture. I. Title.
LB2395.V6 1970 378.1'7'0281 77-118595
70[r79]rev MARC

Zur Didaktik der Studieneinführung : Ausbildungsmaterial für d. student. Mitarb. an d. Informationswochen für Studienanfänger d. OE Sozialwesen WS 76/77 / hrsg. von Dieter Oelschlägel. — Kassel : Gesamthochschul-Bibliothek, 1976.
108 p. ; 21 cm. — (Materialien - [Gesamthochschule Kassel] ; 8)
GFR77-B5
At head of title: Modellversuch "Soziale Studiengänge" an der GhK.
Bibliography: p. 13-14.
ISBN 3-88122-001-1
1. Gesamthochschule Kassel. 2. College student orientation. I. Oelschlägel, Dieter. II. Series: Gesamthochschule Kassel. Materialien - Gesamthochschule Kassel ; 8.
LF3194.K3Z86 79-352667
79 MARC

—HANDBOOKS, MANUALS, ETC.

Edelhart, Michael.
College knowledge / Michael Edelhart. — 1st ed. — Garden City, N.Y. : Anchor Books, 1979.
383 p. : ill. ; 28 cm.
ISBN 0-385-13386-3 : $7.95
1. College student orientation—Handbooks, manuals, etc. I. Title.
LB2343.3.E33 378.1'98 78-1193
79 MARC

—STATISTICS

Saarbrücken. Universität des Saarlandes. Arbeitsgruppe für Empirische Studienforschung.
Orientierungsprobleme und Erfolgsbeeinträchtigung bei Studierenden / Arbeitsgruppe für Empirische Studienforschung der Universität Saarlandes, Universität des Saarlandes Saarbrücken. — Saarbrücken : Universität des Saarlandes, 1977.
3 v. ; 30 cm. — (Saarbrücker Studien zur Hochschulentwicklung ; 22, 28)
GFR78-B3
Summary also in English.
Includes bibliographies.
CONTENTS: T. A. Fragestellung und Durchführung der Untersuchung.—T. B. Ergebnisse in Tabellenform.—T. C. Analyse der Studienprobleme von Abbruchgefährdeten, Fachwechslern und Langzeitstudierenden.
1. Saarbrücken. Universität des Saarlandes—Students—Statistics. 2. College students—Attitudes—Statistics. 3. College student orientation—Statistics. 4. Educational research—Germany, West—Statistics. I. Title. II. Series.
LC148.S2 no. 22, etc 78-385850
[LF3194.S3] 78 MARC

COLLEGE STUDENT SELF-GOVERNMENT see Self-government (in higher education)

COLLEGE STUDENTS

see also College freshmen; Women college students

Hasegawa, Nyozekan, 1875–1969.
(Daigaku oyobi daigakusei)
大學及び大學生 / 長谷川如是閑著. — 東京 : 明治書院, 昭和24 ⌜1949⌝
166 p. ; 19 cm.

1. Universities and colleges—History. 2. Education, Higher—Aims and objectives. 3. College students. I. Title.
LA175.H37 79–800939

(Kokusai shakai e no taiō)
国際社会への対応 : 大学教育の一視点 / 東海大学学生生活研究所, ⌜綿貫達也⌝編. — 東京 : 東海大学出版会, 1979.
vi, 200 p. ; 19 cm. Ja 79-5373
¥1500

1. International education. 2. Intercultural education. 3. College students. I. Watanuki, Tatsuya. II. Tōkai Daigaku. Gakusei Seikatsu Kenkyūjo.
LC1090.K583 79–808788

—ADDRESSES, ESSAYS, LECTURES

(Konnichi no gakusei no tame ni)
今日の學生のために / ⌜著者 小林秀雄 ... et al.⌝. — 東京 : 創元社, 昭和25 ⌜1950⌝
3, 186 p. ; 20 cm. — (創元社學生シリーズ)
Series romanized: Sōgensha gakusei shirizu.

1. College students—Japan—Addresses, essays, lectures. 2. College students—Addresses, essays, lectures. 3. Students—Addresses, essays, lectures. I. Kobayashi, Hideo, 1902–
LA1318.7.K66 79–809301

Yu, Chin-o, 1906–
(Chŏlmun nal ŭi chahwasang)
젊은날의自畵像 / 兪鎭午著. — 서울 : 博英社, 1976.
273 p. : port. ; 18 cm. — (博英文庫 ; 115)
Series romanized: Pagyŏng mun'go.
₩400
1. Yu, Chin-o, 1906– —Biography—Youth—Addresses, essays, lectures. 2. Authors, Korean—20th century—Biography—Addresses, essays, lectures. 3. College students—Addresses, essays, lectures. 4. Books and reading—Addresses, essays, lectures. I. Title.
PL991.94.C5Z464 79–825212

—ANECDOTES, FACETIAE, SATIRE, ETC.

The Harvard lampoon big book of college life / edited by Steven G. Crist and George Meyer ; staff writers, Allan Arffa ... ⌜et al.⌝ ; staff artist, Mel J. Horan. — 1st ed. — Garden City, N.Y. : Doubleday, 1978.
191 p. : ill. ; 28 cm. — (A Dolphin book)
ISBN 0-385-13446-0 : $5.95
1. Education, Higher—Anecdotes, facetiae, satire, etc. 2. College students—Anecdotes, facetiae, satire, etc. I. Crist, Steven G. II. Meyer, George. III. Harvard lampoon.
LB2326.H37 378.1'98'0207 77-15165
00 MARC

—ASPIRATIONS see Student aspirations

—ATTITUDES

Fukuda, Nobuko.
Value choices of education and business administration majors / by Nobuko Fukuda. — Hilo : Social Science Faculty, University of Hawaii, Hilo College, 1978.
28, ⌜1⌝ leaves ; 28 cm. — (Social science research notes ; 7, no. 4)
Bibliography: leaf ⌜29⌝
1. College students—Attitudes. 2. Education—Study and teaching. 3. Student teachers. 4. Business education. I. Title. II. Series.
LB3609.F85 378.1'98'1 78-112475
79 MARC

—ATTITUDES—STATISTICS

Saarbrücken. Universität des Saarlandes. Arbeitsgruppe für Empirische Studienforschung.
Orientierungsprobleme und Erfolgsbeeinträchtigung bei Studierenden / Arbeitsgruppe für Empirische Studienforschung der Universität Saarlandes, Universität des Saarlandes Saarbrüchken. — Saarbrücken : Universität des Saarlandes, 1977.
3 v. ; 30 cm. — (Saarbrücker Studien zur Hochschulentwicklung ; 22, 28) GFR78-B3

Summary also in English.
Includes bibliographies.
CONTENTS: T. A. Fragestellung und Durchführung der Untersuchung.—T. B. Ergebnisse in Tabellenform.—T. C. Analyse der Studienprobleme von Abbruchgefährdeten, Fachwechslern und Langzeitstudierenden.
1. Saarbrücken. Universität des Saarlandes—Students—Statistics. 2. College students—Attitudes—Statistics. 3. College student orientation—Statistics. 4. Educational research—Germany, West—Statistics. I. Title. II. Series.
LC148.S2 no. 22, etc 78-385850
⌜LF3194.S3⌝ 78 MARC

—EMPLOYMENT

see also Education, Cooperative

—EMPLOYMENT—MANITOBA

Manitoba. Youth Secretariat. Employment Services Office.
Provincial summer employment : a study of the Employment Services Office program, 1976. — ⌜Winnipeg : Youth Secretariat⌝, 1976.
21, 3 p. ; 28 cm. C***
1. Youth—Employment—Manitoba. 2. Summer employment—Manitoba. 3. College students—Employment—Manitoba. I. Title.
HD6276.C3M36 1976 331.3'4'097127 79-313232
79 MARC

—EMPLOYMENT—NIGERIA

Education & employment aspirations of Nigerian University students. — Lagos : University of Lagos, Human Resources Research Unit, 1973.
iii, 28 leaves ; 26 cm. — (Research bulletin - Human Resources Research Unit, University of Lagos ; no. 1/002)
Cover title.
"Research project no. 1: Employment aspirations and prospects of Nigerian University undergraduates & graduates."
1. College students—Employment—Nigeria. 2. College graduates—Employment—Nigeria. I. Series: Lagos (Federal Territory). University. Human Resources Research Unit. Research bulletin - Human Resources Research Unit, University of Lagos ; no. 1/002.
HD6276.52.N5E38 331.1'1 79-308979
79 MARC

—EMPLOYMENT—SWEDEN

Sweden. Statistiska centralbyrån. Prognosinstitutet.
Elevströmmarna genom utbildningsväsendet och utflödet till arbetsmarknaden : framskrivning till 1990 = Student flows through the educational system and number of students entering the labour market : projection up to 1990 / ⌜Statistiska centralbyrån, Prognosinstitutet⌝. — Stockholm : Statistiska centralbyrån : Liber distribution, 1978.
181 p. : tables ; 24 cm. — (Information i prognosfrågor ; 1977:4) (Sveriges officiella statistik) S78-6
Table of contents and summary in English.
kr23.00
1. College students—Employment—Sweden—Statistics. 2. College students—Employment—Sweden. 3. Employment forecasting—Sweden—Statistics. I. Title. II. Title: Student flows through the educational system ... III. Series: Sweden. Statistiska centralbyrån. Information i prognosfrågor ; 1977:4. IV. Series: Sweden. Statistiska centralbyrån. Sveriges officiella statistik.
LB2833.S9 1977:4 78-390251
⌜HD6276.52.S8⌝ 79 MARC

—EMPLOYMENT—SWEDEN—STATISTICS

Sweden. Statistiska centralbyrån. Prognosinstitutet.
Elevströmmarna genom utbildningsväsendet och utflödet till arbetsmarknaden : framskrivning till 1990 = Student flows through the educational system and number of students entering the labour market : projection up to 1990 / ⌜Statistiska centralbyrån, Prognosinstitutet⌝. — Stockholm : Statistiska centralbyrån : Liber distribution, 1978.
181 p. : tables ; 24 cm. — (Information i prognosfrågor ; 1977:4) (Sveriges officiella statistik) S78-6
Table of contents and summary in English.
kr23.00
1. College students—Employment—Sweden—Statistics. 2. College students—Employment—Sweden. 3. Employment forecasting—Sweden—Statistics. I. Title. II. Title: Student flows through the educational system ... III. Series: Sweden. Statistiska centralbyrån. Information i prognosfrågor ; 1977:4. IV. Series: Sweden. Statistiska centralbyrån. Sveriges officiella statistik.
LB2833.S9 1977:4 78-390251
⌜HD6276.52.S8⌝ 79 MARC

—HEALTH AND HYGIENE

Nonnamaker, Eldon R
Student health programs : planning, management, and assessing effectiveness / by Eldon R. Nonnamaker and Katherine E. White. — Washington : University Associates, c1978.
vii, 26 p. ; 22 cm. — (The Institutional development series ; 4)
Bibliography: p. 24-26.
1. College students—Health and hygiene. I. White, Katherine E., joint author. II. Title. III. Series.
LB3497.N66 378.1'97'12 78-62717
78 MARC

—LEGAL STATUS, LAWS, ETC.—UNITED STATES

Millington, William G
The law and the college student : justice in evolution / by William G. Millington. — St. Paul : West Pub. Co., 1979.
xlvi, 629 p. ; 24 cm.
Bibliography: p. 613-616.
Includes indexes.
ISBN 0-8299-2047-1 : $15.95
1. College students—Legal status, laws, etc.—United States. 2. Universities and colleges—Law and legislation—United States. I. Title. II. Title: Justice in evolution.
KF4243.M54 344'.73'079 79-14211
79 MARC

—LEGAL STATUS, LAWS, ETC.—UNITED STATES—ADDRESSES, ESSAYS, LECTURES

Sims, Olyn Suthern, 1938-
New directions in campus law enforcement; a handbook for administrators. Edited by O. Suthern Sims, Jr. Athens, University of Georgia Center for Continuing Education ⌜1971⌝
ix, 79 p. 28 cm.
Includes bibliographies.
1. Campus police—Addresses, essays, lectures. 2. College students—Legal status, laws, etc.—United States—Addresses, essays, lectures. 3. College discipline—United States—Addresses, essays, lectures. I. Georgia. University. Georgia Center for Continuing Education. II. Title.
HV8290.S55 363.2 70-634693
74⌜r79⌝rev MARC

—LEGAL STATUS, LAWS, ETC.—UNITED STATES—CASES

Edwards, Harry T
Higher education and the law / by Harry T. Edwards and Virginia Davis Nordin. — Cambridge : Institute for Educational Management, Harvard University, c1979.
xl, 844 p. ; 25 cm.
Includes bibliographical references and index.
1. Universities and colleges—Law and legislation—United States—Cases. 2. College teachers—Legal status, laws, etc.—United States—Cases. 3. College students—Legal status, laws, etc.—United States—Cases. I. Nordin, Virginia Davis, joint author. II. Title.
KF4225.A7E35 344'.73'074 79-88195
79 MARC

—LIBRARY ORIENTATION

Library user instruction in the framework of project studies : teaching humanities students literature search during the basic studies programme at Roskilde University Centre / ⌜by⌝ Tamar Bermann ... ⌜et al.⌝. — Roskilde (Postbox 269, 4000) : Institut for sprogbrugs- socialisations- og videnskabsforskning, Roskilde Universitetscenter, 1977.
44 leaves ; 30 cm. D78-19
Free.
1. College students—Library orientation. 2. Searching, Bibliographical—Study and teaching. 3. Libraries, University and college. 4. Project method in teaching. 5. Roskilde universitetscenter. I. Bermann, Tamar, 1943-
Z675.U5L524 027.7 78-326604
79 MARC

COLLEGE STUDENTS (Continued)

—LIBRARY ORIENTATION —AUDIO-VISUAL AIDS

Hardesty, Larry L
Use of slide/tape presentations in academic libraries / by Larry L. Hardesty ; special section, "Sound/slide presentations: 6 faults" by John Murphy. — New York : J. Norton Publishers, c1978.
vii, 222 p. : ill. ; 27 cm.
Bibliography: pp. 183-192.
Includes index.
ISBN 0-88432-006-5 : $8.95
1. College students—Library orientation—Audio-visual aids. I. Title.
Z675.U5H26 027.7 77-9215
77 MARC

—LIBRARY ORIENTATION—CASE STUDIES

Course-related library and literature-use instruction : working models for programs in undergraduate science education / compiled by Thomas Kirk. — New York : J. Norton Publishers, c1978.
31 p. ; 24 cm. & microfiche (11 sheets (ca. 900 p. in various pagings : ill.) ; 11 x 15 cm.) in pocket.
Bibliography: p. 21-22.
Includes index.
"A micropapers edition."
ISBN 0-88432-013-8
1. College students—Library orientation—Case studies. 2. Library orientation—United States—Case studies. I. Kirk, Thomas, 1943-
Z675.U5C78 027.7 79-64352
00 MARC

—LIBRARY ORIENTATION—CONGRESSES

Conference on Library Orientation, 5th, Eastern Michigan University, 1975.
Faculty involvement in library instruction : their views on participation in and support of academic library use instruction : papers and summaries from the fifth annual Conference on Library Orientation for Academic Libraries, held at Eastern Michigan University, May 15-17, 1975 / edited by Hannelore B. Rader. — Ann Arbor : Published for the Center of Educational Resources, Eastern Michigan University by Pierian Press, 1976.
viii, 119 p. ; 24 cm. — (Library orientation series ; no. 6)
Includes index.
ISBN 0-87650-070-X
1. College students—Library orientation—Congresses. 2. Libraries. University and college—Relations with faculty and curriculum—Congresses. I. Rader, Hannelore B. II. Eastern Michigan University. Center of Educational Resources. III. Title. IV. Series.
Z675.U5C748 1975 027.7 76-21914
76[r79]rev MARC

Conference on Library Orientation, 6th, Eastern Michigan University, 1976.
Library instructions in the seventies : state of the art : papers presented at the sixth annual Conference on Library Orientation for Academic Libraries held at Eastern Michigan University, May 13-14, 1976 / edited by Hannelore B. Rader. — Ann Arbor, Mich. : Published for the Center of Educational Resources, Eastern Michigan University by Pierian Press, 1977.
xi, 130 p. ; 24 cm. — (Library orientation series ; no. 7)
ISBN 0-87650-078-5
1. College students—Library orientation—Congresses. 2. Library orientation—Congresses. I. Hannelore, Rader. II. Eastern Michigan University. Center of Educational Resources. III. Title. IV. Series.
Z675.U5C748 1976 027.7 77-75678
78 MARC

Conference on Library Orientation for Academic Libraries, 7th, Eastern Michigan University, 1977.
Putting library instruction in its place : in the library and in the library school : papers presented at the Seventh Annual Conference on Library Orientation for Academic Libraries, held at Eastern Michigan University, May 12-13, 1977 / edited by Carolyn A. Kirkendall. — Ann Arbor, Mich. : Published for the Center of Educational Resources, Eastern Michigan University by Pierian Press, 1978.
x, 108 p. ; 24 cm. — (Library orientation series ; no. 8)
Includes bibliographies.
ISBN 0-87650-092-0
1. College students—Library orientation—Congresses. 2. Library orientation—Congresses. I. Kirkendall, Carolyn A. II. Title. III. Series.
Z675.U5C748 1977 012 78-53996
79 MARC

—MENTAL HEALTH SERVICES

Hanfmann, Eugenia, 1905-
Effective therapy for college students / Eugenia Hanfmann. — 1st ed. — San Francisco : Jossey-Bass, 1978.
xviii, 347 p. ; 24 cm. — (The Jossey-Bass series in Social and behavioral science & in Higher education)
Bibliography: p. 339-342.
Includes index.
ISBN 0-87589-371-6
1. College students—Mental health services. 2. Adolescent psychotherapy. I. Title.
RC451.4.S7H36 362.2 78-1147
79 MARC

—MISCELLANEA

'Awaḍ, Aḥmad Ḥāfiẓ.
(Min wālid ilá waladih)
من والد الى ولده : وهى رسائل فى التربية والتعليم والآداب كان يبعث بها الى ولده من وقت لآخر / احمد حافظ عوض. — الطبعة 2. — [القاهرة] : [s. n.], 1926
17, 152 p. ; 24 cm.
Includes bibliographical references.
1. College students—Miscellanea. 2. Education, Humanistic—Miscellanea. I. Title.
LB3607.A9 1926 77-960451

—ORIENTATION see College student orientation

—PSYCHOLOGY

Anderson, Robert Orlie, 1945-
Students as real people : interpersonal communication and education / Rob Anderson. — Rochelle Park, N.J. : Hayden Book Co., c1979.
134 p. ; 23 cm.
Includes bibliographical references and index.
ISBN 0-8104-5764-4
1. Interaction analysis in education. 2. Communication in education. 3. College students—Psychology. I. Title.
LB3609.A53 301.5'6 78-11978
78 MARC

Applying new developmental findings / Lee Knefelkamp, Carole Widick, Clyde A. Parker, guest editors. — San Francisco : Jossey-Bass, 1978.
xvi, 125 p. ; 24 cm. — (New directions for student services ; no. 4)
Bibliography: p. 117-119.
Includes index.
1. College students—Psychology. I. Knefelkamp, Lee. II. Widick, Carole. III. Parker, Clyde Alvin. IV. Series.
BF724.A63 378.1'94'019 78-56393
79 MARC

Nagasawa, Kōshichi.
(Joshidaisei no seikatsu to shinri)
女子大生の生活と心理 / 永沢幸七著. — 東京：大日本図書, 1979.
230 p. : ill. ; 17 cm. — (現代心理学ブックス) Ja 79-4987
Series romanized: Gendai shinrigaku bukkusu.
Bibliography: p. 230.
¥550
1. Women college students—Japan. 2. College students—Psychology. I. Title.
LC2366.N32 79-807531

Noah, Berndt Gustav.
Comparative analysis of personality patterns among college students in open and closed housing situations as exhibited on the POI (Personal orientation inventory) / by Berndt Gustav Noah. — [Coral Gables? Fla.] : Noah, 1975.
vi, 171 leaves : ill. ; 29 cm.
Thesis, University of Sarasota.
Bibliography: leaves 125-133.
1. College students—Psychology. I. Title.
LB1117.N6 378.1'98'1 78-324212
79 MARC

Sarason, Irwin G
Coping with academic stressors : a pilot study / Irwin G. Sarason and James H. Johnson ; prepared for Office of Naval Research. — Arlington, Va. : The Office, 1976.
41 leaves ; 29 cm.
Cover title.
"Report SCS-CS-001."
"Sponsored by the Organizational Effectiveness Research Program, Office of Naval Research ... under contract [sic] no. N00014-75-C-0905, NR 170-804."
"Part of the overall project, 'Life stress and coping skills in relation to performance and organizational effectiveness.'"
Bibliography: leaves 20-21.
1. College students—Psychology. 2. Stress (Physiology) I. Johnson, James Harmon, 1943- joint author. II. Title.
LB3609.S27 370.15'8 79-620938
79 MARC

—RECREATION—ADDRESSES, ESSAYS, LECTURES

Sysler, Barry L
Life-time sports for the college student : a behavioral objective approach / Barry L. Sysler, Eugene R. Fox. — 3d ed. — Dubuque, Iowa : Kendall/Hunt Pub. Co., c1978.
viii, 295 p. : ill. ; 23 cm.
Sysler's and Fox's names appear in inverse order on earlier ed.
Includes bibliographies.
ISBN 0-8403-0462-5
1. Sports—Addresses, essays, lectures. 2. College students—Recreation—Addresses, essays, lectures. I. Fox, Eugene R., joint author. II. Title.
GV707.S95 1978 796 78-108756
78 MARC

—SOCIOECONOMIC STATUS

Moos, Rudolf H 1934-
Evaluating educational environments / Rudolf H. Moos. — 1st ed. — San Francisco : Jossey-Bass Publishers, 1979.
xvi, 334 p. : ill. ; 24 cm. — (Jossey-Bass higher education) (Social and behavioral science series)
Bibliography: p. 229-324.
Includes index.
ISBN 0-87589-401-1
1. College environment. 2. College students—Socio-economic status. 3. Universities and colleges—Administration. I. Title.
LB2343.3.M66 378.1'98 79-83568
79 MARC

--STRIKES see Student strikes

—ALASKA

Kleinfeld, Judith.
Alaska native students and college success / by Judith Kleinfeld. — Fairbanks : Institute of Social and Economic Research, University of Alaska, 1978.
iii, 22 p. ; 22 cm. — (ISER occasional papers ; no. 14)
Includes bibliographical references.
$1.00
1. Indians of North America—Alaska—Education. 2. Eskimos—Alaska—Education. 3. College students—Alaska. 4. Student adjustment. I. Title. II. Series: University of Alaska, Fairbanks. Institute of Social and Economic Research. ISER occasional paper ; no. 14.
E97.65.A4K54 378.1'98'09798 78-620025
79 MARC

—ALGERIA

Glasman, Dominique.
Essai sur l'université et les cadres en Algérie : une technocratie sans technologie? / par Dominique Glasman et Jean Kremer. — Paris : Éditions du Centre national de la recherche scientifique, 1978.
252 p. ; 21 cm. — (Les Cahiers du C.R.E.S.M. ; 8 ISSN 0336-5042)
F79-480
Includes bibliographical references and indexes.
ISBN 2-222-02350-5 : 40.00F
1. Education, Higher—Algeria. 2. College students—Algeria. 3. Elite (Social sciences)—Algeria. 4. Social classes—Algeria. I. Kremer, Jean, joint author. II. Title. III. Series: Centre de recherches et d'études sur les sociétés méditerranéennes. Les Cahiers du C.R.E.S.M. ; 8.
LA1673.G55 378.65 79-368008
79 MARC

—ASIA, SOUTHEASTERN

(Tōnan Ajia ryūgakusei no Nihon sentaku)
東南アジア留学生の日本選択 / 川野重任編. — 東京：アジア経済研究所, 1976.
202 p. ; 25 cm. — (研究参考資料 - アジア経済研究所 ; 250) Ja***
Includes bibliographical references.
¥1400
1. Students, Foreign—Japan. 2. College students—Asia, Southeastern. 3. College, Choice of. I. Kawano, Shigetō, 1911- II. Series: Ajia Keizai Kenkyūjo, Tokyo. Kenkyū sankō shiryō ; 250.
LB2376.6.J3T66 79-808325

COLLEGE STUDENTS (Continued)

—AUSTRALIA

Lynch, Robert L
Student participation in physical activities and goal attainment in an Australian university : a report of findings and recommendations for future planning / research consultants, Robert L. Lynch, Geoffrey G. Watson. — ⌜Nedlands, W.A.⌝ : Community Recreation Council of Western Australia, 1978.
vii, 176 p. ; 30 cm. Aus***
Presented to the Community Recreation Council of Western Australia.
Bibliography: p. 126-131.
1. Physical education and training—Australia. 2. College sports—Australia. 3. Academic achievement. 4. College students—Australia. I. Watson, Geoffrey Grainge, joint author. II. Community Recreation Council of Western Australia. III. Title.
GV315.L96 613.7'07'1194 79-311117
79 MARC

—CALIFORNIA—STATISTICS

Fink, Ira Stephen.
1970 census data book : population and housing characteristics of the University of California campus environs / Ira Stephen Fink and Albert Sukoff. — Berkeley : University of California, Office of the President, Assistant Vice President—Physical Planning, 1974.
vii, 339 p. : maps ; 24 cm. — (⌜Community planning publications⌝)
Bibliography: p. 339.
$5.00
1. California—Census, 1970. 2. California—Statistics. 3. College students—California—Statistics. I. Sukoff, Albert, joint author. II. Title. III. Series.
HA266.F56 312'.09794 74-620065
76⌜r79⌝rev MARC

—CALIFORNIA—TRANSPORTATION

Fink, Ira Stephen.
To and from campus : changing student transportation patterns / Ira Stephen Fink. — Berkeley : University of California, Office of the President, Assistant Vice President, Physical Planning, 1974.
xx, 298 p. : ill. ; 24 cm.
Bibliography: p. 295-298.
1. School children—California—Transportation. 2. College students—California—Transportation. I. Title.
LB2864.F52 378.1'98'709794 74-620131
79 MARC

—DENMARK—POLITICAL ACTIVITY—CONGRESSES

Danmarks kommunistiske studenter.
Til kamp for egne interesser—i forbund med arbejderklassen : program vedtaget af KOMM. S. 4. kongres 17.-19. nov. 1978. — København : Danmarks Kommunistiske Studenter : ⌜eksp. Tiden⌝, 1979.
112 p. : ill. ; 24 cm. D79-20
Includes index.
kr10.00
1. College students—Denmark—Political activity—Congresses. 2. Danmarks kommunistiske studenter—Congresses. 3. Student movements—Denmark—Congresses. 4. Higher education and state—Denmark—Congresses. I. Title.
LA878.7.D35 1979 79-383718
79 MARC

—EGYPT—POLITICAL ACTIVITY

Erlikh, Ḥagai.
(Sṭudenṭim be-mishṭar Sadat, 1970-1977)
הסטודנטים במשטר סאדאת, 1977–1970 / חגי ארליך. — תל
אביב : מכון שילוח לחקר המזרח התיכון ואפריקה, אוניברסיטת
תל־אביב, 1978.
46 leaves, ⌜1⌝ leaf of plates : map ; 27 cm. — (סקירות - מכון שילוח לחקר המזרח התיכון ואפריקה, אוניברסיטת תל־אביב ; 60)
Includes bibliographical references.
1. College students—Egypt—Political activity. 2. Egypt—Politics and government—1970- I. Title. II. Series: Mekhon Shiloaḥ le-ḥeḳer ha-Mizraḥ ha-tikhon ṿe-Afriḳah. Seḳirot - Mekhon Shiloaḥ le-ḥeḳer ha-Mizraḥ ha-tikhon ṿe-Afriḳah ; 60.
LA1648.7.E74 79-951720

—FRANCE—ADDRESSES, ESSAYS, LECTURES

Endō, Shūsaku, 1923-
(Furansu no daigakusei)
フランスの大学生 / 遠藤周作. — 東京：角川書店, 昭和52 ⌜1977⌝
210 p. ; 15 cm. — (角川文庫 ; 4007) Ja 77-26945
Series romanized: Kadokawa bunko.
¥220
1. Endō, Shūsaku, 1923- —Journeys—France—Addresses, essays, lectures. 2. France—Description and travel—1945-1974—Addresses, essays, lectures. 3. College students—France—Addresses, essays, lectures. 4. Authors, Japanese—20th century—Biography—Addresses, essays, lectures. I. Title.
PL849.N4Z464 78-816315

—FRANCE—POLITICAL ACTIVITY

Salmon, Jean Marc, 1943-
Hôtel de l'avenir / Jean-Marc Salmon. — Paris (26, rue de Condé, 75006) : Presses d'aujourd'hui, 1978.
294 p. ; 22 cm. F78-14148
52.00F
1. Paris—Riot, 1968. 2. College students—France—Political activity. I. Title.
DC420.S24 944'.361 79-359509
79 MARC

—GERMANY

Diehl, Carl, 1947-
Americans and German scholarship, 1770-1870 / Carl Diehl. — New Haven : Yale University Press, 1978.
xi, 194 p. ; 22 cm. — (Yale historical publications : Miscellany ; 115)
Based on the author's thesis, Yale.
Bibliography: p. 183-189
Includes index.
ISBN 0-300-02079-1
1. American students in Germany—History. 2. College students—United States. 3. College students—Germany. I. Title. II. Series.
LA728.7.D5 378.1'98 77-12931
77⌜r79⌝rev MARC

—GERMANY—POLITICAL ACTIVITY

Bohrmann, Hans.
Strukturwandel der deutschen Studentenpresse : Studentenpolitik und Studentenzeitschriften 1848-1974 / Hans Bohrmann. — München : Verlag Dokumentation, 1975.
337 p. ; 22 cm. — (Kommunikation und Politik ; Bd. 4) GFR***
Originally presented as the author's thesis, Freie Universität, Berlin.
"Bibliographie der deutschen Studentenzeitschriften in den Westzonen und in der Bundesrepublik Deutschland seit 1945": p. 292-327. Bibliography: p. 190-291.
Includes index.
ISBN 3-7940-4020-1
1. College and school journalism—Germany—History. 2. College students—Germany—Political activity. I. Title. II. Series.
LA728.7.B6 378.1'98'970943 75-520231
75⌜r79⌝rev MARC

—GERMANY, WEST—POLITICAL ACTIVITY

Allerbeck, Klaus, 1944-
Soziale Bedingungen für studentischen Radikalismus; eine vergleichende Untersuchung in der Bundesrepublik Deutschland und den Vereinigten Staaten, von Klaus R. Allerbeck. ⌜Köln, 1971?⌝
vii, 423 p. 21 cm. GDB***
Vita.
Bibliography: p. 403-423.
1. College students—Germany, West—Political activity. 2. Conflict of generations. I. Title.
LA728.7.A43 76-858213
79⌜r79⌝rev MARC

—GERMANY, WEST—POLITICAL ACTIVITY—ADDRESSES, ESSAYS, LECTURES

APO und Gewerkschaften : von d. Kooperation zum Bruch / hrsg. von Gudrun Küsel ; ⌜mit e. Vorw. von Theo Pirker⌝. — Berlin : Olle und Wolter, 1978.
188 p. ; 21 cm. GFR78-A
ISBN (invalid) 0-392-24143-X : DM14.80
1. Germany, West—Politics and government—Addresses, essays, lectures. 2. Opposition (Political science)—Addresses, essays, lectures. 3. Trade-unions—Germany, West—Political activity—Addresses, essays, lectures. 4. College students—Germany, West—Political activity—Addresses, essays, lectures. I. Küsel, Gudrun, 1942-
JN3971.A979A66 320.9'43'087 79-377681
79 MARC

—GERMANY, WEST—STATISTICS—PERIODICALS

Germany (Federal Republic, 1949-). Statistisches Bundesamt.
Bevölkerung und Kultur. Reihe 10: Bildungswesen. V. Hochschulen: Studenten an Hochschulen.
—Wintersemester 1974/75.
Stuttgart, W. Kohlhammer.
v. 30 cm. semiannual. (Fachserie - Statistisches Bundesamt, A)
Continued by: Germany (Federal Republic, 1949-). Statistisches Bundesamt. Bildung und Kultur. Reihe 4.1: Studenten an Hochschulen.
1. College attendance—Germany, West—Statistics—Periodicals. 2. College students—Germany, West—Statistics—Periodicals. I. Germany (Federal Republic, 1949-). Statistisches Bundesamt. Studenten an Hochschulen. II. Title. III. Series: Germany (Federal Republic, 1949-). Statistisches Bundesamt. Fachserie.
LC148.G48b 78-648556
MARC-S

—GERMANY, WEST—KIEL—ATTITUDES

Thiele, Ralf.
Einstellung Kieler Geographiestudenten zu ihrem Studienfach : eine exemplarische Untersuchung zu studentischem Verhalten / von Ralf Thiele. — Kiel : R. Thiele, 1978.
vi, 42 p. ; 20 cm. GFR***
"Ergebnis einer Befragung von Studenten am Geographischen Institut der Universität Kiel im Wintersemester 1977/78."
Bibliography: p. 42.
1. Geography—Study and teaching (Higher)—Germany, West—Kiel. 2. Kiel. Universität. Geographisches Institut. 3. College students—Germany, West—Kiel—Attitudes. I. Title.
G76.5.G4T44 910'.7'1143512 79-358368
79 MARC

—HUNGARY

Szentirmai, László, 1932-
Die Tätigkeitsstruktur der Lebensweise der Studenten / László Szentirmai. — Szeged : ⌜s. n.⌝, 1976.
55 p. ; 24 cm. — (Acta Bibliothecaria ; t. 8, fasc. 4 ISSN 0001-7175) H***
Summary in English.
Includes bibliographical references.
1. College students—Hungary. I. Title. II. Series.
LA687.7.S93 78-375816
78 MARC

—HUNGARY—POLITICAL ACTIVITY—HISTORY

A Haladó egyetemi ifjúság mozgalmai Magyarországon, 1918-1945 / ⌜főszerkesztő, Vass Henrik, szerkesztő, Szabó Ágnes ; kiadja a MSZMP KB Párttörténeti Intézete, Felsőoktatási Pedagógiai Központ⌝. — ⌜Budapest⌝ : Kossuth Könyvkiadó, 1978.
379 p. ; 20 cm. H78
Includes bibliographical references.
ISBN 9630910101 : 42.00Ft
1. College students—Hungary—Political activity—History. 2. Socialism and youth—Hungary—History. I. Vass, Henrik. II. Szabó, Ágnes. III. Magyar Szocialista Munkáspárt. Központi Bizottság. Párttörténeti Intézet. IV. Felsőoktatási Pedagógiai Kutatóközpont.
LA683.7.H34 79-343761
*79 MARC

—INDIA—SOUTH INDIA—RELIGIOUS LIFE—STATISTICS

Rajamanickam, M
A psychological study of religious and related attitudes of the student and professional groups in South India / M. Rajamanickam. — 1st ed. — Annamalainagar : Annamalai University, 1976.
xii, 380 p. : ill. ; 22 cm.
Bibliography: p. ⌜366⌝-380.
Rs20.00
1. College students—India—South India—Religious life—Statistics. 2. South India—Religious life and customs—Statistics. I. Title.
BV4531.2.R26 200'.1'9 77-905137
78 MARC

—INDONESIA

Hull, Valerie J
Pembentukan keluarga di kalangan universitas : hasil-hasil sementara dari suatu studi kasus = Family formation in the university community : preliminary results of a case study / Valerie J. Hull, Kodiran, Irawati Singarimbun. — Yogyakarta : Lembaga Kependudukan, Universitas Gadjah Mada, 1976.
ii, 15, ⌜78⌝ leaves ; 33 cm. — (Seri laporan - Lembaga Kependudukan, Universitas Gadjah Mada ; no. 9)
English and Indonesian.
1. College students—Indonesia. 2. Family—Indonesia. I. Kodiran, joint author. II. Singarimbun, Irawati, joint author. III. Title. IV. Title: Family formation in the university community. V. Series: Universitas Gadjah Mada. Lembaga Kependudukan. Seri laporan - Lembaga Kependudukan, Universitas Gadjah Mada ; no. 9.
LA1273.7.H84 76-941448
78 MARC

COLLEGE STUDENTS
—INDONESIA (Continued)

Nasution, Abdul Haris, 1918-
Tanya-jawab mahasiswa, taruna : sekitar sewindu orde baru / oleh A. H. Nasution. — ₍Bandung : Ganaco₎, 1974.
94 p. ; 21 cm.
1. College students—Indonesia. I. Title.
LA1273.7.N37 77-940653
79 MARC

—INDONESIA—STATISTICS
—COLLECTED WORKS

Indonesia. Biro Pusat Statistik. Seksi Statistik Pengadjaran.
Pengadjaran perguruan tinggi.
₍Djakarta₎ Biro Pusat Statistik, Seksi Statistik Pengadjaran₎
v. 20 x 29 cm.
1. Universities and colleges — Indonesia — Statistics — Collected works. 2. College students—Indonesia—Statistics—Collected works. I. Title.
LA1273.I 43a 78-640084
MARC-S

—JAPAN—ADDRESSES, ESSAYS, LECTURES

Amano, Teiyū, 1884-
₍Wakaki hitotachi e₎
學生論 / 天野貞祐. — 東京 : 河出書房, 昭和27 ₍1952₎
353 p. : port. ; 19 cm.
Originally published as v. 3 of Amano Teiyū chosaku shū.
1. College students—Japan—Addresses, essays, lectures. 2. Students—Japan—Addresses, essays, lectures. I. Title.
Title romanized: Gakuseiron.
LA1318.7.A62 1952 79-805787

(Daigakusei, kono kangaeru ashi)
大学生 この考える葦 / 臼井吉見, 河盛好蔵編. — 東京 : 潮文社, 昭和33 ₍1958₎, 昭和41 ₍1966₎ printing.
234 p. ; 18 cm. — (潮文社新書)
Series romanized: Chōbunsha sinsho.
1. College students—Japan—Addresses, essays, lectures. 2. Education, Higher—Japan—1945- —Addresses, essays, lectures. I. Usui, Yoshimi, 1905- II. Kawamori, Yoshizō, 1902-
LA1318.7.D35 78-814091

(Konnichi no gakusei no tame ni)
今日の學生のために / ₍著者 小林秀雄 ... et al.₎. — 東京 : 創元社, 昭和25 ₍1950₎
3, 186 p. ; 20 cm. — (創元社學生シリーズ)
Series romanized: Sōgensha gakusei shirizu.
1. College students—Japan—Addresses, essays, lectures. 2. College students—Addresses, essays, lectures. 3. Students—Addresses, essays, lectures. I. Kobayashi, Hideo, 1902-
LA1318.7.K66 79-809301

—JAPAN—BIOGRAPHY

Tsuchiya, Shukurō, 1908-
(Kurenai moyuru)
紅萌ゆる : 昭和初年の青春 / 土屋祝郎著. — 東京 : 岩波書店, 1978.
ii, 226 p. ; 18 cm. — (岩波新書 ; 47) Ja 78-15002
Series romanized: Iwanami shinsho.
¥280
1. Daisan Kōtō Gakkō—Students. 2. Tsuchiya, Shukurō, 1908- 3. College students—Japan—Biography. I. Title.
LG247.T78 78-813280

—JAPAN—CONDUCT OF LIFE

Matsubara, Tatsuya, 1930-
(Gendai daigakusei katagi)
現代大学生かたぎ : 未来への志向を探る / 松原達哉著. — 東京 : 産心社, 昭和52 ₍1977₎
233 p. : ill. ; 18 cm. — (産心ブックス) Ja 77B-3431
Series romanized: Sanshin bukkusu.
¥680
1. College students—Japan—Conduct of life. 2. College students—Japan—Psychology. I. Title.
LA1318.7.M377 78-811405

—JAPAN—POLITICAL ACTIVITY

Tōdai Gakusei Undō Kenkyūkai.
(Nihon no gakusei undō)
日本の学生運動 : その理論と歴史 / 東大学生運動研究会. — 東京 : 新興出版社, 1956.
340 p. ; 19 cm.
1. Student movements—Japan. 2. College students—Japan—Political activity. I. Title.
LA1318.7.T6 1956 78-812366

—JAPAN—POLITICAL ACTIVITY
—ADDRESSES, ESSAYS, LECTURES

(Zenkyōtō kaitai to genzai)
全共闘 解体と現在 / 田中吉六 ... ₍et al.₎. — 東京 : 田畑書店, 1978.
238 p. ; 20 cm. Ja 78-26972
¥1500
1. Student movements—Japan—Addresses, essays, lectures. 2. College students—Japan—Political activity—Addresses, essays, lectures. 3. Zenkoku Zenkyōtō Rengō—Addresses, essays, lectures. I. Tanaka, Kichiroku, 1907-
LA1318.7.Z42 79-806821

—JAPAN—POLITICAL ACTIVITY
—HISTORY

Kikukawa, Tadao, 1901-1954.
(Gakusei shakai undō shi)
學生社會運動史 / 菊川忠雄著. — 東京 : 中央公論社, 昭和6 ₍1931₎
8, 584 p. ; 20 cm.
1. Student movements—Japan—History. 2. College students—Japan—Political activity—History. 3. High school students—Japan—Political activity—History. I. Title.
LA1318.7.K55 1931 79-801767

Kikukawa, Tadao, 1901-1954.
(Gakusei shakai undō shi)
學生社會運動史 / 菊川忠雄著. — 東京 : 海口書店, 昭和22 ₍1947₎
5, 8, 475, 9 p. ; 19 cm.
1. Student movements—Japan—History. 2. College students—Japan—Political activity—History. 3. High school students—Japan—Political activity—History. I. Title.
LA1318.7.K55 1947 79-801768

—JAPAN—PSYCHOLOGY

Matsubara, Tatsuya, 1930-
(Gendai daigakusei katagi)
現代大学生かたぎ : 未来への志向を探る / 松原達哉著. — 東京 : 産心社, 昭和52 ₍1977₎
233 p. : ill. ; 18 cm. — (産心ブックス) Ja 77B-3431
Series romanized: Sanshin bukkusu.
¥680
1. College students—Japan—Conduct of life. 2. College students—Japan—Psychology. I. Title.
LA1318.7.M377 78-811405

—KOREA—PERIODICALS

(Han saebŏl)
한새벌
₍부산₎ 부산교육대학학생회
v. ill. 21 cm.
1. Teaching—Periodicals. 2. Teachers—Periodicals. 3. College students—Korea—Periodicals. I. Pusan Kyoyuk Taehak. Haksaenghoe.
LH7.P85H35 78-646409
(MARC-S)

—LEBANON—POLITICAL ACTIVITY
—ADDRESSES, ESSAYS, LECTURES

Ṣa'b, Najīb.
(al-Ṭullāb fī Lubnān)
الطلاب في لبنان : من الفوضى الى النظام / نجيب صعب. — الطبعة 1. — بيروت : مؤسسة الدراسات والابحاث اللبنانية، 1977.
64 p. ; 17 cm.
1. College students — Lebanon — Political activity — Addresses, essays, lectures. 2. Lebanon—Politics and government—Addresses, essays, lectures. I. Title.
LA1463.7.S2 79-971177

—NEW YORK (STATE)—BROOKLYN—
RELIGIOUS LIFE

Benjamin, Barbara.
The impossible community : a story of hardship & hope at Brooklyn College in New York / Barbara Benjamin. — Downers Grove, Ill. : InterVarsity Press, c1978.
128 p. ; 18 cm.
ISBN 0-87784-653-7 : $2.50
1. College students—New York (State)—Brooklyn—Religious life. 2. Inter-varsity Christian Fellowship of the United States of America. 3. Brooklyn College. 4. Benjamin, Barbara. I. Title.
BV4531.2.B46 267'.61 77-27741
78 MARC

—POLAND—POZNAN
—SOCIOECONOMIC STATUS

Zandecki, Aleksander.
Społeczno-zawodowe pochodzenie studentów a ich wyniki w nauce / Aleksander Zandecki. — Wyd. 1. — Warszawa : Państwowe Wydawn. Naukowe, 1978.
218 p. : ill. ; 20 cm. — (Raporty z badań)
At head of title: Ministerstwo Nauki, Szkolnictwa Wyższego i Techniki. Instytut Polityki Naukowej, Postępu Technicznego i Szkolnictwa Wyższego.
Summary in English and Russian; table of contents also in English and Russian.
Includes bibliographical references.
zł40.00
1. College students—Poland—Poznan—Socioeconomic status. I. Title.
LC208.8.Z36 378.1'98'094384 79-359056
79 MARC

—RUSSIA

(Student i ego dei͡atel'nost')
Студент и его деятельность / ₍И. Я. Писаренко, В. Я. Кочергин, Т. Н. Гацукевич ... et al. ; Под ред. Г. П. Давидюка ... et al.₎. — Минск : Изд-во БГУ, 1978.
159 p. ; 20 cm. USSR 78
Includes bibliographical references.
1.00rub
1. College students—Russia. 2. Communist education—Russia. I. Pisarenko, I. I͡A. II. Davidzi͡uk, Heorhiĭ Pi͡atrovich.
LA838.7.S77 79-408024

—SENEGAL—BIOGRAPHY

Senegal. Ministère de l'information et des relations avec les assemblées.
Livre blanc sur le suicide d'Oumar Blondin Diop. ₍Dakar, 1973₎
31 p. illus. 24 cm.
1. Diop, Oumar Blondin, 1946-1973. 2. Senegal—Politics and government. 3. College students—Senegal—Political activity. 4. Political crimes and offenses—Senegal. 5. College students—Senegal—Biography. 6. Political prisoners—Senegal—Biography. I. Title.
DT549.82.D56S46 1973 77-477799
79 MARC

COLLEGE STUDENTS (Continued)

—SENEGAL—POLITICAL ACTIVITY

Senegal. Ministère de l'information et des relations avec les assemblées.
Livre blanc sur le suicide d'Oumar Blondin Diop. ⌜Dakar, 1973⌝
31 p. illus. 24 cm.
1. Diop, Oumar Blondin, 1946-1973. 2. Senegal—Politics and government. 3. College students—Senegal—Political activity. 4. Political crimes and offenses—Senegal. 5. College students—Senegal—Biography. 6. Political prisoners—Senegal—Biography. I. Title.
DT549.82.D56S46 1973 77-477799
79 MARC

—SOUTH AFRICA

Du Toit, Brian M 1935-
Drug use and South African students / by Brian M. Du Toit. — ⌜Athens⌝ : Ohio University Center for International Studies, Africa Program, 1978.
127 p. : ill. ; 28 cm. — (Papers in international studies : Africa series ; no. 35)
Bibliography: p. 126-127.
ISBN 0-89680-076-8 : $6.50
1. Drugs and youth—South Africa. 2. High school students—South Arica. 3. College students—South Africa. 4. Cannabis. I. Title. II. Series.
DT1.P33 no. 35 960 s 78-21910
⌜HV5824.Y68⌝ 78 MARC

—SWEDEN—CONDUCT OF LIFE —HISTORY

Jonsson, Jonas, 1908-
Jacobi Filius : en kämpe utan vapenlycka : Scener ur studentlifvet / Jonas Jonsson. — ⌜s.l. : s.n.⌝, 1976 (Malmö : Cron-tryck) S***
xlviii, 95 p. : ill. ; 22 cm.
Includes reprint of the 1833 ed. of Scener ur student-lifvet by C. F. Jacobsson, printed by Elmén & Granberg, Stockholm.
ISBN 9172600799
1. College students—Sweden—History. 2. College students—Sweden—Conduct of life—History. I. Jacobsson, Christopher Fritz, 1807-1875. Scener ur student-lifvet. 1976.
LA908.7.J66 78-372415
78 MARC

—SWEDEN—HISTORY

Jonsson, Jonas, 1908-
Jacobi Filius : en kämpe utan vapenlycka : Scener ur studentlifvet / Jonas Jonsson. — ⌜s.l. : s.n.⌝, 1976 (Malmö : Cron-tryck) S***
xlviii, 95 p. : ill. ; 22 cm.
Includes reprint of the 1833 ed. of Scener ur student-lifvet by C. F. Jacobsson, printed by Elmén & Granberg, Stockholm.
ISBN 9172600799
1. College students—Sweden—History. 2. College students—Sweden—Conduct of life—History. I. Jacobsson, Christopher Fritz, 1807-1875. Scener ur student-lifvet. 1976.
LA908.7.J66 78-372415
78 MARC

—TAIWAN—ADDRESSES, ESSAYS, LECTURES

(Shui pien chung ti hsin i tai)
蛻變中的新一代 / 丁庭宇 ... ⌜et al.⌝ 合著. — 第1版. — 臺北：四季出版事業有限公司, 民國67 ⌜1978⌝
2, 3, 176 p. ; 20 cm. — (四季文萃 ; 18)
Title in English on cover p. ⌜3⌝: The Emerging of a new generation.
In colophon: Chu che Chang Ying-hua.
Series romanized: Ssu chi wen ts'ui.
NT$40.00
1. College students—Taiwan—Addresses, essays, lectures. I. Ting, T'ing-yü. II. Chang, Ying-hua. III. Title. IV. Title: The Emerging of a new generation.
LA1138.7.S55 79-836848

—TAIWAN—POLITICAL ACTIVITY

Ch'en, I-yen, 1942-
(T'ai-wan ti ch'ü ta hsüeh sheng cheng chih she hui hua chih yen chiu)
臺灣地區大學生政治社會化之研究 / 陳義彥撰. — ⌜臺北⌝：政治大學, 民國66 ⌜1977⌝
432 p. (on double leaves) ; 26 cm.
Cover title.
Reproduced from typescript.
"國立政治大學政治研究所博士論文"
Bibliography: p. 383-401.
1. Political socialization—Taiwan. 2. College students—Taiwan—Political activity. I. Title.
JA76.C49 1977 79-840441

—UNITED STATES

Diehl, Carl, 1947-
Americans and German scholarship, 1770-1870 / Carl Diehl. — New Haven : Yale University Press, 1978.
xi, 194 p. ; 22 cm. — (Yale historical publications : Miscellany ; 115)
Based on the author's thesis, Yale.
Bibliography: p. 183-189
Includes index.
ISBN 0-300-02079-1
1. American students in Germany—History. 2. College students—United States. 3. College students—Germany. I. Title. II. Series.
LA728.7.D5 378.1'98 77-12931
77⌜r79⌝rev MARC

The Early college in theory and practice / ⌜edited by Paul E. Kaylor⌝. — Great Barrington, MA : Simon's Rock Early College, c1978.
94 p. ; 23 cm.
Based in part on the proceedings of a consultation held in Oct. 1977 and sponsored by Simon's Rock Early College and the Association of American Colleges.
Includes bibliographical references.
1. Education, Higher—United States. 2. College students—United States. 3. Adolescence. I. Kaylor, Paul E. II. Simon's Rock Early College. III. Association of American Colleges.
LB2331.7.E23 378.73 79-102413
79 MARC

Lamont, Lansing.
Campus shock : a firsthand report on college life today / Lansing Lamont. — 1st ed. — New York : Dutton, c1979.
xii, 144 p. ; 22 cm.
Bibliography: p. 133-136.
Includes index.
ISBN 0-525-07309-4 : $8.95
1. College students—United States. 2. Education, Higher—United States—1965- I. Title.
LA229.L28 1979 378.73 79-608
79 MARC

O'Brien, Robert William, 1907-
The college nisei / Robert W. O'Brien. — New York : Arno Press, 1978 ⌜c1949⌝
viii, 165 p. : ill. ; 24 cm. — (The Asian experience in North America : Chinese and Japanese)
Reprint of the ed. published by Pacific Books, Palo Alto, Calif.
Includes bibliographical references and index.
ISBN 0-405-11286-6
1. Japanese Americans—Evacuation and relocation, 1942-1945. 2. National Japanese American Student Relocation Council. 3. College students—United States. I. Title. II. Series: Asian experience in the North America : Chinese and Japanese.
D769.8.A6O27 1978 940.54'72'73 78-54829
78 MARC

—UNITED STATES—ADDRESSES, ESSAYS, LECTURES

College student development revisited : programs, issues, and practices / editors, Roy F. Giroux ... ⌜et al.⌝. — Rev. ed. — ⌜Washington⌝ : American Personnel and Guidance Association, 1979.
364 p. : ill. ; 23 cm.
Includes bibliographies.
1. College students—United States—Addresses, essays, lectures. I. Giroux, Roy F. II. American Personnel and Guidance Association.
LA229.C634 1979 378.1'98'0973 79-116065
79 MARC

—UNITED STATES—CONDUCT OF LIFE

Carnegie Council on Policy Studies in Higher Education.
Fair practices in higher education : rights and responsibilities of students and their colleges in a period of intensified competition for enrollments : a report of the Carnegie Council on Policy Studies in Higher Education. — 1st ed. — San Francisco : Jossey-Bass Publishers, 1979.
xiii, 91 p. ; 24 cm. — (The Carnegie Council series)
Bibliography: p. ⌜81⌝-86.
Includes index.
ISBN 0-87589-415-1
1. Universities and colleges—United States—Admission. 2. Universities and colleges—United States—Accreditation. 3. Universities and colleges—United States—Moral and religious aspects. 4. College students—United States—Conduct of life. I. Title.
LB2351.C28 1979 378.73 79-84232
79 MARC

Flumiani, Carlo Maria.
The deepening moral degeneration of our colleges & universities / C. M. Flumiani. — ⌜Albuqerque, N.M.⌝ : American Classical College Press, ⌜1978⌝
29 leaves ; 28 cm. — (A Human development series library book)
ISBN 0-89266-102-X
1. Universities and colleges—United States. 2. College students—United States—Conduct of life. 3. College students' socio-economic status—United States. I. Title. II. Series.
⌜LA227.3.F59⌝ 378.73 78-24293
78 MARC

—UNITED STATES—POLITICAL ACTIVITY—HISTORY—20TH CENTURY

Horn, Max, 1920-
The Intercollegiate Socialist Society, 1905-1921 : origins of the modern American student movement / Max Horn. — Boulder, Colo. : Westview Press, 1979.
xv, 259 p. ; 24 cm. — (Westview replica editions)
Bibliography: p. 241-251.
Includes index.
ISBN 0-89158-584-2
1. Student movements—United States. 2. College students—United States—Political activity—History—20th century. 3. League for Industrial Democracy—History. 4. Right and left (Political science) I. Title.
LA229.H56 378.1'98'10973 79-7404
79 MARC

—UNITED STATES—RELIGIOUS LIFE —PERIODICALS

National Institute for Campus Ministries.
The NICM journal for Jews and Christians in higher education. v. 1- winter 1976-
⌜Newton Centre, Mass., National Institute for Campus Ministries⌝
v. 23 cm. quarterly.
Key title: The NICM journal for Jews and Christians in higher education, ISSN 0362-0794
1. College students—United States—Religious life—Periodicals. 2. Church and college—United States—Periodicals. I. National Institute for Campus Ministries. Journal for Jews and Christians in higher education. II. Title.
LB3609.N33a 378.1'98'0973 79-642695
MARC-S

—UNITED STATES—STATISTICS

United States. Dept. of Health, Education, and Welfare. Office for Civil Rights.
Racial, ethnic, and sex enrollment data from institutions of higher education, fall 1976 / Department of Health, Education, and Welfare, Office for Civil Rights. — ⌜Washington⌝ : OCR, 1978.
1356 p. : forms ; 28 cm.
Tables.
1. Minorities—Education (Higher)—United States—Statistics. 2. College attendance—United States—Statistics. 3. College students—United States—Statistics. I. Title.
LC3731.U53 1978 378.1'05'973 79-600842
79 MARC

—YUGOSLAVIA

Jun-lipanj 1968. Dokumenti. ⌜Zbornik dokumenata o studentskim zbivanjima u Jugoslaviji u junskim danima 1968. godine. Zagreb, Hrvatsko filozofsko društvo, 1971⌝ Yu***
xxiii, 518 p. 24 cm. 35.00Din
Issued as Praxis, no. 1/2, 1969, with special title overprinted on cover.
1. College students—Yugoslavia. I. Hrvatsko filozofsko društvo. II. Praxis. Jugoslavensko izdanje.
LA1008.7.J86 73-977903
79⌜r79⌝rev MARC

—YUGOSLAVIA—POLITICAL ACTIVITY

Pervan, Ralph, 1938-
Tito and the students : the university and the university student in self-managing Yugoslavia / ⌜by⌝ Ralph Pervan. — Nedlands, W.A. : University of Western Australia Press ; Forest Grove, Or. : ⌜sold by⌝ International Scholarly Book Services, 1978.
xvi, 239 p. ; 24 cm. Aus
Bibliography: p. 221-234.
Includes index.
ISBN 0-85564-131-2
1. Tito, Josip Broz, Pres. Yugoslavia, 1892- . 2. Education—Yugoslavia—History. 3. Student movements—Yugoslavia. 4. College students—Yugoslavia—Political activity. I. Title.
LA1001.8.P47 378.497 79-314829
79 MARC

COLLEGE STUDENTS, AFRO-AMERICAN
see Afro-American college students

COLLEGE STUDENTS, BLACK

—SOUTH AFRICA—STATISTICS

Van Rensburg, Leon Stephanus Janse.
Onderwystendense : statistiek sedert 1910 : A2 kleurling-studente aan universiteite / L. S. J. van Rensburg. — Pretoria : Instituut vir Statistiese Navorsing, Raad vir Geesteswetenskaplike Navorsing, 1977.
vi, 46 p. : ill. ; 30 cm. — (Verslag - Suid-Afrikaanse Raad vir Geesteswetenskaplike Navorsing, Instituut vir Statistiese Navorsing ; nr. WS-21) SA78
Summary in English.
Includes bibliographical references.
ISBN 0-86965-465-9 : R2.80
1. College students, Black—South Africa—Statistics. I. Title. II. Series: Human Sciences Research Council. Institute for Statistical Research. Verslag WS ; nr. 21.
LC2781.V35 378.1'98'1 79-373223
79 MARC

COLLEGE STUDENTS' SOCIO-ECONOMIC STATUS

—AUSTRALIA

Thomson, Norman J
Tertiary education assistance scheme : means test : a report / by Norman J. Thomson. — Canberra : Govt. Printer of Australia, 1975.
vii, 69 p. : ill. ; 25 cm. — (Parliamentary paper ; no. 105) Aus**
Bibliography: p. 69.
$1.95
1. Student aid—Australia. 2. College students' socio-economic status—Australia. I. Title. II. Series: Australia. Parliament. The records of the proceedings and the printed papers ; 1975, no. 105.
J905.L3 1975, no. 105 328.94'01 s 79-310204
[LB2337.6.A8] 79 MARC

—BRAZIL

Santa Maria, Brazil (Rio Grande do Sul). Universidade Federal. Comissão Permanente de Vestibular.
Análise das características sócio-econômicas-culturais e de desempenho dos candidatos, nas diferentes provas, do concurso vestibular da U.F.S.M., 1977 / Universidade Federal de Santa Maria, Pró-Reitoria de Ensino e Pesquisa, Comisão Permanente de Vestibular. — Santa Maria : A. Comissão, 1977-
v. : ill. ; 24 cm.
Bibliography: v. 1, p. [239]-241.
1. Santa Maria, Brazil (Rio Grande do Sul). Universidade Federal—Curricula—Statistics. 2. College students—Socio-economic status—Brazil. I. Title.
LE31.S24A68 78-376724
78 MARC

—BRAZIL—ESPIRITO SANTO (STATE)

Espírito Santo, Brazil (State). Universidade Federal. Comissão Coordenadora do Vestibular.
Análise sócio-econômica e cultural do vestibular de 1975 na UFES / por Stelio Dias ... [et al.]. — [Vitória] : Universidade Federal do Espírito Santo, Comissão Coordenadora do Vestibular, [1975]
110 p. ; 30 cm.
1. College students' socio-economic status—Brazil—Espirito Santo (State) I. Dias, Stelio. II. Title.
LC208.8.E86 1975 79-379429
79 MARC

—ITALY

Orviati, Silvio.
La popolazione universitaria di Trieste : aspetti socio-economici e previsioni di sviluppo / Silvio Orviati. — 1. ed. — Trieste : CLUET, 1977.
123 p. ; 21 cm. — (Contributi e ricerche scienze politiche, Trieste ; 7) It***
Bibliography: p. 118.
1. Triesta. Università—Students. 2. College students' socio-economic status—Italy. I. Title. II. Series.
LF3893.T7O73 78-378644
*78 MARC

—ITALY—TURIN

Bonet, Luciano.
Gli impiegati-studenti : il caso di scienze politiche / Luciano Bonet. — Milano : F. Angeli, [c1978]
183 p. ; 22 cm. — (Saggi e ricerche dell'istituto di scienze politiche Gioele Solari, Universita di Torino ; n. 1) It77-Dec
L4000
1. Turin. Università. Istituto di scienze politiche— Students. 2. College students' socio-economic status—Italy—Turin. I. Title. II. Series: Turin. Università. Istituto di scienze politiche. Saggi e ricerche dell'istituto di scienze politiche Gioele Solari ; n. 1.
JF1338.T87B66 78-384014
79 MARC

—NEW YORK (CITY)

Kramer, Rena.
Characteristics of enrollees and non enrollees among freshmen, 1972 / Rena Kramer, Barry Kaufman, and Lawrence Podell. — New York : Office of Program and Policy Research, City University of New York, 1974.
19 leaves ; 28 cm.
1. College students' socio-economic status—New York (City) I. Kaufman, Barry, joint author. II. Podell, Lawrence, joint author. III. New York (City). City University of New York. Office of Program and Policy Research. IV. Title.
LC205.8.N5K7 378.1'98'1097471 79-312856
79 MARC

—NIGERIA

University of Lagos. Human Resources Research Unit.
Socio-economic background of Nigerian university students / Human Resources Research Unit (School of Social Studies). — Lagos : University of Lagos, [1973?]
iii, 17 leaves ; 26 cm. — (Research bulletin - Human Resources Research Unit, University of Lagos ; no. 1/001)
Cover title.
"Research project no. 1: Employment aspirations and prospects of Nigerian university undergraduates & graduates."
1. University of Lagos—Students. 2. College students' socio-economic status—Nigeria. I. Title. II. Series: University of Lagos. Human Resources Research Unit. Research bulletin - Human Resources Research Unit, University of Lagos ; no. 1/001.
LG481.L2A55 378.1'98'096691 79-303802
79 MARC

—UNITED STATES

Flumiani, Carlo Maria.
The deepening moral degeneration of our colleges & universities / C. M. Flumiani. — [Albuquerque, N.M.] : American Classical College Press, [1978]
29 leaves ; 28 cm. — (A Human development series library book)
ISBN 0-89266-102-X
1. Universities and colleges—United States. 2. College students—United States—Conduct of life. 3. College students' socio-economic status—United States. I. Title. II. Series.
[LA227.3.F59] 378.73 78-24293
78 MARC

London, Howard B
The culture of a community college / Howard B. London. — New York : Praeger, 1978.
xvii, 181 p. ; 24 cm.
bibliography: p. 175-181.
ISBN 0-03-044701-1
1. Community colleges—United States. 2. United States—Intellectual life. 3. Education, Humanistic—United States. 4. College students' socio-economic status—United States. I. Title.
LB2328.L58 378.1'543'0973 78-8697
78 MARC

COLLEGE TEACHERS

see also Graduate teaching assistants; Medical colleges—Faculty

Comparative perspectives on the academic profession / edited by Philip G. Altbach. — New York : Praeger, 1977.
xii, 214 p. ; 24 cm. — (Praeger special studies in international economics and development)
Includes bibliographies.
ISBN 0-03-040781-8
1. College teachers. 2. Comparative education. I. Altbach, Philip G.
LB2331.7.C65 378.1'2 77-83481
77[r79]rev MARC

Wilson, Logan, 1907-
American academics : then and now / Logan Wilson. — New York : Oxford University Press, 1979.
309 p. ; 22 cm.
Continues the author's earlier work, The academic man.
Includes bibliographical references and index.
ISBN 0-19-502482-6
1. College teachers. I. Title.
LB1778.W52 378.1'2 78-12118
78 MARC

—LEGAL STATUS, LAWS, ETC.—UNITED STATES—CASES

Edwards, Harry T
Higher education and the law / by Harry T. Edwards and Virginia Davis Nordin. — Cambridge : Institute for Educational Management, Harvard University, c1979.
xi, 844 p. ; 25 cm.
Includes bibliographical references and index.
1. Universities and colleges—Law and legislation—United States—Cases. 2. College teachers—Legal status, laws, etc.—United States—Cases. 3. College students—Legal status, laws, etc.—United States—Cases. I. Nordin, Virginia Davis, joint author. II. Title.
KF4225.A7E35 344'.73'074 79-88195
79 MARC

—RETIREMENT

Patton, Carl V
Academia in transition : mid-career change or early retirement / Carl V. Patton. — Cambridge, Mass. : Abt Books, c1979.
xv, 212 p. ; 24 cm.
Includes bibliographical references and index.
ISBN 0-89011-514-1
1. College teaching—Vocational guidance. 2. College teachers—Retirement. 3. Professions—Vocational guidance. 4. Occupational mobility. I. Title.
LB2331.P364 331.2'52 78-66694
79 MARC

—SUPPLY AND DEMAND—CANADA

Cottam, K Jean.
Canadian universities : American takeover of the mind? / K. Jean Cottam. — Toronto : Gall Publications ; distributed by Saannes Publications, 1974.
39 p. ; 22 cm. C***
Includes bibliographical references.
1. College teachers—Supply and demand—Canada. I. Title.
LB2832.4.C2C67 331.1'33 75-311716
75[r79]rev MARC

—TAXATION—UNITED STATES

American Accounting Association.
Tax planning for educators. — Sarasota, Fla. : American Accounting Association, 1979.
ix, 86 p. ; 22 cm.
Published in 1973 under title: Faculty benefits and tax planning.
Includes bibliographical references.
1. College teachers—Taxation—United States. 2. Tax planning—United States. I. Title.
KF6289.8.E3A47 343'.73'0402437 79-108515
79 MARC

—TENURE

Stenger, Alfred, Ph. D.
McCarthyism in Ontario? A submission in form of a case study. Ottawa, 1968.
xii, 150 l. 28 cm. C***
1. Algonquin College—Faculty. 2. College teachers—Tenure. 3. Industrial relations—Study and teaching. I. Title.
T173.A38S74 331.1'33 78-595137
71[r79]rev MARC

—TENURE—PENNSYLVANIA—PHILADELPHIA

Mihram, G Arthur.
An epistle to Dr. Benjamin Franklin / from G. Arthur Mihram. — 1st ed. — Hicksville, N.Y. : Exposition Press, c1975.
ix, 37 p. : port. ; 22 cm. — (An Exposition-university book)
ISBN 0-682-48365-6
1. Pennsylvania. University—Faculty. 2. Pennsylvania. University—Administration. 3. College teachers—Tenure—Pennsylvania—Philadelphia. 4. Privacy, Right of—Pennsylvania—Philadelphia. I. Title.
LD4517.5.M53 378.748'11 79-315383
00 MARC

—TENURE—UNITED STATES

The Hidden professoriate : credentialism, professionalism, and the tenure crisis / edited by Arthur S. Wilke. — Westport, Conn. : Greenwood Press, 1979.
xviii, 290 p. ; 23 cm. — (Contributions in sociology ; no. 29 ISSN 0084-9278)
Bibliography: p. [261]-283.
Includes index.
ISBN 0-8371-9886-0 : $22.95
1. College teachers—Tenure—United States. 2. Higher education and state—United States. 3. Sociologists—United States. I. Wilke, Arthur S. II. Series.
LB2335.7.H5 378.1'22 77-84774
78 MARC

COLLEGE TEACHERS —TENURE—UNITED STATES (Continued)

O'Toole, James.
Tenure, three views / James O'Toole, William W. Van Alstyne, Richard Chait. — ₍New Rochelle, N.Y.₎ : Change Magazine Press, c1979.
55 p. ; 23 cm.
ISBN 0-915390-21-3 : $3.95
1. College teachers—Tenure—United States. I. Van Alstyne, William W., joint author. II. Chait, Richard, joint author. III. Title.
LB2335.7.O86 378.1'22'0973 79-51330
00 MARC

—ALASKA—STATISTICS

Van Houten, John.
Educational institutions occupational employment statistics, Alaska, 1977 / prepared by John O. Van Houten. — ₍Juneau₎ : Alaska Dept. of Labor, 1978.
15 p. : ill. ; 28 cm.
1. Teachers—Alaska—Statistics. 2. School employees—Alaska—Statistics. 3. College teachers—Alaska—Statistics. 4. Universities and colleges—Alaska—Employees—Statistics. I. Alaska. Dept. of Labor. II. Title.
LB2832.3.A4V36 371.1'009798 79-622464
79 MARC

—AUSTRALIA—LEAVES OF ABSENCE

Australia. Tertiary Education Commission.
Study leave in universities and colleges of advanced education : report of the Tertiary Education Commision. — Canberra : Tertiary Education Commission, 1978.
iv, 34 p. ; 25 cm. Aus
1. College teachers—Australia—Leaves of absence. I. Title.
LB2335.A8A9 1978 378.1'21'0994 79-309357
79 MARC

—BELGIUM

Creyf, R
Professoren en het katholiek karakter van hun universiteit : een sociologisch onderzoek naar definiëringsprocessen terzake aan de K. U. Leuven / door R. Creyf, K. Dobbelaere, J. Vanhoutvinck ; m.m.v. A. De Neve. — Leuven : Katholieke Universiteit Leuven, Sociologisch Onderzoeksinstituut, Afdeling Godsdienstsociologie, 1977.
112 leaves ; 29 cm. & 2 suppls. — (Rapport - Katholieke Universiteit Leuven, Sociologisch Onderzoeksinstituut ; 1977/5-7) Be77-9
"F.K.F.O. project 920: Secularisering, een proces van sociale verandering."
Includes bibliographies.
1. Catholic universities and colleges—Social aspects—Belgium. 2. College teachers—Belgium. 3. Sociological research—Belgium. I. Dobbelaere, K., joint author. II. Vanhoutvinck, J., joint author. III. Title. IV. Series: Katholieke Universiteit te Leuven. Sociologisch Onderzoeksinstituut. Rapport ; 1977/5-7.
LC506.B4C73 79-355014
*79 MARC

—CHINA—BIOGRAPHY

Professor Wang Kuang, his life and his works / compiled by China Maritime Institute. — T'ai-pei : China Maritime Institute, 1972.
2, 280 p., ₍7₎ leaves of plates : ill. ; 22 cm.
Cover title.
Title also in Chinese: Wang Kuang chiao shou ch'i jen ch'i shu.
"Professor Wang Kuang's books on transportation science": p. 233-263.
1. Wang, Juang. 2. Merchant marine—China. 3. College teachers—China—Biography. I. Chung-hua min kuo hang yün hsüeh hui. II. Title: Wang Kuang chiao shou chi'i jen ch'i shu.
HE569.W36P76 387.5'0951 79-100054
79 MARC

—ENGLAND—BIOGRAPHY—JUVENILE LITERATURE

Sheffe, Norman.
Goldwin Smith / Norman Sheffe. — Don Mills, Ont. : Fitzhenry & Whiteside, c1976.
63, ₍1₎ p. : ill. ; 22 cm. — (The Canadians) C***
Bibliography: p. ₍64₎
ISBN 0-88902-224-0
1. Smith, Goldwin, 1823-1910—Juvenile literature. 2. College teachers—England—Biography—Juvenile literature. 3. Journalists—England—Biography—Juvenile literature. 4. Journalists—Ontario—Toronto—Biography—Juvenile literature. 5. Toronto—Biography—Juvenile literature. I. Series.
CT788.S592S53 070'.92'4 79-306234
79 MARC

—FRANCE

Klausner, Lucien, 1930-
Où vont les professeurs? / Par Lucien Klausner. — ₍Paris₎ : Casterman, c1979.
164 p. ; 20 cm. — (Collection Orientations E 3) F***
ISBN 2-203-20244-0 : 39.00F
1. College teachers—France. I. Title.
LB2331.7.K55 378.1'2'0944 79-383901
79 MARC

—GERMANY, WEST

Organisation und Reform des Studiums : e. Hochschullehrerbefragung / Christoph Oehler ... ₍et al.₎. — 1. Aufl. — München ; New York : Verlag Dokumentation, 1978.
xi, 102, 15 p. : graphs ; 21 cm. — (Hochschulplanung ; Bd. 29) GFR78-A
"Fortsetzung und Ergänzung der in der Reihe HIS-Hochschulplanung Band 25 veröffentlichten Untersuchung 'Studienplanung und Organisation der Lehre.'"
ISBN 3-7940-3129-6
1. Education, Higher—Germany, West—Curricula. 2. Curriculum planning—Germany, West. 3. College teachers—Germany, West. I. Oehler, Christoph. II. Series.
LB2305.H65 Bd. 29 78-399869
₍LB2362.G3₎ 79 MARC

—GREAT BRITAIN—BIOGRAPHY

Battiscombe, Georgina.
Reluctant pioneer : a life of Elizabeth Wordsworth / ₍by₎ Georgina Battiscombe. — London : Constable, 1978.
320 p., leaf of plate, ₍12₎ p. of plates : ill., ports. ; 23 cm. GB78-18189
"Elizabeth Wordsworth, by E. M. Jamison," p. ₍223₎-311.
Includes bibliographical references and index.
ISBN 0-09-461200-5 : £6.95
1. Wordsworth, Elizabeth, Dame, 1840-1932. 2. College teachers—Great Britain—Biography. I. Jamison, Evelyn Mary, 1877-1972. Elizabeth Wordsworth. 1978. II. Title.
LA2375.G72W673 1978 378.1'1'0924 79-311336
79 MARC

—GREAT BRITAIN—DIRECTORIES

Mruck, Martin L
Handbook of Germanists in Great Britain and Ireland / ₍compiled and edited by Martin L. Mruck. — ₍2nd ed₎. — London : Deutscher Akademischer Austauschdienst, 1976.
x, 135 p. ; 24 cm. GB78-15543
Includes index.
1. German philology—Study and teaching (Higher)—Great Britain—Directories. 2. College teachers—Great Britain—Directories. I. Title.
PF3068.G8M7 1976 430'.7'1141 79-309742
79 MARC

—INDIA—ANDHRA PRADESH—SALARIES, PENSIONS, ETC.

Andhra Pradesh, India. Committee on the Implementation of U.G.C. Scales of Pay to the College and University Teachers in Andhra Pradesh.
Report of the Committee on the Implementation of U.G.C. Pay Scales. — ₍Hyderabad₎ : The Committee, 1976.
iv, 40, xxxii p. ; 25 cm.
Chairman: P. Jaganmohan Reddy.
Rs3.00
1. College teachers—India—Andhra Pradesh—Salaries, pensions, etc. I. Jaganmohan Reddy, Pingle, 1910- . II. Title.
LB2335.I4A52 1976 78-904597
331.2'81'37812095484
79 MARC

—INDIA—MEERUT (CITY) —RELIGIOUS LIFE

Gopal, Krishna, 1937-
The religion of the middle class / K. Gopal. — Meerut City : Anu Prakashan, ₍1977₎
xv, 98 p. ; 23 cm.
On label: D.K. Publishers' Distributors, New Delhi.
Includes bibliographical references.
Rs22.00 ($6.00 U.S.)
1. College teachers—India—Meerut (City)—Religious life. 2. Meerut, India (City)—Religious life and customs. 3. Hindus in Meerut, India (City) I. Title.
BL1228.G66 301.5'8 77-905891
79 MARC

—IRAN—DIRECTORIES

Iran. Daftar-i Bar'rasīhā va Hamāhangi-i Umūr-i Pizhūhishī.
(Rāhnamā-yi a'ẓā-yi hay'at-i 'ilmī-i dānishgāh'hā va mu'assasāt-i āmūzish-i 'ālī-i kishvar, sāl-i taḥṣīlī-i 2534–35)
راهنمای اعضای هیأت علمی دانشگاهها و موسسات آموزش عالی کشور ، سال تحصیلی ۳۵ـ۲۵۳۴ / ₍تهیه‌کننده دفتر بررسیها و هماهنگی امور پژوهشی ؛ به کوشش علی اکبر بیهقی₎. — ₍تهران₎ : شورای پژوهشهای علمی کشور ، 2535 ₍1976₎.
535 p., ₍1₎ leaf of plates : ill. ; 27 cm.
At head of title: Vizārat-i 'Ulūm va Āmūzish-i 'Ālī.
In Persian.
1. Scientists—Iran—Directories. 2. College teachers—Iran—Directories. I. Bayhaqī, 'Alī Akbar. II. Iran. Vizārat-i 'Ulūm va Āmūzish-i 'Ālī. III. Title.
Q149.I 7 I 73 1976 78-971084

—JAPAN

Honda, Akira, 1898-
(Daigaku kyōju)
大学教授：知識人の地獄極楽. — 本多顯彰著. — 東京：光文社, 昭和31 ₍1956₎
184 p. ; 18 cm.
1. College teachers—Japan. I. Title.
LB1778.H58 78-816087

Ikeda, Shin'ichi, 1936-
(Sengo Nihon no daigaku gakumon o dame ni shita kyōju hyaku-ichinin)
戦後日本の大学·学問をダメにした教授101人 / 池田信一. — 東京：山手書房, 1978.
334 p. ; 19 cm. Ja 79-6608
¥980
1. College teachers, Professional ethics for. 2. College teachers—Japan. 3. College presidents—Japan. I. Title.
LB1778.I35 79-808995

—JAPAN—BIOGRAPHY

Miyauchi, Kō, 1937-
(Fūkei o ute)
風景を撃て：大学一九七〇一七五：宮内康建築論集. — 東京：相模書房, 昭和51 ₍1976₎
350 p., ₍1₎ leaf of plates : ill. ; 20 cm. Ja 77-2740
¥1700
1. Miyauchi, Kō, 1937- 2. Architects—Japan—Biography. 3. College teachers—Japan—Biography. 4. Tōkyō Rika Daigaku. Kenchikugakka—Faculty. 5. Architecture—Addresses, essays, lectures. I. Title.
NA1559.M58A2 1976 79-803304

—KOREA—BIOGRAPHY

Zŏng, In-sŏb, 1905-
(Mottahan insaeng)
못다한人生 / 鄭寅燮著. — 서울：徽文出版社, 1978.
426 p. : ill. ; 22 cm. — (나의人生觀)
₩2670
1. Zŏng, In-sŏb, 1905- 2. College teachers—Korea—Biography. 3. Philologists—Korea—Biography. 4. Colleges and universities—Addresses, essays, lectures. 5. Philology—Addresses, essays, lectures. I. Title. II. Series: Na ŭi insaenggwan.
PL991.98.I5Z478 79-825107

—MASSACHUSETTS—WORKLOAD

Massachusetts. Post Audit and Oversight Bureau.
State College System faculty instructional workload. — ₍Boston₎ : Massachusetts General Court, Joint Committee on Post Audit and Oversight, Post Audit and Oversight Bureau, 1977.
73 leaves in various foliations ; 28 cm.
Cover title.
Includes bibliographical references.
1. College teachers—Massachusetts—Workload. 2. Massachusetts State College System—Faculty. I. Title.
LB2331.M37 1977 378.1'22'09744 79-622316
79 MARC

—POLAND

Towarzystwo Uniwersytetów Ludowych Rzeczypospolitej Polskiej.
Uniwersytety ludowe w Polse : sprawozdanie z pierwszego walnego zgromadzenia członków T.U.L.R.P. w Pabianicach 11, 12 i 13 października 1945 roku / ₍komitet redakcyjny, Władysław Radwan ... et al.₎. — Warszawa : Towarzysto Uniwersytetów Ludowych R.P., 1946.
244 p. ; 22 cm.
Bibliography: p. 229-244.
1. College teachers—Poland. 2. Towarzystwo Uniwersyteton Ludowych Rzeczypospolitej Polskiej. I. Radwan, Władysław. II. Title.
LB2331.7.T7 1946 79-351291
79 MARC

COLLEGE TEACHERS (Continued)

—TAIWAN

Wei, Wei-ch'i.
(Hua-kang chiao shou ch'ün chi shu)
華岡教授羣記述 / 魏偉琦著 ; [編輯者中華學術院]. — 台北 : 華岡出版有限公司, 民國67-[1978-
v. : ports. ; 22 cm.
Title in English in colophon: A galaxy of Hwakang professors.
NT$75.00 (v. 1)
1. Chung-kuo wen hua hsüeh yüan—Faculty. 2. College teachers—Taiwan. I. Chung-hua hsüeh shu yüan. II. Title. III. Title: A galaxy of Hwakang professors.
LG55.Y35W44 79-835294

—UNITED STATES—BIOGRAPHY

Altschuler, Glenn C
Andrew D. White, educator, historian, diplomat / by Glenn C. Altschuler. — Ithaca, N.Y. : Cornell University Press, 1979.
300 p. ; 27 cm.
Bibliography: p. [286]-294.
Includes index.
ISBN 0-8014-1156-4
1. White, Andrew Dickson, 1832-1918. 2. Diplomats—United States—Biography. 3. Historians—United States—Biography. 4. College teachers—United States—Biography. I. Title.
E664.W58A57 973.8'092'4 78-58065
78 MARC

Schwarz, Egon, 1922-
Keine Zeit für Eichendorff : Chronik unfreiwilliger Wanderjahre / Egon Schwarz. — Königstein/Ts. : Athenäum, 1979.
190 p. ; 22 cm. GFR79-A
ISBN 3-7610-8046-8 : DM28.00
1. Schwarz, Egon, 1922- —Biography. 2. Critics—Germany—Biography. 3. College teachers—United States—Biography. I. Title.
PT67.S36A34 79-383862
79 MARC

—UNITED STATES—PERIODICALS

Academe. v. 65- Feb. 1979-
[Washington] American Association of University Professors.
v. ill. 28 cm. monthly (except Jan., June, July, Aug.)
"Bulletin of the AAUP."
Continues: AAUP bulletin, ISSN 0001-026X.
Absorbed: Academe (Washington), ISSN 0001-3749 in 1979.
Key title: Academe (Washington. 1979), ISSN 0190-2946.
1. College teachers—United States—Periodicals. 2. American Association of University Professors. I. American Association of University Professors.
LB2301.A3 378.1'2'06273 79-642918
MARC-S

—UNITED STATES—SALARIES, PENSIONS, ETC.

Salary equity : detecting sex bias in salaries among college and university professors / edited by Thomas R. Pezzullo, Barbara E. Brittingham. — Lexington, Mass. : Lexington Book, [1979]
x, 162 p. : ill. ; 24 cm.
Includes bibliographies.
Includes index.
ISBN 0-669-02770-7
1. College teachers—United States—Salaries, pensions, etc. 2. Equal pay for equal work. I. Pezzullo, Thomas R. II. Brittingham, Barbara E.
LB2334.S23 331.2'81'378120973 78-24634
79 MARC

—VIRGINIA—BIOGRAPHY

Dillard, Elizabeth Nash.
Coming hither—going hence : a memoir of Herbert Nash Dillard, 1913-1976 : a biography / by Elizabeth Nash Dillard. — Verona, Va. : McClure Press, c1979.
329 p. : ill. ; 21 cm.
1. Dillard, Herbert Nash, 1913-1976. 2. College teachers—Virginia—Biography. 3. Literature teachers—Virginia—Biography. 4. Virginia. Military Institute, Lexington—Faculty—Biography. I. Title.
U430.V8D543 378.1'2'0924 79-83692
00 MARC

—WASHINGTON (STATE)—SALARIES, PENSIONS, ETC.

Washington (State). Council for Postsecondary Education.
Faculty salaries in Washington public higher education, 1977-79 / Council for Postsecondary Education ; project officers, Denis Curry, Norman Fischer. — [Olympia, Wash.] : The Council, 1978.
v, 41 p. : graph ; 28 cm. — (Report - Council for Postsecondary Education, State of Washington ; 78-11)
1. College teachers—Washington (State)—Salaries, pensions, etc. I. Curry, Denis J. II. Fischer, Norman. III. Title. IV. Series: Washington (State). Council for Postsecondary Education. Report - Council for Postsecondary Education, State of Washington ; 78-11.
LB2334.W335 1978 78-623318
331.2'81'3781209797
79 MARC

—YUGOSLAVIA

Zdravković, Silvija.
Univerzitetski nastavnici u savremenom jugoslovenskom društvu / Silvija Zdravković. — Beograd : Savremena administracija, 1977.
iii, 99, [1] p. ; 24 cm. Yu***
Bibliography: p. 97-[100]
1. College teachers—Yugoslavia. I. Title.
LB2331.7.Z37 78-390295
79 MARC

COLLEGE TEACHERS, PART-TIME

—UNITED STATES—ADDRESSES, ESSAYS, LECTURES

Employing part-time faculty / David W. Leslie, issue editor. — San Francisco : Jossey-Bass Inc., 1978.
vii, 111 p. ; 24 cm. — (New directions for institutional research ; no. 18)
Includes bibliographical references.
1. College teachers, Part-time—United States—Addresses, essays, lectures. I. Leslie, David W. II. Series.
LB2335.35.E46 331.7'61'37812 78-56411
79 MARC

COLLEGE TEACHERS, PROFESSIONAL ETHICS FOR

Ikeda, Shin'ichi, 1936-
(Sengo Nihon no daigaku gakumon o dame ni shita kyōju hyaku-ichinin)
戦後日本の大学・学問をダメにした教授101人 / 池田信一. — 東京 : 山手書房, 1978.
334 p. ; 19 cm. Ja 79-6608
¥980
1. College teachers, Professional ethics for. 2. College teachers—Japan. 3. College presidents—Japan. I. Title.
LB1778.I35 79-808995

Kouda, Tsutomu.
(Daigaku kyōju o kiru)
大学教授を斬る : 知識売人になり下った偽善者たち : 教授能力調査レポート / 鴻生田努著. — 東京 : 日新報道, 1978.
217 p. ; 19 cm. Ja 78-25338
¥890
1. College teachers, Professional ethics for. 2. College teachers, Rating of—Japan. I. Title.
LB1779.K68 79-809176

COLLEGE TEACHERS, RATING OF

—JAPAN

Kouda, Tsutomu.
(Daigaku kyōju o kiru)
大学教授を斬る : 知識売人になり下った偽善者たち : 教授能力調査レポート / 鴻生田努著. — 東京 : 日新報道, 1978.
217 p. ; 19 cm. Ja 78-25338
¥890
1. College teachers, Professional ethics for. 2. College teachers, Rating of—Japan. I. Title.
LB1779.K68 79-809176

—UNITED STATES

Drew, David E
Evaluating science departments : a new index / by David E. Drew and Ronald S. Karpf. — Santa Monica : Rand Corp., 1975.
37 p. : graphs ; 28 cm. — (The Rand paper series ; P-5521)
Cover title.
Includes bibliographical references.
1. Science—Study and teaching (Higher)—United States—Evaluation. 2. Universities and colleges—United States—Evaluation. 3. College teachers, Rating of—United States. I. Karpf, Ronald S., joint author. II. Science departments. III. Title. IV. Series.
AS36.R28 no. 5521 081 s 78-323247
[Q183.3.A1] 79 MARC

—UNITED STATES—ADDRESSES, ESSAYS, LECTURES

Evaluating faculty peformance and vitality / Wayne R. Kirschling, guest editor. — San Francisco : Jossey-Bass, 1978.
x, 106 p. ; 24 cm. — (New directions for institutional research ; no. 20)
Includes bibliographies and index.
1. College teacher, Rating of—United States—Addresses, essays, lectures. I. Kirschling, Wayne R. II. Series.
LB1778.E92 378.1'2 78-56413
79 MARC

COLLEGE TEACHERS, TRAINING OF

—BULGARIA

Sokolov, Nikolaĭ Konstantinovich.
(Partiĭnoe rukovodstvo podgotovkoĭ i vospitaniem kadrov vyssheĭ shkoly BSSR)
Партийное руководство подготовкой и воспитанием кадров высшей школы БССР / Н. К. Соколов. — Минск : Изд-во БГУ, 1978.
143 p. ; 20 cm. USSR 78
Includes bibliographical references.
1.20rub
1. College teachers, Training of—Bulgaria. 2. Communist education—Bulgaria. I. Title.
LB1738.S65 78-421906

—UNITED STATES

O'Banion, Terry, 1936-
Organizing staff development programs that work / Terry O'Banion. — Washington : American Association of Community and Junior Colleges, c1978.
32 p. ; 23 cm.
Bibliography: p. 32.
$5.00
1. College teachers, Training of—United States. I. Title.
LB1738.O2 378.1'2 78-113322
79 MARC

COLLEGE TEACHERS' UNIONS

—UNITED STATES—DIRECTORIES

Garfin, Molly.
Directory of faculty contracts and bargaining agents in institutions [sic] of higher education / [compiled by Molly Garfin]. — New York : National Center for the Study of Collective Bargaining in Higher Education, Baruch College, CUNY, 1979.
iii, 47 p. ; 28 cm.
$5.00
1. Collection bargaining—College teachers—United States. 2. College teachers' unions—United States—Directories. 3. Collective labor agreements—Education, Higher—United States—Directories.
LB2335.885.U6G37 79-113060
331.89'041'378120973
79 MARC

COLLEGE TEACHING

see also Seminars

Beard, Ruth Mary.
Research into teaching methods in higher education, mainly in British universities. — 4th ed. / by Ruth M. Beard. Donald A. Bligh and Alan G. Harding. — Guildford : Society for Research into Higher Education, 1978.
vi, 150 p. : 1 ill. ; 21 cm. — (Research into higher education monographs ; 2 ISSN 0085-1523) GB78-21004
Bibliography: p. 102-147.
Includes index.
ISBN 0-900868-61-9 : £4.50
1. College teaching. 2. Education, Higher—Aims and objectives. 3. Education, Higher—Research. I. Bligh, Donald A., joint author. II. Harding, Alan George, joint author. III. Title. IV. Series.
LB2331.B38 1978 378.1'7'072041 78-326706
79 MARC

COLLEGE TEACHING (Continued)

Bogusz, Jan.
Metody aktywizujące studentów w procesie dydaktycznym szkoły wyższej / Jan Bogusz. — Wyd. 1. — Warszawa : Państwowe Wydawn. Naukowe, 1978.
176 p. : ill. ; 22 cm. — (Biblioteka Nauczyciela akademickiego : Seria Monografie i studia) P***
At head of title: Ministerstwo Nauki, Szkolnictwa Wyższego i Techniki. Instytut Polityki Naukowej, Postępu Technicznego i Szkolnictwa Wyższego.
Summary in English and Russian; table of contents also in English and Russian.
Bibliography: p. 172-[173]
zł25.00
1. Motivation in education. 2. College teaching. I. Title. II. Series.
LB1067.B63 78-392882
79 MARC

Cahn, Steven M
Education and the democratic ideal / Steven M. Cahn. — Chicago : Nelson-Hall, c1979.
xiii, 111 p. ; 23 cm.
Includes bibliographical references and index.
ISBN 0-88229-589-6
1. Education, Higher—Aims and objectives. 2. College teaching. 3. Universities and colleges—Examinations. I. Title.
LB2324.C33 378'.01 78-27155
78 MARC

Cockburn, Barbara.
Participatory discussion / [by] Barbara Cockburn, Alec Ross. — Lancaster : School of Education, University of Lancaster, [1978]
[5], iv, 34 p. ; 21 cm. — (Teaching in higher education series ; 4 ISSN 0309-3565) GB78-19075
Bibliography: p. 32.
ISBN 0-901699-51-9 : £0.80
1. College teaching. 2. Forums (Discussion and debate) I. Ross, Alec M., joint author. II. Title. III. Series.
LB2331.C5236 378.1'7'94 78-324296
79 MARC

Cockburn, Barbara.
Patterns and procedures / [by] Barbara Cockburn, Alec Ross. — Lancaster : School of Education, University of Lancaster, [1978]
[5], v, 57 p. ; 21 cm. — (Teaching in higher education series ; 6 ISSN 0309-3565) GB78-19074
Bibliography: p. 55.
ISBN 0-901699-53-5 : £0.80
1. College teaching. 2. Forums (Discussion and debate) I. Ross, Alec M., joint author. II. Title. III. Series.
LB2331.C5237 378.1'7 78-324045
79 MARC

D'íàchenko, Mikhail Ivanovich.
(Psikhologiía vyssheĭ shkoly)
Психология высшей школы : Особенности деятельности студентов и преподавателей вуза / М. И. Дьяченко, Л. А. Кандыбович. — Минск : Изд-во БГУ, 1978.
319 p. ; 20 cm. USSR 78
Bibliography: p. 311-[317]
1.10rub
1. Educational psychology. 2. College teaching. I. Kandybovich, Lev Aleksandrovich, joint author. II. Title.
LB1051.D5123 78-423086

Didattica universitaria e professione : esperienze in una facoltà umanistica (1964-65-1974-75) / a cura di Giuseppe Galli. — Padova : Antenore, 1976.
206 p., [5] fold. leaves of plates : ill. ; 24 cm. — (Pubblicazioni della Facoltà di lettere e filosofia, Università di Macerata ; 4) It78-July
Includes bibliographical references.
L6000
1. College teaching. 2. Education, Humanistic. I. Galli, Giuseppe. II. Series: Macerata, Italy. Università. Facoltà di lettere e filosofia. Pubblicazioni ; 4.
LB2331.D497 79-352278
*79 MARC

Giesen, Frank.
Hochschulunterricht im Medienverbund : Evaluation d. Reformprojekts "Einführung in die Kommunikationswissenschaft" / Frank Giesen. — München : Minerva-Publikation, 1978.
162 p. ; 21 cm. — (Minerva-Fachserie Wirtschafts- und Sozialwissenschaften) GFR78-A
Originally presented as the author's thesis, Munich, 1977.
Bibliography: p. 102-108.
ISBN 3-597-10008-2 : DM32.00
1. College teaching. 2. Audio-visual education. 3. Mass media—Study and teaching (Higher) I. Title.
LB2331.G53 1978 78-381589
79 MARC

Kallós, Daniel, 1937-
The study of teaching methods in higher education : some lessons from research / Daniel Kallós. — Lund : Dept. of Education, University of Lund, 1976.
18 p. ; 21 cm. — (Pedagogical bulletin ; 1976:4) S***
"Paper presented at the second congress of the European Association of Research and Development in Higher Education in Louvain-La-Neuve, August 30 to September 3, 1976."
Bibliography: p. 16-18.
1. College teaching. 2. Education, Higher—Research—Europe. I. Title. II. Series.
LB2331.K26 378.1'2 ·79-300588
79 MARC

Kozma, Robert B
Instructional techniques in higher education / Robert B. Kozma, Lawrence W. Belle, George W. Williams ; researcher, Norma Pellikka. — Englewood Cliffs, N.J. : Educational Technology Publications, c1978.
xiv, 419 p. ; 25 cm.
Bibliography: p. 375-401.
Includes indexes.
ISBN 0-87778-126-5 : $14.95
1. College teaching. I. Belle, Lawrence W., joint author. II. Williams, George Warner, 1939- joint author. III. Title.
LB2331.K66 378.1'2 78-18384
78 MARC

Layton, David, 1925-
University teaching in transition, edited by David Layton. Edinburgh, Oliver & Boyd, 1968.
ix, 161 p. 23 cm. 40/- B68-23447
Includes bibliographies.
ISBN 0-05-001755-1
1. College teaching. 2. Education, Higher—1945- I. Title.
LB2331.L3 378.1'2 76-365048
70[r79]rev MARC

Milton, Ohmer.
On college teaching / Ohmer Milton and associates. — 1st ed. — San Francisco : Jossey-Bass Publishers, 1978.
xvii, 404 p. : ill. ; 24 cm. — (The Jossey-Bass series in social and behavioral science & in higher education)
Bibliography: p. 383-397.
Includes index.
ISBN 0-87589-377-5
1. College teaching. I. Title.
LB2331.M484 378.1'2 78-50892
78 MARC

Pujol Balcells, Jaime.
Los métodos en la enseñanza universitaria / Jaime Pujol Balcells, José Luis Fons Martín ; prólogo y conclusiones de Rafael Gómez Pérez. — Pamplona : Ediciones Universidad de Navarra, 1978.
195 p. ; 22 cm. — (Colección ICE ; 23) Sp***
Bibliography: p. 187-195.
ISBN 8431305053
1. College teaching. I. Fons Martín, José Luis, joint author. II. Title. III. Series: Pamplona, Spain. Universidad de Navarra. Instituto de Ciencias de la Educación. Colección ICE ; 23.
LB2331.P79 78-384377
78 MARC

—ADDRESSES, ESSAYS, LECTURES

Gruppenarbeit und Tutorenausbildung in den Wirtschaftswissenschaften / Peter Diepold u. Judy Ritter (Hrsg.). — Göttingen : Seminar für Hochschuldidaktik d. Wirtschaftswiss. ; Frankfurt am Main : Verein zur Förderung d. Hochschuldidaktik d. Wirtschaftswiss. e. V. [Vertrieb], 1975.
i, 211 p. ; 21 cm. — (Berichte aus dem Seminar zur Hochschuldidaktik der Wirtschaftswissenschaften ; Bd. 8) GFR76-B15
Second ed. published in 1975 under title: Gruppenarbeit und Tutorenausbildung.
Includes bibliographies.
1. Group work in education—Addresses, essays, lectures. 2. Tutors and tutoring—Addresses, essays, lectures. 3. College teaching—Addresses, essays, lectures. 4. Economics—Study and teaching (Higher)—Addresses, essays, lectures. I. Diepold, Peter. II. Ritter, Judy. III. Series: Göttingen. Universität. Seminar zur Hochschuldidaktik der Wirtschaftswissenschaften. Berichte aus dem Seminar zur Hochschuldidaktik der Wirtschaftswissenschaften ; Bd. 8.
LB1032.G774 79-355081
79 MARC

Institutional renewal through the improvement of teaching / Jerry G. Gaff, guest editor ; [cover design by Willi Baum]. — San Francisco : Jossey-Bass, 1978.
xi, 102 p. ; 24 cm. — (New directions for higher education ; no. 24)
Title on spine: Institutional renewal of teaching.
Includes bibliographies and index.
1. College teaching—Addresses, essays, lectures. I. Gaff, Jerry G. II. Title: Institutional renewal of teaching. III. Series.
LB2331.I58 378.1'2 78-56388
79 MARC

Metodyka ćwiczeń na studiach uniwersyteckich / [redaktor naukowy Leon Leja]. — Wyd. 1. — Poznań : Wydawn. Naukowe Uniwersytetu im. Adama Mickiewicza, 1975.
258 p., [1] fold. leaf : ill. ; 24 cm. — (Seria Psychologia—pedagogika ; nr 35) P***
At head of title: Uniwersytet im. Adama Mickiewicza w Poznaniu. Wydział Filozoficzno -Historyczny.
Includes bibliographies.
zł53.00
1. College teaching—Addresses, essays, lectures. I. Leja, Leon. II. Uniwersytet im. Adama Mickiewicza w Poznaniu. Wydział Filozoficzno-Historyczny. III. Series.
LB2331.M45 78-399367
79 MARC

(Nauchno-metodicheskie osnovy sovershenstvovaniía uchebnogo protšessa v vuze)
Научно-методические основы совершенствования учебного процесса в вузе : Темат. сборник / [Ред. коллегия ... проф. Н. Г. Рассохин (гл. ред.) ... et al.]. — Москва : МЭИ, 1977.
131 p. ; 19 cm. — (Труды Московского энергетического института ; вып. 324) USSR 77
At head of title: Ministerstvo vysshego i srednego spetšial'nogo obrazovaniía SSSR.
Includes bibliographies.
0.58rub
1. College teaching—Addresses, essays, lectures. I. Rassokhin, Nikolaĭ Georgievich. II. Series: Moscow. Ėnergeticheskiĭ institut. Trudy ; vyp. 324.
TJ7.M63 vyp. 324 78-421266
[LB2331]

—ADDRESSES, ESSAYS, LECTURES—COLLECTED WORKS

Auswertung, Rückmeldung, Kritik im Hochschulunterricht / Ludwig Huber ... [et al.], (Hrsg.). — Hamburg : Arbeitsgemeinschaft für Hochschuldidaktik, c1978.
v. (338 p.) : ill. ; 21 cm. — (Blickpunkt Hochschuldidaktik; Heft -51) GFR***
Includes bibliographies.
CONTENTS:
—Bd. 2. Erfahrungen und Folgerungen.

ISBN 3-88334-081-2 (v. 2)
1. College teaching—Addresses, essays, lectures—Collected works. I. Huber, Ludwig, 1937- . II. Arbeitskreis für Hochschuldidaktik. III. Series.
LB1778.B5 Heft 51 79-357991
[LB2331] 79 ·MARC

—CASE STUDIES

Groepswerk en projektonderwijs aan de universiteit / Walter Leirman (ed.). — 3000 Leuven (Tiensestr. 134-136) : Acco, 1977.
v, 130 p. : ill. ; 25 cm. Be78-5
Dutch or German.
Includes bibliographies.
215F
1. College teaching—Case studies. 2. Group work in education—Case studies. 3. Project method in teaching—Case studies. I. Leirman, Walter.
LB2331.G74 78-382324
*79 MARC

—CONGRESSES

Konferencija Technikos mokslų išvystimas respublikoje ir jų rezultatų panaudojimas, Kaunas, 1975.
Aukštojo mokslo pedagogika : Technikos mokslų išvystymo respublikoje ir jų rezultatų panaudojimo konferencijos medžiaga : 1975 m. sausio 31 - vasario 8 d. / redagavo A. Novodvorskis. — Kaunas : Kauno Antano Sniečkaus politechnikos in-tas, 1976.
174 p. : ill. ; 20 cm. USSR***
At head of title: Lietuvos TSR Aukštojo ir specialiojo vidurinio mokslo ministerija.
Added t.p.: Pedagogika vysshego obrazovaniía.
Lithuanian or Russian with summaries in the other language.
Includes bibliographies.
1. Technical education—Lithuania—Congresses. 2. College teaching—Congresses. I. Novodvorskis, Andrius. II. Kaunas. Politechnikos institutas. III. Title. IV. Title: Pedagogika vysshego obrazovaniía.
LC1081.K66 1975 79-362354
79 MARC

Praxisbezug im Studium : Dokumentation des Kongresses vom 20.-22. November 1974 im Congress Centrum Hamburg / hrsg. von Otto Herz. — Hamburg : [Arbeitsgemeinschaft für Hochschuldidaktik], 1975.
457 p. ; 21 cm. — (Blickpunkt Hochschuldidaktik ; 35) GFR***
1. College teaching—Congresses. I. Herz, Otto. II. Arbeitskreis für Hochschuldidaktik. III. Series.
LB1778.B5 Heft 35 79-377988
[LB2331] 79 MARC

—HANDBOOKS, MANUALS, ETC.

Hoover, Kenneth H
College teaching today : a handbook for post-secondary instruction / Kenneth H. Hoover. — Boston : Allyn and Bacon, c1980.
xiv, 405 p. ; 24 cm.
"Outgrowth of a similar work prepared for secondary teachers entitled The professional teachers handbook."
Bibliography: p. 394.
Includes index.
ISBN 0-250-06662-3 : $16.95
1. College teaching—Handbooks, manuals, etc. I. Title.
LB2331.H63 378.1'2 79-12932
79 MARC

COLLEGE TEACHING (Continued)

—VOCATIONAL GUIDANCE

Patton, Carl V
Academia in transition : mid-career change or early retirement / Carl V. Patton. — Cambridge, Mass. : Abt Books, c1979.
xv, 212 p. ; 24 cm.
Includes bibliographical references and index.
ISBN 0-89011-514-1
1. College teaching—Vocational guidance. 2. College teachers—Retirement. 3. Professions—Vocational guidance. 4. Occupational mobility. I. Title.
LB2331.P364 331.2'52 78-66694
79 MARC

—GERMANY, WEST

Lernen und Verändern : zur Theorie u. Praxis d. Projektstudiums / Sabine Hering, Harry Hermanns (Hrsg.). — Hamburg : Arbeitsgemeinschaft für Hochschuldidaktik, 1978.
180 p. ; 21 cm. — (Blickpunkt Hochschuldidaktik ; 49) GFR78-B1
Includes bibliographical references.
ISBN 3-88334-075-8 : DM19.80
1. College teaching—Germany, West. 2. Educational research—Germany, West. I. Hering, Sabine, 1947- . II. Hermanns, Harry. III. Series.
LB1778.B5 Heft 49 79-373952
79 MARC

COLLEGE TEACHING ETHICS see College teachers, Professional ethics for

COLLEGE THEATER

see also College and school drama

—ITALY

Giampaola, Giuseppe.
Scuola e teatro / Giuseppe Giampaola. — [Firenze] : Nuovedizioni E. Vallecchi, [1975?]
253 p. ; 19 cm. It76-Oct
Includes bibliographical references.
L2000
1. Drama in education. 2. Children's plays—Presentation, etc.—Italy. 3. College theater—Italy. 4. Theater and society—Italy. I. Title.
PN3191.I8G5 792'.0945 79-360237
*79 MARC

COLLEGE THEATRICALS see College and school drama

COLLEGE UNIONS see Student unions

COLLEGE VERSE, AMERICAN

see also School verse, American

—COLLECTED WORKS

Studies in poetry. v. 1-
[Lubbock] Dept. of English, Tex. Tech University, 1976-
v. 23 cm.
"A graduate journal."
Key title: Studies in poetry, ISSN 0160-7030
1. College verse, American—Collected works. 2. English poetry—Explication—Collected works. I. Texas Tech University. Dept. of English.
PS591.C65S8 811'.5'408 78-640905
MARC-S

—NEW YORK (STATE)—SYRACUSE

Syracuse poems and stories. 1979-
Syracuse, N. Y., Dept. of English, Syracuse University.
v. 24 cm. annual.
Continues: Syracuse stories and poems, ISSN 0192-9704.
Vols. for 1979- published in limited editions.
Key title: Syracuse poems and stories (1979), ISSN 0194-1003.
1. College verse, American—New York (State)—Syracuse. 2. College prose, American—New York (State)—Syracuse. 3. Syracuse University. I. Syracuse University. Dept. of English. II. Key title.
PN6110.C7S84 810'.8'0054 79-643822
MARC-S

Syracuse stories & poems. 1978-
[Syracuse, N. Y., Dept. of English, Syracuse University]
v. ill. 28 cm. annual.
Continues: Syracuse poems and stories, ISSN 0161-0686.
1978- published in limited editions.
Key title: Syracuse stories & poems, ISSN 0192-9704.
1. College verse, American—New York (State)—Syracuse. 2. College prose, American — New York (State) — Syracuse. 3. Syracuse University. I. Syracuse University. Dept. of English.
PN6110.C7S84 810'.8'0054 79-643211
MARC-S

COLLEGE VERSE, CANADIAN

—ONTARIO—OTTAWA

University of Ottawa. Dept. of English. Writing Workshop.
Choice parts / Writing Workshop, University of Ottawa ; selected by Seymour Mayne. — [Ottawa : The workshop] ; Oakville, Ont. : distribution, Mosaic Press/Valley Ed., c1976.
69 p. ; 26 cm. C***
Selected writings.
1. College verse, Canadian—Ontario—Ottawa. 2. College prose, Canadian—Ontario—Ottawa. I. Mayne, Seymour, 1944- . II. Title.
PR9194.5.C6U5 1976 810'.8'0054 78-321695
78 MARC

University of Ottawa. Dept. of English. Writing Workshop.
Splices / Writing Workshop, University of Ottawa. — [Ottawa : The Workshop ; Oakville, Ont. : distribution, Valley Ed.-/Mosaic Press, c1975]
55 p. : ill. ; 22 cm. C***
Selected writings.
1. College verse, Canadian—Ontario—Ottawa. 2. College prose, Canadian—Ontario—Ottawa. I. Title.
PR9194.5.C6U5 1975 78-322957
79 MARC

—ONTARIO—TORONTO

Shark tank : York Poetry Workshop anthology / edited by Irving Layton. — Toronto : York Poetry Workshop, c1977.
98 p. ; 23 cm. C77-1580-7
ISBN 0-9690730-0-3 : $2.00
1. College verse, Canadian—Ontario—Toronto. 2. Canadian poetry—20th century. 3. York University, Toronto. I. Layton, Irving, 1912- . II. Title.
PR9195.35.C65S5 811'.5'408 79-310680
79 MARC

COLLEGE VERSE, FRENCH-CANADIAN

—QUEBEC—SHERBROOKE

Notes ... / les étudiants du collégial, Séminaire de Sherbrooke. — Sherbrooke : [s.n.], 1976.
[104] leaves ; 18 cm. C***
1. College verse, French-Canadian—Quebec—Sherbrooke. I. Séminaire de Sherbrooke.
PQ3914.N58 841 78-381616
78 MARC

COLLEGE VERSE, LATIN (MEDIEVAL AND MODERN)

—SWITZERLAND—BASEL—HISTORY AND CRITICISM

Buxtorf, Peter, 1913-
Alma mater poetica. Basel, Helbing & Lichtenhahn, 1960.
177 p. 24 cm. (Studien zur Geschichte der Wissenschaften in Basel, 9)
"Verse von und über Basler Professoren, Doktoren und Studenten."
Some text in Latin.
Includes bibliographical references and index in literature.
1. College vverse, Latin (Medieval and modern)—Switzerland—Basel—History and criticism. 2. Basel. Universität in literature. I. Title. II. Series.
PA8065.C65B8 871'.04'0932 a 60-5543
[r79]rev MARC

COLLEGE VERSE, NIGERIAN (ENGLISH)

—PERIODICALS

Omabe. no. 13- Feb. 1975-
Naukka, Dept. of English, University of Nigeria.
no. 21 cm.
Continues: Omaba.
In English.
1. Nigerian poetry (English)—20th century—Periodicals. 2. College verse, Nigerian (English)—Periodicals. I. Nigeria. University, Naukka. Dept. of English.
PR9387.45.O4 821 78-645705
MARC-S

COLLEGE VERSE, PAPUA NEW GUINEA (ENGLISH)

—GOROKA TEACHERS COLLEGE

Flowers of Inonggomang. — Goroka : Expressive Arts Dept., Teachers College, 1974.
49 p. : ill. ; 17 cm.
1. College verse, Papua New Guinea (English)—Goroka Teachers College. I. Goroka Teachers College. Expressive Arts Dept.
PR9655.6.F5 821 79-303192
79 MARC

COLLEGE VERSE, SIERRA LEONE (ENGLISH)

—FOURAH BAY COLLEGE

Poems, Fourah Bay College, 1973 / compiled by M. J. Salt. — [Freetown, Sierra Leone] : R. Butz, [1973?]
15 leaves ; 33 cm.
1. College verse, Sierra Leone (English)—Fourah Bay College. I. Salt, M. J. II. Freetown, Sierra Leone. Fourah Bay College.
PR9393.8.C65P6 821 79-304761
79 MARC

COLLEGE VERSE, SPANISH

—SPAIN—ARAGON

Poesía universitaria. — Zaragoza : Universidad, Rectorado, 1975.
142 p. ; 20 cm. Sp75-Oct
ISBN 8460065847
1. College verse, Spanish—Spain—Aragon. I. Saragossa. Universidad. Rectorado.
PQ6178.C65P6 79-385142
*79 MARC

COLLEGES see Universites and colleges

COLLEGES, AFRO-AMERICAN
see Afro-American universities and colleges

COLLEGES, SMALL see Small colleges

COLLEGES AND UNIVERSITIES

—ADDRESSES, ESSAYS, LECTURES

Zŏng, In-sŏb, 1905-
(Mottahan insaeng)
못다한人生 / 鄭寅燮著. — 서울 : 徽文出版社, 1978.
426 p. : ill. ; 22 cm. — (나의人生觀)
₩2670
1. Zŏng, In-sŏb, 1905- 2. College teachers—Korea—Biography. 3. Philologists—Korea—Biography. 4. Colleges and universities—Addresses, essays, lectures. 5. Philology—Addresses, essays, lectures. I. Title. II. Series: Na ŭi insaenggwan.
PL991.98.I5Z478 79-825107

COLLEGES AND UNIVERSITIES (Continued)

—BUILDINGS—CONGRESSES—COLLECTED WORKS

Internationale Tagung Reproduktion der Materiell-Technischen Basis im Hochschulwesen, 2d, Technische Universität Dresden, 1975.
Reproduktion der materiell-technischen Basis im Hochschulwesen : Referate u. Diskussionsbeitr. : ‹II. internat. Tagung d. Inst. für Hoch- u. Fachschulbau, 9.-10. April 1975 in Dresden / Redaktion, Rudolf Rothe ... et al.›. — Dresden : Inst. für Hoch- u. Fachschulbau an d. Techn. Univ., ‹1976-
v. : numerous ill. ; 30 cm. — (Schriftenreihe Hoch- und Fachschulbau ;-11) GFR77-B4
CONTENTS:
—T. 2. Grundlagen für die Investitionsvorbereitung und Projektierung von Lehr- und Forschungsgebäuden.
1. Colleges and universities—Buildings—Congresses—Collected works. I. Rothe, Rudolf. II. Dresden. Technische Universität. Institut für Hoch- und Fachschulbau. III. Title. IV. Series.
LB3223.I438 1975 79-366419
79 MARC

COLLEGIATE ASSOCIATION FOR THE RESEARCH OF PRINCIPLES

Jetter, Rainer.
Getarnter Rechtsradikalismus : oder, Was will die C.A.R.P.? / Rainer Jetter. — 1. Aufl. — Berlin : DVK Verlag, 1978.
123 p. : ill. ; 21 cm. GFR***
Cover title.
ISBN 3-88107-018-4
1. Radicalism—Germany, West. 2. Right and left (Political science) 3. Youth—Germany, West—Religious life. 4. Moon, Sun Myung. 5. Collegiate Association for the Research of Principles. I. Title.
HN460.R3J47 301.43'15'0943 79-368942
79 MARC

COLLEGIATE SCHOOL, NEW HAVEN see Yale University

COLLEGIO URBANO DI PROPAGANDA FIDE

Piras, Giuseppe.
La Congregazione e il Collegio di Propaganda fide di J. B. Vives, G. Leonardi e M. de Funes / Giuseppe Piras. — Roma : Università gregoriana, 1976.
xi, 154 p. ; 25 cm. — (Documenta missionalia ; 10) It***
Italian or Latin.
1. Catholic Church. Congregatio de Propaganda Fide. 2. Collegio urbano di Propaganda fide. 3. Funes, Martín de, 1560-1611. 4. Leonardi, Giovanni, 1541-1609. 5. Vives, Juan Bautista, 1545-1632. I. Title. II. Series.
BX1837.P57 79-357111
79 MARC

COLLEGIUM BORROMAEUM. SEMINARKIRCHE

Seminarkirche des Collegium Borromaeum, Stadt Freiburg im Breisgau / ‹hrsg. von Robert Zollitsch u. Wolfgang Sauer unter Mitarb. von Karl-Heinz Braun ... et al.›. — 1. Aufl. — München ; Zürich : Schnell und Steiner, 1977.
22, ‹1› p. : ill. (some col.) ; 17 cm. — (Kunstführer ; Nr. 1113) GFR78-A
Caption title.
Bibliography: p. ‹23›
1. Collegium Borromaeum. Seminarkirche. I. Zollitsch, Robert. II. Sauer, Wolfgang. III. Series: Kleine deutsche Kirchenführer ; Nr. 1113.
NA5564.K53 Nr. 1113 914.3'04'8708 s 79-344654
‹NA5586.F7› 79 MARC

COLLEONI, BARTOLOMEO, 1400-1475

—HOMES—ITALY—MALPAGA

Polli, Vittorio.
Il Castello del Colleoni a Malpaga e i suoi affreschi : opera promossa dal Luogo pio della Pietà Istituto Bartolomeo Colleoni in occasione del quinto centenario della morte del condottiero / Vittorio Polli ; fotografie Domenico Lucchetti. — Bergamo : Luogo pio della Pietà Bartolomeo Colleoni, 1975.
68, ‹5› p. : ill. ; 35 cm. — (Monumenta Longobardica ; 7) It77-Nov
Bibliography: p. ‹71›-‹72›
1. Mural painting and decoration, Italian—Italy—Malpaga. 2. Mural painting and decoration, Renaissance—Italy—Malpaga. 3. Malpaga (Castle) 4. Colleoni, Bartolomeo, 1400-1475—Homes—Italy—Malpaga. I. Title. II. Series.
ND2757.M34P64 79-376157
*79 MARC

COLLET, PAUL

Collet, Paul.
I pagani mi hanno parlato di Dio. Presentazione di Carlo Carretto. Assisi, Cittadella editrice, 1971.
102 p. 19 cm. (Quaderni di spiritualità, 8) L800 It71-July
1. Christianity and other religions. 2. Salvation outside the Catholic Church. 3. Collet, Paul. I. Title.
BR127.C5616 79-384567
79 MARC

COLLIANDER, TITO, 1904-

—BIOGRAPHY—YOUTH

Colliander, Tito, 1904-
Bevarat ; Gripen / Tito Colliander. — Helsingfors : Schildt, 1978.
330, ‹1› p. ; 22 cm. — (Finländskt bibliotek) Fi78-6
ISBN 9515001633. ISBN 9515001625 pbk.
1. Colliander, Tito, 1904- —Biography—Youth. 2. Authors, Swedish—20th century—Biography. I. Colliander, Tito, 1904- Gripen. 1978. II. Title.
PT9875.C54Z497 79-347130
79 MARC

—PHILOSOPHY

Colliander, Tito, 1904-
Ihmisen ääni / Tito Colliander ; suom. Kyllikki Härkäpää. — 4.p., ‹13. tuhat›. — Porvoo ; Helsinki ; Juva : WSOY, 1978 (Porvoo : WSOY:n graffiset laitokset)
215, ‹1› p. ; 21 cm. Fi78-8
ISBN 9510083003 : Fmk46.00
1. Colliander, Tito, 1904- —Philosophy. I. Title.
PT9875.C54Z522 79-350750
79 MARC

COLLIEN, PETER

—CATALOGS

Collien, Peter.
Peter Collien : Werkverz. d. Radierungen u. Kupferstiche, 1961-1978. — Berlin : Anabis Verlag, 1978.
‹88› p. : chiefly ill. (some col.) ; 18 x 23 cm. — (Kunst der Gegenwart ; Bd. 1) GFR78-A
Bibliography: p. ‹84›
ISBN 3-921483-14-X : DM45.00
1. Collien, Peter—Catalogs. I. Series: Kunst der Gegenwart (Berlin, 1978-) ; Bd. 1.
NE654.C6A4 1978 769'.92'4 79-364421
79 MARC

COLLIES

see also Border collies

COLLIES, WORKING see Border collies

COLLIN, MATTHÄUS, 1779-1824

—CHARACTERS

Gutkas, Karl.
Kuenringer und Babenberger in der Geschichte und in den Dramen von Matthäus von Collin / Karl Gutkas, Franz Trischler. Zwettl : Schwarz, 1975.
27 p. : geneal. table ; 21 cm. (Bausteine zur Heimatkunde des Hohen Waldviertels ; 2) Au76-4
S35.00
1. Collin, Matthäus, 1779-1824—Characters. 2. Babenberg, House of, in fiction, drama, poetry, etc. 3. Kuenringer family, in fiction, drama, poetry, etc. I. Trischler, Franz. II. Title: Kuenringer und Babenberger ... III. Series.
PT1838.C4Z67 79-382793
79 MARC

COLLIN CO., TEX.

—CENSUS, 1850

United States. Census Office. 7th census, 1850.
Collin County, Texas, 1850 census and mortality schedule. — Richardson, Tex. : J. Bland, c1979.
42 p. ; 28 cm.
Copied by J. B. Bland and R. B. Roose.
Includes index.
1. Collin Co., Tex.—Census, 1850. 2. Collin Co., Tex.—Genealogy. 3. Registers of births, etc.—Texas—Collin Co. I. Bland, Jeanette Bickley. II. Roose, Rita Bickley. III. Title.
F392.C56U53 1850 929'.3764'556 79-115474
79 MARC

—GENEALOGY

Ingmire, Frances Terry.
Collin County, Texas land titles / compiled by Frances Terry Ingmire & Robert Lee Thompson. — St. Louis : Ingmire, c1979.
34 p. ; 29 cm.
Includes index.
1. Collin Co., Tex.—Genealogy. 2. Land titles—Texas—Collin Co. I. Thompson, Robert Lee, joint author. II. Title.
F392.C56I53 929'.3764'556 79-113337
79 MARC

Pitts, Alice.
Collin County (Texas) cemetery inscriptions / compiled by Alice Pitts, Wanda O'Roark, Doris Posey. — McKinney, Tex. : POP Publications, 1975-1978.
2 v. : ill. ; 23 cm.
Includes index.
1. Collin Co., Tex.—Genealogy. 2. Registers of births, etc.—Texas—Collin Co. 3. Inscriptions—Texas—Collin Co. 4. Cemeteries—Texas—Collin Co. I. O'Roark, Wanda, joint author. II. Posey, Doris, joint author. III. Title.
F392.C56P57 929'.3764'556 76-354193
76‹r79›rev MARC

Pitts, Alice.
Marriages, Collin County, Texas. — McKinney, Tex. : Pitts, 1978.
2 v. in 1 ; 29 cm.
Includes index.
CONTENTS: v. 1. 1846-1867.—v. 2. 1867-1876.
1. Collin Co., Tex.—Genealogy. 2. Registers of births, etc.—Texas—Collin Co. I. Title.
F392.C56P58 929'.3764'556 79-111475
00 MARC

United States. Census Office. 7th census, 1850.
Collin County, Texas, 1850 census and mortality schedule. — Richardson, Tex. : J. Bland, c1979.
42 p. ; 28 cm.
Copied by J. B. Bland and R. B. Roose.
Includes index.
1. Collin Co., Tex.—Census, 1850. 2. Collin Co., Tex.—Genealogy. 3. Registers of births, etc.—Texas—Collin Co. I. Bland, Jeanette Bickley. II. Roose, Rita Bickley. III. Title.
F392.C56U53 1850 929'.3764'556 79-115474
79 MARC

COLLINGWOOD, ROBIN GEORGE, 1889-1943

Collingwood, Robin George, 1889-1943.
An autobiography / by R. G. Collingwood. — ‹1st ed.› reprinted ; with a new introduction by Stephen Toulmin. — Oxford ‹etc.› : Oxford University Press, 1978.
xix, 172 p. ; 20 cm. GB79-04809
Includes index.
ISBN 0-19-281247-5 : £2.25
1. Collingwood, Robin George, 1889-1943. 2. Philosophers—England—Biography.
B1618.C74A3 1978 192 79-310127
79 MARC

Greppi Olivetti, Alessandra.
Due saggi su R. G. Collingwood / Alessandra Greppi Olivetti ; con un'appendice di lettere inedite di Collingwood a G. De Ruggiero. — Padova : Liviana, 1977.
x, 108 p. ; 23 cm. — (Collana di testi e saggi ; 8) It78-Aug
English or Italian.
Bibliography: p. 105-106.
Includes index.
L4000
1. Collingwood, Robin George, 1889-1943. 2. History—Philosophy. I. Collingwood, Robin George, 1889-1943. II. Title.
B1618.C74G73 192 79-339544
*79 MARC

COLLINGWOOD COUNTY, N. Z.

—POLITICS AND GOVERNMENT

Newport, J N W
Golden Bay : one hundred years of local government / written by J. N. W. Newport. — ₍Takaka₎ : Golden Bay County Council, 1975.
256 p. : ill., group ports., maps ; 23 cm. NZ76-1
Bibliography: p. 242.
Includes index.
1. Golden Bay County, N.Z.—Politics and government. 2. Collingwood County, N.Z.—Politics and government. 3. Takaka County, N.Z.—Politics and government. I. Title.
JS8350.G64N48 320.9'9315'3 78-319631
79 MARC

COLLINS, ANTHONY, 1676-1729

I Liberi pensatori inglesi del Settecento : Toland, Collins, Tindal : antologia di testi / ₍a cura di Alfredo Sabetti₎. — 1. ed. — Firenze : La nuova Italia, 1978.
ciii, 167 p. ; 20 cm. — (Pensatori antichi e moderni ; 104) It78-June
Includes bibliographical references and index.
L2700
1. Toland, John, 1670-1722. 2. Collins, Anthony, 1676-1729. 3. Tindal, Matthew, 1653?-1733. I. Sabetti, Alfredo.
B1393.Z7L5 78-378138
*78 MARC

COLLINS, DAVID, 1756-1810

Collins, David, 1756-1810.
An account of the English colony in New South Wales : with remarks on the dispositions, customs, manners, etc., of the native inhabitants of that country / by David Collins; edited by Brian H. Fletcher. — Sydney : A. H. & A. W. Reed, 1975.
2 v. : ill. ; 24 cm. Aus***
Bibliography: v. 1, p. 615-628; v. 2, p. 285-298.
Includes indexes.
ISBN 0-589-07167-X
1. New South Wales—Description and travel. 2. Collins, David, 1756-1810. I. Fletcher, Brian H. II. Title.
DU162.C64 1975 994.4'02 78-320094
79 MARC

COLLINS, FLOYD, 1890-1925

Murray, Robert K
Trapped! : The story of the struggle to rescue Floyd Collins from a Kentucky cave in 1925 ... / by Robert K. Murray and Roger W. Brucker. — New York : Putnam, c1979.
335 p., ₍4₎ leaves of plates : ill. ; 24 cm.
Includes bibliographical references and index.
ISBN (invalid) 0-399-12320-2
1. Collins, Floyd, 1890-1925. 2. Speleologists—United States—Biography. 3. Mammoth Cave, Ky. I. Brucker, Roger W., joint author. II. Title.
GB601.6.C64M87 1979 796.5'25'0924 79-10840
79 MARC

COLLINS, MICHAEL, 1890-1922

Joannon, Pierre.
Michael Collins, la naissance de l'I.R.A. / Pierre Joannon ; préf. de Michel Déon. — Paris : La Table ronde, c1978.
297 p. ; 21 cm. F***
Bibliography: p. 289-295.
49.00F
1. Collins, Michael, 1890-1922. 2. Revolutionists—Ireland—Biography. 3. Ireland—History—1910-1921. 4. Irish Republican Army. I. Title.
DA965.C6J62 941.5082'1'0924 79-338736
79 MARC

COLLINS, PETER, 1954-

Collins, Peter, 1954-
The Peter Collins speedway book no. 2 / edited by Richard Bott. — London : S. Paul, 1978.
112 p. : ill., ports. ; 24 cm. GB***
ISBN 0-09-132860-8 : £2.95
1. Collins, Peter, 1954- . 2. Motorcyclists—Great Britain—Biography. 3. Speedway motorcycle racing. I. Bott, Richard. II. Title.
GV1060.2.C64A364 796.7'5'0924 79-312012
79 MARC

COLLINS, WILKIE, 1824-1889

—CRITICISM AND INTERPRETATION

Andrew, Ray Vernon.
Wilkie Collins, a critical survey of his prose fiction, with a bibliography / by R. V. Andrew. — New York : Garland Pub., 1979.
vii, 358 p., ₍4₎ leaves of plates : ill. ; 29 cm. — (The Fiction of popular culture)
Originally presented as the author's thesis, Potchefstroom University for C. H. E., South Africa, 1959.
"Works of Wilkie Collins" : p.
Bibliography: p.
ISBN 0-8240-9667-3 : $35.00
1. Collins, Wilkie, 1824-1889—Criticism and interpretation. I. Title. II. Series: Fiction of popular culture.
PR4497.A5 1979 823'.8 78-60801
79 MARC

Phillips, Walter Clarke, 1881-
Dickens, Reade, and Collins : sensation novelists / by Walter C. Phillips. — New York : Garland Pub., 1979 ₍c1919₎
ix, 230 p. : 19 cm. — (The Fiction of popular culture ; 10)
Reprint of the ed. published by Columbia University Press, New York, originally presented as the author's thesis, Columbia, 1918.
Bibliography: p. 223-230.
ISBN 0-8240-9658-4 : $24.00
1. English fiction—19th century—History and Criticism. 2. Dickens, Charles, 1812-1870—Influence. 3. Reade, Charles, 1814-1884—Criticism and interpretation. 4. Collins, Wilkie, 1824-1889—Criticism and interpretation. 5. Popular literature—Great Britain. 6. Publishers and publishing—Great Britain. I. Title. II. Title: Sensation novelists. III. Series: Fiction of popular culture ; 10.
₍PR871.P5 1979₎ 823'.03 78-60808
78 MARC

COLLINS, WILLIAM, 1721-1759

—CRITICISM AND INTERPRETATION

Garrod, Heathcote William, 1878-1960.
Collins / by H. W. Garrod. — Norwood, Pa. : Norwood Editions, 1977.
123 p. ; 22 cm.
Reprint of the 1928 ed. published by Clarendon Press, Oxford.
Includes bibliographical references.
ISBN 0-8482-0991-5 lib. bdg. : $15.00
1. Collins, William, 1721-1759—Criticism and interpretation.
₍PR3354.G29 1977₎ 821'.5 77-27917
77 MARC

COLLINS FAMILY

The Collins family history : a volume of family history and autobiographical sketches / by Burt Harrison Collins ... ₍et al.₎. — ₍Albuquerque, N.M. : Collins, 1978₎
537, 88 p., ₍14₎ leaves of plates : ill. ; 24 cm.
Cover title: Collins family history, 1740 to 1977.
1. Collins family. I. Collins, Burt Harrison, 1921- .
CS71.C712 1978 929'.2'0973 78-106244
79 MARC

COLLINSON, MOUNT, HONGKONG

—CLIMATE

Lau, Robert.
Climatology of Cape Collinson, 1968-1973 / by R. Lau & W. C. Poon. — Hong Kong : Royal Observatory, 1976.
60 leaves : ill. ; 30 cm. — (Climatological note ; no. 2)
Bibliography: leaf 10.
1. Collinson, Mount, Hongkong—Climate. 2. Meteorological stations—Hongkong—Collinson, Mount. I. Poon, W. C., joint author. II. Hongkong. Royal Observatory. III. Title. IV. Series.
QC990.H64C644 551.6'9'5125 78-321498
79 MARC

COLLISION BROADENING

see also Doppler effect

COLLISIONS, AIRCRAFT see Aeronautics —Accidents

COLLISIONS, AUTOMOBILE see Traffic accidents

COLLISIONS, RAILROAD see Railroads —Accidents

COLLISIONS (NUCLEAR PHYSICS)

see also Charge transfer; Cross sections (Nuclear physics); Electron-molecule scattering; Heavy ion colisions; Inelastic cross sections; Ionization; Nuclear reactions; Radiation; Secondary ion emission; Sputtering (Physics); Stopping power (Nuclear physics); and names of particles, e. g. Electrons, Neutrons, Protons

Atom-molecule collision theory : a guide for the experimentalist / edited by Richard B. Bernstein. — New York : Plenum Press, c1979.
xx, 779 p. : ill. ; 24 cm. — (Physics of atoms and molecules)
Includes bibliographies and index.
ISBN 0-306-40121-5
1. Collisions (Nuclear physics) I. Bernstein, Richard Barry, 1923-
QC794.6.C6A82 539.7'54 78-27380
78 MARC

Hvelplund, P
Accelerated particle interaction with static gases : experimental studies of stopping power and charge-changing collisions / Preben Hvelplund. — ₍Aarhus₎ : Institute of Physics, University of Aarhus, 1977.
90 p. : ill. ; 30 cm. D***
Thesis—University of Aarhus.
Summary in Danish.
Bibliography: p. 83-88.
ISBN 8787548089
1. Stopping power (Nuclear physics) 2. Straggling (Nuclear physics) 3. Collisions (Nuclear physics) 4. Electrons—Measurement. 5. Gases. I. Title.
QC794.6.S8H94 539.7'54 79-300055
79 MARC

Joachain, Charles J
Quantum collision theory / Charles J. Joachain. — Paperback ed. — Amsterdam ; New York : North-Holland Pub. Co. ; New York : sole distributors for the U.S.A. and Canada, Elsevier North-Holland, 1979, c1975.
2 v. (703 p. in various pagings) : ill. ; 22 cm.
Outgrowth of lectures which the author has given at the universities of Berkeley, Calif., Brussels, and Louvain.
Includes bibliographical references and indexes.
ISBN 0-444-85235-2 (Elsevier)
1. Collisions (Nuclear physics) 2. Potential scattering. 3. Scattering (Physics) 4. Quantum theory. I. Title.
QC794.6.C6J6 1979 539.7'54 78-24542
78 MARC

Massey, Harrie Stewart Wilson, Sir.
Atomic and molecular collisions / Sir H. Massey. — London : Taylor & Francis ; New York : Wiley, 1979.
xviii, 309 p. : ill. ; 24 cm.
Includes index.
ISBN 0-470-26742-9
1. Collisions (Nuclear physics) 2. Atoms. 3. Molecules. I. Title.
QC794.6.C6M36 539.7'54 79-11716
79 MARC

Peterkops, Raimonds, 1930-
Theory of ionization of atoms by electron impact / R. K. Peterkop ; translation edited by D. G. Hummer from a draft translation by Elliot Aronson. — Boulder : Colorado Associated University Press, c1977.
vi, 263 p. : graphs ; 24 cm.
Translation of Teoriia ionizatsii atomov élektronnym udarom.
Includes bibliographical references.
ISBN 0-87081-105-3
1. Ionization. 2. Collisions (Nuclear physics) I. Title.
QC702.P4313 539.7'54 77-81310
77₍r79₎rev MARC

White, Sebastian Nicolas.
Direct production of electrons in very high energy proton-proton collisions / Sebastian Nicolas White. — Irvington-on-Hudson, N.Y. : Nevis Laboratories, Columbia University, Physics Dept., 1976.
154 p. : ill. ; 28 cm.
"R-1049, CU-319, Nevis-213."
"National Science Foundation, NSF MPS 75-06955."
Thesis—Columbia University.
Includes bibliographical references.
1. Electrons. 2. Positrons. 3. Pions—Decay. 4. Collisions (Nuclear physics) I. Nevis Laboratories. II. Title.
QC793.5.E622W45 539.7'2112 79-302236
79 MARC

COLLISIONS (NULCEAR PHYSICS) (Continued)

—BIBLIOGRAPHY

(Fizika ėlektronnykh i atomnykh stolknovenii)
Физика электронных и атомных столкновений. Сост. Е. С. Соловьев, Ю. М. Бург, С. Е. Гун и др. род ред. канд. физ.-мат. наук Е. С. Соловьева. Ленинград, 1969–
v. ; 21–26 cm. USSR 69–VKP
At head of title, v. 1–2. Akademii͡a nauk SSSR. Biblioteka Akademii nauk SSSR. Fiziko-tekhnicheskiĭ institut im. A. F. Ioffe.
CONTENTS : ₍вып. 1₎ Библиографический указатель работ советских авторов 1918–1967.—вып. 2, ч. 1. Библиографический указатель 1967–1973 гг.
0.42rub (v. 1) varies
1. Collisions (Nuclear physics)—Bibliography. I. Solov'ev, Evgeniĭ Sergeevich. II. Akademii͡a nauk SSSR. Biblioteka. III. Akademii͡a nauk SSSR. Fiziko-tekhnicheskiĭ institut.
Z7144.N8F58 70–454988
[QC794.6.C6]

COLLISIONS (PHYSICS)

see also Collisions (Nuclear physics); Impact; Scattering (Physics)

COLLISIONS AT SEA

see also Rule of the road at sea; Salvage; Shipwrecks; and names of ships

—PREVENTION

Anglo-French Safety of Navigation Group.
Safety in the Channel : a report / by the Anglo-French Safety of Navigation Group. — London : Department of Trade, 1978.
₍4₎ leaves, 22 p. : 2 charts ; 30 cm. GB78-26119
Errata slip inserted.
Bibliography: p. 20-22.
1. English Channel—Navigation—Safety measures. 2. Collisions at sea—Prevention. I. Great Britain. Dept. of Trade. II. Title.
VK371.A53 1978 623.88'84'0916336 79-302931
79 MARC

—JAPAN—PREVENTION

Kaijō Kōtsū Hōrei Kenkyūkai.
(Kaijō kōtsū anzen hō no kaisetsu)
海上交通安全法の解説 / ₍著者 海上交通法令研究会₎ ; 監修 海上保安庁. — 改訂初版. — 東京 : 海文堂出版, 1978.
3, 6, 199, 49 p. : ill. ; 21 cm. Ja 78–20964
Appendices (p. 1–23, 4th group): 海上衝突予防法—海上交通安全法
¥2500
1. Collisions at sea—Japan—Prevention. 2. Ships—Safety regulations—Japan. I. Japan. Kaijō Hoanchō. II. Japan. Laws, statutes, etc. Kaijō kōtsū anzen hō. 1978. III. Japan. Laws, statutes, etc. Kaijō shōtotsu yobō hō. 1978. IV. Japan. Laws, statutes, etc. Kaijō kōtsū anzen hō. 1978.
78–816434

Makihata, Takeo.
(Meikai kōkai hōki)
明解·航海法規 : 海上衝突予防法·海上交通安全法·港則法 / 巻幡竹夫著. — 東京 : 成山堂書店, 1979.
5, 251 p. : ill. ; 22 cm. Ja 79–6158
¥2800
1. Collisions at sea—Japan—Prevention. 2. Ships—Safety regulations—Japan. 3. Harbors—Japan—Regulations. I. Japan. Laws, statutes, etc. Kaijō shōtotsu yobō hō. 1979. II. Japan. Laws, statutes, etc. Kaijō kōtsū anzen hō. 1979. III. Japan. Laws, statutes, etc. Kōsokuhō. 1979. IV. Title. V. Title: Kōkai hōki.
79–810159

COLLOGEN see Collagen

COLLOIDAL CONCRETE

Brux, Gunther.
Das Colcrete-Verfahren und seine Anwendungsgebiete. Düsseldorf, Beton Verlag ₍1961₎
57 p. illus 21 cm. (Fortschritte des Betonbaues, 6)
Bibliography: p. 55-57.
1. Colloidal concrete. 2. Grouting. 3. Gunite. I. Title.
TA443.C64B78 78-391018
79 MARC

COLLOIDS

see also Particles

Iler, Ralph K
The chemistry of silica : solubility, polymerization, colloid and surface properties, and biochemistry / Ralph K. Iler. — New York : Wiley, c1979.
xxiv, 866 p. : ill. ; 24 cm.
"A Wiley-Interscience publication."
Includes bibliographical references and indexes.
ISBN 0-471-02404-X
1. Silica. 2. Silicates. 3. Colloids. I. Title.
QD181.S6I44 546'.683'2 78-23960
78 MARC

Mândru, Ilie.
Chimia macromoleculelor şi a coloizilor / Ilie Mândru, Minodora Leca. — Bucureşti : Editura Didactică şi Pedagogică, 1977.
361 p. : ill. ; 25 cm. R78-898
At head of title: Ministerul Educaţiei şi Învăţămîntului.
Includes bibliographies.
lei16.40
1. Macromolecules. 2. Colloids. I. Leca, Minodora, joint author. II. Title.
QD381.M372 78-368725
*78 MARC

Mysels, Karol J
Introduction to colloid chemistry / Karol J. Mysels. — Huntington, N.Y. : Krieger, 1978, c1959.
xv, 475 p. : ill. ; 23 cm.
Reprint of the ed. published by Interscience Publishers, New York; with corrections.
Includes bibliographical references and index.
ISBN 0-88275-628-1
1. Colloids. I. Title. II. Title: Colloid chemistry.
QD549.M9 1978 541'.3451 77-13916
77 MARC

Ravich-Shcherbo, Mikhail Iosifovich.
(Fizicheskai͡a i kolloidnai͡a khimii͡a)
Физическая и коллоидная химия : ₍учебник для мед. ин-тов₎ / М. И. Равич-Щербо, В. В. Новиков. — Изд. 3-е, испр. и доп. — Москва : Высш. школа, 1975.
254 p. : ill. ; 23 cm. USSR 75
Bibliography : p. ₍245₎
Includes index.
0.62rub
1. Chemistry, Physical and theoretical. 2. Colloids. I. Novikov, Viktorin Vasil'evich, joint author. II. Title.
QD453.R3 1975 78–420885

Yariv, Shmuel, 1934-
Geochemistry of colloid systems for earth scientists / S. Yariv, H. Cross. — Berlin; New York : Springer-Verlag, 1979.
xii, 450 p. : ill. ; 25 cm.
Includes bibliographies and index.
ISBN 0-387-08980-2
1. Colloids. 2. Geochemistry. I. Cross, Harold, 1930- joint author. II. Title.
QE515.Y37 551.9 78-11407
78 MARC

—BIBLIOGRAPHY—COLLECTED WORKS

(Fiziko-khimicheskai͡a mekhanika i liofil'nost' dispersnykh sistem)
Физико-химическая механика и лиофильность дисперсных систем.
Киев, Академия наук Укр. ССР, Ин-т коллоидной химии и химии воды.
v. 20 cm.
"Библиография работ сотрудников Института коллоидной химии и химии воды АН УССР."
1. Colloids—Bibliography—Collected works. I. Akademii͡a nauk URSR, Kiev. Instytut koloïdnoï khimiï i khimiï vody.
Z5524.C7F54 79–646425
[QD549] (MARC–S)

—COLLECTED WORKS

Kargin, Valentin Alekseevich.
(Kolloidnye sistemy i rastvory polimerov)
Коллоидные системы и растворы полимеров : Избр. тр. / В. А. Каргин ; ₍Биогр. очерк П. В. Козлова₎. — Москва : Наука, 1978.
330 p., ₍5₎ leaves of plates : ill. ; 25 cm. USSR 78
On leaf preceding t. p.: Akademii͡a nauk SSSR. Nauchnyĭ sovet po vysokomolekuli͡arnym soedinenii͡am. Institut neftekhimicheskogo sinteza im. A. V. Topchieva.
Includes bibliographies.
2.90rub
1. Colloids—Collected works. 2. Polymers and polymerization—Collected works. I. Title.
QD549.K36 78–423244

Progress in colloid & polymer science.
Darmstadt, D. Steinkopff,
v 27 cm.
Continues: Fortschrittsberichte über Kolloide & ₍i. e. und₎ Polymere.
English, French, or German.
1. Colloids—Collected works. 2. Polymers and polymerization—Collected works.
QD549.K922 541'.3451 79–645919
MARC-S

—CONGRESSES

Emulsions, latices, and dispersions / edited by Paul Becher, Marvin N. Yudenfreund. — New York : M. Dekker, c1978.
x, 329 p. : ill. ; 24 cm.
Selection of papers presented at a symposium held as part of the 51st Colloid and Surface Science Symposium, held at Grand Isle, N.Y. in June 1977.
Includes bibliographical references and indexes.
ISBN 0-8247-6797-7
1. Colloids—Congresses. 2. Emulsions—Congresses. I. Becher, Paul. II. Yudenfreund, Marvin N., 1943-
QD549.E49 541'.3451 78-11192
78 MARC

International Meeting on the Small Particles and Inorganic Clusters, Lyons, 1976.
Conférence internationale sur les petites particules et amas inorganiques = International meeting on the small particles and inorganic clusters : Lyon-Villeurbanne, France, 13-17 septembre 1976. — ₍Paris₎ : Journal de physique, 1977.
viii, 238 p. : ill. ; 30 cm. F***
"Journal de physique, tome 38, colloque C-2, supplément au no 7, juillet 1977."
"Europhysics journal."
French or English.
Includes bibliographical references and index.
140F
1. Colloids—Congresses. 2. Particles—Congresses. I. Journal de physique.
QD549.I62 1976 531'.16 79-347436
79 MARC

Kolloid-Gesellschaft. 27. Hauptversammlung, Darmstadt, 1975.
Kolloidchemie heute : Vorträge auf der 27. Hauptversammlung in Darmstadt vom 30. September bis 3. Oktober 1975 / hrsg. von H. W. Kohlschütter, A. Weiss. — Darmstadt : D. Steinkopff, 1976-
v. : ill., graphs ; 27 cm. — (Progress in colloid & polymer science ; vol. 59) GFR***
English or German.
Summaries also in English.
Includes bibliographical references.
1. Colloids—Congresses. I. Kohlschütter, Hans Wolfgang, 1902- . II. Weiss, Armin. III. Title. IV. Series.
QD549.K922 vol. 59, etc. 79-376465
541'.3451'08 s
79 MARC

—ELECTRIC PROPERTIES—CONGRESSES

Conference on Electro- Optic and Dielectric Studies on Macromolecules and Colloids, Brunel University, 1978.
Electro-optics and dielectrics of macromolecules and colloids : ₍proceedings of the conference in Electro-optic and Dielectric Studies on Macromolecules and colloids, held at Brunel University, Uxbridge, United Kingdom, April 12-14, 1978₎ / edited by Barry R. Jennings. — New York : Plenum Press, c1979.
xv, 408 p. : ill. ; 26 cm.
Includes bibliographical references and index.
ISBN 0-306-40169-X
1. Colloids—Electric properties—Congresses. 2. Colloids—Optical properties—Congresses. 3. Macromolecules—Electric properties—Congresses. 4. Macromolecules—Optical properties—Congresses. I. Jennings, Barry Randall, 1939- . II. Title.
QD549.C66 1978 547'.7 79-10117
79 MARC

COLLOIDS (Continued)

—OPTICAL PROPERTIES—CONGRESSES

Conference on Electro- Optic and Dielectric Studies on Macromolecules and Colloids, Brunel University, 1978.
Electro-optics and dielectrics of macromolecules and colloids : ⌈proceedings of the conference in Electro-optic and Dielectric Studies on Macromolecules and colloids, held at Brunel University, Uxbridge, United Kingdom, April 12-14, 1978⌉ / edited by Barry R. Jennings. — New York : Plenum Press, c1979.
xv, 408 p. : ill. ; 26 cm.
Includes bibliographical references and index.
ISBN 0-306-40169-X
1. Colloids—Electric properties—Congresses. 2. Colloids—Optical properties—Congresses. 3. Macromolecules—Electric properties—Congresses. 4. Macromolecules—Optical properties—Congresses. I. Jennings, Barry Randall, 1939- . II. Title.
QD549.C66 1978 547'.7 79-10117
79 MARC

COLLOQUIAL BULLS see Bulls, Colloquial

COLLOQUIAL ENGLISH see English language—Spoken English; and subdivision Conversation and phrase books and Spoken English, [French, etc.] under names of languages, e.g. English language—Spoken English

COLLOQUIAL FRENCH see French language--Conversation and phrase books

COLLOQUIAL JAPANESE see Japanese language--Spoken Japanese

COLLOQUIES, RELIGIOUS see Disputations, Religious

COLLUM, CHARLES R.

Collum, Charles R
Dallas nude : a photographic essay / by Charles R. Collum. — 1st ed. — Dallas : Collum Studios, 1977.
96 p. : all ill. ; 36 cm.
$28.50
1. Photography of the nude. 2. Collum, Charles R. I. Title.
TR675.C6 779'.21'0924 77-85236
00 MARC

COLMAR, BATTLE OF, 1945

—ADDRESSES ESSAYS, LECTURES

Société d'histoire et d'archéologie de Colmar.
La Bataille et la libération de Colmar / Société d'histoire et d'archéologie de Colmar. — ⌈Colmar⌉ (Archives municipales, rue des Clefs, 68000) : Société d'histoire et d'archéologie de Colmar, 1975.
163 p. : ill. ; 26 cm. F75-14582
On cover: 1945.
Errata slip inserted.
50.00F
1. Colmar, Battle of, 1945—Addresses essays, lectures. I. Title.
D762.C6S63 1975 940.54'21 79-345564
79 MARC

COLNE, ENG.

—MAPS—BIBLIOGRAPHY—CATALOGS

Colne, Eng. Public Library.
Maps of Colne and district : a classified catalogue of the collection held for reference at Colne Library / ⌈compiled by⌉ Mildred Ellis. — ⌈Nelson (Central Library, Market Sq., Nelson, Lancs.)⌉ : ⌈The Lancashire Library, Pendle District⌉, 1977.
⌈33⌉ leaves ; 33 cm. — (Colne Library publication ; no. 5) GB78-09959
Cover title.
Includes index.
1. Colne, Eng.—Maps—Bibliography—Catalogs. 2. Colne, Eng. Public Library. I. Ellis, Mildred. II. Title. III. Series: Colne, Eng. Public Library. Colne Library publication ; no. 5.
Z6027.G7C64 1977 016.912'427'645 79-302529
⌈GA795.C64⌉ 79 MARC

COLNE, ENG. PUBLIC LIBRARY

Colne, Eng. Public Library.
Maps of Colne and district : a classified catalogue of the collection held for reference at Colne Library / ⌈compiled by⌉ Mildred Ellis. — ⌈Nelson (Central Library, Market Sq., Nelson, Lancs.)⌉ : ⌈The Lancashire Library, Pendle District⌉, 1977.
⌈33⌉ leaves ; 33 cm. — (Colne Library publication ; no. 5) GB78-09959
Cover title.
Includes index.
1. Colne, Eng.—Maps—Bibliography—Catalogs. 2. Colne, Eng. Public Library. I. Ellis, Mildred. II. Title. III. Series: Colne, Eng. Public Library. Colne Library publication ; no. 5.
Z6027.G7C64 1977 016.912'427'645 79-302529
⌈GA795.C64⌉ 79 MARC

COLNE, ENG. (LANCASHIRE)

—GENEALOGY

Colne, Eng. (Lancashire). Parish.
Parochial chapelry of Colne, marriages from 1654-1754 / transcribed by Gladys Whit⌈t⌉aker and Doreen Crowther ; edited by Wilfred M. Spencer. — Burnley : Colne Register Transcriptions, 1975.
ix, 146 p. ; 27 cm. GB76-12290
ISBN 0-9500820-4-X
1. Registers of births, etc.—England—Colne (Lancashire) 2. Colne, Eng. (Lancashire)—Genealogy. 3. Colne, Eng. (Lancashire). Parish. I. Whittaker, Gladys. II. Crowther, Doreen. III. Spencer, Wilfred Micah. IV. Title.
CS436.C72C64 1975 929'.3427'645 79-310926
79 MARC

COLNE, ENG. (LANCASHIRE). PARISH

Colne, Eng. (Lancashire). Parish.
Parochial chapelry of Colne, marriages from 1654-1754 / transcribed by Gladys Whit⌈t⌉aker and Doreen Crowther ; edited by Wilfred M. Spencer. — Burnley : Colne Register Transcriptions, 1975.
ix, 146 p. ; 27 cm. GB76-12290
ISBN 0-9500820-4-X
1. Registers of births, etc.—England—Colne (Lancashire) 2. Colne, Eng. (Lancashire)—Genealogy. 3. Colne, Eng. (Lancashire). Parish. I. Whittaker, Gladys. II. Crowther, Doreen. III. Spencer, Wilfred Micah. IV. Title.
CS436.C72C64 1975 929'.3427'645 79-310926
79 MARC

COLNE VALLEY RAILWAY

The Colne Valley Railway : a pictorial survey. — Halstead : Colne Valley Railway Preservation Society, ⌈1977?⌉
⌈44⌉ p. : chiefly ill., map, plans ; 28 cm. GB78-26702
Cover title.
"A guide to the Colne Valley Railway, Castle Hedingham Station, nr. Halstead, Essex:" ⌈8⌉ p. in pocket.
£1.25
1. Colne Valley Railway. I. Colne Valley Railway Preservation Society.
TF64.C58C64 79-316029
79 MARC

COLOGNE

Alt-Köln-Jubiläums-Kalender : zum 75jährigen Bestehen d. Heimatvereins Alt-Köln 1977 / Red., Peter Joseph Hasenberg. — Köln : Heimatverein Alt-Köln e.V. : Marzellusbuchh. Bachem ⌈Vertrieb⌉, 1976.
192 p. : ill. ; 23 cm. — (Beiträge zur kölnischen Geschichte, Sprache und Eigenart ; Bd. 58) GFR77-B16
1. Cologne. I. Hasenberg, Peter Joseph. II. Heimatverein "Alt-Köln." III. Series: Beiträge zur kölnischen Geschichte, Sprache, Eigenart ; Bd. 58.
DD901.C74A6 943'.55 79-354041
79 MARC

—CITY PLANNING—CITIZEN PARTICIPATION

Wohnungsbaugenossenschaft Urbanes Wohnen Köln.
Bürgerinitiative bie der Planung von Wohnung, Wohnumwelt und Stadt : dargestellt am Modellfall Genossenschaft Urbanes Wohnen Köln : Teilprojekt, Gruppendynamische Phänomene bei der Mitarbeit von Bürgern an Planungsprozessen und ihre Auswirkungen auf die bauliche und städtebauliche Realisierung : Forschungsauftrag BMBau RS II 6 B-70 41 02-99 (1972) / im Auftrag des Bundesministers für Raumordnung, Bauwesen und Städtebau, Bonn-Bad Godesberg, Forschungsbeauftrage, Bürgerinitiative und Wohnungsbaugesnossenschaft Urbanes Wohnen Köln GmbH. — Bonn-Bad Godesberg : Bundesminister für Raumordnung, Bauwesen und Städtebau, 1975.
119 p. : ill., maps ; 30 cm. — (Schriftenreihe städtebauliche Forschung des Bundesministers für Raumordnung, Bauwesen und Städtebau ; 03.039) GFR***
Bibliography: p. 109-114.
Includes indexes.
1. Cologne—City planning—Citizen participation. 2. City planning—Germany, West—Citizen participation. I. Title. II. Series: Schriftenreihe städtebauliche Forschung ; 03.039.
NA9200.C6W63 1975 79-359198
79 MARC

—DESCRIPTION

Kier, Hiltrud.
Die Kölner Neustadt : Planung, Entstehung, Nutzung / von Hiltrud Kier. — 1. Aufl. — Düsseldorf : Schwann, 1978.
240 p., ⌈115⌉ leaves of plates : ill. ; 26 cm. & atlas (32 leaves of plates : plans. — (Beiträge zu den Bau- und Kunstdenkmälern im Rheinland : Bd. 23) GFR78-A
Bibliography: p. 231-240.
ISBN 3-590-29023-4
1. Cologne—Description. I. Title. II. Series.
DD901.C74K46 943'.55 79-352250
79 MARC

—DESCRIPTION—TOURS

Signon, Helmut.
Kölner Spaziergänge / Helmut Signon. — 2., überarb. Aufl. — Köln : Greven, 1978.
247 p. : ill., maps ; 18 cm. GFR78-A
Includes index.
ISBN 3-7743-0148-4 : DM19.80
1. Cologne—Description—Tours. 2. Cologne—History. I. Title.
DD901.C74S56 1978 78-400669
79 MARC

—DESCRIPTION—VIEWS

Fuchs, Peter, 1921-
Köln damals, gestern, heute / Peter Fuchs. — 7. , auf. d. neuesten Stand gebrachte Aufl. — Köln : Greven, 1978.
122 p. : chiefly ill. ; 31 cm. GFR78-A
ISBN 3-7743-0164-6
1. Cologne—Description—Views. I. Title.
DD901.C74F8 1978 943'.55 79-355196
79 MARC

Historische Museen der Stadt Köln.
Jakob und Wilhelm Scheiner : Bilder zur Kölner Stadtentwicklung zwischen 1872 und 1922 : eine Ausstellung der Historischen Museen der Stadt Köln und des Aussenreferats der Kölner Museen im Kölnischen Stadtmuseum, 2. Juni bis 30. September 1978 ⌈Katalog⌉ / ⌈Text, U. Westfehling ; Redaktion, H. Steuer⌉. — Köln : Historische Museen, 1978.
159 p. : chiefly ill. ; 20 cm. GFR***
Bibliography: p. 31-33.
1. Cologne—Description—Views. I. Scheiner, Jakob, 1820-1911. II. Scheiner, Wilhelm, 1853-1922. III. Westfehling, Uwe. IV. Cologne. Kölnisches Stadtmuseum im Zeughaus. V. Title.
DD901.C74H57 1978 78-381276
79 MARC

—ECONOMIC CONDITIONS—18TH CENTURY

Looz-Corswarem, Clemens, Graf von.
Das Finanzwesen der Stadt Köln im 18. Jahrhundert : Beitr. zur Verwaltungsgeschichte e. Reichsstadt / Clemens Graf von Looz-Corswarem. — Köln : Wamper, 1978.
476 p. : ill. ; 24 cm. (Veröffentlichungen des Kölnischen Geschichtsvereins ; 34) GFR79-A8
Originally presented as the author's thesis, Cologne, 1976.
Bibliography: p. 351-371.
Includes indexes.
1. Finance, Public—Germany, West—Cologne—History—18th century. 2. Cologne—Economic conditions—18th century. I. Title. II. Series: Kölnischer Geschichtsverein. Veröffentlichungen ; 34.
HJ9484.C6L66 1978 79-361984
79 MARC

—FORTIFICATIONS, MILITARY INSTALLATIONS, ETC.—HISTORY

Schlegel, Klaus.
Köln und seine preussischen Soldaten : d. Geschichte d. Garnison u. Festung Köln von 1814 bis 1914 / Klaus Schlegel. — 1. Aufl. — Köln : Bachem, 1979.
215 p. : ill. ; 25 cm. GFR79-A
Bibliography: p. 212-215.
Includes index.
ISBN 3-7616-0496-3 : DM38.00
1. Garrisons—Germany, West—Cologne—History. 2. Cologne—Fortifications, military installations, etc.—History. 3. Germany. Heer—History. I. Title.
U375.G3S34 79-378006
79 MARC

COLOGNE (Continued)

—GENEALOGY—PERIODICALS

Kölner genealogische Blätter.
Köln ₍Ger.₎ Bezirksgruppe Köln, Westdeutsche Gesellschaft für Familienkunde.
v. 30 cm.
1. Cologne—Genealogy—Periodicals. I. Westdeutsche Gesellschaft für Familienkunde. Bezirksgruppe Köln.
CS627.C58K6 929'.1'094355 76-643274
MARC-S

—HISTORY

Signon, Helmut.
Kölner Spaziergänge / Helmut Signon. — 2., überarb. Aufl. — Köln : Greven, 1978.
247 p. : ill., maps ; 18 cm. GFR78-A
Includes index.
ISBN 3-7743-0148-4 : DM19.80
1. Cologne—Description—Tours. 2. Cologne—History. I. Title.
DD901.C74S56 1978 78-400669
79 MARC

Wegener, Walter.
Erinnerungen an das alte Köln / Zeichn., Walter Wegener, Texte, Rudolf Spiegel. — Köln : Greven, 1979.
157 p. : ill. ; 25 cm. GFR79-A
ISBN 3-7743-0166-2
1. Wegener, Walter. 2. Cologne in art. 3. Cologne—History. I. Spiegel, Rudolf. II. Title.
NC251.W366A4 1979 79-382601
79 MARC

—HISTORY—SOURCES

Ennen, Leonard, 1820-1880, ed.
Quellen zur Geschichte der Stadt Köln ... Herausgegeben von Dr. Leonard Ennen ... Köln, M. DuMont-Schauberg, 1860-79.
6 v. 16 pl. 25 cm.
On t.-p. of v. 1-2: Hrsg. von Dr. Leonard Ennen ... und Dr. Gottfried Eckertz ...
1. Cologne—History—Sources. I. Eckertz, Gottfried, b. 1817, joint ed. II. Title.
DD901.C72E6 943'.55 14-16839
₍r79₎rev3 MARC

—MUSEUMS—PERIODICALS

Museen der Stadt Köln.
₍Köln, s. n.₎
v. ill. 22-25 cm. monthly.
Supersedes: Museen in Köln.
1. Cologne—Museums—Periodicals.
AM51.C64M87 79-641532
MARC-S

—SOCIAL LIFE AND CUSTOMS—PICTORIAL WORKS

Kroh, Heinz, 1881-1972.
Skizzen aus dem Kölner Milieu / von Heinz Kroh ; mit e. Einf. hrsg. von Hella Rafflenbeul-Kroh. — 1. Aufl., 1.-2. Tsd. — Köln : Bachem, 1978.
151, ₍1₎ p. : chiefly ill. ; 25 cm. GFR78-A
Bibliography: p. ₍152₎
ISBN 3-7616-0469-6 : DM28.00
1. Kroh, Heinz. 2. Cologne in art. 3. Cologne—Social life and customs—Pictorial works. I. Rafflenbeul-Kroh, Hella. II. Title.
NC251.K77A4 1978 79-344841
79 MARC

COLOGNE. KUNSTGEWERBEMUSEUM

—CATALOGS

Cologne. Kunstgewerbemuseum.
Verborgene Schätze aus sieben Jahrhunderten : ausgewählte Werke aus dem Kunstgewerbemuseum der Stadt Köln / ₍Text₎, Brigitte Klesse ; ₍Fotos₎, Hans Mayr. — Köln : Das Museum, 1977.
269 p. : col. ill. ; 26 cm. GFR***
Bibliography: p. 263-268.
1. Art objects—Germany, West—Catalogs. 2. Cologne. Kunstgewerbemuseum—Catalogs. I. Klese, Brigitte, 1929- . II. Mayr, Hans, 1926- . III. Title.
NK480.C73C64 1977 745'.094'0740355 78-365758
78₍r79₎rev MARC

COLOGNE. RÖMISCH-GERMANISCHES MUSEUM

Precht, Gundolf.
Das Grabmal des Lucius Poblicius : Rekonstruktion und Aufbau / Gundolf Precht. — Köln : Römisch-Germanisches Museum der Stadt Köln, 1975.
88 p., ₍18₎ leaves of plates : ill. ; 19 cm. GFR***
Bibliography: p. 86.
1. Poblicius, Lucius—Tomb. 2. Tombs—Rome. 3. Cologne. Römisch-Germanisches Museum. I. Title.
DG103.P73 78-386200
79 MARC

COLOGNE. WALLRAF-RICHARTZ-MUSEUM

—PERIODICALS

Cologne. Museen der Stadt Köln.
Jahresbericht des Wallraf-Richartz-Museum und Museums Ludwig.
₍Köln, Generaldirektor der Museen der Stadt Köln₎
v. ill. 22 cm.
1. Cologne. Wallraf-Richartz-Museum—Periodicals. 2. Museum Ludwig—Periodicals. I. Cologne. Wallraf-Richartz-Museum. II. Museum Ludwig. III. Title.
N2265.C64a 79-642215
MARC-S

COLOGNE (ARCHDIOCESE)

—HISTORY

Penning, Wolf-D
Die weltlichen Zentralbehörden im Erzstift Köln von der ersten Hälfte des 15. bis zum Beginn des 17. Jahrhunderts / von Wolf-Dietrich Penning. — 1. Aufl. — Bonn : Röhrscheid, 1977.
155 p. ; 23 cm. — (Veröffentlichungen des Historischen Vereins für den Niederrhein, insbesondere das alte Erzbistum Köln ; 14) GFR78-A
Originally presented as the author's thesis, Bonn, 1975.
Bibliography: p. 150-55.
ISBN 3-7928-0406-9
1. Cologne (Archdiocese)—History. I. Title. II. Series: Historischer Verein für den Niederrhein, insbesondere die alte Erzdiözese Köln. Veröffentlichungen ; 14.
BX1538.C6P46 1977 78-389259
79 MARC

COLOGNE IN ART

Kroh, Heinz, 1881-1972.
Skizzen aus dem Kölner Milieu / von Heinz Kroh ; mit e. Einf. hrsg. von Hella Rafflenbeul-Kroh. — 1. Aufl., 1.-2. Tsd. — Köln : Bachem, 1978.
151, ₍1₎ p. : chiefly ill. ; 25 cm. GFR78-A
Bibliography: p. ₍152₎
ISBN 3-7616-0469-6 : DM28.00
1. Kroh, Heinz. 2. Cologne in art. 3. Cologne—Social life and customs—Pictorial works. I. Rafflenbeul-Kroh, Hella. II. Title.
NC251.K77A4 1978 79-344841
79 MARC

Wegener, Walter.
Erinnerungen an das alte Köln / Zeichn., Walter Wegener, Texte, Rudolf Spiegel. — Köln : Greven, 1979.
157 p. : ill. ; 25 cm. GFR79-A
ISBN 3-7743-0166-2
1. Wegener, Walter. 2. Cologne in art. 3. Cologne—History. I. Spiegel, Rudolf. II. Title.
NC251.W366A4 1979 79-382601
79 MARC

COLOGNE REGION

—DESCRIPTION AND TRAVEL—TOURS

Hagemann, Christoph Friedel
Köln : Ville und mittlere Erft, Berg. Land, Solingen u. Leverkusen : e. Taschenbuch für d. Freizeit im Grünen mit Wanderparkplätzen, von denen Rundwanderwege ausgehen, Haupt- u. Bezirkswanderwegen d. SGV / ₍d. Texte wurden nach eigenen Erkundungen von C. F. Hagemann erarb. ; die Kt. u. Vignetten zeichnete C. F. Hagemann₎. — 1. Aufl. — Iserlohn : Sauerland-Verlag, 1977.
137, ₍18₎ p. : 100 ill. (some col.), maps ; 21 cm. GFR78-A
"Parken und Wandern."
Pages also numbered 96-231 continuing the paging of the preceding number.
Includes index.
ISBN 3-87695-216-6 : DM12.80
1. Cologne region—Description and travel—Tours. 2. Bergisches Land—Description and travel—Tours. 3. Erft Valley—Description and travel—Tours. 4. Wupper Valley—Description and travel—Tours. I. Title.
DD491.R49568H33 914.3'55 78-370187
78 MARC

COLOGNY, SWITZERLAND. BIBLIOTHECA BODMERIANA

Crum, Margaret C
English and American autographs in the Bodmeriana : catalogue / compiled by Margaret Crum. — Cologny-Genève : Foundation Martin Bodmer, 1977.
104 p. : facsims. ; 25 cm. — (Bibliotheca Bodmeriana : catalogues ; 4)
Sw77-6155
Includes indexes.
ISBN 3-85682-017-X
1. Manuscripts, English—Switzerland—Cologny—Catalogs. 2. Manuscripts, American—Switzerland—Cologny—Catalogs. 3. Autographs—Collections. 4. Cologny, Switzerland. Bibliotheca Bodmeriana. I. Title. II. Series.
Z6623.B64C78 018'.2'09494 78-313790
*78 MARC

Gagnebin, Bernard.
La Fondation Martin Bodmer : une source capitale pour la recherche à Genève / par Bernard Gagnebin. — ₍Genève₎ : Bibliotheca Bodmeriana, ₍1973?₎
54 p. : ill. ; 25 cm. Sw***
"Extrait de Genava, n.s., tome XX, 1972."
1. Cologny, Switzerland. Bibliotheca Bodmeriana.
Z838.C6G34 79-383948
79 MARC

—EXHIBITIONS

Ernst, Fritz, 1889-1958.
Von Zürich nach Weimar; hundert Jahre geistiges Wachstum 1732-1832. ₍Ausstellung₎ Mit einer Bibliographie und 32 Bildtafeln. Einführungen von Martin Bodmer und Carl J. Burckhardt. ₍Zürich, Artemis Verlag, 1953₎
101 p. front., 32 plates (facsims.) 25 cm. (Bibliotheca Bodmeriana, 1)
"Ansprachen anlässlich der Einweihung der Bibliotheca Bodmeriana, Cologny-Genf, 6. Oktober 1951": p. ₍9₎-33.
—— —— Another issue. "Von den 100 Exemplaren der Vorzugsausgabe sind die Nummern 1-80 im Handel, die Nummern I-XX unverkäuflich ... Dieses Exemplar trägt die Nummer 10."
Z838.C6E71953a Rosenwald Coll.
1. Cologny, Switzerland. Biblioteca Bodmerianan—Exhibitions. I. Cologny, Switzerland. Bibliotheca Bodmeriana. II. Title. III. Series.
Z838.C6E7 1953 a 55-5035
₍r79₎rev MARC

COLOMBANI, HELÍ

—BIOGRAPHY

Colombani, Helí.
Memorias precipitadas : experiencia de campo para un hombre de ciudad / Helí Colombani. — Caracas : Ediciones Centauro, ₍19₎78.
231 p. ; 18 cm.
"Obras del autor": p. ₍227₎-231.
Bs12.00
1. Colombani, Helí—Biography. 2. Authors, Venezuelan—20th century—Biography. 3. Country life—Venezuela. I. Title.
PQ8550.13.O44Z463 79-101281
79 MARC

COLOMBIA

—ABSTRACTS

Programa Orinoquia-Amazonia. Comisión Bibliográfica.
Orinoquia-Amazonia, reseñas analíticas. — Bogotá : Comisión Bibliográfica, Programa Orinoquia-Amazonia, Universidad Nacional de Colombia, 1977.
x, 107 p. ; 25 cm. — (Série Bibliografía y documentación ; v. 2)
Cover title.
1. Colombia—Abstracts. 2. Orinoco Valley—Abstracts. 3. Amazon Valley—Abstracts. I. Title. II. Series.
F2252.P76 1977 986.1 79-106289
79 MARC

—ADDRESSES, ESSAYS, LECTURES

Piedrahita Arango, Rubén.
Puntos de vista / Rubén Piedrahita Arango. — Bogotá : ₍s.n.₎, 19
v ; 23 cm.
1. Colombia—Addresses, essays, lectures. 2. South America—Addresses, essays, lectures. I. Title.
F2253.P53 76-483769
76₍r78₎rev MARC

COLOMBIA

—ADDRESSES, ESSAYS, LECTURES (Continued)

Revista de las Indias, 1936-1950 / selección, Alvaro Miranda. — ⌊Bogotá⌋ : Instituto Colombiano de Cultura, Subdirección de Comunicaciones Culturales, División de Publicaciones, c1978.
469 p. ; 21 cm. — (Biblioteca colombiana de cultura) (Colección Autores nacionales) (Serie Las Revistas ; 28)
1. Colombian essays. 2. Colombia—Addresses, essays, lectures. 3. Revista de las Indias. I. Miranda, Alvaro.
PQ8176.E8R4 868 79-104876
79 MARC

—ANTIQUITIES

Chaves Mendoza, Alvaro.
Tierradentro / Alvaro Chaves Mendoza, Mauricio Puerta Restrepo. — Bogotá : Ediciones Zazacuabi, ⌊1976⌋
33 p. : ill. ; 23 cm. — (Colombia, arte y cultura)
1. Indians of South America—Colombia—Tierra Adentro—Antiquities. 2. Tierra Adentro, Colombia—Antiquities. 3. Colombia—Antiquities. I. Puerta Restrepo, Mauricio, joint author. II. Title. III. Series.
F2269.1.T54C47 986.1'5 78-103651
78 MARC

Correal Urrego, Gonzalo.
Investigaciones arqueológicas en los abrigos rocosos del Tequendama : 12.000 años de historia del hombre y su medio ambiente en la altiplanicie de Bogotá : with an extensive English abstract, Archaeological investigations in the Tequendama roch shelters : 12,000 years of history of man and his environment on the high plain of Bogotá / Gonzalo Correal Urrego, Thomas van der Hammen, con la colaboración de Henry E. Coomans ... ⌊et al.⌋. — Bogotá : Fondo de Promoción de la Cultura del Banco Popular, 1977.
194 p. : ill. ; 33 cm. — (Premios de arqueología ; v. 1) (Biblioteca Banco Popular)
"Aparece también como El Cuaternario de Colombia, volumen especial I."
Bibliography: p. 175-176.

1. Tequendama site, Colombia. 2. Indians of South America—Columbia—Cundinamarca (Province)—Antiquities. 3. Cundinamarca, Colombia (Province)—Antiquities. 4. Colombia—Antiquities. I. Hammen, Thomas van der, joint author. II. Title. III. Series.
F2269.1.C95C67 986.1'4 79-101490
79 MARC

—APPROPRIATIONS AND EXPENDITURES

Selowsky, Marcelo.
Who benefits from government expenditure? : A case study of Colombia / Marcelo Selowsky. — New York : Published for the World Bank ⌊by⌋ Oxford University Press, c1979.
xv, 186 p. ; 23 cm.
Bibliography: p. 180-181.
Includes index.
ISBN 0-19-520098-5 : $12.50
1. Colombia—Appropriations and expenditures. 2. Income distribution—Colombia. I. International Bank for Reconstruction and Development. II. Title.
HJ2083.S44 336.861 79-16384
79 MARC

—BIO-BIBLIOGRAPHY

Sánchez López, Luis María.
Diccionario de escritores colombianos. — ⌊Bogotá⌋ : Plaza & Janés, c1978.
547 p. : ill. ; 22 cm.
Bibliography: p. 9-12.
ISBN 8439972407
1. Colombia—Bio-bibliography. I. Title.
Z1740.S26 016.92'00861 79-104308
⌊F2258⌋ 79 MARC

—BIOGRAPHY

Duperly Angueyra, Oswaldo.
Lo que se hereda no se hurta : memorias / Oswaldo Duperly Angueyra. — 1. ed. — Bogotá : Ediciones Tercer Mundo, 1978.
177 p. : ill. ; 23 cm.
1. Duperly Angueyra, Oswaldo. 2. Colombia—Biography. I. Title.
CT708.D86A34 984'.05'0924 79-100989
79 MARC

—CENSUS, 1973

Colombia. Departamento Administrativo Nacional de Estadística.
XIV ⌊i.e. Decimocuarto⌋ censo nacional de población y III de vivienda, 24 de octubre, 1973 : muestra de avance, resumen de los departamentos. — Bogotá : Departamento Administrativo Nacional de Estadística, ⌊1977⌋
58 p. : ill. ; 21 x 22 cm.
"Esta separata reproduce la información de la muestra de avance del XIV censo nacional de población ... dada a conocer en las entregas nos. 289 y 302 del Boletín mensual de estadística, correspondientes a agosto de 1975 y septiembre de 1976."
1. Colombia—Census, 1973. 2. Housing—Colombia—Statistics. I. Title.
HA1011.5 1973.D46 1977 78-108980
312'.09861
79 MARC

—CHURCH HISTORY

Bronx, Humberto.
Golconda y origen de ideas marxistas en minorías del clero colombiano y Templos de Medellín / Humberto Bronx. — ⌊s.l. : s.n., 1978⌋ (Medellín : Editorial A. Salazar)
82 p. ; 17 cm.
Cover title.
Spine title: Golconda e ideas marxistas en el clero.
1. Catholic church in Colombia—History. 2. Communism and christianity—Catholic Church—Colombia. 3. Clergy—Colombia. 4. Medellín, Colombia—Churches. 5. Liberation theology. 6. Colombia—Church history. I. Bronx, Humberto. Templos de Medellín. 1978. II. Title. III. Title: Golconda e ideas marxistas en el clero.
BX1470.2.B76 1978 79-100902
79 MARC

Wängborg, Sune, 1949-
Evangelisation för samhällsförändring : radikala kristna grupper i Colombia / Sune Wängborg. — ⌊Stockholm⌋ : Verbum, 1976.
58, ⌊1⌋ p., ⌊2⌋ leaves of plates : ill. ; 19 cm. — (Kyrkan i Latinamerika)
S77-1/3
Bibliography: p. ⌊59⌋
kr18.85
1. Evangelistic work—Colombia. 2. Christianity and politics. 3. Colombia—Church history. 4. Catholic Church in Colombia—History. I. Title. II. Series.
BV3777.C7W33 78-354186
78⌊r79⌋rev MARC

—CIVILIZATION—PERIODICALS

Nueva frontera. Documentos. 1–
feb. 1976–
⌊Bogotá, Nueva Frontera⌋
no. ill. 34 cm. monthly.
Other title, Febr. 1976– : Documentos Nueva frontera.
Issued as supplement to: Nueva frontera.
1. Colombia—Economic conditions—1970– —Periodicals. 2. Colombia—Civilization—Periodicals. 3. Civilizaton, Modern—1950– —Perodicals. I. Nueva frontera. II. Title: Documentos Nueva frontera.
HC196.N84 330.9'861 78–648733
MARC-S

—CLIMATE—COLLLECTED WORKS

Instituto Colombiano de Hidrología, Meteorología y Adecuación de Tierras.
Boletín climatológico mensual. año 6–
no. 061– enero 1975–
⌊Bogotá⌋ HIMAT.
v. ill. 28 cm.
Continues: Servicio Colombiano de Meteorología e Hidrología. Boletín climatológico mensual.
1. Colombia—Climate—Collected works. I. Title.
QC988.C8S47b 551.6'9'86105 78–646741
MARC-S

—COLLECTED WORKS

Ospina Londoño, Jorge.
Historia, ideología y política / Jorge Ospina Londoño. — 1. ed. — ⌊s.l. : s.n.⌋, 1978 (Medellín, Colombia : Editorial Bedout)
xviii, 535 p. ; 24 cm.
1. Colombia—Collected works. I. Title.
F2253.O85 986.1 79-100983
79 MARC

—COMMERCE

Instituto Colombiano de Comercio Exterior. División Agregados Comerciales.
Análisis comparativo de las exportaciones colombianas por países y productos, enero-junio, 1974-1975 / preparado por División Agregados Comerciales, Instituto Colombiano de Comercio Exterior. — Bogotá : INCOMEX, 1975.
145 leaves : ill. ; 34 cm.
Cover title.
"60.21.agosto.75.1.427"
On t.p.: SE/DAC.
1. Colombia—Commerce. 2. Commercial products—Colombia. I. Title.
HF3426.5.I57 1975 79-338104
79 MARC

—COMMERCE—BIBLIOGRAPHY

Bogotá. Cámara de Comercio. Centro de Informática Económica. Unidad de Documentación.
Obras básicas para bibliotecas de Cámaras de Comercio : una bibliografía / preparada por la Unidad de Documentación del CIEB. — ⌊Bogotá⌋ : Cámara de Comercio de Bogotá, 1973.
30, 8 p. ; 28 cm.
"Compilada por Angela Hernández de Caldas y María Teresa Ramírez Vargas."
"S.D. No. 31."
1. Colombia—Commerce—Bibliography. 2. Commerce—Bibliography. 3. Bibliography—Best books—Commerce. I. Hernández de Caldas, Angela. II. Ramírez Vargas, María Teresa. III. Title.
Z7165.C64B64 1973 78-399215
⌊HF3426.5⌋ 79 MARC

—COMMERCE—INFORMATION SERVICES

Instituto Colombiano de Comercio Exterior.
Circulares del INCOMEX, 1972-1975 / Instituto Colombiano de Comercio Exterior. — ⌊Bogotá⌋ : INCOMEX, 1976.
111 p. ; 29 cm. — (Documentación INCOMEX ; no. 3)
"180.14.1.76.31."
Includes index.
1. Colombia—Commerce—Information services. I. Title. II. Series: Instituto Colombiano de Comercio Exterior. Documentación INCOMEX ; no. 3.
Z7165.C64I57 1976 016.3801'09861 78-399737
⌊HF3424⌋ 79 MARC

—COMMERCE—PERIODICALS

Asociación Latinoamericana de Libre Comercio.
Colombia : estadísticas de comercio exterior, exportaciones. 1961/69–
⌊Montevideo⌋ Asociación Latinoamericana de Libre Comercio.
v. 28 cm. (Serie A - Asociación Latinoamericana de Libre Comercio)
1. Colombia—Commerce—Periodicals. I. Asociación Latinoamericana de Libre Comercio. Estadísticas de comercio exterior, exportacions. II. Title. III. Series: Asociación Latinoamericana de Libre Comercio. Serie A - Asociación Latinoamericana de Libre Comercio (Colombia)
HF3421.A77a 79–645731
MARC-S

Asociación Nacional de Industriales.
Revista ANDI.
⌊Medellín, Colombia, ANDI⌋
no. ill. 24 cm. bimonthly.
Continues: Asociación Nacional de Industriales. ANDI, revista bimestral.
1. Colombia—Economic conditions—1970– —Periodicals. 2. Colombia—Industries—Periodicals. 3. Colombia—Commerce—Periodicals. I. Title.
HC196.A92a 79–649958
MARC-S

—COMMERCIAL POLICY

Simposio sobre Financiamiento Externo, 2d, Medellín, Colombia, 1978.
Política económica externa de Colombia, 1978 / Eduardo Wiesner, editor. — 1. ed. — ⌊Bogota?⌋ : Asociación Bancaria de Colombia, 1978.
349 p. : graphs ; 24 cm. — (Biblioteca Asociación Bancaria de Colombia)
Includes bibliographies.
1. Colombia—Foreign economic relations. 2. Colombia—Commercial policy. 3. Foreign exchange administration—Colombia. I. Wiesner Durán, Eduardo. II. Asociación Bancaria de Colombia. III. Title. IV. Series: Asociación Bancaria de Colombia. Biblioteca Asociación Bancaria de Colombia.
HF1517.S57 1978 382.1'09861 79-101234
79 MARC

COLOMBIA (Continued)

—CONSTITUTIONAL HISTORY

Rivadeneira Vargas, Antonio José.
Historia constitucional de Colombia, 1510-1978 / Antonio José Rivadeneira Vargas. — Bogotá : Editorial Horizontes, 1978.
411, ₍14₎ p. ; 24 cm.
Bibliography: p. ₍413₎-₍419₎
1. Colombia—Constitutional history. 2. Colombia—Politics and government. I. Title.
JL2811.R52 320.9'861 79-114512
79 MARC

—CONSTITUTIONAL LAW

Camargo, Pedro Pablo.
Trayectoria y características del derecho constitucional colombiano / Pedro Pablo Camargo. — 1. ed. — Bogotá : Sociedad Colombiana de Abogados, 1974.
100 p. ; 21 cm. — (Textos jurídicos ; no. 3)
Bibliography: p. ₍89₎-94.
Includes indexes.
1. Colombia—Constitutional law. I. Title. II. Series.
77-155550
79 MARC

Colombia. Laws, statutes, etc.
₍Ley 4 de 1913 (agosto 20)₎
Código de régimen político y municipal : Ley 4a. de 1913 y normas legales complementarias / compilado por Jorge Ortega Torres. — 6. ed. actualizada. — Bogotá : Editorial Temis, 1978.
917 p. ; 21 cm.
Includes index.
1. Administrative law—Colombia. 2. Colombia—Constitutional law. 3. Municipal corporations—Colombia. I. Ortega Torres, Jorge, ed. II. Title.
342'.861 79-103626
79 MARC

Saa Velasco, Ernesto.
Teoría constitucional colombiana / Ernesto Saa Velasco. — Popayán : Universidad del Cauca, 1976.
313 p. ; 25 cm.
Includes bibliographical references and index.
1. Colombia—Constitutional law. I. Title.
342'.861 79-367740
79 MARC

Sarriá, Eustorgio.
Guarda de la Constitución / Eustorgio Sarría. — Bogotá : C.E.I.D.A., ₍1975 or 1976₎
453 p. ; 24 cm.
Includes indexes.
1. Colombia—Constitutional law. I. Title.
342'.861 77-154643
79 MARC

—CONSTITUTIONAL LAW—ADDRESSES, ESSAYS, LECTURES

La Constituyente : itinerario de una propuesta / Jaime Betancur Cuartas, ₍compilador₎. — 1. ed. — Bogotá : Ediciones Tercer Mundo, 1978.
xv, 500 p. ; 23 cm.
Includes bibliographical references.
1. Colombia—Constitutional law—Addresses, essays, lectures. 2. Colombia. Asamblea Constitucional, 1978-1979—Addresses, essays, lectures. I. Betancur Cuartas, Jaime, 1930-
342'.861'029 78-108728
78 MARC

—CONSTITUTIONAL LAW—CASES

Betancur Cuartas, Jaime, 1930-
Jurisprudencia constitucional de la Corte Suprema de Justicia y del Consejo de Estado, años 1969 a 1977 / Jaime Betancur Cuartas, María Teresa Garcés de Angel. — ₍Bogotá₎ : Ediciones Librería del Profesional, ₍1977?₎-
v. ; 24 cm.
1. Colombia—Constitutional law—Cases. I. Colombia. Corte Suprema de Justicia. II. Colombia. Consejo de Estado. III. Title.
342'.861'029 78-107298
79 MARC

—DESCRIPTION AND TRAVEL

Viajeros colombianos por Colombia / prólogo de Gabriel Giraldo Jaramillo ; ilus. de la Comisión Corográfica y de Le Tour du Monde. — Bogotá : Fondo Cultural Cafetero, 1977.
xv, 276 p. : ill. (8 col.) ; 35 cm.
CONTENTS: Posada, E. En vacaciones.—Cordovez Moure, J. M. Un viaje de recreo.—Groot, J. M. Nos fuimos a Ubaque.—Camacho Roldán, S. Algo sobre tierra caliente.—Pombo, M. Bajando el Dagua.—Cordovez Moure, J. M. De Bogotá a Villeta.—Kastos, E. La tierra caliente.—Rivera y Garrido, L. Por Popayán y Pasto.—Posada, E. En el Agua-Nueva.—Samper, J. M. Paseo al Salto de Tequendama.—Posada Gutiérrez, J. El último viaje del Libertador.—Uribe Angel, M. Recuerdo de un viaje de Medellín a Bogotá.—Ancízar, M. De Los Santos a Bucaramanga.—Reyes, R. De Pasto al Amazonas.—Restrepo, E. De Bogotá a Villavicencio.
1. Colombia—Description and travel. 2. Colombia—Social life and customs. I. Giraldo Jaramillo, Gabriel. II. Fondo Cultural Cafetero.
F2263.V5 78-100322
79 MARC

—DESCRIPTION AND TRAVEL—1951-

Burchard, Przemysław,
Dżungla za progiem / Przemysław Burchard. — Wyd. 2. — Warszawa : Czytelnik, 1978 (Krak. : DW)
279, ₍4₎ p., ₍24₎ leaves of plates : ill. ; 22 cm. P78-5996
zł50.00
1. Indians of South America—Colombia. 2. Colombia—Description and travel—1951- . 3. Burchard, Przemysław. I. Title.
F2270.B87 1978 79-347012
79 MARC

Civet, Gérard.
Les trésors de la Colombie / Gérard Civet ; ₍photos. de Gérard Civet, André Chastel et Alain Coutois₎. — Paris : Presses de la Cité, c1979.
121 p. : ill. (some col.) : 27 cm. — (Connaissance du monde ISSN 0182-3671)
F***
Bibliography: p. 121.
ISBN 2-258-00493-4 : 50.00F
1. Colombia—Description and travel—1951- . 2. Civet, Gérard. I. Title.
F2264.C56 986.1'063 79-374301
79 MARC

—DESCRIPTION AND TRAVEL—1951-—GUIDE-BOOKS

Bradt, Hilary.
Backpacking in Venezuela, Colombia, and Ecuador : treks in the Northern Andes / Hilary & George Bradt ; with contributions by P. Alexander ... ₍et al.₎ ; maps, drawings, and photos by Hilary. — Boston : Bradt Enterprises, c1979.
126 p. : ill. ; 22 cm. — (Backpacking guide series)
Bibliography: p. 123-125.
ISBN (invalid) 0-9505797-0-5
1. Backpacking—Venezuela—Guide-books. 2. Backpacking—Colombia—Guide-books. 3. Backpacking—Ecuador—Guide-books. 4. Venezuela—Description and travel—1951- —Guide-books. 5. Colombia—Description and travel—1951- —Guide-books. 6. Ecuador—Description and travel—1951- —Guide-books. I. Bradt, George, joint author. II. Title. III. Series.
GV199.44.V4B72 918.6 79-51623
79 MARC

—ECONOMIC CONDITIONS

Bergquist, Charles W
Coffee and conflict in Colombia, 1886-1910 / Charles W. Bergquist. — Durham, N.C. : Duke University Press, 1978.
x, 277 p. : ill. ; 25 cm.
Bibliography: p. ₍263₎-269.
Includes index.
ISBN 0-8223-0418-X : $16.75
1. Colombia—Politics and government—1886-1903. 2. Colombia—Politics and government—1903-1930. 3. Colombia—Economic conditions. 4. Coffee trade—Colombia—History. I. Title.
F2276.5.B47 320.9'861'061 78-59581
79 MARC

Davenport, Robert Wesley, 1922-
The transfer of technology to Colombian industry through industrial consulting and related services / by Robert W. Davenport, Miguel Bermudez Portocarrero ; prepared for Fundacion Para el Fomento de la Investigacion Cientifica y Technologica ₍sic₎ (FICITEC). — Menlo Park, Calif. : Stanford Research Institute, 1971.
iv, 91 p. ; 27 cm.
"SRI project 8804."
Includes bibliographical references.
1. Technology transfer. 2. Colombia—Economic conditions. I. Bermúdez Portocarrero, Miguel, joint author. II. Fundación para el Fomento de la Investigación Científica y Tecnológica. III. Title.
T174.3.D38 338.91'861 79-307554
79 MARC

Escorcia, José.
Historia económica y social de Colombia, siglo XX / ₍José Escorcia₎. — ₍Bogotá₎ : Editorial Presencia, ₍1978₎
198 p. ; 17 cm.
Includes bibliographical references.
1. Colombia—Economic conditions. 2. Colombia—Social conditions.
HC197.E83 309.1'861'063 79-101128
79 MARC

Galat, José.
Un país prestado / José Galat. — 1. ed. — Bogotá : Tercer Mundo, 1978.
289 p. ; 21 cm. — (Colección Tribuna libre)
Includes bibliographical references.
1. Colombia—Economic conditions. 2. Colombia—Politics and government. I. Title.
HC197.G19 78-107700
79 MARC

Orígenes de las guerras civiles en Colombia. — ₍s.l. : Ediciones Historia y Sociedad, 19—₎.
70 p. ; 17 cm.
On cover: Un análisis marxista.
1. Colombia—History—Colombia—19th century. 2. Civil war. 3. Colombia—Economic conditions.
F2273.074 986.1 78-110011
79₍r79₎rev MARC

Zambrano Pantoja, Fabio.
La economía colombiana en la primera mitad del siglo 19, 1820-1850 / Fabio Zambrano Pantoja. — Medellín, ₍Colombia₎ : Centro de Investigaciones Económicas, Facultad de Ciencias Económicas, Universidad de Antioquia, 1977.
65 p. ; 32 cm.
Cover title: La economía colombiano en el siglo 19.
Includes bibliographical references.
1. Colombia—Economic conditions. I. Title. II. Title: La economía colombiano en el siglo 19.
HC197.Z35 78-105762
78 MARC

—ECONOMIC CONDITIONS—1918-1970

Betancur, Belisario, 1923-
Despiérta, Colombia! / Belisario Betancur. — 1. ed. — Bogotá : Ediciones Tercer Mundo, 1970.
329 p. ; 20 cm. — (Colección Documentos políticos)
1. Colombia—Social conditions. 2. Colombia—Economic conditions—1918-1970. 3. Colombia—Politics and government—1946- . I. Title.
HN303.5.B39 309.1'861 77-475389
77₍r79₎rev MARC

—ECONOMIC CONDITIONS—1918-1970—ADDRESSES, ESSAYS, LECTURES

Bejarano, Jesús A
Ensayos de interpretación de la economía colombiana / Jesús Antonio Bejarano. — Bogotá : Editorial La Carreta, 1978.
291 p. ; 17 cm.
Includes bibliographical references.
CONTENTS: Presentación.—La economía colombiana desde 1950.—Desarrollo clásico y desarrollo dependiente.—Currie.—Contribución al debate sobre el problema agrario.
1. Colombia—Economic conditions—1918-1970—Addresses, essays, lectures. 2. Colombia—Economic conditions—1970- —Addresses, essays, lectures. I. Title.
HC197.B37 330.9'861'063 79-349842
79 MARC

—ECONOMIC CONDITIONS—1970-

Parra Peña, Isidro, 1936-
Capitalismo colombiano : el modelo y su estilo / Isidro Parra-Peña. — 1. ed. — Bogotá : Ediciones Tercer Mundo, 1978.
90 p. ; 21 cm. — (Colección Tribuna libre)
Includes bibliographical references.
1. Capital—Colombia. 2. Colombia—Economic conditions—1970- . 3. Investments, Foreign—Colombia. I. Title.
HC200.C3P37 332'.041'09861 78-107728
79 MARC

Sŏnggyun'gwan Taehakkyo, Seoul, Korea. Han'guk Sanŏp Yŏn'guso.
(Chung-Nammi ŭi chŏngch'i kyŏngje kyoyuk chedo e kwanhan yŏn'gu)
中南美의政治經濟教育制度에關한研究：콜롬비아와페루를中心으로. — ₍서울₎：成均館大學校附設韓國產業研究所, 1972.
x, 258 p. : maps ; 27 cm.
Bibliography: p. 257-258.
1. Colombia—Politics and government—1946- . 2. Colombia—Economic conditions—1970- 3. Peru—Politics and government—1968- 4. Peru—Economic conditions—1968- I. Title.
F2278 S59 1972 78-826771

COLOMBIA
—ECONOMIC CONDITIONS
—1970- (Continued)

Stirnimann, Franz X
Die Rolle des Aktienmarktes in der langfristigen wirtschaftlichen Entwicklung Kolumbiens : eine eimpirische Analyse / Franz X. Stirnimann. — Bern ; Stuttgart : Paul Haupt, c1974.
vi, 267 p. : 4 graphs ; 23 cm. — (Veröffentlichung des Lateinamerikanischen Instituts an der Hochschule St. Gallen für Wirtschafts- und Sozialwissenschaften ; Bd. 8) Sw75-A-109
Originally presented as the author's thesis, Hochschule für Wirtschafts- und Sozialwissenschaften, St. Gall, Switzerland.
Bibliography: p. 261-267.
ISBN 3-258-01159-1 : 38.00F
1. Stock-exchange—Colombia. 2. Colombia—Economic conditions—1970- . I. Title. II. Series: St. Gall, Switzerland. Hochschule für Wirtschafts- und Sozialwissenschaften. Lateinamerikanisches Institut. Veröffentlichungen ; Bd. 8.
HG5352.S74 1974 79-376848
*79 MARC

—ECONOMIC CONDITIONS—1970-
—ADDRESSES, ESSAYS, LECTURES

Bejarano, Jesús A
Ensayos de interpretación de la economía colombiana / Jesús Antonio Bejarano. — Bogotá : Editorial La Carreta, 1978.
291 p. ; 17 cm.
Includes bibliographical references.
CONTENTS: Presentación.—La economía colombiana desde 1950.—Desarrollo clásico y desarrollo dependiente.—Currie.—Contribución al debate sobre el problema agrario.
1. Colombia—Economic conditions—1918-1970—Addresses, essays, lectures. 2. Colombia—Economic conditions—1970- —Addresses, essays, lectures. I. Title.
HC197.B37 330.9'861'063 79-349842
79 MARC

Comentarios económicos. — ⌈Bogotá⌉ : Banco de América Latina, ⌈1977 or 1978-
v. ; 22 cm.
Vol. 1: Principal author of the articles, Alvaro Daza Roa.
1. Colombia—Economic conditions—1970- —Addresses, essays, lectures. 2. Colombia—Economic policy—Addresses, essays, lectures. 3. Finance—Colombia—Addresses, essays, lectures. I. Daza Roa, Alvaro. II. Banco de América Latina.
HC197.C64 78-106070
78 MARC

Metodología y desarrollo en las ciencias sociales : efectos del crecimiento dependiente sobre la estructura social colombiana. — Bogotá : Centro de Estudios sobre Desarrollo Económico, Facultad de Economía, Universidad de los Andes, 1977.
614 p. : graphs ; 18 cm.
Includes bibliographies.
CONTENTS: Havens, A. E. y Parra Sandoval, R. Introducción.—Havens, A. E. Aspectos metrodológicos en el estudio del desarrollo.—Flinn, W. L. Notas sobre las familias de tres asentamientos urbanos. Los movimientos políticos en tres barrios bogotanos.—Havens, A. E. y Flinn, W. L. La tecnología de la revolución verde y el desarrollo de las comunicades.—Havens, A. E. La difusión de nuevas variedades de semillas y sus consecuencias.—Parra Sandoval, R. Empresa multinacional, urbanización y cambios en la estructura ocupacional.—Cepeda Ullos, F. y Mitchell, C. La tendecia tecnocrática.—Havens, A. E. y Flinn, W. L. Determinantes estructurales de la selectividad de la migración de una región de economía en deteriodo.—Havens, A. E., Flinn, W. L. y Lastarria Conhill, S. La reforma agraria y el Frente Nacional.—Havens, A. E. y Flinn, W. L. La clase campesina y las pautas de emigracion.—Uricoechea, F. Formación y expansión del estado burocrático-patrimonial en Colombia y Brasil.
1. Colombia—Social conditions—1970- —Addresses, essays, lectures. 2. Colombia—Economic conditions—1970- —Addresses, essays, lectures. 3. Colombia—Politics and government—1946- —Addresses, essays, lectures. I. Bogota. Universidad de los Andes. Centro de Estudios sobre Desarrollo Económico.
HN303.5.M47 309.1'861'063 79-349270
79 MARC

—ECONOMIC CONDITIONS—1970-
—CONGRESSES

Convención Bancaria y de Instituciones Financieras, 15th, Cartagena, Colombia, 1977.
Banca 77 ⌈i.e. setenta y siete⌉ : memoria de la XV Convención Bancaria y de Instituciones Financieras, Cartagena, octubre 27, 28, y 29 de 1977. — ⌈Bogotá?⌉ : Redactores Asociados, Poligrupo Comunicación, ⌈1977 or 1978⌉
104 p. ; 24 cm.
Sponsored by the Asociación Bancaria de Colombia.
1. Banks and banking—Colombia—Congresses. 2. Colombia—Economic conditions—1970- —Congresses. 3. Coffee trade—Colombia—Congresses. 4. International economic relations—Congresses. I. Asociación Bancaria de Colombia. II. Title.
HG2904.C66 1977 79-115425
79 MARC

—ECONOMIC CONDITIONS—1970-
—PERIODICALS

Asociación Nacional de Industriales.
Revista ANDI.
⌈Medellín, Colombia, ANDI⌉
no. ill. 24 cm. bimonthly.
Continues: Asociación Nacional de Industriales. ANDI, revista bimestral.
1. Colombia—Economic conditions—1970- —Periodicals. 2. Colombia—Industries—Periodicals. 3. Colombia—Commerce—Periodicals. I. Title.
HC196.A92a 79-649958
MARC-S

Nueva frontera. Documentos. 1- feb. 1976-
⌈Bogotá, Nueva Frontera⌉
no. ill. 34 cm. monthly.
Other title, Febr. 1976- : Documentos Nueva frontera.
Issued as supplement to: Nueva frontera.
1. Colombia — Economic conditions — 1970- — Periodicals. 2. Colombia—Civilization—Periodicals. 3. Civilizaton, Modern—1950- —Perodicals. I. Nueva frontera. II. Title: Documentos Nueva frontera.
HC196.N84 330.9'861 78-648733
MARC-S

Parra Peña, Isidro, 1936-
El desarrollo y el ahorro personal / Isidro Parra Peña. — Bogotá : La Asesoría del Departamento de Comunicaciones de la Caja de Crédito Agrario, Industrial y Minero, 1976.
164 p. ; 23 cm.
Includes related legislation.
Includes bibliographical references.
1. Saving and investment—Colombia. 2. Colombia—Economic conditions —1970- . I. Title.
HC200.S3P34 78-107353
78 MARC

—ECONOMIC CONDITIONS—
ADDRESSES, ESSAYS, LECTURES

Camacho Roldán, Salvador, 1827-1900.
Escritos sobre economía y política / Salvador Camacho Roldán ; selección, notas y prólogo de Jesús Antonio Bejarano. — ⌈Bogotá⌉ : Instituto Colombiano de Cultura, Subdirección de Comunicaciones Culturales, 1976.
268, ⌈3⌉ p. ; 21 cm. — (Biblioteca básica colombiana ; 13)
"Bibliografía sobre Salvador Camacho Roldán": p. ⌈269⌉
CONTENTS: Nuestro sistema tributario.—Desamortización de bienes de manos muertas.—Nuestro provenir industrial.—Consideraciones sobre fomento.—La agricultura en Colombia.—Nuestra situación industrial.—Valor trascendental del Río Magdalena.—La nueva administración.—Apelación al pueblo.—El pensamiento del Partido Liberal.—?Hasta dónde se puede llevar la guerra civil?—La unión liberal.—Manuela.—La Convención de Rionegro.
1. Colombia—Economic conditions—Addresses, essays, lectures. 2. Colombia—Politics and government—Addresses, essays, lectures. I. Title. II. Series.
HC197.C355 1976 330.9'861'063 77-551253
77⌈r79⌉rev MARC

Colombia, hoy / por Mario Arrubia ... ⌈et al.⌉. — 1. ed. — Bogotá : Siglo Veintiuno Editores, 1978.
403 p. ; 19 cm. — (Historia inmediata)
Includes bibliographies.
1. Colombia—Politics and government—1810- —Addresses, essays, lectures. 2. Colombia—Economic conditions—Addresses, essays, lectures. 3. Colombian literature—20th century—History and criticism—Addresses, essays, lectures. I. Arrubla, Mario.
F2273.C743 309.1'861 78-376710
78 MARC

El Siglo XIX en Colombia visto por historiadores norteamericanos / compilación, prólogo y notas de Jesús Antonio Bejarano ; ⌈traducción de Luz Jaramillo⌉. — 1. ed. — Bogotá : Editorial La Carreta, 1977.
389 p. : maps ; 17 cm.
Includes bibliographical references.
CONTENTS: Knowlton, R. J. Expropiación de los bienes de la Iglesia en el siglo XIX en Méjico y Colombia.—Harrison, J. P. La evolución de la comercialización del tabaco colombiano hasta 1875.—Bushnell, D. Dos etapas de la política arancelaria colombiana.—Bergquist, C. Una década de regeneración, 1886-1896.—Gilmore, R. L. y Harrison, J. P. Juan Bernardo Elbers y la introducción de la navegación a vapor en el Río Magdalena.—Rippy, F. J. Los comienzos de la era ferroviaria en Colombia.—Beyer, R. C. El transporte y la industria del café en Colombia.—Barnhart, D. S. Auge y fracaso.—Bierck, H. A., Jr. Las pugnas por la abolición de la esclavitud en la Gran Colombia.—Delpar, H. Aspectos del faccionalismo liberal en Colombia, 1875-1885
1. Colombia—Economic conditions—Addresses, essays, lectures. 2. Transportation—Colombia—History—Addresses, essays, lectures. 3. Colombia Politics and government—1863-1885—Addresses, essays, lectures. I. Bejarano, Jesús A.
HC197.S54 78-104827
78 MARC

—ECONOMIC CONDITIONS—
INFORMATION SERVICES—
COLOMBIA

Hernández de Caldas, Angela.
La información socio económica colombiana : dónde se genera, dónde se almacena, cómo se difunde : una contribución a la Reunion de Correspondants del Service de References-Developpment, celebrada en París en noviembre de 1974, con el auspicio del Banco Internacional de Reconstrucción y Desarrollo / Angela Hernández de Caldas. — Bogotá : Cámara de Comercio de Bogotá, Centro de Informática Económica, 1974.
41 leaves ; 30 cm.
Bibliography: leaves 38-41.
1. Colombia—Economic conditions—Information services—Colombia. I. Bogotá. Cámara de Comercio. Centro de Informática Económica. II. Title.
HC197.H43 330.9'861'063 79-341792
79 MARC

—ECONOMIC POLICY

Colombia. Presidente, 1970-1974 (Pastrana Borrero)
Una política social para el desarrollo; mensaje del presidente de la República de Colombia Misael Pastrana Borrero al Congreso Nacional, 20 de julio de 1972. ⌈Bogotá⌉ Tall. Gráf. del Banco de la República ⌈1972⌉
229 p. 28 cm. LACAP73-0717
1. Colombia—Social policy. 2. Colombia—Economic policy. 3. Agriculture and state—Colombia. I. Title.
HN303.5.C64 1972 309.1'861 73-329880
79⌈r79⌉rev MARC

Romero, Manuel.
La política de los comunistas colombianos : texto de estudio de la táctica del Partido Comunista de Colombia / Manuel Romero, Yira Castro. — Bogotá : Ediciones Suramérica, 1978.
162 p. ; 20 cm.
Includes bibliographical references.
1. Partido Comunista de Colombia. 2. Colombia—Economic policy. 3. Colombia—Politics and government—1946- . I. Castro, Yira, joint author. II. Title.
JL2898.C6R65 79-101162
79 MARC

—ECONOMIC POLICY—ADDRESSES,
ESSAYS, LECTURES

Colombia. Ministerio de Hacienda y Crédito Público.
Política fiscal ⌈por⌉ Carlos Villaveces R. Bogotá, Impr. Nacional, 1953.
42 p. 21 cm.
1. Colombia—Economic policy—Addresses, essays, lectures. 2. Fiscal policy—Colombia—Addresses, essays, lectures. I. Villaveces R., Carlos. II. Title.
HC196.A4 1953 55-57868
⌈r79⌉rev MARC

Colombia. Presidente, 1970-1974 (Pastrana Borrero)
Desarrollo y realización de una política social : mensaje del señor presidente de la República de Colombia doctor Misael Pastrana Borrero al Congreso Nacional, 20 de julio de 1973. — ⌈Bogotá⌉ : Tall. Gráf. del Banco de la República, ⌈1973⌉
222 P. ; 28 cm.
1. Colombia—Politics and government—1946- —Addresses, essays, lectures. 2. Colombia—Economic policy—Addresses, essays, lectures. 3. Colombia—Social policy—Addresses, essays, lectures. I. Title.
J220.N327 1973 309.1'861'063 79-352917
79 MARC

Comentarios económicos. — ⌈Bogotá⌉ : Banco de América Latina, ⌈1977 or 1978-
v. ; 22 cm.
Vol. 1: Principal author of the articles, Alvaro Daza Roa.
1. Colombia—Economic conditions—1970- —Addresses, essays, lectures. 2. Colombia—Economic policy—Addresses, essays, lectures. 3. Finance—Colombia—Addresses, essays, lectures. I. Daza Roa, Alvaro. II. Banco de América Latina.
HC197.C64 78-106070
78 MARC

—ECONOMIC POLICY—CONGRESSES

Congreso Nacional de Economistas, 6th, Cali, Colombia, 1977.
Políticas y estrategias del desarrollo económico y social colombiano en un mundo cambiante : informe / VI Congreso Nacional de Economistas, Cali, 6 a 9 de diciembre, 1977. — ⌈Bogotá⌉ : Sociedad Colombiana de Economistas, ⌈1977 or 1978⌉
283 p. ; 23 cm.
1. Colombia—Economic policy—Congresses. 2. Colombia—Social policy—Congresses. I. Sociedad Colombiana de Economistas. II. Title.
HC197.C764 1977 79-101018
79 MARC

—FOREIGN ECONOMIC RELATIONS

Simposio sobre Financiamiento Externo, 2d, Medellín, Colombia, 1978.
Política económica externa de Colombia, 1978 / Eduardo Wiesner, editor. — 1. ed. — ⌈Bogota?⌉ : Asociación Bancaria de Colombia, 1978.
349 p. : graphs ; 24 cm. — (Biblioteca Asociación Bancaria de Colombia)
Includes bibliographies.
1. Colombia—Foreign economic relations. 2. Colombia—Commercial policy. 3. Foreign exchange administration—Colombia. I. Wiesner Durán, Eduardo. II. Asociación Bancaria de Colombia. III. Title. IV. Series: Asociación Bancaria de Colombia. Biblioteca Asociación Bancaria de Colombia.
HF1517.S57 1978 382.1'09861 79-101234
79 MARC

COLOMBIA (Continued)

—FOREIGN RELATIONS—TREATIES

Cavelier, Germán.
Memoria histórico-jurídica sobre el asunto de Los Monjes / Germán Cavelier. — Bogotá : Editorial Kelly, 1977.
543 p., ₍1₎ leaf of plates : ill. ; 25 cm.
Bibliography: p. 503-514.
Includes index.
1. Colombia—Foreign relations—Venezuela. 2. Venezuela—Foreign relations—Colombia. 3. Los Monjes (Islands) 4. Colombia—Foreign relations—Treaties. I. Title.
JX1535.Z7V43 78-381060
78 MARC

—FOREIGN RELATIONS—VENEZUELA

Cavelier, Germán.
Memoria histórico-jurídica sobre el asunto de Los Monjes / Germán Cavelier. — Bogotá : Editorial Kelly, 1977.
543 p., ₍1₎ leaf of plates : ill. ; 25 cm.
Bibliography: p. 503-514.
Includes index.
1. Colombia—Foreign relations—Venezuela. 2. Venezuela—Foreign relations—Colombia. 3. Los Monjes (Islands) 4. Colombia—Foreign relations—Treaties. I. Title.
JX1535.Z7V43 78-381060
78 MARC

—HISTORY—TO 1810

Caballero, Enrique.
América una equivocación / Enrique Caballero. — ₍Bogotá₎ : Editorial Hispana, ₍1978₎
314 p. ; 22 cm.
Bibliography: p. ₍311₎-314.
1. Colombia—History—To 1810. 2. Indians of South America—Colombia. I. Title.
F2272.C2 986.1 79-108636
79 MARC

Friede, Juan.
Los chibchas bajo la dominación española / Juan Friede. — Bogotá : La Carreta, ₍1974₎
286 p., ₍2₎ leaves of plates : maps ; 22 cm. — (Serie Libros de bolsillo de La Carreta)
Published in 1960 under title: Descubrimiento del Nuevo Reino de Granada y fundación de Bogotá; in 1966 under title: Invasión del país de los chibchas.
Includes bibliographical references.
1. Chibcha Indians. 2. Indians, Treatment of—Colombia. 3. Colombia—History—To 1800. 4. Bogotá—History. I. Title.
F2270.2.C4F73 980'.004'98 79-383006
79 MARC

Kuethe, Allan J 1940-
Military reform and society in New Granada, 1773-1808 / Allan J. Kuethe. — Gainesville : University Presses of Florida, 1978.
234 p. : map ; 24 cm. — (Latin American monographs ; 2d ser., 22)
"A University of Florida book."
Bibliography: p. 219-226.
ISBN 0-8130-0570-1 : $10.00
1. Spain. Ejército—Colonial forces—History. 2. Colombia—History, Military. 3. Colombia—History—To 1810. I. Title. II. Series: Florida. University, Gainesville. Center for Latin American Studies. Latin American monographs ; 2d ser., 22.
UA789.K83 355.3'1'0946 77-21908
77 MARC

Rodríguez Fresle, Juan, b. 1566.
El carnero / Juan Rodríguez Freyle ; con notas explicativas del doctor Miguel Aguilera. — ₍Medellín₎ : Editorial Bedout, ₍1976₎
376 p. ; 18 cm. — (Bolsilibros Bedout ; v. 23)
Bibliography: p. 45-46.
1. Colombia—History—To 1810. 2. Bogotá—History. I. Aguilera, Miguel. II. Title.
F2272.R69 1976 986.1 78-103762
78 MARC

—HISTORY—TO 1810—SOURCES

Friede, Juan.
El adelantado, don Gonzalo Jiménez de Quesada / Juan Friede. — 1. ed. — Bogotá : C. Valencia Editores, 1979.
2 v. : ill. ; 21 cm.
Includes bibliographical references.
CONTENTS: I. Estudio biográfico.—II. Documentos.
ISBN 8482770071
1. Jiménez de Quesada, Gonzalo, d. 1579. 2. Explorers—Colombia—Biography. 3. Explorers—Spain—Biography. 4. Colombia—History—To 1810—Sources. I. Title.
F2272.J55F74 986.1'01'0924 79-116374
79 MARC

—HISTORY—INSURRECTION OF THE COMUNEROS, 1781

Gómez Rodríguez, Ramiro.
Revolución de los comuneros / Ramiro Gómez R. — ₍s.l. : s.n., 1978 or 1979₎ (Bucaramanga : Editorial M. A. Gómez)
141 p. ; 20 cm.
Includes bibliographies.
1. Colombia—History—Insurrection of the Comuneros, 1781. I. Title.
F2272.G58 986.1'02 79-108235
79 MARC

—HISTORY—19TH CENTURY

Orígenes de las guerras civiles en Colombia. — ₍s.l. : Ediciones Historia y Sociedad, 19—₎.
70 p. ; 17 cm.
On cover: Un análisis marxista.
1. Colombia—History—Colombia—19th century. 2. Civil war. 3. Colombia—Economic conditions.
F2273.074 986.1 78-110011
79₍r79₎rev MARC

—HISTORY—1810-

Tirado Mejía, Alvaro.
Aspectos sociales de las guerras civiles en Colombia / Alvaro Tirado Mejía. — ₍Bogotá₎ : Instituto Colombiano de Cultura, Subdirección de Comunicaciones Culturales, ₍1976₎
493 p. : ill. ; 21 cm. — (Biblioteca básica colombiana ; 20)
Bibliography: p. 489-493.
1. Colombia—History—1810- . 2. Colombia—History—Sources. I. Title. II. Series.
F2273.T5 986.1 77-550412
77₍r79₎rev MARC

—HISTORY—WAR OF INDEPENDENCE, 1810-1822

Castellanos V , Rafael Ramón.
Pedro Gual, ideólogo de la libertad / Rafael Ramón Castellanos. — Caracas : Italgráfica, 1978.
34 p. ; 27 cm.
Bibliography: p. 32-34.
1. Gual, Pedro, 1783-1862. 2. Venezuela—History—War of Independence, 1810-1823. 3. Colombia—History—War of Independence, 1810-1822. I. Title.
F2324.G77C35 78-113669
79 MARC

—HISTORY—REVOLUTION, 1899-1903

Villegas, Jorge.
La guerra de los mil días / Jorge Villegas, José Yunis. — 1. ed. — Bogotá : C. Valencia, 1978.
323 p., ₍8₎ leaves of plates : ill. ; 21 cm.
"Documentos y cronología": p. ₍131₎-318.
Includes index.
1. Colombia—History—Revolution, 1899-1903. 2. Colombia—History—Revolution, 1899-1903—Sources. I. Yunis, José, joint author. II. Title.
F2276.5.V54 986.1'062 79-348778
79 MARC

—HISTORY—REVOLUTION, 1899-1903—SOURCES

Villegas, Jorge.
La guerra de los mil días / Jorge Villegas, José Yunis. — 1. ed. — Bogotá : C. Valencia, 1978.
323 p., ₍8₎ leaves of plates : ill. ; 21 cm.
"Documentos y cronología": p. ₍131₎-318.
Includes index.
1. Colombia—History—Revolution, 1899-1903. 2. Colombia—History—Revolution, 1899-1903—Sources. I. Yunis, José, joint author. II. Title.
F2276.5.V54 986.1'062 79-348778
79 MARC

—HISTORY—COUP D'ÉTAT, 1953

Abella, Arturo.
Así fue el 13 ₍i.e. trece₎ de junio / por Arturo Abella. — Bogotá : Ediciones Aquí Bogotá, 1973.
80 p. : ill. ; 28 cm. — (Aquí Bogotá ; no. 2)
1. Colombia—History—Coup d'état, 1953. I. Title. II. Series.
F2278.A22 986.1'063 79-372860
79 MARC

—HISTORY—ADDRESSES, ESSAYS, LECTURES

Las Maravillas de Colombia : sorprendentes y poco conocidas / colección dirigida por Enrique Congrains Martín ; investigador asociado, Pedro Pablo Tibaquirá. — Bogotá : Editorial Forja, ₍1979₎
4 v. : ill. ; 24 cm.
CONTENTS: t. 1. Santa Fé de Bogotá. La Sabana. Las tierras frías de Cundinamarca y Boyacá. Santander.—t. 2. Los Llanos Orientales. Ríos Orinoco, Meta y Casanare. Amazonia Colombiana.—t. 3. El Río Magdalena. La Sierra Nevada de Santa Marta. Cartagena y Santa Marta.—t. 4. El Chocó. Antioquia y Medellín. El "Viejo Caldas." El Río Cauca, Cartago y Cali. Popayán. San Agustín.
1. Colombia—History—Addresses, essays, lectures. I. Congrains Martín, Enrique. II. Tibaquirá, Pedro Pablo.
F2271.M3 986.1 79-117115
79 MARC

—HISTORY—COLLECTED WORKS

Jaramillo Uribe, Jaime.
La personalidad histórica de Colombia y otros ensayos / Jaime Jaramillo Uribe. — ₍Bogotá₎ : Instituto Colombiano de Cultura, Subdirección de Comunicaciones Culturales, 1977.
270 p. ; 21 cm. — (Biblioteca basica colombiana ; 28)
Includes bibliographical references.
CONTENTS: Caro y Alberdi.—Etapas de la filosofía en la historia colombiana.—Valoración e influencia de Rousseau en Colombia.—El conflicto entre la conciencia religiosa y la ciencia moderna.—Tres etapas de la historia intelectual de Colombia.—Algunos aspectos de la personalidad histórica de Colombia.—La sociedad colombiana en la época de la visita de Humboldt.—Cambios demográficos y aspectos de la política social española en el Nuevo Reino de Granada.—La influencia de los románticos franceses y de la revolución de 1848 en el pensamiento político colombiano del siglo XIX.—Las Sociedades Democráticas de Artesanos y la coyuntura política y social colombiana de 1848.—El problema de la esclavitud en la historia y en la historiografía de los Estados Unidos.—Discriminación y tolerancia racial en la sociedad cartagenera.—Esquema histórico de la universidad colombiana.
1. Colombia—History—Collected works. I. Title. II. Series.
F2253.J37 986.1 78-111469
79 MARC

Rodríguez Plata, Horacio.
Temas históricos / Horacio Rodríguez Plata. — ₍Medellín, Colombia₎ : Ediciones Fondo Cultural Cafetero, ₍19
v. ; 19 cm.
Includes bibliographical references.
1. Colombia—History—Collected works. I. Title.
F2253.R62 986.1 79-101019
79 MARC

—HISTORY—SOURCES

Tirado Mejía, Alvaro.
Aspectos sociales de las guerras civiles en Colombia / Alvaro Tirado Mejía. — ₍Bogotá₎ : Instituto Colombiano de Cultura, Subdirección de Comunicaciones Culturales, ₍1976₎
493 p. : ill. ; 21 cm. — (Biblioteca básica colombiana ; 20)
Bibliography: p. 489-493.
1. Colombia—History—1810- . 2. Colombia—History—Sources. I. Title. II. Series.
F2273.T5 986.1 77-550412
77₍r79₎rev MARC

—HISTORY, MILITARY

Kuethe, Allan J 1940-
Military reform and society in New Granada, 1773-1808 / Allan J. Kuethe. — Gainesville : University Presses of Florida, 1978.
234 p. : map ; 24 cm. — (Latin American monographs ; 2d ser., 22)
"A University of Florida book."
Bibliography: p. 219-226.
ISBN 0-8130-0570-1 : $10.00
1. Spain. Ejército—Colonial forces—History. 2. Colombia—History, Military. 3. Colombia—History—To 1810. I. Title. II. Series: Florida. University, Gainesville. Center for Latin American Studies. Latin American monograph. ; 2d ser., 22.
UA789.K83 355.3'1'0946 77-21908
77 MARC

—INDUSTRIES

Wogart, Jan Peter.
Industrialization in Colombia : policies, patterns, perspectives / Jan Peter Wogart. — Tübingen : Mohr, 1978.
xiii, 176 p. ; 24 cm. — (Kieler Studien ; 153 ISSN 0304-6989)
GFR78-A
Bibliography: p. 164-176.
ISBN 3-16-340542-8. ISBN 3-16-340541-X pbk.
1. Colombia—Industries. 2. Industry and state—Colombia. 3. Export sales—Colombia. 4. Import substitution—Colombia. I. Title. II. Series.
HC197.W63 338'.09861 78-321765
78 MARC

COLOMBIA (Continued)

—INDUSTRIES—PERIODICALS

Asociación Nacional de Industriales.
Revista ANDI.
[Medellín, Colombia, ANDI]
no. ill. 24 cm. bimonthly.
Continues: Asociación Nacional de Industriales. ANDI, revista bimestral.
1. Colombia—Economic conditions—1970- —Periodicals. 2. Colombia—Industries—Periodicals. 3. Colombia—Commerce—Periodicals. I. Title.
HC196.A92a 79–649958
MARC–S

—LANGUAGES

see also Barasana del Sur language; Ingano language; Tucanoan languages

—MAPS

Colombia. Instituto Geográfico Agustín Codazzi.
Atlas de Colombia. — 3. ed., rev. y aumentada. — Bogotá : Instituto Geografico Agustín Codazzi, 1977.
283 (1 fold.), [3] p. : ill. (some col.), col. maps ; 49 cm.
Planeó y dirigió la obra: Eduardo Acevedo Latorre.
Bibliografía: p. [285]-[286]
CONTENTS: Cosmografía, astronomía y astronáutica.—Geografía general.—Mapa de Colombia.—Geografía física de Colombia.—Geografía humana de Colombia.—Geografía económica de Colombia.—Aspectos administrativos.—Cartografía de las secciones administrativas.—Cartografía histórica.
1. Colombia—Maps. I. Acevedo Latorre, Eduardo. II. Title.
G1730.C65 1977 912'.861 79-354218
79 MARC

Colombia. Instituto Geográfico Agustín Codazzi. Subdirección de Investigación y Divulgación Geográfica.
Atlas básico de Colombia / Ministerio de Hacienda y Credito Publico, Instituto Geográfico "Agustín Codazzi", Sub-dirección de Investigación y Divulgación Geográfica. — 2. ed. — Bogotá : El Instituto, 1978.
141 p. (2 fold.) : ill. (some col.), col. maps ; 35 cm.
CONTENTS: Colombia actual.—La formación del território nacionál.—El medio ambiente.—La población colombiana.—Actividades del hombre.—Divisiones político-administrativas.—Glosario de términos.
1. Colombia—Maps. I. Title.
G1730.C66 1978 912'.861 79-383488
79 MARC

—NATIVE RACES—MAPS

Arango M , Francisco.
Colombia : atlas indigenista / Francisco Arango Montoya. — Bogotá, Colombia : Litografía Arco, 1977.
264, [103] p., [12] leaves of plates : col. ill., 95 maps (some col.) ; 35 cm.
8 col. maps inserted.
Bibliography: p. 221-247.
Includes indexes.
CONTENTS: Lista alfabetica de los grupos indigenas de Colombia.—Vicariatos apostolicos.—Prefecturas apostolicas.—Otras juridicciones eclesisticas.—Cuadros estadisticos.—Bibliografía.—Apendice de las principales leyes indigenas de Colombia.—Estadisticas educativas de los territorios misionales Diciembre 1975.
1. Indigenous church administration. 2. Colombia—Native races—Maps. I. Title.
G1731.E1A7 1977 912'.1'282861 78-379315
78 MARC

—OCCUPATIONS

Servicio Nacional de Aprendizaje.
Lista ampliada de denominaciones : diccionario de sinónimos / República de Colombia, Ministerio de Trabajo y Seguridad Social, Servicio Nacional de Aprendizaje, SENA. — Bogotá : SENA, 1972.
269 p. ; 30 cm.
At head of title: Clasificación nacional de ocupaciones (CNO), edición preliminar.
1. Colombia—Occupations. 2. Occupations—Classification. I. Title.
HB2657.S47 1972 79-338124
79 MARC

—OFFICIALS AND EMPLOYEES

Buitrago C , Luis A
Régimen jurídico de los empleados oficiales / Luis A. Buitrago C. — Bucaramanga [Colombia] : Editorial Nueva Jornada, 1974.
399 p. ; 24 cm.
Includes index.
1. Colombia—Officials and employees. I. Title.
342'.861'068 78-105477
79 MARC

—OFFICIALS AND EMPLOYEES—BIOGRAPHY

Lecompte Luna, Alvaro.
Castillo y Rada, el grancolombiano / Alvaro Lecompte Luna ; prólogo de Lucio Pabón Núñez. — Bogotá : Instituto Caro y Cuervo, 1977.
xi, 170 p., [10] leaves of plates : 11 ill. ; 21 cm. — (Serie La Granada entreabierta ; 19)
Bibliography: p. [141]-142.
1. Castillo y Rada, José María del, 1776-1835. 2. Colombia—Officials and employees—Biography. 3. Politicians—Colombia—Biography. 4. Lawyers—Colombia—Biography. I. Title. II. Series.
F2273.C33L42 986.1'04'0924 78-375543
78 MARC

—POLITICS AND GOVERNMENT

Galat, José.
Un país prestado / José Galat. — 1. ed. — Bogotá : Tercer Mundo, 1978.
289 p. ; 21 cm. — (Colección Tribuna libre)
Includes bibliographical references.
1. Colombia—Economic conditions. 2. Colombia—Politics and government. I. Title.
HC197.G19 78-107700
79 MARC

Guillén Martínez, Fernando, 1925-
El poder : los modelos estructurales del poder político en Colombia / Fernando Guillén Martínez. — Bogotá : Universidad Nacional de Colombia, Centro de Investigaciones para el Desarrollo, 1973.
iii, 391 p. ; 27 cm. & appendix (74 p.)
"Notas, fuentes y referencias": p. [284]-391.
1. Colombia—Politics and government. 2. Power (Social sciences) I. Colombia. Universidad, Bogotá. Centro de Investigaciones para el Desarrollo. II. Title.
JL2831.G84 320.9'861 74-233813
79[r79]rev MARC

Oquist, Paul H
Violencia, conflicto y política en Colombia / Paul Oquist. — Bogotá : Instituto de Estudios Colombianos, 1978.
339 p. ; 24 cm. — (Biblioteca Banco Popular : Divulgación económica y social)
On spine: 1.
"El Instituto de Estudios Colombianos ... inicia con este volumen su serie de ediciones especiales."
Originally presented as the author's thesis, University of California, Berkeley.
Includes bibliographical references.
1. Violence—Colombia—History. 2. Social conflict—History. 3. Colombia—Politics and government. I. Title.
HN310.Z9V56 1978 301.6'33'09861 79-101337
79 MARC

Rivadeneira Vargas, Antonio José.
Historia constitucional de Colombia, 1510-1978 / Antonio José Rivadeneira Vargas. — Bogotá : Editorial Horizontes, 1978.
411, [14] p. ; 24 cm.
Bibliography: p. [413]-[419]
1. Colombia—Constitutional history. 2. Colombia—Politics and government. I. Title.
JL2811.R52 320.9'861 79-114512
79 MARC

Santa, Eduardo, 1928-
Instituciones políticas de Colombia / Eduardo Santa. — Bogotá : Editorial Temis Librería, 1978.
xv, 191 p. ; 24 cm.
Bibliography: p. [186]-191.
ISBN 8482720597
1. Political parties—Colombia. 2. Colombia—Politics and government. 3. Local government—Colombia. I. Title.
JL2898.A1S33 329.9'861 79-101147
79 MARC

—POLITICS AND GOVERNMENT—1810- —ADDRESSES, ESSAYS, LECTURES

Colombia, hoy / por Mario Arrubla ... [et al.]. — 1. ed. — Bogotá : Siglo Veintiuno Editores, 1978.
403 p. ; 19 cm. — (Historia inmediata)
Includes bibliographies.
1. Colombia—Politics and government—1810- —Addresses, essays, lectures. 2. Colombia—Economic conditions—Addresses, essays, lectures. 3. Colombian literature—20th century—History and criticism—Addresses, essays, lectures. I. Arrubla, Mario.
F2273.C743 309.1'861 78-376710
78 MARC

—POLITICS AND GOVERNMENT—1832-1886

Liévano Aguirre, Indalecio.
Rafael Núñez / Indalecio Liévano Aguirre. — 3. ed. — [Bogotá] : Instituto Colombiano de Cultura, Subdirección de Comunicaciones Culturales, División de Publicaciones, c1977.
521 p. ; 22 cm. — (Colección Historia viva ; 2) (Biblioteca colombiana de cultura)
Bibliography: p. 517-518.
1. Núñez, Rafael, Pres. Colombia, 1825-1894. 2. Colombia—Politics and government—1832-1886. 3. Colombia—Politics and government—1886-1903. 4. Colombia—Presidents—Biography.
F2276.N862L53 1977 986.1'061'0924 79-104037
79 MARC

—POLITICS AND GOVERNMENT—1863-1885

Espriella, Ramiro de la.
Acercamiento a Núñez / Ramiro de la Espriella. — 1. ed. — Bogotá : Tercer Mundo, 1978.
xiv, 122 p. ; 21 cm. — (Colección Tribuna libre)
Bibliography: p. 121-122.
1. Núñez, Rafael, Pres. Colombia, 1825-1894. 2. Colombia—Politics and government—1863-1885. 3. Colombia—Politics and government—1886-1903. 4. Colombia—Presidents—Biography. I. Title.
F2276.N862E86 986.1'061'0924 79-101886
79 MARC

—POLITICS AND GOVERNMENT—1863-1885—ADDRESSES, ESSAYS, LECTURES

El Siglo XIX en Colombia visto por historiadores norteamericanos / compilación, prólogo y notas de Jesús Antonio Bejarano ; [traducción de Luz Jaramillo]. — 1. ed. — Bogotá : Editorial La Carreta, 1977.
389 p. : maps ; 17 cm.
Includes bibliographical references.
CONTENTS: Knowlton, R. J. Expropiación de los bienes de la Iglesia en el siglo XIX en Méjico y Colombia.—Harrison, J. P. La evolución de la comercialización del tabaco colombiano hasta 1875.—Bushnell, D. Dos etapas de la política arancelaria colombiana.—Bergquist, C. Una década de regeneración, 1886-1896.—Gilmore, R. L. y Harrison, J. P. Juan Bernardo Elbers y la introducción de la navegación a vapor en el Río Magdalena.—Rippy, F. J. Los comienzos de la era ferroviaria en Colombia.—Beyer, R. C. El transporte y la industria del café en Colombia.—Barnhart, D. S. Auge y fracaso.—Bierck, H. A., Jr. Las pugnas por la abolición de la esclavitud en la Gran Colombia.—Delpar, H. Aspectos del faccionalismo liberal en Colombia, 1875-1885.
1. Colombia—Economic conditions—Addresses, essays, lectures. 2. Transportation—Colombia—History—Addresses, essays, lectures. 3. Colombia—Politics and government—1863-1885—Addresses, essays, lectures. I. Bejarano, Jesús A.
HC197.S54 78-104827
78 MARC

—POLITICS AND GOVERNMENT—1886-1903

Bergquist, Charles W
Coffee and conflict in Colombia, 1886-1910 / Charles W. Bergquist. — Durham, N.C. : Duke University Press, 1978.
x, 277 p. : ill. ; 25 cm.
Bibliography: p. [263]-269.
Includes index.
ISBN 0-8223-0418-X : $16.75
1. Colombia—Politics and government—1886-1903. 2. Colombia—Politics and government—1903-1930. 3. Colombia—Economic conditions. 4. Coffee trade—Colombia—History. I. Title.
F2276.5.B47 320.9'861'061 78-59581
79 MARC

Espriella, Ramiro de la.
Acercamiento a Núñez / Ramiro de la Espriella. — 1. ed. — Bogotá : Tercer Mundo, 1978.
xiv, 122 p. ; 21 cm. — (Colección Tribuna libre)
Bibliography: p. 121-122.
1. Núñez, Rafael, Pres. Colombia, 1825-1894. 2. Colombia—Politics and government—1863-1885. 3. Colombia—Politics and government—1886-1903. 4. Colombia—Presidents—Biography. I. Title.
F2276.N862E86 986.1'061'0924 79-101886
79 MARC

Liévano Aguirre, Indalecio.
Rafael Núñez / Indalecio Liévano Aguirre. — 3. ed. — [Bogotá] : Instituto Colombiano de Cultura, Subdirección de Comunicaciones Culturales, División de Publicaciones, c1977.
521 p. ; 22 cm. — (Colección Historia viva ; 2) (Biblioteca colombiana de cultura)
Bibliography: p. 517-518.
1. Núñez, Rafael, Pres. Colombia, 1825-1894. 2. Colombia—Politics and government—1832-1886. 3. Colombia—Politics and government—1886-1903. 4. Colombia—Presidents—Biography.
F2276.N862L53 1977 986.1'061'0924 79-104037
79 MARC

COLOMBIA (Continued)

—POLITICS AND GOVERNMENT
—1903-1930

Bergquist, Charles W
Coffee and conflict in Colombia, 1886-1910 / Charles W. Bergquist. — Durham, N.C. : Duke University Press, 1978.
x, 277 p. : ill. ; 25 cm.
Bibliography: p. [263]-269.
Includes index.
ISBN 0-8223-0418-X : $16.75
1. Colombia—Politics and government—1886-1903. 2. Colombia—Politics and government—1903-1930. 3. Colombia—Economic conditions. 4. Coffee trade—Colombia—History. I. Title.
F2276.5.B47 320.9'861'061 78-59581
79 MARC

—POLITICS AND GOVERNMENT—1930-1946

Ortiz Márquez, Julio, 1910-
El hombre que fue un pueblo / Julio Ortiz Márquez. — 1. ed. — Bogotá : Carlos Valencia Editores, 1978.
263 p. ; 21 cm.
1. Gaitán, Jorge Eliécer, 1898-1948. 2. Colombia—Politics and government—1930-1946. 3. Colombia—Politics and government—1946- . I. Title.
F2277.G24O77 986.1'063'0924 79-361010
79 MARC

—POLITICS AND GOVERNMENT—1930-1946—COLLECTED WORKS

Zuleta Ferrer, Juan.
La historia contra la pared : selección de ensayos y editoriales, El Colombiano, 1930-1978 / Juan Zuleta Ferrer. — Medellín : [Biblioteca Pública Piloto, 1978]
403 p. ; 23 cm. — (Colección Biblioteca Pública Piloto ; v. 1)
1. Colombia—Politics and government—1930-1946—Collected works. 2. Colombia—Politics and government—1946- —Collected works. 3. Antioquia, Colombia (Dept.)—Politics and government—Collected works. I. El Colombiano, Medellín, Colombia. II. Title. III. Series: Medellín, Colombia. Biblioteca Pública Piloto para la América Latina. Colección Biblioteca Pública Piloto ; v. 1.
F2277.Z84 320.9'861'063 79-100127
79 MARC

—POLITICS AND GOVERNMENT—1946-

Betancur, Belisario, 1923-
Despiérta, Colombia! / Belisario Betancur. — 1. ed. — Bogotá : Ediciones Tercer Mundo, 1970.
329 p. ; 20 cm. — (Colección Documentos políticos)
1. Colombia—Social conditions. 2. Colombia—Economic conditions—1918-1970. 3. Colombia—Politics and government—1946- . I. Title.
HN303.5.B39 309.1'861 77-475389
77[r79]rev MARC

Gallón Restrepo, Octavio.
Ruptura histórica / Octavio Gallón Restrepo. — 1. ed. — Bogotá : Ediciones Tercer Mundo, 1978.
218 p. ; 21 cm.
Includes bibliographical references.
1. Colombia—Politics and government—1946- . 2. Betancur, Belisario, 1923- . 3. Europe—Politics and government—1945- . I. Title.
F2278.G34 320.9'861'063 79-108077
79 MARC

Gómez Buendía, Hernando.
Alfonso López Michelsen, un examen crítico de su pensamiento y de su obra de gobierno / Hernando Gómez Buendía. — 1. ed. — Bogotá : Tercer Mundo, 1978.
367 p. ; 21 cm. — (Colección Tribuna libre)
Includes bibliographical references.
1. López Michelsen, Alfonso, 1913- . 2. Colombia—Politics and government—1946- . I. Title.
F2278.L63G65 986.1'063'0924 79-357536
79 MARC

La Lucha por el poder en 1978 : planteamientos de los candidatos y resultados electorales / editado por Ivan Botero-Páramo. — 1. ed. — Bogotá : Causa Común, 1978.
219 p. : ill. ; 21 cm. — (Colección Documentos políticos)
CONTENTS: Julio César Turbay.—Carlos Lleras Restrepo.—Hernando Agudelo Villa.—Carlos Holmes Trujillo.—Jaime García-Parra.—Belisario Betancur.—Jaime Piedrahita.—Socorro Ramírez.—Julio César Pernía.—Alvaro Valencia Tovar.
1. Elections—Colombia. 2. Political parties—Colombia. 3. Colombia—Politics and government—1946- . I. Botero-Páramo, Iván. II. Series: Colección Documentos políticos (Bogotá)
JL2818.L8 79-102685
79 MARC

Mandato claro : López, 74-78. — [s.l. : s.n., 1974] (Bogotá : Canal Ramírez-Antares)
[45] p. : ill. ; 23 cm.
Cover title: Presidente, 74-78, López.
1. Colombia—Presidents—Election. 2. López Michelsen, Alfonso, 1913- . 3. Colombia—Politics and government—1946- . I. Title: López, setenta y cuatro-setenta y ocho.
JL2841.M35 329'.023'861063 79-338583
79 MARC

Ortiz Márquez, Julio, 1910-
El hombre que fue un pueblo / Julio Ortiz Márquez. — 1. ed. — Bogotá : Carlos Valencia Editores, 1978.
263 p. ; 21 cm.
1. Gaitán, Jorge Eliécer, 1898-1948. 2. Colombia—Politics and government—1930-1946. 3. Colombia—Politics and government—1946- . I. Title.
F2277.G24O77 986.1'063'0924 79-361010
79 MARC

Romero, Manuel.
La política de los comunistas colombianos : texto de estudio de la táctica del Partido Comunista de Colombia / Manuel Romero, Yira Castro. — Bogotá : Ediciones Suramérica, 1978.
162 p. ; 20 cm.
Includes bibliographical references.
1. Partido Comunista de Colombia. 2. Colombia—Economic policy. 3. Colombia—Politics and government—1946- . I. Castro, Yira, joint author. II. Title.
JL2898.C6R65 79-101162
79 MARC

Sŏnggyun'gwan Taehakkyo, Seoul, Korea. Han'guk Sanŏp Yŏn'guso.
(Chung-Nammi ŭi chŏngch'i kyŏngje kyoyuk chedo e kwanhan yŏn'gu)
中南美의政治經濟教育制度에關한研究 : 콜롬비아와페루를中心으로. — [서울] : 成均館大學校附設韓國產業研究所, 1972.
x, 258 p. : maps ; 27 cm
Bibliography: p. 257-258.
1. Colombia—Politics and government—1946- 2. Colombia—Economic conditions—1970- 3. Peru—Politics and government—1968- 4. Peru—Economic conditions—1968- I. Title.
F2278.S59 1972 78-826771

Torres, Mauro.
La tragedia del hombre en Colombia / Mauro Torres. — Bogotá : Ediciones Tercer Mundo, [1978]
36 p. ; 21 cm. — (Temas ; 1)
1. Colombia—Social conditions. 2. Colombia—Politics and government—1946- . I. Title.
HN303.5.T64 78-105053
78 MARC

—POLITICS AND GOVERNMENT—1946- —ADDRESSES, ESSAYS, LECTURES

Araoz, Santiago.
Historia del Frente Nacional y otros ensayos / Santiago Araoz. — [Bogotá? : Editorial Presencia, 1977]
133 p. : ill. ; 17 cm.
Includes bibliographical references.
1. Political participation—Colombia—Addresses, essays, lectures. 2. Colombia—Politics and government—1946- —Addresses, essays, lectures. I. Title.
JL2883.A73 320.9'861'064 78-101313
79 MARC

Charry Samper, Héctor.
El liberalismo, entre la realidad y la utopía / Héctor Charry Samper. — 1. ed. — Bogotá : Ediciones Tercer Mundo, 1977.
xiii, 107 p. ; 21 cm. — (Colección Tribuna libre)
CONTENTS: El liberalismo, exigencias y desafíos.—La ciencia política y la acción de los partidos en Colombia.—El estado de emergencia.—Miguel Samper, el individualismo liberal.—América Latina y el encuentro de dos siglos, el XX y el XXI.
1. War and emergency powers—Colombia—Addresses, essays, lectures. 2. Colombia—Politics and government—1946- —Addresses, essays, lectures. 3. Samper, Miguel, 1825-1899—Addresses, essays, lectures. 4. Liberalism—Addresses, essays, lectures. I. Title.
JL2841.C45 320.9'861'063 79-344491
79 MARC

Colombia. Presidente, 1970-1974 (Pastrana Borrero)
Desarrollo y realización de una política social : mensaje del señor presidente de la República de Colombia doctor Misael Pastrana Borrero al Congreso Nacional, 20 de julio de 1973. — [Bogotá] : Tall. Gráf. del Banco de la República, [1973]
222 P. ; 28 cm.
1. Colombia—Politics and government—1946- —Addresses, essays, lectures. 2. Colombia—Economic policy—Addresses, essays, lectures. 3. Colombia—Social policy—Addresses, essays, lectures. I. Title.
J220.N327 1973 309.1'861'063 79-352917
79 MARC

Colombia. Presidente, 1970-1974 (Pastrana Borrero)
Mensaje del señor presidente de la República de Colombia doctor Misael Pastrana Borrero al Congreso Nacional, 20 de julio de 1971. [Bogotá] Tall. Gráf. del Banco de la República [1971]
117 p. 28 cm. LACAP72-2295
1. Colombia—Politics and government—1946- —Addresses, essays, lectures. I. Title.
J220.N327 1971 354'.861'035 72-335773
79[r79]rev MARC

Gerlein Echeverría, Roberto.
La estructura del poder en Colombia / Roberto Gerlein Echeverría. — 1. ed. — Bogotá : Ediciones Tercer Mundo, 1978.
170 p. ; 21 cm.
CONTENTS: La estructura del poder en Colombia y el conservatismo.—Tesis política.—La situación económica mundial.—Evaluación de los resultados de la Tercera Conferencia de Comercio y Desarrollo (UNCTAD III).—Un programa para el Capitolio.—Conflicto estudiantil.—Ponencia sobre el Proyecto de Ley no. 55.—Ponencia para el primer debate sobre el Proyecto de ley, Acto Legislativo no. 14 de 1975.—Ponencia para el primer debate sobre el Proyecto de Ley no 46 de 1975.—Proyecto de Acto Legislativo no. 65.
1. Colombia—Politics and government—1946- —Addresses, essays, lectures. 2. Power (Social sciences)—Addresses, essays, lectures. I. Title.
JL2815 1978.G47 320.9'861'063 78-107718
79 MARC

López Michelsen, Alfonso, 1913-
Administración López : documentos, discursos. — [Bogotá] : Banco de la República, [1976-
v. ; 23 cm.
Binder's title.
1. Colombia—Politics and government—1946- —Addresses, essays, lectures. I. Banco de la República, Bogotá. II. Title.
J220.N3 1976 354'.861'035 78-107302
79 MARC

López Michelsen, Alfonso, 1913-
El gobierno del mandato claro / Alfonso López Michelsen. — [s.l. : s.n., 1975-1977]
4 v. : ill. ; 25 cm. — ([Bogotá] : Talleres Editoriales de la Impr. Nacional de Colombia)
CONTENTS: t. 1. 7 de agosto de 1974, 6 de febrero de 1975.—t. 2. 7 de febrero de 1975, 7 de agosto de 1975.—t. 3. 7 de agosto de 1975, 6 de febrero de 1976.—t. 4. 7 de febrero a 6 de agosto de 1976.
Includes indexes.
1. Colombia—Politics and government—1946- —Addresses, essays, lectures. I. Title.
F2278.L622 320.9'861'063 77-480848
77[r79]rev MARC

Metodología y desarrollo en las ciencias sociales : efectos del crecimiento dependiente sobre la estructura social colombiana. — Bogotá : Centro de Estudios sobre Desarrollo Económico, Facultad de Economía, Universidad de los Andes, 1977.
614 p. : graphs ; 18 cm.
Includes bibliographies.
CONTENTS: Havens, A. E. y Parra Sandoval, R. Introducción.—Havens, A. E. Aspectos metrodológicos en el estudio del desarrollo.—Flinn, W. L. Notas sobre las familias de tres asentamientos urbanos. Los movimientos políticos en tres barrios bogotanos.—Havens, A. E. y Flinn, W. L. La tecnología de la revolución verde y el desarrollo de las comunicades.—Havens, A. E. La difusión de nuevas variedades de semillas y sus consecuencias.—Parra Sandoval, R. Empresa multinacional, urbanización y cambios en la estructura ocupacional.—Cepeda Ulloa, F. y Mitchell, C. La tendecia tecnocrática.—Havens, A. E. y Flinn, W. L. Determinantes estructurales de la selectividad de la migración de una región de economía en deteriodo.—Havens, A. E., Flinn, W. L. y Lastarria Conhill, S. La reforma agraria y el Frente Nacional.—Havens, A. E. y Flinn, W. L. La clase campesina y las pautas de emigracion.—Uricoechea, F. Formación y expansión del estado burocrático-patrimonial en Colombia y Brasil.
1. Colombia—Social conditions—1970- —Addresses, essays, lectures. 2. Colombia—Economic conditions—1970- —Addresses, essays, lectures. 3. Colombia—Politics and government—1946- —Addresses, essays, lectures. I. Bogota. Universidad de los Andes. Centro de Estudios sobre Desarrollo Económico.
HN303.5.M47 309.1'861'063 79-349270
79 MARC

Socarrás, José Francisco.
La crisis de los valores en Colombia / José Francisco Socarrás. — 1. ed. — Bogotá : Ediciones Tercer Mundo, 1978.
107 p. ; 21 cm. — (Colección El Dedo en la herida)
1. Colombia—Politics and government—1946- —Addresses, essays, lectures. 2. Colombia—Social conditions—1970- —Addresses, essays, lectures. I. Title.
F2278.S58 309.1'861'063 78-107975
79 MARC

—POLITICS AND GOVERNMENT—1946- —COLLECTED WORKS

Zuleta Ferrer, Juan.
La historia contra la pared : selección de ensayos y editoriales, El Colombiano, 1930-1978 / Juan Zuleta Ferrer. — Medellín : [Biblioteca Pública Piloto, 1978]
403 p. ; 23 cm. — (Colección Biblioteca Pública Piloto ; v. 1)
1. Colombia—Politics and government—1930-1946—Collected works. 2. Colombia—Politics and government—1946- —Collected works. 3. Antioquia, Colombia (Dept.)—Politics and government—Collected works. I. El Colombiano, Medellín, Colombia. II. Title. III. Series: Medellín, Colombia. Biblioteca Pública Piloto para la América Latina. Colección Biblioteca Pública Piloto ; v. 1.
F2277.Z84 320.9'861'063 79-100127
79 MARC

COLOMBIA (Continued)

—POLITICS AND GOVERNMENT—ADDRESSES, ESSAYS, LECTURES

Camacho Roldán, Salvador, 1827-1900.
Escritos sobre economía y política / Salvador Camacho Roldán ; selección, notas y prólogo de Jesús Antonio Bejarano. — ₍Bogotá₎ : Instituto Colombiano de Cultura, Subdirección de Comunicaciones Culturales, 1976.
268, ₍3₎ p. ; 21 cm. — (Biblioteca básica colombiana ; 13)
"Bibliografía sobre Salvador Camacho Roldán": p. ₍269₎
CONTENTS: Nuestro sistema tributario.—Desamortización de bienes de manos muertas.—Nuestro provenir industrial.—Consideraciones sobre fomento.—La agricultura en Colombia.—Nuestra situación industrial.—Valor trascendental del Río Magdalena.—La nueva administración.—Apelación al pueblo.—El pensamiento del Partido Liberal.—?Hasta dónde se puede llevar la guerra civil?—La unión liberal.—Manuela.—La Convención de Rionegro.
1. Colombia—Economic conditions—Addresses, essays, lectures. 2. Colombia—Politics and government—Addresses, essays, lectures. I. Title. II. Series.
HC197.C355 1976 330.9'861'063 77-551253
77₍r79₎rev MARC

Vanegas Montoya, Alfredo.
Juicios : ecología, demografía, historia, política, economía / Alfredo Vanegas Montoya. — Medellín : Vanser Editores, 1976.
101 p. ; 17 cm.
Bibliography: p. 77-78.
CONTENTS: Ecología y demografía.—Bolívar y el federalismo.—La salida de Lleras.—Colombia, país de congresos.—La consigna de Antioquia.—The big payoff.—Envigado ciudad con más de tres siglos.—La libertad agresiva de Fernando González.—Los Estados Unidos, 200 años de independencia de un microcosmos.
1. Environmental policy—Addresses, essays, lectures. 2. Colombia—Politics and government—Addresses, essays, lectures. 3. United States—Politics and government—Addresses, essays, lectures. I. Title.
HC79.E5V27 301.31 78-105054
79 MARC

—POPULATION

García N , Carlos.
Características de los inmigrantes en cinco ciudades colombianas : Bogotá, Bucaramanga, Manizales, Medellín y Popayán / Carlos García N. — Bogotá : Centro de Estudios sobre Desarrollo Económico, Universidad de los Andes, Facultad de Economía, 1970.
xiii, 177 leaves ; 28 cm. — (Serie P - CEDE, Universidad de los Andes ; no. 6)
Errata slip inserted.
Bibliography: leaves ₍176₎-177.
1. Colombia—Population. 2. Rural-urban migration—Colombia. 3. Migration, Internal—Colombia. I. Title. II. Series: Bogotá. Universidad de los Andes. Centro de Estudios sobre Desarrollo Económico. Serie P - Universidad de los Andes, Centro de Estudios sobre Desarrollo Económico ; no. 6.
HB2237.G37 79-366579
79 MARC

—PRESIDENTS—BIOGRAPHY

Espriella, Ramiro de la.
Acercamiento a Núñez / Ramiro de la Espriella. — 1. ed. — Bogotá : Tercer Mundo, 1978.
xiv, 122 p. ; 21 cm. — (Colección Tribuna libre)
Bibliography: p. 121-122.
1. Núñez, Rafael, Pres. Colombia, 1825-1894. 2. Colombia—Politics and government—1863-1885. 3. Colombia—Politics and government—1886-1903. 4. Colombia—Presidents—Biography. I. Title.
F2276.N862E86 986.1'061'0924 79-101886
79 MARC

Liévano Aguirre, Indalecio.
Rafael Núñez / Indalecio Liévano Aguirre. — 3. ed. — ₍Bogotá₎ : Instituto Colombiano de Cultura, Subdirección de Comunicaciones Culturales, División de Publicaciones, c1977.
521 p. ; 22 cm. — (Colección Historia viva ; 2) (Biblioteca colombiana de cultura)
Bibliography: p. 517-518.
1. Núñez, Rafael, Pres. Colombia, 1825-1894. 2. Colombia—Politics and government—1832-1886. 3. Colombia—Politics and government—1886-1903. 4. Colombia—Presidents—Biography.
F2276.N862L53 1977 986.1'061'0924 79-104037
79 MARC

Martínez Zulaica, Antonio.
Cólicos republicanos : patobiografía del general Santander / Antonio Martínez Zulaica. — 1. ed. — Tunja : Universidad Pedagógica y Tecnológica de Colombia, Secretaría de Investigaciones y Extensión Universitaria, 1978.
241 p. ; 25 cm. — (Ediciones "La Rana y el águila")
Bibliography: p. 237-241.
1. Santander, Francisco de Paula, Pres. New Granada, 1792-1840—Personality. 2. Colombia—Presidents—Biography. I. Title.
F2273.S26M37 986.1'05'0924 79-101617
00 MARC

Sanín Echeverri, Jaime.
Ospina supo esperar / Jaime Sanín Echeverri. — ₍s.l. : s.n.₎, 1978 (₍Bogotá₎ : Editorial Andes)
267 p., ₍2₎ leaves of plates : ports. ; 21 cm.
1. Ospina Pérez, Mariano, Pres. Colombia, 1891- . 2. Colombia—Presidents—Biography. 3. Businessmen—Colombia—Biography. I. Title.
F2278.O872S26 986.1'063'0924 79-101276
79 MARC

—PRESIDENTS—ELECTION

Mandato claro : López, 74-78. — ₍s.l. : s.n., 1974₎ (Bogotá : Canal Ramírez-Antares)
₍45₎ p. : ill. ; 23 cm.
Cover title: Presidente, 74-78, López.
1. Colombia—Presidents—Election. 2. López Michelsen, Alfonso, 1913- . 3. Colombia—Politics and government—1946- . I. Title: López, setenta y cuatro-setenta y ocho.
JL2841.M35 329'.023'861063 79-338583
79 MARC

—PRESIDENTS—HISTORY

Vázquez Carrizosa, Alfredo.
El poder presidencial en Colombia : la crisis permanente del derecho constitucional / Alfredo Vázquez Carrizosa. — 1. ed. — ₍Bogotá₎ : E. Dobry, Editor, 1979.
437 p. ; 21 cm. — (Colección Ciencia política)
Includes bibliographical references.
1. Executive power—Colombia—History. 2. Colombia—Presidents—History. I. Title.
JL2841.V39 354'.861'032 79-115978
79 MARC

—PUBLIC WORKS—PERIODICALS

Colombia. Ministerio de Obras Públicas y Transporte.
Informe al Congreso Nacional.
₍Bogotá, Ministerio de Obras Públicas y Transporte₎
v. 28 cm.
1. Colombia. Ministerio de Obras Públicas y Transporte—Periodicals. 2. Transportation—Colombia—Periodicals. 3. Colombia—Public works—Periodicals.
HE50.A3M56a 380.5'09861 78-646994
MARC-S

—RELATIONS (GENERAL) WITH FOREIGN COUNTRIES

Dunér, Bertil, 1942-
Cultural dependence in Latin America : the case of Colombian advanced education / by Bertil Dunér. — Rev. ed. — Uppsala : Uppsala University, Dept. of Peace and Conflict Research, 1977.
99 p. ; 21 cm. — (Report - Uppsala University, Department of Peace and Conflict Research ; no. 12) S***
Bibliography: p. 93-99.
ISBN 9150600508 : kr10.00
1. Education, Higher—Colombia. 2. Colombia—Relations (general) with foreign countries. I. Title. II. Series: Uppsala. Universitet. Fredsforskningsseminariet. Report ; no. 12.
LA568.D86 1977 378.861 79-312701
79 MARC

—RURAL CONDITIONS

Rural electrification and development : social and economic impact in Costa Rica and Colombia / John Saunders ... ₍et al.₎. — Boulder, Colo. : Westview Press, 1978.
xxiv, 180 p. : ill. ; 24 cm. — (A Westview replica edition)
Bibliography: p. 177-180.
ISBN 0-89158-274-6 : $17.00
1. Rural electrification—Costa Rica. 2. Rural electrification—Colombia. 3. Costa Rica—Rural conditions. 4. Colombia—Rural conditions. I. Saunders, John Van Dyke, 1930- .
HD9688.C82R87 78-19839
338.4'7'621393097286
78 MARC

—SOCIAL CONDITIONS

Betancur, Belisario, 1923-
Despiérta, Colombia! / Belisario Betancur. — 1. ed. — Bogotá : Ediciones Tercer Mundo, 1970.
329 p. ; 20 cm. — (Colección Documentos políticos)
1. Colombia—Social conditions. 2. Colombia—Economic conditions—1918-1970. 3. Colombia—Politics and government—1946- . I. Title.
HN303.5.B39 309.1'861 77-475389
77₍r79₎rev MARC

Escorcia, José.
Historia económica y social de Colombia, siglo XX / ₍José Escorcia₎. — ₍Bogotá₎ : Editorial Presencia, ₍1978₎
198 p. ; 17 cm.
Includes bibliographical references.
1. Colombia—Economic conditions. 2. Colombia—Social conditions.
HC197.E83 309.1'861'063 79-101128
79 MARC

Torres, Mauro.
La tragedia del hombre en Colombia / Mauro Torres. — Bogotá : Ediciones Tercer Mundo, ₍1978₎
36 p. ; 21 cm. — (Temas ; 1)
1. Colombia—Social conditions. 2. Colombia—Politics and government—1946- . I. Title.
HN303.5.T64 78-105053
78 MARC

—SOCIAL CONDITIONS—1970- —ADDRESSES, ESSAYS, LECTURES

Metodología y desarrollo en las ciencias sociales : efectos del crecimiento dependiente sobre la estructura social colombiana. — Bogotá : Centro de Estudios sobre Desarrollo Económico, Facultad de Economía, Universidad de los Andes, 1977.
614 p. : graphs ; 18 cm.
Includes bibliographies.
CONTENTS: Havens, A. E. y Parra Sandoval, R. Introducción.—Havens, A. E. Aspectos metrodológicos en el estudio del desarrollo.—Flinn, W. L. Notas sobre las familias de tres asentamientos urbanos. Los movimientos políticos en tres barrios bogotanos.—Havens, A. E. y Flinn, W. L. La tecnología de la revolución verde y el desarrollo de las comunicades.—Havens, A. E. La difusión de nuevas variedades de semillas y sus consecuencias.—Parra Sandoval, R. Empresa multinacional, urbanización y cambios en la estructura ocupacional.—Cepeda Ulloa, F. y Mitchell, C. La tendecia tecnocrática.—Havens, A. E. y Flinn, W. L. Determinantes estructurales de la selectividad de la migración de una región de economía en deteriodo.—Havens, A. E., Flinn, W. L. y Lastarria Conhill, S. La reforma agraria y el Frente Nacional.—Havens, A. E. y Flinn, W. L. La clase campesina y las pautas de emigracion.—Uricoechea, F. Formación y expansión del estado burocrático-patrimonial en Colombia y Brasil.
1. Colombia—Social conditions—1970- —Addresses, essays, lectures. 2. Colombia—Economic conditions—1970- —Addresses, essays, lectures. 3. Colombia—Politics and government—1946- —Addresses, essays, lectures. I. Bogota. Universidad de los Andes. Centro de Estudios sobre Desarrollo Económico.
HN303.5.M47 309.1'861'063 79-349270
79 MARC

Socarrás, José Francisco.
La crisis de los valores en Colombia / José Francisco Socarrás. — 1. ed. — Bogotá : Ediciones Tercer Mundo, 1978.
107 p. ; 21 cm. — (Colección El Dedo en la herida)
1. Colombia—Politics and government—1946- —Addresses, essays, lectures. 2. Colombia—Social conditions—1970- —Addresses, essays, lectures. I. Title.
F2278.S58 309.1'861'063 78-107975
79 MARC

—SOCIAL CONDITIONS—1970- —PICTORIAL WORKS

Myrman, Björn, 1942-
Bilder från Colombia / Björn Myrman, Maith Håkansson, Anders Kjellkvist. — Helsingborg : Fyra förläggare, 1978.
119, ₍1₎ p. : ill. ; 29 cm. — (Aktuell fotolitteratur ; 12) S78-29
Bibliography: p. ₍120₎
ISBN 918524628X
1. Colombia—Social conditions—1970- —Pictorial works. 2. Poor—Colombia—Pictorial works. I. Håkansson, Maith, 1937- joint author. II. Kjellkvist, Anders, 1948- joint author. III. Title.
HN303.5.M97 79-343756
79 MARC

—SOCIAL CONDITIONS—ADDRESSES, ESSAYS, LECTURES

Samper, Miguel, 1825-1899.
Selección de escritos / Miguel Samper ; selección y prólogo de Héctor Charry Samper y Santiago Samper Trainer. — Bogotá : Instituto Colombiano de Cultura, Subdirección de Comunicaciones Culturales, c1977.
340 p. ; 21 cm. — (Biblioteca básica colombiana ; 22)
Works by and about Miguel Samper: p. 337-338.
1. Colombia—Politics and government—1832-1886—Addresses, essays, lectures. 2. Colombia—Social conditions—Addresses, essays, lectures. 3. Colombia—Economic conditions—Addresses, essays, lectures. I. Series.
F2276.S224 1977 309.1'861'05 78-110224
79 MARC

—SOCIAL LIFE AND CUSTOMS

Abadía M , Guillermo.
Compendio general de folklore colombiano / Guillermo Abadía Morales. — 3. ed., corr. y aumentada. — ₍Bogotá₎ : Instituto Colombiano de Cultura, Subdirección de Comunicaciones Culturales, ₍1977₎
557 p., ₍25₎ leaves of plates : ill. ; 21 cm. — (Biblioteca básica colombiana ; 24)
Bibliography: p. 543-557.
1. Folk-lore—Colombia. 2. Colombia—Social life and customs. I. Title. II. Series.
GR133.C7A2 1977 78-110570
79 MARC

Viajeros colombianos por Colombia / prólogo de Gabriel Giraldo Jaramillo ; ilus. de la Comisión Corográfica y de Le Tour du Monde. — Bogotá : Fondo Cultural Cafetero, 1977.
xv, 276 p. : ill. (8 col.) ; 35 cm.
CONTENTS: Posada, E. En vacaciones.—Cordovez Moure, J. M. Un viaje de recreo.—Groot, J. M. Nos fuimos a Ubaque.—Camacho Roldán, S. Algo sobre tierra caliente.—Pombo, M. Bajando el Dagua.—Cordovez Moure, J. M. De Bogotá a Villeta.—Kastos, E. La tierra caliente.—Rivera y Garrido, L. Por Popayán y Pasto.—Posada, E. En el Agua-Nueva.—Samper, J. M. Paseo al Salto de Tequendama.—Posada Gutiérrez, J. El último viaje del Libertador.—Uribe Angel, M. Recuerdo de un viaje de Medellín a Bogotá.—Ancízar, M. De Los Santos a Bucaramanga.—Reyes, R. De Pasto al Amazonas.—Restrepo, E. De Bogotá a Villavicencio.
1. Colombia—Description and travel. 2. Colombia—Social life and customs. I. Giraldo Jaramillo, Gabriel. II. Fondo Cultural Cafetero.
F2263.V5 78-100322
79 MARC

COLOMBIA (Continued)

—SOCIAL POLICY

Colombia. Presidente, 1970-1974 (Pastrana Borrero)
Una política social para el desarrollo; mensaje del presidente de la República de Colombia Misael Pastrana Borrero al Congreso Nacional, 20 de julio de 1972. ₍Bogotá₎ Tall. Gráf. del Banco de la República ₍1972₎
229 p. 28 cm. LACAP73-0717
1. Colombia—Social policy. 2. Colombia—Economic policy. 3. Agriculture and state—Colombia. I. Title.
HN303.5.C64 1972 309.1'861 73-329880
79₍r79₎rev MARC

—SOCIAL POLICY—ADDRESSES, ESSAYS, LECTURES

Colombia. Presidente, 1970-1974 (Pastrana Borrero)
Desarrollo y realización de una política social : mensaje del señor presidente de la República de Colombia doctor Misael Pastrana Borrero al Congreso Nacional, 20 de julio de 1973. — ₍Bogotá₎ : Tall. Gráf. del Banco de la República, ₍1973₎
222 P. ; 28 cm.
1. Colombia—Politics and government—1946- —Addresses, essays, lectures. 2. Colombia—Economic policy—Addresses, essays, lectures. 3. Colombia—Social policy—Addresses, essays, lectures. I. Title.
J220.N327 1973 309.1'861'063 79-352917
79 MARC

—SOCIAL POLICY—CONGRESSES

Congreso Nacional de Economistas, 6th, Cali, Colombia, 1977.
Políticas y estrategias del desarrollo económico y social colombiano en un mundo cambiante : informe / VI Congreso Nacional de Economistas, Cali, 6 a 9 de diciembre, 1977. — ₍Bogotá₎ : Sociedad Colombiana de Economistas, ₍1977 or 1978₎
283 p. ; 23 cm.
1. Colombia—Economic policy—Congresses. 2. Colombia—Social policy—Congresses. I. Sociedad Colombiana de Economistas. II. Title.
HC197.C764 1977 79-101018
79 MARC

—STATISTICS—PERIODICALS

Germany (Federal Republic, 1949-). Statistisches Bundesamt.
Länderkurzberichte: Kolumbien.
Stuttgart, W. Kohlhammer.
v. 30 cm. (19 -76: Allgemeine Statistik des Auslandes) (1977- : Statistik des Auslandes)
Title varies slightly.
1. Colombia—Statistics—Periodicals. I. Title. II. Series: Statistik des Auslandes.
HA1015.G46 318.61 70-617708
MARC-S

—VICE-PRESIDENTS—BIOGRAPHY

Echavarría M , Guillermo.
Camilo C. Restrepo / Guillermo Echavarría M. — Medellín, Colombia : Servigráficas, 1979.
97 p., ₍4₎ leaves of plates : ill. ; 21 cm.
Includes bibliographical references.
1. Restrepo, Camilo C., 1864-1933. 2. Colombia—Vice-Presidents—Biography. 3. Antioquia, Colombia—Governors—Biography.
F2277.R385E24 986.1'062'0924 79-117308
79 MARC

COLOMBIA. ASAMBLEA CONSTITUCIONAL, 1978-1979

—ADDRESSES, ESSAYS, LECTURES

La Constituyente : itinerario de una propuesta / Jaime Betancur Cuartas, ₍compilador₎. — 1. ed. — Bogotá : Ediciones Tercer Mundo, 1978.
xv, 500 p. ; 23 cm.
Includes bibliographical references.
1. Colombia—Constitutional law—Addresses, essays, lectures. 2. Colombia. Asamblea Constitucional, 1978-1979—Addresses, essays, lectures. I. Betancur Cuartas, Jaime, 1930-
342'.861'029 78-108728
78 MARC

COLOMBIA. CONGRESO GENERAL, 1821

—CONGRESSES

Venezuela en Lima y en el Rosario de Cúcuta. — Caracas : Instituto Panamericano de Geografía e Historia, 1974.
179 p. : coat of arms ; 23 cm. — (Publicación - Instituto Panamericano de Geografía e Historia, Comisión de Historia, Comité Orígenes de la Emancipación ; no. 17)
Papers presented by members of the Academia Nacional de la Historia at the Congreso Internacional de Historia de América, 5th, Lima, 1971, and at the Congreso Gran Colombiano de Historia, Rosario de Cúcuta, Colombia, 1971.
Includes bibliographical references.
1. South America—History—Wars of Independence, 1806-1830—Congresses. 2. Colombia. Congreso General, 1821—Congresses. 3. Latin American federation—History—Congresses. I. Academia Nacional de la Historia, Caracas. II. Congreso Internacional de Historia de América, 5th, Lima, 1971. III. Congreso Gran Colombiano de Historia, Rosario de Cúcuta, Colombia, 1971. IV. Series: Pan American Institute of Geography and History. Commission on History. Comité de Orígenes de la Emancipación. Publicación ; no. 17.
F2235.V45 980'.02 78-104510
78 MARC

COLOMBIA. DIRECCIÓN GENERAL DEL PRESUPUESTO

Colombia. Laws, statutes, etc.
₍Decreto no. 77 de 1977 (enero 19)₎
Decreto no. 077 (19 de enero 1977). — Bogotá : Ministerio de Hacienda y Crédito Público, Dirección General del Presupuesto, ₍1977 or 1978₎
19 p. ; 28 cm. — (Publicaciones Minhacienda)
1. Colombia. Dirección General del Presupuesto. 2. Budget—Colombia. I. Series: Colombia. Ministerio de Hacienda y Crédito Público. Publicaciones Minhacienda.
343'.861'034 79-104699
79 MARC

COLOMBIA. EJÉRCITO

Echeverri U , Alvaro.
El poder y los militares : un análisis de los ejércitos del continente y Colombia / Alvaro Echeverri U. — 1. ed. — Bogotá : Fondo Editorial Suramérica, ₍1978₎
139 p. ; 23 cm. — (Colección Autores colombianos)
Bibliography: p. 136-139.
1. Latin America—Politics and government—1948- . 2. Latin America—Armed Forces—Political activity. 3. Sociology, Military—Latin America. 4. Colombia. Ejército. I. Title. II. Series.
F1414.2.E25 322'.5'098 79-109110
79 MARC

COLOMBIA. INSTITUTO GEOGRÁFICO AGUSTÍN CODAZZI

Colombia. Ministerio de Hacienda y Crédito Público.
Régimen interno de las dependencias del catastro nacional; Instituto Geográfico Militar y Catastral, Oficinas Seccionales de Catastro, 1943. Bogotá, Impr. Nacional, 1943.
2 p.l., 7-46 p. 16 cm.
At head of title: República de Colombia. Ministerio de Hacienda y Crédito Público.
1. Cadasters—Colombia. 2. Colombia. Instituto Geográfico Agustín Codazzi. I. Title.
HD518.C64 1943 43-14185
₍r79₎rev MARC

COLOMBIA. MINISTERIO DE DESARROLLO ECONÓMICO

El Ministerio de Desarrollo Económico y sus entidades adscritas y vinculadas. — ₍Bogotá₎ : Ministerio de Desarrollo Económico, Oficina de Divulgación, 1977-
v. ; 24 cm.
Includes legislation.
CONTENTS: t. 1. Organización del Ministerio. Superintendencia de Industria y Comercio. Superintendencia de Sociedades.
1. Colombia. Ministerio de Desarrollo Económico. 2. Colombia. Superintendencia de Industria y Comercio. 3. Colombia. Superintendencia de Sociedades. I. Colombia. Ministerio de Desarrollo Económico.
HC197.M56 78-107362
00 MARC

COLOMBIA. MINISTERIO DE OBRAS PÚBLICAS Y TRANSPORTE

—PERIODICALS

Colombia. Ministerio de Obras Públicas y Transporte.
Informe al Congreso Nacional.
₍Bogotá, Ministerio de Obras Públicas y Transporte₎
v. 28 cm.
1. Colombia. Ministerio de Obras Públicas y Transporte—Periodicals. 2. Transportation—Colombia—Periodicals. 3. Colombia—Public works—Periodicals.
HE50.A3M56a 380.5'09861 78-646994
MARC-S

COLOMBIA. SUPERINTENDENCIA DE INDUSTRIA Y COMERCIO

El Ministerio de Desarrollo Económico y sus entidades adscritas y vinculadas. — ₍Bogotá₎ : Ministerio de Desarrollo Económico, Oficina de Divulgación, 1977-
v. ; 24 cm.
Includes legislation.
CONTENTS: t. 1. Organización del Ministerio. Superintendencia de Industria y Comercio. Superintendencia de Sociedades.
1. Colombia. Ministerio de Desarrollo Económico. 2. Colombia. Superintendencia de Industria y Comercio. 3. Colombia. Superintendencia de Sociedades. I. Colombia. Ministerio de Desarrollo Económico.
HC197.M56 78-107362
00 MARC

COLOMBIA. SUPERINTENDENCIA DE SOCIEDADES

El Ministerio de Desarrollo Económico y sus entidades adscritas y vinculadas. — ₍Bogotá₎ : Ministerio de Desarrollo Económico, Oficina de Divulgación, 1977-
v. ; 24 cm.
Includes legislation.
CONTENTS: t. 1. Organización del Ministerio. Superintendencia de Industria y Comercio. Superintendencia de Sociedades.
1. Colombia. Ministerio de Desarrollo Económico. 2. Colombia. Superintendencia de Industria y Comercio. 3. Colombia. Superintendencia de Sociedades. I. Colombia. Ministerio de Desarrollo Económico.
HC197.M56 78-107362
00 MARC

COLOMBIA. UNIVERSIDAD. BOGOTA

—BIBLIOGRAPHY—CATALOGS

Colombia. Universidad. Bogotá. Biblioteca Central.
Catálogo de las publicaciones de la Universidad Nacional de Colombia : suplemento / por Gladys M. Gacharná G. — Bogotá : Universidad Nacional de Colombia, Departamento de Bibliotecas, Biblioteca Central, 1977-
v. ; 27 cm.
1. Colombia. Universidad. Bogota—Bibliography—Catalogs. 2. Colombia. Universidad. Bogota. Biblioteca Central—Catalogs. I. Gacharná García, Gladys M. II. Title.
Z5816.C64C65 1976, suppl. 79-101417
₍LB2391.C5₎ 79 MARC

COLOMBIA. UNIVERSIDAD. BOGOTA. BIBLIOTECA CENTRAL

—CATALOGS

Colombia. Universidad. Bogotá. Biblioteca Central.
Catálogo de las publicaciones de la Universidad Nacional de Colombia : suplemento / por Gladys M. Gacharná G. — Bogotá : Universidad Nacional de Colombia, Departamento de Bibliotecas, Biblioteca Central, 1977-
v. ; 27 cm.
1. Colombia. Universidad. Bogota—Bibliography—Catalogs. 2. Colombia. Universidad. Bogota. Biblioteca Central—Catalogs. I. Gacharná García, Gladys M. II. Title.
Z5816.C64C65 1976, suppl. 79-101417
₍LB2391.C5₎ 79 MARC

COLOMBIAN AUTHORS see Authors, Colombian

COLOMBIAN DRAMA

Buenaventura, Enrique.
Teatro / Enrique Buenaventura. — ₍Bogota₎ : Instituto Colombiano de Cultura, Subdirección de Comunicaciones Culturales, División de Publicaciones, c1977.
270 p. ; 17 cm. — (Biblioteca colombiana de cultura : Colección popular ; 21)
Bibliography: p. 259-261.
CONTENTS: Los papeles del infierno.—El menú.—La orgía.—Soldados.—A la diestra de Dios Padre.
$10.00
PQ8180.12.U38A19 1977 79-103539
79 MARC

López Pérez, Heriberto.
Teatro / Heriberto López Pérez. — ₍1. ed.₎. — ₍Bogotá : ECOE Ediciones, 1977₎
xiv, 152 p. ; 17 cm.
Cover title.
CONTENTS: La última batalla de los vencidos en un día de terror y de miseria.—La leyenda de la calle de las mujeres.—Conversación en el parque.—Feliz entierro, canto de Juancito Verdades y la negra Flora.
PQ8180.22.O64T4 862 78-108094
79 MARC

COLOMBIAN ESSAYS

Revista de las Indias, 1936-1950 / selección, Alvaro Miranda. — ₍Bogotá₎ : Instituto Colombiano de Cultura, Subdirección de Comunicaciones Culturales, División de Publicaciones, c1978.
469 p. ; 21 cm. — (Biblioteca colombiana de cultura) (Colección Autores nacionales) (Serie Las Revistas ; 28)
1. Colombian essays. 2. Colombia—Addresses, essays, lectures. 3. Revista de las Indias. I. Miranda, Alvaro.
PQ8176.E8R4 868 79-104876
79 MARC

COLOMBIAN FICTION

Under this heading, without subdivision, are entered individual works and collections by one author

Alba, Laureano.
El barrio de las flores : novela / Laureano Alba ; ₍portada, Departamento de Arte de Editorial Andes₎. — Bogotá : Plaza y Janés, c1978.
126 p. ; 19 cm. — (Rotativa)
I. Title.
PQ8180.1.L28B3 79-116718
79 MARC

Alonso Villegas, Oscar.
El pecado de los inocentes : novela / Oscar Alonso Villegas. — Medellín, Colombia : Editorial Lealon, 1979.
243 p. ; 17 cm.
I. Title.
PQ8180.1.L6P4 79-116838
79 MARC

Arévalo, Milcíades, 1948-
A la orilla del trópico : relatos / Milcíades Arévalo. — 1. ed. — Bogotá : Ediciones Puesto de Combate, 1978.
118 p. ; 19 cm. — (Colección Línea de fuego)
CONTENTS: La morada radiante.—Otra parte de la soledad.—Trampa para espantar el miedo.—La llave de los secretos.—El fabricante de la risa.—Iohana, Iohana.—El reino circular.—El pan de Pablo.—Entre dos fuegos.—Erasmo Coronel.—El último rebelde.—De profundis domine.—Memorias del día.—Un ave de otro corral.—La orilla del trópico.
I. Title.
PQ8180.1.R4A62 863 79-101277
79 MARC

Atienza y Bermejo, Angel Cayo.
Además ... Urabá de los katíos : novela misional / Angel Cayo Atienza y Bermejo. — Bogotá : Padres Carmelitas, 1977.
239 p. ; 20 cm.
I. Title.
PQ8180.1.T54A66 863 79-108156
79 MARC

Bastidas Padilla, Carlos.
Las raíces de la ira / Carlos Bastidas Padilla. — 2. ed. — ₍Bogotá₎ : Instituto Colombiano de Cultura, Subdirección de Comunicaciones Culturales, División de Publicaciones, c1978.
135 p. ; 17 cm. — (Biblioteca colombiana de cultura : Colección popular ; 28)
I. Title.
PQ8180.12.A76R3 1978 863 79-103925
79 MARC

Bedoya Martínez, Jaime, 1931-
El pijao rebelde : novela / Jaime Bedoya Martínez. — ₍Manizales, Colombia₎ : Editorial Quin-Gráficas, 1978.
428 p., ₍14₎ leaves of plates : ill. ; 22 cm.
I. Title.
PQ8180.12.E33P5 79-116185
79 MARC

Caicedo, Andrés.
Berenice ; El atravesado ; Maternidad ; El tiempo de la ciénaga / Andrés Caicedo Estela. — Bogotá : Plaza y Janés, c1978.
136 p. ; 19 cm. — (Rotativa : Novela)
I. Title.
PQ8180.13.A35B47 863 79-101284
79 MARC

Caicedo, Andrés.
Qué viva la música : novela / Andrés Caicedo. — ₍Bogotá₎ : Instituto Colombiano de Cultura, Subdirección de Comunicaciones Culturales, División de Publicaciones, c1977.
198 p. ; 17 cm. — (Biblioteca colombiana de cultura : Colección popular ; 19)
Discography: p. 189-191.
I. Title.
PQ8180.13.A35Q4 79-103733
79 MARC

Castellanos, Dora.
Año dosmil contigo / Dora Castellanos. — ₍s.l. : s.n., 1977₎ (Caracas : Editorial Arte)
168 p., ₍1₎ leaf of plates : ill. ; 23 cm.
I. Title.
PQ8180.13.A8A8 861 78-108138
79 MARC

Collazos, Oscar, 1942-
Memoria compartida : novela / Oscar Collazos. — Caracas : Monte Avila Editores, ₍1978₎
149 p. ; 18 cm. — (Colección Continentes)
Bs16.00
I. Title.
PQ8180.13.O4M45 863 78-111228
79 MARC

Crespo Paniagua, Renato.
La isla de José Miguel : novela / Renato Crespo Paniagua. — Cochabamba ₍Bolivia₎ : Editorial Los Amigos del libro, 1978.
198 p. ; 21 cm.
I. Title.
PQ8180.13.R43I8 863 79-112012
79 MARC

Dulcey, Arcadio.
Mi lucha / Arcadio Dulcey. — Bogotá : Editorial Kelly, 1978.
287 p. ; 22 cm.
I. Title.
PQ8180.14.U5M5 863 79-101165
79 MARC

Duque López, Alberto, 1943-
Mi revólver es más largo que el tuyo / Alberto Duque López. — ₍Bogotá₎ : Instituto Colombiano de Cultura, Subdirección de Comunicaciones Culturales, División de Publicaciones, ₍1977₎
264 p. : ports. ; 17 cm. — (Biblioteca colombiana de cultura : Colección popular ; 28)
I. Title.
PQ8180.14.U63M5 863 79-103922
79 MARC

Espriella Arango, Alvaro de la.
Los cangrejos no caminan en la nieve / Alvaro de la Espriella Arango. — 1. ed. — Bogotá : Ediciones Tercer Mundo, 1977.
197 p. ; 17 cm. — (Colección Tribuna libre : Serie menor ; 13)
I. Title.
PQ8180.15.S6C3 863 78-104829
79 MARC

García, Eligio.
Para matar el tiempo / Eligio García. — 1. ed. — Bogotá : C. Valencia Editores, 1978.
160 p. ; 21 cm.
ISBN 8482770020
I. Title.
PQ8180.17.A68P3 79-115983
79 MARC

García Márquez, Gabriel, 1928-
Cuatro cuentos / Gabriel García Márquez. — México : Comunidad Latinoamericana de Escritores, c1974.
124 p. ; 15 cm. — (Minilibros CLE ; 1)
CONTENTS: Monólogo de Isabel viendo llover en Macondo.—En este pueblo no hay ladrones.—Los funerales de Mamá Grande.—Un hombre muy viejo con unas alas enormes.
I. Title. II. Series: Comunidad Latinoamericana de Escritores. Minilibros CLE ; 1.
PQ8180.17.A73C8 79-112682
79 MARC

García Márquez, Gabriel, 1928-
La increíble y triste historia de la cándida Eréndira y su abuela desalmada / Gabriel García Márquez. — 1. ed. — Barcelona : Bruguera, 1978.
152 p. ; 18 cm. — (Libro amigo ; 563) Sp***
CONTENTS: Un señor muy viejo con unas alas enormes.—El mar del tiempo perdido.—El ahogado más hermoso del mundo.—Muerte constante más allá del amor.—El último viaje del buque fantasma.—Blacamán el bueno vendedor de milagros.—La increíble y triste historia de la cándida Eréndira y de su abuela desalmada.
ISBN 8402056083 : 125ptas
I. Title.
PQ8180.17.A73I5 1978 863 78-385798
78 MARC

Girón G , Silvio.
Ningunaotraparte : cuentos / Silvio Girón. — ₍s.l. : s.n., 1978₎ (Pereira ₍Colombia₎ : Editorial Gráficas Olímpica)
124 p. : port. ; 17 cm.
CONTENTS: La hermana.—Ningunaotraparte.—La venganza.—La grandota.—Pasos en el infierno.—Pesadilla.—Rebelde con causa.—Angel sodomita.—Deshaciendo los pasos.—La petite puta.
I. Title.
PQ8180.17.I7N5 863 79-102402
79 MARC

Gómez Ortiz, Hernán.
Las diez mil tarántulas / por Hernán Gómez Ortiz. — ₍Bucaramanga?₎ : Ediciones Beneficencia de Santander, ₍1972₎
165 p. ; 17 cm.
CONTENTS: Un avión perdido en el espacio.—Las diez mil tarántulas.—La grieta.—Nuestro duende personal.—Un centavo en la acera.—El policía y las zagalejas.—La bombita.—Un hombre sin pasado.—La trespisos.—La flauta y la cuerda.—Una semilla de mostaza.—El primer paciente.—Tres botijas de barro.
I. Title.
PQ8180.17.O39D5 863 78-389706
78 MARC

Gutiérrez Zambrano, Efraín.
La falsa y violenta historia del verdadero y pacífico general Bululú de la Cruz Olivares / Efraín Gutiérrez Zambrano. — Bogotá : ₍s.n.₎, 1978.
113 p. ; 17 cm.
I. Bululú. II. Title.
PQ8180.17.U86F34 863 79-101610
79 MARC

Hincapié Velásquez, Julio A
El poder del corazón / Julio A. Hincapié Velásquez. — 1. ed. — Bogotá : Ediciones Tercer Mundo, 1978.
142 p. ; 21 cm.
I. Title.
PQ8180.18.I43P6 863 79-108078
79 MARC

Hinestrosa Isaza, Guillermo.
El día que llegó la maldad : obra en dos partes / Guillermo Hinestrosa Isaza. — ₍s.l. : s.n., 1978₎ (₍s.l.₎ : Impresos Super y Cosmográficas)
355 p., ₍1₎ leaf of plates : port. ; 17 cm.
I. Title.
PQ8180.18.I45D5 863 79-101611
79 MARC

Holte-Castelo, Jorge Daniel de, 1928-
Aventuras de un neurasténico / por Jorge Daniel de Holte-Castelo. — ₍1. ed.₎. — ₍Madrid₎ : Sedmay : ₍distribuidora, Maydi, 1977₎
179 p. ; 21 cm. Sp***
ISBN 8473802268
I. Title.
PQ8180.18.O45A95 868 79-342832
79 MARC

Hoyos, Hernán.
Aventuras de un impotente : novela / Hernán Hoyos. — ₍Cali, Colombia : Ediciones Exclusivas, 19₎76.
209 p. ; 17 cm. — (Serie picaresca)
I. Title. II. Series.
PQ8180.18.O8A94 79-115927
79 MARC

Hoyos, Hernán.
Inocentes y perversos : novela / Hernán Hoyos. — 1. ed. — Cali, Colombia : Hoyos, 1979.
171 p. ; 17 cm.
$50.00
I. Title.
PQ8180.18.O8I5 79-116593
79 MARC

Lobo Serna, Ciro Alfonso, 1933-
La ciudad y la culpa / Ciro Alfonso Lobo Serna. — ₍s.l. : s.n., 1978₎ (Bogotá : Tip. Sterner)
190 p. ; 19 cm.
I. Title.
PQ8180.22.O25C5 863 79-109218
79 MARC

Mercado, José Ramón.
Perros de presa / José Ramón Mercado. — Bogotá : Ediciones Nueva Narrativa Colombiana, c1978.
111 p. ; 17 cm.
CONTENTS: Perros de presa.—El bobo de la yuca.—Toro domado.—El extraño visitante.—Un drama de amor y de dolor.—Más intenso que la propia vida.—Tiempo de fiesta.—El día menos pensado.—La noche del nocaut.—Mai broder.—El cerco.
I. Title.
PQ8180.23.E7P47 863 79-102243
79 MARC

Mutis, Alvaro.
La mansión de Araucaíma / Alvaro Mutis. — 1. ed. — Barcelona : Seix Barral, 1978.
139 p. ; 20 cm. — (Biblioteca breve ; 425 : Relatos) Sp***
CONTENTS: La mansión de Araucaíma.— Antes de que cante el gallo.—El último rostro (fragmento).—Sharaya.—La muerte del estratega.
I. Title.
PQ8180.23.U8M3 1978 863 78-395730
79 MARC

Nossa Monroy, Carlos.
Tizones del dolor : de los ardidos leños del dolor quedan tizones que no se extinguen y cuyas candelas arden perennemente / Carlos Nossa Monroy. — ₍s.l. : s.n.₎, 1977 (Bogotá : Editorial Presencia)
124 p. ; 17 cm. — (Leyendas y cuentos de Colombia)
CONTENTS: María del Pilar.—Combate singular.—Tiba.—Historia amorosa de un chigüiro.—Aviso inopinado.—La sombra del destino.—El médico guerrillero.—Cárcel y espanto.—La frontera del odio.—Hija sin nombre.
I. Title. II. Series.
PQ8180.24.O8T5 863 78-105021
78 MARC

Ortiz Betancur, Dario.
El Arenal / Dario Ortiz Betancur. — 1. ed. — ₍Medellín? : L. H. Donado Tovar, Editor, 1979.
234 p. ; 17 cm.
I. Title.
PQ8180.25.R85A89 79-116856
79 MARC

COLOMBIAN FICTION (Continued)

Pardo, Jorge Eliécer.
El jardín de las Hartmann : novela / Jorge Eliécer Pardo. — Bogotá : Plaza y Janés Editores, c1978.
115 p. ; 19 cm. — (Rotativa)
I. Title.
PQ8180.26.A66J3 863 79-101533
79 MARC

Pérez Botero, Luis, 1922-
Ulises en Sabanilla y otros cuentos / Luis Pérez Botero. — Madrid : E.M.E.S.A., c1978.
145 p. : port. ; 18 cm. — (Novelas y cuentos ; 226 : Sección Literatura : Serie Literatura hispanoamericana : Narraciones, siglo XX) Sp***
ISBN 8426572278
I. Title.
PQ8180.26.E713U4 863 79-377435
79 MARC

Pérez Medina, Julián.
Cuentos del ocio / Julián Pérez Medina. — 1. ed. — Medellín : Publicaciones Técnicas, 1978.
68 p. : ill. ; 20 cm.
CONTENTS: Juan, el minero.—Noventa minutos de insensibilidad.—Don Pastor y la culebra.—Historias de Pedro Peludo.—Examen de admisión.—El Hospital de las Glorias.—Los tomines de doña María.—El duelo de Villa Roca. —Isaías en su segunda vida.—Mano Clemente.
I. Title.
PQ8180.26.E715C78 863 79-101030
79 MARC

Pinzón Prieto, José María.
Pedro Pestoso / José María Pinzón P. — 1. ed. — [s.l. : s.n., between 1955 and 1979] [s.l. : Impr. Meridiano)
509 p. ; 17 cm.
I. Title.
PQ8180.26.I59P4 79-116204
79 MARC

Rocha Ochoa, Cesáreo.
Tierra buena : cuentos y relatos / Cesáreo Rocha Ochoa. — [Ibagué, Colombia : Fondo de Cultura de la Beneficencia del Tolima, 1976]
133 p. ; 21 cm.
I. Title.
PQ8180.28.O27T5 863 77-154566
78 MARC

Ruiz Gómez, Darío.
Hojas en el patio / Darío Ruiz Gómez. — [Bogotá] : Instituto Colombiano de Cultura, Subdirección de Comunicaciones Culturales, División de Publicaciones, c1978.
399 p. ; 17 cm. — (Biblioteca colombiana de cultura : Colección popular ; 26)
$20.00
I. Title.
PQ8180.28.U53H6 863 79-103354
79 MARC

Sánchez, Héctor.
Se acabó la casa : relatos / Héctor Sánchez. — [Bogotá] : Instituto Colombiano de Cultura, Subdirección de Comunicaciones Culturales, División de Publicaciones, c1978.
155 p. ; 21 cm. — (Colección Autores nacionales ; 31) (Biblioteca colombiana de cultura)
CONTENTS: La mala renta.—El basural.—Balancín.—Apuesta a gotas.—Las ofrendas.—Las treguas.—Al contacto de los caminos.—El relevo.—Los días sin vuelta.
$80.00
I. Title.
PQ8180.29.A45S4 79-103745
79 MARC

Sánchez, Héctor.
El tejemaneje : novela / Héctor Sánchez ; [portada, Departamento de Arte, Plaza y Janés]. — Bogotá : Plaza y Janés, c1979.
435 p. ; 19 cm. — (Narrativa colombiana) (Novelistas del día)
I. Title. II. Series: Narrativa colombiana (Bogotá.
PQ8180.29.A45T4 79-116133
79 MARC

Sánchez Juliao, David.
El arca de Noé / David Sánchez Juliao. — 1. ed. — Bogotá : Ediciones Tercer Mundo, 1976.
153 p. ; 17 cm.
I. Title.
PQ8180.29.A46A88 868 79-365946
79 MARC

Sánchez Juliao, David.
Nadie es profeta en Lorica / David Sánchez Juliao. — 1. ed. — [Bogotá] : Ediciones Universidades Simón Bolívar, Libre de Pereira y Medellín, 1979.
73 p., [1] leaf of plates : ill. ; 17 cm. — (Colección Universidad y pueblo ; 34)
"Trabajo de creación presentado a la Universidad Simón Bolívar de Barranquilla para recibir el título de doctor honoris causa, Bogotá, diciembre de 1978."
CONTENTS: Nadie es profeta en Lorica.—La fiesta de Joe.—La rebelión de las iguanas.—El rincón de la mujer de blanco.—Acab Nomar y las serpientes. —Pedrito (modelo para armar).—El hombre de la lengua larga.—Fulanito de Brigard, ha muerto.
I. Title.
PQ8180.29.A46N3 79-114548
79 MARC

Santa, Eduardo, 1928-
Los espejos del tiempo / Eduardo Santa ; [ilus., Jorge Elías Triana]. — [Bogotá?] : Ediciones Príncipe, [1978]
177 p. ; 17 cm.
CONTENTS: El viento de Colbo.—La muerte del viejo Oñote.—Esperando al capitán.—Los que nunca volvieron.—Vendedores de sangre.—El defensor.—La noche también es roja.—La caja de embolar.—Una hora de su vida.—El enigma de los espejos.—Muerte por imprevisión.—Los mausoleos familiares.
I. Title.
PQ8179.S385E8 79-114379
79 MARC

Santa, Eduardo, 1928-
El pastor y las estrellas : una filosofía de la vida / Eduardo Santa. — 1. ed. — Bogotá : Ediciones Príncipe, 1977.
169 p. : ill. ; 22 cm.
I. Title.
PQ8179.S385P3 863 78-105129
78 MARC

Santamaría, Germán, 1950-
Morir último / Germán Santamaría. — 1. ed. — Bogotá : C. Valencia Editores, 1978.
117 p. ; 21 cm.
CONTENTS: Una mujer para la segunda madrugada.—Amorcito Case.—Las nubes del porvenir.—Morir último.—Una luz en la noche.—Tu sangre, muchacho, tu sangre.
ISBN 8482770012
I. Title.
PQ8180.29.A476M6 863 79-101559
79 MARC

Solís y Valenzuela, Pedro de.
El desierto prodigioso y prodigio del desierto / Pedro de Solís y Valenzuela ; ed. de Rubén Páez Patiño ; introd., estudios y notas de Jorge Páramo Pomareda, Manuel Briceño Jáuregui, Rubén Páez Patiño. — Bogotá : [Instituto Caro y Cuervo], 1977-
v. : ill. ; 23 cm. — (Publicaciones del Instituto Caro y Cuervo ; 45)
Includes bibliographical references.
I. Páez Patiño, Rubén. II. Páramo Pomareda, Jorge. III. Briceño Jáuregui, Manuel. IV. Title. V. Series: Colombia. Instituto Caro y Cuervo. Publicaciones ; 45.
PQ8179.S59D4 79-113846
79 MARC

Soto Aparicio, Fernando, 1933-
Los funerales de América : novela / Fernando Soto Aparicio. — [Bogotá?] : Ediciones Jaguar, c1978.
376 p. ; 18 cm.
I. Title.
PQ8179.S68F8 79-101266
79 MARC

Suescún, Nicolás.
El retorno a casa : relatos / Nicolás Suescún. — [Bogotá] : Instituto Colombiano de Cultura, Subdirección de Comunicaciones Culturales, División de Publicaciones, c1978.
165 p. ; 17 cm. — (Biblioteca colombiana de cultura : Colección popular ; 24)
CONTENTS: La otra.—Retrato de novios.—En mi pieza.—Un nuevo día.—Un profeta.—La vecina.—Idilio a dos voces.—El amigo de Mario.—Depronto uno despierta.—El retorno a casa.
$20.00
I. Title.
PQ8180.29.U4R4 863 79-103347
79 MARC

Suescún, Nicolás.
El último escalón / Nicolás Suescún. — Bogotá : Editorial Pluma, 1977.
107 p. ; 17 cm.
I. Title.
PQ8180.29.U4U55 863 78-105015
78 MARC

Tafur, Humberto.
La última noticia : novela / Humberto Tafur Charry. — Bogotá : Plaza y Janes, c1979.
117 p. ; 19 cm. — (Rota tiva)
I. Title. II. Series.
PQ8180.3.A34U4 79-115853
79 MARC

Vargas Vila, José María, 1860-1933.
Laureles rojos / J. M. Vargas Vila. — [Medellín] : Beta, [1976 or 1977]
241 p. ; 17 cm. — (His Obras completas ; v. 49)
I. Title.
PQ8179.V3 1976 vol. 49 863 s 78-396484
79 MARC

—20TH CENTURY—HISTORY AND CRITICISM

Williams, Raymond L
La novela colombiana contemporánea / texto, Raimond L. Williams. — Bogotá : Plaza y Janés Editores Colombia, c1976.
93 p. : ports. ; 17 cm. — (Enciclopedia popular ilustrada ; 10)
Bibliography: p. 90-93.
1. Colombian fiction—20th century—History and criticism. I. Title.
PQ8172.W5 863 78-344589
78[r79]rev MARC

—HISTORY AND CRITICISM

Menton, Seymour.
La novela colombiana : planetas y satélites / de Seymour Menton. — Bogotá : Plaza y Janés, c1978.
394 p. ; 19 cm. — (Crítica literaria)
Includes bibliographical references.
1. Colombian fiction—History and criticism. I. Title.
PQ8172.M4 79-101557
79 MARC

—TRANSLATIONS INTO ENGLISH

García Márquez, Gabriel, 1928-
Innocent Eréndira, and other stories / Gabriel García Márquez ; translated from the Spanish by Gregory Rabassa. — 1st ed. — New York : Harper & Row, c1978.
183 p. ; 22 cm.
CONTENTS: The incredible and sad tale of innocent Eréndira and her heartless grandmother.—The sea of lost time.—Death constant beyond love.—The third resignation.—The other side of death.—Eva is inside her cat.—Dialogue with the mirror.—Bitterness for three sleepwalkers.—Eyes of a blue dog. —The woman who came at six o'clock.—Someone has been disarranging these roses.—The night of the curlews.
ISBN 0-06-011416-9 : $8.95
I. Title.
PZ4.G2164 In 1978 863 74-15873
[PQ8180.17.A73] 78 MARC

COLOMBIAN HUMOROUS POETRY see
Humorous poetry, Colombian

COLOMBIAN LITERATURE

Achury Valenzuela, Darío.
Palabras con azar : glosas / Darío Achury Valenzuela. — Bogotá : Instituto Caro y Cuervo, 1975.
185 p. ; 21 cm. — (Serie La Granada entreabierta ; 8)
Includes bibliographical references.
CONTENTS: Azorín. Duelos y quebrantos.—Una incógnita del "Quijote": Duelos y quebrantos".—?"Recamado viso" o "recamado biso"?—Por el mundo de la geofagia.—Ronda alrededor del anglicismo "serendipity".
I. Title. II. Series.
PQ8179.A4P34 868 79-354378
79 MARC

Cepeda Samudio, Alvaro.
Antología / Alvaro Cepeda Samudio ; selección y prólogo de Daniel Samper Pizano. — [Bogotá] : Instituto Colombiano de Cultura, Subdirección de Comunicaciones Culturales, División de Publicaciones, c1977.
482 p. ; 17 cm. — (Biblioteca colombiana de cultura : Colección popular ; 22)
I. Samper Pizano, Daniel.
PQ8180.13.E6A6 1977 868 79-103868
79 MARC

Chávez Vásquez, Gloria, 1949-
Las termitas : fábula, cuentos y ensayos / Gloria Chávez Vásquez. — Bogotá : Unión de Escritores Colombianos, [1978]
94 p. ; 20 cm.
"1."
CONTENTS: Sor Orfelina.—Tatuza.—La letra más linda del alfabeto.—Regálame un canario.—La leyenda americana de la creación del cerebro.—Las termitas.—Mr. Walter.—Letanía inconclusa.—Diario de una neurosis.
I. Title.
PQ8180.13.H36T47 863 79-349797
79 MARC

Collazos, Oscar, 1942-
Textos al margen / Oscar Collazos. — [Bogotá] : Instituto Colombiano de Cultura, Subdirección de Comunicaciones Culturales, División de Publicaciones, c1978.
189 p. ; 22 cm. — (Colección Autores nacionales ; 26) (Biblioteca colombiana de cultura)
Includes bibliographical references.
$80.00
I. Title.
PQ8180.13.O4T4 79-103430
79 MARC

Hoyos, Hernán.
La película / Hernán Hoyos. — [Cali, Colombia : Ediciones Exclusivas, 19]76.
153 p. ; 17 cm. — (Serie social)
I. Title.
PQ8180.18.O8P4 79-115664
79 MARC

Jácome, J J
Fuego de septiembre : jornada en el basurero histórico / J. J. Jácome. — Bogotá : [s.n.], 1978.
132 p. : ill. ; 19 cm.
I. Title.
PQ8180.2.A3F8 79-117730
00 MARC

COLOMBIAN LITERATURE (Continued)

Quessep, Giovanni.
Libro del encantado / Giovanni Quessep. — ₍Bogotá₎ : Instituto Colombiano de Cultura, Subdirección de Comunicaciones Culturales, División de Publicaciones, c1978.
112 p. ; 21 cm. — (Colección Autores nacionales ; 33) (Biblioteca colombiana de cultura)
I. Title.
PQ8180.27.U3L5 79-103935
79 MARC

Sanín Cano, Baldomero, 1861-1957.
Escritos / Baldomero Sanín Cano ; selección y prólogo, J. G. Cobo Borda. — ₍Bogotá₎ : Instituto Colombiano de Cultura, Subdirección de Comunicaciones Culturales, ₍1977₎
789 p. : port. ; 21 cm. — (Biblioteca básica colombiana ; 23)
"Sanín Cano, el oficio de lector, por J. G. Cobo Borda" (p. 17-33) includes bibliographical references.
I. Cobo Borda, J. G., 1948- . II. Series.
PQ8179.S37A6 1977 868 78-111716
79 MARC

Tejada, Luis, 1898-1924.
Gotas de tinta / Luis Tejada ; obra a cargo de Hernando Mejía Arias. — ₍Bogotá : Instituto Colombiano de Cultura₎ Subdirección de Comunicaciones Culturales, División de Publicaciones, c1977.
420 p. : ill. ; 21 cm. — (Biblioteca básica colombiana ; 26)
I. Mejía Arias, Hernando. II. Title. III. Series.
PQ8179.T36G6 1977 868 78-111468
79 MARC

Villegas, Silvio, 1902-1972.
La canción del caminante. ₍Bogotá, Organización Continental de los Festivales del Libro, 1960?₎
163 p. 17 cm. (Biblioteca básica de cultura colombiana. 2. ser., 18)
Biblioteca básica de cultura latino-americana. 2° Festival del libro colombiano.
I. Title.
PQ8179.V56C3 1960 61-36079
₍r78₎rev MARC

Volkening, Ernesto.
Los paseos de Lodovico / por Ernesto Volkening. — 1. ed. — Monterrey, México : Librería Cosmos para Ediciones Sierra Madre, 1974.
viii, 160 p. ; 21 cm. — (Poesía en el mundo : Volumen Extra ; 4)
I. Title.
PQ8179.V64P3 868 79-109973
79 MARC

—20TH CENTURY—HISTORY AND CRITICISM—ADDRESSES, ESSAYS, LECTURES

Colombia, hoy / por Mario Arrubla ... ₍et al.₎. — 1. ed. — Bogotá : Siglo Veintiuno Editores, 1978.
403 p. ; 19 cm. — (Historia inmediata)
Includes bibliographies.
1. Colombia—Politics and government—1810- —Addresses, essays, lectures. 2. Colombia—Economic conditions—Addresses, essays, lectures. 3. Colombian literature—20th century—History and criticism—Addresses, essays, lectures. I. Arrubla, Mario.
F2273.C743 309.1'861 78-376710
78 MARC

—HISTORY AND CRITICISM

Arango Ferrer, Javier.
Horas de literatura colombiana / Javier Arango Ferrer. — ₍Bogotá₎ : Instituto Colombiano de Cultura, Subdirección de Comunicaciones Culturales, División de Publicaciones, c1978.
380 p. ; 17 cm. — (Colección Popular ; 25) (Biblioteca colombiana de cultura)
$20.00
1. Colombian literature—History and criticism. I. Title.
PQ8161.A65 860 79-103742
79 MARC

Vélez de Piedrahita, Rocío.
Comentarios sobre la vida y la obra de algunos autores colombianos / Rocío Vélez de Piedrahita. — 1. ed. — Medellín : Editorial Gamma, ₍1977-
v. ; 22 cm.
Includes bibliographies.
1. Colombian literature—History and criticism. I. Title.
PQ8161.V28 860'.9 78-105022
79 MARC

—HISTORY AND CRITICISM—COLLECTED WORKS

Camacho Guizado, Eduardo.
Sobre literatura colombiana e hispanoamericana / Eduardo Camacho Guizado. — Bogotá : Instituto Colombiano de Cultura, Subdirección de Comunicaciones Culturales, División de Publicaciones, 1978.
407 p. ; 21 cm. — (Colección Autores nacionales ; 27) (Biblioteca colombiana de cultura)
Includes bibliographical references.
1. Colombian literature—History and criticism—Collected works. 2. Spanish literature—20th century—History and criticism—Collected works. I. Title.
PQ8163.C3 860 79-103276
79 MARC

COLOMBIAN LOVE POETRY see Love poetry, Colombian

COLOMBIAN POETRY

Under this heading, without subdivision, are entered individual works and collections by one author

see also Humorous poetry, Colombian; Love poetry, Colombian

Arcángel, Arturo.
13 ₍i.e. Trece₎ poemas a quemarropa / Arturo Arcángel. — ₍s.l. : s.n.₎, 1977.
81 p. : port. ; 23 cm.
I. Title.
PQ8180.1.R25T7 861 79-101252
79 MARC

Artel, Jorge.
Antología poética / Jorge Artel ; ₍carátula, Luz Stella García₎. — 1. ed. — Bogotá : Ecoe, 1979.
xiv, 97 p. ; 21 cm.
I. Title.
PQ8180.1.R77A17 1979 79-115877
79 MARC

Carranza, Eduardo, 1913-
Los pasos cantados (El corazón escrito) Poesía en verso (1935-1968). Madrid, Ed. Cultura Hispánica, 1970.
301 p., 6 l. 20 cm. (Colección La Encina y el mar, 44) Sp71-Jan
I. Title.
PQ8179.C26P3 861 72-311169
00₍r79₎rev MARC

Carranza, Eduardo, 1913-
Los pasos cantados : el corazón escrito (antología-creación, 1935-1975) / Eduardo Carranza. — ₍Bogotá₎ : Instituto Colombiano de Cultura, Subdirección de Comunicaciones Culturales, División de Publicaciones, ₍1975₎.
403 p. ; 21 cm. — (Colección Autores nacionales; 8) (Biblioteca colombiana de cultura)
I. Title.
PQ8179.C26P3 1975 861 79-103265
79 MARC

Castaño A , Jaime León, 1946-
Piel al sol : antiparábolas / Jaime León Castaño A. ; ₍dibujos y diseño de carátula por Luis Fernando Mejía J.₎. — ₍s.l. : s.n., 1978₎ (₍Medellín₎ : Duplicar)
56 p. : ill. ; 23 cm.
I. Title.
PQ8180.13.A746P5 861 78-388189
78 MARC

Cobo Borda, J G 1948-
Consejos para sobrevivir / J. G. Cobo Borda. — 1. ed. — Bogotá : Ediciones La Soga al Cuello, 1974.
78 p. : port. ; 21 cm.
"Consejos y confesiones," by H. Valencia Goelkel: p. 71-74.
I. Title.
PQ8180.13.O24C66 861 78-105350
78 MARC

Echavarría, Rogelio.
El transeúnte / Rogelio Echavarría. — ₍Bogotá₎ : Instituto Colombiano de Cultura, Subdirección de Comunicaciones Culturales, División de Publicaciones, c1977.
115 p. ; 21 cm. — (Colección Autores nacionales ; 23) (Biblioteca colombiana de cultura)
Includes bibliographical references.
I. Title.
PQ8180.15.C35T7 861 79-103732
79 MARC

Echeverri Mejía, Oscar, 1918-
Arte poética : antología, 1942-1976 / Oscar Echeverri Mejía. — ₍s.l. : s.n., 1978₎ (Cali, Colombia : Talleres de la Impr. Departamental)
177 p. : ill. ; 24 cm.
Errata slip inserted.
CONTENTS: Destino de la voz.—Canciones sin palabras.—La rosa sobre el muro.—Cielo de poesía.—La llama y el espejo.—Viaje a la niebla.—Mar de fondo.—España vertebrada.—Las cuatro estaciones.—Humo del tiempo.—La patria ilimite.—Duelos y quebrantos.—Escrito en el agua.
I. Title.
PQ8179.E44A9 861 78-108379
79 MARC

Escobar, Eduardo, 1943-
Antología poética, 1959-1977 ₍i.e. mil novecientos cincuenta y nueve-mil novecientos setenta y siete₎ / Eduardo Escobar. — ₍Bogotá₎ : Instituto Colombiano de Cultura, Subdirección de Comunicaciones Culturales, División de Publicaciones, ₍1978₎
195 p. ; 21 cm. — (Colección autores nacionales ; 35) (Biblioteca colombiana de cultura)
I. Title.
PQ8180.15.S36A17 1978 79-103738
861
79 MARC

Franco Vélez, Jorge, 1922-
Palabras del transeúnte / Jorge Franco Vélez. — 1. ed. — ₍Medellín₎ : Secretaría de Educación y Cultura del Departamento de Antioquia, Dirección de Extensión Cultural, 1978.
419 p. ; 23 cm.
I. Title.
PQ8180.16.R3P3 861 79-103243
79 MARC

Gómez Mejía, Eladio.
Letras de oro y tierra / Eladio Gómez Mejía. — ₍s.l. : s.n.₎, 1977 (Medellín : Editorial Bedout).
135 p. ; 19 cm.
I. Title.
PQ8180.17.O382L4 861 78-105024
78₍r79₎rev MARC

González Mejía, Luis Carlos, 1908-
Fototipias de Urbano Cañarte (Luis Carlos González M.) ; a control remoto? ₍Bogotá, 1969?₎
56 p. 12 x 14 cm. LACAP69-5627
Cover title.
Poems previously published in the periodical Variedades.
1. Pereira, Colombia—Biography—Poetry. I. Title.
PQ8180.17.O5F6 861 72-467529
79₍r79₎rev MARC

González Mejía, Luis Carlos, 1908-
Fototipias de Urbano Cañarte (Luis Carlos González M.) : a control remoto? — 1. ed. — ₍Bogotá₎ : Ediciones Universidades Simón Bolívar, Medellín y Libre de Pereira, 1978.
114 p. ; 17 cm. — (Colección Universidad y pueblo ; 27)
1. Pereira, Colombia—Biography—Poetry. I. Title.
PQ8180.17.O5F6 1978 861 79-101385
79 MARC

Gutiérrez Zambrano, Efraín.
Pueblo natal / Efraín Gutiérrez Zambrano. — Bogotá : Edicolda, ₍1978?₎
64 p. ; 17 cm.
I. Title.
PQ8180.17.U86P8 861 79-101574
79 MARC

Hernández, Javier, 1952-
Vía / Javier Hernández. — 1. ed. — Bogotá : C. Valencia Editores, 1978.
85 p. ; 19 cm.
Errata slip inserted.
I. Title.
PQ8180.18.E63V5 861 79-101116
79 MARC

Leonardo, Diego.
Síquicos e ilógicos córceles / Diego Leonardo. — ₍Medellín₎ : Editorial Sealón, 1977.
150 p. ; 17 cm.
I. Title.
PQ8180.22.E58S57 861 79-349796
79 MARC

COLOMBIAN POETRY (Continued)

Lleras de la Fuente, Fernando.
Silencio de secretos pasadizos / Fernando Lleras de la Fuente. — 1. ed. — ⌈s.l. : s.n.⌉, 1978 (Bogotá : Ediciones Tercer Mundo)
⌈81⌉ p. ; 23 cm.
I. Title.
PQ8180.22.L38S5 861 78-107727
78 MARC

Moyano Ortiz, Juan Carlos.
Espectos : poemas / Juan Carlos Moyano Ortiz ; carátula, fotografía de Carlos Sandino. — 1. ed. — Bogotá : Ediciones Puesto de Combate, 1979.
57 p. ; 17 cm. — (Colección Caballo verde)
I. Title.
PQ8180.23.O9E8 79-116226
79 MARC

Núñez, Rafael, Pres. Colombia, 1825-1894.
Poesías / Rafael Núñez ; prólogo de Ramón de Zubiría ; estudios de Carlos Valderrama Andrade, Nicolás del Castillo Mathieu. — Bogotá : Instituto Caro y Cuervo, 1977.
lii, 402 p. : ill. ; 22 cm. — (Biblioteca colombiana ; 13)
Includes bibliographical references.
I. Valderrama Andrade, Carlos. II. Castillo Mathieu, Nicolás del. III. Title. IV. Series.
PQ8179.N8P6 1977 861 78-105383
79 MARC

Ortiz, Omar, 1950-
La tierra y el éter : poemas / Omar Ortiz ; ⌈carátula y diagramación, Rubén Rueda⌉. — 1. ed. — ⌈s.l. : s.n.⌉, 1979 (Bogotá : Ediciones Alcaraván)
129 p. : ill. ; 21 cm.
I. Title.
PQ8180.25.R82T5 79-117372
79 MARC

Rincón Zapata, Carlos.
Cambio de palabras (1970-1977) / Carlos Rincón Zapata. — 1. ed. — Bogotá : Editorial ABC, 1978.
130 p. ; 22 cm.
I. Title.
PQ8180.28.I53C3 861 79-101334
79 MARC

Rojas, Jorge.
Suma poética, 1939-1976 / Jorge Rojas. — ⌈Bogotá⌉ : Instituto Colombiano de Cultura, Subdirección de Comunicaciones Culturales, División de Publicaciones, 1977.
511 p. ; 21 cm. — (Colección Autores nacionales ; 20) (Biblioteca colombiana de cultura)
Bibliography: p. 507-508.
I. Title.
PQ8179.R56S8 861 79-103266
79 MARC

Ruiz Gómez, Darío.
Geografía : poemas / Darío Ruiz Gómez. — ⌈s.l. : s.n., 1977 or 1978⌉
87 leaves ; 22 cm.
I. Title.
PQ8180.28.U53G4 861 79-101130
79 MARC

Sanders, Teresa.
Huellas en el viento / ⌈Teresa Sanders⌉. — ⌈s.l. : s.n., 1978⌉ (Bogotá : Ediciones Paulinas)
220 p., ⌈1⌉ leaf of plates : ill. ; 17 x 18 cm.
I. Title.
PQ8180.29.A4716H8 861 79-108815
79 MARC

Sossa de Jaramillo, Oliva.
Cuando pasa la brisa : poemas / Oliva Sossa de Jaramillo. — Medellín : Editorial Letras, ⌈1978⌉
101 p. ; 22 cm.
I. Title.
PQ8180.29.O75C8 79-116716
79 MARC

Urraza, Rafael.
Diario : 19 poemas cortos de Rafal Urraza. — 2. ed. — ⌈s.l. : s.n.⌉, 1978 (⌈Bogotá⌉ : Publicaciones Cultural)
⌈45⌉ p. : ill. ; 16 x 22 cm.
PQ8180.31.R7D5 1978 861 79-101087
79 MARC

Valenzuela Sánchez, Germán, 1950-
Poesía social a ras de pueblo / Germán Valenzuela Sánchez. — 1. ed. — Bucaramanga : Impr. Nueva Jornada, 19
v. ; 16 cm.
I. Title.
PQ8180.32.A426P6 861'.6'4 79-101564
79 MARC

—20TH CENTURY

Oficio de poeta : poesía en Bogotá / ⌈compilación de⌉ Rosa Jaramillo. — Bogotá : Universidad de San Buenaventura, 1978.
298 p. : ports. ; 20 cm.
1. Colombian poetry—Colombia—Bogotá. 2. Colombian poetry—20th century. I. Jaramillo, Rosa.
PQ8178.B62O35 861 79-101117
79 MARC

Poemas / ⌈Centro Poético Colombiano⌉. — Bogotá : El Centro, ⌈1977?⌉
226 p. ; 24 cm.
1. Colombian poetry—20th century. I. Centro Poético Colombiano.
PQ8174.P52 861 78-109722
79 MARC

—COLOMBIA—BOGOTÁ

Oficio de poeta : poesía en Bogotá / ⌈compilación de⌉ Rosa Jaramillo. — Bogotá : Universidad de San Buenaventura, 1978.
298 p. : ports. ; 20 cm.
1. Colombian poetry—Colombia—Bogotá. 2. Colombian poetry—20th century. I. Jaramillo, Rosa.
PQ8178.B62O35 861 79-101117
79 MARC

COLOMBIAN WIT AND HUMOR

see also Humorous poetry, Colombian

—COLOMBIA—SALAMINA—HISTORY AND CRITICISM

Amaya, Leonidas, 1927-
Apuntes y cachos salamineños / Leonidas Amaya. — Manizales : Impr. Departamental de Caldas, 1975.
273 p. : ill. ; 20 cm. — (Biblioteca de autores caldenses ; 7. época, v. 44)
1. Salamina, Colombia—Miscellanea. 2. Humorous poetry, Colombian—Colombia—Salamina. 3. Colombian wit and humor—Colombia—Salamina—History and criticism. I. Title. II. Series.
F2291.S16A46 986.1'3 79-352959
79 MARC

COLOMBO, CRISTOFORO

Català i Roca, Pere.
Entorn de Cristòfor Colom / Pere Català i Roca. — Barcelona : R. Dalmau, c1978.
61 p. ; 17 cm. — (Episodis de la història ; 229) Sp***
Includes bibliographical references.
ISBN 8423201287
1. Colombo, Cristoforo. 2. Explorers—Spain—Biography. I. Title.
E120.C37 78-398265
79 MARC

Català i Roca, Pere.
Quatre germans Colom, el 1462 / Pere Català i Roca. — Barcelona : R. Dalmau, c1978.
63 p. ; 17 cm. — (Episodis de la història ; 228) Sp***
Includes bibliographical references.
ISBN 8423201260
1. Colombo family. 2. Colombo, Cristoforo. 3. Catalonia—Biography. I. Title.
E113.C27 78-393694
79 MARC

Colombo, Cristoforo.
Diario de Colón / comentario preliminar por Carlos Sanz. — Madrid : Revista Geográfica Española, ⌈1977?⌉
2 v. : map ; 23 cm. — (Carlos Sanz. La huella de España en el mundo ; 4-5) Sp***
ISBN 8485018028 (v. 1)
1. Colombo, Cristoforo. 2. America—Discovery and exploration—Spanish. 3. Explorers—Spain—Biography. I. Sanz López, Carlos. II. Title. III. Series: Sanz López, Carlos. La huella de España en el mundo ; 4-5.
G288.S24 vol. 4-5 970.01 s 79-350113
⌈E118⌉ 79 MARC

Colombo, Cristoforo.
Diario del descubrimiento / ⌈por⌉ Cristóbal Colón ; estudios, ediciones y notas por Manuel Alvar. — Madrid : La Muralla, 1976.
2 v. ; 24 cm. — (Ediciones del Excmo Cabildo Insular de Gran Canaria : 1, Lengua y literatura) Sp77-Jan
Bibliography: v. 2, p. 245-256.
ISBN 8471332167 : 2500ptas
1. America—Discovery and exploration—Spanish. 2. Colombo, Cristoforo. I. Alvar López, Manuel. II. Title.
E118.C78 1976 970'.01'50924 78-368520
78 MARC

Mahieu, Jacques de, 1915-
L'imposture de Christophe Colomb : la géographie secrète de l'Amérique / par Jacques de Mahieu. — ⌈Paris⌉ : Copernic, 1979.
175 p., ⌈8⌉ leaves of plates : ill. ; 22 cm. F***
Includes bibliographical references.
ISBN 2-85984-025-7 : 44.50F
1. America—Discovery and exploration—Pre-Columbian. 2. Colombo, Cristoforo. I. Title.
E103.M27 970.01 79-369100
79 MARC

Morison, Samuel Eliot, 1887-
Admiral of the ocean sea : a life of Christopher Columbus / Samuel Eliot Morison ; maps by Erwin Raisz, drawings by Bertram Greene. — A limited ed. — Franklin Center, Pa. : Franklin Library, 1978, c1942.
xxiv, 688 p., ⌈2⌉ fold. leaves of plates : ill. ; 25 cm. — (The 100 greatest masterpieces of American literature)
"Notes from the editors" (22 p.) inserted.
Includes index.
1. Colombo, Cristoforo. 2. Explorers—Spain—Biography. 3. America—Discovery and exploration—Spanish. I. Title. II. Series.
E111.M86 1978 970.01'5'0924 78-113100
79 MARC

—CONGRESSES

Convegno internazionale di studi colombiani, Genoa, 1973.
Atti del Convegno internazionale di studi colombiani : 13 e 14 ottobre 1973 / ⌈a cura di Renzo Lagomarsino⌉. — Genova : Civico istituto colombiano, 1974.
174 p. ; 25 cm. It76-Aug
Includes bibliographical references.
1. America—Discovery and exploration—Spanish—Congresses. 2. Colombo, Cristoforo—Congresses. 3. Europe—Exploring expeditions—Congresses. I. Title.
E123.C74 1973 970.01'5 79-361704
*79 MARC

—DRAMA

Chiari, Joseph.
Christopher Columbus : a play / by Joseph Chiari. — New York : Gordian Press, 1979.
x, 82 p. ; 23 cm.
ISBN 0-87752-216-2
1. Colombo, Cristoforo—Drama. I. Title.
PR9105.9.C5C5 822 78-23741
78 MARC

Kazantzak̄es, Nikos, 1883-1957.
Christopher Columbus : ⌈play⌉ / Nikos Kazantzakis. — Kentfield, Calif. : Allen Press, 1972.
79 p. ; 34 cm.
Issued in a case.
"One hundred forty copies."
1. Colombo, Cristoforo—Drama. I. Title.
PA5610.K39C4413 1972 Rosenwald Coll. 78-325375
78 MARC

—EXHIBITIONS

Spain. Instituto de Cultura Hispánica.
Colón y su tiempo : exposición / presentada por el Instituto de Cultura Hispánica de Madrid, como contribución de la Comisión Española para el Bicentenario de los Estados Unidos y la colaboración de la Smithsonian Institution. — Madrid : Ediciones Cultura Hispánica, 1976.
79 p. : ill. ; 23 cm. Sp***
English and Spanish.
1. Colombo, Cristoforo—Exhibitions. 2. America—Discovery and exploration—Spanish—Exhibitions. 3. Explorers—Spain—Biography—Exhibitions. I. Smithsonian Institution. II. Title.
E112.S735 1976 970.01'4'0924 79-368060
79 MARC

—FICTION

Blasco Ibáñez, Vicente, 1867-1928.
En busca del Gran Kan / V. Blasco Ibáñez. — 1. ed. — Esplugas de Llobregat : Plaza & Janés, 1978.
336 p. ; 20 cm. — (Obra de V. Blasco Ibañez) Sp***
ISBN 8401805805
1. Colombo, Cristoforo—Fiction. I. Title. II. Series: Blasco Ibañez, Vicente, 1867-1928. Works. 1976-
PQ6603.L2E4 1978 863'.6'2 79-361838
79 MARC

—JUVENILE LITERATURE

Ceserani, Gian Paolo, 1939-
Il viaggio di Colombo / Piero Ventura ; testo di Gian Paolo Ceserani. — 1. ed. — Milano : A. Mondadori, 1977.
⌈38⌉ p. : ill. ; 30 cm. — (Le pietre preziose) It78-Oct
L3000
1. Colombo, Cristoforo—Juvenile literature. 2. America—Discovery and exploration—Spanish—Juvenile literature. 3. Explorers—Spain—Biography—Juvenile literature. I. Ventura, Piero. II. Title.
E111.C4 970.01'5 79-379430
*79 MARC

COLOMBO, CRISTOFORO
—JUVENILE LITERATURE
(Continued)

Erdődy, János.
Így élt Kolumbusz / Erdődy János. — Bratislava : Madách, 1978.
189 p. : ill. ; 20 cm. — (Így élt ..) CzS***
Kčs15.00
1. Colombo, Cristoforo—Juvenile literature. 2. America—Discovery and exploration—Spanish—Juvenile literature. 3. Explorers—Spain—Biography—Juvenile literature. I. Title.
E111.E78 79-365877
79 MARC

Richards, Dorothy Fay, 1915-
Christopher Columbus, who sailed on! / By Dorothy Fay Richards ; illustrated by John Nelson. — Elgin, Ill. : Child's World ; Chicago : distributed by Childrens Press, c1978.
31 p. : ill. (some col.) ; 25 cm.
SUMMARY: Relates the attempts of Christopher Columbus to find a short sea route to the Indies.
ISBN 0-89565-032-0 : $5.50
1. Columbus, Cristoforo—Juvenile literature. 2. Explorers—Spain—Biography—Juvenile literature. [1. Columbus, Christopher. 2. Explorers] I. Nelson, John, 1928- . II. Title.
E111.R5 970'.1'50924 78-7664
78 MARC

Ventura, Piero.
Christopher Columbus / Piero Ventura ; based on the text by Gian Paolo Ceserani. — New York : Random House, c1978.
[32] p. : col. ill. ; 21 cm.
Based on Il viaggio Colombo.
SUMMARY: A brief biography of Christopher Columbus highlighting his ocean voyages and discovery of new lands.
ISBN 0-394-83907-2
1. Colombo, Cristoforo—Juvenile literature. 2. America—Discovery and exploration—Spanish—Juvenile literature. 3. Explorers—Spain—Biography—Juvenile literature. [1. Columbus, Christopher. 2. Explorers. 3. America—Discovery and exploration—Spanish] I. Ceserani, Gian Paolo, 1939- Il viaggio di Colombo. II. Title.
E111.V47 970.01'5'0924 77-86146
78 MARC

—RELATIONS WITH JEWS

Kayserling, Meyer, 1829-1905.
Christopher Columbus and the participation of the Jews in the Spanish and Portuguese discoveries / by M. Kayserling ; translated from the author's manuscript with his sanction and revision by Charles Gross. — Folcroft, Pa. : Folcroft Library Editions, 1978, c1894.
xv, 189 p. ; 24 cm.
Reprint of the ed. published by Longmans, Green, New York.
English, Spanish, and Latin.
Includes bibliographical references and index.
ISBN 0-8414-5478-7 (lib. bdg.) : $25.00
1. Colombo, Cristoforo—Relations with Jews. 2. America—Discovery and exploration—Spanish. 3. Jews in Spain—History. 4. Maranos. 5. Santángel, Luis de, d. 1505. 6. Santangel family. 7. Spain—History—Ferdinand Isabella, 1479-1516. I. Title.
E111.K23 1978 970'.01'504924 78-26172
78 MARC

COLOMBO, JOHN ROBERT, 1936-

—BIBLIOGRAPHY

Morrow, J David.
A catalogue to the literary papers of John Robert Colombo and working archives of The Tamarack review / compiled by J. David Morrow. — [Hamilton, Ont. : McMaster University Press], 1971.
32 p. ; 27 cm. — (Library research news ; v. 1, no. 5)
1. Colombo, John Robert, 1936- —Manuscripts—Catalogs. 2. Colombo, John Robert, 1936- —Bibliography. 3. The Tamarack review—Archives. I. The Tamarack review. II. Title: A catalogue to the literary papers of John Robert Colombo ... III. Series.
PR9199.3.C635Z459 016.811'5'4 76-152075
78 MARC

—MANUSCRIPTS—CATALOGS

Morrow, J David.
A catalogue to the literary papers of John Robert Colombo and working archives of The Tamarack review / compiled by J. David Morrow. — [Hamilton, Ont. : McMaster University Press], 1971.
32 p. ; 27 cm. — (Library research news ; v. 1, no. 5)
1. Colombo, John Robert, 1936- —Manuscripts—Catalogs. 2. Colombo, John Robert, 1936- —Bibliography. 3. The Tamarack review—Archives. I. The Tamarack review. II. Title: A catalogue to the literary papers of John Robert Colombo ... III. Series.
PR9199.3.C635Z459 016.811'5'4 76-152075
78 MARC

COLOMBO, LEO, 1915-1936

Durando, G
Leo Colombo, modello di figlio, di parrocchiano, di giovane di A. C. e di militare / G. Durando. — 2. ed. — Alba : Istituto missionario Pia società s. Paolo, [1942]
128 p. ; 19 cm. — (In montem sanctum ; 2) (La Vetta)
1. Colombo, Leo, 1915-1936. 2. Catholics—Italy—Biography. I. Title.
BX4705.C6893D87 1942 79-382702
79 MARC

COLOMBO, SRI LANKA

—HARBOR

Ceylon. Varāya Saṃyukta Maṇḍalaya. Publicity and Public Relations Bureau.
Port of Colombo : official handbook, 1968 / [compiled and edited by the Publicity and Public relations Bureau, Port (Cargo) Corporation]. — [s.l. : s.n., 1968?] (Colombo : M. D. Gunasena)
104 p. : ill. ; 28 cm.
1. Colombo, Sri Lanka—Harbor. 2. Shipping—Sri Lanka—Colombo. I. Title.
HE560.C6C49 1968 387.1'09549'3 78-984008
79 MARC

Koḷamba Varāya Komisama.
(Koḷamba Varāya Komisamē Varāya Komasārisgē pālana vārtāva)
කොළඹ වරාය කොමිසමේ වරාය කොමසාරිස්ගේ පාලන වාර්තාව. கொழும்பு கப்பற்றுறை ஆணைக்குழு, கப்பற்றுறை ஆணையாளரின் நிருவாக அறிக்கை. Administration report of the Port Commissioner, Colombo Port Commission.
කොළඹ.
v. 25 cm. annual. (Ceylon. Administration reports. Pt. I: Civil (D))
English, Sinhalese, and Tamil.

1. Colombo, Sri Lanka—Harbor. I. Koḷamba Varāya Komisama. Koḷumpu Kappaṟṟuṟai Āṇaikkuḻu, Kappaṟṟuṟai Āṇaiyāḷariṉ niruvāka aṟikkai. II. Koḷamba Varāya Komisama. Administration report of the Port Commissioner, Colombo Port Commission.

HE560.C6K64a S A 68-19444
MARC-S
[r78]rev 2 PL 480: C-36-8

—MAPS

Sri Lanka. Minindōra Depārtamēntuva.
A to Z atlas of Colombo City, featuring house numbers along main roads, railways, and bus routes. — [Colombo] : Survey Dept., 1972.
iii, 48 p. : maps ; 18 cm.
Cover title: A to Z Colombo atlas & street index.
Scale of sectional maps ca. 1:12,750.
Published under the direction of R. A. Goonewardene, Actg. Surveyor General, Ceylon.
Includes indexes.
Rs1.50
1. Colombo, Sri Lanka—Maps. I. Goonewardene, R. A. II Title. III. Title: A to Z Colombo atlas & street index.
G2294.C6S7 1972 912'.549'3 79-306769
79 MARC

Sri Lanka. Minindōru Depārtamēntuva.
A to Z atlas of Colombo City, featuring house numbers along main roads, railways, and bus routes. — Rev. 1975. — [Colombo] : Survey Dept., 1976.
iii, 50 p., [1] fold. leaf of plates : maps ; 18 cm.
Cover title: A to Z Colombo atlas & street index.
Scale of sectional maps 1:12,750.
Published under the direction of S. J. Munasinghe, Surveyor General of Sri Lanka.
Includes index.
Rs5.00
1. Colombo, Sri Lanka—Maps. I. Munasinghe, S. J. II. Title. III. Title: A to Z Colombo atlas & street index.
G2294.C6S7 1976 912'.549'3 78-324700
79 MARC

COLOMBO CLUB

—HISTORY

Colombo Club.
The first hundred years of the Colombo Club. — Colombo : The Club, [1977?]
41 p., [15] leaves of plates : ill. ; 23 cm.
"1871-1971."
1. Colombo Club—History. I. Title.
HS3115.S725C6332 367'.9549'3 78-908327
79 MARC

COLOMBO FAMILY

Català i Roca, Pere.
Quatre germans Colom, el 1462 / Pere Català i Roca. — Barcelona : R. Dalmau, c1978.
63 p. ; 17 cm. — (Episodis de la història ; 228) Sp***
Includes bibliographical references.
ISBN 8423201260
1. Colombo family. 2. Colombo, Cristoforo. 3. Catalonia—Biography. I. Title.
E113.C27 78-393694
79 MARC

COLOMBO MUSEUM

—CATALOGS

Colombo Museum.
Colombo National Museum : illustrated guide through the exhibited collections / by P. H. D. H. de Silva, director. Dept. of National Museums, Sri Lanka. — [Colombo] : National Museums of Sri Lanka, 1976-
v. : ill. (some col.) ; 19 x 26 cm.
Cover title: Illustrated guide to the Colombo Museum.
Errata slip inserted in v. 1.
Rs20.00 (v. 1)
1. Colombo Museum—Catalogs. I. De Silva, Pilippu Hewa Don Hemasiri, 1927- . II. Title. III. Title: Illustrated guide to the Colombo Museum.
AM101.C69A44 069'.09549'3 77-902378
79 MARC

COLOMBO MUSEUM. LIBRARY

—CATALOGS

Colombo Museum. Library.
Catalogue of palm leaf manuscripts in the Library of the Colombo Museum / by W. A. de Silva. — [Colombo] : Colombo Museum, 1938-
v. : ill. ; 32 cm. — (Memoirs of the Colombo Museum ; series A, no. 4)
English, Pali, Sanskrit (in Sinhalese script) and Sinhalese.
Includes indexes.
1. Buddhist literature—Bibliography—Catalogs. 2. Colombo Museum. Library—Catalogs. I. De Silva, Wilmot Arthur. II. Title. III. Series: Colombo Museum. Memoirs of the Colombo Museum ; series A, no. 4.
Z7865.5.C64 1938 016.2943 77-985639
[BQ1020] 79 MARC

COLON, RAMON

Colon, Ramon.
Another spic got screwed : a case report to the American people / Ramon Colon. — New York : Plus Ultra Educational Publishers, c1976.
143 p. ; 22 cm. — (Colección Arco iris ; 18)
Includes bibliographical references.
ISBN 0-915534-23-1
1. Colon, Ramon. 2. New York (State). State Division of Human Rights. 3. Discrimination in employment—Law and legislation—New York (State) 4. Puerto Ricans in New York (State)—Civil rights. I. Title.
KF228.C64C64 342'.73'085 78-107180
78 MARC

COLÓN, URUGUAY

—HISTORY

Martínez Montero, Homero B
Villa Colón : historia, evolución, desarrollo / Homero Martínez Montero. — Montevideo : Junta de Vecinos de Montevideo, Biblioteca José Artigas, 1976.
268 p., [4] fold. leaves of plates : ill. ; 20 cm.
1. Colón, Uruguay—History. I. Biblioteca José Artigas.
F2791.C48M37 989.5'1 78-102515
78 MARC

COLON, IRRITABLE see Irritable colon

COLON (ANATOMY)

see also Taenia coli

COLON (ANATOMY) (Continued)

—CANCER

Miller, Roscoe E
Examination of the colon / Roscoe E. Miller. — Chicago : Year Book Medical Publishers, 1975.
40 p. : ill. ; 23 cm. — (Current problems in radiology ; v. 5, no. 2)
"References": p. 38-40.
ISBN 0-8151-9912-0 : $5.00
1. Colon (Anatomy)—Cancer—Diagnosis. 2. Colon (Anatomy)—Radiography. I. Title. II. Series.
RC78.A1C87 vol. 5 no. 2 616.07'57'08 s 78-310175
[RC280.C6] 78 MARC

Morson, B C
The pathogenesis of colorectal cancer / Basil C. Morson. — Philadelphia : Saunders, 1978.
xi, 164 p. : ill. ; 25 cm. — (Major problems in pathology)
Includes bibliographies and index.
ISBN 0-7216-6558-6
1. Colon (Anatomy)—Cancer. 2. Rectum—Cancer. I. Title.
RC280.C6M67 616.9'94'347 78-1792
78 MARC

—CANCER—DIAGNOSIS

Miller, Roscoe E
Examination of the colon / Roscoe E. Miller. — Chicago : Year Book Medical Publishers, 1975.
40 p. : ill. ; 23 cm. — (Current problems in radiology ; v. 5, no. 2)
"References": p. 38-40.
ISBN 0-8151-9912-0 : $5.00
1. Colon (Anatomy)—Cancer—Diagnosis. 2. Colon (Anatomy)—Radiography. I. Title. II. Series.
RC78.A1C87 vol. 5 no. 2 616.07'57'08 s 78-310175
[RC280.C6] 78 MARC

—CONGRESSES

Current status of barium enema and colonoscopy in diagnosis of colonic disease : symposium proceedings / cosponsors, Yale Affiliated Gastroenterology Program, Connecticut Society of Gastrointestinal Endoscopy ; edited by Howard M. Spiro, Vincent A. DeLuca, Jr. — Norwalk, Conn. : [Purdue Frederick Co.], c1978.
vii, 109 p. : ill. ; 25 cm.
Title on spine: Barium enema and colonoscopy in diagnosis of colonic disease.
Includes bibliographical references.
1. Colonoscopy—Congresses. 2. Colon (Anatomy)—Diseases—Diagnosis—Congresses. 3. Barium enema—Congresses. 4. Colon (Anatomy)—Congresses. I. Spiro, Howard M. II. DeLuca, Vincent A. III. Yale Affiliated Gastroenterology Program. IV. Connecticut Society of Gastrointestinal Endoscopy. V. Title: Barium enema and colonoscopy in diagnosis of colonic disease.
RC804.C64C87 617'.5547'0754 78-61839
79 MARC

—DISEASES

see also Megacolon

—DISEASES—DIAGNOSIS—CONGRESSES

Current status of barium enema and colonoscopy in diagnosis of colonic disease : symposium proceedings / cosponsors, Yale Affiliated Gastroenterology Program, Connecticut Society of Gastrointestinal Endoscopy ; edited by Howard M. Spiro, Vincent A. DeLuca, Jr. — Norwalk, Conn. : [Purdue Frederick Co.], c1978.
vii, 109 p. : ill. ; 25 cm.
Title on spine: Barium enema and colonoscopy in diagnosis of colonic disease.
Includes bibliographical references.
1. Colonoscopy—Congresses. 2. Colon (Anatomy)—Diseases—Diagnosis—Congresses. 3. Barium enema—Congresses. 4. Colon (Anatomy)—Congresses. I. Spiro, Howard M. II. DeLuca, Vincent A. III. Yale Affiliated Gastroenterology Program. IV. Connecticut Society of Gastrointestinal Endoscopy. V. Title: Barium enema and colonoscopy in diagnosis of colonic disease.
RC804.C64C87 617'.5547'0754 78-61839
79 MARC

—EXAMINATION

see also Colonoscopy

—SURGERY

Symposium on colon and anorectal surgery / James A. Ferguson, guest editor. — Philadelphia : Saunders, 1978.
x p., p. 457-656 : ill. ; 24 cm. — (The Surgical clinics of North America ; v. 58, no. 3)
Cover title: Colon and anorectal surgery.
Includes bibliographies and index.
1. Colon (Anatomy)—Surgery. 2. Rectum—Surgery. 3. Anus—Surgery. I. Ferguson, James A. II. Title: Colon an anorectal surgery. III. Series.
[DNLM: 1. Colon—Surgery. 2. Anus—Surgery. 3. Rectum—Surgery. SU746L v. 58 no. 4 / WI520 S989]
RD34.S85 vol. 58, no. 3 617'.008 s 78-6199
[RD544] 78 MARC

—SURGERY—PERIODICALS

American journal of proctology, gastroenterology & colon & rectal surgery. v. 29–
Jan./Feb. 1978–
[Georgetown, Conn., McMahon Pub. Co.]
v. ill. 29 cm. bimonthly.
"Official publication of the International Academy of Proctology."
Continues: The American journal of proctology, ISSN 0002-9521.
Key title: American journal of proctology, gastroenterology & colon & rectal surgery, ISSN 0162-6566.
1. Gastroenterology—Periodicals. 2. Proctology—Periodicals. 3. Colon (Anatomy)—Surgery—Periodicals. 4. Rectum—Surgery—Periodicals. I. International Academy of Proctology.
616.3'005 78-645416
MARC-S

COLÓN DE TOLEDO, LUIS

—FICTION

Rueda, Manuel.
La prisonera del Alcázar : leyenda de carácter histórico / Manuel Rueda ; ilus., Gilberto Hernández Ortega. — [s.l. : s.n., 1976] ([s.l.] : Editora Amigo del Hogar)
57 p. : ill. ; 24 cm.
1. Colón de Toledo, Luis—Fiction. I. Title.
PQ7409.R8P7 79-115723
79 MARC

COLONA, FRANCESCO, D. 1527

—KNOWLEDGE—ARCHITECTURE

Schmidt, Dorothea.
Untersuchungen zu den Architekturekphrasen in der Hypnerotomachia poliphili : die Beschreibung des Venus-Tempels / Dorothea Schmidt. — Frankfurt am Main : R. G. Fischer, 1978.
215 p. : ill. ; 21 cm. GFR79-A
Bibliography: p. 4-8.
ISBN 3-88323-031-6 : DM28.00
1. Colonna, Francesco, d. 1527. Hypnerotomachia Poliphili. 2. Colona, Francesco, d. 1527—Knowledge—Architecture. 3. Architecture in literature. 4. Architecture, Renaissance—Themes, motives. I. Title.
PQ4619.C9Z88 79-374293
79 MARC

COLONATE

see also Serfdom

COLONIA CECILIA, BRAZIL

Schmidt, Afonso, 1890-1964.
Colônia Cecília : uma aventura anarquista na América, 1889 a 1893 / Afonso Schmidt. — São Paulo, Brasil : Editora Anchieta, 1942.
142, [8] p., [1] leaf of plates : port. ; 19 cm.
"Opiniões dos mais abalisados jornais do país sôbre os livros de Afonso Schmidt 'A marcha' e 'Tesouro de Cananeia'": p. [145-150]
1. Colonia Cecilia, Brazil. 2. Rossi, Giovanni, b. 1855. 3. Anarchism and anarchists—Brazil—Colonia Cecilia. I. Title.
HX875.C6S35 1942b 79-359245
79 MARC

COLONIAL ADMINISTRATORS

—AFRICA—BIOGRAPHY

African proconsuls : European governors in Africa / L. H. Gann and Peter Duignan, editors. — New York : Free Press, c1978.
xi, 548 p., [4] leaves of plates : ill. ; 26 cm.
"Published under the auspices of the Hoover Institution."
Includes bibliographical references and index.
ISBN 0-02-911190-0
1. Africa—Politics and government. 2. Colonial administrators—Africa—Biography. 3. Europe—Colonies—Africa—Administration. I. Gann, Lewis H., 1924- . II. Duignan, Peter.
JQ1875.A1A35 325'.31'0922 78-57054
78 MARC

Chenevix Trench, Charles Pocklington, 1914-
Charley Gordon : an eminent Victorian reassessed / [by] Charles Chenevix Trench. — Lonon : Allen Lane, 1978.
320 p., [16] p. of plates : ill., maps, ports. ; 24 cm.
Bibliography: p. [295]-310.
Includes index.
ISBN 0-7139-0895-5 : £8.95
1. Gordon, Charles George, 1833-1885. 2. Great Britain. Army—Biography. 3. Generals—Great Britain—Biography. 4. Colonial administrators—Africa—Biography. 5. Egypt—History—British occupation, 1882-1936. 6. Sudan—History—1862-1899. 7. China—History—T'ung-chih, 1861-1875.
DA68.32.G6C45 962.4'03'0924 79-301345
79 MARC

—ANGOLA—BIOGRAPHY

Carvalho, João Augusto de Noronha Dias de.
Henrique de Carvalho, uma vida ao serviço da pátria / João Augusto de Noronha Dias de Carvalho. — Lisboa : [s.n.], 1975.
351 p., [3] leaves of plates : ill. ; 24 cm.
Includes bibliographical references.
1. Dias de Carvalho, Henrique Augusto, 1843-1909. 2. Portugal—Colonies—Africa. 3. Angola—Discovery and exploration. 4. Colonial administrators—Portugal—Biography. 5. Colonial administrators—Guinea-Bissau—Biography. 6. Colonial administrators—Angola—Biography. 7. Explorers—Angola—Biography. 8. Explorers—Portugal—Biography. I. Title.
DT611.72.D53C37 325'.31'0924 77-155507
79 MARC

—AUSTRALIA—NEW SOUTH WALES—BIOGRAPHY

Foster, Stephen Glynn, 1948-
Colonial improver : Edward Deas Thomson, 1800-1879 / S. G. Foster. — Carlton [Australia] : Melbourne University Press ; Forest Grove, Or. : exclusive distributor, ISBS, 1978.
xv, 209 p., [2] leaves of plates : ill. ; 23 cm. Aus***
"Based on a thesis which was presented to the University of New England."
Bibliography: p. 187-201.
Includes index.
ISBN 0-522-84136-8
1. Thomson, Edward Deas, 1800-1879. 2. New South Wales—Politics and government. 3. Colonial administrators—Australia—New South Wales—Biography. I. Title.
DU172.T45F67 994.4'02'0924 78-670160
78 MARC

—GREAT BRITAIN—BIOGRAPHY

Douglas-Home, Charles, 1937-
Evelyn Baring : the last proconsul / Charles Douglas-Home. — London : Collins, 1978.
344 p., [4] leaves of plates : ports., maps ; 23 cm. GB***
Bibliography: p. 336-337.
Includes index.
ISBN 0-00-216457-4 : £7.50
1. Baring, Evelyn, Sir, 1903-1973. 2. Colonial administrators—Great Britain—Biography.
JV1009.B37D68 325'.31'0924 79-305791
79 MARC

Flexner, James Thomas, 1908-
Lord of the Mohawks : a biography of Sir William Johnson / James Thomas Flexner ; newly illustrated and with a new foreword. — Boston : Little, Brown, c1979.
xviii, 400 p., [15] leaves of plates : ill. ; 24 cm.
On dust jacket: "A revised edition."
First ed. published in 1959 under title: Mohawk baronet.
Bibliography: p. 361-368.
Includes index.
ISBN 0-316-28609-5
1. Johnson, William, Sir, bart., 1715-1774. 2. United States—History—French and Indian War, 1755-1763. 3. Indians of North America—Government relations—To 1789. 4. Indians of North America—Wars—1750-1815. 5. Soldiers—Great Britain—Biography. 6. Colonial administrators—Great Britain—Biography. 7. Colonial administrators—United States—Biography. I. Title.
E195.J63F57 1979 973.2'6'0924 79-14843
79 MARC

COLONIAL ADMINISTRATORS —GREAT BRITAIN—BIOGRAPHY (Continued)

Lewin, Thomas Herbert, 1839-1916.
A fly on the wheel : or, How I helped to govern India / by Thomas H. Lewin. — Calcutta : Firma KLM, on behalf of Tribal Research Institute, Aizawl, 1977.
318 p., ₍1₎ leaf of plates : ill. ; 22 cm.
Reprint of the 1912 ed., published by Constable, London.
1. Lewin, Thomas Herbert, 1839-1916. 2. India—Description and travel—1859-1900. 3. India—Politics and government—19th century. 4. Lusheis. 5. Colonial administrators—India—Biography. 6. Colonial administrators—Great Britain—Biography. I. Title.
DS479.1.L48A33 1977 954.03'092'4 78-911616
78 MARC

O'Brien, Terence Henry.
Milner : Viscount Milner of St. James's and Cape Town, 1854-1925 / Terence H. O'Brien. — London : Constable, 1979.
447 p., ₍6₎ leaves of plates : ports. ; 24 cm. GB***
Bibliography: p. ₍421₎-426.
Includes index.
ISBN 0-09-457880-X : £10.00
1. Milner, Alfred Milner, 1st Viscount, 1854-1925. 2. South Africa—History—1836-1909. 3. Colonial administrators—South Africa—Biography. 4. Colonial administrators—Great Britain—Biography. 5. Statesmen—Great Britain—Biography. 6. Great Britain—Politics and government—1837-1901. 7. Great Britain—Politics and government—1901-1936.
DT776.M6O27 1979 941.083'092'4 79-315722
79 MARC

Parveen Shaukat Ali.
Pillars of British imperialism : a case study of the political ideas of Sir Alfred Lyall, 1873-1903 / by Parveen Shaukat Ali. — Lahore : Aziz Publishers, 1976.
xix, 176 p. ; 22 cm.
Bibliography: p. 162-168.
Includes index.
Rs40.00 ($7.00 U.S.)
1. Lyall, Alfred Comyn, Sir, 1835-1911. 2. India—Politics and government—1765-1947. 3. Colonial administrators—India—Biography. 4. Colonial administrators—Great Britain—Biography. I. Title.
DS479.1.L9P37 325'.31'0924 77-930237
79 MARC

Raffles, Thomas Stamford, Sir, 1781-1826.
Statement of the services of Sir Stamford Raffles / with an introd. by John Bastin. — Kuala Lumpur ; New York : Oxford University Press, 1978.
xi, 72 p. ; 29 cm. — (Oxford in Asia historical reprints)
ISBN 0-19-580318-3 : $17.50
1. Raffles, Thomas Stamford, Sir, 1781-1826. 2. Indonesia—Politics and government—1798-1942. 3. Singapore—Politics and government. 4. East India Company (English)—History. 5. Colonial administrators—Indonesia—Biography. 6. Colonial administrators—Singapore—Biography. 7. Colonial administrators—Great Britain—Biography.
DS598.S7R267 941.07'3'0924 79-118509
79 MARC

Thornton, Thomas Henry, 1832-1913.
Colonel Sir Robert Sandeman / by Thomas Henry Thornton. — 1st ed. in Pakistan. — Quetta : Gosha-e-Adab : sole distributors Nisa Traders, 1977.
xxiii, 392 p., ₍7₎ leaves of plates : ill. ; 18 cm.
Reprint of 1895 ed. published by John Murray, London.
250 copies printed.
Includes index.
Rs110.00 ($17.00 U.S.)
1. Sandeman, Robert Groves, Sir, 1835-1892. 2. Baluchistan—History. 3. Colonial administrators—Great Britain—Biography. 4. Colonial administrators—Pakistan—Baluchistan—Biography. I. Title.
DS392.B28S257 1977 954'.9'15'0350924 77-930582
79 MARC

Winstone, Harry Victor Frederick.
Gertrude Bell / H. V. F. Winstone. — London : Cape, 1978.
xii, 322 p., ₍8₎ leaves of plates : ill., map (on lining papers), ports. ; 23 cm.
GB78-13967
Bibliography: p. ₍293₎-312.
Includes index.
ISBN 0-224-01432-3 : £3.50
1. Bell, Gerturde Lowthian, 1868-1926. 2. Orientalists—Biography. 3. Colonial administrators—Great Britain—Biography. 4. Colonial administrators—Near East—Biography.
DS61.7.B37W56 915'.04'41 79-315828
79 MARC

—GUINEA-BISSAU—BIOGRAPHY

Carvalho, João Augusto de Noronha Dias de.
Henrique de Carvalho, uma vida ao serviço da pátria / João Augusto de Noronha Dias de Carvalho. — Lisboa : ₍s.n.₎, 1975.
351 p., ₍3₎ leaves of plates : ill. ; 24 cm.
Includes bibliographical references.
1. Dias de Carvalho, Henrique Augusto, 1843-1909. 2. Portugal—Colonies—Africa. 3. Angola—Discovery and exploration. 4. Colonial administrators—Portugal—Biography. 5. Colonial administrators—Guinea-Bissau—Biography. 6. Colonial administrators—Angola—Biography. 7. Explorers—Angola—Biography. 8. Explorers—Portugal—Biography. I. Title.
DT611.72.D53C37 325'.31'0924 77-155507
79 MARC

—INDIA—BIOGRAPHY

Lewin, Thomas Herbert, 1839-1916.
A fly on the wheel : or, How I helped to govern India / by Thomas H. Lewin. — Calcutta : Firma KLM, on behalf of Tribal Research Institute, Aizawl, 1977.
318 p., ₍1₎ leaf of plates : ill. ; 22 cm.
Reprint of the 1912 ed., published by Constable, London.
1. Lewin, Thomas Herbert, 1839-1916. 2. India—Description and travel—1859-1900. 3. India—Politics and government—19th century. 4. Lusheis. 5. Colonial administrators—India—Biography. 6. Colonial administrators—Great Britain—Biography. I. Title.
DS479.1.L48A33 1977 954.03'092'4 78-911616
78 MARC

Parveen Shaukat Ali.
Pillars of British imperialism : a case study of the political ideas of Sir Alfred Lyall, 1873-1903 / by Parveen Shaukat Ali. — Lahore : Aziz Publishers, 1976.
xix, 176 p. ; 22 cm.
Bibliography: p. 162-168.
Includes index.
Rs40.00 ($7.00 U.S.)
1. Lyall, Alfred Comyn, Sir, 1835-1911. 2. India—Politics and government—1765-1947. 3. Colonial administrators—India—Biography. 4. Colonial administrators—Great Britain—Biography. I. Title.
DS479.1.L9P37 325'.31'0924 77-930237
79 MARC

—INDONESIA—BIOGRAPHY

Besturen overzee : herinneringen van oud-ambtenaren bij het binnenlands bestuur in Nederlandsch-Indië / met een inleiding uitg. door S. L. van der Wal ; ₍met bijdragen van J. van Baal ... et al.₎. — Franeker : T. Wever, c1977.
329 p., ₍8₎ leaves of plates : ill. ; 23 cm. Ne***
Includes bibliographical references.
ISBN 9061352533
1. Indonesia—History—20th century—Biography. 2. Indonesia—Biography. 3. Colonial administrators—Indonesia—Biography. 4. Colonial administrators—Netherlands—Biography. I. Baal, Jan van.
DS643.2.B47 79-356804
79 MARC

Raffles, Thomas Stamford, Sir, 1781-1826.
Statement of the services of Sir Stamford Raffles / with an introd. by John Bastin. — Kuala Lumpur ; New York : Oxford University Press, 1978.
xi, 72 p. ; 29 cm. — (Oxford in Asia historical reprints)
ISBN 0-19-580318-3 : $17.50
1. Raffles, Thomas Stamford, Sir, 1781-1826. 2. Indonesia—Politics and government—1798-1942. 3. Singapore—Politics and government. 4. East India Company (English)—History. 5. Colonial administrators—Indonesia—Biography. 6. Colonial administrators—Singapore—Biography. 7. Colonial administrators—Great Britain—Biography.
DS598.S7R267 941.07'3'0924 79-118509
79 MARC

—ITALY—BIOGRAPHY

Gavello, Anna Maria.
Ugo Ferrandi, esploratore novarese / Anna Maria Gavello. — Novara : Tip. La cupola, 1975.
139 p. : ill. ; 24 cm. It77-July
At head of title: Società storica novarese.
"Lettere di Ugo Ferrandi": p. ₍93₎-132.
Includes bibliographical references and index.
1. Ferrandi, Ugo, 1852-1928. 2. Somalia—Discovery and exploration. 3. Somalia—Colonization. 4. Explorers—Somalia—Biography. 5. Explorers—Italy—Biography. 6. Colonial administrators—Somalia—Biography. 7. Colonial administrators—Italy—Biography.
DT405.3.F47G38 916.7'73'0430924 78-398381
*79 MARC

—MAINE—BIOGRAPHY

Gorges, Thomas, 1618-1670.
The letters of Thomas Gorges, Deputy Governor of the Province of Maine, 1640-1643 / edited by Robert E. Moody. — Portland : Maine Historical Society, 1978.
xxii, 148 p. : ill. ; 26 cm.
Bibliography: p. xv-xx.
1. Gorges, Thomas, 1618-1670. 2. Maine—History—Colonial period, ca. 1600-1775—Sources. 3. Colonial administrators—Maine—Biography. 4. Maine—Governors. I. Moody, Robert Earle, 1901- . II. Title.
F23.G69A4 1978 974.1'02'0924 78-52565
78 MARC

—NEAR EAST—BIOGRAPHY

Winstone, Harry Victor Frederick.
Gertrude Bell / H. V. F. Winstone. — London : Cape, 1978.
xii, 322 p., ₍8₎ leaves of plates : ill., map (on lining papers), ports. ; 23 cm.
GB78-13967
Bibliography: p. ₍293₎-312.
Includes index.
ISBN 0-224-01432-3 : £3.50
1. Bell, Gerturde Lowthian, 1868-1926. 2. Orientalists—Biography. 3. Colonial administrators—Great Britain—Biography. 4. Colonial administrators—Near East—Biography.
DS61.7.B37W56 915'.04'41 79-315828
79 MARC

—NETHERLANDS—BIOGRAPHY

Besturen overzee : herinneringen van oud-ambtenaren bij het binnenlands bestuur in Nederlandsch-Indië / met een inleiding uitg. door S. L. van der Wal ; ₍met bijdragen van J. van Baal ... et al.₎. — Franeker : T. Wever, c1977.
329 p., ₍8₎ leaves of plates : ill. ; 23 cm. Ne***
Includes bibliographical references.
ISBN 9061352533
1. Indonesia—History—20th century—Biography. 2. Indonesia—Biography. 3. Colonial administrators—Indonesia—Biography. 4. Colonial administrators—Netherlands—Biography. I. Baal, Jan van.
DS643.2.B47 79-356804
79 MARC

—NIGERIA—BIOGRAPHY

Muffett, D J M
Empire builder extraordinary, Sir George Goldie : his philosophy of government and Empire / D. J. M. Muffett. — Douglas Isle of Man : Shearwater Press, 1978.
334 p. : ill., ports. ; 22 cm. GB***
Bibliography: p. 333-334.
ISBN 0-904980-18-9 : £8.75
1. Goldie, George Dashwood Taubman, Sir, 1846-1925. 2. Nigeria—Colonization. 3. Royal Niger Company. 4. Colonial administrators—Nigeria—Biography. 5. Capitalists and financiers—Great Britain—Biography. I. Title.
DT515.72.G6M83 966.9'03'0924 79-307164
79 MARC

—PAKISTAN—BALUCHISTAN—BIOGRAPHY

Thornton, Thomas Henry, 1832-1913.
Colonel Sir Robert Sandeman / by Thomas Henry Thornton. — 1st ed. in Pakistan. — Quetta : Gosha-e-Adab : sole distributors Nisa Traders, 1977.
xxiii, 392 p., ₍7₎ leaves of plates : ill. ; 18 cm.
Reprint of 1895 ed. published by John Murray, London.
250 copies printed.
Includes index.
Rs110.00 ($17.00 U.S.)
1. Sandeman, Robert Groves, Sir, 1835-1892. 2. Baluchistan—History. 3. Colonial administrators—Great Britain—Biography. 4. Colonial administrators—Pakistan—Baluchistan—Biography. I. Title.
DS392.B28S257 1977 954'.9'15'0350924 77-930582
79 MARC

—PAKISTAN—NORTH-WEST FRONTIER PROVINCE—BIOGRAPHY

Bruce, Richard Isaac.
The forward policy and its results / by Richard Issaq Bruce. — 1st. ed. in Pakistan. — Quetta : Gosha-e- Adab, 1977.
xxviii, 382 p., ₍27₎ leaves of plates : fold. map, ill. ; 23 cm.
Reprint of the 1900 ed. published by Longmans, Green, London.
Includes index.
Rs125.00 ($20.00 U.S.)
1. North-west Frontier Province, Pakistan—Politics and government. 2. Bruce, Richard Isaac. 3. Colonial administrators—Pakistan—North-west Frontier Province—Biography. 4. North-west Frontier Province, Pakistan—Biography. I. Title.
DS392.N67B75 1977 954.9'12 77-930584
79 MARC

—PORTUGAL—BIOGRAPHY

Carvalho, João Augusto de Noronha Dias de.
Henrique de Carvalho, uma vida ao serviço da pátria / João Augusto de Noronha Dias de Carvalho. — Lisboa : ₍s.n.₎, 1975.
351 p., ₍3₎ leaves of plates : ill. ; 24 cm.
Includes bibliographical references.
1. Dias de Carvalho, Henrique Augusto, 1843-1909. 2. Portugal—Colonies—Africa. 3. Angola—Discovery and exploration. 4. Colonial administrators—Portugal—Biography. 5. Colonial administrators—Guinea-Bissau—Biography. 6. Colonial administrators—Angola—Biography. 7. Explorers—Angola—Biography. 8. Explorers—Portugal—Biography. I. Title.
DT611.72.D53C37 325'.31'0924 77-155507
79 MARC

COLONIAL ADMINISTRATORS (Continued)

—SINGAPORE—BIOGRAPHY

Raffles, Thomas Stamford, Sir, 1781-1826.
Statement of the services of Sir Stamford Raffles / with an introd. by John Bastin. — Kuala Lumpur ; New York : Oxford University Press, 1978.
xi, 72 p. ; 29 cm. — (Oxford in Asia historical ⌐reprints)
ISBN 0-19-580318-3 : $17.50
1. Raffles, Thomas Stamford, Sir, 1781-1826. 2. Indonesia—Politics and government—1798-1942. 3. Singapore—Politics and government. 4. East India Company (English)—History. 5. Colonial administrators—Indonesia—Biography. 6. Colonial administrators—Singapore—Biography. 7. Colonial administrators—Great Britain—Biography.
DS598.S7R267 941.07'3'0924 79-118509
79 MARC

—SOMALIA—BIOGRAPHY

Gavello, Anna Maria.
Ugo Ferrandi, esploratore novarese / Anna Maria Gavello. — Novara : Tip. La cupola, 1975.
139 p. : ill. ; 24 cm. It77-July
At head of title: Società storica novarese.
"Lettere di Ugo Ferrandi": p. ⌐93⌐-132.
Includes bibliographical references and index.
1. Ferrandi, Ugo, 1852-1928. 2. Somalia—Discovery and exploration. 3. Somalia—Colonization. 4. Explorers—Somalia—Biography. 5. Explorers—Italy—Biography. 6. Colonial administrators—Somalia—Biography. 7. Colonial administrators—Italy—Biography.
DT405.3.F47G38 916.7'73'0430924 78-398381
*79 MARC

—SOUTH AFRICA—BIOGRAPHY

O'Brien, Terence Henry.
Milner : Viscount Milner of St. James's and Cape Town, 1854-1925 / Terence H. O'Brien. — London : Constable, 1979.
447 p., ⌐6⌐ leaves of plates : ports. ; 24 cm. GB***
Bibliography: p. ⌐421⌐-426.
Includes index.
ISBN 0-09-457880-X : £10.00
1. Milner, Alfred Milner, 1st Viscount, 1854-1925. 2. South Africa—History—1836-1909. 3. Colonial administrators—South Africa—Biography. 4. Colonial administrators—Great Britain—Biography. 5. Statesmen—Great Britain—Biography. 6. Great Britain—Politics and government—1837-1901. 7. Great Britain—Politics and government—1901-1936.
DT776.M6O27 1979 941.083'092'4 79-315722
79 MARC

—UNITED STATES—BIOGRAPHY

Flexner, James Thomas, 1908-
Lord of the Mohawks : a biography of Sir William Johnson / James Thomas Flexner ; newly illustrated and with a new foreword. — Boston : Little, Brown, c1979.
xviii, 400 p., ⌐15⌐ leaves of plates : ill. ; 24 cm.
On dust jacket: "A revised edition."
First ed. published in 1959 under title: Mohawk baronet.
Bibliography: p. 361-368.
Includes index.
ISBN 0-316-28609-5
1. Johnson, William, Sir, bart., 1715-1774. 2. United States—History—French and Indian War, 1755-1763. 3. Indians of North America—Government relations—To 1789. 4. Indians of North America—Wars—1750-1815. 5. Soldiers—Great Britain—Biography. 6. Colonial administrators—Great Britain—Biography. 7. Colonial administrators—United States—Biography. I. Title.
E195.J63F57 1979 973.2'6'0924 79-14843
79 MARC

COLONIAL ADMINISTRATORS' WIVES

—GREAT BRITAIN—BIOGRAPHY

Russell, Elnor, 1907-1976.
Bush life in Nigeria : an account of the experiences of the wife of an administrative officer in Northern Nigeria from 1935 to 1944 / by Elnor Russell. — ⌐Yeovil⌐ : ⌐Sir Kenneth Maddocks⌐, ⌐1978⌐.
⌐6⌐, 136 p., ⌐2⌐ leaves of plates (1 fold.), ⌐8⌐ p. of plates : ill., map, ports. ; 21 cm. GB78-31130
Errata slip inserted.
1. Russell, Elnor, 1907-1976. 2. Northern Nigeria—Social life and customs. 3. British in Northern Nigeria—Biography. 4. Colonial administrators' wives—Nigeria—Northern Nigeria—Biography. 5. Colonial administrators' wives—Great Britain—Biography. I. Title.
DT515.9.N5R87 1978 966.9'5'030924 79-307077
79 MARC

—NIGERIA—NORTHERN NIGERIA—BIOGRAPHY

Russell, Elnor, 1907-1976.
Bush life in Nigeria : an account of the experiences of the wife of an administrative officer in Northern Nigeria from 1935 to 1944 / by Elnor Russell. — ⌐Yeovil⌐ : ⌐Sir Kenneth Maddocks⌐, ⌐1978⌐.
⌐6⌐, 136 p., ⌐2⌐ leaves of plates (1 fold.), ⌐8⌐ p. of plates : ill., map, ports. ; 21 cm. GB78-31130
Errata slip inserted.
1. Russell, Elnor, 1907-1976. 2. Northern Nigeria—Social life and customs. 3. British in Northern Nigeria—Biography. 4. Colonial administrators' wives—Nigeria—Northern Nigeria—Biography. 5. Colonial administrators' wives—Great Britain—Biography. I. Title.
DT515.9.N5R87 1978 966.9'5'030924 79-307077
79 MARC

COLONIAL ARCHITECTURE
see Architecture, Colonial

COLONIAL ART see Art, Colonial

COLONIAL COMPANIES

Lutumba-Lu-Vilu na Wundu.
Les sociétés commerciales tributaires du Comité spécial du Shaba ⌐(ex Katanga) par⌐ Lutumba-Lu-Vilu) (Joseph). Kinshasa, Editions Okapi ⌐1972⌐
155 p. 22 cm.
Includes bibliographies references.
1. Comité special du Katanga. 2. Colonial companies. I. Title.
HC591.C6L87 75-590880
79⌐r79⌐rev MARC

—HISTORY

Babudieri, Fulvio.
L'espansione mercantile austriaca nei territori d'oltremare nel XVIII secolo e suoi riflessi politici ed economici / Fulvio Babudieri. — Milano : A. Giuffrè, 1978.
192 p., ⌐5⌐ leaves of plates : ill., facsims. ; 24 cm. — (⌐Pubblicazioni⌐ - Università di Trieste, Facoltà di scienze politiche ; 11) It***
Bibliography: p. ⌐179⌐-183.
L6000
1. Austria—Commerce—History. 2. Mercantile system—Austria—History. 3. Colonial companies—History. I. Title. II. Series: Trieste. Università. Facoltà di scienze politiche. ⌐Pubblicazioni⌐ - Università di Trieste, Facoltà di scienze politiche ; 11.
HF3545.B3 79-368875
79 MARC

COLONIAL FURNITURE see Furniture, Colonial

COLONIAL NATIONAL HISTORICAL PARK

Kranbuehl, Sarah R
In and around Williamsburg with children / by Sarah R. Kranbuehl ; edited by Virginia A. Carey ; illustrated by Kathryn Blue. — ⌐Williamsburg, Va.⌐ : Kranbuehl, c1978.
60 p. : ill. ; 23 cm.
Includes index.
1. Williamsburg region, Va.—Description and travel—Guide-books. 2. Colonial National Historical Park. 3. Colonial Williamsburg Foundation. 4. Williamsburg, Va.—Description—Guide-books. I. Carey, Virginia A. II. Title.
F234.W7K7 917.55'4252'044 78-108800
78 MARC

COLONIAL PAINTING see Painting, Colonial

COLONIAL POTTERY see Pottery, Colonial

COLONIAL WARS see United States—History—Colonial period, ca. 1600-1775

COLONIAL WILLIAMSBURG, INC.

—ARCHIVES—CATALOGS

Colonial Williamsburg, inc.
Guide to the manuscript collections of Colonial Williamsburg / compiled by Marylee G. McGregor. — 2d ed. — Williamsburg, Va. : Colonial Williamsburg, 1969.
⌐74⌐ p. ; 24 cm.
Includes bibliographical references and index.
1. Virginia—History—Colonial period, ca. 1600-1775—Manuscripts—Catalogs. 2. Virginia—History—Colonial period, ca. 1600-1775—Sources—Bibliography—Catalogs. 3. Virginia—History—Sources—Bibliography—Catalogs. 4. Virginia—History—Manuscripts—Catalogs. 5. Colonial Williamsburg, inc.—Archives—Catalogs. 6. Manuscripts—Virginia—Catalogs. I. McGregor, Marlee G. II. Title.
Z1345.C64 1969 016.9755 79-107956
⌐F229⌐ 79 MARC

COLONIAL WILLIAMSBURG FOUNDATION

Kranbuehl, Sarah R
In and around Williamsburg with children / by Sarah R. Kranbuehl ; edited by Virginia A. Carey ; illustrated by Kathryn Blue. — ⌐Williamsburg, Va.⌐ : Kranbuehl, c1978.
60 p. : ill. ; 23 cm.
Includes index.
1. Williamsburg region, Va.—Description and travel—Guide-books. 2. Colonial National Historical Park. 3. Colonial Williamsburg Foundation. 4. Williamsburg, Va.—Description—Guide-books. I. Carey, Virginia A. II. Title.
F234.W7K7 917.55'4252'044 78-108800
78 MARC

COLONIC DISEASES, FUNCTIONAL
see Irritable colon

COLONIES

see also Agricultural colonies; Colonization; Emigration and immigration; Land settlement

Bertoni, Enrica.
Le cause del colonialismo imperialistico / Enrica Bertoni. — 1. ed. — Milano : ISEDI, 1978.
163 p. : ill. ; 20 cm. — (I Grandi eventi storici e le loro cause ; 8) It78-Nov
Bibliography: p. 161-163.
L2000
1. Imperialism. 2. Colonies. 3. Europe—History—1789-1900. I. Title. II. Series.
JC359.B4 325'.3 79-344274
*79 MARC

Fontenot, Chester J
Frantz Fanon : language as the god gone astray in the flesh / Chester J. Fontenot, Jr. — Lincoln : University of Nebraska, 1979.
vi, 72 p. ; 23 cm. — (University of Nebraska studies ; new ser., no. 60 ISSN 0077-6386)
Bibliography: p. 65-70.
1. Fanon, Frantz, 1925-1961. 2. Colonies. 3. Racism. I. Series: University of Nebraska—Lincoln. University of Nebraska studies ; new ser., no. 60.
JC273.F36F65 325'.3'0924 79-65093
79 MARC

Gabbrielli, Gabriele.
Imperialismi moderni / Gabriele Gabbrielli. — Roma : Istituto nazionale fascista di cultura, 1934.
97 p. ; 22 cm.
Includes bibliographical references.
1. Imperialism. 2. Colonies. I. Title.
JC359.G235 79-364759
79 MARC

Gutiérrez, Guillermo, fl. 1973-
Colonialismo e imperialismo en la historia / Guillermo Gutiérrez. — 1. ed. — Buenos Aires : El Cid Editor : distribución, Capital Federal, Rubbo, 1974.
63 p. : ill. ; 18 cm. — (Colección Liberación o dependencia ; 1)
1. Economic history. 2. Colonies. 3. Imperialism. 4. Latin America—Economic conditions. I. Title.
HC25.G84 330.9 75-586246
79⌐r79⌐rev MARC

Leone, Enrico.
Espansionismo e colonie / Enrico Leone. — Roma : Tip. editrice nazionale, 1911.
235 p. ; 22 cm.
Includes bibliographical references.
1. Colonies. 2. Imperialism. 3. Commercial policy. I. Title.
JV185.L457 79-341486
79 MARC

COLONIES (Continued)

Sargsyan, Harowt'yown Nersesi.
(Neogaghowt'atirowt'yownĕ ew gershahowyt'i gaghowt'ayin aghbyowrnerĕ)
Նեոգաղութատիրությունը և գերշահույթի գաղութային աղբյուրները / Հ. Ն. Սարգսյան. — Երևան : Հայաստանի Պետական Հրատարակչություն, 1963.
52 p. ; 21 cm. — (Օգնություն ՍՄԿՊ ծրագիրն ուսումնասիրողներին)
Includes bibliographical references.
1. Colonies. 2. International economic relations. I. Title. II. Series: Ōgnowt'yown SMKP tsragirn owsowmnasiroghnerin.
JV151.S27 79–970878

Shpirt, Aleksandr I͡Ulianovich.
Tekhnicheskiĭ progress i kolonializm : Konet͡s XVIII—seredina XX v. / A. I͡U. Shpirt. — Moskva : Nauka, 1978.
216 p. ; 20 cm. USSR78
At head of title: Akademii͡a nauk SSSR. Institut vostokovedenii͡a.
Bibliography: p. 206-[215]
0.90rub
1. Underdeveloped areas. 2. Colonies. 3. Technology and civilization. I. Title.
HC59.7.S465 79-375933
79 MARC

Volkov, Maĭ I͡Akovlevich.
(Raspad kolonial'noĭ sistemy imperializma)
Распад колониальной системы империализма / М. Я. Волков. — Москва : Знание, 1978.
64 p. ; 20 cm. — (Новое в жизни, науке, технике : Серия Международная ; № 10, 1978 г.) USSR 78
Includes bibliographical references.
0.11rub
1. World politics—20th century. 2. Colonies. 3. Imperialism. 4. Underdeveloped areas—Nationalism. I. Title. II. Series: Novoe v zhizni, nauke, tekhnike : Serii͡a Mezhdunarodnai͡a ; 1978, no. 10.
D839.N64 1978, no. 10 79–413377
[D443]

—ADDRESSES, ESSAYS, LECTURES

Patriotisme, colonisation / avec une préf. d'Élisée Reclus. — Paris : Temps nouveaux, 1903.
vii, 442 p. ; 24 cm. — (Bibliothèque documentaire ; t. 2)
Bibliographie: p. [429]-436.
Includes index.
1. Colonies—Addresses, essays, lectures. 2. Patriotism—Addresses, essays, lectures.
JV61.P34 79-342647
79 MARC

—CASE STUDIES

Adas, Michael, 1943-
Prophets of rebellion : millenarian protest movements against the European colonial order / by Michael Adas. — Chapel Hill : University of North Carolina Press, c1979.
xxvii, 243 p. : ill. ; 24 cm.
Bibliography: p. 219-234.
Includes index.
ISBN 0-8078-1353-2
1. Colonies—Case studies. 2. Native races—Case studies. 3. Government, Resistence to—Case studies. 4. Millennialism—Case studies. I. Title.
JV305.A3 325'.34'09 78-26775
78 MARC

—HISTORY

Christmann, Helmut.
Kolonialgeschichte / Helmut Christmann. — München : Ehrenwirth, 1975.
186, [1] p. ; 20 cm. — (Materialien und Modelle für den Gymnasialunterricht) (Materialien und Modelle für Studium und Unterricht) GFR***
Bibliography: p. 183-[187]
ISBN 3-431-01713-4
1. Colonies—History. 2. Colonization—History. I. Title.
JV61.C47 325'.309 78-391515
79 MARC

Hammer, Karl, 1936-
Weltmission und Kolonialismus : Sendungsideen d. 19. Jh. im Konflikt / Karl Hammer. — 1. Aufl. — München : Kösel, 1978.
348 p. ; 23 cm. GFR78-A
Bibliography: p. 337-[346]
Includes index.
ISBN 3-466-20180-2 : DM55.00
1. Missions—History. 2. Colonies—History. I. Title.
BV2105.H35 79-354925
79 MARC

Siméon, Michel.
L'économiste et le sauvage : essai / Michel Siméon. — Paris : Hallier, c1978.
278 p. ; 22 cm. — (Essai) F***
Bibliography: p. [247]-262.
1. Colonies—History. 2. Imperialism—History. 3. International economic relations. 4. Underdeveloped areas. I. Title.
JV341.S53 325'.309 79-353065
79 MARC

—HISTORY—ADDRESSES, ESSAYS, LECTURES

Kolonialismus, Imperialismus, Dritte Welt / Alfred Bergmiller, Peter Feldbauer (Hg.) ; mit Beitr. v. Alfred Bergmiller ... [et al.]. — Salzburg : W. Neugebauer, c1977-1978.
2 v. ; 24 cm. — (Geschichte und Sozialkunde ; Bd. 2 : Reihe Lehrbehelfe) Au78-10-264
Vol. 2 has individual title: Materialien.
Vol. 2 edited by J.-P. Lehners.
Bibliography: v. 1, p. 338-342; v. 2, p. 243-245.
Includes index.
ISBN 3-85376-010-4 (v. 1) : S200.00 per vol.
1. Imperialism—History—Addresses, essays, lectures. 2. Colonies—History—Addresses, essays, lectures. 3. Underdeveloped areas—History—Addresses, essays, lectures. I. Bergmiller, Alfred. II. Feldbauer, Peter, 1945- . III. Lehners, Jean-Paul. IV. Series.
JC359.K64 325'.3 78-353959
*78[r79]rev MARC

COLONIES, ARTIST see Artist colonies

COLONIES, SPACE see Space colonies

COLONIES (BIOLOGY)

see also Animal populations

COLONIES IN AFRICA

—LITERARY COLLECTIONS

La Société coloniale vue par les Africains. — [s.l. : s.n., 197-?] (Novo, Dahomey : C.R.A.P.)
52 p. ; 30 cm. — (Littérature africaine : Classe de première ; 1)
1. African literature (French) 2. Colonies in Africa—Literary collections. 3. France—Colonies—Literary collections. I. Series.
PQ3985.S6 840 78-385874
78 MARC

COLONIES IN LITERATURE

Griffiths, Gareth, 1943-
A double exile : African and West Indian writing between two cultures / [by] Gareth Griffiths. — London : Boyars : Distributed by Calder and Boyars, 1978.
205 p. ; 23 cm. — (Critical appraisals series) GB78-21455
Bibliography: p. [194]-202.
Includes index.
ISBN 0-7145-2622-3 : £6.95. ISBN 0-7145-2623-1 pbk.
1. English literature—Commonwealth of Nations authors—History and criticism. 2. African literature (English)—History and criticism. 3. West Indian literature (English)—History and criticism. 4. Colonies in literature. 5. Biculturalism. I. Title.
PR9080.G7 820'.996 78-326174
79 MARC

Laine, Pierre, assistant en lettres à l'Université du Tchad.
L'Afrique centrale et la littérature française des années 20 : Maran, Gide, Celine : actes d'accusation du système colonial / Pierre Laine. — N'Djaména : [Université du Tchad], 1974.
111 leaves ; 29 cm. — (Série Lettres, langues vivantes et sciences humaines ; no 1) (Annales de l'Université du Tchad)
Bibliography: leaves 109-111.
1. French literature—20th century—History and criticism. 2. Africa, Central, in literature. 3. Colonies in literature. 4. Maran, René, 1887-1960. Batouala. 5. Gide, André Paul Guillaume, 1869-1951—Journeys—Africa, Central. 6. Destouches, Louis Ferdinand, 1894-1961. Voyage au bout de la nuit. I. Title. II. Series.
PQ307.A45L3 840'.9'0091 77-980935
79 MARC

—ADDRESSES, ESSAYS, LECTURES

Licari, Anita.
Letteratura, esotismo, colonialismo / [di] Anita Licari, Roberta Maccagnani, Lina Zecchi ; saggio introduttivo di Gianni Celati. — Bologna : Cappelli, 1978.
149 p. ; 21 cm. — (NUC studio ; 4) It78-May
Bibliography: p. 143-145.
L3000
1. French literature—History and criticism—Addresses, essays, lectures. 2. Exoticism in literature—Addresses, essays, lectures. 3. Colonies in literature—Addresses, essays, lectures. I. Maccagnani, Roberta, joint author. II. Zecchi, Lina, joint author. III. Title.
PQ145.7.A2L5 78-374835
*78 MARC

COLONIES IN OCEANICA

—HISTORY

Hempenstall, Peter J
Pacific Islanders under German rule : a study in the meaning of colonial resistance / Peter J. Hempenstall. — Canberra ; Norwalk, Conn. : Australian National University Press, 1978.
xii, 264 p., [2] leaves of plates : ill. ; 23 cm. Aus***
Bibliography: p. 239-246.
Includes index.
ISBN 0-7081-1350-8
1. Colonies in Oceanica—History. 2. Germany—Colonies—Oceanica—Administration. I. Title.
DU60.H55 325'.343'0996 78-58251
79 MARC

COLONIES IN SOUTHEASTERN ASIA

Lapian, A B
Kolonialisme di Asia Tenggara / A. B. Lapian. — Jakarta : Lembaga Research Kebudayaan Nasional, L.I.P.I., 1975.
ii, 39 leaves ; 31 cm. — (Seri studi wilayah ; no. 2)
At head of title: Manusia dan kebudayaan di Asia Tenggara.
Includes bibliographical references.
1. Asia, Southeastern—History. 2. Colonies in Southeastern Asia. I. Title. II. Title: Manusia dan kebudayaan di Asia Tenggara. III. Series.
DS514.3.L36 77-940206
79 MARC

—HISTORY

Furnivall, John Sydenham.
Progress of welfare in southeast Asia : a comparison of colonial policy and practice / by J. S. Furnivall. — 1st AMS ed. — New York : AMS Press, 1978.
84 p. ; 23 cm.
Reprint of the 1941 ed. published by the Secretariat, Institute of Pacific Relations, New York, in series: International research series.
Includes bibliographical references.
ISBN 0-404-59525-1
1. Colonies in Southeastern Asia—History. I. Title. II. Series: Institute of Pacific Relations. International research series.
JV241.F9 1978 325'.31'0959 75-30055
78 MARC

COLONIZATION

see also Agricultural colonies; Colonial companies; Emigration and immigration; Land grants; Space colonies; and subdivision Colonization under names of countries, regions, etc., e.g. Africa—Colonization; and under names of classes of persons, e.g. Epileptics—Colonization; Afro-Americans—Colonization; Mennonites—Colonization

Telkamp, Gerard J
Urban history and European expansion : a review of recent literature concerning colonial cities and a preliminary bibliography / by Gerard J. Telkamp. — [Leiden] : Leiden Centre for the History of European Expansion, 1978.
76 p. : ill. ; 30 cm. — (Intercontinenta ; no. 1 ISSN 0165-2850) Ne***
Includes bibliographies.
1. Cities and towns—Growth. 2. Cities and towns—Growth—Bibliography. 3. Colonization. 4. Colonization—Bibliography. I. Title. II. Series.
HT371.T44 301.36'094 79-301978
79 MARC

La Valorizzazione agraria e la colonizzazione. — Roma : Mondadori, [1940]
316 p., [16] leaves of plates : ill. ; 26 cm.
Cover title.
"Estratto da Gli annali dell'Africa italiana, anno II, v. 3, 1939."
1. Agriculture and state—Italy—Colonies. 2. Agriculture—Africa, Northeast. 3. Colonization.
HD1972 1940.V34 338.1'0961 79-348689
79 MARC

—ADDRESSES, ESSAYS, LECTURES

Jabbour, George.
(al-Ṭabī'ah al-'unṣurīyah lil-isti'mār al-istīṭānī wa-al-masā'il al-qānūnīyah al-nājimah 'anhā)
الطبيعة العنصرية للاستعمار الاستيطاني والمسائل القانونية الناجمة عنها / بقلم جورج جبور. — القاهرة : الامانة العامة لاتحاد المحامين العرب ، [1976]
26 p. ; 25 cm.
«تحتوى الصفحات التالية على نص محاضرة كنت القيتها في ١٩٧٢/١/١٢ بدعوة من نقابة محامى دمشق.»
On cover: al-Mu'tamar al-Thālith 'Ashar li-Ittiḥād al-Muḥāmīn al-'Arab.
£E0.25
1. Colonization—Addresses, essays, lectures. 2. Palestine—Colonization—Addresses, essays, lectures. I. Title: al-Ṭabī'ah al-'unṣurīyah lil-isti'mār al-istīṭānī ...
JV381.J32 77–960745

COLONIZATION (Continued)

—BIBLIOGRAPHY

Telkamp, Gerard J
Urban history and European expansion : a review of recent literature concerning colonial cities and a preliminary bibliography / by Gerard J. Telkamp. — ⌈Leiden⌉ : Leiden Centre for the History of European Expansion, 1978.
76 p. : ill. ; 30 cm. — (Intercontinenta ; no. 1 ISSN 0165-2850) Ne***
Includes bibliographies.
1. Cities and towns—Growth. 2. Cities and towns—Growth—Bibliography. 3. Colonization. 4. Colonization—Bibliography. I. Title. II. Series.
HT371.T44 301.36'094 79-301978
79 MARC

—HISTORY

Besold, Christoph, 1577-1638.
Discursus politicus de incrementis imperiorum, eorumque amplitudine procurandâ. Cvi inserta est dissertatio singularis, De Novo Orbe. Auctore, Christophoro Besoldo. Argentorati, Impensis haeredum Lazari Zetzneri, 1623.
65 p. 19 cm.
"Caput IV. Continens conjectanea de Novo Orbe": p. 26-45.
1. Europe—History—17th century. 2. Colonization—History. 3. America—Discovery and exploration. 4. Ecclesiastical geography—America. 5. Rome—History—Empire, 30 B.C.-284 A.D. I. Besold, Christoph, 1577-1638. De Novo Orbe. 1623. II. Title. III. Title: De Novo Orbe.
D247.B38 79-348875
79 MARC

Christmann, Helmut.
Kolonialgeschichte / Helmut Christmann. — München : Ehrenwirth, 1975.
186, ⌈1⌉ p. ; 20 cm. — (Materialien und Modelle für den Gymnasialunterricht) (Materialien und Modelle für Studium und Unterricht) GFR***
Bibliography: p. 183-⌈187⌉
ISBN 3-431-01713-4
1. Colonies—History. 2. Colonization—History. I. Title.
JV61.C47 325'.309 78-391515
79 MARC

—GHANA

Howard, Rhoda.
Colonialism and underdevelopment in Ghana / Rhoda Howard. — New York : Africana Pub. Co., 1978.
244 p. : map ; 22 cm.
Bibliography: p. 232-237.
Includes index.
ISBN 0-8419-0387-5
1. Ghana—Economic conditions. 2. Ghana—Foreign economic relations. 3. Colonization—Ghana. I. Title.
HC517.G6H68 1978 330.9'667'05 78-5773
78 MARC

COLONIZATION BY AFRO-AMERICANS see Afro-Americans—Colonization

COLONNA, FRANCESCO, D. 1527. HYPNEROTOMACHIA POLIPHILI

Schmidt, Dorothea.
Untersuchungen zu den Architekturekphrasen in der Hypnerotomachia poliphili : die Beschreibung des Venus-Tempels / Dorothea Schmidt. — Frankfurt am Main : R. G. Fischer, 1978.
215 p. : ill. ; 21 cm. GFR79-A
Bibliography: p. 4-8.
ISBN 3-88323-031-6 : DM28.00
1. Colonna, Francesco, d. 1527. Hypnerotomachia Poliphili. 2. Colona, Francesco, d. 1527—Knowledge—Architecture. 3. Architecture in literature. 4. Architecture, Renaissance—Themes, motives. I. Title.
PQ4619.C9Z88 79-374293
79 MARC

COLONNA, FRANCESCO MARIA POMPEO, D. 1726

Costa, Gustavo.
Un collaboratore italiano del conte di Boulainviller: Francesco Maria Pompeo Colonna (1644-1726). Firenze, L. S. Olschki ⌈1965⌉
208-295 p. 25 cm.
At head of title: Accademia toscana di scienze e lettere La Colombaria.
"Dagli Atti e memorie dell'Accademia toscana di scienze e lettere La Colombaria, 29, 1964."
Includes bibliographical references.
1. Colonna, Francesco Maria Pompeo, d. 1726. 2. Boulainvilliers, Henri, comte de, 1658-1722. 3. Scientists—Italy—Biography. 4. Science—History. I. Title.
Q143.C62C67 78-396685
79 MARC

COLONOSCOPY

—CONGRESSES

Current status of barium enema and colonoscopy in diagnosis of colonic disease : symposium proceedings / cosponsors, Yale Affiliated Gastroenterology Program, Connecticut Society of Gastrointestinal Endoscopy ; edited by Howard M. Spiro, Vincent A. DeLuca, Jr. — Norwalk, Conn. : ⌈Purdue Frederick Co.⌉, c1978.
vii, 109 p. : ill. ; 25 cm.
Title on spine: Barium enema and colonoscopy in diagnosis of colonic disease.
Includes bibliographical references.
1. Colonoscopy—Congresses. 2. Colon (Anatomy)—Diseases—Diagnosis—Congresses. 3. Barium enema—Congresses. 4. Colon (Anatomy)—Congresses. I. Spiro, Howard M. II. DeLuca, Vincent A. III. Yale Affiliated Gastroenterology Program. IV. Connecticut Society of Gastrointestinal Endoscopy. V. Title: Barium enema and colonoscopy in diagnosis of colonic disease.
RC804.C64C87 617'.5547'0754 78-61839
79 MARC

COLOPHONS

see also Incipits; Title-page

COLOR

see also Colors; Dyes and dyeing; and names of colors, e. g. Purple, Yellow

Knowles, Edmond.
The essence of color / by Edmond Knowles. — 1st ed. — ⌈Howell⌉ N.J. : E & E Enterprises, c1978.
67 p. : ill. (some col.) ; 28 cm.
Bibliography: p. 65.
Includes index.
ISBN 0-917954-06-8 : $10.50
1. Color. I. Title.
QC495.K73 818'.54'07 78-54111
78 MARC

The Rainbow book : being a collection of essays & illustrations devoted to rainbows in particular & spectral sequences in general focusing on the meaning of color (physically & metaphysically) from ancient to modern times. — Rev. ed. — New York : Vintage Books, 1979.
199 p., ⌈4⌉ leaves of plates : ill. ; 31 cm.
Published on the occasion of an exhibition called the Rainbow Show, held Mar. 21-June 22, 1975, at the M. H. de Young Memorial Museum, et al.
Bibliography: p. 197.
ISBN 0-394-72365-1
1. Rainbow. 2. Color.
⌈QC976.R2R34 1979⌉ 535.6 77-3490
77 MARC

—ATLASES

Küppers, Harald.
DuMonts Farben-Atlas : 5500 Farbnuancen mit Kennzeichnung u. Mischanleitung / Harald Küppers. — Erstveröff. — Köln : DuMont, 1978.
162 p. : ill. (some col.) ; 21 cm. — (DuMont Dokumente) GFR78-A
Includes bibliographical references.
ISBN 3-7701-1058-7 : DM19.80
1. Color—Atlases. I. Title. II. Title: Farben-Atlas.
QC495.2.K83 535.6'022'2 78-365833
78 MARC

—CONGRESSES

International Colour Association.
AIC color 77 : proceedings of the third congress of the International Colour Association, Troy, New York, 10-15 July 1977 = AIC couleur 77 : compte-rendu de la troisième journée de l'Association internationale de la couleur = AIC Farbe 77 : Tagungsbericht von der Dritten Tagung der Internationalen Vereinigung für die Farbe / edited by Fred W. Billmeyer and Gunter Wyszecki. — Bristol ⌈Eng.⌉ : A. Hilger, c1978.
xx, 563 p. : ill. ; 23 cm. GB***
Spine title: Color 77.
English, French, or German.
Organized by the Canadian Society for Color and the Inter-Society Color Council, Inc.
Includes full texts of the 8 invited papers and extended abstracts of the 91 contributed papers.
Includes bibliographies and indexes.
ISBN 0-85274-352-1 : £23.00
1. Color—Congresses. I. Billmeyer, Fred W. II. Wyszecki, Günter. III. Canadian Society for Colour. IV. Inter-society Color Council. V. Title. VI. Title: AIC couleur 77. VII. Title: AIC Farbe 77. VIII. Title: Color 77.
QC494.I57 1978 535.6 78-382652
79 MARC

Praktische Anwendung farbwissenschaftlicher Erkenntnisse : Zusammenfassungen der Vorträge = Applications pratiques de la science de la couleur : résumés des communications = Practical applications of colour science : abstracts of the lectures. — ⌈s.l. : s.n., 1961⌉
⌈83⌉ leaves ; 21 cm.
Cover title.
At head of title: Internatonale Farbtagung. Congrès international de la couleur. International Colour Meeting (5èmes Journées internationales de la couleur) 23-26 Mai 1961, Düsseldorf.
English, French, and German.
1. Color—Congresses. I. International Colour Meeting, 5th, Düsseldorf, 1961. II. Title: Applications pratique de la science de la couleur. III. Title: Practical applications of colour science.
QC494.P72 79-304299
79 MARC

Színdinamikai Konferencia '76, Budapest, 1976.
Színdinamikai Konferencia előadásai, Budapest, 1976. június 8-11 / ⌈szerkesztő lektor, Klausz Csaba ; rend. a Magyar Színbizottság, Magyar Elektrotechnikai Egyesület⌉. — ⌈Budapest⌉ : Magyar Elektrotechnikai Egyesület, ⌈1977⌉
640 p. : ill. ; 24 cm. H77
On cover: AIC.
Cover title: Színdinamika '76 — Colour dynamics '76.
Includes bibliographies.
ISBN 9638141115
1. Color—Congresses. I. Klausz, Csaba. II. Magyar Színbizottság. III. Magyar Elektrotechnikai Egyesület. IV. International Colour Association. V. Title: Színdinamika hetvenhat. VI. Title: Colour dynamics '76.
QC494.S98 1976 79-339077
*79 MARC

—JUVENILE LITERATURE

Anderson, Louis Wilmer, 1933-
Light and color / words by L. W. Anderson. — Milwaukee : Raintree Childrens Books, c1978.
46 p. : col. ill. ; 24 cm. — (Read about)
Bibliography: p. 46.
Includes index.
SUMMARY: Easy-to-read text and examples from daily life introduce the concepts of light and color.
ISBN 0-8393-0077-8 lib. bdg. : $6.60
1. Light—Juvenile literature. 2. Color—Juvenile literature. ⌈1. Light. 2. Color⌉ I. Title.
QC360.A52 535'.013 77-27460
77 MARC

Branley, Franklyn Mansfield, 1915-
Color, from rainbows to lasers / Franklyn M. Branley ; illustrated by Henry Roth. — New York : Crowell, c1978.
87 p. : ill. (some col.) ; 24 cm.
Bibliography: p. 81.
Includes index.
SUMMARY: Explains the basic composition of color, how it is created and used, and its psychological effects.
ISBN 0-690-01256-X : $8.95. ISBN 0-690-03847-X lib. bdg.
1. Color—Juvenile literature. ⌈1. Color⌉ I. Roth, Henry, 1933- II. Title.
QC495.5.B7 1978 535'.6 76-46304
76 MARC

Kirkpatrick, Rena K
Look at rainbow colors / words by Rena K. Kirkpatrick ; pictures by Anna Barnard. — Milwaukee : Raintree Childrens Books, c1978.
32 p. : col. ill. ; 24 cm.
Includes index.
SUMMARY: Easy-to-read text and illustrations introduce the colors of the rainbow.
ISBN 0-8393-0064-6 lib. bdg. : $5.49
1. Rainbow—Juvenile literature. 2. Color—Juvenile literature. ⌈1. Rainbow. 2. Color⌉ I. Barnard, Anna. II. Title.
QC976.R2K57 551.5'67 77-27593
77 MARC

Miller, John Parr, 1913-
Do you know colors? / By J. P. Miller and Katherine Howard. — New York : Random House, c1978.
⌈32⌉ p. : col. ill. ; 21 cm. — (A Random House picture back)
SUMMARY: Reveals how some colors are made from blending others and depicts objects in each color.
ISBN 0-394-83957-9. ISBN 0-394-93957-3 lib. bdg. ISBN 0-394-83956-0 best book club : $0.95
1. Color—Juvenile literature. ⌈1. Color⌉ I. Howard, Katherine, joint author. II. Title.
QC495.5.M54 535.6 78-1133
78 MARC

—MEASUREMENT

see also Colorimetry

—MISCELLANEA

Aléjos, José.
Définition de certains termes appliqués sur la couleur "energiecons" / José Aléjos. — Monte -Carlo : Éditions Regain, c1977.
41 p. ; 19 cm. Mo***
40.00F
1. Color—Miscellanea. I. Title.
BF1999.A6326 78-397623
79 MARC

COLOR (Continued)

—PHYSIOLOGICAL EFFECT

Babey-Brooke, A M
Color therapy : healing with color / Anna M. Babey-Brooke and R. B. Amber. — Rev. and enl. ed. — New York : Santa Barbara Press, 1979.
viii, 271 p. ; 23 cm.
Edition of 1965 entered under R. B. Amber.
Bibliography: p. 215-228.
Includes index.
1. Color—Therapeutic use. 2. Color—Physiological effect. I. Amber, R. B., joint author. II. Title.
RM840.A4 1979 615'.831 78-64591
00 MARC

Birren, Faber, 1900-
Color & human response : aspects of light and color bearing on the reactions of living things and the welfare of human beings / Faber Birren. — New York : Van Nostrand Reinhold Co., 1978.
vi, 141 p., [12] p. : ill. ; 24 cm.
Bibliography: p. 127-134.
Includes index.
ISBN 0-442-20787-5 : $11.95
1. Color—Physiological effect. 2. Color vision. 3. Color—Psychological aspects. 4. Visual perception. I. Title.
QP483.B57 155.9'11'45 77-12505
77 MARC

—PSYCHOLOGICAL ASPECTS

see also Color vision

Birren, Faber, 1900-
Color & human response : aspects of light and color bearing on the reactions of living things and the welfare of human beings / Faber Birren. — New York : Van Nostrand Reinhold Co., 1978.
vi, 141 p., [12] p. : ill. ; 24 cm.
Bibliography: p. 127-134.
Includes index.
ISBN 0-442-20787-5 : $11.95
1. Color—Physiological effect. 2. Color vision. 3. Color—Psychological aspects. 4. Visual perception. I. Title.
QP483.B57 155.9'11'45 77-12505
77 MARC

García-Bermejo Pizarro, Sergio.
El color en el arte infantil : psicología de la percepción, didáctica de la imagen / Sergio García-Bermejo Pizarro. — Madrid : Ciencias de la Educación Preescolar y Especial, [1978?]
179 p., [10] leaves of plates : ill. ; 24 cm. — (Colección Educación preescolar) Sp***
Bibliography: p. 177-178.
ISBN 8485252314 : 875ptas
1. Color vision in children. 2. Color—Psychological aspects. I. Title.
BF723.P36G37 155.4'1 79-346233
79 MARC

Kurokawa, Hitoshi, 1931-
(Chiteki josei wa shikisai shinri ni akarui)
知的女性は色彩心理に明るい：教養としての色彩学入門 / 黒川鴻著. — 東京：こう書房, 昭和52 [1977]
206 p., [4] leaves of plates : ill. (some col.) ; 19 cm. — (Kou books) Ja 77-16742
¥880
1. Color—Psychological aspects. 2. Color-sense. 3. Colors—Analysis. I. Title.
BF789.C7K87 78-813149

Lüscher, Max.
The 4 color person / Max Lüscher ; translated from the German by Joachim Neugroschel. — New York : Simon and Schuster, c1979.
192 p., [1] leaf of plates : ill. ; 22 cm.
Translation of Der 4-Farben-Mensch.
Includes index.
ISBN 0-671-24232-6
1. Color—Psychological aspects. 2. Personality. I. Title.
BF789.C7L8313 1979 155.2'6 78-27360
78 MARC

University of the Trees.
You are a rainbow : original insights into the work of Christopher Hills by researchers practicing his theory of Nuclear Evolution / editor, Norah Hills. — 1st ed. — Boulder Creek, CA : University of the Trees Press, c1979.
125 p. : ill. ; 21 cm.
Includes bibliographical references.
ISBN 0-916438-25-2 : $2.95
1. Personality. 2. Color—Psychological aspects. 3. Aura. I. Hills, Norah. II. Title.
[BF698.U49 1979] 131'.32 79-13393
79 MARC

Wittgenstein, Ludwig, 1889-1951.
Remarks on colour / Ludwig Wittgenstein ; edited by G. E. M. Anscombe ; translated by Linda L. McAlister and Margarete Schättle. — Oxford [Eng.] : B. Blackwell, 1978, c1977.
63, 63 p. ; 23 cm. GB***
Added t.p.: Bemerkungen über die Farben.
English and German.
Opposite pages numbered in duplicate.
ISBN 0-631-17310-2 : £6.95
1. Color—Psychological aspects. I. Anscombe, Gertrude Elizabeth Margaret. II. Title. III. Title: Bemerkungen über die Farben.
BF789.C7W513 1978 110 79-344747
79 MARC

—PSYCHOLOGY

Asari, Atsushi, 1911-
(Jidōga to sekkusu)
児童画とセックス：赤と緑の秘密 / 浅利篤著. — 東京：黎明書房, 昭和34 [1959], 昭和35 [1960] printing.
191 p. : ill. (some col.) ; 19 cm.
Colophon inserted.
1. Drawing, Psychology of. 2. Color—Psychology. 3. Sex in art. 4. Child analysis. I. Title.
BF723.D7A78 79-805183

—TABLES

Daylight illumination : color-contrast tables for full-form objects / M. R. Nagel ... [et al.]. — New York : Academic Press, 1978.
ix, 494 p. : ill. ; 29 cm.
Chiefly tables.
Bibliography: p. 44-45.
ISBN 0-12-513750-8
1. Solar radiation—Tables, etc. 2. Daylight—Tables, etc. 3. Color—Tables. I. Nagel, Max R., 1909-
QC911.D48 551.5'6 77-11218
78 MARC

—TERMINOLOGY

Dürbeck, Helmut.
Zur Charakteristik der griechischen Farbenbezeichnungen / Helmut Dürbeck. — Bonn : Habelt, 1977.
x, 321 p. ; 20 cm. — (Habelts Dissertationsdrucke : Reihe klassische Philologie ; Heft 27) GFR77-A
Originally presented as the author's thesis, Erlangen.
Bibliography: p. iv-x.
Includes indexes.
ISBN 3-7749-1427-3
1. Greek language—Semantics. 2. Color—Terminology. I. Title.
PA427.D8 1977 482 79-340257
79 MARC

Kornerup, Andreas.
Methuen handbook of colour / A. Kornerup and J. H. Wanscher. — 3d ed. / introduced and rev. by Don Pavey. — London : E. Methuen, 1978.
252 p., [60] leaves of col. plates : ill. ; 18 cm. GB***
Translation of Farver i farver.
ISBN 0-413-33400-7 : £7.50
1. Colors. 2. Color—Terminology. I. Wanscher, Johan Henrik, joint author. II. Title.
ND1492.K6713 1978 667 78-320120
78 MARC

—THERAPEUTIC USE

Babey-Brooke, A M
Color therapy : healing with color / Anna M. Babey-Brooke and R. B. Amber. — Rev. and enl. ed. — New York : Santa Barbara Press, 1979.
viii, 271 p. ; 23 cm.
Edition of 1965 entered under R. B. Amber.
Bibliography: p. 215-228.
Includes index.
1. Color—Therapeutic use. 2. Color—Physiological effect. I. Amber, R. B., joint author. II. Title.
RM840.A4 1979 615'.831 78-64591
00 MARC

MacIvor, Virginia.
Vibrations, healing through color, homoeopathy, and radionics / by Virginia MacIvor and Sandra LaForest. — New York : S. Weiser, 1979.
xii, 179 p. : ill. ; 21 cm.
Bibliography: p. [169]-175.
ISBN 0-87728-393-1 : $5.95
1. Color—Therapeutic use. 2. Homeopathy. 3. Radionics. 4. MacNaughton, Wallace Finne, 1886- . I. LaForest, Sandra, joint author. II. Title.
RM840.M32 615'.831 79-114570
79 MARC

COLOR CENTERS

Bójko, Ignacy.
Elektronowy rezonans paramagnetyczny w halogenkach metali alkalicznych zawierających defekty molekularne / Ignacy Bójko. — Opole : WSP, 1977 (Katow. : UŚ)
51, [1] p. : ill. ; 24 cm. — (Zeszyty naukowe Wyższej Szkoły Pedagogicznej im. Powstańców Śląskich w Opolu : Seria B, studia i monografie ; Nr 58) P77-7135
Summary also in English.
Bibliography: p. 50-51.
zł20.00
1. Alkali metal halide crystals—Spectra. 2. Electron paramagnetic resonance spectroscopy. 3. Color centers. I. Title.
QC464.A43B64 78-395103
79 MARC

Marfunin, Arnol'd Sergeevich.
Spectroscopy, luminescence and radiation centers in minerals / A. S. Marfunin ; translated by V. V. Schiffer. — Berlin ; New York : Springer-Verlag, 1979.
xii, 352 p. : ill. ; 25 cm.
Translation of Spektroskopiia, liuminestsentsiia i radiatsionnye tsentry v mineralakh.
Bibliography: p. [304]-345.
Includes index.
ISBN (invalid) 3-540-09070-3 (Berlin)
1. Mineralogy, Determinative. 2. Spectrum analysis. 3. Luminescence. 4. Holes (Electron deficiencies) 5. Color centers. I. Title.
QE369.S65M3713 549'.125 78-26367
78 MARC

Niklas, Andrzej.
Centra barwne w kryształach rubinu / Andrzej Niklas. — Wyd. 1. — Warszawa : Państwowe Wydawn. Naukowe, 1975.
82 p. : ill. ; 24 cm. — (Prace Opolskiego Towarzystwa Przyjaciół Nauk, Wydział IV, Nauk Technicznych i Ścisłych) P***
Summary in English.
Bibliography: p. [80]-82.
zł20.00
1. Ruby crystals. 2. Color centers. I. Title. II. Series: Opolskie Towarzystwo Przyjaciół Nauk. Wydział IV. Nauk Technicznych i Ścisłych. Prace.
QD941.N54 78-399379
79 MARC

COLOR CINEMATOGRAPHY

Limbacher, James L
Four aspects of the film / James L. Limbacher. — New York : Arno Press, 1978, c1969.
386 p., [8] leaves of plates : ill. ; 24 cm. — (Aspects of film)
Reprint of the ed. published by Brussel & Brussel, New York.
Bibliography: p. [255]-256.
Includes index.
ISBN 0-405-11138-X
1. Wide-screen processes (Cinematography) 2. Color cinematography. I. Title. II. Series.
[TR855.L55 1978] 778.5'3 77-11379
78 MARC

COLOR DISCRIMINATION see Color vision

COLOR ETCHINGS see Color prints

COLOR-FIELD PAINTING

—CANADA—EXHIBITIONS

Edmonton Art Gallery.
Canada x ten : David Bolduc, D. T. Chester, Robert Christie, Ann Clarke Darrah, Harold Feist, K. M. Graham, Paul Hutner, Graham Peacock, Milly Ristvedt, Daniel Solomon : an exhibition / organized by the Edmonton Art Gallery. — [s.l. : s.n., 1973]
[32] p. : col. ill. ; 21 cm. C75-6397-1
Includes bibliographies.
$4.00
1. Painting, Abstract—Canada—Exhibitions. 2. Color-field painting—Canada—Exhibitions. 3. Painting, Modern—20th century—Canada—Exhibitions. I. Title.
ND245.5.A28E35 1973 759.11'074'011233 78-369116
78 MARC

☆ U.S. GOVERNMENT PRINTING OFFICE : 1980 O - 318-486 (Vol. 4)